LAXTON'S

BUILDING PRICE BOOK

MAJOR & SMALL WORKS

Edited by **V.B. JOHNSON LLP**

Chartered Quantity Surveyors • Construction Cost Consultants • Project Managers

ONE HUNDRED & SEVENTY EIGHTH EDITION

2006

LAXTON'S

An imprint of Elsevier

AMSTERDAM BOSTON LONDON NEW YORK OXFORD PARIS SAN DIEGO SAN FRANCISCO SINGAPORE SYDNEY TOKYO

www.laxtonsprices.com

Laxton's
An imprint of Elsevier
Linacre House, Jordan Hill, Oxford OX2 8DP
30 Corporate Drive, Burlington, MA 01803

First published 2005

British Library Cataloguing in Publication Data
A catalogue record for this book is available from the British Library

ISBN 0 7506 6560 2

For information on all Elsevier publications
visit our website at www.books.elsevier.com

Printed and bound in Italy

INTRODUCTION

Computers are part of the construction process and V B Johnson have utilized these to the full in the production of Laxton's to the extent that individual material prices shown in the Basic Prices of Materials sections are linked directly with the Measurement Rates, the Composite Prices for the Approximate Estimating and the Example Cost Plans. Any adjustments to material prices received late in the production process can be entered and all sections updated.

Laxton's Measurement Rates and prepared utilizing "first principles" practice and the results clearly detailed within the Measurement Rates pages with all prices being updated.

Laxton's is used by Quantity Surveyors, Contractors, Sub-contractors, Architects, Developers, Students, Government and Public Authorities and those interested in the cost and price of Building Works.

Users should understand the limitations of any price book and should familiarise themselves with these introductory sections including "How to Use", "Essential Information", "Regional Variations" and the "Preliminaries Section".

Laxton's has three levels of pricing – 'Major Works' – for works of a value of £500,000 to £1,500,000 with adjustments for tender values from £250,000 to £5,000,000 – 'Small Works' for works of £50,000 - £100,000 with adjustments for tender values from £15,000 to £250,000 – 'Alterations, Repairs and Maintenance Works' for works up to the value of £15,000.

The Blue section - 'Alterations, Repairs and Maintenance Works' - is intended for use by the smaller builder and others carrying out this type of work, collecting materials from merchants as required and not achieving the usual discounts, very little site set up and often working from the back of a van.

The Composite Prices section has the prices linked to those included in the Major and Small works priced sections and is useful for those who wish to produce a quick estimate as a guide perhaps for budget purposes, although care must be taken to include items not described.

The Basic Prices of Materials sections have been repriced and these prices are linked to the built up prices in the delivered to site column of the priced sections.

The analysis of materials as included herein has shown that an item may consist of a number of different materials with different waste factors. The waste % indicated within the book is for the predominant material.

Building costs rose in the year to 2nd quarter 2005 by 7% to include an increase of 60% in the raw material price of steel and considerable oil price rises.

Regional Variations in construction costs are shown at the start of the measured rates sections with Laxton's based upon national average prices of 1.00, however readers should also note that there are Regional Price differentials in the Retail Price Index with London some 9.7% higher than the UK average. Further detail is available from National Statistics online at www.statistics.gov.uk.

The National agreed wage award this year came into force from 27th June 2005. The award represents the final part of a three year agreement adding 9.5% to the rates for craftsmen and General Building operatives, this is reflected within the prices herein.

The Employers contribution to National Insurance remains this year at 12.8% with an adjustment to threshold levels.

The Aggregates Levy was introduced on the 1st April 2002 and is charged at a single flat rate of £1.60p per tonne (£4.00m3 approx). The levy applies to any sand, gravel or crushed rock commercially exploited in the UK or its territorial waters or imported into the UK. The levy will not apply to coal, clay, slate, metals and metal ores, gemstones or semi precious stones, industrial minerals and recycled aggregates.
www.hmce.gov.uk

As in previous years the "Build-up" of the "All-in" labour rates has been re-calculated; adjustment being made to changes in Labour Rates and in National Insurance contributions.

A section is included in Laxton's setting out the requirements of the Construction (Design and Management) Regulations 1994. This is now law and it is a criminal offence not to observe the requirements of these regulations. Clients, Designers (including Quantity Surveyors) and Contractors please note.

As with the previous editions, the maximum amount of pricing information is given in an easy to read format. Price rates are analysed under ten headings to show the complete estimating picture of materials, labour constants, labour rates and sundries all on one page.

Laxton's Building Price Book is structured generally in accordance with the Seventh Edition Revised 1998 of the Standard Method of Measurement of Building Works (SMM7) which follows the format of the 'Common Arrangement of Work Sections for Building Works 2nd edition published to encourage Project Information to be co-ordinated throughout the industry.

Uniclass table J is based on the Common Arrangement and is used for organising information in specifications and bills of quantities and for classifying information on particular types of construction operations.

The user is urged to read the relevant sections of SMM7 which requires certain measurement rules, definition rules and coverage rules to be observed. Certain labours are deemed to be included in the measured items (and prices) without mention and care should be taken to ascertain precisely the extent of these labours by reference to the Coverage Rules in SMM7.

The index on the front cover and the aligned markers on the edge of pages will assist the reader to find the sections in the book. The contents list and comprehensive index will further assist in locating specific items and subjects.

The Landfill Tax (Qualifying Material) Order 1996 gives a lower rate of £2.00 per tonne for inactive (or inert) waste for materials listed and a standard rate for other waste.

Current standard rates are:-

01.04.02 to 31.03.03 – £13.00 per tonne
01.04.03 to 31.03.04 – £14.00 per tonne
01.04.04 to 31.03.05 – £15.00 per tonne
from 01.04.05 - £18.00 per tonne

Subsequently to be increased in years thereafter, to a medium/long term rate of £35.00 per tonne.
www.hmce.gov.uk

Users of Laxton's should note that the Landfill Tax has been included within Disposal items with alternatives for Inert, Active and Contaminated materials. In this edition the standard rate of tax included is £18.00

There are exemptions including Dredgings, Mining and quarrying, Pet cemeteries, Contaminated land, US Forces, Restoration of Landfill sites and filling of quarries subject to qualification.

The Arbitration Act received Royal Assent on 17th June 1996 and came into operation on 31st January 1997. The Arbitration Act 1996 is reproduced in part in the General Information Section.

The Housing Grants Construction and Regeneration Act 1996 came into force on 1st May 1998, it includes mandatory provisions for disputes in construction contracts to be referred to adjudication, including any contracts which do not comply with these requirements. Also included are agreements made between clients and their consultants, further details are included at the end of the Arbitration Section.

The Contracts (Rights of Third Parties) Bill is now in place and provides rights to third parties automatically under a contract unless parties specifically agree to other constraints. JCT and other forms of contract now generally include a clause to restrict the rights of third parties.

The London Congestion Charge and other local congestion charges are not included within Laxton's. Where applicable due allowances should be made to preliminaries and rates, these costs could be considerable.

Laxton's Approximate Estimating section has been costed to enable complete buildings to be priced following the format of the Standard Form of Cost Analysis as published by the Building Cost Information Service of the Royal Institute of Chartered Surveyors. The section gives examples of a detailed analysis for a detached house and an office block together with alternative composite prices enabling quick and comprehensive estimates to be produced and amended as detail design continues. This section has been tailored to the needs of the Contractor and Quantity Surveyor for approximate estimating data in detail but in a simpler format than can be obtained from other sources.

Descriptions in the approximate estimating section have been abbreviated and care must be taken to understand the extent of works included Allowances should be added for any changes from the norm.

Basic Prices of Materials are included in both the Major and Small Work sections. The materials prices in the analysed rates are taken from these sections.

The pricing level of Laxton's is based on an average National level with indicative regional variations for overall pricing shown on the regional factor maps.

All-in labour rates are shown at the head of each page in each of the Work sections. The build-up of these rates is illustrated in the Preliminaries/General Conditions pages.

Prices take account of the wage rates which became operative on 28th June 2005. Material prices used are those current during the second quarter of 2005.

The Company Information Section contains details of internet links to many company and useful websites these may also be accessed via V B Johnson Website and will enable the reader to find further information on materials, products and services and where to source, search or obtain further details.
www.vbjohnson.co.uk

Every endeavour has been made to ensure the accuracy of the information printed in the book but the Publishers and Editors cannot accept liability for any loss occasioned by the use of the information given.

The section headings are as follows:

Introduction
How to use
Essential Information
Contents
Index
Regional Variations
Preliminaries/General Conditions

MAJOR WORKS

C Existing Site/Buildings/Services
D Groundwork
E In-situ concrete/Large precast concrete
F Masonry
G Structural/Carcassing metal/timber
H Cladding/Covering
J Waterproofing
K Linings/Sheathing/Dry partitioning
L Windows/Doors/Stairs
M Surface finishes
N Furniture/Equipment
P Building fabric sundries
Q Paving/Planting/Fencing/Site furniture
R Disposal systems
S Piped supply systems
T Mechanical heating/cooling/ refrigeration systems
U Ventilation/Air conditioning systems
V Electrical supply/power/lighting systems
W Communications/Security/ Control systems
X Transport systems

Basic Prices of Materials
Composite Prices for Approximate Estimating: -
Cost Plan for an office development
Cost Plan for a detached house
Composite Prices

SMALL WORKS – Yellow Pages

C Existing Site/Buildings/Services
D Groundwork
E In-situ concrete/Large precast concrete
F Masonry
G Structural/Carcassing metal/timber
H Cladding/Covering
J Waterproofing
K Linings/Sheathing/Dry partitioning
L Windows/Doors/Stairs
M Surface finishes
N Furniture/Equipment
P Building fabric sundries
Q Paving/Planting/Fencing/Site furniture
R Disposal systems
S Piped supply systems
T Mechanical heating/cooling/ refrigeration systems
U Ventilation/Air conditioning systems
V Electrical supply/power/lighting systems
W Communications/Security/ Control systems
X Transport systems

Basic Prices of Materials
Composite Prices for Approximate Estimating:-
Cost Plan for an office development
Cost Plan for a detached house
Composite Prices

GENERAL INFORMATION
Standard Rates of Wages
Builders' and Contractors Plant
Guide Prices to Building Types per Square Metre.
Cost Allowances
National Working Rule Agreement
Daywork Charges
Fees
Construction (Design and Management) Regulations 1994
Arbitration
Tables and Memoranda
Metric System

BRANDS AND TRADE NAMES
COMPANY INFORMATION
PRODUCTS AND SERVICES

ALTERATIONS REPAIRS AND MAINTENANCE WORKS – Blue Pages

C Existing Site/Buildings/Services
D Groundwork
E In-situ concrete/Large precast concrete
F Masonry
G Structural/Carcassing metal/timber
H Cladding/Covering
J Waterproofing
K Linings/Sheathing/Dry partitioning
L Windows/Doors/Stairs
M Surface finishes
N Furniture/Equipment
P Building fabric sundries
Q Paving/Planting/Fencing/Site furniture
R Disposal systems
S Piped supply systems
T Mechanical heating/cooling/ refrigeration systems
U Ventilation/Air conditioning systems
V Electrical supply/power/lighting systems

Index

The Editor wishes to thank the Professional bodies, Trade Organisations and firms who have kindly given help and provided information for this edition. For addresses, websites and details see company information section.

ACO Technologies PLC
Air Diffusion Ltd.
Alumasc Interior Building Products
Ancon CCL
Angle Ring Co. Ltd.
Ariel Plastics
Birtley Building Products Ltd.
Blucher UK Ltd.
Boddingtons Ltd.
BRC Ltd.
British Gypsum Ltd.
British Sisalkraft Ltd.
Burlington Slate
CAMAS Building Materials
Caradon Terrain Ltd.
Carter Concrete Ltd.
Celuform Ltd.
Cementation Foundations SKANSKA Ltd.
Civil Engineering Contractors Association.
Civil Engineering Developments Ltd.
Clay Pipe Development Association Ltd.
Construction Confederation
Construction Industry Joint Council
Cordek Ltd
Corus Group Plc
Crescent of Cambridge Ltd.
Crittall Steel Windows Ltd.
Crowthorne Fencing
Dow Construction Products
Drainage Systems
Durable Berkeley.
Electrical Contractors' Association
Eternit Building Materials
Fixatrad Ltd.
Fosroc Ltd.
Grace Construction Products Ltd.
Grass Concrete Ltd.
Halfen Ltd
Hanson Brick
Hanson Concrete Products
Heating and Ventilating Contractors Association
Hepworth Building Products Ltd.
Housing Corporation
H.S.S. Hire Shops
Ibstock Brick Ltd
Jacksons H.S. and Son (Fencing) Ltd.
Jeld-wen UK Ltd
Johnston Pipes Ltd.
Joint Council for the Building and Civil Engineering Industry (Northern Ireland)
Joint Industry Board for the Electrical Contracting Industry
Keller Ground Engineering
Kee Klamps Ltd.
Kingston Craftsmen Structural Timber
Kirkstone
Klargester Environmental Engineers Ltd
Kufa Plastics Ltd.
Lamps and Tubes Ltd
Light Alloy Ltd.
Mandor Engineering Ltd.
Marley Buildings Ltd.
Marley Floors Ltd.
Marshalls Mono Ltd.
Masterbill Micro Systems
Mastic Asphalt Council & Employers Federation Ltd.
Milton Pipes Ltd.
National Domelight Company 01276 451555
National Federation of Terrazzo, Marble and Mosaic Specialists
National Joint Council for the Laying Side of the Mastic Asphalt Industry
Natural Stone Products Ltd.
Oswestry Reinforced Plastics Ltd.
Owens Corning Building Products
Painting & Decorating Association
PreFormed Components Ltd.
Promat Fire Protection
Ramsay & Sons(Forfar) Ltd.
Ranalah Gates Ltd.
R.C. Cutting & Co
Rentokil Ltd.
Rockwool Ltd.
Royal Institute of British Architects
Royal Institution of Chartered Surveyors
Ruberoid Building Products Ltd.
Sadolin(UK) Ltd.
Saint-Gobain Pipe Systems
Sealmaster
Solaglas Ltd.
Stoakes Systems Ltd.
Stowell Concrete Ltd.
Swish Building Products
Syston Rolling Shutters Ltd.
Tarmac (UK) Ltd.
Tarmac Topblock Ltd.
Thermalite
Thorn Lighting Ltd.
Timloc Expamet Building Products
Townscape Products Ltd.
Tremco Ltd.
Velux Co.Ltd.
Wavin Plastics Ltd.
Welconstruct Co Ltd.

V.B.JOHNSON & PARTNERS,
Chartered Quantity Surveyors,
St John's House
304-310 St Albans Road,
Watford
Herts
WD24 6PW

Telephone : 01923 227236
Facsimile : 01923 231134
E Mail : office@ vbjw.demon.co.uk
www.vbjw.demon.co.uk
www.vbjohnson.co.uk

June 2005

HOW TO USE

Laxton's Building Price Book is divided into 6 main sections; the Small Works section is printed on Yellow paper; the Alteration, Repairs and Maintenance section is printed on Blue paper.
The sections are:

SECTIONS

1 INTRODUCTION AND PRELIMINARIES

Important information to be used in conjunction with the price book, including how to Use, allowances to be made for Regional Factors, Tender Values and Preliminary items.

2 MAJOR WORKS

Measured rates are given for all descriptions of building work generally in accordance with SMM7 Revised 1998. A complete breakdown is shown for all measured items, under 10 headings. This approach provides the reader with the price build-up at a glance enabling adjustments to be made easily. Net and gross rates are given for each item.

Details of Basic Prices of materials, Composite Prices for Approximate Estimating and examples of a cost plan for an Office Development and a Detached House are included.

Prices reflect national average costs and the Major Works section is intended to apply to a building contract for works within the range £500,000 to £1,500,000, with adjustment factors for tender values from £250,000 to £5,000,000.

3 SMALL WORKS (Yellow paper)

This section follows an identical pattern to Major Works but is intended to apply to a contract in the range £50,000 to £100,000, with adjustment factors for tender values from £15,000 to £250,000.

The type of contract envisaged would involve one of the following :- (a) a small new build project, (b) a conversion and repair of existing dwelling house with small extensions for additional accommodation, (c) the conversion, alteration and repair of existing office, factory or other type of building with a certain amount of internal reconstruction or renovation.

4 ALTERATIONS, REPAIRS AND MAINTENANCE WORKS (Blue paper)

This section follows a similar approach adopted by the Major and Small Works sections with measured rates given for descriptions of building work generally in accordance with SMM7 with items specifically identified as may be required in works of alterations, repair and maintenance.

Works envisaged would be of minor nature up to the value of £15,000. Additional allowances for disproportionate waste factors, preliminary items such as travelling, vehicles, plant and tools, access, out of hours working and abortive journeys met on works of this nature must be added. A separate index is provided for this section.

5 GENERAL INFORMATION

A useful reference section covering: Standard Rates of Wages, Builders' and Contractors' Plant, Guide Prices to Building Types, Cost Allowances, National Working Rule Agreement, Daywork Charges, Construction (Design and Management) Regulations 1994, Fees, Arbitration, Tables and Memoranda and Metric System.

6 BRANDS AND TRADE NAMES - COMPANY INFORMATION – PRODUCTS AND SERVICES

A unique list of Brands and Trade Names with the name and address of the manufacturer given in the adjacent Company Information Section. The Product and Services section provides details of manufacturers and suppliers. These lists are useful for locating a particular branded item. Every company listed has been written to and amendments received entered within the database. The majority of the companies now have websites and links to these websites is on our website at **vbjohnson.co.uk**, and a useful route should you wish to make frequent visits to these company websites. Simply add our website to your list of favourites.

RATES GENERALLY

Rates given are within the context of a complete contract of construction where all trades work is involved. When pricing sub-contract work careful consideration must be given to the special circumstances pertaining to the pricing of such work and the very different labour outputs that can be achieved by specialists.

In total the Major and Small Works sections of Laxton's contain over 250,000 price elements and constants. The comprehensive index will take the reader straight to any particular measured item.

SCHEDULES OF RATES

In the event that Laxton's is to be used as a Schedule of Rates the employer should at tender stage clearly indicate whether rates are to be adjusted for regional factors, tender values and for overheads and profit.

It is suggested that the employer should state that works are to be valued in accordance with Laxton's Building Price Book 2006 Major Works or Small Works or Alterations, Repairs and Maintenance sections as appropriate. Net Rate column with the addition of a percentage to be stated by the contractor to include for regional factor adjustment, tender value adjustment, overheads, profit and preliminary items not covered elsewhere. A procedure for adjustment of fluctuation should be stated (eg by using future editions of Laxton's or by the use of indices) and dates applicable.

Contractors must recognise that the net rates indicated within Laxton's works sections are national average costs for average projects in the ranges stated. Minor items of maintenance works may require considerable adjustment.

ESSENTIAL INFORMATION

PRICING AND FORMAT

Materials

The delivered to site price is given together with a percentage allowance for waste on site, where appropriate, the resultant materials cost being shown in the shaded column. The prices used are those included in the Basic Prices of Materials section.

Labour

Hours for craft operative and labourer (General Building Operative) are given, these being applied to the labour rates shown in the top left hand shaded panel, the resultant total labour cost being shown in the shaded column.

Sundries

These are incidental costs not included in the main materials or labour cost columns, Many items previously considered as sundries are now included with materials.

Mechanical Plant

Items of Mechanical Plant not included in Preliminaries/General conditions have been allowed for in the measured rates.

Plant and Transport costs are indicated separately in the measured rates for items where these are a major factor and material costs are insignificant as in the case of some groundwork items.

Rates

Net rates shown in the shaded column are exclusive of overheads and profit.

Gross rates are inclusive of 7.5% Main Contractor's overheads and profit in the Major Works section, 12.5% in the Small Works section and 20% in the Alterations, Repairs and Maintenance section. Readers can, of course, adjust the rates for any other percentages that they may wish to apply.

PERCENTAGE ADJUSTMENTS FOR TENDER VALUES

MAJOR WORKS

The measured rates in the Major Works section are based on new work contracts valued in the range of £500,000 to £1,500,000. As a guide to pricing works of larger value and for cost planning or budgetary purposes, the following adjustments may be applied to overall contract values

Contract value

£1,500,000 to £2,500,000..............................deduct 2.5%
£2,500,000 to £5,000,000..............................deduct 5.0%

SMALL WORKS

The measured rates in the Small Works section are based on new work contracts valued in the range of £50,000 to £100,000. As a guide to pricing works of lower or larger values and for cost planning or budgetary purposes, the following adjustments may be applied to overall contract values

Contract value

£ 15,000 to £ 25,000...add 10.0%
£ 25,000 to £ 55,000....................... ....add 5.0%
£ 50,000 to £100,000.................................rates as shown
£100,000 to £200,000.....................................deduct 5.0%
£200,000 to £300,000.....................................deduct 7.5%
£300,000 to £500,000...................................deduct 10.0%

ALTERATIONS, REPAIRS AND MAINTENANCE

The measured rates in the Alterations, Repairs and Maintenance Works section are based on order values in the range of £5,000 to £15,000. The rates may be used for orders of lower value however due allowances must be added for disproportionate waste factors, working in occupied premises and preliminary items.

RENOVATION WORKS AND WORK IN EXISTING UNOCCUPIED PREMISES

The rates in any of the various Works sections may be used as appropriate for this type of work however care must be taken to ensure adequate allowance is made for preparation works, difficulty of access and any double handling.

WORKING IN OCCUPIED PREMISES

Nearly all work of this nature is more costly in execution in that it has to be organised to suit the specific working conditions due to occupation and allowances should be made for the following:

(a) reduction in output arising there from.

(b) moving tradesmen on and off site as the progress of the contract demands.

(c) suppression of noise and dust.

As a guide, the extra labour involved in carrying out work in occupied premises could add between 50% and 100% to the labour cost of the same work in unoccupied premises.

REGIONAL VARIATIONS

Laxton's Building Price Book is based upon national average prices. The Regional Variations section and maps indicate factors that may be applied to adjust overall pricing to regional tender levels from data provided by the B.C.I.S.

USE OF FACTORS AND ADJUSTMENTS

Regional factors and tender value adjustments should only be applied to adjust overall pricing or total contract values. They should not be applied to individual rates or trades.

VALUE ADDED TAX

Prices in all sections exclude Value Added Tax.

CONTENTS

The greatest care has been taken to ensure accuracy but the publishers can accept no responsibility for errors or omission.

Textile Conservation Centre
University of Southampton

WANTED

Information on garments deliberately concealed in buildings

There is an international folk tradition, from at least the Middle Ages onwards, of concealing old garments and other objects in houses and other buildings as they were being constructed or altered. These deliberately concealed objects are usually thought to be tokens of good luck and to have a protective function. The objects are varied and include bottles, trade tokens, shoes and other types of garment.

There has been no consistent recording of such garment finds and little research into the traditions behind the practice. **The Textile Conservation Centre** is undertaking research that will investigate the extent and significance of this tradition.

Several concealed garments have been brought to the Textile Conservation Centre for advice and treatment. The most notable examples include a mediaeval hat found in a church buttress and the ragged remains of two doublets dated c.1600. The doublets are very rare examples of working class dress of the period: one is a coarse linen fibre doublet c.1600 found in a house in Reigate, Surrey and the other a wool doublet c.1620 found in an attic in Abingdon, Oxfordshire (which was found in association with an 18th century pocket, a 19th century baby's cap and various tokens and hops).

The building trade, through its work, is often familiar with the practice, either by finding historically concealed objects during work, or because the tradition of concealing items in the fabric of buildings continues.

The Textile Conservation Centre would like to hear from anyone who knows of this tradition or anyone who has discovered concealed items. We particularly want to record garment finds. Even if the finds are very ragged or very dirty, we'd still like to know about them. For more information please contact:

The Administrator, Deliberately Concealed Garments
The Textile Conservation Centre
University of Southampton
Winchester Campus
Park Avenue
Winchester SO23 8DL
Tel: 023 8059 7100
Fax: 023 8059 7101
Email: tccuk@soton.ac.uk

The Textile Conservation Centre is an internationally renowned Centre of Excellence for conservation education, research and practice.

This project is funded by the Arts and Humanities Research Board

INDEX

C (Cont'd)

D

F (Cont'd)

G

P (Cont'd)

R (Cont'd)

REGIONAL VARIATIONS

The maps of Regional Factors are adapted from the BCIS Quarterly Review of Building Prices and are current as at the second quarter of 2005. A fuller analysis giving county factors is given in the Review. The information gives an approximate guide to the levels of overall pricing for the various regions of the United Kingdom. Prices in this book are based on a national average (1.00) and by multiplying by the factors indicated, comparative prices for regions may be obtained.

Considerable variations may also occur as a result of specific local factors, in particular within inner city areas and the reader should make allowance for these, further detail may be obtained from BCIS.

Introduction

The cost of a building is affected by its location. Many localised variables combine to produce a unique cost, including market factors such as demand and supply of labour and materials, workload, taxation and grants. The physical characteristics of a particular site, its size, accessibility and topography also contribute. Not even identical buildings built at the same time but in different localities will obtain identical tenders.

While all these factors are particular to a time and place, certain areas of the country tend to have different tender levels than others. The location factors given on the map are an attempt to identify some of these general differences using information derived from the BCIS Tender Price Index. The regions chosen are administrative areas and are not significant cost boundaries as far as the building industry is concerned.

It should be stressed that even within counties or large conurbations great variations in tender levels are evident and that in many cases these will outweigh the effect of general regional factors.

Calculation of Regional/County Factors

The location factors are based upon statistical analysis of the projects included in the BCIS All-in Tender Price Index.

Regional trends are calculated to show the differential movement between each region's tender levels and the national trend.

The individual job factors have been grouped into regions and counties. The Regional/County Factors have been calculated from the mean of the groups. The factors have been subject to statistical analysis that adjusts the individual job factors to reflect the regional trend.

Regions

The information has been analysed by county in England and Wales and by region in Scotland with the exception that Greater London has been further sub-divided into Outer London and London Postal Districts. These locations have been grouped into larger regions to give greater statistical reliability. The larger regions used are the Department of the Environment's Standard Statistical Regions except for the South East Region for which smaller groups are used, the South West Regions which has been divided, The Channel Islands, The Isle of Man, The Shetland Islands and Northern Ireland added.

The regions are defined as follows:

ENGLAND

Northern:
Cleveland, Cumbria, Durham, Northumberland, Tyne and Wear.

Yorkshire and Humberside:
Humberside, North Yorkshire, South Yorkshire, West Yorkshire.

East Midlands:
Derbyshire, Leicestershire, Lincolnshire, Northamptonshire, Nottinghamshire.

East Anglia:
Cambridgeshire, Norfolk, Suffolk.

South East:
Bedfordshire, Essex, Hertfordshire, Kent, Surrey, East Sussex and West Sussex, Berkshire, Buckinghamshire, Hampshire, Isle of Wight, Oxfordshire.

Greater London:
The area of the former Greater London Council.

South West:
Avon, Dorset, Gloucestershire, Wiltshire.
Cornwall, Devon Somerset.

West Midlands:
Hereford and Worcester, Shropshire, Staffordshire, Warwickshire, West Midlands.

North West:
Cheshire, Greater Manchester, Lancashire, Merseyside.

WALES
SCOTLAND
NORTHERN IRELAND
THE CHANNEL ISLANDS
THE ISLE OF MAN
THE SHETLAND ISLANDS

Note:
The Regional/County Factors have been shown to represent fairly reliable average differences between locations. They are however averages and any individual project is unlikely to coincide exactly with the average result: the factors provide useful general guidance but on no account should they be used to adjust individual item prices.

This information is abstracted from a more detailed study of location factors which is included in the BCIS Quarterly Review of Building Prices available from BCIS, 12, Great George Street, Parliament Square, London, SW1P 3AD. www.bcis.co.uk

REGIONAL FACTORS MAP

Channel Islands 1.43
Isle of Man 1.32

Laxton's Building Price Book is based upon national average prices (=1.00). Indicative levels of pricing in regions at the second quarter 2005 are given on this map and the factors shown may be applied to adjust overall pricing.

MAJOR WORKS - TENDER VALUES

The measured rates are based on contracts valued in the range of £500,000 to £1,500,000.
As a guide to pricing works of larger value and for cost planning or budgetary purposes, the following adjustments may be made to overall project values:

Project value:
£1,500,000 to £2,500,000 ... deduct 2.5%
£2,500,000 to £5,000,000 ... deduct 5%

SMALL WORKS - TENDER VALUES

The measured rates are based on contracts valued in the range of £50,000 to £100,000.
As a guide to pricing works of smaller or larger value and for cost planning or budgetary purposes, the following adjustments may be made to overall project values:

Project value:
£15,000 to £25,000........... add 10 %
£25,000 to £50,000........... add 5%
£50,000 to £100,000......... rates as shown
£100,000 to £200,000....... deduct 5%
£200,000 to £500,000....... deduct 7.5%

ALTERATIONS/REPAIRS/MAINTENANCE

The measured rates are based on contracts valued in the range of £5,000 to £15,000.
The rates may be used for orders of lower value however due allowances must be added for disproportionate waste factors, working in occupied premises and preliminary items.

PRELIMINARIES/GENERAL CONDITIONS

GENERALLY

The prices throughout this section offer in a convenient form the means of arriving at approximate rates for the various operations commonly met in connection with normal types of buildings. The basis on which the rates are estimated is given below and should be examined and understood before making any adjustments necessary to adapt the prices to a specific contract or to rates of wages other than those given.

BASIC RATES OF WAGES – BUILDING WORKERS ALL-IN LABOUR RATES

Effective Dates of rates used:
Wages..June 27th, 2005
National Insurances..April 6th, 2005
Holidays with Pay (Template)......................... June 30th, 2004
EasyBuild Stakeholder Pension........................... June 30th, 2004
For earlier effective dates of the above see Standard Rates of Wages Section

The basic rates of wages used throughout are those which came into force on June 27th 2005.

BUILDING

Calculation of labour rates used throughout the book:-

		Craft Operative £		General Building Operative £
Guaranteed minimum weekly earnings		351.00		264.03
- 39 hours		9.00		6.77
2,023 hours (includes inclement weather allowance)	at £9.00	18,207.00	at £6.77	13,695.71
Productivity Payments		1,820.70		1,369.57
Non-Productive Overtime 131 hours	at £9.00	1,179.00	at £6.77	886.87
Sick Pay as WR.20 (one week).		93.60		93.60
		21,300.30		16,045.75
National Insurance (Earnings threshold 46.2 weeks at £94.00)....12.8%	of £16,957.50	2,170.56	of £11,702.95	1,497.98
Holidays with Pay (Template) - 226 hours	at £9.00	2,034.00	at £6.77	1,530.02
Training 0.50% of Payroll (CITB Levy)	of £23,333.30	116.67	of £17,575.77	87.88
Retirement Benefit - 52 weeks	at £10.90	566.80	at £10.90	566.88
		26,188.33		19,728.43
Severance Pay 1.5% (Including loss of production during notice period absenteeism and turnover of labour)		392.82		295.93
		26,581.16		20,024.35
Employers Liability and Third Party Insurance .	2.5%	664.54	2.5%	500.61
Trade Supervision	3%	797.43	3%	600.73
Cost per annum		28,043.12		21,125.69
Cost per hour - 1,973 working hours		**£ 14.21**		**£ 10.71**

Calculation of other skill rates as above would be as follows:-

Skilled Operative Rate:-	Skill Rate 1.	£13.55
	Skill Rate 2.	£13.05
	Skill Rate 3.	£12.22
	Skill Rate 4.	£11.52

Note: Public Holidays now included with Holidays with Pay

BASIC RATES OF WAGES (Cont'd)

Calculation of Hours Worked used above.

			Hours
Summer: based on average 45 hours per week working			
	40 weeks		1800
Less Holidays:			
Summer	2 weeks		(90)
Easter (includes Easter Monday)	1 week		(45)
	37 weeks	at 45 hours	1665
Less Bank Holidays (Good Friday, May, Spring and Summer Bank holidays)	4 days at 8 hours.		(32)
			1633

		Hours
Winter: based on average 43 hours per week working		
	12 weeks	516
Less Holidays: Winter(includes Christmas, Boxing and New Years days)	2 weeks	(86)
	10 weeks at 43 hours	430
Less Sick leave	5 days at 8 hours	(40)
		390
Add Summer Hours		1633
Total Hours		2023
(Less Inclement Weather time)		50
Total Actual Hours Worked		1973

Calculation of Non-Productive Overtime hours included above

Based on a 45 hour working week in Summer and a 43 hour working week in Winter, the calculation is as follows:-

				Hours
Summer	6 hours overtime per week at time and a half	=	3 hours x 37 weeks	111
Winter	4 hours overtime per week at time and a half	=	2 hours x 10 weeks	20
Non-Productive hours per annum				131

Holidays with Pay as per W.R. 18

Bank Holidays	63 hours	(8 days)
Winter, Spring, and Summer Holidays	163 hours	(21 days)
	226 hours	(29 days)

PLUMBING

The rates of wages used in the Plumbing and Mechanical Engineering Installations Section are those approved by The Joint Industry Board for Plumbing and Mechanical Engineering Services in England and Wales.

The All-in Labour rate calculations are as follows:

Effective dates of rates used:

Wages .... January 3rd 2005
National Insurance .... April 6th 2005
Holiday and Welfare contributions...............August 23rd 2004

		Trained Plumber £		Advanced Plumber £		Technical Plumber £
46 weeks at 37.5 hours = 1725 hours	at £9.28	16,008.00	at £10.82	18,664.50	at £12.02	20,734.50
Welding Supplement 1725 hours (Gas or Arc)		-	at £0.27	465.75	at £0.27	465.75
Travel time		-		-		-
		16.008.00		19,130.25		21,200.25
Allowance for Incentive Pay	15%	2,401.20	15%	2,869.54	15%	3,180.04
	(a)	18,409.20		21,999.79		24,380.29
National Insurance 12.8% of	14,085.20	1,802.91	17,765.79	2,262.50	20,056.29	2,567.20
(Earnings Threshold 46.2 weeks at £89.01 = £4112.26)						
less rebate on earnings LEL to UEL 3.50%	14,085.20	-492.98	17,765.79	-618.65	20,056.29	-701.97
Pension Contribution of (a)...	7.50%	1,380.69	7.50%	1,649.98	7.50%	1,828.52
Holiday and Welfare Benefit stamps 52 weeks	at £36.10	1,877.20	at £42.10	2,189.20	at £46.80	2,433.60
		22,977.01		27,482.82		30,507.64
Redundancy Payments	1.5%	344.66	1.50%	412.24	1.50%	457.61
Employer's Liability Insurance	2.5%	574.43	2.50%	687.07	2.50%	762.69
Fares say 4720 miles @ 30p		1,416.00		1,416.00		1,416.00
Cost per annum		25,312.09		29,998.13		33,143.95
Cost per hour – 1725 hours		14.67		17.39		19.21
Inclement Weather time	1%	0.15	1%	0.17	1%	0.19
		£ 14.82		**£ 17.56**		**£ 19.41**

The Plumber's all-in wage rate used throughout the book is an average of 1 Trained, 3 Advanced and 1 Technical Plumbers rates giving an average rate of **£17.38** per hour.

Note :
1. Total Weekly Stamp Value is the sum of Holiday Contribution and Welfare Credit.
2. Travel time – Discontinued for distances up to 20 miles
3. Tool Allowance – Discontinued

BASIC RATES OF WAGES (Cont'd)

ELECTRICAL

The rates of wages used in the Electrical Engineering Installation Section are those approved by the Joint Industry Board for the Electrical Contracting Industry. www.jib.org.uk

The All-in Labour rate calculations using Job reporting operatives with own transport is as follows:

Effective dates of rates used:

Wages....................................... January 10th 2005
National Insurance.................... April 6th 2005

				APPROVED + 50p		APPROVED		ELECT		TECH.	
	days	weeks	hours	rate		rate		rate		rate	
HOURS WORKED (LESS 1 WK SICK)		45.20	45.00	11.96	24326.64	11.46	23309.64	10.57	21499.38	12.91	26258.94
STATUTORY SICK PAY (NOT PAYABLE FOR FIRST THREE DAYS now includes bereavement leave)	2.00		68.20	13.64	27.28	13.64	27.28	13.64	27.28	13.64	27.28
NPOT		45.20	3.50	11.96	1892.07	11.46	1812.97	10.57	1672.17	12.91	2042.36
INCENTIVE PAYMENT	15.00%			15.00%	3649.00	15.00%	3496.45	15.00%	3224.91	15.00%	3938.84
TRAVEL TIME– up to 15 miles - Nil											
TRAVEL ALLOWANCE. – up to 15 miles - Nil											
STAT HOLS	8.00		7.50	11.96	717.60	11.46	687.60	10.57	634.20	12.91	774.60
		sub total A			30612.59		29333.94		27057.94		33042.02
NATIONAL INS				12.80%	3918.41	12.80%	3754.74	12.80%	3463.42	12.80%	4229.38
(Lower earnings allowance)		45.20		£94.00	-543.85	£94.00	-543.85	£94.00	-543.85	£94.00	-543.85
		sub total B			33987.15		32544.84		29977.51		36727.56
TRAINING	2.50%of B			2.50%	765.31	2.50%	733.35	2.50%	676.45	2.50%	826.05
TRAVEL ALLOWANCE– up to 15 miles - Nil. (TA TAX & NI EXEMPT ELEMENT)					0.00		0.00		0.00		0.00
JIB STAMP		52.00		44.66	2322.32	44.66	2322.32	41.93	2180.36	49.10	2553.20
			sub total		37074.79		35600.50		32834.32		40106.81
SEVERANCE PAY	2.00%			2.00%	741.50	2.00%	712.01	2.00%	656.69	2.00%	802.14
			sub total		37816.28		36312.51		33491.01		40908.94
EMP. LIAB & 3rd PARTY				2.50%	945.41	2.50%	907.81	2.50%	837.28	2.50%	1022.72
COST PER YEAR					38761.69		37220.33		34328.28		41931.67

GANG RATE						INDIVIDUAL RATES
TECH.	41931.67	x	1.00	NON PRODUCTIVE	41931.67	£20.62
APPROVED ELEC. +	38761.69	x	1.00	100%PRODUCTIVE	38761.69	£19.06
APPROVED ELEC.	37220.33	x	2.00	100%PRODUCTIVE	74440.65	£18.30
ELEC.	34328.28	x	4.00	100%PRODUCTIVE	137313.12	£16.88
					292447.13	

AVERAGE MAN HOURS	45.20	x	45.00	2034	hrs
AVERAGE COST PER MAN			7.00	Men working	**£20.54**

BASIC RATES OF WAGES (Cont'd)

ELECTRICAL (Cont'd)

Note:

Based on National JIB rates

a Basic week= 37.5hrs

b Hours to be worked before overtime paid= 38

c The above JIB hourly rates are from the 5/1/2004.

d Hours remain at 45 as this seems to be the minimum paid in the industry at the moment

The Electricians all in wage rate used throughout the book is an average 1 Approved Electrician +, 2 Approved Electricians and 4 Electricians rates giving an average rate of **£20.54** per hour.

JIB Categories for Operatives Hourly Pay

All-in wage costs

1. Shop Reporting
   - payable to an operative who is required to start and finish at the shop:
     - Technician £18.46
     - Approved Electrician Plus £16.94
     - Approved Electrician £16.18
     - Electrician £14.76
   - Average Cost per Man £18.11b (Technician Non Productive)

2. Job Reporting (transport provided but not included within these calculations)
   - payable to an operative who is required to start and finish at normal starting and finishing times on jobs, travelling in their own time in transport provided by the employer. The operative shall also be entitled to payment for Travel Time, where eligible as detailed in the appropriate scale:
     - Technician £19.68
     - Approved Electrician Plus £18.12
     - Approved Electrician £17.36
     - Electrician £15.92
   - Average Cost per Man £19.46 (Technician Non Productive)

3. Job Reporting - (own transport) as used in the example above
   - payable to an operative who is required to start and finish at normal starting and finishing times on jobs, travelling in their own time by their own means; the operative shall be entitled to payment for travel time and travel allowance where eligible, as detailed in the appropriate scale.

SITE PRELIMINARIES

GENERAL NOTES

The calculation of the costs to be included for Site Preliminaries requires careful consideration of the specific conditions relating to the particular project. No pre-conceived percentage addition can be adopted with any degree of accuracy. The cost of Site Preliminaries is dependent on the type of contract under consideration, its value, its planned programme time, its location and also on the anticipated requirements for the administration and management of the actual site.

Accurate assessment of the contractor's site costs are equally important in small works and major works contracts.

Two factors which must be established prior to commencement of the detailed calculations are (i) the value of the main contractor's work (from which is derived the value of the contractor's labour force) and (ii) the programme time for completion of the Works. If the latter is not stated in the contract conditions, it must be decided by the contractor. If it is stated in the contract conditions, it must be checked and confirmed by the contractor as being reasonable.

For works of Alterations, Repairs and Maintenance examination of the Notes and Specimen Preliminaries given for Small and Major works will enable the user to build up an appropriate addition for this type of work.

NOTES ON SITE PRELIMINARIES

Section A40 - Management and Staff

The numbers, types and grades of Site Staff to be included are matters for judgement by the Contractor.

A selection from the following list will provide the range of staff required to administer a normal building project:

Site Agent or Manager	General Foreman
Engineer	Checker
Storeman	Cashier/Wages Clerk
Materials Purchasing Clerk	Plant Clerk
Quantity Surveyor	Resources Manager
Productivity Controller	Safety Officer
Plant Fitter	Administrative Manager
Typist/Telephonist	

Allowance for Trades Supervision has been made in the Gross Wage build up.

On small contracts, individual staff members may combine the duties of several of the foregoing grades. On larger contracts more than one member may be required for certain duties. For example, two or more Engineers may be required in the initial stages for setting out the Works and for checking the frame construction.

Since 2nd September 1996 The Construction (Health and Welfare) Regulations 1996 apply to the health, safety and welfare of workers on construction sites - the associated costs of which are not specifically identified in this section and acquaintance with these regulations is necessary.

Section A41-1 - Site Accommodation

The size of the office required to accommodate the contractor's site staff will depend on the number of staff envisaged. An allowance of 10m2 per staff member plus 25% for circulation space and toilets is normal.

The overall cost will vary according to the contract period if the hutting is to be hired. Otherwise a use and waste allowance can be made if the hutting is supplied from the contractors own resources. The cost should include for erection and eventual dismantling and, in the case of contracts extending over several years, for labour and materials in maintenance and redecoration and any ancillary accommodation which may be required for the Architect, Clerk of Works or Resident Engineer.

In works of renovation and rehabilitation, office accommodation and space for storage of materials is often available in the existing premises. Otherwise a simple lock up hut is all that is normally required for small works contracts.

Allowance should be made for heating, lighting, furniture and any attendance required. Accommodation priced in this Section could include Offices, Stores, Site Latrines, Canteen, Drying Rooms and Sanitary Facilities, First Aid Cabin, Plant Fitters Sheds, Reinforcement Bending Sheds.

Section A42-2 - Lighting

This item provides for any lighting required inside the buildings during construction or for external floodlighting installed during the winter months to maintain progress. To the cost of wiring, lights and other equipment, should be added the charge for electricity and electrician's time for the installation and removal of the system.

On small works contracts the contractor, by agreement with the client, may be permitted to use existing services. A nominal amount should be included for the use of electricity.

Section A42-4 - Water

Water Company charges

Charges made by various Water Companies for the provision of water for the works differ considerably throughout the country and the actual rate to be paid should be ascertained regionally. Some water companies meter the water required for building purposes. Charges to be considered include:

Water infrastructure charges are £245 for each new domestic property with their own individual supply.

Sewerage infrastructure charges are £245 for each new property

FIRE HYDRANTS

- Installation of new hydrant to existing main in footpath, verge or tarmac road (excluding marker post and plate) £630
- Installation of new hydrant on run of new main (excluding marker post and plate) £340
- Installation of new hydrant at the end of a new main (excluding marker post and plate) £224

BUILDING WATER SUPPLY AND WATER USED DURING CONSTRUCTION

Option 1
The service connection is metered. (£18.50 per year for a 25mm domestic meters) and £0.6004 pence per cubic metre of water used.

Option 2
One off payment for building water
£40 for each house or bungalow.
£20 for each flat or maisonette.
£60 for each small commercial unit (i.e. pub, prefabricated unit etc.)
£130 for each large commercial unit (i.e. hotel, school, supermarket, shops, leisure centre etc.)
Complex developments must be metered (shopping malls, stadiums, factories, etc.)

TEMPORARY CONNECTIONS for site offices, toilets or canteen.

Connected to the building water supply and therefore a separate connection is not required, but in the event that the supply is not metered an additional charge of £2 per property (minimum £30) will be levied to cover the cost of water used.

NOTES ON SITE PRELIMINARIES (Cont'd)
TEMPORARY CONNECTIONS (Cont'd)

Separate connections can sometimes be required across the site during construction and can be charged in two ways.

Option 1
The connection is metered. (£18.50 per year for a 25mm domestic meter) and £0.6004 pence per cubic metre of water used.

Option 2
One off payment
£2 per property (minimum £30

NEW ON-SITE CONNECTIONS (connected to a new main recently laid to serve the development)

The costs for services assume that service ducts are installed in the road. On a site where contamination has been identified, add 10%. For site connections greater than 25mm, quotations will be required

For a new 25mm diameter supply on a greenfield site – no ground contamination

Excavation and backfilling of trench included	Cost per premise	Cost per premise where 6 or more connections to the main can be installed in a single visit to site
1 connection to the water main serving 1 premise	£264	£259
1 connection to the water main serving two premises via a manifold	£213	£207
1 connection to the water main serving 3 or 4 premises via a manifold	£249	£243
1 connection to the water main serving 5 or 6 premises via a manifold	£228	£223

Excavation and backfilling of trench excluded	Cost per premise	Cost per premise where 6 or more connections to the main can be installed in a single visit to site
1 connection to the water main serving 1 premise	£239	£232
1 connection to the water main serving two premises via a manifold	£190	£186
1 connection to the water main serving 3 or 4 premises via a manifold	£227	£223
1 connection to the water main serving 5 or 6 premises via a manifold	£207	£202

For a new 25mm diameter supply to an 'in/on the wall' mounted box

Installation to in/on wall box	Cost per premise	Cost per premise where 6 or more connections to the main can be installed in a single visit to site
connection to the water main (box not included)	£196	£183

NEW AD-HOC CONNECTIONS (Where a new service is connected to an existing main situated in the public highway).

The cost of an ad-hoc connection is determined by the type of road or footway surface and any traffic management scheme and calculated on an individual basis

ON COSTS

To these costs must be added the cost of all temporary plumbing required on the site such as trenching and builder's work in connection, distribution piping, standpipes, hoses and tanks, alterations, repairs and removals.

Any reinstatement costs would also need to be added.

Section A42-6 - Safety, Health and Welfare

Protective helmets, boots, goggles and protective clothing for operatives; wet weather clothing, drying sheds; provision of site latrines and items of a similar nature.

Section A42-7 - Storage of Materials

Consideration should be given to the provision of the following items:

Cement storage shed.
Aggregate storage bins.
Use and waste of tarpaulins and plastic sheeting.
Lockups and toolboxes for tradesmens' tools.

Section A42-8 - Rubbish disposal

Removal of rubbish during construction including clearing out after subcontract trades.

NOTES ON SITE PRELIMINARIES (Cont'd)

Section A42-9 - Cleaning

This Section includes keeping the site tidy throughout the construction period, clearing out after subcontract trades and the final cleaning up on completion of the Works prior to handover.

Section A42-10 - Drying Out

Allowance should be made in this Section for drying out the building by the use of blow heaters or, if permitted, by the provision of fuel to the heating system as installed.

Section A42-11 - Protection of Work

The protection of completed work during the contract period is usually priced in the relevant bill rates. Types of work requiring special protection include expensive stonework or precast concrete features, stair treads, floor finishes, wall linings, high class joinery and glazing to windows.

Section A42-12 - Security

Security of Site - Allow for the provision of a watchman if required, or for periodic visits by a security organization; electronic surveillance, and protection of scaffolds. Some allowance may also be required in particular localities for loss of materials due to theft.

Section A42-14 - Small plant and tools

Small plant and tools comprise ladders, wheelbarrows, trestles, picks, shovels, spades, sledgehammers, cold chisels and similar items. The cost of these minor items is usually allowed for by an addition to the overall cost of labour of between 1% and 2%.

Section A42-15 - Others

a) Insurance of Works

The Contract Bills of Quantities will normally state whether Fire Insurance is the responsibility of the Contractor or of the Employer. Rates of insurance vary according to the nature of the work but are usually in the range £0.10% to £0.30% of the total value to be insured. The total insurable value should include for full reinstatement of any damage caused, all cost escalation, any demolition of damaged work required and an appropriate percentage addition for professional fees.

Note: allowance should be made for Terrorism insurance if required.

b) Travelling Time and Fares

Payments for travelling time and fares can be a major expense and since all contracts, and particularly the JCT Conditions of Contract, place on the Contractor the onus of including in the tender sum the cost of importing labour, investigation should be made into the prevailing conditions regarding the availability of local labour. When all enquiries have been made, the calculations of anticipated expenditure must be based on the Contractor's judgement of the labour position. Rates of payment for travelling time and fares are laid down in the relevant national and local Working Rule Agreements.

c) Miscellaneous

(1) Rates and Taxes on Temporary Buildings. The charges made by the Local Authority for rates on site accommodation should be ascertained locally.

(2) Signboards and notices.

(3) Glass breakage during unloading or after fixing.

(4) Sample panels.

(5) Attendance on site - provision of chainmen and any special attendance required by the Clerk of Works.

(6) Setting out equipment including the use of theodolites, levels or other items and provision of pegs, paint etc.

(7) Provision for winter working. This can range from heating aggregates in the stockpiles to the provision of polythene sheeting to protect the operatives and/or the Works during construction.

(8) Allowance for the cost of labour and materials in rectifying minor defects prior to handover of the Works and during the Defects Liability Period.

(9) Sureties or bonds. The cost of any surety or performance bond for due completion of the Works required by the Contract Documents. Rates for this vary depending on the financial status of the Contractor.

(10) Special Conditions of Contract. These can refer to any particular point required by the Employer, and a careful perusal of the Conditions and the Preambles to the Bills of Quantities or Specification is necessary to ensure that no requirement is overlooked. Typical items are: the suppression of noisy mechanical equipment; regulations dealing with the restricted use of tower cranes; or limitations to ground disturbance during piling operations.

Section A42-16 - General Attendance on Nominated Subcontractors

General Attendance on Nominated Subcontractors by the Main Contractor is deemed to include: the use of the Contractor's temporary roads, pavings and paths; standing scaffolding not required to be altered or retained; standing power-operated hoisting plant; the provision of temporary lighting and water supplies; clearing away rubbish; provision of space for the subcontractor's own offices; the storage of his plant and materials; and the use of messrooms, sanitary accommodation and welfare facilities provided by the Contractor for his own use.

The estimated cost of such attendance is generally added by the Contractor in the Bills of Quantities as a percentage of the Prime Cost Sum shown for each Nominated Subcontractor.

Alternatively it can be included as a lump sum according to the Contractor's experience and judgement. The actual percentage can vary from as little as 0.1%, up to 2.5% or more, depending on the anticipated or known requirements of the Nominated Subcontractor.

Section A43 - Mechanical Plant

This section comprises the cost of all mechanical plant which has not been priced in the Bills of Quantities. It includes Tower Cranes, Mobile Cranes, Goods and Passenger hoists and Pumping plant for dewatering.

The apportioning of the cost of tower cranes and hoists to separate items of the Bills of Quantities is made difficult and unreliable by the fact that very often they deal with various materials in differing trades. As such, the correct time allocation to the individual trades is virtually impossible. Moreover, although not always working to capacity, they cannot be removed from site entirely and a certain proportion of their time is wasted in standing idle. It is therefore normal practice to view the contract as a whole, decide the overall period for which the particular tower crane or hoist will be required, and include the whole cost in the Preliminaries.

NOTES ON SITE PRELIMINARIES (Cont'd)

Section A43 - Mechanical Plant

Tower Cranes - there is a large variety of cranes available in the market and the most suitable crane for each particular contract must be selected. The transport of the crane from the plant yard to the site can be a costly item, and the installation of track, the erection, testing and final dismantling of the crane can, on occasion, constitute a greater expenditure than the hire charge for the crane.

Mobile Tower Crane - A typical build up of cost of mobile tower crane is as follows:

Installation		£
Prepare ground and lay crane track		2,500
Supply of track materials		2,000
Transport erect & dismantle		9,500
		14,000
Running Costs per week	£	
Crane hire & driver	1,900	
Running costs, electricity	110	
Banksman	500	
	2,510	
x 36 weeks		90,360
		£104,360

Static Tower Crane - A typical build up of cost of small static tower crane is as follows:

Installation		£
Transport, erect & dismantle		7,000
Tying-in materials use and waste		600
		7,600
Running Costs per week	£	
Crane hire & driver	1600	
Running costs, electricity	50	
Banksman	500	
	2,150	
x 15 weeks		32,250
		£39,850

Mobile Cranes - A mobile crane may be required during the frame construction for lifting heavy items such as precast concrete floors or cladding units. The weekly hire rates will vary according to the lifting capacity of the crane required but a typical build up of cost is as follows:

	£
Crane hire & driver	1,400
Banksman	500
	1,900
x 4 weeks	7,600
Transporting to and from site	350
	£7,950

Pumping of ground water - Provision for the pumping of ground water (as opposed to water from normal rainfall) is entirely dependent upon the conditions applying to the particular site in question. Under the JCT Standard Form of Contract this is usually covered by a Provisional or Prime Cost sum, but in contracts where the onus is placed on the contractor it is most important that a visit should be paid during the tendering period to the site of the proposed works, to ascertain as far as possible the nature of the ground and the likelihood of encountering water. When, as a result of such an inspection, it is decided that pumping will be necessary, say during excavation of foundations and until the foundation walls have reached a certain height, the calculations will be based upon the length of time it is thought necessary to keep pumps working. A typical build up of cost is as follows:

Assuming one 75mm pump will be kept running continuously for five weeks and for three weeks a second 75mm pump will be required during working hours with a standby pump of 50mm diameter available for emergencies:

		£
(a)	Hire of 75mm pump & hose 5 weeks at £99.00....	495.00
	Operator's time attending, fuelling, etc. –	
	28 hours at £11.52 per hour = £322.56 x 5 weeks	1,612.00
	Hire of extra hose 5 weeks at £14.00..................	70.00
(b)	Hire of 75mm pump & hose 3 weeks at £99.00....	297.00
	Operator's time attending, fuelling, etc. –	
	12 hours at £11.52 per hour = £138.24 x 3 weeks	414.72
	Hire of extra hose 3 weeks at £14.00..................	42.00
(c)	Standby pump 5 weeks at £66.00........................	330.00
	Total cost of pumping	£3,261.52

Section A43-4 - Site Transport

This section includes all requisite site transport not priced in the rates, such as site lorries, dumpers, vans, forklift trucks, tractors and trailers. The cost of transporting mechanical plant, scaffolding and temporary accommodation, is also included in this section. The cost of double handling materials during unloading may also be considered for inclusion in this section.

Section A44-1 - Temporary Roads

This section includes providing access to the site for heavy excavating equipment and for site roads and hardstandings constructed in ash, hardcore or timber sleepers. Allowance should be made for maintenance over the contract period and for eventual breaking up and removal. Crossovers, planked footways and any automatic stop/go lights required should be included. If the base courses of permanent roads can be utilised as temporary roads, an allowance should be made for making good the base prior to final surfacing.

Allow also for any cleaning and maintenance which may be required on Public or Private roads, especially if surplus excavated material has to be carted off site.

Section A44-3 - Access scaffolding

The cost of scaffolding varies not only with the superficial area of the structure but also with the length of time the scaffolding is required to stand and the nature of the building. When calculating the cost, the girth of the building should be multiplied by the height and the area priced at the appropriate rate as shown below, to which should be added the additional charge for standing time exceeding the basic 12 weeks as shown. Provision must also be made for any internal scaffolding required to lift and stair wells and also for fixing suspended ceilings. Debris netting along with boards and ladder access are included with the rates.

Erected

	Small Works		Major Works	
	Putlog	Independent	Putlog	Independent
	£	£	£	£
Scaffolding for brickwork average 6m high m2	8.00	12.50	7.30	10.20
Scaffolding for brickwork average 12m high m2	9.20	13.00	7.60	10.60
Scaffolding for brickwork average 18m high m2	9.40	13.60	7.80	11.10
Add for each additional week..............................m2	0.50	0.55	0.40	0.45

NOTES ON SITE PRELIMINARIES (Cont'd)

Hire Charges

	Small Works				Major Works			
Height	2.20m	4.20m	6.20m	8.20m	2.20m	4.20m	6.20m	8.20m
	£	£	£	£	£	£	£	£
Mobile towers for one week	88.00	124.00	158.00	190.00	79.20	111.60	142.20	171.00

Section A44-4 - Support Scaffolding

Hire Charges

		Small Works	Major Works
		£	£
Scaffolding for suspended ceilings for one week	m2	7.00	6.30
Lift shaft tower to 20m high for one week		96.00	86.40

Section A44-5 - Hoardings, fans and fencing

The nature of these items is so dependent upon the site, type of contract, and the frontage of the building area on the public highway that it is difficult to offer any general method of calculating the anticipated cost. Consideration must be given to the means of stabilizing hoardings on sites where basements or other deep excavations come right up to the line of the hoarding, and gates must be sited to give easy access from the road. Gantries may require artificial lighting throughout their length in certain conditions, particularly in narrow streets in city centres. Any police requirements for offloading materials must be ascertained. Allowance must also be made for giving notices and paying any licence fees required by Local Authorities.

Hire Charges

		Small Works	Major Works
		£	£
Mesh fence 2m high per week – minimum 13 weeks	m	1.50	1.30

SPECIMEN SITE PRELIMINARIES

The examples of the build up of Site Preliminaries which follow are based on building contract values of £75,000 and £1,200,000 (excluding Preliminaries) for Small Works and Major Works respectively made up as follows:

	Small Works	Major Works
	£	£
Nominated Subcontractors	-	300,000
Nominated Suppliers	-	75,000
Provisional Sums...	10,000	75,000
Main Contractors Labour	25,000	300,000
Main Contractors Materials	20,000	300,000
Subcontractors	20,000	150,000
	£75,000	£1,200,0

The Contract Periods have been programmed as 24 weeks (Small Works) and 48 weeks (Major Works).
The Contract Document is the Standard Form of Building Contract 1999 Edition.

Section A40 - Management and Staff

	Small Works			Major Works		
	Per Week £	No. of Weeks	£	Per Week £	No. of Weeks	£
Project Manager/Site Agent	-	-	-	750	24	18,000
General Foreman	600 x 35%	24	5,040	600	36	21,600
Site Engineer	-	-	-	700	4	2,800
Wages Clerk/Productivity Controller	-	-	-	420	24	10,080
Typist/Telephonist..	-	-	-	320	24	7,680
		Carried to Summary	£5,040			£60,160.00

SPECIMEN SITE PRELIMINARIES (Cont'd)

Section A41 - Site Accommodation

	Small Works	Major Works
	£	£
Contractors Office: 50m^2 at £26.00 per m^2	-	1,300
Office furniture and equipment	225	425
Clerk of Works Office: 15m^2 at £26.00 per m^2	-	390
Stores: 30m2 at £16.00 per m^2	-	480
Canteen/Drying Room: 50m^2 at £26.00 per m^2	-	1,300
Canteen equipment	-	365
Cabins or site huts	-	225
Lockups and toolboxes	-	130
Office: Use of existing accommodation. Allow for making good on completion	150	-
Stores: Use existing. Allow for shelving and lockups for materials storage	225	-
Carried to Summary	£600	£4,615

Section A42 - 1 and 2 - Power and Lighting

	Small Works		Major Works	
	£	£	£	£
Connection to mains	-		225	
Distribution unit, cables, floodlights, lamps, plugs and sundry materials	-		1,250	
Electrician in attendance	-		1,350	
Cost of electricity: 48 weeks at £21 per week	-		1,008	
Heating and lighting offices: 48 weeks at £18.00 per week	-		864	
Power and lighting: 24 weeks at £13.00 per week	312	312	-	4,727

Section A42 - 4 - Water

	Small Works		Major Works	
Site Water connection	40		500	
Water Company charges	-		130	
Service piping: 100m at £6.00 per m	-		600	
Stand pipes: 3 no. at £30	-		90	
Hoses tanks and sundry materials	100	140	200	1,520

Section A42 - 5 - Telephone and Administration

	Small Works		Major Works	
Installation	-		180	
Rental	-		225	
Calls: 48 weeks at £25 per week	-		1,200	
Telephones: 24 weeks at £20.00 per week	480	480	-	1,605

Section A42 - 6 - Safety, Health and Welfare

	Small Works		Major Works	
Site Latrines: 2 no. at £275.00	-		550	
Latrine materials: 48 weeks at £4.00 per week	-		192	
Latrine materials 24 weeks at £2.00 per week	48		-	
Sewer connection	-		250	
Drainage; installation and removal	-		600	
First Aid	-		200	
Protective clothing	120		750	
CDM	100	268	500	3042

Section A42 – 8 and 9 - Rubbish Disposal and Cleaning Site

	Small Works		Major Works	
Site clearance during construction: 24 weeks at average £30 per week	720		-	
Site clearance during construction: 48 weeks at average £35 per week	-		1,680	
Final cleaning	200	920	400	2,080

Section A42 - 10 - Drying Out

	Small Works		Major Works	
Heaters and Fuel	100	100	500	500

Section A42 - 12 - Security

	Small Works		Major Works	
Weekend site supervision: 48 weeks at average £30 per week	-		1,440	
Security: allow for loss of materials	200	200	-	14,914
Carried Forward		£2,420		£14,914

SPECIMEN SITE PRELIMINARIES (Cont'd)

		Small Works			Major Works	
		£	£		£	£
Brought Forward			£2,420			£14,914
Section A42 – 14 - Small Plant and Tools						
Labour value £300,000 x 1.5%		-			4,500	
Labour value £ 25,000 x 1.5%		375	375		-	4,500
Section A42 - 15 - Insurance of works						
Contract Sum say		75,000			1,300,000	
Allow for part demolition and clearing up		1,600			25,000	
		76,600			1,325,000	
Increased costs over contract period: say 3%		2,298			-	
Increased costs over contract period: say 6%		-			79,500	
		78,898			1,404,500	
Increased costs over reconstruction period: say 3%		2,367			-	
Increased costs over reconstruction period: say 6%		-			84,270	
		81,265			1,488,770	
Professional fees: 15%		12,190			223,316	
		£ 93,455			£ 1,712,086	
Cost of Insurance: £0.30%			280	£0.20%		3,424
Section A42 - 15 - Travelling Time and Fares						
Main Contractor's Labour	300,000					
Labour in Preliminaries........	40,000					
	£340,000					
- Average £500 per week = 680 man/weeks in contract.						
680 man/weeks x 50% of men local		-			-	
x 20% of men receive £10.00 per week........		-			1,360	
x 20% of men receive £15.00 per week		-			2,040	
x 10% of men receive £20.00 per week		-			1,360	
Travelling time and fares:						
Allow 50 man weeks at average £15.00 per week		750	750		-	4,760
Section A42 - 15 - Miscellaneous						
Rates and Taxes........		225			1,000	
Contractors signboards		-			200	
Chainman attending Engineer		-			400	
Winter working		-			1,250	
Sample panels........		-			300	
Testing concrete cubes		-			300	
Labour attendance in offices		-			1,000	
Defects Liability period:						
Allow 2 tradesmen and 2 labourers for 1 week........		-			2,000	
Materials........		-			400	
Handover and defects liability:						
Labour		400			-	
Materials		105	730		-	6,850
Carried to Summary			£4,555			£34,448

SPECIMEN SITE PRELIMINARIES (Cont'd)

Section A43-1 - Mechanical Plant

	Small Works			Major Works	
Hoists: Assuming a 300kg hoist is required for 12 weeks:					
Hire of hoist per week	-		150		
Fuel, oil and grease	-		30		
Operator (part-time)	-		120		
	-		300		
x 12 hoist weeks				3,600	
Scaffold Hoist: Allow 4 weeks at £95 per week	380			-	
Pumping and Dewatering. It is assumed that a Provisional Sum is included in the Bills of Quantities:					
A typical build up is shown in the notes	-			-	
Pumps: Allow 50mm pump at £80 per week x 2 weeks	160	540		-	3,600

Section A43-4 - Transport

	Small Works		Major Works	
Site tractor and trailer	-		750	
Site van and driver	-		1,000	
Transport Plant to Site:				
Machine excavator	-		270	
Mechanical equipment including concrete mixers, hoists, and pumps	-		300	
Site offices and storage hutting	-		270	
Scaffolding	-		270	
Sundries	-		300	
Double handling materials on site	235		-	
Transport scaffolding and small plant to site	265	500	-	3,160
Carried to summary		£1,040		£6,760

Section A44-1 and 6 - Temporary Roads and Hardstandings

	Small Works		Major Works	
Temporary Roads and Hardstandings:				
Hardcore road: 100m2 at £12.00 per m2	-		1,200	
Hardstandings: 100m2 at £12.00 per m2	-		1,200	
Temporary Roads - allow for access to site	200	200	-	2,400

Section A44-3 and 4 - Scaffolding

Assuming a three storey office building 60m x 10m x 10m high in traditional brick construction:

	Small Works		Major Works	
External putlog scaffolding				
1400m2 at £7.30 per m2	-		10,220	
Mobile tower 8 weeks at £175 each including erection and dismantling	-		1,400	
Internal scaffolding for suspended ceilings				
1800m2 at £6.30 per m2	-		11,340	
Mobile towers				
4 weeks at £180 each including erection and dismantling.	720	720	-	22,960

Section A44-5 - Hoardings, Fans, Fencing etc

	Small Works		Major Works	
Hoarding 2m high: 100m at £25 per metre	-		2500	
Gate	-	-	200	2700
Carried to summary		£920		£28,060

SITE PRELIMINARIES - SUMMARY

SMM7 Section		Small Works £	£	Major Works £	£
A40	Management and Staff	5,040	5,040	60,160	60,160
A41	Site Accommodation	600	600	4,615	4,615
A42	Power and Lighting	312		4,727	
	Water	140		1,520	
	Telephone	480		1,605	
	Safety, Health, and Welfare	268		3,042	
	Rubbish Disposal and Cleaning Site	920		2,080	
	Drying out Building	100		500	
	Security	200		1,440	
	Small Plant and Tools	375		4,500	
	Insurance of Works	280		3,424	
	Travelling Time and Fares	750		4,760	
	Miscellaneous	730	4,555	6,850	34,448
A43	Mechanical Plant	540		3,600	
	Transport	500	1,040	3,160	6,760
A44	Temporary Roads	175		2,400	
	Scaffolding	720		22,960	
	Hoardings	-	920	2,700	28,060
			£12,155		£134,043
	Overheads and Profit: 12.5%		£1,519	7.5%	£10,53
	Total of Site Preliminaries		**£ 13,674**		**£ 144,096**

FIRM PRICE ADDITION (Major Works)

The amount to be added for a firm price tender is dependent upon the overall contract period of the project, the known or anticipated dates of wage awards and the estimated percentage amount of the award. Separate calculations must be carried out for anticipated labour increased costs, materials increases, plant increases, staff salary awards and the increased costs required by direct subcontractors, if any. The total of these represents the addition to be made for a firm price tender.

A typical calculation is as follows:

	£	£
Labour:		
Assume a 5% wage award half way through the 48 week contract period		
The value of labour affected by the award is calculated as £150,000		
The firm price addition is 5% on £150,000		7,500
Materials:		
The value of materials affected by the award is estimated as £150,000		
Firm price addition for materials: allow average 2.5% on £150,000		3750
Plant: Required mainly at the start of project: allow		500
Staff: Allow 4% increase on £25,000		1200
Subcontractors:		
Net value affected by increased costs, say £150,000		
Allow labour 5% on £75,000	3,750	
Allow materials 2.5% on £75000	1,875	5625
Net addition for firm price:		£18,575

FIRM PRICE ADDITION (Small Works)

Firm Price: Most small works contracts are based on firm price tenders. This is due in the main to the relatively short contract periods normally involved. It is therefore the usual practice to make a lump sum allowance in the Preliminaries for any anticipated cost increases in labour and materials. (or maybe ignored altogether should the contractor consider adequate provision is included within the rates)

	£
Allowance for firm price	200
Net addition for firm price:	£200

LANDFILL TAX (Major and Small Works)

The Landfill Tax is included as indicated within the Disposal items of the Groundworks Section within Laxton's rates in this edition.

The two rates of Tax applicable from 01/04/2005 and as used in this edition are:-

£2.00 per tonne for inactive waste.
£18.00 per tonne for all other waste.

Additional allowances should also be added for contaminated waste and specific quotations obtained as required.

The cost of excavation will be affected by the nature of the soil and its subsequent bulking factor in handling and disposal.

A typical conversion rate is 2.10 tonnes per cubic meter. This results in an addition of £4.20 per cubic meter for inactive waste and £37.80 per cubic meter for active waste to basic disposal costs.

Allowances for disposal and Landfill tax must be added to all items of demolitions, alterations and rubbish disposal as appropriate.

AGGREGATES LEVY (Major and Small Works)

The Aggregates Levy came into effect on 1st April 2002. It applies to any sand, gravel or crushed rock commercially exploited in the UK or its territorial waters, or imported into the UK.

The levy does not apply to coal, clay and slate, metals and metal ores, gemstones or semi-precious stones, industrial minerals and recycled aggregates.

The levy is charged at a single flat rate of £1.60 per tonne (£4 per m^3 approx). The effect of this levy on building materials has generally been included within the rates.

Introduction to Health and Safety in Construction

Phil Hughes & Ed Ferrett

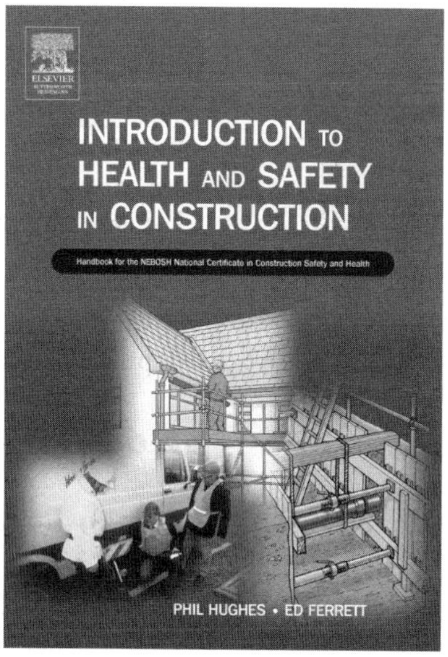

The construction industry is working hard to improve its health and safety record. It is now essential for employees to be aware of the health and safety issues that concern them and demand for qualifications in this area is increasing.

In the UK, NEBOSH (National Examination Board in Occupational Safety and Health) is offering a certificate course in Construction Safety and Health for which this book was written. The NEBOSH course is designed to cover general health and safety issues within a construction context as well as placing emphasis on the specific construction issues. As the basic concepts are the same worldwide, this book is also suitable for international courses in Construction Design and Management, Construction Safety and Health, and the Build Environment.

It is supported with practice questions for the NEBOSH examination and also contains checklists, report forms and record sheets for use at work, making the book a valuable reference tool for Building and Civil Engineering professionals.

£39.99 **Paperback October 2004** **480pp** **210x297 mm** **ISBN 075066343X**

Five Ways to Order

ONLINE
http://books.elsevier.com/construction

E-MAIL
directorders@elsevier.com

FAX
+44 (0) 1865 474011

PHONE
+44 (0) 1865 474010

MAIL
Customer Services, Elsevier Limited
Linacre House, Jordan Hill, Oxford OX2 8DP

MAJOR WORKS

The measured rates in this section are intended to apply to Contracts within the price range £500,000 to £1,500,000 based on National Average prices.

Adjustment to overall contract values for Regional Variations and contract values from £250,000 to £5,000,000 can be made as noted in the "Essential Information" section.

REGIONAL VARIATIONS

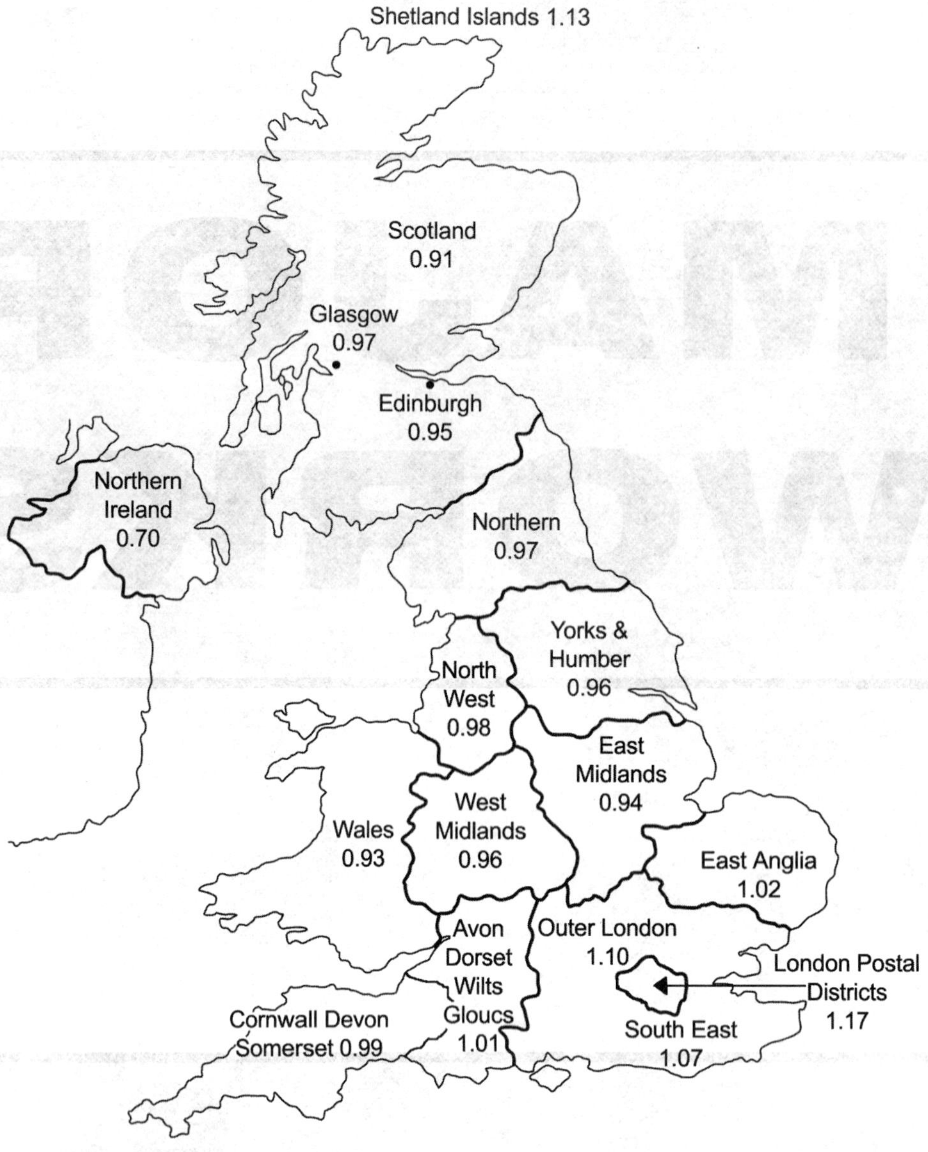

Laxton's Building Price Book is based upon national average prices (=1.00). Indicative levels of pricing in regions at the second quarter 2005 are given on this map and the factors shown may be applied to adjust overall pricing.

MAJOR WORKS - TENDER VALUES

The measured rates are based on contracts valued in the range of £500,000 to £1,500,000.

As a guide to pricing works of larger value and for cost planning or budgetary purposes, the following adjustments may be made to overall project values:

Project value:
£1,500,000 to £2,500,000 ... deduct 2.5%
£2,500,000 to £5,000,000 ... deduct 5%

SMALL WORKS - TENDER VALUES

The measured rates are based on contracts valued in the range of £50,000 to £100,000.

As a guide to pricing works of smaller or larger value and for cost planning or budgetary purposes, the following adjustments may be made to overall project values:

Project value:
£15,000 to £25,000........... add 10 %
£25,000 to £50,000........... add 5%
£50,000 to £100,000......... rates as shown
£100,000 to £200,000....... deduct 5%
£200,000 to £500,000....... deduct 7.5%

ALTERATIONS/REPAIRS/MAINTENANCE

The measured rates are based on contracts valued in the range of £5,000 to £15,000.
The rates may be used for orders of lower value however due allowances must be added for disproportionate waste factors, working in occupied premises and preliminary items.

EXISTING SITE/BUILDINGS/SERVICES

Labour hourly rates: (except Specialists) Craft Operatives £14.21 Labourer £10.71 Rates are national average prices. Refer to REGIONAL VARIATIONS for indicative levels of overall pricing in regions	MATERIALS			LABOUR				RATES		
	Del to Site £	Waste %	Material Cost £	Craft Optve Hrs	Lab Hrs	Labour Cost £	Sunds £	Nett Rate £	Unit	Gross Rate (+7.5%) £
C20: DEMOLITION										
Pulling down outbuildings										
Demolishing individual structures										
timber outbuilding 2.50 x 2.00 x 3.00m maximum high	-	-	-	-	11.00	**117.81**	70.51	**188.32**	nr	**202.44**
outbuilding with half brick walls, one brick piers and felt covered timber roof, 2.50 x 2.00 x 3.00m maximum high	-	-	-	-	22.00	**235.62**	141.02	**376.64**	nr	**404.88**
greenhouse with half brick dwarf walls and timber framing to upper walls and roof, 3.00 x 2.00 x 3.00m maximum high	-	-	-	-	18.00	**192.78**	120.87	**313.65**	nr	**337.17**
Demolishing individual structures; setting aside materials for re-use										
metal framed greenhouse, 3.00 x 2.00 x 3.00m maximum high	-	-	-	4.00	7.00	**131.81**	-	**131.81**	nr	**141.70**
prefabricated concrete garage, 5.40 x 2.60 x 2.40m maximum high	-	-	-	6.00	11.00	**203.07**	-	**203.07**	nr	**218.30**
Demolishing individual structures; making good structures										
timber framed lean-to outbuilding and remove flashings and make good facing brick wall at ridge and vertical abutments, 3.00 x 1.90 x 2.40m maximum high	-	-	-	1.50	12.00	**149.84**	74.66	**224.49**	nr	**241.33**
Unit rates for pricing the above and similar work										
pull down building 3.00 x 1.90 x 2.40m maximum high	-	-	-	-	10.00	**107.10**	70.51	**177.61**	nr	**190.93**
remove flashing	-	-	-	-	0.12	**1.29**	0.40	**1.69**	m	**1.81**
make good facing brick wall at ridge	-	-	-	0.20	0.20	**4.98**	0.37	**5.35**	m	**5.75**
make good facing brick wall at vertical abutment	-	-	-	0.17	0.17	**4.24**	0.37	**4.60**	m	**4.95**
make good rendered wall at ridge	-	-	-	0.20	0.20	**4.98**	0.85	**5.84**	m	**6.27**
make good rendered wall at vertical abutment	-	-	-	0.20	0.20	**4.98**	0.85	**5.84**	m	**6.27**
Demolishing individual structures; making good structures										
lean-to outbuilding with half brick walls and slate covered timber roof and remove flashings, hack off plaster to house wall and make good with rendering to match existing, 1.50 x 1.00 x 2.00m maximum high	-	-	-	4.00	12.50	**190.72**	47.45	**238.16**	nr	**256.02**
Unit rates for pricing the above and similar work										
pull down building with half brick walls, 1.50 x 1.00 x 2.00m maximum high	-	-	-	-	5.30	**56.76**	32.23	**89.00**	nr	**95.67**
pull down building with one brick walls, 1.50 x 1.00 x 2.00m maximum high	-	-	-	-	10.00	**107.10**	60.44	**167.54**	nr	**180.10**
hack off plaster or rendering and make good to match existing brick facings	-	-	-	1.00	2.00	**35.63**	1.21	**36.84**	m2	**39.60**
hack off plaster or rendering and make good to match existing rendering	-	-	-	1.30	2.30	**43.11**	4.87	**47.98**	m2	**51.57**
Removal of old work										
Demolishing parts of structures										
in-situ plain concrete bed										
75mm thick	-	-	-	-	0.90	**9.64**	3.02	**12.66**	m2	**13.61**
100mm thick	-	-	-	-	1.20	**12.85**	4.03	**16.88**	m2	**18.15**
150mm thick	-	-	-	-	1.80	**19.28**	6.04	**25.32**	m2	**27.22**
200mm thick	-	-	-	-	2.40	**25.70**	8.06	**33.76**	m2	**36.29**
in-situ reinforced concrete flat roof										
100mm thick	-	-	-	-	1.90	**20.35**	4.03	**24.38**	m2	**26.21**
150mm thick	-	-	-	-	2.85	**30.52**	6.04	**36.57**	m2	**39.31**
200mm thick	-	-	-	-	3.80	**40.70**	8.06	**48.76**	m2	**52.41**
225mm thick	-	-	-	-	4.28	**45.84**	9.07	**54.90**	m2	**59.02**
in-situ reinforced concrete upper floor										
100mm thick	-	-	-	-	1.90	**20.35**	4.03	**24.38**	m2	**26.21**
150mm thick	-	-	-	-	2.85	**30.52**	6.04	**36.57**	m2	**39.31**

Labour hourly rates: (except Specialists) Craft Operatives £14.21 Labourer £10.71 Rates are national average prices. Refer to REGIONAL VARIATIONS for indicative levels of overall pricing in regions	MATERIALS			LABOUR				RATES		
	Del to Site £	Waste %	Material Cost £	Craft Optve Hrs	Lab Hrs	Labour Cost £	Sunds £	Nett Rate £	Unit	Gross Rate (+7.5%) £
C20: DEMOLITION Cont.										
Removal of old work Cont.										
Demolishing parts of structures Cont.										
in-situ reinforced concrete upper floor Cont.										
200mm thick	-	-	-	-	3.80	**40.70**	8.06	**48.76**	m2	**52.41**
225mm thick	-	-	-	-	4.28	**45.84**	9.07	**54.90**	m2	**59.02**
in-situ reinforced concrete beam	-	-	-	-	19.00	**203.49**	40.29	**243.78**	m3	**262.06**
in-situ reinforced concrete column	-	-	-	-	18.00	**192.78**	40.29	**233.07**	m3	**250.55**
in-situ reinforced concrete wall										
100mm thick	-	-	-	-	1.80	**19.28**	4.03	**23.31**	m2	**25.06**
150mm thick	-	-	-	-	2.70	**28.92**	6.04	**34.96**	m2	**37.58**
200mm thick	-	-	-	-	3.60	**38.56**	8.06	**46.61**	m2	**50.11**
225mm thick	-	-	-	-	4.05	**43.38**	9.07	**52.44**	m2	**56.37**
in-situ reinforced concrete casing to beam	-	-	-	-	16.00	**171.36**	40.29	**211.65**	m3	**227.52**
in-situ reinforced concrete casing to column	-	-	-	-	15.25	**163.33**	40.29	**203.62**	m3	**218.89**
brick internal walls in lime mortar										
102mm thick	-	-	-	-	0.63	**6.75**	4.11	**10.86**	m2	**11.67**
215mm thick	-	-	-	-	1.30	**13.92**	8.66	**22.59**	m2	**24.28**
327mm thick	-	-	-	-	2.00	**21.42**	13.18	**34.60**	m2	**37.19**
brick internal walls in cement mortar										
102mm thick	-	-	-	-	0.95	**10.17**	4.11	**14.29**	m2	**15.36**
215mm thick	-	-	-	-	1.94	**20.78**	8.66	**29.44**	m2	**31.65**
327mm thick	-	-	-	-	2.98	**31.92**	13.18	**45.09**	m2	**48.47**
reinforced brick internal walls in cement lime mortar										
102mm thick	-	-	-	-	0.95	**10.17**	4.11	**14.29**	m2	**15.36**
215mm thick	-	-	-	-	1.94	**20.78**	8.66	**29.44**	m2	**31.65**
hollow clay block internal walls in cement-lime mortar										
50mm thick	-	-	-	-	0.30	**3.21**	2.01	**5.23**	m2	**5.62**
75mm thick	-	-	-	-	0.40	**4.28**	3.02	**7.31**	m2	**7.85**
100mm thick	-	-	-	-	0.52	**5.57**	4.03	**9.60**	m2	**10.32**
hollow clay block internal walls in cement mortar										
50mm thick	-	-	-	-	0.32	**3.43**	2.01	**5.44**	m2	**5.85**
75mm thick	-	-	-	-	0.43	**4.61**	3.02	**7.63**	m2	**8.20**
100mm thick	-	-	-	-	0.57	**6.10**	4.03	**10.13**	m2	**10.89**
concrete block internal walls in cement lime mortar										
75mm thick	-	-	-	-	0.35	**3.75**	3.02	**6.77**	m2	**7.28**
100mm thick	-	-	-	-	0.53	**5.68**	4.03	**9.71**	m2	**10.43**
190mm thick	-	-	-	-	0.70	**7.50**	7.66	**15.15**	m2	**16.29**
215mm thick	-	-	-	-	1.41	**15.10**	8.66	**23.76**	m2	**25.55**
concrete block internal walls in cement mortar										
75mm thick	-	-	-	-	0.56	**6.00**	3.02	**9.02**	m2	**9.70**
100mm thick	-	-	-	-	0.73	**7.82**	4.03	**11.85**	m2	**12.74**
190mm thick	-	-	-	-	0.91	**9.75**	7.66	**17.40**	m2	**18.71**
215mm thick	-	-	-	-	1.67	**17.89**	8.66	**26.55**	m2	**28.54**
if internal walls plastered, add per side	-	-	-	-	0.12	**1.29**	0.81	**2.09**	m2	**2.25**
if internal walls rendered, add per side	-	-	-	-	0.18	**1.93**	0.81	**2.73**	m2	**2.94**
brick external walls in lime mortar										
102mm thick	-	-	-	-	0.48	**5.14**	4.11	**9.25**	m2	**9.95**
215mm thick	-	-	-	-	0.98	**10.50**	8.66	**19.16**	m2	**20.60**
327mm thick	-	-	-	-	1.51	**16.17**	13.18	**29.35**	m2	**31.55**
brick external walls in cement mortar										
102mm thick	-	-	-	-	0.63	**6.75**	4.11	**10.86**	m2	**11.67**
215mm thick	-	-	-	-	1.30	**13.92**	8.66	**22.59**	m2	**24.28**
327mm thick	-	-	-	-	2.00	**21.42**	13.18	**34.60**	m2	**37.19**
reinforced brick external walls in cement lime mortar										
102mm thick	-	-	-	-	0.63	**6.75**	4.11	**10.86**	m2	**11.67**
215mm thick	-	-	-	-	1.30	**13.92**	8.66	**22.59**	m2	**24.28**
if external walls plastered, add per side	-	-	-	-	0.09	**0.96**	0.81	**1.77**	m2	**1.90**
if external walls rendered or rough cast, add per side	-	-	-	-	0.12	**1.29**	0.81	**2.09**	m2	**2.25**
clean old bricks in lime mortar and stack for re-use, per thousand	-	-	-	-	15.00	**160.65**	-	**160.65**	nr	**172.70**
clean old bricks in cement mortar and stack for re-use, per thousand	-	-	-	-	25.00	**267.75**	-	**267.75**	nr	**287.83**
rough rubble walling 600mm thick, in lime mortar	-	-	-	-	3.75	**40.16**	24.17	**64.34**	m2	**69.16**
random rubble walling 350mm thick, in lime mortar	-	-	-	-	2.20	**23.56**	14.10	**37.66**	m2	**40.49**
random rubble walling 500mm thick, in lime mortar	-	-	-	-	3.10	**33.20**	20.15	**53.35**	m2	**57.35**
dressed stone walling 100mm thick, in gauged mortar	-	-	-	2.00	2.30	**53.05**	4.03	**57.08**	m2	**61.36**
dressed stone walling 200mm thick, in gauged mortar	-	-	-	3.90	4.50	**103.61**	8.06	**111.67**	m2	**120.05**
stone copings 300 x 50mm	-	-	-	0.20	0.26	**5.63**	0.60	**6.23**	m	**6.70**
stone copings 375 x 100mm	-	-	-	0.40	0.52	**11.25**	1.52	**12.77**	m	**13.73**
stone staircases	-	-	-	-	15.00	**160.65**	40.29	**200.94**	m3	**216.01**
stone steps	-	-	-	-	14.75	**157.97**	40.29	**198.26**	m3	**213.13**
structural timbers										
50 x 75mm	-	-	-	-	0.13	**1.39**	0.15	**1.55**	m	**1.66**
50 x 100mm	-	-	-	-	0.15	**1.61**	0.20	**1.81**	m	**1.94**
50 x 150mm	-	-	-	-	0.18	**1.93**	0.31	**2.24**	m	**2.40**
50 x 225mm	-	-	-	-	0.24	**2.57**	0.45	**3.02**	m	**3.25**
75 x 100mm	-	-	-	-	0.18	**1.93**	0.31	**2.24**	m	**2.40**
75 x 150mm	-	-	-	-	0.20	**2.14**	0.45	**2.59**	m	**2.79**
75 x 225mm	-	-	-	-	0.31	**3.32**	0.68	**4.00**	m	**4.30**
steelwork										
steel beams, joists and lintels not exceeding 10 kg/m	-	-	-	50.00	50.00	**1246.00**	40.29	**1286.29**	t	**1382.76**
steel beams, joists and lintels 10 - 20 kg/m	-	-	-	46.00	46.00	**1146.32**	40.29	**1186.61**	t	**1275.61**

Labour hourly rates: (except Specialists) Craft Operatives £14.21 Labourer £10.71 Rates are national average prices. Refer to REGIONAL VARIATIONS for indicative levels of overall pricing in regions	MATERIALS			LABOUR				RATES		
	Del to Site £	Waste %	Material Cost £	Craft Optve Hrs	Lab Hrs	Labour Cost £	Sunds £	Nett Rate £	Unit	Gross Rate (+7.5%) £
C20: DEMOLITION Cont.										
Removal of old work Cont.										
Demolishing parts of structures Cont.										
steelwork Cont.										
steel beams, joists and lintels 20 - 50 kg/m .	-	-	-	40.00	40.00	**996.80**	40.29	**1037.09**	t	**1114.87**
steel columns and stanchions not exceeding 10 kg/m	-	-	-	60.00	60.00	**1495.20**	40.29	**1535.49**	t	**1650.65**
steel columns and stanchions 10 - 20 kg/m	-	-	-	56.00	56.00	**1395.52**	40.29	**1435.81**	t	**1543.50**
steel columns and stanchions 20 - 50 kg/m	-	-	-	50.00	50.00	**1246.00**	40.29	**1286.29**	t	**1382.76**
steel purlins and rails not exceeding 10 kg/m	-	-	-	50.00	50.00	**1246.00**	40.29	**1286.29**	t	**1382.76**
steel purlins and rails 10 - 20 kg/m	-	-	-	50.00	50.00	**1246.00**	40.29	**1286.29**	t	**1382.76**
C30: SHORING/FACADE RETENTION										
Support of structures not to be demolished										
Provide and erect timber raking shore complete with sole piece, cleats, needles and 25mm brace boarding with										
two 150 x 150mm rakers (total length 8.00m) and 50 x 175mm wall piece	122.00	10.00	**134.20**	4.00	4.00	**99.68**	22.53	**256.41**	nr	**275.64**
weekly cost of maintaining last.................	-	-	-	1.00	1.00	**24.92**	1.13	**26.05**	nr	**28.00**
three 225 x 225mm rakers (total length 21.00m) and 50 x 250mm wall piece	461.00	10.00	**507.10**	23.20	23.20	**578.14**	97.94	**1183.18**	nr	**1271.92**
weekly cost of maintaining last.................	-	-	-	1.50	1.50	**37.38**	4.84	**42.22**	nr	**45.38**
Provide and erect timber flying shore of 100 x 150mm main member, 50 x 175mm wall pieces and 50 x 100mm straining pieces and 75 x 100mm struts, distance between wall faces										
4.00m	124.00	10.00	**136.40**	15.00	15.00	**373.80**	4.84	**515.04**	nr	**553.66**
weekly cost of maintaining last.................	-	-	-	2.50	2.50	**62.30**	1.13	**63.43**	nr	**68.19**
5.00m	144.00	10.00	**158.40**	17.80	17.80	**443.58**	5.63	**607.61**	nr	**653.18**
weekly cost of maintaining last.................	-	-	-	3.00	3.00	**74.76**	1.29	**76.05**	nr	**81.76**
6.00m	157.00	10.00	**172.70**	5.10	20.60	**293.10**	6.24	**472.04**	nr	**507.44**
weekly cost of maintaining last.................	-	-	-	3.50	3.50	**87.22**	1.48	**88.70**	nr	**95.35**
C40: CLEANING MASONRY/CONCRETE										
Cleaning surfaces										
Thoroughly clean existing concrete surfaces prior to applying damp proof membrane										
floors	-	-	-	-	0.19	**2.03**	0.02	**2.06**	m2	**2.21**
walls	-	-	-	-	0.21	**2.25**	0.02	**2.27**	m2	**2.44**
Thoroughly clean existing concrete surfaces, fill in nail holes and small surface imperfections and leave smooth										
walls	-	-	-	0.35	0.20	**7.12**	0.05	**7.16**	m2	**7.70**
soffits	-	-	-	0.50	0.29	**10.21**	0.05	**10.26**	m2	**11.03**
Thoroughly clean existing brick or block surfaces prior to applying damp proof membrane										
walls	-	-	-	-	0.25	**2.68**	-	**2.68**	m2	**2.88**
Clean out air brick										
225 x 150mm	-	-	-	-	0.20	**2.14**	-	**2.14**	nr	**2.30**
C41: REPAIRING/RENOVATING/CONSERVING MASONRY										
Cut out decayed bricks and replace with new bricks in gauged mortar (1:2:9) and make good surrounding work										
Fair faced common bricks										
singly	0.22	10.00	**0.24**	0.17	0.17	**4.24**	0.15	**4.63**	nr	**4.98**
small patches	13.20	10.00	**14.52**	5.80	5.80	**144.54**	6.43	**165.48**	m2	**177.90**
Picked stock facing bricks										
singly	0.46	10.00	**0.51**	0.17	0.17	**4.24**	0.15	**4.89**	nr	**5.26**
small patches	27.60	10.00	**30.36**	5.80	5.80	**144.54**	6.43	**181.32**	m2	**194.92**
Cut out defective wall and re-build in gauged mortar (1:2:9) and tooth and bond to surrounding work (25% new bricks allowed)										
Half brick wall; stretcher bond										
common bricks	3.30	10.00	**3.63**	4.00	4.00	**99.68**	3.69	**107.00**	m2	**115.02**
One brick wall; English bond										
common bricks	6.60	10.00	**7.26**	7.90	7.90	**196.87**	7.92	**212.05**	m2	**227.95**
common bricks faced one side with picked stock facings	12.00	10.00	**13.20**	8.30	8.30	**206.84**	7.92	**227.96**	m2	**245.06**
picked stock facing bricks faced both sides......	13.80	10.00	**15.18**	8.80	8.80	**219.30**	7.92	**242.40**	m2	**260.58**
One brick wall; Flemish bond										
common bricks	6.60	10.00	**7.26**	7.90	7.90	**196.87**	3.69	**207.82**	m2	**223.40**
common bricks faced one side with picked stock facings	12.00	10.00	**13.20**	8.30	8.30	**206.84**	7.92	**227.96**	m2	**245.06**
picked stock facing bricks faced both sides......	13.80	10.00	**15.18**	8.80	8.80	**219.30**	7.92	**242.40**	m2	**260.58**
275mm hollow wall; skins in stretcher bond										
two half brick skins in common bricks............	6.60	10.00	**7.26**	8.16	8.16	**203.35**	8.07	**218.68**	m2	**235.08**

Labour hourly rates: (except Specialists) Craft Operatives £14.21 Labourer £10.71 Rates are national average prices. Refer to REGIONAL VARIATIONS for indicative levels of overall pricing in regions	MATERIALS			LABOUR				RATES		
	Del to Site £	Waste %	Material Cost £	Craft Optve Hrs	Lab Hrs	Labour Cost £	Sunds £	Nett Rate £	Unit	Gross Rate (+7.5%) £
C41: REPAIRING/RENOVATING/CONSERVING MASONRY Cont.										
Cut out defective wall and re-build in gauged mortar (1:2:9) and tooth and bond to surrounding work (25% new bricks allowed) Cont.										
275mm hollow wall; skins in stretcher bond Cont.										
one half brick skin in common bricks and one half brick skin in picked stock facings..............	10.20	10.00	**11.22**	8.60	8.60	**214.31**	8.07	**233.61**	m2	**251.13**
one 100mm skin in concrete blocks and one half brick skin in picked stock facings..............	9.43	10.00	**10.37**	8.26	8.26	**205.84**	7.55	**223.75**	m2	**240.53**
Cut out crack in brickwork and stitch across with new bricks in gauged mortar(1:2:9); average 450mm wide										
Half brick wall										
common bricks	5.72	10.00	**6.29**	1.90	1.90	**47.35**	3.58	**57.22**	m	**61.51**
common bricks fair faced one side................	5.72	10.00	**6.29**	2.30	2.30	**57.32**	3.58	**67.19**	m	**72.23**
picked stock facing bricks faced both sides......	11.96	10.00	**13.16**	2.40	2.40	**59.81**	3.58	**76.54**	m	**82.28**
One brick wall										
common bricks	11.44	10.00	**12.58**	3.90	3.90	**97.19**	7.71	**117.48**	m	**126.29**
common bricks fair faced one side................	11.44	10.00	**12.58**	4.40	4.40	**109.65**	7.71	**129.94**	m	**139.69**
common bricks faced one side with picked stock facings	20.80	10.00	**22.88**	4.60	4.60	**114.63**	7.71	**145.22**	m	**156.11**
picked stock facing bricks faced both sides......	23.92	10.00	**26.31**	4.70	4.70	**117.12**	7.71	**151.14**	m	**162.48**
Cut out defective arch; re-build in picked stock facing bricks in gauged mortar (1:2:9) including centering										
Brick-on-edge flat arch										
102mm on soffit	3.22	10.00	**3.54**	1.00	1.00	**24.92**	0.90	**29.37**	m	**31.57**
215mm on soffit	6.13	10.00	**6.74**	0.95	0.95	**23.67**	1.58	**32.00**	m	**34.40**
Brick-on-end flat arch										
102mm on soffit	6.13	10.00	**6.74**	0.95	0.95	**23.67**	1.58	**32.00**	m	**34.40**
215mm on soffit	12.26	10.00	**13.49**	1.90	1.90	**47.35**	3.16	**64.00**	m	**68.80**
Segmental arch in two half brick rings										
102mm on soffit	6.13	10.00	**6.74**	2.90	2.90	**72.27**	1.58	**80.59**	m	**86.64**
215mm on soffit	12.26	10.00	**13.49**	2.90	2.90	**72.27**	3.16	**88.92**	m	**95.59**
Semi-circular arch in two half brick rings										
102mm on soffit	6.13	10.00	**6.74**	2.90	2.90	**72.27**	1.58	**80.59**	m	**86.64**
215mm on soffit	12.26	10.00	**13.49**	2.90	2.90	**72.27**	3.16	**88.92**	m	**95.59**
Cut out defective arch; replace with precast concrete, BS 5328, designed mix C20, 20mm aggregate lintel; wedge and pin up to brickwork over with slates in cement mortar (1:3) and extend external rendering over										
Remove defective brick-on-end flat arch										
replace with 102 x 215mm precast concrete lintel reinforced with one 12mm mild steel bar..........	27.27	10.00	**30.00**	1.90	1.90	**47.35**	2.48	**79.83**	m	**85.82**
Cut out defective external sill and re-build										
Shaped brick-on-edge sill in picked stock facing bricks in gauged mortar (1:2:9)										
225mm wide	6.13	10.00	**6.74**	1.00	1.00	**24.92**	2.20	**33.87**	m	**36.41**
Roofing tile sill set weathering and projecting in cement mortar (1:3)										
two courses	6.14	10.00	**6.76**	1.20	1.20	**29.90**	1.14	**37.80**	m	**40.64**
Take down defective brick-on-edge coping and re-build in picked stock facing bricks in gauged mortar (1:2:9)										
Coping										
to one brick wall	6.13	10.00	**6.74**	0.60	0.60	**14.95**	2.20	**23.90**	m	**25.69**
Coping with cement fillets both sides										
to one brick wall; with oversailing course.......	9.20	10.00	**10.12**	0.67	0.67	**16.70**	3.31	**30.13**	m	**32.39**
to one brick wall; with single course tile creasing	9.20	10.00	**10.12**	1.09	1.09	**27.16**	2.77	**40.06**	m	**43.07**
to one brick wall; with double course tile creasing	12.28	10.00	**13.50**	1.38	1.38	**34.39**	3.91	**51.81**	m	**55.69**
Cut out defective air brick and replace with new; bed and point in gauged mortar (1:2:9)										
Cast iron, light, square hole										
225 x 75mm	3.95	-	**3.95**	0.45	0.45	**11.21**	0.17	**15.34**	nr	**16.49**
225 x 150mm	7.25	-	**7.25**	0.45	0.45	**11.21**	0.28	**18.74**	nr	**20.15**
Terra cotta										
215 x 65mm	1.33	-	**1.33**	0.45	0.45	**11.21**	0.17	**12.72**	nr	**13.67**
215 x 140mm	1.85	-	**1.85**	0.45	0.45	**11.21**	0.28	**13.34**	nr	**14.34**
Rake out joints of old brickwork 20mm deep and re-point in cement mortar (1:3)										
For turned in edge of lead flashing										
flush pointing; horizontal......................	-	-	**-**	0.50	0.50	**12.46**	0.26	**12.72**	m	**13.67**
flush pointing; stepped	-	-	**-**	0.85	0.85	**21.18**	0.37	**21.55**	m	**23.16**

Labour hourly rates: (except Specialists) Craft Operatives £14.21 Labourer £10.71 Rates are national average prices. Refer to REGIONAL VARIATIONS for indicative levels of overall pricing in regions	MATERIALS			LABOUR				RATES		
	Del to Site £	Waste %	Material Cost £	Craft Optve Hrs	Lab Hrs	Labour Cost £	Sunds £	Nett Rate £	Unit	Gross Rate (+7.5%) £
C41: REPAIRING/RENOVATING/CONSERVING MASONRY Cont.										
Rake out joints of old brickwork 20mm deep and re-point in cement mortar (1:3) Cont.										
For turned in edge of lead flashing Cont.										
weathered pointing; horizontal	-	-	-	0.55	0.55	**13.71**	0.26	**13.96**	m	**15.01**
weathered pointing; stepped	-	-	-	0.90	0.90	**22.43**	0.37	**22.79**	m	**24.50**
ironed in pointing; horizontal	-	-	-	0.55	0.55	**13.71**	0.26	**13.96**	m	**15.01**
ironed in pointing; stepped	-	-	-	0.90	0.90	**22.43**	0.37	**22.79**	m	**24.50**
Rake out joints of old brickwork 20mm deep and re-point in cement-lime mortar(1:1:6)										
Generally; English bond										
flush pointing	-	-	-	1.00	0.66	**21.28**	0.74	**22.02**	m2	**23.67**
weathered pointing	-	-	-	1.00	0.66	**21.28**	0.74	**22.02**	m2	**23.67**
ironed in pointing	-	-	-	1.20	0.80	**25.62**	0.74	**26.36**	m2	**28.34**
Generally; Flemish bond										
flush pointing	-	-	-	1.00	0.66	**21.28**	0.74	**22.02**	m2	**23.67**
weathered pointing	-	-	-	1.00	0.66	**21.28**	0.74	**22.02**	m2	**23.67**
ironed in pointing	-	-	-	1.20	0.80	**25.62**	0.74	**26.36**	m2	**28.34**
Isolated areas not exceeding 1.00m2; English bond										
flush pointing	-	-	-	1.50	1.00	**32.03**	0.74	**32.77**	m2	**35.22**
weathered pointing	-	-	-	1.50	1.00	**32.03**	0.74	**32.77**	m2	**35.22**
ironed in pointing	-	-	-	1.80	1.20	**38.43**	0.74	**39.17**	m2	**42.11**
Isolated areas not exceeding 1.00m2; Flemish bond										
flush pointing	-	-	-	1.50	1.00	**32.03**	0.74	**32.77**	m2	**35.22**
weathered pointing	-	-	-	1.50	1.00	**32.03**	0.74	**32.77**	m2	**35.22**
ironed in pointing	-	-	-	1.80	1.20	**38.43**	0.74	**39.17**	m2	**42.11**
For using coloured mortar										
add	-	-	-	-	-	-	0.54	**0.54**	m2	**0.58**
Rake out joints of old brickwork 20mm deep and re-point in cement mortar (1:3)										
Generally; English bond										
flush pointing	-	-	-	1.20	0.80	**25.62**	0.74	**26.36**	m2	**28.34**
weathered pointing	-	-	-	1.20	0.80	**25.62**	0.74	**26.36**	m2	**28.34**
Generally; Flemish bond										
flush pointing	-	-	-	1.20	0.80	**25.62**	0.74	**26.36**	m2	**28.34**
weathered pointing	-	-	-	1.20	0.80	**25.62**	0.74	**26.36**	m2	**28.34**
Isolated areas not exceeding 1.00m2; English bond										
flush pointing	-	-	-	1.80	1.20	**38.43**	0.74	**39.17**	m2	**42.11**
weathered pointing	-	-	-	1.80	1.20	**38.43**	0.74	**39.17**	m2	**42.11**
Isolated areas not exceeding 1.00m2; Flemish bond										
flush pointing	-	-	-	1.80	1.20	**38.43**	0.74	**39.17**	m2	**42.11**
weathered pointing	-	-	-	1.80	1.20	**38.43**	0.74	**39.17**	m2	**42.11**
Repairing/Renovating stone										
Note										
restoration and repair of stonework is work for a specialist; prices being obtained for specific projects. Some firms that specialise in this class of work are included within the list at the end of section F-Masonry										
Repairs in Portland stone work and set in gauged mortar (1:2:9) and point to match existing										
Cut out decayed stones in facings of walls, piers or the like, prepare for and set new 75mm thick stone facings										
single stone	357.00	-	**357.00**	15.00	15.00	**373.80**	6.93	**737.73**	m2	**793.05**
areas of two or more adjacent stones	357.00	-	**357.00**	12.00	12.00	**299.04**	6.93	**662.97**	m2	**712.69**
Take out decayed stone sill, prepare for and set new sunk weathered and throated sill										
175 x 100mm	191.00	-	**191.00**	1.80	1.80	**44.86**	0.93	**236.79**	m	**254.55**
250 x 125mm	260.00	-	**260.00**	2.50	2.50	**62.30**	1.58	**323.88**	m	**348.17**
Take out sections of decayed stone coping, prepare for and set new weathered and twice throated coping										
300 x 75mm	232.00	-	**232.00**	1.40	1.40	**34.89**	1.17	**268.06**	m	**288.16**
300 x 100mm	265.00	-	**265.00**	1.70	1.70	**42.36**	1.55	**308.91**	m	**332.08**
Repairs in York stone work and set in gauged mortar (1:2:9) and point to match existing										
Cut out decayed stones in facings of walls, piers or the like, prepare for and set new 75mm thick stone facings										
single stone	268.00	-	**268.00**	17.00	17.00	**423.64**	6.93	**698.57**	m2	**750.96**
areas of two or more adjacent stones	268.00	-	**268.00**	13.75	13.75	**342.65**	6.93	**617.58**	m2	**663.89**
Take out decayed stone sill, prepare for and set new sunk weathered and throated sill										
175 x 100mm	119.00	-	**119.00**	1.80	1.80	**44.86**	1.17	**165.03**	m	**177.40**
250 x 125mm	185.00	-	**185.00**	2.50	2.50	**62.30**	1.55	**248.85**	m	**267.51**

Labour hourly rates: (except Specialists) Craft Operatives £14.21 Labourer £10.71 Rates are national average prices. Refer to REGIONAL VARIATIONS for indicative levels of overall pricing in regions	MATERIALS			LABOUR				RATES		
	Del to Site £	Waste %	Material Cost £	Craft Optve Hrs	Lab Hrs	Labour Cost £	Sunds £	Nett Rate £	Unit	Gross Rate (+7.5%) £
C41: REPAIRING/RENOVATING/CONSERVING MASONRY Cont.										
Repairs in York stone work and set in gauged mortar (1:2:9) and point to match existing Cont.										
Take out sections of decayed stone coping, prepare for and set new weathered and twice throated coping										
300 x 75mm	130.00	-	**130.00**	1.40	1.40	**34.89**	1.17	**166.06**	m	**178.51**
300 x 100mm	154.00	-	**154.00**	1.70	1.70	**42.36**	1.55	**197.91**	m	**212.76**
Rake out decayed mortar joints and re-point in gauged mortar (1:2:9)										
Re-point rubble walling										
coursed	-	-	**-**	1.25	1.25	**31.15**	1.15	**32.30**	m2	**34.72**
uncoursed	-	-	**-**	1.32	1.32	**32.89**	1.15	**34.04**	m2	**36.60**
squared	-	-	**-**	1.25	1.25	**31.15**	1.15	**32.30**	m2	**34.72**
Re-point stonework										
ashlar	-	-	**-**	1.20	1.20	**29.90**	1.15	**31.05**	m2	**33.38**
coursed block	-	-	**-**	1.20	1.20	**29.90**	1.15	**31.05**	m2	**33.38**
C42: REPAIRING/ RENOVATING/CONSERVING CONCRETE										
Re-bed loose copings										
Take off loose concrete coping and clean and remove old bedding mortar										
re-bed, joint and point in cement mortar (1:3)	-	-	**-**	1.00	1.00	**24.92**	1.22	**26.14**	m	**28.10**
Repairing cracks in concrete										
Repair cracks; clean out all dust and debris and fill with mortar mixed with a bonding agent										
up to 5mm wide	-	-	**-**	0.50	0.50	**12.46**	0.59	**13.05**	m	**14.03**
Repair cracks; cutting out to form groove, treat with bonding agent and fill with fine concrete mixed with a bonding agent										
25 x 25mm deep	-	-	**-**	0.50	0.50	**12.46**	1.06	**13.52**	m	**14.54**
40 x 40mm deep	-	-	**-**	0.60	0.60	**14.95**	2.52	**17.47**	m	**18.78**
C45: DAMP PROOF COURSE RENEWAL/INSERTION										
Insert damp proof course in old wall by cutting out one course of brickwork in alternate lengths not exceeding 1.00m and replace old bricks with 50mm bricks in cement mortar (1:3)										
B.S.6398 Class D, bitumen with hessian base laminated with lead in										
one brick wall	3.85	10.00	**4.24**	3.00	3.00	**74.76**	5.94	**84.94**	m	**91.32**
one and a half brick wall	5.79	10.00	**6.37**	4.00	4.00	**99.68**	8.93	**114.98**	m	**123.61**
254mm hollow wall	3.85	10.00	**4.24**	3.50	3.50	**87.22**	5.94	**97.40**	m	**104.71**
Hyload, pitch polymer in										
one brick wall	1.85	10.00	**2.03**	3.00	3.00	**74.76**	5.94	**82.74**	m	**88.94**
one and a half brick wall	2.77	10.00	**3.05**	4.00	4.00	**99.68**	8.93	**111.67**	m	**120.04**
254mm hollow wall	1.85	10.00	**2.03**	3.50	3.50	**87.22**	5.94	**95.20**	m	**102.34**
Two courses slates in cement mortar (1:3) in										
one brick wall	6.04	10.00	**6.64**	3.00	3.00	**74.76**	5.94	**87.35**	m	**93.90**
one and a half brick wall	9.06	10.00	**9.97**	4.00	4.00	**99.68**	8.93	**118.58**	m	**127.47**
254mm hollow wall	6.04	10.00	**6.64**	3.50	3.50	**87.22**	5.94	**99.81**	m	**107.29**
0.6mm bitumen coated copper in										
one brick wall	11.44	10.00	**12.58**	3.00	3.00	**74.76**	5.94	**93.29**	m	**100.28**
one and a half brick wall	17.18	10.00	**18.90**	4.00	4.00	**99.68**	8.93	**127.51**	m	**137.08**
254mm hollow wall	11.44	10.00	**12.58**	3.50	3.50	**87.22**	5.94	**105.75**	m	**113.68**
Insert damp proof course in old wall by hand sawing in 600mm lengths										
B.S.6398 Class D, bitumen with hessian base laminated with lead in										
one brick wall	3.85	10.00	**4.24**	2.50	2.50	**62.30**	8.89	**75.43**	m	**81.09**
one and a half brick wall	5.79	10.00	**6.37**	3.50	3.50	**87.22**	13.34	**106.93**	m	**114.95**
254mm hollow wall	3.85	10.00	**4.24**	3.00	3.00	**74.76**	8.89	**87.89**	m	**94.48**
Hyload, pitch polymer in										
one brick wall	1.85	10.00	**2.03**	2.50	2.50	**62.30**	8.89	**73.22**	m	**78.71**
one and a half brick wall	2.77	10.00	**3.05**	3.50	3.50	**87.22**	13.34	**103.61**	m	**111.38**
254mm hollow wall	1.85	10.00	**2.03**	3.00	3.00	**74.76**	8.89	**85.68**	m	**92.11**
Two courses slates in cement mortar (1:3) in										
one brick wall	6.04	10.00	**6.64**	2.50	2.50	**62.30**	8.89	**77.83**	m	**83.67**
one and a half brick wall	9.06	10.00	**9.97**	3.50	3.50	**87.22**	13.34	**110.53**	m	**118.82**
254mm hollow wall	3.02	10.00	**3.32**	3.00	3.00	**74.76**	8.89	**86.97**	m	**93.50**
0.6mm bitumen coated copper in										
one brick wall	11.44	10.00	**12.58**	2.50	2.50	**62.30**	8.89	**83.77**	m	**90.05**
one and a half brick wall	17.18	10.00	**18.90**	3.50	3.50	**87.22**	13.34	**119.46**	m	**128.42**
254mm hollow wall	11.44	10.00	**12.58**	3.00	3.00	**74.76**	8.89	**96.23**	m	**103.45**

Labour hourly rates: (except Specialists) Craft Operatives £14.21 Labourer £10.71 Rates are national average prices. Refer to REGIONAL VARIATIONS for indicative levels of overall pricing in regions	MATERIALS			LABOUR				RATES		
	Del to Site £	Waste %	Material Cost £	Craft Optve Hrs	Lab Hrs	Labour Cost £	Sunds £	Nett Rate £	Unit	Gross Rate (+7.5%) £
C45: DAMP PROOF COURSE RENEWAL/INSERTION Cont.										
Insert damp proof course in old wall by machine sawing in 600mm lengths										
B.S.6398 Class D, bitumen with hessian base laminated with lead in										
one brick wall	3.85	10.00	**4.24**	2.00	2.00	**49.84**	11.80	**65.88**	m	**70.82**
one and a half brick wall	5.79	10.00	**6.37**	3.00	3.00	**74.76**	17.69	**98.82**	m	**106.24**
254mm hollow wall	3.85	10.00	**4.24**	2.50	2.50	**62.30**	11.80	**78.34**	m	**84.22**
Hyload, pitch polymer in										
one brick wall	1.85	10.00	**2.03**	2.00	2.00	**49.84**	11.80	**63.68**	m	**68.45**
one and a half brick wall	2.77	10.00	**3.05**	3.00	3.00	**74.76**	17.69	**95.51**	m	**102.67**
254mm hollow wall	1.85	10.00	**2.03**	2.50	2.50	**62.30**	11.80	**76.14**	m	**81.85**
Two courses slates in cement mortar (1:3) in										
one brick wall	6.04	10.00	**6.64**	2.00	2.00	**49.84**	11.80	**68.29**	m	**73.41**
one and a half brick wall	9.06	10.00	**9.97**	3.00	3.00	**74.76**	17.69	**102.42**	m	**110.10**
254mm hollow wall	3.02	10.00	**3.32**	2.50	2.50	**62.30**	11.80	**77.43**	m	**83.23**
0.6mm bitumen coated copper in										
one brick wall	11.44	10.00	**12.58**	2.00	2.00	**49.84**	11.80	**74.22**	m	**79.79**
one and a half brick wall	17.18	10.00	**18.90**	3.00	3.00	**74.76**	17.69	**111.35**	m	**119.70**
254mm hollow wall	11.44	10.00	**12.58**	2.50	2.50	**62.30**	11.80	**86.68**	m	**93.19**
Insert cavity tray in old wall by cutting out by hand in short lengths; insert individual trays and make good with picked stock facing bricks in gauged mortar (1:2:9) including pinning up with slates										
Type E Cavitray by Cavity Trays Ltd in										
half brick skin of cavity wall	9.06	10.00	**9.96**	1.75	1.75	**43.61**	7.55	**61.12**	m	**65.70**
external angle	6.25	10.00	**6.88**	1.00	1.00	**24.92**	6.93	**38.73**	nr	**41.63**
internal angle	6.25	10.00	**6.88**	1.25	1.25	**31.15**	6.93	**44.96**	nr	**48.33**
Type X Cavity Trays Ltd to suit 40 degree pitched roof complete with attached code 4 lead flashing and dress over tiles in										
half brick skin of cavity wall	37.66	10.00	**41.42**	4.25	4.25	**105.91**	15.10	**162.43**	m	**174.62**
ridge tray	9.35	10.00	**10.29**	1.00	1.00	**24.92**	3.52	**38.72**	nr	**41.62**
catchment tray	4.43	10.00	**4.87**	0.50	0.50	**12.46**	1.85	**19.18**	nr	**20.62**
corner catchment tray	9.77	10.00	**10.75**	1.00	1.00	**24.92**	3.52	**39.18**	nr	**42.12**
Damp proofing old brick wall by silicone injection method (excluding removal and reinstatement of plaster, etc.)										
Damp proofing										
one brick wall	7.98	-	**7.98**	1.00	1.00	**24.92**	0.81	**33.71**	m	**36.23**
one and a half brick wall	11.92	-	**11.92**	1.50	1.50	**37.38**	1.20	**50.50**	m	**54.29**
254mm hollow wall	7.98	-	**7.98**	1.35	1.35	**33.64**	0.81	**42.43**	m	**45.61**
C50: REPAIRING/RENOVATING/CONSERVING METAL										
Replace defective arch bar										
Take out defective arch bar including cutting away brickwork as necessary, replace with new mild steel arch bar primed all round and make good all work disturbed										
32 x 6mm flat arch bar	1.55	-	**1.55**	1.03	1.03	**25.67**	0.75	**27.96**	m	**30.06**
50 x 6mm flat arch bar	2.40	-	**2.40**	1.12	1.12	**27.91**	1.09	**31.40**	m	**33.76**
50 x 50 x 6mm angle arch bar	4.45	-	**4.45**	1.27	1.27	**31.65**	1.94	**38.04**	m	**40.89**
75 x 50 x 6mm angle arch bar	5.55	-	**5.55**	1.44	1.44	**35.88**	2.45	**43.88**	m	**47.17**
Replace defective baluster										
Cut out defective baluster, clean core and bottom rails, provide new mild steel baluster and weld on										
13mm diameter x 686mm long	0.90	-	**0.90**	0.92	0.92	**22.93**	1.55	**25.38**	nr	**27.28**
13mm diameter x 762mm long	0.90	-	**0.90**	0.94	0.94	**23.42**	1.55	**25.88**	nr	**27.82**
13 x 13 x 686mm long	1.20	-	**1.20**	0.94	0.94	**23.42**	1.84	**26.46**	nr	**28.45**
13 x 13 x 762mm long	1.25	-	**1.25**	0.95	0.95	**23.67**	1.84	**26.76**	nr	**28.77**
Repair damaged weld										
Clean off damaged weld including removing remains of weld and cleaning off surrounding paint back to bright metal and re-weld connection between 10mm rail and										
13mm diameter rod	-	-	**-**	0.50	0.50	**12.46**	2.29	**14.75**	nr	**15.86**
13 x 13mm bar	-	-	**-**	0.53	0.53	**13.21**	2.72	**15.93**	nr	**17.12**
25 x 25mm bar	-	-	**-**	0.60	0.60	**14.95**	3.41	**18.36**	nr	**19.74**
C51: REPAIRING/RENOVATING/CONSERVING TIMBER										
Cut out rotten or infected structural timber members including cutting away wall, etc. and shoring up adjacent work and replace with new pressure impregnated timber										
Plates and bed in cement mortar (1:3)										
50 x 75mm	1.26	10.00	**1.39**	0.30	0.30	**7.48**	0.17	**9.03**	m	**9.71**
75 x 100mm	2.70	10.00	**2.97**	0.50	0.50	**12.46**	0.33	**15.76**	m	**16.94**

Labour hourly rates: (except Specialists) Craft Operatives £14.21 Labourer £10.71 Rates are national average prices. Refer to REGIONAL VARIATIONS for indicative levels of overall pricing in regions	MATERIALS			LABOUR				RATES		
	Del to Site £	Waste %	Material Cost £	Craft Optve Hrs	Lab Hrs	Labour Cost £	Sunds £	Nett Rate £	Unit	Gross Rate (+7.5%) £
C51: REPAIRING/RENOVATING/CONSERVING TIMBER Cont.										
Cut out rotten or infected structural timber members including cutting away wall, etc. and shoring up adjacent work and replace with new pressure impregnated timber Cont.										
Floor or roof joists										
50 x 175mm	2.93	10.00	**3.22**	0.57	0.57	**14.20**	0.38	**17.81**	m	**19.14**
50 x 225mm	3.77	10.00	**4.15**	0.67	0.67	**16.70**	0.49	**21.33**	m	**22.93**
75 x 175mm	4.74	10.00	**5.21**	0.79	0.79	**19.69**	0.58	**25.48**	m	**27.39**
Rafters										
38 x 100mm	1.64	10.00	**1.80**	0.35	0.35	**8.72**	0.17	**10.70**	m	**11.50**
50 x 100mm	1.67	10.00	**1.84**	0.38	0.38	**9.47**	0.22	**11.53**	m	**12.39**
Ceiling joists and collars										
50 x 100mm	1.67	10.00	**1.84**	0.38	0.38	**9.47**	0.22	**11.53**	m	**12.39**
50 x 150mm	2.51	10.00	**2.76**	0.52	0.52	**12.96**	0.33	**16.05**	m	**17.25**
Purlins, ceiling beams and struts										
75 x 100mm	2.70	10.00	**2.97**	0.55	0.55	**13.71**	0.33	**17.01**	m	**18.28**
100 x 225mm	8.52	10.00	**9.37**	1.23	1.23	**30.65**	0.99	**41.01**	m	**44.09**
Roof trusses										
75 x 100mm	2.70	10.00	**2.97**	0.92	0.92	**22.93**	0.33	**26.23**	m	**28.19**
75 x 150mm	4.05	10.00	**4.46**	1.37	1.37	**34.14**	0.49	**39.09**	m	**42.02**
Hangers in roof										
25 x 75mm	0.78	10.00	**0.86**	0.30	0.30	**7.48**	0.09	**8.42**	m	**9.06**
38 x 100mm	1.64	10.00	**1.80**	0.45	0.45	**11.21**	0.17	**13.19**	m	**14.18**
Cut out rotten or infected timber and prepare timber under to receive new										
Roof boarding; provide new 25mm pressure impregnated softwood boarding										
flat boarding; areas not exceeding 0.50m2	11.00	10.00	**12.10**	1.96	1.96	**48.84**	1.09	**62.03**	m2	**66.69**
flat boarding; areas 0.50 - 5.00m2	11.00	10.00	**12.10**	1.63	1.63	**40.62**	1.09	**53.81**	m2	**57.85**
sloping boarding; areas not exceeding 0.50m2	11.00	10.00	**12.10**	2.34	2.34	**58.31**	1.09	**71.50**	m2	**76.87**
sloping boarding; areas 0.50 - 5.00m2	11.00	10.00	**12.10**	1.89	1.89	**47.10**	1.09	**60.29**	m2	**64.81**
Gutter boarding and bearers; provide new 25mm pressure impregnated softwood boarding and 50 x 50mm bearers										
over 300mm wide	14.36	10.00	**15.80**	2.06	2.06	**51.34**	1.31	**68.44**	m2	**73.57**
200mm	3.88	10.00	**4.27**	0.55	0.55	**13.71**	0.44	**18.41**	m	**19.80**
225mm	4.16	10.00	**4.57**	0.62	0.62	**15.45**	0.48	**20.50**	m	**22.04**
Gutter sides; provide new 19mm pressure impregnated softwood boarding										
150mm	1.27	15.00	**1.46**	0.46	0.46	**11.46**	0.12	**13.04**	m	**14.02**
200mm	1.76	15.00	**2.02**	0.50	0.50	**12.46**	0.17	**14.65**	m	**15.75**
225mm	2.15	15.00	**2.47**	0.57	0.57	**14.20**	0.19	**16.87**	m	**18.13**
Eaves or verge soffit and bearers and provide new 25mm pressure impregnated softwood boarding and 25 x 50mm bearers										
over 300mm wide	15.81	15.00	**18.13**	2.10	2.10	**52.33**	1.31	**71.78**	m2	**77.16**
225mm	4.37	10.00	**4.81**	0.67	0.67	**16.70**	0.48	**21.98**	m	**23.63**
Fascia; provide new pressure impregnated softwood										
25 x 150mm	2.28	10.00	**2.51**	0.46	0.46	**11.46**	0.17	**14.14**	m	**15.20**
25 x 200mm	3.00	10.00	**3.30**	0.51	0.51	**12.71**	0.22	**16.23**	m	**17.45**
Barge board; provide new pressure impregnated softwood										
25 x 225mm	3.41	10.00	**3.75**	0.55	0.55	**13.71**	0.25	**17.71**	m	**19.04**
25 x 250mm	3.75	10.00	**4.13**	0.57	0.57	**14.20**	0.28	**18.61**	m	**20.01**
Take up floor boarding, remove all nails and clean joists under, re-lay boarding, clear away and replace damaged boards and sand surface (20% new softwood boards allowed)										
Square edged boarding										
25mm	2.70	15.00	**3.11**	1.73	1.73	**43.11**	0.44	**46.66**	m2	**50.16**
32mm	3.58	15.00	**4.12**	1.76	1.76	**43.86**	0.55	**48.53**	m2	**52.17**
Tongued and grooved boarding										
25mm	2.86	20.00	**3.43**	1.87	1.87	**46.60**	0.44	**50.47**	m2	**54.26**
32mm	3.79	20.00	**4.55**	1.89	1.89	**47.10**	0.55	**52.20**	m2	**56.12**
Take up floor boarding, remove all nails and clean joists under, re-lay boarding, destroy and replace infected boards with new pressure impregnated boards and sand surface (20% new softwood boards allowed)										
Square edged boarding										
25mm	2.97	15.00	**3.42**	1.81	1.81	**45.11**	0.44	**48.96**	m2	**52.63**
32mm	3.94	15.00	**4.53**	1.83	1.83	**45.60**	0.55	**50.68**	m2	**54.49**
Tongued and grooved boarding										
25mm	3.15	20.00	**3.78**	1.94	1.94	**48.34**	0.44	**52.56**	m2	**56.51**
32mm	4.17	20.00	**5.00**	1.95	1.95	**48.59**	0.55	**54.15**	m2	**58.21**

Labour hourly rates: (except Specialists) Craft Operatives £14.21 Labourer £10.71 Rates are national average prices. Refer to REGIONAL VARIATIONS for indicative levels of overall pricing in regions	MATERIALS			LABOUR				RATES		
	Del to Site £	Waste %	Material Cost £	Craft Optve Hrs	Lab Hrs	Labour Cost £	Sunds £	Nett Rate £	Unit	Gross Rate (+7.5%) £
C51: REPAIRING/RENOVATING/CONSERVING TIMBER Cont.										
Take up worn or damaged floor boarding, remove all nails and clean joists under and replace with new softwood boarding										
Square edged boarding										
25mm in areas not exceeding 0.50m2	13.50	15.00	15.53	1.90	1.90	47.35	1.09	63.96	m2	68.76
32mm in areas not exceeding 0.50m2	17.90	15.00	20.59	1.94	1.94	48.34	1.39	70.32	m2	75.59
25mm in areas 0.50 - 3.00m2	13.50	15.00	15.53	1.57	1.57	39.12	1.09	55.74	m2	59.92
32mm in areas 0.50 - 3.00m2	17.90	15.00	20.59	1.61	1.61	40.12	1.39	62.10	m2	66.75
Tongued and grooved boarding										
25mm in areas not exceeding 0.50m2	14.31	20.00	17.17	2.00	2.00	49.84	1.09	68.10	m2	73.21
32mm in areas not exceeding 0.50m2	18.97	20.00	22.76	2.06	2.06	51.34	1.39	75.49	m2	81.15
25mm in areas 0.50 - 3.00m2	14.31	20.00	17.17	1.82	1.82	45.35	1.09	63.62	m2	68.39
32mm in areas 0.50 - 3.00m2	18.97	20.00	22.76	1.87	1.87	46.60	1.39	70.75	m2	76.06
Punch down all nails, remove all tacks, etc. and fill nail holes										
Surface of existing flooring										
generally	-	-	-	0.50	0.10	8.18	0.03	8.21	m2	8.82
generally and machine sand in addition	-	-	-	1.00	0.30	17.42	1.18	18.60	m2	20.00
Note										
removal of old floor coverings included elsewhere										
Easing and overhauling doors										
Ease and adjust door and oil ironmongery										
762 x 1981mm	-	-	-	0.33	0.06	5.33	0.02	5.35	nr	5.75
914 x 2134mm	-	-	-	0.39	0.06	6.18	0.02	6.20	nr	6.67
Take off door, shave 12mm off bottom edge and re-hang										
762 x 1981mm	-	-	-	0.66	0.11	10.56	-	10.56	nr	11.35
914 x 2134mm	-	-	-	0.79	0.13	12.62	-	12.62	nr	13.56
Adapting doors										
Take off door and re-hang on opposite hand, piece out as necessary and oil ironmongery										
762 x 1981mm	-	-	-	2.50	0.25	38.20	0.10	38.30	nr	41.18
914 x 2134mm	-	-	-	2.65	0.44	42.37	0.10	42.47	nr	45.65
Refixing removed doors and frames										
Doors										
762 x 1981mm	-	-	-	2.00	0.34	32.06	0.14	32.20	nr	34.62
914 x 2134mm	-	-	-	2.25	0.38	36.04	0.14	36.18	nr	38.90
Doors and frames; oil existing ironmongery										
762 x 1981mm	-	-	-	4.00	0.66	63.91	0.35	64.26	nr	69.08
914 x 2134mm	-	-	-	4.50	0.75	71.98	0.35	72.33	nr	77.75
Repairing doors and frames										
Piece in door where ironmongery removed										
rim lock	-	-	-	1.00	0.15	15.82	0.16	15.98	nr	17.17
rim lock or latch and furniture	-	-	-	1.50	0.25	23.99	0.39	24.38	nr	26.21
mortice lock	-	-	-	1.50	0.25	23.99	0.39	24.38	nr	26.21
mortice lock or latch and furniture	-	-	-	2.00	0.32	31.85	0.46	32.31	nr	34.73
butt hinge	-	-	-	0.50	0.08	7.96	0.22	8.18	nr	8.80
Piece in frame where ironmongery removed										
lock or latch keep	-	-	-	0.50	0.08	7.96	0.22	8.18	nr	8.80
butt hinge	-	-	-	0.50	0.08	7.96	0.22	8.18	nr	8.80
Take off door, dismantle as necessary, cramp and re-wedge with new glued wedges, pin and re-hang										
762 x 1981mm, panelled	-	-	-	5.00	0.84	80.05	0.16	80.21	nr	86.22
914 x 2134mm, panelled	-	-	-	6.00	1.00	95.97	0.16	96.13	nr	103.34
762 x 1981mm, glazed panelled	-	-	-	5.50	0.90	87.79	0.16	87.95	nr	94.55
914 x 2134mm, glazed panelled	-	-	-	6.50	1.08	103.93	0.16	104.09	nr	111.90
Take off softwood panelled door 762 x 1981mm, dismantle as necessary and replace the following damaged members and re-hang										
38 x 125mm top rail	5.39	10.00	5.93	2.15	0.36	34.41	0.45	40.78	nr	43.84
38 x 125mm hanging stile	14.01	10.00	15.41	2.36	0.39	37.71	1.14	54.26	nr	58.33
38 x 125mm locking stile	14.01	10.00	15.41	2.36	0.39	37.71	1.14	54.26	nr	58.33
38 x 200mm middle rail	8.42	10.00	9.26	4.23	0.71	67.71	0.74	77.71	nr	83.54
38 x 200mm bottom rail	8.42	10.00	9.26	2.20	0.37	35.22	0.74	45.23	nr	48.62
Take off afromosia panelled door 762 x 1981mm, dismantle as necessary and replace the following damaged members and re-hang										
38 x 125mm top rail	27.75	10.00	30.53	5.20	0.87	83.21	0.45	114.19	nr	122.75
38 x 125mm hanging stile	72.15	10.00	79.36	5.36	0.89	85.70	1.14	166.20	nr	178.67
38 x 125mm locking stile	72.15	10.00	79.36	5.36	0.89	85.70	1.14	166.20	nr	178.67
38 x 200mm middle rail	43.93	10.00	48.32	10.40	1.73	166.31	0.74	215.37	nr	231.53
38 x 200mm bottom rail	43.93	10.00	48.32	5.30	0.88	84.74	0.74	133.80	nr	143.84
Take off damaged softwood stop and replace with new										
12 x 38mm	0.46	10.00	0.51	0.35	0.06	5.62	0.10	6.22	m	6.69
25 (fin) x 50mm glue and screw on	0.69	10.00	0.76	0.45	0.08	7.25	0.12	8.13	m	8.74

Labour hourly rates: (except Specialists) Craft Operatives £14.21 Labourer £10.71 Rates are national average prices. Refer to REGIONAL VARIATIONS for indicative levels of overall pricing in regions	MATERIALS			LABOUR				RATES		
	Del to Site £	Waste %	Material Cost £	Craft Optve Hrs	Lab Hrs	Labour Cost £	Sunds £	Nett Rate £	Unit	Gross Rate (+7.5%) £
C51: REPAIRING/RENOVATING/CONSERVING TIMBER Cont.										
Repairing doors and frames Cont.										
Take off damaged afromosia stop and replace with new										
12 x 38mm	1.85	10.00	**2.04**	0.45	0.08	**7.25**	0.10	**9.39**	m	**10.09**
25 (fin) x 50mm glue and screw on	3.70	10.00	**4.07**	0.69	0.12	**11.09**	0.12	**15.28**	m	**16.43**
Flush facing old doors										
Take off panelled door, remove ironmongery, cover both sides with 3.2mm hardboard; refix ironmongery, adjust stops and re-hang										
762 x 1981mm	7.20	10.00	**7.92**	3.10	0.50	**49.41**	0.25	**57.58**	nr	**61.89**
838 x 1981mm	7.20	10.00	**7.92**	3.15	0.51	**50.22**	0.25	**58.39**	nr	**62.77**
Take off panelled door, remove ironmongery, cover one side with 3.2mm hardboard and other side with 6mm fire resistant sheeting screwed on; refix ironmongery, adjust stops and re-hang										
762 x 1981mm	44.09	10.00	**48.50**	4.00	0.67	**64.02**	0.82	**113.33**	nr	**121.83**
838 x 1981mm	44.09	10.00	**48.50**	4.05	0.68	**64.83**	0.82	**114.15**	nr	**122.71**
Take off panelled door, remove ironmongery, cover one side with 3.2mm hardboard, fill panels on other side with 6mm fire resistant sheeting and cover with 6mm fire resistant sheeting screwed on; refix ironmongery, adjust stops and re-hang										
762 x 1981mm	64.34	10.00	**70.77**	5.00	0.83	**79.94**	1.05	**151.76**	nr	**163.14**
838 x 1981mm	64.34	10.00	**70.77**	5.05	0.84	**80.76**	1.05	**152.58**	nr	**164.02**
Take off flush door, remove ironmongery, cover one side with 6mm fire resistant sheeting screwed on; refix ironmongery and re-hang										
762 x 1981mm	40.49	10.00	**44.54**	2.50	0.42	**40.02**	0.53	**85.09**	nr	**91.47**
838 x 1981mm	40.49	10.00	**44.54**	2.50	0.43	**40.13**	0.53	**85.20**	nr	**91.59**
Easing and overhauling windows										
Ease and adjust and oil ironmongery										
casement opening light	-	-	**-**	0.41	0.07	**6.58**	0.02	**6.60**	nr	**7.09**
double hung sash and renew sash lines with best quality flax cord	1.18	10.00	**1.30**	1.64	0.27	**26.20**	0.05	**27.55**	nr	**29.61**
Refixing removed windows										
Windows										
casement opening light and oil ironmongery	-	-	**-**	1.03	0.17	**16.46**	0.04	**16.50**	nr	**17.73**
double hung sash and provide new best quality flax sash cord	1.18	10.00	**1.30**	1.70	0.28	**27.16**	0.05	**28.51**	nr	**30.65**
Single or multi-light timber casements and frames										
440 x 920mm	-	-	**-**	4.00	0.66	**63.91**	0.04	**63.95**	nr	**68.74**
1225 x 1070mm	-	-	**-**	6.00	1.00	**95.97**	0.05	**96.02**	nr	**103.22**
2395 x 1225mm	-	-	**-**	8.00	1.33	**127.92**	0.06	**127.98**	nr	**137.58**
Repairing windows										
Remove damaged members of cased frame and replace with new softwood										
parting bead	0.58	10.00	**0.64**	1.16	0.19	**18.52**	0.05	**19.21**	m	**20.65**
pulley stile	2.25	10.00	**2.48**	1.68	0.28	**26.87**	0.14	**29.49**	m	**31.70**
stop bead	0.75	10.00	**0.83**	0.62	0.10	**9.88**	0.05	**10.76**	m	**11.56**
Take out sash not exceeding 50mm thick, dismantle as necessary, replace the following damaged members in softwood and re-hang with new best quality flax sash cord										
glazing bar	2.80	10.00	**3.08**	3.14	0.52	**50.19**	0.05	**53.32**	m	**57.32**
stile	3.43	10.00	**3.78**	3.03	0.50	**48.41**	0.09	**52.28**	m	**56.20**
rail	5.22	10.00	**5.75**	2.80	0.45	**44.61**	0.25	**50.60**	m	**54.40**
Take out casement opening light not exceeding 50mm thick, dismantle as necessary, replace the following damaged members in softwood and re-hang										
glazing bar	1.62	10.00	**1.78**	1.97	0.32	**31.42**	0.05	**33.25**	m	**35.75**
hanging stile	2.25	10.00	**2.48**	1.85	0.30	**29.50**	0.10	**32.08**	m	**34.48**
shutting stile	2.25	10.00	**2.48**	1.85	0.30	**29.50**	0.10	**32.08**	m	**34.48**
rail	4.04	10.00	**4.44**	1.61	0.27	**25.77**	0.26	**30.47**	m	**32.76**
Take out sash, cramp and re-wedge with new glued wedges, pin and re-hang										
sliding sash	-	-	**-**	5.65	0.95	**90.46**	0.87	**91.33**	nr	**98.18**
opening casement light	-	-	**-**	5.00	0.85	**80.15**	0.87	**81.02**	nr	**87.10**
Repair corner of sash with 100 x 100mm angle repair plate										
let in flush	0.38	-	**0.38**	1.00	0.17	**16.03**	0.17	**16.58**	nr	**17.82**
Carefully rake out and dry crack in sill and fill with two part wood repair system	-	-	**-**	0.50	0.08	**7.96**	1.50	**9.46**	m	**10.17**

Labour hourly rates: (except Specialists) Craft Operatives £14.21 Labourer £10.71 Rates are national average prices. Refer to REGIONAL VARIATIONS for indicative levels of overall pricing in regions	MATERIALS			LABOUR				RATES		
	Del to Site £	Waste %	Material Cost £	Craft Optve Hrs	Lab Hrs	Labour Cost £	Sunds £	Nett Rate £	Unit	Gross Rate (+7.5%) £
C51: REPAIRING/RENOVATING/CONSERVING TIMBER Cont.										
Repairs to softwood skirtings, picture rails, architraves, etc.										
Take off and refix										
skirting	-	-	-	0.40	0.07	**6.43**	0.04	**6.47**	m	**6.96**
dado or picture rail	-	-	-	0.35	0.06	**5.62**	0.03	**5.65**	m	**6.07**
architrave	-	-	-	0.26	0.05	**4.23**	0.03	**4.26**	m	**4.58**
Cut out damaged section and piece in new										
19 x 100mm square skirting	1.04	10.00	**1.14**	0.69	0.11	**10.98**	0.20	**12.33**	m	**13.25**
25 x 150mm square skirting	2.03	10.00	**2.23**	0.71	0.12	**11.37**	0.24	**13.85**	m	**14.89**
19 x 125mm moulded skirting	4.60	10.00	**5.06**	0.70	0.11	**11.13**	0.23	**16.42**	m	**17.65**
19 x 50mm moulded dado or picture rail	2.47	10.00	**2.72**	0.74	0.12	**11.80**	0.10	**14.62**	m	**15.71**
25 x 50mm moulded or splayed architrave	3.33	10.00	**3.66**	0.51	0.09	**8.21**	0.12	**11.99**	m	**12.89**
38 x 75mm moulded or splayed architrave	4.79	10.00	**5.27**	0.52	0.09	**8.35**	0.23	**13.85**	m	**14.89**
Cut out rotten or infected skirting and grounds and replace with new treated softwood										
19 x 100mm square skirting	1.48	10.00	**1.63**	1.00	0.17	**16.03**	0.27	**17.93**	m	**19.27**
25 x 150mm square skirting	2.47	10.00	**2.72**	1.07	0.18	**17.13**	0.37	**20.22**	m	**21.74**
19 x 125mm moulded skirting	5.04	10.00	**5.54**	1.03	0.17	**16.46**	0.31	**22.31**	m	**23.98**
Repairs to staircase members in softwood										
Cut out damaged members of staircase and piece in new										
32 x 275mm tread	7.89	10.00	**8.68**	4.33	0.72	**69.24**	0.46	**78.38**	m	**84.26**
38 x 275mm tread	9.38	10.00	**10.32**	4.85	0.81	**77.59**	0.53	**88.44**	m	**95.07**
19 x 200mm riser	3.84	10.00	**4.22**	2.25	0.38	**36.04**	0.18	**40.45**	m	**43.48**
25 x 200mm riser	4.95	10.00	**5.45**	3.03	0.51	**48.52**	0.27	**54.23**	m	**58.30**
Cut back old tread for a width of 125mm and provide and fix new nosing with glued and dowelled joints to tread										
32mm with rounded nosing	4.24	10.00	**4.66**	2.16	0.36	**34.55**	0.76	**39.97**	m	**42.97**
38mm with rounded nosing	4.90	10.00	**5.39**	2.55	0.43	**40.84**	0.85	**47.08**	m	**50.61**
32mm with moulded nosing	5.04	10.00	**5.54**	2.54	0.42	**40.59**	0.76	**46.90**	m	**50.41**
38mm with moulded nosing	5.72	10.00	**6.29**	3.06	0.51	**48.94**	0.90	**56.14**	m	**60.35**
Replace missing square bar baluster 914mm high; clean out old mortices										
25 x 25mm	0.75	-	**0.75**	0.56	0.09	**8.92**	-	**9.67**	nr	**10.40**
38 x 38mm	1.36	-	**1.36**	0.75	0.13	**12.05**	-	**13.41**	nr	**14.42**
Cut out damaged square bar baluster 914mm high and piece in new										
25 x 25mm	0.75	-	**0.75**	0.75	0.03	**10.98**	0.03	**11.76**	nr	**12.64**
38 x 38mm	1.36	-	**1.36**	0.91	0.15	**14.54**	0.05	**15.95**	nr	**17.14**
Cut out damaged handrail and piece in new										
50mm mopstick	25.21	10.00	**27.73**	1.95	0.33	**31.24**	3.20	**62.17**	m	**66.84**
75 x 100mm moulded	42.04	10.00	**46.24**	3.97	0.66	**63.48**	3.47	**113.20**	m	**121.69**
Refixing removed handrail to new balustrade	-	-	-	0.69	0.12	**11.09**	-	**11.09**	m	**11.92**
Repairs to staircase members in oak										
Cut out damaged members of staircase and piece in new										
32 x 275mm tread	45.40	10.00	**49.94**	6.09	1.02	**97.46**	0.46	**147.86**	m	**158.95**
38 x 275mm tread	54.71	10.00	**60.18**	7.18	1.20	**114.88**	0.53	**175.59**	m	**188.76**
19 x 200mm riser	24.45	10.00	**26.90**	3.34	0.56	**53.46**	0.18	**80.53**	m	**86.57**
25 x 200mm riser	30.27	10.00	**33.30**	4.56	0.76	**72.94**	0.29	**106.52**	m	**114.51**
Cut back old tread for a width of 125mm and provide and fix new nosing with glued and dowelled joints to tread										
32mm with rounded nosing	26.76	10.00	**29.44**	2.23	0.37	**35.65**	0.76	**65.85**	m	**70.79**
38mm with rounded nosing	30.27	10.00	**33.30**	2.92	0.49	**46.74**	0.90	**80.94**	m	**87.01**
32mm with moulded nosing	27.94	10.00	**30.73**	3.20	0.53	**51.15**	0.76	**82.64**	m	**88.84**
38mm with moulded nosing	31.43	10.00	**34.57**	3.95	0.66	**63.20**	0.90	**98.67**	m	**106.07**
Replace missing square bar baluster 914mm high; clean out old mortices										
25 x 25mm	3.95	-	**3.95**	0.94	0.16	**15.07**	-	**19.02**	nr	**20.45**
38 x 38mm	8.03	-	**8.03**	0.99	0.17	**15.89**	-	**23.92**	nr	**25.71**
Cut out damaged square bar baluster 914mm high and piece in new										
25 x 25mm	3.95	-	**3.95**	1.21	0.20	**19.34**	0.03	**23.32**	nr	**25.06**
38 x 38mm	8.03	-	**8.03**	1.32	0.22	**21.11**	0.05	**29.19**	nr	**31.38**
Cut out damaged handrail and piece in new										
50mm mopstick	55.49	10.00	**61.04**	3.64	0.61	**58.26**	3.20	**122.50**	m	**131.68**
75 x 100mm moulded	92.47	10.00	**101.72**	5.50	0.92	**88.01**	3.47	**193.20**	m	**207.68**
Refixing removed handrail to new balustrade	-	-	-	1.35	0.23	**21.65**	-	**21.65**	m	**23.27**
Remove broken ironmongery and fix only new to softwood										
Door ironmongery										
75mm butt hinges	-	-	-	0.23	0.04	**3.70**	0.01	**3.71**	nr	**3.98**
100mm butt hinges	-	-	-	0.23	0.04	**3.70**	0.01	**3.71**	nr	**3.98**
tee hinges up to 300mm	-	-	-	0.62	0.10	**9.88**	0.01	**9.89**	nr	**10.63**

Labour hourly rates: (except Specialists) Craft Operatives £14.21 Labourer £10.71 Rates are national average prices. Refer to REGIONAL VARIATIONS for indicative levels of overall pricing in regions	MATERIALS			LABOUR			RATES			
	Del to Site £	Waste %	Material Cost £	Craft Optve Hrs	Lab Hrs	Labour Cost £	Sunds £	Nett Rate £	Unit	Gross Rate (+7.5%) £
C51: REPAIRING/RENOVATING/CONSERVING TIMBER Cont.										
Remove broken ironmongery and fix only new to softwood Cont.										
Door ironmongery Cont.										
single action floor spring hinges	-	-	-	3.00	0.50	**47.99**	1.30	**49.29**	nr	**52.98**
barrel bolt up to 300mm	-	-	-	0.41	0.07	**6.58**	0.01	**6.59**	nr	**7.08**
lock or latch furniture	-	-	-	0.44	0.07	**7.00**	0.01	**7.01**	nr	**7.54**
rim lock	-	-	-	0.80	0.13	**12.76**	0.01	**12.77**	nr	**13.73**
rim latch	-	-	-	0.80	0.13	**12.76**	0.01	**12.77**	nr	**13.73**
mortice lock	-	-	-	1.22	0.20	**19.48**	0.01	**19.49**	nr	**20.95**
mortice latch	-	-	-	1.08	0.18	**17.27**	0.01	**17.28**	nr	**18.58**
pull handle	-	-	-	0.40	0.07	**6.43**	0.03	**6.46**	nr	**6.95**
Window ironmongery										
sash fastener	-	-	-	1.09	0.18	**17.42**	0.01	**17.43**	nr	**18.73**
sash lift	-	-	-	0.23	0.04	**3.70**	0.01	**3.71**	nr	**3.98**
casement stay	-	-	-	0.40	0.07	**6.43**	0.01	**6.44**	nr	**6.93**
casement fastener	-	-	-	0.40	0.07	**6.43**	0.01	**6.44**	nr	**6.93**
Sundry ironmongery										
hat and coat hook	-	-	-	0.29	0.05	**4.66**	0.01	**4.67**	nr	**5.02**
toilet roll holder	-	-	-	0.29	0.05	**4.66**	0.01	**4.67**	nr	**5.02**
Remove broken ironmongery and fix only new to hardwood										
Door ironmongery										
75mm butt hinges	-	-	-	0.30	0.05	**4.80**	0.01	**4.81**	nr	**5.17**
100mm butt hinges	-	-	-	0.30	0.05	**4.80**	0.01	**4.81**	nr	**5.17**
tee hinges up to 300mm	-	-	-	0.86	0.14	**13.72**	0.01	**13.73**	nr	**14.76**
single action floor spring hinges	-	-	-	4.20	0.70	**67.18**	1.30	**68.48**	nr	**73.61**
barrel bolt up to 300mm	-	-	-	0.60	0.10	**9.60**	0.01	**9.61**	nr	**10.33**
lock or latch furniture	-	-	-	0.61	0.10	**9.74**	0.01	**9.75**	nr	**10.48**
rim lock	-	-	-	1.15	0.19	**18.38**	0.01	**18.39**	nr	**19.77**
rim latch	-	-	-	1.15	0.19	**18.38**	0.01	**18.39**	nr	**19.77**
mortice lock	-	-	-	1.77	0.30	**28.36**	0.01	**28.37**	nr	**30.50**
mortice latch	-	-	-	1.58	0.26	**25.24**	0.01	**25.25**	nr	**27.14**
pull handle	-	-	-	0.57	0.10	**9.17**	0.03	**9.20**	nr	**9.89**
Window ironmongery										
sash fastener	-	-	-	1.57	0.26	**25.09**	0.01	**25.10**	nr	**26.99**
sash lift	-	-	-	0.31	0.05	**4.94**	0.01	**4.95**	nr	**5.32**
casement stay	-	-	-	0.56	0.09	**8.92**	0.01	**8.93**	nr	**9.60**
casement fastener	-	-	-	0.56	0.09	**8.92**	0.01	**8.93**	nr	**9.60**
Sundry ironmongery										
hat and coat hook	-	-	-	0.42	0.07	**6.72**	0.01	**6.73**	nr	**7.23**
toilet roll holder	-	-	-	0.42	0.07	**6.72**	0.01	**6.73**	nr	**7.23**
C52: FUNGUS/BEETLE ERADICATION										
Protective treatment of existing timbers										
Treat with two coats of spray applied preservative										
boarding	0.60	10.00	**0.66**	-	0.15	**1.61**	0.04	**2.30**	m2	**2.47**
structural timbers	0.89	10.00	**0.98**	-	0.20	**2.14**	0.05	**3.17**	m2	**3.41**
Treat with two coats of brush applied preservative										
boarding	0.89	10.00	**0.98**	-	0.26	**2.78**	0.03	**3.80**	m2	**4.08**
structural timbers	1.37	10.00	**1.51**	-	0.80	**8.57**	0.05	**10.12**	m2	**10.88**
Insecticide treatment of existing timbers										
Treat worm infected timbers with spray applied proprietary insecticide										
boarding	0.89	10.00	**0.98**	-	0.24	**2.57**	0.05	**3.60**	m2	**3.87**
structural timbers	1.19	10.00	**1.31**	-	0.28	**3.00**	0.06	**4.37**	m2	**4.70**
joinery timbers	1.19	10.00	**1.31**	-	0.32	**3.43**	0.06	**4.80**	m2	**5.16**
Treat worm infected timbers with brush applied proprietary insecticide										
boarding	1.19	10.00	**1.31**	-	0.36	**3.86**	0.03	**5.20**	m2	**5.59**
structural timbers	1.43	10.00	**1.57**	-	0.42	**4.50**	0.05	**6.12**	m2	**6.58**
joinery timbers	1.43	10.00	**1.57**	-	0.48	**5.14**	0.05	**6.76**	m2	**7.27**
Treatment of wall surfaces										
Treat surfaces of concrete and brickwork adjoining areas where infected timbers removed										
with a blow lamp	-	-	-	-	0.25	**2.68**	0.03	**2.71**	m2	**2.91**
C90: ALTERATIONS - SPOT ITEMS										
Removal of fittings and fixtures										
Removing fittings and fixtures										
cupboard with doors, frames, architraves, etc	-	-	-	-	0.59	**6.32**	1.23	**7.55**	m2	**8.12**
worktop with legs and bearers	-	-	-	-	0.56	**6.00**	1.65	**7.64**	m2	**8.22**
shelving exceeding 300mm wide with bearers and brackets	-	-	-	-	0.45	**4.82**	1.04	**5.86**	m2	**6.30**
shelving not exceeding 300mm wide with bearers and brackets	-	-	-	-	0.20	**2.14**	0.33	**2.47**	m	**2.66**
draining board and bearers										
500 x 600mm	-	-	-	-	0.28	**3.00**	0.63	**3.63**	nr	**3.90**
600 x 900mm	-	-	-	-	0.37	**3.96**	1.11	**5.08**	nr	**5.46**

Labour hourly rates: (except Specialists) Craft Operatives £14.21 Labourer £10.71 Rates are national average prices. Refer to REGIONAL VARIATIONS for indicative levels of overall pricing in regions	MATERIALS			LABOUR				RATES		
	Del to Site £	Waste %	Material Cost £	Craft Optve Hrs	Lab Hrs	Labour Cost £	Sunds £	Nett Rate £	Unit	Gross Rate (+7.5%) £
C90: ALTERATIONS - SPOT ITEMS Cont.										
Removal of fittings and fixtures Cont.										
Removing fittings and fixtures Cont.										
wall cupboard unit										
510 x 305 x 305mm	-	-	-	-	0.47	5.03	1.00	6.03	nr	6.48
510 x 305 x 610mm	-	-	-	-	0.65	6.96	1.93	8.89	nr	9.56
1020 x 305 x 610mm	-	-	-	-	0.94	10.07	3.92	13.99	nr	15.04
1220 x 305 x 610mm	-	-	-	-	1.15	12.32	4.67	16.99	nr	18.26
floor cupboard unit										
510 x 510 x 915mm	-	-	-	-	0.92	9.85	4.91	14.76	nr	15.87
610 x 510 x 915mm	-	-	-	-	1.21	12.96	5.85	18.81	nr	20.22
1020 x 510 x 915mm	-	-	-	-	1.65	17.67	9.81	27.48	nr	29.54
1220 x 510 x 915mm	-	-	-	-	1.93	20.67	11.76	32.43	nr	34.86
sink base unit										
585 x 510 x 895mm	-	-	-	-	0.92	9.85	5.53	15.39	nr	16.54
1070 x 510 x 895mm	-	-	-	-	1.65	17.67	10.07	27.74	nr	29.82
tall cupboard unit										
510 x 510 x 1505mm	-	-	-	-	1.56	16.71	8.08	24.79	nr	26.65
610 x 510 x 1505mm	-	-	-	-	1.77	18.96	9.66	28.61	nr	30.76
bath panel and bearers	-	-	-	-	0.47	5.03	1.04	6.08	nr	6.53
steel or concrete clothes line post with concrete base	-	-	-	-	0.50	5.36	2.06	7.42	nr	7.97

This section continues on the next page

Labour hourly rates: (except Specialists) Craft Operatives £17.38 Labourer £10.71 Rates are national average prices. Refer to REGIONAL VARIATIONS for indicative levels of overall pricing in regions	MATERIALS			LABOUR				RATES		
	Del to Site £	Waste %	Material Cost £	Craft Optve Hrs	Lab Hrs	Labour Cost £	Sunds £	Nett Rate £	Unit	Gross Rate (+7.5%) £
C90: ALTERATIONS - SPOT ITEMS										
Removal of plumbing and electrical installations										
Removing plumbing and electrical installations										
100mm cast iron rainwater gutters	-	-	-	0.20	-	**3.48**	0.20	**3.68**	m	**3.95**
75mm cast iron rainwater pipes	-	-	-	0.25	-	**4.35**	0.23	**4.57**	m	**4.91**
100mm cast iron rainwater pipes	-	-	-	0.25	-	**4.35**	0.38	**4.72**	m	**5.08**
cast iron rainwater head	-	-	-	0.30	-	**5.21**	0.63	**5.84**	nr	**6.28**
cast iron soil and vent pipe with caulked joints										
50mm	-	-	-	0.35	-	**6.08**	0.12	**6.20**	m	**6.67**
75mm	-	-	-	0.35	-	**6.08**	0.23	**6.31**	m	**6.78**
100mm	-	-	-	0.35	-	**6.08**	0.38	**6.46**	m	**6.95**
150mm	-	-	-	0.50	-	**8.69**	0.90	**9.59**	m	**10.31**
100mm asbestos cement soil and vent pipe	-	-	-	0.25	-	**4.35**	0.38	**4.72**	m	**5.08**
150mm asbestos cement soil and vent pipe	-	-	-	0.25	-	**4.35**	0.90	**5.25**	m	**5.64**
100mm p.v.c. soil and vent pipe	-	-	-	0.25	-	**4.35**	0.38	**4.72**	m	**5.08**
150mm p.v.c. soil and vent pipe	-	-	-	0.25	-	**4.35**	0.90	**5.25**	m	**5.64**
100mm lead soil and vent pipe	-	-	-	0.35	-	**6.08**	0.38	**6.46**	m	**6.95**
150mm lead soil and vent pipe	-	-	-	0.35	-	**6.08**	0.90	**6.98**	m	**7.51**
w.c. suites	-	-	-	1.55	-	**26.94**	3.86	**30.80**	nr	**33.11**
lavatory basins	-	-	-	1.35	-	**23.46**	2.32	**25.79**	nr	**27.72**
baths	-	-	-	1.55	-	**26.94**	7.74	**34.68**	nr	**37.28**
glazed ware sinks	-	-	-	1.35	-	**23.46**	3.86	**27.33**	nr	**29.38**
stainless steel sinks	-	-	-	1.35	-	**23.46**	2.32	**25.79**	nr	**27.72**
stainless steel sinks with single drainer	-	-	-	1.35	-	**23.46**	3.10	**26.57**	nr	**28.56**
stainless steel sinks with double drainer	-	-	-	1.35	-	**23.46**	3.86	**27.33**	nr	**29.38**
hot and cold water, waste pipes, etc. 8 - 25mm diameter										
copper	-	-	-	0.15	-	**2.61**	0.05	**2.65**	m	**2.85**
lead	-	-	-	0.15	-	**2.61**	0.05	**2.65**	m	**2.85**
polythene or p.v.c.	-	-	-	0.15	-	**2.61**	0.05	**2.65**	m	**2.85**
stainless steel	-	-	-	0.25	-	**4.35**	0.05	**4.39**	m	**4.72**
steel	-	-	-	0.25	-	**4.35**	0.05	**4.39**	m	**4.72**
hot and cold water, waste pipes, etc. 32 - 50mm diameter										
copper	-	-	-	0.20	-	**3.48**	0.12	**3.59**	m	**3.86**
lead	-	-	-	0.20	-	**3.48**	0.12	**3.59**	m	**3.86**
polythene or p.v.c.	-	-	-	0.20	-	**3.48**	0.12	**3.59**	m	**3.86**
stainless steel	-	-	-	0.30	-	**5.21**	0.12	**5.33**	m	**5.73**
steel	-	-	-	0.30	-	**5.21**	0.12	**5.33**	m	**5.73**
cold water cisterns up to 454 litres capacity	-	-	-	2.00	-	**34.76**	17.56	**52.32**	nr	**56.25**
hot water tanks or cylinders up to 227 litre capacity	-	-	-	2.00	-	**34.76**	8.77	**43.53**	nr	**46.79**
solid fuel boilers up to 66000 BTU output	-	-	-	3.75	-	**65.18**	-	**65.18**	nr	**70.06**
gas or oil fired boilers up to 66000 BTU output	-	-	-	3.75	-	**65.18**	-	**65.18**	nr	**70.06**
wall mounted water heaters	-	-	-	2.00	-	**34.76**	-	**34.76**	nr	**37.37**
wall type radiators up to 914mm long, up to 610mm high	-	-	-	1.00	-	**17.38**	0.78	**18.16**	nr	**19.52**
wall type radiators up to 914mm long, 610 - 914mm high	-	-	-	1.35	-	**23.46**	1.55	**25.02**	nr	**26.89**
wall type radiators 914 - 1829mm long, up to 610mm high	-	-	-	1.55	-	**26.94**	2.37	**29.31**	nr	**31.51**
wall type radiators 914 - 1829mm long, 610 - 914mm high	-	-	-	1.95	-	**33.89**	3.10	**37.00**	nr	**39.77**
column type radiators up to 914mm long, up to 610mm high	-	-	-	1.00	-	**17.38**	4.06	**21.44**	nr	**23.05**
column type radiators up to 914mm long, 610 - 914mm high	-	-	-	1.35	-	**23.46**	8.12	**31.58**	nr	**33.95**
column type radiators 914 - 1829mm long, up to 610mm high	-	-	-	1.55	-	**26.94**	12.18	**39.12**	nr	**42.05**
column type radiators 914 - 1829mm long, 610 - 914mm high	-	-	-	1.95	-	**33.89**	16.23	**50.13**	nr	**53.88**
pipe insulation from pipes 8 - 25mm diameter	-	-	-	0.10	-	**1.74**	0.12	**1.86**	m	**2.00**
pipe insulation from pipes 32 - 50mm diameter	-	-	-	0.13	-	**2.26**	0.23	**2.48**	m	**2.67**
water tank or calorifier insulation from tanks up to 227 litres	-	-	-	0.50	-	**8.69**	8.77	**17.46**	nr	**18.77**
water tank or calorifier insulation from tanks 227 - 454 litres	-	-	-	0.75	-	**13.04**	17.56	**30.60**	nr	**32.89**
Removing plumbing and electrical installations; setting aside for re-use										
cast iron soil and vent pipe with caulked joints										
50mm	-	-	-	0.55	-	**9.56**	-	**9.56**	m	**10.28**
75mm	-	-	-	0.60	-	**10.43**	-	**10.43**	m	**11.21**
100mm	-	-	-	0.80	-	**13.90**	-	**13.90**	m	**14.95**
150mm	-	-	-	1.00	-	**17.38**	-	**17.38**	m	**18.68**

This section continues on the next page

Labour hourly rates: (except Specialists) Craft Operatives £20.54 Labourer £10.71 Rates are national average prices. Refer to REGIONAL VARIATIONS for indicative levels of overall pricing in regions	MATERIALS			LABOUR				RATES		
	Del to Site £	Waste %	Material Cost £	Craft Optve Hrs	Lab Hrs	Labour Cost £	Sunds £	Nett Rate £	Unit	Gross Rate (+7.5%) £
C90: ALTERATIONS - SPOT ITEMS Cont.										
Removal of plumbing and electrical installations Cont.										
Removing plumbing and electrical installations										
wall mounted electric fire	-	-	-	2.00	-	**41.08**	2.06	**43.14**	nr	**46.38**
1.5 kW night storage heater	-	-	-	2.50	-	**51.35**	6.19	**57.54**	nr	**61.85**
2 KW night storage heater	-	-	-	2.50	-	**51.35**	8.25	**59.60**	nr	**64.07**
3 KW night storage heater	-	-	-	2.50	-	**51.35**	12.37	**63.72**	nr	**68.50**
Removing plumbing and electrical installations; extending and making good finishings										
lighting points	-	-	-	1.25	-	**25.68**	1.68	**27.36**	nr	**29.41**
flush type switches	-	-	-	1.25	-	**25.68**	1.68	**27.36**	nr	**29.41**
surface mounted type switches	-	-	-	1.00	-	**20.54**	1.35	**21.89**	nr	**23.53**
flush type socket outlets	-	-	-	1.25	-	**25.68**	1.68	**27.36**	nr	**29.41**
surface mounted type socket outlets	-	-	-	1.00	-	**20.54**	1.35	**21.89**	nr	**23.53**
flush type fitting points	-	-	-	1.25	-	**25.68**	1.68	**27.36**	nr	**29.41**
surface mounted type fitting points	-	-	-	1.00	-	**20.54**	1.35	**21.89**	nr	**23.53**
surface mounted p.v.c. insulated and sheathed cables	-	-	-	0.04	-	**0.82**	0.07	**0.89**	m	**0.96**
surface mounted mineral insulated copper sheathed cables	-	-	-	0.10	-	**2.05**	0.14	**2.20**	m	**2.36**
conduits up to 25mm diameter with junction boxes	-	-	-	0.15	-	**3.08**	0.20	**3.28**	m	**3.53**
surface mounted cable trunking up to 100 x 100mm	-	-	-	0.55	-	**11.30**	0.73	**12.03**	m	**12.93**

This section continues on the next page

Labour hourly rates: (except Specialists) Craft Operatives £14.21 Labourer £10.71 Rates are national average prices. Refer to REGIONAL VARIATIONS for indicative levels of overall pricing in regions	MATERIALS			LABOUR				RATES		
	Del to Site £	Waste %	Material Cost £	Craft Optve Hrs	Lab Hrs	Labour Cost £	Sunds £	Nett Rate £	Unit	Gross Rate (+7.5%) £
C90: ALTERATIONS - SPOT ITEMS Cont.										
Removal of floor, wall and ceiling finishings										
Removing finishings										
cement and sand to floors	-	-	-	-	0.86	**9.21**	2.06	**11.27**	m2	**12.12**
granolithic to floors	-	-	-	-	0.92	**9.85**	2.06	**11.92**	m2	**12.81**
granolithic to treads and risers	-	-	-	-	1.43	**15.32**	2.06	**17.38**	m2	**18.68**
plastic or similar tiles to floors	-	-	-	-	0.20	**2.14**	0.21	**2.36**	m2	**2.53**
plastic or similar tiles to floors; cleaning off for new	-	-	-	-	0.47	**5.03**	0.21	**5.25**	m2	**5.64**
plastic or similar tiles to floors and screed under	-	-	-	-	1.05	**11.25**	2.28	**13.52**	m2	**14.53**
linoleum and underlay to floors	-	-	-	-	0.07	**0.75**	2.06	**2.81**	m2	**3.02**
ceramic or quarry tiles to floors	-	-	-	-	1.21	**12.96**	1.23	**14.19**	m2	**15.26**
ceramic or quarry tiles to floors and screed under	-	-	-	-	2.39	**25.60**	3.29	**28.89**	m2	**31.06**
extra; cleaning and keying surface of concrete under	-	-	-	-	0.53	**5.68**	0.05	**5.72**	m2	**6.15**
asphalt to floors on loose underlay	-	-	-	-	0.35	**3.75**	0.83	**4.58**	m2	**4.92**
asphalt to floors keyed to concrete or screed	-	-	-	-	0.76	**8.14**	0.83	**8.97**	m2	**9.64**
granolithic skirtings	-	-	-	-	0.33	**3.53**	0.21	**3.75**	m	**4.03**
plastic or similar skirtings; cleaning off for new	-	-	-	-	0.13	**1.39**	0.05	**1.44**	m	**1.55**
timber skirtings	-	-	-	-	0.12	**1.29**	0.21	**1.50**	m	**1.61**
ceramic or quarry tile skirtings	-	-	-	-	0.43	**4.61**	0.24	**4.84**	m	**5.21**
plaster to walls	-	-	-	-	0.60	**6.43**	0.83	**7.26**	m2	**7.80**
rendering to walls	-	-	-	-	0.87	**9.32**	0.83	**10.15**	m2	**10.91**
rough cast to walls	-	-	-	-	0.90	**9.64**	1.04	**10.68**	m2	**11.48**
match boarding linings to walls with battens	-	-	-	-	0.38	**4.07**	1.23	**5.30**	m2	**5.70**
plywood or similar sheet linings to walls with battens	-	-	-	-	0.31	**3.32**	1.23	**4.55**	m2	**4.89**
insulating board linings to walls with battens	-	-	-	-	0.31	**3.32**	1.23	**4.55**	m2	**4.89**
plasterboard to walls	-	-	-	-	0.31	**3.32**	0.63	**3.95**	m2	**4.24**
plasterboard dry linings to walls	-	-	-	-	0.50	**5.36**	0.63	**5.98**	m2	**6.43**
plasterboard and skim to walls	-	-	-	-	0.60	**6.43**	0.83	**7.26**	m2	**7.80**
lath and plaster to walls	-	-	-	-	0.44	**4.71**	1.04	**5.76**	m2	**6.19**
metal lath and plaster to walls	-	-	-	-	0.33	**3.53**	1.04	**4.58**	m2	**4.92**
ceramic tiles to walls	-	-	-	-	0.75	**8.03**	0.63	**8.66**	m2	**9.31**
ceramic tiles to walls and backing under	-	-	-	-	1.11	**11.89**	1.04	**12.93**	m2	**13.90**
asphalt coverings to walls keyed to concrete or brickwork	-	-	-	-	0.66	**7.07**	0.83	**7.90**	m2	**8.49**
plaster to ceilings	-	-	-	-	0.74	**7.93**	0.83	**8.75**	m2	**9.41**
match boarding linings to ceilings with battens	-	-	-	-	0.54	**5.78**	1.23	**7.02**	m2	**7.54**
plywood or similar sheet linings to ceilings with battens	-	-	-	-	0.38	**4.07**	1.23	**5.30**	m2	**5.70**
insulating board linings to ceilings with battens	-	-	-	-	0.26	**2.78**	1.23	**4.02**	m2	**4.32**
plasterboard to ceilings	-	-	-	-	0.33	**3.53**	0.63	**4.16**	m2	**4.47**
plasterboard and skim to ceilings	-	-	-	-	0.40	**4.28**	0.83	**5.11**	m2	**5.50**
lath and plaster to ceilings	-	-	-	-	0.59	**6.32**	1.04	**7.36**	m2	**7.91**
metal lath and plaster to ceilings	-	-	-	-	0.47	**5.03**	1.04	**6.08**	m2	**6.53**
Hacking surfaces of concrete as key for new finishings										
ceiling	-	-	-	-	0.67	**7.18**	0.05	**7.22**	m2	**7.76**
floor	-	-	-	-	0.53	**5.68**	0.05	**5.72**	m2	**6.15**
wall	-	-	-	-	0.53	**5.68**	0.05	**5.72**	m2	**6.15**
hacking surfaces of brick wall and raking out joints as key for new finishings	-	-	-	-	0.60	**6.43**	0.05	**6.47**	m2	**6.96**
Hacking surfaces of brick wall and raking out joints as key for new finishings										
wall	-	-	-	-	0.60	**6.43**	0.05	**6.47**	m2	**6.96**
Removing finishings; extending and making good finishings										
plaster cornices up to 100mm girth on face	0.35	10.00	**0.38**	0.78	0.78	**19.44**	0.08	**19.90**	m	**21.39**
plaster cornices 100 - 200mm girth on face	0.57	10.00	**0.62**	0.88	0.88	**21.93**	0.21	**22.77**	m	**24.47**
plaster cornices 200 - 300mm girth on face	1.00	10.00	**1.10**	0.97	0.97	**24.17**	0.63	**25.90**	m	**27.84**
plaster ceiling roses up to 300mm diameter	0.17	10.00	**0.19**	0.41	0.41	**10.22**	0.17	**10.57**	nr	**11.37**
plaster ceiling roses 300 - 450mm diameter	0.35	10.00	**0.38**	0.64	0.64	**15.95**	0.33	**16.66**	nr	**17.91**
plaster ceiling roses 450 - 600mm diameter	0.68	10.00	**0.75**	0.90	0.90	**22.43**	0.65	**23.83**	nr	**25.61**
Removing finishings; carefully handling and disposing toxic or other special waste by approved method										
asbestos cement sheet linings to walls	-	-	-	2.00	2.00	**49.84**	3.29	**53.13**	m2	**57.12**
asbestos cement sheet linings to ceilings	-	-	-	3.00	3.00	**74.76**	3.29	**78.05**	m2	**83.91**
Removal of roof coverings										
Removing coverings										
felt to roofs	-	-	-	-	0.28	**3.00**	0.83	**3.83**	m2	**4.12**
felt skirtings to roofs	-	-	-	-	0.33	**3.53**	0.08	**3.62**	m	**3.89**
asphalt to roofs	-	-	-	-	0.35	**3.75**	1.04	**4.79**	m2	**5.15**
asphalt skirtings to roofs; on expanded metal reinforcement	-	-	-	-	0.20	**2.14**	0.17	**2.31**	m	**2.48**
asphalt skirtings to roofs; keyed to concrete, brickwork, etc.	-	-	-	-	0.30	**3.21**	0.17	**3.38**	m	**3.63**
asphalt coverings to roofs; on expanded metal reinforcement; per 100mm of width	-	-	-	-	0.13	**1.39**	0.13	**1.52**	m	**1.64**
asphalt coverings to roofs; keyed to concrete, brickwork, etc.; per 100mm of width	-	-	-	-	0.20	**2.14**	0.13	**2.27**	m	**2.44**
slate to roofs	-	-	-	-	0.34	**3.64**	0.83	**4.47**	m2	**4.81**
tiles to roofs	-	-	-	-	0.31	**3.32**	1.23	**4.55**	m2	**4.89**
extra; removing battens	-	-	-	-	0.28	**3.00**	0.83	**3.83**	m2	**4.12**
extra; removing counter battens	-	-	-	-	0.10	**1.07**	0.17	**1.24**	m2	**1.33**
extra; removing underfelt	-	-	-	-	0.20	**2.14**	0.33	**2.47**	m2	**2.66**
lead to roofs	-	-	-	-	0.48	**5.14**	0.83	**5.97**	m2	**6.42**
lead flashings to roofs; per 25mm of girth	-	-	-	-	0.05	**0.54**	0.05	**0.58**	m	**0.63**

Labour hourly rates: (except Specialists) Craft Operatives £14.21 Labourer £10.71 Rates are national average prices. Refer to REGIONAL VARIATIONS for indicative levels of overall pricing in regions	MATERIALS			LABOUR				RATES		
	Del to Site £	Waste %	Material Cost £	Craft Optve Hrs	Lab Hrs	Labour Cost £	Sunds £	Nett Rate £	Unit	Gross Rate (+7.5%) £
C90: ALTERATIONS - SPOT ITEMS Cont.										
Removal of roof coverings Cont.										
Removing coverings Cont.										
zinc to roofs	-	-	-	-	0.34	3.64	0.83	4.47	m2	4.81
zinc flashings to roofs; per 25mm of girth	-	-	-	-	0.03	0.32	0.05	0.37	m	0.40
copper to roofs	-	-	-	-	0.34	3.64	0.83	4.47	m2	4.81
copper flashings to roofs; per 25mm of girth	-	-	-	-	0.03	0.32	0.05	0.37	m	0.40
corrugated metal sheeting to roofs	-	-	-	-	0.31	3.32	1.23	4.55	m2	4.89
corrugated translucent sheeting to roofs	-	-	-	-	0.31	3.32	1.23	4.55	m2	4.89
board roof decking	-	-	-	-	0.31	3.32	1.23	4.55	m2	4.89
woodwool roof decking	-	-	-	-	0.39	4.18	1.23	5.41	m2	5.81
Removing coverings; setting aside for re-use										
slate to roofs	-	-	-	-	0.60	6.43	-	6.43	m2	6.91
tiles to roofs	-	-	-	-	0.60	6.43	-	6.43	m2	6.91
clean and stack 405 x 205mm slates; per 100	-	-	-	-	1.50	16.07	-	16.07	nr	17.27
clean and stack 510 x 255mm slates; per 100	-	-	-	-	1.89	20.24	-	20.24	nr	21.76
clean and stack 610 x 305mm slates; per 100	-	-	-	-	2.24	23.99	-	23.99	nr	25.79
clean and stack concrete tiles; per 100	-	-	-	-	2.24	23.99	-	23.99	nr	25.79
clean and stack plain tiles; per 100	-	-	-	-	1.50	16.07	-	16.07	nr	17.27
Removing coverings; carefully handling and disposing toxic or other special waste by approved method										
asbestos cement sheeting to roofs	-	-	-	2.00	2.00	49.84	3.29	53.13	m2	57.12
asbestos cement roof decking	-	-	-	2.00	2.00	49.84	3.29	53.13	m2	57.12
Removal of woodwork										
Removing										
stud partitions plastered both sides	-	-	-	-	0.66	7.07	3.10	10.17	m2	10.94
roof boarding	-	-	-	-	0.19	2.03	1.04	3.08	m2	3.31
roof boarding; prepare joists for new	-	-	-	-	0.45	4.82	1.04	5.86	m2	6.30
gutter boarding and the like	-	-	-	-	0.19	2.03	1.04	3.08	m2	3.31
weather boarding and battens	-	-	-	-	0.24	2.57	1.23	3.80	m2	4.09
tilting fillets, angle fillets and the like	-	-	-	-	0.09	0.96	0.08	1.05	m	1.13
fascia boards 150 mm wide	-	-	-	-	0.10	1.07	0.17	1.24	m	1.33
barge boards 200mm wide	-	-	-	-	0.10	1.07	0.21	1.28	m	1.38
soffit boards 300mm wide	-	-	-	-	0.10	1.07	0.33	1.40	m	1.51
floor boarding	-	-	-	-	0.33	3.53	1.04	4.58	m2	4.92
handrails and brackets	-	-	-	-	0.10	1.07	0.65	1.72	m	1.85
balustrades complete down to and with cappings to aprons or strings	-	-	-	0.58	0.10	9.31	1.65	10.96	m	11.78
newel posts; cut off flush with landing or string	-	-	-	0.48	0.08	7.68	0.41	8.09	nr	8.70
ends of treads projecting beyond face of cut outer string including scotia under; make good treads where balusters removed	-	-	-	0.90	0.15	14.40	1.23	15.63	m	16.80
dado or picture rails with grounds	-	-	-	-	0.10	1.07	0.17	1.24	m	1.33
architrave	-	-	-	-	0.06	0.64	0.17	0.81	m	0.87
window boards and bearers	-	-	-	-	0.12	1.29	0.33	1.62	m	1.74
Removing; setting aside for reuse										
handrails and brackets	-	-	-	0.25	0.04	3.98	-	3.98	m	4.28
Removal of windows and doors										
Removing										
metal windows with internal and external sills; in conjunction with demolition										
997 x 923mm	-	-	-	-	0.40	4.28	1.98	6.26	nr	6.73
1486 x 923mm	-	-	-	-	0.50	5.36	2.84	8.20	nr	8.81
1486 x 1513mm	-	-	-	-	0.60	6.43	4.67	11.09	nr	11.93
1994 x 1513mm	-	-	-	-	0.70	7.50	6.23	13.73	nr	14.76
metal windows with internal and external sills; preparatory to filling openings; cut out lugs										
997 x 923mm	-	-	-	-	1.30	13.92	1.98	15.90	nr	17.09
1486 x 923mm	-	-	-	-	1.60	17.14	2.84	19.98	nr	21.48
1486 x 1513mm	-	-	-	-	2.00	21.42	4.67	26.09	nr	28.05
1994 x 1513mm	-	-	-	-	2.30	24.63	6.23	30.87	nr	33.18
wood single or multi-light casements and frames with internal and external sills; in conjunction with demolition										
440 x 920mm	-	-	-	-	0.28	3.00	1.28	4.28	nr	4.60
1225 x 1070mm	-	-	-	-	0.29	3.11	4.04	7.15	nr	7.68
2395 x 1225mm	-	-	-	-	0.30	3.21	9.08	12.29	nr	13.21
wood single or multi-light casements and frames with internal and external sills; preparatory to filling openings; cut out fixing cramps										
440 x 920mm	-	-	-	-	1.25	13.39	1.28	14.67	nr	15.77
1225 x 1070mm	-	-	-	-	1.70	18.21	4.04	22.25	nr	23.92
2395 x 1225mm	-	-	-	-	2.52	26.99	9.08	36.07	nr	38.77
wood cased frames and sashes with internal and external sills complete with accessories and weights; in conjunction with demolition										
610 x 1225mm	-	-	-	-	0.61	6.53	3.10	9.64	nr	10.36
915 x 1525mm	-	-	-	-	0.66	7.07	5.77	12.84	nr	13.80
1370 x 1525mm	-	-	-	-	0.75	8.03	8.61	16.65	nr	17.90
wood cased frames and sashes with internal and external sills complete with accessories and weights; preparatory to filling openings; cut out fixing cramps										
610 x 1225mm	-	-	-	-	1.79	19.17	3.10	22.28	nr	23.95
915 x 1525mm	-	-	-	-	2.56	27.42	5.77	33.19	nr	35.68
1370 x 1525mm	-	-	-	-	4.11	44.02	8.61	52.63	nr	56.58

Labour hourly rates: (except Specialists) Craft Operatives £14.21 Labourer £10.71 Rates are national average prices. Refer to REGIONAL VARIATIONS for indicative levels of overall pricing in regions	MATERIALS			LABOUR				RATES		
	Del to Site £	Waste %	Material Cost £	Craft Optve Hrs	Lab Hrs	Labour Cost £	Sunds £	Nett Rate £	Unit	Gross Rate (+7.5%) £
C90: ALTERATIONS - SPOT ITEMS Cont.										
Removal of windows and doors Cont.										
Removing Cont.										
wood doors										
single internal doors	-	-	-	-	0.23	**2.46**	2.48	**4.94**	nr	**5.31**
single internal doors and frames; in conjunction with demolition	-	-	-	-	0.18	**1.93**	5.57	**7.50**	nr	**8.06**
single internal doors and frames; preparatory to filling openings; cut out fixing cramps	-	-	-	-	1.00	**10.71**	5.57	**16.28**	nr	**17.50**
double internal doors	-	-	-	-	0.47	**5.03**	4.95	**9.99**	nr	**10.74**
double internal doors and frames; in conjunction with demolition	-	-	-	-	0.19	**2.03**	8.46	**10.50**	nr	**11.28**
double internal doors and frames; preparatory to filling openings; cut out fixing cramps	-	-	-	-	1.20	**12.85**	8.46	**21.31**	nr	**22.91**
single external doors	-	-	-	-	0.25	**2.68**	4.12	**6.80**	nr	**7.31**
single external doors and frames; in conjunction with demolition	-	-	-	-	0.18	**1.93**	7.42	**9.35**	nr	**10.05**
single external doors and frames; preparatory to filling openings; cut out fixing cramps	-	-	-	-	1.00	**10.71**	7.42	**18.13**	nr	**19.49**
double external doors	-	-	-	-	0.47	**5.03**	4.95	**9.99**	nr	**10.74**
double external doors and frames; in conjunction with demolition	-	-	-	-	0.19	**2.03**	8.46	**10.50**	nr	**11.28**
double external doors and frames; preparatory to filling openings; cut out fixing cramps	-	-	-	-	1.20	**12.85**	8.46	**21.31**	nr	**22.91**
single door frames; in conjunction with demolition	-	-	-	-	0.17	**1.82**	3.10	**4.93**	nr	**5.29**
single door frames; preparatory to filling openings; cut out fixing cramps	-	-	-	-	0.77	**8.25**	3.10	**11.35**	nr	**12.20**
double door frames; in conjunction with demolition	-	-	-	-	0.18	**1.93**	3.51	**5.44**	nr	**5.84**
double door frames; preparatory to filling openings; cut out fixing cramps	-	-	-	-	0.82	**8.78**	3.51	**12.29**	nr	**13.21**
Removing; setting aside for reuse										
metal windows with internal and external sills; in conjunction with demolition										
997 x 923mm	-	-	-	0.70	0.50	**15.30**	-	**15.30**	nr	**16.45**
1486 x 923mm	-	-	-	0.85	0.65	**19.04**	-	**19.04**	nr	**20.47**
1486 x 1513mm	-	-	-	0.90	0.75	**20.82**	-	**20.82**	nr	**22.38**
1994 x 1513mm	-	-	-	1.00	0.85	**23.31**	-	**23.31**	nr	**25.06**
metal windows with internal and external sills; preparatory to filling openings; cut out lugs										
997 x 923mm	-	-	-	0.70	1.40	**24.94**	-	**24.94**	nr	**26.81**
1486 x 923mm	-	-	-	0.85	1.75	**30.82**	-	**30.82**	nr	**33.13**
1486 x 1513mm	-	-	-	0.90	2.15	**35.82**	-	**35.82**	nr	**38.50**
1994 x 1513mm	-	-	-	1.00	2.45	**40.45**	-	**40.45**	nr	**43.48**
wood single or multi-light casements and frames with internal and external sills; in conjunction with demolition										
440 x 920mm	-	-	-	0.50	0.36	**10.96**	-	**10.96**	nr	**11.78**
1225 x 1070mm	-	-	-	0.83	0.43	**16.40**	-	**16.40**	nr	**17.63**
2395 x 1225mm	-	-	-	1.00	0.46	**19.14**	-	**19.14**	nr	**20.57**
wood single or multi-light casements and frames with internal and external sills; preparatory to filling openings; cut out fixing cramps										
440 x 920mm	-	-	-	0.50	1.33	**21.35**	-	**21.35**	nr	**22.95**
1225 x 1070mm	-	-	-	0.83	1.84	**31.50**	-	**31.50**	nr	**33.86**
2395 x 1225mm	-	-	-	1.00	2.68	**42.91**	-	**42.91**	nr	**46.13**
wood cased frames and sashes with internal and external sills complete with accessories and weights; in conjunction with demolition										
610 x 1225mm	-	-	-	0.97	0.78	**22.14**	-	**22.14**	nr	**23.80**
915 x 1525mm	-	-	-	1.63	0.93	**33.12**	-	**33.12**	nr	**35.61**
1370 x 1525mm	-	-	-	1.94	1.07	**39.03**	-	**39.03**	nr	**41.95**
wood cased frames and sashes with internal and external sills complete with accessories and weights; preparatory to filling openings; cut out fixing cramps										
610 x 1225mm	-	-	-	0.97	1.96	**34.78**	-	**34.78**	nr	**37.38**
915 x 1525mm	-	-	-	1.63	2.83	**53.47**	-	**53.47**	nr	**57.48**
1370 x 1525mm	-	-	-	1.94	4.43	**75.01**	-	**75.01**	nr	**80.64**
note: the above rates assume that windows and sashes will be re-glazed when re-used										
wood doors										
single internal doors	-	-	-	0.50	0.31	**10.43**	-	**10.43**	nr	**11.21**
single internal doors and frames; in conjunction with demolition	-	-	-	1.00	0.34	**17.85**	-	**17.85**	nr	**19.19**
single internal doors and frames; preparatory to filling openings; cut out fixing cramps	-	-	-	1.00	1.16	**26.63**	-	**26.63**	nr	**28.63**
double internal doors	-	-	-	1.00	0.63	**20.96**	-	**20.96**	nr	**22.53**
double internal doors and frames; in conjunction with demolition	-	-	-	1.47	0.43	**25.49**	-	**25.49**	nr	**27.41**
double internal doors and frames; preparatory to filling openings; cut out fixing cramps	-	-	-	1.47	1.45	**36.42**	-	**36.42**	nr	**39.15**
single external doors	-	-	-	0.50	0.33	**10.64**	-	**10.64**	nr	**11.44**
single external doors and frames; in conjunction with demolition	-	-	-	1.00	0.34	**17.85**	-	**17.85**	nr	**19.19**
single external doors and frames; preparatory to filling openings; cut out fixing cramps	-	-	-	1.00	1.16	**26.63**	-	**26.63**	nr	**28.63**
double external doors	-	-	-	1.00	0.63	**20.96**	-	**20.96**	nr	**22.53**
double external doors and frames; in conjunction with demolition	-	-	-	1.47	0.44	**25.60**	-	**25.60**	nr	**27.52**
double external doors and frames; preparatory to filling openings; cut out fixing cramps	-	-	-	1.47	1.45	**36.42**	-	**36.42**	nr	**39.15**

Labour hourly rates: (except Specialists) Craft Operatives £14.21 Labourer £10.71 Rates are national average prices. Refer to REGIONAL VARIATIONS for indicative levels of overall pricing in regions	MATERIALS			LABOUR				RATES		
	Del to Site £	Waste %	Material Cost £	Craft Optve Hrs	Lab Hrs	Labour Cost £	Sunds £	Nett Rate £	Unit	Gross Rate (+7.5%) £
C90: ALTERATIONS - SPOT ITEMS Cont.										
Removal of windows and doors Cont.										
Removing; setting aside for reuse Cont.										
wood doors Cont.										
single door frames; in conjunction with demolition	-	-	-	0.50	0.25	9.78	-	9.78	nr	10.52
single door frames; preparatory to filling openings; cut out fixing cramps	-	-	-	0.50	0.85	16.21	-	16.21	nr	17.42
double door frames; in conjunction with demolition	-	-	-	0.50	0.26	9.89	-	9.89	nr	10.63
double door frames; preparatory to filling openings; cut out fixing cramps	-	-	-	0.50	0.90	16.74	-	16.74	nr	18.00
Removal of ironmongery										
Removing door ironmongery (piecing in doors and frames included elsewhere)										
butt hinges	-	-	-	-	0.06	0.64	0.01	0.65	nr	0.70
tee hinges up to 300mm	-	-	-	-	0.20	2.14	0.01	2.15	nr	2.32
floor spring hinges and top centres	-	-	-	-	0.92	9.85	0.24	10.09	nr	10.85
barrel bolts up to 200mm	-	-	-	-	0.13	1.39	0.01	1.40	nr	1.51
flush bolts up to 200mm	-	-	-	-	0.17	1.82	0.01	1.83	nr	1.97
indicating bolts	-	-	-	-	0.23	2.46	0.01	2.48	nr	2.66
double panic bolts	-	-	-	-	0.69	7.39	0.13	7.52	nr	8.08
single panic bolts	-	-	-	-	0.58	6.21	0.08	6.29	nr	6.77
Norfolk or Suffolk latches	-	-	-	-	0.23	2.46	0.01	2.48	nr	2.66
cylinder rim night latches	-	-	-	-	0.23	2.46	0.01	2.48	nr	2.66
rim locks or latches and furniture	-	-	-	-	0.23	2.46	0.01	2.48	nr	2.66
mortice dead locks	-	-	-	-	0.17	1.82	0.01	1.83	nr	1.97
mortice locks or latches and furniture	-	-	-	-	0.29	3.11	0.01	3.12	nr	3.35
Bales catches	-	-	-	-	0.12	1.29	0.01	1.30	nr	1.39
overhead door closers, surface fixed	-	-	-	-	0.29	3.11	0.13	3.24	nr	3.48
pull handles	-	-	-	-	0.12	1.29	0.01	1.30	nr	1.39
push plates	-	-	-	-	0.12	1.29	0.01	1.30	nr	1.39
kicking plates	-	-	-	-	0.17	1.82	0.05	1.87	nr	2.01
letter plates	-	-	-	-	0.12	1.29	0.01	1.30	nr	1.39
Removing window ironmongery										
sash centres	-	-	-	-	0.12	1.29	0.01	1.30	nr	1.39
sash fasteners	-	-	-	-	0.17	1.82	0.01	1.83	nr	1.97
sash lifts	-	-	-	-	0.07	0.75	0.01	0.76	nr	0.82
sash screws	-	-	-	-	0.17	1.82	0.01	1.83	nr	1.97
casement fasteners	-	-	-	-	0.12	1.29	0.01	1.30	nr	1.39
casement stays	-	-	-	-	0.12	1.29	0.01	1.30	nr	1.39
quadrant stays	-	-	-	-	0.12	1.29	0.01	1.30	nr	1.39
fanlight catches	-	-	-	-	0.13	1.39	0.01	1.40	nr	1.51
curtain tracks	-	-	-	-	0.15	1.61	0.08	1.69	m	1.82
Removing sundry ironmongery										
hat and coat hooks	-	-	-	-	0.07	0.75	0.01	0.76	nr	0.82
cabin hooks and eyes	-	-	-	-	0.12	1.29	0.01	1.30	nr	1.39
shelf brackets	-	-	-	-	0.07	0.75	0.05	0.80	nr	0.86
toilet roll holders	-	-	-	-	0.09	0.96	0.05	1.01	nr	1.09
towel rollers	-	-	-	-	0.12	1.29	0.08	1.37	nr	1.47
Removal of metalwork										
Removing										
balustrades 1067mm high	-	-	-	-	0.81	8.68	1.07	9.74	m	10.47
wire mesh screens with timber or metal beads screwed on										
generally	-	-	-	-	0.62	6.64	1.04	7.68	m2	8.26
225 x 225mm	-	-	-	-	0.20	2.14	0.08	2.22	nr	2.39
305 x 305mm	-	-	-	-	0.22	2.36	0.13	2.49	nr	2.67
guard bars of vertical bars at 100mm centres welded to horizontal fixing bars at 450mm centres, fixed with bolts	-	-	-	-	0.62	6.64	1.04	7.68	m2	8.26
guard bars of vertical bars at 100mm centres welded to horizontal fixing bars at 450mm centres, fixed with screws	-	-	-	-	0.39	4.18	1.04	5.22	m2	5.61
extra; piecing in softwood after removal of bolts	-	-	-	0.25	0.04	3.98	0.11	4.09	nr	4.39
small bracket	-	-	-	-	0.18	1.93	0.08	2.01	nr	2.16
Removing; making good finishings										
balustrades 1067mm high; making good mortices in treads	-	-	-	0.97	0.97	24.17	1.35	25.52	m	27.44
guard bars of vertical bars at 100mm centres welded to horizontal fixing bars at 450mm centres fixed in mortices; making good concrete or brickwork and plaster	-	-	-	0.60	1.42	23.73	1.41	25.14	m2	27.03
small bracket built or cast in; making good concrete or brickwork and plaster	-	-	-	0.25	0.25	6.23	0.56	6.79	nr	7.30
Removal of manholes										
Removing										
manholes overall size; remove cover and frame; break up brick sides and concrete bottom; fill in void with hardcore										
overall size 914 x 1067mm and 600mm deep to invert	2.00	30.00	2.60	-	10.00	107.10	4.12	113.82	nr	122.36
overall size 914 x 1219mm and 900mm deep to invert	4.00	30.00	5.20	-	11.50	123.17	4.54	132.90	nr	142.87

Labour hourly rates: (except Specialists) Craft Operatives £14.21 Labourer £10.71 Rates are national average prices. Refer to REGIONAL VARIATIONS for indicative levels of overall pricing in regions	MATERIALS			LABOUR				RATES		
	Del to Site £	Waste %	Material Cost £	Craft Optve Hrs	Lab Hrs	Labour Cost £	Sunds £	Nett Rate £	Unit	Gross Rate (+7.5%) £
C90: ALTERATIONS - SPOT ITEMS Cont.										
Removal of manholes Cont.										
Removing Cont.										
manholes overall size; remove cover and frame; break up brick sides and concrete bottom; fill in void with hardcore Cont.										
manhole covers and frames; clean off brickwork	-	-	-	-	0.50	**5.36**	0.73	**6.09**	nr	**6.55**
fresh air inlets; clean out pipe socket	-	-	-	-	0.50	**5.36**	0.24	**5.59**	nr	**6.01**
Removal of fencing										
Removing										
chestnut pale fencing with posts										
610mm high	-	-	-	-	0.30	**3.21**	0.65	**3.86**	m	**4.15**
914mm high	-	-	-	-	0.34	**3.64**	1.00	**4.64**	m	**4.98**
1219mm high	-	-	-	-	0.38	**4.07**	1.33	**5.40**	m	**5.80**
close boarded fencing with posts										
1219mm high	-	-	-	-	0.68	**7.28**	1.52	**8.80**	m	**9.46**
1524mm high	-	-	-	-	0.81	**8.68**	1.90	**10.57**	m	**11.36**
1829mm high	-	-	-	-	0.91	**9.75**	2.28	**12.02**	m	**12.92**
chain link fencing with posts										
914mm high	-	-	-	-	0.29	**3.11**	0.73	**3.84**	m	**4.13**
1219mm high	-	-	-	-	0.34	**3.64**	3.06	**6.70**	m	**7.20**
1524mm high	-	-	-	-	0.38	**4.07**	1.23	**5.30**	m	**5.70**
1829mm high	-	-	-	-	0.42	**4.50**	1.52	**6.02**	m	**6.47**
timber gate and posts	-	-	-	-	0.98	**10.50**	3.10	**13.60**	nr	**14.62**
Strutting for forming openings										
Strutting generally										
the cost of strutting in connection with forming openings is to be added to the cost of forming openings; the following are examples of costs for strutting for various sizes and locations of openings. Costs are based on three uses of timber										
Strutting for forming small openings in internal load bearing walls										
opening size 900 x 2000mm	8.91	5.00	**9.36**	2.00	0.33	**31.95**	0.78	**42.09**	nr	**45.25**
opening size 2000 x 2300mm	19.80	5.00	**20.79**	5.00	0.83	**79.94**	1.69	**102.42**	nr	**110.11**
Strutting for forming openings in external one brick walls										
opening size 900 x 2000mm	13.37	5.00	**14.04**	3.00	0.50	**47.99**	1.35	**63.37**	nr	**68.13**
opening size 1500 x 1200mm	13.37	5.00	**14.04**	3.00	0.50	**47.99**	1.35	**63.37**	nr	**68.13**
Strutting for forming openings in external 280mm brick cavity walls										
opening size 900 x 2000mm	18.32	5.00	**19.24**	4.00	0.66	**63.91**	1.52	**84.66**	nr	**91.01**
opening size 1500 x 1200mm	18.32	5.00	**19.24**	4.00	0.66	**63.91**	1.52	**84.66**	nr	**91.01**
Strutting for forming large openings in internal load bearing walls on the ground floor of an average two storey building with timber floors and pitched roof; with load bearing surface 450mm below ground floor; including cutting holes and making good										
opening size 3700 x 2300mm	123.75	5.00	**129.94**	20.00	20.00	**498.40**	17.96	**646.30**	nr	**694.77**
Unit rates for pricing the above and similar work										
plates, struts, braces and wedges in supports to floor and roof	1.73	5.00	**1.82**	0.28	0.28	**6.98**	0.26	**9.05**	m	**9.73**
dead shore and needle, sole plates, braces and wedges including cutting holes and making good...	62.37	5.00	**65.49**	10.00	10.00	**249.20**	8.95	**323.64**	nr	**347.91**
Strutting for forming large openings in external one brick walls on the ground floor of an average two storey building as described above										
opening size 6000 x 2500mm	250.47	5.00	**262.99**	40.00	40.00	**996.80**	35.63	**1295.43**	nr	**1392.58**
Unit rates for pricing the above and similar work										
strutting to window opening over new opening.....	12.47	5.00	**13.09**	2.00	2.00	**49.84**	1.77	**64.70**	nr	**69.55**
plates, struts, braces and wedges in supports to floor and roof	1.68	5.00	**1.76**	0.28	0.28	**6.98**	0.26	**9.00**	m	**9.68**
dead shore and needle, sole plates, braces and wedges including cutting holes and making good...	98.11	5.00	**103.02**	15.70	15.70	**391.24**	14.04	**508.30**	nr	**546.42**
set of two raking shores with 50mm wall piece, wedges and dogs	25.05	5.00	**26.30**	4.00	4.00	**99.68**	3.54	**129.53**	nr	**139.24**
Strutting for forming large openings in external 280mm brick cavity walls on the ground floor of an average two storey building as above described										
opening size 6000 x 2500mm	274.53	5.00	**288.26**	43.90	43.90	**1093.99**	39.18	**1421.42**	nr	**1528.03**
Unit rates for pricing the above and similar work										
strutting to window opening over new opening.....	24.95	5.00	**26.20**	4.00	4.00	**99.68**	3.54	**129.42**	nr	**139.13**
plates, struts, braces and wedges in supports to floor and roof	1.68	5.00	**1.76**	0.28	0.28	**6.98**	0.26	**9.00**	m	**9.68**
dead shore and needle, sole plates, braces and wedges including cutting holes and making good...	103.95	5.00	**109.15**	16.70	16.70	**416.16**	14.97	**540.28**	nr	**580.80**
set of two raking shores with 50mm wall piece, wedges and dogs	25.34	5.00	**26.61**	4.00	4.00	**99.68**	3.54	**129.83**	nr	**139.57**

Labour hourly rates: (except Specialists) Craft Operatives £14.21 Labourer £10.71 Rates are national average prices. Refer to REGIONAL VARIATIONS for indicative levels of overall pricing in regions	MATERIALS			LABOUR				RATES		
	Del to Site £	Waste %	Material Cost £	Craft Optve Hrs	Lab Hrs	Labour Cost £	Sunds £	Nett Rate £	Unit	Gross Rate (+7.5%) £
C90: ALTERATIONS - SPOT ITEMS Cont.										
Spot items										
Spot items generally										
the following items are usually specified as spot items where all work in all trades is included in one item; to assist in adapting these items for varying circumstances, unit rates have been given for the component operation where possible while these items are not in accordance with SMM7 either by description or measurement, it is felt that the pricing information given will be of value to the reader										
Breaking up and reinstatement of concrete floors for excavations										
Break up concrete bed and 150mm hardcore bed under for a width of 760mm for excavation of trench and reinstate with new hardcore and in-situ concrete, BS 5328, designed mix C20, 20mm aggregate; make good up to new wall and existing concrete bed both side										
with 100mm plain concrete bed	10.83	10.00	**12.20**	-	2.70	**28.92**	7.66	**48.77**	m	**52.43**
with 150mm plain concrete bed	15.53	10.00	**17.37**	-	5.00	**53.55**	9.27	**80.18**	m	**86.20**
with 100mm concrete bed reinforced with steel fabric to BS 4483, Reference A193	12.23	10.00	**13.74**	-	4.05	**43.38**	7.66	**64.77**	m	**69.63**
with 150mm concrete bed reinforced with steel fabric to BS 4483, Reference A193	16.94	10.00	**18.91**	-	7.33	**78.50**	9.27	**106.69**	m	**114.69**
Openings through concrete										
Form opening for door frame through 150mm reinforced concrete wall plastered both sides and with skirting both sides; take off skirtings, cut opening through wall, square up reveals, make good existing plaster up to new frame both sides and form fitted ends on existing skirtings up to new frame; extend cement and sand floor screed and vinyl floor covering through opening and make good up to existing										
838 x 2032mm	18.80	-	**18.80**	22.50	28.00	**619.61**	35.79	**674.19**	nr	**724.76**
Unit rates for pricing the above and similar work										
cut opening through wall										
100mm thick	-	-	-	-	5.30	**56.76**	8.43	**65.19**	m2	**70.08**
150mm thick	-	-	-	-	7.90	**84.61**	12.60	**97.21**	m2	**104.50**
200mm thick	-	-	-	-	10.50	**112.46**	16.77	**129.22**	m2	**138.91**
make good fair face around opening	-	-	-	-	0.80	**8.57**	0.59	**9.16**	m	**9.85**
square up reveals to opening										
100mm wide	-	-	-	0.75	0.75	**18.69**	1.47	**20.16**	m	**21.67**
150mm wide	-	-	-	1.07	1.07	**26.66**	2.07	**28.74**	m	**30.89**
200mm wide	-	-	-	1.38	1.38	**34.39**	3.29	**37.68**	m	**40.51**
make good existing plaster up to new frame	0.32	10.00	**0.35**	0.55	0.55	**13.71**	-	**14.06**	m	**15.11**
13mm two coat hardwall plaster to reveal not exceeding 300mm wide	0.64	10.00	**0.70**	0.64	0.64	**15.95**	-	**16.65**	m	**17.90**
take off old skirting	-	-	-	0.22	-	**3.13**	0.12	**3.24**	m	**3.49**
form fitted end on existing skirting up to new frame	-	-	-	0.38	-	**5.40**	-	**5.40**	nr	**5.80**
short length of old skirting up to 100mm long with mitre with existing one end	-	-	-	0.58	-	**8.24**	0.05	**8.29**	nr	**8.91**
short length of old skirting up to 200mm long with mitres with existing both ends	-	-	-	1.15	-	**16.34**	0.07	**16.41**	nr	**17.64**
38mm cement and sand (1:3) screeded bed in opening	4.93	10.00	**5.43**	1.54	1.54	**38.38**	-	**43.80**	m2	**47.09**
vinyl or similar floor covering to match existing, fixed with adhesive, not exceeding 300mm wide	6.44	10.00	**7.09**	0.15	0.15	**3.74**	0.23	**11.05**	m	**11.88**
iroko threshold; twice oiled; plugged and screwed										
15 x 100mm	5.01	10.00	**5.51**	0.90	-	**12.79**	0.06	**18.36**	m	**19.74**
15 x 150mm	7.33	10.00	**8.06**	1.20	-	**17.05**	0.07	**25.19**	m	**27.08**
15 x 200mm	8.86	10.00	**9.75**	1.55	-	**22.03**	0.11	**31.88**	m	**34.27**
Form opening for staircase through 150mm reinforced concrete suspended floor plastered on soffit and with screed and vinyl floor covering on top; take up floor covering and screed, cut opening through floor, square up edges of slab, make good ceiling plaster up to lining and make good screed and floor covering to lining										
900 x 2134mm	73.99	-	**73.99**	19.00	39.00	**687.68**	46.22	**807.89**	nr	**868.48**
Unit rates for pricing the above and similar work										
take up vinyl or similar floor covering	-	-	-	-	0.55	**5.89**	0.17	**6.06**	m2	**6.51**
hack up screed	-	-	-	-	1.50	**16.07**	1.53	**17.59**	m2	**18.91**
cut opening through 150mm thick slab	-	-	-	-	7.90	**84.61**	12.60	**97.21**	m2	**104.50**
cut opening through 200mm thick slab	-	-	-	-	10.50	**112.46**	16.77	**129.22**	m2	**138.91**
make good fair face around opening	-	-	-	-	0.80	**8.57**	0.56	**9.12**	m	**9.81**
square up edges of 150mm thick slab	-	-	-	1.07	1.07	**26.66**	2.19	**28.86**	m	**31.02**
square up edges of 200mm thick slab	-	-	-	1.38	1.38	**34.39**	2.93	**37.32**	m	**40.12**
make good existing plaster up to new lining	0.32	10.00	**0.35**	0.55	0.55	**13.71**	-	**14.06**	m	**15.11**
make good existing 38mm floor screed up to new lining	0.78	10.00	**0.86**	0.45	0.45	**11.21**	-	**12.07**	m	**12.98**
make good existing vinyl or similar floor covering up to new lining	8.59	10.00	**9.45**	0.67	0.67	**16.70**	0.36	**26.50**	m	**28.49**

Labour hourly rates: (except Specialists) Craft Operatives £14.21 Labourer £10.71 Rates are national average prices. Refer to REGIONAL VARIATIONS for indicative levels of overall pricing in regions	MATERIALS			LABOUR				RATES		
	Del to Site £	Waste %	Material Cost £	Craft Optve Hrs	Lab Hrs	Labour Cost £	Sunds £	Nett Rate £	Unit	Gross Rate (+7.5%) £
C90: ALTERATIONS - SPOT ITEMS Cont.										
Filling holes in concrete										
Fill holes in concrete structure where pipes removed with concrete to BS 5328, designed mix C20, 20mm aggregate; formwork										
wall; thickness 100mm; pipe diameter										
50mm	-	-	-	0.60	0.30	**11.74**	1.81	**13.55**	nr	**14.57**
100mm	-	-	-	0.78	0.39	**15.26**	2.39	**17.65**	nr	**18.98**
150mm	-	-	-	0.98	0.49	**19.17**	2.88	**22.05**	nr	**23.71**
wall; thickness 200mm; pipe diameter										
50mm	-	-	-	0.78	0.39	**15.26**	2.39	**17.65**	nr	**18.98**
100mm	-	-	-	0.98	0.49	**19.17**	2.88	**22.05**	nr	**23.71**
150mm	-	-	-	1.00	0.77	**22.46**	3.70	**26.15**	nr	**28.12**
floor; thickness 200mm; pipe diameter										
50mm	-	-	-	0.74	0.37	**14.48**	2.17	**16.65**	nr	**17.90**
100mm	-	-	-	0.93	0.46	**18.14**	2.77	**20.91**	nr	**22.48**
150mm	-	-	-	0.95	0.48	**18.64**	3.51	**22.15**	nr	**23.81**
Make good fair face to filling and surrounding work										
to one side of wall; pipe diameter										
50mm	-	-	-	0.10	0.10	**2.49**	0.25	**2.74**	nr	**2.95**
100mm	-	-	-	0.20	0.20	**4.98**	0.38	**5.36**	nr	**5.77**
150mm	-	-	-	0.25	0.25	**6.23**	0.46	**6.69**	nr	**7.19**
to soffit of floor; pipe diameter										
50mm	-	-	-	0.10	0.10	**2.49**	0.25	**2.74**	nr	**2.95**
100mm	-	-	-	0.20	0.20	**4.98**	0.38	**5.36**	nr	**5.77**
150mm	-	-	-	0.25	0.25	**6.23**	0.46	**6.69**	nr	**7.19**
Make good plaster to filling and surrounding work										
to one side of wall; pipe diameter										
50mm	-	-	-	0.19	0.19	**4.73**	0.25	**4.98**	nr	**5.36**
100mm	-	-	-	0.36	0.36	**8.97**	0.38	**9.35**	nr	**10.05**
150mm	-	-	-	0.45	0.45	**11.21**	0.46	**11.68**	nr	**12.55**
to soffit of floor; pipe diameter										
50mm	-	-	-	0.20	0.20	**4.98**	0.25	**5.23**	nr	**5.63**
100mm	-	-	-	0.40	0.40	**9.97**	0.38	**10.35**	nr	**11.12**
150mm	-	-	-	0.50	0.50	**12.46**	0.46	**12.92**	nr	**13.89**
Make good floor screed and surrounding work										
pipe diameter 50mm	-	-	-	0.19	0.19	**4.73**	0.11	**4.84**	nr	**5.20**
pipe diameter 100mm	-	-	-	0.36	0.36	**8.97**	0.23	**9.20**	nr	**9.89**
pipe diameter 150mm	-	-	-	0.45	0.45	**11.21**	0.31	**11.52**	nr	**12.39**
Make good vinyl or similar floor covering and surrounding work										
pipe diameter 50mm	-	-	-	0.50	0.50	**12.46**	1.85	**14.31**	nr	**15.38**
pipe diameter 100mm	-	-	-	1.00	1.00	**24.92**	2.48	**27.40**	nr	**29.45**
pipe diameter 150mm	-	-	-	1.20	1.20	**29.90**	3.47	**33.38**	nr	**35.88**
Fill holes in concrete structure where metal sections removed with concrete to BS 5328, designed mix C20, 20mm aggregate; formwork										
wall; thickness 100mm; section not exceeding 250mm deep	-	-	-	0.60	0.60	**14.95**	2.20	**17.16**	nr	**18.44**
wall; thickness 200mm; section not exceeding 250mm deep	-	-	-	0.75	0.75	**18.69**	3.21	**21.90**	nr	**23.54**
wall; thickness 100mm; section 250 - 500mm deep	-	-	-	0.75	0.75	**18.69**	3.21	**21.90**	nr	**23.54**
wall; thickness 200mm; section 250 - 500mm deep	-	-	-	0.90	0.90	**22.43**	3.87	**26.30**	nr	**28.28**
Make good fair face to filling and surrounding work										
to one side of wall; section not exceeding 250mm deep	-	-	-	0.22	0.22	**5.48**	0.21	**5.70**	nr	**6.12**
to one side of wall; section 250 - 500mm deep	-	-	-	0.26	0.26	**6.48**	0.33	**6.81**	nr	**7.32**
Make good plaster to filling and surrounding work										
to one side of wall; section not exceeding 250mm deep	-	-	-	0.35	0.35	**8.72**	0.58	**9.30**	nr	**10.00**
to one side of wall; section 250 - 500mm deep	-	-	-	0.43	0.43	**10.72**	0.88	**11.59**	nr	**12.46**
Filling openings in concrete										
Fill opening where door removed in 150mm reinforced concrete wall with concrete reinforced with 12mm mild steel bars at 300mm centres both ways tied to existing structure; hack off plaster to reveals, hack up floor covering and screed in opening, prepare edges of opening to form joint with filling, drill edges of opening and tie in new reinforcement, plaster filling both sides and extended skirting both sides; making good junction of new and existing plaster and skirtings and make good floor screed and vinyl covering up to filling both sides										
838 x 2032mm	124.08	-	**124.08**	22.50	34.50	**689.22**	11.41	**824.71**	nr	**886.56**
Unit rates for pricing the above and similar work										
hack off plaster to reveal not exceeding 100mm wide	-	-	-	-	0.24	**2.57**	0.08	**2.65**	m	**2.85**
hack off plaster to reveal 100 - 200mm wide	-	-	-	-	0.33	**3.53**	0.17	**3.70**	m	**3.98**
take up vinyl or similar floor covering and screed	-	-	-	-	2.00	**21.42**	1.69	**23.11**	m2	**24.85**
prepare edge of opening 100mm wide to form joint with filling	-	-	-	0.25	0.25	**6.23**	-	**6.23**	m	**6.70**
prepare edge of opening 200mm wide to form joint with filling	-	-	-	0.35	0.35	**8.72**	-	**8.72**	m	**9.38**

Labour hourly rates: (except Specialists) Craft Operatives £14.21 Labourer £10.71 Rates are national average prices. Refer to REGIONAL VARIATIONS for indicative levels of overall pricing in regions	MATERIALS			LABOUR				RATES		
	Del to Site £	Waste %	Material Cost £	Craft Optve Hrs	Lab Hrs	Labour Cost £	Sunds £	Nett Rate £	Unit	Gross Rate (+7.5%) £
C90: ALTERATIONS - SPOT ITEMS Cont.										
Filling openings in concrete Cont.										
Unit rates for pricing the above and similar work Cont.										
concrete to BS 5328, designed mix C20, 20mm aggregate in walls 150 - 450m thick.............	110.82	5.00	**116.36**	-	12.00	**128.52**	0.49	**245.37**	m3	**263.77**
concrete to BS 5328, designed mix C20, 20mm aggregate in walls not exceeding 150mm thick.....	110.82	5.00	**116.36**	-	12.00	**128.52**	0.49	**245.37**	m3	**263.77**
12mm mild steel bar reinforcement; straight......	968.00	10.00	**1064.80**	85.00	-	**1207.85**	5.34	**2277.99**	t	**2448.84**
drill edge of opening and tie in new reinforcement	-	-	**-**	-	1.00	**10.71**	0.49	**11.20**	m	**12.04**
formwork and basic finish to wall, per side......	8.55	10.00	**9.41**	2.50	2.50	**62.30**	1.16	**72.87**	m2	**78.34**
formwork and fine formed finish to wall, per side	8.55	10.00	**9.41**	2.75	2.75	**68.53**	1.30	**79.24**	m2	**85.19**
junction between new and existing fair face......	-	-	**-**	-	0.30	**3.21**	-	**3.21**	m	**3.45**
13mm two coat hardwall plaster over 300mm wide...	2.14	10.00	**2.36**	1.30	1.30	**32.40**	-	**34.75**	m2	**37.36**
25 x 100mm softwood chamfered skirting, primed all round	2.00	10.00	**2.20**	0.56	0.09	**8.92**	0.13	**11.25**	m	**12.10**
junction with existing	-	-	**-**	0.50	0.08	**7.96**	-	**7.96**	nr	**8.56**
make good floor screed and vinyl or similar floor covering up to filling not exceeding 300mm wide..	7.87	10.00	**8.66**	1.12	1.12	**27.91**	0.38	**36.95**	m	**39.72**
Fill opening where staircase removed in 150mm reinforced concrete floor with concrete reinforced with 12mm mild steel bars at 225mm centres both ways tied to existing structure; remove timber lining, prepare edges of opening to form joint with filling, drill edges of opening tie in new reinforcement, plaster filling on soffit and extend cement and sand screed and vinyl floor covering to match existing; make good junction of new and existing plaster and vinyl floor covering										
900 x 2134mm	140.49	-	**140.49**	14.75	22.30	**448.43**	9.48	**598.40**	nr	**643.28**
Unit rates for pricing the above and similar work										
take off timber lining	-	-	**-**	-	0.24	**2.57**	0.12	**2.69**	m	**2.89**
prepare edge of opening 150mm wide to form joint with filling	-	-	**-**	0.30	0.30	**7.48**	-	**7.48**	m	**8.04**
prepare edge of opening 200mm wide to form joint with filling	-	-	**-**	0.35	0.35	**8.72**	-	**8.72**	m	**9.38**
concrete to BS 5328, designed mix C20, 20mm aggregate in slabs 150 - 450mm thick............	110.82	5.00	**116.36**	-	12.00	**128.52**	0.49	**245.37**	m3	**263.77**
concrete to BS 5328, designed mix C20, 20mm aggregate in slabs not exceeding 150mm thick.....	110.82	5.00	**116.36**	-	12.00	**128.52**	0.49	**245.37**	m3	**263.77**
12mm mild steel bar reinforcement; straight......	968.00	10.00	**1064.80**	85.00	-	**1207.85**	5.34	**2277.99**	t	**2448.84**
drill edge of opening and tie in new reinforcement	-	-	**-**	-	1.00	**10.71**	0.49	**11.20**	m	**12.04**
formwork and basic finish to soffit, slab thickness not exceeding 200mm; height to soffit 1.50 - 3.00m	9.50	10.00	**10.45**	2.25	2.25	**56.07**	1.15	**67.67**	m2	**72.74**
formwork and basic finish to soffit, slab thickness 200 - 300mm; height to soffit 1.50 - 3.00m	10.50	10.00	**11.55**	2.50	2.50	**62.30**	1.30	**75.15**	m2	**80.79**
formwork and fine formed finish to soffit, slab thickness not exceeding 200mm; height to soffit 1.50 - 3.00m	9.50	10.00	**10.45**	2.50	2.50	**62.30**	1.33	**74.08**	m2	**79.63**
formwork and fine formed finish to soffit, slab thickness 200 - 300mm; height to soffit 1.50 - 3.00m	10.50	10.00	**11.55**	2.75	2.75	**68.53**	1.37	**81.45**	m2	**87.56**
junction between new and existing fair face......	-	-	**-**	-	0.30	**3.21**	-	**3.21**	m	**3.45**
10mm two coat lightweight plaster over 300mm wide	2.14	10.00	**2.36**	1.30	1.30	**32.40**	-	**34.75**	m2	**37.36**
38mm cement and sand (1:3) trowelled bed.........	4.93	10.00	**5.43**	0.63	0.63	**15.70**	-	**21.13**	m2	**22.71**
vinyl or similar floor covering to match existing fixed with adhesive exceeding 300mm wide.........	21.47	10.00	**23.62**	1.25	0.50	**23.12**	0.66	**47.40**	m2	**50.95**
Openings through brickwork or blockwork										
Form openings for door frames through 100mm block partition plastered both sides and with skirting both sides; take off skirtings, cut opening through partition, insert 100 x 150mm precast concrete lintel and wedge and pin up over, quoin up jambs, extend plaster to faces of lintel and make good junction with existing plaster; make good existing plaster up to new frame both sides and form fitted ends on existing skirtings up to new frame; extend softwood board flooring through opening on and with bearers and make good up to existing										
838 x 2032mm	22.07	-	**22.07**	20.00	17.20	**468.41**	15.05	**505.53**	nr	**543.45**
Unit rates for pricing the above and similar work										
cut opening through 75mm block partition plastered both sides	-	-	**-**	1.10	1.30	**29.55**	4.63	**34.19**	m2	**36.75**
cut opening through 100mm block partition plastered both sides	-	-	**-**	1.30	1.56	**35.18**	5.44	**40.62**	m2	**43.67**
make good fair face around opening...............	-	-	**-**	0.25	0.25	**6.23**	0.12	**6.35**	m	**6.82**
precast concrete, BS 5328, designed mix C20, 20mm aggregate lintel 75 x 150 x 1200mm, reinforced with 1.01 kg of 12mm mild steel bars............	8.70	10.00	**9.57**	0.80	0.80	**19.94**	0.30	**29.80**	nr	**32.03**
precast concrete, BS 5328, designed mix C20, 20mm aggregate lintel 100 x 150 x 1200mm, reinforced with 1.01 kg of 12mm mild steel bars............	17.17	10.00	**18.89**	1.00	1.00	**24.92**	0.39	**44.20**	nr	**47.51**
precast concrete, BS 5328 designed mix C20, 20mm aggregate lintel 75 x 150 x 1200mm, reinforced with 1.01 kg of 12mm mild steel bars, fair finish all faces	8.70	10.00	**9.57**	0.80	0.80	**19.94**	0.30	**29.80**	nr	**32.04**

Labour hourly rates: (except Specialists) Craft Operatives £14.21 Labourer £10.71 Rates are national average prices. Refer to REGIONAL VARIATIONS for indicative levels of overall pricing in regions	MATERIALS			LABOUR				RATES		
	Del to Site £	Waste %	Material Cost £	Craft Optve Hrs	Lab Hrs	Labour Cost £	Sunds £	Nett Rate £	Unit	Gross Rate (+7.5%) £
C90: ALTERATIONS - SPOT ITEMS Cont.										
Openings through brickwork or blockwork Cont.										
Unit rates for pricing the above and similar work Cont.										
precast concrete, BS 5328 designed mix C20, 20mm aggregate lintel 100 x 150 x 1200mm, reinforced with 1.01 kg of 12mm mild steel bars, fair finish all faces	17.17	10.00	**18.89**	1.00	1.00	**24.92**	0.39	**44.20**	nr	**47.51**
wedge and pin up over lintel 75mm wide	-	-	-	0.50	0.50	**12.46**	0.73	**13.19**	m	**14.18**
wedge and pin up over lintel 100mm wide	-	-	-	0.66	0.66	**16.45**	0.97	**17.42**	m	**18.73**
quoin up 75mm wide jambs	-	-	-	0.52	0.52	**12.96**	0.73	**13.69**	m	**14.72**
quoin up 100mm wide jambs	-	-	-	0.68	0.68	**16.95**	0.97	**17.92**	m	**19.26**
extend 13mm hardwall plaster to face of lintel	2.14	10.00	**2.36**	2.13	2.13	**53.08**	0.07	**55.51**	m2	**59.67**
make good existing plaster up to new frame	0.32	10.00	**0.35**	0.55	0.55	**13.71**	-	**14.06**	m	**15.11**
take off old skirting	-	-	-	0.22	-	**3.13**	0.12	**3.24**	m	**3.49**
form fitted end on existing skirting up to new frame	-	-	-	0.38	-	**5.40**	-	**5.40**	nr	**5.80**
38mm cement and sand (1:3) screeded bed in opening	4.93	10.00	**5.43**	1.54	1.54	**38.38**	-	**43.80**	m2	**47.09**
extend 25mm softwood board flooring through opening on and with bearers	31.11	20.00	**35.65**	16.20	2.70	**259.12**	0.64	**295.41**	m2	**317.57**
iroko threshold; twice oiled; plugged and screwed										
19 x 75mm	4.43	10.00	**4.87**	0.80	-	**11.37**	0.08	**16.32**	m	**17.55**
19 x 100mm	5.01	10.00	**5.51**	0.90	-	**12.79**	0.09	**18.39**	m	**19.77**
Form openings for door frames through one brick wall plastered one side and with external rendering other side; take off skirting, cut opening through wall, insert 215 x 150mm precast concrete lintel and wedge and pin up over; quoin up jambs; extend plaster to face of lintel and make good junction with existing plaster; make good existing plaster up to new frame one side; extend external rendering to face of lintel and to reveals and make good up to existing rendering and new frame the other side; form fitted ends on existing skirting up to new frame										
914 x 2082mm	36.18	-	**36.18**	28.00	25.40	**669.91**	33.30	**739.39**	nr	**794.85**
Unit rates for pricing the above and similar work										
cut opening through half brick wall	-	-	-	1.95	2.40	**53.41**	5.64	**59.05**	m2	**63.48**
cut opening through one brick wall	-	-	-	3.80	4.70	**104.34**	10.68	**115.01**	m2	**123.64**
make good fair face around opening	-	-	-	0.50	0.50	**12.46**	0.73	**13.19**	m	**14.18**
make good facings to match existing around opening	1.61	10.00	**1.77**	0.50	0.50	**12.46**	0.73	**14.97**	m	**16.09**
precast concrete, BS 5328, designed mix C20, 20mm aggregate lintel 102 x 150 x 1200mm, reinforced with 1.07 kg of 12mm mild steel bars	17.17	10.00	**18.89**	1.00	1.00	**24.92**	0.43	**44.23**	nr	**47.55**
precast concrete, BS 5328, designed mix C20, 20mm aggregate lintel 215 x 150 x 1200mm, reinforced with 2.13 kg of 12mm mild steel bars	35.18	10.00	**38.70**	1.50	1.50	**37.38**	0.88	**76.95**	nr	**82.73**
precast concrete, BS 5328, designed mix C20, 20mm aggregate lintel 102 x 150 x 1200mm, reinforced with 1.07 kg of 12mm mild steel bars, fair finish all faces	17.17	10.00	**18.89**	1.00	1.00	**24.92**	0.43	**44.23**	nr	**47.55**
precast concrete, BS 5328, designed mix C20, 20mm aggregate lintel 215 x 150 x 1200mm, reinforced with 2.13 kg of 12mm mild steel bars, fair finish all faces	37.14	10.00	**40.85**	1.50	1.50	**37.38**	0.85	**79.09**	nr	**85.02**
wedge and pin up over lintel 102mm wide	-	-	-	0.66	0.66	**16.45**	0.97	**17.42**	m	**18.73**
wedge and pin up over lintel 215mm wide	-	-	-	1.00	1.00	**24.92**	2.07	**26.99**	m	**29.02**
quoin up half brick jambs	-	-	-	0.70	0.70	**17.44**	0.97	**18.42**	m	**19.80**
quoin up one brick jambs	-	-	-	1.27	1.27	**31.65**	2.07	**33.72**	m	**36.25**
facings to match existing to margin	1.61	10.00	**1.77**	0.67	0.67	**16.70**	0.97	**19.44**	m	**20.90**
make good existing plaster up to new frame	0.32	10.00	**0.35**	0.55	0.55	**13.71**	-	**14.06**	m	**15.11**
13mm two coat hardwall plaster to reveal not exceeding 300mm wide	0.64	10.00	**0.70**	0.78	0.78	**19.44**	0.02	**20.17**	m	**21.68**
make good existing external rendering up to new frame	0.24	10.00	**0.27**	0.47	0.47	**11.71**	-	**11.98**	m	**12.88**
13mm two coat cement and sand (1:3) external rendering to reveal not exceeding 300mm wide	0.49	10.00	**0.54**	0.74	0.74	**18.44**	0.02	**19.00**	m	**20.43**
take off old skirting	-	-	-	0.23	-	**3.27**	0.11	**3.37**	m	**3.63**
short length of old skirting up to 100mm long with mitre with existing one end	-	-	-	0.58	-	**8.24**	0.05	**8.29**	nr	**8.91**
short length of old skirting up to 300mm long with mitres with existing both ends	-	-	-	1.15	-	**16.34**	0.06	**16.40**	nr	**17.63**
38mm cement and sand (1:3) screeded bed in opening	4.93	10.00	**5.43**	1.54	1.54	**38.38**	-	**43.80**	m2	**47.09**
extend 25mm softwood board flooring through opening on and with bearers	6.06	15.00	**6.80**	3.65	0.61	**58.40**	0.17	**65.37**	nr	**70.27**
iroko threshold, twice oiled, plugged and screwed										
15 x 113mm	5.14	10.00	**5.65**	0.90	-	**12.79**	0.06	**18.50**	m	**19.89**
15 x 225mm	9.50	10.00	**10.45**	1.65	-	**23.45**	0.11	**34.00**	m	**36.55**
Form opening for window through 275mm hollow wall with two half brick skins, plastered one side and faced with picked stock facings the other; cut opening through wall, insert 275 x 225mm precast concrete boot lintel and wedge and pin up over; insert damp proof course and cavity gutter; quoin up jambs, close cavity at jambs with brickwork bonded to existing and with vertical damp proof course and close cavity at sill with one course of slates; face margin externally to match existing and extend plaster to face of lintel and to reveals and make good up to existing plaster and new frame										
900 x 900mm	55.40	-	**55.40**	15.90	16.80	**405.87**	17.30	**478.57**	nr	**514.46**

Labour hourly rates: (except Specialists) Craft Operatives £14.21 Labourer £10.71 Rates are national average prices. Refer to REGIONAL VARIATIONS for indicative levels of overall pricing in regions	MATERIALS			LABOUR				RATES		
	Del to Site £	Waste %	Material Cost £	Craft Optve Hrs	Lab Hrs	Labour Cost £	Sunds £	Nett Rate £	Unit	Gross Rate (+7.5%) £
C90: ALTERATIONS - SPOT ITEMS Cont.										
Openings through brickwork or blockwork Cont.										
Unit rates for pricing the above and similar work										
cut opening through wall with two half brick skins	-	-	-	3.90	4.80	**106.83**	9.67	**116.50**	m2	**125.23**
cut opening through wall with one half brick and one 100mm block skins	-	-	-	3.25	3.96	**88.59**	9.67	**98.26**	m2	**105.63**
precast concrete, BS 5328, designed mix C20, 20mm aggregate boot lintel 275 x 225 x 1200mm long reinforced with 3.60 kg of 12mm mild steel bars and 0.67 kg of 6mm mild steel links.............	57.19	10.00	**62.91**	2.30	2.30	**57.32**	0.97	**121.20**	nr	**130.29**
close cavity at jambs with blockwork bonded to existing and with 112.5mm wide bituminous hessian based damp proof course, BS 6398, Class A........	2.42	10.00	**2.67**	0.50	0.50	**12.46**	0.25	**15.37**	m	**16.53**
close cavity at jambs with brickwork bonded to existing and with 112.5mm wide bituminous hessian based damp proof course, BS 6398, Class A........	2.45	10.00	**2.69**	0.70	0.70	**17.44**	0.97	**21.11**	m	**22.69**
close cavity at jambs with slates in cement mortar (1:3) set vertically	3.02	10.00	**3.32**	0.44	0.44	**10.96**	0.25	**14.54**	m	**15.63**
close cavity at sill with one course of slates in cement mortar (1:3)	3.02	10.00	**3.32**	0.44	0.44	**10.96**	0.25	**14.54**	m	**15.63**
lead cored bituminous hessian based damp proof course and cavity tray, BS 6398, Class F, width exceeding 225mm	17.13	10.00	**18.84**	0.80	0.80	**19.94**	0.25	**39.03**	m2	**41.95**
precast concrete, BS 5328, designed mix C20, 20mm aggregate lintel 160 x 225 x 1200mm long reinforced with 2.13 kg of 12mm mild steel bars..	39.27	10.00	**43.20**	1.50	1.50	**37.38**	0.73	**81.31**	nr	**87.41**
76 x 76 x 6mm mild steel angle arch bar, primed..	6.70	5.00	**7.04**	0.50	0.50	**12.46**	0.73	**20.23**	m	**21.75**
take out three courses of facing bricks and build brick-on-end flat arch in picked stock facing bricks	6.13	10.00	**6.74**	1.00	1.00	**24.92**	2.14	**33.81**	m	**36.35**
For unit rates for finishes, quoining up jambs, etc see previous items										
Form opening for window where door removed, with window head at same level as old door head, through one brick wall plastered one side and faced the other with picked stock facings; remove old lintel and flat arch, cut away for and insert 102 x 225mm precast concrete lintel and wedge and pin up over; build brick-on-end flat arch in facing bricks on 76 x 76 x 6mm mild steel angle arch bar and point to match existing; cut away jambs of old door opening to extend width; quoin up jambs and face margins externally to match existing; fill in lower part of old opening with brickwork in gauged mortar (1:2:9), bond to existing and face externally and point to match existing										
1200 x 900mm; old door opening 914 x 2082mm; extend plaster to face of lintel, reveals and filling and make good up to existing plaster and new frame; extend skirting one side and make good junction with existing	119.28	-	**119.28**	22.00	21.50	**542.89**	29.63	**691.79**	nr	**743.67**
Unit rates for pricing the above and similar work										
take out old lintel	-	-	-	1.00	1.30	**28.13**	4.03	**32.16**	m	**34.57**
cut away one brick wall to increase width of opening	-	-	-	3.80	4.70	**104.34**	10.07	**114.41**	m2	**122.99**
cut away 275mm hollow wall of two half brick skins to increase width of opening....................	-	-	-	3.90	4.80	**106.83**	10.07	**116.90**	m2	**125.67**
For other unit rates for pricing the above and similar work see previous items and items in 'Filling openings in Brickwork and Blockwork' section										
Form opening for door frame where old window removed, with door head at same level as old window head, through one brick wall plastered one side and faced the other with picked stock facings; remove old lintel and flat arch, cut away for and insert 102 x 225mm precast concrete lintel and wedge and pin up over; build brick-on-end flat arch in facing bricks on 76 x 76 x 6mm mild steel angle arch bar and point, fill old opening at sides of new opening to reduce width with brickwork in gauged mortar (1:2:9), bond to existing, wedge and pin up at soffit, face externally and to margins and point to match existing; take off skirtings, cut away wall below sill of old opening, quion up jambs and face margins externally										
914 x 2082mm, old window opening 1200 x 900mm; extend plaster to face of lintel and filling and make good up to existing plaster and new frame; form fitted ends on existing skirting up to new frame	72.12	-	**72.12**	24.00	23.80	**595.94**	33.77	**701.83**	nr	**754.47**
For unit rates for pricing the above and similar work see previous items and items in 'Filling openings in Brickwork and Blockwork' section										
Air bricks in old walls										
Cut opening through old one brick wall, render all round in cement mortar (1:3); build in clay air brick externally and fibrous plaster ventilator internally and make good facings and plaster										
225 x 150mm	9.17	-	**9.17**	1.75	1.75	**43.61**	0.65	**53.43**	nr	**57.44**
225 x 225mm	7.48	-	**7.48**	2.33	2.33	**58.06**	0.82	**66.36**	nr	**71.34**

Labour hourly rates: (except Specialists) Craft Operatives £14.21 Labourer £10.71 Rates are national average prices. Refer to REGIONAL VARIATIONS for indicative levels of overall pricing in regions	MATERIALS			LABOUR				RATES		
	Del to Site £	Waste %	Material Cost £	Craft Optve Hrs	Lab Hrs	Labour Cost £	Sunds £	Nett Rate £	Unit	Gross Rate (+7.5%) £
C90: ALTERATIONS - SPOT ITEMS Cont.										
Air bricks in old walls Cont.										
Cut opening through old 275mm hollow brick wall and seal cavity with slates; build in clay air brick externally and fibrous plaster ventilator internally and make good facings and plaster										
225 x 150mm	12.19	10.00	**12.49**	2.14	2.14	**53.33**	0.65	**66.47**	nr	**71.46**
225 x 225mm	10.50	10.00	**10.80**	2.81	2.81	**70.03**	0.82	**81.64**	nr	**87.77**
Openings through rubble walling										
Form opening for door frame through 600mm rough rubble wall faced both sides; cut opening through wall, insert 600 x 150mm precast concrete lintel finished fair on all exposed faces and wedge and pin up over; quoin and face up jambs and extend softwood board flooring through opening on and with bearers and make good up to existing										
838 x 2032mm	82.62	-	**82.62**	41.30	43.70	**1054.90**	72.40	**1209.92**	nr	**1300.67**
Unit rates for pricing the above and similar work										
cut opening through 350mm random rubble wall	-	-	**-**	6.00	7.00	**160.23**	14.10	**174.33**	m2	**187.41**
cut opening through 600mm rough rubble wall	-	-	**-**	10.00	12.00	**270.62**	24.17	**294.79**	m2	**316.90**
precast concrete, BS 5328, designed mix C20, 20mm aggregate lintel 350 x 150 x 1140mm reinforced with 3.04 kg of 12mm mild steel bars; fair finish all faces	57.43	10.00	**63.17**	2.25	2.25	**56.07**	0.85	**120.10**	nr	**129.10**
precast concrete, BS 5328, designed mix C20, 20mm aggregate lintel 600 x 150 x 1140mm reinforced with 5.00 kg of 12mm mild steel bars; fair finish all faces	98.45	10.00	**108.30**	3.65	3.65	**90.96**	1.34	**200.59**	nr	**215.64**
wedge and pin up over lintel 350mm wide	-	-	**-**	2.00	2.00	**49.84**	2.44	**52.28**	m	**56.20**
wedge and pin up over lintel 600mm wide	-	-	**-**	3.00	3.00	**74.76**	4.28	**79.04**	m	**84.97**
quoin up 350mm wide jambs	-	-	**-**	2.00	2.00	**49.84**	2.44	**52.28**	m	**56.20**
quoin up 600mm wide jambs	-	-	**-**	3.00	3.00	**74.76**	4.28	**79.04**	m	**84.97**
extend 25mm softwood board flooring through opening on and with bearers	15.15	15.00	**17.00**	9.70	1.60	**154.97**	0.46	**172.44**	nr	**185.37**
Openings through stone faced walls										
Form opening for window through wall of half brick backing, plastered, and 100mm stone facing; cut opening through wall, insert 102 x 150mm precast concrete lintel and wedge and pin up over and build in 76 x 76 x 6mm mild steel angle arch bar to support stone facing, clean soffit of stone and point; quoin up jambs and face up and point externally; extend plaster to face of lintel and to reveals and make good up to existing plaster and new frame										
900 x 900mm	27.20	-	**27.20**	19.20	19.80	**484.89**	22.40	**534.49**	nr	**574.57**
Unit rates for pricing the above and similar work										
cut opening through wall	-	-	**-**	5.15	5.85	**135.84**	9.67	**145.50**	m2	**156.42**
precast concrete, BS 5328, designed mix C20, 20mm aggregate lintel 102 x 150 x 1200mm reinforced with 1.01 kg of 12mm mild steel bars	17.17	10.00	**18.89**	1.05	1.05	**26.17**	0.37	**45.42**	nr	**48.83**
76 x 76 x 6mm mild steel angle arch bar, primed	6.70	5.00	**7.04**	0.50	0.50	**12.46**	0.73	**20.23**	m	**21.75**
quoin up jambs	-	-	**-**	0.70	0.70	**17.44**	1.10	**18.55**	m	**19.94**
face stone jambs and point	-	-	**-**	2.10	2.10	**52.33**	3.06	**55.39**	m	**59.54**
clean stone head and point	-	-	**-**	2.10	2.10	**52.33**	0.37	**52.70**	m	**56.65**
For unit rates for finishes, etc. see previous items										
Filling openings in brickwork or blockwork										
Fill opening where door removed in 100mm block wall with concrete blocks in gauged mortar (1:2:9); provide 50mm preservative treated softwood sole plate on timber floor in opening, bond to blockwork at jambs and wedge and pin up at head; plaster filling both sides and extend skirting both sides; make good junction of new and existing plaster and skirtings										
838 x 2032mm	58.78	-	**58.78**	14.10	8.15	**287.65**	10.05	**356.48**	nr	**383.21**
Unit rates for pricing the above and similar work										
75mm blockwork in filling	8.42	10.00	**9.26**	1.00	1.25	**27.60**	1.34	**38.20**	m2	**41.06**
100mm blockwork in filling	10.10	10.00	**11.11**	1.20	1.50	**33.12**	1.84	**46.06**	m2	**49.52**
75mm blockwork in filling and fair face and flush smooth pointing one side	8.42	10.00	**9.26**	1.25	1.50	**33.83**	1.59	**44.68**	m2	**48.03**
100mm blockwork in filling and fair face and flush smooth pointing one side	10.10	10.00	**11.11**	1.45	1.75	**39.35**	1.96	**52.41**	m2	**56.34**
preservative treated softwood sole plate										
50 x 75mm	1.26	10.00	**1.39**	0.30	0.05	**4.80**	0.25	**6.43**	m	**6.92**
50 x 100mm	1.67	10.00	**1.84**	0.33	0.06	**5.33**	0.33	**7.50**	m	**8.06**
lead cored bituminous hessian based damp proof course, BS 6398, Class F, width not exceeding 225mm	17.13	10.00	**18.84**	0.80	0.80	**19.94**	0.49	**39.26**	m2	**42.21**
wedge and pin up at head 75mm wide	-	-	**-**	0.10	0.10	**2.49**	0.49	**2.98**	m	**3.20**
wedge and pin up at head 100mm wide	-	-	**-**	0.12	0.12	**2.99**	0.73	**3.73**	m	**4.00**
cut pockets and bond 75mm filling to existing blockwork at jambs	-	-	**-**	0.40	0.40	**9.97**	0.85	**10.82**	m	**11.63**
cut pockets and bond 100mm filling to existing blockwork at jambs	-	-	**-**	0.50	0.50	**12.46**	1.21	**13.67**	m	**14.69**

Labour hourly rates: (except Specialists) Craft Operatives £14.21 Labourer £10.71 Rates are national average prices. Refer to REGIONAL VARIATIONS for indicative levels of overall pricing in regions	MATERIALS			LABOUR				RATES		
	Del to Site £	Waste %	Material Cost £	Craft Optve Hrs	Lab Hrs	Labour Cost £	Sunds £	Nett Rate £	Unit	Gross Rate (+7.5%) £
C90: ALTERATIONS - SPOT ITEMS Cont.										
Filling openings in brickwork or blockwork Cont.										
Unit rates for pricing the above and similar work Cont.										
preservative treated softwood sole plate Cont.										
12mm (average) cement and sand dubbing over 300mm wide on filling	1.56	10.00	**1.71**	0.54	0.54	**13.46**	-	**15.17**	m2	**16.31**
13mm two coat lightweight plaster over 300mm wide on filling	2.14	10.00	**2.36**	1.27	1.27	**31.65**	0.02	**34.03**	m2	**36.58**
19 x 100mm softwood chamfered skirting, primed all round	1.63	10.00	**1.79**	0.50	0.08	**7.96**	0.13	**9.89**	m	**10.63**
junction with existing	-	-	**-**	0.50	0.08	**7.96**	-	**7.96**	nr	**8.56**
make good vinyl or similar floor covering up to filling	7.15	10.00	**7.86**	0.50	0.50	**12.46**	0.17	**20.49**	m	**22.03**
Fill opening where door removed in half brick wall with brickwork in gauged mortar (1:2:9); provide lead cored damp proof course in opening, lapped with existing, bond to existing brickwork at jambs and wedge and pin up at head; plaster filling both sides and extend skirting both sides; make good junction of new and existing plaster and skirtings										
838 x 2032mm	51.75	-	**51.75**	15.70	12.75	**359.65**	12.32	**423.72**	nr	**455.50**
Unit rates for pricing the above and similar work										
half brick filling in common bricks	13.20	10.00	**14.52**	2.00	2.00	**49.84**	3.06	**67.42**	m2	**72.47**
half brick filling in common bricks fair faced and flush pointed one side	13.20	10.00	**14.52**	2.33	2.33	**58.06**	3.29	**75.88**	m2	**81.57**
50 x 100mm softwood sole plate	1.67	10.00	**1.84**	0.33	0.06	**5.33**	0.30	**7.47**	m	**8.03**
lead cored bituminous hessian based damp proof course, BS 6398, Class F, width not exceeding 225mm	17.13	10.00	**18.84**	0.80	0.80	**19.94**	0.49	**39.26**	m2	**42.21**
wedge and pin up at head 102mm wide	-	-	**-**	0.12	0.12	**2.99**	0.73	**3.73**	m	**4.00**
cut pockets and bond half brick filling to existing brickwork at jambs	-	-	**-**	0.50	0.50	**12.46**	1.22	**13.68**	m	**14.71**
For unit rates for finishes, etc see previous items										
Fill opening where door removed in one brick external wall with brickwork in gauged mortar (1:2:9) faced externally with picked stock facings pointed to match existing; remove old lintel and arch, provide lead cored damp proof course in opening lapped with existing, bond to existing brickwork at jambs and wedge and pin up at head; plaster filling one side and extend skirting; make good junction of new and existing plaster and skirting										
914 x 2082mm	169.93	-	**169.93**	19.65	18.85	**481.11**	30.45	**681.49**	nr	**732.61**
Unit rates for pricing the above and similar work										
remove old lintel and arch	-	-	**-**	1.46	1.64	**38.31**	2.42	**40.73**	m	**43.78**
one brick filling in common bricks	26.40	10.00	**29.04**	4.00	4.00	**99.68**	7.32	**136.04**	m2	**146.25**
one brick filling in common bricks faced one side with picked stock facings and point to match existing	48.00	10.00	**52.80**	5.00	5.00	**124.60**	7.32	**184.72**	m2	**198.58**
one brick filling in common bricks fair faced and flush smooth pointing one side	26.40	10.00	**29.04**	4.33	4.33	**107.90**	7.32	**144.27**	m2	**155.09**
two course slate damp proof course, width not exceeding 225mm	24.16	10.00	**26.58**	1.02	1.02	**25.42**	0.85	**52.85**	m2	**56.81**
lead cored bituminous hessian based damp proof course, BS 6398, Class F, width not exceeding 225mm	17.13	10.00	**18.84**	0.80	0.80	**19.94**	0.37	**39.15**	m2	**42.08**
wedge and pin up at head 215mm wide	-	-	**-**	0.20	0.20	**4.98**	1.71	**6.69**	m	**7.19**
cut pockets and bond one brick filling to existing brickwork at jambs	-	-	**-**	1.00	1.00	**24.92**	2.81	**27.73**	m	**29.81**
12mm cement and sand (1:3) two coat external rendering over 300mm wide on filling	1.56	10.00	**1.71**	1.25	1.25	**31.15**	0.05	**32.91**	m2	**35.38**
For unit rates for internal finishes, etc. see previous items										
Fill opening where door removed in 275mm external hollow wall with inner skin in common bricks and outer skin in picked stock facings pointed to match existing, in gauged mortar (1:2:9); remove old lintel and arch, provide lead cored combined damp proof course and cavity gutter lapped with existing; form cavity with ties; bond to existing brickwork at jambs and wedge and pin up at head; plaster filling one side and extend skirting; make good junction of new and existing plaster and skirting										
914 x 2082mm	117.77	-	**117.77**	19.90	19.10	**487.34**	31.11	**636.22**	nr	**683.93**
Unit rates for pricing the above and similar work										
half brick filling in common bricks	13.20	10.00	**14.52**	2.00	2.00	**49.84**	3.06	**67.42**	m2	**72.47**
half brick filling in common bricks and fair face and flush smooth pointing one side	13.20	10.00	**14.52**	2.33	2.33	**58.06**	3.29	**75.88**	m2	**81.57**
half brick filling in picked stock facings and point to match existing	27.60	10.00	**30.36**	2.95	2.95	**73.51**	3.29	**107.17**	m2	**115.21**
form 50mm wide cavity with BS 1243 Fig. 3 galvanised steel twisted wall ties built in	1.16	10.00	**1.27**	0.20	0.20	**4.98**	-	**6.26**	m2	**6.73**
lead cored bituminous hessian based damp proof course and cavity tray, BS 6398, Class F, width exceeding 225mm	17.13	10.00	**18.84**	0.80	0.80	**19.94**	0.37	**39.15**	m2	**42.08**
wedge and pin up at head 102mm wide	-	-	**-**	0.12	0.12	**2.99**	0.73	**3.73**	m	**4.00**

Labour hourly rates: (except Specialists) Craft Operatives £14.21 Labourer £10.71 Rates are national average prices. Refer to REGIONAL VARIATIONS for indicative levels of overall pricing in regions	MATERIALS			LABOUR				RATES		
	Del to Site £	Waste %	Material Cost £	Craft Optve Hrs	Lab Hrs	Labour Cost £	Sunds £	Nett Rate £	Unit	Gross Rate (+7.5%) £
C90: ALTERATIONS - SPOT ITEMS Cont.										
Filling openings in brickwork or blockwork Cont.										
Unit rates for pricing the above and similar work Cont.										
cut pockets and bond half brick wall in common bricks to existing brickwork at jambs	-	-	-	0.50	0.50	**12.46**	1.22	**13.68**	m	**14.71**
cut pockets and bond half brick wall in facing bricks to existing brickwork at jambs and make good facings	-	-	-	0.60	0.60	**14.95**	2.56	**17.51**	m	**18.82**
For unit rates for internal finishes, etc. see previous items										
Fillings openings in rubble walling										
Fill opening where window removed in 600mm rough rubble wall with rubble walling in lime mortar (1:3) pointed to match existing; bond to existing walling at jambs and wedge and pin up at head; plaster filling one side and make good junction of new and existing plaster										
900 x 900mm	117.26	-	**117.26**	12.00	12.00	**299.04**	18.84	**435.14**	nr	**467.78**
Unit rates for pricing the above and similar work										
rough rubble walling in filling										
600mm thick	121.00	10.00	**133.10**	6.00	6.00	**149.52**	14.65	**297.27**	m2	**319.56**
random rubble walling in filling										
300mm thick	88.00	10.00	**96.80**	5.80	5.80	**144.54**	7.32	**248.66**	m2	**267.31**
500mm thick	130.00	10.00	**143.00**	10.90	10.90	**271.63**	12.21	**426.83**	m2	**458.85**
wedge and pin up at head										
300mm wide	-	-	-	0.28	0.28	**6.98**	2.44	**9.42**	m	**10.13**
500mm wide	-	-	-	0.47	0.47	**11.71**	4.15	**15.86**	m	**17.05**
600mm wide	-	-	-	0.56	0.56	**13.96**	4.88	**18.84**	m	**20.25**
cut pockets and bond walling to existing at jambs										
300mm	6.60	10.00	**7.26**	0.83	0.83	**20.68**	0.73	**28.68**	m	**30.83**
500mm	9.75	10.00	**10.73**	1.38	1.38	**34.39**	1.10	**46.22**	m	**49.68**
600mm	9.08	10.00	**9.98**	1.75	1.75	**43.61**	1.34	**54.93**	m	**59.05**
For unit rates for internal finishes, etc. see previous items										
Work to chimney stacks										
SEAL AND VENTILATE TOP OF FLUE - remove chimney pot and flaunching and provide 50mm precast concrete, BS 5328, designed mix C20, 20mm aggregate, weathered and throated capping and bed in cement mortar (1:3) to seal top of stack; cut opening through brick side of stack and build in 225 x 225mm clay air brick to ventilate flue and make good facings										
stack size 600 x 600mm with one flue	21.26	10.00	**22.88**	4.90	4.90	**122.11**	5.12	**150.10**	nr	**161.36**
stack size 825 x 600mm with two flues	32.42	10.00	**34.64**	9.00	9.00	**224.28**	9.43	**268.35**	nr	**288.48**
stack size 1050 x 600mm with three flues	43.58	10.00	**46.41**	13.10	13.10	**326.45**	12.93	**385.79**	nr	**414.72**
RENEW DEFECTIVE CHIMNEY POT - remove chimney pot and provide new clay chimney pot set and flaunched in cement mortar (1:3)										
150mm diameter x 600mm high	25.90	-	**25.90**	2.50	2.50	**62.30**	4.63	**92.83**	nr	**99.80**
REBUILD DEFECTIVE CHIMNEY STACK - pull down defective stack to below roof level for a height of 2.00m; prepare for raising and rebuild in common brickwork faced with picked stock facings in gauged mortar (1:2:9) pointed to match existing; parge and core flues; provide No. 4 lead flashings and soakers; provide 150mm diameter clay chimney pots 600mm high set and flaunched in cement mortar (1:3); make good roof tiling and all other work disturbed										
600 x 600mm with one flue	280.77	10.00	**306.26**	35.30	37.55	**903.77**	69.92	**1279.94**	nr	**1375.94**
825 x 600mm with two flues	367.90	10.00	**399.51**	47.15	47.90	**1183.01**	97.17	**1679.69**	nr	**1805.67**
REMOVE DEFECTIVE CHIMNEY STACK - pull down stack to below roof level; remove all flashings and soakers; piece in 50 x 100mm rafters; extend roof tiling with machine made plain tiles to 100mm gauge nailed every fourth course with galvanised nails to and with 25 x 19mm battens; bond new tiling to existing										
850 x 600mm for a height of 2.00m	36.02	10.00	**39.62**	26.00	22.30	**608.29**	42.66	**690.58**	nr	**742.37**
Work to chimney breasts and fireplaces										
TAKE OUT FIREPLACE AND FILL IN OPENING - remove fire surround, fire back and hearth; hack up screed to hearth and extend 25mm tongued and grooved softwood floor boarding over hearth on and with bearers; fill in opening where fireplace removed with 75mm concrete blocks in gauged mortar (1:2:9) bonded to existing at jambs and wedged and pinned up to soffits; form 225 x 225mm opening and provide and build in 225 x 225mm fibrous plaster louvered air vent; plaster filling with 12mm two coat lightweight plaster on 12mm (average) dubbing plaster; extend 19 x 100mm softwood chamfered skirting over filling and join with existing; make good all new finishings to existing										
fireplace opening 570 x 685mm, hearth 1030 x 405mm	39.79	-	**39.79**	8.00	7.45	**193.47**	12.21	**245.47**	nr	**263.87**
fireplace opening 800 x 760mm, hearth 1260 x 405mm	48.64	-	**48.64**	9.65	8.95	**232.98**	14.22	**295.84**	nr	**318.03**

Labour hourly rates: (except Specialists) Craft Operatives £14.21 Labourer £10.71 Rates are national average prices. Refer to REGIONAL VARIATIONS for indicative levels of overall pricing in regions	MATERIALS			LABOUR				RATES		
	Del to Site £	Waste %	Material Cost £	Craft Optve Hrs	Lab Hrs	Labour Cost £	Sunds £	Nett Rate £	Unit	Gross Rate (+7.5%) £
C90: ALTERATIONS - SPOT ITEMS Cont.										
Work to chimney breasts and fireplaces Cont.										
REMOVE CHIMNEY BREAST - pull down chimney breast for full height from ground to roof level (two storeys) including removing two fire surrounds and hearths complete; make out brickwork where flues removed and make out brickwork where breasts removed; extend 50 x 175mm floor joists and 25mm tongued and grooved softwood floor boarding (ground and first floors); extend 50 x 100mm ceiling joists (first floor); plaster walls with 12mm two coat lightweight plaster on 12mm (average) plaster dubbing; extend ceiling plaster with expanded metal lathing and 19mm three coat lightweight plaster; run 19 x 100mm softwood chamfered skirting to walls to join with existing; make good all new finishings up to existing										
chimney breast 2.00m wide and 7.50m high including gathering in roof space	163.90	-	**163.90**	62.00	84.00	**1780.66**	170.64	**2115.20**	nr	**2273.84**
Form trap door in ceiling										
Form new trap door in ceiling with 100mm deep joists and lath and plaster finish; cut away and trim ceiling joists around opening and insert new trimming and trimmer joists; provide 32x 127mm softwood lining with 15 x 25mm stop and 20 x 70mm architrave; provide and place in position 18mm blockboard trap door, lipped all round; make good existing ceiling plaster around openings										
600 x 900mm	43.60	-	**43.60**	21.85	5.60	**370.46**	10.78	**424.85**	nr	**456.71**
Unit rates for pricing the above and similar work										
cut away plastered plasterboard ceiling	-	-	-	1.50	0.33	**24.85**	1.01	**25.86**	m2	**27.80**
cut away lath and plaster ceiling	-	-	-	2.00	0.41	**32.81**	1.21	**34.02**	m2	**36.57**
cut away and trim existing 100mm deep joists around opening										
664 x 964mm	-	-	-	0.95	0.16	**15.21**	0.84	**16.05**	nr	**17.26**
964 x 1444mm	-	-	-	1.90	0.33	**30.53**	1.69	**32.23**	nr	**34.64**
cut away and trim existing 150mm deep joists around opening										
664 x 964mm	-	-	-	1.43	0.24	**22.89**	1.29	**24.18**	nr	**26.00**
964 x 1444mm	-	-	-	2.86	0.48	**45.78**	2.54	**48.32**	nr	**51.94**
softwood trimming or trimmer joists										
75 x 100mm	2.70	10.00	**2.97**	0.95	0.16	**15.21**	0.43	**18.61**	m	**20.01**
75 x 150mm	4.05	10.00	**4.46**	1.00	0.17	**16.03**	0.46	**20.95**	m	**22.52**
softwood lining										
32 x 125mm	2.10	10.00	**2.31**	1.50	0.25	**23.99**	0.13	**26.43**	m	**28.42**
32 x 150mm	2.18	10.00	**2.40**	1.60	0.27	**25.63**	0.09	**28.12**	m	**30.23**
15 x 25mm softwood stop	0.45	10.00	**0.50**	0.33	0.05	**5.22**	0.06	**5.78**	m	**6.21**
19 x 50mm softwood twice rounded architrave to Patt. 18	0.82	10.00	**0.90**	0.33	0.05	**5.22**	0.06	**6.19**	m	**6.65**
18mm blockboard trap door lipped all round										
600 x 900mm	12.94	10.00	**14.23**	3.00	0.50	**47.99**	0.20	**62.42**	nr	**67.10**
900 x 1280mm	29.60	10.00	**32.56**	6.00	1.00	**95.97**	0.36	**128.89**	nr	**138.55**
75mm steel butt hinge to softwood	0.35	-	**0.35**	0.66	0.11	**10.56**	0.20	**11.11**	nr	**11.94**
make good plastered plasterboard ceiling around opening	-	-	-	0.75	0.75	**18.69**	1.46	**20.15**	m	**21.66**
make good lath and plaster ceiling around opening	-	-	-	0.83	0.83	**20.68**	1.79	**22.47**	m	**24.16**
Openings through stud partitions										
Form opening for door frame through lath and plaster finished 50 x 100mm stud partition; take off skirtings, cut away plaster, cut away and trim studding around opening and insert new studding at head and jambs of opening; make good existing lath and plaster up to new frame both sides and form fitted ends on existing skirtings up to new frame; extend softwood board flooring through opening on and with bearers and make good up to existing										
838 x 2032mm	13.94	-	**13.94**	18.10	6.10	**322.53**	19.08	**355.55**	nr	**382.22**
Unit rates for pricing the above and similar work										
cut away plastered plasterboard	-	-	-	1.00	0.25	**16.89**	1.01	**17.89**	m2	**19.24**
cut away lath and plaster	-	-	-	1.10	0.26	**18.42**	1.21	**19.62**	m2	**21.10**
cut away and trim existing 75mm studding around opening										
838 x 2032mm	-	-	-	2.05	0.36	**32.99**	2.58	**35.57**	nr	**38.24**
914 x 2032mm	-	-	-	2.05	0.36	**32.99**	2.58	**35.57**	nr	**38.24**
cut away and trim existing 100mm studding around opening										
838 x 2032mm	-	-	-	2.55	0.45	**41.06**	3.38	**44.43**	nr	**47.76**
914 x 2032mm	-	-	-	2.55	0.45	**41.06**	3.38	**44.43**	nr	**47.76**
50 x 75mm softwood studding	1.26	10.00	**1.39**	0.50	0.08	**7.96**	0.13	**9.48**	m	**10.19**
50 x 100mm softwood studding	1.67	10.00	**1.84**	0.60	0.10	**9.60**	0.14	**11.58**	m	**12.44**
make good plastered plasterboard around opening	-	-	-	0.75	0.75	**18.69**	1.46	**20.15**	m	**21.66**
make good lath and plaster around opening	-	-	-	0.83	0.83	**20.68**	1.79	**22.47**	m	**24.16**
take off skirting	-	-	-	0.21	-	**2.98**	0.21	**3.20**	m	**3.44**
form fitted end on existing skirting	-	-	-	0.57	-	**8.10**	-	**8.10**	nr	**8.71**
extend 25mm softwood tongued and grooved board flooring through opening on and with bearers	3.11	20.00	**3.57**	1.50	0.25	**23.99**	0.07	**27.63**	nr	**29.70**

Labour hourly rates: (except Specialists) Craft Operatives £14.21 Labourer £10.71 Rates are national average prices. Refer to REGIONAL VARIATIONS for indicative levels of overall pricing in regions	MATERIALS			LABOUR				RATES		
	Del to Site £	Waste %	Material Cost £	Craft Optve Hrs	Lab Hrs	Labour Cost £	Sunds £	Nett Rate £	Unit	Gross Rate (+7.5%) £
C90: ALTERATIONS - SPOT ITEMS Cont.										
Filling openings in stud partitions										
Fill opening where door removed in stud partition with 50 x 100mm studding and sole plate, covered on both sides with 9.5mm plasterboard baseboard with 5mm one coat board finish plaster; extend skirting on both sides; make good junction of new and existing plaster and skirtings										
838 x 2032mm	31.98	-	**31.98**	16.00	5.36	**284.77**	3.56	**320.30**	nr	**344.32**
Unit rates for pricing the above and similar work										
50 x 75mm softwood sole plate....................	1.26	10.00	**1.39**	0.50	0.08	**7.96**	0.13	**9.48**	m	**10.19**
50 x 100mm softwood sole plate...................	1.67	10.00	**1.84**	0.60	0.10	**9.60**	0.17	**11.60**	m	**12.47**
50 x 75mm softwood studding......................	1.26	10.00	**1.39**	0.50	0.08	**7.96**	0.13	**9.48**	m	**10.19**
50 x 100mm softwood studding.....................	1.67	10.00	**1.84**	0.60	0.10	**9.60**	0.17	**11.60**	m	**12.47**
9.5mm plasterboard baseboard to filling..........	2.17	10.00	**2.39**	0.50	0.10	**8.18**	0.33	**10.89**	m2	**11.71**
9.5mm plasterboard baseboard to filling including packing out not exceeding 10mm.............	2.17	10.00	**2.39**	1.00	0.17	**16.03**	0.44	**18.86**	m2	**20.27**
5mm one coat board finish plaster on plasterboard to filling	1.14	10.00	**1.25**	0.91	0.91	**22.68**	0.04	**23.97**	m2	**25.76**
10mm two coat hardwall plaster on plasterboard to filling	1.76	10.00	**1.93**	1.27	1.27	**31.65**	0.04	**33.61**	m2	**36.14**
19 x 100mm softwood chamfered skirting, primed all round	1.63	10.00	**1.79**	0.54	0.09	**8.64**	0.21	**10.64**	m	**11.44**
junction with existing	-	-	**-**	0.57	-	**8.10**	-	**8.10**	nr	**8.71**
25 x 150mm softwood chamfered skirting, primed all round	2.92	10.00	**3.21**	0.55	0.09	**8.78**	0.28	**12.28**	m	**13.20**
junction with existing	-	-	**-**	0.57	-	**8.10**	-	**8.10**	nr	**8.71**
Work to existing drains										
BREAK UP CONCRETE PAVING FOR EXCAVATING DRAIN TRENCH AND REINSTATE - break up 100mm concrete paving and 150mm hardcore bed under for excavation of drain trench and reinstate with new hardcore and concrete, BS 5328, designed mix C20, 20mm aggregate and make good up to existing paving both sides										
400mm wide	5.70	30.00	**6.42**	-	3.27	**35.02**	4.03	**45.47**	m	**48.88**
TAKE UP FLAG PAVING FOR EXCAVATING DRAIN TRENCH AND REINSTATE - carefully take up 50mm precast concrete flag paving and set aside for re-use and break up 150mm hardcore bed under for excavation of drain trench and reinstate with new hardcore and re-lay salvaged flag paving on and with 25mm sand bed and grout in lime sand (1:3) and make good up to existing paving on both sides										
400mm wide	1.98	30.00	**2.55**	1.47	1.47	**36.63**	3.71	**42.89**	m	**46.11**
INSERT NEW JUNCTION IN EXISTING DRAIN -excavate for and trace and expose existing drain, break into glazed vitrified clay drain and insert junction with 100mm branch, short length of new pipe and double collar and joint to existing drain; support earthwork, make good concrete bed and haunching and backfill										
existing 100mm diameter drain, invert depth										
600mm..	29.75	10.00	**32.73**	3.00	12.00	**171.15**	10.25	**214.13**	nr	**230.18**
750mm..	29.75	10.00	**32.73**	3.00	14.25	**195.25**	11.08	**239.05**	nr	**256.98**
900mm..	29.75	10.00	**32.73**	3.00	16.50	**219.35**	12.15	**264.22**	nr	**284.03**
existing 150mm diameter drain invert depth										
750mm..	54.13	10.00	**59.54**	4.50	15.75	**232.63**	12.80	**304.97**	nr	**327.84**
900mm..	54.13	10.00	**59.54**	4.50	18.00	**256.73**	14.52	**330.78**	nr	**355.59**
1050mm...	54.13	10.00	**59.54**	4.50	20.25	**280.82**	15.88	**356.24**	nr	**382.96**
1200mm...	54.13	10.00	**59.54**	4.50	22.50	**304.92**	17.36	**381.82**	nr	**410.46**
REPAIR DEFECTIVE DRAIN - excavate for and trace and expose existing drain, break out fractured glazed vitrified clay drain pipe, replace with new pipe and double collars and joint to existing drain; support earthwork, make good concrete bed and haunching and backfill										
existing 100mm diameter drain; single pipe length; invert depth										
450mm..	25.11	10.00	**27.62**	3.00	11.50	**165.80**	7.76	**201.18**	nr	**216.27**
600mm..	25.11	10.00	**27.62**	3.00	13.50	**187.22**	9.36	**224.20**	nr	**241.01**
750mm..	25.11	10.00	**27.62**	3.00	16.00	**213.99**	11.44	**253.05**	nr	**272.02**
add for each additional pipe length; invert depth										
450mm..	16.23	10.00	**17.85**	2.00	6.75	**100.71**	7.41	**125.97**	nr	**135.42**
600mm..	16.23	10.00	**17.85**	2.00	7.60	**109.82**	8.95	**136.62**	nr	**146.86**
750mm..	16.23	10.00	**17.85**	2.00	8.80	**122.67**	10.72	**151.25**	nr	**162.59**
Work to existing manholes										
RAISE TOP OF EXISTING MANHOLE - take off cover and frame and set aside for re-use, prepare level bed on existing one brick sides of manhole internal size 610 x 457mm and raise with Class B engineering brickwork in cement mortar (1:3) finished with a fair face and flush pointed; refix salvaged cover and frame, bed frame in cement mortar and cover in grease and sand										
raising 150mm high	16.74	10.00	**18.41**	5.30	5.30	**132.08**	3.76	**154.25**	nr	**165.81**
raising 225mm high	25.11	10.00	**27.62**	6.75	6.75	**168.21**	5.57	**201.40**	nr	**216.51**
raising 300mm high	33.48	10.00	**36.83**	8.50	8.50	**211.82**	7.13	**255.78**	nr	**274.97**

Labour hourly rates: (except Specialists) Craft Operatives £14.21 Labourer £10.71 Rates are national average prices. Refer to REGIONAL VARIATIONS for indicative levels of overall pricing in regions	MATERIALS			LABOUR				RATES		
	Del to Site £	Waste %	Material Cost £	Craft Optve Hrs	Lab Hrs	Labour Cost £	Sunds £	Nett Rate £	Unit	Gross Rate (+7.5%) £
C90: ALTERATIONS - SPOT ITEMS Cont.										
Work to existing manholes Cont.										
INSERT NEW BRANCH BEND IN BOTTOM OF EXISTING MANHOLE - break into bottom and one brick side of manhole, insert new glazed vitrified clay three quarter section branch bend to discharge over existing main channel, build in end of new drain and make good benching and side of manhole to match existing										
100mm diameter branch bend	10.58	10.00	**11.64**	5.75	5.75	**143.29**	2.69	**157.62**	nr	**169.44**
EXTEND EXISTING MANHOLE - take off cover and frame and set aside; break up one brick end wall of 457mm wide (internal) manhole and excavate for and extend manhole 225mm; support earthwork, level and compact bottom, part backfill and remove surplus spoil from site; extend bottom with 150mm concrete, BS 5328, ordinary prescribed mix C15P, 20mmaggregate bed; extend one brick sides and end with Class B engineering brickwork in cement mortar (1:3) fair faced and flush pointed and bond to existing; extend 100mm main channel and insert 100mm three quarter section branch channel bend; extend benching and build in end of new 100mm drain to one brick side, corbel over end of manhole manholes of the following invert depths; refix salvaged cover and frame, bed frame in cement mortar and cover in grease and sand										
450mm	51.06	10.00	**56.17**	17.60	17.90	**441.81**	10.38	**508.35**	nr	**546.48**
600mm	56.21	10.00	**61.83**	19.80	20.20	**497.70**	13.34	**572.87**	nr	**615.84**
750mm	61.53	10.00	**67.68**	22.00	22.50	**553.60**	16.31	**637.58**	nr	**685.40**
900mm	66.75	10.00	**73.43**	24.20	24.80	**609.49**	19.27	**702.18**	nr	**754.85**
Temporary weatherproof coverings										
Providing and erecting and clearing away on completion; for scaffold tube framework see preliminaries										
corrugated iron sheeting	6.32	10.00	**6.95**	0.15	0.50	**7.49**	0.94	**15.37**	m2	**16.53**
flexible reinforced plastic sheeting	2.73	10.00	**3.00**	0.10	0.25	**4.10**	0.76	**7.86**	m2	**8.45**
tarpaulins	5.80	10.00	**6.38**	0.10	0.25	**4.10**	0.76	**11.24**	m2	**12.08**
Temporary internal screens										
Providing and erecting and clearing away on completion; temporary screen of 50 x 50mm softwood framing and cover with										
reinforced building paper	2.60	10.00	**2.86**	0.43	0.30	**9.32**	0.26	**12.44**	m2	**13.38**
heavy duty polythene sheeting	1.78	10.00	**1.96**	0.43	0.30	**9.32**	0.24	**11.52**	m2	**12.38**
3mm hardboard	2.68	10.00	**2.95**	0.53	0.32	**10.96**	0.39	**14.30**	m2	**15.37**
Providing and erecting and clearing away on completion; temporary dustproof screen of 50 x 75mm softwood framing securely fixed to walls, floor and ceiling, the joints and edges of lining sealed with masking tape and cover with										
3mm hardboard lining one side	3.45	10.00	**3.80**	1.10	0.18	**17.56**	0.52	**21.88**	m2	**23.52**
3mm hardboard lining both sides	4.76	10.00	**5.24**	1.30	0.22	**20.83**	0.65	**26.72**	m2	**28.72**
6mm plywood lining one side	6.74	10.00	**7.41**	1.15	0.19	**18.38**	0.58	**26.37**	m2	**28.35**
6mm plywood lining both sides	11.28	10.00	**12.41**	1.40	0.23	**22.36**	0.91	**35.68**	m2	**38.35**
providing 35mm hardboard faced flush door size 838 x 1981mm in dustproof screen with 25 x 87mm softwood frame, 18 x 25mm softwood stop, pair of 100mm butts, pull handle and ball catch	137.36	10.00	**151.10**	1.50	0.25	**23.99**	3.29	**178.38**	nr	**191.76**
Temporary dustproof corridors										
Note										
for prices of walls and doors, see dustproof screens above										
Providing and erecting and clearing away on completion; ceiling of 50 x 75mm softwood joists at 450mm centres, the joints and edges sealed with masking tape and cover with										
3mm hardboard lining to soffit	3.72	10.00	**4.09**	1.10	0.18	**17.56**	0.52	**22.17**	m2	**23.84**
6mm plywood lining to soffit	6.98	10.00	**7.68**	1.15	0.19	**18.38**	0.65	**26.71**	m2	**28.71**
Temporary timber balustrades										
Providing and erecting and clearing away on completion; temporary softwood balustrade consisting of 50 x 75mm plate, 50 x 50mm standards at 900mm centres, four 25 x 150mm intermediate rails and 50 x 50mm handrail										
1150mm high	6.83	10.00	**7.51**	1.00	0.26	**16.99**	1.49	**26.00**	m	**27.95**
Temporary steel balustrades										
Providing and erecting and clearing away on completion; temporary steel balustrade constructed of 50mm diameter galvanised scaffold tubing and fittings with standards at 900mm centres intermediate rail, handrail, plastic mesh infill and with plates on ends of standards fixed to floor										
1150mm high	6.80	-	**6.80**	0.18	0.90	**12.20**	0.36	**19.35**	m	**20.80**
weekly cost of hire and maintenance	0.67	-	**0.67**	0.02	0.10	**1.36**	0.05	**2.07**	m	**2.23**

Labour hourly rates: (except Specialists) Craft Operatives £14.21 Labourer £10.71 Rates are national average prices. Refer to REGIONAL VARIATIONS for indicative levels of overall pricing in regions	MATERIALS			LABOUR				RATES		
	Del to Site £	Waste %	Material Cost £	Craft Optve Hrs	Lab Hrs	Labour Cost £	Sunds £	Nett Rate £	Unit	Gross Rate (+7.5%) £
C90: ALTERATIONS - SPOT ITEMS Cont.										
Temporary fillings to openings in external walls										
Providing and erecting and clearing away on completion; temporary filling to window or door opening of 50 x 100mm softwood framing and cover with										
corrugated iron sheeting	13.92	10.00	**15.31**	0.45	0.45	**11.21**	1.55	**28.08**	m2	**30.18**
25mm softwood boarding	12.42	10.00	**13.66**	0.60	0.60	**14.95**	1.81	**30.43**	m2	**32.71**
12mm external quality plywood	11.32	10.00	**12.45**	0.50	0.50	**12.46**	1.17	**26.09**	m2	**28.04**
C90: REPAIRING/RENOVATING ASPHALT WORK										
Cut out crack in old covering and make good with new material to match existing										
Floor or tanking										
20mm thick in two coats, horizontal	-	-	-	0.66	0.66	**16.45**	0.59	**17.04**	m	**18.32**
30mm thick in three coats, horizontal	-	-	-	0.90	0.90	**22.43**	0.89	**23.32**	m	**25.07**
30mm thick in three coats, vertical	-	-	-	1.19	1.19	**29.65**	0.89	**30.54**	m	**32.83**
Cut out detached blister in old covering and make good with new material to match existing										
Floor or tanking										
20mm thick in two coats, horizontal	-	-	-	0.30	0.30	**7.48**	0.30	**7.77**	nr	**8.36**
30mm thick in three coats, horizontal	-	-	-	0.35	0.35	**8.72**	0.41	**9.14**	nr	**9.82**
30mm thick in three coats, vertical	-	-	-	0.48	0.48	**11.96**	0.41	**12.38**	nr	**13.30**
Cut out crack in old covering and make good with new material to match existing										
Roof covering										
20mm thick in two coats	-	-	-	0.66	0.66	**16.45**	0.59	**17.04**	m	**18.32**
Cut out detached blister in old covering and make good with new material to match existing										
20mm thick in two coats	-	-	-	0.30	0.30	**7.48**	0.30	**7.77**	nr	**8.36**
C90: REPAIRING/RENOVATING ROOF COVERINGS										
Slate roofing repairs to roofs covered with 610 x 300mm slates										
Remove damaged slates and replace with new; sloping or vertical										
one slate	5.73	5.00	**6.02**	0.32	0.32	**7.97**	0.11	**14.10**	nr	**15.15**
patch of 10 slates	57.30	5.00	**60.17**	1.10	1.10	**27.41**	0.83	**88.41**	nr	**95.04**
patch of slates 1.00 - 3.00m2	70.19	5.00	**73.70**	1.25	1.25	**31.15**	0.90	**105.75**	m2	**113.68**
Examine battens, remove defective and provide 20% new										
19 x 38mm	0.21	10.00	**0.23**	0.09	0.09	**2.24**	0.06	**2.53**	m2	**2.72**
25 x 50mm	0.36	10.00	**0.40**	0.09	0.09	**2.24**	0.08	**2.72**	m2	**2.93**
Re-cover roof with slates previously removed and stacked and fix with slate nails										
with 75mm lap	-	-	-	0.77	0.77	**19.19**	0.11	**19.30**	m2	**20.74**
extra for providing 20% new slates	14.04	5.00	**14.74**	-	-	-	0.21	**14.95**	m2	**16.08**
Remove double course at eaves and										
refix	-	-	-	0.24	0.24	**5.98**	0.37	**6.35**	m	**6.82**
extra for providing 20% new slates	4.87	5.00	**5.11**	-	-	-	0.11	**5.22**	m	**5.61**
Remove double course at verge and replace with new slates and bed and point in mortar	21.43	5.00	**22.50**	0.29	0.29	**7.23**	0.69	**30.42**	m	**32.70**
Slate roofing repairs for roofs covered with 510 x 250mm slates										
Remove damaged slates and replace with new; sloping or vertical										
one slate	2.76	5.00	**2.90**	0.32	0.32	**7.97**	0.11	**10.98**	nr	**11.80**
patch of 10 slates	27.60	5.00	**28.98**	1.00	1.00	**24.92**	0.70	**54.60**	nr	**58.69**
patch of slates 1.00 - 3.00m2	49.68	5.00	**52.16**	1.49	1.49	**37.13**	0.89	**90.18**	m2	**96.95**
Examine battens, remove defective and provide 20% new										
19 x 38mm	0.26	10.00	**0.28**	0.11	0.11	**2.74**	0.08	**3.11**	m2	**3.34**
25 x 50mm	0.44	10.00	**0.49**	0.11	0.11	**2.74**	0.11	**3.33**	m2	**3.58**
Re-cover roof with slates previously removed and stacked and fix with slate nails										
with 75mm lap	-	-	-	1.00	1.00	**24.92**	0.15	**25.07**	m2	**26.95**
extra for providing 20% new slates	9.94	5.00	**10.43**	-	-	-	0.21	**10.65**	m2	**11.44**
Remove double course at eaves and										
refix	-	-	-	0.25	0.25	**6.23**	0.63	**6.86**	m	**7.37**
extra for providing 20% new slates	3.09	5.00	**3.25**	-	-	-	0.09	**3.34**	m	**3.59**
Remove double course at verge and replace with new slates and bed and point in mortar	12.70	5.00	**13.33**	0.29	0.29	**7.23**	0.68	**21.23**	m	**22.83**

Labour hourly rates: (except Specialists) Craft Operatives £14.21 Labourer £10.71 Rates are national average prices. Refer to REGIONAL VARIATIONS for indicative levels of overall pricing in regions	MATERIALS			LABOUR				RATES		
	Del to Site £	Waste %	Material Cost £	Craft Optve Hrs	Lab Hrs	Labour Cost £	Sunds £	Nett Rate £	Unit	Gross Rate (+7.5%) £
C90: REPAIRING/RENOVATING ROOF COVERINGS Cont.										
Slate roofing repairs for roofs covered with 405 x 200mm slates										
Remove damaged slates and replace with new; sloping or vertical										
one slate	1.36	5.00	**1.43**	0.32	0.32	**7.97**	0.09	**9.50**	nr	**10.21**
patch of 10 slates	13.60	5.00	**14.28**	0.90	0.90	**22.43**	0.60	**37.31**	nr	**40.11**
patch of slates 1.00 - 3.00m2	40.80	5.00	**42.84**	2.07	2.07	**51.58**	0.98	**95.41**	m2	**102.56**
Examine battens, remove defective and provide 20% new										
19 x 38mm	0.34	10.00	**0.38**	0.13	0.13	**3.24**	0.09	**3.71**	m2	**3.99**
25 x 50mm	0.59	10.00	**0.65**	0.13	0.13	**3.24**	0.11	**4.00**	m2	**4.30**
Re-cover roof with slates previously removed and stacked and fix with slate nails										
with 75mm lap	-	-	**-**	1.14	1.14	**28.41**	0.23	**28.63**	m2	**30.78**
extra for providing 20% new slates	8.16	5.00	**8.57**	-	-	**-**	0.21	**8.78**	m2	**9.44**
Remove double course at eaves and										
refix	-	-	**-**	0.26	0.26	**6.48**	0.62	**7.10**	m	**7.63**
extra for providing 20% new slates	1.85	5.00	**1.94**	-	-	**-**	0.09	**2.04**	m	**2.19**
Remove double course at verge and replace with new slates and bed and point in mortar	7.93	5.00	**8.33**	0.30	0.30	**7.48**	0.75	**16.55**	m	**17.79**
Tile roofing repairs for roofs covered with machine made clay plain tiles laid to 100mm gauge										
Remove damaged tiles and replace with new, sloping or vertical										
one tile	0.51	10.00	**0.56**	0.32	0.32	**7.97**	0.04	**8.57**	nr	**9.22**
patch of 10 tiles	5.12	10.00	**5.63**	0.90	0.90	**22.43**	0.30	**28.36**	nr	**30.48**
patch of tiles 1.00 - 3.00m2	31.23	10.00	**34.36**	1.72	1.72	**42.86**	1.90	**79.11**	m2	**85.05**
Examine battens, remove defective and provide 20% new										
19 x 25mm	0.46	10.00	**0.51**	0.13	0.13	**3.24**	0.09	**3.84**	m2	**4.13**
19 x 38mm	0.56	10.00	**0.62**	0.14	0.14	**3.49**	0.11	**4.21**	m2	**4.53**
19 x 50mm	0.84	10.00	**0.92**	0.14	0.14	**3.49**	0.13	**4.54**	m2	**4.88**
Re-cover roof with tiles previously removed and stacked										
nail every fourth course with galvanised nails	-	-	**-**	1.00	1.00	**24.92**	0.20	**25.12**	m2	**27.01**
extra for nailing every course	-	-	**-**	0.20	0.20	**4.98**	0.53	**5.52**	m2	**5.93**
extra for providing 20% new tiles	6.25	10.00	**6.87**	-	-	**-**	0.21	**7.08**	m2	**7.62**
Tile roofing repairs to roofs covered with machine made clay plain tiles laid to 90mm gauge										
Remove damaged tiles and replace with new, sloping or vertical										
one tile	0.51	10.00	**0.56**	0.32	0.32	**7.97**	0.04	**8.57**	nr	**9.22**
patch of 10 tiles	5.12	10.00	**5.63**	0.90	0.90	**22.43**	0.30	**28.36**	nr	**30.48**
patch of tiles 1.00 - 3.00m2	34.82	10.00	**38.30**	2.05	2.05	**51.09**	2.11	**91.49**	m2	**98.35**
Examine battens, remove defective and provide 20% new										
19 x 25mm	0.51	10.00	**0.56**	0.14	0.14	**3.49**	0.11	**4.16**	m2	**4.47**
19 x 38mm	0.62	10.00	**0.68**	0.15	0.15	**3.74**	0.13	**4.55**	m2	**4.89**
19 x 50mm	0.93	10.00	**1.03**	0.15	0.15	**3.74**	0.14	**4.91**	m2	**5.27**
Re-cover roof with tiles previously removed and stacked										
nail every fourth course with galvanised nails	-	-	**-**	1.15	1.15	**28.66**	0.23	**28.88**	m2	**31.05**
extra for nailing every course	-	-	**-**	0.23	0.23	**5.73**	0.56	**6.29**	m2	**6.76**
extra for providing 20% new tiles	6.96	10.00	**7.66**	-	-	**-**	0.21	**7.87**	m2	**8.46**
Tile roofing repairs for roofs covered with 381 x 227mm concrete interlocking tiles laid to 75mm laps										
Remove damaged tiles and replace with new, sloping or vertical										
one tile	0.62	10.00	**0.69**	0.32	0.32	**7.97**	0.08	**8.74**	nr	**9.40**
patch of 10 tiles	6.24	10.00	**6.86**	0.90	0.90	**22.43**	0.87	**30.16**	nr	**32.42**
patch of tiles 1.00 - 3.00m2	10.61	10.00	**11.67**	0.85	0.85	**21.18**	1.37	**34.23**	m2	**36.79**
Examine battens, remove defective and provide 20% new										
19 x 25mm	0.17	10.00	**0.18**	0.05	0.05	**1.25**	0.08	**1.51**	m2	**1.62**
19 x 38mm	0.20	10.00	**0.22**	0.06	0.06	**1.50**	0.08	**1.80**	m2	**1.93**
19 x 50mm	0.30	10.00	**0.33**	0.07	0.07	**1.74**	0.09	**2.17**	m2	**2.33**
Re-cover roof with tiles previously removed and stacked										
nail every fourth course with galvanised nails	-	-	**-**	0.50	0.50	**12.46**	0.02	**12.48**	m2	**13.42**
extra for nailing every course	-	-	**-**	0.05	0.05	**1.25**	0.07	**1.32**	m2	**1.42**
extra for providing 20% new tiles	2.50	10.00	**2.75**	-	-	**-**	0.26	**3.01**	m2	**3.23**
Tile roofing repairs generally										
Remove damaged clay tiles and provide new										
half round ridge or hip and bed in mortar	19.73	10.00	**21.71**	0.60	0.60	**14.95**	1.29	**37.95**	m	**40.80**
bonnet hip	24.50	5.00	**25.73**	0.75	0.75	**18.69**	0.36	**44.77**	m	**48.13**
trough valley	24.50	5.00	**25.73**	0.75	0.75	**18.69**	0.36	**44.77**	m	**48.13**
vertical angle	24.50	10.00	**26.95**	0.75	0.75	**18.69**	0.36	**46.00**	m	**49.45**

Labour hourly rates: (except Specialists) Craft Operatives £14.21 Labourer £10.71 Rates are national average prices. Refer to REGIONAL VARIATIONS for indicative levels of overall pricing in regions	MATERIALS			LABOUR				RATES		
	Del to Site £	Waste %	Material Cost £	Craft Optve Hrs	Lab Hrs	Labour Cost £	Sunds £	Nett Rate £	Unit	Gross Rate (+7.5%) £
C90: REPAIRING/RENOVATING ROOF COVERINGS Cont.										
Tile roofing repairs generally Cont.										
Remove double course and replace with new clay tiles and bed and point in mortar										
at eaves	6.14	10.00	**6.76**	0.23	0.23	**5.73**	0.53	**13.02**	m	**14.00**
at verge	12.80	10.00	**14.08**	0.38	0.38	**9.47**	1.65	**25.20**	m	**27.09**
Rake out defective pointing and re-point in cement mortar (1:3)										
ridge or hip tiles	-	-	**-**	0.34	0.34	**8.47**	0.20	**8.67**	m	**9.32**
Hack off defective cement mortar fillet and renew in cement mortar (1:3)	-	-	**-**	0.34	0.34	**8.47**	0.40	**8.88**	m	**9.54**
Remove defective hip hook and replace with new	1.84	5.00	**1.93**	0.34	0.34	**8.47**	0.08	**10.49**	nr	**11.27**
Corrugated fibre cement roofing repairs										
Remove damaged sheets and provide and fix new with screws and washers to timber purlins and ease and adjust edges of adjoining sheets as necessary										
one sheet; 1825mm long Profile 3 natural grey	16.51	10.00	**18.16**	1.23	1.23	**30.65**	2.36	**51.17**	nr	**55.01**
one sheet; 1825mm long Profile 3 standard colour	19.81	10.00	**21.79**	1.23	1.23	**30.65**	2.36	**54.80**	nr	**58.91**
one sheet; 1825mm long Profile 6 natural grey	20.72	10.00	**22.79**	1.78	1.78	**44.36**	3.48	**70.63**	nr	**75.93**
one sheet; 1825mm long Profile 6 standard colour	24.87	10.00	**27.36**	1.78	1.78	**44.36**	3.48	**75.20**	nr	**80.84**
over one sheet; Profile 3 natural grey	15.01	10.00	**16.51**	0.89	0.89	**22.18**	2.23	**40.92**	m2	**43.99**
over one sheet; Profile 3 standard colour	18.01	10.00	**19.81**	0.89	0.89	**22.18**	2.23	**44.22**	m2	**47.53**
over one sheet; Profile 6 natural grey	12.19	10.00	**13.41**	0.89	0.89	**22.18**	3.28	**38.87**	m2	**41.79**
over one sheet; Profile 6 standard colour	14.63	10.00	**16.09**	0.89	0.89	**22.18**	3.28	**41.55**	m2	**44.67**
Remove damaged sheets and provide and fix new with hook bolts and washers to steel purlins and ease and adjust edges of adjoining sheets as necessary										
one sheet; 1825mm long Profile 3 natural grey	16.51	10.00	**18.16**	1.37	1.37	**34.14**	2.36	**54.66**	nr	**58.76**
one sheet; 1825mm long Profile 3 standard colour	19.81	10.00	**21.79**	1.37	1.37	**34.14**	2.36	**58.29**	nr	**62.66**
one sheet; 1825mm long Profile 6 natural grey	20.72	10.00	**22.79**	1.98	1.98	**49.34**	3.48	**75.62**	nr	**81.29**
one sheet; 1825mm long Profile 6 standard colour	24.87	10.00	**27.36**	1.98	1.98	**49.34**	3.48	**80.18**	nr	**86.20**
over one sheet; Profile 3 natural grey	15.01	10.00	**16.51**	1.03	1.03	**25.67**	2.23	**44.41**	m2	**47.74**
over one sheet; Profile 3 standard colour	18.01	10.00	**19.81**	1.03	1.03	**25.67**	2.23	**47.71**	m2	**51.28**
over one sheet; Profile 6 natural grey	12.19	10.00	**13.41**	1.03	1.03	**25.67**	3.28	**42.36**	m2	**45.54**
over one sheet; Profile 6 standard colour	14.63	10.00	**16.09**	1.03	1.03	**25.67**	3.28	**45.04**	m2	**48.42**
Felt roofing repairs for roofs covered with bituminous felt roofing, BS 747										
Sweep clean, cut out defective layer, re-bond adjacent felt and cover with one layer of felt bonded in hot bitumen; patches not exceeding 1.00m2										
fibre based roofing felt type 1B	1.42	15.00	**1.63**	0.88	0.88	**21.93**	2.82	**26.38**	m2	**28.36**
for each defective underlayer cut out and replaced, add	1.42	15.00	**1.63**	0.88	0.88	**21.93**	2.82	**26.38**	m2	**28.36**
fibre based mineral surfaced roofing felt type 1E	2.70	15.00	**3.10**	1.05	1.05	**26.17**	2.82	**32.09**	m2	**34.49**
high performance polyester based roofing felt type 5B	5.50	15.00	**6.33**	1.02	1.02	**25.42**	2.82	**34.56**	m2	**37.16**
for each defective underlayer cut out and replaced, add	5.50	15.00	**6.33**	1.02	1.02	**25.42**	2.82	**34.56**	m2	**37.16**
high performance polyester based mineral surfaced roofing felt type 5E	3.90	15.00	**4.49**	1.13	1.13	**28.16**	2.82	**35.46**	m2	**38.12**
Sweep clean, cut out defective layer, re-bond adjacent felt and cover with one layer of felt bonded in hot bitumen; patches 1.00 - 3.00m2										
fibre based roofing felt type 1B	1.42	15.00	**1.63**	0.43	0.43	**10.72**	2.82	**15.17**	m2	**16.31**
for each defective underlayer cut out and replaced, add	1.42	15.00	**1.63**	0.43	0.43	**10.72**	2.82	**15.17**	m2	**16.31**
fibre based mineral surfaced roofing felt type 1E	2.70	15.00	**3.10**	0.56	0.56	**13.96**	2.82	**19.88**	m2	**21.37**
high performance polyester based roofing felt type 5B	5.50	15.00	**6.33**	0.53	0.53	**13.21**	2.82	**22.35**	m2	**24.03**
for each defective underlayer cut out and replaced, add	5.50	15.00	**6.33**	0.53	0.53	**13.21**	2.82	**22.35**	m2	**24.03**
high performance polyester based mineral surfaced roofing felt type 5E	3.90	15.00	**4.49**	0.65	0.65	**16.20**	2.82	**23.50**	m2	**25.27**
Sweep clean surface of existing roof, prime with hot bitumen and dress with										
13mm layer of limestone or granite chippings	1.80	10.00	**1.98**	-	0.23	**2.46**	1.58	**6.02**	m2	**6.47**
13mm layer of pea shingle	1.25	10.00	**1.38**	-	0.36	**3.86**	1.58	**6.81**	m2	**7.32**
extra; removing existing chippings or shingle	-	-	**-**	-	0.24	**2.57**	0.47	**3.04**	m2	**3.27**
Liquid bitumen proofing (black) brush applied on old roof covering including cleaning old covering										
Corrugated asbestos roofing										
one coat	2.37	10.00	**2.61**	-	0.21	**2.25**	-	**4.86**	m2	**5.22**
two coats	4.42	10.00	**4.86**	-	0.25	**2.68**	-	**7.54**	m2	**8.10**
Felt roofing										
one coat	2.05	10.00	**2.26**	-	0.17	**1.82**	-	**4.08**	m2	**4.38**
two coats	3.29	10.00	**3.62**	-	0.21	**2.25**	-	**5.87**	m2	**6.31**
For top coat in green in lieu black										
add	0.63	10.00	**0.69**	-	-	**-**	-	**0.69**	m2	**0.74**

Labour hourly rates: (except Specialists) Craft Operatives £17.38 Labourer £10.71 Rates are national average prices. Refer to REGIONAL VARIATIONS for indicative levels of overall pricing in regions	MATERIALS			LABOUR				RATES		
	Del to Site £	Waste %	Material Cost £	Craft Optve Hrs	Lab Hrs	Labour Cost £	Sunds £	Nett Rate £	Unit	Gross Rate (+7.5%) £
C90: REPAIRING/RENOVATING ROOF COVERINGS Cont.										
Repairs to lead roofing										
Repair crack clean out and fill with copper bit solder........	-	-	-	0.80	-	**13.90**	0.36	**14.26**	m	**15.33**
Turn back flashing and re-dress	-	-	-	0.48	-	**8.34**	-	**8.34**	m	**8.97**
Repairs to lead flashings, etc.; replace with new lead										
Remove flashing and provide new code Nr4 lead; wedge into groove										
150mm girth	2.99	5.00	**3.14**	1.06	-	**18.42**	0.31	**21.87**	m	**23.51**
240mm girth	4.78	5.00	**5.02**	1.49	-	**25.90**	0.31	**31.22**	m	**33.57**
300mm girth	5.98	5.00	**6.28**	1.82	-	**31.63**	0.31	**38.22**	m	**41.09**
Remove stepped flashing and provide new code Nr 4 lead; wedge into groove										
180mm girth	3.59	5.00	**3.77**	1.54	-	**26.77**	0.31	**30.84**	m	**33.16**
240mm girth	4.78	5.00	**5.02**	1.94	-	**33.72**	0.31	**39.04**	m	**41.97**
300mm girth	5.98	5.00	**6.28**	2.34	-	**40.67**	0.31	**47.26**	m	**50.80**
Remove apron flashing and provide new code Nr 5 lead; wedge into groove										
150mm girth	3.72	5.00	**3.91**	1.06	-	**18.42**	0.31	**22.64**	m	**24.33**
300mm girth	7.44	5.00	**7.81**	1.82	-	**31.63**	0.31	**39.75**	m	**42.73**
450mm girth	11.16	5.00	**11.72**	2.58	-	**44.84**	0.47	**57.03**	m	**61.31**
Remove lining to valley gutter and provide new code Nr 5 lead										
240mm girth	5.95	5.00	**6.25**	1.49	-	**25.90**	-	**32.14**	m	**34.55**
300mm girth	7.44	5.00	**7.81**	1.82	-	**31.63**	-	**39.44**	m	**42.40**
450mm girth	11.16	5.00	**11.72**	2.58	-	**44.84**	-	**56.56**	m	**60.80**
Remove gutter lining and provide new code Nr 6 lead										
360mm girth	10.58	5.00	**11.11**	2.12	-	**36.85**	-	**47.95**	m	**51.55**
420mm girth	12.34	5.00	**12.96**	2.43	-	**42.23**	-	**55.19**	m	**59.33**
450mm girth	13.22	5.00	**13.88**	2.58	-	**44.84**	-	**58.72**	m	**63.13**
Repairs to zinc flashings, etc.; replace with 0.8mm zinc										
Remove flashing and provide new; wedge into groove										
150mm girth	2.47	5.00	**2.59**	1.12	-	**19.47**	0.21	**22.27**	m	**23.94**
240mm girth	3.96	5.00	**4.16**	1.60	-	**27.81**	0.21	**32.18**	m	**34.59**
300mm girth	4.95	5.00	**5.20**	1.94	-	**33.72**	0.31	**39.22**	m	**42.16**
Remove stepped flashing and provide new; wedge into groove										
180mm girth	2.97	5.00	**3.12**	1.65	-	**28.68**	0.21	**32.01**	m	**34.41**
240mm girth	3.96	5.00	**4.16**	2.08	-	**36.15**	0.21	**40.52**	m	**43.56**
300mm girth	4.95	5.00	**5.20**	2.52	-	**43.80**	0.31	**49.30**	m	**53.00**
Remove apron flashing and provide new; wedge into groove										
150mm girth	2.47	5.00	**2.59**	1.12	-	**19.47**	0.21	**22.27**	m	**23.94**
300mm girth	4.95	5.00	**5.20**	1.94	-	**33.72**	0.31	**39.22**	m	**42.16**
450mm girth	7.42	5.00	**7.79**	2.76	-	**47.97**	0.38	**56.14**	m	**60.35**
Remove lining to valley gutter and provide new										
240mm girth	3.96	5.00	**4.16**	1.60	-	**27.81**	-	**31.97**	m	**34.36**
300mm girth	4.95	5.00	**5.20**	1.94	-	**33.72**	-	**38.91**	m	**41.83**
450mm girth	7.42	5.00	**7.79**	2.76	-	**47.97**	-	**55.76**	m	**59.94**
Remove gutter lining and provide new										
360mm girth	5.94	5.00	**6.24**	2.27	-	**39.45**	-	**45.69**	m	**49.12**
420mm girth	6.93	5.00	**7.28**	2.61	-	**45.36**	-	**52.64**	m	**56.59**
450mm girth	7.42	5.00	**7.79**	2.76	-	**47.97**	-	**55.76**	m	**59.94**
Repairs to copper flashings, etc.; replace with 0.55mm copper										
Remove flashing and provide new; wedge into groove										
150mm girth	3.60	5.00	**3.78**	1.06	-	**18.42**	0.43	**22.63**	m	**24.33**
240mm girth	5.76	5.00	**6.05**	1.49	-	**25.90**	0.43	**32.37**	m	**34.80**
300mm girth	7.20	5.00	**7.56**	1.82	-	**31.63**	0.43	**39.62**	m	**42.59**
Remove stepped flashing and provide new; wedge into groove										
180mm girth	4.32	5.00	**4.54**	1.54	-	**26.77**	0.43	**31.73**	m	**34.11**
240mm girth	5.76	5.00	**6.05**	1.94	-	**33.72**	0.43	**40.19**	m	**43.21**
300mm girth	7.20	5.00	**7.56**	2.34	-	**40.67**	0.43	**48.66**	m	**52.30**
Remove apron flashing and provide new; wedge into groove										
150mm girth	3.60	5.00	**3.78**	1.06	-	**18.42**	0.43	**22.63**	m	**24.33**
300mm girth	7.20	5.00	**7.56**	1.82	-	**31.63**	0.43	**39.62**	m	**42.59**
450mm girth	10.80	5.00	**11.34**	2.58	-	**44.84**	0.64	**56.82**	m	**61.08**
Remove lining to valley gutter and provide new										
240mm girth	5.76	5.00	**6.05**	1.49	-	**25.90**	-	**31.94**	m	**34.34**
300mm girth	7.20	5.00	**7.56**	1.82	-	**31.63**	-	**39.19**	m	**42.13**
450mm girth	10.80	5.00	**11.34**	2.58	-	**44.84**	-	**56.18**	m	**60.39**

Labour hourly rates: (except Specialists) Craft Operatives £17.38 Labourer £10.71 Rates are national average prices. Refer to REGIONAL VARIATIONS for indicative levels of overall pricing in regions	MATERIALS			LABOUR				RATES		
	Del to Site £	Waste %	Material Cost £	Craft Optve Hrs	Lab Hrs	Labour Cost £	Sunds £	Nett Rate £	Unit	Gross Rate (+7.5%) £
C90: REPAIRING/RENOVATING ROOF COVERINGS Cont.										
Repairs to copper flashings, etc.; replace with 0.55mm copper Cont.										
Remove gutter lining and provide new										
360mm girth	8.64	5.00	**9.07**	2.12	-	**36.85**	-	**45.92**	m	**49.36**
420mm girth	10.08	5.00	**10.58**	2.43	-	**42.23**	-	**52.82**	m	**56.78**
450mm girth	10.80	5.00	**11.34**	2.58	-	**44.84**	-	**56.18**	m	**60.39**
Repairs to aluminium flashings etc; replace with 0.60mm aluminium										
Remove flashings and provide new; wedge into groove										
150mm girth	2.26	5.00	**2.37**	1.06	-	**18.42**	0.19	**20.99**	m	**22.56**
240mm girth	3.50	5.00	**3.68**	1.49	-	**25.90**	0.19	**29.76**	m	**31.99**
300mm girth	3.50	5.00	**3.68**	1.82	-	**31.63**	0.19	**35.50**	m	**38.16**
Remove stepped flashing and provide new; wedge into groove										
180mm girth	3.05	5.00	**3.20**	1.54	-	**26.77**	0.19	**30.16**	m	**32.42**
240mm girth	3.50	5.00	**3.68**	1.94	-	**33.72**	0.19	**37.58**	m	**40.40**
300mm girth	3.50	5.00	**3.68**	2.34	-	**40.67**	0.19	**44.53**	m	**47.87**
Remove apron flashing and provide new; wedge into groove										
150mm girth	2.26	5.00	**2.37**	1.06	-	**18.42**	0.19	**20.99**	m	**22.56**
300mm girth	3.50	5.00	**3.68**	1.82	-	**31.63**	0.19	**35.50**	m	**38.16**
450mm girth	5.52	5.00	**5.80**	2.56	-	**44.49**	3.56	**53.84**	m	**57.88**
Remove lining to valley gutter and provide new										
240mm girth	3.50	5.00	**3.68**	1.49	-	**25.90**	-	**29.57**	m	**31.79**
300mm girth	3.50	5.00	**3.68**	1.82	-	**31.63**	-	**35.31**	m	**37.95**
450mm girth	5.52	5.00	**5.80**	2.58	-	**44.84**	-	**50.64**	m	**54.43**
Remove gutter lining and provide new										
360mm girth	4.62	5.00	**4.85**	2.12	-	**36.85**	-	**41.70**	m	**44.82**
420mm girth	4.62	5.00	**4.85**	2.43	-	**42.23**	-	**47.08**	m	**50.62**
450mm girth	4.62	5.00	**4.85**	2.58	-	**44.84**	-	**49.69**	m	**53.42**
C90: REPAIRING/RENOVATING PLUMBING										
Repairs to rainwater goods										
Clean out and make good loose fixings										
eaves gutter	-	-	-	0.40	-	**6.95**	-	**6.95**	m	**7.47**
rainwater pipe	-	-	-	0.40	-	**6.95**	-	**6.95**	m	**7.47**
rainwater head	-	-	-	0.75	-	**13.04**	-	**13.04**	nr	**14.01**
Clean out defective joint to existing eaves gutter and re-make joint										
with jointing compound and new bolt	1.00	-	**1.00**	0.75	-	**13.04**	-	**14.04**	nr	**15.09**
Remove remains of old balloon grating and provide new plastic grating in outlet or end of pipe										
50mm diameter	1.96	2.00	**2.00**	0.50	-	**8.69**	-	**10.69**	nr	**11.49**
82mm diameter	1.99	2.00	**2.03**	0.50	-	**8.69**	-	**10.72**	nr	**11.52**
110mm diameter	2.01	2.00	**2.05**	0.50	-	**8.69**	-	**10.74**	nr	**11.55**
Repair or replacement of stopcocks, valves, etc.										
Turn off water supply drain down as necessary and re-new washer to the following up to 19mm										
main stopcock	0.25	-	**0.25**	2.00	-	**34.76**	-	**35.01**	nr	**37.64**
service stopcock	0.25	-	**0.25**	2.00	-	**34.76**	-	**35.01**	nr	**37.64**
bib or pillar cock	0.25	-	**0.25**	2.00	-	**34.76**	-	**35.01**	nr	**37.64**
supatap	0.25	-	**0.25**	2.00	-	**34.76**	-	**35.01**	nr	**37.64**
draining tap	0.25	-	**0.25**	2.00	-	**34.76**	-	**35.01**	nr	**37.64**
Turn off water supply, drain down as necessary, take out old valves and re-new with joints to copper										
13mm brass stopcock (BS 1010)	3.04	2.00	**3.10**	2.00	-	**34.76**	-	**37.86**	nr	**40.70**
13mm chromium plated sink pillar cock (BS 1010)	12.75	2.00	**13.01**	2.00	-	**34.76**	-	**47.77**	nr	**51.35**
13mm brass bib cock (BS 1010)	7.65	2.00	**7.80**	2.00	-	**34.76**	-	**42.56**	nr	**45.76**
13mm drain tap (BS 2879) type 2	1.80	2.00	**1.84**	2.00	-	**34.76**	-	**36.60**	nr	**39.34**
19mm drain tap (BS 2879) type 2	14.52	2.00	**14.81**	2.00	-	**34.76**	-	**49.57**	nr	**53.29**
13mm high pressure ball valve (BS 1212) piston type with copper ball	5.04	2.00	**5.14**	2.00	-	**34.76**	-	**39.90**	nr	**42.89**
13mm high pressure ball valve (BS 1212) piston type with plastic ball	2.99	2.00	**3.05**	2.00	-	**34.76**	-	**37.81**	nr	**40.65**
13mm low pressure ball valve (BS 1212) piston type with copper ball	6.29	2.00	**6.42**	2.00	-	**34.76**	-	**41.18**	nr	**44.26**
13mm low pressure ball valve (BS 1212) piston type with plastic ball	4.24	2.00	**4.32**	2.00	-	**34.76**	-	**39.08**	nr	**42.02**
19mm high pressure ball valve (BS 1212) piston type with copper ball	14.76	2.00	**15.06**	2.50	-	**43.45**	-	**58.51**	nr	**62.89**
19mm high pressure ball valve (BS 1212) piston type with plastic ball	14.29	2.00	**14.58**	2.50	-	**43.45**	-	**58.03**	nr	**62.38**
Replacement of traps										
Take out old plastic trap and replace with new 76mm seal trap (BS EN 274) with 'O'' ring joint outlet										
36mm diameter P trap	3.34	2.00	**3.41**	1.25	-	**21.73**	-	**25.13**	nr	**27.02**
42mm diameter P trap	3.86	2.00	**3.94**	1.50	-	**26.07**	-	**30.01**	nr	**32.26**
36mm diameter S trap	4.23	2.00	**4.31**	1.25	-	**21.73**	-	**26.04**	nr	**27.99**

Labour hourly rates: (except Specialists) Craft Operatives £17.38 Labourer £10.71 Rates are national average prices. Refer to REGIONAL VARIATIONS for indicative levels of overall pricing in regions	MATERIALS			LABOUR				RATES		
	Del to Site £	Waste %	Material Cost £	Craft Optve Hrs	Lab Hrs	Labour Cost £	Sunds £	Nett Rate £	Unit	Gross Rate (+7.5%) £
C90: REPAIRING/RENOVATING PLUMBING Cont.										
Replacement of traps Cont.										
Take out old plastic trap and replace with new 76mm seal trap (BS EN 274) with 'O'' ring joint outlet Cont.										
42mm diameter S trap	4.95	2.00	**5.05**	1.50	-	**26.07**	-	**31.12**	nr	**33.45**
42mm diameter bath trap with overflow connection	7.15	2.00	**7.29**	1.75	-	**30.42**	-	**37.71**	nr	**40.54**
Take out old copper trap and replace with new 76mm seal two piece trap (BS 1184) with compression joint										
35mm diameter P trap	17.40	2.00	**17.75**	1.25	-	**21.73**	-	**39.47**	nr	**42.43**
42mm diameter P trap	20.50	2.00	**20.91**	1.50	-	**26.07**	-	**46.98**	nr	**50.50**
35mm diameter S trap	18.44	2.00	**18.81**	1.25	-	**21.73**	-	**40.53**	nr	**43.57**
42mm diameter S trap	22.18	2.00	**22.62**	1.50	-	**26.07**	-	**48.69**	nr	**52.35**
42mm diameter bath trap with male iron overflow connection	29.10	2.00	**29.68**	1.75	-	**30.42**	-	**60.10**	nr	**64.60**
Preparation of pipe for insertion of new pipe or fitting										
Cut into cast iron pipe and take out a length up to 1.00m for insertion of new pipe or fitting										
50mm diameter	-	-	-	3.70	-	**64.31**	-	**64.31**	nr	**69.13**
64 and 76mm diameter	-	-	-	4.30	-	**74.73**	-	**74.73**	nr	**80.34**
89 and 102mm diameter	-	-	-	4.90	-	**85.16**	-	**85.16**	nr	**91.55**
Cut into cast iron pipe with caulked joint and take out a length up to 1.00m for insertion of new pipe or fitting										
50mm diameter	-	-	-	3.70	-	**64.31**	-	**64.31**	nr	**69.13**
64 and 76mm diameter	-	-	-	4.30	-	**74.73**	-	**74.73**	nr	**80.34**
89 and 102mm diameter	-	-	-	4.90	-	**85.16**	-	**85.16**	nr	**91.55**
Cut into asbestos cement pipe and take out a length up to 1.00m for insertion of new pipe or fitting										
50mm diameter	-	-	-	3.15	-	**54.75**	-	**54.75**	nr	**58.85**
64 and 76mm diameter	-	-	-	3.75	-	**65.18**	-	**65.18**	nr	**70.06**
89 and 102mm diameter	-	-	-	4.35	-	**75.60**	-	**75.60**	nr	**81.27**
Cut into polythene or p.v.c. pipe and take out a length up to 1.00m for insertion of new pipe or fitting										
up to 25mm diameter	-	-	-	2.50	-	**43.45**	-	**43.45**	nr	**46.71**
32 - 50mm diameter	-	-	-	3.15	-	**54.75**	-	**54.75**	nr	**58.85**
64 and 76mm diameter	-	-	-	3.75	-	**65.18**	-	**65.18**	nr	**70.06**
89 and 102mm diameter	-	-	-	4.40	-	**76.47**	-	**76.47**	nr	**82.21**
Cut into steel pipe and take out a length up to 1.00m for insertion of new pipe or fitting										
up to 25mm diameter	-	-	-	3.05	-	**53.01**	-	**53.01**	nr	**56.98**
32 - 50mm diameter	-	-	-	3.70	-	**64.31**	-	**64.31**	nr	**69.13**
64 and 76mm diameter	-	-	-	4.30	-	**74.73**	-	**74.73**	nr	**80.34**
89 and 102mm diameter	-	-	-	4.90	-	**85.16**	-	**85.16**	nr	**91.55**
Cut into stainless steel pipe and take out a length up to 1.00m for insertion of new pipe or fitting										
up to 25mm diameter	-	-	-	2.90	-	**50.40**	-	**50.40**	nr	**54.18**
32 - 50mm diameter	-	-	-	3.50	-	**60.83**	-	**60.83**	nr	**65.39**
64 and 76mm diameter	-	-	-	4.10	-	**71.26**	-	**71.26**	nr	**76.60**
89 and 102mm diameter	-	-	-	4.75	-	**82.56**	-	**82.56**	nr	**88.75**
Cut into copper pipe and take out a length up to 1.00m for insertion of new pipe or fitting										
up to 25mm diameter	-	-	-	2.90	-	**50.40**	-	**50.40**	nr	**54.18**
32 - 50mm diameter	-	-	-	3.50	-	**60.83**	-	**60.83**	nr	**65.39**
64 and 76mm diameter	-	-	-	4.10	-	**71.26**	-	**71.26**	nr	**76.60**
89 and 102mm diameter	-	-	-	4.75	-	**82.56**	-	**82.56**	nr	**88.75**
Cut into lead pipe and take out a length up to 1.00m for insertion of new pipe or fitting										
up to 25mm diameter	-	-	-	3.05	-	**53.01**	-	**53.01**	nr	**56.98**
32 - 50mm diameter	-	-	-	3.70	-	**64.31**	-	**64.31**	nr	**69.13**
64 and 76mm diameter	-	-	-	4.30	-	**74.73**	-	**74.73**	nr	**80.34**
89 and 102mm diameter	-	-	-	4.90	-	**85.16**	-	**85.16**	nr	**91.55**
Removal of traps, valves, etc., for re-use										
Take out the following, including unscrewing or uncoupling, for re-use and including draining down as necessary										
trap up to 25mm diameter	-	-	-	0.70	-	**12.17**	-	**12.17**	nr	**13.08**
trap 32 - 50mm diameter	-	-	-	0.85	-	**14.77**	-	**14.77**	nr	**15.88**
stop valve up to 25mm diameter	-	-	-	2.00	-	**34.76**	-	**34.76**	nr	**37.37**
stop valve 32 - 50mm diameter	-	-	-	2.25	-	**39.11**	-	**39.11**	nr	**42.04**
radiator valve up to 25mm diameter	-	-	-	2.00	-	**34.76**	-	**34.76**	nr	**37.37**
radiator valve 32 - 50mm diameter	-	-	-	2.25	-	**39.11**	-	**39.11**	nr	**42.04**
tap up to 25mm diameter	-	-	-	2.00	-	**34.76**	-	**34.76**	nr	**37.37**
tap 32 - 50mm diameter	-	-	-	2.25	-	**39.11**	-	**39.11**	nr	**42.04**
extra; burning out one soldered joint up to 25mm diameter	-	-	-	0.50	-	**8.69**	-	**8.69**	nr	**9.34**
extra; burning out one soldered joint 32 - 50mm diameter	-	-	-	0.80	-	**13.90**	-	**13.90**	nr	**14.95**

Labour hourly rates: (except Specialists) Craft Operatives £17.38 Labourer £10.71 Rates are national average prices. Refer to REGIONAL VARIATIONS for indicative levels of overall pricing in regions	MATERIALS			LABOUR				RATES		
	Del to Site £	Waste %	Material Cost £	Craft Optve Hrs	Lab Hrs	Labour Cost £	Sunds £	Nett Rate £	Unit	Gross Rate (+7.5%) £
C90: REPAIRING/RENOVATING PLUMBING Cont.										
Replacement of old sanitary fittings										
Disconnect trap, valves and services pipes, remove old fitting and provide and fix new including re-connecting trap, valves and services pipes										
560 x 405mm vitreous china lavatory basin with 32mm chromium plated waste outlet, plug, chain and stay and pair of cast iron brackets.............	47.00	5.00	**49.35**	6.00	-	**104.28**	-	**153.63**	nr	**165.15**
600 x 1000mm stainless steel sink with single drainer with 38mm chromium plated waste outlet, plug, chain and stay	79.75	5.00	**83.74**	6.00	-	**104.28**	-	**188.02**	nr	**202.12**
600 x 1500mm stainless steel sink with double drainer with 38mm chromium plated waste outlet, overflow plug, chain and stay..................	97.50	5.00	**102.38**	6.50	-	**112.97**	-	**215.35**	nr	**231.50**
1500mm pressed steel vitreous enamelled rectangular top bath with cradles and with 38mm chromium plated waste outlet, plug, chain and stay and 32mm chromium plated overflow and including removing old cast iron bath.....................	84.50	5.00	**88.73**	8.00	-	**139.04**	-	**227.77**	nr	**244.85**
1700mm pressed steel vitreous enamelled rectangular top bath with cradles and with 38mm chromium plated waste outlet, plug, chain and stay and 32mm chromium plated overflow and including removing old cast iron bath.....................	84.50	5.00	**88.73**	8.00	-	**139.04**	-	**227.77**	nr	**244.85**
Take off old W.C. seat and provide and fix new ring pattern black plastic seat and cover........	10.85	5.00	**11.39**	1.50	-	**26.07**	-	**37.46**	nr	**40.27**
Disconnect supply pipe, overflow and flush pipe and take out old W.C. cistern and provide and fix new 9 litre black plastic W.C. cistern with cover and connect pipe and ball valve										
low level	41.50	5.00	**43.58**	5.00	-	**86.90**	-	**130.48**	nr	**140.26**
high level	50.75	5.00	**53.29**	5.00	-	**86.90**	-	**140.19**	nr	**150.70**
Disconnect flush pipe, take off seat and remove old W.C. pan, provide and fix new white china pan, connect flush pipe, form cement joint with drain and refix old seat										
S or P trap pan	24.25	5.00	**25.46**	5.00	-	**86.90**	-	**112.36**	nr	**120.79**
Disconnect and remove old high level W.C. suite complete and provide and fix new suite with P or S trap pan, plastic ring seat and cover, plastic flush pipe or bend, 9 litre plastic W.C. cistern with cover, chain and pull or lever handle and 13mm low pressure ball valve; adapt supply pipe and overflow as necessary and connect to new suite and cement joint to drain										
new high level suite	88.75	5.00	**93.19**	8.50	-	**147.73**	-	**240.92**	nr	**258.99**
new low level suite (vitreous china cistern).....	83.00	5.00	**87.15**	9.00	-	**156.42**	-	**243.57**	nr	**261.84**
Disconnect and remove old low level W.C. suite complete and provide and fix new suite with P or S trap pan, plastic ring seat and cover, plastic flush bend, 9 litre vitreous china W.C. cistern with cover, lever handle and 13mm low pressure ball valve with plastic ball; re-connect supply pipe and overflow and cement joint to drain										
new low level suite	83.00	5.00	**87.15**	8.50	-	**147.73**	-	**234.88**	nr	**252.50**
Replacement of old water tanks										
Disconnect all pipework, set aside ball valve and take out cold water storage cistern in roof space; replace with new GRP water storage tank (BS EN 13280); reconnect all pipework and ball valve; remove metal filings and clean inside of tank										
type SC70 size 915 x 610 x 580mm (227 litre).....	198.00	2.00	**201.96**	9.00	-	**156.42**	0.51	**358.89**	nr	**385.81**

This section continues on the next page

Labour hourly rates: (except Specialists) Craft Operatives £20.54 Labourer £10.71 Rates are national average prices. Refer to REGIONAL VARIATIONS for indicative levels of overall pricing in regions	MATERIALS			LABOUR				RATES		
	Del to Site £	Waste %	Material Cost £	Craft Optve Hrs	Lab Hrs	Labour Cost £	Sunds £	Nett Rate £	Unit	Gross Rate (+7.5%) £
C90: REPAIRING/RENOVATING ELECTRICAL INSTALLATIONS										
Generally										
Note the following approximate estimates of costs are dependent on the number and disposition of the points. Lamps and fittings together with cutting and making good are excluded										
Strip out and re-wire complete										
Provide new lamp holders, flush switches, flush socket outlets, fitting outlets and consumer unit with circuit breakers; a three bedroom house with ten lighting points and associated switches, sixteen 13 amp socket outlets, three 13 amp fitting outlets, one 30 amp cooker control panel, three earth bonding points and one eight way consumer unit										
PVC cables in existing conduit	894.28	7.50	**961.35**	52.85	-	**1085.54**	30.00	**2076.89**	nr	**2232.66**
PVC cables and conduit	989.07	7.50	**1063.25**	99.00	-	**2033.46**	30.00	**3126.71**	nr	**3361.21**
Strip out and re-wire with PVC insulated and sheathed cable in existing conduit										
Point to junction box, re-using existing lamp holder, socket outlet etc										
5 amp socket outlet	6.00	7.50	**6.45**	0.85	-	**17.46**	0.45	**24.36**	nr	**26.19**
13 amp socket or fitting outlet	8.00	7.50	**8.60**	0.85	-	**17.46**	0.45	**26.51**	nr	**28.50**
lighting point or switch	4.80	7.50	**5.16**	0.98	-	**20.13**	0.45	**25.74**	nr	**27.67**
Strip out and re-wire with PVC insulated and sheathed cable and new conduit										
Point to junction box, re-using existing lamp holder, socket outlet etc										
5 amp socket outlet	6.97	7.50	**7.49**	1.20	-	**24.65**	0.50	**32.64**	nr	**35.09**
13 amp socket outlet	9.35	7.50	**10.05**	1.20	-	**24.65**	0.50	**35.20**	nr	**37.84**
lighting point or switch	6.63	7.50	**7.13**	1.30	-	**26.70**	0.30	**34.13**	nr	**36.69**
Luminaires and accessories										
Take out damaged and renew										
flourescent lamp fitting 1500mm, batten type, single tube	11.00	2.50	**11.28**	0.56	-	**11.50**	0.25	**23.03**	nr	**24.75**
flourescent lamp fitting 1500mm, batten type, twin tube	18.56	2.50	**19.02**	0.56	-	**11.50**	0.25	**30.78**	nr	**33.08**
flourescent lamp fitting 1500mm, plastic diffused type, twin tube	37.16	2.50	**38.09**	0.56	-	**11.50**	0.25	**49.84**	nr	**53.58**
flourescent lamp fitting 1500mm, plastic diffused type, twin tube	75.23	2.50	**77.11**	0.56	-	**11.50**	0.25	**88.86**	nr	**95.53**
lamp holder, batten type	2.30	7.50	**2.47**	0.56	-	**11.50**	0.25	**14.22**	nr	**15.29**
lamp holder, pendant type with rose	1.15	7.50	**1.24**	1.06	-	**21.77**	0.25	**23.26**	nr	**25.00**
switch, flush type	1.00	7.50	**1.08**	0.75	-	**15.41**	0.45	**16.93**	nr	**18.20**
switch, surface mounted type	1.65	7.50	**1.77**	0.75	-	**15.41**	0.45	**17.63**	nr	**18.95**
13 amp socket outlet, flush type	2.00	7.50	**2.15**	0.75	-	**15.41**	0.45	**18.01**	nr	**19.36**
13 amp socket outlet, surface mounted type	2.90	7.50	**3.12**	0.88	-	**18.08**	0.45	**21.64**	nr	**23.27**
5 amp socket outlet, flush type	4.60	7.50	**4.95**	0.70	-	**14.38**	0.45	**19.77**	nr	**21.26**
5 amp socket outlet, surface mounted type	5.10	7.50	**5.48**	0.88	-	**18.08**	0.45	**24.01**	nr	**25.81**

This section continues on the next page

Labour hourly rates: (except Specialists)
Craft Operatives £14.21 Labourer £10.71
Rates are national average prices.
Refer to REGIONAL VARIATIONS for indicative levels of overall pricing in regions

	MATERIALS			LABOUR				RATES		
	Del to Site £	Waste %	Material Cost £	Craft Optve Hrs	Lab Hrs	Labour Cost £	Sunds £	Nett Rate £	Unit	Gross Rate (+7.5%) £
C90: REPAIRING/RENOVATING FINISHINGS										
Repairs to old wall and ceiling finishings										
Cut out damaged two coat plaster to wall in patches and make out in new plaster to match existing										
under 1.00m2	2.70	10.00	**2.98**	2.13	2.13	**53.08**	0.27	**56.32**	m2	**60.54**
1.00 - 2.00m2	2.70	10.00	**2.98**	1.76	1.76	**43.86**	0.27	**47.10**	m2	**50.63**
2.00 - 4.00m2	2.70	10.00	**2.98**	1.39	1.39	**34.64**	0.27	**37.88**	m2	**40.72**
Cut out damaged three coat plaster to wall in patches and make out in new plaster to match existing										
under 1.00m2	3.59	10.00	**3.95**	2.50	2.50	**62.30**	0.41	**66.66**	m2	**71.66**
1.00 - 2.00m2	3.59	10.00	**3.95**	2.06	2.06	**51.34**	0.41	**55.70**	m2	**59.87**
2.00 - 4.00m2	3.59	10.00	**3.95**	1.62	1.62	**40.37**	0.41	**44.73**	m2	**48.09**
Cut out damaged two coat rendering to wall in patches and make out in new rendering to match existing										
under 1.00m2	1.58	10.00	**1.74**	1.93	1.93	**48.10**	0.27	**50.10**	m2	**53.86**
1.00 - 2.00m2	1.58	10.00	**1.74**	1.60	1.60	**39.87**	0.27	**41.88**	m2	**45.02**
2.00 - 4.00m2	1.58	10.00	**1.74**	1.26	1.26	**31.40**	0.27	**33.41**	m2	**35.91**
Cut out damaged three coat rendering to wall in patches and make out in new rendering to match existing										
under 1.00m2	2.44	10.00	**2.68**	2.28	2.28	**56.82**	0.41	**59.91**	m2	**64.40**
1.00 - 2.00m2	2.44	10.00	**2.68**	1.87	1.87	**46.60**	0.41	**49.69**	m2	**53.42**
2.00 - 4.00m2	2.44	10.00	**2.68**	1.47	1.47	**36.63**	0.41	**39.72**	m2	**42.70**
Cut out damaged two coat plaster to ceiling in patches and make out in new plaster to match existing										
under 1.00m2	2.70	10.00	**2.98**	2.38	2.38	**59.31**	0.27	**62.55**	m2	**67.24**
1.00 - 2.00m2	2.70	10.00	**2.98**	1.96	1.96	**48.84**	0.27	**52.08**	m2	**55.99**
2.00 - 4.00m2	2.70	10.00	**2.98**	1.54	1.54	**38.38**	0.27	**41.62**	m2	**44.74**
Cut out damaged three coat plaster to ceiling in patches and make out in new plaster to match existing										
under 1.00m2	3.59	10.00	**3.95**	2.79	2.79	**69.53**	0.41	**73.89**	m2	**79.43**
1.00 - 2.00m2	3.59	10.00	**3.95**	2.30	2.30	**57.32**	0.41	**61.68**	m2	**66.30**
2.00 - 4.00m2	3.59	10.00	**3.95**	1.81	1.81	**45.11**	0.41	**49.47**	m2	**53.18**
Cut out damaged 9.5mm plasterboard and two coat plaster to wall in patches and make out in new plasterboard and plaster to match existing										
under 1.00m2	4.51	10.00	**4.96**	2.23	2.23	**55.57**	0.45	**60.99**	m2	**65.56**
1.00 - 2.00m2	4.51	10.00	**4.96**	1.78	1.78	**44.36**	0.45	**49.77**	m2	**53.51**
2.00 - 4.00m2	4.51	10.00	**4.96**	1.36	1.36	**33.89**	0.45	**39.31**	m2	**42.25**
Cut out damaged 12.5mm plasterboard and two coat plaster to wall in patches and make out in new plasterboard and plaster to match existing										
under 1.00m2	4.70	10.00	**5.17**	2.28	2.28	**56.82**	0.50	**62.49**	m2	**67.18**
1.00 - 2.00m2	4.70	10.00	**5.17**	1.83	1.83	**45.60**	0.50	**51.28**	m2	**55.12**
2.00 - 4.00m2	4.70	10.00	**5.17**	1.38	1.38	**34.39**	0.50	**40.06**	m2	**43.07**
Cut out damaged 9.5mm plasterboard and two coat plaster to ceiling in patches and make out in new plasterboard and plaster to match existing										
under 1.00m2	4.51	10.00	**4.96**	2.54	2.54	**63.30**	0.45	**68.71**	m2	**73.87**
1.00 - 2.00m2	4.51	10.00	**4.96**	2.03	2.03	**50.59**	0.45	**56.00**	m2	**60.20**
2.00 - 4.00m2	4.51	10.00	**4.96**	1.55	1.55	**38.63**	0.45	**44.04**	m2	**47.34**
Cut out damaged 12.5mm plasterboard and two coat plaster to ceiling in patches and make out in new plasterboard and plaster to match existing										
under 1.00m2	4.70	10.00	**5.17**	2.56	2.56	**63.80**	0.50	**69.47**	m2	**74.68**
1.00 - 2.00m2	4.70	10.00	**5.17**	2.06	2.06	**51.34**	0.50	**57.01**	m2	**61.29**
2.00 - 4.00m2	4.70	10.00	**5.17**	1.57	1.57	**39.12**	0.50	**44.80**	m2	**48.16**
Cut out damaged lath and plaster to wall in patches and make out with new metal lathing and three coat plaster to match existing										
under 1.00m2	6.99	10.00	**7.69**	2.53	2.53	**63.05**	0.50	**71.25**	m2	**76.59**
1.00 - 2.00m2	6.99	10.00	**7.69**	2.14	2.14	**53.33**	0.50	**61.53**	m2	**66.14**
2.00 - 4.00m2	6.99	10.00	**7.69**	1.73	1.73	**43.11**	0.50	**51.31**	m2	**55.16**
Cut out damaged lath and plaster to ceiling in patches and make out with new metal lathing and three coat plaster to match existing										
under 1.00m2	6.99	10.00	**7.69**	2.87	2.87	**71.52**	0.50	**79.72**	m2	**85.70**
1.00 - 2.00m2	6.99	10.00	**7.69**	2.43	2.43	**60.56**	0.50	**68.75**	m2	**73.91**
2.00 - 4.00m2	6.99	10.00	**7.69**	1.95	1.95	**48.59**	0.50	**56.79**	m2	**61.05**
Cut out crack in plaster, form dovetailed key and make good with plaster										
not exceeding 50mm wide	0.13	10.00	**0.14**	0.20	0.20	**4.98**	0.02	**5.14**	m	**5.53**
Make good plaster where the following removed										
small pipe	-	-	-	0.16	0.16	**3.99**	0.02	**4.00**	nr	**4.30**
small steel section	-	-	-	0.21	0.21	**5.23**	0.02	**5.25**	nr	**5.65**
Cut out damaged plaster moulding or cornice and make out with new to match existing										
up to 100mm girth on face	0.84	10.00	**0.92**	1.25	0.63	**24.51**	0.11	**25.54**	m	**27.46**
100 - 150mm girth on face	1.40	10.00	**1.54**	1.86	0.93	**36.39**	0.21	**38.14**	m	**41.00**
150 - 200mm girth on face	2.80	10.00	**3.08**	2.25	1.13	**44.07**	0.38	**47.53**	m	**51.10**

Labour hourly rates: (except Specialists) Craft Operatives £14.21 Labourer £10.71 Rates are national average prices. Refer to REGIONAL VARIATIONS for indicative levels of overall pricing in regions	MATERIALS			LABOUR				RATES		
	Del to Site £	Waste %	Material Cost £	Craft Optve Hrs	Lab Hrs	Labour Cost £	Sunds £	Nett Rate £	Unit	Gross Rate (+7.5%) £
C90: REPAIRING/RENOVATING FINISHINGS Cont.										
Repairs to old wall and ceiling finishings Cont.										
Remove damaged fibrous plaster louvered ventilator, replace with new and make good										
229 x 75mm	1.60	5.00	**1.68**	0.50	0.50	**12.46**	0.05	**14.19**	nr	**15.26**
229 x 150mm	1.92	-	**1.92**	0.60	0.60	**14.95**	0.09	**16.96**	nr	**18.23**
Take out damaged 152 x 152 x 5.5mm white glazed wall tiles and renew to match existing										
isolated tile	0.45	5.00	**0.47**	0.32	0.32	**7.97**	0.03	**8.48**	nr	**9.11**
patch of 5 tiles	2.24	5.00	**2.35**	0.60	0.60	**14.95**	0.15	**17.46**	nr	**18.77**
patch 0.50 - 1.00m2	19.71	5.00	**20.70**	2.32	2.32	**57.81**	1.20	**79.71**	m2	**85.69**
patch 1.00 - 2.00m2	19.71	5.00	**20.70**	1.66	1.66	**41.37**	1.20	**63.26**	m2	**68.01**
Repairs to old floor finishings										
Take out damaged 150 x 150 x 12.5mm clay floor tiles and renew to match existing										
isolated tile	0.32	5.00	**0.34**	0.32	0.21	**6.80**	0.03	**7.16**	nr	**7.70**
patch of 5 tiles	1.60	5.00	**1.68**	0.65	0.45	**14.06**	0.15	**15.89**	nr	**17.08**
patch 0.50 - 1.00m2	14.40	5.00	**15.12**	2.40	1.61	**51.35**	1.20	**67.67**	m2	**72.74**
patch 1.00 - 2.00m2	14.40	5.00	**15.12**	1.90	1.27	**40.60**	1.20	**56.92**	m2	**61.19**
Cut out damaged 25mm granolithic paving in patches and make out in new granolithic to match existing										
under 1.00m2	7.18	5.00	**7.53**	1.50	1.50	**37.38**	0.56	**45.47**	m2	**48.89**
1.00 - 2.00m2	7.18	5.00	**7.53**	1.25	1.25	**31.15**	0.56	**39.24**	m2	**42.19**
2.00 - 4.00m2	7.18	5.00	**7.53**	1.07	1.07	**26.66**	0.56	**34.76**	m2	**37.37**
Cut out damaged 32mm granolithic paving in patches and make out in new granolithic to match existing										
under 1.00m2	9.18	5.00	**9.64**	1.68	1.68	**41.87**	0.66	**52.17**	m2	**56.08**
1.00 - 2.00m2	9.18	5.00	**9.64**	1.38	1.38	**34.39**	0.66	**44.69**	m2	**48.04**
2.00 - 4.00m2	9.18	5.00	**9.64**	1.15	1.15	**28.66**	0.66	**38.96**	m2	**41.88**
Cut out damaged 38mm granolithic paving in patches and make out in new granolithic to match existing										
under 1.00m2	10.91	5.00	**11.45**	1.80	1.80	**44.86**	0.78	**57.08**	m2	**61.36**
1.00 - 2.00m2	10.91	5.00	**11.45**	1.50	1.50	**37.38**	0.78	**49.61**	m2	**53.33**
2.00 - 4.00m2	10.91	5.00	**11.45**	1.25	1.25	**31.15**	0.78	**43.38**	m2	**46.63**
Cut out damaged 50mm granolithic paving in patches and make out in new granolithic to match existing										
under 1.00m2	14.35	5.00	**15.07**	2.12	2.12	**52.83**	0.94	**68.84**	m2	**74.00**
1.00 - 2.00m2	14.35	5.00	**15.07**	1.73	1.73	**43.11**	0.94	**59.12**	m2	**63.56**
2.00 - 4.00m2	14.35	5.00	**15.07**	1.45	1.45	**36.13**	0.94	**52.15**	m2	**56.06**
Cut out damaged granolithic paving to tread 275mm wide and make out in new granolithic to match existing										
25mm thick	2.01	5.00	**2.11**	0.44	0.44	**10.96**	0.16	**13.23**	m	**14.22**
32mm thick	2.58	5.00	**2.71**	0.48	0.48	**11.96**	0.18	**14.85**	m	**15.97**
38mm thick	2.87	5.00	**3.01**	0.50	0.50	**12.46**	0.21	**15.68**	m	**16.86**
50mm thick	4.02	5.00	**4.22**	0.59	0.59	**14.70**	0.26	**19.18**	m	**20.62**
Cut out damaged granolithic covering to plain riser 175mm wide and make out in new granolithic to match existing										
13mm thick	0.57	5.00	**0.60**	0.80	0.80	**19.94**	1.44	**21.98**	m	**23.63**
19mm thick	0.86	5.00	**0.90**	0.84	0.84	**20.93**	1.45	**23.29**	m	**25.03**
25mm thick	1.15	5.00	**1.21**	0.88	0.88	**21.93**	1.46	**24.60**	m	**26.44**
32mm thick	1.72	5.00	**1.81**	0.93	0.93	**23.18**	1.47	**26.46**	m	**28.44**
Cut out damaged granolithic covering to undercut riser 175mm wide and make out in new granolithic to match existing										
13mm thick	0.57	5.00	**0.60**	0.87	0.87	**21.68**	1.44	**23.72**	m	**25.50**
19mm thick	0.86	5.00	**0.90**	0.93	0.93	**23.18**	1.45	**25.53**	m	**27.44**
25mm thick	1.15	5.00	**1.21**	0.98	0.98	**24.42**	1.46	**27.09**	m	**29.12**
32mm thick	1.72	5.00	**1.81**	1.04	1.04	**25.92**	1.47	**29.20**	m	**31.39**
Cut out damaged granolithic skirting 150mm wide and make out in new granolithic to match existing										
13mm thick	0.57	5.00	**0.60**	0.72	0.72	**17.94**	1.23	**19.77**	m	**21.26**
19mm thick	0.86	5.00	**0.90**	0.76	0.76	**18.94**	1.24	**21.09**	m	**22.67**
25mm thick	1.15	5.00	**1.21**	0.80	0.80	**19.94**	1.25	**22.40**	m	**24.08**
32mm thick	1.44	5.00	**1.51**	0.82	0.82	**20.43**	1.27	**23.21**	m	**24.95**
Cut out crack in granolithic paving not exceeding 50mm wide, form dovetailed key and make good to match existing										
25mm thick	0.29	5.00	**0.30**	0.60	0.60	**14.95**	0.03	**15.28**	m	**16.42**
32mm thick	0.57	5.00	**0.60**	0.66	0.66	**16.45**	0.03	**17.08**	m	**18.36**
38mm thick	0.57	5.00	**0.60**	0.69	0.69	**17.19**	0.04	**17.84**	m	**19.18**
50mm thick	0.86	5.00	**0.90**	0.79	0.79	**19.69**	0.05	**20.64**	m	**22.18**
C90: REPAIRING GLAZING										
Resecure glass										
Remove decayed putties, paint one coat on edge of rebate and re-putty										
wood window	0.13	10.00	**0.15**	0.69	-	**9.80**	0.02	**9.97**	m	**10.72**
metal window	0.13	10.00	**0.15**	0.76	-	**10.80**	0.02	**10.97**	m	**11.79**

Labour hourly rates: (except Specialists) Craft Operatives £14.21 Labourer £10.71 Rates are national average prices. Refer to REGIONAL VARIATIONS for indicative levels of overall pricing in regions	MATERIALS			LABOUR				RATES		
	Del to Site £	Waste %	Material Cost £	Craft Optve Hrs	Lab Hrs	Labour Cost £	Sunds £	Nett Rate £	Unit	Gross Rate (+7.5%) £
C90: REPAIRING GLAZING Cont.										
Resecure glass Cont.										
Remove beads, remove decayed bedding materials, paint one coat on edge of rebate, re-bed glass and re-fix beads										
wood window	0.13	10.00	**0.15**	1.22	-	**17.34**	0.02	**17.51**	m	**18.82**
metal window	0.13	10.00	**0.15**	1.32	-	**18.76**	0.02	**18.93**	m	**20.35**
C90: REPAIRING/RENOVATING FENCING										
Repairs to softwood fencing										
Remove defective timber fence post and replace with new 127 x 102mm post; letting post 450mm into ground; backfilling around in concrete mix C20; securing ends of existing arris rails and gravel boards to new post										
to suit fencing 914mm high	13.50	10.00	**14.05**	2.04	2.04	**50.84**	0.41	**65.30**	nr	**70.20**
to suit fencing 1219mm high	14.36	10.00	**14.91**	2.06	2.06	**51.34**	0.47	**66.72**	nr	**71.72**
to suit fencing 1524mm high	15.26	10.00	**15.81**	2.08	2.08	**51.83**	0.55	**68.20**	nr	**73.31**
to suit fencing 1829mm high	16.62	10.00	**17.17**	2.10	2.10	**52.33**	0.64	**70.15**	nr	**75.41**
Remove defective pale from palisade fence and replace with new 19 x 75mm pointed pale; fixing with galvanised nails										
764mm long	0.72	-	**0.72**	0.55	0.09	**8.78**	0.18	**9.68**	nr	**10.41**
1069mm long	1.01	-	**1.01**	0.56	0.09	**8.92**	0.22	**10.15**	nr	**10.91**
1374mm long	1.30	-	**1.30**	0.58	0.10	**9.31**	0.25	**10.86**	nr	**11.68**
Remove defective pales from close boarded fence and replace with new 100mm wide feather edged pales; fixing with galvanised nails										
singly	1.63	-	**1.63**	0.25	0.04	**3.98**	0.21	**5.82**	m	**6.26**
areas up to 1.00m2	9.94	20.00	**11.93**	1.32	0.22	**21.11**	1.75	**34.79**	m2	**37.40**
areas over 1.00m2	9.94	20.00	**11.93**	0.98	0.16	**15.64**	1.75	**29.32**	m2	**31.52**
Remove defective gravel board and centre stump and replace with new 25 x 150mm gravel board and 50 x 50mm centre stump; fixing with galvanised nails										
securing to fence posts	2.44	5.00	**2.56**	0.53	0.09	**8.50**	0.21	**11.27**	m	**12.11**
Repairs to oak fencing										
Remove defective timber fence post and replace with new 127 x 102mm post; letting post 450mm into ground; backfilling around in concrete mix C20; securing ends of existing arris rails and gravel boards to new post										
to suit fencing 914mm high	23.28	10.00	**23.83**	2.17	2.17	**54.08**	0.41	**78.32**	nr	**84.20**
to suit fencing 1219mm high	27.09	10.00	**27.64**	2.21	2.21	**55.07**	0.47	**83.19**	nr	**89.43**
to suit fencing 1524mm high	30.89	10.00	**31.44**	2.25	2.25	**56.07**	0.55	**88.06**	nr	**94.67**
to suit fencing 1829mm high	34.70	10.00	**35.25**	2.29	2.29	**57.07**	0.64	**92.96**	nr	**99.93**
Remove defective pale from palisade fence and replace with new 19 x 75mm pointed pale; fixing with galvanised nails										
764mm long	1.90	-	**1.90**	0.70	0.12	**11.23**	0.18	**13.31**	nr	**14.31**
1069mm long	2.64	-	**2.64**	0.76	0.13	**12.19**	0.22	**15.05**	nr	**16.18**
1374mm long	3.41	-	**3.41**	0.83	0.14	**13.29**	0.25	**16.95**	nr	**18.23**
Remove defective pales from close boarded fence and replace with new 100mm wide feather edged pales; fixing with galvanised nails										
singly	4.27	-	**4.27**	0.27	0.05	**4.37**	0.21	**8.85**	m	**9.52**
areas up to 1.00m2	26.05	20.00	**31.26**	2.56	0.43	**40.98**	1.75	**73.99**	m2	**79.54**
areas over 1.00m2	26.05	20.00	**31.26**	2.07	0.35	**33.16**	1.75	**66.17**	m2	**71.14**
Remove defective gravel board and centre stump and replace with new 25 x 150mm gravel board and 50 x 50mm centre stump; fixing with galvanised nails										
securing to fence posts	6.36	5.00	**6.68**	1.10	0.18	**17.56**	0.21	**24.45**	m	**26.28**
Repairs to posts										
100 x 100mm precast concrete spur 1219mm long; setting into ground; backfilling around in concrete mix C20										
bolting to existing timber post	14.02	10.00	**14.57**	0.76	0.76	**18.94**	1.66	**35.17**	nr	**37.81**
C90: REPAIRING/RENOVATING EXTERNAL PAVINGS										
Repairs to precast concrete flag paving										
Take up uneven paving and set aside for re-use; level up and consolidate existing hardcore and relay salvaged paving										
bedding on 25mm sand bed; grouting in lime mortar (1:3)	0.94	33.00	**1.22**	0.67	0.93	**19.48**	0.12	**20.82**	m2	**22.39**
Repairs to York stone paving										
Take up old paving; re-square; relay random in random sizes										
bedding, jointing and pointing in cement mortar (1:3)	4.39	10.00	**4.83**	1.10	1.10	**27.41**	0.90	**33.14**	m2	**35.62**

Labour hourly rates: (except Specialists) Craft Operatives £14.21 Labourer £10.71 Rates are national average prices. Refer to REGIONAL VARIATIONS for indicative levels of overall pricing in regions	MATERIALS			LABOUR				RATES		
	Del to Site £	Waste %	Material Cost £	Craft Optve Hrs	Lab Hrs	Labour Cost £	Sunds £	Nett Rate £	Unit	Gross Rate (+7.5%) £
C90: REPAIRING/RENOVATING EXTERNAL PAVINGS Cont.										
Repairs to granite sett paving										
Take up, clean and stack old paving	-	-	**-**	-	0.90	**9.64**	-	**9.64**	m2	**10.36**
Take up old paving, clean and re-lay bedding and grouting in cement mortar (1:3)	-	-	**-**	1.65	2.55	**50.76**	1.20	**51.96**	m2	**55.85**
Clean out joints of old paving and re-grout in cement mortar (1:3)	-	-	**-**	-	0.55	**5.89**	0.52	**6.41**	m2	**6.89**
Repairs to bitumen paving										
Cut out damaged 50mm paving to footpath and fill with bitumen macadam under 1.00m2	32.00	10.00	**35.20**	1.65	2.55	**50.76**	1.20	**87.16**	m2	**93.69**
Cut out damaged 75mm paving to road fill with macadam and seal all around with hot bitumen under 1.00m2	48.00	10.00	**52.80**	1.65	2.55	**50.76**	1.20	**104.76**	m2	**112.61**

GROUNDWORK

Labour hourly rates: (except Specialists) Craft Operatives £14.21 Labourer £10.71 Rates are national average prices. Refer to REGIONAL VARIATIONS for indicative levels of overall pricing in regions	PLANT & TRANSPORT			LABOUR				RATES		
	Plant Cost £	Trans Cost £	P & T Cost £	Craft Optve Hrs	Lab Hrs	Labour Cost £	Sunds £	Nett Rate £	Unit	Gross Rate (+7.5%) £
D20: EXCAVATING AND FILLING										
Site preparation										
Removing trees; filling voids left by removal of roots with selected material arising from excavation										
girth 0.60 - 1.50m	66.86	-	**66.86**	-	10.00	**107.10**	-	**173.96**	nr	**187.00**
girth 1.50 - 3.00m	122.57	-	**122.57**	-	20.00	**214.20**	-	**336.77**	nr	**362.03**
girth exceeding 3.00m	178.29	-	**178.29**	-	30.00	**321.30**	-	**499.59**	nr	**537.06**
Clearing site vegetation; filling voids left by removal of roots with selected material arising from excavation										
bushes, scrub, undergrowth, hedges, trees and tree stumps not exceeding 600mm girth	0.53	-	**0.53**	-	-	-	-	**0.53**	m2	**0.57**
Lifting turf for preservation										
stacking on site average 100m distant for future use; watering	-	-	-	-	0.55	**5.89**	-	**5.89**	m2	**6.33**
Excavating - by machine										
Top soil for preservation										
average 150mm deep	0.28	-	**0.28**	-	-	-	-	**0.28**	m2	**0.30**
To reduce levels										
maximum depth not exceeding										
0.25m	2.12	-	**2.12**	-	-	-	-	**2.12**	m3	**2.28**
1.00m	1.11	-	**1.11**	-	-	-	-	**1.11**	m3	**1.20**
2.00m	1.28	-	**1.28**	-	-	-	-	**1.28**	m3	**1.38**
4.00m	1.24	-	**1.24**	-	-	-	-	**1.24**	m3	**1.34**
6.00m	1.53	-	**1.53**	-	-	-	-	**1.53**	m3	**1.65**
Basements and the like										
maximum depth not exceeding										
0.25m	2.12	-	**2.12**	-	0.39	**4.18**	-	**6.29**	m3	**6.77**
1.00m	1.34	-	**1.34**	-	0.41	**4.39**	-	**5.73**	m3	**6.16**
2.00m	1.56	-	**1.56**	-	0.47	**5.03**	-	**6.59**	m3	**7.09**
4.00m	1.92	-	**1.92**	-	0.60	**6.43**	-	**8.34**	m3	**8.97**
6.00m	2.68	-	**2.68**	-	0.74	**7.93**	-	**10.61**	m3	**11.40**
Trenches width not exceeding 0.30m										
maximum depth not exceeding										
0.25m	5.46	-	**5.46**	-	1.71	**18.31**	-	**23.77**	m3	**25.56**
1.00m	5.68	-	**5.68**	-	1.80	**19.28**	-	**24.96**	m3	**26.83**
Trenches width exceeding 0.30m										
maximum depth not exceeding										
0.25m	5.46	-	**5.46**	-	0.57	**6.10**	-	**11.56**	m3	**12.43**
1.00m	4.68	-	**4.68**	-	0.60	**6.43**	-	**11.11**	m3	**11.94**
2.00m	4.68	-	**4.68**	-	0.70	**7.50**	-	**12.18**	m3	**13.09**
4.00m	4.79	-	**4.79**	-	0.87	**9.32**	-	**14.11**	m3	**15.17**
6.00m	4.79	-	**4.79**	-	1.05	**11.25**	-	**16.04**	m3	**17.24**
commencing 1.50m below existing ground level, maximum depth not exceeding										
0.25m	4.68	-	**4.68**	-	0.85	**9.10**	-	**13.78**	m3	**14.82**
1.00m	4.68	-	**4.68**	-	0.90	**9.64**	-	**14.32**	m3	**15.39**
2.00m	4.02	-	**4.02**	-	1.13	**12.10**	-	**16.13**	m3	**17.34**
commencing 3.00m below existing ground level, maximum depth not exceeding										
0.25m	4.02	-	**4.02**	-	1.27	**13.60**	-	**17.63**	m3	**18.95**
1.00m	4.79	-	**4.79**	-	1.33	**14.24**	-	**19.04**	m3	**20.46**
2.00m	4.79	-	**4.79**	-	1.51	**16.17**	-	**20.96**	m3	**22.54**
For pile caps and ground beams between piles										
maximum depth not exceeding										
0.25m	5.68	-	**5.68**	-	0.84	**9.00**	-	**14.68**	m3	**15.78**
1.00m	5.57	-	**5.57**	-	0.88	**9.42**	-	**15.00**	m3	**16.12**
2.00m	5.57	-	**5.57**	-	1.00	**10.71**	-	**16.28**	m3	**17.50**

Labour hourly rates: (except Specialists) Craft Operatives £14.21 Labourer £10.71 Rates are national average prices. Refer to REGIONAL VARIATIONS for indicative levels of overall pricing in regions	PLANT & TRANSPORT			LABOUR				RATES		
	Plant Cost £	Trans Cost £	P & T Cost £	Craft Optve Hrs	Lab Hrs	Labour Cost £	Sunds £	Nett Rate £	Unit	Gross Rate (+7.5%) £
D20: EXCAVATING AND FILLING Cont.										
Excavating - by machine Cont.										
Pits										
maximum depth not exceeding										
0.25m	4.79	-	**4.79**	-	0.84	**9.00**	-	**13.79**	m3	**14.82**
1.00m	5.91	-	**5.91**	-	0.88	**9.42**	-	**15.33**	m3	**16.48**
2.00m	5.91	-	**5.91**	-	1.00	**10.71**	-	**16.62**	m3	**17.86**
4.00m	5.75	-	**5.75**	-	1.27	**13.60**	-	**19.35**	m3	**20.80**
Extra over any types of excavating irrespective of depth										
excavating below ground water level	4.81	-	**4.81**	-	-	**-**	-	**4.81**	m3	**5.17**
excavating in running silt, running sand or liquid mud	8.27	-	**8.27**	-	-	**-**	-	**8.27**	m3	**8.89**
excavating below ground water level in running silt, running sand or liquid mud	11.40	-	**11.40**	-	-	**-**	-	**11.40**	m3	**12.25**
excavating in heavy soil or clay	1.05	-	**1.05**	-	-	**-**	-	**1.05**	m3	**1.13**
excavating in gravel	1.65	-	**1.65**	-	-	**-**	-	**1.65**	m3	**1.77**
excavating in brash, loose rock or chalk	2.08	-	**2.08**	-	-	**-**	-	**2.08**	m3	**2.24**
Breaking out existing materials; extra over any types of excavating irrespective of depth										
sandstone	5.65	-	**5.65**	-	-	**-**	-	**5.65**	m3	**6.07**
hard rock	8.50	-	**8.50**	-	-	**-**	-	**8.50**	m3	**9.14**
concrete	7.06	-	**7.06**	-	-	**-**	-	**7.06**	m3	**7.59**
reinforced concrete	10.61	-	**10.61**	-	-	**-**	-	**10.61**	m3	**11.40**
brickwork, blockwork or stonework	4.71	-	**4.71**	-	-	**-**	-	**4.71**	m3	**5.07**
drain and concrete bed under										
100mm diameter	0.66	-	**0.66**	-	-	**-**	-	**0.66**	m	**0.71**
150mm diameter	0.79	-	**0.79**	-	-	**-**	-	**0.79**	m	**0.85**
225mm diameter	0.93	-	**0.93**	-	-	**-**	-	**0.93**	m	**1.00**
Breaking out existing hard pavings; extra over any types of excavating irrespective of depth										
concrete 150mm thick	1.11	-	**1.11**	-	-	**-**	-	**1.11**	m2	**1.20**
reinforced concrete 200mm thick	2.23	-	**2.23**	-	-	**-**	-	**2.23**	m2	**2.40**
coated macadam or asphalt 75mm thick	0.19	-	**0.19**	-	-	**-**	-	**0.19**	m2	**0.21**
Working space allowance to excavation; including additional earthwork support, disposal (excluding landfill tax) and backfilling										
reduce levels, basements and the like	4.90	-	**4.90**	-	0.28	**3.00**	-	**7.90**	m2	**8.49**
pits	5.57	-	**5.57**	-	0.60	**6.43**	-	**12.00**	m2	**12.90**
trenches	4.90	-	**4.90**	-	0.42	**4.50**	-	**9.40**	m2	**10.11**
Compacting										
bottoms of excavations	0.19	-	**0.19**	-	-	**-**	-	**0.19**	m2	**0.21**
Compacting with 680 kg vibratory roller										
ground	0.20	-	**0.20**	-	-	**-**	-	**0.20**	m2	**0.22**
Compacting with 6-8 tonnes smooth wheeled roller										
ground	0.31	-	**0.31**	-	-	**-**	-	**0.31**	m2	**0.33**
Trimming										
sides of cuttings; vertical or battered	0.45	-	**0.45**	-	-	**-**	-	**0.45**	m2	**0.48**
sides of embankments; vertical or battered	0.45	-	**0.45**	-	-	**-**	-	**0.45**	m2	**0.48**
Disposal of excavated material										
depositing on site in temporary spoil heaps where directed										
25m	-	1.77	**1.77**	-	0.10	**1.07**	-	**2.84**	m3	**3.05**
50m	-	1.86	**1.86**	-	0.10	**1.07**	-	**2.93**	m3	**3.15**
100m	-	2.02	**2.02**	-	0.10	**1.07**	-	**3.09**	m3	**3.32**
400m	-	2.39	**2.39**	-	0.10	**1.07**	-	**3.46**	m3	**3.72**
800m	-	3.12	**3.12**	-	0.10	**1.07**	-	**4.19**	m3	**4.50**
1200m	-	3.32	**3.32**	-	0.10	**1.07**	-	**4.39**	m3	**4.72**
1600m	-	3.52	**3.52**	-	0.10	**1.07**	-	**4.60**	m3	**4.94**
extra for each additional 1600m	-	0.55	**0.55**	-	-	**-**	-	**0.55**	m3	**0.59**
removing from site to tip a) Inert	-	16.00	**16.00**	-	-	**-**	-	**16.00**	m3	**17.20**
removing from site to tip b) Active	-	37.00	**37.00**	-	-	**-**	-	**37.00**	m3	**39.78**
removing from site to tip c) Contaminated (Guide price - always seek a quotation for specialist disposal costs.)	-	80.00	**80.00**	-	-	**-**	-	**80.00**	m3	**86.00**
Disposal of preserved top soil										
depositing on site in temporary spoil heaps where directed										
25m	-	1.77	**1.77**	-	0.10	**1.07**	-	**2.84**	m3	**3.05**
50m	-	1.86	**1.86**	-	0.10	**1.07**	-	**2.93**	m3	**3.15**
100m	-	2.02	**2.02**	-	0.10	**1.07**	-	**3.09**	m3	**3.32**
Excavating - by hand										
Top soil for preservation										
average 150mm deep	-	-	**-**	-	0.30	**3.21**	-	**3.21**	m2	**3.45**
To reduce levels										
maximum depth not exceeding										
0.25m	-	-	**-**	-	2.50	**26.78**	-	**26.78**	m3	**28.78**
1.00m	-	-	**-**	-	2.65	**28.38**	-	**28.38**	m3	**30.51**
2.00m	-	-	**-**	-	3.00	**32.13**	-	**32.13**	m3	**34.54**
4.00m	-	-	**-**	-	3.90	**41.77**	-	**41.77**	m3	**44.90**
6.00m	-	-	**-**	-	4.85	**51.94**	-	**51.94**	m3	**55.84**

Labour hourly rates: (except Specialists) Craft Operatives £14.21 Labourer £10.71 Rates are national average prices. Refer to REGIONAL VARIATIONS for indicative levels of overall pricing in regions	PLANT & TRANSPORT			LABOUR				RATES		
	Plant Cost £	Trans Cost £	P & T Cost £	Craft Optve Hrs	Lab Hrs	Labour Cost £	Sunds £	Nett Rate £	Unit	Gross Rate (+7.5%) £
D20: EXCAVATING AND FILLING Cont.										
Excavating - by hand Cont.										
Basements and the like										
maximum depth not exceeding										
0.25m	-	-	-	-	2.60	**27.85**	-	**27.85**	m3	**29.93**
1.00m	-	-	-	-	2.75	**29.45**	-	**29.45**	m3	**31.66**
2.00m	-	-	-	-	3.10	**33.20**	-	**33.20**	m3	**35.69**
4.00m	-	-	-	-	4.00	**42.84**	-	**42.84**	m3	**46.05**
6.00m	-	-	-	-	4.85	**51.94**	-	**51.94**	m3	**55.84**
Trenches width not exceeding 0.30m										
maximum depth not exceeding										
0.25m	-	-	-	-	4.25	**45.52**	-	**45.52**	m3	**48.93**
1.00m	-	-	-	-	4.50	**48.20**	-	**48.20**	m3	**51.81**
Trenches width exceeding 0.30m										
maximum depth not exceeding										
0.25m	-	-	-	-	2.85	**30.52**	-	**30.52**	m3	**32.81**
1.00m	-	-	-	-	3.00	**32.13**	-	**32.13**	m3	**34.54**
2.00m	-	-	-	-	3.50	**37.49**	-	**37.49**	m3	**40.30**
4.00m	-	-	-	-	4.35	**46.59**	-	**46.59**	m3	**50.08**
6.00m	-	-	-	-	5.50	**58.91**	-	**58.91**	m3	**63.32**
commencing 1.50m below existing ground level, maximum depth not exceeding										
0.25m	-	-	-	-	4.25	**45.52**	-	**45.52**	m3	**48.93**
1.00m	-	-	-	-	4.50	**48.20**	-	**48.20**	m3	**51.81**
2.00m	-	-	-	-	5.65	**60.51**	-	**60.51**	m3	**65.05**
commencing 3.00m below existing ground level, maximum depth not exceeding										
0.25m	-	-	-	-	6.35	**68.01**	-	**68.01**	m3	**73.11**
1.00m	-	-	-	-	6.65	**71.22**	-	**71.22**	m3	**76.56**
2.00m	-	-	-	-	7.65	**81.93**	-	**81.93**	m3	**88.08**
For pile caps and ground beams between piles										
maximum depth not exceeding										
0.25m	-	-	-	-	3.35	**35.88**	-	**35.88**	m3	**38.57**
1.00m	-	-	-	-	3.50	**37.49**	-	**37.49**	m3	**40.30**
2.00m	-	-	-	-	4.00	**42.84**	-	**42.84**	m3	**46.05**
Pits										
maximum depth not exceeding										
0.25m	-	-	-	-	3.35	**35.88**	-	**35.88**	m3	**38.57**
1.00m	-	-	-	-	3.50	**37.49**	-	**37.49**	m3	**40.30**
2.00m	-	-	-	-	4.00	**42.84**	-	**42.84**	m3	**46.05**
4.00m	-	-	-	-	5.10	**54.62**	-	**54.62**	m3	**58.72**
Extra over any types of excavating irrespective of depth										
excavating below ground water level	-	-	-	-	1.50	**16.07**	-	**16.07**	m3	**17.27**
excavating in running silt, running sand or liquid mud	-	-	-	-	4.50	**48.20**	-	**48.20**	m3	**51.81**
excavating below ground water level in running silt, running sand or liquid mud	-	-	-	-	6.00	**64.26**	-	**64.26**	m3	**69.08**
excavating in heavy soil or clay	-	-	-	-	0.65	**6.96**	-	**6.96**	m3	**7.48**
excavating in gravel	-	-	-	-	1.65	**17.67**	-	**17.67**	m3	**19.00**
excavating in brash, loose rock or chalk	-	-	-	-	3.35	**35.88**	-	**35.88**	m3	**38.57**
Breaking out existing materials; extra over any types of excavating irrespective of depth										
sandstone	-	-	-	-	6.65	**71.22**	-	**71.22**	m3	**76.56**
hard rock	-	-	-	-	13.35	**142.98**	-	**142.98**	m3	**153.70**
concrete	-	-	-	-	10.00	**107.10**	-	**107.10**	m3	**115.13**
reinforced concrete	-	-	-	-	15.00	**160.65**	-	**160.65**	m3	**172.70**
brickwork, blockwork or stonework	-	-	-	-	7.00	**74.97**	-	**74.97**	m3	**80.59**
drain and concrete bed under										
100mm diameter	-	-	-	-	0.80	**8.57**	-	**8.57**	m	**9.21**
150mm diameter	-	-	-	-	1.00	**10.71**	-	**10.71**	m	**11.51**
225mm diameter	-	-	-	-	1.15	**12.32**	-	**12.32**	m	**13.24**
Breaking out existing hard pavings; extra over any types of excavating irrespective of depth										
concrete 150mm thick	-	-	-	-	1.50	**16.07**	-	**16.07**	m2	**17.27**
reinforced concrete 200mm thick	-	-	-	-	3.00	**32.13**	-	**32.13**	m2	**34.54**
coated macadam or asphalt 75mm thick	-	-	-	-	0.25	**2.68**	-	**2.68**	m2	**2.88**
Working space allowance to excavation; including additional earthwork support, disposal (excluding landfill tax) and backfilling										
reduce levels, basements and the like	-	-	-	-	2.60	**27.85**	-	**27.85**	m2	**29.93**
pits	-	-	-	-	3.12	**33.42**	-	**33.42**	m2	**35.92**
trenches	-	-	-	-	2.80	**29.99**	-	**29.99**	m2	**32.24**
Compacting										
bottoms of excavations	-	-	-	-	0.12	**1.29**	-	**1.29**	m2	**1.38**
Trimming										
sides of cuttings; vertical or battered	-	-	-	-	0.20	**2.14**	-	**2.14**	m2	**2.30**
sides of embankments; vertical or battered	-	-	-	-	0.20	**2.14**	-	**2.14**	m2	**2.30**
Disposal of excavated material										
depositing on site in temporary spoil heaps where directed										
25m	-	-	-	-	1.25	**13.39**	-	**13.39**	m3	**14.39**
50m	-	-	-	-	1.50	**16.07**	-	**16.07**	m3	**17.27**

Labour hourly rates: (except Specialists) Craft Operatives £14.21 Labourer £10.71 Rates are national average prices. Refer to REGIONAL VARIATIONS for indicative levels of overall pricing in regions	PLANT & TRANSPORT			LABOUR				RATES		
	Plant Cost £	Trans Cost £	P & T Cost £	Craft Optve Hrs	Lab Hrs	Labour Cost £	Sunds £	Nett Rate £	Unit	Gross Rate (+7.5%) £
D20: EXCAVATING AND FILLING Cont.										
Excavating - by hand Cont.										
Disposal of excavated material Cont.										
depositing on site in temporary spoil heaps where directed Cont.										
100m	-	-	-	-	2.00	**21.42**	-	**21.42**	m3	**23.03**
400m	-	2.39	**2.39**	-	1.50	**16.07**	-	**18.46**	m3	**19.84**
800m	-	3.12	**3.12**	-	1.50	**16.07**	-	**19.18**	m3	**20.62**
1200m	-	3.32	**3.32**	-	1.50	**16.07**	-	**19.38**	m3	**20.83**
1600m	-	3.52	**3.52**	-	1.50	**16.07**	-	**19.59**	m3	**21.06**
extra for each additional 1600m	-	0.55	**0.55**	-	-	-	-	**0.55**	m3	**0.59**
removing from site to tip a) Inert	-	16.00	**16.00**	-	1.35	**14.46**	-	**30.46**	m3	**32.74**
removing from site to tip b) Active	-	37.00	**37.00**	-	1.35	**14.46**	-	**51.46**	m3	**55.32**
removing from site to tip c) Contaminated (Guide price - always seek a quotation for specialist disposal costs.)	-	80.00	**80.00**	-	1.35	**14.46**	-	**94.46**	m3	**101.54**
Disposal of preserved top soil										
depositing on site in temporary spoil heaps where directed										
25m	-	-	-	-	1.25	**13.39**	-	**13.39**	m3	**14.39**
50m	-	-	-	-	1.50	**16.07**	-	**16.07**	m3	**17.27**
100m	-	-	-	-	2.00	**21.42**	-	**21.42**	m3	**23.03**
Excavating inside existing building - by hand										
To reduce levels										
maximum depth not exceeding										
0.25m	-	-	-	-	3.50	**37.49**	-	**37.49**	m3	**40.30**
1.00m	-	-	-	-	3.75	**40.16**	-	**40.16**	m3	**43.17**
2.00m	-	-	-	-	4.20	**44.98**	-	**44.98**	m3	**48.36**
4.00m	-	-	-	-	5.40	**57.83**	-	**57.83**	m3	**62.17**
Trenches width not exceeding 0.30m										
maximum depth not exceeding										
0.25m	-	-	-	-	6.00	**64.26**	-	**64.26**	m3	**69.08**
1.00m	-	-	-	-	6.25	**66.94**	-	**66.94**	m3	**71.96**
Trenches width exceeding 0.30m										
maximum depth not exceeding										
0.25m	-	-	-	-	4.00	**42.84**	-	**42.84**	m3	**46.05**
1.00m	-	-	-	-	4.25	**45.52**	-	**45.52**	m3	**48.93**
2.00m	-	-	-	-	5.00	**53.55**	-	**53.55**	m3	**57.57**
4.00m	-	-	-	-	6.15	**65.87**	-	**65.87**	m3	**70.81**
Pits										
maximum depth not exceeding										
0.25m	-	-	-	-	4.75	**50.87**	-	**50.87**	m3	**54.69**
1.00m	-	-	-	-	4.95	**53.01**	-	**53.01**	m3	**56.99**
2.00m	-	-	-	-	5.65	**60.51**	-	**60.51**	m3	**65.05**
4.00m	-	-	-	-	7.15	**76.58**	-	**76.58**	m3	**82.32**
Extra over any types of excavating irrespective of depth										
excavating below ground water level	-	-	-	-	2.15	**23.03**	-	**23.03**	m3	**24.75**
Breaking out existing materials; extra over any types of excavating irrespective of depth										
concrete	-	-	-	-	11.00	**117.81**	-	**117.81**	m3	**126.65**
reinforced concrete	-	-	-	-	15.50	**166.01**	-	**166.01**	m3	**178.46**
brickwork, blockwork or stonework	-	-	-	-	7.00	**74.97**	-	**74.97**	m3	**80.59**
drain and concrete bed under										
100mm diameter	-	-	-	-	1.15	**12.32**	-	**12.32**	m	**13.24**
150mm diameter	-	-	-	-	1.35	**14.46**	-	**14.46**	m	**15.54**
Breaking out existing hard pavings; extra over any types of excavating irrespective of depth										
concrete 150mm thick	-	-	-	-	1.50	**16.07**	-	**16.07**	m2	**17.27**
reinforced concrete 200mm thick	-	-	-	-	3.00	**32.13**	-	**32.13**	m2	**34.54**
Working space allowance to excavation; including additional earthwork support, disposal (excluding landfill tax) and backfilling										
pits	-	-	-	-	4.00	**42.84**	-	**42.84**	m2	**46.05**
trenches	-	-	-	-	3.70	**39.63**	-	**39.63**	m2	**42.60**
Compacting										
bottoms of excavations	-	-	-	-	0.14	**1.50**	-	**1.50**	m2	**1.61**
Disposal of excavated material										
depositing on site in temporary spoil heaps where directed										
25m	-	-	-	-	1.75	**18.74**	-	**18.74**	m3	**20.15**
50m	-	-	-	-	1.95	**20.88**	-	**20.88**	m3	**22.45**
removing from site to tip a) Inert	-	16.00	**16.00**	-	2.25	**24.10**	-	**40.10**	m3	**43.10**
removing from site to tip b) Active	-	37.00	**37.00**	-	2.25	**24.10**	-	**61.10**	m3	**65.68**
removing from site to tip c) Contaminated (Guide price - always seek a quotation for specialist disposal costs.)	-	80.00	**80.00**	-	2.25	**24.10**	-	**104.10**	m3	**111.90**

Labour hourly rates: (except Specialists) Craft Operatives £14.21 Labourer £10.71 Rates are national average prices. Refer to REGIONAL VARIATIONS for indicative levels of overall pricing in regions	MATERIALS			LABOUR				RATES		
	Del to Site £	Waste %	Material Cost £	Craft Optve Hrs	Lab Hrs	Labour Cost £	Sunds £	Nett Rate £	Unit	Gross Rate (+7.5%) £
D20: EXCAVATING AND FILLING Cont.										
Earthwork support										
Earthwork support										
distance between opposing faces not exceeding 2.00m; maximum depth not exceeding										
1.00m	0.68	-	**0.68**	-	0.20	**2.14**	0.02	**2.84**	m2	**3.06**
2.00m	0.89	-	**0.89**	-	0.25	**2.68**	0.03	**3.60**	m2	**3.87**
4.00m	1.05	-	**1.05**	-	0.30	**3.21**	0.03	**4.30**	m2	**4.62**
6.00m	1.16	-	**1.16**	-	0.35	**3.75**	0.04	**4.95**	m2	**5.32**
distance between opposing faces 2.00 - 4.00m; maximum depth not exceeding										
1.00m	0.80	-	**0.80**	-	0.21	**2.25**	0.02	**3.07**	m2	**3.30**
2.00m	0.98	-	**0.98**	-	0.26	**2.78**	0.03	**3.80**	m2	**4.09**
4.00m	1.12	-	**1.12**	-	0.32	**3.43**	0.03	**4.58**	m2	**4.93**
6.00m	1.31	-	**1.31**	-	0.37	**3.96**	0.04	**5.32**	m2	**5.72**
distance between opposing faces exceeding 4.00m; maximum depth not exceeding										
1.00m	0.89	-	**0.89**	-	0.22	**2.36**	0.03	**3.28**	m2	**3.52**
2.00m	1.06	-	**1.06**	-	0.28	**3.00**	0.03	**4.10**	m2	**4.40**
4.00m	1.23	-	**1.23**	-	0.33	**3.53**	0.04	**4.81**	m2	**5.17**
6.00m	1.42	-	**1.42**	-	0.39	**4.18**	0.04	**5.64**	m2	**6.06**
Earthwork support; unstable ground										
distance between opposing faces not exceeding 2.00m; maximum depth not exceeding										
1.00m	1.23	-	**1.23**	-	0.33	**3.53**	0.04	**4.81**	m2	**5.17**
2.00m	1.49	-	**1.49**	-	0.42	**4.50**	0.04	**6.03**	m2	**6.48**
4.00m	1.74	-	**1.74**	-	0.50	**5.36**	0.05	**7.15**	m2	**7.69**
6.00m	1.95	-	**1.95**	-	0.59	**6.32**	0.07	**8.34**	m2	**8.96**
distance between opposing faces 2.00 - 4.00m; maximum depth not exceeding										
1.00m	1.38	-	**1.38**	-	0.35	**3.75**	0.04	**5.17**	m2	**5.56**
2.00m	1.64	-	**1.64**	-	0.43	**4.61**	0.05	**6.30**	m2	**6.77**
4.00m	1.88	-	**1.88**	-	0.53	**5.68**	0.07	**7.63**	m2	**8.20**
6.00m	2.18	-	**2.18**	-	0.62	**6.64**	0.07	**8.88**	m2	**9.55**
distance between opposing faces exceeding 4.00m; maximum depth not exceeding										
1.00m	1.49	-	**1.49**	-	0.37	**3.96**	0.04	**5.49**	m2	**5.90**
2.00m	1.78	-	**1.78**	-	0.47	**5.03**	0.05	**6.87**	m2	**7.38**
4.00m	2.08	-	**2.08**	-	0.55	**5.89**	0.07	**8.04**	m2	**8.64**
6.00m	2.39	-	**2.39**	-	0.65	**6.96**	0.08	**9.42**	m2	**10.13**
Earthwork support; next to roadways										
distance between opposing faces not exceeding 2.00m; maximum depth not exceeding										
1.00m	1.45	-	**1.45**	-	0.40	**4.28**	0.04	**5.78**	m2	**6.21**
2.00m	1.78	-	**1.78**	-	0.50	**5.36**	0.05	**7.19**	m2	**7.73**
4.00m	2.08	-	**2.08**	-	0.60	**6.43**	0.07	**8.57**	m2	**9.22**
6.00m	2.39	-	**2.39**	-	0.70	**7.50**	0.08	**9.96**	m2	**10.71**
distance between opposing faces 2.00 - 4.00m; maximum depth not exceeding										
1.00m	1.64	-	**1.64**	-	0.42	**4.50**	0.05	**6.19**	m2	**6.66**
2.00m	1.94	-	**1.94**	-	0.52	**5.57**	0.07	**7.58**	m2	**8.15**
4.00m	2.28	-	**2.28**	-	0.64	**6.85**	0.08	**9.21**	m2	**9.90**
6.00m	2.60	-	**2.60**	-	0.74	**7.93**	0.09	**10.61**	m2	**11.41**
distance between opposing faces exceeding 4.00m; maximum depth not exceeding										
1.00m	1.74	-	**1.74**	-	0.44	**4.71**	0.05	**6.51**	m2	**7.00**
2.00m	2.15	-	**2.15**	-	0.56	**6.00**	0.07	**8.22**	m2	**8.83**
4.00m	2.50	-	**2.50**	-	0.66	**7.07**	0.08	**9.65**	m2	**10.37**
6.00m	2.85	-	**2.85**	-	0.78	**8.35**	0.09	**11.30**	m2	**12.14**
Earthwork support; left in										
distance between opposing faces not exceeding 2.00m; maximum depth not exceeding										
1.00m	3.47	-	**3.47**	-	0.16	**1.71**	0.04	**5.23**	m2	**5.62**
2.00m	3.47	-	**3.47**	-	0.20	**2.14**	0.05	**5.67**	m2	**6.10**
4.00m	3.47	-	**3.47**	-	0.24	**2.57**	0.05	**6.10**	m2	**6.56**
6.00m	3.47	-	**3.47**	-	0.28	**3.00**	0.07	**6.54**	m2	**7.03**
distance between opposing faces 2.00 - 4.00m; maximum depth not exceeding										
1.00m	3.47	-	**3.47**	-	0.17	**1.82**	0.04	**5.34**	m2	**5.74**
2.00m	3.47	-	**3.47**	-	0.21	**2.25**	0.05	**5.78**	m2	**6.21**
4.00m	3.47	-	**3.47**	-	0.26	**2.78**	0.07	**6.33**	m2	**6.80**
6.00m	3.47	-	**3.47**	-	0.30	**3.21**	0.08	**6.76**	m2	**7.27**
distance between opposing faces exceeding 4.00m; maximum depth not exceeding										
1.00m	4.63	-	**4.63**	-	0.18	**1.93**	0.05	**6.62**	m2	**7.11**
2.00m	4.63	-	**4.63**	-	0.22	**2.36**	0.05	**7.04**	m2	**7.57**
4.00m	4.63	-	**4.63**	-	0.26	**2.78**	0.07	**7.48**	m2	**8.04**
6.00m	4.63	-	**4.63**	-	0.31	**3.32**	0.08	**8.03**	m2	**8.63**
Earthwork support; next to roadways; left in										
distance between opposing faces not exceeding 2.00m; maximum depth not exceeding										
1.00m	11.58	-	**11.58**	-	0.32	**3.43**	0.09	**15.10**	m2	**16.23**
2.00m	11.58	-	**11.58**	-	0.40	**4.28**	0.10	**15.97**	m2	**17.16**
4.00m	11.58	-	**11.58**	-	0.49	**5.25**	0.12	**16.95**	m2	**18.22**
6.00m	11.58	-	**11.58**	-	0.56	**6.00**	0.13	**17.71**	m2	**19.04**
distance between opposing faces 2.00 - 4.00m; maximum depth not exceeding										
1.00m	13.90	-	**13.90**	-	0.34	**3.64**	0.09	**17.63**	m2	**18.95**
2.00m	13.90	-	**13.90**	-	0.42	**4.50**	0.12	**18.52**	m2	**19.91**
4.00m	13.90	-	**13.90**	-	0.51	**5.46**	0.14	**19.50**	m2	**20.97**

Labour hourly rates: (except Specialists) Craft Operatives £14.21 Labourer £10.71 Rates are national average prices. Refer to REGIONAL VARIATIONS for indicative levels of overall pricing in regions	MATERIALS			LABOUR				RATES		
	Del to Site £	Waste %	Material Cost £	Craft Optve Hrs	Lab Hrs	Labour Cost £	Sunds £	Nett Rate £	Unit	Gross Rate (+7.5%) £
D20: EXCAVATING AND FILLING Cont.										
Earthwork support Cont.										
Earthwork support; next to roadways; left in Cont. distance between opposing faces 2.00 - 4.00m; maximum depth not exceeding Cont.										
6.00m	13.90	-	**13.90**	-	0.59	**6.32**	0.16	**20.38**	m2	**21.91**
distance between opposing faces exceeding 4.00m; maximum depth not exceeding										
1.00m	17.37	-	**17.37**	-	0.35	**3.75**	0.10	**21.22**	m2	**22.81**
2.00m	17.37	-	**17.37**	-	0.45	**4.82**	0.13	**22.33**	m2	**24.00**
4.00m	17.37	-	**17.37**	-	0.53	**5.68**	0.15	**23.20**	m2	**24.94**
6.00m	17.37	-	**17.37**	-	0.62	**6.64**	0.19	**24.20**	m2	**26.02**
Earthwork support; unstable ground; left in distance between opposing faces not exceeding 2.00m; maximum depth not exceeding										
1.00m	8.11	-	**8.11**	-	0.26	**2.78**	0.07	**10.96**	m2	**11.78**
2.00m	8.11	-	**8.11**	-	0.34	**3.64**	0.09	**11.84**	m2	**12.72**
4.00m	8.11	-	**8.11**	-	0.40	**4.28**	0.10	**12.49**	m2	**13.43**
6.00m	8.11	-	**8.11**	-	0.47	**5.03**	0.11	**13.25**	m2	**14.25**
distance between opposing faces 2.00 - 4.00m; maximum depth not exceeding										
1.00m	9.27	-	**9.27**	-	0.28	**3.00**	0.08	**12.34**	m2	**13.27**
2.00m	9.27	-	**9.27**	-	0.34	**3.64**	0.09	**13.00**	m2	**13.97**
4.00m	9.27	-	**9.27**	-	0.42	**4.50**	0.10	**13.86**	m2	**14.90**
6.00m	9.27	-	**9.27**	-	0.50	**5.36**	0.12	**14.74**	m2	**15.85**
distance between opposing faces exceeding 4.00m; maximum depth not exceeding										
1.00m	11.58	-	**11.58**	-	0.30	**3.21**	0.09	**14.88**	m2	**16.00**
2.00m	11.58	-	**11.58**	-	0.38	**4.07**	0.10	**15.75**	m2	**16.93**
4.00m	11.58	-	**11.58**	-	0.44	**4.71**	0.12	**16.42**	m2	**17.65**
6.00m	11.58	-	**11.58**	-	0.52	**5.57**	0.13	**17.28**	m2	**18.58**
Earthwork support; inside existing building distance between opposing faces not exceeding 2.00m; maximum depth not exceeding										
1.00m	0.76	-	**0.76**	-	0.22	**2.36**	0.02	**3.14**	m2	**3.37**
2.00m	0.96	-	**0.96**	-	0.28	**3.00**	0.03	**3.99**	m2	**4.29**
4.00m	1.09	-	**1.09**	-	0.30	**3.21**	0.03	**4.33**	m2	**4.66**
distance between opposing faces 2.00 - 4.00m; maximum depth not exceeding										
1.00m	1.05	-	**1.05**	-	0.23	**2.46**	0.02	**3.54**	m2	**3.80**
2.00m	1.09	-	**1.09**	-	0.29	**3.11**	0.03	**4.23**	m2	**4.54**
4.00m	1.16	-	**1.16**	-	0.35	**3.75**	0.04	**4.95**	m2	**5.32**
distance between opposing faces exceeding 4.00m; maximum depth not exceeding										
1.00m	0.95	-	**0.95**	-	0.24	**2.57**	0.03	**3.55**	m2	**3.82**
2.00m	1.10	-	**1.10**	-	0.31	**3.32**	0.03	**4.45**	m2	**4.79**
4.00m	1.32	-	**1.32**	-	0.36	**3.86**	0.04	**5.22**	m2	**5.61**
Earthwork support; unstable ground; inside existing building distance between opposing faces not exceeding 2.00m; maximum depth not exceeding										
1.00m	1.32	-	**1.32**	-	0.36	**3.86**	0.04	**5.22**	m2	**5.61**
2.00m	1.57	-	**1.57**	-	0.46	**4.93**	0.05	**6.55**	m2	**7.04**
4.00m	1.83	-	**1.83**	-	0.55	**5.89**	0.05	**7.77**	m2	**8.35**
distance between opposing faces 2.00 - 4.00m; maximum depth not exceeding										
1.00m	1.43	-	**1.43**	-	0.39	**4.18**	0.04	**5.65**	m2	**6.07**
2.00m	1.73	-	**1.73**	-	0.47	**5.03**	0.05	**6.82**	m2	**7.33**
4.00m	2.02	-	**2.02**	-	0.58	**6.21**	0.07	**8.30**	m2	**8.92**
distance between opposing faces exceeding 4.00m; maximum depth not exceeding										
1.00m	1.57	-	**1.57**	-	0.41	**4.39**	0.05	**6.01**	m2	**6.46**
2.00m	1.86	-	**1.86**	-	0.52	**5.57**	0.05	**7.48**	m2	**8.05**
4.00m	2.16	-	**2.16**	-	0.61	**6.53**	0.07	**8.76**	m2	**9.42**
Excavated material arising from excavation - by machine										
Filling to excavation average thickness exceeding 0.25m	-	-	**-**	-	-	**-**	3.11	**3.11**	m3	**3.35**
Compacting filling	-	-	**-**	-	-	**-**	0.34	**0.34**	m2	**0.37**
Selected excavated material obtained from on site spoil heaps 25m distant - by machine										
Filling to make up levels; depositing in layers 150mm maximum thickness										
average thickness not exceeding 0.25m	-	-	**-**	-	-	**-**	4.79	**4.79**	m3	**5.15**
average thickness exceeding 0.25m	-	-	**-**	-	-	**-**	3.53	**3.53**	m3	**3.80**
Compacting with 680 kg vibratory roller filling	-	-	**-**	-	-	**-**	0.22	**0.22**	m2	**0.24**
Compacting with 6 - 8 tonnes smooth wheeled roller filling	-	-	**-**	-	-	**-**	0.43	**0.43**	m2	**0.46**
Trimming										
sides of cuttings; vertical or battered	-	-	**-**	-	-	**-**	0.54	**0.54**	m2	**0.58**
sides of embankments; vertical or battered	-	-	**-**	-	-	**-**	0.54	**0.54**	m2	**0.58**

Labour hourly rates: (except Specialists) Craft Operatives £14.21 Labourer £10.71 Rates are national average prices. Refer to REGIONAL VARIATIONS for indicative levels of overall pricing in regions	MATERIALS			LABOUR				RATES		
	Del to Site £	Waste %	Material Cost £	Craft Optve Hrs	Lab Hrs	Labour Cost £	Sunds £	Nett Rate £	Unit	Gross Rate (+7.5%) £
D20: EXCAVATING AND FILLING Cont.										
Selected excavated material obtained from on site spoil heaps 50m distant - by machine										
Filling to make up levels; depositing in layers 150mm maximum thickness										
average thickness not exceeding 0.25m............	-	-	**-**	-	-	-	4.68	**4.68**	m3	**5.03**
average thickness exceeding 0.25m................	-	-	**-**	-	-	-	4.68	**4.68**	m3	**5.03**
Compacting with 680 kg vibratory roller										
filling	-	-	**-**	-	-	-	0.22	**0.22**	m2	**0.24**
Compacting with 6 - 8 tonnes smooth wheeled roller										
filling	-	-	**-**	-	-	-	0.43	**0.43**	m2	**0.46**
Trimming										
sides of cuttings; vertical or battered..........	-	-	**-**	-	-	-	0.54	**0.54**	m2	**0.58**
sides of embankments; vertical or battered.......	-	-	**-**	-	-	-	0.54	**0.54**	m2	**0.58**
Preserved topsoil obtained from on site spoil heaps not exceeding 100m distant - by machine										
Filling to make up levels										
average thickness not exceeding 0.25m............	-	-	**-**	-	-	-	6.71	**6.71**	m3	**7.21**
Imported topsoil; wheeling not exceeding 50m - by machine										
Filling to make up levels										
average thickness not exceeding 0.25m............	15.00	10.00	**16.50**	-	-	-	6.08	**22.58**	m3	**24.27**
Hard, dry, broken brick or stone to be obtained off site; wheeling not exceeding 25m - by machine										
Filling to make up levels; depositing in layers 150mm maximum thickness										
average thickness not exceeding 0.25m............	17.00	25.00	**21.25**	-	-	-	3.59	**24.84**	m3	**26.71**
average thickness exceeding 0.25m................	17.00	25.00	**21.25**	-	-	-	2.98	**24.23**	m3	**26.05**
Surface packing to filling										
to vertical or battered faces....................	-	-	**-**	-	-	-	0.61	**0.61**	m2	**0.66**
Compacting with 680 kg vibratory roller										
filling; blinding with sand, ashes or similar fine material	0.15	40.00	**0.21**	-	0.05	**0.54**	0.22	**0.97**	m2	**1.04**
Compacting with 6 - 8 tonnes smooth wheeled roller										
filling; blinding with sand, ashes or similar fine material	0.15	40.00	**0.21**	-	0.05	**0.54**	0.43	**1.17**	m2	**1.26**
Hard, dry, broken brick or stone to be obtained off site; wheeling not exceeding 50m - by machine										
Filling to make up levels; depositing in layers 150mm maximum thickness										
average thickness not exceeding 0.25m............	17.00	25.00	**21.25**	-	-	-	4.18	**25.43**	m3	**27.34**
average thickness exceeding 0.25m................	17.00	25.00	**21.25**	-	-	-	3.59	**24.84**	m3	**26.71**
Surface packing to filling										
to vertical or battered faces....................	-	-	**-**	-	-	-	0.61	**0.61**	m2	**0.66**
Compacting with 680 kg vibratory roller										
filling; blinding with sand, ashes or similar fine material	0.15	40.00	**0.21**	-	0.05	**0.54**	0.22	**0.97**	m2	**1.04**
Compacting with 6 - 8 tonnes smooth wheeled roller										
filling; blinding with sand, ashes or similar fine material	0.15	40.00	**0.21**	-	0.05	**0.54**	0.39	**1.14**	m2	**1.22**
MOT Type 1 to be obtained off site; wheeling not exceeding 25m - by machine										
Filling to make up levels; depositing in layers 150mm maximum thickness										
average thickness not exceeding 0.25m............	19.95	25.00	**24.94**	-	-	-	4.18	**29.12**	m3	**31.30**
average thickness exceeding 0.25m................	19.95	25.00	**24.94**	-	-	-	3.47	**28.41**	m3	**30.54**
Surface packing to filling										
to vertical or battered faces....................	-	-	**-**	-	-	-	0.61	**0.61**	m2	**0.66**
Compacting with 680 kg vibratory roller										
filling; blinding with sand, ashes or similar fine material	0.15	40.00	**0.21**	-	0.05	**0.54**	0.22	**0.97**	m2	**1.04**
Compacting with 6 - 8 tonnes smooth wheeled roller										
filling; blinding with sand, ashes or similar fine material	0.15	40.00	**0.21**	-	0.05	**0.54**	0.43	**1.17**	m2	**1.26**
MOT Type 1 to be obtained off site; wheeling not exceeding 50m - by machine										
Filling to make up levels; depositing in layers 150mm maximum thickness										
average thickness not exceeding 0.25m............	19.95	25.00	**24.94**	-	-	-	4.18	**29.12**	m3	**31.30**
average thickness exceeding 0.25m................	19.95	25.00	**24.94**	-	-	-	3.59	**28.53**	m3	**30.67**

Labour hourly rates: (except Specialists) Craft Operatives £14.21 Labourer £10.71 Rates are national average prices. Refer to REGIONAL VARIATIONS for indicative levels of overall pricing in regions	MATERIALS			LABOUR				RATES		
	Del to Site £	Waste %	Material Cost £	Craft Optve Hrs	Lab Hrs	Labour Cost £	Sunds £	Nett Rate £	Unit	Gross Rate (+7.5%) £
D20: EXCAVATING AND FILLING Cont.										
MOT Type 1 to be obtained off site; wheeling not exceeding 50m - by machine Cont.										
Surface packing to filling to vertical or battered faces....................	-	-	-	-	-	-	0.61	**0.61**	m2	**0.66**
Compacting with 680 kg vibratory roller filling; blinding with sand, ashes or similar fine material	0.15	40.00	**0.21**	-	0.05	**0.54**	0.22	**0.97**	m2	**1.04**
Compacting with 6 - 8 tonnes smooth wheeled roller filling; blinding with sand, ashes or similar fine material	0.15	40.00	**0.21**	-	0.05	**0.54**	0.43	**1.17**	m2	**1.26**
MOT Type 2 to be obtained off site; wheeling not exceeding 25m - by machine										
Filling to make up levels; depositing in layers 150mm maximum thickness										
average thickness not exceeding 0.25m...........	18.48	25.00	**23.10**	-	-	-	3.59	**26.69**	m3	**28.69**
average thickness exceeding 0.25m................	18.48	25.00	**23.10**	-	-	-	2.98	**26.08**	m3	**28.04**
Surface packing to filling to vertical or battered faces....................	-	-	-	-	-	-	0.61	**0.61**	m2	**0.66**
Compacting with 680 kg vibratory roller filling; blinding with sand, ashes or similar fine material	0.15	40.00	**0.21**	-	0.05	**0.54**	0.22	**0.97**	m2	**1.04**
Compacting with 6 - 8 tonnes smooth wheeled roller filling; blinding with sand, ashes or similar fine material	0.15	40.00	**0.21**	-	0.05	**0.54**	0.43	**1.17**	m2	**1.26**
MOT Type 2 to be obtained off site; wheeling not exceeding 50m - by machine										
Filling to make up levels; depositing in layers 150mm maximum thickness										
average thickness not exceeding 0.25m...........	18.48	25.00	**23.10**	-	-	-	4.18	**27.28**	m3	**29.33**
average thickness exceeding 0.25m................	18.48	25.00	**23.10**	-	-	-	3.59	**26.69**	m3	**28.69**
Surface packing to filling to vertical or battered faces....................	-	-	-	-	-	-	0.61	**0.61**	m2	**0.66**
Compacting with 680 kg vibratory roller filling; blinding with sand, ashes or similar fine material	0.15	40.00	**0.21**	-	0.05	**0.54**	0.22	**0.97**	m2	**1.04**
Compacting with 6 - 8 tonnes smooth wheeled roller filling; blinding with sand, ashes or similar fine material	0.15	40.00	**0.21**	-	0.05	**0.54**	0.43	**1.17**	m2	**1.26**
Ashes to be obtained off site; wheeling not exceeding 25m - by machine										
Filling to make up levels; depositing in layers 150mm maximum thickness										
average thickness not exceeding 0.25m...........	9.99	40.00	**13.98**	-	-	-	3.80	**17.78**	m3	**19.11**
Compacting with 680 kg vibratory roller filling	-	-	-	-	-	-	0.22	**0.22**	m2	**0.24**
Compacting with 6 - 8 tonnes smooth wheeled roller filling	-	-	-	-	-	-	0.43	**0.43**	m2	**0.46**
Ashes to be obtained off site; wheeling not exceeding 50m - by machine										
Filling to make up levels; depositing in layers 150mm maximum thickness										
average thickness not exceeding 0.25m...........	9.99	40.00	**13.98**	-	-	-	5.14	**19.12**	m3	**20.55**
Compacting with 680 kg vibratory roller filling	-	-	-	-	-	-	0.22	**0.22**	m2	**0.24**
Compacting with 6 - 8 tonnes smooth wheeled roller filling	-	-	-	-	-	-	0.43	**0.43**	m2	**0.46**
Sand to be obtained off site; wheeling not exceeding 25m - by machine										
Filling to make up levels; depositing in layers 150mm maximum thickness										
average thickness not exceeding 0.25m...........	14.58	33.00	**19.39**	-	-	-	4.18	**23.57**	m3	**25.34**
average thickness exceeding 0.25m................	14.58	33.00	**19.39**	-	-	-	3.59	**22.98**	m3	**24.71**
Compacting with 680 kg vibratory roller filling	-	-	-	-	-	-	0.22	**0.22**	m2	**0.24**
Compacting with 6 - 8 tonnes smooth wheeled roller filling	-	-	-	-	-	-	0.43	**0.43**	m2	**0.46**

Labour hourly rates: (except Specialists) Craft Operatives £14.21 Labourer £10.71 Rates are national average prices. Refer to REGIONAL VARIATIONS for indicative levels of overall pricing in regions	MATERIALS			LABOUR				RATES		
	Del to Site £	Waste %	Material Cost £	Craft Optve Hrs	Lab Hrs	Labour Cost £	Sunds £	Nett Rate £	Unit	Gross Rate (+7.5%) £
D20: EXCAVATING AND FILLING Cont.										
Sand to be obtained off site; wheeling not exceeding 50m - by machine										
Filling to make up levels; depositing in layers 150mm maximum thickness										
average thickness not exceeding 0.25m............	14.58	33.00	**19.39**	-	-	-	4.77	**24.16**	m3	**25.97**
average thickness exceeding 0.25m................	14.58	33.00	**19.39**	-	-	-	4.18	**23.57**	m3	**25.34**
Compacting with 680 kg vibratory roller										
filling	-	-	-	-	-	-	0.22	**0.22**	m2	**0.24**
Compacting with 6 - 8 tonnes smooth wheeled roller										
filling	-	-	-	-	-	-	0.43	**0.43**	m2	**0.46**
Hoggin to be obtained off site; wheeling not exceeding 25m - by machine										
Filling to make up levels; depositing in layers 150mm maximum thickness										
average thickness not exceeding 0.25m............	16.00	33.00	**21.28**	-	-	-	4.17	**25.45**	m3	**27.36**
average thickness exceeding 0.25m................	16.00	33.00	**21.28**	-	-	-	3.47	**24.75**	m3	**26.61**
Compacting with 680 kg vibratory roller										
filling	-	-	-	-	-	-	0.22	**0.22**	m2	**0.24**
Compacting with 6 - 8 tonnes smooth wheeled roller										
filling	-	-	-	-	-	-	0.43	**0.43**	m2	**0.46**
Hoggin to be obtained off site; wheeling not exceeding 50m - by machine										
Filling to make up levels; depositing in layers 150mm maximum thickness										
average thickness not exceeding 0.25m............	16.00	33.00	**21.28**	-	-	-	3.59	**24.87**	m3	**26.74**
average thickness exceeding 0.25m................	16.00	33.00	**21.28**	-	-	-	2.98	**24.26**	m3	**26.08**
Compacting with 680 kg vibratory roller										
filling	-	-	-	-	-	-	0.22	**0.22**	m2	**0.24**
Compacting with 6 - 8 tonnes smooth wheeled roller										
filling	-	-	-	-	-	-	0.43	**0.43**	m2	**0.46**
Excavated material arising from excavation - by hand										
Filling to excavation										
average thickness exceeding 0.25m................	-	-	-	-	1.33	**14.24**	-	**14.24**	m3	**15.31**
Compacting										
filling	-	-	-	-	-	-	0.36	**0.36**	m2	**0.38**
Selected excavated material obtained from on site spoil heaps 25m distant - by hand										
Filling to make up levels; depositing in layers 150mm maximum thickness										
average thickness not exceeding 0.25m............	-	-	-	-	1.75	**18.74**	-	**18.74**	m3	**20.15**
average thickness exceeding 0.25m................	-	-	-	-	1.50	**16.07**	-	**16.07**	m3	**17.27**
Compacting with 680 kg vibratory roller										
filling	-	-	-	-	-	-	0.22	**0.22**	m2	**0.24**
Compacting with 6 - 8 tonnes smooth wheeled roller										
filling	-	-	-	-	-	-	0.43	**0.43**	m2	**0.46**
Trimming										
sides of cuttings; vertical or battered..........	-	-	-	-	0.20	**2.14**	-	**2.14**	m2	**2.30**
sides of embankments; vertical or battered.......	-	-	-	-	0.20	**2.14**	-	**2.14**	m2	**2.30**
Selected excavated material obtained from on site spoil heaps 50m distant - by hand										
Filling to make up levels; depositing in layers 150mm maximum thickness										
average thickness not exceeding 0.25m............	-	-	-	-	2.00	**21.42**	-	**21.42**	m3	**23.03**
average thickness exceeding 0.25m................	-	-	-	-	1.85	**19.81**	-	**19.81**	m3	**21.30**
Compacting with 680 kg vibratory roller										
filling	-	-	-	-	-	-	0.22	**0.22**	m2	**0.24**
Compacting with 6 - 8 tonnes smooth wheeled roller										
filling	-	-	-	-	-	-	0.43	**0.43**	m2	**0.46**
Trimming										
sides of cuttings; vertical or battered..........	-	-	-	-	0.20	**2.14**	-	**2.14**	m2	**2.30**
sides of embankments; vertical or battered.......	-	-	-	-	0.20	**2.14**	-	**2.14**	m2	**2.30**
Preserved topsoil obtained from on site spoil heaps not exceeding 100m distant - by hand										
Filling to make up levels										
average thickness not exceeding 0.25m............	-	-	-	-	4.00	**42.84**	-	**42.84**	m3	**46.05**
Filling to external planters										
average thickness not exceeding 0.25m............	-	-	-	-	5.00	**53.55**	-	**53.55**	m3	**57.57**

Labour hourly rates: (except Specialists) Craft Operatives £14.21 Labourer £10.71 Rates are national average prices. Refer to REGIONAL VARIATIONS for indicative levels of overall pricing in regions	MATERIALS			LABOUR				RATES		
	Del to Site £	Waste %	Material Cost £	Craft Optve Hrs	Lab Hrs	Labour Cost £	Sunds £	Nett Rate £	Unit	Gross Rate (+7.5%) £
D20: EXCAVATING AND FILLING Cont.										
Imported topsoil; wheeling not exceeding 50m - by hand										
Filling to make up levels										
average thickness not exceeding 0.25m...........	15.00	10.00	**16.50**	-	3.50	**37.49**	-	**53.99**	m3	**58.03**
Filling to external planters										
average thickness exceeding 0.25m...............	15.00	10.00	**16.50**	-	4.50	**48.20**	-	**64.70**	m3	**69.55**
Hard, dry, broken brick or stone to be obtained off site; wheeling not exceeding 25m - by hand										
Filling to make up levels; depositing in layers 150mm maximum thickness										
average thickness not exceeding 0.25m...........	17.00	25.00	**21.25**	-	1.50	**16.07**	-	**37.32**	m3	**40.11**
average thickness exceeding 0.25m...............	17.00	25.00	**21.25**	-	1.25	**13.39**	-	**34.64**	m3	**37.24**
Surface packing to filling										
to vertical or battered faces...................	-	-	**-**	-	0.20	**2.14**	-	**2.14**	m2	**2.30**
Compacting with 680 kg vibratory roller										
filling; blinding with sand, ashes or similar fine material	0.15	40.00	**0.21**	-	0.05	**0.54**	0.22	**0.97**	m2	**1.04**
Compacting with 6 - 8 tonnes smooth wheeled roller										
filling; blinding with sand, ashes or similar fine material	0.15	40.00	**0.21**	-	0.05	**0.54**	0.43	**1.17**	m2	**1.26**
Hard, dry, broken brick or stone to be obtained off site; wheeling not exceeding 50m - by hand										
Filling to make up levels; depositing in layers 150mm maximum thickness										
average thickness not exceeding 0.25m...........	17.00	25.00	**21.25**	-	1.75	**18.74**	-	**39.99**	m3	**42.99**
average thickness exceeding 0.25m...............	17.00	25.00	**21.25**	-	1.50	**16.07**	-	**37.32**	m3	**40.11**
Surface packing to filling										
to vertical or battered faces...................	-	-	**-**	-	0.20	**2.14**	-	**2.14**	m2	**2.30**
Compacting with 680 kg vibratory roller										
filling; blinding with sand, ashes or similar fine material	0.15	40.00	**0.21**	-	0.05	**0.54**	0.22	**0.97**	m2	**1.04**
Compacting with 6 - 8 tonnes smooth wheeled roller										
filling; blinding with sand, ashes or similar fine material	0.15	40.00	**0.21**	-	0.05	**0.54**	0.43	**1.17**	m2	**1.26**
MOT Type 1 to be obtained off site; wheeling not exceeding 25m - by hand										
Filling to make up levels; depositing in layers 150mm maximum thickness										
average thickness not exceeding 0.25m...........	19.95	25.00	**24.94**	-	1.50	**16.07**	-	**41.00**	m3	**44.08**
average thickness exceeding 0.25m...............	19.95	25.00	**24.94**	-	1.25	**13.39**	-	**38.33**	m3	**41.20**
Surface packing to filling										
to vertical or battered faces...................	-	-	**-**	-	0.20	**2.14**	-	**2.14**	m2	**2.30**
Compacting with 680 kg vibratory roller										
filling; blinding with sand, ashes or similar fine material	0.15	40.00	**0.21**	-	0.05	**0.54**	0.22	**0.97**	m2	**1.04**
Compacting with 6 - 8 tonnes smooth wheeled roller										
filling; blinding with sand, ashes or similar fine material	0.15	40.00	**0.21**	-	0.05	**0.54**	0.43	**1.17**	m2	**1.26**
MOT Type 1 to be obtained off site; wheeling not exceeding 50m - by hand										
Filling to make up levels; depositing in layers 150mm maximum thickness										
average thickness not exceeding 0.25m...........	19.95	25.00	**24.94**	-	1.75	**18.74**	-	**43.68**	m3	**46.96**
average thickness exceeding 0.25m...............	19.95	25.00	**24.94**	-	1.50	**16.07**	-	**41.00**	m3	**44.08**
Surface packing to filling										
to vertical or battered faces...................	-	-	**-**	-	0.20	**2.14**	-	**2.14**	m2	**2.30**
Compacting with 680 kg vibratory roller										
filling; blinding with sand, ashes or similar fine material	0.15	40.00	**0.21**	-	0.05	**0.54**	0.22	**0.97**	m2	**1.04**
Compacting with 6 - 8 tonnes smooth wheeled roller										
filling; blinding with sand, ashes or similar fine material	0.15	40.00	**0.21**	-	0.05	**0.54**	0.43	**1.17**	m2	**1.26**
MOT Type 2 to be obtained off site; wheeling not exceeding 25m - by hand										
Filling to make up levels; depositing in layers 150mm maximum thickness										
average thickness not exceeding 0.25m...........	18.48	25.00	**23.10**	-	1.50	**16.07**	-	**39.17**	m3	**42.10**
average thickness exceeding 0.25m...............	18.48	25.00	**23.10**	-	1.25	**13.39**	-	**36.49**	m3	**39.22**
Surface packing to filling										
to vertical or battered faces...................	-	-	**-**	-	0.20	**2.14**	-	**2.14**	m2	**2.30**

Labour hourly rates: (except Specialists) Craft Operatives £14.21 Labourer £10.71 Rates are national average prices. Refer to REGIONAL VARIATIONS for indicative levels of overall pricing in regions	MATERIALS			LABOUR				RATES		
	Del to Site £	Waste %	Material Cost £	Craft Optve Hrs	Lab Hrs	Labour Cost £	Sunds £	Nett Rate £	Unit	Gross Rate (+7.5%) £
D20: EXCAVATING AND FILLING Cont.										
MOT Type 2 to be obtained off site; wheeling not exceeding 25m - by hand Cont.										
Compacting with 680 kg vibratory roller filling; blinding with sand, ashes or similar fine material	0.15	40.00	**0.21**	-	0.05	**0.54**	0.22	**0.97**	m2	**1.04**
Compacting with 6 - 8 tonnes smooth wheeled roller filling; blinding with sand, ashes or similar fine material	0.15	40.00	**0.21**	-	0.05	**0.54**	0.43	**1.17**	m2	**1.26**
MOT Type 2 to be obtained off site; wheeling not exceeding 50m - by hand										
Filling to make up levels; depositing in layers 150mm maximum thickness										
average thickness not exceeding 0.25m	18.48	25.00	**23.10**	-	1.75	**18.74**	-	**41.84**	m3	**44.98**
average thickness exceeding 0.25m	18.48	25.00	**23.10**	-	1.50	**16.07**	-	**39.17**	m3	**42.10**
Surface packing to filling to vertical or battered faces	-	-	**-**	-	0.20	**2.14**	-	**2.14**	m2	**2.30**
Compacting with 680 kg vibratory roller filling; blinding with sand, ashes or similar fine material	0.15	40.00	**0.21**	-	0.05	**0.54**	0.22	**0.97**	m2	**1.04**
Compacting with 6 - 8 tonnes smooth wheeled roller filling; blinding with sand, ashes or similar fine material	0.15	40.00	**0.21**	-	0.05	**0.54**	0.43	**1.17**	m2	**1.26**
Ashes to be obtained off site; wheeling not exceeding 25m - by hand										
Filling to make up levels; depositing in layers 150mm maximum thickness										
average thickness not exceeding 0.25m	9.99	40.00	**13.98**	-	1.60	**17.14**	-	**31.12**	m3	**33.45**
Compacting with 680 kg vibratory roller filling	-	-	**-**	-	-	**-**	0.22	**0.22**	m2	**0.24**
Compacting with 6 - 8 tonnes smooth wheeled roller filling	-	-	**-**	-	-	**-**	0.43	**0.43**	m2	**0.46**
Ashes to be obtained off site; wheeling not exceeding 50m - by hand										
Filling to make up levels; depositing in layers 150mm maximum thickness										
average thickness not exceeding 0.25m	9.99	40.00	**13.98**	-	1.85	**19.81**	-	**33.79**	m3	**36.33**
Compacting with 680 kg vibratory roller filling	-	-	**-**	-	-	**-**	0.22	**0.22**	m2	**0.24**
Compacting with 6 - 8 tonnes smooth wheeled roller filling	-	-	**-**	-	-	**-**	0.43	**0.43**	m2	**0.46**
Sand to be obtained off site; wheeling not exceeding 25m - by hand										
Filling to make up levels; depositing in layers 150mm maximum thickness										
average thickness not exceeding 0.25m	14.58	33.00	**19.39**	-	1.75	**18.74**	-	**38.13**	m3	**40.99**
average thickness exceeding 0.25m	14.58	33.00	**19.39**	-	1.50	**16.07**	-	**35.46**	m3	**38.12**
Compacting with 680 kg vibratory roller filling	-	-	**-**	-	-	**-**	0.22	**0.22**	m2	**0.24**
Compacting with 6 - 8 tonnes smooth wheeled roller filling	-	-	**-**	-	-	**-**	0.43	**0.43**	m2	**0.46**
Sand to be obtained off site; wheeling not exceeding 50m - by hand										
Filling to make up levels; depositing in layers 150mm maximum thickness										
average thickness not exceeding 0.25m	14.58	33.00	**19.39**	-	2.00	**21.42**	-	**40.81**	m3	**43.87**
average thickness exceeding 0.25m	14.58	33.00	**19.39**	-	1.75	**18.74**	-	**38.13**	m3	**40.99**
Compacting with 680 kg vibratory roller filling	-	-	**-**	-	-	**-**	0.22	**0.22**	m2	**0.24**
Compacting with 6 - 8 tonnes smooth wheeled roller filling	-	-	**-**	-	-	**-**	0.43	**0.43**	m2	**0.46**
Hoggin to be obtained off site; wheeling not exceeding 25m - by hand										
Filling to make up levels; depositing in layers 150mm maximum thickness										
average thickness not exceeding 0.25m	16.00	33.00	**21.28**	-	1.50	**16.07**	-	**37.35**	m3	**40.15**
average thickness exceeding 0.25m	16.00	33.00	**21.28**	-	1.25	**13.39**	-	**34.67**	m3	**37.27**
Compacting with 680 kg vibratory roller filling	-	-	**-**	-	-	**-**	0.22	**0.22**	m2	**0.24**

Labour hourly rates: (except Specialists) Craft Operatives £14.21 Labourer £10.71 Rates are national average prices. Refer to REGIONAL VARIATIONS for indicative levels of overall pricing in regions	MATERIALS			LABOUR				RATES		
	Del to Site £	Waste %	Material Cost £	Craft Optve Hrs	Lab Hrs	Labour Cost £	Sunds £	Nett Rate £	Unit	Gross Rate (+7.5%) £
D20: EXCAVATING AND FILLING Cont.										
Hoggin to be obtained off site; wheeling not exceeding 25m - by hand Cont.										
Compacting with 6 - 8 tonnes smooth wheeled roller										
filling	-	-	-	-	-	-	0.43	**0.43**	m2	**0.46**
Hoggin to be obtained off site; wheeling not exceeding 50m - by hand										
Filling to make up levels; depositing in layers 150mm maximum thickness										
average thickness not exceeding 0.25m	16.00	33.00	**21.28**	-	1.75	**18.74**	-	**40.02**	m3	**43.02**
average thickness exceeding 0.25m	16.00	33.00	**21.28**	-	1.50	**16.07**	-	**37.35**	m3	**40.15**
Compacting with 680 kg vibratory roller										
filling	-	-	-	-	-	-	0.22	**0.22**	m2	**0.24**
Compacting with 6 - 8 tonnes smooth wheeled roller										
filling	-	-	-	-	-	-	0.43	**0.43**	m2	**0.46**
Excavated material arising from excavation; work inside existing building - by hand										
Filling to excavation										
average thickness exceeding 0.25m	-	-	-	-	2.67	**28.60**	-	**28.60**	m3	**30.74**
Compacting										
filling	-	-	-	-	-	-	0.22	**0.22**	m2	**0.24**
Selected excavated material obtained from on site spoil heaps 25m distant; work inside existing building - by hand										
Filling to make up levels; depositing in layers 150mm maximum thickness										
average thickness not exceeding 0.25m	-	-	-	-	3.10	**33.20**	-	**33.20**	m3	**35.69**
average thickness exceeding 0.25m	-	-	-	-	2.93	**31.38**	-	**31.38**	m3	**33.73**
Compacting										
filling	-	-	-	-	-	-	0.22	**0.22**	m2	**0.24**
Selected excavated material obtained from on site spoil heaps 50m distant; work inside existing building - by hand										
Filling to make up levels; depositing in layers 150mm maximum thickness										
average thickness not exceeding 0.25m	-	-	-	-	3.35	**35.88**	-	**35.88**	m3	**38.57**
average thickness exceeding 0.25m	-	-	-	-	3.20	**34.27**	-	**34.27**	m3	**36.84**
Compacting										
filling	-	-	-	-	-	-	0.22	**0.22**	m2	**0.24**
Hard, dry, broken brick or stone to be obtained off site; wheeling not exceeding 25m; work inside existing building - by hand										
Filling to make up levels; depositing in layers 150mm maximum thickness										
average thickness not exceeding 0.25m	17.00	25.00	**21.25**	-	2.85	**30.52**	-	**51.77**	m3	**55.66**
average thickness exceeding 0.25m	17.00	25.00	**21.25**	-	2.68	**28.70**	-	**49.95**	m3	**53.70**
Surface packing to filling										
to vertical or battered faces	-	-	-	-	0.20	**2.14**	-	**2.14**	m2	**2.30**
Compacting										
filling; blinding with sand, ashes or similar fine material	0.15	40.00	**0.21**	-	0.05	**0.54**	0.22	**0.97**	m2	**1.04**
Hard, dry, broken brick or stone to be obtained off site; wheeling not exceeding 50m; work inside existing building - by hand										
Filling to make up levels; depositing in layers 150mm maximum thickness										
average thickness not exceeding 0.25m	17.00	25.00	**21.25**	-	3.10	**33.20**	-	**54.45**	m3	**58.53**
average thickness exceeding 0.25m	17.00	25.00	**21.25**	-	2.85	**30.52**	-	**51.77**	m3	**55.66**
Surface packing to filling										
to vertical or battered faces	-	-	-	-	0.20	**2.14**	-	**2.14**	m2	**2.30**
Compacting										
filling; blinding with sand, ashes or similar fine material	0.15	40.00	**0.21**	-	0.05	**0.54**	0.22	**0.97**	m2	**1.04**
Ashes to be obtained off site; wheeling not exceeding 25m; work inside existing building - by hand										
Filling to make up levels; depositing in layers 150mm maximum thickness										
average thickness not exceeding 0.25m	9.99	40.00	**13.98**	-	2.95	**31.59**	-	**45.57**	m3	**48.99**
Compacting										
filling	-	-	-	-	-	-	0.22	**0.22**	m2	**0.24**

Labour hourly rates: (except Specialists) Craft Operatives £14.21 Labourer £10.71 Rates are national average prices. Refer to REGIONAL VARIATIONS for indicative levels of overall pricing in regions	MATERIALS			LABOUR				RATES		
	Del to Site £	Waste %	Material Cost £	Craft Optve Hrs	Lab Hrs	Labour Cost £	Sunds £	Nett Rate £	Unit	Gross Rate (+7.5%) £
D20: EXCAVATING AND FILLING Cont.										
Ashes to be obtained off site; wheeling not exceeding 50m; work inside existing building - by hand										
Filling to make up levels; depositing in layers 150mm maximum thickness										
average thickness not exceeding 0.25m............	9.99	40.00	**13.98**	-	3.20	**34.27**	-	**48.25**	m3	**51.87**
Compacting										
filling	-	-	-	-	-	-	0.22	**0.22**	m2	**0.24**
Sand to be obtained off site; wheeling not exceeding 25m; work inside existing building - by hand										
Filling to make up levels; depositing in layers 150mm maximum thickness										
average thickness not exceeding 0.25m............	14.58	33.00	**19.39**	-	3.10	**33.20**	-	**52.59**	m3	**56.54**
average thickness exceeding 0.25m................	14.58	33.00	**19.39**	-	2.85	**30.52**	-	**49.91**	m3	**53.66**
Compacting										
filling	-	-	-	-	-	-	0.22	**0.22**	m2	**0.24**
Sand to be obtained off site; wheeling not exceeding 50m; work inside existing building - by hand										
Filling to make up levels; depositing in layers 150mm maximum thickness										
average thickness not exceeding 0.25m............	14.58	33.00	**19.39**	-	3.35	**35.88**	-	**55.27**	m3	**59.42**
average thickness exceeding 0.25m................	14.58	33.00	**19.39**	-	3.10	**33.20**	-	**52.59**	m3	**56.54**
Compacting										
filling	-	-	-	-	-	-	0.22	**0.22**	m2	**0.24**
D30: CAST IN PLACE PILING										
Prices include for a 21 N/mm2 concrete mix, nominal reinforcement and a minimum number of 50 piles on any one contract. The working loads sizes and lengths given below will depend on the nature of the soils in which the piles will be founded as well as structure to be supported (KELLER GROUND ENGINEERING)										
On/Off site charge										
in addition to the following prices add approximately	-	-	**Spclist**	-	-	**Spclist**	-	**4735.33**	Sm	**5090.48**
Short auger piles										
up to 6m long; 450mm nominal diameter; 8 tonnes normal working load	-	-	**Spclist**	-	-	**Spclist**	-	**27.74**	m	**29.82**
up to 6m long; 610mm nominal diameter; 30 tonnes normal working load	-	-	**Spclist**	-	-	**Spclist**	-	**41.94**	m	**45.09**
Bored cast in-situ piles										
up to 15m long; 450mm nominal diameter; 40 tonnes normal working load	-	-	**Spclist**	-	-	**Spclist**	-	**25.60**	m	**27.52**
up to 15m long; 610mm nominal diameter; 120 tonnes normal working load	-	-	**Spclist**	-	-	**Spclist**	-	**39.81**	m	**42.80**
Auger piles										
up to 20m long; 450mm nominal diameter...........	-	-	**Spclist**	-	-	**Spclist**	-	**25.96**	m	**27.90**
up to 30m long; 1200mm nominal diameter..........	-	-	**Spclist**	-	-	**Spclist**	-	**128.70**	m	**138.35**
Large diameter auger piles (plain shaft)										
up to 20m long; 610mm nominal diameter; 150 tonnes normal working load	-	-	**Spclist**	-	-	**Spclist**	-	**39.48**	m	**42.44**
up to 30m long; 1525mm nominal diameter; 600 tonnes normal working load.......................	-	-	**Spclist**	-	-	**Spclist**	-	**200.20**	m	**215.22**
Large diameter auger piles (belled base)										
up to 20m long; 610mm nominal diameter; 150 tonnes normal working load	-	-	**Spclist**	-	-	**Spclist**	-	**54.67**	m	**58.77**
up to 30m long; 1525mm nominal diameter; 1000 tonnes normal working load.......................	-	-	**Spclist**	-	-	**Spclist**	-	**257.46**	m	**276.77**
Boring through obstructions, rock like formations, etc undertaken on a time basis at a rate per piling rig per hour										
Cutting off tops of piles; including preparation and integration of reinforcement into pile cap or ground beam and disposal										
450-610mm nominal diameter.......................	-	-	**Spclist**	-	-	**Spclist**	-	**9.50**	m	**10.21**
610-1525mm nominal diameter......................	-	-	**Spclist**	-	-	**Spclist**	-	**35.94**	m	**38.63**

Labour hourly rates: (except Specialists) Craft Operatives £14.21 Labourer £10.71 Rates are national average prices. Refer to REGIONAL VARIATIONS for indicative levels of overall pricing in regions	MATERIALS			LABOUR				RATES		
	Del to Site £	Waste %	Material Cost £	Craft Optve Hrs	Lab Hrs	Labour Cost £	Sunds £	Nett Rate £	Unit	Gross Rate (+7.5%) £
D32: STEEL PILING										
Mild steel Universal Bearing Piles 'H' section, to BS EN 10025 Grade S275 and high yield steel Grade S275JR in lengths 9 - 15m supplied, handled, pitched and driven vertically with landbased plant (KELLER GROUND ENGINEERING)										
Note; the following prices are based on quantities of 25 - 150 tonnes										
On/Off site charge										
in addition to the following prices, add approximately	-	-	Spclist	-	-	Spclist	-	4067.66	Sm	4372.73
Mild steel piles										
203 x 203 x 45 kg/m, SWL 40 tonnes	-	-	Spclist	-	-	Spclist	-	39.30	m	42.24
203 x 203 x 54 kg/m, SWL 50 tonnes	-	-	Spclist	-	-	Spclist	-	44.31	m	47.63
254 x 254 x 63 kg/m, SWL 60 tonnes	-	-	Spclist	-	-	Spclist	-	50.08	m	53.83
254 x 254 x 71 kg/m, SWL 70 tonnes	-	-	Spclist	-	-	Spclist	-	53.94	m	57.98
305 x 305 x 79 kg/m, SWL 75 tonnes	-	-	Spclist	-	-	Spclist	-	59.39	m	63.84
254 x 254 x 85 kg/m, SWL 80 tonnes	-	-	Spclist	-	-	Spclist	-	63.03	m	67.76
305 x 305 x 88 kg/m, SWL 85 tonnes	-	-	Spclist	-	-	Spclist	-	68.85	m	74.02
305 x 305 x 95 kg/m, SWL 90 tonnes	-	-	Spclist	-	-	Spclist	-	73.20	m	78.70
356 x 368 x 109 kg/m, SWL 105 tonnes	-	-	Spclist	-	-	Spclist	-	87.73	m	94.31
305 x 305 x 110 kg/m, SWL 105 tonnes	-	-	Spclist	-	-	Spclist	-	84.98	m	91.35
305 x 305 x 126 kg/m, SWL 120 tonnes	-	-	Spclist	-	-	Spclist	-	94.14	m	101.20
305 x 305 x 149 kg/m, SWL 140 tonnes	-	-	Spclist	-	-	Spclist	-	103.90	m	111.69
356 x 368 x 133 kg/m, SWL 130 tonnes	-	-	Spclist	-	-	Spclist	-	108.74	m	116.90
356 x 368 x 152 kg/m, SWL 140 tonnes	-	-	Spclist	-	-	Spclist	-	115.42	m	124.07
356 x 368 x 174 kg/m, SWL 165 tonnes	-	-	Spclist	-	-	Spclist	-	132.73	m	142.69
305 x 305 x 186 kg/m, SWL 175 tonnes	-	-	Spclist	-	-	Spclist	-	140.78	m	151.34
305 x 305 x 223 kg/m, SWL 210 tonnes	-	-	Spclist	-	-	Spclist	-	157.95	m	169.80
Extra for high yield steel										
add	-	-	Spclist	-	-	Spclist	-	42.61	T	45.81
For quiet piling, add										
to terminal charge, lump sum	-	-	Spclist	-	-	Spclist	-	852.27	Sm	916.19
plus for all piles	-	-	Spclist	-	-	Spclist	-	2.11	m	2.26
Mild steel sheet piling to BS EN 10025 Grade S275 or high yield steel Grade S275JR in lengths 4.5-15m supplied, handled, pitched and driven by land based plant in one visit (KELLER GROUND ENGINEERING)										
On/Off site charge										
in addition to the following prices, add approximately	-	-	Spclist	-	-	Spclist	-	3719.00	Sm	3997.93
Mild steel sheet piling										
Larssen Section 6W	-	-	Spclist	-	-	Spclist	-	85.12	m2	91.51
extra for high yield steel	-	-	Spclist	-	-	Spclist	-	4.66	m2	5.01
extra for one coat L.B.V.(Lowca Varnish to BS 1070 Type 2) before driving	-	-	Spclist	-	-	Spclist	-	3.41	m2	3.66
extra for corners	-	-	Spclist	-	-	Spclist	-	34.07	m	36.62
extra for junctions	-	-	Spclist	-	-	Spclist	-	46.45	m	49.93
Frodingham Section 1N and Larssen Section 9W	-	-	Spclist	-	-	Spclist	-	94.11	m2	101.17
extra for high yield steel	-	-	Spclist	-	-	Spclist	-	3.87	m2	4.16
extra for one coat L.B.V.(Lowca Varnish to BS 1070 Type 2) before driving	-	-	Spclist	-	-	Spclist	-	3.41	m2	3.66
extra for corners	-	-	Spclist	-	-	Spclist	-	34.07	m	36.62
extra for junctions	-	-	Spclist	-	-	Spclist	-	46.45	m	49.93
Frodingham Section 2N and Larssen Section 12W	-	-	Spclist	-	-	Spclist	-	104.76	m2	112.62
extra for high yield steel	-	-	Spclist	-	-	Spclist	-	4.96	m2	5.33
extra for one coat L.B.V.(Lowca Varnish to BS 1070 Type 2) before driving	-	-	Spclist	-	-	Spclist	-	3.41	m2	3.66
extra for corners	-	-	Spclist	-	-	Spclist	-	34.07	m	36.62
extra for junctions	-	-	Spclist	-	-	Spclist	-	46.45	m	49.93
Frodingham Section 1BXN	-	-	Spclist	-	-	Spclist	-	116.12	m2	124.82
extra for high yield steel	-	-	Spclist	-	-	Spclist	-	7.90	m2	8.50
extra for one coat L.B.V.(Lowca Varnish to BS 1070 Type 2) before driving	-	-	Spclist	-	-	Spclist	-	3.10	m2	3.33
extra for corners	-	-	Spclist	-	-	Spclist	-	37.16	m	39.95
extra for junctions	-	-	Spclist	-	-	Spclist	-	46.45	m	49.93
Frodingham Section 3N and Larssen Section 16W	-	-	Spclist	-	-	Spclist	-	117.15	m2	125.94
extra for high yield steel	-	-	Spclist	-	-	Spclist	-	6.34	m2	6.82
extra for one coat L.B.V.(Lowca Varnish to BS 1070 Type 2) before driving	-	-	Spclist	-	-	Spclist	-	3.10	m2	3.33
extra for corners	-	-	Spclist	-	-	Spclist	-	37.16	m	39.95
extra for junctions	-	-	Spclist	-	-	Spclist	-	46.45	m	49.93
Larssen Section 20W	-	-	Spclist	-	-	Spclist	-	132.55	m2	142.50
extra for high yield steel	-	-	Spclist	-	-	Spclist	-	6.97	m2	7.50
extra for one coat L.B.V.(Lowca Varnish to BS 1070 Type 2) before driving	-	-	Spclist	-	-	Spclist	-	3.49	m2	3.75
extra for corners	-	-	Spclist	-	-	Spclist	-	40.94	m	44.01
extra for junctions	-	-	Spclist	-	-	Spclist	-	51.36	m	55.21
Frodingham Section 4N and Larssen Section 25W	-	-	Spclist	-	-	Spclist	-	146.78	m2	157.79
extra for high yield steel	-	-	Spclist	-	-	Spclist	-	7.59	m2	8.16
extra for one coat L.B.V.(Lowca Varnish to BS 1070 Type 2) before driving	-	-	Spclist	-	-	Spclist	-	3.87	m2	4.16
extra for corners	-	-	Spclist	-	-	Spclist	-	43.35	m	46.60
extra for junctions	-	-	Spclist	-	-	Spclist	-	54.18	m	58.25
Larssen Section 32W	-	-	Spclist	-	-	Spclist	-	173.30	m2	186.29
extra for high yield steel	-	-	Spclist	-	-	Spclist	-	8.43	m2	9.06
extra for one coat L.B.V.(Lowca Varnish to BS 1070 Type 2) before driving	-	-	Spclist	-	-	Spclist	-	4.18	m2	4.50

Labour hourly rates: (except Specialists) Craft Operatives £14.21 Labourer £10.71 Rates are national average prices. Refer to REGIONAL VARIATIONS for indicative levels of overall pricing in regions	MATERIALS			LABOUR				RATES		
	Del to Site £	Waste %	Material Cost £	Craft Optve Hrs	Lab Hrs	Labour Cost £	Sunds £	Nett Rate £	Unit	Gross Rate (+7.5%) £
D32: STEEL PILING Cont.										
Mild steel sheet piling to BS EN 10025 Grade S275 or high yield steel Grade S275JR in lengths 4.5-15m supplied, handled, pitched and driven by land based plant in one visit (KELLER GROUND ENGINEERING) Cont.										
Mild steel sheet piling Cont.										
extra for corners	-	-	Spclist	-	-	Spclist	-	46.45	m	49.93
extra for junctions	-	-	Spclist	-	-	Spclist	-	57.28	m	61.58
Frodingham Section 5	-	-	Spclist	-	-	Spclist	-	195.56	m2	210.22
extra for high yield steel	-	-	Spclist	-	-	Spclist	-	11.14	m2	11.98
extra for one coat L.B.V.(Lowca Varnish to BS 1070 Type 2) before driving	-	-	Spclist	-	-	Spclist	-	7.27	m2	7.82
extra for corners	-	-	Spclist	-	-	Spclist	-	46.45	m	49.93
extra for junctions	-	-	Spclist	-	-	Spclist	-	61.92	m	66.56
For quiet piling, add										
lump sum	-	-	Spclist	-	-	Spclist	-	774.79	Sm	832.90
plus for all sections, from	-	-	Spclist	-	-	Spclist	-	3.62	m2	3.89
to	-	-	Spclist	-	-	Spclist	-	6.00	m2	6.46
D40: EMBEDDED RETAINING WALLS										
Embedded retaining walls; contiguous panel construction; panel lengths not exceeding 5m. Note:- the following prices are indicative only: firm quotations should always be obtained (CEMENTATION FOUNDATIONS SKANSKA LIMITED)										
On/Off site charge in addition to the following prices, add for bringing plant to site, erecting and dismantling, maintaining and removing from site, approximately	-	-	Spclist	-	-	Spclist	-	76149.90	Sm	81861.14
Excavation and Bentonite slurry and disposal										
600mm thick wall; maximum depth										
5m	-	-	Spclist	-	-	Spclist	-	228.45	m3	245.58
10m	-	-	Spclist	-	-	Spclist	-	228.45	m3	245.58
15m	-	-	Spclist	-	-	Spclist	-	228.45	m3	245.58
20m	-	-	Spclist	-	-	Spclist	-	217.62	m3	233.94
800mm thick wall; maximum depth										
5m	-	-	Spclist	-	-	Spclist	-	190.37	m3	204.65
10m	-	-	Spclist	-	-	Spclist	-	190.37	m3	204.65
15m	-	-	Spclist	-	-	Spclist	-	190.37	m3	204.65
20m	-	-	Spclist	-	-	Spclist	-	190.37	m3	204.65
1000mm thick wall; maximum depth										
5m	-	-	Spclist	-	-	Spclist	-	152.30	m3	163.72
10m	-	-	Spclist	-	-	Spclist	-	152.30	m3	163.72
15m	-	-	Spclist	-	-	Spclist	-	152.30	m3	163.72
20m	-	-	Spclist	-	-	Spclist	-	152.18	m3	163.59
Excavating through obstructions, rock like formations, etc undertaken on a time basis at a rate per rig per hour	-	-	Spclist	-	-	Spclist	-	444.21	hr	477.52
Reinforced concrete; BS 5328, designed mix C25, 20mm aggregate, minimum cement content 400 kg/m3										
600mm thick wall	-	-	Spclist	-	-	Spclist	-	95.19	m3	102.33
800mm thick wall	-	-	Spclist	-	-	Spclist	-	95.19	m3	102.33
1000mm thick wall	-	-	Spclist	-	-	Spclist	-	95.19	m3	102.33
Reinforced in-situ concrete; sulphate resisting; BS 5328, designed mix C25, 20mm aggregate, minimum cement content 400 kg/m3										
600mm thick wall	-	-	Spclist	-	-	Spclist	-	107.88	m3	115.97
800mm thick wall	-	-	Spclist	-	-	Spclist	-	107.88	m3	115.97
1000mm thick wall	-	-	Spclist	-	-	Spclist	-	107.88	m3	115.97
Reinforcement bars; BS 4449 hot rolled plain round mild steel; including hooks tying wire, and spacers and chairs which are at the discretion of the Contractor										
16mm; straight	-	-	Spclist	-	-	Spclist	-	679.00	t	729.93
20mm; straight	-	-	Spclist	-	-	Spclist	-	679.00	t	729.93
25mm; straight	-	-	Spclist	-	-	Spclist	-	679.00	t	729.93
32mm; straight	-	-	Spclist	-	-	Spclist	-	679.00	t	729.93
40mm; straight	-	-	Spclist	-	-	Spclist	-	679.00	t	729.93
16mm; bent	-	-	Spclist	-	-	Spclist	-	679.00	t	729.93
20mm; bent	-	-	Spclist	-	-	Spclist	-	679.00	t	729.93
25mm; bent	-	-	Spclist	-	-	Spclist	-	679.00	t	729.93
32mm; bent	-	-	Spclist	-	-	Spclist	-	679.00	t	729.93
40mm; bent	-	-	Spclist	-	-	Spclist	-	679.00	t	729.93
Reinforcement bars; BS 4449 hot rolled deformed high yield steel; including hooks tying wire, and spacers and chairs which are at the discretion of the Contractor										
16mm; straight	-	-	Spclist	-	-	Spclist	-	679.00	t	729.93
20mm; straight	-	-	Spclist	-	-	Spclist	-	679.00	t	729.93
25mm; straight	-	-	Spclist	-	-	Spclist	-	679.00	t	729.93
32mm; straight	-	-	Spclist	-	-	Spclist	-	679.00	t	729.93
40mm; straight	-	-	Spclist	-	-	Spclist	-	679.00	t	729.93
16mm; bent	-	-	Spclist	-	-	Spclist	-	679.00	t	729.93
20mm; bent	-	-	Spclist	-	-	Spclist	-	679.00	t	729.93

Labour hourly rates: (except Specialists) Craft Operatives £14.21 Labourer £10.71 Rates are national average prices. Refer to REGIONAL VARIATIONS for indicative levels of overall pricing in regions	MATERIALS			LABOUR				RATES		
	Del to Site £	Waste %	Material Cost £	Craft Optve Hrs	Lab Hrs	Labour Cost £	Sunds £	Nett Rate £	Unit	Gross Rate (+7.5%) £
D40: EMBEDDED RETAINING WALLS Cont.										
Embedded retaining walls; contiguous panel construction; panel lengths not exceeding 5m. Note:- the following prices are indicative only: firm quotations should always be obtained (CEMENTATION FOUNDATIONS SKANSKA LIMITED) Cont.										
Reinforcement bars; BS 4449 hot rolled deformed high yield steel; including hooks tying wire, and spacers and chairs which are at the discretion of the Contractor Cont.										
25mm; bent	-	-	Spclist	-	-	Spclist	-	679.00	t	729.93
32mm; bent	-	-	Spclist	-	-	Spclist	-	679.00	t	729.93
40mm; bent	-	-	Spclist	-	-	Spclist	-	679.00	t	729.93
Guide walls; excavation, disposal and support; reinforced in-situ concrete; BS 5328, designed mix C25, 20mm aggregate, minimum cement content 290 kg/m3; reinforced with one layer fabric BS 4483 reference A252, 3.95 kg/m2 including laps, tying wire, all cutting and bending, and spacers and chairs which are at the discretion of the Contractor; formwork both sides										
both sides; 1000mm high, 600mm apart; propped top and bottom at 2000mm centres	-	-	Spclist	-	-	Spclist	-	234.80	m	252.41
both sides; 1000mm high, 800mm apart; propped top and bottom at 2000mm centres	-	-	Spclist	-	-	Spclist	-	234.80	m	252.41
both sides; 1000mm high, 1000mm apart; propped top and bottom at 2000mm centres	-	-	Spclist	-	-	Spclist	-	234.80	m	252.41
both sides; 1500mm high, 600mm apart; propped top and bottom at 2000mm centres	-	-	Spclist	-	-	Spclist	-	361.71	m	388.84
both sides; 1500mm high, 800mm apart; propped top and bottom at 2000mm centres	-	-	Spclist	-	-	Spclist	-	361.71	m	388.84
both sides; 1500mm high, 1000mm apart; propped top and bottom at 2000mm centres	-	-	Spclist	-	-	Spclist	-	361.71	m	388.84
D41: CRIB WALLS/GABIONS/REINFORCED EARTH										
Retaining Walls										
Betoflor precast concrete landscape retaining walls including soil filling to pockets but excluding excavation, concrete foundations, stone backfill to rear of wall and planting which are all deemed measured separately										
Betoflor interlocking units 500mm long x 250mm wide x 200mm modular deep in wall 250mm wide	42.01	5.00	44.11	-	3.50	37.49	-	81.60	m2	87.72
Extra over for colours	4.02	5.00	4.22	-	0.50	5.36	-	9.58	m2	10.30
Betoatlas interlocking units 250mm long x 500mm wide x 200mm modular deep in wall 500mm wide	58.12	5.00	61.02	-	5.00	53.55	-	114.57	m2	123.17
Extra over for colours	5.25	5.00	5.51	-	0.10	1.07	-	6.58	m2	7.08
Betonap 150/50 woven mesh geotextile as reinforcement to backfill	3.63	5.00	3.81	-	0.20	2.14	-	5.96	m2	6.40
graded stone filling behind Betoflor or Betoatlas.	25.20	5.00	26.46	-	3.00	32.13	-	58.59	m3	62.98
concrete haunching to base of blocks	4.40	5.00	4.62	-	0.36	3.86	-	8.47	m	9.11
Betoflor range; Betojard precast concrete vertical acoustic wall including soil filling to pockets but excluding excavation, concrete foundations, stone backfill to rear of wall and planting which are all deemed measured separately										
Betojard standard 500mm long x 250mm wide x 200mm deep units in wall 250mm wide	52.42	5.00	55.04	-	4.00	42.84	-	97.88	m2	105.23
Extra over for colours	1.74	5.00	1.82	-	0.05	0.54	-	2.36	m2	2.54
Betojard pivot 290mm x 250mm wide x 200mm deep	52.42	5.00	55.04	-	5.00	53.55	-	108.59	m2	116.74
Extra over for colours	1.74	5.00	1.82	-	0.05	0.54	-	2.36	m2	2.54
Extra over fill core to pivot block with reinforced concrete with 4 No 10mm diameter vertical mild steel bars and 6mm links	5.46	5.00	5.84	-	1.50	16.07	-	21.91	m	23.55
Extra over dowelled connection between pivot and standard block with concrete infill to standard block and 1 No 10mm diameter x 200mm long galvanized dowel debonded one end at 300mm centres vertically	7.94	5.00	8.45	-	0.75	8.03	-	16.48	m	17.72
concrete haunching to base of blocks	4.40	5.00	4.62	-	0.36	3.86	-	8.47	m	9.11
D50: UNDERPINNING										
Information										
The work of underpinning in this section comprises work to be carried out in short lengths; prices are exclusive of shoring and other temporary supports										
Excavating										
Preliminary trenches										
maximum depth not exceeding										
1.00m	-	-	-	-	4.00	42.84	-	42.84	m3	46.05
2.00m	-	-	-	-	5.00	53.55	-	53.55	m3	57.57
4.00m	-	-	-	-	6.00	64.26	-	64.26	m3	69.08
Underpinning pits										
maximum depth not exceeding										
1.00m	-	-	-	-	5.20	55.69	-	55.69	m3	59.87

Labour hourly rates: (except Specialists) Craft Operatives £14.21 Labourer £10.71 Rates are national average prices. Refer to REGIONAL VARIATIONS for indicative levels of overall pricing in regions	MATERIALS			LABOUR				RATES		
	Del to Site £	Waste %	Material Cost £	Craft Optve Hrs	Lab Hrs	Labour Cost £	Sunds £	Nett Rate £	Unit	Gross Rate (+7.5%) £
D50: UNDERPINNING Cont.										
Excavating Cont.										
Underpinning pits Cont. maximum depth not exceeding Cont.										
2.00m	-	-	-	-	6.50	69.62	-	69.62	m3	74.84
4.00m	-	-	-	-	7.80	83.54	-	83.54	m3	89.80
Earthwork support										
preliminary trenches; distance between opposing faces not exceeding 2.00m; maximum depth not exceeding										
1.00m	1.28	-	1.28	-	0.44	4.71	0.02	6.01	m2	6.46
2.00m	1.72	-	1.72	-	0.56	6.00	0.03	7.75	m2	8.33
4.00m	1.99	-	1.99	-	0.60	6.43	0.03	8.45	m2	9.08
preliminary trenches; distance between opposing faces 2.00 - 4.00m; maximum depth not exceeding										
1.00m	1.51	-	1.51	-	0.46	4.93	0.02	6.46	m2	6.94
2.00m	1.90	-	1.90	-	0.58	6.21	0.03	8.14	m2	8.75
4.00m	2.07	-	2.07	-	0.70	7.50	0.04	9.61	m2	10.33
underpinning pits; distance between opposing faces not exceeding 2.00m; maximum depth not exceeding										
1.00m	1.60	-	1.60	-	0.51	5.46	0.02	7.09	m2	7.62
2.00m	2.01	-	2.01	-	0.64	6.85	0.03	8.90	m2	9.57
4.00m	2.25	-	2.25	-	0.69	7.39	0.04	9.68	m2	10.41
underpinning pits; distance between opposing faces 2.00 - 4.00m; maximum depth not exceeding										
1.00m	1.71	-	1.71	-	0.53	5.68	0.02	7.41	m2	7.96
2.00m	2.16	-	2.16	-	0.67	7.18	0.03	9.37	m2	10.08
4.00m	2.49	-	2.49	-	0.81	8.68	0.04	11.21	m2	12.05
Cutting away existing projecting foundations										
masonry; maximum width 103mm; maximum depth 150mm	-	-	-	-	0.80	8.57	-	8.57	m	9.21
masonry; maximum width 154mm; maximum depth 225mm	-	-	-	-	0.95	10.17	-	10.17	m	10.94
concrete; maximum width 253mm; maximum depth 190mm	-	-	-	-	1.15	12.32	-	12.32	m	13.24
concrete; maximum width 304mm; maximum depth 300mm	-	-	-	-	1.35	14.46	-	14.46	m	15.54
Preparing the underside of the existing work to receive the pinning up of the new work										
350mm wide	-	-	-	-	0.40	4.28	-	4.28	m	4.61
500mm wide	-	-	-	-	0.50	5.36	-	5.36	m	5.76
1000mm wide	-	-	-	-	1.00	10.71	-	10.71	m	11.51
Compacting										
bottoms of excavations	-	-	-	-	0.40	4.28	-	4.28	m2	4.61
Disposal of excavated material										
removing from site to tip (including tipping charges but excluding landfill tax)	-	-	-	-	3.00	32.13	11.61	43.74	m3	47.02
Excavated material arising from excavations										
Filling to excavations										
average thickness exceeding 0.25m	-	-	-	-	2.00	21.42	-	21.42	m3	23.03
Compacting										
filling	-	-	-	-	0.20	2.14	-	2.14	m2	2.30
Plain in-situ concrete; BS 5328, ordinary prescribed mix ST3, 20mm aggregate										
Foundations; poured on or against earth or unblinded hardcore										
generally	58.80	7.50	63.21	-	4.75	50.87	-	114.09	m3	122.64
Plain in-situ concrete; BS 5328, ordinary prescribed mix ST4, 20mm aggregate										
Foundations; poured on or against earth or unblinded hardcore										
generally	60.03	7.50	64.53	-	4.75	50.87	-	115.40	m3	124.05
Formwork and basic finish										
Sides of foundations; plain vertical										
height exceeding 1.00m	6.96	10.00	7.65	3.50	0.70	57.23	1.79	66.68	m2	71.68
height not exceeding 250mm	2.08	10.00	2.28	1.06	0.22	17.42	0.54	20.24	m	21.76
height 250 - 500mm	3.82	10.00	4.20	1.92	0.38	31.35	0.97	36.53	m	39.27
height 0.50m - 1.00m	7.02	10.00	7.72	3.58	0.72	58.58	1.83	68.13	m	73.24
Common bricks, BS EN 772, Category M, 215 x 102.5 x 65mm, compressive strength not less than 5.2 N/mm2; in cement mortar (1:3)										
Walls; vertical										
215mm thick; English bond	22.02	5.00	23.12	4.70	3.40	103.20	6.53	132.84	m2	142.81
327mm thick; English bond	33.12	5.00	34.77	5.70	4.20	125.98	10.26	171.01	m2	183.83
440mm thick; English bond	44.03	5.00	46.23	6.30	4.75	140.40	13.99	200.61	m2	215.66
Bonding to existing including extra material										
thickness of new work 215mm	2.41	5.00	2.53	1.32	0.88	28.18	0.70	31.41	m	33.76
thickness of new work 327mm	3.70	5.00	3.89	1.90	1.27	40.60	1.05	45.54	m	48.95
thickness of new work 440mm	4.81	5.00	5.05	2.50	1.67	53.41	1.41	59.87	m	64.36

Labour hourly rates: (except Specialists) Craft Operatives £14.21 Labourer £10.71 Rates are national average prices. Refer to REGIONAL VARIATIONS for indicative levels of overall pricing in regions	MATERIALS			LABOUR				RATES		
	Del to Site £	Waste %	Material Cost £	Craft Optve Hrs	Lab Hrs	Labour Cost £	Sunds £	Nett Rate £	Unit	Gross Rate (+7.5%) £
D50: UNDERPINNING Cont.										
Second Hard Stock bricks, BS EN 772, Category M, 215 x 102.5 x 65mm, in cement mortar (1:3)										
Walls; vertical										
215mm thick; English bond	43.44	5.00	**45.61**	4.70	3.40	**103.20**	6.53	**155.33**	m2	**166.98**
327mm thick; English bond	65.34	5.00	**68.60**	5.70	4.20	**125.98**	10.26	**204.84**	m2	**220.20**
440mm thick; English bond	86.87	5.00	**91.21**	6.30	4.75	**140.40**	13.99	**245.59**	m2	**264.01**
Bonding to existing including extra material										
thickness of new work 215mm	4.75	5.00	**4.98**	1.32	0.88	**28.18**	0.70	**33.86**	m	**36.40**
thickness of new work 327mm	7.30	5.00	**7.67**	1.90	1.27	**40.60**	1.05	**49.32**	m	**53.02**
thickness of new work 440mm	9.49	5.00	**9.96**	2.50	1.67	**53.41**	1.41	**64.78**	m	**69.64**
Engineering bricks, BS EN 772, Category F, 215 x 102.5 x 65mm, class A; in cement mortar (1:3)										
Walls; vertical										
215mm thick; English bond	32.73	5.00	**34.36**	5.20	3.75	**114.05**	6.53	**154.94**	m2	**166.56**
327mm thick; English bond	49.23	5.00	**51.69**	6.30	4.60	**138.79**	10.26	**200.73**	m2	**215.79**
440mm thick; English bond	65.45	5.00	**68.72**	6.95	5.25	**154.99**	13.99	**237.70**	m2	**255.52**
Bonding to existing including extra material										
thickness of new work 215mm	3.58	5.00	**3.75**	1.45	0.97	**30.99**	0.70	**35.45**	m	**38.10**
thickness of new work 327mm	5.50	5.00	**5.78**	2.10	1.40	**44.84**	1.05	**51.66**	m	**55.54**
thickness of new work 440mm	7.15	5.00	**7.51**	2.75	1.83	**58.68**	1.41	**67.59**	m	**72.66**
Engineering bricks, BS EN 772, Category F, 215 x 102.5 x 65mm, Class B; in cement mortar (1:3)										
Walls; vertical										
215mm thick; English bond	23.21	5.00	**24.37**	4.33	3.75	**101.69**	6.53	**132.58**	m2	**142.53**
327mm thick; English bond	34.91	5.00	**36.65**	6.30	4.60	**138.79**	10.26	**185.69**	m2	**199.62**
440mm thick; English bond	46.41	5.00	**48.73**	6.95	5.25	**154.99**	13.99	**217.70**	m2	**234.03**
Bonding to existing including extra material										
thickness of new work 215mm	2.54	5.00	**2.66**	1.45	0.97	**30.99**	0.70	**34.35**	m	**36.93**
thickness of new work 327mm	3.90	5.00	**4.10**	2.10	1.40	**44.84**	1.05	**49.98**	m	**53.73**
thickness of new work 440mm	5.07	5.00	**5.32**	2.75	1.83	**58.68**	1.41	**65.41**	m	**70.31**
Sundry items										
Wedging and pinning up to underside of existing construction with two courses slates in cement mortar (1:3)										
215mm walls	7.77	15.00	**8.94**	0.60	0.40	**12.81**	0.18	**21.93**	m	**23.57**
327mm walls	11.65	15.00	**13.40**	0.87	0.58	**18.57**	0.22	**32.19**	m	**34.61**
440mm walls	15.65	15.00	**17.99**	1.14	0.76	**24.34**	0.25	**42.58**	m	**45.77**

IN-SITU CONCRETE/LARGE PRECAST CONCRETE

Labour hourly rates: (except Specialists) Craft Operatives £14.21 Labourer £10.71 Rates are national average prices. Refer to REGIONAL VARIATIONS for indicative levels of overall pricing in regions	MATERIALS			LABOUR				RATES		
	Del to Site £	Waste %	Material Cost £	Craft Optve Hrs	Lab Hrs	Labour Cost £	Sunds £	Nett Rate £	Unit	Gross Rate (+7.5%) £
E10: MIXING/CASTING/CURING IN-SITU CONCRETE - (READY MIXED)										
Plain in-situ concrete; BS 5328, ordinary prescribed mix ST3, 20mm aggregate										
Foundations poured on or against earth or unblinded hardcore										
generally	58.80	7.50	**63.21**	-	2.10	**22.49**	-	**85.71**	m3	**92.13**
Isolated foundations; poured on or against earth or unblinded hardcore										
generally	58.80	7.50	**63.21**	-	2.50	**26.78**	-	**89.99**	m3	**96.74**
Beds; poured on or against earth or unblinded hardcore										
thickness exceeding 450mm	58.80	7.50	**63.21**	-	2.00	**21.42**	-	**84.63**	m3	**90.98**
thickness 150 - 450mm	58.80	7.50	**63.21**	-	2.25	**24.10**	-	**87.31**	m3	**93.86**
thickness not exceeding 150mm	58.80	7.50	**63.21**	-	2.85	**30.52**	-	**93.74**	m3	**100.77**
Filling hollow walls										
thickness not exceeding 150mm	58.80	5.00	**61.74**	-	5.00	**53.55**	-	**115.29**	m3	**123.94**
Plain in-situ concrete; BS 5328, ordinary prescribed mix ST4, 20mm aggregate										
Foundations poured on or against earth or unblinded hardcore										
generally	60.03	7.50	**64.53**	-	2.10	**22.49**	-	**87.02**	m3	**93.54**
Isolated foundations; poured on or against earth or unblinded hardcore										
generally	60.03	7.50	**64.53**	-	2.50	**26.78**	-	**91.30**	m3	**98.15**
Beds; poured on or against earth or unblinded hardcore										
thickness exceeding 450mm	60.03	7.50	**64.53**	-	2.00	**21.42**	-	**85.95**	m3	**92.39**
thickness 150 - 450mm	60.03	7.50	**64.53**	-	2.25	**24.10**	-	**88.62**	m3	**95.27**
thickness not exceeding 150mm	60.03	7.50	**64.53**	-	2.85	**30.52**	-	**95.05**	m3	**102.18**
Filling hollow walls										
thickness not exceeding 150mm	60.03	5.00	**63.03**	-	5.00	**53.55**	-	**116.58**	m3	**125.32**
Plain in-situ concrete; BS 5328, ordinary prescribed mix ST5, 20mm aggregate										
Foundations poured on or against earth or unblinded hardcore										
generally	62.08	7.50	**66.73**	-	2.10	**22.49**	-	**89.22**	m3	**95.91**
Isolated foundations; poured on or against earth or unblinded hardcore										
generally	62.08	7.50	**66.73**	-	2.50	**26.78**	-	**93.51**	m3	**100.52**
Beds; poured on or against earth or unblinded hardcore										
thickness exceeding 450mm	62.08	7.50	**66.73**	-	2.00	**21.42**	-	**88.15**	m3	**94.76**
thickness 150 - 450mm	62.08	7.50	**66.73**	-	2.25	**24.10**	-	**90.83**	m3	**97.64**
thickness not exceeding 150mm	62.08	7.50	**66.73**	-	2.85	**30.52**	-	**97.26**	m3	**104.55**
Filling hollow walls										
thickness not exceeding 150mm	62.08	5.00	**65.18**	-	5.00	**53.55**	-	**118.73**	m3	**127.63**
Reinforced in-situ concrete; BS 5328, designed mix C15, 20mm aggregate, minimum cement content 220 kg/m3; vibrated										
Foundations										
generally	60.24	5.00	**63.25**	-	2.35	**25.17**	0.61	**89.03**	m3	**95.71**
Ground beams										
generally	60.24	5.00	**63.25**	-	3.25	**34.81**	0.61	**98.67**	m3	**106.07**

Labour hourly rates: (except Specialists) Craft Operatives £14.21 Labourer £10.71 Rates are national average prices. Refer to REGIONAL VARIATIONS for indicative levels of overall pricing in regions	MATERIALS			LABOUR				RATES		
	Del to Site £	Waste %	Material Cost £	Craft Optve Hrs	Lab Hrs	Labour Cost £	Sunds £	Nett Rate £	Unit	Gross Rate (+7.5%) £
E10: MIXING/CASTING/CURING IN-SITU CONCRETE - (READY MIXED) Cont.										
Reinforced in-situ concrete; BS 5328, designed mix C15, 20mm aggregate, minimum cement content 220 kg/m3; vibrated Cont.										
Isolated foundations										
generally	60.24	5.00	**63.25**	-	2.75	**29.45**	0.61	**93.31**	m3	**100.31**
Beds										
thickness exceeding 450mm	60.24	5.00	**63.25**	-	2.25	**24.10**	0.61	**87.96**	m3	**94.55**
thickness 150 - 450mm	60.24	5.00	**63.25**	-	2.50	**26.78**	0.61	**90.63**	m3	**97.43**
thickness not exceeding 150mm	60.24	5.00	**63.25**	-	3.10	**33.20**	0.61	**97.06**	m3	**104.34**
Slabs										
thickness exceeding 450mm	60.24	2.50	**61.74**	-	2.90	**31.06**	0.61	**93.41**	m3	**100.42**
thickness 150 - 450mm	60.24	2.50	**61.74**	-	3.15	**33.74**	0.61	**96.09**	m3	**103.30**
thickness not exceeding 150mm	60.24	2.50	**61.74**	-	4.15	**44.45**	0.61	**106.80**	m3	**114.81**
Walls										
thickness exceeding 450mm	60.24	2.50	**61.74**	-	3.20	**34.27**	0.61	**96.63**	m3	**103.87**
thickness 150 - 450mm	60.24	2.50	**61.74**	-	3.50	**37.49**	0.61	**99.84**	m3	**107.33**
thickness not exceeding 150mm	60.24	2.50	**61.74**	-	4.50	**48.20**	0.61	**110.55**	m3	**118.84**
Beams										
isolated	60.24	2.50	**61.74**	-	4.25	**45.52**	0.61	**107.87**	m3	**115.96**
isolated deep	60.24	2.50	**61.74**	-	4.40	**47.12**	0.61	**109.48**	m3	**117.69**
attached deep	60.24	2.50	**61.74**	-	4.25	**45.52**	0.61	**107.87**	m3	**115.96**
Beam casings										
isolated	60.24	2.50	**61.74**	-	4.50	**48.20**	0.61	**110.55**	m3	**118.84**
isolated deep	60.24	2.50	**61.74**	-	4.65	**49.80**	0.61	**112.16**	m3	**120.57**
attached deep	60.24	2.50	**61.74**	-	4.50	**48.20**	0.61	**110.55**	m3	**118.84**
Columns										
generally	60.24	2.50	**61.74**	-	5.75	**61.58**	0.61	**123.94**	m3	**133.23**
Column casings										
generally	60.24	2.50	**61.74**	-	6.00	**64.26**	0.61	**126.61**	m3	**136.11**
Staircases										
generally	60.24	2.50	**61.74**	-	5.00	**53.55**	0.61	**115.90**	m3	**124.60**
Upstands										
generally	60.24	2.50	**61.74**	-	7.00	**74.97**	0.61	**137.32**	m3	**147.62**
Reinforced in-situ concrete; BS 5328, designed mix C20, 20mm aggregate, minimum cement content 240 kg/m3; vibrated										
Foundations										
generally	61.25	5.00	**64.31**	-	2.35	**25.17**	0.61	**90.09**	m3	**96.85**
Ground beams										
generally	61.25	5.00	**64.31**	-	3.25	**34.81**	0.61	**99.73**	m3	**107.21**
Isolated foundations										
generally	61.25	5.00	**64.31**	-	2.75	**29.45**	0.61	**94.38**	m3	**101.45**
Beds										
thickness exceeding 450mm	61.25	5.00	**64.31**	-	2.25	**24.10**	0.61	**89.02**	m3	**95.70**
thickness 150 - 450mm	61.25	5.00	**64.31**	-	2.50	**26.78**	0.61	**91.70**	m3	**98.58**
thickness not exceeding 150mm	61.25	5.00	**64.31**	-	3.10	**33.20**	0.61	**98.12**	m3	**105.48**
Slabs										
thickness exceeding 450mm	61.25	2.50	**62.78**	-	2.90	**31.06**	0.61	**94.45**	m3	**101.54**
thickness 150 - 450mm	61.25	2.50	**62.78**	-	3.15	**33.74**	0.61	**97.13**	m3	**104.41**
thickness not exceeding 150mm	61.25	2.50	**62.78**	-	4.15	**44.45**	0.61	**107.84**	m3	**115.93**
Walls										
thickness exceeding 450mm	61.25	2.50	**62.78**	-	3.20	**34.27**	0.61	**97.66**	m3	**104.99**
thickness 150 - 450mm	61.25	2.50	**62.78**	-	3.50	**37.49**	0.61	**100.88**	m3	**108.44**
thickness not exceeding 150mm	61.25	2.50	**62.78**	-	4.50	**48.20**	0.61	**111.59**	m3	**119.96**
Beams										
isolated	61.25	2.50	**62.78**	-	4.25	**45.52**	0.61	**108.91**	m3	**117.08**
isolated deep	61.25	2.50	**62.78**	-	4.40	**47.12**	0.61	**110.52**	m3	**118.80**
attached deep	61.25	2.50	**62.78**	-	4.25	**45.52**	0.61	**108.91**	m3	**117.08**
Beam casings										
isolated	61.25	2.50	**62.78**	-	4.50	**48.20**	0.61	**111.59**	m3	**119.96**
isolated deep	61.25	2.50	**62.78**	-	4.65	**49.80**	0.61	**113.19**	m3	**121.68**
attached deep	61.25	2.50	**62.78**	-	4.50	**48.20**	0.61	**111.59**	m3	**119.96**
Columns										
generally	61.25	2.50	**62.78**	-	5.75	**61.58**	0.61	**124.97**	m3	**134.35**
Column casings										
generally	61.25	2.50	**62.78**	-	6.00	**64.26**	0.61	**127.65**	m3	**137.23**
Staircases										
generally	61.25	2.50	**62.78**	-	5.00	**53.55**	0.61	**116.94**	m3	**125.71**
Upstands										
generally	61.25	2.50	**62.78**	-	7.00	**74.97**	0.61	**138.36**	m3	**148.74**

Labour hourly rates: (except Specialists) Craft Operatives £14.21 Labourer £10.71 Rates are national average prices. Refer to REGIONAL VARIATIONS for indicative levels of overall pricing in regions	MATERIALS			LABOUR				RATES		
	Del to Site £	Waste %	Material Cost £	Craft Optve Hrs	Lab Hrs	Labour Cost £	Sunds £	Nett Rate £	Unit	Gross Rate (+7.5%) £
E10: MIXING/CASTING/CURING IN-SITU CONCRETE - (READY MIXED) Cont.										
Reinforced in-situ concrete; BS 5328, designed mix C25, 20mm aggregate, minimum cement content 290 kg/m3; vibrated										
Foundations										
generally	63.45	5.00	**66.62**	-	2.35	**25.17**	0.61	**92.40**	m3	**99.33**
Ground beams										
generally	63.45	5.00	**66.62**	-	3.25	**34.81**	0.61	**102.04**	m3	**109.70**
Isolated foundations										
generally	63.45	5.00	**66.62**	-	2.75	**29.45**	0.61	**96.69**	m3	**103.94**
Beds										
thickness exceeding 450mm	63.45	5.00	**66.62**	-	2.25	**24.10**	0.61	**91.33**	m3	**98.18**
thickness 150 - 450mm	63.45	5.00	**66.62**	-	2.50	**26.78**	0.61	**94.01**	m3	**101.06**
thickness not exceeding 150mm	63.45	5.00	**66.62**	-	3.10	**33.20**	0.61	**100.44**	m3	**107.97**
Slabs										
thickness exceeding 450mm	63.45	2.50	**65.04**	-	2.90	**31.06**	0.61	**96.71**	m3	**103.96**
thickness 150 - 450mm	63.45	2.50	**65.04**	-	3.15	**33.74**	0.61	**99.38**	m3	**106.84**
thickness not exceeding 150mm	63.45	2.50	**65.04**	-	4.15	**44.45**	0.61	**110.09**	m3	**118.35**
Walls										
thickness exceeding 450mm	63.45	2.50	**65.04**	-	3.20	**34.27**	0.61	**99.92**	m3	**107.41**
thickness 150 - 450mm	63.45	2.50	**65.04**	-	3.50	**37.49**	0.61	**103.13**	m3	**110.87**
thickness not exceeding 150mm	63.45	2.50	**65.04**	-	4.50	**48.20**	0.61	**113.84**	m3	**122.38**
Beams										
isolated	63.45	2.50	**65.04**	-	4.25	**45.52**	0.61	**111.17**	m3	**119.50**
isolated deep	63.45	2.50	**65.04**	-	4.40	**47.12**	0.61	**112.77**	m3	**121.23**
attached deep	63.45	2.50	**65.04**	-	4.25	**45.52**	0.61	**111.17**	m3	**119.50**
Beam casings										
isolated	63.45	2.50	**65.04**	-	4.50	**48.20**	0.61	**113.84**	m3	**122.38**
isolated deep	63.45	2.50	**65.04**	-	4.65	**49.80**	0.61	**115.45**	m3	**124.11**
attached deep	63.45	2.50	**65.04**	-	4.50	**48.20**	0.61	**113.84**	m3	**122.38**
Columns										
generally	63.45	2.50	**65.04**	-	5.75	**61.58**	0.61	**127.23**	m3	**136.77**
Column casings										
generally	63.45	2.50	**65.04**	-	6.00	**64.26**	0.61	**129.91**	m3	**139.65**
Staircases										
generally	63.45	2.50	**65.04**	-	5.00	**53.55**	0.61	**119.20**	m3	**128.14**
Upstands										
generally	63.45	2.50	**65.04**	-	7.00	**74.97**	0.61	**140.62**	m3	**151.16**
Reinforced in-situ concrete; BS 5328, designed mix C30, 20mm aggregate, minimum cement content 290 kg/m3; vibrated										
Foundations										
generally	64.91	5.00	**68.15**	-	2.35	**25.17**	0.61	**93.93**	m3	**100.98**
Ground beams										
generally	64.91	5.00	**68.15**	-	3.25	**34.81**	0.61	**103.57**	m3	**111.34**
Isolated foundations										
generally	64.91	5.00	**68.15**	-	2.75	**29.45**	0.61	**98.21**	m3	**105.58**
Beds										
thickness exceeding 450mm	64.91	5.00	**68.15**	-	2.25	**24.10**	0.61	**92.86**	m3	**99.82**
thickness 150 - 450mm	64.91	5.00	**68.15**	-	2.50	**26.78**	0.61	**95.54**	m3	**102.70**
thickness not exceeding 150mm	64.91	5.00	**68.15**	-	3.10	**33.20**	0.61	**101.96**	m3	**109.61**
Slabs										
thickness exceeding 450mm	64.91	2.50	**66.53**	-	2.90	**31.06**	0.61	**98.20**	m3	**105.56**
thickness 150 - 450mm	64.91	2.50	**66.53**	-	3.15	**33.74**	0.61	**100.88**	m3	**108.44**
thickness not exceeding 150mm	64.91	2.50	**66.53**	-	4.15	**44.45**	0.61	**111.59**	m3	**119.96**
Walls										
thickness exceeding 450mm	64.91	2.50	**66.53**	-	3.20	**34.27**	0.61	**101.41**	m3	**109.02**
thickness 150 - 450mm	64.91	2.50	**66.53**	-	3.50	**37.49**	0.61	**104.62**	m3	**112.47**
thickness not exceeding 150mm	64.91	2.50	**66.53**	-	4.50	**48.20**	0.61	**115.33**	m3	**123.98**
Beams										
isolated	64.91	2.50	**66.53**	-	4.25	**45.52**	0.61	**112.66**	m3	**121.11**
isolated deep	64.91	2.50	**66.53**	-	4.40	**47.12**	0.61	**114.26**	m3	**122.83**
attached deep	64.91	2.50	**66.53**	-	4.25	**45.52**	0.61	**112.66**	m3	**121.11**
Beam casings										
isolated	64.91	2.50	**66.53**	-	4.50	**48.20**	0.61	**115.33**	m3	**123.98**
isolated deep	64.91	2.50	**66.53**	-	4.65	**49.80**	0.61	**116.94**	m3	**125.71**
attached deep	64.91	2.50	**66.53**	-	4.50	**48.20**	0.61	**115.33**	m3	**123.98**
Columns										
generally	64.91	2.50	**66.53**	-	5.75	**61.58**	0.61	**128.72**	m3	**138.38**
Column casings										
generally	64.91	2.50	**66.53**	-	6.00	**64.26**	0.61	**131.40**	m3	**141.25**

Labour hourly rates: (except Specialists) Craft Operatives £14.21 Labourer £10.71 Rates are national average prices. Refer to REGIONAL VARIATIONS for indicative levels of overall pricing in regions	MATERIALS			LABOUR				RATES		
	Del to Site	Waste	Material Cost	Craft Optve	Lab	Labour Cost	Sunds	Nett Rate	Unit	Gross Rate (+7.5%)
	£	%	£	Hrs	Hrs	£	£	£		£
E10: MIXING/CASTING/CURING IN-SITU CONCRETE - (READY MIXED) Cont.										
Reinforced in-situ concrete; BS 5328, designed mix C30, 20mm aggregate, minimum cement content 290 kg/m3; vibrated Cont.										
Upstands										
generally	64.91	2.50	**66.53**	-	7.00	**74.97**	0.61	**142.11**	m3	**152.77**
E10: MIXING/CASTING/CURING IN-SITU CONCRETE - (SITE MIXED)										
Plain in-situ concrete; mix 1:6, all in aggregate										
Foundations poured on or against earth or unblinded hardcore										
generally	77.37	7.50	**83.17**	-	3.10	**33.20**	1.83	**118.20**	m3	**127.07**
Isolated foundations; poured on or against earth or unblinded hardcore										
generally	77.37	7.50	**83.17**	-	3.50	**37.49**	1.83	**122.49**	m3	**131.67**
Beds; poured on or against earth or unblinded hardcore										
thickness 150 - 450mm	77.37	7.50	**83.17**	-	3.25	**34.81**	1.83	**119.81**	m3	**128.79**
thickness not exceeding 150mm	77.37	7.50	**83.17**	-	3.85	**41.23**	1.83	**126.23**	m3	**135.70**
Beds; sloping not exceeding 15 degrees; poured on or against earth or unblinded hardcore										
thickness 150 - 450mm	77.37	7.50	**83.17**	-	3.45	**36.95**	1.83	**121.95**	m3	**131.10**
thickness not exceeding 150mm	77.37	7.50	**83.17**	-	4.05	**43.38**	1.83	**128.38**	m3	**138.01**
Beds; sloping exceeding 15 degrees; poured on or against earth or unblinded hardcore										
thickness 150 - 450mm	77.37	7.50	**83.17**	-	3.65	**39.09**	1.83	**124.09**	m3	**133.40**
thickness not exceeding 150mm	77.37	7.50	**83.17**	-	4.25	**45.52**	1.83	**130.52**	m3	**140.31**
Plain in-situ concrete; mix 1:8, all in aggregate										
Foundations poured on or against earth or unblinded hardcore										
generally	69.50	7.50	**74.71**	-	3.10	**33.20**	1.83	**109.75**	m3	**117.98**
Isolated foundations; poured on or against earth or unblinded hardcore										
generally	69.50	7.50	**74.71**	-	3.50	**37.49**	1.83	**114.03**	m3	**122.59**
Beds; poured on or against earth or unblinded hardcore										
thickness 150 - 450mm	69.50	7.50	**74.71**	-	3.25	**34.81**	1.83	**111.36**	m3	**119.71**
thickness not exceeding 150mm	69.50	7.50	**74.71**	-	3.85	**41.23**	1.83	**117.78**	m3	**126.61**
Beds; sloping not exceeding 15 degrees; poured on or against earth or unblinded hardcore										
thickness 150 - 450mm	69.50	7.50	**74.71**	-	3.45	**36.95**	1.83	**113.50**	m3	**122.01**
thickness not exceeding 150mm	69.50	7.50	**74.71**	-	4.05	**43.38**	1.83	**119.92**	m3	**128.92**
Beds; sloping exceeding 15 degrees; poured on or against earth or unblinded hardcore										
thickness 150 - 450mm	69.50	7.50	**74.71**	-	3.65	**39.09**	1.83	**115.64**	m3	**124.31**
thickness not exceeding 150mm	69.50	7.50	**74.71**	-	4.25	**45.52**	1.83	**122.07**	m3	**131.22**
Plain in-situ concrete; mix 1:12, all in aggregate										
Foundations poured on or against earth or unblinded hardcore										
generally	61.61	7.50	**66.23**	-	3.10	**33.20**	1.83	**101.27**	m3	**108.86**
Isolated foundations; poured on or against earth or unblinded hardcore										
generally	61.61	7.50	**66.23**	-	3.50	**37.49**	1.83	**105.55**	m3	**113.47**
Beds; poured on or against earth or unblinded hardcore										
thickness 150 - 450mm	61.61	7.50	**66.23**	-	3.25	**34.81**	1.83	**102.87**	m3	**110.59**
thickness not exceeding 150mm	61.61	7.50	**66.23**	-	3.85	**41.23**	1.83	**109.30**	m3	**117.50**
Beds; sloping not exceeding 15 degrees; poured on or against earth or unblinded hardcore										
thickness 150 - 450mm	61.61	7.50	**66.23**	-	3.45	**36.95**	1.83	**105.01**	m3	**112.89**
thickness not exceeding 150mm	61.61	7.50	**66.23**	-	4.05	**43.38**	1.83	**111.44**	m3	**119.80**
Beds; sloping exceeding 15 degrees; poured on or against earth or unblinded hardcore										
thickness 150 - 450mm	61.61	7.50	**66.23**	-	3.65	**39.09**	1.83	**107.16**	m3	**115.19**
thickness not exceeding 150mm	61.61	7.50	**66.23**	-	4.25	**45.52**	1.83	**113.58**	m3	**122.10**
Plain in-situ concrete; BS 5328, ordinary prescribed mix C15P, 20mm aggregate										
Foundations poured on or against earth or unblinded hardcore										
generally	78.78	7.50	**84.69**	-	3.10	**33.20**	1.83	**119.73**	m3	**128.71**
Isolated foundations; poured on or against earth or unblinded hardcore										
generally	78.78	7.50	**84.69**	-	3.50	**37.49**	1.83	**124.01**	m3	**133.31**

Labour hourly rates: (except Specialists) Craft Operatives £14.21 Labourer £10.71 Rates are national average prices. Refer to REGIONAL VARIATIONS for indicative levels of overall pricing in regions	MATERIALS			LABOUR				RATES		
	Del to Site £	Waste %	Material Cost £	Craft Optve Hrs	Lab Hrs	Labour Cost £	Sunds £	Nett Rate £	Unit	Gross Rate (+7.5%) £
E10: MIXING/CASTING/CURING IN-SITU CONCRETE - (SITE MIXED) Cont.										
Plain in-situ concrete; BS 5328, ordinary prescribed mix C15P, 20mm aggregate Cont.										
Beds; poured on or against earth or unblinded hardcore										
thickness exceeding 450mm	78.78	7.50	**84.69**	-	3.00	**32.13**	1.83	**118.65**	m3	**127.55**
thickness 150 - 450mm	78.78	7.50	**84.69**	-	3.25	**34.81**	1.83	**121.33**	m3	**130.43**
thickness not exceeding 150mm	78.78	7.50	**84.69**	-	3.85	**41.23**	1.83	**127.76**	m3	**137.34**
Filling hollow walls										
thickness not exceeding 150mm	78.78	7.50	**84.69**	-	6.00	**64.26**	1.83	**150.78**	m3	**162.09**
Plain in-situ concrete; BS 5328, ordinary prescribed mix C20P, 20mm aggregate										
Foundations poured on or against earth or unblinded hardcore										
generally	81.09	7.50	**87.17**	-	3.10	**33.20**	1.83	**122.21**	m3	**131.37**
Isolated foundations; poured on or against earth or unblinded hardcore										
generally	81.09	7.50	**87.17**	-	3.50	**37.49**	1.83	**126.49**	m3	**135.98**
Beds; poured on or against earth or unblinded hardcore										
thickness exceeding 450mm	81.09	7.50	**87.17**	-	3.00	**32.13**	1.83	**121.14**	m3	**130.22**
thickness 150 - 450mm	81.09	7.50	**87.17**	-	3.25	**34.81**	1.83	**123.81**	m3	**133.10**
thickness not exceeding 150mm	81.09	7.50	**87.17**	-	3.85	**41.23**	1.83	**130.24**	m3	**140.01**
Filling hollow walls										
thickness not exceeding 150mm	81.09	7.50	**87.17**	-	6.00	**64.26**	1.83	**153.27**	m3	**164.76**
Plain in-situ concrete; BS 5328, ordinary prescribed mix C25P, 20mm aggregate										
Foundations poured on or against earth or unblinded hardcore										
generally	84.46	7.50	**90.80**	-	3.10	**33.20**	1.83	**125.83**	m3	**135.27**
Isolated foundations; poured on or against earth or unblinded hardcore										
generally	84.46	7.50	**90.80**	-	3.50	**37.49**	1.83	**130.11**	m3	**139.87**
Beds; poured on or against earth or unblinded hardcore										
thickness exceeding 450mm	84.46	7.50	**90.80**	-	3.00	**32.13**	1.83	**124.76**	m3	**134.12**
thickness 150 - 450mm	84.46	7.50	**90.80**	-	3.25	**34.81**	1.83	**127.44**	m3	**136.99**
thickness not exceeding 150mm	84.46	7.50	**90.80**	-	3.85	**41.23**	1.83	**133.86**	m3	**143.90**
Filling hollow walls										
thickness not exceeding 150mm	84.46	7.50	**90.80**	-	6.00	**64.26**	1.83	**156.89**	m3	**168.66**
Reinforced in-situ concrete; BS 5328, designed mix C15, 20mm aggregate, minimum cement content 220 kg/m3; vibrated										
Foundations										
generally	73.64	5.00	**77.32**	-	3.35	**35.88**	2.44	**115.65**	m3	**124.32**
Ground beams										
generally	73.64	5.00	**77.32**	-	4.25	**45.52**	2.44	**125.29**	m3	**134.68**
Isolated foundations										
generally	73.64	5.00	**77.32**	-	3.75	**40.16**	2.44	**119.93**	m3	**128.93**
Beds										
thickness exceeding 450mm	73.64	5.00	**77.32**	-	3.25	**34.81**	2.44	**114.58**	m3	**123.17**
thickness 150 - 450mm	73.64	5.00	**77.32**	-	3.50	**37.49**	2.44	**117.25**	m3	**126.05**
thickness not exceeding 150mm	73.64	5.00	**77.32**	-	4.10	**43.91**	2.44	**123.68**	m3	**132.96**
Reinforced in-situ concrete; BS 5328, designed mix C20, 20mm aggregate, minimum cement content 240 kg/m3; vibrated										
Foundations										
generally	74.94	5.00	**78.68**	-	3.35	**35.88**	2.44	**117.01**	m3	**125.78**
Ground beams										
generally	74.94	5.00	**78.68**	-	4.25	**45.52**	2.44	**126.64**	m3	**136.14**
Isolated foundations										
generally	74.94	5.00	**78.68**	-	3.75	**40.16**	2.44	**121.29**	m3	**130.39**
Beds										
thickness exceeding 450mm	74.94	5.00	**78.68**	-	3.25	**34.81**	2.44	**115.93**	m3	**124.63**
thickness 150 - 450mm	74.94	5.00	**78.68**	-	3.50	**37.49**	2.44	**118.61**	m3	**127.51**
thickness not exceeding 150mm	74.94	5.00	**78.68**	-	4.10	**43.91**	2.44	**125.04**	m3	**134.42**
Reinforced in-situ concrete; BS 5328, designed mix C25, 20mm aggregate, minimum cement content 290 kg/m3; vibrated										
Foundations										
generally	79.32	5.00	**83.29**	-	3.35	**35.88**	2.44	**121.61**	m3	**130.73**

Labour hourly rates: (except Specialists) Craft Operatives £14.21 Labourer £10.71 Rates are national average prices. Refer to REGIONAL VARIATIONS for indicative levels of overall pricing in regions	MATERIALS			LABOUR				RATES		
	Del to Site £	Waste %	Material Cost £	Craft Optve Hrs	Lab Hrs	Labour Cost £	Sunds £	Nett Rate £	Unit	Gross Rate (+7.5%) £
E10: MIXING/CASTING/CURING IN-SITU CONCRETE - (SITE MIXED) Cont.										
Reinforced in-situ concrete; BS 5328, designed mix C25, 20mm aggregate, minimum cement content 290 kg/m3; vibrated Cont.										
Ground beams										
generally	79.32	5.00	**83.29**	-	4.25	**45.52**	2.44	**131.25**	m3	**141.09**
Isolated foundations										
generally	79.32	5.00	**83.29**	-	3.75	**40.16**	2.44	**125.89**	m3	**135.34**
Beds										
thickness exceeding 450mm	79.32	5.00	**83.29**	-	3.25	**34.81**	2.44	**120.54**	m3	**129.58**
thickness 150 - 450mm	79.32	5.00	**83.29**	-	3.50	**37.49**	2.44	**123.22**	m3	**132.46**
thickness not exceeding 150mm	79.32	5.00	**83.29**	-	4.10	**43.91**	2.44	**129.64**	m3	**139.37**
Cement and sand (1:3); for grouting and the like										
Grouting.										
stanchion bases.	2.19	10.00	**2.41**	-	0.40	**4.28**	1.10	**7.79**	nr	**8.38**
E10: MIXING/CASTING/CURING IN-SITU CONCRETE - (SUNDRIES)										
The foregoing concrete is based on the use of Portland cement. For other cements and waterproofers ADD as follows										
Concrete 1:6										
rapid hardening cement	3.96	-	**3.96**	-	-	**-**	-	**3.96**	m3	**4.26**
sulphate resisting cement	4.97	-	**4.97**	-	-	**-**	-	**4.97**	m3	**5.34**
waterproofing powder	11.45	-	**11.45**	-	-	**-**	-	**11.45**	m3	**12.31**
waterproofing liquid	5.84	-	**5.84**	-	-	**-**	-	**5.84**	m3	**6.27**
Concrete 1:8										
rapid hardening cement	2.98	-	**2.98**	-	-	**-**	-	**2.98**	m3	**3.20**
sulphate resisting cement	3.73	-	**3.73**	-	-	**-**	-	**3.73**	m3	**4.01**
waterproofing powder	8.60	-	**8.60**	-	-	**-**	-	**8.60**	m3	**9.24**
waterproofing liquid	5.84	-	**5.84**	-	-	**-**	-	**5.84**	m3	**6.27**
Concrete 1:12										
rapid hardening cement	1.99	-	**1.99**	-	-	**-**	-	**1.99**	m3	**2.14**
sulphate resisting cement	2.48	-	**2.48**	-	-	**-**	-	**2.48**	m3	**2.66**
waterproofing powder	5.72	-	**5.72**	-	-	**-**	-	**5.72**	m3	**6.15**
waterproofing liquid	5.84	-	**5.84**	-	-	**-**	-	**5.84**	m3	**6.27**
Concrete ST3 and C15										
rapid hardening cement	2.65	-	**2.65**	-	-	**-**	-	**2.65**	m3	**2.85**
sulphate resisting cement	3.30	-	**3.30**	-	-	**-**	-	**3.30**	m3	**3.55**
waterproofing powder	7.64	-	**7.64**	-	-	**-**	-	**7.64**	m3	**8.21**
waterproofing liquid	5.84	-	**5.84**	-	-	**-**	-	**5.84**	m3	**6.27**
Concrete ST4 and C20										
rapid hardening cement	3.92	-	**3.92**	-	-	**-**	-	**3.92**	m3	**4.21**
sulphate resisting cement	4.91	-	**4.91**	-	-	**-**	-	**4.91**	m3	**5.28**
waterproofing powder	11.29	-	**11.29**	-	-	**-**	-	**11.29**	m3	**12.14**
waterproofing liquid	5.84	-	**5.84**	-	-	**-**	-	**5.84**	m3	**6.27**
Concrete ST5 and C25										
rapid hardening cement	5.29	-	**5.29**	-	-	**-**	-	**5.29**	m3	**5.68**
sulphate resisting cement	6.61	-	**6.61**	-	-	**-**	-	**6.61**	m3	**7.11**
waterproofing powder	15.26	-	**15.26**	-	-	**-**	-	**15.26**	m3	**16.40**
waterproofing liquid	5.84	-	**5.84**	-	-	**-**	-	**5.84**	m3	**6.27**
Reinforced in-situ lightweight concrete; 20.5 N/mm2; vibrated										
Walls										
thickness exceeding 450mm	85.44	2.50	**87.57**	-	4.40	**47.12**	2.44	**137.14**	m3	**147.43**
thickness 150 - 450mm	85.44	2.50	**87.57**	-	4.65	**49.80**	2.44	**139.82**	m3	**150.31**
thickness not exceeding 150mm	85.44	2.50	**87.57**	-	5.10	**54.62**	2.44	**144.64**	m3	**155.49**
Beam casings										
isolated	85.44	2.50	**87.57**	-	5.50	**58.91**	2.44	**148.92**	m3	**160.09**
Reinforced in-situ lightweight concrete; 26.0 N/mm2; vibrated										
Slabs										
thickness exceeding 450mm	89.55	2.50	**91.79**	-	4.00	**42.84**	2.44	**137.07**	m3	**147.35**
thickness 150 - 450mm	89.55	2.50	**91.79**	-	4.10	**43.91**	2.44	**138.14**	m3	**148.50**
thickness not exceeding 150mm	89.55	2.50	**91.79**	-	4.65	**49.80**	2.44	**144.03**	m3	**154.83**
Reinforced in-situ lightweight concrete; 41.5 N/mm2; vibrated										
Beams										
isolated	97.77	2.50	**100.21**	-	5.00	**53.55**	2.44	**156.21**	m3	**167.92**
Polythene sheeting as temporary protection to surface of concrete (use and waste)										
Polythene sheeting and laying										
125mu	0.35	10.00	**0.38**	-	0.03	**0.32**	-	**0.70**	m2	**0.76**

Labour hourly rates: (except Specialists) Craft Operatives £14.21 Labourer £10.71 Rates are national average prices. Refer to REGIONAL VARIATIONS for indicative levels of overall pricing in regions	MATERIALS			LABOUR				RATES		
	Del to Site £	Waste %	Material Cost £	Craft Optve Hrs	Lab Hrs	Labour Cost £	Sunds £	Nett Rate £	Unit	Gross Rate (+7.5%) £
E10: MIXING/CASTING/CURING IN-SITU CONCRETE - (SUNDRIES) Cont.										
Waterproof building paper, BS 1521; laying on hardcore to receive concrete										
Waterproof building paper and laying										
Grade B1F	0.34	25.00	**0.43**	-	0.05	**0.54**	-	**0.96**	m2	**1.04**
Grade B2	1.21	25.00	**1.51**	-	0.05	**0.54**	-	**2.05**	m2	**2.20**
Polythene sheeting, laying on hardcore to receive concrete										
Polythene sheeting and laying										
300mu	0.50	20.00	**0.60**	-	0.05	**0.54**	-	**1.14**	m2	**1.22**
E20: FORMWORK FOR IN-SITU CONCRETE										
Formwork and basic finish										
Sides of foundations; plain vertical										
height exceeding 1.00m	3.32	12.50	**3.74**	1.75	0.35	**28.62**	0.83	**33.18**	m2	**35.67**
height not exceeding 250mm	1.03	12.50	**1.15**	0.53	0.11	**8.71**	0.26	**10.12**	m	**10.88**
height 250 - 500mm	1.81	12.50	**2.04**	0.96	0.19	**15.68**	0.48	**18.19**	m	**19.56**
height 0.50m - 1.00m	3.41	12.50	**3.84**	1.79	0.36	**29.29**	0.83	**33.96**	m	**36.51**
Sides of foundations; plain vertical; left in										
height exceeding 1.00m	14.95	-	**14.95**	1.00	0.20	**16.35**	0.50	**31.80**	m2	**34.19**
height not exceeding 250mm	4.58	-	**4.58**	0.30	0.06	**4.91**	0.18	**9.66**	m	**10.39**
height 250 - 500mm	8.23	-	**8.23**	0.55	0.11	**8.99**	0.28	**17.51**	m	**18.82**
height 0.50m - 1.00m	15.31	-	**15.31**	1.03	0.21	**16.89**	0.54	**32.73**	m	**35.18**
Sides of ground beams and edges of beds; plain vertical										
height exceeding 1.00m	3.32	12.50	**3.74**	1.75	0.35	**28.62**	0.83	**33.18**	m2	**35.67**
height not exceeding 250mm	1.03	12.50	**1.15**	0.53	0.11	**8.71**	0.26	**10.12**	m	**10.88**
height 250 - 500mm	1.81	12.50	**2.04**	0.96	0.19	**15.68**	0.48	**18.19**	m	**19.56**
height 0.50m - 1.00m	3.41	12.50	**3.84**	1.79	0.36	**29.29**	0.83	**33.96**	m	**36.51**
Sides of ground beams and edges of beds; plain vertical; left in										
height exceeding 1.00m	14.95	-	**14.95**	1.00	0.20	**16.35**	0.48	**31.78**	m2	**34.16**
height not exceeding 250mm	4.58	-	**4.58**	0.30	0.06	**4.91**	0.18	**9.66**	m	**10.39**
height 250 - 500mm	8.23	-	**8.23**	0.55	0.11	**8.99**	0.28	**17.51**	m	**18.82**
height 0.50m - 1.00m	15.31	-	**15.31**	1.03	0.21	**16.89**	0.54	**32.73**	m	**35.18**
Edges of suspended slabs; plain vertical										
height not exceeding 250mm	1.26	12.50	**1.41**	0.66	0.13	**10.77**	0.29	**12.48**	m	**13.41**
height 250 - 500mm	2.31	12.50	**2.60**	1.20	0.24	**19.62**	0.57	**22.79**	m	**24.50**
extra; recesses; plain rectangular										
12 x 25mm	0.01	-	**0.01**	0.05	-	**0.71**	-	**0.72**	m	**0.77**
25 x 50mm	0.03	-	**0.03**	0.07	-	**0.99**	-	**1.03**	m	**1.11**
50 x 75mm	0.16	-	**0.16**	0.09	-	**1.28**	-	**1.44**	m	**1.55**
50 x 100mm	0.24	-	**0.24**	0.11	-	**1.56**	-	**1.80**	m	**1.94**
extra; rebates; plain rectangular										
12 x 25mm	0.01	-	**0.01**	0.07	-	**0.99**	-	**1.00**	m	**1.08**
25 x 50mm	0.03	-	**0.03**	0.08	-	**1.14**	-	**1.17**	m	**1.26**
50 x 75mm	0.16	-	**0.16**	0.10	-	**1.42**	-	**1.58**	m	**1.70**
50 x 100mm	0.24	-	**0.24**	0.13	-	**1.85**	-	**2.09**	m	**2.24**
extra; chamfers										
60mm wide	0.07	-	**0.07**	0.05	-	**0.71**	-	**0.78**	m	**0.84**
75mm wide	0.08	-	**0.08**	0.07	-	**0.99**	-	**1.07**	m	**1.16**
100mm wide	0.21	-	**0.21**	0.11	-	**1.56**	-	**1.77**	m	**1.90**
Edges of suspended slabs; plain vertical; curved 10m radius										
height not exceeding 250mm	1.84	12.50	**2.06**	1.00	0.20	**16.35**	0.50	**18.92**	m	**20.34**
Edges of suspended slabs; plain vertical; curved 1m radius										
height not exceeding 250mm	3.13	12.50	**3.52**	1.65	0.33	**26.98**	0.80	**31.29**	m	**33.64**
Edges of suspended slabs; plain vertical; curved 20m radius										
height not exceeding 250mm	1.40	12.50	**1.57**	0.74	0.15	**12.12**	0.33	**14.02**	m	**15.07**
Sides of upstands; plain vertical										
height not exceeding 250mm	1.45	12.50	**1.64**	0.75	0.15	**12.26**	0.33	**14.23**	m	**15.29**
height 250 - 500mm	1.72	12.50	**1.93**	0.88	0.18	**14.43**	0.43	**16.80**	m	**18.06**
extra; recesses; plain rectangular										
12 x 25mm	0.01	-	**0.01**	0.05	-	**0.71**	-	**0.72**	m	**0.77**
25 x 50mm	0.03	-	**0.03**	0.07	-	**0.99**	-	**1.03**	m	**1.11**
50 x 75mm	0.16	-	**0.16**	0.09	-	**1.28**	-	**1.44**	m	**1.55**
50 x 100mm	0.24	-	**0.24**	0.11	-	**1.56**	-	**1.80**	m	**1.94**
extra; rebates; plain rectangular										
12 x 25mm	0.01	-	**0.01**	0.07	-	**0.99**	-	**1.00**	m	**1.08**
25 x 50mm	0.03	-	**0.03**	0.08	-	**1.14**	-	**1.17**	m	**1.26**
50 x 75mm	0.16	-	**0.16**	0.10	-	**1.42**	-	**1.58**	m	**1.70**
50 x 100mm	0.24	-	**0.24**	0.13	-	**1.85**	-	**2.09**	m	**2.24**
extra; chamfers										
60mm wide	0.07	-	**0.07**	0.05	-	**0.71**	-	**0.78**	m	**0.84**
75mm wide	0.08	-	**0.08**	0.07	-	**0.99**	-	**1.07**	m	**1.16**
100mm wide	0.21	-	**0.21**	0.11	-	**1.56**	-	**1.77**	m	**1.90**
Machine bases and plinths; plain vertical										
height exceeding 1.00m	3.63	12.50	**4.09**	1.93	0.38	**31.50**	0.91	**36.50**	m2	**39.23**

Labour hourly rates: (except Specialists) Craft Operatives £14.21 Labourer £10.71 Rates are national average prices. Refer to REGIONAL VARIATIONS for indicative levels of overall pricing in regions	MATERIALS			LABOUR				RATES		
	Del to Site £	Waste %	Material Cost £	Craft Optve Hrs	Lab Hrs	Labour Cost £	Sunds £	Nett Rate £	Unit	Gross Rate (+7.5%) £
E20: FORMWORK FOR IN-SITU CONCRETE Cont.										
Formwork and basic finish Cont.										
Machine bases and plinths; plain vertical Cont.										
height not exceeding 250mm	1.08	12.50	**1.22**	0.58	0.11	**9.42**	0.28	**10.92**	m	**11.74**
height 250 - 500mm	1.98	12.50	**2.23**	1.06	0.20	**17.20**	0.51	**19.95**	m	**21.45**
height 0.50m - 1.00m	3.70	12.50	**4.16**	1.98	0.39	**32.31**	0.98	**37.46**	m	**40.27**
Soffits of slabs; horizontal										
slab thickness not exceeding 200mm; height to soffit										
1.50 - 3.00m	10.02	10.00	**11.02**	1.50	0.30	**24.53**	1.17	**36.72**	m2	**39.48**
3.00 - 4.50m	10.02	10.00	**11.02**	1.72	0.35	**28.19**	1.60	**40.82**	m2	**43.88**
4.50 - 6.00m	10.02	10.00	**11.02**	1.94	0.40	**31.85**	2.00	**44.88**	m2	**48.24**
slab thickness 200 - 300mm; height to soffit										
1.50 - 3.00m	11.06	10.00	**12.17**	1.65	0.33	**26.98**	1.31	**40.46**	m2	**43.49**
3.00 - 4.50m	11.06	10.00	**12.17**	1.87	0.38	**30.64**	1.72	**44.53**	m2	**47.87**
4.50 - 6.00m	11.06	10.00	**12.17**	2.09	0.43	**34.30**	2.11	**48.58**	m2	**52.22**
Soffits of slabs; horizontal; left in										
slab thickness not exceeding 200mm; height to soffit										
not exceeding 1.50m	26.88	-	**26.88**	1.00	0.20	**16.35**	0.75	**43.98**	m2	**47.28**
1.50 - 3.00m	26.88	-	**26.88**	1.00	0.20	**16.35**	0.83	**44.06**	m2	**47.36**
Soffits of slabs; horizontal; with frequent uses of prefabricated panels										
slab thickness not exceeding 200mm; height to soffit										
1.50 - 3.00m	3.02	15.00	**3.48**	1.50	0.30	**24.53**	1.17	**29.17**	m2	**31.36**
3.00 - 4.50m	3.02	15.00	**3.48**	1.72	0.35	**28.19**	1.60	**33.27**	m2	**35.76**
4.50 - 6.00m	3.02	15.00	**3.48**	1.94	0.40	**31.85**	2.00	**37.33**	m2	**40.13**
slab thickness 200 - 300mm; height to soffit										
1.50 - 3.00m	3.29	15.00	**3.78**	1.65	0.33	**26.98**	1.28	**32.04**	m2	**34.44**
3.00 - 4.50m	3.29	15.00	**3.78**	1.87	0.38	**30.64**	1.72	**36.14**	m2	**38.85**
4.50 - 6.00m	3.29	15.00	**3.78**	2.09	0.43	**34.30**	2.11	**40.19**	m2	**43.20**
Soffits of slabs; sloping not exceeding 15 degrees										
slab thickness not exceeding 200mm; height to soffit										
1.50 - 3.00m	11.07	10.00	**12.18**	1.65	0.33	**26.98**	1.26	**40.42**	m2	**43.46**
3.00 - 4.50m	11.07	10.00	**12.18**	1.87	0.38	**30.64**	1.68	**44.51**	m2	**47.85**
4.50 - 6.00m	11.07	10.00	**12.18**	2.09	0.43	**34.30**	2.11	**48.59**	m2	**52.23**
Soffits of slabs; sloping exceeding 15 degrees										
slab thickness not exceeding 200mm; height to soffit										
1.50 - 3.00m	11.53	10.00	**12.69**	1.73	0.35	**28.33**	1.49	**42.51**	m2	**45.69**
3.00 - 4.50m	11.53	10.00	**12.69**	1.95	0.40	**31.99**	1.90	**46.58**	m2	**50.07**
4.50 - 6.00m	11.53	10.00	**12.69**	2.17	0.45	**35.66**	2.32	**50.66**	m2	**54.46**
Soffits of landings; horizontal										
slab thickness not exceeding 200mm; height to soffit										
not exceeding 1.50m	10.02	10.00	**11.02**	1.80	0.36	**29.43**	1.17	**41.63**	m2	**44.75**
1.50 - 3.00m	10.02	10.00	**11.02**	2.02	0.41	**33.10**	1.60	**45.72**	m2	**49.15**
3.00 - 4.50m	10.02	10.00	**11.02**	2.24	0.46	**36.76**	2.00	**49.78**	m2	**53.52**
Top formwork										
sloping exceeding 15 degrees	3.77	10.00	**4.15**	1.50	0.30	**24.53**	0.67	**29.34**	m2	**31.54**
Walls; vertical										
plain	6.66	15.00	**7.65**	1.60	0.32	**26.16**	1.08	**34.89**	m2	**37.51**
plain; height exceeding 3.00m above floor level	7.48	15.00	**8.60**	1.60	0.32	**26.16**	1.21	**35.97**	m2	**38.67**
interrupted	7.35	15.00	**8.45**	1.75	0.35	**28.62**	1.18	**38.25**	m2	**41.11**
interrupted; height exceeding 3.00m above floor level	7.85	15.00	**9.03**	1.75	0.35	**28.62**	1.31	**38.96**	m2	**41.88**
extra; recesses; plain rectangular										
12 x 25mm	0.01	-	**0.01**	0.05	-	**0.71**	-	**0.72**	m	**0.78**
25 x 50mm	0.04	-	**0.04**	0.07	-	**0.99**	-	**1.03**	m	**1.11**
50 x 75mm	0.19	-	**0.19**	0.09	-	**1.28**	-	**1.46**	m	**1.57**
50 x 100mm	0.28	-	**0.28**	0.11	-	**1.56**	-	**1.84**	m	**1.98**
Walls; vertical; curved 1m radius										
plain	16.61	15.00	**19.10**	4.00	0.80	**65.41**	2.68	**87.19**	m2	**93.73**
Walls; vertical; curved 10m radius										
plain	9.98	15.00	**11.48**	2.40	0.48	**39.24**	1.64	**52.36**	m2	**56.29**
Walls; vertical; curved 20m radius										
plain	7.47	15.00	**8.58**	1.80	0.36	**29.43**	1.21	**39.22**	m2	**42.17**
Walls; battered										
plain	7.75	15.00	**8.91**	1.92	0.38	**31.35**	1.31	**41.57**	m2	**44.69**
Beams; attached to slabs										
regular shaped; rectangular; height to soffit										
1.50 - 3.00m	6.97	10.00	**7.67**	2.25	0.45	**36.79**	1.03	**45.49**	m2	**48.90**
3.00 - 4.50m	6.97	10.00	**7.67**	2.25	0.45	**36.79**	1.25	**45.72**	m2	**49.14**
regular shaped; rectangular; with 50mm wide chamfers; height to soffit										
1.50 - 3.00m	7.11	10.00	**7.82**	2.31	0.45	**37.64**	1.03	**46.49**	m2	**49.98**
3.00 - 4.50m	7.11	10.00	**7.82**	2.31	0.45	**37.64**	1.25	**46.71**	m2	**50.22**

Labour hourly rates: (except Specialists) Craft Operatives £14.21 Labourer £10.71 Rates are national average prices. Refer to REGIONAL VARIATIONS for indicative levels of overall pricing in regions	MATERIALS			LABOUR				RATES		
	Del to Site £	Waste %	Material Cost £	Craft Optve Hrs	Lab Hrs	Labour Cost £	Sunds £	Nett Rate £	Unit	Gross Rate (+7.5%) £
E20: FORMWORK FOR IN-SITU CONCRETE Cont.										
Formwork and basic finish Cont.										
Beams; isolated										
regular shaped; rectangular; height to soffit										
1.50 - 3.00m	7.20	10.00	**7.92**	2.35	0.47	**38.43**	1.03	**47.38**	m2	**50.93**
3.00 - 4.50m	7.20	10.00	**7.92**	2.35	0.47	**38.43**	1.25	**47.60**	m2	**51.17**
regular shaped; rectangular; with 50mm wide chamfers; height to soffit										
1.50 - 3.00m	7.36	10.00	**8.10**	2.41	0.47	**39.28**	1.03	**48.41**	m2	**52.04**
3.00 - 4.50m	7.36	10.00	**8.10**	2.41	0.47	**39.28**	1.25	**48.63**	m2	**52.27**
Columns; attached to walls										
regular shaped; rectangular	4.84	10.00	**5.33**	2.30	0.46	**37.61**	0.97	**43.91**	m2	**47.20**
regular shaped; rectangular, height exceeding 3.00m above floor level	4.53	10.00	**4.98**	2.40	0.46	**39.03**	1.17	**45.18**	m2	**48.57**
Columns; isolated										
regular shaped; rectangular	4.84	10.00	**5.33**	2.25	0.45	**36.79**	0.97	**43.09**	m2	**46.32**
regular shaped; rectangular, height exceeding 3.00m above floor level	4.84	10.00	**5.33**	2.30	0.45	**37.50**	1.17	**44.00**	m2	**47.30**
regular shaped; rectangular; with 50mm wide chamfers	5.23	10.00	**5.75**	2.39	0.45	**38.78**	1.03	**45.57**	m2	**48.98**
regular shaped; rectangular; with 50mm wide chamfers; height exceeding 3.00m above floor level	5.23	10.00	**5.75**	2.50	0.45	**40.34**	1.25	**47.35**	m2	**50.90**
Column casings; attached to walls										
regular shaped; rectangular	4.84	10.00	**5.33**	2.30	0.46	**37.61**	0.97	**43.91**	m2	**47.20**
regular shaped; rectangular, height exceeding 3.00m above floor level	4.84	10.00	**5.33**	2.40	0.46	**39.03**	1.17	**45.53**	m2	**48.94**
Column casings; isolated										
regular shaped; rectangular	4.84	10.00	**5.33**	2.25	0.45	**36.79**	0.97	**43.09**	m2	**46.32**
regular shaped; rectangular, height exceeding 3.00m above floor level	4.84	10.00	**5.33**	2.30	0.45	**37.50**	1.17	**44.00**	m2	**47.30**
regular shaped; rectangular; with 50mm wide chamfers	5.23	10.00	**5.75**	2.39	0.45	**38.78**	1.03	**45.57**	m2	**48.98**
regular shaped; rectangular; with 50mm wide chamfers; height exceeding 3.00m above floor level	5.23	10.00	**5.75**	2.50	0.45	**40.34**	1.17	**47.27**	m2	**50.81**
Extra over a basic finish for a fine formed finish										
slabs	-	-	**-**	-	0.35	**3.75**	-	**3.75**	m2	**4.03**
walls	-	-	**-**	-	0.35	**3.75**	-	**3.75**	m2	**4.03**
beams	-	-	**-**	-	0.35	**3.75**	-	**3.75**	m2	**4.03**
columns	-	-	**-**	-	0.35	**3.75**	-	**3.75**	m2	**4.03**
Wall kickers										
straight	0.91	15.00	**1.05**	0.25	0.05	**4.09**	0.21	**5.35**	m	**5.75**
curved 2m radius	2.35	15.00	**2.71**	0.66	0.15	**10.99**	0.42	**14.11**	m	**15.17**
curved 10m radius	1.45	15.00	**1.67**	0.40	0.08	**6.54**	0.27	**8.48**	m	**9.12**
curved 20m radius	1.08	15.00	**1.25**	0.30	0.06	**4.91**	0.23	**6.39**	m	**6.87**
Suspended wall kickers										
straight	1.02	15.00	**1.17**	0.28	0.06	**4.62**	0.22	**6.01**	m	**6.46**
curved 2m radius	2.60	15.00	**2.98**	0.73	0.17	**12.19**	0.46	**15.63**	m	**16.81**
curved 10m radius	1.58	15.00	**1.82**	0.44	0.09	**7.22**	0.29	**9.33**	m	**10.03**
curved 20m radius	1.21	15.00	**1.39**	0.33	0.07	**5.44**	0.26	**7.09**	m	**7.62**
Wall ends, soffits and steps in walls; plain										
width not exceeding 250mm	1.96	15.00	**2.26**	0.53	0.11	**8.71**	0.33	**11.29**	m	**12.14**
Openings in walls; plain										
width not exceeding 250mm	1.96	15.00	**2.26**	0.53	0.11	**8.71**	0.33	**11.29**	m	**12.14**
Stairflights; 1000mm wide; 155mm thick waist; 178mm risers; includes formwork to soffits, risers and strings										
strings 300mm wide	14.96	10.00	**16.46**	6.05	1.21	**98.93**	2.43	**117.82**	m	**126.66**
string 300mm wide; junction with wall	14.96	10.00	**16.46**	6.05	1.21	**98.93**	2.43	**117.82**	m	**126.66**
Stairflights; 1500mm wide; 180mm thick waist; 178mm risers; includes formwork to soffits, risers and strings										
strings 325mm wide	18.71	10.00	**20.58**	7.56	1.51	**123.60**	3.04	**147.22**	m	**158.26**
string 325mm wide; junction with wall	18.71	10.00	**20.58**	7.56	1.51	**123.60**	3.04	**147.22**	m	**158.26**
Stairflights; 1000mm wide; 155mm thick waist; 178mm undercut risers; includes formwork to soffits, risers and strings										
strings 300mm wide	14.96	10.00	**16.46**	6.05	1.21	**98.93**	2.43	**117.82**	m	**126.66**
string 300mm wide; junction with wall	14.96	10.00	**16.46**	6.05	1.21	**98.93**	2.43	**117.82**	m	**126.66**
Stairflights; 1500mm wide; 180mm thick waist; 178mm undercut risers; includes formwork to soffits, risers and strings										
strings 325mm wide	18.71	10.00	**20.58**	7.56	1.51	**123.60**	3.04	**147.22**	m	**158.26**
string 325mm wide; junction with wall	18.71	10.00	**20.58**	7.56	1.51	**123.60**	3.04	**147.22**	m	**158.26**
Mortices; rectangular										
girth not exceeding 500mm; depth not exceeding 250mm	0.89	15.00	**1.02**	0.48	0.10	**7.89**	0.23	**9.15**	nr	**9.83**
girth 0.50 - 1.00m; depth 250 - 500mm	2.95	15.00	**3.40**	1.54	0.31	**25.20**	0.74	**29.34**	nr	**31.54**
girth 0.50 - 1.00m; depth 0.50 - 1.00m	5.71	15.00	**6.57**	3.08	0.62	**50.41**	1.45	**58.43**	nr	**62.81**

Labour hourly rates: (except Specialists) Craft Operatives £14.21 Labourer £10.71 Rates are national average prices. Refer to REGIONAL VARIATIONS for indicative levels of overall pricing in regions	MATERIALS			LABOUR				RATES		
	Del to Site £	Waste %	Material Cost £	Craft Optve Hrs	Lab Hrs	Labour Cost £	Sunds £	Nett Rate £	Unit	Gross Rate (+7.5%) £
E20: FORMWORK FOR IN-SITU CONCRETE Cont.										
Formwork and basic finish Cont.										
Mortices; circular										
girth not exceeding 500mm; depth not exceeding 250mm	1.81	15.00	**2.08**	0.96	0.20	**15.78**	0.46	**18.32**	nr	**19.70**
Holes; rectangular										
girth not exceeding 500mm; depth not exceeding 250mm	0.89	15.00	**1.02**	0.48	0.10	**7.89**	0.23	**9.15**	nr	**9.83**
girth not exceeding 500mm; depth 250 - 500mm	1.81	15.00	**2.08**	0.88	0.18	**14.43**	0.42	**16.94**	nr	**18.21**
girth 0.50 - 1.00m; depth not exceeding 250mm	1.56	15.00	**1.79**	0.85	0.17	**13.90**	0.40	**16.09**	nr	**17.29**
girth 0.50 - 1.00m; depth 250 - 500mm	2.86	15.00	**3.29**	1.54	0.31	**25.20**	0.74	**29.23**	nr	**31.42**
girth 1.00m - 2.00m; depth not exceeding 250mm	2.25	15.00	**2.59**	1.21	0.24	**19.76**	0.57	**22.93**	nr	**24.64**
girth 2.00m - 3.00m; depth not exceeding 250mm	2.93	15.00	**3.37**	1.57	0.31	**25.63**	0.75	**29.75**	nr	**31.98**
girth 3.00m - 4.00m; depth not exceeding 250mm	3.65	15.00	**4.19**	1.93	0.39	**31.60**	0.91	**36.71**	nr	**39.46**
girth 1.00m - 2.00m; depth 250 - 500mm	4.07	15.00	**4.68**	2.20	0.44	**35.97**	1.03	**41.69**	nr	**44.81**
girth 2.00m - 3.00m; depth 250 - 500mm	5.31	15.00	**6.10**	2.85	0.57	**46.60**	1.36	**54.06**	nr	**58.12**
girth 3.00m - 4.00m; depth 250 - 500mm	6.63	15.00	**7.63**	3.50	0.70	**57.23**	1.65	**66.51**	nr	**71.50**
Holes; circular										
girth not exceeding 500mm; depth not exceeding 250mm	1.81	15.00	**2.08**	0.96	0.20	**15.78**	0.46	**18.32**	nr	**19.70**
girth not exceeding 500mm; depth 250 - 500mm	3.25	15.00	**3.74**	1.76	0.36	**28.87**	0.82	**33.42**	nr	**35.93**
girth 0.50 - 1.00m; depth not exceeding 250mm	3.14	15.00	**3.61**	1.70	0.34	**27.80**	0.80	**32.20**	nr	**34.62**
girth 0.50 - 1.00m; depth 250 - 500mm	5.71	15.00	**6.57**	3.08	0.62	**50.41**	1.45	**58.43**	nr	**62.81**
girth 1.00m - 2.00m; depth not exceeding 250mm	4.48	15.00	**5.15**	2.42	0.48	**39.53**	1.15	**45.82**	nr	**49.26**
girth 2.00m - 3.00m; depth not exceeding 250mm	5.83	15.00	**6.70**	3.14	0.62	**51.26**	1.49	**59.44**	nr	**63.90**
girth 3.00m - 4.00m; depth not exceeding 250mm	7.29	15.00	**8.38**	3.86	0.78	**63.20**	1.80	**73.39**	nr	**78.89**
girth 1.00m - 2.00m; depth 250 - 500mm	8.14	15.00	**9.37**	4.40	0.88	**71.95**	2.07	**83.39**	nr	**89.64**
girth 2.00m - 3.00m; depth 250 - 500mm	10.62	15.00	**12.22**	5.70	1.14	**93.21**	2.68	**108.10**	nr	**116.21**
girth 3.00m - 4.00m; depth 250 - 500mm	13.29	15.00	**15.28**	7.00	1.40	**114.46**	3.30	**133.04**	nr	**143.02**
Formwork and basic finish; coating with retarding agent										
Sides of foundations; plain vertical										
height exceeding 1.00m	4.00	12.50	**4.50**	1.75	0.59	**31.19**	0.83	**36.52**	m2	**39.26**
height not exceeding 250mm	1.11	12.50	**1.25**	0.53	0.14	**9.03**	0.26	**10.53**	m	**11.32**
height 250 - 500mm	2.09	12.50	**2.35**	0.96	0.28	**16.64**	0.48	**19.47**	m	**20.93**
height 0.50m - 1.00m	3.91	12.50	**4.40**	1.79	0.54	**31.22**	0.83	**36.45**	m	**39.18**
Sides of ground beams and edges of beds; plain vertical										
height exceeding 1.00m	4.00	12.50	**4.50**	1.75	0.59	**31.19**	0.83	**36.52**	m2	**39.26**
height not exceeding 250mm	1.11	12.50	**1.25**	0.53	0.14	**9.03**	0.26	**10.53**	m	**11.32**
height 250 - 500mm	2.09	12.50	**2.35**	0.96	0.28	**16.64**	0.48	**19.47**	m	**20.93**
height 0.50m - 1.00m	3.91	12.50	**4.40**	1.79	0.54	**31.22**	0.83	**36.45**	m	**39.18**
Edges of suspended slabs; plain vertical										
height not exceeding 250mm	1.30	12.50	**1.47**	0.66	0.16	**11.09**	0.29	**12.85**	m	**13.82**
height 250 - 500mm	2.60	12.50	**2.92**	1.20	0.33	**20.59**	0.57	**24.08**	m	**25.88**
Edges of suspended slabs; plain vertical; curved 10m radius										
height not exceeding 250mm	3.18	12.50	**3.58**	1.65	0.36	**27.30**	0.80	**31.68**	m	**34.05**
Edges of suspended slabs; plain vertical; curved 1m radius										
height not exceeding 250mm	1.94	12.50	**2.18**	1.00	0.23	**16.67**	0.50	**19.36**	m	**20.81**
Edges of suspended slabs; plain vertical; curved 20m radius										
height not exceeding 250mm	1.48	12.50	**1.66**	0.74	0.18	**12.44**	0.32	**14.42**	m	**15.50**
Sides of upstands; plain vertical										
height not exceeding 250mm	1.55	12.50	**1.74**	0.75	0.18	**12.59**	0.32	**14.64**	m	**15.74**
height 250 - 500mm	1.97	12.50	**2.22**	0.88	0.27	**15.40**	0.42	**18.04**	m	**19.39**
Machine bases and plinths; plain vertical										
height exceeding 1.00m	4.35	12.50	**4.89**	1.93	0.62	**34.07**	0.91	**39.87**	m2	**42.86**
height not exceeding 250mm	1.15	12.50	**1.30**	0.58	0.14	**9.74**	0.28	**11.32**	m	**12.17**
height 250 - 500mm	2.26	12.50	**2.54**	1.06	0.29	**18.17**	0.51	**21.23**	m	**22.82**
height 0.50m - 1.00m	4.23	12.50	**4.76**	1.98	0.57	**34.24**	0.96	**39.96**	m	**42.96**
Soffits of slabs; horizontal										
slab thickness not exceeding 200mm; height to soffit										
1.50 - 3.00m	10.74	10.00	**11.81**	1.50	0.54	**27.10**	1.17	**40.08**	m2	**43.09**
3.00 - 4.50m	10.74	10.00	**11.81**	1.72	0.59	**30.76**	1.60	**44.18**	m2	**47.49**
4.50 - 6.00m	10.74	10.00	**11.81**	1.94	0.64	**34.42**	2.00	**48.24**	m2	**51.85**
slab thickness 200 - 300mm; height to soffit										
1.50 - 3.00m	11.74	10.00	**12.92**	1.65	0.57	**29.55**	1.28	**43.74**	m2	**47.02**
3.00 - 4.50m	11.74	10.00	**12.92**	1.87	0.62	**33.21**	1.72	**47.85**	m2	**51.44**
4.50 - 6.00m	11.74	10.00	**12.92**	2.09	0.67	**36.87**	2.11	**51.90**	m2	**55.79**
Soffits of slabs; horizontal; with frequent uses of prefabricated panels										
slab thickness not exceeding 200mm; height to soffit										
1.50 - 3.00m	3.69	15.00	**4.24**	1.50	0.54	**27.10**	1.17	**32.51**	m2	**34.95**
3.00 - 4.50m	3.69	15.00	**4.24**	1.72	0.59	**30.76**	1.60	**36.61**	m2	**39.35**
4.50 - 6.00m	3.69	15.00	**4.24**	1.94	0.64	**34.42**	2.00	**40.67**	m2	**43.72**
slab thickness 200 - 300mm; height to soffit										
1.50 - 3.00m	4.38	15.00	**5.04**	1.65	0.78	**31.80**	1.28	**38.12**	m2	**40.98**

Labour hourly rates: (except Specialists) Craft Operatives £14.21 Labourer £10.71 Rates are national average prices. Refer to REGIONAL VARIATIONS for indicative levels of overall pricing in regions	MATERIALS			LABOUR				RATES		
	Del to Site £	Waste %	Material Cost £	Craft Optve Hrs	Lab Hrs	Labour Cost £	Sunds £	Nett Rate £	Unit	Gross Rate (+7.5%) £
E20: FORMWORK FOR IN-SITU CONCRETE Cont.										
Formwork and basic finish; coating with retarding agent Cont.										
Soffits of slabs; horizontal; with frequent uses of prefabricated panels Cont.										
slab thickness 200 - 300mm; height to soffit Cont.										
3.00 - 4.50m	4.38	15.00	**5.04**	1.87	0.83	**35.46**	1.72	**42.22**	m2	**45.39**
4.50 - 6.00m	4.38	15.00	**5.04**	2.09	0.88	**39.12**	2.11	**46.27**	m2	**49.74**
Soffits of slabs; sloping not exceeding 15 degrees										
slab thickness not exceeding 200mm; height to soffit										
1.50 - 3.00m	11.74	15.00	**13.50**	1.65	0.57	**29.55**	1.28	**44.33**	m2	**47.65**
3.00 - 4.50m	11.74	15.00	**13.50**	1.87	0.62	**33.21**	1.72	**48.43**	m2	**52.07**
4.50 - 6.00m	11.74	15.00	**13.50**	2.09	0.67	**36.87**	2.11	**52.48**	m2	**56.42**
Soffits of slabs; sloping exceeding 15 degrees										
slab thickness not exceeding 200mm; height to soffit										
1.50 - 3.00m	12.22	15.00	**14.05**	1.73	0.59	**30.90**	1.49	**46.44**	m2	**49.92**
3.00 - 4.50m	12.22	15.00	**14.05**	1.95	0.64	**34.56**	1.90	**50.51**	m2	**54.29**
4.50 - 6.00m	12.22	15.00	**14.05**	2.17	0.69	**38.23**	2.32	**54.59**	m2	**58.68**
Soffits of landings; horizontal										
slab thickness not exceeding 200mm; height to soffit										
not exceeding 1.50m	10.74	15.00	**12.35**	1.80	0.60	**32.00**	1.17	**45.52**	m2	**48.94**
1.50 - 3.00m	10.74	15.00	**12.35**	2.02	0.65	**35.67**	1.60	**49.62**	m2	**53.34**
3.00 - 4.50m	10.74	15.00	**12.35**	2.24	0.70	**39.33**	2.00	**53.68**	m2	**57.70**
Top formwork										
sloping exceeding 15 degrees	4.45	15.00	**5.12**	1.50	0.54	**27.10**	0.67	**32.89**	m2	**35.35**
Walls; vertical										
plain	7.47	15.00	**8.58**	1.60	0.56	**28.73**	1.08	**38.39**	m2	**41.27**
plain; height exceeding 3.00m above floor level	8.26	15.00	**9.50**	1.60	0.56	**28.73**	1.21	**39.44**	m2	**42.40**
interrupted	8.11	15.00	**9.32**	1.75	0.59	**31.19**	1.18	**41.69**	m2	**44.82**
interrupted; height exceeding 3.00m above floor level	8.67	15.00	**9.96**	1.75	0.59	**31.19**	1.31	**42.46**	m2	**45.65**
Walls; vertical; curved 1m radius										
plain	17.38	15.00	**19.99**	4.00	1.04	**67.98**	2.68	**90.65**	m2	**97.44**
Walls; vertical; curved 10m radius										
plain	10.75	15.00	**12.37**	2.40	0.72	**41.82**	1.64	**55.82**	m2	**60.01**
Walls; vertical; curved 20m radius										
plain	8.22	15.00	**9.46**	1.80	0.60	**32.00**	1.21	**42.67**	m2	**45.87**
Walls; battered										
plain	8.77	15.00	**10.09**	1.92	0.62	**33.92**	1.31	**45.32**	m2	**48.72**
Beams; attached to slabs										
regular shaped; rectangular; height to soffit										
1.50 - 3.00m	7.77	10.00	**8.55**	2.25	0.69	**39.36**	1.03	**48.94**	m2	**52.61**
3.00 - 4.50m	7.77	10.00	**8.55**	2.25	0.69	**39.36**	1.25	**49.17**	m2	**52.85**
regular shaped; rectangular; with 50mm wide chamfers; height to soffit										
1.50 - 3.00m	7.95	10.00	**8.74**	2.31	0.69	**40.22**	1.03	**49.98**	m2	**53.73**
3.00 - 4.50m	7.95	10.00	**8.74**	2.31	0.69	**40.22**	1.25	**50.21**	m2	**53.97**
Beams; isolated										
regular shaped; rectangular; height to soffit										
1.50 - 3.00m	8.12	10.00	**8.93**	2.35	0.71	**41.00**	1.03	**50.96**	m2	**54.78**
3.00 - 4.50m	8.12	10.00	**8.93**	2.35	0.71	**41.00**	1.25	**51.18**	m2	**55.02**
regular shaped; rectangular; with 50mm wide chamfers; height to soffit										
1.50 - 3.00m	8.37	10.00	**9.21**	2.41	0.71	**41.85**	1.03	**52.09**	m2	**55.99**
3.00 - 4.50m	8.37	10.00	**9.21**	2.41	0.71	**41.85**	1.25	**52.31**	m2	**56.23**
Columns; attached to walls										
regular shaped; rectangular	5.64	10.00	**6.21**	2.30	0.70	**40.18**	0.97	**47.36**	m2	**50.91**
regular shaped; rectangular, height exceeding 3.00m above floor level	5.64	10.00	**6.21**	2.43	0.70	**42.03**	1.17	**49.40**	m2	**53.11**
Columns; isolated										
regular shaped; rectangular	5.64	10.00	**6.21**	2.25	0.70	**39.47**	0.97	**46.65**	m2	**50.15**
regular shaped; rectangular, height exceeding 3.00m above floor level	5.64	10.00	**6.21**	2.48	0.70	**42.74**	1.17	**50.11**	m2	**53.87**
regular shaped; rectangular; with 50mm wide chamfers	6.06	10.00	**6.66**	2.43	0.69	**41.92**	1.03	**49.61**	m2	**53.33**
regular shaped; rectangular; with 50mm wide chamfers; height exceeding 3.00m above floor level	6.06	10.00	**6.66**	2.50	0.70	**43.02**	1.25	**50.94**	m2	**54.76**
Column casings; attached to walls										
regular shaped; rectangular	5.64	10.00	**6.21**	2.30	0.70	**40.18**	1.17	**47.56**	m2	**51.12**
regular shaped; rectangular, height exceeding 3.00m above floor level	5.64	10.00	**6.21**	2.40	0.70	**41.60**	0.97	**48.78**	m2	**52.44**
Column casings; isolated										
regular shaped; rectangular	5.64	10.00	**6.21**	2.38	0.69	**41.21**	0.97	**48.39**	m2	**52.02**
regular shaped; rectangular, height exceeding 3.00m above floor level	5.64	10.00	**6.21**	2.48	0.69	**42.63**	1.17	**50.01**	m2	**53.76**

Labour hourly rates: (except Specialists) Craft Operatives £14.21 Labourer £10.71 Rates are national average prices. Refer to REGIONAL VARIATIONS for indicative levels of overall pricing in regions	MATERIALS			LABOUR				RATES		
	Del to Site £	Waste %	Material Cost £	Craft Optve Hrs	Lab Hrs	Labour Cost £	Sunds £	Nett Rate £	Unit	Gross Rate (+7.5%) £
E20: FORMWORK FOR IN-SITU CONCRETE Cont.										
Formwork and basic finish; coating with retarding agent Cont.										
Column casings; isolated Cont.										
regular shaped; rectangular; with 50mm wide chamfers	6.06	10.00	**6.66**	2.53	0.69	**43.34**	1.03	**51.03**	m2	**54.86**
regular shaped; rectangular; with 50mm wide chamfers; height exceeding 3.00m above floor level	6.06	10.00	**6.66**	2.63	0.69	**44.76**	1.25	**52.68**	m2	**56.63**
Extra over a basic finish for a fine formed finish										
slabs	-	-	**-**	-	0.35	**3.75**	-	**3.75**	m2	**4.03**
walls	-	-	**-**	-	0.35	**3.75**	-	**3.75**	m2	**4.03**
beams	-	-	**-**	-	0.35	**3.75**	-	**3.75**	m2	**4.03**
columns	-	-	**-**	-	0.35	**3.75**	-	**3.75**	m2	**4.03**
Wall kickers										
straight	1.06	15.00	**1.22**	0.25	0.09	**4.52**	0.21	**5.95**	m	**6.39**
curved 2m radius	2.48	15.00	**2.85**	0.66	0.17	**11.20**	0.42	**14.47**	m	**15.56**
curved 10m radius	1.53	15.00	**1.76**	0.40	0.12	**6.97**	0.27	**9.00**	m	**9.68**
curved 20m radius	1.18	15.00	**1.35**	0.30	0.10	**5.33**	0.23	**6.92**	m	**7.44**
Suspended wall kickers										
straight	1.18	15.00	**1.35**	0.28	0.10	**5.05**	0.23	**6.64**	m	**7.13**
curved 2m radius	2.70	15.00	**3.10**	0.73	0.19	**12.41**	0.47	**15.98**	m	**17.18**
curved 10m radius	1.73	15.00	**1.99**	0.44	0.13	**7.64**	0.29	**9.93**	m	**10.67**
curved 20m radius	1.30	15.00	**1.50**	0.33	0.11	**5.87**	0.26	**7.62**	m	**8.20**
Wall ends, soffits and steps in walls; plain										
width not exceeding 250mm	2.15	15.00	**2.47**	0.53	0.17	**9.35**	0.33	**12.15**	m	**13.06**
Openings in walls; plain										
width not exceeding 250mm	2.15	15.00	**2.47**	0.53	0.17	**9.35**	0.33	**12.15**	m	**13.06**
Stairflights; 1000mm wide; 155mm thick waist; 178mm risers; includes formwork to soffits, risers and strings										
strings 300mm wide	16.52	10.00	**18.17**	6.05	1.74	**104.61**	2.43	**125.21**	m	**134.60**
string 300mm wide; junction with wall	16.52	10.00	**18.17**	6.05	1.74	**104.61**	2.43	**125.21**	m	**134.60**
Stairflights; 1500mm wide; 180mm thick waist; 178mm risers; includes formwork to soffits, risers and strings										
strings 325mm wide	20.63	10.00	**22.70**	7.56	2.17	**130.67**	3.04	**156.41**	m	**168.14**
string 325mm wide; junction with wall	20.63	10.00	**22.70**	7.56	2.17	**130.67**	3.04	**156.41**	m	**168.14**
Stairflights; 1000mm wide; 155mm thick waist; 178mm undercut risers; includes formwork to soffits, risers and strings										
strings 300mm wide	16.52	10.00	**18.17**	6.05	1.74	**104.61**	2.43	**125.21**	m	**134.60**
string 300mm wide; junction with wall	16.52	10.00	**18.17**	6.05	1.74	**104.61**	2.43	**125.21**	m	**134.60**
Stairflights; 1500mm wide; 180mm thick waist; 178mm undercut risers; includes formwork to soffits, risers and strings										
strings 325mm wide	20.63	10.00	**22.70**	7.56	2.17	**130.67**	3.04	**156.41**	m	**168.14**
string 325mm wide; junction with wall	20.63	10.00	**22.70**	7.56	2.17	**130.67**	3.04	**156.41**	m	**168.14**
Claymaster low density expanded polystyrene permanent formwork; fixing with Clayfix hooks - 3/m2										
Sides of foundations; plain vertical 75mm thick										
height exceeding 1.00m	12.35	5.00	**12.97**	-	0.06	**0.64**	0.74	**14.35**	m2	**15.42**
height not exceeding 250mm	3.09	5.00	**3.24**	-	0.02	**0.21**	0.25	**3.70**	m	**3.98**
height 250-500mm	6.18	5.00	**6.48**	-	0.04	**0.43**	0.25	**7.16**	m	**7.69**
height 0.50m-1.00m	9.26	5.00	**9.73**	-	0.07	**0.75**	0.49	**10.97**	m	**11.79**
Sides of foundations; plain vertical 100mm thick										
height exceeding 1.00m	20.71	5.00	**21.75**	-	0.08	**0.86**	0.74	**23.34**	m2	**25.09**
height not exceeding 250mm	5.18	5.00	**5.44**	-	0.03	**0.32**	0.25	**6.00**	m	**6.45**
height 250-500mm	10.36	5.00	**10.87**	-	0.05	**0.54**	0.25	**11.65**	m	**12.53**
height 0.50m-1.00m	15.53	5.00	**16.31**	-	0.09	**0.96**	0.49	**17.76**	m	**19.10**
Sides of foundations; plain vertical 150mm thick										
height exceeding 1.00m	31.06	5.00	**32.61**	-	0.10	**1.07**	0.74	**34.42**	m2	**37.00**
height not exceeding 250mm	7.77	5.00	**8.15**	-	0.03	**0.32**	0.25	**8.72**	m	**9.37**
height 250-500mm	15.53	5.00	**16.31**	-	0.06	**0.64**	0.25	**17.19**	m	**18.48**
height 0.50m-1.00m	23.30	5.00	**24.46**	-	0.11	**1.18**	0.49	**26.13**	m	**28.09**
To underside of foundations; laid on earth or hardcore; 75mm thick										
horizontal	12.35	5.00	**12.97**	-	0.05	**0.54**	-	**13.50**	m2	**14.52**
To underside of foundations; laid on earth or hardcore; 100mm thick										
horizontal	20.71	5.00	**21.75**	-	0.06	**0.64**	-	**22.39**	m2	**24.07**
To underside of foundations; laid on earth or hardcore; 150mm thick										
horizontal	31.06	5.00	**32.61**	-	0.07	**0.75**	-	**33.36**	m2	**35.86**

Labour hourly rates: (except Specialists) Craft Operatives £14.21 Labourer £10.71 Rates are national average prices. Refer to REGIONAL VARIATIONS for indicative levels of overall pricing in regions	MATERIALS			LABOUR				RATES		
	Del to Site	Waste	Material Cost	Craft Optve	Lab	Labour Cost	Sunds	Nett Rate	Unit	Gross Rate (+7.5%)
	£	%	£	Hrs	Hrs	£	£	£		£
E20: FORMWORK FOR IN-SITU CONCRETE Cont.										
Expamet Hy-rib permanent shuttering and reinforcement										
Reference 2611, 4.86 kg/m2 to soffits of slabs; horizontal; one rib side laps; 150mm end laps										
slab thickness 75mm; strutting and supports at 750mm centres; height to soffit										
1.50 - 3.00m	15.50	5.00	**16.28**	1.19	0.30	**20.12**	1.25	**37.65**	m2	**40.47**
3.00 - 4.50m	15.50	5.00	**16.28**	1.37	0.34	**23.11**	1.54	**40.93**	m2	**44.00**
slab thickness 100mm; strutting and supports at 650mm centres; height to soffit										
1.50 - 3.00m	15.50	5.00	**16.28**	1.21	0.30	**20.41**	1.42	**38.10**	m2	**40.96**
3.00 - 4.50m	15.50	5.00	**16.28**	1.39	0.35	**23.50**	1.91	**41.68**	m2	**44.81**
slab thickness 125mm; strutting and supports at 550mm centres; height to soffit										
1.50 - 3.00m	15.50	5.00	**16.28**	1.23	0.31	**20.80**	1.59	**38.67**	m2	**41.57**
3.00 - 4.50m	15.50	5.00	**16.28**	1.41	0.35	**23.78**	2.30	**42.37**	m2	**45.54**
slab thickness 150mm; strutting and supports at 450mm centres; height to soffit										
1.50 - 3.00m	15.50	5.00	**16.28**	1.25	0.31	**21.08**	1.79	**39.15**	m2	**42.08**
3.00 - 4.50m	15.50	5.00	**16.28**	1.44	0.36	**24.32**	2.68	**43.27**	m2	**46.52**
Reference 2411, 6.34 kg/m2 to soffits of slabs; horizontal; one rib side laps; 150mm end laps										
slab thickness 75mm; strutting and supports at 850mm centres; height to soffit										
1.50 - 3.00m	19.80	5.00	**20.79**	1.22	0.31	**20.66**	1.25	**42.70**	m2	**45.90**
3.00 - 4.50m	19.80	5.00	**20.79**	1.40	0.35	**23.64**	1.54	**45.98**	m2	**49.43**
slab thickness 100mm; strutting and supports at 750mm centres; height to soffit										
1.50 - 3.00m	19.80	5.00	**20.79**	1.24	0.31	**20.94**	1.42	**43.15**	m2	**46.38**
3.00 - 4.50m	19.80	5.00	**20.79**	1.43	0.36	**24.18**	1.91	**46.88**	m2	**50.39**
slab thickness 125mm; strutting and supports at 650mm centres; height to soffit										
1.50 - 3.00m	19.80	5.00	**20.79**	1.25	0.32	**21.19**	1.59	**43.57**	m2	**46.84**
3.00 - 4.50m	19.80	5.00	**20.79**	1.45	0.36	**24.46**	2.30	**47.56**	m2	**51.12**
slab thickness 150mm; strutting and supports at 550mm centres; height to soffit										
1.50 - 3.00m	19.80	5.00	**20.79**	1.28	0.32	**21.62**	1.79	**44.20**	m2	**47.51**
3.00 - 4.50m	19.80	5.00	**20.79**	1.47	0.37	**24.85**	2.68	**48.32**	m2	**51.95**
slab thickness 175mm; strutting and supports at 450mm centres; height to soffit										
1.50 - 3.00m	19.80	5.00	**20.79**	1.31	0.33	**22.15**	2.06	**45.00**	m2	**48.38**
3.00 - 4.50m	19.80	5.00	**20.79**	1.51	0.38	**25.53**	3.19	**49.51**	m2	**53.23**
slab thickness 200mm; strutting and supports at 350mm centres; height to soffit										
1.50 - 3.00m	19.80	5.00	**20.79**	1.34	0.34	**22.68**	2.32	**45.79**	m2	**49.23**
3.00 - 4.50m	19.80	5.00	**20.79**	1.54	0.39	**26.06**	3.72	**50.57**	m2	**54.37**
Reference 2611, 4.86 kg/m2 to soffits of arched slabs; one rib side laps										
900mm span; 75mm rise; height to soffit										
1.50 - 3.00m	15.50	5.00	**16.28**	1.31	0.33	**22.15**	0.92	**39.35**	m2	**42.30**
1200mm span; 75mm rise; one row strutting and supports per span; height to soffit										
1.50 - 3.00m	15.50	5.00	**16.28**	1.31	0.33	**22.15**	1.25	**39.68**	m2	**42.65**
3.00 - 4.50m	15.50	5.00	**16.28**	1.49	0.37	**25.14**	1.54	**42.96**	m2	**46.18**
Reference 2411, 6.34 kg/m2 to soffits of arched slabs; one rib side laps										
1500mm span; 100mm rise; two rows strutting and supports per span; height to soffit										
1.50 - 3.00m	19.80	5.00	**20.79**	1.62	0.40	**27.30**	2.67	**50.76**	m2	**54.57**
3.00 - 4.50m	19.80	5.00	**20.79**	1.84	0.46	**31.07**	3.57	**55.43**	m2	**59.59**
1800mm span; 150mm rise; two rows strutting and supports per span; height to soffit										
1.50 - 3.00m	19.80	5.00	**20.79**	1.62	0.40	**27.30**	2.67	**50.76**	m2	**54.57**
3.00 - 4.50m	19.80	5.00	**20.79**	1.84	0.46	**31.07**	3.57	**55.43**	m2	**59.59**
E30: REINFORCEMENT FOR IN-SITU CONCRETE										
Reinforcement bars; BS 4449, hot rolled plain round mild steel including hooks and tying wire, and spacers and chairs which are at the discretion of the Contractor										
Straight										
6mm	420.00	5.00	**441.00**	46.00	4.00	**696.50**	25.24	**1162.74**	t	**1249.94**
8mm	405.00	5.00	**425.25**	40.00	4.00	**611.24**	22.40	**1058.89**	t	**1138.31**
10mm	395.00	5.00	**414.75**	33.00	4.00	**511.77**	18.13	**944.65**	t	**1015.50**
12mm	390.00	5.00	**409.50**	28.00	4.00	**440.72**	15.46	**865.68**	t	**930.60**
16mm	385.00	5.00	**404.25**	22.00	4.00	**355.46**	12.71	**772.42**	t	**830.35**
20mm	385.00	5.00	**404.25**	18.00	4.00	**298.62**	7.03	**709.90**	t	**763.14**
25mm	385.00	5.00	**404.25**	18.00	4.00	**298.62**	7.03	**709.90**	t	**763.14**
32mm	380.00	5.00	**399.00**	18.00	4.00	**298.62**	7.03	**704.65**	t	**757.50**
Bent										
6mm	420.00	5.00	**441.00**	46.00	4.00	**696.50**	25.24	**1162.74**	t	**1249.94**
8mm	405.00	5.00	**425.25**	40.00	4.00	**611.24**	22.40	**1058.89**	t	**1138.31**
10mm	395.00	5.00	**414.75**	33.00	4.00	**511.77**	18.13	**944.65**	t	**1015.50**
12mm	390.00	5.00	**409.50**	28.00	4.00	**440.72**	15.46	**865.68**	t	**930.60**
16mm	385.00	5.00	**404.25**	22.00	4.00	**355.46**	12.71	**772.42**	t	**830.35**
20mm	385.00	5.00	**404.25**	18.00	4.00	**298.62**	7.03	**709.90**	t	**763.14**
25mm	385.00	5.00	**404.25**	18.00	4.00	**298.62**	7.03	**709.90**	t	**763.14**
32mm	380.00	5.00	**399.00**	18.00	4.00	**298.62**	7.03	**704.65**	t	**757.50**

Labour hourly rates: (except Specialists) Craft Operatives £14.21 Labourer £10.71 Rates are national average prices. Refer to REGIONAL VARIATIONS for indicative levels of overall pricing in regions	MATERIALS			LABOUR			RATES			
	Del to Site £	Waste %	Material Cost £	Craft Optve Hrs	Lab Hrs	Labour Cost £	Sunds £	Nett Rate £	Unit	Gross Rate (+7.5%) £
E30: REINFORCEMENT FOR IN-SITU CONCRETE Cont.										
Reinforcement bars; BS 4449, hot rolled plain round mild steel including hooks and tying wire, and spacers and chairs which are at the discretion of the Contractor Cont.										
Links										
6mm	420.00	5.00	**441.00**	60.00	4.00	**895.44**	25.24	**1361.68**	t	**1463.80**
8mm	405.00	5.00	**425.25**	60.00	4.00	**895.44**	22.40	**1343.09**	t	**1443.83**
Reinforcement bars; BS 4449, hot rolled deformed high yield steel including hooks and tying wire, and spacers and chairs which are at the discretion of the Contractor										
Straight										
6mm	430.00	5.00	**451.50**	46.00	4.00	**696.50**	25.24	**1173.24**	t	**1261.23**
8mm	410.00	5.00	**430.50**	40.00	4.00	**611.24**	22.40	**1064.14**	t	**1143.95**
10mm	400.00	5.00	**420.00**	33.00	4.00	**511.77**	18.13	**949.90**	t	**1021.15**
12mm	400.00	5.00	**420.00**	28.00	4.00	**440.72**	15.46	**876.18**	t	**941.89**
16mm	390.00	5.00	**409.50**	22.00	4.00	**355.46**	12.71	**777.67**	t	**835.99**
20mm	390.00	5.00	**409.50**	18.00	4.00	**298.62**	7.03	**715.15**	t	**768.79**
25mm	390.00	5.00	**409.50**	18.00	4.00	**298.62**	7.03	**715.15**	t	**768.79**
32mm	390.00	5.00	**409.50**	18.00	4.00	**298.62**	7.03	**715.15**	t	**768.79**
Bent										
6mm	430.00	5.00	**451.50**	46.00	4.00	**696.50**	25.24	**1173.24**	t	**1261.23**
8mm	410.00	5.00	**430.50**	40.00	4.00	**611.24**	22.40	**1064.14**	t	**1143.95**
10mm	400.00	5.00	**420.00**	33.00	4.00	**511.77**	18.13	**949.90**	t	**1021.15**
12mm	400.00	5.00	**420.00**	28.00	4.00	**440.72**	15.46	**876.18**	t	**941.89**
16mm	390.00	5.00	**409.50**	22.00	4.00	**355.46**	12.71	**777.67**	t	**835.99**
20mm	390.00	5.00	**409.50**	18.00	4.00	**298.62**	7.03	**715.15**	t	**768.79**
25mm	390.00	5.00	**409.50**	18.00	4.00	**298.62**	7.03	**715.15**	t	**768.79**
32mm	390.00	5.00	**409.50**	18.00	4.00	**298.62**	7.03	**715.15**	t	**768.79**
Links										
6mm	430.00	5.00	**451.50**	60.00	4.00	**895.44**	25.24	**1372.18**	t	**1475.09**
8mm	410.00	5.00	**430.50**	60.00	4.00	**895.44**	22.40	**1348.34**	t	**1449.47**
Take delivery, cut, bend and fix reinforcing rods; including hooks and tying wire, and spacers and chairs which are at the discretion of the Contractor										
Straight										
6mm	-	-	-	50.00	4.00	**753.34**	25.24	**778.58**	t	**836.97**
8mm	-	-	-	44.00	4.00	**668.08**	22.40	**690.48**	t	**742.27**
10mm	-	-	-	36.00	4.00	**554.40**	18.13	**572.53**	t	**615.47**
12mm	-	-	-	31.00	4.00	**483.35**	15.46	**498.81**	t	**536.22**
16mm	-	-	-	24.00	4.00	**383.88**	12.71	**396.59**	t	**426.33**
20mm	-	-	-	20.00	4.00	**327.04**	7.03	**334.07**	t	**359.13**
25mm	-	-	-	20.00	4.00	**327.04**	7.03	**334.07**	t	**359.13**
32mm	-	-	-	20.00	4.00	**327.04**	7.03	**334.07**	t	**359.13**
Bent										
6mm	-	-	-	68.00	4.00	**1009.12**	25.24	**1034.36**	t	**1111.93**
8mm	-	-	-	60.00	4.00	**895.44**	22.40	**917.84**	t	**986.68**
10mm	-	-	-	50.00	4.00	**753.34**	18.13	**771.47**	t	**829.33**
12mm	-	-	-	43.00	4.00	**653.87**	15.46	**669.33**	t	**719.53**
16mm	-	-	-	34.00	4.00	**525.98**	12.71	**538.69**	t	**579.09**
20mm	-	-	-	28.00	4.00	**440.72**	7.03	**447.75**	t	**481.33**
25mm	-	-	-	26.00	4.00	**412.30**	7.03	**419.33**	t	**450.78**
32mm	-	-	-	26.00	4.00	**412.30**	7.03	**419.33**	t	**450.78**
Links										
6mm	-	-	-	90.00	4.00	**1321.74**	25.24	**1346.98**	t	**1448.00**
8mm	-	-	-	90.00	4.00	**1321.74**	22.40	**1344.14**	t	**1444.95**
Take delivery and fix reinforcing rods supplied cut to length and bent; including tying wire, and spacers and chairs which are at the discretion of the Contractor										
Straight										
6mm	-	-	-	46.00	4.00	**696.50**	25.24	**721.74**	t	**775.87**
8mm	-	-	-	40.00	4.00	**611.24**	22.40	**633.64**	t	**681.17**
10mm	-	-	-	33.00	4.00	**511.77**	18.13	**529.90**	t	**569.65**
12mm	-	-	-	28.00	4.00	**440.72**	15.46	**456.18**	t	**490.39**
16mm	-	-	-	22.00	4.00	**355.46**	12.71	**368.17**	t	**395.78**
20mm	-	-	-	18.00	4.00	**298.62**	7.03	**305.65**	t	**328.57**
25mm	-	-	-	18.00	4.00	**298.62**	7.03	**305.65**	t	**328.57**
32mm	-	-	-	18.00	4.00	**298.62**	7.03	**305.65**	t	**328.57**
Bent										
6mm	-	-	-	46.00	4.00	**696.50**	25.24	**721.74**	t	**775.87**
8mm	-	-	-	40.00	4.00	**611.24**	22.40	**633.64**	t	**681.17**
10mm	-	-	-	33.00	4.00	**511.77**	18.13	**529.90**	t	**569.65**
12mm	-	-	-	28.00	4.00	**440.72**	15.46	**456.18**	t	**490.39**
16mm	-	-	-	22.00	4.00	**355.46**	12.71	**368.17**	t	**395.78**
20mm	-	-	-	18.00	4.00	**298.62**	7.03	**305.65**	t	**328.57**
25mm	-	-	-	18.00	4.00	**298.62**	7.03	**305.65**	t	**328.57**
Links										
6mm	-	-	-	60.00	4.00	**895.44**	25.24	**920.68**	t	**989.73**
8mm	-	-	-	60.00	4.00	**895.44**	22.40	**917.84**	t	**986.68**

Labour hourly rates: (except Specialists) Craft Operatives £14.21 Labourer £10.71 Rates are national average prices. Refer to REGIONAL VARIATIONS for indicative levels of overall pricing in regions	MATERIALS			LABOUR				RATES		
	Del to Site £	Waste %	Material Cost £	Craft Optve Hrs	Lab Hrs	Labour Cost £	Sunds £	Nett Rate £	Unit	Gross Rate (+7.5%) £
E30: REINFORCEMENT FOR IN-SITU CONCRETE Cont.										
Reinforcement bars; high yield stainless steel BS 6744, Type 2 (minimum yield stress 460 N/mm2); including hooks and tying wire, and spacers and chairs which are at the discretion of the Contractor										
Straight										
8mm	4100.00	5.00	**4305.00**	75.00	9.00	**1162.14**	92.99	**5560.13**	t	**5977.14**
10mm	4100.00	5.00	**4305.00**	63.00	9.00	**991.62**	77.50	**5374.12**	t	**5777.17**
12mm	4100.00	5.00	**4305.00**	54.00	9.00	**863.73**	65.10	**5233.83**	t	**5626.36**
16mm	4100.00	5.00	**4305.00**	45.00	9.00	**735.84**	54.25	**5095.09**	t	**5477.22**
20mm	4100.00	5.00	**4305.00**	37.50	9.00	**629.27**	43.40	**4977.66**	t	**5350.99**
25mm	4100.00	5.00	**4305.00**	30.00	9.00	**522.69**	29.84	**4857.53**	t	**5221.85**
Bent										
8mm	4100.00	5.00	**4305.00**	75.00	9.00	**1162.14**	92.99	**5560.13**	t	**5977.14**
10mm	4100.00	5.00	**4305.00**	63.00	9.00	**991.62**	77.50	**5374.12**	t	**5777.17**
12mm	4100.00	5.00	**4305.00**	54.00	9.00	**863.73**	65.10	**5233.83**	t	**5626.36**
16mm	4100.00	5.00	**4305.00**	45.00	9.00	**735.84**	54.25	**5095.09**	t	**5477.22**
20mm	4100.00	5.00	**4305.00**	37.50	9.00	**629.27**	43.40	**4977.66**	t	**5350.99**
25mm	4100.00	5.00	**4305.00**	30.00	9.00	**522.69**	29.84	**4857.53**	t	**5221.85**
Links										
6mm	4100.00	5.00	**4305.00**	120.00	9.00	**1801.59**	105.39	**6211.98**	t	**6677.88**
8mm	4100.00	5.00	**4305.00**	105.00	9.00	**1588.44**	92.99	**5986.43**	t	**6435.42**
Reinforcement fabric; BS 4483, hard drawn plain round steel; welded; including laps, tying wire, all cutting and bending, and spacers and chairs which are at the discretion of the Contractor										
Reference A98, 1.54 kg/m2; 200mm side laps; 200mm end laps										
generally	0.80	15.00	**0.92**	0.06	0.01	**0.96**	0.06	**1.93**	m2	**2.08**
strips in one width										
750mm wide	0.80	15.00	**0.92**	0.09	0.01	**1.39**	0.06	**2.36**	m2	**2.53**
900mm wide	0.80	15.00	**0.92**	0.08	0.01	**1.24**	0.06	**2.22**	m2	**2.38**
1050mm wide	0.80	15.00	**0.92**	0.08	0.01	**1.24**	0.06	**2.22**	m2	**2.38**
1200mm wide	0.80	15.00	**0.92**	0.07	0.01	**1.10**	0.06	**2.07**	m2	**2.23**
Reference A142, 2.22kg/m2; 200mm side laps; 200mm end laps										
generally	1.11	15.00	**1.28**	0.07	0.02	**1.21**	0.08	**2.57**	m2	**2.77**
strips in one width										
750mm wide	1.11	15.00	**1.28**	0.11	0.02	**1.78**	0.08	**3.14**	m2	**3.38**
900mm wide	1.11	15.00	**1.28**	0.10	0.02	**1.64**	0.08	**3.00**	m2	**3.22**
1050mm wide	1.11	15.00	**1.28**	0.09	0.02	**1.49**	0.08	**2.86**	m2	**3.07**
1200mm wide	1.11	15.00	**1.28**	0.08	0.02	**1.35**	0.08	**2.71**	m2	**2.92**
Reference A193, 3.02kg/m2; 200mm side laps; 200mm end laps										
generally	1.47	15.00	**1.69**	0.07	0.02	**1.21**	0.11	**3.01**	m2	**3.23**
strips in one width										
750mm wide	1.47	15.00	**1.69**	0.11	0.02	**1.78**	0.11	**3.58**	m2	**3.84**
900mm wide	1.47	15.00	**1.69**	0.10	0.02	**1.64**	0.11	**3.43**	m2	**3.69**
1050mm wide	1.47	15.00	**1.69**	0.09	0.02	**1.49**	0.11	**3.29**	m2	**3.54**
1200mm wide	1.47	15.00	**1.69**	0.08	0.02	**1.35**	0.11	**3.15**	m2	**3.39**
Reference A252, 3.95kg/m2; 200mm side laps; 200mm end laps										
generally	2.01	15.00	**2.31**	0.08	0.02	**1.35**	0.14	**3.80**	m2	**4.09**
strips in one width										
750mm wide	2.01	15.00	**2.31**	0.12	0.02	**1.92**	0.14	**4.37**	m2	**4.70**
900mm wide	2.01	15.00	**2.31**	0.11	0.02	**1.78**	0.14	**4.23**	m2	**4.55**
1050mm wide	2.01	15.00	**2.31**	0.10	0.02	**1.64**	0.14	**4.09**	m2	**4.40**
1200mm wide	2.01	15.00	**2.31**	0.10	0.02	**1.64**	0.14	**4.09**	m2	**4.40**
Reference A393, 6.16 kg/m2; 200mm side laps; 200mm end laps										
generally	2.97	15.00	**3.41**	0.09	0.02	**1.49**	0.21	**5.12**	m2	**5.50**
strips in one width										
750mm wide	2.97	15.00	**3.41**	0.14	0.02	**2.20**	0.21	**5.83**	m2	**6.27**
900mm wide	2.97	15.00	**3.41**	0.13	0.02	**2.06**	0.21	**5.69**	m2	**6.11**
1050mm wide	2.97	15.00	**3.41**	0.12	0.02	**1.92**	0.21	**5.54**	m2	**5.96**
1200mm wide	2.97	15.00	**3.41**	0.11	0.02	**1.78**	0.21	**5.40**	m2	**5.81**
Reference B1131, 10.90 kg/m2; 100mm side laps; 200mm end laps										
generally	5.35	15.00	**6.16**	0.12	0.03	**2.03**	0.38	**8.56**	m2	**9.21**
strips in one width										
750mm wide	5.35	15.00	**6.16**	0.18	0.03	**2.88**	0.38	**9.42**	m2	**10.12**
900mm wide	5.35	15.00	**6.16**	0.17	0.03	**2.74**	0.38	**9.27**	m2	**9.97**
1050mm wide	5.35	15.00	**6.16**	0.16	0.03	**2.59**	0.38	**9.13**	m2	**9.82**
1200mm wide	5.35	15.00	**6.16**	0.14	0.03	**2.31**	0.38	**8.85**	m2	**9.51**
Reference B196, 3.05 kg/m2; 100mm side laps; 200mm end laps										
generally	1.50	15.00	**1.73**	0.07	0.02	**1.21**	0.21	**3.15**	m2	**3.39**
strips in one width										
750mm wide	1.50	15.00	**1.73**	0.11	0.02	**1.78**	0.21	**3.72**	m2	**4.00**
900mm wide	1.50	15.00	**1.73**	0.10	0.02	**1.64**	0.21	**3.58**	m2	**3.84**
1050mm wide	1.50	15.00	**1.73**	0.09	0.02	**1.49**	0.21	**3.43**	m2	**3.69**
1200mm wide	1.50	15.00	**1.73**	0.08	0.02	**1.35**	0.21	**3.29**	m2	**3.54**

Labour hourly rates: (except Specialists)
Craft Operatives £14.21 Labourer £10.71
Rates are national average prices.
Refer to REGIONAL VARIATIONS for indicative levels of overall pricing in regions

	MATERIALS			LABOUR				RATES		
	Del to Site £	Waste %	Material Cost £	Craft Optve Hrs	Lab Hrs	Labour Cost £	Sunds £	Nett Rate £	Unit	Gross Rate (+7.5%) £
E30: REINFORCEMENT FOR IN-SITU CONCRETE Cont.										
Reinforcement fabric; BS 4483, hard drawn plain round steel; welded; including laps, tying wire, all cutting and bending, and spacers and chairs which are at the discretion of the Contractor Cont.										
Reference B283, 3.73 kg/m2; 100mm side laps; 200mm end laps										
generally	1.83	15.00	**2.10**	0.07	0.02	**1.21**	0.13	**3.43**	m2	**3.69**
strips in one width										
750mm wide	1.83	15.00	**2.10**	0.11	0.02	**1.78**	0.13	**4.00**	m2	**4.30**
900mm wide	1.83	15.00	**2.10**	0.10	0.02	**1.64**	0.13	**3.86**	m2	**4.15**
1050mm wide	1.83	15.00	**2.10**	0.09	0.02	**1.49**	0.13	**3.72**	m2	**4.00**
1200mm wide	1.83	15.00	**2.10**	0.08	0.02	**1.35**	0.13	**3.58**	m2	**3.84**
Reference B385, 4.53 kg/m2; 100mm side laps; 200mm end laps										
generally	2.22	15.00	**2.56**	0.08	0.02	**1.35**	0.15	**4.06**	m2	**4.37**
strips in one width										
750mm wide	2.22	15.00	**2.56**	0.12	0.02	**1.92**	0.15	**4.63**	m2	**4.98**
900mm wide	2.22	15.00	**2.56**	0.11	0.02	**1.78**	0.15	**4.49**	m2	**4.83**
1050mm wide	2.22	15.00	**2.56**	0.10	0.02	**1.64**	0.15	**4.35**	m2	**4.67**
1200mm wide	2.22	15.00	**2.56**	0.10	0.02	**1.64**	0.15	**4.35**	m2	**4.67**
Reference B503, 5.93 kg/m2; 100mm side laps; 200mm end laps										
generally	2.88	15.00	**3.31**	0.09	0.02	**1.49**	0.21	**5.02**	m2	**5.40**
strips in one width										
750mm wide	2.88	15.00	**3.31**	0.14	0.02	**2.20**	0.21	**5.73**	m2	**6.16**
900mm wide	2.88	15.00	**3.31**	0.13	0.02	**2.06**	0.21	**5.59**	m2	**6.01**
1050mm wide	2.88	15.00	**3.31**	0.12	0.02	**1.92**	0.21	**5.45**	m2	**5.85**
1200mm wide	2.88	15.00	**3.31**	0.11	0.02	**1.78**	0.21	**5.30**	m2	**5.70**
Reference B785, 8.14 kg/m2; 100mm side laps; 200mm end laps										
generally	3.96	15.00	**4.56**	0.10	0.03	**1.74**	0.28	**6.58**	m2	**7.07**
strips in one width										
750mm wide	3.96	15.00	**4.56**	0.15	0.03	**2.45**	0.28	**7.29**	m2	**7.84**
900mm wide	3.96	15.00	**4.56**	0.14	0.03	**2.31**	0.28	**7.15**	m2	**7.68**
1050mm wide	3.96	15.00	**4.56**	0.13	0.03	**2.17**	0.28	**7.01**	m2	**7.53**
1200mm wide	3.96	15.00	**4.56**	0.12	0.03	**2.03**	0.28	**6.86**	m2	**7.38**
Reference C283, 2.61 kg/m2; 100mm side laps; 400mm end laps										
generally	1.27	15.00	**1.46**	0.07	0.02	**1.21**	0.10	**2.76**	m2	**2.97**
strips in one width										
750mm wide	1.27	15.00	**1.46**	0.11	0.02	**1.78**	0.10	**3.33**	m2	**3.58**
900mm wide	1.27	15.00	**1.46**	0.11	0.02	**1.78**	0.10	**3.33**	m2	**3.58**
1050mm wide	1.27	15.00	**1.46**	0.09	0.02	**1.49**	0.10	**3.05**	m2	**3.28**
1200mm wide	1.27	15.00	**1.46**	0.08	0.02	**1.35**	0.10	**2.91**	m2	**3.12**
Reference C385, 3.41 kg/m2; 100mm side laps; 400mm end laps										
generally	1.65	15.00	**1.90**	0.07	0.02	**1.21**	0.13	**3.24**	m2	**3.48**
strips in one width										
750mm wide	1.65	15.00	**1.90**	0.11	0.02	**1.78**	0.13	**3.80**	m2	**4.09**
900mm wide	1.65	15.00	**1.90**	0.10	0.02	**1.64**	0.13	**3.66**	m2	**3.94**
1050mm wide	1.65	15.00	**1.90**	0.09	0.02	**1.49**	0.13	**3.52**	m2	**3.78**
1200mm wide	1.65	15.00	**1.90**	0.08	0.02	**1.35**	0.13	**3.38**	m2	**3.63**
Reference C503, 4.34 kg/m2; 100mm side laps; 400mm end laps										
generally	2.04	15.00	**2.34**	0.08	0.02	**1.35**	0.15	**3.85**	m2	**4.14**
strips in one width										
750mm wide	2.04	15.00	**2.34**	0.12	0.02	**1.92**	0.15	**4.42**	m2	**4.75**
900mm wide	2.04	15.00	**2.34**	0.11	0.02	**1.78**	0.15	**4.27**	m2	**4.59**
1050mm wide	2.04	15.00	**2.34**	0.10	0.02	**1.64**	0.15	**4.13**	m2	**4.44**
1200mm wide	2.04	15.00	**2.34**	0.10	0.02	**1.64**	0.15	**4.13**	m2	**4.44**
Reference C636, 5.55 kg/m2; 100mm side laps; 400mm end laps										
generally	2.72	15.00	**3.13**	0.09	0.02	**1.49**	0.20	**4.82**	m2	**5.18**
strips in one width										
750mm wide	2.72	15.00	**3.13**	0.14	0.02	**2.20**	0.20	**5.53**	m2	**5.94**
900mm wide	2.72	15.00	**3.13**	0.13	0.02	**2.06**	0.20	**5.39**	m2	**5.79**
1050mm wide	2.72	15.00	**3.13**	0.12	0.02	**1.92**	0.20	**5.24**	m2	**5.64**
1200mm wide	2.72	15.00	**3.13**	0.11	0.02	**1.78**	0.20	**5.10**	m2	**5.48**
Reference C785, 6.72 kg/m2; 100mm side laps; 400mm end laps										
generally	3.28	15.00	**3.77**	0.09	0.02	**1.49**	0.24	**5.50**	m2	**5.92**
strips in one width										
750mm wide	3.28	15.00	**3.77**	0.14	0.02	**2.20**	0.24	**6.21**	m2	**6.68**
900mm wide	3.28	15.00	**3.77**	0.13	0.02	**2.06**	0.24	**6.07**	m2	**6.53**
1050mm wide	3.28	15.00	**3.77**	0.12	0.02	**1.92**	0.24	**5.93**	m2	**6.37**
1200mm wide	3.28	15.00	**3.77**	0.11	0.02	**1.78**	0.24	**5.79**	m2	**6.22**
Reference D49, 0.77 kg/m2; 100mm side laps; 100mm end laps										
bent	0.77	15.00	**0.88**	0.03	0.01	**0.53**	0.08	**1.50**	m2	**1.61**
Reference D98, 1.54 kg/m2; 200mm side laps; 200mm end laps										
bent	0.72	15.00	**0.82**	0.03	0.01	**0.53**	0.08	**1.44**	m2	**1.55**

Labour hourly rates: (except Specialists) Craft Operatives £14.21 Labourer £10.71 Rates are national average prices. Refer to REGIONAL VARIATIONS for indicative levels of overall pricing in regions	MATERIALS			LABOUR				RATES		
	Del to Site £	Waste %	Material Cost £	Craft Optve Hrs	Lab Hrs	Labour Cost £	Sunds £	Nett Rate £	Unit	Gross Rate (+7.5%) £
E30: REINFORCEMENT FOR IN-SITU CONCRETE Cont.										
Expamet 76mm mesh expanded steel reinforcement (uncoated); including laps, tying wire, all cutting and bending, and spacers and chairs which are at the discretion of the Contractor										
Reference 8, 3.80 kg/m2; 150mm side laps; 150mm end laps										
generally	11.04	15.00	**12.70**	0.07	0.02	**1.21**	0.59	**14.50**	m2	**15.59**
strips in one width										
750mm wide	11.04	15.00	**12.70**	0.11	0.02	**1.78**	0.59	**15.07**	m2	**16.20**
900mm wide	11.04	15.00	**12.70**	0.10	0.02	**1.64**	0.59	**14.93**	m2	**16.04**
1050mm wide	11.04	15.00	**12.70**	0.09	0.02	**1.49**	0.59	**14.78**	m2	**15.89**
1200mm wide	11.04	15.00	**12.70**	0.08	0.02	**1.35**	0.59	**14.64**	m2	**15.74**
Expamet 'Securilath' flattened security mesh expanded steel reinforcement (uncoated); including laps										
Reference 2073F; 150mm side laps; 150mm end laps										
generally	24.51	15.00	**28.19**	0.10	0.02	**1.64**	0.28	**30.10**	m2	**32.36**
E40: DESIGNED JOINTS IN IN-SITU CONCRETE										
Formed joints										
Incorporating 10mm thick Korkpak, Grace Construction Products Limited - Servicised; formwork; reinforcement laid continuously across joint										
in concrete, depth not exceeding 150mm; horizontal	3.60	5.00	**3.78**	-	0.07	**0.75**	4.96	**9.49**	m	**10.20**
in concrete, depth 150 - 300mm; horizontal	6.30	5.00	**6.61**	-	0.08	**0.86**	9.13	**16.60**	m	**17.84**
in concrete, depth 300 - 450mm; horizontal	9.00	5.00	**9.44**	-	0.09	**0.96**	13.28	**23.69**	m	**25.47**
Incorporating 13mm thick Korkpak, Grace Construction Products Limited - Servicised; formwork; reinforcement laid continuously across joint										
in concrete, depth not exceeding 150mm; horizontal	3.68	5.00	**3.86**	-	0.07	**0.75**	4.96	**9.57**	m	**10.29**
in concrete, depth 150 - 300mm; horizontal	6.43	5.00	**6.75**	-	0.08	**0.86**	9.13	**16.74**	m	**18.00**
in concrete, depth 300 - 450mm; horizontal	9.19	5.00	**9.65**	-	0.09	**0.96**	13.28	**23.90**	m	**25.69**
Incorporating 19mm thick Korkpak, Grace Construction Products Limited - Servicised; formwork; reinforcement laid continuously across joint										
in concrete, depth not exceeding 150mm; horizontal	5.18	5.00	**5.43**	-	0.07	**0.75**	4.96	**11.15**	m	**11.98**
in concrete, depth 150 - 300mm; horizontal	9.06	5.00	**9.51**	-	0.08	**0.86**	9.13	**19.50**	m	**20.96**
in concrete, depth 300 - 450mm; horizontal	12.94	5.00	**13.59**	-	0.09	**0.96**	13.28	**27.83**	m	**29.92**
Incorporating 195mm wide 'Serviseal 195', external face type p.v.c waterstop, Grace Construction Products Limited - Servicised; heat welded joints; formwork; reinforcement laid continuously across joint										
in concrete, depth not exceeding 150mm; horizontal	5.83	10.00	**6.41**	-	0.40	**4.28**	4.96	**15.66**	m	**16.83**
in concrete, depth 150 - 300mm; horizontal	5.83	10.00	**6.41**	-	0.40	**4.28**	9.13	**19.83**	m	**21.31**
in concrete, depth 300 - 450mm; horizontal	5.83	10.00	**6.41**	-	0.40	**4.28**	13.28	**23.98**	m	**25.78**
vertical L piece	15.59	10.00	**17.15**	-	0.20	**2.14**	-	**19.29**	nr	**20.74**
flat L piece	8.68	10.00	**9.55**	-	0.20	**2.14**	-	**11.69**	nr	**12.57**
flat T piece	14.89	10.00	**16.38**	-	0.25	**2.68**	-	**19.06**	nr	**20.49**
flat X piece	22.46	10.00	**24.71**	-	0.30	**3.21**	-	**27.92**	nr	**30.01**
in concrete, width not exceeding 150mm; vertical	5.83	10.00	**6.41**	-	0.50	**5.36**	4.96	**16.73**	m	**17.98**
in concrete, width 150 - 300mm; vertical	5.83	10.00	**6.41**	-	0.50	**5.36**	9.13	**20.90**	m	**22.46**
in concrete, width 300 - 450mm; vertical	5.83	10.00	**6.41**	-	0.50	**5.36**	13.28	**25.05**	m	**26.93**
flat L piece	8.68	10.00	**9.55**	-	0.20	**2.14**	-	**11.69**	nr	**12.57**
flat T piece	14.89	10.00	**16.38**	-	0.25	**2.68**	-	**19.06**	nr	**20.49**
flat X piece	22.46	10.00	**24.71**	-	0.30	**3.21**	-	**27.92**	nr	**30.01**
Incorporating 240mm wide 'Serviseal 240', heavy duty section external face type p.v.c. waterstop, Grace Construction Products Limited - Servicised; heat welded joints; formwork; reinforcement laid continuously across joint										
in concrete, depth not exceeding 150mm; horizontal	7.12	10.00	**7.83**	-	0.45	**4.82**	4.96	**17.61**	m	**18.93**
in concrete, depth 150 - 300mm; horizontal	7.12	10.00	**7.83**	-	0.45	**4.82**	9.13	**21.78**	m	**23.41**
in concrete, depth 300 - 450mm; horizontal	7.12	10.00	**7.83**	-	0.45	**4.82**	13.28	**25.93**	m	**27.88**
vertical L piece	16.30	10.00	**17.93**	-	0.20	**2.14**	-	**20.07**	nr	**21.58**
flat L piece	9.87	10.00	**10.86**	-	0.20	**2.14**	-	**13.00**	nr	**13.97**
flat T piece	16.69	10.00	**18.36**	-	0.25	**2.68**	-	**21.04**	nr	**22.61**
flat X piece	24.85	10.00	**27.34**	-	0.30	**3.21**	-	**30.55**	nr	**32.84**
in concrete, width not exceeding 150mm; vertical	7.12	10.00	**7.83**	-	0.55	**5.89**	4.96	**18.68**	m	**20.09**
in concrete, width 150 - 300mm; vertical	7.12	10.00	**7.83**	-	0.55	**5.89**	9.13	**22.85**	m	**24.56**
in concrete, width 300 - 450mm; vertical	7.12	10.00	**7.83**	-	0.55	**5.89**	13.28	**27.01**	m	**29.03**
flat L piece	9.87	10.00	**10.86**	-	0.20	**2.14**	-	**13.00**	nr	**13.97**
flat T piece	16.69	10.00	**18.36**	-	0.25	**2.68**	-	**21.04**	nr	**22.61**
flat X piece	24.85	10.00	**27.34**	-	0.30	**3.21**	-	**30.55**	nr	**32.84**
Incorporating 320mm wide 'Serviseal Kicker 320', external face type p.v.c. kickerjoint waterstop, Grace Construction Products Limited - Servicised; heat welded joints; formwork; reinforcement laid continuously across joint										
in concrete, depth not exceeding 150mm; vertical	9.96	10.00	**10.96**	-	0.60	**6.43**	4.96	**22.34**	m	**24.02**
in concrete, depth 150 - 300mm; vertical	9.96	10.00	**10.96**	-	0.60	**6.43**	9.13	**26.51**	m	**28.50**
in concrete, depth 300 - 450mm; vertical	9.96	10.00	**10.96**	-	0.60	**6.43**	13.28	**30.67**	m	**32.97**
vertical L piece	18.92	10.00	**20.81**	-	0.25	**2.68**	-	**23.49**	nr	**25.25**
flat L piece	22.23	10.00	**24.45**	-	0.25	**2.68**	-	**27.13**	nr	**29.17**
flat T piece	32.55	10.00	**35.81**	-	0.30	**3.21**	-	**39.02**	nr	**41.94**
flat X piece	40.54	10.00	**44.59**	-	0.35	**3.75**	-	**48.34**	nr	**51.97**

Labour hourly rates: (except Specialists) Craft Operatives £14.21 Labourer £10.71 Rates are national average prices. Refer to REGIONAL VARIATIONS for indicative levels of overall pricing in regions	MATERIALS			LABOUR				RATES		
	Del to Site £	Waste %	Material Cost £	Craft Optve Hrs	Lab Hrs	Labour Cost £	Sunds £	Nett Rate £	Unit	Gross Rate (+7.5%) £
E40: DESIGNED JOINTS IN IN-SITU CONCRETE Cont.										
Formed joints Cont.										
Incorporating 210mm wide 'PVC Edgetie 210' centre bulb type internally placed p.v.c. waterstop, Grace Construction Products Limited - Servicised; heat welded joints; formwork; reinforcement laid continuously across joint										
in concrete, depth not exceeding 150mm; horizontal	5.10	10.00	**5.61**	-	0.40	**4.28**	4.96	**14.86**	m	**15.97**
in concrete, depth 150 - 300mm; horizontal	5.10	10.00	**5.61**	-	0.40	**4.28**	9.13	**19.02**	m	**20.45**
in concrete, depth 300 - 450mm; horizontal	5.10	10.00	**5.61**	-	0.40	**4.28**	13.28	**23.18**	m	**24.92**
flat L piece	10.98	10.00	**12.08**	-	0.20	**2.14**	-	**14.22**	nr	**15.29**
flat T piece	14.51	10.00	**15.96**	-	0.25	**2.68**	-	**18.64**	nr	**20.04**
flat X piece	21.95	10.00	**24.15**	-	0.30	**3.21**	-	**27.36**	nr	**29.41**
in concrete, width not exceeding 150mm; vertical	5.10	10.00	**5.61**	-	0.50	**5.36**	4.96	**15.93**	m	**17.12**
in concrete, width 150 - 300mm; vertical	5.10	10.00	**5.61**	-	0.50	**5.36**	9.13	**20.09**	m	**21.60**
in concrete, width 300 - 450mm; vertical	5.10	10.00	**5.61**	-	0.50	**5.36**	13.28	**24.25**	m	**26.07**
vertical L piece	14.15	10.00	**15.57**	-	0.20	**2.14**	-	**17.71**	nr	**19.04**
flat L piece	10.98	10.00	**12.08**	-	0.20	**2.14**	-	**14.22**	nr	**15.29**
vertical T piece	22.13	10.00	**24.34**	-	0.25	**2.68**	-	**27.02**	nr	**29.05**
flat T piece	14.51	10.00	**15.96**	-	0.25	**2.68**	-	**18.64**	nr	**20.04**
flat X piece	21.95	10.00	**24.15**	-	0.30	**3.21**	-	**27.36**	nr	**29.41**
Incorporating 260mm wide 'PVC Edgetie 260' centre bulb type internally placed p.v.c. waterstop, Grace Construction Products Limited - Servicised; heat welded joints; formwork; reinforcement laid continuously across joint										
in concrete, depth not exceeding 150mm; horizontal	5.84	10.00	**6.42**	-	0.45	**4.82**	4.96	**16.20**	m	**17.42**
in concrete, depth 150 - 300mm; horizontal	5.84	10.00	**6.42**	-	0.45	**4.82**	9.13	**20.37**	m	**21.90**
in concrete, depth 300 - 450mm; horizontal	5.84	10.00	**6.42**	-	0.45	**4.82**	13.28	**24.53**	m	**26.37**
flat L piece	12.36	10.00	**13.60**	-	0.20	**2.14**	-	**15.74**	nr	**16.92**
flat T piece	19.00	10.00	**20.90**	-	0.25	**2.68**	-	**23.58**	nr	**25.35**
flat X piece	26.68	10.00	**29.35**	-	0.30	**3.21**	-	**32.56**	nr	**35.00**
in concrete, width not exceeding 150mm; vertical	5.84	10.00	**6.42**	-	0.55	**5.89**	4.96	**17.28**	m	**18.57**
in concrete, width 150 - 300mm; vertical	5.84	10.00	**6.42**	-	0.55	**5.89**	9.13	**21.44**	m	**23.05**
in concrete, width 300 - 450mm; vertical	5.84	10.00	**6.42**	-	0.55	**5.89**	13.28	**25.60**	m	**27.52**
vertical L piece	19.64	10.00	**21.60**	-	0.20	**2.14**	-	**23.75**	nr	**25.53**
flat L piece	12.36	10.00	**13.60**	-	0.20	**2.14**	-	**15.74**	nr	**16.92**
vertical T piece	26.50	10.00	**29.15**	-	0.25	**2.68**	-	**31.83**	nr	**34.21**
flat T piece	19.00	10.00	**20.90**	-	0.25	**2.68**	-	**23.58**	nr	**25.35**
flat X piece	26.68	10.00	**29.35**	-	0.30	**3.21**	-	**32.56**	nr	**35.00**
Incorporating 445mm wide Reference 2411, 6.34 kg/m2 Expamet Hy-rib permanent shuttering and reinforcement; 150mm end laps; temporary supports; formwork laid continuously across joint										
in concrete, depth not exceeding 150mm; horizontal	3.11	5.00	**3.26**	0.23	0.06	**3.91**	5.50	**12.67**	m	**13.62**
in concrete, depth 150 - 300mm; horizontal	6.21	5.00	**6.52**	0.23	0.06	**3.91**	9.59	**20.03**	m	**21.53**
in concrete, width not exceeding 150mm; vertical	3.11	5.00	**3.26**	0.35	0.09	**5.94**	5.50	**14.70**	m	**15.80**
in concrete, width 150 - 300mm; vertical	6.21	5.00	**6.52**	0.35	0.09	**5.94**	9.59	**22.05**	m	**23.71**
Incorporating 445mm wide Reference 2611, 4.86 kg/m2 Expamet Hy-rib permanent shuttering and reinforcement; 150mm end laps; temporary supports; formwork laid continuously across joint										
in concrete, depth not exceeding 150mm; horizontal	1.85	5.00	**1.95**	0.23	0.06	**3.91**	5.50	**11.36**	m	**12.21**
in concrete, depth 150 - 300mm; horizontal	3.71	5.00	**3.89**	0.23	0.06	**3.91**	9.59	**17.39**	m	**18.70**
in concrete, width not exceeding 150mm; vertical	1.85	5.00	**1.95**	0.35	0.09	**5.94**	5.50	**13.38**	m	**14.39**
in concrete, width 150 - 300mm; vertical	3.71	5.00	**3.89**	0.35	0.09	**5.94**	9.59	**19.42**	m	**20.88**
Incorporating 445mm wide Reference 2811, 3.39 kg/m2 Expamet Hy-rib permanent shuttering and reinforcement; 150mm end laps; temporary supports; formwork laid continuously across joint										
in concrete, depth not exceeding 150mm; horizontal	4.66	5.00	**4.89**	0.23	0.06	**3.91**	5.50	**14.30**	m	**15.37**
in concrete, depth 150 - 300mm; horizontal	9.32	5.00	**9.78**	0.23	0.06	**3.91**	9.59	**23.29**	m	**25.03**
in concrete, width not exceeding 150mm; vertical	4.66	5.00	**4.89**	0.35	0.09	**5.94**	5.50	**16.33**	m	**17.55**
in concrete, width 150 - 300mm; vertical	9.32	5.00	**9.78**	0.35	0.09	**5.94**	9.59	**25.31**	m	**27.21**
Sealant to joint; Grace Construction Products Limited - Servicised. 'Servigard DW' moisture tolerant 2 part dark grey compound; including preparation, cleaning and priming with water, mixing sealant with cement										
10 x 25mm; horizontal	3.58	5.00	**3.76**	-	0.15	**1.61**	-	**5.37**	m	**5.77**
13 x 25mm; horizontal	4.58	5.00	**4.81**	-	0.20	**2.14**	-	**6.95**	m	**7.47**
19 x 25mm; horizontal	6.58	5.00	**6.91**	-	0.25	**2.68**	-	**9.58**	m	**10.30**
Sealant to joint; Grace Construction Products Limited - Servicised. 'Paraseal' pouring grade 2 part grey polysulphide sealant ; including preparation, cleaners, primers and sealant										
10 x 25mm; horizontal	2.45	5.00	**2.57**	-	0.10	**1.07**	-	**3.64**	m	**3.91**
13 x 25mm; horizontal	3.18	5.00	**3.34**	-	0.11	**1.18**	-	**4.52**	m	**4.86**
19 x 25mm; horizontal	4.65	5.00	**4.88**	-	0.12	**1.29**	-	**6.17**	m	**6.63**
E41: WORKED FINISHES/ CUTTING ON IN-SITU CONCRETE										
Worked finishes on in-situ concrete										
Vacuum dewatering; power floating and power trowelling surfaces	-	-	-	-	0.20	**2.14**	0.27	**2.41**	m2	**2.59**
Tamping unset concrete surfaces	-	-	-	-	0.20	**2.14**	0.33	**2.47**	m2	**2.66**

Labour hourly rates: (except Specialists) Craft Operatives £14.21 Labourer £10.71 Rates are national average prices. Refer to REGIONAL VARIATIONS for indicative levels of overall pricing in regions	MATERIALS			LABOUR				RATES		
	Del to Site £	Waste %	Material Cost £	Craft Optve Hrs	Lab Hrs	Labour Cost £	Sunds £	Nett Rate £	Unit	Gross Rate (+7.5%) £
E41: WORKED FINISHES/ CUTTING ON IN-SITU CONCRETE Cont.										
Worked finishes on in-situ concrete Cont.										
Tamping unset concrete Cont.										
surfaces to falls	-	-	-	-	0.40	**4.28**	0.42	**4.70**	m2	**5.05**
Trowelling										
surfaces	-	-	-	-	0.33	**3.53**	-	**3.53**	m2	**3.80**
surfaces to falls	-	-	-	-	0.35	**3.75**	-	**3.75**	m2	**4.03**
Hacking; by hand										
surfaces	-	-	-	-	0.50	**5.36**	-	**5.36**	m2	**5.76**
to soffits	-	-	-	-	0.65	**6.96**	-	**6.96**	m2	**7.48**
Hacking; by machine										
surfaces	-	-	-	-	0.20	**2.14**	0.27	**2.41**	m2	**2.59**
to soffits	-	-	-	-	0.25	**2.68**	0.27	**2.95**	m2	**3.17**
Bush hammering										
surfaces	-	-	-	1.30	-	**18.47**	0.76	**19.23**	m2	**20.67**
to soffits	-	-	-	2.00	-	**28.42**	1.00	**29.42**	m2	**31.63**
Cutting on in-situ concrete										
Cutting chases										
depth not exceeding 50mm	-	-	-	-	0.17	**1.82**	-	**1.82**	m	**1.96**
Cutting chases; making good										
depth not exceeding 50mm	-	-	-	-	0.20	**2.14**	0.24	**2.39**	m	**2.57**
Cutting mortices										
50 x 50mm; depth not exceeding 100mm	-	-	-	-	0.50	**5.36**	-	**5.36**	nr	**5.76**
75 x 75mm; depth not exceeding 100mm	-	-	-	-	0.50	**5.36**	-	**5.36**	nr	**5.76**
100 x 100mm; depth 100 - 200mm	-	-	-	-	0.60	**6.43**	-	**6.43**	nr	**6.91**
38mm diameter; depth not exceeding 100mm	-	-	-	-	0.35	**3.75**	-	**3.75**	nr	**4.03**
Cutting holes										
225 x 150mm; depth not exceeding 100mm	-	-	-	-	1.32	**14.14**	0.35	**14.49**	nr	**15.58**
300 x 150mm; depth not exceeding 100mm	-	-	-	-	1.45	**15.53**	0.39	**15.92**	nr	**17.11**
300 x 300mm; depth not exceeding 100mm	-	-	-	-	1.45	**15.53**	0.39	**15.92**	nr	**17.11**
225 x 150mm; depth 100 - 200mm	-	-	-	-	1.92	**20.56**	0.56	**21.13**	nr	**22.71**
300 x 150mm; depth 100 - 200mm	-	-	-	-	2.11	**22.60**	0.57	**23.17**	nr	**24.91**
300 x 300mm; depth 100 - 200mm	-	-	-	-	2.11	**22.60**	0.57	**23.17**	nr	**24.91**
225 x 150mm; depth 200 - 300mm	-	-	-	-	2.21	**23.67**	0.59	**24.26**	nr	**26.07**
300 x 150mm; depth 200 - 300mm	-	-	-	-	2.41	**25.81**	0.65	**26.46**	nr	**28.44**
300 x 300mm; depth 200 - 300mm	-	-	-	-	2.41	**25.81**	0.65	**26.46**	nr	**28.44**
50mm diameter; depth not exceeding 100mm	-	-	-	-	0.85	**9.10**	0.22	**9.32**	nr	**10.02**
100mm diameter; depth not exceeding 100mm	-	-	-	-	0.94	**10.07**	0.28	**10.35**	nr	**11.12**
150mm diameter; depth not exceeding 100mm	-	-	-	-	0.94	**10.07**	0.28	**10.35**	nr	**11.12**
50mm diameter; depth 100 - 200mm	-	-	-	-	1.37	**14.67**	0.37	**15.04**	nr	**16.17**
100mm diameter; depth 100 - 200mm	-	-	-	-	1.47	**15.74**	0.39	**16.13**	nr	**17.34**
150mm diameter; depth 100 - 200mm	-	-	-	-	1.47	**15.74**	0.39	**16.13**	nr	**17.34**
50mm diameter; depth 200 - 300mm	-	-	-	-	1.63	**17.46**	0.45	**17.91**	nr	**19.25**
100mm diameter; depth 200 - 300mm	-	-	-	-	1.79	**19.17**	0.50	**19.67**	nr	**21.15**
150mm diameter; depth 200 - 300mm	-	-	-	-	1.79	**19.17**	0.50	**19.67**	nr	**21.15**
Grouting into mortices with cement mortar (1:1); around steel										
50 x 50mm; depth not exceeding 100mm	-	-	-	-	0.12	**1.29**	0.55	**1.84**	nr	**1.97**
75 x 75mm; depth not exceeding 100mm	-	-	-	-	0.12	**1.29**	0.55	**1.84**	nr	**1.97**
100 x 100mm; depth 100 - 200mm	-	-	-	-	0.15	**1.61**	0.71	**2.32**	nr	**2.49**
38mm diameter; depth not exceeding 100mm	-	-	-	-	0.08	**0.86**	0.39	**1.25**	nr	**1.34**
Cutting on reinforced in-situ concrete										
Cutting holes										
225 x 150mm; depth not exceeding 100mm	-	-	-	-	1.64	**17.56**	0.45	**18.02**	nr	**19.37**
300 x 150mm; depth not exceeding 100mm	-	-	-	-	1.81	**19.39**	0.51	**19.90**	nr	**21.39**
300 x 300mm; depth not exceeding 100mm	-	-	-	-	1.81	**19.39**	0.51	**19.90**	nr	**21.39**
225 x 150mm; depth 100 - 200mm	-	-	-	-	2.39	**25.60**	0.65	**26.24**	nr	**28.21**
300 x 150mm; depth 100 - 200mm	-	-	-	-	2.63	**28.17**	0.71	**28.88**	nr	**31.04**
300 x 300mm; depth 100 - 200mm	-	-	-	-	2.63	**28.17**	0.65	**28.81**	nr	**30.98**
225 x 150mm; depth 200 - 300mm	-	-	-	-	2.78	**29.77**	0.76	**30.53**	nr	**32.82**
300 x 150mm; depth 200 - 300mm	-	-	-	-	3.03	**32.45**	0.83	**33.28**	nr	**35.78**
300 x 300mm; depth 200 - 300mm	-	-	-	-	3.03	**32.45**	0.83	**33.28**	nr	**35.78**
50mm diameter; depth not exceeding 100mm	-	-	-	-	1.05	**11.25**	0.23	**11.48**	nr	**12.34**
100mm diameter; depth not exceeding 100mm	-	-	-	-	1.18	**12.64**	0.34	**12.98**	nr	**13.95**
150mm diameter; depth not exceeding 100mm	-	-	-	-	1.18	**12.64**	0.34	**12.98**	nr	**13.95**
50mm diameter; depth 100 - 200mm	-	-	-	-	1.72	**18.42**	0.48	**18.90**	nr	**20.32**
100mm diameter; depth 100 - 200mm	-	-	-	-	1.84	**19.71**	0.51	**20.22**	nr	**21.74**
150mm diameter; depth 100 - 200mm	-	-	-	-	1.84	**19.71**	0.51	**20.22**	nr	**21.74**
50mm diameter; depth 200 - 300mm	-	-	-	-	2.04	**21.85**	0.56	**22.41**	nr	**24.09**
100mm diameter; depth 200 - 300mm	-	-	-	-	2.24	**23.99**	0.60	**24.59**	nr	**26.43**
150mm diameter; depth 200 - 300mm	-	-	-	-	2.24	**23.99**	0.60	**24.59**	nr	**26.43**
E42: ACCESSORIES CAST INTO IN-SITU CONCRETE										
Cast in accessories										
Galvanized steel dowels										
12mm diameter x 150mm long	0.58	10.00	**0.64**	-	0.08	**0.86**	-	**1.50**	nr	**1.61**
12mm diameter x 300mm long	0.86	10.00	**0.95**	-	0.08	**0.86**	-	**1.81**	nr	**1.94**
12mm diameter x 600mm long	1.30	10.00	**1.43**	-	0.12	**1.29**	-	**2.72**	nr	**2.92**

Labour hourly rates: (except Specialists) Craft Operatives £14.21 Labourer £10.71 Rates are national average prices. Refer to REGIONAL VARIATIONS for indicative levels of overall pricing in regions	MATERIALS			LABOUR				RATES		
	Del to Site £	Waste %	Material Cost £	Craft Optve Hrs	Lab Hrs	Labour Cost £	Sunds £	Nett Rate £	Unit	Gross Rate (+7.5%) £
E42: ACCESSORIES CAST INTO IN-SITU CONCRETE Cont.										
Cast in accessories Cont.										
Galvanized expanded metal tie										
300 x 50mm; bending and temporarily fixing to formwork	0.70	10.00	**0.77**	-	0.12	**1.29**	-	**2.05**	nr	**2.21**
Galvanized steel dovetailed masonry slots with 3mm thick twisted tie to suit 50mm cavity										
100mm long; temporarily fixing to formwork	0.71	10.00	**0.78**	0.10	-	**1.42**	-	**2.20**	nr	**2.37**
Stainless steel channel with anchors at 250mm centres										
28 x 15mm; temporarily fixing to formwork	11.44	5.00	**12.01**	0.28	-	**3.98**	-	**15.99**	m	**17.18**
28 x 15 x 150mm long; temporarily fixing to formwork	1.98	5.00	**2.08**	0.12	-	**1.71**	-	**3.78**	nr	**4.07**
M10 bolt 50mm long with `T' head nut and washers	3.91	5.00	**4.10**	0.05	-	**0.71**	-	**4.81**	nr	**5.17**
fishtailed tie to suit 50mm cavity	1.65	5.00	**1.73**	0.06	-	**0.85**	-	**2.58**	nr	**2.78**
Stainless steel angle drilled at 450mm centres										
70 x 90 x 5mm	28.68	5.00	**30.12**	0.80	-	**11.37**	-	**41.49**	m	**44.60**
Copper dovetailed masonry slots with 3mm thick twisted tie to suit 50mm cavity										
100mm long; temporarily fixing to formwork	8.37	10.00	**9.21**	0.10	-	**1.42**	-	**10.63**	nr	**11.42**
Bolt boxes; The Expanded Metal Co. Ltd										
reference 220 for use with poured concrete; 75mm diameter										
150mm long	1.28	10.00	**1.41**	0.25	-	**3.55**	0.68	**5.65**	nr	**6.07**
225mm long	1.54	10.00	**1.70**	0.30	-	**4.26**	0.68	**6.64**	nr	**7.14**
300mm long	1.82	10.00	**2.00**	0.35	-	**4.97**	0.68	**7.66**	nr	**8.23**
reference 220 for use with poured concrete; 100mm diameter										
375mm long	2.93	10.00	**3.23**	0.40	-	**5.68**	0.68	**9.59**	nr	**10.31**
450mm long	3.30	10.00	**3.63**	0.45	-	**6.39**	0.68	**10.71**	nr	**11.51**
600mm long	3.66	10.00	**4.02**	0.50	-	**7.11**	0.68	**11.81**	nr	**12.70**
reference 220 for use with vibrated concrete; wrap with single layer of thin polythene sheet; 75mm diameter										
150mm long	1.44	10.00	**1.58**	0.25	-	**3.55**	0.82	**5.95**	nr	**6.40**
225mm long	1.72	10.00	**1.89**	0.30	-	**4.26**	0.82	**6.97**	nr	**7.50**
300mm long	1.99	10.00	**2.19**	0.35	-	**4.97**	0.82	**7.99**	nr	**8.58**
reference 220 for use with vibrated concrete; wrap with single layer of thin polythene sheet; 100mm diameter										
375mm long	3.11	10.00	**3.42**	0.40	-	**5.68**	0.82	**9.92**	nr	**10.67**
450mm long	3.51	10.00	**3.87**	0.45	-	**6.39**	0.82	**11.08**	nr	**11.91**
600mm long	3.86	10.00	**4.24**	0.50	-	**7.11**	0.82	**12.17**	nr	**13.08**
Steel rag bolts										
M 10 x 100mm long	1.28	5.00	**1.34**	0.09	-	**1.28**	-	**2.62**	nr	**2.82**
M 10 x 160mm long	1.38	5.00	**1.44**	0.10	-	**1.42**	-	**2.86**	nr	**3.08**
M 12 x 100mm long	1.46	5.00	**1.53**	0.09	-	**1.28**	-	**2.81**	nr	**3.02**
M 12 x 160mm long	1.64	5.00	**1.72**	0.10	-	**1.42**	-	**3.14**	nr	**3.38**
M 12 x 200mm long	1.76	5.00	**1.85**	0.12	-	**1.71**	-	**3.56**	nr	**3.82**
M 16 x 120mm long	2.19	5.00	**2.30**	0.09	-	**1.28**	-	**3.58**	nr	**3.85**
M 16 x 160mm long	2.50	5.00	**2.62**	0.10	-	**1.42**	-	**4.04**	nr	**4.35**
M 16 x 200mm long	2.80	5.00	**2.94**	0.12	-	**1.71**	-	**4.65**	nr	**5.00**
E60: PRECAST/COMPOSITE CONCRETE DECKING										
Prestressed concrete beams and 100mm building blocks; hoist bed and grout										
155mm thick floors (Bison 'House Beam')										
beams at 510mm centres	12.09	2.50	**12.40**	-	1.05	**11.25**	0.49	**24.13**	m2	**25.94**
beams at 510mm and 285mm centres	14.06	2.50	**14.41**	-	0.95	**10.17**	0.49	**25.08**	m2	**26.96**
double beams at 624mm centres	16.54	2.50	**16.95**	-	0.90	**9.64**	0.49	**27.08**	m2	**29.11**
beams at 285mm centres	17.61	2.50	**18.05**	-	1.05	**11.25**	0.49	**29.78**	m2	**32.01**
double beams at 399mm centres	23.27	2.50	**23.85**	-	0.95	**10.17**	0.49	**34.51**	m2	**37.10**
beams at 114mm centres with concrete between	39.08	2.50	**40.05**	-	0.90	**9.64**	0.49	**50.18**	m2	**53.94**
225mm thick floors (Bison 'Deep Beam')										
beams at 530mm centres	16.51	2.50	**16.93**	-	1.05	**11.25**	0.49	**28.66**	m2	**30.81**
beams at 305mm centres	24.93	2.50	**25.55**	-	0.95	**10.17**	0.49	**36.21**	m2	**38.93**
beams at 140mm centres with concrete between	50.63	2.50	**51.90**	-	0.90	**9.64**	0.49	**62.03**	m2	**66.68**
Precast concrete hollowcore floors and hoist bed and grout (HANSON CONCRETE PRODUCTS LTD)										
150 mm thick planks 750 wide										
span between supports 7.6m; UDL not exceeding 2KN/m2	35.22	2.50	**36.10**	-	-	**-**	-	**36.10**	m2	**38.81**
span between supports 6.6m; UDL not exceeding 4KN/m2	35.22	2.50	**36.10**	-	-	**-**	-	**36.10**	m2	**38.81**
span between supports 5.9m; UDL not exceeding 6KN/m2	35.22	2.50	**36.10**	-	-	**-**	-	**36.10**	m2	**38.81**
span between supports 5.4m; UDL not exceeding 8KN/m2	35.22	2.50	**36.10**	-	-	**-**	-	**36.10**	m2	**38.81**
span between supports 5.0m; UDL not exceeding 10KN/m2	35.22	2.50	**36.10**	-	-	**-**	-	**36.10**	m2	**38.81**
200 mm thick planks 750 wide										
span between supports 9.3m; UDL not exceeding 2KN/m2	38.91	2.50	**39.88**	-	-	**-**	-	**39.88**	m2	**42.87**

Labour hourly rates: (except Specialists) Craft Operatives £14.21 Labourer £10.71 Rates are national average prices. Refer to REGIONAL VARIATIONS for indicative levels of overall pricing in regions	MATERIALS			LABOUR				RATES		
	Del to Site £	Waste %	Material Cost £	Craft Optve Hrs	Lab Hrs	Labour Cost £	Sunds £	Nett Rate £	Unit	Gross Rate (+7.5%) £
E60: PRECAST/COMPOSITE CONCRETE DECKING Cont.										
Precast concrete hollowcore floors and hoist bed and grout (HANSON CONCRETE PRODUCTS LTD) Cont.										
200 mm thick planks 750 wide Cont.										
span between supports 8.1m; UDL not exceeding 4KN/m2	38.91	2.50	**39.88**	-	-	-	-	**39.88**	m2	**42.87**
span between supports 7.3m; UDL not exceeding 6KN/m2	38.91	2.50	**39.88**	-	-	-	-	**39.88**	m2	**42.87**
span between supports 6.7m; UDL not exceeding 8KN/m2	38.91	2.50	**39.88**	-	-	-	-	**39.88**	m2	**42.87**
span between supports 6.2m; UDL not exceeding 10KN/m2	38.91	2.50	**39.88**	-	-	-	-	**39.88**	m2	**42.87**
250 mm thick planks 750 wide										
span between supports 10.6m; UDL not exceeding 2KN/m2	42.06	2.50	**43.11**	-	-	-	-	**43.11**	m2	**46.34**
span between supports 9.3m; UDL not exceeding 4KN/m2	42.06	2.50	**43.11**	-	-	-	-	**43.11**	m2	**46.34**
span between supports 8.5m; UDL not exceeding 6KN/m2	42.06	2.50	**43.11**	-	-	-	-	**43.11**	m2	**46.34**
span between supports 7.8m; UDL not exceeding 8KN/m2	42.06	2.50	**43.11**	-	-	-	-	**43.11**	m2	**46.34**
span between supports 7.2m; UDL not exceeding 10KN/m2	42.06	2.50	**43.11**	-	-	-	-	**43.11**	m2	**46.34**
300 mm thick planks 750 wide										
span between supports 11.5m; UDL not exceeding 2KN/m2	45.55	2.50	**46.69**	-	-	-	-	**46.69**	m2	**50.19**
span between supports 10.3m; UDL not exceeding 4KN/m2	45.55	2.50	**46.69**	-	-	-	-	**46.69**	m2	**50.19**
span between supports 9.4m; UDL not exceeding 6KN/m2	45.55	2.50	**46.69**	-	-	-	-	**46.69**	m2	**50.19**
span between supports 8.7m; UDL not exceeding 8KN/m2	45.55	2.50	**46.69**	-	-	-	-	**46.69**	m2	**50.19**
span between supports 8.1m; UDL not exceeding 10KN/m2	45.55	2.50	**46.69**	-	-	-	-	**46.69**	m2	**50.19**
Passive fall arrest provision										
Add to the foregoing rates for providing safety decking, airbags or netting as appropriate to the works										
total area less than 100 m2	5.01	2.50	**5.13**	-	-	-	-	**5.13**	m2	**5.52**
total area over 100m2	3.80	2.50	**3.90**	-	-	-	-	**3.90**	m2	**4.19**

MASONRY

Labour hourly rates: (except Specialists) Craft Operatives £14.21 Labourer £10.71 Rates are national average prices. Refer to REGIONAL VARIATIONS for indicative levels of overall pricing in regions	MATERIALS			LABOUR				RATES		
	Del to Site £	Waste %	Material Cost £	Craft Optve Hrs	Lab Hrs	Labour Cost £	Sunds £	Nett Rate £	Unit	Gross Rate (+7.5%) £
F10: BRICK/BLOCK WALLING										
Common bricks, BS EN 772, Category M, 215 x 102.5 x 65mm, compressive strength 20.5 N/mm2; in cement-lime mortar (1:2:9)										
Walls; vertical										
102mm thick; stretcher bond	10.92	5.00	**11.46**	1.45	1.10	**32.39**	2.63	**46.48**	m2	**49.97**
215mm thick; English bond	22.02	5.00	**23.12**	2.35	1.85	**53.21**	6.15	**82.47**	m2	**88.66**
327mm thick; English bond	33.12	5.00	**34.77**	2.85	2.30	**65.13**	9.66	**109.57**	m2	**117.78**
extra; grooved bricks; 50% wall surface keyed	0.16	5.00	**0.17**	-	-	**-**	-	**0.17**	m2	**0.18**
extra; grooved bricks; 100% wall surface keyed	0.31	5.00	**0.33**	-	-	**-**	-	**0.33**	m2	**0.35**
Walls; building against concrete (ties measured separately); vertical										
102mm thick; stretcher bond	10.92	5.00	**11.46**	1.60	1.20	**35.59**	2.63	**49.68**	m2	**53.41**
Walls; building against old brickwork; tie new to old with 3mm diameter galvanized twisted wire ties - 6/m2; vertical										
102mm thick; stretcher bond	11.65	10.00	**12.27**	1.60	1.20	**35.59**	2.63	**50.49**	m2	**54.28**
215mm thick; English bond	22.75	10.00	**23.93**	2.65	2.05	**59.61**	6.15	**89.69**	m2	**96.42**
Walls; bonding to stonework; including extra material; vertical										
102mm thick; stretcher bond	11.53	5.00	**12.11**	1.60	1.20	**35.59**	2.63	**50.33**	m2	**54.10**
215mm thick; English bond	23.25	5.00	**24.41**	2.65	2.05	**59.61**	6.15	**90.17**	m2	**96.94**
327mm thick; English bond	34.35	5.00	**36.07**	3.15	2.50	**71.54**	9.66	**117.27**	m2	**126.06**
440mm thick; English bond	45.26	5.00	**47.53**	3.70	3.00	**84.71**	13.18	**145.41**	m2	**156.32**
Walls; bonding to old brickwork; cutting pockets; including extra material; vertical										
102mm thick; stretcher bond	11.53	5.00	**12.11**	3.65	2.55	**79.18**	2.63	**93.92**	m2	**100.96**
215mm thick; English bond	23.25	5.00	**24.41**	4.60	3.30	**100.71**	6.15	**131.27**	m2	**141.11**
Walls; curved 2m radius; including extra material; vertical										
102mm thick; stretcher bond	10.92	5.00	**11.46**	2.85	2.05	**62.45**	2.63	**76.55**	m2	**82.29**
215mm thick; English bond	22.02	5.00	**23.12**	5.20	3.75	**114.05**	6.15	**143.32**	m2	**154.07**
327mm thick; English bond	33.12	5.00	**34.77**	7.10	5.10	**155.51**	9.66	**199.95**	m2	**214.94**
Walls; curved 6m radius; including extra material; vertical										
102mm thick; stretcher bond	10.92	5.00	**11.46**	2.15	1.55	**47.15**	2.63	**61.25**	m2	**65.84**
215mm thick; English bond	22.02	5.00	**23.12**	3.75	2.80	**83.28**	6.15	**112.54**	m2	**120.98**
327mm thick; English bond	33.12	5.00	**34.77**	4.95	3.70	**109.97**	9.66	**154.40**	m2	**165.98**
Isolated piers; vertical										
215mm thick; English bond	22.02	5.00	**23.12**	4.70	3.40	**103.20**	6.15	**132.47**	m2	**142.40**
327mm thick; English bond	33.12	5.00	**34.77**	5.70	4.20	**125.98**	9.66	**170.41**	m2	**183.20**
440mm thick; English bond	44.03	5.00	**46.23**	6.60	4.95	**146.80**	13.18	**206.21**	m2	**221.68**
Chimney stacks; vertical										
440mm thick; English bond	44.03	5.00	**46.23**	6.60	5.45	**152.16**	13.18	**211.57**	m2	**227.43**
890mm thick; English bond	88.06	5.00	**92.46**	10.40	8.80	**242.03**	26.36	**360.85**	m2	**387.92**
Projections; vertical										
215mm wide x 112mm projection	2.47	5.00	**2.59**	0.30	0.23	**6.73**	0.70	**10.01**	m	**10.77**
215mm wide x 215mm projection	4.93	5.00	**5.18**	0.50	0.39	**11.28**	1.41	**17.87**	m	**19.21**
327mm wide x 112mm projection	3.70	5.00	**3.89**	0.45	0.35	**10.14**	1.05	**15.08**	m	**16.21**
327mm wide x 215mm projection	7.40	5.00	**7.77**	0.75	0.59	**16.98**	2.11	**26.85**	m	**28.87**
Projections; horizontal										
215mm wide x 112mm projection	2.47	5.00	**2.59**	0.30	0.23	**6.73**	0.70	**10.01**	m	**10.77**
215mm wide x 215mm projection	4.93	5.00	**5.18**	0.50	0.39	**11.28**	1.41	**17.87**	m	**19.21**
327mm wide x 112mm projection	3.70	5.00	**3.89**	0.45	0.35	**10.14**	1.05	**15.08**	m	**16.21**
327mm wide x 215mm projection	7.40	5.00	**7.77**	0.75	0.59	**16.98**	2.11	**26.85**	m	**28.87**
Projections; bonding to old brickwork; cutting pockets; including extra material; vertical										
215mm wide x 112mm projection	4.11	5.00	**4.31**	0.93	0.65	**20.18**	1.17	**25.66**	m	**27.59**

Labour hourly rates: (except Specialists) Craft Operatives £14.21 Labourer £10.71 Rates are national average prices. Refer to REGIONAL VARIATIONS for indicative levels of overall pricing in regions	MATERIALS			LABOUR			RATES			
	Del to Site £	Waste %	Material Cost £	Craft Optve Hrs	Lab Hrs	Labour Cost £	Sunds £	Nett Rate £	Unit	Gross Rate (+7.5%) £
F10: BRICK/BLOCK WALLING Cont.										
Common bricks, BS EN 772, Category M, 215 x 102.5 x 65mm, compressive strength 20.5 N/mm2; in cement-lime mortar (1:2:9) Cont.										
Projections; bonding to old brickwork; cutting pockets; including extra material; vertical Cont.										
215mm wide x 215mm projection	6.57	5.00	**6.90**	1.13	0.81	**24.73**	1.87	**33.51**	m	**36.02**
327mm wide x 112mm projection	6.17	5.00	**6.47**	1.40	0.98	**30.39**	1.75	**38.62**	m	**41.51**
327mm wide x 215mm projection	9.04	5.00	**9.49**	1.70	1.22	**37.22**	2.58	**49.30**	m	**53.00**
Projections; bonding to old brickwork; cutting pockets; including extra material; horizontal										
215mm wide x 112mm projection	4.11	5.00	**4.31**	0.93	0.65	**20.18**	1.17	**25.66**	m	**27.59**
215mm wide x 215mm projection	6.57	5.00	**6.90**	1.13	0.81	**24.73**	1.87	**33.51**	m	**36.02**
327mm wide x 112mm projection	6.17	5.00	**6.47**	1.40	0.98	**30.39**	1.75	**38.62**	m	**41.51**
327mm wide x 215mm projection	9.04	5.00	**9.49**	1.70	1.22	**37.22**	2.58	**49.30**	m	**53.00**
Arches including centering; flat										
112mm high on face; 215mm thick; width of exposed soffit 215mm; bricks-on-edge	2.47	5.00	**2.59**	1.50	1.00	**32.03**	0.70	**35.31**	m	**37.96**
215mm high on face; 112mm thick; width of exposed soffit 112mm; bricks-on-end	2.47	5.00	**2.59**	1.45	1.00	**31.31**	0.70	**34.60**	m	**37.20**
Arches including centering; semi-circular										
112mm high on face; 215mm thick; width of exposed soffit 215mm; bricks-on-edge	3.87	5.00	**4.07**	2.15	1.35	**45.01**	1.11	**50.18**	m	**53.95**
215mm high on face; 215mm thick; width of exposed soffit 215mm; bricks-on-edge	7.75	5.00	**8.13**	2.95	1.95	**62.80**	2.20	**73.14**	m	**78.63**
Closing cavities; vertical										
50mm wide with brickwork 102mm thick	1.20	5.00	**1.26**	0.40	0.27	**8.58**	0.34	**10.18**	m	**10.95**
75mm wide with brickwork 102mm thick	1.20	5.00	**1.26**	0.40	0.27	**8.58**	0.34	**10.18**	m	**10.95**
Closing cavities; horizontal										
50mm wide with slates	1.92	5.00	**2.02**	0.66	0.44	**14.09**	0.34	**16.45**	m	**17.68**
75mm wide with slates	1.92	5.00	**2.02**	0.66	0.44	**14.09**	0.34	**16.45**	m	**17.68**
50mm wide with brickwork 102mm thick	1.20	5.00	**1.26**	0.40	0.27	**8.58**	0.34	**10.18**	m	**10.95**
75mm wide with brickwork 102mm thick	1.20	5.00	**1.26**	0.40	0.27	**8.58**	0.34	**10.18**	m	**10.95**
Bonding ends to existing common brickwork; cutting pockets; extra material										
walls; bonding every third course										
102mm thick	0.82	5.00	**0.86**	0.37	0.25	**7.94**	0.24	**9.03**	m	**9.71**
215mm thick	1.64	5.00	**1.72**	0.66	0.44	**14.09**	0.47	**16.29**	m	**17.51**
327mm thick	2.47	5.00	**2.59**	0.95	0.63	**20.25**	0.70	**23.53**	m	**25.30**
Second hard stock bricks, BS EN 772, Category M, 215 x 102.5 x 65mm; in cement-lime mortar (1:2:9)										
Walls; vertical										
102mm thick; stretcher bond	21.54	5.00	**22.61**	1.45	1.10	**32.39**	2.63	**57.63**	m2	**61.95**
215mm thick; English bond	43.44	5.00	**45.61**	2.35	1.85	**53.21**	6.15	**104.96**	m2	**112.83**
327mm thick; English bond	65.34	5.00	**68.60**	2.85	2.30	**65.13**	9.66	**143.40**	m2	**154.15**
Walls; building against concrete (ties measured separately); vertical										
102mm thick; stretcher bond	21.54	5.00	**22.61**	1.60	1.20	**35.59**	2.63	**60.83**	m2	**65.40**
Walls; building against old brickwork; tie new to old with 3mm diameter galvanized twisted wire ties - 6/m2; vertical										
102mm thick; stretcher bond	22.27	10.00	**23.42**	1.60	1.20	**35.59**	2.63	**61.65**	m2	**66.27**
215mm thick; English bond	44.17	10.00	**46.42**	2.65	2.05	**59.61**	6.15	**112.18**	m2	**120.59**
Walls; bonding to stonework; including extra material; vertical										
102mm thick; stretcher bond	22.75	5.00	**23.89**	1.60	1.20	**35.59**	2.78	**62.26**	m2	**66.93**
215mm thick; English bond	45.87	5.00	**48.16**	2.65	2.05	**59.61**	6.50	**114.28**	m2	**122.85**
327mm thick; English bond	67.77	5.00	**71.16**	3.15	2.50	**71.54**	10.02	**152.71**	m2	**164.17**
440mm thick; English bond	89.30	5.00	**93.77**	3.70	3.00	**84.71**	13.55	**192.02**	m2	**206.42**
Walls; bonding to old brickwork; cutting pockets; including extra material; vertical										
102mm thick; stretcher bond	22.75	5.00	**23.89**	3.65	2.55	**79.18**	2.78	**105.85**	m2	**113.79**
215mm thick; English bond	45.87	5.00	**48.16**	4.60	3.30	**100.71**	6.50	**155.38**	m2	**167.03**
Walls; curved 2m radius; including extra material; vertical										
102mm thick; stretcher bond	21.54	5.00	**22.61**	2.85	2.05	**62.45**	2.63	**87.70**	m2	**94.28**
215mm thick; English bond	43.44	5.00	**45.61**	5.20	3.75	**114.05**	6.15	**165.81**	m2	**178.25**
327mm thick; English bond	65.34	5.00	**68.60**	7.10	5.10	**155.51**	9.66	**233.78**	m2	**251.31**
Walls; curved 6m radius; including extra material; vertical										
102mm thick; stretcher bond	21.54	5.00	**22.61**	2.15	1.55	**47.15**	2.63	**72.40**	m2	**77.83**
215mm thick; English bond	43.44	5.00	**45.61**	3.75	2.80	**83.28**	6.15	**135.03**	m2	**145.16**
327mm thick; English bond	65.34	5.00	**68.60**	4.95	3.70	**109.97**	9.66	**188.23**	m2	**202.35**
Isolated piers; vertical										
215mm thick; English bond	43.44	5.00	**45.61**	4.70	3.40	**103.20**	6.15	**154.96**	m2	**166.58**
327mm thick; English bond	65.34	5.00	**68.60**	5.70	4.20	**125.98**	9.66	**204.25**	m2	**219.56**
440mm thick; English bond	86.87	5.00	**91.21**	6.60	4.95	**146.80**	13.18	**251.19**	m2	**270.03**

Labour hourly rates: (except Specialists) Craft Operatives £14.21 Labourer £10.71 Rates are national average prices. Refer to REGIONAL VARIATIONS for indicative levels of overall pricing in regions	MATERIALS			LABOUR				RATES		
	Del to Site £	Waste %	Material Cost £	Craft Optve Hrs	Lab Hrs	Labour Cost £	Sunds £	Nett Rate £	Unit	Gross Rate (+7.5%) £
F10: BRICK/BLOCK WALLING Cont.										
Second hard stock bricks, BS EN 772, Category M, 215 x 102.5 x 65mm; in cement-lime mortar (1:2:9) Cont.										
Chimney stacks; vertical										
440mm thick; English bond	86.87	5.00	**91.21**	6.60	5.45	**152.16**	13.18	**256.55**	m2	**275.79**
890mm thick; English bond	173.74	5.00	**182.43**	10.40	8.80	**242.03**	26.36	**450.82**	m2	**484.63**
Projections; vertical										
215mm wide x 112mm projection..................	4.87	5.00	**5.11**	0.30	0.23	**6.73**	0.70	**12.53**	m	**13.47**
215mm wide x 215mm projection..................	9.73	5.00	**10.22**	0.50	0.39	**11.28**	1.41	**22.91**	m	**24.63**
327mm wide x 112mm projection..................	7.30	5.00	**7.67**	0.45	0.35	**10.14**	1.05	**18.86**	m	**20.28**
327mm wide x 215mm projection..................	14.60	5.00	**15.33**	0.75	0.59	**16.98**	2.11	**34.41**	m	**36.99**
Projections; horizontal										
215mm wide x 112mm projection..................	4.87	5.00	**5.11**	0.30	0.23	**6.73**	0.70	**12.53**	m	**13.47**
215mm wide x 215mm projection..................	9.73	5.00	**10.22**	0.50	0.39	**11.28**	1.41	**22.91**	m	**24.63**
327mm wide x 112mm projection..................	7.30	5.00	**7.67**	0.45	0.35	**10.14**	1.05	**18.86**	m	**20.28**
327mm wide x 215mm projection..................	14.60	5.00	**15.33**	0.75	0.59	**16.98**	2.11	**34.41**	m	**36.99**
Projections; bonding to old brickwork; cutting pockets; including extra material; vertical										
215mm wide x 112mm projection..................	8.11	5.00	**8.51**	0.93	0.65	**20.18**	1.17	**29.86**	m	**32.10**
215mm wide x 215mm projection..................	12.97	5.00	**13.62**	1.13	0.81	**24.73**	1.87	**40.22**	m	**43.24**
327mm wide x 112mm projection..................	12.17	5.00	**12.77**	1.40	0.98	**30.39**	1.75	**44.92**	m	**48.28**
327mm wide x 215mm projection..................	17.84	5.00	**18.73**	1.70	1.22	**37.22**	2.58	**58.54**	m	**62.93**
Projections; bonding to old brickwork; cutting pockets; including extra material; horizontal										
215mm wide x 112mm projection..................	8.11	5.00	**8.51**	0.93	0.65	**20.18**	1.17	**29.86**	m	**32.10**
215mm wide x 215mm projection..................	12.97	5.00	**13.62**	1.13	0.81	**24.73**	1.87	**40.22**	m	**43.24**
327mm wide x 112mm projection..................	12.17	5.00	**12.77**	1.40	0.98	**30.39**	1.75	**44.92**	m	**48.28**
327mm wide x 215mm projection..................	17.84	5.00	**18.73**	1.70	1.22	**37.22**	2.58	**58.54**	m	**62.93**
Arches including centering; flat										
112mm high on face; 215mm thick; width of exposed soffit 215mm; bricks-on-edge..................	4.87	5.00	**5.11**	1.50	1.00	**32.03**	0.70	**37.83**	m	**40.67**
215mm high on face; 112mm thick; width of exposed soffit 112mm; bricks-on-end..................	4.87	5.00	**5.11**	1.45	1.00	**31.31**	0.70	**37.12**	m	**39.91**
Arches including centering; semi-circular										
112mm high on face; 215mm thick; width of exposed soffit 215mm; bricks-on-edge..................	7.64	5.00	**8.02**	2.15	1.35	**45.01**	1.11	**54.14**	m	**58.20**
215mm high on face; 215mm thick; width of exposed soffit 215mm; bricks-on-edge..................	15.28	5.00	**16.05**	2.95	1.95	**62.80**	2.20	**81.05**	m	**87.13**
Closing cavities; vertical										
50mm wide with brickwork 102mm thick............	2.37	5.00	**2.49**	0.40	0.27	**8.58**	0.34	**11.41**	m	**12.27**
75mm wide with brickwork 102mm thick............	2.37	5.00	**2.49**	0.40	0.27	**8.58**	0.34	**11.41**	m	**12.27**
Closing cavities; horizontal										
50mm wide with slates	1.92	5.00	**2.02**	0.66	0.44	**14.09**	0.34	**16.45**	m	**17.68**
75mm wide with slates	1.92	5.00	**2.02**	0.66	0.44	**14.09**	0.34	**16.45**	m	**17.68**
50mm wide with brickwork 102mm thick............	2.37	5.00	**2.49**	0.40	0.27	**8.58**	0.34	**11.41**	m	**12.27**
75mm wide with brickwork 102mm thick............	2.37	5.00	**2.49**	0.40	0.27	**8.58**	0.34	**11.41**	m	**12.27**
Bonding ends to existing common brickwork; cutting pockets; extra material										
walls; bonding every third course										
102mm thick..................................	1.62	5.00	**1.70**	0.37	0.25	**7.94**	0.24	**9.87**	m	**10.61**
215mm thick..................................	3.24	5.00	**3.40**	0.66	0.44	**14.09**	0.47	**17.97**	m	**19.31**
327mm thick..................................	4.87	5.00	**5.11**	0.95	0.63	**20.25**	0.70	**26.05**	m	**28.01**
Engineering bricks, BS EN 772, Category F, 215 x 102.5 x 65, Class B; in cement-lime mortar (1:2:9)										
Walls; vertical										
102mm thick; stretcher bond....................	11.51	5.00	**12.08**	1.60	1.20	**35.59**	2.63	**50.30**	m2	**54.07**
215mm thick; English bond	23.21	5.00	**24.37**	2.60	2.05	**58.90**	6.15	**89.42**	m2	**96.12**
327mm thick; English bond	34.91	5.00	**36.65**	3.15	2.55	**72.07**	9.66	**118.39**	m2	**127.27**
Walls; building against concrete (ties measured separately); vertical										
102mm thick; stretcher bond....................	11.51	5.00	**12.08**	1.75	1.30	**38.79**	2.63	**53.50**	m2	**57.52**
Walls; building against old brickwork; tie new to old with 3mm diameter galvanized twisted wire ties - 6/m2; vertical										
102mm thick; stretcher bond....................	12.24	10.00	**12.89**	1.75	1.30	**38.79**	2.63	**54.32**	m2	**58.39**
215mm thick; English bond	23.94	10.00	**25.18**	2.90	2.25	**65.31**	6.15	**96.63**	m2	**103.88**
Walls; bonding to stonework; including extra material; vertical										
102mm thick; stretcher bond....................	12.15	5.00	**12.76**	1.75	1.30	**38.79**	2.78	**54.34**	m2	**58.41**
215mm thick; English bond	24.51	5.00	**25.73**	2.90	2.25	**65.31**	6.50	**97.54**	m2	**104.86**
327mm thick; English bond	36.21	5.00	**38.02**	3.45	2.75	**78.48**	10.02	**126.51**	m2	**136.00**
440mm thick; English bond	47.71	5.00	**50.10**	4.05	3.30	**92.89**	13.55	**156.53**	m2	**168.27**
Walls; bonding to old brickwork; cutting pockets; including extra material; vertical										
102mm thick; stretcher bond....................	12.15	5.00	**12.76**	4.00	2.80	**86.83**	2.78	**102.37**	m2	**110.05**
215mm thick; English bond	24.51	5.00	**25.73**	5.05	3.65	**110.85**	6.50	**143.09**	m2	**153.82**

Labour hourly rates: (except Specialists) Craft Operatives £14.21 Labourer £10.71 Rates are national average prices. Refer to REGIONAL VARIATIONS for indicative levels of overall pricing in regions	MATERIALS			LABOUR				RATES		
	Del to Site £	Waste %	Material Cost £	Craft Optve Hrs	Lab Hrs	Labour Cost £	Sunds £	Nett Rate £	Unit	Gross Rate (+7.5%) £
F10: BRICK/BLOCK WALLING Cont.										
Engineering bricks, BS EN 772, Category F, 215 x 102.5 x 65, Class B; in cement-lime mortar (1:2:9) Cont.										
Walls; curved 2m radius; including extra material; vertical										
102mm thick; stretcher bond	11.51	5.00	**12.08**	3.15	2.25	**68.86**	2.63	**83.57**	m2	**89.84**
215mm thick; English bond	23.21	5.00	**24.37**	5.70	4.10	**124.91**	6.15	**155.42**	m2	**167.08**
327mm thick; English bond	34.91	5.00	**36.65**	7.80	5.60	**170.81**	9.66	**217.13**	m2	**233.41**
Walls; curved 6m radius; including extra material; vertical										
102mm thick; stretcher bond	11.51	5.00	**12.08**	2.35	1.70	**51.60**	2.63	**66.31**	m2	**71.29**
215mm thick; English bond	23.21	5.00	**24.37**	4.10	3.10	**91.46**	6.15	**121.98**	m2	**131.12**
327mm thick; English bond	34.91	5.00	**36.65**	5.45	4.05	**120.82**	9.66	**167.13**	m2	**179.67**
Isolated piers; vertical										
215mm thick; English bond	23.21	5.00	**24.37**	5.15	3.75	**113.34**	6.15	**143.86**	m2	**154.65**
327mm thick; English bond	34.91	5.00	**36.65**	6.25	4.60	**138.08**	9.66	**184.39**	m2	**198.22**
440mm thick; English bond	46.41	5.00	**48.73**	7.25	5.45	**161.39**	13.18	**223.30**	m2	**240.05**
Chimney stacks; vertical										
440mm thick; English bond	46.41	5.00	**48.73**	7.25	5.45	**161.39**	13.18	**223.30**	m2	**240.05**
890mm thick; English bond	92.82	5.00	**97.46**	11.45	8.80	**256.95**	26.36	**380.77**	m2	**409.33**
Projections; vertical										
215mm wide x 112mm projection	2.60	5.00	**2.73**	0.33	0.25	**7.37**	0.70	**10.79**	m	**11.60**
215mm wide x 215mm projection	5.20	5.00	**5.46**	0.55	0.42	**12.31**	1.41	**19.18**	m	**20.62**
327mm wide x 112mm projection	3.90	5.00	**4.10**	0.50	0.38	**11.17**	1.05	**16.32**	m	**17.55**
327mm wide x 215mm projection	7.80	5.00	**8.19**	0.83	0.63	**18.54**	2.11	**28.84**	m	**31.00**
Projections; horizontal										
215mm wide x 112mm projection	2.60	5.00	**2.73**	0.33	0.25	**7.37**	0.70	**10.79**	m	**11.60**
215mm wide x 215mm projection	5.20	5.00	**5.46**	0.55	0.42	**12.31**	1.41	**19.18**	m	**20.62**
327mm wide x 112mm projection	3.90	5.00	**4.10**	0.50	0.38	**11.17**	1.05	**16.32**	m	**17.55**
327mm wide x 215mm projection	7.80	5.00	**8.19**	0.83	0.63	**18.54**	2.11	**28.84**	m	**31.00**
Projections; bonding to old brickwork; cutting pockets; including extra material; vertical										
215mm wide x 112mm projection	4.33	5.00	**4.55**	1.02	0.72	**22.21**	1.17	**27.92**	m	**30.02**
215mm wide x 215mm projection	6.93	5.00	**7.28**	1.24	0.89	**27.15**	1.87	**36.30**	m	**39.02**
327mm wide x 112mm projection	6.50	5.00	**6.82**	1.54	1.08	**33.45**	1.75	**42.03**	m	**45.18**
327mm wide x 215mm projection	9.53	5.00	**10.01**	1.87	1.34	**40.92**	2.58	**53.51**	m	**57.53**
Projections; bonding to old brickwork; cutting pockets; including extra material; horizontal										
215mm wide x 112mm projection	4.33	5.00	**4.55**	1.02	0.72	**22.21**	1.17	**27.92**	m	**30.02**
215mm wide x 215mm projection	6.93	5.00	**7.28**	1.24	0.89	**27.15**	1.87	**36.30**	m	**39.02**
327mm wide x 112mm projection	6.50	5.00	**6.82**	1.54	1.08	**33.45**	1.75	**42.03**	m	**45.18**
327mm wide x 215mm projection	9.53	5.00	**10.01**	1.87	1.34	**40.92**	2.58	**53.51**	m	**57.53**
Arches including centering; flat										
112mm high on face; 215mm thick; width of exposed soffit 215mm; bricks-on-edge	2.60	5.00	**2.73**	1.60	1.05	**33.98**	0.70	**37.41**	m	**40.22**
215mm high on face; 112mm thick; width of exposed soffit 112mm; bricks-on-end	2.60	5.00	**2.73**	1.55	1.10	**33.81**	0.70	**37.23**	m	**40.03**
Arches including centering; semi-circular										
112mm high on face; 215mm thick; width of exposed soffit 215mm; bricks-on-edge	4.08	5.00	**4.29**	2.25	1.40	**46.97**	1.11	**52.36**	m	**56.29**
215mm high on face; 215mm thick; width of exposed soffit 215mm; bricks-on-edge	8.16	5.00	**8.57**	3.10	2.05	**66.01**	2.20	**76.78**	m	**82.54**
Closing cavities; vertical										
50mm wide with brickwork 102mm thick	1.27	5.00	**1.33**	0.44	0.30	**9.47**	0.34	**11.14**	m	**11.98**
75mm wide with brickwork 102mm thick	1.27	5.00	**1.33**	0.44	0.30	**9.47**	0.34	**11.14**	m	**11.98**
Closing cavities; horizontal										
50mm wide with slates	1.92	5.00	**2.02**	0.66	0.44	**14.09**	0.34	**16.45**	m	**17.68**
75mm wide with slates	1.92	5.00	**2.02**	0.66	0.44	**14.09**	0.34	**16.45**	m	**17.68**
50mm wide with brickwork 102mm thick	1.27	5.00	**1.33**	0.44	0.30	**9.47**	0.34	**11.14**	m	**11.98**
75mm wide with brickwork 102mm thick	1.27	5.00	**1.33**	0.44	0.30	**9.47**	0.34	**11.14**	m	**11.98**
Bonding ends to existing common brickwork; cutting pockets; extra material										
walls; bonding every third course										
102mm thick	0.87	5.00	**0.91**	0.41	0.28	**8.82**	0.24	**9.97**	m	**10.72**
215mm thick	1.73	5.00	**1.82**	0.73	0.48	**15.51**	0.47	**17.81**	m	**19.14**
327mm thick	2.60	5.00	**2.73**	1.05	0.69	**22.31**	0.70	**25.74**	m	**27.67**
Engineering bricks, BS EN 772, Category F, 215 x 102.5 x 65mm, Class A; in cement-lime mortar (1:2:9)										
Walls; vertical										
102mm thick; stretcher bond	16.23	5.00	**17.04**	1.60	1.20	**35.59**	2.63	**55.26**	m2	**59.40**
215mm thick; English bond	32.73	5.00	**34.36**	2.60	2.05	**58.90**	6.15	**99.41**	m2	**106.87**
327mm thick; English bond	49.23	5.00	**51.69**	3.15	2.55	**72.07**	9.66	**133.42**	m2	**143.43**
Walls; building against concrete (ties measured separately); vertical										
102mm thick; stretcher bond	16.23	5.00	**17.04**	1.75	1.30	**38.79**	2.63	**58.46**	m2	**62.85**

Labour hourly rates: (except Specialists) Craft Operatives £14.21 Labourer £10.71 Rates are national average prices. Refer to REGIONAL VARIATIONS for indicative levels of overall pricing in regions	MATERIALS			LABOUR				RATES		
	Del to Site £	Waste %	Material Cost £	Craft Optve Hrs	Lab Hrs	Labour Cost £	Sunds £	Nett Rate £	Unit	Gross Rate (+7.5%) £
F10: BRICK/BLOCK WALLING Cont.										
Engineering bricks, BS EN 772, Category F, 215 x 102.5 x 65mm, Class A; in cement-lime mortar (1:2:9) Cont.										
Walls; building against old brickwork; tie new to old with 3mm diameter galvanized twisted wire ties - 6/m2; vertical										
102mm thick; stretcher bond	16.96	10.00	**17.85**	1.75	1.30	**38.79**	2.63	**59.27**	m2	**63.72**
215mm thick; English bond	33.46	10.00	**35.17**	2.90	2.25	**65.31**	6.15	**106.63**	m2	**114.63**
Walls; bonding to stonework; including extra material; vertical										
102mm thick; stretcher bond	17.14	5.00	**18.00**	1.75	1.30	**38.79**	2.78	**59.57**	m2	**64.04**
215mm thick; English bond	34.56	5.00	**36.29**	2.90	2.25	**65.31**	6.50	**108.10**	m2	**116.20**
327mm thick; English bond	51.06	5.00	**53.61**	3.45	2.75	**78.48**	10.02	**142.11**	m2	**152.77**
440mm thick; English bond	67.28	5.00	**70.65**	4.05	3.30	**92.89**	13.55	**177.09**	m2	**190.37**
Walls; bonding to old brickwork; cutting pockets; including extra material; vertical										
102mm thick; stretcher bond	17.14	5.00	**18.00**	4.00	2.80	**86.83**	2.78	**107.61**	m2	**115.68**
215mm thick; English bond	34.56	5.00	**36.29**	5.05	3.65	**110.85**	6.50	**153.64**	m2	**165.17**
Walls; curved 2m radius; including extra material; vertical										
102mm thick; stretcher bond	16.23	5.00	**17.04**	3.15	2.25	**68.86**	2.63	**88.53**	m2	**95.17**
215mm thick; English bond	32.73	5.00	**34.36**	5.70	4.10	**124.91**	6.15	**165.42**	m2	**177.82**
327mm thick; English bond	49.23	5.00	**51.69**	7.80	5.60	**170.81**	9.66	**232.16**	m2	**249.58**
Walls; curved 6m radius; including extra material; vertical										
102mm thick; stretcher bond	16.23	5.00	**17.04**	2.35	1.70	**51.60**	2.63	**71.27**	m2	**76.62**
215mm thick; English bond	32.73	5.00	**34.36**	4.10	3.10	**91.46**	6.15	**131.97**	m2	**141.87**
327mm thick; English bond	49.23	5.00	**51.69**	5.45	4.05	**120.82**	9.66	**182.17**	m2	**195.83**
Isolated piers; vertical										
215mm thick; English bond	32.73	5.00	**34.36**	5.15	3.75	**113.34**	6.15	**153.85**	m2	**165.39**
327mm thick; English bond	49.23	5.00	**51.69**	6.25	4.60	**138.08**	9.66	**199.43**	m2	**214.39**
440mm thick; English bond	65.45	5.00	**68.72**	7.25	5.45	**161.39**	13.18	**243.29**	m2	**261.54**
Chimney stacks; vertical										
440mm thick; English bond	65.45	5.00	**68.72**	7.25	5.45	**161.39**	13.18	**243.29**	m2	**261.54**
890mm thick; English bond	130.90	5.00	**137.45**	11.45	8.80	**256.95**	26.36	**420.76**	m2	**452.31**
Projections; vertical										
215mm wide x 112mm projection	3.67	5.00	**3.85**	0.33	0.25	**7.37**	0.70	**11.91**	m	**12.81**
215mm wide x 215mm projection	7.33	5.00	**7.70**	0.55	0.42	**12.31**	1.41	**21.42**	m	**23.03**
327mm wide x 112mm projection	5.50	5.00	**5.78**	0.50	0.38	**11.17**	1.05	**18.00**	m	**19.35**
327mm wide x 215mm projection	11.00	5.00	**11.55**	0.83	0.63	**18.54**	2.11	**32.20**	m	**34.61**
Projections; horizontal										
215mm wide x 112mm projection	3.67	5.00	**3.85**	0.33	0.25	**7.37**	0.70	**11.91**	m	**12.81**
215mm wide x 215mm projection	7.33	5.00	**7.70**	0.55	0.42	**12.31**	1.41	**21.42**	m	**23.03**
327mm wide x 112mm projection	5.50	5.00	**5.78**	0.50	0.38	**11.17**	1.05	**18.00**	m	**19.35**
327mm wide x 215mm projection	11.00	5.00	**11.55**	0.83	0.63	**18.54**	2.11	**32.20**	m	**34.61**
Projections; bonding to old brickwork; cutting pockets; including extra material; vertical										
215mm wide x 112mm projection	6.11	5.00	**6.41**	1.02	0.72	**22.21**	1.17	**29.79**	m	**32.02**
215mm wide x 215mm projection	9.77	5.00	**10.26**	1.24	0.89	**27.15**	1.87	**39.28**	m	**42.23**
327mm wide x 112mm projection	9.17	5.00	**9.62**	1.54	1.08	**33.45**	1.75	**44.83**	m	**48.19**
327mm wide x 215mm projection	13.44	5.00	**14.11**	1.87	1.34	**40.92**	2.58	**57.62**	m	**61.94**
Projections; bonding to old brickwork; cutting pockets; including extra material; horizontal										
215mm wide x 112mm projection	6.11	5.00	**6.41**	1.02	0.72	**22.21**	1.17	**29.79**	m	**32.02**
215mm wide x 215mm projection	9.77	5.00	**10.26**	1.24	0.89	**27.15**	1.87	**39.28**	m	**42.23**
327mm wide x 112mm projection	9.17	5.00	**9.62**	1.54	1.08	**33.45**	1.75	**44.83**	m	**48.19**
327mm wide x 215mm projection	13.44	5.00	**14.11**	1.87	1.34	**40.92**	2.58	**57.62**	m	**61.94**
Arches including centering; flat										
112mm high on face; 215mm thick; width of exposed soffit 215mm; bricks-on-edge	3.67	5.00	**3.85**	1.60	1.05	**33.98**	0.70	**38.53**	m	**41.42**
215mm high on face; 112mm thick; width of exposed soffit 112mm; bricks-on-end	3.67	5.00	**3.85**	1.55	1.10	**33.81**	0.70	**38.35**	m	**41.23**
Arches including centering; semi-circular										
112mm high on face; 215mm thick; width of exposed soffit 215mm; bricks-on-edge	5.76	5.00	**6.04**	2.25	1.40	**46.97**	1.11	**54.12**	m	**58.18**
215mm high on face; 215mm thick; width of exposed soffit 215mm; bricks-on-edge	11.51	5.00	**12.09**	3.10	2.05	**66.01**	2.20	**80.30**	m	**86.32**
Closing cavities; vertical										
50mm wide with brickwork 102mm thick	1.79	5.00	**1.88**	0.44	0.30	**9.47**	0.34	**11.69**	m	**12.56**
75mm wide with brickwork 102mm thick	1.79	5.00	**1.88**	0.44	0.30	**9.47**	0.34	**11.69**	m	**12.56**
Closing cavities; horizontal										
50mm wide with slates	1.92	5.00	**2.02**	0.66	0.44	**14.09**	0.34	**16.45**	m	**17.68**
75mm wide with slates	1.92	5.00	**2.02**	0.66	0.44	**14.09**	0.34	**16.45**	m	**17.68**
50mm wide with brickwork 102mm thick	1.79	5.00	**1.88**	0.44	0.30	**9.47**	0.34	**11.69**	m	**12.56**
75mm wide with brickwork 102mm thick	1.79	5.00	**1.88**	0.44	0.30	**9.47**	0.34	**11.69**	m	**12.56**
Bonding ends to existing common brickwork; cutting pockets; extra material										
walls; bonding every third course										
102mm thick	1.22	5.00	**1.28**	0.41	0.28	**8.82**	0.24	**10.34**	m	**11.12**

Labour hourly rates: (except Specialists) Craft Operatives £14.21 Labourer £10.71 Rates are national average prices. Refer to REGIONAL VARIATIONS for indicative levels of overall pricing in regions	MATERIALS			LABOUR				RATES		
	Del to Site £	Waste %	Material Cost £	Craft Optve Hrs	Lab Hrs	Labour Cost £	Sunds £	Nett Rate £	Unit	Gross Rate (+7.5%) £
F10: BRICK/BLOCK WALLING Cont.										
Engineering bricks, BS EN 772, Category F, 215 x 102.5 x 65mm, Class A; in cement-lime mortar (1:2:9) Cont.										
Bonding ends to existing common brickwork; cutting pockets; extra material Cont.										
walls; bonding every third course Cont.										
215mm thick	2.44	5.00	**2.56**	0.73	0.48	**15.51**	0.47	**18.55**	m	**19.94**
327mm thick	3.67	5.00	**3.85**	1.05	0.69	**22.31**	0.70	**26.86**	m	**28.87**
For other mortar mixes, ADD or DEDUCT as follows:										
For each half brick thickness										
1:1:6 cement-lime mortar	-	-	**-**	-	-	**-**	0.39	**0.39**	m2	**0.42**
1:4 cement mortar	-	-	**-**	-	-	**-**	0.12	**0.12**	m2	**0.13**
1:3 cement mortar	-	-	**-**	-	-	**-**	0.16	**0.16**	m2	**0.17**
For using Sulphate Resisting cement in lieu of Portland cement, ADD as follows										
For each half brick thickness										
1:2:9 cement-lime mortar	-	-	**-**	-	-	**-**	0.11	**0.11**	m2	**0.12**
1:1:6 cement-lime mortar	-	-	**-**	-	-	**-**	0.15	**0.15**	m2	**0.16**
1:4 cement mortar	-	-	**-**	-	-	**-**	0.25	**0.25**	m2	**0.27**
1:3 cement mortar	-	-	**-**	-	-	**-**	0.31	**0.31**	m2	**0.34**
Staffordshire wirecut bricks, BS EN 772, Category F, 215 x 102.5 x 65mm; in cement mortar (1:3)										
Walls; vertical										
215mm thick; English bond	42.25	5.00	**44.36**	2.60	2.05	**58.90**	6.53	**109.78**	m2	**118.02**
Staffordshire pressed bricks, BS EN 772, Category F, 215 x 102.5 x 65mm; in cement mortar (1:3)										
Walls; vertical										
215mm thick; English bond	49.39	5.00	**51.85**	2.60	2.05	**58.90**	6.53	**117.28**	m2	**126.08**
Composite walling of Staffordshire bricks, BS EN 772, Category F, 215 x 102.5 x 65mm; in cement mortar (1:3); wirecut bricks backing; pressed bricks facing; weather struck pointing as work proceeds										
Walls; vertical										
English bond; facework one side	47.23	5.00	**49.59**	2.90	2.25	**65.31**	6.53	**121.42**	m2	**130.53**
Flemish bond; facework one side	46.63	5.00	**48.96**	2.90	2.25	**65.31**	6.53	**120.79**	m2	**129.85**
Common bricks, BS EN 772, Category M, 215 x 102.5 x 65mm, compressive strength 20.5 N/mm2; in cement-lime mortar (1:2:9); flush smooth pointing as work proceeds										
Walls; vertical										
stretcher bond; facework one side										
102mm thick	10.92	5.00	**11.46**	1.75	1.30	**38.79**	2.63	**52.89**	m2	**56.85**
stretcher bond; facework both sides										
102mm thick	10.92	5.00	**11.46**	2.05	1.50	**45.20**	2.63	**59.29**	m2	**63.74**
English bond; facework one side										
215mm thick	22.02	5.00	**23.12**	2.65	2.05	**59.61**	6.15	**88.88**	m2	**95.54**
English bond; facework both sides										
215mm thick	22.02	5.00	**23.12**	2.95	2.25	**66.02**	6.15	**95.28**	m2	**102.43**
Engineering bricks, BS EN 772, Category F, 215 x 102.5 x 65mm, Class B; in cement-lime mortar (1:2:9); flush smooth pointing as work proceeds										
Walls; vertical										
stretcher bond; facework one side										
102mm thick	11.51	5.00	**12.08**	1.90	1.40	**41.99**	2.63	**56.71**	m2	**60.96**
stretcher bond; facework both sides										
102mm thick	11.51	5.00	**12.08**	2.20	1.60	**48.40**	2.63	**63.11**	m2	**67.85**
English bond; facework one side										
215mm thick	23.21	5.00	**24.37**	2.90	2.25	**65.31**	6.15	**95.82**	m2	**103.01**
English bond; facework both sides										
215mm thick	23.21	5.00	**24.37**	3.20	2.45	**71.71**	6.15	**102.23**	m2	**109.89**
Engineering bricks, BS EN 772, Category F, 215 x 102.5 x 65mm, Class A; in cement-lime mortar (1:2:9); flush smooth pointing as work proceeds										
Walls; vertical										
stretcher bond; facework one side										
102mm thick	16.23	5.00	**17.04**	1.90	1.40	**41.99**	2.63	**61.66**	m2	**66.29**
stretcher bond; facework both sides										
102mm thick	16.23	5.00	**17.04**	2.20	1.60	**48.40**	2.63	**68.07**	m2	**73.17**
English bond; facework one side										
215mm thick	32.73	5.00	**34.36**	2.90	2.25	**65.31**	6.15	**105.82**	m2	**113.75**
English bond; facework both sides										
215mm thick	32.73	5.00	**34.36**	3.20	2.45	**71.71**	6.15	**112.22**	m2	**120.64**

Labour hourly rates: (except Specialists) Craft Operatives £14.21 Labourer £10.71 Rates are national average prices. Refer to REGIONAL VARIATIONS for indicative levels of overall pricing in regions	MATERIALS			LABOUR				RATES		
	Del to Site £	Waste %	Material Cost £	Craft Optve Hrs	Lab Hrs	Labour Cost £	Sunds £	Nett Rate £	Unit	Gross Rate (+7.5%) £
F10: BRICK/BLOCK WALLING Cont.										
Facing bricks, second hard stocks, BS EN 772, Category M, 215 x 102.5 x 65mm; in cement-lime mortar (1:2:9); flush smooth pointing as work proceeds										
Walls; vertical										
stretcher bond; facework one side										
102mm thick	21.54	5.00	**22.61**	1.75	1.30	**38.79**	2.63	**64.04**	m2	**68.84**
stretcher bond; facework both sides										
102mm thick	21.54	5.00	**22.61**	2.05	1.50	**45.20**	2.63	**70.44**	m2	**75.72**
English bond; facework one side										
215mm thick	43.44	5.00	**45.61**	2.65	2.05	**59.61**	6.15	**111.37**	m2	**119.72**
English bond; facework both sides										
215mm thick	43.44	5.00	**45.61**	2.95	2.25	**66.02**	6.15	**117.77**	m2	**126.61**
Facing bricks p.c. £150.00 per 1000, 215 x 102.5 x 65mm; in cement-lime mortar (1:2:9); flush smooth pointing as work proceeds										
Walls; vertical										
stretcher bond; facework one side										
102mm thick	8.85	5.00	**9.29**	1.75	1.30	**38.79**	2.63	**50.72**	m2	**54.52**
stretcher bond; facework both sides										
102mm thick	8.85	5.00	**9.29**	2.05	1.50	**45.20**	2.63	**57.12**	m2	**61.41**
English bond; facework both sides										
215mm thick	17.85	5.00	**18.74**	2.65	2.05	**59.61**	6.15	**84.50**	m2	**90.84**
Flemish bond; facework both sides										
215mm thick	17.85	5.00	**18.74**	2.95	2.25	**66.02**	6.15	**90.91**	m2	**97.73**
extra; special bricks; vertical angles; squint; BS 4729, type AN.1.1.	23.33	5.00	**24.49**	0.25	0.15	**5.16**	-	**29.65**	m	**31.88**
extra; special bricks; intersections; birdsmouth; BS 4729, type AN.4.1.	69.98	5.00	**73.48**	0.25	0.15	**5.16**	-	**78.64**	m	**84.54**
Walls; building overhand; vertical										
stretcher bond; facework one side										
102mm thick	8.85	5.00	**9.29**	2.15	1.55	**47.15**	2.63	**59.08**	m2	**63.51**
stretcher bond; facework both sides										
102mm thick	8.85	5.00	**9.29**	2.45	1.75	**53.56**	2.63	**65.48**	m2	**70.39**
English bond; facework both sides										
215mm thick	17.85	5.00	**18.74**	3.05	2.30	**67.97**	6.15	**92.87**	m2	**99.83**
Flemish bond; facework both sides										
215mm thick	17.85	5.00	**18.74**	3.35	2.50	**74.38**	6.15	**99.27**	m2	**106.72**
Arches including centering; flat										
112mm high on face; 215mm thick; width of exposed soffit 112mm; bricks-on-edge	2.00	5.00	**2.10**	1.05	0.80	**23.49**	0.70	**26.29**	m	**28.26**
112mm high on face; 215mm thick; width of exposed soffit 215mm; bricks-on-edge	2.00	5.00	**2.10**	1.50	1.00	**32.03**	0.70	**34.82**	m	**37.43**
215mm high on face; 112mm thick; width of exposed soffit 112mm; bricks-on-end	2.00	5.00	**2.10**	1.45	1.00	**31.31**	0.70	**34.11**	m	**36.67**
Arches including centering; segmental										
215mm high on face; 215mm thick; width of exposed soffit 112mm; bricks-on-edge	4.00	5.00	**4.20**	2.40	1.65	**51.78**	1.41	**57.38**	m	**61.69**
215mm high on face; 215mm thick; width of exposed soffit 215mm; bricks-on-edge	4.00	5.00	**4.20**	2.95	1.95	**62.80**	1.41	**68.41**	m	**73.54**
Arches including centering; semi-circular										
215mm high on face; 215mm thick; width of exposed soffit 112mm; bricks-on-edge	4.00	5.00	**4.20**	2.40	1.65	**51.78**	1.41	**57.38**	m	**61.69**
215mm high on face; 215mm thick; width of exposed soffit 215mm; bricks-on-edge	4.00	5.00	**4.20**	2.95	1.95	**62.80**	1.41	**68.41**	m	**73.54**
Facework copings; bricks-on-edge; pointing top and each side										
215 x 102.5mm; horizontal	2.00	5.00	**2.10**	0.70	0.50	**15.30**	0.70	**18.10**	m	**19.46**
215 x 102.5mm; with oversailing course and cement fillets both sides; horizontal	2.00	5.00	**2.10**	1.10	0.75	**23.66**	0.70	**26.46**	m	**28.45**
327 x 102.5mm; horizontal	3.00	5.00	**3.15**	1.05	0.75	**22.95**	1.05	**27.16**	m	**29.19**
327 x 102.5mm; with oversailing course and cement fillets both sides; horizontal	3.00	5.00	**3.15**	1.50	1.05	**32.56**	1.05	**36.76**	m	**39.52**
extra; galvanized iron coping cramp at angle or end	2.12	5.00	**2.23**	0.04	0.03	**0.89**	-	**3.12**	nr	**3.35**
Facing bricks p.c. £200.00 per 1000, 215 x 102.5 x 65mm; in cement-lime mortar (1:2:9); flush smooth pointing as work proceeds										
Walls; vertical										
stretcher bond; facework one side										
102mm thick	11.80	5.00	**12.39**	1.75	1.30	**38.79**	2.63	**53.81**	m2	**57.85**
stretcher bond; facework both sides										
102mm thick	11.80	5.00	**12.39**	2.05	1.50	**45.20**	2.63	**60.22**	m2	**64.74**
English bond; facework both sides										
215mm thick	23.80	5.00	**24.99**	2.65	2.05	**59.61**	6.15	**90.75**	m2	**97.56**
Flemish bond; facework both sides										
215mm thick	23.80	5.00	**24.99**	2.95	2.25	**66.02**	6.15	**97.16**	m2	**104.44**
extra; special bricks; vertical angles; squint; BS 4729, type AN.1.1.	23.33	5.00	**24.49**	0.25	0.15	**5.16**	-	**29.65**	m	**31.88**
extra; special bricks; intersections; birdsmouth; BS 4729, type AN.4.1.	69.98	5.00	**73.48**	0.25	0.15	**5.16**	-	**78.64**	m	**84.54**
Walls; building overhand; vertical										
stretcher bond; facework one side										
102mm thick	11.80	5.00	**12.39**	2.15	1.55	**47.15**	2.63	**62.18**	m2	**66.84**

Labour hourly rates: (except Specialists) Craft Operatives £14.21 Labourer £10.71 Rates are national average prices. Refer to REGIONAL VARIATIONS for indicative levels of overall pricing in regions	MATERIALS			LABOUR				RATES		
	Del to Site £	Waste %	Material Cost £	Craft Optve Hrs	Lab Hrs	Labour Cost £	Sunds £	Nett Rate £	Unit	Gross Rate (+7.5%) £
F10: BRICK/BLOCK WALLING Cont.										
Facing bricks p.c. £200.00 per 1000, 215 x 102.5 x 65mm; in cement-lime mortar (1:2:9); flush smooth pointing as work proceeds Cont.										
Walls; building overhand; vertical Cont.										
stretcher bond; facework both sides										
102mm thick	11.80	5.00	**12.39**	2.45	1.75	**53.56**	2.63	**68.58**	m2	**73.72**
English bond; facework both sides										
215mm thick	23.80	5.00	**24.99**	3.05	2.30	**67.97**	6.15	**99.11**	m2	**106.55**
Flemish bond; facework both sides										
215mm thick	23.80	5.00	**24.99**	3.35	2.50	**74.38**	6.15	**105.52**	m2	**113.43**
Arches including centering; flat										
112mm high on face; 215mm thick; width of exposed soffit 112mm; bricks-on-edge	2.67	5.00	**2.80**	1.05	0.80	**23.49**	0.70	**26.99**	m	**29.01**
112mm high on face; 215mm thick; width of exposed soffit 215mm; bricks-on-edge	2.67	5.00	**2.80**	1.50	1.00	**32.03**	0.70	**35.52**	m	**38.19**
215mm high on face; 112mm thick; width of exposed soffit 112mm; bricks-on-end	2.67	5.00	**2.80**	1.45	1.00	**31.31**	0.70	**34.81**	m	**37.42**
Arches including centering; segmental										
215mm high on face; 215mm thick; width of exposed soffit 112mm; bricks-on-edge	5.33	5.00	**5.60**	2.40	1.65	**51.78**	1.41	**58.78**	m	**63.19**
215mm high on face; 215mm thick; width of exposed soffit 215mm; bricks-on-edge	5.33	5.00	**5.60**	2.95	1.95	**62.80**	1.41	**69.81**	m	**75.05**
Arches including centering; semi-circular										
215mm high on face; 215mm thick; width of exposed soffit 112mm; bricks-on-edge	5.33	5.00	**5.60**	2.40	1.65	**51.78**	1.41	**58.78**	m	**63.19**
215mm high on face; 215mm thick; width of exposed soffit 215mm; bricks-on-edge	5.33	5.00	**5.60**	2.95	1.95	**62.80**	1.41	**69.81**	m	**75.05**
Facework copings; bricks-on-edge; pointing top and each side										
215 x 102.5mm; horizontal	2.67	5.00	**2.80**	0.70	0.50	**15.30**	0.70	**18.80**	m	**20.21**
215 x 102.5mm; with oversailing course and cement fillets both sides; horizontal	2.67	5.00	**2.80**	1.10	0.75	**23.66**	0.70	**27.16**	m	**29.20**
327 x 102.5mm; horizontal	4.00	5.00	**4.20**	1.05	0.75	**22.95**	1.05	**28.21**	m	**30.32**
327 x 102.5mm; with oversailing course and cement fillets both sides; horizontal	4.00	5.00	**4.20**	1.50	1.05	**32.56**	1.05	**37.81**	m	**40.65**
extra; galvanized iron coping cramp at angle or end	2.12	5.00	**2.23**	0.04	0.03	**0.89**	-	**3.12**	nr	**3.35**
Facing bricks p.c. £250.00 per 1000, 215 x 102.5 x 65mm; in cement-lime mortar (1:2:9); flush smooth pointing as work proceeds										
Walls; vertical										
stretcher bond; facework one side										
102mm thick	14.75	5.00	**15.49**	1.75	1.30	**38.79**	2.63	**56.91**	m2	**61.18**
stretcher bond; facework both sides										
102mm thick	14.75	5.00	**15.49**	2.05	1.50	**45.20**	2.63	**63.32**	m2	**68.07**
English bond; facework both sides										
215mm thick	29.75	5.00	**31.24**	2.65	2.05	**59.61**	6.15	**97.00**	m2	**104.27**
Flemish bond; facework both sides										
215mm thick	29.75	5.00	**31.24**	2.95	2.25	**66.02**	6.15	**103.40**	m2	**111.16**
extra; special bricks; vertical angles; squint; BS 4729, type AN.1.1.	23.33	5.00	**24.49**	0.25	0.15	**5.16**	-	**29.65**	m	**31.88**
extra; special bricks; intersections; birdsmouth; BS 4729, type AN.4.1.	69.98	5.00	**73.48**	0.25	0.15	**5.16**	-	**78.64**	m	**84.54**
Walls; building overhand; vertical										
stretcher bond; facework one side										
102mm thick	14.75	5.00	**15.49**	2.15	1.55	**47.15**	2.63	**65.27**	m2	**70.17**
stretcher bond; facework both sides										
102mm thick	14.75	5.00	**15.49**	2.45	1.75	**53.56**	2.63	**71.68**	m2	**77.05**
English bond; facework both sides										
215mm thick	29.75	5.00	**31.24**	3.05	2.30	**67.97**	6.15	**105.36**	m2	**113.26**
Flemish bond; facework both sides										
215mm thick	29.75	5.00	**31.24**	3.35	2.50	**74.38**	6.15	**111.77**	m2	**120.15**
Arches including centering; flat										
112mm high on face; 215mm thick; width of exposed soffit 112mm; bricks-on-edge	3.33	5.00	**3.50**	1.05	0.80	**23.49**	0.70	**27.69**	m	**29.76**
112mm high on face; 215mm thick; width of exposed soffit 215mm; bricks-on-edge	3.33	5.00	**3.50**	1.50	1.00	**32.03**	0.70	**36.22**	m	**38.94**
215mm high on face; 112mm thick; width of exposed soffit 112mm; bricks-on-end	3.33	5.00	**3.50**	1.45	1.00	**31.31**	0.70	**35.51**	m	**38.18**
Arches including centering; segmental										
215mm high on face; 215mm thick; width of exposed soffit 112mm; bricks-on-edge	6.67	5.00	**7.00**	2.40	1.65	**51.78**	1.41	**60.18**	m	**64.70**
215mm high on face; 215mm thick; width of exposed soffit 215mm; bricks-on-edge	6.67	5.00	**7.00**	2.95	1.95	**62.80**	1.41	**71.21**	m	**76.55**
Arches including centering; semi-circular										
215mm high on face; 215mm thick; width of exposed soffit 112mm; bricks-on-edge	6.67	5.00	**7.00**	2.40	1.65	**51.78**	1.41	**60.18**	m	**64.70**
215mm high on face; 215mm thick; width of exposed soffit 215mm; bricks-on-edge	6.67	5.00	**7.00**	2.95	1.95	**62.80**	1.41	**71.21**	m	**76.55**
Facework copings; bricks-on-edge; pointing top and each side										
215 x 102.5mm; horizontal	3.33	5.00	**3.50**	0.70	0.50	**15.30**	0.70	**19.50**	m	**20.96**

Labour hourly rates: (except Specialists) Craft Operatives £14.21 Labourer £10.71 Rates are national average prices. Refer to REGIONAL VARIATIONS for indicative levels of overall pricing in regions	MATERIALS			LABOUR				RATES		
	Del to Site £	Waste %	Material Cost £	Craft Optve Hrs	Lab Hrs	Labour Cost £	Sunds £	Nett Rate £	Unit	Gross Rate (+7.5%) £
F10: BRICK/BLOCK WALLING Cont.										
Facing bricks p.c. £250.00 per 1000, 215 x 102.5 x 65mm; in cement-lime mortar (1:2:9); flush smooth pointing as work proceeds Cont.										
Facework copings; bricks-on-edge; pointing top and each side Cont.										
215 x 102.5mm; with oversailing course and cement fillets both sides; horizontal	3.33	5.00	**3.50**	1.10	0.75	**23.66**	0.70	**27.86**	m	**29.95**
327 x 102.5mm; horizontal	5.00	5.00	**5.25**	1.05	0.75	**22.95**	1.05	**29.26**	m	**31.45**
327 x 102.5mm; with oversailing course and cement fillets both sides; horizontal	5.00	5.00	**5.25**	1.50	1.05	**32.56**	1.05	**38.86**	m	**41.78**
extra; galvanized iron coping cramp at angle or end	2.12	5.00	**2.23**	0.04	0.03	**0.89**	-	**3.12**	nr	**3.35**
Facing bricks p.c. £300.00 per 1000, 215 x 102.5 x 65mm; in cement-lime mortar (1:2:9); flush smooth pointing as work proceeds										
Walls; vertical										
stretcher bond; facework one side										
102mm thick	17.70	5.00	**18.59**	1.75	1.30	**38.79**	2.63	**60.01**	m2	**64.51**
stretcher bond; facework both sides										
102mm thick	17.70	5.00	**18.59**	2.05	1.50	**45.20**	2.63	**66.41**	m2	**71.40**
English bond; facework both sides										
215mm thick	35.70	5.00	**37.49**	2.65	2.05	**59.61**	6.15	**103.25**	m2	**110.99**
Flemish bond; facework both sides										
215mm thick	35.70	5.00	**37.49**	2.95	2.25	**66.02**	6.15	**109.65**	m2	**117.87**
extra; special bricks; vertical angles; squint; BS 4729, type AN.1.1.	23.33	5.00	**24.49**	0.25	0.15	**5.16**	-	**29.65**	m	**31.88**
extra; special bricks; intersections; birdsmouth; BS 4729, type AN.4.1.	69.98	5.00	**73.48**	0.25	0.15	**5.16**	-	**78.64**	m	**84.54**
Walls; building overhand; vertical										
stretcher bond; facework one side										
102mm thick	17.70	5.00	**18.59**	2.15	1.55	**47.15**	2.63	**68.37**	m2	**73.50**
stretcher bond; facework both sides										
102mm thick	17.70	5.00	**18.59**	2.45	1.75	**53.56**	2.63	**74.78**	m2	**80.38**
English bond; facework both sides										
215mm thick	35.70	5.00	**37.49**	3.05	2.30	**67.97**	6.15	**111.61**	m2	**119.98**
Flemish bond; facework both sides										
215mm thick	35.70	5.00	**37.49**	3.35	2.50	**74.38**	6.15	**118.01**	m2	**126.86**
Arches including centering; flat										
112mm high on face; 215mm thick; width of exposed soffit 112mm; bricks-on-edge	4.00	5.00	**4.20**	1.05	0.80	**23.49**	0.70	**28.39**	m	**30.52**
112mm high on face; 215mm thick; width of exposed soffit 215mm; bricks-on-edge	4.00	5.00	**4.20**	1.50	1.00	**32.03**	0.70	**36.92**	m	**39.69**
215mm high on face; 112mm thick; width of exposed soffit 112mm; bricks-on-end	4.00	5.00	**4.20**	1.45	1.00	**31.31**	0.70	**36.21**	m	**38.93**
Arches including centering; segmental										
215mm high on face; 215mm thick; width of exposed soffit 112mm; bricks-on-edge	8.00	5.00	**8.40**	2.40	1.65	**51.78**	1.41	**61.58**	m	**66.20**
215mm high on face; 215mm thick; width of exposed soffit 215mm; bricks-on-edge	8.00	5.00	**8.40**	2.95	1.95	**62.80**	1.41	**72.61**	m	**78.06**
Arches including centering; semi-circular										
215mm high on face; 215mm thick; width of exposed soffit 112mm; bricks-on-edge	8.00	5.00	**8.40**	2.40	1.65	**51.78**	1.41	**61.58**	m	**66.20**
215mm high on face; 215mm thick; width of exposed soffit 215mm; bricks-on-edge	8.00	5.00	**8.40**	2.95	1.95	**62.80**	1.41	**72.61**	m	**78.06**
Facework copings; bricks-on-edge; pointing top and each side										
215 x 102.5mm; horizontal	4.00	5.00	**4.20**	0.70	0.50	**15.30**	0.70	**20.20**	m	**21.71**
215 x 102.5mm; with oversailing course and cement fillets both sides; horizontal	4.00	5.00	**4.20**	1.10	0.75	**23.66**	0.70	**28.56**	m	**30.70**
327 x 102.5mm; horizontal	6.00	5.00	**6.30**	1.05	0.75	**22.95**	1.05	**30.31**	m	**32.58**
327 x 102.5mm; with oversailing course and cement fillets both sides; horizontal	6.00	5.00	**6.30**	1.50	1.05	**32.56**	1.05	**39.91**	m	**42.91**
extra; galvanized iron coping cramp at angle or end	2.12	5.00	**2.23**	0.04	0.03	**0.89**	-	**3.12**	nr	**3.35**
Facing bricks p.c. £350.00 per 1000, 215 x 102.5 x 65mm; in cement-lime mortar (1:2:9); flush smooth pointing as work proceeds										
Walls; vertical										
stretcher bond; facework one side										
102mm thick	20.65	5.00	**21.68**	1.75	1.30	**38.79**	2.63	**63.11**	m2	**67.84**
stretcher bond; facework both sides										
102mm thick	20.65	5.00	**21.68**	2.05	1.50	**45.20**	2.63	**69.51**	m2	**74.73**
English bond; facework both sides										
215mm thick	41.65	5.00	**43.73**	2.65	2.05	**59.61**	6.15	**109.49**	m2	**117.71**
Flemish bond; facework both sides										
215mm thick	41.65	5.00	**43.73**	2.95	2.25	**66.02**	6.15	**115.90**	m2	**124.59**
extra; special bricks; vertical angles; squint; BS 4729, type AN.1.1.	23.33	5.00	**24.49**	0.25	0.15	**5.16**	-	**29.65**	m	**31.88**
extra; special bricks; intersections; birdsmouth; BS 4729, type AN.4.1.	69.98	5.00	**73.48**	0.25	0.15	**5.16**	-	**78.64**	m	**84.54**
Walls; building overhand; vertical										
stretcher bond; facework one side										
102mm thick	20.65	5.00	**21.68**	2.15	1.55	**47.15**	2.63	**71.47**	m2	**76.83**

Labour hourly rates: (except Specialists) Craft Operatives £14.21 Labourer £10.71 Rates are national average prices. Refer to REGIONAL VARIATIONS for indicative levels of overall pricing in regions	MATERIALS			LABOUR				RATES		
	Del to Site £	Waste %	Material Cost £	Craft Optve Hrs	Lab Hrs	Labour Cost £	Sunds £	Nett Rate £	Unit	Gross Rate (+7.5%) £
F10: BRICK/BLOCK WALLING Cont.										
Facing bricks p.c. £350.00 per 1000, 215 x 102.5 x 65mm; in cement-lime mortar (1:2:9); flush smooth pointing as work proceeds Cont.										
Walls; building overhand; vertical Cont.										
stretcher bond; facework both sides										
102mm thick	20.65	5.00	**21.68**	2.45	1.75	**53.56**	2.63	**77.87**	m2	**83.71**
English bond; facework both sides										
215mm thick	41.65	5.00	**43.73**	3.05	2.30	**67.97**	6.15	**117.86**	m2	**126.69**
Flemish bond; facework both sides										
215mm thick	41.65	5.00	**43.73**	3.35	2.50	**74.38**	6.15	**124.26**	m2	**133.58**
Arches including centering; flat										
112mm high on face; 215mm thick; width of exposed soffit 112mm; bricks-on-edge	4.67	5.00	**4.90**	1.05	0.80	**23.49**	0.70	**29.09**	m	**31.27**
112mm high on face; 215mm thick; width of exposed soffit 215mm; bricks-on-edge	4.67	5.00	**4.90**	1.50	1.00	**32.03**	0.70	**37.62**	m	**40.44**
215mm high on face; 112mm thick; width of exposed soffit 112mm; bricks-on-end	4.67	5.00	**4.90**	1.45	1.00	**31.31**	0.70	**36.91**	m	**39.68**
Arches including centering; segmental										
215mm high on face; 215mm thick; width of exposed soffit 112mm; bricks-on-edge	9.33	5.00	**9.80**	2.40	1.65	**51.78**	1.41	**62.98**	m	**67.70**
215mm high on face; 215mm thick; width of exposed soffit 215mm; bricks-on-edge	9.33	5.00	**9.80**	2.95	1.95	**62.80**	1.41	**74.01**	m	**79.56**
Arches including centering; semi-circular										
215mm high on face; 215mm thick; width of exposed soffit 112mm; bricks-on-edge	9.33	5.00	**9.80**	2.40	1.65	**51.78**	1.41	**62.98**	m	**67.70**
215mm high on face; 215mm thick; width of exposed soffit 215mm; bricks-on-edge	9.33	5.00	**9.80**	2.95	1.95	**62.80**	1.41	**74.01**	m	**79.56**
Facework copings; bricks-on-edge; pointing top and each side										
215 x 102.5mm; horizontal	4.67	5.00	**4.90**	0.70	0.50	**15.30**	0.70	**20.90**	m	**22.47**
215 x 102.5mm; with oversailing course and cement fillets both sides; horizontal	4.67	5.00	**4.90**	1.10	0.75	**23.66**	0.70	**29.26**	m	**31.46**
327 x 102.5mm; horizontal	7.00	5.00	**7.35**	1.05	0.75	**22.95**	1.05	**31.36**	m	**33.71**
327 x 102.5mm; with oversailing course and cement fillets both sides; horizontal	7.00	5.00	**7.35**	1.50	1.05	**32.56**	1.05	**40.96**	m	**44.04**
extra; galvanized iron coping cramp at angle or end	2.12	5.00	**2.23**	0.04	0.03	**0.89**	-	**3.12**	nr	**3.35**
Composite walling of bricks 215 x 102.5x 65mm; in cement-lime mortar (1:2:9); common bricks BS EN 772 Category M backing, compressive strength 20.5 N/mm2; facing bricks p.c. £150.00 per 1000; flush smooth pointing as work proceeds										
Walls; vertical										
English bond; facework one side										
215mm thick	18.72	5.00	**19.65**	2.65	2.05	**59.61**	6.15	**85.41**	m2	**91.82**
327mm thick	29.63	5.00	**31.11**	3.15	2.50	**71.54**	9.66	**112.31**	m2	**120.74**
Flemish bond; facework one side										
215mm thick	19.07	5.00	**20.02**	2.65	2.05	**59.61**	6.15	**85.78**	m2	**92.21**
327mm thick	29.98	5.00	**31.48**	3.15	2.50	**71.54**	9.66	**112.68**	m2	**121.13**
Walls; building overhand; vertical										
English bond; facework one side										
215mm thick	18.72	5.00	**19.65**	3.05	2.35	**68.51**	6.15	**94.31**	m2	**101.38**
327mm thick	29.63	5.00	**31.11**	3.55	2.80	**80.43**	9.66	**121.21**	m2	**130.30**
Flemish bond; facework one side										
215mm thick	19.07	5.00	**20.02**	3.05	2.35	**68.51**	6.15	**94.68**	m2	**101.78**
327mm thick	29.98	5.00	**31.48**	3.55	2.80	**80.43**	9.66	**121.58**	m2	**130.70**
Composite walling of bricks 215 x 102.5x 65mm; in cement-lime mortar (1:2:9); common bricks BS EN 772 Category M backing, compressive strength 20.5 N/mm2; facing bricks p.c. £200.00 per 1000; flush smooth pointing as work proceeds										
Walls; vertical										
English bond; facework one side										
215mm thick	23.17	5.00	**24.32**	2.65	2.05	**59.61**	6.15	**90.08**	m2	**96.84**
327mm thick	34.08	5.00	**35.78**	3.15	2.50	**71.54**	9.66	**116.98**	m2	**125.76**
Flemish bond; facework one side										
215mm thick	23.02	5.00	**24.17**	2.65	2.05	**59.61**	6.15	**89.93**	m2	**96.67**
327mm thick	33.93	5.00	**35.63**	3.15	2.50	**71.54**	9.66	**116.83**	m2	**125.59**
Walls; building overhand; vertical										
English bond; facework one side										
215mm thick	23.17	5.00	**24.32**	3.05	2.35	**68.51**	6.15	**98.98**	m2	**106.40**
327mm thick	34.08	5.00	**35.78**	3.55	2.80	**80.43**	9.66	**125.88**	m2	**135.32**
Flemish bond; facework one side										
215mm thick	23.02	5.00	**24.17**	3.05	2.35	**68.51**	6.15	**98.82**	m2	**106.24**
327mm thick	33.93	5.00	**35.63**	3.55	2.80	**80.43**	9.66	**125.72**	m2	**135.15**

Labour hourly rates: (except Specialists) Craft Operatives £14.21 Labourer £10.71 Rates are national average prices. Refer to REGIONAL VARIATIONS for indicative levels of overall pricing in regions	MATERIALS			LABOUR				RATES		
	Del to Site £	Waste %	Material Cost £	Craft Optve Hrs	Lab Hrs	Labour Cost £	Sunds £	Nett Rate £	Unit	Gross Rate (+7.5%) £
F10: BRICK/BLOCK WALLING Cont.										
Composite walling of bricks 215 x 102.5 x 65mm; in cement-lime mortar (1:2:9); common bricks BS EN 772 Category M backing, compressive strength 20.5 N/mm2; facing bricks p.c. £250.00 per 1000; flush smooth pointing as work proceeds										
Walls; vertical										
English bond; facework one side										
215mm thick	27.62	5.00	**29.00**	2.65	2.05	**59.61**	6.15	**94.76**	m2	**101.86**
327mm thick	38.53	5.00	**40.46**	3.15	2.50	**71.54**	9.66	**121.66**	m2	**130.78**
Flemish bond; facework one side										
215mm thick	26.97	5.00	**28.31**	2.65	2.05	**59.61**	6.15	**94.07**	m2	**101.13**
327mm thick	37.88	5.00	**39.77**	3.15	2.50	**71.54**	9.66	**120.97**	m2	**130.05**
Walls; building overhand; vertical										
English bond; facework one side										
215mm thick	27.62	5.00	**29.00**	3.05	2.35	**68.51**	6.15	**103.65**	m2	**111.43**
327mm thick	38.53	5.00	**40.46**	3.55	2.80	**80.43**	9.66	**130.55**	m2	**140.35**
Flemish bond; facework one side										
215mm thick	26.97	5.00	**28.31**	3.05	2.35	**68.51**	6.15	**102.97**	m2	**110.69**
327mm thick	37.88	5.00	**39.77**	3.55	2.80	**80.43**	9.66	**129.87**	m2	**139.61**
Composite walling of bricks 215 x 102.5 x 65mm; in cement-lime mortar (1:2:9); common bricks BS EN 772 Category M backing, compressive strength 20.5 N/mm2; facing bricks p.c. £300.00 per 1000; flush smooth pointing as work proceeds										
Walls; vertical										
English bond; facework one side										
215mm thick	32.07	5.00	**33.67**	2.65	2.05	**59.61**	6.15	**99.43**	m2	**106.89**
327mm thick	42.98	5.00	**45.13**	3.15	2.50	**71.54**	9.66	**126.33**	m2	**135.80**
Flemish bond; facework one side										
215mm thick	30.92	5.00	**32.46**	2.65	2.05	**59.61**	6.15	**98.22**	m2	**105.59**
327mm thick	41.83	5.00	**43.92**	3.15	2.50	**71.54**	9.66	**125.12**	m2	**134.51**
Walls; building overhand; vertical										
English bond; facework one side										
215mm thick	32.07	5.00	**33.67**	3.05	2.35	**68.51**	6.15	**108.33**	m2	**116.45**
327mm thick	42.98	5.00	**45.13**	3.55	2.80	**80.43**	9.66	**135.23**	m2	**145.37**
Flemish bond; facework one side										
215mm thick	30.92	5.00	**32.46**	3.05	2.35	**68.51**	6.15	**107.12**	m2	**115.15**
327mm thick	41.83	5.00	**43.92**	3.55	2.80	**80.43**	9.66	**134.02**	m2	**144.07**
Composite walling of bricks 215 x 102.5 x 65mm; in cement-lime mortar (1:2:9); common bricks BS EN 772 Category M backing, compressive strength 20.5 N/mm2; facing bricks p.c. £350.00 per 1000; flush smooth pointing as work proceeds										
Walls; vertical										
English bond; facework one side										
215mm thick	36.52	5.00	**38.34**	2.65	2.05	**59.61**	6.15	**104.10**	m2	**111.91**
327mm thick	47.43	5.00	**49.80**	3.15	2.50	**71.54**	9.66	**131.00**	m2	**140.83**
Flemish bond; facework one side										
215mm thick	34.87	5.00	**36.61**	2.65	2.05	**59.61**	6.15	**102.37**	m2	**110.05**
327mm thick	45.78	5.00	**48.07**	3.15	2.50	**71.54**	9.66	**129.27**	m2	**138.96**
Walls; building overhand; vertical										
English bond; facework one side										
215mm thick	36.52	5.00	**38.34**	3.05	2.35	**68.51**	6.15	**113.00**	m2	**121.47**
327mm thick	47.43	5.00	**49.80**	3.55	2.80	**80.43**	9.66	**139.90**	m2	**150.39**
Flemish bond; facework one side										
215mm thick	34.87	5.00	**36.61**	3.05	2.35	**68.51**	6.15	**111.27**	m2	**119.61**
327mm thick	45.78	5.00	**48.07**	3.55	2.80	**80.43**	9.66	**138.17**	m2	**148.53**
For each £10.00 difference in cost of 1000 facing bricks, add or deduct the following										
Wall in facing bricks										
102mm thick; English bond; facework one side	0.89	5.00	**0.93**	-	-	**-**	-	**0.93**	m2	**1.00**
215mm thick; facework both sides	1.19	5.00	**1.25**	-	-	**-**	-	**1.25**	m2	**1.34**
Composite wall; common brick backing; facing bricks facework										
215mm thick; English bond; facework one side	0.89	5.00	**0.93**	-	-	**-**	-	**0.93**	m2	**1.00**
215mm thick; Flemish bond; facework one side	0.79	5.00	**0.83**	-	-	**-**	-	**0.83**	m2	**0.89**
Extra over flush smooth pointing for the following types of pointing										
Pointing as work proceeds										
ironing in joints	-	-	**-**	0.03	0.02	**0.64**	-	**0.64**	m2	**0.69**
weathered pointing	-	-	**-**	0.03	0.02	**0.64**	-	**0.64**	m2	**0.69**
recessed pointing	-	-	**-**	0.04	0.03	**0.89**	-	**0.89**	m2	**0.96**
Raking out joints and pointing on completion										
flush smooth pointing with cement-lime mortar (1:2:9)	-	-	**-**	0.75	0.50	**16.01**	0.17	**16.18**	m2	**17.40**
flush smooth pointing with coloured cement-lime mortar (1:2:9)	-	-	**-**	0.75	0.50	**16.01**	0.24	**16.25**	m2	**17.47**
weathered pointing with cement-lime mortar (1:2:9)	-	-	**-**	0.83	0.55	**17.68**	0.17	**17.86**	m2	**19.20**
weathered pointing with coloured cement-lime mortar (1:2:9)	-	-	**-**	0.83	0.55	**17.68**	0.24	**17.92**	m2	**19.27**

Labour hourly rates: (except Specialists) Craft Operatives £14.21 Labourer £10.71 Rates are national average prices. Refer to REGIONAL VARIATIONS for indicative levels of overall pricing in regions	MATERIALS			LABOUR				RATES		
	Del to Site	Waste	Material Cost	Craft Optve	Lab	Labour Cost	Sunds	Nett Rate	Unit	Gross Rate (+7.5%)
	£	%	£	Hrs	Hrs	£	£	£		£
F10: BRICK/BLOCK WALLING Cont.										
Extra over cement-lime mortar (1:2:9) for bedding and pointing for using coloured cement-lime mortar (1:2:9) for the following										
Wall										
102mm thick; pointing both sides	-	-	-	-	-	-	0.89	**0.89**	m2	**0.96**
215mm thick; pointing both sides	-	-	-	-	-	-	2.07	**2.07**	m2	**2.23**
Aerated concrete blocks, Thermalite, 440 x 215mm, Shield blocks (4.0 N/mm2);in cement-lime mortar (1:1:6)										
Walls; vertical										
75mm thick; stretcher bond	6.56	5.00	**6.89**	0.60	0.50	**13.88**	0.81	**21.58**	m2	**23.19**
90mm thick; stretcher bond	7.80	5.00	**8.19**	0.80	0.65	**18.33**	1.01	**27.53**	m2	**29.59**
100mm thick; stretcher bond	8.63	5.00	**9.06**	0.80	0.65	**18.33**	1.01	**28.40**	m2	**30.53**
140mm thick; stretcher bond	11.96	5.00	**12.56**	0.90	0.75	**20.82**	1.52	**34.90**	m2	**37.51**
190mm thick; stretcher bond	16.10	5.00	**16.91**	0.95	0.80	**22.07**	2.01	**40.98**	m2	**44.06**
Closing cavities; vertical										
50mm wide with blockwork 100mm thick	2.16	5.00	**2.27**	0.20	0.10	**3.91**	0.22	**6.39**	m	**6.87**
75mm wide with blockwork 100mm thick	2.16	5.00	**2.27**	0.20	0.10	**3.91**	0.22	**6.39**	m	**6.87**
Closing cavities; horizontal										
50mm wide with blockwork 100mm thick	2.16	5.00	**2.27**	0.20	0.10	**3.91**	0.22	**6.39**	m	**6.87**
75mm wide with blockwork 100mm thick	2.16	5.00	**2.27**	0.20	0.10	**3.91**	0.22	**6.39**	m	**6.87**
Bonding ends to common brickwork; forming pockets; extra material										
walls; bonding every third course										
75mm	1.64	5.00	**1.72**	0.15	0.10	**3.20**	-	**4.92**	m	**5.29**
90mm	1.95	5.00	**2.05**	0.25	0.20	**5.69**	-	**7.74**	m	**8.32**
100mm	2.16	5.00	**2.27**	0.25	0.20	**5.69**	-	**7.96**	m	**8.56**
140mm	2.99	5.00	**3.14**	0.35	0.25	**7.65**	-	**10.79**	m	**11.60**
190mm	4.03	5.00	**4.23**	0.40	0.30	**8.90**	-	**13.12**	m	**14.11**
Bonding ends to existing common brickwork; cutting pockets; extra material										
walls; bonding every third course										
75mm	1.64	5.00	**1.72**	0.30	0.20	**6.41**	-	**8.13**	m	**8.74**
90mm	1.95	5.00	**2.05**	0.50	0.40	**11.39**	-	**13.44**	m	**14.44**
100mm	2.16	5.00	**2.27**	0.50	0.40	**11.39**	-	**13.65**	m	**14.68**
140mm	2.99	5.00	**3.14**	0.70	0.50	**15.30**	-	**18.44**	m	**19.82**
190mm	4.03	5.00	**4.23**	0.80	0.60	**17.79**	-	**22.02**	m	**23.67**
Bonding ends to existing concrete blockwork; cutting pockets; extra material										
walls; bonding every third course										
75mm	1.64	5.00	**1.72**	0.25	0.20	**5.69**	-	**7.42**	m	**7.97**
90mm	1.95	5.00	**2.05**	0.35	0.30	**8.19**	-	**10.23**	m	**11.00**
100mm	2.16	5.00	**2.27**	0.40	0.30	**8.90**	-	**11.16**	m	**12.00**
140mm	2.99	5.00	**3.14**	0.55	0.40	**12.10**	-	**15.24**	m	**16.38**
190mm	4.03	5.00	**4.23**	0.65	0.50	**14.59**	-	**18.82**	m	**20.23**
Aerated concrete blocks, Thermalite, 440 x 215mm, Turbo blocks (2.8 N/mm2); in cement-lime mortar (1:1:6)										
Walls; vertical										
100mm thick; stretcher bond	8.77	5.00	**9.21**	0.80	0.65	**18.33**	1.01	**28.55**	m2	**30.69**
115mm thick; stretcher bond	10.05	5.00	**10.55**	0.80	0.65	**18.33**	1.01	**29.89**	m2	**32.13**
125mm thick; stretcher bond	10.91	5.00	**11.46**	0.85	0.70	**19.58**	1.52	**32.55**	m2	**34.99**
130mm thick; stretcher bond	11.34	5.00	**11.91**	0.85	0.70	**19.58**	1.52	**33.00**	m2	**35.47**
150mm thick; stretcher bond	13.05	5.00	**13.70**	0.90	0.75	**20.82**	1.52	**36.04**	m2	**38.74**
190mm thick; stretcher bond	16.47	5.00	**17.29**	0.95	0.80	**22.07**	2.01	**41.37**	m2	**44.47**
200mm thick; stretcher bond	17.32	5.00	**18.19**	1.00	0.85	**23.31**	2.01	**43.51**	m2	**46.77**
215mm thick; stretcher bond	18.61	5.00	**19.54**	1.00	0.85	**23.31**	2.52	**45.37**	m2	**48.77**
Closing cavities; vertical										
50mm wide with blockwork 100mm thick	2.19	5.00	**2.30**	0.20	0.10	**3.91**	0.22	**6.43**	m	**6.91**
75mm wide with blockwork 100mm thick	2.19	5.00	**2.30**	0.20	0.10	**3.91**	0.22	**6.43**	m	**6.91**
Closing cavities; horizontal										
50mm wide with blockwork 100mm thick	2.19	5.00	**2.30**	0.20	0.10	**3.91**	0.22	**6.43**	m	**6.91**
75mm wide with blockwork 100mm thick	2.19	5.00	**2.30**	0.20	0.10	**3.91**	0.22	**6.43**	m	**6.91**
Aerated concrete blocks, Thermalite, 440 x 215mm, Turbo blocks (2.8 N/mm2); in thin joint mortar										
Walls; vertical										
115mm thick; stretcher bond	11.31	5.00	**11.88**	0.64	0.72	**16.81**	1.09	**29.77**	m2	**32.00**
Aerated concrete blocks, Thermalite, (Large Format) Turbo blocks (2.8 N/mm2); in thin joint mortar										
Walls; vertical										
115mm thick; stretcher bond	11.31	5.00	**11.88**	0.52	0.59	**13.65**	1.09	**26.62**	m2	**28.61**
Aerated concrete blocks, Thermalite, 440 x 215mm, Party wall blocks (4.0 N/mm2); in cement-lime mortar (1:1:6)										
Walls; vertical										
215mm thick; stretcher bond	18.34	5.00	**19.26**	1.00	0.85	**23.31**	2.52	**45.09**	m2	**48.47**

Labour hourly rates: (except Specialists) Craft Operatives £14.21 Labourer £10.71 Rates are national average prices. Refer to REGIONAL VARIATIONS for indicative levels of overall pricing in regions	MATERIALS			LABOUR				RATES		
	Del to Site £	Waste %	Material Cost £	Craft Optve Hrs	Lab Hrs	Labour Cost £	Sunds £	Nett Rate £	Unit	Gross Rate (+7.5%) £
F10: BRICK/BLOCK WALLING Cont.										
Aerated concrete blocks, Thermalite, 440 x 215mm, Hi-Strength 7 blocks (7.0 N/mm2); in cement-lime mortar (1:1:6)										
Walls; vertical										
100mm thick; stretcher bond	10.78	5.00	**11.32**	0.80	0.65	**18.33**	1.01	**30.66**	m2	**32.96**
140mm thick; stretcher bond	14.94	5.00	**15.69**	0.90	0.75	**20.82**	1.52	**38.02**	m2	**40.88**
150mm thick; stretcher bond	15.98	5.00	**16.78**	0.90	0.75	**20.82**	1.52	**39.12**	m2	**42.05**
190mm thick; stretcher bond	20.14	5.00	**21.15**	0.95	0.80	**22.07**	2.01	**45.22**	m2	**48.62**
200mm thick; stretcher bond	21.18	5.00	**22.24**	1.00	0.85	**23.31**	2.01	**47.56**	m2	**51.13**
215mm thick; stretcher bond	22.73	5.00	**23.87**	1.00	0.85	**23.31**	2.52	**49.70**	m2	**53.42**
Closing cavities; vertical										
50mm wide with blockwork 100mm thick	2.70	5.00	**2.83**	0.20	0.10	**3.91**	0.22	**6.96**	m	**7.48**
75mm wide with blockwork 100mm thick	2.70	5.00	**2.83**	0.20	0.10	**3.91**	0.22	**6.96**	m	**7.48**
Closing cavities; horizontal										
50mm wide with blockwork 100mm thick	2.70	5.00	**2.83**	0.20	0.10	**3.91**	0.22	**6.96**	m	**7.48**
75mm wide with blockwork 100mm thick	2.70	5.00	**2.83**	0.20	0.10	**3.91**	0.22	**6.96**	m	**7.48**
Aerated concrete blocks, Thermalite, 440 x 215mm, Trenchblocks (4.0 N/mm2); in cement mortar (1:4)										
Walls; vertical										
255mm thick; stretcher bond	21.70	5.00	**22.79**	1.30	1.05	**29.72**	2.10	**54.60**	m2	**58.69**
275mm thick; stretcher bond	23.37	5.00	**24.54**	1.45	1.15	**32.92**	2.52	**59.98**	m2	**64.47**
305mm thick; stretcher bond	25.89	5.00	**27.18**	1.65	1.25	**36.83**	2.52	**66.53**	m2	**71.52**
355mm thick; stretcher bond	30.08	5.00	**31.58**	1.95	1.50	**43.77**	2.92	**78.28**	m2	**84.15**
Aerated concrete blocks, Tarmac Toplite foundation blocks; 440 x 215mm, (3.5 N/mm2); in cement mortar (1:4)										
Walls; vertical										
260mm thick; stretcher bond	18.33	5.00	**19.25**	1.40	0.95	**30.07**	2.52	**51.83**	m2	**55.72**
275mm thick; stretcher bond	19.39	5.00	**20.36**	1.60	1.05	**33.98**	2.52	**56.86**	m2	**61.12**
300mm thick; stretcher bond	21.15	5.00	**22.21**	1.70	1.10	**35.94**	2.52	**60.66**	m2	**65.21**
Tarmac Topblock, medium density concrete block, Hemelite, 440 x 215mm, solid Standard blocks (3.5 N/mm2); in cement-lime mortar (1:1:6)										
Walls; vertical										
75mm thick; stretcher bond	5.38	5.00	**5.65**	0.75	0.65	**17.62**	0.81	**24.07**	m2	**25.88**
90mm thick; stretcher bond	5.98	5.00	**6.28**	0.90	0.75	**20.82**	1.01	**28.11**	m2	**30.22**
100mm thick; stretcher bond	6.50	5.00	**6.83**	0.90	0.75	**20.82**	1.01	**28.66**	m2	**30.81**
140mm thick; stretcher bond	9.56	5.00	**10.04**	1.00	0.85	**23.31**	1.52	**34.87**	m2	**37.48**
190mm thick; stretcher bond	12.97	5.00	**13.62**	1.05	0.90	**24.56**	2.01	**40.19**	m2	**43.20**
215mm thick; stretcher bond	14.67	5.00	**15.40**	1.10	0.95	**25.81**	2.52	**43.72**	m2	**47.00**
Closing cavities; vertical										
50mm wide with blockwork 100mm thick	1.63	5.00	**1.71**	0.20	0.10	**3.91**	0.22	**5.83**	m	**6.27**
75mm wide with blockwork 100mm thick	1.63	5.00	**1.71**	0.20	0.10	**3.91**	0.22	**5.83**	m	**6.27**
Closing cavities; horizontal										
50mm wide with blockwork 100mm thick	1.63	5.00	**1.71**	0.20	0.10	**3.91**	0.22	**5.83**	m	**6.27**
75mm wide with blockwork 100mm thick	1.63	5.00	**1.71**	0.20	0.10	**3.91**	0.22	**5.83**	m	**6.27**
Bonding ends to common brickwork; forming pockets; extra material										
walls; bonding every third course										
75mm	1.35	5.00	**1.41**	0.25	0.20	**5.69**	-	**7.11**	m	**7.64**
90mm	1.50	5.00	**1.57**	0.30	0.25	**6.94**	-	**8.51**	m	**9.15**
100mm	1.63	5.00	**1.71**	0.30	0.25	**6.94**	-	**8.65**	m	**9.30**
140mm	2.39	5.00	**2.51**	0.40	0.30	**8.90**	-	**11.41**	m	**12.26**
190mm	3.24	5.00	**3.40**	0.45	0.35	**10.14**	-	**13.55**	m	**14.56**
215mm	3.67	5.00	**3.85**	0.55	0.40	**12.10**	-	**15.95**	m	**17.15**
Bonding ends to existing common brickwork; cutting pockets; extra material										
walls; bonding every third course										
75mm	1.35	5.00	**1.41**	0.50	0.40	**11.39**	-	**12.80**	m	**13.76**
90mm	1.50	5.00	**1.57**	0.60	0.50	**13.88**	-	**15.45**	m	**16.61**
100mm	1.63	5.00	**1.71**	0.60	0.50	**13.88**	-	**15.59**	m	**16.76**
140mm	2.39	5.00	**2.51**	0.80	0.60	**17.79**	-	**20.30**	m	**21.83**
190mm	3.24	5.00	**3.40**	0.90	0.70	**20.29**	-	**23.69**	m	**25.47**
215mm	3.67	5.00	**3.85**	1.10	0.80	**24.20**	-	**28.05**	m	**30.15**
Bonding ends to existing concrete blockwork; cutting pockets; extra material										
walls; bonding every third course										
75mm	1.35	5.00	**1.41**	0.40	0.30	**8.90**	-	**10.31**	m	**11.08**
90mm	1.50	5.00	**1.57**	0.45	0.35	**10.14**	-	**11.71**	m	**12.59**
100mm	1.63	5.00	**1.71**	0.45	0.35	**10.14**	-	**11.85**	m	**12.74**
140mm	2.39	5.00	**2.51**	0.60	0.45	**13.35**	-	**15.86**	m	**17.04**
190mm	3.24	5.00	**3.40**	0.70	0.55	**15.84**	-	**19.24**	m	**20.69**
215mm	3.67	5.00	**3.85**	0.85	0.60	**18.50**	-	**22.36**	m	**24.03**
Tarmac Topblock, medium density concrete block, Hemelite, 440 x 215mm, solid Standard blocks (7.0 N/mm2); in cement-lime mortar (1:1:6)										
Walls; vertical										
90mm thick; stretcher bond	6.28	5.00	**6.59**	0.90	0.75	**20.82**	1.01	**28.43**	m2	**30.56**

Labour hourly rates: (except Specialists) Craft Operatives £14.21 Labourer £10.71 Rates are national average prices. Refer to REGIONAL VARIATIONS for indicative levels of overall pricing in regions	MATERIALS Del to Site £	 Waste %	 Material Cost £	LABOUR Craft Optve Hrs	 Lab Hrs	 Labour Cost £	 Sunds £	RATES Nett Rate £	 Unit	 Gross Rate (+7.5%) £
F10: BRICK/BLOCK WALLING Cont.										
Tarmac Topblock, medium density concrete block, Hemelite, 440 x 215mm, solid Standard blocks (7.0 N/mm2); in cement-lime mortar (1:1:6) Cont.										
Walls; vertical Cont.										
100mm thick; stretcher bond	6.90	5.00	**7.25**	0.90	0.75	**20.82**	1.01	**29.08**	m2	**31.26**
140mm thick; stretcher bond	10.14	5.00	**10.65**	1.00	0.85	**23.31**	1.52	**35.48**	m2	**38.14**
190mm thick; stretcher bond	13.77	5.00	**14.46**	1.05	0.90	**24.56**	2.01	**41.03**	m2	**44.11**
215mm thick; stretcher bond	15.58	5.00	**16.36**	1.10	0.95	**25.81**	2.52	**44.68**	m2	**48.03**
Closing cavities; vertical										
50mm wide with blockwork 100mm thick	1.73	5.00	**1.81**	0.20	0.10	**3.91**	0.22	**5.94**	m	**6.38**
75mm wide with blockwork 100mm thick	1.73	5.00	**1.81**	0.20	0.10	**3.91**	0.22	**5.94**	m	**6.38**
Closing cavities; horizontal										
50mm wide with blockwork 100mm thick	1.73	5.00	**1.81**	0.20	0.10	**3.91**	0.22	**5.94**	m	**6.38**
75mm wide with blockwork 100mm thick	1.73	5.00	**1.81**	0.20	0.10	**3.91**	0.22	**5.94**	m	**6.38**
Tarmac Topblock, fair face concrete blocks, Lignacite, 440 x 215mm, solid Standard blocks (7.0 N/mm2); in cement-lime mortar (1:1:6)										
Walls; vertical										
100mm thick; stretcher bond	12.13	5.00	**12.74**	0.95	0.80	**22.07**	1.01	**35.81**	m2	**38.50**
140mm thick; stretcher bond	17.86	5.00	**18.75**	1.10	0.90	**25.27**	1.52	**45.54**	m2	**48.95**
190mm thick; stretcher bond	24.23	5.00	**25.44**	1.15	0.95	**26.52**	2.01	**53.97**	m2	**58.02**
Closing cavities; vertical										
50mm wide with blockwork 100mm thick	3.03	5.00	**3.18**	0.20	0.10	**3.91**	0.22	**7.31**	m	**7.86**
75mm wide with blockwork 100mm thick	3.03	5.00	**3.18**	0.20	0.10	**3.91**	0.22	**7.31**	m	**7.86**
Closing cavities; horizontal										
50mm wide with blockwork 100mm thick	3.03	5.00	**3.18**	0.20	0.10	**3.91**	0.22	**7.31**	m	**7.86**
75mm wide with blockwork 100mm thick	3.03	5.00	**3.18**	0.20	0.10	**3.91**	0.22	**7.31**	m	**7.86**
Tarmac Topblock, dense concrete blocks, Topcrete, 440 x 215mm, solid Standard blocks (7.0 N/mm2); in cement-lime mortar (1:1:6)										
Walls; vertical										
90mm thick; stretcher bond	7.62	5.00	**8.00**	1.15	0.95	**26.52**	1.01	**35.53**	m2	**38.19**
100mm thick; stretcher bond	8.47	5.00	**8.89**	1.20	1.00	**27.76**	1.01	**37.67**	m2	**40.49**
140mm thick; stretcher bond	11.86	5.00	**12.45**	1.30	1.10	**30.25**	1.52	**44.22**	m2	**47.54**
190mm thick; stretcher bond	16.09	5.00	**16.89**	1.40	1.15	**32.21**	2.01	**51.12**	m2	**54.95**
215mm thick; stretcher bond	18.21	5.00	**19.12**	1.50	1.20	**34.17**	2.52	**55.80**	m2	**59.99**
Closing cavities; vertical										
50mm wide with blockwork 100mm thick	2.12	5.00	**2.22**	0.20	0.10	**3.91**	0.22	**6.35**	m	**6.83**
75mm wide with blockwork 100mm thick	2.12	5.00	**2.22**	0.20	0.10	**3.91**	0.22	**6.35**	m	**6.83**
Closing cavities; horizontal										
50mm wide with blockwork 100mm thick	2.12	5.00	**2.22**	0.20	0.10	**3.91**	0.22	**6.35**	m	**6.83**
75mm wide with blockwork 100mm thick	2.12	5.00	**2.22**	0.20	0.10	**3.91**	0.22	**6.35**	m	**6.83**
Tarmac Topblock, dense concrete blocks, Topcrete, 440 x 215mm hollow standard blocks (7.0 N/mm2); in cement-lime mortar (1:1:6)										
Walls; vertical										
215mm thick; stretcher bond	12.68	5.00	**13.31**	1.50	0.81	**29.99**	2.52	**45.82**	m2	**49.26**
Aerated concrete blocks, Thermalite, 440 x 215mm, Smooth Face blocks (4.0 N/mm2); in cement-lime mortar (1:1:6); flush smooth pointing as work proceeds										
Walls; vertical										
stretcher bond; facework one side										
100mm thick	14.43	5.00	**15.15**	1.00	0.80	**22.78**	1.01	**38.94**	m2	**41.86**
140mm thick	20.07	5.00	**21.07**	1.10	0.90	**25.27**	1.52	**47.86**	m2	**51.45**
190mm thick	27.12	5.00	**28.48**	1.15	0.95	**26.52**	2.01	**57.00**	m2	**61.28**
215mm thick	30.65	5.00	**32.18**	1.20	1.00	**27.76**	2.52	**62.46**	m2	**67.14**
stretcher bond; facework both sides										
100mm thick	14.43	5.00	**15.15**	1.20	0.95	**27.23**	1.01	**43.39**	m2	**46.64**
140mm thick	20.07	5.00	**21.07**	1.30	1.05	**29.72**	1.52	**52.31**	m2	**56.23**
190mm thick	27.12	5.00	**28.48**	1.35	1.10	**30.96**	2.01	**61.45**	m2	**66.06**
215mm thick	30.65	5.00	**32.18**	1.40	1.15	**32.21**	2.52	**66.91**	m2	**71.93**
Tarmac Topblock, fair face concrete blocks, Lignacite, 440 x 215mm, solid Standard blocks (7.0 N/mm2); in cement-lime mortar (1:1:6); flush smooth pointing as work proceeds										
Walls; vertical										
stretcher bond; facework one side										
100mm thick	12.13	5.00	**12.74**	1.15	0.95	**26.52**	1.01	**40.26**	m2	**43.28**
140mm thick	17.86	5.00	**18.75**	1.30	1.05	**29.72**	1.52	**49.99**	m2	**53.74**
190mm thick	24.23	5.00	**25.44**	1.35	1.10	**30.96**	2.01	**58.42**	m2	**62.80**
stretcher bond; facework both sides										
100mm thick	12.13	5.00	**12.74**	1.35	1.10	**30.96**	1.01	**44.71**	m2	**48.06**
140mm thick	17.86	5.00	**18.75**	1.50	1.20	**34.17**	1.52	**54.44**	m2	**58.52**
190mm thick	24.23	5.00	**25.44**	1.55	1.25	**35.41**	2.01	**62.86**	m2	**67.58**

Labour hourly rates: (except Specialists) Craft Operatives £14.21 Labourer £10.71 Rates are national average prices. Refer to REGIONAL VARIATIONS for indicative levels of overall pricing in regions	MATERIALS			LABOUR				RATES		
	Del to Site £	Waste %	Material Cost £	Craft Optve Hrs	Lab Hrs	Labour Cost £	Sunds £	Nett Rate £	Unit	Gross Rate (+7.5%) £
F10: BRICK/BLOCK WALLING Cont.										
Reconstructed stone blocks, Marshalls Mono Ltd. Cromwell coursed random length Split Face buff walling blocks in cement-lime mortar (1:1:6); flat recessed pointing as work proceeds										
Walls; vertical										
100mm thick; alternate courses 102mm and 140mm high blocks; stretcher bond; facework one side...	17.41	5.00	**18.28**	1.90	1.40	**41.99**	1.52	**61.79**	m2	**66.42**
100mm thick; one course 102mm high blocks and two courses 140mm high blocks; stretcher bond; facework one side	17.41	5.00	**18.28**	1.90	1.40	**41.99**	1.52	**61.79**	m2	**66.42**
Reconstructed stone blocks, Marshalls Mono Ltd. Cromwell coursed random length Pitched Face buff walling blocks in cement-lime mortar (1:1:6); flat recessed pointing as work proceeds										
Walls; vertical										
90mm thick; alternate courses 102mm and 140mm high blocks; stretcher bond; facework one side........	19.73	5.00	**20.72**	1.90	1.40	**41.99**	1.52	**64.23**	m2	**69.04**
90mm thick; one course 102mm high blocks and two courses 140mm high blocks; stretcher bond; facework one side	19.73	5.00	**20.72**	1.90	1.40	**41.99**	1.52	**64.23**	m2	**69.04**
Firebricks, 215 x 102.5 x 65mm; in fire cement										
Flue linings; bonding to surrounding brickwork with headers -4/m2										
112mm thick; stretcher bond.....................	100.83	5.00	**105.87**	1.90	1.40	**41.99**	7.20	**155.07**	m2	**166.70**
Flue linings; built clear of main brickwork but with one header in each course set projecting to contact main work										
112mm thick; stretcher bond.....................	100.83	5.00	**105.87**	1.90	1.40	**41.99**	7.20	**155.07**	m2	**166.70**
112mm thick; stretcher bond; in segmental top to flue	100.83	5.00	**105.87**	2.50	1.80	**54.80**	7.20	**167.88**	m2	**180.47**
HR Supra Flue lining bricks, 230 x 114 x 76mm, Hepworth Refractories, Sheffield, MPK 21 mortar										
Flue linings; bonding to surrounding brickwork with headers -4/m2										
114mm thick; stretcher bond.....................	40.71	5.00	**42.75**	1.90	1.40	**41.99**	4.35	**89.09**	m2	**95.77**
230mm thick; English bond	82.11	5.00	**86.22**	3.15	2.35	**69.93**	7.78	**163.93**	m2	**176.22**
F11: GLASS BLOCK WALLING - SCREENS AND PANELS										
Hollow glass blocks, white, cross ribbed, in cement mortar (1:3); continuous joints; flat recessed pointing as work proceeds										
Screens or panels; facework both sides										
190 x 190mm blocks; vertical; 80mm thick.........	149.75	10.00	**164.73**	3.00	3.00	**74.76**	53.50	**292.99**	m2	**314.96**
240 x 240mm blocks; vertical; 80mm thick.........	121.28	10.00	**133.41**	2.50	2.50	**62.30**	43.06	**238.77**	m2	**256.68**
F20: NATURAL STONE RUBBLE WALLING										
Stone rubble work; random stones of Yorkshire limestone; bedding and jointing in lime mortar (1:3); uncoursed										
Walls; tapering both sides; including extra material										
500mm thick	50.00	15.00	**57.50**	3.75	3.75	**93.45**	8.46	**159.41**	m2	**171.37**
600mm thick	55.00	15.00	**63.25**	4.50	4.50	**112.14**	10.16	**185.55**	m2	**199.46**
Stone rubble work; squared rubble face stones of Yorkshire limestone; bedding and jointing in lime mortar (1:3); irregular coursed; courses average 150mm high										
Walls; vertical; face stones 100 - 150mm on bed; bonding to brickwork; including extra material; scappled or axed face; weather struck pointing as work proceeds										
150mm thick; faced one side......................	40.00	15.00	**46.00**	3.00	3.00	**74.76**	2.82	**123.58**	m2	**132.85**
Walls; vertical; face stones 100 - 150mm on bed; bonding to brickwork; including extra material; hammer dressed face; weather struck pointing as work proceeds										
150mm thick; faced one side......................	40.00	15.00	**46.00**	3.30	3.30	**82.24**	2.82	**131.06**	m2	**140.89**
Walls; vertical; face stones 100 - 150mm on bed; bonding to brickwork; including extra material; rock worked face; pointing with a parallel joint as work proceeds										
150mm thick; faced one side.....................	42.00	15.00	**48.30**	3.60	3.60	**89.71**	2.82	**140.83**	m2	**151.40**

Labour hourly rates: (except Specialists) Craft Operatives £14.21 Labourer £10.71 Rates are national average prices. Refer to REGIONAL VARIATIONS for indicative levels of overall pricing in regions	MATERIALS			LABOUR				RATES		
	Del to Site £	Waste %	Material Cost £	Craft Optve Hrs	Lab Hrs	Labour Cost £	Sunds £	Nett Rate £	Unit	Gross Rate (+7.5%) £
F20: NATURAL STONE RUBBLE WALLING Cont.										
Stone rubble work; squared rubble face stones of Yorkshire limestone; bedding and jointing in lime mortar (1:3); regular coursed; courses average 150mm high										
Walls; vertical; face stones 100 - 150mm on bed; bonding to brickwork; including extra material; scappled or axed face; weather struck pointing as work proceeds										
150mm thick; faced one side	55.00	15.00	**63.25**	3.20	3.20	**79.74**	2.82	**145.82**	m2	**156.75**
Walls; vertical; face stones 100 - 150mm on bed; bonding to brickwork; including extra material; hammer dressed face; weather struck pointing as work proceeds										
150mm thick; faced one side	55.00	15.00	**63.25**	3.50	3.50	**87.22**	2.82	**153.29**	m2	**164.79**
Walls; vertical; face stones 100 - 150mm on bed; bonding to brickwork; including extra material; rock worked face; pointing with a parallel joint as work proceeds										
150mm thick; faced one side	57.00	15.00	**65.55**	3.80	3.80	**94.70**	2.82	**163.07**	m2	**175.30**
Stone rubble work; random rubble backing and squared rubble face stones of Yorkshire limestone; bedding and jointing in lime mortar (1:3); irregular coursed; courses average 150mm high										
Walls; vertical; face stones 100 - 150mm on bed; scappled or axed face; weather struck pointing as work proceeds										
350mm thick; faced one side	64.00	15.00	**73.60**	4.00	4.00	**99.68**	5.90	**179.18**	m2	**192.62**
500mm thick; faced one side	81.00	15.00	**93.15**	4.75	4.75	**118.37**	8.46	**219.98**	m2	**236.48**
Walls; vertical; face stones 100 - 150mm on bed; hammer dressed face; weather struck pointing as work proceeds										
350mm thick; faced one side	64.00	15.00	**73.60**	4.30	4.30	**107.16**	5.90	**186.66**	m2	**200.65**
500mm thick; faced one side	81.00	15.00	**93.15**	5.05	5.05	**125.85**	8.46	**227.46**	m2	**244.52**
Walls; vertical; face stones 100 - 150mm on bed; rock worked face; pointing with a parallel joint as work proceeds										
350mm thick; faced one side	64.00	15.00	**73.60**	4.60	4.60	**114.63**	5.90	**194.13**	m2	**208.69**
500mm thick; faced one side	81.00	15.00	**93.15**	5.35	5.35	**133.32**	8.46	**234.94**	m2	**252.56**
Quoin stones; scappled or axed face; attached faced two adjacent faces										
250 x 200 x 350mm	26.00	5.00	**27.30**	0.50	0.25	**9.78**	0.29	**37.37**	nr	**40.17**
250 x 200 x 500mm	37.25	5.00	**39.11**	0.80	0.40	**15.65**	0.41	**55.17**	nr	**59.31**
250 x 250 x 350mm	33.50	5.00	**35.18**	0.65	0.35	**12.99**	0.36	**48.52**	nr	**52.16**
250 x 250 x 500mm	46.50	5.00	**48.83**	0.90	0.45	**17.61**	0.51	**66.95**	nr	**71.97**
380 x 200 x 350mm	39.00	5.00	**40.95**	0.80	0.40	**15.65**	0.46	**57.06**	nr	**61.34**
380 x 200 x 500mm	55.00	5.00	**57.75**	1.00	0.50	**19.57**	0.62	**77.93**	nr	**83.78**
Quoin stones; rock worked face; attached faced two adjacent faces										
250 x 200 x 350mm	24.50	5.00	**25.73**	0.55	0.25	**10.49**	0.29	**36.51**	nr	**39.24**
250 x 200 x 500mm	33.00	5.00	**34.65**	0.85	0.40	**16.36**	0.41	**51.42**	nr	**55.28**
250 x 250 x 350mm	28.50	5.00	**29.93**	0.70	0.35	**13.70**	0.36	**43.98**	nr	**47.28**
250 x 250 x 500mm	40.50	5.00	**42.53**	0.95	0.45	**18.32**	0.51	**61.36**	nr	**65.96**
380 x 200 x 350mm	35.00	5.00	**36.75**	0.85	0.40	**16.36**	0.46	**53.57**	nr	**57.59**
380 x 200 x 500mm	49.00	5.00	**51.45**	1.10	0.50	**20.99**	0.62	**73.05**	nr	**78.53**
Arches; relieving										
225mm high on face, 180mm wide on soffit	95.00	5.00	**99.75**	1.50	0.75	**29.35**	0.67	**129.76**	m	**139.50**
225mm high on face, 250mm wide on soffit	130.00	5.00	**136.50**	2.00	1.00	**39.13**	0.92	**176.55**	m	**189.79**
Grooves										
12 x 25mm	-	-	**-**	0.25	-	**3.55**	-	**3.55**	m	**3.82**
12 x 38mm	-	-	**-**	0.30	-	**4.26**	-	**4.26**	m	**4.58**
Stone rubble work; random rubble backing and squared rubble face stones of Yorkshire limestone; bedding and jointing in lime mortar (1:3); regular coursed; courses average 150mm high										
Walls; vertical; face stones 100 - 150mm on bed; scappled or axed face; weather struck pointing as work proceeds										
350mm thick; faced one side	81.00	15.00	**93.15**	4.20	4.20	**104.66**	5.90	**203.71**	m2	**218.99**
Walls; vertical; face stones 125 - 200mm on bed; scappled or axed face; weather struck pointing as work proceeds										
500mm thick; faced one side	99.00	15.00	**113.85**	4.95	4.95	**123.35**	8.46	**245.67**	m2	**264.09**
Walls; vertical; face stones 100 - 150mm on bed; hammer dressed face; weather struck pointing as work proceeds										
350mm thick; faced one side	81.00	15.00	**93.15**	4.50	4.50	**112.14**	5.90	**211.19**	m2	**227.03**
500mm thick; faced one side	99.00	15.00	**113.85**	5.25	5.25	**130.83**	8.46	**253.14**	m2	**272.13**

Labour hourly rates: (except Specialists) Craft Operatives £14.21 Labourer £10.71 Rates are national average prices. Refer to REGIONAL VARIATIONS for indicative levels of overall pricing in regions	MATERIALS			LABOUR				RATES		
	Del to Site £	Waste %	Material Cost £	Craft Optve Hrs	Lab Hrs	Labour Cost £	Sunds £	Nett Rate £	Unit	Gross Rate (+7.5%) £
F20: NATURAL STONE RUBBLE WALLING Cont.										
Stone rubble work; random rubble backing and squared rubble face stones of Yorkshire limestone; bedding and jointing in lime mortar (1:3); regular coursed; courses average 150mm high Cont.										
Walls; vertical; face stones 100 - 150mm on bed; rock worked face; pointing with a parallel joint as work proceeds										
350mm thick; faced one side	81.00	15.00	**93.15**	4.80	4.80	**119.62**	5.90	**218.67**	m2	**235.07**
500mm thick; faced one side	99.00	15.00	**113.85**	5.55	5.55	**138.31**	8.46	**260.62**	m2	**280.17**
F21: NATURAL STONE ASHLAR WALLING/DRESSINGS										
Natural stonework										
Note stonework is work for a specialist; prices being obtained for specific projects. Some firms that specialise in this class of work are included within the list at the end of this section										
Natural stonework dressings; natural Dorset limestone; Portland Whitbed; bedding and jointing in mason's mortar (1:3:12); flush smooth pointing as work proceeds; slurrying with weak lime mortar and cleaning down on completion										
Walls; vertical; building against brickwork; BS 1243 Fig 1 specification 3.5 wall ties - 4/m2 built in										
50mm thick; plain and rubbed one side	136.00	2.50	**139.40**	4.00	2.60	**84.69**	5.64	**229.73**	m2	**246.96**
75mm thick; plain and rubbed one side	187.50	2.50	**192.19**	4.50	3.15	**97.68**	6.16	**296.03**	m2	**318.23**
100mm thick; plain and rubbed one side	245.00	2.50	**251.13**	5.00	3.70	**110.68**	6.67	**368.47**	m2	**396.11**
Natural stonework; natural Dorset limestone, Portland Whitbed; bedding and jointing in cement-lime mortar (1:2:9); flush smooth pointing as work proceeds										
Lintels plain and rubbed faces -3; splayed and rubbed faces -1										
200 x 100mm	60.00	2.50	**61.50**	1.25	0.85	**26.87**	1.28	**89.65**	m	**96.37**
225 x 125mm	77.00	2.50	**78.93**	1.50	1.10	**33.10**	1.64	**113.66**	m	**122.19**
Sills plain and rubbed faces -3; sunk weathered and rubbed faces -1; grooves -1; throats -1										
200 x 75mm	80.00	2.50	**82.00**	1.00	0.65	**21.17**	1.95	**105.12**	m	**113.00**
250 x 75mm	97.00	2.50	**99.43**	1.20	0.80	**25.62**	2.26	**127.30**	m	**136.85**
300 x 75mm	112.00	2.50	**114.80**	1.40	1.00	**30.60**	2.51	**147.92**	m	**159.01**
Jamb stones; attached plain and rubbed faces -3; splayed and rubbed faces -1; rebates -1; grooves -1										
175 x 75mm	62.00	2.50	**63.55**	1.00	0.65	**21.17**	1.69	**86.41**	m	**92.90**
200 x 100mm	67.00	2.50	**68.68**	1.25	1.00	**28.47**	2.21	**99.35**	m	**106.80**
Band courses; moulded; horizontal										
225 x 125mm	140.00	2.50	**143.50**	1.60	1.15	**35.05**	3.08	**181.63**	m	**195.25**
250 x 150mm	168.00	2.50	**172.20**	1.80	1.35	**40.04**	3.59	**215.83**	m	**232.01**
300 x 150mm	198.00	2.50	**202.95**	2.00	1.55	**45.02**	4.21	**252.18**	m	**271.09**
Copings; horizontal plain and rubbed faces -2; weathered and rubbed faces -1; throats -2; cramped joints with stainless steel cramps										
300 x 50mm	97.00	2.50	**99.43**	0.75	0.50	**16.01**	4.31	**119.75**	m	**128.73**
300 x 75mm	113.00	2.50	**115.83**	0.90	0.70	**20.29**	4.82	**140.93**	m	**151.50**
375 x 100mm	156.00	2.50	**159.90**	1.35	1.00	**29.89**	5.80	**195.59**	m	**210.26**
Natural stonework, natural Yorkshire sandstone, Bolton Wood; bedding and jointing in cement-lime mortar (1:2:9); flush smooth pointing as work proceeds										
Sills plain and rubbed faces -3; sunk weathered and rubbed faces -1; throats -1										
175 x 100mm	58.00	2.50	**59.45**	1.25	0.85	**26.87**	1.28	**87.60**	m	**94.17**
250 x 125mm	90.00	2.50	**92.25**	1.40	1.00	**30.60**	1.69	**124.55**	m	**133.89**
Copings; horizontal plain and rubbed faces -2; weathered and rubbed faces -2; throats -2; cramped joints with stainless steel cramps										
300 x 75mm	63.50	2.50	**65.09**	0.90	0.70	**20.29**	3.08	**88.45**	m	**95.09**
300 x 100mm	75.00	2.50	**76.88**	1.10	0.80	**24.20**	3.39	**104.46**	m	**112.29**
Kerbs; horizontal plain and sawn faces - 4										
150 x 150mm	39.00	2.50	**39.98**	0.85	0.65	**19.04**	1.13	**60.14**	m	**64.65**
225 x 150mm	59.00	2.50	**60.48**	1.20	1.00	**27.76**	1.69	**89.93**	m	**96.67**
Cover stones										
75 x 300; rough edges -2; plain and sawn face -1.	33.75	2.50	**34.59**	0.85	0.65	**19.04**	2.26	**55.89**	m	**60.08**

Labour hourly rates: (except Specialists) Craft Operatives £14.21 Labourer £10.71 Rates are national average prices. Refer to REGIONAL VARIATIONS for indicative levels of overall pricing in regions	MATERIALS			LABOUR				RATES		
	Del to Site £	Waste %	Material Cost £	Craft Optve Hrs	Lab Hrs	Labour Cost £	Sunds £	Nett Rate £	Unit	Gross Rate (+7.5%) £
F21: NATURAL STONE ASHLAR WALLING/DRESSINGS Cont.										
Natural stonework, natural Yorkshire sandstone, Bolton Wood; bedding and jointing in cement-lime mortar (1:2:9); flush smooth pointing as work proceeds Cont.										
Templates										
150 x 300; rough edges -2; plain and sawn face -1	67.50	2.50	**69.19**	1.50	1.20	**34.17**	2.82	**106.18**	m	**114.14**
Steps; plain										
plain and rubbed top and front										
225 x 75mm	27.00	2.50	**27.68**	0.75	0.55	**16.55**	2.26	**46.48**	m	**49.97**
225 x 150mm	54.00	2.50	**55.35**	1.30	1.00	**29.18**	2.82	**87.35**	m	**93.91**
300 x 150mm	72.00	2.50	**73.80**	1.60	1.30	**36.66**	3.39	**113.84**	m	**122.38**
Landings										
75 x 900 x 900mm; sawn edges -4; plain and rubbed face -1	99.00	2.50	**101.48**	2.00	1.75	**47.16**	3.39	**152.02**	nr	**163.43**
F22: CAST STONE ASHLAR WALLING/DRESSINGS										
Cast stonework dressings; simulated Dorset limestone; Portland Whitbed; bedding and jointing in cement-lime mortar (1:2:9); flush smooth pointing as work proceeds										
Walls; vertical building against brickwork; BS 1243 Fig 1 specification 3.5 wall ties -4/m2 built in										
100mm thick; plain and rubbed one side	80.00	5.00	**84.00**	4.00	2.50	**83.62**	5.64	**173.26**	m2	**186.25**
Fair raking cutting										
100 thick	16.00	5.00	**16.80**	1.00	0.50	**19.57**	-	**36.37**	m	**39.09**
Cast stonework; simulated Dorset limestone; Portland Whitbed; bedding and jointing in cement-lime mortar (1:2:9); flush smooth pointing as work proceeds										
Lintels										
plain and rubbed faces -3; splayed and rubbed faces -1										
200 x 100mm	23.00	5.00	**24.15**	1.00	0.60	**20.64**	0.82	**45.61**	m	**49.03**
225 x 125mm	34.00	5.00	**35.70**	1.25	0.75	**25.80**	1.13	**62.62**	m	**67.32**
Sills										
plain and rubbed faces -3; sunk weathered and rubbed faces -1; grooves -1; throats -1										
200 x 75mm	21.00	5.00	**22.05**	0.90	0.50	**18.14**	1.13	**41.32**	m	**44.42**
extra; stoolings	5.00	5.00	**5.25**	-	-	**-**	-	**5.25**	nr	**5.64**
250 x 75mm	23.00	5.00	**24.15**	1.00	0.60	**20.64**	1.39	**46.17**	m	**49.63**
extra; stoolings	5.00	5.00	**5.25**	-	-	**-**	-	**5.25**	nr	**5.64**
300 x 75mm	27.00	5.00	**28.35**	1.25	0.75	**25.80**	1.69	**55.84**	m	**60.03**
extra; stoolings	5.00	5.00	**5.25**	-	-	**-**	-	**5.25**	nr	**5.64**
Jamb stones; attached										
plain and rubbed faces -3; splayed and rubbed faces -1; rebates -1; grooves -1										
175 x 75mm	22.50	5.00	**23.63**	0.90	0.50	**18.14**	0.87	**42.64**	m	**45.84**
200 x 100mm	29.00	5.00	**30.45**	1.00	0.60	**20.64**	1.69	**52.78**	m	**56.74**
Band courses; plain; horizontal										
225 x 125mm	22.50	5.00	**23.63**	1.40	1.00	**30.60**	1.39	**55.61**	m	**59.79**
extra; external return	16.00	5.00	**16.80**	-	-	**-**	-	**16.80**	nr	**18.06**
250 x 150mm	30.00	5.00	**31.50**	1.50	1.10	**33.10**	1.69	**66.29**	m	**71.26**
extra; external return	16.00	5.00	**16.80**	-	-	**-**	-	**16.80**	nr	**18.06**
300 x 150mm	33.00	5.00	**34.65**	1.60	1.20	**35.59**	1.95	**72.19**	m	**77.60**
extra; external return	16.00	5.00	**16.80**	-	-	**-**	-	**16.80**	nr	**18.06**
Copings; horizontal										
plain and rubbed faces -2; weathered and rubbed faces -2; throats -2; cramped joints with stainless steel cramps										
300 x 50mm	17.50	5.00	**18.38**	0.50	0.30	**10.32**	2.57	**31.26**	m	**33.60**
extra; internal angles	16.00	5.00	**16.80**	-	-	**-**	-	**16.80**	nr	**18.06**
extra; external angles	16.00	5.00	**16.80**	-	-	**-**	-	**16.80**	nr	**18.06**
300 x 75mm	22.50	5.00	**23.63**	0.60	0.35	**12.27**	2.82	**38.72**	m	**41.63**
extra; internal angles	16.00	5.00	**16.80**	-	-	**-**	-	**16.80**	nr	**18.06**
extra; external angles	16.00	5.00	**16.80**	-	-	**-**	-	**16.80**	nr	**18.06**
375 x 100mm	37.50	5.00	**39.38**	0.75	0.50	**16.01**	3.39	**58.77**	m	**63.18**
extra; internal angles	16.00	5.00	**16.80**	-	-	**-**	-	**16.80**	nr	**18.06**
extra; external angles	16.00	5.00	**16.80**	-	-	**-**	-	**16.80**	nr	**18.06**
Steps; plain										
300 x 150mm; plain and rubbed tread and riser	36.00	5.00	**37.80**	1.25	0.90	**27.40**	2.26	**67.46**	m	**72.52**
Steps; spandril										
250mm wide tread; 180mm high riser; plain and rubbed tread and riser; carborundum finish to tread	45.00	5.00	**47.25**	2.00	1.25	**41.81**	1.69	**90.75**	m	**97.56**
Landings										
150 x 900 x 900mm; plain and rubbed top surface	80.00	5.00	**84.00**	1.50	1.00	**32.03**	2.26	**118.28**	nr	**127.15**

Labour hourly rates: (except Specialists) Craft Operatives £14.21 Labourer £10.71 Rates are national average prices. Refer to REGIONAL VARIATIONS for indicative levels of overall pricing in regions	MATERIALS			LABOUR				RATES		
	Del to Site £	Waste %	Material Cost £	Craft Optve Hrs	Lab Hrs	Labour Cost £	Sunds £	Nett Rate £	Unit	Gross Rate (+7.5%) £
F30: ACCESSORIES/SUNDRY ITEMS FOR BRICK/BLOCK/STONE WALLING										
Forming cavities in hollow walls										
Cavity with BS 1243 Fig 1 stainless steel wire butterfly wall ties, -4/m2 built in										
width of cavity 50mm	0.49	10.00	**0.54**	0.10	0.07	**2.17**	-	**2.71**	m2	**2.92**
Cavity with Staifix HRT4; 200mm Housing stainless steel wall ties -4/m2 built in; 50 fibreglass resin bonded slab cavity insulation										
width of cavity 50mm	3.75	10.00	**4.12**	0.35	0.25	**7.65**	1.16	**12.93**	m2	**13.90**
Cavity with BS 1243 Fig 3 stainless steel twisted wall ties -4/m2 built in; 25 Celotex CW035 tuff-R Zero cavity insulation boards										
width of cavity 50mm	7.44	10.00	**8.19**	0.35	0.35	**8.72**	1.16	**18.07**	m2	**19.42**
Cavity with 200mm stainless steel vertical twisted wall ties -4/m2 built in; 50 fibreglass resin bonded slab cavity insulation										
width of cavity 50mm	5.58	10.00	**6.14**	0.35	0.25	**7.65**	1.16	**14.95**	m2	**16.08**
Cavity with Staifix HRT4; 200mm Housing stainless steel wall ties -4/m2 built in; 75 fibreglass resin bonded slab cavity insulation										
width of cavity 75mm	5.04	10.00	**5.54**	0.40	0.25	**8.36**	1.16	**15.06**	m2	**16.19**
Cavity with stainless steel vertical twisted wall ties -4/m2 built in; 75 fibreglass resin bonded slab cavity insulation										
width of cavity 75mm	6.87	10.00	**7.56**	0.40	0.25	**8.36**	1.16	**17.08**	m2	**18.37**
Damp proof courses										
BS 6398 Class A, bitumen with hessian base; 100mm laps; bedding in cement mortar (1:3); no allowance made for laps										
width not exceeding 225mm; vertical	6.91	5.00	**7.26**	0.45	-	**6.39**	-	**13.65**	m2	**14.67**
width exceeding 225mm; vertical	6.91	5.00	**7.26**	0.40	-	**5.68**	-	**12.94**	m2	**13.91**
width not exceeding 225mm; horizontal	6.91	5.00	**7.26**	0.30	-	**4.26**	-	**11.52**	m2	**12.38**
width exceeding 225mm; horizontal	6.91	5.00	**7.26**	0.25	-	**3.55**	-	**10.81**	m2	**11.62**
cavity trays; width not exceeding 225mm; horizontal	6.91	5.00	**7.26**	0.50	-	**7.11**	-	**14.36**	m2	**15.44**
cavity trays; width exceeding 225mm; horizontal	6.91	5.00	**7.26**	0.45	-	**6.39**	-	**13.65**	m2	**14.67**
BS 6398 Class B, bitumen with fibre base; 100mm laps; bedding in cement mortar (1:3); no allowance made for laps										
width not exceeding 225mm; vertical	4.13	5.00	**4.34**	0.45	-	**6.39**	-	**10.73**	m2	**11.54**
width exceeding 225mm; vertical	4.13	5.00	**4.34**	0.40	-	**5.68**	-	**10.02**	m2	**10.77**
width not exceeding 225mm; horizontal	4.13	5.00	**4.34**	0.30	-	**4.26**	-	**8.60**	m2	**9.24**
width exceeding 225mm; horizontal	4.13	5.00	**4.34**	0.25	-	**3.55**	-	**7.89**	m2	**8.48**
BS 6398 Class D, bitumen with hessian base laminated with lead; 100mm laps; bedding in cement mortar (1:3); no allowance made for laps										
width not exceeding 225mm; vertical	13.70	5.00	**14.39**	0.45	-	**6.39**	-	**20.78**	m2	**22.34**
width exceeding 225mm; vertical	13.70	5.00	**14.39**	0.40	-	**5.68**	-	**20.07**	m2	**21.57**
width not exceeding 225mm; horizontal	13.70	5.00	**14.39**	0.30	-	**4.26**	-	**18.65**	m2	**20.05**
width exceeding 225mm; horizontal	13.70	5.00	**14.39**	0.25	-	**3.55**	-	**17.94**	m2	**19.28**
BS 6398 Class E, bitumen with fibre base laminated with lead; 100mm laps; bedding in cement mortar (1:3); no allowance made for laps										
width not exceeding 225mm; vertical	15.22	5.00	**15.98**	0.45	-	**6.39**	-	**22.38**	m2	**24.05**
width exceeding 225mm; vertical	15.22	5.00	**15.98**	0.40	-	**5.68**	-	**21.67**	m2	**23.29**
width not exceeding 225mm; horizontal	15.22	5.00	**15.98**	0.30	-	**4.26**	-	**20.24**	m2	**21.76**
width exceeding 225mm; horizontal	15.22	5.00	**15.98**	0.25	-	**3.55**	-	**19.53**	m2	**21.00**
Hyload, pitch polymer; 100mm laps; bedding in cement lime mortar (1:1:6); no allowance made for laps										
width not exceeding 225mm; vertical	5.52	5.00	**5.80**	0.45	-	**6.39**	-	**12.19**	m2	**13.10**
width exceeding 225mm; vertical	5.52	5.00	**5.80**	0.40	-	**5.68**	-	**11.48**	m2	**12.34**
width not exceeding 225mm; horizontal	5.52	5.00	**5.80**	0.30	-	**4.26**	-	**10.06**	m2	**10.81**
width exceeding 225mm; horizontal	5.52	5.00	**5.80**	0.25	-	**3.55**	-	**9.35**	m2	**10.05**
Synthaprufe bituminous latex emulsion; two coats brushed on; blinded with sand										
width not exceeding 225mm; vertical	4.68	5.00	**4.92**	-	0.50	**5.36**	0.15	**10.42**	m2	**11.20**
width exceeding 225mm; vertical	4.68	5.00	**4.92**	-	0.30	**3.21**	0.15	**8.28**	m2	**8.90**
Synthaprufe bituminous latex emulsion; three coats brushed on; blinded with sand										
width not exceeding 225mm; vertical	7.04	5.00	**7.39**	-	0.70	**7.50**	0.17	**15.06**	m2	**16.19**
width exceeding 225mm; vertical	7.04	5.00	**7.39**	-	0.42	**4.50**	0.17	**12.06**	m2	**12.97**
Bituminous emulsion; two coats brushed on										
width not exceeding 225mm; vertical	3.52	5.00	**3.70**	-	0.42	**4.50**	0.15	**8.34**	m2	**8.97**
width exceeding 225mm; vertical	3.52	5.00	**3.70**	-	0.26	**2.78**	0.15	**6.63**	m2	**7.13**
Bituminous emulsion; three coats brushed on										
width not exceeding 225mm; vertical	4.89	5.00	**5.14**	-	0.60	**6.43**	0.17	**11.74**	m2	**12.62**
width exceeding 225mm; vertical	4.89	5.00	**5.14**	-	0.36	**3.86**	0.17	**9.17**	m2	**9.85**

Labour hourly rates: (except Specialists) Craft Operatives £14.21 Labourer £10.71 Rates are national average prices. Refer to REGIONAL VARIATIONS for indicative levels of overall pricing in regions	MATERIALS			LABOUR				RATES		
	Del to Site £	Waste %	Material Cost £	Craft Optve Hrs	Lab Hrs	Labour Cost £	Sunds £	Nett Rate £	Unit	Gross Rate (+7.5%) £
F30: ACCESSORIES/SUNDRY ITEMS FOR BRICK/BLOCK/STONE WALLING Cont.										
Damp proof courses Cont.										
One course slates in cement mortar (1:3)										
width not exceeding 225mm; vertical	16.02	15.00	**18.42**	1.30	0.90	**28.11**	-	**46.54**	m2	**50.03**
width exceeding 225mm; vertical	16.02	15.00	**18.42**	1.20	0.80	**25.62**	-	**44.04**	m2	**47.35**
width not exceeding 225mm; horizontal	16.02	15.00	**18.42**	0.80	0.55	**17.26**	-	**35.68**	m2	**38.36**
width exceeding 225mm; horizontal	16.02	15.00	**18.42**	0.70	0.50	**15.30**	-	**33.73**	m2	**36.25**
Two courses slates in cement mortar (1:3)										
width not exceeding 225mm; vertical	32.04	15.00	**36.85**	2.20	1.50	**47.33**	-	**84.17**	m2	**90.49**
width exceeding 225mm; vertical	32.04	15.00	**36.85**	2.00	1.35	**42.88**	-	**79.72**	m2	**85.70**
width not exceeding 225mm; horizontal	32.04	15.00	**36.85**	1.30	0.90	**28.11**	-	**64.96**	m2	**69.83**
width exceeding 225mm; horizontal	32.04	15.00	**36.85**	1.20	0.80	**25.62**	-	**62.47**	m2	**67.15**
Cavity trays										
Type G Cavitray by Cavity Trays Ltd in										
half brick skin of cavity wall	9.05	5.00	**9.50**	0.20	-	**2.84**	-	**12.34**	m	**13.27**
external angle	7.22	5.00	**7.58**	0.10	-	**1.42**	-	**9.00**	nr	**9.68**
internal angle	7.22	5.00	**7.58**	0.10	-	**1.42**	-	**9.00**	nr	**9.68**
Type X Cavitray by Cavity Trays Ltd to suit 40 degree pitched roof complete with attached code 4 lead flashing and dress over tiles in										
half brick skin of cavity wall	37.66	5.00	**39.54**	0.20	-	**2.84**	-	**42.38**	m	**45.56**
ridge tray	9.35	5.00	**9.82**	0.10	-	**1.42**	-	**11.24**	nr	**12.08**
catchment tray	4.43	5.00	**4.65**	0.10	-	**1.42**	-	**6.07**	nr	**6.53**
corner catchment angle tray	9.77	5.00	**10.26**	0.10	-	**1.42**	-	**11.68**	nr	**12.56**
Type W Cavity weep/ventilator by Cavity Trays Ltd in										
half brick skin of cavity wall	0.50	5.00	**0.52**	0.10	-	**1.42**	-	**1.95**	nr	**2.09**
extension duct	0.76	5.00	**0.80**	0.10	-	**1.42**	-	**2.22**	nr	**2.39**
Joint reinforcement										
Expamet grade 304/S15 Exmet reinforcement, stainless steel; 150mm laps;										
65mm wide	1.04	5.00	**1.09**	0.04	0.02	**0.78**	0.03	**1.91**	m	**2.05**
115mm wide	1.58	5.00	**1.66**	0.05	0.03	**1.03**	0.03	**2.72**	m	**2.93**
175mm wide	2.73	5.00	**2.87**	0.06	0.03	**1.17**	0.04	**4.08**	m	**4.39**
225mm wide	3.57	5.00	**3.75**	0.07	0.04	**1.42**	0.05	**5.23**	m	**5.62**
Bed Joint Reinforcement (UK) Ltd Netspan reinforcement, stainless steel grade 304/S15; 150mm laps;										
61mm wide	1.04	5.00	**1.09**	0.04	0.02	**0.78**	0.03	**1.91**	m	**2.05**
112mm wide	1.58	5.00	**1.66**	0.05	0.03	**1.03**	0.03	**2.72**	m	**2.93**
178mm wide	2.73	5.00	**2.87**	0.05	0.03	**1.03**	0.03	**3.93**	m	**4.23**
228mm wide	3.57	5.00	**3.75**	0.06	0.03	**1.17**	0.04	**4.97**	m	**5.34**
Bed Joint Reinforcement (UK) Ltd grade 304/S15 Wallspan reinforcement, stainless steel; 150mm laps;										
210mm wide	6.15	5.00	**6.46**	0.06	0.03	**1.17**	0.04	**7.67**	m	**8.25**
222.5mm wide	6.15	5.00	**6.46**	0.06	0.03	**1.17**	0.04	**7.67**	m	**8.25**
235mm wide	6.15	5.00	**6.46**	0.07	0.04	**1.42**	0.05	**7.93**	m	**8.53**
250mm wide	6.15	5.00	**6.46**	0.07	0.04	**1.42**	0.05	**7.93**	m	**8.53**
275mm wide	6.15	5.00	**6.46**	0.07	0.04	**1.42**	0.05	**7.93**	m	**8.53**
Brickforce , galvanised steel; 150mm laps;										
60mm wide ref GBF30W60	0.38	5.00	**0.40**	0.04	0.02	**0.78**	0.03	**1.22**	m	**1.31**
60mm wide ref GBF35W60	0.52	5.00	**0.55**	0.04	0.02	**0.78**	0.03	**1.36**	m	**1.46**
60mm wide ref GBF40W60	0.63	5.00	**0.66**	0.04	0.02	**0.78**	0.03	**1.47**	m	**1.58**
60mm wide ref GBF45W60	0.84	5.00	**0.88**	0.04	0.02	**0.78**	0.03	**1.69**	m	**1.82**
60mm wide ref GBF50W60	1.04	5.00	**1.09**	0.04	0.02	**0.78**	0.03	**1.91**	m	**2.05**
160mm wide ref GBF30W160	0.55	5.00	**0.58**	0.05	0.03	**1.03**	0.03	**1.64**	m	**1.77**
160mm wide ref GBF35W160	0.75	5.00	**0.79**	0.05	0.03	**1.03**	0.03	**1.85**	m	**1.99**
160mm wide ref GBF40W160	0.90	5.00	**0.95**	0.05	0.03	**1.03**	0.03	**2.01**	m	**2.16**
160mm wide ref GBF45W160	1.22	5.00	**1.28**	0.05	0.03	**1.03**	0.03	**2.34**	m	**2.52**
160mm wide ref GBF50W160	1.51	5.00	**1.58**	0.05	0.03	**1.03**	0.03	**2.64**	m	**2.84**
Brickforce , stainless steel; 150mm laps;										
60mm wide ref SBF30W60	1.01	5.00	**1.06**	0.04	0.02	**0.78**	0.03	**1.87**	m	**2.01**
60mm wide ref SBF35W60	1.38	5.00	**1.45**	0.04	0.02	**0.78**	0.03	**2.26**	m	**2.43**
60mm wide ref SBF40W60	1.81	5.00	**1.90**	0.04	0.02	**0.78**	0.03	**2.72**	m	**2.92**
60mm wide ref SBF45W60	2.23	5.00	**2.34**	0.04	0.02	**0.78**	0.03	**3.15**	m	**3.39**
60mm wide ref SBF50W60	2.75	5.00	**2.88**	0.04	0.02	**0.78**	0.03	**3.70**	m	**3.97**
160mm wide ref SBF30W160	1.06	5.00	**1.11**	0.05	0.03	**1.03**	0.03	**2.18**	m	**2.34**
160mm wide ref SBF35W160	1.44	5.00	**1.51**	0.05	0.03	**1.03**	0.03	**2.58**	m	**2.77**
160mm wide ref SBF40W160	1.91	5.00	**2.00**	0.05	0.03	**1.03**	0.03	**3.07**	m	**3.30**
160mm wide ref SBF45W160	2.33	5.00	**2.45**	0.05	0.03	**1.03**	0.03	**3.51**	m	**3.78**
160mm wide ref SBF50W160	2.88	5.00	**3.03**	0.05	0.03	**1.03**	0.03	**4.09**	m	**4.40**
Pointing in flashings										
Cement mortar (1:3)										
horizontal	-	-	**-**	0.40	-	**5.68**	0.24	**5.92**	m	**6.36**
horizontal; in old wall	-	-	**-**	0.45	-	**6.39**	0.24	**6.63**	m	**7.13**
stepped	-	-	**-**	0.65	-	**9.24**	0.38	**9.61**	m	**10.33**
stepped; in old wall	-	-	**-**	0.70	-	**9.95**	0.38	**10.32**	m	**11.10**

Labour hourly rates: (except Specialists) Craft Operatives £14.21 Labourer £10.71 Rates are national average prices. Refer to REGIONAL VARIATIONS for indicative levels of overall pricing in regions	MATERIALS			LABOUR				RATES		
	Del to Site £	Waste %	Material Cost £	Craft Optve Hrs	Lab Hrs	Labour Cost £	Sunds £	Nett Rate £	Unit	Gross Rate (+7.5%) £
F30: ACCESSORIES/SUNDRY ITEMS FOR BRICK/BLOCK/STONE WALLING Cont.										
Wedging and pinning										
Two courses slates in cement mortar (1:3)										
width of wall 215mm	8.01	15.00	**9.21**	0.50	0.25	**9.78**	0.14	**19.13**	m	**20.57**
Joints										
Expansion joints in facing brickwork 13mm wide; vertical; filling with Servicised Ltd, Fibrepack filler, Vertiseal compound pointing one side; including preparation, cleaners, primers and sealers										
102mm thick wall	3.13	7.50	**3.36**	0.09	0.18	**3.21**	-	**6.57**	m	**7.06**
215mm thick wall	4.96	7.50	**5.33**	0.10	0.20	**3.56**	-	**8.90**	m	**9.56**
Expansion joints in blockwork 13mm wide; vertical; filling with Servicised Ltd, Fibrepack filler, Vertiseal compound pointing one side; including preparation, cleaners, primers and sealers										
100mm thick wall	3.13	7.50	**3.36**	0.09	0.18	**3.21**	-	**6.57**	m	**7.06**
Expansion joints in glass blockwork 10mm wide; vertical; in filling with compressible material, polysulphide sealant both sides including preparation, cleaners, primers and sealers										
80mm thick wall	4.16	7.50	**4.47**	0.09	0.18	**3.21**	-	**7.68**	m	**8.25**
Slates and tiles for creasing										
Nibless creasing tiles, red, machine made, 265 x 165 x 10mm; in cement-lime mortar (1:1:6)										
one course 253mm wide	4.38	5.00	**4.60**	0.50	0.35	**10.85**	0.25	**15.70**	m	**16.88**
one course 365mm wide	7.11	5.00	**7.47**	0.70	0.55	**15.84**	0.35	**23.66**	m	**25.43**
two courses 253mm wide	8.77	5.00	**9.21**	0.85	0.55	**17.97**	0.51	**27.68**	m	**29.76**
two courses 365mm wide	14.22	5.00	**14.93**	1.15	0.85	**25.45**	0.71	**41.09**	m	**44.17**
Slate and tile sills										
Clay plain roofing tiles, red, BS EN 1304, machine made, 265 x 165 x 13mm; in cement-lime mortar										
one course 150mm wide; set weathering	1.58	5.00	**1.66**	0.45	0.30	**9.61**	0.24	**11.50**	m	**12.37**
two courses 150mm wide; set weathering	3.16	5.00	**3.32**	0.75	0.50	**16.01**	0.46	**19.79**	m	**21.28**
Fires and fire parts										
Solid one piece or two piece firebacks; BS 1251; bed and joint in fire cement; concrete filling at back										
fire size 400mm	31.85	2.50	**32.65**	1.00	0.67	**21.39**	1.16	**55.19**	nr	**59.33**
fire size 450mm	34.21	2.50	**35.07**	1.10	0.73	**23.45**	1.29	**59.80**	nr	**64.29**
Solid one piece or two piece firebacks with cut out for boiler; BS 1251; bed and joint in fire cement; concrete filling at back										
fire size 400mm	38.58	2.50	**39.54**	1.00	0.67	**21.39**	1.16	**62.09**	nr	**66.75**
fire size 450mm	44.94	2.50	**46.06**	1.10	0.73	**23.45**	1.29	**70.80**	nr	**76.11**
Frets and stools; black vitreous enamelled; place in position										
fire size 400mm	26.87	-	**26.87**	0.33	0.22	**7.05**	-	**33.92**	nr	**36.46**
fire size 450mm	41.99	-	**41.99**	0.36	0.24	**7.69**	-	**49.68**	nr	**53.40**
Frets and stools; lustre finish; place in position										
fire size 400mm	43.74	-	**43.74**	0.33	0.22	**7.05**	-	**50.79**	nr	**54.59**
fire size 450mm	28.66	-	**28.66**	0.36	0.24	**7.69**	-	**36.35**	nr	**39.07**
Continuous burning open fire; vitreous finish; self contained open fire; plugging and screwing; sealing to opening										
fire size 400mm	134.49	-	**134.49**	3.50	2.33	**74.69**	2.38	**211.56**	nr	**227.42**
fire size 450mm	153.62	-	**153.62**	3.75	2.50	**80.06**	4.60	**238.28**	nr	**256.15**
Gas ignited smokeless fuel fires; assembling; plugging and screwing; bedding and jointing in fire cement; sealing to opening										
grate only; black										
fire size 400mm	87.30	-	**87.30**	2.50	1.67	**53.41**	-	**140.71**	nr	**151.26**
fire size 450mm	88.48	-	**88.48**	2.50	1.67	**53.41**	-	**141.89**	nr	**152.53**
grate only; colours										
fire size 400mm	88.48	-	**88.48**	2.50	1.67	**53.41**	-	**141.89**	nr	**152.53**
fire size 450mm	89.64	-	**89.64**	2.50	1.67	**53.41**	-	**143.05**	nr	**153.78**
boiler and self contained flue										
fire size 400mm	202.52	-	**202.52**	4.50	3.00	**96.08**	2.38	**300.97**	nr	**323.54**
fire size 450mm	215.37	-	**215.37**	4.50	3.00	**96.08**	4.60	**316.05**	nr	**339.75**
threefold brick sets										
fire size 400mm	26.89	-	**26.89**	1.00	0.67	**21.39**	1.16	**49.44**	nr	**53.14**
fire size 450mm	26.89	-	**26.89**	1.00	0.67	**21.39**	1.29	**49.57**	nr	**53.28**
extra; chrome finish to fire front; fire size 400 or 450mm	26.89	-	**26.89**	-	-	**-**	-	**26.89**	nr	**28.91**
extra; back boiler unit with Bower Barffed rustless boiler; fire size 400 or 450mm	209.52	-	**209.52**	-	-	**-**	-	**209.52**	nr	**225.23**

Labour hourly rates: (except Specialists) Craft Operatives £14.21 Labourer £10.71 Rates are national average prices. Refer to REGIONAL VARIATIONS for indicative levels of overall pricing in regions	MATERIALS			LABOUR				RATES		
	Del to Site £	Waste %	Material Cost £	Craft Optve Hrs	Lab Hrs	Labour Cost £	Sunds £	Nett Rate £	Unit	Gross Rate (+7.5%) £
F30: ACCESSORIES/SUNDRY ITEMS FOR BRICK/BLOCK/STONE WALLING Cont.										
Fires and fire parts Cont.										
Gas ignited smokeless fuel fires; assembling; plugging and screwing; bedding and jointing in fire cement; sealing to opening Cont.										
threefold brick sets Cont.										
extra; back boiler unit with copper boiler; 400 or 450mm	320.08	-	**320.08**	-	-	**-**	-	**320.08**	nr	**344.09**
Flue linings										
Clay flue linings, BS 1181; rebated joints; jointed in cement mortar (1:3)										
150mm diameter, Type 2	18.47	5.00	**19.39**	0.45	0.35	**10.14**	0.31	**29.85**	m	**32.09**
terminal Type 6F	17.66	5.00	**18.54**	0.45	0.35	**10.14**	0.16	**28.85**	nr	**31.01**
185mm diameter, Type 2	23.67	5.00	**24.85**	0.50	0.40	**11.39**	0.41	**36.65**	m	**39.40**
terminal Type 6F	19.21	5.00	**20.17**	0.50	0.40	**11.39**	0.18	**31.74**	nr	**34.12**
225mm diameter, Type 2	39.46	5.00	**41.43**	0.55	0.45	**12.64**	0.44	**54.51**	m	**58.60**
terminal Type 6F	20.67	5.00	**21.70**	0.55	0.45	**12.64**	0.24	**34.58**	nr	**37.17**
185 x 185mm, Type 1	25.42	5.00	**26.69**	0.55	0.45	**12.64**	0.41	**39.73**	m	**42.71**
terminal Type 4D	28.76	5.00	**30.20**	0.55	0.45	**12.64**	0.18	**43.02**	nr	**46.24**
Chimney pots										
Clay chimney pots; set and flaunched in cement mortar (1:3)										
tapered roll top; 600mm high	35.40	5.00	**37.17**	0.75	0.63	**17.40**	0.41	**54.98**	nr	**59.11**
tapered roll top; 750mm high	48.20	5.00	**50.61**	0.88	0.75	**20.54**	0.52	**71.66**	nr	**77.04**
tapered roll top; 900mm high	63.57	5.00	**66.75**	1.00	0.93	**24.17**	0.61	**91.53**	nr	**98.40**
Air bricks										
Clay air bricks, BS 493, square hole pattern; opening with slate lintel over										
common brick wall 102mm thick; opening size										
225 x 75mm	1.90	-	**1.90**	0.20	0.10	**3.91**	0.61	**6.43**	nr	**6.91**
225 x 150mm	3.62	-	**3.62**	0.25	0.13	**4.94**	0.65	**9.21**	nr	**9.90**
225 x 225mm	6.35	-	**6.35**	0.30	0.15	**5.87**	0.71	**12.93**	nr	**13.90**
common brick wall 215mm thick; opening size										
225 x 75mm	1.90	-	**1.90**	0.30	0.15	**5.87**	1.16	**8.93**	nr	**9.60**
225 x 150mm	3.62	-	**3.62**	0.38	0.19	**7.43**	1.21	**12.27**	nr	**13.19**
225 x 225mm	6.35	-	**6.35**	0.45	0.23	**8.86**	1.30	**16.51**	nr	**17.75**
common brick wall 215mm thick; facework one side; opening size										
225 x 75mm	1.90	-	**1.90**	0.40	0.20	**7.83**	1.16	**10.89**	nr	**11.70**
225 x 150mm	3.62	-	**3.62**	0.50	0.25	**9.78**	1.21	**14.62**	nr	**15.71**
225 x 225mm	6.35	-	**6.35**	0.60	0.30	**11.74**	1.30	**19.39**	nr	**20.84**
common brick wall 327mm thick; facework one side; opening size										
225 x 75mm	1.90	-	**1.90**	0.60	0.30	**11.74**	1.87	**15.51**	nr	**16.67**
225 x 150mm	3.62	-	**3.62**	0.70	0.35	**13.70**	1.97	**19.28**	nr	**20.73**
225 x 225mm	6.35	-	**6.35**	0.80	0.40	**15.65**	2.03	**24.03**	nr	**25.84**
common brick wall 440mm thick; facework one side; opening size										
225 x 75mm	1.90	-	**1.90**	0.80	0.40	**15.65**	2.49	**20.05**	nr	**21.55**
225 x 150mm	3.62	-	**3.62**	0.90	0.45	**17.61**	2.58	**23.81**	nr	**25.59**
225 x 225mm	6.35	-	**6.35**	1.00	0.50	**19.57**	2.67	**28.58**	nr	**30.72**
cavity wall 252mm thick with 102mm facing brick outer skin, 100mm block inner skin and 50mm cavity; sealing cavity with slates in cement mortar (1:3); rendering all round with cement mortar (1:3); opening size										
225 x 75mm	1.90	-	**1.90**	0.60	0.30	**11.74**	2.25	**15.89**	nr	**17.08**
225 x 150mm	3.62	-	**3.62**	0.70	0.35	**13.70**	2.76	**20.08**	nr	**21.58**
225 x 225mm	6.35	-	**6.35**	0.80	0.40	**15.65**	3.24	**25.24**	nr	**27.13**
Gas flue blocks										
Marflex HP system precast refractory concrete gas flue blocks; bedding and jointing in Flue joint refractory mortar										
recess block reference HP1, 405 x 140 x 222mm	1.90	5.00	**2.00**	0.26	-	**3.69**	0.61	**6.30**	nr	**6.78**
cover block reference HP2, 385 x 140 x 222mm	4.80	5.00	**5.04**	0.45	-	**6.39**	0.61	**12.05**	nr	**12.95**
Standard block reference HP3, 355 x 140 x 72mm	2.40	5.00	**2.52**	0.30	-	**4.26**	0.61	**7.40**	nr	**7.95**
Standard block reference HP3, 355 x 140 x 112mm	2.50	5.00	**2.63**	0.30	-	**4.26**	0.61	**7.50**	nr	**8.06**
Standard block reference HP3, 355 x 140 x 222mm	3.30	5.00	**3.47**	0.30	-	**4.26**	0.61	**8.34**	nr	**8.97**
Vent unit for bathroom/kitchen extract fans reference HP1 and HP2	5.36	5.00	**5.63**	0.30	-	**4.26**	0.61	**10.50**	nr	**11.29**
standard block reference HP4, 280 x 140 x 72mm	2.50	5.00	**2.63**	0.25	-	**3.55**	0.54	**6.72**	nr	**7.22**
standard block reference HP4, 280 x 140 x 112mm	2.60	5.00	**2.73**	0.25	-	**3.55**	0.54	**6.82**	nr	**7.33**
standard block reference HP4, 280 x 140 x 222mm	3.30	5.00	**3.47**	0.25	-	**3.55**	0.54	**7.56**	nr	**8.12**
120mm side offset block reference HP5, 400 x 140 x 222mm	3.60	5.00	**3.78**	0.37	-	**5.26**	0.78	**9.82**	nr	**10.56**
70mm back offset block reference HP6, 280 x 210 x 222mm	12.50	5.00	**13.13**	0.35	-	**4.97**	0.78	**18.88**	nr	**20.30**
vertical exit block reference HP7, 280 x 181 x 222mm	7.25	5.00	**7.61**	0.35	-	**4.97**	0.61	**13.20**	nr	**14.19**
angled entry/exit block reference HP8, 280 x 140 x 230mm	7.25	5.00	**7.61**	0.40	-	**5.68**	0.84	**14.14**	nr	**15.20**
double rebate block used when in conjunction with a boiler reference HP9, 280 x 140 x 222mm	5.20	5.00	**5.46**	0.40	-	**5.68**	0.54	**11.68**	nr	**12.56**
corbel block reference HP10	7.19	5.00	**7.55**	0.40	-	**5.68**	0.61	**13.85**	nr	**14.88**

Labour hourly rates: (except Specialists) Craft Operatives £14.21 Labourer £10.71 Rates are national average prices. Refer to REGIONAL VARIATIONS for indicative levels of overall pricing in regions	MATERIALS			LABOUR				RATES		
	Del to Site £	Waste %	Material Cost £	Craft Optve Hrs	Lab Hrs	Labour Cost £	Sunds £	Nett Rate £	Unit	Gross Rate (+7.5%) £
F30: ACCESSORIES/SUNDRY ITEMS FOR BRICK/BLOCK/STONE WALLING Cont.										
Gas flue blocks Cont.										
Marflex HP system precast refractory concrete gas flue blocks; bedding and jointing in Flue joint refractory mortar Cont.										
conversion unit for back boiler recess units reference HP25, 280 x 262 x 222mm	10.72	5.00	**11.26**	0.40	-	**5.68**	0.54	**17.48**	nr	18.79
Arch bars										
Steel flat arch bar										
30 x 6mm	2.06	-	**2.06**	0.70	-	**9.95**	-	**12.01**	m	12.91
50 x 6mm	3.18	-	**3.18**	0.70	-	**9.95**	-	**13.13**	m	14.11
Steel angle arch bar										
50 x 50 x 6mm	5.94	-	**5.94**	0.70	-	**9.95**	-	**15.89**	m	17.08
75 x 50 x 6mm	7.53	-	**7.53**	0.80	-	**11.37**	-	**18.90**	m	20.32
Building in										
Building in metal windows; building in lugs; bedding in cement mortar (1:3), pointing with Secomastic standard mastic one side										
200mm high; lugs to brick jambs; plugging and screwing to brick sill and concrete head										
500mm wide	-	-	-	0.53	0.22	**9.89**	1.00	**10.89**	nr	11.70
600mm wide	-	-	-	0.58	0.24	**10.81**	1.25	**12.06**	nr	12.96
900mm wide	-	-	-	0.74	0.29	**13.62**	1.71	**15.33**	nr	16.48
1200mm wide	-	-	-	0.90	0.34	**16.43**	2.20	**18.63**	nr	20.03
1500mm wide	-	-	-	1.06	0.39	**19.24**	2.67	**21.91**	nr	23.55
1800mm wide	-	-	-	1.22	0.44	**22.05**	3.13	**25.18**	nr	27.07
500mm high; lugs to brick jambs; plugging and screwing to brick sill and concrete head										
500mm wide	-	-	-	0.64	0.25	**11.77**	1.55	**13.32**	nr	14.32
600mm wide	-	-	-	0.69	0.27	**12.70**	1.70	**14.40**	nr	15.47
900mm wide	-	-	-	0.85	0.32	**15.51**	2.16	**17.67**	nr	18.99
1200mm wide	-	-	-	1.02	0.37	**18.46**	2.63	**21.09**	nr	22.67
1500mm wide	-	-	-	1.18	0.43	**21.37**	3.13	**24.50**	nr	26.34
1800mm wide	-	-	-	1.34	0.48	**24.18**	3.56	**27.74**	nr	29.82
700mm high; lugs to brick jambs; plugging and screwing to brick sill and concrete head										
500mm wide	-	-	-	0.75	0.28	**13.66**	1.85	**15.51**	nr	16.67
600mm wide	-	-	-	0.80	0.30	**14.58**	2.01	**16.59**	nr	17.84
900mm wide	-	-	-	0.97	0.35	**17.53**	2.49	**20.03**	nr	21.53
1200mm wide	-	-	-	1.13	0.41	**20.45**	2.96	**23.40**	nr	25.16
1500mm wide	-	-	-	1.30	0.46	**23.40**	3.42	**26.82**	nr	28.83
1800mm wide	-	-	-	1.46	0.52	**26.32**	3.87	**30.19**	nr	32.45
900mm high; lugs to brick jambs; plugging and screwing to brick sill and concrete head										
500mm wide	-	-	-	0.85	0.30	**15.29**	2.14	**17.43**	nr	18.74
600mm wide	-	-	-	0.91	0.32	**16.36**	2.26	**18.62**	nr	20.01
900mm wide	-	-	-	1.08	0.38	**19.42**	2.77	**22.19**	nr	23.85
1200mm wide	-	-	-	1.25	0.44	**22.47**	3.24	**25.71**	nr	27.64
1500mm wide	-	-	-	1.42	0.50	**25.53**	3.71	**29.24**	nr	31.44
1800mm wide	-	-	-	1.58	0.55	**28.34**	4.15	**32.49**	nr	34.93
1100mm high; lugs to brick jambs; plugging and screwing to brick sill and concrete head										
500mm wide	-	-	-	0.96	0.33	**17.18**	2.48	**19.66**	nr	21.13
600mm wide	-	-	-	1.02	0.35	**18.24**	2.59	**20.83**	nr	22.40
900mm wide	-	-	-	1.19	0.41	**21.30**	3.07	**24.38**	nr	26.20
1200mm wide	-	-	-	1.36	0.47	**24.36**	3.55	**27.91**	nr	30.00
1500mm wide	-	-	-	1.54	0.53	**27.56**	4.02	**31.58**	nr	33.95
1800mm wide	-	-	-	1.71	0.59	**30.62**	4.49	**35.11**	nr	37.74
1300mm high; lugs to brick jambs; plugging and screwing to brick sill and concrete head										
500mm wide	-	-	-	1.07	0.36	**19.06**	2.74	**21.80**	nr	23.44
600mm wide	-	-	-	1.12	0.38	**19.99**	2.90	**22.89**	nr	24.60
900mm wide	-	-	-	1.30	0.44	**23.19**	3.38	**26.56**	nr	28.55
1200mm wide	-	-	-	1.48	0.50	**26.39**	3.85	**30.23**	nr	32.50
1500mm wide	-	-	-	1.65	0.57	**29.55**	4.33	**33.88**	nr	36.42
1800mm wide	-	-	-	1.83	0.63	**32.75**	4.79	**37.55**	nr	40.36
1500mm high; lugs to brick jambs; plugging and screwing to brick sill and concrete head										
500mm wide	-	-	-	1.18	0.39	**20.94**	3.06	**24.01**	nr	25.81
600mm wide	-	-	-	1.23	0.41	**21.87**	3.19	**25.06**	nr	26.94
900mm wide	-	-	-	1.42	0.47	**25.21**	3.68	**28.89**	nr	31.06
1200mm wide	-	-	-	1.59	0.54	**28.38**	4.14	**32.52**	nr	34.95
1500mm wide	-	-	-	1.77	0.60	**31.58**	4.59	**36.17**	nr	38.88
1800mm wide	-	-	-	1.95	0.67	**34.89**	5.10	**39.98**	nr	42.98
2100mm high; lugs to brick jambs; plugging and screwing to brick sill and concrete head										
500mm wide	-	-	-	1.39	0.44	**24.46**	3.96	**28.42**	nr	30.55
600mm wide	-	-	-	1.45	0.46	**25.53**	4.12	**29.65**	nr	31.87
900mm wide	-	-	-	1.64	0.53	**28.98**	4.58	**33.56**	nr	36.08
1200mm wide	-	-	-	1.82	0.60	**32.29**	5.10	**37.38**	nr	40.19
1500mm wide	-	-	-	2.01	0.67	**35.74**	5.53	**41.26**	nr	44.36
1800mm wide	-	-	-	2.19	0.74	**39.05**	6.02	**45.07**	nr	48.45

Labour hourly rates: (except Specialists) Craft Operatives £14.21 Labourer £10.71 Rates are national average prices. Refer to REGIONAL VARIATIONS for indicative levels of overall pricing in regions	MATERIALS			LABOUR				RATES		
	Del to Site	Waste	Material Cost	Craft Optve	Lab	Labour Cost	Sunds	Nett Rate	Unit	Gross Rate (+7.5%)
	£	%	£	Hrs	Hrs	£	£	£		£
F30: ACCESSORIES/SUNDRY ITEMS FOR BRICK/BLOCK/STONE WALLING Cont.										
Building in Cont.										
Building in factory glazed metal windows; screwing with galvanized screws; bedding in cement mortar (1:3), pointing with Secomastic standard mastic one side										
300mm high; plugging and screwing lugs to brick jambs and sill and concrete head										
600mm wide	-	-	-	0.70	0.30	**13.16**	1.39	**14.55**	nr	**15.64**
900mm wide	-	-	-	0.84	0.34	**15.58**	1.87	**17.45**	nr	**18.76**
1200mm wide	-	-	-	0.98	0.38	**18.00**	2.33	**20.33**	nr	**21.85**
1500mm wide	-	-	-	1.12	0.41	**20.31**	2.81	**23.11**	nr	**24.85**
1800mm wide	-	-	-	1.26	0.45	**22.72**	3.26	**25.98**	nr	**27.93**
2400mm wide	-	-	-	1.54	0.53	**27.56**	4.21	**31.77**	nr	**34.16**
3000mm wide	-	-	-	1.82	0.60	**32.29**	5.14	**37.43**	nr	**40.23**
700mm high; plugging and screwing lugs to brick jambs and sill and concrete head										
600mm wide	-	-	-	0.96	0.40	**17.93**	2.01	**19.94**	nr	**21.43**
900mm wide	-	-	-	1.17	0.48	**21.77**	2.49	**24.26**	nr	**26.08**
1200mm wide	-	-	-	1.37	0.57	**25.57**	2.99	**28.56**	nr	**30.70**
1500mm wide	-	-	-	1.58	0.64	**29.31**	3.42	**32.72**	nr	**35.18**
1800mm wide	-	-	-	1.79	0.73	**33.25**	3.87	**37.12**	nr	**39.91**
2400mm wide	-	-	-	2.21	0.90	**41.04**	4.83	**45.87**	nr	**49.31**
3000mm wide	-	-	-	2.62	1.06	**48.58**	5.78	**54.37**	nr	**58.44**
900mm high; plugging and screwing lugs to brick jambs and sill and concrete head										
600mm wide	-	-	-	1.09	0.45	**20.31**	2.31	**22.62**	nr	**24.32**
900mm wide	-	-	-	1.33	0.55	**24.79**	2.77	**27.56**	nr	**29.63**
1200mm wide	-	-	-	1.57	0.66	**29.38**	3.24	**32.61**	nr	**35.06**
1500mm wide	-	-	-	1.81	0.76	**33.86**	3.71	**37.57**	nr	**40.39**
1800mm wide	-	-	-	2.06	0.87	**38.59**	4.15	**42.74**	nr	**45.95**
2400mm wide	-	-	-	2.54	1.08	**47.66**	5.12	**52.78**	nr	**56.74**
3000mm wide	-	-	-	3.02	1.29	**56.73**	6.06	**62.79**	nr	**67.50**
1100mm high; plugging and screwing lugs to brick jambs and sill and concrete head										
600mm wide	-	-	-	1.22	0.50	**22.69**	2.60	**25.29**	nr	**27.19**
900mm wide	-	-	-	1.49	0.62	**27.81**	3.07	**30.89**	nr	**33.20**
1200mm wide	-	-	-	1.77	0.75	**33.18**	3.54	**36.72**	nr	**39.48**
1500mm wide	-	-	-	2.04	0.88	**38.41**	4.02	**42.43**	nr	**45.62**
1800mm wide	-	-	-	2.32	1.01	**43.78**	4.49	**48.28**	nr	**51.90**
2400mm wide	-	-	-	2.87	1.27	**54.38**	5.42	**59.80**	nr	**64.29**
3000mm wide	-	-	-	3.42	1.52	**64.88**	6.37	**71.25**	nr	**76.60**
1300mm high; plugging and screwing lugs to brick jambs and sill and concrete head										
600mm wide	-	-	-	1.34	0.54	**24.82**	3.07	**27.90**	nr	**29.99**
900mm wide	-	-	-	1.66	0.70	**31.09**	3.38	**34.46**	nr	**37.05**
1200mm wide	-	-	-	1.96	0.85	**36.96**	3.86	**40.81**	nr	**43.88**
1500mm wide	-	-	-	2.28	0.99	**43.00**	4.33	**47.33**	nr	**50.88**
1800mm wide	-	-	-	2.59	1.15	**49.12**	4.79	**53.91**	nr	**57.96**
2400mm wide	-	-	-	3.21	1.45	**61.14**	5.70	**66.84**	nr	**71.85**
3000mm wide	-	-	-	3.83	1.75	**73.17**	6.63	**79.80**	nr	**85.78**
1500mm high; plugging and screwing lugs to brick jambs and sill and concrete head										
600mm wide	-	-	-	1.47	0.59	**27.21**	3.19	**30.40**	nr	**32.68**
900mm wide	-	-	-	1.82	0.77	**34.11**	3.68	**37.79**	nr	**40.62**
1200mm wide	-	-	-	2.16	0.94	**40.76**	4.14	**44.90**	nr	**48.27**
1500mm wide	-	-	-	2.51	1.11	**47.56**	4.63	**52.19**	nr	**56.10**
1800mm wide	-	-	-	2.85	1.29	**54.31**	5.10	**59.41**	nr	**63.87**
2400mm wide	-	-	-	3.54	1.64	**67.87**	6.03	**73.90**	nr	**79.44**
3000mm wide	-	-	-	4.23	1.98	**81.31**	6.96	**88.27**	nr	**94.89**
2100mm high; plugging and screwing lugs to brick jambs and sill and concrete head										
600mm wide	-	-	-	1.86	0.74	**34.36**	4.12	**38.47**	nr	**41.36**
900mm wide	-	-	-	2.31	0.98	**43.32**	4.58	**47.90**	nr	**51.49**
1200mm wide	-	-	-	2.75	1.22	**52.14**	5.08	**57.23**	nr	**61.52**
1500mm wide	-	-	-	3.20	1.46	**61.11**	5.53	**66.63**	nr	**71.63**
1800mm wide	-	-	-	3.65	1.71	**70.18**	5.99	**76.17**	nr	**81.88**
2400mm wide	-	-	-	4.54	2.19	**87.97**	6.93	**94.90**	nr	**102.02**
3000mm wide	-	-	-	5.43	2.67	**105.76**	7.86	**113.61**	nr	**122.14**
Building in wood windows; building in lugs; bedding in cement mortar (1:3), pointing with Secomastic standard mastic one side										
768mm high; lugs to brick jambs										
438mm wide	-	-	-	1.43	0.09	**21.28**	1.87	**23.15**	nr	**24.89**
641mm wide	-	-	-	1.64	0.09	**24.27**	2.20	**26.47**	nr	**28.46**
1225mm wide	-	-	-	2.24	0.10	**32.90**	3.14	**36.04**	nr	**38.74**
1809mm wide	-	-	-	2.86	0.11	**41.82**	4.10	**45.91**	nr	**49.36**
2394mm wide	-	-	-	3.46	0.12	**50.45**	5.05	**55.50**	nr	**59.67**
920mm high; lugs to brick jambs										
438mm wide	-	-	-	1.61	0.09	**23.84**	2.09	**25.93**	nr	**27.87**
641mm wide	-	-	-	1.82	0.10	**26.93**	2.40	**29.33**	nr	**31.53**
1225mm wide	-	-	-	2.45	0.11	**35.99**	3.38	**39.37**	nr	**42.32**
1809mm wide	-	-	-	3.07	0.13	**45.02**	4.33	**49.35**	nr	**53.05**
2394mm wide	-	-	-	3.70	0.14	**54.08**	5.27	**59.34**	nr	**63.79**
1073mm high; lugs to brick jambs										
438mm wide	-	-	-	1.80	0.10	**26.65**	2.33	**28.98**	nr	**31.16**
641mm wide	-	-	-	2.01	0.11	**29.74**	2.67	**32.41**	nr	**34.84**
1225mm wide	-	-	-	2.66	0.12	**39.08**	3.62	**42.71**	nr	**45.91**
1809mm wide	-	-	-	3.30	0.14	**48.39**	4.55	**52.94**	nr	**56.91**
2394mm wide	-	-	-	3.95	0.16	**57.84**	5.51	**63.36**	nr	**68.11**
1225mm high; lugs to brick jambs										
438mm wide	-	-	-	1.98	0.10	**29.21**	2.58	**31.79**	nr	**34.17**

Labour hourly rates: (except Specialists)
Craft Operatives £14.21 Labourer £10.71
Rates are national average prices.
Refer to REGIONAL VARIATIONS for indicative levels of overall pricing in regions

	MATERIALS			LABOUR				RATES		
	Del to Site £	Waste %	Material Cost £	Craft Optve Hrs	Lab Hrs	Labour Cost £	Sunds £	Nett Rate £	Unit	Gross Rate (+7.5%) £
F30: ACCESSORIES/SUNDRY ITEMS FOR BRICK/BLOCK/STONE WALLING Cont.										
Building in Cont.										
Building in wood windows; building in lugs; bedding in cement mortar (1:3), pointing with Secomastic standard mastic one side Cont.										
1225mm high; lugs to brick jambs Cont.										
641mm wide	-	-	-	2.20	0.11	**32.44**	2.88	**35.32**	nr	**37.97**
1225mm wide	-	-	-	2.87	0.13	**42.18**	3.85	**46.02**	nr	**49.48**
1809mm wide	-	-	-	3.52	0.16	**51.73**	4.75	**56.48**	nr	**60.72**
2394mm wide	-	-	-	4.19	0.18	**61.47**	5.72	**67.19**	nr	**72.23**
1378mm high; lugs to brick jambs										
438mm wide	-	-	-	2.16	0.10	**31.76**	2.80	**34.56**	nr	**37.15**
641mm wide	-	-	-	2.39	0.11	**35.14**	3.13	**38.27**	nr	**41.14**
1225mm wide	-	-	-	3.07	0.14	**45.12**	4.09	**49.21**	nr	**52.90**
1809mm wide	-	-	-	3.75	0.17	**55.11**	5.02	**60.13**	nr	**64.64**
2394mm wide	-	-	-	4.43	0.20	**65.09**	5.98	**71.07**	nr	**76.40**
Building in fireplace interior with slabbed tile surround and loose hearth tiles for 400mm opening; bed and joint interior in fire cement and fill with concrete at back; plug and screw on surround lugs; bed hearth tiles in cement mortar (1:3) and point in										
with firebrick back	-	-	-	5.00	3.50	**108.54**	4.75	**113.29**	nr	**121.78**
with back boiler unit and self contained flue	-	-	-	6.00	4.00	**128.10**	5.72	**133.82**	nr	**143.86**
Proprietary items										
Precast prestressed concrete lintels; Stressline Ltd; Roughcast; bedding in cement-lime mortar (1:1:6)										
100 x 70mm										
900mm long	3.23	2.50	**3.31**	0.24	0.12	**4.70**	0.26	**8.26**	nr	**8.88**
1050mm long	3.77	2.50	**3.86**	0.27	0.14	**5.34**	0.28	**9.48**	nr	**10.19**
1200mm long	4.30	2.50	**4.41**	0.31	0.16	**6.12**	0.33	**10.86**	nr	**11.67**
1500mm long	5.38	2.50	**5.51**	0.38	0.19	**7.43**	0.40	**13.35**	nr	**14.35**
1800mm long	6.46	2.50	**6.62**	0.44	0.22	**8.61**	0.44	**15.67**	nr	**16.85**
2100mm long	7.53	2.50	**7.72**	0.50	0.25	**9.78**	0.51	**18.01**	nr	**19.36**
2400mm long	8.61	2.50	**8.83**	0.55	0.28	**10.81**	0.58	**20.22**	nr	**21.74**
2700mm long	9.68	2.50	**9.92**	0.60	0.30	**11.74**	0.62	**22.28**	nr	**23.96**
3000mm long	10.76	2.50	**11.03**	0.66	0.33	**12.91**	0.70	**24.64**	nr	**26.49**
150 x 70mm										
900mm long	4.39	2.50	**4.50**	0.25	0.13	**4.94**	0.26	**9.70**	nr	**10.43**
1050mm long	5.12	2.50	**5.25**	0.28	0.14	**5.48**	0.30	**11.03**	nr	**11.85**
1200mm long	5.84	2.50	**5.99**	0.32	0.16	**6.26**	0.33	**12.58**	nr	**13.52**
1500mm long	7.31	2.50	**7.49**	0.39	0.20	**7.68**	0.41	**15.59**	nr	**16.75**
1800mm long	8.78	2.50	**9.00**	0.45	0.23	**8.86**	0.45	**18.31**	nr	**19.68**
2100mm long	10.23	2.50	**10.49**	0.51	0.26	**10.03**	0.54	**21.05**	nr	**22.63**
2400mm long	11.70	2.50	**11.99**	0.57	0.29	**11.21**	0.58	**23.78**	nr	**25.56**
2700mm long	13.16	2.50	**13.49**	0.62	0.31	**12.13**	0.61	**26.23**	nr	**28.20**
3000mm long	14.62	2.50	**14.99**	0.68	0.34	**13.30**	0.72	**29.01**	nr	**31.19**
225 x 70mm										
900mm long	6.76	2.50	**6.93**	0.26	0.13	**5.09**	0.26	**12.27**	nr	**13.19**
1050mm long	7.89	2.50	**8.09**	0.29	0.15	**5.73**	0.30	**14.12**	nr	**15.17**
1200mm long	9.01	2.50	**9.24**	0.33	0.17	**6.51**	0.33	**16.08**	nr	**17.28**
1500mm long	11.27	2.50	**11.55**	0.40	0.20	**7.83**	0.41	**19.79**	nr	**21.27**
1800mm long	13.53	2.50	**13.87**	0.46	0.23	**9.00**	0.45	**23.32**	nr	**25.07**
2100mm long	15.77	2.50	**16.16**	0.52	0.26	**10.17**	0.54	**26.88**	nr	**28.89**
2400mm long	18.03	2.50	**18.48**	0.58	0.29	**11.35**	0.61	**30.44**	nr	**32.72**
2700mm long	20.28	2.50	**20.79**	0.63	0.32	**12.38**	0.69	**33.85**	nr	**36.39**
3000mm long	22.53	2.50	**23.09**	0.69	0.35	**13.55**	0.72	**37.37**	nr	**40.17**
255 x 70mm										
900mm long	7.20	2.50	**7.38**	0.26	0.13	**5.09**	0.29	**12.76**	nr	**13.71**
1050mm long	8.40	2.50	**8.61**	0.30	0.15	**5.87**	0.31	**14.79**	nr	**15.90**
1200mm long	9.60	2.50	**9.84**	0.33	0.17	**6.51**	0.35	**16.70**	nr	**17.96**
1500mm long	12.00	2.50	**12.30**	0.40	0.20	**7.83**	0.41	**20.53**	nr	**22.07**
1800mm long	14.39	2.50	**14.75**	0.47	0.24	**9.25**	0.48	**24.48**	nr	**26.32**
2100mm long	16.80	2.50	**17.22**	0.53	0.27	**10.42**	0.56	**28.20**	nr	**30.32**
2400mm long	19.19	2.50	**19.67**	0.60	0.30	**11.74**	0.60	**32.01**	nr	**34.41**
2700mm long	21.60	2.50	**22.14**	0.64	0.32	**12.52**	0.69	**35.35**	nr	**38.00**
3000mm long	23.99	2.50	**24.59**	0.70	0.35	**13.70**	0.72	**39.01**	nr	**41.93**
100 x 145mm										
900mm long	6.53	2.50	**6.69**	0.26	0.13	**5.09**	0.26	**12.04**	nr	**12.94**
1050mm long	7.62	2.50	**7.81**	0.29	0.15	**5.73**	0.30	**13.84**	nr	**14.88**
1200mm long	8.71	2.50	**8.93**	0.33	0.17	**6.51**	0.35	**15.79**	nr	**16.98**
1500mm long	10.89	2.50	**11.16**	0.40	0.20	**7.83**	0.41	**19.40**	nr	**20.85**
1800mm long	13.06	2.50	**13.39**	0.46	0.23	**9.00**	0.48	**22.87**	nr	**24.59**
2100mm long	15.24	2.50	**15.62**	0.52	0.26	**10.17**	0.54	**26.33**	nr	**28.31**
2400mm long	17.43	2.50	**17.87**	0.58	0.29	**11.35**	0.61	**29.83**	nr	**32.06**
2700mm long	19.60	2.50	**20.09**	0.63	0.32	**12.38**	0.69	**33.16**	nr	**35.64**
3000mm long	21.78	2.50	**22.32**	0.69	0.35	**13.55**	0.72	**36.60**	nr	**39.34**
Galvanized steel lintels; SUPERGALV (BIRTLEY) lintels reference CB 50; bedding in cement-lime mortar (1:1:6)										
125mm deep x 750mm long	20.97	2.50	**21.49**	0.28	0.28	**6.98**	-	**28.47**	nr	**30.61**
125mm deep x 1200mm long	33.07	2.50	**33.90**	0.38	0.38	**9.47**	-	**43.37**	nr	**46.62**
125mm deep x 1350mm long	39.03	2.50	**40.01**	0.41	0.41	**10.22**	-	**50.22**	nr	**53.99**
125mm deep x 1500mm long	43.12	2.50	**44.20**	0.44	0.44	**10.96**	-	**55.16**	nr	**59.30**
125mm deep x 1650mm long	49.16	2.50	**50.39**	0.47	0.47	**11.71**	-	**62.10**	nr	**66.76**
165mm deep x 1800mm long	53.19	2.50	**54.52**	0.50	0.50	**12.46**	-	**66.98**	nr	**72.00**
165mm deep x 1950mm long	58.37	2.50	**59.83**	0.53	0.53	**13.21**	-	**73.04**	nr	**78.51**
165mm deep x 2100mm long	61.10	2.50	**62.63**	0.56	0.56	**13.96**	-	**76.58**	nr	**82.33**
165mm deep x 2250mm long	70.02	2.50	**71.77**	0.60	0.60	**14.95**	-	**86.72**	nr	**93.23**
165mm deep x 2400mm long	74.06	2.50	**75.91**	0.63	0.63	**15.70**	-	**91.61**	nr	**98.48**
215mm deep x 2550mm long	83.48	2.50	**85.57**	0.67	0.67	**16.70**	-	**102.26**	nr	**109.93**

Labour hourly rates: (except Specialists) Craft Operatives £14.21 Labourer £10.71 Rates are national average prices. Refer to REGIONAL VARIATIONS for indicative levels of overall pricing in regions	MATERIALS Del to Site £	Waste %	Material Cost £	LABOUR Craft Optve Hrs	Lab Hrs	Labour Cost £	Sunds £	RATES Nett Rate £	Unit	Gross Rate (+7.5%) £
F30: ACCESSORIES/SUNDRY ITEMS FOR BRICK/BLOCK/STONE WALLING Cont.										
Proprietary items Cont.										
Galvanized steel lintels; SUPERGALV (BIRTLEY) lintels reference CB 50; bedding in cement-lime mortar (1:1:6) Cont.										
215mm deep x 2850mm long	113.76	2.50	**116.60**	0.71	0.71	**17.69**	-	**134.30**	nr	**144.37**
215mm deep x 3000mm long	121.13	2.50	**124.16**	0.75	0.75	**18.69**	-	**142.85**	nr	**153.56**
215mm deep x 3300mm long	139.65	2.50	**143.14**	0.81	0.81	**20.19**	-	**163.33**	nr	**175.58**
215mm deep x 3600mm long	160.01	2.50	**164.01**	0.88	0.88	**21.93**	-	**185.94**	nr	**199.89**
215mm deep x 3900mm long	195.26	2.50	**200.14**	0.94	0.94	**23.42**	-	**223.57**	nr	**240.33**
Galvanized steel lintels; SUPERGALV (BIRTLEY) lintels reference CB 50 H.D.; bedding in cement-lime mortar (1:1:6)										
165mm deep x 750mm long	24.75	2.50	**25.37**	0.28	0.28	**6.98**	-	**32.35**	nr	**34.77**
165mm deep x 1050mm long	32.59	2.50	**33.40**	0.34	0.34	**8.47**	-	**41.88**	nr	**45.02**
165mm deep x 1200mm long	37.13	2.50	**38.06**	0.38	0.38	**9.47**	-	**47.53**	nr	**51.09**
165mm deep x 1350mm long	45.14	2.50	**46.27**	0.41	0.41	**10.22**	-	**56.49**	nr	**60.72**
165mm deep x 1500mm long	49.76	2.50	**51.00**	0.44	0.44	**10.96**	-	**61.97**	nr	**66.62**
215mm deep x 1650mm long	59.54	2.50	**61.03**	0.47	0.47	**11.71**	-	**72.74**	nr	**78.20**
215mm deep x 1800mm long	65.00	2.50	**66.63**	0.50	0.50	**12.46**	-	**79.09**	nr	**85.02**
215mm deep x 2100mm long	76.66	2.50	**78.58**	0.56	0.56	**13.96**	-	**92.53**	nr	**99.47**
215mm deep x 2250mm long	96.39	2.50	**98.80**	0.60	0.60	**14.95**	-	**113.75**	nr	**122.28**
215mm deep x 2400mm long	113.11	2.50	**115.94**	0.63	0.63	**15.70**	-	**131.64**	nr	**141.51**
215mm deep x 2550mm long	122.89	2.50	**125.96**	0.67	0.67	**16.70**	-	**142.66**	nr	**153.36**
215mm deep x 3000mm long	185.39	2.50	**190.02**	0.78	0.78	**19.44**	-	**209.46**	nr	**225.17**
Galvanized steel lintels; SUPERGALV (BIRTLEY) lintels reference CB 70; bedding in cement-lime mortar (1:1:6)										
120mm deep x 750mm long	20.81	2.50	**21.33**	0.29	0.29	**7.23**	-	**28.56**	nr	**30.70**
120mm deep x 1200mm long	33.06	2.50	**33.89**	0.40	0.40	**9.97**	-	**43.85**	nr	**47.14**
120mm deep x 1350mm long	38.84	2.50	**39.81**	0.43	0.43	**10.72**	-	**50.53**	nr	**54.32**
120mm deep x 1500mm long	43.13	2.50	**44.21**	0.46	0.46	**11.46**	-	**55.67**	nr	**59.85**
120mm deep x 1650mm long	48.86	2.50	**50.08**	0.49	0.49	**12.21**	-	**62.29**	nr	**66.96**
160mm deep x 1800mm long	53.05	2.50	**54.38**	0.53	0.53	**13.21**	-	**67.58**	nr	**72.65**
160mm deep x 1950mm long	59.24	2.50	**60.72**	0.56	0.56	**13.96**	-	**74.68**	nr	**80.28**
160mm deep x 2100mm long	61.88	2.50	**63.43**	0.59	0.59	**14.70**	-	**78.13**	nr	**83.99**
215mm deep x 2250mm long	69.44	2.50	**71.18**	0.63	0.63	**15.70**	-	**86.88**	nr	**93.39**
215mm deep x 2400mm long	74.06	2.50	**75.91**	0.66	0.66	**16.45**	-	**92.36**	nr	**99.29**
215mm deep x 2550mm long	85.12	2.50	**87.25**	0.70	0.70	**17.44**	-	**104.69**	nr	**112.54**
215mm deep x 2850mm long	112.44	2.50	**115.25**	0.75	0.75	**18.69**	-	**133.94**	nr	**143.99**
215mm deep x 3000mm long	120.78	2.50	**123.80**	0.79	0.79	**19.69**	-	**143.49**	nr	**154.25**
215mm deep x 3300mm long	139.67	2.50	**143.16**	0.85	0.85	**21.18**	-	**164.34**	nr	**176.67**
215mm deep x 3600mm long	153.53	2.50	**157.37**	0.92	0.92	**22.93**	-	**180.29**	nr	**193.82**
215mm deep x 3900mm long	187.43	2.50	**192.12**	0.99	0.99	**24.67**	-	**216.79**	nr	**233.05**
Galvanized steel lintels; SUPERGALV (BIRTLEY) lintels reference CB 70 H.D.; bedding in cement-lime mortar (1:1:6)										
160mm deep x 750mm long	24.77	2.50	**25.39**	0.29	0.29	**7.23**	-	**32.62**	nr	**35.06**
160mm deep x 1050mm long	33.14	2.50	**33.97**	0.36	0.36	**8.97**	-	**42.94**	nr	**46.16**
160mm deep x 1200mm long	37.13	2.50	**38.06**	0.40	0.40	**9.97**	-	**48.03**	nr	**51.63**
160mm deep x 1350mm long	45.76	2.50	**46.90**	0.43	0.43	**10.72**	-	**57.62**	nr	**61.94**
160mm deep x 1500mm long	49.77	2.50	**51.01**	0.46	0.46	**11.46**	-	**62.48**	nr	**67.16**
215mm deep x 1650mm long	65.33	2.50	**66.96**	0.49	0.49	**12.21**	-	**79.17**	nr	**85.11**
215mm deep x 1800mm long	69.51	2.50	**71.25**	0.53	0.53	**13.21**	-	**84.46**	nr	**90.79**
215mm deep x 2100mm long	79.94	2.50	**81.94**	0.59	0.59	**14.70**	-	**96.64**	nr	**103.89**
215mm deep x 2250mm long	98.60	2.50	**101.07**	0.63	0.63	**15.70**	-	**116.76**	nr	**125.52**
215mm deep x 2400mm long	113.14	2.50	**115.97**	0.66	0.66	**16.45**	-	**132.42**	nr	**142.35**
215mm deep x 2550mm long	124.55	2.50	**127.66**	0.70	0.70	**17.44**	-	**145.11**	nr	**155.99**
215mm deep x 3300mm long	185.44	2.50	**190.08**	0.82	0.82	**20.43**	-	**210.51**	nr	**226.30**
Galvanized steel lintels; SUPERGALV (BIRTLEY) lintels reference CB 50/130; bedding in cement-lime mortar (1:1:6)										
120mm deep x 750mm long	21.06	2.50	**21.59**	0.31	0.31	**7.73**	-	**29.31**	nr	**31.51**
120mm deep x 1200mm long	33.47	2.50	**34.31**	0.42	0.42	**10.47**	-	**44.77**	nr	**48.13**
120mm deep x 1350mm long	38.36	2.50	**39.32**	0.45	0.45	**11.21**	-	**50.53**	nr	**54.32**
120mm deep x 1500mm long	42.42	2.50	**43.48**	0.48	0.48	**11.96**	-	**55.44**	nr	**59.60**
120mm deep x 1650mm long	49.34	2.50	**50.57**	0.52	0.52	**12.96**	-	**63.53**	nr	**68.30**
165mm deep x 1800mm long	53.15	2.50	**54.48**	0.55	0.55	**13.71**	-	**68.18**	nr	**73.30**
165mm deep x 1950mm long	58.64	2.50	**60.11**	0.58	0.58	**14.45**	-	**74.56**	nr	**80.15**
165mm deep x 2100mm long	63.35	2.50	**64.93**	0.62	0.62	**15.45**	-	**80.38**	nr	**86.41**
165mm deep x 2250mm long	72.61	2.50	**74.43**	0.66	0.66	**16.45**	-	**90.87**	nr	**97.69**
165mm deep x 2400mm long	77.47	2.50	**79.41**	0.69	0.69	**17.19**	-	**96.60**	nr	**103.85**
215mm deep x 2550mm long	87.79	2.50	**89.98**	0.74	0.74	**18.44**	-	**108.43**	nr	**116.56**
215mm deep x 2850mm long	123.71	2.50	**126.80**	0.78	0.78	**19.44**	-	**146.24**	nr	**157.21**
215mm deep x 3000mm long	127.88	2.50	**131.08**	0.83	0.83	**20.68**	-	**151.76**	nr	**163.14**
215mm deep x 3300mm long	142.82	2.50	**146.39**	0.89	0.89	**22.18**	-	**168.57**	nr	**181.21**
215mm deep x 3600mm long	157.28	2.50	**161.21**	0.97	0.97	**24.17**	-	**185.38**	nr	**199.29**
215mm deep x 3900mm long	230.48	2.50	**236.24**	1.03	1.03	**25.67**	-	**261.91**	nr	**281.55**
Galvanized steel lintels; SUPERGALV (BIRTLEY) lintels reference CB 50/130 H.D.; bedding in cement-lime mortar (1:1:6)										
160mm deep x 750mm long	30.09	2.50	**30.84**	0.31	0.31	**7.73**	-	**38.57**	nr	**41.46**
160mm deep x 1200mm long	45.04	2.50	**46.17**	0.42	0.42	**10.47**	-	**56.63**	nr	**60.88**
160mm deep x 1350mm long	56.36	2.50	**57.77**	0.45	0.45	**11.21**	-	**68.98**	nr	**74.16**
160mm deep x 1500mm long	62.09	2.50	**63.64**	0.48	0.48	**11.96**	-	**75.60**	nr	**81.27**
215mm deep x 1650mm long	71.12	2.50	**72.90**	0.52	0.52	**12.96**	-	**85.86**	nr	**92.30**
215mm deep x 1800mm long	76.98	2.50	**78.90**	0.55	0.55	**13.71**	-	**92.61**	nr	**99.56**
215mm deep x 1950mm long	95.84	2.50	**98.24**	0.58	0.58	**14.45**	-	**112.69**	nr	**121.14**
215mm deep x 2250mm long	110.27	2.50	**113.03**	0.66	0.66	**16.45**	-	**129.47**	nr	**139.18**

Labour hourly rates: (except Specialists) Craft Operatives £14.21 Labourer £10.71 Rates are national average prices. Refer to REGIONAL VARIATIONS for indicative levels of overall pricing in regions	MATERIALS			LABOUR				RATES		
	Del to Site £	Waste %	Material Cost £	Craft Optve Hrs	Lab Hrs	Labour Cost £	Sunds £	Nett Rate £	Unit	Gross Rate (+7.5%) £
F30: ACCESSORIES/SUNDRY ITEMS FOR BRICK/BLOCK/STONE WALLING Cont.										
Proprietary items Cont.										
Galvanized steel lintels; SUPERGALV (BIRTLEY) lintels reference CB 50/130 H.D.; bedding in cement-lime mortar (1:1:6) Cont.										
215mm deep x 2400mm long	119.19	2.50	**122.17**	0.69	0.69	**17.19**	-	**139.36**	nr	**149.82**
215mm deep x 2550mm long	130.67	2.50	**133.94**	0.74	0.74	**18.44**	-	**152.38**	nr	**163.81**
215mm deep x 2700mm long	138.03	2.50	**141.48**	0.80	0.80	**19.94**	-	**161.42**	nr	**173.52**
215mm deep x 3300mm long	189.18	2.50	**193.91**	0.86	0.86	**21.43**	-	**215.34**	nr	**231.49**
Galvanized steel lintels; SUPERGALV (BIRTLEY) lintels reference CB 70/130; bedding in cement-lime mortar (1:1:6)										
115mm deep x 750mm long	25.49	2.50	**26.13**	0.31	0.31	**7.73**	-	**33.85**	nr	**36.39**
115mm deep x 1200mm long	41.99	2.50	**43.04**	0.42	0.42	**10.47**	-	**53.51**	nr	**57.52**
115mm deep x 1350mm long	48.34	2.50	**49.55**	0.45	0.45	**11.21**	-	**60.76**	nr	**65.32**
115mm deep x 1500mm long	53.04	2.50	**54.37**	0.48	0.48	**11.96**	-	**66.33**	nr	**71.30**
115mm deep x 1650mm long	65.41	2.50	**67.05**	0.52	0.52	**12.96**	-	**80.00**	nr	**86.00**
155mm deep x 1800mm long	72.96	2.50	**74.78**	0.55	0.55	**13.71**	-	**88.49**	nr	**95.13**
155mm deep x 1950mm long	79.04	2.50	**81.02**	0.58	0.58	**14.45**	-	**95.47**	nr	**102.63**
155mm deep x 2100mm long	84.02	2.50	**86.12**	0.62	0.62	**15.45**	-	**101.57**	nr	**109.19**
155mm deep x 2250mm long	113.66	2.50	**116.50**	0.66	0.66	**16.45**	-	**132.95**	nr	**142.92**
155mm deep x 2400mm long	121.24	2.50	**124.27**	0.69	0.69	**17.19**	-	**141.47**	nr	**152.08**
210mm deep x 2550mm long	134.37	2.50	**137.73**	0.74	0.74	**18.44**	-	**156.17**	nr	**167.88**
210mm deep x 2850mm long	179.18	2.50	**183.66**	0.78	0.78	**19.44**	-	**203.10**	nr	**218.33**
210mm deep x 3000mm long	200.09	2.50	**205.09**	0.83	0.83	**20.68**	-	**225.78**	nr	**242.71**
210mm deep x 3300mm long	220.09	2.50	**225.59**	0.89	0.89	**22.18**	-	**247.77**	nr	**266.35**
Galvanized steel lintels; SUPERGALV (BIRTLEY) lintels reference HD50; bedding in cement-lime mortar (1:1:6)										
225mm deep x 3900mm long	229.23	2.50	**234.96**	1.02	1.02	**25.42**	-	**260.38**	nr	**279.91**
225mm deep x 4200mm long	271.10	2.50	**277.88**	1.05	1.05	**26.17**	-	**304.04**	nr	**326.85**
225mm deep x 4500mm long	299.66	2.50	**307.15**	1.09	1.09	**27.16**	-	**334.31**	nr	**359.39**
225mm deep x 4800mm long	319.61	2.50	**327.60**	1.12	1.12	**27.91**	-	**355.51**	nr	**382.17**
Galvanized steel lintels; SUPERGALV (BIRTLEY) lintels reference HD 70; bedding in cement-lime mortar (1:1:6)										
225mm deep x 3000mm long	186.14	2.50	**190.79**	0.85	0.85	**21.18**	-	**211.98**	nr	**227.87**
225mm deep x 3300mm long	204.74	2.50	**209.86**	0.89	0.89	**22.18**	-	**232.04**	nr	**249.44**
225mm deep x 3600mm long	224.39	2.50	**230.00**	0.92	0.92	**22.93**	-	**252.93**	nr	**271.90**
Galvanized steel lintels; SUPERGALV (BIRTLEY) lintels reference SB100; bedding in cement-lime mortar (1:1:6)										
75mm deep x 750mm long	13.96	2.50	**14.31**	0.45	0.45	**11.21**	-	**25.52**	nr	**27.44**
75mm deep x 900mm long	16.30	2.50	**16.71**	0.48	0.48	**11.96**	-	**28.67**	nr	**30.82**
75mm deep x 1050mm long	19.31	2.50	**19.79**	0.51	0.51	**12.71**	-	**32.50**	nr	**34.94**
75mm deep x 1200mm long	21.47	2.50	**22.01**	0.54	0.54	**13.46**	-	**35.46**	nr	**38.12**
75mm deep x 1500mm long	26.88	2.50	**27.55**	0.60	0.60	**14.95**	-	**42.50**	nr	**45.69**
140mm deep x 1800mm long	41.12	2.50	**42.15**	0.65	0.65	**16.20**	-	**58.35**	nr	**62.72**
140mm deep x 1950mm long	46.79	2.50	**47.96**	0.69	0.69	**17.19**	-	**65.15**	nr	**70.04**
140mm deep x 2100mm long	49.69	2.50	**50.93**	0.73	0.73	**18.19**	-	**69.12**	nr	**74.31**
140mm deep x 2250mm long	53.45	2.50	**54.79**	0.77	0.77	**19.19**	-	**73.97**	nr	**79.52**
140mm deep x 2400mm long	57.61	2.50	**59.05**	0.82	0.82	**20.43**	-	**79.48**	nr	**85.45**
140mm deep x 2550mm long	62.15	2.50	**63.70**	0.87	0.87	**21.68**	-	**85.38**	nr	**91.79**
140mm deep x 2700mm long	66.05	2.50	**67.70**	0.92	0.92	**22.93**	-	**90.63**	nr	**97.42**
215mm deep x 2850mm long	99.92	2.50	**102.42**	0.93	0.93	**23.18**	-	**125.59**	nr	**135.01**
215mm deep x 3000mm long	105.91	2.50	**108.56**	0.95	0.95	**23.67**	-	**132.23**	nr	**142.15**
Galvanized steel lintels; SUPERGALV (BIRTLEY) lintels reference SB140; bedding in cement-lime mortar (1:1:6)										
140mm deep x 750mm long	17.64	2.50	**18.08**	0.45	0.45	**11.21**	-	**29.30**	nr	**31.49**
140mm deep x 900mm long	21.29	2.50	**21.82**	0.48	0.48	**11.96**	-	**33.78**	nr	**36.32**
140mm deep x 1050mm long	25.10	2.50	**25.72**	0.51	0.51	**12.71**	-	**38.43**	nr	**41.31**
140mm deep x 1200mm long	28.70	2.50	**29.42**	0.54	0.54	**13.46**	-	**42.88**	nr	**46.09**
140mm deep x 1350mm long	32.56	2.50	**33.37**	0.57	0.57	**14.20**	-	**47.58**	nr	**51.14**
140mm deep x 1500mm long	36.98	2.50	**37.91**	0.60	0.60	**14.95**	-	**52.86**	nr	**56.82**
140mm deep x 1650mm long	44.42	2.50	**45.53**	0.62	0.62	**15.45**	-	**60.98**	nr	**65.55**
140mm deep x 1800mm long	48.28	2.50	**49.48**	0.65	0.65	**16.20**	-	**65.68**	nr	**70.61**
140mm deep x 1950mm long	54.56	2.50	**55.93**	0.69	0.69	**17.19**	-	**73.12**	nr	**78.61**
140mm deep x 2100mm long	59.49	2.50	**60.98**	0.73	0.73	**18.19**	-	**79.17**	nr	**85.11**
140mm deep x 2250mm long	65.60	2.50	**67.23**	0.77	0.77	**19.19**	-	**86.42**	nr	**92.91**
140mm deep x 2400mm long	68.88	2.50	**70.60**	0.82	0.82	**20.43**	-	**91.04**	nr	**97.86**
140mm deep x 2550mm long	73.00	2.50	**74.82**	0.87	0.87	**21.68**	-	**96.50**	nr	**103.74**
140mm deep x 2700mm long	77.28	2.50	**79.21**	0.92	0.92	**22.93**	-	**102.14**	nr	**109.80**
215mm deep x 2850mm long	112.03	2.50	**114.83**	0.95	0.95	**23.67**	-	**138.50**	nr	**148.89**
215mm deep x 3000mm long	116.63	2.50	**119.54**	0.97	0.97	**24.17**	-	**143.71**	nr	**154.49**
Galvanized steel lintels; SUPERGALV (BIRTLEY) lintels reference SBL200; bedding in cement-lime mortar (1:1:6)										
142mm deep x 750mm long	24.39	2.50	**25.00**	0.45	0.45	**11.21**	-	**36.21**	nr	**38.93**
142mm deep x 900mm long	29.24	2.50	**29.97**	0.48	0.48	**11.96**	-	**41.93**	nr	**45.07**
142mm deep x 1050mm long	33.90	2.50	**34.75**	0.51	0.51	**12.71**	-	**47.46**	nr	**51.02**
142mm deep x 1200mm long	38.35	2.50	**39.31**	0.54	0.54	**13.46**	-	**52.76**	nr	**56.72**
142mm deep x 1350mm long	43.42	2.50	**44.50**	0.60	0.60	**14.95**	-	**59.46**	nr	**63.91**
142mm deep x 1500mm long	48.97	2.50	**50.19**	0.65	0.65	**16.20**	-	**66.39**	nr	**71.37**
142mm deep x 1650mm long	55.53	2.50	**56.92**	0.67	0.67	**16.70**	-	**73.61**	nr	**79.14**
142mm deep x 1800mm long	60.61	2.50	**62.12**	0.72	0.72	**17.94**	-	**80.07**	nr	**86.07**
142mm deep x 1950mm long	66.26	2.50	**67.92**	0.76	0.76	**18.94**	-	**86.86**	nr	**93.37**
142mm deep x 2100mm long	70.94	2.50	**72.72**	0.80	0.77	**19.61**	-	**92.33**	nr	**99.26**
142mm deep x 2250mm long	81.33	2.50	**83.36**	0.77	0.80	**19.51**	-	**102.87**	nr	**110.59**
142mm deep x 2400mm long	85.45	2.50	**87.58**	0.82	0.82	**20.43**	-	**108.02**	nr	**116.12**
142mm deep x 2550mm long	92.51	2.50	**94.82**	0.87	0.87	**21.68**	-	**116.50**	nr	**125.24**

Labour hourly rates: (except Specialists) Craft Operatives £14.21 Labourer £10.71 Rates are national average prices. Refer to REGIONAL VARIATIONS for indicative levels of overall pricing in regions	MATERIALS			LABOUR				RATES		
	Del to Site £	Waste %	Material Cost £	Craft Optve Hrs	Lab Hrs	Labour Cost £	Sunds £	Nett Rate £	Unit	Gross Rate (+7.5%) £
F30: ACCESSORIES/SUNDRY ITEMS FOR BRICK/BLOCK/STONE WALLING Cont.										
Proprietary items Cont.										
Galvanized steel lintels; SUPERGALV (BIRTLEY) lintels reference SBL200; bedding in cement-lime mortar (1:1:6) Cont.										
142mm deep x 2700mm long	95.81	2.50	**98.20**	0.92	0.92	**22.93**	-	**121.13**	nr	**130.21**
218mm deep x 2850mm long	135.87	2.50	**139.27**	0.93	0.93	**23.18**	-	**162.44**	nr	**174.63**
218mm deep x 3000mm long	144.12	2.50	**147.72**	0.95	0.95	**23.67**	-	**171.40**	nr	**184.25**
Galvanized steel lintels; SUPERGALV (BIRTLEY) internal door lintels reference INT100; bedding in cement-lime mortar (1:1:6)										
100mm wide, 900mm long	11.08	2.50	**11.35**	0.33	0.33	**8.22**	-	**19.58**	nr	**21.05**
100mm wide, 1050mm long	12.95	2.50	**13.27**	0.36	0.36	**8.97**	-	**22.24**	nr	**23.91**
100mm wide, 1200mm long	14.81	2.50	**15.18**	0.38	0.38	**9.47**	-	**24.65**	nr	**26.50**
Stainless steel Furfix wall extension profiles; single flange; plugging and screwing to brickwork and building ties in to joints of new walls										
100mm	10.88	5.00	**11.42**	0.30	-	**4.26**	-	**15.69**	m	**16.86**
150mm	11.73	5.00	**12.32**	0.35	-	**4.97**	-	**17.29**	m	**18.59**
215mm	21.21	5.00	**22.27**	0.40	-	**5.68**	-	**27.95**	m	**30.05**
Harris and Edgar Ltd; Hemax restraint channel ref. 36/8; fixing to steelwork with Hemax M8 screws complete with plate washer at 450mm centres; incorporating ties at 450mm centres										
100mm long ties reference BP36 fishtail	9.77	5.00	**10.26**	0.50	0.25	**9.78**	0.68	**20.72**	m	**22.27**
125mm long ties reference BP36 fishtail	9.81	5.00	**10.30**	0.50	0.25	**9.78**	0.80	**20.88**	m	**22.44**
150mm long ties reference BP36 fishtail	9.87	5.00	**10.36**	0.50	0.25	**9.78**	0.90	**21.05**	m	**22.63**
Thermabate insulated cavity closers; fixing to timber with nails										
Thermabate 50; vertical	4.21	5.00	**4.42**	0.33	0.16	**6.40**	-	**10.82**	m	**11.64**
Thermabate 75; vertical	4.40	5.00	**4.62**	0.33	0.16	**6.40**	-	**11.02**	m	**11.85**
Thermabate 50; horizontal	4.21	5.00	**4.42**	0.33	0.16	**6.40**	-	**10.82**	m	**11.64**
Thermabate 75; horizontal	4.40	5.00	**4.62**	0.33	0.16	**6.40**	-	**11.02**	m	**11.85**
Thermabate insulated cavity closers; fixing to masonry with PVC-U ties at 225mm centres										
Thermabate 50; vertical	4.21	5.00	**4.42**	0.33	0.16	**6.40**	-	**10.82**	m	**11.64**
Thermabate 75; vertical	4.40	5.00	**4.62**	0.33	0.16	**6.40**	-	**11.02**	m	**11.85**
Thermabate 50; horizontal	4.21	5.00	**4.42**	0.33	0.16	**6.40**	-	**10.82**	m	**11.64**
Thermabate 75; horizontal	4.40	5.00	**4.62**	0.33	0.16	**6.40**	-	**11.02**	m	**11.85**
F31: PRECAST CONCRETE SILLS/LINTELS/COPINGS/FEATURES										
Precast concrete sills; BS 5642 Part 1; bedding in cement lime mortar (1:1:6)										
Sills; figure 2 or figure 4										
50 x 150 x 300mm splayed and grooved	2.08	2.50	**2.13**	0.15	0.15	**3.74**	0.12	**5.99**	nr	**6.44**
50 x 150 x 400mm splayed and grooved	2.77	2.50	**2.84**	0.20	0.20	**4.98**	0.14	**7.96**	nr	**8.56**
50 x 150 x 700mm splayed and grooved	4.85	2.50	**4.97**	0.30	0.30	**7.48**	0.20	**12.65**	nr	**13.60**
50 x 150 x 1300mm splayed and grooved	9.01	2.50	**9.23**	0.50	0.50	**12.46**	0.30	**21.99**	nr	**23.64**
extra for stooled end	2.35	-	**2.35**	-	-	**-**	-	**2.35**	nr	**2.53**
75 x 150 x 300mm splayed and grooved	3.12	2.50	**3.19**	0.17	0.17	**4.24**	0.12	**7.55**	nr	**8.11**
75 x 150 x 400mm splayed and grooved	4.16	2.50	**4.26**	0.22	0.22	**5.48**	0.14	**9.88**	nr	**10.62**
75 x 150 x 700mm splayed and grooved	7.27	2.50	**7.46**	0.35	0.35	**8.72**	0.20	**16.38**	nr	**17.61**
75 x 150 x 1300mm splayed and grooved	13.51	2.50	**13.85**	0.55	0.55	**13.71**	0.30	**27.85**	nr	**29.94**
extra for stooled end	2.94	-	**2.94**	-	-	**-**	-	**2.94**	nr	**3.16**
100 x 150 x 300mm splayed and grooved	4.16	2.50	**4.26**	0.20	0.20	**4.98**	0.12	**9.36**	nr	**10.07**
100 x 150 x 400mm splayed and grooved	5.54	2.50	**5.68**	0.25	0.25	**6.23**	0.14	**12.05**	nr	**12.95**
100 x 150 x 700mm splayed and grooved	9.70	2.50	**9.94**	0.40	0.40	**9.97**	0.20	**20.11**	nr	**21.62**
100 x 150 x 1300mm splayed and grooved	18.01	2.50	**18.46**	0.60	0.60	**14.95**	0.30	**33.72**	nr	**36.24**
extra for stooled end	3.53	-	**3.53**	-	-	**-**	-	**3.53**	nr	**3.79**
Precast concrete copings BS 5642 Part 2; bedding in cement-lime mortar (1:1:6)										
Copings; figure 1; horizontal										
75 x 200mm; splayed; rebated joints	10.39	2.50	**10.65**	0.25	0.13	**4.94**	0.17	**15.77**	m	**16.95**
extra; stopped ends	1.18	-	**1.18**	-	-	**-**	-	**1.18**	nr	**1.27**
extra; internal angles	2.35	-	**2.35**	-	-	**-**	-	**2.35**	nr	**2.53**
100 x 300mm; splayed; rebated joints	20.78	2.50	**21.30**	0.50	0.25	**9.78**	0.27	**31.35**	m	**33.70**
extra; stopped ends	1.77	-	**1.77**	-	-	**-**	-	**1.77**	nr	**1.90**
extra; internal angles	2.94	-	**2.94**	-	-	**-**	-	**2.94**	nr	**3.16**
75 x 200mm; saddleback; rebated joints	11.08	2.50	**11.36**	0.25	0.13	**4.94**	0.17	**16.48**	m	**17.71**
extra; hipped ends	1.18	-	**1.18**	-	-	**-**	-	**1.18**	nr	**1.27**
extra; internal angles	2.35	-	**2.35**	-	-	**-**	-	**2.35**	nr	**2.53**
100 x 300mm; saddleback; rebated joints	22.17	2.50	**22.72**	0.50	0.25	**9.78**	0.27	**32.77**	m	**35.23**
extra; hipped ends	1.77	-	**1.77**	-	-	**-**	-	**1.77**	nr	**1.90**
extra; internal angles	2.94	-	**2.94**	-	-	**-**	-	**2.94**	nr	**3.16**
Precast concrete lintels; BS 5328, designed mix C25, 20mm aggregate minimum cement content 360 kg/m3; vibrated; reinforcement bars BS 4449 hot rolled plain round mild steel; bedding in cement-lime mortar (1:1:6)										
Lintels										
75 x 150 x 1200mm; reinforced with 1.20 kg of 12mm bars	6.41	2.50	**6.57**	0.18	0.09	**3.52**	0.12	**10.21**	nr	**10.98**

Labour hourly rates: (except Specialists) Craft Operatives £14.21 Labourer £10.71 Rates are national average prices. Refer to REGIONAL VARIATIONS for indicative levels of overall pricing in regions	MATERIALS			LABOUR				RATES		
	Del to Site £	Waste %	Material Cost £	Craft Optve Hrs	Lab Hrs	Labour Cost £	Sunds £	Nett Rate £	Unit	Gross Rate (+7.5%) £
F31: PRECAST CONCRETE SILLS/LINTELS/COPINGS/FEATURES Cont.										
Precast concrete lintels; BS 5328, designed mix C25, 20mm aggregate minimum cement content 360 kg/m3; vibrated; reinforcement bars BS 4449 hot rolled plain round mild steel; bedding in cement-lime mortar (1:1:6) Cont.										
Lintels Cont.										
75 x 150 x 1650mm; reinforced with 1.60 kg of 12mm bars	8.81	2.50	9.03	0.25	0.13	4.94	0.15	14.13	nr	15.19
102 x 150 x 1200mm; reinforced with 1.20 kg of 12mm bars	8.72	2.50	8.93	0.30	0.15	5.87	0.13	14.93	nr	16.05
102 x 150 x 1650mm; reinforced with 1.60 kg of 12mm bars	11.99	2.50	12.29	0.41	0.20	7.97	0.15	20.40	nr	21.93
215 x 150 x 1200mm; reinforced with 2.40 kg of 12mm bars	18.37	2.50	18.83	0.60	0.30	11.74	0.19	30.76	nr	33.07
215 x 150 x 1650mm; reinforced with 3.20 kg of 12mm bars	25.26	2.50	25.89	0.83	0.41	16.19	0.26	42.34	nr	45.51
215 x 225 x 1200mm; reinforced with 2.40 kg of 12mm bars	27.56	2.50	28.25	0.78	0.39	15.26	0.20	43.71	nr	46.99
215 x 225 x 1650mm; reinforced with 3.20 kg of 12mm bars	37.89	2.50	38.84	1.07	0.53	20.88	0.27	59.99	nr	64.49
327 x 150 x 1200mm; reinforced with 3.60 kg of 12mm bars	27.94	2.50	28.64	0.78	0.39	15.26	0.20	44.11	nr	47.42
327 x 150 x 1650mm; reinforced with 4.80 kg of 12mm bars	38.42	2.50	39.38	1.07	0.54	20.99	0.27	60.64	nr	65.19
327 x 225 x 1200mm; reinforced with 3.60 kg of 12mm bars	41.92	2.50	42.96	1.20	0.60	23.48	0.25	66.69	nr	71.69
327 x 225 x 1650mm; reinforced with 4.80 kg of 12mm bars	57.64	2.50	59.08	1.65	0.83	32.34	0.31	91.72	nr	98.60
75 x 150 x 1200mm; reinforced with 1.20 kg of 12mm bars; fair finish two faces	7.88	2.50	8.08	0.18	0.09	3.52	0.12	11.72	nr	12.60
75 x 150 x 1650mm; reinforced with 1.60 kg of 12mm bars; fair finish two faces	10.84	2.50	11.11	0.25	0.13	4.94	0.15	16.21	nr	17.42
102 x 150 x 1200mm; reinforced with 1.20 kg of 12mm bars; fair finish two faces	10.72	2.50	10.99	0.30	0.15	5.87	0.12	16.98	nr	18.25
102 x 150 x 1650mm; reinforced with 1.60 kg of 12mm bars; fair finish two faces	14.74	2.50	15.11	0.41	0.20	7.97	0.15	23.23	nr	24.97
215 x 150 x 1200mm; reinforced with 2.40 kg of 12mm bars; fair finish two faces	22.60	2.50	23.16	0.60	0.30	11.74	0.19	35.09	nr	37.73
215 x 150 x 1650mm; reinforced with 3.20 kg of 12mm bars; fair finish two faces	31.07	2.50	31.85	0.83	0.41	16.19	0.26	48.29	nr	51.91
215 x 225 x 1200mm; reinforced with 2.40 kg of 12mm bars; fair finish two faces	33.90	2.50	34.74	0.78	0.39	15.26	0.20	50.21	nr	53.97
215 x 225 x 1650mm; reinforced with 3.20 kg of 12mm bars; fair finish two faces	46.61	2.50	47.77	1.07	0.53	20.88	0.27	68.92	nr	74.09
327 x 150 x 1200mm; reinforced with 3.60 kg of 12mm bars; fair finish two faces	34.37	2.50	35.23	0.78	0.39	15.26	0.20	50.69	nr	54.49
327 x 150 x 1650mm; reinforced with 4.80 kg of 12mm bars; fair finish two faces	47.26	2.50	48.44	1.07	0.54	20.99	0.27	69.70	nr	74.92
327 x 225 x 1200mm; reinforced with 3.60 kg of 12mm bars; fair finish two faces	51.55	2.50	52.84	1.20	0.60	23.48	0.24	76.56	nr	82.30
327 x 225 x 1650mm; reinforced with 4.80 kg of 12mm bars; fair finish two faces	70.88	2.50	72.66	1.65	0.83	32.34	0.35	105.35	nr	113.25
Boot lintels										
215 x 150 x 1200mm; reinforced with 3.60 kg of 12mm bars and 0.51 kg of 6mm links	22.60	2.50	23.16	0.60	0.30	11.74	0.18	35.08	nr	37.71
215 x 150 x 1800mm; reinforced with 5.19 kg of 12mm bars and 0.77 kg of 6mm links	33.90	2.50	34.74	0.90	0.45	17.61	0.27	52.62	nr	56.57
215 x 225 x 1200mm; reinforced with 3.60 kg of 12mm bars and 0.62 kg of 6mm links	33.90	2.50	34.74	0.78	0.39	15.26	0.19	50.20	nr	53.96
215 x 225 x 1800mm; reinforced with 5.19 kg of 12mm bars and 0.93 kg of 6mm links	50.84	2.50	52.11	1.18	0.59	23.09	0.28	75.48	nr	81.14
252 x 150 x 1200mm; reinforced with 3.60 kg of 12mm bars and 0.56 kg of 6mm links	26.49	2.50	27.15	0.72	0.36	14.09	0.24	41.47	nr	44.58
252 x 150 x 1800mm; reinforced with 5.19 kg of 12mm bars and 0.83 kg of 6mm links	39.73	2.50	40.72	1.08	0.54	21.13	0.32	62.17	nr	66.84
252 x 225 x 1200mm; reinforced with 3.60 kg of 12mm bars and 0.67 kg of 6mm links	39.73	2.50	40.72	0.90	0.45	17.61	0.24	58.57	nr	62.96
252 x 225 x 1800mm; reinforced with 5.19 kg of 12mm bars and 1.00 kg of 6mm links	59.59	2.50	61.08	1.35	0.67	26.36	0.33	87.77	nr	94.36
Precast concrete padstones; BS 5328, designed mix C25, 20mm aggregate, minimum cement content 360 kg/m3; vibrated; reinforced at contractor's discretion; bedding in cement-lime mortar (1:1:6)										
Padstones										
215 x 215 x 75mm	2.08	2.50	2.13	0.06	0.03	1.17	0.12	3.42	nr	3.68
215 x 215 x 150mm	4.16	2.50	4.27	0.08	0.04	1.57	0.12	5.95	nr	6.39
327 x 215 x 150mm	6.33	2.50	6.49	0.11	0.05	2.10	0.13	8.71	nr	9.37
327 x 327 x 150mm	9.63	2.50	9.87	0.17	0.08	3.27	0.14	13.28	nr	14.27
440 x 215 x 150mm	8.52	2.50	8.73	0.17	0.08	3.27	0.15	12.15	nr	13.06
440 x 327 x 150mm	12.95	2.50	13.27	0.20	0.10	3.91	0.16	17.35	nr	18.65
440 x 440 x 150mm	17.43	2.50	17.86	0.20	0.10	3.91	0.17	21.95	nr	23.59

NATURAL STONEWORK

Natural stonework including restoration and repair is work for the specialist and the cost depends on many factors, prices should be obtained for specific projects.

There are a number of reputable firms who specialize in this class of work and who will submit quotations. Among these are the following:-

STONEWEST LTD.
Lamberts Place, St James Road, Croydon, Surrey, CR9 2HX.
Tel: 020 8684 6646
Fax: 020 8684 9323

A. F. JONES (STONEMASONS).
33 Bedford Road, Reading, Berkshire, RG1 7EX.
Tel: 0118 957 3537
Fax: 0118 957 4334

DIESPEKER MARBLE & TERRAZZO LTD.
132-136 Ormside Street, Peckham, London, SE15 1TF
Tel: 020 7358 0160
Fax: 020 7369 1695

C.W.O. LIMITED.
Terminus Road, Chichester, West Sussex, PO19 2TX.
Tel: 01243 784225
Fax: 01243 813700

LAING STONEMASONRY
Dalston Road, Carlisle, Cumbria, CA2 5NR.
Tel: 01228 521404
Fax: 01228 511306

MACC STONE CO.
196-198 New Street, Biddulph Moor, Staffordshire, Moorlands, ST8 7NW.
Tel: 01782 514353
Fax: 01782 516783

STONEGUARD (LONDON) LTD.
St Martins House, The Runway, South Ruislip, Middlesex, H04 6SG.
Tel: 020 8839 9977
Fax: 020 8839 9988

APS MASONRY LTD.
Osney Mead, Oxford, OX2 0EQ.
Tel: 01865 246677
Fax: 01865 254617

DUNHOUSE QUARRY CO. LTD.
Staine Drop, Darlington, Co. Durham DL2 3QU
Tel: 01833 660208
Fax: 01833 660748

STRUCTURAL/CARCASSING METAL/TIMBER

Labour hourly rates: (except Specialists) Craft Operatives £14.21 Labourer £10.71 Rates are national average prices. Refer to REGIONAL VARIATIONS for indicative levels of overall pricing in regions	MATERIALS			LABOUR				RATES		
	Del to Site £	Waste %	Material Cost £	Craft Optve Hrs	Lab Hrs	Labour Cost £	Sunds £	Nett Rate £	Unit	Gross Rate (+7.5%) £
G10: STRUCTURAL STEEL FRAMING										
Prices generally for fabricated steelwork										
Note prices can vary considerably dependent upon the character of the work; the following are average prices for work fabricated and delivered to site										
Framing, fabrication; including shop and site black bolts, nuts and washers for structural framing to structural framing connections										
Weldable steel, BS EN 10025 : 1993 Grade S275JR (formerly BS 7668 Grade 43B), hot rolled sections BS 4-1 (Euronorm 54); welded fabrication in accordance with BS 5950-1										
Columns										
weight less than 40 Kg/m	1030.00	-	**1030.00**	-	-	-	-	**1030.00**	t	**1107.25**
weight 40-100 Kg/m	1030.00	-	**1030.00**	-	-	-	-	**1030.00**	t	**1107.25**
weight exceeding 100 Kg/m	1030.00	-	**1030.00**	-	-	-	-	**1030.00**	t	**1107.25**
Beams										
weight less than 40 Kg/m	1030.00	-	**1030.00**	-	-	-	-	**1030.00**	t	**1107.25**
weight 40-100 Kg/m	1030.00	-	**1030.00**	-	-	-	-	**1030.00**	t	**1107.25**
weight exceeding 100 Kg/m	1030.00	-	**1030.00**	-	-	-	-	**1030.00**	t	**1107.25**
Beams, castellated										
weight less than 40 Kg/m	1235.00	-	**1235.00**	-	-	-	-	**1235.00**	t	**1327.63**
weight 40-100 Kg/m	1235.00	-	**1235.00**	-	-	-	-	**1235.00**	t	**1327.63**
weight exceeding 100 Kg/m	1235.00	-	**1235.00**	-	-	-	-	**1235.00**	t	**1327.63**
Beams, curved										
weight less than 40 Kg/m	1500.00	-	**1500.00**	-	-	-	-	**1500.00**	t	**1612.50**
weight 40-100 Kg/m	1500.00	-	**1500.00**	-	-	-	-	**1500.00**	t	**1612.50**
weight exceeding 100 Kg/m	1500.00	-	**1500.00**	-	-	-	-	**1500.00**	t	**1612.50**
Bracings, tubular										
weight less than 40 Kg/m	1080.00	-	**1080.00**	-	-	-	-	**1080.00**	t	**1161.00**
weight 40-100 Kg/m	1080.00	-	**1080.00**	-	-	-	-	**1080.00**	t	**1161.00**
weight exceeding 100 Kg/m	1080.00	-	**1080.00**	-	-	-	-	**1080.00**	t	**1161.00**
Purlins and cladding rails										
weight less than 40 Kg/m	1045.00	-	**1045.00**	-	-	-	-	**1045.00**	t	**1123.38**
weight 40-100 Kg/m	1045.00	-	**1045.00**	-	-	-	-	**1045.00**	t	**1123.38**
weight exceeding 100 Kg/m	1045.00	-	**1045.00**	-	-	-	-	**1045.00**	t	**1123.38**
portal frames	975.00	-	**975.00**	-	-	-	-	**975.00**	t	**1048.13**
trusses 12m to 18m span	1550.00	-	**1550.00**	-	-	-	-	**1550.00**	t	**1666.25**
trusses 6m to 12m span	1605.00	-	**1605.00**	-	-	-	-	**1605.00**	t	**1725.38**
trusses up to 6m span	1710.00	-	**1710.00**	-	-	-	-	**1710.00**	t	**1838.25**
Weldable steel BS EN 10210 : 1994 Grade S275JOH (formerly BS 7668 Grade 43C), hot rolled sections BS EN 10210-2 (Euronorm 57); welded fabrication in accordance with BS 5950-1										
column, square hollow section										
weight less than 40 Kg/m	1175.00	-	**1175.00**	-	-	-	-	**1175.00**	t	**1263.13**
weight 40-100 Kg/m	1175.00	-	**1175.00**	-	-	-	-	**1175.00**	t	**1263.13**
weight exceeding 100 Kg/m	1175.00	-	**1175.00**	-	-	-	-	**1175.00**	t	**1263.13**
column, rectangular hollow section										
weight less than 40 Kg/m	1190.00	-	**1190.00**	-	-	-	-	**1190.00**	t	**1279.25**
weight 40-100 Kg/m	1190.00	-	**1190.00**	-	-	-	-	**1190.00**	t	**1279.25**
weight exceeding 100 Kg/m	1190.00	-	**1190.00**	-	-	-	-	**1190.00**	t	**1279.25**
Holding down bolts or assemblies; mild steel										
rag or indented bolts; M 10 with nuts and washers										
100mm long	1.21	5.00	**1.27**	0.10	-	**1.42**	-	**2.69**	nr	**2.90**
160mm long	1.26	5.00	**1.32**	0.11	-	**1.56**	-	**2.89**	nr	**3.10**
rag or indented bolts; M 12 with nuts and washers										
100mm long	1.31	5.00	**1.37**	0.10	-	**1.42**	-	**2.79**	nr	**3.00**
160mm long	1.52	5.00	**1.59**	0.13	-	**1.85**	-	**3.44**	nr	**3.70**
200mm long	1.70	5.00	**1.79**	0.14	-	**1.99**	-	**3.78**	nr	**4.06**
rag or indented bolts; M 16 with nuts and washers										
120mm long	1.97	5.00	**2.07**	0.11	-	**1.56**	-	**3.63**	nr	**3.90**
160mm long	2.26	5.00	**2.37**	0.14	-	**1.99**	-	**4.36**	nr	**4.69**
200mm long	2.63	5.00	**2.77**	0.16	-	**2.27**	-	**5.04**	nr	**5.42**

Labour hourly rates: (except Specialists) Craft Operatives £14.21 Labourer £10.71 Rates are national average prices. Refer to REGIONAL VARIATIONS for indicative levels of overall pricing in regions	MATERIALS			LABOUR				RATES		
	Del to Site £	Waste %	Material Cost £	Craft Optve Hrs	Lab Hrs	Labour Cost £	Sunds £	Nett Rate £	Unit	Gross Rate (+7.5%) £
G10: STRUCTURAL STEEL FRAMING Cont.										
Framing, fabrication; including shop and site black bolts, nuts and washers for structural framing to structural framing connections Cont.										
Holding down bolts or assemblies; mild steel Cont.										
holding down bolt assembly; M 20 bolt with 100 x 100 x 10mm plate washer tack welded to head; with nuts and washers										
300mm long	4.55	5.00	**4.77**	0.28	-	**3.98**	-	**8.75**	nr	**9.41**
350mm long	5.19	5.00	**5.45**	0.32	-	**4.55**	-	**9.99**	nr	**10.74**
400mm long	5.89	5.00	**6.18**	0.35	-	**4.97**	-	**11.15**	nr	**11.99**
450mm long	6.74	5.00	**7.07**	0.55	-	**7.82**	-	**14.89**	nr	**16.01**
High strength friction grip bolts; BS 4395 Part 1 - general grade										
M 16; with nuts and washers										
50mm long	0.74	5.00	**0.77**	0.08	-	**1.14**	-	**1.91**	nr	**2.05**
60mm long	0.81	5.00	**0.85**	0.08	-	**1.14**	-	**1.98**	nr	**2.13**
70mm long	0.86	5.00	**0.91**	0.08	-	**1.14**	-	**2.04**	nr	**2.20**
80mm long	0.93	5.00	**0.98**	0.09	-	**1.28**	-	**2.26**	nr	**2.43**
90mm long	0.99	5.00	**1.04**	0.09	-	**1.28**	-	**2.32**	nr	**2.49**
100mm long	1.03	5.00	**1.08**	0.10	-	**1.42**	-	**2.50**	nr	**2.69**
120mm long	1.13	5.00	**1.19**	0.10	-	**1.42**	-	**2.61**	nr	**2.80**
140mm long	1.28	5.00	**1.35**	0.11	-	**1.56**	-	**2.91**	nr	**3.13**
150mm long	1.29	5.00	**1.36**	0.12	-	**1.71**	-	**3.06**	nr	**3.29**
160mm long	1.43	5.00	**1.51**	0.13	-	**1.85**	-	**3.35**	nr	**3.60**
180mm long	1.62	5.00	**1.70**	0.13	-	**1.85**	-	**3.55**	nr	**3.82**
200mm long	4.12	5.00	**4.32**	0.14	-	**1.99**	-	**6.31**	nr	**6.78**
220mm long	4.37	5.00	**4.59**	0.14	-	**1.99**	-	**6.58**	nr	**7.07**
M 20; with nuts and washers										
60mm long	1.40	5.00	**1.47**	0.08	-	**1.14**	-	**2.61**	nr	**2.80**
70mm long	1.49	5.00	**1.57**	0.08	-	**1.14**	-	**2.70**	nr	**2.91**
80mm long	1.55	5.00	**1.63**	0.09	-	**1.28**	-	**2.91**	nr	**3.13**
90mm long	1.63	5.00	**1.71**	0.10	-	**1.42**	-	**3.13**	nr	**3.37**
100mm long	1.69	5.00	**1.78**	0.10	-	**1.42**	-	**3.20**	nr	**3.44**
120mm long	1.89	5.00	**1.98**	0.11	-	**1.56**	-	**3.55**	nr	**3.81**
140mm long	2.12	5.00	**2.23**	0.12	-	**1.71**	-	**3.93**	nr	**4.23**
150mm long	2.35	5.00	**2.47**	0.13	-	**1.85**	-	**4.32**	nr	**4.64**
160mm long	2.41	5.00	**2.53**	0.14	-	**1.99**	-	**4.52**	nr	**4.86**
180mm long	2.76	5.00	**2.90**	0.14	-	**1.99**	-	**4.89**	nr	**5.26**
200mm long	4.78	5.00	**5.02**	0.15	-	**2.13**	-	**7.15**	nr	**7.69**
220mm long	5.16	5.00	**5.42**	0.17	-	**2.42**	-	**7.84**	nr	**8.43**
240mm long	5.57	5.00	**5.85**	0.18	-	**2.56**	-	**8.41**	nr	**9.04**
260mm long	5.96	5.00	**6.25**	0.20	-	**2.84**	-	**9.10**	nr	**9.78**
280mm long	6.36	5.00	**6.68**	0.21	-	**2.98**	-	**9.67**	nr	**10.39**
300mm long	6.77	5.00	**7.11**	0.22	-	**3.13**	-	**10.24**	nr	**11.00**
M 24; with nuts and washers										
70mm long	2.49	5.00	**2.62**	0.09	-	**1.28**	-	**3.90**	nr	**4.19**
80mm long	2.82	5.00	**2.96**	0.11	-	**1.56**	-	**4.53**	nr	**4.86**
90mm long	2.87	5.00	**3.01**	0.12	-	**1.71**	-	**4.72**	nr	**5.07**
100mm long	3.01	5.00	**3.16**	0.12	-	**1.71**	-	**4.86**	nr	**5.23**
120mm long	3.39	5.00	**3.56**	0.13	-	**1.85**	-	**5.41**	nr	**5.81**
140mm long	3.85	5.00	**4.04**	0.15	-	**2.13**	-	**6.17**	nr	**6.63**
150mm long	3.96	5.00	**4.16**	0.15	-	**2.13**	-	**6.29**	nr	**6.76**
160mm long	4.26	5.00	**4.47**	0.16	-	**2.27**	-	**6.74**	nr	**7.25**
180mm long	4.84	5.00	**5.08**	0.17	-	**2.42**	-	**7.49**	nr	**8.06**
200mm long	6.94	5.00	**7.28**	0.18	-	**2.56**	-	**9.84**	nr	**10.58**
220mm long	7.55	5.00	**7.93**	0.20	-	**2.84**	-	**10.77**	nr	**11.58**
240mm long	8.14	5.00	**8.54**	0.21	-	**2.98**	-	**11.53**	nr	**12.39**
260mm long	8.74	5.00	**9.18**	0.22	-	**3.13**	-	**12.30**	nr	**13.23**
280mm long	9.32	5.00	**9.79**	0.24	-	**3.41**	-	**13.20**	nr	**14.19**
300mm long	9.95	5.00	**10.45**	0.25	-	**3.55**	-	**14.00**	nr	**15.05**
Framing, erection; in accordance with BS 5950-1										
Note the following prices for erection include for the site to be reasonably clear, ease of access and the erection carried out during normal working hours										
Permanent erection of fabricated steelwork on site with bolted connections										
framing	-	-	**-**	23.00	-	**326.83**	-	**326.83**	t	**351.34**
Cold rolled zed purlins and cladding rails; Metsec, galvanized steel										
Purlin sleeved system (section only); fixing to cleats on frame members at 6000mm centres; hoist and fix 3.00m above ground level; fixing with bolts										
reference 14214; 3.03 kg/m	14.79	2.50	**15.16**	0.20	0.03	**3.16**	-	**18.32**	m	**19.70**
reference 14216; 3.47 kg/m	16.70	2.50	**17.12**	0.20	0.03	**3.16**	-	**20.28**	m	**21.80**
reference 17214; 3.66 kg/m	17.50	2.50	**17.94**	0.25	0.03	**3.87**	-	**21.81**	m	**23.45**
reference 17216; 4.11 kg/m	19.88	2.50	**20.38**	0.25	0.03	**3.87**	-	**24.25**	m	**26.07**
reference 20216; 4.49 kg/m	21.71	2.50	**22.25**	0.30	0.04	**4.69**	-	**26.94**	m	**28.96**
reference 20218; 5.03 kg/m	24.48	2.50	**25.09**	0.30	0.04	**4.69**	-	**29.78**	m	**32.02**
reference 20220; 5.57 kg/m	26.89	2.50	**27.56**	0.30	0.04	**4.69**	-	**32.25**	m	**34.67**
reference 23218; 5.73 kg/m	27.79	2.50	**28.48**	0.33	0.04	**5.12**	-	**33.60**	m	**36.12**
reference 23220; 6.34 kg/m	30.82	2.50	**31.59**	0.33	0.04	**5.12**	-	**36.71**	m	**39.46**
Purlin sleeved system (section only); fixing to cleats on frame members at 6000mm centres; hoist and fix 6.00m above ground level; fixing with bolts										
reference 14214; 3.03 kg/m	14.79	2.50	**15.16**	0.22	0.03	**3.45**	-	**18.61**	m	**20.00**

Labour hourly rates: (except Specialists) Craft Operatives £14.21 Labourer £10.71 Rates are national average prices. Refer to REGIONAL VARIATIONS for indicative levels of overall pricing in regions	MATERIALS			LABOUR				RATES		
	Del to Site £	Waste %	Material Cost £	Craft Optve Hrs	Lab Hrs	Labour Cost £	Sunds £	Nett Rate £	Unit	Gross Rate (+7.5%) £
G10: STRUCTURAL STEEL FRAMING Cont.										
Cold rolled zed purlins and cladding rails; Metsec, galvanized steel Cont.										
Purlin sleeved system (section only); fixing to cleats on frame members at 6000mm centres; hoist and fix 6.00m above ground level; fixing with bolts Cont.										
reference 14216; 3.47 kg/m......................	16.70	2.50	**17.12**	0.22	0.03	**3.45**	-	**20.57**	m	**22.11**
reference 17214; 3.66 kg/m......................	17.50	2.50	**17.94**	0.28	0.04	**4.41**	-	**22.34**	m	**24.02**
reference 17216; 4.11 kg/m......................	19.88	2.50	**20.38**	0.28	0.04	**4.41**	-	**24.78**	m	**26.64**
reference 20216; 4.49 kg/m......................	21.71	2.50	**22.25**	0.33	0.04	**5.12**	-	**27.37**	m	**29.42**
reference 20218; 5.03 kg/m......................	24.48	2.50	**25.09**	0.33	0.04	**5.12**	-	**30.21**	m	**32.48**
reference 20220; 5.57 kg/m......................	26.89	2.50	**27.56**	0.33	0.04	**5.12**	-	**32.68**	m	**35.13**
reference 23218; 5.73 kg/m......................	27.79	2.50	**28.48**	0.36	0.05	**5.65**	-	**34.14**	m	**36.70**
reference 23220; 6.34 kg/m......................	30.82	2.50	**31.59**	0.36	0.05	**5.65**	-	**37.24**	m	**40.03**
Purlin overlap system (cleat only); fixing to cleats on frame members at 6000mm centres; hoist and fix 3.00m above ground level; fixing with bolts galvanized										
reference 142	9.94	2.50	**10.19**	0.20	0.03	**3.16**	-	**13.35**	m	**14.35**
reference 172	10.57	2.50	**10.83**	0.25	0.03	**3.87**	-	**14.70**	m	**15.81**
reference 202	11.51	2.50	**11.80**	0.30	0.04	**4.69**	-	**16.49**	m	**17.72**
reference 232	12.45	2.50	**12.76**	0.33	0.04	**5.12**	-	**17.88**	m	**19.22**
Purlin overlap system (cleat only); fixing to cleats on frame members at 6000mm centres; hoist and fix 6.00m above ground level; fixing with bolts galvanized										
reference 142	9.94	2.50	**10.19**	0.22	0.03	**3.45**	-	**13.63**	m	**14.66**
reference 172	10.57	2.50	**10.83**	0.28	0.04	**4.41**	-	**15.24**	m	**16.38**
reference 202	11.51	2.50	**11.80**	0.33	0.04	**5.12**	-	**16.91**	m	**18.18**
reference 232	12.45	2.50	**12.76**	0.36	0.05	**5.65**	-	**18.41**	m	**19.79**
Purlin Non-Continuous system (section only) ; fixing to cleats on frame members at 6000mm centres; hoist and fix 3.00m above ground level; fixing with bolts										
reference 17215; 3.85 kg/m......................	18.61	2.50	**19.08**	0.25	0.03	**3.87**	-	**22.95**	m	**24.67**
reference 17216; 4.11 kg/m......................	19.88	2.50	**20.38**	0.25	0.03	**3.87**	-	**24.25**	m	**26.07**
reference 20216; 4.49 kg/m......................	21.71	2.50	**22.25**	0.30	0.04	**4.69**	-	**26.94**	m	**28.96**
reference 20218; 5.03 kg/m......................	24.48	2.50	**25.09**	0.30	0.04	**4.69**	-	**29.78**	m	**32.02**
reference 20220; 5.57 kg/m......................	26.89	2.50	**27.56**	0.30	0.04	**4.69**	-	**32.25**	m	**34.67**
reference 23218; 5.73 kg/m......................	27.79	2.50	**28.48**	0.33	0.04	**5.12**	-	**33.60**	m	**36.12**
reference 23223; 7.26 kg/m......................	35.61	2.50	**36.50**	0.33	0.04	**5.12**	-	**41.62**	m	**44.74**
reference 26223; 7.92 kg/m......................	38.78	2.50	**39.75**	0.40	0.05	**6.22**	-	**45.97**	m	**49.42**
reference 26229; 9.88 kg/m......................	48.09	2.50	**49.29**	0.40	0.05	**6.22**	-	**55.51**	m	**59.68**
Purlin Non-Continuous system (section only) ; fixing to cleats on frame members at 6000mm centres; hoist and fix 6.00m above ground level; fixing with bolts										
reference 17215; 3.85 kg/m......................	18.61	2.50	**19.08**	0.28	0.04	**4.41**	-	**23.48**	m	**25.24**
reference 17216; 4.11 kg/m......................	19.88	2.50	**20.38**	0.28	0.04	**4.41**	-	**24.78**	m	**26.64**
reference 20216; 4.49 kg/m......................	21.71	2.50	**22.25**	0.33	0.04	**5.12**	-	**27.37**	m	**29.42**
reference 20218; 5.03 kg/m......................	24.48	2.50	**25.09**	0.33	0.04	**5.12**	-	**30.21**	m	**32.48**
reference 20220; 5.57 kg/m......................	26.89	2.50	**27.56**	0.33	0.04	**5.12**	-	**32.68**	m	**35.13**
reference 23218; 5.73 kg/m......................	27.79	2.50	**28.48**	0.36	0.05	**5.65**	-	**34.14**	m	**36.70**
reference 23223; 7.26 kg/m......................	35.61	2.50	**36.50**	0.36	0.05	**5.65**	-	**42.15**	m	**45.31**
reference 26223; 7.92 kg/m......................	38.78	2.50	**39.75**	0.44	0.06	**6.90**	-	**46.64**	m	**50.14**
reference 26229; 9.88 kg/m......................	48.09	2.50	**49.29**	0.44	0.06	**6.90**	-	**56.19**	m	**60.40**
Galvanized purlin cleats; weld on										
Reference 142 for 142mm deep purlins............	5.17	2.50	**5.30**	-	-	**-**	-	**5.30**	nr	**5.70**
Reference 172 for 172mm deep purlins............	5.89	2.50	**6.04**	-	-	**-**	-	**6.04**	nr	**6.49**
Reference 202 for 202mm deep purlins............	6.43	2.50	**6.59**	-	-	**-**	-	**6.59**	nr	**7.09**
Reference 232 for 232mm deep purlins............	7.08	2.50	**7.26**	-	-	**-**	-	**7.26**	nr	**7.80**
Reference 262 for 262mm deep purlins............	7.73	2.50	**7.92**	-	-	**-**	-	**7.92**	nr	**8.52**
Round Lok anti-sag rods; push fit to purlins										
purlins at 1150mm centres	5.20	2.50	**5.33**	0.25	0.03	**3.87**	-	**9.20**	nr	**9.89**
purlins at 1350mm centres	6.04	2.50	**6.19**	0.25	0.03	**3.87**	-	**10.06**	nr	**10.82**
purlins at 1550mm centres	6.83	2.50	**7.00**	0.33	0.04	**5.12**	-	**12.12**	nr	**13.03**
purlins at 1700mm centres	7.69	2.50	**7.88**	0.33	0.04	**5.12**	-	**13.00**	nr	**13.97**
purlins at 1950mm centres	8.44	2.50	**8.65**	0.33	0.04	**5.12**	-	**13.77**	nr	**14.80**
Side rail sleeved system (section only); fixing to cleats on stanchions at 6000mm centres; hoist and fix 3.00m above ground level; fixing with bolts										
reference 14214; 3.03 kg/m......................	14.79	2.50	**15.16**	0.20	0.03	**3.16**	-	**18.32**	m	**19.70**
reference 14216; 3.47 kg/m......................	16.70	2.50	**17.12**	0.20	0.03	**3.16**	-	**20.28**	m	**21.80**
reference 17215; 3.85 kg/m......................	18.61	2.50	**19.08**	0.25	0.03	**3.87**	-	**22.95**	m	**24.67**
reference 17216; 4.11 kg/m......................	19.88	2.50	**20.38**	0.25	0.03	**3.87**	-	**24.25**	m	**26.07**
reference 20216; 4.49 kg/m......................	21.71	2.50	**22.25**	0.30	0.04	**4.69**	-	**26.94**	m	**28.96**
reference 20218; 5.03 kg/m......................	24.48	2.50	**25.09**	0.30	0.04	**4.69**	-	**29.78**	m	**32.02**
reference 20220; 5.57 kg/m......................	26.89	2.50	**27.56**	0.30	0.04	**4.69**	-	**32.25**	m	**34.67**
reference 23218; 5.73 kg/m......................	27.79	2.50	**28.48**	0.33	0.04	**5.12**	-	**33.60**	m	**36.12**
reference 23223; 7.26 kg/m......................	35.61	2.50	**36.50**	0.33	0.04	**5.12**	-	**41.62**	m	**44.74**
Side rail sleeved system (section only); fixing to cleats on stanchions at 6000mm centres; hoist and fix 6.00m above ground level; fixing with bolts										
reference 14214; 3.03 kg/m......................	14.79	2.50	**15.16**	0.22	0.03	**3.45**	-	**18.61**	m	**20.00**
reference 14216; 3.47 kg/m......................	16.70	2.50	**17.12**	0.22	0.03	**3.45**	-	**20.57**	m	**22.11**
reference 17215; 3.85 kg/m......................	18.61	2.50	**19.08**	0.28	0.04	**4.41**	-	**23.48**	m	**25.24**
reference 17216; 4.11 kg/m......................	19.88	2.50	**20.38**	0.28	0.04	**4.41**	-	**24.78**	m	**26.64**
reference 20216; 4.49 kg/m......................	21.71	2.50	**22.25**	0.33	0.04	**5.12**	-	**27.37**	m	**29.42**
reference 20218; 5.03 kg/m......................	24.48	2.50	**25.09**	0.33	0.04	**5.12**	-	**30.21**	m	**32.48**
reference 20220; 5.57 kg/m......................	26.89	2.50	**27.56**	0.33	0.04	**5.12**	-	**32.68**	m	**35.13**

Labour hourly rates: (except Specialists) Craft Operatives £14.21 Labourer £10.71 Rates are national average prices. Refer to REGIONAL VARIATIONS for indicative levels of overall pricing in regions	MATERIALS			LABOUR			RATES			
	Del to Site £	Waste %	Material Cost £	Craft Optve Hrs	Lab Hrs	Labour Cost £	Sunds £	Nett Rate £	Unit	Gross Rate (+7.5%) £
G10: STRUCTURAL STEEL FRAMING Cont.										
Cold rolled zed purlins and cladding rails; Metsec, galvanized steel Cont.										
Side rail sleeved system (section only); fixing to cleats on stanchions at 6000mm centres; hoist and fix 6.00m above ground level; fixing with bolts Cont.										
reference 23218; 5.73 kg/m	27.79	2.50	**28.48**	0.36	0.05	**5.65**	-	**34.14**	m	**36.70**
reference 23223; 7.26 kg/m	35.61	2.50	**36.50**	0.36	0.05	**5.65**	-	**42.15**	m	**45.31**
Side rail single span system (section only) ; fixing to cleats on stanchions at 6000mm centres; hoist and fix 3.00m above ground level; fixing with bolts										
reference 14214; 3.03 kg/m	14.79	2.50	**15.16**	0.20	0.03	**3.16**	-	**18.32**	m	**19.70**
reference 14216; 3.47 kg/m	16.70	2.50	**17.12**	0.20	0.03	**3.16**	-	**20.28**	m	**21.80**
reference 17215; 3.85 kg/m	18.61	2.50	**19.08**	0.25	0.03	**3.87**	-	**22.95**	m	**24.67**
reference 17216; 4.11 kg/m	19.88	2.50	**20.38**	0.25	0.03	**3.87**	-	**24.25**	m	**26.07**
reference 20216; 4.49 kg/m	21.71	2.50	**22.25**	0.30	0.04	**4.69**	-	**26.94**	m	**28.96**
reference 20218; 5.03 kg/m	24.48	2.50	**25.09**	0.30	0.04	**4.69**	-	**29.78**	m	**32.02**
reference 20220; 5.57 kg/m	26.89	2.50	**27.56**	0.30	0.04	**4.69**	-	**32.25**	m	**34.67**
reference 23218; 5.73 kg/m	27.79	2.50	**28.48**	0.33	0.04	**5.12**	-	**33.60**	m	**36.12**
reference 23223; 7.26 kg/m	35.61	2.50	**36.50**	0.33	0.04	**5.12**	-	**41.62**	m	**44.74**
Side rail single span system (section only) ; fixing to cleats on stanchions at 6000mm centres; hoist and fix 6.00m above ground level; fixing with bolts										
reference 14214; 3.03 kg/m	14.79	2.50	**15.16**	0.22	0.03	**3.45**	-	**18.61**	m	**20.00**
reference 14216; 3.47 kg/m	16.70	2.50	**17.12**	0.22	0.03	**3.45**	-	**20.57**	m	**22.11**
reference 17215; 3.85 kg/m	18.61	2.50	**19.08**	0.28	0.04	**4.41**	-	**23.48**	m	**25.24**
reference 17216; 4.11 kg/m	19.88	2.50	**20.38**	0.28	0.04	**4.41**	-	**24.78**	m	**26.64**
reference 20216; 4.49 kg/m	21.71	2.50	**22.25**	0.33	0.04	**5.12**	-	**27.37**	m	**29.42**
reference 20218; 5.03 kg/m	24.48	2.50	**25.09**	0.33	0.04	**5.12**	-	**30.21**	m	**32.48**
reference 20220; 5.57 kg/m	26.89	2.50	**27.56**	0.33	0.04	**5.12**	-	**32.68**	m	**35.13**
reference 23218; 5.73 kg/m	27.79	2.50	**28.48**	0.36	0.05	**5.65**	-	**34.14**	m	**36.70**
reference 23223; 7.26 kg/m	35.61	2.50	**36.50**	0.36	0.05	**5.65**	-	**42.15**	m	**45.31**
Galvanized side rail cleats; weld on										
Reference 142 for 142mm wide rails	5.18	2.50	**5.31**	-	-	**-**	-	**5.31**	nr	**5.71**
Reference 172 for 172mm wide rails	7.05	2.50	**7.23**	-	-	**-**	-	**7.23**	nr	**7.77**
Reference 202 for 202mm wide rails	8.68	2.50	**8.90**	-	-	**-**	-	**8.90**	nr	**9.56**
Reference 232 for 232mm wide rails	10.09	2.50	**10.34**	-	-	**-**	-	**10.34**	nr	**11.12**
Galvanized side rail supports; 122-262 series; weld on										
rail 1000mm long	15.20	2.50	**15.58**	0.35	0.04	**5.40**	-	**20.98**	nr	**22.56**
rail 1400mm long	16.59	2.50	**17.00**	0.35	0.04	**5.40**	-	**22.41**	nr	**24.09**
rail 1600mm long	18.40	2.50	**18.86**	0.45	0.06	**7.04**	-	**25.90**	nr	**27.84**
rail 1800mm long	22.14	2.50	**22.69**	0.45	0.06	**7.04**	-	**29.73**	nr	**31.96**
Diagonal tie wire ropes (assembled with end brackets); bolt on										
1700mm long	29.66	2.50	**30.40**	0.30	0.04	**4.69**	-	**35.09**	nr	**37.72**
2200mm long	33.18	2.50	**34.01**	0.30	0.04	**4.69**	-	**38.70**	nr	**41.60**
2600mm long	36.70	2.50	**37.62**	0.40	0.05	**6.22**	-	**43.84**	nr	**47.12**
3600mm long	45.31	2.50	**46.44**	0.40	0.05	**6.22**	-	**52.66**	nr	**56.61**
Surface preparation at works										
Note notwithstanding the requirement of SMM7 to measure painting on structural steelwork in m2, painting off site has been given in tonnes of structural steelwork in accordance with normal steelwork contractors practice										
Blast cleaning										
surfaces of steelwork	-	-	**-**	-	-	**-**	-	**82.32**	t	**88.49**
Surface treatment at works										
One coat micaceous oxide primer, 75 microns										
surfaces of steelwork	-	-	**-**	-	-	**-**	-	**88.20**	t	**94.82**
Two coats micaceous oxide primer, 150 microns										
surfaces of steelwork	-	-	**-**	-	-	**-**	-	**164.64**	t	**176.99**
G12: ISOLATED STRUCTURAL METAL MEMBERS										
Isolated unfabricated structural members										
Weldable steel, BS EN 10025 : 1993 Grade S275 (formerly BS 7668 Grade 43A), hot rolled sections										
plain member; beam										
weight less than 40 Kg/m	950.00	-	**950.00**	44.12	-	**626.95**	-	**1576.95**	t	**1695.22**
weight 40-100 Kg/m	950.00	-	**950.00**	41.18	-	**585.17**	-	**1535.17**	t	**1650.31**
weight exceeding 100 Kg/m	950.00	-	**950.00**	36.75	-	**522.22**	-	**1472.22**	t	**1582.63**
Steel short span lattice joists primed at works; treated timber inserts in top and bottom chords; full depth end seatings for bolting to supporting structure; hoist and fix 3.00m above ground level										
200mm deep, 8.2 kg/m	41.52	2.50	**42.55**	0.66	0.08	**10.24**	-	**52.79**	m	**56.75**
250mm deep, 8.5 kg/m	43.26	2.50	**44.34**	0.68	0.09	**10.63**	-	**54.97**	m	**59.09**
300mm deep, 10.7 kg/m	54.65	2.50	**56.01**	0.75	0.09	**11.62**	-	**67.64**	m	**72.71**
350mm deep, 11.6 kg/m	58.75	2.50	**60.22**	0.81	0.10	**12.58**	-	**72.80**	m	**78.26**
350mm deep, 12.8 kg/m	64.21	2.50	**65.82**	0.90	0.11	**13.97**	-	**79.78**	m	**85.77**

Labour hourly rates: (except Specialists) Craft Operatives £14.21 Labourer £10.71 Rates are national average prices. Refer to REGIONAL VARIATIONS for indicative levels of overall pricing in regions	MATERIALS			LABOUR				RATES		
	Del to Site £	Waste %	Material Cost £	Craft Optve Hrs	Lab Hrs	Labour Cost £	Sunds £	Nett Rate £	Unit	Gross Rate (+7.5%) £
G12: ISOLATED STRUCTURAL METAL MEMBERS Cont.										
Isolated unfabricated structural members Cont.										
Steel short span lattice joists primed at works; treated timber inserts in top and bottom chords; full depth end seatings for bolting to supporting structure; hoist and fix 6.00m above ground level										
200mm deep, 8.2 kg/m	41.52	2.50	**42.55**	0.72	0.09	**11.20**	-	**53.75**	m	**57.78**
250mm deep, 8.5 kg/m	43.26	2.50	**44.34**	0.75	0.09	**11.62**	-	**55.97**	m	**60.16**
300mm deep, 10.7 kg/m	54.65	2.50	**56.01**	0.82	0.10	**12.72**	-	**68.74**	m	**73.89**
350mm deep, 11.6 kg/m	58.75	2.50	**60.22**	0.89	0.11	**13.83**	-	**74.04**	m	**79.59**
350mm deep, 12.8 kg/m	64.21	2.50	**65.82**	0.98	0.12	**15.21**	-	**81.03**	m	**87.10**
Steel intermediate span lattice joists primed at works; treated timber inserts in top and bottom chords; full depth end seatings for bolting to supporting structure; hoist and fix 3.00m above ground level										
450mm deep, 12.2 kg/m	56.17	2.50	**57.57**	0.85	0.11	**13.26**	-	**70.83**	m	**76.14**
500mm deep, 15.8 kg/m	69.45	2.50	**71.19**	1.03	0.13	**16.03**	-	**87.21**	m	**93.75**
550mm deep, 19.4 kg/m	82.12	2.50	**84.18**	1.26	0.16	**19.62**	-	**103.80**	m	**111.58**
600mm deep, 22.5 kg/m	93.96	2.50	**96.31**	1.35	0.17	**21.00**	-	**117.32**	m	**126.12**
650mm deep, 29.7 kg/m	123.79	2.50	**126.89**	1.78	0.22	**27.65**	-	**154.54**	m	**166.13**
Steel intermediate span lattice joists primed at works; treated timber inserts in top and bottom chords; full depth end seatings for bolting to supporting structure; hoist and fix 6.00m above ground level										
450mm deep, 12.2 kg/m	56.17	2.50	**57.57**	0.94	0.12	**14.64**	-	**72.21**	m	**77.63**
500mm deep, 15.8 kg/m	69.45	2.50	**71.19**	1.13	0.14	**17.56**	-	**88.74**	m	**95.40**
550mm deep, 19.4 kg/m	82.12	2.50	**84.18**	1.39	0.17	**21.57**	-	**105.75**	m	**113.68**
600mm deep, 22.5 kg/m	93.96	2.50	**96.31**	1.49	0.19	**23.21**	-	**119.52**	m	**128.48**
650mm deep, 29.7 kg/m	123.79	2.50	**126.89**	1.96	0.25	**30.53**	-	**157.42**	m	**169.22**
Steel long span lattice joists primed at works; treated timber inserts in top and bottom chords; full depth end seatings for bolting to supporting structure; hoist and fix 3.00m above ground level										
700mm deep, 39.2 kg/m	160.15	2.50	**164.15**	1.96	0.25	**30.53**	-	**194.68**	m	**209.28**
800mm deep, 44.1 kg/m	171.38	2.50	**175.67**	2.21	0.28	**34.40**	-	**210.07**	m	**225.82**
900mm deep, 45.3 kg/m	172.37	2.50	**176.68**	2.27	0.28	**35.26**	-	**211.93**	m	**227.83**
1000mm deep, 46.1 kg/m	176.09	2.50	**180.49**	2.31	0.29	**35.93**	-	**216.42**	m	**232.65**
1500mm deep, 54.2 kg/m	208.27	2.50	**213.48**	2.40	0.30	**37.32**	-	**250.79**	m	**269.60**
Steel long span lattice joists primed at works; treated timber inserts in top and bottom chords; full depth end seatings for bolting to supporting structure; hoist and fix 6.00m above ground level										
700mm deep, 39.2 kg/m	160.15	2.50	**164.15**	2.16	0.27	**33.59**	-	**197.74**	m	**212.57**
800mm deep, 44.1 kg/m	171.38	2.50	**175.67**	2.43	0.30	**37.74**	-	**213.41**	m	**229.42**
900mm deep, 45.3 kg/m	172.37	2.50	**176.68**	2.50	0.31	**38.85**	-	**215.52**	m	**231.69**
1000mm deep, 46.1 kg/m	176.09	2.50	**180.49**	2.54	0.32	**39.52**	-	**220.01**	m	**236.51**
1500mm deep, 54.2 kg/m	208.27	2.50	**213.48**	2.64	0.33	**41.05**	-	**254.53**	m	**273.61**
Fixing bolts; steel										
Black bolts, BS 4190 grade 4.6										
M 8; with nuts and washers										
30mm long	0.13	5.00	**0.14**	0.06	-	**0.85**	-	**0.99**	nr	**1.06**
40mm long	0.14	5.00	**0.15**	0.06	-	**0.85**	-	**1.00**	nr	**1.07**
50mm long	0.15	5.00	**0.16**	0.06	-	**0.85**	-	**1.01**	nr	**1.09**
60mm long	0.16	5.00	**0.17**	0.06	-	**0.85**	-	**1.02**	nr	**1.10**
70mm long	0.18	5.00	**0.18**	0.07	-	**0.99**	-	**1.18**	nr	**1.27**
80mm long	0.20	5.00	**0.21**	0.07	-	**0.99**	-	**1.20**	nr	**1.29**
90mm long	0.22	5.00	**0.23**	0.07	-	**0.99**	-	**1.23**	nr	**1.32**
100mm long	0.25	5.00	**0.26**	0.07	-	**0.99**	-	**1.25**	nr	**1.35**
120mm long	0.29	5.00	**0.31**	0.08	-	**1.14**	-	**1.44**	nr	**1.55**
M 10; with nuts and washers										
40mm long	0.20	5.00	**0.21**	0.06	-	**0.85**	-	**1.06**	nr	**1.14**
50mm long	0.21	5.00	**0.22**	0.06	-	**0.85**	-	**1.07**	nr	**1.15**
60mm long	0.23	5.00	**0.24**	0.06	-	**0.85**	-	**1.10**	nr	**1.18**
70mm long	0.25	5.00	**0.26**	0.07	-	**0.99**	-	**1.25**	nr	**1.35**
80mm long	0.30	5.00	**0.32**	0.07	-	**0.99**	-	**1.31**	nr	**1.41**
90mm long	0.33	5.00	**0.34**	0.07	-	**0.99**	-	**1.34**	nr	**1.44**
100mm long	0.37	5.00	**0.39**	0.07	-	**0.99**	-	**1.39**	nr	**1.49**
120mm long	0.40	5.00	**0.42**	0.08	-	**1.14**	-	**1.55**	nr	**1.67**
140mm long	0.51	5.00	**0.54**	0.09	-	**1.28**	-	**1.82**	nr	**1.95**
150mm long	0.58	5.00	**0.61**	0.10	-	**1.42**	-	**2.03**	nr	**2.19**
M 12; with nuts and washers										
40mm long	0.28	5.00	**0.29**	0.07	-	**0.99**	-	**1.29**	nr	**1.39**
50mm long	0.29	5.00	**0.31**	0.07	-	**0.99**	-	**1.30**	nr	**1.40**
60mm long	0.32	5.00	**0.33**	0.07	-	**0.99**	-	**1.33**	nr	**1.42**
70mm long	0.36	5.00	**0.38**	0.07	-	**0.99**	-	**1.37**	nr	**1.48**
80mm long	0.40	5.00	**0.42**	0.07	-	**0.99**	-	**1.41**	nr	**1.52**
90mm long	0.41	5.00	**0.43**	0.07	-	**0.99**	-	**1.42**	nr	**1.53**
100mm long	0.44	5.00	**0.47**	0.08	-	**1.14**	-	**1.60**	nr	**1.72**
120mm long	0.54	5.00	**0.56**	0.08	-	**1.14**	-	**1.70**	nr	**1.83**
140mm long	0.67	5.00	**0.70**	0.10	-	**1.42**	-	**2.12**	nr	**2.28**
150mm long	0.84	5.00	**0.88**	0.10	-	**1.42**	-	**2.30**	nr	**2.47**
160mm long	1.20	5.00	**1.26**	0.10	-	**1.42**	-	**2.68**	nr	**2.88**
180mm long	1.26	5.00	**1.32**	0.11	-	**1.56**	-	**2.89**	nr	**3.10**
200mm long	1.32	5.00	**1.38**	0.12	-	**1.71**	-	**3.09**	nr	**3.32**
220mm long	1.39	5.00	**1.46**	0.13	-	**1.85**	-	**3.30**	nr	**3.55**
240mm long	1.45	5.00	**1.52**	0.14	-	**1.99**	-	**3.51**	nr	**3.77**

Labour hourly rates: (except Specialists) Craft Operatives £14.21 Labourer £10.71 Rates are national average prices. Refer to REGIONAL VARIATIONS for indicative levels of overall pricing in regions	MATERIALS			LABOUR				RATES		
	Del to Site £	Waste %	Material Cost £	Craft Optve Hrs	Lab Hrs	Labour Cost £	Sunds £	Nett Rate £	Unit	Gross Rate (+7.5%) £
G12: ISOLATED STRUCTURAL METAL MEMBERS Cont.										
Fixing bolts; steel Cont.										
Black bolts, BS 4190 grade 4.6 Cont.										
M 12; with nuts and washers Cont.										
260mm long	1.52	5.00	**1.59**	0.15	-	**2.13**	-	**3.72**	nr	**4.00**
280mm long	1.60	5.00	**1.68**	0.16	-	**2.27**	-	**3.95**	nr	**4.25**
300mm long	1.66	5.00	**1.74**	0.17	-	**2.42**	-	**4.15**	nr	**4.46**
M 16; with nuts and washers										
50mm long	0.51	5.00	**0.54**	0.07	-	**0.99**	-	**1.53**	nr	**1.65**
60mm long	0.57	5.00	**0.60**	0.07	-	**0.99**	-	**1.60**	nr	**1.71**
70mm long	0.63	5.00	**0.66**	0.07	-	**0.99**	-	**1.66**	nr	**1.78**
80mm long	0.69	5.00	**0.72**	0.08	-	**1.14**	-	**1.86**	nr	**2.00**
90mm long	0.74	5.00	**0.77**	0.08	-	**1.14**	-	**1.91**	nr	**2.05**
100mm long	0.86	5.00	**0.91**	0.09	-	**1.28**	-	**2.18**	nr	**2.35**
120mm long	1.00	5.00	**1.05**	0.09	-	**1.28**	-	**2.33**	nr	**2.51**
140mm long	1.12	5.00	**1.17**	0.10	-	**1.42**	-	**2.60**	nr	**2.79**
150mm long	1.18	5.00	**1.24**	0.11	-	**1.56**	-	**2.80**	nr	**3.01**
160mm long	1.63	5.00	**1.71**	0.12	-	**1.71**	-	**3.42**	nr	**3.68**
180mm long	1.70	5.00	**1.79**	0.12	-	**1.71**	-	**3.49**	nr	**3.75**
200mm long	1.81	5.00	**1.90**	0.13	-	**1.85**	-	**3.74**	nr	**4.03**
220mm long	1.91	5.00	**2.01**	0.13	-	**1.85**	-	**3.85**	nr	**4.14**
M 20; with nuts and washers										
60mm long	0.97	5.00	**1.02**	0.07	-	**0.99**	-	**2.01**	nr	**2.16**
70mm long	1.07	5.00	**1.13**	0.07	-	**0.99**	-	**2.12**	nr	**2.28**
80mm long	1.21	5.00	**1.27**	0.08	-	**1.14**	-	**2.41**	nr	**2.59**
90mm long	1.22	5.00	**1.29**	0.09	-	**1.28**	-	**2.56**	nr	**2.76**
100mm long	1.40	5.00	**1.47**	0.09	-	**1.28**	-	**2.75**	nr	**2.95**
120mm long	1.64	5.00	**1.73**	0.10	-	**1.42**	-	**3.15**	nr	**3.38**
140mm long	1.91	5.00	**2.01**	0.11	-	**1.56**	-	**3.57**	nr	**3.84**
150mm long	1.96	5.00	**2.06**	0.12	-	**1.71**	-	**3.76**	nr	**4.04**
160mm long	2.41	5.00	**2.53**	0.13	-	**1.85**	-	**4.38**	nr	**4.71**
180mm long	2.54	5.00	**2.67**	0.13	-	**1.85**	-	**4.52**	nr	**4.85**
200mm long	2.72	5.00	**2.85**	0.14	-	**1.99**	-	**4.84**	nr	**5.20**
220mm long	2.82	5.00	**2.96**	0.15	-	**2.13**	-	**5.09**	nr	**5.48**
240mm long	2.98	5.00	**3.13**	0.16	-	**2.27**	-	**5.41**	nr	**5.81**
260mm long	3.07	5.00	**3.22**	0.17	-	**2.42**	-	**5.64**	nr	**6.06**
280mm long	3.15	5.00	**3.30**	0.18	-	**2.56**	-	**5.86**	nr	**6.30**
300mm long	3.33	5.00	**3.50**	0.20	-	**2.84**	-	**6.34**	nr	**6.82**
M 24; with nuts and washers										
70mm long	2.44	5.00	**2.56**	0.09	-	**1.28**	-	**3.84**	nr	**4.12**
80mm long	2.54	5.00	**2.67**	0.09	-	**1.28**	-	**3.95**	nr	**4.24**
90mm long	2.62	5.00	**2.75**	0.10	-	**1.42**	-	**4.18**	nr	**4.49**
100mm long	2.72	5.00	**2.85**	0.10	-	**1.42**	-	**4.27**	nr	**4.59**
120mm long	2.90	5.00	**3.05**	0.10	-	**1.42**	-	**4.47**	nr	**4.80**
140mm long	3.26	5.00	**3.43**	0.11	-	**1.56**	-	**4.99**	nr	**5.36**
150mm long	3.46	5.00	**3.64**	0.13	-	**1.85**	-	**5.48**	nr	**5.89**
160mm long	3.58	5.00	**3.76**	0.13	-	**1.85**	-	**5.61**	nr	**6.03**
180mm long	3.67	5.00	**3.86**	0.13	-	**1.85**	-	**5.70**	nr	**6.13**
200mm long	3.85	5.00	**4.04**	0.14	-	**1.99**	-	**6.03**	nr	**6.48**
220mm long	4.24	5.00	**4.45**	0.15	-	**2.13**	-	**6.59**	nr	**7.08**
240mm long	4.36	5.00	**4.58**	0.17	-	**2.42**	-	**6.99**	nr	**7.52**
260mm long	4.65	5.00	**4.88**	0.18	-	**2.56**	-	**7.44**	nr	**8.00**
280mm long	4.80	5.00	**5.04**	0.22	-	**3.13**	-	**8.17**	nr	**8.78**
300mm long	4.94	5.00	**5.19**	0.24	-	**3.41**	-	**8.60**	nr	**9.24**
Black bolts, BS 4190 grade 4.6; galvanized										
M 8; with nuts and washers										
30mm long	0.18	5.00	**0.18**	0.06	-	**0.85**	-	**1.04**	nr	**1.11**
40mm long	0.20	5.00	**0.21**	0.06	-	**0.85**	-	**1.06**	nr	**1.14**
50mm long	0.21	5.00	**0.22**	0.06	-	**0.85**	-	**1.07**	nr	**1.15**
60mm long	0.23	5.00	**0.24**	0.06	-	**0.85**	-	**1.10**	nr	**1.18**
70mm long	0.26	5.00	**0.27**	0.07	-	**0.99**	-	**1.26**	nr	**1.36**
80mm long	0.27	5.00	**0.28**	0.07	-	**0.99**	-	**1.28**	nr	**1.37**
90mm long	0.30	5.00	**0.32**	0.07	-	**0.99**	-	**1.31**	nr	**1.41**
100mm long	0.37	5.00	**0.39**	0.07	-	**0.99**	-	**1.39**	nr	**1.49**
120mm long	0.44	5.00	**0.47**	0.08	-	**1.14**	-	**1.60**	nr	**1.72**
M 10; with nuts and washers										
40mm long	0.27	5.00	**0.28**	0.06	-	**0.85**	-	**1.14**	nr	**1.22**
50mm long	0.30	5.00	**0.32**	0.06	-	**0.85**	-	**1.17**	nr	**1.26**
60mm long	0.34	5.00	**0.36**	0.06	-	**0.85**	-	**1.21**	nr	**1.30**
70mm long	0.37	5.00	**0.39**	0.07	-	**0.99**	-	**1.39**	nr	**1.49**
80mm long	0.46	5.00	**0.48**	0.07	-	**0.99**	-	**1.47**	nr	**1.58**
90mm long	0.50	5.00	**0.53**	0.07	-	**0.99**	-	**1.52**	nr	**1.64**
100mm long	0.55	5.00	**0.58**	0.07	-	**0.99**	-	**1.57**	nr	**1.69**
120mm long	0.58	5.00	**0.61**	0.08	-	**1.14**	-	**1.75**	nr	**1.88**
140mm long	0.76	5.00	**0.80**	0.09	-	**1.28**	-	**2.07**	nr	**2.23**
150mm long	0.88	5.00	**0.92**	0.10	-	**1.42**	-	**2.34**	nr	**2.52**
M 12; with nuts and washers										
40mm long	0.46	5.00	**0.48**	0.07	-	**0.99**	-	**1.47**	nr	**1.58**
50mm long	0.48	5.00	**0.50**	0.07	-	**0.99**	-	**1.50**	nr	**1.61**
60mm long	0.51	5.00	**0.54**	0.07	-	**0.99**	-	**1.53**	nr	**1.65**
70mm long	0.57	5.00	**0.60**	0.07	-	**0.99**	-	**1.60**	nr	**1.71**
80mm long	0.62	5.00	**0.65**	0.07	-	**0.99**	-	**1.64**	nr	**1.77**
90mm long	0.64	5.00	**0.67**	0.07	-	**0.99**	-	**1.67**	nr	**1.79**
100mm long	0.70	5.00	**0.73**	0.08	-	**1.14**	-	**1.87**	nr	**2.01**
120mm long	0.86	5.00	**0.91**	0.08	-	**1.14**	-	**2.04**	nr	**2.20**
140mm long	1.02	5.00	**1.07**	0.10	-	**1.42**	-	**2.49**	nr	**2.67**
150mm long	1.28	5.00	**1.35**	0.10	-	**1.42**	-	**2.77**	nr	**2.97**
160mm long	1.81	5.00	**1.90**	0.10	-	**1.42**	-	**3.32**	nr	**3.57**
180mm long	1.90	5.00	**2.00**	0.11	-	**1.56**	-	**3.56**	nr	**3.82**
200mm long	1.98	5.00	**2.08**	0.12	-	**1.71**	-	**3.79**	nr	**4.07**
220mm long	2.09	5.00	**2.19**	0.13	-	**1.85**	-	**4.04**	nr	**4.34**
240mm long	2.17	5.00	**2.28**	0.14	-	**1.99**	-	**4.27**	nr	**4.59**
260mm long	2.30	5.00	**2.41**	0.15	-	**2.13**	-	**4.54**	nr	**4.88**

Labour hourly rates: (except Specialists) Craft Operatives £14.21 Labourer £10.71 Rates are national average prices. Refer to REGIONAL VARIATIONS for indicative levels of overall pricing in regions	MATERIALS			LABOUR				RATES		
	Del to Site £	Waste %	Material Cost £	Craft Optve Hrs	Lab Hrs	Labour Cost £	Sunds £	Nett Rate £	Unit	Gross Rate (+7.5%) £
G12: ISOLATED STRUCTURAL METAL MEMBERS Cont.										
Fixing bolts; steel Cont.										
Black bolts, BS 4190 grade 4.6; galvanized Cont.										
M 12; with nuts and washers Cont.										
280mm long	2.37	5.00	**2.49**	0.16	-	**2.27**	-	**4.76**	nr	**5.12**
300mm long	2.47	5.00	**2.59**	0.17	-	**2.42**	-	**5.01**	nr	**5.39**
M 16; with nuts and washers										
50mm long	0.83	5.00	**0.87**	0.07	-	**0.99**	-	**1.86**	nr	**2.00**
60mm long	0.90	5.00	**0.94**	0.07	-	**0.99**	-	**1.94**	nr	**2.08**
70mm long	0.97	5.00	**1.02**	0.07	-	**0.99**	-	**2.01**	nr	**2.16**
80mm long	1.05	5.00	**1.10**	0.08	-	**1.14**	-	**2.24**	nr	**2.41**
90mm long	1.11	5.00	**1.16**	0.08	-	**1.14**	-	**2.30**	nr	**2.47**
100mm long	1.17	5.00	**1.22**	0.09	-	**1.28**	-	**2.50**	nr	**2.69**
120mm long	1.52	5.00	**1.59**	0.09	-	**1.28**	-	**2.87**	nr	**3.08**
140mm long	1.69	5.00	**1.78**	0.10	-	**1.42**	-	**3.20**	nr	**3.44**
150mm long	1.80	5.00	**1.88**	0.11	-	**1.56**	-	**3.45**	nr	**3.71**
160mm long	2.42	5.00	**2.55**	0.12	-	**1.71**	-	**4.25**	nr	**4.57**
180mm long	2.58	5.00	**2.70**	0.12	-	**1.71**	-	**4.41**	nr	**4.74**
200mm long	2.74	5.00	**2.88**	0.13	-	**1.85**	-	**4.72**	nr	**5.08**
220mm long	2.86	5.00	**3.00**	0.13	-	**1.85**	-	**4.85**	nr	**5.21**
M 20; with nuts and washers										
60mm long	1.49	5.00	**1.57**	0.07	-	**0.99**	-	**2.56**	nr	**2.75**
70mm long	1.64	5.00	**1.73**	0.07	-	**0.99**	-	**2.72**	nr	**2.92**
80mm long	1.85	5.00	**1.95**	0.08	-	**1.14**	-	**3.08**	nr	**3.31**
90mm long	1.87	5.00	**1.96**	0.09	-	**1.28**	-	**3.24**	nr	**3.48**
100mm long	2.13	5.00	**2.24**	0.09	-	**1.28**	-	**3.52**	nr	**3.78**
120mm long	2.47	5.00	**2.59**	0.10	-	**1.42**	-	**4.02**	nr	**4.32**
140mm long	2.88	5.00	**3.02**	0.11	-	**1.56**	-	**4.59**	nr	**4.93**
150mm long	2.95	5.00	**3.10**	0.12	-	**1.71**	-	**4.80**	nr	**5.16**
160mm long	3.65	5.00	**3.83**	0.13	-	**1.85**	-	**5.68**	nr	**6.10**
180mm long	3.77	5.00	**3.95**	0.13	-	**1.85**	-	**5.80**	nr	**6.24**
200mm long	4.01	5.00	**4.21**	0.14	-	**1.99**	-	**6.20**	nr	**6.66**
220mm long	4.17	5.00	**4.38**	0.15	-	**2.13**	-	**6.51**	nr	**7.00**
240mm long	4.43	5.00	**4.65**	0.16	-	**2.27**	-	**6.92**	nr	**7.44**
260mm long	4.53	5.00	**4.76**	0.17	-	**2.42**	-	**7.18**	nr	**7.71**
280mm long	4.65	5.00	**4.88**	0.18	-	**2.56**	-	**7.44**	nr	**8.00**
300mm long	4.93	5.00	**5.18**	0.20	-	**2.84**	-	**8.02**	nr	**8.62**
M 24; with nuts and washers										
70mm long	3.60	5.00	**3.78**	0.09	-	**1.28**	-	**5.06**	nr	**5.44**
80mm long	3.77	5.00	**3.95**	0.09	-	**1.28**	-	**5.23**	nr	**5.62**
90mm long	3.88	5.00	**4.08**	0.10	-	**1.42**	-	**5.50**	nr	**5.91**
100mm long	4.03	5.00	**4.23**	0.10	-	**1.42**	-	**5.66**	nr	**6.08**
120mm long	4.26	5.00	**4.47**	0.10	-	**1.42**	-	**5.89**	nr	**6.33**
140mm long	4.83	5.00	**5.07**	0.11	-	**1.56**	-	**6.63**	nr	**7.13**
150mm long	5.13	5.00	**5.38**	0.13	-	**1.85**	-	**7.23**	nr	**7.77**
160mm long	5.29	5.00	**5.56**	0.13	-	**1.85**	-	**7.40**	nr	**7.96**
180mm long	5.42	5.00	**5.69**	0.13	-	**1.85**	-	**7.54**	nr	**8.10**
200mm long	5.48	5.00	**5.75**	0.14	-	**1.99**	-	**7.74**	nr	**8.32**
220mm long	5.98	5.00	**6.28**	0.15	-	**2.13**	-	**8.41**	nr	**9.04**
240mm long	6.11	5.00	**6.41**	0.17	-	**2.42**	-	**8.83**	nr	**9.49**
260mm long	6.42	5.00	**6.74**	0.18	-	**2.56**	-	**9.30**	nr	**10.00**
280mm long	6.59	5.00	**6.92**	0.22	-	**3.13**	-	**10.04**	nr	**10.79**
300mm long	6.80	5.00	**7.13**	0.24	-	**3.41**	-	**10.55**	nr	**11.34**
G20: CARPENTRY/TIMBER FRAMING/FIRST FIXING										
Sasco Trussed Rafters in softwood, sawn, BS 4978, GS grade, impregnated; note:- prices are guide prices only, based on contracts within 30 mile radius of Widnes and are subject to adjustment for quantity: firm quotations should always be obtained										
Trussed rafters; Hydro-Nail plated joints										
22.5 degree Standard duo pitch; 450mm overhangs; fixing with clips (included elsewhere); span over wall plates										
5000mm	33.34	2.50	**34.18**	1.10	0.14	**17.13**	1.82	**53.12**	nr	**57.11**
6000mm	38.37	2.50	**39.33**	1.20	0.15	**18.66**	1.98	**59.97**	nr	**64.47**
7000mm	42.94	2.50	**44.01**	1.30	0.16	**20.19**	2.15	**66.34**	nr	**71.32**
8000mm	50.83	2.50	**52.10**	1.40	0.17	**21.71**	2.31	**76.13**	nr	**81.84**
9000mm	58.84	2.50	**60.31**	1.50	0.19	**23.35**	2.48	**86.14**	nr	**92.60**
10000mm	71.24	2.50	**73.02**	1.60	0.20	**24.88**	2.64	**100.54**	nr	**108.08**
35 degree Standard duo pitch; 450mm overhangs; fixing with clips (included elsewhere); span over wall plates										
5000mm	34.66	2.50	**35.53**	1.10	0.14	**17.13**	1.82	**54.47**	nr	**58.56**
6000mm	40.06	2.50	**41.06**	1.20	0.15	**18.66**	1.98	**61.70**	nr	**66.32**
7000mm	44.64	2.50	**45.75**	1.30	0.16	**20.19**	2.15	**68.08**	nr	**73.19**
8000mm	53.58	2.50	**54.92**	1.40	0.17	**21.71**	2.31	**78.94**	nr	**84.86**
9000mm	59.64	2.50	**61.13**	1.50	0.19	**23.35**	2.48	**86.96**	nr	**93.48**
45 degree Standard duo pitch; 450mm overhangs; fixing with clips (included elsewhere); span over wall plates										
5000mm	54.41	2.50	**55.77**	1.10	0.14	**17.13**	1.82	**74.71**	nr	**80.32**
6000mm	65.11	2.50	**66.74**	1.20	0.15	**18.66**	1.98	**87.37**	nr	**93.93**
7000mm	75.94	2.50	**77.84**	1.30	0.16	**20.19**	2.15	**100.17**	nr	**107.69**
8000mm	86.84	2.50	**89.01**	1.40	0.17	**21.71**	2.31	**113.03**	nr	**121.51**
9000mm	97.85	2.50	**100.29**	1.50	0.19	**23.35**	2.48	**126.12**	nr	**135.58**
22.5 degree Bobtail duo pitch; 450mm overhangs; fixing with clips (included elsewhere); span over wall plates										
4000mm	36.66	2.50	**37.57**	1.00	0.13	**15.60**	1.65	**54.82**	nr	**58.94**
5000mm	45.69	2.50	**46.83**	1.10	0.14	**17.13**	1.82	**65.77**	nr	**70.71**
6000mm	54.88	2.50	**56.25**	1.20	0.15	**18.66**	1.98	**76.89**	nr	**82.66**

Labour hourly rates: (except Specialists) Craft Operatives £14.21 Labourer £10.71 Rates are national average prices. Refer to REGIONAL VARIATIONS for indicative levels of overall pricing in regions	MATERIALS			LABOUR				RATES		
	Del to Site £	Waste %	Material Cost £	Craft Optve Hrs	Lab Hrs	Labour Cost £	Sunds £	Nett Rate £	Unit	Gross Rate (+7.5%) £
G20: CARPENTRY/TIMBER FRAMING/FIRST FIXING Cont.										
Sasco Trussed Rafters in softwood, sawn, BS 4978, GS grade, impregnated; note:- prices are guide prices only, based on contracts within 30 mile radius of Widnes and are subject to adjustment for quantity: firm quotations should always be obtained Cont.										
Trussed rafters; Hydro-Nail plated joints Cont.										
22.5 degree Bobtail duo pitch; 450mm overhangs; fixing with clips (included elsewhere); span over wall plates Cont.										
7000mm	64.10	2.50	**65.71**	1.30	0.16	**20.19**	2.15	**88.04**	nr	**94.64**
8000mm	73.31	2.50	**75.14**	1.40	0.17	**21.71**	2.31	**99.17**	nr	**106.61**
9000mm	82.61	2.50	**84.67**	1.50	0.19	**23.35**	2.48	**110.50**	nr	**118.79**
35 degree Bobtail duo pitch; 450mm overhangs; fixing with clips (included elsewhere); span over wall plates										
4000mm	38.24	2.50	**39.19**	1.00	0.13	**15.60**	1.65	**56.45**	nr	**60.68**
5000mm	47.65	2.50	**48.84**	1.10	0.14	**17.13**	1.82	**67.79**	nr	**72.87**
6000mm	57.33	2.50	**58.77**	1.20	0.15	**18.66**	1.98	**79.41**	nr	**85.36**
7000mm	66.63	2.50	**68.30**	1.30	0.16	**20.19**	2.15	**90.63**	nr	**97.42**
8000mm	76.21	2.50	**78.11**	1.40	0.17	**21.71**	2.31	**102.14**	nr	**109.80**
9000mm	85.64	2.50	**87.78**	1.50	0.19	**23.35**	2.48	**113.60**	nr	**122.12**
45 degree Bobtail duo pitch; 450mm overhangs; fixing with clips (included elsewhere); span over wall plates										
4000mm	59.31	2.50	**60.80**	1.00	0.13	**15.60**	1.65	**78.05**	nr	**83.90**
5000mm	74.08	2.50	**75.93**	1.10	0.14	**17.13**	1.82	**94.88**	nr	**101.99**
6000mm	89.01	2.50	**91.23**	1.20	0.15	**18.66**	1.98	**111.87**	nr	**120.26**
7000mm	103.85	2.50	**106.44**	1.30	0.16	**20.19**	2.15	**128.78**	nr	**138.43**
8000mm	128.09	2.50	**131.29**	1.40	0.17	**21.71**	2.31	**155.31**	nr	**166.96**
9000mm	133.56	2.50	**136.90**	1.50	0.19	**23.35**	2.48	**162.72**	nr	**174.92**
22.5 degree Monopitch; 450mm overhangs; fixing with clips (included elsewhere); span over wall plates										
2000mm	17.94	2.50	**18.39**	0.80	0.10	**12.44**	1.32	**32.15**	nr	**34.56**
3000mm	23.77	2.50	**24.36**	0.90	0.11	**13.97**	1.49	**39.81**	nr	**42.80**
4000mm	29.59	2.50	**30.33**	1.00	0.13	**15.60**	1.65	**47.58**	nr	**51.15**
5000mm	36.17	2.50	**37.07**	1.10	0.14	**17.13**	1.82	**56.02**	nr	**60.22**
6000mm	46.71	2.50	**47.87**	1.20	0.15	**18.66**	1.98	**68.51**	nr	**73.65**
35 degree Monopitch; 450mm overhangs; fixing with clips (included elsewhere); span over wall plates										
2000mm	19.19	2.50	**19.66**	0.80	0.10	**12.44**	1.32	**33.42**	nr	**35.93**
3000mm	25.27	2.50	**25.91**	0.90	0.11	**13.97**	1.49	**41.36**	nr	**44.46**
4000mm	31.48	2.50	**32.27**	1.00	0.13	**15.60**	1.65	**49.52**	nr	**53.24**
5000mm	38.83	2.50	**39.80**	1.10	0.14	**17.13**	1.82	**58.74**	nr	**63.15**
6000mm	50.02	2.50	**51.27**	1.20	0.15	**18.66**	1.98	**71.91**	nr	**77.30**
45 degree Monopitch; 450mm overhangs; fixing with clips (included elsewhere); span over wall plates										
2000mm	20.49	2.50	**21.00**	0.80	0.10	**12.44**	1.32	**34.76**	nr	**37.36**
3000mm	39.02	2.50	**40.00**	0.90	0.11	**13.97**	1.49	**55.45**	nr	**59.61**
4000mm	46.14	2.50	**47.30**	1.00	0.13	**15.60**	1.65	**64.55**	nr	**69.39**
22.5 degree duo pitch Girder Truss; fixing with clips (included elsewhere); span over wall plates										
5000mm	134.67	2.50	**138.03**	1.10	0.14	**17.13**	1.82	**156.98**	nr	**168.75**
6000mm	152.43	2.50	**156.24**	1.20	0.15	**18.66**	1.98	**176.88**	nr	**190.15**
7000mm	165.22	2.50	**169.35**	1.30	0.16	**20.19**	2.15	**191.68**	nr	**206.05**
8000mm	191.63	2.50	**196.42**	1.40	0.17	**21.71**	2.31	**220.44**	nr	**236.98**
9000mm	213.05	2.50	**218.37**	1.50	0.19	**23.35**	2.48	**244.20**	nr	**262.51**
10000mm	251.05	2.50	**257.32**	1.60	0.20	**24.88**	2.64	**284.84**	nr	**306.20**
35 degree duo pitch Girder Truss; fixing with clips (included elsewhere); span over wall plates										
5000mm	141.07	2.50	**144.59**	1.10	0.14	**17.13**	1.82	**163.54**	nr	**175.80**
6000mm	157.13	2.50	**161.06**	1.20	0.15	**18.66**	1.98	**181.70**	nr	**195.33**
7000mm	171.81	2.50	**176.10**	1.30	0.16	**20.19**	2.15	**198.44**	nr	**213.32**
8000mm	198.22	2.50	**203.18**	1.40	0.17	**21.71**	2.31	**227.20**	nr	**244.24**
9000mm	219.62	2.50	**225.11**	1.50	0.19	**23.35**	2.48	**250.94**	nr	**269.76**
10000mm	257.49	2.50	**263.93**	1.60	0.20	**24.88**	2.64	**291.45**	nr	**313.30**
45 degree duo pitch Girder Truss; fixing with clips (included elsewhere); span over wall plates										
5000mm	190.96	2.50	**195.74**	1.10	0.14	**17.13**	1.82	**214.68**	nr	**230.78**
6000mm	232.14	2.50	**237.95**	1.20	0.15	**18.66**	1.98	**258.58**	nr	**277.98**
7000mm	265.96	2.50	**272.61**	1.30	0.16	**20.19**	2.15	**294.94**	nr	**317.06**
8000mm	297.92	2.50	**305.36**	1.40	0.17	**21.71**	2.31	**329.39**	nr	**354.09**
9000mm	330.82	2.50	**339.09**	1.50	0.19	**23.35**	2.48	**364.91**	nr	**392.28**
22.5 degree pitch gable ladder; 450mm overhang; span over wall plate										
5000mm	17.77	2.50	**18.21**	1.10	0.14	**17.13**	1.82	**37.16**	nr	**39.94**
6000mm	20.29	2.50	**20.80**	1.20	0.15	**18.66**	1.98	**41.44**	nr	**44.55**
7000mm	22.75	2.50	**23.32**	1.30	0.16	**20.19**	2.15	**45.65**	nr	**49.07**
8000mm	27.27	2.50	**27.95**	1.40	0.17	**21.71**	2.31	**51.98**	nr	**55.87**
9000mm	31.23	2.50	**32.01**	1.50	0.19	**23.35**	2.48	**57.84**	nr	**62.17**
10000mm	37.88	2.50	**38.83**	1.60	0.20	**24.88**	2.64	**66.35**	nr	**71.32**
35 degree pitch gable ladder; 450mm overhang; span over wall plate										
5000mm	18.43	2.50	**18.89**	1.10	0.14	**17.13**	1.82	**37.84**	nr	**40.68**
6000mm	21.33	2.50	**21.86**	1.20	0.15	**18.66**	1.98	**42.50**	nr	**45.69**
7000mm	23.38	2.50	**23.97**	1.30	0.16	**20.19**	2.15	**46.30**	nr	**49.77**
8000mm	28.38	2.50	**29.09**	1.40	0.17	**21.71**	2.31	**53.11**	nr	**57.10**
9000mm	32.41	2.50	**33.22**	1.50	0.19	**23.35**	2.48	**59.05**	nr	**63.48**
10000mm	39.29	2.50	**40.27**	1.60	0.20	**24.88**	2.64	**67.79**	nr	**72.87**
45 degree pitch gable ladder; 450mm overhang; span over wall plate										
5000mm	28.95	2.50	**29.68**	1.10	0.14	**17.13**	1.82	**48.62**	nr	**52.27**
6000mm	34.66	2.50	**35.53**	1.20	0.15	**18.66**	1.98	**56.17**	nr	**60.38**

Labour hourly rates: (except Specialists) Craft Operatives £14.21 Labourer £10.71 Rates are national average prices. Refer to REGIONAL VARIATIONS for indicative levels of overall pricing in regions	MATERIALS			LABOUR				RATES		
	Del to Site £	Waste %	Material Cost £	Craft Optve Hrs	Lab Hrs	Labour Cost £	Sunds £	Nett Rate £	Unit	Gross Rate (+7.5%) £
G20: CARPENTRY/TIMBER FRAMING/FIRST FIXING Cont.										
Sasco Trussed Rafters in softwood, sawn, BS 4978, GS grade, impregnated; note:- prices are guide prices only, based on contracts within 30 mile radius of Widnes and are subject to adjustment for quantity: firm quotations should always be obtained Cont.										
Trussed rafters; Hydro-Nail plated joints Cont.										
45 degree pitch gable ladder; 450mm overhang; span over wall plate Cont.										
7000mm	40.31	2.50	**41.31**	1.30	0.16	**20.19**	2.15	**63.65**	nr	**68.42**
8000mm	45.32	2.50	**46.45**	1.40	0.17	**21.71**	2.31	**70.47**	nr	**75.76**
9000mm	51.52	2.50	**52.81**	1.50	0.19	**23.35**	2.48	**78.64**	nr	**84.54**
10000mm	56.79	2.50	**58.21**	1.60	0.20	**24.88**	2.64	**85.72**	nr	**92.15**
Softwood, wrot, BS 4978, GS grade - laminated beams										
Glued laminated beams; BS EN 386										
65 x 150 x 4000mm	54.23	2.50	**55.58**	0.50	0.06	**7.75**	1.06	**64.39**	nr	**69.22**
65 x 175 x 4000mm	63.16	2.50	**64.74**	0.60	0.07	**9.28**	1.24	**75.26**	nr	**80.90**
65 x 200 x 4000mm	72.08	2.50	**73.88**	0.60	0.07	**9.28**	1.40	**84.56**	nr	**90.90**
65 x 225 x 4000mm	81.42	2.50	**83.45**	0.80	0.10	**12.44**	1.57	**97.46**	nr	**104.77**
65 x 250 x 4000mm	90.77	2.50	**93.04**	0.90	0.11	**13.97**	1.75	**108.76**	nr	**116.91**
65 x 250 x 6000mm	136.08	2.50	**139.49**	1.10	0.14	**17.13**	1.95	**158.56**	nr	**170.46**
65 x 275 x 4000mm	99.63	2.50	**102.12**	1.00	0.13	**15.60**	1.95	**119.67**	nr	**128.64**
65 x 275 x 6000mm	151.38	2.50	**155.16**	1.20	0.15	**18.66**	2.11	**175.93**	nr	**189.13**
65 x 300 x 4000mm	107.42	2.50	**110.11**	1.20	0.15	**18.66**	2.11	**130.88**	nr	**140.70**
65 x 300 x 6000mm	162.07	2.50	**166.12**	1.30	0.16	**20.19**	2.28	**188.58**	nr	**202.73**
65 x 325 x 4000mm	119.32	2.50	**122.31**	1.30	0.16	**20.19**	2.28	**144.77**	nr	**155.63**
65 x 325 x 6000mm	175.45	2.50	**179.84**	1.40	0.18	**21.82**	2.46	**204.12**	nr	**219.42**
65 x 325 x 8000mm	234.69	2.50	**240.56**	1.50	0.19	**23.35**	2.64	**266.55**	nr	**286.54**
90 x 150 x 4000mm	68.76	2.50	**70.48**	0.70	0.09	**10.91**	1.40	**82.79**	nr	**89.00**
90 x 175 x 4000mm	81.03	2.50	**83.06**	0.80	0.10	**12.44**	1.57	**97.07**	nr	**104.35**
90 x 200 x 4000mm	92.19	2.50	**94.49**	0.90	0.11	**13.97**	1.75	**110.21**	nr	**118.47**
90 x 225 x 4000mm	102.96	2.50	**105.53**	1.00	0.13	**15.60**	1.95	**123.08**	nr	**132.31**
90 x 250 x 4000mm	116.26	2.50	**119.17**	1.10	0.14	**17.13**	2.11	**138.41**	nr	**148.79**
90 x 250 x 6000mm	171.75	2.50	**176.04**	1.30	0.16	**20.19**	2.28	**198.51**	nr	**213.40**
90 x 275 x 4000mm	127.13	2.50	**130.31**	1.20	0.15	**18.66**	2.28	**151.24**	nr	**162.58**
90 x 275 x 6000mm	191.44	2.50	**196.23**	1.40	0.18	**21.82**	2.46	**220.51**	nr	**237.05**
90 x 300 x 4000mm	137.53	2.50	**140.97**	1.40	0.18	**21.82**	2.46	**165.25**	nr	**177.65**
90 x 300 x 6000mm	209.64	2.50	**214.88**	1.50	0.19	**23.35**	2.66	**240.89**	nr	**258.96**
90 x 325 x 4000mm	152.02	2.50	**155.82**	1.50	0.19	**23.35**	2.66	**181.83**	nr	**195.47**
90 x 325 x 6000mm	226.00	2.50	**231.65**	1.60	0.20	**24.88**	2.82	**259.35**	nr	**278.80**
90 x 325 x 8000mm	299.23	2.50	**306.71**	1.70	0.21	**26.41**	3.00	**336.12**	nr	**361.33**
90 x 350 x 4000mm	160.59	2.50	**164.60**	1.60	0.20	**24.88**	2.82	**192.30**	nr	**206.72**
90 x 350 x 6000mm	243.85	2.50	**249.94**	1.70	0.21	**26.41**	3.00	**279.35**	nr	**300.30**
90 x 350 x 8000mm	325.63	2.50	**333.77**	1.80	0.22	**27.93**	3.17	**364.87**	nr	**392.24**
90 x 375 x 4000mm	171.75	2.50	**176.04**	1.70	0.21	**26.41**	3.00	**205.45**	nr	**220.86**
90 x 375 x 6000mm	258.71	2.50	**265.18**	1.80	0.22	**27.93**	3.17	**296.28**	nr	**318.50**
90 x 375 x 8000mm	346.81	2.50	**355.48**	1.90	0.24	**29.57**	3.37	**388.42**	nr	**417.55**
90 x 375 x 10000mm	433.44	2.50	**444.27**	2.00	0.25	**31.10**	3.51	**478.88**	nr	**514.80**
90 x 400 x 4000mm	186.60	2.50	**191.27**	1.80	0.22	**27.93**	3.17	**222.37**	nr	**239.05**
90 x 400 x 6000mm	276.56	2.50	**283.47**	1.90	0.24	**29.57**	3.37	**316.41**	nr	**340.14**
90 x 400 x 8000mm	369.49	2.50	**378.73**	2.00	0.25	**31.10**	3.51	**413.34**	nr	**444.34**
90 x 400 x 10000mm	461.30	2.50	**472.83**	2.10	0.26	**32.63**	3.70	**509.15**	nr	**547.34**
90 x 425 x 6000mm	309.27	2.50	**317.00**	2.00	0.25	**31.10**	3.51	**351.62**	nr	**377.99**
90 x 425 x 8000mm	389.56	2.50	**399.30**	2.10	0.26	**32.63**	3.70	**435.62**	nr	**468.29**
90 x 425 x 10000mm	485.47	2.50	**497.60**	2.20	0.28	**34.26**	3.86	**535.73**	nr	**575.91**
90 x 425 x 12000mm	582.85	2.50	**597.42**	2.30	0.29	**35.79**	4.04	**637.25**	nr	**685.04**
90 x 450 x 6000mm	312.25	2.50	**320.05**	2.10	0.26	**32.63**	3.70	**356.37**	nr	**383.10**
90 x 450 x 8000mm	415.59	2.50	**425.98**	2.20	0.28	**34.26**	3.86	**464.10**	nr	**498.91**
90 x 450 x 10000mm	518.17	2.50	**531.12**	2.30	0.29	**35.79**	4.04	**570.95**	nr	**613.78**
90 x 450 x 12000mm	623.00	2.50	**638.58**	2.40	0.30	**37.32**	4.22	**680.12**	nr	**731.13**
115 x 250 x 4000mm	146.83	2.50	**150.50**	1.30	0.16	**20.19**	2.46	**173.15**	nr	**186.13**
115 x 250 x 6000mm	217.16	2.50	**222.59**	1.50	0.19	**23.35**	2.66	**248.60**	nr	**267.24**
115 x 275 x 4000mm	159.47	2.50	**163.45**	1.40	0.18	**21.82**	2.66	**187.93**	nr	**202.03**
115 x 275 x 6000mm	243.85	2.50	**249.94**	1.60	0.20	**24.88**	2.82	**277.64**	nr	**298.47**
115 x 300 x 4000mm	176.57	2.50	**180.98**	1.60	0.20	**24.88**	2.82	**208.68**	nr	**224.33**
115 x 300 x 6000mm	265.03	2.50	**271.66**	1.70	0.21	**26.41**	3.00	**301.07**	nr	**323.65**
115 x 325 x 4000mm	191.44	2.50	**196.23**	1.70	0.21	**26.41**	3.00	**225.64**	nr	**242.56**
115 x 325 x 6000mm	287.71	2.50	**294.91**	1.80	0.22	**27.93**	3.17	**326.01**	nr	**350.46**
115 x 325 x 8000mm	384.33	2.50	**393.94**	1.90	0.24	**29.57**	3.37	**426.88**	nr	**458.89**
115 x 350 x 4000mm	209.27	2.50	**214.50**	1.80	0.22	**27.93**	3.17	**245.60**	nr	**264.02**
115 x 350 x 6000mm	312.25	2.50	**320.05**	1.90	0.24	**29.57**	3.37	**352.99**	nr	**379.46**
115 x 350 x 8000mm	415.59	2.50	**425.98**	2.00	0.25	**31.10**	3.51	**460.59**	nr	**495.13**
115 x 375 x 4000mm	224.14	2.50	**229.75**	1.90	0.24	**29.57**	3.37	**262.68**	nr	**282.38**
115 x 375 x 6000mm	333.81	2.50	**342.15**	2.00	0.25	**31.10**	3.51	**376.76**	nr	**405.02**
115 x 375 x 8000mm	444.92	2.50	**456.04**	2.10	0.26	**32.63**	3.70	**492.36**	nr	**529.29**
115 x 400 x 4000mm	237.16	2.50	**243.09**	2.00	0.25	**31.10**	3.51	**277.70**	nr	**298.53**
115 x 400 x 6000mm	356.85	2.50	**365.77**	2.10	0.26	**32.63**	3.70	**402.09**	nr	**432.25**
115 x 400 x 8000mm	475.80	2.50	**487.69**	2.20	0.28	**34.26**	3.86	**525.81**	nr	**565.25**
115 x 400 x 10000mm	595.49	2.50	**610.38**	2.30	0.29	**35.79**	4.04	**650.21**	nr	**698.98**
115 x 425 x 6000mm	369.49	2.50	**378.73**	2.20	0.28	**34.26**	3.86	**416.85**	nr	**448.12**
115 x 425 x 8000mm	494.00	2.50	**506.35**	2.30	0.29	**35.79**	4.04	**546.18**	nr	**587.14**
115 x 425 x 10000mm	614.82	2.50	**630.19**	2.40	0.30	**37.32**	4.22	**671.73**	nr	**722.12**
115 x 425 x 12000mm	736.00	2.50	**754.40**	2.50	0.31	**38.85**	4.41	**797.65**	nr	**857.48**
115 x 450 x 6000mm	391.04	2.50	**400.82**	2.30	0.29	**35.79**	4.04	**440.65**	nr	**473.70**
115 x 450 x 8000mm	521.89	2.50	**534.94**	2.40	0.30	**37.32**	4.22	**576.48**	nr	**619.71**
115 x 450 x 10000mm	654.22	2.50	**670.58**	2.50	0.31	**38.85**	4.41	**713.83**	nr	**767.36**
115 x 450 x 12000mm	785.07	2.50	**804.70**	2.60	0.33	**40.48**	4.57	**849.75**	nr	**913.48**
115 x 475 x 6000mm	415.59	2.50	**425.98**	2.40	0.30	**37.32**	4.22	**467.52**	nr	**502.58**
115 x 475 x 8000mm	554.59	2.50	**568.46**	2.50	0.31	**38.85**	4.41	**611.71**	nr	**657.58**
115 x 475 x 10000mm	693.76	2.50	**711.10**	2.60	0.33	**40.48**	4.57	**756.16**	nr	**812.87**
115 x 475 x 12000mm	828.93	2.50	**849.66**	2.70	0.34	**42.01**	4.77	**896.43**	nr	**963.67**
115 x 475 x 14000mm	966.47	2.50	**990.63**	2.80	0.35	**43.54**	4.93	**1039.10**	nr	**1117.03**

Labour hourly rates: (except Specialists) Craft Operatives £14.21 Labourer £10.71 Rates are national average prices. Refer to REGIONAL VARIATIONS for indicative levels of overall pricing in regions	MATERIALS			LABOUR				RATES		
	Del to Site £	Waste %	Material Cost £	Craft Optve Hrs	Lab Hrs	Labour Cost £	Sunds £	Nett Rate £	Unit	Gross Rate (+7.5%) £
G20: CARPENTRY/TIMBER FRAMING/FIRST FIXING Cont.										
Softwood, wrot, BS 4978, GS grade - laminated beams Cont.										
Glued laminated beams; BS EN 386 Cont.										
115 x 500 x 6000mm	438.63	2.50	**449.59**	2.50	0.31	**38.85**	4.41	**492.84**	nr	**529.81**
115 x 500 x 8000mm	582.11	2.50	**596.66**	2.60	0.33	**40.48**	4.57	**641.71**	nr	**689.84**
115 x 500 x 10000mm	731.17	2.50	**749.45**	2.70	0.34	**42.01**	4.77	**796.23**	nr	**855.94**
115 x 500 x 12000mm	876.47	2.50	**898.38**	2.80	0.35	**43.54**	4.93	**946.85**	nr	**1017.86**
115 x 500 x 14000mm	1020.74	2.50	**1046.26**	2.90	0.36	**45.06**	5.12	**1096.44**	nr	**1178.67**
140 x 350 x 4000mm	251.65	2.50	**257.94**	2.00	0.25	**31.10**	3.51	**292.56**	nr	**314.50**
140 x 350 x 6000mm	381.01	2.50	**390.53**	2.10	0.26	**32.63**	3.70	**426.86**	nr	**458.87**
140 x 350 x 8000mm	507.02	2.50	**519.69**	2.20	0.28	**34.26**	3.86	**557.81**	nr	**599.65**
140 x 375 x 4000mm	268.00	2.50	**274.70**	2.10	0.26	**32.63**	3.70	**311.02**	nr	**334.35**
140 x 375 x 6000mm	420.42	2.50	**430.93**	2.20	0.28	**34.26**	3.86	**469.05**	nr	**504.23**
140 x 375 x 8000mm	538.25	2.50	**551.70**	2.30	0.29	**35.79**	4.04	**591.53**	nr	**635.90**
140 x 375 x 10000mm	672.07	2.50	**688.87**	2.40	0.30	**37.32**	4.22	**730.41**	nr	**785.19**
140 x 400 x 6000mm	425.24	2.50	**435.87**	2.30	0.29	**35.79**	4.04	**475.70**	nr	**511.38**
140 x 400 x 8000mm	570.96	2.50	**585.23**	2.40	0.30	**37.32**	4.22	**626.77**	nr	**673.78**
140 x 400 x 10000mm	711.47	2.50	**729.25**	2.50	0.31	**38.85**	4.41	**772.50**	nr	**830.44**
140 x 425 x 6000mm	459.44	2.50	**470.93**	2.40	0.30	**37.32**	4.22	**512.47**	nr	**550.90**
140 x 425 x 8000mm	614.82	2.50	**630.19**	2.50	0.31	**38.85**	4.41	**673.44**	nr	**723.95**
140 x 425 x 10000mm	768.70	2.50	**787.92**	2.60	0.33	**40.48**	4.57	**832.97**	nr	**895.44**
140 x 425 x 12000mm	922.60	2.50	**945.67**	2.70	0.34	**42.01**	4.77	**992.44**	nr	**1066.88**
140 x 450 x 6000mm	483.97	2.50	**496.07**	2.50	0.31	**38.85**	4.41	**539.32**	nr	**579.77**
140 x 450 x 8000mm	646.05	2.50	**662.20**	2.60	0.33	**40.48**	4.57	**707.26**	nr	**760.30**
140 x 450 x 10000mm	805.88	2.50	**826.03**	2.70	0.34	**42.01**	4.77	**872.81**	nr	**938.27**
140 x 450 x 12000mm	966.47	2.50	**990.63**	2.80	0.35	**43.54**	4.93	**1039.10**	nr	**1117.03**
140 x 475 x 6000mm	507.02	2.50	**519.69**	2.60	0.33	**40.48**	4.57	**564.74**	nr	**607.10**
140 x 475 x 8000mm	675.41	2.50	**692.29**	2.70	0.34	**42.01**	4.77	**739.07**	nr	**794.50**
140 x 475 x 10000mm	845.66	2.50	**866.81**	2.80	0.35	**43.54**	4.93	**915.28**	nr	**983.92**
140 x 475 x 12000mm	1015.54	2.50	**1040.92**	2.90	0.36	**45.06**	5.12	**1091.10**	nr	**1172.94**
140 x 475 x 14000mm	1179.09	2.50	**1208.56**	3.00	0.38	**46.70**	5.28	**1260.54**	nr	**1355.08**
165 x 425 x 6000mm	541.22	2.50	**554.75**	2.60	0.33	**40.48**	4.57	**599.80**	nr	**644.79**
165 x 425 x 8000mm	722.61	2.50	**740.67**	2.70	0.34	**42.01**	4.77	**787.45**	nr	**846.51**
165 x 425 x 10000mm	903.10	2.50	**925.67**	2.80	0.35	**43.54**	4.93	**974.14**	nr	**1047.20**
165 x 425 x 12000mm	1081.87	2.50	**1108.92**	2.90	0.36	**45.06**	5.12	**1159.10**	nr	**1246.03**
165 x 450 x 6000mm	564.27	2.50	**578.38**	2.70	0.34	**42.01**	4.77	**625.16**	nr	**672.04**
165 x 450 x 8000mm	753.85	2.50	**772.70**	2.80	0.35	**43.54**	4.93	**821.17**	nr	**882.75**
165 x 450 x 10000mm	941.93	2.50	**965.48**	2.90	0.36	**45.06**	5.12	**1015.66**	nr	**1091.84**
165 x 450 x 12000mm	1131.51	2.50	**1159.80**	3.00	0.38	**46.70**	5.28	**1211.78**	nr	**1302.66**
165 x 475 x 6000mm	598.48	2.50	**613.44**	2.80	0.35	**43.54**	4.93	**661.91**	nr	**711.55**
165 x 475 x 8000mm	799.93	2.50	**819.93**	2.90	0.36	**45.06**	5.12	**870.11**	nr	**935.37**
165 x 475 x 10000mm	997.69	2.50	**1022.63**	3.00	0.38	**46.70**	5.28	**1074.61**	nr	**1155.20**
165 x 475 x 12000mm	1200.27	2.50	**1230.28**	3.10	0.39	**48.23**	5.46	**1283.97**	nr	**1380.27**
165 x 475 x 14000mm	1398.40	2.50	**1433.36**	3.20	0.40	**49.76**	5.64	**1488.76**	nr	**1600.42**
190 x 475 x 6000mm	693.26	2.50	**710.59**	3.00	0.38	**46.70**	5.28	**762.57**	nr	**819.76**
190 x 475 x 8000mm	922.22	2.50	**945.27**	3.10	0.39	**48.23**	5.46	**998.96**	nr	**1073.89**
190 x 475 x 10000mm	1153.07	2.50	**1181.90**	3.20	0.40	**49.76**	5.64	**1237.30**	nr	**1330.09**
190 x 475 x 12000mm	1383.54	2.50	**1418.12**	3.30	0.41	**51.28**	5.81	**1475.22**	nr	**1585.86**
190 x 475 x 14000mm	1614.00	2.50	**1654.35**	3.40	0.42	**52.81**	5.97	**1713.14**	nr	**1841.62**
Softwood, sawn, BS 4978, GS grade - carcassing										
Floor members										
38 x 75mm	0.57	10.00	**0.63**	0.12	0.01	**1.81**	0.01	**2.45**	m	**2.63**
38 x 100mm	0.76	10.00	**0.84**	0.14	0.02	**2.20**	0.01	**3.05**	m	**3.27**
50 x 75mm	0.73	10.00	**0.80**	0.14	0.02	**2.20**	0.01	**3.01**	m	**3.24**
50 x 100mm	0.98	10.00	**1.07**	0.16	0.02	**2.49**	0.01	**3.57**	m	**3.84**
50 x 150mm	1.46	10.00	**1.61**	0.20	0.03	**3.11**	0.01	**4.73**	m	**5.09**
50 x 175mm	1.71	10.00	**1.88**	0.24	0.03	**3.73**	0.01	**5.62**	m	**6.04**
50 x 200mm	1.95	10.00	**2.15**	0.27	0.04	**4.27**	0.01	**6.42**	m	**6.91**
50 x 225mm	2.19	10.00	**2.41**	0.29	0.04	**4.55**	0.02	**6.98**	m	**7.50**
50 x 250mm	2.44	10.00	**2.68**	0.32	0.04	**4.98**	0.02	**7.67**	m	**8.25**
75 x 150mm	2.25	10.00	**2.48**	0.29	0.04	**4.55**	0.02	**7.04**	m	**7.57**
75 x 175mm	2.63	10.00	**2.89**	0.33	0.04	**5.12**	0.02	**8.02**	m	**8.62**
75 x 200mm	3.00	10.00	**3.30**	0.35	0.05	**5.51**	0.02	**8.83**	m	**9.49**
75 x 225mm	3.38	10.00	**3.71**	0.40	0.05	**6.22**	0.02	**9.95**	m	**10.70**
75 x 250mm	3.75	10.00	**4.13**	0.45	0.06	**7.04**	0.02	**11.19**	m	**12.02**
Wall or partition members										
38 x 75mm	0.57	10.00	**0.63**	0.18	0.02	**2.77**	0.01	**3.41**	m	**3.66**
38 x 100mm	0.76	10.00	**0.84**	0.21	0.03	**3.31**	0.01	**4.15**	m	**4.46**
50 x 75mm	0.73	10.00	**0.80**	0.21	0.03	**3.31**	0.01	**4.12**	m	**4.43**
50 x 100mm	0.98	10.00	**1.07**	0.24	0.03	**3.73**	0.01	**4.82**	m	**5.18**
75 x 100mm	1.50	10.00	**1.65**	0.30	0.04	**4.69**	0.02	**6.36**	m	**6.83**
38 x 75mm; fixing to masonry	0.57	10.00	**0.63**	0.36	0.04	**5.54**	0.02	**6.19**	m	**6.66**
38 x 100mm; fixing to masonry	0.76	10.00	**0.84**	0.39	0.05	**6.08**	0.02	**6.93**	m	**7.45**
50 x 75mm; fixing to masonry	0.73	10.00	**0.80**	0.39	0.05	**6.08**	0.02	**6.90**	m	**7.42**
50 x 100mm; fixing to masonry	0.98	10.00	**1.07**	0.43	0.05	**6.65**	0.02	**7.74**	m	**8.32**
75 x 100mm; fixing to masonry	1.50	10.00	**1.65**	0.50	0.06	**7.75**	0.03	**9.42**	m	**10.13**
Plates										
38 x 75mm	0.57	10.00	**0.63**	0.18	0.02	**2.77**	0.01	**3.41**	m	**3.66**
38 x 100mm	0.76	10.00	**0.84**	0.21	0.03	**3.31**	0.01	**4.15**	m	**4.46**
50 x 75mm	0.73	10.00	**0.80**	0.21	0.03	**3.31**	0.01	**4.12**	m	**4.43**
50 x 100mm	0.98	10.00	**1.07**	0.24	0.03	**3.73**	0.01	**4.82**	m	**5.18**
75 x 150mm	1.50	10.00	**1.65**	0.40	0.05	**6.22**	0.02	**7.89**	m	**8.48**
38 x 75mm; fixing by bolting	0.57	10.00	**0.63**	0.22	0.03	**3.45**	-	**4.07**	m	**4.38**
38 x 100mm; fixing by bolting	0.76	10.00	**0.84**	0.25	0.03	**3.87**	-	**4.71**	m	**5.06**
50 x 75mm; fixing by bolting	0.73	10.00	**0.80**	0.25	0.03	**3.87**	-	**4.68**	m	**5.03**
50 x 100mm; fixing by bolting	0.98	10.00	**1.07**	0.29	0.04	**4.55**	-	**5.62**	m	**6.04**
75 x 150mm; fixing by bolting	1.50	10.00	**1.65**	0.48	0.06	**7.46**	-	**9.11**	m	**9.80**

Labour hourly rates: (except Specialists) Craft Operatives £14.21 Labourer £10.71 Rates are national average prices. Refer to REGIONAL VARIATIONS for indicative levels of overall pricing in regions	MATERIALS			LABOUR				RATES		
	Del to Site £	Waste %	Material Cost £	Craft Optve Hrs	Lab Hrs	Labour Cost £	Sunds £	Nett Rate £	Unit	Gross Rate (+7.5%) £
G20: CARPENTRY/TIMBER FRAMING/FIRST FIXING Cont.										
Softwood, sawn, BS 4978, GS grade - carcassing Cont.										
Roof members; flat										
50 x 175mm	1.71	10.00	**1.88**	0.24	0.03	**3.73**	0.01	**5.62**	m	6.04
50 x 200mm	1.95	10.00	**2.15**	0.27	0.04	**4.27**	0.01	**6.42**	m	6.91
50 x 225mm	2.19	10.00	**2.41**	0.29	0.04	**4.55**	0.02	**6.98**	m	7.50
50 x 250mm	2.44	10.00	**2.68**	0.32	0.04	**4.98**	0.02	**7.67**	m	8.25
75 x 175mm	2.63	10.00	**2.89**	0.33	0.04	**5.12**	0.02	**8.02**	m	8.62
75 x 200mm	3.00	10.00	**3.30**	0.35	0.05	**5.51**	0.02	**8.83**	m	9.49
75 x 225mm	3.38	10.00	**3.71**	0.40	0.05	**6.22**	0.02	**9.95**	m	10.70
75 x 250mm	3.75	10.00	**4.13**	0.45	0.06	**7.04**	0.02	**11.19**	m	12.02
Roof members; pitched										
25 x 75mm	0.38	10.00	**0.41**	0.14	0.02	**2.20**	0.01	**2.62**	m	2.82
25 x 125mm	0.63	10.00	**0.69**	0.16	0.02	**2.49**	0.01	**3.19**	m	3.42
32 x 175mm	1.12	10.00	**1.23**	0.22	0.03	**3.45**	0.01	**4.69**	m	5.04
32 x 225mm	1.44	10.00	**1.58**	0.25	0.03	**3.87**	0.01	**5.47**	m	5.88
38 x 75mm	0.57	10.00	**0.63**	0.15	0.02	**2.35**	0.01	**2.98**	m	3.20
38 x 100mm	0.76	10.00	**0.84**	0.18	0.02	**2.77**	0.01	**3.62**	m	3.89
50 x 75mm	0.73	10.00	**0.80**	0.18	0.02	**2.77**	0.01	**3.59**	m	3.85
50 x 100mm	0.98	10.00	**1.07**	0.20	0.03	**3.16**	0.01	**4.25**	m	4.56
50 x 125mm	1.22	10.00	**1.34**	0.23	0.03	**3.59**	0.01	**4.94**	m	5.31
50 x 150mm	1.46	10.00	**1.61**	0.26	0.03	**4.02**	0.01	**5.64**	m	6.06
50 x 225mm	2.19	10.00	**2.41**	0.36	0.05	**5.65**	0.02	**8.08**	m	8.69
50 x 250mm	2.44	10.00	**2.68**	0.40	0.05	**6.22**	0.02	**8.92**	m	9.59
75 x 100mm	1.50	10.00	**1.65**	0.26	0.03	**4.02**	0.01	**5.68**	m	6.11
75 x 150mm	2.25	10.00	**2.48**	0.35	0.05	**5.51**	0.02	**8.00**	m	8.60
100 x 150mm	3.00	10.00	**3.30**	0.45	0.06	**7.04**	0.02	**10.36**	m	11.14
100 x 225mm	4.50	10.00	**4.95**	0.60	0.08	**9.38**	0.03	**14.37**	m	15.44
Joist strutting; herringbone; depth of joist 175mm 50 x 50mm	0.88	10.00	**0.97**	0.65	0.08	**10.09**	0.04	**11.10**	m	11.93
Joist strutting; herringbone; depth of joist 200mm 50 x 50mm	1.02	10.00	**1.13**	0.65	0.08	**10.09**	0.04	**11.26**	m	12.10
Joist strutting; herringbone; depth of joist 225mm 50 x 50mm	1.17	10.00	**1.29**	0.65	0.08	**10.09**	0.04	**11.42**	m	12.27
Joist strutting; herringbone; depth of joist 250mm 50 x 50mm	1.32	10.00	**1.45**	0.65	0.08	**10.09**	0.04	**11.58**	m	12.45
Joist strutting; block; depth of joist 100mm 50 x 100mm	0.98	10.00	**1.07**	0.65	0.08	**10.09**	0.04	**11.20**	m	12.04
Joist strutting; block; depth of joist 150mm 50 x 150mm	1.46	10.00	**1.61**	0.65	0.08	**10.09**	0.04	**11.74**	m	12.62
Joist strutting; block; depth of joist 175mm 50 x 175mm	1.71	10.00	**1.88**	0.70	0.09	**10.91**	0.04	**12.82**	m	13.79
Joist strutting; block; depth of joist 200mm 50 x 200mm	1.95	10.00	**2.15**	0.70	0.09	**10.91**	0.04	**13.09**	m	14.07
Joist strutting; block; depth of joist 225mm 50 x 225mm	2.19	10.00	**2.41**	0.75	0.09	**11.62**	0.04	**14.07**	m	15.13
Joist strutting; block; depth of joist 250mm 50 x 250mm	2.44	10.00	**2.68**	0.75	0.09	**11.62**	0.04	**14.34**	m	15.42
Noggings to joists										
50 x 50mm	0.49	10.00	**0.54**	0.28	0.04	**4.41**	0.02	**4.96**	m	5.33
50 x 75mm	0.73	10.00	**0.80**	0.33	0.04	**5.12**	0.02	**5.94**	m	6.38
Wrot surfaces										
plain; 50mm wide	-	-	-	0.15	-	**2.13**	-	**2.13**	m	2.29
plain; 75mm wide	-	-	-	0.20	-	**2.84**	-	**2.84**	m	3.06
plain; 100mm wide	-	-	-	0.25	-	**3.55**	-	**3.55**	m	3.82
plain; 150mm wide	-	-	-	0.30	-	**4.26**	-	**4.26**	m	4.58
plain; 200mm wide	-	-	-	0.35	-	**4.97**	-	**4.97**	m	5.35
Softwood, sawn, BS 4978, GS grade, impregnated - carcassing										
Floor members										
38 x 75mm	0.64	10.00	**0.71**	0.12	0.01	**1.81**	0.01	**2.52**	m	2.71
38 x 100mm	0.86	10.00	**0.94**	0.14	0.02	**2.20**	0.01	**3.15**	m	3.39
50 x 75mm	0.83	10.00	**0.91**	0.14	0.02	**2.20**	0.01	**3.12**	m	3.35
50 x 100mm	1.10	10.00	**1.21**	0.16	0.02	**2.49**	0.01	**3.71**	m	3.99
50 x 150mm	1.65	10.00	**1.82**	0.20	0.03	**3.11**	0.01	**4.94**	m	5.31
50 x 175mm	1.93	10.00	**2.12**	0.24	0.03	**3.73**	0.01	**5.86**	m	6.30
50 x 200mm	2.20	10.00	**2.42**	0.27	0.04	**4.27**	0.01	**6.70**	m	7.20
50 x 225mm	2.48	10.00	**2.72**	0.29	0.04	**4.55**	0.02	**7.29**	m	7.83
50 x 250mm	2.75	10.00	**3.03**	0.32	0.04	**4.98**	0.02	**8.02**	m	8.62
75 x 150mm	2.53	10.00	**2.78**	0.29	0.04	**4.55**	0.02	**7.35**	m	7.90
75 x 175mm	2.95	10.00	**3.25**	0.33	0.04	**5.12**	0.02	**8.38**	m	9.01
75 x 200mm	3.38	10.00	**3.71**	0.35	0.05	**5.51**	0.02	**9.24**	m	9.93
75 x 225mm	3.80	10.00	**4.18**	0.40	0.05	**6.22**	0.02	**10.42**	m	11.20
75 x 250mm	4.22	10.00	**4.64**	0.45	0.06	**7.04**	0.02	**11.70**	m	12.58
Wall or partition members										
38 x 75mm	0.64	10.00	**0.71**	0.18	0.02	**2.77**	0.01	**3.49**	m	3.75
38 x 100mm	0.86	10.00	**0.94**	0.21	0.03	**3.31**	0.01	**4.26**	m	4.57
50 x 75mm	0.83	10.00	**0.91**	0.21	0.03	**3.31**	0.01	**4.22**	m	4.54

STRUCTURAL/CARCASSING METAL/TIMBER

Labour hourly rates: (except Specialists) Craft Operatives £14.21 Labourer £10.71 Rates are national average prices. Refer to REGIONAL VARIATIONS for indicative levels of overall pricing in regions	MATERIALS			LABOUR				RATES		
	Del to Site £	Waste %	Material Cost £	Craft Optve Hrs	Lab Hrs	Labour Cost £	Sunds £	Nett Rate £	Unit	Gross Rate (+7.5%) £
G20: CARPENTRY/TIMBER FRAMING/FIRST FIXING Cont.										
Softwood, sawn, BS 4978, GS grade, impregnated - carcassing Cont.										
Wall or partition members Cont.										
50 x 100mm	1.10	10.00	**1.21**	0.24	0.03	**3.73**	0.01	**4.95**	m	**5.33**
75 x 100mm	1.69	10.00	**1.86**	0.30	0.04	**4.69**	0.02	**6.56**	m	**7.06**
38 x 75mm; fixing to masonry	0.64	10.00	**0.71**	0.36	0.04	**5.54**	0.02	**6.27**	m	**6.74**
38 x 100mm; fixing to masonry	0.86	10.00	**0.94**	0.39	0.05	**6.08**	0.02	**7.04**	m	**7.57**
50 x 75mm; fixing to masonry	0.83	10.00	**0.91**	0.39	0.05	**6.08**	0.02	**7.00**	m	**7.53**
50 x 100mm; fixing to masonry	1.10	10.00	**1.21**	0.43	0.05	**6.65**	0.02	**7.88**	m	**8.47**
75 x 100mm; fixing to masonry	1.69	10.00	**1.86**	0.50	0.06	**7.75**	0.03	**9.63**	m	**10.35**
Plates										
38 x 75mm	0.64	10.00	**0.71**	0.18	0.02	**2.77**	0.01	**3.49**	m	**3.75**
38 x 100mm	0.86	10.00	**0.94**	0.21	0.03	**3.31**	0.01	**4.26**	m	**4.57**
50 x 75mm	0.83	10.00	**0.91**	0.21	0.03	**3.31**	0.01	**4.22**	m	**4.54**
50 x 100mm	1.10	10.00	**1.21**	0.24	0.03	**3.73**	0.01	**4.95**	m	**5.33**
75 x 150mm	1.69	10.00	**1.86**	0.40	0.05	**6.22**	0.02	**8.10**	m	**8.70**
38 x 75mm; fixing by bolting	0.64	10.00	**0.71**	0.22	0.03	**3.45**	-	**4.15**	m	**4.46**
38 x 100mm; fixing by bolting	0.86	10.00	**0.94**	0.25	0.03	**3.87**	-	**4.81**	m	**5.18**
50 x 75mm; fixing by bolting	0.83	10.00	**0.91**	0.25	0.03	**3.87**	-	**4.78**	m	**5.14**
50 x 100mm; fixing by bolting	1.10	10.00	**1.21**	0.29	0.04	**4.55**	-	**5.76**	m	**6.19**
75 x 150mm; fixing by bolting	1.69	10.00	**1.86**	0.48	0.06	**7.46**	-	**9.32**	m	**10.02**
Roof members; flat										
50 x 100mm	1.10	10.00	**1.21**	0.20	0.03	**3.11**	0.01	**4.33**	m	**4.66**
50 x 175mm	1.93	10.00	**2.12**	0.24	0.03	**3.73**	0.01	**5.86**	m	**6.30**
50 x 200mm	2.20	10.00	**2.42**	0.27	0.04	**4.27**	0.01	**6.70**	m	**7.20**
50 x 225mm	2.48	10.00	**2.72**	0.29	0.04	**4.55**	0.02	**7.29**	m	**7.83**
50 x 250mm	2.75	10.00	**3.03**	0.32	0.04	**4.98**	0.02	**8.02**	m	**8.62**
75 x 175mm	2.95	10.00	**3.25**	0.33	0.04	**5.12**	0.02	**8.38**	m	**9.01**
75 x 200mm	3.38	10.00	**3.71**	0.35	0.05	**5.51**	0.02	**9.24**	m	**9.93**
75 x 225mm	3.80	10.00	**4.18**	0.40	0.05	**6.22**	0.02	**10.42**	m	**11.20**
75 x 250mm	4.22	10.00	**4.64**	0.45	0.06	**7.04**	0.02	**11.70**	m	**12.58**
Roof members; pitched										
25 x 75mm	0.42	10.00	**0.46**	0.14	0.02	**2.20**	0.01	**2.67**	m	**2.87**
25 x 125mm	0.70	10.00	**0.77**	0.16	0.02	**2.49**	0.01	**3.27**	m	**3.52**
32 x 175mm	1.26	10.00	**1.39**	0.22	0.03	**3.45**	0.01	**4.85**	m	**5.21**
32 x 225mm	1.62	10.00	**1.78**	0.25	0.03	**3.87**	0.01	**5.67**	m	**6.09**
38 x 75mm	0.64	10.00	**0.71**	0.15	0.02	**2.35**	0.01	**3.06**	m	**3.29**
38 x 100mm	0.86	10.00	**0.94**	0.18	0.02	**2.77**	0.01	**3.72**	m	**4.00**
50 x 75mm	0.83	10.00	**0.91**	0.18	0.02	**2.77**	0.01	**3.69**	m	**3.97**
50 x 100mm	1.10	10.00	**1.21**	0.20	0.03	**3.16**	0.01	**4.38**	m	**4.71**
50 x 125mm	1.38	10.00	**1.51**	0.23	0.03	**3.59**	0.01	**5.12**	m	**5.50**
50 x 150mm	1.65	10.00	**1.82**	0.26	0.03	**4.02**	0.01	**5.84**	m	**6.28**
50 x 225mm	2.48	10.00	**2.72**	0.36	0.05	**5.65**	0.02	**8.39**	m	**9.02**
50 x 250mm	2.75	10.00	**3.03**	0.40	0.05	**6.22**	0.02	**9.26**	m	**9.96**
75 x 100mm	1.69	10.00	**1.86**	0.26	0.03	**4.02**	0.01	**5.89**	m	**6.33**
75 x 150mm	2.53	10.00	**2.78**	0.35	0.05	**5.51**	0.02	**8.31**	m	**8.94**
100 x 150mm	3.38	10.00	**3.71**	0.45	0.06	**7.04**	0.02	**10.77**	m	**11.58**
100 x 225mm	5.06	10.00	**5.57**	0.60	0.08	**9.38**	0.03	**14.99**	m	**16.11**
Joist strutting; herringbone; depth of joist 175mm 50 x 50mm	0.99	10.00	**1.09**	0.65	0.08	**10.09**	0.04	**11.22**	m	**12.06**
Joist strutting; herringbone; depth of joist 200mm 50 x 50mm	1.16	10.00	**1.27**	0.65	0.08	**10.09**	0.04	**11.40**	m	**12.26**
Joist strutting; herringbone; depth of joist 225mm 50 x 50mm	1.32	10.00	**1.45**	0.65	0.08	**10.09**	0.04	**11.58**	m	**12.45**
Joist strutting; herringbone; depth of joist 250mm 50 x 50mm	1.49	10.00	**1.63**	0.65	0.08	**10.09**	0.04	**11.76**	m	**12.65**
Joist strutting; block; depth of joist 150mm 50 x 150mm	1.65	10.00	**1.82**	0.65	0.08	**10.09**	0.04	**11.94**	m	**12.84**
Joist strutting; block; depth of joist 175mm 50 x 175mm	1.93	10.00	**2.12**	0.70	0.09	**10.91**	0.04	**13.06**	m	**14.04**
Joist strutting; block; depth of joist 200mm 50 x 200mm	2.20	10.00	**2.42**	0.70	0.09	**10.91**	0.04	**13.37**	m	**14.37**
Joist strutting; block; depth of joist 225mm 50 x 225mm	2.48	10.00	**2.72**	0.75	0.09	**11.62**	0.04	**14.38**	m	**15.46**
Joist strutting; block; depth of joist 250mm 50 x 250mm	2.75	10.00	**3.03**	0.75	0.09	**11.62**	0.04	**14.69**	m	**15.79**
Noggings to joists										
50 x 50mm	0.55	10.00	**0.61**	0.28	0.04	**4.41**	0.02	**5.03**	m	**5.41**
50 x 75mm	0.83	10.00	**0.91**	0.33	0.04	**5.12**	0.02	**6.04**	m	**6.49**
Wrot surfaces										
plain; 50mm wide	-	-	**-**	0.15	-	**2.13**	-	**2.13**	m	**2.29**
plain; 75mm wide	-	-	**-**	0.20	-	**2.84**	-	**2.84**	m	**3.06**
plain; 100mm wide	-	-	**-**	0.25	-	**3.55**	-	**3.55**	m	**3.82**
plain; 150mm wide	-	-	**-**	0.30	-	**4.26**	-	**4.26**	m	**4.58**
plain; 200mm wide	-	-	**-**	0.35	-	**4.97**	-	**4.97**	m	**5.35**

Labour hourly rates: (except Specialists) Craft Operatives £14.21 Labourer £10.71 Rates are national average prices. Refer to REGIONAL VARIATIONS for indicative levels of overall pricing in regions	MATERIALS			LABOUR				RATES		
	Del to Site £	Waste %	Material Cost £	Craft Optve Hrs	Lab Hrs	Labour Cost £	Sunds £	Nett Rate £	Unit	Gross Rate (+7.5%) £
G20: CARPENTRY/TIMBER FRAMING/FIRST FIXING Cont.										
Oak, sawn - carcassing										
Floor members										
38 x 75mm	4.27	10.00	**4.70**	0.21	0.03	**3.31**	0.01	**8.01**	m	**8.61**
38 x 100mm	5.70	10.00	**6.27**	0.24	0.03	**3.73**	0.01	**10.01**	m	**10.77**
50 x 75mm	5.62	10.00	**6.18**	0.24	0.03	**3.73**	0.01	**9.93**	m	**10.67**
50 x 100mm	7.50	10.00	**8.25**	0.28	0.04	**4.41**	0.02	**12.67**	m	**13.62**
50 x 175mm	13.12	10.00	**14.43**	0.42	0.05	**6.50**	0.02	**20.96**	m	**22.53**
50 x 200mm	15.00	10.00	**16.50**	0.47	0.06	**7.32**	0.03	**23.85**	m	**25.64**
50 x 225mm	16.87	10.00	**18.56**	0.51	0.06	**7.89**	0.03	**26.47**	m	**28.46**
50 x 250mm	18.75	10.00	**20.62**	0.56	0.07	**8.71**	0.03	**29.36**	m	**31.56**
75 x 150mm	16.87	10.00	**18.56**	0.51	0.06	**7.89**	0.03	**26.47**	m	**28.46**
75 x 175mm	19.70	10.00	**21.67**	0.58	0.07	**8.99**	0.03	**30.69**	m	**32.99**
75 x 200mm	22.50	10.00	**24.75**	0.61	0.08	**9.52**	0.03	**34.31**	m	**36.88**
75 x 225mm	25.33	10.00	**27.86**	0.70	0.09	**10.91**	0.04	**38.81**	m	**41.72**
75 x 250mm	28.12	10.00	**30.93**	0.79	0.10	**12.30**	0.04	**43.27**	m	**46.52**
Wall or partition members										
38 x 75mm	4.27	10.00	**4.70**	0.32	0.04	**4.98**	0.02	**9.69**	m	**10.41**
38 x 100mm	5.70	10.00	**6.27**	0.37	0.05	**5.79**	0.02	**12.08**	m	**12.99**
50 x 75mm	5.62	10.00	**6.18**	0.37	0.05	**5.79**	0.02	**11.99**	m	**12.89**
50 x 100mm	7.50	10.00	**8.25**	0.42	0.05	**6.50**	0.02	**14.78**	m	**15.89**
75 x 100mm	11.25	10.00	**12.37**	0.52	0.06	**8.03**	0.03	**20.43**	m	**21.97**
38 x 75mm; fixing to masonry	4.27	10.00	**4.70**	0.53	0.07	**8.28**	0.03	**13.01**	m	**13.98**
38 x 100mm; fixing to masonry	5.70	10.00	**6.27**	0.59	0.07	**9.13**	0.03	**15.44**	m	**16.59**
50 x 75mm; fixing to masonry	5.62	10.00	**6.18**	0.59	0.07	**9.13**	0.03	**15.35**	m	**16.50**
50 x 100mm; fixing to masonry	7.50	10.00	**8.25**	0.65	0.08	**10.09**	0.04	**18.38**	m	**19.76**
75 x 100mm; fixing to masonry	11.25	10.00	**12.37**	0.78	0.10	**12.15**	0.04	**24.57**	m	**26.42**
Plates										
38 x 75mm	4.27	10.00	**4.70**	0.32	0.04	**4.98**	0.02	**9.69**	m	**10.41**
38 x 100mm	5.70	10.00	**6.27**	0.37	0.05	**5.79**	0.02	**12.08**	m	**12.99**
50 x 75mm	5.62	10.00	**6.18**	0.37	0.05	**5.79**	0.02	**11.99**	m	**12.89**
50 x 100mm	7.50	10.00	**8.25**	0.42	0.05	**6.50**	0.02	**14.78**	m	**15.89**
75 x 150mm	11.25	10.00	**12.37**	0.52	0.06	**8.03**	0.03	**20.43**	m	**21.97**
38 x 75mm; fixing by bolting	4.27	10.00	**4.70**	0.38	0.05	**5.94**	-	**10.63**	m	**11.43**
38 x 100mm; fixing by bolting	5.70	10.00	**6.27**	0.44	0.05	**6.79**	-	**13.06**	m	**14.04**
50 x 75mm; fixing by bolting	5.62	10.00	**6.18**	0.44	0.06	**6.90**	-	**13.08**	m	**14.06**
50 x 100mm; fixing by bolting	7.50	10.00	**8.25**	0.50	0.06	**7.75**	-	**16.00**	m	**17.20**
75 x 150mm; fixing by bolting	11.25	10.00	**12.37**	0.62	0.07	**9.56**	-	**21.93**	m	**23.58**
Roof members; flat										
50 x 175mm	13.12	10.00	**14.43**	0.42	0.05	**6.50**	0.02	**20.96**	m	**22.53**
50 x 200mm	15.00	10.00	**16.50**	0.47	0.06	**7.32**	0.03	**23.85**	m	**25.64**
50 x 225mm	16.87	10.00	**18.56**	0.51	0.06	**7.89**	0.03	**26.47**	m	**28.46**
50 x 250mm	18.75	10.00	**20.62**	0.56	0.07	**8.71**	0.03	**29.36**	m	**31.56**
75 x 175mm	19.70	10.00	**21.67**	0.58	0.07	**8.99**	0.03	**30.69**	m	**32.99**
75 x 200mm	22.50	10.00	**24.75**	0.61	0.08	**9.52**	0.03	**34.31**	m	**36.88**
75 x 225mm	25.33	10.00	**27.86**	0.70	0.09	**10.91**	0.04	**38.81**	m	**41.72**
75 x 250mm	28.12	10.00	**30.93**	0.79	0.10	**12.30**	0.04	**43.27**	m	**46.52**
Roof members; pitched										
25 x 75mm	2.83	10.00	**3.11**	0.25	0.03	**3.87**	0.01	**7.00**	m	**7.52**
25 x 125mm	4.70	10.00	**5.17**	0.28	0.04	**4.41**	0.02	**9.59**	m	**10.31**
32 x 175mm	8.40	10.00	**9.24**	0.39	0.05	**6.08**	0.02	**15.34**	m	**16.49**
32 x 225mm	10.80	10.00	**11.88**	0.44	0.06	**6.90**	0.02	**18.80**	m	**20.21**
38 x 75mm	4.27	10.00	**4.70**	0.26	0.03	**4.02**	0.01	**8.73**	m	**9.38**
38 x 100mm	5.70	10.00	**6.27**	0.32	0.04	**4.98**	0.02	**11.26**	m	**12.11**
50 x 75mm	5.62	10.00	**6.18**	0.32	0.04	**4.98**	0.02	**11.17**	m	**12.01**
50 x 100mm	7.50	10.00	**8.25**	0.35	0.04	**5.40**	0.02	**13.67**	m	**14.70**
50 x 125mm	9.37	10.00	**10.31**	0.40	0.05	**6.22**	0.02	**16.55**	m	**17.79**
50 x 150mm	11.25	10.00	**12.37**	0.46	0.06	**7.18**	0.03	**19.58**	m	**21.05**
50 x 225mm	16.87	10.00	**18.56**	0.63	0.08	**9.81**	0.03	**28.40**	m	**30.53**
50 x 250mm	18.75	10.00	**20.62**	0.70	0.09	**10.91**	0.04	**31.57**	m	**33.94**
75 x 100mm	11.25	10.00	**12.37**	0.46	0.06	**7.18**	0.03	**19.58**	m	**21.05**
75 x 150mm	16.87	10.00	**18.56**	0.61	0.08	**9.52**	0.03	**28.11**	m	**30.22**
100 x 150mm	22.50	10.00	**24.75**	0.79	0.10	**12.30**	0.04	**37.09**	m	**39.87**
100 x 225mm	33.75	10.00	**37.12**	1.05	0.13	**16.31**	0.06	**53.49**	m	**57.51**
Joist strutting; herringbone; depth of joist 175mm 50 x 50mm	6.75	10.00	**7.42**	1.10	0.14	**17.13**	0.06	**24.61**	m	**26.46**
Joist strutting; herringbone; depth of joist 200mm 50 x 50mm	7.87	10.00	**8.66**	1.10	0.14	**17.13**	0.06	**25.85**	m	**27.79**
Joist strutting; herringbone; depth of joist 225mm 50 x 50mm	9.00	10.00	**9.90**	1.10	0.14	**17.13**	0.06	**27.09**	m	**29.12**
Joist strutting; herringbone; depth of joist 250mm 50 x 50mm	10.12	10.00	**11.14**	1.10	0.14	**17.13**	0.06	**28.33**	m	**30.45**
Joist strutting; block; depth of joist 150mm 50 x 150mm	11.25	10.00	**12.37**	1.10	0.14	**17.13**	0.06	**29.56**	m	**31.78**
Joist strutting; block; depth of joist 175mm 50 x 175mm	13.12	10.00	**14.43**	1.20	0.15	**18.66**	0.07	**33.16**	m	**35.64**
Joist strutting; block; depth of joist 200mm 50 x 200mm	15.00	10.00	**16.50**	1.20	0.15	**18.66**	0.07	**35.22**	m	**37.87**
Joist strutting; block; depth of joist 225mm 50 x 225mm	16.87	10.00	**18.56**	1.30	0.16	**20.19**	0.07	**38.81**	m	**41.72**

Labour hourly rates: (except Specialists) Craft Operatives £14.21 Labourer £10.71 Rates are national average prices. Refer to REGIONAL VARIATIONS for indicative levels of overall pricing in regions	MATERIALS			LABOUR				RATES		
	Del to Site	Waste	Material Cost	Craft Optve	Lab	Labour Cost	Sunds	Nett Rate	Unit	Gross Rate (+7.5%)
	£	%	£	Hrs	Hrs	£	£	£		£
G20: CARPENTRY/TIMBER FRAMING/FIRST FIXING Cont.										
Oak, sawn - carcassing Cont.										
Joist strutting; block; depth of joist 250mm										
50 x 250mm	18.75	10.00	**20.62**	1.30	0.16	**20.19**	0.07	**40.88**	m	**43.95**
Noggings to joists										
50 x 50mm	3.75	10.00	**4.12**	0.50	0.06	**7.75**	0.03	**11.90**	m	**12.79**
50 x 75mm	5.62	10.00	**6.18**	0.60	0.08	**9.38**	0.03	**15.60**	m	**16.77**
Wrot surfaces										
plain; 50mm wide	-	-	**-**	0.30	-	**4.26**	-	**4.26**	m	**4.58**
plain; 75mm wide	-	-	**-**	0.40	-	**5.68**	-	**5.68**	m	**6.11**
plain; 100mm wide	-	-	**-**	0.50	-	**7.11**	-	**7.11**	m	**7.64**
plain; 150mm wide	-	-	**-**	0.60	-	**8.53**	-	**8.53**	m	**9.17**
plain; 200mm wide	-	-	**-**	0.70	-	**9.95**	-	**9.95**	m	**10.69**
Stress grading softwood										
General Structural (GS) grade included in rates										
Special Structural (SS) grade add 10% to materials prices										
Flame proofing treatment to softwood										
For timbers treated with proofing process to Class 1 add 100% to materials prices										
Softwood, sawn - supports										
Butt jointed supports										
width exceeding 300mm; 19 x 38mm at 300mm centres; fixing to masonry	0.48	10.00	**0.53**	1.00	0.13	**15.60**	0.05	**16.18**	m2	**17.40**
width exceeding 300mm; 25 x 50mm at 300mm centres; fixing to masonry	0.83	10.00	**0.92**	1.10	0.14	**17.13**	0.06	**18.11**	m2	**19.46**
Framed supports										
width exceeding 300mm; 38 x 50mm members at 400mm centres one way; 38 x 50mm subsidiary members at 400mm centres one way	1.90	10.00	**2.09**	1.00	0.13	**15.60**	0.05	**17.75**	m2	**19.08**
width exceeding 300mm; 38 x 50mm members at 500mm centres one way; 38 x 50mm subsidiary members at 400mm centres one way	1.71	10.00	**1.88**	0.80	0.10	**12.44**	0.04	**14.36**	m2	**15.44**
width exceeding 300mm; 50 x 50mm members at 400mm centres one way; 50 x 50mm subsidiary members at 400mm centres one way	2.50	10.00	**2.75**	1.10	0.14	**17.13**	0.06	**19.94**	m2	**21.44**
width exceeding 300mm; 50 x 50mm members at 500mm centres one way; 50 x 50mm subsidiary members at 400mm centres one way	2.25	10.00	**2.48**	0.90	0.11	**13.97**	0.05	**16.49**	m2	**17.73**
width exceeding 300mm; 19 x 38mm members at 300mm centres one way; 19 x 38mm subsidiary members at 300mm centres one way; fixing to masonry	0.96	10.00	**1.05**	2.25	0.28	**34.97**	0.12	**36.15**	m2	**38.86**
width exceeding 300mm; 25 x 50mm members at 300mm centres one way; 25 x 50mm subsidiary members at 300mm centres one way; fixing to masonry	1.67	10.00	**1.83**	2.35	0.29	**36.50**	0.13	**38.46**	m2	**41.34**
width not exceeding 300mm; 38 x 50mm members longitudinally; 38 x 50mm subsidiary members at 400mm centres laterally	1.90	10.00	**2.09**	0.40	0.05	**6.22**	0.02	**8.33**	m	**8.95**
width not exceeding 300mm; 50 x 50mm members longitudinally; 50 x 50mm subsidiary members at 400mm centres laterally	2.50	10.00	**2.75**	0.45	0.06	**7.04**	0.02	**9.81**	m	**10.55**
Individual supports										
6 x 38mm	0.08	10.00	**0.09**	0.08	0.01	**1.24**	0.00	**1.34**	m	**1.44**
12 x 38mm	0.12	10.00	**0.13**	0.08	0.01	**1.24**	0.00	**1.38**	m	**1.48**
19 x 38mm	0.14	10.00	**0.16**	0.09	0.01	**1.39**	0.00	**1.55**	m	**1.66**
25 x 50mm	0.25	10.00	**0.28**	0.10	0.01	**1.53**	0.01	**1.81**	m	**1.95**
38 x 50mm	0.38	10.00	**0.42**	0.11	0.01	**1.67**	0.01	**2.09**	m	**2.25**
50 x 50mm	0.50	10.00	**0.55**	0.12	0.02	**1.92**	0.01	**2.48**	m	**2.66**
50 x 150mm	1.50	10.00	**1.65**	0.17	0.02	**2.63**	0.01	**4.29**	m	**4.61**
50 x 300mm	3.00	10.00	**3.30**	0.23	0.03	**3.59**	0.01	**6.90**	m	**7.42**
6 x 38mm; fixing to masonry	0.08	10.00	**0.09**	0.23	0.03	**3.59**	0.01	**3.69**	m	**3.97**
12 x 38mm; fixing to masonry	0.12	10.00	**0.13**	0.23	0.03	**3.59**	0.01	**3.73**	m	**4.01**
19 x 38mm; fixing to masonry	0.14	10.00	**0.16**	0.24	0.03	**3.73**	0.01	**3.90**	m	**4.20**
25 x 50mm; fixing to masonry	0.25	10.00	**0.28**	0.25	0.03	**3.87**	0.01	**4.16**	m	**4.47**
38 x 50mm; fixing to masonry	0.38	10.00	**0.42**	0.27	0.03	**4.16**	0.01	**4.59**	m	**4.93**
50 x 50mm; fixing to masonry	0.50	10.00	**0.55**	0.28	0.04	**4.41**	0.02	**4.97**	m	**5.35**
Softwood, sawn, impregnated - supports										
Butt jointed supports										
width exceeding 300mm; 50mm wide x 25mm average depth at 450mm centres	1.25	10.00	**1.37**	0.30	0.04	**4.69**	0.02	**6.08**	m2	**6.54**
width exceeding 300mm; 50mm wide x 25mm average depth at 600mm centres	0.93	10.00	**1.03**	0.23	0.03	**3.59**	0.01	**4.63**	m2	**4.98**
width exceeding 300mm; 50mm wide x 50mm average depth at 450mm centres	2.50	10.00	**2.75**	0.36	0.05	**5.65**	0.02	**8.42**	m2	**9.05**
width exceeding 300mm; 50mm wide x 50mm average depth at 600mm centres	1.87	10.00	**2.05**	0.27	0.03	**4.16**	0.01	**6.23**	m2	**6.69**
width exceeding 300mm; 50mm wide x 63mm average depth at 450mm centres	3.15	10.00	**3.46**	0.40	0.05	**6.22**	0.02	**9.70**	m2	**10.43**
width exceeding 300mm; 50mm wide x 63mm average depth at 600mm centres	2.35	10.00	**2.59**	0.30	0.04	**4.69**	0.02	**7.30**	m2	**7.84**
width exceeding 300mm; 50mm wide x 75mm average depth at 450mm centres	3.75	10.00	**4.12**	0.43	0.05	**6.65**	0.02	**10.79**	m2	**11.60**

Labour hourly rates: (except Specialists) Craft Operatives £14.21 Labourer £10.71 Rates are national average prices. Refer to REGIONAL VARIATIONS for indicative levels of overall pricing in regions	MATERIALS			LABOUR				RATES		
	Del to Site £	Waste %	Material Cost £	Craft Optve Hrs	Lab Hrs	Labour Cost £	Sunds £	Nett Rate £	Unit	Gross Rate (+7.5%) £
G20: CARPENTRY/TIMBER FRAMING/FIRST FIXING Cont.										
Softwood, sawn, impregnated - supports Cont.										
Butt jointed supports Cont.										
width exceeding 300mm; 50mm wide x 75mm average depth at 600mm centres	2.80	10.00	**3.08**	0.33	0.04	**5.12**	0.02	**8.22**	m2	**8.83**
Individual supports										
25 x 25mm; 1 labours	0.18	10.00	**0.19**	0.11	0.01	**1.67**	0.01	**1.87**	m	**2.01**
25 x 25mm; 2 labours	0.21	10.00	**0.23**	0.11	0.01	**1.67**	0.01	**1.91**	m	**2.05**
25 x 38mm; 1 labours	0.27	10.00	**0.29**	0.11	0.01	**1.67**	0.01	**1.97**	m	**2.12**
25 x 38mm; 2 labours	0.32	10.00	**0.35**	0.11	0.01	**1.67**	0.01	**2.03**	m	**2.18**
triangular; extreme dimensions										
25 x 25mm	0.18	10.00	**0.19**	0.11	0.01	**1.67**	0.01	**1.87**	m	**2.01**
38 x 75mm	0.80	10.00	**0.88**	0.13	0.02	**2.06**	0.01	**2.95**	m	**3.17**
50 x 75mm	1.06	10.00	**1.16**	0.15	0.02	**2.35**	0.01	**3.51**	m	**3.78**
50 x 100mm	1.41	10.00	**1.55**	0.17	0.02	**2.63**	0.01	**4.19**	m	**4.50**
75 x 100mm	2.11	10.00	**2.32**	0.19	0.02	**2.91**	0.01	**5.24**	m	**5.64**
rounded roll for lead										
50 x 50mm	1.69	10.00	**1.86**	0.22	0.03	**3.45**	0.01	**5.32**	m	**5.72**
50 x 75mm	2.53	10.00	**2.78**	0.25	0.03	**3.87**	0.01	**6.67**	m	**7.17**
rounded roll for lead; birdsmouthed on to ridge or hip										
50 x 50mm	1.69	10.00	**1.86**	0.38	0.05	**5.94**	0.02	**7.81**	m	**8.40**
50 x 75mm	2.53	10.00	**2.78**	0.41	0.05	**6.36**	0.02	**9.17**	m	**9.86**
roll for zinc; 32 x 44mm	0.95	10.00	**1.04**	0.16	0.02	**2.49**	0.01	**3.54**	m	**3.80**
Softwood, wrot, impregnated - gutter boarding										
Gutter boards including sides, tongued and grooved joints										
width exceeding 300mm; 19mm thick	7.04	10.00	**7.74**	2.00	0.25	**31.10**	0.11	**38.95**	m2	**41.87**
width exceeding 300mm; 25mm thick	9.28	10.00	**10.21**	2.25	0.28	**34.97**	0.12	**45.30**	m2	**48.69**
width not exceeding 300mm; 19mm thick										
150mm wide	1.06	10.00	**1.16**	0.30	0.04	**4.69**	0.02	**5.87**	m	**6.31**
225mm wide	1.58	10.00	**1.74**	0.40	0.05	**6.22**	0.02	**7.98**	m	**8.58**
width not exceeding 300mm; 25mm thick										
150mm wide	1.39	10.00	**1.53**	0.33	0.04	**5.12**	0.02	**6.67**	m	**7.17**
225mm wide	2.09	10.00	**2.30**	0.44	0.05	**6.79**	0.02	**9.11**	m	**9.79**
cesspool 225 x 225 x 150mm deep	1.72	10.00	**1.89**	1.50	0.19	**23.35**	0.08	**25.32**	nr	**27.22**
Chimney gutter boards, butt joints										
width not exceeding 300mm; 25mm thick x 100mm average wide	0.93	10.00	**1.02**	0.50	0.06	**7.75**	0.03	**8.79**	m	**9.45**
gusset end	0.32	10.00	**0.35**	0.35	0.04	**5.40**	0.02	**5.77**	nr	**6.21**
width not exceeding 300mm; 25mm thick x 175mm average wide	1.62	10.00	**1.79**	0.65	0.08	**10.09**	0.04	**11.92**	m	**12.81**
gusset end	0.45	10.00	**0.49**	0.40	0.05	**6.22**	0.02	**6.73**	nr	**7.24**
Lier boards; tongued and grooved joints										
width exceeding 300mm; 19mm thick	7.04	10.00	**7.74**	0.90	0.11	**13.97**	0.05	**21.76**	m2	**23.39**
width exceeding 300mm; 25mm thick	9.28	10.00	**10.21**	1.00	0.12	**15.50**	0.05	**25.76**	m2	**27.69**
Valley sole boards; butt joints										
width not exceeding 300mm; 25mm thick										
100mm wide	0.93	10.00	**1.02**	0.45	0.06	**7.04**	0.02	**8.08**	m	**8.69**
150mm wide	1.39	10.00	**1.53**	0.50	0.06	**7.75**	0.03	**9.31**	m	**10.00**
Plywood BS EN 636, II/III grade, WBP bonded, butt joints - gutter boarding										
Gutter boards including sides, butt joints										
width exceeding 300mm; 18mm thick	10.19	10.00	**11.21**	1.50	0.19	**23.35**	0.08	**34.64**	m2	**37.24**
width exceeding 300mm; 25mm thick	14.16	10.00	**15.58**	1.65	0.21	**25.70**	0.09	**41.36**	m2	**44.46**
width not exceeding 300mm; 18mm thick										
150mm wide	1.53	10.00	**1.68**	0.20	0.03	**3.16**	0.01	**4.85**	m	**5.22**
225mm wide	2.29	10.00	**2.52**	0.22	0.03	**3.45**	0.01	**5.98**	m	**6.43**
width not exceeding 300mm; 25mm thick										
150mm wide	2.12	10.00	**2.34**	0.23	0.03	**3.59**	0.01	**5.94**	m	**6.38**
225mm wide	3.19	10.00	**3.50**	0.25	0.03	**3.87**	0.01	**7.39**	m	**7.95**
cesspool 225 x 225 x 150mm deep	2.62	10.00	**2.88**	1.50	0.19	**23.35**	0.08	**26.31**	nr	**28.28**
Softwood, wrot, impregnated - eaves and verge boarding										
Fascia and barge boards										
width not exceeding 300mm; 25mm thick										
150mm wide	1.39	10.00	**1.53**	0.22	0.03	**3.45**	0.01	**4.99**	m	**5.37**
200mm wide	1.86	10.00	**2.04**	0.23	0.03	**3.59**	0.01	**5.64**	m	**6.07**
Fascia and barge boards, tongued, grooved and veed joints										
width not exceeding 300mm; 25mm thick										
225mm wide	2.09	10.00	**2.30**	0.30	0.04	**4.69**	0.02	**7.00**	m	**7.53**
250mm wide	2.32	10.00	**2.55**	0.32	0.04	**4.98**	0.02	**7.54**	m	**8.11**
Eaves or verge soffit boards										
width exceeding 300mm; 25mm thick	9.28	10.00	**10.21**	1.00	0.12	**15.50**	0.05	**25.76**	m2	**27.69**
width not exceeding 300mm; 25mm thick										
225mm wide	2.09	10.00	**2.30**	0.30	0.04	**4.69**	0.02	**7.00**	m	**7.53**

Labour hourly rates: (except Specialists) Craft Operatives £14.21 Labourer £10.71 Rates are national average prices. Refer to REGIONAL VARIATIONS for indicative levels of overall pricing in regions	MATERIALS			LABOUR				RATES		
	Del to Site £	Waste %	Material Cost £	Craft Optve Hrs	Lab Hrs	Labour Cost £	Sunds £	Nett Rate £	Unit	Gross Rate (+7.5%) £
G20: CARPENTRY/TIMBER FRAMING/FIRST FIXING Cont.										
Plywood BS EN 636, II/III grade, WBP bonded, butt joints - eaves and verge boarding										
Fascia and barge boards										
width not exceeding 300mm; 18mm thick										
150mm wide	1.53	10.00	**1.68**	0.22	0.03	**3.45**	0.01	**5.14**	m	**5.53**
225mm wide	2.29	10.00	**2.52**	0.23	0.03	**3.59**	0.01	**6.12**	m	**6.58**
250mm wide	2.55	10.00	**2.80**	0.24	0.03	**3.73**	0.01	**6.55**	m	**7.04**
Eaves or verge soffit boards										
width exceeding 300mm; 18mm thick	10.19	10.00	**11.21**	0.75	0.09	**11.62**	0.04	**22.87**	m2	**24.59**
width not exceeding 300mm; 18mm thick										
225mm wide	2.29	10.00	**2.52**	0.25	0.03	**3.87**	0.01	**6.41**	m	**6.89**
Fascia and bargeboards; Caradon Celuform Ltd										
Fascia reference 4247; butt joint with joint trims reference 4560; fixing to timber with Polytop screws										
150mm wide	11.13	5.00	**11.69**	0.20	0.03	**3.16**	0.16	**15.01**	m	**16.13**
200mm wide	13.39	5.00	**14.06**	0.20	0.03	**3.16**	0.16	**17.38**	m	**18.69**
250mm wide	17.60	5.00	**18.48**	0.20	0.03	**3.16**	0.32	**21.96**	m	**23.61**
Elite solid fascia board reference 4245; butt joint with joint trims reference 4560; fixing to timber with Polytop screws										
150mm wide	11.19	5.00	**11.75**	0.20	0.03	**3.16**	0.16	**15.07**	m	**16.20**
200mm wide	13.77	5.00	**14.46**	0.20	0.03	**3.16**	0.16	**17.78**	m	**19.11**
250mm wide	18.58	5.00	**19.51**	0.20	0.03	**3.16**	0.32	**22.99**	m	**24.71**
Supaliner fascia board reference 4270; butt joint with joint trims reference 4586; fixing to timber with Polytop screws										
150mm wide	9.01	5.00	**9.46**	0.30	0.04	**4.69**	0.16	**14.31**	m	**15.39**
170mm wide	10.38	5.00	**10.90**	0.30	0.04	**4.69**	0.16	**15.75**	m	**16.93**
225mm wide	13.17	5.00	**13.83**	0.30	0.04	**4.69**	0.16	**18.68**	m	**20.08**
250mm wide	14.55	5.00	**15.28**	0.35	0.04	**5.40**	0.32	**21.00**	m	**22.58**
Celuvent multi purpose/soffite board reference 4258; butt joint with panel joint trim reference 4570; fixing to timber with stainless steel nails; fixing soffite board channel reference 4572 to masonry with Polytop screws										
100mm wide	10.93	5.00	**11.48**	0.40	0.06	**6.33**	0.16	**17.97**	m	**19.31**
225mm wide	18.32	5.00	**19.24**	0.40	0.06	**6.33**	0.32	**25.88**	m	**27.82**
325mm wide	24.17	5.00	**25.37**	0.60	0.08	**9.38**	0.48	**35.23**	m	**37.88**
Multi purpose/soffite board reference 4558; butt joint with panel joint trim reference 4570; fixing to timber with stainless steel nails; fixing soffite board channel reference 4572 to masonry with Polytop screws										
125mm wide	12.25	5.00	**12.87**	0.40	0.06	**6.33**	0.16	**19.35**	m	**20.80**
175mm wide	14.92	5.00	**15.67**	0.40	0.06	**6.33**	0.16	**22.15**	m	**23.81**
200mm wide	15.54	5.00	**16.32**	0.40	0.06	**6.33**	0.32	**22.96**	m	**24.69**
300mm wide	22.63	5.00	**23.76**	0.60	0.08	**9.38**	0.48	**33.61**	m	**36.14**
Steel - tie rods and straps										
Tie rods										
19mm diameter; threaded	5.48	2.50	**5.62**	0.25	0.03	**3.87**	-	**9.49**	m	**10.21**
extra; with nut and washer	0.27	2.50	**0.27**	0.20	0.03	**3.16**	-	**3.44**	nr	**3.70**
25mm diameter; threaded	7.41	2.50	**7.60**	0.30	0.04	**4.69**	-	**12.29**	m	**13.21**
extra; with nut and washer	0.48	2.50	**0.49**	0.30	0.04	**4.69**	-	**5.19**	nr	**5.57**
Straps										
3 x 38; holes -4; fixing with bolts (bolts included elsewhere)										
500mm long	1.09	2.50	**1.12**	0.13	0.02	**2.06**	-	**3.18**	nr	**3.42**
750mm long	1.62	2.50	**1.66**	0.19	0.02	**2.91**	-	**4.57**	nr	**4.91**
1000mm long	2.20	2.50	**2.25**	0.25	0.03	**3.87**	-	**6.12**	nr	**6.58**
6 x 50; holes -4; fixing with bolts (bolts included elsewhere)										
500mm long	2.40	2.50	**2.46**	0.16	0.02	**2.49**	-	**4.95**	nr	**5.32**
750mm long	3.55	2.50	**3.63**	0.23	0.03	**3.59**	-	**7.22**	nr	**7.76**
1000mm long	4.77	2.50	**4.89**	0.30	0.04	**4.69**	-	**9.58**	nr	**10.29**
Steel; galvanized - straps										
Straps and clips; BAT Building Products Ltd.										
30 x 2.5mm standard strapping; fixing with nails										
600mm long	0.74	5.00	**0.78**	0.15	0.02	**2.35**	0.01	**3.13**	nr	**3.36**
800mm long	1.08	5.00	**1.14**	0.17	0.02	**2.63**	0.01	**3.78**	nr	**4.06**
1000mm long	1.39	5.00	**1.46**	0.18	0.02	**2.77**	0.01	**4.24**	nr	**4.56**
1200mm long	1.68	5.00	**1.77**	0.19	0.02	**2.91**	0.01	**4.69**	nr	**5.04**
1600mm long	2.18	5.00	**2.29**	0.20	0.02	**3.06**	0.01	**5.36**	nr	**5.76**
30 x 2.5mm standard strapping; twists -1; fixing with nails										
600mm long	0.81	5.00	**0.85**	0.15	0.02	**2.35**	0.01	**3.21**	nr	**3.45**
800mm long	1.17	5.00	**1.23**	0.17	0.02	**2.63**	0.01	**3.87**	nr	**4.16**
1000mm long	1.47	5.00	**1.54**	0.18	0.02	**2.77**	0.01	**4.32**	nr	**4.65**
1200mm long	1.75	5.00	**1.83**	0.19	0.02	**2.91**	0.01	**4.76**	nr	**5.11**
1600mm long	2.26	5.00	**2.37**	0.20	0.02	**3.06**	0.01	**5.44**	nr	**5.85**
30 x 5mm M 305 strapping; bends -1; nailing one end (building in included elsewhere)										
600mm long	1.56	5.00	**1.64**	0.08	0.02	**1.35**	0.01	**3.00**	nr	**3.22**

Labour hourly rates: (except Specialists) Craft Operatives £14.21 Labourer £10.71 Rates are national average prices. Refer to REGIONAL VARIATIONS for indicative levels of overall pricing in regions	MATERIALS			LABOUR				RATES		
	Del to Site £	Waste %	Material Cost £	Craft Optve Hrs	Lab Hrs	Labour Cost £	Sunds £	Nett Rate £	Unit	Gross Rate (+7.5%) £
G20: CARPENTRY/TIMBER FRAMING/FIRST FIXING Cont.										
Steel; galvanized - straps Cont.										
Straps and clips; BAT Building Products Ltd. Cont.										
30 x 5mm M 305 strapping; bends -1; nailing one end (building in included elsewhere) Cont.										
800mm long	2.00	5.00	**2.10**	0.08	0.02	**1.35**	0.01	**3.46**	nr	**3.72**
1000mm long	2.55	5.00	**2.68**	0.09	0.02	**1.49**	0.01	**4.18**	nr	**4.49**
1200mm long	3.06	5.00	**3.22**	0.09	0.02	**1.49**	0.01	**4.72**	nr	**5.07**
1600mm long	4.04	5.00	**4.24**	0.10	0.02	**1.64**	0.01	**5.88**	nr	**6.32**
30 x 5mm M305 strapping; bends -1; twists -1; nailing one end (building in included elsewhere)										
600mm long	1.63	5.00	**1.71**	0.08	0.02	**1.35**	0.01	**3.07**	nr	**3.30**
800mm long	2.05	5.00	**2.15**	0.08	0.02	**1.35**	0.01	**3.51**	nr	**3.77**
1000mm long	2.61	5.00	**2.74**	0.09	0.02	**1.49**	0.01	**4.24**	nr	**4.56**
1200mm long	3.14	5.00	**3.29**	0.09	0.02	**1.49**	0.01	**4.79**	nr	**5.15**
1600mm long	4.11	5.00	**4.32**	0.10	0.02	**1.64**	0.01	**5.96**	nr	**6.41**
truss clips; fixing with nails										
for 38mm thick members	0.29	5.00	**0.30**	0.20	0.02	**3.06**	0.01	**3.37**	nr	**3.62**
for 50mm thick members	0.31	5.00	**0.33**	0.20	0.02	**3.06**	0.01	**3.39**	nr	**3.65**
Steel; galvanized - joist hangers										
Joist hangers; BAT Building Products Ltd; (building in where required included elsewhere)										
SPH type S, for 50 x 100mm joist	1.55	2.50	**1.59**	-	-	**-**	-	**1.59**	nr	**1.71**
SPH type S, for 50 x 125mm joist	1.59	2.50	**1.62**	-	-	**-**	-	**1.62**	nr	**1.75**
SPH type S, for 50 x 150mm joist	1.59	2.50	**1.62**	-	-	**-**	-	**1.62**	nr	**1.75**
SPH type S, for 50 x 175mm joist	1.66	2.50	**1.70**	-	-	**-**	-	**1.70**	nr	**1.83**
SPH type S, for 50 x 200mm joist	1.84	2.50	**1.89**	-	-	**-**	-	**1.89**	nr	**2.03**
SPH type S, for 50 x 225mm joist	1.97	2.50	**2.02**	-	-	**-**	-	**2.02**	nr	**2.17**
SPH type S, for 50 x 250mm joist	2.52	2.50	**2.58**	-	-	**-**	-	**2.58**	nr	**2.77**
SPH type S, for 63 x 100mm joist	1.98	2.50	**2.03**	-	-	**-**	-	**2.03**	nr	**2.18**
SPH type S, for 63 x 125mm joist	2.01	2.50	**2.06**	-	-	**-**	-	**2.06**	nr	**2.22**
SPH type S, for 63 x 150mm joist	2.30	2.50	**2.36**	-	-	**-**	-	**2.36**	nr	**2.54**
SPH type S, for 63 x 175mm joist	2.16	2.50	**2.22**	-	-	**-**	-	**2.22**	nr	**2.38**
SPH type S, for 63 x 200mm joist	2.30	2.50	**2.36**	-	-	**-**	-	**2.36**	nr	**2.54**
SPH type S, for 63 x 225mm joist	2.46	2.50	**2.52**	-	-	**-**	-	**2.52**	nr	**2.71**
SPH type S, for 63 x 250mm joist	2.63	2.50	**2.70**	-	-	**-**	-	**2.70**	nr	**2.90**
SPH type R, for 50 x 100mm joist	2.35	2.50	**2.40**	-	-	**-**	-	**2.40**	nr	**2.58**
SPH type R, for 50 x 125mm joist	2.39	2.50	**2.45**	-	-	**-**	-	**2.45**	nr	**2.63**
SPH type R, for 50 x 150mm joist	2.46	2.50	**2.52**	-	-	**-**	-	**2.52**	nr	**2.71**
SPH type R, for 50 x 175mm joist	2.56	2.50	**2.62**	-	-	**-**	-	**2.62**	nr	**2.82**
SPH type R, for 50 x 200mm joist	2.72	2.50	**2.79**	-	-	**-**	-	**2.79**	nr	**3.00**
SPH type R, for 50 x 225mm joist	2.98	2.50	**3.05**	-	-	**-**	-	**3.05**	nr	**3.28**
SPH type R, for 50 x 250mm joist	3.53	2.50	**3.62**	-	-	**-**	-	**3.62**	nr	**3.89**
SPH type R, for 63 x 100mm joist	2.87	2.50	**2.94**	-	-	**-**	-	**2.94**	nr	**3.16**
SPH type R, for 63 x 125mm joist	2.91	2.50	**2.99**	-	-	**-**	-	**2.99**	nr	**3.21**
SPH type R, for 63 x 150mm joist	3.23	2.50	**3.32**	-	-	**-**	-	**3.32**	nr	**3.56**
SPH type R, for 63 x 175mm joist	3.04	2.50	**3.12**	-	-	**-**	-	**3.12**	nr	**3.35**
SPH type R, for 63 x 200mm joist	3.23	2.50	**3.32**	-	-	**-**	-	**3.32**	nr	**3.56**
SPH type R, for 63 x 225mm joist	3.46	2.50	**3.55**	-	-	**-**	-	**3.55**	nr	**3.81**
SPH type R, for 63 x 250mm joist	3.65	2.50	**3.74**	-	-	**-**	-	**3.74**	nr	**4.02**
SPH type ST for 50 x 100mm joist	4.11	2.50	**4.22**	-	-	**-**	-	**4.22**	nr	**4.53**
SPH type ST for 50 x 125mm joist	4.23	2.50	**4.34**	-	-	**-**	-	**4.34**	nr	**4.66**
SPH type ST for 50 x 150mm joist	4.03	2.50	**4.13**	-	-	**-**	-	**4.13**	nr	**4.44**
SPH type ST for 50 x 175mm joist	4.59	2.50	**4.71**	-	-	**-**	-	**4.71**	nr	**5.06**
SPH type ST for 50 x 200mm joist	4.82	2.50	**4.94**	-	-	**-**	-	**4.94**	nr	**5.31**
SPH type ST for 50 x 225mm joist	5.27	2.50	**5.40**	-	-	**-**	-	**5.40**	nr	**5.81**
SPH type ST for 50 x 250mm joist	6.39	2.50	**6.55**	-	-	**-**	-	**6.55**	nr	**7.05**
Speedy Minor type for the following size joists; fixing with nails										
38 x 100mm	0.44	2.50	**0.45**	0.15	0.02	**2.35**	0.01	**2.80**	nr	**3.01**
50 x 100mm	0.44	2.50	**0.45**	0.15	0.02	**2.35**	0.01	**2.80**	nr	**3.01**
Speedy short leg type for the following size joists fixing with nails										
38 x 100mm	0.80	2.50	**0.82**	0.15	0.02	**2.35**	0.01	**3.18**	nr	**3.41**
50 x 175mm	0.80	2.50	**0.82**	0.15	0.02	**2.35**	0.01	**3.18**	nr	**3.41**
Speedy Standard Leg type for the following size joists; fixing with nails										
38 x 100mm	0.88	2.50	**0.90**	0.15	0.02	**2.35**	0.01	**3.25**	nr	**3.50**
50 x 175mm	0.88	2.50	**0.90**	0.15	0.02	**2.35**	0.01	**3.25**	nr	**3.50**
63 x 225mm	0.91	2.50	**0.93**	0.15	0.02	**2.35**	0.01	**3.29**	nr	**3.53**
75 x 225mm	0.91	2.50	**0.93**	0.17	0.02	**2.63**	0.01	**3.57**	nr	**3.84**
100 x 225mm	1.01	2.50	**1.03**	0.17	0.02	**2.63**	0.01	**3.67**	nr	**3.95**
Steel; galvanized - joist struts										
Herringbone joist struts; BAT Building Products Ltd. to suit joists at the following centres; fixing with nails										
400mm	0.31	5.00	**0.33**	0.25	0.03	**3.87**	0.01	**4.21**	m	**4.53**
450mm	0.34	5.00	**0.36**	0.25	0.03	**3.87**	0.01	**4.25**	m	**4.57**
600mm	0.40	5.00	**0.42**	0.20	0.03	**3.16**	0.01	**3.59**	m	**3.86**
Steel; galvanized - truss plates and framing anchors										
Truss plates; BAT Building Products Ltd. fixing with nails										
51 x 114mm	0.16	5.00	**0.17**	0.35	0.04	**5.40**	0.02	**5.59**	nr	**6.01**
76 x 254mm	0.60	5.00	**0.63**	1.00	0.13	**15.60**	0.05	**16.28**	nr	**17.51**
114 x 152mm	0.48	5.00	**0.51**	1.00	0.13	**15.60**	0.05	**16.16**	nr	**17.37**

Labour hourly rates: (except Specialists) Craft Operatives £14.21 Labourer £10.71 Rates are national average prices. Refer to REGIONAL VARIATIONS for indicative levels of overall pricing in regions.	MATERIALS			LABOUR				RATES		
	Del to Site £	Waste %	Material Cost £	Craft Optve Hrs	Lab Hrs	Labour Cost £	Sunds £	Nett Rate £	Unit	Gross Rate (+7.5%) £
G20: CARPENTRY/TIMBER FRAMING/FIRST FIXING Cont.										
Steel; galvanized - truss plates and framing anchors Cont.										
Truss plates; BAT Building Products Ltd. Cont.										
fixing with nails Cont.										
114 x 254mm	0.82	5.00	**0.87**	1.50	0.19	**23.35**	0.08	**24.30**	nr	**26.12**
152 x 152mm	0.66	5.00	**0.70**	1.60	0.20	**24.88**	0.09	**25.66**	nr	**27.59**
Framing anchors; BAT Building Products Ltd.										
type A; fixing with nails	0.41	5.00	**0.43**	0.20	0.02	**3.06**	0.01	**3.49**	nr	**3.76**
type B; fixing with nails	0.41	5.00	**0.43**	0.20	0.02	**3.06**	0.01	**3.49**	nr	**3.76**
type C; fixing with nails	0.41	5.00	**0.43**	0.20	0.02	**3.06**	0.01	**3.49**	nr	**3.76**
Steel - timber connectors										
Connectors BS EN 912										
split ring connectors table 1; 64mm diameter	1.07	5.00	**1.12**	0.08	0.01	**1.24**	-	**2.37**	nr	**2.55**
split ring connectors table 1; 100 mm diameter	2.95	5.00	**3.09**	0.08	0.01	**1.24**	-	**4.34**	nr	**4.66**
shear plate connectors, table 2; 67mm diameter	1.76	5.00	**1.84**	0.08	0.01	**1.24**	-	**3.09**	nr	**3.32**
single sided round toothed-plate connectors, table 4										
38mm diameter	0.18	5.00	**0.19**	0.08	0.01	**1.24**	-	**1.44**	nr	**1.54**
51mm diameter	0.25	5.00	**0.26**	0.08	0.01	**1.24**	-	**1.50**	nr	**1.62**
64mm diameter	0.33	5.00	**0.35**	0.08	0.01	**1.24**	-	**1.59**	nr	**1.71**
76mm diameter	0.47	5.00	**0.49**	0.08	0.01	**1.24**	-	**1.74**	nr	**1.87**
double sided round toothed-plate connector, table 4										
38mm diameter	0.25	5.00	**0.26**	0.08	0.01	**1.24**	-	**1.50**	nr	**1.62**
51mm diameter	0.28	5.00	**0.29**	0.08	0.01	**1.24**	-	**1.54**	nr	**1.65**
64mm diameter	0.35	5.00	**0.37**	0.08	0.01	**1.24**	-	**1.62**	nr	**1.74**
76mm diameter	0.51	5.00	**0.54**	0.08	0.01	**1.24**	-	**1.78**	nr	**1.92**
Cast iron - connectors										
Connectors BS EN 912										
shear plate connectors, table 2; 102mm diameter	6.25	5.00	**6.56**	0.12	0.01	**1.81**	-	**8.37**	nr	**9.00**
Steel - bolts and nuts										
Black bolts, BS 4190 grade 4.6, hexagon head										
M 12; with nuts and washers										
100mm long	0.48	5.00	**0.50**	0.09	0.01	**1.39**	-	**1.89**	nr	**2.03**
120mm long	0.58	5.00	**0.61**	0.11	0.01	**1.67**	-	**2.28**	nr	**2.45**
160mm long	1.24	5.00	**1.30**	0.11	0.01	**1.67**	-	**2.97**	nr	**3.19**
180mm long	1.29	5.00	**1.36**	0.12	0.01	**1.81**	-	**3.17**	nr	**3.41**
200mm long	1.35	5.00	**1.42**	0.14	0.02	**2.20**	-	**3.62**	nr	**3.89**
M 16; with nuts and washers										
100mm long	0.67	5.00	**0.70**	0.09	0.01	**1.39**	-	**2.08**	nr	**2.24**
120mm long	1.03	5.00	**1.08**	0.11	0.01	**1.67**	-	**2.75**	nr	**2.95**
160mm long	1.67	5.00	**1.75**	0.12	0.01	**1.81**	-	**3.56**	nr	**3.83**
180mm long	1.76	5.00	**1.85**	0.13	0.02	**2.06**	-	**3.91**	nr	**4.20**
200mm long	1.87	5.00	**1.96**	0.14	0.02	**2.20**	-	**4.16**	nr	**4.47**
M 20; with nuts and washers										
100mm long	1.45	5.00	**1.52**	0.09	0.01	**1.39**	-	**2.90**	nr	**3.12**
120mm long	1.97	5.00	**2.07**	0.11	0.01	**1.67**	-	**3.74**	nr	**4.02**
160mm long	2.48	5.00	**2.61**	0.14	0.02	**2.20**	-	**4.81**	nr	**5.17**
180mm long	2.51	5.00	**2.63**	0.14	0.02	**2.20**	-	**4.83**	nr	**5.20**
200mm long	2.76	5.00	**2.90**	0.14	0.02	**2.20**	-	**5.10**	nr	**5.49**
M 12; with nuts and 38 x 38 x 3mm square plate washers										
100mm long	0.49	5.00	**0.51**	0.09	0.01	**1.39**	-	**1.90**	nr	**2.04**
120mm long	0.61	5.00	**0.64**	0.11	0.01	**1.67**	-	**2.31**	nr	**2.48**
160mm long	1.25	5.00	**1.31**	0.11	0.01	**1.67**	-	**2.98**	nr	**3.20**
180mm long	1.34	5.00	**1.41**	0.12	0.01	**1.81**	-	**3.22**	nr	**3.46**
200mm long	1.38	5.00	**1.44**	0.14	0.02	**2.20**	-	**3.65**	nr	**3.92**
M 12; with nuts and 50 x 50 x 3mm square plate washers										
100mm long	0.53	5.00	**0.55**	0.09	0.01	**1.39**	-	**1.94**	nr	**2.08**
120mm long	0.62	5.00	**0.65**	0.11	0.01	**1.67**	-	**2.32**	nr	**2.49**
160mm long	1.28	5.00	**1.35**	0.11	0.01	**1.67**	-	**3.02**	nr	**3.24**
180mm long	1.34	5.00	**1.41**	0.12	0.01	**1.81**	-	**3.22**	nr	**3.46**
200mm long	1.41	5.00	**1.48**	0.14	0.02	**2.20**	-	**3.69**	nr	**3.96**
Black bolts, BS 4933 Grade 4.6, cup head, square neck										
M 6; with nuts and washers										
25mm long	0.08	5.00	**0.08**	0.07	0.01	**1.10**	-	**1.18**	nr	**1.27**
50mm long	0.09	5.00	**0.09**	0.07	0.01	**1.10**	-	**1.20**	nr	**1.28**
75mm long	0.12	5.00	**0.12**	0.08	0.01	**1.24**	-	**1.37**	nr	**1.47**
100mm long	0.16	5.00	**0.17**	0.08	0.01	**1.24**	-	**1.42**	nr	**1.52**
150mm long	0.23	5.00	**0.24**	0.09	0.01	**1.39**	-	**1.63**	nr	**1.75**
M 8; with nuts and washers										
25mm long	0.12	5.00	**0.12**	0.07	0.01	**1.10**	-	**1.22**	nr	**1.32**
50mm long	0.13	5.00	**0.14**	0.07	0.01	**1.10**	-	**1.24**	nr	**1.33**
75mm long	0.16	5.00	**0.17**	0.08	0.01	**1.24**	-	**1.42**	nr	**1.52**
100mm long	0.25	5.00	**0.26**	0.08	0.01	**1.24**	-	**1.50**	nr	**1.61**
150mm long	0.35	5.00	**0.37**	0.09	0.01	**1.39**	-	**1.75**	nr	**1.89**
M 10; with nuts and washers										
50mm long	0.21	5.00	**0.22**	0.07	0.01	**1.10**	-	**1.32**	nr	**1.42**
75mm long	0.23	5.00	**0.24**	0.08	0.01	**1.24**	-	**1.49**	nr	**1.60**
100mm long	0.35	5.00	**0.37**	0.08	0.01	**1.24**	-	**1.61**	nr	**1.73**
150mm long	0.50	5.00	**0.53**	0.10	0.01	**1.53**	-	**2.06**	nr	**2.21**
M 12; with nuts and washers										
50mm long	0.30	5.00	**0.32**	0.08	0.01	**1.24**	-	**1.56**	nr	**1.68**
75mm long	0.40	5.00	**0.42**	0.08	0.01	**1.24**	-	**1.66**	nr	**1.78**

Labour hourly rates: (except Specialists) Craft Operatives £14.21 Labourer £10.71 Rates are national average prices. Refer to REGIONAL VARIATIONS for indicative levels of overall pricing in regions	MATERIALS			LABOUR				RATES		
	Del to Site £	Waste %	Material Cost £	Craft Optve Hrs	Lab Hrs	Labour Cost £	Sunds £	Nett Rate £	Unit	Gross Rate (+7.5%) £
G20: CARPENTRY/TIMBER FRAMING/FIRST FIXING Cont.										
Steel - bolts and nuts Cont.										
Black bolts, BS 4933 Grade 4.6, cup head, square neck Cont.										
M 12; with nuts and washers Cont.										
100mm long	0.49	5.00	**0.51**	0.09	0.01	**1.39**	-	**1.90**	nr	**2.04**
150mm long	0.68	5.00	**0.71**	0.11	0.01	**1.67**	-	**2.38**	nr	**2.56**
Steel; galvanized - bolts and nuts										
Black bolts, BS 4190 grade 4.6, hexagon head										
M 12; with nuts and washers										
100mm long	0.80	5.00	**0.84**	0.09	0.01	**1.39**	-	**2.22**	nr	**2.39**
120mm long	0.99	5.00	**1.04**	0.11	0.01	**1.67**	-	**2.71**	nr	**2.91**
160mm long	2.07	5.00	**2.17**	0.11	0.01	**1.67**	-	**3.84**	nr	**4.13**
180mm long	2.18	5.00	**2.28**	0.12	0.01	**1.81**	-	**4.10**	nr	**4.40**
200mm long	2.26	5.00	**2.38**	0.14	0.02	**2.20**	-	**4.58**	nr	**4.92**
M 16; with nuts and washers										
100mm long	1.33	5.00	**1.40**	0.09	0.01	**1.39**	-	**2.78**	nr	**2.99**
120mm long	1.73	5.00	**1.82**	0.11	0.01	**1.67**	-	**3.49**	nr	**3.75**
160mm long	2.78	5.00	**2.91**	0.12	0.01	**1.81**	-	**4.73**	nr	**5.08**
180mm long	2.94	5.00	**3.09**	0.13	0.02	**2.06**	-	**5.15**	nr	**5.54**
200mm long	3.13	5.00	**3.29**	0.14	0.02	**2.20**	-	**5.49**	nr	**5.90**
M 20; with nuts and washers										
100mm long	2.44	5.00	**2.56**	0.10	0.01	**1.53**	-	**4.09**	nr	**4.40**
120mm long	3.29	5.00	**3.45**	0.11	0.01	**1.67**	-	**5.12**	nr	**5.50**
160mm long	4.15	5.00	**4.36**	0.14	0.02	**2.20**	-	**6.56**	nr	**7.05**
180mm long	4.31	5.00	**4.52**	0.14	0.02	**2.20**	-	**6.73**	nr	**7.23**
200mm long	4.58	5.00	**4.81**	0.14	0.02	**2.20**	-	**7.02**	nr	**7.54**
M 12; with nuts and 38 x 38 x 3mm square plate washers										
100mm long	0.88	5.00	**0.92**	0.09	0.01	**1.39**	-	**2.31**	nr	**2.48**
120mm long	1.06	5.00	**1.11**	0.11	0.01	**1.67**	-	**2.78**	nr	**2.99**
160mm long	2.14	5.00	**2.25**	0.11	0.01	**1.67**	-	**3.92**	nr	**4.21**
180mm long	2.25	5.00	**2.37**	0.12	0.01	**1.81**	-	**4.18**	nr	**4.49**
200mm long	2.35	5.00	**2.47**	0.14	0.02	**2.20**	-	**4.67**	nr	**5.02**
M 12; with nuts and 50 x 50 x 3mm square plate washers										
100mm long	0.93	5.00	**0.98**	0.09	0.01	**1.39**	-	**2.36**	nr	**2.54**
120mm long	1.10	5.00	**1.15**	0.11	0.01	**1.67**	-	**2.82**	nr	**3.04**
160mm long	2.20	5.00	**2.31**	0.11	0.01	**1.67**	-	**3.98**	nr	**4.28**
180mm long	2.32	5.00	**2.44**	0.12	0.01	**1.81**	-	**4.25**	nr	**4.57**
200mm long	2.48	5.00	**2.60**	0.14	0.02	**2.20**	-	**4.80**	nr	**5.16**
Steel - expanding bolts										
Expanding bolts; bolt projecting Rawlbolts; drilling masonry										
with nuts and washers; reference										
44 - 505 (M6 10P)	0.94	5.00	**0.98**	0.20	0.02	**3.06**	-	**4.04**	nr	**4.34**
44 - 510 (M6 25P)	1.05	5.00	**1.10**	0.20	0.02	**3.06**	-	**4.16**	nr	**4.47**
44 - 515 (M6 60P)	1.10	5.00	**1.16**	0.20	0.02	**3.06**	-	**4.21**	nr	**4.53**
44 - 555 (M8 10P)	1.18	5.00	**1.24**	0.22	0.03	**3.45**	-	**4.69**	nr	**5.04**
44 - 560 (M8 25P)	1.25	5.00	**1.31**	0.22	0.03	**3.45**	-	**4.75**	nr	**5.11**
44 - 565 (M8 60P)	1.33	5.00	**1.40**	0.22	0.03	**3.45**	-	**4.84**	nr	**5.21**
44 - 565 (M8 60P)	1.62	5.00	**1.70**	0.25	0.03	**3.87**	-	**5.57**	nr	**5.99**
44 - 610 (M10 30P)	1.70	5.00	**1.78**	0.25	0.03	**3.87**	-	**5.66**	nr	**6.08**
44 - 615 (M10 60P)	1.77	5.00	**1.86**	0.25	0.03	**3.87**	-	**5.73**	nr	**6.16**
44 - 655 (M12 15P)	2.58	5.00	**2.71**	0.27	0.03	**4.16**	-	**6.87**	nr	**7.38**
44 - 660 (M12 30P)	2.77	5.00	**2.91**	0.27	0.03	**4.16**	-	**7.07**	nr	**7.60**
44 - 665 (M12 75P)	3.45	5.00	**3.63**	0.27	0.03	**4.16**	-	**7.78**	nr	**8.37**
44 - 705 (M16 15P)	6.10	5.00	**6.41**	0.30	0.04	**4.69**	-	**11.10**	nr	**11.93**
44 - 710 (M16 35P)	6.60	5.00	**6.93**	0.30	0.04	**4.69**	-	**11.62**	nr	**12.49**
44 - 715 (M16 75P)	6.92	5.00	**7.26**	0.30	0.04	**4.69**	-	**11.95**	nr	**12.85**
44 - 755 (M20 15P)	9.19	5.00	**9.65**	0.33	0.04	**5.12**	-	**14.77**	nr	**15.87**
44 - 760 (M20 30P)	9.93	5.00	**10.42**	0.33	0.04	**5.12**	-	**15.54**	nr	**16.70**
44 - 765 (M20 100P)	11.29	5.00	**11.85**	0.33	0.04	**5.12**	-	**16.97**	nr	**18.25**
Expanding bolts; Sleeve Anchor (Rawlok); bolt projecting type; drilling masonry										
with nuts and washers; reference										
69 - 504 (M4.5 9P)	0.27	5.00	**0.28**	0.16	0.02	**2.49**	-	**2.77**	nr	**2.97**
69 - 506 (M4.5 27P)	0.33	5.00	**0.34**	0.16	0.02	**2.49**	-	**2.83**	nr	**3.04**
69 - 508 (M6 9P)	0.34	5.00	**0.36**	0.18	0.02	**2.77**	-	**3.13**	nr	**3.37**
69 - 510 (M6 35P)	0.43	5.00	**0.45**	0.18	0.02	**2.77**	-	**3.22**	nr	**3.46**
69 - 512 (M6 60P)	0.57	5.00	**0.60**	0.18	0.02	**2.77**	-	**3.37**	nr	**3.62**
69 - 514 (M8 9P)	0.49	5.00	**0.52**	0.20	0.03	**3.16**	-	**3.68**	nr	**3.96**
69 - 516 (M8 36P)	0.63	5.00	**0.66**	0.20	0.03	**3.16**	-	**3.83**	nr	**4.12**
69 - 518 (M8 60P)	0.82	5.00	**0.86**	0.20	0.03	**3.16**	-	**4.03**	nr	**4.33**
69 - 520 (M10 9P)	0.75	5.00	**0.79**	0.22	0.03	**3.45**	-	**4.24**	nr	**4.55**
69 - 522 (M10 22P)	0.84	5.00	**0.88**	0.22	0.03	**3.45**	-	**4.33**	nr	**4.65**
69 - 524 (M10 50P)	1.16	5.00	**1.22**	0.22	0.03	**3.45**	-	**4.67**	nr	**5.02**
69 - 525 (M10 80P)	1.44	5.00	**1.52**	0.22	0.03	**3.45**	-	**4.96**	nr	**5.34**
69 - 526 (M12 13P)	1.53	5.00	**1.61**	0.25	0.03	**3.87**	-	**5.48**	nr	**5.89**
69 - 528 (M12 55P)	1.80	5.00	**1.89**	0.25	0.03	**3.87**	-	**5.76**	nr	**6.20**
69 - 530 (M12 90P)	2.58	5.00	**2.71**	0.25	0.03	**3.87**	-	**6.58**	nr	**7.08**
69 - 533 (M16 25P)	2.60	5.00	**2.73**	0.27	0.03	**4.16**	-	**6.88**	nr	**7.40**
69 - 534 (M16 57P)	3.35	5.00	**3.51**	0.27	0.03	**4.16**	-	**7.67**	nr	**8.25**
69 - 536 (M16 100P)	4.03	5.00	**4.24**	0.27	0.03	**4.16**	-	**8.39**	nr	**9.02**
Expanding bolts; loose bolt Rawlbolts; drilling masonry										
with washers; reference										
44 - 015 (M6 10L)	0.94	5.00	**0.99**	0.20	0.02	**3.06**	-	**4.04**	nr	**4.35**

Labour hourly rates: (except Specialists) Craft Operatives £14.21 Labourer £10.71 Rates are national average prices. Refer to REGIONAL VARIATIONS for indicative levels of overall pricing in regions	MATERIALS			LABOUR				RATES		
	Del to Site £	Waste %	Material Cost £	Craft Optve Hrs	Lab Hrs	Labour Cost £	Sunds £	Nett Rate £	Unit	Gross Rate (+7.5%) £
G20: CARPENTRY/TIMBER FRAMING/FIRST FIXING Cont.										
Steel - expanding bolts Cont.										
Expanding bolts; loose bolt Rawlbolts; drilling masonry Cont.										
with washers; reference Cont.										
44 - 020 (M6 25L)	1.00	5.00	**1.04**	0.20	0.02	**3.06**	-	**4.10**	nr	**4.41**
44 - 025 (M6 40L)	1.00	5.00	**1.05**	0.20	0.02	**3.06**	-	**4.10**	nr	**4.41**
44 - 055 (M8 10L)	1.17	5.00	**1.23**	0.22	0.03	**3.45**	-	**4.68**	nr	**5.03**
44 - 060 (M8 25L)	1.22	5.00	**1.28**	0.22	0.03	**3.45**	-	**4.73**	nr	**5.08**
44 - 065 (M8 40L)	1.30	5.00	**1.36**	0.22	0.03	**3.45**	-	**4.81**	nr	**5.17**
44 - 105 (M10 10L)	1.56	5.00	**1.64**	0.25	0.03	**3.87**	-	**5.52**	nr	**5.93**
44 - 110 (M10 25L)	1.61	5.00	**1.69**	0.25	0.03	**3.87**	-	**5.56**	nr	**5.98**
44 - 115 (M10 50L)	1.70	5.00	**1.79**	0.25	0.03	**3.87**	-	**5.66**	nr	**6.08**
44 - 120 (M10 75L)	1.76	5.00	**1.85**	0.25	0.03	**3.87**	-	**5.72**	nr	**6.15**
44 - 155 (M12 10L)	2.34	5.00	**2.45**	0.27	0.03	**4.16**	-	**6.61**	nr	**7.11**
44 - 160 (M12 25L)	2.58	5.00	**2.71**	0.27	0.03	**4.16**	-	**6.87**	nr	**7.38**
44 - 165 (M12 40L)	2.69	5.00	**2.83**	0.27	0.03	**4.16**	-	**6.98**	nr	**7.51**
44 - 170 (M12 60L)	2.84	5.00	**2.98**	0.27	0.03	**4.16**	-	**7.13**	nr	**7.67**
44 - 205 (M16 15L)	5.44	5.00	**5.71**	0.30	0.04	**4.69**	-	**10.40**	nr	**11.18**
44 - 210 (M16 30L)	6.27	5.00	**6.59**	0.30	0.04	**4.69**	-	**11.28**	nr	**12.12**
44 - 215 (M16 60L)	6.79	5.00	**7.13**	0.30	0.04	**4.69**	-	**11.82**	nr	**12.70**
44 - 255 (M20 60L)	10.65	5.00	**11.18**	0.33	0.04	**5.12**	-	**16.30**	nr	**17.52**
44 - 260 (M20 100L)	11.05	5.00	**11.60**	0.33	0.04	**5.12**	-	**16.72**	nr	**17.97**
Spit Mega high performance safety anchors; loose nut version; drilling masonry										
with nuts and washers, reference										
E10 - 15/20	2.37	5.00	**2.49**	0.25	0.03	**3.87**	-	**6.37**	nr	**6.84**
E10 - 15/45	2.63	5.00	**2.76**	0.25	0.03	**3.87**	-	**6.63**	nr	**7.13**
E10 - 15/65	2.79	5.00	**2.93**	0.25	0.03	**3.87**	-	**6.81**	nr	**7.32**
E12 - 18/25	3.35	5.00	**3.51**	0.27	0.03	**4.16**	-	**7.67**	nr	**8.25**
E12 - 18/45	3.48	5.00	**3.65**	0.27	0.03	**4.16**	-	**7.81**	nr	**8.40**
E12 - 18/65	3.72	5.00	**3.91**	0.27	0.03	**4.16**	-	**8.07**	nr	**8.67**
E16 - 24/25	9.43	5.00	**9.90**	0.30	0.03	**4.58**	-	**14.48**	nr	**15.57**
E16 - 24/45	10.00	5.00	**10.49**	0.30	0.04	**4.69**	-	**15.19**	nr	**16.33**
E16 - 24/95	12.28	5.00	**12.90**	0.30	0.04	**4.69**	-	**17.59**	nr	**18.91**
E20 - 28/25	13.78	5.00	**14.46**	0.33	0.04	**5.12**	-	**19.58**	nr	**21.05**
E20 - 28/45	16.49	5.00	**17.31**	0.33	0.04	**5.12**	-	**22.43**	nr	**24.11**
E20 - 28/95	16.58	5.00	**17.41**	0.33	0.04	**5.12**	-	**22.53**	nr	**24.22**
Spit Mega high performance safety anchors; bolt head version; drilling masonry										
with washers, reference										
V10 - 15/20	2.37	5.00	**2.49**	0.25	0.03	**3.87**	-	**6.37**	nr	**6.84**
V12 - 18/25	3.36	5.00	**3.53**	0.27	0.03	**4.16**	-	**7.68**	nr	**8.26**
V16 - 24/25	7.26	5.00	**7.62**	0.30	0.04	**4.69**	-	**12.32**	nr	**13.24**
V20 - 28/25	13.76	5.00	**14.45**	0.33	0.04	**5.12**	-	**19.57**	nr	**21.04**
Spit Fix high performance through bolt BZP anchors; drilling masonry										
with nuts and washers, reference										
6/10/45	0.34	5.00	**0.35**	0.18	0.03	**2.88**	-	**3.23**	nr	**3.48**
6/20/55	0.35	5.00	**0.37**	0.18	0.03	**2.88**	-	**3.25**	nr	**3.49**
6/50/85	0.39	5.00	**0.41**	0.18	0.03	**2.88**	-	**3.28**	nr	**3.53**
8/10/50	0.41	5.00	**0.43**	0.20	0.03	**3.16**	-	**3.59**	nr	**3.86**
8/50/90	0.53	5.00	**0.56**	0.20	0.03	**3.16**	-	**3.72**	nr	**4.00**
8/90/130	0.61	5.00	**0.64**	0.20	0.03	**3.16**	-	**3.81**	nr	**4.09**
10/10/60	0.63	5.00	**0.66**	0.22	0.03	**3.45**	-	**4.11**	nr	**4.41**
10/25/75	0.64	5.00	**0.67**	0.22	0.03	**3.45**	-	**4.12**	nr	**4.43**
10/45/95	0.69	5.00	**0.72**	0.22	0.03	**3.45**	-	**4.17**	nr	**4.48**
12/10/70	0.83	5.00	**0.87**	0.25	0.03	**3.87**	-	**4.75**	nr	**5.10**
12/40/100	1.13	5.00	**1.19**	0.25	0.03	**3.87**	-	**5.06**	nr	**5.44**
12/80/140	1.31	5.00	**1.38**	0.25	0.03	**3.87**	-	**5.25**	nr	**5.65**
12/120/180	1.66	5.00	**1.74**	0.25	0.03	**3.87**	-	**5.62**	nr	**6.04**
12/160/220	2.84	5.00	**2.98**	0.25	0.03	**3.87**	-	**6.86**	nr	**7.37**
16/10/90	2.23	5.00	**2.34**	0.27	0.03	**4.16**	-	**6.50**	nr	**6.98**
16/45/125	2.26	5.00	**2.38**	0.27	0.03	**4.16**	-	**6.53**	nr	**7.02**
16/95/175	2.31	5.00	**2.43**	0.27	0.03	**4.16**	-	**6.59**	nr	**7.08**
20/20/120	2.42	5.00	**2.54**	0.30	0.03	**4.58**	-	**7.13**	nr	**7.66**
20/60/160	2.66	5.00	**2.80**	0.30	0.04	**4.69**	-	**7.49**	nr	**8.05**
20/115/215	2.95	5.00	**3.10**	0.30	0.04	**4.69**	-	**7.79**	nr	**8.37**
Steel; plated - expanding bolts										
Expanding bolts; hook Rawlbolts; drilling masonry										
with nuts and washers; reference										
44 - 401 (M6)	1.26	5.00	**1.32**	0.20	0.03	**3.16**	-	**4.49**	nr	**4.82**
44 - 406 (M8)	1.54	5.00	**1.62**	0.22	0.03	**3.45**	-	**5.06**	nr	**5.44**
44 - 411 (M10)	2.48	5.00	**2.61**	0.25	0.03	**3.87**	-	**6.48**	nr	**6.96**
44 - 416 (M12)	3.92	5.00	**4.12**	0.27	0.03	**4.16**	-	**8.27**	nr	**8.89**
Expanding bolts; eye Rawlbolts; drilling masonry										
with nuts and washers; reference										
44 - 432 (M6)	2.79	5.00	**2.93**	0.20	0.03	**3.16**	-	**6.09**	nr	**6.55**
44 - 437 (M8)	3.18	5.00	**3.34**	0.22	0.03	**3.45**	-	**6.79**	nr	**7.30**
44 - 442 (M10)	4.26	5.00	**4.47**	0.25	0.03	**3.87**	-	**8.35**	nr	**8.97**
44 - 447 (M12)	6.37	5.00	**6.68**	0.27	0.03	**4.16**	-	**10.84**	nr	**11.66**
Expanding bolts; bolt projecting Through Bolts; drilling masonry										
with nuts and washers; reference										
56 - 104 (M6 x 55mm)	0.41	5.00	**0.43**	0.18	0.02	**2.77**	-	**3.20**	nr	**3.44**
56 - 110 (M6 x 85mm)	0.53	5.00	**0.55**	0.18	0.02	**2.77**	-	**3.32**	nr	**3.57**
56 - 114 (M8 x 50mm)	0.57	5.00	**0.60**	0.20	0.03	**3.16**	-	**3.76**	nr	**4.04**

Labour hourly rates: (except Specialists) Craft Operatives £14.21 Labourer £10.71 Rates are national average prices. Refer to REGIONAL VARIATIONS for indicative levels of overall pricing in regions	MATERIALS			LABOUR				RATES		
	Del to Site £	Waste %	Material Cost £	Craft Optve Hrs	Lab Hrs	Labour Cost £	Sunds £	Nett Rate £	Unit	Gross Rate (+7.5%) £
G20: CARPENTRY/TIMBER FRAMING/FIRST FIXING Cont.										
Steel; plated - expanding bolts Cont.										
Expanding bolts; bolt projecting Through Bolts; drilling masonry Cont.										
with nuts and washers; reference Cont.										
56 - 116 (M8 x 65mm)	0.59	5.00	**0.62**	0.20	0.03	**3.16**	-	**3.78**	nr	**4.07**
56 - 120 (M8 x 80mm)	0.62	5.00	**0.65**	0.20	0.03	**3.16**	-	**3.81**	nr	**4.10**
56 - 124 (M8 x 105mm)	0.66	5.00	**0.70**	0.20	0.03	**3.16**	-	**3.86**	nr	**4.15**
56 - 125 (M8 x 115mm)	0.73	5.00	**0.76**	0.20	0.03	**3.16**	-	**3.93**	nr	**4.22**
56 - 128 (M10 x 80mm)	0.76	5.00	**0.80**	0.22	0.03	**3.45**	-	**4.24**	nr	**4.56**
56 - 129 (M10 x 65mm)	0.72	5.00	**0.76**	0.22	0.03	**3.45**	-	**4.20**	nr	**4.52**
56 - 132 (M10 x 95mm)	0.80	5.00	**0.84**	0.22	0.03	**3.45**	-	**4.28**	nr	**4.61**
56 - 136 (M10 x 115mm)	0.86	5.00	**0.90**	0.22	0.03	**3.45**	-	**4.35**	nr	**4.67**
56 - 138 (M10 x 130mm)	0.99	5.00	**1.04**	0.22	0.03	**3.45**	-	**4.49**	nr	**4.82**
56 - 139 (M12 x 80mm)	1.14	5.00	**1.19**	0.25	0.03	**3.87**	-	**5.07**	nr	**5.45**
56 - 140 (M12 x 100mm)	1.15	5.00	**1.21**	0.25	0.03	**3.87**	-	**5.08**	nr	**5.46**
56 - 144 (M12 x 120mm)	1.24	5.00	**1.30**	0.25	0.03	**3.87**	-	**5.18**	nr	**5.56**
56 - 148 (M12 x 135mm)	1.32	5.00	**1.39**	0.25	0.03	**3.87**	-	**5.26**	nr	**5.65**
56 - 150 (M12 x 150mm)	1.39	5.00	**1.46**	0.25	0.03	**3.87**	-	**5.33**	nr	**5.73**
56 - 152 (M16 x 140mm)	2.07	5.00	**2.17**	0.27	0.03	**4.16**	-	**6.33**	nr	**6.80**
56 - 153 (M16 x 105mm)	1.85	5.00	**1.94**	0.27	0.03	**4.16**	-	**6.10**	nr	**6.55**
56 - 156 (M16 x 180mm)	2.51	5.00	**2.64**	0.27	0.03	**4.16**	-	**6.79**	nr	**7.30**
56 - 158 (M16 x 220mm)	3.25	5.00	**3.41**	0.27	0.03	**4.16**	-	**7.57**	nr	**8.14**
56 - 159 (M20 x 125mm)	3.26	5.00	**3.42**	0.30	0.04	**4.69**	-	**8.11**	nr	**8.72**
56 - 160 (M20 x 160mm)	3.55	5.00	**3.73**	0.30	0.04	**4.69**	-	**8.42**	nr	**9.05**
56 - 164 (M20 x 200mm)	4.53	5.00	**4.75**	0.30	0.04	**4.69**	-	**9.44**	nr	**10.15**
56 - 166 (M20 x 300mm)	5.65	5.00	**5.94**	0.30	0.04	**4.69**	-	**10.63**	nr	**11.42**
56 - 168 (M24 x 180mm)	6.49	5.00	**6.82**	0.33	0.04	**5.12**	-	**11.94**	nr	**12.83**
56 - 172 (M24 x 260mm)	7.97	5.00	**8.37**	0.33	0.04	**5.12**	-	**13.49**	nr	**14.50**
Stainless steel grade 316 - expanding bolts										
Expanding bolts; bolt projecting Through Bolts; drilling masonry										
with nuts and washers; reference										
56 - 604 (M6 x 55mm)	1.92	5.00	**2.01**	0.18	0.02	**2.77**	-	**4.78**	nr	**5.14**
56 - 616 (M8 x 65mm)	2.34	5.00	**2.46**	0.20	0.02	**3.06**	-	**5.52**	nr	**5.93**
56 - 624 (M8 x 105mm)	2.90	5.00	**3.04**	0.20	0.02	**3.06**	-	**6.10**	nr	**6.55**
56 - 628 (M10 x 80mm)	3.02	5.00	**3.17**	0.22	0.03	**3.45**	-	**6.62**	nr	**7.11**
56 - 636 (M10 x 115mm)	3.57	5.00	**3.74**	0.22	0.03	**3.45**	-	**7.19**	nr	**7.73**
56 - 638 (M10 x 130mm)	3.99	5.00	**4.19**	0.22	0.03	**3.45**	-	**7.63**	nr	**8.21**
56 - 640 (M12 x 100mm)	5.48	5.00	**5.75**	0.25	0.03	**3.87**	-	**9.63**	nr	**10.35**
56 - 648 (M12 x 135mm)	6.60	5.00	**6.93**	0.25	0.03	**3.87**	-	**10.80**	nr	**11.61**
56 - 650 (M12 x 150mm)	7.34	5.00	**7.71**	0.25	0.03	**3.87**	-	**11.58**	nr	**12.45**
56 - 652 (M16 x 140mm)	9.44	5.00	**9.91**	0.27	0.03	**4.16**	-	**14.07**	nr	**15.13**
56 - 658 (M16 x 220mm)	13.64	5.00	**14.32**	0.27	0.03	**4.16**	-	**18.48**	nr	**19.86**
56 - 660 (M20 x 160mm)	18.49	5.00	**19.42**	0.30	0.04	**4.69**	-	**24.11**	nr	**25.91**
56 - 666 (M20 x 300mm)	27.42	5.00	**28.79**	0.30	0.04	**4.69**	-	**33.49**	nr	**36.00**
56 - 672 (M24 x 260mm)	37.84	5.00	**39.73**	0.33	0.04	**5.12**	-	**44.85**	nr	**48.21**
Chemical anchors										
Chemical anchors; Kemfix capsules and standard studs; drilling masonry										
capsule reference 60-428; stud reference 60-649; with nuts and washers (M8 x 110mm)	1.91	5.00	**2.01**	0.24	0.03	**3.73**	-	**5.74**	nr	**6.17**
capsule reference 60-430; stud reference 60-653; with nuts and washers (M10 x 130mm)	2.14	5.00	**2.25**	0.27	0.03	**4.16**	-	**6.41**	nr	**6.89**
capsule reference 60-432; stud reference 60-661; with nuts and washers (M12 x 160mm)	2.63	5.00	**2.76**	0.30	0.04	**4.69**	-	**7.45**	nr	**8.01**
Chemical anchors; Kemfix capsules and stainless steel studs; drilling masonry										
capsule reference 60-428; stud reference 60-906; with nuts and washers (M8 x 110mm)	3.29	5.00	**3.46**	0.24	0.03	**3.73**	-	**7.19**	nr	**7.73**
capsule reference 60-430; stud reference 60-911; with nuts and washers (M10 x 130mm)	4.40	5.00	**4.61**	0.27	0.03	**4.16**	-	**8.77**	nr	**9.43**
capsule reference 60-432; stud reference 60-916; with nuts and washers (M12 x 160mm)	6.04	5.00	**6.34**	0.30	0.04	**4.69**	-	**11.03**	nr	**11.86**
capsule reference 60-436; stud reference 60-921; with nuts and washers (M16 x 190mm)	10.29	5.00	**10.80**	0.33	0.04	**5.12**	-	**15.92**	nr	**17.12**
capsule reference 60-442; stud reference 60-926; with nuts and washers (M20 x 240mm)	16.18	5.00	**16.99**	0.35	0.04	**5.40**	-	**22.39**	nr	**24.07**
capsule reference 60-446; stud reference 60-931; with nuts and washers (M24 x 295mm)	26.37	5.00	**27.68**	0.38	0.05	**5.94**	-	**33.62**	nr	**36.14**
Chemical anchors; Kemfix capsules and standard internal threaded sockets; drilling masonry										
capsule reference 60-428; socket reference 60-622 (M8 x 75mm)	2.57	5.00	**2.70**	0.30	0.04	**4.69**	-	**7.39**	nr	**7.94**
capsule reference 60-430; socket reference 60-624 (M10 x 75mm)	2.96	5.00	**3.10**	0.33	0.04	**5.12**	-	**8.22**	nr	**8.84**
capsule reference 60-432; socket reference 60-628 (M12 x 100mm)	4.01	5.00	**4.21**	0.35	0.04	**5.40**	-	**9.61**	nr	**10.33**
capsule reference 60-436; socket reference 60-630 (M16 x 125mm)	7.40	5.00	**7.77**	0.38	0.05	**5.94**	-	**13.70**	nr	**14.73**
Chemcial anchors; Kemfix capsules and stainless steel internal threaded sockets; drilling masonry										
capsule reference 60-428; socket reference 60-987	4.42	5.00	**4.64**	0.30	0.04	**4.69**	-	**9.33**	nr	**10.03**
capsule reference 60-430; socket reference 60-989	5.72	5.00	**6.01**	0.33	0.04	**5.12**	-	**11.13**	nr	**11.96**
capsule reference 60-432; socket reference 60-993	7.35	5.00	**7.72**	0.35	0.04	**5.40**	-	**13.12**	nr	**14.10**
capsule reference 60-436; socket reference 60-995	14.52	5.00	**15.24**	0.38	0.05	**5.94**	-	**21.18**	nr	**22.77**

Labour hourly rates: (except Specialists) Craft Operatives £14.21 Labourer £10.71 Rates are national average prices. Refer to REGIONAL VARIATIONS for indicative levels of overall pricing in regions	MATERIALS			LABOUR				RATES		
	Del to Site £	Waste %	Material Cost £	Craft Optve Hrs	Lab Hrs	Labour Cost £	Sunds £	Nett Rate £	Unit	Gross Rate (+7.5%) £
G20: CARPENTRY/TIMBER FRAMING/FIRST FIXING Cont.										
Chemical anchors Cont.										
Chemical anchors in low density material; Kemfix capsules, mesh sleeves and standard studs; drilling masonry										
capsule reference 60-428; sleeve reference 60-805; stud reference 60-805; with nuts and washers.....	2.69	5.00	**2.82**	0.24	0.03	**3.73**	-	**6.56**	nr	**7.05**
capsule reference 60-430; sleeve reference 60-807; stud reference 60-807; with nuts and washers.....	3.21	5.00	**3.37**	0.27	0.03	**4.16**	-	**7.52**	nr	**8.09**
capsule reference 60-432; sleeve reference 60-550; stud reference 60-113; with nuts and washers.....	3.78	5.00	**3.97**	0.30	0.04	**4.69**	-	**8.66**	nr	**9.31**
capsule reference 60-436; sleeve reference 60-562; stud reference 60-117; with nuts and washers.....	5.29	5.00	**5.56**	0.35	0.04	**5.40**	-	**10.96**	nr	**11.78**
Chemical anchors in low density material; Kemfix capsules, mesh sleeves and stainless steel studs; drilling masonry										
capsule reference 60-428; sleeve reference 60-805; stud reference 60-905; with nuts and washers.....	4.12	5.00	**4.32**	0.24	0.03	**3.73**	-	**8.06**	nr	**8.66**
capsule reference 60-430; sleeve reference 60-807; stud reference 60-910; with nuts and washers.....	5.51	5.00	**5.78**	0.27	0.03	**4.16**	-	**9.94**	nr	**10.68**
capsule reference 60-432; sleeve reference 60-113; stud reference 60-915; with nuts and washers.....	7.27	5.00	**7.63**	0.30	0.04	**4.69**	-	**12.32**	nr	**13.25**
capsule reference 60-436; sleeve reference 60-117; stud reference 60-920; with nuts and washers.....	11.57	5.00	**12.15**	0.35	0.04	**5.40**	-	**17.55**	nr	**18.86**
Chemical anchors in low density material; Kemfix capsules, mesh sleeves and standard internal threaded sockets; drilling masonry										
capsule reference 60-428; sleeve reference 60-805; socket reference 60-662	3.39	5.00	**3.56**	0.24	0.03	**3.73**	-	**7.30**	nr	**7.84**
capsule reference 60-430; sleeve reference 60-807; socket reference 60-668	4.07	5.00	**4.27**	0.27	0.03	**4.16**	-	**8.43**	nr	**9.06**
capsule reference 60-432; sleeve reference 60-113; socket reference 60-674	5.24	5.00	**5.50**	0.30	0.04	**4.69**	-	**10.19**	nr	**10.95**
capsule reference 60-436; sleeve reference 60-117; socket reference 60-676	8.68	5.00	**9.11**	0.33	0.04	**5.12**	-	**14.23**	nr	**15.30**
Chemical anchors in low density material; Kemfix capsules, mesh sleeves and stainless steel internal threaded sockets; drilling masonry										
capsule reference 60-428; sleeve reference 60-805; socket reference 60-947	5.25	5.00	**5.51**	0.24	0.03	**3.73**	-	**9.24**	nr	**9.93**
capsule reference 60-430; sleeve reference 60-807; socket reference 60-949	6.83	5.00	**7.17**	0.27	0.03	**4.16**	-	**11.33**	nr	**12.18**
capsule reference 60-436; sleeve reference 60-117; socket reference 60-955	15.80	5.00	**16.59**	0.33	0.04	**5.12**	-	**21.70**	nr	**23.33**
Spit Maxi high performance chemical anchors; capsules and zinc coated steelstuds; drilling masonry										
capsule reference M8 stud reference SM8; with nuts and washers	1.57	5.00	**1.64**	0.25	0.03	**3.87**	-	**5.52**	nr	**5.93**
capsule reference M10; stud reference SM10; with nuts and washers	1.90	5.00	**2.00**	0.27	0.03	**4.16**	-	**6.16**	nr	**6.62**
capsule reference M12; stud reference SM12; with nuts and washers	2.43	5.00	**2.55**	0.25	0.03	**3.87**	-	**6.43**	nr	**6.91**
capsule reference M16; stud reference SM16; with nuts and washers	3.37	5.00	**3.54**	0.27	0.03	**4.16**	-	**7.70**	nr	**8.28**
capsule reference M20; stud reference SM20; with nuts and washers	5.76	5.00	**6.04**	0.30	0.04	**4.69**	-	**10.74**	nr	**11.54**
capsule reference M24; stud reference SM24; with nuts and washers	10.12	5.00	**10.62**	0.35	0.04	**5.40**	-	**16.02**	nr	**17.22**
Spit Maxi high performance chemical anchors; capsules and stainless steel studs Grade 316 (A4); drilling masonry										
capsule reference M8 stud reference SM8i; with nuts and washers	1.78	5.00	**1.87**	0.25	0.03	**3.87**	-	**5.74**	nr	**6.18**
capsule reference M10; stud reference SM10i; with nuts and washers	2.37	5.00	**2.49**	0.27	0.03	**4.16**	-	**6.65**	nr	**7.15**
capsule reference M12; stud reference SM12i; with nuts and washers	3.40	5.00	**3.57**	0.25	0.03	**3.87**	-	**7.44**	nr	**8.00**
capsule reference M16; stud reference SM16i; with nuts and washers	5.39	5.00	**5.66**	0.27	0.03	**4.16**	-	**9.82**	nr	**10.56**
capsule reference M20; stud reference SM20i; with nuts and washers	9.71	5.00	**10.19**	0.30	0.04	**4.69**	-	**14.88**	nr	**16.00**
capsule reference M24; stud reference SM24i; with nuts and washers	14.17	5.00	**14.88**	0.35	0.04	**5.40**	-	**20.28**	nr	**21.81**
G30: METAL PROFILED SHEET DECKING										
Galvanized steel troughed decking; 0.7mm thick metal, 35mm overall depth; natural soffit and top surface; 150mm end laps and one corrugation side laps										
Decking; fixed to steel rails at 1200mm centres with self tapping screws; drilling holes										
sloping; 10 degrees pitch	-	-	**Spclist**	-	-	**Spclist**	-	**22.08**	m2	**23.74**
Extra over decking for										
raking cutting	-	-	**Spclist**	-	-	**Spclist**	-	**4.33**	m	**4.65**
holes 50mm diameter; formed on site.............	-	-	**Spclist**	-	-	**Spclist**	-	**11.88**	nr	**12.77**

Labour hourly rates: (except Specialists) Craft Operatives £14.21 Labourer £10.71 Rates are national average prices. Refer to REGIONAL VARIATIONS for indicative levels of overall pricing in regions	MATERIALS			LABOUR				RATES		
	Del to Site £	Waste %	Material Cost £	Craft Optve Hrs	Lab Hrs	Labour Cost £	Sunds £	Nett Rate £	Unit	Gross Rate (+7.5%) £
G30: METAL PROFILED SHEET DECKING Cont.										
Galvanized steel troughed decking; 0.7mm thick metal, 48mm overall depth; natural soffit and top surface; 150mm end laps and one corrugation side laps										
Decking; fixed to steel rails at 1500mm centres with self tapping screws; drilling holes										
sloping; 10 degrees pitch	-	-	Spclist	-	-	Spclist	-	22.37	m2	24.05
Extra over decking for										
raking cutting	-	-	Spclist	-	-	Spclist	-	4.33	m	4.65
holes 50mm diameter; formed on site	-	-	Spclist	-	-	Spclist	-	11.88	nr	12.77
Galvanized steel troughed decking; 0.7mm thick metal, 63mm overall depth; natural soffit and top surface; 150mm end laps and one corrugation side laps										
Decking; fixed to steel rails at 2000mm centres with self tapping screws; drilling holes										
sloping; 10 degrees pitch	-	-	Spclist	-	-	Spclist	-	23.82	m2	25.61
Extra over decking for										
raking cutting	-	-	Spclist	-	-	Spclist	-	4.33	m	4.65
holes 50mm diameter; formed on site	-	-	Spclist	-	-	Spclist	-	11.88	nr	12.77
Galvanized steel troughed decking; 0.7mm thick metal, 100mm overall depth; natural soffit and top surface; 150mm end laps and one corrugation side laps										
Decking; fixed to steel rails at 3500mm centres with self tapping screws; drilling holes										
sloping; 10 degrees pitch	-	-	Spclist	-	-	Spclist	-	25.44	m2	27.35
Extra over decking for										
raking cutting	-	-	Spclist	-	-	Spclist	-	4.33	m	4.65
holes 50mm diameter; formed on site	-	-	Spclist	-	-	Spclist	-	11.88	nr	12.77
Galvanized steel troughed decking; 0.9mm thick metal, 35mm overall depth; natural soffit and top surface; 150mm end laps and one corrugation side laps										
Decking; fixed to steel rails at 1500mm centres with self tapping screws; drilling holes										
sloping; 10 degrees pitch	-	-	Spclist	-	-	Spclist	-	25.10	m2	26.98
Extra over decking for										
raking cutting	-	-	Spclist	-	-	Spclist	-	4.33	m	4.65
holes 50mm diameter; formed on site	-	-	Spclist	-	-	Spclist	-	11.88	nr	12.77
Galvanized steel troughed decking; 0.9mm thick metal, 48mm overall depth; natural soffit and top surface; 150mm end laps and one corrugation side laps										
Decking; fixing to steel rails at 2000mm centres with self tapping screws; drilling holes										
sloping; 10 degrees pitch	-	-	Spclist	-	-	Spclist	-	25.50	m2	27.41
Extra over decking for										
raking cutting	-	-	Spclist	-	-	Spclist	-	4.33	m	4.65
holes 50mm diameter; formed on site	-	-	Spclist	-	-	Spclist	-	11.88	nr	12.77
Galvanized steel troughed decking; 0.9mm thick metal, 63mm overall depth; natural soffit and top surface; 150mm end laps and one corrugation side laps										
Decking; fixing to steel rails at 2500mm centres with self tapping screws; drilling holes										
sloping; 10 degrees pitch	-	-	Spclist	-	-	Spclist	-	26.73	m2	28.73
Extra over decking for										
raking cutting	-	-	Spclist	-	-	Spclist	-	4.33	m	4.65
holes 50mm diameter; formed on site	-	-	Spclist	-	-	Spclist	-	11.88	nr	12.77
Galvanized steel troughed decking; 0.9mm thick metal, 100mm overall depth; natural soffit and top surface; 150mm end laps and one corrugation side laps										
Decking; fixing to steel rails at 4000mm centres with self tapping screws; drilling holes										
sloping; 10 degrees pitch	-	-	Spclist	-	-	Spclist	-	28.76	m2	30.92
Extra over decking for										
raking cutting	-	-	Spclist	-	-	Spclist	-	4.33	m	4.65
holes 50mm diameter; formed on site	-	-	Spclist	-	-	Spclist	-	11.88	nr	12.77
Galvanized steel troughed decking; 1.2mm thick metal, 35mm overall depth; natural soffit and top surface; 150mm end laps and one corrugation side laps										
Decking; fixing to steel rails at 2000mm centres with self tapping screws; drilling holes										
sloping; 10 degrees pitch	-	-	Spclist	-	-	Spclist	-	28.91	m2	31.08
Extra over decking for										
raking cutting	-	-	Spclist	-	-	Spclist	-	4.33	m	4.65

Labour hourly rates: (except Specialists) Craft Operatives £14.21 Labourer £10.71 Rates are national average prices. Refer to REGIONAL VARIATIONS for indicative levels of overall pricing in regions	MATERIALS			LABOUR				RATES		
	Del to Site £	Waste %	Material Cost £	Craft Optve Hrs	Lab Hrs	Labour Cost £	Sunds £	Nett Rate £	Unit	Gross Rate (+7.5%) £
G30: METAL PROFILED SHEET DECKING Cont.										
Galvanized steel troughed decking; 1.2mm thick metal, 35mm overall depth; natural soffit and top surface; 150mm end laps and one corrugation side laps Cont.										
Extra over decking for Cont.										
holes 50mm diameter; formed on site	-	-	Spclist	-	-	Spclist	-	11.88	nr	12.77
Galvanized steel troughed decking; 1.2mm thick metal, 48mm overall depth; natural soffit and top surface; 150mm end laps and one corrugation side laps										
Decking; fixing to steel rails at 2500mm centres with self tapping screws; drilling holes										
sloping; 10 degrees pitch	-	-	Spclist	-	-	Spclist	-	29.38	m2	31.58
Extra over decking for										
raking cutting	-	-	Spclist	-	-	Spclist	-	4.33	m	4.65
holes 50mm diameter; formed on site	-	-	Spclist	-	-	Spclist	-	11.88	nr	12.77
Galvanized steel troughed decking; 1.2mm thick metal, 63mm overall depth; natural soffit and top surface; 150mm end laps and one corrugation side laps										
Decking; fixing to steel rails at 3000mm centres with self tapping screws; drilling holes										
sloping; 10 degrees pitch	-	-	Spclist	-	-	Spclist	-	30.97	m2	33.29
Extra over decking for										
raking cutting	-	-	Spclist	-	-	Spclist	-	4.33	m	4.65
holes 50mm diameter; formed on site	-	-	Spclist	-	-	Spclist	-	11.88	nr	12.77
Galvanized steel troughed decking; 1.2mm thick metal, 100mm overall depth; natural soffit and top surface; 150mm end laps and one corrugation side laps										
Decking; fixing to steel rails at 4500mm centres with self tapping screws; drilling holes										
sloping; 10 degrees pitch	-	-	Spclist	-	-	Spclist	-	33.63	m2	36.15
Extra over decking for										
raking cutting	-	-	Spclist	-	-	Spclist	-	4.33	m	4.65
holes 50mm diameter; formed on site	-	-	Spclist	-	-	Spclist	-	11.88	nr	12.77
Aluminium troughed decking; 0.9mm thick metal 35mm overall depth; natural soffit and top surface; 150mm end laps and one corrugation side laps										
Decking; fixing to steel rails at 900mm centres with self tapping screws; drilling holes										
sloping; 10 degrees pitch	-	-	Spclist	-	-	Spclist	-	27.81	m2	29.90
Extra over decking for										
raking cutting	-	-	Spclist	-	-	Spclist	-	4.33	m	4.65
holes 50mm diameter; formed on site	-	-	Spclist	-	-	Spclist	-	11.88	nr	12.77
Aluminium troughed decking; 0.9mm thick metal, 48mm overall depth; natural soffit and top surface; 150mm end laps and one corrugation side laps										
Decking; fixing to steel rails at 1200mm centres with self tapping screws; drilling holes										
sloping; 10 degrees pitch	-	-	Spclist	-	-	Spclist	-	28.37	m2	30.50
Extra over decking for										
raking cutting	-	-	Spclist	-	-	Spclist	-	4.33	m	4.65
holes 50mm diameter; formed on site	-	-	Spclist	-	-	Spclist	-	11.88	nr	12.77
Aluminium troughed decking; 0.9mm thick metal, 63mm overall depth; natural soffit and top surface; 150mm end laps and one corrugation side laps										
Decking; fixing to steel rails at 1500mm centres with self tapping screws; drilling holes										
sloping; 10 degrees pitch	-	-	Spclist	-	-	Spclist	-	30.30	m2	32.57
Extra over decking for										
raking cutting	-	-	Spclist	-	-	Spclist	-	4.33	m	4.65
holes 50mm diameter; formed on site	-	-	Spclist	-	-	Spclist	-	11.88	nr	12.77
Aluminium troughed decking; 0.9mm thick metal, 100mm overall depth; natural soffit and top surface; 150mm end laps and one corrugation side laps										
Decking; fixing to steel rails at 2900mm centres with self tapping screws; drilling holes										
sloping; 10 degrees pitch	-	-	Spclist	-	-	Spclist	-	33.26	m2	35.75
Extra over decking for										
raking cutting	-	-	Spclist	-	-	Spclist	-	4.33	m	4.65
holes 50mm diameter; formed on site	-	-	Spclist	-	-	Spclist	-	11.88	nr	12.77

Labour hourly rates: (except Specialists) Craft Operatives £14.21 Labourer £10.71 Rates are national average prices. Refer to REGIONAL VARIATIONS for indicative levels of overall pricing in regions	MATERIALS			LABOUR				RATES		
	Del to Site £	Waste %	Material Cost £	Craft Optve Hrs	Lab Hrs	Labour Cost £	Sunds £	Nett Rate £	Unit	Gross Rate (+7.5%) £
G30: METAL PROFILED SHEET DECKING Cont.										
Aluminium troughed decking; 1.2mm thick metal, 35mm overall depth; natural soffit and top surface; 150mm end laps and one corrugation side laps										
Decking; fixing to steel rails at 1200mm centres with self tapping screws; drilling holes										
sloping; 10 degrees pitch	-	-	Spclist	-	-	Spclist	-	31.69	m2	34.07
Extra over decking for										
raking cutting	-	-	Spclist	-	-	Spclist	-	4.33	m	4.65
holes 50mm diameter; formed on site	-	-	Spclist	-	-	Spclist	-	11.88	nr	12.77
Aluminium troughed decking; 1.2mm thick metal, 48mm overall depth; natural soffit and top surface; 150mm end laps and one corrugation side laps										
Decking; fixing to steel rails at 1700mm centres with self tapping screws; drilling holes										
sloping; 10 degrees pitch	-	-	Spclist	-	-	Spclist	-	32.22	m2	34.64
Extra over decking for										
raking cutting	-	-	Spclist	-	-	Spclist	-	4.33	m	4.65
holes 50mm diameter; formed on site	-	-	Spclist	-	-	Spclist	-	11.88	nr	12.77
Aluminium troughed decking; 1.2mm thick metal, 63mm overall depth; natural soffit and top surface; 150mm end laps and one corrugation side laps										
Decking; fixing to steel rails at 2000mm centres with self tapping screws; drilling holes										
sloping; 10 degrees pitch	-	-	Spclist	-	-	Spclist	-	34.18	m2	36.74
Extra over decking for										
raking cutting	-	-	Spclist	-	-	Spclist	-	4.33	m	4.65
holes 50mm diameter; formed on site	-	-	Spclist	-	-	Spclist	-	11.88	nr	12.77
Aluminium troughed decking; 1.2mm thick metal, 100mm overall depth; natural soffit and top surface; 150mm end laps and one corrugation side laps										
Decking; fixing to steel rails at 3000mm centres with self tapping screws; drilling holes										
sloping; 10 degrees pitch	-	-	Spclist	-	-	Spclist	-	37.53	m2	40.34
Extra over decking for										
raking cutting	-	-	Spclist	-	-	Spclist	-	4.33	m	4.65
holes 50mm diameter; formed on site	-	-	Spclist	-	-	Spclist	-	11.88	nr	12.77

CLADDING/COVERING

Labour hourly rates: (except Specialists) Craft Operatives £14.21 Labourer £10.71 Rates are national average prices. Refer to REGIONAL VARIATIONS for indicative levels of overall pricing in regions	MATERIALS			LABOUR				RATES		
	Del to Site £	Waste %	Material Cost £	Craft Optve Hrs	Lab Hrs	Labour Cost £	Sunds £	Nett Rate £	Unit	Gross Rate (+7.5%) £
H10: PATENT GLAZING										
Patent glazing with aluminium alloy bars 2000mm long with aluminium wings and seatings for glass, spaced at approximately 600mm centres, glazed with BS 952 Georgian wired cast, 7mm thick										
Roof areas										
single tier	108.78	5.00	**114.22**	1.50	1.50	**37.38**	2.40	**154.00**	m2	**165.55**
multi-tier	116.76	5.00	**122.60**	1.70	1.70	**42.36**	2.40	**167.36**	m2	**179.91**
Vertical surfaces										
single tier	121.17	5.00	**127.23**	1.50	1.50	**37.38**	2.40	**167.01**	m2	**179.53**
multi-tier	129.15	5.00	**135.61**	1.70	1.70	**42.36**	2.40	**180.37**	m2	**193.90**
Patent glazing with aluminium alloy bars 2000mm long with aluminium wings and seatings for glass, spaced at approximately 600mm centres, glazed with BS 952 Georgian wired polished, 6mm thick										
Roof areas										
single tier	170.10	5.00	**178.61**	1.65	1.65	**41.12**	2.40	**222.12**	m2	**238.78**
multi-tier	182.81	5.00	**191.95**	1.90	1.90	**47.35**	2.40	**241.70**	m2	**259.83**
Vertical surfaces										
single tier	182.49	5.00	**191.61**	1.65	1.65	**41.12**	2.40	**235.13**	m2	**252.77**
multi-tier	194.15	5.00	**203.86**	1.90	1.90	**47.35**	2.40	**253.61**	m2	**272.63**
Patent glazing with aluminium alloy bars 3000mm long with aluminium wings and seatings for glass, spaced at approximately 600mm centres, glazed with BS 952 Georgian wired cast, 7mm thick										
Roof areas										
single tier	129.26	5.00	**135.72**	1.50	1.50	**37.38**	2.40	**175.50**	m2	**188.67**
multi-tier	137.66	5.00	**144.54**	1.70	1.70	**42.36**	2.40	**189.31**	m2	**203.51**
Vertical surfaces										
single tier	140.39	5.00	**147.41**	1.50	1.50	**37.38**	2.40	**187.19**	m2	**201.23**
multi-tier	150.05	5.00	**157.55**	1.70	1.70	**42.36**	2.40	**202.32**	m2	**217.49**
Patent glazing with aluminium alloy bars 3000mm long with aluminium wings and seatings for glass, spaced at approximately 600mm centres, glazed with BS 952 Georgian wired polished, 6mm thick										
Roof areas										
single tier	189.53	5.00	**199.01**	1.65	1.65	**41.12**	2.40	**242.52**	m2	**260.71**
multi-tier	203.70	5.00	**213.89**	1.90	1.90	**47.35**	2.40	**263.63**	m2	**283.41**
Vertical surfaces										
single tier	201.92	5.00	**212.02**	1.65	1.65	**41.12**	2.40	**255.53**	m2	**274.70**
multi-tier	216.20	5.00	**227.01**	1.90	1.90	**47.35**	2.40	**276.76**	m2	**297.51**
Extra over patent glazing in roof areas with aluminium alloy bars 2000mm long with aluminium wings and seatings for glass, spaced at approximately 600mm centres, glazed with BS 952 Georgian wired polished, 6mm thick										
Opening lights including opening gear										
600 x 600mm	230.41	5.00	**241.93**	4.40	4.40	**109.65**	-	**351.58**	nr	**377.95**
600 x 900mm	279.85	5.00	**293.84**	5.50	5.50	**137.06**	-	**430.90**	nr	**463.22**

Labour hourly rates: (except Specialists) Craft Operatives £14.21 Labourer £10.71 Rates are national average prices. Refer to REGIONAL VARIATIONS for indicative levels of overall pricing in regions	MATERIALS			LABOUR				RATES		
	Del to Site £	Waste %	Material Cost £	Craft Optve Hrs	Lab Hrs	Labour Cost £	Sunds £	Nett Rate £	Unit	Gross Rate (+7.5%) £
H10: PATENT GLAZING Cont.										
Extra over patent glazing in roof areas with aluminium alloy bars 2000mm long with aluminium wings and seatings for glass, spaced at approximately 600mm centres, glazed with BS 952 Georgian wired cast, 7mm thick										
Opening lights including opening gear										
600 x 600mm	230.41	5.00	**241.93**	4.00	4.00	**99.68**	-	**341.61**	nr	**367.23**
600 x 900mm	279.85	5.00	**293.84**	5.00	5.00	**124.60**	-	**418.44**	nr	**449.83**
Extra over patent glazing in vertical surfaces with aluminium alloy bars 3000mm long with aluminium wings and seatings for glass, spaced at approximately 600mm centres, glazed with BS 952 Georgian wired cast, 7mm thick										
Opening lights including opening gear										
600 x 600mm	336.30	5.00	**353.12**	2.00	2.00	**49.84**	-	**402.96**	nr	**433.18**
600 x 900mm	408.91	5.00	**429.36**	3.00	3.00	**74.76**	-	**504.12**	nr	**541.92**
Extra over patent glazing to vertical surfaces with aluminium alloy bars 3000mm long with aluminium wings and seatings for glass, spaced at approximately 600mm centres, glazed with BS 952 Georgian wired polished, 6mm thick										
Opening lights including opening gear										
600 x 600mm	336.30	5.00	**353.12**	2.20	2.20	**54.82**	-	**407.94**	nr	**438.53**
600 x 900mm	408.91	5.00	**429.36**	3.30	3.30	**82.24**	-	**511.59**	nr	**549.96**
H21: TIMBER WEATHERBOARDING										
Softwood, wrought, impregnated										
Boarding to walls, shiplapped joints; 19mm thick, 150mm wide boards										
width exceeding 300mm; fixing to timber	7.30	10.00	**8.03**	0.70	0.09	**10.91**	0.11	**19.05**	m2	**20.48**
Boarding to walls, shiplapped joints; 25mm thick, 150mm wide boards										
width exceeding 300mm; fixing to timber	9.60	10.00	**10.56**	0.75	0.10	**11.73**	0.12	**22.41**	m2	**24.09**
Western Red Cedar										
Boarding to walls, shiplapped joints; 19mm thick, 150mm wide boards										
width exceeding 300mm; fixing to timber	25.08	10.00	**27.59**	0.75	0.10	**11.73**	0.12	**39.44**	m2	**42.39**
Boarding to walls, shiplapped joints; 25mm thick, 150mm wide boards										
width exceeding 300mm; fixing to timber	33.00	10.00	**36.30**	0.80	0.10	**12.44**	0.13	**48.87**	m2	**52.53**
Abutments										
19mm thick softwood boarding										
raking cutting	0.73	10.00	**0.80**	0.06	-	**0.85**	-	**1.66**	m	**1.78**
curved cutting	1.10	10.00	**1.20**	0.09	-	**1.28**	-	**2.48**	m	**2.67**
25mm thick softwood boarding										
raking cutting	0.96	10.00	**1.06**	0.07	-	**0.99**	-	**2.05**	m	**2.20**
curved cutting	1.44	10.00	**1.58**	0.10	-	**1.42**	-	**3.01**	m	**3.23**
19mm thick Western Red Cedar boarding										
raking cutting	2.51	10.00	**2.76**	0.08	-	**1.14**	-	**3.90**	m	**4.19**
curved cutting	3.76	10.00	**4.14**	0.12	-	**1.71**	-	**5.84**	m	**6.28**
25mm thick Western Red Cedar boarding										
raking cutting	3.30	10.00	**3.63**	0.09	-	**1.28**	-	**4.91**	m	**5.28**
curved cutting	4.95	10.00	**5.45**	0.13	-	**1.85**	-	**7.29**	m	**7.84**
H30: FIBRE CEMENT PROFILED SHEET CLADDING/COVERING/SIDING										
Roof coverings; corrugated reinforced cement Profile 3 sheeting, standard grey colour; lapped one and a half corrugations at sides and 150mm at ends										
Coverings; fixing to timber joists at 900mm general spacing with galvanized mild steel drive screws and washers; drilling holes										
pitch 30 degrees from horizontal	10.92	15.00	**12.56**	0.25	0.25	**6.23**	0.44	**19.23**	m2	**20.67**
Coverings; fixing to steel purlins at 900mm general spacing with galvanized hook bolts and washers; drilling holes										
pitch 30 degrees from horizontal	10.92	15.00	**12.56**	0.33	0.33	**8.22**	0.87	**21.65**	m2	**23.28**
Eaves										
eaves filler pieces	9.36	10.00	**10.30**	0.10	0.10	**2.49**	-	**12.79**	m	**13.75**
Ridges										
two piece plain angular adjustable ridge tiles	13.54	10.00	**14.89**	0.20	0.20	**4.98**	-	**19.88**	m	**21.37**

Labour hourly rates: (except Specialists) Craft Operatives £14.21 Labourer £10.71 Rates are national average prices. Refer to REGIONAL VARIATIONS for indicative levels of overall pricing in regions	MATERIALS			LABOUR				RATES		
	Del to Site £	Waste %	Material Cost £	Craft Optve Hrs	Lab Hrs	Labour Cost £	Sunds £	Nett Rate £	Unit	Gross Rate (+7.5%) £
H30: FIBRE CEMENT PROFILED SHEET CLADDING/COVERING/SIDING Cont.										
Roof coverings; corrugated reinforced cement Profile 3 sheeting, standard grey colour; lapped one and a half corrugations at sides and 150mm at ends Cont.										
Barge boards										
standard barge boards	7.67	10.00	**8.44**	0.10	0.10	**2.49**	-	**10.93**	m	**11.75**
Aprons/sills										
apron flashings	9.40	10.00	**10.34**	0.13	0.13	**3.24**	-	**13.58**	m	**14.60**
Finials										
standard ridge cap finials	8.47	10.00	**9.32**	0.15	0.15	**3.74**	-	**13.06**	nr	**14.03**
Cutting										
raking	2.18	15.00	**2.51**	0.25	0.25	**6.23**	-	**8.74**	m	**9.40**
Holes										
for pipes, standards or the like	-	-	**-**	0.50	-	**7.11**	-	**7.11**	nr	**7.64**
Roof coverings; corrugated reinforced cement Profile 3 sheeting, standard coloured; lapped one and a half corrugations at sides and 150mm at ends										
Coverings; fixing to timber joists at 900mm general spacing with galvanized mild steel drive screws and washers; drilling holes										
pitch 30 degrees from horizontal	13.10	15.00	**15.07**	0.25	0.25	**6.23**	0.56	**21.86**	m2	**23.49**
Coverings; fixing to steel purlins at 900mm general spacing with galvanized hook bolts and washers; drilling holes										
pitch 30 degrees from horizontal	13.10	15.00	**15.07**	0.33	0.33	**8.22**	1.12	**24.41**	m2	**26.24**
Eaves										
eaves filler pieces	11.71	10.00	**12.88**	0.10	0.10	**2.49**	-	**15.37**	m	**16.53**
Ridges										
two piece plain angular adjustable ridge tiles	16.33	10.00	**17.96**	0.20	0.20	**4.98**	-	**22.95**	m	**24.67**
Barge boards										
standard barge boards	9.59	10.00	**10.55**	0.10	0.10	**2.49**	-	**13.04**	m	**14.02**
Aprons/sills										
apron flashings	11.74	10.00	**12.91**	0.13	0.13	**3.24**	-	**16.15**	m	**17.37**
Finials										
standard ridge cap finials	9.63	10.00	**10.59**	0.15	0.15	**3.74**	-	**14.33**	nr	**15.41**
Cutting										
raking	2.62	15.00	**3.01**	0.25	0.25	**6.23**	-	**9.24**	m	**9.94**
Holes										
for pipes, standards or the like	-	-	**-**	0.50	-	**7.11**	-	**7.11**	nr	**7.64**
Roof coverings; corrugated reinforced cement Profile 6 sheeting, standard grey colour; lapped half a corrugation at sides and 150mm at ends										
Coverings; fixing to timber joists at 900mm general spacing with galvanized mild steel drive screws and washers; drilling holes										
pitch 30 degrees from horizontal	8.69	15.00	**9.99**	0.25	0.25	**6.23**	0.45	**16.67**	m2	**17.92**
Coverings; fixing to steel purlins at 900mm general spacing with galvanized hook bolts and washers; drilling holes										
pitch 30 degrees from horizontal	8.69	15.00	**9.99**	0.33	0.33	**8.22**	0.91	**19.13**	m2	**20.56**
Eaves										
eaves filler pieces	6.21	10.00	**6.83**	0.10	0.10	**2.49**	-	**9.32**	m	**10.02**
Ridges										
two piece plain angular adjustable ridge tiles	11.42	10.00	**12.56**	0.20	0.20	**4.98**	-	**17.55**	m	**18.86**
Barge boards										
standard barge boards	7.67	10.00	**8.44**	0.10	0.10	**2.49**	-	**10.93**	m	**11.75**
Aprons/sills										
apron flashings	6.69	10.00	**7.36**	0.13	0.13	**3.24**	-	**10.60**	m	**11.39**
Finials										
standard ridge cap finials	7.56	10.00	**8.32**	0.15	0.15	**3.74**	-	**12.05**	nr	**12.96**
Cutting										
raking	1.74	15.00	**2.00**	0.25	0.25	**6.23**	-	**8.23**	m	**8.85**
Holes										
for pipes, standards or the like	-	-	**-**	0.50	-	**7.11**	-	**7.11**	nr	**7.64**

Labour hourly rates: (except Specialists) Craft Operatives £14.21 Labourer £10.71 Rates are national average prices. Refer to REGIONAL VARIATIONS for indicative levels of overall pricing in regions	MATERIALS			LABOUR				RATES		
	Del to Site £	Waste %	Material Cost £	Craft Optve Hrs	Lab Hrs	Labour Cost £	Sunds £	Nett Rate £	Unit	Gross Rate (+7.5%) £
H30: FIBRE CEMENT PROFILED SHEET CLADDING/COVERING/SIDING Cont.										
Roof coverings; corrugated reinforced cement Profile 6 sheeting, standard coloured; lapped half a corrugation at sides and 150mm at ends										
Coverings; fixing to timber joists at 900mm general spacing with galvanized mild steel drive screws and washers; drilling holes										
pitch 30 degrees from horizontal	10.43	15.00	**11.99**	0.25	0.25	**6.23**	0.56	**18.78**	m2	**20.19**
Coverings; fixing to steel purlins at 900mm general spacing with galvanized hook bolts and washers; drilling holes										
pitch 30 degrees from horizontal	10.43	15.00	**11.99**	0.33	0.33	**8.22**	1.12	**21.34**	m2	**22.94**
Eaves										
eaves filler pieces	7.76	10.00	**8.54**	0.10	0.10	**2.49**	-	**11.03**	m	**11.86**
Ridges										
two piece plain angular adjustable ridge tiles	14.28	10.00	**15.71**	0.20	0.20	**4.98**	-	**20.69**	m	**22.24**
Barge boards										
standard barge boards	9.59	10.00	**10.55**	0.10	0.10	**2.49**	-	**13.04**	m	**14.02**
Aprons/sills										
apron flashings	8.38	10.00	**9.22**	0.13	0.13	**3.24**	-	**12.46**	m	**13.39**
Finials										
standard ridge cap finials	9.45	10.00	**10.40**	0.15	0.15	**3.74**	-	**14.13**	nr	**15.19**
Cutting										
raking	2.09	15.00	**2.40**	0.25	0.25	**6.23**	-	**8.63**	m	**9.28**
Holes										
for pipes, standards or the like	-	-	**-**	0.50	-	**7.11**	-	**7.11**	nr	**7.64**
H31: METAL PROFILED/FLAT SHEET CLADDING/COVERING/SIDING										
Wall cladding; PVC colour coated both sides galvanized steel profiled sheeting 0.70mm thick and with sheets secured at seams and laps										
Coverings; fixing to steel rails at 900mm general spacing with galvanized hook bolts and washers; drilling holes										
pitch 90 degrees from horizontal	14.40	5.00	**15.12**	0.35	0.35	**8.72**	1.06	**24.90**	m2	**26.77**
Vertical angles										
vertical corner flashings	5.38	5.00	**5.65**	0.15	0.15	**3.74**	-	**9.39**	m	**10.09**
Filler blocks										
polyethylene	0.61	5.00	**0.64**	0.10	0.10	**2.49**	-	**3.13**	m	**3.37**
PVC	1.71	5.00	**1.80**	0.10	0.10	**2.49**	-	**4.29**	m	**4.61**
black synthetic rubber	1.90	5.00	**2.00**	0.10	0.10	**2.49**	-	**4.49**	m	**4.82**
Cutting										
raking	2.88	5.00	**3.02**	0.25	0.25	**6.23**	-	**9.25**	m	**9.95**
Roof coverings; PVC colour coated both sides galvanized steel profiled sheeting 0.70mm thick and with sheets secured at seams and laps										
Coverings; fixing to steel purlins at 900mm general spacing with galvanized hook bolts and washers; drilling holes										
pitch 30 degrees from horizontal	14.40	5.00	**15.12**	0.30	0.30	**7.48**	1.08	**23.68**	m2	**25.45**
Ridges										
ridge cappings	5.38	5.00	**5.65**	0.15	0.15	**3.74**	-	**9.39**	m	**10.09**
Flashings										
gable flashings	5.38	5.00	**5.65**	0.15	0.15	**3.74**	-	**9.39**	m	**10.09**
eaves flashings	3.58	5.00	**3.76**	0.15	0.15	**3.74**	-	**7.50**	m	**8.06**
Filler blocks										
polyethylene	0.61	5.00	**0.64**	0.10	0.10	**2.49**	-	**3.13**	m	**3.37**
PVC	1.71	5.00	**1.80**	0.10	0.10	**2.49**	-	**4.29**	m	**4.61**
black synthetic rubber	1.90	5.00	**2.00**	0.10	0.10	**2.49**	-	**4.49**	m	**4.82**
Cutting										
raking	2.88	5.00	**3.02**	0.25	0.25	**6.23**	-	**9.25**	m	**9.95**
H32: PLASTICS PROFILED SHEET CLADDING/ COVERING/SIDING										
Roof coverings; standard corrugated glass fibre reinforced translucent sheeting; lapped one and a half corrugations at sides and 150mm at ends										
Coverings; fixing to timber joists at 900mm general spacing with galvanized mild steel drive screws and washers; drilling holes										
pitch 30 degrees from horizontal	13.27	5.00	**13.93**	0.25	0.25	**6.23**	0.30	**20.46**	m2	**22.00**

Labour hourly rates: (except Specialists) Craft Operatives £14.21 Labourer £10.71 Rates are national average prices. Refer to REGIONAL VARIATIONS for indicative levels of overall pricing in regions	MATERIALS			LABOUR				RATES		
	Del to Site £	Waste %	Material Cost £	Craft Optve Hrs	Lab Hrs	Labour Cost £	Sunds £	Nett Rate £	Unit	Gross Rate (+7.5%) £
H32: PLASTICS PROFILED SHEET CLADDING/ COVERING/SIDING Cont.										
Roof coverings; standard corrugated glass fibre reinforced translucent sheeting; lapped one and a half corrugations at sides and 150mm at ends Cont.										
Coverings; fixing to steel purlins at 900mm general spacing with galvanized hook bolts and washers; drilling holes										
pitch 30 degrees from horizontal	13.27	5.00	**13.93**	0.30	0.30	**7.48**	0.37	**21.78**	m2	**23.41**
Cutting										
raking	2.65	5.00	**2.79**	0.25	0.25	**6.23**	-	**9.02**	m	**9.69**
Roof coverings; fire resisting corrugated glass fibre reinforced translucent sheeting; lapped one and a half corrugations at sides and 150mm at ends										
Coverings; fixing to timber joists at 900mm general spacing with galvanized mild steel drive screws and washers; drilling holes										
pitch 30 degrees from horizontal	15.59	5.00	**16.37**	0.25	0.25	**6.23**	0.30	**22.90**	m2	**24.62**
Coverings; fixing to steel purlins at 900mm general spacing with galvanized hook bolts and washers; drilling holes										
pitch 30 degrees from horizontal	15.59	5.00	**16.37**	0.30	0.30	**7.48**	0.46	**24.31**	m2	**26.13**
Cutting										
raking	3.12	5.00	**3.27**	0.25	0.25	**6.23**	-	**9.50**	m	**10.22**
Roof coverings; standard vinyl corrugated sheeting; lapped one and a half corrugations at sides and 150mm at ends										
Coverings; fixing to timber joists at 900mm general spacing with galvanized mild steel drive screws and washers; drilling holes										
pitch 30 degrees from horizontal	17.65	5.00	**18.53**	0.25	0.25	**6.23**	0.46	**25.22**	m2	**27.11**
Coverings; fixing to steel purlins at 900mm general spacing with galvanized hook bolts and washers; drilling holes										
pitch 30 degrees from horizontal	17.65	5.00	**18.53**	0.30	0.30	**7.48**	0.46	**26.47**	m2	**28.45**
Cutting										
raking	3.53	5.00	**3.71**	0.25	0.25	**6.23**	-	**9.94**	m	**10.68**
Wall cladding; Swish Products high impact rigid uPVC profiled sections; colour white; secured with starter sections and clips										
Coverings with shiplap profile code C002 giving 150mm cover; fixing to timber										
vertical cladding										
sections applied horizontally	35.75	5.00	**37.54**	0.38	0.38	**9.47**	1.75	**48.76**	m2	**52.41**
sections applied vertically	35.75	5.00	**37.54**	0.38	0.38	**9.47**	1.75	**48.76**	m2	**52.41**
Coverings with open V profile code C003 giving 150mm cover; fixing to timber										
vertical cladding										
sections applied horizontally	35.75	5.00	**37.54**	0.40	0.40	**9.97**	2.50	**50.01**	m2	**53.76**
sections applied vertically	35.75	5.00	**37.54**	0.40	0.40	**9.97**	2.50	**50.01**	m2	**53.76**
Coverings with open V profile code C269 giving 100mm cover; fixing to timber										
vertical cladding										
sections applied horizontally	40.36	5.00	**42.38**	0.35	0.35	**8.72**	1.75	**52.85**	m2	**56.81**
sections applied vertically	40.36	5.00	**42.38**	0.35	0.35	**8.72**	1.75	**52.85**	m2	**56.81**
Vertical angles										
section C030 for vertically applied section	4.09	5.00	**4.29**	0.10	0.10	**2.49**	0.14	**6.93**	m	**7.45**
H51: NATURAL STONE SLAB CLADDING/FEATURES										
English blue/grey slate facings (Best Quality); natural riven finish; bedding, jointing and pointing in gauged mortar (1:2:9)										
450 x 600 x 20mm units to walls on brickwork or blockwork base										
plain, width exceeding 300mm	84.89	2.50	**87.01**	2.00	2.00	**49.84**	0.45	**137.30**	m2	**147.60**
450 x 600 x 30mm units to walls on brickwork or blockwork base										
plain, width exceeding 300mm	111.35	2.50	**114.13**	2.25	2.25	**56.07**	0.51	**170.71**	m2	**183.52**
450 x 600 x 40mm units to walls on brickwork or blockwork base										
plain, width exceeding 300mm	122.38	2.50	**125.44**	2.50	2.50	**62.30**	0.70	**188.44**	m2	**202.57**
450 x 600 x 50mm units to walls on brickwork or blockwork base										
plain, width exceeding 300mm	151.04	2.50	**154.82**	3.00	3.00	**74.76**	0.86	**230.44**	m2	**247.72**

Labour hourly rates: (except Specialists) Craft Operatives £14.21 Labourer £10.71 Rates are national average prices. Refer to REGIONAL VARIATIONS for indicative levels of overall pricing in regions	MATERIALS			LABOUR				RATES		
	Del to Site £	Waste %	Material Cost £	Craft Optve Hrs	Lab Hrs	Labour Cost £	Sunds £	Nett Rate £	Unit	Gross Rate (+7.5%) £
H51: NATURAL STONE SLAB CLADDING/FEATURES Cont.										
English blue/grey slate facings (Best quality); fine rubbed finish, one face; bedding, jointing and pointing in gauged mortar (1:2:9)										
750 x 1200 x 20mm units to walls on brickwork or blockwork base										
plain, width exceeding 300mm....................	212.00	2.50	**217.30**	2.00	2.00	**49.84**	1.31	**268.45**	m2	**288.58**
extra; rubbed square edges......................	17.00	-	**17.00**	-	-	**-**	-	**17.00**	m	**18.28**
extra; half rounded edges	52.10	-	**52.10**	-	-	**-**	-	**52.10**	m	**56.01**
extra; full rounded edges	52.10	-	**52.10**	-	-	**-**	-	**52.10**	m	**56.01**
extra; rebated joints	24.62	-	**24.62**	-	-	**-**	-	**24.62**	m	**26.47**
750 x 1200 x 30mm units to walls on brickwork or blockwork base										
plain, width exceeding 300mm....................	253.34	2.50	**259.67**	2.25	2.25	**56.07**	1.39	**317.13**	m2	**340.92**
extra; rubbed square edges......................	21.45	-	**21.45**	-	-	**-**	-	**21.45**	m	**23.06**
extra; half rounded edges	59.50	-	**59.50**	-	-	**-**	-	**59.50**	m	**63.96**
extra; full rounded edges	59.50	-	**59.50**	-	-	**-**	-	**59.50**	m	**63.96**
extra; rebated joints	29.26	-	**29.26**	-	-	**-**	-	**29.26**	m	**31.45**
750 x 1200 x 40mm units to walls on brickwork or blockwork base										
plain, width exceeding 300mm....................	259.70	2.50	**266.19**	2.50	2.50	**62.30**	1.56	**330.05**	m2	**354.81**
extra; rubbed square edges......................	21.45	-	**21.45**	-	-	**-**	-	**21.45**	m	**23.06**
extra; half rounded edges	59.50	-	**59.50**	-	-	**-**	-	**59.50**	m	**63.96**
extra; full rounded edges	59.50	-	**59.50**	-	-	**-**	-	**59.50**	m	**63.96**
extra; rebated joints	29.26	-	**29.26**	-	-	**-**	-	**29.26**	m	**31.45**
750 x 1200 x 50mm units to walls on brickwork or blockwork base										
plain, width exceeding 300mm....................	275.60	2.50	**282.49**	3.00	3.00	**74.76**	1.73	**358.98**	m2	**385.90**
extra; rubbed square edges......................	21.45	-	**21.45**	-	-	**-**	-	**21.45**	m	**23.06**
extra; half rounded edges	59.50	-	**59.50**	-	-	**-**	-	**59.50**	m	**63.96**
extra; full rounded edges	59.50	-	**59.50**	-	-	**-**	-	**59.50**	m	**63.96**
extra; rebated joints	29.26	-	**29.26**	-	-	**-**	-	**29.26**	m	**31.45**
English blue/grey slate facings (Best quality); fine rubbed finish, both faces; bedding, jointing and pointing in gauged mortar (1:2:9)										
750 x 1200 x 20mm units to walls on brickwork or blockwork base										
plain, width exceeding 300mm....................	247.00	2.50	**253.18**	2.50	2.50	**62.30**	2.17	**317.65**	m2	**341.47**
750 x 1200 x 30mm units to walls on brickwork or blockwork base										
plain, width exceeding 300mm....................	288.34	2.50	**295.55**	2.75	2.75	**68.53**	2.28	**366.36**	m2	**393.84**
750 x 1200 x 40mm units to walls on brickwork or blockwork base										
plain, width exceeding 300mm....................	294.70	2.50	**302.07**	3.00	3.00	**74.76**	2.46	**379.29**	m2	**407.73**
750 x 1200 x 50mm units to walls on brickwork or blockwork base										
plain, width exceeding 300mm....................	310.60	2.50	**318.37**	3.50	3.50	**87.22**	2.58	**408.17**	m2	**438.78**
Kirkstone green slate; bedding, jointing and pointing in gauged mortar (1:2:9)										
30mm thick units to sills on brickwork or blockwork base										
width 125mm; weathered, throated and grooved.....	60.10	2.50	**61.60**	0.55	0.55	**13.71**	0.57	**75.88**	m	**81.57**
width 190mm; weathered, throated and grooved.....	76.88	2.50	**78.80**	0.60	0.60	**14.95**	0.79	**94.54**	m	**101.63**
40mm thick units to sills on brickwork or blockwork base										
width 125mm	67.45	2.50	**69.14**	0.60	0.60	**14.95**	0.57	**84.66**	m	**91.01**
extra; stooling for jambs	22.26	-	**22.26**	-	-	**-**	-	**22.26**	nr	**23.93**
extra; notching for jambs and mullions...........	22.26	-	**22.26**	-	-	**-**	-	**22.26**	nr	**23.93**
width 190mm	88.52	2.50	**90.73**	0.75	0.75	**18.69**	0.74	**110.16**	m	**118.43**
extra; stooling for jambs	22.26	-	**22.26**	-	-	**-**	-	**22.26**	nr	**23.93**
extra; notching for jambs and mullions...........	22.26	-	**22.26**	-	-	**-**	-	**22.26**	nr	**23.93**
50mm thick units to sills on brickwork or blockwork base										
width 125mm	79.86	2.50	**81.86**	0.75	0.75	**18.69**	0.57	**101.12**	m	**108.70**
extra; stooling for jambs	22.26	-	**22.26**	-	-	**-**	-	**22.26**	nr	**23.93**
extra; notching for jambs and mullions...........	22.26	-	**22.26**	-	-	**-**	-	**22.26**	nr	**23.93**
width 190mm	106.91	2.50	**109.58**	1.05	1.05	**26.17**	0.74	**136.49**	m	**146.73**
extra; stooling for jambs	22.26	-	**22.26**	-	-	**-**	-	**22.26**	nr	**23.93**
extra; notching for jambs and mullions...........	22.26	-	**22.26**	-	-	**-**	-	**22.26**	nr	**23.93**
30mm units to window boards on brickwork or blockwork base										
width 230mm	76.09	2.50	**77.99**	0.75	0.75	**18.69**	0.93	**97.61**	m	**104.93**
25mm units to combined sills and window boards on brickwork or blockwork base										
width 360mm	109.66	2.50	**112.40**	1.00	1.00	**24.92**	1.31	**138.63**	m	**149.03**

Labour hourly rates: (except Specialists) Craft Operatives £14.21 Labourer £10.71 Rates are national average prices. Refer to REGIONAL VARIATIONS for indicative levels of overall pricing in regions	MATERIALS			LABOUR				RATES		
	Del to Site £	Waste %	Material Cost £	Craft Optve Hrs	Lab Hrs	Labour Cost £	Sunds £	Nett Rate £	Unit	Gross Rate (+7.5%) £
H60: PLAIN ROOF TILING										
Quantities required for 1m2 of tiling										
Plain tiles to 65mm lap tiles, 60nr battens, 10m										
Plain tiles to 85mm lap tiles, 70nr battens, 11m										
Roof coverings; clayware Machine made plain tiles BS EN 1304, red, 265 x 165mm; fixing every fourth course with two galvanized nails per tile to 65mm lap; 38 x 19mm pressure impregnated softwood battens fixed with galvanized nails; underlay BS 747 type 1F reinforced bitumen felt; 150mm laps; fixing with galvanized steel clout nails										
Pitched 50 degrees from horizontal										
generally	30.95	5.00	**32.68**	0.62	0.31	**12.13**	1.39	**46.20**	m2	**49.67**
holes	-	-	**-**	0.50	0.25	**9.78**	-	**9.78**	nr	**10.52**
abutments; square	2.77	5.00	**2.91**	0.25	0.13	**4.94**	-	**7.86**	m	**8.44**
abutments; raking	3.93	5.00	**4.12**	0.38	0.19	**7.43**	-	**11.56**	m	**12.42**
double course at eaves; Machine made eaves tile	2.88	5.00	**3.02**	0.30	0.15	**5.87**	0.49	**9.38**	m	**10.09**
verges; bed and point in coloured cement-lime mortar (1:1:6)	2.77	5.00	**2.91**	0.35	0.18	**6.90**	0.37	**10.18**	m	**10.95**
verges; single extra undercloak course plain tiles; bed and point in coloured cement-lime mortar (1:1:6)	3.70	5.00	**3.88**	0.50	0.25	**9.78**	0.37	**14.03**	m	**15.09**
Machine made ridge tiles, half round; in 300mm effective lengths; butt jointed; bedding and pointing in coloured cement-lime mortar (1:1:6)	16.32	5.00	**17.13**	0.30	0.15	**5.87**	0.37	**23.37**	m	**25.13**
Machine made hip tiles, half round; in 300mm effective lengths; butt jointed; bedding and pointing in coloured cement-lime mortar (1:1:6)	16.32	5.00	**17.13**	0.30	0.15	**5.87**	0.37	**23.37**	m	**25.13**
Machine made hip tiles; angular	16.32	5.00	**17.13**	0.30	0.15	**5.87**	0.37	**23.37**	m	**25.13**
Machine made hip tiles; bonnet pattern; bedding and pointing in coloured cement-lime mortar (1:1:6)	16.32	5.00	**17.13**	0.30	0.15	**5.87**	0.37	**23.37**	m	**25.13**
Machine made valley tiles; angular	16.32	5.00	**17.13**	0.45	0.23	**8.86**	0.37	**26.36**	m	**28.34**
Roof coverings; clayware Machine made plain tiles BS EN 1304, red, 265 x 165mm; fixing every fourth course with two galvanized nails per tile to 65mm lap; 50 x 19mm pressure impregnated softwood battens fixed with galvanized nails; underlay BS 747 type 1F reinforced bitumen felt; 150mm laps; fixing with galvanized steel clout nails										
Pitched 50 degrees from horizontal										
generally	32.15	5.00	**34.00**	0.62	0.31	**12.13**	1.39	**47.52**	m2	**51.09**
Roof coverings; clayware Machine made plain tiles BS EN 1304, red, 265 x 165mm; fixing every course with two galvanized nails per tile to 65mm lap; 38 x 19mm pressure impregnated softwood battens fixed with galvanized nails; underlay BS 747 type 1F reinforced bitumen felt; 150mm laps; fixing with galvanized steel clout nails										
Pitched 50 degrees from horizontal										
generally	30.95	5.00	**32.68**	0.74	0.37	**14.48**	2.67	**49.83**	m2	**53.57**
Roof coverings; clayware Machine made plain tiles BS EN 1304, red, 265 x 165mm; fixing every fourth course with two copper nails per tile to 65mm lap; 38 x19mm pressure impregnated softwood battens fixed with galvanized nails; underlay BS 747 type 1F reinforced bitumen felt; 150mm laps; fixing with galvanized steel clout nails										
Pitched 50 degrees from horizontal										
generally	30.95	5.00	**32.68**	0.62	0.31	**12.13**	2.13	**46.94**	m2	**50.46**
Roof coverings; clayware Machine made plain tiles BS EN 1304, red, 265 x 165mm; fixing every course with two copper nails per tile to 65mm lap; 38 x 19mm pressure impregnated softwood battens fixed with galvanized nails; underlay BS 747 type 1F reinforced bitumen felt; 150mm laps; fixing with galvanized steel clout nails										
Pitched 50 degrees from horizontal										
generally	30.95	5.00	**32.68**	0.74	0.37	**14.48**	4.89	**52.05**	m2	**55.95**
Roof coverings; clayware Machine made plain tiles BS EN 1304, red, 265 x 165mm; fixing every fourth course with two galvanized nails per tile to 85mm lap; 38 x 19mm pressure impregnated softwood battens fixed with galvanized nails; underlay BS 747 type 1F reinforced bitumen felt; 150mm laps; fixing with galvanized steel clout nails										
Pitched 40 degrees from horizontal										
generally	35.84	5.00	**37.83**	0.70	0.35	**13.70**	1.39	**52.92**	m2	**56.88**

Labour hourly rates: (except Specialists) Craft Operatives £14.21 Labourer £10.71 Rates are national average prices. Refer to REGIONAL VARIATIONS for indicative levels of overall pricing in regions	MATERIALS			LABOUR				RATES		
	Del to Site £	Waste %	Material Cost £	Craft Optve Hrs	Lab Hrs	Labour Cost £	Sunds £	Nett Rate £	Unit	Gross Rate (+7.5%) £
H60: PLAIN ROOF TILING Cont.										
Wall coverings; clayware Machine made plain tiles BS EN 1304, red, 265 x 165mm;fixing every course with two galvanized nails per tile to 38mm lap; 38 x 19mm pressure impregnated softwood battens fixed with galvanized nails; underlay BS 747 type 1F reinforced bitumen felt; 150mm laps; fixing with galvanized steel clout nails										
Vertical										
generally	27.57	5.00	**29.14**	0.75	0.38	**14.73**	2.03	**45.90**	m2	**49.34**
Roof coverings; clayware Hand made plain tiles BS EN 1304, red, 265 x 165mm; fixing every fourth course with two galvanized nails per tile to 65mm lap; 38 x 19mm pressure impregnated softwood battens fixed with galvanized nails; underlay BS 747 type 1F reinforced bitumen felt; 150mm laps; fixing with galvanized steel clout nails										
Pitched 50 degrees from horizontal										
generally	45.13	5.00	**47.57**	0.62	0.31	**12.13**	1.39	**61.09**	m2	**65.67**
holes	-	-	**-**	0.50	0.25	**9.78**	-	**9.78**	nr	**10.52**
abutments; square	4.19	5.00	**4.40**	0.25	0.13	**4.94**	-	**9.34**	m	**10.04**
abutments; raking	5.94	5.00	**6.23**	0.38	0.19	**7.43**	-	**13.67**	m	**14.69**
double course at eaves; Hand made eaves tile	4.80	5.00	**5.04**	0.30	0.15	**5.87**	0.49	**11.40**	m	**12.25**
verges; bed and point in coloured cement-lime mortar (1:1:6)	4.19	5.00	**4.40**	0.35	0.18	**6.90**	0.37	**11.67**	m	**12.55**
verges; single extra undercloak course plain tiles; bed and point in coloured cement-lime mortar (1:1:6)	5.59	5.00	**5.87**	0.50	0.25	**9.78**	0.37	**16.02**	m	**17.22**
Hand made ridge tiles, half round; in 300mm effective lengths; butt jointed; bedding and pointing in coloured cement-lime mortar (1:1:6)	14.69	5.00	**15.42**	0.30	0.15	**5.87**	0.37	**21.66**	m	**23.28**
Hand made hip tiles, half round; in 300mm effective lengths; butt jointed; bedding and pointing in coloured cement-lime mortar (1:1:6)	14.69	5.00	**15.42**	0.30	0.15	**5.87**	0.37	**21.66**	m	**23.28**
Hand made hip tiles; angular	20.85	5.00	**21.89**	0.30	0.15	**5.87**	0.37	**28.13**	m	**30.24**
Hand made hip tiles; bonnet pattern; bedding and pointing in coloured cement-lime mortar (1:1:6)	16.32	5.00	**17.13**	0.30	0.15	**5.87**	0.37	**23.37**	m	**25.13**
Hand made valley tiles; angular	15.42	5.00	**16.19**	0.45	0.23	**8.86**	0.37	**25.42**	m	**27.32**
Roof coverings; clayware Hand made plain tiles BS EN 1304, red, 265 x 165mm; fixing every fourth course with two galvanized nails per tile to 85mm lap; 38 x 19mm pressure impregnated softwood battens fixed with galvanized nails; underlay BS 747 type 1F reinforced bitumen felt; 150mm laps; fixing with galvanized steel clout nails										
Pitched 40 degrees from horizontal										
generally	52.38	5.00	**55.20**	0.70	0.35	**13.70**	1.36	**70.25**	m2	**75.52**
Fittings										
Hip irons										
galvanized mild steel; 32 x 3 x 380mm girth; scrolled; fixing with galvanized steel screws to timber	1.84	5.00	**1.93**	0.10	0.05	**1.96**	-	**3.89**	nr	**4.18**
Lead soakers										
fixing only	-	-	**-**	0.30	0.15	**5.87**	-	**5.87**	nr	**6.31**
Roof coverings; Redland Plain granular faced tiles, 268 x 165mm; fixing every fifth course with two aluminium nails to each tile to 65mm lap; 38 x 19mm pressure impregnated softwood battens fixed with galvanized nails; underlay BS 747 type 1F bitumen reinforced felt; 150mm laps; fixing with galvanized steel clout nails										
Pitched 40 degrees from horizontal										
generally	20.73	5.00	**21.93**	0.62	0.31	**12.13**	1.02	**35.08**	m2	**37.71**
double course at eaves and nailing each tile with two aluminium nails	1.80	5.00	**1.89**	0.30	0.15	**5.87**	0.45	**8.21**	m	**8.83**
verges; Redland Dry Verge system with plain tiles and tile-and-a-half tiles in alternate courses, clips and aluminium nails	10.44	5.00	**10.96**	0.25	0.12	**4.84**	0.96	**16.76**	m	**18.02**
verges; plain tile undercloak; tiles and undercloak bedded and pointed in tinted cement mortar (1:3) and with plain tiles and tile-and-a-half tiles in alternate courses and aluminium nails	0.90	5.00	**0.95**	0.25	0.12	**4.84**	0.40	**6.18**	m	**6.65**
double course at top edges with clips and nailing each tile with two aluminium nails	1.82	5.00	**1.91**	0.30	0.15	**5.87**	3.39	**11.17**	m	**12.01**
ridge or hip tiles, Redland Plain angle; butt jointed; bedding and pointing in tinted cement mortar (1:3)	16.02	5.00	**16.82**	0.30	0.15	**5.87**	0.37	**23.06**	m	**24.79**
ridge or hip tiles, Redland half round; butt jointed; bedding and pointing in tinted cement mortar (1:3)	6.67	5.00	**7.00**	0.30	0.15	**5.87**	0.37	**13.24**	m	**14.23**
valley tiles , Redland Plain valley tiles	8.53	5.00	**8.96**	0.30	0.15	**5.87**	0.37	**15.20**	m	**16.34**

Labour hourly rates: (except Specialists) Craft Operatives £14.21 Labourer £10.71 Rates are national average prices. Refer to REGIONAL VARIATIONS for indicative levels of overall pricing in regions	MATERIALS			LABOUR				RATES		
	Del to Site £	Waste %	Material Cost £	Craft Optve Hrs	Lab Hrs	Labour Cost £	Sunds £	Nett Rate £	Unit	Gross Rate (+7.5%) £
H60: PLAIN ROOF TILING Cont.										
Roof coverings; Redland Plain granular faced tiles, 268 x 165mm; fixing every fifth course with two aluminium nails to each tile to 65mm lap; 50 x 25mm pressure impregnated softwood battens fixed with galvanized nails; underlay; BS 747 type 1F bitumen reinforced felt; 150mm laps; fixing with galvanized steel clout nails										
Pitched 40 degrees from horizontal										
generally	20.73	5.00	**21.93**	0.62	0.31	**12.13**	1.02	**35.08**	m2	**37.71**
abutments; square	1.80	5.00	**1.89**	0.25	0.13	**4.94**	-	**6.83**	m	**7.35**
abutments; raking	2.58	5.00	**2.71**	0.38	0.19	**7.43**	-	**10.14**	m	**10.90**
abutments; curved to 3000mm radius	3.60	5.00	**3.78**	0.50	0.25	**9.78**	-	**13.56**	m	**14.58**
Roof coverings; Redland Plain granular faced tiles, 268 x 165mm; fixing each tile with two aluminium nails to 65mm lap; 38 x 19mm pressure impregnated softwood battens fixed with galvanized nails; underlay; BS 747 type 1F bitumen reinforced felt; 150mm laps; fixing with galvanized steel clout nails										
Pitched 40 degrees from horizontal										
generally	20.73	5.00	**21.93**	0.74	0.37	**14.48**	1.28	**37.69**	m2	**40.51**
Roof coverings; Plain through coloured tiles, 268 x 165mm; fixing every fifth course with two aluminium nails to each tile to 65mm lap; 38 x 19mm pressure impregnated softwood battens fixed with galvanized nails; underlay; BS 747 type 1F bitumen reinforced felt; 150mm laps; fixing with galvanized steel clout nails										
Pitched 40 degrees from horizontal										
generally	21.23	5.00	**22.48**	0.62	0.31	**12.13**	1.02	**35.63**	m2	**38.30**
double course at eaves and nailing each tile with two aluminium nails	1.80	5.00	**1.89**	0.30	0.15	**5.87**	0.45	**8.21**	m	**8.83**
verges; Redland Dry Verge system with plain tiles and tile-and-a-half tiles in alternate courses, clips and aluminium nails	10.44	5.00	**10.96**	0.25	0.12	**4.84**	0.96	**16.76**	m	**18.02**
verges; plain tile undercloak; tiles and undercloak bedded and pointed in tinted cement mortar (1:3) and with plain tiles and tile-and-a-half tiles in alternate courses and aluminium nails	1.80	5.00	**1.89**	0.25	0.12	**4.84**	0.40	**7.13**	m	**7.66**
double course at top edges with clips and nailing each tile with two aluminium nails	1.80	5.00	**1.89**	0.30	0.15	**5.87**	3.39	**11.15**	m	**11.99**
ridge or hip tiles, Redland Plain angle; butt jointed; bedding and pointing in tinted cement mortar (1:3)	16.02	5.00	**16.82**	0.30	0.15	**5.87**	0.37	**23.06**	m	**24.79**
ridge or hip tiles, Redland half round; butt jointed; bedding and pointing in tinted cement mortar (1:3)	6.66	5.00	**6.99**	0.30	0.15	**5.87**	0.37	**13.23**	m	**14.22**
valley tiles, Redland Plain valley tiles	8.52	5.00	**8.95**	0.30	0.15	**5.87**	0.37	**15.19**	m	**16.33**
Roof coverings; Redland Downland plain granular faced tiles, 268 x 165mm; fixing every fifth course with two aluminium nails to each tile to 65mm lap; 38 x 19mm pressure impregnated softwood battens fixed with galvanized nails; underlay; BS 747 type 1F bitumen reinforced felt; 150mm laps; fixing with galvanized steel clout nails										
Pitched 40 degrees from horizontal										
generally	23.33	5.00	**24.68**	0.62	0.31	**12.13**	1.02	**37.83**	m2	**40.67**
double course at eaves and nailing each tile with two aluminium nails	2.01	5.00	**2.11**	0.30	0.15	**5.87**	0.45	**8.43**	m	**9.06**
verges; Redland Dry Verge system with plain tiles and tile-and-a-half tiles in alternate courses, clips and aluminium nails	10.55	5.00	**11.07**	0.25	0.12	**4.84**	0.96	**16.87**	m	**18.14**
verges; plain tile undercloak; tiles and undercloak bedded and pointed in tinted cement mortar (1:3) and with plain tiles and tile-and-a-half tiles in alternate courses and aluminium nails	2.01	5.00	**2.11**	0.25	0.12	**4.84**	0.40	**7.35**	m	**7.90**
double course at top edges with clips and nailing each tile with two aluminium nails	2.01	5.00	**2.11**	0.30	0.15	**5.87**	3.39	**11.37**	m	**12.22**
ridge or hip tiles, Redland Plain angle; butt jointed; bedding and pointing in tinted cement mortar (1:3)	16.02	5.00	**16.82**	0.25	0.15	**5.16**	0.37	**22.35**	m	**24.02**
ridge or hip tiles, Redland half round; butt jointed; bedding and pointing in tinted cement mortar (1:3)	6.66	5.00	**6.99**	0.30	0.15	**5.87**	0.37	**13.23**	m	**14.22**
valley tiles, Redland Plain valley tiles	9.82	5.00	**10.31**	0.30	0.15	**5.87**	0.37	**16.55**	m	**17.80**
Roof coverings; Redland Downland plain through coloured tiles, 268 x 165mm; fixing every fifth course with two aluminium nails to each tile to 65mm lap; 38 x 19mm pressure impregnated softwood battens fixed with galvanized nails; underlay; BS 747 type 1F bitumen reinforced felt; 150mm laps; fixing with galvanized steel clout nails										
Pitched 40 degrees from horizontal										
generally	23.33	5.00	**24.68**	0.62	0.31	**12.13**	1.02	**37.83**	m2	**40.67**

Labour hourly rates: (except Specialists) Craft Operatives £14.21 Labourer £10.71 Rates are national average prices. Refer to REGIONAL VARIATIONS for indicative levels of overall pricing in regions	MATERIALS			LABOUR				RATES		
	Del to Site £	Waste %	Material Cost £	Craft Optve Hrs	Lab Hrs	Labour Cost £	Sunds £	Nett Rate £	Unit	Gross Rate (+7.5%) £
H60: PLAIN ROOF TILING Cont.										
Roof coverings; Redland Downland plain through coloured tiles, 268 x 165mm; fixing every fifth course with two aluminium nails to each tile to 65mm lap; 38 x 19mm pressure impregnated softwood battens fixed with galvanized nails; underlay; BS 747 type 1F bitumen reinforced felt; 150mm laps; fixing with galvanized steel clout nails Cont.										
Pitched 40 degrees from horizontal Cont.										
double course at eaves and nailing each tile with two aluminium nails	2.01	5.00	**2.11**	0.30	0.15	**5.87**	0.45	**8.43**	m	**9.06**
verges; Redland Dry Verge system with plain tiles and tile-and-a-half tiles in alternate courses, clips and aluminium nails	10.55	5.00	**11.07**	0.25	0.12	**4.84**	0.96	**16.87**	m	**18.14**
verges; plain tile undercloak; tiles and undercloak bedded and pointed in tinted cement mortar (1:3) and with plain tiles and tile-and-a-half tiles in alternate courses and aluminium nails	2.01	5.00	**2.11**	0.25	0.12	**4.84**	0.40	**7.35**	m	**7.90**
double course at top edges with clips and nailing each tile with two aluminium nails	2.01	5.00	**2.11**	0.30	0.15	**5.87**	3.39	**11.37**	m	**12.22**
ridge or hip tiles, Redland Plain angle; butt jointed; bedding and pointing in tinted cement mortar (1:3)	16.02	5.00	**16.82**	0.30	0.15	**5.87**	0.37	**23.06**	m	**24.79**
ridge or hip tiles, Redland half round; butt jointed; bedding and pointing in tinted cement mortar (1:3)	9.99	5.00	**10.49**	0.30	0.15	**5.87**	0.37	**16.73**	m	**17.98**
valley tiles, Redland Plain valley tiles	9.82	5.00	**10.31**	0.30	0.15	**5.87**	0.37	**16.55**	m	**17.80**
Fittings										
Ventilator tiles										
Red Vent ridge ventilation terminal 450mm long for										
half round ridge	45.20	5.00	**47.46**	1.30	0.65	**25.43**	0.59	**73.48**	nr	**79.00**
universal angle ridge	45.20	5.00	**47.46**	1.30	0.65	**25.43**	0.59	**73.48**	nr	**79.00**
Redvent eaves ventilators fixing with aluminium nails	8.16	5.00	**8.57**	0.50	0.25	**9.78**	0.03	**18.38**	m	**19.76**
Red Vent Thruvent tiles for										
Stonewold slates	34.50	5.00	**36.23**	0.50	0.25	**9.78**	0.22	**46.23**	nr	**49.69**
Regent tiles	34.50	5.00	**36.23**	0.50	0.25	**9.78**	0.22	**46.23**	nr	**49.69**
Grovebury double pantiles	34.50	5.00	**36.23**	0.50	0.25	**9.78**	0.22	**46.23**	nr	**49.69**
Norfolk pantiles	34.50	5.00	**36.23**	0.50	0.25	**9.78**	0.22	**46.23**	nr	**49.69**
Redland 49 tiles	34.50	5.00	**36.23**	0.50	0.25	**9.78**	0.22	**46.23**	nr	**49.69**
Renown tiles	34.50	5.00	**36.23**	0.50	0.25	**9.78**	0.22	**46.23**	nr	**49.69**
Redland 50 double roman tiles	34.50	5.00	**36.23**	0.50	0.25	**9.78**	0.22	**46.23**	nr	**49.69**
Redland plain tiles	40.10	5.00	**42.11**	0.50	0.25	**9.78**	0.22	**52.11**	nr	**56.02**
Redland Downland plain tiles	42.40	5.00	**44.52**	0.50	0.25	**9.78**	0.22	**54.52**	nr	**58.61**
Red line ventilation tile complete with underlay seal and fixing clips for										
Stonewold slates	43.00	5.00	**45.15**	1.00	0.50	**19.57**	0.22	**64.94**	nr	**69.81**
Regent tiles	43.00	5.00	**45.15**	1.00	0.50	**19.57**	0.22	**64.94**	nr	**69.81**
Grovebury double pantiles	43.00	5.00	**45.15**	1.00	0.50	**19.57**	0.22	**64.94**	nr	**69.81**
Renown tiles	43.00	5.00	**45.15**	1.00	0.50	**19.57**	0.22	**64.94**	nr	**69.81**
Redland 50 double roman tiles	43.00	5.00	**45.15**	1.00	0.50	**19.57**	0.22	**64.94**	nr	**69.81**
Gas terminals										
Gas Flue ridge terminal Mark III, 450mm long with sealing gasket and fixing brackets										
half round ridge	44.30	5.00	**46.52**	1.50	0.75	**29.35**	0.59	**76.45**	nr	**82.19**
universal angle ridge	48.70	5.00	**51.14**	1.50	0.75	**29.35**	0.59	**81.07**	nr	**87.15**
Gas Flue ridge terminal Mark III, 450mm long with sealing gasket and fixing brackets with 150mm R type adaptor for										
half round ridge	61.00	5.00	**64.05**	1.50	0.75	**29.35**	0.59	**93.99**	nr	**101.04**
universal angle ridge	65.66	5.00	**68.94**	1.50	2.56	**48.73**	0.59	**118.27**	nr	**127.14**
extra for extension adaptor and gasket	16.70	5.00	**17.54**	0.30	0.15	**5.87**	0.22	**23.62**	nr	**25.40**
Hip irons										
galvanized mild steel; 32 x 3 x 380mm girth; scrolled; fixing with galvanized steel screws to timber	1.84	5.00	**1.93**	0.10	0.05	**1.96**	-	**3.89**	nr	**4.18**
Lead soakers										
fixing only	-	-	**-**	0.30	0.15	**5.87**	-	**5.87**	nr	**6.31**
H61: FIBRE CEMENT SLATING										
Quantities required for 1m2 of slating										
Slates to 70mm lap										
400 x 240mm, 25.3nr										
Slates to 102mm lap										
600 x 300mm, 13nr										
500 x 250mm, 19.5nr										
400 x 200mm, 32nr										
Roof coverings; asbestos-free cement slates, 600 x 300mm; centre fixing with copper nails and copper disc rivets to 102mm lap; 38 x 19mm pressure impregnated softwood battens fixed with galvanized nails; underlay; BS 747 type 1F bitumen reinforced felt; 150mm laps; fixing with galvanized steel clout nails										
Pitched 30 degrees from horizontal										
generally	14.40	5.00	**15.23**	0.52	0.26	**10.17**	1.42	**26.82**	m2	**28.83**

Labour hourly rates: (except Specialists) Craft Operatives £14.21 Labourer £10.71 Rates are national average prices. Refer to REGIONAL VARIATIONS for indicative levels of overall pricing in regions	MATERIALS			LABOUR				RATES		
	Del to Site £	Waste %	Material Cost £	Craft Optve Hrs	Lab Hrs	Labour Cost £	Sunds £	Nett Rate £	Unit	Gross Rate (+7.5%) £
H61: FIBRE CEMENT SLATING Cont.										
Roof coverings; asbestos-free cement slates, 600 x 300mm; centre fixing with copper nails and copper disc rivets to 102mm lap; 38 x 19mm pressure impregnated softwood battens fixed with galvanized nails; underlay; BS 747 type 1F bitumen reinforced felt; 150mm laps; fixing with galvanized steel clout nails Cont.										
Pitched 30 degrees from horizontal Cont.										
holes	-	-	**-**	0.50	0.25	**9.78**	-	**9.78**	nr	**10.52**
abutments; square	1.28	5.00	**1.34**	0.15	0.08	**2.99**	-	**4.33**	m	**4.66**
abutments; raking	1.92	5.00	**2.01**	2.70	0.12	**39.65**	-	**41.67**	m	**44.79**
abutments; curved to 3000mm radius	2.56	5.00	**2.69**	0.30	0.15	**5.87**	-	**8.56**	m	**9.20**
double course at eaves	3.44	5.00	**3.62**	0.15	0.08	**2.99**	0.65	**7.25**	m	**7.80**
verges; slate undercloak and point in cement mortar (1:3)	1.97	5.00	**2.07**	0.08	0.04	**1.57**	0.53	**4.16**	m	**4.47**
ridges or hips; asbestos free cement; fixing with nails	23.66	5.00	**24.84**	0.20	0.10	**3.91**	0.37	**29.13**	m	**31.31**
Roof coverings; asbestos-free cement slates, 600 x 300mm; centre fixing with copper nails and copper disc rivets to 102mm lap; 50 x 25mm pressure impregnated softwood battens fixed with galvanized nails; underlay; BS 747 type 1F bitumen reinforced felt; 150mm laps; fixing with galvanized steel clout nails										
Pitched 30 degrees from horizontal										
generally	15.24	5.00	**16.15**	0.56	0.28	**10.96**	1.42	**28.53**	m2	**30.67**
Roof coverings; asbestos-free cement slates, 600 x 300mm; centre fixing with copper nails and copper disc rivets to 102mm lap (close boarding on rafters included elsewhere); underlay; BS 747 type 1B underslating felt; 150mm laps; fixing with galvanized steel clout nails										
Pitched 30 degrees from horizontal										
generally	13.32	5.00	**14.04**	0.42	0.21	**8.22**	1.04	**23.30**	m2	**25.05**
Roof coverings; asbestos-free cement slates, 500 x 250mm; centre fixing with copper nails and copper disc rivets to 102mm lap; 38 x 19mm pressure impregnated softwood battens fixed with galvanized nails; underlay; BS 747 type 1F bitumen reinforced felt; 150mm laps; fixing with galvanized steel clout nails										
Pitched 30 degrees from horizontal										
generally	17.32	5.00	**18.31**	0.58	0.29	**11.35**	1.94	**31.60**	m2	**33.97**
holes	-	-	**-**	0.50	0.25	**9.78**	-	**9.78**	nr	**10.52**
abutments; square	1.54	5.00	**1.62**	0.20	0.10	**3.91**	-	**5.53**	m	**5.95**
abutments; raking	2.32	5.00	**2.44**	0.30	0.15	**5.87**	-	**8.31**	m	**8.93**
abutments; curved to 3000mm radius	3.09	5.00	**3.24**	0.40	0.20	**7.83**	-	**11.07**	m	**11.90**
double course at eaves	3.17	5.00	**3.33**	0.18	0.09	**3.52**	0.69	**7.54**	m	**8.10**
verges; slate undercloak and point in cement mortar (1:3)	1.98	5.00	**2.08**	0.10	0.05	**1.96**	0.57	**4.61**	m	**4.95**
ridges or hips; asbestos free cement; fixing with nails	23.66	5.00	**24.84**	0.20	0.10	**3.91**	0.37	**29.13**	m	**31.31**
Roof coverings; asbestos-free cement slates, 500 x 250mm; centre fixing with copper nails and copper disc rivets to 102mm lap; 50 x 25mm pressure impregnated softwood battens fixed with galvanized nails; underlay; BS 747 type 1F bitumen reinforced felt; 150mm laps; fixing with galvanized steel clout nails										
Pitched 30 degrees from horizontal										
generally	18.37	5.00	**19.47**	0.63	0.32	**12.38**	1.94	**33.79**	m2	**36.32**
Roof coverings; asbestos-free cement slates, 400 x 200mm; centre fixing with copper nails and copper disc rivets to 102mm lap; 38 x 19mm pressure impregnated softwood battens fixed with galvanized nails; underlay; BS 747 type 1F bitumen reinforced felt; 150mm laps; fixing with galvanized steel clout nails										
Pitched 30 degrees from horizontal										
generally	20.45	5.00	**21.61**	0.62	0.31	**12.13**	2.94	**36.68**	m2	**39.43**
holes	-	-	**-**	0.50	0.25	**9.78**	-	**9.78**	nr	**10.52**
abutments; square	1.83	5.00	**1.92**	0.25	0.13	**4.94**	-	**6.87**	m	**7.38**
abutments; raking	2.75	5.00	**2.88**	0.38	0.19	**7.43**	-	**10.32**	m	**11.09**
abutments; curved to 3000mm radius	3.66	5.00	**3.84**	0.50	0.25	**9.78**	-	**13.63**	m	**14.65**
double course at eaves	2.86	5.00	**3.00**	0.20	0.10	**3.91**	0.77	**7.69**	m	**8.26**
verges; slate undercloak and point in cement mortar (1:3)	1.72	5.00	**1.80**	0.12	0.06	**2.35**	0.61	**4.76**	m	**5.12**
ridges or hips; asbestos free cement; fixing with nails	23.66	5.00	**24.84**	0.20	0.10	**3.91**	0.37	**29.13**	m	**31.31**

Labour hourly rates: (except Specialists) Craft Operatives £14.21 Labourer £10.71 Rates are national average prices. Refer to REGIONAL VARIATIONS for indicative levels of overall pricing in regions	MATERIALS			LABOUR				RATES		
	Del to Site £	Waste %	Material Cost £	Craft Optve Hrs	Lab Hrs	Labour Cost £	Sunds £	Nett Rate £	Unit	Gross Rate (+7.5%) £
H61: FIBRE CEMENT SLATING Cont.										
Roof coverings; asbestos-free cement slates, 400 x 200mm; centre fixing with copper nails and copper disc rivets to 102mm lap; 50 x 25mm pressure impregnated softwood battens fixed with galvanized nails; underlay; BS 747 type 1F bitumen reinforced felt; 150mm laps; fixing with galvanized steel clout nails										
Pitched 30 degrees from horizontal										
generally	21.71	5.00	**23.00**	0.70	0.35	**13.70**	2.94	**39.63**	m2	**42.60**
Fittings										
Hip irons										
galvanized mild steel; 32 x 3 x 380mm girth; scrolled; fixing with galvanized steel screws to timber	1.84	5.00	**1.93**	0.10	0.05	**1.96**	-	**3.89**	nr	**4.18**
Lead soakers										
fixing only	-	-	**-**	0.30	0.15	**5.87**	-	**5.87**	nr	**6.31**
H62: NATURAL SLATING										
Quantities required for 1m2 of slating										
Slates to 75mm lap										
600 x 300mm, 12.65nr										
500 x 250mm, 18.9nr										
400 x 200mm, 31nr										
Roof coverings; blue/grey slates, 600 x 300mm, 6.5mm thick; fixing with slate nails to 75mm lap; 38 x 19mm pressure impregnated softwood battens fixed with galvanized nails; underlay; BS 747 type 1F bitumen reinforced felt; 150mm laps; fixing with galvanized steel clout nails										
Pitched 30 degrees from horizontal										
generally	66.43	5.00	**69.86**	0.50	0.25	**9.78**	0.50	**80.14**	m2	**86.15**
holes	-	-	**-**	0.50	0.25	**9.78**	-	**9.78**	nr	**10.52**
abutments; square	6.30	5.00	**6.62**	0.15	0.08	**2.99**	-	**9.61**	m	**10.33**
abutments; raking	9.48	5.00	**9.95**	0.23	0.12	**4.55**	-	**14.51**	m	**15.59**
double course at eaves	17.93	5.00	**18.83**	0.14	0.07	**2.74**	0.44	**22.01**	m	**23.66**
verges; slate undercloak and point in cement mortar (1:3)	10.25	5.00	**10.76**	0.08	0.04	**1.57**	0.41	**12.74**	m	**13.69**
Roof coverings; blue/grey slates, 600 x 300mm, 6.5mm thick; fixing with aluminium nails to 75mm lap; 38 x 19mm pressure impregnated softwood battens fixed with galvanized nails; underlay; BS 747 type 1F bitumen reinforced felt; 150mm laps; fixing with galvanized steel clout nails										
Pitched 30 degrees from horizontal										
generally	66.43	5.00	**69.86**	0.50	0.25	**9.78**	0.99	**80.63**	m2	**86.68**
Roof coverings; blue/grey slates, 600 x 300mm, 6.5mm thick; fixing with copper nails to 75mm laps; 38 x 19mm pressure impregnated softwood battens fixed with galvanized nails; underlay; BS 747 type 1F bitumen reinforced felt; 150mm laps; fixing with galvanized steel clout nails										
Pitched 30 degrees from horizontal										
generally	66.43	5.00	**69.86**	0.50	0.25	**9.78**	1.23	**80.87**	m2	**86.93**
Roof coverings; blue/grey slates, 600 x 300mm, 6.5mm thick; fixing with slate nails to 75mm lap; 50 x 25mm pressure impregnated softwood battens fixed with galvanized nails; underlay; BS 747 type 1F bitumen reinforced felt; 150mm laps; fixing with galvanized steel clout nails										
Pitched 30 degrees from horizontal										
generally	67.27	5.00	**70.78**	0.54	0.27	**10.57**	0.50	**81.85**	m2	**87.98**
Roof coverings; blue/grey slates, 600 x 300mm, 6.5mm thick; fixing with slate nails to 75mm lap; 50 x 25mm pressure impregnated softwood counterbattens at 1067mm centres and 38 x 19mm pressure impregnated softwood battens fixed with galvanized nails; underlay; BS 747 type 1B underslating felt; 150mm laps; fixing with galvanized steel clout nails										
Pitched 30 degrees from horizontal										
generally	67.27	5.00	**70.78**	0.54	0.27	**10.57**	1.01	**82.36**	m2	**88.53**
Roof coverings; blue/grey slates, 500 x 250mm, 6.5mm thick; fixing with slate nails to 75mm lap; 38 x 19mm pressure impregnated softwood battens fixed with galvanized nails; underlay; BS 747 type 1F bitumen reinforced felt; 150mm laps; fixing with galvanized steel clout nails										
Pitched 30 degrees from horizontal										
generally	48.71	5.00	**51.27**	0.58	0.29	**11.35**	0.74	**63.36**	m2	**68.11**

Labour hourly rates: (except Specialists) Craft Operatives £14.21 Labourer £10.71 Rates are national average prices. Refer to REGIONAL VARIATIONS for indicative levels of overall pricing in regions	MATERIALS			LABOUR				RATES		
	Del to Site £	Waste %	Material Cost £	Craft Optve Hrs	Lab Hrs	Labour Cost £	Sunds £	Nett Rate £	Unit	Gross Rate (+7.5%) £
H62: NATURAL SLATING Cont.										
Roof coverings; blue/grey slates, 500 x 250mm, 6.5mm thick; fixing with slate nails to 75mm lap; 38 x 19mm pressure impregnated softwood battens fixed with galvanized nails; underlay; BS 747 type 1F bitumen reinforced felt; 150mm laps; fixing with galvanized steel clout nails Cont.										
Pitched 30 degrees from horizontal Cont.										
holes	-	-	-	0.50	0.25	9.78	-	9.78	nr	10.52
abutments; square	4.51	5.00	4.74	0.20	0.10	3.91	-	8.65	m	9.30
abutments; raking	6.76	5.00	7.10	0.30	0.15	5.87	-	12.97	m	13.95
double course at eaves	9.91	5.00	10.41	0.17	0.09	3.38	0.71	14.50	m	15.58
verges; slate undercloak and point in cement mortar (1:3)	6.20	5.00	6.50	0.10	0.05	1.96	0.42	8.88	m	9.55
Roof coverings; blue/grey slates, 500 x 250mm, 6.5mm thick; fixing with aluminium nails to 75mm lap; 38 x 19mm pressure impregnated softwood battens fixed with galvanized nails; underlay; BS 747 type 1F bitumen reinforced felt; 150mm laps; fixing with galvanized steel clout nails										
Pitched 30 degrees from horizontal										
generally	48.71	5.00	51.27	0.58	0.29	11.35	1.47	64.09	m2	68.89
Roof coverings; blue/grey slates, 500 x 250mm, 6.5mm thick; fixing with copper nails to 75mm lap; 38 x 19mm pressure impregnated softwood battens fixed with galvanized nails; underlay; BS 747 type 1F bitumen reinforced felt; 150mm laps; fixing with galvanized steel clout nails										
Pitched 30 degrees from horizontal										
generally	48.71	5.00	51.27	0.58	0.29	11.35	1.83	64.45	m2	69.28
Roof coverings; blue/grey slates, 500 x 250mm, 6.5mm thick; fixing with slate nails to 75mm lap; 50 x 25mm pressure impregnated softwood battens fixed with galvanized nails; underlay; BS 747 type 1F bitumen reinforced felt; 150mm laps; fixing with galvanized steel clout nails										
Pitched 30 degrees from horizontal										
generally	49.76	5.00	52.43	0.63	0.32	12.38	0.74	65.54	m2	70.46
Roof coverings; blue/grey slates, 400 x 200mm, 6.5mm thick; fixing with slate nails to 75mm lap; 38 x 19mm pressure impregnated softwood battens fixed with galvanized nails; underlay; BS 747 type 1F bitumen reinforced felt; 150mm laps; fixing with galvanized steel clout nails										
Pitched 30 degrees from horizontal										
generally	37.30	5.00	39.30	0.70	0.35	13.70	1.23	54.23	m2	58.30
holes	-	-	-	0.50	0.25	9.78	-	9.78	nr	10.52
abutments; square	3.40	5.00	3.57	0.25	0.13	4.94	-	8.52	m	9.16
abutments; raking	5.10	5.00	5.36	0.38	0.19	7.43	-	12.79	m	13.75
double course at eaves	6.80	5.00	7.14	0.20	0.10	3.91	0.36	11.42	m	12.27
verges; slate undercloak and point in cement mortar (1:3)	3.40	5.00	3.57	0.12	0.06	2.35	0.43	6.35	m	6.83
Roof coverings; blue/grey slates, 400 x 200mm, 6.5mm thick; fixing with aluminium nails to 75mm lap; 38 x 19mm pressure impregnated softwood battens fixed with galvanized nails; underlay; BS 747 type 1F bitumen reinforced felt; 150mm laps; fixing with galvanized steel clout nails										
Pitched 30 degrees from horizontal										
generally	37.30	5.00	39.30	0.70	0.35	13.70	2.43	55.43	m2	59.59
Roof coverings; blue/grey slates, 400 x200mm, 6.5mm thick; fixing with copper nails to 75mmlap; 38 x 19mm pressure impregnated softwood battens fixed with galvanized nails; underlay; BS 747 type 1F bitumen reinforced felt; 150mm laps; fixing with galvanized steel clout nails										
Pitched 30 degrees from horizontal										
generally	37.30	5.00	39.30	0.70	0.35	13.70	3.03	56.03	m2	60.23
Roof coverings; blue/grey slates, 400 x 200mm, 6.5mm thick; fixing with slate nails to 75mm lap; 50 x 25mm pressure impregnated softwood battens fixed with galvanized nails; underlay; BS 747 type 1F bitumen reinforced felt; 150mm laps; fixing with galvanized steel clout nails										
Pitched 30 degrees from horizontal										
generally	38.56	5.00	40.69	0.76	0.38	14.87	1.23	56.79	m2	61.05

Labour hourly rates: (except Specialists) Craft Operatives £14.21 Labourer £10.71 Rates are national average prices. Refer to REGIONAL VARIATIONS for indicative levels of overall pricing in regions	MATERIALS			LABOUR				RATES		
	Del to Site £	Waste %	Material Cost £	Craft Optve Hrs	Lab Hrs	Labour Cost £	Sunds £	Nett Rate £	Unit	Gross Rate (+7.5%) £
H62: NATURAL SLATING Cont.										
Roof coverings; Burlington green slates, best random fixed in diminishing courses with alloy nails to 75mm lap; 38 x 19mm pressure impregnated softwood battens fixed with galvanized nails; underlay; BS 747 type 1F bitumen reinforced felt; 150mm laps; fixing with galvanized steel clout nails										
Pitched 30 degrees from horizontal										
generally	110.17	5.00	**115.81**	0.75	0.38	**14.73**	1.80	**132.34**	m2	**142.27**
abutments; square	10.80	5.00	**11.34**	0.20	0.10	**3.91**	-	**15.26**	m	**16.40**
abutments; raking	16.20	5.00	**17.01**	0.30	0.15	**5.87**	-	**22.88**	m	**24.60**
double course at eaves	23.57	5.00	**24.75**	0.20	0.10	**3.91**	0.90	**29.56**	m	**31.78**
verges; slate undercloak and point in cement mortar (1:3)	15.71	5.00	**16.50**	0.20	0.10	**3.91**	0.55	**20.96**	m	**22.53**
hips, valleys and angles; close mitred	17.68	5.00	**18.56**	0.35	0.18	**6.90**	0.58	**26.04**	m	**27.99**
Fittings										
Hip irons										
galvanized mild steel; 32 x 3 x 380mm girth; scrolled; fixing with galvanized steel screws to timber	1.84	5.00	**1.93**	0.10	0.05	**1.96**	-	**3.89**	nr	**4.18**
Lead soakers										
fixing only	-	-	**-**	0.30	0.15	**5.87**	-	**5.87**	nr	**6.31**
H63: RECONSTRUCTED STONE SLATING/ TILING										
Roof coverings; Redland Cambrian interlocking riven textured slates, 300 x 336mm; fixing every slate with two stainless steel nails and one stainless steel clip to 50mm laps; 38 x 19mm pressure impregnated softwood battens fixed with galvanized nails; underlay; BS 747 type 1F bitumen reinforced felt; 150mm laps; fixing with galvanized steel clout nails										
Pitched 40 degrees from horizontal										
generally	24.43	5.00	**25.76**	0.48	0.24	**9.39**	1.74	**36.89**	m2	**39.66**
abutments; square	2.28	5.00	**2.39**	0.10	0.05	**1.96**	-	**4.35**	m	**4.67**
abutments; raking	3.42	5.00	**3.60**	0.15	0.08	**2.99**	-	**6.58**	m	**7.08**
abutments; curved to 3000mm radius	4.55	5.00	**4.78**	0.20	0.10	**3.91**	-	**8.69**	m	**9.35**
supplementary fixing at eaves with stainless steel eaves clip to each slate	0.48	5.00	**0.50**	0.04	0.02	**0.78**	-	**1.29**	m	**1.38**
verges; 150 x 6mm fibre cement undercloak; slates and undercloak bedded and pointed in tinted cement mortar (1:3) with full slates and slate and a half slates in alternate courses and stainless steel verge clips	7.39	5.00	**7.76**	0.20	0.10	**3.91**	2.02	**13.69**	m	**14.71**
verges; 150 x 6mm fibre cement undercloak; slates and undercloak bedded and pointed in tinted cement mortar (1:3) with verge slates and slate and a half verge slates in alternate courses and stainless steel verge clips	7.39	5.00	**7.76**	0.20	0.10	**3.91**	2.02	**13.69**	m	**14.71**
ridge or hip tiles, Redland third round; butt jointed; bedding and pointing in tinted cement mortar (1:3)	6.90	5.00	**7.25**	0.20	0.10	**3.91**	0.37	**11.53**	m	**12.39**
ridge or hip tiles, Redland half round; butt jointed; bedding and pointing in tinted cement mortar (1:3)	6.90	5.00	**7.25**	0.20	0.10	**3.91**	0.37	**11.53**	m	**12.39**
ridge or hip tiles, Redland universal angle; butt jointed; bedding and pointing in tinted cement mortar (1:3)	7.50	5.00	**7.87**	0.20	0.10	**3.91**	0.37	**12.16**	m	**13.07**
ridge tiles, Redland Dry Ridge system with half round ridge tiles; fixing with stainless steel batten straps and ring shanked fixing nails with neoprene washers and sleeves; polypropylene ridge seals and uPVC profile filler units	23.51	5.00	**24.69**	0.40	0.20	**7.83**	-	**32.51**	m	**34.95**
ridge tiles, Redland Dry Ridge system with universal angle ridge tiles; fixing with stainless steel batten straps and ring shanked fixing nails with neoprene washers and sleeves; polypropylene ridge seals and uPVC profile filler units	24.11	5.00	**25.32**	0.40	0.20	**7.83**	-	**33.14**	m	**35.63**
ridge tiles, Redland Dry Vent Ridge system with half round ridge tiles; fixing with stainless steel batten straps and ring shanked nails with neoprene washers and sleeves; polypropylene ridge seals, PVC air flow control units and uPVC ventilated profile filler units	23.51	5.00	**24.69**	0.40	0.20	**7.83**	-	**32.51**	m	**34.95**
ridge tiles, Redland Dry Vent Ridge system with Universal angle ridge tiles; fixing with stainless steel batten straps and ring shanked nails with neoprene washers and sleeves; polypropylene ridge seals, PVC air flow control units and uPVC ventilated profile filler units	24.11	5.00	**25.32**	0.40	0.20	**7.83**	-	**33.14**	m	**35.63**
Monoridge filler units in conjunction with top tile, bed tile in nonsetting mastic sealant and screw on through filler unit with screws, washers and caps	34.51	5.00	**36.24**	0.20	0.10	**3.91**	-	**40.15**	m	**43.16**
valley tiles, Redland Universal valley troughs; laid with 100mm laps	24.48	5.00	**25.70**	0.50	0.25	**9.78**	0.37	**35.86**	m	**38.55**

Labour hourly rates: (except Specialists) Craft Operatives £14.21 Labourer £10.71 Rates are national average prices. Refer to REGIONAL VARIATIONS for indicative levels of overall pricing in regions	MATERIALS			LABOUR				RATES		
	Del to Site £	Waste %	Material Cost £	Craft Optve Hrs	Lab Hrs	Labour Cost £	Sunds £	Nett Rate £	Unit	Gross Rate (+7.5%) £
H63: RECONSTRUCTED STONE SLATING/ TILING Cont.										
Roof coverings; Redland Cambrian interlocking riven textured slates, 300 x 336mm; fixing every slate with two stainless steel nails and one stainless steel clip to 90mm lap; 38 x 25mm pressure impregnated softwood battens fixed with galvanized nails; underlay; BS 747 type 1F bitumen reinforced felt; 150mm laps; fixing with galvanized steel clout nails										
Pitched 40 degrees from horizontal										
generally	29.10	5.00	**30.68**	0.55	0.28	**10.81**	2.25	**43.74**	m2	**47.02**
abutments; square	2.72	5.00	**2.86**	0.10	0.05	**1.96**	-	**4.81**	m	**5.18**
abutments; raking	4.07	5.00	**4.28**	0.15	0.08	**2.99**	-	**7.27**	m	**7.81**
abutments; curved to 3000mm radius	5.44	5.00	**5.72**	0.20	0.10	**3.91**	-	**9.63**	m	**10.35**
Fittings										
Ventilator tiles										
Red Vent ridge ventilation terminal 450mm long for half round ridge	44.30	5.00	**46.52**	1.30	0.65	**25.43**	-	**71.95**	nr	**77.35**
universal angle ridge	44.30	5.00	**46.52**	1.30	0.65	**25.43**	-	**71.95**	nr	**77.35**
Red Vent eaves ventilators; fixing with stainless steel nails	8.17	5.00	**8.58**	0.40	0.20	**7.83**	-	**16.40**	m	**17.63**
Redland Cambrian Thruvent interlocking slate complete with weather cap, underlay seal and fixing clips	36.20	5.00	**38.01**	0.50	0.25	**9.78**	-	**47.79**	nr	**51.38**
Gas terminals										
Gas Flue ridge terminal Mark III, 450mm long with sealing gasket and fixing brackets half round ridge	44.30	5.00	**46.52**	1.50	0.75	**29.35**	-	**75.86**	nr	**81.55**
Gas Flue ridge terminal Mark III, 450mm long with sealing gasket and fixing brackets with 150mm R type adaptor for half round ridge	44.30	5.00	**46.52**	1.50	0.75	**29.35**	28.50	**104.36**	nr	**112.19**
Hip irons										
galvanized mild steel; 32 x 3 x 380mm girth; scrolled; fixing with galvanized steel screws to timber	1.84	5.00	**1.93**	0.10	0.05	**1.96**	-	**3.89**	nr	**4.18**
Lead soakers										
fixing only	-	-	**-**	0.30	0.15	**5.87**	-	**5.87**	nr	**6.31**
Fittings										
Timloc uPVC Mark 3 eaves ventilators; reference 1126/40 fixing with nails to timber at 400mm centres 330mm wide	3.68	5.00	**3.86**	0.25	0.12	**4.84**	0.27	**8.97**	m	**9.64**
Timloc uPVC Mark 3 eaves ventilators; reference 1123 fixing with nails to timber at 600mm centres 330mm wide	2.46	5.00	**2.58**	0.20	0.10	**3.91**	0.27	**6.77**	m	**7.27**
Timlock uPVC soffit ventilators; reference 1137, 10mm airflow; fixing with screws to timber type C	1.97	5.00	**2.07**	0.15	0.05	**2.67**	0.10	**4.84**	m	**5.20**
Timlock uPVC Mark 2 eaves ventilators; reference 1122; fitting between trusses at 600mm centres 300mm girth	2.46	5.00	**2.58**	0.20	0.10	**3.91**	0.27	**6.77**	m	**7.27**
Timlock polypropylene over-fascia ventilators; reference 3011; fixing with screws to timber to top of fascia	3.50	5.00	**3.68**	0.20	0.10	**3.91**	0.16	**7.75**	m	**8.33**
H65: SINGLE LAP ROOF TILING										
Roof coverings; Redland Stonewold through coloured slates, 430 x 380mm; fixing every fourth course with galvanized nails to 75mm lap; 38 x 19mm pressure impregnated softwood battens fixed with galvanized nails; underlay; BS 747 type 1F reinforced bitumen felt; 150mm laps; fixing with galvanized steel clout nails										
Pitched 40 degrees from horizontal										
generally	12.66	5.00	**13.38**	0.42	0.21	**8.22**	0.72	**22.32**	m2	**23.99**
abutments; square	1.13	5.00	**1.19**	0.10	0.05	**1.96**	-	**3.14**	m	**3.38**
abutments; raking	1.70	5.00	**1.78**	0.15	0.08	**2.99**	-	**4.77**	m	**5.13**
abutments; curved to 3000mm radius	2.26	5.00	**2.38**	0.20	0.10	**3.91**	-	**6.29**	m	**6.76**
supplementary fixing eaves course with one clip per slate	0.33	5.00	**0.35**	0.03	0.02	**0.64**	-	**0.99**	m	**1.06**
verges; Redland Dry Verge system with half slates, full slates and clips	13.68	5.00	**14.36**	0.25	0.12	**4.84**	1.71	**20.91**	m	**22.48**
verges; Redland Dry Verge system with half slates, verge slates and clips	14.09	5.00	**14.80**	0.25	0.12	**4.84**	1.71	**21.35**	m	**22.95**
verges; 150 x 6mm fibre cement undercloak, slates and undercloak bedded and pointed in tinted cement mortar (1:3) and with half slates, full slates and clips	15.80	5.00	**16.59**	0.20	0.10	**3.91**	2.02	**22.53**	m	**24.22**
verges; 150 x 6mm fibre cement undercloak, slates and undercloak bedded and pointed in tinted cement mortar (1:3) and with half slates, verge slates and clips	6.39	5.00	**6.71**	0.20	0.10	**3.91**	2.02	**12.64**	m	**13.59**

Labour hourly rates: (except Specialists) Craft Operatives £14.21 Labourer £10.71 Rates are national average prices. Refer to REGIONAL VARIATIONS for indicative levels of overall pricing in regions	MATERIALS			LABOUR				RATES		
	Del to Site £	Waste %	Material Cost £	Craft Optve Hrs	Lab Hrs	Labour Cost £	Sunds £	Nett Rate £	Unit	Gross Rate (+7.5%) £
H65: SINGLE LAP ROOF TILING Cont.										
Roof coverings; Redland Stonewold through coloured slates, 430 x 380mm; fixing every fourth course with galvanized nails to 75mm lap; 38 x 19mm pressure impregnated softwood battens fixed with galvanized nails; underlay; BS 747 type 1F reinforced bitumen felt; 150mm laps; fixing with galvanized steel clout nails Cont.										
Pitched 40 degrees from horizontal Cont.										
ridge or hip tiles, Redland third round; butt jointed; bedding and pointing in tinted cement mortar (1:3)	6.90	5.00	**7.25**	0.20	0.10	**3.91**	0.37	**11.53**	m	**12.39**
ridge or hip tiles, Redland half round; butt jointed; bedding and pointing in tinted cement mortar (1:3)	6.90	5.00	**7.25**	0.20	0.10	**3.91**	0.37	**11.53**	m	**12.39**
ridge tiles, Redland universal angle; butt jointed; bedding and pointing in tinted cement mortar (1:3)	7.50	5.00	**7.87**	0.20	0.10	**3.91**	0.37	**12.16**	m	**13.07**
ridge tiles, Redland universal Stonewold type Monopitch; butt jointed; fixing with aluminium nails and bedding and pointing in tinted cement mortar (1:3)	18.03	5.00	**18.93**	0.20	0.10	**3.91**	0.37	**23.21**	m	**24.96**
ridge tiles, Redland universal half round type Monopitch; butt jointed; fixing with aluminium nails and bedding and pointing in tinted cement mortar (1:3)	18.03	5.00	**18.93**	0.20	0.10	**3.91**	0.37	**23.21**	m	**24.96**
ridge tiles, Redland Dry Ridge system with half round ridge tiles; fixing with stainless steel batten straps and ring shanked fixing nails with neoprene washers and sleeves; polypropylene ridge seals and uPVC profile filler units	24.79	5.00	**26.03**	0.40	0.20	**7.83**	-	**33.86**	m	**36.39**
ridge tiles, Redland Dry Ridge system with universal angle ridge tiles; fixing with stainless steel batten straps and ring shanked fixing nails with neoprene washers and sleeves; polypropylene ridge seals and uPVC profile filler units	24.30	5.00	**25.52**	0.40	0.20	**7.83**	-	**33.34**	m	**35.84**
ridge tiles, Redland Dry Vent Ridge system with half round ridge tiles; fixing with stainless steel batten straps and ring shanked nails with neoprene washers and sleeves; polypropylene ridge seals, PVC air flow control units and uPVC ventilated profile filler units	25.75	5.00	**27.04**	0.40	0.20	**7.83**	-	**34.86**	m	**37.48**
ridge tiles, Redland Dry Vent Ridge system with Universal angle ridge tiles; fixing with stainless steel batten straps and ring shanked nails with neoprene washers and sleeves; polypropylene ridge seals, PVC air flow control units and uPVC ventilated profile filler units	24.30	5.00	**25.52**	0.40	0.20	**7.83**	-	**33.34**	m	**35.84**
Monoridge filler units in conjunction with top tile, bed tile in nonsetting mastic sealant and screw on through filler unit with screws, washers and caps	31.49	5.00	**33.06**	0.20	0.10	**3.91**	0.37	**37.35**	m	**40.15**
valley tiles, Redland Universal valley troughs; laid with 100mm laps	24.57	5.00	**25.80**	0.50	0.25	**9.78**	0.37	**35.95**	m	**38.65**
Roof coverings; Redland Stonewold through coloured slates, 430 x 380mm; fixing every fourth course with galvanized nails to 75mm lap; 38 x 25mm pressure impregnated softwood battens fixed with galvanized nails; underlay; BS 747 type 1F reinforced bitumen felt; 150mm laps; fixing with galvanized steel clout nails										
Pitched 40 degrees from horizontal										
generally	12.81	5.00	**13.55**	0.42	0.21	**8.22**	0.72	**22.48**	m2	**24.17**
Roof coverings; Redland Stonewold through coloured slates, 430 x 380mm; fixing each course with galvanized nails to 75mm; 38 x 25mm pressure impregnated softwood battens fixed with galvanized nails; underlay; B.S 747 type 1F reinforced bitumen felt; 150mm laps; fixing with galvanized steel clout nails										
Pitched 40 degrees from horizontal										
generally	12.81	5.00	**13.55**	0.46	0.23	**9.00**	2.08	**24.63**	m2	**26.47**
Roof coverings; Redland Stonewold through coloured slates, 430 x 380mm; fixing every fourth course with galvanized nails to 75mm lap; 38 x 25mm pressure impregnated softwood battens fixed with galvanized nails; underlay, BS 747 type 1F aluminium foil surfaced reinforced bitumen felt; 150mm laps; fixing with galvanized steel clout nails										
Pitched 40 degrees from horizontal										
generally	12.81	5.00	**13.55**	0.43	0.21	**8.36**	0.72	**22.63**	m2	**24.32**

Labour hourly rates: (except Specialists) Craft Operatives £14.21 Labourer £10.71 Rates are national average prices. Refer to REGIONAL VARIATIONS for indicative levels of overall pricing in regions	MATERIALS			LABOUR				RATES		
	Del to Site £	Waste %	Material Cost £	Craft Optve Hrs	Lab Hrs	Labour Cost £	Sunds £	Nett Rate £	Unit	Gross Rate (+7.5%) £
H65: SINGLE LAP ROOF TILING Cont.										
Roof coverings; Redland Stonewold through coloured slates, 430 x 380mm; fixing every fourth course with galvanized nails to 75mm lap; 38 x 25mm pressure impregnated softwood battens fixed with galvanized nails; underlay, BS 747 type 1F reinforced bitumen felt with 50mm glass fibre insulation bonded on; 150mm laps; fixing with galvanized steel clout nails										
Pitched 40 degrees from horizontal										
generally	12.81	5.00	**13.55**	0.52	0.26	**10.17**	0.72	**24.44**	m2	**26.27**
Roof coverings; Redland Regent granular faced tiles, 418 x 332mm; fixing every fourth course with galvanized nails to 75mm lap; 38 x 25mm pressure impregnated softwood battens fixed with galvanized nails; underlay; BS 747 type 1F bitumen reinforced felt; 150mm laps; fixing with galvanized steel clout nails										
Pitched 40 degrees from horizontal										
generally	8.51	5.00	**9.04**	0.42	0.21	**8.22**	0.81	**18.07**	m2	**19.42**
reform eaves filler unit and eaves clip to each eaves tile	0.33	5.00	**0.35**	0.06	0.03	**1.17**	0.63	**2.15**	m	**2.31**
verges; cloaked verge tiles and aluminium nails	8.85	5.00	**9.29**	0.15	0.08	**2.99**	0.06	**12.34**	m	**13.27**
verges; half tiles, cloaked verge tiles and aluminium nails	10.30	5.00	**10.81**	0.20	0.10	**3.91**	0.10	**14.83**	m	**15.94**
verges; 150 x 6mm fibre cement undercloak; tiles and undercloak bedded and pointed in tinted cement mortar (1:3) and with standard tiles and clips	4.42	5.00	**4.64**	0.20	0.10	**3.91**	2.02	**10.58**	m	**11.37**
verges; 150 x 6mm fibre cement undercloak; tiles and undercloak bedded and pointed in tinted cement mortar (1:3) and with verge tiles and clips	4.42	5.00	**4.64**	0.20	0.10	**3.91**	2.02	**10.58**	m	**11.37**
verges; 150 x 6mm fibre cement undercloak; tiles and undercloak bedded and pointed in tinted cement mortar (1:3) and with half tiles, standard tiles and clips	7.90	5.00	**8.30**	0.20	0.10	**3.91**	2.02	**14.23**	m	**15.30**
verges; 150 x 6mm fibre cement undercloak; tiles and undercloak bedded and pointed in tinted cement mortar (1:3) and with half tiles, verge tiles and clips	7.90	5.00	**8.30**	0.20	0.10	**3.91**	2.02	**14.23**	m	**15.30**
ridge or hip tiles, Redland third round; butt jointed; bedding and pointing in tinted cement mortar (1:3); dentil slips in pan of each tile set in bedding	6.60	5.00	**6.93**	0.30	0.15	**5.87**	1.45	**14.25**	m	**15.32**
ridge or hip tiles, Redland half round; butt jointed; bedding and pointing in tinted cement mortar (1:3); dentil slips in pan of each tile set in bedding	6.60	5.00	**6.93**	0.30	0.15	**5.87**	1.45	**14.25**	m	**15.32**
ridge tiles, Redland universal half round type Monopitch; butt jointed; fixing with aluminium nails and bedding and pointing in tinted cement mortar (1:3); dentil slips in pan of each tile set in bedding	18.03	5.00	**18.93**	0.30	0.15	**5.87**	1.45	**26.25**	m	**28.22**
valley tiles, Redland universal valley troughs; laid with 100mm laps	24.57	5.00	**25.80**	0.30	0.15	**5.87**	0.37	**32.04**	m	**34.44**
Roof coverings; Redland Regent granular faced tiles, 418 x 332mm; fixing every fourth course with aluminium nails to 75mm lap; 38 x 25mm pressure impregnated softwood battens fixed with galvanized nails; underlay; BS 747 type 1F bitumen reinforced felt; 150mm laps; fixing with galvanized steel clout nails										
Pitched 40 degrees from horizontal										
generally	8.51	5.00	**9.04**	0.42	0.21	**8.22**	0.95	**18.21**	m2	**19.57**
Roof coverings; Redland Regent granular faced tiles, 418 x 332mm; fixing every fourth course with galvanized nails to 100mm lap; 38 x 25mm pressure impregnated softwood battens fixed with galvanized nails; underlay; BS 747 type 1F bitumen reinforced felt; 150mm laps; fixing with galvanized steel clout nails										
Pitched 40 degrees from horizontal										
generally	9.09	5.00	**9.65**	0.42	0.21	**8.22**	0.58	**18.44**	m2	**19.83**
abutments; square	0.76	5.00	**0.80**	0.10	0.05	**1.96**	-	**2.75**	m	**2.96**
abutments; raking	1.14	5.00	**1.19**	0.15	0.08	**2.99**	-	**4.18**	m	**4.50**
abutments; curved to 3000mm radius	1.52	5.00	**1.60**	0.20	0.10	**3.91**	-	**5.51**	m	**5.92**
Roof coverings; Redland Regent through coloured tiles, 418 x 332mm; fixing every fourth course with galvanized nails to 75mm lap; 38 x 25mm pressure impregnated softwood battens fixed with galvanized nails; underlay; BS 747 type 1F bitumen reinforced felt; 150mm laps; fixing with galvanized steel clout nails										
Pitched 40 degrees from horizontal										
generally	8.51	5.00	**9.04**	0.42	0.21	**8.22**	0.81	**18.07**	m2	**19.42**
reform eaves filler unit and eaves clip to each eaves tile	0.33	5.00	**0.35**	0.06	0.03	**1.17**	0.63	**2.15**	m	**2.31**
verges; cloaked verge tiles and aluminium nails	8.85	5.00	**9.29**	0.15	0.08	**2.99**	0.06	**12.34**	m	**13.27**

Labour hourly rates: (except Specialists) Craft Operatives £14.21 Labourer £10.71 Rates are national average prices. Refer to REGIONAL VARIATIONS for indicative levels of overall pricing in regions	MATERIALS			LABOUR				RATES		
	Del to Site £	Waste %	Material Cost £	Craft Optve Hrs	Lab Hrs	Labour Cost £	Sunds £	Nett Rate £	Unit	Gross Rate (+7.5%) £
H65: SINGLE LAP ROOF TILING Cont.										
Roof coverings; Redland Regent through coloured tiles, 418 x 332mm; fixing every fourth course with galvanized nails to 75mm lap; 38 x 25mm pressure impregnated softwood battens fixed with galvanized nails; underlay; BS 747 type 1F bitumen reinforced felt; 150mm laps; fixing with galvanized steel clout nails Cont.										
Pitched 40 degrees from horizontal Cont.										
verges; 150 x 6mm fibre cement undercloak; tiles and undercloak bedded and pointed in tinted cement mortar (1:3) and with standard tiles and clips...	4.42	5.00	**4.64**	0.20	0.10	**3.91**	2.02	**10.58**	m	**11.37**
verges; 150 x 6mm fibre cement undercloak; tiles and undercloak bedded and pointed in tinted cement mortar (1:3) and with verge tiles and clips......	4.42	5.00	**4.64**	0.20	0.10	**3.91**	2.02	**10.58**	m	**11.37**
verges; 150 x 6mm fibre cement undercloak; tiles and undercloak bedded and pointed in tinted cement mortar (1:3) and with half tiles, standard tiles and clips	7.90	5.00	**8.30**	0.20	0.10	**3.91**	2.02	**14.23**	m	**15.30**
verges; 150 x 6mm fibre cement undercloak; tiles and undercloak bedded and pointed in tinted cement mortar (1:3) and with half tiles, verge tiles and clips	7.90	5.00	**8.30**	0.20	0.10	**3.91**	2.02	**14.23**	m	**15.30**
valley tiles, Redland universal valley troughs; laid with 100mm laps	24.57	5.00	**25.80**	0.30	0.15	**5.87**	0.37	**32.04**	m	**34.44**
Roof coverings; Redland Regent through coloured tiles, 418 x 332mm; fixing every fourth course with galvanized nails to 100mm lap; 38 x 25mm pressure impregnated softwood battens fixed with galvanized nails; underlay; BS 747 type 1F bitumen reinforced felt; 150mm laps; fixing with galvanized steel clout nails										
Pitched 40 degrees from horizontal										
generally	9.09	5.00	**9.65**	0.42	0.21	**8.22**	0.81	**18.67**	m2	**20.08**
abutments; square	0.76	5.00	**0.80**	0.10	0.05	**1.96**	-	**2.75**	m	**2.96**
abutments; raking	1.14	5.00	**1.19**	0.15	0.08	**2.99**	-	**4.18**	m	**4.50**
abutments; curved to 3000mm radius..............	1.52	5.00	**1.60**	0.20	0.10	**3.91**	-	**5.51**	m	**5.92**
Roof coverings; Redland Grovebury granular faced double pantiles, 418 x 332mm; fixing every fourth course with galvanized nails to 75mm lap; 38 x 22mm pressure impregnated softwood battens fixed with galvanized nails; underlay; BS 747 type 1F bitumen reinforced felt; 150mm laps; fixing with galvanized steel clout nails										
Pitched 40 degrees from horizontal										
generally	8.51	5.00	**9.04**	0.42	0.21	**8.22**	0.81	**18.07**	m2	**19.42**
reform eaves filler unit and eaves clip to each eaves tile	0.33	5.00	**0.35**	0.06	0.03	**1.17**	0.63	**2.15**	m	**2.31**
verges; cloaked verge tiles and aluminium nails..	8.85	5.00	**9.29**	0.15	0.08	**2.99**	0.06	**12.34**	m	**13.27**
verges; half tiles, cloaked verge tiles and aluminium nails	10.30	5.00	**10.81**	0.20	0.10	**3.91**	0.10	**14.83**	m	**15.94**
verges; 150 x 6mm fibre cement undercloak; tiles and undercloak bedded and pointed in tinted cement mortar (1:3) and with standard tiles and clips...	4.42	5.00	**4.64**	0.20	0.10	**3.91**	2.02	**10.58**	m	**11.37**
verges; 150 x 6mm fibre cement undercloak; tiles and undercloak bedded and pointed in tinted cement mortar (1:3) and with treble roll verge tiles and clips	4.42	5.00	**4.64**	0.20	0.10	**3.91**	2.02	**10.58**	m	**11.37**
ridge or hip tiles, Redland third round; butt jointed; bedding and pointing in tinted cement mortar (1:3)	6.60	5.00	**6.93**	0.30	0.15	**5.87**	2.02	**14.82**	m	**15.93**
ridge or hip tiles, Redland half round; butt jointed; bedding and pointing in tinted cement mortar (1:3)	6.60	5.00	**6.93**	0.30	0.15	**5.87**	2.02	**14.82**	m	**15.93**
valley tiles, Redland Universal valley troughs; laid with 100mm laps	24.57	5.00	**25.80**	0.30	0.15	**5.87**	0.37	**32.04**	m	**34.44**
Roof coverings; Redland Grovebury through coloured double pantiles, 418 x 332mm; fixing every fourth course with galvanized nails to 75mm lap; 38 x 22mm pressure impregnated softwood battens fixed with galvanized nails; underlay; BS 747 type 1F bitumen reinforced felt; 150mm laps; fixing with galvanized steel clout nails										
Pitched 40 degrees from horizontal										
generally	8.51	5.00	**9.04**	0.42	0.21	**8.22**	0.81	**18.07**	m2	**19.42**
reform eaves filler unit and eaves clip to each eaves tile	0.33	5.00	**0.35**	0.06	0.03	**1.17**	0.63	**2.15**	m	**2.31**
verges; cloaked verge tiles and aluminium nails..	8.85	5.00	**9.29**	0.15	0.08	**2.99**	0.06	**12.34**	m	**13.27**
verges; half tiles, cloaked verge tiles and aluminium nails	10.30	5.00	**10.81**	0.20	0.10	**3.91**	0.10	**14.83**	m	**15.94**
verges; 150 x 6mm fibre cement undercloak; tiles and undercloak bedded and pointed in tinted cement mortar (1:3) and with standard tiles and clips...	4.42	5.00	**4.64**	0.20	0.10	**3.91**	2.02	**10.58**	m	**11.37**
verges; 150 x 6mm fibre cement undercloak; tiles and undercloak bedded and pointed in tinted cement mortar (1:3) and with treble roll verge tiles and clips	4.42	5.00	**4.64**	0.20	0.10	**3.91**	2.02	**10.58**	m	**11.37**

Labour hourly rates: (except Specialists) Craft Operatives £14.21 Labourer £10.71 Rates are national average prices. Refer to REGIONAL VARIATIONS for indicative levels of overall pricing in regions	MATERIALS			LABOUR				RATES		
	Del to Site £	Waste %	Material Cost £	Craft Optve Hrs	Lab Hrs	Labour Cost £	Sunds £	Nett Rate £	Unit	Gross Rate (+7.5%) £
H65: SINGLE LAP ROOF TILING Cont.										
Roof coverings; Redland Grovebury through coloured double pantiles, 418 x 332mm; fixing every fourth course with galvanized nails to 75mm lap; 38 x 22mm pressure impregnated softwood battens fixed with galvanized nails; underlay; BS 747 type 1F bitumen reinforced felt; 150mm laps; fixing with galvanized steel clout nails Cont.										
Pitched 40 degrees from horizontal Cont.										
ridge or hip tiles, Redland third round; butt jointed; bedding and pointing in tinted cement mortar (1:3)	6.60	5.00	**6.93**	0.30	0.15	**5.87**	0.37	**13.17**	m	**14.16**
ridge or hip tiles, Redland half round; butt jointed ; bedding and pointing in tinted cement mortar (1:3)	6.60	5.00	**6.93**	0.30	0.15	**5.87**	0.37	**13.17**	m	**14.16**
valley tiles, Redland Universal valley troughs; laid with 100mm laps	24.57	5.00	**25.80**	0.30	0.15	**5.87**	0.37	**32.04**	m	**34.44**
Roof coverings; Redland Norfolk through coloured pantiles, 381 x 227mm; fixing every fourth course with galvanized nails to 75mm lap; 38 x 25mm pressure impregnated softwood battens fixed with galvanized nails; underlay; BS 747 type 1F bitumen reinforced felt; 150mm laps; fixing with galvanized steel clout nails										
Pitched 40 degrees from horizontal										
generally	10.14	5.00	**10.76**	0.54	0.27	**10.57**	1.17	**22.50**	m2	**24.19**
abutments; square	0.83	5.00	**0.87**	0.10	0.05	**1.96**	-	**2.83**	m	**3.04**
abutments; raking	1.25	5.00	**1.31**	0.15	0.08	**2.99**	-	**4.30**	m	**4.62**
abutments; curved to 3000mm radius	1.67	5.00	**1.75**	0.20	0.10	**3.91**	-	**5.66**	m	**6.09**
reform eaves filler unit and aluminium nails to each eaves tile	0.33	5.00	**0.35**	0.06	0.03	**1.17**	0.63	**2.15**	m	**2.31**
verges; plain tile undercloak; tiles and undercloak bedded and pointed in tinted cement mortar (1:3) and with standard tiles and aluminium nails	1.53	5.00	**1.61**	0.10	0.05	**1.96**	0.40	**3.97**	m	**4.26**
verges; plain tile undercloak; tiles and undercloak bedded and pointed in tinted cement mortar (1:3) and with standard tiles and clips	1.53	5.00	**1.61**	0.10	0.05	**1.96**	3.07	**6.64**	m	**7.13**
verges; 150 x 6mm fibre cement undercloak; tiles and undercloak bedded and pointed in tinted cement mortar (1:3) and with standard tiles and aluminium nails	3.78	5.00	**3.97**	0.15	0.08	**2.99**	0.43	**7.39**	m	**7.94**
verges; 150 x 6mm fibre cement undercloak; tiles and undercloak bedded and pointed in tinted cement mortar (1:3) and with standard tiles and clips	3.78	5.00	**3.97**	0.15	0.08	**2.99**	3.07	**10.03**	m	**10.78**
ridge or hip tiles, Redland third round; butt jointed; bedding and pointing in tinted cement mortar (1:3)	6.90	5.00	**7.25**	0.30	0.15	**5.87**	0.37	**13.48**	m	**14.50**
ridge or hip tiles, Redland half round; butt jointed; bedding and pointing in tinted cement mortar (1:3)	6.90	5.00	**7.25**	0.30	0.15	**5.87**	0.37	**13.48**	m	**14.50**
valley tiles, Redland Universal valley troughs; laid with 100mm laps	24.57	5.00	**25.80**	0.30	0.15	**5.87**	0.37	**32.04**	m	**34.44**
Roof coverings; Redland Norfolk through coloured pantiles, 381 x 227mm; fixing every fourth course with aluminium nails to 75mm lap; 38 x 25mm pressure impregnated softwood battens fixed with galvanized nails; underlay; BS 747 type 1F bitumen reinforced felt; 150mm laps; fixing with galvanized steel clout nails										
Pitched 40 degrees from horizontal										
generally	10.14	5.00	**10.76**	0.54	0.27	**10.57**	1.17	**22.50**	m2	**24.19**
Roof coverings; Redland Norfolk through coloured pantiles, 381 x 227mm; fixing every fourth course with galvanized nails to 100mm lap; 38 x 25mm pressure impregnated softwood battens fixed with galvanized nails; underlay; BS 747 type 1F bitumen reinforced felt; 150mm laps; fixing with galvanized steel clout nails										
Pitched 40 degrees from horizontal										
generally	10.91	5.00	**11.57**	0.59	0.30	**11.60**	1.25	**24.41**	m2	**26.25**
Roof coverings; Redland 49 granular faced tiles, 381 x 227mm; fixing every fourth course with galvanized nails to 75mm lap; 38 x 25mm pressure impregnated softwood battens fixed with galvanized nails; underlay BS 747 type 1F bitumen reinforced felt; 150mm laps; fixing with galvanized steel clout nails										
Pitched 40 degrees from horizontal										
generally	9.62	5.00	**10.21**	0.54	0.27	**10.57**	1.17	**21.94**	m2	**23.59**
supplementary fixing eaves course with one aluminium nail per tile	-	-	**-**	0.02	0.01	**0.39**	0.03	**0.42**	m	**0.45**
verges; Redland Dry Verge system with standard tiles and clips	11.04	5.00	**11.59**	0.25	0.12	**4.84**	1.71	**18.14**	m	**19.50**
verges; Redland Dry Verge system with verge tiles and clips	11.04	5.00	**11.59**	0.25	0.12	**4.84**	1.71	**18.14**	m	**19.50**

Labour hourly rates: (except Specialists) Craft Operatives £14.21 Labourer £10.71 Rates are national average prices. Refer to REGIONAL VARIATIONS for indicative levels of overall pricing in regions	MATERIALS			LABOUR				RATES		
	Del to Site £	Waste %	Material Cost £	Craft Optve Hrs	Lab Hrs	Labour Cost £	Sunds £	Nett Rate £	Unit	Gross Rate (+7.5%) £
H65: SINGLE LAP ROOF TILING Cont.										
Roof coverings; Redland 49 granular faced tiles, 381 x 227mm; fixing every fourth course with galvanized nails to 75mm lap; 38 x 25mm pressure impregnated softwood battens fixed with galvanized nails; underlay BS 747 type 1F bitumen reinforced felt; 150mm laps; fixing with galvanized steel clout nails Cont.										
Pitched 40 degrees from horizontal Cont.										
verges; plain tile undercloak; tiles and undercloak bedded and pointed in tinted cement mortar (1:3) and with standard tiles and aluminium nails	1.50	5.00	**1.57**	0.10	0.05	**1.96**	0.43	**3.96**	m	**4.26**
verges; plain tile undercloak; tiles and undercloak bedded and pointed in tinted cement mortar (1:3) and with verge tiles and aluminium nails	1.50	5.00	**1.57**	0.10	0.05	**1.96**	0.43	**3.96**	m	**4.26**
verges; 150 x 6mm fibre cement undercloak; tiles and undercloak bedded and pointed in tinted cement mortar (1:3) and with standard tiles and aluminium nails	3.75	5.00	**3.93**	0.15	0.08	**2.99**	0.43	**7.35**	m	**7.90**
verges; 150 x 6mm fibre cement undercloak; tiles and undercloak bedded and pointed in tinted cement mortar (1:3) and with verge tiles and aluminium nails	3.75	5.00	**3.93**	0.15	0.08	**2.99**	0.43	**7.35**	m	**7.90**
ridge or hip tiles, Redland third round; butt jointed; bedding and pointing in tinted cement mortar (1:3)	6.90	5.00	**7.25**	0.30	0.15	**5.87**	0.37	**13.48**	m	**14.50**
ridge or hip tiles, Redland half round; butt jointed; bedding and pointing in tinted cement mortar (1:3)	6.90	5.00	**7.25**	0.30	0.15	**5.87**	0.37	**13.48**	m	**14.50**
valley tiles, Redland Universal valley troughs; laid with 100mm laps	24.57	5.00	**25.80**	0.30	0.15	**5.87**	0.37	**32.04**	m	**34.44**
Roof coverings; Redland 49 granular faced tiles, 381 x 227mm; fixing every fourth course with galvanized nails to 100mm lap; 38 x 25mm pressure impregnated softwood battens fixed with galvanized nails; underlay; BS 747 type 1F bitumen reinforced felt; 150mm laps; fixing with galvanized steel clout nails										
Pitched 40 degrees from horizontal										
generally	10.37	5.00	**10.99**	0.59	0.30	**11.60**	1.25	**23.84**	m2	**25.63**
abutments; square	0.89	5.00	**0.93**	0.10	0.05	**1.96**	-	**2.89**	m	**3.11**
abutments; raking	1.33	5.00	**1.40**	0.15	0.08	**2.99**	-	**4.39**	m	**4.72**
abutments; curved to 3000mm radius	1.41	5.00	**1.48**	0.20	0.10	**3.91**	-	**5.39**	m	**5.79**
Roof coverings; Redland 49 through coloured tiles, 381 x 227mm; fixing every fourth course with galvanized nails to 75mm lap; 38 x 25mm pressure impregnated softwood battens fixed with galvanized nails; underlay; BS 747 type 1F bitumen reinforced felt; 150mm laps; fixing with galvanized steel clout nails										
Pitched 40 degrees from horizontal										
generally	9.62	5.00	**10.21**	0.54	0.27	**10.57**	1.17	**21.94**	m2	**23.59**
abutments; square	0.81	5.00	**0.85**	0.10	0.05	**1.96**	-	**2.81**	m	**3.02**
abutments; raking	1.22	5.00	**1.28**	0.15	0.08	**2.99**	-	**4.27**	m	**4.59**
abutments; curved to 3000mm radius	1.65	5.00	**1.73**	0.20	0.10	**3.91**	-	**5.64**	m	**6.07**
supplementary fixing eaves course with one aluminium nail per tile	-	-	**-**	0.02	0.01	**0.39**	0.03	**0.42**	m	**0.45**
verges; Redland Dry Verge system with standard tiles and clips	11.04	5.00	**11.59**	0.25	0.12	**4.84**	1.71	**18.14**	m	**19.50**
verges; Redland Dry Verge system with verge tiles and clips	11.04	5.00	**11.59**	0.25	0.12	**4.84**	1.71	**18.14**	m	**19.50**
verges; plain tile undercloak; tiles and undercloak bedded and pointed in tinted cement mortar (1:3) and with standard tiles and aluminium nails	1.50	5.00	**1.57**	0.10	0.05	**1.96**	0.43	**3.96**	m	**4.26**
verges; plain tile undercloak; tiles and undercloak bedded and pointed in tinted cement mortar (1:3) and with verge tiles and aluminium nails	1.50	5.00	**1.57**	0.10	0.05	**1.96**	0.43	**3.96**	m	**4.26**
verges; 150 x 6mm fibre cement undercloak; tiles and undercloak bedded and pointed in tinted cement mortar (1:3) and with standard tiles and aluminium nails	3.75	5.00	**3.93**	0.15	0.08	**2.99**	0.43	**7.35**	m	**7.90**
verges; 150 x 6mm fibre cement undercloak; tiles and undercloak bedded and pointed in tinted cement mortar (1:3) and with verge tiles and aluminium nails	3.75	5.00	**3.93**	0.15	0.08	**2.99**	0.43	**7.35**	m	**7.90**
ridge or hip tiles, Redland third round; butt jointed; bedding and pointing in tinted cement mortar (1:3)	6.90	5.00	**7.25**	0.30	0.15	**5.87**	0.37	**13.48**	m	**14.50**
ridge or hip tiles, Redland half round; butt jointed; bedding and pointing in tinted cement mortar (1:3)	6.90	5.00	**7.25**	0.30	0.15	**5.87**	0.37	**13.48**	m	**14.50**
valley tiles, Redland Universal valley troughs; laid with 100mm laps	24.57	5.00	**25.80**	0.30	0.15	**5.87**	0.37	**32.04**	m	**34.44**

Labour hourly rates: (except Specialists) Craft Operatives £14.21 Labourer £10.71 Rates are national average prices. Refer to REGIONAL VARIATIONS for indicative levels of overall pricing in regions	MATERIALS			LABOUR				RATES		
	Del to Site £	Waste %	Material Cost £	Craft Optve Hrs	Lab Hrs	Labour Cost £	Sunds £	Nett Rate £	Unit	Gross Rate (+7.5%) £
H65: SINGLE LAP ROOF TILING Cont.										
Roof coverings; Redland 49 through coloured tiles, 381 x 227mm; fixing every fourth course with galvanized nails to 100mm lap; 38 x 25mm pressure impregnated softwood battens fixed with galvanized nails; underlay; BS 747 type 1F bitumen reinforced felt; 150mm laps; fixing with galvanized steel clout nails										
Pitched 40 degrees from horizontal										
generally	10.37	5.00	**10.99**	0.59	0.30	**11.60**	1.25	**23.84**	m2	**25.63**
Roof coverings; Redland Renown granular faced tiles, 418 x 330mm; fixing every fourth course with galvanized nails to 75mm lap; 38 x 25mm pressure impregnated softwood battens fixed with galvanized nails; underlay; BS 747 type 1F bitumen reinforced felt; 150mm laps; fixing with galvanized steel clout nails										
Pitched 40 degrees from horizontal										
generally	8.28	5.00	**8.80**	0.42	0.21	**8.22**	0.51	**17.52**	m2	**18.84**
abutments; square	0.68	5.00	**0.71**	0.10	0.05	**1.96**	-	**2.67**	m	**2.87**
abutments; raking	1.02	5.00	**1.07**	0.15	0.08	**2.99**	-	**4.06**	m	**4.37**
abutments; curved to 3000mm radius	1.36	5.00	**1.43**	0.20	0.10	**3.91**	-	**5.34**	m	**5.74**
reform eaves filler unit and aluminium nails to each eaves tile	0.33	5.00	**0.35**	0.06	0.03	**1.17**	0.63	**2.15**	m	**2.31**
verges; cloaked verge tiles and aluminium nails	2.31	5.00	**2.43**	0.15	0.08	**2.99**	0.10	**5.51**	m	**5.93**
verges; half tiles, cloaked verge tiles and aluminium nails	10.95	5.00	**11.50**	0.15	0.08	**2.99**	0.43	**14.92**	m	**16.03**
verges; plain tile undercloak; tiles and undercloak bedded and pointed in tinted cement mortar (1:3) and with standard tiles and aluminium nails	2.10	5.00	**2.21**	0.10	0.05	**1.96**	0.43	**4.59**	m	**4.94**
verges; plain tile undercloak; tiles and undercloak bedded and pointed in tinted cement mortar (1:3) and with verge tiles and aluminium nails	2.10	5.00	**2.21**	0.10	0.05	**1.96**	0.43	**4.59**	m	**4.94**
verges; 150 x 6mm fibre cement undercloak; tiles and undercloak bedded and pointed in tinted cement mortar (1:3) and with standard tiles and aluminium nails	4.35	5.00	**4.57**	0.15	0.08	**2.99**	0.43	**7.99**	m	**8.58**
verges; 150 x 6mm fibre cement undercloak; tiles and undercloak bedded and pointed in tinted cement mortar (1:3) and with verge tiles and aluminium nails	4.35	5.00	**4.57**	0.15	0.08	**2.99**	0.43	**7.99**	m	**8.58**
ridge or hip tiles, Redland third round; butt jointed; bedding and pointing in tinted cement mortar (1:3)	6.90	5.00	**7.25**	0.30	0.15	**5.87**	0.37	**13.48**	m	**14.50**
ridge or hip tiles, Redland half round; butt jointed; bedding and pointing in tinted cement mortar (1:3)	6.90	5.00	**7.25**	0.30	0.15	**5.87**	0.37	**13.48**	m	**14.50**
valley tiles, Redland Universal valley troughs; laid with 100mm laps	18.84	5.00	**19.78**	0.30	0.15	**5.87**	0.37	**26.02**	m	**27.97**
Roof coverings; Redland 50 granular faced double roman tiles, 418 x 330mm; fixing every fourth course with galvanized nails to 75mm lap; 38 x 25mm pressure impregnated softwood battens fixed with galvanized nails; underlay; BS 747 type 1F bitumen reinforced felt; 150mm laps; fixing with galvanized steel clout nails										
Pitched 40 degrees from horizontal										
generally	8.28	5.00	**8.80**	0.42	0.21	**8.22**	0.61	**17.62**	m2	**18.94**
reform eaves filler unit and aluminium nails to each eaves tile	0.33	5.00	**0.35**	0.06	0.03	**1.17**	0.63	**2.15**	m	**2.31**
verges; cloaked verge tiles and aluminium nails	10.95	5.00	**11.50**	0.15	0.08	**2.99**	0.06	**14.55**	m	**15.64**
verges; half tiles, cloaked verge tiles and aluminium nails	10.25	5.00	**10.76**	0.20	0.10	**3.91**	0.06	**14.74**	m	**15.84**
verges; plain tile undercloak; tiles and undercloak bedded and pointed in tinted cement mortar (1:3) and with standard tiles and aluminium nails	2.10	5.00	**2.21**	0.10	0.05	**1.96**	0.43	**4.59**	m	**4.94**
verges; plain tile undercloak; tiles and undercloak bedded and pointed in tinted cement mortar (1:3) and with treble roll verge tiles and aluminium nails	2.10	5.00	**2.21**	0.10	0.05	**1.96**	0.43	**4.59**	m	**4.94**
verges; 150 x 6mm fibre cement undercloak; tiles and undercloak bedded and pointed in tinted cement mortar (1:3) and with standard tiles and aluminium nails	4.35	5.00	**4.57**	0.15	0.08	**2.99**	0.43	**7.99**	m	**8.58**
verges; 150 x 6mm fibre cement undercloak; tiles and undercloak bedded and pointed in tinted cement mortar (1:3) and with treble roll verge tiles and aluminium nails	4.35	5.00	**4.57**	0.15	0.08	**2.99**	0.43	**7.99**	m	**8.58**
ridge or hip tiles, Redland third round; butt jointed; bedding and pointing in tinted cement mortar (1:3)	6.90	5.00	**7.25**	0.30	0.15	**5.87**	0.37	**13.48**	m	**14.50**
ridge or hip tiles, Redland half round; butt jointed; bedding and pointing in tinted cement mortar (1:3)	6.90	5.00	**7.25**	0.30	0.15	**5.87**	0.37	**13.48**	m	**14.50**
valley tiles, Redland Universal valley troughs; laid with 100mm laps	24.57	5.00	**25.80**	0.30	0.15	**5.87**	0.37	**32.04**	m	**34.44**

Labour hourly rates: (except Specialists) Craft Operatives £14.21 Labourer £10.71 Rates are national average prices. Refer to REGIONAL VARIATIONS for indicative levels of overall pricing in regions	MATERIALS			LABOUR				RATES		
	Del to Site £	Waste %	Material Cost £	Craft Optve Hrs	Lab Hrs	Labour Cost £	Sunds £	Nett Rate £	Unit	Gross Rate (+7.5%) £
H65: SINGLE LAP ROOF TILING Cont.										
Roof coverings; Redland 50 through coloured double roman tiles, 418 x 330mm; fixing every fourth course with galvanized nails to 75mm lap; 38 x 25mm pressure impregnated softwood battens fixed with galvanized nails; underlay; BS 747 type 1F bitumen reinforced felt; 150mm laps; fixing with galvanized steel clout nails										
Pitched 40 degrees from horizontal										
generally	8.28	5.00	**8.80**	0.42	0.21	**8.22**	0.61	**17.62**	m2	**18.94**
reform eaves filler unit and aluminium nails to each eaves tile	0.33	5.00	**0.35**	0.06	0.03	**1.17**	0.63	**2.15**	m	**2.31**
verges; cloaked verge tiles and aluminium nails	10.95	5.00	**11.50**	0.15	0.08	**2.99**	0.06	**14.55**	m	**15.64**
verges; half tiles, cloaked verge tiles and aluminium nails	10.95	5.00	**11.50**	0.15	0.08	**2.99**	0.06	**14.55**	m	**15.64**
verges; plain tile undercloak; tiles and undercloak bedded and pointed in tinted cement mortar (1:3) and with standard tiles and aluminium nails	2.10	5.00	**2.21**	0.10	0.05	**1.96**	0.43	**4.59**	m	**4.94**
verges; plain tile undercloak; tiles and undercloak bedded and pointed in tinted cement mortar (1:3) and with treble roll verge tiles and aluminium nails	2.10	5.00	**2.21**	0.10	0.05	**1.96**	0.43	**4.59**	m	**4.94**
verges; 150 x 6mm fibre cement undercloak; tiles and undercloak bedded and pointed in tinted cement mortar (1:3) and with standard tiles and aluminium nails	4.35	5.00	**4.57**	0.15	0.08	**2.99**	0.43	**7.99**	m	**8.58**
verges; 150 x 6mm fibre cement undercloak; tiles and undercloak bedded and pointed in tinted cement mortar (1:3) and with treble roll verge tiles and aluminium nails	4.35	5.00	**4.57**	0.15	0.08	**2.99**	0.43	**7.99**	m	**8.58**
ridge or hip tiles, Redland third round; butt jointed; bedding and pointing in tinted cement mortar (1:3)	6.90	5.00	**7.25**	0.30	0.15	**5.87**	0.37	**13.48**	m	**14.50**
ridge or hip tiles, Redland half round; butt jointed; bedding and pointing in tinted cement mortar (1:3)	6.90	5.00	**7.25**	0.30	0.15	**5.87**	0.37	**13.48**	m	**14.50**
valley tiles, Redland Universal valley troughs; laid with 100mm laps	24.57	5.00	**25.80**	0.30	0.15	**5.87**	0.37	**32.04**	m	**34.44**
Fittings										
Ventilator tiles										
Red Vent ridge ventilation terminal 450mm long for										
half round ridge	45.20	5.00	**47.46**	1.30	0.65	**25.43**	0.59	**73.48**	nr	**79.00**
universal angle ridge	45.20	5.00	**47.46**	1.30	0.65	**25.43**	0.59	**73.48**	nr	**79.00**
Redvent eaves ventilators fixing with aluminium nails	8.17	5.00	**8.58**	0.50	0.25	**9.78**	0.03	**18.39**	m	**19.77**
Red Vent Thruvent tiles for										
Stonewold slates	34.50	5.00	**36.23**	0.50	0.25	**9.78**	0.22	**46.23**	nr	**49.69**
Regent tiles	34.50	5.00	**36.23**	0.50	0.25	**9.78**	0.22	**46.23**	nr	**49.69**
Grovebury double pantiles	34.50	5.00	**36.23**	0.50	0.25	**9.78**	0.22	**46.23**	nr	**49.69**
Norfolk pantiles	34.50	5.00	**36.23**	0.50	0.25	**9.78**	0.22	**46.23**	nr	**49.69**
Redland 49 tiles	34.50	5.00	**36.23**	0.50	0.25	**9.78**	0.22	**46.23**	nr	**49.69**
Renown tiles	34.50	5.00	**36.23**	0.50	0.25	**9.78**	0.22	**46.23**	nr	**49.69**
Redland 50 double roman tiles	34.50	5.00	**36.23**	0.50	0.25	**9.78**	0.22	**46.23**	nr	**49.69**
Red line ventilation tile complete with underlay seal and fixing clips for										
Stonewold slates	42.75	5.00	**44.89**	1.00	0.50	**19.57**	0.22	**64.67**	nr	**69.52**
Regent tiles	42.75	5.00	**44.89**	1.00	0.50	**19.57**	0.22	**64.67**	nr	**69.52**
Grovebury double pantiles	42.75	5.00	**44.89**	1.00	0.50	**19.57**	0.22	**64.67**	nr	**69.52**
Renown tiles	42.75	5.00	**44.89**	1.00	0.50	**19.57**	0.22	**64.67**	nr	**69.52**
Redland 50 double roman tiles	42.75	5.00	**44.89**	1.00	0.50	**19.57**	0.22	**64.67**	nr	**69.52**
Gas terminals										
Gas Flue ridge terminal Mark III, 450mm long with sealing gasket and fixing brackets										
half round ridge	44.30	5.00	**46.52**	1.50	0.75	**29.35**	0.59	**76.45**	nr	**82.19**
extra for 150mm extension adaptor	16.70	5.00	**17.54**	0.30	0.15	**5.87**	0.22	**23.62**	nr	**25.40**
Hip irons										
galvanized mild steel; 32 x 3 x 380mm girth; scrolled; fixing with galvanized steel screws to timber	1.84	5.00	**1.93**	0.10	0.05	**1.96**	-	**3.89**	nr	**4.18**
Lead soakers										
fixing only	-	-	**-**	0.30	0.15	**5.87**	-	**5.87**	nr	**6.31**

This section continues on the next page

Labour hourly rates: (except Specialists) Craft Operatives £17.38 Labourer £10.71 Rates are national average prices. Refer to REGIONAL VARIATIONS for indicative levels of overall pricing in regions	MATERIALS			LABOUR				RATES		
	Del to Site £	Waste %	Material Cost £	Craft Optve Hrs	Lab Hrs	Labour Cost £	Sunds £	Nett Rate £	Unit	Gross Rate (+7.5%) £
H71: LEAD SHEET COVERINGS/FLASHINGS										
Sheet lead										
Technical data										
Code 3, 1.32mm thick, 14.97 kg/m2, colour code green										
Code 4, 1.80mm thick, 20.41 kg/m2, colour code blue										
Code 5, 2.24mm thick, 25.40 kg/m2, colour code red										
Code 6, 2.65mm thick, 30.05 kg/m2, colour code black										
Code 7, 3.15mm thick, 35.72 kg/m2, colour code white										
Code 8, 3.55mm thick, 40.26 kg/m2, colour code orange										
Milled lead sheet, BS EN 12588										
Nr 3 roof coverings; fixing to timber with milled lead cleats and galvanized screws										
pitch 7.5 degrees from horizontal	14.64	5.00	**15.37**	3.85	-	**66.91**	0.20	**82.49**	m2	**88.68**
pitch 40 degrees from horizontal	14.64	5.00	**15.37**	4.40	-	**76.47**	0.20	**92.05**	m2	**98.95**
pitch 75 degrees from horizontal	14.64	5.00	**15.37**	4.95	-	**86.03**	0.20	**101.61**	m2	**109.23**
Nr 4 roof coverings; fixing to timber with milled lead cleats and galvanized screws										
pitch 7.5 degrees from horizontal	19.92	5.00	**20.92**	4.40	-	**76.47**	0.27	**97.65**	m2	**104.98**
pitch 40 degrees from horizontal	19.92	5.00	**20.92**	4.95	-	**86.03**	0.27	**107.21**	m2	**115.25**
pitch 75 degrees from horizontal	19.92	5.00	**20.92**	5.50	-	**95.59**	0.27	**116.77**	m2	**125.53**
Nr 5 roof coverings; fixing to timber with milled lead cleats and galvanized screws										
pitch 7.5 degrees from horizontal	24.79	5.00	**26.03**	4.95	-	**86.03**	0.33	**112.39**	m2	**120.82**
pitch 40 degrees from horizontal	24.79	5.00	**26.03**	5.50	-	**95.59**	0.33	**121.95**	m2	**131.09**
pitch 75 degrees from horizontal	24.79	5.00	**26.03**	6.05	-	**105.15**	0.33	**131.50**	m2	**141.37**
extra; oil patination to surfaces	1.60	10.00	**1.76**	0.20	-	**3.48**	-	**5.24**	m2	**5.63**
Nr 3 wall coverings; fixing to timber with milled lead cleats and galvanized screws										
vertical	14.64	5.00	**15.37**	6.95	-	**120.79**	0.20	**136.37**	m2	**146.59**
Nr 4 wall coverings; fixing to timber with milled lead cleats and galvanized screws										
vertical	19.92	5.00	**20.92**	7.50	-	**130.35**	0.27	**151.53**	m2	**162.90**
Nr 5 wall coverings; fixing to timber with milled lead cleats and galvanized screws										
vertical	24.79	5.00	**26.03**	8.05	-	**139.91**	0.33	**166.26**	m2	**178.73**
Flashings; horizontal										
Nr 3; 150mm lapped joints; fixing to masonry with milled lead clips and lead wedges										
150mm girth	2.20	5.00	**2.31**	0.40	-	**6.95**	0.20	**9.47**	m	**10.18**
240mm girth	3.51	5.00	**3.69**	0.64	-	**11.12**	0.20	**15.01**	m	**16.14**
300mm girth	4.39	5.00	**4.61**	0.80	-	**13.90**	0.20	**18.72**	m	**20.12**
Nr 4; 150mm lapped joints; fixing to masonry with milled lead clips and lead wedges										
150mm girth	2.99	5.00	**3.14**	0.44	-	**7.65**	0.27	**11.05**	m	**11.88**
240mm girth	4.78	5.00	**5.02**	0.71	-	**12.34**	0.27	**17.62**	m	**18.95**
300mm girth	5.98	5.00	**6.28**	0.89	-	**15.47**	0.27	**22.01**	m	**23.66**
Nr 5; 150mm lapped joints; fixing to masonry with milled lead clips and lead wedges										
150mm girth	3.72	5.00	**3.91**	0.50	-	**8.69**	0.33	**12.92**	m	**13.89**
240mm girth	5.95	5.00	**6.25**	0.80	-	**13.90**	0.33	**20.48**	m	**22.01**
300mm girth	7.44	5.00	**7.81**	1.00	-	**17.38**	0.33	**25.52**	m	**27.43**
Nr 3; 150mm lapped joints; fixing to timber with copper nails										
150mm girth	2.20	5.00	**2.31**	0.40	-	**6.95**	0.26	**9.52**	m	**10.23**
240mm girth	3.51	5.00	**3.69**	0.64	-	**11.12**	0.26	**15.06**	m	**16.19**
300mm girth	4.39	5.00	**4.61**	0.80	-	**13.90**	0.26	**18.77**	m	**20.18**
Nr 4; 150mm lapped joints; fixing to timber with copper nails										
150mm girth	2.99	5.00	**3.14**	0.44	-	**7.65**	0.26	**11.04**	m	**11.87**
240mm girth	4.78	5.00	**5.02**	0.71	-	**12.34**	0.26	**17.61**	m	**18.93**
300mm girth	5.98	5.00	**6.28**	0.89	-	**15.47**	0.26	**22.00**	m	**23.65**
Nr 5; 150mm lapped joints; fixing to timber with copper nails										
150mm girth	3.72	5.00	**3.91**	0.50	-	**8.69**	0.26	**12.85**	m	**13.81**
240mm girth	5.95	5.00	**6.25**	0.80	-	**13.90**	0.26	**20.41**	m	**21.94**
300mm girth	7.44	5.00	**7.81**	1.00	-	**17.38**	0.26	**25.45**	m	**27.36**
Flashings; stepped										
Nr 3; 150mm lapped joints; fixing to masonry with milled lead clips and lead wedges										
180mm girth	2.64	5.00	**2.77**	0.64	-	**11.12**	0.24	**14.14**	m	**15.20**
240mm girth	3.51	5.00	**3.69**	0.86	-	**14.95**	0.24	**18.88**	m	**20.29**
300mm girth	4.39	5.00	**4.61**	1.08	-	**18.77**	0.24	**23.62**	m	**25.40**

Labour hourly rates: (except Specialists)
Craft Operatives £17.38 Labourer £10.71
Rates are national average prices.
Refer to REGIONAL VARIATIONS for indicative levels of overall pricing in regions

	MATERIALS			LABOUR				RATES		
	Del to Site £	Waste %	Material Cost £	Craft Optve Hrs	Lab Hrs	Labour Cost £	Sunds £	Nett Rate £	Unit	Gross Rate (+7.5%) £
H71: LEAD SHEET COVERINGS/FLASHINGS Cont.										
Flashings; stepped Cont.										
Nr 4; 150mm lapped joints; fixing to masonry with milled lead clips and lead wedges										
180mm girth	3.59	5.00	**3.77**	0.71	-	**12.34**	0.41	**16.52**	m	**17.76**
240mm girth	4.78	5.00	**5.02**	0.95	-	**16.51**	0.41	**21.94**	m	**23.58**
300mm girth	5.98	5.00	**6.28**	1.19	-	**20.68**	0.41	**27.37**	m	**29.42**
Nr 5; 150mm lapped joints; fixing to masonry with milled lead clips and lead wedges										
180mm girth	4.46	5.00	**4.68**	0.80	-	**13.90**	0.49	**19.08**	m	**20.51**
240mm girth	5.95	5.00	**6.25**	1.07	-	**18.60**	0.49	**25.33**	m	**27.23**
300mm girth	7.44	5.00	**7.81**	1.33	-	**23.12**	0.49	**31.42**	m	**33.77**
Aprons; horizontal										
Nr 3; 150mm lapped joints; fixing to masonry with milled lead clips and lead wedges										
150mm girth	2.20	5.00	**2.31**	0.40	-	**6.95**	0.20	**9.47**	m	**10.18**
240mm girth	3.51	5.00	**3.69**	0.64	-	**11.12**	0.20	**15.01**	m	**16.14**
300mm girth	4.39	5.00	**4.61**	0.80	-	**13.90**	0.20	**18.72**	m	**20.12**
450mm girth	6.59	5.00	**6.92**	1.20	-	**20.86**	0.31	**28.08**	m	**30.19**
Nr 4; 150mm lapped joints; fixing to masonry with milled lead clips and lead wedges										
150mm girth	2.99	5.00	**3.14**	0.44	-	**7.65**	0.27	**11.05**	m	**11.88**
240mm girth	4.78	5.00	**5.02**	0.71	-	**12.34**	0.27	**17.62**	m	**18.95**
300mm girth	5.98	5.00	**6.28**	0.89	-	**15.47**	0.27	**22.01**	m	**23.66**
450mm girth	8.96	5.00	**9.41**	1.33	-	**23.12**	0.41	**32.93**	m	**35.40**
Nr 5; 150mm lapped joints; fixing to masonry with milled lead clips and lead wedges										
150mm girth	3.72	5.00	**3.91**	0.50	-	**8.69**	0.33	**12.92**	m	**13.89**
240mm girth	5.95	5.00	**6.25**	0.80	-	**13.90**	0.33	**20.48**	m	**22.01**
300mm girth	7.44	5.00	**7.81**	1.00	-	**17.38**	0.33	**25.52**	m	**27.43**
450mm girth	11.16	5.00	**11.72**	1.50	-	**26.07**	0.49	**38.28**	m	**41.15**
Hips; sloping; dressing over slating and tiling										
Nr 3; 150mm lapped joints; fixing to timber with milled lead clips										
240mm girth	3.51	5.00	**3.69**	0.64	-	**11.12**	0.20	**15.01**	m	**16.14**
300mm girth	4.39	5.00	**4.61**	0.80	-	**13.90**	0.20	**18.72**	m	**20.12**
450mm girth	6.59	5.00	**6.92**	1.20	-	**20.86**	0.31	**28.08**	m	**30.19**
Nr 4; 150mm lapped joints; fixing to timber with milled lead clips										
240mm girth	4.78	5.00	**5.02**	0.71	-	**12.34**	0.27	**17.62**	m	**18.95**
300mm girth	5.98	5.00	**6.28**	0.89	-	**15.47**	0.27	**22.01**	m	**23.66**
450mm girth	8.96	5.00	**9.41**	1.33	-	**23.12**	0.41	**32.93**	m	**35.40**
Nr 5; 150mm lapped joints; fixing to timber with milled lead clips										
240mm girth	5.95	5.00	**6.25**	0.80	-	**13.90**	0.33	**20.48**	m	**22.01**
300mm girth	7.44	5.00	**7.81**	1.00	-	**17.38**	0.33	**25.52**	m	**27.43**
450mm girth	11.16	5.00	**11.72**	1.50	-	**26.07**	0.49	**38.28**	m	**41.15**
Kerbs; horizontal										
Nr 3; 150mm lapped joints; fixing to timber with copper nails										
240mm girth	3.51	5.00	**3.69**	0.64	-	**11.12**	0.20	**15.01**	m	**16.14**
300mm girth	4.39	5.00	**4.61**	0.80	-	**13.90**	0.20	**18.72**	m	**20.12**
450mm girth	6.59	5.00	**6.92**	1.20	-	**20.86**	0.31	**28.08**	m	**30.19**
Nr 4; 150mm lapped joints; fixing to timber with copper nails										
240mm girth	4.78	5.00	**5.02**	0.71	-	**12.34**	0.27	**17.62**	m	**18.95**
300mm girth	5.98	5.00	**6.28**	0.89	-	**15.47**	0.27	**22.01**	m	**23.66**
450mm girth	8.96	5.00	**9.41**	1.33	-	**23.12**	0.41	**32.93**	m	**35.40**
Nr 5; 150mm lapped joints; fixing to timber with copper nails										
240mm girth	5.95	5.00	**6.25**	0.80	-	**13.90**	0.33	**20.48**	m	**22.01**
300mm girth	7.44	5.00	**7.81**	1.00	-	**17.38**	0.33	**25.52**	m	**27.43**
450mm girth	11.16	5.00	**11.72**	1.50	-	**26.07**	0.49	**38.28**	m	**41.15**
Ridges; horizontal; dressing over slating and tiling										
Nr 3; 150mm lapped joints; fixing to timber with milled lead clips										
240mm girth	3.51	5.00	**3.69**	0.64	-	**11.12**	0.20	**15.01**	m	**16.14**
300mm girth	4.39	5.00	**4.61**	0.80	-	**13.90**	0.20	**18.72**	m	**20.12**
450mm girth	6.59	5.00	**6.92**	1.20	-	**20.86**	0.31	**28.08**	m	**30.19**
Nr 4; 150mm lapped joints; fixing to timber with milled lead clips										
240mm girth	4.78	5.00	**5.02**	0.71	-	**12.34**	0.27	**17.62**	m	**18.95**
300mm girth	5.98	5.00	**6.28**	0.89	-	**15.47**	0.27	**22.01**	m	**23.66**
450mm girth	8.96	5.00	**9.41**	1.33	-	**23.12**	0.41	**32.93**	m	**35.40**
Nr 5; 150mm lapped joints; fixing to timber with milled lead clips										
240mm girth	5.95	5.00	**6.25**	0.80	-	**13.90**	0.33	**20.48**	m	**22.01**

Labour hourly rates: (except Specialists) Craft Operatives £17.38 Labourer £10.71 Rates are national average prices. Refer to REGIONAL VARIATIONS for indicative levels of overall pricing in regions	MATERIALS			LABOUR				RATES		
	Del to Site £	Waste %	Material Cost £	Craft Optve Hrs	Lab Hrs	Labour Cost £	Sunds £	Nett Rate £	Unit	Gross Rate (+7.5%) £
H71: LEAD SHEET COVERINGS/FLASHINGS Cont.										
Ridges; horizontal; dressing over slating and tiling Cont.										
Nr 5; 150mm lapped joints; fixing to timber with milled lead clips Cont.										
300mm girth	7.44	5.00	**7.81**	1.00	-	**17.38**	0.33	**25.52**	m	**27.43**
450mm girth	11.16	5.00	**11.72**	1.50	-	**26.07**	0.49	**38.28**	m	**41.15**
Valleys; sloping										
Nr 3; dressing over tilting fillets -1; 150mm lapped joints; fixing to timber with copper nails										
240mm girth	3.51	5.00	**3.69**	0.64	-	**11.12**	0.20	**15.01**	m	**16.14**
300mm girth	4.39	5.00	**4.61**	0.80	-	**13.90**	0.20	**18.72**	m	**20.12**
450mm girth	6.59	5.00	**6.92**	1.20	-	**20.86**	0.31	**28.08**	m	**30.19**
Nr 4; dressing over tilting fillets -1; 150mm lapped joints; fixing to timber with copper nails										
240mm girth	4.78	5.00	**5.02**	0.71	-	**12.34**	0.27	**17.62**	m	**18.95**
300mm girth	5.98	5.00	**6.28**	0.89	-	**15.47**	0.27	**22.01**	m	**23.66**
450mm girth	8.96	5.00	**9.41**	1.33	-	**23.12**	0.41	**32.93**	m	**35.40**
Nr 5; dressing over tilting fillets -1; 150mm lapped joints; fixing to timber with copper nails										
240mm girth	5.95	5.00	**6.25**	0.80	-	**13.90**	0.33	**20.48**	m	**22.01**
300mm girth	7.44	5.00	**7.81**	1.00	-	**17.38**	0.33	**25.52**	m	**27.43**
450mm girth	11.16	5.00	**11.72**	1.50	-	**26.07**	0.49	**38.28**	m	**41.15**
Cavity gutters										
Nr 3; 150mm lapped joints; bedding in cement mortar (1:3)										
225mm girth	3.51	5.00	**3.69**	0.64	-	**11.12**	0.41	**15.22**	m	**16.36**
345mm girth	5.27	5.00	**5.53**	0.96	-	**16.68**	0.61	**22.83**	m	**24.54**
Nr 3 edges										
welted	-	-	**-**	0.17	-	**2.95**	-	**2.95**	m	**3.18**
beaded	-	-	**-**	0.17	-	**2.95**	-	**2.95**	m	**3.18**
Nr 4 edges										
welted	-	-	**-**	0.17	-	**2.95**	-	**2.95**	m	**3.18**
beaded	-	-	**-**	0.17	-	**2.95**	-	**2.95**	m	**3.18**
Nr 5 edges										
welted	-	-	**-**	0.17	-	**2.95**	-	**2.95**	m	**3.18**
beaded	-	-	**-**	0.17	-	**2.95**	-	**2.95**	m	**3.18**
Nr 3 seams										
leadburned	0.85	-	**0.85**	0.20	-	**3.48**	-	**4.33**	m	**4.65**
Nr 4 seams										
leadburned	1.05	-	**1.05**	0.20	-	**3.48**	-	**4.53**	m	**4.87**
Nr 5 seams										
leadburned	1.35	-	**1.35**	0.20	-	**3.48**	-	**4.83**	m	**5.19**
Nr 3 dressings										
corrugated roofing; fibre cement; down corrugations	-	-	**-**	0.25	-	**4.35**	-	**4.35**	m	**4.67**
corrugated roofing; fibre cement; across corrugations	-	-	**-**	0.25	-	**4.35**	-	**4.35**	m	**4.67**
glass and glazing bars; timber	-	-	**-**	0.25	-	**4.35**	-	**4.35**	m	**4.67**
Nr 4 dressings										
corrugated roofing; fibre cement; down corrugations	-	-	**-**	0.25	-	**4.35**	-	**4.35**	m	**4.67**
corrugated roofing; fibre cement; across corrugations	-	-	**-**	0.25	-	**4.35**	-	**4.35**	m	**4.67**
glass and glazing bars; timber	-	-	**-**	0.25	-	**4.35**	-	**4.35**	m	**4.67**
Nr 5 dressings										
corrugated roofing; fibre cement; down corrugations	-	-	**-**	0.25	-	**4.35**	-	**4.35**	m	**4.67**
corrugated roofing; fibre cement; across corrugations	-	-	**-**	0.25	-	**4.35**	-	**4.35**	m	**4.67**
glass and glazing bars; timber	-	-	**-**	0.25	-	**4.35**	-	**4.35**	m	**4.67**
Nr 3 soakers and slates handed to others for fixing										
180 x 180mm	0.48	5.00	**0.50**	0.20	-	**3.48**	-	**3.98**	nr	**4.28**
180 x 300mm	0.79	5.00	**0.83**	0.20	-	**3.48**	-	**4.31**	nr	**4.63**
450 x 450mm	2.97	5.00	**3.12**	0.20	-	**3.48**	-	**6.59**	nr	**7.09**
Nr 4 soakers and slates handed to others for fixing										
180 x 180mm	0.65	5.00	**0.68**	0.20	-	**3.48**	-	**4.16**	nr	**4.47**
180 x 300mm	1.08	5.00	**1.13**	0.20	-	**3.48**	-	**4.61**	nr	**4.96**
450 x 450mm	4.03	5.00	**4.23**	0.20	-	**3.48**	-	**7.71**	nr	**8.29**
Nr 5 soakers and slates handed to others for fixing										
180 x 180mm	0.80	5.00	**0.84**	0.20	-	**3.48**	-	**4.32**	nr	**4.64**
180 x 300mm	1.34	5.00	**1.41**	0.20	-	**3.48**	-	**4.88**	nr	**5.25**
450 x 450mm	5.02	5.00	**5.27**	0.20	-	**3.48**	-	**8.75**	nr	**9.40**

Labour hourly rates: (except Specialists)
Craft Operatives £17.38 Labourer £10.71
Rates are national average prices.
Refer to REGIONAL VARIATIONS for indicative levels of overall pricing in regions

	MATERIALS			LABOUR				RATES		
	Del to Site £	Waste %	Material Cost £	Craft Optve Hrs	Lab Hrs	Labour Cost £	Sunds £	Nett Rate £	Unit	Gross Rate (+7.5%) £
H71: LEAD SHEET COVERINGS/FLASHINGS Cont.										
Cavity gutters Cont.										
Nr 3 collars around pipes, standards and the like 150mm long; soldered joints to metal covering										
50mm diameter	3.80	5.00	**3.99**	1.00	-	**17.38**	-	**21.37**	nr	**22.97**
100mm diameter	4.28	5.00	**4.49**	1.00	-	**17.38**	-	**21.87**	nr	**23.51**
Nr 4 collars around pipes, standards and the like 150mm long; soldered joints to metal covering										
50mm diameter	4.88	5.00	**5.12**	1.00	-	**17.38**	-	**22.50**	nr	**24.19**
100mm diameter	5.44	5.00	**5.71**	1.00	-	**17.38**	-	**23.09**	nr	**24.82**
Nr 5 collars around pipes, standards and the like 150mm long; soldered joints to metal covering										
50mm diameter	6.08	5.00	**6.38**	1.00	-	**17.38**	-	**23.76**	nr	**25.55**
100mm diameter	6.79	5.00	**7.13**	1.00	-	**17.38**	-	**24.51**	nr	**26.35**
Nr 3 dots										
cast lead	0.65	5.00	**0.68**	0.67	-	**11.64**	-	**12.33**	nr	**13.25**
soldered	6.20	5.00	**6.51**	0.75	-	**13.04**	-	**19.55**	nr	**21.01**
Nr 4 dots										
cast lead	0.90	5.00	**0.95**	0.67	-	**11.64**	-	**12.59**	nr	**13.53**
soldered	6.20	5.00	**6.51**	0.75	-	**13.04**	-	**19.55**	nr	**21.01**
Nr 5 dots										
cast lead	1.05	5.00	**1.10**	0.67	-	**11.64**	-	**12.75**	nr	**13.70**
soldered	6.20	5.00	**6.51**	0.75	-	**13.04**	-	**19.55**	nr	**21.01**
H72: ALUMINIUM SHEET STRIP COVERINGS/FLASHINGS										
Aluminium sheet, BS EN 485 grade S1BO, commercial purity										
0.60mm thick roof coverings; fixing to timber with aluminium cleats and aluminium alloy screws										
pitch 7.5 degrees from horizontal	9.13	5.00	**9.59**	4.40	-	**76.47**	0.12	**86.18**	m2	**92.64**
pitch 40 degrees from horizontal	9.13	5.00	**9.59**	4.95	-	**86.03**	0.12	**95.74**	m2	**102.92**
pitch 75 degrees from horizontal	9.13	5.00	**9.59**	5.50	-	**95.59**	0.12	**105.30**	m2	**113.20**
Flashings; horizontal 0.60mm thick; 150mm lapped joints; fixing to masonry with aluminium clips and wedges										
150mm girth	2.26	5.00	**2.37**	0.44	-	**7.65**	0.12	**10.14**	m	**10.90**
240mm girth	3.50	5.00	**3.68**	0.71	-	**12.34**	0.12	**16.14**	m	**17.35**
300mm girth	3.50	5.00	**3.68**	0.89	-	**15.47**	0.12	**19.27**	m	**20.71**
Flashings; stepped 0.60mm thick; 150mm lapped joints; fixing to masonry with aluminium clips and wedges										
180mm girth	3.05	5.00	**3.20**	0.71	-	**12.34**	0.18	**15.73**	m	**16.91**
240mm girth	3.50	5.00	**3.68**	0.95	-	**16.51**	0.18	**20.37**	m	**21.90**
300mm girth	3.50	5.00	**3.68**	1.19	-	**20.68**	0.18	**24.54**	m	**26.38**
Aprons; horizontal 0.60mm thick; 150mm lapped joints; fixing to masonry with aluminium clips and wedges										
150mm girth	2.26	5.00	**2.37**	0.44	-	**7.65**	0.12	**10.14**	m	**10.90**
180mm girth	3.05	5.00	**3.20**	0.71	-	**12.34**	0.12	**15.66**	m	**16.84**
240mm girth	3.50	5.00	**3.68**	0.89	-	**15.47**	0.12	**19.27**	m	**20.71**
450mm girth	5.52	5.00	**5.80**	1.33	-	**23.12**	0.18	**29.10**	m	**31.28**
Hips; sloping 0.60mm thick; 150mm lapped joints; fixing to timber with aluminium clips										
240mm girth	3.50	5.00	**3.68**	0.71	-	**12.34**	0.12	**16.14**	m	**17.35**
300mm girth	3.50	5.00	**3.68**	0.89	-	**15.47**	0.12	**19.27**	m	**20.71**
450mm girth	5.52	5.00	**5.80**	1.33	-	**23.12**	0.18	**29.10**	m	**31.28**
Kerbs; horizontal 0.60mm thick; 150mm lapped joints; fixing to timber with aluminium clips										
240mm girth	3.50	5.00	**3.68**	0.71	-	**12.34**	0.12	**16.14**	m	**17.35**
300mm girth	3.50	5.00	**3.68**	0.89	-	**15.47**	0.12	**19.27**	m	**20.71**
450mm girth	5.52	5.00	**5.80**	1.33	-	**23.12**	0.18	**29.10**	m	**31.28**
Ridges; horizontal 0.60mm thick; 150mm lapped joints; fixing to timber with aluminium clips										
240mm girth	3.50	5.00	**3.68**	0.71	-	**12.34**	0.12	**16.14**	m	**17.35**
300mm girth	3.50	5.00	**3.68**	0.89	-	**15.47**	0.12	**19.27**	m	**20.71**
450mm girth	5.52	5.00	**5.80**	1.33	-	**23.12**	0.18	**29.10**	m	**31.28**
Valleys; sloping 0.60mm thick; 150mm lapped joints; fixing to timber with aluminium clips										
240mm girth	3.50	5.00	**3.68**	0.71	-	**12.34**	0.12	**16.14**	m	**17.35**
300mm girth	3.50	5.00	**3.68**	0.89	-	**15.47**	0.12	**19.27**	m	**20.71**
450mm girth	5.52	5.00	**5.80**	1.33	-	**23.12**	0.18	**29.10**	m	**31.28**
0.60mm thick edges										
welted	-	-	**-**	0.17	-	**2.95**	-	**2.95**	m	**3.18**
beaded	-	-	**-**	0.17	-	**2.95**	-	**2.95**	m	**3.18**

Labour hourly rates: (except Specialists) Craft Operatives £17.38 Labourer £10.71 Rates are national average prices. Refer to REGIONAL VARIATIONS for indicative levels of overall pricing in regions	MATERIALS			LABOUR				RATES		
	Del to Site £	Waste %	Material Cost £	Craft Optve Hrs	Lab Hrs	Labour Cost £	Sunds £	Nett Rate £	Unit	Gross Rate (+7.5%) £
H72: ALUMINIUM SHEET STRIP COVERINGS/FLASHINGS Cont.										
Aluminium sheet, BS EN 485 grade S1BO, commercial purity Cont.										
0.60mm thick soakers and slates handed to others for fixing										
180 x 180mm	0.55	5.00	**0.58**	0.20	-	**3.48**	-	**4.05**	nr	**4.36**
180 x 300mm	0.92	5.00	**0.97**	0.20	-	**3.48**	-	**4.44**	nr	**4.78**
H73: COPPER STRIP SHEET COVERINGS/FLASHINGS										
Copper sheet; BS EN 1172										
0.60mm thick roof coverings; fixing to timber with copper cleats and copper nails										
pitch 7.5 degrees from horizontal	24.00	5.00	**25.20**	4.40	-	**76.47**	0.37	**102.04**	m2	**109.69**
pitch 40 degrees from horizontal	24.00	5.00	**25.20**	4.95	-	**86.03**	0.37	**111.60**	m2	**119.97**
pitch 75 degrees from horizontal	24.00	5.00	**25.20**	5.50	-	**95.59**	0.37	**121.16**	m2	**130.24**
0.70mm thick roof coverings; fixing to timber with copper cleats and copper nails										
pitch 7.5 degrees from horizontal	30.00	5.00	**31.50**	4.95	-	**86.03**	0.46	**117.99**	m2	**126.84**
pitch 40 degrees from horizontal	30.00	5.00	**31.50**	5.50	-	**95.59**	0.46	**127.55**	m2	**137.12**
pitch 75 degrees from horizontal	30.00	5.00	**31.50**	6.05	-	**105.15**	0.46	**137.11**	m2	**147.39**
Flashings; horizontal										
0.60mm thick; 150mm lapped joints; fixing to masonry with copper clips and wedges										
150mm girth	3.60	5.00	**3.78**	0.44	-	**7.65**	0.37	**11.79**	m	**12.68**
240mm girth	5.76	5.00	**6.05**	0.71	-	**12.34**	0.37	**18.76**	m	**20.16**
300mm girth	7.20	5.00	**7.56**	0.89	-	**15.47**	0.37	**23.40**	m	**25.15**
0.70mm thick; 150mm lapped joints; fixing to masonry with copper clips and wedges										
150mm girth	4.50	5.00	**4.73**	0.50	-	**8.69**	0.46	**13.87**	m	**14.91**
240mm girth	7.20	5.00	**7.56**	0.80	-	**13.90**	0.46	**21.92**	m	**23.57**
300mm girth	9.00	5.00	**9.45**	1.00	-	**17.38**	0.46	**27.29**	m	**29.34**
Flashings; stepped										
0.60mm thick; 150mm lapped joints; fixing to masonry with copper clips and wedges										
180mm girth	4.32	5.00	**4.54**	0.71	-	**12.34**	0.55	**17.43**	m	**18.73**
240mm girth	5.76	5.00	**6.05**	0.95	-	**16.51**	0.55	**23.11**	m	**24.84**
300mm girth	7.20	5.00	**7.56**	1.19	-	**20.68**	0.55	**28.79**	m	**30.95**
0.70mm thick; 150mm lapped joints; fixing to masonry with copper clips and wedges										
180mm girth	5.40	5.00	**5.67**	0.80	-	**13.90**	0.69	**20.27**	m	**21.79**
240mm girth	7.20	5.00	**7.56**	1.07	-	**18.60**	0.69	**26.85**	m	**28.86**
300mm girth	9.00	5.00	**9.45**	1.33	-	**23.12**	0.69	**33.26**	m	**35.75**
Aprons; horizontal										
0.60mm thick; 150mm lapped joints; fixing to timber with copper clips										
150mm girth	3.60	5.00	**3.78**	0.44	-	**7.65**	0.37	**11.79**	m	**12.68**
240mm girth	5.76	5.00	**6.05**	0.71	-	**12.34**	0.37	**18.76**	m	**20.16**
300mm girth	7.20	5.00	**7.56**	0.89	-	**15.47**	0.37	**23.40**	m	**25.15**
450mm girth	10.80	5.00	**11.34**	1.33	-	**23.12**	0.55	**35.01**	m	**37.63**
0.70mm thick; 150mm lapped joints; fixing to timber with copper clips										
150mm girth	4.50	5.00	**4.73**	0.50	-	**8.69**	0.46	**13.87**	m	**14.91**
240mm girth	7.20	5.00	**7.56**	0.80	-	**13.90**	0.46	**21.92**	m	**23.57**
300mm girth	9.00	5.00	**9.45**	1.00	-	**17.38**	0.46	**27.29**	m	**29.34**
450mm girth	13.50	5.00	**14.18**	1.50	-	**26.07**	0.69	**40.94**	m	**44.01**
Hips; sloping										
0.60mm thick; 150mm lapped joints; fixing to timber with copper clips										
240mm girth	5.76	5.00	**6.05**	0.71	-	**12.34**	0.37	**18.76**	m	**20.16**
300mm girth	7.20	5.00	**7.56**	0.89	-	**15.47**	0.37	**23.40**	m	**25.15**
450mm girth	10.80	5.00	**11.34**	1.33	-	**23.12**	0.55	**35.01**	m	**37.63**
0.70mm thick; 150mm lapped joints; fixing to timber with copper clips										
240mm girth	7.20	5.00	**7.56**	0.80	-	**13.90**	0.46	**21.92**	m	**23.57**
300mm girth	9.00	5.00	**9.45**	1.00	-	**17.38**	0.46	**27.29**	m	**29.34**
450mm girth	13.50	5.00	**14.18**	1.50	-	**26.07**	0.69	**40.94**	m	**44.01**
Kerbs; horizontal										
0.60mm thick; 150mm lapped joints; fixing to timber with copper clips										
240mm girth	5.76	5.00	**6.05**	0.71	-	**12.34**	0.37	**18.76**	m	**20.16**
300mm girth	7.20	5.00	**7.56**	0.89	-	**15.47**	0.37	**23.40**	m	**25.15**
450mm girth	10.80	5.00	**11.34**	1.33	-	**23.12**	0.55	**35.01**	m	**37.63**
0.70mm thick; 150mm lapped joints; fixing to timber with copper clips										
240mm girth	7.20	5.00	**7.56**	0.80	-	**13.90**	0.46	**21.92**	m	**23.57**
300mm girth	9.00	5.00	**9.45**	1.00	-	**17.38**	0.46	**27.29**	m	**29.34**
450mm girth	13.50	5.00	**14.18**	1.50	-	**26.07**	0.69	**40.94**	m	**44.01**
Ridges; horizontal										
0.60mm thick; 150mm lapped joints; fixing to timber with copper clips										
240mm girth	5.76	5.00	**6.05**	0.71	-	**12.34**	0.37	**18.76**	m	**20.16**
300mm girth	7.20	5.00	**7.56**	0.89	-	**15.47**	0.37	**23.40**	m	**25.15**
450mm girth	10.80	5.00	**11.34**	1.33	-	**23.12**	0.55	**35.01**	m	**37.63**

Labour hourly rates: (except Specialists) Craft Operatives £17.38 Labourer £10.71 Rates are national average prices. Refer to REGIONAL VARIATIONS for indicative levels of overall pricing in regions	MATERIALS			LABOUR				RATES		
	Del to Site £	Waste %	Material Cost £	Craft Optve Hrs	Lab Hrs	Labour Cost £	Sunds £	Nett Rate £	Unit	Gross Rate (+7.5%) £
H73: COPPER STRIP SHEET COVERINGS/FLASHINGS Cont.										
Copper sheet; BS EN 1172 Cont.										
Ridges; horizontal Cont.										
0.70mm thick; 150mm lapped joints; fixing to timber with copper clips										
240mm girth	7.20	5.00	**7.56**	0.80	-	**13.90**	0.46	**21.92**	m	**23.57**
300mm girth	9.00	5.00	**9.45**	1.00	-	**17.38**	0.46	**27.29**	m	**29.34**
450mm girth	13.50	5.00	**14.18**	1.50	-	**26.07**	0.69	**40.94**	m	**44.01**
Valleys; sloping										
0.60mm thick; 150mm lapped joints; fixing to timber with copper clips										
240mm girth	5.76	5.00	**6.05**	0.71	-	**12.34**	0.37	**18.76**	m	**20.16**
300mm girth	7.20	5.00	**7.56**	0.89	-	**15.47**	0.37	**23.40**	m	**25.15**
450mm girth	10.80	5.00	**11.34**	1.33	-	**23.12**	0.55	**35.01**	m	**37.63**
0.70mm thick; 150mm lapped joints; fixing to timber with copper clips										
240mm girth	7.20	5.00	**7.56**	0.80	-	**13.90**	0.46	**21.92**	m	**23.57**
300mm girth	9.00	5.00	**9.45**	1.07	-	**18.60**	0.46	**28.51**	m	**30.64**
450mm girth	13.50	5.00	**14.18**	1.33	-	**23.12**	0.69	**37.98**	m	**40.83**
0.60mm thick edges										
welted	-	-	**-**	0.17	-	**2.95**	-	**2.95**	m	**3.18**
beaded	-	-	**-**	0.17	-	**2.95**	-	**2.95**	m	**3.18**
0.70mm thick edges										
welted	-	-	**-**	0.17	-	**2.95**	-	**2.95**	m	**3.18**
beaded	-	-	**-**	0.17	-	**2.95**	-	**2.95**	m	**3.18**
H74: ZINC STRIP SHEET COVERINGS/FLASHINGS										
Zinc alloy sheet; BS EN 988										
0.8mm thick roof coverings; roll cap; fixing to timber with zinc clips										
pitch 7.5 degrees from horizontal	16.50	5.00	**17.33**	4.95	-	**86.03**	0.18	**103.54**	m2	**111.31**
pitch 40 degrees from horizontal	16.50	5.00	**17.33**	5.50	-	**95.59**	0.18	**113.10**	m2	**121.58**
pitch 75 degrees from horizontal	16.50	5.00	**17.33**	6.05	-	**105.15**	0.20	**122.68**	m2	**131.88**
Flashings; horizontal										
0.8mm thick; 150mm lapped joints; fixing to masonry with zinc clips and wedges										
150mm girth	2.47	5.00	**2.59**	0.44	-	**7.65**	0.13	**10.37**	m	**11.15**
240mm girth	3.96	5.00	**4.16**	0.71	-	**12.34**	0.13	**16.63**	m	**17.88**
300mm girth	4.95	5.00	**5.20**	0.89	-	**15.47**	0.13	**20.80**	m	**22.36**
Flashings; stepped										
0.8mm thick; 150mm lapped joints; fixing to masonry with zinc clips and wedges										
180mm girth	2.97	5.00	**3.12**	0.71	-	**12.34**	0.20	**15.66**	m	**16.84**
240mm girth	3.96	5.00	**4.16**	0.95	-	**16.51**	0.20	**20.87**	m	**22.44**
300mm girth	4.95	5.00	**5.20**	1.19	-	**20.68**	0.20	**26.08**	m	**28.04**
Aprons; horizontal										
0.8mm thick; 150mm lapped joints; fixing to masonry with zinc clips and wedges										
150mm girth	2.47	5.00	**2.59**	0.44	-	**7.65**	0.13	**10.37**	m	**11.15**
240mm girth	3.96	5.00	**4.16**	0.71	-	**12.34**	0.13	**16.63**	m	**17.88**
300mm girth	4.95	5.00	**5.20**	0.89	-	**15.47**	0.13	**20.80**	m	**22.36**
450mm girth	7.42	5.00	**7.79**	1.33	-	**23.12**	0.20	**31.11**	m	**33.44**
Hips; sloping										
0.8mm thick; 150mm lapped joints; fixing to timber with zinc clips										
240mm girth	3.96	5.00	**4.16**	0.71	-	**12.34**	0.13	**16.63**	m	**17.88**
300mm girth	4.95	5.00	**5.20**	0.89	-	**15.47**	0.13	**20.80**	m	**22.36**
450mm girth	7.42	5.00	**7.79**	1.33	-	**23.12**	0.20	**31.11**	m	**33.44**
Kerbs; horizontal										
0.8mm thick; 150mm lapped joints; fixing to timber with zinc clips										
240mm girth	3.96	5.00	**4.16**	0.71	-	**12.34**	0.13	**16.63**	m	**17.88**
300mm girth	4.95	5.00	**5.20**	0.89	-	**15.47**	0.13	**20.80**	m	**22.36**
450mm girth	7.42	5.00	**7.79**	1.33	-	**23.12**	0.20	**31.11**	m	**33.44**
Ridges; horizontal										
0.8mm thick; 150mm lapped joints; fixing to timber with zinc clips										
240mm girth	3.96	5.00	**4.16**	0.71	-	**12.34**	0.13	**16.63**	m	**17.88**
300mm girth	4.95	5.00	**5.20**	0.89	-	**15.47**	0.13	**20.80**	m	**22.36**
450mm girth	7.42	5.00	**7.79**	1.33	-	**23.12**	0.20	**31.11**	m	**33.44**
Valleys; sloping										
0.8mm thick; 150mm lapped joints; fixing to timber with zinc clips										
240mm girth	3.96	5.00	**4.16**	0.71	-	**12.34**	0.13	**16.63**	m	**17.88**
300mm girth	4.95	5.00	**5.20**	0.89	-	**15.47**	0.13	**20.80**	m	**22.36**
450mm girth	7.42	5.00	**7.79**	1.33	-	**23.12**	0.20	**31.11**	m	**33.44**
0.65mm thick edges										
welted	-	-	**-**	0.17	-	**2.95**	-	**2.95**	m	**3.18**
beaded	-	-	**-**	0.17	-	**2.95**	-	**2.95**	m	**3.18**

Labour hourly rates: (except Specialists) Craft Operatives £17.38 Labourer £10.71 Rates are national average prices. Refer to REGIONAL VARIATIONS for indicative levels of overall pricing in regions	MATERIALS			LABOUR				RATES		
	Del to Site £	Waste %	Material Cost £	Craft Optve Hrs	Lab Hrs	Labour Cost £	Sunds £	Nett Rate £	Unit	Gross Rate (+7.5%) £
H74: ZINC STRIP SHEET COVERINGS/FLASHINGS Cont.										
Zinc alloy sheet; BS EN 988 Cont.										
0.8mm thick edges										
welted	-	-	-	0.17	-	**2.95**	-	**2.95**	m	**3.18**
beaded	-	-	-	0.17	-	**2.95**	-	**2.95**	m	**3.18**
0.8mm thick soakers and slates - handed to others for fixing										
180 x 180mm	0.54	5.00	**0.57**	0.20	-	**3.48**	-	**4.04**	nr	**4.35**
180 x 300mm	0.89	5.00	**0.93**	0.20	-	**3.48**	-	**4.41**	nr	**4.74**
450 x 450mm	3.34	5.00	**3.51**	0.20	-	**3.48**	-	**6.98**	nr	**7.51**
0.8mm thick collars around pipes, standards and the like										
150mm long; soldered joints to metal covering										
50mm diameter	3.20	5.00	**3.36**	1.00	-	**17.38**	-	**20.74**	nr	**22.30**
100mm diameter	3.57	5.00	**3.75**	1.00	-	**17.38**	-	**21.13**	nr	**22.71**

WATERPROOFING

Labour hourly rates: (except Specialists) Craft Operatives £14.21 Labourer £10.71 Rates are national average prices. Refer to REGIONAL VARIATIONS for indicative levels of overall pricing in regions	MATERIALS			LABOUR				RATES		
	Del to Site £	Waste %	Material Cost £	Craft Optve Hrs	Lab Hrs	Labour Cost £	Sunds £	Nett Rate £	Unit	Gross Rate (+7.5%) £
J20: MASTIC ASPHALT TANKING/ DAMP PROOF MEMBRANES										
Tanking and damp proofing BS 6925 (limestone aggregate)										
Tanking and damp proofing 13mm thick; coats of asphalt -1; to concrete base; horizontal work subsequently covered										
width not exceeding 150mm	-	-	Spclist	-	-	Spclist	-	5.57	lm	5.99
width 150 - 225mm	-	-	Spclist	-	-	Spclist	-	6.82	lm	7.33
width 225 - 300mm	-	-	Spclist	-	-	Spclist	-	8.31	lm	8.93
width exceeding 300mm	-	-	Spclist	-	-	Spclist	-	14.03	m2	15.08
Tanking and damp proofing 20mm thick; coats of asphalt -2; to concrete base; horizontal work subsequently covered										
width not exceeding 150mm	-	-	Spclist	-	-	Spclist	-	6.27	lm	6.74
width 150 - 225mm	-	-	Spclist	-	-	Spclist	-	7.87	lm	8.46
width 225 - 300mm	-	-	Spclist	-	-	Spclist	-	9.71	lm	10.44
width exceeding 300mm	-	-	Spclist	-	-	Spclist	-	20.45	m2	21.98
Tanking and damp proofing 30mm thick; coats of asphalt -3; to concrete base; horizontal work subsequently covered										
width not exceeding 150mm	-	-	Spclist	-	-	Spclist	-	9.41	lm	10.12
width 150 - 225mm	-	-	Spclist	-	-	Spclist	-	11.78	lm	12.66
width 225 - 300mm	-	-	Spclist	-	-	Spclist	-	14.57	lm	15.66
width exceeding 300mm	-	-	Spclist	-	-	Spclist	-	29.13	m2	31.31
Tanking and damp proofing 20mm thick; coats of asphalt -2; to concrete base; vertical work subsequently covered										
width not exceeding 150mm	-	-	Spclist	-	-	Spclist	-	12.34	lm	13.27
width 150 - 225mm	-	-	Spclist	-	-	Spclist	-	14.43	lm,	15.51
width 225 - 300mm	-	-	Spclist	-	-	Spclist	-	16.45	lm	17.68
width exceeding 300mm	-	-	Spclist	-	-	Spclist	-	42.51	m2	45.70
Tanking and damp proofing 20mm thick; coats of asphalt -3; to concrete base; vertical work subsequently covered										
width not exceeding 150mm	-	-	Spclist	-	-	Spclist	-	18.55	lm	19.94
width 150 - 225mm	-	-	Spclist	-	-	Spclist	-	21.72	lm	23.35
width 225 - 300mm	-	-	Spclist	-	-	Spclist	-	24.77	lm	26.63
width exceeding 300mm	-	-	Spclist	-	-	Spclist	-	59.02	m2	63.45
Internal angle fillets to concrete base; priming base with bitumen										
two coats; work subsequently covered	-	-	Spclist	-	-	Spclist	-	5.52	m	5.93
J21: MASTIC ASPHALT ROOFING/INSULATION/FINISHES										
Roofing; BS 6925 (limestone aggregate); sand rubbing; BS 747 type 4A sheathing felt isolating membrane, butt joints										
Roofing 20mm thick; coats of asphalt -2; to concrete base; pitch not exceeding 6 degrees from horizontal										
width exceeding 300mm	-	-	Spclist	-	-	Spclist	-	22.03	m2	23.68
extra; solid water check roll	-	-	Spclist	-	-	Spclist	-	11.05	m	11.88
Roofing 20mm thick; coats of asphalt -2; to concrete base; pitch 30 degrees from horizontal										
width exceeding 300mm	-	-	Spclist	-	-	Spclist	-	33.69	m2	36.22
Roofing 20mm thick; coats of asphalt -2; to concrete base; pitch 45 degrees from horizontal										
width exceeding 300mm	-	-	Spclist	-	-	Spclist	-	43.11	m2	46.34
Paint two coats of Solar reflective roof paint on surfaces of asphalt										
width exceeding 300mm	-	-	Spclist	-	-	Spclist	-	6.00	m2	6.45

Labour hourly rates: (except Specialists) Craft Operatives £14.21 Labourer £10.71 Rates are national average prices. Refer to REGIONAL VARIATIONS for indicative levels of overall pricing in regions	MATERIALS			LABOUR				RATES		
	Del to Site £	Waste %	Material Cost £	Craft Optve Hrs	Lab Hrs	Labour Cost £	Sunds £	Nett Rate £	Unit	Gross Rate (+7.5%) £
J21: MASTIC ASPHALT ROOFING/INSULATION/FINISHES Cont.										
Roofing; BS 6925 (limestone aggregate); sand rubbing; BS 747 type 4A sheathing felt isolating membrane, butt joints Cont.										
Skirtings 13mm thick; coats of asphalt -2; to brickwork base										
girth not exceeding 150mm	-	-	**Spclist**	-	-	**Spclist**	-	**13.08**	m	**14.06**
Coverings to kerbs 20mm thick; coats of asphalt -2; to concrete base										
girth 600mm	-	-	**Spclist**	-	-	**Spclist**	-	**27.74**	m	**29.82**
Jointing new roofing to existing										
20mm thick	-	-	**Spclist**	-	-	**Spclist**	-	**16.58**	m	**17.82**
Edge trim; aluminium; silver anodized priming with bituminous primer; butt joints with internal jointing sleeves; fixing with aluminium alloy screws to timber; working two coat asphalt into grooves -1										
63.5mm wide x 44.4mm face depth	7.48	5.00	**7.85**	0.36	-	**5.12**	1.25	**14.22**	m	**15.29**
63.5mm wide x 76.2mm face depth	8.76	5.00	**9.20**	0.36	-	**5.12**	1.31	**15.62**	m	**16.80**
104.8mm wide x 44.4mm face depth	8.38	5.00	**8.80**	0.36	-	**5.12**	1.31	**15.22**	m	**16.37**
104.8mm wide x 76.2mm face depth	8.78	5.00	**9.22**	0.44	-	**6.25**	1.49	**16.96**	m	**18.23**
76.2mm fixing arm at 10 degrees x 38.1mm face depth	8.86	5.00	**9.30**	0.36	-	**5.12**	1.31	**15.73**	m	**16.91**
Edge trim; glass fibre reinforced butt joints; fixing with stainless steel screws to timber; working two coat asphalt into grooves -1.	10.31	5.00	**10.83**	0.30	-	**4.26**	1.49	**16.58**	m	**17.82**
Roof Ventilators										
plastic; setting in position	7.40	2.50	**7.59**	0.75	-	**10.66**	0.49	**18.73**	nr	**20.14**
aluminium; setting in position	8.21	2.50	**8.42**	0.75	-	**10.66**	0.49	**19.56**	nr	**21.03**
Roofing; BS 6925 (limestone aggregate); covering with 13mm white spar chippings in hot bitumen; BS 747 type 4A sheathing felt isolating membrane; butt joints										
Roofing 20mm thick; coats of asphalt -2; to concrete base; pitch not exceeding 6 degrees from horizontal										
width exceeding 300mm	-	-	-	-	-	-	-	**26.48**	m2	**28.47**
Roofing 20mm thick; coats of asphalt -2; to concrete base; pitch 30 degrees from horizontal										
width exceeding 300mm	-	-	-	-	-	-	-	**38.12**	m2	**40.98**
J30: LIQUID APPLIED TANKING/DAMP PROOF MEMBRANES										
Synthaprufe bituminous emulsion; blinding with sand										
Coverings, coats -2; to concrete base; horizontal										
width not exceeding 150mm	4.72	10.00	**5.19**	-	0.40	**4.28**	0.13	**9.60**	m2	**10.32**
width 150 - 225mm	4.72	10.00	**5.19**	-	0.33	**3.53**	0.13	**8.85**	m2	**9.52**
width 225 - 300mm	4.72	10.00	**5.19**	-	0.26	**2.78**	0.13	**8.10**	m2	**8.71**
width exceeding 300mm	4.72	10.00	**5.19**	-	0.20	**2.14**	0.13	**7.46**	m2	**8.02**
Coverings, coats -3; to concrete base; horizontal										
width not exceeding 150mm	7.04	10.00	**7.74**	-	0.56	**6.00**	0.13	**13.87**	m2	**14.91**
width 150 - 225mm	7.04	10.00	**7.74**	-	0.46	**4.93**	0.13	**12.80**	m2	**13.76**
width 225 - 300mm	7.04	10.00	**7.74**	-	0.36	**3.86**	0.13	**11.73**	m2	**12.61**
width exceeding 300mm	7.04	10.00	**7.74**	-	0.28	**3.00**	0.13	**10.87**	m2	**11.69**
Coverings; coats -2; to concrete base; vertical										
width not exceeding 150mm	4.72	10.00	**5.19**	-	0.60	**6.43**	0.13	**11.74**	m2	**12.63**
width 150 - 225mm	4.72	10.00	**5.19**	-	0.50	**5.36**	0.13	**10.67**	m2	**11.47**
width 225 - 300mm	4.72	10.00	**5.19**	-	0.40	**4.28**	0.13	**9.60**	m2	**10.32**
width exceeding 300mm	4.72	10.00	**5.19**	-	0.30	**3.21**	0.13	**8.53**	m2	**9.17**
Coverings; coats -3; to concrete base; vertical										
width not exceeding 150mm	7.04	10.00	**7.74**	-	0.84	**9.00**	0.13	**16.87**	m2	**18.14**
width 150 - 225mm	7.04	10.00	**7.74**	-	0.70	**7.50**	0.13	**15.37**	m2	**16.52**
width 225 - 300mm	7.04	10.00	**7.74**	-	0.56	**6.00**	0.13	**13.87**	m2	**14.91**
width exceeding 300mm	7.04	10.00	**7.74**	-	0.42	**4.50**	0.13	**12.37**	m2	**13.30**
Bituminous emulsion										
Coverings, coats -2; to concrete base; horizontal										
width not exceeding 150mm	3.52	10.00	**3.87**	-	0.34	**3.64**	0.13	**7.64**	m2	**8.22**
width 150 - 225mm	3.52	10.00	**3.87**	-	0.28	**3.00**	0.13	**7.00**	m2	**7.53**
width 225 - 300mm	3.52	10.00	**3.87**	-	0.22	**2.36**	0.13	**6.36**	m2	**6.84**
width exceeding 300mm	3.52	10.00	**3.87**	-	0.17	**1.82**	0.13	**5.82**	m2	**6.26**
Coverings, coats -3; to concrete base; horizontal										
width not exceeding 150mm	4.93	10.00	**5.42**	-	0.48	**5.14**	0.13	**10.69**	m2	**11.49**
width 150 - 225mm	4.93	10.00	**5.42**	-	0.40	**4.28**	0.13	**9.83**	m2	**10.57**
width 225 - 300mm	4.93	10.00	**5.42**	-	0.32	**3.43**	0.13	**8.98**	m2	**9.65**
width exceeding 300mm	4.93	10.00	**5.42**	-	0.24	**2.57**	0.13	**8.12**	m2	**8.73**
Coverings; coats -2; to concrete base; vertical										
width not exceeding 150mm	3.52	10.00	**3.87**	-	0.51	**5.46**	0.13	**9.46**	m2	**10.17**
width 150 - 225mm	3.52	10.00	**3.87**	-	0.42	**4.50**	0.13	**8.50**	m2	**9.14**
width 225 - 300mm	3.52	10.00	**3.87**	-	0.33	**3.53**	0.13	**7.54**	m2	**8.10**
width exceeding 300mm	3.52	10.00	**3.87**	-	0.26	**2.78**	0.13	**6.79**	m2	**7.30**

Labour hourly rates: (except Specialists) Craft Operatives £14.21 Labourer £10.71 Rates are national average prices. Refer to REGIONAL VARIATIONS for indicative levels of overall pricing in regions	MATERIALS			LABOUR				RATES		
	Del to Site £	Waste %	Material Cost £	Craft Optve Hrs	Lab Hrs	Labour Cost £	Sunds £	Nett Rate £	Unit	Gross Rate (+7.5%) £
J30: LIQUID APPLIED TANKING/DAMP PROOF MEMBRANES Cont.										
Bituminous emulsion Cont.										
Coverings; coats -3; to concrete base; vertical										
width not exceeding 150mm	4.93	10.00	**5.42**	-	0.72	**7.71**	0.13	**13.26**	m2	**14.26**
width 150 - 225mm	4.93	10.00	**5.42**	-	0.60	**6.43**	0.13	**11.98**	m2	**12.88**
width 225 - 300mm	4.93	10.00	**5.42**	-	0.48	**5.14**	0.13	**10.69**	m2	**11.49**
width exceeding 300mm	4.93	10.00	**5.42**	-	0.36	**3.86**	0.13	**9.41**	m2	**10.11**
R.I.W. liquid asphaltic composition										
Coverings, coats -2; to concrete base; horizontal										
width not exceeding 150mm	7.30	10.00	**8.03**	-	0.40	**4.28**	0.13	**12.45**	m2	**13.38**
width 150 - 225mm	7.30	10.00	**8.03**	-	0.33	**3.53**	0.13	**11.70**	m2	**12.58**
width 225 - 300mm	7.30	10.00	**8.03**	-	0.26	**2.78**	0.13	**10.95**	m2	**11.77**
width exceeding 300mm	7.30	10.00	**8.03**	-	0.20	**2.14**	0.13	**10.31**	m2	**11.08**
Coverings, coats -3; to concrete base; horizontal										
width not exceeding 150mm	11.00	10.00	**12.10**	-	0.56	**6.00**	0.13	**18.22**	m2	**19.59**
width 150 - 225mm	11.00	10.00	**12.10**	-	0.46	**4.93**	0.13	**17.15**	m2	**18.44**
width 225 - 300mm	11.00	10.00	**12.10**	-	0.36	**3.86**	0.13	**16.08**	m2	**17.29**
width exceeding 300mm	11.00	10.00	**12.10**	-	0.28	**3.00**	0.13	**15.23**	m2	**16.37**
Coverings; coats -2; to concrete base; vertical										
width not exceeding 150mm	7.30	10.00	**8.03**	-	0.60	**6.43**	0.13	**14.59**	m2	**15.68**
width 150 - 225mm	7.30	10.00	**8.03**	-	0.50	**5.36**	0.13	**13.52**	m2	**14.53**
width 225 - 300mm	7.30	10.00	**8.03**	-	0.40	**4.28**	0.13	**12.45**	m2	**13.38**
width exceeding 300mm	7.30	10.00	**8.03**	-	0.30	**3.21**	0.13	**11.38**	m2	**12.23**
Coverings; coats -3; to concrete base; vertical										
width not exceeding 150mm	11.00	10.00	**12.10**	-	0.84	**9.00**	0.13	**21.22**	m2	**22.82**
width 150 - 225mm	11.00	10.00	**12.10**	-	0.70	**7.50**	0.13	**19.72**	m2	**21.20**
width 225 - 300mm	11.00	10.00	**12.10**	-	0.56	**6.00**	0.13	**18.22**	m2	**19.59**
width exceeding 300mm	11.00	10.00	**12.10**	-	0.42	**4.50**	0.13	**16.73**	m2	**17.98**
J40: FLEXIBLE SHEET TANKING/DAMP PROOFING MEMBRANES										
Polythene sheeting; 100mm welted laps										
Tanking and damp proofing; 125mu										
horizontal; on concrete base	0.33	20.00	**0.40**	-	0.05	**0.54**	-	**0.93**	m2	**1.00**
vertical; on concrete base	0.33	20.00	**0.40**	-	0.08	**0.86**	-	**1.25**	m2	**1.35**
Tanking and damp proofing; 250mu										
horizontal; on concrete base	0.38	20.00	**0.46**	-	0.08	**0.86**	-	**1.31**	m2	**1.41**
vertical; on concrete base	0.38	20.00	**0.46**	-	0.12	**1.29**	-	**1.74**	m2	**1.87**
Tanking and damp proofing; 300mu										
horizontal; on concrete base	0.50	20.00	**0.60**	-	0.09	**0.96**	-	**1.56**	m2	**1.68**
vertical; on concrete base	0.50	20.00	**0.60**	-	0.14	**1.50**	-	**2.10**	m2	**2.26**
Bituthene 500X self adhesive damp proof membrane; 25mm lapped joints										
Tanking and damp proofing										
horizontal; on concrete base	4.20	20.00	**5.04**	-	0.15	**1.61**	-	**6.65**	m2	**7.14**
vertical; on concrete base	4.20	20.00	**5.04**	-	0.25	**2.68**	-	**7.72**	m2	**8.30**
Bituthene 500X self adhesive damp proof membrane; 75mm lapped joints										
Tanking and damp proofing										
horizontal; on concrete base	4.60	20.00	**5.52**	-	0.20	**2.14**	-	**7.66**	m2	**8.24**
vertical; on concrete base; priming with Servicised B primer	4.60	20.00	**5.52**	-	0.30	**3.21**	0.24	**8.97**	m2	**9.65**
extra; Bituthene internal angle fillet; 40 x 40mm	5.18	10.00	**5.70**	-	0.10	**1.07**	-	**6.77**	m	**7.28**
British Sisalkraft Ltd 728 damp proof membrane; 150mm laps sealed with tape										
Tanking and damp proofing										
horizontal; on concrete base	1.30	20.00	**1.56**	-	0.04	**0.43**	0.10	**2.09**	m2	**2.25**
Visqueen 1200 super damp proof membrane; 150mm laps sealed with tape and mastic										
Tanking and damp proofing										
horizontal; on concrete base	0.50	20.00	**0.60**	-	0.04	**0.43**	0.10	**1.13**	m2	**1.21**
J41: BUILT UP FELT ROOF COVERINGS										
Roofing; felt BS 747, comprising 3G Rubervent underlay 2.6 kg/m2, 1 layer Ruberfort HP180 underlay 3.4 kg/m2, 1 layer Ruberfort HP180 mineral surface top sheet 3.6 kg/m2										
Roof coverings; layers of felt -3; bonding with hot bitumen compound to timber base										
pitch 7.5 degrees from horizontal	9.19	15.00	**10.71**	0.40	0.05	**6.22**	-	**16.93**	m2	**18.20**
pitch 40 degrees from horizontal	9.19	15.00	**10.71**	0.67	0.08	**10.38**	-	**21.09**	m2	**22.67**
pitch 75 degrees from horizontal	9.19	15.00	**10.71**	0.80	0.10	**12.44**	-	**23.15**	m2	**24.88**
Roof coverings; layers of felt -3; bonding with hot bitumen compound to cement and sand or concrete base										
pitch 7.5 degrees from horizontal	9.19	15.00	**10.71**	0.40	0.05	**6.22**	-	**16.93**	m2	**18.20**

Labour hourly rates: (except Specialists) Craft Operatives £14.21 Labourer £10.71 Rates are national average prices. Refer to REGIONAL VARIATIONS for indicative levels of overall pricing in regions	MATERIALS			LABOUR				RATES		
	Del to Site £	Waste %	Material Cost £	Craft Optve Hrs	Lab Hrs	Labour Cost £	Sunds £	Nett Rate £	Unit	Gross Rate (+7.5%) £
J41: BUILT UP FELT ROOF COVERINGS Cont.										
Roofing; felt BS 747, comprising 3G Rubervent underlay 2.6 kg/m2, 1 layer Ruberfort HP180 underlay 3.4 kg/m2, 1 layer Ruberfort HP180 mineral surface top sheet 3.6 kg/m2 Cont.										
Roof coverings; layers of felt -3; bonding with hot bitumen compound to cement and sand or concrete base Cont.										
pitch 40 degrees from horizontal	9.19	15.00	**10.71**	0.67	0.08	**10.38**	-	**21.09**	m2	**22.67**
pitch 75 degrees from horizontal	9.19	15.00	**10.71**	0.80	0.10	**12.44**	-	**23.15**	m2	**24.88**
Skirtings; layers of felt - 3; bonding with hot bitumen compound to brickwork base										
girth 0 - 200mm	1.84	15.00	**2.14**	0.25	0.03	**3.87**	-	**6.02**	m	**6.47**
girth 200 - 400mm	3.68	15.00	**4.28**	0.40	0.05	**6.22**	-	**10.50**	m	**11.29**
Collars around pipes; standard and the like; layers of felt -1; bonding with hot bitumen compound to metal base										
50mm diameter x 150mm long; hole in 3 layer covering	0.77	15.00	**0.89**	0.15	0.02	**2.35**	-	**3.24**	nr	**3.48**
100mm diameter x 150mm long; hole in 3 layer covering	0.98	15.00	**1.14**	0.20	0.03	**3.16**	-	**4.30**	nr	**4.62**
Eaves trim; aluminium; silver anodized butt joints over matching sleeve pieces 200mm long; fixing to timber base with aluminium alloy screws; bedding in mastic										
63.5mm wide x 44.4mm face depth	7.48	5.00	**7.85**	0.36	-	**5.12**	1.37	**14.34**	m	**15.42**
63.5mm wide x 76.2mm face depth	8.76	5.00	**9.20**	0.36	-	**5.12**	1.44	**15.75**	m	**16.94**
104.8mm wide x 44.4mm face depth	8.38	5.00	**8.80**	0.36	-	**5.12**	1.44	**15.35**	m	**16.51**
104.8mm wide x 76.2mm face depth	8.78	5.00	**9.22**	0.44	-	**6.25**	1.61	**17.08**	m	**18.36**
Eaves trim; glass fibre reinforced butt joints; fixing to timber base with stainless steel screws; bedding in mastic	10.31	5.00	**10.83**	0.30	-	**4.26**	1.61	**16.70**	m	**17.95**
Roof ventilators										
plastic; setting in position	7.40	2.50	**7.59**	0.75	-	**10.66**	0.50	**18.74**	nr	**20.15**
aluminium; setting in position	8.21	2.50	**8.42**	0.75	-	**10.66**	0.50	**19.57**	nr	**21.04**
Roofing; BS 747, all layers type 3B -1.8 kg/m2, 75mm laps; fully bonding layers with hot bitumen bonding compound										
Roof coverings; layers of felt -2; bonding with hot bitumen compound to timber base										
pitch 7.5 degrees from horizontal	3.54	15.00	**4.17**	0.33	0.04	**5.12**	-	**9.28**	m2	**9.98**
pitch 40 degrees from horizontal	3.54	15.00	**4.17**	0.55	0.07	**8.57**	-	**12.73**	m2	**13.68**
pitch 75 degrees from horizontal	3.54	15.00	**4.17**	0.66	0.08	**10.24**	-	**14.40**	m2	**15.48**
Roof coverings; layers of felt -2; bonding with hot bitumen compound to cement and sand or concrete base										
pitch 7.5 degrees from horizontal	3.54	15.00	**4.17**	0.33	0.04	**5.12**	-	**9.28**	m2	**9.98**
pitch 40 degrees from horizontal	3.54	15.00	**4.17**	0.55	0.07	**8.57**	-	**12.73**	m2	**13.68**
pitch 75 degrees from horizontal	3.54	15.00	**4.17**	0.66	0.08	**10.24**	-	**14.40**	m2	**15.48**
Roof coverings; layers of felt -3; bonding with hot bitumen compound to timber base										
pitch 7.5 degrees from horizontal	5.31	15.00	**6.25**	0.48	0.06	**7.46**	-	**13.71**	m2	**14.74**
pitch 40 degrees from horizontal	5.31	15.00	**6.25**	0.80	0.10	**12.44**	-	**18.69**	m2	**20.09**
pitch 75 degrees from horizontal	5.31	15.00	**6.25**	0.96	0.12	**14.93**	-	**21.17**	m2	**22.76**
Roof coverings; layers of felt -3; bonding with hot bitumen compound to cement and sand or concrete base										
pitch 7.5 degrees from horizontal	5.31	15.00	**6.25**	0.48	0.06	**7.46**	-	**13.71**	m2	**14.74**
pitch 40 degrees from horizontal	5.31	15.00	**6.25**	0.80	0.10	**12.44**	-	**18.69**	m2	**20.09**
pitch 75 degrees from horizontal	5.31	15.00	**6.25**	0.96	0.12	**14.93**	-	**21.17**	m2	**22.76**
Roofing; felt BS 747, bottom and intermediate layers type 3B -1.8 kg/m2, top layer type 3E -2.8 kg/m2, 75mm laps; fully bonding layers with hot bitumen bonding compound										
Roof coverings; layers of felt -2; bonding with hot bitumen compound to timber base										
pitch 7.5 degrees from horizontal	4.38	15.00	**5.13**	0.33	0.04	**5.12**	-	**10.25**	m2	**11.02**
pitch 40 degrees from horizontal	4.38	15.00	**5.13**	0.55	0.07	**8.57**	-	**13.70**	m2	**14.72**
pitch 75 degrees from horizontal	4.38	15.00	**5.13**	0.66	0.08	**10.24**	-	**15.37**	m2	**16.52**
Roof coverings; layers of felt -2; bonding with hot bitumen compound to cement and sand or concrete base										
pitch 7.5 degrees from horizontal	4.38	15.00	**5.13**	0.33	0.04	**5.12**	-	**10.25**	m2	**11.02**
pitch 40 degrees from horizontal	4.38	15.00	**5.13**	0.55	0.07	**8.57**	-	**13.70**	m2	**14.72**
pitch 75 degrees from horizontal	4.38	15.00	**5.13**	0.66	0.08	**10.24**	-	**15.37**	m2	**16.52**
Roof coverings; layers of felt -3; bonding with hot bitumen compound to timber base										
pitch 7.5 degrees from horizontal	6.15	15.00	**7.21**	0.48	0.06	**7.46**	-	**14.68**	m2	**15.78**
pitch 40 degrees from horizontal	6.15	15.00	**7.21**	0.80	0.10	**12.44**	-	**19.65**	m2	**21.13**
pitch 75 degrees from horizontal	6.15	15.00	**7.21**	0.96	0.12	**14.93**	-	**22.14**	m2	**23.80**

Labour hourly rates: (except Specialists) Craft Operatives £14.21 Labourer £10.71 Rates are national average prices. Refer to REGIONAL VARIATIONS for indicative levels of overall pricing in regions	MATERIALS			LABOUR				RATES		
	Del to Site £	Waste %	Material Cost £	Craft Optve Hrs	Lab Hrs	Labour Cost £	Sunds £	Nett Rate £	Unit	Gross Rate (+7.5%) £
J41: BUILT UP FELT ROOF COVERINGS Cont.										
Roofing; felt BS 747, bottom and intermediate layers type 3B -1.8 kg/m2, top layer type 3E -2.8 kg/m2, 75mm laps; fully bonding layers with hot bitumen bonding compound Cont.										
Roof coverings; layers of felt -3; bonding with hot bitumen compound to cement and sand or concrete base										
pitch 7.5 degrees from horizontal................	6.15	15.00	**7.21**	0.48	0.06	**7.46**	-	**14.68**	m2	**15.78**
pitch 40 degrees from horizontal................	6.15	15.00	**7.21**	0.80	0.10	**12.44**	-	**19.65**	m2	**21.13**
pitch 75 degrees from horizontal.................	6.15	15.00	**7.21**	0.96	0.12	**14.93**	-	**22.14**	m2	**23.80**
Skirtings; layers of felt - 3; bonding with hot bitumen compound to brickwork base										
girth 0 - 200mm	1.23	15.00	**1.44**	0.25	0.03	**3.87**	-	**5.32**	m	**5.72**
girth 200 - 400mm	2.46	15.00	**2.89**	0.40	0.05	**6.22**	-	**9.10**	m	**9.79**
Collars around pipes; standard and the like; layers of felt -1; bonding with hot bitumen compound to metal base										
50mm diameter x 150mm long; hole in 3 layer covering	0.49	15.00	**0.56**	0.15	0.02	**2.35**	-	**2.91**	nr	**3.13**
100mm diameter x 150mm long; hole in 3 layer covering	0.64	15.00	**0.75**	0.20	0.03	**3.16**	-	**3.91**	nr	**4.20**
Roofing; sheet, Ruberoid Building Products Superflex Ultrabond system, bottom layer Rubervent BS 747 type 3G; intermediate layer Superbase Ultrabond; top layer Superflex Ultrabond slate surfaced cap sheet, 50mm side laps, 75mm end laps; partially bonding bottom layer, fully bonding other layers in Ruberoid 95/25 hot bonding bitumen										
Roof coverings; layers of sheet -3; bonding with hot bitumen compound to timber base										
pitch not exceeding 5.0 degrees from horizontal..	14.20	15.00	**16.47**	0.48	0.24	**9.39**	-	**25.86**	m2	**27.80**
Roof coverings; layers of sheet -3; bonding with hot bitumen compound to cement and sand or concrete base										
pitch not exceeding 5.0 degrees from horizontal..	14.20	15.00	**16.47**	0.48	0.24	**9.39**	-	**25.86**	m2	**27.80**
Roof coverings; layers of sheet -3; bonding with hot bitumen compound to metal base										
pitch not exceeding 5.0 degrees from horizontal..	14.20	15.00	**16.47**	0.52	0.24	**9.96**	-	**26.43**	m2	**28.41**
Skirtings; layers of sheet -2; bonding with hot bitumen compound to brickwork base										
girth 0 - 200mm	1.74	15.00	**2.02**	0.25	0.03	**3.87**	-	**5.89**	m	**6.33**
girth 200 - 400mm	3.48	15.00	**4.04**	0.40	0.05	**6.22**	-	**10.25**	m	**11.02**
Roofing; sheet, Ruberoid Building Products Superflex Ultrabond system, bottom layer Rubervent BS 747 type 3G; intermediate layer Superbase Ultrabond; top layer Superflex Ultrabond slate surfaced cap sheet, 50mm side laps, 75mm end laps; partially bonding bottom layer, fully bonding other layers in Ruberoid 95/25 hot bonding bitumen; 25mm Kingspan Thermaroof TR21 (PIR) Board										
Roof coverings; layers of sheet -3; bonding with hot bitumen compound to timber base										
pitch not exceeding 5.0 degrees from horizontal..	22.33	15.00	**25.82**	0.96	0.31	**16.96**	1.01	**43.79**	m2	**47.07**
Roof coverings; layers of sheet -3; bonding with hot bitumen compound to cement and sand or concrete base										
pitch not exceeding 5.0 degrees from horizontal..	22.33	15.00	**25.82**	0.96	0.31	**16.96**	1.01	**43.79**	m2	**47.07**
Roof coverings; layers of sheet -3; bonding with hot bitumen compound to metal base										
pitch not exceeding 5.0 degrees from horizontal..	22.33	15.00	**25.82**	1.00	0.31	**17.53**	1.01	**44.36**	m2	**47.69**
Roofing; sheet, Ruberoid Building Products Superflex Ultrabond system, bottom layer Rubervent BS 747 type 3G; intermediate layer Superbase Ultrabond; top layer Superflex Ultrabond slate surfaced cap sheet, 50mm side laps, 75mm end laps; partially bonding bottom layer, fully bonding other layers in Ruberoid 95/25 hot bonding bitumen; 30mm Kingspan Thermaroof TR21 (PIR) Board										
Roof coverings; layers of sheet -3; bonding with hot bitumen compound to timber base										
pitch not exceeding 5.0 degrees from horizontal..	23.52	15.00	**27.19**	0.98	0.31	**17.25**	-	**44.43**	m2	**47.77**

Labour hourly rates: (except Specialists) Craft Operatives £14.21 Labourer £10.71 Rates are national average prices. Refer to REGIONAL VARIATIONS for indicative levels of overall pricing in regions	MATERIALS			LABOUR				RATES		
	Del to Site £	Waste %	Material Cost £	Craft Optve Hrs	Lab Hrs	Labour Cost £	Sunds £	Nett Rate £	Unit	Gross Rate (+7.5%) £
J41: BUILT UP FELT ROOF COVERINGS Cont.										
Roofing; sheet, Ruberoid Building Products Superflex Ultrabond system, bottom layer Rubervent BS 747 type 3G; intermediate layer Superbase Ultrabond; top layer Superflex Ultrabond slate surfaced cap sheet, 50mm side laps, 75mm end laps; partially bonding bottom layer, fully bonding other layers in Ruberoid 95/25 hot bonding bitumen; 40mm Kingspan Thermaroof TR21 (PIR) Board										
Roof coverings; layers of sheet -3; bonding with hot bitumen compound to timber base pitch not exceeding 5.0 degrees from horizontal ..	24.97	15.00	**28.86**	1.02	0.31	**17.81**	-	**46.67**	m2	**50.17**
Roofing; sheet, Ruberoid Building Products Superflex Ultrabond system, bottom layer Rubervent BS 747 type 3G; intermediate layer Superbase Ultrabond; top layer Superflex Ultrabond slate surfaced cap sheet, 50mm side laps, 75mm end laps; partially bonding bottom layer, fully bonding other layers in Ruberoid 95/25 hot bonding bitumen; 50mm Kingspan Thermaboard TR21 (PIR) Board										
Roof coverings; layers of sheet -3; bonding with hot bitumen compound to timber base pitch not exceeding 5.0 degrees from horizontal ..	26.50	15.00	**30.62**	1.05	0.31	**18.24**	-	**48.86**	m2	**52.52**
Roofing; sheet, Ruberoid Building Products Superflex Ultratorch system, bottom layer Superbase Ultratorch; top layer Superflex Ultratorch slate surfaced cap sheet, 75mm side and end laps; fully bonding both layers (by torching); SuperRock Torch, mineral wool insulation with bitumen primed surface; Superbar vapour control layer										
Roof coverings; layers of sheet -3; bonding to timber base pitch not exceeding 5.0 degrees from horizontal ..	35.86	15.00	**41.24**	0.48	0.24	**9.39**	2.61	**53.24**	m2	**57.23**
Roof coverings; layers of sheet -3; bonding to cement and sand or concrete base pitch not exceeding 5.0 degrees from horizontal ..	35.86	15.00	**41.24**	0.48	0.24	**9.39**	2.61	**53.24**	m2	**57.23**
Roof coverings; layers of sheet -3; bonding to metal base pitch not exceeding 5.0 degrees from horizontal ..	35.86	15.00	**41.24**	0.52	0.24	**9.96**	2.61	**53.81**	m2	**57.84**
Skirtings; layers of sheet -2; bonding with hot bitumen compound to brickwork base										
girth 0 - 200mm	2.69	15.00	**3.10**	0.25	0.03	**3.87**	0.53	**7.50**	m	**8.06**
girth 200 - 400mm	5.39	15.00	**6.20**	0.40	0.05	**6.22**	1.05	**13.47**	m	**14.48**
Roofing; sheet, Ruberoid Building Products Superflex Ultratorch system, bottom layer Superbase Ultratorch; top layer Superflex Ultratorch slate surfaced cap sheet, 75mm side and end laps; fully bonding both layers (by torching); SuperRock Torch, 25mm Kingspan Thermaroof TR21 (PIR Board) insulation with bitumen primed surface; Superbar vapour control layer										
Roof coverings; layers of sheet -3; bonding to timber base pitch not exceeding 5.0 degrees from horizontal ..	43.99	15.00	**50.59**	0.96	0.31	**16.96**	2.61	**70.16**	m2	**75.42**
Roof coverings; layers of sheet -3; bonding to cement and sand or concrete base pitch not exceeding 5.0 degrees from horizontal ..	43.99	15.00	**50.59**	1.02	0.31	**17.81**	2.61	**71.01**	m2	**76.34**
Roof coverings; layers of sheet -3; bonding to metal base pitch not exceeding 5.0 degrees from horizontal ..	43.99	15.00	**50.59**	1.00	0.31	**17.53**	2.61	**70.73**	m2	**76.03**
Roofing; sheet, Ruberoid Building Products Superflex Ultratorch system, bottom layer Superbase Ultratorch; top layer Superflex Ultratorch slate surfaced cap sheet, 75mm side and end laps; fully bonding both layers (by torching); SuperRock Torch, 30mm Kingspan Thermaroof TR21 (PIR Board) insulation with bitumen primed surface; Superbar vapour control layer										
Roof coverings; layers of sheet -3; bonding to timber base pitch not exceeding 5.0 degrees from horizontal ..	45.18	15.00	**51.96**	0.98	0.31	**17.25**	2.61	**71.81**	m2	**77.20**
Roofing; sheet, Ruberoid Building Products Superflex Ultratorch system, bottom layer Superbase Ultratorch; top layer Superflex Ultratorch slate surfaced cap sheet, 75mm side and end laps; fully bonding both layers (by torching); SuperRock Torch, 40mm Kingspan Thermaroof TR21 (PIR Board) insulation with bitumen primed surface; Superbar vapour control layer										
Roof coverings; layers of sheet -3; bonding to timber base pitch not exceeding 5.0 degrees from horizontal ..	46.63	15.00	**53.62**	1.02	0.31	**17.81**	2.61	**74.05**	m2	**79.60**

LININGS/SHEATHING/DRY PARTITIONING

Labour hourly rates: (except Specialists) Craft Operatives £14.21 Labourer £10.71 Rates are national average prices. Refer to REGIONAL VARIATIONS for indicative levels of overall pricing in regions	MATERIALS			LABOUR				RATES		
	Del to Site £	Waste %	Material Cost £	Craft Optve Hrs	Lab Hrs	Labour Cost £	Sunds £	Nett Rate £	Unit	Gross Rate (+7.5%) £
K10: PLASTERBOARD DRY LINING/PARTITIONS/CEILINGS										
Linings; tapered edge sheeting of one layer 9.5mm thick Gyproc wall board for direct decoration, BS 1230, butt joints; fixing with galvanized nails to timber base; butt joints filled with joint filler tape and joint finish, spot filling										
Walls										
height 2100 - 2400mm	4.69	5.00	**4.93**	0.84	0.42	**16.43**	-	**21.36**	m	**22.97**
height 2400 - 2700mm	5.28	5.00	**5.55**	0.95	0.48	**18.64**	-	**24.19**	m	**26.00**
height 2700 - 3000mm	5.80	5.00	**6.09**	1.05	0.53	**20.60**	-	**26.68**	m	**28.68**
Beams; 3nr faces										
total girth 600 - 1200mm	2.38	10.00	**2.60**	0.88	0.21	**14.75**	-	**17.35**	m	**18.65**
Columns; 4nr faces										
total girth 600 - 1200mm	2.44	10.00	**2.66**	0.88	0.21	**14.75**	-	**17.41**	m	**18.72**
total girth 1200 - 1800mm	3.61	10.00	**3.93**	1.31	0.32	**22.04**	-	**25.97**	m	**27.92**
Reveals and soffits of openings and recesses										
width not exceeding 300mm	0.64	10.00	**0.70**	0.21	0.05	**3.52**	-	**4.21**	m	**4.53**
Ceilings										
generally	1.97	5.00	**2.07**	0.40	0.18	**7.61**	-	**9.68**	m2	**10.40**
Linings; tapered edge sheeting of one layer 12.5mm thick Gyproc wallboard for direct decoration, BS 1230 butt joints; fixing with galvanized nails to timber base; butt joints filled with joint filler tape and joint finish, spot filling										
Walls										
height 2100 - 2400mm	4.93	5.00	**5.18**	0.96	0.48	**18.78**	-	**23.96**	m	**25.76**
height 2400 - 2700mm	5.55	5.00	**5.83**	1.08	0.54	**21.13**	-	**26.96**	m	**28.98**
height 2700 - 3000mm	6.17	5.00	**6.48**	1.20	0.60	**23.48**	-	**29.95**	m	**32.20**
Beams; 3nr faces										
total girth 600 - 1200mm	2.50	10.00	**2.73**	1.02	0.24	**17.06**	-	**19.79**	m	**21.28**
Columns; 4nr faces										
total girth 600 - 1200mm	2.56	10.00	**2.79**	1.02	0.24	**17.06**	-	**19.85**	m	**21.34**
total girth 1200 - 1800mm	3.79	10.00	**4.13**	1.53	0.36	**25.60**	-	**29.73**	m	**31.95**
Reveals and soffits of openings and recesses										
width not exceeding 300mm	0.67	10.00	**0.73**	0.21	0.06	**3.63**	-	**4.36**	m	**4.68**
Ceilings										
generally	2.07	5.00	**2.17**	0.46	0.20	**8.68**	-	**10.85**	m2	**11.66**
Linings; tapered edge sheeting of one layer 15mm thick Gyproc wall board for direct decoration, BS 1230, butt joints; fixing with galvanized nails to timber base; butt joints filled with joint filler tape and joint finish, spot filling										
Walls										
height 2100 - 2400mm	7.41	5.00	**7.78**	1.08	1.23	**28.52**	-	**36.30**	m	**39.02**
height 2400 - 2700mm	8.33	5.00	**8.75**	1.21	0.61	**23.73**	-	**32.48**	m	**34.91**
height 2700 - 3000mm	9.18	5.00	**9.64**	1.35	0.68	**26.47**	-	**36.11**	m	**38.82**
Beams; 3nr faces										
total girth 600 - 1200mm	3.74	10.00	**4.09**	1.16	0.27	**19.38**	-	**23.47**	m	**25.22**
Columns; 4nr faces										
total girth 600 - 1200mm	3.80	10.00	**4.15**	1.16	0.27	**19.38**	-	**23.52**	m	**25.29**
total girth 1200 - 1800mm	5.64	10.00	**6.17**	1.74	0.41	**29.12**	-	**35.28**	m	**37.93**
Reveals and soffits of openings and recesses										
width not exceeding 300mm	0.98	10.00	**1.07**	0.22	0.06	**3.77**	-	**4.84**	m	**5.20**

Labour hourly rates: (except Specialists) Craft Operatives £14.21 Labourer £10.71 Rates are national average prices. Refer to REGIONAL VARIATIONS for indicative levels of overall pricing in regions	MATERIALS			LABOUR				RATES		
	Del to Site £	Waste %	Material Cost £	Craft Optve Hrs	Lab Hrs	Labour Cost £	Sunds £	Nett Rate £	Unit	Gross Rate (+7.5%) £
K10: PLASTERBOARD DRY LINING/PARTITIONS/CEILINGS Cont.										
Linings; tapered edge sheeting of one layer 15mm thick Gyproc wall board for direct decoration, BS 1230, butt joints; fixing with galvanized nails to timber base; butt joints filled with joint filler tape and joint finish, spot filling Cont.										
Ceilings										
generally	3.10	5.00	**3.25**	0.52	0.22	**9.75**	-	**13.00**	m2	**13.97**
Linings; tapered edge sheeting of one layer 9.5mm thick Gyproc Duplex wallboard for direct decoration, BS 1230 butt joints; fixing with galvanized nails to timber base; butt joints filled with joint filler tape and joint finish, spot filling										
Walls										
height 2100 - 2400mm	6.01	5.00	**6.32**	0.84	0.42	**16.43**	-	**22.75**	m	**24.46**
height 2400 - 2700mm	6.77	5.00	**7.11**	0.95	0.48	**18.64**	-	**25.75**	m	**27.68**
height 2700 - 3000mm	7.44	5.00	**7.82**	1.05	0.53	**20.60**	-	**28.41**	m	**30.54**
Beams; 3nr faces										
total girth 600 - 1200mm	3.04	10.00	**3.32**	0.88	0.21	**14.75**	-	**18.08**	m	**19.43**
Columns; 4nr faces										
total girth 600 - 1200mm	3.10	10.00	**3.38**	0.88	0.21	**14.75**	-	**18.14**	m	**19.50**
total girth 1200 - 1800mm	4.60	10.00	**5.02**	1.31	0.32	**22.04**	-	**27.06**	m	**29.09**
Reveals and soffits of openings and recesses										
width not exceeding 300mm	0.80	10.00	**0.88**	0.21	0.05	**3.52**	-	**4.40**	m	**4.73**
Ceilings										
generally	2.52	5.00	**2.64**	0.40	0.18	**7.61**	-	**10.26**	m2	**11.03**
Linings; tapered edge sheeting of one layer 12.5mm thick Gyproc Duplex wallboard for direct decoration, BS 1230, butt joints; fixing with galvanized nails to timber base; butt joints filled with joint filler tape and joint finish, spot filling										
Walls										
height 2100 - 2400mm	6.33	5.00	**6.64**	0.96	0.48	**18.78**	-	**25.43**	m	**27.33**
height 2400 - 2700mm	7.12	5.00	**7.47**	1.08	0.54	**21.13**	-	**28.60**	m	**30.75**
height 2700 - 3000mm	7.91	5.00	**8.30**	1.20	0.60	**23.48**	-	**31.78**	m	**34.16**
Beams; 3nr faces										
total girth 600 - 1200mm	3.20	10.00	**3.50**	1.02	0.24	**17.06**	-	**20.56**	m	**22.10**
Columns; 4nr faces										
total girth 600 - 1200mm	3.26	10.00	**3.55**	1.02	0.24	**17.06**	-	**20.62**	m	**22.17**
total girth 1200 - 1800mm	4.83	10.00	**5.28**	1.53	0.36	**25.60**	-	**30.87**	m	**33.19**
Reveals and soffits of openings and recesses										
width not exceeding 300mm	0.84	10.00	**0.92**	0.21	0.06	**3.63**	-	**4.55**	m	**4.89**
Ceilings										
generally	2.65	5.00	**2.78**	0.46	0.20	**8.68**	-	**11.46**	m2	**12.32**
Linings; tapered edge sheeting of one layer 15mm thick Gyproc Duplex wallboard for direct decoration, BS 1230, butt joints; fixing with galvanized nails to timber base; butt joints filled with joint filler tape and joint finish, spot filling										
Walls										
height 2100 - 2400mm	9.64	5.00	**10.12**	1.08	0.54	**21.13**	-	**31.25**	m	**33.59**
height 2400 - 2700mm	10.84	5.00	**11.39**	1.21	0.61	**23.73**	-	**35.11**	m	**37.75**
height 2700 - 3000mm	11.94	5.00	**12.54**	1.35	0.68	**26.47**	-	**39.00**	m	**41.93**
Beams; 3nr faces										
total girth 600 - 1200mm	4.86	10.00	**5.32**	1.16	0.27	**19.38**	-	**24.69**	m	**26.54**
Columns; 4nr faces										
total girth 600 - 1200mm	4.91	10.00	**5.38**	1.16	0.27	**19.38**	-	**24.75**	m	**26.61**
total girth 1200 - 1800mm	7.31	10.00	**8.01**	1.74	0.41	**29.12**	-	**37.13**	m	**39.91**
Reveals and soffits of openings and recesses										
width not exceeding 300mm	1.26	10.00	**1.38**	0.22	0.06	**3.77**	-	**5.14**	m	**5.53**
Ceilings										
generally	4.03	5.00	**4.23**	0.52	0.22	**9.75**	-	**13.98**	m2	**15.02**
Linings; tapered edge sheeting of one layer 19mm thick Gyproc plank for direct decoration, BS 1230, butt joints; fixing with galvanized nails to timber base; butt joints filled with joint filler tape and joint finish, spot filling										
Walls										
height 2100 - 2400mm	10.45	5.00	**10.98**	1.20	0.60	**23.48**	-	**34.45**	m	**37.04**
height 2400 - 2700mm	11.76	5.00	**12.35**	1.35	0.68	**26.47**	-	**38.81**	m	**41.72**
height 2700 - 3000mm	13.00	5.00	**13.65**	1.50	0.75	**29.35**	-	**43.00**	m	**46.22**
Beams; 3nr faces										
total girth 600 - 1200mm	5.23	10.00	**5.73**	1.50	0.30	**24.53**	-	**30.26**	m	**32.53**

Labour hourly rates: (except Specialists) Craft Operatives £14.21 Labourer £10.71 Rates are national average prices. Refer to REGIONAL VARIATIONS for indicative levels of overall pricing in regions	MATERIALS			LABOUR				RATES		
	Del to Site £	Waste %	Material Cost £	Craft Optve Hrs	Lab Hrs	Labour Cost £	Sunds £	Nett Rate £	Unit	Gross Rate (+7.5%) £
K10: PLASTERBOARD DRY LINING/PARTITIONS/CEILINGS Cont.										
Linings; tapered edge sheeting of one layer 19mm thick Gyproc plank for direct decoration, BS 1230, butt joints; fixing with galvanized nails to timber base; butt joints filled with joint filler tape and joint finish, spot filling Cont.										
Columns; 4nr faces										
total girth 600 - 1200mm	5.29	10.00	**5.79**	1.50	0.30	**24.53**	-	**30.32**	m	**32.59**
total girth 1200 - 1800mm	7.88	10.00	**8.63**	2.25	0.45	**36.79**	-	**45.42**	m	**48.82**
Reveals and soffits of openings and recesses										
width not exceeding 300mm	1.35	10.00	**1.48**	0.30	0.08	**5.12**	-	**6.60**	m	**7.09**
Ceilings										
generally	4.37	5.00	**4.59**	0.60	0.25	**11.20**	-	**15.79**	m2	**16.97**
Linings; two layers of Gypsum wallboard, first layer square edge sheeting 9.5mm thick, second layer tapered edge sheeting 9.5mm thick for direct decoration, BS 1230, butt joints; fixing with galvanized nails to timber base; butt joints of second layer filled with joint filler tape and joint finish, spot filling										
Walls										
height 2100 - 2400mm	8.92	5.00	**9.37**	1.44	0.72	**28.17**	-	**37.54**	m	**40.36**
height 2400 - 2700mm	10.04	5.00	**10.54**	1.62	0.81	**31.70**	-	**42.24**	m	**45.40**
height 2700 - 3000mm	11.08	5.00	**11.63**	1.80	0.90	**35.22**	-	**46.85**	m	**50.36**
Beams; 3nr faces										
total girth 600 - 1200mm	2.59	10.00	**2.81**	1.62	0.36	**26.88**	-	**29.69**	m	**31.92**
Columns; 4nr faces										
total girth 600 - 1200mm	4.55	10.00	**4.97**	1.62	0.36	**26.88**	-	**31.85**	m	**34.24**
total girth 1200 - 1800mm	6.78	10.00	**7.40**	2.43	0.54	**40.31**	-	**47.72**	m	**51.30**
Reveals and soffits of openings and recesses										
width not exceeding 300mm	1.17	10.00	**1.27**	0.33	0.09	**5.65**	-	**6.93**	m	**7.45**
Ceilings										
generally	3.73	5.00	**3.92**	0.70	0.30	**13.16**	-	**17.08**	m2	**18.36**
Linings; two layers of Gypsum wallboard, first layer square edge sheeting 9.5mm thick, second layer tapered edge sheeting 12.5mm thick for direct decoration, BS 1230, butt joints; fixing with galvanized nails to timber base; butt joints of second layer filled with joint filler tape and joint finish, spot filling										
Walls										
height 2100 - 2400mm	9.16	5.00	**9.62**	1.56	0.78	**30.52**	-	**40.14**	m	**43.15**
height 2400 - 2700mm	10.31	5.00	**10.82**	1.76	0.88	**34.43**	-	**45.26**	m	**48.65**
height 2700 - 3000mm	11.38	5.00	**11.95**	1.95	0.98	**38.21**	-	**50.15**	m	**53.92**
Beams; 3nr faces										
total girth 600 - 1200mm	4.62	10.00	**4.95**	1.78	0.39	**29.47**	-	**34.42**	m	**37.00**
Columns; 4nr faces										
total girth 600 - 1200mm	4.67	10.00	**5.01**	1.78	0.39	**29.47**	-	**34.48**	m	**37.07**
total girth 1200 - 1800mm	6.96	10.00	**7.46**	2.66	0.59	**44.12**	-	**51.58**	m	**55.44**
Reveals and soffits of openings and recesses										
width not exceeding 300mm	1.20	10.00	**1.28**	0.36	0.10	**6.19**	-	**7.47**	m	**8.03**
Ceilings										
generally	3.83	5.00	**4.02**	0.76	0.33	**14.33**	-	**18.36**	m2	**19.73**
Linings; two layers of Gypsum wallboard, first layer square edge sheeting 12.5mm thick, second layer tapered edge sheeting 12.5mm thick for direct decoration, BS 1230, butt joints; fixing with galvanized nails to timber base; butt joints of second layer filled with joint filler tape and joint finish, spot filling										
Walls										
height 2100 - 2400mm	9.40	5.00	**9.87**	1.68	0.84	**32.87**	-	**42.74**	m	**45.95**
height 2400 - 2700mm	10.58	5.00	**11.11**	1.89	0.95	**37.03**	-	**48.14**	m	**51.75**
height 2700 - 3000mm	11.68	5.00	**12.26**	2.10	1.05	**41.09**	-	**53.35**	m	**57.35**
Beams; 3nr faces										
total girth 600 - 1200mm	4.74	10.00	**5.18**	1.92	0.42	**31.78**	-	**36.96**	m	**39.73**
Columns; 4nr faces										
total girth 600 - 1200mm	4.79	10.00	**5.24**	1.92	0.42	**31.78**	-	**37.02**	m	**39.79**
total girth 1200 - 1800mm	7.14	10.00	**7.80**	2.88	0.63	**47.67**	-	**55.47**	m	**59.63**
Reveals and soffits of openings and recesses										
width not exceeding 300mm	1.23	10.00	**1.34**	0.39	0.11	**6.72**	-	**8.06**	m	**8.66**
Ceilings										
generally	3.93	5.00	**4.13**	0.82	0.35	**15.40**	-	**19.53**	m2	**20.99**

Labour hourly rates: (except Specialists) Craft Operatives £14.21 Labourer £10.71 Rates are national average prices. Refer to REGIONAL VARIATIONS for indicative levels of overall pricing in regions	MATERIALS			LABOUR				RATES		
	Del to Site £	Waste %	Material Cost £	Craft Optve Hrs	Lab Hrs	Labour Cost £	Sunds £	Nett Rate £	Unit	Gross Rate (+7.5%) £
K10: PLASTERBOARD DRY LINING/PARTITIONS/CEILINGS Cont.										
Linings; two layers of Gypsum wallboard, first layer square edge sheeting 15mm thick, second layer tapered edge sheeting 15mm thick for direct decoration, BS 1230, butt joints; fixing with galvanized nails to timber base; butt joints of second layer filled with joint filler tape and joint finish, spot filling										
Walls										
height 2100 - 2400mm	13.61	5.00	**14.29**	1.80	0.90	**35.22**	-	**49.50**	m	**53.22**
height 2400 - 2700mm	16.14	5.00	**16.95**	2.02	1.01	**39.52**	-	**56.47**	m	**60.70**
height 2700 - 3000mm	17.86	5.00	**18.75**	2.25	1.13	**44.07**	-	**62.83**	m	**67.54**
Beams; 3nr faces										
total girth 600 - 1200mm	7.21	10.00	**7.90**	1.68	0.45	**28.69**	-	**36.59**	m	**39.33**
Columns; 4nr faces										
total girth 600 - 1200mm	7.27	10.00	**7.96**	1.68	0.45	**28.69**	-	**36.65**	m	**39.40**
total girth 1200 - 1800mm	10.85	10.00	**11.88**	2.52	0.68	**43.09**	-	**54.97**	m	**59.09**
Reveals and soffits of openings and recesses										
width not exceeding 300mm	1.85	10.00	**2.02**	0.45	0.11	**7.57**	-	**9.59**	m	**10.31**
Ceilings										
generally	5.99	5.00	**6.29**	0.86	0.38	**16.29**	-	**22.58**	m2	**24.27**
Linings; two layers of Gypsum wallboard, first layer Duplex square edge sheeting 9.5mm thick, second layer tapered edge sheeting 9.5mm thick for direct decoration, BS 1230, butt joints; fixing with galvanized nails to timber base; butt joints of second layer filled with joint filler tape and joint finish, spot filling										
Walls										
height 2100 - 2400mm	10.24	5.00	**10.76**	1.44	0.72	**28.17**	-	**38.93**	m	**41.85**
height 2400 - 2700mm	11.52	5.00	**12.10**	1.62	0.81	**31.70**	-	**43.80**	m	**47.08**
height 2700 - 3000mm	12.73	5.00	**13.37**	1.80	0.90	**35.22**	-	**48.58**	m	**52.23**
Beams; 3nr faces										
total girth 600 - 1200mm	5.16	10.00	**5.51**	1.62	0.36	**26.88**	-	**32.39**	m	**34.82**
Columns; 4nr faces										
total girth 600 - 1200mm	5.21	10.00	**5.57**	1.62	0.36	**26.88**	-	**32.45**	m	**34.88**
total girth 1200 - 1800mm	7.77	10.00	**8.30**	2.43	0.54	**40.31**	-	**48.61**	m	**52.26**
Reveals and soffits of openings and recesses										
width not exceeding 300mm	1.33	10.00	**1.42**	0.33	0.09	**5.65**	-	**7.08**	m	**7.61**
Ceilings										
generally	5.87	5.00	**6.16**	0.70	0.30	**13.16**	-	**19.32**	m2	**20.77**
Linings; two layers of Gypsum wallboard, first layer Duplex square edge sheeting 9.5mm thick, second layer tapered edge sheeting 12.5mm thick for direct decoration, BS 1230, butt joints; fixing with galvanized nails to timber base; butt joints of second layer filled with joint filler tape and joint finish, spot filling										
Walls										
height 2100 - 2400mm	10.48	5.00	**11.01**	1.56	0.78	**30.52**	-	**41.53**	m	**44.64**
height 2400 - 2700mm	11.79	5.00	**12.38**	1.76	0.88	**34.43**	-	**46.82**	m	**50.33**
height 2700 - 3000mm	13.03	5.00	**13.68**	1.95	0.98	**38.21**	-	**51.89**	m	**55.78**
Beams; 3nr faces										
total girth 600 - 1200mm	5.28	10.00	**5.64**	1.78	0.39	**29.47**	-	**35.11**	m	**37.75**
Columns; 4nr faces										
total girth 600 - 1200mm	5.33	10.00	**5.70**	1.78	0.39	**29.47**	-	**35.17**	m	**37.81**
total girth 1200 - 1800mm	7.95	10.00	**8.50**	2.66	0.59	**44.12**	-	**52.62**	m	**56.56**
Reveals and soffits of openings and recesses										
width not exceeding 300mm	1.36	10.00	**1.46**	0.36	0.10	**6.19**	-	**7.64**	m	**8.22**
Ceilings										
generally	4.38	5.00	**4.60**	0.76	0.33	**14.33**	-	**18.93**	m2	**20.35**
Linings; two layers of Gypsum wallboard, first layer Duplex square edge sheeting 12.5mm thick, second layer tapered edge sheeting 12.5mm thick for direct decoration, BS 1230, butt joints; fixing with galvanized nails to timber base; butt joints of second layer filled with joint filler tape and joint finish, spot filling										
Walls										
height 2100 - 2400mm	10.80	5.00	**11.34**	1.68	0.84	**32.87**	-	**44.20**	m	**47.52**
height 2400 - 2700mm	12.15	5.00	**12.75**	1.89	0.95	**37.03**	-	**49.78**	m	**53.52**
height 2700 - 3000mm	13.42	5.00	**14.09**	2.10	1.05	**41.09**	-	**55.18**	m	**59.31**
Beams; 3nr faces										
total girth 600 - 1200mm	5.43	10.00	**5.81**	1.92	0.42	**31.78**	-	**37.59**	m	**40.41**

Labour hourly rates: (except Specialists) Craft Operatives £14.21 Labourer £10.71 Rates are national average prices. Refer to REGIONAL VARIATIONS for indicative levels of overall pricing in regions	MATERIALS			LABOUR				RATES		
	Del to Site £	Waste %	Material Cost £	Craft Optve Hrs	Lab Hrs	Labour Cost £	Sunds £	Nett Rate £	Unit	Gross Rate (+7.5%) £
K10: PLASTERBOARD DRY LINING/PARTITIONS/CEILINGS Cont.										
Linings; two layers of Gypsum wallboard, first layer Duplex square edge sheeting 12.5mm thick, second layer tapered edge sheeting 12.5mm thick for direct decoration, BS 1230, butt joints; fixing with galvanized nails to timber base; butt joints of second layer filled with joint filler tape and joint finish, spot filling Cont.										
Columns; 4nr faces										
total girth 600 - 1200mm	5.49	10.00	**5.87**	1.92	0.42	**31.78**	-	**37.65**	m	**40.47**
total girth 1200 - 1800mm	8.18	10.00	**8.74**	2.88	0.63	**47.67**	-	**56.42**	m	**60.65**
Reveals and soffits of openings and recesses										
width not exceeding 300mm	1.40	10.00	**1.50**	0.39	0.11	**6.72**	-	**8.22**	m	**8.83**
Ceilings										
generally	4.51	5.00	**4.74**	0.82	0.35	**15.40**	-	**20.14**	m2	**21.65**
Linings; two layers of Gypsum wallboard, first layer Duplex square edge sheeting 15mm thick, second layer tapered edge sheeting 15mm thick for direct decoration, BS 1230, butt joints; fixing with galvanized nails to timber base; butt joints of second layer filled with joint filler tape and joint finish, spot filling										
Walls										
height 2100 - 2400mm	15.84	5.00	**16.63**	1.80	0.90	**35.22**	-	**51.85**	m	**55.74**
height 2400 - 2700mm	18.65	5.00	**19.58**	2.02	1.01	**39.52**	-	**59.11**	m	**63.54**
height 2700 - 3000mm	20.65	5.00	**21.68**	2.25	1.13	**44.07**	-	**65.76**	m	**70.69**
Beams; 3nr faces										
total girth 600 - 1200mm	8.33	10.00	**8.91**	1.68	0.45	**28.69**	-	**37.60**	m	**40.42**
Columns; 4nr faces										
total girth 600 - 1200mm	8.38	10.00	**8.96**	1.68	0.45	**28.69**	-	**37.66**	m	**40.48**
total girth 1200 - 1800mm	12.52	10.00	**13.39**	2.52	0.68	**43.09**	-	**56.48**	m	**60.72**
Reveals and soffits of openings and recesses										
width not exceeding 300mm	2.13	10.00	**2.27**	0.45	0.11	**7.57**	-	**9.84**	m	**10.58**
Ceilings										
generally	6.92	5.00	**7.27**	0.86	0.38	**16.29**	-	**23.56**	m2	**25.32**
K10: PLASTERBOARD DRY LINING/PARTITIONS/CEILINGS: PLASTIC FACED PLASTERBOARD										
Linings; square edge sheeting of one layer 12.5mm thick Gyproc Industrial Grade plastic faced plasterboard, BS 1230, butt joints; fixing with galvanized nails to timber base; butt joints sealed with self adhesive white tape										
Walls										
height 2100 - 2400mm	9.08	5.00	**9.58**	0.70	0.35	**13.70**	-	**23.28**	m	**25.02**
height 2400 - 2700mm	10.22	5.00	**10.78**	0.78	0.39	**15.26**	-	**26.04**	m	**27.99**
height 2700 - 3000mm	11.35	5.00	**11.98**	0.86	0.43	**16.83**	-	**28.80**	m	**30.96**
Beams; 3nr faces										
total girth 600 - 1200mm	4.58	10.00	**5.03**	0.87	0.17	**14.18**	-	**19.22**	m	**20.66**
Columns; 4nr faces										
total girth 600 - 1200mm	4.98	10.00	**5.47**	0.87	0.17	**14.18**	-	**19.66**	m	**21.13**
total girth 1200 - 1800mm	7.07	10.00	**7.77**	1.30	0.26	**21.26**	-	**29.03**	m	**31.20**
Reveals and soffits of openings and recesses										
width not exceeding 300mm	1.66	10.00	**1.82**	0.17	0.05	**2.95**	-	**4.77**	m	**5.13**
Ceilings										
generally	3.86	5.00	**4.07**	0.35	0.15	**6.58**	-	**10.65**	m2	**11.45**
3.50 - 5.00m above floor	3.86	5.00	**4.07**	0.38	0.15	**7.01**	-	**11.08**	m2	**11.91**
K10: PLASTERBOARD DRY LINING/PARTITIONS/CEILING - THERMAL BOARD										
Linings; tapered edged sheeting of one layer 22mm thick Gyproc Thermaline BASIC board vapour check grade for direct decoration, butt joints; fixing with galvanized nails to timber base; butt joints filled with joint filler tape and joint finish, spot filling										
Walls										
height 2100 - 2400mm	15.10	5.00	**15.85**	1.44	0.72	**28.17**	-	**44.02**	m	**47.33**
height 2400 - 2700mm	16.98	5.00	**17.83**	1.62	0.81	**31.70**	-	**49.53**	m	**53.24**
height 2700 - 3000mm	18.79	5.00	**19.73**	1.80	0.90	**35.22**	-	**54.95**	m	**59.07**
Reveals and soffits of openings and recesses										
width not exceeding 300mm	1.94	10.00	**2.12**	0.36	0.09	**6.08**	-	**8.20**	m	**8.82**
Ceilings										
generally	6.30	5.00	**6.62**	0.72	0.30	**13.44**	-	**20.06**	m2	**21.57**

Labour hourly rates: (except Specialists) Craft Operatives £14.21 Labourer £10.71 Rates are national average prices. Refer to REGIONAL VARIATIONS for indicative levels of overall pricing in regions	MATERIALS			LABOUR				RATES		
	Del to Site £	Waste %	Material Cost £	Craft Optve Hrs	Lab Hrs	Labour Cost £	Sunds £	Nett Rate £	Unit	Gross Rate (+7.5%) £
K10: PLASTERBOARD DRY LINING/PARTITIONS/CEILING - THERMAL BOARD Cont.										
Linings; tapered edged sheeting of one layer 30mm thick Gyproc thermaline BASIC board vapour check grade for direct decoration, butt joints; fixing with galvanized nails to timber base; butt joints filled with joint filler tape and joint finish, spot filling										
Walls										
height 2100 - 2400mm	15.48	5.00	**16.25**	1.68	0.84	**32.87**	-	**49.12**	m	**52.81**
height 2400 - 2700mm	17.41	5.00	**18.28**	1.89	0.95	**37.03**	-	**55.32**	m	**59.46**
height 2700 - 3000mm	19.27	5.00	**20.24**	2.10	1.05	**41.09**	-	**61.32**	m	**65.92**
Reveals and soffits of openings and recesses										
width not exceeding 300mm	1.99	10.00	**2.18**	0.42	0.11	**7.15**	-	**9.32**	m	**10.02**
Ceilings										
generally	6.46	5.00	**6.79**	0.84	0.36	**15.79**	-	**22.58**	m2	**24.27**
Linings; tapered edged sheeting of one layer 40mm thick Gyproc Thermaline BASIC board vapour check grade for direct decoration, butt joints; fixing with galvanized nails to timber base; butt joints filled with joint filler tape and joint finish, spot filling										
Walls										
height 2100 - 2400mm	18.22	5.00	**19.13**	1.92	0.96	**37.56**	-	**56.69**	m	**60.94**
height 2400 - 2700mm	20.49	5.00	**21.52**	2.16	1.08	**42.26**	-	**63.78**	m	**68.56**
height 2700 - 3000mm	22.69	5.00	**23.83**	2.40	1.20	**46.96**	-	**70.79**	m	**76.09**
Reveals and soffits of openings and recesses										
width not exceeding 300mm	2.33	10.00	**2.55**	0.48	0.12	**8.11**	-	**10.66**	m	**11.46**
Ceilings										
generally	7.60	5.00	**7.98**	0.96	0.42	**18.14**	-	**26.12**	m2	**28.08**
Linings; tapered edged sheeting of one layer 50mm thick Gyproc Thermaline BASIC board vapour check grade for direct decoration, butt joints; fixing with galvanized nails to timber base; butt joints filled with joint filler tape and joint finish, spot filling										
Walls										
height 2100 - 2400mm	22.80	5.00	**23.94**	2.16	1.08	**42.26**	-	**66.20**	m	**71.17**
height 2400 - 2700mm	25.65	5.00	**26.93**	2.43	1.22	**47.60**	-	**74.53**	m	**80.12**
height 2700 - 3000mm	28.42	5.00	**29.85**	2.70	1.35	**52.83**	-	**82.67**	m	**88.87**
Reveals and soffits of openings and recesses										
width not exceeding 300mm	2.90	10.00	**3.18**	0.54	0.14	**9.17**	-	**12.36**	m	**13.28**
Ceilings										
generally	9.51	5.00	**9.99**	1.08	0.48	**20.49**	-	**30.48**	m2	**32.76**
K10: PLASTERBOARD DRY LINING/PARTITIONS/CEILING - CELLULAR PARTITIONS										
Proprietary partitions; Panelwall PDP single leaf, tapered edge panels for direct decoration, butt joints; fixing with galvanized nails to 19 x 37mm (f sizes) impregnated wrot softwood sole plate with 19 x 37 x 300mm (f sizes) locating blocks at 900mm centres and 19x 37mm (f sizes) impregnated wrought softwood wall and ceiling battens; applying continuous beads of Gyproc acoustical sealant to perimeters of construction both sides; butt joints filled with joint filler tape and joint finish, spot filling; fixing to masonry with cartridge fired nails										
Height 2100 - 2400mm, 57mm thick										
boarded both sides	32.82	5.00	**34.46**	2.60	1.30	**50.87**	-	**85.33**	m	**91.73**
Height 2400 - 2700mm, 57mm thick										
boarded both sides	36.26	5.00	**38.07**	2.85	1.43	**55.81**	-	**93.89**	m	**100.93**
Angles to partitions; 37 x 19mm softwood batten, 37 x 37mm softwood batten, fixing together through partitions; one panel rebated by cutting back one face by full panel thickness; joints filled with joint filler tape and joint finish										
Plain										
57mm thick partitions	0.73	5.00	**0.77**	0.25	0.13	**4.94**	-	**5.71**	m	**6.14**
Tee junctions to partitions; 37 x 19mm softwood batten, 37 x 37mm softwood batten, fixing together through partitions; joints filled with joint filler tape and joint finish										
Plain										
57mm thick partitions	0.73	5.00	**0.77**	0.15	0.08	**2.99**	-	**3.75**	m	**4.04**

Labour hourly rates: (except Specialists) Craft Operatives £14.21 Labourer £10.71 Rates are national average prices. Refer to REGIONAL VARIATIONS for indicative levels of overall pricing in regions	MATERIALS			LABOUR				RATES		
	Del to Site £	Waste %	Material Cost £	Craft Optve Hrs	Lab Hrs	Labour Cost £	Sunds £	Nett Rate £	Unit	Gross Rate (+7.5%) £
K10: PLASTERBOARD DRY LINING/PARTITIONS/CEILING - CELLULAR PARTITIONS Cont.										
Fair ends to partitions										
57 mm thick partitions										
facing with 9.5mm Gyproc wallboard; fixing to 37 x 37mm softwood batten with galvanized nails.......	0.50	10.00	**0.53**	0.10	0.05	**1.96**	-	**2.49**	m	**2.67**
Fixings for heavy fittings										
Sinks										
insert 37 x 37mm softwood fixing block into core prior to erection of panels; not exceeding 150mm long	0.06	5.00	**0.06**	0.10	-	**1.42**	-	**1.49**	nr	**1.60**
insert 37 x 37mm softwood fixing block into core prior to erection panels; 150 - 300mm long.......	0.12	5.00	**0.13**	0.10	-	**1.42**	-	**1.55**	nr	**1.67**
Radiators										
insert 37 x 37mm softwood fixing block into core prior to erection of panels; not exceeding 150mm long	0.06	5.00	**0.06**	0.10	-	**1.42**	-	**1.49**	nr	**1.60**
insert 37 x 37mm softwood fixing block into core prior to erection panels; 150 - 300mm long.......	0.12	5.00	**0.13**	0.10	-	**1.42**	-	**1.55**	nr	**1.67**
K10: PLASTERBOARD DRY LINING/PARTITIONS/CEILINGS: LAMINATED PARTITIONS										
Proprietary partitions; Gyproc SolidWall laminated partition; tapered edge sheeting of one layer 12.5mm thick Gyproc wall board each side for direct decoration, BS 1230, butt joints; applying continuous beads of Gyproc acoustical sealant to perimeters of construction on both sides, fixing with nails to timber frame comprising 25 x 38mm (f sizes) perimeter members with one layer 19mm thick Gyproc plank infill; butt joints filled with joint filler tape and joint finish, spot filling; fixing to masonry with cartridge fired nails										
Height 2100 - 2400mm, 50mm thick boarded both sides	24.65	5.00	**25.88**	2.90	1.45	**56.74**	-	**82.62**	m	**88.81**
Height 2400 - 2700mm, 50mm thick boarded both sides	27.12	5.00	**28.48**	3.20	1.60	**62.61**	-	**91.09**	m	**97.92**
Proprietary partitions; Gyproc SolidWall laminated partition; tapered edge sheeting of one layer 19mm thick Gyproc plank for direct decoration, B.S 1230, butt joints; applying continuous beads of Gyproc acoustical sealant to perimeters of construction on both sides, fixing with nails to timber frame comprising 25 x 38mm (f sizes) perimeter members with one layer 19mm thick Gyproc plank infill; butt joints filled with joint filler tape and joint finish, spot filling; fixing to masonry with cartridge fired nails										
Height 2100 - 2400mm, 65mm thick boarded both sides	35.30	5.00	**37.07**	3.15	1.58	**61.68**	-	**98.75**	m	**106.16**
Height 2400 - 2700mm, 65mm thick boarded both sides	39.11	5.00	**41.07**	3.45	1.73	**67.55**	-	**108.62**	m	**116.77**
Height 2700 - 3000mm, 65mm thick boarded both sides	42.92	5.00	**45.07**	3.80	1.90	**74.35**	-	**119.41**	m	**128.37**
Angles to partitions; 38 x 25mm softwood batten, fixing together through partitions; each panel rebated by cutting back core and inner layers; joints filled with joint filler tape and joint finish										
Plain										
50mm thick partitions	0.29	5.00	**0.30**	0.25	-	**3.55**	-	**3.86**	m	**4.15**
65mm thick partitions	0.29	5.00	**0.30**	0.25	-	**3.55**	-	**3.86**	m	**4.15**
Tee junctions to partitions; forming groove by cutting back one face layer; bedding end in groove in bonding compound										
Plain										
50mm thick partitions	-	-	**-**	0.20	-	**2.84**	-	**2.84**	m	**3.06**
65mm thick partitions	-	-	**-**	0.20	-	**2.84**	-	**2.84**	m	**3.06**

Labour hourly rates: (except Specialists) Craft Operatives £14.21 Labourer £10.71 Rates are national average prices. Refer to REGIONAL VARIATIONS for indicative levels of overall pricing in regions	MATERIALS			LABOUR				RATES		
	Del to Site £	Waste %	Material Cost £	Craft Optve Hrs	Lab Hrs	Labour Cost £	Sunds £	Nett Rate £	Unit	Gross Rate (+7.5%) £
K10: PLASTERBOARD DRY LINING/PARTITIONS/CEILINGS: METAL STUD PARTITIONS										
Proprietary partitions; Gyproc Gypwall metal stud; tapered edge sheeting of one layer 12.5mm thick Gyproc wallboard for direct decoration, BS 1230, butt joints; applying continuous beads of Gyproc acoustical sealant to perimeters of construction both sides, fixing with Pozidriv head screws to steel frame comprising 50mm head and floor channels, 48mm vertical studs at 600mm centres; butt joints filled with joint filler tape and joint finish, spot filling; fixing to masonry with cartridge fired nails										
Height 2100 - 2400mm, 75mm thick boarded both sides	22.12	5.00	**23.22**	2.50	1.25	**48.91**	-	**72.14**	m	**77.55**
Proprietary partitions; Gyproc Gypwall metal stud; tapered edge sheeting of two layers 12.5mm thick Gyproc wallboard for direct decoration, BS 1230, butt joints; applying continuous beads of Gyproc acoustical sealant to perimeters of construction both sides, fixing with Pozidriv head screws to steel frame comprising 50mm head and floor channels, 48mm vertical studs at 600mm centres; butt joints filled with joint filler tape and joint finish, spot filling; fixing to masonry with cartridge fired nails										
Height 2100 - 2400mm, 100mm thick boarded both sides	31.06	5.00	**32.62**	2.80	1.40	**54.78**	-	**87.40**	m	**93.95**
Height 2400 - 2700mm, 100mm thick boarded both sides	34.05	5.00	**35.76**	3.15	1.58	**61.68**	-	**97.44**	m	**104.75**
Height 2700 - 3000mm, 100mm thick boarded both sides	37.05	5.00	**38.90**	3.50	1.75	**68.48**	-	**107.38**	m	**115.43**
Angles to partitions; fixing end studs together through partition; extending sheeting to one side across end of partition; joints filled with joint filler tape and joint finish										
Plain										
75mm thick partitions	1.28	10.00	**1.35**	0.10	-	**1.42**	-	**2.77**	m	**2.98**
100mm thick partitions	1.32	10.00	**1.39**	0.10	-	**1.42**	-	**2.81**	m	**3.03**
K10: PLASTERBOARD DRY LINING/PARTITIONS/CEILINGS: STAIRCASE AREAS										
Linings; tapered edge sheeting of one layer 9.5mm thick Gyproc wall board for direct decoration, BS 1230, butt joints; fixing with galvanized nails to timber base; butt joints filled with joint filler tape and joint finish, spot filling										
Walls										
height 2100 - 2400mm	4.69	5.00	**4.93**	0.98	0.42	**18.42**	-	**23.35**	m	**25.10**
height 2400 - 2700mm	5.28	5.00	**5.55**	1.11	0.48	**20.91**	-	**26.46**	m	**28.44**
height 2700 - 3000mm	5.80	5.00	**6.09**	1.23	0.53	**23.15**	-	**29.24**	m	**31.43**
Reveals and soffits of openings and recesses width not exceeding 300mm	0.64	10.00	**0.70**	0.20	0.05	**3.38**	-	**4.07**	m	**4.38**
Ceilings generally	1.97	5.00	**2.07**	0.44	0.18	**8.18**	-	**10.25**	m2	**11.02**
Linings; tapered edge sheeting of one layer 12.5mm thick Gyproc wallboard for direct decoration, BS 1230 butt joints; fixing with galvanized nails to timber base; butt joints filled with joint filler tape and joint finish, spot filling										
Walls										
height 2100 - 2400mm	4.93	5.00	**5.18**	1.15	0.48	**21.48**	-	**26.66**	m	**28.66**
height 2400 - 2700mm	5.55	5.00	**5.83**	1.30	0.54	**24.26**	-	**30.09**	m	**32.34**
height 2700 - 3000mm	6.17	5.00	**6.48**	1.44	0.60	**26.89**	-	**33.36**	m	**35.87**
Reveals and soffits of openings and recesses width not exceeding 300mm	0.67	10.00	**0.73**	0.23	0.06	**3.91**	-	**4.64**	m	**4.99**
Ceilings generally	2.07	5.00	**2.17**	0.51	0.20	**9.39**	-	**11.56**	m2	**12.43**
Linings; tapered edge sheeting of one layer 15mm thick Gyproc wallboard for direct decoration, BS 1230 butt joints; fixing with galvanized nails to timber base; butt joints filled with joint filler tape and joint finish, spot filling										
Walls										
height 2100 - 2400mm	7.41	5.00	**7.78**	1.24	0.54	**23.40**	-	**31.18**	m	**33.52**
height 2400 - 2700mm	8.33	5.00	**8.75**	1.39	0.61	**26.29**	-	**35.03**	m	**37.66**
height 2700 - 3000mm	9.18	5.00	**9.64**	1.55	0.68	**29.31**	-	**38.95**	m	**41.87**
Reveals and soffits of openings and recesses width not exceeding 300mm	0.98	10.00	**1.07**	0.25	0.06	**4.20**	-	**5.26**	m	**5.66**

Labour hourly rates: (except Specialists) Craft Operatives £14.21 Labourer £10.71 Rates are national average prices. Refer to REGIONAL VARIATIONS for indicative levels of overall pricing in regions	MATERIALS			LABOUR				RATES		
	Del to Site £	Waste %	Material Cost £	Craft Optve Hrs	Lab Hrs	Labour Cost £	Sunds £	Nett Rate £	Unit	Gross Rate (+7.5%) £
K10: PLASTERBOARD DRY LINING/PARTITIONS/CEILINGS: STAIRCASE AREAS Cont.										
Linings; tapered edge sheeting of one layer 15mm thick Gyproc wallboard for direct decoration, BS 1230 butt joints; fixing with galvanized nails to timber base; butt joints filled with joint filler tape and joint finish, spot filling Cont.										
Ceilings										
generally	3.10	5.00	**3.25**	0.57	0.22	**10.46**	-	**13.71**	m2	**14.74**
Linings; tapered edge sheeting of one layer 9.5mm thick Gyproc Duplex wallboard for direct decoration, BS 1230 butt joints; fixing with galvanized nails to timber base; butt joints filled with joint filler tape and joint finish, spot filling										
Walls										
height 2100 - 2400mm	6.01	10.00	**6.57**	0.98	0.42	**18.42**	-	**25.00**	m	**26.87**
height 2400 - 2700mm	6.77	5.00	**7.11**	1.11	0.48	**20.91**	-	**28.02**	m	**30.12**
height 2700 - 3000mm	7.44	5.00	**7.82**	1.23	0.53	**23.15**	-	**30.97**	m	**33.29**
Reveals and soffits of openings and recesses										
width not exceeding 300mm	0.80	10.00	**0.88**	0.20	0.05	**3.38**	-	**4.25**	m	**4.57**
Ceilings										
generally	2.52	5.00	**2.64**	0.44	0.18	**8.18**	-	**10.82**	m2	**11.64**
Linings; tapered edge sheeting of one layer 12.5mm thick Gyproc Duplex wallboard for direct decoration, BS 1230, butt joints; fixing with galvanized nails to timber base; butt joints filled with joint filler tape and joint finish, spot filling										
Walls										
height 2100 - 2400mm	6.33	5.00	**6.64**	1.15	0.48	**21.48**	-	**28.13**	m	**30.23**
height 2400 - 2700mm	7.12	5.00	**7.47**	1.30	0.54	**24.26**	-	**31.73**	m	**34.11**
height 2700 - 3000mm	7.91	5.00	**8.30**	1.44	0.60	**26.89**	-	**35.19**	m	**37.83**
Reveals and soffits of openings and recesses										
width not exceeding 300mm	0.84	10.00	**0.92**	0.23	0.06	**3.91**	-	**4.83**	m	**5.19**
Ceilings										
generally	2.65	5.00	**2.78**	0.51	0.20	**9.39**	-	**12.17**	m2	**13.08**
Linings; tapered edge sheeting of one layer 15mm thick Gyproc Duplex wallboard for direct decoration, BS 1230 butt joints; fixing with galvanized nails to timber base; butt joints filled with joint filler tape and joint finish, spot filling										
Walls										
height 2100 - 2400mm	9.64	5.00	**10.12**	1.24	0.54	**23.40**	-	**33.52**	m	**36.04**
height 2400 - 2700mm	10.84	5.00	**11.39**	1.39	0.61	**26.29**	-	**37.67**	m	**40.50**
height 2700 - 3000mm	11.94	5.00	**12.54**	1.55	0.68	**29.31**	-	**41.84**	m	**44.98**
Reveals and soffits of openings and recesses										
width not exceeding 300mm	1.26	10.00	**1.38**	0.25	0.06	**4.20**	-	**5.57**	m	**5.99**
Ceilings										
generally	4.03	5.00	**4.23**	0.57	0.22	**10.46**	-	**14.69**	m2	**15.79**
Linings; tapered edge sheeting of one layer 19mm thick Gyproc plank for direct decoration, BS 1230, butt joints; fixing with galvanized nails to timber base; butt joints filled with joint filler tape and joint finish, spot filling										
Walls										
height 2100 - 2400mm	10.45	5.00	**10.98**	1.51	0.60	**27.88**	-	**38.86**	m	**41.77**
height 2400 - 2700mm	11.76	5.00	**12.35**	1.70	0.68	**31.44**	-	**43.79**	m	**47.07**
height 2700 - 3000mm	13.00	5.00	**13.65**	1.89	0.75	**34.89**	-	**48.54**	m	**52.18**
Reveals and soffits of openings and recesses										
width not exceeding 300mm	1.35	10.00	**1.48**	0.34	0.08	**5.69**	-	**7.16**	m	**7.70**
Ceilings										
generally	4.37	5.00	**4.59**	0.68	0.25	**12.34**	-	**16.93**	m2	**18.20**
Linings; two layers of Gypsum wallboard, first layer square edge sheeting 9.5mm thick, second layer tapered edge sheeting 9.5mm thick for direct decoration, BS 1230, butt joints; fixing with galvanized nails to timber base; butt joints of second layer filled with joint filler tape and joint finish, spot filling										
Walls										
height 2100 - 2400mm	8.92	5.00	**9.37**	1.75	0.72	**32.58**	-	**41.95**	m	**45.09**
height 2400 - 2700mm	10.04	5.00	**10.54**	1.97	0.81	**36.67**	-	**47.21**	m	**50.75**
height 2700 - 3000mm	11.08	5.00	**11.63**	2.19	0.90	**40.76**	-	**52.39**	m	**56.32**
Reveals and soffits of openings and recesses										
width not exceeding 300mm	1.17	10.00	**1.27**	0.37	0.09	**6.22**	-	**7.50**	m	**8.06**

Labour hourly rates: (except Specialists) Craft Operatives £14.21 Labourer £10.71 Rates are national average prices. Refer to REGIONAL VARIATIONS for indicative levels of overall pricing in regions	MATERIALS			LABOUR				RATES		
	Del to Site £	Waste %	Material Cost £	Craft Optve Hrs	Lab Hrs	Labour Cost £	Sunds £	Nett Rate £	Unit	Gross Rate (+7.5%) £
K10: PLASTERBOARD DRY LINING/PARTITIONS/CEILINGS: STAIRCASE AREAS Cont.										
Linings; two layers of Gypsum wallboard, first layer square edge sheeting 9.5mm thick, second layer tapered edge sheeting 9.5mm thick for direct decoration, BS 1230, butt joints; fixing with galvanized nails to timber base; butt joints of second layer filled with joint filler tape and joint finish, spot filling Cont.										
Ceilings										
generally	3.73	5.00	**3.92**	0.78	0.30	**14.30**	-	**18.21**	m2	**19.58**
Linings; two layers of Gypsum wallboard, first layer square edge sheeting 9.5mm thick, second layer tapered edge sheeting 12.5mm thick for direct decoration, BS 1230, butt joints; fixing with galvanized nails to timber base; butt joints of second layer filled with joint filler tape and joint finish, spot filling										
Walls										
height 2100 - 2400mm	9.16	5.00	**9.62**	1.90	0.78	**35.35**	-	**44.97**	m	**48.35**
height 2400 - 2700mm	10.31	5.00	**10.82**	2.13	0.88	**39.69**	-	**50.52**	m	**54.31**
height 2700 - 3000mm	11.38	5.00	**11.95**	2.37	0.98	**44.17**	-	**56.12**	m	**60.33**
Reveals and soffits of openings and recesses										
width not exceeding 300mm	1.20	10.00	**1.28**	0.40	0.10	**6.76**	-	**8.04**	m	**8.64**
Ceilings										
generally	3.83	5.00	**4.02**	0.84	0.33	**15.47**	-	**19.49**	m2	**20.95**
Linings; two layers of Gypsum wallboard, first layer square edge sheeting 12.5mm thick, second layer tapered edge sheeting 12.5mm thick for direct decoration, BS 1230, butt joints; fixing with galvanized nails to timber base; butt joints of second layer filled with joint filler tape and joint finish, spot filling										
Walls										
height 2100 - 2400mm	9.40	5.00	**9.87**	2.04	0.84	**37.98**	-	**47.86**	m	**51.45**
height 2400 - 2700mm	10.58	5.00	**11.11**	2.30	0.95	**42.86**	-	**53.97**	m	**58.01**
height 2700 - 3000mm	11.68	5.00	**12.26**	2.55	1.05	**47.48**	-	**59.74**	m	**64.23**
Reveals and soffits of openings and recesses										
width not exceeding 300mm	1.23	10.00	**1.34**	0.44	0.11	**7.43**	-	**8.77**	m	**9.43**
Ceilings										
generally	3.93	5.00	**4.13**	0.91	0.35	**16.68**	-	**20.81**	m2	**22.37**
Linings; two layers of Gypsum wallboard, first layer square edge sheeting 15mm thick, second layer tapered edge sheeting 15mm thick for direct decoration, BS 1230, butt joints; fixing with galvanized nails to timber base; butt joints of second layer filled with joint filler tape and joint finish, spot filling										
Walls										
height 2100 - 2400mm	13.61	5.00	**14.29**	2.16	0.90	**40.33**	-	**54.62**	m	**58.72**
height 2400 - 2700mm	16.14	5.00	**16.95**	2.43	1.01	**45.35**	-	**62.30**	m	**66.97**
height 2700 - 3000mm	17.86	5.00	**18.75**	2.70	1.13	**50.47**	-	**69.22**	m	**74.41**
Reveals and soffits of openings and recesses										
width not exceeding 300mm	1.85	10.00	**2.02**	0.50	0.11	**8.28**	-	**10.30**	m	**11.08**
Ceilings										
generally	5.99	5.00	**6.29**	0.95	0.38	**17.57**	-	**23.86**	m2	**25.65**
Linings; two layers of Gypsum wallboard, first layer Duplex square edge sheeting 9.5mm thick, second layer tapered edge sheeting 9.5mm thick for direct decoration, BS 1230, butt joints; finish with galvanized nails to timber base; butt joint on second layer filled with joint filler tape and joint finish, spot filling										
Walls										
height 2100 - 2400mm	10.24	5.00	**10.76**	1.75	0.72	**32.58**	-	**43.33**	m	**46.58**
height 2400 - 2700mm	11.52	5.00	**12.10**	1.97	0.81	**36.67**	-	**48.77**	m	**52.43**
height 2700 - 3000mm	12.73	5.00	**13.37**	2.19	0.90	**40.76**	-	**54.12**	m	**58.18**
Reveals and soffits of openings and recesses										
width not exceeding 300mm	1.33	10.00	**1.42**	0.37	0.09	**6.22**	-	**7.65**	m	**8.22**
Ceilings										
generally	4.28	5.00	**4.49**	0.78	0.30	**14.30**	-	**18.79**	m2	**20.20**

Labour hourly rates: (except Specialists) Craft Operatives £14.21 Labourer £10.71 Rates are national average prices. Refer to REGIONAL VARIATIONS for indicative levels of overall pricing in regions	MATERIALS			LABOUR				RATES		
	Del to Site £	Waste %	Material Cost £	Craft Optve Hrs	Lab Hrs	Labour Cost £	Sunds £	Nett Rate £	Unit	Gross Rate (+7.5%) £
K10: PLASTERBOARD DRY LINING/PARTITIONS/CEILINGS: STAIRCASE AREAS Cont.										
Linings; two layers of Gypsum wallboard, first layer Duplex square edge sheeting 9.5mm thick, second layer tapered edge sheeting 12.5mm thick for direct decoration, BS 1230, butt joints; fixing with galvanized nails to timber base; butt joints of second layer filled with joint filler tape and joint finish, spot filling										
Walls										
height 2100 - 2400mm	10.48	5.00	**11.01**	1.90	0.78	**35.35**	-	**46.36**	m	**49.84**
height 2400 - 2700mm	11.79	5.00	**12.38**	2.13	0.88	**39.69**	-	**52.08**	m	**55.98**
height 2700 - 3000mm	13.03	5.00	**13.68**	2.37	0.98	**44.17**	-	**57.85**	m	**62.19**
Reveals and soffits of openings and recesses										
width not exceeding 300mm	1.36	10.00	**1.46**	0.40	0.10	**6.76**	-	**8.21**	m	**8.83**
Ceilings										
generally	4.38	5.00	**4.60**	0.84	0.33	**15.47**	-	**20.07**	m2	**21.58**
Linings; two layers of Gypsum wallboard, first layer Duplex square edge sheeting 12.5mm thick, second layer tapered edge sheeting 12.5mm thick for direct decoration, BS 1230, butt joints; fixing with galvanized nails to timber base; butt joints of second layer filled with joint filler tape and joint finish, spot filling										
Walls										
height 2100 - 2400mm	10.80	5.00	**11.34**	2.04	0.84	**37.98**	-	**49.32**	m	**53.02**
height 2400 - 2700mm	12.15	5.00	**12.75**	2.30	0.95	**42.86**	-	**55.61**	m	**59.78**
height 2700 - 3000mm	13.42	5.00	**14.09**	2.55	1.05	**47.48**	-	**61.57**	m	**66.19**
Reveals and soffits of openings and recesses										
width not exceeding 300mm	1.40	10.00	**1.50**	0.44	0.11	**7.43**	-	**8.93**	m	**9.60**
Ceilings										
generally	4.51	5.00	**4.74**	0.91	0.35	**16.68**	-	**21.42**	m2	**23.02**
Linings; two layers of Gypsum wallboard, first layer Duplex square edge sheeting 15mm thick, second layer tapered edge sheeting 15mm thick for direct decoration, BS 1230, butt joints; fixing with galvanized nails to timber base; butt joints of second layer filled with joint filler tape and joint finish, spot filling										
Walls										
height 2100 - 2400mm	15.84	5.00	**16.63**	2.16	0.90	**40.33**	-	**56.96**	m	**61.24**
height 2400 - 2700mm	18.65	5.00	**19.58**	2.43	1.01	**45.35**	-	**64.93**	m	**69.80**
height 2700 - 3000mm	20.65	5.00	**21.68**	2.70	1.13	**50.47**	-	**72.15**	m	**77.56**
Reveals and soffits of openings and recesses										
width not exceeding 300mm	2.13	10.00	**2.27**	0.50	0.11	**8.28**	-	**10.56**	m	**11.35**
Ceilings										
generally	6.92	5.00	**7.27**	0.95	0.38	**17.57**	-	**24.84**	m2	**26.70**
K10: PLASTERBOARD DRY LINING/PARTITIONS/CEILING: STAIRCASE AREAS: PLASTIC FACED										
Linings; square edge sheeting of one layer 12.5mm thick Gyproc Industrial grade plastic faced plasterboard, BS 1230, butt joints; fixing with galvanized nails to timber base; butt joints sealed with self adhesive white tape										
Walls										
height 2100 - 2400mm	9.08	5.00	**9.58**	0.85	0.35	**15.83**	-	**25.41**	m	**27.31**
height 2400 - 2700mm	10.22	5.00	**10.78**	0.97	0.39	**17.96**	-	**28.74**	m	**30.90**
height 2700 - 3000mm	11.36	5.00	**11.98**	1.07	0.44	**19.92**	-	**31.90**	m	**34.29**
Reveals and soffits of openings and recesses										
width not exceeding 300mm	1.66	10.00	**1.82**	0.21	0.05	**3.52**	-	**5.34**	m	**5.74**
Ceilings										
generally	3.86	5.00	**4.07**	0.39	0.15	**7.15**	-	**11.22**	m2	**12.06**
K11: RIGID SHEET FLOORING/SHEATHING/LININGS/CASINGS										
Douglas Fir plywood, unsanded select sheathing quality, WBP bonded; tongued and grooved joints long edges										
15mm thick sheeting to floors										
width exceeding 300mm	6.23	7.50	**6.70**	0.29	0.04	**4.55**	0.03	**11.28**	m2	**12.12**
18mm thick sheeting to floors										
width exceeding 300mm	7.89	7.50	**8.49**	0.30	0.04	**4.69**	0.03	**13.21**	m2	**14.20**

Labour hourly rates: (except Specialists) Craft Operatives £14.21 Labourer £10.71 Rates are national average prices. Refer to REGIONAL VARIATIONS for indicative levels of overall pricing in regions	MATERIALS			LABOUR				RATES		
	Del to Site £	Waste %	Material Cost £	Craft Optve Hrs	Lab Hrs	Labour Cost £	Sunds £	Nett Rate £	Unit	Gross Rate (+7.5%) £
K11: RIGID SHEET FLOORING/SHEATHING/LININGS/CASINGS Cont.										
Birch faced plywood, BB quality, WBP bonded; tongued and grooved joints all edges										
12mm thick sheeting to floors										
width exceeding 300mm	18.71	7.50	**20.11**	0.27	0.04	**4.27**	0.03	**24.41**	m2	**26.24**
15mm thick sheeting to floors										
width exceeding 300mm	22.46	7.50	**24.15**	0.29	0.04	**4.55**	0.03	**28.73**	m2	**30.88**
18mm thick sheeting to floors										
width exceeding 300mm	27.24	7.50	**29.29**	0.30	0.04	**4.69**	0.03	**34.01**	m2	**36.56**
Wood chipboard, BS EN 312, type II flooring, square edges; butt joints										
18mm thick sheeting to floors										
width exceeding 300mm	2.45	7.50	**2.63**	0.30	0.04	**4.69**	0.03	**7.36**	m2	**7.91**
22mm thick sheeting to floors										
width exceeding 300mm	2.91	7.50	**3.13**	0.30	0.04	**4.69**	0.03	**7.85**	m2	**8.44**
Wood chipboard, BS EN 312, type II flooring; tongued and grooved joints										
18mm thick sheeting to floors										
width exceeding 300mm	3.85	7.50	**4.14**	0.30	0.04	**4.69**	0.03	**8.86**	m2	**9.53**
22mm thick sheeting to floors										
width exceeding 300mm	5.28	7.50	**5.68**	0.30	0.04	**4.69**	0.03	**10.40**	m2	**11.18**
Abutments										
12mm thick plywood linings										
raking cutting	1.87	7.50	**2.01**	0.03	-	**0.43**	-	**2.44**	m	**2.62**
curved cutting	2.81	7.50	**3.02**	0.05	-	**0.71**	-	**3.73**	m	**4.01**
15mm thick plywood linings										
raking cutting	2.25	7.50	**2.41**	0.04	-	**0.57**	-	**2.98**	m	**3.21**
curved cutting	3.37	7.50	**3.62**	0.06	-	**0.85**	-	**4.47**	m	**4.81**
18mm thick plywood linings										
raking cutting	2.72	7.50	**2.93**	0.04	-	**0.57**	-	**3.50**	m	**3.76**
curved cutting	4.09	7.50	**4.39**	0.06	-	**0.85**	-	**5.25**	m	**5.64**
18mm thick wood chipboard linings										
raking cutting	0.25	7.50	**0.26**	0.03	-	**0.43**	-	**0.69**	m	**0.74**
curved cutting	0.37	7.50	**0.40**	0.05	-	**0.71**	-	**1.11**	m	**1.19**
22mm thick wood chipboard linings										
raking cutting	0.29	7.50	**0.31**	0.04	-	**0.57**	-	**0.88**	m	**0.95**
curved cutting	0.44	7.50	**0.47**	0.06	-	**0.85**	-	**1.32**	m	**1.42**
K11: RIGID SHEET FLOORING/SHEATHING/LININGS/CASINGS: SOUND INSULATED FLOORS										
Foam backed softwood battens covered with tongued and grooved chipboard or plywood panels										
19mm chipboard panels to level sub-floors										
width exceeding 300mm; battens at 450mm centres; elevation										
50mm	7.62	10.00	**8.29**	0.27	0.03	**4.16**	0.09	**12.54**	m2	**13.48**
75mm	8.21	10.00	**8.91**	0.27	0.03	**4.16**	0.09	**13.16**	m2	**14.15**
width exceeding 300mm; battens at 600mm centres; elevation										
50mm	7.18	10.00	**7.83**	0.27	0.03	**4.16**	0.07	**12.06**	m2	**12.96**
75mm	7.61	10.00	**8.29**	0.27	0.03	**4.16**	0.08	**12.52**	m2	**13.46**
19mm chipboard panels with improved moisture resistance to level sub-floors										
width exceeding 300mm; battens at 450mm centres; elevation										
50mm	4.68	10.00	**5.06**	0.27	0.03	**4.16**	0.09	**9.30**	m2	**10.00**
75mm	5.26	10.00	**5.67**	0.27	0.03	**4.16**	0.09	**9.92**	m2	**10.67**
width exceeding 300mm; battens at 600mm centres; elevation										
50mm	4.23	10.00	**4.59**	0.27	0.03	**4.16**	0.07	**8.82**	m2	**9.48**
75mm	4.67	10.00	**5.05**	0.27	0.03	**4.16**	0.08	**9.29**	m2	**9.98**
19mm plywood panels to level sub-floors										
width exceeding 300mm; battens at 450mm centres; elevation										
50mm	9.65	10.00	**10.53**	0.27	0.03	**4.16**	0.09	**14.77**	m2	**15.88**
75mm	10.24	10.00	**11.15**	0.27	0.03	**4.16**	0.09	**15.40**	m2	**16.55**
K11: RIGID SHEET FLOORING/SHEATHING/LININGS/CASINGS: FIBREBOARDS AND HARDBOARDS										
Standard hardboard, BS EN 622; butt joints										
3.2mm thick linings to walls										
width exceeding 300mm	1.25	5.00	**1.31**	0.33	0.04	**5.12**	0.11	**6.54**	m2	**7.03**
width not exceeding 300mm	0.37	10.00	**0.41**	0.20	0.01	**2.95**	0.07	**3.43**	m	**3.68**

Labour hourly rates: (except Specialists) Craft Operatives £14.21 Labourer £10.71 Rates are national average prices. Refer to REGIONAL VARIATIONS for indicative levels of overall pricing in regions	MATERIALS			LABOUR				RATES		
	Del to Site £	Waste %	Material Cost £	Craft Optve Hrs	Lab Hrs	Labour Cost £	Sunds £	Nett Rate £	Unit	Gross Rate (+7.5%) £
K11: RIGID SHEET FLOORING/SHEATHING/LININGS/CASINGS: FIBREBOARDS AND HARDBOARDS Cont.										
Standard hardboard, BS EN 622; butt joints Cont.										
4.8mm thick linings to walls										
width exceeding 300mm	2.27	5.00	**2.38**	0.34	0.04	**5.26**	0.11	**7.76**	m2	**8.34**
width not exceeding 300mm	0.68	10.00	**0.75**	0.20	0.01	**2.95**	0.07	**3.76**	m	**4.05**
6.4mm thick linings to walls										
width exceeding 300mm	2.84	5.00	**2.98**	0.35	0.04	**5.40**	0.12	**8.50**	m2	**9.14**
width not exceeding 300mm	0.85	10.00	**0.94**	0.21	0.01	**3.09**	0.07	**4.10**	m	**4.40**
3.2mm thick sheeting to floors										
width exceeding 300mm	1.25	5.00	**1.31**	0.17	0.02	**2.63**	0.06	**4.00**	m2	**4.30**
3.2mm thick linings to ceilings										
width exceeding 300mm	1.25	5.00	**1.31**	0.41	0.05	**6.36**	0.14	**7.81**	m2	**8.39**
width not exceeding 300mm	0.37	10.00	**0.41**	0.25	0.02	**3.77**	0.08	**4.26**	m	**4.58**
4.8mm thick linings to ceilings										
width exceeding 300mm	2.27	5.00	**2.38**	0.43	0.05	**6.65**	0.14	**9.17**	m2	**9.86**
width not exceeding 300mm	0.68	10.00	**0.75**	0.26	0.02	**3.91**	0.09	**4.74**	m	**5.10**
6.4mm thick linings to ceilings										
width exceeding 300mm	2.84	5.00	**2.98**	0.44	0.05	**6.79**	0.15	**9.91**	m2	**10.66**
width not exceeding 300mm	0.85	10.00	**0.94**	0.26	0.02	**3.91**	0.09	**4.93**	m	**5.30**
Tempered hardboard, BS EN 622; butt joints										
3.2mm thick linings to walls										
width exceeding 300mm	1.79	5.00	**1.88**	0.33	0.04	**5.12**	0.11	**7.11**	m2	**7.64**
width not exceeding 300mm	0.54	10.00	**0.59**	0.20	0.01	**2.95**	0.07	**3.61**	m	**3.88**
4.8mm thick linings to walls										
width exceeding 300mm	2.61	5.00	**2.74**	0.34	0.04	**5.26**	0.11	**8.11**	m2	**8.72**
width not exceeding 300mm	0.78	10.00	**0.86**	0.20	0.01	**2.95**	0.07	**3.88**	m	**4.17**
6.4mm thick linings to walls										
width exceeding 300mm	3.30	5.00	**3.47**	0.35	0.04	**5.40**	0.12	**8.99**	m2	**9.66**
width not exceeding 300mm	0.99	10.00	**1.09**	0.21	0.01	**3.09**	0.07	**4.25**	m	**4.57**
3.2mm thick linings to ceilings										
width exceeding 300mm	1.79	5.00	**1.88**	0.41	0.05	**6.36**	0.14	**8.38**	m2	**9.01**
width not exceeding 300mm	0.54	10.00	**0.59**	0.25	0.02	**3.77**	0.08	**4.44**	m	**4.77**
4.8mm thick linings to ceilings										
width exceeding 300mm	2.61	5.00	**2.74**	0.43	0.05	**6.65**	0.14	**9.53**	m2	**10.24**
width not exceeding 300mm	0.78	10.00	**0.86**	0.26	0.02	**3.91**	0.09	**4.86**	m	**5.22**
6.4mm thick linings to ceilings										
width exceeding 300mm	3.30	5.00	**3.47**	0.44	0.05	**6.79**	0.15	**10.40**	m2	**11.18**
width not exceeding 300mm	0.99	10.00	**1.09**	0.26	0.02	**3.91**	0.09	**5.09**	m	**5.47**
Flame retardant hardboard, BS EN 622 tested to BS 476 Part 7 Class 1; butt joints										
3.2mm thick linings to walls										
width exceeding 300mm	2.84	5.00	**2.98**	0.33	0.04	**5.12**	0.11	**8.21**	m2	**8.82**
width not exceeding 300mm	0.85	10.00	**0.94**	0.20	0.01	**2.95**	0.07	**3.95**	m	**4.25**
4.8mm thick linings to walls										
width exceeding 300mm	5.68	5.00	**5.96**	0.34	0.04	**5.26**	0.11	**11.33**	m2	**12.18**
width not exceeding 300mm	1.70	10.00	**1.87**	0.20	0.01	**2.95**	0.07	**4.89**	m	**5.26**
6.4mm thick linings to walls										
width exceeding 300mm	6.25	5.00	**6.56**	0.35	0.04	**5.40**	0.12	**12.08**	m2	**12.98**
width not exceeding 300mm	1.87	10.00	**2.06**	0.21	0.01	**3.09**	0.07	**5.22**	m	**5.61**
3.2mm thick linings to ceilings										
width exceeding 300mm	2.84	5.00	**2.98**	0.41	0.05	**6.36**	0.14	**9.48**	m2	**10.19**
width not exceeding 300mm	0.85	10.00	**0.94**	0.25	0.02	**3.77**	0.08	**4.79**	m	**5.15**
4.8mm thick linings to ceilings										
width exceeding 300mm	5.68	5.00	**5.96**	0.43	0.05	**6.65**	0.14	**12.75**	m2	**13.71**
width not exceeding 300mm	1.70	10.00	**1.87**	0.26	0.02	**3.91**	0.09	**5.87**	m	**6.31**
6.4mm thick linings to ceilings										
width exceeding 300mm	6.25	5.00	**6.56**	0.44	0.06	**6.90**	0.15	**13.60**	m2	**14.62**
width not exceeding 300mm	1.87	10.00	**2.06**	0.26	0.02	**3.91**	0.09	**6.06**	m	**6.51**
Low density medium board, BS EN 622, type LME; butt joints										
6mm thick linings to walls										
width exceeding 300mm	7.81	5.00	**8.20**	0.35	0.04	**5.40**	0.12	**13.71**	m2	**14.74**
width not exceeding 300mm	2.34	10.00	**2.58**	0.21	0.01	**3.09**	0.07	**5.74**	m	**6.17**
9mm thick linings to walls										
width exceeding 300mm	9.99	5.00	**10.49**	0.36	0.05	**5.65**	0.12	**16.26**	m2	**17.48**
width not exceeding 300mm	3.00	10.00	**3.30**	0.22	0.02	**3.34**	0.07	**6.71**	m	**7.21**
12mm thick linings to walls										
width exceeding 300mm	12.74	5.00	**13.38**	0.38	0.05	**5.94**	0.13	**19.44**	m2	**20.90**
width not exceeding 300mm	3.82	10.00	**4.20**	0.23	0.02	**3.48**	0.08	**7.76**	m	**8.35**

Labour hourly rates: (except Specialists) Craft Operatives £14.21 Labourer £10.71 Rates are national average prices. Refer to REGIONAL VARIATIONS for indicative levels of overall pricing in regions	MATERIALS			LABOUR				RATES		
	Del to Site £	Waste %	Material Cost £	Craft Optve Hrs	Lab Hrs	Labour Cost £	Sunds £	Nett Rate £	Unit	Gross Rate (+7.5%) £
K11: RIGID SHEET FLOORING/SHEATHING/LININGS/CASINGS: FIBREBOARDS AND HARDBOARDS Cont.										
Low density medium board, BS EN 622, type LME; butt joints Cont.										
6mm thick linings to ceilings										
width exceeding 300mm	7.81	5.00	**8.20**	0.44	0.06	**6.90**	0.15	**15.24**	m2	**16.38**
width not exceeding 300mm	2.34	10.00	**2.58**	0.26	0.02	**3.91**	0.09	**6.57**	m	**7.06**
9mm thick linings to ceilings										
width exceeding 300mm	9.99	5.00	**10.49**	0.45	0.06	**7.04**	0.15	**17.68**	m2	**19.00**
width not exceeding 300mm	3.00	10.00	**3.30**	0.27	0.02	**4.05**	0.09	**7.44**	m	**8.00**
12mm thick linings to ceilings										
width exceeding 300mm	12.74	5.00	**13.38**	0.48	0.06	**7.46**	0.16	**21.00**	m2	**22.58**
width not exceeding 300mm	3.82	10.00	**4.20**	0.29	0.02	**4.34**	0.10	**8.64**	m	**9.28**
Low density medium board, BS EN 622, type LMN; butt joints										
6mm thick linings to walls										
width exceeding 300mm	6.68	5.00	**7.02**	0.35	0.04	**5.40**	0.12	**12.53**	m2	**13.47**
width not exceeding 300mm	2.00	10.00	**2.21**	0.21	0.01	**3.09**	0.07	**5.37**	m	**5.77**
12mm thick linings to walls										
width exceeding 300mm	8.31	5.00	**8.72**	0.36	0.05	**5.65**	0.12	**14.49**	m2	**15.58**
width not exceeding 300mm	2.49	10.00	**2.74**	0.22	0.02	**3.34**	0.07	**6.15**	m	**6.62**
6mm thick linings to ceilings										
width exceeding 300mm	6.68	5.00	**7.02**	0.44	0.06	**6.90**	0.15	**14.06**	m2	**15.11**
width not exceeding 300mm	2.00	10.00	**2.21**	0.26	0.02	**3.91**	0.09	**6.20**	m	**6.66**
12mm thick linings to ceilings										
width exceeding 300mm	8.31	5.00	**8.72**	0.45	0.06	**7.04**	0.15	**15.91**	m2	**17.10**
width not exceeding 300mm	2.49	10.00	**2.74**	0.27	0.02	**4.05**	0.09	**6.88**	m	**7.40**
Medium density fibreboard, BS EN 622, type MDF; butt joints										
9mm thick linings to walls										
width exceeding 300mm	3.28	5.00	**3.44**	0.35	0.04	**5.40**	0.12	**8.96**	m2	**9.63**
width not exceeding 300mm	0.98	10.00	**1.08**	0.21	0.01	**3.09**	0.07	**4.24**	m	**4.56**
12mm thick linings to walls										
width exceeding 300mm	4.03	5.00	**4.23**	0.38	0.05	**5.94**	0.13	**10.29**	m2	**11.06**
width not exceeding 300mm	1.21	10.00	**1.33**	0.23	0.02	**3.48**	0.08	**4.89**	m	**5.25**
18mm thick linings to walls										
width exceeding 300mm	5.29	5.00	**5.55**	0.41	0.05	**6.36**	0.14	**12.05**	m2	**12.96**
width not exceeding 300mm	1.59	10.00	**1.75**	0.24	0.02	**3.62**	0.08	**5.45**	m	**5.86**
25mm thick linings to walls										
width exceeding 300mm	7.31	5.00	**7.68**	0.44	0.06	**6.90**	0.15	**14.72**	m2	**15.82**
width not exceeding 300mm	2.19	10.00	**2.41**	0.26	0.02	**3.91**	0.09	**6.41**	m	**6.89**
9mm thick linings to ceilings										
width exceeding 300mm	3.28	5.00	**3.44**	0.44	0.06	**6.90**	0.15	**10.48**	m2	**11.27**
width not exceeding 300mm	0.98	10.00	**1.08**	0.26	0.02	**3.91**	0.09	**5.08**	m	**5.46**
12mm thick linings to ceilings										
width exceeding 300mm	4.03	5.00	**4.23**	0.48	0.06	**7.46**	0.16	**11.85**	m2	**12.74**
width not exceeding 300mm	1.21	10.00	**1.33**	0.29	0.02	**4.34**	0.10	**5.76**	m	**6.19**
18mm thick linings to ceilings										
width exceeding 300mm	5.29	5.00	**5.55**	0.52	0.07	**8.14**	0.17	**13.87**	m2	**14.90**
width not exceeding 300mm	1.59	10.00	**1.75**	0.31	0.02	**4.62**	0.10	**6.47**	m	**6.95**
25mm thick linings to ceilings										
width exceeding 300mm	7.31	5.00	**7.68**	0.56	0.07	**8.71**	0.18	**16.57**	m2	**17.81**
width not exceeding 300mm	2.19	10.00	**2.41**	0.34	0.02	**5.05**	0.11	**7.57**	m	**8.14**
Insulating softboard, BS EN 622; butt joints										
13mm thick linings to walls										
width exceeding 300mm	1.95	5.00	**2.04**	0.38	0.05	**5.94**	0.13	**8.10**	m2	**8.71**
width not exceeding 300mm	0.58	10.00	**0.64**	0.23	0.02	**3.48**	0.08	**4.20**	m	**4.52**
13mm thick linings to ceilings										
width exceeding 300mm	1.95	5.00	**2.04**	0.48	0.06	**7.46**	0.16	**9.67**	m2	**10.39**
width not exceeding 300mm	0.58	10.00	**0.64**	0.29	0.02	**4.34**	0.10	**5.07**	m	**5.45**
Bitumen impregnated insulating board, BS EN 622 ; butt joints										
15mm thick linings to walls										
width exceeding 300mm	2.25	5.00	**2.36**	0.38	0.05	**5.94**	0.13	**8.42**	m2	**9.05**
width not exceeding 300mm	0.67	10.00	**0.74**	0.23	0.02	**3.48**	0.08	**4.30**	m	**4.62**
15mm thick linings to ceilings										
width exceeding 300mm	2.25	5.00	**2.36**	0.48	0.06	**7.46**	0.16	**9.98**	m2	**10.73**
width not exceeding 300mm	0.67	10.00	**0.74**	0.29	0.02	**4.34**	0.10	**5.17**	m	**5.56**

Labour hourly rates: (except Specialists) Craft Operatives £14.21 Labourer £10.71 Rates are national average prices. Refer to REGIONAL VARIATIONS for indicative levels of overall pricing in regions	MATERIALS			LABOUR				RATES		
	Del to Site £	Waste %	Material Cost £	Craft Optve Hrs	Lab Hrs	Labour Cost £	Sunds £	Nett Rate £	Unit	Gross Rate (+7.5%) £
K11: RIGID SHEET FLOORING/SHEATHING/LININGS/CASINGS: ASBESTOS FREE BOARD										
Masterboard Class 0 fire resisting boards; butt joints										
6mm thick linings to walls										
width exceeding 300mm	10.28	5.00	**10.79**	0.45	0.06	**7.04**	0.15	**17.98**	m2	**19.33**
width not exceeding 300mm........................	3.08	10.00	**3.39**	0.27	0.02	**4.05**	0.09	**7.53**	m	**8.10**
9mm thick linings to walls										
width exceeding 300mm	21.72	5.00	**22.81**	0.49	0.06	**7.61**	0.16	**30.57**	m2	**32.87**
width not exceeding 300mm........................	6.52	10.00	**7.17**	0.29	0.02	**4.34**	0.10	**11.60**	m	**12.47**
12mm thick linings to walls										
width exceeding 300mm	28.64	5.00	**30.07**	0.53	0.07	**8.28**	0.17	**38.53**	m2	**41.42**
width not exceeding 300mm........................	8.59	10.00	**9.45**	0.32	0.02	**4.76**	0.11	**14.32**	m	**15.39**
6mm thick linings to ceilings										
width exceeding 300mm	10.28	5.00	**10.79**	0.57	0.07	**8.85**	0.19	**19.83**	m2	**21.32**
width not exceeding 300mm........................	3.08	10.00	**3.39**	0.34	0.02	**5.05**	0.11	**8.55**	m	**9.19**
9mm thick linings to ceilings										
width exceeding 300mm	21.72	5.00	**22.81**	0.61	0.08	**9.52**	0.20	**32.53**	m2	**34.97**
width not exceeding 300mm........................	6.52	10.00	**7.17**	0.37	0.02	**5.47**	0.12	**12.76**	m	**13.72**
12mm thick linings to ceilings										
width exceeding 300mm	28.64	5.00	**30.07**	0.67	0.08	**10.38**	0.22	**40.67**	m2	**43.72**
width not exceeding 300mm........................	8.59	10.00	**9.45**	0.40	0.02	**5.90**	0.13	**15.48**	m	**16.64**
Supalux fire resisting boards, Sanded; butt joints; fixing to timber with screws, countersinking										
6mm thick linings to walls										
width exceeding 300mm	14.24	5.00	**14.95**	0.45	0.06	**7.04**	0.15	**22.14**	m2	**23.80**
width not exceeding 300mm........................	4.27	10.00	**4.70**	0.27	0.02	**4.05**	0.09	**8.84**	m	**9.50**
9mm thick linings to walls										
width exceeding 300mm	24.24	5.00	**25.45**	0.49	0.06	**7.61**	0.16	**33.22**	m2	**35.71**
width not exceeding 300mm........................	7.27	10.00	**8.00**	0.29	0.02	**4.34**	0.10	**12.43**	m	**13.36**
12mm thick linings to walls										
width exceeding 300mm	32.08	5.00	**33.68**	0.53	0.07	**8.28**	0.17	**42.14**	m2	**45.30**
width not exceeding 300mm........................	9.62	10.00	**10.59**	0.32	0.02	**4.76**	0.11	**15.45**	m	**16.61**
6mm thick linings to ceilings										
width exceeding 300mm	14.24	5.00	**14.95**	0.57	0.07	**8.85**	0.19	**23.99**	m2	**25.79**
width not exceeding 300mm........................	4.27	10.00	**4.70**	0.34	0.02	**5.05**	0.11	**9.86**	m	**10.60**
9mm thick linings to ceilings										
width exceeding 300mm	24.24	5.00	**25.45**	0.61	0.08	**9.52**	0.20	**35.18**	m2	**37.82**
width not exceeding 300mm........................	7.27	10.00	**8.00**	0.37	0.02	**5.47**	0.12	**13.59**	m	**14.61**
12mm thick linings to ceilings										
width exceeding 300mm	32.08	5.00	**33.68**	0.67	0.08	**10.38**	0.22	**44.28**	m2	**47.60**
width not exceeding 300mm........................	9.62	10.00	**10.59**	0.40	0.02	**5.90**	0.13	**16.62**	m	**17.86**
6mm thick casings to isolated beams or the like										
total girth not exceeding 600mm.................	14.24	10.00	**15.66**	1.13	0.14	**17.56**	0.37	**33.59**	m2	**36.11**
total girth 600 - 1200mm.........................	14.24	5.00	**14.95**	0.68	0.09	**10.63**	0.22	**25.80**	m2	**27.74**
9mm thick casings to isolated beams or the like										
total girth not exceeding 600mm.................	24.24	10.00	**26.66**	1.21	0.15	**18.80**	0.40	**45.86**	m2	**49.30**
total girth 600 - 1200mm.........................	24.24	5.00	**25.45**	0.74	0.09	**11.48**	0.24	**37.18**	m2	**39.96**
12mm thick casings to isolated beams or the like										
total girth not exceeding 600mm.................	32.08	10.00	**35.29**	1.33	0.17	**20.72**	0.44	**56.45**	m2	**60.68**
total girth 600 - 1200mm.........................	32.08	5.00	**33.68**	0.80	0.10	**12.44**	0.26	**46.39**	m2	**49.87**
6mm thick casings to isolated columns or the like										
total girth 600 - 1200mm.........................	14.24	10.00	**15.66**	1.13	0.14	**17.56**	0.37	**33.59**	m2	**36.11**
total girth 1200 - 1800mm........................	14.24	5.00	**14.95**	0.68	0.09	**10.63**	0.22	**25.80**	m2	**27.74**
9mm thick casings to isolated columns or the like										
total girth 600 - 1200mm.........................	24.24	10.00	**26.66**	1.21	0.15	**18.80**	0.40	**45.86**	m2	**49.30**
total girth 1200 - 1800mm........................	24.24	5.00	**25.45**	0.74	0.09	**11.48**	0.24	**37.18**	m2	**39.96**
12mm thick casings to isolated columns or the like										
total girth 600 - 1200mm.........................	32.08	10.00	**35.29**	1.33	0.17	**20.72**	0.44	**56.45**	m2	**60.68**
total girth 1200 - 1800mm........................	32.08	5.00	**33.68**	0.80	0.10	**12.44**	0.26	**46.39**	m2	**49.87**
Vermiculux fire resisting boards; butt joints; fixing to steel with self tapping screws, countersinking										
20mm thick casings to isolated beams or the like										
total girth not exceeding 600mm.................	21.96	10.00	**24.15**	1.50	0.19	**23.35**	0.50	**48.00**	m2	**51.60**
total girth 600 - 1200mm.........................	21.96	5.00	**23.06**	0.90	0.11	**13.97**	0.30	**37.32**	m2	**40.12**
30mm thick casings to isolated beams or the like										
total girth not exceeding 600mm.................	35.91	10.00	**39.50**	1.65	0.21	**25.70**	0.54	**65.74**	m2	**70.67**
total girth 600 - 1200mm.........................	35.93	5.00	**37.72**	1.00	0.13	**15.60**	0.33	**53.65**	m2	**57.68**
40mm thick casings to isolated beams or the like										
total girth not exceeding 600mm.................	54.45	10.00	**59.90**	1.80	0.23	**28.04**	0.59	**88.53**	m2	**95.17**
total girth 600 - 1200mm.........................	54.45	5.00	**57.17**	1.08	0.14	**16.85**	0.36	**74.38**	m2	**79.95**

Labour hourly rates: (except Specialists) Craft Operatives £14.21 Labourer £10.71 Rates are national average prices. Refer to REGIONAL VARIATIONS for indicative levels of overall pricing in regions	MATERIALS			LABOUR				RATES		
	Del to Site £	Waste %	Material Cost £	Craft Optve Hrs	Lab Hrs	Labour Cost £	Sunds £	Nett Rate £	Unit	Gross Rate (+7.5%) £
K11: RIGID SHEET FLOORING/SHEATHING/LININGS/CASINGS: ASBESTOS FREE BOARD Cont.										
Vermiculux fire resisting boards; butt joints; fixing to steel with self tapping screws, countersinking Cont.										
50mm thick casings to isolated beams or the like										
total girth not exceeding 600mm	72.51	10.00	**79.76**	1.95	0.24	**30.28**	0.64	**110.68**	m2	**118.98**
total girth 600 - 1200mm	72.51	5.00	**76.13**	1.17	0.15	**18.23**	0.39	**94.75**	m2	**101.86**
60mm thick casings to isolated beams or the like										
total girth not exceeding 600mm	86.46	10.00	**95.11**	2.10	0.26	**32.63**	0.69	**128.43**	m2	**138.06**
total girth 600 - 1200mm	86.46	5.00	**90.79**	1.26	0.16	**19.62**	0.42	**110.82**	m2	**119.13**
20mm thick casings to isolated columns or the like										
total girth 600 - 1200mm	21.96	10.00	**24.15**	1.50	0.19	**23.35**	0.50	**48.00**	m2	**51.60**
total girth 1200 - 1800mm	21.96	5.00	**23.06**	0.90	0.11	**13.97**	0.30	**37.32**	m2	**40.12**
30mm thick casings to isolated columns or the like										
total girth 600 - 1200mm	35.91	10.00	**39.50**	1.65	0.21	**25.70**	0.54	**65.74**	m2	**70.67**
total girth 1200 - 1800mm	35.93	5.00	**37.72**	1.00	0.13	**15.60**	0.33	**53.65**	m2	**57.68**
40mm thick casings to isolated columns or the like										
total girth 600 - 1200mm	54.45	10.00	**59.90**	1.80	0.23	**28.04**	0.59	**88.53**	m2	**95.17**
total girth 1200 - 1800mm	54.45	5.00	**57.17**	1.08	0.14	**16.85**	0.36	**74.38**	m2	**79.95**
50mm thick casings to isolated columns or the like										
total girth 600 - 1200mm	72.51	10.00	**79.76**	1.95	0.24	**30.28**	0.64	**110.68**	m2	**118.98**
total girth 1200 - 1800mm	72.51	5.00	**76.13**	1.17	0.15	**18.23**	0.39	**94.75**	m2	**101.86**
60mm thick casings to isolated columns or the like										
total girth 600 - 1200mm	86.46	10.00	**95.11**	2.10	0.26	**32.63**	0.69	**128.43**	m2	**138.06**
total girth 1200 - 1800mm	86.46	5.00	**90.79**	1.26	0.16	**19.62**	0.42	**110.82**	m2	**119.13**
50 x 25mm x 20g mild steel fixing angle										
fixing to masonry	1.71	5.00	**1.80**	0.40	0.05	**6.22**	0.13	**8.15**	m	**8.76**
fixing to steel	1.71	5.00	**1.80**	0.40	0.05	**6.22**	0.13	**8.15**	m	**8.76**
Tacboard Class 0 fire resisting boards; butt joints										
6mm thick linings to walls										
width exceeding 300mm	8.85	5.00	**9.29**	0.45	0.06	**7.04**	0.15	**16.48**	m2	**17.72**
width not exceeding 300mm	2.66	10.00	**2.92**	0.27	0.02	**4.05**	0.09	**7.06**	m	**7.59**
9mm thick linings to walls										
width exceeding 300mm	16.31	5.00	**17.12**	0.49	0.06	**7.61**	0.16	**24.89**	m2	**26.76**
width not exceeding 300mm	4.89	10.00	**5.38**	0.29	0.02	**4.34**	0.10	**9.81**	m	**10.55**
12mm thick linings to walls										
width exceeding 300mm	21.23	5.00	**22.29**	0.53	0.07	**8.28**	0.17	**30.75**	m2	**33.06**
width not exceeding 300mm	6.37	10.00	**7.01**	0.32	0.02	**4.76**	0.11	**11.87**	m	**12.76**
6mm thick linings to ceilings										
width exceeding 300mm	8.85	5.00	**9.29**	0.57	0.07	**8.85**	0.19	**18.33**	m2	**19.71**
width not exceeding 300mm	2.66	10.00	**2.92**	0.34	0.02	**5.05**	0.11	**8.08**	m	**8.68**
New Tacfire Class 1 fire resisting boards; butt joints										
6mm thick linings to walls										
width exceeding 300mm	11.83	5.00	**12.43**	0.45	0.06	**7.04**	0.15	**19.61**	m2	**21.08**
width not exceeding 300mm	3.55	10.00	**3.91**	0.27	0.02	**4.05**	0.09	**8.05**	m	**8.65**
9mm thick linings to walls										
width exceeding 300mm	18.13	5.00	**19.03**	0.49	0.06	**7.61**	0.16	**26.80**	m2	**28.81**
width not exceeding 300mm	5.44	10.00	**5.98**	0.29	0.02	**4.34**	0.10	**10.41**	m	**11.19**
12mm thick linings to walls										
width exceeding 300mm	23.90	5.00	**25.10**	0.53	0.07	**8.28**	0.17	**33.55**	m2	**36.07**
width not exceeding 300mm	7.17	10.00	**7.89**	0.32	0.02	**4.76**	0.11	**12.75**	m	**13.71**
6mm thick linings to ceilings; 6mm cover fillets										
width exceeding 300mm	13.15	5.00	**13.74**	0.57	0.07	**8.85**	0.19	**22.78**	m2	**24.48**
width not exceeding 300mm	3.94	10.00	**4.30**	0.34	0.02	**5.05**	0.11	**9.46**	m	**10.17**
9mm thick linings to ceilings; 9mm cover fillets										
width exceeding 300mm	20.56	5.00	**21.47**	0.61	0.08	**9.52**	0.20	**31.20**	m2	**33.54**
width not exceeding 300mm	6.17	10.00	**6.71**	0.37	0.02	**5.47**	0.12	**12.31**	m	**13.23**
12mm thick linings to ceilings; 9mm cover fillets										
width exceeding 300mm	26.34	5.00	**27.53**	0.67	0.08	**10.38**	0.22	**38.13**	m2	**40.99**
width not exceeding 300mm	7.90	10.00	**8.62**	0.40	0.02	**5.90**	0.13	**14.65**	m	**15.75**
Viculad fire resisting boards; butt joints; fixing with screws; countersinking										
18mm thick casings to isolated beams or the like; including noggins										
total girth not exceeding 600mm	19.41	10.00	**21.28**	1.42	0.18	**22.11**	0.47	**43.85**	m2	**47.14**
total girth 600 - 1200mm	19.41	5.00	**20.39**	0.85	0.11	**13.26**	0.28	**33.92**	m2	**36.47**
25mm thick casings to isolated beams or the like; including noggins										
total girth not exceeding 600mm	22.07	10.00	**24.19**	1.58	0.20	**24.59**	0.52	**49.30**	m2	**53.00**
total girth 600 - 1200mm	22.07	5.00	**23.17**	0.95	0.12	**14.78**	0.31	**38.27**	m2	**41.14**

Labour hourly rates: (except Specialists) Craft Operatives £14.21 Labourer £10.71 Rates are national average prices. Refer to REGIONAL VARIATIONS for indicative levels of overall pricing in regions	MATERIALS			LABOUR				RATES		
	Del to Site £	Waste %	Material Cost £	Craft Optve Hrs	Lab Hrs	Labour Cost £	Sunds £	Nett Rate £	Unit	Gross Rate (+7.5%) £
K11: RIGID SHEET FLOORING/SHEATHING/LININGS/CASINGS: ASBESTOS FREE BOARD Cont.										
Vicuclad fire resisting boards; butt joints; fixing with screws; countersinking Cont.										
35mm thick casings to isolated beams or the like; including noggins										
total girth not exceeding 600mm..................	26.72	10.00	**29.29**	1.65	0.21	**25.70**	0.54	**55.53**	m2	**59.69**
total girth 600 - 1200mm..........................	26.70	5.00	**28.03**	1.00	0.13	**15.60**	0.33	**43.96**	m2	**47.26**
18mm thick casings to isolated columns or the like										
total girth 600 - 1200mm..........................	18.49	10.00	**20.31**	1.42	0.18	**22.11**	0.47	**42.89**	m2	**46.10**
total girth 1200 - 1800mm........................	18.49	5.00	**19.42**	0.85	0.11	**13.26**	0.28	**32.96**	m2	**35.43**
25mm thick casings to isolated columns or the like										
total girth 600 - 1200mm..........................	21.15	10.00	**23.22**	1.58	0.20	**24.59**	0.52	**48.34**	m2	**51.96**
total girth 1200 - 1800mm........................	21.15	5.00	**22.21**	0.95	0.12	**14.78**	0.31	**37.31**	m2	**40.10**
35mm thick casings to isolated columns or the like										
total girth 600 - 1200mm..........................	25.80	10.00	**28.32**	1.65	0.21	**25.70**	0.54	**54.56**	m2	**58.65**
total girth 1200 - 1800mm........................	25.80	5.00	**27.09**	1.00	0.13	**15.60**	0.33	**43.02**	m2	**46.25**
K11: RIGID SHEET FLOORING/SHEATHING/LININGS/CASINGS: DUCT FRONTS										
Plywood BS EN 636, II/III grade, MR bonded; butt joints; fixing to timber with screws, countersinking										
12mm thick linings to walls										
width exceeding 300mm	12.58	20.00	**15.10**	1.00	0.13	**15.60**	0.05	**30.75**	m2	**33.06**
width exceeding 300mm; fixing to timber with brass screws and cups	12.58	20.00	**15.10**	1.50	0.19	**23.35**	0.19	**38.64**	m2	**41.54**
Blockboard, BS 3444, 2/2 grade, MR bonded; butt joints										
18mm thick linings to walls										
width exceeding 300mm	9.17	20.00	**11.00**	1.20	0.15	**18.66**	0.07	**29.73**	m2	**31.96**
width exceeding 300mm; fixing to timber with brass screws and cups	9.17	20.00	**11.00**	1.75	0.22	**27.22**	0.20	**38.43**	m2	**41.31**
Extra over plywood BS EN 636, II/III grade, MR bonded; butt joints; 12mm thick linings to walls; width exceeding 300mm										
Access panels										
300 x 600mm	-	-	**-**	0.50	0.05	**7.64**	0.03	**7.67**	nr	**8.24**
600 x 600mm	-	-	**-**	0.65	0.07	**9.99**	0.03	**10.02**	nr	**10.77**
600 x 900mm	-	-	**-**	0.80	0.08	**12.22**	0.04	**12.27**	nr	**13.19**
Extra over blockboard, BS 3444, 2/2 grade, MR bonded, butt joints; 18mm thick linings to walls; width exceeding 300mm										
Access panels										
300 x 600mm	-	-	**-**	0.60	0.07	**9.28**	0.03	**9.31**	nr	**10.01**
600 x 600mm	-	-	**-**	0.80	0.08	**12.22**	0.04	**12.27**	nr	**13.19**
600 x 900mm	-	-	**-**	1.10	0.11	**16.81**	0.06	**16.87**	nr	**18.13**
K11: RIGID SHEET FLOORING/SHEATHING/LININGS/CASINGS: PIPE CASINGS										
Standard hardboard, BS EN 622, type SHA; butt joints										
3mm thick linings to pipe ducts and casings										
width exceeding 300mm	1.19	20.00	**1.43**	0.80	0.10	**12.44**	0.04	**13.91**	m2	**14.95**
width not exceeding 300mm........................	0.36	20.00	**0.43**	0.25	0.03	**3.87**	0.01	**4.32**	m	**4.64**
Plywood BS EN 636, II/III grade, MR bonded; butt joints; fixing to timber with screws, countersinking										
12mm thick linings to pipe ducts and casings										
width exceeding 300mm	12.58	20.00	**15.10**	1.50	0.19	**23.35**	0.08	**38.53**	m2	**41.41**
width not exceeding 300mm........................	3.77	20.00	**4.53**	0.50	0.06	**7.75**	0.03	**12.30**	m	**13.23**
Supalux Class O fire resisting boards, sanded; butt joints; fixing to timber with screws, countersinking										
9mm thick linings to pipe ducts and casings										
width exceeding 300mm	24.24	20.00	**29.09**	1.50	0.19	**23.35**	0.08	**52.52**	m2	**56.46**
width not exceeding 300mm........................	7.27	20.00	**8.73**	0.50	0.06	**7.75**	0.03	**16.50**	m	**17.74**
Pendock Profiles Ltd preformed plywood casings; white melamine finish; butt joints; fixing to timber with polytop white screws, countersinking										
5mm thick casings; horizontal										
reference TK45/110, 45 x 110mm...................	6.72	20.00	**8.07**	0.40	0.05	**6.22**	0.07	**14.35**	m	**15.43**
extra; internal corner	6.47	2.50	**6.63**	0.25	0.03	**3.87**	0.07	**10.57**	nr	**11.37**
extra; external corner	8.43	2.50	**8.64**	0.25	0.03	**3.87**	0.07	**12.58**	nr	**13.53**
extra; stop end	5.85	2.50	**5.99**	0.15	0.02	**2.35**	0.03	**8.37**	nr	**9.00**
reference TK45/150, 45 x 150mm...................	8.49	20.00	**10.19**	0.42	0.05	**6.50**	0.07	**16.76**	m	**18.02**
extra; internal corner	7.65	2.50	**7.84**	0.25	0.03	**3.87**	0.07	**11.78**	nr	**12.66**
extra; external corner	9.69	2.50	**9.94**	0.25	0.03	**3.87**	0.07	**13.88**	nr	**14.92**
extra; stop end	8.18	2.50	**8.39**	0.15	0.02	**2.35**	0.03	**10.77**	nr	**11.57**
reference TK45/190, 45 x 190mm...................	11.16	20.00	**13.39**	0.44	0.05	**6.79**	0.07	**20.24**	m	**21.76**

Labour hourly rates: (except Specialists) Craft Operatives £14.21 Labourer £10.71 Rates are national average prices. Refer to REGIONAL VARIATIONS for indicative levels of overall pricing in regions	MATERIALS			LABOUR				RATES		
	Del to Site £	Waste %	Material Cost £	Craft Optve Hrs	Lab Hrs	Labour Cost £	Sunds £	Nett Rate £	Unit	Gross Rate (+7.5%) £
K11: RIGID SHEET FLOORING/SHEATHING/LININGS/CASINGS: PIPE CASINGS Cont.										
Pendock Profiles Ltd preformed plywood casings; white melamine finish; butt joints; fixing to timber with polytop white screws, countersinking Cont.										
5mm thick casings; horizontal Cont.										
extra; internal corner	7.70	2.50	**7.89**	0.25	0.03	**3.87**	0.07	**11.83**	nr	**12.71**
extra; external corner	9.74	2.50	**9.99**	0.25	0.03	**3.87**	0.07	**13.93**	nr	**14.97**
extra; stop end	9.74	2.50	**9.99**	0.15	0.02	**2.35**	0.03	**12.37**	nr	**13.29**
reference TK25/125, 25 x 125mm	6.71	20.00	**8.05**	0.40	0.05	**6.22**	0.07	**14.34**	m	**15.41**
extra; internal corner	6.48	2.50	**6.65**	0.25	0.03	**3.87**	0.07	**10.59**	nr	**11.38**
extra; external corner	8.43	2.50	**8.64**	0.25	0.03	**3.87**	0.07	**12.58**	nr	**13.53**
extra; stop end	5.85	2.50	**5.99**	0.15	0.02	**2.35**	0.03	**8.37**	nr	**9.00**
5mm thick casings; vertical										
reference TK45/110, 45 x 110mm	6.12	20.00	**7.35**	0.40	0.05	**6.22**	0.07	**13.63**	m	**14.65**
reference TK45/135, 45 x 135mm	8.93	20.00	**10.72**	0.41	0.05	**6.36**	0.07	**17.15**	m	**18.43**
reference TK45/150, 45 x 150mm	7.83	20.00	**9.40**	0.42	0.05	**6.50**	0.07	**15.97**	m	**17.17**
reference TK45/190, 45 x 190mm	10.48	20.00	**12.58**	0.44	0.05	**6.79**	0.07	**19.43**	m	**20.89**
5mm thick casings; horizontal										
reference MX 100/100, 100 x 100mm	11.21	20.00	**13.45**	0.42	0.05	**6.50**	0.07	**20.02**	m	**21.52**
extra; stop end	9.72	2.50	**9.96**	0.15	0.02	**2.35**	0.03	**12.34**	nr	**13.27**
8mm thick casings; horizontal										
reference MX 150/150, 150 x 150mm	16.96	20.00	**20.36**	0.45	0.06	**7.04**	0.07	**27.46**	m	**29.52**
extra; internal corner	12.08	2.50	**12.38**	0.25	0.03	**3.87**	0.07	**16.32**	nr	**17.55**
extra; external corner	18.21	2.50	**18.67**	0.25	0.03	**3.87**	0.07	**22.61**	nr	**24.30**
extra; stop end	9.72	2.50	**9.96**	0.15	0.02	**2.35**	0.03	**12.34**	nr	**13.27**
reference MX 150/200, 200 x 150mm	18.80	20.00	**22.56**	0.46	0.06	**7.18**	0.07	**29.81**	m	**32.04**
extra; stop end	9.72	2.50	**9.96**	0.15	0.02	**2.35**	0.03	**12.34**	nr	**13.27**
reference MX 200/200, 200 x 200mm	20.50	20.00	**24.60**	0.47	0.06	**7.32**	0.07	**31.99**	m	**34.39**
extra; internal corner	12.18	2.50	**12.48**	0.25	0.03	**3.87**	0.07	**16.42**	nr	**17.66**
extra; external corner	18.41	2.50	**18.87**	0.25	0.03	**3.87**	0.07	**22.81**	nr	**24.53**
extra; stop end	9.72	2.50	**9.96**	0.15	0.02	**2.35**	0.03	**12.34**	nr	**13.27**
reference MX 200/300, 200 x 300mm	24.06	20.00	**28.87**	0.48	0.06	**7.46**	0.07	**36.40**	m	**39.13**
extra; stop end	10.67	2.50	**10.94**	0.15	0.02	**2.35**	0.03	**13.31**	nr	**14.31**
reference MX 150/300, 300 x 150mm	22.75	20.00	**27.30**	0.48	0.06	**7.46**	0.07	**34.83**	m	**37.44**
extra; internal corner	12.93	-	**12.93**	0.25	0.03	**3.87**	0.07	**16.87**	nr	**18.13**
extra; external corner	19.21	-	**19.21**	0.25	0.03	**3.87**	0.07	**23.15**	nr	**24.89**
extra; stop end	10.67	2.50	**10.94**	0.15	0.02	**2.35**	0.03	**13.31**	nr	**14.31**
reference MX 300/300, 300 x 300mm	27.43	20.00	**32.92**	0.50	0.06	**7.75**	0.07	**40.73**	m	**43.79**
extra; internal corner	13.79	2.50	**14.14**	0.25	0.03	**3.87**	0.07	**18.08**	nr	**19.43**
extra; external corner	20.09	2.50	**20.59**	0.25	0.03	**3.87**	0.07	**24.53**	nr	**26.37**
extra; stop end	13.25	2.50	**13.59**	0.15	0.02	**2.35**	0.03	**15.96**	nr	**17.16**
5mm thick casings; vertical										
reference MX 100/100, 100 x 100mm	9.38	20.00	**11.26**	0.42	0.05	**6.50**	0.07	**17.83**	m	**19.17**
8mm thick casings; vertical										
reference MX 150/150, 150 x 150mm	15.12	20.00	**18.14**	0.45	0.06	**7.04**	0.07	**25.24**	m	**27.14**
reference MX 150/200, 200 x 150mm	16.98	20.00	**20.37**	0.46	0.06	**7.18**	0.07	**27.62**	m	**29.69**
reference MX 200/200, 200 x 200mm	18.69	20.00	**22.43**	0.47	0.06	**7.32**	0.07	**29.81**	m	**32.05**
reference MX 200/300, 200 x 300mm	22.24	20.00	**26.68**	0.48	0.06	**7.46**	0.07	**34.21**	m	**36.78**
reference MX 150/300, 300 x 150mm	20.92	20.00	**25.11**	0.49	0.06	**7.61**	0.07	**32.78**	m	**35.24**
reference MX 300/300, 300 x 300mm	25.60	20.00	**30.72**	0.50	0.06	**7.75**	0.07	**38.53**	m	**41.42**
K11: RIGID SHEET FLOORING/SHEATHING/LININGS/CASINGS: ROOF BOARDING										
Plywood BS EN 636, II/III grade, WBP bonded; butt joints										
18mm thick sheeting to roofs; external										
width exceeding 300mm	10.19	10.00	**11.21**	0.30	0.04	**4.69**	0.02	**15.92**	m2	**17.11**
extra; cesspool 225 x 225 x 150mm	2.04	10.00	**2.24**	1.40	0.18	**21.82**	0.08	**24.14**	nr	**25.95**
extra; cesspool 300 x 300 x 150mm	2.80	10.00	**3.08**	1.50	0.19	**23.35**	0.08	**26.51**	nr	**28.50**
width exceeding 300mm; sloping	10.19	10.00	**11.21**	0.40	0.05	**6.22**	0.02	**17.45**	m2	**18.76**
24mm thick sheeting to roofs; external										
width exceeding 300mm	14.16	10.00	**15.58**	0.35	0.04	**5.40**	0.02	**20.99**	m2	**22.57**
extra; cesspool 225 x 225 x 150mm	2.83	10.00	**3.12**	1.50	0.19	**23.35**	0.08	**26.54**	nr	**28.54**
extra; cesspool 300 x 300 x 150mm	3.89	10.00	**4.28**	1.60	0.20	**24.88**	0.09	**29.25**	nr	**31.44**
width exceeding 300mm; sloping	14.16	10.00	**15.58**	0.45	0.06	**7.04**	0.02	**22.64**	m2	**24.33**
Wood chipboard, BS EN 312, type 1 standard; butt joints										
12mm thick sheeting to roofs; external										
width exceeding 300mm	2.15	10.00	**2.37**	0.25	0.03	**3.87**	0.01	**6.25**	m2	**6.72**
extra; cesspool 225 x 225 x 150mm	0.43	10.00	**0.47**	1.30	0.16	**20.19**	0.07	**20.73**	nr	**22.28**
extra; cesspool 300 x 300 x 150mm	0.59	10.00	**0.65**	1.40	0.18	**21.82**	0.08	**22.55**	nr	**24.24**
width exceeding 300mm; sloping	2.15	10.00	**2.37**	0.35	0.04	**5.40**	0.02	**7.79**	m2	**8.37**
18mm thick sheeting to roofs; external										
width exceeding 300mm	2.92	10.00	**3.21**	0.30	0.04	**4.69**	0.02	**7.92**	m2	**8.51**
extra; cesspool 225 x 225 x 150mm	0.58	10.00	**0.64**	1.40	0.18	**21.82**	0.08	**22.54**	nr	**24.23**
extra; cesspool 300 x 300 x 150mm	0.80	10.00	**0.88**	1.50	0.19	**23.35**	0.08	**24.31**	nr	**26.14**
width exceeding 300mm; sloping	2.92	10.00	**3.21**	0.40	0.05	**6.22**	0.02	**9.45**	m2	**10.16**
Abutments										
18mm thick plywood linings										
raking cutting	1.02	10.00	**1.12**	0.30	-	**4.26**	-	**5.38**	m	**5.79**

Labour hourly rates: (except Specialists) Craft Operatives £14.21 Labourer £10.71 Rates are national average prices. Refer to REGIONAL VARIATIONS for indicative levels of overall pricing in regions	MATERIALS			LABOUR				RATES		
	Del to Site £	Waste %	Material Cost £	Craft Optve Hrs	Lab Hrs	Labour Cost £	Sunds £	Nett Rate £	Unit	Gross Rate (+7.5%) £
K11: RIGID SHEET FLOORING/SHEATHING/LININGS/CASINGS: ROOF BOARDING Cont.										
Abutments Cont.										
18mm thick plywood linings Cont.										
curved cutting	1.53	10.00	**1.68**	0.45	-	**6.39**	-	**8.08**	m	**8.68**
rebates	-	-	-	0.18	-	**2.56**	-	**2.56**	m	**2.75**
grooves	-	-	-	0.18	-	**2.56**	-	**2.56**	m	**2.75**
chamfers	-	-	-	0.13	-	**1.85**	-	**1.85**	m	**1.99**
24mm thick plywood linings										
raking cutting	1.42	10.00	**1.56**	0.40	-	**5.68**	-	**7.24**	m	**7.78**
curved cutting	2.12	10.00	**2.34**	0.60	-	**8.53**	-	**10.86**	m	**11.68**
rebates	-	-	-	0.18	-	**2.56**	-	**2.56**	m	**2.75**
grooves	-	-	-	0.18	-	**2.56**	-	**2.56**	m	**2.75**
chamfers	-	-	-	0.13	-	**1.85**	-	**1.85**	m	**1.99**
12mm thick wood chipboard linings										
raking cutting	0.22	10.00	**0.24**	0.25	-	**3.55**	-	**3.79**	m	**4.07**
curved cutting	0.32	10.00	**0.35**	0.35	-	**4.97**	-	**5.33**	m	**5.73**
rebates	-	-	-	0.18	-	**2.56**	-	**2.56**	m	**2.75**
grooves	-	-	-	0.18	-	**2.56**	-	**2.56**	m	**2.75**
chamfers	-	-	-	0.13	-	**1.85**	-	**1.85**	m	**1.99**
18mm thick wood chipboard linings										
raking cutting	0.29	10.00	**0.32**	0.30	-	**4.26**	-	**4.58**	m	**4.93**
curved cutting	0.44	10.00	**0.48**	0.45	-	**6.39**	-	**6.88**	m	**7.39**
rebates	-	-	-	0.18	-	**2.56**	-	**2.56**	m	**2.75**
grooves	-	-	-	0.18	-	**2.56**	-	**2.56**	m	**2.75**
chamfers	-	-	-	0.13	-	**1.85**	-	**1.85**	m	**1.99**
K11: RIGID SHEET FLOORING/SHEATHING/ LININGS/CASINGS; STAIRCASE AREAS - ASBESTOS FREE BOARD										
Masterboard Class 0 fire resisting boards; butt joints										
6mm thick linings to walls										
width exceeding 300mm	10.28	5.00	**10.79**	0.57	0.07	**8.85**	0.19	**19.83**	m2	**21.32**
width not exceeding 300mm	3.08	10.00	**3.39**	0.34	0.02	**5.05**	0.11	**8.55**	m	**9.19**
9mm thick linings to walls										
width exceeding 300mm	21.72	5.00	**22.81**	0.61	0.08	**9.52**	0.20	**32.53**	m2	**34.97**
width not exceeding 300mm	6.52	10.00	**7.17**	0.37	0.02	**5.47**	0.12	**12.76**	m	**13.72**
12mm thick linings to walls										
width exceeding 300mm	28.64	5.00	**30.07**	0.67	0.08	**10.38**	0.22	**40.67**	m2	**43.72**
width not exceeding 300mm	8.59	10.00	**9.45**	0.40	0.02	**5.90**	0.13	**15.48**	m	**16.64**
6mm thick linings to ceilings										
width exceeding 300mm	10.28	5.00	**10.79**	0.61	0.08	**9.52**	0.20	**20.52**	m2	**22.06**
width not exceeding 300mm	3.08	10.00	**3.39**	0.37	0.02	**5.47**	0.12	**8.99**	m	**9.66**
9mm thick linings to ceilings										
width exceeding 300mm	21.72	5.00	**22.81**	0.67	0.08	**10.38**	0.22	**33.40**	m2	**35.91**
width not exceeding 300mm	6.52	10.00	**7.17**	0.40	0.02	**5.90**	0.13	**13.20**	m	**14.19**
12mm thick linings to ceilings										
width exceeding 300mm	28.64	5.00	**30.07**	0.72	0.09	**11.20**	0.24	**41.50**	m2	**44.62**
width not exceeding 300mm	8.59	10.00	**9.45**	0.43	0.03	**6.43**	0.14	**16.02**	m	**17.23**
Supalux fire resisting boards, sanded; butt joints; fixing to timber with screws, countersinking										
6mm thick linings to walls										
width exceeding 300mm	14.24	5.00	**14.95**	0.57	0.07	**8.85**	0.19	**23.99**	m2	**25.79**
width not exceeding 300mm	4.27	10.00	**4.70**	0.34	0.02	**5.05**	0.11	**9.86**	m	**10.60**
9mm thick linings to walls										
width exceeding 300mm	24.24	5.00	**25.45**	0.61	0.08	**9.52**	0.20	**35.18**	m2	**37.82**
width not exceeding 300mm	7.27	10.00	**8.00**	0.37	0.02	**5.47**	0.12	**13.59**	m	**14.61**
12mm thick linings to walls										
width exceeding 300mm	32.08	5.00	**33.68**	0.67	0.08	**10.38**	0.22	**44.28**	m2	**47.60**
width not exceeding 300mm	9.62	10.00	**10.59**	0.40	0.02	**5.90**	0.13	**16.62**	m	**17.86**
6mm thick linings to ceilings										
width exceeding 300mm	14.24	5.00	**14.95**	0.61	0.08	**9.52**	0.20	**24.68**	m2	**26.53**
width not exceeding 300mm	4.27	10.00	**4.70**	0.37	0.02	**5.47**	0.12	**10.29**	m	**11.07**
9mm thick linings to ceilings										
width exceeding 300mm	24.24	5.00	**25.45**	0.67	0.08	**10.38**	0.22	**36.05**	m2	**38.75**
width not exceeding 300mm	7.27	10.00	**8.00**	0.40	0.02	**5.90**	0.13	**14.03**	m	**15.08**
12mm thick linings to ceilings										
width exceeding 300mm	32.08	5.00	**33.68**	0.72	0.09	**11.20**	0.24	**45.12**	m2	**48.50**
width not exceeding 300mm	9.62	10.00	**10.59**	0.43	0.03	**6.43**	0.14	**17.16**	m	**18.45**

Labour hourly rates: (except Specialists) Craft Operatives £14.21 Labourer £10.71 Rates are national average prices. Refer to REGIONAL VARIATIONS for indicative levels of overall pricing in regions	MATERIALS			LABOUR				RATES		
	Del to Site £	Waste %	Material Cost £	Craft Optve Hrs	Lab Hrs	Labour Cost £	Sunds £	Nett Rate £	Unit	Gross Rate (+7.5%) £
K12: UNDER PURLIN/INSIDE RAIL PANEL LININGS: PLASTERBOARD										
Gypliner metal grid fixing system; square edge sheeting of one layer 12.5mm thick Gyproc wallboard, BS 1230, butt joints;screwing to mild steel grid.										
12.5mm thick linings to walls										
width exceeding 300mm	5.72	5.00	**6.03**	0.80	0.40	**15.65**	0.26	**21.94**	m2	**23.59**
width not exceeding 300mm	5.72	5.00	**6.00**	0.48	0.12	**8.11**	0.26	**14.37**	m	**15.45**
12.5mm thick linings to ceilings										
width exceeding 300mm	5.65	5.00	**5.96**	1.00	0.50	**19.57**	0.26	**25.79**	m2	**27.73**
width not exceeding 300mm	5.72	5.00	**6.00**	0.60	0.15	**10.13**	0.26	**16.40**	m	**17.63**
K13: RIGID SHEET FINE LININGS/PANELLING										
Softwood, wrot - wall panelling										
19mm thick panelled linings to walls; square framed; including grounds										
width exceeding 300mm	68.27	2.50	**69.97**	1.60	0.20	**24.88**	0.09	**94.94**	m2	**102.06**
width exceeding 300mm; fixing to masonry with screws	68.27	2.50	**69.97**	2.60	0.32	**40.37**	0.15	**110.49**	m2	**118.78**
25mm thick panelled linings to walls; square framed; including grounds										
width exceeding 300mm	69.19	2.50	**70.92**	1.70	0.21	**26.41**	0.10	**97.42**	m2	**104.73**
width exceeding 300mm; fixing to masonry with screws	69.19	2.50	**70.92**	2.70	0.34	**42.01**	0.16	**113.08**	m2	**121.56**
19mm thick panelled linings to walls; square framed; obstructed by integral services; including grounds										
width exceeding 300mm	68.27	2.50	**69.97**	2.10	0.26	**32.63**	0.12	**102.72**	m2	**110.42**
width exceeding 300mm; fixing to masonry with screws	68.27	2.50	**69.97**	3.10	0.39	**48.23**	0.17	**118.37**	m2	**127.25**
25mm thick panelled linings to walls; square framed; obstructed by integral services; including grounds										
width exceeding 300mm	69.19	2.50	**70.92**	2.20	0.27	**34.15**	0.13	**105.20**	m2	**113.09**
width exceeding 300mm; fixing to masonry with screws	69.19	2.50	**70.92**	3.20	0.40	**49.76**	0.18	**120.85**	m2	**129.92**
19mm thick panelled linings to walls; moulded; including grounds										
width exceeding 300mm	74.72	2.50	**76.59**	1.70	0.21	**26.41**	0.10	**103.09**	m2	**110.82**
width exceeding 300mm; fixing to masonry with screws	74.72	2.50	**76.59**	2.70	0.34	**42.01**	0.16	**118.75**	m2	**127.66**
25mm thick panelled linings to walls; moulded; including grounds										
width exceeding 300mm	77.49	2.50	**79.43**	1.80	0.22	**27.93**	0.10	**107.46**	m2	**115.52**
width exceeding 300mm; fixing to masonry with screws	77.49	2.50	**79.43**	2.80	0.35	**43.54**	0.16	**123.12**	m2	**132.36**
19mm thick panelled linings to walls; moulded; obstructed by integral services; including grounds										
width exceeding 300mm	74.72	2.50	**76.59**	2.20	0.27	**34.15**	0.13	**110.87**	m2	**119.18**
width exceeding 300mm; fixing to masonry with screws	74.72	2.50	**76.59**	3.20	0.40	**49.76**	0.18	**126.53**	m2	**136.02**
25mm thick panelled linings to walls; moulded; obstructed by integral services; including grounds										
width exceeding 300mm	77.49	2.50	**79.43**	2.30	0.29	**35.79**	0.13	**115.34**	m2	**124.00**
width exceeding 300mm; fixing to masonry with screws	77.49	2.50	**79.43**	3.30	0.41	**51.28**	0.19	**130.90**	m2	**140.72**
Afromosia, wrot, selected for transparent finish - wall panelling										
19mm thick panelled linings to walls; square framed; including grounds										
width exceeding 300mm	194.40	2.50	**199.26**	2.25	0.28	**34.97**	0.13	**234.36**	m2	**251.94**
width exceeding 300mm; fixing to masonry with screws	194.40	2.50	**199.26**	3.25	0.41	**50.57**	0.18	**250.02**	m2	**268.77**
25mm thick panelled linings to walls; square framed; including grounds										
width exceeding 300mm	199.80	2.50	**204.80**	2.35	0.29	**36.50**	0.13	**241.43**	m2	**259.53**
width exceeding 300mm; fixing to masonry with screws	199.80	2.50	**204.80**	3.35	0.42	**52.10**	0.19	**257.09**	m2	**276.37**
19mm thick panelled linings to walls; square framed; obstructed by integral services; including grounds										
width exceeding 300mm	194.40	2.50	**199.26**	2.75	0.34	**42.72**	0.16	**242.13**	m2	**260.29**
width exceeding 300mm; fixing to masonry with screws	194.40	2.50	**199.26**	3.75	0.47	**58.32**	0.21	**257.79**	m2	**277.13**
25mm thick panelled linings to walls; square framed; obstructed by integral services; including grounds										
width exceeding 300mm	199.80	2.50	**204.80**	2.85	0.36	**44.35**	0.16	**249.31**	m2	**268.01**
width exceeding 300mm; fixing to masonry with screws	199.80	2.50	**204.80**	3.85	0.48	**59.85**	0.22	**264.86**	m2	**284.73**

Labour hourly rates: (except Specialists) Craft Operatives £14.21 Labourer £10.71 Rates are national average prices. Refer to REGIONAL VARIATIONS for indicative levels of overall pricing in regions	MATERIALS			LABOUR				RATES		
	Del to Site £	Waste %	Material Cost £	Craft Optve Hrs	Lab Hrs	Labour Cost £	Sunds £	Nett Rate £	Unit	Gross Rate (+7.5%) £
K13: RIGID SHEET FINE LININGS/PANELLING Cont.										
Afromosia, wrot, selected for transparent finish - wall panelling Cont.										
19mm thick panelled linings to walls; moulded; including grounds										
width exceeding 300mm	208.80	2.50	**214.02**	2.40	0.30	**37.32**	0.14	**251.47**	m2	**270.33**
width exceeding 300mm; fixing to masonry with screws	208.80	2.50	**214.02**	3.40	0.42	**52.81**	0.19	**267.02**	m2	**287.05**
25mm thick panelled linings to walls; moulded; including grounds										
width exceeding 300mm	222.30	2.50	**227.86**	2.50	0.31	**38.85**	0.14	**266.84**	m2	**286.86**
width exceeding 300mm; fixing to masonry with screws	222.30	2.50	**227.86**	3.50	0.44	**54.45**	0.20	**282.50**	m2	**303.69**
19mm thick panelled linings to walls; moulded; obstructed by integral services; including grounds										
width exceeding 300mm	208.80	2.50	**214.02**	2.90	0.36	**45.06**	0.16	**259.25**	m2	**278.69**
width exceeding 300mm; fixing to masonry with screws	208.80	2.50	**214.02**	3.90	0.49	**60.67**	0.22	**274.90**	m2	**295.52**
25mm thick panelled linings to walls; moulded; obstructed by integral services; including grounds										
width exceeding 300mm	222.30	2.50	**227.86**	3.00	0.37	**46.59**	0.17	**274.62**	m2	**295.21**
width exceeding 300mm; fixing to masonry with screws	222.30	2.50	**227.86**	4.00	0.50	**62.20**	0.22	**290.28**	m2	**312.05**
Sapele, wrot, selected for transparent finish - wall panelling										
19mm thick panelled linings to walls; square framed; including grounds										
width exceeding 300mm	156.60	2.50	**160.52**	2.05	0.26	**31.92**	0.12	**192.55**	m2	**206.99**
width exceeding 300mm; fixing to masonry with screws	156.60	2.50	**160.52**	3.05	0.38	**47.41**	0.17	**208.10**	m2	**223.71**
25mm thick panelled linings to walls; square framed; including grounds										
width exceeding 300mm	164.20	2.50	**168.31**	2.15	0.27	**33.44**	0.12	**201.87**	m2	**217.01**
width exceeding 300mm; fixing to masonry with screws	164.20	2.50	**168.31**	3.15	0.39	**48.94**	0.18	**217.42**	m2	**233.73**
19mm thick panelled linings to walls; square framed; obstructed by integral services; including grounds										
width exceeding 300mm	156.60	2.50	**160.52**	2.55	0.32	**39.66**	0.15	**200.32**	m2	**215.35**
width exceeding 300mm; fixing to masonry with screws	156.60	2.50	**160.52**	3.55	0.44	**55.16**	0.20	**215.87**	m2	**232.06**
25mm thick panelled linings to walls; square framed; obstructed by integral services; including grounds										
width exceeding 300mm	164.20	2.50	**168.31**	2.65	0.33	**41.19**	0.15	**209.64**	m2	**225.37**
width exceeding 300mm; fixing to masonry with screws	164.20	2.50	**168.31**	3.65	0.46	**56.79**	0.20	**225.30**	m2	**242.20**
19mm thick panelled linings to walls; moulded; including grounds										
width exceeding 300mm	172.80	2.50	**177.12**	2.15	0.27	**33.44**	0.12	**210.69**	m2	**226.49**
width exceeding 300mm; fixing to masonry with screws	172.80	2.50	**177.12**	3.15	0.39	**48.94**	0.18	**226.24**	m2	**243.20**
25mm thick panelled linings to walls; moulded; including grounds										
width exceeding 300mm	179.30	2.50	**183.78**	2.30	0.29	**35.79**	0.13	**219.70**	m2	**236.18**
width exceeding 300mm; fixing to masonry with screws	179.30	2.50	**183.78**	3.30	0.41	**51.28**	0.19	**235.25**	m2	**252.90**
19mm thick panelled linings to walls; moulded; obstructed by integral services; including grounds										
width exceeding 300mm	172.80	2.50	**177.12**	2.65	0.33	**41.19**	0.15	**218.46**	m2	**234.84**
width exceeding 300mm; fixing to masonry with screws	172.80	2.50	**177.12**	3.65	0.46	**56.79**	0.20	**234.12**	m2	**251.68**
25mm thick panelled linings to walls; moulded; obstructed by integral services; including grounds										
width exceeding 300mm	179.30	2.50	**183.78**	2.80	0.35	**43.54**	0.16	**227.48**	m2	**244.54**
width exceeding 300mm; fixing to masonry with screws	179.30	2.50	**183.78**	3.80	0.47	**59.03**	0.21	**243.03**	m2	**261.26**
K13: RIGID SHEET FINE LININGS/PANELLING: VENEERED PLYWOOD PANELLING										
Pre-finished, decorative veneers; butt joints - wall panelling. Note. prices are for panelling with Afromosia, Ash, Beech, Cherry, Elm, Oak, Knotted Pine, Sapele or Teak faced veneers										
4mm thick linings to walls										
width exceeding 300mm	9.56	15.00	**10.99**	0.80	0.10	**12.44**	0.04	**23.48**	m2	**25.24**
width exceeding 300 mm; fixing to timber with adhesive	9.56	15.00	**10.99**	0.95	0.12	**14.78**	0.66	**26.44**	m2	**28.42**
width exceeding 300mm; fixing to plaster with adhesive	9.56	15.00	**10.99**	0.90	0.11	**13.97**	0.66	**25.62**	m2	**27.54**

Labour hourly rates: (except Specialists) Craft Operatives £14.21 Labourer £10.71 Rates are national average prices. Refer to REGIONAL VARIATIONS for indicative levels of overall pricing in regions	MATERIALS			LABOUR				RATES		
	Del to Site £	Waste %	Material Cost £	Craft Optve Hrs	Lab Hrs	Labour Cost £	Sunds £	Nett Rate £	Unit	Gross Rate (+7.5%) £
K13: RIGID SHEET FINE LININGS/PANELLING: VENEERED PLYWOOD PANELLING Cont.										
Flame retardant, pre-finished, decorative, veneers not matched; random V-grooves on face; butt joints - wall panelling. Note. prices are for panelling with Afromosia, Ash, Beech, Cherry, Elm, Oak, Knotted Pine, Sapele or Teak faced veneers										
4mm thick linings to walls										
width exceeding 300mm	25.65	15.00	**29.50**	0.80	0.10	**12.44**	0.04	**41.98**	m2	**45.13**
width exceeding 300 mm; fixing to timber with adhesive	25.65	15.00	**29.50**	0.95	0.12	**14.78**	0.66	**44.94**	m2	**48.31**
width exceeding 300mm; fixing to plaster with adhesive	25.65	15.00	**29.50**	0.90	0.11	**13.97**	0.66	**44.12**	m2	**47.43**
K20: TIMBER BOARD FLOORING/SHEATHING/LININGS/CASINGS										
Softwood, sawn; fixing to timber - boarded flooring										
Boarding to floors, square edges; 19mm thick, 75mm wide boards										
width exceeding 300mm	4.03	7.50	**4.33**	0.60	0.08	**9.38**	0.03	**13.75**	m2	**14.78**
Boarding to floors, square edges; 25mm thick, 125mm wide boards										
width exceeding 300mm	5.05	7.50	**5.42**	0.55	0.07	**8.57**	0.03	**14.02**	m2	**15.07**
Boarding to floors, square edges; 32mm thick, 150mm wide boards										
width exceeding 300mm	6.41	7.50	**6.89**	0.55	0.07	**8.57**	0.03	**15.49**	m2	**16.65**
Softwood, wrought; fixing to timber - boarded flooring										
Boarding to floors, square edges; 19mm thick, 75mm wide boards										
width exceeding 300mm	6.72	7.50	**7.22**	0.70	0.09	**10.91**	0.04	**18.17**	m2	**19.53**
Boarding to floors, square edges; 25mm thick, 125mm wide boards										
width exceeding 300mm	8.40	7.50	**9.04**	0.65	0.08	**10.09**	0.04	**19.16**	m2	**20.60**
Boarding to floors, square edges; 32mm thick, 150mm wide boards										
width exceeding 300mm	10.69	7.50	**11.49**	0.65	0.08	**10.09**	0.04	**21.62**	m2	**23.24**
Boarding to floors, tongued and grooved joints; 19mm thick, 75mm wide boards										
width exceeding 300mm	7.68	7.50	**8.26**	0.80	0.10	**12.44**	0.04	**20.74**	m2	**22.29**
Boarding to floors, tongued and grooved joints; 25mm thick, 125mm wide boards										
width exceeding 300mm	9.20	7.50	**9.90**	0.75	0.10	**11.73**	0.04	**21.66**	m2	**23.29**
Boarding to floors, tongued and grooved joints; 32mm thick, 150mm wide boards										
width exceeding 300mm	11.52	7.50	**12.38**	0.75	0.10	**11.73**	0.04	**24.15**	m2	**25.96**
Abutments										
19mm thick softwood boarding										
raking cutting	0.67	7.50	**0.72**	0.06	-	**0.85**	-	**1.57**	m	**1.69**
curved cutting	1.01	7.50	**1.08**	0.09	-	**1.28**	-	**2.36**	m	**2.54**
25mm thick softwood boarding										
raking cutting	0.84	7.50	**0.90**	0.06	-	**0.85**	-	**1.76**	m	**1.89**
curved cutting	1.26	7.50	**1.36**	0.09	-	**1.28**	-	**2.63**	m	**2.83**
32mm thick softwood boarding										
raking cutting	1.07	7.50	**1.15**	0.06	-	**0.85**	-	**2.00**	m	**2.15**
curved cutting	1.60	7.50	**1.72**	0.09	-	**1.28**	-	**3.00**	m	**3.23**
K20: TIMBER BOARD FLOORING/SHEATHING/LININGS/CASINGS: SURFACE TREATMENT OF EXISTING WOOD FLOORING										
Sanding and sealing existing wood flooring										
Machine sanding										
width exceeding 300mm	-	-	**-**	0.50	-	**7.11**	0.66	**7.77**	m2	**8.35**
Prepare, one priming coat and one finish coat of seal										
width exceeding 300mm	-	-	**-**	0.16	-	**2.27**	0.33	**2.60**	m2	**2.80**
K20: TIMBER BOARD FLOORING/SHEATHING/LININGS/CASINGS: WALL AND CEILING BOARDING										
Softwood, wrought; fixing to timber										
Boarding to walls, tongued, grooved and veed joints; 19mm thick, 100mm wide boards										
width exceeding 300mm	11.04	10.00	**12.14**	1.00	0.13	**15.60**	0.05	**27.80**	m2	**29.88**
Boarding to walls, tongued, grooved and veed joints; 19mm thick, 150mm wide boards										
width exceeding 300mm	10.08	10.00	**11.09**	0.70	0.09	**10.91**	0.04	**22.03**	m2	**23.69**

Labour hourly rates: (except Specialists) Craft Operatives £14.21 Labourer £10.71 Rates are national average prices. Refer to REGIONAL VARIATIONS for indicative levels of overall pricing in regions	MATERIALS			LABOUR				RATES		
	Del to Site £	Waste %	Material Cost £	Craft Optve Hrs	Lab Hrs	Labour Cost £	Sunds £	Nett Rate £	Unit	Gross Rate (+7.5%) £
K20: TIMBER BOARD FLOORING/SHEATHING/LININGS/CASINGS: WALL AND CEILING BOARDING Cont.										
Softwood, wrought; fixing to timber Cont.										
Boarding to walls, tongued, grooved and veed joints; 25mm thick, 100mm wide boards width exceeding 300mm	14.40	10.00	**15.84**	1.05	0.13	**16.31**	0.06	**32.21**	m2	**34.62**
Boarding to walls, tongued, grooved and veed joints; 25mm thick, 150mm wide boards width exceeding 300mm	13.44	10.00	**14.78**	0.75	0.09	**11.62**	0.04	**26.44**	m2	**28.43**
Boarding to ceilings, tongued, grooved and veed joints; 19mm thick, 100mm wide boards width exceeding 300mm	11.04	10.00	**12.14**	1.50	0.19	**23.35**	0.08	**35.57**	m2	**38.24**
Boarding to ceilings, tongued, grooved and veed joints; 19mm thick, 150mm wide boards width exceeding 300mm	10.08	10.00	**11.09**	1.00	0.13	**15.60**	0.05	**26.74**	m2	**28.75**
Boarding to ceilings, tongued, grooved and veed joints; 25mm thick, 100mm wide boards width exceeding 300mm	14.40	10.00	**15.84**	1.55	0.19	**24.06**	0.08	**39.98**	m2	**42.98**
Boarding to ceilings, tongued, grooved and veed joints; 25mm thick, 150mm wide boards width exceeding 300mm	13.44	10.00	**14.78**	1.05	0.13	**16.31**	0.06	**31.15**	m2	**33.49**
Western Red Cedar, wrought; fixing to timber										
Boarding to walls, tongued, grooved and veed joints; 19mm thick, 100mm wide boards width exceeding 300mm	42.70	10.00	**46.97**	1.05	0.13	**16.31**	0.06	**63.34**	m2	**68.09**
Boarding to walls, tongued, grooved and veed joints; 19mm thick, 150mm wide boards width exceeding 300mm	36.42	10.00	**40.06**	0.75	0.10	**11.73**	0.04	**51.83**	m2	**55.72**
Boarding to walls, tongued, grooved and veed joints; 25mm thick, 100mm wide boards width exceeding 300mm	53.60	10.00	**58.96**	1.10	0.14	**17.13**	0.06	**76.15**	m2	**81.86**
Boarding to walls, tongued, grooved and veed joints; 25mm thick, 150mm wide boards width exceeding 300mm	48.30	10.00	**53.13**	0.80	0.10	**12.44**	0.04	**65.61**	m2	**70.53**
Knotty Pine, wrought, selected for transparent finish; fixing to timber										
Boarding to walls, tongued, grooved and veed joints; 19mm thick, 100mm wide boards width exceeding 300mm	32.81	10.00	**36.09**	1.30	0.16	**20.19**	0.07	**56.35**	m2	**60.57**
Boarding to walls, tongued, grooved and veed joints; 19mm thick, 150mm wide boards width exceeding 300mm	28.02	10.00	**30.82**	0.90	0.11	**13.97**	0.05	**44.84**	m2	**48.20**
Boarding to walls, tongued, grooved and veed joints; 25mm thick, 100mm wide boards width exceeding 300mm	41.27	10.00	**45.40**	1.35	0.17	**21.00**	0.07	**66.47**	m2	**71.46**
Boarding to walls, tongued, grooved and veed joints; 25mm thick, 150mm wide boards width exceeding 300mm	37.17	10.00	**40.89**	0.95	0.12	**14.78**	0.05	**55.72**	m2	**59.90**
Boarding to ceilings, tongued, grooved and veed joints; 19mm thick, 100mm wide boards width exceeding 300mm	32.81	10.00	**36.09**	1.85	0.23	**28.75**	0.10	**64.94**	m2	**69.81**
Boarding to ceilings, tongued, grooved and veed joints; 19mm thick, 150mm wide boards width exceeding 300mm	28.02	10.00	**30.82**	1.35	0.17	**21.00**	0.07	**51.90**	m2	**55.79**
Boarding to ceilings, tongued, grooved and veed joints; 25mm thick, 100mm wide boards width exceeding 300mm	41.27	10.00	**45.40**	1.90	0.24	**29.57**	0.10	**75.07**	m2	**80.70**
Boarding to ceilings, tongued, grooved and veed joints; 25mm thick, 150mm wide boards width exceeding 300mm	37.17	10.00	**40.89**	1.40	0.18	**21.82**	0.08	**62.78**	m2	**67.49**
Sapele, wrought, selected for transparent finish; fixing to timber										
Boarding to walls, tongued, grooved and veed joints; 19mm thick, 100mm wide boards width exceeding 300mm	55.43	10.00	**60.97**	1.45	0.18	**22.53**	0.08	**83.58**	m2	**89.85**
Boarding to walls, tongued, grooved and veed joints; 19mm thick, 150mm wide boards width exceeding 300mm	46.36	10.00	**51.00**	1.00	0.12	**15.50**	0.05	**66.54**	m2	**71.53**
Boarding to walls, tongued, grooved and veed joints; 25mm thick, 100mm wide boards width exceeding 300mm	67.45	10.00	**74.20**	1.50	0.19	**23.35**	0.08	**97.62**	m2	**104.95**

Labour hourly rates: (except Specialists) Craft Operatives £14.21 Labourer £10.71 Rates are national average prices. Refer to REGIONAL VARIATIONS for indicative levels of overall pricing in regions	MATERIALS			LABOUR				RATES		
	Del to Site £	Waste %	Material Cost £	Craft Optve Hrs	Lab Hrs	Labour Cost £	Sunds £	Nett Rate £	Unit	Gross Rate (+7.5%) £
K20: TIMBER BOARD FLOORING/SHEATHING/LININGS/CASINGS: WALL AND CEILING BOARDING Cont.										
Sapele, wrought, selected for transparent finish; fixing to timber Cont.										
Boarding to walls, tongued, grooved and veed joints; 25mm thick, 150mm wide boards										
width exceeding 300mm	60.66	10.00	**66.73**	1.05	0.13	**16.31**	0.06	**83.09**	m2	**89.33**
Boarding to ceilings, tongued, grooved and veed joints; 19mm thick, 100mm wide boards										
width exceeding 300mm	55.43	10.00	**60.97**	2.00	0.25	**31.10**	0.11	**92.18**	m2	**99.09**
Boarding to ceilings, tongued, grooved and veed joints; 19mm thick, 150mm wide boards										
width exceeding 300mm	46.36	10.00	**51.00**	1.50	0.19	**23.35**	0.08	**74.43**	m2	**80.01**
Boarding to ceilings, tongued, grooved and veed joints; 25mm thick, 100mm wide boards										
width exceeding 300mm	67.45	10.00	**74.20**	2.05	0.26	**31.92**	0.11	**106.22**	m2	**114.19**
Boarding to ceilings, tongued, grooved and veed joints; 25mm thick, 150mm wide boards										
width exceeding 300mm	60.66	10.00	**66.73**	1.55	0.19	**24.06**	0.08	**90.87**	m2	**97.68**
Abutments										
19mm thick softwood boarding										
raking cutting	1.10	10.00	**1.21**	0.06	-	**0.85**	-	**2.07**	m	**2.22**
curved cutting	1.66	10.00	**1.82**	0.09	-	**1.28**	-	**3.10**	m	**3.33**
25mm thick softwood boarding										
raking cutting	1.44	10.00	**1.58**	0.07	-	**0.99**	-	**2.58**	m	**2.77**
curved cutting	2.16	10.00	**2.38**	0.10	-	**1.42**	-	**3.80**	m	**4.08**
19mm thick Western Red Cedar boarding										
raking cutting	4.27	10.00	**4.70**	0.07	-	**0.99**	-	**5.69**	m	**6.12**
curved cutting	6.41	10.00	**7.05**	0.10	-	**1.42**	-	**8.47**	m	**9.10**
25mm thick Western Red Cedar boarding										
raking cutting	5.36	10.00	**5.90**	0.08	-	**1.14**	-	**7.03**	m	**7.56**
curved cutting	8.04	10.00	**8.84**	0.11	-	**1.56**	-	**10.41**	m	**11.19**
19 mm thick hardwood boarding										
raking cutting	5.54	10.00	**6.10**	0.14	-	**1.99**	-	**8.09**	m	**8.69**
curved cutting	8.31	10.00	**9.15**	0.21	-	**2.98**	-	**12.13**	m	**13.04**
25 mm thick hardwood boarding										
raking cutting	6.75	10.00	**7.42**	0.16	-	**2.27**	-	**9.69**	m	**10.42**
curved cutting	10.12	10.00	**11.13**	0.24	-	**3.41**	-	**14.54**	m	**15.63**
Finished angles										
External; 19mm thick softwood boarding										
tongued and mitred	-	-	**-**	0.67	-	**9.52**	-	**9.52**	m	**10.23**
External; 25mm thick softwood boarding										
tongued and mitred	-	-	**-**	0.70	-	**9.95**	-	**9.95**	m	**10.69**
External; 19mm thick hardwood boarding										
tongued and mitred	-	-	**-**	0.95	-	**13.50**	-	**13.50**	m	**14.51**
External; 25mm thick hardwood boarding										
tongued and mitred	-	-	**-**	1.00	-	**14.21**	-	**14.21**	m	**15.28**
K20: TIMBER BOARD FLOORING/SHEATHING/LININGS/CASINGS: ROOF BOARDING										
Softwood, sawn, impregnated; fixing to timber										
Boarding to roofs, butt joints; 19mm thick, 75mm wide boards; external										
width exceeding 300mm	4.16	7.50	**4.47**	0.60	0.07	**9.28**	0.03	**13.78**	m2	**14.81**
width exceeding 300mm; sloping	4.16	7.50	**4.47**	0.70	0.09	**10.91**	0.04	**15.42**	m2	**16.58**
Boarding to roofs, butt joints; 25mm thick, 125mm wide boards; external										
width exceeding 300mm	5.47	7.50	**5.88**	0.55	0.07	**8.57**	0.03	**14.48**	m2	**15.56**
width exceeding 300mm; sloping	5.47	7.50	**5.88**	0.65	0.08	**10.09**	0.04	**16.01**	m2	**17.21**
Softwood, wrought, impregnated; fixing to timber										
Boarding to roofs, tongued and grooved joints; 19mm thick, 75mm wide boards external										
width exceeding 300mm	8.32	7.50	**8.94**	0.80	0.10	**12.44**	0.04	**21.43**	m2	**23.03**
extra; cross rebated and rounded drip; 50mm wide	0.86	10.00	**0.95**	0.20	0.03	**3.16**	0.01	**4.12**	m	**4.43**
extra; dovetailed cesspool 225 x 225 x 150mm	1.39	10.00	**1.53**	1.40	0.17	**21.71**	0.08	**23.32**	nr	**25.06**
extra; dovetailed cesspool 300 x 300 x 150mm	1.74	10.00	**1.91**	1.60	0.20	**24.88**	0.09	**26.88**	nr	**28.89**
width exceeding 300mm; sloping	8.32	7.50	**8.94**	0.90	0.11	**13.97**	0.05	**22.96**	m2	**24.68**
Boarding to roofs, tongued and grooved joints; 25mm thick, 125mm wide boards external										
width exceeding 300mm	9.96	7.50	**10.71**	0.75	0.09	**11.62**	0.04	**22.37**	m2	**24.05**
extra; cross rebated and rounded drip; 50mm wide	0.86	10.00	**0.95**	0.20	0.03	**3.16**	0.01	**4.12**	m	**4.43**
extra; dovetailed cesspool 225 x 225 x 150mm	1.74	10.00	**1.91**	1.50	0.19	**23.35**	0.08	**25.34**	nr	**27.24**

Labour hourly rates: (except Specialists) Craft Operatives £14.21 Labourer £10.71 Rates are national average prices. Refer to REGIONAL VARIATIONS for indicative levels of overall pricing in regions	MATERIALS			LABOUR				RATES		
	Del to Site £	Waste %	Material Cost £	Craft Optve Hrs	Lab Hrs	Labour Cost £	Sunds £	Nett Rate £	Unit	Gross Rate (+7.5%) £
K20: TIMBER BOARD FLOORING/SHEATHING/LININGS/CASINGS: ROOF BOARDING Cont.										
Softwood, wrought, impregnated; fixing to timber Cont.										
Boarding to roofs, tongued and grooved joints; 25mm thick, 125mm wide boards external Cont.										
extra; dovetailed cesspool 300 x 300 x 150mm.....	2.34	10.00	**2.57**	1.70	0.21	**26.41**	0.09	**29.07**	nr	**31.25**
width exceeding 300mm; sloping..................	9.96	7.50	**10.71**	0.85	0.11	**13.26**	0.05	**24.01**	m2	**25.81**
Abutments										
19mm thick softwood boarding										
raking cutting	0.83	7.50	**0.89**	0.35	-	**4.97**	-	**5.87**	m	**6.31**
curved cutting	1.25	7.50	**1.34**	0.55	-	**7.82**	-	**9.16**	m	**9.84**
rebates	-	-	**-**	0.18	-	**2.56**	-	**2.56**	m	**2.75**
grooves	-	-	**-**	0.18	-	**2.56**	-	**2.56**	m	**2.75**
chamfers	-	-	**-**	0.13	-	**1.85**	-	**1.85**	m	**1.99**
25mm thick softwood boarding										
raking cutting	0.98	7.50	**1.06**	0.45	-	**6.39**	-	**7.45**	m	**8.01**
curved cutting	1.48	7.50	**1.59**	0.65	-	**9.24**	-	**10.82**	m	**11.64**
rebates	-	-	**-**	0.18	-	**2.56**	-	**2.56**	m	**2.75**
grooves	-	-	**-**	0.18	-	**2.56**	-	**2.56**	m	**2.75**
chamfers	-	-	**-**	0.13	-	**1.85**	-	**1.85**	m	**1.99**
K21: TIMBER STRIP/BOARD FINE FLOORING/LININGS: STRIP FLOORING										
Hardwood, wrought, selected for transparent finish; sanded, two coats sealer finish										
Boarding to floors, tongued and grooved joints; 22mm thick, 75mm wide boards; fixing on and with 25 x 50mm impregnated softwood battens to masonry										
width exceeding 300 mm; secret fixing to timber..	36.71	7.50	**39.44**	1.25	0.15	**19.37**	0.07	**58.88**	m2	**63.29**
Boarding to floors, tongued and grooved joints; 12mm thick, 75mm wide boards; overlay										
width exceeding 300 mm; secret fixing to timber..	29.93	7.50	**32.18**	1.00	0.13	**15.60**	0.05	**47.83**	m2	**51.42**
K21: TIMBER STRIP/BOARD FINE FLOORING/LININGS: SEMI-SPRUNG FLOORS										
Foam backed softwood battens covered with hardwood tongued and grooved strip flooring										
22mm Standard Beech strip flooring to level sub-floors width exceeding 300mm; battens at approximately 400mm centres; elevation										
75mm..	-	-	**-**	-	-	**-**	-	**49.10**	m2	**52.78**
22mm 'Sylva Squash' Beech strip flooring to level sub-floors width exceeding 300mm; battens at approximately 430mm centres; elevation										
75mm..	-	-	**-**	-	-	**-**	-	**52.50**	m2	**56.44**
20mm Prime Maple strip flooring to level sub-floors width exceeding 300mm; battens at approximately 300mm centres; elevation										
75mm..	-	-	**-**	-	-	**-**	-	**39.95**	m2	**42.95**
20mm First Grade Maple strip flooring to level sub-floors width exceeding 300mm; battens at approximately 300mm centres; elevation										
75mm..	-	-	**-**	-	-	**-**	-	**43.37**	m2	**46.62**
K40: DEMOUNTABLE SUSPENDED CEILINGS										
Suspended ceilings; 1200 x 600mm bevelled, grooved and rebated asbestos-free fire resisting tiles; laying in position in metal suspension system of main channel members on wire or rod hangers										
Depth of suspension 150 - 500mm										
9mm thick linings; fixing hangers to masonry.....	15.80	5.00	**16.59**	0.50	0.25	**9.78**	4.35	**30.72**	m2	**33.03**
9mm thick linings not exceeding 300mm wide; fixing hangers to masonry	4.74	10.00	**5.21**	0.30	0.08	**5.07**	4.35	**14.63**	m	**15.73**
Suspended ceilings; 600 x 600mm bevelled, grooved and rebated asbestos-free fire resisting tiles; laying in position in metal suspension system of main channel members on wire or rod hangers										
Depth of suspension 150 - 500mm										
9mm thick linings; fixing hangers to masonry.....	15.45	5.00	**16.22**	0.60	0.30	**11.74**	4.35	**32.31**	m2	**34.73**
9mm thick linings not exceeding 300mm wide; fixing hangers to masonry	4.64	10.00	**5.10**	0.36	0.09	**6.08**	4.35	**15.53**	m	**16.69**

Labour hourly rates: (except Specialists) Craft Operatives £14.21 Labourer £10.71 Rates are national average prices. Refer to REGIONAL VARIATIONS for indicative levels of overall pricing in regions	MATERIALS			LABOUR				RATES		
	Del to Site £	Waste %	Material Cost £	Craft Optve Hrs	Lab Hrs	Labour Cost £	Sunds £	Nett Rate £	Unit	Gross Rate (+7.5%) £
K40: DEMOUNTABLE SUSPENDED CEILINGS Cont.										
Suspended ceilings; 1200 x 600mm bevelled, grooved and rebated plain mineral fibre tiles; laying in position in metal suspension system of main channel members on wire or rod hangers										
Depth of suspension 150 - 500mm										
15.8mm thick linings; fixing hangers to masonry..	9.60	5.00	**10.08**	0.50	0.25	**9.78**	4.35	**24.21**	m2	**26.03**
15.8mm thick linings not exceeding 300mm wide; fixing hangers to masonry.......................	2.88	10.00	**3.17**	0.30	0.08	**5.07**	2.60	**10.83**	m	**11.65**
Suspended ceilings; 600 x 600mm bevelled, grooved and rebated plain mineral fibre tiles; laying in position in metal suspension system of main channel members on wire or rod hangers										
Depth of suspension 150 - 500mm										
15.8mm thick linings; fixing hangers to masonry..	9.60	5.00	**10.08**	0.60	0.30	**11.74**	2.65	**24.47**	m2	**26.30**
15.8mm thick linings not exceeding 300mm wide; fixing hangers to masonry.......................	3.60	10.00	**3.96**	0.36	0.09	**6.08**	2.60	**12.64**	m	**13.59**
Suspended ceilings; 1200 x 600mm bevelled, grooved and rebated textured mineral fibre tiles; laying in position in metal suspension system of main channel members on wire or rod hangers										
Depth of suspension 150 - 500mm										
15.8mm thick linings; fixing hangers to masonry..	12.00	5.00	**12.60**	0.50	0.25	**9.78**	4.35	**26.73**	m2	**28.74**
15.8mm thick linings not exceeding 300mm wide; fixing hangers to masonry.......................	3.60	10.00	**3.96**	0.30	0.08	**5.07**	2.60	**11.63**	m	**12.50**
Suspended ceilings; 600 x 600mm bevelled, grooved and rebated regular drilled mineral fibre tiles; laying in position in metal suspension system of main channel members on wire or rod hangers										
Depth of suspension 150 - 500mm										
15.8mm thick linings; fixing hangers to masonry..	12.00	5.00	**12.60**	0.60	0.30	**11.74**	2.65	**26.99**	m2	**29.01**
15.8mm thick linings not exceeding 300mm wide; fixing hangers to masonry.......................	3.56	10.00	**3.91**	0.36	0.09	**6.08**	1.60	**11.59**	m	**12.46**
Suspended ceilings; 1200 x 600mm bevelled, grooved and rebated patterned tiles p.c. £15.00/m2; laying in position in metal suspension system of main channel members on wire or rod hangers										
Depth of suspension 150 - 500mm										
15.8mm thick linings; fixing hangers to masonry..	15.75	5.00	**16.54**	0.50	0.25	**9.78**	4.35	**30.67**	m2	**32.97**
15.8mm thick linings not exceeding 300mm wide; fixing hangers to masonry.......................	4.95	10.00	**5.45**	0.30	0.08	**5.07**	2.60	**13.11**	m	**14.09**
Suspended ceilings; 600 x 600mm bevelled, grooved and rebated patterned tiles p.c. £15.00/m2; laying in position in metal suspension system of main channel members on wire or rod hangers										
Depth of suspension 150 - 500mm										
15.8mm thick linings; fixing hangers to masonry..	15.75	5.00	**16.54**	0.60	0.30	**11.74**	2.65	**30.93**	m2	**33.25**
15.8mm thick linings not exceeding 300mm wide; fixing hangers to masonry.......................	4.95	10.00	**5.45**	0.36	0.09	**6.08**	1.60	**13.12**	m	**14.11**
Suspended ceiling; British Gypsum Ltd, 1200 x 600mm Glasroc GRG tiles; fixing into Casoline MF grid system										
Depth of suspension 150-500mm										
10mm thick linings; fixing hangers to masonry....	5.15	5.00	**5.41**	0.60	0.30	**11.74**	4.35	**21.50**	m2	**23.11**
10mm thick linings not exceeding 300mm wide; fixing hangers to masonry.......................	1.55	10.00	**1.70**	0.36	0.09	**6.08**	4.35	**12.13**	m	**13.04**
Edge trims										
Plain										
angle section; white stove enamelled aluminium; fixing to masonry with screws at 450mm centres...	0.95	5.00	**1.00**	0.40	0.20	**7.83**	-	**8.82**	m	**9.49**
K41: RAISED ACCESS FLOORS										
Flooring; Microfloor 'Bonded 600' light grade full access system; 600 x 600mm high density particle board panels, BS EN 312 and DIN 68761; 100 x 100mm precast lightweight concrete pedestals at 600mm centres fixed to sub-floor with epoxy resin adhesive										
Thickness of panel 30mm										
finished floor height										
50mm.......................................	-	-	**-**	-	-	**-**	-	**16.73**	m2	**17.98**
75mm.......................................	-	-	**-**	-	-	**-**	-	**16.85**	m2	**18.11**
100mm......................................	-	-	**-**	-	-	**-**	-	**17.06**	m2	**18.34**
125mm......................................	-	-	**-**	-	-	**-**	-	**17.24**	m2	**18.53**
150mm......................................	-	-	**-**	-	-	**-**	-	**17.59**	m2	**18.91**
175mm......................................	-	-	**-**	-	-	**-**	-	**17.87**	m2	**19.21**
200mm......................................	-	-	**-**	-	-	**-**	-	**18.03**	m2	**19.38**

Labour hourly rates: (except Specialists) Craft Operatives £14.21 Labourer £10.71 Rates are national average prices. Refer to REGIONAL VARIATIONS for indicative levels of overall pricing in regions	MATERIALS			LABOUR				RATES		
	Del to Site £	Waste %	Material Cost £	Craft Optve Hrs	Lab Hrs	Labour Cost £	Sunds £	Nett Rate £	Unit	Gross Rate (+7.5%) £
K41: RAISED ACCESS FLOORS Cont.										
Flooring; Microfloor 'Bonded 600' medium grade full access system; 600 x 600mm high density particle board panels, BS EN 312 and DIN 68761; 100 x 100mm precast lightweight concrete pedestals at 600mm centres fixed to sub-floor with epoxy resin adhesive										
Thickness of panel 38mm										
finished floor height										
50mm	-	-	-	-	-	-	-	**19.03**	m2	**20.46**
75mm	-	-	-	-	-	-	-	**19.03**	m2	**20.46**
100mm	-	-	-	-	-	-	-	**19.39**	m2	**20.84**
125mm	-	-	-	-	-	-	-	**19.56**	m2	**21.03**
150mm	-	-	-	-	-	-	-	**20.19**	m2	**21.70**
175mm	-	-	-	-	-	-	-	**20.36**	m2	**21.89**
200mm	-	-	-	-	-	-	-	**20.51**	m2	**22.05**
Flooring; Microfloor 'Bonded 600' office loadings grade full access system; 600 x 600mm high density particle board panels, BS EN 634 and DIN 68761; 100 x 100mm precast lightweight concrete pedestals at 600mm centres fixed to sub-floor with epoxy resin adhesive										
Thickness of panel 30mm										
finished floor height										
50mm	-	-	-	-	-	-	-	**13.03**	m2	**14.01**
75mm	-	-	-	-	-	-	-	**13.10**	m2	**14.08**
100mm	-	-	-	-	-	-	-	**13.39**	m2	**14.39**
125mm	-	-	-	-	-	-	-	**13.56**	m2	**14.58**
150mm	-	-	-	-	-	-	-	**13.89**	m2	**14.93**
175mm	-	-	-	-	-	-	-	**14.20**	m2	**15.27**
200mm	-	-	-	-	-	-	-	**14.35**	m2	**15.43**

WINDOWS/DOORS/STAIRS

Labour hourly rates: (except Specialists) Craft Operatives £14.21 Labourer £10.71 Rates are national average prices. Refer to REGIONAL VARIATIONS for indicative levels of overall pricing in regions	MATERIALS			LABOUR				RATES		
	Del to Site £	Waste %	Material Cost £	Craft Optve Hrs	Lab Hrs	Labour Cost £	Sunds £	Nett Rate £	Unit	Gross Rate (+7.5%) £
L10: WINDOWS/ROOFLIGHTS/SCREENS/LOUVRES: TIMBER WINDOWS										
Casements in softwood, base coat stain by manufacturer										
Magnet Trade; softwood Statesman Plus casement windows without bars; softwood sub-sills; hinges; fasteners										
488 x 900mm overall; N09V	73.95	2.50	**75.80**	1.20	0.60	**23.48**	1.40	**100.68**	nr	**108.23**
631 x 750mm overall; 107V	86.05	2.50	**88.20**	1.20	0.60	**23.48**	1.40	**113.08**	nr	**121.56**
631 x 900mm overall; 109V	89.73	2.50	**91.98**	1.30	0.65	**25.43**	1.40	**118.81**	nr	**127.72**
631 x 1050mm overall; 110V	93.14	2.50	**95.46**	1.50	0.75	**29.35**	1.40	**126.21**	nr	**135.68**
631 x 1200mm overall; 112v	96.73	2.50	**99.14**	1.60	0.80	**31.30**	1.40	**131.85**	nr	**141.74**
915 x 900mm overall; 2N09W	111.94	2.50	**114.74**	1.70	0.85	**33.26**	1.40	**149.40**	nr	**160.60**
915 x 1050mm overall; 2N10W	109.77	2.50	**112.51**	1.70	0.85	**33.26**	1.40	**147.17**	nr	**158.21**
915 x 1200mm overall; 2N12W	119.41	2.50	**122.39**	1.80	0.90	**35.22**	1.40	**159.01**	nr	**170.93**
1200 x 900mm overall; 209W	130.94	2.50	**134.21**	1.70	0.85	**33.26**	1.40	**168.87**	nr	**181.53**
1200 x 1050mm overall; 210CV	158.72	2.50	**162.69**	1.90	0.95	**37.17**	1.40	**201.26**	nr	**216.35**
1200 x 1050mm overall; 210W	133.39	2.50	**136.73**	1.90	0.95	**37.17**	1.40	**175.30**	nr	**188.45**
1200 x 1200mm overall; 212CV	158.53	2.50	**162.49**	2.00	1.00	**39.13**	1.40	**203.02**	nr	**218.25**
1200 x 1200mm overall; 212W	141.24	2.50	**144.77**	2.00	1.00	**39.13**	1.40	**185.30**	nr	**199.19**
1524 x 1050mm overall; 3NN10WW	171.29	2.50	**175.57**	2.10	1.05	**41.09**	1.40	**218.06**	nr	**234.41**
1769 x 1050mm overall; 310CVC	229.78	2.50	**235.53**	2.40	1.20	**46.96**	1.40	**283.88**	nr	**305.17**
1769 x 1200mm overall; 312CVC	257.47	2.50	**263.91**	2.50	1.25	**48.91**	1.40	**314.22**	nr	**337.79**
Fully reversible windows in softwood, wrought; base coat stain by manufacturer										
Magnet Trade Hi-Profile; softwood horizontal tilt windows; locking catch; high security espagnolette bolt; fully weatherstripped and Factory Double Glazed; clear low E glass										
600 x 900mm overall; HP0906	269.80	2.50	**276.54**	1.35	0.67	**26.36**	1.40	**304.30**	nr	**327.12**
900 x 900mm overall; HP0909	298.41	2.50	**305.87**	1.60	0.80	**31.30**	1.40	**338.57**	nr	**363.97**
900 x 1050mm overall; HP0910	310.60	2.50	**318.37**	1.70	0.85	**33.26**	1.40	**353.03**	nr	**379.50**
900 x 1200mm overall; HP0912	324.52	2.50	**332.64**	1.80	0.90	**35.22**	1.40	**369.25**	nr	**396.95**
1200 x 900mm overall; HP1209	335.24	2.50	**343.63**	1.80	0.90	**35.22**	1.40	**380.24**	nr	**408.76**
1200 x 1050mm overall; HP1210	354.73	2.50	**363.60**	1.90	0.95	**37.17**	1.40	**402.17**	nr	**432.34**
1200 x 1200mm overall; HP1212	371.87	2.50	**381.16**	2.00	1.00	**39.13**	1.40	**421.69**	nr	**453.32**
Sash windows in softwood, wrought; base coat stain by manufacturer										
Magnet Victorian Statesman plus non bar sash softwood windows; glazing beads; fasteners; base coat stain by manufacturer										
1200 x 900mm overall; VBS209B	178.73	2.50	**183.20**	2.40	1.20	**46.96**	1.40	**231.56**	nr	**248.92**
1200 x 1050mm overall; VBS210B	187.16	2.50	**191.83**	2.55	1.27	**49.84**	1.40	**243.07**	nr	**261.30**
1200 x 1200mm overall; VBS212B	187.16	2.50	**191.83**	2.75	1.37	**53.75**	1.40	**246.98**	nr	**265.51**
1200 x 1350mm overall; VBS213B	193.71	2.50	**198.55**	3.00	1.50	**58.70**	1.40	**258.64**	nr	**278.04**
1769 x 1050mm overall; VBS310BB	337.85	2.50	**346.30**	3.10	1.55	**60.65**	1.40	**408.35**	nr	**438.97**
1769 x 1200mm overall; VBS312BB	277.26	2.50	**284.19**	3.40	1.70	**66.52**	1.40	**352.12**	nr	**378.52**
1769 x 1350mm overall; VBS313BB	348.57	2.50	**357.28**	3.60	1.80	**70.43**	1.40	**429.12**	nr	**461.30**
1769 x 1050mm overall; VBN310BBMOE	552.92	2.50	**566.75**	3.10	1.55	**60.65**	1.40	**628.80**	nr	**675.96**
1769 x 1200mm overall; VBN312BBMOE	569.03	2.50	**583.26**	3.40	1.70	**66.52**	1.40	**651.18**	nr	**700.02**
1769 x 1350mm overall; VBN313BBMOE	580.54	2.50	**595.05**	3.60	1.80	**70.43**	1.40	**666.88**	nr	**716.90**
Windows in softwood, base coat stained by manufacturer										
Jeld-Wen Ltd. Softwood Sovereign windows plain casement non bar type side hung; glazing beads; weatherstripping, hinges; fasteners; base coat stain by manufacturer										
630 x 750mm overall; W107C	94.46	2.50	**96.82**	1.60	0.80	**31.30**	1.40	**129.53**	nr	**139.24**
630 x 900mm overall; W109C	100.23	2.50	**102.74**	1.85	0.92	**36.14**	1.40	**140.28**	nr	**150.80**
630 x 1050mm overall; W110C	100.64	2.50	**103.16**	2.05	1.02	**40.05**	1.40	**144.61**	nr	**155.46**
630 x 1200mm overall; W112C	104.39	2.50	**107.00**	2.20	1.10	**43.04**	1.40	**151.44**	nr	**162.80**
630 x 1350mm overall; W113C	117.04	2.50	**119.97**	2.40	1.20	**46.96**	1.40	**168.32**	nr	**180.95**
1200 x 750mm overall; W207C	123.32	2.50	**126.40**	2.15	1.07	**42.01**	1.40	**169.81**	nr	**182.55**
1200 x 900mm overall; W209C	126.17	2.50	**129.32**	2.40	1.20	**46.96**	1.40	**177.68**	nr	**191.01**
1200 x 1050mm overall; W210C	128.68	2.50	**131.90**	2.55	1.27	**49.84**	1.40	**183.13**	nr	**196.87**
1200 x 1200mm overall; W212C	131.19	2.50	**134.47**	2.75	1.37	**53.75**	1.40	**189.62**	nr	**203.84**

Labour hourly rates: (except Specialists) Craft Operatives £14.21 Labourer £10.71 Rates are national average prices. Refer to REGIONAL VARIATIONS for indicative levels of overall pricing in regions	MATERIALS			LABOUR				RATES		
	Del to Site £	Waste %	Material Cost £	Craft Optve Hrs	Lab Hrs	Labour Cost £	Sunds £	Nett Rate £	Unit	Gross Rate (+7.5%) £
L10: WINDOWS/ROOFLIGHTS/SCREENS/LOUVRES: TIMBER WINDOWS Cont.										
Windows in softwood, base coat stained by manufacturer Cont.										
Jeld-Wen Ltd. Softwood Sovereign windows plain casement non bar type side hung; glazing beads; weatherstripping, hinges; fasteners; base coat stain by manufacturer Cont.										
1200 x 1350mm overall; W213C	147.05	2.50	**150.73**	2.95	1.47	**57.66**	1.40	**209.79**	nr	**225.52**
1770 x 750mm overall; W307CC	181.54	2.50	**186.08**	2.75	1.37	**53.75**	1.40	**241.23**	nr	**259.32**
1770 x 900mm overall; W309CC	179.70	2.50	**184.19**	2.95	1.47	**57.66**	1.40	**243.26**	nr	**261.50**
1770 x 1050mm overall; W310C; Landscape	151.27	2.50	**155.05**	3.10	1.55	**60.65**	1.40	**217.10**	nr	**233.38**
1770 x 1050mm overall; W310CC	186.61	2.50	**191.28**	3.10	1.55	**60.65**	1.40	**253.33**	nr	**272.33**
1770 x 1200mm overall; W312C; Landscape	193.40	2.50	**198.24**	3.35	1.67	**65.49**	1.40	**265.12**	nr	**285.01**
1770 x 1200mm overall; W312CC	186.94	2.50	**191.61**	3.35	1.67	**65.49**	1.40	**258.50**	nr	**277.89**
1770 x 1350mm overall; W313C; Landscape	163.57	2.50	**167.66**	3.60	1.80	**70.43**	1.40	**239.49**	nr	**257.46**
1770 x 1350mm overall; W313CC	218.14	2.50	**223.59**	3.60	1.80	**70.43**	1.40	**295.43**	nr	**317.58**
2339 x 900mm overall; W409CMC	232.36	2.50	**238.17**	3.60	1.80	**70.43**	1.40	**310.00**	nr	**333.25**
2339 x 1050mm overall; W410CMC	242.22	2.50	**248.28**	3.85	1.92	**75.27**	1.40	**324.95**	nr	**349.32**
2339 x 1200mm overall; W412CMC	252.44	2.50	**258.75**	4.05	2.02	**79.18**	1.40	**339.34**	nr	**364.79**
2339 x 1350mm overall; W413CMC	268.53	2.50	**275.24**	4.25	2.12	**83.10**	1.40	**359.74**	nr	**386.72**
Jeld-Wen Ltd. Softwood Alpha Sovereign windows casement with vents top hung type; glazing beads; weatherstripping, hinges; fasteners; base coat stain by manufacturer										
630 x 750mm overall; W107V	92.59	2.50	**94.90**	1.50	0.75	**29.35**	1.40	**125.65**	nr	**135.08**
630 x 900mm overall; W109V	92.89	2.50	**95.21**	1.60	0.80	**31.30**	1.40	**127.92**	nr	**137.51**
630 x 1050mm overall; W110V	95.08	2.50	**97.46**	1.85	0.92	**36.14**	1.40	**135.00**	nr	**145.12**
630 x 1200mm overall; W112V	102.57	2.50	**105.13**	2.05	1.02	**40.05**	1.40	**146.59**	nr	**157.58**
630 x 1350mm overall; W113V	105.48	2.50	**108.12**	2.20	1.10	**43.04**	1.40	**152.56**	nr	**164.00**
915 x 750mm overall; W2N07CV	106.81	2.50	**109.48**	1.80	0.90	**35.22**	1.40	**146.10**	nr	**157.06**
915 x 900mm overall; W2N09CV	109.59	2.50	**112.33**	1.95	0.97	**38.10**	1.40	**151.83**	nr	**163.22**
915 x 1050mm overall; W2N10CV	111.08	2.50	**113.86**	2.15	1.07	**42.01**	1.40	**157.27**	nr	**169.07**
915 x 1200mm overall; W2N12CV	119.86	2.50	**122.86**	2.35	1.17	**45.92**	1.40	**170.18**	nr	**182.95**
1200 x 900mm overall; W209T	155.69	2.50	**159.58**	2.15	1.07	**42.01**	1.40	**202.99**	nr	**218.22**
1200 x 1050mm overall; W210T	155.89	2.50	**159.79**	2.40	1.20	**46.96**	1.40	**208.14**	nr	**223.75**
1200 x 1200mm overall; W212T	160.83	2.50	**164.85**	2.55	1.27	**49.84**	1.40	**216.09**	nr	**232.29**
1200 x 1350mm overall; W213T	178.08	2.50	**182.53**	2.75	1.37	**53.75**	1.40	**237.68**	nr	**255.51**
1200 x 1500mm overall; W215T	220.75	2.50	**226.27**	2.65	1.32	**51.79**	1.40	**279.46**	nr	**300.42**
1770 x 1050mm overall; W310TXT	227.95	2.50	**233.65**	2.75	1.37	**53.75**	1.40	**288.80**	nr	**310.46**
1770 x 1350mm overall; W313TXT	249.59	2.50	**255.83**	3.10	1.55	**60.65**	1.40	**317.88**	nr	**341.72**
1770 x 1500mm overall; W315TXT	221.87	2.50	**227.41**	3.35	1.67	**65.49**	1.40	**294.30**	nr	**316.38**
Windows in hardwood, base coat stained by manufacturer										
Jeld-Wen Ltd. Hardwood Sovereign windows plain casement non bar type side hung; glazing beads; weatherstripping, hinges; fasteners; base coat stain by manufacturer										
630 x 750mm overall; W107CH	176.81	2.50	**181.23**	2.05	1.02	**40.05**	1.40	**222.68**	nr	**239.39**
630 x 900mm overall; W109CH	184.43	2.50	**189.04**	2.20	1.10	**43.04**	1.40	**233.48**	nr	**251.00**
630 x 1050mm overall; W110CH	192.42	2.50	**197.23**	2.20	1.10	**43.04**	1.40	**241.67**	nr	**259.80**
630 x 1200mm overall; W112CH	200.35	2.50	**205.36**	2.35	1.15	**45.71**	1.40	**252.47**	nr	**271.40**
630 x 1350mm overall; W113CH	214.93	2.50	**220.30**	2.35	1.15	**45.71**	1.40	**267.41**	nr	**287.47**
1200 x 750mm overall; W207CH	227.45	2.50	**233.14**	2.15	1.07	**42.01**	1.40	**276.55**	nr	**297.29**
1200 x 900mm overall; W209CH	238.77	2.50	**244.74**	2.40	1.20	**46.96**	1.40	**293.10**	nr	**315.08**
1200 x 1050mm overall; W210CH	247.41	2.50	**253.60**	2.55	1.37	**50.91**	1.40	**305.90**	nr	**328.85**
1200 x 1200mm overall; W212CH	257.88	2.50	**264.33**	2.75	1.37	**53.75**	1.40	**319.48**	nr	**343.44**
1200 x 1350mm overall; W213CH	275.44	2.50	**282.33**	2.95	1.47	**57.66**	1.40	**341.39**	nr	**366.99**
1770 x 750mm overall; W307CCH	345.13	2.50	**353.76**	2.75	1.37	**53.75**	1.40	**408.91**	nr	**439.58**
1770 x 900mm overall; W309CCH	349.82	2.50	**358.57**	2.95	1.47	**57.66**	1.40	**417.63**	nr	**448.95**
1770 x 1050mm overall; W310CH; Landscape	281.35	2.50	**288.39**	3.10	1.55	**60.65**	1.40	**350.44**	nr	**376.72**
1770 x 1050mm overall; W310CCH	376.41	2.50	**385.82**	3.10	1.55	**60.65**	1.40	**447.87**	nr	**481.46**
1770 x 1200mm overall; W312CH; Landscape	391.23	2.50	**401.01**	3.35	1.67	**65.49**	1.40	**467.90**	nr	**502.99**
1770 x 1200mm overall; W312CCH	378.15	2.50	**387.60**	3.35	1.67	**65.49**	1.40	**454.49**	nr	**488.58**
1770 x 1350mm overall; W313CH; Landscape	296.88	2.50	**304.30**	3.60	1.80	**70.43**	1.40	**376.14**	nr	**404.35**
1770 x 1350mm overall; W313CCH	419.23	2.50	**429.71**	3.60	1.80	**70.43**	1.40	**501.54**	nr	**539.16**
2339 x 900mm overall; W409CMCH	456.75	2.50	**468.17**	3.75	1.85	**73.10**	1.40	**542.67**	nr	**583.37**
2339 x 1050mm overall; W410CMCH	467.63	2.50	**479.32**	3.85	1.92	**75.27**	1.40	**555.99**	nr	**597.69**
2339 x 1200mm overall; W412CMCH	488.40	2.50	**500.61**	4.05	2.02	**79.18**	1.40	**581.19**	nr	**624.78**
2339 x 1350mm overall; W413CMCH	528.87	2.50	**542.09**	4.25	2.12	**83.10**	1.40	**626.59**	nr	**673.58**
Jeld-Wen Ltd. Hardwood Sovereign windows casement with vents; glazing beads; weatherstripping, hinges; fasteners; base coat stain by manufacturer										
630 x 750mm overall; W107VH	169.80	2.50	**174.05**	1.65	0.80	**32.01**	1.40	**207.46**	nr	**223.02**
630 x 900mm overall; W109VH	175.77	2.50	**180.16**	1.85	0.92	**36.14**	1.40	**217.71**	nr	**234.03**
630 x 1050mm overall; W110VH	180.27	2.50	**184.78**	2.05	1.02	**40.05**	1.40	**226.23**	nr	**243.20**
630 x 1200mm overall; W112VH	185.35	2.50	**189.98**	2.20	1.10	**43.04**	1.40	**234.43**	nr	**252.01**
630 x 1350mm overall; W113VH	191.46	2.50	**196.25**	2.40	1.20	**46.96**	1.40	**244.60**	nr	**262.95**
915 x 750mm overall; W2N07CVH	204.75	2.50	**209.87**	2.05	1.02	**40.05**	1.40	**251.33**	nr	**270.18**
915 x 900mm overall; W2N09CVH	212.34	2.50	**217.65**	2.15	1.07	**42.01**	1.40	**261.06**	nr	**280.64**
915 x 1050mm overall; W2N10CVH	215.76	2.50	**221.16**	2.35	1.17	**45.92**	1.40	**268.48**	nr	**288.62**
915 x 1200mm overall; W2N12CVH	221.18	2.50	**226.71**	2.50	1.25	**48.91**	1.40	**277.02**	nr	**297.80**
1200 x 900mm overall; W209TH	301.43	2.50	**308.97**	2.40	1.20	**46.96**	1.40	**357.32**	nr	**384.12**
1200 x 1050mm overall; W210TH	310.70	2.50	**318.47**	2.55	1.27	**49.84**	1.40	**369.70**	nr	**397.43**
1200 x 1200mm overall; W212TH	321.66	2.50	**329.70**	2.75	1.37	**53.75**	1.40	**384.85**	nr	**413.72**
1200 x 1350mm overall; W213TH	338.29	2.50	**346.75**	3.00	1.50	**58.70**	1.40	**406.84**	nr	**437.36**
1200 x 1500mm overall; W215TH	424.32	2.50	**434.93**	3.00	1.50	**58.70**	1.40	**495.02**	nr	**532.15**
1770 x 1050mm overall; W310TXTH	438.84	2.50	**449.81**	3.10	1.55	**60.65**	1.40	**511.86**	nr	**550.25**
1770 x 1350mm overall; W313TXTH	482.77	2.50	**494.84**	3.60	1.80	**70.43**	1.40	**566.67**	nr	**609.17**
1770 x 1500mm overall; W315TXTH	414.76	2.50	**425.13**	3.80	1.90	**74.35**	1.40	**500.88**	nr	**538.44**

Labour hourly rates: (except Specialists) Craft Operatives £14.21 Labourer £10.71 Rates are national average prices. Refer to REGIONAL VARIATIONS for indicative levels of overall pricing in regions	MATERIALS			LABOUR				RATES		
	Del to Site £	Waste %	Material Cost £	Craft Optve Hrs	Lab Hrs	Labour Cost £	Sunds £	Nett Rate £	Unit	Gross Rate (+7.5%) £
L10: WINDOWS/ROOFLIGHTS/SCREENS/LOUVRES: TIMBER WINDOWS Cont.										
Windows in hardwood, base coat stained by manufacturer Cont.										
Jeld-Wen Ltd. Hardwood Sovereign windows, horizontal glazing bar type; glazing beads; weatherstripping, hinges; fasteners; base coat stain by manufacturer										
630 x 900mm overall; WH109CH	111.11	2.50	113.89	1.85	0.92	36.14	1.40	151.43	nr	162.79
630 x 1050mm overall; WH110CH	114.62	2.50	117.49	2.05	1.02	40.05	1.40	158.94	nr	170.86
630 x 1200mm overall; WH112CH	118.38	2.50	121.34	2.20	1.10	43.04	1.40	165.78	nr	178.22
630 x 1350mm overall; WH113CH	125.68	2.50	128.82	2.40	1.20	46.96	1.40	177.18	nr	190.47
1200 x 900mm overall; WH209CH	194.66	2.50	199.53	2.40	1.20	46.96	1.40	247.88	nr	266.47
1200 x 1050mm overall; WH210CH	201.50	2.50	206.54	2.55	1.27	49.84	1.40	257.77	nr	277.11
1200 x 1200mm overall; WH212CH	208.88	2.50	214.10	2.75	1.37	53.75	1.40	269.25	nr	289.45
1200 x 1350mm overall; WH213CH	223.95	2.50	229.55	3.00	1.50	58.70	1.40	289.64	nr	311.37
1770 x 1050mm overall; WH310CCH	224.13	2.50	229.73	3.10	1.55	60.65	1.40	291.78	nr	313.67
1770 x 1200mm overall; WH312CCH	231.63	2.50	237.42	3.40	1.70	66.52	1.40	305.34	nr	328.24
1770 x 1350mm overall; WH313CCH	245.60	2.50	251.74	3.60	1.80	70.43	1.40	323.57	nr	347.84
2339 x 1050mm overall; WH410CMCH	279.42	2.50	286.41	3.85	1.77	73.67	1.40	361.47	nr	388.58
2339 x 1200mm overall; WH412CMCH	289.16	2.50	296.39	4.05	2.02	79.18	1.40	376.97	nr	405.25
2339 x 1350mm overall; WH413CMCH	305.79	2.50	313.43	4.30	2.15	84.13	1.40	398.96	nr	428.89
Purpose made windows in softwood, wrought										
38mm moulded casements or fanlights										
in one pane	92.30	2.50	94.61	1.50	0.19	23.35	0.24	118.20	m2	127.06
divided into panes 0.10 - 0.50m2	106.15	2.50	108.80	1.50	0.19	23.35	0.24	132.39	m2	142.32
divided into panes not exceeding 0.10m2	119.99	2.50	122.99	1.50	0.19	23.35	0.24	146.58	m2	157.57
50mm moulded casements or fanlights										
in one pane	106.15	2.50	108.80	1.60	0.20	24.88	0.26	133.94	m2	143.98
divided into panes 0.10 - 0.50m2	119.99	2.50	122.99	1.60	0.20	24.88	0.26	148.13	m2	159.24
divided into panes not exceeding 0.10m2	133.84	2.50	137.18	1.60	0.20	24.88	0.26	162.32	m2	174.49
38mm moulded casements with semi-circular heads (measured square)										
in one pane	184.60	2.50	189.22	1.65	0.21	25.70	0.27	215.18	m2	231.32
divided into panes 0.10 - 0.50m2	198.45	2.50	203.41	1.65	0.21	25.70	0.27	229.37	m2	246.57
divided into panes not exceeding 0.10m2	212.29	2.50	217.60	1.65	0.21	25.70	0.27	243.56	m2	261.83
50mm moulded casements with semi-circular heads (measured square)										
in one pane	212.29	2.50	217.60	1.75	0.22	27.22	0.28	245.10	m2	263.48
divided into panes 0.10 - 0.50m2	226.14	2.50	231.79	1.75	0.22	27.22	0.28	259.29	m2	278.74
divided into panes not exceeding 0.10m2	239.98	2.50	245.98	1.75	0.22	27.22	0.28	273.48	m2	293.99
38mm bullseye casements										
457mm diameter in one pane	138.45	2.50	141.91	1.00	0.12	15.50	0.16	157.57	nr	169.38
762mm diameter in one pane	184.60	2.50	189.22	1.50	0.19	23.35	0.24	212.80	nr	228.77
50mm bullseye casements										
457mm diameter in one pane	147.68	2.50	151.37	1.10	0.14	17.13	0.18	168.68	nr	181.33
762mm diameter in one pane	198.45	2.50	203.41	1.65	0.21	25.70	0.27	229.37	nr	246.57
Labours										
check throated edge	0.47	2.50	0.48	-	-	-	-	0.48	m	0.52
rebated and splayed bottom rail	0.47	2.50	0.48	-	-	-	-	0.48	m	0.52
rebated and beaded meeting stile	0.65	2.50	0.67	-	-	-	-	0.67	m	0.72
fitting and hanging casement or fanlight on butts (included elsewhere)										
38mm	-	-	-	0.67	0.08	10.38	-	10.38	nr	11.16
50mm	-	-	-	0.74	0.09	11.48	-	11.48	nr	12.34
fitting and hanging casement or fanlight on sash centres (included elsewhere)										
38mm	-	-	-	1.50	0.19	23.35	-	23.35	nr	25.10
50mm	-	-	-	1.65	0.21	25.70	-	25.70	nr	27.62
Purpose made windows in Afrormosia, wrought										
38mm moulded casements or fanlights										
in one pane	178.00	2.50	182.45	2.25	0.28	34.97	0.37	217.79	m2	234.13
divided into panes 0.10 - 0.50m2	204.70	2.50	209.82	2.25	0.28	34.97	0.37	245.16	m2	263.55
divided into panes not exceeding 0.10m2	231.40	2.50	237.19	2.25	0.28	34.97	0.37	272.53	m2	292.97
50mm moulded casements or fanlights										
in one pane	204.70	2.50	209.82	2.40	0.30	37.32	0.39	247.52	m2	266.09
divided into panes 0.10 - 0.50m2	231.40	2.50	237.19	2.40	0.30	37.32	0.39	274.89	m2	295.51
divided into panes not exceeding 0.10m2	258.10	2.50	264.55	2.40	0.30	37.32	0.39	302.26	m2	324.93
38mm moulded casements with semi-circular heads (measured square)										
in one pane	356.00	2.50	364.90	2.45	0.31	38.13	0.40	403.43	m2	433.69
divided into panes 0.10 - 0.50m2	382.70	2.50	392.27	2.45	0.31	38.13	0.40	430.80	m2	463.11
divided into panes not exceeding 0.10m2	409.40	2.50	419.64	2.45	0.31	38.13	0.40	458.17	m2	492.53
50mm moulded casements with semi-circular heads (measured square)										
in one pane	409.40	2.50	419.64	2.60	0.32	40.37	0.42	460.43	m2	494.96
divided into panes 0.10 - 0.50m2	436.10	2.50	447.00	2.60	0.32	40.37	0.42	487.80	m2	524.38
divided into panes not exceeding 0.10m2	462.80	2.50	474.37	2.60	0.32	40.37	0.42	515.16	m2	553.80
38mm bullseye casements										
457mm diameter in one pane	267.00	2.50	273.68	1.50	0.19	23.35	0.24	297.26	nr	319.56
762mm diameter in one pane	356.00	2.50	364.90	2.25	0.28	34.97	0.37	400.24	nr	430.26

Labour hourly rates: (except Specialists) Craft Operatives £14.21 Labourer £10.71 Rates are national average prices. Refer to REGIONAL VARIATIONS for indicative levels of overall pricing in regions	MATERIALS			LABOUR				RATES		
	Del to Site £	Waste %	Material Cost £	Craft Optve Hrs	Lab Hrs	Labour Cost £	Sunds £	Nett Rate £	Unit	Gross Rate (+7.5%) £
L10: WINDOWS/ROOFLIGHTS/SCREENS/LOUVRES: TIMBER WINDOWS Cont.										
Purpose made windows in Afrormosia, wrought Cont.										
50mm bullseye casements										
457mm diameter in one pane	284.80	2.50	**291.92**	1.65	0.21	**25.70**	0.27	**317.89**	nr	**341.73**
762mm diameter in one pane	382.70	2.50	**392.27**	2.45	0.31	**38.13**	0.40	**430.80**	nr	**463.11**
Labours										
check throated edge	0.89	2.50	**0.91**	-	-	**-**	-	**0.91**	m	**0.98**
rebated and splayed bottom rail	0.89	2.50	**0.91**	-	-	**-**	-	**0.91**	m	**0.98**
rebated and beaded meeting stile	1.25	2.50	**1.28**	-	-	**-**	-	**1.28**	m	**1.38**
fitting and hanging casement or fanlight on butts (included elsewhere)										
38mm	-	-	**-**	1.00	0.12	**15.50**	-	**15.50**	nr	**16.66**
50mm	-	-	**-**	1.10	0.14	**17.13**	-	**17.13**	nr	**18.42**
fitting and hanging casement or fanlight on sash centres (included elsewhere)										
38mm	-	-	**-**	2.25	0.28	**34.97**	-	**34.97**	nr	**37.59**
50mm	-	-	**-**	2.45	0.31	**38.13**	-	**38.13**	nr	**40.99**
Purpose made windows in European Oak, wrought										
38mm moulded casements or fanlights										
in one pane	234.30	2.50	**240.16**	3.00	0.37	**46.59**	0.49	**287.24**	m2	**308.78**
divided into panes 0.10 - 0.50m2	269.45	2.50	**276.18**	3.00	0.37	**46.59**	0.49	**323.26**	m2	**347.51**
divided into panes not exceeding 0.10m2	304.59	2.50	**312.20**	3.00	0.37	**46.59**	0.49	**359.29**	m2	**386.23**
50mm moulded casements or fanlights										
in one pane	269.45	2.50	**276.18**	3.20	0.40	**49.76**	0.52	**326.46**	m2	**350.94**
divided into panes 0.10 - 0.50m2	304.59	2.50	**312.20**	3.20	0.40	**49.76**	0.52	**362.48**	m2	**389.67**
divided into panes not exceeding 0.10m2	339.74	2.50	**348.23**	3.20	0.40	**49.76**	0.52	**398.50**	m2	**428.39**
38mm moulded casements with semi-circular heads (measured square)										
in one pane	468.60	2.50	**480.32**	3.30	0.41	**51.28**	0.54	**532.14**	m2	**572.05**
divided into panes 0.10 - 0.50m2	503.75	2.50	**516.34**	3.30	0.41	**51.28**	0.54	**568.16**	m2	**610.77**
divided into panes not exceeding 0.10m2	538.89	2.50	**552.36**	3.30	0.41	**51.28**	0.54	**604.19**	m2	**649.50**
50mm moulded casements with semi-circular heads (measured square)										
in one pane	538.89	2.50	**552.36**	3.50	0.44	**54.45**	0.57	**607.38**	m2	**652.93**
divided into panes 0.10 - 0.50m2	574.04	2.50	**588.39**	3.50	0.44	**54.45**	0.57	**643.40**	m2	**691.66**
divided into panes not exceeding 0.10m2	609.18	2.50	**624.41**	3.50	0.44	**54.45**	0.57	**679.43**	m2	**730.38**
38mm bullseye casements										
457mm diameter in one pane	351.45	2.50	**360.24**	2.00	0.25	**31.10**	0.33	**391.66**	nr	**421.04**
762mm diameter in one pane	468.60	2.50	**480.32**	3.00	0.37	**46.59**	0.49	**527.40**	nr	**566.95**
50mm bullseye casements										
457mm diameter in one pane	374.88	2.50	**384.25**	2.20	0.27	**34.15**	0.36	**418.77**	nr	**450.17**
762mm diameter in one pane	503.75	2.50	**516.34**	3.30	0.41	**51.28**	0.54	**568.16**	nr	**610.77**
Labours										
check throated edge	1.18	2.50	**1.21**	-	-	**-**	-	**1.21**	m	**1.30**
rebated and splayed bottom rail	1.18	2.50	**1.21**	-	-	**-**	-	**1.21**	m	**1.30**
rebated and beaded meeting stile	1.64	2.50	**1.68**	-	-	**-**	-	**1.68**	m	**1.81**
fitting and hanging casement or fanlight on butts (included elsewhere)										
38mm	-	-	**-**	1.30	0.16	**20.19**	-	**20.19**	nr	**21.70**
50mm	-	-	**-**	1.45	0.18	**22.53**	-	**22.53**	nr	**24.22**
fitting and hanging casement or fanlight on sash centres (included elsewhere)										
38mm	-	-	**-**	3.00	0.37	**46.59**	-	**46.59**	nr	**50.09**
50mm	-	-	**-**	3.30	0.41	**51.28**	-	**51.28**	nr	**55.13**
Sash windows in softwood, wrought, preservative treated; priming or staining by manufacturer										
Magnet Trade; weather stripping; spiral balances; brass plated fittings; fixing to masonry with galvanized steel cramps -4 nr, 25 x 3 x 150mm girth, flat section holes -2										
825 x 1094mm overall; SS3	295.00	2.50	**302.38**	3.30	1.65	**64.56**	1.40	**368.34**	nr	**395.97**
825 x 1394mm overall; SS4	291.20	2.50	**298.48**	3.60	1.80	**70.43**	1.40	**370.31**	nr	**398.09**
920 x 1225mm overall; SS2	297.18	2.50	**304.61**	3.60	1.80	**70.43**	1.40	**376.44**	nr	**404.67**
1051 x 1094mm overall; SSW3	292.83	2.50	**300.15**	3.60	1.80	**70.43**	1.40	**371.99**	nr	**399.88**
1051 x 1394mm overall; SSW4	305.45	2.50	**313.09**	3.90	1.95	**76.30**	1.40	**390.79**	nr	**420.10**
1051 x 1694mm overall; SSW5	361.30	2.50	**370.33**	4.20	2.10	**82.17**	1.40	**453.90**	nr	**487.94**
Magnet Trade; Georgian, open rebated for glass; weatherstripping; spiral balances; brass plated fittings; fixing to masonry with galvanized steel cramps -4 nr, 25 x 3 x 150mm girth, flat section, holes -2										
825 x 1094mm overall; GS3	336.18	2.50	**344.59**	2.25	1.12	**43.97**	1.40	**389.95**	nr	**419.20**
825 x 1394mm overall; GS4	344.12	2.50	**352.72**	2.60	1.30	**50.87**	1.40	**404.99**	nr	**435.36**
1051 x 1394mm overall; GSW4	381.54	2.50	**391.08**	2.95	1.47	**57.66**	1.40	**450.14**	nr	**483.90**

Labour hourly rates: (except Specialists) Craft Operatives £14.21 Labourer £10.71 Rates are national average prices. Refer to REGIONAL VARIATIONS for indicative levels of overall pricing in regions	MATERIALS			LABOUR				RATES		
	Del to Site £	Waste %	Material Cost £	Craft Optve Hrs	Lab Hrs	Labour Cost £	Sunds £	Nett Rate £	Unit	Gross Rate (+7.5%) £
L10: WINDOWS/ROOFLIGHTS/SCREENS/LOUVRES: TIMBER WINDOWS Cont.										
Sash windows mainly in softwood, wrought										
Cased frames; 25mm inside and outside linings; 32mm pulley stiles and head; 10mm beads, back linings, etc; 76mm oak sunk weathered and throated sills; 38mm moulded sashes										
in one pane	150.50	2.50	**154.26**	2.30	0.29	**35.79**	0.37	**190.42**	m2	**204.70**
divided into panes 0.50 - 1.00m2	158.03	2.50	**161.98**	2.30	0.29	**35.79**	0.37	**198.13**	m2	**212.99**
divided into panes 0.10 - 0.50m2	173.08	2.50	**177.40**	2.30	0.29	**35.79**	0.37	**213.56**	m2	**229.58**
divided into panes not exceeding 0.10m2	195.65	2.50	**200.54**	2.30	0.29	**35.79**	0.37	**236.70**	m2	**254.45**
Cased frames; 25mm inside and outside linings; 32mm pulley stiles and head; 10mm beads, back linings, etc; 76mm oak sunk weathered and throated sills; 50mm moulded sashes										
in one pane	173.08	2.50	**177.40**	2.50	0.31	**38.85**	0.41	**216.66**	m2	**232.91**
divided into panes 0.50 - 1.00m2	180.60	2.50	**185.12**	2.50	0.31	**38.85**	0.41	**224.37**	m2	**241.20**
divided into panes 0.10 - 0.50m2	195.65	2.50	**200.54**	2.50	0.31	**38.85**	0.41	**239.80**	m2	**257.78**
divided into panes not exceeding 0.10m2	218.23	2.50	**223.68**	2.50	0.31	**38.85**	0.41	**262.94**	m2	**282.66**
extra; windows in three lights with boxed mullions	30.10	2.50	**30.85**	0.70	0.09	**10.91**	0.11	**41.87**	nr	**45.01**
extra; windows with moulded horns	37.63	2.50	**38.57**	-	-	**-**	-	**38.57**	nr	**41.46**
extra; deep bottom rails and draught beads	3.76	2.50	**3.86**	-	-	**-**	-	**3.86**	m	**4.15**
Labours										
throats	0.39	2.50	**0.40**	-	-	**-**	-	**0.40**	m	**0.43**
grooves for jamb linings	0.45	2.50	**0.46**	-	-	**-**	-	**0.46**	m	**0.50**
fitting and hanging sashes in double hung windows; providing brass faced iron pulleys, best flax cords and iron weights										
38mm thick weighing 14 lbs per sash	40.00	2.50	**41.00**	1.40	0.17	**21.71**	-	**62.71**	nr	**67.42**
50mm thick weighing 20 lbs per sash	45.00	2.50	**46.13**	1.40	0.17	**21.71**	-	**67.84**	nr	**72.93**
Note										
for small frames and sashes, i.e. 1.25m2 and under, add 20% to the foregoing prices										
Sash windows in Afrormosia, wrought										
Cased frames; 25mm inside and outside linings; 32mm pulley stiles and head; 10mm beads, back linings, etc; 76mm oak sunk weathered and throated sills; 38mm moulded sashes										
in one pane	349.99	2.50	**358.74**	4.00	0.50	**62.20**	0.65	**421.58**	m2	**453.20**
divided into panes 0.50 - 1.00m2	367.49	2.50	**376.68**	4.00	0.50	**62.20**	0.65	**439.52**	m2	**472.49**
divided into panes 0.10 - 0.50m2	402.49	2.50	**412.55**	4.00	0.50	**62.20**	0.65	**475.40**	m2	**511.05**
divided into panes not exceeding 0.10m2	454.99	2.50	**466.36**	4.00	0.50	**62.20**	0.65	**529.21**	m2	**568.90**
Cased frames; 25mm inside and outside linings; 32mm pulley stiles and head; 10mm beads, back linings, etc; 76mm oak sunk weathered and throated sills; 50mm moulded sashes										
in one pane	402.49	2.50	**412.55**	4.40	0.55	**68.41**	0.72	**481.69**	m2	**517.81**
divided into panes 0.50 - 1.00m2	419.99	2.50	**430.49**	4.40	0.55	**68.41**	0.72	**499.62**	m2	**537.10**
divided into panes 0.10 - 0.50m2	454.99	2.50	**466.36**	4.40	0.55	**68.41**	0.72	**535.50**	m2	**575.66**
divided into panes not exceeding 0.10m2	507.49	2.50	**520.18**	4.40	0.55	**68.41**	0.72	**589.31**	m2	**633.51**
extra; windows in three lights with boxed mullions	70.00	2.50	**71.75**	1.20	0.15	**18.66**	0.20	**90.61**	nr	**97.40**
extra; windows with moulded horns	87.50	2.50	**89.69**	-	-	**-**	-	**89.69**	nr	**96.41**
extra; deep bottom rails and draught beads	8.75	2.50	**8.97**	-	-	**-**	-	**8.97**	m	**9.64**
Labours										
throats	0.88	2.50	**0.90**	-	-	**-**	-	**0.90**	m	**0.97**
grooves for jamb linings	1.05	2.50	**1.08**	-	-	**-**	-	**1.08**	m	**1.16**
fitting and hanging sashes in double hung windows; providing brass faced iron pulleys, best flax cords and iron weights										
38mm thick weighing 14 lbs per sash	40.00	2.50	**41.00**	2.50	0.31	**38.85**	-	**79.85**	nr	**85.83**
50mm thick weighing 20 lbs per sash	45.00	2.50	**46.13**	2.50	0.31	**38.85**	-	**84.97**	nr	**91.34**
Note										
for small frames and sashes, i.e. 1.25m2 and under, add 20% to the foregoing prices										
Sash windows in European Oak, wrought										
Cased frames; 25mm inside and outside linings; 32mm pulley stiles and head; 10mm beads, back linings, etc; 76mm oak sunk weathered and throated sills; 38mm moulded sashes										
in one pane	462.04	2.50	**473.60**	4.60	0.57	**71.47**	0.75	**545.82**	m2	**586.75**
divided into panes 0.50 - 1.00m2	485.14	2.50	**497.27**	4.60	0.57	**71.47**	0.75	**569.49**	m2	**612.21**
divided into panes 0.10 - 0.50m2	531.34	2.50	**544.63**	4.60	0.57	**71.47**	0.75	**616.85**	m2	**663.11**
divided into panes not exceeding 0.10m2	600.66	2.50	**615.67**	4.60	0.57	**71.47**	0.75	**687.89**	m2	**739.48**
Cased frames; 25mm inside and outside linings; 32mm pulley stiles and head; 10mm beads, back linings, etc; 76mm oak sunk weathered and throated sills; 50mm moulded sashes										
in one pane	531.34	2.50	**544.63**	5.00	0.62	**77.69**	0.81	**623.13**	m2	**669.86**
divided into panes 0.50 - 1.00m2	554.44	2.50	**568.31**	5.00	0.62	**77.69**	0.81	**646.81**	m2	**695.32**
divided into panes 0.10 - 0.50m2	600.66	2.50	**615.67**	5.00	0.62	**77.69**	0.81	**694.17**	m2	**746.23**
divided into panes not exceeding 0.10m2	669.96	2.50	**686.70**	5.00	0.62	**77.69**	0.81	**765.20**	m2	**822.59**
extra; windows in three lights with boxed mullions	118.81	2.50	**121.78**	1.40	0.17	**21.71**	0.23	**143.73**	nr	**154.51**
extra; windows with moulded horns	115.51	2.50	**118.40**	-	-	**-**	-	**118.40**	nr	**127.28**
extra; deep bottom rails and draught beads	11.56	2.50	**11.85**	-	-	**-**	-	**11.85**	m	**12.74**

Labour hourly rates: (except Specialists) Craft Operatives £14.21 Labourer £10.71 Rates are national average prices. Refer to REGIONAL VARIATIONS for indicative levels of overall pricing in regions	MATERIALS			LABOUR				RATES		
	Del to Site £	Waste %	Material Cost £	Craft Optve Hrs	Lab Hrs	Labour Cost £	Sunds £	Nett Rate £	Unit	Gross Rate (+7.5%) £
L10: WINDOWS/ROOFLIGHTS/SCREENS/LOUVRES: TIMBER WINDOWS Cont.										
Sash windows in European Oak, wrought Cont.										
Labours										
throats	1.16	2.50	**1.18**	-	-	**-**	-	**1.18**	m	**1.27**
grooves for jamb linings	1.16	2.50	**1.18**	-	-	**-**	-	**1.18**	m	**1.27**
fitting and hanging sashes in double hung windows; providing brass faced iron pulleys, best flax cords and iron weights										
38mm thick weighing 14 lbs per sash	40.00	2.50	**41.00**	2.80	0.35	**43.54**	-	**84.54**	nr	**90.88**
50mm thick weighing 20 lbs per sash	45.00	2.50	**46.13**	2.80	0.35	**43.54**	-	**89.66**	nr	**96.39**
Note										
for small frames and sashes, i.e. 1.25m2 and under, add 20% to the foregoing prices										
L10: WINDOWS/ROOFLIGHTS/SCREENS/LOUVRES: TIMBER ROOFLIGHTS										
Skylights in softwood, wrought										
Chamfered, straight bar										
38mm	72.51	2.50	**74.32**	1.30	0.16	**20.19**	0.21	**94.72**	m2	**101.82**
50mm	83.96	2.50	**86.06**	1.50	0.19	**23.35**	0.24	**109.65**	m2	**117.87**
63mm	95.41	2.50	**97.79**	1.80	0.22	**27.93**	0.29	**126.02**	m2	**135.47**
Moulded, straight bar										
38mm	76.33	2.50	**78.23**	1.30	0.16	**20.19**	0.21	**98.63**	m2	**106.03**
50mm	87.77	2.50	**89.97**	1.50	0.19	**23.35**	0.24	**113.56**	m2	**122.07**
63mm	99.22	2.50	**101.70**	1.80	0.22	**27.93**	0.29	**129.93**	m2	**139.67**
Skylights in Oak, wrought										
Chamfered, straight bar										
38mm	222.59	2.50	**228.15**	2.00	0.25	**31.10**	0.33	**259.58**	m2	**279.05**
50mm	257.73	2.50	**264.17**	2.25	0.28	**34.97**	0.37	**299.51**	m2	**321.98**
63mm	292.88	2.50	**300.20**	2.70	0.34	**42.01**	0.44	**342.65**	m2	**368.34**
Moulded, straight bar										
38mm	234.30	2.50	**240.16**	2.00	0.25	**31.10**	0.33	**271.59**	m2	**291.95**
50mm	269.45	2.50	**276.18**	2.25	0.28	**34.97**	0.37	**311.52**	m2	**334.89**
63mm	304.59	2.50	**312.20**	2.70	0.34	**42.01**	0.44	**354.65**	m2	**381.25**
Skylight kerbs in softwood, wrought										
Kerbs; dovetailed at angles										
38 x 225mm	9.66	2.50	**9.90**	0.27	0.03	**4.16**	0.04	**14.10**	m	**15.16**
50 x 225mm	11.58	2.50	**11.87**	0.30	0.04	**4.69**	0.05	**16.61**	m	**17.86**
38 x 225mm; chamfers -1 nr	10.13	2.50	**10.39**	0.27	0.03	**4.16**	0.04	**14.58**	m	**15.68**
50 x 225mm; chamfers -1 nr	12.06	2.50	**12.37**	0.30	0.04	**4.69**	0.05	**17.11**	m	**18.39**
Kerbs; in two thicknesses to circular skylights										
38 x 225mm	28.94	2.50	**29.66**	0.41	0.05	**6.36**	0.07	**36.09**	m	**38.80**
50 x 225mm	34.69	2.50	**35.56**	0.46	0.06	**7.18**	0.08	**42.82**	m	**46.03**
38 x 225mm; chamfers -1 nr	29.91	2.50	**30.66**	0.41	0.05	**6.36**	0.07	**37.09**	m	**39.88**
50 x 225mm; chamfers -1 nr	35.70	2.50	**36.59**	0.46	0.06	**7.18**	0.08	**43.85**	m	**47.14**
Skylight kerbs in Oak, wrought										
Kerbs; dovetailed at angles										
38 x 225mm	47.30	2.50	**48.48**	0.54	0.07	**8.42**	0.09	**57.00**	m	**61.27**
50 x 225mm	56.76	2.50	**58.18**	0.60	0.08	**9.38**	0.10	**67.66**	m	**72.74**
38 x 225mm; chamfers -1 nr	49.67	2.50	**50.91**	0.54	0.07	**8.42**	0.09	**59.42**	m	**63.88**
50 x 225mm; chamfers -1 nr	59.13	2.50	**60.60**	0.60	0.08	**9.38**	0.10	**70.09**	m	**75.34**
Kerbs; in two thicknesses to circular skylights										
38 x 225mm	141.90	2.50	**145.45**	0.82	0.10	**12.72**	0.13	**158.30**	m	**170.17**
50 x 225mm	170.28	2.50	**174.54**	0.92	0.12	**14.36**	0.15	**189.05**	m	**203.22**
38 x 225mm; chamfers -1 nr	146.63	2.50	**150.30**	0.82	0.10	**12.72**	0.13	**163.15**	m	**175.39**
50 x 225mm; chamfers -1 nr	175.01	2.50	**179.39**	0.92	0.12	**14.36**	0.15	**193.89**	m	**208.44**
Roof windows in Nordic red pine, wrought, treated										
Velux roof windows GGL/GHL 3000 Range, factory glazed clear float double glazed sealed unit; type EDL 0000; fixing to timber with screws; dressing flashings										
778 x 550mm overall, C02 GGL Range	151.01	2.50	**154.79**	4.50	2.25	**88.04**	-	**242.83**	nr	**261.04**
978 x 550mm overall, C04 GGL Range	168.12	2.50	**172.32**	5.25	2.62	**102.66**	-	**274.99**	nr	**295.61**
1178 x 660mm overall, F06 GGL Range	198.74	2.50	**203.71**	5.25	2.62	**102.66**	-	**306.37**	nr	**329.35**
978 x 780mm overall, M04 GGL Range	190.21	2.50	**194.97**	5.00	2.50	**97.83**	-	**292.79**	nr	**314.75**
1178 x 780mm overall, M06 GGL Range	218.41	2.50	**223.87**	5.00	2.50	**97.83**	-	**321.69**	nr	**345.82**
1398 x 780mm overall, M08 GGL Range	227.44	2.50	**233.13**	5.50	2.75	**107.61**	-	**340.73**	nr	**366.29**
1600 x 942mm overall, P10 GGL Range	284.64	2.50	**291.76**	5.50	2.75	**107.61**	-	**399.36**	nr	**429.32**
1178 x 1140mm overall, S06 GGL Range	253.31	2.50	**259.65**	6.00	3.00	**117.39**	-	**377.04**	nr	**405.31**
778 x 550mm overall, C02 GHL Range	151.01	2.50	**154.79**	5.75	2.87	**112.45**	-	**267.23**	nr	**287.27**
978 x 550mm overall, C04 GHL Range	231.27	2.50	**237.05**	5.75	2.87	**112.45**	-	**349.49**	nr	**375.71**
1178 x 660mm overall, F06 GHL Range	264.86	2.50	**271.48**	5.75	2.87	**112.45**	-	**383.93**	nr	**412.72**
978 x 780mm overall, M04 GHL Range	276.79	2.50	**283.71**	5.75	2.87	**112.45**	-	**396.15**	nr	**425.86**
1178 x 780mm overall, M06 GHL Range	279.69	2.50	**286.68**	6.50	3.25	**127.17**	-	**413.86**	nr	**444.90**
1398 x 780mm overall, M08 GHL Range	298.46	2.50	**305.92**	6.50	3.25	**127.17**	-	**433.10**	nr	**465.58**
1600 x 942mm overall, P10 GHL Range	284.64	2.50	**291.76**	6.50	3.25	**127.17**	-	**418.93**	nr	**450.35**
1178 x 1140mm overall, S06 GHL Range	321.21	2.50	**329.24**	6.50	3.25	**127.17**	-	**456.41**	nr	**490.64**

Labour hourly rates: (except Specialists) Craft Operatives £14.21 Labourer £10.71 Rates are national average prices. Refer to REGIONAL VARIATIONS for indicative levels of overall pricing in regions	MATERIALS			LABOUR				RATES		
	Del to Site £	Waste %	Material Cost £	Craft Optve Hrs	Lab Hrs	Labour Cost £	Sunds £	Nett Rate £	Unit	Gross Rate (+7.5%) £
L10: WINDOWS/ROOFLIGHTS/SCREENS/LOUVRES: TIMBER ROOFLIGHTS Cont.										
Roof windows in Nordic red pine, wrought, treated Cont.										
Velux roof windows GGL/GHL 3000 Range, factory glazed clear float double glazed sealed unit; type EDZ 0000; fixing to timber with screws; dressing flashings										
778 x 550mm overall, C02 GGL Range	154.71	2.50	**158.58**	4.50	2.25	**88.04**	-	**246.62**	nr	265.12
978 x 550mm overall, C04 GGL Range	171.07	2.50	**175.35**	5.25	2.62	**102.66**	-	**278.01**	nr	298.86
1178 x 660mm overall, F06 GGL Range	201.57	2.50	**206.61**	5.25	2.62	**102.66**	-	**309.27**	nr	332.46
978 x 780mm overall, M04 GGL Range	194.04	2.50	**198.89**	5.00	2.50	**97.83**	-	**296.71**	nr	318.97
1178 x 780mm overall, M06 GGL Range	223.36	2.50	**228.94**	5.00	2.50	**97.83**	-	**326.77**	nr	351.27
1398 x 780mm overall, M08 GGL Range	233.12	2.50	**238.95**	5.50	2.75	**107.61**	-	**346.55**	nr	372.54
1600 x 942mm overall, P10 GGL Range	289.59	2.50	**296.83**	5.50	2.75	**107.61**	-	**404.44**	nr	434.77
1178 x 1140mm overall, S06 GGL Range	257.93	2.50	**264.37**	6.00	3.00	**117.39**	-	**381.76**	nr	410.40
778 x 550mm overall, C02 GHL Range	154.71	2.50	**158.58**	5.75	2.87	**112.45**	-	**271.03**	nr	291.35
978 x 550mm overall, C04 GHL Range	234.22	2.50	**240.07**	5.75	2.87	**112.45**	-	**352.52**	nr	378.96
1178 x 660mm overall, F06 GHL Range	267.72	2.50	**274.41**	5.75	2.87	**112.45**	-	**386.86**	nr	415.87
978 x 780mm overall, M04 GHL Range	257.87	2.50	**264.32**	5.75	2.87	**112.45**	-	**376.76**	nr	405.02
1178 x 780mm overall, M06 GHL Range	284.64	2.50	**291.76**	6.50	3.25	**127.17**	-	**418.93**	nr	450.35
1398 x 780mm overall, M08 GHL Range	304.14	2.50	**311.74**	6.50	3.25	**127.17**	-	**438.92**	nr	471.84
1600 x 942mm overall, P10 GHL Range	289.59	2.50	**296.83**	6.50	3.25	**127.17**	-	**424.00**	nr	455.80
1178 x 1140mm overall, S06 GHL Range	325.71	2.50	**333.85**	6.50	3.25	**127.17**	-	**461.02**	nr	495.60
Velux roof windows GGL/GHL 3073 Range, factory glazed clear float/heat absorbing glass double glazed sealed unit; type EDZ 0000; fixing to timber with screws; dressing flashings										
778 x 550mm overall, C02 GGL Range	195.92	2.50	**200.82**	4.50	2.25	**88.04**	-	**288.86**	nr	310.53
978 x 550mm overall, C04 GGL Range	216.43	2.50	**221.84**	5.25	2.62	**102.66**	-	**324.51**	nr	348.85
1178 x 660mm overall, F06 GGL Range	254.97	2.50	**261.35**	5.25	2.62	**102.66**	-	**364.01**	nr	391.31
978 x 780mm overall, M04 GGL Range	245.15	2.50	**251.27**	5.00	2.50	**97.83**	-	**349.10**	nr	375.28
1178 x 780mm overall, M06 GGL Range	269.81	2.50	**276.55**	5.00	2.50	**97.83**	-	**374.38**	nr	402.46
1398 x 780mm overall, M08 GGL Range	294.50	2.50	**301.87**	5.50	2.75	**107.61**	-	**409.47**	nr	440.18
1600 x 942mm overall, P10 GGL Range	358.76	2.50	**367.73**	5.50	2.75	**107.61**	-	**475.34**	nr	510.99
1178 x 1140mm overall, S06 GGL Range	329.12	2.50	**337.34**	6.00	3.00	**117.39**	-	**454.73**	nr	488.84
778 x 550mm overall, C02 GHL Range	195.92	2.50	**200.82**	5.75	2.87	**112.45**	-	**313.27**	nr	336.76
978 x 550mm overall, C04 GHL Range	271.78	2.50	**278.58**	5.75	2.87	**112.45**	-	**391.02**	nr	420.35
1178 x 660mm overall, F06 GHL Range	314.28	2.50	**322.14**	5.75	2.87	**112.45**	-	**434.59**	nr	467.18
978 x 780mm overall, M04 GHL Range	301.48	2.50	**309.02**	5.75	2.87	**112.45**	-	**421.46**	nr	453.07
1178 x 780mm overall, M06 GHL Range	328.59	2.50	**336.80**	6.50	3.25	**127.17**	-	**463.98**	nr	498.77
1398 x 780mm overall, M08 GHL Range	364.68	2.50	**373.80**	6.50	3.25	**127.17**	-	**500.97**	nr	538.55
1600 x 942mm overall, P10 GHL Range	358.76	2.50	**367.73**	6.50	3.25	**127.17**	-	**494.90**	nr	532.02
1178 x 1140mm overall, S06 GHL Range	395.33	2.50	**405.21**	6.50	3.25	**127.17**	-	**532.38**	nr	572.31
Velux roof windows GGL/GHL 3073 Range, factory glazed clear float/heat absorbing glass double glazed sealed unit; type EDL 0000; fixing to timber with screws; dressing flashings										
778 x 550mm overall, C02 GGL Range	199.63	2.50	**204.62**	4.50	2.25	**88.04**	-	**292.66**	nr	314.61
978 x 550mm overall, C04 GGL Range	219.38	2.50	**224.87**	5.25	2.62	**102.66**	-	**327.53**	nr	352.10
1178 x 660mm overall, F06 GGL Range	257.80	2.50	**264.25**	5.25	2.62	**102.66**	-	**366.91**	nr	394.43
978 x 780mm overall, M04 GGL Range	248.97	2.50	**255.20**	5.00	2.50	**97.83**	-	**353.02**	nr	379.50
1178 x 780mm overall, M06 GGL Range	274.76	2.50	**281.62**	5.00	2.50	**97.83**	-	**379.45**	nr	407.91
1398 x 780mm overall, M08 GGL Range	300.18	2.50	**307.68**	5.50	2.75	**107.61**	-	**415.29**	nr	446.44
1600 x 942mm overall, P10 GGL Range	363.71	2.50	**372.80**	5.50	2.75	**107.61**	-	**480.41**	nr	516.44
1178 x 1140mm overall, S06 GGL Range	333.73	2.50	**342.07**	6.00	3.00	**117.39**	-	**459.46**	nr	493.92
778 x 550mm overall, C02 GHL Range	199.63	2.50	**204.62**	5.75	2.87	**112.45**	-	**317.06**	nr	340.84
978 x 550mm overall, C04 GHL Range	274.73	2.50	**281.60**	5.75	2.87	**112.45**	-	**394.05**	nr	423.60
1178 x 660mm overall, F06 GHL Range	317.11	2.50	**325.04**	5.75	2.87	**112.45**	-	**437.48**	nr	470.30
978 x 780mm overall, M04 GHL Range	305.31	2.50	**312.94**	5.75	2.87	**112.45**	-	**425.39**	nr	457.29
1178 x 780mm overall, M06 GHL Range	333.31	2.50	**341.65**	6.50	3.25	**127.17**	-	**468.82**	nr	503.98
1398 x 780mm overall, M08 GHL Range	364.68	2.50	**373.80**	6.50	3.25	**127.17**	-	**500.97**	nr	538.55
1600 x 942mm overall, P10 GHL Range	358.76	2.50	**367.73**	6.50	3.25	**127.17**	-	**494.90**	nr	532.02
1178 x 1140mm overall, S06 GHL Range	395.33	2.50	**405.21**	6.50	3.25	**127.17**	-	**532.38**	nr	572.31
Bedding and pointing frames										
Bedding in cement mortar (1:3); pointing with Secomastic standard mastic one side										
wood frames	0.72	5.00	**0.78**	0.18	-	**2.56**	-	**3.34**	m	3.59
Bedding in cement mortar (1:3); pointing with coloured two part polysulphide mastic one side										
wood frames	0.75	5.00	**0.81**	0.18	-	**2.56**	-	**3.37**	m	3.62
L10: WINDOWS/ROOFLIGHTS/SCREENS/LOUVRES: METAL WINDOWS										
Windows in galvanized steel										
Crittall Windows Ltd. Duralife Homelight; weatherstripping; fixing to masonry with lugs										
fixed lights										
508 x 292mm, NG5	20.08	2.50	**20.58**	0.53	0.22	**9.89**	-	**30.46**	nr	32.75
508 x 923mm, NC5	27.53	2.50	**28.21**	0.85	0.30	**15.29**	-	**43.51**	nr	46.77
508 x 1218mm, ND5	31.43	2.50	**32.21**	1.07	0.36	**19.06**	-	**51.27**	nr	55.12
997 x 628mm, NE13	36.34	2.50	**37.25**	0.97	0.35	**17.53**	-	**54.78**	nr	58.89
997 x 923mm, NC13	42.39	2.50	**43.45**	1.08	0.38	**19.42**	-	**62.87**	nr	67.58
997 x 1218mm, ND13	46.94	2.50	**48.11**	1.30	0.44	**23.19**	-	**71.29**	nr	76.64
1486 x 628mm, NE14	43.49	2.50	**44.58**	1.30	0.46	**23.40**	-	**67.98**	nr	73.08
1486 x 923mm, NC14	48.89	2.50	**50.12**	1.42	0.50	**25.53**	-	**75.65**	nr	81.32
1486 x 1218mm, ND14	55.33	2.50	**56.71**	1.65	0.57	**29.55**	-	**86.26**	nr	92.73
top hung lights										
508 x 292mm, NG1	67.37	2.50	**69.05**	0.53	0.22	**9.89**	-	**78.94**	nr	84.86

Labour hourly rates: (except Specialists)
Craft Operatives £14.21 Labourer £10.71
Rates are national average prices.
Refer to REGIONAL VARIATIONS for indicative levels of overall pricing in regions

	MATERIALS			LABOUR				RATES		
	Del to Site £	Waste %	Material Cost £	Craft Optve Hrs	Lab Hrs	Labour Cost £	Sunds £	Nett Rate £	Unit	Gross Rate (+7.5%) £
L10: WINDOWS/ROOFLIGHTS/SCREENS/LOUVRES: METAL WINDOWS Cont.										
Windows in galvanized steel Cont.										
Crittall Windows Ltd. Duralife Homelight; weatherstripping; fixing to masonry with lugs Cont.										
top hung lights Cont.										
508 x 457mm, NH1	73.44	2.50	**75.28**	0.64	0.25	**11.77**	-	**87.05**	nr	**93.58**
508 x 628mm, NE1	79.01	2.50	**80.98**	0.75	0.28	**13.66**	-	**94.64**	nr	**101.73**
997 x 628mm, NE13E	110.85	2.50	**113.62**	0.97	0.35	**17.53**	-	**131.15**	nr	**140.98**
997 x 923mm, NC13C	119.30	2.50	**122.29**	1.08	0.38	**19.42**	-	**141.70**	nr	**152.33**
bottom hung lights										
508 x 628mm, NL1	102.99	2.50	**105.57**	0.75	0.28	**13.66**	-	**119.23**	nr	**128.17**
side hung lights										
508 x 628mm, NES1	86.03	2.50	**88.18**	0.75	0.28	**13.66**	-	**101.84**	nr	**109.48**
508 x 923mm, NC1	92.44	2.50	**94.75**	0.85	0.30	**15.29**	-	**110.04**	nr	**118.30**
508 x 1067mm, NC01	97.99	2.50	**100.44**	0.96	0.33	**17.18**	-	**117.62**	nr	**126.44**
508 x 1218mm, ND1	104.30	2.50	**106.91**	1.07	0.36	**19.06**	-	**125.97**	nr	**135.41**
mixed lights										
279 x 923mm, NC6F	70.87	2.50	**72.64**	0.85	0.30	**15.29**	-	**87.93**	nr	**94.52**
508 x 923mm, NC5F	81.79	2.50	**83.83**	0.85	0.30	**15.29**	-	**99.12**	nr	**106.56**
508 x 1067mm, NC05F	83.73	2.50	**85.83**	0.96	0.33	**17.18**	-	**103.00**	nr	**110.73**
997 x 292mm, NG2	82.81	2.50	**84.88**	0.74	0.29	**13.62**	-	**98.50**	nr	**105.88**
997 x 457mm, NH2	90.52	2.50	**92.78**	0.85	0.32	**15.51**	-	**108.29**	nr	**116.41**
997 x 628mm, NE2	98.34	2.50	**100.79**	0.97	0.35	**17.53**	-	**118.33**	nr	**127.20**
997 x 628mm, NES2	116.68	2.50	**119.60**	0.97	0.35	**17.53**	-	**137.13**	nr	**147.42**
997 x 923mm, NC2	127.43	2.50	**130.61**	1.08	0.38	**19.42**	-	**150.03**	nr	**161.28**
997 x 923mm, NC2F	169.15	2.50	**173.37**	1.08	0.38	**19.42**	-	**192.79**	nr	**207.25**
997 x 1067mm, NC02	138.59	2.50	**142.05**	1.19	0.41	**21.30**	-	**163.35**	nr	**175.61**
997 x 1067mm, NC02F	184.46	2.50	**189.07**	1.19	0.41	**21.30**	-	**210.37**	nr	**226.15**
997 x 1218mm, ND2	147.65	2.50	**151.34**	1.30	0.44	**23.19**	-	**174.53**	nr	**187.62**
997 x 1218mm, ND2F	186.50	2.50	**191.16**	1.30	0.44	**23.19**	-	**214.34**	nr	**230.42**
997 x 1513mm, NDV2FSB	214.94	2.50	**220.31**	1.42	0.47	**25.21**	-	**245.52**	nr	**263.94**
1486 x 628mm, NE3	123.87	2.50	**126.97**	1.30	0.46	**23.40**	-	**150.37**	nr	**161.64**
1486 x 923mm, NC4	210.73	2.50	**215.99**	1.42	0.50	**25.53**	-	**241.53**	nr	**259.64**
1486 x 923mm, NC4F	257.95	2.50	**264.40**	1.42	0.50	**25.53**	-	**289.93**	nr	**311.67**
1486 x 1067mm, NC04	226.97	2.50	**232.64**	1.54	0.53	**27.56**	-	**260.20**	nr	**279.71**
1486 x 1067mm, NC04F	272.66	2.50	**279.47**	1.54	0.53	**27.56**	-	**307.03**	nr	**330.06**
1486 x 1218mm, ND4	242.17	2.50	**248.22**	1.65	0.57	**29.55**	-	**277.77**	nr	**298.60**
1486 x 1218mm, ND4F	287.90	2.50	**295.10**	1.65	0.57	**29.55**	-	**324.65**	nr	**349.00**
1486 x 1513mm, NDV4FSB	332.89	2.50	**341.21**	1.77	0.60	**31.58**	-	**372.79**	nr	**400.75**
1994 x 923mm, NC11F	279.75	2.50	**286.74**	1.75	0.60	**31.29**	-	**318.04**	nr	**341.89**
1994 x 1218mm, ND11F	302.76	2.50	**310.32**	2.00	0.70	**35.92**	-	**346.24**	nr	**372.21**
1994 x 1513mm, NDV11FSB	353.85	2.50	**362.69**	2.15	0.75	**38.58**	-	**401.28**	nr	**431.37**
Extra for										
mullions										
292mm high	11.69	2.50	**11.98**	0.20	0.10	**3.91**	-	**15.89**	nr	**17.08**
457mm high	16.34	2.50	**16.75**	0.30	0.15	**5.87**	-	**22.62**	nr	**24.32**
628mm high	21.11	2.50	**21.63**	0.40	0.20	**7.83**	-	**29.46**	nr	**31.67**
923mm high	29.35	2.50	**30.08**	0.50	0.25	**9.78**	-	**39.86**	nr	**42.85**
1067mm high	33.98	2.50	**34.83**	0.60	0.30	**11.74**	-	**46.57**	nr	**50.06**
1218mm high	38.24	2.50	**39.19**	0.80	0.40	**15.65**	-	**54.85**	nr	**58.96**
1513mm high	46.47	2.50	**47.63**	1.00	0.50	**19.57**	-	**67.19**	nr	**72.23**
transoms										
279mm wide	12.12	2.50	**12.42**	0.20	0.10	**3.91**	-	**16.34**	nr	**17.56**
508mm wide	18.68	2.50	**19.14**	0.30	0.15	**5.87**	-	**25.01**	nr	**26.89**
997mm wide	48.71	2.50	**49.93**	0.60	0.30	**11.74**	-	**61.67**	nr	**66.29**
1143mm wide	54.10	2.50	**55.45**	1.00	0.50	**19.57**	-	**75.02**	nr	**80.64**
Controlair Ventilators, permanent, mill finish										
508mm wide	31.20	2.50	**31.98**	-	-	**-**	-	**31.98**	nr	**34.38**
997mm wide	41.90	2.50	**42.95**	-	-	**-**	-	**42.95**	nr	**46.17**
1486mm wide	94.39	2.50	**96.75**	-	-	**-**	-	**96.75**	nr	**104.00**
Controlair Ventilators, permanent, mill finish, flyscreen										
508mm wide	37.08	2.50	**38.01**	-	-	**-**	-	**38.01**	nr	**40.86**
997mm wide	52.83	2.50	**54.15**	-	-	**-**	-	**54.15**	nr	**58.21**
1486mm wide	109.25	2.50	**111.99**	-	-	**-**	-	**111.99**	nr	**120.38**
Controlair Ventilators, adjustable										
508mm wide 1-LT	37.19	2.50	**38.11**	-	-	**-**	-	**38.11**	nr	**40.97**
997mm wide 2-LT	47.02	2.50	**48.19**	-	-	**-**	-	**48.19**	nr	**51.81**
Controlair Ventilators, adjustable, flyscreen										
508mm wide 1-LT	42.84	2.50	**43.91**	-	-	**-**	-	**43.91**	nr	**47.20**
997mm wide 2-LT	57.95	2.50	**59.40**	-	-	**-**	-	**59.40**	nr	**63.85**
locks with Parkes locking handles for										
side hung lights	38.58	2.50	**39.55**	-	-	**-**	-	**39.55**	nr	**42.51**
horizontally pivoted lights	38.58	2.50	**39.55**	-	-	**-**	-	**39.55**	nr	**42.51**
Windows in galvanized steel; polyester powder coated; white matt										
Crittall Windows Ltd. Duralife Homelight; weatherstripping; fixing to masonry with lugs										
fixed lights										
508 x 292mm, NG5	23.57	2.50	**24.16**	0.53	0.22	**9.89**	-	**34.04**	nr	**36.60**
508 x 923mm, NC5	32.29	2.50	**33.09**	0.85	0.30	**15.29**	-	**48.39**	nr	**52.01**
508 x 1218mm, ND5	36.88	2.50	**37.80**	1.07	0.36	**19.06**	-	**56.86**	nr	**61.12**
997 x 628mm, NE13	42.66	2.50	**43.72**	0.97	0.35	**17.53**	-	**61.25**	nr	**65.85**
997 x 923mm, NC13	49.94	2.50	**51.18**	1.08	0.38	**19.42**	-	**70.60**	nr	**75.89**
997 x 1218mm, ND13	57.89	2.50	**59.34**	1.30	0.44	**23.19**	-	**82.52**	nr	**88.71**
1486 x 628mm, NE14	51.19	2.50	**52.47**	1.30	0.46	**23.40**	-	**75.87**	nr	**81.56**
1486 x 923mm, NC14	57.36	2.50	**58.80**	1.42	0.50	**25.53**	-	**84.33**	nr	**90.65**
1486 x 1218mm, ND14	64.92	2.50	**66.54**	1.65	0.57	**29.55**	-	**96.09**	nr	**103.30**
top hung lights										
508 x 292mm, NG1	75.05	2.50	**76.92**	0.53	0.22	**9.89**	-	**86.81**	nr	**93.32**

Labour hourly rates: (except Specialists) Craft Operatives £14.21 Labourer £10.71 Rates are national average prices. Refer to REGIONAL VARIATIONS for indicative levels of overall pricing in regions	MATERIALS			LABOUR				RATES		
	Del to Site £	Waste %	Material Cost £	Craft Optve Hrs	Lab Hrs	Labour Cost £	Sunds £	Nett Rate £	Unit	Gross Rate (+7.5%) £
L10: WINDOWS/ROOFLIGHTS/SCREENS/LOUVRES: METAL WINDOWS Cont.										
Windows in galvanized steel; polyester powder coated; white matt Cont.										
Crittall Windows Ltd. Duralife Homelight; weatherstripping; fixing to masonry with lugs Cont.										
top hung lights Cont.										
508 x 457mm, NH1	83.92	2.50	**86.01**	0.64	0.25	**11.77**	-	**97.78**	nr	**105.12**
508 x 628mm, NE1	90.04	2.50	**92.29**	0.75	0.28	**13.66**	-	**105.95**	nr	**113.89**
997 x 628mm, NE13E	124.90	2.50	**128.02**	0.97	0.35	**17.53**	-	**145.55**	nr	**156.47**
997 x 923mm, NC13C	135.54	2.50	**138.93**	1.08	0.38	**19.42**	-	**158.35**	nr	**170.23**
bottom hung lights										
508 x 628mm, NL1	127.72	2.50	**130.91**	0.75	0.28	**13.66**	-	**144.56**	nr	**155.41**
side hung lights										
508 x 628mm, NES1	96.78	2.50	**99.20**	0.75	0.28	**13.66**	-	**112.85**	nr	**121.32**
508 x 923mm, NC1	104.09	2.50	**106.70**	0.85	0.30	**15.29**	-	**121.99**	nr	**131.14**
508 x 1067mm, NC01	110.34	2.50	**113.10**	0.96	0.33	**17.18**	-	**130.28**	nr	**140.05**
508 x 1218mm, ND1	117.62	2.50	**120.56**	1.07	0.36	**19.06**	-	**139.62**	nr	**150.09**
mixed lights										
279 x 923mm, NC6F	81.25	2.50	**83.28**	0.85	0.30	**15.29**	-	**98.57**	nr	**105.96**
508 x 923mm, NC5F	91.92	2.50	**94.21**	0.85	0.30	**15.29**	-	**109.50**	nr	**117.72**
508 x 1067mm, NC05F	94.19	2.50	**96.55**	0.96	0.33	**17.18**	-	**113.72**	nr	**122.25**
997 x 292mm, NG2	97.60	2.50	**100.04**	0.74	0.29	**13.62**	-	**113.66**	nr	**122.19**
997 x 457mm, NH2	102.77	2.50	**105.33**	0.85	0.32	**15.51**	-	**120.84**	nr	**129.90**
997 x 628mm, NE2	113.25	2.50	**116.08**	0.97	0.35	**17.53**	-	**133.61**	nr	**143.63**
997 x 628mm, NES2	131.92	2.50	**135.21**	0.97	0.35	**17.53**	-	**152.75**	nr	**164.20**
997 x 923mm, NC2	143.88	2.50	**147.47**	1.08	0.38	**19.42**	-	**166.89**	nr	**179.41**
997 x 923mm, NC2F	189.14	2.50	**193.87**	1.08	0.38	**19.42**	-	**213.29**	nr	**229.28**
997 x 1067mm, NC02	155.70	2.50	**159.59**	1.19	0.41	**21.30**	-	**180.89**	nr	**194.46**
997 x 1067mm, NC02F	206.41	2.50	**211.57**	1.19	0.41	**21.30**	-	**232.87**	nr	**250.34**
997 x 1218mm, ND2	168.25	2.50	**172.46**	1.30	0.44	**23.19**	-	**195.64**	nr	**210.32**
997 x 1218mm, ND2F	208.67	2.50	**213.88**	1.30	0.44	**23.19**	-	**237.07**	nr	**254.85**
997 x 1513mm, NDV2FSB	242.03	2.50	**248.08**	1.42	0.47	**25.21**	-	**273.29**	nr	**293.79**
1486 x 628mm, NE3	147.20	2.50	**150.87**	1.30	0.46	**23.40**	-	**174.27**	nr	**187.35**
1486 x 923mm, NC4	235.60	2.50	**241.49**	1.42	0.50	**25.53**	-	**267.02**	nr	**287.05**
1486 x 923mm, NC4F	288.01	2.50	**295.21**	1.42	0.50	**25.53**	-	**320.75**	nr	**344.80**
1486 x 1067mm, NC04	265.88	2.50	**272.53**	1.54	0.53	**27.56**	-	**300.09**	nr	**322.59**
1486 x 1067mm, NC04F	307.72	2.50	**315.42**	1.54	0.53	**27.56**	-	**342.97**	nr	**368.70**
1486 x 1218mm, ND4	286.73	2.50	**293.90**	1.65	0.57	**29.55**	-	**323.45**	nr	**347.71**
1486 x 1218mm, ND4F	324.43	2.50	**332.54**	1.65	0.57	**29.55**	-	**362.09**	nr	**389.25**
1486 x 1513mm, NDV4FSB	395.04	2.50	**404.91**	1.77	0.60	**31.58**	-	**436.49**	nr	**469.23**
1994 x 923mm, NC11F	315.00	2.50	**322.88**	1.75	0.60	**31.29**	-	**354.17**	nr	**380.73**
1994 x 1218mm, ND11F	342.13	2.50	**350.68**	2.00	0.70	**35.92**	-	**386.60**	nr	**415.59**
1994 x 1513mm, NDV11FSB	398.05	2.50	**408.00**	2.15	0.75	**38.58**	-	**446.58**	nr	**480.07**
Extra for										
mullions										
292mm high	14.24	2.50	**14.59**	0.20	0.10	**3.91**	-	**18.51**	nr	**19.89**
457mm high	18.91	2.50	**19.38**	0.30	0.15	**5.87**	-	**25.25**	nr	**27.14**
628mm high	23.84	2.50	**24.44**	0.40	0.20	**7.83**	-	**32.26**	nr	**34.68**
923mm high	32.16	2.50	**32.96**	0.50	0.25	**9.78**	-	**42.75**	nr	**45.95**
1067mm high	36.98	2.50	**37.90**	0.60	0.30	**11.74**	-	**49.64**	nr	**53.37**
1218mm high	41.23	2.50	**42.26**	0.80	0.40	**15.65**	-	**57.91**	nr	**62.25**
1513mm high	49.62	2.50	**50.86**	1.00	0.50	**19.57**	-	**70.42**	nr	**75.70**
transoms										
279mm wide	15.12	2.50	**15.50**	0.20	0.10	**3.91**	-	**19.41**	nr	**20.87**
508mm wide	21.88	2.50	**22.43**	0.30	0.15	**5.87**	-	**28.30**	nr	**30.42**
997mm wide	59.64	2.50	**61.13**	0.60	0.30	**11.74**	-	**72.87**	nr	**78.34**
1143mm wide	65.32	2.50	**66.95**	1.00	0.50	**19.57**	-	**86.51**	nr	**93.00**
Controlair Ventilators, permanent, mill finish										
508mm wide	31.20	2.50	**31.98**	-	-	**-**	-	**31.98**	nr	**34.38**
997mm wide	41.90	2.50	**42.95**	-	-	**-**	-	**42.95**	nr	**46.17**
1486mm wide	94.39	2.50	**96.75**	-	-	**-**	-	**96.75**	nr	**104.00**
Controlair Ventilators, permanent, mill finish, flyscreen										
508mm wide	37.08	2.50	**38.01**	-	-	**-**	-	**38.01**	nr	**40.86**
997mm wide	52.83	2.50	**54.15**	-	-	**-**	-	**54.15**	nr	**58.21**
1486mm wide	109.25	2.50	**111.99**	-	-	**-**	-	**111.99**	nr	**120.38**
Controlair Ventilators, adjustable										
508mm wide 1-LT	37.19	2.50	**38.11**	-	-	**-**	-	**38.11**	nr	**40.97**
997mm wide 2-LT	47.02	2.50	**48.19**	-	-	**-**	-	**48.19**	nr	**51.81**
Controlair Ventilators, adjustable, flyscreen										
508mm wide 1-LT	42.84	2.50	**43.91**	-	-	**-**	-	**43.91**	nr	**47.20**
997mm wide 2-LT	57.95	2.50	**59.40**	-	-	**-**	-	**59.40**	nr	**63.85**
locks with Parkes locking handles for										
side hung lights	38.58	2.50	**39.55**	-	-	**-**	-	**39.55**	nr	**42.51**
horizontally pivoted lights	38.58	2.50	**39.55**	-	-	**-**	-	**39.55**	nr	**42.51**
Windows in aluminium; polyester powder finish; white matt										
Kawneer Products 102 Casement range; factory glazed, 4mm, one pane clear glass; fixing to masonry with lugs										
fixed lights										
600 x 400mm	82.89	2.50	**84.96**	0.70	0.30	**13.16**	-	**98.12**	nr	**105.48**
600 x 800mm	101.18	2.50	**103.71**	1.09	0.45	**20.31**	-	**124.01**	nr	**133.32**
600 x 1000mm	113.37	2.50	**116.20**	1.22	0.50	**22.69**	-	**138.89**	nr	**149.31**
600 x 1200mm	121.90	2.50	**124.95**	1.47	0.59	**27.21**	-	**152.16**	nr	**163.57**
600 x 1600mm	145.06	2.50	**148.69**	1.47	0.59	**27.21**	-	**175.90**	nr	**189.09**
800 x 400mm	90.21	2.50	**92.46**	0.84	0.34	**15.58**	-	**108.04**	nr	**116.14**
800 x 800mm	113.37	2.50	**116.20**	1.33	0.55	**24.79**	-	**140.99**	nr	**151.57**
800 x 1000mm	124.34	2.50	**127.45**	1.49	0.62	**27.81**	-	**155.26**	nr	**166.90**
800 x 1200mm	137.75	2.50	**141.19**	1.82	0.77	**34.11**	-	**175.30**	nr	**188.45**
800 x 1600mm	163.35	2.50	**167.43**	1.82	0.77	**34.11**	-	**201.54**	nr	**216.65**

Labour hourly rates: (except Specialists) Craft Operatives £14.21 Labourer £10.71 Rates are national average prices. Refer to REGIONAL VARIATIONS for indicative levels of overall pricing in regions	MATERIALS			LABOUR				RATES		
	Del to Site £	Waste %	Material Cost £	Craft Optve Hrs	Lab Hrs	Labour Cost £	Sunds £	Nett Rate £	Unit	Gross Rate (+7.5%) £
L10: WINDOWS/ROOFLIGHTS/SCREENS/LOUVRES: METAL WINDOWS Cont.										
Windows in aluminium; polyester powder finish; white matt Cont.										
Kawneer Products 102 Casement range; factory glazed, 4mm, one pane clear glass; fixing to masonry with lugs Cont.										
fixed lights Cont.										
1200 x 400mm	108.49	2.50	111.20	0.98	0.38	18.00	-	129.20	nr	138.89
1200 x 800mm	128.00	2.50	131.19	1.57	0.66	29.38	-	160.57	nr	172.62
1200 x 1000mm	152.38	2.50	156.18	1.77	0.75	33.18	-	189.37	nr	203.57
1200 x 1200mm	167.00	2.50	171.18	2.16	0.94	40.76	-	211.94	nr	227.83
1200 x 1600mm	196.26	2.50	201.17	2.16	0.94	40.76	-	241.93	nr	260.07
1400 x 800mm	148.72	2.50	152.44	1.81	0.76	33.86	-	186.30	nr	200.27
1400 x 1000mm	165.78	2.50	169.93	2.04	0.88	38.41	-	208.34	nr	223.97
1400 x 1400mm	197.48	2.50	202.41	2.51	1.11	47.56	-	249.97	nr	268.72
1400 x 1600mm	236.49	2.50	242.40	2.51	1.11	47.56	-	289.95	nr	311.70
top hung casement										
600 x 800mm	216.98	2.50	222.41	1.09	0.45	20.31	-	242.71	nr	260.92
600 x 1000mm	235.27	2.50	241.15	1.22	0.50	22.69	-	263.84	nr	283.63
600 x 1200mm	249.90	2.50	256.14	1.47	0.59	27.21	-	283.35	nr	304.60
800 x 800mm	235.27	2.50	241.15	1.33	0.55	24.79	-	265.94	nr	285.88
800 x 1000mm	253.55	2.50	259.89	1.49	0.62	27.81	-	287.70	nr	309.28
800 x 1200mm	270.62	2.50	277.38	1.82	0.77	34.11	-	311.49	nr	334.85
top hung fanlights										
600 x400mm	215.76	2.50	221.16	0.70	0.30	13.16	-	234.32	nr	251.89
800 x 600mm	240.14	2.50	246.15	0.84	0.34	15.58	-	261.72	nr	281.35
1200 x 600mm	260.87	2.50	267.39	0.98	0.38	18.00	-	285.38	nr	306.79
Kawneer Products 102 Casement range; factory glazed, 4mm, one pane obscure glass; fixing to masonry with lugs										
fixed lights										
600 x 400mm	82.89	2.50	84.96	0.70	0.30	13.16	-	98.12	nr	105.48
600 x 800mm	102.40	2.50	104.96	1.09	0.45	20.31	-	125.26	nr	134.66
600 x 1000mm	113.37	2.50	116.20	1.22	0.50	22.69	-	138.89	nr	149.31
600 x 1200mm	124.34	2.50	127.45	1.47	0.59	27.21	-	154.65	nr	166.25
600 x 1600mm	148.72	2.50	152.44	1.47	0.59	27.21	-	179.64	nr	193.12
800 x 400mm	90.21	2.50	92.46	0.84	0.34	15.58	-	108.04	nr	116.14
800 x 800mm	114.59	2.50	117.45	1.33	0.55	24.79	-	142.24	nr	152.91
800 x 1000mm	126.78	2.50	129.95	1.49	0.62	27.81	-	157.76	nr	169.59
1200 x 400mm	109.71	2.50	112.45	0.98	0.38	18.00	-	130.45	nr	140.23
1200 x 800mm	140.19	2.50	143.69	1.57	0.66	29.38	-	173.07	nr	186.05
top hung casement										
600 x 800mm	218.20	2.50	223.66	1.09	0.45	20.31	-	243.96	nr	262.26
600 x 1000mm	236.49	2.50	242.40	1.22	0.50	22.69	-	265.09	nr	284.97
600 x 1200mm	252.33	2.50	258.64	1.47	0.59	27.21	-	285.85	nr	307.29
800 x 800mm	236.49	2.50	242.40	1.33	0.55	24.79	-	267.19	nr	287.23
800 x 1000mm	254.77	2.50	261.14	1.49	0.62	27.81	-	288.95	nr	310.62
800 x 1200mm	271.84	2.50	278.63	1.82	0.77	34.11	-	312.74	nr	336.20
top hung fanlights										
600 x400mm	218.20	2.50	223.66	0.70	0.30	13.16	-	236.82	nr	254.58
800 x 600mm	242.58	2.50	248.65	0.84	0.34	15.58	-	264.22	nr	284.04
1200 x 600mm	266.96	2.50	273.64	0.98	0.38	18.00	-	291.63	nr	313.50
Sliding windows in aluminium; polyester powder finish; white matt										
Kawneer Products Kingsley Equal series; factory double glazed, 24mm - 4/16/4 units with clear 'Pilkington K Glass' glass; fixing to masonry with lugs										
vertical sliders										
600 x 800mm	355.05	2.50	363.92	1.09	0.45	20.31	-	384.23	nr	413.05
600 x 1000mm	416.26	2.50	426.67	1.22	0.50	22.69	-	449.36	nr	483.06
600 x 1200mm	487.49	2.50	499.68	1.34	0.54	24.82	-	524.51	nr	563.84
800 x 800mm	603.25	2.50	618.33	1.33	0.55	24.79	-	643.12	nr	691.35
800 x 1000mm	703.42	2.50	721.00	1.49	0.62	27.81	-	748.81	nr	804.98
800 x 1200mm	786.89	2.50	806.56	1.66	0.70	31.09	-	837.65	nr	900.47
1200 x 800mm	1259.92	2.50	1291.41	1.77	0.75	33.18	-	1324.60	nr	1423.94
1200 x 1000mm	1341.17	2.50	1374.69	1.96	0.85	36.96	-	1411.65	nr	1517.52
1200 x 1200mm	1454.69	2.50	1491.06	2.16	0.94	40.76	-	1531.82	nr	1646.71
1400 x 800mm	1594.93	2.50	1634.80	2.28	1.11	44.29	-	1679.09	nr	1805.02
1400 x 1000mm	1699.55	2.50	1742.04	2.51	1.11	47.56	-	1789.59	nr	1923.81
1400 x 1200mm	1777.46	2.50	1821.90	2.51	1.11	47.56	-	1869.45	nr	2009.66
horizontal sliders										
1200 x 800mm	331.67	2.50	339.97	1.57	0.66	29.38	-	369.34	nr	397.04
1200 x 1000mm	349.48	2.50	358.22	1.77	0.75	33.18	-	391.40	nr	420.76
1200 x 1400mm	394.00	2.50	403.85	1.96	0.85	36.96	-	440.81	nr	473.87
1400 x 800mm	348.37	2.50	357.08	1.81	0.76	33.86	-	390.94	nr	420.26
1400 x 1000mm	371.74	2.50	381.04	2.04	0.88	38.41	-	419.45	nr	450.91
1400 x 1400mm	411.81	2.50	422.11	2.28	0.99	43.00	-	465.11	nr	499.99
1400 x 1600mm	432.96	2.50	443.78	2.51	1.11	47.56	-	491.34	nr	528.19
1600 x 800mm	369.52	2.50	378.75	2.06	0.87	38.59	-	417.34	nr	448.65
1600 x 1000mm	392.89	2.50	402.71	2.32	1.01	43.78	-	446.50	nr	479.98
1600 x 1600mm	457.44	2.50	468.88	2.85	1.29	54.31	-	523.19	nr	562.43
Window sills in galvanized pressed steel BS 6510; fixing to metal										
for opening										
508mm wide; AWA	23.36	2.50	23.95	0.50	0.25	9.78	-	33.73	nr	36.26
508mm wide; AWB	25.85	2.50	26.50	0.50	0.25	9.78	-	36.28	nr	39.00
508mm wide; PS	26.79	2.50	27.46	0.50	0.25	9.78	-	37.24	nr	40.03
508mm wide; RPS	26.79	2.50	27.46	0.50	0.25	9.78	-	37.24	nr	40.03
628mm wide; AWA	25.95	2.50	26.60	0.60	0.30	11.74	-	38.34	nr	41.21

Labour hourly rates: (except Specialists) Craft Operatives £14.21 Labourer £10.71 Rates are national average prices. Refer to REGIONAL VARIATIONS for indicative levels of overall pricing in regions	MATERIALS			LABOUR				RATES		
	Del to Site £	Waste %	Material Cost £	Craft Optve Hrs	Lab Hrs	Labour Cost £	Sunds £	Nett Rate £	Unit	Gross Rate (+7.5%) £
L10: WINDOWS/ROOFLIGHTS/SCREENS/LOUVRES: METAL WINDOWS Cont.										
Sliding windows in aluminium; polyester powder finish; white matt Cont.										
Window sills in galvanized pressed steel BS 6510; fixing to metal Cont.										
for opening Cont.										
628mm wide; AWB	28.22	2.50	**28.92**	0.60	0.30	**11.74**	-	**40.66**	nr	**43.71**
628mm wide; PS	31.84	2.50	**32.64**	0.60	0.30	**11.74**	-	**44.38**	nr	**47.71**
628mm wide; RPS	31.84	2.50	**32.64**	0.60	0.30	**11.74**	-	**44.38**	nr	**47.71**
997mm wide; AWA	36.46	2.50	**37.38**	0.70	0.35	**13.70**	-	**51.07**	nr	**54.90**
997mm wide; AWB	41.13	2.50	**42.16**	0.70	0.35	**13.70**	-	**55.85**	nr	**60.04**
997mm wide; PS	36.79	2.50	**37.71**	0.70	0.35	**13.70**	-	**51.41**	nr	**55.26**
997mm wide; RPS	37.90	2.50	**38.84**	0.70	0.35	**13.70**	-	**52.54**	nr	**56.48**
1237mm wide; AWA	41.13	2.50	**42.16**	0.80	0.40	**15.65**	-	**57.81**	nr	**62.14**
1237mm wide; AWB	42.46	2.50	**43.53**	0.80	0.40	**15.65**	-	**59.18**	nr	**63.62**
1237mm wide; PS	38.37	2.50	**39.33**	0.80	0.40	**15.65**	-	**54.98**	nr	**59.11**
1237mm wide; RPS	39.77	2.50	**40.77**	0.80	0.40	**15.65**	-	**56.42**	nr	**60.65**
1486mm wide; AWA	46.00	2.50	**47.15**	0.90	0.45	**17.61**	-	**64.76**	nr	**69.62**
1486mm wide; AWB	47.46	2.50	**48.64**	0.90	0.45	**17.61**	-	**66.25**	nr	**71.22**
1486mm wide; PS	42.52	2.50	**43.58**	0.90	0.45	**17.61**	-	**61.19**	nr	**65.78**
1486mm wide; RPS	44.18	2.50	**45.29**	0.90	0.45	**17.61**	-	**62.89**	nr	**67.61**
1846mm wide; AWA	49.27	2.50	**50.50**	1.00	0.50	**19.57**	-	**70.07**	nr	**75.32**
1846mm wide; AWB	51.07	2.50	**52.35**	1.00	0.50	**19.57**	-	**71.91**	nr	**77.31**
1846mm wide; PS	44.90	2.50	**46.02**	1.00	0.50	**19.57**	-	**65.59**	nr	**70.51**
1846mm wide; RPS	46.96	2.50	**48.13**	1.00	0.50	**19.57**	-	**67.70**	nr	**72.77**
L10: WINDOWS/ROOFLIGHTS/SCREENS/LOUVRES: METAL ROOFLIGHTS										
Rooflights, mainly in mill finish aluminium										
Rooflights; factory glazed with one layer 6mm thick Georgian wired cast glass										
non-ventilating base frame; fixing to masonry with screws										
600 x 600mm	159.50	2.50	**163.49**	1.80	0.90	**35.22**	-	**198.70**	nr	**213.61**
600 x 900mm	195.50	2.50	**200.39**	2.00	1.00	**39.13**	-	**239.52**	nr	**257.48**
900 x 900mm	262.50	2.50	**269.06**	2.00	1.00	**39.13**	-	**308.19**	nr	**331.31**
900 x 1200mm	311.50	2.50	**319.29**	2.20	1.10	**43.04**	-	**362.33**	nr	**389.51**
1200 x 1200mm	378.50	2.50	**387.96**	2.50	1.25	**48.91**	-	**436.88**	nr	**469.64**
ventilating base frame; fixing to masonry with screws										
600 x 600mm	235.46	2.50	**241.35**	1.80	0.90	**35.22**	-	**276.56**	nr	**297.31**
600 x 900mm	296.50	2.50	**303.91**	2.00	1.00	**39.13**	-	**343.04**	nr	**368.77**
900 x 900mm	357.54	2.50	**366.48**	2.00	1.00	**39.13**	-	**405.61**	nr	**436.03**
900 x 1200mm	438.66	2.50	**449.63**	2.20	1.10	**43.04**	-	**492.67**	nr	**529.62**
1200 x 1200mm	519.78	2.50	**532.77**	2.50	1.25	**48.91**	-	**581.69**	nr	**625.31**
ventilating base frame and kerb; fixing to masonry with screws										
600 x 600mm	279.16	2.50	**286.14**	1.80	0.90	**35.22**	-	**321.36**	nr	**345.46**
600 x 900mm	340.50	2.50	**349.01**	2.00	1.00	**39.13**	-	**388.14**	nr	**417.25**
900 x 900mm	401.54	2.50	**411.58**	2.00	1.00	**39.13**	-	**450.71**	nr	**484.51**
900 x 1200mm	467.17	2.50	**478.85**	2.20	1.10	**43.04**	-	**521.89**	nr	**561.03**
1200 x 1200mm	532.80	2.50	**546.12**	2.50	1.25	**48.91**	-	**595.03**	nr	**639.66**
L10: WINDOWS/ROOFLIGHTS/SCREENS/LOUVRES: METAL SCREENS, BORROWED LIGHTS, FRAMES AND GRILLES										
Screens in mild steel										
Screens; 6 x 50 x 50mm angle framing to perimeter; 75 x 75 x 3mm mesh infill; welded connections										
2000 x 2000mm overall; fixing to masonry with screws	345.52	2.50	**354.16**	6.00	3.00	**117.39**	-	**471.55**	nr	**506.92**
3000 x 2000mm overall; fixing to masonry with screws	427.57	2.50	**438.26**	8.00	4.00	**156.52**	-	**594.78**	nr	**639.39**
Screens; 6 x 50 x 50mm angle framing to perimeter; 75 x 75 x 5mm mesh infill; welded connections										
2000 x 2000mm overall; fixing to masonry with screws	358.24	2.50	**367.19**	6.00	3.00	**117.39**	-	**484.58**	nr	**520.93**
Screens; 6 x 50 x 75mm angle framing to perimeter; 75 x 75 x 3mm mesh infill; welded connections										
3000 x 2000mm overall; fixing to masonry with screws	525.80	2.50	**538.94**	8.00	4.00	**156.52**	-	**695.46**	nr	**747.62**
Screens; 6 x 50 x 50mm angle framing to perimeter; 6 x 51 x 102mm tee mullion; 75 x 75 x 3mm mesh infill; welded connections										
2000 x 4000mm overall; mullions -1 nr; fixing to masonry with screws	901.37	2.50	**923.90**	8.00	4.00	**156.52**	-	**1080.42**	nr	**1161.45**
Screens; 38.1 x 38.1 x 3.2mm hollow section framing to perimeter; 75 x 75 x 3mm mesh infill; welded connections										
2000 x 2000mm overall; fixing to masonry with screws	462.24	2.50	**473.80**	6.00	3.00	**117.39**	-	**591.19**	nr	**635.52**

Labour hourly rates: (except Specialists) Craft Operatives £14.21 Labourer £10.71 Rates are national average prices. Refer to REGIONAL VARIATIONS for indicative levels of overall pricing in regions	MATERIALS			LABOUR				RATES		
	Del to Site £	Waste %	Material Cost £	Craft Optve Hrs	Lab Hrs	Labour Cost £	Sunds £	Nett Rate £	Unit	Gross Rate (+7.5%) £
L10: WINDOWS/ROOFLIGHTS/SCREENS/LOUVRES: METAL SCREENS, BORROWED LIGHTS, FRAMES AND GRILLES Cont.										
Screens in mild steel Cont.										
Screens; 38.1 x 38.1 x 3.2mm hollow section framing to perimeter; 75 x 75 x 5mm mesh infill; welded connections										
2000 x 2000mm overall; fixing to masonry with screws	496.91	2.50	**509.33**	6.00	3.00	**117.39**	-	**626.72**	nr	**673.72**
Screens; 76.2 x 38.1 x 4mm hollow section framing to perimeter; 75 x 75 x 3mm mesh infill; welded connections										
2000 x 2000mm overall; fixing to masonry with screws	502.69	2.50	**515.25**	6.00	3.00	**117.39**	-	**632.64**	nr	**680.09**
Grilles in mild steel										
Grilles; 13 x 51mm flat bar framing to perimeter; 13mm diameter vertical infill bars at 150mm centres; welded connections										
1000 x 1000mm overall; fixing to masonry with screws	352.46	2.50	**361.27**	6.00	3.00	**117.39**	-	**478.66**	nr	**514.56**
2000 x 1000mm overall; fixing to masonry with screws	473.80	2.50	**485.64**	8.00	4.00	**156.52**	-	**642.16**	nr	**690.32**
Grilles; 13 x 51mm flat bar framing to perimeter; 13mm diameter vertical infill bars at 300mm centres; welded connections										
1000 x 1000mm overall; fixing to masonry with screws	335.12	2.50	**343.50**	4.00	2.00	**78.26**	-	**421.76**	nr	**453.39**
Grilles; 13 x 51mm flat bar framing to perimeter; 18mm diameter vertical infill bars at 150mm centres; welded connections										
1000 x 1000mm overall; fixing to masonry with screws	342.06	2.50	**350.61**	4.00	2.00	**78.26**	-	**428.87**	nr	**461.03**
Glazing frames in mild steel										
Glazing frames; 13 x 38 x 3mm angle framing to perimeter; welded connections; 13 x 13mm glazing beads, fixed with screws										
500 x 500mm overall; fixing to timber with screws	208.01	2.50	**213.21**	3.00	1.50	**58.70**	-	**271.90**	nr	**292.30**
Glazing frames; 15 x 21 x 3mm angle framing to perimeter; welded connections; 13 x 13mm glazing beads, fixed with screws										
500 x 1000mm overall; fixing to timber with screws	335.12	2.50	**343.50**	3.00	1.50	**58.70**	-	**402.20**	nr	**432.36**
Glazing frames; 18 x 25 x 3mm angle framing to perimeter; welded connections; 9 x 25mm glazing beads, fixed with screws										
1000 x 1000mm overall; fixing to timber with screws	404.46	2.50	**414.57**	4.00	2.00	**78.26**	-	**492.83**	nr	**529.79**
Glazing frames; 18 x 25 x 3mm angle framing to perimeter; welded connections; 13 x 13mm glazing beads, fixed with screws										
1000 x 1000mm overall; fixing to timber with screws	436.82	2.50	**447.74**	4.00	2.00	**78.26**	-	**526.00**	nr	**565.45**
Glazing frames; 18 x 38 x 3mm angle framing to perimeter; welded connections; 9 x 25mm glazing beads, fixed with screws										
1000 x 1000mm overall; fixing to timber with screws	543.13	2.50	**556.71**	3.00	1.50	**58.70**	-	**615.41**	nr	**661.56**
Glazing frames; 18 x 25 x 3mm angle framing to perimeter; welded connections; 15 x 15 x 3mm channel glazing beads; fixed with screws										
1000 x 1000mm overall; fixing to timber with screws	543.13	2.50	**556.71**	3.00	1.50	**58.70**	-	**615.41**	nr	**661.56**
Glazing frames; 18 x 38 x 3mm angle framing to perimeter; welded connections; 15 x 15 x 3mm channel glazing beads; fixed with screws										
1000 x 1000mm overall; fixing to timber with screws	543.13	2.50	**556.71**	3.00	1.50	**58.70**	-	**615.41**	nr	**661.56**
Glazing frames in aluminium										
Glazing frames; 13 x 38 x 3mm angle framing to perimeter; welded connections; 13 x 13mm glazing beads, fixed with screws										
500 x 500mm overall; fixing to timber with screws	354.77	2.50	**363.64**	3.00	1.50	**58.70**	-	**422.33**	nr	**454.01**
Glazing frames; 15 x 21 x 3mm angle framing to perimeter; welded connections; 13 x 13mm glazing beads, fixed with screws										
500 x 1000mm overall; fixing to timber with screws	354.77	2.50	**363.64**	3.00	1.50	**58.70**	-	**422.33**	nr	**454.01**

Labour hourly rates: (except Specialists) Craft Operatives £14.21 Labourer £10.71 Rates are national average prices. Refer to REGIONAL VARIATIONS for indicative levels of overall pricing in regions	MATERIALS			LABOUR				RATES		
	Del to Site £	Waste %	Material Cost £	Craft Optve Hrs	Lab Hrs	Labour Cost £	Sunds £	Nett Rate £	Unit	Gross Rate (+7.5%) £
L10: WINDOWS/ROOFLIGHTS/SCREENS/LOUVRES: METAL SCREENS, BORROWED LIGHTS, FRAMES AND GRILLES Cont.										
Glazing frames in aluminium Cont.										
Glazing frames; 18 x 25 x 3mm angle framing to perimeter; welded connections; 9 x 25mm glazing beads, fixed with screws										
1000 x 1000mm overall; fixing to timber with screws	381.35	2.50	**390.88**	4.00	2.00	**78.26**	-	**469.14**	nr	**504.33**
Glazing frames; 18 x 25 x 3mm angle framing to perimeter; welded connections; 13 x 13mm glazing beads, fixed with screws										
1000 x 1000mm overall; fixing to timber with screws	392.90	2.50	**402.73**	4.00	2.00	**78.26**	-	**480.99**	nr	**517.06**
Glazing frames; 18 x 38 x 3mm angle framing to perimeter; welded connections; 9 x 25mm glazing beads, fixed with screws										
1000 x 1000mm overall; fixing to timber with screws	444.91	2.50	**456.03**	4.00	2.00	**78.26**	-	**534.29**	nr	**574.36**
Glazing frames; 18 x 25 x 3mm angle framing to perimeter; welded connections; 15 x 15 x 3mm channel glazing beads; fixed with screws										
1000 x 1000mm overall; fixing to timber with screws	444.91	2.50	**456.03**	4.00	2.00	**78.26**	-	**534.29**	nr	**574.36**
Glazing frames; 18 x 38 x 3mm angle framing to perimeter; welded connections; 15 x 15 x 3mm channel glazing beads; fixed with screws										
1000 x 1000mm overall; fixing to timber with screws	502.69	2.50	**515.25**	4.00	2.00	**78.26**	-	**593.51**	nr	**638.03**
Bedding and pointing frames										
Pointing with Secomastic										
standard metal frames one side	0.73	10.00	**0.80**	0.12	-	**1.71**	-	**2.51**	m	**2.70**
Bedding in cement mortar (1:3); pointing with Secomastic standard mastic one side										
metal frames	0.87	5.00	**0.94**	0.18	-	**2.56**	-	**3.50**	m	**3.76**
L10 - WINDOWS/ROOFLIGHTS/SCREENS/LOUVRES: PLASTICS WINDOWS										
Windows in uPVC; white										
Windows; factory glazed 20mm double glazed units; hinges; fastenings										
fixed light; fixing to masonry with cleats and screws; overall size										
600 x 600mm	144.71	2.50	**148.33**	2.40	1.20	**46.96**	1.40	**196.69**	nr	**211.44**
600 x 900mm	155.08	2.50	**158.96**	3.00	1.50	**58.70**	1.40	**219.05**	nr	**235.48**
600 x 1050mm	165.41	2.50	**169.54**	3.30	1.65	**64.56**	1.40	**235.51**	nr	**253.17**
600 x 1200mm	178.44	2.50	**182.90**	3.25	1.62	**63.53**	1.40	**247.83**	nr	**266.42**
600 x 1500mm	209.43	2.50	**214.67**	3.80	1.90	**74.35**	1.40	**290.41**	nr	**312.20**
750 x 600mm	149.88	2.50	**153.62**	2.45	1.22	**47.88**	1.40	**202.90**	nr	**218.12**
750 x 900mm	175.75	2.50	**180.14**	3.00	1.50	**58.70**	1.40	**240.23**	nr	**258.25**
750 x 1050mm	186.07	2.50	**190.73**	3.25	1.62	**63.53**	1.40	**255.66**	nr	**274.83**
750 x 1200mm	196.41	2.50	**201.32**	3.50	1.75	**68.48**	1.40	**271.20**	nr	**291.54**
750 x 1500mm	230.00	2.50	**235.75**	4.05	2.02	**79.18**	1.40	**316.33**	nr	**340.06**
900 x 600mm	162.82	2.50	**166.89**	2.70	1.35	**52.83**	1.40	**221.11**	nr	**237.70**
900 x 900mm	214.49	2.50	**219.86**	3.25	1.62	**63.53**	1.40	**284.79**	nr	**306.15**
900 x 1050mm	206.74	2.50	**211.91**	3.40	1.70	**66.52**	1.40	**279.83**	nr	**300.82**
900 x 1200mm	222.25	2.50	**227.80**	3.80	1.90	**74.35**	1.40	**303.55**	nr	**326.31**
900 x 1500mm	261.13	2.50	**267.66**	4.30	2.15	**84.13**	1.40	**353.19**	nr	**379.68**
1200 x 600mm	178.44	2.50	**182.90**	3.05	1.52	**59.62**	1.40	**243.92**	nr	**262.21**
1200 x 900mm	214.60	2.50	**219.96**	3.60	1.80	**70.43**	1.40	**291.79**	nr	**313.68**
1200 x 1050mm	235.28	2.50	**241.16**	3.85	1.92	**75.27**	1.40	**317.83**	nr	**341.67**
1200 x 1200mm	255.96	2.50	**262.36**	4.15	2.07	**81.14**	1.40	**344.90**	nr	**370.76**
1200 x 1500mm	304.95	2.50	**312.58**	4.75	2.37	**92.88**	1.40	**406.86**	nr	**437.37**
fixed light with fixed light; fixing to masonry with cleats and screws; overall size										
600 x 1350mm	258.44	2.50	**264.90**	3.50	1.75	**68.48**	1.40	**334.78**	nr	**359.89**
600 x 1500mm	279.11	2.50	**286.09**	3.80	1.90	**74.35**	1.40	**361.84**	nr	**388.97**
600 x 1800mm	307.64	2.50	**315.34**	4.05	2.02	**79.18**	1.40	**395.92**	nr	**425.61**
600 x 2100mm	335.95	2.50	**344.35**	4.30	2.15	**84.13**	1.40	**429.88**	nr	**462.12**
750 x 1350mm	279.11	2.50	**286.09**	3.80	1.90	**74.35**	1.40	**361.84**	nr	**388.97**
750 x 1500mm	297.30	2.50	**304.74**	4.05	2.02	**79.18**	1.40	**385.32**	nr	**414.22**
750 x 1800mm	335.95	2.50	**344.35**	4.30	2.15	**84.13**	1.40	**429.88**	nr	**462.12**
750 x 2100mm	377.31	2.50	**386.74**	4.60	2.30	**90.00**	1.40	**478.14**	nr	**514.00**
900 x 1350mm	302.48	2.50	**310.04**	4.05	2.02	**79.18**	1.40	**390.63**	nr	**419.92**
900 x 1500mm	320.46	2.50	**328.47**	4.30	2.15	**84.13**	1.40	**414.00**	nr	**445.05**
900 x 1800mm	359.33	2.50	**368.32**	4.50	2.25	**88.04**	1.40	**457.76**	nr	**492.09**
900 x 2100mm	397.77	2.50	**407.72**	4.85	2.42	**94.84**	1.40	**503.96**	nr	**541.75**
1200 x 1350mm	341.13	2.50	**349.65**	4.50	2.25	**88.04**	1.40	**439.10**	nr	**472.03**
1200 x 1500mm	364.51	2.50	**373.62**	4.75	2.37	**92.88**	1.40	**467.90**	nr	**503.00**
1200 x 1800mm	411.01	2.50	**421.28**	5.05	2.52	**98.75**	1.40	**521.43**	nr	**560.54**
1200 x 2100mm	454.83	2.50	**466.20**	5.30	2.65	**103.69**	1.40	**571.30**	nr	**614.14**
overall, tilt/turn; fixing to masonry with cleats and screws; overall size										
600 x 600mm	421.34	2.50	**431.87**	2.15	1.07	**42.01**	1.40	**475.28**	nr	**510.93**
600 x 900mm	465.18	2.50	**476.81**	2.70	1.35	**52.83**	1.40	**531.04**	nr	**570.87**
600 x 1050mm	501.14	2.50	**513.67**	3.00	1.50	**58.70**	1.40	**573.76**	nr	**616.79**

Labour hourly rates: (except Specialists) Craft Operatives £14.21 Labourer £10.71 Rates are national average prices. Refer to REGIONAL VARIATIONS for indicative levels of overall pricing in regions	MATERIALS			LABOUR				RATES		
	Del to Site £	Waste %	Material Cost £	Craft Optve Hrs	Lab Hrs	Labour Cost £	Sunds £	Nett Rate £	Unit	Gross Rate (+7.5%) £
L10 - WINDOWS/ROOFLIGHTS/SCREENS/LOUVRES: PLASTICS WINDOWS Cont.										
Windows in uPVC; white Cont.										
Windows; factory glazed 20mm double glazed units; hinges; fastenings Cont.										
overall, tilt/turn; fixing to masonry with cleats and screws; overall size Cont.										
600 x 1200mm	516.86	2.50	**529.78**	3.25	1.62	**63.53**	1.40	**594.71**	nr	**639.32**
600 x 1350mm	552.81	2.50	**566.63**	3.50	1.75	**68.48**	1.40	**636.51**	nr	**684.25**
600 x 1500mm	581.58	2.50	**596.12**	3.80	1.90	**74.35**	1.40	**671.86**	nr	**722.25**
750 x 600mm	437.96	2.50	**448.91**	2.45	1.22	**47.88**	1.40	**498.19**	nr	**535.56**
750 x 900mm	488.55	2.50	**500.77**	3.00	1.50	**58.70**	1.40	**560.86**	nr	**602.93**
750 x 1050mm	517.33	2.50	**530.26**	3.25	1.62	**63.53**	1.40	**595.20**	nr	**639.84**
750 x 1200mm	545.38	2.50	**559.02**	3.50	1.75	**68.48**	1.40	**628.89**	nr	**676.06**
750 x 1350mm	573.72	2.50	**588.07**	3.80	1.90	**74.35**	1.40	**663.81**	nr	**713.60**
750 x 1500mm	599.56	2.50	**614.54**	4.05	2.02	**79.18**	1.40	**695.13**	nr	**747.26**
900 x 900mm	516.86	2.50	**529.78**	3.25	1.62	**63.53**	1.40	**594.71**	nr	**639.32**
900 x 1050mm	547.87	2.50	**561.56**	3.40	1.70	**66.52**	1.40	**629.49**	nr	**676.70**
900 x 1200mm	576.41	2.50	**590.83**	3.80	1.90	**74.35**	1.40	**666.57**	nr	**716.57**
900 x 1350mm	604.72	2.50	**619.84**	4.05	2.02	**79.18**	1.40	**700.42**	nr	**752.95**
900 x 1500mm	640.89	2.50	**656.91**	4.30	2.15	**84.13**	1.40	**742.44**	nr	**798.13**
1200 x 1050mm	615.06	2.50	**630.44**	3.85	1.92	**75.27**	1.40	**707.11**	nr	**760.14**
1200 x 1200mm	651.24	2.50	**667.53**	4.15	2.07	**81.14**	1.40	**750.07**	nr	**806.32**
1200 x 1350mm	679.78	2.50	**696.77**	4.50	2.25	**88.04**	1.40	**786.22**	nr	**845.18**
1200 x 1500mm	715.96	2.50	**733.86**	4.75	2.37	**92.88**	1.40	**828.14**	nr	**890.25**
tilt/turn with fixed lights; fixing to masonry with cleats and screws; overall size										
600 x 1350mm	540.22	2.50	**553.72**	3.50	1.75	**68.48**	1.40	**623.60**	nr	**670.37**
600 x 1500mm	568.56	2.50	**582.77**	3.80	1.90	**74.35**	1.40	**658.52**	nr	**707.91**
600 x 2100mm	656.41	2.50	**672.82**	4.30	2.15	**84.13**	1.40	**758.35**	nr	**815.22**
750 x 1350mm	576.41	2.50	**590.83**	3.80	1.90	**74.35**	1.40	**666.57**	nr	**716.57**
750 x 1500mm	602.25	2.50	**617.30**	4.05	2.02	**79.18**	1.40	**697.89**	nr	**750.23**
750 x 2100mm	687.65	2.50	**704.84**	4.60	2.30	**90.00**	1.40	**796.24**	nr	**855.95**
900 x 1350mm	617.99	2.50	**633.44**	4.05	2.02	**79.18**	1.40	**714.03**	nr	**767.58**
900 x 1500mm	640.89	2.50	**656.91**	4.30	2.15	**84.13**	1.40	**742.44**	nr	**798.13**
900 x 2100mm	733.94	2.50	**752.29**	4.85	2.42	**94.84**	1.40	**848.53**	nr	**912.17**
1200 x 1350mm	697.76	2.50	**715.20**	4.50	2.25	**88.04**	1.40	**804.64**	nr	**864.99**
1200 x 1500mm	726.29	2.50	**744.45**	4.75	2.37	**92.88**	1.40	**838.73**	nr	**901.63**
1200 x 2100mm	803.58	2.50	**823.67**	5.30	2.65	**103.69**	1.40	**928.77**	nr	**998.42**
tilt/turn sash with fixed side light; fixing to masonry with cleats and screws; overall size										
1800 x 900mm	687.65	2.50	**704.84**	4.40	2.20	**86.09**	1.40	**792.32**	nr	**851.75**
1800 x 1050mm	733.94	2.50	**752.29**	4.70	2.35	**91.96**	1.40	**845.65**	nr	**909.07**
1800 x 1200mm	780.47	2.50	**799.98**	5.05	2.52	**98.75**	1.40	**900.13**	nr	**967.64**
1800 x 1350mm	826.97	2.50	**847.64**	5.40	2.70	**105.65**	1.40	**954.69**	nr	**1026.29**
1800 x 1500mm	873.48	2.50	**895.31**	5.65	2.82	**110.49**	1.40	**1007.20**	nr	**1082.74**
tilt/turn sash with centre fixed light; fixing to masonry with cleats and screws; overall size										
2400 x 900mm	1085.41	2.50	**1112.54**	4.40	2.20	**86.09**	1.40	**1200.03**	nr	**1290.03**
2400 x 1050mm	1162.91	2.50	**1191.99**	5.75	2.87	**112.45**	1.40	**1305.83**	nr	**1403.77**
2400 x 1200mm	1240.45	2.50	**1271.46**	6.10	3.05	**119.35**	1.40	**1392.21**	nr	**1496.62**
2400 x 1350mm	1305.17	2.50	**1337.80**	6.40	3.20	**125.22**	1.40	**1464.41**	nr	**1574.24**
2400 x 1500mm	678.43	2.50	**695.39**	6.65	3.32	**130.05**	1.40	**826.84**	nr	**888.86**
L10 - WINDOWS/ROOFLIGHTS/SCREENS/LOUVRES: PLASTIC ROOFLIGHTS										
Roof domelights in glass reinforced plastics										
Domelights, translucent										
plain; fixing to masonry with screws										
610 x 610mm	84.67	2.50	**86.78**	1.00	1.00	**24.92**	0.27	**111.97**	nr	**120.37**
914 x 914mm	108.79	2.50	**111.51**	2.00	2.00	**49.84**	0.54	**161.89**	nr	**174.03**
1219 x 1219mm	144.65	2.50	**148.27**	3.00	3.00	**74.76**	0.81	**223.84**	nr	**240.63**
fixed ventilator; fixing to masonry with screws										
610 x 610mm	150.53	2.50	**154.30**	1.00	1.00	**24.92**	0.27	**179.49**	nr	**192.95**
914 x 914mm	181.10	2.50	**185.63**	2.00	2.00	**49.84**	0.54	**236.01**	nr	**253.71**
1219 x 1219mm	223.46	2.50	**229.04**	3.00	3.00	**74.76**	0.81	**304.61**	nr	**327.46**
controlled ventilator; fixing to masonry with screws										
610 x 610mm	168.77	2.50	**172.98**	1.00	1.00	**24.92**	0.27	**198.17**	nr	**213.04**
914 x 914mm	229.32	2.50	**235.05**	2.00	2.00	**49.84**	0.54	**285.43**	nr	**306.84**
1219 x 1219mm	307.60	2.50	**315.29**	3.00	3.00	**74.76**	0.81	**390.86**	nr	**420.17**
Bedding and pointing frames										
Pointing with Secomastic standard mastic plastics frames one side	1.31	10.00	**1.45**	0.12	-	**1.71**	-	**3.15**	m	**3.39**
Bedding in cement mortar (1:3); pointing with Secomastic standard mastic one side plastics frames	1.82	5.00	**1.99**	0.18	-	**2.56**	-	**4.54**	m	**4.88**
L20: DOORS/SHUTTERS/HATCHES: TIMBER DOORS										
Doors in softwood, wrought										
Internal panelled doors; SA; Magnet Trade 762 x 1981 x 34mm (f sizes)	116.05	2.50	**118.95**	1.20	0.15	**18.66**	-	**137.61**	nr	**147.93**
Internal panelled doors; SA; glazed with bevelled glass; Magnet Trade 762 x 1981 x 34mm (f sizes)	147.39	2.50	**151.08**	4.20	0.53	**65.36**	-	**216.44**	nr	**232.67**

Labour hourly rates: (except Specialists) Craft Operatives £14.21 Labourer £10.71 Rates are national average prices. Refer to REGIONAL VARIATIONS for indicative levels of overall pricing in regions	MATERIALS			LABOUR				RATES		
	Del to Site £	Waste %	Material Cost £	Craft Optve Hrs	Lab Hrs	Labour Cost £	Sunds £	Nett Rate £	Unit	Gross Rate (+7.5%) £
L20: DOORS/SHUTTERS/HATCHES: TIMBER DOORS Cont.										
Doors in softwood, wrought, preservative treated										
External ledged and braced doors; L & B Magnet Trade										
686 x 1981 x 44mm (f sizes)......................	72.80	2.50	**74.62**	1.10	0.14	**17.13**	-	**91.75**	nr	**98.63**
762 x 1981 x 44mm (f sizes)......................	67.72	2.50	**69.41**	1.30	0.16	**20.19**	-	**89.60**	nr	**96.31**
838 x 1981 x 44mm (f sizes)......................	79.58	2.50	**81.57**	1.50	0.19	**23.35**	-	**104.92**	nr	**112.79**
External framed, ledged and braced doors; YX; Magnet Trade										
686 x 1981 x 44mm (f sizes)......................	96.54	2.50	**98.95**	1.10	0.14	**17.13**	-	**116.08**	nr	**124.79**
762 x 1981 x 44mm (f sizes)......................	67.72	2.50	**69.41**	1.30	0.16	**20.19**	-	**89.60**	nr	**96.31**
813 x 1981 x 44mm (f sizes)......................	112.33	2.50	**115.14**	1.50	0.19	**23.35**	-	**138.49**	nr	**148.88**
838 x 1981 x 44mm (f sizes)......................	93.15	2.50	**95.48**	1.50	0.19	**23.35**	-	**118.83**	nr	**127.74**
External panelled doors; Richmond; Magnet Trade										
762 x 1981 x 44mm (f sizes)......................	160.80	2.50	**164.82**	1.30	0.16	**20.19**	-	**185.00**	nr	**198.88**
838 x 1981 x 44mm (f sizes)......................	164.15	2.50	**168.26**	1.50	0.19	**23.35**	-	**191.61**	nr	**205.98**
External panelled doors; KXT; Magnet Trade										
762 x 1981 x 44mm (f sizes)......................	97.08	2.50	**99.50**	1.30	0.16	**20.19**	-	**119.69**	nr	**128.67**
External panelled doors; Stable; softwood; Magnet Trade										
762 x 1981 x 44mm (f sizes)......................	144.00	2.50	**147.60**	2.40	0.30	**37.32**	-	**184.92**	nr	**198.79**
838 x 1981 x 44mm (f sizes)......................	153.53	2.50	**157.37**	2.60	0.33	**40.48**	-	**197.85**	nr	**212.69**
External panelled doors; Stable upper door glazed with Geneva glazing panel, Magnet Trade										
762 x 1981 x 44mm (f sizes)......................	202.95	2.50	**208.02**	2.70	0.34	**42.01**	-	**250.03**	nr	**268.78**
838 x 1981 x 44mm (f sizes)......................	213.99	2.50	**219.34**	2.90	0.36	**45.06**	-	**264.40**	nr	**284.23**
External panelled doors; 2XG; Magnet Trade										
762 x 1981 x 44mm (f sizes)......................	59.24	2.50	**60.72**	1.30	0.16	**20.19**	-	**80.91**	nr	**86.98**
813 x 2032 x 44mm (f sizes)......................	62.63	2.50	**64.20**	1.50	0.19	**23.35**	-	**87.55**	nr	**94.11**
838 x 1981 x 44mm (f sizes)......................	60.94	2.50	**62.47**	1.50	0.19	**23.35**	-	**85.82**	nr	**92.25**
External panelled doors; 2XGG; Magnet Trade										
762 x 1981 x 44mm (f sizes)......................	63.52	2.50	**65.10**	1.30	0.16	**20.19**	-	**85.29**	nr	**91.69**
838 x 1981 x 44mm (f sizes)......................	63.52	2.50	**65.10**	1.50	0.19	**23.35**	-	**88.45**	nr	**95.09**
External panelled doors; 2XG, glazed with 4mm clear toughend glass; Magnet Trade										
762 x 1981 x 44mm (f sizes)......................	81.06	2.50	**83.09**	1.70	0.21	**26.41**	-	**109.49**	nr	**117.70**
813 x 2032 x 44mm (f sizes)......................	93.09	2.50	**95.42**	1.90	0.24	**29.57**	-	**124.99**	nr	**134.36**
838 x 1981 x 44mm (f sizes)......................	87.84	2.50	**90.04**	1.90	0.24	**29.57**	-	**119.61**	nr	**128.58**
External panelled doors; 2XG, glazed with 4mm obscure toughend glass; Magnet Trade										
762 x 1981 x 44mm (f sizes)......................	84.69	2.50	**86.81**	1.70	0.21	**26.41**	-	**113.21**	nr	**121.70**
813 x 2032 x 44mm (f sizes)......................	90.71	2.50	**92.98**	1.90	0.24	**29.57**	-	**122.55**	nr	**131.74**
838 x 1981 x 44mm (f sizes)......................	89.02	2.50	**91.25**	1.90	0.24	**29.57**	-	**120.82**	nr	**129.88**
External panelled doors; 2XGG, glazed with 4mm obscure toughend glass; Magnet Trade										
762 x 1981 x 44mm (f sizes)......................	104.63	2.50	**107.25**	2.00	0.25	**31.10**	-	**138.35**	nr	**148.72**
838 x 1981 x 44mm (f sizes)......................	108.53	2.50	**111.24**	2.20	0.28	**34.26**	-	**145.50**	nr	**156.42**
Garage doors; MFL, pair, side hung; Magnet Trade										
1981 x 2134 x 44mm (f sizes) overall............	284.67	2.50	**291.79**	3.50	0.44	**54.45**	-	**346.24**	nr	**372.21**
2134 x 2134 x 44mm (f sizes) overall............	292.81	2.50	**300.13**	3.65	0.46	**56.79**	-	**356.92**	nr	**383.69**
Garage doors; 301, pair, side hung; Magnet Trade										
1981 x 2134 x 44mm (f sizes) overall............	335.43	2.50	**343.81**	3.50	0.44	**54.45**	-	**398.26**	nr	**428.13**
2134 x 2134 x 44mm (f sizes) overall............	343.56	2.50	**352.15**	3.65	0.46	**56.79**	-	**408.94**	nr	**439.61**
Doors in Hemlock, wrought										
Internal panelled doors; 2G; Magnet Trade										
762 x 1981 x 34mm (f sizes)......................	86.91	2.50	**89.09**	1.20	0.15	**18.66**	-	**107.74**	nr	**115.83**
Internal panelled doors; 2GG; Magnet Trade										
762 x 1981 x 34mm (f sizes)......................	81.83	2.50	**83.87**	1.20	0.15	**18.66**	-	**102.53**	nr	**110.22**
Internal panelled doors; SA; Magnet Trade										
762 x 1981 x 34mm (f sizes)......................	116.05	2.50	**118.95**	1.20	0.15	**18.66**	-	**137.61**	nr	**147.93**
Internal panelled doors; 10; Magnet Trade										
762 x 1981 x 34mm (f sizes)......................	64.33	2.50	**65.94**	1.20	0.15	**18.66**	-	**84.60**	nr	**90.95**
Internal panelled doors; 2G, glazed with 4mm clear toughend glass; Magnet Trade										
762 x 1981 x 34mm (f sizes)......................	111.56	2.50	**114.35**	1.60	0.20	**24.88**	-	**139.23**	nr	**149.67**
Internal panelled doors; 2GG, glazed with 4mm clear toughend glass; Magnet Trade										
762 x 1981 x 34mm (f sizes)......................	122.75	2.50	**125.82**	1.90	0.24	**29.57**	-	**155.39**	nr	**167.04**
Internal panelled doors; 10, glazed with 4mm clear toughend glass; Magnet Trade										
762 x 1981 x 34mm (f sizes)......................	106.27	2.50	**108.93**	1.70	0.22	**26.51**	-	**135.44**	nr	**145.60**
Internal panelled doors; 2G, glazed with 4mm obscure toughend glass; Magnet Trade										
762 x 1981 x 34mm (f sizes)......................	115.19	2.50	**118.07**	1.60	0.20	**24.88**	-	**142.94**	nr	**153.67**

Labour hourly rates: (except Specialists) Craft Operatives £14.21 Labourer £10.71 Rates are national average prices. Refer to REGIONAL VARIATIONS for indicative levels of overall pricing in regions	MATERIALS			LABOUR				RATES		
	Del to Site £	Waste %	Material Cost £	Craft Optve Hrs	Lab Hrs	Labour Cost £	Sunds £	Nett Rate £	Unit	Gross Rate (+7.5%) £
L20: DOORS/SHUTTERS/HATCHES: TIMBER DOORS Cont.										
Doors in Hemlock, wrought Cont.										
Internal panelled doors; 2GG, glazed with 4mm obscure toughend glass; Magnet Trade										
762 x 1981 x 34mm (f sizes)......................	125.78	2.50	**128.92**	1.90	0.24	**29.57**	-	**158.49**	nr	**170.38**
Internal panelled doors; 10, glazed with 4mm obscure toughend glass; Magnet Trade										
762 x 1981 x 34mm (f sizes)......................	107.60	2.50	**110.29**	1.70	0.22	**26.51**	-	**136.81**	nr	**147.07**
Internal panelled doors; SA, glazed with bevelled glass; Magnet Trade										
762 x 1981 x 34mm (f sizes)......................	147.39	2.50	**151.08**	4.20	0.53	**65.36**	-	**216.44**	nr	**232.67**
External panelled doors; 2XG; Magnet Trade										
762 x 1981 x 44mm (f sizes)......................	67.71	2.50	**69.41**	1.30	0.16	**20.19**	-	**89.59**	nr	**96.31**
813 x 2032 x 44mm (f sizes)......................	82.97	2.50	**85.04**	1.50	0.19	**23.35**	-	**108.39**	nr	**116.52**
838 x 1981 x 44mm (f sizes)......................	76.20	2.50	**78.11**	1.50	0.19	**23.35**	-	**101.46**	nr	**109.06**
External panelled doors; 2XGG; Magnet Trade										
762 x 1981 x 44mm (f sizes)......................	79.59	2.50	**81.58**	1.30	0.16	**20.19**	-	**101.77**	nr	**109.40**
813 x 2032 x 44mm (f sizes)......................	72.17	2.50	**73.97**	1.50	0.19	**23.35**	-	**97.32**	nr	**104.62**
838 x 1981 x 44mm (f sizes)......................	86.37	2.50	**88.53**	1.50	0.19	**23.35**	-	**111.88**	nr	**120.27**
External panelled doors; KXT; Magnet Trade										
762 x 1981 x 44mm (f sizes)......................	142.31	2.50	**145.87**	1.30	0.16	**20.19**	-	**166.05**	nr	**178.51**
838 x 1981 x 44mm (f sizes)......................	144.00	2.50	**147.60**	1.50	0.19	**23.35**	-	**170.95**	nr	**183.78**
External panelled doors; SA; Magnet Trade										
762 x 1981 x 44mm (f sizes)......................	110.10	2.50	**112.85**	1.30	0.16	**20.19**	-	**133.04**	nr	**143.02**
838 x 1981 x 44mm (f sizes)......................	163.70	2.50	**167.79**	1.50	0.19	**23.35**	-	**191.14**	nr	**205.48**
External panelled doors; 10; Magnet Trade										
762 x 1981 x 44mm (f sizes)......................	74.51	2.50	**76.37**	1.30	0.16	**20.19**	-	**96.55**	nr	**103.80**
External panelled doors; 2XG, glazed with 4mm clear toughend glass; Magnet Trade										
762 x 1981 x 44mm (f sizes)......................	89.53	2.50	**91.77**	1.70	0.21	**26.41**	-	**118.18**	nr	**127.04**
813 x 2032 x 44mm (f sizes)......................	113.44	2.50	**116.27**	1.90	0.24	**29.57**	-	**145.84**	nr	**156.78**
838 x 1981 x 44mm (f sizes)......................	103.10	2.50	**105.67**	1.90	0.24	**29.57**	-	**135.24**	nr	**145.39**
External panelled doors; 2XGG, glazed with 4mm clear toughend glass; Magnet Trade										
762 x 1981 x 44mm (f sizes)......................	117.63	2.50	**120.57**	2.00	0.25	**31.10**	-	**151.67**	nr	**163.04**
813 x 2032 x 44mm (f sizes)......................	119.36	2.50	**122.34**	2.20	0.28	**34.26**	-	**156.60**	nr	**168.35**
838 x 1981 x 44mm (f sizes)......................	130.00	2.50	**133.25**	2.20	0.28	**34.26**	-	**167.51**	nr	**180.08**
External panelled doors; 10, glazed with 4mm clear toughend glass; Magnet Trade										
762 x 1981 x 44mm (f sizes)......................	114.11	2.50	**116.97**	1.80	0.23	**28.04**	-	**145.01**	nr	**155.88**
External panelled doors; 2XG, glazed with 4mm obscure toughend glass; Magnet Trade										
762 x 1981 x 44mm (f sizes)......................	93.16	2.50	**95.49**	1.70	0.21	**26.41**	-	**121.90**	nr	**131.04**
813 x 2032 x 44mm (f sizes)......................	111.05	2.50	**113.83**	1.90	0.24	**29.57**	-	**143.40**	nr	**154.15**
838 x 1981 x 44mm (f sizes)......................	104.28	2.50	**106.89**	1.90	0.24	**29.57**	-	**136.46**	nr	**146.69**
External panelled doors; 2XGG, glazed with 4mm obscure toughend glass; Magnet Trade										
762 x 1981 x 44mm (f sizes)......................	120.70	2.50	**123.72**	2.00	0.25	**31.10**	-	**154.82**	nr	**166.43**
813 x 2032 x 44mm (f sizes)......................	117.13	2.50	**120.06**	2.20	0.28	**34.26**	-	**154.32**	nr	**165.89**
838 x 1981 x 44mm (f sizes)......................	131.38	2.50	**134.67**	2.00	0.28	**31.42**	-	**166.08**	nr	**178.54**
External panelled doors; SA, glazed with 6mm bevelled glass; Magnet Trade										
762 x 1981 x 44mm (f sizes)......................	221.92	2.50	**227.47**	4.30	0.54	**66.89**	-	**294.36**	nr	**316.43**
838 x 1981 x 44mm (f sizes)......................	282.64	2.50	**289.71**	4.50	0.56	**69.94**	-	**359.65**	nr	**386.62**
Doors in Hardwood, wrought										
External panelled doors; Alicante; Magnet Trade										
813 x 2032 x 44mm (f sizes)......................	129.63	2.50	**132.87**	2.30	0.29	**35.79**	-	**168.66**	nr	**181.31**
838 x 1981 x 44mm (f sizes)......................	129.63	2.50	**132.87**	2.30	0.29	**35.79**	-	**168.66**	nr	**181.31**
External panelled doors; Carolina, glazed in clear glass; Magnet Trade										
813 x 2032 x 44mm (f sizes)......................	156.67	2.50	**160.59**	2.30	0.29	**35.79**	-	**196.38**	nr	**211.11**
838 x 1981 x 44mm (f sizes)......................	102.82	2.50	**105.39**	2.30	0.29	**35.79**	-	**141.18**	nr	**151.77**
External panelled doors; Alicante, double glazed with clear glass; Magnet Trade										
813 x 2032 x 44mm (f sizes)......................	198.22	2.50	**203.18**	4.10	0.51	**63.72**	-	**266.90**	nr	**286.92**
838 x 1981 x 44mm (f sizes)......................	198.22	2.50	**203.18**	4.10	0.51	**63.72**	-	**266.90**	nr	**286.92**
Doors in Hardwood (solid, laminated or veneered), wrought										
External panelled doors, fire resisting; FD30 2XGG; Magnet Trade										
838 x 1981 x 44mm (f sizes)......................	158.62	2.50	**162.58**	2.20	0.28	**34.26**	-	**196.84**	nr	**211.60**
External panelled doors; Kendall; Magnet Trade										
762 x 1981 x 44mm (f sizes)......................	132.17	2.50	**135.48**	2.10	0.26	**32.63**	-	**168.10**	nr	**180.71**
838 x 1981 x 44mm (f sizes)......................	132.17	2.50	**135.48**	2.30	0.29	**35.79**	-	**171.27**	nr	**184.11**

Labour hourly rates: (except Specialists) Craft Operatives £14.21 Labourer £10.71 Rates are national average prices. Refer to REGIONAL VARIATIONS for indicative levels of overall pricing in regions	MATERIALS			LABOUR				RATES		
	Del to Site £	Waste %	Material Cost £	Craft Optve Hrs	Lab Hrs	Labour Cost £	Sunds £	Nett Rate £	Unit	Gross Rate (+7.5%) £
L20: DOORS/SHUTTERS/HATCHES: TIMBER DOORS Cont.										
Doors in Hardwood (solid, laminated or veneered), wrought Cont.										
External panelled doors; Steeton; Magnet Trade										
838 x 1981 x 44mm (f sizes)......................	178.96	2.50	**183.43**	2.30	0.29	**35.79**	-	**219.22**	nr	235.66
External panelled doors; Conway; Magnet Trade										
762 x 1981 x 44mm (f sizes)......................	168.79	2.50	**173.01**	2.10	0.26	**32.63**	-	**205.63**	nr	221.05
813 x 2032 x 44mm (f sizes)......................	168.79	2.50	**173.01**	2.30	0.29	**35.79**	-	**208.79**	nr	224.45
838 x 1981 x 44mm (f sizes)......................	168.79	2.50	**173.01**	2.30	0.29	**35.79**	-	**208.79**	nr	224.45
External panelled doors; Elizabethan; Magnet Trade										
813 x 2032 x 44mm (f sizes)......................	159.63	2.50	**163.62**	2.30	0.29	**35.79**	-	**199.41**	nr	214.37
838 x 1981 x 44mm (f sizes)......................	142.31	2.50	**145.87**	2.30	0.29	**35.79**	-	**181.66**	nr	195.28
External panelled doors; Manilla; Magnet Trade										
762 x 1981 x 44mm (f sizes)......................	203.33	2.50	**208.42**	2.10	0.26	**32.63**	-	**241.04**	nr	259.12
813 x 2032 x 44mm (f sizes)......................	203.33	2.50	**208.42**	2.30	0.29	**35.79**	-	**244.20**	nr	262.52
838 x 1981 x 44mm (f sizes)......................	164.15	2.50	**168.26**	2.30	0.29	**35.79**	-	**204.05**	nr	219.35
External panelled doors; Richmond; Magnet Trade										
838 x 1981 x 44mm (f sizes)......................	188.62	2.50	**193.33**	2.30	0.29	**35.79**	-	**229.12**	nr	246.31
External panelled doors; Rutland; Magnet Trade										
762 x 1981 x 44mm (f sizes)......................	128.10	2.50	**131.31**	2.10	0.26	**32.63**	-	**163.93**	nr	176.23
838 x 1981 x 44mm (f sizes)......................	128.10	2.50	**131.31**	2.30	0.29	**35.79**	-	**167.10**	nr	179.63
External panelled doors; Stable door; Magnet Trade										
762 x 1981 x 44mm (f sizes)......................	199.81	2.50	**204.80**	3.60	0.45	**55.98**	-	**260.78**	nr	280.34
838 x 1981 x 44mm (f sizes)......................	208.96	2.50	**214.18**	3.85	0.48	**59.85**	-	**274.03**	nr	294.59
External panelled doors; Stourbridge; Magnet Trade										
838 x 1981 x 44mm (f sizes)......................	321.34	2.50	**329.38**	2.30	0.29	**35.79**	-	**365.17**	nr	392.55
External panelled doors; Stuart; Magnet Trade										
762 x 1981 x 44mm (f sizes)......................	132.17	2.50	**135.48**	2.10	0.26	**32.63**	-	**168.10**	nr	180.71
838 x 1981 x 44mm (f sizes)......................	132.17	2.50	**135.48**	2.30	0.29	**35.79**	-	**171.27**	nr	184.11
External panelled doors; Kendall, Double glazed with clear glass; Magnet Trade										
762 x 1981 x 44mm (f sizes)......................	209.41	2.50	**214.65**	2.50	0.31	**38.85**	-	**253.49**	nr	272.50
838 x 1981 x 44mm (f sizes)......................	208.39	2.50	**213.60**	2.70	0.34	**42.01**	-	**255.61**	nr	274.78
External panelled doors; Kendall, Double glazed with obscure glass; Magnet Trade										
762 x 1981 x 44mm (f sizes)......................	216.02	2.50	**221.42**	2.50	0.31	**38.85**	-	**260.27**	nr	279.79
838 x 1981 x 44mm (f sizes)......................	217.55	2.50	**222.99**	2.70	0.34	**42.01**	-	**265.00**	nr	284.87
External panelled doors; Rutland, Double glazed with clear glass; Magnet Trade										
762 x 1981 x 44mm (f sizes)......................	248.01	2.50	**254.22**	2.80	0.35	**43.54**	-	**297.75**	nr	320.08
838 x 1981 x 44mm (f sizes)......................	247.00	2.50	**253.17**	3.00	0.38	**46.70**	-	**299.87**	nr	322.36
External panelled doors; Rutland, Double glazed with obscure glass; Magnet Trade										
838 x 1981 x 44mm (f sizes)......................	258.19	2.50	**264.64**	3.00	0.38	**46.70**	-	**311.34**	nr	334.69
External panelled doors; Stourbridge, Double glazed with clear glass; Magnet Trade										
838 x 1981 x 44mm (f sizes)......................	535.89	2.50	**549.28**	5.30	0.66	**82.38**	-	**631.67**	nr	679.04
External panelled doors; Stuart, glazed with 6mm bevelled glass; Magnet Trade										
762 x 1981 x 44mm (f sizes)......................	238.91	2.50	**244.88**	5.10	0.64	**79.33**	-	**324.20**	nr	348.52
838 x 1981 x 44mm (f sizes)......................	249.08	2.50	**255.30**	5.30	0.66	**82.38**	-	**337.69**	nr	363.01
External panelled doors; Stable door, upper door glazed with decorative leaded obscured glazing; Magnet Trade										
762 x 1981 x 44mm (f sizes)......................	258.74	2.50	**265.21**	5.00	0.63	**77.80**	-	**343.00**	nr	368.73
838 x 1981 x 44mm (f sizes)......................	269.42	2.50	**276.15**	5.25	0.66	**81.67**	-	**357.82**	nr	384.66
External panelled doors; Steeton, glazed with clear glass double glazing unit; Magnet Trade										
838 x 1981 x 44mm (f sizes)......................	239.92	2.50	**245.92**	2.60	0.33	**40.48**	-	**286.40**	nr	307.88
External panelled doors; Richmond, glazed with decorative leaded obscured double glazing; Magnet Trade										
838 x 1981 x 44mm (f sizes)......................	367.06	2.50	**376.23**	2.90	0.36	**45.06**	-	**421.30**	nr	452.89
External panelled doors; Rutland, glazed with clear glass double glazing unit; Magnet Trade										
762 x 1981 x 44mm (f sizes)......................	248.01	2.50	**254.22**	2.80	0.35	**43.54**	-	**297.75**	nr	320.08
838 x 1981 x 44mm (f sizes)......................	247.00	2.50	**253.17**	3.00	0.38	**46.70**	-	**299.87**	nr	322.36
Doors in Hardwood (solid, laminated or fibreboard core and/or veneered), wrought										
Internal panelled doors; SA; Magnet Trade										
686 x 1981 x 34mm (f sizes)......................	76.18	2.50	**78.08**	1.90	0.24	**29.57**	-	**107.65**	nr	115.73
762 x 1981 x 34mm (f sizes)......................	76.18	2.50	**78.08**	2.10	0.26	**32.63**	-	**110.71**	nr	119.01

Labour hourly rates: (except Specialists) Craft Operatives £14.21 Labourer £10.71 Rates are national average prices. Refer to REGIONAL VARIATIONS for indicative levels of overall pricing in regions	MATERIALS			LABOUR				RATES		
	Del to Site £	Waste %	Material Cost £	Craft Optve Hrs	Lab Hrs	Labour Cost £	Sunds £	Nett Rate £	Unit	Gross Rate (+7.5%) £
L20: DOORS/SHUTTERS/HATCHES: TIMBER DOORS Cont.										
Doors in Hardwood (solid, laminated or fibreboard core and/or veneered), wrought Cont.										
Internal panelled doors; Victorian; Magnet Trade										
686 x 1981 x 34mm (f sizes)	119.46	2.50	**122.45**	1.90	0.24	**29.57**	-	**152.01**	nr	**163.42**
762 x 1981 x 34mm (f sizes)	121.49	2.50	**124.53**	2.10	0.26	**32.63**	-	**157.16**	nr	**168.94**
Internal panelled doors; Louis; Magnet Trade										
686 x 1981 x 34mm (f sizes)	301.61	2.50	**309.15**	1.90	0.24	**29.57**	-	**338.72**	nr	**364.13**
762 x 1981 x 34mm (f sizes)	301.61	2.50	**309.15**	2.10	0.26	**32.63**	-	**341.78**	nr	**367.41**
Internal panelled doors; Windsor; Magnet Trade										
686 x 1981 x 34mm (f sizes)	122.51	2.50	**125.57**	1.90	0.24	**29.57**	-	**155.14**	nr	**166.78**
762 x 1981 x 34mm (f sizes)	123.53	2.50	**126.62**	2.10	0.26	**32.63**	-	**159.24**	nr	**171.18**
Internal panelled doors; 10; Magnet Trade										
762 x 1981 x 34mm (f sizes)	72.67	2.50	**74.49**	2.10	0.26	**32.63**	-	**107.12**	nr	**115.15**
Internal panelled doors; 10, glazed with 4mm clear toughend glass; Magnet Trade										
762 x 1981 x 34mm (f sizes)	111.77	2.50	**114.57**	2.60	0.33	**40.48**	-	**155.05**	nr	**166.68**
Internal panelled doors; 10, glazed with 4mm obscure toughend glass; Magnet Trade										
762 x 1981 x 34mm (f sizes)	113.11	2.50	**115.94**	2.60	0.33	**40.48**	-	**156.42**	nr	**168.15**
Internal panelled doors; Malaga; Magnet Trade										
762 x 1981 x 34mm (f sizes)	142.34	2.50	**145.90**	2.10	0.26	**32.63**	-	**178.53**	nr	**191.92**
Internal panelled doors; SA, glazed with bevelled glass; Magnet Trade										
762 x 1981 x 34mm (f sizes)	135.53	2.50	**138.92**	5.10	0.64	**79.33**	-	**218.24**	nr	**234.61**
Doors in solid or laminated construction, Oak facings										
Internal panelled doors; Louis; Magnet Trade										
762 x 1981 x 34mm (f sizes)	135.53	2.50	**138.92**	2.10	0.26	**32.63**	-	**171.55**	nr	**184.41**
Doors in hardboard, embossed										
Internal panelled doors; Colonist; Magnet Trade										
610 x 1981 x 34mm (f sizes)	25.35	2.50	**25.98**	1.60	0.20	**24.88**	-	**50.86**	nr	**54.67**
686 x 1981 x 34mm (f sizes)	25.35	2.50	**25.98**	1.70	0.21	**26.41**	-	**52.39**	nr	**56.32**
762 x 1981 x 34mm (f sizes)	25.35	2.50	**25.98**	1.90	0.24	**29.57**	-	**55.55**	nr	**59.72**
Internal panelled doors; Sentinel; Magnet Trade										
686 x 1981 x 34mm (f sizes)	47.37	2.50	**48.56**	1.70	0.21	**26.41**	-	**74.96**	nr	**80.58**
762 x 1981 x 34mm (f sizes)	37.20	2.50	**38.13**	1.90	0.24	**29.57**	-	**67.70**	nr	**72.78**
Flush doors										
Internal; hardboard facings; Magnet Trade										
305 x 1981 x 34mm	27.92	2.50	**28.62**	1.00	0.13	**15.60**	-	**44.22**	nr	**47.54**
381 x 1981 x 34mm	26.40	2.50	**27.06**	1.10	0.14	**17.13**	-	**44.19**	nr	**47.50**
457 x 1981 x 34mm	26.40	2.50	**27.06**	1.20	0.15	**18.66**	-	**45.72**	nr	**49.15**
533 x 1981 x 34mm	25.35	2.50	**25.98**	1.30	0.16	**20.19**	-	**46.17**	nr	**49.63**
610 x 1981 x 34mm	25.35	2.50	**25.98**	1.40	0.18	**21.82**	-	**47.80**	nr	**51.39**
686 x 1981 x 34mm	25.35	2.50	**25.98**	1.50	0.19	**23.35**	-	**49.33**	nr	**53.03**
711 x 1981 x 34mm	27.42	2.50	**28.10**	1.60	0.20	**24.88**	-	**52.98**	nr	**56.95**
762 x 1981 x 34mm	25.35	2.50	**25.98**	1.70	0.21	**26.41**	-	**52.39**	nr	**56.32**
813 x 2032 x 34mm	30.47	2.50	**31.23**	1.90	0.24	**29.57**	-	**60.80**	nr	**65.36**
838 x 1981 x 34mm	26.19	2.50	**26.85**	1.90	0.24	**29.57**	-	**56.42**	nr	**60.65**
Internal; Magnaseal; Magnet Trade										
610 x 1981 x 34mm	21.37	2.50	**21.90**	1.80	0.23	**28.04**	-	**49.94**	nr	**53.69**
686 x 1981 x 34mm	21.37	2.50	**21.90**	1.70	0.21	**26.41**	-	**48.31**	nr	**51.93**
762 x 1981 x 34mm	19.72	2.50	**20.21**	1.90	0.24	**29.57**	-	**49.78**	nr	**53.51**
838 x 1981 x 34mm	27.99	2.50	**28.69**	2.20	0.28	**34.26**	-	**62.95**	nr	**67.67**
Internal; plywood facings; Magnet Trade										
381 x 1981 x 34mm	32.12	2.50	**32.92**	1.10	0.14	**17.13**	-	**50.05**	nr	**53.80**
457 x 1981 x 34mm	32.12	2.50	**32.92**	1.20	0.15	**18.66**	-	**51.58**	nr	**55.45**
533 x 1981 x 34mm	32.98	2.50	**33.80**	1.30	0.16	**20.19**	-	**53.99**	nr	**58.04**
610 x 1981 x 34mm	32.12	2.50	**32.92**	1.40	0.18	**21.82**	-	**54.74**	nr	**58.85**
686 x 1981 x 34mm	25.35	2.50	**25.98**	1.50	0.19	**23.35**	-	**49.33**	nr	**53.03**
711 x 1981 x 34mm	32.12	2.50	**32.92**	1.60	0.20	**24.88**	-	**57.80**	nr	**62.13**
762 x 1981 x 34mm	25.35	2.50	**25.98**	1.70	0.21	**26.41**	-	**52.39**	nr	**56.32**
813 x 2032 x 34mm	35.52	2.50	**36.41**	1.90	0.24	**29.57**	-	**65.98**	nr	**70.92**
838 x 1981 x 34mm	33.82	2.50	**34.67**	1.90	0.24	**29.57**	-	**64.24**	nr	**69.06**
Internal; Sapele Showpiece; Magnet Trade										
457 x 1981 x 34mm	46.74	2.50	**47.91**	1.40	0.18	**21.82**	-	**69.73**	nr	**74.96**
533 x 1981 x 34mm	46.74	2.50	**47.91**	1.50	0.19	**23.35**	-	**71.26**	nr	**76.60**
610 x 1981 x 34mm	40.60	2.50	**41.62**	1.60	0.20	**24.88**	-	**66.50**	nr	**71.48**
686 x 1981 x 34mm	33.82	2.50	**34.67**	1.70	0.21	**26.41**	-	**61.07**	nr	**65.66**
762 x 1981 x 34mm	28.74	2.50	**29.46**	2.10	0.26	**32.63**	-	**62.08**	nr	**66.74**
838 x 1981 x 34mm	39.74	2.50	**40.74**	2.10	0.26	**32.63**	-	**73.36**	nr	**78.87**
External; MF1X, plywood facings; Magnet Trade										
762 x 1981 x 44mm	60.94	2.50	**62.47**	1.80	0.23	**28.04**	-	**90.51**	nr	**97.30**
838 x 1981 x 44mm	62.63	2.50	**64.19**	2.00	0.25	**31.10**	-	**95.29**	nr	**102.44**
External; MF2X, plywood facings; Magnet Trade										
762 x 1981 x 44mm; glazing aperture 457 x 457mm	83.35	2.50	**85.44**	1.80	0.23	**28.04**	-	**113.48**	nr	**121.99**

Labour hourly rates: (except Specialists) Craft Operatives £14.21 Labourer £10.71 Rates are national average prices. Refer to REGIONAL VARIATIONS for indicative levels of overall pricing in regions	MATERIALS			LABOUR				RATES		
	Del to Site £	Waste %	Material Cost £	Craft Optve Hrs	Lab Hrs	Labour Cost £	Sunds £	Nett Rate £	Unit	Gross Rate (+7.5%) £
L20: DOORS/SHUTTERS/HATCHES: TIMBER DOORS Cont.										
Flush doors Cont.										
External; MF2X, plywood facings; Magnet Trade Cont.										
838 x 1981 x 44mm; glazing aperture 457 x 457mm..	83.35	2.50	**85.44**	2.00	0.25	**31.10**	-	**116.53**	nr	**125.27**
External; MF4X, plywood facings; Magnet Trade										
762 x 1981 x 44mm; glazing aperture 559 x 864mm..	82.34	2.50	**84.39**	1.80	0.23	**28.04**	-	**112.44**	nr	**120.87**
838 x 1981 x 44mm; glazing aperture 635 x 864mm..	85.90	2.50	**88.04**	2.00	0.25	**31.10**	-	**119.14**	nr	**128.08**
Internal; fire resisting; Magnaseal; FD30; Magnet Trade										
813 x 2032 x 44mm	59.76	2.50	**61.25**	2.20	0.28	**34.26**	-	**95.51**	nr	**102.68**
Internal; fire resisting; Plywood facings; FD30; Magnet Trade										
686 x 1981 x 44mm	59.76	2.50	**61.25**	1.70	0.21	**26.41**	-	**87.66**	nr	**94.23**
762 x 1981 x 44mm	59.76	2.50	**61.25**	1.90	0.24	**29.57**	-	**90.82**	nr	**97.63**
813 x 2032 x 44mm	62.40	2.50	**63.96**	2.20	0.28	**34.26**	-	**98.22**	nr	**105.59**
838 x 1981 x 44mm	62.40	2.50	**63.96**	2.20	0.28	**34.26**	-	**98.22**	nr	**105.59**
Internal; fire resisting; plywood facings; FDG30; Magnet Trade										
762 x 1981 x 44mm; glazing aperture 508 x 508mm..	74.51	2.50	**76.37**	1.90	0.24	**29.57**	-	**105.94**	nr	**113.88**
813 x 2032 x 44mm; glazing aperture 559 x 559mm..	88.44	2.50	**90.65**	2.20	0.29	**34.37**	-	**125.02**	nr	**134.39**
838 x 1981 x 44mm; glazing aperture 584 x 584mm..	76.20	2.50	**78.11**	2.20	0.29	**34.37**	-	**112.47**	nr	**120.91**
External; fire resisting; plywood facings; FD30; Magnet Trade										
762 x 1981 x 44mm	27.03	2.50	**27.71**	1.90	0.24	**29.57**	-	**57.28**	nr	**61.57**
838 x 1981 x 44mm	27.03	2.50	**27.71**	2.30	0.29	**35.79**	-	**63.50**	nr	**68.26**
Internal; fire resisting; Sapele facings; FD30; Magnet Trade										
762 x 1981 x 44mm	85.39	2.50	**87.52**	2.20	0.29	**34.37**	-	**121.89**	nr	**131.03**
838 x 1981 x 44mm	92.00	2.50	**94.30**	2.40	0.30	**37.32**	-	**131.62**	nr	**141.49**
Trap doors in softwood, wrought										
19mm matchboarding on 25 x 75mm ledges										
457 x 610mm	23.29	2.50	**23.87**	0.18	0.02	**2.77**	-	**26.64**	nr	**28.64**
610 x 610mm	28.80	2.50	**29.52**	0.19	0.02	**2.91**	-	**32.43**	nr	**34.87**
762 x 610mm	34.92	2.50	**35.79**	0.20	0.02	**3.06**	-	**38.85**	nr	**41.76**
Trap doors in B.C. Pine, wrought										
19mm matchboarding on 25 x 75mm ledges										
457 x 610mm	28.84	2.50	**29.56**	0.18	0.02	**2.77**	-	**32.33**	nr	**34.76**
610 x 610mm	40.78	2.50	**41.80**	0.19	0.02	**2.91**	-	**44.71**	nr	**48.06**
762 x 610mm	52.13	2.50	**53.43**	0.20	0.02	**3.06**	-	**56.49**	nr	**60.72**
Panelled doors										
Note										
the following prices for panelled doors are for doors to detail in moderate numbers: doors in large numbers to one pattern would cost considerably less										
Panelled doors in softwood, wrought										
38mm square framed (or chamfered or moulded one or both sides)										
two panel	65.87	2.50	**67.52**	0.85	0.11	**13.26**	-	**80.77**	m2	**86.83**
four panel	73.25	2.50	**75.08**	0.85	0.11	**13.26**	-	**88.34**	m2	**94.97**
six panel	80.91	2.50	**82.93**	0.85	0.11	**13.26**	-	**96.19**	m2	**103.40**
add if upper panels open moulded in small squares for glass	15.90	2.50	**16.30**	-	-	**-**	-	**16.30**	m2	**17.52**
50mm square framed (or chamfered or moulded one or both sides)										
two panel	73.25	2.50	**75.08**	1.00	0.11	**15.39**	-	**90.47**	m2	**97.26**
four panel	80.91	2.50	**82.93**	1.00	0.11	**15.39**	-	**98.32**	m2	**105.69**
six panel	88.58	2.50	**90.79**	1.00	0.11	**15.39**	-	**106.18**	m2	**114.15**
add if upper panels open moulded in small squares for glass	15.90	2.50	**16.30**	-	-	**-**	-	**16.30**	m2	**17.52**
Panelled doors in Afromosia, wrought										
38mm square framed (or chamfered or moulded one or both sides)										
two panel	189.10	2.50	**193.83**	1.50	0.19	**23.35**	-	**217.18**	m2	**233.47**
four panel	201.72	2.50	**206.76**	1.50	0.19	**23.35**	-	**230.11**	m2	**247.37**
six panel	214.33	2.50	**219.69**	1.50	0.19	**23.35**	-	**243.04**	m2	**261.27**
add if upper panels open moulded in small squares for glass	35.93	2.50	**36.83**	-	-	**-**	-	**36.83**	m2	**39.59**
50mm square framed (or chamfered or moulded one or both sides)										
two panel	208.03	2.50	**213.23**	1.75	0.22	**27.22**	-	**240.45**	m2	**258.49**
four panel	220.63	2.50	**226.15**	1.75	0.22	**27.22**	-	**253.37**	m2	**272.37**
six panel	233.24	2.50	**239.07**	1.75	0.22	**27.22**	-	**266.29**	m2	**286.27**
add if upper panels open moulded in small squares for glass	35.93	2.50	**36.83**	-	-	**-**	-	**36.83**	m2	**39.59**

Labour hourly rates: (except Specialists) Craft Operatives £14.21 Labourer £10.71 Rates are national average prices. Refer to REGIONAL VARIATIONS for indicative levels of overall pricing in regions	MATERIALS			LABOUR				RATES		
	Del to Site £	Waste %	Material Cost £	Craft Optve Hrs	Lab Hrs	Labour Cost £	Sunds £	Nett Rate £	Unit	Gross Rate (+7.5%) £
L20: DOORS/SHUTTERS/HATCHES: TIMBER DOORS Cont.										
Panelled doors in Sapele, wrought										
38mm square framed (or chamfered or moulded one or both sides)										
two panel	165.31	2.50	**169.44**	1.25	0.16	**19.48**	-	**188.92**	m2	**203.09**
four panel	178.22	2.50	**182.68**	1.25	0.16	**19.48**	-	**202.16**	m2	**217.32**
six panel	191.15	2.50	**195.93**	1.25	0.16	**19.48**	-	**215.40**	m2	**231.56**
add if upper panels open moulded in small squares for glass	31.00	2.50	**31.77**	-	-	**-**	-	**31.77**	m2	**34.15**
50mm square framed (or chamfered or moulded one or both sides)										
two panel	178.22	2.50	**182.68**	1.50	0.19	**23.35**	-	**206.03**	m2	**221.48**
four panel	191.15	2.50	**195.93**	1.50	0.19	**23.35**	-	**219.28**	m2	**235.72**
six panel	204.06	2.50	**209.16**	1.50	0.19	**23.35**	-	**232.51**	m2	**249.95**
add if upper panels open moulded in small squares for glass	31.00	2.50	**31.77**	-	-	**-**	-	**31.77**	m2	**34.15**
Garage doors in softwood, wrought										
Side hung; framed, tongued and grooved boarded										
2134 x 1981 x 44mm overall, pair	290.66	2.50	**297.92**	4.10	0.51	**63.72**	-	**361.65**	nr	**388.77**
2134 x 2134 x 44mm overall, pair	302.28	2.50	**309.84**	4.25	0.53	**66.07**	-	**375.91**	nr	**404.10**
Up and over; framed, tongued and grooved boarded; up and over door gear										
2134 x 1981 x 44mm	406.91	2.50	**417.08**	8.00	1.00	**124.39**	-	**541.47**	nr	**582.09**
Garage doors in Western Red Cedar, wrought										
Side hung; framed, tongued and grooved boarded										
2134 x 1981 x 44mm overall, pair	407.93	2.50	**418.13**	4.50	0.56	**69.94**	-	**488.07**	nr	**524.68**
2134 x 2134 x 44mm overall, pair	418.95	2.50	**429.42**	4.75	0.59	**73.82**	-	**503.24**	nr	**540.98**
Up and over; framed, tongued and grooved boarded; up and over door gear										
2134 x 1981 x 44mm	518.18	2.50	**531.13**	8.40	1.05	**130.61**	-	**661.74**	nr	**711.37**
Garage doors in Oak, wrought										
Side hung; framed, tongued and grooved boarded										
2134 x 1981 x 44mm overall, pair	1090.85	2.50	**1118.12**	7.15	0.90	**111.24**	-	**1229.36**	nr	**1321.56**
2134 x 2134 x 44mm overall, pair	1151.46	2.50	**1180.25**	7.40	0.95	**115.33**	-	**1295.58**	nr	**1392.74**
Up and over; framed, tongued and grooved boarded; up and over door gear										
2134 x 1981 x 44mm	1272.67	2.50	**1304.49**	12.00	1.50	**186.59**	-	**1491.07**	nr	**1602.90**
Patio doors in hardwood, wrought										
Double glazed clear toughened glass, Magnet Trade, one opening leaf, doors and frames treated with base coat stains, fixing frame to masonry with screws										
2073 x 1805mm overall, HP 6	1272.67	2.50	**1304.49**	12.00	1.50	**186.59**	-	**1491.07**	nr	**1602.90**
2073 X 2387mm overall, HP 8	1272.67	2.50	**1304.49**	12.00	1.50	**186.59**	-	**1491.07**	nr	**1602.90**
extra; ventilator head, for HP 6	20.00	2.50	**20.50**	-	-	**-**	-	**20.50**	nr	**22.04**
extra; ventilator head, for HP 8	20.00	2.50	**20.50**	-	-	**-**	-	**20.50**	nr	**22.04**
Doorsets mainly in softwood, wrought										
Doorsets; 28mm thick jambs, head and transome with 12mm thick stop to suit 100mm thick wall; 14mm thick hardwood threshold; honeycomb core plywood faced flush door lipped two long edges; 6mm plywood transome panel fixed with pinned beads; 65mm snap in hinges; 57mm backset mortice latch; fixing frame to masonry with screws										
526 x 2040 x 40mm flush door -1 nr; basic dimensions										
600 x 2100mm	120.70	2.50	**123.72**	1.25	0.16	**19.48**	-	**143.19**	nr	**153.93**
600 x 2400mm	147.58	2.50	**151.27**	1.35	0.17	**21.00**	-	**172.27**	nr	**185.19**
626 x 2040 x 40mm flush door -1 nr; basic dimensions										
700 x 2100mm	120.70	2.50	**123.72**	1.25	0.16	**19.48**	-	**143.19**	nr	**153.93**
700 x 2400mm	147.58	2.50	**151.27**	1.35	0.17	**21.00**	-	**172.27**	nr	**185.19**
726 x 2040 x 40mm flush door -1 nr; basic dimensions										
800 x 2100mm	120.70	2.50	**123.72**	1.25	0.16	**19.48**	-	**143.19**	nr	**153.93**
800 x 2400mm	147.58	2.50	**151.27**	1.35	0.17	**21.00**	-	**172.27**	nr	**185.19**
826 x 2040 x 40mm flush door -1 nr; basic dimensions										
900 x 2100mm	125.95	2.50	**129.10**	1.25	0.16	**19.48**	-	**148.57**	nr	**159.72**
900 x 2400mm	150.73	2.50	**154.50**	1.35	0.17	**21.00**	-	**175.50**	nr	**188.67**

Labour hourly rates: (except Specialists) Craft Operatives £14.21 Labourer £10.71 Rates are national average prices. Refer to REGIONAL VARIATIONS for indicative levels of overall pricing in regions	MATERIALS			LABOUR				RATES		
	Del to Site £	Waste %	Material Cost £	Craft Optve Hrs	Lab Hrs	Labour Cost £	Sunds £	Nett Rate £	Unit	Gross Rate (+7.5%) £
L20: DOORS/SHUTTERS/HATCHES: TIMBER DOORS Cont.										
Doorsets mainly in softwood, wrought Cont.										
Note										
actual frame size is 14mm narrower and 14mm shorter than basic size dimensions										
L20: DOORS/SHUTTERS/HATCHES: ROLLER SHUTTERS										
Roller shutters in wood										
Note										
roller shutters are always purpose made to order and the following prices are indicative only. Firm quotations should always be obtained										
Pole and hook operation										
2134 x 2134mm	1500.00	-	1500.00	12.00	12.00	299.04	25.00	1824.04	nr	1960.84
L20: DOORS/SHUTTERS/HATCHES: LININGS										
Door frames and door lining sets in softwood, wrought										
Sets										
linings; fixing to masonry with screws										
32 x 113mm	4.99	2.50	5.11	0.40	0.05	6.22	0.07	11.40	m	12.25
32 x 150mm	5.98	2.50	6.13	0.40	0.05	6.22	0.07	12.42	m	13.36
32 x 225mm	8.95	2.50	9.17	0.60	0.07	9.28	0.10	18.55	m	19.94
32 x 330mm	14.06	2.50	14.41	0.60	0.07	9.28	0.10	23.79	m	25.57
38 x 113mm	5.78	2.50	5.92	0.40	0.05	6.22	0.07	12.21	m	13.13
38 x 150mm	6.99	2.50	7.17	0.40	0.05	6.22	0.07	13.46	m	14.47
38 x 225mm	10.29	2.50	10.55	0.60	0.07	9.28	0.10	19.92	m	21.42
38 x 330mm	16.62	2.50	17.04	0.60	0.07	9.28	0.10	26.42	m	28.40
linings										
32 x 113mm; labours -1	5.12	2.50	5.25	0.20	0.02	3.06	0.04	8.34	m	8.97
32 x 113mm; labours -2	5.19	2.50	5.32	0.20	0.02	3.06	0.04	8.42	m	9.05
32 x 113mm; labours -3	5.24	2.50	5.37	0.20	0.02	3.06	0.04	8.47	m	9.11
32 x 113mm; labours -4	5.31	2.50	5.45	0.20	0.02	3.06	0.04	8.54	m	9.18
38 x 113mm; labours -1	5.92	2.50	6.07	0.20	0.02	3.06	0.04	9.17	m	9.85
38 x 113mm; labours -2	5.98	2.50	6.13	0.20	0.02	3.06	0.04	9.23	m	9.92
38 x 113mm; labours -3	6.06	2.50	6.21	0.20	0.02	3.06	0.04	9.30	m	10.00
38 x 113mm; labours -4	6.12	2.50	6.27	0.20	0.02	3.06	0.04	9.37	m	10.07
Internal door lining sets; supplied unassembled; fixing to masonry with screws										
32 x 115mm rebated linings; assembling										
for 610 x 1981mm doors	22.22	2.50	22.77	2.00	0.25	31.10	0.32	54.19	nr	58.25
for 686 x 1981mm doors	22.22	2.50	22.77	2.00	0.25	31.10	0.32	54.19	nr	58.25
for 762 x 1981mm doors	22.22	2.50	22.77	2.00	0.25	31.10	0.32	54.19	nr	58.25
32 x 140mm rebated linings; assembling										
for 610 x 1981mm doors	24.24	2.50	24.84	2.00	0.25	31.10	0.32	56.26	nr	60.48
for 686 x 1981mm doors	24.24	2.50	24.84	2.00	0.25	31.10	0.32	56.26	nr	60.48
for 762 x 1981mm doors	24.24	2.50	24.84	2.00	0.25	31.10	0.32	56.26	nr	60.48
Door frames and door lining sets in Afromosia, wrought, selected for transparent finish										
Sets										
linings; fixing to masonry with screws										
32 x 113mm	12.03	2.50	12.33	0.60	0.08	9.38	0.10	21.81	m	23.45
32 x 150mm	14.56	2.50	14.92	0.60	0.08	9.38	0.10	24.41	m	26.24
32 x 225mm	21.57	2.50	22.11	0.90	0.11	13.97	0.15	36.23	m	38.94
32 x 330mm	33.59	2.50	34.43	0.90	0.11	13.97	0.15	48.55	m	52.19
38 x 113mm	14.03	2.50	14.38	0.60	0.08	9.38	0.10	23.86	m	25.65
38 x 150mm	16.80	2.50	17.22	0.60	0.08	9.38	0.10	26.70	m	28.71
38 x 225mm	24.63	2.50	25.25	0.90	0.11	13.97	0.15	39.36	m	42.32
38 x 330mm	39.19	2.50	40.17	0.90	0.11	13.97	0.15	54.29	m	58.36
linings										
32 x 113mm; labours -1	12.20	2.50	12.51	0.30	0.04	4.69	0.06	17.26	m	18.55
32 x 113mm; labours -2	12.32	2.50	12.63	0.30	0.04	4.69	0.06	17.38	m	18.68
32 x 113mm; labours -3	12.43	2.50	12.74	0.30	0.04	4.69	0.06	17.49	m	18.80
32 x 113mm; labours -4	12.56	2.50	12.87	0.30	0.04	4.69	0.06	17.63	m	18.95
38 x 113mm; labours -1	14.20	2.50	14.56	0.30	0.04	4.69	0.06	19.31	m	20.75
38 x 113mm; labours -2	14.32	2.50	14.68	0.30	0.04	4.69	0.06	19.43	m	20.89
38 x 113mm; labours -3	14.44	2.50	14.80	0.30	0.04	4.69	0.06	19.55	m	21.02
38 x 113mm; labours -4	14.56	2.50	14.92	0.30	0.04	4.69	0.06	19.68	m	21.15
Door frames and door lining sets in Sapele, wrought, selected for transparent finish										
Sets										
linings; fixing to masonry with screws										
32 x 113mm	10.48	2.50	10.74	0.55	0.07	8.57	0.09	19.39	m	20.85
32 x 150mm	12.57	2.50	12.89	0.55	0.07	8.57	0.09	21.54	m	23.16
32 x 225mm	18.86	2.50	19.33	0.80	0.08	12.22	0.14	31.69	m	34.07
32 x 330mm	29.33	2.50	30.07	0.80	0.08	12.22	0.14	42.43	m	45.61
38 x 113mm	12.27	2.50	12.57	0.55	0.07	8.57	0.09	21.23	m	22.82
38 x 150mm	14.66	2.50	15.03	0.55	0.07	8.57	0.09	23.68	m	25.46
38 x 225mm	21.55	2.50	22.08	0.80	0.08	12.22	0.14	34.45	m	37.03
38 x 330mm	34.71	2.50	35.58	0.80	0.08	12.22	0.14	47.95	m	51.54
linings										
32 x 113mm; labours -1	10.66	2.50	10.93	0.27	0.03	4.16	0.04	15.12	m	16.26
32 x 113mm; labours -2	10.77	2.50	11.04	0.27	0.03	4.16	0.04	15.24	m	16.38
32 x 113mm; labours -3	10.90	2.50	11.17	0.27	0.03	4.16	0.04	15.37	m	16.52
32 x 113mm; labours -4	11.01	2.50	11.29	0.27	0.03	4.16	0.04	15.49	m	16.65
38 x 113mm; labours -1	12.45	2.50	12.76	0.27	0.03	4.16	0.04	16.96	m	18.23
38 x 113mm; labours -2	12.57	2.50	12.89	0.27	0.03	4.16	0.04	17.09	m	18.37

Labour hourly rates: (except Specialists)
Craft Operatives £14.21 Labourer £10.71
Rates are national average prices.
Refer to REGIONAL VARIATIONS for indicative levels of overall pricing in regions

	MATERIALS			LABOUR				RATES		
	Del to Site £	Waste %	Material Cost £	Craft Optve Hrs	Lab Hrs	Labour Cost £	Sunds £	Nett Rate £	Unit	Gross Rate (+7.5%) £
L20: DOORS/SHUTTERS/HATCHES: LININGS Cont.										
Door frames and door lining sets in Sapele, wrought, selected for transparent finish Cont.										
Sets Cont.										
linings Cont.										
38 x 113mm; labours -3	12.69	2.50	**13.01**	0.27	0.03	**4.16**	0.04	**17.20**	m	**18.49**
38 x 113mm; labours -4	12.81	2.50	**13.13**	0.27	0.03	**4.16**	0.04	**17.33**	m	**18.63**
L20: DOORS/SHUTTERS/HATCHES: FRAMES										
Door frames and door lining sets in softwood, wrought - frames										
Sets										
38 x 75mm jambs	4.67	2.50	**4.78**	0.18	0.02	**2.77**	0.03	**7.58**	m	**8.15**
38 x 100mm jambs	5.40	2.50	**5.53**	0.18	0.02	**2.77**	0.03	**8.33**	m	**8.96**
50 x 75mm jambs	5.40	2.50	**5.53**	0.20	0.03	**3.16**	0.03	**8.73**	m	**9.38**
50 x 100mm jambs	6.26	2.50	**6.42**	0.20	0.03	**3.16**	0.03	**9.61**	m	**10.33**
50 x 125mm jambs	6.99	2.50	**7.17**	0.20	0.03	**3.16**	0.03	**10.36**	m	**11.14**
63 x 100mm jambs	6.99	2.50	**7.17**	0.20	0.03	**3.16**	0.03	**10.36**	m	**11.14**
75 x 100mm jambs	8.95	2.50	**9.17**	0.22	0.03	**3.45**	0.04	**12.66**	m	**13.61**
75 x 113mm jambs	10.23	2.50	**10.48**	0.22	0.03	**3.45**	0.04	**13.97**	m	**15.02**
100 x 100mm jambs	11.30	2.50	**11.58**	0.25	0.03	**3.87**	0.04	**15.50**	m	**16.66**
100 x 113mm jambs	12.99	2.50	**13.31**	0.27	0.03	**4.16**	0.04	**17.51**	m	**18.82**
100 x 125mm jambs	14.68	2.50	**15.04**	0.27	0.03	**4.16**	0.04	**19.24**	m	**20.69**
113 x 113mm jambs	14.68	2.50	**15.04**	0.27	0.03	**4.16**	0.04	**19.24**	m	**20.69**
113 x 125mm jambs	16.27	2.50	**16.68**	0.30	0.04	**4.69**	0.05	**21.42**	m	**23.03**
113 x 150mm jambs	17.91	2.50	**18.36**	0.30	0.04	**4.69**	0.05	**23.10**	m	**24.83**
38 x 75mm jambs; labours -1	4.78	2.50	**4.90**	0.18	0.02	**2.77**	0.03	**7.70**	m	**8.28**
38 x 75mm jambs; labours -2	4.83	2.50	**4.95**	0.18	0.02	**2.77**	0.03	**7.75**	m	**8.33**
38 x 75mm jambs; labours -3	4.92	2.50	**5.05**	0.18	0.02	**2.77**	0.03	**7.85**	m	**8.44**
38 x 75mm jambs; labours -4	4.99	2.50	**5.11**	0.18	0.02	**2.77**	0.03	**7.91**	m	**8.51**
38 x 75mm heads	4.67	2.50	**4.78**	0.18	0.02	**2.77**	0.03	**7.58**	m	**8.15**
38 x 100mm heads	5.40	2.50	**5.53**	0.18	0.02	**2.77**	0.03	**8.33**	m	**8.96**
50 x 75mm heads	5.40	2.50	**5.53**	0.20	0.03	**3.16**	0.03	**8.73**	m	**9.38**
50 x 100mm heads	6.26	2.50	**6.42**	0.20	0.03	**3.16**	0.03	**9.61**	m	**10.33**
50 x 125mm heads	6.99	2.50	**7.17**	0.20	0.03	**3.16**	0.03	**10.36**	m	**11.14**
63 x 100mm heads	6.99	2.50	**7.17**	0.20	0.03	**3.16**	0.03	**10.36**	m	**11.14**
75 x 100mm heads	8.95	2.50	**9.17**	0.22	0.03	**3.45**	0.04	**12.66**	m	**13.61**
75 x 113mm heads	10.23	2.50	**10.48**	0.22	0.03	**3.45**	0.04	**13.97**	m	**15.02**
100 x 100mm heads	11.30	2.50	**11.58**	0.25	0.03	**3.87**	0.04	**15.50**	m	**16.66**
100 x 113mm heads	12.99	2.50	**13.31**	0.27	0.03	**4.16**	0.04	**17.51**	m	**18.82**
100 x 125mm heads	14.69	2.50	**15.05**	0.27	0.03	**4.16**	0.04	**19.25**	m	**20.70**
113 x 113mm heads	14.69	2.50	**15.05**	0.27	0.03	**4.16**	0.04	**19.25**	m	**20.70**
113 x 125mm heads	16.27	2.50	**16.68**	0.30	0.04	**4.69**	0.05	**21.42**	m	**23.03**
113 x 150mm heads	17.91	2.50	**18.36**	0.30	0.04	**4.69**	0.05	**23.10**	m	**24.83**
38 x 75mm heads; labours -1	4.78	2.50	**4.90**	0.18	0.02	**2.77**	0.03	**7.70**	m	**8.28**
38 x 75mm heads; labours -2	4.83	2.50	**4.95**	0.18	0.02	**2.77**	0.03	**7.75**	m	**8.33**
38 x 75mm heads; labours -3	4.92	2.50	**5.05**	0.18	0.02	**2.77**	0.03	**7.85**	m	**8.44**
38 x 75mm heads; labours -4	4.99	2.50	**5.11**	0.18	0.02	**2.77**	0.03	**7.91**	m	**8.51**
63 x 125mm sills	10.63	2.50	**10.90**	0.23	0.03	**3.59**	0.04	**14.52**	m	**15.61**
75 x 125mm sills	12.11	2.50	**12.42**	0.24	0.03	**3.73**	0.04	**16.19**	m	**17.40**
75 x 150mm sills	13.60	2.50	**13.94**	0.25	0.03	**3.94**	0.04	**17.85**	m	**19.19**
63 x 125mm sills; labours -1	10.76	2.50	**11.03**	0.23	0.03	**3.59**	0.04	**14.66**	m	**15.76**
63 x 125mm sills; labours -2	10.82	2.50	**11.09**	0.23	0.03	**3.59**	0.04	**14.71**	m	**15.82**
63 x 125mm sills; labours -3	10.91	2.50	**11.18**	0.23	0.03	**3.59**	0.04	**14.81**	m	**15.92**
63 x 125mm sills; labours -4	10.96	2.50	**11.23**	0.23	0.03	**3.59**	0.04	**14.86**	m	**15.98**
63 x 100mm mullions	6.99	2.50	**7.17**	0.05	0.01	**0.82**	-	**7.99**	m	**8.59**
75 x 100mm mullions	8.95	2.50	**9.17**	0.05	0.01	**0.82**	-	**9.99**	m	**10.74**
75 x 113mm mullions	10.23	2.50	**10.48**	0.05	0.01	**0.82**	-	**11.30**	m	**12.15**
100 x 100mm mullions	11.30	2.50	**11.58**	0.05	0.01	**0.82**	-	**12.40**	m	**13.33**
100 x 125mm mullions	14.69	2.50	**15.05**	0.05	0.01	**0.82**	-	**15.87**	m	**17.06**
63 x 100mm mullions; labours -1	7.14	2.50	**7.32**	0.05	0.01	**0.82**	-	**8.13**	m	**8.74**
63 x 100mm mullions; labours -2	7.20	2.50	**7.38**	0.05	0.01	**0.82**	-	**8.20**	m	**8.81**
63 x 100mm mullions; labours -3	7.28	2.50	**7.46**	0.05	0.01	**0.82**	-	**8.28**	m	**8.90**
63 x 100mm mullions; labours -4	7.33	2.50	**7.52**	0.05	0.01	**0.82**	-	**8.33**	m	**8.96**
63 x 100mm transoms	6.99	2.50	**7.17**	0.05	0.01	**0.82**	-	**7.99**	m	**8.59**
75 x 100mm transoms	8.95	2.50	**9.17**	0.05	0.01	**0.82**	-	**9.99**	m	**10.74**
75 x 113mm transoms	10.23	2.50	**10.48**	0.05	0.01	**0.82**	-	**11.30**	m	**12.15**
100 x 100mm transoms	11.30	2.50	**11.58**	0.05	0.01	**0.82**	-	**12.40**	m	**13.33**
100 x 125mm transoms	14.69	2.50	**15.05**	0.05	0.01	**0.82**	-	**15.87**	m	**17.06**
63 x 100mm transoms; labours -1	7.14	2.50	**7.32**	0.05	0.01	**0.82**	-	**8.13**	m	**8.74**
63 x 100mm transoms; labours -2	7.20	2.50	**7.38**	0.05	0.01	**0.82**	-	**8.20**	m	**8.81**
63 x 100mm transoms; labours -3	7.28	2.50	**7.46**	0.05	0.01	**0.82**	-	**8.28**	m	**8.90**
63 x 100mm transoms; labours -4	7.33	2.50	**7.52**	0.05	0.01	**0.82**	-	**8.33**	m	**8.96**
Internal door frame sets; supplied unassembled										
38 x 63mm jambs and head; stops (supplied loose); assembling										
for 686 x 1981mm doors	20.20	2.50	**20.70**	1.00	0.13	**15.60**	0.16	**36.47**	nr	**39.20**
for 762 x 1981mm doors	20.20	2.50	**20.70**	1.00	0.13	**15.60**	0.16	**36.47**	nr	**39.20**
38 x 75mm jambs and head; stops (supplied loose); assembling										
for 686 x 1981mm doors	21.54	2.50	**22.08**	1.00	0.13	**15.60**	0.16	**37.84**	nr	**40.68**
for 762 x 1981mm doors	21.54	2.50	**22.08**	1.00	0.13	**15.60**	0.16	**37.84**	nr	**40.68**
50 x 100mm jambs and head; stops (supplied loose); assembling										
for 686 x 1981mm doors	33.65	2.50	**34.49**	1.00	0.13	**15.60**	0.16	**50.25**	nr	**54.02**
for 762 x 1981mm doors	33.65	2.50	**34.49**	1.00	0.13	**15.60**	0.16	**50.25**	nr	**54.02**
50 x 113mm jambs and head; stops (supplied loose); assembling										
for 686 x 1981mm doors	41.74	2.50	**42.78**	1.00	0.13	**15.60**	0.16	**58.54**	nr	**62.93**
for 762 x 1981mm doors	41.74	2.50	**42.78**	1.00	0.13	**15.60**	0.16	**58.54**	nr	**62.93**

Labour hourly rates: (except Specialists) Craft Operatives £14.21 Labourer £10.71 Rates are national average prices. Refer to REGIONAL VARIATIONS for indicative levels of overall pricing in regions	MATERIALS			LABOUR				RATES		
	Del to Site £	Waste %	Material Cost £	Craft Optve Hrs	Lab Hrs	Labour Cost £	Sunds £	Nett Rate £	Unit	Gross Rate (+7.5%) £
L20: DOORS/SHUTTERS/HATCHES: FRAMES Cont.										
Door frames and door lining sets in softwood, wrought - frames Cont.										
External door frame sets; one coat external primer before delivery to site										
63 x 75mm jambs and head; rebates -1										
for 762 x 1981 x 50mm doors	36.34	2.50	**37.25**	1.00	0.13	**15.60**	0.16	**53.01**	nr	**56.98**
for 838 x 1981 x 50mm doors	37.69	2.50	**38.63**	1.00	0.13	**15.60**	0.16	**54.39**	nr	**58.47**
63 x 88mm jambs and head; rebates -1										
for 762 x 1981 x 50mm doors	37.69	2.50	**38.63**	1.00	0.13	**15.60**	0.16	**54.39**	nr	**58.47**
for 838 x 1981 x 50mm doors	39.05	2.50	**40.02**	1.00	0.13	**15.60**	0.16	**55.79**	nr	**59.97**
63 x 75mm jambs and head, rebates -1 nr; hardwood sill										
for 762 x 1981 x 50mm doors	80.76	2.50	**82.78**	1.25	0.16	**19.48**	0.20	**102.46**	nr	**110.14**
for 838 x 1981 x 50mm doors	82.11	2.50	**84.16**	1.25	0.16	**19.48**	0.20	**103.84**	nr	**111.63**
63 x 88mm jambs and head, rebates -1 nr; hardwood sill										
for 762 x 1981 x 50mm doors	83.44	2.50	**85.53**	1.25	0.16	**19.48**	0.20	**105.20**	nr	**113.09**
for 838 x 1981 x 50mm doors	84.80	2.50	**86.92**	1.25	0.16	**19.48**	0.20	**106.60**	nr	**114.59**
for 1168 x 1981 x 50mm doors	86.13	2.50	**88.28**	1.50	0.19	**23.35**	0.24	**111.87**	nr	**120.26**
External garage door frame sets; one coat external primer before delivery to site; supplied unassembled										
75 x 100mm jambs and head; assembling										
for 2134 x 1981mm side hung doors	59.41	2.50	**60.90**	2.25	0.28	**34.97**	0.37	**96.24**	nr	**103.45**
for 2134 x 2134mm side hung doors	63.36	2.50	**64.94**	2.25	0.28	**34.97**	0.37	**100.28**	nr	**107.80**
75 x 75mm jambs and head; assembling										
for 2134 x 1981mm up and over doors	55.18	2.50	**56.56**	2.25	0.28	**34.97**	0.37	**91.90**	nr	**98.79**
Door frames and door lining sets in Afromosia, wrought, selected for transparent finish - frames										
Sets										
38 x 75mm jambs	11.20	2.50	**11.48**	0.31	0.04	**4.83**	0.05	**16.36**	m	**17.59**
38 x 100mm jambs	12.90	2.50	**13.22**	0.31	0.04	**4.83**	0.05	**18.11**	m	**19.46**
50 x 75mm jambs	12.90	2.50	**13.22**	0.35	0.04	**5.40**	0.06	**18.68**	m	**20.09**
50 x 100mm jambs	15.15	2.50	**15.53**	0.35	0.04	**5.40**	0.06	**20.99**	m	**22.56**
50 x 125mm jambs	16.80	2.50	**17.22**	0.35	0.04	**5.40**	0.06	**22.68**	m	**24.38**
63 x 100mm jambs	16.80	2.50	**17.22**	0.35	0.04	**5.40**	0.06	**22.68**	m	**24.38**
75 x 100mm jambs	21.57	2.50	**22.11**	0.38	0.05	**5.94**	0.06	**28.10**	m	**30.21**
75 x 113mm jambs	24.63	2.50	**25.25**	0.38	0.05	**5.94**	0.06	**31.24**	m	**33.58**
100 x 100mm jambs	27.46	2.50	**28.15**	0.44	0.05	**6.79**	0.07	**35.00**	m	**37.63**
100 x 113mm jambs	31.36	2.50	**32.14**	0.47	0.06	**7.32**	0.08	**39.55**	m	**42.51**
100 x 125mm jambs	35.31	2.50	**36.19**	0.47	0.06	**7.32**	0.08	**43.59**	m	**46.86**
113 x 113mm jambs	35.31	2.50	**36.19**	0.47	0.06	**7.32**	0.08	**43.59**	m	**46.86**
113 x 125mm jambs	39.19	2.50	**40.17**	0.52	0.07	**8.14**	0.09	**48.40**	m	**52.03**
113 x 150mm jambs	43.15	2.50	**44.23**	0.52	0.07	**8.14**	0.09	**52.46**	m	**56.39**
38 x 75mm jambs; labours -1	11.37	2.50	**11.65**	0.31	0.04	**4.83**	0.05	**16.54**	m	**17.78**
38 x 75mm jambs; labours -2	11.49	2.50	**11.78**	0.31	0.04	**4.83**	0.05	**16.66**	m	**17.91**
38 x 75mm jambs; labours -3	11.67	2.50	**11.96**	0.31	0.04	**4.83**	0.05	**16.85**	m	**18.11**
38 x 75mm jambs; labours -4	11.85	2.50	**12.15**	0.31	0.04	**4.83**	0.05	**17.03**	m	**18.31**
38 x 75mm heads	11.20	2.50	**11.48**	0.31	0.04	**4.83**	0.05	**16.36**	m	**17.59**
38 x 100mm heads	12.90	2.50	**13.22**	0.31	0.04	**4.83**	0.05	**18.11**	m	**19.46**
50 x 75mm heads	12.90	2.50	**13.22**	0.31	0.04	**4.83**	0.05	**18.11**	m	**19.46**
50 x 100mm heads	15.15	2.50	**15.53**	0.35	0.04	**5.40**	0.06	**20.99**	m	**22.56**
50 x 125mm heads	16.80	2.50	**17.22**	0.35	0.04	**5.40**	0.06	**22.68**	m	**24.38**
63 x 100mm heads	16.80	2.50	**17.22**	0.35	0.04	**5.40**	0.06	**22.68**	m	**24.38**
75 x 100mm heads	21.57	2.50	**22.11**	0.38	0.05	**5.94**	0.06	**28.10**	m	**30.21**
75 x 113mm heads	24.63	2.50	**25.25**	0.38	0.05	**5.94**	0.06	**31.24**	m	**33.58**
100 x 100mm heads	27.46	2.50	**28.15**	0.44	0.05	**6.79**	0.07	**35.00**	m	**37.63**
100 x 113mm heads	31.36	2.50	**32.14**	0.47	0.06	**7.32**	0.08	**39.55**	m	**42.51**
100 x 125mm heads	35.31	2.50	**36.19**	0.47	0.06	**7.32**	0.08	**43.59**	m	**46.86**
113 x 113mm heads	35.31	2.50	**36.19**	0.47	0.06	**7.32**	0.08	**43.59**	m	**46.86**
113 x 125mm heads	39.19	2.50	**40.17**	0.52	0.07	**8.14**	0.09	**48.40**	m	**52.03**
113 x 150mm heads	43.15	2.50	**44.23**	0.52	0.07	**8.14**	0.09	**52.46**	m	**56.39**
38 x 75mm heads; labours -1	11.37	2.50	**11.65**	0.31	0.04	**4.83**	0.05	**16.54**	m	**17.78**
38 x 75mm heads; labours -2	11.49	2.50	**11.78**	0.31	0.04	**4.83**	0.05	**16.66**	m	**17.91**
38 x 75mm heads; labours -3	11.67	2.50	**11.96**	0.31	0.04	**4.83**	0.05	**16.85**	m	**18.11**
38 x 75mm heads; labours -4	11.85	2.50	**12.15**	0.31	0.04	**4.83**	0.05	**17.03**	m	**18.31**
63 x 125mm sills	25.46	2.50	**26.10**	0.40	0.05	**6.22**	0.06	**32.38**	m	**34.80**
75 x 125mm sills	29.12	2.50	**29.85**	0.44	0.06	**6.90**	0.07	**36.81**	m	**39.57**
75 x 150mm sills	32.78	2.50	**33.60**	0.45	0.06	**7.04**	0.07	**40.71**	m	**43.76**
63 x 125mm sills; labours -1	25.76	2.50	**26.40**	0.40	0.05	**6.22**	0.06	**32.68**	m	**35.13**
63 x 125mm sills; labours -2	25.88	2.50	**26.53**	0.40	0.05	**6.22**	0.06	**32.81**	m	**35.27**
63 x 125mm sills; labours -3	25.99	2.50	**26.64**	0.40	0.05	**6.22**	0.06	**32.92**	m	**35.39**
63 x 125mm sills; labours -4	26.11	2.50	**26.76**	0.40	0.05	**6.22**	0.06	**33.04**	m	**35.52**
63 x 100mm mullions	16.80	2.50	**17.22**	0.05	0.01	**0.82**	-	**18.04**	m	**19.39**
75 x 100mm mullions	21.57	2.50	**22.11**	0.05	0.01	**0.82**	-	**22.93**	m	**24.65**
75 x 113mm mullions	24.63	2.50	**25.25**	0.05	0.01	**0.82**	-	**26.06**	m	**28.02**
100 x 100mm mullions	27.46	2.50	**28.15**	0.05	0.01	**0.82**	-	**28.96**	m	**31.14**
100 x 125mm mullions	35.31	2.50	**36.19**	0.05	0.01	**0.82**	-	**37.01**	m	**39.79**
63 x 100mm mullions; labours -1	16.97	2.50	**17.39**	0.05	0.01	**0.82**	-	**18.21**	m	**19.58**
63 x 100mm mullions; labours -2	17.09	2.50	**17.52**	0.05	0.01	**0.82**	-	**18.33**	m	**19.71**
63 x 100mm mullions; labours -3	17.21	2.50	**17.64**	0.05	0.01	**0.82**	-	**18.46**	m	**19.84**
63 x 100mm mullions; labours -4	17.33	2.50	**17.76**	0.05	0.01	**0.82**	-	**18.58**	m	**19.97**
63 x 100mm transoms	16.80	2.50	**17.22**	0.05	0.01	**0.82**	-	**18.04**	m	**19.39**
75 x 100mm transoms	21.57	2.50	**22.11**	0.05	0.01	**0.82**	-	**22.93**	m	**24.65**
75 x 113mm transoms	24.63	2.50	**25.25**	0.05	0.01	**0.82**	-	**26.06**	m	**28.02**
100 x 100mm transoms	27.46	2.50	**28.15**	0.05	0.01	**0.82**	-	**28.96**	m	**31.14**
100 x 125mm transoms	35.31	2.50	**36.19**	0.05	0.01	**0.82**	-	**37.01**	m	**39.79**
63 x 100mm transoms; labours -1	16.97	2.50	**17.39**	0.05	0.01	**0.82**	-	**18.21**	m	**19.58**
63 x 100mm transoms; labours -2	17.09	2.50	**17.52**	0.05	0.01	**0.82**	-	**18.33**	m	**19.71**
63 x 100mm transoms; labours -3	17.21	2.50	**17.64**	0.05	0.01	**0.82**	-	**18.46**	m	**19.84**
63 x 100mm transoms; labours -4	17.33	2.50	**17.76**	0.05	0.01	**0.82**	-	**18.58**	m	**19.97**

Labour hourly rates: (except Specialists) Craft Operatives £14.21 Labourer £10.71 Rates are national average prices. Refer to REGIONAL VARIATIONS for indicative levels of overall pricing in regions	MATERIALS			LABOUR				RATES		
	Del to Site £	Waste %	Material Cost £	Craft Optve Hrs	Lab Hrs	Labour Cost £	Sunds £	Nett Rate £	Unit	Gross Rate (+7.5%) £
L20: DOORS/SHUTTERS/HATCHES: FRAMES Cont.										
Door frames and door lining sets in Sapele, wrought, selected for transparent finish - frames										
Sets										
38 x 75mm jambs	9.82	2.50	**10.06**	0.27	0.03	**4.16**	0.04	**14.26**	m	**15.33**
38 x 100mm jambs	11.31	2.50	**11.59**	0.27	0.03	**4.16**	0.04	**15.79**	m	**16.97**
50 x 75mm jambs	11.31	2.50	**11.59**	0.30	0.04	**4.69**	0.05	**16.33**	m	**17.56**
50 x 100mm jambs	13.17	2.50	**13.50**	0.30	0.04	**4.69**	0.05	**18.24**	m	**19.61**
50 x 125mm jambs	14.66	2.50	**15.03**	0.30	0.04	**4.69**	0.05	**19.77**	m	**21.25**
63 x 100mm jambs	14.66	2.50	**15.03**	0.30	0.04	**4.69**	0.05	**19.77**	m	**21.25**
75 x 100mm jambs	18.86	2.50	**19.33**	0.33	0.04	**5.12**	0.05	**24.49**	m	**26.33**
75 x 113mm jambs	21.55	2.50	**22.08**	0.33	0.04	**5.12**	0.05	**27.25**	m	**29.30**
100 x 100mm jambs	23.83	2.50	**24.42**	0.37	0.05	**5.79**	0.06	**30.27**	m	**32.55**
100 x 113mm jambs	27.29	2.50	**27.97**	0.40	0.05	**6.22**	0.07	**34.26**	m	**36.83**
100 x 125mm jambs	30.83	2.50	**31.60**	0.40	0.05	**6.22**	0.07	**37.89**	m	**40.73**
113 x 113mm jambs	30.83	2.50	**31.60**	0.40	0.05	**6.22**	0.07	**37.89**	m	**40.73**
113 x 125mm jambs	34.36	2.50	**35.22**	0.45	0.06	**7.04**	0.07	**42.33**	m	**45.50**
113 x 150mm jambs	37.71	2.50	**38.65**	0.45	0.06	**7.04**	0.07	**45.76**	m	**49.19**
38 x 75mm jambs; labours -1	10.00	2.50	**10.25**	0.27	0.03	**4.16**	0.04	**14.45**	m	**15.53**
38 x 75mm jambs; labours -2	10.11	2.50	**10.36**	0.27	0.03	**4.16**	0.04	**14.56**	m	**15.65**
38 x 75mm jambs; labours -3	10.24	2.50	**10.49**	0.27	0.03	**4.16**	0.04	**14.69**	m	**15.79**
38 x 75mm jambs; labours -4	10.35	2.50	**10.61**	0.27	0.03	**4.16**	0.04	**14.81**	m	**15.92**
38 x 75mm heads	9.82	2.50	**10.06**	0.27	0.03	**4.16**	0.04	**14.26**	m	**15.33**
38 x 100mm heads	11.31	2.50	**11.59**	0.27	0.03	**4.16**	0.04	**15.79**	m	**16.97**
50 x 75mm heads	11.31	2.50	**11.59**	0.30	0.04	**4.69**	0.05	**16.33**	m	**17.56**
50 x 100mm heads	13.17	2.50	**13.50**	0.30	0.04	**4.69**	0.05	**18.24**	m	**19.61**
50 x 125mm heads	14.66	2.50	**15.03**	0.30	0.04	**4.69**	0.05	**19.77**	m	**21.25**
63 x 100mm heads	14.66	2.50	**15.03**	0.30	0.04	**4.69**	0.05	**19.77**	m	**21.25**
75 x 100mm heads	15.26	2.50	**15.65**	0.33	0.04	**5.12**	0.05	**20.81**	m	**22.37**
75 x 113mm heads	21.55	2.50	**22.08**	0.33	0.04	**5.12**	0.05	**27.25**	m	**29.30**
100 x 100mm heads	23.83	2.50	**24.42**	0.37	0.05	**5.79**	0.06	**30.27**	m	**32.55**
100 x 113mm heads	27.29	2.50	**27.97**	0.40	0.05	**6.22**	0.07	**34.26**	m	**36.83**
100 x 125mm heads	30.83	2.50	**31.60**	0.40	0.05	**6.22**	0.07	**37.89**	m	**40.73**
113 x 113mm heads	30.83	2.50	**31.60**	0.40	0.05	**6.22**	0.07	**37.89**	m	**40.73**
113 x 125mm heads	34.36	2.50	**35.22**	0.45	0.06	**7.04**	0.07	**42.33**	m	**45.50**
113 x 150mm heads	37.71	2.50	**38.65**	0.45	0.06	**7.04**	0.07	**45.76**	m	**49.19**
38 x 75mm heads; labours -1	10.00	2.50	**10.25**	0.27	0.03	**4.16**	0.04	**14.45**	m	**15.53**
38 x 75mm heads; labours -2	10.00	2.50	**10.25**	0.27	0.03	**4.16**	0.04	**14.45**	m	**15.53**
38 x 75mm heads; labours -3	10.24	2.50	**10.49**	0.27	0.03	**4.16**	0.04	**14.69**	m	**15.79**
38 x 75mm heads; labours -4	10.35	2.50	**10.61**	0.27	0.03	**4.16**	0.04	**14.81**	m	**15.92**
63 x 125mm sills	22.39	2.50	**22.95**	0.35	0.04	**5.40**	0.06	**28.41**	m	**30.54**
75 x 125mm sills	25.43	2.50	**26.07**	0.37	0.05	**5.79**	0.06	**31.92**	m	**34.32**
75 x 150mm sills	28.73	2.50	**29.45**	0.37	0.05	**5.79**	0.06	**35.30**	m	**37.95**
63 x 125mm sills; labours -1	22.56	2.50	**23.12**	0.35	0.04	**5.40**	0.06	**28.59**	m	**30.73**
63 x 125mm sills; labours -2	22.69	2.50	**23.25**	0.35	0.04	**5.40**	0.06	**28.72**	m	**30.87**
63 x 125mm sills; labours -3	22.80	2.50	**23.37**	0.35	0.04	**5.40**	0.06	**28.83**	m	**30.99**
63 x 125mm sills; labours -4	22.93	2.50	**23.50**	0.35	0.04	**5.40**	0.06	**28.96**	m	**31.13**
63 x 100mm mullions	14.66	2.50	**15.03**	0.05	0.01	**0.82**	-	**15.84**	m	**17.03**
75 x 100mm mullions	15.26	2.50	**15.65**	0.05	0.01	**0.82**	-	**16.46**	m	**17.70**
75 x 113mm mullions	21.55	2.50	**22.08**	0.05	0.01	**0.82**	-	**22.90**	m	**24.62**
100 x 100mm mullions	23.83	2.50	**24.42**	0.05	0.01	**0.82**	-	**25.24**	m	**27.13**
100 x 125mm mullions	30.83	2.50	**31.60**	0.05	0.01	**0.82**	-	**32.41**	m	**34.84**
63 x 100mm mullions; labours -1	14.84	2.50	**15.21**	0.05	0.01	**0.82**	-	**16.03**	m	**17.23**
63 x 100mm mullions; labours -2	14.97	2.50	**15.34**	0.05	0.01	**0.82**	-	**16.16**	m	**17.37**
63 x 100mm mullions; labours -3	15.08	2.50	**15.46**	0.05	0.01	**0.82**	-	**16.28**	m	**17.50**
63 x 100mm mullions; labours -4	15.21	2.50	**15.59**	0.05	0.01	**0.82**	-	**16.41**	m	**17.64**
63 x 100mm transoms	14.66	2.50	**15.03**	0.05	0.01	**0.82**	-	**15.84**	m	**17.03**
75 x 100mm transoms	15.26	2.50	**15.65**	0.05	0.01	**0.82**	-	**16.46**	m	**17.70**
75 x 113mm transoms	21.55	2.50	**22.08**	0.05	0.01	**0.82**	-	**22.90**	m	**24.62**
100 x 100mm transoms	23.83	2.50	**24.42**	0.05	0.01	**0.82**	-	**25.24**	m	**27.13**
100 x 125mm transoms	30.83	2.50	**31.60**	0.05	0.01	**0.82**	-	**32.41**	m	**34.84**
63 x 100mm transoms; labours -1	14.84	2.50	**15.21**	0.05	0.01	**0.82**	-	**16.03**	m	**17.23**
63 x 100mm transoms; labours -2	14.97	2.50	**15.34**	0.05	0.01	**0.82**	-	**16.16**	m	**17.37**
63 x 100mm transoms; labours -3	15.08	2.50	**15.46**	0.05	0.01	**0.82**	-	**16.28**	m	**17.50**
63 x 100mm transoms; labours -4	15.21	2.50	**15.59**	0.05	0.01	**0.82**	-	**16.41**	m	**17.64**
Intumescent strips and smoke seals										
Albi-Flex, white PVC sleeved self adhesive; Rentokil Ltd.										
12 x 4mm intumescent strip, half hour application; setting into groove in timber frame or door	3.68	10.00	**4.05**	0.15	-	**2.13**	-	**6.18**	m	**6.64**
18 x 4mm intumescent strip, one hour application; setting into groove in timber frame or door	5.15	10.00	**5.67**	0.15	-	**2.13**	-	**7.80**	m	**8.39**
22 x 4mm intumescent strip with integral cold smoke seal, half hour application; setting into groove in timber frame or door	7.80	10.00	**8.58**	0.15	-	**2.13**	-	**10.72**	m	**11.52**
28 x 4mm intumescent strip with integral cold smoke seal, one hour application; setting into groove in timber frame or door	9.05	10.00	**9.96**	0.15	-	**2.13**	-	**12.09**	m	**12.99**
12 x 4mm intumescent strip, half hour application; fixing to both sides of glass behind glazing beads	7.37	10.00	**8.11**	0.30	-	**4.26**	-	**12.37**	m	**13.30**
25 x 4mm intumescent strip, one hour application; fixing to both sides of glass behind glazing beads	13.24	10.00	**14.56**	0.30	-	**4.26**	-	**18.82**	m	**20.24**
10 x 4mm cold smoke seal; setting into groove in timber frame or door	4.04	10.00	**4.45**	0.15	-	**2.13**	-	**6.58**	m	**7.07**
Lorient Polyproducts Ltd; System 36										
15 x 12mm glazing channel reference LG1512; fitting over the edge of the pane	1.72	10.00	**1.89**	0.15	-	**2.13**	-	**4.02**	m	**4.32**
Lorient Polyproducts Ltd; System 90										
intumescent lining reference LX4402 to suit 44mm thick doors; fixing to timber with adhesive	3.74	10.00	**4.11**	0.30	-	**4.26**	-	**8.38**	m	**9.00**

Labour hourly rates: (except Specialists) Craft Operatives £14.21 Labourer £10.71 Rates are national average prices. Refer to REGIONAL VARIATIONS for indicative levels of overall pricing in regions	MATERIALS			LABOUR				RATES		
	Del to Site £	Waste %	Material Cost £	Craft Optve Hrs	Lab Hrs	Labour Cost £	Sunds £	Nett Rate £	Unit	Gross Rate (+7.5%) £
L20: DOORS/SHUTTERS/HATCHES: FRAMES Cont.										
Intumescent strips and smoke seals Cont.										
Lorient Polyproducts Ltd; System 90 Cont.										
intumescent lining reference LX5402 to suit 54mm thick doors; fixing to timber with adhesive......	4.19	10.00	**4.61**	0.30	-	**4.26**	-	**8.87**	m	**9.53**
Mann McGowan Fabrications Ltd										
Pyroglaze 30 half hour application; fixing to both sides of glass behind glazing beads.............	3.03	10.00	**3.33**	0.35	-	**4.97**	-	**8.30**	m	**8.92**
Pyroglaze 60 one hour application; fixing to both sides of glass behind glazing beads.............	4.74	10.00	**5.22**	0.35	-	**4.97**	-	**10.19**	m	**10.96**
Bedding and pointing frames										
Bedding in cement mortar (1:3); pointing one side wood frames	0.29	5.00	**0.30**	0.12	-	**1.71**	-	**2.00**	m	**2.15**
Bedding in cement mortar (1:3); pointing each side wood frames	0.38	5.00	**0.40**	0.18	-	**2.56**	-	**2.96**	m	**3.18**
Bedding in cement mortar (1:3); pointing with Secomastic standard mastic one side wood frames one side	1.16	5.00	**1.26**	0.18	-	**2.56**	-	**3.82**	m	**4.11**
L20: DOORS/SHUTTERS/HATCHES: METAL DOORS										
Doors and sidelights in galvanized steel										
Crittall Windows Ltd. Duralife Homelight; weatherstripping; fixing to masonry with lugs										
761 x 2056mm doors NA15.........................	695.03	2.50	**712.41**	2.80	1.40	**54.78**	-	**767.19**	nr	**824.73**
997 x 2056mm doors NA2	1046.15	2.50	**1072.31**	3.00	1.50	**58.70**	-	**1131.00**	nr	**1215.83**
1143 x 2056mm doors NA25........................	1065.83	2.50	**1092.47**	3.20	1.60	**62.61**	-	**1155.08**	nr	**1241.71**
279 x 2056mm sidelights, NA6....................	165.42	2.50	**169.55**	1.20	0.60	**23.48**	-	**193.03**	nr	**207.51**
508 x 2056mm sidelights, NA5....................	201.04	2.50	**206.07**	1.50	0.75	**29.35**	-	**235.42**	nr	**253.07**
997 x 2056mm sidelights, NA13F..................	382.48	2.50	**392.04**	1.80	0.90	**35.22**	-	**427.26**	nr	**459.30**
Doors and sidelights in galvanized steel; polyester powder coated; white matt										
Crittall Windows Ltd. Duralife Homelight; weatherstripping; fixing to masonry with lugs										
761 x 2056mm doors NA15.........................	875.13	2.50	**897.01**	2.50	1.40	**50.52**	-	**947.53**	nr	**1018.59**
997 x 2056mm doors NA2	1316.67	2.50	**1349.58**	3.00	1.50	**58.70**	-	**1408.28**	nr	**1513.90**
1143 x 2056mm doors NA25........................	1341.81	2.50	**1375.36**	3.20	1.60	**62.61**	-	**1437.97**	nr	**1545.81**
279 x 2056mm sidelights, NA6....................	205.14	2.50	**210.27**	1.20	0.60	**23.48**	-	**233.75**	nr	**251.28**
508 x 2056mm sidelights, NA5....................	249.33	2.50	**255.56**	1.50	0.75	**29.35**	-	**284.91**	nr	**306.28**
997 x 2056mm sidelights, NA13F..................	465.07	2.50	**476.69**	1.80	0.90	**35.22**	-	**511.91**	nr	**550.31**
Garage doors; Henderson Durham Castle in galvanized steel; one coat primer by manufacturer										
Overhead garage doors; tensioning device Tracked spring counterbalanced										
fixing to timber with screws; for opening size										
2134 x 1981mm	265.00	2.50	**271.63**	4.00	4.00	**99.68**	-	**371.31**	nr	**399.15**
2134 x 2134mm	302.00	2.50	**309.55**	4.00	4.00	**99.68**	-	**409.23**	nr	**439.92**
Overhead garage doors; counterbalanced by springs										
4267 x 1981mm	924.00	2.50	**947.10**	6.00	6.00	**149.52**	-	**1096.62**	nr	**1178.87**
4267 x 2134mm	924.00	2.50	**947.10**	6.00	6.00	**149.52**	-	**1096.62**	nr	**1178.87**
L20: DOORS/SHUTTERS/HATCHES: ROLLER SHUTTERS										
Roller shutters in steel, galvanized										
Note roller shutters are always purpose made to order and the following prices are indicative only. Firm quotations should always be obtained										
Crank handle operation										
4572 x 3658mm overall; fixing to masonry with screws	1247.09	2.50	**1278.26**	16.00	16.00	**398.72**	-	**1676.98**	nr	**1802.76**
Endless hand chain operation										
5486 x 4267mm overall; fixing to masonry with screws	1306.47	2.50	**1339.13**	20.00	20.00	**498.40**	-	**1837.53**	nr	**1975.35**
Electric motor operation										
6096 x 6096mm overall; fixing to masonry with screws	2434.79	2.50	**2495.65**	30.00	30.00	**747.60**	-	**3243.25**	nr	**3486.50**
Roller shutters in aluminium										
Endless hand chain operation										
3553 x 3048mm overall; fixing to masonry with screws	1134.68	2.50	**1163.04**	20.00	20.00	**498.40**	-	**1661.44**	nr	**1786.05**
Bedding and pointing frames										
Pointing with Secomastic standard mastic metal frames one side	0.88	10.00	**0.96**	0.12	0.06	**2.35**	-	**3.31**	m	**3.56**

Labour hourly rates: (except Specialists) Craft Operatives £14.21 Labourer £10.71 Rates are national average prices. Refer to REGIONAL VARIATIONS for indicative levels of overall pricing in regions	MATERIALS			LABOUR				RATES		
	Del to Site £	Waste %	Material Cost £	Craft Optve Hrs	Lab Hrs	Labour Cost £	Sunds £	Nett Rate £	Unit	Gross Rate (+7.5%) £
L20: DOORS/SHUTTERS/HATCHES: ROLLER SHUTTERS Cont.										
Bedding and pointing frames Cont.										
Bedding in cement mortar (1:3); pointing with Secomastic standard mastic one side										
metal frames one side	1.02	5.00	**1.10**	0.12	0.10	**2.78**	-	**3.88**	m	**4.17**
L20: DOORS/SHUTTERS/HATCHES: PLASTICS/RUBBER DOORS										
Flexible doors mainly in plastics and rubber										
Mancuna standard doors; Mandor Engineering Ltd; 43mm diameter steel tube frame; top and bottom plates with pivoting arrangements and spring unit; 8mm rubber panel doors with triangular perspex windows in each leaf										
fixing to masonry; for opening size										
1800 x 2100mm	892.50	2.50	**914.81**	4.00	4.00	**99.68**	-	**1014.49**	nr	**1090.58**
2440 x 2400mm	997.50	2.50	**1022.44**	5.00	5.00	**124.60**	-	**1147.04**	nr	**1233.07**
Mancuna standard doors; Mandor Engineering Ltd; 43mm diameter steel tube frame; top and bottom plates with pivoting arrangements and spring unit; 12mm rubber panel doors with triangular perspex windows in each leaf										
fixing to masonry; for opening size										
1800 x 2100mm	997.50	2.50	**1022.44**	4.50	4.50	**112.14**	-	**1134.58**	nr	**1219.67**
2440 x 2400mm	1139.25	2.50	**1167.73**	5.50	5.50	**137.06**	-	**1304.79**	nr	**1402.65**
Mancuna standard doors; Mandor Engineering Ltd; 43mm diameter steel tube frame; top and bottom plates with pivoting arrangements and spring unit; 6mm clear flexible panel doors										
fixing to masonry; for opening size										
1800 x 2100mm	756.00	2.50	**774.90**	3.50	3.50	**87.22**	-	**862.12**	nr	**926.78**
2440 x 2400mm	803.25	2.50	**823.33**	4.00	4.00	**99.68**	-	**923.01**	nr	**992.24**
Mancuna standard doors; Mandor Engineering Ltd; 43mm diameter steel tube frame; top and bottom plates with pivoting arrangements and spring unit; 9mm clear flexible panel doors										
fixing to masonry; for opening size										
1800 x 2100mm	840.00	2.50	**861.00**	3.75	3.75	**93.45**	-	**954.45**	nr	**1026.03**
2440 x 2400mm	903.00	2.50	**925.58**	4.25	4.25	**105.91**	-	**1031.49**	nr	**1108.85**
Mandor heavy duty doors; Mandor Engineering Ltd; 63mm diameter steel tube frame; top and bottom plates with pivoting arrangements and spring unit; 12mm rubber panel doors with oval panel windows in each leaf										
fixing to masonry; for opening size										
2440 x 3050mm	1522.50	2.50	**1560.56**	8.00	8.00	**199.36**	-	**1759.92**	nr	**1891.92**
2740 x 3350mm	1785.00	2.50	**1829.63**	9.00	9.00	**224.28**	-	**2053.91**	nr	**2207.95**
3050 x 3660mm	1879.50	2.50	**1926.49**	10.00	10.00	**249.20**	-	**2175.69**	nr	**2338.86**
Mandor heavy duty doors; Mandor Engineering Ltd; 63mm diameter steel tube frame; top and bottom plates with pivoting arrangements and spring unit; 9mm clear flexible panel doors										
fixing to masonry; for opening size										
2440 x 3050mm	1417.50	2.50	**1452.94**	6.00	6.00	**149.52**	-	**1602.46**	nr	**1722.64**
2740 x 3350mm	1648.50	2.50	**1689.71**	7.00	7.00	**174.44**	-	**1864.15**	nr	**2003.96**
3050 x 3660mm	1932.00	2.50	**1980.30**	8.00	8.00	**199.36**	-	**2179.66**	nr	**2343.13**
Flexible strip curtains mainly in pvc										
Mandor strip curtains; Mandor Engineering Ltd; 200mm wide x 2.0mm thick clear pvc strip curtains, suspended from 80 x 80 x 6mm steel angle; fixing angle to masonry with bolts										
double overlap; for opening size										
1800 x 2100mm	115.50	2.50	**118.39**	2.00	2.00	**49.84**	-	**168.23**	nr	**180.84**
2100 x 2440mm	162.75	2.50	**166.82**	2.30	2.30	**57.32**	-	**224.13**	nr	**240.94**
single overlap; for opening size										
1800 x 2100mm	93.45	2.50	**95.79**	1.80	1.80	**44.86**	-	**140.64**	nr	**151.19**
2100 x 2440mm	129.15	2.50	**132.38**	2.10	2.10	**52.33**	-	**184.71**	nr	**198.56**
Mandor strip curtains; Mandor Engineering Ltd; 300mm wide x 2.5mm thick clear pvc strip curtains, suspended from 80 x 80 x 6mm steel angle; fixing angle to masonry with bolts										
double overlap; for opening size										
2440 x 2740mm	226.80	2.50	**232.47**	2.64	2.64	**65.79**	-	**298.26**	nr	**320.63**
2740 x 2900mm	267.75	2.50	**274.44**	2.94	2.94	**73.26**	-	**347.71**	nr	**373.79**
single overlap; for opening size										
2440 x 2740mm	175.35	2.50	**179.73**	2.44	2.44	**60.80**	-	**240.54**	nr	**258.58**
2740 x 2900mm	207.90	2.50	**213.10**	2.74	2.74	**68.28**	-	**281.38**	nr	**302.48**

Labour hourly rates: (except Specialists) Craft Operatives £14.21 Labourer £10.71 Rates are national average prices. Refer to REGIONAL VARIATIONS for indicative levels of overall pricing in regions	MATERIALS			LABOUR				RATES		
	Del to Site £	Waste %	Material Cost £	Craft Optve Hrs	Lab Hrs	Labour Cost £	Sunds £	Nett Rate £	Unit	Gross Rate (+7.5%) £
L30 STAIRS/WALKWAYS/BALUSTRADES: TIMBER STAIRS										
Stairs in softwood, wrought										
Straight flight staircase and balustrade; 25mm treads; 19mm risers, glued; wedged and blocked; 25mm wall string; 38mm outer string; 75 x 75mm newels; 25 x 25mm balusters; 38 x 75mm hardwood handrail										
864mm wide x 2600mm rise overall; balustrade to one side; fixing to masonry with screws	666.75	2.50	**683.42**	12.00	1.50	**186.59**	4.25	**874.25**	nr	**939.82**
864mm wide x 2600mm rise overall with 3 nr winders at bottom; balustrade to one side; fixing to masonry with screws	800.10	2.50	**820.10**	16.00	2.00	**248.78**	4.50	**1073.38**	nr	**1153.89**
864mm wide x 2600mm rise overall with 3 nr winders at top; balustrade to one side; fixing to masonry with screws	800.10	2.50	**820.10**	16.00	2.00	**248.78**	4.50	**1073.38**	nr	**1153.89**
Balustrades in softwood, wrought										
Isolated balustrades; 25 x 25mm balusters at 150mm centres, housed construction (handrail included elsewhere)										
914mm high; fixing to timber with screws	35.02	2.50	**35.89**	0.65	0.08	**10.09**	0.11	**46.10**	m	**49.55**
Isolated balustrades; 38 x 38mm balusters at 150mm centres, housed construction (handrail included elsewhere)										
914mm high; fixing to timber with screws	51.32	2.50	**52.61**	0.95	0.12	**14.78**	0.15	**67.54**	m	**72.61**
Isolated balustrades; 50 x 50mm balusters at 150mm centres, housed construction (handrail included elsewhere)										
914mm high; fixing to timber with screws	67.43	2.50	**69.12**	1.10	0.14	**17.13**	0.18	**86.43**	m	**92.91**
Isolated balustrades; 25 x 25mm balusters at 150mm centres, 50mm mopstick handrail, housed construction										
914mm high; fixing to timber with screws	58.70	2.50	**60.16**	1.00	0.12	**15.50**	0.16	**75.82**	m	**81.50**
extra; ramps	36.03	2.50	**36.93**	0.90	0.11	**13.97**	0.15	**51.04**	nr	**54.87**
extra; wreaths	72.04	2.50	**73.84**	1.80	0.23	**28.04**	0.29	**102.17**	nr	**109.84**
extra; bends	36.03	2.50	**36.93**	0.60	0.08	**9.38**	0.10	**46.41**	nr	**49.89**
Isolated balustrades; 38 x 38mm balusters at 150mm centres, 50mm mopstick handrail, housed construction										
914mm high; fixing to timber with screws	74.71	2.50	**76.58**	1.30	0.16	**20.19**	0.21	**96.97**	m	**104.24**
extra; ramps	36.03	2.50	**36.93**	0.90	0.11	**13.97**	0.15	**51.04**	nr	**54.87**
extra; wreaths	72.04	2.50	**73.84**	1.80	0.23	**28.04**	0.29	**102.17**	nr	**109.84**
extra; bends	36.03	2.50	**36.93**	0.60	0.08	**9.38**	0.10	**46.41**	nr	**49.89**
Isolated balustrades; 50 x 50mm balusters at 150mm centres, 50mm mopstick handrail, housed construction										
914mm high; fixing to timber with screws	90.71	2.50	**92.98**	1.45	0.18	**22.53**	0.24	**115.75**	m	**124.43**
extra; ramps	36.03	2.50	**36.93**	0.90	0.11	**13.97**	0.15	**51.04**	nr	**54.87**
extra; wreaths	72.04	2.50	**73.84**	1.80	0.23	**28.04**	0.29	**102.17**	nr	**109.84**
extra; bends	36.03	2.50	**36.93**	0.60	0.08	**9.38**	0.10	**46.41**	nr	**49.89**
Isolated balustrades; 25 x 25mm balusters at 150mm centres, 75 x 100mm moulded handrail; housed construction										
914mm high; fixing to timber with screws	72.04	2.50	**73.84**	1.05	0.13	**16.31**	0.17	**90.32**	m	**97.10**
extra; ramps	53.36	2.50	**54.70**	2.00	0.25	**31.10**	0.33	**86.12**	nr	**92.58**
extra; wreaths	106.72	2.50	**109.39**	4.00	0.50	**62.20**	0.65	**172.24**	nr	**185.15**
extra; bends	53.36	2.50	**54.70**	1.20	0.15	**18.66**	0.20	**73.55**	nr	**79.07**
Isolated balustrades; 38 x 38mm balusters at 150mm centres, 75 x 100mm moulded handrail; housed construction										
914mm high; fixing to timber with screws	88.04	2.50	**90.24**	1.35	0.17	**21.00**	0.22	**111.47**	m	**119.83**
extra; ramps	53.36	2.50	**54.70**	2.00	0.25	**31.10**	0.33	**86.12**	nr	**92.58**
extra; wreaths	106.72	2.50	**109.39**	4.00	0.50	**62.20**	0.65	**172.24**	nr	**185.15**
extra; bends	53.36	2.50	**54.70**	1.20	0.15	**18.66**	0.20	**73.55**	nr	**79.07**
Isolated balustrades; 50 x 50mm balusters at 150mm centres, 75 x 100mm moulded handrail; housed construction										
914mm high; fixing to timber with screws	104.06	2.50	**106.66**	1.50	0.19	**23.35**	0.24	**130.25**	m	**140.01**
extra; ramps	53.36	2.50	**54.70**	2.00	0.25	**31.10**	0.33	**86.12**	nr	**92.58**
extra; wreaths	106.72	2.50	**109.39**	4.00	0.50	**62.20**	0.65	**172.24**	nr	**185.15**
extra; bends	53.36	2.50	**54.70**	1.20	0.15	**18.66**	0.20	**73.55**	nr	**79.07**
Balustrades in European Oak, wrought, selected for transparent finish										
Isolated balustrades; 25 x 25mm balusters at 150mm centres, 50mm mopstick handrail, housed construction										
914mm high; fixing to timber with screws	129.13	2.50	**132.36**	1.70	0.21	**26.41**	0.28	**159.04**	m	**170.97**
extra; ramps	79.26	2.50	**81.24**	1.50	0.19	**23.35**	0.24	**104.83**	nr	**112.69**
extra; wreaths	158.48	2.50	**162.44**	3.00	0.38	**46.70**	0.49	**209.63**	nr	**225.35**
extra; bends	79.26	2.50	**81.24**	1.10	0.14	**17.13**	0.18	**98.55**	nr	**105.94**
Isolated balustrades; 38 x 38mm balusters at 150mm centres, 50mm mopstick handrail, housed construction										
914mm high; fixing to timber with screws	164.36	2.50	**168.47**	2.15	0.27	**33.44**	0.35	**202.26**	m	**217.43**
extra; ramps	79.26	2.50	**81.24**	1.50	0.19	**23.35**	0.24	**104.83**	nr	**112.69**
extra; wreaths	158.48	2.50	**162.44**	3.00	0.38	**46.70**	0.49	**209.63**	nr	**225.35**
extra; bends	79.26	2.50	**81.24**	1.10	0.14	**17.13**	0.18	**98.55**	nr	**105.94**

Labour hourly rates: (except Specialists) Craft Operatives £14.21 Labourer £10.71 Rates are national average prices. Refer to REGIONAL VARIATIONS for indicative levels of overall pricing in regions	MATERIALS			LABOUR				RATES		
	Del to Site £	Waste %	Material Cost £	Craft Optve Hrs	Lab Hrs	Labour Cost £	Sunds £	Nett Rate £	Unit	Gross Rate (+7.5%) £
L30 STAIRS/WALKWAYS/BALUSTRADES: TIMBER STAIRS Cont.										
Balustrades in European Oak, wrought, selected for transparent finish Cont.										
Isolated balustrades; 50 x 50mm balusters at 150mm centres, 50mm mopstick handrail, housed construction										
914mm high; fixing to timber with screws	199.57	2.50	**204.56**	2.50	0.31	**38.85**	0.41	**243.82**	m	**262.10**
extra; ramps	79.26	2.50	**81.24**	1.50	0.19	**23.35**	0.24	**104.83**	nr	**112.69**
extra; wreaths	158.48	2.50	**162.44**	3.00	0.38	**46.70**	0.49	**209.63**	nr	**225.35**
extra; bends	79.26	2.50	**81.24**	1.10	0.14	**17.13**	0.18	**98.55**	nr	**105.94**
Isolated balustrades; 25 x 25mm balusters at 150mm centres, 75 x 100mm moulded handrail; housed construction										
914mm high; fixing to timber with screws	158.48	2.50	**162.44**	1.80	0.23	**28.04**	0.29	**190.77**	m	**205.08**
extra; ramps	117.39	2.50	**120.33**	3.10	0.39	**48.23**	0.50	**169.06**	nr	**181.74**
extra; wreaths	234.79	2.50	**240.66**	6.20	0.78	**96.46**	1.00	**338.11**	nr	**363.47**
extra; bends	117.39	2.50	**120.33**	2.15	0.27	**33.44**	0.35	**154.12**	nr	**165.68**
Isolated balustrades; 38 x 38mm balusters at 150mm centres, 75 x 100mm moulded handrail; housed construction										
914mm high; fixing to timber with screws	193.71	2.50	**198.55**	2.25	0.28	**34.97**	0.37	**233.89**	m	**251.43**
extra; ramps	117.39	2.50	**120.33**	3.10	0.39	**48.23**	0.50	**169.06**	nr	**181.74**
extra; wreaths	234.79	2.50	**240.66**	6.20	0.78	**96.46**	1.00	**338.11**	nr	**363.47**
extra; bends	117.39	2.50	**120.33**	2.15	0.27	**33.44**	0.35	**154.12**	nr	**165.68**
Isolated balustrades; 50 x 50mm balusters at 150mm centres, 75 x 100mm moulded handrail; housed construction										
914mm high; fixing to timber with screws	228.92	2.50	**234.64**	2.60	0.33	**40.48**	0.42	**275.54**	m	**296.21**
extra; ramps	117.39	2.50	**120.33**	3.10	0.39	**48.23**	0.50	**169.06**	nr	**181.74**
extra; wreaths	234.79	2.50	**240.66**	6.20	0.78	**96.46**	1.00	**338.11**	nr	**363.47**
extra; bends	117.39	2.50	**120.33**	2.15	0.27	**33.44**	0.35	**154.12**	nr	**165.68**
Handrails in softwood, wrought										
Associated handrails										
50mm mopstick; fixing through metal backgrounds with screws	20.01	5.00	**21.01**	0.35	0.04	**5.40**	0.06	**26.48**	m	**28.46**
extra; ramps	36.03	5.00	**37.83**	1.00	0.13	**15.60**	0.16	**53.59**	nr	**57.61**
extra; wreaths	72.04	5.00	**75.64**	2.00	0.25	**31.10**	0.33	**107.07**	nr	**115.10**
extra; bends	36.03	5.00	**37.83**	1.00	0.13	**15.60**	0.16	**53.59**	nr	**57.61**
75 x 100mm; moulded; fixing through metal backgrounds with screws	33.36	5.00	**35.03**	0.40	0.05	**6.22**	0.07	**41.32**	m	**44.41**
extra; ramps	53.36	5.00	**56.03**	1.20	0.15	**18.66**	0.20	**74.89**	nr	**80.50**
extra; wreaths	106.72	5.00	**112.06**	2.40	0.30	**37.32**	0.39	**149.77**	nr	**161.00**
extra; bends	53.36	5.00	**56.03**	1.20	0.15	**18.66**	0.20	**74.89**	nr	**80.50**
Handrails in African Mahogany, wrought, selected for transparent finish										
Associated handrails										
50mm mopstick; fixing through metal backgrounds with screws	25.02	5.00	**26.27**	0.50	0.06	**7.75**	0.08	**34.10**	m	**36.66**
extra; ramps	45.03	5.00	**47.29**	1.50	0.19	**23.35**	0.24	**70.88**	nr	**76.19**
extra; wreaths	90.05	5.00	**94.55**	3.00	0.38	**46.70**	0.49	**141.74**	nr	**152.37**
extra; bends	45.03	5.00	**47.29**	1.50	0.19	**23.35**	0.24	**70.88**	nr	**76.19**
75 x 100mm; moulded; fixing through metal backgrounds with screws	41.70	5.00	**43.78**	0.60	0.08	**9.38**	0.10	**53.26**	m	**57.26**
extra; ramps	66.71	5.00	**70.04**	1.80	0.23	**28.04**	0.29	**98.37**	nr	**105.75**
extra; wreaths	133.40	5.00	**140.07**	3.60	0.45	**55.98**	0.59	**196.64**	nr	**211.39**
extra; bends	66.71	5.00	**70.04**	1.80	0.73	**33.40**	0.36	**103.80**	nr	**111.58**
Handrails in European Oak, wrought, selected for transparent finish										
Associated handrails										
50mm mopstick; fixing through metal backgrounds with screws	44.04	5.00	**46.24**	0.50	0.06	**7.75**	0.08	**54.07**	m	**58.12**
extra; ramps	79.26	5.00	**83.22**	1.50	0.19	**23.35**	0.24	**106.81**	nr	**114.82**
extra; wreaths	158.48	5.00	**166.40**	3.00	0.38	**46.70**	0.49	**213.59**	nr	**229.61**
extra; bends	79.26	5.00	**83.22**	1.50	0.19	**23.35**	0.24	**106.81**	nr	**114.82**
75 x 100mm; moulded; fixing through metal backgrounds with screws	73.39	5.00	**77.06**	0.60	0.08	**9.38**	0.10	**86.54**	m	**93.03**
extra; ramps	117.39	5.00	**123.26**	1.80	0.23	**28.04**	0.29	**151.60**	nr	**162.96**
extra; wreaths	234.79	5.00	**246.53**	3.60	0.45	**55.98**	0.59	**303.09**	nr	**325.83**
extra; bends	117.39	5.00	**123.26**	1.80	0.23	**28.04**	0.36	**151.67**	nr	**163.04**
L30: STAIRS/WALKWAYS/BALUSTRADES: METAL STAIRS AND BALUSTRADES										
Staircases in steel										
Straight flight staircases; 180 x 10mm flat stringers, shaped ends to top and bottom; 6 x 250mm on plain raised pattern plate treads, with 50 x 50 x 6mm shelf angles and 40 x 40 x 6mm stiffening bars, bolted to stringers; welded, cleated and bolted connections										
770mm wide x 3000mm going x 2600mm rise overall; fixing to masonry with 4 nr Rawlbolts	-	-	-	-	-	-	-	**2087.25**	nr	**2243.79**
920mm wide x 3000mm going x 2600mm rise overall; fixing to masonry with 4 nr Rawlbolts	-	-	-	-	-	-	-	**2208.00**	nr	**2373.60**

Labour hourly rates: (except Specialists) Craft Operatives £14.21 Labourer £10.71 Rates are national average prices. Refer to REGIONAL VARIATIONS for indicative levels of overall pricing in regions	MATERIALS			LABOUR				RATES		
	Del to Site £	Waste %	Material Cost £	Craft Optve Hrs	Lab Hrs	Labour Cost £	Sunds £	Nett Rate £	Unit	Gross Rate (+7.5%) £
L30: STAIRS/WALKWAYS/BALUSTRADES: METAL STAIRS AND BALUSTRADES Cont.										
Staircases in steel Cont.										
Straight flight staircases, 180 x 10mm flat stringers, shaped ends to top and bottom; 6 x 250mm tray treads, with three 6mm diameter reinforcing bars welded to inside, and 50 x 50 x 6mm shelf angles, bolted to stringers; welded, cleated and bolted connections										
770mm wide x 3000mm going x 2600mm rise overall; fixing to masonry with 4 nr Rawlbolts...........	-	-	-	-	-	-	-	**1978.00**	nr	**2126.35**
920mm wide x 3000mm going x 2600mm rise overall; fixing to masonry with 4 nr Rawlbolts...........	-	-	-	-	-	-	-	**2047.00**	nr	**2200.53**
Straight flight staircases; 178 x 76mm channel stringers, shaped ends to top and bottom; 6 x 250mm on plain raised pattern plate treads, with 50 x 50 x 6mm shelf angles and 40 x 40 x 6mm stiffening bars, bolted to stringers; welded, cleated and bolted connections										
770mm wide x 3000mm going x 2600mm rise overall; fixing to masonry with 4 nr Rawlbolts...........	-	-	-	-	-	-	-	**2259.75**	nr	**2429.23**
920mm wide x 3000mm going x 2600mm rise overall; fixing to masonry with 4 nr Rawlbolts...........	-	-	-	-	-	-	-	**2576.00**	nr	**2769.20**
Straight flight staircases and balustrades; 180 x 10mm flat stringers, shaped ends to top and bottom; 6 x 250mm on plain raised pattern plate treads with 50 x 50 x 6mm shelf angles and 40 x 40 x 6mm stiffening bars, bolted to stringers; 915mm high balustrade to both sides consisting of 25mm diameter solid bar handrail and 32mm diameter solid bar standards at 250mm centres with base plate welded on and bolted to face of stringer, and ball type joints at intersections; welded, cleated and bolted connections										
770mm wide x 3000mm going x 2600mm rise overall; fixing to masonry with 4 nr Rawlbolts...........	-	-	-	-	-	-	-	**3036.00**	nr	**3263.70**
Straight flight staircases and balustrades; 180 x 10mm flat stringers, shaped ends to top and bottom; 6 x 250mm on plain raised pattern plate treads with 50 x 50 x 6mm shelf angles and 40 x 40 x 6mm stiffening bars, bolted to stringers; 915mm high balustrade to both sides consisting of 25mm diameter solid bar handrail and 32mm diameter solid bar standards at 250mm centres with base plate welded on and bolted to top of stringer, and ball type joints at intersections; welded, cleated and bolted connections										
770mm wide x 3000mm going x 2600mm rise overall; fixing to masonry with 4 nr Rawlbolts...........	-	-	-	-	-	-	-	**3036.00**	nr	**3263.70**
Straight flight staircases and balustrades; 180 x 10mm flat stringers, shaped ends to top and bottom; 6 x 250mm on plain raised pattern plate treads with 50 x 50 x 6mm shelf angles and 40 x 40 x 6mm stiffening bars, bolted to stringers; 915mm high balustrade to one side consisting of 25mm diameter solid bar handrail and 32mm diameter solid bar standards at 250mm centres with base plate welded on and bolted to face of stringer, and ball type joints at intersections; welded, cleated and bolted connections										
770mm wide x 3000mm going x 2600mm rise overall; fixing to masonry with 4 nr Rawlbolts...........	-	-	-	-	-	-	-	**2829.00**	nr	**3041.18**
Straight flight staircases and balustrades; 180 x 10mm flat stringers, shaped ends to top and bottom; 6 x 250mm on plain raised pattern plate treads with 50 x 50 x 6mm shelf angles and 40 x 40 x 6mm stiffening bars, bolted to stringers; 915mm high balustrade to one side consisting of 25mm diameter solid bar handrail and 32mm diameter solid bar standards at 250mm centres with base plate welded on and bolted to top of stringer, and ball type joints at intersections; welded, cleated and bolted connections										
770mm wide x 3000mm going x 2600mm rise overall; fixing to masonry with 4 nr Rawlbolts...........	-	-	-	-	-	-	-	**2829.00**	nr	**3041.18**
Straight flight staircases and balustrades; 180 x 10mm flat stringers, shaped ends to top and bottom; 6 x 250mm on plain raised pattern plate treads with 50 x 50 x 6mm shelf angles and 40 x 40 x 6mm stiffening bars, bolted to stringers; 1070mm high balustrade to both sides consisting of 25mm diameter solid bar handrail and 32mm diameter solid bar standards at 250mm centres with base plate welded on and bolted to top of stringer, and ball type joints at intersections; welded, cleated and bolted connections										
770mm wide x 3000mm going x 2600mm rise overall; fixing to masonry with 4 nr Rawlbolts...........	-	-	-	-	-	-	-	**3064.75**	nr	**3294.61**

Labour hourly rates: (except Specialists) Craft Operatives £14.21 Labourer £10.71 Rates are national average prices. Refer to REGIONAL VARIATIONS for indicative levels of overall pricing in regions	MATERIALS			LABOUR				RATES		
	Del to Site £	Waste %	Material Cost £	Craft Optve Hrs	Lab Hrs	Labour Cost £	Sunds £	Nett Rate £	Unit	Gross Rate (+7.5%) £
L30: STAIRS/WALKWAYS/BALUSTRADES: METAL STAIRS AND BALUSTRADES Cont.										
Staircases in steel Cont.										
Straight flight staircases and balustrades; 180 x 10mm flat stringers, shaped ends to top and bottom; 6 x 250mm on plain raised pattern plate treads with 50 x 50 x 6mm shelf angles and 40 x 40 x 6mm stiffening bars, bolted to stringers; 1070mm high balustrade to one side consisting of 25mm diameter solid bar handrail and 32mm diameter solid bar standards at 250mm centres with base plate welded on and bolted to top of stringer, and ball type joints at intersections; welded, cleated and bolted connections										
770mm wide x 3000mm going x 2600mm rise overall; fixing to masonry with 4 nr Rawlbolts	-	-	-	-	-	-	-	**2857.75**	nr	**3072.08**
Straight flight staircases and balustrades; 180 x 10mm flat stringers, shaped ends to top and bottom; 6 x 250mm on plain raised pattern plate treads with 50 x 50 x 6mm shelf angles and 40 x 40 x 6mm stiffening bars, bolted to stringers; 915mm high balustrade to one side consisting of 25mm diameter solid bar handrail and intermediate rail, and 32mm diameter solid bar standards at 250mm centres with base plate welded on and bolted to top of stringer, and ball type joints at intersections										
770mm wide x 3000mm going x 2600mm rise overall; fixing to masonry with 4 nr Rawlbolts	-	-	-	-	-	-	-	**2834.75**	nr	**3047.36**
Straight flight staircases and balustrades; 180 x 10mm flat stringers, shaped ends to top and bottom; 6 x 250mm on plain raised pattern plate treads with 50 x 50 x 6mm shelf angles and 40 x 40 x 6mm stiffening bars, bolted to stringers; 1070mm high balustrade to one side consisting of 25mm diameter solid bar handrail and intermediate rail, and 32mm diameter solid bar standards at 250mm centres with base plate welded on and bolted to top of stringer, and ball type joints at intersections										
770mm wide x 3000mm going x 2600mm rise overall; fixing to masonry with 4 nr Rawlbolts	-	-	-	-	-	-	-	**2857.75**	nr	**3072.08**
Quarter landing staircases, in two flights; 180 x 10mm flat stringers, shaped ends to top and bottom; 6 x 250mm on plain raised pattern plate treads, with 50 x 50 x 6mm shelf angles and 40 x 40 x 6mm stiffening bars, bolted to stringers; 6mm on plain raised pattern plate landing, welded on; 100 x 8mm flat kicking plates welded on; welded, cleated and bolted connections										
770mm wide x 2000mm going first flight excluding landing x 1000mm going second flight excluding landing x 2600mm rise overall; 770 x 770mm landing overall; fixing to masonry with 4 nr Rawlbolts	-	-	-	-	-	-	-	**4565.50**	nr	**4907.91**
920mm wide x 2000mm going first flight excluding landing x 1000mm going second flight excluding landing x 2600mm rise overall; 920 x 920mm landing overall; fixing to masonry with 4 nr Rawlbolts	-	-	-	-	-	-	-	**4795.50**	nr	**5155.16**
Half landing staircases, in two flights; 180 x 10mm flat stringers, shaped ends to top and bottom; 6 x 250mm on plain raised pattern plate treads, with 50 x 50 x 6mm shelf angles and 40 x 40 x 6mm stiffening bars, bolted to stringers; 6mm on plain raised pattern plate landing, welded on; 100 x 8mm flat kicking plates welded on; welded, cleated and bolted connections										
770mm wide x 2000mm going first flight excluding landing x 1000mm going second flight excluding landing x 2600mm rise overall; 770 x 1640mm landing overall; fixing to masonry with 4 nr Rawlbolts	-	-	-	-	-	-	-	**5117.50**	nr	**5501.31**
920mm wide x 2000mm going first flight excluding landing x 1000mm going second flight excluding landing x 2600mm rise overall; 920 x 1940mm landing overall; fixing to masonry with 4 nr Rawlbolts	-	-	-	-	-	-	-	**5807.50**	nr	**6243.06**
Balustrades in steel										
Isolated balustrades; 6 x 38mm flat core rail; 13mm diameter balusters at 250mm centres; welded fabrication ground to smooth finish; casting into mortices in concrete; wedging in position; temporary wedges										
838mm high; level	139.83	2.50	**143.32**	1.00	0.50	**19.57**	-	**162.89**	m	**175.11**
extra; ramps	23.92	2.50	**24.52**	1.00	0.50	**19.57**	-	**44.08**	nr	**47.39**
extra; wreaths	39.87	2.50	**40.86**	2.00	1.00	**39.13**	-	**79.99**	nr	**85.99**
extra; bends	21.73	2.50	**22.27**	1.00	0.50	**19.57**	-	**41.83**	nr	**44.97**
838mm high; raking	144.45	2.50	**148.06**	1.25	0.65	**24.72**	-	**172.79**	m	**185.74**
914mm high; level	153.52	2.50	**157.36**	1.00	0.50	**19.57**	-	**176.93**	m	**190.19**
extra; ramps	23.92	2.50	**24.52**	1.00	0.50	**19.57**	-	**44.08**	nr	**47.39**
extra; wreaths	39.87	2.50	**40.86**	2.00	1.00	**39.13**	-	**79.99**	nr	**85.99**
extra; bends	21.73	2.50	**22.27**	1.00	0.50	**19.57**	-	**41.83**	nr	**44.97**

Labour hourly rates: (except Specialists) Craft Operatives £14.21 Labourer £10.71 Rates are national average prices. Refer to REGIONAL VARIATIONS for indicative levels of overall pricing in regions	MATERIALS			LABOUR				RATES		
	Del to Site £	Waste %	Material Cost £	Craft Optve Hrs	Lab Hrs	Labour Cost £	Sunds £	Nett Rate £	Unit	Gross Rate (+7.5%) £
L30: STAIRS/WALKWAYS/BALUSTRADES: METAL STAIRS AND BALUSTRADES Cont.										
Balustrades in steel Cont.										
Isolated balustrades; 6 x 38mm flat core rail; 13mm diameter balusters at 250mm centres; welded fabrication ground to smooth finish; casting into mortices in concrete; wedging in position; temporary wedges Cont.										
914mm high; raking	162.25	2.50	**166.30**	1.25	0.65	**24.72**	-	**191.03**	m	**205.35**
Isolated balustrades; 6 x 38mm flat core rail; 13 x 13mm balusters at 250mm centres; welded fabrication ground to smooth finish; casting into mortices in concrete; wedging in position; temporary wedges										
838mm high; level	146.07	2.50	**149.72**	1.00	0.50	**19.57**	-	**169.28**	m	**181.98**
838mm high; raking	151.04	2.50	**154.81**	1.00	0.50	**19.57**	-	**174.38**	m	**187.46**
914mm high; level	158.55	2.50	**162.51**	1.00	0.50	**19.57**	-	**182.08**	m	**195.73**
914mm high; raking	168.72	2.50	**172.94**	1.00	0.50	**19.57**	-	**192.50**	m	**206.94**
Isolated balustrades; 13 x 51mm rounded handrail; 13mm diameter balusters at 250mm centres; welded fabrication ground to smooth finish; casting into mortices in concrete; wedging in position; temporary wedges										
838mm high; level	168.72	2.50	**172.94**	1.00	0.50	**19.57**	-	**192.50**	m	**206.94**
extra; ramps	26.23	2.50	**26.89**	1.00	0.50	**19.57**	-	**46.45**	nr	**49.94**
extra; wreaths	43.68	2.50	**44.77**	2.00	1.00	**39.13**	-	**83.90**	nr	**90.20**
extra; bends	23.75	2.50	**24.34**	1.00	0.50	**19.57**	-	**43.91**	nr	**47.20**
838mm high; raking	172.18	2.50	**176.49**	1.00	0.50	**19.57**	-	**196.05**	m	**210.76**
914mm high; level	179.70	2.50	**184.19**	1.00	0.50	**19.57**	-	**203.75**	m	**219.03**
extra; ramps	26.23	2.50	**26.89**	1.00	0.50	**19.57**	-	**46.45**	nr	**49.94**
extra; wreaths	43.79	2.50	**44.88**	2.00	1.00	**39.13**	-	**84.01**	nr	**90.31**
extra; bends	23.75	2.50	**24.34**	1.00	0.50	**19.57**	-	**43.91**	nr	**47.20**
914mm high; raking	187.21	2.50	**191.89**	1.00	0.50	**19.57**	-	**211.45**	m	**227.31**
Isolated balustrades; 6 x 38mm flat core rail; 10 x 51mm flat bottom rail; 13mm diameter infill balusters at 250mm centres; 25 x 25mm standards at 3000mm centres; welded fabrication ground to smooth finish; casting into mortices in concrete; wedging in position; temporary wedges										
838mm high; level	187.21	2.50	**191.89**	1.00	0.50	**19.57**	-	**211.45**	m	**227.31**
extra; ramps	44.95	2.50	**46.08**	1.00	0.50	**19.57**	-	**65.64**	nr	**70.56**
extra; wreaths	56.16	2.50	**57.57**	2.00	1.00	**39.13**	-	**96.70**	nr	**103.95**
extra; bends	39.87	2.50	**40.86**	1.00	0.50	**19.57**	-	**60.43**	nr	**64.96**
838mm high; raking	193.45	2.50	**198.28**	1.00	0.50	**19.57**	-	**217.85**	m	**234.19**
914mm high; level	205.93	2.50	**211.08**	1.00	0.50	**19.57**	-	**230.64**	m	**247.94**
extra; ramps	44.95	2.50	**46.08**	1.00	0.50	**19.57**	-	**65.64**	nr	**70.56**
extra; wreaths	56.16	2.50	**57.57**	2.00	1.00	**39.13**	-	**96.70**	nr	**103.95**
extra; bends	39.87	2.50	**40.86**	1.00	0.50	**19.57**	-	**60.43**	nr	**64.96**
Balustrades in steel tubing, BS 1387, medium grade; galvanized after fabrication										
Isolated balustrades; 40mm diameter handrails; 40mm diameter standards at 1500mm centres; welded fabrication ground to smooth finish; casting into mortices in concrete; wedging in position; temporary wedges										
838mm high; level	87.36	2.50	**89.55**	0.75	0.40	**14.94**	-	**104.49**	m	**112.33**
838mm high; raking	93.60	2.50	**95.94**	0.75	0.40	**14.94**	-	**110.89**	m	**119.20**
914mm high; level	93.60	2.50	**95.94**	0.75	0.40	**14.94**	-	**110.89**	m	**119.20**
914mm high; raking	99.84	2.50	**102.34**	0.75	0.40	**14.94**	-	**117.28**	m	**126.08**
Isolated balustrades; 40mm diameter handrails; 40mm diameter standards at 1500mm centres, with 6 x 75mm diameter fixing plates welded on to end, holes -3 nr; welded fabrication ground to smooth finish; fixing to masonry with Rawlbolts										
838mm high; level	99.84	2.50	**102.34**	1.00	0.50	**19.57**	-	**121.91**	m	**131.05**
838mm high; raking	106.08	2.50	**108.74**	1.00	0.50	**19.57**	-	**128.30**	m	**137.92**
914mm high; level	118.57	2.50	**121.53**	1.00	0.50	**19.57**	-	**141.09**	m	**151.68**
914mm high; raking	131.05	2.50	**134.32**	1.00	0.50	**19.57**	-	**153.89**	m	**165.43**
Spiral staircases in steel; one coat red oxide primer before delivery to site										
Crescent 'H' range domestic spiral staircases; model 15 'H'; Crescent of Cambridge Ltd.; comprising tread modules complete with centre column, tread support, wooden tread, tread baluster, handrail section, PVC handrail cover, baluster wicket infill panel and tread riser bar, and with attachment brackets for fixing to floor and newel extending from centre column; assembling and bolting together										
1568mm diameter; for floor to floor height units between 2160mm and 2640mm; plywood treads with hardwood edgings; fixing base and ground plates to timber	1179.78	2.50	**1209.27**	30.00	15.00	**586.95**	-	**1796.22**	nr	**1930.94**
extra; additional tread modules, 180 - 220mm per rise	86.81	2.50	**88.98**	1.50	0.75	**29.35**	-	**118.33**	nr	**127.21**
extra; mini landing to match treads, at upper floor level	217.04	2.50	**222.46**	2.00	1.00	**39.13**	-	**261.59**	nr	**281.21**

Labour hourly rates: (except Specialists) Craft Operatives £14.21 Labourer £10.71 Rates are national average prices. Refer to REGIONAL VARIATIONS for indicative levels of overall pricing in regions	MATERIALS			LABOUR				RATES		
	Del to Site £	Waste %	Material Cost £	Craft Optve Hrs	Lab Hrs	Labour Cost £	Sunds £	Nett Rate £	Unit	Gross Rate (+7.5%) £
L30: STAIRS/WALKWAYS/BALUSTRADES: METAL STAIRS AND BALUSTRADES Cont.										
Spiral staircases in steel; one coat red oxide primer before delivery to site Cont.										
Crescent 'H' range domestic spiral staircases; model 15 'H'; Crescent of Cambridge Ltd.; comprising tread modules complete with centre column, tread support, wooden tread, tread baluster, handrail section, PVC handrail cover, baluster wicket infill panel and tread riser bar, and with attachment brackets for fixing to floor and newel extending from centre column; assembling and bolting together Cont.										
1568mm diameter; for floor to floor height units between 2160mm and 2640mm; plywood treads with carpet recess; fixing base and ground plates to timber	1346.73	2.50	**1380.40**	30.00	15.00	**586.95**	-	**1967.35**	nr	**2114.90**
extra; additional tread modules, 180 - 220mm per rise	103.51	2.50	**106.10**	2.00	1.00	**39.13**	-	**145.23**	nr	**156.12**
1568mm diameter; for floor to floor height units between 2160mm and 2640mm; hardwood treads, plain or with hardwood surround providing carpet recess; fixing base and ground plates to timber	1530.38	2.50	**1568.63**	30.00	15.00	**586.95**	-	**2155.58**	nr	**2317.25**
extra; additional tread modules, 180 - 220mm per rise	117.98	2.50	**120.93**	2.00	1.00	**39.13**	-	**160.06**	nr	**172.06**
extra; mini landing to match treads, at upper floor level	260.44	2.50	**266.95**	2.00	1.00	**39.13**	-	**306.08**	nr	**329.04**
stairwell balustrading with straight bar infill and PVC handrail cover	96.83	2.50	**99.25**	1.75	0.90	**34.51**	-	**133.76**	m	**143.79**
stairwell balustrading to match staircase including PVC handrail cover	158.05	2.50	**162.00**	1.75	0.90	**34.51**	-	**196.50**	m	**211.24**
Crescent 'H' range domestic spiral staircases; model 18 'H'; Crescent of Cambridge Ltd.; comprising tread modules complete with centre column, tread support, wooden tread, tread baluster, handrail section, PVC handrail cover, baluster wicket infill panel and tread riser bar, and with attachment brackets for fixing to floor and newel extending from centre column; assembling and bolting together										
1956 diameter; for floor to floor height units between 2160mm and 2640mm; plywood treads with hardwood edgings; fixing base and ground plates to timber	1218.74	2.50	**1249.20**	30.00	15.00	**586.95**	-	**1836.15**	nr	**1973.86**
extra; additional tread modules. 180-220mm per rise	91.27	2.50	**93.55**	1.50	0.75	**29.35**	-	**122.90**	nr	**132.11**
extra; mini landing to match treads, at upper floor level	224.83	2.50	**230.45**	2.00	1.00	**39.13**	-	**269.58**	nr	**289.79**
1956 diameter; for floor to floor height units between 2160mm and 2640mm; plywood treads with carpet recess fixing base and ground plates to timber	1391.25	2.50	**1426.03**	30.00	15.00	**586.95**	-	**2012.98**	nr	**2163.95**
extra; additional tread modules, 180-220mm per rise	104.62	2.50	**107.24**	1.50	0.75	**29.35**	-	**136.59**	nr	**146.83**
1956 diameter; for floor to floor height units between 2160mm and 2640mm; hardwood treads, plain or with hardwood surround providing carpet recess; fixing base and ground plates to timber	1558.20	2.50	**1597.16**	30.00	15.00	**586.95**	-	**2184.11**	nr	**2347.91**
extra; additional tread modules, 180-220mm per rise	119.09	2.50	**122.07**	1.50	0.75	**29.35**	-	**151.42**	nr	**162.77**
extra; mini landing to match treads, at upper floor level	267.12	2.50	**273.80**	2.00	1.00	**39.13**	-	**312.93**	nr	**336.40**
stairwell balustrading with straight bar infill and PVC handrail cover	96.83	2.50	**99.25**	1.75	0.90	**34.51**	-	**133.76**	m	**143.79**
stairwell balustrading to match staircase including PVC handrail cover	158.05	2.50	**162.00**	1.75	0.90	**34.51**	-	**196.50**	nr	**211.24**
Crescent 'H' range domestic spiral staircases; model 20 'H'; Crescent of Cambridge Ltd.; comprising tread modules complete with centre column, tread support, wooden tread, tread baluster, handrail section, PVC handrail cover, baluster wicket infill panel and tread riser bar, and with attachment brackets for fixing to floor and newel extending from centre column; assembling and bolting together										
2092mm diameter; for floor to floor height units between 2160mm and 2640mm; plywood treads with hardwood edgings; fixing base and ground plates to timber	1268.82	2.50	**1300.54**	30.00	15.00	**586.95**	-	**1887.49**	nr	**2029.05**
extra; additional tread modules, 180 - 220mm per rise	93.49	2.50	**95.83**	1.50	0.75	**29.35**	-	**125.18**	nr	**134.57**
extra; mini landing to match treads, at upper floor level	233.73	2.50	**239.57**	2.00	1.00	**39.13**	-	**278.70**	nr	**299.61**
2092mm diameter; for floor to floor height units between 2160mm and 2640mm; plywood treads with carpet recess; fixing base and ground plates to timber	1435.77	2.50	**1471.66**	30.00	15.00	**586.95**	-	**2058.61**	nr	**2213.01**
extra; additional tread modules, 180 - 220mm per rise	105.74	2.50	**108.38**	1.50	0.75	**29.35**	-	**137.73**	nr	**148.06**
2092mm diameter; for floor to floor height units between 2160mm and 2640mm; hardwood treads, plain or with hardwood surround providing carpet recess; fixing base and ground plates to timber	1586.03	2.50	**1625.68**	31.50	16.00	**618.98**	-	**2244.65**	nr	**2413.00**
extra; additional tread modules, 180 - 220mm per rise	122.43	2.50	**125.49**	1.50	0.75	**29.35**	-	**154.84**	nr	**166.45**

Labour hourly rates: (except Specialists) Craft Operatives £14.21 Labourer £10.71 Rates are national average prices. Refer to REGIONAL VARIATIONS for indicative levels of overall pricing in regions	MATERIALS			LABOUR				RATES		
	Del to Site £	Waste %	Material Cost £	Craft Optve Hrs	Lab Hrs	Labour Cost £	Sunds £	Nett Rate £	Unit	Gross Rate (+7.5%) £
L30: STAIRS/WALKWAYS/BALUSTRADES: METAL STAIRS AND BALUSTRADES Cont.										
Spiral staircases in steel; one coat red oxide primer before delivery to site Cont.										
Crescent 'H' range domestic spiral staircases; model 20 'H'; Crescent of Cambridge Ltd.; comprising tread modules complete with centre column, tread support, wooden tread, tread baluster, handrail section, PVC handrail cover, baluster wicket infill panel and tread riser bar, and with attachment brackets for fixing to floor and newel extending from centre column; assembling and bolting together Cont.										
extra; mini landing to match treads, at upper floor level	274.91	2.50	**281.78**	2.00	1.00	**39.13**	-	**320.91**	nr	**344.98**
stairwell balustrading with straight bar infill and PVC handrail cover	96.83	2.50	**99.25**	1.75	0.90	**34.51**	-	**133.76**	m	**143.79**
stairwell balustrading to match staircase including PVC handrail cover	158.05	2.50	**162.00**	1.75	0.90	**34.51**	-	**196.50**	m	**211.24**
Crescent 'S' range shop, office, factory and domestic spiral staircases; model 15 'S'; Crescent of Cambridge Ltd.; comprising tread modules complete with tread/collar section, baluster bar and section of handrail, and with upper floor attachment details with flanged column tube section, newel and top handrail baluster; assembling and bolting together										
1548mm diameter; for floor to floor height units between 2520mm and 3080mm; plain or ribbed treads; fixing base and ground plates and upper floor attachment to masonry with expanding bolts	1303.32	2.50	**1335.91**	35.00	17.50	**684.78**	-	**2020.68**	nr	**2172.23**
extra; additional tread modules, 180 - 220mm per rise	86.81	2.50	**88.98**	1.50	0.75	**29.35**	-	**118.33**	nr	**127.21**
1548mm diameter; for floor to floor height units between 2520mm and 3080mm; recessed treads; fixing base and ground plates and upper floor attachment to masonry with expanding bolts	1391.25	2.50	**1426.03**	36.50	19.00	**722.16**	-	**2148.19**	nr	**2309.30**
extra; additional tread modules, 180 - 220mm per rise	92.38	2.50	**94.69**	1.50	0.75	**29.35**	-	**124.04**	nr	**133.34**
1548mm diameter; for floor to floor height units between 2520mm and 3080mm; recessed treads with studded rubber tread profile covers; fixing base and ground plates and upper floor attachment to masonry with expanding bolts	178.08	2.50	**182.53**	36.50	19.00	**722.16**	-	**904.69**	nr	**972.54**
extra; additional tread modules, 180 - 220mm per rise	119.09	2.50	**122.07**	1.50	0.75	**29.35**	-	**151.42**	nr	**162.77**
extra; 825 x 825mm plain or Durbar plate landing at upper floor level, complete with column attachment section, column newel, top handrail, baluster and handrail section, and balustrading complete to one side	395.12	2.50	**404.99**	3.00	1.50	**58.70**	-	**463.69**	nr	**498.46**
extra; 825 x 825mm recessed landing at upper floor level complete with column attachment section, column newel, top handrail, baluster and handrail section, and balustrading complete to one side	430.73	2.50	**441.50**	3.00	1.50	**58.70**	-	**500.19**	nr	**537.71**
extra; stairwell balustrade including handrail balusters and ground rail	82.36	2.50	**84.42**	1.75	0.90	**34.51**	-	**118.93**	m	**127.85**
extra; additional tread balusters	8.90	2.50	**9.13**	0.80	0.40	**15.65**	-	**24.78**	nr	**26.64**
extra; riser bars	7.35	2.50	**7.53**	0.80	0.40	**15.65**	-	**23.18**	nr	**24.92**
extra; PVC handrail cover	4.45	2.50	**4.56**	0.30	0.15	**5.87**	-	**10.43**	m	**11.22**
Crescent 'S' range shop, office, factory and domestic spiral staircases; model 20 'S'; Crescent of Cambridge Ltd.; comprising tread modules complete with tread/collar section, baluster bar and section of handrail, and with upper floor attachment details with flanged column tube section, newel and top handrail baluster; assembling and bolting together										
2072mm diameter; for floor to floor height units between 2520mm and 3080mm; plain or ribbed treads; fixing base and ground plates and upper floor attachment to masonry with expanding bolts	1446.90	2.50	**1483.07**	36.50	19.00	**722.16**	-	**2205.23**	nr	**2370.62**
extra; additional tread modules, 180 - 220mm per rise	96.83	2.50	**99.25**	1.50	0.75	**29.35**	-	**128.60**	nr	**138.24**
2072mm diameter; for floor to floor height units between 2520mm and 3080mm; recessed treads; fixing base and ground plates and upper floor attachment to masonry with expanding bolts	1524.81	2.50	**1562.93**	36.50	19.00	**722.16**	-	**2285.09**	nr	**2456.47**
extra; additional tread modules, 180 - 220mm per rise	102.40	2.50	**104.96**	1.50	0.75	**29.35**	-	**134.30**	nr	**144.38**
2072mm diameter; for floor to floor height units between 2520mm and 3080mm; recessed treads with studded rubber tread profile covers; fixing base and ground plates and upper floor attachment to masonry with expanding bolts	2029.00	2.50	**2079.72**	36.50	19.00	**722.16**	-	**2801.88**	nr	**3012.02**
extra; additional tread modules, 180 - 220mm per rise	138.01	2.50	**141.46**	1.50	0.75	**29.35**	-	**170.81**	nr	**183.62**
extra; 1100 x 1100mm plain or Durbar plate landing at upper floor level, complete with column attachment section, column newel, top handrail, baluster and handrail section, and balustrading complete to one side	471.91	2.50	**483.71**	3.00	1.50	**58.70**	-	**542.40**	nr	**583.09**

Labour hourly rates: (except Specialists) Craft Operatives £14.21 Labourer £10.71 Rates are national average prices. Refer to REGIONAL VARIATIONS for indicative levels of overall pricing in regions	MATERIALS			LABOUR				RATES		
	Del to Site £	Waste %	Material Cost £	Craft Optve Hrs	Lab Hrs	Labour Cost £	Sunds £	Nett Rate £	Unit	Gross Rate (+7.5%) £
L30: STAIRS/WALKWAYS/BALUSTRADES: METAL STAIRS AND BALUSTRADES Cont.										
Spiral staircases in steel; one coat red oxide primer before delivery to site Cont.										
Crescent 'S' range shop, office, factory and domestic spiral staircases; model 20 'S'; Crescent of Cambridge Ltd.; comprising tread modules complete with tread/collar section, baluster bar and section of handrail, and with upper floor attachment details with flanged column tube section, newel and top handrail baluster; assembling and bolting together Cont.										
extra; 1100 x 1100mm recessed landing at upper floor level complete with column attachment section, column newel, top handrail, baluster and handrail section, and balustrading complete to one side	511.98	2.50	**524.78**	3.50	1.75	**68.48**	-	**593.26**	nr	**637.75**
extra; stairwell balustrade including handrail balusters and ground rail	80.14	2.50	**82.14**	1.75	0.90	**34.51**	-	**116.65**	m	**125.39**
extra; additional tread balusters	8.90	2.50	**9.13**	0.80	0.40	**15.65**	-	**24.78**	nr	**26.64**
extra; riser bars	7.35	2.50	**7.53**	0.80	0.40	**15.65**	-	**23.18**	nr	**24.92**
extra; PVC handrail cover	4.45	2.50	**4.56**	0.30	0.15	**5.87**	-	**10.43**	m	**11.22**
Spiral staircases in steel; hot dip galvanized finish										
Crescent 'S' range shop, office, factory and domestic spiral staircases; model 15 'S'; Crescent of Cambridge Ltd.; comprising tread modules complete with tread/collar section, baluster bar and section of handrail, and with upper floor attachment details with flanged column tube section, newel and top handrail baluster; assembling and bolting together										
1548mm diameter; for floor to floor height units between 2520mm and 3080mm; plain or ribbed treads; fixing base and ground plates and upper floor attachment to masonry with expanding bolts	1391.25	2.50	**1426.03**	40.00	20.00	**782.60**	-	**2208.63**	nr	**2374.28**
extra; additional tread modules, 180 - 220mm per rise	93.49	2.50	**95.83**	1.50	0.75	**29.35**	-	**125.18**	nr	**134.57**
1548mm diameter; for floor to floor height units between 2520mm and 3080mm; recessed treads; fixing base and ground plates and upper floor attachment to masonry with expanding bolts	1480.29	2.50	**1517.30**	36.50	19.00	**722.16**	-	**2239.45**	nr	**2407.41**
extra; additional tread modules, 180 - 220mm per rise	100.17	2.50	**102.67**	1.50	0.75	**29.35**	-	**132.02**	nr	**141.92**
extra; 825 x 825mm plain or Durbar plate landing at upper floor level, complete with column attachment section, column newel, top handrail, baluster and handrail section, and balustrading complete to one side	440.75	2.50	**451.77**	3.00	1.50	**58.70**	-	**510.46**	nr	**548.75**
extra; 825 x 825mm recessed landing at upper floor level complete with column attachment section, column newel, top handrail, baluster and handrail section, and balustrading complete to one side	474.14	2.50	**485.99**	3.50	1.75	**68.48**	-	**554.47**	nr	**596.05**
extra; stairwell balustrade including handrail balusters and ground rail	89.71	2.50	**91.95**	1.75	0.90	**34.51**	-	**126.46**	m	**135.94**
extra; additional tread balusters	10.24	2.50	**10.50**	0.80	0.40	**15.65**	-	**26.15**	nr	**28.11**
extra; riser bars	8.96	2.50	**9.18**	0.80	0.40	**15.65**	-	**24.84**	nr	**26.70**
Crescent 'S' range shop, office, factory and domestic spiral staircases; model 20 'S'; Crescent of Cambridge Ltd.; comprising tread modules complete with tread/collar section, baluster bar and section of handrail, and with upper floor attachment details with flanged column tube section, newel and top handrail baluster; assembling and bolting together										
2072mm diameter; for floor to floor height units between 2520mm and 3080mm; plain or ribbed treads; fixing base and ground plates and upper floor attachment to masonry with expanding bolts	1530.38	2.50	**1568.63**	36.50	19.00	**722.16**	-	**2290.79**	nr	**2462.60**
extra; additional tread modules, 180 - 220mm per rise	104.62	2.50	**107.24**	1.50	0.75	**29.35**	-	**136.59**	nr	**146.83**
2072mm diameter; for floor to floor height units between 2520mm and 3080mm; recessed treads; fixing base and ground plates and upper floor attachment to masonry with expanding bolts	1604.95	2.50	**1645.07**	36.50	19.00	**722.16**	-	**2367.22**	nr	**2544.77**
extra; additional tread modules, 180 - 220mm per rise	110.19	2.50	**112.94**	1.50	0.75	**29.35**	-	**142.29**	nr	**152.96**
extra; 1100 x 1100mm plain or Durbar plate landing at upper floor level, complete with column attachment section, column newel, top handrail, baluster and handrail section, and balustrading complete to one side	536.47	2.50	**549.88**	3.00	1.50	**58.70**	-	**608.57**	nr	**654.22**
extra; 1100 x 1100mm recessed landing at upper floor level complete with column attachment section, column newel, top handrail, baluster and handrail section, and balustrading complete to one side	584.33	2.50	**598.93**	3.50	1.75	**68.48**	-	**667.41**	nr	**717.47**
extra; stairwell balustrade including handrail balusters and ground rail	89.71	2.50	**91.95**	1.75	0.90	**34.51**	-	**126.46**	m	**135.94**
extra; additional tread balusters	10.24	2.50	**10.50**	0.80	0.40	**15.65**	-	**26.15**	nr	**28.11**
extra; riser bars	8.96	2.50	**9.18**	0.80	0.40	**15.65**	-	**24.84**	nr	**26.70**

Labour hourly rates: (except Specialists) Craft Operatives £14.21 Labourer £10.71 Rates are national average prices. Refer to REGIONAL VARIATIONS for indicative levels of overall pricing in regions	MATERIALS			LABOUR				RATES		
	Del to Site £	Waste %	Material Cost £	Craft Optve Hrs	Lab Hrs	Labour Cost £	Sunds £	Nett Rate £	Unit	Gross Rate (+7.5%) £
L30: STAIRS/WALKWAYS/BALUSTRADES: METAL STAIRS AND BALUSTRADES Cont.										
Spiral staircases in steel; hot dip galvanized finish Cont.										
Crescent modular spiral fire escape staircases; model 15; Crescent of Cambridge Ltd.; comprising tread modules with plain or ribbed treads, and complete with centre core, spacers, baluster bar and section of handrail, tread riser bars and standard landing with balustrade one side, bars at 115mm centres; assembling and bolting together										
1548mm diameter; ground to first floor assembly, with base and ground plates; for floor to floor heights between 2520mm and 3080mm; fixing base plates and landing to masonry with expanding bolts	2270.52	2.50	**2327.28**	40.00	20.00	**782.60**	-	**3109.88**	nr	**3343.12**
extra; base extension	146.92	2.50	**150.59**	2.00	1.00	**39.13**	-	**189.72**	nr	**203.95**
1548mm diameter; intermediate floor assembly; for floor to floor heights between 2520mm and 3080mm; fixing landing to masonry with expanding bolts...	2183.71	2.50	**2238.30**	40.00	20.00	**782.60**	-	**3020.90**	nr	**3247.47**
1548mm diameter; top floor assembly with two additional lengths of balustrade to landing; for floor to floor heights between 2520mm and 3080mm; fixing landing to masonry with expanding bolts...	2534.30	2.50	**2597.66**	30.00	15.00	**586.95**	-	**3184.61**	nr	**3423.45**
extra; additional tread modules 180 - 220mm per rise	117.98	2.50	**120.93**	1.50	0.75	**29.35**	-	**150.27**	nr	**161.55**
extra; shield type infill panel in lieu of additional balusters to tread module............	173.63	2.50	**177.97**	1.00	0.50	**19.57**	-	**197.53**	nr	**212.35**
extra; 'wicket bar' type infill panel in lieu of additional balusters to tread module............	267.12	2.50	**273.80**	1.00	0.50	**19.57**	-	**293.36**	nr	**315.37**
extra; PVC handrail cover.......................	4.50	2.50	**4.61**	0.30	0.15	**5.87**	-	**10.48**	m	**11.26**
Crescent modular spiral fire escape staircases; model 20; Crescent of Cambridge Ltd.; comprising tread modules with plain or ribbed treads, and complete with centre core, spacers, baluster bar and section of handrail, tread riser bars and standard landing with balustrade one side, bars at 115mm centres; assembling and bolting together										
2072mm diameter; ground to first floor assembly, with base and ground plates; for floor to floor heights between 2520mm and 3080mm; fixing base plates and landing to masonry with expanding bolts	2699.03	2.50	**2766.50**	40.00	20.00	**782.60**	-	**3549.10**	nr	**3815.28**
extra; base extension	162.50	2.50	**166.56**	2.00	1.00	**39.13**	-	**205.69**	nr	**221.12**
2072mm diameter; intermediate floor assembly; for floor to floor heights between 2520mm and 3080mm; fixing landing to masonry with expanding bolts...	2609.99	2.50	**2675.23**	30.00	15.00	**586.95**	-	**3262.18**	nr	**3506.85**
2072mm diameter; top floor assembly with two additional lengths of balustrade to landing; for floor to floor heights between 2520mm and 3080mm; fixing landing to masonry with expanding bolts...	2960.58	2.50	**3034.59**	30.00	15.00	**586.95**	-	**3621.54**	nr	**3893.16**
extra; additional tread modules 180 - 220mm per rise	116.87	2.50	**119.79**	1.50	0.75	**29.35**	-	**149.13**	nr	**160.32**
extra; 'wicket bar' type infill panel in lieu of additional balusters to tread module............	146.92	2.50	**150.59**	1.00	0.50	**19.57**	-	**170.15**	nr	**182.92**
extra; PVC handrail cover.......................	4.50	2.50	**4.61**	0.30	0.15	**5.87**	-	**10.48**	m	**11.26**
Ladders in steel										
Vertical ladders; 65 x 10mm flat stringers; 65 x 10 x 250mm girth stringer brackets, bent once, bolted to stringers; 20mm diameter solid bar rungs welded to stringers; fixing to masonry with expanding bolts										
395mm wide x 3000mm long overall; brackets -6 nr; rungs -7 nr	525.80	2.50	**538.94**	10.00	10.00	**249.20**	-	**788.14**	nr	**847.25**
395mm wide x 5000mm long overall; brackets -8 nr; rungs -14 nr	525.80	2.50	**538.94**	12.00	12.00	**299.04**	-	**837.98**	nr	**900.83**
470mm wide x 3000mm long overall; brackets -6 nr; rungs -7 nr	444.91	2.50	**456.03**	10.00	10.00	**249.20**	-	**705.23**	nr	**758.12**
470mm wide x 5000mm long overall; brackets -8 nr; rungs -14 nr	780.03	2.50	**799.53**	12.00	12.00	**299.04**	-	**1098.57**	nr	**1180.96**
395mm wide x 3000mm long overall, stringers rising and returning 900mm above top fixing points; brackets -6 nr; rungs -7 nr....................	444.91	2.50	**456.03**	10.00	10.00	**249.20**	-	**705.23**	nr	**758.12**
395mm wide x 5000mm long overall, stringers rising and returning 900mm above top fixing points; brackets -8 nr; rungs -14 nr...................	780.03	2.50	**799.53**	12.00	12.00	**299.04**	-	**1098.57**	nr	**1180.96**
470mm wide x 3000mm long overall, stringers rising and returning 900mm above top fixing points; brackets -6 nr; rungs -7 nr....................	780.03	2.50	**799.53**	10.00	10.00	**249.20**	-	**1048.73**	nr	**1127.39**
470mm wide x 5000mm long overall, stringers rising and returning 900mm above top fixing points; brackets -8 nr; rungs -14 nr....................	947.59	2.50	**971.28**	12.00	12.00	**299.04**	-	**1270.32**	nr	**1365.60**
Vertical ladders; 65 x 10mm flat stringers; 65 x 10 x 250mm girth stringer brackets, bent once, bolted to stringers; 20mm diameter solid bar rungs welded to stringers; 65 x 10mm flat back hoops 900mm diameter, welded to stringers										
395mm wide x 3000mm long overall; brackets -6 nr; rungs -7 nr; hoops -7 nr.......................	947.59	2.50	**971.28**	15.00	15.00	**373.80**	-	**1345.08**	nr	**1445.96**
395mm wide x 5000mm long overall; brackets -8 nr; rungs -14 nr; hoops -14 nr.....................	1144.04	2.50	**1172.65**	20.00	20.00	**498.40**	-	**1671.05**	nr	**1796.37**
470mm wide x 3000mm long overall; brackets -6 nr; rungs -7 nr; hoops -7 nr.......................	1080.49	2.50	**1107.50**	15.00	15.00	**373.80**	-	**1481.30**	nr	**1592.40**

Labour hourly rates: (except Specialists) Craft Operatives £14.21 Labourer £10.71 Rates are national average prices. Refer to REGIONAL VARIATIONS for indicative levels of overall pricing in regions	MATERIALS			LABOUR				RATES		
	Del to Site £	Waste %	Material Cost £	Craft Optve Hrs	Lab Hrs	Labour Cost £	Sunds £	Nett Rate £	Unit	Gross Rate (+7.5%) £
L30: STAIRS/WALKWAYS/BALUSTRADES: METAL STAIRS AND BALUSTRADES Cont.										
Ladders in steel Cont.										
Vertical ladders; 65 x 10mm flat stringers; 65 x 10 x 250mm girth stringer brackets, bent once, bolted to stringers; 20mm diameter solid bar rungs welded to stringers; 65 x 10mm flat back hoops 900mm diameter, welded to stringers Cont.										
470mm wide x 5000mm long overall; brackets -8 nr; rungs -14 nr; hoops -14 nr	1271.16	2.50	**1302.94**	20.00	20.00	**498.40**	-	**1801.34**	nr	**1936.44**
395mm wide x 3000mm long overall, stringers rising and returning 900mm above top fixing points; brackets -6 nr; rungs -7 nr; hoops -7 nr	907.15	2.50	**929.82**	15.00	15.00	**373.80**	-	**1303.62**	nr	**1401.40**
395mm wide x 5000mm long overall, stringers rising and returning 900mm above top fixing points; brackets -8 nr; rungs -14 nr; hoops -14 nr	1126.71	2.50	**1154.88**	20.00	20.00	**498.40**	-	**1653.28**	nr	**1777.27**
470mm wide x 3000mm long overall, stringers rising and returning 900mm above top fixing points; brackets -6 nr; rungs -7 nr; hoops -7 nr	1028.48	2.50	**1054.20**	15.00	15.00	**373.80**	-	**1428.00**	nr	**1535.10**
470mm wide x 5000mm long overall, stringers rising and returning 900mm above top fixing points; brackets -8 nr; rungs -14 nr; hoops -14 nr	1340.50	2.50	**1374.01**	20.00	20.00	**498.40**	-	**1872.41**	nr	**2012.84**
Ship type ladders; 75 x 10mm flat stringers, shaped and cleated ends to tops and bottoms; 65 x 10 x 350mm girth stringer brackets, bent once, bolted to stringers; 8 x 75mm on plain raised pattern plate treads with 50 x 50 x 6mm shelf angles, bolted to stringers; 25mm diameter solid bar handrails to both sides, with flattened ends bolted to stringers; 32mm diameter x 175mm long solid bar standards with three way ball type joint at intersection with handrail, and flattened ends bolted to stringers; fixing to masonry with expanding bolts										
420mm wide x 3000mm long overall, handrails rising and returning 900mm above top of stringers; brackets -4 nr; treads -12 nr; standards -6 nr	751.14	2.50	**769.92**	10.00	10.00	**249.20**	-	**1019.12**	nr	**1095.55**
420mm wide x 5000mm long overall, handrails rising and returning 900mm above top of stringers; brackets -4 nr; treads -21 nr; standards -10 nr	1063.15	2.50	**1089.73**	12.00	12.00	**299.04**	-	**1388.77**	nr	**1492.93**
470mm wide x 3000mm long overall, handrails rising and returning 900mm above top of stringers; brackets -4 nr; treads -12 nr; standards -6 nr	872.48	2.50	**894.29**	10.00	10.00	**249.20**	-	**1143.49**	nr	**1229.25**
470mm wide x 5000mm long overall, handrails rising and returning 900mm above top of stringers; brackets -4 nr; treads -21 nr; standards -10 nr	1178.71	2.50	**1208.18**	12.00	12.00	**299.04**	-	**1507.22**	nr	**1620.26**
Ship type ladders; 75 x 10mm flat stringers, shaped and cleated ends to tops and bottoms; 65 x 10 x 350mm girth stringer brackets, bent once, bolted to stringers; 8 x 75mm on plain raised pattern plate treads with 50 x 50 x 6mm shelf angles, bolted to stringers; 25mm diameter solid bar handrails to both sides, with flattened ends bolted to stringers; 32mm diameter x 330mm long solid bar standards with three way ball type joint at intersection with handrail, and flattened ends bolted to stringers; fixing to masonry with expanding bolts										
420mm wide x 3000mm long overall, handrails rising and returning 900mm above top of stringers; brackets -4 nr; treads -12 nr; standards -6 nr	751.14	2.50	**769.92**	10.00	10.00	**249.20**	-	**1019.12**	nr	**1095.55**
420mm wide x 5000mm long overall, handrails rising and returning 900mm above top of stringers; brackets -4 nr; treads -21 nr; standards -10 nr	1080.49	2.50	**1107.50**	12.00	12.00	**299.04**	-	**1406.54**	nr	**1512.03**
470mm wide x 3000mm long overall, handrails rising and returning 900mm above top of stringers; brackets -4 nr; treads -12 nr; standards -6 nr	936.04	2.50	**959.44**	10.00	10.00	**249.20**	-	**1208.64**	nr	**1299.28**
470mm wide x 5000mm long overall, handrails rising and returning 900mm above top of stringers; brackets -4 nr; treads -21 nr; standards -10 nr	1196.05	2.50	**1225.95**	12.00	12.00	**299.04**	-	**1524.99**	nr	**1639.36**
L40: GENERAL GLAZING										
Glass; BS 952, clear float										
3mm thick to wood rebates with BS 544 putty										
not exceeding 0.15m2	37.63	5.00	**39.51**	0.80	0.08	**12.22**	0.30	**52.03**	m2	**55.94**
0.15 - 4.00m2	37.63	5.00	**39.51**	0.40	0.04	**6.11**	0.14	**45.77**	m2	**49.20**
4mm thick to wood rebates with BS 544 putty										
not exceeding 0.15m2	37.63	5.00	**39.51**	0.80	0.08	**12.22**	0.30	**52.03**	m2	**55.94**
0.15 - 4.00m2	37.63	5.00	**39.51**	0.40	0.04	**6.11**	0.14	**45.77**	m2	**49.20**
5mm thick to wood rebates with BS 544 putty										
not exceeding 0.15m2	55.01	5.00	**57.76**	0.90	0.08	**13.65**	0.30	**71.70**	m2	**77.08**
0.15 - 4.00m2	55.01	5.00	**57.76**	0.45	0.04	**6.82**	0.14	**64.73**	m2	**69.58**
6mm thick to wood rebates with BS 544 putty										
not exceeding 0.15m2	55.01	5.00	**57.76**	0.90	0.08	**13.65**	0.30	**71.70**	m2	**77.08**
0.15 - 4.00m2	55.01	5.00	**57.76**	0.45	0.04	**6.82**	0.14	**64.73**	m2	**69.58**
3mm thick to wood rebates with bradded wood beads (included elsewhere) and BS 544 putty										
not exceeding 0.15m2	37.63	5.00	**39.51**	1.00	0.08	**15.07**	0.30	**54.88**	m2	**58.99**

Labour hourly rates: (except Specialists) Craft Operatives £14.21 Labourer £10.71 Rates are national average prices. Refer to REGIONAL VARIATIONS for indicative levels of overall pricing in regions	MATERIALS			LABOUR				RATES		
	Del to Site £	Waste %	Material Cost £	Craft Optve Hrs	Lab Hrs	Labour Cost £	Sunds £	Nett Rate £	Unit	Gross Rate (+7.5%) £
L40: GENERAL GLAZING Cont.										
Glass; BS 952, clear float Cont.										
3mm thick to wood rebates with bradded wood beads (included elsewhere) and BS 544 putty Cont.										
0.15 - 4.00m2	37.63	5.00	**39.51**	0.50	0.04	**7.53**	0.14	**47.19**	m2	**50.73**
4mm thick to wood rebates with bradded wood beads (included elsewhere) and BS 544 putty										
not exceeding 0.15m2	37.63	5.00	**39.51**	1.00	0.08	**15.07**	0.30	**54.88**	m2	**58.99**
0.15 - 4.00m2	37.63	5.00	**39.51**	0.50	0.04	**7.53**	0.14	**47.19**	m2	**50.73**
5mm thick to wood rebates with bradded wood beads (included elsewhere) and BS 544 putty										
not exceeding 0.15m2	55.01	5.00	**57.76**	1.10	0.08	**16.49**	0.30	**74.55**	m2	**80.14**
0.15 - 4.00m2	55.01	5.00	**57.76**	0.55	0.04	**8.24**	0.14	**66.15**	m2	**71.11**
6mm thick to wood rebates with bradded wood beads (included elsewhere) and BS 544 putty										
not exceeding 0.15m2	55.01	5.00	**57.76**	1.10	0.08	**16.49**	0.30	**74.55**	m2	**80.14**
0.15 - 4.00m2	55.01	5.00	**57.76**	0.55	0.04	**8.24**	0.14	**66.15**	m2	**71.11**
3mm thick to wood rebates with screwed wood beads and glazing strip (included elsewhere)										
not exceeding 0.15m2	37.63	5.00	**39.51**	1.40	0.08	**20.75**	0.30	**60.56**	m2	**65.10**
0.15 - 4.00m2	37.63	5.00	**39.51**	0.70	0.04	**10.38**	0.14	**50.03**	m2	**53.78**
4mm thick to wood rebates with screwed wood beads and glazing strip (included elsewhere)										
not exceeding 0.15m2	37.63	5.00	**39.51**	1.40	0.08	**20.75**	0.30	**60.56**	m2	**65.10**
0.15 - 4.00m2	37.63	5.00	**39.51**	0.70	0.04	**10.38**	0.14	**50.03**	m2	**53.78**
5mm thick to wood rebates with screwed wood beads and glazing strip (included elsewhere)										
not exceeding 0.15m2	55.01	5.00	**57.76**	1.50	0.08	**22.17**	0.30	**80.23**	m2	**86.25**
0.15 - 4.00m2	55.01	5.00	**57.76**	0.75	0.04	**11.09**	0.14	**68.99**	m2	**74.16**
6mm thick to wood rebates with screwed wood beads and glazing strip (included elsewhere)										
not exceeding 0.15m2	55.01	5.00	**57.76**	1.50	0.08	**22.17**	0.30	**80.23**	m2	**86.25**
0.15 - 4.00m2	55.01	5.00	**57.76**	0.75	0.04	**11.09**	0.14	**68.99**	m2	**74.16**
3mm thick to metal rebates with metal casement glazing compound										
not exceeding 0.15m2	37.63	5.00	**39.51**	0.90	0.08	**13.65**	0.34	**53.50**	m2	**57.51**
0.15 - 4.00m2	37.63	5.00	**39.51**	0.45	0.04	**6.82**	0.17	**46.50**	m2	**49.99**
4mm thick to metal rebates with metal casement glazing compound										
not exceeding 0.15m2	37.63	5.00	**39.51**	0.90	0.08	**13.65**	0.34	**53.50**	m2	**57.51**
0.15 - 4.00m2	37.63	5.00	**39.51**	0.45	0.04	**6.82**	0.17	**46.50**	m2	**49.99**
5mm thick to metal rebates with metal casement glazing compound										
not exceeding 0.15m2	55.01	5.00	**57.76**	1.00	0.08	**15.07**	0.34	**73.17**	m2	**78.66**
0.15 - 4.00m2	55.01	5.00	**57.76**	0.50	0.04	**7.53**	0.17	**65.46**	m2	**70.37**
6mm thick to metal rebates with metal casement glazing compound										
not exceeding 0.15m2	55.01	5.00	**57.76**	1.00	0.08	**15.07**	0.34	**73.17**	m2	**78.66**
0.15 - 4.00m2	55.01	5.00	**57.76**	0.50	0.04	**7.53**	0.17	**65.46**	m2	**70.37**
3mm thick to metal rebates with clipped metal beads and gaskets (included elsewhere)										
not exceeding 0.15m2	37.63	5.00	**39.51**	1.20	0.08	**17.91**	0.30	**57.72**	m2	**62.05**
0.15 - 4.00m2	37.63	5.00	**39.51**	0.60	0.04	**8.95**	0.14	**48.61**	m2	**52.25**
4mm thick to metal rebates with clipped metal beads and gaskets (included elsewhere)										
not exceeding 0.15m2	37.63	5.00	**39.51**	1.20	0.08	**17.91**	0.30	**57.72**	m2	**62.05**
0.15 - 4.00m2	37.63	5.00	**39.51**	0.60	0.04	**8.95**	0.14	**48.61**	m2	**52.25**
5mm thick to metal rebates with clipped metal beads and gaskets (included elsewhere)										
not exceeding 0.15m2	55.01	5.00	**57.76**	1.30	0.08	**19.33**	0.30	**77.39**	m2	**83.19**
0.15 - 4.00m2	55.01	5.00	**57.76**	0.65	0.04	**9.66**	0.14	**67.57**	m2	**72.64**
6mm thick to metal rebates with clipped metal beads and gaskets (included elsewhere)										
not exceeding 0.15m2	55.01	5.00	**57.76**	1.30	0.08	**19.33**	0.30	**77.39**	m2	**83.19**
0.15 - 4.00m2	55.01	5.00	**57.76**	0.65	0.04	**9.66**	0.14	**67.57**	m2	**72.64**
Glass; BS 952, rough cast										
6mm thick to wood rebates with BS 544 putty										
not exceeding 0.15m2	49.46	5.00	**51.93**	0.90	0.08	**13.65**	0.30	**65.88**	m2	**70.82**
0.15 - 4.00m2	49.46	5.00	**51.93**	0.45	0.04	**6.82**	0.14	**58.90**	m2	**63.32**
6mm thick to wood rebates with bradded wood beads (included elsewhere) and BS 544 putty										
not exceeding 0.15m2	49.46	5.00	**51.93**	1.10	0.08	**16.49**	0.30	**68.72**	m2	**73.87**
0.15 - 4.00m2	49.46	5.00	**51.93**	0.55	0.04	**8.24**	0.14	**60.32**	m2	**64.84**

Labour hourly rates: (except Specialists) Craft Operatives £14.21 Labourer £10.71 Rates are national average prices. Refer to REGIONAL VARIATIONS for indicative levels of overall pricing in regions	MATERIALS			LABOUR				RATES		
	Del to Site £	Waste %	Material Cost £	Craft Optve Hrs	Lab Hrs	Labour Cost £	Sunds £	Nett Rate £	Unit	Gross Rate (+7.5%) £
L40: GENERAL GLAZING Cont.										
Glass; BS 952, rough cast Cont.										
6mm thick to wood rebates with screwed wood beads and glazing strip (included elsewhere)										
not exceeding 0.15m2	49.46	5.00	**51.93**	1.50	0.08	**22.17**	0.30	**74.40**	m2	**79.98**
0.15 - 4.00m2	49.46	5.00	**51.93**	0.75	0.04	**11.09**	0.14	**63.16**	m2	**67.90**
6mm thick to metal rebates with metal casement glazing compound										
not exceeding 0.15m2	49.46	5.00	**51.93**	1.00	0.08	**15.07**	0.34	**67.34**	m2	**72.39**
0.15 - 4.00m2	49.46	5.00	**51.93**	0.50	0.04	**7.53**	0.17	**59.63**	m2	**64.10**
6mm thick to metal rebates with clipped metal beads and gaskets (included elsewhere)										
not exceeding 0.15m2	49.46	5.00	**51.93**	1.30	0.08	**19.33**	0.34	**71.60**	m2	**76.97**
0.15 - 4.00m2	49.46	5.00	**51.93**	0.65	0.04	**9.66**	0.17	**61.76**	m2	**66.40**
Glass; BS 952, Pyroshield clear										
6mm thick to wood rebates with screwed beads and intumescent glazing strip (included elsewhere)										
not exceeding 0.15m2	89.03	5.00	**93.48**	1.80	0.08	**26.43**	-	**119.92**	m2	**128.91**
not exceeding 0.15m2; aligning panes with adjacent panes	89.03	8.50	**96.60**	2.00	0.08	**29.28**	-	**125.87**	m2	**135.31**
0.15 - 4.00m2	89.03	5.00	**93.48**	0.90	0.04	**13.22**	-	**106.70**	m2	**114.70**
0.15 - 4.00m2; aligning panes with adjacent panes	89.03	8.50	**96.60**	1.00	0.04	**14.64**	-	**111.24**	m2	**119.58**
6mm thick to wood rebates with screwed beads and Pyroglazing strip (included elsewhere)										
not exceeding 0.15m2	89.03	5.00	**93.48**	1.80	0.08	**26.43**	-	**119.92**	m2	**128.91**
not exceeding 0.15m2; aligning panes with adjacent panes	89.03	8.50	**96.60**	2.00	0.08	**29.28**	-	**125.87**	m2	**135.31**
0.15 - 4.00m2	89.03	5.00	**93.48**	0.90	0.04	**13.22**	-	**106.70**	m2	**114.70**
0.15 - 4.00m2; aligning panes with adjacent panes	89.03	8.50	**96.60**	1.00	0.04	**14.64**	-	**111.24**	m2	**119.58**
6mm thick to wood rebates with screwed beads and Lorient System 90 (included elsewhere)										
not exceeding 0.15m2	89.03	5.00	**93.48**	2.00	0.08	**29.28**	-	**122.76**	m2	**131.97**
not exceeding 0.15m2; aligning panes with adjacent panes	89.03	8.50	**96.60**	2.20	0.08	**32.12**	-	**128.72**	m2	**138.37**
0.15 - 4.00m2	89.03	5.00	**93.48**	1.00	0.04	**14.64**	-	**108.12**	m2	**116.23**
0.15 - 4.00m2; aligning panes with adjacent panes	89.03	8.50	**96.60**	1.10	0.04	**16.06**	-	**112.66**	m2	**121.11**
6mm thick to metal rebates with screwed metal beads and mild steel flat strips (included elsewhere)										
not exceeding 0.15m2	89.03	5.00	**93.48**	1.80	0.08	**26.43**	-	**119.92**	m2	**128.91**
not exceeding 0.15m2; aligning panes with adjacent panes	89.03	8.50	**96.60**	2.00	0.08	**29.28**	-	**125.87**	m2	**135.31**
0.15 - 4.00m2	89.03	5.00	**93.48**	0.90	0.04	**13.22**	-	**106.70**	m2	**114.70**
0.15 - 4.00m2; aligning panes with adjacent panes	89.03	8.50	**96.60**	1.00	0.04	**14.64**	-	**111.24**	m2	**119.58**
Glass; BS 952, Pyroshield safety clear										
6mm thick to wood rebates with screwed beads and intumescent glazing strip (included elsewhere)										
not exceeding 0.15m2	104.84	5.00	**110.08**	1.80	0.08	**26.43**	-	**136.52**	m2	**146.76**
not exceeding 0.15m2; aligning panes with adjacent panes	104.84	8.50	**113.75**	2.00	0.08	**29.28**	-	**143.03**	m2	**153.76**
0.15 - 4.00m2	104.84	5.00	**110.08**	0.90	0.04	**13.22**	-	**123.30**	m2	**132.55**
0.15 - 4.00m2; aligning panes with adjacent panes	104.84	8.50	**113.75**	1.00	0.04	**14.64**	-	**128.39**	m2	**138.02**
6mm thick to wood rebates with screwed beads and Pyroglazing strip (included elsewhere)										
not exceeding 0.15m2	104.84	5.00	**110.08**	1.80	0.08	**26.43**	-	**136.52**	m2	**146.76**
not exceeding 0.15m2; aligning panes with adjacent panes	104.84	8.50	**113.75**	2.00	0.08	**29.28**	-	**143.03**	m2	**153.76**
0.15 - 4.00m2	104.84	5.00	**110.08**	0.90	0.04	**13.22**	-	**123.30**	m2	**132.55**
0.15 - 4.00m2; aligning panes with adjacent panes	104.84	8.50	**113.75**	1.00	0.04	**14.64**	-	**128.39**	m2	**138.02**
6mm thick to wood rebates with screwed beads and Lorient System 90 (included elsewhere)										
not exceeding 0.15m2	104.84	5.00	**110.08**	2.00	0.08	**29.28**	-	**139.36**	m2	**149.81**
not exceeding 0.15m2; aligning panes with adjacent panes	104.84	8.50	**113.75**	2.20	0.08	**32.12**	-	**145.87**	m2	**156.81**
0.15 - 4.00m2	104.84	5.00	**110.08**	1.00	0.04	**14.64**	-	**124.72**	m2	**134.07**
0.15 - 4.00m2; aligning panes with adjacent panes	104.84	8.50	**113.75**	1.10	0.04	**16.06**	-	**129.81**	m2	**139.55**
6mm thick to metal rebates with screwed metal beads and mild steel flat strips (included elsewhere)										
not exceeding 0.15m2	104.84	5.00	**110.08**	1.80	0.08	**26.43**	-	**136.52**	m2	**146.76**
not exceeding 0.15m2; aligning panes with adjacent panes	104.84	8.50	**113.75**	2.00	0.08	**29.28**	-	**143.03**	m2	**153.76**
0.15 - 4.00m2	104.84	5.00	**110.08**	0.90	0.04	**13.22**	-	**123.30**	m2	**132.55**
0.15 - 4.00m2; aligning panes with adjacent panes	104.84	8.50	**113.75**	1.00	0.04	**14.64**	-	**128.39**	m2	**138.02**
Glass; BS 952, Pyroshield texture										
7mm thick to wood rebates with screwed beads and intumescent glazing strip (included elsewhere)										
not exceeding 0.15m2	41.19	5.00	**43.25**	1.80	0.08	**26.43**	-	**69.68**	m2	**74.91**
not exceeding 0.15m2; aligning panes with adjacent panes	41.19	8.50	**44.69**	2.00	0.08	**29.28**	-	**73.97**	m2	**79.52**
0.15 - 4.00m2	41.19	5.00	**43.25**	0.90	0.04	**13.22**	-	**56.47**	m2	**60.70**
0.15 - 4.00m2; aligning panes with adjacent panes	41.19	8.50	**44.69**	1.00	0.04	**14.64**	-	**59.33**	m2	**63.78**

Labour hourly rates: (except Specialists) Craft Operatives £14.21 Labourer £10.71 Rates are national average prices. Refer to REGIONAL VARIATIONS for indicative levels of overall pricing in regions	MATERIALS			LABOUR				RATES		
	Del to Site £	Waste %	Material Cost £	Craft Optve Hrs	Lab Hrs	Labour Cost £	Sunds £	Nett Rate £	Unit	Gross Rate (+7.5%) £
L40: GENERAL GLAZING Cont.										
Glass; BS 952, Pyroshield texture Cont.										
7mm thick to wood rebates with screwed beads and Pyroglazing strip (included elsewhere)										
not exceeding 0.15m2	41.19	5.00	**43.25**	1.80	0.08	**26.43**	-	**69.68**	m2	**74.91**
not exceeding 0.15m2; aligning panes with adjacent panes	41.19	8.50	**44.69**	2.00	0.08	**29.28**	-	**73.97**	m2	**79.52**
0.15 - 4.00m2	41.19	5.00	**43.25**	0.90	0.04	**13.22**	-	**56.47**	m2	**60.70**
0.15 - 4.00m2; aligning panes with adjacent panes	41.19	8.50	**44.69**	1.00	0.04	**14.64**	-	**59.33**	m2	**63.78**
7mm thick to wood rebates with screwed beads and Lorient System 90 (included elsewhere)										
not exceeding 0.15m2	41.19	5.00	**43.25**	2.00	0.08	**29.28**	-	**72.53**	m2	**77.97**
not exceeding 0.15m2; aligning panes with adjacent panes	41.19	8.50	**44.69**	2.20	0.08	**32.12**	-	**76.81**	m2	**82.57**
0.15 - 4.00m2	41.19	5.00	**43.25**	1.00	0.04	**14.64**	-	**57.89**	m2	**62.23**
0.15 - 4.00m2; aligning panes with adjacent panes	41.19	8.50	**44.69**	1.10	0.04	**16.06**	-	**60.75**	m2	**65.31**
7mm thick to metal rebates with screwed metal beads and mild steel flat strips (included elsewhere)										
not exceeding 0.15m2	41.19	5.00	**43.25**	1.80	0.08	**26.43**	-	**69.68**	m2	**74.91**
not exceeding 0.15m2; aligning panes with adjacent panes	41.19	8.50	**44.69**	2.00	0.08	**29.28**	-	**73.97**	m2	**79.52**
0.15 - 4.00m2	41.19	5.00	**43.25**	0.90	0.04	**13.22**	-	**56.47**	m2	**60.70**
0.15 - 4.00m2; aligning panes with adjacent panes	41.19	8.50	**44.69**	1.00	0.04	**14.64**	-	**59.33**	m2	**63.78**
Glass; BS 952, Pyroshield safety texture										
7mm thick to wood rebates with screwed beads and intumescent glazing strip (included elsewhere)										
not exceeding 0.15m2	53.59	5.00	**56.27**	1.80	0.08	**26.43**	-	**82.70**	m2	**88.91**
not exceeding 0.15m2; aligning panes with adjacent panes	53.59	8.50	**58.15**	2.00	0.08	**29.28**	-	**87.42**	m2	**93.98**
0.15 - 4.00m2	53.59	5.00	**56.27**	0.90	0.04	**13.22**	-	**69.49**	m2	**74.70**
0.15 - 4.00m2; aligning panes with adjacent panes	53.59	8.50	**58.15**	1.00	0.04	**14.64**	-	**72.78**	m2	**78.24**
7mm thick to wood rebates with screwed beads and Pyroglazing strip (included elsewhere)										
not exceeding 0.15m2	53.59	5.00	**56.27**	1.80	0.08	**26.43**	-	**82.70**	m2	**88.91**
not exceeding 0.15m2; aligning panes with adjacent panes	53.59	8.50	**58.15**	2.00	0.08	**29.28**	-	**87.42**	m2	**93.98**
0.15 - 4.00m2	53.59	5.00	**56.27**	0.90	0.04	**13.22**	-	**69.49**	m2	**74.70**
0.15 - 4.00m2; aligning panes with adjacent panes	53.59	8.50	**58.15**	1.00	0.04	**14.64**	-	**72.78**	m2	**78.24**
7mm thick to wood rebates with screwed beads and Lorient System 90 (included elsewhere)										
not exceeding 0.15m2	53.59	5.00	**56.27**	2.00	0.08	**29.28**	-	**85.55**	m2	**91.96**
not exceeding 0.15m2; aligning panes with adjacent panes	53.59	8.50	**58.15**	2.20	0.08	**32.12**	-	**90.26**	m2	**97.03**
0.15 - 4.00m2	53.59	5.00	**56.27**	1.00	0.04	**14.64**	-	**70.91**	m2	**76.23**
0.15 - 4.00m2; aligning panes with adjacent panes	53.59	8.50	**58.15**	1.10	0.04	**16.06**	-	**74.20**	m2	**79.77**
7mm thick to metal rebates with screwed metal beads and mild steel flat strips (included elsewhere)										
not exceeding 0.15m2	53.59	5.00	**56.27**	1.80	0.08	**26.43**	-	**82.70**	m2	**88.91**
not exceeding 0.15m2; aligning panes with adjacent panes	53.59	8.50	**58.15**	2.00	0.08	**29.28**	-	**87.42**	m2	**93.98**
0.15 - 4.00m2	53.59	5.00	**56.27**	0.90	0.04	**13.22**	-	**69.49**	m2	**74.70**
0.15 - 4.00m2; aligning panes with adjacent panes	53.59	8.50	**58.15**	1.00	0.04	**14.64**	-	**72.78**	m2	**78.24**
Glass; BS 952, white patterned										
4mm thick to wood rebates with BS 544 putty										
not exceeding 0.15m2	34.90	5.00	**36.65**	0.80	0.08	**12.22**	0.30	**49.17**	m2	**52.85**
0.15 - 4.00m2	34.90	5.00	**36.65**	0.40	0.04	**6.11**	0.14	**42.90**	m2	**46.12**
6mm thick to wood rebates with BS 544 putty										
not exceeding 0.15m2	52.53	5.00	**55.16**	0.90	0.08	**13.65**	0.30	**69.10**	m2	**74.28**
0.15 - 4.00m2	52.53	5.00	**55.16**	0.45	0.04	**6.82**	0.14	**62.12**	m2	**66.78**
4mm thick to wood rebates with bradded wood beads (included elsewhere) and BS 544 putty										
not exceeding 0.15m2	34.90	5.00	**36.65**	1.00	0.08	**15.07**	0.30	**52.01**	m2	**55.91**
0.15 - 4.00m2	34.90	5.00	**36.65**	0.50	0.04	**7.53**	0.14	**44.32**	m2	**47.65**
6mm thick to wood rebates with bradded wood beads (included elsewhere) and BS 544 putty										
not exceeding 0.15m2	52.53	5.00	**55.16**	1.10	0.08	**16.49**	0.30	**71.94**	m2	**77.34**
0.15 - 4.00m2	52.53	5.00	**55.16**	0.55	0.04	**8.24**	0.14	**63.54**	m2	**68.31**
4mm thick to wood rebates with screwed wood beads and glazing strip (included elsewhere)										
not exceeding 0.15m2	34.90	5.00	**36.65**	1.40	0.08	**20.75**	0.30	**57.69**	m2	**62.02**
0.15 - 4.00m2	34.90	5.00	**36.65**	0.70	0.04	**10.38**	0.14	**47.16**	m2	**50.70**
6mm thick to wood rebates with screwed wood beads and glazing strip (included elsewhere)										
not exceeding 0.15m2	52.53	5.00	**55.16**	1.50	0.08	**22.17**	0.30	**77.63**	m2	**83.45**
0.15 - 4.00m2	52.53	5.00	**55.16**	0.75	0.04	**11.09**	0.14	**66.39**	m2	**71.36**
4mm thick to metal rebates with metal casement glazing compound										
not exceeding 0.15m2	34.90	5.00	**36.65**	0.90	0.08	**13.65**	0.34	**50.63**	m2	**54.43**
0.15 - 4.00m2	34.90	5.00	**36.65**	0.45	0.04	**6.82**	0.17	**43.63**	m2	**46.91**

Labour hourly rates: (except Specialists) Craft Operatives £14.21 Labourer £10.71 Rates are national average prices. Refer to REGIONAL VARIATIONS for indicative levels of overall pricing in regions	MATERIALS			LABOUR				RATES		
	Del to Site £	Waste %	Material Cost £	Craft Optve Hrs	Lab Hrs	Labour Cost £	Sunds £	Nett Rate £	Unit	Gross Rate (+7.5%) £
L40: GENERAL GLAZING Cont.										
Glass; BS 952, white patterned Cont.										
6mm thick to metal rebates with metal casement glazing compound										
not exceeding 0.15m2	52.53	5.00	**55.16**	1.00	0.08	**15.07**	0.34	**70.56**	m2	**75.86**
0.15 - 4.00m2	52.53	5.00	**55.16**	0.50	0.04	**7.53**	0.17	**62.85**	m2	**67.57**
4mm thick to metal rebates with clipped metal beads and gaskets (included elsewhere)										
not exceeding 0.15m2	34.90	5.00	**36.65**	1.20	0.08	**17.91**	0.30	**54.85**	m2	**58.96**
0.15 - 4.00m2	34.90	5.00	**36.65**	0.60	0.04	**8.95**	0.14	**45.74**	m2	**49.17**
6mm thick to metal rebates with clipped metal beads and gaskets (included elsewhere)										
not exceeding 0.15m2	52.53	5.00	**55.16**	1.30	0.08	**19.33**	0.30	**74.78**	m2	**80.39**
0.15 - 4.00m2	52.53	5.00	**55.16**	0.65	0.04	**9.66**	0.14	**64.96**	m2	**69.84**
Glass; BS 952, tinted patterned										
4mm thick to wood rebates with BS 544 putty										
not exceeding 0.15m2	51.54	5.00	**54.12**	0.80	0.08	**12.22**	0.30	**66.64**	m2	**71.64**
0.15 - 4.00m2	51.54	5.00	**54.12**	0.40	0.04	**6.11**	0.14	**60.37**	m2	**64.90**
6mm thick to wood rebates with BS 544 putty										
not exceeding 0.15m2	58.32	5.00	**61.24**	0.90	0.08	**13.65**	0.30	**75.18**	m2	**80.82**
0.15 - 4.00m2	58.32	5.00	**61.24**	0.45	0.04	**6.82**	0.14	**68.20**	m2	**73.32**
4mm thick to wood rebates with bradded wood beads (included elsewhere) and BS 544 putty										
not exceeding 0.15m2	51.54	5.00	**54.12**	1.00	0.08	**15.07**	0.30	**69.48**	m2	**74.69**
0.15 - 4.00m2	51.54	5.00	**54.12**	0.50	0.04	**7.53**	0.14	**61.79**	m2	**66.43**
6mm thick to wood rebates with bradded wood beads (included elsewhere) and BS 544 putty										
not exceeding 0.15m2	58.32	5.00	**61.24**	1.10	0.08	**16.49**	0.30	**78.02**	m2	**83.87**
0.15 - 4.00m2	58.32	5.00	**61.24**	0.55	0.04	**8.24**	0.14	**69.62**	m2	**74.84**
4mm thick to wood rebates with screwed wood beads and glazing strip (included elsewhere)										
not exceeding 0.15m2	51.54	5.00	**54.12**	1.40	0.08	**20.75**	0.30	**75.16**	m2	**80.80**
0.15 - 4.00m2	51.54	5.00	**54.12**	0.70	0.04	**10.38**	0.14	**64.64**	m2	**69.48**
6mm thick to wood rebates with screwed wood beads and glazing strip (included elsewhere)										
not exceeding 0.15m2	58.32	5.00	**61.24**	1.50	0.08	**22.17**	0.30	**83.70**	m2	**89.98**
0.15 - 4.00m2	58.32	5.00	**61.24**	0.75	0.04	**11.09**	0.14	**72.46**	m2	**77.90**
4mm thick to metal rebates with metal casement glazing compound										
not exceeding 0.15m2	51.54	5.00	**54.12**	0.90	0.08	**13.65**	0.34	**68.10**	m2	**73.21**
0.15 - 4.00m2	51.54	5.00	**54.12**	0.45	0.04	**6.82**	0.17	**61.10**	m2	**65.69**
6mm thick to metal rebates with metal casement glazing compound										
not exceeding 0.15m2	58.32	5.00	**61.24**	1.00	0.08	**15.07**	0.34	**76.64**	m2	**82.39**
0.15 - 4.00m2	58.32	5.00	**61.24**	0.50	0.04	**7.53**	0.17	**68.93**	m2	**74.10**
4mm thick to metal rebates with clipped metal beads and gaskets (included elsewhere)										
not exceeding 0.15m2	51.54	5.00	**54.12**	1.20	0.08	**17.91**	0.30	**72.32**	m2	**77.75**
0.15 - 4.00m2	51.54	5.00	**54.12**	0.60	0.04	**8.95**	0.14	**63.21**	m2	**67.96**
6mm thick to metal rebates with clipped metal beads and gaskets (included elsewhere)										
not exceeding 0.15m2	58.32	5.00	**61.24**	1.30	0.08	**19.33**	0.30	**80.86**	m2	**86.93**
0.15 - 4.00m2	58.32	5.00	**61.24**	0.65	0.04	**9.66**	0.14	**71.04**	m2	**76.37**
Patterns available										
4mm Bronze tint										
Autumn; Cotswold; Everglade; Sycamore										
4mm white										
Arctic										
6mm white										
Deep Flemish										
4 and 6mm white										
Autumn; Cotswold; Driftwood; Everglade; Flemish; Linkon; Mayflower; Reeded; Stippolyte										
Glass; BS 952, antisun float; grey										
4mm thick to wood rebates with BS 544 putty										
not exceeding 2400 x 1200mm	71.87	5.00	**75.46**	0.60	0.10	**9.60**	0.21	**85.27**	m2	**91.66**
6mm thick to wood rebates with BS 544 putty										
not exceeding 5950 x 3150mm	103.66	5.00	**108.84**	0.65	0.10	**10.31**	0.21	**119.36**	m2	**128.31**
10mm thick to wood rebates with BS 544 putty										
not exceeding 5950 x 3150mm	185.34	5.00	**194.61**	1.50	0.10	**22.39**	0.26	**217.26**	m2	**233.55**
12mm thick to wood rebates with BS 544 putty										
not exceeding 5950 x 3150mm	256.78	5.00	**269.62**	2.00	0.10	**29.49**	0.33	**299.44**	m2	**321.90**

Labour hourly rates: (except Specialists) Craft Operatives £14.21 Labourer £10.71 Rates are national average prices. Refer to REGIONAL VARIATIONS for indicative levels of overall pricing in regions	MATERIALS			LABOUR				RATES		
	Del to Site £	Waste %	Material Cost £	Craft Optve Hrs	Lab Hrs	Labour Cost £	Sunds £	Nett Rate £	Unit	Gross Rate (+7.5%) £
L40: GENERAL GLAZING Cont.										
Glass; BS 952, antisun float; grey Cont.										
4mm thick to wood rebates with bradded wood beads (included elsewhere) and BS 544 putty not exceeding 2400 x 1200mm	71.87	5.00	**75.46**	0.70	0.10	**11.02**	0.21	**86.69**	m2	**93.19**
6mm thick to wood rebates with bradded wood beads (included elsewhere) and BS 544 putty not exceeding 5950 x 3150mm	103.66	5.00	**108.84**	0.75	0.10	**11.73**	0.21	**120.78**	m2	**129.84**
10mm thick to wood rebates with bradded wood beads (included elsewhere) and BS 544 putty not exceeding 5950 x 3150mm	185.34	5.00	**194.61**	1.60	0.10	**23.81**	0.26	**218.68**	m2	**235.08**
12mm thick to wood rebates with bradded wood beads (included elsewhere) and BS 544 putty not exceeding 5950 x 3150mm	256.78	5.00	**269.62**	2.10	0.10	**30.91**	0.33	**300.86**	m2	**323.43**
4mm thick to wood rebates with screwed wood beads and glazing strip (included elsewhere) not exceeding 2400 x 1200mm	71.87	5.00	**75.46**	0.90	0.10	**13.86**	-	**89.32**	m2	**96.02**
6mm thick to wood rebates with screwed wood beads and glazing strip (included elsewhere) not exceeding 5950 x 3150mm	103.66	5.00	**108.84**	0.95	0.10	**14.57**	-	**123.41**	m2	**132.67**
10mm thick to wood rebates with screwed wood beads and glazing strip (included elsewhere) not exceeding 5950 x 3150mm	185.34	5.00	**194.61**	1.70	0.10	**25.23**	-	**219.84**	m2	**236.32**
12mm thick to wood rebates with screwed wood beads and glazing strip (included elsewhere) not exceeding 5950 x 3150mm	256.78	5.00	**269.62**	2.20	0.10	**32.33**	-	**301.95**	m2	**324.60**
Glass; BS 952, antisun float; bronze										
4mm thick to wood rebates with BS 544 putty not exceeding 2400 x 1200mm	71.87	5.00	**75.46**	0.60	0.10	**9.60**	0.21	**85.27**	m2	**91.66**
6mm thick to wood rebates with BS 544 putty not exceeding 5950 x 3150mm	103.66	5.00	**108.84**	0.65	0.10	**10.31**	0.21	**119.36**	m2	**128.31**
10mm thick to wood rebates with BS 544 putty not exceeding 5950 x 3150mm	185.34	5.00	**194.61**	1.50	0.10	**22.39**	0.26	**217.26**	m2	**233.55**
12mm thick to wood rebates with BS 544 putty not exceeding 5950 x 3150mm	255.83	5.00	**268.62**	2.00	0.10	**29.49**	0.33	**298.44**	m2	**320.83**
4mm thick to wood rebates with bradded wood beads (included elsewhere) and BS 544 putty not exceeding 2400 x 1200mm	71.87	5.00	**75.46**	0.70	0.10	**11.02**	0.21	**86.69**	m2	**93.19**
6mm thick to wood rebates with bradded wood beads (included elsewhere) and BS 544 putty not exceeding 5950 x 3150mm	103.66	5.00	**108.84**	0.75	0.10	**11.73**	0.21	**120.78**	m2	**129.84**
10mm thick to wood rebates with bradded wood beads (included elsewhere) and BS 544 putty not exceeding 5950 x 3150mm	185.34	5.00	**194.61**	1.60	0.10	**23.81**	0.26	**218.68**	m2	**235.08**
12mm thick to wood rebates with bradded wood beads (included elsewhere) and BS 544 putty not exceeding 5950 x 3150mm	255.83	5.00	**268.62**	2.10	0.10	**30.91**	0.33	**299.86**	m2	**322.35**
4mm thick to wood rebates with screwed wood beads and glazing strip (included elsewhere) not exceeding 2400 x 1200mm	71.87	5.00	**75.46**	0.90	0.10	**13.86**	-	**89.32**	m2	**96.02**
6mm thick to wood rebates with screwed wood beads and glazing strip (included elsewhere) not exceeding 5950 x 3150mm	103.66	5.00	**108.84**	0.95	0.10	**14.57**	-	**123.41**	m2	**132.67**
10mm thick to wood rebates with screwed wood beads and glazing strip (included elsewhere) not exceeding 5950 x 3150mm	185.34	5.00	**194.61**	1.70	0.10	**25.23**	-	**219.84**	m2	**236.32**
12mm thick to wood rebates with screwed wood beads and glazing strip (included elsewhere) not exceeding 5950 x 3150mm	255.83	5.00	**268.62**	2.20	0.10	**32.33**	-	**300.95**	m2	**323.53**
Glass; BS 952, antisun float; green										
6mm thick to wood rebates with BS 544 putty not exceeding 3150 x 2050mm	132.88	5.00	**139.52**	0.65	0.10	**10.31**	0.21	**150.04**	m2	**161.29**
6mm thick to wood rebates with bradded wood beads (included elsewhere) and BS 544 putty not exceeding 3150 x 2050mm	132.88	5.00	**139.52**	0.75	0.10	**11.73**	0.21	**151.46**	m2	**162.82**
6mm thick to wood rebates with screwed wood beads and glazing strip (included elsewhere) not exceeding 3150 x 2050mm	132.88	5.00	**139.52**	0.95	0.10	**14.57**	-	**154.09**	m2	**165.65**

Labour hourly rates: (except Specialists) Craft Operatives £14.21 Labourer £10.71 Rates are national average prices. Refer to REGIONAL VARIATIONS for indicative levels of overall pricing in regions	MATERIALS			LABOUR				RATES		
	Del to Site £	Waste %	Material Cost £	Craft Optve Hrs	Lab Hrs	Labour Cost £	Sunds £	Nett Rate £	Unit	Gross Rate (+7.5%) £
L40: GENERAL GLAZING Cont.										
Glass; BS 952, clear float										
10mm thick to wood rebates with screwed wood beads and glazing strip (included elsewhere) not exceeding 5950 x 3150mm	111.43	5.00	**117.00**	2.25	0.10	**33.04**	-	**150.05**	m2	**161.30**
12mm thick to wood rebates with screwed wood beads and glazing strip (included elsewhere) not exceeding 5950 x 3150mm	128.47	5.00	**134.89**	2.75	0.10	**40.15**	-	**175.04**	m2	**188.17**
15mm thick to wood rebates with screwed wood beads and glazing strip (included elsewhere) not exceeding 2950 x 2000mm	213.74	5.00	**224.43**	3.50	0.10	**50.81**	-	**275.23**	m2	**295.88**
19mm thick to wood rebates with screwed wood beads and glazing strip (included elsewhere) not exceeding 2950 x 2000mm	302.76	5.00	**317.90**	4.00	0.10	**57.91**	-	**375.81**	m2	**403.99**
25mm thick to wood rebates with screwed wood beads and glazing strip (included elsewhere) not exceeding 2950 x 2000mm	480.61	5.00	**504.64**	4.50	0.10	**65.02**	-	**569.66**	m2	**612.38**
Glass; BS 952, toughened clear float										
4mm thick to metal rebates with screwed metal beads and gaskets (included elsewhere) not exceeding 2400 x 1300mm	52.97	5.00	**55.62**	0.60	0.10	**9.60**	-	**65.22**	m2	**70.11**
5mm thick to metal rebates with screwed metal beads and gaskets (included elsewhere) not exceeding 2500 x 1520mm	79.49	5.00	**83.46**	0.60	0.10	**9.60**	-	**93.06**	m2	**100.04**
6mm thick to metal rebates with screwed metal beads and gaskets (included elsewhere) not exceeding 2500 x 1520mm	79.49	5.00	**83.46**	0.65	0.10	**10.31**	-	**93.77**	m2	**100.80**
10mm thick to metal rebates with screwed metal beads and gaskets (included elsewhere) not exceeding 2500 x 1520mm	126.08	5.00	**132.38**	1.50	0.10	**22.39**	-	**154.77**	m2	**166.38**
12mm thick to metal rebates with screwed metal beads and gaskets (included elsewhere) not exceeding 2500 x 1520mm	230.87	5.00	**242.41**	2.00	0.10	**29.49**	-	**271.90**	m2	**292.30**
Glass; BS 952, toughened white patterned										
4mm thick to metal rebates with screwed metal beads and gaskets (included elsewhere) not exceeding 2100 x 1300mm	106.78	5.00	**112.12**	0.60	0.10	**9.60**	-	**121.72**	m2	**130.84**
6mm thick to metal rebates with screwed metal beads and gaskets (included elsewhere) not exceeding 2100 x 1300mm	111.48	5.00	**117.05**	0.65	0.10	**10.31**	-	**127.36**	m2	**136.91**
Glass; BS 952, toughened tinted patterned										
4mm thick to metal rebates with screwed metal beads and gaskets (included elsewhere) not exceeding 2100 x 1300mm	106.79	5.00	**112.13**	0.60	0.10	**9.60**	-	**121.73**	m2	**130.86**
6mm thick to metal rebates with screwed metal beads and gaskets (included elsewhere) not exceeding 2100 x 1300mm	111.48	5.00	**117.05**	0.65	0.10	**10.31**	-	**127.36**	m2	**136.91**
Patterns available										
4mm tinted Everglade										
4mm white Reeded										
4mm tinted Autumn; Cotswold										
4 and 6mm white Autumn; Cotswold; Driftwood; Everglade; Flemish; Linkon; Mayflower; Stippolyte; Sycamore										
Glass; BS 952, toughened antisun float; grey										
4mm thick to metal rebates with screwed metal beads and gaskets (included elsewhere) not exceeding 2100 x 1250mm	97.21	5.00	**102.07**	0.60	0.10	**9.60**	-	**111.67**	m2	**120.04**
6mm thick to metal rebates with screwed metal beads and gaskets (included elsewhere) not exceeding 2500 x 1520mm	101.34	5.00	**106.41**	0.65	0.10	**10.31**	-	**116.71**	m2	**125.47**
10mm thick to metal rebates with screwed metal beads and gaskets (included elsewhere) not exceeding 2500 x 1520mm	200.58	5.00	**210.61**	1.50	0.10	**22.39**	-	**233.00**	m2	**250.47**
12mm thick to metal rebates with screwed metal beads and gaskets (included elsewhere) not exceeding 2500 x 1520mm	263.93	5.00	**277.13**	2.00	0.10	**29.49**	-	**306.62**	m2	**329.61**

Labour hourly rates: (except Specialists) Craft Operatives £14.21 Labourer £10.71 Rates are national average prices. Refer to REGIONAL VARIATIONS for indicative levels of overall pricing in regions	MATERIALS			LABOUR				RATES		
	Del to Site £	Waste %	Material Cost £	Craft Optve Hrs	Lab Hrs	Labour Cost £	Sunds £	Nett Rate £	Unit	Gross Rate (+7.5%) £
L40: GENERAL GLAZING Cont.										
Glass; BS 952, toughened antisun float; bronze										
4mm thick to metal rebates with screwed metal beads and gaskets (included elsewhere) not exceeding 2100 x 1250mm	97.13	5.00	**101.99**	0.60	0.10	**9.60**	-	**111.58**	m2	**119.95**
6mm thick to metal rebates with screwed metal beads and gaskets (included elsewhere) not exceeding 2500 x 1520mm	101.34	5.00	**106.41**	0.65	0.10	**10.31**	-	**116.71**	m2	**125.47**
10mm thick to metal rebates with screwed metal beads and gaskets (included elsewhere) not exceeding 2500 x 1520mm	201.92	5.00	**212.02**	1.50	0.10	**22.39**	-	**234.40**	m2	**251.98**
12mm thick to metal rebates with screwed metal beads and gaskets (included elsewhere) not exceeding 2500 x 1520mm	263.93	5.00	**277.13**	2.00	0.10	**29.49**	-	**306.62**	m2	**329.61**
Glass; BS 952, laminated safety, clear float										
4.4mm thick to metal rebates with screwed metal beads and gaskets (included elsewhere) not exceeding 2100 x 1200mm	70.02	5.00	**73.52**	1.60	0.10	**23.81**	-	**97.33**	m2	**104.63**
6.4mm thick to metal rebates with screwed metal beads and gaskets (included elsewhere) not exceeding 3210 x 2000mm	58.71	5.00	**61.65**	2.00	0.10	**29.49**	-	**91.14**	m2	**97.97**
Glass; BS 952, laminated anti-bandit, clear float										
7.5mm thick to metal rebates with screwed metal beads and gaskets (included elsewhere) not exceeding 3210 x 2000mm	122.51	5.00	**128.64**	2.50	0.10	**36.60**	-	**165.23**	m2	**177.62**
9.5mm thick to metal rebates with screwed metal beads and gaskets (included elsewhere) not exceeding 3600 x 2500mm	143.77	5.00	**150.96**	2.75	0.10	**40.15**	-	**191.11**	m2	**205.44**
11.5mm thick to metal rebates with screwed metal beads and gaskets (included elsewhere) not exceeding 3180 x 2000mm	164.14	5.00	**172.35**	3.25	0.10	**47.25**	-	**219.60**	m2	**236.07**
not exceeding 4500 x 2500mm	197.04	5.00	**206.89**	3.25	0.10	**47.25**	-	**254.15**	m2	**273.21**
Materials resembling glass; UVA stabilised polycarbonate sheet latex paper masked both sides										
3mm thick to metal rebates with screwed metal beads and gaskets (included elsewhere) standard grade	38.07	5.00	**39.97**	1.00	0.10	**15.28**	-	**55.25**	m2	**59.40**
4mm thick to metal rebates with screwed metal beads and gaskets (included elsewhere) standard grade	50.77	5.00	**53.31**	1.00	0.10	**15.28**	-	**68.59**	m2	**73.73**
5mm thick to metal rebates with screwed metal beads and gaskets (included elsewhere) standard grade	63.46	5.00	**66.63**	1.00	0.10	**15.28**	-	**81.91**	m2	**88.06**
6mm thick to metal rebates with screwed metal beads and gaskets (included elsewhere) standard grade	76.19	5.00	**80.00**	1.15	0.10	**17.41**	-	**97.41**	m2	**104.72**
8mm thick to metal rebates with screwed metal beads and gaskets (included elsewhere) standard grade	101.54	5.00	**106.62**	1.35	0.10	**20.25**	-	**126.87**	m2	**136.39**
9.5mm thick to metal rebates with screwed metal beads and gaskets (included elsewhere) standard grade	120.58	5.00	**126.61**	1.50	0.10	**22.39**	-	**149.00**	m2	**160.17**
3mm thick to metal rebates with screwed metal beads and gaskets (included elsewhere) abrasion resistant hard coated grade	98.13	5.00	**103.04**	1.00	0.10	**15.28**	-	**118.32**	m2	**127.19**
4mm thick to metal rebates with screwed metal beads and gaskets (included elsewhere) abrasion resistant hard coated grade	115.01	5.00	**120.76**	1.00	0.10	**15.28**	-	**136.04**	m2	**146.24**
5mm thick to metal rebates with screwed metal beads and gaskets (included elsewhere) abrasion resistant hard coated grade	132.59	5.00	**139.22**	1.00	0.10	**15.28**	-	**154.50**	m2	**166.09**
6mm thick to metal rebates with screwed metal beads and gaskets (included elsewhere) abrasion resistant hard coated grade	155.36	5.00	**163.13**	1.15	0.10	**17.41**	-	**180.54**	m2	**194.08**
8mm thick to metal rebates with screwed metal beads and gaskets (included elsewhere) abrasion resistant hard coated grade	200.41	5.00	**210.43**	1.35	0.10	**20.25**	-	**230.69**	m2	**247.99**
9.5mm thick to metal rebates with screwed metal beads and gaskets (included elsewhere) abrasion resistant hard coated grade	234.54	5.00	**246.27**	1.50	0.10	**22.39**	-	**268.65**	m2	**288.80**

Labour hourly rates: (except Specialists) Craft Operatives £14.21 Labourer £10.71 Rates are national average prices. Refer to REGIONAL VARIATIONS for indicative levels of overall pricing in regions	MATERIALS			LABOUR				RATES		
	Del to Site £	Waste %	Material Cost £	Craft Optve Hrs	Lab Hrs	Labour Cost £	Sunds £	Nett Rate £	Unit	Gross Rate (+7.5%) £
L40: GENERAL GLAZING Cont.										
Materials resembling glass; Meshlite vandal resistant glazing; Georgian, clear, smooth both faces or crinkle one face										
3mm thick to wood rebates with screwed wood beads (included elsewhere) and BS 544 putty										
general purpose grade	28.52	5.00	**29.95**	1.00	0.10	**15.28**	0.15	**45.38**	m2	**48.78**
fire retardant grade, class 2	34.40	5.00	**36.12**	1.00	0.10	**15.28**	0.15	**51.56**	m2	**55.42**
fire retardant grade, class 0	40.40	5.00	**42.42**	1.00	0.10	**15.28**	0.15	**57.86**	m2	**62.19**
4mm thick to wood rebates with screwed wood beads (included elsewhere) and BS 544 putty										
general purpose grade	36.85	5.00	**38.69**	1.00	0.10	**15.28**	0.15	**54.13**	m2	**58.19**
fire retardant grade, class 2	44.49	5.00	**46.71**	1.00	0.10	**15.28**	0.15	**62.15**	m2	**66.81**
fire retardant grade, class 0	52.38	5.00	**55.00**	1.00	0.10	**15.28**	0.15	**70.43**	m2	**75.72**
6mm thick to wood rebates with screwed wood beads (included elsewhere) and BS 544 putty										
general purpose grade	49.26	5.00	**51.72**	1.15	0.10	**17.41**	0.15	**69.29**	m2	**74.49**
fire retardant grade, class 2	62.42	5.00	**65.54**	1.15	0.10	**17.41**	0.15	**83.11**	m2	**89.34**
fire retardant grade, class 0	75.80	5.00	**79.59**	1.15	0.10	**17.41**	0.15	**97.16**	m2	**104.44**
3mm thick to metal rebates with screwed metal beads and gaskets (included elsewhere)										
general purpose grade	28.52	5.00	**29.95**	1.00	0.10	**15.28**	-	**45.23**	m2	**48.62**
fire retardant grade, class 2	34.40	5.00	**36.12**	1.00	0.10	**15.28**	-	**51.40**	m2	**55.26**
fire retardant grade, class 0	40.40	5.00	**42.42**	1.00	0.10	**15.28**	-	**57.70**	m2	**62.03**
4mm thick to metal rebates with screwed metal beads and gaskets (included elsewhere)										
general purpose grade	36.85	5.00	**38.69**	1.00	0.10	**15.28**	-	**53.97**	m2	**58.02**
fire retardant grade, class 2	44.49	5.00	**46.71**	1.00	0.10	**15.28**	-	**62.00**	m2	**66.65**
fire retardant grade, class 0	52.38	5.00	**55.00**	1.00	0.10	**15.28**	-	**70.28**	m2	**75.55**
6mm thick to metal rebates with screwed metal beads and gaskets (included elsewhere)										
general purpose grade	49.26	5.00	**51.72**	1.15	0.10	**17.41**	-	**69.14**	m2	**74.32**
fire retardant grade, class 2	62.42	5.00	**65.54**	1.15	0.10	**17.41**	-	**82.95**	m2	**89.18**
fire retardant grade, class 0	75.80	5.00	**79.59**	1.15	0.10	**17.41**	-	**97.00**	m2	**104.28**
Materials resembling glass; Meshlite Vandal resistant glazing; plain, opaque colours or clear, smooth both faces or crinkle one face										
2mm thick to wood rebates with screwed wood beads (included elsewhere) and BS 544 putty										
general purpose grade	21.60	5.00	**22.68**	1.00	0.10	**15.28**	0.15	**38.12**	m2	**40.97**
3mm thick to wood rebates with screwed wood beads (included elsewhere) and BS 544 putty										
general purpose grade	25.85	5.00	**27.14**	1.00	0.10	**15.28**	0.15	**42.58**	m2	**45.77**
fire retardant grade, class 2	32.51	5.00	**34.14**	1.00	0.10	**15.28**	0.15	**49.57**	m2	**53.29**
fire retardant grade, class 0	39.16	5.00	**41.12**	1.00	0.10	**15.28**	0.15	**56.55**	m2	**60.79**
4mm thick to wood rebates with screwed wood beads (included elsewhere) and BS 544 putty										
general purpose grade	34.40	5.00	**36.12**	1.00	0.10	**15.28**	0.15	**51.56**	m2	**55.42**
fire retardant grade, class 2	42.25	5.00	**44.36**	1.00	0.10	**15.28**	0.15	**59.80**	m2	**64.28**
fire retardant grade, class 0	51.10	5.00	**53.66**	1.00	0.10	**15.28**	0.15	**69.09**	m2	**74.27**
6mm thick to wood rebates with screwed wood beads (included elsewhere) and BS 544 putty										
general purpose grade	45.42	5.00	**47.69**	1.15	0.10	**17.41**	0.15	**65.26**	m2	**70.15**
fire retardant grade, class 2	59.89	5.00	**62.88**	1.15	0.10	**17.41**	0.15	**80.45**	m2	**86.48**
fire retardant grade, class 0	74.36	5.00	**78.08**	1.15	0.10	**17.41**	0.15	**95.64**	m2	**102.82**
2mm thick to metal rebates with screwed metal beads and gaskets (included elsewhere)										
general purpose grade; plain	21.60	5.00	**22.68**	1.00	0.10	**15.28**	-	**37.96**	m2	**40.81**
3mm thick to metal rebates with screwed metal beads and gaskets (included elsewhere)										
general purpose grade	25.85	5.00	**27.14**	1.00	0.10	**15.28**	-	**42.42**	m2	**45.61**
fire retardant grade, class 2	32.51	5.00	**34.14**	1.00	0.10	**15.28**	-	**49.42**	m2	**53.12**
fire retardant grade, class 0	39.16	5.00	**41.12**	1.00	0.10	**15.28**	-	**56.40**	m2	**60.63**
4mm thick to metal rebates with screwed metal beads and gaskets (included elsewhere)										
general purpose grade	34.40	5.00	**36.12**	1.00	0.10	**15.28**	-	**51.40**	m2	**55.26**
fire retardant grade, class 2	42.25	5.00	**44.36**	1.00	0.10	**15.28**	-	**59.64**	m2	**64.12**
fire retardant grade, class 0	51.10	5.00	**53.66**	1.00	0.10	**15.28**	-	**68.94**	m2	**74.11**
6mm thick to metal rebates with screwed metal beads and gaskets (included elsewhere)										
general purpose grade	45.42	5.00	**47.69**	1.15	0.10	**17.41**	-	**65.10**	m2	**69.99**
fire retardant grade, class 2	59.89	5.00	**62.88**	1.15	0.10	**17.41**	-	**80.30**	m2	**86.32**
fire retardant grade, class 0	74.36	5.00	**78.08**	1.15	0.10	**17.41**	-	**95.49**	m2	**102.65**

Labour hourly rates: (except Specialists) Craft Operatives £14.21 Labourer £10.71 Rates are national average prices. Refer to REGIONAL VARIATIONS for indicative levels of overall pricing in regions	MATERIALS			LABOUR				RATES		
	Del to Site £	Waste %	Material Cost £	Craft Optve Hrs	Lab Hrs	Labour Cost £	Sunds £	Nett Rate £	Unit	Gross Rate (+7.5%) £
L40: GENERAL GLAZING Cont.										
Materials resembling glass; Meshlite Vandal resistant glazing; diamond, clear, smooth both faces or crinkle one face										
3mm thick to wood rebates with screwed wood beads (included elsewhere) and BS 544 putty										
general purpose grade	29.06	5.00	**30.51**	1.00	0.10	**15.28**	0.15	**45.95**	m2	**49.39**
fire retardant grade, class 2	34.37	5.00	**36.09**	1.00	0.10	**15.28**	0.15	**51.52**	m2	**55.39**
fire retardant grade, class 0	39.80	5.00	**41.79**	1.00	0.10	**15.28**	0.15	**57.23**	m2	**61.52**
4mm thick to wood rebates with screwed wood beads (included elsewhere) and BS 544 putty										
general purpose grade	38.04	5.00	**39.94**	1.00	0.10	**15.28**	0.15	**55.38**	m2	**59.53**
fire retardant grade, class 2	45.01	5.00	**47.26**	1.00	0.10	**15.28**	0.15	**62.70**	m2	**67.40**
fire retardant grade, class 0	51.98	5.00	**54.58**	1.00	0.10	**15.28**	0.15	**70.01**	m2	**75.27**
6mm thick to wood rebates with screwed wood beads (included elsewhere) and BS 544 putty										
general purpose grade	50.02	5.00	**52.52**	1.15	0.10	**17.41**	0.15	**70.09**	m2	**75.34**
fire retardant grade, class 2	62.79	5.00	**65.93**	1.15	0.10	**17.41**	0.15	**83.50**	m2	**89.76**
fire retardant grade, class 0	75.28	5.00	**79.04**	1.15	0.10	**17.41**	0.15	**96.61**	m2	**103.86**
3mm thick to metal rebates with screwed metal beads and gaskets (included elsewhere)										
general purpose grade	29.06	5.00	**30.51**	1.00	0.10	**15.28**	-	**45.79**	m2	**49.23**
fire retardant grade, class 2	34.37	5.00	**36.09**	1.00	0.10	**15.28**	-	**51.37**	m2	**55.22**
fire retardant grade, class 0	39.80	5.00	**41.79**	1.00	0.10	**15.28**	-	**57.07**	m2	**61.35**
4mm thick to metal rebates with screwed metal beads and gaskets (included elsewhere)										
general purpose grade	38.04	5.00	**39.94**	1.00	0.10	**15.28**	-	**55.22**	m2	**59.36**
fire retardant grade, class 2	45.01	5.00	**47.26**	1.00	0.10	**15.28**	-	**62.54**	m2	**67.23**
fire retardant grade, class 0	51.98	5.00	**54.58**	1.00	0.10	**15.28**	-	**69.86**	m2	**75.10**
6mm thick to metal rebates with screwed metal beads and gaskets (included elsewhere)										
general purpose grade	50.02	5.00	**52.52**	1.15	0.10	**17.41**	-	**69.93**	m2	**75.18**
fire retardant grade, class 2	62.79	5.00	**65.93**	1.15	0.10	**17.41**	-	**83.34**	m2	**89.59**
fire retardant grade, class 0	75.28	5.00	**79.04**	1.15	0.10	**17.41**	-	**96.46**	m2	**103.69**
Factory made double glazed hermetically sealed units										
Two panes BS 952, clear float 3 or 4mm thick; to metal rebates with screwed metal beads and gaskets (included elsewhere)										
521mm wide x 421mm high	17.75	5.00	**18.64**	1.00	-	**14.21**	-	**32.85**	nr	**35.31**
521mm wide x 621mm high	26.18	5.00	**27.49**	1.25	-	**17.76**	-	**45.25**	nr	**48.65**
740mm wide x 740mm high	44.30	5.00	**46.52**	1.75	-	**24.87**	-	**71.38**	nr	**76.74**
848mm wide x 848mm high	58.18	5.00	**61.09**	2.00	-	**28.42**	-	**89.51**	nr	**96.22**
1048mm wide x 1048mm high	88.87	5.00	**93.31**	2.50	-	**35.53**	-	**128.84**	nr	**138.50**
1148mm wide x 1248mm high	115.92	5.00	**121.72**	3.00	-	**42.63**	-	**164.35**	nr	**176.67**
Two panes BS 952, clear float 5 or 6mm thick; to metal rebates with screwed metal beads and gaskets (included elsewhere)										
521mm wide x 421mm high	26.14	5.00	**27.45**	1.10	-	**15.63**	-	**43.08**	nr	**46.31**
521mm wide x 621mm high	38.56	5.00	**40.49**	1.40	-	**19.89**	-	**60.38**	nr	**64.91**
740mm wide x 740mm high	65.27	5.00	**68.53**	1.90	-	**27.00**	-	**95.53**	nr	**102.70**
848mm wide x 848mm high	85.71	5.00	**90.00**	2.20	-	**31.26**	-	**121.26**	nr	**130.35**
1048mm wide x 1048mm high	105.92	5.00	**111.22**	2.75	-	**39.08**	-	**150.29**	nr	**161.57**
1148mm wide x 1248mm high	157.08	5.00	**164.93**	3.30	-	**46.89**	-	**211.83**	nr	**227.71**
Inner pane BS 952, clear float 4mm thick; outer pane BS 952, white patterned 4mm thick; to metal rebates with screwed metal beads and gaskets (included elsewhere)										
521mm wide x 421mm high	21.80	5.00	**22.89**	1.00	-	**14.21**	-	**37.10**	nr	**39.88**
521mm wide x 621mm high	32.00	5.00	**33.60**	1.25	-	**17.76**	-	**51.36**	nr	**55.21**
740mm wide x 740mm high	54.16	5.00	**56.87**	1.75	-	**24.87**	-	**81.74**	nr	**87.87**
848mm wide x 848mm high	71.11	5.00	**74.67**	2.00	-	**28.42**	-	**103.09**	nr	**110.82**
1048mm wide x 1048mm high	87.89	5.00	**92.28**	2.50	-	**35.53**	-	**127.81**	nr	**137.40**
1148mm wide x 1248mm high	141.69	5.00	**148.77**	3.00	-	**42.63**	-	**191.40**	nr	**205.76**
Inner pane BS 952, clear float 4mm thick; outer pane BS 952, white patterned 6mm thick; to metal rebates with screwed metal beads and gaskets (included elsewhere)										
521mm wide x 421mm high	25.98	5.00	**27.28**	1.10	-	**15.63**	-	**42.91**	nr	**46.13**
521mm wide x 621mm high	38.31	5.00	**40.23**	1.40	-	**19.89**	-	**60.12**	nr	**64.63**
740mm wide x 740mm high	64.84	5.00	**68.08**	1.90	-	**27.00**	-	**95.08**	nr	**102.21**
848mm wide x 848mm high	85.15	5.00	**89.41**	2.20	-	**31.26**	-	**120.67**	nr	**129.72**
1048mm wide x 1048mm high	105.23	5.00	**110.49**	2.75	-	**39.08**	-	**149.57**	nr	**160.79**
1148mm wide x 1248mm high	169.65	5.00	**178.13**	3.30	-	**46.89**	-	**225.03**	nr	**241.90**
Polyester window films; Durable Berkeley Company Ltd.,										
3M Scotchshield Safety films; applying to glass										
type SH4CLL, optically clear	-	-	**-**	-	-	**-**	-	**26.90**	m2	**28.92**
type SH4S1L, combination solar/safety	-	-	**-**	-	-	**-**	-	**32.28**	m2	**34.70**
3M Scotchtint Solar Control films; applying to glass										
type P18, silver	-	-	**-**	-	-	**-**	-	**29.59**	m2	**31.81**
type RE15S1X, external	-	-	**-**	-	-	**-**	-	**37.66**	m2	**40.48**
type RE35NEARL, neutral	-	-	**-**	-	-	**-**	-	**29.59**	m2	**31.81**

Labour hourly rates: (except Specialists) Craft Operatives £14.21 Labourer £10.71 Rates are national average prices. Refer to REGIONAL VARIATIONS for indicative levels of overall pricing in regions	MATERIALS			LABOUR				RATES		
	Del to Site £	Waste %	Material Cost £	Craft Optve Hrs	Lab Hrs	Labour Cost £	Sunds £	Nett Rate £	Unit	Gross Rate (+7.5%) £
L40: GENERAL GLAZING Cont.										
Polyester window films; Durable Berkeley Company Ltd., Cont.										
3M Scotchtint Plus All Seasons insulating films; applying to glass										
type LE20 S1AR (silver)	-	-	-	-	-	-	-	37.66	m2	40.48
type LE35 AMARL (bronze)	-	-	-	-	-	-	-	37.66	m2	40.48
type LE50 AMARL (bronze)	-	-	-	-	-	-	-	37.66	m2	40.48
Glass doors; glass, BS 952, toughened clear, polished plate; fittings finish BMA, satin chrome or polished chrome; prices include the provision of floor springs but exclude handles										
12mm thick										
750mm wide x 2150mm high	1484.00	-	1484.00	8.00	8.00	199.36	-	1683.36	nr	1809.61
762mm wide x 2134mm high	1401.00	-	1401.00	8.00	8.00	199.36	-	1600.36	nr	1720.39
800mm wide x 2150mm high	1534.00	-	1534.00	8.00	8.00	199.36	-	1733.36	nr	1863.36
838mm wide x 2134mm high	1473.00	-	1473.00	8.00	8.00	199.36	-	1672.36	nr	1797.79
850mm wide x 2150mm high	1579.00	-	1579.00	8.00	8.00	199.36	-	1778.36	nr	1911.74
900mm wide x 2150mm high	1632.00	-	1632.00	8.00	8.00	199.36	-	1831.36	nr	1968.71
914mm wide x 2134mm high	1533.00	-	1533.00	8.00	8.00	199.36	-	1732.36	nr	1862.29
950mm wide x 2150mm high	1733.00	-	1733.00	9.00	9.00	224.28	-	1957.28	nr	2104.08
1000mm wide x 2150mm high	1839.00	-	1839.00	9.00	9.00	224.28	-	2063.28	nr	2218.03
1100mm wide x 2150mm high	2052.00	-	2052.00	10.00	10.00	249.20	-	2301.20	nr	2473.79
1200mm wide x 2150mm high	2148.00	-	2148.00	10.00	10.00	249.20	-	2397.20	nr	2576.99
Drilling										
Hole through sheet or float glass; not exceeding 6mm thick										
6 - 15mm diameter	3.71	-	3.71	-	-	-	-	3.71	nr	3.98
16 - 38mm diameter	5.23	-	5.23	-	-	-	-	5.23	nr	5.63
exceeding 38mm diameter	10.56	-	10.56	-	-	-	-	10.56	nr	11.35
Hole through sheet or float glass; not exceeding 10mm thick										
6 - 15mm diameter	4.79	-	4.79	-	-	-	-	4.79	nr	5.15
16 - 38mm diameter	7.05	-	7.05	-	-	-	-	7.05	nr	7.57
exceeding 38mm diameter	12.81	-	12.81	-	-	-	-	12.81	nr	13.77
Hole through sheet or float glass; not exceeding 12mm thick										
6 - 15mm diameter	5.94	-	5.94	-	-	-	-	5.94	nr	6.38
16 - 38mm diameter	8.40	-	8.40	-	-	-	-	8.40	nr	9.03
exceeding 38mm diameter	15.20	-	15.20	-	-	-	-	15.20	nr	16.34
Hole through sheet or float glass; not exceeding 19mm thick										
6 - 15mm diameter	7.42	-	7.42	-	-	-	-	7.42	nr	7.98
16 - 38mm diameter	10.56	-	10.56	-	-	-	-	10.56	nr	11.35
exceeding 38mm diameter	18.71	-	18.71	-	-	-	-	18.71	nr	20.11
Hole through sheet or float glass; not exceeding 25mm thick										
6 - 15mm diameter	9.34	-	9.34	-	-	-	-	9.34	nr	10.04
16 - 38mm diameter	13.22	-	13.22	-	-	-	-	13.22	nr	14.22
exceeding 38mm diameter	23.31	-	23.31	-	-	-	-	23.31	nr	25.06
For wired and laminated glass add 50%										
For countersunk holes add 33 1/3%										
Bedding edges of panes										
Wash leather strips										
to edges of 3mm thick glass or the like	0.54	5.00	0.57	0.05	-	0.71	-	1.28	m	1.37
to edges of 6mm thick glass or the like	0.67	5.00	0.70	0.06	-	0.85	-	1.56	m	1.67
Rubber glazing strips										
to edges of 3mm thick glass or the like	0.50	5.00	0.53	0.05	-	0.71	-	1.24	m	1.33
to edges of 6mm thick glass or the like	0.60	5.00	0.63	0.06	-	0.85	-	1.48	m	1.59
Hacking out existing glass; preparing for re-glazing										
Float glass										
wood rebates	-	-	-	0.38	0.38	9.47	0.18	9.65	m	10.37
wood rebates and screwed wood beads; storing beads for re-use	-	-	-	0.45	0.45	11.21	0.18	11.39	m	12.24
metal rebates	-	-	-	0.40	0.40	9.97	0.18	10.14	m	10.90
metal rebates and screwed metal beads; storing beads for re-use	-	-	-	0.48	0.48	11.96	0.18	12.14	m	13.05
Float glass behind guard bars in position										
wood rebates and screwed wood beads; storing beads for re-use	-	-	-	0.75	0.75	18.69	0.18	18.87	m	20.28
metal rebates and screwed metal beads; storing beads for re-use	-	-	-	0.80	0.80	19.94	0.18	20.11	m	21.62

SURFACE FINISHES

Labour hourly rates: (except Specialists) Craft Operatives £14.21 Labourer £10.71 Rates are national average prices. Refer to REGIONAL VARIATIONS for indicative levels of overall pricing in regions	MATERIALS			LABOUR				RATES		
	Del to Site £	Waste %	Material Cost £	Craft Optve Hrs	Lab Hrs	Labour Cost £	Sunds £	Nett Rate £	Unit	Gross Rate (+7.5%) £
M10: CEMENT: SAND/CONCRETE SCREEDS/TOPPINGS										
Mortar, cement and sand (1:3) - screeds										
13mm work to walls on brickwork or blockwork base; one coat; screeded										
width exceeding 300mm	1.46	5.00	**1.53**	0.40	0.20	**7.83**	-	**9.35**	m2	**10.06**
width not exceeding 300mm	0.45	10.00	**0.49**	0.24	0.06	**4.05**	-	**4.55**	m	**4.89**
13mm work to walls on brickwork or blockwork base; one coat; troweled										
width exceeding 300mm	1.46	5.00	**1.53**	0.52	0.20	**9.53**	-	**11.06**	m2	**11.89**
width not exceeding 300mm	0.45	10.00	**0.49**	0.31	0.06	**5.05**	-	**5.54**	m	**5.96**
13mm work to walls on brickwork or blockwork base; one coat; floated										
width exceeding 300mm	1.46	5.00	**1.53**	0.50	0.20	**9.25**	-	**10.78**	m2	**11.58**
width not exceeding 300mm	0.45	10.00	**0.49**	0.30	0.06	**4.91**	-	**5.40**	m	**5.80**
19mm work to floors on concrete base; one coat; screeded										
level and to falls only not exceeding 15 degrees from horizontal	2.28	5.00	**2.39**	0.24	0.15	**5.02**	-	**7.40**	m2	**7.96**
to falls and crossfalls and to slopes not exceeding 15 degrees from horizontal	2.28	5.00	**2.39**	0.34	0.15	**6.44**	-	**8.82**	m2	**9.49**
25mm work to floors on concrete base; one coat; screeded										
level and to falls only not exceeding 15 degrees from horizontal	2.95	5.00	**3.09**	0.26	0.16	**5.41**	-	**8.50**	m2	**9.14**
to falls and crossfalls and to slopes not exceeding 15 degrees from horizontal	2.95	5.00	**3.09**	0.36	0.16	**6.83**	-	**9.92**	m2	**10.67**
32mm work to floors on concrete base; one coat; screeded										
level and to falls only not exceeding 15 degrees from horizontal	3.74	5.00	**3.92**	0.29	0.18	**6.05**	-	**9.96**	m2	**10.71**
to falls and crossfalls and to slopes not exceeding 15 degrees from horizontal	3.74	5.00	**3.92**	0.39	0.18	**7.47**	-	**11.39**	m2	**12.24**
38mm work to floors on concrete base; one coat; screeded										
level and to falls only not exceeding 15 degrees from horizontal	4.41	5.00	**4.62**	0.31	0.19	**6.44**	-	**11.06**	m2	**11.89**
to falls and crossfalls and to slopes not exceeding 15 degrees from horizontal	4.41	5.00	**4.62**	0.41	0.19	**7.86**	-	**12.48**	m2	**13.42**
50mm work to floors on concrete base; one coat; screeded										
level and to falls only not exceeding 15 degrees from horizontal	5.75	5.00	**6.03**	0.36	0.21	**7.36**	-	**13.40**	m2	**14.40**
to falls and crossfalls and to slopes not exceeding 15 degrees from horizontal	5.75	5.00	**6.03**	0.46	0.21	**8.79**	-	**14.82**	m2	**15.93**
19mm work to floors on concrete base; one coat; troweled										
level and to falls only not exceeding 15 degrees from horizontal	2.28	5.00	**2.39**	0.36	0.15	**6.72**	-	**9.11**	m2	**9.79**
to falls and crossfalls and to slopes not exceeding 15 degrees from horizontal	2.28	5.00	**2.39**	0.46	0.15	**8.14**	-	**10.53**	m2	**11.32**
25mm work to floors on concrete base; one coat; troweled										
level and to falls only not exceeding 15 degrees from horizontal	2.95	5.00	**3.09**	0.38	0.16	**7.11**	-	**10.21**	m2	**10.97**
to falls and crossfalls and to slopes not exceeding 15 degrees from horizontal	2.95	5.00	**3.09**	0.48	0.16	**8.53**	-	**11.63**	m2	**12.50**

Labour hourly rates: (except Specialists) Craft Operatives £14.21 Labourer £10.71 Rates are national average prices. Refer to REGIONAL VARIATIONS for indicative levels of overall pricing in regions	MATERIALS			LABOUR				RATES		
	Del to Site £	Waste %	Material Cost £	Craft Optve Hrs	Lab Hrs	Labour Cost £	Sunds £	Nett Rate £	Unit	Gross Rate (+7.5%) £
M10: CEMENT: SAND/CONCRETE SCREEDS/TOPPINGS Cont.										
Mortar, cement and sand (1:3) - screeds Cont.										
32mm work to floors on concrete base; one coat; troweled										
level and to falls only not exceeding 15 degrees from horizontal	3.74	5.00	**3.92**	0.41	0.18	**7.75**	-	**11.67**	m2	**12.54**
to falls and crossfalls and to slopes not exceeding 15 degrees from horizontal	3.74	5.00	**3.92**	0.51	0.18	**9.17**	-	**13.09**	m2	**14.07**
38mm work to floors on concrete base; one coat; troweled										
level and to falls only not exceeding 15 degrees from horizontal	4.41	5.00	**4.62**	0.43	0.19	**8.15**	-	**12.77**	m2	**13.72**
to falls and crossfalls and to slopes not exceeding 15 degrees from horizontal	4.41	5.00	**4.62**	0.53	0.19	**9.57**	-	**14.19**	m2	**15.25**
50mm work to floors on concrete base; one coat; troweled										
level and to falls only not exceeding 15 degrees from horizontal	5.75	5.00	**6.03**	0.48	0.21	**9.07**	-	**15.10**	m2	**16.24**
to falls and crossfalls and to slopes not exceeding 15 degrees from horizontal	5.75	5.00	**6.03**	0.58	0.21	**10.49**	-	**16.52**	m2	**17.76**
19mm work to floors on concrete base; one coat; floated										
level and to falls only not exceeding 15 degrees from horizontal	2.28	5.00	**2.39**	0.34	0.15	**6.44**	-	**8.82**	m2	**9.49**
to falls and crossfalls and to slopes not exceeding 15 degrees from horizontal	2.28	5.00	**2.39**	0.44	0.15	**7.86**	-	**10.25**	m2	**11.01**
25mm work to floors on concrete base; one coat; floated										
level and to falls only not exceeding 15 degrees from horizontal	2.95	5.00	**3.09**	0.36	0.16	**6.83**	-	**9.92**	m2	**10.67**
to falls and crossfalls and to slopes not exceeding 15 degrees from horizontal	2.95	5.00	**3.09**	0.46	0.16	**8.25**	-	**11.34**	m2	**12.19**
32mm work to floors on concrete base; one coat; floated										
level and to falls only not exceeding 15 degrees from horizontal	3.74	5.00	**3.92**	0.39	0.18	**7.47**	-	**11.39**	m2	**12.24**
to falls and crossfalls and to slopes not exceeding 15 degrees from horizontal	3.74	5.00	**3.92**	0.49	0.18	**8.89**	-	**12.81**	m2	**13.77**
38mm work to floors on concrete base; one coat; floated										
level and to falls only not exceeding 15 degrees from horizontal	4.41	5.00	**4.62**	0.41	0.19	**7.86**	-	**12.48**	m2	**13.42**
to falls and crossfalls and to slopes not exceeding 15 degrees from horizontal	4.41	5.00	**4.62**	0.51	0.19	**9.28**	-	**13.90**	m2	**14.95**
50mm work to floors on concrete base; one coat; floated										
level and to falls only not exceeding 15 degrees from horizontal	5.75	5.00	**6.03**	0.46	0.21	**8.79**	-	**14.82**	m2	**15.93**
to falls and crossfalls and to slopes not exceeding 15 degrees from horizontal	5.75	5.00	**6.03**	0.56	0.21	**10.21**	-	**16.24**	m2	**17.46**
Mortar, cement and sand (1:3) - paving										
25mm work to floors on concrete base; one coat; troweled										
level and to falls only not exceeding 15 degrees from horizontal	2.95	5.00	**3.09**	0.38	0.16	**7.11**	-	**10.21**	m2	**10.97**
to falls and crossfalls and to slopes not exceeding 15 degrees from horizontal	2.95	5.00	**3.09**	0.48	0.16	**8.53**	-	**11.63**	m2	**12.50**
to slopes exceeding 15 degrees from horizontal	2.95	5.00	**3.09**	0.53	0.16	**9.24**	-	**12.34**	m2	**13.26**
32mm work to floors on concrete base; one coat; troweled										
level and to falls only not exceeding 15 degrees from horizontal	3.74	5.00	**3.92**	0.41	0.18	**7.75**	-	**11.67**	m2	**12.54**
to falls and crossfalls and to slopes not exceeding 15 degrees from horizontal	3.74	5.00	**3.92**	0.51	0.18	**9.17**	-	**13.09**	m2	**14.07**
to slopes exceeding 15 degrees from horizontal	3.74	5.00	**3.92**	0.56	0.18	**9.89**	-	**13.80**	m2	**14.84**
38mm work to floors on concrete base; one coat; troweled										
level and to falls only not exceeding 15 degrees from horizontal	4.41	5.00	**4.62**	0.43	0.19	**8.15**	-	**12.77**	m2	**13.72**
to falls and crossfalls and to slopes not exceeding 15 degrees from horizontal	4.41	5.00	**4.62**	0.53	0.19	**9.57**	-	**14.19**	m2	**15.25**
to slopes exceeding 15 degrees from horizontal	4.41	5.00	**4.62**	0.58	0.19	**10.28**	-	**14.90**	m2	**16.02**
50mm work to floors on concrete base; one coat; troweled										
level and to falls only not exceeding 15 degrees from horizontal	5.75	5.00	**6.03**	0.48	0.21	**9.07**	-	**15.10**	m2	**16.24**
to falls and crossfalls and to slopes not exceeding 15 degrees from horizontal	5.75	5.00	**6.03**	0.58	0.21	**10.49**	-	**16.52**	m2	**17.76**
to slopes exceeding 15 degrees from horizontal	5.75	5.00	**6.03**	0.63	0.21	**11.20**	-	**17.23**	m2	**18.53**
If paving oil proofed, add										
25mm work	0.19	10.00	**0.21**	-	0.03	**0.32**	-	**0.53**	m2	**0.57**

Labour hourly rates: (except Specialists) Craft Operatives £14.21 Labourer £10.71 Rates are national average prices. Refer to REGIONAL VARIATIONS for indicative levels of overall pricing in regions	MATERIALS			LABOUR				RATES		
	Del to Site £	Waste %	Material Cost £	Craft Optve Hrs	Lab Hrs	Labour Cost £	Sunds £	Nett Rate £	Unit	Gross Rate (+7.5%) £
M10: CEMENT: SAND/CONCRETE SCREEDS/TOPPINGS Cont.										
Mortar, cement and sand (1:3) - paving Cont.										
If paving oil proofed, add Cont.										
32mm work	0.25	10.00	**0.27**	-	0.03	**0.32**	-	**0.59**	m2	**0.64**
38mm work	0.29	10.00	**0.32**	-	0.04	**0.43**	-	**0.75**	m2	**0.81**
50mm work	0.38	10.00	**0.42**	-	0.05	**0.54**	-	**0.96**	m2	**1.03**
19mm work to skirtings on brickwork or blockwork base										
height 100mm	0.34	10.00	**0.37**	0.70	0.05	**10.48**	-	**10.85**	m	**11.67**
height 150mm	0.45	10.00	**0.49**	0.80	0.05	**11.90**	-	**12.40**	m	**13.33**
25mm work to skirtings on brickwork or blockwork base										
height 100mm	0.45	10.00	**0.49**	0.75	0.05	**11.19**	-	**11.69**	m	**12.56**
height 150mm	0.67	10.00	**0.74**	0.85	0.05	**12.61**	-	**13.35**	m	**14.35**
Granolithic screed; steel troweled										
25mm work to floors on concrete base; one coat										
level and to falls only not exceeding 15 degrees from horizontal	3.98	5.00	**4.16**	0.42	0.18	**7.90**	-	**12.06**	m2	**12.96**
to falls and crossfalls and to slopes not exceeding 15 degrees from horizontal	3.98	5.00	**4.16**	0.52	0.18	**9.32**	-	**13.48**	m2	**14.49**
to slopes exceeding 15 degrees from horizontal	3.98	5.00	**4.16**	0.57	0.18	**10.03**	-	**14.19**	m2	**15.26**
32mm work to floors on concrete base; one coat										
level and to falls only not exceeding 15 degrees from horizontal	5.01	5.00	**5.24**	0.45	0.20	**8.54**	-	**13.78**	m2	**14.81**
to falls and crossfalls and to slopes not exceeding 15 degrees from horizontal	5.01	5.00	**5.24**	0.55	0.20	**9.96**	-	**15.20**	m2	**16.34**
to slopes exceeding 15 degrees from horizontal	5.01	5.00	**5.24**	0.61	0.20	**10.81**	-	**16.05**	m2	**17.26**
38mm work to floors on concrete base; one coat										
level and to falls only not exceeding 15 degrees from horizontal	5.89	5.00	**6.17**	0.48	0.21	**9.07**	-	**15.24**	m2	**16.38**
to falls and crossfalls and to slopes not exceeding 15 degrees from horizontal	5.89	5.00	**6.17**	0.58	0.21	**10.49**	-	**16.66**	m2	**17.91**
to slopes exceeding 15 degrees from horizontal	5.89	5.00	**6.17**	0.63	0.21	**11.20**	-	**17.37**	m2	**18.67**
50mm work to floors on concrete base; one coat										
level and to falls only not exceeding 15 degrees from horizontal	7.66	5.00	**8.02**	0.54	0.24	**10.24**	-	**18.27**	m2	**19.64**
to falls and crossfalls and to slopes not exceeding 15 degrees from horizontal	7.66	5.00	**8.02**	0.64	0.24	**11.66**	-	**19.69**	m2	**21.16**
to slopes exceeding 15 degrees from horizontal	7.66	5.00	**8.02**	0.69	0.24	**12.38**	-	**20.40**	m2	**21.93**
If paving tinted, add according to tint										
25mm work	1.70	10.00	**1.86**	-	0.03	**0.32**	-	**2.19**	m2	**2.35**
32mm work	2.17	10.00	**2.39**	-	0.03	**0.32**	-	**2.71**	m2	**2.91**
38mm work	2.58	10.00	**2.83**	-	0.04	**0.43**	-	**3.26**	m2	**3.51**
50mm work	3.39	10.00	**3.73**	-	0.05	**0.54**	-	**4.26**	m2	**4.58**
If carborundum troweled into surface										
add	3.45	10.00	**3.80**	0.08	0.04	**1.57**	-	**5.36**	m2	**5.76**
25mm linings to channels on concrete base										
150mm girth on face; to falls	0.92	10.00	**1.01**	0.20	0.05	**3.38**	-	**4.39**	m	**4.71**
225mm girth on face; to falls	1.25	10.00	**1.37**	0.25	0.05	**4.09**	-	**5.46**	m	**5.87**
32mm linings to channels on concrete base										
150mm girth on face; to falls	1.07	10.00	**1.17**	0.23	0.05	**3.80**	-	**4.97**	m	**5.35**
225mm girth on face; to falls	1.69	10.00	**1.86**	0.28	0.05	**4.51**	-	**6.37**	m	**6.85**
38mm linings to channels on concrete base										
150mm girth on face; to falls	1.36	10.00	**1.49**	0.25	0.05	**4.09**	-	**5.58**	m	**6.00**
225mm girth on face; to falls	1.99	10.00	**2.18**	0.30	0.05	**4.80**	-	**6.98**	m	**7.50**
50mm linings to channels on concrete base										
150mm girth on face; to falls	1.66	10.00	**1.82**	0.30	0.05	**4.80**	-	**6.62**	m	**7.11**
225mm girth on face; to falls	2.58	10.00	**2.83**	0.35	0.05	**5.51**	-	**8.33**	m	**8.96**
13mm work to skirtings on concrete base										
height 150mm	0.44	10.00	**0.49**	0.75	0.05	**11.19**	-	**11.68**	m	**12.55**
19mm work to skirtings on concrete base										
height 150mm	0.59	10.00	**0.65**	0.80	0.05	**11.90**	-	**12.55**	m	**13.49**
25mm work to skirtings on concrete base										
height 150mm	0.88	10.00	**0.97**	0.85	0.05	**12.61**	-	**13.58**	m	**14.60**
32mm work to skirtings on concrete base										
height 150mm	1.03	10.00	**1.13**	0.90	0.05	**13.32**	-	**14.46**	m	**15.54**
Rounded angles and intersections										
10 - 100mm radius	-	-	**-**	0.18	-	**2.56**	-	**2.56**	m	**2.75**
Surface hardeners on paving										
Proprietary surface hardener (Epoxy Resin Based)										
two coats	1.49	10.00	**1.63**	-	0.20	**2.14**	-	**3.78**	m2	**4.06**
three coats	2.03	10.00	**2.24**	-	0.25	**2.68**	-	**4.92**	m2	**5.28**
Polyurethane floor sealer										
two coats	1.92	10.00	**2.11**	-	0.20	**2.14**	-	**4.26**	m2	**4.58**
three coats	2.48	10.00	**2.73**	-	0.25	**2.68**	-	**5.41**	m2	**5.81**

Labour hourly rates: (except Specialists) Craft Operatives £14.21 Labourer £10.71 Rates are national average prices. Refer to REGIONAL VARIATIONS for indicative levels of overall pricing in regions	MATERIALS			LABOUR				RATES		
	Del to Site £	Waste %	Material Cost £	Craft Optve Hrs	Lab Hrs	Labour Cost £	Sunds £	Nett Rate £	Unit	Gross Rate (+7.5%) £
M10: CEMENT: SAND/CONCRETE SCREEDS/TOPPINGS Cont.										
Surface hardeners on paving Cont.										
Nitoflor Lithurin concrete surface dressing										
one coat	0.36	10.00	**0.40**	-	0.17	**1.82**	-	**2.22**	m2	**2.39**
two coats	0.73	10.00	**0.80**	-	0.24	**2.57**	-	**3.37**	m2	**3.62**
If on old floors										
add for cleaning and degreasing	0.32	5.00	**0.34**	-	0.35	**3.75**	-	**4.09**	m2	**4.39**
Vermiculite screed consisting of cement and Vermiculite aggregate finished with 20mm cement and sand (1:4) screeded bed										
45mm work to roofs on concrete base; two coats to falls and crossfalls and to slopes not exceeding 15 degrees from horizontal	6.62	5.00	**6.95**	0.45	0.48	**11.54**	-	**18.48**	m2	**19.87**
60mm work to roofs on concrete base; two coats to falls and crossfalls and to slopes not exceeding 15 degrees from horizontal	9.29	5.00	**9.75**	0.48	0.51	**12.28**	-	**22.03**	m2	**23.68**
70mm work to roofs on concrete base; two coats to falls and crossfalls and to slopes not exceeding 15 degrees from horizontal	11.07	5.00	**11.62**	0.51	0.54	**13.03**	-	**24.65**	m2	**26.50**
80mm work to roofs on concrete base; two coats to falls and crossfalls and to slopes not exceeding 15 degrees from horizontal	12.85	5.00	**13.49**	0.53	0.56	**13.53**	-	**27.02**	m2	**29.04**
Lightweight concrete screed consisting of cement and lightweight aggregate, medium grade, 10 - 5 gauge, 800 kg/m3 (1:10) finished with 15mm cement and sand (1:4) troweled bed										
50mm work to roofs on concrete base; two coats to falls and crossfalls and to slopes not exceeding 15 degrees from horizontal	5.48	5.00	**5.75**	0.60	0.51	**13.99**	-	**19.74**	m2	**21.22**
75mm work to roofs on concrete base; two coats to falls and crossfalls and to slopes not exceeding 15 degrees from horizontal	8.21	5.00	**8.61**	0.65	0.56	**15.23**	-	**23.84**	m2	**25.63**
100mm work to roofs on concrete base; two coats to falls and crossfalls and to slopes not exceeding 15 degrees from horizontal	10.93	5.00	**11.47**	0.70	0.61	**16.48**	-	**27.95**	m2	**30.05**
Screed reinforcement										
Reinforcement; galvanized wire netting, BS EN 10223-2, 13mm mesh, 22 gauge wire; 150mm laps; placing in position										
floors	2.22	5.00	**2.33**	0.05	-	**0.71**	-	**3.04**	m2	**3.27**
Reinforcement; galvanized wire netting, BS EN 10223-2, 25mm mesh, 20 gauge wire; 150mm laps; placing in position										
floors	1.55	5.00	**1.63**	0.05	-	**0.71**	-	**2.34**	m2	**2.51**
Reinforcement; galvanized wire netting, BS EN 10223-2, 38mm mesh, 19 gauge wire; 150mm laps; placing in position										
floors	1.29	5.00	**1.35**	0.05	-	**0.71**	-	**2.07**	m2	**2.22**
Reinforcement; galvanized wire netting, BS EN 10223-2, 50mm mesh, 19 gauge wire; 150mm laps; placing in position										
floors	0.95	5.00	**1.00**	0.05	-	**0.71**	-	**1.71**	m2	**1.84**
Division strips										
Aluminium dividing strips; setting in bed										
5 x 25mm; flat section	4.85	5.00	**5.09**	0.15	-	**2.13**	-	**7.22**	m	**7.77**
Brass dividing strips; setting in bed										
5 x 25mm; flat section	6.10	5.00	**6.41**	0.15	-	**2.13**	-	**8.54**	m	**9.18**
M10: CEMENT: SAND/CONCRETE SCREEDS/TOPPINGS; STAIRCASE AREAS										
Mortar, cement and sand (1:3) - paving										
13mm work to walls on brickwork or blockwork base; one coat; screeded										
width exceeding 300mm	1.46	5.00	**1.53**	0.60	0.20	**10.67**	-	**12.20**	m2	**13.11**
width not exceeding 300mm	0.45	10.00	**0.49**	0.36	0.06	**5.76**	-	**6.25**	m	**6.72**
13mm work to walls on brickwork or blockwork base; one coat; troweled										
width exceeding 300mm	1.46	5.00	**1.53**	0.78	0.20	**13.23**	-	**14.75**	m2	**15.86**
width not exceeding 300mm	0.45	10.00	**0.49**	0.47	0.06	**7.32**	-	**7.81**	m	**8.40**
13mm work to walls on brickwork or blockwork base; one coat; floated										
width exceeding 300mm	1.46	5.00	**1.53**	0.75	0.20	**12.80**	-	**14.33**	m2	**15.40**

Labour hourly rates: (except Specialists) Craft Operatives £14.21 Labourer £10.71 Rates are national average prices. Refer to REGIONAL VARIATIONS for indicative levels of overall pricing in regions	MATERIALS			LABOUR				RATES		
	Del to Site £	Waste %	Material Cost £	Craft Optve Hrs	Lab Hrs	Labour Cost £	Sunds £	Nett Rate £	Unit	Gross Rate (+7.5%) £
M10: CEMENT: SAND/CONCRETE SCREEDS/TOPPINGS; STAIRCASE AREAS Cont.										
Mortar, cement and sand (1:3) - paving Cont.										
13mm work to walls on brickwork or blockwork base; one coat; floated Cont. width not exceeding 300mm	0.45	10.00	**0.49**	0.45	0.06	**7.04**	-	**7.53**	m	**8.09**
19mm work to floors on concrete base; one coat; screeded level and to falls only not exceeding 15 degrees from horizontal	2.28	5.00	**2.39**	0.36	0.15	**6.72**	-	**9.11**	m2	**9.79**
25mm work to floors on concrete base; one coat; screeded level and to falls only not exceeding 15 degrees from horizontal	2.95	5.00	**3.09**	0.39	0.16	**7.26**	-	**10.35**	m2	**11.12**
32mm work to floors on concrete base; one coat; screeded level and to falls only not exceeding 15 degrees from horizontal	3.74	5.00	**3.92**	0.44	0.18	**8.18**	-	**12.10**	m2	**13.00**
38mm work to floors on concrete base; one coat; screeded level and to falls only not exceeding 15 degrees from horizontal	4.41	5.00	**4.62**	0.47	0.19	**8.71**	-	**13.34**	m2	**14.34**
50mm work to floors on concrete base; one coat; screeded level and to falls only not exceeding 15 degrees from horizontal	5.75	5.00	**6.03**	0.54	0.21	**9.92**	-	**15.96**	m2	**17.15**
19mm work to floors on concrete base; one coat; troweled level and to falls only not exceeding 15 degrees from horizontal	2.28	5.00	**2.39**	0.54	0.15	**9.28**	-	**11.67**	m2	**12.54**
25mm work to floors on concrete base; one coat; troweled level and to falls only not exceeding 15 degrees from horizontal	2.95	5.00	**3.09**	0.57	0.16	**9.81**	-	**12.91**	m2	**13.87**
32mm work to floors on concrete base; one coat; troweled level and to falls only not exceeding 15 degrees from horizontal	3.74	5.00	**3.92**	0.62	0.18	**10.74**	-	**14.65**	m2	**15.75**
38mm work to floors on concrete base; one coat; troweled level and to falls only not exceeding 15 degrees from horizontal	4.41	5.00	**4.62**	0.65	0.19	**11.27**	-	**15.89**	m2	**17.08**
50mm work to floors on concrete base; one coat; troweled level and to falls only not exceeding 15 degrees from horizontal	5.75	5.00	**6.03**	0.72	0.21	**12.48**	-	**18.51**	m2	**19.90**
19mm work to floors on concrete base; one coat; floated level and to falls only not exceeding 15 degrees from horizontal	2.28	5.00	**2.39**	0.51	0.15	**8.85**	-	**11.24**	m2	**12.08**
25mm work to floors on concrete base; one coat; floated level and to falls only not exceeding 15 degrees from horizontal	2.95	5.00	**3.09**	0.54	0.16	**9.39**	-	**12.48**	m2	**13.42**
32mm work to floors on concrete base; one coat; floated level and to falls only not exceeding 15 degrees from horizontal	3.74	5.00	**3.92**	0.59	0.18	**10.31**	-	**14.23**	m2	**15.29**
38mm work to floors on concrete base; one coat; floated level and to falls only not exceeding 15 degrees from horizontal	4.41	5.00	**4.62**	0.62	0.19	**10.85**	-	**15.47**	m2	**16.63**
50mm work to floors on concrete base; one coat; floated level and to falls only not exceeding 15 degrees from horizontal	5.75	5.00	**6.03**	0.69	0.21	**12.05**	-	**18.09**	m2	**19.44**
19mm work to treads on concrete base; screeded width 275mm	0.93	10.00	**1.02**	0.15	0.04	**2.56**	-	**3.58**	m	**3.85**
25mm work to treads on concrete base; screeded width 275mm	1.16	10.00	**1.27**	0.16	0.04	**2.70**	-	**3.97**	m	**4.27**
32mm work to treads on concrete base; screeded width 275mm	1.49	10.00	**1.64**	0.18	0.05	**3.09**	-	**4.73**	m	**5.09**
38mm work to treads on concrete base; screeded width 275mm	1.83	10.00	**2.01**	0.20	0.05	**3.38**	-	**5.39**	m	**5.79**

Labour hourly rates: (except Specialists) Craft Operatives £14.21 Labourer £10.71 Rates are national average prices. Refer to REGIONAL VARIATIONS for indicative levels of overall pricing in regions	MATERIALS			LABOUR				RATES		
	Del to Site £	Waste %	Material Cost £	Craft Optve Hrs	Lab Hrs	Labour Cost £	Sunds £	Nett Rate £	Unit	Gross Rate (+7.5%) £
M10: CEMENT: SAND/CONCRETE SCREEDS/TOPPINGS; STAIRCASE AREAS Cont.										
Mortar, cement and sand (1:3) - paving Cont.										
19mm work to treads on concrete base; floated width 275mm	0.93	10.00	**1.02**	0.20	0.04	**3.27**	-	**4.29**	m	**4.62**
25mm work to treads on concrete base; floated width 275mm	1.16	10.00	**1.27**	0.21	0.04	**3.41**	-	**4.68**	m	**5.03**
32mm work to treads on concrete base; floated width 275mm	1.49	10.00	**1.64**	0.23	0.05	**3.80**	-	**5.44**	m	**5.85**
38mm work to treads on concrete base; floated width 275mm	1.83	10.00	**2.01**	0.25	0.05	**4.09**	-	**6.10**	m	**6.55**
19mm work to treads on concrete base; troweled width 275mm	0.93	10.00	**1.02**	0.25	0.04	**3.98**	-	**5.00**	m	**5.38**
25mm work to treads on concrete base; troweled width 275mm	1.16	10.00	**1.27**	0.26	0.04	**4.12**	-	**5.39**	m	**5.80**
32mm work to treads on concrete base; troweled width 275mm	1.49	10.00	**1.64**	0.28	0.05	**4.51**	-	**6.15**	m	**6.62**
38mm work to treads on concrete base; troweled width 275mm	1.83	10.00	**2.01**	0.30	0.05	**4.80**	-	**6.81**	m	**7.32**
19mm work to plain risers on concrete base; keyed height 175mm	0.56	10.00	**0.62**	0.30	0.03	**4.58**	-	**5.20**	m	**5.59**
25mm work to plain risers on concrete base; keyed height 175mm	0.78	10.00	**0.86**	0.35	0.03	**5.29**	-	**6.16**	m	**6.62**
32mm work to plain risers on concrete base; keyed height 175mm	0.90	10.00	**0.99**	0.40	0.03	**6.01**	-	**6.99**	m	**7.52**
19mm work to plain risers on concrete base; troweled height 175mm	0.56	10.00	**0.62**	0.50	0.03	**7.43**	-	**8.04**	m	**8.65**
25mm work to plain risers on concrete base; troweled height 175mm	0.78	10.00	**0.86**	0.55	0.03	**8.14**	-	**9.00**	m	**9.67**
32mm work to plain risers on concrete base; troweled height 175mm	0.90	10.00	**0.99**	0.60	0.03	**8.85**	-	**9.83**	m	**10.57**
19mm work to undercut risers on concrete base; keyed height 175mm	0.56	10.00	**0.62**	0.35	0.03	**5.29**	-	**5.91**	m	**6.35**
25mm work to undercut risers on concrete base; keyed height 175mm	0.78	10.00	**0.86**	0.40	0.03	**6.01**	-	**6.87**	m	**7.38**
32mm work to undercut risers on concrete base; keyed height 175mm	0.90	10.00	**0.99**	0.45	0.03	**6.72**	-	**7.70**	m	**8.28**
19mm work to undercut risers on concrete base; troweled height 175mm	0.56	10.00	**0.62**	0.55	0.03	**8.14**	-	**8.75**	m	**9.41**
25mm work to undercut risers on concrete base; troweled height 175mm	0.78	10.00	**0.86**	0.60	0.03	**8.85**	-	**9.71**	m	**10.44**
32mm work to undercut risers on concrete base; troweled height 175mm	0.90	10.00	**0.99**	0.65	0.03	**9.56**	-	**10.54**	m	**11.33**
Granolithic screed; steel troweled										
25mm work to floors on concrete base; one coat level and to falls only not exceeding 15 degrees from horizontal	3.98	5.00	**4.16**	0.63	0.18	**10.88**	-	**15.04**	m2	**16.17**
32mm work to floors on concrete base; one coat level and to falls only not exceeding 15 degrees from horizontal	5.01	5.00	**5.24**	0.68	0.20	**11.80**	-	**17.05**	m2	**18.33**
38mm work to floors on concrete base; one coat level and to falls only not exceeding 15 degrees from horizontal	5.89	5.00	**6.17**	0.72	0.21	**12.48**	-	**18.65**	m2	**20.05**
50mm work to floors on concrete base; one coat level and to falls only not exceeding 15 degrees from horizontal	7.66	5.00	**8.02**	0.81	0.24	**14.08**	-	**22.10**	m2	**23.76**
25mm work to treads on concrete base width 275mm	1.55	10.00	**1.69**	0.25	0.05	**4.09**	-	**5.78**	m	**6.21**
32mm work to treads on concrete base width 275mm	1.99	10.00	**2.18**	0.30	0.06	**4.91**	-	**7.08**	m	**7.62**
38mm work to treads on concrete base width 275mm	2.43	10.00	**2.66**	0.35	0.06	**5.62**	-	**8.28**	m	**8.90**

Labour hourly rates: (except Specialists) Craft Operatives £14.21 Labourer £10.71 Rates are national average prices. Refer to REGIONAL VARIATIONS for indicative levels of overall pricing in regions	MATERIALS			LABOUR				RATES		
	Del to Site £	Waste %	Material Cost £	Craft Optve Hrs	Lab Hrs	Labour Cost £	Sunds £	Nett Rate £	Unit	Gross Rate (+7.5%) £
M10: CEMENT: SAND/CONCRETE SCREEDS/TOPPINGS; STAIRCASE AREAS Cont.										
Granolithic screed; steel troweled Cont.										
50mm work to treads on concrete base width 275mm	3.16	10.00	**3.47**	0.40	0.07	**6.43**	-	**9.91**	m	**10.65**
13mm work to plain risers on concrete base height 175mm	0.44	10.00	**0.49**	0.50	0.05	**7.64**	-	**8.13**	m	**8.74**
19mm work to plain risers on concrete base height 175mm	0.74	10.00	**0.81**	0.55	0.05	**8.35**	-	**9.16**	m	**9.85**
25mm work to plain risers on concrete base height 175mm	1.03	10.00	**1.13**	0.60	0.05	**9.06**	-	**10.19**	m	**10.96**
32mm work to plain risers on concrete base height 175mm	1.18	10.00	**1.29**	0.65	0.05	**9.77**	-	**11.07**	m	**11.90**
13mm work to undercut risers on concrete base height 175mm	0.44	10.00	**0.49**	0.55	0.05	**8.35**	-	**8.84**	m	**9.50**
19mm work to undercut risers on concrete base height 175mm	0.74	10.00	**0.81**	0.60	0.05	**9.06**	-	**9.87**	m	**10.61**
25mm work to undercut risers on concrete base height 175mm	1.03	10.00	**1.13**	0.65	0.05	**9.77**	-	**10.90**	m	**11.72**
32mm work to undercut risers on concrete base height 175mm	1.18	10.00	**1.29**	0.70	0.05	**10.48**	-	**11.78**	m	**12.66**
13mm work to strings on concrete base height 300mm	0.88	10.00	**0.97**	0.70	0.05	**10.48**	-	**11.45**	m	**12.31**
19mm work to strings on concrete base height 300mm	1.32	10.00	**1.46**	0.75	0.05	**11.19**	-	**12.65**	m	**13.60**
25mm work to strings on concrete base height 300mm	1.62	10.00	**1.78**	0.80	0.05	**11.90**	-	**13.68**	m	**14.71**
32mm work to strings on concrete base height 300mm	2.06	10.00	**2.26**	0.85	0.06	**12.72**	-	**14.98**	m	**16.11**
13mm work to strings on brickwork or blockwork base height 275mm	0.74	10.00	**0.81**	0.65	0.05	**9.77**	-	**10.58**	m	**11.37**
19mm work to strings on brickwork or blockwork base height 275mm	1.18	10.00	**1.29**	0.70	0.05	**10.48**	-	**11.78**	m	**12.66**
25mm work to strings on brickwork or blockwork base height 275mm	1.47	10.00	**1.62**	0.75	0.05	**11.19**	-	**12.81**	m	**13.77**
32mm work to strings on brickwork or blockwork base height 275mm	1.91	10.00	**2.10**	0.80	0.06	**12.01**	-	**14.11**	m	**15.17**
13mm work to aprons on concrete base height 150mm	0.44	10.00	**0.49**	0.50	0.05	**7.64**	-	**8.13**	m	**8.74**
19mm work to aprons on concrete base height 150mm	0.59	10.00	**0.65**	0.55	0.05	**8.35**	-	**9.00**	m	**9.67**
25mm work to aprons on concrete base height 150mm	0.88	10.00	**0.97**	0.60	0.05	**9.06**	-	**10.03**	m	**10.78**
32mm work to aprons on concrete base height 150mm	1.03	10.00	**1.13**	0.65	0.05	**9.77**	-	**10.90**	m	**11.72**
M11: MASTIC ASPHALT FLOORING/FLOOR UNDERLAYS										
Paving; BS 6925 (limestone aggregate) , brown, smooth floated finish										
Flooring and underlay 15mm thick; coats of asphalt - 1; to concrete base; flat										
width not exceeding 150mm	-	-	**-**	-	-	**-**	-	**24.75**	m2	**26.61**
width 150 - 225mm	-	-	**-**	-	-	**-**	-	**20.25**	m2	**21.77**
width 225 - 300mm	-	-	**-**	-	-	**-**	-	**16.10**	m2	**17.31**
width exceeding 300mm	-	-	**-**	-	-	**-**	-	**12.20**	m2	**13.12**
Skirtings 15mm thick; coats of asphalt - 1; no underlay; to brickwork base girth not exceeding 150mm	-	-	**-**	-	-	**-**	-	**7.80**	m	**8.39**
Paving; BS 6925 (limestone aggregate) , red, smooth floated finish										
Flooring and underlay 15mm thick; coats of asphalt - 1; to concrete base; flat										
width not exceeding 150mm	-	-	**-**	-	-	**-**	-	**26.25**	m2	**28.22**
width 150 - 225mm	-	-	**-**	-	-	**-**	-	**21.80**	m2	**23.44**
width 225 - 300mm	-	-	**-**	-	-	**-**	-	**17.50**	m2	**18.81**
width exceeding 300mm	-	-	**-**	-	-	**-**	-	**13.10**	m2	**14.08**

Labour hourly rates: (except Specialists) Craft Operatives £14.21 Labourer £10.71 Rates are national average prices. Refer to REGIONAL VARIATIONS for indicative levels of overall pricing in regions	MATERIALS			LABOUR				RATES		
	Del to Site £	Waste %	Material Cost £	Craft Optve Hrs	Lab Hrs	Labour Cost £	Sunds £	Nett Rate £	Unit	Gross Rate (+7.5%) £
M11: MASTIC ASPHALT FLOORING/FLOOR UNDERLAYS Cont.										
Paving; BS 6925 (limestone aggregate) , red, smooth floated finish Cont.										
Skirtings 15mm thick; coats of asphalt - 1; no underlay; to brickwork base										
girth not exceeding 150mm	-	-	-	-	-	-	-	**8.95**	m	**9.62**
M12: TROWELED BITUMEN/RESIN/RUBBER LATEX FLOORING										
One coat levelling screeds										
3mm work to floors on concrete base; one coat level and to falls only not exceeding 15 degrees from horizontal	2.66	5.00	**2.80**	0.40	0.05	**6.22**	-	**9.02**	m2	**9.69**
3mm work to floors on existing timber boarded base; one coat level and to falls only not exceeding 15 degrees from horizontal	2.66	5.00	**2.80**	0.50	0.06	**7.75**	-	**10.54**	m2	**11.34**
M20: PLASTERED/RENDERED/ROUGHCAST COATINGS - PLASTERBOARD										
Baseboarding; Gyproc Handiboard; 5mm joints, filled with plaster and scrimmed; fixing with Galvanized nails, 30mm long at 400mm ccs; internal										
9.5mm work to walls on timber base										
width exceeding 300mm	2.67	5.00	**2.81**	0.25	0.13	**4.94**	-	**7.75**	m2	**8.33**
width not exceeding 300mm	0.87	10.00	**0.95**	0.15	0.04	**2.56**	-	**3.51**	m	**3.77**
9.5mm work to ceilings on timber base										
width exceeding 300mm	2.67	5.00	**2.81**	0.30	0.13	**5.66**	-	**8.46**	m2	**9.10**
width not exceeding 300mm	0.92	10.00	**1.00**	0.18	0.04	**2.99**	-	**3.98**	m	**4.28**
9.5mm work to isolated beams on timber base										
width exceeding 300mm	2.67	5.00	**2.81**	0.38	0.13	**6.79**	-	**9.60**	m2	**10.32**
width not exceeding 300mm	0.92	10.00	**1.00**	0.23	0.04	**3.70**	-	**4.69**	m	**5.05**
9.5mm work to isolated columns on timber base										
width exceeding 300mm	2.67	5.00	**2.81**	0.38	0.13	**6.79**	-	**9.60**	m2	**10.32**
width not exceeding 300mm	0.92	10.00	**1.00**	0.23	0.04	**3.70**	-	**4.69**	m	**5.05**
Baseboarding; Gyproc Handiboard; 5mm joints, filled with plaster and scrimmed; fixing with adhesive dabs; internal										
9.5mm work to walls on masonry base										
width exceeding 300mm	2.64	5.00	**2.78**	0.25	0.13	**4.94**	-	**7.73**	m2	**8.31**
width not exceeding 300mm	0.85	10.00	**0.93**	0.15	0.04	**2.56**	-	**3.49**	m	**3.76**
Baseboarding; Gyproc Handiboard Duplex; 5mm joints filled with plaster and scrimmed; fixing with Galvanized nails, 30mm long at 400mm ccs; internal										
9.5mm work to walls on timber base										
width exceeding 300mm	3.10	5.00	**3.26**	0.25	0.13	**4.94**	-	**8.20**	m2	**8.82**
width not exceeding 300mm	1.00	10.00	**1.09**	0.15	0.04	**2.56**	-	**3.65**	m	**3.93**
9.5mm work to ceilings on timber base										
width exceeding 300mm	3.10	5.00	**3.26**	0.30	0.13	**5.66**	-	**8.92**	m2	**9.58**
width not exceeding 300mm	1.05	10.00	**1.14**	0.18	0.04	**2.99**	-	**4.13**	m	**4.43**
9.5mm work to isolated beams on timber base										
width exceeding 300mm	3.10	5.00	**3.26**	0.38	0.13	**6.79**	-	**10.05**	m2	**10.81**
width not exceeding 300mm	1.05	10.00	**1.14**	0.23	0.04	**3.70**	-	**4.84**	m	**5.20**
9.5mm work to isolated columns on timber base										
width exceeding 300mm	3.10	5.00	**3.26**	0.38	0.13	**6.79**	-	**10.05**	m2	**10.81**
width not exceeding 300mm	1.05	10.00	**1.14**	0.23	0.04	**3.70**	-	**4.84**	m	**5.20**
Baseboarding; Gyproc Handiboard; 3mm joints, filled with plaster; fixing with Galvanized nails, 30mm long at 400mm ccs; internal										
12.5mm work to walls on timber base										
width exceeding 300mm	3.77	5.00	**3.96**	0.30	0.15	**5.87**	-	**9.83**	m2	**10.57**
width not exceeding 300mm	1.20	10.00	**1.31**	0.18	0.05	**3.09**	-	**4.41**	m	**4.74**
12.5mm work to ceilings on timber base										
width exceeding 300mm	3.60	5.00	**3.78**	0.36	0.15	**6.72**	-	**10.50**	m2	**11.29**
width not exceeding 300mm	1.20	10.00	**1.31**	0.22	0.05	**3.66**	-	**4.97**	m	**5.35**
12.5mm work to isolated beams on timber base										
width exceeding 300mm	3.77	5.00	**3.96**	0.45	0.15	**8.00**	-	**11.96**	m2	**12.86**
width not exceeding 300mm	1.20	10.00	**1.31**	0.27	0.05	**4.37**	-	**5.69**	m	**6.11**
12.5mm work to isolated columns on timber base										
width exceeding 300mm	3.77	5.00	**3.96**	0.45	0.15	**8.00**	-	**11.96**	m2	**12.86**
width not exceeding 300mm	1.20	10.00	**1.31**	0.27	0.05	**4.37**	-	**5.69**	m	**6.11**

Labour hourly rates: (except Specialists) Craft Operatives £14.21 Labourer £10.71 Rates are national average prices. Refer to REGIONAL VARIATIONS for indicative levels of overall pricing in regions	MATERIALS			LABOUR				RATES		
	Del to Site £	Waste %	Material Cost £	Craft Optve Hrs	Lab Hrs	Labour Cost £	Sunds £	Nett Rate £	Unit	Gross Rate (+7.5%) £
M20: PLASTERED/RENDERED/ROUGHCAST COATINGS - PLASTERBOARD Cont.										
Baseboarding; Gyproc plank, square edge; 5mm joints, filled with plaster and scrimmed; internal										
19mm work to walls on timber base										
width exceeding 300mm	4.41	5.00	**4.64**	0.40	0.20	**7.83**	-	**12.46**	m2	**13.40**
width not exceeding 300mm	1.41	10.00	**1.54**	0.24	0.06	**4.05**	-	**5.59**	m	**6.01**
19mm work to ceilings on timber base										
width exceeding 300mm	4.41	5.00	**4.64**	0.48	0.20	**8.96**	-	**13.60**	m2	**14.62**
width not exceeding 300mm	1.41	10.00	**1.54**	0.29	0.06	**4.76**	-	**6.30**	m	**6.78**
19mm work to isolated beams on timber base										
width exceeding 300mm	4.41	5.00	**4.64**	0.60	0.20	**10.67**	-	**15.30**	m2	**16.45**
width not exceeding 300mm	1.41	10.00	**1.54**	0.36	0.04	**5.54**	-	**7.08**	m	**7.62**
19mm work to isolated columns on timber base										
width exceeding 300mm	4.41	5.00	**4.64**	0.60	0.20	**10.67**	-	**15.30**	m2	**16.45**
width not exceeding 300mm	1.41	10.00	**1.54**	0.36	0.04	**5.54**	-	**7.08**	m	**7.62**
Baseboarding; Gyproc square edge wallboard; 3mm joints filled with plaster and scrimmed; fixing with Galvanized nails, 30mm long at 400mm ccs; internal										
9.5mm work to walls on timber base										
width exceeding 300mm	1.93	5.00	**2.03**	0.25	0.13	**4.94**	-	**6.98**	m2	**7.50**
width not exceeding 300mm	0.65	10.00	**0.71**	0.15	0.04	**2.56**	-	**3.27**	m	**3.51**
12.5mm work to walls on timber base										
width exceeding 300mm	2.03	5.00	**2.14**	0.30	0.15	**5.87**	-	**8.01**	m2	**8.61**
width not exceeding 300mm	0.68	10.00	**0.74**	0.18	0.05	**3.09**	-	**3.83**	m	**4.12**
15mm work to walls on timber base										
width exceeding 300mm	3.06	5.00	**3.22**	0.35	0.18	**6.90**	-	**10.12**	m2	**10.88**
width not exceeding 300mm	0.99	10.00	**1.08**	0.21	0.05	**3.52**	-	**4.60**	m	**4.94**
9.5mm work to ceilings on timber base										
width exceeding 300mm	1.93	5.00	**2.03**	0.30	0.13	**5.66**	-	**7.69**	m2	**8.26**
width not exceeding 300mm	0.65	10.00	**0.71**	0.18	0.04	**2.99**	-	**3.69**	m	**3.97**
12.5mm work to ceilings on timber base										
width exceeding 300mm	2.03	5.00	**2.14**	0.36	0.15	**6.72**	-	**8.86**	m2	**9.52**
width not exceeding 300mm	0.68	10.00	**0.74**	0.22	0.05	**3.66**	-	**4.40**	m	**4.73**
15mm work to ceilings on timber base										
width exceeding 300mm	3.06	5.00	**3.22**	0.42	0.18	**7.90**	-	**11.11**	m2	**11.95**
width not exceeding 300mm	0.99	10.00	**1.08**	0.26	0.05	**4.23**	-	**5.31**	m	**5.71**
9.5mm work to isolated beams on timber base										
width exceeding 300mm	1.93	5.00	**2.03**	0.38	0.13	**6.79**	-	**8.82**	m2	**9.49**
width not exceeding 300mm	0.65	10.00	**0.71**	0.23	0.04	**3.70**	-	**4.40**	m	**4.73**
12.5mm work to isolated beams on timber base										
width exceeding 300mm	2.03	5.00	**2.14**	0.45	0.15	**8.00**	-	**10.14**	m2	**10.90**
width not exceeding 300mm	0.68	10.00	**0.74**	0.27	0.05	**4.37**	-	**5.11**	m	**5.49**
15mm work to isolated beams on timber base										
width exceeding 300mm	3.06	5.00	**3.22**	0.52	0.18	**9.32**	-	**12.54**	m2	**13.48**
width not exceeding 300mm	0.99	10.00	**1.08**	0.31	0.05	**4.94**	-	**6.02**	m	**6.47**
9.5mm work to isolated columns on timber base										
width exceeding 300mm	1.93	5.00	**2.03**	0.38	0.13	**6.79**	-	**8.82**	m2	**9.49**
width not exceeding 300mm	0.65	10.00	**0.71**	0.23	0.04	**3.70**	-	**4.40**	m	**4.73**
12.5mm work to isolated columns on timber base										
width exceeding 300mm	2.03	5.00	**2.14**	0.45	0.15	**8.00**	-	**10.14**	m2	**10.90**
width not exceeding 300mm	0.68	10.00	**0.74**	0.27	0.05	**4.37**	-	**5.11**	m	**5.49**
15mm work to isolated columns on timber base										
width exceeding 300mm	3.06	5.00	**3.22**	0.52	0.18	**9.32**	-	**12.54**	m2	**13.48**
width not exceeding 300mm	0.99	10.00	**1.08**	0.31	0.05	**4.94**	-	**6.02**	m	**6.47**
Baseboarding; Gyproc square edge Duplex wallboard; 3mm joints filled with plaster and scrimmed; fixing with Galvanized nails, 30mm long at 400mm ccs; internal										
9.5mm work to walls on timber base										
width exceeding 300mm	2.48	5.00	**2.61**	0.25	0.13	**4.94**	-	**7.55**	m2	**8.12**
width not exceeding 300mm	0.81	10.00	**0.89**	0.15	0.04	**2.56**	-	**3.45**	m	**3.71**
12.5mm work to walls on timber base										
width exceeding 300mm	2.61	5.00	**2.75**	0.30	0.15	**5.87**	-	**8.62**	m2	**9.26**
width not exceeding 300mm	0.85	10.00	**0.93**	0.18	0.05	**3.09**	-	**4.02**	m	**4.33**
15mm work to walls on timber base										
width exceeding 300mm	3.99	5.00	**4.19**	0.35	0.18	**6.90**	-	**11.10**	m2	**11.93**
width not exceeding 300mm	1.27	10.00	**1.39**	0.21	0.05	**3.52**	-	**4.91**	m	**5.27**
9.5mm work to ceilings on timber base										
width exceeding 300mm	2.48	5.00	**2.61**	0.30	0.13	**5.66**	-	**8.26**	m2	**8.88**
width not exceeding 300mm	0.81	10.00	**0.89**	0.18	0.04	**2.99**	-	**3.87**	m	**4.16**
12.5mm work to ceilings on timber base										
width exceeding 300mm	2.61	5.00	**2.75**	0.36	0.15	**6.72**	-	**9.47**	m2	**10.18**

Labour hourly rates: (except Specialists) Craft Operatives £14.21 Labourer £10.71 Rates are national average prices. Refer to REGIONAL VARIATIONS for indicative levels of overall pricing in regions	MATERIALS			LABOUR				RATES		
	Del to Site £	Waste %	Material Cost £	Craft Optve Hrs	Lab Hrs	Labour Cost £	Sunds £	Nett Rate £	Unit	Gross Rate (+7.5%) £
M20: PLASTERED/RENDERED/ROUGHCAST COATINGS - PLASTERBOARD Cont.										
Baseboarding; Gyproc square edge Duplex wallboard; 3mm joints filled with plaster and scrimmed; fixing with Galvanized nails, 30mm long at 400mm ccs; internal Cont.										
12.5mm work to ceilings on timber base Cont.										
width not exceeding 300mm	0.85	10.00	**0.93**	0.22	0.05	**3.66**	-	**4.59**	m	**4.94**
15mm work to ceilings on timber base										
width exceeding 300mm	3.99	5.00	**4.19**	0.42	0.18	**7.90**	-	**12.09**	m2	**13.00**
width not exceeding 300mm	1.27	10.00	**1.39**	0.26	0.05	**4.23**	-	**5.62**	m	**6.04**
9.5mm work to isolated beams on timber base										
width exceeding 300mm	2.48	5.00	**2.61**	0.38	0.13	**6.79**	-	**9.40**	m2	**10.11**
width not exceeding 300mm	0.81	10.00	**0.89**	0.23	0.04	**3.70**	-	**4.58**	m	**4.93**
12.5mm work to isolated beams on timber base										
width exceeding 300mm	2.61	5.00	**2.75**	0.45	0.15	**8.00**	-	**10.75**	m2	**11.55**
width not exceeding 300mm	0.85	10.00	**0.93**	0.27	0.05	**4.37**	-	**5.30**	m	**5.70**
15mm work to isolated beams on timber base										
width exceeding 300mm	3.99	5.00	**4.19**	0.52	0.18	**9.32**	-	**13.51**	m2	**14.52**
width not exceeding 300mm	1.27	10.00	**1.39**	0.31	0.05	**4.94**	-	**6.33**	m	**6.80**
9.5mm work to isolated columns on timber base										
width exceeding 300mm	2.48	5.00	**2.61**	0.38	0.13	**6.79**	-	**9.40**	m2	**10.11**
width not exceeding 300mm	0.81	10.00	**0.89**	0.23	0.04	**3.70**	-	**4.58**	m	**4.93**
12.5mm work to isolated columns on timber base										
width exceeding 300mm	2.61	5.00	**2.75**	0.45	0.15	**8.00**	-	**10.75**	m2	**11.55**
width not exceeding 300mm	0.85	10.00	**0.93**	0.27	0.05	**4.37**	-	**5.30**	m	**5.70**
15mm work to isolated columns on timber base										
width exceeding 300mm	3.99	5.00	**4.19**	0.52	0.18	**9.32**	-	**13.51**	m2	**14.52**
width not exceeding 300mm	1.27	10.00	**1.39**	0.31	0.05	**4.94**	-	**6.33**	m	**6.80**
Baseboarding; two layers 9.5mm Gyproc square edge wallboard; second layer with 3mm joints filled with plaster and scrimmed; fixing with Galvanized nails, 30mm long at 400mm ccs; internal										
19mm work to walls on timber base										
width exceeding 300mm	3.69	5.00	**3.87**	0.50	0.25	**9.78**	-	**13.66**	m2	**14.68**
width not exceeding 300mm	1.23	10.00	**1.33**	0.30	0.08	**5.12**	-	**6.45**	m	**6.94**
19mm work to ceilings on timber base										
width exceeding 300mm	3.69	5.00	**3.87**	0.60	0.25	**11.20**	-	**15.08**	m2	**16.21**
width not exceeding 300mm	1.23	10.00	**1.33**	0.36	0.08	**5.97**	-	**7.31**	m	**7.85**
19mm work to isolated beams on timber base										
width exceeding 300mm	3.69	5.00	**3.87**	0.75	0.25	**13.34**	-	**17.21**	m2	**18.50**
width not exceeding 300mm	1.23	10.00	**1.33**	0.45	0.08	**7.25**	-	**8.58**	m	**9.23**
19mm work to isolated columns on timber base										
width exceeding 300mm	3.69	5.00	**3.87**	0.75	0.25	**13.34**	-	**17.21**	m2	**18.50**
width not exceeding 300mm	1.23	10.00	**1.33**	0.45	0.08	**7.25**	-	**8.58**	m	**9.23**
Baseboarding; one layer 9.5mm and one layer 12.5mm Gyproc square edge wallboard; second layer with 3mm joints filled with plaster and scrimmed; fixing with Galvanized nails, 30mm long at 400mm ccs, internal										
22mm work to walls on timber base										
width exceeding 300mm	3.79	5.00	**3.98**	0.55	0.28	**10.81**	-	**14.79**	m2	**15.90**
width not exceeding 300mm	1.26	10.00	**1.37**	0.33	0.08	**5.55**	-	**6.91**	m	**7.43**
22mm work to ceilings on timber base										
width exceeding 300mm	3.79	5.00	**3.98**	0.66	0.28	**12.38**	-	**16.36**	m2	**17.58**
width not exceeding 300mm	1.26	10.00	**1.37**	0.40	0.08	**6.54**	-	**7.91**	m	**8.50**
22mm work to isolated beams on timber base										
width exceeding 300mm	3.79	5.00	**3.98**	0.83	0.28	**14.79**	-	**18.77**	m2	**20.18**
width not exceeding 300mm	1.26	10.00	**1.37**	0.50	0.08	**7.96**	-	**9.33**	m	**10.03**
22mm work to isolated columns on timber base										
width exceeding 300mm	3.79	5.00	**3.98**	0.83	0.28	**14.79**	-	**18.77**	m2	**20.18**
width not exceeding 300mm	1.26	10.00	**1.37**	0.50	0.08	**7.96**	-	**9.33**	m	**10.03**
Baseboarding; two layers 12.5mm Gyproc square edge wallboard; second layer with 3mm joints filled with plaster and scrimmed; fixing with Galvanized nails, 30mm long at 400mm ccs; internal										
25mm work to walls on timber base										
width exceeding 300mm	3.89	5.00	**4.08**	0.60	0.30	**11.74**	-	**15.82**	m2	**17.01**
width not exceeding 300mm	1.29	10.00	**1.40**	0.36	0.09	**6.08**	-	**7.48**	m	**8.04**
25mm work to ceilings on timber base										
width exceeding 300mm	3.89	5.00	**4.08**	0.72	0.30	**13.44**	-	**17.53**	m2	**18.84**
width not exceeding 300mm	1.29	10.00	**1.40**	0.43	0.09	**7.07**	-	**8.47**	m	**9.11**
25mm work to isolated beams on timber base										
width exceeding 300mm	3.89	5.00	**4.08**	0.90	0.30	**16.00**	-	**20.09**	m2	**21.59**

Labour hourly rates: (except Specialists) Craft Operatives £14.21 Labourer £10.71 Rates are national average prices. Refer to REGIONAL VARIATIONS for indicative levels of overall pricing in regions	MATERIALS			LABOUR				RATES		
	Del to Site £	Waste %	Material Cost £	Craft Optve Hrs	Lab Hrs	Labour Cost £	Sunds £	Nett Rate £	Unit	Gross Rate (+7.5%) £
M20: PLASTERED/RENDERED/ROUGHCAST COATINGS - PLASTERBOARD Cont.										
Baseboarding; two layers 12.5mm Gyproc square edge wallboard; second layer with 3mm joints filled with plaster and scrimmed; fixing with Galvanized nails, 30mm long at 400mm ccs; internal Cont.										
25mm work to isolated beams on timber base Cont.										
width not exceeding 300mm........................	1.29	10.00	**1.40**	0.54	0.09	**8.64**	-	**10.04**	m	**10.79**
25mm work to isolated columns on timber base										
width exceeding 300mm	3.89	5.00	**4.08**	0.90	0.30	**16.00**	-	**20.09**	m2	**21.59**
width not exceeding 300mm........................	1.29	10.00	**1.40**	0.54	0.09	**8.64**	-	**10.04**	m	**10.79**
Baseboarding; two layers 15mm Gyproc square edge wallboard; second layer with 3mm joints filled with plaster and scrimmed; fixing with Galvanized nails, 30mm long at 400mm ccs; internal										
30mm work to walls on timber base										
width exceeding 300mm	5.95	5.00	**6.25**	0.65	0.33	**12.77**	-	**19.02**	m2	**20.44**
width not exceeding 300mm........................	1.91	10.00	**2.08**	0.39	0.10	**6.61**	-	**8.69**	m	**9.34**
30mm work to ceilings on timber base										
width exceeding 300mm	5.95	5.00	**6.25**	0.78	0.33	**14.62**	-	**20.86**	m2	**22.43**
width not exceeding 300mm........................	1.91	10.00	**2.08**	0.47	0.10	**7.75**	-	**9.83**	m	**10.57**
30mm work to isolated beams on timber base										
width exceeding 300mm	5.95	5.00	**6.25**	0.98	0.33	**17.46**	-	**23.71**	m2	**25.48**
width not exceeding 300mm........................	1.91	10.00	**2.08**	0.59	0.10	**9.45**	-	**11.53**	m	**12.40**
30mm work to isolated columns on timber base										
width exceeding 300mm	5.95	5.00	**6.25**	0.98	0.33	**17.46**	-	**23.71**	m2	**25.48**
width not exceeding 300mm........................	1.91	10.00	**2.08**	0.59	0.10	**9.45**	-	**11.53**	m	**12.40**
Baseboarding; two layers; first layer 9.5mm Gyproc Duplex wallboard, second layer 9.5mm Gyproc wallboard both layers with square edge boards; second layer with 3mm joints filled with plaster and scrimmed; fixing with Galvanized nails, 30mm long at 400mm ccs; internal										
19mm work to walls on timber base										
width exceeding 300mm	4.24	5.00	**4.45**	0.50	0.25	**9.78**	-	**14.23**	m2	**15.30**
width not exceeding 300mm........................	1.39	10.00	**1.52**	0.30	0.08	**5.12**	-	**6.63**	m	**7.13**
19mm work to ceilings on timber base										
width exceeding 300mm	4.24	5.00	**4.45**	0.60	0.25	**11.20**	-	**15.65**	m2	**16.83**
width not exceeding 300mm........................	1.39	10.00	**1.52**	0.36	0.08	**5.97**	-	**7.49**	m	**8.05**
19mm work to isolated beams on timber base										
width exceeding 300mm	4.24	5.00	**4.45**	0.75	0.25	**13.34**	-	**17.79**	m2	**19.12**
width not exceeding 300mm........................	1.39	10.00	**1.52**	0.45	0.08	**7.25**	-	**8.77**	m	**9.42**
19mm work to isolated columns on timber base										
width exceeding 300mm	4.24	5.00	**4.45**	0.75	0.25	**13.34**	-	**17.79**	m2	**19.12**
width not exceeding 300mm........................	1.39	10.00	**1.52**	0.45	0.08	**7.25**	-	**8.77**	m	**9.42**
Baseboarding; two layers; first layer 12.5mm Gyproc Duplex wallboard, second layer 9.5mm Gyproc wallboard, both layers with square edge boards; second layer with 3mm joints filled with plaster and scrimmed; fixing with Galvanized nails, 30mm long at 400mm ccs; internal										
22mm work to walls on timber base										
width exceeding 300mm	4.37	5.00	**4.59**	0.55	0.28	**10.81**	-	**15.40**	m2	**16.56**
width not exceeding 300mm........................	1.43	10.00	**1.56**	0.33	0.08	**5.55**	-	**7.10**	m	**7.64**
22mm work to ceilings on timber base										
width exceeding 300mm	4.37	5.00	**4.59**	0.66	0.28	**12.38**	-	**16.96**	m2	**18.24**
width not exceeding 300mm........................	1.43	10.00	**1.56**	0.40	0.08	**6.54**	-	**8.10**	m	**8.71**
22mm work to isolated beams on timber base										
width exceeding 300mm	4.37	5.00	**4.59**	0.83	0.28	**14.79**	-	**19.38**	m2	**20.83**
width not exceeding 300mm........................	1.43	10.00	**1.56**	0.50	0.08	**7.96**	-	**9.52**	m	**10.23**
22mm work to isolated columns on timber base										
width exceeding 300mm	4.37	5.00	**4.59**	0.83	0.28	**14.79**	-	**19.38**	m2	**20.83**
width not exceeding 300mm........................	1.43	10.00	**1.56**	0.50	0.08	**7.96**	-	**9.52**	m	**10.23**
Baseboarding; two layers; first layer 12.5mm Gyproc Duplex wallboard, second layer 12.5mm Gyproc wallboard, both layers with square edge boards; second layer with 3mm joints filled with plaster and scrimmed; fixing with Galvanized nails, 30mm long at 400mm ccs; internal										
25mm work to walls on timber base										
width exceeding 300mm	4.47	5.00	**4.69**	0.60	0.30	**11.74**	-	**16.43**	m2	**17.66**
width not exceeding 300mm........................	1.46	10.00	**1.59**	0.36	0.09	**6.08**	-	**7.67**	m	**8.25**
25mm work to ceilings on timber base										
width exceeding 300mm	4.47	5.00	**4.69**	0.72	0.30	**13.44**	-	**18.14**	m2	**19.50**
width not exceeding 300mm........................	1.46	10.00	**1.59**	0.43	0.09	**7.07**	-	**8.67**	m	**9.32**

Labour hourly rates: (except Specialists) Craft Operatives £14.21 Labourer £10.71 Rates are national average prices. Refer to REGIONAL VARIATIONS for indicative levels of overall pricing in regions	MATERIALS			LABOUR				RATES		
	Del to Site £	Waste %	Material Cost £	Craft Optve Hrs	Lab Hrs	Labour Cost £	Sunds £	Nett Rate £	Unit	Gross Rate (+7.5%) £
M20: PLASTERED/RENDERED/ROUGHCAST COATINGS - PLASTERBOARD Cont.										
Baseboarding; two layers; first layer 12.5mm Gyproc Duplex wallboard, second layer 12.5mm Gyproc wallboard, both layers with square edge boards; second layer with 3mm joints filled with plaster and scrimmed; fixing with Galvanized nails, 30mm long at 400mm ccs; internal Cont.										
25mm work to isolated beams on timber base										
width exceeding 300mm	4.47	5.00	**4.69**	0.90	0.30	**16.00**	-	**20.69**	m2	**22.25**
width not exceeding 300mm	1.46	10.00	**1.59**	0.54	0.09	**8.64**	-	**10.23**	m	**11.00**
25mm work to isolated columns on timber base										
width exceeding 300mm	4.47	5.00	**4.69**	0.90	0.30	**16.00**	-	**20.69**	m2	**22.25**
width not exceeding 300mm	1.46	10.00	**1.59**	0.54	0.09	**8.64**	-	**10.23**	m	**11.00**
Baseboarding; two layers; first layer 15mm Gyproc Duplex wallboard, second layer 15mm Gyproc wallboard, both layers with square edge boards; second layer with 3mm joints filled with plaster and scrimmed; fixing with Galvanized nails, 30mm long at 400mm ccs; internal										
30mm work to walls on timber base										
width exceeding 300mm	6.88	5.00	**7.22**	0.65	0.33	**12.77**	-	**19.99**	m2	**21.49**
width not exceeding 300mm	2.19	10.00	**2.39**	0.39	0.10	**6.61**	-	**9.00**	m	**9.67**
30mm work to ceilings on timber base										
width exceeding 300mm	6.88	5.00	**7.22**	0.78	0.33	**14.62**	-	**21.84**	m2	**23.48**
width not exceeding 300mm	2.19	10.00	**2.39**	0.47	0.10	**7.75**	-	**10.14**	m	**10.90**
30mm work to isolated beams on timber base										
width exceeding 300mm	6.88	5.00	**7.22**	0.98	0.33	**17.46**	-	**24.68**	m2	**26.53**
width not exceeding 300mm	2.19	10.00	**2.39**	0.59	0.10	**9.45**	-	**11.84**	m	**12.73**
30mm work to isolated columns on timber base										
width exceeding 300mm	6.88	5.00	**7.22**	0.98	0.33	**17.46**	-	**24.68**	m2	**26.53**
width not exceeding 300mm	2.19	10.00	**2.39**	0.59	0.10	**9.45**	-	**11.84**	m	**12.73**
M20: PLASTERED/RENDERED/ROUGHCAST COATINGS: DAMP WALL TREATMENT										
Newlath; fixing with masonry nails; internal										
Work to walls on brickwork or blockwork base										
width exceeding 300mm	12.00	5.00	**12.48**	0.36	0.18	**7.04**	-	**19.52**	m2	**20.98**
width not exceeding 300mm	3.60	5.00	**3.74**	0.22	0.05	**3.66**	-	**7.40**	m	**7.96**
M20: PLASTERED/RENDERED/ROUGHCAST COATINGS: PORTLAND CEMENT WORK										
Plain face, first and finishing coats cement and sand (1:3), total 13mm thick; wood floated; internal										
13mm work to walls on brickwork or blockwork base										
width exceeding 300mm	1.46	5.00	**1.53**	0.45	0.23	**8.86**	-	**10.39**	m2	**11.17**
width exceeding 300mm; dubbing average 6mm thick	2.13	5.00	**2.23**	0.68	0.44	**14.38**	-	**16.61**	m2	**17.86**
width exceeding 300mm; dubbing average 12mm thick	2.80	5.00	**2.94**	0.68	0.44	**14.38**	-	**17.32**	m2	**18.61**
width exceeding 300mm; dubbing average 19mm thick	3.58	5.00	**3.76**	0.70	0.57	**16.05**	-	**19.81**	m2	**21.30**
width exceeding 300mm; curved to 3000mm radius	1.46	5.00	**1.53**	0.60	0.23	**10.99**	-	**12.52**	m2	**13.46**
width not exceeding 300mm	0.45	10.00	**0.49**	0.27	0.07	**4.59**	-	**5.08**	m	**5.46**
width not exceeding 300mm; dubbing average 6mm thick	0.67	10.00	**0.74**	0.41	0.13	**7.22**	-	**7.96**	m	**8.55**
width not exceeding 300mm; dubbing average 12mm thick	0.90	10.00	**0.99**	0.41	0.13	**7.22**	-	**8.20**	m	**8.82**
width not exceeding 300mm; dubbing average 19mm thick	1.12	10.00	**1.23**	0.42	0.17	**7.79**	-	**9.02**	m	**9.70**
width not exceeding 300mm; curved to 3000mm radius	0.45	10.00	**0.49**	0.36	0.07	**5.87**	-	**6.36**	m	**6.83**
If plain face waterproofed										
add	0.29	10.00	**0.32**	-	0.02	**0.21**	-	**0.54**	m2	**0.58**
Rounded angles										
radius 10 - 100mm	-	-	-	0.24	-	**3.41**	-	**3.41**	m	**3.67**
Work forming flush skirting; 13mm thick										
height 150mm	0.22	50.00	**0.34**	0.50	0.05	**7.64**	-	**7.98**	m	**8.57**
height 225mm	0.34	50.00	**0.50**	0.55	0.05	**8.35**	-	**8.86**	m	**9.52**
height 150mm; curved to 3000mm radius	0.22	50.00	**0.34**	0.67	0.05	**10.06**	-	**10.39**	m	**11.17**
height 225mm; curved to 3000mm radius	0.34	50.00	**0.50**	0.73	0.05	**10.91**	-	**11.41**	m	**12.27**
Work forming projecting skirting; 13mm projection										
height 150mm	0.22	50.00	**0.34**	0.75	0.05	**11.19**	-	**11.53**	m	**12.39**
height 225mm	0.34	50.00	**0.50**	0.80	0.05	**11.90**	-	**12.41**	m	**13.34**
height 150mm; curved to 3000mm radius	0.22	50.00	**0.34**	1.00	0.05	**14.75**	-	**15.08**	m	**16.21**
height 225mm; curved to 3000mm radius	0.34	50.00	**0.50**	1.07	0.05	**15.74**	-	**16.24**	m	**17.46**

Labour hourly rates: (except Specialists) Craft Operatives £14.21 Labourer £10.71 Rates are national average prices. Refer to REGIONAL VARIATIONS for indicative levels of overall pricing in regions	MATERIALS			LABOUR				RATES		
	Del to Site £	Waste %	Material Cost £	Craft Optve Hrs	Lab Hrs	Labour Cost £	Sunds £	Nett Rate £	Unit	Gross Rate (+7.5%) £
M20: PLASTERED/RENDERED/ROUGHCAST COATINGS: ROUGH CAST										
Render and dry dash; first and finishing coats cement and sand (1:3), total 15mm thick; wood floated; dry dash of pea shingle; external										
15mm work to walls on brickwork or blockwork base										
width exceeding 300mm	1.83	5.00	**1.93**	0.75	0.38	**14.73**	-	**16.66**	m2	**17.90**
extra; spatterdash coat	0.75	10.00	**0.82**	0.20	0.10	**3.91**	-	**4.73**	m2	**5.09**
width exceeding 300mm; curved to 3000mm radius	1.83	5.00	**1.93**	1.00	0.38	**18.28**	-	**20.21**	m2	**21.72**
width not exceeding 300mm	0.40	10.00	**0.44**	0.45	0.11	**7.57**	-	**8.01**	m	**8.61**
extra; spatterdash coat	0.02	10.00	**0.02**	0.12	0.03	**2.03**	-	**2.05**	m	**2.20**
width not exceeding 300mm; curved to 3000mm radius	0.40	10.00	**0.44**	0.60	0.11	**9.70**	-	**10.14**	m	**10.90**
Render and wet dash; first and finishing coats cement and sand (1:3), total 15mm thick; wood floated; wet dash of crushed stone or shingle and cement slurry										
15mm work to walls on brickwork or blockwork base										
width exceeding 300mm	2.51	5.00	**2.68**	0.85	0.43	**16.68**	-	**19.37**	m2	**20.82**
extra; spatterdash coat	0.46	10.00	**0.50**	0.25	0.13	**4.94**	-	**5.45**	m2	**5.85**
width exceeding 300mm; curved to 3000mm radius	2.51	5.00	**2.68**	1.13	0.43	**20.66**	-	**23.34**	m2	**25.10**
width not exceeding 300mm	0.79	10.00	**0.87**	0.51	0.13	**8.64**	-	**9.51**	m	**10.22**
extra; spatterdash coat	0.22	10.00	**0.24**	0.15	0.04	**2.56**	-	**2.80**	m	**3.01**
width not exceeding 300mm; curved to 3000mm radius	1.39	10.00	**1.53**	0.68	0.13	**11.06**	-	**12.59**	m	**13.53**
M20: PLASTERED/RENDERED/ROUGHCAST COATINGS: TYROLEAN FINISH										
Render and Tyrolean finish; first and finishing coats cement and sand (1:3), total 15mm thick; wood floated; Tyrolean finish of 'Cullamix' mixture applied by machine										
15mm work to walls on brickwork or blockwork base										
width exceeding 300mm	2.29	5.00	**2.40**	0.90	0.25	**15.47**	-	**18.43**	m2	**19.81**
width not exceeding 300mm	0.76	10.00	**0.84**	0.54	0.08	**8.53**	-	**9.71**	m	**10.43**
M20: PLASTERED/RENDERED/ROUGHCAST COATINGS: HARDWALL PLASTERING										
Plaster, BS 1191, Part 1, Class B; finishing coat of board finish, 3mm thick; steel troweled; internal										
3mm work to walls on concrete or plasterboard base										
width exceeding 300mm	0.49	5.00	**0.52**	0.25	0.13	**4.94**	0.01	**5.48**	m2	**5.89**
width exceeding 300mm; curved to 3000mm radius	0.49	5.00	**0.52**	0.33	0.13	**6.08**	0.01	**6.61**	m2	**7.11**
width not exceeding 300mm	0.26	10.00	**0.28**	0.15	0.04	**2.56**	0.00	**2.85**	m	**3.06**
width not exceeding 300mm; curved to 3000mm radius	0.26	10.00	**0.28**	0.20	0.04	**3.27**	0.00	**3.56**	m	**3.82**
3mm work to ceilings on concrete or plasterboard base										
width exceeding 300mm	0.49	5.00	**0.52**	0.30	0.13	**5.66**	0.01	**6.19**	m2	**6.65**
width exceeding 300mm; 3.50 - 5.00m above floor	0.49	5.00	**0.52**	0.33	0.13	**6.08**	0.01	**6.61**	m2	**7.11**
width not exceeding 300mm	0.26	10.00	**0.28**	0.18	0.04	**2.99**	0.00	**3.27**	m	**3.52**
width not exceeding 300mm; 3.50 - 5.00m above floor	0.26	10.00	**0.28**	0.20	0.04	**3.27**	0.00	**3.56**	m	**3.82**
3mm work to isolated beams on concrete or plasterboard base										
width exceeding 300mm	0.49	5.00	**0.52**	0.38	0.13	**6.79**	0.01	**7.32**	m2	**7.87**
width not exceeding 300mm	0.26	10.00	**0.28**	0.23	0.04	**3.70**	0.00	**3.98**	m	**4.28**
3mm work to isolated columns on concrete or plasterboard base										
width exceeding 300mm	0.49	5.00	**0.52**	0.38	0.13	**6.79**	0.01	**7.32**	m2	**7.87**
width not exceeding 300mm	0.26	10.00	**0.28**	0.23	0.04	**3.70**	0.00	**3.98**	m	**4.28**
Plaster; first coat of hardwall, 11mm thick; finishing coat of multi-finish 2mm thick; steel troweled; internal										
13mm work to walls on brickwork or blockwork base										
width exceeding 300mm	1.95	5.00	**2.05**	0.45	0.23	**8.86**	0.05	**10.95**	m2	**11.78**
width exceeding 300mm; curved to 3000mm radius	1.95	5.00	**2.05**	0.60	0.23	**10.99**	0.05	**13.09**	m2	**14.07**
width not exceeding 300mm	0.88	10.00	**0.97**	0.27	0.07	**4.59**	0.02	**5.57**	m	**5.99**
width not exceeding 300mm; curved to 3000mm radius	0.88	10.00	**0.97**	0.36	0.07	**5.87**	0.02	**6.85**	m	**7.37**
13mm work to ceilings on concrete base										
width exceeding 300mm	1.95	5.00	**2.05**	0.54	0.23	**10.14**	0.05	**12.23**	m2	**13.15**
width exceeding 300mm; 3.50 - 5.00m above floor	1.95	5.00	**2.05**	0.59	0.23	**10.85**	0.05	**12.94**	m2	**13.91**
width not exceeding 300mm	0.88	10.00	**0.97**	0.32	0.07	**5.30**	0.02	**6.28**	m	**6.75**
width not exceeding 300mm; 3.50 - 5.00m above floor	0.88	10.00	**0.97**	0.35	0.07	**5.72**	0.02	**6.71**	m	**7.21**
13mm work to isolated beams on concrete base										
width exceeding 300mm	1.95	5.00	**2.05**	0.68	0.23	**12.13**	0.05	**14.22**	m2	**15.29**
width not exceeding 300mm	0.88	10.00	**0.97**	0.41	0.07	**6.58**	0.02	**7.56**	m	**8.13**
13mm work to isolated columns on concrete base										
width exceeding 300mm	1.95	5.00	**2.05**	0.68	0.23	**12.13**	0.05	**14.22**	m2	**15.29**
width not exceeding 300mm	0.88	10.00	**0.97**	0.41	0.07	**6.58**	0.02	**7.56**	m	**8.13**
Rounded angles										
radius 10 - 100mm	-	-	**-**	0.24	-	**3.41**	-	**3.41**	m	**3.67**

Labour hourly rates: (except Specialists) Craft Operatives £14.21 Labourer £10.71 Rates are national average prices. Refer to REGIONAL VARIATIONS for indicative levels of overall pricing in regions	MATERIALS			LABOUR				RATES		
	Del to Site £	Waste %	Material Cost £	Craft Optve Hrs	Lab Hrs	Labour Cost £	Sunds £	Nett Rate £	Unit	Gross Rate (+7.5%) £
M20: PLASTERED/RENDERED/ROUGHCAST COATINGS: HARDWALL PLASTERING Cont.										
Plaster; Thistle Universal one coat plaster, 13mm thick; steel troweled; internal										
Note: the thickness is from the face of the metal lathing										
13mm work to walls on metal lathing base										
width exceeding 300mm	3.13	5.00	**3.28**	0.35	0.18	**6.90**	0.08	**10.26**	m2	**11.03**
width exceeding 300mm; curved to 3000mm radius	3.13	5.00	**3.28**	0.47	0.18	**8.61**	0.08	**11.97**	m2	**12.87**
width not exceeding 300mm	1.30	10.00	**1.43**	0.21	0.05	**3.52**	0.03	**4.98**	m	**5.35**
width not exceeding 300mm; curved to 3000mm radius	1.30	10.00	**1.43**	0.28	0.05	**4.51**	0.03	**5.97**	m	**6.42**
13mm work to ceilings on metal lathing base										
width exceeding 300mm	3.13	5.00	**3.28**	0.42	0.18	**7.90**	0.08	**11.26**	m2	**12.10**
width exceeding 300mm; 3.50 - 5.00m above floor	3.13	5.00	**3.28**	0.46	0.18	**8.46**	0.08	**11.83**	m2	**12.71**
width not exceeding 300mm	1.30	10.00	**1.43**	0.25	0.05	**4.09**	0.03	**5.55**	m	**5.96**
width not exceeding 300mm; 3.50 - 5.00m above floor	1.30	10.00	**1.43**	0.28	0.05	**4.51**	0.03	**5.97**	m	**6.42**
13mm work to isolated beams on metal lathing base										
width exceeding 300mm	3.13	5.00	**3.28**	0.53	0.18	**9.46**	0.08	**12.82**	m2	**13.78**
width not exceeding 300mm	1.30	10.00	**1.43**	0.32	0.05	**5.08**	0.03	**6.54**	m	**7.03**
13mm work to isolated columns on metal lathing base										
width exceeding 300mm	3.13	5.00	**3.28**	0.53	0.18	**9.46**	0.08	**12.82**	m2	**13.78**
width not exceeding 300mm	1.30	10.00	**1.43**	0.32	0.05	**5.08**	0.03	**6.54**	m	**7.03**
Rounded angles										
radius 10 - 100mm	-	-	**-**	0.18	-	**2.56**	-	**2.56**	m	**2.75**
M20: PLASTERED/RENDERED/ROUGHCAST COATINGS: LIGHTWEIGHT PLASTERING										
Plaster, Thistle; pre-mixed; floating coat of browning, 11mm thick; finishing coat of finish, 2mm thick; steel troweled; internal										
13mm work to walls on brickwork or blockwork base										
width exceeding 300mm	1.43	5.00	**1.50**	0.40	0.20	**7.83**	0.04	**9.36**	m2	**10.07**
width exceeding 300mm; curved to 3000mm radius	1.43	5.00	**1.50**	0.53	0.20	**9.67**	0.04	**11.21**	m2	**12.05**
width exceeding 300mm; dubbing average 6mm thick	2.12	5.00	**2.22**	0.63	0.38	**13.02**	0.05	**15.30**	m2	**16.45**
width exceeding 300mm; dubbing average 6mm thick; curved to 3000mm radius	2.12	5.00	**2.22**	0.84	0.38	**16.01**	0.05	**18.28**	m2	**19.65**
width exceeding 300mm; dubbing average 12mm thick	2.63	5.00	**2.76**	0.63	0.38	**13.02**	0.07	**15.86**	m2	**17.04**
width exceeding 300mm; dubbing average 12mm thick; curved to 3000mm radius	2.63	5.00	**2.76**	0.84	0.38	**16.01**	0.07	**18.84**	m2	**20.25**
width exceeding 300mm; dubbing average 19mm thick	3.32	5.00	**3.49**	0.65	0.49	**14.48**	0.07	**18.04**	m2	**19.39**
width exceeding 300mm; dubbing average 19mm thick; curved to 3000mm radius	3.49	5.00	**3.67**	0.87	0.49	**17.61**	0.07	**21.34**	m2	**22.95**
width not exceeding 300mm	0.63	10.00	**0.69**	0.24	0.06	**4.05**	0.01	**4.76**	m	**5.11**
width not exceeding 300mm; curved to 3000mm radius	0.63	10.00	**0.69**	0.32	0.06	**5.19**	0.01	**5.89**	m	**6.34**
width not exceeding 300mm; dubbing average 6mm thick	0.97	10.00	**1.07**	0.36	0.09	**6.08**	0.02	**7.17**	m	**7.70**
width not exceeding 300mm; dubbing average 6mm thick; curved to 3000mm radius	0.97	10.00	**1.07**	0.48	0.09	**7.78**	0.02	**8.87**	m	**9.54**
width not exceeding 300mm; dubbing average 12mm thick	1.14	10.00	**1.26**	0.38	0.09	**6.36**	0.02	**7.64**	m	**8.22**
width not exceeding 300mm; dubbing average 12mm thick; curved to 3000mm radius	1.14	10.00	**1.26**	0.51	0.09	**8.21**	0.02	**9.49**	m	**10.20**
width not exceeding 300mm; dubbing average 19mm thick	1.49	10.00	**1.64**	0.39	0.15	**7.15**	0.02	**8.81**	m	**9.47**
width not exceeding 300mm; dubbing average 19mm thick; curved to 3000mm radius	1.49	10.00	**1.64**	0.52	0.15	**9.00**	0.02	**10.65**	m	**11.45**
Rounded angles										
radius 10 - 100mm	-	-	**-**	0.24	-	**3.41**	-	**3.41**	m	**3.67**
Plaster, Thistle; pre-mixed; floating coat of bonding 8mm thick; finishing coat of finish, 2mm thick; steel troweled; internal										
10mm work to walls on concrete or plasterboard base										
width exceeding 300mm	1.43	5.00	**1.50**	0.40	0.20	**7.83**	0.04	**9.36**	m2	**10.07**
width exceeding 300mm; curved to 3000mm radius	1.43	5.00	**1.50**	0.53	0.20	**9.67**	0.04	**11.21**	m2	**12.05**
width not exceeding 300mm	0.63	10.00	**0.69**	0.24	0.06	**4.05**	0.01	**4.76**	m	**5.11**
width not exceeding 300mm; curved to 3000mm radius	0.63	10.00	**0.69**	0.32	0.06	**5.19**	0.01	**5.89**	m	**6.34**
10mm work to ceilings on concrete or plasterboard base										
width exceeding 300mm	1.43	5.00	**1.50**	0.48	0.20	**8.96**	0.04	**10.50**	m2	**11.29**
width exceeding 300mm; 3.50 - 5.00m above floor	1.43	5.00	**1.50**	0.52	0.20	**9.53**	0.04	**11.07**	m2	**11.90**
width not exceeding 300mm	0.63	10.00	**0.69**	0.29	0.06	**4.76**	0.01	**5.47**	m	**5.88**
width not exceeding 300mm; 3.50 - 5.00m above floor	0.63	10.00	**0.69**	0.31	0.06	**5.05**	0.01	**5.75**	m	**6.18**
10mm work to isolated beams on concrete or plasterboard base										
width exceeding 300mm	1.43	5.00	**1.50**	0.60	0.20	**10.67**	0.04	**12.21**	m2	**13.12**
width not exceeding 300mm	0.63	10.00	**0.69**	0.36	0.06	**5.76**	0.01	**6.46**	m	**6.95**
10mm work to isolated columns on concrete or plasterboard base										
width exceeding 300mm	1.43	5.00	**1.50**	0.60	0.20	**10.67**	0.04	**12.21**	m2	**13.12**
width not exceeding 300mm	0.63	10.00	**0.69**	0.36	0.06	**5.76**	0.01	**6.46**	m	**6.95**

Labour hourly rates: (except Specialists) Craft Operatives £14.21 Labourer £10.71 Rates are national average prices. Refer to REGIONAL VARIATIONS for indicative levels of overall pricing in regions	MATERIALS			LABOUR				RATES		
	Del to Site £	Waste %	Material Cost £	Craft Optve Hrs	Lab Hrs	Labour Cost £	Sunds £	Nett Rate £	Unit	Gross Rate (+7.5%) £
M20: PLASTERED/RENDERED/ROUGHCAST COATINGS: LIGHTWEIGHT PLASTERING Cont.										
Plaster, Thistle; pre-mixed; floating coat of bonding 8mm thick; finishing coat of finish, 2mm thick; steel troweled; internal Cont.										
Rounded angles										
radius 10 - 100mm	-	-	-	0.24	-	3.41	-	3.41	m	3.67
Plaster, Thistle; pre-mixed; floating coat of bonding 11mm thick; finishing coat of finish, 2mm thick; steel troweled; internal										
13mm work to ceilings on precast concrete beam and infill blocks base										
width exceeding 300mm	1.95	5.00	2.04	0.48	0.20	8.96	0.05	11.06	m2	11.88
width not exceeding 300mm	0.97	10.00	1.07	0.29	0.06	4.76	0.02	5.85	m	6.29
Plaster, Thistle; tough coat, 11mm thick;finish 2mm thick steel troweled; internal										
Note: the thickness is from the face of the metal lathing										
13mm work to walls on metal lathing base										
width exceeding 300mm	1.46	5.00	1.54	0.55	0.28	10.81	0.04	12.39	m2	13.32
width exceeding 300mm; curved to 3000mm radius	1.46	5.00	1.54	0.73	0.28	13.37	0.04	14.94	m2	16.07
width not exceeding 300mm	0.64	10.00	0.71	0.33	0.08	5.55	0.01	6.27	m	6.74
width not exceeding 300mm; curved to 3000mm radius	0.64	10.00	0.71	0.44	0.08	7.11	0.01	7.83	m	8.42
13mm work to ceilings on metal lathing base										
width exceeding 300mm	1.46	5.00	1.54	0.66	0.28	12.38	0.04	13.95	m2	15.00
width exceeding 300mm; 3.50 - 5.00m above floor	1.46	5.00	1.54	0.72	0.28	13.23	0.04	14.80	m2	15.91
width not exceeding 300mm	0.64	10.00	0.71	0.40	0.08	6.54	0.01	7.26	m	7.80
width not exceeding 300mm; 3.50 - 5.00m above floor	0.64	10.00	0.71	0.43	0.08	6.97	0.01	7.69	m	8.26
13mm work to isolated beams on metal lathing base										
width exceeding 300mm	1.46	5.00	1.54	0.83	0.28	14.79	0.04	16.37	m2	17.59
width not exceeding 300mm	0.64	10.00	0.71	0.50	0.08	7.96	0.01	8.68	m	9.33
13mm work to isolated columns on metal lathing base										
width exceeding 300mm	1.46	5.00	1.54	0.83	0.28	14.79	0.04	16.37	m2	17.59
width not exceeding 300mm	0.64	10.00	0.71	0.50	0.08	7.96	0.01	8.68	m	9.33
Rounded angles										
radius 10 - 100mm	-	-	-	0.30	-	4.26	-	4.26	m	4.58
M20: PLASTERED/RENDERED/ROUGHCAST COATINGS: RENOVATING PLASTER										
Renovating plaster, Thistle; undercoat 11mm thick; finishing coat 2mm thick; steel troweled; internal										
13mm work to walls on existing brickwork or blockwork base										
width exceeding 300mm	2.22	5.00	2.33	0.50	0.25	9.78	0.06	12.17	m2	13.08
width exceeding 300mm; curved to 3000mm radius	2.22	5.00	2.33	0.67	0.25	12.20	0.06	14.59	m2	15.68
width exceeding 300mm; dubbing average 6mm thick	3.05	5.00	3.20	0.70	0.37	13.91	0.08	17.19	m2	18.48
width exceeding 300mm; dubbing average 6mm thick; curved to 3000mm radius	3.05	5.00	3.20	0.93	0.37	17.18	0.08	20.46	m2	21.99
width exceeding 300mm; dubbing average 12mm thick	4.08	5.00	4.28	0.73	0.48	15.51	0.11	19.90	m2	21.40
width exceeding 300mm; dubbing average 12mm thick; curved to 3000mm radius	4.08	5.00	4.28	0.97	0.48	18.92	0.11	23.31	m2	25.06
width not exceeding 300mm	1.01	10.00	1.11	0.30	0.08	5.12	0.02	6.25	m	6.71
width not exceeding 300mm; curved to 3000mm radius	1.01	10.00	1.11	0.40	0.08	6.54	0.02	7.67	m	8.24
width not exceeding 300mm; dubbing average 6mm thick	1.42	10.00	1.56	0.42	0.12	7.25	0.02	8.84	m	9.50
width not exceeding 300mm; dubbing average 6mm thick; curved to 3000mm radius	1.42	10.00	1.56	0.56	0.12	9.24	0.02	10.83	m	11.64
width not exceeding 300mm; dubbing average 12mm thick	1.83	10.00	2.02	0.44	0.15	7.86	0.03	9.91	m	10.65
width not exceeding 300mm; dubbing average 12mm thick; curved to 3000mm radius	1.83	10.00	2.02	0.59	0.15	9.99	0.03	12.04	m	12.94
Rounded angles										
radius 10 - 100mm	-	-	-	0.30	-	4.26	-	4.26	m	4.58
M20: PLASTERED/RENDERED/ROUGHCAST COATINGS: PLASTER BEADS AND STOPS										
Accessories										
galvanized steel angle beads										
Expamet fixing to brickwork or blockwork with masonry nails										
Reference 550	0.57	10.00	0.62	0.08	-	1.14	-	1.76	m	1.89
Reference 553 for 3mm	0.57	10.00	0.62	0.08	-	1.14	-	1.76	m	1.89
Reference 554 for 6mm	0.73	10.00	0.80	0.08	-	1.14	-	1.94	m	2.08
galvanized steel stop beads										
Expamet fixing to brickwork or blockwork with masonry nails										
Reference 560 for 3mm	0.76	10.00	0.83	0.08	-	1.14	-	1.97	m	2.11
Reference 561for 6mm	0.76	10.00	0.83	0.08	-	1.14	-	1.97	m	2.11
Reference 562 for 10mm	0.71	10.00	0.77	0.08	-	1.14	-	1.91	m	2.05

Labour hourly rates: (except Specialists) Craft Operatives £14.21 Labourer £10.71 Rates are national average prices. Refer to REGIONAL VARIATIONS for indicative levels of overall pricing in regions	MATERIALS			LABOUR				RATES		
	Del to Site £	Waste %	Material Cost £	Craft Optve Hrs	Lab Hrs	Labour Cost £	Sunds £	Nett Rate £	Unit	Gross Rate (+7.5%) £
M20: PLASTERED/RENDERED/ROUGHCAST COATINGS: PLASTER BEADS AND STOPS Cont.										
Accessories Cont.										
galvanized steel stop beads Cont. Expamet fixing to brickwork or blockwork with masonry nails Cont.										
Reference 563 for 13mm	0.71	10.00	**0.77**	0.08	-	**1.14**	-	**1.91**	m	**2.05**
Reference 565 for 16mm	0.90	10.00	**0.98**	0.08	-	**1.14**	-	**2.12**	m	**2.28**
Reference 566 for 19mm	0.90	10.00	**0.98**	0.08	-	**1.14**	-	**2.12**	m	**2.28**
galvanized steel plasterboard edging beads Expamet fixing to timber with galvanized nails, 30mm long at 400mm ccs										
Reference 567 for 10mm	7.58	10.00	**7.69**	0.08	-	**1.14**	-	**8.83**	m	**9.49**
Reference 568 for 13mm	7.58	10.00	**7.69**	0.08	-	**1.14**	-	**8.83**	m	**9.49**
Reference 576 for 15mm	7.58	10.00	**7.69**	0.08	-	**1.14**	-	**8.83**	m	**9.49**
Accessories; external										
galvanized steel stop beads Expamet fixing to brickwork or blockwork with masonry nails										
Reference 570	0.70	10.00	**0.76**	0.08	-	**1.14**	-	**1.90**	m	**2.04**
Stainless steel angle beads Expamet fixing to brickwork or blockwork with masonry nails										
Reference 545	1.88	10.00	**2.07**	0.08	-	**1.14**	-	**3.20**	m	**3.44**
Stainless steel stop beads Expamet fixing to brickwork or blockwork with masonry nails										
Reference 547	1.82	10.00	**2.00**	0.08	-	**1.14**	-	**3.13**	m	**3.37**
M20: PLASTERED/RENDERED/ROUGHCAST COATINGS: ANTI-CRACK STRIPS										
Accessories										
Lathing; BB galvanized expanded metal lath, 9mm mesh X 0.500mm thick x 1.11 kg/m2; butt joints fixing with nails										
100mm wide to walls; one edge to timber; one edge to brickwork or blockwork	0.48	5.00	**0.50**	0.08	-	**1.14**	0.03	**1.66**	m	**1.79**
100mm wide to walls; one edge to timber; one edge to concrete	0.48	5.00	**0.50**	0.10	-	**1.42**	0.03	**1.95**	m	**2.10**
100mm wide to walls; one edge to brickwork or blockwork; one edge to concrete	0.48	5.00	**0.50**	0.12	-	**1.71**	0.04	**2.24**	m	**2.41**
Lathing; BB galvanized expanded metal lath, 9mm mesh x 0.725mm thick x 1.61 kg/m2; butt joints; fixing with nails										
100mm wide to walls; one edge to timber; one edge to brickwork or blockwork	0.56	5.00	**0.58**	0.08	-	**1.14**	0.03	**1.75**	m	**1.88**
100mm wide to walls; one edge to timber; one edge to concrete	0.56	5.00	**0.58**	0.10	-	**1.42**	0.03	**2.04**	m	**2.19**
100mm wide to walls; one edge to brickwork or blockwork; one edge to concrete	0.56	5.00	**0.58**	0.12	-	**1.71**	0.04	**2.33**	m	**2.50**
M20: PLASTERED/RENDERED/ROUGHCAST COATINGS: BONDING FLUID										
Prepare and apply bonding fluid to receive plaster or cement rendering										
Work to walls on existing cement and sand base width exceeding 300mm	0.49	10.00	**0.54**	0.20	-	**2.84**	-	**3.38**	m2	**3.64**
Work to walls on existing glazed tiling base width exceeding 300mm	0.37	10.00	**0.41**	0.15	-	**2.13**	-	**2.54**	m2	**2.73**
Work to walls on existing painted surface width exceeding 300mm	0.44	10.00	**0.49**	0.18	-	**2.56**	-	**3.05**	m2	**3.27**
Work to walls on existing concrete base width exceeding 300mm	0.49	10.00	**0.54**	0.20	-	**2.84**	-	**3.38**	m2	**3.64**
Work to ceilings on existing cement and sand base width exceeding 300mm	0.49	10.00	**0.54**	0.24	-	**3.41**	-	**3.95**	m2	**4.25**
Work to ceilings on existing painted surface width exceeding 300mm	0.44	10.00	**0.49**	0.22	-	**3.13**	-	**3.61**	m2	**3.89**
Work to ceilings on existing concrete base width exceeding 300mm	0.49	10.00	**0.54**	0.24	-	**3.41**	-	**3.95**	m2	**4.25**

Labour hourly rates: (except Specialists) Craft Operatives £14.21 Labourer £10.71 Rates are national average prices. Refer to REGIONAL VARIATIONS for indicative levels of overall pricing in regions	MATERIALS			LABOUR				RATES		
	Del to Site £	Waste %	Material Cost £	Craft Optve Hrs	Lab Hrs	Labour Cost £	Sunds £	Nett Rate £	Unit	Gross Rate (+7.5%) £
M20: PLASTERED/RENDERED/ROUGHCAST COATINGS; STAIRCASE AREAS - PLASTERBOARD										
Baseboarding; Gyproc Handiboard; 5mm joints, filled with plaster and scrimmed; fixing with galvanized nails, 30mm long at 400mm ccs; internal										
9.5mm work to walls on timber base										
width exceeding 300mm	2.67	5.00	**2.81**	0.31	0.13	**5.80**	-	**8.61**	m2	**9.25**
width not exceeding 300mm	0.87	10.00	**0.95**	0.19	0.04	**3.13**	-	**4.08**	m	**4.38**
9.5mm work to ceilings on timber base										
width exceeding 300mm	2.67	5.00	**2.81**	0.38	0.13	**6.79**	-	**9.60**	m2	**10.32**
width not exceeding 300mm	0.92	10.00	**1.00**	0.23	0.04	**3.70**	-	**4.69**	m	**5.05**
Baseboarding; Gyproc Handiboard Duplex; 5mm joints filled with plaster and scrimmed; fixing with galvanized nails, 30mm long at 400mm ccs; internal										
9.5mm work to walls on timber base										
width exceeding 300mm	3.10	5.00	**3.26**	0.31	0.13	**5.80**	-	**9.06**	m2	**9.74**
width not exceeding 300mm	1.00	10.00	**1.09**	0.19	0.04	**3.13**	-	**4.22**	m	**4.54**
9.5mm work to ceilings on timber base										
width exceeding 300mm	3.10	5.00	**3.26**	0.38	0.13	**6.79**	-	**10.05**	m2	**10.81**
width not exceeding 300mm	1.05	10.00	**1.14**	0.23	0.04	**3.70**	-	**4.84**	m	**5.20**
Baseboarding; Gyproc Handiboard; 3mm joints, filled with plaster; fixing with galvanized nails, 30mm long at 400mm ccs; internal										
12.5mm work to walls on timber base										
width exceeding 300mm	3.77	5.00	**3.96**	0.38	0.15	**7.01**	-	**10.97**	m2	**11.79**
width not exceeding 300mm	1.20	10.00	**1.31**	0.23	0.05	**3.80**	-	**5.12**	m	**5.50**
12.5mm work to ceilings on timber base										
width exceeding 300mm	3.60	5.00	**3.78**	0.45	0.15	**8.00**	-	**11.78**	m2	**12.67**
width not exceeding 300mm	1.20	10.00	**1.31**	0.27	0.05	**4.37**	-	**5.69**	m	**6.11**
Baseboarding; Gyproc plank, square edge; 5mm joints, filled with plaster and scrimmed; internal										
19mm work to walls on timber base										
width exceeding 300mm	4.41	5.00	**4.64**	0.50	0.20	**9.25**	-	**13.88**	m2	**14.93**
width not exceeding 300mm	1.41	10.00	**1.54**	0.30	0.06	**4.91**	-	**6.45**	m	**6.93**
19mm work to ceilings on timber base										
width exceeding 300mm	4.41	5.00	**4.64**	0.60	0.20	**10.67**	-	**15.30**	m2	**16.45**
width not exceeding 300mm	1.41	10.00	**1.54**	0.36	0.06	**5.76**	-	**7.30**	m	**7.85**
Baseboarding; Gyproc square edge wallboard; 3mm joints filled with plaster and scrimmed; fixing with galvanized nails, 30mm long at 400mm ccs; internal										
9.5mm work to walls on timber base										
width exceeding 300mm	1.93	5.00	**2.03**	0.31	0.13	**5.80**	-	**7.83**	m2	**8.42**
width not exceeding 300mm	0.65	10.00	**0.71**	0.19	0.04	**3.13**	-	**3.83**	m	**4.12**
12.5mm work to walls on timber base										
width exceeding 300mm	2.03	5.00	**2.14**	0.38	0.15	**7.01**	-	**9.14**	m2	**9.83**
width not exceeding 300mm	0.68	10.00	**0.74**	0.23	0.05	**3.80**	-	**4.54**	m	**4.88**
15mm work to walls on timber base										
width exceeding 300mm	3.06	5.00	**3.22**	0.45	0.18	**8.32**	-	**11.54**	m2	**12.41**
width not exceeding 300mm	0.99	10.00	**1.08**	0.27	0.05	**4.37**	-	**5.45**	m	**5.86**
9.5mm work to ceilings on timber base										
width exceeding 300mm	1.93	5.00	**2.03**	0.38	0.13	**6.79**	-	**8.82**	m2	**9.49**
width not exceeding 300mm	0.65	10.00	**0.71**	0.23	0.04	**3.70**	-	**4.40**	m	**4.73**
12.5mm work to ceilings on timber base										
width exceeding 300mm	2.03	5.00	**2.14**	0.45	0.15	**8.00**	-	**10.14**	m2	**10.90**
width not exceeding 300mm	0.68	10.00	**0.74**	0.27	0.05	**4.37**	-	**5.11**	m	**5.49**
15mm work to ceilings on timber base										
width exceeding 300mm	3.06	5.00	**3.22**	0.52	0.18	**9.32**	-	**12.54**	m2	**13.48**
width not exceeding 300mm	0.99	10.00	**1.08**	0.31	0.05	**4.94**	-	**6.02**	m	**6.47**
Baseboarding; Gyproc square edge Duplex wallboard; 3mm joints filled with plaster and scrimmed; fixing with galvanized nails, 30mm long at 400mm ccs; internal										
9.5mm work to walls on timber base										
width exceeding 300mm	2.48	5.00	**2.61**	0.31	0.13	**5.80**	-	**8.41**	m2	**9.04**
width not exceeding 300mm	0.81	10.00	**0.89**	0.19	0.04	**3.13**	-	**4.02**	m	**4.32**
12.5mm work to walls on timber base										
width exceeding 300mm	2.61	5.00	**2.75**	0.38	0.15	**7.01**	-	**9.75**	m2	**10.48**
width not exceeding 300mm	0.85	10.00	**0.93**	0.23	0.05	**3.80**	-	**4.73**	m	**5.09**
15mm work to walls on timber base										
width exceeding 300mm	3.99	5.00	**4.19**	0.45	0.18	**8.32**	-	**12.52**	m2	**13.46**
width not exceeding 300mm	1.27	10.00	**1.39**	0.27	0.05	**4.37**	-	**5.76**	m	**6.19**
9.5mm work to ceilings on timber base										
width exceeding 300mm	2.48	5.00	**2.61**	0.38	0.13	**6.79**	-	**9.40**	m2	**10.11**

Labour hourly rates: (except Specialists) Craft Operatives £14.21 Labourer £10.71 Rates are national average prices. Refer to REGIONAL VARIATIONS for indicative levels of overall pricing in regions	MATERIALS			LABOUR				RATES		
	Del to Site £	Waste %	Material Cost £	Craft Optve Hrs	Lab Hrs	Labour Cost £	Sunds £	Nett Rate £	Unit	Gross Rate (+7.5%) £
M20: PLASTERED/RENDERED/ROUGHCAST COATINGS; STAIRCASE AREAS - PLASTERBOARD Cont.										
Baseboarding; Gyproc square edge Duplex wallboard; 3mm joints filled with plaster and scrimmed; fixing with galvanized nails, 30mm long at 400mm ccs; internal Cont.										
9.5mm work to ceilings on timber base Cont.										
width not exceeding 300mm	0.81	10.00	**0.89**	0.23	0.04	**3.70**	-	**4.58**	m	**4.93**
12.5mm work to ceilings on timber base										
width exceeding 300mm	2.61	5.00	**2.75**	0.45	0.15	**8.00**	-	**10.75**	m2	**11.55**
width not exceeding 300mm	0.85	10.00	**0.93**	0.27	0.05	**4.37**	-	**5.30**	m	**5.70**
15mm work to ceilings on timber base										
width exceeding 300mm	3.99	5.00	**4.19**	0.52	0.18	**9.32**	-	**13.51**	m2	**14.52**
width not exceeding 300mm	1.27	10.00	**1.39**	0.31	0.05	**4.94**	-	**6.33**	m	**6.80**
Baseboarding; two layers 9.5mm Gyproc square edge wallboard; second layer with 3mm joints filled with plaster and scrimmed; fixing with galvanized nails, 30mm long at 400mm ccs; internal										
19mm work to walls on timber base										
width exceeding 300mm	3.69	5.00	**3.87**	0.63	0.25	**11.63**	-	**15.50**	m2	**16.67**
width not exceeding 300mm	1.23	10.00	**1.33**	0.38	0.08	**6.26**	-	**7.59**	m	**8.16**
19mm work to ceilings on timber base										
width exceeding 300mm	3.69	5.00	**3.87**	0.75	0.25	**13.34**	-	**17.21**	m2	**18.50**
width not exceeding 300mm	1.23	10.00	**1.33**	0.45	0.08	**7.25**	-	**8.58**	m	**9.23**
Baseboarding; one layer 9.5mm and one layer 12.5mm Gyproc square edge wallboard; second layer with 3mm joints filled with plaster and scrimmed; fixing with Galvanized nails, 30mm long at 400mm ccs, internal										
22mm work to walls on timber base										
width exceeding 300mm	3.79	5.00	**3.98**	0.69	0.28	**12.80**	-	**16.78**	m2	**18.04**
width not exceeding 300mm	1.26	10.00	**1.37**	0.41	0.08	**6.68**	-	**8.05**	m	**8.65**
22mm work to ceilings on timber base										
width exceeding 300mm	3.79	5.00	**3.98**	0.83	0.28	**14.79**	-	**18.77**	m2	**20.18**
width not exceeding 300mm	1.26	10.00	**1.37**	0.50	0.08	**7.96**	-	**9.33**	m	**10.03**
Baseboarding; two layers 12.5mm Gyproc square edge wallboard; second layer with 3mm joints filled with plaster and scrimmed; fixing with galvanized nails, 30mm long at 400mm ccs; internal										
25mm work to walls on timber base										
width exceeding 300mm	3.89	5.00	**4.08**	0.75	0.30	**13.87**	-	**17.95**	m2	**19.30**
width not exceeding 300mm	1.29	10.00	**1.40**	0.45	0.09	**7.36**	-	**8.76**	m	**9.41**
25mm work to ceilings on timber base										
width exceeding 300mm	3.89	5.00	**4.08**	0.90	0.30	**16.00**	-	**20.09**	m2	**21.59**
width not exceeding 300mm	1.29	10.00	**1.40**	0.54	0.09	**8.64**	-	**10.04**	m	**10.79**
Baseboarding; two layers 15mm Gyproc square edge wallboard; second layer with 3mm joints filled with plaster and scrimmed; fixing with galvanized nails, 30mm long at 400mm ccs; internal										
30mm work to walls on timber base										
width exceeding 300mm	5.95	5.00	**6.25**	0.81	0.33	**15.04**	-	**21.29**	m2	**22.89**
width not exceeding 300mm	1.91	10.00	**2.08**	0.49	0.10	**8.03**	-	**10.11**	m	**10.87**
30mm work to ceilings on timber base										
width exceeding 300mm	5.95	5.00	**6.25**	0.97	0.33	**17.32**	-	**23.56**	m2	**25.33**
width not exceeding 300mm	1.91	10.00	**2.08**	0.58	0.10	**9.31**	-	**11.39**	m	**12.25**
Baseboarding; two layers; first layer 9.5mm Gyproc Duplex wallboard, second layer 9.5mm Gyproc wallboard both layers with square edge boards; second layer with 3mm joints filled with plaster and scrimmed; fixing with galvanized nails, 30mm long at 400mm ccs; internal										
19mm work to walls on timber base										
width exceeding 300mm	4.24	5.00	**4.45**	0.63	0.25	**11.63**	-	**16.08**	m2	**17.29**
width not exceeding 300mm	1.39	10.00	**1.52**	0.38	0.08	**6.26**	-	**7.77**	m	**8.35**
19mm work to ceilings on timber base										
width exceeding 300mm	4.24	5.00	**4.45**	0.75	0.25	**13.34**	-	**17.79**	m2	**19.12**
width not exceeding 300mm	1.39	10.00	**1.52**	0.45	0.08	**7.25**	-	**8.77**	m	**9.42**
Baseboarding; two layers; first layer 12.5mm Gyproc Duplex wallboard, second layer 9.5mm Gyproc wallboard, both layers with square edge boards; second layer with 3mm joints filled with plaster and scrimmed; fixing with galvanized nails, 30mm long at 400mm ccs; internal										
22mm work to walls on timber base										
width exceeding 300mm	4.37	5.00	**4.59**	0.69	0.28	**12.80**	-	**17.39**	m2	**18.70**
width not exceeding 300mm	1.43	10.00	**1.56**	0.41	0.08	**6.68**	-	**8.24**	m	**8.86**

Labour hourly rates: (except Specialists) Craft Operatives £14.21 Labourer £10.71 Rates are national average prices. Refer to REGIONAL VARIATIONS for indicative levels of overall pricing in regions	MATERIALS			LABOUR				RATES		
	Del to Site £	Waste %	Material Cost £	Craft Optve Hrs	Lab Hrs	Labour Cost £	Sunds £	Nett Rate £	Unit	Gross Rate (+7.5%) £
M20: PLASTERED/RENDERED/ROUGHCAST COATINGS; STAIRCASE AREAS - PLASTERBOARD Cont.										
Baseboarding; two layers; first layer 12.5mm Gyproc Duplex wallboard, second layer 9.5mm Gyproc wallboard, both layers with square edge boards; second layer with 3mm joints filled with plaster and scrimmed; fixing with galvanized nails, 30mm long at 400mm ccs; internal Cont.										
22mm work to ceilings on timber base										
width exceeding 300mm	4.37	5.00	**4.59**	0.83	0.28	**14.79**	-	**19.38**	m2	**20.83**
width not exceeding 300mm	1.43	10.00	**1.56**	0.50	0.08	**7.96**	-	**9.52**	m	**10.23**
Baseboarding; two layers; first layer 12.5mm Gyproc Duplex wallboard, second layer 12.5mm Gyproc wallboard, both layers with square edge boards; second layer with 3mm joints filled with plaster and scrimmed; fixing with galvanized nails, 30mm long at 400mm ccs; internal										
25mm work to walls on timber base										
width exceeding 300mm	4.47	5.00	**4.69**	0.75	0.30	**13.87**	-	**18.56**	m2	**19.95**
width not exceeding 300mm	1.46	10.00	**1.59**	0.45	0.09	**7.36**	-	**8.95**	m	**9.62**
25mm work to ceilings on timber base										
width exceeding 300mm	4.47	5.00	**4.69**	0.90	0.30	**16.00**	-	**20.69**	m2	**22.25**
width not exceeding 300mm	1.46	10.00	**1.59**	0.54	0.09	**8.64**	-	**10.23**	m	**11.00**
Baseboarding; two layers; first layer 15mm Gyproc Duplex wallboard, second layer 15mm Gyproc wallboard, both layers with square edge boards; second layer with 3mm joints filled with plaster and scrimmed; fixing with galvanized nails, 30mm long at 400mm ccs; internal										
30mm work to walls on timber base										
width exceeding 300mm	6.88	5.00	**7.22**	0.81	0.33	**15.04**	-	**22.27**	m2	**23.94**
width not exceeding 300mm	2.19	10.00	**2.39**	0.49	0.10	**8.03**	-	**10.42**	m	**11.20**
30mm work to ceilings on timber base										
width exceeding 300mm	6.88	5.00	**7.22**	0.97	0.33	**17.32**	-	**24.54**	m2	**26.38**
width not exceeding 300mm	2.19	10.00	**2.39**	0.58	0.10	**9.31**	-	**11.70**	m	**12.58**
M20: PLASTERED/RENDERED/ROUGHCAST COATINGS: STAIRCASE AREAS - HARDWALL PLASTERING										
Plaster, BS 1191, Part 1, Class B; finishing coat of board finish, 3mm thick; steel troweled; internal										
3mm work to walls on concrete or plasterboard base										
width exceeding 300mm	0.49	5.00	**0.52**	0.31	0.13	**5.80**	0.01	**6.33**	m2	**6.80**
width not exceeding 300mm	0.26	10.00	**0.28**	0.19	0.04	**3.13**	0.00	**3.41**	m	**3.67**
3mm work to ceilings on concrete or plasterboard base										
width exceeding 300mm	0.49	5.00	**0.52**	0.38	0.13	**6.79**	0.01	**7.32**	m2	**7.87**
width not exceeding 300mm	0.26	10.00	**0.28**	0.23	0.04	**3.70**	0.00	**3.98**	m	**4.28**
Plaster; first coat of hardwall, 11mm thick; finishing coat of multi-finish 2mm thick; steel troweled; internal										
13mm work to walls on brickwork or blockwork base										
width exceeding 300mm	1.95	5.00	**2.05**	0.56	0.23	**10.42**	0.05	**12.52**	m2	**13.46**
width not exceeding 300mm	0.88	10.00	**0.97**	0.34	0.07	**5.58**	0.02	**6.57**	m	**7.06**
13mm work to ceilings on concrete base										
width exceeding 300mm	1.95	5.00	**2.05**	0.68	0.23	**12.13**	0.05	**14.22**	m2	**15.29**
width not exceeding 300mm	0.88	10.00	**0.97**	0.41	0.07	**6.58**	0.02	**7.56**	m	**8.13**
Plaster; Thistle Universal one coat plaster, 13mm thick; steel troweled; internal										
Note: the thickness is from the face of the metal lathing										
13mm work to walls on metal lathing base										
width exceeding 300mm	3.13	5.00	**3.28**	0.44	0.18	**8.18**	0.08	**11.54**	m2	**12.41**
width not exceeding 300mm	1.30	10.00	**1.43**	0.26	0.05	**4.23**	0.03	**5.69**	m	**6.12**
13mm work to ceilings on metal lathing base										
width exceeding 300mm	3.13	5.00	**3.28**	0.53	0.18	**9.46**	0.08	**12.82**	m2	**13.78**
width not exceeding 300mm	1.30	10.00	**1.43**	0.32	0.05	**5.08**	0.03	**6.54**	m	**7.03**
M20: PLASTERED/RENDERED/ROUGHCAST COATINGS: STAIRCASE AREAS - LIGHTWEIGHT PLASTERING										
Plaster, Thistle; pre-mixed; floating coat of browning, 11mm thick; finishing coat of finish, 2mm thick; steel troweled; internal										
13mm work to walls on brickwork or blockwork base										
width exceeding 300mm	1.43	5.00	**1.50**	0.50	0.20	**9.25**	0.04	**10.79**	m2	**11.59**
width not exceeding 300mm	0.46	10.00	**0.50**	0.30	0.06	**4.91**	0.01	**5.42**	m	**5.83**

Labour hourly rates: (except Specialists) Craft Operatives £14.21 Labourer £10.71 Rates are national average prices. Refer to REGIONAL VARIATIONS for indicative levels of overall pricing in regions	MATERIALS			LABOUR				RATES		
	Del to Site £	Waste %	Material Cost £	Craft Optve Hrs	Lab Hrs	Labour Cost £	Sunds £	Nett Rate £	Unit	Gross Rate (+7.5%) £
M20: PLASTERED/RENDERED/ROUGHCAST COATINGS: STAIRCASE AREAS - LIGHTWEIGHT PLASTERING Cont.										
Plaster, Thistle; pre-mixed; floating coat of bonding 8mm thick; finishing coat of finish, 2mm thick; steel troweled; internal										
10mm work to walls on concrete or plasterboard base										
width exceeding 300mm	1.43	5.00	**1.50**	0.50	0.20	**9.25**	0.04	**10.79**	m2	**11.59**
width not exceeding 300mm	0.63	10.00	**0.69**	0.30	0.06	**4.91**	0.01	**5.61**	m	**6.03**
10mm work to ceilings on concrete or plasterboard base										
width exceeding 300mm	1.43	5.00	**1.50**	0.60	0.20	**10.67**	0.04	**12.21**	m2	**13.12**
width not exceeding 300mm	0.63	10.00	**0.69**	0.36	0.06	**5.76**	0.01	**6.46**	m	**6.95**
Plaster, Thistle; tough coat, 11mm thick;finish 2mm thick steel troweled; internal										
Note: the thickness is from the face of the metal lathing										
13mm work to walls on metal lathing base										
width exceeding 300mm	1.46	5.00	**1.54**	0.69	0.28	**12.80**	0.04	**14.38**	m2	**15.45**
width not exceeding 300mm	0.64	10.00	**0.71**	0.41	0.08	**6.68**	0.01	**7.40**	m	**7.96**
13mm work to ceilings on metal lathing base										
width exceeding 300mm	1.46	5.00	**1.54**	0.83	0.28	**14.79**	0.04	**16.37**	m2	**17.59**
width not exceeding 300mm	0.64	10.00	**0.71**	0.50	0.08	**7.96**	0.01	**8.68**	m	**9.33**
M30: METAL MESH LATHING/ANCHORED REINFORCEMENT FOR PLASTERED COATINGS										
Lathing; BB galvanized expanded metal lath, 9mm mesh x 0.500mm thick x 1.11 kg/m2; butt joints; fixing with galvanized staples										
Work to walls										
width exceeding 300mm	4.75	5.00	**4.99**	0.15	0.08	**2.99**	0.05	**8.03**	m2	**8.63**
width not exceeding 300mm	1.66	10.00	**1.83**	0.09	0.02	**1.49**	0.03	**3.35**	m	**3.60**
Work to ceilings										
width exceeding 300mm	4.75	5.00	**4.99**	0.18	0.08	**3.41**	0.06	**8.46**	m2	**9.10**
width not exceeding 300mm	1.66	10.00	**1.83**	0.11	0.02	**1.78**	0.04	**3.64**	m	**3.92**
Work to isolated beams										
width exceeding 300mm	4.75	5.00	**4.99**	0.23	0.08	**4.13**	0.08	**9.19**	m2	**9.88**
width not exceeding 300mm	1.66	10.00	**1.83**	0.14	0.02	**2.20**	0.05	**4.08**	m	**4.38**
Work to isolated columns										
width exceeding 300mm	4.75	5.00	**4.99**	0.23	0.08	**4.13**	0.08	**9.19**	m2	**9.88**
width not exceeding 300mm	1.66	10.00	**1.83**	0.14	0.02	**2.20**	0.05	**4.08**	m	**4.38**
Lathing; BB galvanized expanded metal lath 9mm mesh x 0.725mm thick x 1.61 kg/m2; butt joints; fixing with galvanized staples										
Work to walls										
width exceeding 300mm	5.55	5.00	**5.83**	0.15	0.08	**2.99**	0.05	**8.86**	m2	**9.53**
width not exceeding 300mm	1.94	10.00	**2.14**	0.09	0.02	**1.49**	0.05	**3.68**	m	**3.95**
Work to ceilings										
width exceeding 300mm	5.55	5.00	**5.83**	0.18	0.08	**3.41**	0.05	**9.29**	m2	**9.98**
width not exceeding 300mm	1.94	10.00	**2.14**	0.11	0.02	**1.78**	0.05	**3.96**	m	**4.26**
Work to isolated beams										
width exceeding 300mm	5.55	5.00	**5.83**	0.23	0.08	**4.13**	0.05	**10.00**	m2	**10.75**
width not exceeding 300mm	1.94	10.00	**2.14**	0.14	0.02	**2.20**	0.05	**4.39**	m	**4.72**
Work to isolated columns										
width exceeding 300mm	5.55	5.00	**5.83**	0.23	0.08	**4.13**	0.05	**10.00**	m2	**10.75**
width not exceeding 300mm	1.94	10.00	**2.14**	0.14	0.02	**2.20**	0.05	**4.39**	m	**4.72**
Lathing; galvanized expanded metal lath 9mm mesh x 0.950mm thick x 2.50 kg/m2; butt joints; fixing with galvanized staples										
Work to walls										
width exceeding 300mm	9.00	5.00	**9.45**	0.15	0.08	**2.99**	0.05	**12.48**	m2	**13.42**
width not exceeding 300mm	3.15	10.00	**3.47**	0.09	0.02	**1.49**	0.05	**5.00**	m	**5.38**
Work to ceilings										
width exceeding 300mm	9.00	5.00	**9.45**	0.18	0.08	**3.41**	0.05	**12.91**	m2	**13.88**
width not exceeding 300mm	3.15	10.00	**3.47**	0.11	0.02	**1.78**	0.05	**5.29**	m	**5.69**
Work to isolated beams										
width exceeding 300mm	9.00	5.00	**9.45**	0.23	0.08	**4.13**	0.05	**13.62**	m2	**14.64**
width not exceeding 300mm	3.15	10.00	**3.47**	0.14	0.02	**2.20**	0.05	**5.71**	m	**6.14**
Work to isolated columns										
width exceeding 300mm	9.00	5.00	**9.45**	0.23	0.08	**4.13**	0.05	**13.62**	m2	**14.64**
width not exceeding 300mm	3.15	10.00	**3.47**	0.14	0.02	**2.20**	0.05	**5.71**	m	**6.14**

Labour hourly rates: (except Specialists) Craft Operatives £14.21 Labourer £10.71 Rates are national average prices. Refer to REGIONAL VARIATIONS for indicative levels of overall pricing in regions	MATERIALS			LABOUR				RATES		
	Del to Site £	Waste %	Material Cost £	Craft Optve Hrs	Lab Hrs	Labour Cost £	Sunds £	Nett Rate £	Unit	Gross Rate (+7.5%) £
M30: METAL MESH LATHING/ANCHORED REINFORCEMENT FOR PLASTERED COATINGS Cont.										
Lathing; Expamet galvanized Rib-Lath, 0.300mm thick x 1.16 kg/m2; butt joints; fixing with galvanized staples										
Work to walls										
width exceeding 300mm	5.90	5.00	**6.20**	0.15	0.08	**2.99**	0.05	**9.23**	m2	**9.92**
width not exceeding 300mm	2.07	10.00	**2.27**	0.09	0.02	**1.49**	0.05	**3.81**	m	**4.10**
Work to ceilings										
width exceeding 300mm	5.90	5.00	**6.20**	0.18	0.08	**3.41**	0.05	**9.66**	m2	**10.38**
width not exceeding 300mm	2.07	10.00	**2.27**	0.11	0.02	**1.78**	0.05	**4.10**	m	**4.40**
Work to isolated beams										
width exceeding 300mm	5.90	5.00	**6.20**	0.23	0.08	**4.13**	0.05	**10.37**	m2	**11.14**
width not exceeding 300mm	2.07	10.00	**2.27**	0.14	0.02	**2.20**	0.05	**4.52**	m	**4.86**
Work to isolated columns										
width exceeding 300mm	5.90	5.00	**6.20**	0.23	0.08	**4.13**	0.05	**10.37**	m2	**11.14**
width not exceeding 300mm	2.07	10.00	**2.27**	0.14	0.02	**2.20**	0.05	**4.52**	m	**4.86**
Lathing; Expamet galvanized Rib-Lath, 0.500mm thick x 1.90 kg/m2; butt joints; fixing with galvanized staples										
Work to walls										
width exceeding 300mm	6.85	5.00	**7.19**	0.15	0.08	**2.99**	0.05	**10.23**	m2	**10.99**
width not exceeding 300mm	2.40	10.00	**2.64**	0.09	0.02	**1.49**	0.05	**4.18**	m	**4.49**
Work to ceilings										
width exceeding 300mm	6.85	5.00	**7.19**	0.18	0.08	**3.41**	0.05	**10.65**	m2	**11.45**
width not exceeding 300mm	2.40	10.00	**2.64**	0.11	0.02	**1.78**	0.05	**4.46**	m	**4.80**
Work to isolated beams										
width exceeding 300mm	6.85	5.00	**7.19**	0.23	0.08	**4.13**	0.05	**11.36**	m2	**12.22**
width not exceeding 300mm	2.40	10.00	**2.64**	0.14	0.02	**2.20**	0.05	**4.89**	m	**5.25**
Work to isolated columns										
width exceeding 300mm	6.85	5.00	**7.19**	0.23	0.08	**4.13**	0.05	**11.36**	m2	**12.22**
width not exceeding 300mm	2.40	10.00	**2.64**	0.14	0.02	**2.20**	0.05	**4.89**	m	**5.25**
Lathing; Expamet stainless steel Rib-Lath, 0.300mm thick x 1.16 kg/m2; butt joints; fixing with stainless steel staples										
Work to walls										
width exceeding 300mm	15.80	5.00	**16.59**	0.15	0.08	**2.99**	0.05	**19.62**	m2	**21.10**
width not exceeding 300mm	5.53	10.00	**6.08**	0.09	0.02	**1.49**	0.05	**7.62**	m	**8.19**
Work to ceilings										
width exceeding 300mm	15.80	5.00	**16.59**	0.18	0.08	**3.41**	0.05	**20.05**	m2	**21.55**
width not exceeding 300mm	5.53	10.00	**6.08**	0.11	0.02	**1.78**	0.05	**7.91**	m	**8.50**
Work to isolated beams										
width exceeding 300mm	15.80	5.00	**16.59**	0.23	0.08	**4.13**	0.05	**20.76**	m2	**22.32**
width not exceeding 300mm	5.53	10.00	**6.08**	0.14	0.02	**2.20**	0.05	**8.33**	m	**8.96**
Work to isolated columns										
width exceeding 300mm	15.80	5.00	**16.59**	0.23	0.08	**4.13**	0.05	**20.76**	m2	**22.32**
width not exceeding 300mm	5.53	10.00	**6.08**	0.14	0.02	**2.20**	0.05	**8.33**	m	**8.96**
Lathing; galvanized expanded metal lath 9.2mm x 28.6mm mesh x 0.460mm thick x 1.15 kg/m2 for render; butt joints; fixing with galvanized staples										
Work to walls										
width exceeding 300mm	8.35	5.00	**8.77**	0.15	0.08	**2.99**	0.05	**11.80**	m2	**12.69**
width not exceeding 300mm	2.92	10.00	**3.21**	0.09	0.02	**1.49**	0.05	**4.75**	m	**5.11**
Work to ceilings										
width exceeding 300mm	8.35	5.00	**8.77**	0.18	0.08	**3.41**	0.05	**12.23**	m2	**13.15**
width not exceeding 300mm	2.92	10.00	**3.21**	0.11	0.02	**1.78**	0.05	**5.04**	m	**5.42**
Work to isolated beams										
width exceeding 300mm	8.35	5.00	**8.77**	0.23	0.08	**4.13**	0.05	**12.94**	m2	**13.91**
width not exceeding 300mm	2.92	10.00	**3.21**	0.14	0.02	**2.20**	0.05	**5.46**	m	**5.87**
Work to isolated columns										
width exceeding 300mm	8.35	5.00	**8.77**	0.23	0.08	**4.13**	0.05	**12.94**	m2	**13.91**
width not exceeding 300mm	2.92	10.00	**3.21**	0.14	0.02	**2.20**	0.05	**5.46**	m	**5.87**
Lathing; galvanized Red-rib lath, 0.500mm thick x 1.91kg/m2, butt joints; fixing with galvanized staples										
Work to walls										
width exceeding 300mm	7.50	5.00	**7.88**	0.15	0.08	**2.99**	0.05	**10.91**	m2	**11.73**
width not exceeding 300mm	2.63	10.00	**2.89**	0.09	0.02	**1.49**	0.05	**4.43**	m	**4.76**
Work to ceilings										
width exceeding 300mm	7.50	5.00	**7.88**	0.18	0.08	**3.41**	0.05	**11.34**	m2	**12.19**
width not exceeding 300mm	2.63	10.00	**2.89**	0.11	0.02	**1.78**	0.05	**4.71**	m	**5.06**
Work to isolated beams										
width exceeding 300mm	7.50	5.00	**7.88**	0.23	0.08	**4.13**	0.05	**12.05**	m2	**12.95**

Labour hourly rates: (except Specialists) Craft Operatives £14.21 Labourer £10.71 Rates are national average prices. Refer to REGIONAL VARIATIONS for indicative levels of overall pricing in regions	MATERIALS			LABOUR				RATES		
	Del to Site £	Waste %	Material Cost £	Craft Optve Hrs	Lab Hrs	Labour Cost £	Sunds £	Nett Rate £	Unit	Gross Rate (+7.5%) £
M30: METAL MESH LATHING/ANCHORED REINFORCEMENT FOR PLASTERED COATINGS Cont.										
Lathing; galvanized Red-rib lath, 0.500mm thick x 1.91kg/m2, butt joints; fixing with galvanized staples Cont.										
Work to isolated beams Cont.										
width not exceeding 300mm	2.63	10.00	**2.89**	0.14	0.02	**2.20**	0.05	**5.14**	m	**5.52**
Work to isolated columns										
width exceeding 300mm	7.50	5.00	**7.88**	0.23	0.08	**4.13**	0.05	**12.05**	m2	**12.95**
width not exceeding 300mm	2.63	10.00	**2.89**	0.14	0.02	**2.20**	0.05	**5.14**	m	**5.52**
Extra cost of fixing lathing to brickwork, blockwork or concrete with cartridge fired nails in lieu of to timber with staples										
Work to walls										
width exceeding 300mm	-	-	**-**	0.10	-	**1.42**	0.13	**1.56**	m2	**1.67**
width not exceeding 300mm	-	-	**-**	0.06	-	**0.85**	0.08	**0.93**	m	**1.00**
Extra cost of fixing lathing to steel with tying wire in lieu of to timber with staples										
Work to walls										
width exceeding 300mm	-	-	**-**	0.15	-	**2.13**	0.15	**2.28**	m2	**2.45**
width not exceeding 300mm	-	-	**-**	0.09	-	**1.28**	0.09	**1.37**	m	**1.47**
Work to ceilings										
width exceeding 300mm	-	-	**-**	0.15	-	**2.13**	0.15	**2.28**	m2	**2.45**
width not exceeding 300mm	-	-	**-**	0.09	-	**1.28**	0.09	**1.37**	m	**1.47**
Work to isolated beams										
width exceeding 300mm	-	-	**-**	0.15	-	**2.13**	0.15	**2.28**	m2	**2.45**
width not exceeding 300mm	-	-	**-**	0.09	-	**1.28**	0.09	**1.37**	m	**1.47**
Work to isolated columns										
width exceeding 300mm	-	-	**-**	0.15	-	**2.13**	0.15	**2.28**	m2	**2.45**
width not exceeding 300mm	-	-	**-**	0.09	-	**1.28**	0.09	**1.37**	m	**1.47**
Expamet galvanized arch formers and fix to brickwork or blockwork with galvanized nails, 30mm long at 400mm ccs										
Arch corners										
372mm radius	13.80	5.00	**14.49**	0.33	0.17	**6.51**	0.14	**21.14**	nr	**22.72**
452mm radius	16.85	5.00	**17.69**	0.33	0.17	**6.51**	0.17	**24.37**	nr	**26.20**
602mm radius	21.00	5.00	**22.05**	0.33	0.17	**6.51**	0.21	**28.77**	nr	**30.93**
752mm radius	27.60	5.00	**28.98**	0.33	0.17	**6.51**	0.28	**35.77**	nr	**38.45**
Semi-circular arches										
372mm radius	27.50	5.00	**28.88**	0.67	0.34	**13.16**	0.28	**42.31**	nr	**45.49**
397mm radius	28.25	5.00	**29.66**	0.67	0.34	**13.16**	0.28	**43.11**	nr	**46.34**
412mm radius	29.20	5.00	**30.66**	0.67	0.34	**13.16**	0.29	**44.11**	nr	**47.42**
452mm radius	33.80	5.00	**35.49**	0.67	0.34	**13.16**	0.34	**48.99**	nr	**52.66**
602mm radius	42.55	5.00	**44.68**	0.67	0.34	**13.16**	0.43	**58.27**	nr	**62.63**
752mm radius	58.60	5.00	**61.53**	0.67	0.34	**13.16**	0.59	**75.28**	nr	**80.92**
Elliptical arches										
1220mm wide x 340mm rise	53.00	5.00	**55.65**	0.67	0.34	**13.16**	0.53	**69.34**	nr	**74.54**
1370mm wide x 360mm rise	54.00	5.00	**56.70**	0.67	0.34	**13.16**	0.54	**70.40**	nr	**75.68**
1520mm wide x 380mm rise	59.55	5.00	**62.53**	0.67	0.34	**13.16**	0.60	**76.29**	nr	**82.01**
1830mm wide x 410mm rise	63.30	5.00	**66.47**	1.00	0.50	**19.57**	0.63	**86.66**	nr	**93.16**
2130mm wide x 430mm rise	68.15	5.00	**71.56**	1.00	0.50	**19.57**	0.68	**91.80**	nr	**98.69**
2440mm wide x 440mm rise	69.50	5.00	**72.98**	1.00	0.50	**19.57**	0.70	**93.24**	nr	**100.23**
3050mm wide x 520mm rise	72.35	5.00	**75.97**	1.00	0.50	**19.57**	0.72	**96.26**	nr	**103.48**
Spandrel arches										
760mm wide x 180mm radius x 220mm rise	35.80	5.00	**37.59**	0.67	0.34	**13.16**	0.36	**51.11**	nr	**54.94**
910mm wide x 180mm radius x 240mm rise	37.90	5.00	**39.80**	0.67	0.34	**13.16**	0.38	**53.34**	nr	**57.34**
1220mm wide x 230mm radius x 290mm rise	48.80	5.00	**51.24**	0.67	0.34	**13.16**	0.49	**64.89**	nr	**69.76**
1520mm wide x 230mm radius x 330mm rise	54.25	5.00	**56.96**	0.67	0.34	**13.16**	0.54	**70.67**	nr	**75.97**
1830mm wide x 230mm radius x 360mm rise	54.85	5.00	**57.59**	0.67	0.34	**13.16**	0.55	**71.30**	nr	**76.65**
2130mm wide x 230mm radius x 370mm rise	65.00	5.00	**68.25**	0.67	0.34	**13.16**	0.65	**82.06**	nr	**88.22**
2440mm wide x 230mm radius x 390mm rise	69.00	5.00	**72.45**	1.33	0.67	**26.08**	0.69	**99.22**	nr	**106.66**
3050mm wide x 230mm radius x 440mm rise	72.85	5.00	**76.49**	1.33	0.67	**26.08**	0.73	**103.30**	nr	**111.04**
Bulls-eyes										
222mm radius	34.45	5.00	**36.17**	0.17	0.09	**3.38**	0.34	**39.90**	nr	**42.89**
Soffit strips										
155mm wide	3.50	5.00	**3.68**	0.09	0.05	**1.81**	0.04	**5.52**	m	**5.94**
Make-up pieces										
600mm long	8.35	5.00	**8.77**	0.17	0.09	**3.38**	0.08	**12.23**	nr	**13.15**
Expamet galvanized circular window formers and fix to brickwork or blockwork with galvanized nails, 30mm long at 400mm ccs										
Circular windows										
594mm diameter	30.00	5.00	**31.50**	1.33	0.67	**26.08**	0.30	**57.88**	nr	**62.22**

Labour hourly rates: (except Specialists) Craft Operatives £14.21 Labourer £10.71 Rates are national average prices. Refer to REGIONAL VARIATIONS for indicative levels of overall pricing in regions	MATERIALS			LABOUR				RATES		
	Del to Site £	Waste %	Material Cost £	Craft Optve Hrs	Lab Hrs	Labour Cost £	Sunds £	Nett Rate £	Unit	Gross Rate (+7.5%) £
M30: METAL MESH LATHING/ANCHORED REINFORCEMENT FOR PLASTERED COATINGS; STAIRCASE AREAS										
Lathing; BB galvanized expanded metal lath, 9mm mesh x 0.500mm thick x 1.11 kg/m2; butt joints; fixing with galvanized staples										
Work to walls										
width exceeding 300mm	4.75	5.00	**4.99**	0.19	0.08	**3.56**	0.06	**8.61**	m2	**9.25**
width not exceeding 300mm	1.66	10.00	**1.83**	0.11	0.02	**1.78**	0.04	**3.64**	m	**3.92**
Work to ceilings										
width exceeding 300mm	4.75	5.00	**4.99**	0.23	0.08	**4.13**	0.08	**9.19**	m2	**9.88**
width not exceeding 300mm	1.66	10.00	**1.83**	0.14	0.02	**2.20**	0.05	**4.08**	m	**4.38**
Lathing; BB galvanized expanded metal lath 9mm mesh x 0.725mm thick x 1.61 kg/m2; butt joints; fixing with galvanized staples										
Work to walls										
width exceeding 300mm	5.55	5.00	**5.83**	0.19	0.08	**3.56**	0.06	**9.45**	m2	**10.16**
width not exceeding 300mm	1.94	10.00	**2.14**	0.11	0.02	**1.78**	0.04	**3.95**	m	**4.25**
Work to ceilings										
width exceeding 300mm	5.55	5.00	**5.83**	0.23	0.08	**4.13**	0.08	**10.03**	m2	**10.78**
width not exceeding 300mm	1.94	10.00	**2.14**	0.14	0.02	**2.20**	0.05	**4.39**	m	**4.72**
Lathing; galvanized expanded metal lath 9mm mesh x 0.950mm thick x 2.50 kg/m2; butt joints; fixing with galvanized staples										
Work to walls										
width exceeding 300mm	9.00	5.00	**9.45**	0.19	0.08	**3.56**	0.06	**13.07**	m2	**14.05**
width not exceeding 300mm	3.15	10.00	**3.47**	0.11	0.02	**1.78**	0.04	**5.28**	m	**5.67**
Work to ceilings										
width exceeding 300mm	9.00	5.00	**9.45**	0.23	0.08	**4.13**	0.08	**13.65**	m2	**14.67**
width not exceeding 300mm	3.15	10.00	**3.47**	0.14	0.02	**2.20**	0.05	**5.71**	m	**6.14**
Lathing; Expamet galvanized Rib-Lath, 0.300mm thick x 1.16 kg/m2; butt joints; fixing with galvanized staples										
Work to walls										
width exceeding 300mm	5.90	5.00	**6.20**	0.19	0.08	**3.56**	0.06	**9.81**	m2	**10.55**
width not exceeding 300mm	2.07	10.00	**2.27**	0.11	0.02	**1.78**	0.04	**4.09**	m	**4.39**
Work to ceilings										
width exceeding 300mm	5.90	5.00	**6.20**	0.23	0.08	**4.13**	0.08	**10.40**	m2	**11.18**
width not exceeding 300mm	2.07	10.00	**2.27**	0.14	0.02	**2.20**	0.05	**4.52**	m	**4.86**
Lathing; Expamet galvanized Rib-Lath, 0.500mm thick x 1.90 kg/m2; butt joints; fixing with galvanized staples										
Work to walls										
width exceeding 300mm	6.85	5.00	**7.19**	0.19	0.08	**3.56**	0.06	**10.81**	m2	**11.62**
width not exceeding 300mm	2.40	10.00	**2.64**	0.11	0.02	**1.78**	0.04	**4.45**	m	**4.78**
Work to ceilings										
width exceeding 300mm	6.85	5.00	**7.19**	0.23	0.08	**4.13**	0.08	**11.39**	m2	**12.25**
width not exceeding 300mm	2.40	10.00	**2.64**	0.14	0.02	**2.20**	0.05	**4.89**	m	**5.25**
Lathing; Expamet stainless steel Rib-Lath, 0.300mm thick x 1.16 kg/m2; butt joints; fixing with stainless steel staples										
Work to walls										
width exceeding 300mm	15.80	5.00	**16.59**	0.19	0.08	**3.56**	0.06	**20.21**	m2	**21.73**
width not exceeding 300mm	5.53	10.00	**6.08**	0.11	0.02	**1.78**	0.04	**7.90**	m	**8.49**
Work to ceilings										
width exceeding 300mm	15.80	5.00	**16.59**	0.23	0.08	**4.13**	0.08	**20.79**	m2	**22.35**
width not exceeding 300mm	5.53	10.00	**6.08**	0.14	0.02	**2.20**	0.05	**8.33**	m	**8.96**
Lathing; galvanized Red-rib lath, 0.500mm thick x 1.91kg/m2, butt joints; fixing with galvanized staples										
Work to walls										
width exceeding 300mm	7.50	5.00	**7.88**	0.19	0.08	**3.56**	0.06	**11.49**	m2	**12.36**
width not exceeding 300mm	2.63	10.00	**2.89**	0.11	0.02	**1.78**	0.04	**4.70**	m	**5.05**
Work to ceilings										
width exceeding 300mm	7.50	5.00	**7.88**	0.23	0.08	**4.13**	0.08	**12.08**	m2	**12.98**
width not exceeding 300mm	2.63	10.00	**2.89**	0.14	0.02	**2.20**	0.05	**5.14**	m	**5.52**
M31: FIBROUS PLASTER: PLASTERBOARD COVE										
Coves										
Gyproc cove; fixing with adhesive										
100mm girth	1.14	10.00	**1.26**	0.15	0.08	**2.99**	-	**4.25**	m	**4.56**
extra; ends	-	-	**-**	0.05	-	**0.71**	-	**0.71**	nr	**0.76**
extra; internal angles	-	-	**-**	0.25	-	**3.55**	-	**3.55**	nr	**3.82**
extra; external angles	-	-	**-**	0.25	-	**3.55**	-	**3.55**	nr	**3.82**
127mm girth	1.02	10.00	**1.13**	0.15	0.08	**2.99**	-	**4.11**	m	**4.42**

Labour hourly rates: (except Specialists) Craft Operatives £14.21 Labourer £10.71 Rates are national average prices. Refer to REGIONAL VARIATIONS for indicative levels of overall pricing in regions	MATERIALS			LABOUR				RATES		
	Del to Site £	Waste %	Material Cost £	Craft Optve Hrs	Lab Hrs	Labour Cost £	Sunds £	Nett Rate £	Unit	Gross Rate (+7.5%) £
M31: FIBROUS PLASTER: PLASTERBOARD COVE Cont.										
Coves Cont.										
Gyproc cove; fixing with adhesive Cont.										
extra; ends	-	-	-	0.05	-	**0.71**	-	**0.71**	nr	**0.76**
extra; internal angles	-	-	-	0.25	-	**3.55**	-	**3.55**	nr	**3.82**
extra; external angles	-	-	-	0.25	-	**3.55**	-	**3.55**	nr	**3.82**
Gyproc cove; fixing with nails to timber										
100mm girth	1.01	10.00	**1.11**	0.10	0.05	**1.96**	0.03	**3.10**	m	**3.33**
extra; ends	-	-	-	0.05	-	**0.71**	-	**0.71**	nr	**0.76**
extra; internal angles	-	-	-	0.25	-	**3.55**	-	**3.55**	nr	**3.82**
extra; external angles	-	-	-	0.25	-	**3.55**	-	**3.55**	nr	**3.82**
127mm girth	0.89	10.00	**0.98**	0.10	0.05	**1.96**	0.03	**2.97**	m	**3.19**
extra; ends	-	-	-	0.05	-	**0.71**	-	**0.71**	nr	**0.76**
extra; internal angles	-	-	-	0.25	-	**3.55**	-	**3.55**	nr	**3.82**
extra; external angles	-	-	-	0.25	-	**3.55**	-	**3.55**	nr	**3.82**
M31: FIBROUS PLASTER: PLASTER VENTILATORS										
Fibrous plaster ventilator; fixing in plastered wall										
Plain										
229 x 79mm	0.99	5.00	**1.04**	0.13	-	**1.85**	-	**2.89**	nr	**3.10**
229 x 152mm	1.19	5.00	**1.25**	0.15	-	**2.13**	-	**3.38**	nr	**3.63**
229 x 229mm	1.47	5.00	**1.54**	0.18	-	**2.56**	-	**4.10**	nr	**4.41**
Flyproof										
229 x 79mm	1.13	5.00	**1.19**	0.18	-	**2.56**	-	**3.74**	nr	**4.03**
229 x 152mm	1.42	5.00	**1.49**	0.20	-	**2.84**	-	**4.33**	nr	**4.66**
229 x 229mm	1.84	5.00	**1.93**	0.23	-	**3.27**	-	**5.20**	nr	**5.59**
M40: STONE/CONCRETE/QUARRY/CERAMIC TILING/MOSAIC: CLAY TILE PAVING										
Clay floor quarries, BS 6431, terracotta; 3mm joints, symmetrical layout; bedding in 10mm cement mortar (1:3); pointing in cement mortar (1:3); on cement and sand base										
150 x 150 x 12.5mm units to floors level or to falls only not exceeding 15 degrees from horizontal plain	15.78	5.00	**16.57**	0.80	0.50	**16.72**	-	**33.30**	m2	**35.79**
150 x 150 x 20mm units to floors level or to falls only not exceeding 15 degrees from horizontal plain	22.84	5.00	**23.98**	1.00	0.60	**20.64**	-	**44.61**	m2	**47.96**
225 x 225 x 25mm units to floors level or to falls only not exceeding 15 degrees from horizontal plain	23.98	5.00	**25.18**	0.70	0.45	**14.77**	-	**39.95**	m2	**42.94**
150 x 150 x 12.5mm units to floors to falls and crossfalls and to slopes not exceeding 15 degrees from horizontal plain	15.78	5.00	**16.57**	0.88	0.50	**17.86**	-	**34.43**	m2	**37.02**
150 x 150 x 20mm units to floors to falls and crossfalls and to slopes not exceeding 15 degrees from horizontal plain	22.84	5.00	**23.98**	1.10	0.60	**22.06**	-	**46.03**	m2	**49.49**
225 x 225 x 25mm units to floors to falls and crossfalls and to slopes not exceeding 15 degrees from horizontal plain	23.98	5.00	**25.18**	0.75	0.45	**15.48**	-	**40.66**	m2	**43.71**
Extra for pointing with tinted mortar										
150 x 150 x 12.5mm tiles	0.85	10.00	**0.93**	0.10	-	**1.42**	-	**2.36**	m2	**2.53**
150 x 150 x 20mm tiles	0.85	10.00	**0.93**	0.10	-	**1.42**	-	**2.36**	m2	**2.53**
225 x 225 x 25mm tiles	0.85	10.00	**0.93**	0.07	-	**0.99**	-	**1.93**	m2	**2.07**
150 x 150 x 12.5mm units to skirtings on brickwork or blockwork base										
height 150mm; square top edge	2.63	10.00	**2.88**	0.30	0.08	**5.12**	-	**8.00**	m	**8.60**
height 150mm; rounded top edge	2.84	10.00	**3.11**	0.30	0.08	**5.12**	-	**8.23**	m	**8.85**
height 150mm; rounded top edge and cove at bottom	4.48	10.00	**4.91**	0.30	0.08	**5.12**	-	**10.03**	m	**10.78**
150 x 150 x 20mm units to skirtings on brickwork or blockwork base										
height 150mm; square top edge	3.78	10.00	**4.14**	0.38	0.09	**6.36**	-	**10.50**	m	**11.29**
height 150mm; rounded top edge	4.35	10.00	**4.76**	0.38	0.09	**6.36**	-	**11.13**	m	**11.96**
Clay floor quarries, BS 6431, blended; 3mm joints symmetrical layout; bedding in 10mm cement mortar (1:3); pointing in cement mortar (1:3);on cement and sand base										
150 x 150 x 12.5mm units to floors level or to falls only not exceeding 15 degrees from horizontal plain	19.91	5.00	**20.91**	0.80	0.50	**16.72**	-	**37.63**	m2	**40.45**
150 x 150 x 20mm units to floors level or to falls only not exceeding 15 degrees from horizontal plain	21.89	5.00	**22.98**	1.00	0.60	**20.64**	-	**43.62**	m2	**46.89**

Labour hourly rates: (except Specialists) Craft Operatives £14.21 Labourer £10.71 Rates are national average prices. Refer to REGIONAL VARIATIONS for indicative levels of overall pricing in regions	MATERIALS			LABOUR				RATES		
	Del to Site £	Waste %	Material Cost £	Craft Optve Hrs	Lab Hrs	Labour Cost £	Sunds £	Nett Rate £	Unit	Gross Rate (+7.5%) £
M40: STONE/CONCRETE/QUARRY/CERAMIC TILING/MOSAIC: CLAY TILE PAVING Cont.										
Clay floor quarries, BS 6431, blended; 3mm joints symmetrical layout; bedding in 10mm cement mortar (1:3); pointing in cement mortar (1:3);on cement and sand base Cont.										
225 x 225 x 25mm units to floors level or to falls only not exceeding 15 degrees from horizontal										
plain	24.70	5.00	**25.94**	0.70	0.45	**14.77**	-	**40.70**	m2	**43.75**
150 x 150 x 12.5mm units to floors to falls and crossfalls and to slopes not exceeding 15 degrees from horizontal										
plain	19.91	5.00	**20.91**	0.88	0.50	**17.86**	-	**38.77**	m2	**41.67**
150 x 150 x 20mm units to floors to falls and crossfalls and to slopes not exceeding 15 degrees from horizontal										
plain	21.89	5.00	**22.98**	1.10	0.60	**22.06**	-	**45.04**	m2	**48.42**
225 x 225 x 25mm units to floors to falls and crossfalls and to slopes not exceeding 15 degrees from horizontal										
plain	24.70	5.00	**25.94**	0.75	0.45	**15.48**	-	**41.41**	m2	**44.52**
Extra for pointing with tinted mortar										
150 x 150 x 12.5mm tiles	0.85	10.00	**0.93**	0.10	-	**1.42**	-	**2.36**	m2	**2.53**
150 x 150 x 20mm tiles	0.85	10.00	**0.93**	0.10	-	**1.42**	-	**2.36**	m2	**2.53**
225 x 225 x 25mm tiles	0.85	10.00	**0.93**	0.07	-	**0.99**	-	**1.93**	m2	**2.07**
150 x 150 x 12.5mm units to skirtings on brickwork or blockwork base										
height 150mm; square top edge	3.30	10.00	**3.62**	0.30	0.08	**5.12**	-	**8.74**	m	**9.39**
height 150mm; rounded top edge	3.72	10.00	**4.07**	0.30	0.08	**5.12**	-	**9.19**	m	**9.88**
height 150mm; rounded top edge and cove at bottom	4.64	10.00	**5.09**	0.30	0.08	**5.12**	-	**10.21**	m	**10.97**
150 x 150 x 20mm units to skirtings on brickwork or blockwork base										
height 150mm; square top edge	3.72	10.00	**4.07**	0.38	0.09	**6.36**	-	**10.44**	m	**11.22**
height 150mm; rounded top edge	4.01	10.00	**4.40**	0.38	0.09	**6.36**	-	**10.76**	m	**11.57**
Clay floor quarries, BS 6431, dark; 3mm joints, symmetrical layout; bedding in 10mm cement mortar (1:3); pointing in cement mortar (1:3); on cement and sand base										
150 x 150 x 12.5mm units to floors level or to falls only not exceeding 15degrees from horizontal										
plain	22.62	5.00	**23.75**	0.80	0.50	**16.72**	-	**40.48**	m2	**43.51**
150 x 150 x 12.5mm units to floors to falls and crossfalls and to slopes not exceeding 15 degrees from horizontal										
plain	22.62	5.00	**23.75**	0.88	0.50	**17.86**	-	**41.61**	m2	**44.73**
150 x 150 x 12.5mm units to skirtings on brickwork or blockwork base										
height 150mm; square top edge	3.67	10.00	**4.02**	0.30	0.08	**5.12**	-	**9.14**	m	**9.82**
height 150mm; rounded top edge	3.97	10.00	**4.35**	0.30	0.08	**5.12**	-	**9.47**	m	**10.18**
height 150mm; rounded top edge and cove at bottom	5.37	10.00	**5.89**	0.30	0.08	**5.12**	-	**11.01**	m	**11.83**
Ceramic floor tiles, BS 6431, fully vitrified, red; 3mm joints, symmetrical layout; bedding in 10mm cement mortar (1:3); pointing in cement mortar (1:3);on cement and sand base										
100 x 100 x 9mm units to floors level or to falls only not exceeding 15 degrees from horizontal										
plain	16.53	5.00	**17.36**	1.60	0.90	**32.38**	-	**49.73**	m2	**53.46**
152 x 152 x 12mm units to floors level or to falls only not exceeding 15 degrees from horizontal										
plain	18.03	5.00	**18.93**	0.90	0.55	**18.68**	-	**37.61**	m2	**40.43**
200 x 200 x 12mm units to floors level or to falls only not exceeding 15 degrees from horizontal										
plain	19.23	5.00	**20.19**	0.60	0.40	**12.81**	-	**33.00**	m2	**35.48**
100 x 100 x 9mm units to floors to falls and crossfalls and to slopes not exceeding 15 degrees from horizontal										
plain	16.53	5.00	**17.36**	1.76	0.90	**34.65**	-	**52.01**	m2	**55.91**
152 x 152 x 12mm units to floors to falls and crossfalls and to slopes not exceeding 15 degrees from horizontal										
plain	18.03	5.00	**18.93**	0.99	0.55	**19.96**	-	**38.89**	m2	**41.81**
200 x 200 x 12mm units to floors to falls and crossfalls and to slopes not exceeding 15 degrees from horizontal										
plain	19.23	5.00	**20.19**	0.66	0.40	**13.66**	-	**33.85**	m2	**36.39**

Labour hourly rates: (except Specialists) Craft Operatives £14.21 Labourer £10.71 Rates are national average prices. Refer to REGIONAL VARIATIONS for indicative levels of overall pricing in regions	MATERIALS			LABOUR				RATES		
	Del to Site £	Waste %	Material Cost £	Craft Optve Hrs	Lab Hrs	Labour Cost £	Sunds £	Nett Rate £	Unit	Gross Rate (+7.5%) £
M40: STONE/CONCRETE/QUARRY/CERAMIC TILING/MOSAIC: CLAY TILE PAVING Cont.										
Ceramic floor tiles, BS 6431, fully vitrified, red; 3mm joints, symmetrical layout; bedding in 10mm cement mortar (1:3); pointing in cement mortar (1:3);on cement and sand base Cont.										
Extra for pointing with tinted mortar										
100 x 100 x 9mm tiles	0.85	10.00	**0.93**	0.15	-	**2.13**	-	**3.07**	m2	**3.30**
152 x 152 x 12mm tiles	0.85	10.00	**0.93**	0.10	-	**1.42**	-	**2.36**	m2	**2.53**
200 x 200 x 12mm tiles	0.85	10.00	**0.93**	0.08	-	**1.14**	-	**2.07**	m2	**2.23**
100 x 100 x 9mm units to skirtings on brickwork or blockwork base										
height 100mm; square top edge	1.71	10.00	**1.87**	0.36	0.09	**6.08**	-	**7.95**	m	**8.54**
height 100mm; rounded top edge	2.20	10.00	**2.41**	0.36	0.09	**6.08**	-	**8.49**	m	**9.13**
152 x 152 x 12mm units to skirtings on brickwork or blockwork base										
height 152mm; square top edge	3.00	10.00	**3.29**	0.33	0.08	**5.55**	-	**8.83**	m	**9.50**
height 152mm; rounded top edge	3.57	10.00	**3.91**	0.33	0.08	**5.55**	-	**9.45**	m	**10.16**
height 152mm; rounded top edge and cove at bottom	5.02	10.00	**5.50**	0.33	0.08	**5.55**	-	**11.05**	m	**11.88**
200 x 200 x 12mm units to skirtings on brickwork or blockwork base										
height 200mm; square top edge	4.07	10.00	**4.45**	0.30	0.08	**5.12**	-	**9.57**	m	**10.29**
Ceramic floor tiles, BS 6431, fully vitrified, cream; 3mm joints, symmetrical layout; bedding in 10mm cement mortar (1:3); pointing in cement mortar (1:3); on cement and sand base										
100 x 100 x 9mm units to floors level or to falls only not exceeding 15 degrees from horizontal										
plain	17.43	5.00	**18.30**	1.60	0.90	**32.38**	-	**50.68**	m2	**54.48**
152 x 152 x 12mm units to floors level or to falls only not exceeding 15 degrees from horizontal										
plain	18.73	5.00	**19.67**	0.90	0.55	**18.68**	-	**38.35**	m2	**41.22**
200 x 200 x 12mm units to floors level or to falls only not exceeding 15 degrees from horizontal										
plain	19.93	5.00	**20.93**	0.60	0.40	**12.81**	-	**33.74**	m2	**36.27**
100 x 100 x 9mm units to floors to falls and crossfalls and to slopes not exceeding 15 degrees from horizontal										
plain	17.43	5.00	**18.30**	1.76	0.90	**34.65**	-	**52.95**	m2	**56.92**
152 x 152 x 12mm units to floors to falls and crossfalls and to slopes not exceeding 15 degrees from horizontal										
plain	18.73	5.00	**19.67**	0.99	0.55	**19.96**	-	**39.62**	m2	**42.60**
200 x 200 x 12mm units to floors to falls and crossfalls and to slopes not exceeding 15 degrees from horizontal										
plain	19.93	5.00	**20.93**	0.66	0.40	**13.66**	-	**34.59**	m2	**37.18**
Extra for pointing with tinted mortar										
100 x 100 x 9mm tiles	0.85	10.00	**0.93**	0.15	-	**2.13**	-	**3.07**	m2	**3.30**
152 x 152 x 12mm tiles	0.85	10.00	**0.93**	0.10	-	**1.42**	-	**2.36**	m2	**2.53**
200 x 200 x 12mm tiles	0.85	10.00	**0.93**	0.08	-	**1.14**	-	**2.07**	m2	**2.23**
100 x 100 x 9mm units to skirtings on brickwork or blockwork base										
height 100mm; square top edge	1.80	10.00	**1.97**	0.36	0.09	**6.08**	-	**8.05**	m	**8.65**
height 100mm; rounded top edge	2.29	10.00	**2.51**	0.36	0.09	**6.08**	-	**8.59**	m	**9.23**
152 x 152 x 12mm units to skirtings on brickwork or blockwork base										
height 152mm; square top edge	2.93	10.00	**3.20**	0.33	0.08	**5.55**	-	**8.75**	m	**9.41**
height 152mm; rounded top edge	3.63	10.00	**3.97**	0.33	0.08	**5.55**	-	**9.52**	m	**10.23**
height 152mm; rounded top edge and cove at bottom	5.02	10.00	**5.50**	0.33	0.08	**5.55**	-	**11.05**	m	**11.88**
200 x 200 x 12mm units to skirtings on brickwork or blockwork base										
height 200mm; square top edge	4.21	10.00	**4.60**	0.30	0.08	**5.12**	-	**9.72**	m	**10.45**
Ceramic floor tiles, BS 6431, fully vitrified, black; 3mm joints, symmetrical layout; bedding in 10mm cement mortar (1:3); pointing in cement mortar (1:3); on cement and sand base										
100 x 100 x 9mm units to floors level or to falls only not exceeding 15 degrees from horizontal										
plain	21.03	5.00	**22.08**	1.60	0.90	**32.38**	-	**54.46**	m2	**58.54**
152 x 152 x 12mm units to floors level or to falls only not exceeding 15 degrees from horizontal										
plain	22.03	5.00	**23.13**	0.90	0.55	**18.68**	-	**41.81**	m2	**44.95**
200 x 200 x 12mm units to floors level or to falls only not exceeding 15 degrees from horizontal										
plain	22.78	5.00	**23.92**	0.60	0.40	**12.81**	-	**36.73**	m2	**39.48**

Labour hourly rates: (except Specialists) Craft Operatives £14.21 Labourer £10.71 Rates are national average prices. Refer to REGIONAL VARIATIONS for indicative levels of overall pricing in regions	MATERIALS			LABOUR				RATES		
	Del to Site £	Waste %	Material Cost £	Craft Optve Hrs	Lab Hrs	Labour Cost £	Sunds £	Nett Rate £	Unit	Gross Rate (+7.5%) £
M40: STONE/CONCRETE/QUARRY/CERAMIC TILING/MOSAIC: CLAY TILE PAVING Cont.										
Ceramic floor tiles, BS 6431, fully vitrified, black; 3mm joints, symmetrical layout; bedding in 10mm cement mortar (1:3); pointing in cement mortar (1:3); on cement and sand base Cont.										
100 x 100 x 9mm units to floors to falls and crossfalls and to slopes not exceeding 15 degrees from horizontal										
plain	21.03	5.00	**22.08**	1.76	0.90	**34.65**	-	**56.73**	m2	**60.98**
152 x 152 x 12mm units to floors to falls and crossfalls and to slopes not exceeding 15 degrees from horizontal										
plain	22.03	5.00	**23.13**	0.99	0.55	**19.96**	-	**43.09**	m2	**46.32**
200 x 200 x 12mm units to floors to falls and crossfalls and to slopes not exceeding 15 degrees from horizontal										
plain	22.78	5.00	**23.92**	0.66	0.40	**13.66**	-	**37.58**	m2	**40.40**
Extra for pointing with tinted mortar										
100 x 100 x 9mm tiles	0.85	10.00	**0.93**	0.15	-	**2.13**	-	**3.07**	m2	**3.30**
152 x 152 x 12mm tiles	0.85	10.00	**0.93**	0.10	-	**1.42**	-	**2.36**	m2	**2.53**
200 x 200 x 12mm tiles	0.85	10.00	**0.93**	0.08	-	**1.14**	-	**2.07**	m2	**2.23**
100 x 100 x 9mm units to skirtings on brickwork or blockwork base										
height 100mm; square top edge	2.16	10.00	**2.36**	0.36	0.09	**6.08**	-	**8.44**	m	**9.08**
height 100mm; rounded top edge	2.66	10.00	**2.91**	0.36	0.09	**6.08**	-	**8.99**	m	**9.67**
152 x 152 x 12mm units to skirtings on brickwork or blockwork base										
height 152mm; square top edge	3.43	10.00	**3.76**	0.33	0.08	**5.55**	-	**9.30**	m	**10.00**
height 152mm; rounded top edge	3.95	10.00	**4.32**	0.33	0.08	**5.55**	-	**9.87**	m	**10.61**
height 152mm; rounded top edge and cove at bottom	5.02	10.00	**5.50**	0.33	0.08	**5.55**	-	**11.05**	m	**11.88**
200 x 200 x 12mm units to skirtings on brickwork or blockwork base										
height 200mm; square top edge	4.78	10.00	**5.23**	0.30	0.08	**5.12**	-	**10.35**	m	**11.13**
M40: STONE/CONCRETE/QUARRY/CERAMIC TILING/MOSAIC: TILE WALL LININGS										
Ceramic tiles, BS 6431, glazed white; 2mm joints, symmetrical layout; bedding in 10mm cement mortar (1:3); pointing with neat white cement; on brickwork or blockwork base										
108 x 108 x 6.5mm units to walls										
plain, width exceeding 300mm	14.94	5.00	**15.70**	1.30	0.75	**26.51**	-	**42.21**	m2	**45.37**
plain, width not exceeding 300mm	4.94	10.00	**5.41**	0.78	0.47	**16.12**	-	**21.53**	m	**23.14**
152 x 152 x 5.5mm units to walls										
plain, width exceeding 300mm	10.41	5.00	**10.94**	1.00	0.60	**20.64**	-	**31.58**	m2	**33.95**
plain, width not exceeding 300mm	3.23	10.00	**3.53**	0.60	0.36	**12.38**	-	**15.91**	m	**17.11**
Ceramic tiles, BS 6431, glazed white; 2mm joints; symmetrical layout; fixing with thin bed adhesive; pointing with neat white cement; on plaster base										
108 x 108 x 6.5mm units to walls										
plain, width exceeding 300mm	15.33	5.00	**16.19**	1.30	0.75	**26.51**	-	**42.69**	m2	**45.90**
plain, width not exceeding 300mm	5.28	10.00	**5.81**	0.78	0.47	**16.12**	-	**21.92**	m	**23.57**
152 x 152 x 5.5mm units to walls										
plain, width exceeding 300mm	10.80	5.00	**11.43**	1.00	0.60	**20.64**	-	**32.06**	m2	**34.47**
plain, width not exceeding 300mm	3.57	10.00	**3.93**	0.60	0.36	**12.38**	-	**16.31**	m	**17.53**
Ceramic tiles, BS 6431, glazed white; 2mm joints, symmetrical layout; fixing with thick bed adhesive; pointing with neat white cement; on cement and sand base										
108 x 108 x 6.5mm units to walls										
plain, width exceeding 300mm	16.09	5.00	**17.02**	1.30	0.75	**26.51**	-	**43.53**	m2	**46.79**
plain, width not exceeding 300mm	5.62	10.00	**6.18**	0.78	0.47	**16.12**	-	**22.30**	m	**23.97**
152 x 152 x 5.5mm units to walls										
plain, width exceeding 300mm	11.56	5.00	**12.26**	1.00	0.60	**20.64**	-	**32.90**	m2	**35.37**
plain, width not exceeding 300mm	3.91	10.00	**4.30**	0.60	0.36	**12.38**	-	**16.68**	m	**17.93**
Ceramic tiles, BS 6431, glazed light colour; 2mm joints symmetrical layout; bedding in 10mm cement mortar (1:3); pointing with neat white cement; on cement and sand base										
108 x 108 x 6.5mm units to walls										
plain, width exceeding 300mm	14.94	5.00	**15.70**	1.30	0.75	**26.51**	-	**42.21**	m2	**45.37**
plain, width not exceeding 300mm	4.94	10.00	**5.41**	0.78	0.47	**16.12**	-	**21.53**	m	**23.14**
152 x 152 x 5.5mm units to walls										
plain, width exceeding 300mm	10.41	5.00	**10.94**	1.00	0.60	**20.64**	-	**31.58**	m2	**33.95**
plain, width not exceeding 300mm	3.23	10.00	**3.53**	0.60	0.36	**12.38**	-	**15.91**	m	**17.11**

Labour hourly rates: (except Specialists) Craft Operatives £14.21 Labourer £10.71 Rates are national average prices. Refer to REGIONAL VARIATIONS for indicative levels of overall pricing in regions	MATERIALS			LABOUR				RATES		
	Del to Site £	Waste %	Material Cost £	Craft Optve Hrs	Lab Hrs	Labour Cost £	Sunds £	Nett Rate £	Unit	Gross Rate (+7.5%) £
M40: STONE/CONCRETE/QUARRY/CERAMIC TILING/MOSAIC: TILE WALL LININGS Cont.										
Ceramic tiles, BS 6431, glazed light colour; 2mm joints, symmetrical layout; fixing with thin bed adhesive; pointing with neat white cement; on plaster base										
108 x 108 x 6.5mm units to walls										
plain, width exceeding 300mm....................	15.33	5.00	**16.19**	1.30	0.75	**26.51**	-	**42.69**	m2	**45.90**
plain, width not exceeding 300mm................	5.28	10.00	**5.81**	0.78	0.47	**16.12**	-	**21.92**	m	**23.57**
152 x 152 x 5.5mm units to walls										
plain, width exceeding 300mm....................	10.80	5.00	**11.43**	1.00	0.60	**20.64**	-	**32.06**	m2	**34.47**
plain, width not exceeding 300mm................	3.57	10.00	**3.93**	0.60	0.36	**12.38**	-	**16.31**	m	**17.53**
Ceramic tiles, BS 6431, glazed light colour; 2mm joints, symmetrical layout; fixing with thick bed adhesive; pointing with neat white cement; on cement and sand base										
108 x 108 x 6.5mm units to walls										
plain, width exceeding 300mm....................	16.09	5.00	**17.02**	1.30	0.75	**26.51**	-	**43.53**	m2	**46.79**
plain, width not exceeding 300mm................	5.62	10.00	**6.18**	0.78	0.47	**16.12**	-	**22.30**	m	**23.97**
152 x 152 x 5.5mm units to walls										
plain, width exceeding 300mm....................	11.56	5.00	**12.26**	1.00	0.60	**20.64**	-	**32.90**	m2	**35.37**
plain, width not exceeding 300mm................	3.91	10.00	**4.30**	0.60	0.36	**12.38**	-	**16.68**	m	**17.93**
Ceramic tiles, BS 6431, glazed dark colour; 2mm joints, symmetrical layout; bedding in 10mm cement mortar (1:3); pointing with neat white cement; on brickwork or blockwork base										
108 x 108 x 6.5mm units to walls										
plain, width exceeding 300mm....................	17.14	5.00	**18.01**	1.30	0.75	**26.51**	-	**44.52**	m2	**47.86**
plain, width not exceeding 300mm................	5.66	10.00	**6.21**	0.78	0.47	**16.12**	-	**22.33**	m	**24.00**
152 x 152 x 5.5mm units to walls										
plain, width exceeding 300mm....................	12.61	5.00	**13.25**	1.00	0.60	**20.64**	-	**33.89**	m2	**36.43**
plain, width not exceeding 300mm................	3.90	10.00	**4.28**	0.60	0.36	**12.38**	-	**16.66**	m	**17.91**
Ceramic tiles, BS 6431, glazed dark colour; 2mm joints, symmetrical layout; fixing with thin bed adhesive; pointing with neat white cement; on plaster base										
108 x 108 x 6.5mm units to walls										
plain, width exceeding 300mm....................	17.53	5.00	**18.50**	1.30	0.75	**26.51**	-	**45.00**	m2	**48.38**
plain, width not exceeding 300mm................	6.01	10.00	**6.61**	0.78	0.47	**16.12**	-	**22.72**	m	**24.43**
152 x 152 x 5.5mm units to walls										
plain, width exceeding 300mm....................	13.00	5.00	**13.74**	1.00	0.60	**20.64**	-	**34.37**	m2	**36.95**
plain, width not exceeding 300mm................	4.25	10.00	**4.67**	0.60	0.36	**12.38**	-	**17.05**	m	**18.33**
Ceramic tiles, BS 6431, glazed dark colour; 2mm joints, symmetrical layout; fixing with thick bed adhesive; pointing with neat white cement; on cement and sand base										
108 x 108 x 6.5mm units to walls										
plain, width exceeding 300mm....................	18.29	5.00	**19.33**	1.30	0.75	**26.51**	-	**45.84**	m2	**49.28**
plain, width not exceeding 300mm................	6.35	10.00	**6.98**	0.78	0.47	**16.12**	-	**23.10**	m	**24.83**
152 x 152 x 5.5mm units to walls										
plain, width exceeding 300mm....................	13.76	5.00	**14.57**	1.00	0.60	**20.64**	-	**35.21**	m2	**37.85**
plain, width not exceeding 300mm................	4.59	10.00	**5.05**	0.60	0.36	**12.38**	-	**17.43**	m	**18.74**
Extra over ceramic tiles, BS 6431, glazed white; 2mm joints, symmetrical layout; bedding or fixing in any material to general surfaces on any base										
Special tiles										
rounded edge	2.97	5.00	**3.12**	-	-	**-**	-	**3.12**	m	**3.35**
external angle	5.94	5.00	**6.24**	-	-	**-**	-	**6.24**	m	**6.70**
internal angle to skirting......................	-	-	**-**	0.07	-	**0.99**	-	**0.99**	nr	**1.07**
external angle to skirting......................	-	-	**-**	0.07	-	**0.99**	-	**0.99**	nr	**1.07**
internal angle	-	-	**-**	0.10	-	**1.42**	-	**1.42**	m	**1.53**
Extra over ceramic tiles, BS 6431, glazed light colour; 2mm joints, symmetrical layout; bedding or fixing in any material to general surfaces on any base										
Special tiles										
rounded edge	2.97	5.00	**3.12**	-	-	**-**	-	**3.12**	m	**3.35**
external angle	5.94	5.00	**6.24**	-	-	**-**	-	**6.24**	m	**6.70**
internal angle to skirting......................	-	-	**-**	0.07	-	**0.99**	-	**0.99**	nr	**1.07**
external angle to skirting......................	-	-	**-**	0.07	-	**0.99**	-	**0.99**	nr	**1.07**
internal angle	-	-	**-**	0.10	-	**1.42**	-	**1.42**	m	**1.53**
Extra over ceramic tiles, BS 6431, glazed dark colour; 2mm joints, symmetrical layout; bedding or fixing in any material to general surfaces on any base										
Special tiles										
rounded edge	3.45	5.00	**3.63**	-	-	**-**	-	**3.63**	m	**3.90**

Labour hourly rates: (except Specialists) Craft Operatives £14.21 Labourer £10.71 Rates are national average prices. Refer to REGIONAL VARIATIONS for indicative levels of overall pricing in regions	MATERIALS			LABOUR				RATES		
	Del to Site £	Waste %	Material Cost £	Craft Optve Hrs	Lab Hrs	Labour Cost £	Sunds £	Nett Rate £	Unit	Gross Rate (+7.5%) £
M40: STONE/CONCRETE/QUARRY/CERAMIC TILING/MOSAIC: TILE WALL LININGS Cont.										
Extra over ceramic tiles, BS 6431, glazed dark colour; 2mm joints, symmetrical layout; bedding or fixing in any material to general surfaces on any base Cont.										
Special tiles Cont.										
external angle bead	3.45	5.00	**3.63**	0.22	-	**3.13**	-	**6.75**	m	**7.26**
internal angle to skirting	-	-	-	0.07	-	**0.99**	-	**0.99**	nr	**1.07**
external angle to skirting	-	-	-	0.07	-	**0.99**	-	**0.99**	nr	**1.07**
internal angle	-	-	-	0.10	-	**1.42**	-	**1.42**	m	**1.53**
M40: STONE/CONCRETE/QUARRY/CERAMIC TILING/MOSAIC: TILE CILLS										
Ceramic tiles, BS 6431, glazed white; 2mm joints, symmetrical layout; bedding in 10mm cement mortar (1:3); pointing with neat white cement; on brickwork or blockwork base - cills										
108 x 108 x 4mm units to cills on brickwork or blockwork base										
width 150mm; rounded angle	2.52	5.00	**2.65**	0.39	0.11	**6.72**	-	**9.37**	m	**10.08**
width 225mm; rounded angle	4.16	5.00	**4.37**	0.54	0.17	**9.49**	-	**13.86**	m	**14.90**
width 300mm; rounded angle	5.16	5.00	**5.42**	0.69	0.23	**12.27**	-	**17.69**	m	**19.02**
152 x 152 x 5.5mm units to cills on brickwork or blockwork base										
width 150mm; rounded angle	2.32	5.00	**2.44**	0.33	0.09	**5.65**	-	**8.09**	m	**8.70**
width 225mm; rounded angle	3.14	5.00	**3.30**	0.44	0.14	**7.75**	-	**11.06**	m	**11.89**
width 300mm; rounded angle	4.08	5.00	**4.29**	0.55	0.18	**9.74**	-	**14.03**	m	**15.09**
Clay floor quarries, BS 6431, terracotta; 3mm joints, symmetrical layout; bedding in 10mm cement mortar (1:3); pointing in cement mortar (1:3); on cement and sand base - cills										
152 x 152 x 12.5mm units to cills on brickwork or blockwork base										
width 152mm; rounded angle	2.84	10.00	**3.11**	0.30	0.08	**5.12**	-	**8.23**	m	**8.85**
width 225mm; rounded angle	4.21	5.00	**4.55**	0.40	0.12	**6.97**	-	**11.52**	m	**12.38**
width 300mm; rounded angle	5.59	5.00	**5.99**	0.51	0.17	**9.07**	-	**15.06**	m	**16.19**
M40: STONE/CONCRETE/QUARRY/CERAMIC TILING/MOSAIC; PLANT ROOMS										
Clay floor quarries, BS 6431, terracotta; 3mm joints, symmetrical layout; bedding in 10mm cement mortar (1:3); pointing in cement mortar (1:3); on cement and sand base										
150 x 150 x 12.5mm units to floors level or to falls only not exceeding 15 degrees from horizontal										
plain	15.78	5.00	**16.57**	0.88	0.50	**17.86**	-	**34.43**	m2	**37.02**
150 x 150 x 20mm units to floors level or to falls only not exceeding 15 degrees from horizontal										
plain	22.84	5.00	**23.98**	1.10	0.60	**22.06**	-	**46.03**	m2	**49.49**
225 x 225 x 25mm units to floors level or to falls only not exceeding 15 degrees from horizontal										
plain	23.98	5.00	**25.18**	0.75	0.45	**15.48**	-	**40.66**	m2	**43.71**
150 x 150 x 12.5mm units to floors to falls and crossfalls and to slopes not exceeding 15 degrees from horizontal										
plain	15.78	5.00	**16.57**	0.96	0.50	**19.00**	-	**35.57**	m2	**38.24**
150 x 150 x 20mm units to floors to falls and crossfalls and to slopes not exceeding 15 degrees from horizontal										
plain	22.84	5.00	**23.98**	1.20	0.60	**23.48**	-	**47.46**	m2	**51.01**
225 x 225 x 25mm units to floors to falls and crossfalls and to slopes not exceeding 15 degrees from horizontal										
plain	23.98	5.00	**25.18**	0.80	0.45	**16.19**	-	**41.37**	m2	**44.47**
Clay floor quarries, BS 6431, blended; 3mm joints symmetrical layout; bedding in 10mm cement mortar (1:3); pointing in cement mortar (1:3);on cement and sand base										
150 x 150 x 12.5mm units to floors level or to falls only not exceeding 15 degrees from horizontal										
plain	19.91	5.00	**20.91**	0.88	0.50	**17.86**	-	**38.77**	m2	**41.67**
150 x 150 x 20mm units to floors level or to falls only not exceeding 15 degrees from horizontal										
plain	21.89	5.00	**22.98**	1.10	0.60	**22.06**	-	**45.04**	m2	**48.42**
225 x 225 x 25mm units to floors level or to falls only not exceeding 15 degrees from horizontal										
plain	24.70	5.00	**25.94**	0.75	0.45	**15.48**	-	**41.41**	m2	**44.52**

Labour hourly rates: (except Specialists) Craft Operatives £14.21 Labourer £10.71 Rates are national average prices. Refer to REGIONAL VARIATIONS for indicative levels of overall pricing in regions	MATERIALS			LABOUR				RATES		
	Del to Site £	Waste %	Material Cost £	Craft Optve Hrs	Lab Hrs	Labour Cost £	Sunds £	Nett Rate £	Unit	Gross Rate (+7.5%) £
M40: STONE/CONCRETE/QUARRY/CERAMIC TILING/MOSAIC; PLANT ROOMS Cont.										
Clay floor quarries, BS 6431, blended; 3mm joints symmetrical layout; bedding in 10mm cement mortar (1:3); pointing in cement mortar (1:3);on cement and sand base Cont.										
150 x 150 x 12.5mm units to floors to falls and crossfalls and to slopes not exceeding 15 degrees from horizontal										
plain	19.91	5.00	**20.91**	0.96	0.50	**19.00**	-	**39.90**	m2	**42.90**
150 x 150 x 20mm units to floors to falls and crossfalls and to slopes not exceeding 15 degrees from horizontal										
plain	21.89	5.00	**22.98**	1.20	0.60	**23.48**	-	**46.46**	m2	**49.95**
225 x 225 x 25mm units to floors to falls and crossfalls and to slopes not exceeding 15 degrees from horizontal										
plain	24.70	5.00	**25.94**	0.80	0.45	**16.19**	-	**42.12**	m2	**45.28**
Clay floor quarries, BS 6431, dark; 3mm joints, symmetrical layout; bedding in 10mm cement mortar (1:3); pointing in cement mortar (1:3); on cement and sand base										
150 x 150 x 12.5mm units to floors level or to falls only not exceeding 15degrees from horizontal										
plain	22.62	5.00	**23.75**	0.88	0.50	**17.86**	-	**41.61**	m2	**44.73**
150 x 150 x 12.5mm units to floors to falls and crossfalls and to slopes not exceeding 15 degrees from horizontal										
plain	22.62	5.00	**23.75**	0.96	0.50	**19.00**	-	**42.75**	m2	**45.95**
M41: TERRAZZO TILING/IN SITU TERRAZZO										
Terrazzo tiles, BS 4131, aggregate size random, ground, grouted and polished, standard colour range; 3mm joints, symmetrical layout; bedding in 40mm cement mortar (1:3); grouting with white cement; in-situ margins										
300 x 300 x 28mm units to floors on concrete base; level or to falls only not exceeding 15 degrees from horizontal										
plain	38.47	5.00	**40.55**	1.10	0.65	**22.59**	-	**63.14**	m2	**67.87**
Terrazzo, white cement and white marble chippings (2:5); polished; on cement and sand base										
6mm work to walls										
width exceeding 300mm	6.00	5.00	**6.30**	3.60	1.80	**70.43**	-	**76.73**	m2	**82.49**
width not exceeding 300mm	3.00	10.00	**3.30**	2.16	0.54	**36.48**	-	**39.78**	m	**42.76**
16mm work to floors; one coat; floated										
level and to falls only not exceeding 15 degrees from horizontal; laid in bays, average size 610 x 610mm	16.00	5.00	**16.80**	2.25	1.13	**44.07**	-	**60.87**	m2	**65.44**
6mm work to skirtings										
height 75mm	1.00	5.00	**1.05**	1.25	0.14	**19.26**	-	**20.31**	m	**21.84**
height 75mm; curved to 3000mm radius	1.00	5.00	**1.05**	1.67	0.14	**25.23**	-	**26.28**	m	**28.25**
16mm work to skirtings on brickwork or blockwork base										
height 150mm	4.00	5.00	**4.20**	1.60	0.27	**25.63**	-	**29.83**	m	**32.06**
height 150mm; curved to 3000mm radius	4.00	5.00	**4.20**	2.13	0.27	**33.16**	-	**37.36**	m	**40.16**
Rounded angles and intersections										
10 - 100mm radius	-	-	**-**	0.18	-	**2.56**	-	**2.56**	m	**2.75**
Coves										
25mm girth	-	-	**-**	0.40	-	**5.68**	-	**5.68**	m	**6.11**
40mm girth	-	-	**-**	0.60	-	**8.53**	-	**8.53**	m	**9.17**
Accessories										
Plastic dividing strips; setting in bed and finishing										
6 x 16mm; flat section	1.95	-	**1.95**	0.25	-	**3.55**	-	**5.50**	m	**5.92**
M41: TERRAZZO TILING/IN SITU TERRAZZO; STAIRCASE AREAS										
Terrazzo tiles, BS 4131, aggregate size random, ground, grouted and polished, standard colour range; 3mm joints, symmetrical layout; bedding in 40mm cement mortar (1:3); grouting with white cement; in-situ margins										
300 x 300 x 28mm units to floors on concrete base; level or to falls only not exceeding 15 degrees from horizontal										
plain	38.47	5.00	**40.55**	1.65	0.65	**30.41**	-	**70.95**	m2	**76.28**

Labour hourly rates: (except Specialists) Craft Operatives £14.21 Labourer £10.71 Rates are national average prices. Refer to REGIONAL VARIATIONS for indicative levels of overall pricing in regions	MATERIALS			LABOUR				RATES		
	Del to Site £	Waste %	Material Cost £	Craft Optve Hrs	Lab Hrs	Labour Cost £	Sunds £	Nett Rate £	Unit	Gross Rate (+7.5%) £
M41: TERRAZZO TILING/IN SITU TERRAZZO; STAIRCASE AREAS Cont.										
Terrazzo tiles, BS 4131, aggregate size random, ground, grouted and polished, standard colour range; 3mm joints, symmetrical layout; bedding in 40mm cement mortar (1:3); grouting with white cement; in-situ margins Cont.										
305 x 305 x 28mm units to treads on concrete base width 292mm; rounded nosing	17.66	5.00	**18.60**	0.75	0.19	**12.69**	-	**31.30**	m	**33.64**
305 x 305 x 28mm units to plain risers on concrete base height 165mm	5.61	5.00	**5.93**	0.50	0.11	**8.28**	-	**14.21**	m	**15.28**
Terrazzo, white cement and white marble chippings (2:5); polished; on cement and sand base										
16mm work to treads; one coat										
width 279mm	7.00	5.00	**7.35**	3.00	0.32	**46.06**	-	**53.41**	m	**57.41**
extra; two line carborundum non-slip inlay	8.10	5.00	**8.51**	0.25	-	**3.55**	-	**12.06**	m	**12.96**
16mm work to undercut risers; one coat height 178mm	4.00	5.00	**4.20**	2.00	0.20	**30.56**	-	**34.76**	m	**37.37**
6mm work to strings; one coat										
height 150mm	1.00	5.00	**1.05**	1.80	0.27	**28.47**	-	**29.52**	m	**31.73**
height 200mm	2.00	5.00	**2.10**	2.50	0.36	**39.38**	-	**41.48**	m	**44.59**
M42: WOOD BLOCK/COMPOSITION BLOCK/ PARQUET FLOORING										
Maple wood blocks, tongued and grooved joints; symmetrical herringbone pattern layout, two block plain borders; fixing with adhesive; sanding, one coat sealer										
70 x 230 x 20mm units to floors on cement and sand base; level or to falls only not exceeding 15 degrees from horizontal plain	29.37	5.00	**30.84**	1.50	1.25	**34.70**	-	**65.54**	m2	**70.46**
Merbau wood blocks, tongued and grooved joints; symmetrical herringbone pattern layout, two block plain borders; fixing with adhesive; sanding, one coat sealer										
70 x 230 x 20mm units to floors on cement and sand base; level or to falls only not exceeding 15 degrees from horizontal plain	24.37	5.00	**25.59**	1.50	1.25	**34.70**	-	**60.29**	m2	**64.81**
Oak wood blocks, tongued and grooved joints; symmetrical herringbone pattern layout, two block plain borders; fixing with adhesive; sanding, one coat sealer										
70 x 230 x 20mm units to floors on cement and sand base; level or to falls only not exceeding 15 degrees from horizontal plain	33.37	5.00	**35.04**	1.50	1.25	**34.70**	-	**69.74**	m2	**74.97**
M50: RUBBER/PLASTICS/CORK/LINO/CARPET TILING/SHEETING: RUBBER FLOORING										
Rubber floor tiles, BS 1711; butt joints, symmetrical layout; fixing with adhesive; on cement and sand base										
610 x 610 x 2mm units to floors level or to falls only not exceeding 15 degrees from horizontal										
width exceeding 300mm	24.20	5.00	**25.52**	0.35	0.23	**7.44**	-	**32.96**	m2	**35.43**
width not exceeding 300mm	7.60	10.00	**8.36**	0.21	0.07	**3.73**	-	**12.09**	m	**13.00**
610 x 610 x 3.0mm units to floors level or to falls only not exceeding 15 degrees from horizontal										
width exceeding 300mm	28.05	5.00	**29.56**	0.35	0.23	**7.44**	-	**37.00**	m2	**39.77**
width not exceeding 300mm	8.75	10.00	**9.63**	0.21	0.07	**3.73**	-	**13.36**	m	**14.36**
Smooth finish rubber matting, BS 1711; butt joints; fixing with adhesive; on cement and sand base										
2mm work to floors level or to falls only not exceeding 15 degrees from horizontal										
width exceeding 300mm	23.30	5.00	**24.58**	0.15	0.13	**3.52**	-	**28.10**	m2	**30.21**
width not exceeding 300mm	7.33	10.00	**8.06**	0.09	0.04	**1.71**	-	**9.77**	m	**10.50**
3mm work to floors level or to falls only not exceeding 15 degrees from horizontal										
width exceeding 300mm	26.75	5.00	**28.20**	0.15	0.13	**3.52**	-	**31.72**	m2	**34.10**
width not exceeding 300mm	8.36	10.00	**9.20**	0.09	0.04	**1.71**	-	**10.91**	m	**11.72**
5mm work to floors level or to falls only not exceeding 15 degrees from horizontal										
width exceeding 300mm	30.20	5.00	**31.82**	0.15	0.13	**3.52**	-	**35.34**	m2	**37.99**
width not exceeding 300mm	9.40	10.00	**10.34**	0.09	0.04	**1.71**	-	**12.04**	m	**12.95**

Labour hourly rates: (except Specialists) Craft Operatives £14.21 Labourer £10.71 Rates are national average prices. Refer to REGIONAL VARIATIONS for indicative levels of overall pricing in regions	MATERIALS			LABOUR				RATES		
	Del to Site £	Waste %	Material Cost £	Craft Optve Hrs	Lab Hrs	Labour Cost £	Sunds £	Nett Rate £	Unit	Gross Rate (+7.5%) £
M50: RUBBER/PLASTICS/CORK/LINO/CARPET TILING/SHEETING: PVC FLOORING										
P.V.C. floor tiles, BS EN 654; butt joints, symmetrical layout; fixing with adhesive; two coats sealer; on cement and sand base										
300 x 300 x 2mm units to floors level or to falls only not exceeding 15 degrees from horizontal										
width exceeding 300mm	10.78	5.00	**11.49**	0.45	0.23	**8.86**	-	**20.34**	m2	**21.87**
width not exceeding 300mm	3.74	10.00	**4.11**	0.27	0.07	**4.59**	-	**8.69**	m	**9.35**
300 x 300 x 2.5mm units to floors level or to falls only not exceeding 15 degrees from horizontal										
width exceeding 300mm	12.43	5.00	**13.22**	0.45	0.23	**8.86**	-	**22.08**	m2	**23.73**
width not exceeding 300mm	4.23	10.00	**4.65**	0.27	0.07	**4.59**	-	**9.24**	m	**9.93**
Fully flexible PVC heavy duty floor tiles, BS EN 649 type A; butt joints, symmetrical layout; fixing with adhesive; two coats sealer; on cement and sand base										
610 x 610 x 2mm units to floors level or to falls only not exceeding 15 degrees from horizontal										
width exceeding 300mm	16.93	5.00	**17.94**	0.45	0.23	**8.86**	-	**26.80**	m2	**28.81**
width not exceeding 300mm	5.58	10.00	**6.14**	0.27	0.07	**4.59**	-	**10.72**	m	**11.53**
610 x 610 x 2.5mm units to floors level or to falls only not exceeding 15 degrees from horizontal										
width exceeding 300mm	19.08	5.00	**20.20**	0.45	0.23	**8.86**	-	**29.06**	m2	**31.24**
width not exceeding 300mm	6.23	10.00	**6.85**	0.27	0.07	**4.59**	-	**11.43**	m	**12.29**
Fully flexible P.V.C. heavy duty sheet,BS EN 649 Type A; butt joints; welded; fixing with adhesive; two coats sealer; on cement and sand base										
2mm work to floors level or to falls only not exceeding 15 degrees from horizontal										
width exceeding 300mm	10.78	5.00	**11.49**	0.30	0.15	**5.87**	-	**17.36**	m2	**18.66**
width not exceeding 300mm	3.74	10.00	**4.11**	0.18	0.05	**3.09**	-	**7.20**	m	**7.74**
2.5mm work to floors level or to falls only not exceeding 15 degrees from horizontal										
width exceeding 300mm	12.43	5.00	**13.22**	0.30	0.15	**5.87**	-	**19.09**	m2	**20.52**
width not exceeding 300mm	4.23	10.00	**4.65**	0.18	0.05	**3.09**	-	**7.75**	m	**8.33**
P.V.C. coved skirting; fixing with adhesive										
Skirtings on plaster base										
height 100mm - Set in	2.10	5.00	**2.23**	0.20	0.01	**2.95**	-	**5.18**	m	**5.57**
height 100mm - Sit on	2.40	5.00	**2.55**	0.20	0.02	**3.06**	-	**5.60**	m	**6.02**
M50: RUBBER/PLASTICS/CORK/LINO/CARPET TILING/SHEETING: CORK FLOORING										
Cork tiles, BS 8203; butt joints; symmetrical layout; fixing with adhesive; two coats seal; on cement and sand base										
300 x 300 x 3.2mm units to floors level or to falls only not exceeding 15 degrees from horizontal										
width exceeding 300mm	14.23	5.00	**15.13**	0.75	0.23	**13.12**	-	**28.26**	m2	**30.37**
width not exceeding 300mm	4.65	10.00	**5.12**	0.45	0.07	**7.14**	-	**12.26**	m	**13.18**
300 x 300 x 4.75mm units to floors level or to falls only not exceeding 15 degrees from horizontal										
width exceeding 300mm	15.98	5.00	**16.97**	0.75	0.23	**13.12**	-	**30.09**	m2	**32.35**
width not exceeding 300mm	5.19	10.00	**5.70**	0.45	0.07	**7.14**	-	**12.85**	m	**13.81**
M50: RUBBER/PLASTICS/CORK/LINO/CARPET TILING/SHEETING: LINOLEUM FLOORING										
Linoleum tiles, BS EN 548 & 670 butt joints, symmetrical layout; fixing with adhesive; on cement and sand base										
500 x 500 x 2mm units to floors level or to falls only not exceeding 15 degrees from horizontal										
width exceeding 300mm	15.52	5.00	**16.34**	0.35	0.23	**7.44**	-	**23.77**	m2	**25.56**
width not exceeding 300mm	4.79	10.00	**5.27**	0.21	0.07	**3.73**	-	**9.01**	m	**9.68**
500 x 500 x 3.2mm units to floors level or to falls only not exceeding 15 degrees from horizontal										
width exceeding 300mm	16.92	5.00	**17.81**	0.35	0.23	**7.44**	-	**25.24**	m2	**27.14**
width not exceeding 300mm	5.21	10.00	**5.73**	0.21	0.07	**3.73**	-	**9.47**	m	**10.18**
Linoleum sheet, BS EN 548 & 670 butt joints; laying loose; on cement and sand base										
2mm work to floors level or to falls only not exceeding 15 degrees from horizontal										
width exceeding 300mm	13.60	5.00	**14.28**	0.15	0.13	**3.52**	-	**17.80**	m2	**19.14**
width not exceeding 300mm	4.08	10.00	**4.49**	0.09	0.04	**1.71**	-	**6.20**	m	**6.66**
3.2mm work to floors level or to falls only not exceeding 15 degrees from horizontal										
width exceeding 300mm	17.10	5.00	**17.96**	0.15	0.13	**3.52**	-	**21.48**	m2	**23.09**

Labour hourly rates: (except Specialists) Craft Operatives £14.21 Labourer £10.71 Rates are national average prices. Refer to REGIONAL VARIATIONS for indicative levels of overall pricing in regions	MATERIALS			LABOUR				RATES		
	Del to Site £	Waste %	Material Cost £	Craft Optve Hrs	Lab Hrs	Labour Cost £	Sunds £	Nett Rate £	Unit	Gross Rate (+7.5%) £
M50: RUBBER/PLASTICS/CORK/LINO/CARPET TILING/SHEETING: LINOLEUM FLOORING Cont.										
Linoleum sheet, BS EN 548 & 670 butt joints; laying loose; on cement and sand base Cont.										
3.2mm work to floors level or to falls only not exceeding 15 degrees from horizontal Cont.										
width not exceeding 300mm	5.13	10.00	**5.64**	0.09	0.04	**1.71**	-	**7.35**	m	**7.90**
M50: RUBBER/PLASTICS/CORK/LINO/CARPET TILING/SHEETING; STAIRCASE AREAS - STAIR NOSINGS AND TREADS										
Accessories										
Heavy duty aluminium alloy stair nosings with anti-slip inserts; 46mm wide with single line insert; fixing to timber with screws										
11.5mm drop	7.81	5.00	**8.20**	0.40	-	**5.68**	0.04	**13.92**	m	**14.97**
22mm drop	9.08	5.00	**9.54**	0.40	-	**5.68**	0.04	**15.26**	m	**16.41**
25mm drop	9.76	5.00	**10.25**	0.40	-	**5.68**	0.04	**15.98**	m	**17.18**
32mm drop	11.05	5.00	**11.60**	0.40	-	**5.68**	0.04	**17.33**	m	**18.63**
33mm drop	11.05	5.00	**11.60**	0.40	-	**5.68**	0.04	**17.33**	m	**18.63**
38mm drop	12.06	5.00	**12.66**	0.40	-	**5.68**	0.04	**18.39**	m	**19.77**
46mm drop	12.70	5.00	**13.33**	0.40	-	**5.68**	0.04	**19.06**	m	**20.49**
Heavy duty aluminium alloy stair nosings with anti-slip inserts; 80mm wide with two line inserts; fixing to timber with screws										
22mm drop	13.01	5.00	**13.66**	0.55	-	**7.82**	0.06	**21.54**	m	**23.15**
25mm drop	16.24	5.00	**17.06**	0.55	-	**7.82**	0.06	**24.93**	m	**26.80**
32mm drop	17.58	5.00	**18.46**	0.55	-	**7.82**	0.06	**26.34**	m	**28.31**
33mm drop	17.58	5.00	**18.46**	0.55	-	**7.82**	0.06	**26.34**	m	**28.31**
51mm drop	18.21	5.00	**19.12**	0.55	-	**7.82**	0.06	**27.00**	m	**29.02**
63mm drop	19.85	5.00	**20.84**	0.55	-	**7.82**	0.06	**28.72**	m	**30.87**
Heavy duty aluminium alloy stair nosings with anti-slip inserts; 46mm wide with single line insert; fixing to masonry with screws										
11.5mm drop	7.81	5.00	**8.20**	0.65	-	**9.24**	0.06	**17.50**	m	**18.81**
22mm drop	9.08	5.00	**9.54**	0.65	-	**9.24**	0.06	**18.84**	m	**20.25**
25mm drop	9.76	5.00	**10.25**	0.65	-	**9.24**	0.06	**19.55**	m	**21.02**
32mm drop	11.05	5.00	**11.60**	0.65	-	**9.24**	0.06	**20.90**	m	**22.47**
33mm drop	11.05	5.00	**11.60**	0.65	-	**9.24**	0.06	**20.90**	m	**22.47**
38mm drop	12.06	5.00	**12.66**	0.65	-	**9.24**	0.06	**21.96**	m	**23.61**
46mm drop	12.70	5.00	**13.33**	0.65	-	**9.24**	0.06	**22.63**	m	**24.33**
Heavy duty aluminium alloy stair nosings with anti-slip inserts; 80mm wide with two line inserts; fixing to masonry with screws										
22mm drop	13.01	5.00	**13.66**	0.80	-	**11.37**	0.09	**25.12**	m	**27.00**
25mm drop	16.24	5.00	**17.06**	0.80	-	**11.37**	0.09	**28.52**	m	**30.65**
32mm drop	17.58	5.00	**18.46**	0.80	-	**11.37**	0.09	**29.92**	m	**32.17**
33mm drop	17.58	5.00	**18.46**	0.80	-	**11.37**	0.09	**29.92**	m	**32.17**
51mm drop	18.21	5.00	**19.12**	0.80	-	**11.37**	0.09	**30.58**	m	**32.87**
63mm drop	19.85	5.00	**20.84**	0.80	-	**11.37**	0.09	**32.30**	m	**34.72**
Aluminium stair nosings with anti-slip inserts; 46mm wide with single line insert; fixing to timber with screws										
22mm drop	6.48	5.00	**6.80**	0.40	-	**5.68**	0.04	**12.53**	m	**13.47**
28mm drop	6.48	5.00	**6.80**	0.40	-	**5.68**	0.04	**12.53**	m	**13.47**
33mm drop	8.44	5.00	**8.86**	0.40	-	**5.68**	0.04	**14.58**	m	**15.68**
Aluminium stair nosings with anti-slip inserts; 66mm wide with single line insert; fixing to timber with screws										
22mm drop	9.08	5.00	**9.54**	0.45	-	**6.39**	0.06	**15.99**	m	**17.19**
25mm drop	9.76	5.00	**10.25**	0.45	-	**6.39**	0.06	**16.71**	m	**17.96**
32mm drop	11.05	5.00	**11.60**	0.45	-	**6.39**	0.06	**18.06**	m	**19.41**
33mm drop	11.67	5.00	**12.25**	0.45	-	**6.39**	0.06	**18.71**	m	**20.11**
Aluminium stair nosings with anti-slip inserts; 80mm wide with two line inserts; fixing to timber with screws										
22mm drop	10.42	5.00	**10.94**	0.55	-	**7.82**	0.06	**18.82**	m	**20.23**
28mm drop	11.67	5.00	**12.25**	0.55	-	**7.82**	0.06	**20.13**	m	**21.64**
33mm drop	12.38	5.00	**13.00**	0.55	-	**7.82**	0.06	**20.88**	m	**22.44**
46mm drop	13.01	5.00	**13.66**	0.55	-	**7.82**	0.06	**21.54**	m	**23.15**
51mm drop	13.66	5.00	**14.34**	0.55	-	**7.82**	0.06	**22.22**	m	**23.88**
Aluminium alloy stair nosings with anti-slip inserts; 46mm wide with single line insert; fixing to masonry with screws										
22mm drop	6.48	5.00	**6.80**	0.65	-	**9.24**	0.06	**16.10**	m	**17.31**
28mm drop	6.48	5.00	**6.80**	0.65	-	**9.24**	0.06	**16.10**	m	**17.31**
33mm drop	8.44	5.00	**8.86**	0.65	-	**9.24**	0.06	**18.16**	m	**19.52**
Aluminium alloy stair nosings with anti-slip inserts; 66mm wide with single line insert; fixing to masonry with screws										
22mm drop	9.08	5.00	**9.54**	0.70	-	**9.95**	0.09	**19.58**	m	**21.05**
25mm drop	9.76	5.00	**10.25**	0.70	-	**9.95**	0.09	**20.29**	m	**21.81**
32mm drop	11.05	5.00	**11.60**	0.70	-	**9.95**	0.09	**21.64**	m	**23.27**
33mm drop	11.67	5.00	**12.25**	0.70	-	**9.95**	0.09	**22.29**	m	**23.96**

Labour hourly rates: (except Specialists) Craft Operatives £14.21 Labourer £10.71 Rates are national average prices. Refer to REGIONAL VARIATIONS for indicative levels of overall pricing in regions	MATERIALS			LABOUR				RATES		
	Del to Site £	Waste %	Material Cost £	Craft Optve Hrs	Lab Hrs	Labour Cost £	Sunds £	Nett Rate £	Unit	Gross Rate (+7.5%) £
M50: RUBBER/PLASTICS/CORK/LINO/CARPET TILING/SHEETING; STAIRCASE AREAS - STAIR NOSINGS AND TREADS Cont.										
Accessories Cont.										
Aluminium alloy stair nosings with anti-slip inserts; 80mm wide with two line inserts; fixing to masonry with screws										
22mm drop	10.42	5.00	**10.94**	0.80	-	**11.37**	0.09	**22.40**	m	**24.08**
28mm drop	11.67	5.00	**12.25**	0.80	-	**11.37**	0.09	**23.71**	m	**25.49**
33mm drop	12.38	5.00	**13.00**	0.80	-	**11.37**	0.09	**24.46**	m	**26.29**
46mm drop	13.01	5.00	**13.66**	0.80	-	**11.37**	0.09	**25.12**	m	**27.00**
51mm drop	13.66	5.00	**14.34**	0.80	-	**11.37**	0.09	**25.80**	m	**27.74**
M51: EDGE FIXED CARPETING										
Underlay to carpeting; Tredaire; fixing with tacks; on timber base										
Work to floors; level or to falls only not exceeding 15 degrees from horizontal										
width exceeding 300mm	1.30	5.00	**1.37**	0.10	0.05	**1.96**	0.13	**3.45**	m2	**3.71**
Fitted carpeting; Tufted wool/nylon, p.c. £20.00 m2; fixing with tackless grippers; on timber base										
Work to floors; level or to falls only not exceeding 15 degrees from horizontal										
width exceeding 300mm	20.00	5.00	**21.00**	0.20	0.13	**4.23**	0.13	**25.36**	m2	**27.26**
Fitted carpeting; Axminster wool/nylon, P.C. £26.00m2; fixing with tackless grippers; on timber base										
Work to floors; level or to falls only not exceeding 15 degrees from horizontal										
width exceeding 300mm	26.00	5.00	**27.30**	0.20	0.13	**4.23**	0.13	**31.66**	m2	**34.04**
Fitted carpeting; Wilton wool/nylon, p.c. £32.5000 m2; fixing with tackless grippers; on timber base										
Work to floors; level or to falls only not exceeding 15 degrees from horizontal										
width exceeding 300mm	32.50	5.00	**34.13**	0.20	0.13	**4.23**	0.13	**38.49**	m2	**41.37**
M51: EDGE FIXED CARPETING; STAIRCASE AREAS - STAIR NOSINGS AND TREADS										
Accessories										
Aluminium alloy carpet nosings with anti-slip inserts; with single line insert; fixing to timber with screws										
65mm wide x 24mm drop	14.78	5.00	**15.52**	0.45	-	**6.39**	0.06	**21.98**	m	**23.62**
116mm wide overall with carpet gripper x 42mm drop	21.12	5.00	**22.17**	0.50	-	**7.11**	0.09	**29.37**	m	**31.57**
Aluminium alloy carpet nosings with anti-slip inserts; with two line inserts; fixing to timber with screws										
80mm wide x 31mm drop	16.89	5.00	**17.74**	0.55	-	**7.82**	0.06	**25.61**	m	**27.53**
Aluminium alloy carpet nosings with anti-slip inserts; with single line insert; fixing to masonry with screws										
65mm wide x 24mm drop	14.78	5.00	**15.52**	0.70	-	**9.95**	0.09	**25.56**	m	**27.48**
116mm wide overall with carpet gripper x 42mm drop	21.12	5.00	**22.17**	0.75	-	**10.66**	0.12	**32.95**	m	**35.42**
Aluminium alloy carpet nosings with anti-slip inserts; with two line inserts; fixing to masonry with screws										
80mm wide x 31mm drop	16.89	5.00	**17.74**	0.80	-	**11.37**	0.09	**29.20**	m	**31.39**
M52: DECORATIVE PAPERS/FABRICS										
Lining paper (prime cost sum for supply and allowance for waste included elsewhere); sizing; applying adhesive; hanging; butt joints										
Plaster walls and columns exceeding 0.50m2	0.04	-	**0.04**	0.12	-	**1.71**	0.03	**1.78**	m2	**1.91**
Plaster ceilings and beams exceeding 0.50m2	0.04	-	**0.04**	0.13	-	**1.85**	0.04	**1.93**	m2	**2.07**
Pulp paper (prime cost sum for supply and allowance for waste included elsewhere); sizing; applying adhesive; hanging; butt joints										
Plaster walls and columns exceeding 0.50m2	0.04	-	**0.04**	0.28	-	**3.98**	0.08	**4.10**	m2	**4.41**
Plaster ceilings and beams exceeding 0.50m2	0.04	-	**0.04**	0.31	-	**4.41**	0.08	**4.53**	m2	**4.87**
Washable paper (prime cost sum for supply and allowance for waste included elsewhere); sizing; applying adhesive; hanging; butt joints										
Plaster walls and columns exceeding 0.50m2	0.06	-	**0.06**	0.22	-	**3.13**	0.06	**3.24**	m2	**3.49**

Labour hourly rates: (except Specialists) Craft Operatives £14.21 Labourer £10.71 Rates are national average prices. Refer to REGIONAL VARIATIONS for indicative levels of overall pricing in regions	MATERIALS			LABOUR				RATES		
	Del to Site £	Waste %	Material Cost £	Craft Optve Hrs	Lab Hrs	Labour Cost £	Sunds £	Nett Rate £	Unit	Gross Rate (+7.5%) £
M52: DECORATIVE PAPERS/FABRICS Cont.										
Washable paper (prime cost sum for supply and allowance for waste included elsewhere); sizing; applying adhesive; hanging; butt joints Cont.										
Plaster ceilings and beams exceeding 0.50m2	0.06	-	**0.06**	0.24	-	**3.41**	0.06	**3.53**	m2	**3.80**
Vinyl coated paper (prime cost sum for supply and allowance for waste included elsewhere); sizing; applying adhesive; hanging; butt joints										
Plaster walls and columns exceeding 0.50m2	0.06	-	**0.06**	0.22	-	**3.13**	0.06	**3.24**	m2	**3.49**
Plaster ceilings and beams exceeding 0.50m2	0.06	-	**0.06**	0.24	-	**3.41**	0.06	**3.53**	m2	**3.79**
Embossed or textured paper (prime cost sum for supply and allowance for waste included elsewhere); sizing; applying adhesive; hanging; butt joints										
Plaster walls and columns exceeding 0.50m2	0.06	-	**0.06**	0.22	-	**3.13**	0.06	**3.24**	m2	**3.49**
Plaster ceilings and beams exceeding 0.50m2	0.12	-	**0.12**	0.24	-	**3.41**	0.06	**3.59**	m2	**3.86**
Woodchip paper (prime cost sum for supply and allowance for waste included elsewhere); sizing; applying adhesive; hanging; butt joints										
Plaster walls and columns exceeding 0.50m2	0.04	-	**0.04**	0.22	-	**3.13**	0.06	**3.23**	m2	**3.47**
Plaster ceilings and beams exceeding 0.50m2	0.04	-	**0.04**	0.24	-	**3.41**	0.06	**3.52**	m2	**3.78**
Paper border strips (prime cost sum for supply and allowance for waste included elsewhere); sizing; applying adhesive; hanging; butt joints										
Border strips 25mm wide to papered walls and columns	0.01	-	**0.01**	0.08	-	**1.14**	0.02	**1.16**	m	**1.25**
75mm wide to papered walls and columns	0.02	-	**0.02**	0.09	-	**1.28**	0.02	**1.32**	m	**1.42**
Lining paper p.c. £1.00 roll; sizing; applying adhesive; hanging; butt joints										
Plaster walls and columns exceeding 0.50m2	0.24	12.00	**0.27**	0.12	-	**1.71**	0.03	**2.01**	m2	**2.16**
Plaster ceilings and beams exceeding 0.50m2	0.24	12.00	**0.27**	0.13	-	**1.85**	0.03	**2.15**	m2	**2.31**
Pulp paper p.c. £3.00 roll; sizing; applying adhesive; hanging; butt joints										
Plaster walls and columns exceeding 0.50m2	0.64	30.00	**0.82**	0.22	-	**3.13**	0.06	**4.01**	m2	**4.31**
Plaster ceilings and beams exceeding 0.50m2	0.64	25.00	**0.79**	0.24	-	**3.41**	0.06	**4.27**	m2	**4.59**
Pulp paper (24' drop pattern match) p.c. £4.00 roll; sizing; applying adhesive; hanging; butt joints										
Plaster walls and columns exceeding 0.50m2	0.84	40.00	**1.16**	0.22	-	**3.13**	0.06	**4.35**	m2	**4.68**
Plaster ceilings and beams exceeding 0.50m2	0.84	35.00	**1.12**	0.24	-	**3.41**	0.06	**4.60**	m2	**4.94**
Washable paper p.c. £4.00 roll; sizing; applying adhesive; hanging; butt joints										
Plaster walls and columns exceeding 0.50m2	0.86	25.00	**1.06**	0.22	-	**3.13**	0.06	**4.24**	m2	**4.56**
Plaster ceilings and beams exceeding 0.50m2	0.86	20.00	**1.02**	0.24	-	**3.41**	0.06	**4.49**	m2	**4.83**
Vinyl coated paper p.c. £6.50 roll; sizing; applying adhesive; hanging; butt joints										
Plaster walls and columns exceeding 0.50m2	1.36	25.00	**1.68**	0.22	-	**3.13**	0.06	**4.87**	m2	**5.23**
Plaster ceilings and beams exceeding 0.50m2	1.36	20.00	**1.62**	0.24	-	**3.41**	0.06	**5.09**	m2	**5.47**

Labour hourly rates: (except Specialists) Craft Operatives £14.21 Labourer £10.71 Rates are national average prices. Refer to REGIONAL VARIATIONS for indicative levels of overall pricing in regions	MATERIALS			LABOUR				RATES		
	Del to Site £	Waste %	Material Cost £	Craft Optve Hrs	Lab Hrs	Labour Cost £	Sunds £	Nett Rate £	Unit	Gross Rate (+7.5%) £
M52: DECORATIVE PAPERS/FABRICS Cont.										
Vinyl coated paper p.c. £7.00 roll; sizing; applying adhesive; hanging; butt joints										
Plaster walls and columns exceeding 0.50m2	1.46	25.00	**1.81**	0.22	-	**3.13**	0.06	**4.99**	m2	**5.37**
Plaster ceilings and beams exceeding 0.50m2	1.46	20.00	**1.74**	0.24	-	**3.41**	0.06	**5.21**	m2	**5.60**
Embossed paper p.c. £4.50 roll; sizing; applying adhesive; hanging; butt joints										
Plaster walls and columns exceeding 0.50m2	1.02	25.00	**1.24**	0.22	-	**3.13**	0.06	**4.43**	m2	**4.76**
Plaster ceilings and beams exceeding 0.50m2	1.02	20.00	**1.20**	0.24	-	**3.41**	0.06	**4.67**	m2	**5.02**
Textured paper p.c. £5.00 roll; sizing; applying adhesive; hanging; butt joints										
Plaster walls and columns exceeding 0.50m2	1.09	25.00	**1.34**	0.22	-	**3.13**	0.06	**4.52**	m2	**4.86**
Plaster ceilings and beams exceeding 0.50m2	1.09	20.00	**1.29**	0.24	-	**3.41**	0.06	**4.76**	m2	**5.12**
Woodchip paper p.c. £1.50 roll; sizing; applying adhesive; hanging; butt joints										
Plaster walls and columns exceeding 0.50m2	0.34	25.00	**0.42**	0.22	-	**3.13**	0.06	**3.60**	m2	**3.87**
Plaster ceilings and beams exceeding 0.50m2	0.34	20.00	**0.40**	0.24	-	**3.41**	0.06	**3.88**	m2	**4.17**
Paper border strips p.c. £0.25m; applying adhesive; hanging; butt joints										
Border strips 25mm wide to papered walls and columns	0.26	10.00	**0.28**	0.08	-	**1.14**	0.02	**1.44**	m	**1.55**
Paper border strips p.c. £0.50m; applying adhesive; hanging; butt joints										
Border strips 75mm wide to papered walls and columns	0.52	10.00	**0.57**	0.09	-	**1.28**	0.02	**1.87**	m	**2.01**
Hessian wall covering (prime cost sum for supply and allowance for waste included elsewhere); sizing; applying adhesive; hanging; butt joints										
Plaster walls and columns exceeding 0.50m2	0.18	-	**0.18**	0.40	-	**5.68**	0.11	**5.97**	m2	**6.41**
Textile hessian paper backed wall covering (prime cost sum for supply and allowance for waste included elsewhere); sizing; applying adhesive; hanging; butt joints										
Plaster walls and columns exceeding 0.50m2	0.18	-	**0.18**	0.33	-	**4.69**	0.09	**4.95**	m2	**5.32**
Hessian wall covering p.c. £4.00m2; sizing; applying adhesive; hanging; butt joints										
Plaster walls and columns exceeding 0.50m2	4.18	30.00	**5.38**	0.40	-	**5.68**	0.11	**11.17**	m2	**12.00**
Textile hessian paper backed wall covering p.c. £7.50m2; sizing; applying adhesive; hanging; butt joints										
Plaster walls and columns exceeding 0.50m2	7.68	25.00	**9.55**	0.33	-	**4.69**	0.09	**14.33**	m2	**15.40**
M52: DECORATIVE PAPERS/FABRICS - REDECORATIONS										
Stripping existing paper; lining paper (prime cost sum for supply and allowance for waste included elsewhere); sizing; applying adhesive hanging; butt joints										
Hard building board walls and columns exceeding 0.50m2	0.04	-	**0.04**	0.29	-	**4.12**	0.19	**4.36**	m2	**4.69**
Plaster walls and columns exceeding 0.50m2	0.04	-	**0.04**	0.25	-	**3.55**	0.17	**3.76**	m2	**4.05**
Hard building board ceilings and beams exceeding 0.50m2	0.04	-	**0.04**	0.32	-	**4.55**	0.21	**4.81**	m2	**5.17**
Plaster ceilings and beams exceeding 0.50m2	0.04	-	**0.04**	0.27	-	**3.84**	0.18	**4.06**	m2	**4.37**

Labour hourly rates: (except Specialists) Craft Operatives £14.21 Labourer £10.71 Rates are national average prices. Refer to REGIONAL VARIATIONS for indicative levels of overall pricing in regions	MATERIALS			LABOUR				RATES		
	Del to Site £	Waste %	Material Cost £	Craft Optve Hrs	Lab Hrs	Labour Cost £	Sunds £	Nett Rate £	Unit	Gross Rate (+7.5%) £
M52: DECORATIVE PAPERS/FABRICS - REDECORATIONS Cont.										
Stripping existing paper; lining paper and cross lining (prime cost sum for supply and allowance for waste included elsewhere); sizing; applying adhesive; hanging; butt joints										
Hard building board walls and columns exceeding 0.50m2	0.09	-	**0.09**	0.41	-	**5.83**	0.27	**6.19**	m2	**6.65**
Plaster walls and columns exceeding 0.50m2	0.09	-	**0.09**	0.37	-	**5.26**	0.25	**5.59**	m2	**6.01**
Hard building board ceilings and beams exceeding 0.50m2	0.09	-	**0.09**	0.45	-	**6.39**	0.30	**6.78**	m2	**7.29**
Plaster ceilings and beams exceeding 0.50m2	0.09	-	**0.09**	0.40	-	**5.68**	0.27	**6.04**	m2	**6.49**
Stripping existing paper; pulp paper (prime cost sum for supply and allowance for waste included elsewhere); applying adhesive; hanging; butt joints										
Hard building board walls and columns exceeding 0.50m2	0.04	-	**0.04**	0.39	-	**5.54**	0.26	**5.85**	m2	**6.29**
Plaster walls and columns exceeding 0.50m2	0.04	-	**0.04**	0.35	-	**4.97**	0.23	**5.25**	m2	**5.65**
Hard building board ceilings and beams exceeding 0.50m2	0.04	-	**0.04**	0.43	-	**6.11**	0.29	**6.44**	m2	**6.93**
Plaster ceilings and beams exceeding 0.50m2	0.04	-	**0.04**	0.38	-	**5.40**	0.25	**5.70**	m2	**6.13**
Stripping existing washable or vinyl coated paper; washable paper (prime cost sum for supply and allowance for waste included elsewhere); applying adhesive; hanging; butt joints										
Hard building board walls and columns exceeding 0.50m2	0.06	-	**0.06**	0.40	-	**5.68**	0.27	**6.01**	m2	**6.46**
Plaster walls and columns exceeding 0.50m2	0.06	-	**0.06**	0.36	-	**5.12**	0.24	**5.41**	m2	**5.82**
Hard building board ceilings and beams exceeding 0.50m2	0.06	-	**0.06**	0.44	-	**6.25**	0.29	**6.60**	m2	**7.10**
Plaster ceilings and beams exceeding 0.50m2	0.06	-	**0.06**	0.39	-	**5.54**	0.26	**5.86**	m2	**6.30**
Stripping existing washable or vinyl coated paper; vinyl coated paper (prime cost sum for supply and allowance for waste included elsewhere); sizing; applying adhesive; hanging; butt joints										
Hard building board walls and columns exceeding 0.50m2	0.06	-	**0.06**	0.40	-	**5.68**	0.27	**6.01**	m2	**6.46**
Plaster walls and columns exceeding 0.50m2	0.06	-	**0.06**	0.36	-	**5.12**	0.24	**5.41**	m2	**5.82**
Hard building board ceilings and beams exceeding 0.50m2	0.06	-	**0.06**	0.44	-	**6.25**	0.29	**6.60**	m2	**7.10**
Plaster ceilings and beams exceeding 0.50m2	0.06	-	**0.06**	0.39	-	**5.54**	0.26	**5.86**	m2	**6.30**
Stripping existing paper; embossed or textured paper; (prime cost sum for supply and allowance for waste included elsewhere); sizing; applying adhesive; hanging; butt joints										
Hard building board walls and columns exceeding 0.50m2	0.12	-	**0.12**	0.40	-	**5.68**	0.27	**6.07**	m2	**6.52**
Plaster walls and columns exceeding 0.50m2	0.12	-	**0.12**	0.36	-	**5.12**	0.24	**5.47**	m2	**5.88**
Hard building board ceilings and beams exceeding 0.50m2	0.12	-	**0.12**	0.44	-	**6.25**	0.29	**6.66**	m2	**7.16**
Plaster ceilings and beams exceeding 0.50m2	0.12	-	**0.12**	0.39	-	**5.54**	0.26	**5.92**	m2	**6.36**
Stripping existing paper; woodchip paper (prime cost sum for supply and allowance for waste included elsewhere); sizing; applying adhesive; hanging; butt joints										
Hard building board walls and columns exceeding 0.50m2	0.04	-	**0.04**	0.40	-	**5.68**	0.27	**6.00**	m2	**6.45**
Plaster walls and columns exceeding 0.50m2	0.04	-	**0.04**	0.36	-	**5.12**	0.24	**5.40**	m2	**5.81**

Labour hourly rates: (except Specialists) Craft Operatives £14.21 Labourer £10.71 Rates are national average prices. Refer to REGIONAL VARIATIONS for indicative levels of overall pricing in regions	MATERIALS			LABOUR				RATES		
	Del to Site £	Waste %	Material Cost £	Craft Optve Hrs	Lab Hrs	Labour Cost £	Sunds £	Nett Rate £	Unit	Gross Rate (+7.5%) £
M52: DECORATIVE PAPERS/FABRICS - REDECORATIONS Cont.										
Stripping existing paper; woodchip paper (prime cost sum for supply and allowance for waste included elsewhere); sizing; applying adhesive; hanging; butt joints Cont.										
Hard building board ceilings and beams exceeding 0.50m2	0.04	-	**0.04**	0.44	-	**6.25**	0.29	**6.59**	m2	**7.09**
Plaster ceilings and beams exceeding 0.50m2	0.04	-	**0.04**	0.39	-	**5.54**	0.26	**5.85**	m2	**6.29**
Stripping existing varnished or painted paper; woodchip paper (prime cost sum for supply and allowance for waste included elsewhere); sizing; applying adhesive; hanging; butt joints										
Hard building board walls and columns exceeding 0.50m2	0.04	-	**0.04**	0.52	-	**7.39**	0.35	**7.78**	m2	**8.36**
Plaster walls and columns exceeding 0.50m2	0.04	-	**0.04**	0.48	-	**6.82**	0.32	**7.19**	m2	**7.73**
Hard building board ceilings and beams exceeding 0.50m2	0.04	-	**0.04**	0.57	-	**8.10**	0.38	**8.53**	m2	**9.16**
Plaster ceilings and beams exceeding 0.50m2	0.04	-	**0.04**	0.53	-	**7.53**	0.36	**7.93**	m2	**8.52**
Stripping existing paper; lining paper p.c. £1.00 roll; sizing; applying adhesive; hanging; butt joints										
Hard building board walls and columns exceeding 0.50m2	0.24	12.00	**0.27**	0.29	-	**4.12**	0.19	**4.58**	m2	**4.93**
Plaster walls and columns exceeding 0.50m2	0.24	12.00	**0.27**	0.25	-	**3.55**	0.17	**3.99**	m2	**4.29**
Hard building board ceilings and beams exceeding 0.50m2	0.24	12.00	**0.27**	0.32	-	**4.55**	0.21	**5.03**	m2	**5.41**
Plaster ceilings and beams exceeding 0.50m2	0.24	12.00	**0.27**	0.27	-	**3.84**	0.18	**4.29**	m2	**4.61**
Stripping existing paper; lining paper p.c. £1.00 roll and cross lining P.C. £1.00 roll; sizing; applying adhesive; hanging; butt joints										
Hard building board walls and columns exceeding 0.50m2	0.49	12.00	**0.54**	0.41	-	**5.83**	0.27	**6.64**	m2	**7.13**
Plaster walls and columns exceeding 0.50m2	0.49	12.00	**0.54**	0.37	-	**5.26**	0.25	**6.04**	m2	**6.49**
Hard building board ceilings and beams exceeding 0.50m2	0.49	12.00	**0.54**	0.45	-	**6.39**	0.30	**7.23**	m2	**7.77**
Plaster ceilings and beams exceeding 0.50m2	0.49	12.00	**0.54**	0.40	-	**5.68**	0.27	**6.49**	m2	**6.97**
Stripping existing paper; pulp paper p.c. £3.00 roll; sizing; applying adhesive; hanging; butt joints										
Hard building board walls and columns exceeding 0.50m2	0.64	30.00	**0.82**	0.39	-	**5.54**	0.26	**6.63**	m2	**7.12**
Plaster walls and columns exceeding 0.50m2	0.64	30.00	**0.82**	0.35	-	**4.97**	0.23	**6.03**	m2	**6.48**
Hard building board ceilings and beams exceeding 0.50m2	0.64	25.00	**0.79**	0.43	-	**6.11**	0.29	**7.19**	m2	**7.73**
Plaster ceilings and beams exceeding 0.50m2	0.64	25.00	**0.79**	0.38	-	**5.40**	0.25	**6.45**	m2	**6.93**
Stripping existing paper; pulp paper with 24' drop pattern; p.c. £4.00 roll; sizing; applying adhesive; hanging; butt joints										
Hard building board walls and columns exceeding 0.50m2	0.84	40.00	**1.16**	0.45	-	**6.39**	0.30	**7.86**	m2	**8.45**
Plaster walls and columns exceeding 0.50m2	0.84	40.00	**1.16**	0.41	-	**5.83**	0.27	**7.26**	m2	**7.81**
Hard building board ceilings and beams exceeding 0.50m2	0.84	35.00	**1.12**	0.49	-	**6.96**	0.33	**8.41**	m2	**9.05**
Plaster ceilings and beams exceeding 0.50m2	0.84	35.00	**1.12**	0.44	-	**6.25**	0.29	**7.67**	m2	**8.25**

Labour hourly rates: (except Specialists) Craft Operatives £14.21 Labourer £10.71 Rates are national average prices. Refer to REGIONAL VARIATIONS for indicative levels of overall pricing in regions	MATERIALS			LABOUR				RATES		
	Del to Site £	Waste %	Material Cost £	Craft Optve Hrs	Lab Hrs	Labour Cost £	Sunds £	Nett Rate £	Unit	Gross Rate (+7.5%) £
M52: DECORATIVE PAPERS/FABRICS - REDECORATIONS Cont.										
Stripping existing washable or vinyl coated paper; washable paper p.c. £4.00 roll; sizing; applying adhesive; hanging; butt joints										
Hard building board walls and columns exceeding 0.50m2	0.86	25.00	**1.06**	0.40	-	**5.68**	0.27	**7.01**	m2	**7.54**
Plaster walls and columns exceeding 0.50m2	0.86	25.00	**1.06**	0.36	-	**5.12**	0.24	**6.41**	m2	**6.90**
Hard building board ceilings and beams exceeding 0.50m2	0.86	20.00	**1.02**	0.44	-	**6.25**	0.29	**7.56**	m2	**8.13**
Plaster ceilings and beams exceeding 0.50m2	0.86	20.00	**1.02**	0.39	-	**5.54**	0.26	**6.82**	m2	**7.33**
Stripping existing washable or vinyl coated paper; vinyl coated paper P.C. £6.50 roll; sizing; applying adhesive; hanging; butt joints										
Hard building board walls and columns exceeding 0.50m2	1.36	25.00	**1.68**	0.40	-	**5.68**	0.27	**7.63**	m2	**8.21**
Plaster walls and columns exceeding 0.50m2	1.36	25.00	**1.68**	0.36	-	**5.12**	0.24	**7.04**	m2	**7.57**
Hard building board ceilings and beams exceeding 0.50m2	1.36	20.00	**1.62**	0.44	-	**6.25**	0.29	**8.16**	m2	**8.78**
Plaster ceilings and beams exceeding 0.50m2	1.36	20.00	**1.62**	0.39	-	**5.54**	0.26	**7.42**	m2	**7.98**
Stripping existing washable or vinyl coated paper; vinyl coated paper P.C. £8.00 roll; sizing; applying adhesive; hanging; butt joints										
Hard building board walls and columns exceeding 0.50m2	1.66	25.00	**2.06**	0.40	-	**5.68**	0.27	**8.01**	m2	**8.61**
Plaster walls and columns exceeding 0.50m2	1.66	25.00	**2.06**	0.36	-	**5.12**	0.24	**7.41**	m2	**7.97**
Hard building board ceilings and beams exceeding 0.50m2	1.66	20.00	**1.98**	0.44	-	**6.25**	0.29	**8.52**	m2	**9.16**
Plaster ceilings and beams exceeding 0.50m2	1.66	20.00	**1.98**	0.39	-	**5.54**	0.26	**7.78**	m2	**8.36**
Stripping existing paper; embossed paper p.c. £4.50 roll; sizing; applying adhesive; hanging; butt joints										
Hard building board walls and columns exceeding 0.50m2	1.02	25.00	**1.24**	0.39	-	**5.54**	0.26	**7.05**	m2	**7.57**
Plaster walls and columns exceeding 0.50m2	1.02	25.00	**1.24**	0.35	-	**4.97**	0.23	**6.45**	m2	**6.93**
Hard building board ceilings and beams exceeding 0.50m2	1.02	20.00	**1.20**	0.43	-	**6.11**	0.29	**7.60**	m2	**8.17**
Plaster ceilings and beams exceeding 0.50m2	1.02	20.00	**1.20**	0.38	-	**5.40**	0.25	**6.85**	m2	**7.37**
Stripping existing paper; textured paper p.c. £5.00 roll; sizing; applying adhesive; hanging; butt joints										
Hard building board walls and columns exceeding 0.50m2	1.09	25.00	**1.34**	0.39	-	**5.54**	0.26	**7.14**	m2	**7.68**
Plaster walls and columns exceeding 0.50m2	1.09	25.00	**1.34**	0.35	-	**4.97**	0.23	**6.55**	m2	**7.04**
Hard building board ceilings and beams exceeding 0.50m2	1.09	20.00	**1.29**	0.43	-	**6.11**	0.29	**7.69**	m2	**8.26**
Plaster ceilings and beams exceeding 0.50m2	1.09	20.00	**1.29**	0.38	-	**5.40**	0.25	**6.94**	m2	**7.46**
Stripping existing paper; woodchip paper p.c. £1.50 roll; sizing; applying adhesive; hanging; butt joints										
Hard building board walls and columns exceeding 0.50m2	0.34	25.00	**0.42**	0.39	-	**5.54**	0.26	**6.22**	m2	**6.69**
Plaster walls and columns exceeding 0.50m2	0.34	25.00	**0.42**	0.35	-	**4.97**	0.23	**5.63**	m2	**6.05**
Hard building board ceilings and beams exceeding 0.50m2	0.34	20.00	**0.40**	0.43	-	**6.11**	0.29	**6.80**	m2	**7.31**
Plaster ceilings and beams exceeding 0.50m2	0.34	20.00	**0.40**	0.38	-	**5.40**	0.25	**6.06**	m2	**6.51**

Labour hourly rates: (except Specialists) Craft Operatives £14.21 Labourer £10.71 Rates are national average prices. Refer to REGIONAL VARIATIONS for indicative levels of overall pricing in regions	MATERIALS			LABOUR				RATES		
	Del to Site £	Waste %	Material Cost £	Craft Optve Hrs	Lab Hrs	Labour Cost £	Sunds £	Nett Rate £	Unit	Gross Rate (+7.5%) £
M52: DECORATIVE PAPERS/FABRICS - REDECORATIONS Cont.										
Washing down old distempered or painted surfaces; lining paper p.c. £1.00 roll; sizing; applying adhesive; hanging; butt joints										
Hard building board walls and columns exceeding 0.50m2	0.24	12.00	**0.27**	0.20	-	**2.84**	0.05	**3.16**	m2	**3.40**
Plaster walls and columns exceeding 0.50m2	0.24	12.00	**0.27**	0.20	-	**2.84**	0.05	**3.16**	m2	**3.40**
Hard building board ceilings and beams exceeding 0.50m2	0.24	12.00	**0.27**	0.21	-	**2.98**	0.06	**3.31**	m2	**3.56**
Plaster ceilings and beams exceeding 0.50m2	0.24	12.00	**0.27**	0.21	-	**2.98**	0.06	**3.31**	m2	**3.56**
Lining papered walls and columns exceeding 0.50m2	0.24	12.00	**0.27**	0.20	-	**2.84**	0.05	**3.16**	m2	**3.40**
Lining papered ceilings and beams exceeding 0.50m2	0.24	12.00	**0.27**	0.21	-	**2.98**	0.06	**3.31**	m2	**3.56**
Washing down old distempered or painted surfaces; lining paper p.c. £1.00 roll; and cross lining p.c. £1.00 roll; sizing; applying adhesive; hanging; butt joints										
Hard building board walls and columns exceeding 0.50m2	0.49	12.00	**0.54**	0.32	-	**4.55**	0.09	**5.17**	m2	**5.56**
Plaster walls and columns exceeding 0.50m2	0.49	12.00	**0.54**	0.32	-	**4.55**	0.09	**5.17**	m2	**5.56**
Hard building board ceilings and beams exceeding 0.50m2	0.49	12.00	**0.54**	0.34	-	**4.83**	0.09	**5.46**	m2	**5.87**
Plaster ceilings and beams exceeding 0.50m2	0.49	12.00	**0.54**	0.34	-	**4.83**	0.09	**5.46**	m2	**5.87**
Washing down old distempered or painted surfaces; pulp paper p.c. £3.00 roll; sizing; applying adhesive; hanging; butt joints										
Hard building board walls and columns exceeding 0.50m2	0.64	30.00	**0.82**	0.30	-	**4.26**	0.08	**5.17**	m2	**5.56**
Plaster walls and columns exceeding 0.50m2	0.64	30.00	**0.82**	0.30	-	**4.26**	0.08	**5.17**	m2	**5.56**
Hard building board ceilings and beams exceeding 0.50m2	0.64	25.00	**0.79**	0.32	-	**4.55**	0.09	**5.43**	m2	**5.83**
Plaster ceilings and beams exceeding 0.50m2	0.64	25.00	**0.79**	0.32	-	**4.55**	0.09	**5.43**	m2	**5.83**
Washing down old distempered or painted surfaces; pulp paper with 24' drop pattern; p.c. £4.00 roll; sizing; applying adhesive; hanging; butt joints										
Hard building board walls and columns exceeding 0.50m2	0.84	40.00	**1.16**	0.36	-	**5.12**	0.10	**6.38**	m2	**6.85**
Plaster walls and columns exceeding 0.50m2	0.84	40.00	**1.16**	0.36	-	**5.12**	0.10	**6.38**	m2	**6.85**
Hard building board ceilings and beams exceeding 0.50m2	0.84	35.00	**1.12**	0.38	-	**5.40**	0.10	**6.63**	m2	**7.12**
Plaster ceilings and beams exceeding 0.50m2	0.84	35.00	**1.12**	0.38	-	**5.40**	0.10	**6.63**	m2	**7.12**
Washing down old distempered or painted surfaces; washable paper p.c. £4.00 roll; sizing; applying adhesive; hanging; butt joints										
Hard building board walls and columns exceeding 0.50m2	0.86	25.00	**1.06**	0.30	-	**4.26**	0.08	**5.40**	m2	**5.81**
Plaster walls and columns exceeding 0.50m2	0.86	25.00	**1.06**	0.30	-	**4.26**	0.08	**5.40**	m2	**5.81**
Hard building board ceilings and beams exceeding 0.50m2	0.86	20.00	**1.02**	0.32	-	**4.55**	0.09	**5.65**	m2	**6.08**
Plaster ceilings and beams exceeding 0.50m2	0.86	20.00	**1.02**	0.32	-	**4.55**	0.09	**5.65**	m2	**6.08**
Washing down old distempered or painted surfaces; vinyl coated paper p.c. £6.50 roll; sizing; applying adhesive; hanging; butt joints										
Hard building board walls and columns exceeding 0.50m2	1.36	25.00	**1.68**	0.30	-	**4.26**	0.08	**6.03**	m2	**6.48**

Labour hourly rates: (except Specialists) Craft Operatives £14.21 Labourer £10.71 Rates are national average prices. Refer to REGIONAL VARIATIONS for indicative levels of overall pricing in regions	MATERIALS			LABOUR				RATES		
	Del to Site £	Waste %	Material Cost £	Craft Optve Hrs	Lab Hrs	Labour Cost £	Sunds £	Nett Rate £	Unit	Gross Rate (+7.5%) £
M52: DECORATIVE PAPERS/FABRICS - REDECORATIONS Cont.										
Washing down old distempered or painted surfaces; vinyl coated paper p.c. £6.50 roll; sizing; applying adhesive; hanging; butt joints Cont.										
Plaster walls and columns exceeding 0.50m2	1.36	25.00	**1.68**	0.30	-	**4.26**	0.08	**6.03**	m2	**6.48**
Hard building board ceilings and beams exceeding 0.50m2	1.36	20.00	**1.62**	0.32	-	**4.55**	0.09	**6.25**	m2	**6.72**
Plaster ceilings and beams exceeding 0.50m2	1.36	20.00	**1.62**	0.32	-	**4.55**	0.09	**6.25**	m2	**6.72**
Washing down old distempered or painted surfaces; vinyl coated paper p.c. £7.00 roll; sizing; applying adhesive; hanging; butt joints										
Hard building board walls and columns exceeding 0.50m2	1.46	25.00	**1.81**	0.30	-	**4.26**	0.08	**6.15**	m2	**6.61**
Plaster walls and columns exceeding 0.50m2	1.46	25.00	**1.81**	0.30	-	**4.26**	0.08	**6.15**	m2	**6.61**
Hard building board ceilings and beams exceeding 0.50m2	1.46	20.00	**1.74**	0.32	-	**4.55**	0.09	**6.37**	m2	**6.85**
Plaster ceilings and beams exceeding 0.50m2	1.46	20.00	**1.74**	0.32	-	**4.55**	0.09	**6.37**	m2	**6.85**
Washing down old distempered or painted surfaces; embossed paper p.c. £4.50 roll; sizing; applying adhesive; hanging; butt joints										
Hard building board walls and columns exceeding 0.50m2	1.02	25.00	**1.24**	0.30	-	**4.26**	0.08	**5.59**	m2	**6.01**
Plaster walls and columns exceeding 0.50m2	1.02	25.00	**1.24**	0.30	-	**4.26**	0.08	**5.59**	m2	**6.01**
Hard building board ceilings and beams exceeding 0.50m2	1.02	20.00	**1.20**	0.32	-	**4.55**	0.09	**5.83**	m2	**6.27**
Plaster ceilings and beams exceeding 0.50m2	1.02	20.00	**1.20**	0.32	-	**4.55**	0.09	**5.83**	m2	**6.27**
Washing down old distempered or painted surfaces; textured paper p.c. £5.00 roll; sizing; applying adhesive; hanging; butt joints										
Hard building board walls and columns exceeding 0.50m2	1.09	25.00	**1.34**	0.30	-	**4.26**	0.08	**5.68**	m2	**6.11**
Plaster walls and columns exceeding 0.50m2	1.09	25.00	**1.34**	0.30	-	**4.26**	0.08	**5.68**	m2	**6.11**
Hard building board ceilings and beams exceeding 0.50m2	1.09	20.00	**1.29**	0.32	-	**4.55**	0.09	**5.92**	m2	**6.37**
Plaster ceilings and beams exceeding 0.50m2	1.09	20.00	**1.29**	0.32	-	**4.55**	0.09	**5.92**	m2	**6.37**
Washing down old distempered or painted surfaces; woodchip paper p.c. £1.50 roll; sizing; applying adhesive; hanging; butt joints										
Hard building board walls and columns exceeding 0.50m2	0.34	25.00	**0.42**	0.30	-	**4.26**	0.08	**4.76**	m2	**5.12**
Plaster walls and columns exceeding 0.50m2	0.34	25.00	**0.42**	0.30	-	**4.26**	0.08	**4.76**	m2	**5.12**
Hard building board ceilings and beams exceeding 0.50m2	0.34	20.00	**0.40**	0.32	-	**4.55**	0.08	**5.03**	m2	**5.41**
Plaster ceilings and beams exceeding 0.50m2	0.34	20.00	**0.40**	0.32	-	**4.55**	0.09	**5.04**	m2	**5.42**
Stripping existing paper; hessian wall covering (prime cost sum for supply and allowance for waste included elsewhere); sizing; applying adhesive; hanging; butt joints										
Hard building board walls and columns exceeding 0.50m2	0.18	-	**0.18**	0.57	-	**8.10**	0.38	**8.66**	m2	**9.31**
Plaster walls and columns exceeding 0.50m2	0.18	-	**0.18**	0.53	-	**7.53**	0.36	**8.06**	m2	**8.67**

Labour hourly rates: (except Specialists) Craft Operatives £14.21 Labourer £10.71 Rates are national average prices. Refer to REGIONAL VARIATIONS for indicative levels of overall pricing in regions	MATERIALS			LABOUR				RATES		
	Del to Site £	Waste %	Material Cost £	Craft Optve Hrs	Lab Hrs	Labour Cost £	Sunds £	Nett Rate £	Unit	Gross Rate (+7.5%) £
M52: DECORATIVE PAPERS/FABRICS - REDECORATIONS Cont.										
Stripping existing paper; textile hessian paper backed wall covering (prime cost sum for supply and allowance for waste included elsewhere); sizing; applying adhesive; hanging; butt joints										
Hard building board walls and columns exceeding 0.50m2	0.18	-	**0.18**	0.50	-	**7.11**	0.34	**7.62**	m2	**8.19**
Plaster walls and columns exceeding 0.50m2	0.18	-	**0.18**	0.46	-	**6.54**	0.31	**7.02**	m2	**7.55**
Stripping existing hessian; hessian surfaced wall paper (prime cost sum for supply and allowance for waste included elsewhere); sizing; applying adhesive; hanging; butt joints										
Hard building board walls and columns exceeding 0.50m2	0.18	-	**0.18**	0.94	-	**13.36**	0.63	**14.16**	m2	**15.22**
Plaster walls and columns exceeding 0.50m2	0.18	-	**0.18**	0.90	-	**12.79**	0.60	**13.57**	m2	**14.58**
Stripping existing hessian; textile hessian paper backed wall covering (prime cost sum for supply and allowance for waste included elsewhere); sizing; applying adhesive; hanging; butt joints										
Hard building board walls and columns exceeding 0.50m2	0.18	-	**0.18**	0.87	-	**12.36**	0.58	**13.12**	m2	**14.10**
Plaster walls and columns exceeding 0.50m2	0.18	-	**0.18**	0.83	-	**11.79**	0.56	**12.53**	m2	**13.46**
Stripping existing hessian, paper backed; hessian surfaced wall paper (prime cost sum for supply and allowance for waste included elsewhere); sizing; applying adhesive; hanging; butt joints										
Hard building board walls and columns exceeding 0.50m2	0.18	-	**0.18**	0.57	-	**8.10**	0.38	**8.66**	m2	**9.31**
Plaster walls and columns exceeding 0.50m2	0.18	-	**0.18**	0.53	-	**7.53**	0.36	**8.06**	m2	**8.67**
Stripping existing hessian, paper backed; textile hessian paper backed wall covering (prime cost sum for supply and allowance for waste included elsewhere); sizing; applying adhesive; hanging; butt joints										
Hard building board walls and columns exceeding 0.50m2	0.18	-	**0.18**	0.50	-	**7.11**	0.34	**7.62**	m2	**8.19**
Plaster walls and columns exceeding 0.50m2	0.18	-	**0.18**	0.46	-	**6.54**	0.31	**7.02**	m2	**7.55**
Stripping existing paper; hessian surfaced wall paper p.c. £4.00m2; sizing; applying adhesive; hanging; butt joints										
Hard building board walls and columns exceeding 0.50m2	4.18	30.00	**5.38**	0.57	-	**8.10**	0.38	**13.86**	m2	**14.90**
Plaster walls and columns exceeding 0.50m2	4.18	30.00	**5.38**	0.53	-	**7.53**	0.36	**13.26**	m2	**14.26**
Stripping existing paper; textile hessian paper backed wall covering P.C. £7.50m2; sizing; applying adhesive; hanging; butt joints										
Hard building board walls and columns exceeding 0.50m2	4.18	30.00	**5.38**	0.50	-	**7.11**	0.34	**12.82**	m2	**13.78**
Plaster walls and columns exceeding 0.50m2	4.18	30.00	**5.38**	0.46	-	**6.54**	0.31	**12.22**	m2	**13.14**
Stripping existing hessian; hessian surfaced wall paper p.c. £4.00m2; sizing; applying adhesive; hanging ; butt joints										
Hard building board walls and columns exceeding 0.50m2	4.18	30.00	**5.38**	0.94	-	**13.36**	0.63	**19.36**	m2	**20.81**
Plaster walls and columns exceeding 0.50m2	4.18	30.00	**5.38**	0.90	-	**12.79**	0.60	**18.77**	m2	**20.17**
Stripping existing hessian; textile hessian paper backed wall covering P.C. £7.5m2; sizing; applying adhesive; hanging; butt joints										
Hard building board walls and columns exceeding 0.50m2	4.18	30.00	**5.38**	0.87	-	**12.36**	0.58	**18.32**	m2	**19.69**
Plaster walls and columns exceeding 0.50m2	4.18	30.00	**5.38**	0.83	-	**11.79**	0.56	**17.73**	m2	**19.05**

Labour hourly rates: (except Specialists) Craft Operatives £14.21 Labourer £10.71 Rates are national average prices. Refer to REGIONAL VARIATIONS for indicative levels of overall pricing in regions	MATERIALS			LABOUR				RATES		
	Del to Site £	Waste %	Material Cost £	Craft Optve Hrs	Lab Hrs	Labour Cost £	Sunds £	Nett Rate £	Unit	Gross Rate (+7.5%) £
M52: DECORATIVE PAPERS/FABRICS - REDECORATIONS Cont.										
Stripping existing hessian, paper backed; hessian surfaced wall paper p.c. £4.00m2; sizing; applying adhesive; hanging; butt joints										
Hard building board walls and columns exceeding 0.50m2	4.18	30.00	**5.38**	0.57	-	**8.10**	0.38	**13.86**	m2	**14.90**
Plaster walls and columns exceeding 0.50m2	4.18	30.00	**5.38**	0.53	-	**7.53**	0.36	**13.26**	m2	**14.26**
Stripping existing hessian, paper backed; textile hessian paper backed wall covering p.c. £7.50m2; sizing; applying adhesive; hanging; butt joints										
Hard building board walls and columns exceeding 0.50m2	7.68	25.00	**9.55**	0.50	-	**7.11**	0.34	**16.99**	m2	**18.26**
Plaster walls and columns exceeding 0.50m2	0.85	25.00	**1.03**	0.46	-	**6.54**	-	**7.57**	m2	**8.14**
Expanded polystyrene sheet 2mm thick; sizing; applying adhesive; hanging; butt joints										
Plaster walls and columns exceeding 0.50m2	0.40	25.00	**0.50**	0.33	-	**4.69**	0.18	**5.37**	m2	**5.77**
Plaster ceilings and beams exceeding 0.50m2	0.40	25.00	**0.50**	0.39	-	**5.54**	0.18	**6.22**	m2	**6.69**
M60: PAINTING/CLEAR FINISHING - EMULSION PAINTING										
Mist coat, one full coat emulsion paint										
Concrete general surfaces girth exceeding 300mm	0.17	10.00	**0.19**	0.15	-	**2.13**	0.04	**2.36**	m2	**2.53**
Concrete general surfaces 3.50 - 5.00m above floor girth exceeding 300mm	0.17	10.00	**0.19**	0.17	-	**2.42**	0.05	**2.65**	m2	**2.85**
Plaster general surfaces girth exceeding 300mm	0.14	10.00	**0.15**	0.13	-	**1.85**	0.04	**2.03**	m2	**2.18**
isolated surfaces, girth not exceeding 300mm	0.41	10.00	**0.45**	0.07	-	**0.99**	0.02	**1.46**	m	**1.57**
Plaster general surfaces 3.50 - 5.00m above floor girth exceeding 300mm	0.14	10.00	**0.15**	0.15	-	**2.13**	0.04	**2.32**	m2	**2.49**
Plasterboard general surfaces girth exceeding 300mm	0.14	10.00	**0.15**	0.13	-	**1.85**	0.04	**2.03**	m2	**2.18**
Plasterboard general surfaces 3.50 - 5.00m above floor girth exceeding 300mm	0.14	10.00	**0.15**	0.15	-	**2.13**	0.04	**2.32**	m2	**2.49**
Brickwork general surfaces girth exceeding 300mm	0.17	10.00	**0.19**	0.17	-	**2.42**	0.05	**2.65**	m2	**2.85**
Paper covered general surfaces girth exceeding 300mm	0.14	10.00	**0.15**	0.14	-	**1.99**	0.04	**2.18**	m2	**2.34**
Paper covered general surfaces 3.50 - 5.00m above floor girth exceeding 300mm	0.14	10.00	**0.15**	0.16	-	**2.27**	0.04	**2.47**	m2	**2.65**
Mist coat, two full coats emulsion paint										
Concrete general surfaces girth exceeding 300mm	0.28	10.00	**0.31**	0.21	-	**2.98**	0.06	**3.35**	m2	**3.60**
Concrete general surfaces 3.50 - 5.00m above floor girth exceeding 300mm	0.28	10.00	**0.31**	0.24	-	**3.41**	0.06	**3.79**	m2	**4.07**
Plaster general surfaces girth exceeding 300mm	0.23	10.00	**0.25**	0.19	-	**2.70**	0.05	**3.00**	m2	**3.23**
isolated surfaces, girth not exceeding 300mm	0.07	10.00	**0.07**	0.10	-	**1.42**	0.03	**1.52**	m	**1.64**
Plaster general surfaces 3.50 - 5.00m above floor girth exceeding 300mm	0.23	10.00	**0.25**	0.22	-	**3.13**	0.06	**3.44**	m2	**3.70**
Plasterboard general surfaces girth exceeding 300mm	0.23	10.00	**0.25**	0.19	-	**2.70**	0.05	**3.00**	m2	**3.23**
Plasterboard general surfaces 3.50 - 5.00m above floor girth exceeding 300mm	0.23	10.00	**0.25**	0.22	-	**3.13**	0.06	**3.44**	m2	**3.70**
Brickwork general surfaces girth exceeding 300mm	0.28	10.00	**0.31**	0.24	-	**3.41**	0.06	**3.79**	m2	**4.07**
Paper covered general surfaces girth exceeding 300mm	0.23	10.00	**0.25**	0.20	-	**2.84**	0.05	**3.15**	m2	**3.38**
Paper covered general surfaces 3.50 - 5.00m above floor girth exceeding 300mm	0.23	10.00	**0.25**	0.30	-	**4.26**	0.08	**4.60**	m2	**4.94**

Labour hourly rates: (except Specialists) Craft Operatives £14.21 Labourer £10.71 Rates are national average prices. Refer to REGIONAL VARIATIONS for indicative levels of overall pricing in regions	MATERIALS			LABOUR				RATES		
	Del to Site £	Waste %	Material Cost £	Craft Optve Hrs	Lab Hrs	Labour Cost £	Sunds £	Nett Rate £	Unit	Gross Rate (+7.5%) £
M60: PAINTING/CLEAR FINISHING - CEMENT PAINTING										
One coat waterproof cement paint; external work										
Cement rendered general surfaces										
girth exceeding 300mm	1.10	10.00	**1.21**	0.11	-	**1.56**	0.03	**2.80**	m2	**3.01**
isolated surfaces, girth not exceeding 300mm	1.10	10.00	**1.21**	0.04	-	**0.57**	0.01	**1.79**	m	**1.92**
Concrete general surfaces										
girth exceeding 300mm	1.10	10.00	**1.21**	0.11	-	**1.56**	0.03	**2.80**	m2	**3.01**
Brickwork general surfaces										
girth exceeding 300mm	1.27	10.00	**1.40**	0.13	-	**1.85**	0.04	**3.28**	m2	**3.53**
Rough cast general surfaces										
girth exceeding 300mm	1.45	10.00	**1.59**	0.17	-	**2.42**	0.05	**4.06**	m2	**4.36**
Two coats waterproof cement paint; external work										
Cement rendered general surfaces										
girth exceeding 300mm	2.90	10.00	**3.19**	0.22	-	**3.13**	0.06	**6.37**	m2	**6.85**
isolated surfaces, girth not exceeding 300mm	0.88	10.00	**0.97**	0.08	-	**1.14**	0.02	**2.12**	m	**2.28**
Concrete general surfaces										
girth exceeding 300mm	2.20	10.00	**2.41**	0.22	-	**3.13**	0.06	**5.60**	m2	**6.02**
Brickwork general surfaces										
girth exceeding 300mm	2.55	10.00	**2.80**	0.24	-	**3.41**	0.06	**6.28**	m2	**6.75**
Rough cast general surfaces										
girth exceeding 300mm	2.90	10.00	**3.19**	0.33	-	**4.69**	0.09	**7.97**	m2	**8.56**
One coat sealer, one coat waterproof cement paint; external work										
Cement rendered general surfaces										
girth exceeding 300mm	2.30	10.00	**2.52**	0.18	-	**2.56**	0.05	**5.13**	m2	**5.52**
isolated surfaces, girth not exceeding 300mm	0.68	10.00	**0.75**	0.06	-	**0.85**	0.02	**1.62**	m	**1.74**
Concrete general surfaces										
girth exceeding 300mm	1.94	10.00	**2.14**	0.18	-	**2.56**	0.05	**4.74**	m2	**5.10**
Brickwork general surfaces										
girth exceeding 300mm	2.12	10.00	**2.33**	0.21	-	**2.98**	0.06	**5.37**	m2	**5.78**
Rough cast general surfaces										
girth exceeding 300mm	2.30	10.00	**2.52**	0.27	-	**3.84**	0.07	**6.43**	m2	**6.92**
Two coats sealer, one coat waterproof cement paint; external work										
Cement rendered general surfaces										
girth exceeding 300mm	2.88	10.00	**3.17**	0.34	-	**4.83**	0.09	**8.09**	m2	**8.70**
isolated surfaces, girth not exceeding 300mm	0.86	10.00	**0.95**	0.11	-	**1.56**	0.03	**2.54**	m	**2.73**
Concrete general surfaces										
girth exceeding 300mm	2.35	10.00	**2.58**	0.34	-	**4.83**	0.09	**7.50**	m2	**8.07**
Brickwork general surfaces										
girth exceeding 300mm	2.70	10.00	**2.97**	0.39	-	**5.54**	0.11	**8.62**	m2	**9.27**
Rough cast general surfaces										
girth exceeding 300mm	2.88	10.00	**3.17**	0.51	-	**7.25**	0.14	**10.55**	m2	**11.34**
One coat textured masonry paint; external work										
Cement rendered general surfaces										
girth exceeding 300mm	0.18	10.00	**0.20**	0.11	-	**1.56**	0.03	**1.79**	m2	**1.92**
isolated surfaces, girth not exceeding 300mm	0.07	10.00	**0.08**	0.04	-	**0.57**	0.01	**0.66**	m	**0.71**
Concrete general surfaces										
girth exceeding 300mm	0.25	10.00	**0.28**	0.11	-	**1.56**	0.03	**1.87**	m2	**2.01**
Brickwork general surfaces										
girth exceeding 300mm	0.45	10.00	**0.49**	0.13	-	**1.85**	0.04	**2.37**	m2	**2.55**
Rough cast general surfaces										
girth exceeding 300mm	0.59	10.00	**0.65**	0.17	-	**2.42**	0.05	**3.11**	m2	**3.34**
Two coats textured masonry paint; external work										
Cement rendered general surfaces										
girth exceeding 300mm	0.50	10.00	**0.55**	0.22	-	**3.13**	0.06	**3.74**	m2	**4.02**
isolated surfaces, girth not exceeding 300mm	0.14	10.00	**0.16**	0.08	-	**1.14**	0.02	**1.32**	m	**1.41**
Concrete general surfaces										
girth exceeding 300mm	0.50	10.00	**0.55**	0.22	-	**3.13**	0.06	**3.74**	m2	**4.02**
Brickwork general surfaces										
girth exceeding 300mm	0.90	10.00	**0.98**	0.24	-	**3.41**	0.06	**4.46**	m2	**4.79**
Rough cast general surfaces										
girth exceeding 300mm	1.18	10.00	**1.30**	0.33	-	**4.69**	0.09	**6.08**	m2	**6.53**

Labour hourly rates: (except Specialists) Craft Operatives £14.21 Labourer £10.71 Rates are national average prices. Refer to REGIONAL VARIATIONS for indicative levels of overall pricing in regions	MATERIALS			LABOUR				RATES		
	Del to Site £	Waste %	Material Cost £	Craft Optve Hrs	Lab Hrs	Labour Cost £	Sunds £	Nett Rate £	Unit	Gross Rate (+7.5%) £
M60: PAINTING/CLEAR FINISHING - CEMENT PAINTING Cont.										
One coat stabilizing solution, one coat textured masonry paint; external work										
Cement rendered general surfaces										
girth exceeding 300mm	0.49	10.00	**0.53**	0.18	-	**2.56**	0.05	**3.14**	m2	**3.38**
isolated surfaces, girth not exceeding 300mm	0.14	10.00	**0.16**	0.06	-	**0.85**	0.02	**1.03**	m	**1.10**
Concrete general surfaces										
girth exceeding 300mm	0.49	10.00	**0.53**	0.18	-	**2.56**	0.05	**3.14**	m2	**3.38**
Brickwork general surfaces										
girth exceeding 300mm	0.68	10.00	**0.75**	0.21	-	**2.98**	0.06	**3.79**	m2	**4.08**
Rough cast general surfaces										
girth exceeding 300mm	0.83	10.00	**0.91**	0.27	-	**3.84**	0.07	**4.82**	m2	**5.18**
One coat stabilizing solution, two coats textured masonry paint; external work										
Cement rendered general surfaces										
girth exceeding 300mm	0.74	10.00	**0.81**	0.34	-	**4.83**	0.09	**5.73**	m2	**6.16**
isolated surfaces, girth not exceeding 300mm	0.21	10.00	**0.24**	0.11	-	**1.56**	0.03	**1.83**	m	**1.97**
Concrete general surfaces										
girth exceeding 300mm	0.74	10.00	**0.81**	0.34	-	**4.83**	0.09	**5.73**	m2	**6.16**
Brickwork general surfaces										
girth exceeding 300mm	1.13	10.00	**1.24**	0.39	-	**5.54**	0.11	**6.89**	m2	**7.41**
Rough cast general surfaces										
girth exceeding 300mm	1.42	10.00	**1.56**	0.51	-	**7.25**	0.14	**8.94**	m2	**9.61**
M60: PAINTING/CLEAR FINISHING - PRESERVATIVE TREATMENT										
One coat creosote BS 144; external work (sawn timber)										
Wood general surfaces										
girth exceeding 300mm	0.26	10.00	**0.29**	0.10	-	**1.42**	0.03	**1.73**	m2	**1.86**
isolated surfaces, girth not exceeding 300mm	0.08	10.00	**0.09**	0.04	-	**0.57**	0.01	**0.67**	m	**0.72**
Two coats creosote BS 144; external work (sawn timber)										
Wood general surfaces										
girth exceeding 300mm	0.53	10.00	**0.58**	0.20	-	**2.84**	0.05	**3.48**	m2	**3.74**
isolated surfaces, girth not exceeding 300mm	0.16	10.00	**0.17**	0.07	-	**0.99**	0.02	**1.19**	m	**1.28**
One coat wood preservative, internal work (sawn timber)										
Wood general surfaces										
girth exceeding 300mm	0.30	10.00	**0.33**	0.11	-	**1.56**	0.03	**1.92**	m2	**2.07**
isolated surfaces, girth not exceeding 300mm	0.09	10.00	**0.10**	0.04	-	**0.57**	0.01	**0.68**	m	**0.73**
One coat wood preservative, internal work (wrought timber)										
Wood general surfaces										
girth exceeding 300mm	0.25	10.00	**0.28**	0.09	-	**1.28**	0.02	**1.58**	m2	**1.70**
isolated surfaces, girth not exceeding 300mm	0.08	10.00	**0.08**	0.03	-	**0.43**	0.01	**0.52**	m	**0.56**
One coat wood preservative, external work (sawn timber)										
Wood general surfaces										
girth exceeding 300mm	0.30	10.00	**0.33**	0.10	-	**1.42**	0.03	**1.78**	m2	**1.91**
isolated surfaces, girth not exceeding 300mm	0.09	10.00	**0.10**	0.04	-	**0.57**	0.01	**0.68**	m	**0.73**
One coat wood preservative, external work (wrought timber)										
Wood general surfaces										
girth exceeding 300mm	0.25	10.00	**0.28**	0.08	-	**1.14**	0.02	**1.43**	m2	**1.54**
isolated surfaces, girth not exceeding 300mm	0.08	10.00	**0.08**	0.02	-	**0.28**	0.01	**0.37**	m	**0.40**
Two coats wood preservative, internal work (sawn timber)										
Wood general surfaces										
girth exceeding 300mm	0.60	10.00	**0.66**	0.22	-	**3.13**	0.06	**3.85**	m2	**4.13**
isolated surfaces, girth not exceeding 300mm	0.19	10.00	**0.21**	0.08	-	**1.14**	0.02	**1.36**	m	**1.47**
Two coats wood preservative, internal work (wrought timber)										
Wood general surfaces										
girth exceeding 300mm	0.50	10.00	**0.55**	0.18	-	**2.56**	0.05	**3.16**	m2	**3.39**
isolated surfaces, girth not exceeding 300mm	0.15	10.00	**0.17**	0.05	-	**0.71**	0.01	**0.89**	m	**0.96**
Two coats wood preservative, external work (sawn timber)										
Wood general surfaces										
girth exceeding 300mm	0.60	10.00	**0.66**	0.20	-	**2.84**	0.05	**3.56**	m2	**3.82**
isolated surfaces, girth not exceeding 300mm	0.19	10.00	**0.21**	0.07	-	**0.99**	0.02	**1.22**	m	**1.31**

Labour hourly rates: (except Specialists) Craft Operatives £14.21 Labourer £10.71 Rates are national average prices. Refer to REGIONAL VARIATIONS for indicative levels of overall pricing in regions	MATERIALS			LABOUR				RATES		
	Del to Site £	Waste %	Material Cost £	Craft Optve Hrs	Lab Hrs	Labour Cost £	Sunds £	Nett Rate £	Unit	Gross Rate (+7.5%) £
M60: PAINTING/CLEAR FINISHING - PRESERVATIVE TREATMENT Cont.										
Two coats wood preservative, external work (wrought timber)										
Wood general surfaces										
girth exceeding 300mm	0.50	10.00	**0.55**	0.16	-	**2.27**	0.04	**2.87**	m2	**3.08**
isolated surfaces, girth not exceeding 300mm	0.15	10.00	**0.17**	0.05	-	**0.71**	0.01	**0.89**	m	**0.96**
Two coats Sadolins High Performance Varnish										
Wood general surfaces										
girth exceeding 300mm	1.14	10.00	**1.26**	0.27	-	**3.84**	0.07	**5.17**	m2	**5.56**
isolated surfaces, girth not exceeding 300mm	0.34	10.00	**0.38**	0.10	-	**1.35**	0.03	**1.75**	m	**1.88**
Wood glazed doors										
girth exceeding 300mm; panes, area not exceeding 0.10m2	0.65	10.00	**0.72**	0.89	-	**12.65**	0.24	**13.61**	m2	**14.63**
girth exceeding 300mm; panes, area 0.10 - 0.50m2	0.46	10.00	**0.51**	0.64	-	**9.09**	0.17	**9.78**	m2	**10.51**
girth exceeding 300mm; panes, area 0.50 - 1.00m2	0.41	10.00	**0.45**	0.54	-	**7.67**	0.15	**8.27**	m2	**8.89**
girth exceeding 300mm; panes, area exceeding 1.00m2	0.41	10.00	**0.45**	0.48	-	**6.82**	0.13	**7.40**	m2	**7.96**
Wood partially glazed doors										
girth exceeding 300mm; panes, area 0.50 - 1.00m2	0.63	10.00	**0.69**	0.54	-	**7.67**	0.15	**8.51**	m2	**9.15**
girth exceeding 300mm; panes, area exceeding 1.00m2	0.63	10.00	**0.69**	0.52	-	**7.39**	0.14	**8.22**	m2	**8.84**
Wood windows and screens										
girth exceeding 300mm; panes, area not exceeding 0.10m2	0.65	10.00	**0.72**	0.98	-	**13.93**	0.26	**14.91**	m2	**16.03**
girth exceeding 300mm; panes, area 0.10 - 0.50m2	0.46	10.00	**0.51**	0.71	-	**10.09**	0.19	**10.79**	m2	**11.60**
girth exceeding 300mm; panes, area 0.50 - 1.00m2	0.41	10.00	**0.45**	0.60	-	**8.53**	0.18	**9.16**	m2	**9.85**
girth exceeding 300mm; panes, area exceeding 1.00m2	0.41	10.00	**0.45**	0.52	-	**7.39**	0.14	**7.98**	m2	**8.58**
One coat Sadolins Classic; two coats Sadolins Extra; external work										
Wood general surfaces										
girth exceeding 300mm	1.78	10.00	**1.95**	0.50	-	**7.11**	0.14	**9.19**	m2	**9.88**
isolated surfaces, girth not exceeding 300mm	0.52	10.00	**0.58**	0.17	-	**2.42**	0.05	**3.04**	m	**3.26**
Wood glazed doors										
girth exceeding 300mm; panes, area not exceeding 0.10m2	0.97	10.00	**1.07**	0.78	-	**11.08**	0.21	**12.36**	m2	**13.29**
girth exceeding 300mm; panes, area 0.10 - 0.50m2	0.70	10.00	**0.77**	0.56	-	**7.96**	0.15	**8.88**	m2	**9.54**
girth exceeding 300mm; panes, area 0.50 - 1.00m2	0.63	10.00	**0.69**	0.47	-	**6.68**	0.13	**7.50**	m2	**8.06**
girth exceeding 300mm; panes, area exceeding 1.00m2	0.63	10.00	**0.69**	0.42	-	**5.97**	0.11	**6.77**	m2	**7.28**
Wood partially glazed doors										
girth exceeding 300mm; panes, area 0.50 - 1.00m2	0.97	10.00	**1.07**	0.47	-	**6.68**	0.13	**7.87**	m2	**8.47**
girth exceeding 300mm; panes, area exceeding 1.00m2	0.97	10.00	**1.07**	0.45	-	**6.39**	0.12	**7.59**	m2	**8.15**
Wood windows and screens										
girth exceeding 300mm; panes, area not exceeding 0.10m2	0.97	10.00	**1.07**	0.89	-	**12.65**	0.24	**13.96**	m2	**15.00**
girth exceeding 300mm; panes, area 0.10 - 0.50m2	0.70	10.00	**0.77**	0.65	-	**9.24**	0.18	**10.18**	m2	**10.94**
girth exceeding 300mm; panes, area 0.50 - 1.00m2	0.63	10.00	**0.69**	0.55	-	**7.82**	0.15	**8.66**	m2	**9.31**
girth exceeding 300mm; panes, area exceeding 1.00m2	0.63	10.00	**0.69**	0.49	-	**6.96**	0.13	**7.79**	m2	**8.37**
Wood railings fences and gates; open type										
girth exceeding 300mm	1.36	10.00	**1.49**	0.84	-	**11.94**	0.23	**13.66**	m2	**14.68**
isolated surfaces, girth not exceeding 300mm	0.52	10.00	**0.58**	0.25	-	**3.55**	0.07	**4.20**	m	**4.51**
Wood railings fences and gates; close type										
girth exceeding 300mm	1.78	10.00	**1.95**	0.84	-	**11.94**	0.23	**14.12**	m2	**15.17**
M60: PAINTING/CLEAR FINISHING - OIL PAINTING WALLS AND CEILINGS										
One coat primer, one undercoat, one coat full gloss finish										
Concrete general surfaces										
girth exceeding 300mm	0.81	10.00	**0.89**	0.30	-	**4.26**	0.08	**5.23**	m2	**5.62**
Concrete general surfaces 3.50 - 5.00m above floor										
girth exceeding 300mm	0.81	10.00	**0.89**	0.33	-	**4.69**	0.09	**5.66**	m2	**6.09**
Brickwork general surfaces										
girth exceeding 300mm	0.94	10.00	**1.03**	0.34	-	**4.83**	0.09	**5.96**	m2	**6.40**
Plasterboard general surfaces										
girth exceeding 300mm	0.74	10.00	**0.81**	0.27	-	**3.84**	0.07	**4.72**	m2	**5.07**
Plasterboard general surfaces 3.50 - 5.00m above floor										
girth exceeding 300mm	0.74	10.00	**0.81**	0.30	-	**4.26**	0.08	**5.16**	m2	**5.54**
Plaster general surfaces										
girth exceeding 300mm	0.74	10.00	**0.81**	0.27	-	**3.84**	0.07	**4.72**	m2	**5.07**
isolated surfaces, girth not exceeding 300mm	1.43	10.00	**1.57**	0.12	-	**1.71**	0.03	**3.31**	m	**3.56**

Labour hourly rates: (except Specialists) Craft Operatives £14.21 Labourer £10.71 Rates are national average prices. Refer to REGIONAL VARIATIONS for indicative levels of overall pricing in regions	MATERIALS			LABOUR				RATES		
	Del to Site £	Waste %	Material Cost £	Craft Optve Hrs	Lab Hrs	Labour Cost £	Sunds £	Nett Rate £	Unit	Gross Rate (+7.5%) £
M60: PAINTING/CLEAR FINISHING - OIL PAINTING WALLS AND CEILINGS Cont.										
One coat primer, one undercoat, one coat full gloss finish Cont.										
Plaster general surfaces 3.50 - 5.00m above floor										
girth exceeding 300mm	0.74	10.00	**0.81**	0.30	-	**4.26**	0.08	**5.16**	m2	**5.54**
One coat primer, two undercoats, one coat full gloss finish										
Concrete general surfaces										
girth exceeding 300mm	0.96	10.00	**1.05**	0.40	-	**5.68**	0.11	**6.85**	m2	**7.36**
Concrete general surfaces 3.50 - 5.00m above floor										
girth exceeding 300mm	0.96	10.00	**1.05**	0.44	-	**6.25**	0.12	**7.42**	m2	**7.98**
Brickwork general surfaces										
girth exceeding 300mm	1.16	10.00	**1.28**	0.45	-	**6.39**	0.12	**7.79**	m2	**8.38**
Plasterboard general surfaces										
girth exceeding 300mm	0.98	10.00	**1.08**	0.36	-	**5.12**	0.10	**6.29**	m2	**6.77**
Plasterboard general surfaces 3.50 - 5.00m above floor										
girth exceeding 300mm	0.89	10.00	**0.98**	0.40	-	**5.68**	0.11	**6.77**	m2	**7.28**
Plaster general surfaces										
girth exceeding 300mm	0.89	10.00	**0.98**	0.36	-	**5.12**	0.10	**6.19**	m2	**6.66**
isolated surfaces, girth not exceeding 300mm	0.27	10.00	**0.30**	0.16	-	**2.27**	0.04	**2.62**	m	**2.81**
Plaster general surfaces 3.50 - 5.00m above floor										
girth exceeding 300mm	0.89	10.00	**0.98**	0.40	-	**5.68**	0.11	**6.77**	m2	**7.28**
One coat primer, one undercoat, one coat eggshell finish										
Concrete general surfaces										
girth exceeding 300mm	0.84	10.00	**0.92**	0.30	-	**4.26**	0.08	**5.27**	m2	**5.66**
Concrete general surfaces 3.50 - 5.00m above floor										
girth exceeding 300mm	0.84	10.00	**0.92**	0.33	-	**4.69**	0.09	**5.70**	m2	**6.13**
Brickwork general surfaces										
girth exceeding 300mm	1.04	10.00	**1.15**	0.34	-	**4.83**	0.09	**6.07**	m2	**6.52**
Plasterboard general surfaces										
girth exceeding 300mm	0.77	10.00	**0.85**	0.27	-	**3.84**	0.07	**4.76**	m2	**5.11**
Plasterboard general surfaces 3.50 - 5.00m above floor										
girth exceeding 300mm	0.77	10.00	**0.85**	0.30	-	**4.26**	0.08	**5.19**	m2	**5.58**
Plaster general surfaces										
girth exceeding 300mm	0.77	10.00	**0.85**	0.27	-	**3.84**	0.07	**4.76**	m2	**5.11**
isolated surfaces, girth not exceeding 300mm	1.44	10.00	**1.58**	0.12	-	**1.71**	0.03	**3.32**	m	**3.57**
Plaster general surfaces 3.50 - 5.00m above floor										
girth exceeding 300mm	0.77	10.00	**0.85**	0.30	-	**4.26**	0.08	**5.19**	m2	**5.58**
One coat primer, two undercoats, one coat eggshell finish										
Concrete general surfaces										
girth exceeding 300mm	0.99	10.00	**1.09**	0.40	-	**5.68**	0.11	**6.88**	m2	**7.40**
Concrete general surfaces 3.50 - 5.00m above floor										
girth exceeding 300mm	0.99	10.00	**1.09**	0.44	-	**6.25**	0.12	**7.46**	m2	**8.02**
Brickwork general surfaces										
girth exceeding 300mm	1.21	10.00	**1.33**	0.45	-	**6.39**	0.12	**7.84**	m2	**8.43**
Plasterboard general surfaces										
girth exceeding 300mm	0.92	10.00	**1.01**	0.36	-	**5.12**	0.10	**6.23**	m2	**6.69**
Plasterboard general surfaces 3.50 - 5.00m above floor										
girth exceeding 300mm	0.92	10.00	**1.01**	0.40	-	**5.68**	0.11	**6.81**	m2	**7.32**
Plaster general surfaces										
girth exceeding 300mm	0.92	10.00	**1.01**	0.36	-	**5.12**	0.10	**6.23**	m2	**6.69**
isolated surfaces, girth not exceeding 300mm	0.28	10.00	**0.31**	0.16	-	**2.27**	0.04	**2.63**	m	**2.83**
Plaster general surfaces 3.50 - 5.00m above floor										
girth exceeding 300mm	0.92	10.00	**1.01**	0.40	-	**5.68**	0.02	**6.72**	m2	**7.22**
M60: PAINTING/CLEAR FINISHING - SPRAY PAINTING										
Spray one coat primer, one undercoat, one coat full gloss finish										
Concrete general surfaces										
girth exceeding 300mm	0.87	10.00	**0.96**	0.26	-	**3.69**	0.07	**4.73**	m2	**5.08**
Brickwork general surfaces										
girth exceeding 300mm	1.07	10.00	**1.17**	0.30	-	**4.26**	0.08	**5.52**	m2	**5.93**

Labour hourly rates: (except Specialists) Craft Operatives £14.21 Labourer £10.71 Rates are national average prices. Refer to REGIONAL VARIATIONS for indicative levels of overall pricing in regions	MATERIALS			LABOUR				RATES		
	Del to Site £	Waste %	Material Cost £	Craft Optve Hrs	Lab Hrs	Labour Cost £	Sunds £	Nett Rate £	Unit	Gross Rate (+7.5%) £
M60: PAINTING/CLEAR FINISHING - SPRAY PAINTING Cont.										
Spray one coat primer, one undercoat, one coat full gloss finish Cont.										
Plaster general surfaces girth exceeding 300mm	0.81	10.00	**0.89**	0.24	-	**3.41**	0.06	**4.36**	m2	**4.69**
Spray one coat primer, two undercoats, one coat full gloss finish										
Concrete general surfaces girth exceeding 300mm	1.05	10.00	**1.15**	0.31	-	**4.41**	0.08	**5.64**	m2	**6.07**
Brickwork general surfaces girth exceeding 300mm	1.30	10.00	**1.43**	0.35	-	**4.97**	0.09	**6.50**	m2	**6.98**
Plaster general surfaces girth exceeding 300mm	0.98	10.00	**1.08**	0.28	-	**3.98**	0.08	**5.13**	m2	**5.52**
Spray one coat primer, one basecoat, one coat multicolour finish										
Concrete general surfaces girth exceeding 300mm	1.37	10.00	**1.50**	0.39	-	**5.54**	0.11	**7.15**	m2	**7.69**
Brickwork general surfaces girth exceeding 300mm	1.64	10.00	**1.81**	0.41	-	**5.83**	0.11	**7.74**	m2	**8.32**
Plaster general surfaces girth exceeding 300mm	1.24	10.00	**1.36**	0.37	-	**5.26**	0.10	**6.72**	m2	**7.22**
M60: PAINTING/CLEAR FINISHING - ACRYLATED RUBBER PAINTING										
One coat primer, two coats acrylated rubber paint										
Concrete general surfaces girth exceeding 300mm	3.31	10.00	**3.64**	0.30	-	**4.26**	0.08	**7.98**	m2	**8.58**
Brickwork general surfaces girth exceeding 300mm	3.95	10.00	**4.34**	0.32	-	**4.55**	0.09	**8.97**	m2	**9.65**
Plasterboard general surfaces girth exceeding 300mm	2.97	10.00	**3.27**	0.29	-	**4.12**	0.08	**7.46**	m2	**8.02**
Plaster general surfaces girth exceeding 300mm	2.97	10.00	**3.27**	0.29	-	**4.12**	0.08	**7.46**	m2	**8.02**
isolated surfaces, girth not exceeding 300mm	0.86	10.00	**0.95**	0.13	-	**1.85**	0.04	**2.83**	m	**3.05**
M60: PAINTING/CLEAR FINISHING - PLASTIC FINISH										
Textured plastic coating - Stippled finish										
Concrete general surfaces girth exceeding 300mm	0.31	10.00	**0.35**	0.25	-	**3.55**	0.07	**3.97**	m2	**4.26**
Concrete general surfaces 3.50 - 5.00m above floor girth exceeding 300mm	0.31	10.00	**0.35**	0.28	-	**3.98**	0.08	**4.40**	m2	**4.73**
Brickwork general surfaces girth exceeding 300mm	0.38	10.00	**0.41**	0.30	-	**4.26**	0.08	**4.76**	m2	**5.11**
Plasterboard general surfaces girth exceeding 300mm	0.28	10.00	**0.31**	0.20	-	**2.84**	0.05	**3.21**	m2	**3.45**
Plasterboard general surfaces 3.50 - 5.00m above floor girth exceeding 300mm	0.28	10.00	**0.31**	0.22	-	**3.13**	0.06	**3.50**	m2	**3.76**
Plaster general surfaces girth exceeding 300mm	0.28	10.00	**0.31**	0.20	-	**2.84**	0.05	**3.21**	m2	**3.45**
Plaster general surfaces 3.50 - 5.00m above floor girth exceeding 300mm	0.28	10.00	**0.31**	0.22	-	**3.13**	0.06	**3.50**	m2	**3.76**
Textured plastic coating - Combed Finish										
Concrete general surfaces girth exceeding 300mm	0.38	10.00	**0.41**	0.30	-	**4.26**	0.08	**4.76**	m2	**5.11**
Concrete general surfaces 3.50 - 5.00m above floor girth exceeding 300mm	0.38	10.00	**0.41**	0.33	-	**4.69**	0.09	**5.19**	m2	**5.58**
Brickwork general surfaces girth exceeding 300mm	0.45	10.00	**0.50**	0.35	-	**4.97**	0.09	**5.56**	m2	**5.98**
Plasterboard general surfaces girth exceeding 300mm	0.34	10.00	**0.37**	0.25	-	**3.55**	0.07	**3.99**	m2	**4.29**
Plasterboard general surfaces 3.50 - 5.00m above floor girth exceeding 300mm	0.34	10.00	**0.37**	0.27	-	**3.84**	0.07	**4.28**	m2	**4.60**
Plaster general surfaces girth exceeding 300mm	0.34	10.00	**0.37**	0.25	-	**3.55**	0.07	**3.99**	m2	**4.29**

Labour hourly rates: (except Specialists)
Craft Operatives £14.21 Labourer £10.71
Rates are national average prices.
Refer to REGIONAL VARIATIONS for indicative levels of overall pricing in regions

	MATERIALS			LABOUR				RATES		
	Del to Site £	Waste %	Material Cost £	Craft Optve Hrs	Lab Hrs	Labour Cost £	Sunds £	Nett Rate £	Unit	Gross Rate (+7.5%) £
M60: PAINTING/CLEAR FINISHING - PLASTIC FINISH Cont.										
Textured plastic coating - Combed Finish Cont.										
Plaster general surfaces 3.50 - 5.00m above floor girth exceeding 300mm	0.34	10.00	**0.37**	0.27	-	**3.84**	0.07	**4.28**	m2	**4.60**
M60: PAINTING/CLEAR FINISHING - OIL PAINTING METALWORK										
One undercoat, one coat full gloss finish on ready primed metal surfaces										
Iron or steel structural work										
girth exceeding 300mm	0.30	10.00	**0.33**	0.28	-	**3.98**	0.08	**4.39**	m2	**4.72**
isolated surfaces, girth not exceeding 300mm	0.09	10.00	**0.10**	0.12	-	**1.71**	0.03	**1.84**	m	**1.98**
Iron or steel structural members of roof trusses, lattice girders, purlins and the like										
girth exceeding 300mm	0.30	10.00	**0.33**	0.36	-	**5.12**	0.10	**5.55**	m2	**5.96**
isolated surfaces, girth not exceeding 300mm	0.09	10.00	**0.10**	0.12	-	**1.71**	0.03	**1.84**	m	**1.98**
Two undercoats, one coat full gloss finish on ready primed metal surfaces										
Iron or steel structural work										
girth exceeding 300mm	0.46	10.00	**0.50**	0.42	-	**5.97**	0.11	**6.58**	m2	**7.08**
isolated surfaces, girth not exceeding 300mm	0.14	10.00	**0.15**	0.14	-	**1.99**	0.04	**2.18**	m	**2.35**
Iron or steel structural members of roof trusses, lattice girders, purlins and the like										
girth exceeding 300mm	0.46	10.00	**0.50**	0.54	-	**7.67**	0.15	**8.32**	m2	**8.95**
isolated surfaces, girth not exceeding 300mm	0.14	10.00	**0.15**	0.18	-	**2.56**	0.05	**2.76**	m	**2.97**
One coat primer, one undercoat, one coat full gloss finish on metal surfaces										
Iron or steel general surfaces										
girth exceeding 300mm	0.85	10.00	**0.94**	0.39	-	**5.54**	0.11	**6.59**	m2	**7.08**
isolated surfaces, girth not exceeding 300mm	0.27	10.00	**0.30**	0.13	-	**1.85**	0.04	**2.18**	m	**2.34**
Galvanized glazed doors, windows or screens										
girth exceeding 300mm; panes, area not exceeding 0.10m2	0.46	10.00	**0.51**	0.72	-	**10.23**	0.19	**10.94**	m2	**11.76**
girth exceeding 300mm; panes, area 0.10 - 0.50m2	0.34	10.00	**0.38**	0.51	-	**7.25**	0.14	**7.76**	m2	**8.34**
girth exceeding 300mm; panes, area 0.50 - 1.00m2	0.29	10.00	**0.32**	0.43	-	**6.11**	0.12	**6.55**	m2	**7.04**
girth exceeding 300mm; panes, area exceeding 1.00m2	0.29	10.00	**0.32**	0.39	-	**5.54**	0.11	**5.97**	m2	**6.42**
Iron or steel structural work										
girth exceeding 300mm	0.85	10.00	**0.94**	0.42	-	**5.97**	0.11	**7.02**	m2	**7.55**
isolated surfaces, girth not exceeding 300mm	0.27	10.00	**0.30**	0.18	-	**2.56**	0.05	**2.90**	m	**3.12**
Iron or steel structural members of roof trusses, lattice girders, purlins and the like										
girth exceeding 300mm	0.85	10.00	**0.94**	0.54	-	**7.67**	0.15	**8.76**	m2	**9.41**
isolated surfaces, girth not exceeding 300mm	0.27	10.00	**0.30**	0.18	-	**2.56**	0.05	**2.90**	m	**3.12**
Iron or steel services										
girth exceeding 300mm	0.85	10.00	**0.94**	0.54	-	**7.67**	0.15	**8.76**	m2	**9.41**
isolated surfaces, girth not exceeding 300mm	0.27	10.00	**0.30**	0.18	-	**2.56**	0.05	**2.90**	m	**3.12**
isolated areas not exceeding 0.50m2 irrespective of girth	0.44	10.00	**0.48**	0.54	-	**7.67**	0.15	**8.30**	nr	**8.92**
Copper services										
girth exceeding 300mm	0.85	10.00	**0.94**	0.54	-	**7.67**	0.15	**8.76**	m2	**9.41**
isolated surfaces, girth not exceeding 300mm	0.27	10.00	**0.30**	0.18	-	**2.56**	0.05	**2.90**	m	**3.12**
Galvanized services										
girth exceeding 300mm	0.85	10.00	**0.94**	0.54	-	**7.67**	0.15	**8.76**	m2	**9.41**
isolated surfaces, girth not exceeding 300mm	0.27	10.00	**0.30**	0.18	-	**2.56**	0.05	**2.90**	m	**3.12**
One coat primer, one undercoat, one coat full gloss finish on metal surfaces ; external work										
Iron or steel general surfaces										
girth exceeding 300mm	0.85	10.00	**0.94**	0.42	-	**5.97**	0.11	**7.02**	m2	**7.55**
isolated surfaces, girth not exceeding 300mm	0.27	10.00	**0.30**	0.16	-	**2.27**	0.04	**2.61**	m	**2.81**
Galvanized glazed doors, windows or screens										
girth exceeding 300mm; panes, area not exceeding 0.10m2	0.46	10.00	**0.51**	0.75	-	**10.66**	0.20	**11.37**	m2	**12.22**
girth exceeding 300mm; panes, area 0.10 - 0.50m2	0.34	10.00	**0.38**	0.54	-	**7.67**	0.15	**8.19**	m2	**8.81**
girth exceeding 300mm; panes, area 0.50 - 1.00m2	0.29	10.00	**0.32**	0.46	-	**6.54**	0.12	**6.98**	m2	**7.51**
girth exceeding 300mm; panes, area exceeding 1.00m2	0.29	10.00	**0.32**	0.42	-	**5.97**	0.11	**6.40**	m2	**6.88**
Iron or steel structural work										
girth exceeding 300mm	0.85	10.00	**0.94**	0.42	-	**5.97**	0.11	**7.02**	m2	**7.55**
isolated surfaces, girth not exceeding 300mm	0.27	10.00	**0.30**	0.18	-	**2.56**	0.05	**2.90**	m	**3.12**
Iron or steel structural members of roof trusses, lattice girders, purlins and the like										
girth exceeding 300mm	0.85	10.00	**0.94**	0.54	-	**7.67**	0.15	**8.76**	m2	**9.41**
isolated surfaces, girth not exceeding 300mm	0.27	10.00	**0.30**	0.18	-	**2.56**	0.05	**2.90**	m	**3.12**

Labour hourly rates: (except Specialists) Craft Operatives £14.21 Labourer £10.71 Rates are national average prices. Refer to REGIONAL VARIATIONS for indicative levels of overall pricing in regions	MATERIALS			LABOUR				RATES		
	Del to Site £	Waste %	Material Cost £	Craft Optve Hrs	Lab Hrs	Labour Cost £	Sunds £	Nett Rate £	Unit	Gross Rate (+7.5%) £
M60: PAINTING/CLEAR FINISHING - OIL PAINTING METALWORK Cont.										
One coat primer, one undercoat, one coat full gloss finish on metal surfaces ; external work Cont.										
Iron or steel railings, fences and gates; plain open type										
girth exceeding 300mm	0.66	10.00	**0.72**	0.39	-	**5.54**	0.11	**6.37**	m2	**6.85**
isolated surfaces, girth not exceeding 300mm	0.27	10.00	**0.30**	0.13	-	**1.85**	0.04	**2.18**	m	**2.34**
Iron or steel railings, fences and gates; close type										
girth exceeding 300mm	0.85	10.00	**0.94**	0.33	-	**4.69**	0.09	**5.72**	m2	**6.15**
Iron or steel railings, fences and gates; ornamental type										
girth exceeding 300mm	0.66	10.00	**0.72**	0.66	-	**9.38**	0.18	**10.28**	m2	**11.05**
Iron or steel eaves gutters										
girth exceeding 300mm	0.85	10.00	**0.94**	0.45	-	**6.39**	0.12	**7.45**	m2	**8.01**
isolated surfaces, girth not exceeding 300mm	0.27	10.00	**0.30**	0.15	-	**2.13**	0.04	**2.47**	m	**2.65**
Galvanized eaves gutters										
girth exceeding 300mm	0.85	10.00	**0.94**	0.45	-	**6.39**	0.12	**7.45**	m2	**8.01**
isolated surfaces, girth not exceeding 300mm	0.27	10.00	**0.30**	0.15	-	**2.13**	0.04	**2.47**	m	**2.65**
Iron or steel services										
girth exceeding 300mm	0.85	10.00	**0.94**	0.54	-	**7.67**	0.15	**8.76**	m2	**9.41**
isolated surfaces, girth not exceeding 300mm	0.27	10.00	**0.30**	0.18	-	**2.56**	0.05	**2.90**	m	**3.12**
isolated areas not exceeding 0.50m2 irrespective of girth	0.44	10.00	**0.48**	0.54	-	**7.67**	0.15	**8.30**	nr	**8.92**
Copper services										
girth exceeding 300mm	0.85	10.00	**0.94**	0.54	-	**7.67**	0.15	**8.76**	m2	**9.41**
isolated surfaces, girth not exceeding 300mm	0.27	10.00	**0.30**	0.18	-	**2.56**	0.05	**2.90**	m	**3.12**
Galvanized services										
girth exceeding 300mm	0.85	10.00	**0.94**	0.54	-	**7.67**	0.15	**8.76**	m2	**9.41**
isolated surfaces, girth not exceeding 300mm	0.27	10.00	**0.30**	0.18	-	**2.56**	0.05	**2.90**	m	**3.12**
One coat primer, two undercoats, one coat full gloss finish on metal surfaces										
Iron or steel general surfaces										
girth exceeding 300mm	1.01	10.00	**1.11**	0.52	-	**7.39**	0.14	**8.64**	m2	**9.28**
isolated surfaces, girth not exceeding 300mm	0.32	10.00	**0.35**	0.17	-	**2.42**	0.05	**2.81**	m	**3.02**
Galvanized glazed doors, windows or screens										
girth exceeding 300mm; panes, area not exceeding 0.10m2	0.55	10.00	**0.60**	0.96	-	**13.64**	0.26	**14.50**	m2	**15.59**
girth exceeding 300mm; panes, area 0.10 - 0.50m2	0.40	10.00	**0.44**	0.68	-	**9.66**	0.18	**10.29**	m2	**11.06**
girth exceeding 300mm; panes, area 0.50 - 1.00m2	0.34	10.00	**0.37**	0.57	-	**8.10**	0.15	**8.63**	m2	**9.27**
girth exceeding 300mm; panes, area exceeding 1.00m2	0.34	10.00	**0.37**	0.52	-	**7.39**	0.14	**7.90**	m2	**8.50**
Iron or steel structural work										
girth exceeding 300mm	1.01	10.00	**1.11**	0.56	-	**7.96**	0.15	**9.21**	m2	**9.91**
isolated surfaces, girth not exceeding 300mm	0.32	10.00	**0.35**	0.22	-	**3.13**	0.06	**3.53**	m	**3.80**
Iron or steel structural members of roof trusses, lattice girders, purlins and the like										
girth exceeding 300mm	1.01	10.00	**1.11**	0.72	-	**10.23**	0.19	**11.53**	m2	**12.40**
isolated surfaces, girth not exceeding 300mm	0.32	10.00	**0.35**	0.22	-	**3.13**	0.06	**3.53**	m	**3.80**
Iron or steel services										
girth exceeding 300mm	1.01	10.00	**1.11**	0.63	-	**8.95**	0.17	**10.23**	m2	**11.00**
isolated surfaces, girth not exceeding 300mm	0.32	10.00	**0.35**	0.21	-	**2.98**	0.06	**3.39**	m	**3.64**
isolated areas not exceeding 0.50m2 irrespective of girth	0.51	10.00	**0.56**	0.63	-	**8.95**	0.17	**9.68**	nr	**10.41**
Copper services										
girth exceeding 300mm	1.01	10.00	**1.11**	0.63	-	**8.95**	0.17	**10.23**	m2	**11.00**
isolated surfaces, girth not exceeding 300mm	0.32	10.00	**0.35**	0.21	-	**2.98**	0.06	**3.39**	m	**3.64**
Galvanized services										
girth exceeding 300mm	1.01	10.00	**1.11**	0.63	-	**8.95**	0.17	**10.23**	m2	**11.00**
isolated surfaces, girth not exceeding 300mm	0.32	10.00	**0.35**	0.21	-	**2.98**	0.06	**3.39**	m	**3.64**
One coat primer, two undercoats, one coat full gloss finish on metal surfaces; external work										
Iron or steel general surfaces										
girth exceeding 300mm	1.01	10.00	**1.11**	0.56	-	**7.96**	0.15	**9.21**	m2	**9.91**
isolated surfaces, girth not exceeding 300mm	0.32	10.00	**0.35**	0.19	-	**2.70**	0.05	**3.10**	m	**3.33**
Galvanized glazed doors, windows or screens										
girth exceeding 300mm; panes, area not exceeding 0.10m2	0.55	10.00	**0.60**	1.00	-	**14.21**	0.27	**15.08**	m2	**16.21**
girth exceeding 300mm; panes, area 0.10 - 0.50m2	0.40	10.00	**0.44**	0.72	-	**10.23**	0.19	**10.87**	m2	**11.68**
girth exceeding 300mm; panes, area 0.50 - 1.00m2	0.34	10.00	**0.37**	0.61	-	**8.67**	0.16	**9.21**	m2	**9.90**
girth exceeding 300mm; panes, area exceeding 1.00m2	0.34	10.00	**0.37**	0.56	-	**7.96**	0.15	**8.48**	m2	**9.12**
Iron or steel structural work										
girth exceeding 300mm	1.01	10.00	**1.11**	0.56	-	**7.96**	0.15	**9.21**	m2	**9.91**
isolated surfaces, girth not exceeding 300mm	0.32	10.00	**0.35**	0.22	-	**3.13**	0.06	**3.53**	m	**3.80**

Labour hourly rates: (except Specialists) Craft Operatives £14.21 Labourer £10.71 Rates are national average prices. Refer to REGIONAL VARIATIONS for indicative levels of overall pricing in regions	MATERIALS			LABOUR				RATES		
	Del to Site £	Waste %	Material Cost £	Craft Optve Hrs	Lab Hrs	Labour Cost £	Sunds £	Nett Rate £	Unit	Gross Rate (+7.5%) £
M60: PAINTING/CLEAR FINISHING - OIL PAINTING METALWORK Cont.										
One coat primer, two undercoats, one coat full gloss finish on metal surfaces; external work Cont.										
Iron or steel structural members of roof trusses, lattice girders, purlins and the like										
girth exceeding 300mm	1.01	10.00	**1.11**	0.72	-	**10.23**	0.19	**11.53**	m2	**12.40**
isolated surfaces, girth not exceeding 300mm	0.32	10.00	**0.35**	0.22	-	**3.13**	0.06	**3.53**	m	**3.80**
Iron or steel railings, fences and gates; plain open type										
girth exceeding 300mm	0.78	10.00	**0.85**	0.52	-	**7.39**	0.14	**8.38**	m2	**9.01**
isolated surfaces, girth not exceeding 300mm	0.32	10.00	**0.35**	0.17	-	**2.42**	0.05	**2.81**	m	**3.02**
Iron or steel railings, fences and gates; close type										
girth exceeding 300mm	1.01	10.00	**1.11**	0.44	-	**6.25**	0.12	**7.48**	m2	**8.04**
Iron or steel railings, fences and gates; ornamental type										
girth exceeding 300mm	0.78	10.00	**0.85**	0.88	-	**12.50**	0.24	**13.60**	m2	**14.61**
Iron or steel eaves gutters										
girth exceeding 300mm	1.01	10.00	**1.11**	0.60	-	**8.53**	0.16	**9.79**	m2	**10.53**
isolated surfaces, girth not exceeding 300mm	0.32	10.00	**0.35**	0.20	-	**2.84**	0.05	**3.24**	m	**3.49**
Galvanized eaves gutters										
girth exceeding 300mm	1.01	10.00	**1.11**	0.60	-	**8.53**	0.16	**9.79**	m2	**10.53**
isolated surfaces, girth not exceeding 300mm	0.32	10.00	**0.35**	0.20	-	**2.84**	0.05	**3.24**	m	**3.49**
Iron or steel services										
girth exceeding 300mm	1.01	10.00	**1.11**	0.63	-	**8.95**	0.17	**10.23**	m2	**11.00**
isolated surfaces, girth not exceeding 300mm	0.32	10.00	**0.35**	0.21	-	**2.98**	0.06	**3.39**	m	**3.64**
isolated areas not exceeding 0.50m2 irrespective of girth	0.52	10.00	**0.57**	0.63	-	**8.95**	0.17	**9.69**	nr	**10.42**
Copper services										
girth exceeding 300mm	1.01	10.00	**1.11**	0.63	-	**8.95**	0.17	**10.23**	m2	**11.00**
isolated surfaces, girth not exceeding 300mm	0.32	10.00	**0.35**	0.21	-	**2.98**	0.06	**3.39**	m	**3.64**
Galvanized services										
girth exceeding 300mm	1.01	10.00	**1.11**	0.63	-	**8.95**	0.17	**10.23**	m2	**11.00**
isolated surfaces, girth not exceeding 300mm	0.32	10.00	**0.35**	0.21	-	**2.98**	0.06	**3.39**	m	**3.64**
One coat primer, one undercoat, two coats full gloss finish on metal surfaces										
Iron or steel general surfaces										
girth exceeding 300mm	1.01	10.00	**1.11**	0.52	-	**7.39**	0.14	**8.64**	m2	**9.28**
isolated surfaces, girth not exceeding 300mm	0.32	10.00	**0.35**	0.17	-	**2.42**	0.05	**2.81**	m	**3.02**
Galvanized glazed doors, windows or screens										
girth exceeding 300mm; panes, area not exceeding 0.10m2	0.55	10.00	**0.60**	0.96	-	**13.64**	0.26	**14.50**	m2	**15.59**
girth exceeding 300mm; panes, area 0.10 - 0.50m2	0.40	10.00	**0.44**	0.68	-	**9.66**	0.18	**10.29**	m2	**11.06**
girth exceeding 300mm; panes, area 0.50 - 1.00m2	0.34	10.00	**0.37**	0.57	-	**8.10**	0.15	**8.63**	m2	**9.27**
girth exceeding 300mm; panes, area exceeding 1.00m2	0.34	10.00	**0.37**	0.52	-	**7.39**	0.14	**7.90**	m2	**8.50**
Iron or steel structural work										
girth exceeding 300mm	1.01	10.00	**1.11**	0.56	-	**7.96**	0.15	**9.21**	m2	**9.91**
isolated surfaces, girth not exceeding 300mm	0.32	10.00	**0.35**	0.22	-	**3.13**	0.06	**3.53**	m	**3.80**
Iron or steel structural members of roof trusses, lattice girders, purlins and the like										
girth exceeding 300mm	1.01	10.00	**1.11**	0.72	-	**10.23**	0.19	**11.53**	m2	**12.40**
isolated surfaces, girth not exceeding 300mm	0.32	10.00	**0.35**	0.22	-	**3.13**	0.06	**3.53**	m	**3.80**
Iron or steel services										
girth exceeding 300mm	1.01	10.00	**1.11**	0.63	-	**8.95**	0.17	**10.23**	m2	**11.00**
isolated surfaces, girth not exceeding 300mm	0.32	10.00	**0.35**	0.21	-	**2.98**	0.06	**3.39**	m	**3.64**
isolated areas not exceeding 0.50m2 irrespective of girth	0.52	10.00	**0.57**	0.63	-	**8.95**	0.17	**9.69**	nr	**10.42**
Copper services										
girth exceeding 300mm	1.01	10.00	**1.11**	0.63	-	**8.95**	0.17	**10.23**	m2	**11.00**
isolated surfaces, girth not exceeding 300mm	0.32	10.00	**0.35**	0.21	-	**2.98**	0.06	**3.39**	m	**3.64**
Galvanized services										
girth exceeding 300mm	1.01	10.00	**1.11**	0.63	-	**8.95**	0.17	**10.23**	m2	**11.00**
isolated surfaces, girth not exceeding 300mm	0.32	10.00	**0.35**	0.21	-	**2.98**	0.06	**3.39**	m	**3.64**
One coat primer, one undercoat, two coats full gloss finish on metal surfaces; external work										
Iron or steel general surfaces										
girth exceeding 300mm	1.01	10.00	**1.11**	0.56	-	**7.96**	0.15	**9.21**	m2	**9.91**
isolated surfaces, girth not exceeding 300mm	0.32	10.00	**0.35**	0.19	-	**2.70**	0.05	**3.10**	m	**3.33**
Galvanized glazed doors, windows or screens										
girth exceeding 300mm; panes, area not exceeding 0.10m2	0.55	10.00	**0.60**	1.00	-	**14.21**	0.27	**15.08**	m2	**16.21**
girth exceeding 300mm; panes, area 0.10 - 0.50m2	0.40	10.00	**0.44**	0.72	-	**10.23**	0.19	**10.87**	m2	**11.68**
girth exceeding 300mm; panes, area 0.50 - 1.00m2	0.34	10.00	**0.37**	0.61	-	**8.67**	0.16	**9.21**	m2	**9.90**

Labour hourly rates: (except Specialists) Craft Operatives £14.21 Labourer £10.71 Rates are national average prices. Refer to REGIONAL VARIATIONS for indicative levels of overall pricing in regions	MATERIALS			LABOUR				RATES		
	Del to Site £	Waste %	Material Cost £	Craft Optve Hrs	Lab Hrs	Labour Cost £	Sunds £	Nett Rate £	Unit	Gross Rate (+7.5%) £
M60: PAINTING/CLEAR FINISHING - OIL PAINTING METALWORK Cont.										
One coat primer, one undercoat, two coats full gloss finish on metal surfaces; external work Cont.										
Galvanized glazed doors, windows or screens Cont.										
girth exceeding 300mm; panes, area exceeding 1.00m2	0.34	10.00	**0.37**	0.56	-	**7.96**	0.15	**8.48**	m2	**9.12**
Iron or steel structural work										
girth exceeding 300mm	1.01	10.00	**1.11**	0.56	-	**7.96**	0.15	**9.21**	m2	**9.91**
isolated surfaces, girth not exceeding 300mm	0.32	10.00	**0.35**	0.22	-	**3.13**	0.06	**3.53**	m2	**3.80**
Iron or steel structural members of roof trusses, lattice girders, purlins and the like										
girth exceeding 300mm	1.01	10.00	**1.11**	0.56	-	**7.96**	0.15	**9.21**	m2	**9.91**
isolated surfaces, girth not exceeding 300mm	0.32	10.00	**0.35**	0.22	-	**3.13**	0.06	**3.53**	m	**3.80**
Iron or steel railings, fences and gates; plain open type										
girth exceeding 300mm	0.78	10.00	**0.85**	0.52	-	**7.39**	0.14	**8.38**	m2	**9.01**
isolated surfaces, girth not exceeding 300mm	0.32	10.00	**0.35**	0.17	-	**2.42**	0.05	**2.81**	m	**3.02**
Iron or steel railings, fences and gates; close type										
girth exceeding 300mm	1.01	10.00	**1.11**	0.44	-	**6.25**	0.12	**7.48**	m2	**8.04**
Iron or steel railings, fences and gates; ornamental type										
girth exceeding 300mm	0.78	10.00	**0.85**	0.88	-	**12.50**	0.24	**13.60**	m2	**14.61**
Iron or steel eaves gutters										
girth exceeding 300mm	1.01	10.00	**1.11**	0.60	-	**8.53**	0.16	**9.79**	m2	**10.53**
isolated surfaces, girth not exceeding 300mm	0.32	10.00	**0.35**	0.20	-	**2.84**	0.05	**3.24**	m	**3.49**
Galvanized eaves gutters										
girth exceeding 300mm	1.01	10.00	**1.11**	0.60	-	**8.53**	0.16	**9.79**	m2	**10.53**
isolated surfaces, girth not exceeding 300mm	0.32	10.00	**0.35**	0.20	-	**2.84**	0.05	**3.24**	m	**3.49**
Iron or steel services										
girth exceeding 300mm	1.01	10.00	**1.11**	0.63	-	**8.95**	0.17	**10.23**	m2	**11.00**
isolated surfaces, girth not exceeding 300mm	0.32	10.00	**0.35**	0.21	-	**2.98**	0.06	**3.39**	m	**3.64**
isolated areas not exceeding 0.50m2 irrespective of girth	0.52	10.00	**0.57**	0.63	-	**8.95**	0.17	**9.69**	nr	**10.42**
Copper services										
girth exceeding 300mm	1.01	10.00	**1.11**	0.63	-	**8.95**	0.17	**10.23**	m2	**11.00**
isolated surfaces, girth not exceeding 300mm	0.32	10.00	**0.35**	0.21	-	**2.98**	0.06	**3.39**	m	**3.64**
Galvanized services										
girth exceeding 300mm	1.01	10.00	**1.11**	0.63	-	**8.95**	0.17	**10.23**	m2	**11.00**
isolated surfaces, girth not exceeding 300mm	0.32	10.00	**0.35**	0.21	-	**2.98**	0.06	**3.39**	m	**3.64**
M60: PAINTING/CLEAR FINISHING - METALLIC PAINTING										
One coat aluminium metallic paint										
Iron or steel radiators; panel type										
girth exceeding 300mm	0.46	10.00	**0.50**	0.14	-	**1.99**	0.04	**2.53**	m2	**2.72**
Iron or steel radiators; column type										
girth exceeding 300mm	0.46	10.00	**0.50**	0.20	-	**2.84**	0.05	**3.40**	m2	**3.65**
Iron or steel services										
isolated surfaces, girth not exceeding 300mm	0.14	10.00	**0.15**	0.05	-	**0.71**	0.01	**0.88**	m	**0.94**
Copper services										
isolated surfaces, girth not exceeding 300mm	0.14	10.00	**0.15**	0.05	-	**0.71**	0.01	**0.88**	m	**0.94**
Galvanized services										
isolated surfaces, girth not exceeding 300mm	0.14	10.00	**0.15**	0.05	-	**0.71**	0.01	**0.88**	m	**0.94**
Two coats aluminium metallic paint										
Iron or steel radiators; panel type										
girth exceeding 300mm	0.91	10.00	**1.00**	0.28	-	**3.98**	0.08	**5.06**	m2	**5.43**
Iron or steel radiators; column type										
girth exceeding 300mm	0.91	10.00	**1.00**	0.40	-	**5.68**	0.11	**6.79**	m2	**7.30**
Iron or steel services										
isolated surfaces, girth not exceeding 300mm	0.28	10.00	**0.31**	0.10	-	**1.42**	0.03	**1.76**	m	**1.89**
Copper services										
isolated surfaces, girth not exceeding 300mm	0.28	10.00	**0.31**	0.10	-	**1.42**	0.03	**1.76**	m	**1.89**
Galvanized services										
isolated surfaces, girth not exceeding 300mm	0.28	10.00	**0.31**	0.10	-	**1.42**	0.03	**1.76**	m	**1.89**
One coat gold or bronze metallic paint										
Iron or steel radiators; panel type										
girth exceeding 300mm	0.90	10.00	**0.99**	0.14	-	**1.99**	0.04	**3.01**	m2	**3.24**
Iron or steel radiators; column type										
girth exceeding 300mm	0.90	10.00	**0.99**	0.20	-	**2.84**	0.05	**3.88**	m2	**4.17**

Labour hourly rates: (except Specialists)
Craft Operatives £14.21 Labourer £10.71
Rates are national average prices.
Refer to REGIONAL VARIATIONS for indicative levels of overall pricing in regions

	MATERIALS			LABOUR				RATES		
	Del to Site £	Waste %	Material Cost £	Craft Optve Hrs	Lab Hrs	Labour Cost £	Sunds £	Nett Rate £	Unit	Gross Rate (+7.5%) £
M60: PAINTING/CLEAR FINISHING - METALLIC PAINTING Cont.										
One coat gold or bronze metallic paint Cont.										
Iron or steel services										
isolated surfaces, girth not exceeding 300mm.....	0.26	10.00	**0.29**	0.05	-	**0.71**	0.01	**1.01**	m	**1.09**
Copper services										
isolated surfaces, girth not exceeding 300mm.....	0.26	10.00	**0.29**	0.05	-	**0.71**	0.01	**1.01**	m	**1.09**
Galvanized services										
isolated surfaces, girth not exceeding 300mm.....	0.26	10.00	**0.29**	0.05	-	**0.71**	0.01	**1.01**	m	**1.09**
Two coats gold or bronze metallic paint										
Iron or steel radiators; panel type										
girth exceeding 300mm	1.79	10.00	**1.97**	0.28	-	**3.98**	0.08	**6.03**	m2	**6.48**
Iron or steel radiators; column type										
girth exceeding 300mm	1.79	10.00	**1.97**	0.40	-	**5.68**	0.11	**7.76**	m2	**8.35**
Iron or steel services										
isolated surfaces, girth not exceeding 300mm.....	0.53	10.00	**0.58**	0.10	-	**1.42**	0.03	**2.03**	m	**2.18**
Copper services										
isolated surfaces, girth not exceeding 300mm.....	0.53	10.00	**0.58**	0.10	-	**1.42**	0.03	**2.03**	m	**2.18**
Galvanized services										
isolated surfaces, girth not exceeding 300mm.....	0.53	10.00	**0.58**	0.10	-	**1.42**	0.03	**2.03**	m	**2.18**
M60: PAINTING/CLEAR FINISHING - BITUMINOUS PAINT										
One coat black bitumen paint; external work										
Iron or steel general surfaces										
girth exceeding 300mm	0.22	10.00	**0.25**	0.13	-	**1.85**	0.04	**2.13**	m2	**2.29**
Iron or steel eaves gutters										
girth exceeding 300mm	0.22	10.00	**0.25**	0.14	-	**1.99**	0.04	**2.27**	m2	**2.44**
Iron or steel services										
girth exceeding 300mm	0.22	10.00	**0.25**	0.15	-	**2.13**	0.04	**2.42**	m2	**2.60**
isolated surfaces, girth not exceeding 300mm.....	0.07	10.00	**0.08**	0.05	-	**0.71**	0.01	**0.80**	m	**0.86**
Two coats black bitumen paint; external work										
Iron or steel general surfaces										
girth exceeding 300mm	0.45	10.00	**0.49**	0.26	-	**3.69**	0.07	**4.26**	m2	**4.58**
Iron or steel eaves gutters										
girth exceeding 300mm	0.45	10.00	**0.49**	0.28	-	**3.98**	0.08	**4.55**	m2	**4.89**
isolated surfaces, girth not exceeding 300mm.....	0.13	10.00	**0.15**	0.09	-	**1.28**	0.02	**1.45**	m	**1.56**
Iron or steel services										
girth exceeding 300mm	0.45	10.00	**0.49**	0.30	-	**4.26**	0.08	**4.84**	m2	**5.20**
isolated surfaces, girth not exceeding 300mm.....	0.13	10.00	**0.15**	0.09	-	**1.28**	0.02	**1.45**	m	**1.56**
M60: PAINTING/CLEAR FINISHING - OIL PAINTING WOODWORK										
One coat primer; carried out on site before fixing members										
Wood general surfaces										
girth exceeding 300mm	0.23	10.00	**0.25**	0.16	-	**2.27**	0.04	**2.57**	m2	**2.76**
isolated surfaces, girth not exceeding 300mm.....	0.07	10.00	**0.07**	0.06	-	**0.85**	0.02	**0.94**	m	**1.01**
One undercoat, one coat full gloss finish on ready primed wood surfaces										
Wood general surfaces										
girth exceeding 300mm	0.30	10.00	**0.33**	0.32	-	**4.55**	0.09	**4.97**	m2	**5.34**
isolated surfaces, girth not exceeding 300mm.....	0.09	10.00	**0.10**	0.11	-	**1.56**	0.03	**1.70**	m	**1.82**
Wood glazed doors										
girth exceeding 300mm; panes, area not exceeding 0.10m2	0.16	10.00	**0.18**	0.54	-	**7.67**	0.15	**8.00**	m2	**8.60**
girth exceeding 300mm; panes, area 0.10 - 0.50m2 .	0.12	10.00	**0.13**	0.38	-	**5.40**	0.10	**5.63**	m2	**6.05**
girth exceeding 300mm; panes, area 0.50 - 1.00m2 .	0.12	10.00	**0.13**	0.32	-	**4.55**	0.09	**4.76**	m2	**5.12**
girth exceeding 300mm; panes, area exceeding 1.00m2	0.12	10.00	**0.13**	0.29	-	**4.12**	0.08	**4.33**	m2	**4.65**
Wood partially glazed doors										
girth exceeding 300mm; panes, area not exceeding 0.10m2	0.23	10.00	**0.26**	0.43	-	**6.11**	0.12	**6.48**	m2	**6.97**
girth exceeding 300mm; panes, area 0.10 - 0.50m2 .	0.19	10.00	**0.21**	0.35	-	**4.97**	0.09	**5.27**	m2	**5.67**
girth exceeding 300mm; panes, area 0.50 - 1.00m2 .	0.16	10.00	**0.18**	0.32	-	**4.55**	0.09	**4.81**	m2	**5.17**
girth exceeding 300mm; panes, area exceeding 1.00m2	0.16	10.00	**0.18**	0.31	-	**4.41**	0.08	**4.67**	m2	**5.02**
Wood windows and screens										
girth exceeding 300mm; panes, area not exceeding 0.10m2	0.23	10.00	**0.26**	0.59	-	**8.38**	0.16	**8.80**	m2	**9.46**
girth exceeding 300mm; panes, area 0.10 - 0.50m2 .	0.19	10.00	**0.21**	0.42	-	**5.97**	0.11	**6.29**	m2	**6.76**
girth exceeding 300mm; panes, area 0.50 - 1.00m2 .	0.16	10.00	**0.18**	0.35	-	**4.97**	0.09	**5.25**	m2	**5.64**

Labour hourly rates: (except Specialists) Craft Operatives £14.21 Labourer £10.71 Rates are national average prices. Refer to REGIONAL VARIATIONS for indicative levels of overall pricing in regions	MATERIALS			LABOUR				RATES		
	Del to Site £	Waste %	Material Cost £	Craft Optve Hrs	Lab Hrs	Labour Cost £	Sunds £	Nett Rate £	Unit	Gross Rate (+7.5%) £
M60: PAINTING/CLEAR FINISHING - OIL PAINTING WOODWORK Cont.										
One undercoat, one coat full gloss finish on ready primed wood surfaces Cont.										
Wood windows and screens Cont.										
girth exceeding 300mm; panes, area exceeding 1.00m2	0.16	10.00	0.18	0.32	-	4.55	0.09	4.81	m2	5.17
One undercoat, one coat full gloss finish on ready primed wood surfaces; external work										
Wood general surfaces										
girth exceeding 300mm	0.30	10.00	0.33	0.34	-	4.83	0.09	5.26	m2	5.65
isolated surfaces, girth not exceeding 300mm	0.09	10.00	0.10	0.12	-	1.71	0.03	1.84	m	1.98
Wood glazed doors										
girth exceeding 300mm; panes, area not exceeding 0.10m2	0.16	10.00	0.18	0.54	-	7.67	0.15	8.00	m2	8.60
girth exceeding 300mm; panes, area 0.10 - 0.50m2	0.12	10.00	0.13	0.38	-	5.40	0.10	5.63	m2	6.05
girth exceeding 300mm; panes, area 0.50 - 1.00m2	0.12	10.00	0.13	0.32	-	4.55	0.09	4.76	m2	5.12
girth exceeding 300mm; panes, area exceeding 1.00m2	0.12	10.00	0.13	0.29	-	4.12	0.08	4.33	m2	4.65
Wood partially glazed doors										
girth exceeding 300mm; panes, area not exceeding 0.10m2	0.23	10.00	0.26	0.43	-	6.11	0.12	6.48	m2	6.97
girth exceeding 300mm; panes, area 0.10 - 0.50m2	0.19	10.00	0.21	0.35	-	4.97	0.09	5.27	m2	5.67
girth exceeding 300mm; panes, area 0.50 - 1.00m2	0.16	10.00	0.18	0.32	-	4.55	0.09	4.81	m2	5.17
girth exceeding 300mm; panes, area exceeding 1.00m2	0.16	10.00	0.18	0.31	-	4.41	0.08	4.67	m2	5.02
Wood windows and screens										
girth exceeding 300mm; panes, area not exceeding 0.10m2	0.23	10.00	0.26	0.61	-	8.67	0.16	9.09	m2	9.77
girth exceeding 300mm; panes, area 0.10 - 0.50m2	0.19	10.00	0.21	0.44	-	6.25	0.12	6.58	m2	7.07
girth exceeding 300mm; panes, area 0.50 - 1.00m2	0.16	10.00	0.18	0.37	-	5.26	0.10	5.54	m2	5.95
girth exceeding 300mm; panes, area exceeding 1.00m2	0.16	10.00	0.18	0.34	-	4.83	0.09	5.10	m2	5.49
Wood railings fences and gates; open type										
girth exceeding 300mm	0.23	10.00	0.26	0.25	-	3.55	0.07	3.88	m2	4.17
isolated surfaces, girth not exceeding 300mm	0.09	10.00	0.10	0.08	-	1.14	0.02	1.26	m	1.36
Wood railings fences and gates; close type										
girth exceeding 300mm	0.30	10.00	0.33	0.22	-	3.13	0.06	3.52	m2	3.78
Two undercoats, one coat full gloss finish on ready primed wood surfaces										
Wood general surfaces										
girth exceeding 300mm	0.46	10.00	0.50	0.46	-	6.54	0.12	7.16	m2	7.70
isolated surfaces, girth not exceeding 300mm	0.14	10.00	0.15	0.15	-	2.13	0.04	2.33	m	2.50
Wood glazed doors										
girth exceeding 300mm; panes, area not exceeding 0.10m2	0.25	10.00	0.27	0.76	-	10.80	0.21	11.28	m2	12.12
girth exceeding 300mm; panes, area 0.10 - 0.50m2	0.18	10.00	0.19	0.55	-	7.82	0.15	8.16	m2	8.77
girth exceeding 300mm; panes, area 0.50 - 1.00m2	0.18	10.00	0.19	0.46	-	6.54	0.12	6.85	m2	7.37
girth exceeding 300mm; panes, area exceeding 1.00m2	0.18	10.00	0.19	0.41	-	5.83	0.11	6.13	m2	6.59
Wood partially glazed doors										
girth exceeding 300mm; panes, area not exceeding 0.10m2	0.35	10.00	0.39	0.61	-	8.67	0.16	9.22	m2	9.91
girth exceeding 300mm; panes, area 0.10 - 0.50m2	0.28	10.00	0.31	0.51	-	7.25	0.14	7.69	m2	8.27
girth exceeding 300mm; panes, area 0.50 - 1.00m2	0.25	10.00	0.27	0.46	-	6.54	0.12	6.93	m2	7.45
girth exceeding 300mm; panes, area exceeding 1.00m2	0.25	10.00	0.27	0.44	-	6.25	0.12	6.64	m2	7.14
Wood windows and screens										
girth exceeding 300mm; panes, area not exceeding 0.10m2	0.35	10.00	0.39	0.84	-	11.94	0.23	12.55	m2	13.49
girth exceeding 300mm; panes, area 0.10 - 0.50m2	0.28	10.00	0.31	0.61	-	8.67	0.16	9.14	m2	9.83
girth exceeding 300mm; panes, area 0.50 - 1.00m2	0.25	10.00	0.27	0.51	-	7.25	0.14	7.66	m2	8.23
girth exceeding 300mm; panes, area exceeding 1.00m2	0.25	10.00	0.27	0.45	-	6.39	0.12	6.79	m2	7.30
Two undercoats, one coat full gloss finish on ready primed wood surfaces; external work										
Wood general surfaces										
girth exceeding 300mm	0.46	10.00	0.50	0.49	-	6.96	0.13	7.60	m2	8.17
isolated surfaces, girth not exceeding 300mm	0.14	10.00	0.15	0.17	-	2.42	0.05	2.62	m	2.81
Wood glazed doors										
girth exceeding 300mm; panes, area not exceeding 0.10m2	0.25	10.00	0.27	0.76	-	10.80	0.21	11.28	m2	12.12
girth exceeding 300mm; panes, area 0.10 - 0.50m2	0.18	10.00	0.19	0.55	-	7.82	0.15	8.16	m2	8.77
girth exceeding 300mm; panes, area 0.50 - 1.00m2	0.18	10.00	0.19	0.46	-	6.54	0.12	6.85	m2	7.37
girth exceeding 300mm; panes, area exceeding 1.00m2	0.18	10.00	0.19	0.41	-	5.83	0.11	6.13	m2	6.59

Labour hourly rates: (except Specialists) Craft Operatives £14.21 Labourer £10.71 Rates are national average prices. Refer to REGIONAL VARIATIONS for indicative levels of overall pricing in regions	MATERIALS			LABOUR				RATES		
	Del to Site £	Waste %	Material Cost £	Craft Optve Hrs	Lab Hrs	Labour Cost £	Sunds £	Nett Rate £	Unit	Gross Rate (+7.5%) £
M60: PAINTING/CLEAR FINISHING - OIL PAINTING WOODWORK Cont.										
Two undercoats, one coat full gloss finish on ready primed wood surfaces; external work Cont.										
Wood partially glazed doors										
girth exceeding 300mm; panes, area not exceeding 0.10m2	0.35	10.00	**0.39**	0.61	-	**8.67**	0.16	**9.22**	m2	**9.91**
girth exceeding 300mm; panes, area 0.10 - 0.50m2	0.28	10.00	**0.31**	0.51	-	**7.25**	0.14	**7.69**	m2	**8.27**
girth exceeding 300mm; panes, area 0.50 - 1.00m2	0.25	10.00	**0.27**	0.46	-	**6.54**	0.12	**6.93**	m2	**7.45**
girth exceeding 300mm; panes, area exceeding 1.00m2	0.25	10.00	**0.27**	0.44	-	**6.25**	0.12	**6.64**	m2	**7.14**
Wood windows and screens										
girth exceeding 300mm; panes, area not exceeding 0.10m2	0.35	10.00	**0.39**	0.87	-	**12.36**	0.23	**12.98**	m2	**13.96**
girth exceeding 300mm; panes, area 0.10 - 0.50m2	0.28	10.00	**0.31**	0.64	-	**9.09**	0.17	**9.58**	m2	**10.29**
girth exceeding 300mm; panes, area 0.50 - 1.00m2	0.25	10.00	**0.27**	0.54	-	**7.67**	0.15	**8.09**	m2	**8.70**
girth exceeding 300mm; panes, area exceeding 1.00m2	0.25	10.00	**0.27**	0.48	-	**6.82**	0.13	**7.22**	m2	**7.76**
Wood railings fences and gates; open type										
girth exceeding 300mm	0.35	10.00	**0.39**	0.38	-	**5.40**	0.10	**5.89**	m2	**6.33**
isolated surfaces, girth not exceeding 300mm	0.14	10.00	**0.15**	0.13	-	**1.85**	0.04	**2.04**	m	**2.19**
Wood railings fences and gates; close type										
girth exceeding 300mm	0.46	10.00	**0.50**	0.33	-	**4.69**	0.09	**5.28**	m2	**5.68**
One coat primer, one undercoat, one coat full gloss finish on wood surfaces										
Wood general surfaces										
girth exceeding 300mm	0.54	10.00	**0.59**	0.46	-	**6.54**	0.12	**7.25**	m2	**7.79**
isolated surfaces, girth not exceeding 300mm	0.16	10.00	**0.18**	0.15	-	**2.13**	0.04	**2.35**	m	**2.52**
Wood glazed doors										
girth exceeding 300mm; panes, area not exceeding 0.10m2	0.30	10.00	**0.33**	0.76	-	**10.80**	0.21	**11.33**	m2	**12.18**
girth exceeding 300mm; panes, area 0.10 - 0.50m2	0.22	10.00	**0.24**	0.55	-	**7.82**	0.15	**8.20**	m2	**8.82**
girth exceeding 300mm; panes, area 0.50 - 1.00m2	0.20	10.00	**0.22**	0.46	-	**6.54**	0.12	**6.88**	m2	**7.40**
girth exceeding 300mm; panes, area exceeding 1.00m2	0.20	10.00	**0.22**	0.41	-	**5.83**	0.11	**6.16**	m2	**6.62**
Wood partially glazed doors										
girth exceeding 300mm; panes, area not exceeding 0.10m2	0.42	10.00	**0.46**	0.61	-	**8.67**	0.16	**9.29**	m2	**9.99**
girth exceeding 300mm; panes, area 0.10 - 0.50m2	0.32	10.00	**0.35**	0.51	-	**7.25**	0.14	**7.74**	m2	**8.32**
girth exceeding 300mm; panes, area 0.50 - 1.00m2	0.30	10.00	**0.33**	0.46	-	**6.54**	0.12	**6.99**	m2	**7.51**
girth exceeding 300mm; panes, area exceeding 1.00m2	0.30	10.00	**0.33**	0.44	-	**6.25**	0.12	**6.70**	m2	**7.20**
Wood windows and screens										
girth exceeding 300mm; panes, area not exceeding 0.10m2	0.42	10.00	**0.46**	0.84	-	**11.94**	0.23	**12.62**	m2	**13.57**
girth exceeding 300mm; panes, area 0.10 - 0.50m2	0.32	10.00	**0.35**	0.61	-	**8.67**	0.16	**9.18**	m2	**9.87**
girth exceeding 300mm; panes, area 0.50 - 1.00m2	0.30	10.00	**0.33**	0.51	-	**7.25**	0.14	**7.71**	m2	**8.29**
girth exceeding 300mm; panes, area exceeding 1.00m2	0.30	10.00	**0.33**	0.45	-	**6.39**	0.12	**6.84**	m2	**7.35**
One coat primer, one undercoat, one coat full gloss finish on wood surfaces; external work										
Wood general surfaces										
girth exceeding 300mm	0.54	10.00	**0.59**	0.49	-	**6.96**	0.13	**7.68**	m2	**8.26**
isolated surfaces, girth not exceeding 300mm	0.16	10.00	**0.18**	0.17	-	**2.42**	0.05	**2.64**	m	**2.84**
Wood glazed doors										
girth exceeding 300mm; panes, area not exceeding 0.10m2	0.30	10.00	**0.33**	0.76	-	**10.80**	0.21	**11.33**	m2	**12.18**
girth exceeding 300mm; panes, area 0.10 - 0.50m2	0.22	10.00	**0.24**	0.55	-	**7.82**	0.15	**8.20**	m2	**8.82**
girth exceeding 300mm; panes, area 0.50 - 1.00m2	0.20	10.00	**0.22**	0.46	-	**6.54**	0.02	**6.78**	m2	**7.29**
girth exceeding 300mm; panes, area exceeding 1.00m2	0.20	10.00	**0.22**	0.41	-	**5.83**	0.11	**6.16**	m2	**6.62**
Wood partially glazed doors										
girth exceeding 300mm; panes, area not exceeding 0.10m2	0.42	10.00	**0.46**	0.61	-	**8.67**	0.16	**9.29**	m2	**9.99**
girth exceeding 300mm; panes, area 0.10 - 0.50m2	0.32	10.00	**0.35**	0.51	-	**7.25**	0.14	**7.74**	m2	**8.32**
girth exceeding 300mm; panes, area 0.50 - 1.00m2	0.30	10.00	**0.33**	0.46	-	**6.54**	0.12	**6.99**	m2	**7.51**
girth exceeding 300mm; panes, area exceeding 1.00m2	0.30	10.00	**0.33**	0.44	-	**6.25**	0.12	**6.70**	m2	**7.20**
Wood windows and screens										
girth exceeding 300mm; panes, area not exceeding 0.10m2	0.42	10.00	**0.46**	0.87	-	**12.36**	0.23	**13.06**	m2	**14.03**
girth exceeding 300mm; panes, area 0.10 - 0.50m2	0.32	10.00	**0.35**	0.64	-	**9.09**	0.17	**9.62**	m2	**10.34**
girth exceeding 300mm; panes, area 0.50 - 1.00m2	0.30	10.00	**0.33**	0.54	-	**7.67**	0.15	**8.15**	m2	**8.76**
girth exceeding 300mm; panes, area exceeding 1.00m2	0.30	10.00	**0.33**	0.48	-	**6.82**	0.13	**7.28**	m2	**7.82**
Wood railings fences and gates; open type										
girth exceeding 300mm	0.42	10.00	**0.46**	0.38	-	**5.40**	0.10	**5.96**	m2	**6.41**
isolated surfaces, girth not exceeding 300mm	0.16	10.00	**0.18**	0.13	-	**1.85**	0.04	**2.06**	m	**2.21**

Labour hourly rates: (except Specialists) Craft Operatives £14.21 Labourer £10.71 Rates are national average prices. Refer to REGIONAL VARIATIONS for indicative levels of overall pricing in regions	MATERIALS			LABOUR				RATES		
	Del to Site £	Waste %	Material Cost £	Craft Optve Hrs	Lab Hrs	Labour Cost £	Sunds £	Nett Rate £	Unit	Gross Rate (+7.5%) £
M60: PAINTING/CLEAR FINISHING - OIL PAINTING WOODWORK Cont.										
One coat primer, one undercoat, one coat full gloss finish on wood surfaces; external work Cont.										
Wood railings fences and gates; close type										
girth exceeding 300mm	0.54	10.00	0.59	0.33	-	4.69	0.09	5.37	m2	5.77
One coat primer, two undercoats, one coat full gloss finish on wood surfaces										
Wood general surfaces										
girth exceeding 300mm	0.69	10.00	0.76	0.60	-	8.53	0.16	9.44	m2	10.15
isolated surfaces, girth not exceeding 300mm.....	0.21	10.00	0.23	0.20	-	2.84	0.05	3.12	m	3.36
Wood glazed doors										
girth exceeding 300mm; panes, area not exceeding 0.10m2	0.38	10.00	0.42	1.00	-	14.21	0.27	14.90	m2	16.01
girth exceeding 300mm; panes, area 0.10 - 0.50m2.	0.27	10.00	0.30	0.72	-	10.23	0.19	10.73	m2	11.53
girth exceeding 300mm; panes, area 0.50 - 1.00m2.	0.26	10.00	0.28	0.60	-	8.53	0.16	8.97	m2	9.64
girth exceeding 300mm; panes, area exceeding 1.00m2	0.26	10.00	0.28	0.54	-	7.67	0.15	8.10	m2	8.71
Wood partially glazed doors										
girth exceeding 300mm; panes, area not exceeding 0.10m2	0.53	10.00	0.59	0.80	-	11.37	0.22	12.17	m2	13.08
girth exceeding 300mm; panes, area 0.10 - 0.50m2.	0.41	10.00	0.45	0.66	-	9.38	0.18	10.01	m2	10.76
girth exceeding 300mm; panes, area 0.50 - 1.00m2.	0.38	10.00	0.42	0.60	-	8.53	0.16	9.10	m2	9.79
girth exceeding 300mm; panes, area exceeding 1.00m2	0.38	10.00	0.42	0.57	-	8.10	0.15	8.67	m2	9.32
Wood windows and screens										
girth exceeding 300mm; panes, area not exceeding 0.10m2	0.53	10.00	0.59	1.10	-	15.63	0.30	16.51	m2	17.75
girth exceeding 300mm; panes, area 0.10 - 0.50m2.	0.41	10.00	0.45	0.79	-	11.23	0.21	11.89	m2	12.79
girth exceeding 300mm; panes, area 0.50 - 1.00m2.	0.38	10.00	0.42	0.66	-	9.38	0.18	9.97	m2	10.72
girth exceeding 300mm; panes, area exceeding 1.00m2	0.38	10.00	0.42	0.59	-	8.38	0.16	8.96	m2	9.63
One coat primer, two undercoats, one coat full gloss finish on wood surfaces; external work										
Wood general surfaces										
girth exceeding 300mm	0.69	10.00	0.76	0.64	-	9.09	0.17	10.02	m2	10.78
isolated surfaces, girth not exceeding 300mm.....	0.21	10.00	0.23	0.22	-	3.13	0.06	3.41	m2	3.67
Wood glazed doors										
girth exceeding 300mm; panes, area not exceeding 0.10m2	0.38	10.00	0.42	1.00	-	14.21	0.27	14.90	m2	16.01
girth exceeding 300mm; panes, area 0.10 - 0.50m2.	0.27	10.00	0.30	0.72	-	10.23	0.19	10.73	m2	11.53
girth exceeding 300mm; panes, area 0.50 - 1.00m2.	0.26	10.00	0.28	0.60	-	8.53	0.16	8.97	m2	9.64
girth exceeding 300mm; panes, area exceeding 1.00m2	0.26	10.00	0.28	0.54	-	7.67	0.15	8.10	m2	8.71
Wood partially glazed doors										
girth exceeding 300mm; panes, area not exceeding 0.10m2	0.53	10.00	0.59	0.80	-	11.37	0.22	12.17	m2	13.08
girth exceeding 300mm; panes, area 0.10 - 0.50m2.	0.41	10.00	0.45	0.66	-	9.38	0.18	10.01	m2	10.76
girth exceeding 300mm; panes, area 0.50 - 1.00m2.	0.38	10.00	0.42	0.60	-	8.53	0.16	9.10	m2	9.79
girth exceeding 300mm; panes, area exceeding 1.00m2	0.38	10.00	0.42	0.57	-	8.10	0.15	8.67	m2	9.32
Wood windows and screens										
girth exceeding 300mm; panes, area not exceeding 0.10m2	0.53	10.00	0.59	1.14	-	16.20	0.31	17.09	m2	18.38
girth exceeding 300mm; panes, area 0.10 - 0.50m2.	0.41	10.00	0.45	0.83	-	11.79	0.22	12.47	m2	13.41
girth exceeding 300mm; panes, area 0.50 - 1.00m2.	0.38	10.00	0.42	0.70	-	9.95	0.19	10.55	m2	11.34
girth exceeding 300mm; panes, area exceeding 1.00m2	0.38	10.00	0.42	0.63	-	8.95	0.17	9.54	m2	10.25
Wood railings fences and gates; open type										
girth exceeding 300mm	0.53	10.00	0.59	0.50	-	7.11	0.14	7.83	m2	8.41
isolated surfaces, girth not exceeding 300mm.....	0.21	10.00	0.23	0.17	-	2.42	0.05	2.69	m	2.89
Wood railings fences and gates; close type										
girth exceeding 300mm	0.69	10.00	0.76	0.44	-	6.25	0.12	7.13	m2	7.66
One coat primer, one undercoat, two coats full gloss finish on wood surfaces										
Wood general surfaces										
girth exceeding 300mm	0.69	10.00	0.76	0.60	-	8.53	0.16	9.44	m2	10.15
isolated surfaces, girth not exceeding 300mm.....	0.21	10.00	0.23	0.20	-	2.84	0.05	3.12	m	3.36
Wood glazed doors										
girth exceeding 300mm; panes, area not exceeding 0.10m2	0.38	10.00	0.42	1.00	-	14.21	0.27	14.90	m2	16.01
girth exceeding 300mm; panes, area 0.10 - 0.50m2.	0.27	10.00	0.30	0.72	-	10.23	0.19	10.73	m2	11.53
girth exceeding 300mm; panes, area 0.50 - 1.00m2.	0.26	10.00	0.28	0.60	-	8.53	0.16	8.97	m2	9.64
girth exceeding 300mm; panes, area exceeding 1.00m2	0.26	10.00	0.28	0.54	-	7.67	0.15	8.10	m2	8.71
Wood partially glazed doors										
girth exceeding 300mm; panes, area not exceeding 0.10m2	0.53	10.00	0.59	0.80	-	11.37	0.22	12.17	m2	13.08

Labour hourly rates: (except Specialists) Craft Operatives £14.21 Labourer £10.71 Rates are national average prices. Refer to REGIONAL VARIATIONS for indicative levels of overall pricing in regions	MATERIALS			LABOUR				RATES		
	Del to Site £	Waste %	Material Cost £	Craft Optve Hrs	Lab Hrs	Labour Cost £	Sunds £	Nett Rate £	Unit	Gross Rate (+7.5%) £
M60: PAINTING/CLEAR FINISHING - OIL PAINTING WOODWORK Cont.										
One coat primer, one undercoat, two coats full gloss finish on wood surfaces Cont.										
Wood partially glazed doors Cont.										
girth exceeding 300mm; panes, area 0.10 - 0.50m2	0.41	10.00	**0.45**	0.66	-	**9.38**	0.18	**10.01**	m2	**10.76**
girth exceeding 300mm; panes, area 0.50 - 1.00m2	0.38	10.00	**0.42**	0.60	-	**8.53**	0.16	**9.10**	m2	**9.79**
girth exceeding 300mm; panes, area exceeding 1.00m2	0.38	10.00	**0.42**	0.57	-	**8.10**	0.15	**8.67**	m2	**9.32**
Wood windows and screens										
girth exceeding 300mm; panes, area not exceeding 0.10m2	0.53	10.00	**0.59**	1.10	-	**15.63**	0.30	**16.51**	m2	**17.75**
girth exceeding 300mm; panes, area 0.10 - 0.50m2	0.41	10.00	**0.45**	0.79	-	**11.23**	0.21	**11.89**	m2	**12.79**
girth exceeding 300mm; panes, area 0.50 - 1.00m2	0.38	10.00	**0.42**	0.66	-	**9.38**	0.18	**9.97**	m2	**10.72**
girth exceeding 300mm; panes, area exceeding 1.00m2	0.38	10.00	**0.42**	0.59	-	**8.38**	0.16	**8.96**	m2	**9.63**
One coat primer, one undercoat, two coats full gloss finish wood surfaces; external work										
Wood general surfaces										
girth exceeding 300mm	0.69	10.00	**0.76**	0.64	-	**9.09**	0.17	**10.02**	m2	**10.78**
isolated surfaces, girth not exceeding 300mm	0.21	10.00	**0.23**	0.22	-	**3.13**	0.06	**3.41**	m	**3.67**
Wood glazed doors										
girth exceeding 300mm; panes, area not exceeding 0.10m2	0.38	10.00	**0.42**	1.00	-	**14.21**	0.27	**14.90**	m2	**16.01**
girth exceeding 300mm; panes, area 0.10 - 0.50m2	0.27	10.00	**0.30**	0.72	-	**10.23**	0.19	**10.73**	m2	**11.53**
girth exceeding 300mm; panes, area 0.50 - 1.00m2	0.26	10.00	**0.28**	0.60	-	**8.53**	0.16	**8.97**	m2	**9.64**
girth exceeding 300mm; panes, area exceeding 1.00m2	0.26	10.00	**0.28**	0.54	-	**7.67**	0.15	**8.10**	m2	**8.71**
Wood partially glazed doors										
girth exceeding 300mm; panes, area not exceeding 0.10m2	0.53	10.00	**0.59**	0.80	-	**11.37**	0.22	**12.17**	m2	**13.08**
girth exceeding 300mm; panes, area 0.10 - 0.50m2	0.41	10.00	**0.45**	0.66	-	**9.38**	0.18	**10.01**	m2	**10.76**
girth exceeding 300mm; panes, area 0.50 - 1.00m2	0.38	10.00	**0.42**	0.60	-	**8.53**	0.16	**9.10**	m2	**9.79**
girth exceeding 300mm; panes, area exceeding 1.00m2	0.38	10.00	**0.42**	0.57	-	**8.10**	0.15	**8.67**	m2	**9.32**
Wood windows and screens										
girth exceeding 300mm; panes, area not exceeding 0.10m2	0.53	10.00	**0.59**	1.14	-	**16.20**	0.31	**17.09**	m2	**18.38**
girth exceeding 300mm; panes, area 0.10 - 0.50m2	0.41	10.00	**0.45**	0.83	-	**11.79**	0.22	**12.47**	m2	**13.41**
girth exceeding 300mm; panes, area 0.50 - 1.00m2	0.38	10.00	**0.42**	0.70	-	**9.95**	0.19	**10.55**	m2	**11.34**
girth exceeding 300mm; panes, area exceeding 1.00m2	0.38	10.00	**0.42**	0.63	-	**8.95**	0.17	**9.54**	m2	**10.25**
Wood railings fences and gates; open type										
girth exceeding 300mm	0.53	10.00	**0.59**	0.50	-	**7.11**	0.14	**7.83**	m2	**8.41**
isolated surfaces, girth not exceeding 300mm	0.21	10.00	**0.23**	0.17	-	**2.42**	0.05	**2.69**	m	**2.89**
Wood railings fences and gates; close type										
girth exceeding 300mm	0.69	10.00	**0.76**	0.44	-	**6.25**	0.12	**7.13**	m2	**7.66**
M60: PAINTING/CLEAR FINISHING - POLYURETHANE LACQUER										
Two coats polyurethane lacquer										
Wood general surfaces										
girth exceeding 300mm	0.77	10.00	**0.85**	0.40	-	**5.68**	0.11	**6.64**	m2	**7.14**
isolated surfaces, girth not exceeding 300mm	0.23	10.00	**0.25**	0.13	-	**1.85**	0.04	**2.14**	m	**2.30**
Wood glazed doors										
girth exceeding 300mm; panes, area not exceeding 0.10m2	0.44	10.00	**0.48**	0.67	-	**9.52**	0.18	**10.18**	m2	**10.94**
girth exceeding 300mm; panes, area 0.10 - 0.50m2	0.31	10.00	**0.34**	0.48	-	**6.82**	0.13	**7.29**	m2	**7.84**
girth exceeding 300mm; panes, area 0.50 - 1.00m2	0.28	10.00	**0.31**	0.40	-	**5.68**	0.11	**6.10**	m2	**6.56**
girth exceeding 300mm; panes, area exceeding 1.00m2	0.28	10.00	**0.31**	0.36	-	**5.12**	0.10	**5.52**	m2	**5.94**
Wood partially glazed doors										
girth exceeding 300mm; panes, area not exceeding 0.10m2	0.62	10.00	**0.68**	0.54	-	**7.67**	0.15	**8.50**	m2	**9.13**
girth exceeding 300mm; panes, area 0.10 - 0.50m2	0.46	10.00	**0.51**	0.44	-	**6.25**	0.12	**6.88**	m2	**7.40**
girth exceeding 300mm; panes, area 0.50 - 1.00m2	0.44	10.00	**0.48**	0.40	-	**5.68**	0.11	**6.27**	m2	**6.74**
girth exceeding 300mm; panes, area exceeding 1.00m2	0.44	10.00	**0.48**	0.38	-	**5.40**	0.10	**5.98**	m2	**6.43**
Wood windows and screens										
girth exceeding 300mm; panes, area not exceeding 0.10m2	0.62	10.00	**0.68**	0.74	-	**10.52**	0.20	**11.39**	m2	**12.25**
girth exceeding 300mm; panes, area 0.10 - 0.50m2	0.46	10.00	**0.51**	0.53	-	**7.53**	0.14	**8.18**	m2	**8.80**
girth exceeding 300mm; panes, area 0.50 - 1.00m2	0.44	10.00	**0.48**	0.44	-	**6.25**	0.12	**6.85**	m2	**7.36**
girth exceeding 300mm; panes, area exceeding 1.00m2	0.44	10.00	**0.48**	0.40	-	**5.68**	0.11	**6.27**	m2	**6.74**
Two coats polyurethane lacquer; external work										
Wood general surfaces										
girth exceeding 300mm	0.77	10.00	**0.85**	0.42	-	**5.97**	0.11	**6.93**	m2	**7.45**
isolated surfaces, girth not exceeding 300mm	0.23	10.00	**0.25**	0.14	-	**1.99**	0.04	**2.28**	m	**2.45**

Labour hourly rates: (except Specialists) Craft Operatives £14.21 Labourer £10.71 Rates are national average prices. Refer to REGIONAL VARIATIONS for indicative levels of overall pricing in regions	MATERIALS			LABOUR				RATES		
	Del to Site £	Waste %	Material Cost £	Craft Optve Hrs	Lab Hrs	Labour Cost £	Sunds £	Nett Rate £	Unit	Gross Rate (+7.5%) £
M60: PAINTING/CLEAR FINISHING - POLYURETHANE LACQUER Cont.										
Two coats polyurethane lacquer; external work Cont.										
Wood glazed doors										
girth exceeding 300mm; panes, area not exceeding 0.10m2	0.44	10.00	**0.48**	0.67	-	**9.52**	0.18	**10.18**	m2	**10.94**
girth exceeding 300mm; panes, area 0.10 - 0.50m2	0.31	10.00	**0.34**	0.48	-	**6.82**	0.13	**7.29**	m2	**7.84**
girth exceeding 300mm; panes, area 0.50 - 1.00m2	0.28	10.00	**0.31**	0.40	-	**5.68**	0.11	**6.10**	m2	**6.56**
girth exceeding 300mm; panes, area exceeding 1.00m2	0.28	10.00	**0.31**	0.36	-	**5.12**	0.10	**5.52**	m2	**5.94**
Wood partially glazed doors										
girth exceeding 300mm; panes, area not exceeding 0.10m2	0.62	10.00	**0.68**	0.54	-	**7.67**	0.15	**8.50**	m2	**9.13**
girth exceeding 300mm; panes, area 0.10 - 0.50m2	0.46	10.00	**0.51**	0.44	-	**6.25**	0.12	**6.88**	m2	**7.40**
girth exceeding 300mm; panes, area 0.50 - 1.00m2	0.44	10.00	**0.48**	0.40	-	**5.68**	0.11	**6.27**	m2	**6.74**
girth exceeding 300mm; panes, area exceeding 1.00m2	0.44	10.00	**0.48**	0.38	-	**5.40**	0.10	**5.98**	m2	**6.43**
Wood windows and screens										
girth exceeding 300mm; panes, area not exceeding 0.10m2	0.62	10.00	**0.68**	0.76	-	**10.80**	0.21	**11.68**	m2	**12.56**
girth exceeding 300mm; panes, area 0.10 - 0.50m2	0.46	10.00	**0.51**	0.55	-	**7.82**	0.15	**8.47**	m2	**9.11**
girth exceeding 300mm; panes, area 0.50 - 1.00m2	0.44	10.00	**0.48**	0.46	-	**6.54**	0.12	**7.14**	m2	**7.68**
girth exceeding 300mm; panes, area exceeding 1.00m2	0.44	10.00	**0.48**	0.42	-	**5.97**	0.11	**6.56**	m2	**7.05**
Three coats polyurethane lacquer										
Wood general surfaces										
girth exceeding 300mm	1.13	10.00	**1.24**	0.47	-	**6.68**	0.13	**8.05**	m2	**8.65**
isolated surfaces, girth not exceeding 300mm	0.33	10.00	**0.37**	0.16	-	**2.27**	0.04	**2.68**	m	**2.88**
Wood glazed doors										
girth exceeding 300mm; panes, area not exceeding 0.10m2	0.64	10.00	**0.71**	0.78	-	**11.08**	0.21	**12.00**	m2	**12.90**
girth exceeding 300mm; panes, area 0.10 - 0.50m2	0.46	10.00	**0.51**	0.56	-	**7.96**	0.15	**8.62**	m2	**9.26**
girth exceeding 300mm; panes, area 0.50 - 1.00m2	0.41	10.00	**0.45**	0.47	-	**6.68**	0.13	**7.26**	m2	**7.80**
girth exceeding 300mm; panes, area exceeding 1.00m2	0.41	10.00	**0.45**	0.42	-	**5.97**	0.11	**6.53**	m2	**7.02**
Wood partially glazed doors										
girth exceeding 300mm; panes, area not exceeding 0.10m2	0.90	10.00	**0.99**	0.63	-	**8.95**	0.17	**10.11**	m2	**10.87**
girth exceeding 300mm; panes, area 0.10 - 0.50m2	0.67	10.00	**0.73**	0.57	-	**8.10**	0.15	**8.99**	m2	**9.66**
girth exceeding 300mm; panes, area 0.50 - 1.00m2	0.62	10.00	**0.68**	0.47	-	**6.68**	0.13	**7.48**	m2	**8.04**
girth exceeding 300mm; panes, area exceeding 1.00m2	0.62	10.00	**0.68**	0.45	-	**6.39**	0.12	**7.19**	m2	**7.73**
Wood windows and screens										
girth exceeding 300mm; panes, area not exceeding 0.10m2	0.64	10.00	**0.71**	0.86	-	**12.22**	0.23	**13.16**	m2	**14.15**
girth exceeding 300mm; panes, area 0.10 - 0.50m2	0.46	10.00	**0.51**	0.62	-	**8.81**	0.17	**9.49**	m2	**10.20**
girth exceeding 300mm; panes, area 0.50 - 1.00m2	0.41	10.00	**0.45**	0.52	-	**7.39**	0.14	**7.98**	m2	**8.58**
girth exceeding 300mm; panes, area exceeding 1.00m2	0.41	10.00	**0.45**	0.46	-	**6.54**	0.12	**7.11**	m2	**7.65**
Three coats polyurethane lacquer; external work										
Wood general surfaces										
girth exceeding 300mm	1.13	10.00	**1.24**	0.50	-	**7.11**	0.14	**8.48**	m2	**9.12**
isolated surfaces, girth not exceeding 300mm	0.33	10.00	**0.37**	0.17	-	**2.42**	0.05	**2.83**	m	**3.04**
Wood glazed doors										
girth exceeding 300mm; panes, area not exceeding 0.10m2	0.64	10.00	**0.71**	0.78	-	**11.08**	0.21	**12.00**	m2	**12.90**
girth exceeding 300mm; panes, area 0.10 - 0.50m2	0.46	10.00	**0.51**	0.56	-	**7.96**	0.15	**8.62**	m2	**9.26**
girth exceeding 300mm; panes, area 0.50 - 1.00m2	0.41	10.00	**0.45**	0.47	-	**6.68**	0.13	**7.26**	m2	**7.80**
girth exceeding 300mm; panes, area exceeding 1.00m2	0.41	10.00	**0.45**	0.42	-	**5.97**	0.11	**6.53**	m2	**7.02**
Wood partially glazed doors										
girth exceeding 300mm; panes, area not exceeding 0.10m2	0.90	10.00	**0.99**	0.63	-	**8.95**	0.17	**10.11**	m2	**10.87**
girth exceeding 300mm; panes, area 0.10 - 0.50m2	0.67	10.00	**0.73**	0.57	-	**8.10**	0.15	**8.99**	m2	**9.66**
girth exceeding 300mm; panes, area 0.50 - 1.00m2	0.62	10.00	**0.68**	0.47	-	**6.68**	0.13	**7.48**	m2	**8.04**
girth exceeding 300mm; panes, area exceeding 1.00m2	0.62	10.00	**0.68**	0.45	-	**6.39**	0.12	**7.19**	m2	**7.73**
Wood windows and screens										
girth exceeding 300mm; panes, area not exceeding 0.10m2	0.64	10.00	**0.71**	0.89	-	**12.65**	0.24	**13.59**	m2	**14.61**
girth exceeding 300mm; panes, area 0.10 - 0.50m2	0.46	10.00	**0.51**	0.65	-	**9.24**	0.18	**9.92**	m2	**10.66**
girth exceeding 300mm; panes, area 0.50 - 1.00m2	0.41	10.00	**0.45**	0.55	-	**7.82**	0.15	**8.42**	m2	**9.05**
girth exceeding 300mm; panes, area exceeding 1.00m2	0.41	10.00	**0.45**	0.49	-	**6.96**	0.13	**7.55**	m2	**8.11**
M60: PAINTING/CLEAR FINISHING - FIRE RETARDANT PAINTS AND VARNISHES										
Two coats fire retardant paint										
Wood general surfaces										
girth exceeding 300mm	3.04	10.00	**3.34**	0.40	-	**5.68**	0.11	**9.13**	m2	**9.82**

Labour hourly rates: (except Specialists) Craft Operatives £14.21 Labourer £10.71 Rates are national average prices. Refer to REGIONAL VARIATIONS for indicative levels of overall pricing in regions	MATERIALS			LABOUR				RATES		
	Del to Site £	Waste %	Material Cost £	Craft Optve Hrs	Lab Hrs	Labour Cost £	Sunds £	Nett Rate £	Unit	Gross Rate (+7.5%) £
M60: PAINTING/CLEAR FINISHING - FIRE RETARDANT PAINTS AND VARNISHES Cont.										
Two coats fire retardant paint Cont.										
Wood general surfaces Cont.										
isolated surfaces, girth not exceeding 300mm.....	0.92	10.00	**1.01**	0.13	-	**1.85**	0.04	**2.89**	m	**3.11**
Two coats fire retardant varnish, one overcoat varnish										
Wood general surfaces										
girth exceeding 300mm	3.23	10.00	**3.55**	0.54	-	**7.67**	0.15	**11.37**	m2	**12.22**
isolated surfaces, girth not exceeding 300mm.....	0.99	10.00	**1.09**	0.18	-	**2.56**	0.05	**3.70**	m	**3.98**
M60: PAINTING/CLEAR FINISHING - OILING HARDWOOD										
Two coats raw linseed oil										
Wood general surfaces										
girth exceeding 300mm	0.86	10.00	**0.94**	0.44	-	**6.25**	0.12	**7.31**	m2	**7.86**
isolated surfaces, girth not exceeding 300mm.....	0.26	10.00	**0.28**	0.15	-	**2.13**	0.04	**2.45**	m	**2.64**
M60: PAINTING/CLEAR FINISHING - FRENCH AND WAX POLISHING										
Stain; two coats white polish										
Wood general surfaces										
girth exceeding 300mm	2.06	10.00	**2.26**	0.51	-	**7.25**	0.14	**9.65**	m2	**10.37**
isolated surfaces, girth not exceeding 300mm.....	0.65	10.00	**0.71**	0.17	-	**2.42**	0.05	**3.17**	m	**3.41**
Open grain French polish										
Wood general surfaces										
girth exceeding 300mm	2.12	10.00	**2.33**	2.40	-	**34.10**	0.65	**37.08**	m2	**39.86**
isolated surfaces, girth not exceeding 300mm.....	0.67	10.00	**0.73**	0.80	-	**11.37**	0.22	**12.32**	m	**13.24**
Stain; body in; fully French polish										
Wood general surfaces										
girth exceeding 300mm	3.52	10.00	**3.87**	3.60	-	**51.16**	0.97	**56.00**	m2	**60.20**
isolated surfaces, girth not exceeding 300mm.....	1.11	10.00	**1.22**	1.20	-	**17.05**	0.32	**18.60**	m	**19.99**
M60: PAINTING/CLEAR FINISHING - ROAD MARKINGS										
One coat solvent based road marking paint; external work										
Concrete general surfaces										
isolated surfaces, 50mm wide....................	1.29	5.00	**1.35**	0.10	-	**1.42**	0.03	**2.80**	m	**3.01**
isolated surfaces, 100mm wide...................	2.58	5.00	**2.70**	0.12	-	**1.71**	0.03	**4.44**	m	**4.77**
Two coats solvent based road marking paint; external work										
Concrete general surfaces										
isolated surfaces, 50mm wide....................	2.58	5.00	**2.70**	0.15	-	**2.13**	0.04	**4.88**	m	**5.24**
isolated surfaces, 100mm wide...................	5.15	5.00	**5.41**	0.18	-	**2.56**	0.05	**8.01**	m	**8.61**
Prime and apply non-reflective self adhesive road marking tape; external work										
Concrete general surfaces										
isolated surfaces, 50mm wide....................	1.64	5.00	**1.72**	0.10	-	**1.42**	0.03	**3.17**	m	**3.40**
isolated surfaces, 100mm wide...................	3.02	5.00	**3.17**	0.13	-	**1.85**	0.04	**5.06**	m	**5.43**
Prime and apply reflective self adhesive road marking tape; external work										
Concrete general surfaces										
isolated surfaces, 50mm wide....................	2.28	5.00	**2.39**	0.10	-	**1.42**	0.03	**3.84**	m	**4.13**
isolated surfaces, 100mm wide...................	4.20	5.00	**4.41**	0.13	-	**1.85**	0.04	**6.29**	m	**6.77**
M60: PAINTING/CLEAR FINISHING (REDECORATIONS) - EMULSION PAINTING										
Generally										
Note: the following rates include for the cost of all preparatory work, e.g. washing down, etc.										
Two coats emulsion paint, existing emulsion painted surfaces										
Concrete general surfaces										
girth exceeding 300mm	0.23	10.00	**0.25**	0.29	-	**4.12**	0.08	**4.45**	m2	**4.79**
Plaster general surfaces										
girth exceeding 300mm	0.18	10.00	**0.20**	0.25	-	**3.55**	0.07	**3.82**	m2	**4.11**
Plasterboard general surfaces										
girth exceeding 300mm	0.18	10.00	**0.20**	0.26	-	**3.69**	0.07	**3.97**	m2	**4.26**
Brickwork general surfaces										
girth exceeding 300mm	0.23	10.00	**0.25**	0.29	-	**4.12**	0.08	**4.45**	m2	**4.79**

Labour hourly rates: (except Specialists) Craft Operatives £14.21 Labourer £10.71 Rates are national average prices. Refer to REGIONAL VARIATIONS for indicative levels of overall pricing in regions	MATERIALS			LABOUR				RATES		
	Del to Site £	Waste %	Material Cost £	Craft Optve Hrs	Lab Hrs	Labour Cost £	Sunds £	Nett Rate £	Unit	Gross Rate (+7.5%) £
M60: PAINTING/CLEAR FINISHING (REDECORATIONS) - EMULSION PAINTING Cont.										
Two coats emulsion paint, existing emulsion painted surfaces Cont.										
Paper covered general surfaces girth exceeding 300mm	0.18	10.00	**0.20**	0.26	-	**3.69**	0.07	**3.97**	m2	**4.26**
Two full coats emulsion paint, existing washable distempered surfaces										
Concrete general surfaces girth exceeding 300mm	0.23	10.00	**0.25**	0.30	-	**4.26**	0.08	**4.60**	m2	**4.94**
Plaster general surfaces girth exceeding 300mm	0.18	10.00	**0.20**	0.26	-	**3.69**	0.07	**3.97**	m2	**4.26**
Plasterboard general surfaces girth exceeding 300mm	0.18	10.00	**0.20**	0.27	-	**3.84**	0.07	**4.11**	m2	**4.42**
Brickwork general surfaces girth exceeding 300mm	0.23	10.00	**0.25**	0.30	-	**4.26**	0.08	**4.60**	m2	**4.94**
Paper covered general surfaces girth exceeding 300mm	0.18	10.00	**0.20**	0.27	-	**3.84**	0.07	**4.11**	m2	**4.42**
Two full coats emulsion paint, existing textured plastic coating surfaces										
Concrete general surfaces girth exceeding 300mm	0.23	10.00	**0.25**	0.32	-	**4.55**	0.09	**4.89**	m2	**5.25**
Plaster general surfaces girth exceeding 300mm	0.18	10.00	**0.20**	0.28	-	**3.98**	0.08	**4.25**	m2	**4.57**
Plasterboard general surfaces girth exceeding 300mm	0.18	10.00	**0.20**	0.29	-	**4.12**	0.08	**4.40**	m2	**4.73**
Brickwork general surfaces girth exceeding 300mm	0.24	10.00	**0.26**	0.32	-	**4.55**	0.09	**4.89**	m2	**5.26**
Paper covered general surfaces girth exceeding 300mm	0.18	10.00	**0.20**	0.29	-	**4.12**	0.08	**4.40**	m2	**4.73**
Mist coat, two full coats emulsion paint, existing non-washable distempered surfaces										
Concrete general surfaces girth exceeding 300mm	0.28	10.00	**0.31**	0.35	-	**4.97**	0.09	**5.38**	m2	**5.78**
Plaster general surfaces girth exceeding 300mm	0.23	10.00	**0.25**	0.31	-	**4.41**	0.08	**4.74**	m2	**5.10**
Plasterboard general surfaces girth exceeding 300mm	0.23	10.00	**0.25**	0.32	-	**4.55**	0.09	**4.89**	m2	**5.25**
Brickwork general surfaces girth exceeding 300mm	0.28	10.00	**0.31**	0.35	-	**4.97**	0.09	**5.38**	m2	**5.78**
Paper covered general surfaces girth exceeding 300mm	0.23	10.00	**0.25**	0.32	-	**4.55**	0.09	**4.89**	m2	**5.25**
Two full coats emulsion paint, existing gloss painted surfaces										
Concrete general surfaces girth exceeding 300mm	0.23	10.00	**0.25**	0.30	-	**4.26**	0.08	**4.60**	m2	**4.94**
Plaster general surfaces girth exceeding 300mm	0.18	10.00	**0.20**	0.26	-	**3.69**	0.07	**3.97**	m2	**4.26**
Plasterboard general surfaces girth exceeding 300mm	0.18	10.00	**0.20**	0.27	-	**3.84**	0.07	**4.11**	m2	**4.42**
Brickwork general surfaces girth exceeding 300mm	0.24	10.00	**0.26**	0.30	-	**4.26**	0.08	**4.60**	m2	**4.95**
Paper covered general surfaces girth exceeding 300mm	0.18	10.00	**0.20**	0.27	-	**3.84**	0.07	**4.11**	m2	**4.42**
M60: PAINTING/CLEAR FINISHING (REDECORATIONS) - CEMENT PAINTING										
Generally										
Note: the following rates include for the cost of all preparatory work, e.g. washing down, etc.										
One coat sealer, one coat waterproof cement paint; external work										
Cement rendered general surfaces girth exceeding 300mm	2.30	10.00	**2.52**	0.32	-	**4.55**	0.09	**7.16**	m2	**7.69**
Concrete general surfaces girth exceeding 300mm	1.94	10.00	**2.14**	0.38	-	**5.40**	0.10	**7.64**	m2	**8.21**

Labour hourly rates: (except Specialists) Craft Operatives £14.21 Labourer £10.71 Rates are national average prices. Refer to REGIONAL VARIATIONS for indicative levels of overall pricing in regions	MATERIALS			LABOUR				RATES		
	Del to Site £	Waste %	Material Cost £	Craft Optve Hrs	Lab Hrs	Labour Cost £	Sunds £	Nett Rate £	Unit	Gross Rate (+7.5%) £
M60: PAINTING/CLEAR FINISHING (REDECORATIONS) - CEMENT PAINTING Cont.										
One coat sealer, one coat waterproof cement paint; external work Cont.										
Brickwork general surfaces girth exceeding 300mm	2.12	10.00	**2.33**	0.38	-	**5.40**	0.10	**7.83**	m2	**8.42**
Rough cast general surfaces girth exceeding 300mm	2.30	10.00	**2.52**	0.47	-	**6.68**	0.13	**9.33**	m2	**10.03**
Two coats sealer, one coat waterproof cement paint; external work										
Cement rendered general surfaces girth exceeding 300mm	2.88	10.00	**3.17**	0.43	-	**6.11**	0.12	**9.39**	m2	**10.10**
Concrete general surfaces girth exceeding 300mm	2.35	10.00	**2.58**	0.52	-	**7.39**	0.14	**10.11**	m2	**10.87**
Brickwork general surfaces girth exceeding 300mm	2.70	10.00	**2.97**	0.52	-	**7.39**	0.14	**10.50**	m2	**11.29**
Rough cast general surfaces girth exceeding 300mm	2.88	10.00	**3.17**	0.64	-	**9.09**	0.17	**12.43**	m2	**13.37**
One coat waterproof cement paint existing cement painted surfaces ; external work										
Cement rendered general surfaces girth exceeding 300mm	1.10	10.00	**1.21**	0.24	-	**3.41**	0.06	**4.68**	m2	**5.03**
Concrete general surfaces girth exceeding 300mm	1.10	10.00	**1.21**	0.30	-	**4.26**	0.08	**5.55**	m2	**5.97**
Brickwork general surfaces girth exceeding 300mm	1.27	10.00	**1.40**	0.30	-	**4.26**	0.08	**5.74**	m2	**6.18**
Rough cast general surfaces girth exceeding 300mm	1.45	10.00	**1.59**	0.39	-	**5.54**	0.11	**7.24**	m2	**7.78**
Two coats waterproof cement paint existing cement painted surfaces ; external work										
Cement rendered general surfaces girth exceeding 300mm	2.90	10.00	**3.19**	0.35	-	**4.97**	0.09	**8.26**	m2	**8.87**
Concrete general surfaces girth exceeding 300mm	2.20	10.00	**2.41**	0.44	-	**6.25**	0.12	**8.79**	m2	**9.44**
Brickwork general surfaces girth exceeding 300mm	2.55	10.00	**2.80**	0.44	-	**6.25**	0.12	**9.17**	m2	**9.86**
Rough cast general surfaces girth exceeding 300mm	2.90	10.00	**3.19**	0.56	-	**7.96**	0.15	**11.30**	m2	**12.14**
One coat stabilizing solution, one coat textured masonry paint, existing undecorated surfaces; external work										
Cement rendered general surfaces girth exceeding 300mm	0.41	10.00	**0.46**	0.32	-	**4.55**	0.09	**5.09**	m2	**5.47**
Concrete general surfaces girth exceeding 300mm	0.49	10.00	**0.53**	0.38	-	**5.40**	0.10	**6.04**	m2	**6.49**
Brickwork general surfaces girth exceeding 300mm	0.68	10.00	**0.75**	0.38	-	**5.40**	0.10	**6.25**	m2	**6.72**
Rough cast general surfaces girth exceeding 300mm	0.83	10.00	**0.91**	0.47	-	**6.68**	0.13	**7.71**	m2	**8.29**
One coat stabilizing solution, two coats textured masonry paint, existing undecorated surfaces; external work										
Cement rendered general surfaces girth exceeding 300mm	0.74	10.00	**0.81**	0.43	-	**6.11**	0.12	**7.04**	m2	**7.56**
Concrete general surfaces girth exceeding 300mm	0.74	10.00	**0.81**	0.52	-	**7.39**	0.14	**8.34**	m2	**8.97**
Brickwork general surfaces girth exceeding 300mm	1.13	10.00	**1.24**	0.52	-	**7.39**	0.14	**8.77**	m2	**9.43**
Rough cast general surfaces girth exceeding 300mm	1.42	10.00	**1.56**	0.64	-	**9.09**	0.17	**10.83**	m2	**11.64**
One coat textured masonry paint, existing cement painted surfaces; external work										
Cement rendered general surfaces girth exceeding 300mm	0.18	10.00	**0.20**	0.24	-	**3.41**	0.06	**3.67**	m2	**3.95**

Labour hourly rates: (except Specialists) Craft Operatives £14.21 Labourer £10.71 Rates are national average prices. Refer to REGIONAL VARIATIONS for indicative levels of overall pricing in regions	MATERIALS			LABOUR				RATES		
	Del to Site £	Waste %	Material Cost £	Craft Optve Hrs	Lab Hrs	Labour Cost £	Sunds £	Nett Rate £	Unit	Gross Rate (+7.5%) £
M60: PAINTING/CLEAR FINISHING (REDECORATIONS) - CEMENT PAINTING Cont.										
One coat textured masonry paint, existing cement painted surfaces; external work Cont.										
Concrete general surfaces girth exceeding 300mm	0.25	10.00	**0.28**	0.30	-	**4.26**	0.08	**4.62**	m2	**4.97**
Brickwork general surfaces girth exceeding 300mm	0.45	10.00	**0.49**	0.30	-	**4.26**	0.08	**4.84**	m2	**5.20**
Rough cast general surfaces girth exceeding 300mm	0.59	10.00	**0.65**	0.39	-	**5.54**	0.11	**6.30**	m2	**6.77**
Two coats textured masonry paint, existing cement painted surfaces; external work										
Cement rendered general surfaces girth exceeding 300mm	0.50	10.00	**0.55**	0.35	-	**4.97**	0.09	**5.62**	m2	**6.04**
Concrete general surfaces girth exceeding 300mm	0.50	10.00	**0.55**	0.44	-	**6.25**	0.12	**6.92**	m2	**7.44**
Brickwork general surfaces girth exceeding 300mm	0.90	10.00	**0.98**	0.44	-	**6.25**	0.12	**7.36**	m2	**7.91**
Rough cast general surfaces girth exceeding 300mm	1.18	10.00	**1.30**	0.56	-	**7.96**	0.15	**9.41**	m2	**10.11**
M60: PAINTING/CLEAR FINISHING (REDECORATIONS) - OIL PAINTING WALLS AND CEILINGS										
Generally										
Note: the following rates include for the cost of all preparatory work, e.g. washing down, etc.										
One undercoat, one coat full gloss finish, existing gloss painted surfaces										
Concrete general surfaces girth exceeding 300mm	0.32	10.00	**0.35**	0.35	-	**4.97**	0.09	**5.42**	m2	**5.82**
Brickwork general surfaces girth exceeding 300mm	0.34	10.00	**0.37**	0.36	-	**5.12**	0.10	**5.59**	m2	**6.00**
Plasterboard general surfaces girth exceeding 300mm	0.29	10.00	**0.32**	0.32	-	**4.55**	0.09	**4.96**	m2	**5.33**
Plaster general surfaces girth exceeding 300mm	0.29	10.00	**0.32**	0.31	-	**4.41**	0.08	**4.81**	m2	**5.17**
One undercoat, one coat eggshell finish, existing gloss painted surfaces										
Concrete general surfaces girth exceeding 300mm	0.35	10.00	**0.39**	0.35	-	**4.97**	0.09	**5.45**	m2	**5.86**
Brickwork general surfaces girth exceeding 300mm	0.44	10.00	**0.49**	0.36	-	**5.12**	0.10	**5.70**	m2	**6.13**
Plasterboard general surfaces girth exceeding 300mm	0.32	10.00	**0.36**	0.32	-	**4.55**	0.09	**4.99**	m2	**5.37**
Plaster general surfaces girth exceeding 300mm	0.32	10.00	**0.36**	0.31	-	**4.41**	0.08	**4.85**	m2	**5.21**
Two undercoats, one coat full gloss finish, existing gloss painted surfaces										
Concrete general surfaces girth exceeding 300mm	0.47	10.00	**0.51**	0.44	-	**6.25**	0.12	**6.89**	m2	**7.40**
Brickwork general surfaces girth exceeding 300mm	0.56	10.00	**0.62**	0.44	-	**6.25**	0.12	**6.99**	m2	**7.51**
Plasterboard general surfaces girth exceeding 300mm	0.44	10.00	**0.49**	0.41	-	**5.83**	0.11	**6.43**	m2	**6.91**
Plaster general surfaces girth exceeding 300mm	0.44	10.00	**0.49**	0.40	-	**5.68**	0.11	**6.28**	m2	**6.75**
Two undercoats, one coat eggshell finish, existing gloss painted surfaces										
Concrete general surfaces girth exceeding 300mm	0.50	10.00	**0.55**	0.44	-	**6.25**	0.12	**6.92**	m2	**7.44**
Brickwork general surfaces girth exceeding 300mm	0.60	10.00	**0.67**	0.44	-	**6.25**	0.12	**7.04**	m2	**7.56**
Plasterboard general surfaces girth exceeding 300mm	0.48	10.00	**0.52**	0.41	-	**5.83**	0.11	**6.46**	m2	**6.95**
Plaster general surfaces girth exceeding 300mm	0.48	10.00	**0.52**	0.40	-	**5.68**	0.11	**6.32**	m2	**6.79**

Labour hourly rates: (except Specialists) Craft Operatives £14.21 Labourer £10.71 Rates are national average prices. Refer to REGIONAL VARIATIONS for indicative levels of overall pricing in regions	MATERIALS			LABOUR				RATES		
	Del to Site £	Waste %	Material Cost £	Craft Optve Hrs	Lab Hrs	Labour Cost £	Sunds £	Nett Rate £	Unit	Gross Rate (+7.5%) £
M60: PAINTING/CLEAR FINISHING (REDECORATIONS) - OIL PAINTING WALLS AND CEILINGS Cont.										
One coat primer, one undercoat, one coat full gloss finish, existing washable distempered or emulsion painted surfaces										
Concrete general surfaces girth exceeding 300mm	0.86	10.00	**0.95**	0.44	-	**6.25**	0.12	**7.32**	m2	**7.87**
Brickwork general surfaces girth exceeding 300mm	1.01	10.00	**1.11**	0.44	-	**6.25**	0.12	**7.49**	m2	**8.05**
Plasterboard general surfaces girth exceeding 300mm	0.79	10.00	**0.87**	0.41	-	**5.83**	0.11	**6.81**	m2	**7.32**
Plaster general surfaces girth exceeding 300mm	0.79	10.00	**0.87**	0.40	-	**5.68**	0.11	**6.66**	m2	**7.16**
One coat primer, one undercoat, one coat eggshell finish, existing washable distempered or emulsion painted surfaces										
Concrete general surfaces girth exceeding 300mm	0.90	10.00	**0.99**	0.44	-	**6.25**	0.12	**7.36**	m2	**7.91**
Brickwork general surfaces girth exceeding 300mm	1.11	10.00	**1.23**	0.44	-	**6.25**	0.12	**7.60**	m2	**8.17**
Plasterboard general surfaces girth exceeding 300mm	0.82	10.00	**0.91**	0.41	-	**5.83**	0.11	**6.84**	m2	**7.36**
Plaster general surfaces girth exceeding 300mm	0.82	10.00	**0.91**	0.40	-	**5.68**	0.11	**6.70**	m2	**7.20**
One coat primer, two undercoats, one coat full gloss finish, existing washable distempered or emulsion painted surfaces										
Concrete general surfaces girth exceeding 300mm	1.02	10.00	**1.12**	0.53	-	**7.53**	0.14	**8.79**	m2	**9.45**
Brickwork general surfaces girth exceeding 300mm	1.24	10.00	**1.36**	0.54	-	**7.67**	0.15	**9.18**	m2	**9.87**
Plasterboard general surfaces girth exceeding 300mm	0.94	10.00	**1.04**	0.50	-	**7.11**	0.14	**8.28**	m2	**8.90**
Plaster general surfaces girth exceeding 300mm	0.94	10.00	**1.04**	0.49	-	**6.96**	0.13	**8.13**	m2	**8.74**
One coat primer, two undercoats, one coat eggshell finish, existing washable distempered or emulsion painted surfaces										
Concrete general surfaces girth exceeding 300mm	1.05	10.00	**1.16**	0.53	-	**7.53**	0.14	**8.83**	m2	**9.49**
Brickwork general surfaces girth exceeding 300mm	1.28	10.00	**1.41**	0.54	-	**7.67**	0.15	**9.23**	m2	**9.92**
Plasterboard general surfaces girth exceeding 300mm	0.98	10.00	**1.07**	0.50	-	**7.11**	0.14	**8.31**	m2	**8.94**
Plaster general surfaces girth exceeding 300mm	0.98	10.00	**1.07**	0.49	-	**6.96**	0.13	**8.17**	m2	**8.78**
One coat primer, one undercoat, one coat full gloss finish, existing non-washable distempered surfaces										
Concrete general surfaces girth exceeding 300mm	0.86	10.00	**0.95**	0.44	-	**6.25**	0.12	**7.32**	m2	**7.87**
Brickwork general surfaces girth exceeding 300mm	1.01	10.00	**1.11**	0.44	-	**6.25**	0.12	**7.49**	m2	**8.05**
Plasterboard general surfaces girth exceeding 300mm	0.79	10.00	**0.87**	0.41	-	**5.83**	0.11	**6.81**	m2	**7.32**
Plaster general surfaces girth exceeding 300mm	0.79	10.00	**0.87**	0.40	-	**5.68**	0.11	**6.66**	m2	**7.16**
One coat primer, one undercoat, one coat eggshell finish, existing non-washable distempered surfaces										
Concrete general surfaces girth exceeding 300mm	0.90	10.00	**0.99**	0.44	-	**6.25**	0.12	**7.36**	m2	**7.91**
Brickwork general surfaces girth exceeding 300mm	1.11	10.00	**1.23**	0.44	-	**6.25**	0.12	**7.60**	m2	**8.17**
Plasterboard general surfaces girth exceeding 300mm	0.82	10.00	**0.91**	0.41	-	**5.83**	0.11	**6.84**	m2	**7.36**
Plaster general surfaces girth exceeding 300mm	0.82	10.00	**0.91**	0.40	-	**5.68**	0.11	**6.70**	m2	**7.20**

Labour hourly rates: (except Specialists) Craft Operatives £14.21 Labourer £10.71 Rates are national average prices. Refer to REGIONAL VARIATIONS for indicative levels of overall pricing in regions	MATERIALS			LABOUR				RATES		
	Del to Site £	Waste %	Material Cost £	Craft Optve Hrs	Lab Hrs	Labour Cost £	Sunds £	Nett Rate £	Unit	Gross Rate (+7.5%) £
M60: PAINTING/CLEAR FINISHING (REDECORATIONS) - OIL PAINTING WALLS AND CEILINGS Cont.										
One coat primer, two undercoats, one coat full gloss finish, existing non-washable distempered surfaces										
Concrete general surfaces girth exceeding 300mm	1.02	10.00	**1.12**	0.53	-	**7.53**	0.14	**8.79**	m2	**9.45**
Brickwork general surfaces girth exceeding 300mm	1.24	10.00	**1.36**	0.54	-	**7.67**	0.15	**9.18**	m2	**9.87**
Plasterboard general surfaces girth exceeding 300mm	0.94	10.00	**1.04**	0.50	-	**7.11**	0.14	**8.28**	m2	**8.90**
Plaster general surfaces girth exceeding 300mm	0.94	10.00	**1.04**	0.49	-	**6.96**	0.13	**8.13**	m2	**8.74**
One coat primer, two undercoats, one coat eggshell finish, existing non-washable distempered surfaces										
Concrete general surfaces girth exceeding 300mm	1.05	10.00	**1.16**	0.53	-	**7.53**	0.14	**8.83**	m2	**9.49**
Brickwork general surfaces girth exceeding 300mm	1.28	10.00	**1.41**	0.54	-	**7.67**	0.15	**9.23**	m2	**9.92**
Plasterboard general surfaces girth exceeding 300mm	0.98	10.00	**1.07**	0.50	-	**7.11**	0.14	**8.31**	m2	**8.94**
Plaster general surfaces girth exceeding 300mm	0.98	10.00	**1.07**	0.49	-	**6.96**	0.13	**8.17**	m2	**8.78**
M60: PAINTING/CLEAR FINISHING (REDECORATIONS) - PLASTIC FINISH										
Generally										
Note: the following rates include for the cost of all preparatory work, e.g. washing down, etc.										
Textured plastic coating - stippled finish, existing washable distempered or emulsion painted surfaces										
Concrete general surfaces girth exceeding 300mm	0.31	10.00	**0.35**	0.36	-	**5.12**	0.10	**5.56**	m2	**5.98**
Brickwork general surfaces girth exceeding 300mm	0.38	10.00	**0.41**	0.41	-	**5.83**	0.11	**6.35**	m2	**6.83**
Plasterboard general surfaces girth exceeding 300mm	0.28	10.00	**0.31**	0.31	-	**4.41**	0.08	**4.80**	m2	**5.16**
Plaster general surfaces girth exceeding 300mm	0.28	10.00	**0.31**	0.31	-	**4.41**	0.08	**4.80**	m2	**5.16**
Textured plastic coating - combed finish, existing washable distempered or emulsion painted surfaces										
Concrete general surfaces girth exceeding 300mm	0.38	10.00	**0.41**	0.41	-	**5.83**	0.11	**6.35**	m2	**6.83**
Brickwork general surfaces girth exceeding 300mm	0.45	10.00	**0.50**	0.46	-	**6.54**	0.12	**7.16**	m2	**7.69**
Plasterboard general surfaces girth exceeding 300mm	0.34	10.00	**0.37**	0.36	-	**5.12**	0.10	**5.59**	m2	**6.00**
Plaster general surfaces girth exceeding 300mm	0.34	10.00	**0.37**	0.36	-	**5.12**	0.10	**5.59**	m2	**6.00**
Textured plastic coating - stippled finish, existing non-washable distempered surfaces										
Concrete general surfaces girth exceeding 300mm	0.31	10.00	**0.35**	0.41	-	**5.83**	0.11	**6.28**	m2	**6.75**
Brickwork general surfaces girth exceeding 300mm	0.38	10.00	**0.41**	0.46	-	**6.54**	0.12	**7.07**	m2	**7.60**
Plasterboard general surfaces girth exceeding 300mm	0.28	10.00	**0.31**	0.36	-	**5.12**	0.10	**5.52**	m2	**5.94**
Plaster general surfaces girth exceeding 300mm	0.28	10.00	**0.31**	0.36	-	**5.12**	0.10	**5.52**	m2	**5.94**
Textured plastic coating - combed finish, existing non-washable distempered surfaces										
Concrete general surfaces girth exceeding 300mm	0.38	10.00	**0.41**	0.46	-	**6.54**	0.12	**7.07**	m2	**7.60**
Brickwork general surfaces girth exceeding 300mm	0.45	10.00	**0.50**	0.51	-	**7.25**	0.14	**7.88**	m2	**8.47**
Plasterboard general surfaces girth exceeding 300mm	0.34	10.00	**0.37**	0.41	-	**5.83**	0.11	**6.31**	m2	**6.78**

Labour hourly rates: (except Specialists) Craft Operatives £14.21 Labourer £10.71 Rates are national average prices. Refer to REGIONAL VARIATIONS for indicative levels of overall pricing in regions	MATERIALS			LABOUR				RATES		
	Del to Site £	Waste %	Material Cost £	Craft Optve Hrs	Lab Hrs	Labour Cost £	Sunds £	Nett Rate £	Unit	Gross Rate (+7.5%) £
M60: PAINTING/CLEAR FINISHING (REDECORATIONS) - PLASTIC FINISH Cont.										
Textured plastic coating - combed finish, existing non-washable distempered surfaces Cont.										
Plaster general surfaces										
girth exceeding 300mm	0.34	10.00	**0.37**	0.41	-	**5.83**	0.11	**6.31**	m2	**6.78**
M60: PAINTING/CLEAR FINISHING (REDECORATIONS) - CLEAN OUT GUTTERS										
Generally										
Note: the following rates include for the cost of all preparatory work, e.g. washing down, etc.										
Clean out gutters prior to repainting, staunch joints with red lead, bituminous compound or mastic										
Iron or steel eaves gutters										
generally	-	-	**-**	0.05	-	**0.71**	0.06	**0.77**	m	**0.83**
M60: PAINTING/CLEAR FINISHING (REDECORATIONS) - OIL PAINTING METALWORK										
Generally										
Note: the following rates include for the cost of all preparatory work, e.g. washing down, etc.										
One undercoat, one coat full gloss finish, existing gloss painted metal surfaces										
Iron or steel general surfaces										
girth exceeding 300mm	0.30	10.00	**0.33**	0.42	-	**5.97**	0.11	**6.42**	m2	**6.90**
isolated surfaces, girth not exceeding 300mm	0.09	10.00	**0.10**	0.14	-	**1.99**	0.04	**2.13**	m	**2.29**
Galvanized glazed doors, windows or screens										
girth exceeding 300mm; panes, area not exceeding 0.10m2	0.16	10.00	**0.18**	0.79	-	**11.23**	0.21	**11.62**	m2	**12.49**
girth exceeding 300mm; panes, area 0.10 - 0.50m2	0.12	10.00	**0.13**	0.58	-	**8.24**	0.16	**8.53**	m2	**9.17**
girth exceeding 300mm; panes, area 0.50 - 1.00m2	0.09	10.00	**0.10**	0.47	-	**6.68**	0.13	**6.91**	m2	**7.43**
girth exceeding 300mm; panes, area exceeding 1.00m2	0.09	10.00	**0.10**	0.42	-	**5.97**	0.11	**6.18**	m2	**6.65**
Iron or steel structural work										
girth exceeding 300mm	0.30	10.00	**0.33**	0.46	-	**6.54**	0.12	**7.00**	m2	**7.52**
isolated surfaces, girth not exceeding 300mm	0.09	10.00	**0.10**	0.19	-	**2.70**	0.05	**2.85**	m	**3.07**
Iron or steel structural members of roof trusses, lattice girders, purlins and the like										
girth exceeding 300mm	0.30	10.00	**0.33**	0.59	-	**8.38**	0.16	**8.88**	m2	**9.54**
isolated surfaces, girth not exceeding 300mm	0.09	10.00	**0.10**	0.20	-	**2.84**	0.05	**3.00**	m	**3.22**
Iron or steel services										
girth exceeding 300mm	0.30	10.00	**0.33**	0.59	-	**8.38**	0.16	**8.88**	m2	**9.54**
isolated surfaces, girth not exceeding 300mm	0.09	10.00	**0.10**	0.20	-	**2.84**	0.05	**3.00**	m	**3.22**
isolated areas not exceeding 0.50m2 irrespective of girth	0.16	10.00	**0.18**	0.59	-	**8.38**	0.16	**8.72**	nr	**9.38**
Copper services										
girth exceeding 300mm	0.30	10.00	**0.33**	0.59	-	**8.38**	0.16	**8.88**	m2	**9.54**
isolated surfaces, girth not exceeding 300mm	0.09	10.00	**0.10**	0.20	-	**2.84**	0.05	**3.00**	m	**3.22**
Galvanized services										
girth exceeding 300mm	0.30	10.00	**0.33**	0.59	-	**8.38**	0.16	**8.88**	m2	**9.54**
isolated surfaces, girth not exceeding 300mm	0.09	10.00	**0.10**	0.20	-	**2.84**	0.05	**3.00**	m	**3.22**
One undercoat, one coat full gloss finish, existing gloss painted metal surfaces; external work										
Iron or steel general surfaces										
girth exceeding 300mm	0.30	10.00	**0.33**	0.45	-	**6.39**	0.12	**6.85**	m2	**7.36**
isolated surfaces, girth not exceeding 300mm	0.09	10.00	**0.10**	0.15	-	**2.13**	0.04	**2.27**	m	**2.45**
Galvanized glazed doors, windows or screens										
girth exceeding 300mm; panes, area not exceeding 0.10m2	0.16	10.00	**0.18**	0.81	-	**11.51**	0.22	**11.91**	m2	**12.80**
girth exceeding 300mm; panes, area 0.10 - 0.50m2	0.12	10.00	**0.13**	0.60	-	**8.53**	0.16	**8.82**	m2	**9.48**
girth exceeding 300mm; panes, area 0.50 - 1.00m2	0.09	10.00	**0.10**	0.50	-	**7.11**	0.14	**7.34**	m2	**7.89**
girth exceeding 300mm; panes, area exceeding 1.00m2	0.09	10.00	**0.10**	0.44	-	**6.25**	0.12	**6.47**	m2	**6.96**
Iron or steel structural work										
girth exceeding 300mm	0.30	10.00	**0.33**	0.46	-	**6.54**	0.12	**7.00**	m2	**7.52**
isolated surfaces, girth not exceeding 300mm	0.09	10.00	**0.10**	0.19	-	**2.70**	0.05	**2.85**	m	**3.07**
Iron or steel structural members of roof trusses, lattice girders, purlins and the like										
girth exceeding 300mm	0.30	10.00	**0.33**	0.59	-	**8.38**	0.16	**8.88**	m2	**9.54**
isolated surfaces, girth not exceeding 300mm	0.09	10.00	**0.10**	0.20	-	**2.84**	0.05	**3.00**	m	**3.22**
Iron or steel railings, fences and gates; plain open type										
girth exceeding 300mm	0.23	10.00	**0.26**	0.44	-	**6.25**	0.12	**6.63**	m2	**7.13**
isolated surfaces, girth not exceeding 300mm	0.09	10.00	**0.10**	0.15	-	**2.13**	0.04	**2.27**	m	**2.45**

Labour hourly rates: (except Specialists) Craft Operatives £14.21 Labourer £10.71 Rates are national average prices. Refer to REGIONAL VARIATIONS for indicative levels of overall pricing in regions	MATERIALS Del to Site £	 Waste %	 Material Cost £	LABOUR Craft Optve Hrs	 Lab Hrs	 Labour Cost £	 Sunds £	RATES Nett Rate £	 Unit	 Gross Rate (+7.5%) £
M60: PAINTING/CLEAR FINISHING (REDECORATIONS) - OIL PAINTING METALWORK Cont.										
One undercoat, one coat full gloss finish, existing gloss painted metal surfaces; external work Cont.										
Iron or steel railings, fences and gates; close type										
girth exceeding 300mm	0.30	10.00	**0.33**	0.40	-	**5.68**	0.11	**6.13**	m2	**6.59**
Iron or steel railings, fences and gates; ornamental type										
girth exceeding 300mm	0.23	10.00	**0.26**	0.80	-	**11.37**	0.22	**11.84**	m2	**12.73**
Iron or steel eaves gutters										
girth exceeding 300mm	0.30	10.00	**0.33**	0.48	-	**6.82**	0.13	**7.29**	m2	**7.83**
isolated surfaces, girth not exceeding 300mm.....	0.09	10.00	**0.10**	0.16	-	**2.27**	0.04	**2.42**	m	**2.60**
Galvanized eaves gutters										
girth exceeding 300mm	0.30	10.00	**0.33**	0.48	-	**6.82**	0.13	**7.29**	m2	**7.83**
isolated surfaces, girth not exceeding 300mm.....	0.09	10.00	**0.10**	0.16	-	**2.27**	0.04	**2.42**	m	**2.60**
Iron or steel services										
girth exceeding 300mm	0.30	10.00	**0.33**	0.59	-	**8.38**	0.16	**8.88**	m2	**9.54**
isolated surfaces, girth not exceeding 300mm.....	0.09	10.00	**0.10**	0.20	-	**2.84**	0.05	**3.00**	m	**3.22**
isolated areas not exceeding 0.50m2 irrespective of girth	0.16	10.00	**0.18**	0.59	-	**8.38**	0.16	**8.72**	nr	**9.38**
Copper services										
girth exceeding 300mm	0.30	10.00	**0.33**	0.59	-	**8.38**	0.16	**8.88**	m2	**9.54**
isolated surfaces, girth not exceeding 300mm.....	0.09	10.00	**0.10**	0.20	-	**2.84**	0.05	**3.00**	m	**3.22**
Galvanized services										
girth exceeding 300mm	0.30	10.00	**0.33**	0.59	-	**8.38**	0.16	**8.88**	m2	**9.54**
isolated surfaces, girth not exceeding 300mm.....	0.09	10.00	**0.10**	0.20	-	**2.84**	0.05	**3.00**	m	**3.22**
Two coats full gloss finish, existing gloss painted metal surfaces										
Iron or steel general surfaces										
girth exceeding 300mm	0.30	10.00	**0.33**	0.43	-	**6.11**	0.12	**6.56**	m2	**7.05**
isolated surfaces, girth not exceeding 300mm.....	0.09	10.00	**0.10**	0.14	-	**1.99**	0.04	**2.13**	m	**2.29**
Galvanized glazed doors, windows or screens										
girth exceeding 300mm; panes, area not exceeding 0.10m2	0.16	10.00	**0.18**	0.79	-	**11.23**	0.21	**11.62**	m2	**12.49**
girth exceeding 300mm; panes, area 0.10 - 0.50m2 .	0.12	10.00	**0.13**	0.58	-	**8.24**	0.16	**8.53**	m2	**9.17**
girth exceeding 300mm; panes, area 0.50 - 1.00m2 .	0.09	10.00	**0.10**	0.48	-	**6.82**	0.13	**7.05**	m2	**7.58**
girth exceeding 300mm; panes, area exceeding 1.00m2	0.09	10.00	**0.10**	0.42	-	**5.97**	0.11	**6.18**	m2	**6.65**
Iron or steel structural work										
girth exceeding 300mm	0.30	10.00	**0.33**	0.46	-	**6.54**	0.12	**7.00**	m2	**7.52**
isolated surfaces, girth not exceeding 300mm.....	0.09	10.00	**0.10**	0.19	-	**2.70**	0.05	**2.85**	m	**3.07**
Iron or steel structural members of roof trusses, lattice girders, purlins and the like										
girth exceeding 300mm	0.30	10.00	**0.33**	0.59	-	**8.38**	0.16	**8.88**	m2	**9.54**
isolated surfaces, girth not exceeding 300mm.....	0.09	10.00	**0.10**	0.20	-	**2.84**	0.05	**3.00**	m	**3.22**
Iron or steel services										
girth exceeding 300mm	0.30	10.00	**0.33**	0.59	-	**8.38**	0.16	**8.88**	m2	**9.54**
isolated surfaces, girth not exceeding 300mm.....	0.09	10.00	**0.10**	0.20	-	**2.84**	0.05	**3.00**	m	**3.22**
isolated areas not exceeding 0.50m2 irrespective of girth	0.15	10.00	**0.17**	0.59	-	**8.38**	0.16	**8.71**	nr	**9.36**
Copper services										
girth exceeding 300mm	0.30	10.00	**0.33**	0.59	-	**8.38**	0.16	**8.88**	m2	**9.54**
isolated surfaces, girth not exceeding 300mm.....	0.09	10.00	**0.10**	0.20	-	**2.84**	0.05	**3.00**	m	**3.22**
Galvanized services										
girth exceeding 300mm	0.30	10.00	**0.33**	0.59	-	**8.38**	0.16	**8.88**	m2	**9.54**
isolated surfaces, girth not exceeding 300mm.....	0.09	10.00	**0.10**	0.20	-	**2.84**	0.05	**3.00**	m	**3.22**
Two coats full gloss finish, existing gloss painted metal surfaces; external work										
Iron or steel general surfaces										
girth exceeding 300mm	0.30	10.00	**0.33**	0.45	-	**6.39**	0.12	**6.85**	m2	**7.36**
isolated surfaces, girth not exceeding 300mm.....	0.09	10.00	**0.10**	0.15	-	**2.13**	0.04	**2.27**	m	**2.45**
Galvanized glazed doors, windows or screens										
girth exceeding 300mm; panes, area not exceeding 0.10m2	0.16	10.00	**0.18**	0.81	-	**11.51**	0.22	**11.91**	m2	**12.80**
girth exceeding 300mm; panes, area 0.10 - 0.50m2 .	0.12	10.00	**0.13**	0.60	-	**8.53**	0.16	**8.82**	m2	**9.48**
girth exceeding 300mm; panes, area 0.50 - 1.00m2 .	0.09	10.00	**0.10**	0.50	-	**7.11**	0.14	**7.34**	m	**7.89**
girth exceeding 300mm; panes, area exceeding 1.00m2	0.09	10.00	**0.10**	0.44	-	**6.25**	0.12	**6.47**	m2	**6.96**
Iron or steel structural work										
girth exceeding 300mm	0.30	10.00	**0.33**	0.46	-	**6.54**	0.12	**7.00**	m2	**7.52**
isolated surfaces, girth not exceeding 300mm.....	0.09	10.00	**0.10**	0.19	-	**2.70**	0.05	**2.85**	m	**3.07**
Iron or steel structural members of roof trusses, lattice girders, purlins and the like										
girth exceeding 300mm	0.30	10.00	**0.33**	0.59	-	**8.38**	0.16	**8.88**	m2	**9.54**
isolated surfaces, girth not exceeding 300mm.....	0.09	10.00	**0.10**	0.20	-	**2.84**	0.05	**3.00**	m	**3.22**

Labour hourly rates: (except Specialists) Craft Operatives £14.21 Labourer £10.71 Rates are national average prices. Refer to REGIONAL VARIATIONS for indicative levels of overall pricing in regions	MATERIALS			LABOUR				RATES		
	Del to Site £	Waste %	Material Cost £	Craft Optve Hrs	Lab Hrs	Labour Cost £	Sunds £	Nett Rate £	Unit	Gross Rate (+7.5%) £
M60: PAINTING/CLEAR FINISHING (REDECORATIONS) - OIL PAINTING METALWORK Cont.										
Two coats full gloss finish, existing gloss painted metal surfaces; external work Cont.										
Iron or steel railings, fences and gates; plain open type										
girth exceeding 300mm	0.23	10.00	**0.26**	0.44	-	**6.25**	0.12	**6.63**	m2	**7.13**
isolated surfaces, girth not exceeding 300mm	0.09	10.00	**0.10**	0.15	-	**2.13**	0.04	**2.27**	m	**2.45**
Iron or steel railings, fences and gates; close type										
girth exceeding 300mm	0.30	10.00	**0.33**	0.40	-	**5.68**	0.11	**6.13**	m2	**6.59**
Iron or steel railings, fences and gates; ornamental type										
girth exceeding 300mm	0.23	10.00	**0.26**	0.67	-	**9.52**	0.18	**9.96**	m2	**10.71**
Iron or steel eaves gutters										
girth exceeding 300mm	0.30	10.00	**0.33**	0.48	-	**6.82**	0.13	**7.29**	m2	**7.83**
isolated surfaces, girth not exceeding 300mm	0.09	10.00	**0.10**	0.16	-	**2.27**	0.04	**2.42**	m	**2.60**
Galvanized eaves gutters										
girth exceeding 300mm	0.30	10.00	**0.33**	0.48	-	**6.82**	0.13	**7.29**	m2	**7.83**
isolated surfaces, girth not exceeding 300mm	0.09	10.00	**0.10**	0.16	-	**2.27**	0.04	**2.42**	m	**2.60**
Iron or steel services										
girth exceeding 300mm	0.30	10.00	**0.33**	0.59	-	**8.38**	0.16	**8.88**	m2	**9.54**
isolated surfaces, girth not exceeding 300mm	0.09	10.00	**0.10**	0.20	-	**2.84**	0.05	**3.00**	m	**3.22**
isolated areas not exceeding 0.50m2 irrespective of girth	0.15	10.00	**0.17**	0.59	-	**8.38**	0.16	**8.71**	nr	**9.36**
Copper services										
girth exceeding 300mm	0.30	10.00	**0.33**	0.59	-	**8.38**	0.16	**8.88**	m2	**9.54**
isolated surfaces, girth not exceeding 300mm	0.09	10.00	**0.10**	0.20	-	**2.84**	0.05	**3.00**	m	**3.22**
Galvanized services										
girth exceeding 300mm	0.30	10.00	**0.33**	0.59	-	**8.38**	0.16	**8.88**	m2	**9.54**
isolated surfaces, girth not exceeding 300mm	0.09	10.00	**0.10**	0.20	-	**2.84**	0.05	**3.00**	m	**3.22**
Two undercoats, one coat full gloss finish, existing gloss painted metal surfaces										
Iron or steel general surfaces										
girth exceeding 300mm	0.46	10.00	**0.50**	0.56	-	**7.96**	0.15	**8.61**	m2	**9.26**
isolated surfaces, girth not exceeding 300mm	0.14	10.00	**0.15**	0.18	-	**2.56**	0.05	**2.76**	m	**2.97**
Galvanized glazed doors, windows or screens										
girth exceeding 300mm; panes, area not exceeding 0.10m2	0.25	10.00	**0.27**	1.03	-	**14.64**	0.28	**15.18**	m2	**16.32**
girth exceeding 300mm; panes, area 0.10 - 0.50m2	0.18	10.00	**0.19**	0.75	-	**10.66**	0.20	**11.05**	m2	**11.88**
girth exceeding 300mm; panes, area 0.50 - 1.00m2	0.14	10.00	**0.15**	0.63	-	**8.95**	0.17	**9.28**	m2	**9.97**
girth exceeding 300mm; panes, area exceeding 1.00m2	0.14	10.00	**0.15**	0.55	-	**7.82**	0.15	**8.12**	m2	**8.73**
Iron or steel structural work										
girth exceeding 300mm	0.46	10.00	**0.50**	0.60	-	**8.53**	0.16	**9.19**	m2	**9.88**
isolated surfaces, girth not exceeding 300mm	0.14	10.00	**0.15**	0.25	-	**3.55**	0.07	**3.77**	m	**4.06**
Iron or steel structural members of roof trusses, lattice girders, purlins and the like										
girth exceeding 300mm	0.46	10.00	**0.50**	0.77	-	**10.94**	0.21	**11.65**	m2	**12.53**
isolated surfaces, girth not exceeding 300mm	0.14	10.00	**0.15**	0.26	-	**3.69**	0.07	**3.92**	m	**4.21**
Iron or steel services										
girth exceeding 300mm	0.46	10.00	**0.50**	0.77	-	**10.94**	0.21	**11.65**	m2	**12.53**
isolated surfaces, girth not exceeding 300mm	0.14	10.00	**0.15**	0.26	-	**3.69**	0.07	**3.92**	m	**4.21**
isolated areas not exceeding 0.50m2 irrespective of girth	0.23	10.00	**0.26**	0.77	-	**10.94**	0.21	**11.41**	nr	**12.26**
Copper services										
girth exceeding 300mm	0.46	10.00	**0.50**	0.77	-	**10.94**	0.21	**11.65**	m2	**12.53**
isolated surfaces, girth not exceeding 300mm	0.14	10.00	**0.15**	0.26	-	**3.69**	0.07	**3.92**	m	**4.21**
Galvanized services										
girth exceeding 300mm	0.46	10.00	**0.50**	0.77	-	**10.94**	0.21	**11.65**	m2	**12.53**
isolated surfaces, girth not exceeding 300mm	0.14	10.00	**0.15**	0.26	-	**3.69**	0.07	**3.92**	m	**4.21**
Two undercoats, one coat full gloss finish, existing gloss painted metal surfaces; external work										
Iron or steel general surfaces										
girth exceeding 300mm	0.46	10.00	**0.50**	0.59	-	**8.38**	0.16	**9.05**	m2	**9.72**
isolated surfaces, girth not exceeding 300mm	0.14	10.00	**0.15**	0.20	-	**2.84**	0.05	**3.05**	m	**3.28**
Galvanized glazed doors, windows or screens										
girth exceeding 300mm; panes, area not exceeding 0.10m2	0.25	10.00	**0.27**	1.06	-	**15.06**	0.29	**15.62**	m2	**16.79**
girth exceeding 300mm; panes, area 0.10 - 0.50m2	0.18	10.00	**0.19**	0.78	-	**11.08**	0.21	**11.49**	m2	**12.35**
girth exceeding 300mm; panes, area 0.50 - 1.00m2	0.14	10.00	**0.15**	0.66	-	**9.38**	0.18	**9.71**	m2	**10.44**
girth exceeding 300mm; panes, area exceeding 1.00m2	0.14	10.00	**0.15**	0.58	-	**8.24**	0.16	**8.55**	m2	**9.19**
Iron or steel structural work										
girth exceeding 300mm	0.46	10.00	**0.50**	0.60	-	**8.53**	0.16	**9.19**	m2	**9.88**
isolated surfaces, girth not exceeding 300mm	0.14	10.00	**0.15**	0.25	-	**3.55**	0.07	**3.77**	m	**4.06**

Labour hourly rates: (except Specialists) Craft Operatives £14.21 Labourer £10.71 Rates are national average prices. Refer to REGIONAL VARIATIONS for indicative levels of overall pricing in regions	MATERIALS			LABOUR				RATES		
	Del to Site £	Waste %	Material Cost £	Craft Optve Hrs	Lab Hrs	Labour Cost £	Sunds £	Nett Rate £	Unit	Gross Rate (+7.5%) £
M60: PAINTING/CLEAR FINISHING (REDECORATIONS) - OIL PAINTING METALWORK Cont.										
Two undercoats, one coat full gloss finish, existing gloss painted metal surfaces; external work Cont.										
Iron or steel structural members of roof trusses, lattice girders, purlins and the like										
girth exceeding 300mm	0.46	10.00	**0.50**	0.77	-	**10.94**	0.21	**11.65**	m2	**12.53**
isolated surfaces, girth not exceeding 300mm	0.14	10.00	**0.15**	0.26	-	**3.69**	0.07	**3.92**	m	**4.21**
Iron or steel railings, fences and gates; plain open type										
girth exceeding 300mm	0.35	10.00	**0.39**	0.57	-	**8.10**	0.15	**8.64**	m2	**9.29**
isolated surfaces, girth not exceeding 300mm	0.14	10.00	**0.15**	0.19	-	**2.70**	0.05	**2.91**	m	**3.12**
Iron or steel railings, fences and gates; close type										
girth exceeding 300mm	0.46	10.00	**0.50**	0.51	-	**7.25**	0.14	**7.89**	m2	**8.48**
Iron or steel railings, fences and gates; ornamental type										
girth exceeding 300mm	0.35	10.00	**0.39**	0.97	-	**13.78**	0.26	**14.43**	m2	**15.51**
Iron or steel eaves gutters										
girth exceeding 300mm	0.46	10.00	**0.50**	0.62	-	**8.81**	0.17	**9.48**	m2	**10.19**
isolated surfaces, girth not exceeding 300mm	0.14	10.00	**0.15**	0.21	-	**2.98**	0.06	**3.20**	m	**3.43**
Galvanized eaves gutters										
girth exceeding 300mm	0.46	10.00	**0.50**	0.62	-	**8.81**	0.17	**9.48**	m2	**10.19**
isolated surfaces, girth not exceeding 300mm	0.14	10.00	**0.15**	0.21	-	**2.98**	0.06	**3.20**	m	**3.43**
Iron or steel services										
girth exceeding 300mm	0.46	10.00	**0.50**	0.77	-	**10.94**	0.21	**11.65**	m2	**12.53**
isolated surfaces, girth not exceeding 300mm	0.14	10.00	**0.15**	0.26	-	**3.69**	0.07	**3.92**	m	**4.21**
isolated areas not exceeding 0.50m2 irrespective of girth	0.23	10.00	**0.26**	0.77	-	**10.94**	0.21	**11.41**	nr	**12.26**
Copper services										
girth exceeding 300mm	0.46	10.00	**0.50**	0.77	-	**10.94**	0.21	**11.65**	m2	**12.53**
isolated surfaces, girth not exceeding 300mm	0.14	10.00	**0.15**	0.26	-	**3.69**	0.07	**3.92**	m	**4.21**
Galvanized services										
girth exceeding 300mm	0.46	10.00	**0.50**	0.77	-	**10.94**	0.21	**11.65**	m2	**12.53**
isolated surfaces, girth not exceeding 300mm	0.14	10.00	**0.15**	0.26	-	**3.69**	0.07	**3.92**	m	**4.21**
One undercoat, two coats full gloss finish, existing gloss painted metal surfaces										
Iron or steel general surfaces										
girth exceeding 300mm	0.46	10.00	**0.50**	0.56	-	**7.96**	0.15	**8.61**	m2	**9.26**
isolated surfaces, girth not exceeding 300mm	0.14	10.00	**0.15**	0.18	-	**2.56**	0.05	**2.76**	m	**2.97**
Galvanized glazed doors, windows or screens										
girth exceeding 300mm; panes, area not exceeding 0.10m2	0.25	10.00	**0.27**	1.03	-	**14.64**	0.28	**15.18**	m2	**16.32**
girth exceeding 300mm; panes, area 0.10 - 0.50m2	0.18	10.00	**0.19**	0.75	-	**10.66**	0.20	**11.05**	m2	**11.88**
girth exceeding 300mm; panes, area 0.50 - 1.00m2	0.14	10.00	**0.15**	0.63	-	**8.95**	0.17	**9.28**	m2	**9.97**
girth exceeding 300mm; panes, area exceeding 1.00m2	0.14	10.00	**0.15**	0.55	-	**7.82**	0.15	**8.12**	m2	**8.73**
Iron or steel structural work										
girth exceeding 300mm	0.46	10.00	**0.50**	0.60	-	**8.53**	0.16	**9.19**	m2	**9.88**
isolated surfaces, girth not exceeding 300mm	0.14	10.00	**0.15**	0.25	-	**3.55**	0.07	**3.77**	m	**4.06**
Iron or steel structural members of roof trusses, lattice girders, purlins and the like										
girth exceeding 300mm	0.46	10.00	**0.50**	0.77	-	**10.94**	0.21	**11.65**	m2	**12.53**
isolated surfaces, girth not exceeding 300mm	0.14	10.00	**0.15**	0.26	-	**3.69**	0.07	**3.92**	m	**4.21**
Iron or steel services										
girth exceeding 300mm	0.46	10.00	**0.50**	0.77	-	**10.94**	0.21	**11.65**	m2	**12.53**
isolated surfaces, girth not exceeding 300mm	0.14	10.00	**0.15**	0.26	-	**3.69**	0.07	**3.92**	m	**4.21**
isolated areas not exceeding 0.50m2 irrespective of girth	0.25	10.00	**0.27**	0.77	-	**10.94**	0.21	**11.42**	nr	**12.28**
Copper services										
girth exceeding 300mm	0.46	10.00	**0.50**	0.77	-	**10.94**	0.21	**11.65**	m2	**12.53**
isolated surfaces, girth not exceeding 300mm	0.14	10.00	**0.15**	0.26	-	**3.69**	0.07	**3.92**	m	**4.21**
Galvanized services										
girth exceeding 300mm	0.46	10.00	**0.50**	0.77	-	**10.94**	0.21	**11.65**	m2	**12.53**
isolated surfaces, girth not exceeding 300mm	0.14	10.00	**0.15**	0.26	-	**3.69**	0.07	**3.92**	m	**4.21**
One undercoat, two coats full gloss finish, existing gloss painted metal surfaces; external work										
Iron or steel general surfaces										
girth exceeding 300mm	0.46	10.00	**0.50**	0.59	-	**8.38**	0.16	**9.05**	m2	**9.72**
isolated surfaces, girth not exceeding 300mm	0.14	10.00	**0.15**	0.20	-	**2.84**	0.05	**3.05**	m	**3.28**
Galvanized glazed doors, windows or screens										
girth exceeding 300mm; panes, area not exceeding 0.10m2	0.25	10.00	**0.27**	1.06	-	**15.06**	0.29	**15.62**	m2	**16.79**
girth exceeding 300mm; panes, area 0.10 - 0.50m2	0.18	10.00	**0.19**	0.78	-	**11.08**	0.21	**11.49**	m2	**12.35**
girth exceeding 300mm; panes, area 0.50 - 1.00m2	0.14	10.00	**0.15**	0.66	-	**9.38**	0.18	**9.71**	m2	**10.44**

Labour hourly rates: (except Specialists) Craft Operatives £14.21 Labourer £10.71 Rates are national average prices. Refer to REGIONAL VARIATIONS for indicative levels of overall pricing in regions	MATERIALS Del to Site £	Waste %	Material Cost £	LABOUR Craft Optve Hrs	Lab Hrs	Labour Cost £	Sunds £	RATES Nett Rate £	Unit	Gross Rate (+7.5%) £
M60: PAINTING/CLEAR FINISHING (REDECORATIONS) - OIL PAINTING METALWORK Cont.										
One undercoat, two coats full gloss finish, existing gloss painted metal surfaces; external work Cont.										
Galvanized glazed doors, windows or screens Cont.										
girth exceeding 300mm; panes, area exceeding 1.00m2	0.14	10.00	**0.15**	0.58	-	**8.24**	0.16	**8.55**	m2	**9.19**
Iron or steel structural work										
girth exceeding 300mm	0.46	10.00	**0.50**	0.60	-	**8.53**	0.16	**9.19**	m2	**9.88**
isolated surfaces, girth not exceeding 300mm	0.14	10.00	**0.15**	0.25	-	**3.55**	0.07	**3.77**	m	**4.06**
Iron or steel structural members of roof trusses, lattice girders, purlins and the like										
girth exceeding 300mm	0.46	10.00	**0.50**	0.77	-	**10.94**	0.21	**11.65**	m2	**12.53**
isolated surfaces, girth not exceeding 300mm	0.14	10.00	**0.15**	0.26	-	**3.69**	0.07	**3.92**	m	**4.21**
Iron or steel railings, fences and gates; plain open type										
girth exceeding 300mm	0.35	10.00	**0.39**	0.57	-	**8.10**	0.15	**8.64**	m2	**9.29**
isolated surfaces, girth not exceeding 300mm	0.14	10.00	**0.15**	0.19	-	**2.70**	0.05	**2.91**	m	**3.12**
Iron or steel railings, fences and gates; close type										
girth exceeding 300mm	0.46	10.00	**0.50**	0.51	-	**7.25**	0.14	**7.89**	m2	**8.48**
Iron or steel railings, fences and gates; ornamental type										
girth exceeding 300mm	0.35	10.00	**0.39**	0.97	-	**13.78**	0.26	**14.43**	m2	**15.51**
Iron or steel eaves gutters										
girth exceeding 300mm	0.46	10.00	**0.50**	0.62	-	**8.81**	0.17	**9.48**	m2	**10.19**
isolated surfaces, girth not exceeding 300mm	0.14	10.00	**0.15**	0.21	-	**2.98**	0.06	**3.20**	m	**3.43**
Galvanized eaves gutters										
girth exceeding 300mm	0.46	10.00	**0.50**	0.62	-	**8.81**	0.17	**9.48**	m2	**10.19**
isolated surfaces, girth not exceeding 300mm	0.14	10.00	**0.15**	0.21	-	**2.98**	0.06	**3.20**	m	**3.43**
Iron or steel services										
girth exceeding 300mm	0.46	10.00	**0.50**	0.77	-	**10.94**	0.21	**11.65**	m2	**12.53**
isolated surfaces, girth not exceeding 300mm	0.14	10.00	**0.15**	0.26	-	**3.69**	0.07	**3.92**	m	**4.21**
isolated areas not exceeding 0.50m2 irrespective of girth	0.25	10.00	**0.27**	0.77	-	**10.94**	0.21	**11.42**	nr	**12.28**
Copper services										
girth exceeding 300mm	0.46	10.00	**0.50**	0.77	-	**10.94**	0.21	**11.65**	m2	**12.53**
isolated surfaces, girth not exceeding 300mm	0.14	10.00	**0.15**	0.26	-	**3.69**	0.07	**3.92**	m	**4.21**
Galvanized services										
girth exceeding 300mm	0.46	10.00	**0.50**	0.77	-	**10.94**	0.21	**11.65**	m2	**12.53**
isolated surfaces, girth not exceeding 300mm	0.14	10.00	**0.15**	0.26	-	**3.69**	0.07	**3.92**	m	**4.21**
M60: PAINTING/CLEAR FINISHING (REDECORATIONS) - OIL PAINTING WOODWORK										
Generally										
Note: the following rates include for the cost of all preparatory work, e.g. washing down, etc.										
One undercoat, one coat full gloss finish, existing gloss painted wood surfaces										
Wood general surfaces										
girth exceeding 300mm	0.30	10.00	**0.33**	0.41	-	**5.83**	0.11	**6.27**	m2	**6.74**
isolated surfaces, girth not exceeding 300mm	0.09	10.00	**0.10**	0.14	-	**1.99**	0.04	**2.13**	m	**2.29**
Wood glazed doors										
girth exceeding 300mm; panes, area not exceeding 0.10m2	0.16	10.00	**0.18**	0.72	-	**10.23**	0.19	**10.61**	m2	**11.40**
girth exceeding 300mm; panes, area 0.10 - 0.50m2	0.12	10.00	**0.13**	0.52	-	**7.39**	0.14	**7.66**	m2	**8.23**
girth exceeding 300mm; panes, area 0.50 - 1.00m2	0.12	10.00	**0.13**	0.43	-	**6.11**	0.12	**6.36**	m2	**6.83**
girth exceeding 300mm; panes, area exceeding 1.00m2	0.12	10.00	**0.13**	0.37	-	**5.26**	0.10	**5.49**	m2	**5.90**
Wood partially glazed doors										
girth exceeding 300mm; panes, area not exceeding 0.10m2	0.23	10.00	**0.26**	0.61	-	**8.67**	0.16	**9.09**	m2	**9.77**
girth exceeding 300mm; panes, area 0.10 - 0.50m2	0.19	10.00	**0.21**	0.49	-	**6.96**	0.13	**7.30**	m2	**7.85**
girth exceeding 300mm; panes, area 0.50 - 1.00m2	0.16	10.00	**0.18**	0.43	-	**6.11**	0.12	**6.41**	m2	**6.89**
girth exceeding 300mm; panes, area exceeding 1.00m2	0.16	10.00	**0.18**	0.39	-	**5.54**	0.11	**5.83**	m2	**6.26**
Wood windows and screens										
girth exceeding 300mm; panes, area not exceeding 0.10m2	0.23	10.00	**0.26**	0.77	-	**10.94**	0.21	**11.41**	m2	**12.26**
girth exceeding 300mm; panes, area 0.10 - 0.50m2	0.19	10.00	**0.21**	0.56	-	**7.96**	0.15	**8.31**	m2	**8.94**
girth exceeding 300mm; panes, area 0.50 - 1.00m2	0.16	10.00	**0.18**	0.46	-	**6.54**	0.12	**6.84**	m2	**7.35**
girth exceeding 300mm; panes, area exceeding 1.00m2	0.16	10.00	**0.18**	0.40	-	**5.68**	0.11	**5.97**	m2	**6.42**
One undercoat, one coat full gloss finish, existing gloss painted wood surfaces; external work										
Wood general surfaces										
girth exceeding 300mm	0.30	10.00	**0.33**	0.44	-	**6.25**	0.12	**6.71**	m2	**7.21**

Labour hourly rates: (except Specialists) Craft Operatives £14.21 Labourer £10.71 Rates are national average prices. Refer to REGIONAL VARIATIONS for indicative levels of overall pricing in regions	MATERIALS			LABOUR				RATES		
	Del to Site £	Waste %	Material Cost £	Craft Optve Hrs	Lab Hrs	Labour Cost £	Sunds £	Nett Rate £	Unit	Gross Rate (+7.5%) £
M60: PAINTING/CLEAR FINISHING (REDECORATIONS) - OIL PAINTING WOODWORK Cont.										
One undercoat, one coat full gloss finish, existing gloss painted wood surfaces; external work Cont.										
Wood general surfaces Cont.										
isolated surfaces, girth not exceeding 300mm.....	0.09	10.00	**0.10**	0.15	-	**2.13**	0.04	**2.27**	m	**2.45**
Wood glazed doors										
girth exceeding 300mm; panes, area not exceeding 0.10m2	0.16	10.00	**0.18**	0.73	-	**10.37**	0.20	**10.75**	m2	**11.56**
girth exceeding 300mm; panes, area 0.10 - 0.50m2 .	0.12	10.00	**0.13**	0.53	-	**7.53**	0.14	**7.80**	m2	**8.39**
girth exceeding 300mm; panes, area 0.50 - 1.00m2 .	0.12	10.00	**0.13**	0.44	-	**6.25**	0.12	**6.50**	m2	**6.99**
girth exceeding 300mm; panes, area exceeding 1.00m2	0.12	10.00	**0.13**	0.38	-	**5.40**	0.10	**5.63**	m2	**6.05**
Wood partially glazed doors										
girth exceeding 300mm; panes, area not exceeding 0.10m2	0.23	10.00	**0.26**	0.62	-	**8.81**	0.17	**9.24**	m2	**9.93**
girth exceeding 300mm; panes, area 0.10 - 0.50m2 .	0.19	10.00	**0.21**	0.50	-	**7.11**	0.14	**7.45**	m2	**8.00**
girth exceeding 300mm; panes, area 0.50 - 1.00m2 .	0.16	10.00	**0.18**	0.44	-	**6.25**	0.12	**6.55**	m2	**7.04**
girth exceeding 300mm; panes, area exceeding 1.00m2	0.16	10.00	**0.18**	0.40	-	**5.68**	0.11	**5.97**	m2	**6.42**
Wood windows and screens										
girth exceeding 300mm; panes, area not exceeding 0.10m2	0.23	10.00	**0.26**	0.80	-	**11.37**	0.22	**11.84**	m2	**12.73**
girth exceeding 300mm; panes, area 0.10 - 0.50m2 .	0.19	10.00	**0.21**	0.59	-	**8.38**	0.16	**8.75**	m2	**9.41**
girth exceeding 300mm; panes, area 0.50 - 1.00m2 .	0.16	10.00	**0.18**	0.49	-	**6.96**	0.13	**7.28**	m2	**7.82**
girth exceeding 300mm; panes, area exceeding 1.00m2	0.16	10.00	**0.18**	0.43	-	**6.11**	0.12	**6.41**	m2	**6.89**
Wood railings fences and gates; open type										
girth exceeding 300mm	0.23	10.00	**0.26**	0.37	-	**5.26**	0.10	**5.62**	m2	**6.04**
isolated surfaces, girth not exceeding 300mm.....	0.09	10.00	**0.10**	0.12	-	**1.71**	0.03	**1.84**	m	**1.98**
Wood railings fences and gates; close type										
girth exceeding 300mm	0.30	10.00	**0.33**	0.34	-	**4.83**	0.09	**5.26**	m2	**5.65**
Two coats full gloss finish, existing gloss painted wood surfaces										
Wood general surfaces										
girth exceeding 300mm	0.30	10.00	**0.33**	0.41	-	**5.83**	0.11	**6.27**	m2	**6.74**
isolated surfaces, girth not exceeding 300mm.....	0.09	10.00	**0.10**	0.14	-	**1.99**	0.04	**2.13**	m	**2.29**
Wood glazed doors										
girth exceeding 300mm; panes, area not exceeding 0.10m2	0.16	10.00	**0.18**	0.72	-	**10.23**	0.19	**10.61**	m2	**11.40**
girth exceeding 300mm; panes, area 0.10 - 0.50m2 .	0.12	10.00	**0.13**	0.52	-	**7.39**	0.14	**7.66**	m2	**8.23**
girth exceeding 300mm; panes, area 0.50 - 1.00m2 .	0.12	10.00	**0.13**	0.43	-	**6.11**	0.12	**6.36**	m2	**6.83**
girth exceeding 300mm; panes, area exceeding 1.00m2	0.12	10.00	**0.13**	0.37	-	**5.26**	0.10	**5.49**	m2	**5.90**
Wood partially glazed doors										
girth exceeding 300mm; panes, area not exceeding 0.10m2	0.23	10.00	**0.26**	0.61	-	**8.67**	0.16	**9.09**	m2	**9.77**
girth exceeding 300mm; panes, area 0.10 - 0.50m2 .	0.19	10.00	**0.21**	0.49	-	**6.96**	0.13	**7.30**	m2	**7.85**
girth exceeding 300mm; panes, area 0.50 - 1.00m2 .	0.16	10.00	**0.18**	0.43	-	**6.11**	0.12	**6.41**	m2	**6.89**
girth exceeding 300mm; panes, area exceeding 1.00m2	0.16	10.00	**0.18**	0.39	-	**5.54**	0.11	**5.83**	m2	**6.26**
Wood windows and screens										
girth exceeding 300mm; panes, area not exceeding 0.10m2	0.23	10.00	**0.26**	0.77	-	**10.94**	0.21	**11.41**	m2	**12.26**
girth exceeding 300mm; panes, area 0.10 - 0.50m2 .	0.19	10.00	**0.21**	0.56	-	**7.96**	0.15	**8.31**	m2	**8.94**
girth exceeding 300mm; panes, area 0.50 - 1.00m2 .	0.16	10.00	**0.18**	0.46	-	**6.54**	0.12	**6.84**	m2	**7.35**
girth exceeding 300mm; panes, area exceeding 1.00m2	0.16	10.00	**0.18**	0.40	-	**5.68**	0.11	**5.97**	m2	**6.42**
Two coats full gloss finish, existing gloss painted wood surfaces; external work										
Wood general surfaces										
girth exceeding 300mm	0.30	10.00	**0.33**	0.44	-	**6.25**	0.12	**6.71**	m2	**7.21**
isolated surfaces, girth not exceeding 300mm.....	0.09	10.00	**0.10**	0.15	-	**2.13**	0.04	**2.27**	m	**2.45**
Wood glazed doors										
girth exceeding 300mm; panes, area not exceeding 0.10m2	0.16	10.00	**0.18**	0.73	-	**10.37**	0.20	**10.75**	m2	**11.56**
girth exceeding 300mm; panes, area 0.10 - 0.50m2 .	0.12	10.00	**0.13**	0.53	-	**7.53**	0.14	**7.80**	m2	**8.39**
girth exceeding 300mm; panes, area 0.50 - 1.00m2 .	0.12	10.00	**0.13**	0.44	-	**6.25**	0.12	**6.50**	m2	**6.99**
girth exceeding 300mm; panes, area exceeding 1.00m2	0.12	10.00	**0.13**	0.38	-	**5.40**	0.10	**5.63**	m2	**6.05**
Wood partially glazed doors										
girth exceeding 300mm; panes, area not exceeding 0.10m2	0.23	10.00	**0.26**	0.62	-	**8.81**	0.17	**9.24**	m2	**9.93**
girth exceeding 300mm; panes, area 0.10 - 0.50m2 .	0.19	10.00	**0.21**	0.50	-	**7.11**	0.14	**7.45**	m2	**8.00**
girth exceeding 300mm; panes, area 0.50 - 1.00m2 .	0.16	10.00	**0.18**	0.44	-	**6.25**	0.12	**6.55**	m2	**7.04**
girth exceeding 300mm; panes, area exceeding 1.00m2	0.16	10.00	**0.18**	0.40	-	**5.68**	0.11	**5.97**	m2	**6.42**

Labour hourly rates: (except Specialists) Craft Operatives £14.21 Labourer £10.71 Rates are national average prices. Refer to REGIONAL VARIATIONS for indicative levels of overall pricing in regions	MATERIALS			LABOUR				RATES		
	Del to Site £	Waste %	Material Cost £	Craft Optve Hrs	Lab Hrs	Labour Cost £	Sunds £	Nett Rate £	Unit	Gross Rate (+7.5%) £
M60: PAINTING/CLEAR FINISHING (REDECORATIONS) - OIL PAINTING WOODWORK Cont.										
Two coats full gloss finish, existing gloss painted wood surfaces; external work Cont.										
Wood windows and screens										
girth exceeding 300mm; panes, area not exceeding 0.10m2	0.23	10.00	**0.26**	0.80	-	**11.37**	0.22	**11.84**	m2	**12.73**
girth exceeding 300mm; panes, area 0.10 - 0.50m2	0.19	10.00	**0.21**	0.59	-	**8.38**	0.16	**8.75**	m2	**9.41**
girth exceeding 300mm; panes, area 0.50 - 1.00m2	0.16	10.00	**0.18**	0.49	-	**6.96**	0.13	**7.28**	m2	**7.82**
girth exceeding 300mm; panes, area exceeding 1.00m2	0.16	10.00	**0.18**	0.43	-	**6.11**	0.12	**6.41**	m2	**6.89**
Wood railings fences and gates; open type										
girth exceeding 300mm	0.23	10.00	**0.26**	0.37	-	**5.26**	0.10	**5.62**	m2	**6.04**
isolated surfaces, girth not exceeding 300mm	0.09	10.00	**0.10**	0.12	-	**1.71**	0.03	**1.84**	m	**1.98**
Wood railings fences and gates; close type										
girth exceeding 300mm	0.30	10.00	**0.33**	0.34	-	**4.83**	0.09	**5.26**	m2	**5.65**
Two undercoats, one coat full gloss finish, existing gloss painted wood surfaces										
Wood general surfaces										
girth exceeding 300mm	0.46	10.00	**0.50**	0.55	-	**7.82**	0.15	**8.47**	m2	**9.10**
isolated surfaces, girth not exceeding 300mm	0.14	10.00	**0.15**	0.18	-	**2.56**	0.05	**2.76**	m	**2.97**
Wood glazed doors										
girth exceeding 300mm; panes, area not exceeding 0.10m2	0.25	10.00	**0.27**	0.94	-	**13.36**	0.25	**13.88**	m2	**14.92**
girth exceeding 300mm; panes, area 0.10 - 0.50m2	0.18	10.00	**0.19**	0.69	-	**9.80**	0.19	**10.18**	m2	**10.95**
girth exceeding 300mm; panes, area 0.50 - 1.00m2	0.18	10.00	**0.19**	0.57	-	**8.10**	0.15	**8.45**	m2	**9.08**
girth exceeding 300mm; panes, area exceeding 1.00m2	0.18	10.00	**0.19**	0.49	-	**6.96**	0.13	**7.29**	m2	**7.83**
Wood partially glazed doors										
girth exceeding 300mm; panes, area not exceeding 0.10m2	0.35	10.00	**0.39**	0.79	-	**11.23**	0.21	**11.83**	m2	**12.71**
girth exceeding 300mm; panes, area 0.10 - 0.50m2	0.28	10.00	**0.31**	0.65	-	**9.24**	0.18	**9.72**	m2	**10.45**
girth exceeding 300mm; panes, area 0.50 - 1.00m2	0.25	10.00	**0.27**	0.57	-	**8.10**	0.15	**8.52**	m2	**9.16**
girth exceeding 300mm; panes, area exceeding 1.00m2	0.25	10.00	**0.27**	0.52	-	**7.39**	0.14	**7.80**	m2	**8.38**
Wood windows and screens										
girth exceeding 300mm; panes, area not exceeding 0.10m2	0.35	10.00	**0.39**	1.02	-	**14.49**	0.28	**15.16**	m2	**16.29**
girth exceeding 300mm; panes, area 0.10 - 0.50m2	0.28	10.00	**0.31**	0.75	-	**10.66**	0.20	**11.17**	m2	**12.01**
girth exceeding 300mm; panes, area 0.50 - 1.00m2	0.25	10.00	**0.27**	0.62	-	**8.81**	0.17	**9.25**	m2	**9.94**
girth exceeding 300mm; panes, area exceeding 1.00m2	0.25	10.00	**0.27**	0.53	-	**7.53**	0.14	**7.94**	m2	**8.54**
Two undercoats, one coat full gloss finish, existing gloss painted wood surfaces; external work										
Wood general surfaces										
girth exceeding 300mm	0.46	10.00	**0.50**	0.59	-	**8.38**	0.16	**9.05**	m2	**9.72**
isolated surfaces, girth not exceeding 300mm	0.14	10.00	**0.15**	0.20	-	**2.84**	0.05	**3.05**	m	**3.28**
Wood glazed doors										
girth exceeding 300mm; panes, area not exceeding 0.10m2	0.25	10.00	**0.27**	0.95	-	**13.50**	0.26	**14.03**	m2	**15.08**
girth exceeding 300mm; panes, area 0.10 - 0.50m2	0.18	10.00	**0.19**	0.70	-	**9.95**	0.19	**10.33**	m2	**11.10**
girth exceeding 300mm; panes, area 0.50 - 1.00m2	0.18	10.00	**0.19**	0.58	-	**8.24**	0.16	**8.59**	m2	**9.24**
girth exceeding 300mm; panes, area exceeding 1.00m2	0.18	10.00	**0.19**	0.50	-	**7.11**	0.14	**7.43**	m2	**7.99**
Wood partially glazed doors										
girth exceeding 300mm; panes, area not exceeding 0.10m2	0.35	10.00	**0.39**	0.80	-	**11.37**	0.22	**11.97**	m2	**12.87**
girth exceeding 300mm; panes, area 0.10 - 0.50m2	0.28	10.00	**0.31**	0.66	-	**9.38**	0.18	**9.87**	m2	**10.61**
girth exceeding 300mm; panes, area 0.50 - 1.00m2	0.25	10.00	**0.27**	0.58	-	**8.24**	0.16	**8.67**	m2	**9.32**
girth exceeding 300mm; panes, area exceeding 1.00m2	0.25	10.00	**0.27**	0.53	-	**7.53**	0.14	**7.94**	m2	**8.54**
Wood windows and screens										
girth exceeding 300mm; panes, area not exceeding 0.10m2	0.35	10.00	**0.39**	1.05	-	**14.92**	0.28	**15.59**	m2	**16.76**
girth exceeding 300mm; panes, area 0.10 - 0.50m2	0.28	10.00	**0.31**	0.79	-	**11.23**	0.21	**11.75**	m2	**12.63**
girth exceeding 300mm; panes, area 0.50 - 1.00m2	0.25	10.00	**0.27**	0.68	-	**9.66**	0.18	**10.12**	m2	**10.88**
girth exceeding 300mm; panes, area exceeding 1.00m2	0.25	10.00	**0.27**	0.57	-	**8.10**	0.15	**8.52**	m2	**9.16**
Wood railings fences and gates; open type										
girth exceeding 300mm	0.35	10.00	**0.39**	0.50	-	**7.11**	0.14	**7.63**	m2	**8.20**
isolated surfaces, girth not exceeding 300mm	0.14	10.00	**0.15**	0.17	-	**2.42**	0.05	**2.62**	m	**2.81**
Wood railings fences and gates; close type										
girth exceeding 300mm	0.46	10.00	**0.50**	0.45	-	**6.39**	0.12	**7.02**	m2	**7.54**
One undercoat, two coats full gloss finish, existing gloss painted wood surfaces										
Wood general surfaces										
girth exceeding 300mm	0.46	10.00	**0.50**	0.57	-	**8.10**	0.15	**8.76**	m2	**9.41**
isolated surfaces, girth not exceeding 300mm	0.14	10.00	**0.15**	0.18	-	**2.56**	0.05	**2.76**	m	**2.97**

Labour hourly rates: (except Specialists) Craft Operatives £14.21 Labourer £10.71 Rates are national average prices. Refer to REGIONAL VARIATIONS for indicative levels of overall pricing in regions	MATERIALS			LABOUR				RATES		
	Del to Site £	Waste %	Material Cost £	Craft Optve Hrs	Lab Hrs	Labour Cost £	Sunds £	Nett Rate £	Unit	Gross Rate (+7.5%) £
M60: PAINTING/CLEAR FINISHING (REDECORATIONS) - OIL PAINTING WOODWORK Cont.										
One undercoat, two coats full gloss finish, existing gloss painted wood surfaces Cont.										
Wood glazed doors										
girth exceeding 300mm; panes, area not exceeding 0.10m2	0.25	10.00	**0.27**	0.94	-	**13.36**	0.25	**13.88**	m2	**14.92**
girth exceeding 300mm; panes, area 0.10 - 0.50m2	0.18	10.00	**0.19**	0.69	-	**9.80**	0.19	**10.18**	m2	**10.95**
girth exceeding 300mm; panes, area 0.50 - 1.00m2	0.18	10.00	**0.19**	0.57	-	**8.10**	0.15	**8.45**	m2	**9.08**
girth exceeding 300mm; panes, area exceeding 1.00m2	0.18	10.00	**0.19**	0.49	-	**6.96**	0.13	**7.29**	m2	**7.83**
Wood partially glazed doors										
girth exceeding 300mm; panes, area not exceeding 0.10m2	0.35	10.00	**0.39**	0.79	-	**11.23**	0.21	**11.83**	m2	**12.71**
girth exceeding 300mm; panes, area 0.10 - 0.50m2	0.28	10.00	**0.31**	0.65	-	**9.24**	0.18	**9.72**	m2	**10.45**
girth exceeding 300mm; panes, area 0.50 - 1.00m2	0.25	10.00	**0.27**	0.57	-	**8.10**	0.15	**8.52**	m2	**9.16**
girth exceeding 300mm; panes, area exceeding 1.00m2	0.25	10.00	**0.27**	0.52	-	**7.39**	0.14	**7.80**	m2	**8.38**
Wood windows and screens										
girth exceeding 300mm; panes, area not exceeding 0.10m2	0.35	10.00	**0.39**	1.02	-	**14.49**	0.28	**15.16**	m2	**16.29**
girth exceeding 300mm; panes, area 0.10 - 0.50m2	0.28	10.00	**0.31**	0.75	-	**10.66**	0.20	**11.17**	m2	**12.01**
girth exceeding 300mm; panes, area 0.50 - 1.00m2	0.25	10.00	**0.27**	0.62	-	**8.81**	0.17	**9.25**	m2	**9.94**
girth exceeding 300mm; panes, area exceeding 1.00m2	0.25	10.00	**0.27**	0.53	-	**7.53**	0.14	**7.94**	m2	**8.54**
One undercoat, two coats full gloss finish, existing gloss painted wood surfaces; external work										
Wood general surfaces										
girth exceeding 300mm	0.46	10.00	**0.50**	0.59	-	**8.38**	0.16	**9.05**	m2	**9.72**
isolated surfaces, girth not exceeding 300mm	0.14	10.00	**0.15**	0.20	-	**2.84**	0.05	**3.05**	m	**3.28**
Wood glazed doors										
girth exceeding 300mm; panes, area not exceeding 0.10m2	0.25	10.00	**0.27**	0.95	-	**13.50**	0.26	**14.03**	m2	**15.08**
girth exceeding 300mm; panes, area 0.10 - 0.50m2	0.18	10.00	**0.19**	0.70	-	**9.95**	0.19	**10.33**	m2	**11.10**
girth exceeding 300mm; panes, area 0.50 - 1.00m2	0.18	10.00	**0.19**	0.58	-	**8.24**	0.16	**8.59**	m2	**9.24**
girth exceeding 300mm; panes, area exceeding 1.00m2	0.18	10.00	**0.19**	0.50	-	**7.11**	0.14	**7.43**	m2	**7.99**
Wood partially glazed doors										
girth exceeding 300mm; panes, area not exceeding 0.10m2	0.35	10.00	**0.39**	0.80	-	**11.37**	0.22	**11.97**	m2	**12.87**
girth exceeding 300mm; panes, area 0.10 - 0.50m2	0.28	10.00	**0.31**	0.66	-	**9.38**	0.18	**9.87**	m2	**10.61**
girth exceeding 300mm; panes, area 0.50 - 1.00m2	0.25	10.00	**0.27**	0.58	-	**8.24**	0.16	**8.67**	m2	**9.32**
girth exceeding 300mm; panes, area exceeding 1.00m2	0.25	10.00	**0.27**	0.53	-	**7.53**	0.14	**7.94**	m2	**8.54**
Wood windows and screens										
girth exceeding 300mm; panes, area not exceeding 0.10m2	0.35	10.00	**0.39**	1.06	-	**15.06**	0.29	**15.73**	m2	**16.92**
girth exceeding 300mm; panes, area 0.10 - 0.50m2	0.28	10.00	**0.31**	0.79	-	**11.23**	0.21	**11.75**	m2	**12.63**
girth exceeding 300mm; panes, area 0.50 - 1.00m2	0.25	10.00	**0.27**	0.66	-	**9.38**	0.18	**9.83**	m2	**10.56**
girth exceeding 300mm; panes, area exceeding 1.00m2	0.25	10.00	**0.27**	0.57	-	**8.10**	0.15	**8.52**	m2	**9.16**
Wood railings fences and gates; open type										
girth exceeding 300mm	0.35	10.00	**0.39**	0.50	-	**7.11**	0.14	**7.63**	m2	**8.20**
isolated surfaces, girth not exceeding 300mm	0.14	10.00	**0.15**	0.17	-	**2.42**	0.05	**2.62**	m	**2.81**
Wood railings fences and gates; close type										
girth exceeding 300mm	0.46	10.00	**0.50**	0.45	-	**6.39**	0.12	**7.02**	m2	**7.54**
Burn off, one coat primer, one undercoat, one coat full gloss finish, existing painted wood surfaces										
Wood general surfaces										
girth exceeding 300mm	0.54	10.00	**0.59**	0.82	-	**11.65**	0.55	**12.79**	m2	**13.75**
isolated surfaces, girth not exceeding 300mm	0.16	10.00	**0.18**	0.24	-	**3.41**	0.16	**3.75**	m	**4.03**
Wood glazed doors										
girth exceeding 300mm; panes, area not exceeding 0.10m2	0.30	10.00	**0.33**	1.36	-	**19.33**	0.91	**20.56**	m2	**22.10**
girth exceeding 300mm; panes, area 0.10 - 0.50m2	0.22	10.00	**0.24**	0.98	-	**13.93**	0.66	**14.82**	m2	**15.93**
girth exceeding 300mm; panes, area 0.50 - 1.00m2	0.20	10.00	**0.22**	0.82	-	**11.65**	0.55	**12.42**	m2	**13.35**
girth exceeding 300mm; panes, area exceeding 1.00m2	0.20	10.00	**0.22**	0.73	-	**10.37**	0.49	**11.08**	m2	**11.91**
Wood partially glazed doors										
girth exceeding 300mm; panes, area not exceeding 0.10m2	0.42	10.00	**0.46**	1.09	-	**15.49**	0.73	**16.68**	m2	**17.93**
girth exceeding 300mm; panes, area 0.10 - 0.50m2	0.32	10.00	**0.35**	0.90	-	**12.79**	0.60	**13.74**	m2	**14.77**
girth exceeding 300mm; panes, area 0.50 - 1.00m2	0.30	10.00	**0.33**	0.82	-	**11.65**	0.55	**12.53**	m2	**13.47**
girth exceeding 300mm; panes, area exceeding 1.00m2	0.30	10.00	**0.33**	0.78	-	**11.08**	0.52	**11.93**	m2	**12.83**
Wood windows and screens										
girth exceeding 300mm; panes, area not exceeding 0.10m2	0.42	10.00	**0.46**	1.50	-	**21.32**	1.01	**22.78**	m2	**24.49**
girth exceeding 300mm; panes, area 0.10 - 0.50m2	0.32	10.00	**0.35**	1.07	-	**15.20**	0.72	**16.27**	m2	**17.49**
girth exceeding 300mm; panes, area 0.50 - 1.00m2	0.30	10.00	**0.33**	0.90	-	**12.79**	0.60	**13.72**	m2	**14.75**

Labour hourly rates: (except Specialists) Craft Operatives £14.21 Labourer £10.71 Rates are national average prices. Refer to REGIONAL VARIATIONS for indicative levels of overall pricing in regions	MATERIALS			LABOUR				RATES		
	Del to Site £	Waste %	Material Cost £	Craft Optve Hrs	Lab Hrs	Labour Cost £	Sunds £	Nett Rate £	Unit	Gross Rate (+7.5%) £
M60: PAINTING/CLEAR FINISHING (REDECORATIONS) - OIL PAINTING WOODWORK Cont.										
Burn off, one coat primer, one undercoat, one coat full gloss finish, existing painted wood surfaces Cont.										
Wood windows and screens Cont.										
girth exceeding 300mm; panes, area exceeding 1.00m2	0.30	10.00	**0.33**	0.80	-	**11.37**	0.54	**12.23**	m2	**13.15**
Burn off, one coat primer, one undercoat, one coat full gloss finish, existing painted wood surfaces; external work										
Wood general surfaces										
girth exceeding 300mm	0.54	10.00	**0.59**	0.86	-	**12.22**	0.58	**13.39**	m2	**14.39**
isolated surfaces, girth not exceeding 300mm	0.16	10.00	**0.18**	0.26	-	**3.69**	0.17	**4.04**	m	**4.35**
Wood glazed doors										
girth exceeding 300mm; panes, area not exceeding 0.10m2	0.30	10.00	**0.33**	1.36	-	**19.33**	0.91	**20.56**	m2	**22.10**
girth exceeding 300mm; panes, area 0.10 - 0.50m2	0.22	10.00	**0.24**	0.98	-	**13.93**	0.66	**14.82**	m2	**15.93**
girth exceeding 300mm; panes, area 0.50 - 1.00m2	0.20	10.00	**0.22**	0.82	-	**11.65**	0.55	**12.42**	m2	**13.35**
girth exceeding 300mm; panes, area exceeding 1.00m2	0.20	10.00	**0.22**	0.73	-	**10.37**	0.49	**11.08**	m2	**11.91**
Wood partially glazed doors										
girth exceeding 300mm; panes, area not exceeding 0.10m2	0.42	10.00	**0.46**	1.09	-	**15.49**	0.73	**16.68**	m2	**17.93**
girth exceeding 300mm; panes, area 0.10 - 0.50m2	0.32	10.00	**0.35**	0.90	-	**12.79**	0.60	**13.74**	m2	**14.77**
girth exceeding 300mm; panes, area 0.50 - 1.00m2	0.30	10.00	**0.33**	0.82	-	**11.65**	0.55	**12.53**	m2	**13.47**
girth exceeding 300mm; panes, area exceeding 1.00m2	0.30	10.00	**0.33**	0.78	-	**11.08**	0.52	**11.93**	m2	**12.83**
Wood windows and screens										
girth exceeding 300mm; panes, area not exceeding 0.10m2	0.42	10.00	**0.46**	1.54	-	**21.88**	1.03	**23.37**	m2	**25.13**
girth exceeding 300mm; panes, area 0.10 - 0.50m2	0.32	10.00	**0.35**	1.11	-	**15.77**	0.74	**16.87**	m2	**18.13**
girth exceeding 300mm; panes, area 0.50 - 1.00m2	0.30	10.00	**0.33**	0.94	-	**13.36**	0.63	**14.31**	m2	**15.39**
girth exceeding 300mm; panes, area exceeding 1.00m2	0.30	10.00	**0.33**	0.84	-	**11.94**	0.56	**12.83**	m2	**13.79**
Wood railings fences and gates; open type										
girth exceeding 300mm	0.42	10.00	**0.46**	0.74	-	**10.52**	0.50	**11.47**	m2	**12.33**
isolated surfaces, girth not exceeding 300mm	0.16	10.00	**0.18**	0.22	-	**3.13**	0.15	**3.45**	m	**3.71**
Wood railings fences and gates; close type										
girth exceeding 300mm	0.54	10.00	**0.59**	0.69	-	**9.80**	0.46	**10.86**	m2	**11.67**
Burn off, one coat primer, two undercoats, one coat full gloss finish, existing painted wood surfaces										
Wood general surfaces										
girth exceeding 300mm	0.69	10.00	**0.76**	0.96	-	**13.64**	0.64	**15.04**	m2	**16.17**
isolated surfaces, girth not exceeding 300mm	0.21	10.00	**0.23**	0.29	-	**4.12**	0.19	**4.54**	m	**4.88**
Wood glazed doors										
girth exceeding 300mm; panes, area not exceeding 0.10m2	0.38	10.00	**0.42**	1.60	-	**22.74**	1.07	**24.22**	m2	**26.04**
girth exceeding 300mm; panes, area 0.10 - 0.50m2	0.27	10.00	**0.30**	1.15	-	**16.34**	0.77	**17.41**	m2	**18.72**
girth exceeding 300mm; panes, area 0.50 - 1.00m2	0.26	10.00	**0.28**	0.96	-	**13.64**	0.64	**14.57**	m2	**15.66**
girth exceeding 300mm; panes, area exceeding 1.00m2	0.26	10.00	**0.28**	0.86	-	**12.22**	0.58	**13.08**	m2	**14.06**
Wood partially glazed doors										
girth exceeding 300mm; panes, area not exceeding 0.10m2	0.53	10.00	**0.59**	1.28	-	**18.19**	0.86	**19.63**	m2	**21.11**
girth exceeding 300mm; panes, area 0.10 - 0.50m2	0.41	10.00	**0.45**	1.06	-	**15.06**	0.71	**16.23**	m2	**17.44**
girth exceeding 300mm; panes, area 0.50 - 1.00m2	0.38	10.00	**0.42**	0.96	-	**13.64**	0.64	**14.70**	m2	**15.80**
girth exceeding 300mm; panes, area exceeding 1.00m2	0.38	10.00	**0.42**	0.91	-	**12.93**	0.61	**13.96**	m2	**15.00**
Wood windows and screens										
girth exceeding 300mm; panes, area not exceeding 0.10m2	0.53	10.00	**0.59**	1.76	-	**25.01**	1.18	**26.78**	m2	**28.78**
girth exceeding 300mm; panes, area 0.10 - 0.50m2	0.41	10.00	**0.45**	1.27	-	**18.05**	0.85	**19.35**	m2	**20.80**
girth exceeding 300mm; panes, area 0.50 - 1.00m2	0.38	10.00	**0.42**	1.06	-	**15.06**	0.71	**16.19**	m2	**17.40**
girth exceeding 300mm; panes, area exceeding 1.00m2	0.38	10.00	**0.42**	0.94	-	**13.36**	0.63	**14.40**	m2	**15.48**
Burn off, one coat primer, two undercoats, one coat full gloss finish, existing painted wood surfaces; external work										
Wood general surfaces										
girth exceeding 300mm	0.69	10.00	**0.76**	1.01	-	**14.35**	0.68	**15.79**	m2	**16.97**
isolated surfaces, girth not exceeding 300mm	0.21	10.00	**0.23**	0.32	-	**4.55**	0.21	**4.99**	m	**5.36**
Wood glazed doors										
girth exceeding 300mm; panes, area not exceeding 0.10m2	0.38	10.00	**0.42**	1.60	-	**22.74**	1.07	**24.22**	m2	**26.04**
girth exceeding 300mm; panes, area 0.10 - 0.50m2	0.27	10.00	**0.30**	1.15	-	**16.34**	0.77	**17.41**	m2	**18.72**
girth exceeding 300mm; panes, area 0.50 - 1.00m2	0.26	10.00	**0.28**	0.96	-	**13.64**	0.64	**14.57**	m2	**15.66**
girth exceeding 300mm; panes, area exceeding 1.00m2	0.26	10.00	**0.28**	0.86	-	**12.22**	0.58	**13.08**	m2	**14.06**

Labour hourly rates: (except Specialists) Craft Operatives £14.21 Labourer £10.71 Rates are national average prices. Refer to REGIONAL VARIATIONS for indicative levels of overall pricing in regions	MATERIALS			LABOUR				RATES		
	Del to Site £	Waste %	Material Cost £	Craft Optve Hrs	Lab Hrs	Labour Cost £	Sunds £	Nett Rate £	Unit	Gross Rate (+7.5%) £
M60: PAINTING/CLEAR FINISHING (REDECORATIONS) - OIL PAINTING WOODWORK Cont.										
Burn off, one coat primer, two undercoats, one coat full gloss finish, existing painted wood surfaces; external work Cont.										
Wood partially glazed doors										
girth exceeding 300mm; panes, area not exceeding 0.10m2	0.53	10.00	**0.59**	1.28	-	**18.19**	0.86	**19.63**	m2	**21.11**
girth exceeding 300mm; panes, area 0.10 - 0.50m2	0.41	10.00	**0.45**	1.06	-	**15.06**	0.71	**16.23**	m2	**17.44**
girth exceeding 300mm; panes, area 0.50 - 1.00m2	0.38	10.00	**0.42**	0.96	-	**13.64**	0.64	**14.70**	m2	**15.80**
girth exceeding 300mm; panes, area exceeding 1.00m2	0.38	10.00	**0.42**	0.91	-	**12.93**	0.61	**13.96**	m2	**15.00**
Wood windows and screens										
girth exceeding 300mm; panes, area not exceeding 0.10m2	0.53	10.00	**0.59**	1.81	-	**25.72**	1.21	**27.52**	m2	**29.58**
girth exceeding 300mm; panes, area 0.10 - 0.50m2	0.41	10.00	**0.45**	1.32	-	**18.76**	0.88	**20.10**	m2	**21.60**
girth exceeding 300mm; panes, area 0.50 - 1.00m2	0.38	10.00	**0.42**	1.11	-	**15.77**	0.74	**16.93**	m2	**18.20**
girth exceeding 300mm; panes, area exceeding 1.00m2	0.38	10.00	**0.42**	0.99	-	**14.07**	0.66	**15.15**	m2	**16.28**
Wood railings fences and gates; open type										
girth exceeding 300mm	0.53	10.00	**0.59**	0.86	-	**12.22**	0.58	**13.38**	m2	**14.39**
isolated surfaces, girth not exceeding 300mm	0.21	10.00	**0.23**	0.26	-	**3.69**	0.17	**4.10**	m	**4.40**
Wood railings fences and gates; close type										
girth exceeding 300mm	0.69	10.00	**0.76**	0.80	-	**11.37**	0.54	**12.66**	m2	**13.61**
Burn off, one coat primer, one undercoat, two coats full gloss finish, existing painted wood surfaces										
Wood general surfaces										
girth exceeding 300mm	0.69	10.00	**0.76**	0.96	-	**13.64**	0.64	**15.04**	m2	**16.17**
isolated surfaces, girth not exceeding 300mm	0.21	10.00	**0.23**	0.29	-	**4.12**	0.19	**4.54**	m	**4.88**
Wood glazed doors										
girth exceeding 300mm; panes, area not exceeding 0.10m2	0.38	10.00	**0.42**	1.60	-	**22.74**	1.07	**24.22**	m2	**26.04**
girth exceeding 300mm; panes, area 0.10 - 0.50m2	0.27	10.00	**0.30**	1.15	-	**16.34**	0.77	**17.41**	m2	**18.72**
girth exceeding 300mm; panes, area 0.50 - 1.00m2	0.26	10.00	**0.28**	0.96	-	**13.64**	0.64	**14.57**	m2	**15.66**
girth exceeding 300mm; panes, area exceeding 1.00m2	0.26	10.00	**0.28**	0.86	-	**12.22**	0.58	**13.08**	m2	**14.06**
Wood partially glazed doors										
girth exceeding 300mm; panes, area not exceeding 0.10m2	0.53	10.00	**0.59**	1.28	-	**18.19**	0.86	**19.63**	m2	**21.11**
girth exceeding 300mm; panes, area 0.10 - 0.50m2	0.41	10.00	**0.45**	1.06	-	**15.06**	0.71	**16.23**	m2	**17.44**
girth exceeding 300mm; panes, area 0.50 - 1.00m2	0.38	10.00	**0.42**	0.96	-	**13.64**	0.64	**14.70**	m2	**15.80**
girth exceeding 300mm; panes, area exceeding 1.00m2	0.38	10.00	**0.42**	0.91	-	**12.93**	0.61	**13.96**	m2	**15.00**
Wood windows and screens										
girth exceeding 300mm; panes, area not exceeding 0.10m2	0.53	10.00	**0.59**	1.76	-	**25.01**	1.18	**26.78**	m2	**28.78**
girth exceeding 300mm; panes, area 0.10 - 0.50m2	0.41	10.00	**0.45**	1.27	-	**18.05**	0.85	**19.35**	m2	**20.80**
girth exceeding 300mm; panes, area 0.50 - 1.00m2	0.38	10.00	**0.42**	1.06	-	**15.06**	0.71	**16.19**	m2	**17.40**
girth exceeding 300mm; panes, area exceeding 1.00m2	0.38	10.00	**0.42**	0.94	-	**13.36**	0.63	**14.40**	m2	**15.48**
Burn off, one coat primer, one undercoat, two coats full gloss finish, existing painted wood surfaces; external work										
Wood general surfaces										
girth exceeding 300mm	0.69	10.00	**0.76**	1.01	-	**14.35**	0.68	**15.79**	m2	**16.97**
isolated surfaces, girth not exceeding 300mm	0.21	10.00	**0.23**	0.32	-	**4.55**	0.21	**4.99**	m	**5.36**
Wood glazed doors										
girth exceeding 300mm; panes, area not exceeding 0.10m2	0.38	10.00	**0.42**	1.60	-	**22.74**	1.07	**24.22**	m2	**26.04**
girth exceeding 300mm; panes, area 0.10 - 0.50m2	0.27	10.00	**0.30**	1.15	-	**16.34**	0.77	**17.41**	m2	**18.72**
girth exceeding 300mm; panes, area 0.50 - 1.00m2	0.26	10.00	**0.28**	0.96	-	**13.64**	0.64	**14.57**	m2	**15.66**
girth exceeding 300mm; panes, area exceeding 1.00m2	0.26	10.00	**0.28**	0.86	-	**12.22**	0.58	**13.08**	m2	**14.06**
Wood partially glazed doors										
girth exceeding 300mm; panes, area not exceeding 0.10m2	0.53	10.00	**0.59**	1.28	-	**18.19**	0.86	**19.63**	m2	**21.11**
girth exceeding 300mm; panes, area 0.10 - 0.50m2	0.41	10.00	**0.45**	1.06	-	**15.06**	0.71	**16.23**	m2	**17.44**
girth exceeding 300mm; panes, area 0.50 - 1.00m2	0.38	10.00	**0.42**	0.96	-	**13.64**	0.64	**14.70**	m2	**15.80**
girth exceeding 300mm; panes, area exceeding 1.00m2	0.38	10.00	**0.42**	0.91	-	**12.93**	0.61	**13.96**	m2	**15.00**
Wood windows and screens										
girth exceeding 300mm; panes, area not exceeding 0.10m2	0.53	10.00	**0.59**	1.81	-	**25.72**	1.21	**27.52**	m2	**29.58**
girth exceeding 300mm; panes, area 0.10 - 0.50m2	0.41	10.00	**0.45**	1.32	-	**18.76**	0.88	**20.10**	m2	**21.60**
girth exceeding 300mm; panes, area 0.50 - 1.00m2	0.38	10.00	**0.42**	1.11	-	**15.77**	0.74	**16.93**	m2	**18.20**
girth exceeding 300mm; panes, area exceeding 1.00m2	0.38	10.00	**0.42**	0.99	-	**14.07**	0.66	**15.15**	m2	**16.28**
Wood railings fences and gates; open type										
girth exceeding 300mm	0.53	10.00	**0.59**	0.86	-	**12.22**	0.58	**13.38**	m2	**14.39**
isolated surfaces, girth not exceeding 300mm	0.21	10.00	**0.23**	0.26	-	**3.69**	0.17	**4.10**	m	**4.40**

Labour hourly rates: (except Specialists) Craft Operatives £14.21 Labourer £10.71 Rates are national average prices. Refer to REGIONAL VARIATIONS for indicative levels of overall pricing in regions	MATERIALS Del to Site £	Waste %	Material Cost £	LABOUR Craft Optve Hrs	Lab Hrs	Labour Cost £	Sunds £	RATES Nett Rate £	Unit	Gross Rate (+7.5%) £
M60: PAINTING/CLEAR FINISHING (REDECORATIONS) - OIL PAINTING WOODWORK Cont.										
Burn off, one coat primer, one undercoat, two coats full gloss finish, existing painted wood surfaces; external work Cont.										
Wood railings fences and gates; close type										
girth exceeding 300mm	0.69	10.00	**0.76**	0.80	-	**11.37**	0.54	**12.66**	m2	**13.61**
M60: PAINTING/CLEAR FINISHING (REDECORATIONS) - POLYURETHANE LACQUER										
Generally										
Note: the following rates include for the cost of all preparatory work, e.g. washing down, etc.										
Two coats polyurethane lacquer, existing lacquered surfaces										
Wood general surfaces										
girth exceeding 300mm	0.77	10.00	**0.85**	0.49	-	**6.96**	0.13	**7.94**	m2	**8.54**
isolated surfaces, girth not exceeding 300mm	0.23	10.00	**0.25**	0.16	-	**2.27**	0.04	**2.57**	m	**2.76**
Wood glazed doors										
girth exceeding 300mm; panes, area not exceeding 0.10m2	0.44	10.00	**0.48**	0.82	-	**11.65**	0.22	**12.35**	m2	**13.28**
girth exceeding 300mm; panes, area 0.10 - 0.50m2	0.31	10.00	**0.34**	0.60	-	**8.53**	0.16	**9.03**	m2	**9.70**
girth exceeding 300mm; panes, area 0.50 - 1.00m2	0.28	10.00	**0.31**	0.50	-	**7.11**	0.14	**7.55**	m2	**8.12**
girth exceeding 300mm; panes, area exceeding 1.00m2	0.28	10.00	**0.31**	0.43	-	**6.11**	0.12	**6.54**	m2	**7.03**
Wood partially glazed doors										
girth exceeding 300mm; panes, area not exceeding 0.10m2	0.62	10.00	**0.68**	0.69	-	**9.80**	0.19	**10.67**	m2	**11.47**
girth exceeding 300mm; panes, area 0.10 - 0.50m2	0.46	10.00	**0.51**	0.56	-	**7.96**	0.15	**8.62**	m2	**9.26**
girth exceeding 300mm; panes, area 0.50 - 1.00m2	0.44	10.00	**0.48**	0.50	-	**7.11**	0.14	**7.72**	m2	**8.30**
girth exceeding 300mm; panes, area exceeding 1.00m2	0.44	10.00	**0.48**	0.45	-	**6.39**	0.12	**7.00**	m2	**7.52**
Wood windows and screens										
girth exceeding 300mm; panes, area not exceeding 0.10m2	0.62	10.00	**0.68**	0.89	-	**12.65**	0.24	**13.56**	m2	**14.58**
girth exceeding 300mm; panes, area 0.10 - 0.50m2	0.46	10.00	**0.51**	0.65	-	**9.24**	0.18	**9.92**	m2	**10.66**
girth exceeding 300mm; panes, area 0.50 - 1.00m2	0.44	10.00	**0.48**	0.54	-	**7.67**	0.15	**8.30**	m2	**8.92**
girth exceeding 300mm; panes, area exceeding 1.00m2	0.44	10.00	**0.48**	0.47	-	**6.68**	0.13	**7.29**	m2	**7.83**
Two coats polyurethane lacquer, existing lacquered surfaces; external work										
Wood general surfaces										
girth exceeding 300mm	0.77	10.00	**0.85**	0.52	-	**7.39**	0.14	**8.38**	m2	**9.00**
isolated surfaces, girth not exceeding 300mm	0.23	10.00	**0.25**	0.18	-	**2.56**	0.05	**2.86**	m	**3.07**
Wood glazed doors										
girth exceeding 300mm; panes, area not exceeding 0.10m2	0.44	10.00	**0.48**	0.85	-	**12.08**	0.23	**12.79**	m2	**13.75**
girth exceeding 300mm; panes, area 0.10 - 0.50m2	0.31	10.00	**0.34**	0.62	-	**8.81**	0.17	**9.32**	m2	**10.01**
girth exceeding 300mm; panes, area 0.50 - 1.00m2	0.28	10.00	**0.31**	0.51	-	**7.25**	0.14	**7.70**	m2	**8.27**
girth exceeding 300mm; panes, area exceeding 1.00m2	0.28	10.00	**0.31**	0.44	-	**6.25**	0.12	**6.68**	m2	**7.18**
Wood partially glazed doors										
girth exceeding 300mm; panes, area not exceeding 0.10m2	0.62	10.00	**0.68**	0.72	-	**10.23**	0.19	**11.10**	m2	**11.94**
girth exceeding 300mm; panes, area 0.10 - 0.50m2	0.46	10.00	**0.51**	0.58	-	**8.24**	0.16	**8.91**	m2	**9.57**
girth exceeding 300mm; panes, area 0.50 - 1.00m2	0.44	10.00	**0.48**	0.51	-	**7.25**	0.14	**7.86**	m2	**8.45**
girth exceeding 300mm; panes, area exceeding 1.00m2	0.44	10.00	**0.48**	0.46	-	**6.54**	0.12	**7.14**	m2	**7.68**
Wood windows and screens										
girth exceeding 300mm; panes, area not exceeding 0.10m2	0.62	10.00	**0.68**	0.94	-	**13.36**	0.25	**14.29**	m2	**15.36**
girth exceeding 300mm; panes, area 0.10 - 0.50m2	0.46	10.00	**0.51**	0.69	-	**9.80**	0.19	**10.50**	m2	**11.29**
girth exceeding 300mm; panes, area 0.50 - 1.00m2	0.44	10.00	**0.48**	0.57	-	**8.10**	0.15	**8.73**	m2	**9.39**
girth exceeding 300mm; panes, area exceeding 1.00m2	0.44	10.00	**0.48**	0.50	-	**7.11**	0.14	**7.72**	m2	**8.30**
Three coats polyurethane lacquer, existing lacquered surfaces										
Wood general surfaces										
girth exceeding 300mm	1.13	10.00	**1.24**	0.56	-	**7.96**	0.15	**9.35**	m2	**10.05**
isolated surfaces, girth not exceeding 300mm	0.33	10.00	**0.37**	0.19	-	**2.70**	0.05	**3.12**	m	**3.35**
Wood glazed doors										
girth exceeding 300mm; panes, area not exceeding 0.10m2	0.64	10.00	**0.71**	0.93	-	**13.22**	0.25	**14.17**	m2	**15.23**
girth exceeding 300mm; panes, area 0.10 - 0.50m2	0.46	10.00	**0.51**	0.68	-	**9.66**	0.18	**10.35**	m2	**11.13**
girth exceeding 300mm; panes, area 0.50 - 1.00m2	0.41	10.00	**0.45**	0.57	-	**8.10**	0.15	**8.71**	m2	**9.36**
girth exceeding 300mm; panes, area exceeding 1.00m2	0.41	10.00	**0.45**	0.49	-	**6.96**	0.13	**7.55**	m2	**8.11**

Labour hourly rates: (except Specialists) Craft Operatives £14.21 Labourer £10.71 Rates are national average prices. Refer to REGIONAL VARIATIONS for indicative levels of overall pricing in regions	MATERIALS			LABOUR				RATES		
	Del to Site £	Waste %	Material Cost £	Craft Optve Hrs	Lab Hrs	Labour Cost £	Sunds £	Nett Rate £	Unit	Gross Rate (+7.5%) £
M60: PAINTING/CLEAR FINISHING (REDECORATIONS) - POLYURETHANE LACQUER Cont.										
Three coats polyurethane lacquer, existing lacquered surfaces Cont.										
Wood partially glazed doors										
girth exceeding 300mm; panes, area not exceeding 0.10m2	0.90	10.00	0.99	0.78	-	11.08	0.21	12.28	m2	13.20
girth exceeding 300mm; panes, area 0.10 - 0.50m2	0.67	10.00	0.73	0.69	-	9.80	0.19	10.72	m2	11.53
girth exceeding 300mm; panes, area 0.50 - 1.00m2	0.62	10.00	0.68	0.57	-	8.10	0.15	8.93	m2	9.60
girth exceeding 300mm; panes, area exceeding 1.00m2	0.62	10.00	0.68	0.51	-	7.25	0.14	8.06	m2	8.67
Wood windows and screens										
girth exceeding 300mm; panes, area not exceeding 0.10m2	0.64	10.00	0.71	1.01	-	14.35	0.27	15.33	m2	16.48
girth exceeding 300mm; panes, area 0.10 - 0.50m2	0.46	10.00	0.51	0.74	-	10.52	0.20	11.22	m2	12.06
girth exceeding 300mm; panes, area 0.50 - 1.00m2	0.41	10.00	0.45	0.62	-	8.81	0.17	9.43	m2	10.14
girth exceeding 300mm; panes, area exceeding 1.00m2	0.41	10.00	0.45	0.53	-	7.53	0.14	8.13	m2	8.74
Three coats polyurethane lacquer, existing lacquered surfaces; external work										
Wood general surfaces										
girth exceeding 300mm	1.13	10.00	1.24	0.60	-	8.53	0.16	9.93	m2	10.67
isolated surfaces, girth not exceeding 300mm	0.33	10.00	0.37	0.20	-	2.84	0.05	3.26	m	3.51
Wood glazed doors										
girth exceeding 300mm; panes, area not exceeding 0.10m2	0.64	10.00	0.71	0.96	-	13.64	0.26	14.61	m2	15.70
girth exceeding 300mm; panes, area 0.10 - 0.50m2	0.46	10.00	0.51	0.70	-	9.95	0.19	10.64	m2	11.44
girth exceeding 300mm; panes, area 0.50 - 1.00m2	0.41	10.00	0.45	0.58	-	8.24	0.16	8.85	m2	9.51
girth exceeding 300mm; panes, area exceeding 1.00m2	0.41	10.00	0.45	0.50	-	7.11	0.14	7.69	m2	8.27
Wood partially glazed doors										
girth exceeding 300mm; panes, area not exceeding 0.10m2	0.90	10.00	0.99	0.81	-	11.51	0.22	12.72	m2	13.67
girth exceeding 300mm; panes, area 0.10 - 0.50m2	0.67	10.00	0.73	0.71	-	10.09	0.19	11.01	m2	11.84
girth exceeding 300mm; panes, area 0.50 - 1.00m2	0.62	10.00	0.68	0.58	-	8.24	0.16	9.08	m2	9.76
girth exceeding 300mm; panes, area exceeding 1.00m2	0.62	10.00	0.68	0.53	-	7.53	0.14	8.35	m2	8.98
Wood windows and screens										
girth exceeding 300mm; panes, area not exceeding 0.10m2	0.64	10.00	0.71	1.07	-	15.20	0.29	16.20	m2	17.41
girth exceeding 300mm; panes, area 0.10 - 0.50m2	0.46	10.00	0.51	0.79	-	11.23	0.21	11.95	m2	12.84
girth exceeding 300mm; panes, area 0.50 - 1.00m2	0.41	10.00	0.45	0.66	-	9.38	0.18	10.01	m2	10.76
girth exceeding 300mm; panes, area exceeding 1.00m2	0.41	10.00	0.45	0.57	-	8.10	0.15	8.71	m2	9.36
M60: PAINTING/CLEAR FINISHING (REDECORATIONS) - FRENCH AND WAX POLISHING										
Generally										
Note: the following rates include for the cost of all preparatory work, e.g. washing down, etc.										
Stain; two coats white polish										
Wood general surfaces										
girth exceeding 300mm	2.05	10.00	2.25	0.40	-	5.68	0.11	8.04	m2	8.65
isolated surfaces, girth not exceeding 300mm	0.64	10.00	0.71	0.13	-	1.85	0.04	2.59	m	2.78
Strip old polish, oil, two coats wax polish, existing polished surfaces										
Wood general surfaces										
girth exceeding 300mm	2.05	10.00	2.25	1.12	-	15.92	0.30	18.47	m2	19.86
isolated surfaces, girth not exceeding 300mm	0.64	10.00	0.71	0.34	-	4.83	0.09	5.63	m	6.05
Strip old polish, oil, stain, two coats wax polish, existing polished surfaces										
Wood general surfaces										
girth exceeding 300mm	2.05	10.00	2.25	1.23	-	17.48	0.33	20.06	m2	21.57
isolated surfaces, girth not exceeding 300mm	0.64	10.00	0.71	0.38	-	5.40	0.10	6.21	m	6.68
Open grain French polish existing polished surfaces										
Wood general surfaces										
girth exceeding 300mm	2.12	10.00	2.33	2.40	-	34.10	0.65	37.08	m2	39.86
isolated surfaces, girth not exceeding 300mm	0.67	10.00	0.73	0.80	-	11.37	0.22	12.32	m	13.24
Fully French polish existing polished surfaces										
Wood general surfaces										
girth exceeding 300mm	3.52	10.00	3.87	1.40	-	19.89	0.38	24.14	m2	25.95
isolated surfaces, girth not exceeding 300mm	1.11	10.00	1.22	0.47	-	6.68	0.13	8.03	m	8.63

Labour hourly rates: (except Specialists) Craft Operatives £14.21 Labourer £10.71 Rates are national average prices. Refer to REGIONAL VARIATIONS for indicative levels of overall pricing in regions	MATERIALS			LABOUR				RATES		
	Del to Site £	Waste %	Material Cost £	Craft Optve Hrs	Lab Hrs	Labour Cost £	Sunds £	Nett Rate £	Unit	Gross Rate (+7.5%) £
M60: PAINTING/CLEAR FINISHING (REDECORATIONS) - FRENCH AND WAX POLISHING Cont.										
Strip old polish, oil, open grain French polish, existing polished surfaces										
Wood general surfaces										
girth exceeding 300mm	2.12	10.00	**2.33**	3.12	-	**44.34**	0.84	**47.51**	m2	**51.07**
isolated surfaces, girth not exceeding 300mm	0.67	10.00	**0.73**	1.04	-	**14.78**	0.28	**15.79**	m	**16.98**
Strip old polish, oil, stain, open grain French polish, existing polished surfaces										
Wood general surfaces										
girth exceeding 300mm	2.12	10.00	**2.33**	3.23	-	**45.90**	0.87	**49.10**	m2	**52.78**
isolated surfaces, girth not exceeding 300mm	0.67	10.00	**0.73**	1.08	-	**15.35**	0.29	**16.37**	m	**17.60**
Strip old polish, oil, body in, fully French polish existing polished surfaces										
Wood general surfaces										
girth exceeding 300mm	3.52	10.00	**3.87**	4.21	-	**59.82**	1.14	**64.83**	m2	**69.69**
isolated surfaces, girth not exceeding 300mm	1.11	10.00	**1.22**	1.40	-	**19.89**	0.38	**21.49**	m	**23.10**
Strip old polish, oil, stain, body in, fully French polish, existing polished surfaces										
Wood general surfaces										
girth exceeding 300mm	3.52	10.00	**3.87**	4.32	-	**61.39**	1.17	**66.42**	m2	**71.41**
isolated surfaces, girth not exceeding 300mm	1.11	10.00	**1.22**	1.44	-	**20.46**	0.39	**22.07**	m	**23.73**
M60: PAINTING/CLEAR FINISHING (REDECORATIONS) - WATER REPELLENT										
One coat solvent based water repellent, existing surfaces										
Cement rendered general surfaces										
girth exceeding 300mm	0.14	5.00	**0.15**	0.11	-	**1.56**	0.03	**1.74**	m2	**1.87**
Stone general surfaces										
girth exceeding 300mm	0.17	5.00	**0.18**	0.12	-	**1.71**	0.03	**1.92**	m2	**2.06**
Brickwork general surfaces										
girth exceeding 300mm	0.17	5.00	**0.18**	0.12	-	**1.71**	0.03	**1.92**	m2	**2.06**
M60: PAINTING/CLEAR FINISHING (REDECORATIONS) - REMOVAL OF MOULD GROWTH										
Apply fungicide to existing decorated and infected surfaces										
Cement rendered general surfaces										
girth exceeding 300mm	0.11	5.00	**0.11**	0.20	-	**2.84**	0.05	**3.01**	m2	**3.23**
Concrete general surfaces										
girth exceeding 300mm	0.13	5.00	**0.14**	0.20	-	**2.84**	0.05	**3.03**	m2	**3.26**
Plaster general surfaces										
girth exceeding 300mm	0.10	5.00	**0.10**	0.20	-	**2.84**	0.05	**3.00**	m2	**3.22**
Brickwork general surfaces										
girth exceeding 300mm	0.13	5.00	**0.14**	0.22	-	**3.13**	0.06	**3.32**	m2	**3.57**
Apply fungicide to existing decorated and infected surfaces; external work										
Cement rendered general surfaces										
girth exceeding 300mm	0.11	5.00	**0.11**	0.22	-	**3.13**	0.06	**3.30**	m2	**3.55**
Concrete general surfaces										
girth exceeding 300mm	0.13	5.00	**0.14**	0.22	-	**3.13**	0.06	**3.32**	m2	**3.57**
Plaster general surfaces										
girth exceeding 300mm	0.10	5.00	**0.10**	0.22	-	**3.13**	0.06	**3.29**	m2	**3.53**
Brickwork general surfaces										
girth exceeding 300mm	0.13	5.00	**0.14**	0.24	-	**3.41**	0.06	**3.61**	m2	**3.88**
M60: PAINTING/CLEAR FINISHING; STAIRCASE AREAS: EMULSION PAINTING										
Mist coat, one full coat emulsion paint										
Concrete general surfaces										
girth exceeding 300mm	0.14	10.00	**0.15**	0.21	-	**2.98**	0.06	**3.19**	m2	**3.43**
Plaster general surfaces										
girth exceeding 300mm	0.11	10.00	**0.13**	0.19	-	**2.70**	0.05	**2.88**	m2	**3.09**
Plasterboard general surfaces										
girth exceeding 300mm	0.11	10.00	**0.13**	0.19	-	**2.70**	0.05	**2.88**	m2	**3.09**
Brickwork general surfaces										
girth exceeding 300mm	0.17	10.00	**0.19**	0.23	-	**3.27**	0.06	**3.52**	m2	**3.78**

Labour hourly rates: (except Specialists) Craft Operatives £14.21 Labourer £10.71 Rates are national average prices. Refer to REGIONAL VARIATIONS for indicative levels of overall pricing in regions	MATERIALS			LABOUR				RATES		
	Del to Site £	Waste %	Material Cost £	Craft Optve Hrs	Lab Hrs	Labour Cost £	Sunds £	Nett Rate £	Unit	Gross Rate (+7.5%) £
M60: PAINTING/CLEAR FINISHING; STAIRCASE AREAS: EMULSION PAINTING Cont.										
Mist coat, one full coat emulsion paint Cont.										
Paper covered general surfaces girth exceeding 300mm	0.12	10.00	**0.13**	0.20	-	**2.84**	0.05	**3.03**	m2	3.26
Mist coat, two full coats emulsion paint										
Concrete general surfaces girth exceeding 300mm	0.24	10.00	**0.27**	0.30	-	**4.26**	0.08	**4.61**	m2	4.96
Plaster general surfaces girth exceeding 300mm	0.20	10.00	**0.22**	0.28	-	**3.98**	0.08	**4.28**	m2	4.60
Plasterboard general surfaces girth exceeding 300mm	0.20	10.00	**0.22**	0.28	-	**3.98**	0.08	**4.28**	m2	4.60
Brickwork general surfaces girth exceeding 300mm	0.30	10.00	**0.33**	0.33	-	**4.69**	0.09	**5.11**	m2	5.50
Paper covered general surfaces girth exceeding 300mm	0.22	10.00	**0.25**	0.29	-	**4.12**	0.08	**4.44**	m2	4.78
M60: PAINTING/CLEAR FINISHING; STAIRCASE AREAS: CEMENT PAINTING										
One coat textured masonry paint										
Cement rendered general surfaces girth exceeding 300mm	0.30	10.00	**0.32**	0.14	-	**1.99**	0.04	**2.35**	m2	2.53
Concrete general surfaces girth exceeding 300mm	0.22	10.00	**0.25**	0.14	-	**1.99**	0.04	**2.27**	m2	2.44
Brickwork general surfaces girth exceeding 300mm	0.26	10.00	**0.29**	0.16	-	**2.27**	0.04	**2.60**	m2	2.80
Rough cast general surfaces girth exceeding 300mm	0.59	10.00	**0.65**	0.20	-	**2.84**	0.05	**3.55**	m2	3.81
Two coats textured masonry paint										
Cement rendered general surfaces girth exceeding 300mm	0.59	10.00	**0.65**	0.28	-	**3.98**	0.08	**4.70**	m2	5.06
Concrete general surfaces girth exceeding 300mm	0.45	10.00	**0.49**	0.28	-	**3.98**	0.08	**4.55**	m2	4.89
Brickwork general surfaces girth exceeding 300mm	0.52	10.00	**0.57**	0.30	-	**4.26**	0.08	**4.92**	m2	5.28
Rough cast general surfaces girth exceeding 300mm	1.18	10.00	**1.30**	0.36	-	**5.12**	0.10	**6.51**	m2	7.00
One coat stabilizing solution, one coat textured masonry paint										
Cement rendered general surfaces girth exceeding 300mm	0.59	10.00	**0.65**	0.24	-	**3.41**	0.06	**4.12**	m2	4.43
Concrete general surfaces girth exceeding 300mm	0.48	10.00	**0.53**	0.24	-	**3.41**	0.06	**4.01**	m2	4.31
Brickwork general surfaces girth exceeding 300mm	0.55	10.00	**0.61**	0.27	-	**3.84**	0.07	**4.52**	m2	4.86
Rough cast general surfaces girth exceeding 300mm	0.89	10.00	**0.97**	0.33	-	**4.69**	0.09	**5.75**	m2	6.18
One coat sealer, two coats waterproof cement paint; external work										
Cement rendered general surfaces girth exceeding 300mm	6.64	10.00	**7.31**	0.38	-	**5.40**	0.10	**12.81**	m2	13.77
Concrete general surfaces girth exceeding 300mm	5.24	10.00	**5.76**	0.38	-	**5.40**	0.10	**11.26**	m2	12.11
Brickwork general surfaces girth exceeding 300mm	3.45	10.00	**3.80**	0.41	-	**5.83**	0.11	**9.74**	m2	10.47
Rough cast general surfaces girth exceeding 300mm	6.64	10.00	**7.31**	0.49	-	**6.96**	0.13	**14.40**	m2	15.48
M60: PAINTING/CLEAR FINISHING; STAIRCASE AREAS: OIL PAINTING										
One coat primer, one undercoat, one coat full gloss finish										
Concrete general surfaces girth exceeding 300mm	0.86	10.00	**0.95**	0.39	-	**5.54**	0.11	**6.60**	m2	7.09

Labour hourly rates: (except Specialists) Craft Operatives £14.21 Labourer £10.71 Rates are national average prices. Refer to REGIONAL VARIATIONS for indicative levels of overall pricing in regions	MATERIALS			LABOUR				RATES		
	Del to Site £	Waste %	Material Cost £	Craft Optve Hrs	Lab Hrs	Labour Cost £	Sunds £	Nett Rate £	Unit	Gross Rate (+7.5%) £
M60: PAINTING/CLEAR FINISHING; STAIRCASE AREAS: OIL PAINTING Cont.										
One coat primer, one undercoat, one coat full gloss finish Cont.										
Brickwork general surfaces girth exceeding 300mm	1.07	10.00	**1.18**	0.43	-	**6.11**	0.12	**7.40**	m2	**7.96**
Plasterboard general surfaces girth exceeding 300mm	0.70	10.00	**0.78**	0.36	-	**5.12**	0.10	**5.99**	m2	**6.44**
Plaster general surfaces girth exceeding 300mm	0.70	10.00	**0.78**	0.36	-	**5.12**	0.10	**5.99**	m2	**6.44**
One coat primer, two undercoats, one coat full gloss finish										
Concrete general surfaces girth exceeding 300mm	1.05	10.00	**1.16**	0.52	-	**7.39**	0.14	**8.69**	m2	**9.34**
Brickwork general surfaces girth exceeding 300mm	1.30	10.00	**1.43**	0.57	-	**8.10**	0.15	**9.69**	m2	**10.41**
Plasterboard general surfaces girth exceeding 300mm	0.86	10.00	**0.94**	0.48	-	**6.82**	0.13	**7.89**	m2	**8.48**
Plaster general surfaces girth exceeding 300mm	0.86	10.00	**0.94**	0.48	-	**6.82**	0.13	**7.89**	m2	**8.48**
One coat primer, one undercoat, one coat eggshell finish										
Concrete general surfaces girth exceeding 300mm	0.91	10.00	**1.00**	0.39	-	**5.54**	0.11	**6.65**	m2	**7.14**
Brickwork general surfaces girth exceeding 300mm	1.12	10.00	**1.24**	0.43	-	**6.11**	0.12	**7.46**	m2	**8.02**
Plasterboard general surfaces girth exceeding 300mm	0.74	10.00	**0.81**	0.36	-	**5.12**	0.10	**6.03**	m2	**6.48**
Plaster general surfaces girth exceeding 300mm	0.74	10.00	**0.81**	0.36	-	**5.12**	0.10	**6.03**	m2	**6.48**
One coat primer, two undercoats, one coat eggshell finish										
Concrete general surfaces girth exceeding 300mm	1.09	10.00	**1.20**	0.52	-	**7.39**	0.14	**8.73**	m2	**9.39**
Brickwork general surfaces girth exceeding 300mm	1.36	10.00	**1.49**	0.57	-	**8.10**	0.15	**9.75**	m2	**10.48**
Plasterboard general surfaces girth exceeding 300mm	0.89	10.00	**0.98**	0.48	-	**6.82**	0.13	**7.93**	m2	**8.53**
Plaster general surfaces girth exceeding 300mm	0.89	10.00	**0.98**	0.48	-	**6.82**	0.13	**7.93**	m2	**8.53**
M60: PAINTING/CLEAR FINISHING; STAIRCASE AREAS: PLASTIC FINISH										
Textured plastic coating - Stippled finish										
Concrete general surfaces girth exceeding 300mm	0.37	10.00	**0.41**	0.34	-	**4.83**	0.09	**5.33**	m2	**5.73**
Brickwork general surfaces girth exceeding 300mm	0.45	10.00	**0.49**	0.44	-	**6.25**	0.12	**6.86**	m2	**7.38**
Plasterboard general surfaces girth exceeding 300mm	0.30	10.00	**0.33**	0.26	-	**3.69**	0.07	**4.10**	m2	**4.40**
Plaster general surfaces girth exceeding 300mm	0.30	10.00	**0.33**	0.26	-	**3.69**	0.07	**4.10**	m2	**4.40**
Textured plastic coating - Combed Finish										
Concrete general surfaces girth exceeding 300mm	0.37	10.00	**0.41**	0.39	-	**5.54**	0.11	**6.05**	m2	**6.51**
Brickwork general surfaces girth exceeding 300mm	0.45	10.00	**0.49**	0.54	-	**7.67**	0.15	**8.31**	m2	**8.93**
Plasterboard general surfaces girth exceeding 300mm	0.30	10.00	**0.33**	0.31	-	**4.41**	0.08	**4.82**	m2	**5.18**
Plaster general surfaces girth exceeding 300mm	0.37	10.00	**0.41**	0.31	-	**4.41**	0.08	**4.90**	m2	**5.26**

Labour hourly rates: (except Specialists) Craft Operatives £14.21 Labourer £10.71 Rates are national average prices. Refer to REGIONAL VARIATIONS for indicative levels of overall pricing in regions	MATERIALS			LABOUR				RATES		
	Del to Site £	Waste %	Material Cost £	Craft Optve Hrs	Lab Hrs	Labour Cost £	Sunds £	Nett Rate £	Unit	Gross Rate (+7.5%) £
M60: PAINTING/CLEAR FINISHING; STAIRCASE AREAS - REDECORATIONS: EMULSION PAINTING										
Generally										
Note: the following rates include for the cost of all preparatory work, e.g. washing down, etc.										
Two coats emulsion paint, existing emulsion painted surfaces										
Concrete general surfaces girth exceeding 300mm	0.22	10.00	**0.24**	0.35	-	**4.97**	0.09	**5.31**	m2	**5.70**
Plaster general surfaces girth exceeding 300mm	0.18	10.00	**0.20**	0.31	-	**4.41**	0.08	**4.69**	m2	**5.04**
Plasterboard general surfaces girth exceeding 300mm	0.18	10.00	**0.20**	0.32	-	**4.55**	0.09	**4.83**	m2	**5.20**
Brickwork general surfaces girth exceeding 300mm	0.27	10.00	**0.30**	0.35	-	**4.97**	0.09	**5.37**	m2	**5.77**
Paper covered general surfaces girth exceeding 300mm	0.20	10.00	**0.22**	0.32	-	**4.55**	0.09	**4.86**	m2	**5.22**
Two full coats emulsion paint, existing washable distempered surfaces										
Concrete general surfaces girth exceeding 300mm	0.22	10.00	**0.24**	0.36	-	**5.12**	0.10	**5.45**	m2	**5.86**
Plaster general surfaces girth exceeding 300mm	0.18	10.00	**0.20**	0.32	-	**4.55**	0.09	**4.83**	m2	**5.20**
Plasterboard general surfaces girth exceeding 300mm	0.18	10.00	**0.20**	0.33	-	**4.69**	0.09	**4.98**	m2	**5.35**
Brickwork general surfaces girth exceeding 300mm	0.27	10.00	**0.30**	0.36	-	**5.12**	0.10	**5.51**	m2	**5.92**
Paper covered general surfaces girth exceeding 300mm	0.20	10.00	**0.22**	0.33	-	**4.69**	0.09	**5.00**	m2	**5.38**
Two full coats emulsion paint, existing textured plastic coated surfaces										
Concrete general surfaces girth exceeding 300mm	0.22	10.00	**0.24**	0.39	-	**5.54**	0.11	**5.88**	m2	**6.33**
Plaster general surfaces girth exceeding 300mm	0.18	10.00	**0.20**	0.34	-	**4.83**	0.09	**5.12**	m2	**5.51**
Plasterboard general surfaces girth exceeding 300mm	0.18	10.00	**0.20**	0.35	-	**4.97**	0.09	**5.27**	m2	**5.66**
Brickwork general surfaces girth exceeding 300mm	0.27	10.00	**0.30**	0.38	-	**5.40**	0.10	**5.80**	m2	**6.23**
Paper covered general surfaces girth exceeding 300mm	0.20	10.00	**0.22**	0.35	-	**4.97**	0.09	**5.29**	m2	**5.69**
Mist coat, two full coats emulsion paint, existing non-washable distempered surfaces										
Concrete general surfaces girth exceeding 300mm	0.24	10.00	**0.27**	0.44	-	**6.25**	0.12	**6.64**	m2	**7.14**
Plaster general surfaces girth exceeding 300mm	0.20	10.00	**0.22**	0.40	-	**5.68**	0.11	**6.01**	m2	**6.47**
Plasterboard general surfaces girth exceeding 300mm	0.20	10.00	**0.22**	0.41	-	**5.83**	0.11	**6.16**	m2	**6.62**
Brickwork general surfaces girth exceeding 300mm	0.30	10.00	**0.33**	0.44	-	**6.25**	0.12	**6.71**	m2	**7.21**
Paper covered general surfaces girth exceeding 300mm	0.22	10.00	**0.25**	0.41	-	**5.83**	0.11	**6.18**	m2	**6.65**
Two full coats emulsion paint, existing gloss painted surfaces										
Concrete general surfaces girth exceeding 300mm	0.22	10.00	**0.24**	0.36	-	**5.12**	0.10	**5.45**	m2	**5.86**
Plaster general surfaces girth exceeding 300mm	0.18	10.00	**0.20**	0.32	-	**4.55**	0.09	**4.83**	m2	**5.20**
Plasterboard general surfaces girth exceeding 300mm	0.18	10.00	**0.20**	0.33	-	**4.69**	0.09	**4.98**	m2	**5.35**
Brickwork general surfaces girth exceeding 300mm	0.27	10.00	**0.30**	0.36	-	**5.12**	0.10	**5.51**	m2	**5.92**
Paper covered general surfaces girth exceeding 300mm	0.20	10.00	**0.22**	0.33	-	**4.69**	0.09	**5.00**	m2	**5.38**

Labour hourly rates: (except Specialists) Craft Operatives £14.21 Labourer £10.71 Rates are national average prices. Refer to REGIONAL VARIATIONS for indicative levels of overall pricing in regions	MATERIALS			LABOUR				RATES		
	Del to Site £	Waste %	Material Cost £	Craft Optve Hrs	Lab Hrs	Labour Cost £	Sunds £	Nett Rate £	Unit	Gross Rate (+7.5%) £
M60: PAINTING/CLEAR FINISHING; STAIRCASE AREAS - REDECORATIONS: OIL PAINTING										
Generally										
Note: the following rates include for the cost of all preparatory work, e.g. washing down, etc.										
One undercoat, one coat full gloss finish, existing gloss painted surfaces										
Concrete general surfaces girth exceeding 300mm	0.32	10.00	**0.35**	0.41	-	**5.83**	0.11	**6.28**	m2	**6.76**
Brickwork general surfaces girth exceeding 300mm	0.34	10.00	**0.37**	0.42	-	**5.97**	0.11	**6.45**	m2	**6.94**
Plasterboard general surfaces girth exceeding 300mm	0.29	10.00	**0.32**	0.38	-	**5.40**	0.10	**5.82**	m2	**6.26**
Plaster general surfaces girth exceeding 300mm	0.29	10.00	**0.32**	0.37	-	**5.26**	0.10	**5.68**	m2	**6.11**
One undercoat, one coat eggshell finish, existing gloss painted surfaces										
Concrete general surfaces girth exceeding 300mm	0.35	10.00	**0.39**	0.41	-	**5.83**	0.11	**6.32**	m2	**6.80**
Brickwork general surfaces girth exceeding 300mm	0.44	10.00	**0.49**	0.42	-	**5.97**	0.11	**6.57**	m2	**7.06**
Plasterboard general surfaces girth exceeding 300mm	0.32	10.00	**0.36**	0.38	-	**5.40**	0.10	**5.86**	m2	**6.30**
Plaster general surfaces girth exceeding 300mm	0.32	10.00	**0.36**	0.32	-	**4.55**	0.09	**4.99**	m2	**5.37**
Two undercoats, one coat full gloss finish, existing gloss painted surfaces										
Concrete general surfaces girth exceeding 300mm	0.47	10.00	**0.51**	0.53	-	**7.53**	0.14	**8.19**	m2	**8.80**
Brickwork general surfaces girth exceeding 300mm	0.56	10.00	**0.62**	0.53	-	**7.53**	0.14	**8.29**	m2	**8.91**
Plasterboard general surfaces girth exceeding 300mm	0.44	10.00	**0.49**	0.50	-	**7.11**	0.14	**7.73**	m2	**8.31**
Plaster general surfaces girth exceeding 300mm	0.44	10.00	**0.49**	0.49	-	**6.96**	0.13	**7.58**	m2	**8.15**
Two undercoats, one coat eggshell finish, existing gloss painted surfaces										
Concrete general surfaces girth exceeding 300mm	0.50	10.00	**0.55**	0.53	-	**7.53**	0.14	**8.23**	m2	**8.84**
Brickwork general surfaces girth exceeding 300mm	0.60	10.00	**0.67**	0.53	-	**7.53**	0.14	**8.34**	m2	**8.97**
Plasterboard general surfaces girth exceeding 300mm	0.48	10.00	**0.52**	0.50	-	**7.11**	0.14	**7.76**	m2	**8.35**
Plaster general surfaces girth exceeding 300mm	0.48	10.00	**0.52**	0.49	-	**6.96**	0.13	**7.62**	m2	**8.19**
One coat primer, one undercoat, one coat full gloss finish, existing washable distempered or emulsion painted surfaces										
Concrete general surfaces girth exceeding 300mm	0.86	10.00	**0.95**	0.53	-	**7.53**	0.14	**8.63**	m2	**9.27**
Brickwork general surfaces girth exceeding 300mm	1.01	10.00	**1.11**	0.53	-	**7.53**	0.14	**8.79**	m2	**9.45**
Plasterboard general surfaces girth exceeding 300mm	0.64	10.00	**0.71**	0.50	-	**7.11**	0.18	**7.99**	m2	**8.59**
Plaster general surfaces girth exceeding 300mm	0.74	10.00	**0.82**	0.49	-	**6.96**	0.13	**7.91**	m2	**8.50**
One coat primer, one undercoat, one coat eggshell finish, existing washable distempered or emulsion painted surfaces										
Concrete general surfaces girth exceeding 300mm	0.90	10.00	**0.99**	0.53	-	**7.53**	0.14	**8.66**	m2	**9.31**
Brickwork general surfaces girth exceeding 300mm	1.11	10.00	**1.23**	0.53	-	**7.53**	0.14	**8.90**	m2	**9.57**
Plasterboard general surfaces girth exceeding 300mm	0.82	10.00	**0.91**	0.50	-	**7.11**	0.14	**8.15**	m2	**8.76**

Labour hourly rates: (except Specialists) Craft Operatives £14.21 Labourer £10.71 Rates are national average prices. Refer to REGIONAL VARIATIONS for indicative levels of overall pricing in regions	MATERIALS			LABOUR				RATES		
	Del to Site £	Waste %	Material Cost £	Craft Optve Hrs	Lab Hrs	Labour Cost £	Sunds £	Nett Rate £	Unit	Gross Rate (+7.5%) £
M60: PAINTING/CLEAR FINISHING; STAIRCASE AREAS - REDECORATIONS: OIL PAINTING Cont.										
One coat primer, one undercoat, one coat eggshell finish, existing washable distempered or emulsion painted surfaces Cont.										
Plaster general surfaces girth exceeding 300mm	0.82	10.00	**0.91**	0.49	-	**6.96**	0.13	**8.00**	m2	**8.60**
One coat primer, two undercoats, one coat full gloss finish, existing washable distempered or emulsion painted surfaces										
Concrete general surfaces girth exceeding 300mm	1.02	10.00	**1.12**	0.65	-	**9.24**	0.18	**10.53**	m2	**11.32**
Brickwork general surfaces girth exceeding 300mm	1.24	10.00	**1.36**	0.66	-	**9.38**	0.18	**10.92**	m2	**11.73**
Plasterboard general surfaces girth exceeding 300mm	1.04	10.00	**1.14**	0.62	-	**8.81**	0.17	**10.12**	m2	**10.88**
Plaster general surfaces girth exceeding 300mm	0.94	10.00	**1.04**	0.61	-	**8.67**	0.16	**9.87**	m2	**10.61**
One coat primer, two undercoats, one coat eggshell finish, existing washable distempered or emulsion painted surfaces										
Concrete general surfaces girth exceeding 300mm	1.05	10.00	**1.16**	0.65	-	**9.24**	0.18	**10.57**	m2	**11.36**
Brickwork general surfaces girth exceeding 300mm	1.28	10.00	**1.41**	0.66	-	**9.38**	0.18	**10.96**	m2	**11.79**
Plasterboard general surfaces girth exceeding 300mm	0.98	10.00	**1.07**	0.62	-	**8.81**	0.17	**10.05**	m2	**10.81**
Plaster general surfaces girth exceeding 300mm	0.98	10.00	**1.07**	0.61	-	**8.67**	0.16	**9.91**	m2	**10.65**
One coat primer, one undercoat, one coat full gloss finish, existing non-washable distempered surfaces										
Concrete general surfaces girth exceeding 300mm	0.86	10.00	**0.95**	0.53	-	**7.53**	0.14	**8.63**	m2	**9.27**
Brickwork general surfaces girth exceeding 300mm	1.01	10.00	**1.11**	0.53	-	**7.53**	0.14	**8.79**	m2	**9.45**
Plasterboard general surfaces girth exceeding 300mm	0.79	10.00	**0.87**	0.50	-	**7.11**	0.14	**8.11**	m2	**8.72**
Plaster general surfaces girth exceeding 300mm	0.79	10.00	**0.87**	0.49	-	**6.96**	0.13	**7.97**	m2	**8.56**
One coat primer, one undercoat, one coat eggshell finish, existing non-washable distempered surfaces										
Concrete general surfaces girth exceeding 300mm	0.90	10.00	**0.99**	0.53	-	**7.53**	0.14	**8.66**	m2	**9.31**
Brickwork general surfaces girth exceeding 300mm	1.11	10.00	**1.23**	0.53	-	**7.53**	0.14	**8.90**	m2	**9.57**
Plasterboard general surfaces girth exceeding 300mm	0.82	10.00	**0.91**	0.50	-	**7.11**	0.14	**8.15**	m2	**8.76**
Plaster general surfaces girth exceeding 300mm	0.82	10.00	**0.91**	0.49	-	**6.96**	0.13	**8.00**	m2	**8.60**
One coat primer, two undercoats, one coat full gloss finish, existing non-washable distempered surfaces										
Concrete general surfaces girth exceeding 300mm	1.02	10.00	**1.12**	0.65	-	**9.24**	0.18	**10.53**	m2	**11.32**
Brickwork general surfaces girth exceeding 300mm	1.24	10.00	**1.36**	0.66	-	**9.38**	0.18	**10.92**	m2	**11.73**
Plasterboard general surfaces girth exceeding 300mm	1.04	10.00	**1.14**	0.62	-	**8.81**	0.17	**10.12**	m2	**10.88**
Plaster general surfaces girth exceeding 300mm	0.94	10.00	**1.04**	0.61	-	**8.67**	0.16	**9.87**	m2	**10.61**
One coat primer, two undercoats, one coat eggshell finish, existing non-washable distempered surfaces										
Concrete general surfaces girth exceeding 300mm	1.05	10.00	**1.16**	0.65	-	**9.24**	0.18	**10.57**	m2	**11.36**
Brickwork general surfaces girth exceeding 300mm	1.28	10.00	**1.41**	0.66	-	**9.38**	0.18	**10.96**	m2	**11.79**

Labour hourly rates: (except Specialists) Craft Operatives £14.21 Labourer £10.71 Rates are national average prices. Refer to REGIONAL VARIATIONS for indicative levels of overall pricing in regions	MATERIALS			LABOUR				RATES		
	Del to Site £	Waste %	Material Cost £	Craft Optve Hrs	Lab Hrs	Labour Cost £	Sunds £	Nett Rate £	Unit	Gross Rate (+7.5%) £
M60: PAINTING/CLEAR FINISHING; STAIRCASE AREAS - REDECORATIONS: OIL PAINTING Cont.										
One coat primer, two undercoats, one coat eggshell finish, existing non-washable distempered surfaces Cont.										
Plasterboard general surfaces girth exceeding 300mm	0.98	10.00	**1.07**	0.62	-	**8.81**	0.17	**10.05**	m2	**10.81**
Plaster general surfaces girth exceeding 300mm	0.98	10.00	**1.07**	0.61	-	**8.67**	0.16	**9.91**	m2	**10.65**
M60: PAINTING/CLEAR FINISHING; STAIRCASE AREAS - REDECORATIONS: PLASTIC FINISH										
Generally										
Note: the following rates include for the cost of all preparatory work, e.g. washing down, etc.										
Textured plastic coating - stippled finish, existing washable distempered or emulsion painted surfaces										
Concrete general surfaces girth exceeding 300mm	0.31	10.00	**0.35**	0.40	-	**5.68**	0.11	**6.14**	m2	**6.60**
Brickwork general surfaces girth exceeding 300mm	0.38	10.00	**0.41**	0.40	-	**5.68**	0.11	**6.21**	m2	**6.67**
Plasterboard general surfaces girth exceeding 300mm	0.28	10.00	**0.31**	0.37	-	**5.26**	0.10	**5.67**	m2	**6.09**
Plaster general surfaces girth exceeding 300mm	0.28	10.00	**0.31**	0.36	-	**5.12**	0.10	**5.52**	m2	**5.94**
Textured plastic coating - combed finish, existing washable distempered or emulsion painted surfaces										
Concrete general surfaces girth exceeding 300mm	0.38	10.00	**0.41**	0.45	-	**6.39**	0.12	**6.93**	m2	**7.45**
Brickwork general surfaces girth exceeding 300mm	0.45	10.00	**0.50**	0.45	-	**6.39**	0.12	**7.01**	m2	**7.54**
Plasterboard general surfaces girth exceeding 300mm	0.34	10.00	**0.37**	0.42	-	**5.97**	0.11	**6.45**	m2	**6.94**
Plaster general surfaces girth exceeding 300mm	0.34	10.00	**0.37**	0.41	-	**5.83**	0.11	**6.31**	m2	**6.78**
Textured plastic coating - stippled finish, existing non-washable distempered surfaces										
Concrete general surfaces girth exceeding 300mm	0.31	10.00	**0.35**	0.51	-	**7.25**	0.14	**7.73**	m2	**8.31**
Brickwork general surfaces girth exceeding 300mm	0.38	10.00	**0.41**	0.51	-	**7.25**	0.14	**7.80**	m2	**8.38**
Plasterboard general surfaces girth exceeding 300mm	0.28	10.00	**0.31**	0.48	-	**6.82**	0.13	**7.26**	m2	**7.81**
Plaster general surfaces girth exceeding 300mm	0.28	10.00	**0.31**	0.47	-	**6.68**	0.13	**7.12**	m2	**7.65**
Textured plastic coating - combed finish, existing non-washable distempered surfaces										
Concrete general surfaces girth exceeding 300mm	0.38	10.00	**0.41**	0.56	-	**7.96**	0.15	**8.52**	m2	**9.16**
Brickwork general surfaces girth exceeding 300mm	0.45	10.00	**0.50**	0.56	-	**7.96**	0.15	**8.61**	m2	**9.25**
Plasterboard general surfaces girth exceeding 300mm	0.34	10.00	**0.37**	0.53	-	**7.53**	0.14	**8.05**	m2	**8.65**
Plaster general surfaces girth exceeding 300mm	0.34	10.00	**0.37**	0.52	-	**7.39**	0.14	**7.90**	m2	**8.49**

FURNITURE/EQUIPMENT

Labour hourly rates: (except Specialists) Craft Operatives £14.21 Labourer £10.71 Rates are national average prices. Refer to REGIONAL VARIATIONS for indicative levels of overall pricing in regions	MATERIALS			LABOUR				RATES		
	Del to Site £	Waste %	Material Cost £	Craft Optve Hrs	Lab Hrs	Labour Cost £	Sunds £	Nett Rate £	Unit	Gross Rate (+7.5%) £
N10: GENERAL FIXTURES/FURNISHING/EQUIPMENT: MIRRORS										
BS 952, clear float, SG, silvered and protected with copper backing										
6mm thick; fixing to masonry with brass screws, chromium plated dome covers, rubber sleeves and washers										
holes 6mm diameter -4; edges polished										
254 x 400mm	25.90	5.00	**27.20**	0.50	-	**7.11**	0.42	**34.72**	nr	**37.32**
300 x 460mm	30.07	5.00	**31.57**	0.50	-	**7.11**	0.42	**39.10**	nr	**42.03**
360 x 500mm	34.65	5.00	**36.38**	0.50	-	**7.11**	0.42	**43.91**	nr	**47.20**
460 x 560mm	42.85	5.00	**44.99**	0.65	-	**9.24**	0.42	**54.65**	nr	**58.75**
460 x 600mm	44.81	5.00	**47.05**	0.65	-	**9.24**	0.42	**56.71**	nr	**60.96**
460 x 900mm	59.56	5.00	**62.54**	0.70	-	**9.95**	0.42	**72.90**	nr	**78.37**
500 x 680mm	51.45	5.00	**54.02**	0.65	-	**9.24**	0.42	**63.68**	nr	**68.45**
600 x 900mm	71.61	5.00	**75.19**	0.75	-	**10.66**	0.42	**86.27**	nr	**92.74**
holes 6mm diameter -4; edges bevelled										
254 x 400mm	32.77	5.00	**34.41**	0.50	-	**7.11**	0.42	**41.93**	nr	**45.08**
300 x 460mm	38.05	5.00	**39.95**	0.50	-	**7.11**	0.42	**47.48**	nr	**51.04**
360 x 500mm	43.68	5.00	**45.86**	0.50	-	**7.11**	0.42	**53.39**	nr	**57.39**
460 x 560mm	53.56	5.00	**56.24**	0.65	-	**9.24**	0.42	**65.89**	nr	**70.83**
460 x 600mm	55.94	5.00	**58.74**	0.65	-	**9.24**	0.42	**68.39**	nr	**73.52**
460 x 900mm	73.84	5.00	**77.53**	0.70	-	**9.95**	0.42	**87.90**	nr	**94.49**
500 x 680mm	63.84	5.00	**67.03**	0.65	-	**9.24**	0.42	**76.69**	nr	**82.44**
600 x 900mm	87.36	5.00	**91.73**	0.75	-	**10.66**	0.42	**102.80**	nr	**110.51**
N10: GENERAL FIXTURES/FURNISHING/EQUIPMENT: CURTAIN TRACKS										
Fixing only curtain track										
Metal or plastic track with fittings										
fixing with screws to softwood	-	-	-	0.40	-	**5.68**	-	**5.68**	m	**6.11**
fixing with screws to hardwood	-	-	-	0.60	-	**8.53**	-	**8.53**	m	**9.17**
N10: GENERAL FIXTURES/FURNISHING/EQUIPMENT: BLINDS										
Internal blinds										
Venetian blinds; stove enamelled aluminium alloy slats 25mm wide; plain colours										
1200mm drop; fixing to timber with screws										
1000mm wide	29.56	2.50	**30.30**	1.00	0.13	**15.60**	0.05	**45.96**	nr	**49.40**
2000mm wide	45.35	2.50	**46.48**	1.50	0.19	**23.35**	0.08	**69.91**	nr	**75.15**
3000mm wide	61.10	2.50	**62.63**	2.00	0.25	**31.10**	0.11	**93.83**	nr	**100.87**
Venetian blinds; stove enamelled aluminium alloy slats 50mm wide; plain colours										
1200mm drop; fixing to timber with screws										
1000mm wide	27.25	2.50	**27.94**	1.15	0.14	**17.84**	0.06	**45.84**	nr	**49.28**
2000mm wide	42.05	2.50	**43.10**	1.70	0.21	**26.41**	0.09	**69.60**	nr	**74.81**
3000mm wide	57.16	2.50	**58.59**	2.25	0.28	**34.97**	0.12	**93.69**	nr	**100.71**
Venetian blinds; stove enamelled aluminium alloy slats 25mm wide; plain colours with single cord control										
1200mm drop; fixing to timber with screws										
1000mm wide	42.05	2.50	**43.10**	1.00	0.13	**15.60**	0.05	**58.75**	nr	**63.16**
2000mm wide	65.69	2.50	**67.33**	1.50	0.19	**23.35**	0.08	**90.76**	nr	**97.57**
3000mm wide	89.02	2.50	**91.25**	2.00	0.25	**31.10**	0.11	**122.45**	nr	**131.64**
Venetian blinds; stove enamelled aluminium alloy slats 50mm wide; plain colours with single cord control										
1200mm drop; fixing to timber with screws										
1000mm wide	38.10	2.50	**39.05**	1.15	0.14	**17.84**	0.06	**56.96**	nr	**61.23**
2000mm wide	57.81	2.50	**59.26**	1.70	0.21	**26.41**	0.09	**85.75**	nr	**92.19**
3000mm wide	79.49	2.50	**81.48**	2.25	0.28	**34.97**	0.12	**116.57**	nr	**125.31**

Labour hourly rates: (except Specialists) Craft Operatives £14.21 Labourer £10.71 Rates are national average prices. Refer to REGIONAL VARIATIONS for indicative levels of overall pricing in regions	MATERIALS			LABOUR				RATES		
	Del to Site £	Waste %	Material Cost £	Craft Optve Hrs	Lab Hrs	Labour Cost £	Sunds £	Nett Rate £	Unit	Gross Rate (+7.5%) £
N10: GENERAL FIXTURES/FURNISHING/EQUIPMENT: BLINDS Cont.										
Internal blinds Cont.										
Venetian blinds; stove enamelled aluminium alloy slats 25mm wide; plain colours with channels for dimout										
1200mm drop; fixing to timber with screws										
1000mm wide	71.61	2.50	**73.40**	2.00	0.25	**31.10**	0.11	**104.60**	nr	**112.45**
2000mm wide	88.03	2.50	**90.23**	2.50	0.31	**38.85**	0.14	**129.21**	nr	**138.90**
3000mm wide	103.79	2.50	**106.39**	3.00	0.38	**46.70**	0.16	**153.25**	nr	**164.74**
Venetian blinds; stove enamelled aluminium alloy slats 50mm wide; plain colours with channels for dimout										
1200mm drop; fixing to timber with screws										
1000mm wide	68.97	2.50	**70.70**	2.25	0.28	**34.97**	0.12	**105.79**	nr	**113.72**
2000mm wide	84.75	2.50	**86.87**	2.80	0.35	**43.54**	0.15	**130.55**	nr	**140.35**
3000mm wide	99.86	2.50	**102.35**	3.35	0.42	**52.10**	0.18	**154.64**	nr	**166.23**
Roller blinds; automatic ratchet action; fire resistant material										
1200mm drop; fixing to timber with screws										
1000mm wide	55.19	2.50	**56.57**	1.00	0.13	**15.60**	0.05	**72.22**	nr	**77.64**
2000mm wide	67.67	2.50	**69.36**	1.50	0.19	**23.35**	0.08	**92.79**	nr	**99.75**
3000mm wide	80.13	2.50	**82.14**	2.00	0.25	**31.10**	0.11	**113.34**	nr	**121.85**
Roller blinds; automatic ratchet action; holland type material										
1200mm drop; fixing to timber with screws										
1000mm wide	53.22	2.50	**54.55**	1.00	0.13	**15.60**	0.05	**70.20**	nr	**75.47**
2000mm wide	63.72	2.50	**65.32**	1.50	0.19	**23.35**	0.08	**88.75**	nr	**95.40**
3000mm wide	74.20	2.50	**76.05**	2.00	0.25	**31.10**	0.11	**107.26**	nr	**115.31**
Roller blinds; self acting roller; blackout material										
1200mm drop; fixing to timber with screws										
1000mm wide	49.27	2.50	**50.50**	1.15	0.14	**17.84**	0.06	**68.41**	nr	**73.54**
2000mm wide	59.79	2.50	**61.28**	1.70	0.21	**26.41**	0.09	**87.78**	nr	**94.37**
3000mm wide	70.29	2.50	**72.05**	2.25	0.28	**34.97**	0.12	**107.15**	nr	**115.18**
Roller blinds; 100% blackout; natural anodised box and channels										
1200mm drop; fixing to timber with screws										
1000mm wide	162.92	2.50	**167.00**	2.00	0.25	**31.10**	0.11	**198.20**	nr	**213.07**
2000mm wide	202.95	2.50	**208.02**	2.75	0.34	**42.72**	0.15	**250.89**	nr	**269.71**
3000mm wide	248.32	2.50	**254.52**	3.50	0.44	**54.45**	0.19	**309.16**	nr	**332.35**
Vertical louvre blinds; 89mm wide louvres in standard material										
1200mm drop; fixing to timber with screws										
1000mm wide	59.12	2.50	**60.60**	0.90	0.11	**13.97**	0.05	**74.62**	nr	**80.21**
2000mm wide	103.79	2.50	**106.39**	1.35	0.17	**21.00**	0.07	**127.46**	nr	**137.02**
3000mm wide	148.47	2.50	**152.18**	1.80	0.23	**28.04**	0.10	**180.32**	nr	**193.84**
Vertical louvre blinds; 127mm wide louvres in standard material										
1200mm drop; fixing to timber with screws										
1000mm wide	53.22	2.50	**54.55**	0.95	0.12	**14.78**	0.05	**69.38**	nr	**74.59**
2000mm wide	92.63	2.50	**94.94**	1.40	0.18	**21.82**	0.08	**116.84**	nr	**125.61**
3000mm wide	132.70	2.50	**136.01**	1.85	0.23	**28.75**	0.10	**164.86**	nr	**177.23**
External blinds; manually operated										
Venetian blinds; stove enamelled aluminium slats 80mm wide; natural anodised side guides; excluding boxing										
1200mm drop; fixing to timber with screws										
1000mm wide	-	-	**Spclist**	-	-	**Spclist**	-	**242.69**	nr	**260.89**
2000mm wide	-	-	**Spclist**	-	-	**Spclist**	-	**361.20**	nr	**388.29**
3000mm wide	-	-	**Spclist**	-	-	**Spclist**	-	**486.50**	nr	**522.99**
Queensland awnings; acrylic material; natural anodised arms and boxing										
1000mm projection										
2000mm long	-	-	**Spclist**	-	-	**Spclist**	-	**568.89**	nr	**611.56**
3000mm long	-	-	**Spclist**	-	-	**Spclist**	-	**853.54**	nr	**917.56**
4000mm long	-	-	**Spclist**	-	-	**Spclist**	-	**1137.78**	nr	**1223.11**
Rollscreen vertical drop roller blinds; mesh material; natural anodised side guides and boxing										
1200mm drop; fixing to timber with screws										
1000mm wide	-	-	**Spclist**	-	-	**Spclist**	-	**237.01**	nr	**254.79**
2000mm wide	-	-	**Spclist**	-	-	**Spclist**	-	**355.56**	nr	**382.23**
3000mm wide	-	-	**Spclist**	-	-	**Spclist**	-	**473.87**	nr	**509.41**
Quandrant canopy; acrylic material; natural anodised frames										
1000mm projection										
2000mm long	-	-	**Spclist**	-	-	**Spclist**	-	**313.79**	nr	**337.32**
3000mm long	-	-	**Spclist**	-	-	**Spclist**	-	**479.72**	nr	**515.70**
4000mm long	-	-	**Spclist**	-	-	**Spclist**	-	**652.42**	nr	**701.35**
Foldaway awning; acrylic material; natural anodised arms and front rail										
2000mm projection										
2000mm long	-	-	**Spclist**	-	-	**Spclist**	-	**509.07**	nr	**547.25**
3000mm long	-	-	**Spclist**	-	-	**Spclist**	-	**747.23**	nr	**803.27**
4000mm long	-	-	**Spclist**	-	-	**Spclist**	-	**971.86**	nr	**1044.75**

Labour hourly rates: (except Specialists) Craft Operatives £14.21 Labourer £10.71 Rates are national average prices. Refer to REGIONAL VARIATIONS for indicative levels of overall pricing in regions	MATERIALS			LABOUR				RATES		
	Del to Site £	Waste %	Material Cost £	Craft Optve Hrs	Lab Hrs	Labour Cost £	Sunds £	Nett Rate £	Unit	Gross Rate (+7.5%) £
N10: GENERAL FIXTURES/FURNISHING/EQUIPMENT: BLINDS Cont.										
External blinds; electrically operated										
Venetian blinds; stove enamelled aluminium slats 80mm wide; natural anodised side guides; excluding boxing										
1200mm drop; fixing to timber with screws										
1000mm wide	-	-	**Spclist**	-	-	**Spclist**	-	**557.60**	nr	**599.42**
2000mm wide	-	-	**Spclist**	-	-	**Spclist**	-	**681.77**	nr	**732.90**
3000mm wide	-	-	**Spclist**	-	-	**Spclist**	-	**799.16**	nr	**859.10**
Queensland awnings; acrylic material; natural anodised arms and boxing										
1000mm projection										
2000mm long	-	-	**Spclist**	-	-	**Spclist**	-	**652.42**	nr	**701.35**
3000mm long	-	-	**Spclist**	-	-	**Spclist**	-	**931.22**	nr	**1001.06**
4000mm long	-	-	**Spclist**	-	-	**Spclist**	-	**1185.19**	nr	**1274.08**
Rollscreen vertical drop roller blinds; mesh material; natural anodised side guides and boxing										
1200mm drop; fixing to timber with screws										
1000mm wide	-	-	**Spclist**	-	-	**Spclist**	-	**451.50**	nr	**485.36**
2000mm wide	-	-	**Spclist**	-	-	**Spclist**	-	**568.89**	nr	**611.56**
3000mm wide	-	-	**Spclist**	-	-	**Spclist**	-	**687.41**	nr	**738.97**
Quandrant canopy; acrylic material; natural anodised frames										
1000mm projection										
2000mm long	-	-	**Spclist**	-	-	**Spclist**	-	**408.61**	nr	**439.26**
3000mm long	-	-	**Spclist**	-	-	**Spclist**	-	**568.89**	nr	**611.56**
4000mm long	-	-	**Spclist**	-	-	**Spclist**	-	**758.52**	nr	**815.41**
Foldaway awning; acrylic material; natural anodised arms and front rail										
2000mm projection										
2000mm long	-	-	**Spclist**	-	-	**Spclist**	-	**746.11**	nr	**802.07**
3000mm long	-	-	**Spclist**	-	-	**Spclist**	-	**977.50**	nr	**1050.81**
4000mm long	-	-	**Spclist**	-	-	**Spclist**	-	**1241.63**	nr	**1334.75**
N10: GENERAL FIXTURES/FURNISHING/EQUIPMENT: STORAGE SYSTEMS										
Vista two tone coloured steel boltless storage systems; Welconstruct Co. Ltd.; units 900mm wide x 1800mm high; assembling										
Open shelving with top, bottom, three shelves and braces; placing in position										
300mm deep code L106-101	126.30	2.50	**129.46**	1.00	0.50	**19.57**	-	**149.02**	nr	**160.20**
450mm deep code L106-102	159.00	2.50	**162.98**	1.20	0.60	**23.48**	-	**186.45**	nr	**200.44**
600mm deep code L106-103	183.50	2.50	**188.09**	1.40	0.70	**27.39**	-	**215.48**	nr	**231.64**
Closed shelving with top, bottom, and three shelves; placing in position										
300mm deep code L106-111	178.00	2.50	**182.45**	1.10	0.55	**21.52**	-	**203.97**	nr	**219.27**
extra for additional shelf code L106-301	12.40	2.50	**12.71**	0.20	0.10	**3.91**	-	**16.62**	nr	**17.87**
450mm deep code L106-112	214.00	2.50	**219.35**	1.30	0.65	**25.43**	-	**244.78**	nr	**263.14**
extra for additional shelf code L106-302	18.90	2.50	**19.37**	0.20	0.10	**3.91**	-	**23.29**	nr	**25.03**
600mm deep code L106-113	241.00	2.50	**247.03**	1.50	0.75	**29.35**	-	**276.37**	nr	**297.10**
extra for additional shelf code L106-303	23.60	2.50	**24.19**	0.20	0.10	**3.91**	-	**28.10**	nr	**30.21**
Extra over unit for										
pair of lockable doors code L106-401	303.00	2.50	**310.58**	0.20	0.10	**3.91**	-	**314.49**	nr	**338.07**
N10: GENERAL FIXTURES/FURNISHING/EQUIPMENT: SHELVING SYSTEMS										
Spanwel wide access boltless steel shelving systems; Welconstruct Co Ltd.; upright frames connected with shelf beams and with shelves; assembling										
Unit 1800mm high with 2 upright frames and three standard duty steel shelves including top and bottom; placing in position										
1200mm long x 650mm deep code L108-201	270.00	2.50	**276.75**	1.40	0.70	**27.39**	-	**304.14**	nr	**326.95**
extra; additional standard duty steel shelf and shelf beam code L108-225	58.90	2.50	**60.37**	0.20	0.10	**3.91**	-	**64.29**	nr	**69.11**
1200mm long x 800mm deep code L108-202	285.00	2.50	**292.13**	1.50	0.75	**29.35**	-	**321.47**	nr	**345.58**
extra; additional standard duty steel shelf and shelf beam L108-226	62.30	2.50	**63.86**	0.20	0.10	**3.91**	-	**67.77**	nr	**72.85**
1200mm long x 950mm deep code L108-203	299.00	2.50	**306.48**	1.60	0.80	**31.30**	-	**337.78**	nr	**363.11**
extra; additional standard duty steel shelf and shelf beam code L108-227	66.10	2.50	**67.75**	0.20	0.10	**3.91**	-	**71.67**	nr	**77.04**
2400mm long x 650mm deep code L108-204	386.00	2.50	**395.65**	1.50	0.75	**29.35**	-	**425.00**	nr	**456.87**
extra; additional standard duty steel shelf and shelf beam code L108-228	97.40	2.50	**99.84**	0.20	0.10	**3.91**	-	**103.75**	nr	**111.53**
2400mm long x 800mm deep code L108-205	410.00	2.50	**420.25**	1.60	0.80	**31.30**	-	**451.55**	nr	**485.42**
extra; additional standard duty steel shelf and shelf beam code L108-229	103.90	2.50	**106.50**	0.20	0.10	**3.91**	-	**110.41**	nr	**118.69**
2400mm long x 950mm deep code L108-206	435.00	2.50	**445.88**	1.70	0.85	**33.26**	-	**479.14**	nr	**515.07**
extra; additional standard duty steel shelf and shelf beam code L108-230	111.30	2.50	**114.08**	0.20	0.10	**3.91**	-	**118.00**	nr	**126.85**

Labour hourly rates: (except Specialists) Craft Operatives £14.21 Labourer £10.71 Rates are national average prices. Refer to REGIONAL VARIATIONS for indicative levels of overall pricing in regions	MATERIALS			LABOUR				RATES		
	Del to Site £	Waste %	Material Cost £	Craft Optve Hrs	Lab Hrs	Labour Cost £	Sunds £	Nett Rate £	Unit	Gross Rate (+7.5%) £
N10: GENERAL FIXTURES/FURNISHING/EQUIPMENT: SHELVING SYSTEMS Cont.										
Spanwel wide access boltless steel shelving systems; Welconstruct Co Ltd.; upright frames connected with shelf beams and with shelves; assembling Cont.										
Unit 2400mm high with 3 upright frames and five standard duty steel shelves including top and bottom; placing in position										
2400mm long x 650mm deep	670.00	2.50	**686.75**	2.80	1.40	**54.78**	-	**741.53**	nr	**797.15**
2400mm long x 800mm deep	707.00	2.50	**724.68**	3.00	1.50	**58.70**	-	**783.37**	nr	**842.12**
2400mm long x 950mm deep	761.00	2.50	**780.03**	3.20	1.60	**62.61**	-	**842.63**	nr	**905.83**
N10: GENERAL FIXTURES/FURNISHING/EQUIPMENT: SHELVING SYSTEMS - SHELVING SUPPORT SYSTEMS										
Spur patent shelf supports in steel with white enamelled finish; Spur Shelving Ltd; Spur Steel-Lok components										
Wall uprights										
type 9011, 430mm long; fixing to masonry with screws	1.65	2.50	**1.69**	0.25	-	**3.55**	-	**5.24**	nr	**5.64**
type 9013, 1000mm long; fixing to masonry with screws	3.25	2.50	**3.33**	0.45	-	**6.39**	-	**9.73**	nr	**10.46**
type 9016, 1600mm long; fixing to masonry with screws	4.95	2.50	**5.07**	0.55	-	**7.82**	-	**12.89**	nr	**13.86**
type 9020, 2400mm long; fixing to masonry with screws	6.95	2.50	**7.12**	0.70	-	**9.95**	-	**17.07**	nr	**18.35**
Straight brackets										
type 9001, 120mm long	0.95	2.50	**0.97**	0.05	-	**0.71**	-	**1.68**	nr	**1.81**
type 9003, 270mm long	1.40	2.50	**1.44**	0.05	-	**0.71**	-	**2.15**	nr	**2.31**
type 9004, 360mm long	1.95	2.50	**2.00**	0.07	-	**0.99**	-	**2.99**	nr	**3.22**
type 9006, 470mm long	3.25	2.50	**3.33**	0.07	-	**0.99**	-	**4.33**	nr	**4.65**
N10: GENERAL FIXTURES/FURNISHING/EQUIPMENT: MAT RIMS										
Matwells in Aluminium										
Mat frames; welded fabrication										
34 x 26 x 6mm angle section; angles mitred; plain lugs -4, welded on; mat space										
610 x 457mm	38.39	2.50	**39.35**	1.00	1.00	**24.92**	-	**64.27**	nr	**69.09**
762 x 457mm	44.01	2.50	**45.11**	1.25	1.25	**31.15**	-	**76.26**	nr	**81.98**
914 x 610mm	55.08	2.50	**56.46**	1.50	1.50	**37.38**	-	**93.84**	nr	**100.87**
Matwells in polished brass										
Mat frames; brazed fabrication										
38 x 38 x 6mm angle section; angles mitred; plain lugs -4, welded on; mat space										
610 x 457mm	118.80	2.50	**121.77**	1.25	1.25	**31.15**	-	**152.92**	nr	**164.39**
762 x 457mm	140.40	2.50	**143.91**	1.50	1.50	**37.38**	-	**181.29**	nr	**194.89**
914 x 610mm	183.60	2.50	**188.19**	1.75	1.75	**43.61**	-	**231.80**	nr	**249.19**
N10: GENERAL FIXTURES/FURNISHING/EQUIPMENT: CLOTHES LOCKERS										
Individual compartment clothes lockers; Welconstruct Co. Ltd.; steel, stove enamelled finish; standard colour solid doors and with cam locks 442 series. 12-19 ordered										
2 compartment; placing in position										
305 x 305 x 1830mm code L442-021	70.50	2.50	**72.26**	-	0.50	**5.36**	-	**77.62**	nr	**83.44**
305 x 460 x 1830mm code L442-022	76.70	2.50	**78.62**	-	0.55	**5.89**	-	**84.51**	nr	**90.85**
380 x 380 x 1830mm code L442-024	78.80	2.50	**80.77**	-	0.60	**6.43**	-	**87.20**	nr	**93.74**
460 x 460 x 1830mm code L442-025	97.40	2.50	**99.84**	-	0.70	**7.50**	-	**107.33**	nr	**115.38**
3 compartment; placing in position										
305 x 305 x 1830mm code L442-026	75.70	2.50	**77.59**	-	0.50	**5.36**	-	**82.95**	nr	**89.17**
305 x 460 x 1830mm code L442-027	85.00	2.50	**87.13**	-	0.55	**5.89**	-	**93.02**	nr	**99.99**
380 x 380 x 1830mm code L442-029	85.00	2.50	**87.13**	-	0.60	**6.43**	-	**93.55**	nr	**100.57**
460 x 460 x 1830mm code L442-030	102.60	2.50	**105.17**	-	0.70	**7.50**	-	**112.66**	nr	**121.11**
4 compartment; placing in position										
305 x 305 x 1830mm code L442-031	81.90	2.50	**83.95**	-	0.50	**5.36**	-	**89.30**	nr	**96.00**
305 x 460 x 1830mm code L442-032	90.20	2.50	**92.46**	-	0.50	**5.36**	-	**97.81**	nr	**105.15**
380 x 380 x 1830mm code L442-034	92.30	2.50	**94.61**	-	0.60	**6.43**	-	**101.03**	nr	**108.61**
460 x 460 x 1830mm code L442-035	113.00	2.50	**115.83**	-	0.70	**7.50**	-	**123.32**	nr	**132.57**
6 compartment; placing in position										
305 x 305 x 1830mm code L442-036	101.60	2.50	**104.14**	-	0.50	**5.36**	-	**109.50**	nr	**117.71**
305 x 460 x 1830mm code L442-037	109.90	2.50	**112.65**	-	0.55	**5.89**	-	**118.54**	nr	**127.43**
380 x 380 x 1830mm code L442-039	109.90	2.50	**112.65**	-	0.60	**6.43**	-	**119.07**	nr	**128.00**
460 x 460 x 1830mm code L442-040	130.60	2.50	**133.87**	-	0.70	**7.50**	-	**141.36**	nr	**151.96**
Swimming bath lockers; Welconstruct Co Ltd.; galvanized steel, stove enamelled finish; standard colour doors with mastered series cam locks 452 series. 20-49 ordered										
1 compartment; placing in position										
305 x 305 x 1830mm code L452-001	100.60	2.50	**103.12**	-	0.50	**5.36**	-	**108.47**	nr	**116.61**

Labour hourly rates: (except Specialists) Craft Operatives £14.21 Labourer £10.71 Rates are national average prices. Refer to REGIONAL VARIATIONS for indicative levels of overall pricing in regions	MATERIALS			LABOUR				RATES		
	Del to Site £	Waste %	Material Cost £	Craft Optve Hrs	Lab Hrs	Labour Cost £	Sunds £	Nett Rate £	Unit	Gross Rate (+7.5%) £
N10: GENERAL FIXTURES/FURNISHING/EQUIPMENT: CLOTHES LOCKERS Cont.										
Swimming bath lockers; Welconstruct Co Ltd.; galvanized steel, stove enamelled finish; standard colour doors with mastered series cam locks 452 series. 20-49 ordered Cont.										
2 compartment; placing in position										
305 x 305 x 1830mm code L452-002	130.40	2.50	**133.66**	-	0.50	**5.36**	-	**139.02**	nr	**149.44**
305 x 460 x 1830mm code L452-003	136.60	2.50	**140.02**	-	0.55	**5.89**	-	**145.91**	nr	**156.85**
3 compartment; placing in position										
305 x 305 x 1830mm code L452-004	153.50	2.50	**157.34**	-	0.50	**5.36**	-	**162.69**	nr	**174.89**
305 x 460 x 1830mm code L452-005	161.50	2.50	**165.54**	-	0.55	**5.89**	-	**171.43**	nr	**184.29**
4 compartment; placing in position										
305 x 305 x 1830mm code L452-006	165.50	2.50	**169.64**	-	0.50	**5.36**	-	**174.99**	nr	**188.12**
305 x 460 x 1830mm code L452-007	177.00	2.50	**181.43**	-	0.55	**5.89**	-	**187.32**	nr	**201.36**
N10: GENERAL FIXTURES/FURNISHING/EQUIPMENT: CLOAK ROOM EQUIPMENT										
Static square tube double sided coat racks; Welconstruct Co. Ltd.; hardwood seats; in assembled units										
Racks 1500mm long x 1675mm high x 610mm deep; placing in position; 40mm square tube										
10 hooks and angle framed steel mesh single shoe tray, reference L456-102 AND 109; placing in position	468.30	2.50	**480.01**	-	0.50	**5.36**	-	**485.36**	nr	**521.76**
10 hooks, reference L456-102; placing in position	405.00	2.50	**415.13**	-	0.50	**5.36**	-	**420.48**	nr	**452.02**
10 hooks and 5 baskets L456-102 and 106	471.70	2.50	**483.49**	-	0.50	**5.36**	-	**488.85**	nr	**525.51**
10 hooks and 10 baskets L456-102 and 107	518.20	2.50	**531.16**	-	0.50	**5.36**	-	**536.51**	nr	**576.75**
Static square tube single sided coat racks; Welconstruct Co. Ltd.; mobile; in assembled units										
Racks 1500mm long x 1825mm high x 600mm deep; placing in position; 40mm square tube										
15 hangers and top tray, reference L456-103	391.00	2.50	**400.78**	-	0.50	**5.36**	-	**406.13**	nr	**436.59**
Free-standing bench seats; Welconstruct Co. Ltd.; steel square tube framing; hardwood seats; placing in position										
Bench seat with shelf 1500mm long x 300mm deep and 450mm high; placing in position; 40mm square tube										
single sided 300mm deep, reference L456-115	189.50	2.50	**194.24**	-	0.50	**5.36**	-	**199.59**	nr	**214.56**
double sided 600mm deep L456-116	305.00	2.50	**312.63**	-	0.50	**5.36**	-	**317.98**	nr	**341.83**
Wall rack Welconstruct Co. Ltd; hardwood										
1500mm long; fixing to masonry										
5 hooks, reference L456-029	45.45	2.50	**46.59**	-	0.60	**6.43**	-	**53.01**	nr	**56.99**
N10: GENERAL FIXTURES/FURNISHING/EQUIPMENT: HAT AND COAT RAILS										
Hat and coat rails										
Rails in softwood, wrought										
25 x 75mm; chamfers -2	1.29	10.00	**1.42**	0.25	0.03	**3.87**	0.01	**5.30**	m	**5.70**
25 x 100mm; chamfers -2	1.73	10.00	**1.90**	0.25	0.03	**3.87**	0.01	**5.79**	m	**6.22**
25 x 75mm; chamfers -2; fixing to masonry with screws	1.29	10.00	**1.42**	0.50	0.06	**7.75**	0.03	**9.19**	m	**9.88**
25 x 100mm; chamfers -2; fixing to masonry with screws	1.73	10.00	**1.90**	0.50	0.06	**7.75**	0.03	**9.68**	m	**10.40**
Rails in Afromosia, wrought, selected for transparent finish										
25 x 75mm; chamfers -2	3.86	10.00	**4.25**	0.50	0.06	**7.75**	0.03	**12.02**	m	**12.92**
25 x 100mm; chamfers -2	5.16	10.00	**5.68**	0.50	0.06	**7.75**	0.03	**13.45**	m	**14.46**
25 x 75mm; chamfers -2; fixing to masonry with screws, countersinking and pellating	3.86	10.00	**4.25**	1.00	0.12	**15.50**	0.05	**19.79**	m	**21.28**
25 x 100mm; chamfers -2; fixing to masonry with screws, countersinking and pellating	5.16	10.00	**5.68**	1.00	0.12	**15.50**	0.05	**21.22**	m	**22.82**
Hat and coat hooks										
B.M.A. finish	2.00	2.50	**2.05**	0.25	-	**3.55**	-	**5.60**	nr	**6.02**
chromium plated	2.00	2.50	**2.05**	0.25	-	**3.55**	-	**5.60**	nr	**6.02**
SAA finish	1.00	2.50	**1.03**	0.25	-	**3.55**	-	**4.58**	nr	**4.92**
N11: DOMESTIC KITCHEN FITTINGS										
Standard melamine finish on chipboard units, with backs										
Wall units										
fixing to masonry with screws										
400 x 300 x 600mm	68.24	2.50	**69.95**	1.75	0.22	**27.22**	-	**97.17**	nr	**104.46**
400 x 300 x 900mm	88.77	2.50	**90.99**	1.85	0.23	**28.75**	-	**119.74**	nr	**128.73**
500 x 300 x 600mm	73.29	2.50	**75.12**	1.90	0.25	**29.68**	-	**104.80**	nr	**112.65**
500 x 300 x 900mm	92.92	2.50	**95.24**	1.95	0.25	**30.39**	-	**125.63**	nr	**135.05**
600 x 300 x 600mm	80.07	2.50	**82.07**	1.95	0.25	**30.39**	-	**112.45**	nr	**120.89**

Labour hourly rates: (except Specialists) Craft Operatives £14.21 Labourer £10.71 Rates are national average prices. Refer to REGIONAL VARIATIONS for indicative levels of overall pricing in regions	MATERIALS			LABOUR				RATES		
	Del to Site £	Waste %	Material Cost £	Craft Optve Hrs	Lab Hrs	Labour Cost £	Sunds £	Nett Rate £	Unit	Gross Rate (+7.5%) £
N11: DOMESTIC KITCHEN FITTINGS Cont.										
Standard melamine finish on chipboard units, with backs Cont.										
Wall units Cont.										
fixing to masonry with screws Cont.										
600 x 300 x 900mm	96.86	2.50	**99.28**	2.05	0.26	**31.92**	-	**131.20**	nr	**141.04**
1000 x 300 x 600mm	128.64	2.50	**131.85**	2.35	0.30	**36.61**	-	**168.46**	nr	**181.09**
1000 x 300 x 900mm	156.78	2.50	**160.70**	2.50	0.31	**38.85**	-	**199.55**	nr	**214.52**
1200 x 300 x 600mm	139.71	2.50	**143.20**	2.50	0.31	**38.85**	-	**182.05**	nr	**195.70**
1200 x 300 x 900mm	159.08	2.50	**163.05**	2.70	0.35	**42.12**	-	**205.17**	nr	**220.56**
Floor units on plinths and with plastic faced worktops; without drawers										
fixing to masonry with screws										
400 x 600 x 900mm	129.16	2.50	**132.39**	2.00	0.25	**31.10**	-	**163.48**	nr	**175.75**
500 x 600 x 900mm	141.88	2.50	**145.43**	2.20	0.28	**34.26**	-	**179.69**	nr	**193.16**
600 x 600 x 900mm	151.90	2.50	**155.69**	2.40	0.30	**37.32**	-	**193.01**	nr	**207.49**
1000 x 600 x 900mm	253.79	2.50	**260.14**	3.20	0.40	**49.76**	-	**309.89**	nr	**333.13**
1200 x 600 x 900mm	275.79	2.50	**282.69**	3.50	0.44	**54.45**	-	**337.13**	nr	**362.42**
Floor units on plinths and with plastic faced worktops; with one drawer										
fixing to masonry with screws										
500 x 600 x 900mm	177.07	2.50	**181.50**	2.20	0.28	**34.26**	-	**215.76**	nr	**231.94**
600 x 600 x 900mm	187.09	2.50	**191.77**	2.40	0.30	**37.32**	-	**229.09**	nr	**246.27**
1000 x 600 x 900mm	304.85	2.50	**312.47**	3.20	0.40	**49.76**	-	**362.23**	nr	**389.39**
1200 x 600 x 900mm	329.43	2.50	**337.67**	3.50	0.44	**54.45**	-	**392.11**	nr	**421.52**
Floor units on plinths and with plastic faced worktops; with four drawers										
fixing to masonry with screws										
500 x 600 x 900mm	246.99	2.50	**253.16**	2.20	0.28	**34.26**	-	**287.42**	nr	**308.98**
600 x 600 x 900mm	261.08	2.50	**267.61**	2.40	0.30	**37.32**	-	**304.92**	nr	**327.79**
Sink units on plinths										
1200 x 600 x 900mm, with one drawer; fixing to masonry with screws	233.94	2.50	**239.79**	3.50	0.44	**54.45**	-	**294.24**	nr	**316.30**
1200 x 600 x 900mm, without drawer; fixing to masonry with screws	178.72	2.50	**183.19**	3.50	0.44	**54.45**	-	**237.64**	nr	**255.46**
1500 x 600 x 900mm, without drawer; fixing to masonry with screws	279.88	2.50	**286.88**	3.75	0.47	**58.32**	-	**345.20**	nr	**371.09**
Store cupboards on plinths										
fixing to masonry with screws										
500 x 600 x 1950mm without shelves	206.07	2.50	**211.22**	3.60	0.45	**55.98**	-	**267.20**	nr	**287.24**
500 x 600 x 1950mm with shelves	220.43	2.50	**225.94**	3.60	0.45	**55.98**	-	**281.92**	nr	**303.06**
600 x 600 x 1950mm without shelves	224.32	2.50	**229.93**	4.00	0.50	**62.20**	-	**292.13**	nr	**314.03**
600 x 600 x 1950mm with shelves	241.92	2.50	**247.97**	4.00	0.50	**62.20**	-	**310.16**	nr	**333.42**
Laminated plastic faced units complete with ironmongery and with backs										
Wall units										
400 x 300 x 600mm, with one door; fixing to masonry with screws	136.42	2.50	**139.83**	1.75	0.22	**27.22**	-	**167.06**	nr	**179.59**
500 x 300 x 600mm, with one door; fixing to masonry with screws	146.55	2.50	**150.21**	1.85	0.23	**28.75**	-	**178.96**	nr	**192.39**
1000 x 300 x 600mm, with two doors; fixing to masonry with screws	257.26	2.50	**263.69**	2.35	0.30	**36.61**	-	**300.30**	nr	**322.82**
1200 x 300 x 600mm, with two doors; fixing to masonry with screws	279.41	2.50	**286.39**	2.50	0.31	**38.85**	-	**325.24**	nr	**349.63**
Floor units on plinths and with worktops										
500 x 600 x 900mm, with one drawer and one cupboard; fixing to masonry with screws	354.16	2.50	**363.01**	2.20	0.28	**34.26**	-	**397.27**	nr	**427.07**
500 x 600 x 900mm, with four drawers; fixing to masonry with screws	493.98	2.50	**506.33**	2.40	0.30	**37.32**	-	**543.64**	nr	**584.42**
1000 x 600 x 900mm, with two drawers and two cupboards; fixing to masonry with screws	609.69	2.50	**624.93**	3.20	0.40	**49.76**	-	**674.68**	nr	**725.29**
1500 x 600 x 900mm, with three drawers and three cupboards; fixing to masonry with screws	787.71	2.50	**807.40**	3.85	0.48	**59.85**	-	**867.25**	nr	**932.29**
Sink units on plinths										
1000 x 600 x 900mm, with one drawer and two cupboards, fixing to masonry with screws	467.67	2.50	**479.36**	3.65	0.47	**56.90**	-	**536.26**	nr	**576.48**
1500 x 600 x 900mm, with two drawers and three cupboards; fixing to masonry with screws	699.54	2.50	**717.03**	3.75	0.50	**58.64**	-	**775.67**	nr	**833.84**
Tall units on plinths										
500 x 600 x 1950mm, broom cupboard; fixing to masonry with screws	430.15	2.50	**440.90**	3.75	0.50	**58.64**	-	**499.55**	nr	**537.01**
600 x 600 x 1950mm, larder unit; fixing to masonry with screws	448.83	2.50	**460.05**	3.75	0.50	**58.64**	-	**518.69**	nr	**557.59**
Hardwood veneered units complete with ironmongery but without backs										
Wardrobe units										
fixing to masonry with screws										
1000 x 600 x 2175mm	497.70	2.50	**510.14**	4.75	0.60	**73.92**	-	**584.07**	nr	**627.87**
1500 x 600 x 2175mm	625.28	2.50	**640.91**	4.75	0.60	**73.92**	-	**714.83**	nr	**768.44**

Labour hourly rates: (except Specialists) Craft Operatives £14.21 Labourer £10.71 Rates are national average prices. Refer to REGIONAL VARIATIONS for indicative levels of overall pricing in regions	MATERIALS			LABOUR				RATES		
	Del to Site £	Waste %	Material Cost £	Craft Optve Hrs	Lab Hrs	Labour Cost £	Sunds £	Nett Rate £	Unit	Gross Rate (+7.5%) £
N11: DOMESTIC KITCHEN FITTINGS Cont.										
Pre-Assembled kitchen units, white melamine finish; Module 500 range; Magnet PLC										
Drawerline units; fixing to masonry with screws										
floor units										
300mm F300	82.71	2.50	**84.78**	1.25	0.31	**21.08**	-	**105.86**	nr	**113.80**
500mm F1	94.54	2.50	**96.91**	1.25	0.34	**21.40**	-	**118.31**	nr	**127.18**
600mm F6	98.76	2.50	**101.23**	1.25	0.34	**21.40**	-	**122.63**	nr	**131.83**
1000mm F2	155.39	2.50	**159.27**	1.75	0.45	**29.69**	-	**188.96**	nr	**203.13**
hob floor units										
1000mm H2	120.68	2.50	**123.69**	1.75	0.44	**29.58**	-	**153.27**	nr	**164.77**
sink units										
1000mm S2	114.80	2.50	**117.67**	1.75	0.44	**29.58**	-	**147.25**	nr	**158.30**
Tall units; fixing to masonry with screws										
floor units										
500mm DF1	74.23	2.50	**76.09**	2.50	0.30	**38.74**	-	**114.82**	nr	**123.44**
Drawer units; fixing to masonry with screws										
four drawer floor units										
500mm FD1	184.18	2.50	**188.79**	1.25	0.46	**22.69**	-	**211.48**	nr	**227.34**
Wall units; fixing to masonry with screws										
standard units 600mm high										
300mm W306	41.30	2.50	**42.33**	2.00	0.32	**31.85**	-	**74.18**	nr	**79.74**
500mm W7	75.92	2.50	**77.81**	2.00	0.35	**32.17**	-	**109.98**	nr	**118.23**
600mm W5	41.30	2.50	**42.33**	2.00	0.37	**32.38**	-	**74.71**	nr	**80.32**
1000mm W2	88.59	2.50	**90.80**	2.60	0.47	**41.98**	-	**132.78**	nr	**142.74**
standard corner units 900mm high										
600 x 600mm CW6	121.52	2.50	**124.56**	3.00	0.46	**47.56**	-	**172.12**	nr	**185.02**
Worktops; fixing with screws										
round front edge										
38mm worktops, round front edge ADEN Range	25.33	25.00	**31.67**	1.50	0.19	**23.35**	-	**55.02**	m	**59.14**
Plinth										
Minster 500	3.67	25.00	**4.59**	0.25	0.03	**3.87**	-	**8.46**	m	**9.10**
Cornice										
Minster 500	6.22	25.00	**7.78**	0.25	0.03	**3.87**	-	**11.65**	m	**12.52**
Pelmet										
Minster 500	4.04	25.00	**5.05**	0.25	0.03	**3.87**	-	**8.92**	m	**9.59**
Worktop trim										
coloured	7.34	25.00	**9.18**	0.25	0.03	**3.87**	-	**13.05**	m	**14.03**
N13: SANITARY APPLIANCES/FITTINGS										
Attendance on sanitary fittings										
Note:										
Sanitary fittings are usually included as p.c. items or provisional sums, the following are the allowances for attendance by the main contractor i.e. unloading, storing and distributing fittings for fixing by the plumber and returning empty cases and packings										
Sinks, fireclay										
610 x 457 x 254mm	-	-	**-**	-	1.00	**10.71**	0.55	**11.26**	nr	**12.11**
762 x 508 x 254mm	-	-	**-**	-	1.25	**13.39**	0.69	**14.08**	nr	**15.14**
add if including tubular stands	-	-	**-**	-	0.75	**8.03**	0.46	**8.49**	nr	**9.13**
Wash basins, 559 x 406mm; earthenware or fireclay										
single	-	-	**-**	-	0.80	**8.57**	0.46	**9.03**	nr	**9.70**
range of 4 with cover overlaps between basins	-	-	**-**	-	2.50	**26.78**	1.24	**28.02**	nr	**30.12**
Combined sinks and drainers; stainless steel										
1067 x 533mm	-	-	**-**	-	1.40	**14.99**	0.82	**15.81**	nr	**17.00**
1600 x 610mm	-	-	**-**	-	1.70	**18.21**	0.97	**19.18**	nr	**20.61**
add if including tubular stands	-	-	**-**	-	0.75	**8.03**	0.46	**8.49**	nr	**9.13**
Combined sinks and drainers; porcelain enamel										
1067 x 533mm	-	-	**-**	-	1.40	**14.99**	0.82	**15.81**	nr	**17.00**
1600 x 610mm	-	-	**-**	-	1.70	**18.21**	0.97	**19.18**	nr	**20.61**
Baths excluding panels; pressed steel										
694 x 1688 x 570mm overall	-	-	**-**	-	2.50	**26.78**	1.20	**27.98**	nr	**30.08**
W.C. suites; china										
complete with WWP, flush pipe etc.,	-	-	**-**	-	1.40	**14.99**	0.82	**15.81**	nr	**17.00**
Block pattern urinals; with glazed ends, back and channel in one piece with separate tread including flushing cistern etc.,										
single stall	-	-	**-**	-	2.75	**29.45**	1.38	**30.83**	nr	**33.14**
range of four with loose overlaps	-	-	**-**	-	6.00	**64.26**	3.21	**67.47**	nr	**72.53**
Wall urinals, bowl type, with flushing cistern, etc.,										
single	-	-	**-**	-	1.40	**14.99**	0.82	**15.81**	nr	**17.00**
range of 3 with two divisions	-	-	**-**	-	3.30	**35.34**	1.65	**37.00**	nr	**39.77**

Labour hourly rates: (except Specialists) Craft Operatives £14.21 Labourer £10.71 Rates are national average prices. Refer to REGIONAL VARIATIONS for indicative levels of overall pricing in regions	MATERIALS			LABOUR				RATES		
	Del to Site £	Waste %	Material Cost £	Craft Optve Hrs	Lab Hrs	Labour Cost £	Sunds £	Nett Rate £	Unit	Gross Rate (+7.5%) £
N13: SANITARY APPLIANCES/FITTINGS Cont.										
Attendance on sanitary fittings Cont.										
Slab urinals with one return end, flushing cistern, sparge pipe, etc.,										
1219mm long x 1067mm high	-	-	-	-	2.75	**29.45**	1.84	**31.29**	nr	**33.64**
1829mm long x 1067mm high	-	-	-	-	3.75	**40.16**	2.75	**42.92**	nr	**46.14**

This section continues on the next page

Labour hourly rates: (except Specialists) Craft Operatives £17.38 Labourer £10.71 Rates are national average prices. Refer to REGIONAL VARIATIONS for indicative levels of overall pricing in regions	MATERIALS			LABOUR				RATES		
	Del to Site £	Waste %	Material Cost £	Craft Optve Hrs	Lab Hrs	Labour Cost £	Sunds £	Nett Rate £	Unit	Gross Rate (+7.5%) £
N13: SANITARY APPLIANCES/FITTINGS Cont.										
Fix only appliances (prime cost sum for supply included elsewhere)										
Sink units, stainless steel; combined overflow and waste outlet, plug and chain; pair pillar taps										
fixing on base unit with metal clips										
1000 x 600mm	-	-	-	3.00	-	**52.14**	0.82	**52.96**	nr	**56.93**
1600 x 600mm	-	-	-	3.50	-	**60.83**	0.82	**61.65**	nr	**66.27**
Sinks; white glazed fireclay; waste outlet, plug, chain and stay; cantilever brackets; pair bib taps										
fixing brackets to masonry with screws; sealing at back with white sealant										
610 x 457 x 254mm	-	-	-	4.00	-	**69.52**	1.33	**70.85**	nr	**76.16**
762 x 508 x 254mm	-	-	-	4.00	-	**69.52**	1.33	**70.85**	nr	**76.16**
Sinks; white glazed fireclay; waste outlet, plug, chain and stay; legs and screw-to-wall bearers; pair bib taps										
fixing legs to masonry with screws; sealing at back with mastic sealant										
610 x 457 x 254mm	-	-	-	4.50	-	**78.21**	1.33	**79.54**	nr	**85.50**
762 x 508 x 254mm	-	-	-	4.50	-	**78.21**	1.33	**79.54**	nr	**85.50**
Wash basins; vitreous china; waste outlet, plug, chain and stay; screw to wall brackets; pair pillar taps										
560 x 406mm; fixing brackets to masonry with screws; sealing at back with mastic sealant	-	-	-	3.00	-	**52.14**	1.33	**53.47**	nr	**57.48**
range of four, each 559 x 406mm with overlap strips; fixing brackets to masonry with screws; sealing overlap strips and at back with mastic sealant	-	-	-	12.75	-	**221.60**	5.05	**226.64**	nr	**243.64**
Wash basins; vitreous china; waste outlet, plug, chain and stay; legs and screw-to-wall bearers										
560 x 406mm; fixing legs to masonry with screws; sealing at back with mastic sealant	-	-	-	3.50	-	**60.83**	1.33	**62.16**	nr	**66.82**
range of 4, each 559 x 406mm with overlap strips; fixing legs to masonry with screws; sealing overlap strips and at back with mastic sealant	-	-	-	13.75	-	**238.98**	5.05	**244.02**	nr	**262.33**
Baths; enamelled pressed steel; combined overflow and waste outlet, plug and chain; metal cradles; pair pillar taps										
700 x 1700 x 570mm overall; sealing at walls with mastic sealant	-	-	-	5.50	-	**95.59**	2.96	**98.55**	nr	**105.94**
High level W.C. suites; vitreous china pan; plastics cistern with valveless fittings, ball valve, chain and pull handle; flush pipe; plastics seat and cover										
fixing pan and cistern brackets to masonry with screws; bedding pan in mastic; jointing pan to drain with cement mortar (1:2) and gaskin joint	-	-	-	4.50	-	**78.21**	1.79	**80.00**	nr	**85.99**
fixing pan and cistern brackets to masonry with screws; bedding pan in mastic; jointing pan to soil pipe with Multikwik connector	-	-	-	4.50	-	**78.21**	1.79	**80.00**	nr	**85.99**
Low level W.C. suites; vitreous china pan; plastics cistern with valveless fittings and ball valve; flush pipe; plastics seat and cover										
fixing pan and cistern brackets to masonry with screws; bedding pan in mastic; jointing pan to drain with cement mortar (1:2) and gaskin joint	-	-	-	4.50	-	**78.21**	1.79	**80.00**	nr	**85.99**
fixing pan and cistern brackets to masonry with screws; bedding pan in mastic; jointing pan to soil pipe with Multikwik connector	-	-	-	4.50	-	**78.21**	1.79	**80.00**	nr	**85.99**
Wall urinals; vitreous china bowl; automatic cistern; spreads and flush pipe; waste outlet										
single bowl; fixing bowl, cistern and pipe brackets to masonry with screws	-	-	-	4.50	-	**78.21**	1.38	**79.59**	nr	**85.56**
range of three bowls; fixing bowls, divisions, cistern and pipe brackets to masonry with screws	-	-	-	11.50	-	**199.87**	2.81	**202.68**	nr	**217.88**
Single stall urinals; white glazed fireclay in one piece; white glazed fireclay tread; plastics automatic cistern; spreader and flush pipe; waste outlet and domed grating										
bedding stall and tread in cement mortar (1:4) and jointing tread with waterproof jointing compound; fixing cistern and pipe brackets to masonry with screws	-	-	-	7.00	-	**121.66**	3.37	**125.03**	nr	**134.40**
Slab urinals; range of four; white glazed fireclay back, ends, divisions, channel and tread; automatic cistern; flush and sparge pipes; waste outlet and domed grating										
bedding back, ends, channel and tread in cement mortar (1:4) and jointing with waterproof jointing compound; fixing divisions, cistern and pipe brackets to masonry with screws	-	-	-	19.00	-	**330.22**	10.71	**340.93**	nr	**366.50**

Labour hourly rates: (except Specialists) Craft Operatives £17.38 Labourer £10.71 Rates are national average prices. Refer to REGIONAL VARIATIONS for indicative levels of overall pricing in regions	MATERIALS			LABOUR				RATES		
	Del to Site	Waste	Material Cost	Craft Optve	Lab	Labour Cost	Sunds	Nett Rate	Unit	Gross Rate (+7.5%)
	£	%	£	Hrs	Hrs	£	£	£		£
N13: SANITARY APPLIANCES/FITTINGS Cont.										
Fix only appliances (prime cost sum for supply included elsewhere) Cont.										
Showers; white glazed fireclay tray 760 x 760 x 180mm; waste outlet; recessed valve and spray head										
bedding tray in cement mortar (1:4); fixing valve head to masonry with screws; sealing at walls with mastic sealant	-	-	-	4.75	-	**82.56**	4.34	**86.89**	nr	**93.41**
Shower curtains with rails, hooks, supports and fittings										
straight, 914mm long x 1829mm drop; fixing supports to masonry with screws	-	-	-	1.00	-	**17.38**	0.34	**17.72**	nr	**19.05**
angled, 1676mm girth x 1829mm drop; fixing supports to masonry with screws	-	-	-	1.50	-	**26.07**	0.46	**26.53**	nr	**28.52**
Supply and fix appliances - sinks										
Note: the following prices include joints to copper services and wastes but do not include traps										
Sink units; single bowl with single drainer and back ledge; stainless steel; BS EN 13310 Type A; 38mm waste, plug and chain to BS EN 274 Part 1 with combined overflow; pair 13mm pillar taps to BS 1010 Part 2										
fixing on base unit with metal clips										
1000 x 600mm	79.75	2.50	**81.74**	3.00	-	**52.14**	0.82	**134.70**	nr	**144.80**
1200 x 600mm	85.50	2.50	**87.64**	3.50	-	**60.83**	0.82	**149.28**	nr	**160.48**
Sink units; single bowl with double drainer and back ledge; stainless steel; BS EN 13310 Type B; 38mm waste, plug and chain to BS EN 274 Part 1 with combined overflow; pair 13mm pillar taps to BS 1010 Part 2										
fixing on base unit with metal clips										
1500 x 600mm	97.50	2.50	**99.94**	3.00	-	**52.14**	1.07	**153.15**	nr	**164.63**
Sinks; white glazed fireclay; BS 1206 with weir overflow; 38mm slotted waste, chain, stay and plug to BS EN 274 Part 1; aluminium alloy cantilever brackets										
screwing chain stay; fixing brackets to masonry with screws; sealing at back with mastic sealant										
455 x 380 x 205mm	83.75	2.50	**85.84**	3.50	-	**60.83**	1.33	**148.00**	nr	**159.10**
610 x 455 x 255mm	116.75	2.50	**119.67**	3.50	-	**60.83**	1.33	**181.82**	nr	**195.46**
760 x 455 x 255mm	168.00	2.50	**172.20**	3.50	-	**60.83**	1.33	**234.36**	nr	**251.93**
Sinks; white glazed fireclay; BS 1206 with wier overflow; 38mm slotted waste, chain, stay and plug to BS EN 274 Part 1; painted legs and screw-to-wall bearers										
screwing chain stay; fixing legs to masonry with screws; sealing at back with mastic sealant										
455 x 380 x 205mm	100.50	2.50	**103.01**	3.50	-	**60.83**	1.33	**165.17**	nr	**177.56**
610 x 455 x 255mm	130.50	2.50	**133.76**	3.50	-	**60.83**	1.33	**195.92**	nr	**210.61**
760 x 455 x 255mm	181.75	2.50	**186.29**	3.50	-	**60.83**	1.33	**248.45**	nr	**267.08**
Supply and fix appliances - lavatory basins										
Wash basins; white vitreous china; BS 5506 Part 3; 32mm slotted waste, chain, stay and plug to BS EN 274 Part 1; painted cantilever brackets; pair 13 pillar taps to BS 1010 Part 2										
sealing at back with mastic sealant										
560 x 406mm	47.00	2.50	**48.18**	3.50	-	**60.83**	1.33	**110.33**	nr	**118.61**
Wash basins; white vitreous china; BS 5506 Part 3; 32mm slotted waste, chain, stay and plug to BS EN 274 Part 1; painted legs and screw-to-wall bearers; pair 13mm taps to BS 1010 Part 2										
fixing legs to masonry with screws; sealing at back with mastic sealant										
560 x 406mm	55.00	2.50	**56.38**	4.10	-	**71.26**	1.33	**128.96**	nr	**138.63**
Wash basins; white vitreous china; BS 5506 Part 3; 32mm slotted waste, chain, stay and plug to BS EN 274 Part 1; painted towel rail brackets; pair 13mm pillar taps to BS 1010 Part 2										
fixing brackets to masonry with screws; sealing at back with mastic sealant										
560 x 406mm	41.25	2.50	**42.28**	4.00	-	**69.52**	1.33	**113.13**	nr	**121.61**
Wash basins; coloured vitreous china; BS 5506 Part 3; 32mm slotted waste, chain, stay and plug to BS EN 274 Part 1; painted build-in brackets; pair 13mm pillar taps to BS 1010 Part 2										
building in brackets to masonry; sealing at back with mastic sealant										
560 x 406mm	68.75	2.50	**70.47**	3.50	-	**60.83**	1.33	**132.62**	nr	**142.57**

Labour hourly rates: (except Specialists) Craft Operatives £17.38 Labourer £10.71 Rates are national average prices. Refer to REGIONAL VARIATIONS for indicative levels of overall pricing in regions	MATERIALS			LABOUR				RATES		
	Del to Site £	Waste %	Material Cost £	Craft Optve Hrs	Lab Hrs	Labour Cost £	Sunds £	Nett Rate £	Unit	Gross Rate (+7.5%) £
N13: SANITARY APPLIANCES/FITTINGS Cont.										
Supply and fix appliances - lavatory basins Cont.										
Wash basins; coloured vitreous china; BS 5506 Part 3; 32mm slotted waste, chain, stay and plug to BS EN 274 Part 1; painted legs and screw-to-wall bearers; pair 13mm pillar taps to BS 1010 Part 2										
fixing legs to masonry with screws; sealing at back with mastic sealant										
560 x 406mm	76.75	2.50	**78.67**	4.10	-	**71.26**	1.33	**151.25**	nr	**162.60**
Wash basins; coloured vitreous china; BS 5506 Part 3; 32mm slotted waste, chain, stay and plug to BS EN 274 Part 1; painted towel rail brackets; pair 13mm pillar taps to BS 1010 Part 2										
fixing brackets to masonry with screws; sealing at back with mastic sealant										
560 x 406mm	63.00	2.50	**64.58**	4.00	-	**69.52**	1.33	**135.42**	nr	**145.58**
Wash basins, angle type; white vitreous china; BS 5506 Part 3; 32mm slotted waste, chain, stay and plug to BS EN 274 Part 1; concealed brackets; pair 13mm pillar taps to BS 1010 Part 2										
fixing brackets to masonry with screws; sealing at back with mastic sealant										
457 x 431mm	62.75	2.50	**64.32**	4.00	-	**69.52**	1.33	**135.16**	nr	**145.30**
Wash basins, angle type; coloured vitreous china; BS 5506 Part 3; 32mm slotted waste, chain, stay and plug to BS EN 274 Part 1; concealed brackets; pair 13mm pillar taps to BS 1010 Part 2										
fixing brackets to masonry with screws; sealing at back with mastic sealant										
457 x 431mm	73.50	2.50	**75.34**	4.00	-	**69.52**	1.33	**146.18**	nr	**157.15**
Wash basins with pedestal; white vitreous china; BS 5506 Part 3; 32mm slotted waste, chain, stay and plug to BS EN 274 Part 1; pair 13mm pillar taps to BS 1010 Part 2										
fixing basin to masonry with screws; sealing at back with mastic sealant										
560 x 406mm	61.75	2.50	**63.29**	4.50	-	**78.21**	1.33	**142.83**	nr	**153.54**
Wash basins with pedestal; coloured vitreous china; BS 5506 Part 3; 32mm slotted waste, chain, stay and plug to BS EN 274 Part 1; pair 13mm pillar taps to BS 1010 Part 2										
fixing basin to masonry with screws; sealing at back with mastic sealant										
560 x 406mm	86.25	2.50	**88.41**	3.75	-	**65.18**	1.33	**154.91**	nr	**166.53**
Supply and fix appliances - baths										
Baths; white enamelled pressed steel; BS EN 232; 32mm overflow with front grid and 38mm waste, chain and plug to BS EN 274 Part 1 with combined overflow; metal cradles with adjustable feet; pair 19mm pillar taps to BS 1010 Part 2										
sealing at walls with mastic sealant										
700 x 1700 x 570mm overall	84.50	2.50	**86.61**	5.50	-	**95.59**	2.96	**185.16**	nr	**199.05**
724 x 1500 x 570mm overall	84.50	2.50	**86.61**	5.50	-	**95.59**	2.96	**185.16**	nr	**199.05**
Baths; coloured enamelled pressed steel; BS EN 232; 32mm overflow with front grid and 38mm waste, chain and plug to BS EN 274 Part 1 with combined overflow; metal cradles with adjustable feet; pair 19mm pillar taps to BS 1010 Part 2										
sealing at walls with mastic sealant										
700 x 1700 x 570mm overall	84.50	2.50	**86.61**	5.50	-	**95.59**	2.96	**185.16**	nr	**199.05**
Baths; white acrylic; BS 4305; 32mm overflow with front grid and 38mm waste, chain and plug to BS EN 274 Part 1 with combined overflow; metal cradles with adjustable feet and wall fixing brackets; pair 19mm pillar taps to BS 1010 Part 2										
700 x 1700 x 570mm overall; fixing brackets to masonry with screws; sealing at walls with mastic sealant	107.75	2.50	**110.44**	5.00	-	**86.90**	2.96	**200.30**	nr	**215.32**
Baths; coloured acrylic; BS 4305; 32mm overflow with front grid and 38mm waste, chain and plug to BS EN 274 Part 1 with combined overflow; metal cradles with adjustable feet and wall fixing brackets; pair 19mm pillar taps to BS 1010 Part 2										
700 x 1700 x 570mm overall; fixing brackets to masonry with screws; sealing at walls with mastic sealant	107.75	2.50	**110.44**	5.00	-	**86.90**	2.96	**200.30**	nr	**215.32**
Supply and fix appliances - W.C. pans										
White vitreous china pans to BS 5503										
pan with horizontal outlet; fixing pan to timber with screws; bedding pan in mastic; jointing pan to soil pipe with Multikwik connector	24.25	2.50	**24.86**	2.50	-	**43.45**	0.71	**69.02**	nr	**74.20**

FURNITURE/EQUIPMENT

Labour hourly rates: (except Specialists) Craft Operatives £17.38 Labourer £10.71 Rates are national average prices. Refer to REGIONAL VARIATIONS for indicative levels of overall pricing in regions	MATERIALS			LABOUR				RATES		
	Del to Site £	Waste %	Material Cost £	Craft Optve Hrs	Lab Hrs	Labour Cost £	Sunds £	Nett Rate £	Unit	Gross Rate (+7.5%) £
N13: SANITARY APPLIANCES/FITTINGS Cont.										
Supply and fix appliances - W.C. pans Cont.										
White vitreous china pans to BS 5503 Cont.										
pan with horizontal outlet; fixing pan to masonry with screws; bedding pan in mastic; jointing pan to soil pipe with Multikwik connector...........	24.25	2.50	**24.86**	3.00	-	**52.14**	1.02	**78.02**	nr	**83.87**
Coloured vitreous china pans to BS 5503										
pan with horizontal outlet; fixing pan to timber with screws; bedding pan in mastic; jointing pan to soil pipe with Multikwik connector...........	47.25	2.50	**48.43**	2.50	-	**43.45**	0.71	**92.60**	nr	**99.54**
pan with horizontal outlet; fixing pan to masonry with screws; bedding pan in mastic; jointing pan to soil pipe with Multikwik connector...........	47.25	2.50	**48.43**	3.00	-	**52.14**	1.02	**101.59**	nr	**109.21**
Plastics W.C. seats; BS 1254, type 1										
black; fixing to pan	7.75	2.50	**7.94**	0.25	-	**4.35**	-	**12.29**	nr	**13.21**
Plastics W.C. seats and covers; BS 1254, type 1										
black; fixing to pan	10.85	2.50	**11.12**	0.25	-	**4.35**	-	**15.47**	nr	**16.63**
coloured; fixing to pan	19.50	2.50	**19.99**	0.25	-	**4.35**	-	**24.33**	nr	**26.16**
white; fixing to pan	17.50	2.50	**17.94**	0.25	-	**4.35**	-	**22.28**	nr	**23.95**
Supply and fix appliances - cisterns										
9 litre black plastics cistern with valveless fittings, chain and pull handle to BS 1125; 13mm piston type high pressure ball valve to BS 1212 Part 1 with 127mm plastics float to BS 2456										
fixing cistern brackets to masonry with screws...	50.75	2.50	**52.02**	1.00	-	**17.38**	0.34	**69.74**	nr	**74.97**
Plastics flush pipe to BS 1125; adjustable										
for back wall fixing	2.40	2.50	**2.46**	0.75	-	**13.04**	0.34	**15.83**	nr	**17.02**
for side wall fixing	2.90	2.50	**2.97**	0.75	-	**13.04**	0.34	**16.34**	nr	**17.57**
Supply and fix appliances - W.C. suites										
High level W.C. suites; white vitreous china pan to BS 5503; 9 litre black plastics cistern with valveless fittings, chain and pull handle and plastics flush pipe to BS 1125; 13mm piston type high pressure ball valve to BS 1212 Part 1 with 127mm plastics float to BS 2456; black plastics seat and cover to BS 1254 type 1										
pan with S or P trap conversion bend to BS 5627; fixing pan to timber with screws; fixing cistern brackets to masonry with screws; bedding pan in mastic; jointing pan to soil pipe with Multikwik connector	88.75	2.50	**90.97**	4.00	-	**69.52**	1.79	**162.27**	nr	**174.44**
pan with S or P trap conversion bend to BS 5627; fixing pan and cistern brackets to masonry with screws; bedding pan in mastic; jointing pan to soil pipe with Multikwik connector..............	88.75	2.50	**90.97**	4.50	-	**78.21**	1.79	**170.96**	nr	**183.79**
Low level W.C. suites; white vitreous china pan to BS 5503; 9 litre white vitreous china cistern with valveless fittings and plastics flush bend to BS 1125; 13mm piston type high pressure ball valve to BS 1212 Part 1 with 127mm plastics float to BS 2456; white plastics seat and cover to BS 1254 type 1										
pan with S or P trap conversion bend to BS 5627; fixing pan to timber with screws; fixing cistern brackets to masonry with screws; bedding pan in mastic; jointing pan to soil pipe with Multikwik connector	83.00	2.50	**85.08**	4.00	-	**69.52**	1.79	**156.38**	nr	**168.11**
pan with S or P trap conversion bend to BS 5627; fixing pan and cistern brackets to masonry with screws; bedding pan in mastic; jointing pan to soil pipe with Multikwik connector..............	83.00	2.50	**85.08**	4.50	-	**78.21**	1.79	**165.07**	nr	**177.45**
Low level W.C. suites; coloured vitreous china pan to BS 5503; 9 litre coloured vitreous china cistern with valveless fittings and plastics flush bend to BS 1125; 13mm piston type high pressure ball valve to BS 1212 Part 1 with 127mm plastics float to BS 2456; coloured plastics seat and cover to BS 1254 type 1										
pan with S or P trap conversion bend to BS 5627; fixing pan to timber with screws; fixing cistern brackets to masonry with screws; bedding pan in mastic; jointing pan to soil pipe with Multikwik connector	134.50	2.50	**137.86**	4.00	-	**69.52**	1.79	**209.17**	nr	**224.86**
pan with S or P trap conversion bend to BS 5627; fixing pan and cistern brackets to masonry with screws; bedding pan in mastic; jointing pan to soil pipe with Multikwik connector..............	134.50	2.50	**137.86**	4.50	-	**78.21**	1.79	**217.86**	nr	**234.20**
Supply and fix appliances - urinals										
Wall urinals; white vitreous china bowl to BS 5520; 4.5 litre white vitreous china automatic cistern to BS 1876; spreader and stainless steel flush pipe; 32mm plastics waste outlet										
fixing bowl, cistern and pipe brackets to masonry with screws	212.00	2.50	**217.30**	4.50	-	**78.21**	1.38	**296.89**	nr	**319.15**

Labour hourly rates: (except Specialists) Craft Operatives £17.38 Labourer £10.71 Rates are national average prices. Refer to REGIONAL VARIATIONS for indicative levels of overall pricing in regions	MATERIALS			LABOUR				RATES		
	Del to Site £	Waste %	Material Cost £	Craft Optve Hrs	Lab Hrs	Labour Cost £	Sunds £	Nett Rate £	Unit	Gross Rate (+7.5%) £
N13: SANITARY APPLIANCES/FITTINGS Cont.										
Supply and fix appliances - urinals Cont.										
Wall urinals; range of 2; white vitreous china bowls to BS 5520; white vitreous china divisions; 9 litre white plastics automatic cistern to BS 1876; stainless steel flush pipes and spreaders; 32mm plastics waste outlets										
fixing bowls, divisions, cistern and pipe brackets to masonry with screws	398.50	2.50	**408.46**	9.00	-	**156.42**	2.04	**566.92**	nr	**609.44**
Wall urinals; range of 3; white vitreous china bowls to BS 5520; white vitreous china divisions; 14 litre white plastics automatic cistern to BS 1876; stainless steel flush pipes and spreaders; 32mm plastics waste outlets										
fixing bowls, divisions, cistern and pipe brackets to masonry with screws	612.50	2.50	**627.81**	13.50	-	**234.63**	2.70	**865.15**	nr	**930.03**
1200mm long stainless steel wall hung trough urinal with plastic; white; automatic cistern to BS 1876; spreader and stainless steel flush pipe; waste outlet and domed grating										
fixing urinal cistern and pipe brackets to masonry with screws	575.00	2.50	**589.38**	5.00	-	**86.90**	3.37	**679.64**	nr	**730.61**
1800mm long stainless steel wall hung trough urinal with plastic; white; automatic cistern to BS 1876; spreader and stainless steel flush pipe; waste outlet and domed grating										
fixing urinal cistern and pipe brackets to masonry with screws	665.00	2.50	**681.63**	7.00	-	**121.66**	5.81	**809.10**	nr	**869.78**
2400mm long stainless steel wall hung trough urinal with plastic; white; automatic cistern to BS 1876; spreader and stainless steel flush pipe; waste outlet and domed grating										
fixing urinal cistern and pipe brackets to masonry with screws	685.00	2.50	**702.13**	10.00	-	**173.80**	8.16	**884.09**	nr	**950.39**
Supply and fix appliances - showers										
Showers; white glazed Armastone tray, 760 x 760 x 180mm; 38mm waste outlet; surface mechanical valve										
bedding tray in cement mortar (1:4); fixing valve and head to masonry with screws; sealing at walls with mastic sealant	240.75	2.50	**246.77**	4.75	-	**82.56**	4.34	**333.66**	nr	**358.68**
Showers; white glazed Armastone tray, 760 x 760 x 180mm; 38mm waste outlet; recessed thermostatic valve and swivel spray head										
bedding tray in cement mortar (1:4); fixing valve and head to masonry with screws; sealing at walls with mastic sealant	370.75	2.50	**380.02**	4.75	-	**82.56**	4.34	**466.91**	nr	**501.93**
Showers; coloured glazed acrylic tray, 760 x 760 x 180mm; 38mm waste outlet; recessed thermostatic valve and swivel spray head										
bedding tray in cement mortar (1:4); fixing valve and head to masonry with screws; sealing at walls with mastic sealant	412.75	2.50	**423.07**	4.75	-	**82.56**	4.34	**509.96**	nr	**548.21**
Showers; white moulded acrylic tray with removable front and side panels, 760 x 760 x 260mm with adjustable metal cradle; 38mm waste outlet; surface fixing mechanical valve, flexible tube hand spray and slide bar										
fixing cradle, valve and slide bar to masonry with screws; sealing at walls with mastic sealant	282.75	2.50	**289.82**	4.75	-	**82.56**	2.91	**375.28**	nr	**403.43**
Showers; white moulded acrylic tray with removable front and side panels, 760 x 760 x 260mm with adjustable metal cradle; 38mm waste outlet; surface fixing thermostatic valve, flexible tube hand spray and slide bar										
fixing cradle, valve and slide bar to masonry with screws; sealing at walls with mastic sealant	412.75	2.50	**423.07**	4.75	-	**82.56**	2.91	**508.53**	nr	**546.67**
Shower curtains; nylon; anodised aluminium rail, glider hooks, end and suspension fittings										
straight, 914mm long x 1829mm drop	32.25	2.50	**33.06**	1.00	-	**17.38**	0.34	**50.77**	nr	**54.58**
angled, 1676mm girth x 1829mm drop	45.50	2.50	**46.64**	1.50	-	**26.07**	0.51	**73.22**	nr	**78.71**
Shower curtains; heavy duty plastic; anodised aluminium rail, glider hooks, end and suspension fitting; fixing supports to masonry with screws										
straight, 914mm long x 1829mm drop	36.50	2.50	**37.41**	1.00	-	**17.38**	0.34	**55.13**	nr	**59.26**
angled, 1676mm girth x 1829mm drop	49.25	2.50	**50.48**	1.50	-	**26.07**	0.51	**77.06**	nr	**82.84**

Labour hourly rates: (except Specialists) Craft Operatives £14.21 Labourer £10.71 Rates are national average prices. Refer to REGIONAL VARIATIONS for indicative levels of overall pricing in regions	MATERIALS			LABOUR				RATES		
	Del to Site £	Waste %	Material Cost £	Craft Optve Hrs	Lab Hrs	Labour Cost £	Sunds £	Nett Rate £	Unit	Gross Rate (+7.5%) £
N15: SIGNS/NOTICES										
Signwriting in gloss paint; one coat										
Letters or numerals, Helvetica medium style, on painted or varnished surfaces										
50mm high	0.01	10.00	**0.01**	0.14	-	**1.99**	0.00	**2.00**	nr	**2.16**
extra; shading	0.01	10.00	**0.01**	0.08	-	**1.14**	0.00	**1.15**	nr	**1.24**
extra; outline	0.01	10.00	**0.01**	0.12	-	**1.71**	0.00	**1.72**	nr	**1.85**
100mm high	0.01	10.00	**0.01**	0.28	-	**3.98**	0.01	**4.00**	nr	**4.30**
extra; shading	0.01	10.00	**0.01**	0.16	-	**2.27**	0.00	**2.29**	nr	**2.46**
extra; outline	0.01	10.00	**0.01**	0.24	-	**3.41**	0.01	**3.43**	nr	**3.69**
150mm high	0.02	10.00	**0.03**	0.42	-	**5.97**	0.01	**6.00**	nr	**6.45**
extra; shading	0.01	10.00	**0.01**	0.24	-	**3.41**	0.01	**3.43**	nr	**3.69**
extra; outline	0.01	10.00	**0.01**	0.36	-	**5.12**	0.01	**5.14**	nr	**5.52**
200mm high	0.02	10.00	**0.03**	0.56	-	**7.96**	0.01	**7.99**	nr	**8.59**
extra; shading	0.01	10.00	**0.01**	0.32	-	**4.55**	0.01	**4.57**	nr	**4.91**
extra; outline	0.01	10.00	**0.01**	0.48	-	**6.82**	0.01	**6.84**	nr	**7.36**
300mm high	0.04	10.00	**0.04**	0.84	-	**11.94**	0.02	**11.99**	nr	**12.89**
extra; shading	0.02	10.00	**0.03**	0.48	-	**6.82**	0.01	**6.86**	nr	**7.37**
extra; outline	0.02	10.00	**0.03**	0.72	-	**10.23**	0.01	**10.27**	nr	**11.04**
stops	-	-	-	0.04	-	**0.57**	-	**0.57**	nr	**0.61**
Signwriting in gloss paint; two coats										
Letters or numerals, Helvetica medium style, on painted or varnished surfaces										
50mm high	0.01	10.00	**0.01**	0.25	-	**3.55**	0.01	**3.57**	nr	**3.84**
100mm high	0.02	10.00	**0.03**	0.50	-	**7.11**	0.01	**7.14**	nr	**7.68**
150mm high	0.02	10.00	**0.03**	0.75	-	**10.66**	0.01	**10.70**	nr	**11.50**
200mm high	0.04	10.00	**0.04**	1.00	-	**14.21**	0.02	**14.27**	nr	**15.34**
250mm high	0.04	10.00	**0.04**	1.25	-	**17.76**	0.02	**17.83**	nr	**19.16**
300mm high	0.05	10.00	**0.05**	1.50	-	**21.32**	0.03	**21.40**	nr	**23.00**
stops	-	-	-	0.07	-	**0.99**	0.00	**1.00**	nr	**1.07**

BUILDING FABRIC SUNDRIES

Labour hourly rates: (except Specialists) Craft Operatives £14.21 Labourer £10.71 Rates are national average prices. Refer to REGIONAL VARIATIONS for indicative levels of overall pricing in regions	MATERIALS			LABOUR				RATES		
	Del to Site £	Waste %	Material Cost £	Craft Optve Hrs	Lab Hrs	Labour Cost £	Sunds £	Nett Rate £	Unit	Gross Rate (+7.5%) £
P10: SUNDRY INSULATION/PROOFING WORK/ FIRE STOPS										
Waterproof reinforced building paper BS 1521; grade A1F; 150mm lapped joints										
Across members at 450mm centres vertical	0.72	5.00	**0.75**	0.17	0.02	**2.63**	0.02	**3.40**	m2	**3.66**
Waterproof reinforced building paper; reflection (thermal) grade, single sided; 150mm lapped joints										
Across members at 450mm centres vertical	1.07	5.00	**1.12**	0.17	0.02	**2.63**	0.02	**3.77**	m2	**4.06**
Waterproof reinforced building paper; reflection (thermal) grade, double sided; 150mm lapped joints										
Across members at 450mm centres vertical	1.59	5.00	**1.67**	0.17	0.02	**2.63**	0.02	**4.32**	m2	**4.64**
British Sisalkraft Ltd Insulex 714 vapour control layer; 150mm laps sealed with tape										
Across members at 450mm centres horizontal; fixing to timber with stainless steel staples	1.60	10.00	**1.76**	0.09	0.01	**1.39**	0.05	**3.19**	m2	**3.43**
Plain areas horizontal; laid loose	1.60	10.00	**1.76**	0.05	0.01	**0.82**	-	**2.58**	m2	**2.77**
British Sisalkraft Ltd SK860 vapour control layer; 100mm laps sealed with tape										
Plain areas horizontal; laid loose	1.23	10.00	**1.36**	0.07	0.01	**1.10**	0.03	**2.49**	m2	**2.68**
Mineral wool insulation quilt; butt joints										
60mm thick; across members at 450mm centres horizontal; laid loose	2.18	15.00	**2.50**	0.09	0.01	**1.39**	-	**3.89**	m2	**4.18**
Paper faced mineral wool insulation quilt; butt joints										
60mm thick; between members at 450mm centres vertical	4.21	15.00	**4.85**	0.20	0.02	**3.06**	0.03	**7.93**	m2	**8.53**
Glass fibre insulation quilt; butt joints										
60mm thick; across members at 450mm centres horizontal; laid loose	2.25	15.00	**2.58**	0.09	0.01	**1.39**	-	**3.97**	m2	**4.27**
80mm thick; across members at 450mm centres horizontal; laid loose	3.15	15.00	**3.62**	0.09	0.01	**1.39**	-	**5.01**	m2	**5.38**
100mm thick; across members at 450mm centres horizontal; laid loose	3.91	15.00	**4.50**	0.09	0.01	**1.39**	-	**5.89**	m2	**6.33**
60mm thick; between members at 450mm centres horizontal; laid loose	2.25	10.00	**2.47**	0.15	0.02	**2.35**	-	**4.82**	m2	**5.18**
80mm thick; between members at 450mm centres horizontal; laid loose	3.15	10.00	**3.46**	0.15	0.02	**2.35**	-	**5.81**	m2	**6.25**
100mm thick; between members at 450mm centres horizontal; laid loose	3.91	10.00	**4.30**	0.15	0.02	**2.35**	-	**6.65**	m2	**7.15**
Paper faced glass fibre insulation quilt; butt joints										
60mm thick; across members at 450mm centres horizontal; laid loose	4.21	15.00	**4.85**	0.09	0.01	**1.39**	-	**6.23**	m2	**6.70**

Labour hourly rates: (except Specialists)
Craft Operatives £14.21 Labourer £10.71
Rates are national average prices.
Refer to REGIONAL VARIATIONS for indicative levels of overall pricing in regions

	MATERIALS			LABOUR				RATES		
	Del to Site £	Waste %	Material Cost £	Craft Optve Hrs	Lab Hrs	Labour Cost £	Sunds £	Nett Rate £	Unit	Gross Rate (+7.5%) £
P10: SUNDRY INSULATION/PROOFING WORK/ FIRE STOPS Cont.										
Paper faced glass fibre insulation quilt; butt joints Cont.										
80mm thick; across members at 450mm centres										
horizontal; laid loose	5.09	15.00	**5.86**	0.09	0.01	**1.39**	-	**7.24**	m2	**7.79**
Owens Corning Building Products (UK) Ltd; Crown wool insulation quilt; butt joints										
100mm thick; between members at 450mm centres										
horizontal; laid loose	2.85	10.00	**3.13**	0.10	0.01	**1.53**	-	**4.66**	m2	**5.01**
150mm thick; between members at 450mm centres										
horizontal; laid loose	4.31	10.00	**4.74**	0.10	0.01	**1.53**	-	**6.27**	m2	**6.73**
200mm thick; between members at 450mm centres										
horizontal; laid loose	6.04	10.00	**6.65**	0.20	0.03	**3.16**	-	**9.81**	m2	**10.55**
100mm thick; between members at 450mm centres; 50mm thick across members at 450mm centres										
horizontal; laid loose	4.31	10.00	**4.74**	0.28	0.05	**4.51**	-	**9.25**	m2	**9.95**
100mm thick; between members at 450mm centres; 100mm thick across members at 450mm centres										
horizontal; laid loose	5.70	10.00	**6.27**	0.36	0.05	**5.65**	-	**11.92**	m2	**12.81**
Rockwool Rollbatts; insulation quilt; butt joints										
100mm thick; between members at 450mm centres										
horizontal; laid loose	2.96	10.00	**3.26**	0.10	0.01	**1.53**	-	**4.79**	m2	**5.15**
150mm thick; between members at 450mm centres										
horizontal; laid loose	4.56	10.00	**5.02**	0.15	0.02	**2.35**	-	**7.36**	m2	**7.92**
100mm thick; between members at 450mm centres; 100mm thick across members at 450mm centres										
horizontal; laid loose	5.93	10.00	**6.52**	0.20	0.03	**3.16**	-	**9.68**	m2	**10.41**
Glass fibre medium density insulation board; butt joints										
30mm thick; plain areas										
horizontal; bedding in bitumen	1.94	10.00	**2.14**	0.20	0.02	**3.06**	1.50	**6.70**	m2	**7.20**
50mm thick; plain areas										
horizontal; bedding in bitumen	3.47	10.00	**3.82**	0.22	0.03	**3.45**	1.50	**8.77**	m2	**9.43**
75mm thick; plain areas										
horizontal; bedding in bitumen	5.07	10.00	**5.58**	0.25	0.03	**3.87**	1.50	**10.95**	m2	**11.77**
30mm thick; across members at 450mm centres										
vertical	1.94	10.00	**2.14**	0.20	0.02	**3.06**	0.03	**5.23**	m2	**5.62**
50mm thick; across members at 450mm centres										
vertical	3.47	10.00	**3.82**	0.22	0.03	**3.45**	0.03	**7.30**	m2	**7.84**
75mm thick; across members at 450mm centres										
vertical	5.07	10.00	**5.58**	0.25	0.03	**3.87**	0.03	**9.48**	m2	**10.19**
Dow Construction Products; Styrofoam Floormate 200; butt joints										
25mm thick; plain areas										
horizontal; laid loose	5.51	10.00	**6.06**	0.10	0.01	**1.53**	-	**7.59**	m2	**8.16**
35mm thick; plain areas										
horizontal; laid loose	7.54	10.00	**8.29**	0.15	0.02	**2.35**	-	**10.64**	m2	**11.43**
50mm thick; plain areas										
horizontal; laid loose	10.57	10.00	**11.63**	0.20	0.03	**3.16**	-	**14.79**	m2	**15.90**
Sempatap latex foam sheeting; butt joints										
5mm thick; plain areas										
soffit; fixing with adhesive	9.27	15.00	**10.66**	1.25	1.25	**31.15**	2.40	**44.21**	m2	**47.53**
vertical; fixing with adhesive	9.27	15.00	**10.66**	1.00	1.00	**24.92**	2.40	**37.98**	m2	**40.83**
Sempafloor SBR latex foam sheeting with coated non woven polyester surface; butt joints										
4.5mm thick; plain areas										
horizontal; fixing with adhesive	10.35	15.00	**11.90**	0.75	0.75	**18.69**	2.05	**32.64**	m2	**35.09**
Expanded polystyrene sheeting; butt joints										
13mm thick; plain areas										
horizontal; laid loose	0.79	5.00	**0.83**	0.08	0.04	**1.57**	-	**2.39**	m2	**2.57**
vertical; fixing with adhesive	0.79	5.00	**0.83**	0.23	0.12	**4.55**	1.00	**6.38**	m2	**6.86**
19mm thick; plain areas										
horizontal; laid loose	1.23	5.00	**1.29**	0.09	0.05	**1.81**	-	**3.10**	m2	**3.34**
vertical; fixing with adhesive	1.23	5.00	**1.29**	0.24	0.12	**4.70**	1.00	**6.98**	m2	**7.51**

Labour hourly rates: (except Specialists) Craft Operatives £14.21 Labourer £10.71 Rates are national average prices. Refer to REGIONAL VARIATIONS for indicative levels of overall pricing in regions	MATERIALS			LABOUR				RATES		
	Del to Site £	Waste %	Material Cost £	Craft Optve Hrs	Lab Hrs	Labour Cost £	Sunds £	Nett Rate £	Unit	Gross Rate (+7.5%) £
P10: SUNDRY INSULATION/PROOFING WORK/ FIRE STOPS Cont.										
Expanded polystyrene sheeting; butt joints Cont.										
25mm thick; plain areas										
horizontal; laid loose	1.55	5.00	**1.63**	0.10	0.05	**1.96**	-	**3.59**	m2	**3.85**
vertical; fixing with adhesive	1.55	5.00	**1.63**	0.25	0.13	**4.94**	1.00	**7.57**	m2	**8.14**
50mm thick; plain areas										
horizontal; laid loose	3.07	5.00	**3.22**	0.12	0.06	**2.35**	-	**5.57**	m2	**5.99**
vertical; fixing with adhesive	3.07	5.00	**3.22**	0.27	0.14	**5.34**	1.00	**9.56**	m2	**10.27**
75mm thick; plain areas										
horizontal; laid loose	4.62	5.00	**4.85**	0.15	0.08	**2.99**	-	**7.84**	m2	**8.43**
vertical; fixing with adhesive	4.62	5.00	**4.85**	0.30	0.15	**5.87**	1.00	**11.72**	m2	**12.60**
Expanded polystyrene sheeting, non-flammable; butt joints										
13mm thick; plain areas										
horizontal; laid loose	1.24	5.00	**1.30**	0.08	0.04	**1.57**	-	**2.87**	m2	**3.08**
vertical; fixing with adhesive	1.24	5.00	**1.30**	0.23	0.12	**4.55**	1.00	**6.85**	m2	**7.37**
19mm thick; plain areas										
horizontal; laid loose	1.47	5.00	**1.54**	0.09	0.05	**1.81**	-	**3.36**	m2	**3.61**
vertical; fixing with adhesive	1.47	5.00	**1.54**	0.24	0.12	**4.70**	1.00	**7.24**	m2	**7.78**
25mm thick; plain areas										
horizontal; laid loose	1.77	5.00	**1.86**	0.10	0.05	**1.96**	-	**3.82**	m2	**4.10**
vertical; fixing with adhesive	1.77	5.00	**1.86**	0.25	0.13	**4.94**	1.00	**7.80**	m2	**8.39**
50mm thick; plain areas										
horizontal; laid loose	4.25	5.00	**4.46**	0.12	0.06	**2.35**	-	**6.81**	m2	**7.32**
vertical; fixing with adhesive	4.25	5.00	**4.46**	0.27	0.14	**5.34**	1.00	**10.80**	m2	**11.61**
75mm thick; plain areas										
horizontal; laid loose	6.39	5.00	**6.71**	0.15	0.08	**2.99**	-	**9.70**	m2	**10.43**
vertical; fixing with adhesive	6.39	5.00	**6.71**	0.30	0.15	**5.87**	1.00	**13.58**	m2	**14.60**
Perforated zinc sheeting; butt joints										
Nr 7 gauge; across members at 450mm centres										
soffit	22.81	20.00	**27.37**	1.25	0.16	**19.48**	0.20	**47.04**	m2	**50.57**
vertical	22.81	20.00	**27.37**	0.75	0.09	**11.62**	0.12	**39.11**	m2	**42.04**
Nr 8 gauge; across members at 450mm centres										
soffit	25.00	20.00	**30.01**	1.25	0.16	**19.48**	0.20	**49.68**	m2	**53.41**
vertical	25.00	20.00	**30.01**	0.75	0.09	**11.62**	0.12	**41.75**	m2	**44.88**
Nr 9 gauge; across members at 450mm centres										
soffit	27.96	20.00	**33.55**	1.25	0.16	**19.48**	0.20	**53.22**	m2	**57.22**
vertical	27.96	20.00	**33.55**	0.75	0.09	**11.62**	0.12	**45.29**	m2	**48.69**
galvanized wire netting, BS EN 10223-2; butt joints										
13mm mesh x Nr 22 gauge; across members at 450mm centres										
soffit	2.31	10.00	**2.54**	-	0.12	**1.29**	0.03	**3.86**	m2	**4.15**
vertical	2.31	10.00	**2.54**	-	0.08	**0.86**	0.03	**3.43**	m2	**3.68**
25mm mesh x Nr 19 gauge; across members at 450mm centres										
soffit	1.85	10.00	**2.04**	-	0.12	**1.29**	0.03	**3.35**	m2	**3.60**
vertical	1.85	10.00	**2.04**	-	0.08	**0.86**	0.03	**2.92**	m2	**3.14**
38mm mesh x Nr 19 gauge; across members at 450mm centres										
soffit	1.33	10.00	**1.46**	-	0.12	**1.29**	0.03	**2.78**	m2	**2.99**
vertical	1.33	10.00	**1.46**	-	0.08	**0.86**	0.03	**2.35**	m2	**2.53**
50mm mesh x Nr 19 gauge; across members at 450mm centres										
soffit	1.05	10.00	**1.16**	-	0.12	**1.29**	0.03	**2.47**	m2	**2.66**
vertical	1.05	10.00	**1.16**	-	0.08	**0.86**	0.03	**2.04**	m2	**2.19**
P20: UNFRAMED ISOLATED TRIMS/SKIRTINGS/SUNDRY ITEMS										
Softwood, wrought - skirtings, architraves, picture rails and cover fillets to BS EN 942										
Skirtings, picture rails, architraves and the like; (finished sizes)										
13 x 45mm; reference 13CA45	0.83	10.00	**0.91**	0.22	0.02	**3.34**	0.03	**4.28**	m	**4.60**
20 x 70mm; reference 20CA70	1.29	10.00	**1.42**	0.24	0.03	**3.73**	0.03	**5.18**	m	**5.57**
13 x 45mm; reference 13CP45	0.83	10.00	**0.91**	0.23	0.03	**3.59**	0.03	**4.53**	m	**4.87**
13 x 70mm; reference 13CP70	1.07	10.00	**1.17**	0.23	0.03	**3.59**	0.03	**4.79**	m	**5.15**
13 x 70mm; reference 13CS70	1.07	10.00	**1.17**	0.25	0.03	**3.87**	0.03	**5.08**	m	**5.46**
13 x 95mm; reference 13CS95	1.37	10.00	**1.51**	0.23	0.03	**3.59**	0.03	**5.13**	m	**5.51**
20 x 120mm; reference 20CS120	2.12	10.00	**2.33**	0.27	0.03	**4.16**	0.04	**6.53**	m	**7.02**
9 x 33mm; reference 9RA33	0.61	10.00	**0.67**	0.20	0.02	**3.06**	0.03	**3.76**	m	**4.04**
13 x 45mm; reference 13RA45	0.83	10.00	**0.91**	0.23	0.03	**3.59**	0.03	**4.53**	m	**4.87**
20 x 70mm; reference 20RA70	1.37	10.00	**1.51**	0.24	0.03	**3.73**	0.03	**5.27**	m	**5.67**
13 x 70mm; reference 13RS70	1.11	10.00	**1.22**	0.25	0.03	**3.87**	0.03	**5.12**	m	**5.50**
13 x 95mm; reference 13RS95	1.43	10.00	**1.57**	0.23	0.03	**3.59**	0.03	**5.19**	m	**5.58**
20 x 120mm; reference 20RS120	2.12	10.00	**2.33**	0.27	0.03	**4.16**	0.04	**6.53**	m	**7.02**
13 x 45mm; reference 13RP45	0.87	10.00	**0.95**	0.23	0.03	**3.59**	0.03	**4.57**	m	**4.92**

Labour hourly rates: (except Specialists) Craft Operatives £14.21 Labourer £10.71 Rates are national average prices. Refer to REGIONAL VARIATIONS for indicative levels of overall pricing in regions	MATERIALS			LABOUR				RATES		
	Del to Site £	Waste %	Material Cost £	Craft Optve Hrs	Lab Hrs	Labour Cost £	Sunds £	Nett Rate £	Unit	Gross Rate (+7.5%) £
P20: UNFRAMED ISOLATED TRIMS/SKIRTINGS/SUNDRY ITEMS Cont.										
Softwood, wrought - skirtings, architraves, picture rails and cover fillets to BS EN 942 Cont.										
Skirtings, picture rails, architraves and the like; (finished sizes) Cont.										
13 x 70mm; reference 13RP70	1.16	10.00	**1.28**	0.25	0.03	**3.87**	0.03	**5.18**	m	**5.57**
Cover fillets, stops, trims, beads, nosings and the like; (finished sizes)										
11 x 33mm; reference C33	0.61	10.00	**0.67**	0.19	0.02	**2.91**	0.02	**3.61**	m	**3.88**
11 x 45mm; reference C45	0.64	10.00	**0.70**	0.19	0.02	**2.91**	0.02	**3.64**	m	**3.91**
13 x 33mm; reference HR33	0.72	10.00	**0.79**	0.19	0.02	**2.91**	0.02	**3.73**	m	**4.01**
20 x 45mm; reference HR45	0.96	10.00	**1.06**	0.21	0.03	**3.31**	0.03	**4.39**	m	**4.72**
11 x 11mm; reference Q11	0.32	10.00	**0.35**	0.19	0.02	**2.91**	0.02	**3.29**	m	**3.53**
13 x 13mm; reference Q13	0.51	10.00	**0.56**	0.19	0.02	**2.91**	0.02	**3.49**	m	**3.75**
20 x 20mm; reference Q20	0.64	10.00	**0.70**	0.19	0.02	**2.91**	0.02	**3.64**	m	**3.91**
13 x 13mm; reference S13	0.51	10.00	**0.56**	0.19	0.02	**2.91**	0.02	**3.49**	m	**3.75**
20 x 20mm; reference S20	0.64	10.00	**0.70**	0.19	0.02	**2.91**	0.02	**3.64**	m	**3.91**
27 x 27mm; reference S27	0.83	10.00	**0.91**	0.21	0.03	**3.31**	0.03	**4.24**	m	**4.56**
12 x 95mm; reference 12HRCS95	1.52	10.00	**1.67**	0.25	0.03	**3.87**	0.04	**5.58**	m	**6.00**
15 x 95mm; reference 15HRCS95	1.61	10.00	**1.77**	0.27	0.03	**4.16**	0.04	**5.97**	m	**6.42**
Softwood, wrought - skirtings, architraves, picture rails and cover fillets										
Skirtings, picture rails, architraves and the like; (finished sizes)										
19 x 100mm	2.03	10.00	**2.23**	0.25	0.03	**3.87**	0.04	**6.14**	m	**6.60**
19 x 150mm	2.84	10.00	**3.12**	0.28	0.04	**4.41**	0.05	**7.58**	m	**8.15**
25 x 100mm	2.59	10.00	**2.84**	0.27	0.03	**4.16**	0.04	**7.04**	m	**7.57**
25 x 150mm	3.64	10.00	**4.00**	0.31	0.04	**4.83**	0.05	**8.88**	m	**9.55**
25 x 50mm; splays -1	1.47	10.00	**1.61**	0.22	0.03	**3.45**	0.04	**5.10**	m	**5.48**
25 x 63mm; splays -1	1.71	10.00	**1.88**	0.24	0.03	**3.73**	0.04	**5.65**	m	**6.07**
25 x 75mm; splays -1	2.19	10.00	**2.40**	0.25	0.03	**3.87**	0.04	**6.32**	m	**6.79**
32 x 100mm; splays -1	3.16	10.00	**3.47**	0.30	0.04	**4.69**	0.05	**8.22**	m	**8.83**
38 x 75mm; splays -1	2.90	10.00	**3.20**	0.28	0.04	**4.41**	0.05	**7.65**	m	**8.23**
15 x 50mm; mouldings -1	0.96	10.00	**1.06**	0.21	0.03	**3.31**	0.04	**4.40**	m	**4.73**
15 x 125mm; mouldings -1	1.79	10.00	**1.96**	0.25	0.03	**3.87**	0.04	**5.88**	m	**6.32**
19 x 50mm; mouldings -1	1.21	10.00	**1.33**	0.22	0.03	**3.45**	0.04	**4.82**	m	**5.18**
19 x 125mm; mouldings -1	2.27	10.00	**2.49**	0.27	0.03	**4.16**	0.04	**6.69**	m	**7.19**
25 x 50mm; mouldings -1	1.47	10.00	**1.61**	0.22	0.03	**3.45**	0.04	**5.10**	m	**5.48**
25 x 63mm; mouldings -1	1.71	10.00	**1.88**	0.24	0.03	**3.73**	0.04	**5.65**	m	**6.07**
25 x 75mm; mouldings -1	2.19	10.00	**2.40**	0.25	0.03	**3.87**	0.04	**6.32**	m	**6.79**
32 x 100mm; mouldings -1	3.16	10.00	**3.47**	0.30	0.04	**4.69**	0.05	**8.22**	m	**8.83**
38 x 75mm; mouldings -1	2.90	10.00	**3.20**	0.28	0.04	**4.41**	0.05	**7.65**	m	**8.23**
15 x 125mm; chamfers -1	1.79	10.00	**1.96**	0.25	0.03	**3.87**	0.04	**5.88**	m	**6.32**
19 x 125mm; chamfers -1	2.27	10.00	**2.49**	0.26	0.03	**4.02**	0.04	**6.55**	m	**7.04**
25 x 75mm; chamfers -2	2.27	10.00	**2.49**	0.25	0.03	**3.87**	0.04	**6.41**	m	**6.89**
25 x 100mm; chamfers -2	2.76	10.00	**3.03**	0.26	0.03	**4.02**	0.04	**7.09**	m	**7.62**
Cover fillets, stops, trims, beads, nosings and the like; (finished sizes)										
12 x 50mm; mouldings -1	0.83	10.00	**0.91**	0.21	0.03	**3.31**	0.04	**4.25**	m	**4.57**
19 x 19mm; mouldings -1	0.64	10.00	**0.70**	0.19	0.02	**2.91**	0.03	**3.65**	m	**3.92**
19 x 50mm; mouldings -1	1.21	10.00	**1.33**	0.22	0.03	**3.45**	0.04	**4.82**	m	**5.18**
25 x 25mm; mouldings -1	0.83	10.00	**0.91**	0.21	0.03	**3.31**	0.04	**4.25**	m	**4.57**
25 x 38mm; mouldings -1	1.15	10.00	**1.26**	0.22	0.03	**3.45**	0.04	**4.75**	m	**5.10**
19 x 19mm; chamfers -1	0.64	10.00	**0.70**	0.19	0.02	**2.91**	0.03	**3.65**	m	**3.92**
25 x 25mm; chamfers -1	0.83	10.00	**0.91**	0.21	0.03	**3.31**	0.04	**4.25**	m	**4.57**
25 x 38mm; chamfers -1	1.15	10.00	**1.26**	0.22	0.03	**3.45**	0.04	**4.75**	m	**5.10**
Afrormosia, wrought - skirtings, architraves, picture rails and cover fillets										
Skirtings, picture rails, architraves and the like; (finished sizes)										
19 x 100mm	6.03	10.00	**6.63**	0.42	0.05	**6.50**	0.10	**13.24**	m	**14.23**
19 x 150mm	7.81	10.00	**8.59**	0.48	0.06	**7.46**	0.10	**16.15**	m	**17.37**
25 x 100mm	7.44	10.00	**8.18**	0.46	0.06	**7.18**	0.10	**15.46**	m	**16.62**
25 x 150mm	10.06	10.00	**11.07**	0.48	0.06	**7.46**	0.10	**18.63**	m	**20.03**
extra; ends	1.54	10.00	**1.69**	0.05	0.01	**0.82**	0.10	**2.61**	nr	**2.81**
extra; angles	2.50	-	**2.50**	0.10	0.01	**1.53**	0.10	**4.13**	nr	**4.44**
extra; mitres	2.50	-	**2.50**	0.10	0.01	**1.53**	0.10	**4.13**	nr	**4.44**
extra; intersections	2.50	-	**2.50**	0.10	0.01	**1.53**	0.10	**4.13**	nr	**4.44**
25 x 50mm; splays -1	4.36	10.00	**4.80**	0.37	0.05	**5.79**	0.10	**10.69**	m	**11.49**
25 x 63mm; splays -1	5.13	10.00	**5.64**	0.41	0.05	**6.36**	0.10	**12.10**	m	**13.01**
25 x 75mm; splays -1	6.34	10.00	**6.97**	0.42	0.05	**6.50**	0.10	**13.58**	m	**14.60**
32 x 100mm; splays -1	8.71	10.00	**9.58**	0.48	0.06	**7.46**	0.10	**17.14**	m	**18.43**
38 x 75mm; splays -1	8.08	10.00	**8.89**	0.48	0.06	**7.46**	0.10	**16.45**	m	**17.69**
extra; ends	1.21	-	**1.21**	0.05	0.01	**0.82**	0.10	**2.13**	nr	**2.29**
extra; angles	2.05	-	**2.05**	0.10	0.01	**1.53**	0.10	**3.68**	nr	**3.95**
extra; mitres	2.05	-	**2.05**	0.10	0.01	**1.53**	0.10	**3.68**	nr	**3.95**
extra; intersections	2.05	-	**2.05**	0.10	0.01	**1.53**	0.10	**3.68**	nr	**3.95**
15 x 50mm; mouldings -1	2.83	10.00	**3.11**	0.36	0.05	**5.65**	0.10	**8.86**	m	**9.53**
15 x 125mm; mouldings -1	5.39	10.00	**5.93**	0.42	0.05	**6.50**	0.10	**12.53**	m	**13.47**
19 x 50mm; mouldings -1	3.53	10.00	**3.88**	0.37	0.05	**5.79**	0.10	**9.78**	m	**10.51**
19 x 125mm; mouldings -1	6.79	10.00	**7.47**	0.46	0.06	**7.18**	0.10	**14.75**	m	**15.85**
25 x 50mm; mouldings -1	4.36	10.00	**4.80**	0.37	0.05	**5.79**	0.10	**10.69**	m	**11.49**
25 x 63mm; mouldings -1	5.13	10.00	**5.64**	0.41	0.05	**6.36**	0.10	**12.10**	m	**13.01**
25 x 75mm; mouldings -1	6.34	10.00	**6.97**	0.42	0.05	**6.50**	0.10	**13.58**	m	**14.60**
32 x 100mm; mouldings -1	8.71	10.00	**9.58**	0.48	0.06	**7.46**	0.10	**17.14**	m	**18.43**
38 x 75mm; mouldings -1	8.08	10.00	**8.89**	0.48	0.06	**7.46**	0.10	**16.45**	m	**17.69**
extra; ends	1.21	-	**1.21**	0.05	0.01	**0.82**	0.10	**2.13**	nr	**2.29**

Labour hourly rates: (except Specialists) Craft Operatives £14.21 Labourer £10.71 Rates are national average prices. Refer to REGIONAL VARIATIONS for indicative levels of overall pricing in regions	MATERIALS Del to Site £	Waste %	Material Cost £	LABOUR Craft Optve Hrs	Lab Hrs	Labour Cost £	Sunds £	RATES Nett Rate £	Unit	Gross Rate (+7.5%) £
P20: UNFRAMED ISOLATED TRIMS/SKIRTINGS/SUNDRY ITEMS Cont.										
Afrormosia, wrought - skirtings, architraves, picture rails and cover fillets Cont.										
Skirtings, picture rails, architraves and the like; (finished sizes) Cont.										
extra; angles	2.05	-	**2.05**	0.10	0.01	**1.53**	0.10	**3.68**	nr	**3.95**
extra; mitres	2.05	-	**2.05**	0.10	0.01	**1.53**	0.10	**3.68**	nr	**3.95**
extra; intersections	2.05	-	**2.05**	0.10	0.01	**1.53**	0.10	**3.68**	nr	**3.95**
15 x 125mm; chamfers -1	5.39	10.00	**5.93**	0.42	0.05	**6.50**	0.10	**12.53**	m	**13.47**
19 x 125mm; chamfers -1	6.79	10.00	**7.47**	0.46	0.06	**7.18**	0.10	**14.75**	m	**15.85**
25 x 75mm; chamfers -2	6.54	10.00	**7.19**	0.42	0.05	**6.50**	0.10	**13.80**	m	**14.83**
25 x 100mm; chamfers -2	7.81	10.00	**8.59**	0.44	0.06	**6.90**	0.10	**15.59**	m	**16.75**
Cover fillets, stops, trims, beads, nosings and the like; (finished sizes)										
12 x 50mm; mouldings -1	2.38	10.00	**2.62**	0.36	0.05	**5.65**	0.08	**8.35**	m	**8.98**
19 x 19mm; mouldings -1	1.60	10.00	**1.76**	0.32	0.04	**4.98**	0.08	**6.82**	m	**7.33**
19 x 50mm; mouldings -1	3.53	10.00	**3.88**	0.37	0.05	**5.79**	0.08	**9.76**	m	**10.49**
25 x 25mm; mouldings -1	2.44	10.00	**2.68**	0.36	0.05	**5.65**	0.08	**8.42**	m	**9.05**
25 x 38mm; mouldings -1	3.26	10.00	**3.59**	0.37	0.05	**5.79**	0.08	**9.46**	m	**10.17**
19 x 19mm; chamfers -1	1.60	10.00	**1.76**	0.32	0.04	**4.98**	0.08	**6.82**	m	**7.33**
25 x 25mm; chamfers -1	2.44	10.00	**2.68**	0.36	0.05	**5.65**	0.08	**8.42**	m	**9.05**
25 x 38mm; chamfers -1	3.26	10.00	**3.59**	0.37	0.05	**5.79**	0.08	**9.46**	m	**10.17**
Medium density fibreboard, BS EN 622, skirtings, architraves, picture rails and cover fillets to BS 1186 Part 3										
Skirtings, picture rails, architraves and the like; (finished sizes)										
13 x 45mm; reference 13CA45	1.17	10.00	**1.28**	0.28	0.04	**4.41**	0.06	**5.75**	m	**6.18**
20 x 70mm; reference 20CA70	1.83	10.00	**2.01**	0.30	0.04	**4.69**	0.06	**6.76**	m	**7.27**
13 x 45mm; reference 13CP45	1.17	10.00	**1.28**	0.29	0.04	**4.55**	0.06	**5.89**	m	**6.34**
13 x 70mm; reference 13CP70	1.47	10.00	**1.62**	0.29	0.04	**4.55**	0.06	**6.23**	m	**6.69**
13 x 70mm; reference 13CS70	1.47	10.00	**1.62**	0.31	0.04	**4.83**	0.06	**6.51**	m	**7.00**
13 x 95mm; reference 13CS95	1.94	10.00	**2.13**	0.29	0.04	**4.55**	0.06	**6.74**	m	**7.24**
20 x 120mm; reference 20CS120	3.00	10.00	**3.30**	0.34	0.04	**5.26**	0.06	**8.62**	m	**9.26**
9 x 33mm; reference 9RA33	0.88	10.00	**0.96**	0.25	0.03	**3.87**	0.06	**4.90**	m	**5.26**
13 x 45mm; reference 13RA45	1.17	10.00	**1.28**	0.29	0.04	**4.55**	0.06	**5.89**	m	**6.34**
20 x 70mm; reference 20RA70	1.94	10.00	**2.13**	0.30	0.04	**4.69**	0.06	**6.88**	m	**7.40**
13 x 70mm; reference 13RS70	1.56	10.00	**1.71**	0.31	0.04	**4.83**	0.06	**6.61**	m	**7.10**
13 x 95mm; reference 13RS95	2.02	10.00	**2.22**	0.29	0.04	**4.55**	0.06	**6.83**	m	**7.35**
20 x 120mm; reference 20RS120	3.00	10.00	**3.30**	0.34	0.04	**5.26**	0.06	**8.62**	m	**9.26**
13 x 45mm; reference 13RP45	1.22	10.00	**1.34**	0.29	0.04	**4.55**	0.06	**5.95**	m	**6.40**
13 x 70mm; reference 13RP70	1.64	10.00	**1.81**	0.31	0.04	**4.83**	0.06	**6.70**	m	**7.20**
Cover fillets, stops, trims, beads, nosings and the like; (finished sizes)										
11 x 33mm; reference C33	0.88	10.00	**0.96**	0.24	0.02	**3.62**	0.05	**4.64**	m	**4.99**
11 x 45mm; reference C45	0.92	10.00	**1.01**	0.24	0.02	**3.62**	0.05	**4.69**	m	**5.04**
13 x 33mm; reference HR33	1.03	10.00	**1.13**	0.24	0.02	**3.62**	0.05	**4.80**	m	**5.17**
20 x 45mm; reference HR45	1.38	10.00	**1.52**	0.26	0.04	**4.12**	0.05	**5.70**	m	**6.12**
11 x 11mm; reference Q11	0.47	10.00	**0.51**	0.24	0.02	**3.62**	0.05	**4.19**	m	**4.50**
13 x 13mm; reference Q13	0.71	10.00	**0.79**	0.24	0.02	**3.62**	0.05	**4.46**	m	**4.79**
20 x 20mm; reference Q20	0.92	10.00	**1.01**	0.24	0.02	**3.62**	0.05	**4.69**	m	**5.04**
13 x 13mm; reference S13	0.71	10.00	**0.79**	0.24	0.02	**3.62**	0.05	**4.46**	m	**4.79**
20 x 20mm; reference S20	0.92	10.00	**1.01**	0.24	0.02	**3.62**	0.05	**4.69**	m	**5.04**
27 x 27mm; reference S27	1.17	10.00	**1.28**	0.26	0.04	**4.12**	0.05	**5.46**	m	**5.87**
12 x 95mm; reference 12HRCS95	2.16	10.00	**2.38**	0.31	0.04	**4.83**	0.05	**7.26**	m	**7.81**
15 x 95mm; reference 15HRCS95	2.27	10.00	**2.50**	0.34	0.04	**5.26**	0.05	**7.81**	m	**8.39**
Softwood, wrought - cappings										
Cover fillets, stops, trims, beads, nosings and the like; (finished sizes)										
25 x 50mm; level; rebates -1; mouldings -1	1.47	10.00	**1.61**	0.22	0.03	**3.45**	0.04	**5.10**	m	**5.48**
50 x 75mm; level; rebates -1; mouldings -1	4.21	10.00	**4.63**	0.31	0.04	**4.83**	0.05	**9.52**	m	**10.23**
50 x 100mm; level; rebates -1; mouldings -1	5.26	10.00	**5.79**	0.34	0.04	**5.26**	0.05	**11.10**	m	**11.93**
25 x 50mm; ramped; rebates -1; mouldings -1	1.47	10.00	**1.61**	0.33	0.04	**5.12**	0.05	**6.78**	m	**7.29**
50 x 75mm; ramped; rebates -1; mouldings -1	4.21	10.00	**4.63**	0.46	0.06	**7.18**	0.08	**11.89**	m	**12.78**
50 x 100mm; ramped; rebates -1; mouldings -1	5.26	10.00	**5.79**	0.51	0.06	**7.89**	0.08	**13.76**	m	**14.79**
Afrormosia, wrought - cappings										
Cover fillets, stops, trims, beads, nosings and the like; (finished sizes)										
25 x 50mm; level; rebates -1; mouldings -1	4.49	10.00	**4.94**	0.37	0.05	**5.79**	0.06	**10.79**	m	**11.60**
50 x 75mm; level; rebates -1; mouldings -1	11.54	10.00	**12.69**	0.53	0.07	**8.28**	0.09	**21.07**	m	**22.64**
extra; ends	1.73	-	**1.73**	0.08	0.01	**1.24**	0.01	**2.98**	nr	**3.21**
extra; angles	2.89	-	**2.89**	0.13	0.02	**2.06**	0.02	**4.97**	nr	**5.34**
extra; mitres	2.89	-	**2.89**	0.13	0.02	**2.06**	0.02	**4.97**	nr	**5.34**
extra; rounded corners not exceeding 300mm girth	11.54	-	**11.54**	0.53	0.07	**8.28**	0.09	**19.91**	nr	**21.40**
50 x 100mm; level; rebates -1; mouldings -1	14.35	10.00	**15.79**	0.58	0.07	**8.99**	0.09	**24.87**	m	**26.73**
extra; ends	2.18	-	**2.18**	0.09	0.01	**1.39**	0.01	**3.58**	nr	**3.84**
extra; angles	3.59	-	**3.59**	0.14	0.02	**2.20**	0.02	**5.81**	nr	**6.25**
extra; mitres	3.59	-	**3.59**	0.14	0.02	**2.20**	0.02	**5.81**	nr	**6.25**
extra; rounded corners not exceeding 300mm girth	14.35	-	**14.35**	0.58	0.07	**8.99**	0.09	**23.43**	nr	**25.19**
25 x 50mm; ramped; rebates -1; mouldings -1	4.49	10.00	**4.94**	0.55	0.07	**8.57**	0.09	**13.59**	m	**14.61**
50 x 75mm; ramped; rebates -1; mouldings -1	11.54	10.00	**12.69**	0.80	0.10	**12.44**	0.13	**25.26**	m	**27.16**
extra; ends	1.73	-	**1.73**	0.12	0.02	**1.92**	0.02	**3.67**	nr	**3.94**
extra; angles	2.89	-	**2.89**	0.20	0.03	**3.16**	0.03	**6.08**	nr	**6.54**
extra; mitres	2.89	-	**2.89**	0.20	0.03	**3.16**	0.03	**6.08**	nr	**6.54**
extra; rounded corners not exceeding 300mm girth	11.54	-	**11.54**	0.80	0.10	**12.44**	0.13	**24.11**	nr	**25.92**

Labour hourly rates: (except Specialists) Craft Operatives £14.21 Labourer £10.71 Rates are national average prices. Refer to REGIONAL VARIATIONS for indicative levels of overall pricing in regions	MATERIALS			LABOUR				RATES		
	Del to Site £	Waste %	Material Cost £	Craft Optve Hrs	Lab Hrs	Labour Cost £	Sunds £	Nett Rate £	Unit	Gross Rate (+7.5%) £
P20: UNFRAMED ISOLATED TRIMS/SKIRTINGS/SUNDRY ITEMS Cont.										
Afrormosia, wrought - cappings Cont.										
Cover fillets, stops, trims, beads, nosings and the like; (finished sizes) Cont.										
50 x 100mm; ramped; rebates -1; mouldings -1.....	14.35	10.00	**15.79**	0.90	0.11	**13.97**	0.15	**29.90**	m	**32.14**
extra; ends	2.18	-	**2.18**	0.13	0.02	**2.06**	0.02	**4.26**	nr	**4.58**
extra; angles	3.59	-	**3.59**	0.22	0.03	**3.45**	0.04	**7.08**	nr	**7.61**
extra; mitres	3.59	-	**3.59**	0.22	0.03	**3.45**	0.04	**7.08**	nr	**7.61**
extra; rounded corners not exceeding 300mm girth.	14.35	-	**14.35**	0.90	0.11	**13.97**	0.15	**28.47**	nr	**30.60**
Sapele; wrought - cappings										
Cover fillets, stops, trims, beads, nosings and the like; (finished sizes)										
25 x 50mm; level; rebates -1; mouldings -1.......	3.72	10.00	**4.09**	0.33	0.04	**5.12**	0.05	**9.26**	m	**9.95**
50 x 75mm; level; rebates -1; mouldings -1.......	9.28	10.00	**10.21**	0.46	0.06	**7.18**	0.08	**17.47**	m	**18.78**
extra; ends	1.38	-	**1.38**	0.07	0.01	**1.10**	0.01	**2.49**	nr	**2.68**
extra; angles	2.34	-	**2.34**	0.12	0.02	**1.92**	0.02	**4.28**	nr	**4.60**
extra; mitres	2.34	-	**2.34**	0.12	0.02	**1.92**	0.02	**4.28**	nr	**4.60**
extra; rounded corners not exceeding 300mm girth.	9.28	-	**9.28**	0.46	0.06	**7.18**	0.08	**16.54**	nr	**17.78**
50 x 100mm; level; rebates -1; mouldings -1......	11.38	10.00	**12.51**	0.50	0.06	**7.75**	0.08	**20.34**	m	**21.87**
extra; ends	1.73	-	**1.73**	0.08	0.01	**1.24**	0.01	**2.99**	nr	**3.21**
extra; angles	2.87	-	**2.87**	0.13	0.02	**2.06**	0.02	**4.95**	nr	**5.33**
extra; mitres	2.87	-	**2.87**	0.13	0.02	**2.06**	0.02	**4.95**	nr	**5.33**
extra; rounded corners not exceeding 300mm girth.	11.38	-	**11.38**	0.50	0.06	**7.75**	0.08	**19.20**	nr	**20.65**
25 x 50mm; ramped; rebates -1; mouldings -1......	3.72	10.00	**4.09**	0.50	0.06	**7.75**	0.08	**11.92**	m	**12.81**
50 x 75mm; ramped; rebates -1; mouldings -1......	9.28	10.00	**10.21**	0.70	0.09	**10.91**	0.11	**21.23**	m	**22.82**
extra; ends	1.38	10.00	**1.52**	0.10	0.01	**1.53**	0.02	**3.07**	nr	**3.30**
extra; angles	2.34	-	**2.34**	0.17	0.02	**2.63**	0.03	**5.00**	nr	**5.37**
extra; mitres	2.34	-	**2.34**	0.17	0.02	**2.63**	0.03	**5.00**	nr	**5.37**
extra; rounded corners not exceeding 300mm girth.	9.28	-	**9.28**	0.70	0.09	**10.91**	0.11	**20.30**	nr	**21.82**
50 x 100mm; ramped; rebates -1; mouldings -1.....	11.38	10.00	**12.51**	0.75	0.09	**11.62**	0.12	**24.26**	m	**26.08**
extra; ends	1.73	-	**1.73**	0.11	0.01	**1.67**	0.02	**3.42**	nr	**3.68**
extra; angles	2.87	-	**2.87**	0.19	0.02	**2.91**	0.03	**5.82**	nr	**6.25**
extra; mitres	2.87	-	**2.87**	0.19	0.02	**2.91**	0.03	**5.82**	nr	**6.25**
extra; rounded corners not exceeding 300mm girth.	11.38	-	**11.38**	0.75	0.09	**11.62**	0.12	**23.12**	nr	**24.85**
Softwood, wrought - window boards										
Cover fillets, stops, trims, beads, nosings and the like; (finished sizes)										
mouldings -1; tongued on										
25 x 50mm	1.71	10.00	**1.88**	0.25	0.03	**3.87**	0.04	**5.79**	m	**6.22**
25 x 75mm	2.76	10.00	**3.03**	0.26	0.03	**4.02**	0.04	**7.09**	m	**7.62**
32 x 50mm	2.12	10.00	**2.33**	0.28	0.04	**4.41**	0.05	**6.79**	m	**7.30**
32 x 75mm	3.25	10.00	**3.58**	0.29	0.04	**4.55**	0.05	**8.18**	m	**8.79**
rounded edges -1; tongued on										
25 x 50mm	1.71	10.00	**1.88**	0.25	0.03	**3.87**	0.04	**5.79**	m	**6.22**
25 x 75mm	2.76	10.00	**3.03**	0.26	0.03	**4.02**	0.04	**7.09**	m	**7.62**
32 x 50mm	2.12	10.00	**2.33**	0.28	0.04	**4.41**	0.05	**6.79**	m	**7.30**
32 x 75mm	3.25	10.00	**3.58**	0.29	0.04	**4.55**	0.05	**8.18**	m	**8.79**
Window boards										
25 x 150mm; mouldings -1	3.73	10.00	**4.10**	0.30	0.04	**4.69**	0.05	**8.85**	m	**9.51**
32 x 150mm; mouldings -1	4.38	10.00	**4.82**	0.33	0.04	**5.12**	0.05	**9.99**	m	**10.74**
25 x 150mm; rounded edges -1	3.73	10.00	**4.10**	0.30	0.04	**4.69**	0.05	**8.85**	m	**9.51**
32 x 150mm; rounded edges -1	4.38	10.00	**4.82**	0.33	0.04	**5.12**	0.05	**9.99**	m	**10.74**
Afrormosia, wrought - window boards										
Cover fillets, stops, trims, beads, nosings and the like; (finished sizes)										
mouldings -1; tongued on										
25 x 50mm	4.16	10.00	**4.58**	0.38	0.05	**5.94**	0.06	**10.57**	m	**11.36**
25 x 75mm	6.46	10.00	**7.11**	0.40	0.05	**6.22**	0.06	**13.39**	m	**14.39**
32 x 50mm	5.01	10.00	**5.51**	0.42	0.05	**6.50**	0.07	**12.08**	m	**12.99**
32 x 75mm	7.67	10.00	**8.44**	0.44	0.06	**6.90**	0.07	**15.40**	m	**16.56**
rounded edges -1; tongued on										
25 x 50mm	4.16	10.00	**4.58**	0.38	0.05	**5.94**	0.06	**10.57**	m	**11.36**
25 x 75mm	6.46	10.00	**7.11**	0.40	0.05	**6.22**	0.06	**13.39**	m	**14.39**
32 x 50mm	5.01	10.00	**5.51**	0.42	0.05	**6.50**	0.07	**12.08**	m	**12.99**
32 x 75mm	7.67	10.00	**8.44**	0.44	0.06	**6.90**	0.07	**15.40**	m	**16.56**
Window boards										
25 x 150mm; mouldings -1	10.27	10.00	**11.30**	0.52	0.07	**8.14**	0.09	**19.53**	m	**20.99**
32 x 150mm; mouldings -1	12.20	10.00	**13.42**	0.58	0.07	**8.99**	0.09	**22.50**	m	**24.19**
25 x 150mm; rounded edges -1	10.27	10.00	**11.30**	0.52	0.07	**8.14**	0.09	**19.53**	m	**20.99**
32 x 150mm; rounded edges -1	12.20	10.00	**13.42**	0.58	0.07	**8.99**	0.09	**22.50**	m	**24.19**
Sapele; wrought - window boards										
Cover fillets, stops, trims, beads, nosings and the like; (finished sizes)										
mouldings -1; tongued on										
25 x 50mm	3.48	10.00	**3.82**	0.37	0.05	**5.79**	0.06	**9.68**	m	**10.40**
25 x 75mm	5.32	10.00	**5.86**	0.39	0.05	**6.08**	0.06	**11.99**	m	**12.89**
32 x 50mm	4.07	10.00	**4.48**	0.41	0.05	**6.36**	0.07	**10.91**	m	**11.73**
32 x 75mm	6.35	10.00	**6.98**	0.43	0.05	**6.65**	0.07	**13.70**	m	**14.73**
rounded edges -1; tongued on										
25 x 50mm	3.48	10.00	**3.82**	0.37	0.05	**5.79**	0.06	**9.68**	m	**10.40**
25 x 75mm	5.32	10.00	**5.86**	0.39	0.05	**6.08**	0.06	**11.99**	m	**12.89**
32 x 50mm	4.07	10.00	**4.48**	0.41	0.05	**6.36**	0.07	**10.91**	m	**11.73**
32 x 75mm	6.35	10.00	**6.98**	0.43	0.05	**6.65**	0.07	**13.70**	m	**14.73**

Labour hourly rates: (except Specialists) Craft Operatives £14.21 Labourer £10.71 Rates are national average prices. Refer to REGIONAL VARIATIONS for indicative levels of overall pricing in regions	MATERIALS			LABOUR				RATES		
	Del to Site £	Waste %	Material Cost £	Craft Optve Hrs	Lab Hrs	Labour Cost £	Sunds £	Nett Rate £	Unit	Gross Rate (+7.5%) £
P20: UNFRAMED ISOLATED TRIMS/SKIRTINGS/SUNDRY ITEMS Cont.										
Sapele; wrought - window boards Cont.										
Window boards										
25 x 150mm; mouldings -1	8.44	10.00	**9.28**	0.45	0.06	**7.04**	0.07	**16.39**	m	**17.62**
32 x 150mm; mouldings -1	10.00	10.00	**11.00**	0.50	0.06	**7.75**	0.08	**18.83**	m	**20.24**
25 x 150mm; rounded edges -1	8.44	10.00	**9.28**	0.45	0.06	**7.04**	0.07	**16.39**	m	**17.62**
32 x 150mm; rounded edges -1	10.00	10.00	**11.00**	0.50	0.06	**7.75**	0.08	**18.83**	m	**20.24**
Moisture resistant medium density fibreboard, BS EN 622										
Window boards										
25 x 225mm; nosed and tongued	5.70	10.00	**6.27**	0.30	0.04	**4.69**	0.05	**11.01**	m	**11.84**
25 x 250mm; nosed and tongued	6.33	10.00	**6.97**	0.32	0.04	**4.98**	0.05	**11.99**	m	**12.89**
Softwood, wrought - isolated shelves and worktops										
Cover fillets, stops, trims, beads, nosings and the like; (finished sizes)										
19 x 12mm	0.34	10.00	**0.37**	0.18	0.35	**6.31**	0.30	**6.98**	m	**7.50**
19 x 16mm	0.41	10.00	**0.45**	0.19	0.02	**2.91**	0.03	**3.40**	m	**3.65**
19 x 18mm	0.47	10.00	**0.52**	0.19	0.02	**2.91**	0.03	**3.46**	m	**3.72**
19 x 22mm	0.52	10.00	**0.57**	0.20	0.03	**3.16**	0.03	**3.77**	m	**4.05**
19 x 25mm	0.58	10.00	**0.64**	0.20	0.03	**3.16**	0.03	**3.83**	m	**4.12**
Isolated shelves and worktops										
25 x 150mm	2.22	10.00	**2.44**	0.20	0.03	**3.16**	0.03	**5.64**	m	**6.06**
25 x 225mm	3.31	10.00	**3.65**	0.25	0.03	**3.87**	0.04	**7.56**	m	**8.13**
25 x 300mm	4.67	10.00	**5.14**	0.30	0.04	**4.69**	0.05	**9.88**	m	**10.62**
32 x 150mm	2.82	10.00	**3.10**	0.22	0.03	**3.45**	0.04	**6.59**	m	**7.08**
32 x 225mm	4.29	10.00	**4.71**	0.27	0.03	**4.16**	0.04	**8.91**	m	**9.58**
32 x 300mm	5.67	10.00	**6.24**	0.33	0.04	**5.12**	0.05	**11.40**	m	**12.26**
25 x 450mm; cross tongued	9.97	2.50	**10.22**	0.35	0.04	**5.40**	0.06	**15.68**	m	**16.85**
25 x 600mm; cross tongued	13.28	2.50	**13.61**	0.45	0.06	**7.04**	0.07	**20.72**	m	**22.28**
32 x 450mm; cross tongued	12.76	2.50	**13.08**	0.40	0.06	**6.33**	0.07	**19.48**	m	**20.94**
32 x 600mm; cross tongued	16.97	2.50	**17.40**	0.50	0.06	**7.75**	0.08	**25.23**	m	**27.12**
38 x 450mm; cross tongued	15.13	2.50	**15.51**	0.45	0.06	**7.04**	0.07	**22.61**	m	**24.31**
38 x 600mm; cross tongued	20.47	2.50	**20.98**	0.55	0.07	**8.57**	0.09	**29.64**	m	**31.86**
25 x 450mm overall; 25 x 50mm slats spaced 25mm apart	5.17	10.00	**5.69**	1.00	0.13	**15.60**	0.16	**21.45**	m	**23.06**
25 x 450mm overall; 25 x 50mm slats spaced 32mm apart	4.43	10.00	**4.87**	0.90	0.11	**13.97**	0.15	**18.99**	m	**20.42**
25 x 600mm overall; 25 x 50mm slats spaced 25mm apart	6.65	10.00	**7.32**	1.25	0.16	**19.48**	0.20	**26.99**	m	**29.02**
25 x 600mm overall; 25 x 50mm slats spaced 32mm apart	5.90	10.00	**6.49**	1.15	0.14	**17.84**	0.19	**24.52**	m	**26.36**
25 x 900mm overall; 25 x 50mm slats spaced 25mm apart	9.60	10.00	**10.56**	1.90	0.24	**29.57**	0.31	**40.44**	m	**43.48**
25 x 900mm overall; 25 x 50mm slats spaced 32mm apart	8.86	10.00	**9.75**	1.80	0.23	**28.04**	0.29	**38.08**	m	**40.94**
Afrormosia, wrought - isolated shelves and worktops										
Cover fillets, stops, trims, beads, nosings and the like; (finished sizes)										
19 x 12mm	1.00	10.00	**1.10**	0.31	0.04	**4.83**	0.05	**5.98**	m	**6.43**
19 x 16mm	1.18	10.00	**1.30**	0.33	0.04	**5.12**	0.05	**6.47**	m	**6.95**
19 x 18mm	1.36	10.00	**1.50**	0.33	0.04	**5.12**	0.05	**6.66**	m	**7.16**
19 x 22mm	1.47	10.00	**1.62**	0.35	0.04	**5.40**	0.06	**7.08**	m	**7.61**
19 x 25mm	1.71	10.00	**1.88**	0.35	0.04	**5.40**	0.06	**7.34**	m	**7.89**
Isolated shelves and worktops										
25 x 150mm	7.66	10.00	**8.43**	0.35	0.04	**5.40**	0.06	**13.89**	m	**14.93**
25 x 225mm	11.50	10.00	**12.65**	0.44	0.06	**6.90**	0.07	**19.62**	m	**21.09**
25 x 300mm	15.33	10.00	**16.86**	0.53	0.07	**8.28**	0.09	**25.23**	m	**27.13**
32 x 150mm	9.73	10.00	**10.70**	0.38	0.05	**5.94**	0.06	**16.70**	m	**17.95**
32 x 225mm	14.44	10.00	**15.88**	0.47	0.06	**7.32**	0.08	**23.29**	m	**25.03**
32 x 300mm	19.45	10.00	**21.40**	0.56	0.07	**8.71**	0.09	**30.19**	m	**32.46**
25 x 450mm; cross tongued	31.83	2.50	**32.63**	0.60	0.08	**9.38**	0.10	**42.11**	m	**45.27**
25 x 600mm; cross tongued	42.44	2.50	**43.50**	0.80	0.10	**12.44**	0.13	**56.07**	m	**60.28**
32 x 450mm; cross tongued	40.66	2.50	**41.68**	0.70	0.09	**10.91**	0.11	**52.70**	m	**56.65**
32 x 600mm; cross tongued	54.22	2.50	**55.58**	0.90	0.11	**13.97**	0.15	**69.69**	m	**74.92**
38 x 450mm; cross tongued	48.33	2.50	**49.54**	0.80	0.10	**12.44**	0.13	**62.11**	m	**66.77**
38 x 600mm; cross tongued	64.84	2.50	**66.46**	1.00	0.13	**15.60**	0.16	**82.22**	m	**88.39**
Plywood BS EN 636, II/III grade, INT bonded; butt joints - isolated shelves and worktops										
Cover fillets, stops, trims, beads, nosings and the like										
50 x 6.5mm	0.24	20.00	**0.29**	0.16	0.02	**2.49**	0.05	**2.83**	m	**3.04**
50 x 12mm	0.34	20.00	**0.41**	0.16	0.02	**2.49**	0.05	**2.94**	m	**3.16**
50 x 15mm	0.42	20.00	**0.50**	0.17	0.02	**2.63**	0.05	**3.18**	m	**3.42**
50 x 19mm	0.49	20.00	**0.59**	0.18	0.02	**2.77**	0.05	**3.41**	m	**3.66**
100 x 6.5mm	0.37	20.00	**0.45**	0.16	0.02	**2.49**	0.05	**2.98**	m	**3.21**
100 x 12mm	0.66	20.00	**0.80**	0.17	0.02	**2.63**	0.05	**3.48**	m	**3.74**
100 x 15mm	0.86	20.00	**1.03**	0.18	0.02	**2.77**	0.05	**3.86**	m	**4.14**
100 x 19mm	0.97	20.00	**1.16**	0.19	0.02	**2.91**	0.05	**4.12**	m	**4.43**
Isolated shelves and worktops										
6.5 x 150mm	0.80	20.00	**0.96**	0.18	0.02	**2.77**	0.05	**3.79**	m	**4.07**
6.5 x 225mm	1.23	20.00	**1.48**	0.23	0.02	**3.48**	0.05	**5.01**	m	**5.39**
6.5 x 300mm	1.64	20.00	**1.97**	0.28	0.04	**4.41**	0.05	**6.43**	m	**6.91**

Labour hourly rates: (except Specialists) Craft Operatives £14.21 Labourer £10.71 Rates are national average prices. Refer to REGIONAL VARIATIONS for indicative levels of overall pricing in regions	MATERIALS			LABOUR				RATES		
	Del to Site £	Waste %	Material Cost £	Craft Optve Hrs	Lab Hrs	Labour Cost £	Sunds £	Nett Rate £	Unit	Gross Rate (+7.5%) £
P20: UNFRAMED ISOLATED TRIMS/SKIRTINGS/SUNDRY ITEMS Cont.										
Plywood BS EN 636, II/III grade, INT bonded; butt joints - isolated shelves and worktops Cont.										
Isolated shelves and worktops Cont.										
6.5 x 450mm	2.44	20.00	**2.93**	0.33	0.04	**5.12**	0.05	**8.10**	m	**8.71**
6.5 x 600mm	3.25	20.00	**3.90**	0.38	0.05	**5.94**	0.05	**9.88**	m	**10.62**
12 x 150mm	1.51	20.00	**1.82**	0.19	0.02	**2.91**	0.05	**4.78**	m	**5.14**
12 x 225mm	2.26	20.00	**2.71**	0.24	0.03	**3.73**	0.05	**6.49**	m	**6.98**
12 x 300mm	3.04	20.00	**3.65**	0.29	0.04	**4.55**	0.05	**8.24**	m	**8.86**
12 x 450mm	4.53	20.00	**5.43**	0.34	0.04	**5.26**	0.05	**10.74**	m	**11.55**
12 x 600mm	6.03	20.00	**7.24**	0.39	0.05	**6.08**	0.05	**13.36**	m	**14.36**
15 x 150mm	1.82	20.00	**2.18**	0.23	0.03	**3.59**	0.05	**5.82**	m	**6.25**
15 x 225mm	2.74	20.00	**3.28**	0.25	0.03	**3.87**	0.05	**7.21**	m	**7.75**
15 x 300mm	3.65	20.00	**4.39**	0.30	0.04	**4.69**	0.05	**9.13**	m	**9.81**
15 x 450mm	5.47	20.00	**6.56**	0.35	0.04	**5.40**	0.05	**12.02**	m	**12.92**
15 x 600mm	7.30	20.00	**8.76**	0.40	0.05	**6.22**	0.05	**15.03**	m	**16.15**
19 x 150mm	2.01	20.00	**2.42**	0.20	0.03	**3.16**	0.05	**5.63**	m	**6.05**
19 x 225mm	3.03	20.00	**3.63**	0.25	0.03	**3.87**	0.05	**7.56**	m	**8.12**
19 x 300mm	4.00	20.00	**4.80**	0.30	0.04	**4.69**	0.05	**9.55**	m	**10.26**
19 x 450mm	6.01	20.00	**7.21**	0.35	0.04	**5.40**	0.05	**12.66**	m	**13.61**
19 x 600mm	8.01	20.00	**9.61**	0.40	0.05	**6.22**	0.05	**15.88**	m	**17.07**
Blockboard, BS 3444, 2/2 grade, INT bonded, butt joints - isolated shelves and worktops										
Isolated shelves and worktops										
16 x 150mm	4.54	20.00	**5.45**	0.20	0.03	**3.16**	0.07	**8.68**	m	**9.34**
16 x 225mm	6.81	20.00	**8.18**	0.25	0.03	**3.87**	0.07	**12.12**	m	**13.03**
16 x 300mm	9.08	20.00	**10.90**	0.30	0.04	**4.69**	0.07	**15.66**	m	**16.84**
16 x 450mm	13.63	20.00	**16.35**	0.35	0.04	**5.40**	0.07	**21.82**	m	**23.46**
16 x 600mm	18.10	20.00	**21.72**	0.40	0.05	**6.22**	0.07	**28.01**	m	**30.11**
18 x 150mm	4.68	20.00	**5.61**	0.20	0.03	**3.16**	0.07	**8.85**	m	**9.51**
18 x 225mm	7.02	20.00	**8.42**	0.25	0.03	**3.87**	0.07	**12.37**	m	**13.29**
18 x 300mm	9.33	20.00	**11.20**	0.30	0.04	**4.69**	0.07	**15.96**	m	**17.16**
18 x 450mm	13.99	20.00	**16.79**	0.35	0.04	**5.40**	0.07	**22.26**	m	**23.93**
18 x 600mm	18.69	20.00	**22.43**	0.40	0.05	**6.22**	0.07	**28.72**	m	**30.87**
22 x 150mm	6.31	20.00	**7.58**	0.20	0.03	**3.16**	0.07	**10.81**	m	**11.62**
22 x 225mm	9.47	20.00	**11.36**	0.25	0.03	**3.87**	0.07	**15.31**	m	**16.46**
22 x 300mm	12.20	20.00	**14.64**	0.30	0.04	**4.69**	0.07	**19.40**	m	**20.85**
22 x 450mm	18.94	20.00	**22.73**	0.35	0.04	**5.40**	0.07	**28.20**	m	**30.32**
22 x 600mm	25.26	20.00	**30.31**	0.40	0.05	**6.22**	0.07	**36.60**	m	**39.34**
25 x 150mm	6.84	20.00	**8.20**	0.20	0.03	**3.16**	0.07	**11.44**	m	**12.29**
25 x 225mm	10.20	20.00	**12.24**	0.25	0.03	**3.87**	0.07	**16.18**	m	**17.39**
25 x 300mm	13.63	20.00	**16.35**	0.30	0.04	**4.69**	0.07	**21.11**	m	**22.70**
25 x 450mm	20.44	20.00	**24.53**	0.35	0.04	**5.40**	0.07	**30.00**	m	**32.25**
25 x 600mm	27.25	20.00	**32.70**	0.40	0.05	**6.22**	0.07	**38.99**	m	**41.92**
Wood chipboard BS EN 312, type I standard; butt joints - isolated shelves and worktops										
Isolated shelves and worktops										
12 x 150mm	0.50	20.00	**0.60**	0.19	0.02	**2.91**	0.05	**3.56**	m	**3.83**
12 x 225mm	0.72	20.00	**0.87**	0.24	0.03	**3.73**	0.05	**4.65**	m	**5.00**
12 x 300mm	0.98	20.00	**1.17**	0.29	0.04	**4.55**	0.05	**5.77**	m	**6.21**
12 x 450mm	1.45	20.00	**1.75**	0.34	0.04	**5.26**	0.05	**7.06**	m	**7.58**
12 x 600mm	1.96	20.00	**2.35**	0.39	0.05	**6.08**	0.05	**8.47**	m	**9.11**
18 x 150mm	0.65	20.00	**0.78**	0.20	0.03	**3.16**	0.05	**4.00**	m	**4.30**
18 x 225mm	0.97	20.00	**1.16**	0.25	0.03	**3.87**	0.05	**5.08**	m	**5.46**
18 x 300mm	1.27	20.00	**1.52**	0.30	0.04	**4.69**	0.05	**6.26**	m	**6.73**
18 x 450mm	1.91	20.00	**2.29**	0.35	0.04	**5.40**	0.05	**7.74**	m	**8.32**
18 x 600mm	2.54	20.00	**3.04**	0.40	0.05	**6.22**	0.05	**9.31**	m	**10.01**
25 x 150mm	1.20	20.00	**1.44**	0.20	0.03	**3.16**	0.05	**4.65**	m	**5.00**
25 x 225mm	1.79	20.00	**2.15**	0.25	0.03	**3.87**	0.05	**6.07**	m	**6.53**
25 x 300mm	2.39	20.00	**2.86**	0.30	0.04	**4.69**	0.05	**7.60**	m	**8.18**
25 x 450mm	3.57	20.00	**4.29**	0.35	0.04	**5.40**	0.05	**9.74**	m	**10.47**
25 x 600mm	4.76	20.00	**5.71**	0.40	0.05	**6.22**	0.05	**11.98**	m	**12.88**
Blockboard, BS 3444, 2/2 grade, INT bonded, faced with 1.5mm laminated plastic sheet, BS EN 438, classified HGS, with 1.2mm laminated plastic sheet balance veneer; butt joints										
Isolated shelves and worktops										
16 x 150mm	28.73	20.00	**34.48**	0.40	0.05	**6.22**	0.09	**40.79**	m	**43.84**
16 x 225mm	38.33	20.00	**45.99**	0.50	0.06	**7.75**	0.09	**53.83**	m	**57.87**
16 x 300mm	47.93	20.00	**57.51**	0.60	0.08	**9.38**	0.09	**66.99**	m	**72.01**
16 x 450mm	65.16	20.00	**78.20**	0.70	0.09	**10.91**	0.09	**89.20**	m	**95.89**
16 x 600mm	76.66	20.00	**91.99**	0.80	0.10	**12.44**	0.09	**104.52**	m	**112.36**
18 x 150mm	28.73	20.00	**34.48**	0.40	0.05	**6.22**	0.09	**40.79**	m	**43.84**
18 x 225mm	38.33	20.00	**45.99**	0.50	0.06	**7.75**	0.09	**53.83**	m	**57.87**
18 x 300mm	47.93	20.00	**57.51**	0.60	0.08	**9.38**	0.09	**66.99**	m	**72.01**
18 x 450mm	65.16	20.00	**78.20**	0.70	0.09	**10.91**	0.09	**89.20**	m	**95.89**
18 x 600mm	76.00	20.00	**91.20**	0.80	0.10	**12.44**	0.09	**103.73**	m	**111.51**
22 x 150mm	30.66	20.00	**36.80**	0.40	0.05	**6.22**	0.09	**43.10**	m	**46.34**
22 x 225mm	41.79	20.00	**50.15**	0.50	0.06	**7.75**	0.09	**57.98**	m	**62.33**
22 x 300mm	49.82	20.00	**59.78**	0.60	0.08	**9.38**	0.09	**69.26**	m	**74.45**
22 x 450mm	67.08	20.00	**80.50**	0.70	0.09	**10.91**	0.09	**91.50**	m	**98.37**
22 x 600mm	78.47	20.00	**94.17**	0.80	0.10	**12.44**	0.09	**106.70**	m	**114.70**
25 x 150mm	32.65	20.00	**39.18**	0.40	0.05	**6.22**	0.09	**45.49**	m	**48.90**
25 x 225mm	42.17	20.00	**50.60**	0.50	0.06	**7.75**	0.09	**58.44**	m	**62.82**
25 x 300mm	51.83	20.00	**62.20**	0.60	0.08	**9.38**	0.09	**71.67**	m	**77.05**
25 x 450mm	69.03	20.00	**82.84**	0.70	0.09	**10.91**	0.09	**93.84**	m	**100.88**
25 x 600mm	80.50	20.00	**96.60**	0.80	0.10	**12.44**	0.09	**109.12**	m	**117.31**

Labour hourly rates: (except Specialists) Craft Operatives £14.21 Labourer £10.71 Rates are national average prices. Refer to REGIONAL VARIATIONS for indicative levels of overall pricing in regions	MATERIALS			LABOUR				RATES		
	Del to Site £	Waste %	Material Cost £	Craft Optve Hrs	Lab Hrs	Labour Cost £	Sunds £	Nett Rate £	Unit	Gross Rate (+7.5%) £
P20: UNFRAMED ISOLATED TRIMS/SKIRTINGS/SUNDRY ITEMS Cont.										
Laminated plastic sheet, BS EN 438 classification HGS										
Cover fillets, stops, trims, beads, nosings and the like; fixing with adhesive										
1.5 x 16mm	0.62	20.00	**0.74**	0.20	0.03	**3.16**	0.06	**3.96**	m	**4.26**
1.5 x 18mm	0.69	20.00	**0.82**	0.21	0.03	**3.31**	0.07	**4.20**	m	**4.51**
1.5 x 22mm	0.75	20.00	**0.90**	0.23	0.03	**3.59**	0.08	**4.57**	m	**4.91**
1.5 x 25mm	0.82	20.00	**0.98**	0.25	0.03	**3.87**	0.09	**4.94**	m	**5.32**
1.5 x 75mm	1.56	20.00	**1.88**	0.30	0.04	**4.69**	0.28	**6.85**	m	**7.36**
Isolated shelves and worktops; fixing with adhesive										
1.5 x 150mm	3.10	20.00	**3.72**	0.35	0.04	**5.40**	0.56	**9.69**	m	**10.41**
1.5 x 225mm	4.64	20.00	**5.57**	0.40	0.05	**6.22**	0.84	**12.63**	m	**13.58**
1.5 x 300mm	6.29	20.00	**7.55**	0.45	0.06	**7.04**	1.12	**15.70**	m	**16.88**
1.5 x 450mm	9.39	20.00	**11.27**	0.53	0.07	**8.28**	1.69	**21.24**	m	**22.84**
1.5 x 600mm	12.50	20.00	**15.00**	0.60	0.08	**9.38**	2.25	**26.63**	m	**28.63**
P21: IRONMONGERY										
Fix only ironmongery (prime cost sum for supply included elsewhere)										
To softwood										
butt hinges; 50mm	-	-	**-**	0.15	-	**2.13**	-	**2.13**	nr	**2.29**
butt hinges; 100mm	-	-	**-**	0.15	-	**2.13**	-	**2.13**	nr	**2.29**
rising hinges; 100mm	-	-	**-**	0.25	-	**3.55**	-	**3.55**	nr	**3.82**
tee hinges; 150mm	-	-	**-**	0.40	-	**5.68**	-	**5.68**	nr	**6.11**
tee hinges; 300mm	-	-	**-**	0.50	-	**7.11**	-	**7.11**	nr	**7.64**
tee hinges; 450mm	-	-	**-**	0.60	-	**8.53**	-	**8.53**	nr	**9.17**
hook and band hinges; 300mm	-	-	**-**	0.80	-	**11.37**	-	**11.37**	nr	**12.22**
hook and band hinges; 450mm	-	-	**-**	1.00	-	**14.21**	-	**14.21**	nr	**15.28**
hook and band hinges; 900mm	-	-	**-**	1.50	-	**21.32**	-	**21.32**	nr	**22.91**
collinge hinges; 600mm	-	-	**-**	0.90	-	**12.79**	-	**12.79**	nr	**13.75**
collinge hinges; 750mm	-	-	**-**	1.15	-	**16.34**	-	**16.34**	nr	**17.57**
collinge hinges; 900mm	-	-	**-**	1.40	-	**19.89**	-	**19.89**	nr	**21.39**
single action floor springs and top centres	-	-	**-**	2.50	-	**35.53**	-	**35.53**	nr	**38.19**
double action floor springs and top centres	-	-	**-**	3.00	-	**42.63**	-	**42.63**	nr	**45.83**
coil springs	-	-	**-**	0.25	-	**3.55**	-	**3.55**	nr	**3.82**
overhead door closers	-	-	**-**	1.50	-	**21.32**	-	**21.32**	nr	**22.91**
concealed overhead door closers	-	-	**-**	2.50	-	**35.53**	-	**35.53**	nr	**38.19**
Perko door closers	-	-	**-**	1.50	-	**21.32**	-	**21.32**	nr	**22.91**
door selectors	-	-	**-**	2.00	0.25	**31.10**	-	**31.10**	nr	**33.43**
cabin hooks	-	-	**-**	0.20	-	**2.84**	-	**2.84**	nr	**3.06**
fanlight catches	-	-	**-**	0.35	-	**4.97**	-	**4.97**	nr	**5.35**
roller catches	-	-	**-**	0.35	-	**4.97**	-	**4.97**	nr	**5.35**
casement fasteners	-	-	**-**	0.35	-	**4.97**	-	**4.97**	nr	**5.35**
sash fasteners	-	-	**-**	1.00	-	**14.21**	-	**14.21**	nr	**15.28**
sash screws	-	-	**-**	0.50	-	**7.11**	-	**7.11**	nr	**7.64**
mortice latches	-	-	**-**	0.65	-	**9.24**	-	**9.24**	nr	**9.93**
night latches	-	-	**-**	1.00	-	**14.21**	-	**14.21**	nr	**15.28**
Norfolk latches	-	-	**-**	0.80	-	**11.37**	-	**11.37**	nr	**12.22**
rim latches	-	-	**-**	0.85	-	**12.08**	-	**12.08**	nr	**12.98**
Suffolk latches	-	-	**-**	0.80	-	**11.37**	-	**11.37**	nr	**12.22**
budget locks	-	-	**-**	0.70	-	**9.95**	-	**9.95**	nr	**10.69**
cylinder locks	-	-	**-**	1.00	-	**14.21**	-	**14.21**	nr	**15.28**
dead locks	-	-	**-**	0.80	-	**11.37**	-	**11.37**	nr	**12.22**
mortice locks	-	-	**-**	1.00	-	**14.21**	-	**14.21**	nr	**15.28**
rim locks	-	-	**-**	0.80	-	**11.37**	-	**11.37**	nr	**12.22**
automatic coin collecting locks	-	-	**-**	2.00	-	**28.42**	-	**28.42**	nr	**30.55**
casement stays	-	-	**-**	0.25	-	**3.55**	-	**3.55**	nr	**3.82**
quadrant stays	-	-	**-**	0.25	-	**3.55**	-	**3.55**	nr	**3.82**
barrel bolts; 150mm	-	-	**-**	0.30	-	**4.26**	-	**4.26**	nr	**4.58**
barrel bolts; 250mm	-	-	**-**	0.40	-	**5.68**	-	**5.68**	nr	**6.11**
door bolts; 150mm	-	-	**-**	0.30	-	**4.26**	-	**4.26**	nr	**4.58**
door bolts; 250mm	-	-	**-**	0.40	-	**5.68**	-	**5.68**	nr	**6.11**
monkey tail bolts										
300mm	-	-	**-**	0.35	-	**4.97**	-	**4.97**	nr	**5.35**
450mm	-	-	**-**	0.40	-	**5.68**	-	**5.68**	nr	**6.11**
600mm	-	-	**-**	0.50	-	**7.11**	-	**7.11**	nr	**7.64**
flush bolts; 200mm	-	-	**-**	0.70	-	**9.95**	-	**9.95**	nr	**10.69**
flush bolts; 450mm	-	-	**-**	1.00	-	**14.21**	-	**14.21**	nr	**15.28**
indicating bolts	-	-	**-**	1.00	-	**14.21**	-	**14.21**	nr	**15.28**
panic bolts; to single door	-	-	**-**	1.50	-	**21.32**	-	**21.32**	nr	**22.91**
panic bolts; to double doors	-	-	**-**	2.00	-	**28.42**	-	**28.42**	nr	**30.55**
knobs	-	-	**-**	0.20	-	**2.84**	-	**2.84**	nr	**3.06**
lever handles	-	-	**-**	0.20	-	**2.84**	-	**2.84**	nr	**3.06**
sash lifts	-	-	**-**	0.20	-	**2.84**	-	**2.84**	nr	**3.06**
pull handles; 150mm	-	-	**-**	0.15	-	**2.13**	-	**2.13**	nr	**2.29**
pull handles; 225mm	-	-	**-**	0.20	-	**2.84**	-	**2.84**	nr	**3.06**
pull handles; 300mm	-	-	**-**	0.25	-	**3.55**	-	**3.55**	nr	**3.82**
back plates	-	-	**-**	0.25	-	**3.55**	-	**3.55**	nr	**3.82**
escutcheon plates	-	-	**-**	0.20	-	**2.84**	-	**2.84**	nr	**3.06**
kicking plates	-	-	**-**	0.50	-	**7.11**	-	**7.11**	nr	**7.64**
letter plates	-	-	**-**	1.50	-	**21.32**	-	**21.32**	nr	**22.91**
push plates; 225mm	-	-	**-**	0.20	-	**2.84**	-	**2.84**	nr	**3.06**
push plates; 300mm	-	-	**-**	0.25	-	**3.55**	-	**3.55**	nr	**3.82**
shelf brackets	-	-	**-**	0.25	-	**3.55**	-	**3.55**	nr	**3.82**
sash cleats	-	-	**-**	0.20	-	**2.84**	-	**2.84**	nr	**3.06**
To softwood and brickwork										
cabin hooks	-	-	**-**	0.25	-	**3.55**	-	**3.55**	nr	**3.82**

Labour hourly rates: (except Specialists) Craft Operatives £14.21 Labourer £10.71 Rates are national average prices. Refer to REGIONAL VARIATIONS for indicative levels of overall pricing in regions	MATERIALS			LABOUR				RATES		
	Del to Site £	Waste %	Material Cost £	Craft Optve Hrs	Lab Hrs	Labour Cost £	Sunds £	Nett Rate £	Unit	Gross Rate (+7.5%) £
P21: IRONMONGERY Cont.										
Fix only ironmongery (prime cost sum for supply included elsewhere)										
To hardwood or the like										
butt hinges; 50mm	-	-	-	0.20	-	2.84	-	2.84	nr	3.06
butt hinges; 100mm	-	-	-	0.20	-	2.84	-	2.84	nr	3.06
rising hinges; 100mm	-	-	-	0.35	-	4.97	-	4.97	nr	5.35
tee hinges; 150mm	-	-	-	0.55	-	7.82	-	7.82	nr	8.40
tee hinges; 300mm	-	-	-	0.70	-	9.95	-	9.95	nr	10.69
tee hinges; 450mm	-	-	-	0.80	-	11.37	-	11.37	nr	12.22
hook and band hinges; 300mm	-	-	-	1.10	-	15.63	-	15.63	nr	16.80
hook and band hinges; 450mm	-	-	-	1.40	-	19.89	-	19.89	nr	21.39
hook and band hinges; 900mm	-	-	-	2.00	-	28.42	-	28.42	nr	30.55
collinge hinges; 600mm	-	-	-	1.30	-	18.47	-	18.47	nr	19.86
collinge hinges; 750mm	-	-	-	1.70	-	24.16	-	24.16	nr	25.97
collinge hinges; 900mm	-	-	-	2.05	-	29.13	-	29.13	nr	31.32
single action floor springs and top centres	-	-	-	3.75	-	53.29	-	53.29	nr	57.28
double action floor springs and top centres	-	-	-	4.50	-	63.95	-	63.95	nr	68.74
coil springs	-	-	-	0.40	-	5.68	-	5.68	nr	6.11
overhead door closers	-	-	-	2.25	-	31.97	-	31.97	nr	34.37
concealed overhead door closers	-	-	-	3.75	-	53.29	-	53.29	nr	57.28
Perko door closers	-	-	-	2.25	-	31.97	-	31.97	nr	34.37
door selectors	-	-	-	3.00	0.38	46.70	-	46.70	nr	50.20
cabin hooks	-	-	-	0.30	-	4.26	-	4.26	nr	4.58
fanlight catches	-	-	-	0.50	-	7.11	-	7.11	nr	7.64
roller catches	-	-	-	0.50	-	7.11	-	7.11	nr	7.64
casement fasteners	-	-	-	0.50	-	7.11	-	7.11	nr	7.64
sash fasteners	-	-	-	1.50	-	21.32	-	21.32	nr	22.91
sash screws	-	-	-	0.75	-	10.66	-	10.66	nr	11.46
mortice latches	-	-	-	1.00	-	14.21	-	14.21	nr	15.28
night latches	-	-	-	1.50	-	21.32	-	21.32	nr	22.91
Norfolk latches	-	-	-	1.20	-	17.05	-	17.05	nr	18.33
rim latches	-	-	-	0.75	-	10.66	-	10.66	nr	11.46
Suffolk latches	-	-	-	1.20	-	17.05	-	17.05	nr	18.33
budget locks	-	-	-	1.00	-	14.21	-	14.21	nr	15.28
cylinder locks	-	-	-	1.50	-	21.32	-	21.32	nr	22.91
dead locks	-	-	-	1.20	-	17.05	-	17.05	nr	18.33
mortice locks	-	-	-	1.50	-	21.32	-	21.32	nr	22.91
rim locks	-	-	-	1.20	-	17.05	-	17.05	nr	18.33
automatic coin collecting locks	-	-	-	3.00	-	42.63	-	42.63	nr	45.83
casement stays	-	-	-	0.35	-	4.97	-	4.97	nr	5.35
quadrant stays	-	-	-	0.35	-	4.97	-	4.97	nr	5.35
barrel bolts; 150mm	-	-	-	0.40	-	5.68	-	5.68	nr	6.11
barrel bolts; 250mm	-	-	-	0.55	-	7.82	-	7.82	nr	8.40
door bolts; 150mm	-	-	-	0.40	-	5.68	-	5.68	nr	6.11
door bolts; 250mm	-	-	-	0.55	-	7.82	-	7.82	nr	8.40
monkey tail bolts										
300mm	-	-	-	0.50	-	7.11	-	7.11	nr	7.64
450mm	-	-	-	0.60	-	8.53	-	8.53	nr	9.17
600mm	-	-	-	0.70	-	9.95	-	9.95	nr	10.69
flush bolts; 200mm	-	-	-	1.00	-	14.21	-	14.21	nr	15.28
flush bolts; 450mm	-	-	-	1.50	-	21.32	-	21.32	nr	22.91
indicating bolts	-	-	-	1.50	-	21.32	-	21.32	nr	22.91
panic bolts; to single door	-	-	-	2.25	-	31.97	-	31.97	nr	34.37
panic bolts; to double doors	-	-	-	3.00	-	42.63	-	42.63	nr	45.83
knobs	-	-	-	0.30	-	4.26	-	4.26	nr	4.58
lever handles	-	-	-	0.30	-	4.26	-	4.26	nr	4.58
sash lifts	-	-	-	0.30	-	4.26	-	4.26	nr	4.58
pull handles; 150mm	-	-	-	0.20	-	2.84	-	2.84	nr	3.06
pull handles; 225mm	-	-	-	0.30	-	4.26	-	4.26	nr	4.58
pull handles; 300mm	-	-	-	0.35	-	4.97	-	4.97	nr	5.35
back plates	-	-	-	0.35	-	4.97	-	4.97	nr	5.35
escutcheon plates	-	-	-	0.30	-	4.26	-	4.26	nr	4.58
kicking plates	-	-	-	0.75	-	10.66	-	10.66	nr	11.46
letter plates	-	-	-	2.25	-	31.97	-	31.97	nr	34.37
push plates; 225mm	-	-	-	0.30	-	4.26	-	4.26	nr	4.58
push plates; 300mm	-	-	-	0.35	-	4.97	-	4.97	nr	5.35
shelf brackets	-	-	-	0.40	-	5.68	-	5.68	nr	6.11
sash cleats	-	-	-	0.30	-	4.26	-	4.26	nr	4.58
To hardwood and brickwork										
cabin hooks	-	-	-	0.35	-	4.97	-	4.97	nr	5.35
Water bars; steel; galvanized										
Water bars; to concrete										
flat section; setting in groove in mastic										
25 x 3 x 900mm long	2.31	5.00	2.43	0.60	0.08	9.38	0.40	12.21	nr	13.12
40 x 3 x 900mm long	3.68	5.00	3.86	0.60	0.08	9.38	0.50	13.74	nr	14.77
40 x 6 x 900mm long	5.28	5.00	5.55	0.60	0.08	9.38	0.50	15.43	nr	16.59
50 x 6 x 900mm long	7.12	5.00	7.47	0.60	0.08	9.38	0.50	17.36	nr	18.66
Water bars; to hardwood										
flat section; setting in groove in mastic										
25 x 3 x 900mm long	2.31	5.00	2.43	0.50	0.06	7.75	0.40	10.57	nr	11.37
40 x 3 x 900mm long	3.68	5.00	3.86	0.50	0.06	7.75	0.50	12.11	nr	13.01
40 x 6 x 900mm long	5.28	5.00	5.55	0.50	0.06	7.75	0.50	13.79	nr	14.83
50 x 6 x 900mm long	7.12	5.00	7.47	0.50	0.06	7.75	0.50	15.72	nr	16.90

Labour hourly rates: (except Specialists) Craft Operatives £14.21 Labourer £10.71 Rates are national average prices. Refer to REGIONAL VARIATIONS for indicative levels of overall pricing in regions	MATERIALS			LABOUR				RATES		
	Del to Site £	Waste %	Material Cost £	Craft Optve Hrs	Lab Hrs	Labour Cost £	Sunds £	Nett Rate £	Unit	Gross Rate (+7.5%) £
P21: IRONMONGERY Cont.										
Sliding door gear, P.C Henderson Ltd.										
To softwood										
interior straight sliding door gear sets for commercial and domestic doors; Senator single door set comprising track, hangers, end stops and bottom guide; for doors 20 - 35mm thick, maximum weight 25kg maximum 900mm wide..................	22.99	2.50	**23.56**	1.50	-	**21.32**	-	**44.88**	nr	**48.25**
extra for pelmet 1855mm long....................	18.27	2.50	**18.72**	0.65	-	**9.24**	-	**27.96**	nr	**30.06**
extra for pelmet end cap	2.00	2.50	**2.05**	0.10	-	**1.42**	-	**3.47**	nr	**3.73**
extra for Doorseal deflector guide..............	8.55	2.50	**8.77**	0.35	-	**4.97**	-	**13.74**	nr	**14.77**
interior straight sliding door gear sets for commercial and domestic doors; Phantom single door set comprising top assembly, hangers, adjustable nylon guide and door stops; for doors 30 - 50mm thick, maximum weight 45kg, 610 - 915mm wide.....	37.08	2.50	**38.01**	1.75	-	**24.87**	-	**62.87**	nr	**67.59**
extra for soffit fixing bracket and bolt.........	1.51	2.50	**1.55**	0.20	-	**2.84**	-	**4.39**	nr	**4.72**
extra for pelmet 1550mm long....................	15.28	2.50	**15.66**	0.55	-	**7.82**	-	**23.47**	nr	**25.23**
extra for pelmet 1855mm long....................	18.27	2.50	**18.72**	0.65	-	**9.24**	-	**27.96**	nr	**30.06**
extra for pelmet end cap	2.00	2.50	**2.05**	0.10	-	**1.42**	-	**3.47**	nr	**3.73**
extra for Doorseal deflector guide..............	8.55	2.50	**8.77**	0.35	-	**4.97**	-	**13.74**	nr	**14.77**
interior straight sliding door gear sets for commercial and domestic doors; Marathon Junior Nr J2 single door set comprising top assembly, hangers, end stops, inverted guide channel and nylon guide; for doors 32 - 50mm thick, maximum weight, 55kg, 400 - 750mm wide; forming groove for guide channel	34.47	2.50	**35.33**	2.40	-	**34.10**	-	**69.44**	nr	**74.65**
interior straight sliding door gear sets for commercial and domestic doors; Marathon Junior Nr J3 single door set comprising top assembly, hangers, end stops, inverted guide channel and nylon guide; for doors 32 - 50mm thick, maximum weight, 55kg, 750 - 900mm wide; forming groove for guide channel	36.38	2.50	**37.28**	2.60	-	**36.95**	-	**74.23**	nr	**79.80**
interior straight sliding door gear sets for commercial and domestic doors; Marathon Junior Nr J4 single door set comprising top assembly, hangers, end stops, inverted guide channel and nylon guide; for doors 32 - 50mm thick, maximum weight, 55kg, 900 - 1050mm wide; forming groove for guide channel	39.38	2.50	**40.36**	2.80	-	**39.79**	-	**80.15**	nr	**86.16**
interior straight sliding door gear sets for commercial and domestic doors; Marathon Junior Nr J5 single door set comprising top assembly, hangers, end stops, inverted guide channel and nylon guide; for doors 32 - 50mm thick, maximum weight, 55kg, 1050 - 1200mm wide; forming groove for guide channel	44.50	2.50	**45.61**	3.00	-	**42.63**	-	**88.24**	nr	**94.86**
interior straight sliding door gear sets for commercial and domestic doors; Marathon Junior Nr J6 single door set comprising top assembly, hangers, end stops, inverted guide channel and nylon guide; for doors 32 - 50mm thick, maximum weight, 55kg, 1200 - 1500mm wide; forming groove for guide channel	73.86	2.50	**75.71**	3.20	-	**45.47**	-	**121.18**	nr	**130.27**
extra for soffit fixing bracket and bolt.........	1.51	2.50	**1.55**	0.20	-	**2.84**	-	**4.39**	nr	**4.72**
extra for pelmet 1550mm long....................	15.28	2.50	**15.66**	0.55	-	**7.82**	-	**23.47**	nr	**25.23**
extra for pelmet 1855mm long....................	18.27	2.50	**18.72**	0.65	-	**9.24**	-	**27.96**	nr	**30.06**
extra for pelmet end cap	2.00	2.50	**2.05**	0.10	-	**1.42**	-	**3.47**	nr	**3.73**
extra for Doorseal deflector guide..............	8.55	2.50	**8.77**	0.35	-	**4.97**	-	**13.74**	nr	**14.77**
interior straight sliding door gear sets for commercial and domestic doors; Marathon Senior Nr S3 single door set comprising top assembly, hangers, end stops, inverted guide channel and nylon guide; for doors 32 - 50mm thick maximum weight 90kg, 750 - 900mm wide; forming groove for guide channel	48.81	2.50	**50.03**	2.80	-	**39.79**	-	**89.82**	nr	**96.55**
interior straight sliding door gear sets for commercial and domestic doors; Marathon Senior Nr S4 single door set comprising top assembly, hangers, end stops, inverted guide channel and nylon guide; for doors 32 - 50mm thick maximum weight 90kg, 900 - 1050mm wide; forming groove for guide channel	54.20	2.50	**55.55**	3.00	-	**42.63**	-	**98.18**	nr	**105.55**
interior straight sliding door gear sets for commercial and domestic doors; Marathon Senior Nr S5 single door set comprising top assembly, hangers, end stops, inverted guide channel and nylon guide; for doors 32 - 50mm thick maximum weight 90kg, 1050 - 1200mm wide; forming groove for guide channel	59.80	2.50	**61.29**	3.15	-	**44.76**	-	**106.05**	nr	**114.01**
interior straight sliding door gear sets for commercial and domestic doors; Marathon Senior Nr S6 single door set comprising top assembly, hangers, end stops, inverted guide channel and nylon guide; for doors 32 - 50mm thick maximum weight 90kg, 1200 - 1500mm wide; forming groove for guide channel	95.89	2.50	**98.29**	3.30	-	**46.89**	-	**145.18**	nr	**156.07**
extra for soffit fixing bracket and bolt.........	1.51	2.50	**1.55**	0.20	-	**2.84**	-	**4.39**	nr	**4.72**
extra for pelmet 1550mm long....................	15.28	2.50	**15.66**	0.55	-	**7.82**	-	**23.47**	nr	**25.23**
extra for pelmet 1855mm long....................	18.27	2.50	**18.72**	0.65	-	**9.24**	-	**27.96**	nr	**30.06**
extra for pelmet end cap	2.00	2.50	**2.05**	0.10	-	**1.42**	-	**3.47**	nr	**3.73**
extra for Doorseal deflector guide..............	8.55	2.50	**8.77**	0.30	-	**4.26**	-	**13.03**	nr	**14.01**

Labour hourly rates: (except Specialists) Craft Operatives £14.21 Labourer £10.71 Rates are national average prices. Refer to REGIONAL VARIATIONS for indicative levels of overall pricing in regions	MATERIALS			LABOUR				RATES		
	Del to Site £	Waste %	Material Cost £	Craft Optve Hrs	Lab Hrs	Labour Cost £	Sunds £	Nett Rate £	Unit	Gross Rate (+7.5%) £
P21: IRONMONGERY Cont.										
Sliding door gear, P.C Henderson Ltd. Cont.										
To softwood Cont.										
interior straight sliding door gear sets for wardrobe and cupboard doors; Single Top Nr ST12 single door set comprising track, hangers, guides and safety stop; for door 16 - 35mm thick, maximum weight 25kg, maximum 900mm wide; one door to 600mm wide opening	14.03	2.50	**14.38**	1.25	-	**17.76**	-	**32.14**	nr	**34.55**
interior straight sliding door gear sets for wardrobe and cupboard doors; Single Top Nr ST15 single door set comprising track, hangers, guides and safety stop; for door 16 - 35mm thick, maximum weight 25kg, maximum 900mm wide; one door to 750mm wide opening	15.81	2.50	**16.20**	1.35	-	**19.18**	-	**35.39**	nr	**38.04**
interior straight sliding door gear sets for wardrobe and cupboard doors; Single Top Nr ST18 single door set comprising track, hangers, guides and safety stop; for door 16 - 35mm thick, maximum weight 25kg, maximum 900mm wide; one door to 900mm wide opening	17.90	2.50	**18.34**	1.45	-	**20.60**	-	**38.95**	nr	**41.87**
interior straight sliding door gear sets for wardrobe and cupboard doors; Double Top Nr W12 bi-passing door set comprising double track section, hangers, guides and safety stop; for doors 16 - 35mm thick, maximum weight 25kg, maximum 900mm wide; two doors in 1200mm wide opening	24.06	2.50	**24.67**	2.00	-	**28.42**	-	**53.09**	nr	**57.07**
interior straight sliding door gear sets for wardrobe and cupboard doors; Double Top Nr W15 bi-passing door set comprising double track section, hangers, guides and safety stop; for doors 16 - 35mm thick, maximum weight 25kg, maximum 900mm wide; two doors in 1500mm wide opening	26.82	2.50	**27.49**	2.10	-	**29.84**	-	**57.33**	nr	**61.63**
interior straight sliding door gear sets for wardrobe and cupboard doors; Double Top Nr W18 bi-passing door set comprising double track section, hangers, guides and safety stop; for doors 16 - 35mm thick, maximum weight 25kg, maximum 900mm wide; two doors in 1800mm wide opening	29.76	2.50	**30.51**	2.20	-	**31.26**	-	**61.77**	nr	**66.40**
interior straight sliding door gear sets for wardrobe and cupboard doors; Double Top Nr W24 bi-passing door set comprising double track section, hangers, guides and safety stop; for doors 16 - 35mm thick, maximum weight 25kg, maximum 900mm wide; three doors in 2400mm wide opening	39.01	2.50	**39.98**	2.40	-	**34.10**	-	**74.09**	nr	**79.64**
interior straight sliding door gear sets for wardrobe and cupboard doors; Bi-Fold Nr B10-2 folding door set comprising top guide track, top and bottom pivots; top guide and hinges; for doors 20 - 35mm thick, maximum weight 14kg each leaf, maximum 530mm wide; two doors in 1065mm wide opening	22.12	2.50	**22.68**	3.00	-	**42.63**	-	**65.31**	nr	**70.21**
interior straight sliding door gear sets for wardrobes and cupboard doors; Bi-Fold Nr B15-4 folding door set comprising top guide track, top and bottom pivots, top guide and hinges; for doors 20 - 35mm thick, maximum weight 14kg each leaf, maximum 530mm wide; four doors in 1525mm wide opening with aligner	37.55	2.50	**38.49**	6.00	-	**85.26**	-	**123.75**	nr	**133.03**
interior straight sliding door gear sets for wardrobes and cupboard doors; Bi-Fold Nr B20-4 folding door set comprising top guide track, top and bottom pivots, top guide and hinges; for doors 20 - 35mm thick, maximum weight 14kg each leaf, maximum 530mm wide; four doors in 2135mm wide opening with aligner	41.10	2.50	**42.12**	6.50	-	**92.37**	-	**134.49**	nr	**144.58**
interior straight sliding door gear sets for built in cupboard doors; Slipper Nr SS4 double passing door set comprising two top tracks, sliders, safety stop, flush pulls and guides; for doors 16 - 30mm thick, maximum weight 9kg, maximum 900mm wide; two doors in 1200mm wide opening	16.96	2.50	**17.38**	1.85	-	**26.29**	-	**43.67**	nr	**46.95**
interior straight sliding door gear sets for built in cupboard doors; Slipper Nr SS5 double passing door set comprising two top tracks, sliders, safety stop, flush pulls and guides; for doors 16 - 30mm thick, maximum weight 9kg, maximum 900mm wide; two doors in 1500mm wide opening	19.96	2.50	**20.46**	2.00	-	**28.42**	-	**48.88**	nr	**52.55**
interior straight sliding door gear sets for built in cupboard doors; Slipper Nr SS6 double passing door set comprising two top tracks, sliders, safety stop, flush pulls and guides; for doors 16 - 30mm thick, maximum weight 9kg, maximum 900mm wide; two doors in 1800mm wide opening	22.30	2.50	**22.85**	2.15	-	**30.55**	-	**53.41**	nr	**57.41**
interior straight sliding door gear sets for cupboards, book cases and cabinet work; Loretto Nr D4 bi-passing door set comprising two top guide channels, two bottom rails, nylon guides and bottom rollers; for doors 20 - 45mm thick; maximum weight 23kg, maximum 900mm wide; two doors in 1200mm wide opening; forming grooves for top guide channels and bottom rails; forming sinkings for bottom rollers	28.75	2.50	**29.47**	2.50	-	**35.53**	-	**64.99**	nr	**69.87**

Labour hourly rates: (except Specialists) Craft Operatives £14.21 Labourer £10.71 Rates are national average prices. Refer to REGIONAL VARIATIONS for indicative levels of overall pricing in regions	MATERIALS			LABOUR				RATES		
	Del to Site £	Waste %	Material Cost £	Craft Optve Hrs	Lab Hrs	Labour Cost £	Sunds £	Nett Rate £	Unit	Gross Rate (+7.5%) £
P21: IRONMONGERY Cont.										
Sliding door gear, P.C Henderson Ltd. Cont.										
To softwood Cont.										
interior straight sliding door gear sets for cupboards, book cases and cabinet work; Loretto Nr D5 bi-passing door set comprising two top guide channels, two bottom rails, nylon guides and bottom rollers; for doors 20 - 45mm thick; maximum weight 23kg, maximum 900mm wide; two doors in 1500mm wide opening; forming grooves for top guide channels and bottom rails; forming sinkings for bottom rollers	31.99	2.50	**32.79**	2.60	-	**36.95**	-	**69.73**	nr	**74.96**
interior straight sliding door gear sets for cupboards, book cases and cabinet work; Loretto Nr D6 bi-passing door set comprising two top guide channels, two bottom rails, nylon guides and bottom rollers; for doors 20 - 45mm thick; maximum weight 23kg, maximum 900mm wide; two doors in 1800mm wide opening; forming grooves for top guide channels and bottom rails; forming sinkings for bottom rollers	36.14	2.50	**37.04**	2.70	-	**38.37**	-	**75.41**	nr	**81.07**
extra for retractable roller guide bolt	6.79	2.50	**6.96**	0.35	-	**4.97**	-	**11.93**	nr	**12.82**
extra for flush pull	2.94	2.50	**3.02**	0.25	-	**3.55**	-	**6.57**	nr	**7.06**
extra for safety stop	0.93	2.50	**0.95**	0.10	-	**1.42**	-	**2.37**	nr	**2.55**
extra for cylinder lock	14.20	2.50	**14.56**	0.60	-	**8.53**	-	**23.08**	nr	**24.81**
interior straight sliding/folding room divider gear sets; Husky Folding Nr HF-2 folding door set comprising top track, brackets, top and bottom pivots, hangers and hinges, bottom channel and roller guides; for doors 20 - 40mm thick, maximum weight 25kg, maximum 600mm wide, maximum 2400mm high; two doors in 1200mm wide opening; forming groove for bottom channel	61.05	2.50	**62.57**	5.80	-	**82.42**	-	**144.99**	nr	**155.86**
interior straight sliding/folding room divider gear sets; Husky Folding Nr HF-4 folding door set comprising top track, brackets, top and bottom pivots, hangers and hinges, bottom channel and roller guides; for doors 20 - 40mm thick, maximum weight 25kg, maximum 600mm wide, maximum 2400mm high; four doors in 2400mm wide opening; forming groove for bottom channel	115.90	2.50	**118.80**	11.50	-	**163.42**	-	**282.21**	nr	**303.38**
interior straight sliding door gear sets for glass panels; Zenith 2 Nr 12 double passing door set comprising double top guide, bottom rail, glass rail with rubber glazing strip, end caps and bottom rollers; for panels 6mm thick, maximum weight 16kg or 1m2 per panel; two panels nr 1200mm wide opening	43.64	2.50	**44.73**	1.50	-	**21.32**	-	**66.05**	nr	**71.00**
extra for dust seal	4.10	2.50	**4.21**	0.30	-	**4.26**	-	**8.47**	m	**9.10**
interior straight sliding door gear sets for glass panels; Zenith 2 Nr 15 double passing door set comprising double top guide, bottom rail, glass rail with rubber glazing strip, end caps and bottom rollers; for panels 6mm thick, maximum weight 16kg or 1m2 per panel; two panels nr 1500mm wide opening	50.27	2.50	**51.52**	1.65	-	**23.45**	-	**74.97**	nr	**80.59**
extra for dust seal	4.10	2.50	**4.21**	0.30	-	**4.26**	-	**8.47**	m	**9.10**
interior straight sliding door gear sets for glass panels; Zenith 2 Nr 18 double passing door set comprising double top guide, bottom rail, glass rail with rubber glazing strip, end caps and bottom rollers; for panels 6mm thick, maximum weight 16kg or 1m2 per panel; two panels nr 1800mm wide opening	56.28	2.50	**57.68**	1.80	-	**25.58**	-	**83.26**	nr	**89.50**
extra for finger pull	2.94	2.50	**3.02**	0.40	-	**5.68**	-	**8.70**	nr	**9.35**
extra for cylinder lock	14.20	2.50	**14.56**	1.00	-	**14.21**	-	**28.77**	nr	**30.92**
extra for dust seal	4.10	2.50	**4.21**	0.30	-	**4.26**	-	**8.47**	m	**9.10**
To softwood and concrete										
exterior straight sliding bottom roller gear; Sterling 225 components comprising top guide reference 900, top guide brackets for face fixing for single run reference 1/900, top guide rollers for face fixing reference 54/900, bottom rollers reference 5 and bottom rail reference 299; for timber doors 44 - 50mm thick, maximum weight 225kg, maximum 3300mm high; single door in 1200mm wide opening; forming notches for bottom rollers; casting bottom rail into concrete	288.83	2.50	**296.05**	6.15	-	**87.39**	-	**383.44**	nr	**412.20**
exterior straight sliding bottom roller gear; Sterling 225 components comprising top guide reference 900, top guide brackets for soffit fixing for single run reference 3/900, top guide rollers for edge fixing reference 203/900, bottom rollers reference 5 and bottom rail reference 299; for timber doors 44 - 50mm thick, maximum weight 225kg, maximum 3300mm high; single door in 1200mm wide opening; forming notches for bottom rollers; casting bottom rail into concrete	312.42	2.50	**320.23**	6.35	-	**90.23**	-	**410.47**	nr	**441.25**

Labour hourly rates: (except Specialists) Craft Operatives £14.21 Labourer £10.71 Rates are national average prices. Refer to REGIONAL VARIATIONS for indicative levels of overall pricing in regions	MATERIALS			LABOUR				RATES		
	Del to Site £	Waste %	Material Cost £	Craft Optve Hrs	Lab Hrs	Labour Cost £	Sunds £	Nett Rate £	Unit	Gross Rate (+7.5%) £
P21: IRONMONGERY Cont.										
Sliding door gear, P.C Henderson Ltd. Cont.										
To softwood and concrete Cont.										
exterior straight sliding bottom roller gear; Sterling 225 components comprising top guide reference 900, top guide brackets for face fixing for single run reference 1/900, top guide rollers for face fixing reference 54/900, bottom rollers reference 5 and bottom rail reference 299; for timber doors 44 - 50mm thick, maximum weight 225kg, maximum 3300mm high; two doors in 2400mm wide opening; forming notches for bottom rollers; casting bottom rails into concrete..............	569.10	2.50	**583.33**	12.00	-	**170.52**	-	**753.85**	nr	**810.39**
exterior straight sliding bottom roller gear; Sterling 225 components comprising top guides reference 900, top guide brackets for face fixing for double run reference 5/900, top guide rollers for face fixing reference 54/900, bottom rollers reference 5 and bottom rails reference 299; for timber doors 44 - 50mm thick, maximum weight 225kg, maximum 3300 high; two doors in 2400mm wide opening; forming notches for bottom rollers; casting bottom rails into concrete..............	647.31	2.50	**663.49**	12.35	-	**175.49**	-	**838.98**	nr	**901.91**
exterior straight sliding bottom roller gear; Sterling 350 components comprising top guide reference 99, top guide brackets for face fixing for single run reference 31 and 31x, top guide rollers reference 53/99, bottom rollers reference 2 and bottom rail reference 299; for timber doors 44 - 50mm thick, maximum weight 350kg, maximum 4000mm high; single door in 2400mm wide opening; forming notches for bottom rollers; casting bottom rail into concrete	640.25	2.50	**656.25**	12.90	-	**183.31**	-	**839.56**	nr	**902.53**
exterior straight sliding bottom roller gear; Sterling 350 components comprising top guide reference 99, top guide brackets for face fixing for single run reference 31 and 31X, top guide rollers reference 53/99, bottom rollers reference 2 and bottom rail reference 299; for timber doors 44 - 50mm thick, maximum weight 350kg, maximum 4000mm high; two doors in 3600mm wide opening; forming notches for bottom rollers; casting bottom rail into concrete	1074.81	2.50	**1101.68**	16.33	-	**232.05**	-	**1333.73**	nr	**1433.76**
exterior straight sliding bottom roller gear; Sterling 800 components comprising top guide reference 99, top guide brackets for face fixing for single run reference 31 and 31X, top guide rollers reference 53/99, bottom rollers reference 3 and bottom rail reference 298; for timber doors 54 - 63mm thick, maximum weight 800kg, maximum 5200mm high; single door in 3600mm wide opening; forming notches for bottom rollers; casting bottom rail into concrete	926.51	2.50	**949.67**	14.65	-	**208.18**	-	**1157.84**	nr	**1244.68**
exterior straight sliding bottom roller gear, Sterling 800 components comprising top guide, reference 99, top guide brackets for face fixing for single run reference 31 and 31X, top guide rollers reference 53/99, bottom rollers reference 3 and bottom rail reference 298; for timber doors 54 - 63mm thick, maximum weight 800kg, maximum 5200mm high; two doors in 5400mm wide opening; forming notches for bottom rollers; casting bottom rail into concrete	1525.43	2.50	**1563.57**	23.55	-	**334.65**	-	**1898.22**	nr	**2040.58**
Hinges; standard quality										
To softwood										
backflap hinges; steel										
25mm..	0.23	2.50	**0.24**	0.15	-	**2.13**	-	**2.37**	nr	**2.54**
38mm..	0.27	2.50	**0.28**	0.15	-	**2.13**	-	**2.41**	nr	**2.59**
50mm..	0.35	2.50	**0.36**	0.15	-	**2.13**	-	**2.49**	nr	**2.68**
63mm..	1.04	2.50	**1.07**	0.15	-	**2.13**	-	**3.20**	nr	**3.44**
75mm..	1.65	2.50	**1.69**	0.15	-	**2.13**	-	**3.82**	nr	**4.11**
butt hinges; steel; light medium pattern										
38mm..	0.14	2.50	**0.14**	0.15	-	**2.13**	-	**2.28**	nr	**2.45**
50mm..	0.14	2.50	**0.14**	0.15	-	**2.13**	-	**2.28**	nr	**2.45**
63mm..	0.16	2.50	**0.16**	0.15	-	**2.13**	-	**2.30**	nr	**2.47**
75mm..	0.16	2.50	**0.16**	0.15	-	**2.13**	-	**2.30**	nr	**2.47**
100mm...	0.33	2.50	**0.34**	0.15	-	**2.13**	-	**2.47**	nr	**2.65**
butt hinges; steel; strong pattern										
75mm..	0.64	2.50	**0.66**	0.15	-	**2.13**	-	**2.79**	nr	**3.00**
100mm...	0.70	2.50	**0.72**	0.15	-	**2.13**	-	**2.85**	nr	**3.06**
butt hinges; cast iron; light										
50mm..	0.84	2.50	**0.86**	0.15	-	**2.13**	-	**2.99**	nr	**3.22**
63mm..	0.94	2.50	**0.96**	0.15	-	**2.13**	-	**3.10**	nr	**3.33**
75mm..	1.04	2.50	**1.07**	0.15	-	**2.13**	-	**3.20**	nr	**3.44**
100mm...	1.13	2.50	**1.16**	0.15	-	**2.13**	-	**3.29**	nr	**3.54**
butt hinges; brass, brass pin										
38mm..	0.79	2.50	**0.81**	0.15	-	**2.13**	-	**2.94**	nr	**3.16**
50mm..	0.91	2.50	**0.93**	0.15	-	**2.13**	-	**3.06**	nr	**3.29**
63mm..	1.07	2.50	**1.10**	0.15	-	**2.13**	-	**3.23**	nr	**3.47**
75mm..	1.16	2.50	**1.19**	0.15	-	**2.13**	-	**3.32**	nr	**3.57**
100mm...	2.68	2.50	**2.75**	0.15	-	**2.13**	-	**4.88**	nr	**5.24**

Labour hourly rates: (except Specialists) Craft Operatives £14.21 Labourer £10.71 Rates are national average prices. Refer to REGIONAL VARIATIONS for indicative levels of overall pricing in regions	MATERIALS			LABOUR				RATES		
	Del to Site £	Waste %	Material Cost £	Craft Optve Hrs	Lab Hrs	Labour Cost £	Sunds £	Nett Rate £	Unit	Gross Rate (+7.5%) £
P21: IRONMONGERY Cont.										
Hinges; standard quality Cont.										
To softwood Cont.										
butt hinges; brass, steel washers, steel pin										
75mm	1.85	2.50	**1.90**	0.15	-	**2.13**	-	**4.03**	nr	**4.33**
100mm	3.17	2.50	**3.25**	0.15	-	**2.13**	-	**5.38**	nr	**5.78**
butt hinges; brass, B.M.A., steel washers, steel pin										
75mm	3.50	2.50	**3.59**	0.15	-	**2.13**	-	**5.72**	nr	**6.15**
100mm	5.69	2.50	**5.83**	0.15	-	**2.13**	-	**7.96**	nr	**8.56**
butt hinges; brass, chromium plated, steel washers, steel pin										
75mm	3.50	2.50	**3.59**	0.15	-	**2.13**	-	**5.72**	nr	**6.15**
100mm	6.12	2.50	**6.27**	0.15	-	**2.13**	-	**8.40**	nr	**9.03**
butt hinges; aluminium, stainless steel pin and washers										
75mm	2.74	2.50	**2.81**	0.15	-	**2.13**	-	**4.94**	nr	**5.31**
100mm	3.39	2.50	**3.47**	0.15	-	**2.13**	-	**5.61**	nr	**6.03**
rising hinges; steel										
75mm	1.02	2.50	**1.05**	0.25	-	**3.55**	-	**4.60**	nr	**4.94**
100mm	1.37	2.50	**1.40**	0.25	-	**3.55**	-	**4.96**	nr	**5.33**
rising hinges; cast iron										
75mm	1.97	2.50	**2.02**	0.25	-	**3.55**	-	**5.57**	nr	**5.99**
100mm	2.89	2.50	**2.96**	0.25	-	**3.55**	-	**6.51**	nr	**7.00**
spring hinges; steel, japanned; single action										
100mm	12.78	2.50	**13.10**	0.50	-	**7.11**	-	**20.20**	nr	**21.72**
125mm	14.37	2.50	**14.73**	0.55	-	**7.82**	-	**22.54**	nr	**24.24**
150mm	15.85	2.50	**16.25**	0.60	-	**8.53**	-	**24.77**	nr	**26.63**
spring hinges; steel, japanned; double action										
75mm	13.82	2.50	**14.17**	0.60	-	**8.53**	-	**22.69**	nr	**24.39**
100mm	15.85	2.50	**16.25**	0.60	-	**8.53**	-	**24.77**	nr	**26.63**
125mm	18.58	2.50	**19.04**	0.65	-	**9.24**	-	**28.28**	nr	**30.40**
150mm	21.31	2.50	**21.84**	0.70	-	**9.95**	-	**31.79**	nr	**34.17**
tee hinges; japanned; light										
150mm	0.35	2.50	**0.36**	0.40	-	**5.68**	-	**6.04**	nr	**6.50**
230mm	0.52	2.50	**0.53**	0.40	-	**5.68**	-	**6.22**	nr	**6.68**
300mm	0.68	2.50	**0.70**	0.45	-	**6.39**	-	**7.09**	nr	**7.62**
375mm	0.97	2.50	**0.99**	0.50	-	**7.11**	-	**8.10**	nr	**8.71**
450mm	1.34	2.50	**1.37**	0.55	-	**7.82**	-	**9.19**	nr	**9.88**
tee hinges; self colour; heavy										
150mm	0.62	2.50	**0.64**	0.45	-	**6.39**	-	**7.03**	nr	**7.56**
230mm	0.84	2.50	**0.86**	0.45	-	**6.39**	-	**7.26**	nr	**7.80**
300mm	1.34	2.50	**1.37**	0.50	-	**7.11**	-	**8.48**	nr	**9.11**
375mm	2.03	2.50	**2.08**	0.55	-	**7.82**	-	**9.90**	nr	**10.64**
450mm	2.56	2.50	**2.62**	0.60	-	**8.53**	-	**11.15**	nr	**11.99**
600mm	4.81	2.50	**4.93**	0.70	-	**9.95**	-	**14.88**	nr	**15.99**
hook and band hinges; on plate; heavy										
300mm	2.46	2.50	**2.52**	0.80	-	**11.37**	-	**13.89**	nr	**14.93**
450mm	4.04	2.50	**4.14**	1.00	-	**14.21**	-	**18.35**	nr	**19.73**
610mm	6.40	2.50	**6.56**	1.20	-	**17.05**	-	**23.61**	nr	**25.38**
914mm	11.21	2.50	**11.49**	1.50	-	**21.32**	-	**32.81**	nr	**35.27**
collinge hinges; cup for wood; best quality										
610mm	30.37	2.50	**31.13**	0.90	-	**12.79**	-	**43.92**	nr	**47.21**
762mm	37.43	2.50	**38.37**	1.15	-	**16.34**	-	**54.71**	nr	**58.81**
914mm	47.80	2.50	**49.00**	1.40	-	**19.89**	-	**68.89**	nr	**74.06**
parliament hinges; steel; 100mm	3.44	2.50	**3.53**	0.25	-	**3.55**	-	**7.08**	nr	**7.61**
cellar flap hinges; wrought, welded, plain joint; 450mm	22.40	2.50	**22.96**	0.50	-	**7.11**	-	**30.07**	nr	**32.32**
To hardwood or the like										
backflap hinges; steel										
25mm	0.23	2.50	**0.24**	0.20	-	**2.84**	-	**3.08**	nr	**3.31**
38mm	0.27	2.50	**0.28**	0.20	-	**2.84**	-	**3.12**	nr	**3.35**
50mm	0.35	2.50	**0.36**	0.20	-	**2.84**	-	**3.20**	nr	**3.44**
63mm	1.04	2.50	**1.07**	0.20	-	**2.84**	-	**3.91**	nr	**4.20**
75mm	1.65	2.50	**1.69**	0.20	-	**2.84**	-	**4.53**	nr	**4.87**
butt hinges; steel; light medium pattern										
38mm	0.14	2.50	**0.14**	0.20	-	**2.84**	-	**2.99**	nr	**3.21**
50mm	0.14	2.50	**0.14**	0.20	-	**2.84**	-	**2.99**	nr	**3.21**
63mm	0.16	2.50	**0.16**	0.20	-	**2.84**	-	**3.01**	nr	**3.23**
75mm	0.16	2.50	**0.16**	0.20	-	**2.84**	-	**3.01**	nr	**3.23**
100mm	0.33	2.50	**0.34**	0.20	-	**2.84**	-	**3.18**	nr	**3.42**
butt hinges; steel; strong pattern										
75mm	0.64	2.50	**0.66**	0.20	-	**2.84**	-	**3.50**	nr	**3.76**
100mm	0.70	2.50	**0.72**	0.20	-	**2.84**	-	**3.56**	nr	**3.83**
butt hinges; cast iron; light										
50mm	0.84	2.50	**0.86**	0.20	-	**2.84**	-	**3.70**	nr	**3.98**
63mm	0.94	2.50	**0.96**	0.20	-	**2.84**	-	**3.81**	nr	**4.09**
75mm	1.04	2.50	**1.07**	0.20	-	**2.84**	-	**3.91**	nr	**4.20**
100mm	1.13	2.50	**1.16**	0.20	-	**2.84**	-	**4.00**	nr	**4.30**
butt hinges; brass, brass pin										
38mm	0.79	2.50	**0.81**	0.20	-	**2.84**	-	**3.65**	nr	**3.93**
50mm	0.91	2.50	**0.93**	0.20	-	**2.84**	-	**3.77**	nr	**4.06**
63mm	1.07	2.50	**1.10**	0.20	-	**2.84**	-	**3.94**	nr	**4.23**
75mm	1.16	2.50	**1.19**	0.20	-	**2.84**	-	**4.03**	nr	**4.33**
100mm	2.68	2.50	**2.75**	0.20	-	**2.84**	-	**5.59**	nr	**6.01**
butt hinges; brass, steel washers, steel pin										
75mm	1.85	2.50	**1.90**	0.20	-	**2.84**	-	**4.74**	nr	**5.09**
100mm	3.17	2.50	**3.25**	0.20	-	**2.84**	-	**6.09**	nr	**6.55**
butt hinges; brass, B.M.A., steel washers, steel pin										
75mm	3.50	2.50	**3.59**	0.20	-	**2.84**	-	**6.43**	nr	**6.91**
100mm	5.69	2.50	**5.83**	0.20	-	**2.84**	-	**8.67**	nr	**9.32**

Labour hourly rates: (except Specialists) Craft Operatives £14.21 Labourer £10.71 Rates are national average prices. Refer to REGIONAL VARIATIONS for indicative levels of overall pricing in regions	MATERIALS			LABOUR				RATES		
	Del to Site £	Waste %	Material Cost £	Craft Optve Hrs	Lab Hrs	Labour Cost £	Sunds £	Nett Rate £	Unit	Gross Rate (+7.5%) £
P21: IRONMONGERY Cont.										
Hinges; standard quality Cont.										
To hardwood or the like Cont.										
butt hinges; brass, chromium plated, steel washers, steel pin										
75mm	3.50	2.50	**3.59**	0.20	-	**2.84**	-	**6.43**	nr	**6.91**
100mm	6.12	2.50	**6.27**	0.20	-	**2.84**	-	**9.12**	nr	**9.80**
butt hinges; aluminium, stainless steel pin and washers										
75mm	2.74	2.50	**2.81**	0.20	-	**2.84**	-	**5.65**	nr	**6.07**
100mm	3.39	2.50	**3.47**	0.20	-	**2.84**	-	**6.32**	nr	**6.79**
rising hinges; steel										
75mm	1.02	2.50	**1.05**	0.35	-	**4.97**	-	**6.02**	nr	**6.47**
100mm	1.37	2.50	**1.40**	0.35	-	**4.97**	-	**6.38**	nr	**6.86**
rising hinges; cast iron										
75mm	1.97	2.50	**2.02**	0.35	-	**4.97**	-	**6.99**	nr	**7.52**
100mm	2.89	2.50	**2.96**	0.35	-	**4.97**	-	**7.94**	nr	**8.53**
spring hinges; steel, japanned; single action										
100mm	12.78	2.50	**13.10**	0.75	-	**10.66**	-	**23.76**	nr	**25.54**
125mm	14.37	2.50	**14.73**	0.80	-	**11.37**	-	**26.10**	nr	**28.05**
150mm	15.85	2.50	**16.25**	0.85	-	**12.08**	-	**28.32**	nr	**30.45**
spring hinges; steel, japanned; double action										
75mm	13.82	2.50	**14.17**	0.80	-	**11.37**	-	**25.53**	nr	**27.45**
100mm	15.85	2.50	**16.25**	0.85	-	**12.08**	-	**28.32**	nr	**30.45**
125mm	18.58	2.50	**19.04**	0.95	-	**13.50**	-	**32.54**	nr	**34.98**
150mm	21.31	2.50	**21.84**	1.00	-	**14.21**	-	**36.05**	nr	**38.76**
tee hinges; japanned; light										
150mm	0.35	2.50	**0.36**	0.55	-	**7.82**	-	**8.17**	nr	**8.79**
230mm	0.52	2.50	**0.53**	0.55	-	**7.82**	-	**8.35**	nr	**8.97**
300mm	0.68	2.50	**0.70**	0.65	-	**9.24**	-	**9.93**	nr	**10.68**
375mm	0.97	2.50	**0.99**	0.75	-	**10.66**	-	**11.65**	nr	**12.53**
450mm	1.34	2.50	**1.37**	0.80	-	**11.37**	-	**12.74**	nr	**13.70**
tee hinges; self colour; heavy										
150mm	0.62	2.50	**0.64**	0.60	-	**8.53**	-	**9.16**	nr	**9.85**
230mm	0.84	2.50	**0.86**	0.60	-	**8.53**	-	**9.39**	nr	**10.09**
300mm	1.34	2.50	**1.37**	0.70	-	**9.95**	-	**11.32**	nr	**12.17**
375mm	2.03	2.50	**2.08**	0.80	-	**11.37**	-	**13.45**	nr	**14.46**
450mm	2.56	2.50	**2.62**	0.85	-	**12.08**	-	**14.70**	nr	**15.81**
600mm	4.81	2.50	**4.93**	0.95	-	**13.50**	-	**18.43**	nr	**19.81**
hook and band hinges; on plate; heavy										
300mm	2.46	2.50	**2.52**	1.10	-	**15.63**	-	**18.15**	nr	**19.51**
450mm	4.04	2.50	**4.14**	1.40	-	**19.89**	-	**24.04**	nr	**25.84**
610mm	6.40	2.50	**6.56**	1.75	-	**24.87**	-	**31.43**	nr	**33.78**
914mm	11.21	2.50	**11.49**	2.00	-	**28.42**	-	**39.91**	nr	**42.90**
collinge hinges; cup for wood; best quality										
610mm	30.37	2.50	**31.13**	1.30	-	**18.47**	-	**49.60**	nr	**53.32**
762mm	37.43	2.50	**38.37**	1.70	-	**24.16**	-	**62.52**	nr	**67.21**
914mm	47.80	2.50	**49.00**	2.05	-	**29.13**	-	**78.13**	nr	**83.98**
parliament hinges; steel; 100mm	3.44	2.50	**3.53**	0.35	-	**4.97**	-	**8.50**	nr	**9.14**
cellar flap hinges; wrought, welded, plain joint; 450mm	22.40	2.50	**22.96**	0.75	-	**10.66**	-	**33.62**	nr	**36.14**
Floor springs; standard quality										
To softwood										
single action floor springs and top centres; B.M.A.	185.76	2.50	**190.40**	2.50	-	**35.53**	-	**225.93**	nr	**242.87**
single action floor springs and top centres; chromium plated	185.76	2.50	**190.40**	2.50	-	**35.53**	-	**225.93**	nr	**242.87**
double action floor springs and top centres; B.M.A.	224.00	2.50	**229.60**	3.00	-	**42.63**	-	**272.23**	nr	**292.65**
double action floor springs and top centres; chromium plated	224.00	2.50	**229.60**	3.00	-	**42.63**	-	**272.23**	nr	**292.65**
To hardwood or the like										
single action floor springs and top centres; B.M.A.	185.76	2.50	**190.40**	3.75	-	**53.29**	-	**243.69**	nr	**261.97**
single action floor springs and top centres; chromium plated	185.76	2.50	**190.40**	3.75	-	**53.29**	-	**243.69**	nr	**261.97**
double action floor springs and top centres; B.M.A.	224.00	2.50	**229.60**	4.50	-	**63.95**	-	**293.55**	nr	**315.56**
double action floor springs and top centres; chromium plated	224.00	2.50	**229.60**	4.50	-	**63.95**	-	**293.55**	nr	**315.56**
Door closers; standard quality										
To softwood										
coil door springs; japanned	3.93	2.50	**4.03**	0.25	-	**3.55**	-	**7.58**	nr	**8.15**
overhead door closers; liquid check and spring; 'Briton' 2000 series; silver; light doors	63.89	2.50	**65.49**	1.50	-	**21.32**	-	**86.80**	nr	**93.31**
overhead door closers; liquid check and spring; 'Briton' 2000 series; silver; medium doors	74.30	2.50	**76.16**	1.50	-	**21.32**	-	**97.47**	nr	**104.78**
overhead door closers; liquid check and spring; 'Briton' 2000 series; silver; heavy doors	95.61	2.50	**98.00**	1.50	-	**21.32**	-	**119.32**	nr	**128.26**
concealed overhead door closers; liquid check and spring; medium doors	106.54	2.50	**109.20**	2.50	-	**35.53**	-	**144.73**	nr	**155.58**
Perko door closers; brass plate	6.66	2.50	**6.83**	1.50	-	**21.32**	-	**28.14**	nr	**30.25**
To hardwood or the like										
coil door springs; japanned	3.93	2.50	**4.03**	0.40	-	**5.68**	-	**9.71**	nr	**10.44**
overhead door closers; liquid check and spring; 'Briton' 2000 series; silver; light doors	63.89	2.50	**65.49**	2.25	-	**31.97**	-	**97.46**	nr	**104.77**
overhead door closers; liquid check and spring; 'Briton' 2000 series; silver; medium doors	74.30	2.50	**76.16**	2.25	-	**31.97**	-	**108.13**	nr	**116.24**

Labour hourly rates: (except Specialists) Craft Operatives £14.21 Labourer £10.71 Rates are national average prices. Refer to REGIONAL VARIATIONS for indicative levels of overall pricing in regions	MATERIALS			LABOUR				RATES		
	Del to Site £	Waste %	Material Cost £	Craft Optve Hrs	Lab Hrs	Labour Cost £	Sunds £	Nett Rate £	Unit	Gross Rate (+7.5%) £
P21: IRONMONGERY Cont.										
Door closers; standard quality Cont.										
To hardwood or the like Cont.										
overhead door closers; liquid check and spring; 'Briton' 2000 series; silver; heavy doors........	95.61	2.50	**98.00**	2.25	-	**31.97**	-	**129.97**	nr	**139.72**
concealed overhead door closers; liquid check and spring; medium doors	106.54	2.50	**109.20**	3.75	-	**53.29**	-	**162.49**	nr	**174.68**
Perko door closers; brass plate.................	6.66	2.50	**6.83**	2.25	-	**31.97**	-	**38.80**	nr	**41.71**
Door selectors; standard quality										
To softwood or the like										
Union 8815 door selector; aluminium anodised silver finish	47.87	2.50	**49.07**	2.00	0.25	**31.10**	-	**80.16**	nr	**86.18**
Close Rite door selector; satin nickel plated....	35.74	2.50	**36.63**	2.25	0.28	**34.97**	-	**71.60**	nr	**76.98**
To hardwood or the like										
Union 8815 door selector; aluminium anodised silver finish	47.87	2.50	**49.07**	3.00	0.38	**46.70**	-	**95.77**	nr	**102.95**
Close Rite door selector; satin nickel plated....	35.74	2.50	**36.63**	3.50	0.44	**54.45**	-	**91.08**	nr	**97.91**
Locks and latches; standard quality										
To softwood										
magnetic catches; 6lb pull......................	0.52	2.50	**0.53**	0.20	-	**2.84**	-	**3.38**	nr	**3.63**
magnetic catches; 9lb pull......................	0.64	2.50	**0.66**	0.20	-	**2.84**	-	**3.50**	nr	**3.76**
magnetic catches; 13lb pull.....................	1.06	2.50	**1.09**	0.20	-	**2.84**	-	**3.93**	nr	**4.22**
roller catches; mortice; nylon; 18mm............	1.06	2.50	**1.09**	0.35	-	**4.97**	-	**6.06**	nr	**6.51**
roller catches; mortice; nylon; 27mm............	1.06	2.50	**1.09**	0.35	-	**4.97**	-	**6.06**	nr	**6.51**
roller catches; surface, adjustable; nylon; 16mm.	1.85	2.50	**1.90**	0.25	-	**3.55**	-	**5.45**	nr	**5.86**
roller catches; mortice, adjustable; satin chrome plated; 25 x 22 x 10mm	4.53	2.50	**4.64**	0.40	-	**5.68**	-	**10.33**	nr	**11.10**
roller catches; double ball; brass; 43mm........	0.69	2.50	**0.71**	0.30	-	**4.26**	-	**4.97**	nr	**5.34**
mortice latches; stamped steel case; 75mm........	1.02	2.50	**1.05**	0.65	-	**9.24**	-	**10.28**	nr	**11.05**
mortice latches; locking; stamped steel case; 75mm	2.03	2.50	**2.08**	0.65	-	**9.24**	-	**11.32**	nr	**12.17**
cylinder mortice latches; 'Union' key operated outside, knob inside, bolt held by slide; B.M.A.; for end set to suit 13mm rebate.................	39.94	2.50	**40.94**	0.75	-	**10.66**	-	**51.60**	nr	**55.47**
cylinder mortice latches; 'Union' key operated outside, knob inside, bolt held by slide; chromium plated; for end set to suit 13mm rebate..........	37.94	2.50	**38.89**	0.75	-	**10.66**	-	**49.55**	nr	**53.26**
cylinder rim night latches; 'Legge' 707; chromium plated	13.01	2.50	**13.34**	1.00	-	**14.21**	-	**27.55**	nr	**29.61**
cylinder rim night latches; 'Union' 1022; chromium plated	15.19	2.50	**15.57**	1.00	-	**14.21**	-	**29.78**	nr	**32.01**
cylinder rim night latches; 'Yale' 88; chromium plated	16.88	2.50	**17.30**	1.00	-	**14.21**	-	**31.51**	nr	**33.88**
Suffolk latches; japanned; medium; size Nr 2.....	2.03	2.50	**2.08**	0.80	-	**11.37**	-	**13.45**	nr	**14.46**
Suffolk latches; japanned; medium; size Nr 3.....	2.62	2.50	**2.69**	0.80	-	**11.37**	-	**14.05**	nr	**15.11**
Suffolk latches; japanned; heavy; size Nr 4......	4.26	2.50	**4.37**	0.80	-	**11.37**	-	**15.73**	nr	**16.91**
Suffolk latches; galvanized; size Nr 4...........	4.53	2.50	**4.64**	0.80	-	**11.37**	-	**16.01**	nr	**17.21**
cupboard locks; 1 lever; japanned; 63mm..........	3.99	2.50	**4.09**	0.75	-	**10.66**	-	**14.75**	nr	**15.85**
cupboard locks; 2 lever; brass; 50mm.............	5.41	2.50	**5.55**	0.75	-	**10.66**	-	**16.20**	nr	**17.42**
cupboard locks; 2 lever; brass; 63mm.............	5.41	2.50	**5.55**	0.75	-	**10.66**	-	**16.20**	nr	**17.42**
cupboard locks; 4 lever; brass; 63mm.............	5.78	2.50	**5.92**	0.75	-	**10.66**	-	**16.58**	nr	**17.83**
rim dead lock; japanned case; 100mm..............	3.44	2.50	**3.53**	0.80	-	**11.37**	-	**14.89**	nr	**16.01**
rim dead lock; japanned case; 125mm..............	3.99	2.50	**4.09**	0.80	-	**11.37**	-	**15.46**	nr	**16.62**
rim dead lock; japanned case; 150mm..............	4.81	2.50	**4.93**	0.80	-	**11.37**	-	**16.30**	nr	**17.52**
mortice dead locks; japanned case; 75mm..........	5.03	2.50	**5.16**	0.80	-	**11.37**	-	**16.52**	nr	**17.76**
mortice locks; three levers......................	3.93	2.50	**4.03**	1.00	-	**14.21**	-	**18.24**	nr	**19.61**
mortice locks; five levers.......................	10.38	2.50	**10.64**	1.00	-	**14.21**	-	**24.85**	nr	**26.71**
rim locks; japanned case; 150 x 100mm............	4.81	2.50	**4.93**	0.80	-	**11.37**	-	**16.30**	nr	**17.52**
rim locks; japanned case; 150 x 75mm.............	5.36	2.50	**5.49**	0.80	-	**11.37**	-	**16.86**	nr	**18.13**
rim locks; japanned case; strong pattern; 150 x 100mm	7.21	2.50	**7.39**	0.80	-	**11.37**	-	**18.76**	nr	**20.17**
padlocks, 'Squire', anti-pilfer bolt, warded spring; zinc plated; 32mm	1.37	2.50	**1.40**	-	-	**-**	-	**1.40**	nr	**1.51**
padlocks, 'Squire', anti-pilfer bolt, warded spring; zinc plated; 38mm	2.13	2.50	**2.18**	-	-	**-**	-	**2.18**	nr	**2.35**
padlocks, 'Squire', anti-pilfer bolt, warded spring; zinc plated; 44mm	2.35	2.50	**2.41**	-	-	**-**	-	**2.41**	nr	**2.59**
padlocks; 'Squire', anti-pilfer bolt, 4 pin tumblers; zinc plated; 32mm.....................	4.32	2.50	**4.43**	-	-	**-**	-	**4.43**	nr	**4.76**
padlocks; 'Squire', anti-pilfer bolt, 4 pin tumblers; zinc plated; 38mm.....................	5.69	2.50	**5.83**	-	-	**-**	-	**5.83**	nr	**6.27**
padlocks; 'Squire', anti-pilfer bolt, 4 pin tumblers; zinc plated; 44mm.....................	6.72	2.50	**6.89**	-	-	**-**	-	**6.89**	nr	**7.40**
padlocks, 'Squire', 4 levers; galvanized case....	3.39	2.50	**3.47**	-	-	**-**	-	**3.47**	nr	**3.74**
padlocks, 'Squire', 4 levers; brass case.........	4.37	2.50	**4.48**	-	-	**-**	-	**4.48**	nr	**4.82**
padlocks, 'Squire', strong pattern; chromium plated	29.28	2.50	**30.01**	-	-	**-**	-	**30.01**	nr	**32.26**
hasps and staples; japanned wire; light; 75mm....	0.56	2.50	**0.57**	0.30	-	**4.26**	-	**4.84**	nr	**5.20**
hasps and staples; japanned wire; light; 100mm...	0.66	2.50	**0.68**	0.30	-	**4.26**	-	**4.94**	nr	**5.31**
hasps and staples; galvanized wire; heavy; 75mm..	1.22	2.50	**1.25**	0.30	-	**4.26**	-	**5.51**	nr	**5.93**
hasps and staples; galvanized wire; heavy; 100mm.	1.32	2.50	**1.35**	0.30	-	**4.26**	-	**5.62**	nr	**6.04**
hasps and staples; galvanized wire; heavy; 150mm.	1.37	2.50	**1.40**	0.30	-	**4.26**	-	**5.67**	nr	**6.09**
locking bars; japanned; 200mm....................	5.14	2.50	**5.27**	0.45	-	**6.39**	-	**11.66**	nr	**12.54**
locking bars; japanned; 250mm....................	5.36	2.50	**5.49**	0.45	-	**6.39**	-	**11.89**	nr	**12.78**
locking bars; japanned; 300mm....................	6.78	2.50	**6.95**	0.45	-	**6.39**	-	**13.34**	nr	**14.34**
locking bars; japanned; 350mm....................	7.98	2.50	**8.18**	0.45	-	**6.39**	-	**14.57**	nr	**15.67**
To hardwood or the like										
magnetic catches; 6lb pull......................	0.52	2.50	**0.53**	0.30	-	**4.26**	-	**4.80**	nr	**5.16**
magnetic catches; 9lb pull......................	0.64	2.50	**0.66**	0.30	-	**4.26**	-	**4.92**	nr	**5.29**

Labour hourly rates: (except Specialists) Craft Operatives £14.21 Labourer £10.71 Rates are national average prices. Refer to REGIONAL VARIATIONS for indicative levels of overall pricing in regions	MATERIALS			LABOUR				RATES		
	Del to Site £	Waste %	Material Cost £	Craft Optve Hrs	Lab Hrs	Labour Cost £	Sunds £	Nett Rate £	Unit	Gross Rate (+7.5%) £
P21: IRONMONGERY Cont.										
Locks and latches; standard quality Cont.										
To hardwood or the like Cont.										
magnetic catches; 13lb pull	1.06	2.50	**1.09**	0.30	-	**4.26**	-	**5.35**	nr	**5.75**
roller catches; mortice; nylon; 18mm	1.06	2.50	**1.09**	0.50	-	**7.11**	-	**8.19**	nr	**8.81**
roller catches; mortice; nylon; 27mm	1.06	2.50	**1.09**	0.50	-	**7.11**	-	**8.19**	nr	**8.81**
roller catches; surface, adjustable; nylon; 16mm	1.85	2.50	**1.90**	0.40	-	**5.68**	-	**7.58**	nr	**8.15**
roller catches; mortice, adjustable; satin chrome plated; 25 x 22 x 10mm	4.53	2.50	**4.64**	0.60	-	**8.53**	-	**13.17**	nr	**14.16**
roller catches; double ball; brass; 43mm	0.69	2.50	**0.71**	0.45	-	**6.39**	-	**7.10**	nr	**7.63**
mortice latches; stamped steel case; 75mm	1.02	2.50	**1.05**	1.00	-	**14.21**	-	**15.26**	nr	**16.40**
mortice latches; locking; stamped steel case; 75mm	2.03	2.50	**2.08**	1.00	-	**14.21**	-	**16.29**	nr	**17.51**
cylinder mortice latches; 'Union' key operated outside, knob inside, bolt held by slide; B.M.A.; for end set to suit 13mm rebate	39.94	2.50	**40.94**	1.10	-	**15.63**	-	**56.57**	nr	**60.81**
cylinder mortice latches; 'Union' key operated outside, knob inside, bolt held by slide; chromium plated; for end set to suit 13mm rebate	37.94	2.50	**38.89**	1.10	-	**15.63**	-	**54.52**	nr	**58.61**
cylinder rim night latches; 'Legge' 707; chromium plated	13.01	2.50	**13.34**	1.50	-	**21.32**	-	**34.65**	nr	**37.25**
cylinder rim night latches; 'Union' 1022; chromium plated	15.19	2.50	**15.57**	1.50	-	**21.32**	-	**36.88**	nr	**39.65**
cylinder rim night latches; 'Yale' 88; chromium plated	16.88	2.50	**17.30**	1.50	-	**21.32**	-	**38.62**	nr	**41.51**
Suffolk latches; japanned; medium; size Nr 2	2.03	2.50	**2.08**	1.20	-	**17.05**	-	**19.13**	nr	**20.57**
Suffolk latches; japanned; medium; size Nr 3	2.62	2.50	**2.69**	1.20	-	**17.05**	-	**19.74**	nr	**21.22**
Suffolk latches; japanned; heavy; size Nr 4	4.26	2.50	**4.37**	1.20	-	**17.05**	-	**21.42**	nr	**23.02**
Suffolk latches; galvanized; size Nr 4	4.53	2.50	**4.64**	1.20	-	**17.05**	-	**21.70**	nr	**23.32**
cupboard locks; 1 lever; japanned; 63mm	3.99	2.50	**4.09**	1.10	-	**15.63**	-	**19.72**	nr	**21.20**
cupboard locks; 2 lever; brass; 50mm	5.41	2.50	**5.55**	1.10	-	**15.63**	-	**21.18**	nr	**22.76**
cupboard locks; 2 lever; brass; 63mm	5.41	2.50	**5.55**	1.10	-	**15.63**	-	**21.18**	nr	**22.76**
cupboard locks; 4 lever; brass; 63mm	5.78	2.50	**5.92**	1.10	-	**15.63**	-	**21.56**	nr	**23.17**
rim dead lock; japanned case; 100mm	3.44	2.50	**3.53**	1.20	-	**17.05**	-	**20.58**	nr	**22.12**
rim dead lock; japanned case; 125mm	3.99	2.50	**4.09**	1.20	-	**17.05**	-	**21.14**	nr	**22.73**
rim dead lock; japanned case; 150mm	4.81	2.50	**4.93**	1.20	-	**17.05**	-	**21.98**	nr	**23.63**
mortice dead locks; japanned case; 75mm	5.03	2.50	**5.16**	1.20	-	**17.05**	-	**22.21**	nr	**23.87**
mortice locks; three levers	3.93	2.50	**4.03**	1.50	-	**21.32**	-	**25.34**	nr	**27.24**
mortice locks; five levers	10.38	2.50	**10.64**	1.50	-	**21.32**	-	**31.95**	nr	**34.35**
rim locks; japanned case; 150 x 100mm	4.81	2.50	**4.93**	1.20	-	**17.05**	-	**21.98**	nr	**23.63**
rim locks; japanned case; 150 x 75mm	5.36	2.50	**5.49**	1.20	-	**17.05**	-	**22.55**	nr	**24.24**
rim locks; japanned case; strong pattern; 150 x 100mm	7.21	2.50	**7.39**	1.20	-	**17.05**	-	**24.44**	nr	**26.28**
padlocks, 'Squire', anti-pilfer bolt, warded spring; zinc plated; 32mm	1.37	2.50	**1.40**	-	-	**-**	-	**1.40**	nr	**1.51**
padlocks, 'Squire', anti-pilfer bolt, warded spring; zinc plated; 38mm	2.13	2.50	**2.18**	-	-	**-**	-	**2.18**	nr	**2.35**
padlocks, 'Squire', anti-pilfer bolt, warded spring; zinc plated; 44mm	2.35	2.50	**2.41**	-	-	**-**	-	**2.41**	nr	**2.59**
padlocks; 'Squire', anti-pilfer bolt, 4 pin tumblers; zinc plated; 32mm	4.32	2.50	**4.43**	-	-	**-**	-	**4.43**	nr	**4.76**
padlocks; 'Squire', anti-pilfer bolt, 4 pin tumblers; zinc plated; 38mm	5.69	2.50	**5.83**	-	-	**-**	-	**5.83**	nr	**6.27**
padlocks; 'Squire', anti-pilfer bolt, 4 pin tumblers; zinc plated; 44mm	6.72	2.50	**6.89**	-	-	**-**	-	**6.89**	nr	**7.40**
padlocks, 'Squire', 4 levers; galvanized case	3.39	2.50	**3.47**	-	-	**-**	-	**3.47**	nr	**3.74**
padlocks, 'Squire', 4 levers; brass case	4.37	2.50	**4.48**	-	-	**-**	-	**4.48**	nr	**4.82**
padlocks; 'Squire', strong pattern; chromium plated	29.28	2.50	**30.01**	-	-	**-**	-	**30.01**	nr	**32.26**
hasps and staples; japanned wire; light; 75mm	0.56	2.50	**0.57**	0.50	-	**7.11**	-	**7.68**	nr	**8.25**
hasps and staples; japanned wire; light; 100mm	0.66	2.50	**0.68**	0.50	-	**7.11**	-	**7.78**	nr	**8.37**
hasps and staples; galvanized wire; heavy; 75mm	1.22	2.50	**1.25**	0.50	-	**7.11**	-	**8.36**	nr	**8.98**
hasps and staples; galvanized wire; heavy; 100mm	1.32	2.50	**1.35**	0.50	-	**7.11**	-	**8.46**	nr	**9.09**
hasps and staples; galvanized wire; heavy; 150mm	1.37	2.50	**1.40**	0.50	-	**7.11**	-	**8.51**	nr	**9.15**
locking bars; japanned; 200mm	5.14	2.50	**5.27**	0.60	-	**8.53**	-	**13.79**	nr	**14.83**
locking bars; japanned; 250mm	5.36	2.50	**5.49**	0.60	-	**8.53**	-	**14.02**	nr	**15.07**
locking bars; japanned; 300mm	6.78	2.50	**6.95**	0.60	-	**8.53**	-	**15.48**	nr	**16.64**
locking bars; japanned; 350mm	7.98	2.50	**8.18**	0.60	-	**8.53**	-	**16.71**	nr	**17.96**
Bolts; standard quality										
To softwood										
barrel bolts; japanned, steel barrel; medium										
100mm	1.04	2.50	**1.07**	0.25	-	**3.55**	-	**4.62**	nr	**4.96**
150mm	1.32	2.50	**1.35**	0.30	-	**4.26**	-	**5.62**	nr	**6.04**
200mm	1.70	2.50	**1.74**	0.35	-	**4.97**	-	**6.72**	nr	**7.22**
barrel bolts; japanned, steel barrel; heavy										
100mm	2.62	2.50	**2.69**	0.25	-	**3.55**	-	**6.24**	nr	**6.71**
150mm	3.17	2.50	**3.25**	0.30	-	**4.26**	-	**7.51**	nr	**8.08**
200mm	3.83	2.50	**3.93**	0.35	-	**4.97**	-	**8.90**	nr	**9.57**
250mm	4.53	2.50	**4.64**	0.40	-	**5.68**	-	**10.33**	nr	**11.10**
300mm	5.36	2.50	**5.49**	0.45	-	**6.39**	-	**11.89**	nr	**12.78**
barrel bolts; extruded brass, round brass shoot; 25mm wide										
75mm	1.04	2.50	**1.07**	0.25	-	**3.55**	-	**4.62**	nr	**4.96**
100mm	1.26	2.50	**1.29**	0.25	-	**3.55**	-	**4.84**	nr	**5.21**
150mm	1.58	2.50	**1.62**	0.30	-	**4.26**	-	**5.88**	nr	**6.32**
barrel bolts; extruded brass; B.M.A., round brass shoot; 25mm wide										
75mm	1.42	2.50	**1.46**	0.25	-	**3.55**	-	**5.01**	nr	**5.38**
100mm	1.70	2.50	**1.74**	0.25	-	**3.55**	-	**5.30**	nr	**5.69**
150mm	2.13	2.50	**2.18**	0.30	-	**4.26**	-	**6.45**	nr	**6.93**
barrel bolts; extruded brass, chromium plated, round brass shoot; 25mm wide										
75mm	1.42	2.50	**1.46**	0.25	-	**3.55**	-	**5.01**	nr	**5.38**

Labour hourly rates: (except Specialists) Craft Operatives £14.21 Labourer £10.71 Rates are national average prices. Refer to REGIONAL VARIATIONS for indicative levels of overall pricing in regions	MATERIALS			LABOUR				RATES		
	Del to Site £	Waste %	Material Cost £	Craft Optve Hrs	Lab Hrs	Labour Cost £	Sunds £	Nett Rate £	Unit	Gross Rate (+7.5%) £
P21: IRONMONGERY Cont.										
Bolts; standard quality Cont.										
To softwood Cont.										
barrel bolts; extruded brass, chromium plated, round brass shoot; 25mm wide Cont.										
100mm	1.70	2.50	**1.74**	0.25	-	**3.55**	-	**5.30**	nr	**5.69**
150mm	2.13	2.50	**2.18**	0.30	-	**4.26**	-	**6.45**	nr	**6.93**
barrel bolts; extruded aluminium, S.A.A., round aluminium shoot; 25mm wide										
75mm	1.04	2.50	**1.07**	0.25	-	**3.55**	-	**4.62**	nr	**4.96**
100mm	1.14	2.50	**1.17**	0.25	-	**3.55**	-	**4.72**	nr	**5.08**
150mm	1.32	2.50	**1.35**	0.30	-	**4.26**	-	**5.62**	nr	**6.04**
monkey tail bolts										
japanned; 300mm	7.98	2.50	**8.18**	0.35	-	**4.97**	-	**13.15**	nr	**14.14**
japanned; 450mm	10.65	2.50	**10.92**	0.40	-	**5.68**	-	**16.60**	nr	**17.85**
japanned; 610mm	12.78	2.50	**13.10**	0.50	-	**7.11**	-	**20.20**	nr	**21.72**
flush bolts; B.M.A.										
100mm	6.88	2.50	**7.05**	0.50	-	**7.11**	-	**14.16**	nr	**15.22**
150mm	9.94	2.50	**10.19**	0.60	-	**8.53**	-	**18.71**	nr	**20.12**
200mm	17.05	2.50	**17.48**	0.70	-	**9.95**	-	**27.42**	nr	**29.48**
flush bolts; S.C.P.										
100mm	7.43	2.50	**7.62**	0.50	-	**7.11**	-	**14.72**	nr	**15.82**
150mm	10.11	2.50	**10.36**	0.60	-	**8.53**	-	**18.89**	nr	**20.31**
200mm	15.95	2.50	**16.35**	0.70	-	**9.95**	-	**26.30**	nr	**28.27**
lever action flush bolts; B.M.A.										
150mm	8.52	2.50	**8.73**	0.60	-	**8.53**	-	**17.26**	nr	**18.55**
200 x 25mm	9.07	2.50	**9.30**	0.70	-	**9.95**	-	**19.24**	nr	**20.69**
lever action flush bolts; S.A.A.										
150mm	6.88	2.50	**7.05**	0.60	-	**8.53**	-	**15.58**	nr	**16.75**
200 x 25mm	7.11	2.50	**7.29**	0.70	-	**9.95**	-	**17.23**	nr	**18.53**
necked bolts; extruded brass, round brass shoot; 25mm wide										
75mm	1.32	2.50	**1.35**	0.25	-	**3.55**	-	**4.91**	nr	**5.27**
100mm	1.43	2.50	**1.47**	0.25	-	**3.55**	-	**5.02**	nr	**5.39**
150mm	1.91	2.50	**1.96**	0.30	-	**4.26**	-	**6.22**	nr	**6.69**
necked bolts; extruded aluminium, S.A.A., round aluminium shoot; 25mm wide										
75mm	0.94	2.50	**0.96**	0.25	-	**3.55**	-	**4.52**	nr	**4.85**
100mm	1.04	2.50	**1.07**	0.25	-	**3.55**	-	**4.62**	nr	**4.96**
150mm	1.32	2.50	**1.35**	0.30	-	**4.26**	-	**5.62**	nr	**6.04**
indicating bolts; S.A.A.	5.69	2.50	**5.83**	1.00	-	**14.21**	-	**20.04**	nr	**21.55**
indicating bolts; B.M.A.	5.41	2.50	**5.55**	1.00	-	**14.21**	-	**19.76**	nr	**21.24**
panic bolts; to single door; iron, bronzed or silver	53.26	2.50	**54.59**	1.50	-	**21.32**	-	**75.91**	nr	**81.60**
panic bolts; to single door; aluminium box, steel shoots and cross rail, anodised silver	52.99	2.50	**54.31**	1.50	-	**21.32**	-	**75.63**	nr	**81.30**
panic bolts; to double doors; iron bronzed or silver	63.38	2.50	**64.96**	2.00	-	**28.42**	-	**93.38**	nr	**100.39**
panic bolts; to double doors; aluminium box, steel shoots and cross rail, anodised silver	63.38	2.50	**64.96**	2.00	-	**28.42**	-	**93.38**	nr	**100.39**
padlock bolts; galvanized; heavy										
150mm	3.12	2.50	**3.20**	0.35	-	**4.97**	-	**8.17**	nr	**8.78**
200mm	3.33	2.50	**3.41**	0.35	-	**4.97**	-	**8.39**	nr	**9.02**
250mm	3.99	2.50	**4.09**	0.50	-	**7.11**	-	**11.19**	nr	**12.03**
300mm	4.92	2.50	**5.04**	0.50	-	**7.11**	-	**12.15**	nr	**13.06**
To hardwood or the like										
barrel bolts; japanned, steel barrel; medium										
100mm	1.04	2.50	**1.07**	0.35	-	**4.97**	-	**6.04**	nr	**6.49**
150mm	1.32	2.50	**1.35**	0.40	-	**5.68**	-	**7.04**	nr	**7.56**
200mm	1.70	2.50	**1.74**	0.50	-	**7.11**	-	**8.85**	nr	**9.51**
barrel bolts; japanned, steel barrel; heavy										
100mm	2.62	2.50	**2.69**	0.35	-	**4.97**	-	**7.66**	nr	**8.23**
150mm	3.17	2.50	**3.25**	0.40	-	**5.68**	-	**8.93**	nr	**9.60**
200mm	3.83	2.50	**3.93**	0.50	-	**7.11**	-	**11.03**	nr	**11.86**
250mm	4.53	2.50	**4.64**	0.55	-	**7.82**	-	**12.46**	nr	**13.39**
300mm	5.36	2.50	**5.49**	0.65	-	**9.24**	-	**14.73**	nr	**15.84**
barrel bolts; extruded brass, round brass shoot; 25mm wide										
75mm	1.04	2.50	**1.07**	0.35	-	**4.97**	-	**6.04**	nr	**6.49**
100mm	1.26	2.50	**1.29**	0.35	-	**4.97**	-	**6.27**	nr	**6.73**
150mm	1.58	2.50	**1.62**	0.40	-	**5.68**	-	**7.30**	nr	**7.85**
barrel bolts; extruded brass; B.M.A., round brass shoot; 25mm wide										
75mm	1.42	2.50	**1.46**	0.35	-	**4.97**	-	**6.43**	nr	**6.91**
100mm	1.70	2.50	**1.74**	0.35	-	**4.97**	-	**6.72**	nr	**7.22**
150mm	2.13	2.50	**2.18**	0.40	-	**5.68**	-	**7.87**	nr	**8.46**
barrel bolts; extruded brass, chromium plated, round brass shoot; 25mm wide										
75mm	1.42	2.50	**1.46**	0.35	-	**4.97**	-	**6.43**	nr	**6.91**
100mm	1.70	2.50	**1.74**	0.35	-	**4.97**	-	**6.72**	nr	**7.22**
150mm	2.13	2.50	**2.18**	0.40	-	**5.68**	-	**7.87**	nr	**8.46**
barrel bolts; extruded aluminium, S.A.A., round aluminium shoot; 25mm wide										
75mm	1.04	2.50	**1.07**	0.35	-	**4.97**	-	**6.04**	nr	**6.49**
100mm	1.14	2.50	**1.17**	0.35	-	**4.97**	-	**6.14**	nr	**6.60**
150mm	1.32	2.50	**1.35**	0.40	-	**5.68**	-	**7.04**	nr	**7.56**
monkey tail bolts										
japanned; 300mm	7.98	2.50	**8.18**	0.50	-	**7.11**	-	**15.28**	nr	**16.43**
japanned; 450mm	10.65	2.50	**10.92**	0.60	-	**8.53**	-	**19.44**	nr	**20.90**
japanned; 610mm	12.78	2.50	**13.10**	0.70	-	**9.95**	-	**23.05**	nr	**24.77**
flush bolts; B.M.A.										
100mm	6.88	2.50	**7.05**	0.75	-	**10.66**	-	**17.71**	nr	**19.04**
150mm	9.94	2.50	**10.19**	0.90	-	**12.79**	-	**22.98**	nr	**24.70**

Labour hourly rates: (except Specialists) Craft Operatives £14.21 Labourer £10.71 Rates are national average prices. Refer to REGIONAL VARIATIONS for indicative levels of overall pricing in regions	MATERIALS			LABOUR				RATES		
	Del to Site £	Waste %	Material Cost £	Craft Optve Hrs	Lab Hrs	Labour Cost £	Sunds £	Nett Rate £	Unit	Gross Rate (+7.5%) £
P21: IRONMONGERY Cont.										
Bolts; standard quality Cont.										
To hardwood or the like Cont.										
flush bolts; B.M.A. Cont.										
200mm	17.05	2.50	**17.48**	1.00	-	**14.21**	-	**31.69**	nr	**34.06**
flush bolts; S.C.P.										
100mm	7.43	2.50	**7.62**	0.75	-	**10.66**	-	**18.27**	nr	**19.64**
150mm	10.11	2.50	**10.36**	0.90	-	**12.79**	-	**23.15**	nr	**24.89**
200mm	15.95	2.50	**16.35**	1.00	-	**14.21**	-	**30.56**	nr	**32.85**
lever action flush bolts; B.M.A.										
150mm	8.52	2.50	**8.73**	0.90	-	**12.79**	-	**21.52**	nr	**23.14**
200 x 25mm	9.07	2.50	**9.30**	1.00	-	**14.21**	-	**23.51**	nr	**25.27**
lever action flush bolts; S.A.A.										
150mm	6.88	2.50	**7.05**	0.90	-	**12.79**	-	**19.84**	nr	**21.33**
200 x 25mm	7.11	2.50	**7.29**	1.00	-	**14.21**	-	**21.50**	nr	**23.11**
necked bolts; extruded brass, round brass shoot; 25mm wide										
75mm	1.32	2.50	**1.35**	0.35	-	**4.97**	-	**6.33**	nr	**6.80**
100mm	1.43	2.50	**1.47**	0.35	-	**4.97**	-	**6.44**	nr	**6.92**
150mm	1.91	2.50	**1.96**	0.40	-	**5.68**	-	**7.64**	nr	**8.21**
necked bolts; extruded aluminium, S.A.A., round aluminium shoot; 25mm wide										
75mm	0.94	2.50	**0.96**	0.35	-	**4.97**	-	**5.94**	nr	**6.38**
100mm	1.04	2.50	**1.07**	0.35	-	**4.97**	-	**6.04**	nr	**6.49**
150mm	1.32	2.50	**1.35**	0.40	-	**5.68**	-	**7.04**	nr	**7.56**
indicating bolts; S.A.A.	5.69	2.50	**5.83**	1.50	-	**21.32**	-	**27.15**	nr	**29.18**
indicating bolts; B.M.A.	5.41	2.50	**5.55**	1.50	-	**21.32**	-	**26.86**	nr	**28.87**
panic bolts; to single door; iron, bronzed or silver	53.26	2.50	**54.59**	2.25	-	**31.97**	-	**86.56**	nr	**93.06**
panic bolts; to single door; aluminium box, steel shoots and cross rail, anodised silver	52.99	2.50	**54.31**	2.25	-	**31.97**	-	**86.29**	nr	**92.76**
panic bolts; to double doors; iron bronzed or silver	63.38	2.50	**64.96**	3.00	-	**42.63**	-	**107.59**	nr	**115.66**
panic bolts; to double doors; aluminium box, steel shoots and cross rail, anodised silver	63.38	2.50	**64.96**	3.00	-	**42.63**	-	**107.59**	nr	**115.66**
padlock bolts; galvanized; heavy										
150mm	3.12	2.50	**3.20**	0.50	-	**7.11**	-	**10.30**	nr	**11.08**
200mm	3.33	2.50	**3.41**	0.50	-	**7.11**	-	**10.52**	nr	**11.31**
250mm	3.99	2.50	**4.09**	0.75	-	**10.66**	-	**14.75**	nr	**15.85**
300mm	4.92	2.50	**5.04**	0.75	-	**10.66**	-	**15.70**	nr	**16.88**
Door handles; standard quality										
To softwood										
lever handles; spring action; B.M.A., best quality; 41 x 150mm	24.60	2.50	**25.22**	0.20	-	**2.84**	-	**28.06**	nr	**30.16**
lever handles; spring action; chromium plated, housing quality; 41 x 150mm	6.40	2.50	**6.56**	0.20	-	**2.84**	-	**9.40**	nr	**10.11**
lever handles; spring action; chromium plated, best quality; 41 x 150mm	22.40	2.50	**22.96**	0.20	-	**2.84**	-	**25.80**	nr	**27.74**
lever handles; spring action; S.A.A. housing quality; 41 x 150mm	3.88	2.50	**3.98**	0.20	-	**2.84**	-	**6.82**	nr	**7.33**
lever handles; spring action; S.A.A., best quality; 41 x 150mm	14.76	2.50	**15.13**	0.20	-	**2.84**	-	**17.97**	nr	**19.32**
lever handles; spring action; plastic/nylon, housing quality; 41 x 150mm	3.12	2.50	**3.20**	0.15	-	**2.13**	-	**5.33**	nr	**5.73**
pull handles; B.M.A.; 150mm	8.96	2.50	**9.18**	0.15	-	**2.13**	-	**11.32**	nr	**12.16**
pull handles; B.M.A.; 225mm	16.79	2.50	**17.21**	0.20	-	**2.84**	-	**20.05**	nr	**21.56**
pull handles; B.M.A.; 300mm	18.90	2.50	**19.37**	0.25	-	**3.55**	-	**22.93**	nr	**24.64**
pull handles; S.A.A.; 150mm	3.44	2.50	**3.53**	0.15	-	**2.13**	-	**5.66**	nr	**6.08**
pull handles; S.A.A.; 225mm	5.36	2.50	**5.49**	0.20	-	**2.84**	-	**8.34**	nr	**8.96**
pull handles; S.A.A.; 300mm	6.88	2.50	**7.05**	0.25	-	**3.55**	-	**10.60**	nr	**11.40**
pull handles on 50 x 250mm plate, lettered; B.M.A.; 225mm	38.80	2.50	**39.77**	0.40	-	**5.68**	-	**45.45**	nr	**48.86**
pull handles on 50 x 250mm plate, lettered; S.A.A.; 225mm	27.16	2.50	**27.84**	0.40	-	**5.68**	-	**33.52**	nr	**36.04**
To hardwood or the like										
lever handles; spring action; B.M.A., best quality; 41 x 150mm	24.60	2.50	**25.22**	0.30	-	**4.26**	-	**29.48**	nr	**31.69**
lever handles; spring action; chromium plated, housing quality; 41 x 150mm	6.40	2.50	**6.56**	0.30	-	**4.26**	-	**10.82**	nr	**11.63**
lever handles; spring action; chromium plated, best quality; 41 x 150mm	22.40	2.50	**22.96**	0.30	-	**4.26**	-	**27.22**	nr	**29.26**
lever handles; spring action; S.A.A. housing quality; 41 x 150mm	3.88	2.50	**3.98**	0.30	-	**4.26**	-	**8.24**	nr	**8.86**
lever handles; spring action; S.A.A., best quality; 41 x 150mm	14.76	2.50	**15.13**	0.30	-	**4.26**	-	**19.39**	nr	**20.85**
lever handles; spring action; plastic/nylon, housing quality; 41 x 150mm	3.12	2.50	**3.20**	0.20	-	**2.84**	-	**6.04**	nr	**6.49**
pull handles; B.M.A.; 150mm	8.96	2.50	**9.18**	0.20	-	**2.84**	-	**12.03**	nr	**12.93**
pull handles; B.M.A.; 225mm	16.79	2.50	**17.21**	0.30	-	**4.26**	-	**21.47**	nr	**23.08**
pull handles; B.M.A.; 300mm	18.90	2.50	**19.37**	0.35	-	**4.97**	-	**24.35**	nr	**26.17**
pull handles; S.A.A.; 150mm	3.44	2.50	**3.53**	0.20	-	**2.84**	-	**6.37**	nr	**6.85**
pull handles; S.A.A.; 225mm	5.36	2.50	**5.49**	0.30	-	**4.26**	-	**9.76**	nr	**10.49**
pull handles; S.A.A.; 300mm	6.88	2.50	**7.05**	0.35	-	**4.97**	-	**12.03**	nr	**12.93**
pull handles on 50 x 250mm plate, lettered; B.M.A.; 225mm	38.80	2.50	**39.77**	0.60	-	**8.53**	-	**48.30**	nr	**51.92**
pull handles on 50 x 250mm plate, lettered; S.A.A.; 225mm	27.16	2.50	**27.84**	0.60	-	**8.53**	-	**36.37**	nr	**39.09**

Labour hourly rates: (except Specialists) Craft Operatives £14.21 Labourer £10.71 Rates are national average prices. Refer to REGIONAL VARIATIONS for indicative levels of overall pricing in regions	MATERIALS			LABOUR				RATES		
	Del to Site £	Waste %	Material Cost £	Craft Optve Hrs	Lab Hrs	Labour Cost £	Sunds £	Nett Rate £	Unit	Gross Rate (+7.5%) £
P21: IRONMONGERY Cont.										
Door furniture; standard quality										
To softwood										
knobs; real B.M.A.; surface fixing	8.25	2.50	**8.46**	0.20	-	**2.84**	-	**11.30**	nr	**12.15**
knobs; S.A.A.; surface fixing	4.81	2.50	**4.93**	0.20	-	**2.84**	-	**7.77**	nr	**8.36**
knobs; real B.M.A.; secret fixing	13.34	2.50	**13.67**	0.30	-	**4.26**	-	**17.94**	nr	**19.28**
knobs; S.A.A.; secret fixing	6.94	2.50	**7.11**	0.30	-	**4.26**	-	**11.38**	nr	**12.23**
To hardwood or the like										
knobs; real B.M.A.; surface fixing	8.25	2.50	**8.46**	0.30	-	**4.26**	-	**12.72**	nr	**13.67**
knobs; S.A.A.; surface fixing	4.81	2.50	**4.93**	0.30	-	**4.26**	-	**9.19**	nr	**9.88**
knobs; real B.M.A.; secret fixing	13.34	2.50	**13.67**	0.45	-	**6.39**	-	**20.07**	nr	**21.57**
knobs; S.A.A.; secret fixing	6.94	2.50	**7.11**	0.45	-	**6.39**	-	**13.51**	nr	**14.52**
Window furniture; standard quality										
To softwood										
fanlight catches; brass	4.92	2.50	**5.04**	0.35	-	**4.97**	-	**10.02**	nr	**10.77**
fanlight catches; B.M.A.	5.36	2.50	**5.49**	0.35	-	**4.97**	-	**10.47**	nr	**11.25**
fanlight catches; chromium plated	5.46	2.50	**5.60**	0.35	-	**4.97**	-	**10.57**	nr	**11.36**
casement fasteners; wedge plate; black malleable iron	1.04	2.50	**1.07**	0.35	-	**4.97**	-	**6.04**	nr	**6.49**
casement fasteners; wedge plate; B.M.A.	2.41	2.50	**2.47**	0.35	-	**4.97**	-	**7.44**	nr	**8.00**
casement fasteners; wedge plate; chromium plated	2.41	2.50	**2.47**	0.35	-	**4.97**	-	**7.44**	nr	**8.00**
casement fasteners; wedge plate; S.A.A.	1.91	2.50	**1.96**	0.35	-	**4.97**	-	**6.93**	nr	**7.45**
sash fasteners; brass; 70mm	5.52	2.50	**5.66**	1.00	-	**14.21**	-	**19.87**	nr	**21.36**
sash fasteners; B.M.A.; 70mm	5.90	2.50	**6.05**	1.00	-	**14.21**	-	**20.26**	nr	**21.78**
sash fasteners; chromium plated; 70mm	6.50	2.50	**6.66**	1.00	-	**14.21**	-	**20.87**	nr	**22.44**
casement stays; two pins; grey malleable iron										
200mm	1.37	2.50	**1.40**	0.35	-	**4.97**	-	**6.38**	nr	**6.86**
250mm	1.47	2.50	**1.51**	0.35	-	**4.97**	-	**6.48**	nr	**6.97**
300mm	1.58	2.50	**1.62**	0.35	-	**4.97**	-	**6.59**	nr	**7.09**
casement stays; two pins; B.M.A.										
200mm	3.17	2.50	**3.25**	0.35	-	**4.97**	-	**8.22**	nr	**8.84**
250mm	3.44	2.50	**3.53**	0.35	-	**4.97**	-	**8.50**	nr	**9.14**
300mm	3.72	2.50	**3.81**	0.35	-	**4.97**	-	**8.79**	nr	**9.45**
casement stays; two pins; chromium plated										
200mm	3.17	2.50	**3.25**	0.35	-	**4.97**	-	**8.22**	nr	**8.84**
250mm	3.44	2.50	**3.53**	0.35	-	**4.97**	-	**8.50**	nr	**9.14**
300mm	3.72	2.50	**3.81**	0.35	-	**4.97**	-	**8.79**	nr	**9.45**
casement stays; two pins; S.A.A.										
200mm	1.75	2.50	**1.79**	0.35	-	**4.97**	-	**6.77**	nr	**7.27**
250mm	1.91	2.50	**1.96**	0.35	-	**4.97**	-	**6.93**	nr	**7.45**
300mm	2.03	2.50	**2.08**	0.35	-	**4.97**	-	**7.05**	nr	**7.58**
sash lifts; polished brass; 50mm	1.09	2.50	**1.12**	0.20	-	**2.84**	-	**3.96**	nr	**4.26**
sash lifts; B.M.A.; 50mm	1.32	2.50	**1.35**	0.20	-	**2.84**	-	**4.20**	nr	**4.51**
sash lifts; chromium plated; 50mm	1.47	2.50	**1.51**	0.20	-	**2.84**	-	**4.35**	nr	**4.67**
flush lifts; brass; 75mm	1.42	2.50	**1.46**	0.50	-	**7.11**	-	**8.56**	nr	**9.20**
flush lifts; B.M.A.; 75mm	1.70	2.50	**1.74**	0.50	-	**7.11**	-	**8.85**	nr	**9.51**
flush lifts; chromium plated; 75mm	1.70	2.50	**1.74**	0.50	-	**7.11**	-	**8.85**	nr	**9.51**
sash cleats; polished brass; 76mm	1.70	2.50	**1.74**	0.20	-	**2.84**	-	**4.58**	nr	**4.93**
sash cleats; B.M.A.; 76mm	1.75	2.50	**1.79**	0.20	-	**2.84**	-	**4.64**	nr	**4.98**
sash cleats; chromium plated; 76mm	2.13	2.50	**2.18**	0.20	-	**2.84**	-	**5.03**	nr	**5.40**
sash pulleys; frame and wheel	2.18	2.50	**2.23**	0.50	-	**7.11**	-	**9.34**	nr	**10.04**
To hardwood or the like										
fanlight catches; brass	4.92	2.50	**5.04**	0.50	-	**7.11**	-	**12.15**	nr	**13.06**
fanlight catches; B.M.A.	5.36	2.50	**5.49**	0.50	-	**7.11**	-	**12.60**	nr	**13.54**
fanlight catches; chromium plated	5.46	2.50	**5.60**	0.50	-	**7.11**	-	**12.70**	nr	**13.65**
casement fasteners; wedge plate; black malleable iron	1.04	2.50	**1.07**	0.50	-	**7.11**	-	**8.17**	nr	**8.78**
casement fasteners; wedge plate; B.M.A.	2.41	2.50	**2.47**	0.50	-	**7.11**	-	**9.58**	nr	**10.29**
casement fasteners; wedge plate; chromium plated	2.41	2.50	**2.47**	0.50	-	**7.11**	-	**9.58**	nr	**10.29**
casement fasteners; wedge plate; S.A.A.	1.91	2.50	**1.96**	0.50	-	**7.11**	-	**9.06**	nr	**9.74**
sash fasteners; brass; 70mm	5.52	2.50	**5.66**	1.50	-	**21.32**	-	**26.97**	nr	**29.00**
sash fasteners; B.M.A.; 70mm	5.90	2.50	**6.05**	1.50	-	**21.32**	-	**27.36**	nr	**29.41**
sash fasteners; chromium plated; 70mm	6.50	2.50	**6.66**	1.50	-	**21.32**	-	**27.98**	nr	**30.08**
casement stays; two pins; grey malleable iron										
200mm	1.37	2.50	**1.40**	0.50	-	**7.11**	-	**8.51**	nr	**9.15**
250mm	1.47	2.50	**1.51**	0.50	-	**7.11**	-	**8.61**	nr	**9.26**
300mm	1.58	2.50	**1.62**	0.50	-	**7.11**	-	**8.72**	nr	**9.38**
casement stays; two pins; B.M.A.										
200mm	3.17	2.50	**3.25**	0.50	-	**7.11**	-	**10.35**	nr	**11.13**
250mm	3.44	2.50	**3.53**	0.50	-	**7.11**	-	**10.63**	nr	**11.43**
300mm	3.72	2.50	**3.81**	0.50	-	**7.11**	-	**10.92**	nr	**11.74**
casement stays; two pins; chromium plated										
200mm	3.17	2.50	**3.25**	0.50	-	**7.11**	-	**10.35**	nr	**11.13**
250mm	3.44	2.50	**3.53**	0.50	-	**7.11**	-	**10.63**	nr	**11.43**
300mm	3.72	2.50	**3.81**	0.50	-	**7.11**	-	**10.92**	nr	**11.74**
casement stays; two pins; S.A.A.										
200mm	1.75	2.50	**1.79**	0.50	-	**7.11**	-	**8.90**	nr	**9.57**
250mm	1.91	2.50	**1.96**	0.50	-	**7.11**	-	**9.06**	nr	**9.74**
300mm	2.03	2.50	**2.08**	0.50	-	**7.11**	-	**9.19**	nr	**9.87**
sash lifts; polished brass; 50mm	1.09	2.50	**1.12**	0.30	-	**4.26**	-	**5.38**	nr	**5.78**
sash lifts; B.M.A.; 50mm	1.32	2.50	**1.35**	0.30	-	**4.26**	-	**5.62**	nr	**6.04**
sash lifts; chromium plated; 50mm	1.47	2.50	**1.51**	0.30	-	**4.26**	-	**5.77**	nr	**6.20**
flush lifts; brass; 75mm	1.42	2.50	**1.46**	0.75	-	**10.66**	-	**12.11**	nr	**13.02**
flush lifts; B.M.A.; 75mm	1.70	2.50	**1.74**	0.75	-	**10.66**	-	**12.40**	nr	**13.33**
flush lifts; chromium plated; 75mm	1.70	2.50	**1.74**	0.75	-	**10.66**	-	**12.40**	nr	**13.33**
sash cleats; polished brass; 76mm	1.70	2.50	**1.74**	0.30	-	**4.26**	-	**6.01**	nr	**6.46**
sash cleats; B.M.A.; 76mm	1.75	2.50	**1.79**	0.30	-	**4.26**	-	**6.06**	nr	**6.51**
sash cleats; chromium plated; 76mm	2.13	2.50	**2.18**	0.30	-	**4.26**	-	**6.45**	nr	**6.93**
sash pulleys; frame and wheel	2.18	2.50	**2.23**	0.75	-	**10.66**	-	**12.89**	nr	**13.86**

Labour hourly rates: (except Specialists) Craft Operatives £14.21 Labourer £10.71 Rates are national average prices. Refer to REGIONAL VARIATIONS for indicative levels of overall pricing in regions	MATERIALS			LABOUR				RATES		
	Del to Site	Waste	Material Cost	Craft Optve	Lab	Labour Cost	Sunds	Nett Rate	Unit	Gross Rate (+7.5%)
	£	%	£	Hrs	Hrs	£	£	£		£
P21: IRONMONGERY Cont.										
Window furniture; Willan Building Services Ltd - to softwood or the like										
Glidevale frame vent; aluminium with epoxy paint finish										
reference TV FV2; 198mm long; fixing with screws and clip on covers	7.73	2.50	**7.92**	0.20	0.03	**3.16**	0.15	**11.23**	nr	**12.08**
reference TV FV4; 358mm long; fixing with screws and clip on covers	9.31	2.50	**9.54**	0.30	0.04	**4.69**	0.18	**14.42**	nr	**15.50**
Glidevale canopy grille; aluminium with epoxy paint finish										
reference TV CG2; 155 x 15mm slot size; fixing with screws and clip on covers	3.86	2.50	**3.96**	0.20	0.03	**3.16**	0.07	**7.19**	nr	**7.73**
reference TV CG4; 308 x 15mm slot size; fixing with screws and clip on covers	5.46	2.50	**5.60**	0.30	0.03	**4.58**	0.11	**10.29**	nr	**11.06**
Glidevale condensation drainage channel; aluminium with powder coated finish										
reference CDN2500; to suit 4 or 6mm glazing; fixing with screws	24.32	2.50	**24.93**	0.40	0.05	**6.22**	0.48	**31.63**	m	**34.00**
Window furniture; Willan Building Services Ltd - to hardwood or the like										
Glidevale frame vent; aluminium with epoxy paint finish										
reference TV FV2; 198mm long; fixing with screws and clip on covers	7.73	2.50	**7.92**	0.30	0.04	**4.69**	0.15	**12.76**	nr	**13.72**
reference TV FV4; 358mm long; fixing with screws and clip on covers	9.31	2.50	**9.54**	0.45	0.06	**7.04**	0.18	**16.76**	nr	**18.02**
Glidevale canopy grille; aluminium with epoxy paint finish										
reference TV CG2; 155 x 15mm slot size; fixing with screws and clip on covers	3.86	2.50	**3.96**	0.30	0.04	**4.69**	0.07	**8.72**	nr	**9.38**
reference TV CG4; 308 x 15mm slot size; fixing with screws and clip on covers	5.46	2.50	**5.60**	0.45	0.06	**7.04**	0.11	**12.74**	nr	**13.70**
Glidevale condensation drainage channel; aluminium with powder coated finish										
reference CDN2500; to suit 4 or 6mm glazing; fixing with screws	24.32	2.50	**24.93**	0.60	0.08	**9.38**	0.48	**34.79**	m	**37.40**
Letter plates; standard quality										
To softwood										
letter plates; plain; real B.M.A.; 356mm wide	43.71	2.50	**44.80**	1.50	-	**21.32**	-	**66.12**	nr	**71.08**
letter plates; plain; anodised silver; 356mm wide	6.88	2.50	**7.05**	1.50	-	**21.32**	-	**28.37**	nr	**30.49**
letter plates; gravity flap; B.M.A.; 254mm wide	8.30	2.50	**8.51**	1.50	-	**21.32**	-	**29.82**	nr	**32.06**
letter plates; gravity flap; chromium plated; 254mm wide	8.30	2.50	**8.51**	1.50	-	**21.32**	-	**29.82**	nr	**32.06**
letter plates; gravity flap, two numerals; anodised silver; 254mm wide	7.21	2.50	**7.39**	1.60	-	**22.74**	-	**30.13**	nr	**32.39**
postal knockers; S.A.A.; frame 260 x 83mm; opening 203 x 44m	9.56	2.50	**9.80**	1.60	-	**22.74**	-	**32.54**	nr	**34.98**
To hardwood or the like										
letter plates; plain; real B.M.A.; 356mm wide	43.71	2.50	**44.80**	2.25	-	**31.97**	-	**76.78**	nr	**82.53**
letter plates; plain; anodised silver; 356mm wide	6.88	2.50	**7.05**	2.25	-	**31.97**	-	**39.02**	nr	**41.95**
letter plates; gravity flap; B.M.A.; 254mm wide	8.30	2.50	**8.51**	2.25	-	**31.97**	-	**40.48**	nr	**43.52**
letter plates; gravity flap; chromium plated; 254mm wide	8.30	2.50	**8.51**	2.25	-	**31.97**	-	**40.48**	nr	**43.52**
letter plates; gravity flap, two numerals; anodised silver; 254mm wide	7.21	2.50	**7.39**	2.35	-	**33.39**	-	**40.78**	nr	**43.84**
postal knockers; S.A.A.; frame 260 x 83mm; opening 203 x 44m	9.56	2.50	**9.80**	2.40	-	**34.10**	-	**43.90**	nr	**47.20**
Security locks; standard quality										
To hardwood or the like										
window catches; locking; polished aluminium	4.53	2.50	**4.64**	1.00	-	**14.21**	-	**18.85**	nr	**20.27**
dual screws; sanded brass	2.08	2.50	**2.13**	0.80	-	**11.37**	-	**13.50**	nr	**14.51**
window stops; locking; brass	4.81	2.50	**4.93**	0.80	-	**11.37**	-	**16.30**	nr	**17.52**
mortice latches; locking	53.00	2.50	**54.33**	3.00	-	**42.63**	-	**96.96**	nr	**104.23**
double cylinder automatic deadlatches; B.M.A.	31.69	2.50	**32.48**	3.00	-	**42.63**	-	**75.11**	nr	**80.75**
double cylinder automatic deadlatches; chromium plated	30.60	2.50	**31.37**	3.00	-	**42.63**	-	**74.00**	nr	**79.54**
mortice deadlocks; Chubb	24.60	2.50	**25.22**	3.00	-	**42.63**	-	**67.85**	nr	**72.93**
mortice locks; two bolt	30.60	2.50	**31.37**	3.00	-	**42.63**	-	**74.00**	nr	**79.54**
metal window locks; white	5.46	2.50	**5.60**	1.25	-	**17.76**	-	**23.36**	nr	**25.11**
security door chains; steel chain; brass	4.43	2.50	**4.54**	0.50	-	**7.11**	-	**11.65**	nr	**12.52**
security door chains; steel chain; chromium plated	4.92	2.50	**5.04**	0.50	-	**7.11**	-	**12.15**	nr	**13.06**
security mortice bolts; loose keys; chromium plated; 16mm diameter	4.15	2.50	**4.25**	1.00	-	**14.21**	-	**18.46**	nr	**19.85**
security mortice bolts; loose keys; chromium plated; 32mm diameter	5.52	2.50	**5.66**	1.10	-	**15.63**	-	**21.29**	nr	**22.89**
security hinge bolts; silver anodised	4.48	2.50	**4.59**	1.50	-	**21.32**	-	**25.91**	nr	**27.85**
door viewers; chromium plated	4.15	2.50	**4.25**	1.25	-	**17.76**	-	**22.02**	nr	**23.67**

Labour hourly rates: (except Specialists) Craft Operatives £14.21 Labourer £10.71 Rates are national average prices. Refer to REGIONAL VARIATIONS for indicative levels of overall pricing in regions	MATERIALS			LABOUR				RATES		
	Del to Site £	Waste %	Material Cost £	Craft Optve Hrs	Lab Hrs	Labour Cost £	Sunds £	Nett Rate £	Unit	Gross Rate (+7.5%) £
P21: IRONMONGERY Cont.										
Shelf brackets; standard quality										
To softwood										
shelf brackets; grey finished										
100 x 75mm	0.16	2.50	**0.16**	0.20	-	**2.84**	-	**3.01**	nr	**3.23**
150 x 125mm	0.22	2.50	**0.23**	0.20	-	**2.84**	-	**3.07**	nr	**3.30**
225 x 175mm	0.35	2.50	**0.36**	0.20	-	**2.84**	-	**3.20**	nr	**3.44**
300 x 250mm	0.52	2.50	**0.53**	0.20	-	**2.84**	-	**3.38**	nr	**3.63**
350 x 300mm	0.68	2.50	**0.70**	0.20	-	**2.84**	-	**3.54**	nr	**3.80**
Cabin hooks; standard quality										
To softwood										
cabin hooks and eyes; black japanned										
100mm	1.04	2.50	**1.07**	0.20	-	**2.84**	-	**3.91**	nr	**4.20**
150mm	1.32	2.50	**1.35**	0.20	-	**2.84**	-	**4.20**	nr	**4.51**
200mm	1.91	2.50	**1.96**	0.20	-	**2.84**	-	**4.80**	nr	**5.16**
250mm	2.13	2.50	**2.18**	0.20	-	**2.84**	-	**5.03**	nr	**5.40**
cabin hooks and eyes; polished brass										
100mm	4.37	2.50	**4.48**	0.20	-	**2.84**	-	**7.32**	nr	**7.87**
150mm	4.92	2.50	**5.04**	0.20	-	**2.84**	-	**7.89**	nr	**8.48**
To hardwood or the like										
cabin hooks and eyes; black japanned										
100mm	1.04	2.50	**1.07**	0.30	-	**4.26**	-	**5.33**	nr	**5.73**
150mm	1.32	2.50	**1.35**	0.30	-	**4.26**	-	**5.62**	nr	**6.04**
200mm	1.91	2.50	**1.96**	0.30	-	**4.26**	-	**6.22**	nr	**6.69**
250mm	2.13	2.50	**2.18**	0.30	-	**4.26**	-	**6.45**	nr	**6.93**
cabin hooks and eyes; polished brass										
100mm	4.37	2.50	**4.48**	0.30	-	**4.26**	-	**8.74**	nr	**9.40**
150mm	4.92	2.50	**5.04**	0.30	-	**4.26**	-	**9.31**	nr	**10.00**
To softwood and brickwork										
cabin hooks and eyes; black japanned										
100mm	1.04	2.50	**1.07**	0.25	-	**3.55**	-	**4.62**	nr	**4.96**
150mm	1.32	2.50	**1.35**	0.25	-	**3.55**	-	**4.91**	nr	**5.27**
200mm	1.91	2.50	**1.96**	0.25	-	**3.55**	-	**5.51**	nr	**5.92**
250mm	2.13	2.50	**2.18**	0.25	-	**3.55**	-	**5.74**	nr	**6.17**
cabin hooks and eyes; polished brass										
100mm	4.37	2.50	**4.48**	0.25	-	**3.55**	-	**8.03**	nr	**8.63**
150mm	4.92	2.50	**5.04**	0.25	-	**3.55**	-	**8.60**	nr	**9.24**
To hardwood and brickwork										
cabin hooks and eyes; black japanned										
100mm	1.04	2.50	**1.07**	0.35	-	**4.97**	-	**6.04**	nr	**6.49**
150mm	1.32	2.50	**1.35**	0.35	-	**4.97**	-	**6.33**	nr	**6.80**
200mm	1.91	2.50	**1.96**	0.35	-	**4.97**	-	**6.93**	nr	**7.45**
250mm	2.13	2.50	**2.18**	0.35	-	**4.97**	-	**7.16**	nr	**7.69**
cabin hooks and eyes; polished brass										
100mm	4.37	2.50	**4.48**	0.35	-	**4.97**	-	**9.45**	nr	**10.16**
150mm	4.92	2.50	**5.04**	0.35	-	**4.97**	-	**10.02**	nr	**10.77**
Draught seals and strips; standard quality										
To softwood										
draught excluders; plastic foam, self-adhesive										
900mm long	0.16	10.00	**0.18**	0.15	-	**2.13**	-	**2.31**	nr	**2.48**
2000mm long	0.33	10.00	**0.36**	0.30	-	**4.26**	-	**4.63**	nr	**4.97**
draught excluders; aluminium section, rubber tubing										
900mm long	6.83	10.00	**7.51**	0.17	0.02	**2.63**	-	**10.14**	nr	**10.90**
2000mm long	15.46	10.00	**17.01**	0.33	0.04	**5.12**	-	**22.12**	nr	**23.78**
draught excluders; aluminium section, vinyl seal										
900mm long	6.83	10.00	**7.51**	0.17	0.02	**2.63**	-	**10.14**	nr	**10.90**
2000mm long	15.46	10.00	**17.01**	0.33	0.04	**5.12**	-	**22.12**	nr	**23.78**
draught excluders; plastic moulding, nylon brush; 900mm long	5.85	10.00	**6.44**	0.17	0.02	**2.63**	-	**9.06**	nr	**9.74**
draught excluders; aluminium base, flexible arch; 900mm long	15.57	10.00	**17.13**	0.25	0.03	**3.87**	-	**21.00**	nr	**22.58**
draught excluders; aluminium threshold, flexible arch; 900mm long	15.57	10.00	**17.13**	0.30	0.04	**4.69**	-	**21.82**	nr	**23.45**
Draught seals and strips; Sealmaster Ltd										
Threshold seals										
reference BDA; fixing to masonry with screws	9.76	2.50	**10.00**	0.40	0.05	**6.22**	0.20	**16.42**	m	**17.66**
reference BDB; fixing to masonry with screws	12.51	2.50	**12.82**	0.40	0.05	**6.22**	0.25	**19.29**	m	**20.74**
reference BDWB (weather board); fixing to timber with screws	15.20	2.50	**15.58**	0.60	0.08	**9.38**	0.30	**25.26**	m	**27.16**
reference TTM (stop seal); fixing to timber with screws	9.23	2.50	**9.46**	0.30	0.04	**4.69**	0.20	**14.35**	m	**15.43**
Drawer pulls; standard quality										
To softwood										
drawer pulls; brass; 100mm	1.42	2.50	**1.46**	0.15	-	**2.13**	-	**3.59**	nr	**3.86**
drawer pulls; B.M.A.; 100mm	1.85	2.50	**1.90**	0.15	-	**2.13**	-	**4.03**	nr	**4.33**
drawer pulls; chromium plated; 100mm	1.80	2.50	**1.85**	0.15	-	**2.13**	-	**3.98**	nr	**4.27**
drawer pulls; S.A.A.; 100mm	1.09	2.50	**1.12**	0.15	-	**2.13**	-	**3.25**	nr	**3.49**
To hardwood or the like										
drawer pulls; brass; 100mm	1.42	2.50	**1.46**	0.25	-	**3.55**	-	**5.01**	nr	**5.38**
drawer pulls; B.M.A.; 100mm	1.85	2.50	**1.90**	0.25	-	**3.55**	-	**5.45**	nr	**5.86**
drawer pulls; chromium plated; 100mm	1.80	2.50	**1.85**	0.25	-	**3.55**	-	**5.40**	nr	**5.80**

Labour hourly rates: (except Specialists) Craft Operatives £14.21 Labourer £10.71 Rates are national average prices. Refer to REGIONAL VARIATIONS for indicative levels of overall pricing in regions	MATERIALS			LABOUR				RATES		
	Del to Site £	Waste %	Material Cost £	Craft Optve Hrs	Lab Hrs	Labour Cost £	Sunds £	Nett Rate £	Unit	Gross Rate (+7.5%) £
P21: IRONMONGERY Cont.										
Drawer pulls; standard quality Cont.										
To hardwood or the like Cont.										
drawer pulls; S.A.A.; 100mm	1.09	2.50	**1.12**	0.25	-	**3.55**	-	**4.67**	nr	**5.02**
Hooks; standard quality										
To softwood										
cup hooks; polished brass; 25mm	0.16	2.50	**0.16**	0.05	-	**0.71**	-	**0.87**	nr	**0.94**
hat and coat hooks; B.M.A.	1.74	2.50	**1.78**	0.25	-	**3.55**	-	**5.34**	nr	**5.74**
hat and coat hooks; chromium plated	1.74	2.50	**1.78**	0.25	-	**3.55**	-	**5.34**	nr	**5.74**
hat and coat hooks; S.A.A.	0.76	2.50	**0.78**	0.25	-	**3.55**	-	**4.33**	nr	**4.66**
To hardwood or the like										
cup hooks; polished brass; 25mm	0.16	2.50	**0.16**	0.07	-	**0.99**	-	**1.16**	nr	**1.25**
hat and coat hooks; B.M.A.	1.74	2.50	**1.78**	0.40	-	**5.68**	-	**7.47**	nr	**8.03**
hat and coat hooks; chromium plated	1.74	2.50	**1.78**	0.40	-	**5.68**	-	**7.47**	nr	**8.03**
hat and coat hooks; S.A.A.	0.76	2.50	**0.78**	0.40	-	**5.68**	-	**6.46**	nr	**6.95**

This section continues on the next page

Labour hourly rates: (except Specialists) Craft Operatives £14.21 Labourer £10.71 Rates are national average prices. Refer to REGIONAL VARIATIONS for indicative levels of overall pricing in regions	PLANT & TRANSPORT			LABOUR				RATES		
	Plant Cost £	Trans Cost £	P & T Cost £	Craft Optve Hrs	Lab Hrs	Labour Cost £	Sunds £	Nett Rate £	Unit	Gross Rate (+7.5%) £
P30: TRENCHES/PIPEWAYS/PITS FOR BURIED ENGINEERING SERVICES										
Site preparation										
Lifting turf for preservation										
stacking on site average 50 metres distance for immediate use; watering	0.00	-	**0.00**	-	0.50	**5.36**	-	**5.36**	m2	**5.76**
Re-laying turf										
taking from stack average 50 metres distance, laying on prepared bed, watering, maintaining	0.00	-	**0.00**	-	0.50	**5.36**	-	**5.36**	m2	**5.76**
Excavating trenches to receive services not exceeding 200mm nominal size - by machine										
Excavations commencing from natural ground level; compacting; backfilling with excavated material										
not exceeding 1m deep; average 750mm deep	1.55	0.48	**2.03**	-	0.20	**2.14**	2.25	**6.43**	m	**6.91**
not exceeding 1m deep; average 750mm deep; next to roadways	1.68	0.48	**2.16**	-	0.20	**2.14**	3.44	**7.74**	m	**8.33**
Excavations commencing from existing ground level; levelling and grading backfilling to receive turf										
not exceeding 1m deep; average 750mm deep	1.68	0.48	**2.16**	-	0.20	**2.14**	3.50	**7.81**	m	**8.39**
Extra over excavating trenches irrespective of depth; breaking out existing materials										
hard rock	17.29	-	**17.29**	-	-	**-**	-	**17.29**	m3	**18.58**
concrete	14.14	-	**14.14**	-	-	**-**	-	**14.14**	m3	**15.20**
reinforced concrete	21.22	-	**21.22**	-	-	**-**	-	**21.22**	m3	**22.81**
brickwork, blockwork or stonework	9.43	-	**9.43**	-	-	**-**	-	**9.43**	m3	**10.14**
Extra over excavating trenches irrespective of depth; breaking out existing hard pavings										
concrete 150mm thick	2.20	-	**2.20**	-	-	**-**	-	**2.20**	m2	**2.37**
reinforced concrete 200mm thick	4.56	-	**4.56**	-	-	**-**	-	**4.56**	m2	**4.90**
concrete 150mm thick; reinstating	2.67	-	**2.67**	-	-	**-**	13.83	**16.50**	m2	**17.74**
macadam paving 75mm thick	0.39	-	**0.39**	-	-	**-**	-	**0.39**	m2	**0.42**
macadam paving 75mm thick; reinstating	0.39	-	**0.39**	-	-	**-**	17.54	**17.93**	m2	**19.27**
concrete flag paving 50mm thick	-	-	**-**	-	0.26	**2.78**	-	**2.78**	m2	**2.99**
concrete flag paving 50mm thick; re-instating	-	-	**-**	0.50	0.76	**15.24**	11.69	**26.93**	m2	**28.95**
Extra over excavating trenches irrespective of depth; excavating next to existing services										
electricity services -1	-	-	**-**	-	5.00	**53.55**	-	**53.55**	m	**57.57**
gas services -1	-	-	**-**	-	5.00	**53.55**	-	**53.55**	m	**57.57**
water services -1	-	-	**-**	-	5.00	**53.55**	-	**53.55**	m	**57.57**
Extra over excavating trenches irrespective of depth; excavating around existing services crossing trench										
electricity services; cables crossing -1	-	-	**-**	-	7.50	**80.33**	-	**80.33**	nr	**86.35**
gas services services crossing -1	-	-	**-**	-	7.50	**80.33**	-	**80.33**	nr	**86.35**
water services services crossing -1	-	-	**-**	-	7.50	**80.33**	-	**80.33**	nr	**86.35**
Site preparation										
Lifting turf for preservation										
stacking on site average 50 metres distance for immediate use; watering	-	-	**-**	-	0.50	**5.36**	-	**5.36**	m2	**5.76**
Re-laying turf										
taking from stack average 50 metres distance, laying on prepared bed, watering, maintaining	-	-	**-**	-	0.50	**5.36**	-	**5.36**	m2	**5.76**
Excavating trenches to receive services not exceeding 200mm nominal size - by hand										
Excavations commencing from natural ground level; compacting; backfilling with excavated material										
not exceeding 1m deep; average 750mm deep	-	-	**-**	-	1.15	**12.32**	7.18	**19.50**	m	**20.96**
not exceeding 1m deep; average 750mm deep; next to roadways	-	-	**-**	-	1.15	**12.32**	7.62	**19.94**	m	**21.44**
Excavations commencing from existing ground level; levelling and grading backfilling to receive turf										
not exceeding 1m deep; average 750mm deep	-	-	**-**	-	1.15	**12.32**	8.60	**20.92**	m	**22.49**
Extra over excavating trenches irrespective of depth; breaking out existing materials										
hard rock	-	-	**-**	-	13.35	**142.98**	-	**142.98**	m3	**153.70**
concrete	-	-	**-**	-	10.00	**107.10**	-	**107.10**	m3	**115.13**
reinforced concrete	-	-	**-**	-	15.00	**160.65**	-	**160.65**	m3	**172.70**
brickwork, blockwork or stonework	-	-	**-**	-	7.00	**74.97**	-	**74.97**	m3	**80.59**
Extra over excavating trenches irrespective of depth; breaking out existing hard pavings										
concrete 150mm thick	-	-	**-**	-	1.50	**16.07**	-	**16.07**	m2	**17.27**
reinforced concrete 150mm thick	-	-	**-**	-	3.00	**32.13**	-	**32.13**	m2	**34.54**
concrete 150mm thick; reinstating	-	-	**-**	-	3.00	**32.13**	16.15	**48.28**	m2	**51.90**
macadam paving 75mm thick	-	-	**-**	-	0.25	**2.68**	-	**2.68**	m2	**2.88**
macadam paving 75mm thick; reinstating	-	-	**-**	-	0.25	**2.68**	20.47	**23.15**	m2	**24.89**
concrete flag paving 50mm thick	-	-	**-**	-	0.26	**2.78**	-	**2.78**	m2	**2.99**
concrete flag paving 50mm thick; re-instating	-	-	**-**	-	0.76	**8.14**	13.65	**21.78**	m2	**23.42**

Labour hourly rates: (except Specialists) Craft Operatives £14.21 Labourer £10.71 Rates are national average prices. Refer to REGIONAL VARIATIONS for indicative levels of overall pricing in regions	PLANT & TRANSPORT			LABOUR				RATES		
	Plant Cost £	Trans Cost £	P & T Cost £	Craft Optve Hrs	Lab Hrs	Labour Cost £	Sunds £	Nett Rate £	Unit	Gross Rate (+7.5%) £
P30: TRENCHES/PIPEWAYS/PITS FOR BURIED ENGINEERING SERVICES Cont.										
Excavating trenches to receive services not exceeding 200mm nominal size - by hand Cont.										
Extra over excavating trenches irrespective of depth; excavating next existing services										
electricity services -1	-	-	-	-	5.00	**53.55**	-	**53.55**	m	**57.57**
gas services -1	-	-	-	-	5.00	**53.55**	-	**53.55**	m	**57.57**
water services -1	-	-	-	-	5.00	**53.55**	-	**53.55**	m	**57.57**
Extra over excavating trenches irrespective of depth; excavating around existing services crossing trench										
electricity services; cables crossing -1	-	-	-	-	7.50	**80.33**	-	**80.33**	nr	**86.35**
gas services services crossing -1	-	-	-	-	7.50	**80.33**	-	**80.33**	nr	**86.35**
water services services crossing -1	-	-	-	-	7.50	**80.33**	-	**80.33**	nr	**86.35**

This section continues on the next page

Labour hourly rates: (except Specialists) Craft Operatives £14.21 Labourer £10.71 Rates are national average prices. Refer to REGIONAL VARIATIONS for indicative levels of overall pricing in regions	MATERIALS			LABOUR				RATES		
	Del to Site £	Waste %	Material Cost £	Craft Optve Hrs	Lab Hrs	Labour Cost £	Sunds £	Nett Rate £	Unit	Gross Rate (+7.5%) £
P30: TRENCHES/PIPEWAYS/PITS FOR BURIED ENGINEERING SERVICES Cont.										
Underground ducts; vitrified clayware, BS 65 extra strength; flexible joints										
Straight										
100mm nominal size single way duct; laid in position in trench	3.20	5.00	**3.36**	0.16	0.02	**2.49**	-	**5.85**	m	**6.29**
100mm nominal size bonded to form 2 way duct; laid in position in trench	6.40	5.00	**6.72**	0.25	0.03	**3.87**	-	**10.59**	m	**11.39**
100mm nominal size bonded to form 4 way duct; laid in position in trench	12.80	5.00	**13.44**	0.35	0.04	**5.40**	-	**18.84**	m	**20.26**
100mm nominal size bonded to form 6 way duct; laid in position in trench	19.20	5.00	**20.16**	0.45	0.06	**7.04**	-	**27.20**	m	**29.24**
extra; providing and laying in position nylon draw wire	0.12	10.00	**0.13**	0.02	-	**0.28**	-	**0.42**	m	**0.45**
Stop cock pits, valve chambers and the like										
Stop cock pits; half brick thick sides of common bricks, BS EN 772, Category M, 215 x 102.5 x 65mm, compressive strength 20.5 N/mm2, in cement mortar (1:3); 100mm thick base and top of plain in-situ concrete, BS 5328, ordinary prescribed mix ST4, 20mm aggregate, formwork; including all excavation, backfilling, disposal of surplus excavated material, earthwork support and compaction of ground										
600mm deep in clear; (surface boxes included elsewhere)										
internal size 225 x 225mm	16.79	5.00	**17.88**	1.50	1.50	**37.38**	0.06	**57.72**	nr	**62.05**
internal size 338 x 338mm	28.38	5.00	**30.67**	2.25	2.25	**56.07**	0.10	**90.51**	nr	**97.30**
aggregate, formwork										
750mm deep in clear; (surface boxes included elsewhere)										
150mm diameter	28.74	5.00	**29.51**	1.15	1.15	**28.66**	0.07	**60.64**	nr	**65.19**
Cast iron surface boxes and covers, coated to BS 4147; bedding in cement mortar (1:3)										
Surface boxes, BS 5834 Part 2 marked S.V.										
light grade, hinged lid; 150 x 150mm overall top size, 75mm deep	9.15	15.00	**9.25**	0.75	0.38	**14.73**	0.25	**24.22**	nr	**26.04**
Surface boxes, BS 750, marked 'FIRE HYDRANT'										
heavy grade; minimum clear opening 230 x 380mm, minimum depth 125mm	40.26	15.00	**40.46**	1.25	0.63	**24.51**	1.21	**66.18**	nr	**71.15**
Surface boxes, BS 5834 Part 2 marked 'W' or WATER										
medium grade; double triangular cover; minimum clear opening 300 x 300mm, minimum depth 100mm	49.00	15.00	**49.21**	1.50	0.75	**29.35**	1.20	**79.75**	nr	**85.74**
Accessories/Sundry items										
Stop cock keys										
tee	18.73	-	**18.73**	-	-	**-**	-	**18.73**	nr	**20.13**
P31: HOLES/CHASES/COVERS/SUPPORTS FOR SERVICES; BUILDERS WORK IN CONNECTION WITH MECHANICAL SERVICES										
Cutting holes for services										
For ducts through concrete; making good										
rectangular ducts not exceeding 1.00m girth										
150mm thick	-	-	-	1.60	-	**22.74**	0.54	**23.27**	nr	**25.02**
200mm thick	-	-	-	1.90	-	**27.00**	0.64	**27.64**	nr	**29.71**
300mm thick	-	-	-	2.20	-	**31.26**	0.75	**32.01**	nr	**34.41**
rectangular ducts 1.00-2.00m girth										
150mm thick	-	-	-	3.20	-	**45.47**	1.19	**46.66**	nr	**50.16**
200mm thick	-	-	-	3.80	-	**54.00**	1.29	**55.29**	nr	**59.44**
300mm thick	-	-	-	4.40	-	**62.52**	1.50	**64.03**	nr	**68.83**
rectangular ducts 2.00-3.00m girth										
150mm thick	-	-	-	4.00	-	**56.84**	1.37	**58.21**	nr	**62.57**
200mm thick	-	-	-	4.75	-	**67.50**	1.62	**69.11**	nr	**74.30**
300mm thick	-	-	-	5.50	-	**78.16**	1.85	**80.00**	nr	**86.00**
For pipes through concrete; making good										
pipes not exceeding 55mm nominal size										
150mm thick	-	-	-	0.90	-	**12.79**	0.34	**13.13**	nr	**14.11**
200mm thick	-	-	-	1.15	-	**16.34**	0.43	**16.77**	nr	**18.03**
300mm thick	-	-	-	1.45	-	**20.60**	0.54	**21.14**	nr	**22.73**
pipes 55 - 110mm nominal size										
150mm thick	-	-	-	1.12	-	**15.92**	0.42	**16.34**	nr	**17.56**
200mm thick	-	-	-	1.36	-	**19.33**	0.49	**19.82**	nr	**21.31**
300mm thick	-	-	-	1.80	-	**25.58**	0.67	**26.25**	nr	**28.22**
pipes exceeding 110mm nominal size										
150mm thick	-	-	-	1.35	-	**19.18**	0.47	**19.66**	nr	**21.13**
200mm thick	-	-	-	1.72	-	**24.44**	0.60	**25.04**	nr	**26.92**
300mm thick	-	-	-	2.18	-	**30.98**	0.78	**31.75**	nr	**34.14**
For ducts through reinforced concrete; making good										
rectangular ducts not exceeding 1.00m girth										
150mm thick	-	-	-	2.40	-	**34.10**	0.78	**34.88**	nr	**37.50**
200mm thick	-	-	-	2.85	-	**40.50**	0.91	**41.41**	nr	**44.52**
300mm thick	-	-	-	3.30	-	**46.89**	1.07	**47.96**	nr	**51.56**

Labour hourly rates: (except Specialists) Craft Operatives £14.21 Labourer £10.71 Rates are national average prices. Refer to REGIONAL VARIATIONS for indicative levels of overall pricing in regions	MATERIALS			LABOUR				RATES		
	Del to Site £	Waste %	Material Cost £	Craft Optve Hrs	Lab Hrs	Labour Cost £	Sunds £	Nett Rate £	Unit	Gross Rate (+7.5%) £
P31: HOLES/CHASES/COVERS/SUPPORTS FOR SERVICES; BUILDERS WORK IN CONNECTION WITH MECHANICAL SERVICES Cont.										
Cutting holes for services Cont.										
For ducts through reinforced concrete; making good Cont.										
rectangular ducts 1.00-2.00m girth										
150mm thick	-	-	-	4.80	-	**68.21**	1.58	**69.78**	nr	**75.02**
200mm thick	-	-	-	5.70	-	**81.00**	1.86	**82.86**	nr	**89.07**
300mm thick	-	-	-	6.60	-	**93.79**	2.16	**95.95**	nr	**103.15**
rectangular ducts 2.00-3.00m girth										
150mm thick	-	-	-	6.00	-	**85.26**	1.95	**87.21**	nr	**93.75**
200mm thick	-	-	-	7.00	-	**99.47**	2.32	**101.79**	nr	**109.42**
300mm thick	-	-	-	8.25	-	**117.23**	2.65	**119.88**	nr	**128.87**
For pipes through reinforced concrete; making good										
pipes not exceeding 55mm nominal size										
150mm thick	-	-	-	1.35	-	**19.18**	0.48	**19.67**	nr	**21.14**
200mm thick	-	-	-	1.75	-	**24.87**	0.63	**25.50**	nr	**27.41**
300mm thick	-	-	-	2.20	-	**31.26**	0.79	**32.05**	nr	**34.45**
pipes 55 - 110mm nominal size										
150mm thick	-	-	-	1.70	-	**24.16**	0.61	**24.77**	nr	**26.62**
200mm thick	-	-	-	2.05	-	**29.13**	0.74	**29.87**	nr	**32.11**
300mm thick	-	-	-	2.70	-	**38.37**	0.97	**39.33**	nr	**42.28**
pipes exceeding 110mm nominal size										
150mm thick	-	-	-	2.00	-	**28.42**	0.71	**29.13**	nr	**31.32**
200mm thick	-	-	-	2.60	-	**36.95**	0.92	**37.87**	nr	**40.71**
300mm thick	-	-	-	3.20	-	**45.47**	1.14	**46.62**	nr	**50.11**
For ducts through brickwork; making good										
rectangular ducts not exceeding 1.00m girth										
102.5mm thick	-	-	-	1.20	-	**17.05**	0.37	**17.42**	nr	**18.73**
215mm thick	-	-	-	1.40	-	**19.89**	0.44	**20.34**	nr	**21.86**
327.5mm thick	-	-	-	1.65	-	**23.45**	0.53	**23.97**	nr	**25.77**
rectangular ducts 1.00-2.00m girth										
102.5mm thick	-	-	-	2.40	-	**34.10**	0.74	**34.84**	nr	**37.45**
215mm thick	-	-	-	2.80	-	**39.79**	0.88	**40.67**	nr	**43.72**
327.5mm thick	-	-	-	3.30	-	**46.89**	1.05	**47.94**	nr	**51.54**
rectangular ducts 2.00-3.00m girth										
102.5mm thick	-	-	-	3.00	-	**42.63**	0.95	**43.58**	nr	**46.84**
215mm thick	-	-	-	3.50	-	**49.74**	1.10	**50.84**	nr	**54.65**
327.5mm thick	-	-	-	4.10	-	**58.26**	1.31	**59.57**	nr	**64.04**
For pipes through brickwork; making good										
pipes not exceeding 55mm nominal size										
102.5mm thick	-	-	-	0.40	-	**5.68**	0.15	**5.83**	nr	**6.27**
215mm thick	-	-	-	0.66	-	**9.38**	0.23	**9.61**	nr	**10.33**
327.5mm thick	-	-	-	1.00	-	**14.21**	0.36	**14.57**	nr	**15.66**
pipes 55 - 110mm nominal size										
102.5mm thick	-	-	-	0.53	-	**7.53**	0.19	**7.72**	nr	**8.30**
215mm thick	-	-	-	0.88	-	**12.50**	0.32	**12.82**	nr	**13.78**
327.5mm thick	-	-	-	1.33	-	**18.90**	0.47	**19.37**	nr	**20.82**
pipes exceeding 110mm nominal size										
102.5mm thick	-	-	-	0.67	-	**9.52**	0.24	**9.76**	nr	**10.49**
215mm thick	-	-	-	1.10	-	**15.63**	0.39	**16.02**	nr	**17.22**
327.5mm thick	-	-	-	1.67	-	**23.73**	0.59	**24.32**	nr	**26.14**
For ducts through brickwork; making good fair face or facings one side										
rectangular ducts not exceeding 1.00m girth										
102.5mm thick	-	-	-	1.45	-	**20.60**	0.58	**21.18**	nr	**22.77**
215mm thick	-	-	-	1.65	-	**23.45**	0.68	**24.13**	nr	**25.94**
327.5mm thick	-	-	-	1.90	-	**27.00**	0.74	**27.73**	nr	**29.81**
rectangular ducts 1.00-2.00m girth										
102.5mm thick	-	-	-	2.70	-	**38.37**	1.16	**39.52**	nr	**42.49**
215mm thick	-	-	-	3.10	-	**44.05**	1.37	**45.42**	nr	**48.82**
327.5mm thick	-	-	-	3.60	-	**51.16**	1.47	**52.63**	nr	**56.57**
rectangular ducts 2.00-3.00m girth										
102.5mm thick	-	-	-	3.40	-	**48.31**	1.47	**49.78**	nr	**53.52**
215mm thick	-	-	-	3.90	-	**55.42**	1.73	**57.15**	nr	**61.44**
327.5mm thick	-	-	-	4.50	-	**63.95**	1.84	**65.78**	nr	**70.72**
For pipes through brickwork; making good fair face or facings one side										
pipes not exceeding 55mm nominal size										
102.5mm thick	-	-	-	0.50	-	**7.11**	0.19	**7.29**	nr	**7.84**
215mm thick	-	-	-	0.76	-	**10.80**	0.27	**11.07**	nr	**11.90**
327.5mm thick	-	-	-	1.10	-	**15.63**	0.39	**16.02**	nr	**17.22**
pipes 55 - 110mm nominal size										
102.5mm thick	-	-	-	0.68	-	**9.66**	0.23	**9.89**	nr	**10.64**
215mm thick	-	-	-	1.03	-	**14.64**	0.35	**14.98**	nr	**16.11**
327.5mm thick	-	-	-	1.48	-	**21.03**	0.51	**21.55**	nr	**23.16**
pipes exceeding 110mm nominal size										
102.5mm thick	-	-	-	0.87	-	**12.36**	0.28	**12.65**	nr	**13.59**
215mm thick	-	-	-	1.30	-	**18.47**	0.42	**18.89**	nr	**20.31**
327.5mm thick	-	-	-	1.87	-	**26.57**	0.62	**27.19**	nr	**29.23**
For ducts through blockwork; making good										
rectangular ducts not exceeding 1.00m girth										
100mm thick	-	-	-	0.90	-	**12.79**	0.37	**13.16**	nr	**14.14**
140mm thick	-	-	-	1.00	-	**14.21**	0.44	**14.65**	nr	**15.75**
190mm thick	-	-	-	1.10	-	**15.63**	0.53	**16.16**	nr	**17.37**
rectangular ducts 1.00-2.00m girth										
100mm thick	-	-	-	1.80	-	**25.58**	0.74	**26.31**	nr	**28.29**

Labour hourly rates: (except Specialists) Craft Operatives £14.21 Labourer £10.71 Rates are national average prices. Refer to REGIONAL VARIATIONS for indicative levels of overall pricing in regions	MATERIALS			LABOUR				RATES		
	Del to Site £	Waste %	Material Cost £	Craft Optve Hrs	Lab Hrs	Labour Cost £	Sunds £	Nett Rate £	Unit	Gross Rate (+7.5%) £
P31: HOLES/CHASES/COVERS/SUPPORTS FOR SERVICES; BUILDERS WORK IN CONNECTION WITH MECHANICAL SERVICES Cont.										
Cutting holes for services Cont.										
For ducts through blockwork; making good Cont.										
rectangular ducts 1.00-2.00m girth Cont.										
140mm thick	-	-	-	1.95	-	27.71	0.88	28.59	nr	30.74
190mm thick	-	-	-	2.10	-	29.84	1.05	30.89	nr	33.21
rectangular ducts 2.00-3.00m girth										
100mm thick	-	-	-	2.25	-	31.97	0.95	32.92	nr	35.39
140mm thick	-	-	-	2.45	-	34.81	1.10	35.92	nr	38.61
190mm thick	-	-	-	2.65	-	37.66	1.31	38.97	nr	41.89
For pipes through blockwork; making good										
pipes not exceeding 55mm nominal size										
100mm thick	-	-	-	0.30	-	4.26	0.14	4.40	nr	4.73
140mm thick	-	-	-	0.40	-	5.68	0.22	5.90	nr	6.35
190mm thick	-	-	-	0.50	-	7.11	0.34	7.44	nr	8.00
pipes 55 - 110mm nominal size										
100mm thick	-	-	-	0.40	-	5.68	0.18	5.86	nr	6.30
140mm thick	-	-	-	0.53	-	7.53	0.29	7.83	nr	8.41
190mm thick	-	-	-	0.66	-	9.38	0.45	9.83	nr	10.57
pipes exceeding 110mm nominal size										
100mm thick	-	-	-	0.50	-	7.11	0.23	7.34	nr	7.89
140mm thick	-	-	-	0.68	-	9.66	0.37	10.03	nr	10.78
190mm thick	-	-	-	0.83	-	11.79	0.57	12.36	nr	13.29
For ducts through existing brickwork with plaster finish; making good each side										
rectangular ducts not exceeding 1.00m girth										
102.5mm thick	-	-	-	1.95	-	27.71	0.63	28.34	nr	30.46
215mm thick	-	-	-	2.25	-	31.97	0.79	32.76	nr	35.22
327.5mm thick	-	-	-	2.55	-	36.24	0.89	37.13	nr	39.91
rectangular ducts 1.00-2.00m girth										
102.5mm thick	-	-	-	3.60	-	51.16	1.26	52.42	nr	56.35
215mm thick	-	-	-	4.20	-	59.68	1.58	61.26	nr	65.85
327.5mm thick	-	-	-	4.80	-	68.21	1.79	69.99	nr	75.24
rectangular ducts 2.00-3.00m girth										
102.5mm thick	-	-	-	4.50	-	63.95	1.89	65.84	nr	70.77
215mm thick	-	-	-	5.25	-	74.60	2.36	76.97	nr	82.74
327.5mm thick	-	-	-	6.05	-	85.97	2.68	88.65	nr	95.30
For ducts through existing brickwork with plaster finish; making good one side										
rectangular ducts not exceeding 1.00m girth										
102.5mm thick	-	-	-	1.75	-	24.87	0.57	25.43	nr	27.34
215mm thick	-	-	-	2.05	-	29.13	0.71	29.84	nr	32.08
327.5mm thick	-	-	-	2.30	-	32.68	0.80	33.48	nr	35.99
rectangular ducts 1.00-2.00m girth										
102.5mm thick	-	-	-	3.25	-	46.18	1.13	47.32	nr	50.87
215mm thick	-	-	-	3.80	-	54.00	1.42	55.42	nr	59.57
327.5mm thick	-	-	-	4.35	-	61.81	1.63	63.44	nr	68.20
rectangular ducts 2.00-3.00m girth										
102.5mm thick	-	-	-	4.05	-	57.55	1.68	59.23	nr	63.67
215mm thick	-	-	-	4.75	-	67.50	2.10	69.60	nr	74.82
327.5mm thick	-	-	-	5.45	-	77.44	2.42	79.86	nr	85.85
For pipes through existing brickwork with plaster finish; making good each side										
pipes not exceeding 55mm nominal size										
102.5mm thick	-	-	-	0.65	-	9.24	0.23	9.47	nr	10.18
215mm thick	-	-	-	1.00	-	14.21	0.33	14.54	nr	15.63
327.5mm thick	-	-	-	1.45	-	20.60	0.46	21.07	nr	22.65
pipes 55 - 110mm nominal size										
102.5mm thick	-	-	-	0.95	-	13.50	0.27	13.77	nr	14.81
215mm thick	-	-	-	1.35	-	19.18	0.42	19.60	nr	21.07
327.5mm thick	-	-	-	1.95	-	27.71	0.62	28.33	nr	30.45
pipes exceeding 110mm nominal size										
102.5mm thick	-	-	-	1.20	-	17.05	0.34	17.39	nr	18.69
215mm thick	-	-	-	1.75	-	24.87	0.50	25.37	nr	27.27
327.5mm thick	-	-	-	2.50	-	35.53	0.75	36.27	nr	38.99
For pipes through existing brickwork with plaster finish; making good one side										
pipes not exceeding 55mm nominal size										
102.5mm thick	-	-	-	0.60	-	8.53	0.19	8.72	nr	9.37
215mm thick	-	-	-	0.90	-	12.79	0.27	13.06	nr	14.04
327.5mm thick	-	-	-	1.30	-	18.47	0.39	18.86	nr	20.28
pipes 55 - 110mm nominal size										
102.5mm thick	-	-	-	0.85	-	12.08	0.23	12.31	nr	13.23
215mm thick	-	-	-	1.20	-	17.05	0.35	17.40	nr	18.70
327.5mm thick	-	-	-	1.75	-	24.87	0.51	25.38	nr	27.29
pipes exceeding 110mm nominal size										
102.5mm thick	-	-	-	1.10	-	15.63	0.28	15.91	nr	17.11
215mm thick	-	-	-	1.60	-	22.74	0.42	23.16	nr	24.89
327.5mm thick	-	-	-	2.25	-	31.97	0.57	32.54	nr	34.98
For pipes through softwood										
pipes not exceeding 55mm nominal size										
19mm thick	-	-	-	0.10	-	1.42	-	1.42	nr	1.53
50mm thick	-	-	-	0.13	-	1.85	-	1.85	nr	1.99
pipes 55 - 110mm nominal size										
19mm thick	-	-	-	0.15	-	2.13	-	2.13	nr	2.29
50mm thick	-	-	-	0.20	-	2.84	-	2.84	nr	3.06

Labour hourly rates: (except Specialists) Craft Operatives £14.21 Labourer £10.71 Rates are national average prices. Refer to REGIONAL VARIATIONS for indicative levels of overall pricing in regions	MATERIALS			LABOUR				RATES		
	Del to Site £	Waste %	Material Cost £	Craft Optve Hrs	Lab Hrs	Labour Cost £	Sunds £	Nett Rate £	Unit	Gross Rate (+7.5%) £
P31: HOLES/CHASES/COVERS/SUPPORTS FOR SERVICES; BUILDERS WORK IN CONNECTION WITH MECHANICAL SERVICES Cont.										
Cutting holes for services Cont.										
For pipes through plywood										
pipes not exceeding 55mm nominal size										
13mm thick	-	-	-	0.10	-	1.42	-	1.42	nr	1.53
19mm thick	-	-	-	0.13	-	1.85	-	1.85	nr	1.99
pipes 55 - 110mm nominal size										
13mm thick	-	-	-	0.15	-	2.13	-	2.13	nr	2.29
19mm thick	-	-	-	0.20	-	2.84	-	2.84	nr	3.06
For pipes through existing plasterboard with skim finish; making good one side										
pipes not exceeding 55mm nominal size										
13mm thick	-	-	-	0.20	-	2.84	-	2.84	nr	3.06
pipes 55 - 110mm nominal size										
13mm thick	-	-	-	0.30	-	4.26	-	4.26	nr	4.58
Cutting chases for services										
In concrete; making good										
15mm nominal size pipes -1	-	-	-	0.30	-	4.26	-	4.26	m	4.58
22mm nominal size pipes -1	-	-	-	0.35	-	4.97	-	4.97	m	5.35
In brickwork										
15mm nominal size pipes -1	-	-	-	0.25	-	3.55	-	3.55	m	3.82
22mm nominal size pipes -1	-	-	-	0.30	-	4.26	-	4.26	m	4.58
In brickwork; making good fair face or facings										
15mm nominal size pipes -1	-	-	-	0.30	-	4.26	0.12	4.38	m	4.71
22mm nominal size pipes -1	-	-	-	0.35	-	4.97	0.18	5.15	m	5.54
In existing brickwork with plaster finish; making good										
15mm nominal size pipes -1	-	-	-	0.40	-	5.68	0.23	5.92	m	6.36
22mm nominal size pipes -1	-	-	-	0.50	-	7.11	0.36	7.46	m	8.02
In blockwork; making good										
15mm nominal size pipes -1	-	-	-	0.20	-	2.84	-	2.84	m	3.06
22mm nominal size pipes -1	-	-	-	0.25	-	3.55	-	3.55	m	3.82
Pipe and duct sleeves										
Steel pipes BS 1387 Table 2; casting into concrete; making good										
100mm long; to the following nominal size steel pipes										
15mm	0.58	5.00	0.61	0.10	0.05	1.96	-	2.56	nr	2.76
20mm	0.64	5.00	0.67	0.12	0.06	2.35	-	3.02	nr	3.25
25mm	0.79	5.00	0.83	0.13	0.06	2.49	-	3.32	nr	3.57
32mm	0.92	5.00	0.97	0.14	0.07	2.74	-	3.71	nr	3.99
40mm	1.06	5.00	1.11	0.16	0.08	3.13	-	4.24	nr	4.56
50mm	1.34	5.00	1.41	0.20	0.10	3.91	-	5.32	nr	5.72
150mm long; to the following nominal size steel pipes										
15mm	0.71	5.00	0.75	0.12	0.06	2.35	-	3.10	nr	3.33
20mm	0.80	5.00	0.84	0.14	0.07	2.74	-	3.58	nr	3.85
25mm	1.00	5.00	1.05	0.15	0.08	2.99	-	4.04	nr	4.34
32mm	1.18	5.00	1.23	0.16	0.08	3.13	-	4.37	nr	4.69
40mm	1.34	5.00	1.41	0.17	0.09	3.38	-	4.79	nr	5.15
50mm	1.74	5.00	1.83	0.21	0.10	4.06	-	5.89	nr	6.33
200mm long; to the following nominal size steel pipes										
15mm	0.84	5.00	0.88	0.13	0.06	2.49	-	3.37	nr	3.62
20mm	0.95	5.00	0.99	0.15	0.08	2.99	-	3.98	nr	4.28
25mm	1.20	5.00	1.26	0.16	0.08	3.13	-	4.39	nr	4.72
32mm	1.43	5.00	1.50	0.17	0.09	3.38	-	4.88	nr	5.24
40mm	1.64	5.00	1.72	0.18	0.09	3.52	-	5.24	nr	5.63
50mm	2.15	5.00	2.26	0.22	0.11	4.30	-	6.56	nr	7.06
250mm long; to the following nominal size steel pipes										
15mm	0.98	5.00	1.03	0.15	0.08	2.99	-	4.01	nr	4.32
20mm	1.09	5.00	1.15	0.17	0.09	3.38	-	4.53	nr	4.87
25mm	1.41	5.00	1.48	0.18	0.09	3.52	-	5.00	nr	5.37
32mm	1.69	5.00	1.78	0.19	0.10	3.77	-	5.55	nr	5.96
40mm	1.93	5.00	2.03	0.20	0.11	4.02	-	6.05	nr	6.50
50mm	2.56	5.00	2.69	0.24	0.12	4.70	-	7.39	nr	7.94
100mm long; to the following nominal size copper pipes										
15mm	0.64	5.00	0.67	0.10	0.05	1.96	-	2.63	nr	2.83
22mm	0.70	5.00	0.74	0.12	0.06	2.35	-	3.09	nr	3.32
28mm	0.87	5.00	0.92	0.13	0.06	2.49	-	3.41	nr	3.66
32mm	1.02	5.00	1.07	0.14	0.07	2.74	-	3.81	nr	4.09
45mm	1.17	5.00	1.22	0.16	0.08	3.13	-	4.35	nr	4.68
54mm	1.48	5.00	1.56	0.20	0.10	3.91	-	5.47	nr	5.88
150mm long; to the following nominal size copper pipes										
15mm	0.79	5.00	0.83	0.12	0.06	2.35	-	3.18	nr	3.41
22mm	0.88	5.00	0.93	0.14	0.07	2.74	-	3.67	nr	3.94
28mm	1.10	5.00	1.16	0.15	0.08	2.99	-	4.15	nr	4.46
32mm	1.29	5.00	1.36	0.16	0.08	3.13	-	4.49	nr	4.82
45mm	1.48	5.00	1.56	0.17	0.09	3.38	-	4.93	nr	5.30
54mm	1.92	5.00	2.02	0.21	0.10	4.06	-	6.07	nr	6.53

Labour hourly rates: (except Specialists)
Craft Operatives £14.21 Labourer £10.71
Rates are national average prices.
Refer to REGIONAL VARIATIONS for indicative levels of overall pricing in regions

	MATERIALS			LABOUR				RATES		
	Del to Site £	Waste %	Material Cost £	Craft Optve Hrs	Lab Hrs	Labour Cost £	Sunds £	Nett Rate £	Unit	Gross Rate (+7.5%) £
P31: HOLES/CHASES/COVERS/SUPPORTS FOR SERVICES; BUILDERS WORK IN CONNECTION WITH MECHANICAL SERVICES Cont.										
Pipe and duct sleeves Cont.										
Steel pipes BS 1387 Table 2; casting into concrete; making good Cont.										
200mm long; to the following nominal size copper pipes										
15mm	0.92	5.00	**0.97**	0.13	0.06	**2.49**	-	**3.46**	nr	**3.72**
22mm	1.04	5.00	**1.09**	0.15	0.08	**2.99**	-	**4.08**	nr	**4.39**
28mm	1.31	5.00	**1.38**	0.16	0.08	**3.13**	-	**4.51**	nr	**4.85**
32mm	1.58	5.00	**1.65**	0.17	0.09	**3.38**	-	**5.03**	nr	**5.41**
45mm	1.81	5.00	**1.90**	0.18	0.09	**3.52**	-	**5.42**	nr	**5.82**
54mm	2.37	5.00	**2.49**	0.22	0.11	**4.30**	-	**6.80**	nr	**7.31**
250mm long; to the following nominal size copper pipes										
15mm	1.08	5.00	**1.14**	0.15	0.08	**2.99**	-	**4.12**	nr	**4.43**
22mm	1.20	5.00	**1.26**	0.17	0.09	**3.38**	-	**4.64**	nr	**4.98**
28mm	1.54	5.00	**1.62**	0.18	0.09	**3.52**	-	**5.14**	nr	**5.53**
32mm	1.86	5.00	**1.95**	0.19	0.10	**3.77**	-	**5.72**	nr	**6.15**
45mm	2.11	5.00	**2.22**	0.20	0.10	**3.91**	-	**6.13**	nr	**6.59**
54mm	2.81	5.00	**2.95**	0.24	0.12	**4.70**	-	**7.65**	nr	**8.22**
Steel pipes BS 1387 Table 2; building into blockwork; bedding and pointing in cement mortar (1:3); making good										
100mm long; to the following nominal size steel pipes										
15mm	0.58	5.00	**0.61**	0.15	0.08	**2.99**	-	**3.60**	nr	**3.86**
20mm	0.64	5.00	**0.67**	0.16	0.08	**3.13**	-	**3.80**	nr	**4.09**
25mm	0.79	5.00	**0.83**	0.17	0.09	**3.38**	-	**4.21**	nr	**4.52**
32mm	0.92	5.00	**0.97**	0.18	0.09	**3.52**	-	**4.49**	nr	**4.83**
40mm	1.06	5.00	**1.11**	0.19	0.10	**3.77**	-	**4.88**	nr	**5.25**
50mm	1.34	5.00	**1.41**	0.20	0.10	**3.91**	-	**5.32**	nr	**5.72**
140mm long; to the following nominal size steel pipes										
15mm	0.68	5.00	**0.72**	0.17	0.09	**3.38**	-	**4.10**	nr	**4.40**
20mm	0.77	5.00	**0.81**	0.18	0.09	**3.52**	-	**4.33**	nr	**4.65**
25mm	0.96	5.00	**1.00**	0.19	0.10	**3.77**	-	**4.77**	nr	**5.13**
32mm	1.12	5.00	**1.18**	0.20	0.10	**3.91**	-	**5.09**	nr	**5.48**
40mm	1.29	5.00	**1.36**	0.21	0.11	**4.16**	-	**5.52**	nr	**5.93**
50mm	1.67	5.00	**1.75**	0.22	0.11	**4.30**	-	**6.06**	nr	**6.51**
190mm long; to the following nominal size steel pipes										
15mm	0.82	5.00	**0.86**	0.19	0.10	**3.77**	-	**4.63**	nr	**4.98**
20mm	0.91	5.00	**0.96**	0.20	0.10	**3.91**	-	**4.87**	nr	**5.24**
25mm	1.17	5.00	**1.22**	0.21	0.11	**4.16**	-	**5.39**	nr	**5.79**
32mm	1.39	5.00	**1.46**	0.22	0.11	**4.30**	-	**5.76**	nr	**6.19**
40mm	1.58	5.00	**1.65**	0.23	0.12	**4.55**	-	**6.21**	nr	**6.67**
50mm	2.07	5.00	**2.17**	0.24	0.12	**4.70**	-	**6.87**	nr	**7.38**
215mm long; to the following nominal size steel pipes										
15mm	0.88	5.00	**0.93**	0.20	0.10	**3.91**	-	**4.84**	nr	**5.20**
20mm	0.99	5.00	**1.04**	0.21	0.11	**4.16**	-	**5.20**	nr	**5.59**
25mm	1.27	5.00	**1.33**	0.22	0.11	**4.30**	-	**5.64**	nr	**6.06**
32mm	1.51	5.00	**1.59**	0.23	0.12	**4.55**	-	**6.14**	nr	**6.60**
40mm	1.72	5.00	**1.81**	0.24	0.12	**4.70**	-	**6.50**	nr	**6.99**
50mm	2.28	5.00	**2.39**	0.25	0.13	**4.94**	-	**7.34**	nr	**7.89**
100mm long; to the following nominal size copper pipes										
15mm	0.64	5.00	**0.67**	0.15	0.08	**2.99**	-	**3.66**	nr	**3.94**
22mm	0.69	5.00	**0.73**	0.16	0.08	**3.13**	-	**3.86**	nr	**4.15**
28mm	0.87	5.00	**0.92**	0.17	0.09	**3.38**	-	**4.30**	nr	**4.62**
32mm	1.02	5.00	**1.07**	0.18	0.09	**3.52**	-	**4.59**	nr	**4.94**
45mm	1.06	5.00	**1.11**	0.19	0.10	**3.77**	-	**4.88**	nr	**5.25**
54mm	1.48	5.00	**1.56**	0.20	0.10	**3.91**	-	**5.47**	nr	**5.88**
140mm long; to the following nominal size copper pipes										
15mm	0.76	5.00	**0.79**	0.17	0.09	**3.38**	-	**4.17**	nr	**4.49**
22mm	0.84	5.00	**0.88**	0.18	0.09	**3.52**	-	**4.40**	nr	**4.73**
28mm	1.06	5.00	**1.11**	0.19	0.10	**3.77**	-	**4.88**	nr	**5.25**
32mm	1.24	5.00	**1.30**	0.20	0.10	**3.91**	-	**5.21**	nr	**5.60**
45mm	1.42	5.00	**1.49**	0.21	0.11	**4.16**	-	**5.65**	nr	**6.07**
54mm	1.84	5.00	**1.93**	0.22	0.11	**4.30**	-	**6.23**	nr	**6.70**
190mm long; to the following nominal size copper pipes										
15mm	0.90	5.00	**0.95**	0.19	0.10	**3.77**	-	**4.72**	nr	**5.07**
22mm	1.01	5.00	**1.06**	0.20	0.10	**3.91**	-	**4.97**	nr	**5.34**
28mm	1.28	5.00	**1.35**	0.21	0.11	**4.16**	-	**5.51**	nr	**5.92**
32mm	1.52	5.00	**1.60**	0.22	0.11	**4.30**	-	**5.90**	nr	**6.35**
45mm	1.73	5.00	**1.82**	0.23	0.12	**4.55**	-	**6.37**	nr	**6.85**
54mm	2.28	5.00	**2.39**	0.24	0.12	**4.70**	-	**7.09**	nr	**7.62**
215mm long; to the following nominal size copper pipes										
15mm	0.97	5.00	**1.01**	0.20	0.10	**3.91**	-	**4.93**	nr	**5.30**
22mm	1.08	5.00	**1.14**	0.21	0.11	**4.16**	-	**5.30**	nr	**5.70**
28mm	1.40	5.00	**1.47**	0.22	0.11	**4.30**	-	**5.77**	nr	**6.20**
32mm	1.66	5.00	**1.74**	0.23	0.12	**4.55**	-	**6.30**	nr	**6.77**
45mm	1.89	5.00	**1.98**	0.24	0.12	**4.70**	-	**6.68**	nr	**7.18**
54mm	2.51	5.00	**2.64**	0.25	0.13	**4.94**	-	**7.58**	nr	**8.15**

Labour hourly rates: (except Specialists) Craft Operatives £14.21 Labourer £10.71 Rates are national average prices. Refer to REGIONAL VARIATIONS for indicative levels of overall pricing in regions	MATERIALS			LABOUR				RATES		
	Del to Site £	Waste %	Material Cost £	Craft Optve Hrs	Lab Hrs	Labour Cost £	Sunds £	Nett Rate £	Unit	Gross Rate (+7.5%) £
P31: HOLES/CHASES/COVERS/SUPPORTS FOR SERVICES; BUILDERS WORK IN CONNECTION WITH MECHANICAL SERVICES Cont.										
Ends of supports for equipment, fittings, appliances and ancillaries										
Fix only; casting into concrete; making good										
holderbat or bracket	-	-	-	0.25	-	**3.55**	0.29	**3.85**	nr	**4.13**
Fix only; building into brickwork										
holderbat or bracket	-	-	-	0.10	-	**1.42**	-	**1.42**	nr	**1.53**
Fix only; cutting and pinning to brickwork; making good										
holderbat or bracket	-	-	-	0.12	-	**1.71**	0.09	**1.80**	nr	**1.93**
Fix only; cutting and pinning to brickwork; making good fair face or facings one side										
holderbat or bracket	-	-	-	0.15	-	**2.13**	0.14	**2.27**	nr	**2.44**
Cavity fixings for ends of supports for pipes and ducts										
Rawlnut Multi-purpose fixings, plated pan head screws; The Rawlplug Co. Ltd; fixing to soft building board, drilling										
M4 with 30mm long screw; product code 09 - 130										
at 300mm centres	1.40	5.00	**1.47**	0.17	-	**2.42**	-	**3.89**	m	**4.18**
at 450mm centres	0.93	5.00	**0.98**	0.11	-	**1.56**	-	**2.54**	m	**2.73**
isolated	0.42	5.00	**0.44**	0.05	-	**0.71**	-	**1.15**	nr	**1.24**
M5 with 30mm long screw; product code 09 - 235										
at 300mm centres	1.40	5.00	**1.47**	0.20	-	**2.84**	-	**4.31**	m	**4.64**
at 450mm centres	0.93	5.00	**0.98**	0.13	-	**1.85**	-	**2.83**	m	**3.04**
isolated	0.42	5.00	**0.44**	0.06	-	**0.85**	-	**1.29**	nr	**1.39**
M5 with 50mm long screw; product code 09 - 317										
at 300mm centres	1.88	5.00	**1.97**	0.20	-	**2.84**	-	**4.82**	m	**5.18**
at 450mm centres	1.25	5.00	**1.32**	0.13	-	**1.85**	-	**3.16**	m	**3.40**
isolated	0.56	5.00	**0.59**	0.06	-	**0.85**	-	**1.44**	nr	**1.55**
Rawlnut Multi-purpose fixings, plated pan head screws; The Rawlplug Co. Ltd; fixing to sheet metal, drilling										
M4 with 30mm long screw; product code 09 - 130										
at 300mm centres	1.40	5.00	**1.47**	0.33	-	**4.69**	-	**6.16**	m	**6.62**
at 450mm centres	0.93	5.00	**0.98**	0.22	-	**3.13**	-	**4.11**	m	**4.41**
isolated	0.42	5.00	**0.44**	0.10	-	**1.42**	-	**1.86**	nr	**2.00**
M5 with 30mm long screw; product code 09 - 235										
at 300mm centres	1.40	5.00	**1.47**	0.40	-	**5.68**	-	**7.15**	m	**7.69**
at 450mm centres	0.93	5.00	**0.98**	0.27	-	**3.84**	-	**4.82**	m	**5.18**
isolated	0.42	5.00	**0.44**	0.12	-	**1.71**	-	**2.15**	nr	**2.31**
M5 with 50mm long screw; product code 09 - 317										
at 300mm centres	1.88	5.00	**1.97**	0.40	-	**5.68**	-	**7.66**	m	**8.23**
at 450mm centres	1.25	5.00	**1.32**	0.27	-	**3.84**	-	**5.15**	m	**5.54**
isolated	0.56	5.00	**0.59**	0.12	-	**1.71**	-	**2.30**	nr	**2.47**
Interset high performance cavity fixings, plated pan head screws; The Rawlplug Co. Ltd; fixing to soft building board, drilling										
M4 with 40mm long screw; product code 41 - 620										
at 300mm centres	1.62	5.00	**1.70**	0.17	-	**2.42**	-	**4.12**	m	**4.43**
at 450mm centres	1.08	5.00	**1.14**	0.11	-	**1.56**	-	**2.70**	m	**2.90**
isolated	0.49	5.00	**0.51**	0.05	-	**0.71**	-	**1.22**	nr	**1.31**
M5 with 40mm long screw; product code 41 - 636										
at 300mm centres	1.94	5.00	**2.04**	0.20	-	**2.84**	-	**4.88**	m	**5.24**
at 450mm centres	1.29	5.00	**1.36**	0.13	-	**1.85**	-	**3.21**	m	**3.45**
isolated	0.58	5.00	**0.61**	0.06	-	**0.85**	-	**1.46**	nr	**1.57**
M5 with 55mm long screw; product code 41 - 652										
at 300mm centres	2.03	5.00	**2.13**	0.20	-	**2.84**	-	**4.97**	m	**5.34**
at 450mm centres	1.35	5.00	**1.42**	0.13	-	**1.85**	-	**3.27**	m	**3.51**
isolated	0.61	5.00	**0.64**	0.06	-	**0.85**	-	**1.49**	nr	**1.60**
Spring toggles plated pan head screws; The Rawlplug Co. Ltd; fixing to soft building board, drilling										
M5 with 80mm long screw; product code 94 - 439										
at 300mm centres	1.34	5.00	**1.41**	0.20	-	**2.84**	-	**4.25**	m	**4.57**
at 450mm centres	0.90	5.00	**0.94**	0.13	-	**1.85**	-	**2.79**	m	**3.00**
isolated	0.40	5.00	**0.42**	0.06	-	**0.85**	-	**1.28**	nr	**1.37**
M6 with 60mm long screw; product code 94 - 442										
at 300mm centres	1.57	5.00	**1.64**	0.20	-	**2.84**	-	**4.49**	m	**4.82**
at 450mm centres	1.04	5.00	**1.10**	0.13	-	**1.85**	-	**2.94**	m	**3.16**
isolated	0.47	5.00	**0.49**	0.06	-	**0.85**	-	**1.35**	nr	**1.45**
M6 with 80mm long screw; product code 94 - 464										
at 300mm centres	1.93	5.00	**2.02**	0.20	-	**2.84**	-	**4.86**	m	**5.23**
at 450mm centres	1.28	5.00	**1.35**	0.13	-	**1.85**	-	**3.20**	m	**3.44**
isolated	0.58	5.00	**0.61**	0.06	-	**0.85**	-	**1.46**	nr	**1.57**
Trench covers and frames										
Duct covers in cast iron, coated; continuous covers in 610mm lengths; steel bearers										
22mm deep, for pedestrian traffic; bedding frames to concrete in cement mortar (1:3); nominal width										
150mm	90.42	-	**90.42**	0.50	0.50	**12.46**	0.68	**103.56**	m	**111.32**
225mm	94.23	-	**94.23**	0.55	0.55	**13.71**	0.68	**108.62**	m	**116.77**
300mm	109.92	-	**109.92**	0.65	0.65	**16.20**	0.68	**126.80**	m	**136.31**
375mm	125.35	-	**125.35**	0.70	0.70	**17.44**	0.68	**143.47**	m	**154.23**

Labour hourly rates: (except Specialists) Craft Operatives £14.21 Labourer £10.71 Rates are national average prices. Refer to REGIONAL VARIATIONS for indicative levels of overall pricing in regions	MATERIALS			LABOUR				RATES		
	Del to Site £	Waste %	Material Cost £	Craft Optve Hrs	Lab Hrs	Labour Cost £	Sunds £	Nett Rate £	Unit	Gross Rate (+7.5%) £
P31: HOLES/CHASES/COVERS/SUPPORTS FOR SERVICES; BUILDERS WORK IN CONNECTION WITH MECHANICAL SERVICES Cont.										
Trench covers and frames Cont.										
Duct covers in cast iron, coated; continuous covers in 610mm lengths; steel bearers Cont.										
22mm deep, for pedestrian traffic; bedding frames to concrete in cement mortar (1:3); nominal width Cont.										
450mm	141.25	-	**141.25**	0.75	0.75	**18.69**	0.68	**160.62**	m	**172.66**
Pendock Profiles Ltd. floor ducting profiles; galvanized mild steel tray section with 12mm plywood cover board; fixing tray section to masonry with nails; fixing cover board with screws, countersinking										
For 50mm screeds										
reference FDT 100/50, 100mm wide	10.15	2.50	**10.41**	0.35	-	**4.97**	0.10	**15.48**	m	**16.64**
extra; stop end	2.89	2.50	**2.96**	0.10	-	**1.42**	-	**4.39**	nr	**4.71**
extra; corner	12.53	2.50	**12.84**	0.25	-	**3.55**	0.15	**16.55**	nr	**17.79**
extra; tee	12.53	2.50	**12.84**	0.30	-	**4.26**	0.15	**17.26**	nr	**18.55**
reference FDT 150/50, 150mm wide	13.42	2.50	**13.76**	0.40	-	**5.68**	0.10	**19.54**	m	**21.01**
extra; stop end	2.89	2.50	**2.96**	0.12	-	**1.71**	-	**4.67**	nr	**5.02**
extra; corner	13.95	2.50	**14.30**	0.30	-	**4.26**	0.15	**18.71**	nr	**20.11**
extra; tee	13.95	2.50	**14.30**	0.35	-	**4.97**	0.15	**19.42**	nr	**20.88**
reference FDT 200/50, 200mm wide	16.34	2.50	**16.75**	0.50	-	**7.11**	0.10	**23.95**	m	**25.75**
extra; stop end	2.89	2.50	**2.96**	0.15	-	**2.13**	-	**5.10**	nr	**5.48**
extra; corner	16.22	2.50	**16.63**	0.35	-	**4.97**	0.15	**21.75**	nr	**23.38**
extra; tee	16.22	2.50	**16.63**	0.40	-	**5.68**	0.15	**22.46**	nr	**24.15**
For 70mm screeds										
reference FDT 100/70, 100mm wide	10.97	2.50	**11.25**	0.35	-	**4.97**	0.10	**16.32**	m	**17.54**
extra; stop end	2.89	2.50	**2.96**	0.10	-	**1.42**	-	**4.39**	nr	**4.71**
extra; corner	12.53	2.50	**12.84**	0.25	-	**3.55**	0.15	**16.55**	nr	**17.79**
extra; tee	12.53	2.50	**12.84**	0.30	-	**4.26**	0.15	**17.26**	nr	**18.55**
reference FDT 150/70, 150mm wide	14.25	2.50	**14.60**	0.40	-	**5.68**	0.10	**20.39**	m	**21.92**
extra; stop end	2.89	2.50	**2.96**	0.12	-	**1.71**	-	**4.67**	nr	**5.02**
extra; corner	13.95	2.50	**14.30**	0.30	-	**4.26**	0.15	**18.71**	nr	**20.11**
extra; tee	13.95	2.50	**14.30**	0.35	-	**4.97**	0.15	**19.42**	nr	**20.88**
reference FDT 200/70, 200mm wide	17.05	2.50	**17.47**	0.50	-	**7.11**	0.10	**24.68**	m	**26.53**
extra; stop end	2.89	2.50	**2.96**	0.15	-	**2.13**	-	**5.10**	nr	**5.48**
extra; corner	16.22	2.50	**16.63**	0.35	-	**4.97**	0.15	**21.75**	nr	**23.38**
extra; tee	16.22	2.50	**16.63**	0.40	-	**5.68**	0.15	**22.46**	nr	**24.15**
Casings, bearers and supports for equipment										
Softwood, sawn										
25mm boarded platforms	5.20	10.00	**5.71**	1.10	0.14	**17.13**	0.15	**22.99**	m2	**24.72**
50 x 50mm bearers	0.47	10.00	**0.52**	0.16	0.02	**2.49**	0.02	**3.02**	m	**3.25**
50 x 75mm bearers	0.67	10.00	**0.74**	0.19	0.02	**2.91**	0.02	**3.67**	m	**3.95**
50 x 100mm bearers	0.94	10.00	**1.03**	0.23	0.03	**3.59**	0.03	**4.65**	m	**5.00**
38 x 50mm bearers; framed	0.38	10.00	**0.41**	0.25	0.03	**3.87**	0.03	**4.32**	m	**4.64**
Softwood, wrought										
25mm boarded platforms; tongued and grooved	10.15	10.00	**11.17**	1.20	0.15	**18.66**	0.17	**30.00**	m2	**32.25**
19mm boarded sides; tongued and grooved	7.62	10.00	**8.38**	1.10	0.14	**17.13**	0.15	**25.66**	m2	**27.59**
extra for holes for pipes not exceeding 55mm nominal size	-	-	**-**	0.25	-	**3.55**	-	**3.55**	nr	**3.82**
19mm boarded cover; tongued and grooved; ledged; sectional	14.59	10.00	**16.05**	1.20	0.15	**18.66**	0.17	**34.88**	m2	**37.50**
Fibreboard										
13mm sides	8.16	10.00	**8.98**	2.10	0.26	**32.63**	0.29	**41.89**	m2	**45.03**
extra for holes for pipes not exceeding 55mm nominal size	-	-	**-**	0.20	-	**2.84**	-	**2.84**	nr	**3.06**
Vermiculite insulation										
packing around tank	62.22	10.00	**68.45**	4.80	0.60	**74.63**	-	**143.08**	m3	**153.81**
Slag wool insulation										
packing around tank	30.97	10.00	**34.06**	6.00	0.75	**93.29**	-	**127.35**	m3	**136.91**
P31: HOLES/CHASES/COVERS/SUPPORTS FOR SERVICES; BUILDERS WORK IN CONNECTION WITH ELECTRICAL SERVICES										
Cutting or forming holes, mortices, sinkings and chases - New buildings										
Concealed steel conduits; making good										
luminaire points	-	-	**-**	0.70	0.35	**13.70**	0.37	**14.06**	nr	**15.12**
socket outlet points	-	-	**-**	0.60	0.30	**11.74**	0.32	**12.05**	nr	**12.96**
fitting outlet points	-	-	**-**	0.60	0.30	**11.74**	0.32	**12.05**	nr	**12.96**
equipment and control gear points	-	-	**-**	0.80	0.40	**15.65**	0.43	**16.08**	nr	**17.29**
Exposed p.v.c. conduits; making good										
luminaire points	-	-	**-**	0.25	0.13	**4.94**	0.13	**5.07**	nr	**5.45**
socket outlet points	-	-	**-**	0.20	0.10	**3.91**	0.11	**4.02**	nr	**4.32**
fitting outlet points	-	-	**-**	0.20	0.10	**3.91**	0.11	**4.02**	nr	**4.32**
equipment and control gear points	-	-	**-**	0.30	0.15	**5.87**	0.15	**6.02**	nr	**6.47**

Labour hourly rates: (except Specialists) Craft Operatives £14.21 Labourer £10.71 Rates are national average prices. Refer to REGIONAL VARIATIONS for indicative levels of overall pricing in regions	MATERIALS			LABOUR				RATES		
	Del to Site £	Waste %	Material Cost £	Craft Optve Hrs	Lab Hrs	Labour Cost £	Sunds £	Nett Rate £	Unit	Gross Rate (+7.5%) £
P31: HOLES/CHASES/COVERS/SUPPORTS FOR SERVICES; BUILDERS WORK IN CONNECTION WITH ELECTRICAL SERVICES Cont.										
Cutting mortices, sinking and the like for services - Existing buildings										
In existing concrete; making good										
75 x 75 x 35mm	-	-	-	0.45	0.13	**7.79**	0.53	**8.31**	nr	**8.94**
150 x 75 x 35mm	-	-	-	0.60	0.15	**10.13**	0.79	**10.92**	nr	**11.74**
In existing brickwork										
75 x 75 x 35mm	-	-	-	0.25	0.10	**4.62**	0.32	**4.94**	nr	**5.31**
150 x 75 x 35mm	-	-	-	0.40	0.13	**7.08**	0.47	**7.55**	nr	**8.11**
In existing brickwork with plaster finish; making good										
75 x 75 x 35mm	-	-	-	0.30	0.18	**6.19**	0.53	**6.72**	nr	**7.22**
150 x 75 x 35mm	-	-	-	0.45	0.20	**8.54**	0.74	**9.27**	nr	**9.97**
In existing blockwork; making good										
75 x 75 x 35mm	-	-	-	0.25	0.08	**4.41**	0.35	**4.76**	nr	**5.11**
150 x 75 x 35mm	-	-	-	0.35	0.10	**6.04**	0.38	**6.42**	nr	**6.90**
In existing blockwork with plaster finish; making good										
75 x 75 x 35mm	-	-	-	0.40	0.13	**7.08**	0.47	**7.55**	nr	**8.11**
150 x 75 x 35mm	-	-	-	0.46	0.15	**8.14**	0.42	**8.56**	nr	**9.21**
Cutting chases for services										
In existing concrete; making good										
20mm nominal size conduits -1	-	-	-	0.40	0.20	**7.83**	0.37	**8.19**	m	**8.81**
20mm nominal size conduits -3	-	-	-	0.50	0.25	**9.78**	0.43	**10.21**	m	**10.98**
20mm nominal size conduits -6	-	-	-	0.60	0.30	**11.74**	0.49	**12.23**	m	**13.15**
In existing brickwork										
20mm nominal size conduits -1	-	-	-	0.30	0.15	**5.87**	0.37	**6.24**	m	**6.70**
20mm nominal size conduits -3	-	-	-	0.40	0.20	**7.83**	0.43	**8.26**	m	**8.88**
20mm nominal size conduits -6	-	-	-	0.50	0.25	**9.78**	0.49	**10.28**	m	**11.05**
In existing brickwork with plaster finish; making good										
20mm nominal size conduits -1	-	-	-	0.45	0.23	**8.86**	0.48	**9.34**	m	**10.04**
20mm nominal size conduits -3	-	-	-	0.55	0.28	**10.81**	0.55	**11.36**	m	**12.21**
20mm nominal size conduits -6	-	-	-	0.65	0.33	**12.77**	0.61	**13.38**	m	**14.38**
In existing blockwork										
20mm nominal size conduits -1	-	-	-	0.23	0.12	**4.55**	0.37	**4.92**	m	**5.29**
20mm nominal size conduits -3	-	-	-	0.30	0.15	**5.87**	0.43	**6.30**	m	**6.77**
20mm nominal size conduits -6	-	-	-	0.38	0.19	**7.43**	0.49	**7.93**	m	**8.52**
In existing blockwork with plaster finish; making good										
20mm nominal size conduits -1	-	-	-	0.38	0.19	**7.43**	0.48	**7.92**	m	**8.51**
20mm nominal size conduits -3	-	-	-	0.45	0.23	**8.86**	0.55	**9.40**	m	**10.11**
20mm nominal size conduits -6	-	-	-	0.53	0.27	**10.42**	0.61	**11.03**	m	**11.86**
Lifting and replacing floorboards										
For cables or conduits										
in groups 1 - 3	-	-	-	0.16	0.08	**3.13**	0.11	**3.24**	m	**3.48**
in groups 3 - 6	-	-	-	0.20	0.10	**3.91**	0.12	**4.03**	m	**4.33**
in groups exceeding 6	-	-	-	0.25	0.13	**4.94**	0.13	**5.07**	m	**5.45**

PAVING/PLANTING/FENCING/SITE FURNITURE

Labour hourly rates: (except Specialists) Craft Operatives £14.21 Labourer £10.71 Rates are national average prices. Refer to REGIONAL VARIATIONS for indicative levels of overall pricing in regions	PLANT & TRANSPORT			LABOUR				RATES		
	Plant Cost £	Trans Cost £	P & T Cost £	Craft Optve Hrs	Lab Hrs	Labour Cost £	Sunds £	Nett Rate £	Unit	Gross Rate (+7.5%) £
Q10: KERBS/EDGINGS/CHANNELS/PAVING ACCESSORIES										
Excavating - by machine										
Trenches width not exceeding 0.30m										
maximum depth not exceeding 0.25m................	3.90	-	**3.90**	-	1.71	**18.31**	-	**22.21**	m3	**23.88**
maximum depth not exceeding 1.00m................	4.06	-	**4.06**	-	1.80	**19.28**	-	**23.34**	m3	**25.09**
Extra over any types of excavating irrespective of depth										
excavating below ground water level.............	3.44	-	**3.44**	-	-	**-**	-	**3.44**	m3	**3.70**
excavating in running silt, running sand or liquid mud	5.91	-	**5.91**	-	-	**-**	-	**5.91**	m3	**6.35**
Breaking out existing materials; extra over any types of excavating irrespective of depth										
hard rock	8.55	-	**8.55**	-	-	**-**	-	**8.55**	m3	**9.19**
concrete	7.10	-	**7.10**	-	-	**-**	-	**7.10**	m3	**7.63**
reinforced concrete	10.66	-	**10.66**	-	-	**-**	-	**10.66**	m3	**11.46**
brickwork, blockwork or stonework...............	4.74	-	**4.74**	-	-	**-**	-	**4.74**	m3	**5.09**
100mm diameter drain and concrete bed under......	0.67	-	**0.67**	-	-	**-**	-	**0.67**	m	**0.72**
Breaking out existing hard pavings; extra over any types of excavating irrespective of depth										
concrete 150mm thick	1.12	-	**1.12**	-	-	**-**	-	**1.12**	m2	**1.20**
reinforced concrete 200mm thick.................	2.24	-	**2.24**	-	-	**-**	-	**2.24**	m2	**2.41**
brickwork, blockwork or stonework 100mm thick....	0.53	-	**0.53**	-	-	**-**	-	**0.53**	m2	**0.57**
coated macadam or asphalt 75mm thick............	0.19	-	**0.19**	-	-	**-**	-	**0.19**	m2	**0.20**
Disposal of excavated material										
depositing on site in temporary spoil heaps where directed										
25m.......................................	-	1.77	**1.77**	-	0.10	**1.07**	-	**2.84**	m3	**3.05**
100m......................................	-	2.02	**2.02**	-	0.10	**1.07**	-	**3.09**	m3	**3.32**
800m......................................	-	3.12	**3.12**	-	0.10	**1.07**	-	**4.19**	m3	**4.50**
1600m.....................................	-	3.52	**3.52**	-	0.10	**1.07**	-	**4.60**	m3	**4.94**
extra for each additional 1600m	-	0.55	**0.55**	-	-	**-**	-	**0.55**	m3	**0.59**
removing from site to tip a) Inert	-	16.00	**16.00**	-	-	**-**	-	**16.00**	m3	**17.20**
removing from site to tip b) Active	-	37.00	**37.00**	-	-	**-**	-	**37.00**	m3	**39.78**
removing from site to tip c) Contaminated (Guide price - always seek a quotation for specialist disposal costs.)	-	80.00	**80.00**	-	-	**-**	-	**80.00**	m3	**86.00**
Disposal of preserved top soil										
depositing on site in temporary spoil heaps where directed										
25m.......................................	-	1.77	**1.77**	-	0.10	**1.07**	-	**2.84**	m3	**3.05**
50m.......................................	-	1.86	**1.86**	-	0.10	**1.07**	-	**2.93**	m3	**3.15**
100m......................................	-	2.02	**2.02**	-	0.10	**1.07**	-	**3.09**	m3	**3.32**
Surface treatments										
Compacting										
bottoms of excavations	0.19	-	**0.19**	-	-	**-**	-	**0.19**	m2	**0.21**
Excavating - by hand										
Trenches width not exceeding 0.30m										
maximum depth not exceeding 0.25m................	-	-	**-**	-	4.25	**45.52**	-	**45.52**	m3	**48.93**
maximum depth not exceeding 1.00m................	-	-	**-**	-	4.50	**48.20**	-	**48.20**	m3	**51.81**
Extra over any types of excavating irrespective of depth										
excavating below ground water level.............	-	-	**-**	-	1.50	**16.07**	-	**16.07**	m3	**17.27**
excavating in running silt, running sand or liquid mud	-	-	**-**	-	4.50	**48.20**	-	**48.20**	m3	**51.81**
Breaking out existing materials; extra over any types of excavating irrespective of depth										
hard rock	-	-	**-**	-	10.00	**107.10**	-	**107.10**	m3	**115.13**
concrete	-	-	**-**	-	7.00	**74.97**	-	**74.97**	m3	**80.59**

Labour hourly rates: (except Specialists) Craft Operatives £14.21 Labourer £10.71 Rates are national average prices. Refer to REGIONAL VARIATIONS for indicative levels of overall pricing in regions	PLANT & TRANSPORT			LABOUR				RATES		
	Plant Cost £	Trans Cost £	P & T Cost £	Craft Optve Hrs	Lab Hrs	Labour Cost £	Sunds £	Nett Rate £	Unit	Gross Rate (+7.5%) £
Q10: KERBS/EDGINGS/CHANNELS/PAVING ACCESSORIES Cont.										
Excavating - by hand Cont.										
Breaking out existing materials; extra over any types of excavating irrespective of depth Cont.										
reinforced concrete	-	-	-	-	12.00	**128.52**	-	**128.52**	m3	**138.16**
brickwork, blockwork or stonework	-	-	-	-	4.00	**42.84**	-	**42.84**	m3	**46.05**
100mm diameter drain and concrete bed under	-	-	-	-	1.00	**10.71**	-	**10.71**	m	**11.51**
Breaking out existing hard pavings; extra over any types of excavating irrespective of depth										
concrete 150mm thick	-	-	-	-	1.50	**16.07**	-	**16.07**	m2	**17.27**
reinforced concrete 200mm thick	-	-	-	-	2.00	**21.42**	-	**21.42**	m2	**23.03**
brickwork, blockwork or stonework 100mm thick	-	-	-	-	0.70	**7.50**	-	**7.50**	m2	**8.06**
coated macadam or asphalt 75mm thick	-	-	-	-	0.25	**2.68**	-	**2.68**	m2	**2.88**
Disposal of excavated material										
depositing on site in temporary spoil heaps where directed										
25m	-	-	-	-	1.25	**13.39**	-	**13.39**	m3	**14.39**
50m	-	-	-	-	1.50	**16.07**	-	**16.07**	m3	**17.27**
100m	-	-	-	-	2.00	**21.42**	-	**21.42**	m3	**23.03**
extra for removing beyond 100m and not exceeding 400m	-	-	-	-	1.55	**16.60**	2.18	**18.78**	m3	**20.19**
extra for removing beyond 100m and not exceeding 800m	-	-	-	-	1.55	**16.60**	2.84	**19.44**	m3	**20.90**
depositing on site in permanent spoil heaps average 25m distant	-	-	-	-	1.25	**13.39**	-	**13.39**	m3	**14.39**
removing from site to tip										
a) Inert	-	-	-	-	1.35	**14.46**	16.00	**30.46**	m3	**32.74**
b) Active	-	-	-	-	1.35	**14.46**	40.70	**55.16**	m3	**59.30**
c) Contaminated (guide price - always seek a quote for specialist disposal cost	-	-	-	-	1.35	**14.46**	86.00	**100.46**	m3	**107.99**
Surface treatments										
Compacting										
bottoms of excavations	-	-	-	-	0.12	**1.29**	-	**1.29**	m2	**1.38**

This section continues on the next page

Labour hourly rates: (except Specialists)
Craft Operatives £14.21 Labourer £10.71
Rates are national average prices.
Refer to REGIONAL VARIATIONS for indicative levels of overall pricing in regions

	MATERIALS			LABOUR				RATES		
	Del to Site £	Waste %	Material Cost £	Craft Optve Hrs	Lab Hrs	Labour Cost £	Sunds £	Nett Rate £	Unit	Gross Rate (+7.5%) £
Q10: KERBS/EDGINGS/CHANNELS/PAVING ACCESSORIES Cont.										
Earthwork support										
Earthwork support										
maximum depth not exceeding 1.00m; distance between opposing faces not exceeding 2.00m.......	0.74	-	**0.74**	-	0.20	**2.14**	0.02	**2.90**	m2	**3.12**
Earthwork support; curved										
Maximum depth not exceeding 1.00m; distance between opposing faces not exceeding 2.00m.......	0.98	-	**0.98**	-	0.30	**3.21**	0.04	**4.24**	m2	**4.56**
Earthwork support; unstable ground										
maximum depth not exceeding 1.00m; distance between opposing faces not exceeding 2.00m.......	1.32	-	**1.32**	-	0.36	**3.86**	0.02	**5.20**	m2	**5.59**
Earthwork support; next to roadways										
maximum depth not exceeding 1.00m; distance between opposing faces not exceeding 2.00m.......	1.45	-	**1.45**	-	0.40	**4.28**	0.04	**5.78**	m2	**6.21**
Precast concrete; standard or stock pattern units; BS 7263 Part 1; bedding, jointing and pointing in cement mortar (1:3); on plain in-situ concrete foundation; BS 5328 ordinary prescribed mix ST4, 20mm aggregate										
Kerbs; Figs. 1, 2, 6 and 7; concrete foundation and haunching; formwork										
125 x 255mm kerb; 400 x 200mm foundation.........	10.54	5.00	**11.21**	-	1.09	**11.67**	-	**22.88**	m	**24.60**
125 x 255mm kerb; 400 x 200mm foundation; curved 10.00m radius	13.94	5.00	**14.78**	-	1.16	**12.42**	-	**27.20**	m	**29.24**
150 x 305mm kerb; 400 x 200mm foundation.........	13.44	5.00	**14.26**	-	1.16	**12.42**	-	**26.68**	m	**28.68**
150 x 305mm kerb; 400 x 200mm foundation; curved 10.00m radius	20.35	5.00	**21.51**	-	1.26	**13.49**	-	**35.01**	m	**37.63**
Edgings; Figs. 10, 11, 12 and 13; concrete foundation and haunching; formwork										
50 x 150mm edging; 300 x 150mm foundation........	6.61	5.00	**7.02**	-	0.78	**8.35**	-	**15.38**	m	**16.53**
50 x 200mm edging; 300 x 150mm foundation........	7.28	5.00	**7.72**	-	0.90	**9.64**	-	**17.36**	m	**18.66**
50 x 250mm edging; 300 x 150mm foundation........	7.56	5.00	**8.02**	-	1.02	**10.92**	-	**18.94**	m	**20.37**
Channels; Fig. 8; concrete foundation and haunching; formwork										
255 x 125mm channel; 450 x 150mm foundation......	11.80	5.00	**12.49**	-	0.86	**9.21**	-	**21.70**	m	**23.33**
255 x 125mm channel; 450 x 150mm foundation; curved 10.00 radius	14.12	5.00	**14.93**	-	0.93	**9.96**	-	**24.90**	m	**26.76**
Quadrants; Fig. 14; concrete foundation and haunching; formwork										
305 x 305 x 150mm quadrant; 500 x 500 x 150mm foundation	11.33	5.00	**11.96**	-	0.61	**6.53**	-	**18.49**	nr	**19.88**
305 x 305 x 255mm quadrant; 500 x 500 x 150mm foundation	11.63	5.00	**12.28**	-	0.69	**7.39**	-	**19.67**	nr	**21.14**
455 x 455 x 150mm quadrant; 650 x 650 x 150mm foundation	14.72	5.00	**15.55**	-	0.85	**9.10**	-	**24.66**	nr	**26.51**
455 x 455 x 255mm quadrant; 650 x 650 x 150mm foundation	15.22	5.00	**16.08**	-	0.94	**10.07**	-	**26.15**	nr	**28.11**
Granite; standard units; BS EN 1343; bedding jointing and pointing in cement mortar (1:3); on plain in-situ concrete foundation; BS 5328 ordinary prescribed mix ST4, 20mm aggregate										
Edge kerb; concrete foundation and haunching; formwork										
150 x 300mm; 300 x 200mm foundation.............	18.68	2.50	**19.42**	0.50	0.65	**14.07**	-	**33.48**	m	**36.00**
150 x 300mm; 300 x 200mm foundation; curved external radius exceeding 1000mm................	29.58	2.50	**30.63**	0.80	0.65	**18.33**	-	**48.96**	m	**52.63**
200 x 300mm; 350 x 200mm foundation.............	29.61	2.50	**30.65**	0.60	0.73	**16.34**	-	**46.99**	m	**50.52**
200 x 300mm; 350 x 200mm foundation; curved external radius exceeding 1000mm................	36.22	2.50	**37.48**	0.90	0.73	**20.61**	-	**58.09**	m	**62.44**
Flat kerb; concrete foundation and haunching; formwork										
300 x 150mm; 450 x 200mm foundation.............	31.95	2.50	**33.11**	0.50	0.73	**14.92**	-	**48.03**	m	**51.64**
300 x 150mm; 450 x 200mm foundation; curved external radius exceeding 1000mm................	41.01	2.50	**42.44**	0.80	0.73	**19.19**	-	**61.63**	m	**66.25**
300 x 200mm; 450 x 200mm foundation.............	40.31	2.50	**41.68**	0.60	0.75	**16.56**	-	**58.24**	m	**62.61**
300 x 200mm; 450 x 200mm foundation; curved external radius exceeding 1000mm................	51.35	2.50	**53.04**	0.90	0.75	**20.82**	-	**73.86**	m	**79.40**
Q20: GRANULAR SUB-BASES TO ROADS/PAVINGS										
Hard, dry, broken brick or stone to be obtained off site; wheeling not exceeding 25m - by machine										
Filling to make up levels; depositing in layers 150mm maximum thickness										
average thickness not exceeding 0.25m...........	17.00	25.00	**21.25**	-	-	**-**	3.59	**24.84**	m3	**26.71**
average thickness exceeding 0.25m...............	17.00	25.00	**21.25**	-	-	**-**	2.98	**24.23**	m3	**26.05**
Surface packing to filling										
to vertical or battered faces....................	-	-	**-**	-	-	**-**	0.61	**0.61**	m2	**0.66**
Compacting with 680 kg vibratory roller										
filling; blinding with sand, ashes or similar fine material	0.15	40.00	**0.21**	-	0.05	**0.54**	0.22	**0.97**	m2	**1.04**

Labour hourly rates: (except Specialists) Craft Operatives £14.21 Labourer £10.71 Rates are national average prices. Refer to REGIONAL VARIATIONS for indicative levels of overall pricing in regions	MATERIALS			LABOUR			RATES			
	Del to Site £	Waste %	Material Cost £	Craft Optve Hrs	Lab Hrs	Labour Cost £	Sunds £	Nett Rate £	Unit	Gross Rate (+7.5%) £
Q20: GRANULAR SUB-BASES TO ROADS/PAVINGS Cont.										
Hard, dry, broken brick or stone to be obtained off site; wheeling not exceeding 25m - by machine Cont.										
Compacting with 6 - 8 tonnes smooth wheeled roller filling; blinding with sand, ashes or similar fine material	0.15	40.00	**0.21**	-	0.05	**0.54**	0.43	**1.17**	m2	**1.26**
Hard, dry, broken brick or stone to be obtained off site; wheeling not exceeding 50m - by machine										
Filling to make up levels; depositing in layers 150mm maximum thickness										
average thickness not exceeding 0.25m	17.00	25.00	**21.25**	-	-	**-**	4.18	**25.43**	m3	**27.34**
average thickness exceeding 0.25m	17.00	25.00	**21.25**	-	-	**-**	3.59	**24.84**	m3	**26.71**
Surface packing to filling										
to vertical or battered faces	-	-	**-**	-	-	**-**	0.61	**0.61**	m2	**0.66**
Compacting with 680 kg vibratory roller filling; blinding with sand, ashes or similar fine material	0.15	40.00	**0.21**	-	0.05	**0.54**	0.22	**0.97**	m2	**1.04**
Compacting with 6 - 8 tonnes smooth wheeled roller filling; blinding with sand, ashes or similar fine material	0.15	40.00	**0.21**	-	0.05	**0.54**	0.39	**1.14**	m2	**1.22**
MOT Type 1 to be obtained off site; wheeling not exceeding 25m - by machine										
Filling to make up levels; depositing in layers 150mm maximum thickness										
average thickness not exceeding 0.25m	19.95	25.00	**24.94**	-	-	**-**	4.18	**29.12**	m3	**31.30**
average thickness exceeding 0.25m	19.95	25.00	**24.94**	-	-	**-**	3.47	**28.41**	m3	**30.54**
Surface packing to filling										
to vertical or battered faces	-	-	**-**	-	-	**-**	0.61	**0.61**	m2	**0.66**
Compacting with 680 kg vibratory roller filling; blinding with sand, ashes or similar fine material	0.15	40.00	**0.21**	-	0.05	**0.54**	0.22	**0.97**	m2	**1.04**
Compacting with 6 - 8 tonnes smooth wheeled roller filling; blinding with sand, ashes or similar fine material	0.15	40.00	**0.21**	-	0.05	**0.54**	0.43	**1.17**	m2	**1.26**
MOT Type 1 to be obtained off site; wheeling not exceeding 50m - by machine										
Filling to make up levels; depositing in layers 150mm maximum thickness										
average thickness not exceeding 0.25m	19.95	25.00	**24.94**	-	-	**-**	4.18	**29.12**	m3	**31.30**
average thickness exceeding 0.25m	19.95	25.00	**24.94**	-	-	**-**	3.59	**28.53**	m3	**30.67**
Surface packing to filling										
to vertical or battered faces	-	-	**-**	-	-	**-**	0.61	**0.61**	m2	**0.66**
Compacting with 680 kg vibratory roller filling; blinding with sand, ashes or similar fine material	0.15	40.00	**0.21**	-	0.05	**0.54**	0.22	**0.97**	m2	**1.04**
Compacting with 6 - 8 tonnes smooth wheeled roller filling; blinding with sand, ashes or similar fine material	0.15	40.00	**0.21**	-	0.05	**0.54**	0.43	**1.17**	m2	**1.26**
MOT Type 2 to be obtained off site; wheeling not exceeding 25m - by machine										
Filling to make up levels; depositing in layers 150mm maximum thickness										
average thickness not exceeding 0.25m	18.48	25.00	**23.10**	-	-	**-**	3.59	**26.69**	m3	**28.69**
average thickness exceeding 0.25m	18.48	25.00	**23.10**	-	-	**-**	2.98	**26.08**	m3	**28.04**
Surface packing to filling										
to vertical or battered faces	-	-	**-**	-	-	**-**	0.61	**0.61**	m2	**0.66**
Compacting with 680 kg vibratory roller filling; blinding with sand, ashes or similar fine material	0.15	40.00	**0.21**	-	0.05	**0.54**	0.22	**0.97**	m2	**1.04**
Compacting with 6 - 8 tonnes smooth wheeled roller filling; blinding with sand, ashes or similar fine material	0.15	40.00	**0.21**	-	0.05	**0.54**	0.43	**1.17**	m2	**1.26**
MOT Type 2 to be obtained off site; wheeling not exceeding 50m - by machine										
Filling to make up levels; depositing in layers 150mm maximum thickness										
average thickness not exceeding 0.25m	18.48	25.00	**23.10**	-	-	**-**	4.18	**27.28**	m3	**29.33**
average thickness exceeding 0.25m	18.48	25.00	**23.10**	-	-	**-**	3.59	**26.69**	m3	**28.69**
Surface packing to filling										
to vertical or battered faces	-	-	**-**	-	-	**-**	0.61	**0.61**	m2	**0.66**

Labour hourly rates: (except Specialists) Craft Operatives £14.21 Labourer £10.71 Rates are national average prices. Refer to REGIONAL VARIATIONS for indicative levels of overall pricing in regions	MATERIALS			LABOUR				RATES		
	Del to Site £	Waste %	Material Cost £	Craft Optve Hrs	Lab Hrs	Labour Cost £	Sunds £	Nett Rate £	Unit	Gross Rate (+7.5%) £
Q20: GRANULAR SUB-BASES TO ROADS/PAVINGS Cont.										
MOT Type 2 to be obtained off site; wheeling not exceeding 50m - by machine Cont.										
Compacting with 680 kg vibratory roller filling; blinding with sand, ashes or similar fine material	0.15	40.00	**0.21**	-	0.05	**0.54**	0.22	**0.97**	m2	**1.04**
Compacting with 6 - 8 tonnes smooth wheeled roller filling; blinding with sand, ashes or similar fine material	0.15	40.00	**0.21**	-	0.05	**0.54**	0.43	**1.17**	m2	**1.26**
Ashes to be obtained off site; wheeling not exceeding 25m - by machine										
Filling to make up levels; depositing in layers 150mm maximum thickness										
average thickness not exceeding 0.25m	9.99	40.00	**13.98**	-	-	**-**	3.80	**17.78**	m3	**19.11**
Compacting with 680 kg vibratory roller filling	-	-	**-**	-	-	**-**	0.22	**0.22**	m2	**0.24**
Compacting with 6 - 8 tonnes smooth wheeled roller filling	-	-	**-**	-	-	**-**	0.43	**0.43**	m2	**0.46**
Ashes to be obtained off site; wheeling not exceeding 50m - by machine										
Filling to make up levels; depositing in layers 150mm maximum thickness										
average thickness not exceeding 0.25m	9.99	40.00	**13.98**	-	-	**-**	5.14	**19.12**	m3	**20.55**
Compacting with 680 kg vibratory roller filling	-	-	**-**	-	-	**-**	0.22	**0.22**	m2	**0.24**
Compacting with 6 - 8 tonnes smooth wheeled roller filling	-	-	**-**	-	-	**-**	0.43	**0.43**	m2	**0.46**
Hoggin to be obtained off site; wheeling not exceeding 25m - by machine										
Filling to make up levels; depositing in layers 150mm maximum thickness										
average thickness not exceeding 0.25m	16.00	33.00	**21.28**	-	-	**-**	4.17	**25.45**	m3	**27.36**
average thickness exceeding 0.25m	16.00	33.00	**21.28**	-	-	**-**	3.47	**24.75**	m3	**26.61**
Compacting with 680 kg vibratory roller filling	-	-	**-**	-	-	**-**	0.22	**0.22**	m2	**0.24**
Compacting with 6 - 8 tonnes smooth wheeled roller filling	-	-	**-**	-	-	**-**	0.43	**0.43**	m2	**0.46**
Hoggin to be obtained off site; wheeling not exceeding 50m - by machine										
Filling to make up levels; depositing in layers 150mm maximum thickness										
average thickness not exceeding 0.25m	16.00	33.00	**21.28**	-	-	**-**	3.59	**24.87**	m3	**26.74**
average thickness exceeding 0.25m	16.00	33.00	**21.28**	-	-	**-**	2.98	**24.26**	m3	**26.08**
Compacting with 680 kg vibratory roller filling	-	-	**-**	-	-	**-**	0.22	**0.22**	m2	**0.24**
Compacting with 6 - 8 tonnes smooth wheeled roller filling	-	-	**-**	-	-	**-**	0.43	**0.43**	m2	**0.46**
Hard, dry, broken brick or stone to be obtained off site; wheeling not exceeding 25m - by hand										
Filling to make up levels; depositing in layers 150mm maximum thickness										
average thickness not exceeding 0.25m	17.00	25.00	**21.25**	-	1.50	**16.07**	-	**37.32**	m3	**40.11**
average thickness exceeding 0.25m	17.00	25.00	**21.25**	-	1.25	**13.39**	-	**34.64**	m3	**37.24**
Compacting with 680 kg vibratory roller filling; blinding with sand, ashes or similar fine material	0.15	40.00	**0.21**	-	0.05	**0.54**	0.22	**0.97**	m2	**1.04**
Compacting with 6 - 8 tonnes smooth wheeled roller filling; blinding with sand, ashes or similar fine material	0.15	40.00	**0.21**	-	0.05	**0.54**	0.43	**1.17**	m2	**1.26**
Hard, dry, broken brick or stone to be obtained off site; wheeling not exceeding 50m - by hand										
Filling to make up levels; depositing in layers 150mm maximum thickness										
average thickness not exceeding 0.25m	17.00	25.00	**21.25**	-	1.75	**18.74**	-	**39.99**	m3	**42.99**
average thickness exceeding 0.25m	17.00	25.00	**21.25**	-	1.50	**16.07**	-	**37.32**	m3	**40.11**
Compacting with 680 kg vibratory roller filling; blinding with sand, ashes or similar fine material	0.15	40.00	**0.21**	-	0.05	**0.54**	0.22	**0.97**	m2	**1.04**

Labour hourly rates: (except Specialists) Craft Operatives £14.21 Labourer £10.71 Rates are national average prices. Refer to REGIONAL VARIATIONS for indicative levels of overall pricing in regions	MATERIALS			LABOUR				RATES		
	Del to Site £	Waste %	Material Cost £	Craft Optve Hrs	Lab Hrs	Labour Cost £	Sunds £	Nett Rate £	Unit	Gross Rate (+7.5%) £
Q20: GRANULAR SUB-BASES TO ROADS/PAVINGS Cont.										
Hard, dry, broken brick or stone to be obtained off site; wheeling not exceeding 50m - by hand Cont.										
Compacting with 6 - 8 tonnes smooth wheeled roller filling; blinding with sand, ashes or similar fine material	0.15	40.00	**0.21**	-	0.05	**0.54**	0.43	**1.17**	m2	**1.26**
MOT Type 1 to be obtained off site; wheeling not exceeding 25m - by hand										
Filling to make up levels; depositing in layers 150mm maximum thickness										
average thickness not exceeding 0.25m	19.95	25.00	**24.94**	-	1.50	**16.07**	-	**41.00**	m3	**44.08**
average thickness exceeding 0.25m	19.95	25.00	**24.94**	-	1.25	**13.39**	-	**38.33**	m3	**41.20**
Compacting with 680 kg vibratory roller filling; blinding with sand, ashes or similar fine material	0.15	40.00	**0.21**	-	0.05	**0.54**	0.22	**0.97**	m2	**1.04**
Compacting with 6 - 8 tonnes smooth wheeled roller filling; blinding with sand, ashes or similar fine material	0.15	40.00	**0.21**	-	0.05	**0.54**	0.43	**1.17**	m2	**1.26**
MOT Type 1 to be obtained off site; wheeling not exceeding 50m - by hand										
Filling to make up levels; depositing in layers 150mm maximum thickness										
average thickness not exceeding 0.25m	19.95	25.00	**24.94**	-	1.75	**18.74**	-	**43.68**	m3	**46.96**
average thickness exceeding 0.25m	19.95	25.00	**24.94**	-	1.50	**16.07**	-	**41.00**	m3	**44.08**
Compacting with 680 kg vibratory roller filling; blinding with sand, ashes or similar fine material	0.15	40.00	**0.21**	-	0.05	**0.54**	0.22	**0.97**	m2	**1.04**
Compacting with 6 - 8 tonnes smooth wheeled roller filling; blinding with sand, ashes or similar fine material	0.15	40.00	**0.21**	-	0.05	**0.54**	0.43	**1.17**	m2	**1.26**
MOT Type 2 to be obtained off site; wheeling not exceeding 25m - by hand										
Filling to make up levels; depositing in layers 150mm maximum thickness										
average thickness not exceeding 0.25m	18.48	25.00	**23.10**	-	1.50	**16.07**	-	**39.17**	m3	**42.10**
average thickness exceeding 0.25m	18.48	25.00	**23.10**	-	1.25	**13.39**	-	**36.49**	m3	**39.22**
Compacting with 680 kg vibratory roller filling; blinding with sand, ashes or similar fine material	0.15	40.00	**0.21**	-	0.05	**0.54**	0.22	**0.97**	m2	**1.04**
Compacting with 6 - 8 tonnes smooth wheeled roller filling; blinding with sand, ashes or similar fine material	0.15	40.00	**0.21**	-	0.05	**0.54**	0.43	**1.17**	m2	**1.26**
MOT Type 2 to be obtained off site; wheeling not exceeding 50m - by hand										
Filling to make up levels; depositing in layers 150mm maximum thickness										
average thickness not exceeding 0.25m	18.48	25.00	**23.10**	-	1.75	**18.74**	-	**41.84**	m3	**44.98**
average thickness exceeding 0.25m	18.48	25.00	**23.10**	-	1.50	**16.07**	-	**39.17**	m3	**42.10**
Compacting with 680 kg vibratory roller filling; blinding with sand, ashes or similar fine material	0.15	40.00	**0.21**	-	0.05	**0.54**	0.22	**0.97**	m2	**1.04**
Compacting with 6 - 8 tonnes smooth wheeled roller filling; blinding with sand, ashes or similar fine material	0.15	40.00	**0.21**	-	0.05	**0.54**	0.43	**1.17**	m2	**1.26**
Ashes to be obtained off site; wheeling not exceeding 25m - by hand										
Filling to make up levels; depositing in layers 150mm maximum thickness										
average thickness not exceeding 0.25m	9.99	40.00	**13.98**	-	1.60	**17.14**	-	**31.12**	m3	**33.45**
Compacting with 680 kg vibratory roller filling	-	-	**-**	-	-	**-**	0.22	**0.22**	m2	**0.24**
Compacting with 6 - 8 tonnes smooth wheeled roller filling	-	-	**-**	-	-	**-**	0.43	**0.43**	m2	**0.46**
Ashes to be obtained off site; wheeling not exceeding 50m - by hand										
Filling to make up levels; depositing in layers 150mm maximum thickness										
average thickness not exceeding 0.25m	9.99	40.00	**13.98**	-	1.85	**19.81**	-	**33.79**	m3	**36.33**
Compacting with 680 kg vibratory roller filling	-	-	**-**	-	-	**-**	0.22	**0.22**	m2	**0.24**

Labour hourly rates: (except Specialists) Craft Operatives £14.21 Labourer £10.71 Rates are national average prices. Refer to REGIONAL VARIATIONS for indicative levels of overall pricing in regions	MATERIALS			LABOUR				RATES		
	Del to Site £	Waste %	Material Cost £	Craft Optve Hrs	Lab Hrs	Labour Cost £	Sunds £	Nett Rate £	Unit	Gross Rate (+7.5%) £
Q20: GRANULAR SUB-BASES TO ROADS/PAVINGS Cont.										
Ashes to be obtained off site; wheeling not exceeding 50m - by hand Cont.										
Compacting with 6 - 8 tonnes smooth wheeled roller filling	-	-	-	-	-	-	0.43	**0.43**	m2	**0.46**
Hoggin to be obtained off site; wheeling not exceeding 25m - by hand										
Filling to make up levels; depositing in layers 150mm maximum thickness										
average thickness not exceeding 0.25m	16.00	33.00	**21.28**	-	1.50	**16.07**	-	**37.35**	m3	**40.15**
average thickness exceeding 0.25m	16.00	33.00	**21.28**	-	1.25	**13.39**	-	**34.67**	m3	**37.27**
Compacting with 680 kg vibratory roller filling	-	-	-	-	-	-	0.22	**0.22**	m2	**0.24**
Compacting with 6 - 8 tonnes smooth wheeled roller filling	-	-	-	-	-	-	0.43	**0.43**	m2	**0.46**
Hoggin to be obtained off site; wheeling not exceeding 50m - by hand										
Filling to make up levels; depositing in layers 150mm maximum thickness										
average thickness not exceeding 0.25m	16.00	33.00	**21.28**	-	1.75	**18.74**	-	**40.02**	m3	**43.02**
average thickness exceeding 0.25m	16.00	33.00	**21.28**	-	1.50	**16.07**	-	**37.35**	m3	**40.15**
Compacting with 680 kg vibratory roller filling	-	-	-	-	-	-	0.22	**0.22**	m2	**0.24**
Compacting with 6 - 8 tonnes smooth wheeled roller filling	-	-	-	-	-	-	0.43	**0.43**	m2	**0.46**
Q21: IN-SITU CONCRETE ROADS/PAVINGS										
Plain in-situ concrete; mix 1:8, all in aggregate										
Beds; poured on or against earth or unblinded hardcore thickness not exceeding 150mm	69.50	7.50	**74.71**	-	4.25	**45.52**	1.50	**121.73**	m3	**130.86**
Plain in-situ concrete; mix 1:6; all in aggregate										
Beds; poured on or against earth or unblinded hardcore										
thickness 150 - 450mm	77.37	7.50	**83.17**	-	4.00	**42.84**	1.50	**127.51**	m3	**137.07**
thickness not exceeding 150mm	77.37	7.50	**83.17**	-	4.25	**45.52**	1.50	**130.19**	m3	**139.95**
Plain in-situ concrete; BS 5328, ordinary prescribed mix ST4, 20mm aggregate										
Beds thickness not exceeding 150mm	60.03	5.00	**63.03**	-	4.25	**45.52**	1.50	**110.04**	m3	**118.30**
Beds; poured on or against earth or unblinded hardcore thickness not exceeding 150mm	60.03	5.00	**63.03**	-	4.25	**45.52**	1.50	**110.04**	m3	**118.30**
Reinforced in-situ concrete; BS 5328, designed mix C20, 20mm aggregate, minimum cement content 240 kg/m3; vibrated										
Beds										
thickness 150 - 450mm	61.25	5.00	**64.31**	-	4.25	**45.52**	0.50	**110.33**	m3	**118.60**
thickness not exceeding 150mm	61.25	5.00	**64.31**	-	4.50	**48.20**	0.50	**113.01**	m3	**121.48**
Formwork and basic finish										
Edges of beds										
height not exceeding 250mm	1.03	12.50	**1.15**	0.53	0.11	**8.71**	0.22	**10.08**	m	**10.84**
height 250 - 500mm	1.81	12.50	**2.04**	0.96	0.19	**15.68**	0.41	**18.12**	m	**19.48**
Formwork steel forms										
Edges of beds										
height not exceeding 250mm	0.65	15.00	**0.75**	0.36	0.07	**5.87**	0.22	**6.84**	m	**7.35**
height 250 - 500mm	1.07	15.00	**1.23**	0.64	0.13	**10.49**	0.41	**12.13**	m	**13.04**
Reinforcement fabric; BS 4483, hard drawn plain round steel; welded; including laps, tying wire, all cutting and bending, and spacers and chairs which are at the discretion of the Contractor										
Reference C283, 2.61 kg/m2; 100mm side laps; 400mm end laps generally	1.27	15.00	**1.46**	0.05	0.02	**0.92**	0.07	**2.45**	m2	**2.64**
Reference C385, 3.41 kg/m2; 100mm side laps; 400mm end laps generally	1.65	15.00	**1.90**	0.05	0.02	**0.92**	0.09	**2.91**	m2	**3.13**
Reference C503, 4.34 kg/m2; 100mm side laps; 400mm end laps generally	2.04	15.00	**2.34**	0.06	0.02	**1.07**	0.11	**3.52**	m2	**3.78**

Labour hourly rates: (except Specialists) Craft Operatives £14.21 Labourer £10.71 Rates are national average prices. Refer to REGIONAL VARIATIONS for indicative levels of overall pricing in regions	MATERIALS			LABOUR				RATES		
	Del to Site £	Waste %	Material Cost £	Craft Optve Hrs	Lab Hrs	Labour Cost £	Sunds £	Nett Rate £	Unit	Gross Rate (+7.5%) £
Q21: IN-SITU CONCRETE ROADS/PAVINGS Cont.										
Reinforcement fabric; BS 4483, hard drawn plain round steel; welded; including laps, tying wire, all cutting and bending, and spacers and chairs which are at the discretion of the Contractor Cont.										
Reference C636, 5.55 kg/m2; 100mm side laps; 400mm end laps										
generally	2.72	15.00	**3.13**	0.06	0.02	**1.07**	0.14	**4.33**	m2	**4.66**
Reference C785, 6.72 kg/m2; 100mm side laps; 400mm end laps										
generally	3.28	15.00	**3.77**	0.07	0.02	**1.21**	0.17	**5.15**	m2	**5.54**
Formed joints										
Sealant to joint; Grace Construction Products. ' Servimastic 96' including preparation, cleaners, primers and sealers										
10 x 25mm	2.45	5.00	**2.57**	-	0.10	**1.07**	-	**3.64**	m	**3.91**
12.5 x 25mm	3.18	5.00	**3.34**	-	0.11	**1.18**	-	**4.52**	m	**4.86**
Worked finish on in-situ concrete										
Tamping by mechanical means										
level surfaces	-	-	**-**	-	0.15	**1.61**	0.27	**1.88**	m2	**2.02**
sloping surfaces	-	-	**-**	-	0.30	**3.21**	0.27	**3.48**	m2	**3.74**
surfaces to falls	-	-	**-**	-	0.40	**4.28**	0.34	**4.62**	m2	**4.97**
Trowelling										
level surfaces	-	-	**-**	-	0.33	**3.53**	-	**3.53**	m2	**3.80**
sloping surfaces	-	-	**-**	-	0.40	**4.28**	-	**4.28**	m2	**4.61**
surfaces to falls	-	-	**-**	-	0.50	**5.36**	-	**5.36**	m2	**5.76**
Rolling with an indenting roller										
level surfaces	-	-	**-**	-	0.10	**1.07**	0.27	**1.34**	m2	**1.44**
sloping surfaces	-	-	**-**	-	0.15	**1.61**	0.27	**1.88**	m2	**2.02**
surfaces to falls	-	-	**-**	-	0.30	**3.21**	0.34	**3.55**	m2	**3.82**
Accessories cast into in-situ concrete										
Mild steel dowels; half coated with bitumen										
16mm diameter x 400mm long	0.31	10.00	**0.34**	-	0.06	**0.64**	0.18	**1.16**	nr	**1.25**
20mm diameter x 500mm long; with plastic compression cap	0.60	10.00	**0.66**	-	0.08	**0.86**	0.18	**1.70**	nr	**1.83**
Q22: COATED MACADAM/ASPHALT ROADS/PAVINGS										
Coated macadam, BS 4987; base course 28mm nominal size aggregate, 50mm thick; wearing course 6mm nominal size aggregate, 20mm thick; rolled with 3 - 4 tonne roller; external										
70mm roads on concrete base; to falls and crossfalls, and slopes not exceeding 15 degrees from horizontal										
generally	-	-	**Spclist**	-	-	**Spclist**	-	**12.62**	m2	**13.57**
70mm pavings on concrete base; to falls and crossfalls, and slopes not exceeding 15 degrees from horizontal										
generally	-	-	**Spclist**	-	-	**Spclist**	-	**14.02**	m2	**15.07**
Coated macadam, BS 4987; base course 28mm nominal size aggregate, 50mm thick; wearing course 6mm nominal size aggregate, 20mm thick; rolled with 3 - 4 tonne roller; dressing surface with coated grit brushed on and lightly rolled; external										
70mm roads on concrete base; to falls and crossfalls, and slopes not exceeding 15 degrees from horizontal										
generally	-	-	**Spclist**	-	-	**Spclist**	-	**15.29**	m2	**16.44**
70mm pavings on concrete base; to falls and crossfalls, and slopes not exceeding 15 degrees from horizontal										
generally	-	-	**Spclist**	-	-	**Spclist**	-	**16.67**	m2	**17.92**
Coated macadam, BS 4987; base course 40mm nominal size aggregate, 65mm thick; wearing course 10mm nominal size aggregate, 25mm thick; rolled with 6 - 8 tonne roller; external										
90mm roads on concrete base; to falls and crossfalls, and slopes not exceeding 15 degrees from horizontal										
generally	-	-	**Spclist**	-	-	**Spclist**	-	**15.29**	m2	**16.44**
90mm pavings on concrete base; to falls and crossfalls, and slopes not exceeding 15 degrees from horizontal										
generally	-	-	**Spclist**	-	-	**Spclist**	-	**16.78**	m2	**18.04**

Labour hourly rates: (except Specialists) Craft Operatives £14.21 Labourer £10.71 Rates are national average prices. Refer to REGIONAL VARIATIONS for indicative levels of overall pricing in regions	MATERIALS			LABOUR				RATES		
	Del to Site £	Waste %	Material Cost £	Craft Optve Hrs	Lab Hrs	Labour Cost £	Sunds £	Nett Rate £	Unit	Gross Rate (+7.5%) £
Q22: COATED MACADAM/ASPHALT ROADS/PAVINGS Cont.										
Coated macadam, BS 4987; base course 40mm nominal size aggregate, 65mm thick; wearing course 10mm nominal size aggregate, 25mm thick; rolled with 6 - 8 tonne roller; dressing surface with coated grit brushed on and lightly rolled; external										
90mm roads on concrete base; to falls and crossfalls, and slopes not exceeding 15 degrees from horizontal generally	-	-	**Spclist**	-	-	**Spclist**	-	**18.22**	m2	**19.59**
90mm pavings on concrete base; to falls and crossfalls, and slopes not exceeding 15 degrees from horizontal generally	-	-	**Spclist**	-	-	**Spclist**	-	**18.99**	m2	**20.41**
Fine cold asphalt, BS 4987; single course 6mm nominal size aggregate; rolled with 3 - 4 tonne roller; external										
12mm roads on concrete base; to falls and crossfalls, and slopes not exceeding 15 degrees from horizontal generally	-	-	**Spclist**	-	-	**Spclist**	-	**5.75**	m2	**6.18**
19mm roads on concrete base; to falls and crossfalls, and slopes not exceeding 15 degrees from horizontal generally	-	-	**Spclist**	-	-	**Spclist**	-	**6.71**	m2	**7.21**
25mm roads on concrete base; to falls and crossfalls, and slopes not exceeding 15 degrees from horizontal generally	-	-	**Spclist**	-	-	**Spclist**	-	**7.90**	m2	**8.49**
12mm pavings on concrete base; to falls and crossfalls, and slopes not exceeding 15 degrees from horizontal generally	-	-	**Spclist**	-	-	**Spclist**	-	**6.10**	m2	**6.56**
19mm pavings on concrete base; to falls and crossfalls, and slopes not exceeding 15 degrees from horizontal generally	-	-	**Spclist**	-	-	**Spclist**	-	**7.66**	m2	**8.23**
25mm pavings on concrete base; to falls and crossfalls, and slopes not exceeding 15 degrees from horizontal generally	-	-	**Spclist**	-	-	**Spclist**	-	**8.73**	m2	**9.38**
Q22: GRAVEL/HOGGIN/WOODCHIP ROADS/PAVINGS										
Gravel paving; first layer clinker 75mm thick; intermediate layer coarse gravel 75mm thick; wearing layer blinding gravel 38mm thick; well water and roll										
188mm roads on blinded hardcore base; to falls and crossfalls and slopes not exceeding 15 degrees from horizontal generally	5.52	10.00	**6.07**	-	0.44	**4.71**	-	**11.19**	m2	**12.03**
Gravel paving; single layer blinding gravel; well water and roll										
50mm pavings on blinded hardcore base; to falls and crossfalls and slopes not exceeding 15 degrees from horizontal generally	2.06	10.00	**2.27**	-	0.13	**1.39**	-	**3.76**	m2	**4.05**
Gravel paving; first layer clinker 50mm thick; wearing layer blinding gravel 38mm thick; well water and roll										
88mm pavings on blinded hardcore base; to falls and crossfalls and slopes not exceeding 15 degrees from horizontal generally	2.64	10.00	**2.90**	-	0.24	**2.57**	-	**5.58**	m2	**5.99**
extra; 'Colas' and 10mm shingle dressing; one coat	1.17	5.00	**1.23**	-	0.15	**1.61**	-	**2.84**	m2	**3.05**
extra; 'Colas' and 10mm shingle dressing; two coats	1.99	5.00	**2.09**	-	0.25	**2.68**	-	**4.77**	m2	**5.12**
Q25: SLAB/BRICK/BLOCK/SETT/COBBLE PAVINGS										
York stone paving; 13mm thick bedding, jointing and pointing in cement mortar (1:3)										
Paving on blinded hardcore base										
50mm thick; to falls and crossfalls and to slopes not exceeding 15 degrees from horizontal	65.63	10.00	**68.90**	0.27	0.27	**6.73**	1.02	**76.65**	m2	**82.40**
75mm thick; to falls and crossfalls and to slopes not exceeding 15 degrees from horizontal	74.44	10.00	**78.16**	0.33	0.33	**8.22**	1.08	**87.47**	m2	**94.03**
extra; rubbed top surface	15.96	5.00	**16.76**	-	-	**-**	-	**16.76**	m2	**18.01**
Treads on concrete base										
50mm thick; 300mm wide	63.05	5.00	**66.20**	0.33	0.33	**8.22**	0.27	**74.70**	m	**80.30**

Labour hourly rates: (except Specialists) Craft Operatives £14.21 Labourer £10.71 Rates are national average prices. Refer to REGIONAL VARIATIONS for indicative levels of overall pricing in regions	MATERIALS			LABOUR				RATES		
	Del to Site £	Waste %	Material Cost £	Craft Optve Hrs	Lab Hrs	Labour Cost £	Sunds £	Nett Rate £	Unit	Gross Rate (+7.5%) £
Q25: SLAB/BRICK/BLOCK/SETT/COBBLE PAVINGS Cont.										
Precast concrete flags, BS 7263 Part 1, natural finish; 6mm joints, symmetrical layout; bedding in 13mm cement mortar (1:3); jointing and pointing with lime and sand (1:2)										
600 x 450 x 50mm units to pavings on sand, granular or blinded hardcore base										
50mm thick; to falls and crossfalls and to slopes not exceeding 15 degrees from horizontal.........	9.67	5.00	**10.10**	0.27	0.27	**6.73**	1.08	**17.91**	m2	**19.25**
600 x 600 x 50mm units to pavings on sand, granular or blinded hardcore base										
50mm thick; to falls and crossfalls and to slopes not exceeding 15 degrees from horizontal.........	8.09	5.00	**8.45**	0.25	0.25	**6.23**	1.01	**15.69**	m2	**16.86**
600 x 750 x 50mm units to pavings on sand, granular or blinded hardcore base										
50mm thick; to falls and crossfalls and to slopes not exceeding 15 degrees from horizontal.........	8.42	5.00	**8.77**	0.23	0.46	**8.19**	1.03	**18.00**	m2	**19.35**
600 x 900 x 50mm units to pavings on sand, granular or blinded hardcore base										
50mm thick; to falls and crossfalls and to slopes not exceeding 15 degrees from horizontal.........	7.58	5.00	**7.89**	0.21	0.42	**7.48**	1.04	**16.41**	m2	**17.64**
600 x 450 x 63mm units to pavings on sand, granular or blinded hardcore base										
63mm thick; to falls and crossfalls and to slopes not exceeding 15 degrees from horizontal.........	12.09	5.00	**12.63**	0.27	0.27	**6.73**	1.06	**20.42**	m2	**21.95**
600 x 600 x 63mm units to pavings on sand, granular or blinded hardcore base										
63mm thick; to falls and crossfalls and to slopes not exceeding 15 degrees from horizontal.........	10.92	5.00	**11.40**	0.25	0.25	**6.23**	1.00	**18.63**	m2	**20.03**
600 x 750 x 63mm units to pavings on sand , granular or blinded hardcore base										
63mm thick; to falls and crossfalls and to slopes not exceeding 15 degrees from horizontal.........	9.44	5.00	**9.85**	0.23	0.46	**8.19**	1.01	**19.05**	m2	**20.48**
600 x 900 x 63mm units to pavings on sand, granular or blinded hardcore base										
63mm thick; to falls and crossfalls and to slopes not exceeding 15 degrees from horizontal.........	10.43	5.00	**10.88**	0.21	0.42	**7.48**	1.02	**19.38**	m2	**20.84**
Precast concrete flags, BS 7263 Part 1, natural finish; 6mm joints, symmetrical layout; bedding in 13mm lime mortar (1:3); jointing and pointing with lime and sand (1:2)										
600 x 450 x 50mm units to pavings on sand, granular or blinded hardcore base										
50mm thick; to falls and crossfalls and to slopes not exceeding 15 degrees from horizontal.........	10.07	5.00	**10.50**	0.27	0.27	**6.73**	0.67	**17.90**	m2	**19.24**
600 x 600 x 50mm units to pavings on sand, granular or blinded hardcore base										
50mm thick; to falls and crossfalls and to slopes not exceeding 15 degrees from horizontal.........	8.50	5.00	**8.85**	0.25	0.25	**6.23**	0.60	**15.68**	m2	**16.85**
600 x 750 x 50mm units to pavings on sand, granular or blinded hardcore base										
50mm thick; to falls and crossfalls and to slopes not exceeding 15 degrees from horizontal.........	8.55	5.00	**8.90**	0.23	0.46	**8.19**	0.61	**17.71**	m2	**19.03**
600 x 900 x 50mm units to pavings on sand, granular or blinded hardcore base										
50mm thick; to falls and crossfalls and to slopes not exceeding 15 degrees from horizontal.........	7.71	5.00	**8.02**	0.21	0.42	**7.48**	0.59	**16.09**	m2	**17.30**
600 x 450 x 63mm units to pavings on sand, granular or blinded hardcore base										
63mm thick; to falls and crossfalls and to slopes not exceeding 15 degrees from horizontal.........	12.22	5.00	**12.76**	0.27	0.27	**6.73**	1.06	**20.54**	m2	**22.09**
600 x 600 x 63mm units to pavings on sand, granular or blinded hardcore base										
63mm thick; to falls and crossfalls and to slopes not exceeding 15 degrees from horizontal.........	11.05	5.00	**11.53**	0.25	0.25	**6.23**	1.00	**18.76**	m2	**20.17**
600 x 750 x 63mm units to pavings on sand, granular or blinded hardcore base										
63mm thick; to falls and crossfalls and to slopes not exceeding 15 degrees from horizontal.........	9.57	5.00	**9.97**	0.23	0.46	**8.19**	1.01	**19.18**	m2	**20.62**
600 x 900 x 63mm units to pavings on sand , granular or blinded hardcore base										
63mm thick; to falls and crossfalls and to slopes not exceeding 15 degrees from horizontal.........	10.55	5.00	**11.01**	0.21	0.42	**7.48**	1.02	**19.51**	m2	**20.97**

Labour hourly rates: (except Specialists) Craft Operatives £14.21 Labourer £10.71 Rates are national average prices. Refer to REGIONAL VARIATIONS for indicative levels of overall pricing in regions	MATERIALS			LABOUR				RATES		
	Del to Site £	Waste %	Material Cost £	Craft Optve Hrs	Lab Hrs	Labour Cost £	Sunds £	Nett Rate £	Unit	Gross Rate (+7.5%) £
Q25: SLAB/BRICK/BLOCK/SETT/COBBLE PAVINGS Cont.										
Precast concrete flags, BS 7263 Part 1, natural finish; 6mm joints, symmetrical layout; bedding in 25mm sand; jointing and pointing with lime and sand (1:2)										
600 x 450 x 50mm units to pavings on sand, granular or blinded hardcore base										
50mm thick; to falls and crossfalls and to slopes not exceeding 15 degrees from horizontal.........	8.82	5.00	**9.25**	0.27	0.27	**6.73**	0.63	**16.60**	m2	**17.85**
600 x 600 x 50mm units to pavings on sand, granular or blinded hardcore base										
50mm thick; to falls and crossfalls and to slopes not exceeding 15 degrees from horizontal.........	7.42	5.00	**7.77**	0.25	0.25	**6.23**	0.54	**14.54**	m2	**15.64**
600 x 750 x 50mm units to pavings on sand, granular or blinded hardcore base										
50mm thick; to falls and crossfalls and to slopes not exceeding 15 degrees from horizontal.........	7.29	5.00	**7.65**	0.23	0.46	**8.19**	0.55	**16.39**	m2	**17.62**
600 x 900 x 50mm units to pavings on sand, granular or blinded hardcore base										
50mm thick; to falls and crossfalls and to slopes not exceeding 15 degrees from horizontal.........	6.45	5.00	**6.76**	0.21	0.42	**7.48**	0.56	**14.80**	m2	**15.91**
Keyblok precast concrete paving blocks; Marshalls Mono Ltd Driveline 50; in standard units; bedding on sand; covering with sand, compacting with plate vibrator, sweeping off surplus										
200 x 100mm units to pavings on 50mm sand base; natural colour										
symmetrical half bond layout; level and to falls only										
50mm thick..................................	8.03	10.00	**8.83**	-	1.05	**11.25**	1.40	**21.48**	m2	**23.09**
laid in straight herringbone pattern; level and to falls only										
50mm thick..................................	8.03	10.00	**8.83**	-	1.15	**12.32**	1.40	**22.55**	m2	**24.24**
Brick paving on concrete base; 13mm thick bedding, jointing and pointing in cement mortar (1:3)										
Paving to falls and crossfalls and to slopes not exceeding 15 degrees from horizontal										
25mm thick Brick paviors P.C. £300.00 per 1000...	18.00	5.00	**18.90**	2.50	1.40	**50.52**	1.01	**70.43**	m2	**75.71**
65mm thick Facing bricks P.C. £210.00 per 1000...	12.60	5.00	**13.23**	2.50	1.40	**50.52**	1.01	**64.76**	m2	**69.62**
50mm thick Staffordshire blue chequered paviors..	18.00	5.00	**18.90**	2.50	1.40	**50.52**	0.96	**70.38**	m2	**75.66**
50mm thick Accrington non slip paviors..........	18.00	5.00	**18.90**	2.50	1.40	**50.52**	0.96	**70.38**	m2	**75.66**
Granite sett paving on concrete base; 13mm thick bedding and grouting in cement mortar (1:3)										
Paving to falls and crossfalls and to slopes not exceeding 15 degrees from horizontal										
125mm thick; new	34.55	2.50	**35.35**	1.65	2.00	**44.87**	2.23	**82.45**	m2	**88.63**
150mm thick; new	41.31	2.50	**42.28**	1.80	2.15	**48.60**	2.55	**93.43**	m2	**100.44**
100mm thick; reclaimed	38.67	2.50	**39.59**	1.50	1.80	**40.59**	1.91	**82.09**	m2	**88.25**
150mm thick; reclaimed	34.28	2.50	**35.08**	1.65	2.00	**44.87**	2.23	**82.18**	m2	**88.34**
200mm thick; reclaimed	45.24	2.50	**46.30**	1.80	2.15	**48.60**	2.55	**97.46**	m2	**104.76**
Cobble paving set in concrete bed (measured separately) tight butted, dry grouted with cement and sand (1:3) watered and brushed										
Paving level and to falls only										
plain	13.44	2.50	**13.78**	1.65	1.65	**41.12**	2.55	**57.45**	m2	**61.75**
set to pattern	13.44	2.50	**13.78**	2.00	2.00	**49.84**	2.55	**66.17**	m2	**71.13**
Crazy paving on blinded hardcore base, bedding in 38mm thick sand; pointing in cement mortar (1:4)										
Paving to falls and crossfalls and to slopes not exceeding 15 degrees from horizontal										
50mm thick precast concrete flag.................	4.65	2.50	**4.77**	0.40	0.40	**9.97**	2.15	**16.88**	m2	**18.15**
38-50mm thick York stone	52.87	2.50	**54.19**	0.60	0.60	**14.95**	3.10	**72.24**	m2	**77.66**
38mm thick Westmorland green slate..............	108.80	2.50	**111.52**	0.60	0.60	**14.95**	2.15	**128.62**	m2	**138.27**
Grass concrete paving; voids filled with vegetable soil and sown with grass seed										
Precast concrete perforated slabs										
120mm thick units on 25mm thick sand bed; level and to falls only	16.70	2.50	**17.30**	0.28	0.28	**6.98**	-	**24.28**	m2	**26.10**
Q30: SEEDING/TURFING										
Cultivating										
Surfaces of natural ground										
digging over one spit deep; removing debris; weeding	-	-	**-**	-	0.25	**2.68**	-	**2.68**	m2	**2.88**
Surfaces of filling										
digging over one spit deep......................	-	-	**-**	-	0.30	**3.21**	-	**3.21**	m2	**3.45**

Labour hourly rates: (except Specialists) Craft Operatives £14.21 Labourer £10.71 Rates are national average prices. Refer to REGIONAL VARIATIONS for indicative levels of overall pricing in regions	MATERIALS			LABOUR				RATES		
	Del to Site £	Waste %	Material Cost £	Craft Optve Hrs	Lab Hrs	Labour Cost £	Sunds £	Nett Rate £	Unit	Gross Rate (+7.5%) £
Q30: SEEDING/TURFING Cont.										
Surface applications										
Bone meal, 0.06 kg/m2; raking in										
general surfaces	0.03	5.00	0.03	-	0.05	0.54	-	0.57	m2	0.61
Selective weedkiller, 0.03 kg/m2; applying by spreader										
general surfaces	0.08	5.00	0.08	-	0.05	0.54	-	0.62	m2	0.67
Pre-seeding fertilizer at the rate of 0.05 kg/m2										
general surfaces	0.08	5.00	0.08	-	0.05	0.54	-	0.62	m2	0.67
General grass fertilizer at the rate of 0.06 kg/m2										
general surfaces	0.09	5.00	0.09	-	0.05	0.54	-	0.63	m2	0.68
Seeding										
Grass seed, 0.07 kg/m2, raking in; rolling; maintaining for 12 months after laying										
general surfaces	0.31	5.00	0.33	-	0.20	2.14	-	2.47	m2	2.65
Stone pick, roll and cut grass with a rotary cutter and remove arisings										
general surfaces	-	-	-	-	0.10	1.07	-	1.07	m2	1.15
Turfing										
Take turf from stack, wheel not exceeding 100m, lay, roll and water, maintaining for 12 months after laying										
general surfaces	-	-	-	-	0.50	5.36	-	5.36	m2	5.76
Rolawn® Medallion turf; cultivated; lay, roll and water; maintaining for 12 months after laying										
general surfaces	2.21	5.00	2.32	-	0.35	3.75	-	6.06	m2	6.52
Q31: PLANTING										
Planting trees, shrubs and hedge plants										
Excavate or form pit, hole or trench, dig over ground in bottom, spread and pack around roots with finely broken soil, refill with top soil with one third by volume of farmyard manure incorporated, water in, remove surplus excavated material and provide labelling										
small tree	0.68	-	0.68	-	0.75	8.03	0.39	9.10	nr	9.79
medium tree	0.86	-	0.86	-	1.50	16.07	0.56	17.49	nr	18.80
large tree	1.08	-	1.08	-	3.00	32.13	0.72	33.93	nr	36.47
shrub	0.51	-	0.51	-	0.50	5.36	0.16	6.03	nr	6.48
hedge plant	0.51	-	0.51	-	0.25	2.68	0.16	3.35	nr	3.60
60mm diameter treated softwood tree stake, pointed and driven into ground and with two PVC tree ties secured around tree and nailed to stake										
2100mm long	3.13	-	3.13	-	0.33	3.53	1.45	8.11	nr	8.72
2400mm long	3.32	-	3.32	-	0.33	3.53	1.45	8.30	nr	8.93
2700mm long	3.67	-	3.67	-	0.33	3.53	1.45	8.65	nr	9.30
Planting herbaceous plants, bulbs, corms and tubers										
Provide planting bed of screened and conditioned Rolawn® topsoil										
150mm thick	3.23	5.00	3.39	-	0.75	8.03	-	11.42	m2	12.28
225mm thick	4.85	5.00	5.09	-	1.12	12.00	-	17.09	m2	18.37
Form hole and plant										
herbaceous plants	0.41	5.00	0.43	-	0.20	2.14	-	2.57	nr	2.77
bulbs, corms and tubers	0.09	5.00	0.09	-	0.10	1.07	-	1.17	nr	1.25
Mulching after planting										
25mm peat	0.89	-	0.89	-	0.10	1.07	-	1.96	m2	2.11
75mm farmyard manure	0.43	-	0.43	-	0.10	1.07	-	1.50	m2	1.61
Q40: FENCING: POST AND WIRE FENCING										
Fencing; strained wire; BS 1722-2; concrete posts and struts; 4.00mm diameter galvanized mild steel line wires; galvanized steel fittings and accessories; backfilling around posts in concrete mix ST4										
900mm high fencing; type SC90; wires -3; posts with rounded tops										
posts at 2743mm centres; 600mm into ground	5.51	2.50	5.61	-	0.75	8.03	-	13.64	m	14.67
extra; end posts with struts -1	20.52	2.50	20.95	-	1.00	10.71	-	31.66	nr	34.04
extra; angle posts with struts -2	33.50	2.50	34.22	-	1.50	16.07	-	50.29	nr	54.06
1050mm high fencing; type SC105A; wires-5; posts rounded tops										
posts at 2743mm centres; 600mm into ground	5.93	2.50	6.04	-	0.80	8.57	-	14.61	m	15.70
extra; end posts with struts -1	22.16	2.50	22.64	-	1.10	11.78	-	34.42	nr	37.00
extra; angle posts with struts -2	36.19	2.50	36.97	-	1.65	17.67	-	54.65	nr	58.74

Labour hourly rates: (except Specialists) Craft Operatives £14.21 Labourer £10.71 Rates are national average prices. Refer to REGIONAL VARIATIONS for indicative levels of overall pricing in regions	MATERIALS			LABOUR				RATES		
	Del to Site £	Waste %	Material Cost £	Craft Optve Hrs	Lab Hrs	Labour Cost £	Sunds £	Nett Rate £	Unit	Gross Rate (+7.5%) £
Q40: FENCING: CHAIN LINK FENCING										
Fencing; chain link; BS 1722-1; concrete posts; galvanized mesh, line and tying wires; galvanized steel fittings and accessories; backfilling around posts in concrete mix ST4										
900mm high fencing; type GLC 90; posts with rounded tops										
posts at 2743mm centres; 600mm into ground.......	11.28	5.00	**11.76**	-	1.00	**10.71**	-	**22.47**	m	**24.16**
extra; end posts with struts -1.................	20.52	5.00	**21.39**	-	1.20	**12.85**	-	**34.24**	nr	**36.81**
extra; angle posts with struts -2...............	30.90	5.00	**32.21**	-	1.60	**17.14**	-	**49.35**	nr	**53.05**
1200mm high fencing; type GLC 120; posts with rounded tops										
posts at 2743mm centres; 600mm into ground.......	12.21	5.00	**12.74**	-	1.10	**11.78**	-	**24.52**	m	**26.36**
extra; end posts with struts -1.................	27.41	5.00	**28.63**	-	1.40	**14.99**	-	**43.62**	nr	**46.89**
extra; angle posts with struts -2...............	44.22	5.00	**46.20**	-	1.80	**19.28**	-	**65.48**	nr	**70.39**
1400mm high fencing; type GLC 140A; posts with rounded tops										
posts at 2743mm centres; 600mm into ground.......	12.68	5.00	**13.24**	-	1.20	**12.85**	-	**26.09**	m	**28.05**
extra; end posts with struts -1.................	27.21	5.00	**28.41**	-	1.55	**16.60**	-	**45.01**	nr	**48.39**
extra; angle posts with struts -2...............	45.14	5.00	**47.17**	-	2.00	**21.42**	-	**68.59**	nr	**73.73**
1800mm high fencing; type GLC 180; posts with rounded tops										
posts at 2743mm centres; 600mm into ground.......	14.77	5.00	**15.43**	-	1.45	**15.53**	-	**30.96**	m	**33.29**
extra; end posts with struts -1.................	33.68	5.00	**35.21**	-	1.65	**17.67**	-	**52.88**	nr	**56.84**
extra; angle posts with struts -2...............	53.61	5.00	**56.06**	-	2.15	**23.03**	-	**79.09**	nr	**85.02**
Fencing; chain link; BS 1722-1; concrete posts; plastics coated Grade A mesh, line and tying wire; galvanized steel fittings and accessories; backfilling around posts in concrete mix ST4										
900mm high fencing; type PLC 90A; posts with rounded tops										
posts at 2743mm centres; 600mm into ground.......	7.30	5.00	**7.59**	-	1.00	**10.71**	-	**18.30**	m	**19.67**
extra; end posts with struts -1.................	20.52	5.00	**21.39**	-	1.20	**12.85**	-	**34.24**	nr	**36.81**
extra; angle posts with struts -2...............	30.90	5.00	**32.21**	-	1.60	**17.14**	-	**49.35**	nr	**53.05**
1200mm high fencing; type PLC 120; posts with rounded tops										
posts at 2743mm centres; 600mm into ground.......	8.36	5.00	**8.70**	-	1.10	**11.78**	-	**20.48**	m	**22.02**
extra; end posts with struts -1.................	27.41	5.00	**28.63**	-	1.40	**14.99**	-	**43.62**	nr	**46.89**
extra; angle posts with struts -2...............	44.22	5.00	**46.20**	-	1.80	**19.28**	-	**65.48**	nr	**70.39**
1400mm high fencing; type PLC 140A; posts with rounded tops										
posts at 2743mm centres; 600mm into ground.......	9.53	5.00	**9.93**	-	1.20	**12.85**	-	**22.78**	m	**24.49**
extra; end posts with struts -1.................	27.21	5.00	**28.41**	-	1.55	**16.60**	-	**45.01**	nr	**48.39**
extra; angle posts with struts -2...............	45.14	5.00	**47.17**	-	2.00	**21.42**	-	**68.59**	nr	**73.73**
1800mm high fencing; type PLC 180; posts with rounded tops										
posts at 2743mm centres; 600mm into ground.......	12.10	5.00	**12.62**	-	1.45	**15.53**	-	**28.15**	m	**30.26**
extra; end posts with struts -1.................	33.68	5.00	**35.21**	-	1.65	**17.67**	-	**52.88**	nr	**56.84**
extra; angle posts with struts -2...............	53.61	5.00	**56.06**	-	2.15	**23.03**	-	**79.09**	nr	**85.02**
Fencing; chain link; BS 1722-1; rolled steel angle posts; galvanized mesh, line and tying wires; galvanized steel fittings and accessories; backfilling around posts in concrete mix ST4										
900mm high fencing; type GLS 90; posts with rounded tops										
posts at 2743mm centres; 600mm into ground.......	7.59	2.50	**7.73**	-	0.65	**6.96**	-	**14.69**	m	**15.79**
extra; end posts with struts -1.................	24.96	2.50	**25.49**	-	1.00	**10.71**	-	**36.20**	nr	**38.92**
extra; angle posts with struts -2...............	38.30	2.50	**39.12**	-	1.60	**17.14**	-	**56.25**	nr	**60.47**
1200mm high fencing; type GLS 120; posts with rounded tops										
posts at 2743mm centres; 600mm into ground.......	8.97	2.50	**9.15**	-	0.75	**8.03**	-	**17.18**	m	**18.47**
extra; end posts with struts -1.................	29.17	2.50	**29.81**	-	1.10	**11.78**	-	**41.59**	nr	**44.71**
extra; angle posts with struts -2...............	45.27	2.50	**46.26**	-	1.75	**18.74**	-	**65.00**	nr	**69.88**
1400mm high fencing; type GLS 140A; posts with rounded tops										
posts at 2743mm centres; 600mm into ground.......	9.63	2.50	**9.82**	-	0.80	**8.57**	-	**18.39**	m	**19.77**
extra; end posts with struts -1.................	31.37	2.50	**32.06**	-	1.50	**16.07**	-	**48.13**	nr	**51.74**
extra; angle posts with struts -2...............	48.70	2.50	**49.78**	-	1.90	**20.35**	-	**70.13**	nr	**75.38**
1800mm high fencing; type GLS 180; posts with rounded tops										
posts at 2743mm centres; 600mm into ground.......	11.45	2.50	**11.69**	-	1.00	**10.71**	-	**22.40**	m	**24.08**
extra; end posts with struts -1.................	40.60	2.50	**41.52**	-	1.60	**17.14**	-	**58.66**	nr	**63.06**
extra; angle posts with struts -2...............	60.79	2.50	**62.17**	-	2.10	**22.49**	-	**84.66**	nr	**91.01**

Labour hourly rates: (except Specialists) Craft Operatives £14.21 Labourer £10.71 Rates are national average prices. Refer to REGIONAL VARIATIONS for indicative levels of overall pricing in regions	MATERIALS			LABOUR				RATES		
	Del to Site £	Waste %	Material Cost £	Craft Optve Hrs	Lab Hrs	Labour Cost £	Sunds £	Nett Rate £	Unit	Gross Rate (+7.5%) £
Q40: FENCING: CHAIN LINK FENCING Cont.										
Fencing; chain link; BS 1722 Part 1; rolled steel angle posts; plastics coated Grade A mesh, line and tying wires; galvanized steel fittings and accessories; backfilling around posts in concrete mix ST4										
900mm high fencing; type PLS 90A; posts with rounded tops										
posts at 2743mm centres; 600mm into ground.......	11.09	5.00	**11.55**	-	0.65	**6.96**	-	**18.51**	m	**19.90**
extra; end posts with struts -1.................	32.70	5.00	**34.15**	-	1.00	**10.71**	-	**44.86**	nr	**48.23**
extra; angle posts with struts -2...............	48.84	5.00	**51.00**	-	1.60	**17.14**	-	**68.14**	nr	**73.25**
1200mm high fencing; type PLS 120; posts with rounded tops										
posts at 2743mm centres; 600mm into ground.......	12.12	5.00	**12.63**	-	0.75	**8.03**	-	**20.66**	m	**22.21**
extra; end posts with struts -1.................	43.57	5.00	**45.56**	-	1.10	**11.78**	-	**57.35**	nr	**61.65**
extra; angle posts with struts -2...............	58.50	5.00	**61.14**	-	1.75	**18.74**	-	**79.89**	nr	**85.88**
1400mm high fencing; type PLS 140A; posts with rounded tops										
posts at 2743mm centres; 600mm into ground.......	13.60	5.00	**14.18**	-	0.80	**8.57**	-	**22.75**	m	**24.46**
extra; end posts with struts -1.................	47.01	5.00	**49.18**	-	1.50	**16.07**	-	**65.24**	nr	**70.13**
extra; angle posts with struts -2...............	62.65	5.00	**65.50**	-	1.90	**20.35**	-	**85.85**	nr	**92.29**
1800mm high fencing; type PLS 180; posts with rounded tops										
posts at 2743mm centres; 600mm into ground.......	16.48	5.00	**17.21**	-	1.00	**10.71**	-	**27.92**	m	**30.02**
extra; end posts with struts -1.................	63.10	5.00	**66.07**	-	1.60	**17.14**	-	**83.21**	nr	**89.45**
extra; angle posts with struts -2...............	79.79	5.00	**83.50**	-	2.10	**22.49**	-	**105.99**	nr	**113.94**
Q40: FENCING: CHESTNUT FENCING										
Fencing; cleft chestnut pale; BS 1722 Part 4; sweet chestnut posts and struts; galvanized accessories; backfilling around posts in concrete mix ST4										
900mm high fencing; type CW90										
posts at 2000mm centres; 600mm into ground.......	9.85	5.00	**10.06**	-	0.25	**2.68**	-	**12.73**	m	**13.69**
extra; end posts with struts - 1................	17.68	5.00	**17.99**	-	0.30	**3.21**	-	**21.20**	nr	**22.79**
extra; angle posts with struts - 2..............	24.80	5.00	**25.18**	-	0.40	**4.28**	-	**29.46**	nr	**31.67**
1200mm high fencing; type CW 120										
posts at 2000mm centres; 600mm into ground.......	10.90	5.00	**11.16**	-	0.17	**1.82**	-	**12.98**	m	**13.95**
extra; end posts with struts - 1................	19.03	5.00	**19.41**	-	0.45	**4.82**	-	**24.22**	nr	**26.04**
extra; angle posts with struts - 2..............	26.55	5.00	**27.02**	-	0.60	**6.43**	-	**33.44**	nr	**35.95**
1500mm high fencing; type CW 150										
posts at 2000mm centres; 600mm into ground.......	13.72	5.00	**14.12**	-	0.22	**2.36**	-	**16.47**	m	**17.71**
extra; end posts with struts - 1................	20.28	5.00	**20.72**	-	0.50	**5.36**	-	**26.07**	nr	**28.03**
extra; angle posts with struts - 2..............	28.20	5.00	**28.75**	-	0.65	**6.96**	-	**35.71**	nr	**38.39**
Q40: FENCING: BOARDED FENCING										
Fencing; close boarded; BS 1722 Part 5; sawn softwood posts, rails, pales, gravel boards and centre stumps, pressure impregnated with preservative; backfilling around posts in concrete mix ST4										
1200mm high fencing; type BW 120										
posts at 3000mm centres; 600mm into ground.......	24.44	5.00	**25.29**	1.25	-	**17.76**	-	**43.05**	m	**46.28**
1500mm high fencing; type BW 150										
posts at 3000mm centres; 750mm into ground.......	29.28	5.00	**30.37**	1.50	-	**21.32**	-	**51.68**	m	**55.56**
1800mm high fencing; type BW 180A										
posts at 3000mm centres; 750mm into ground.......	31.77	5.00	**32.98**	1.67	-	**23.73**	-	**56.71**	m	**60.96**
Fencing; wooden palisade; BS 1722-5; softwood posts, rails, pales and stumps; pressure impregnated with preservative; backfilling around posts in concrete mix ST4										
1050mm high fencing; type WPW 105; 75 x 19mm rectangular pales with pointed tops										
posts at 3000mm centres; 600mm into ground.......	22.50	5.00	**23.25**	1.00	-	**14.21**	-	**37.46**	m	**40.27**
1200mm high fencing; type WPW 120; 75 x 19mm rectangular pales with pointed tops										
posts at 3000mm centres; 600mm into ground.......	24.58	5.00	**25.43**	1.25	-	**17.76**	-	**43.20**	m	**46.44**
Q40: FENCING: STEEL RAILINGS										
Ornamental steel railings; Ranalah Gates Ltd.; primed at works; fixing in brick openings										
Railing										
reference 200, 356mm high	35.47	1.50	**36.01**	0.75	0.38	**14.73**	0.21	**50.94**	m	**54.76**
reference 220, 279mm high	66.29	1.50	**67.29**	0.75	0.38	**14.73**	0.21	**82.23**	m	**88.39**
reference 230, 432mm high	54.18	1.50	**54.99**	0.75	0.38	**14.73**	0.21	**69.93**	m	**75.17**
reference 240, 838mm high	73.10	1.50	**74.20**	1.00	0.50	**19.57**	0.26	**94.02**	m	**101.07**

Labour hourly rates: (except Specialists) Craft Operatives £14.21 Labourer £10.71 Rates are national average prices. Refer to REGIONAL VARIATIONS for indicative levels of overall pricing in regions	MATERIALS			LABOUR				RATES		
	Del to Site £	Waste %	Material Cost £	Craft Optve Hrs	Lab Hrs	Labour Cost £	Sunds £	Nett Rate £	Unit	Gross Rate (+7.5%) £
Q40: FENCING: GUARD RAILS										
galvanized mild steel guard rail of tubing to BS 1387 medium grade with Kee Klamp fittings										
Rail or standard										
40mm diameter ref. 7-2-G	5.92	5.00	**6.22**	0.25	-	**3.55**	0.30	**10.07**	m	**10.82**
extra; flanged end	4.16	5.00	**4.37**	-	-	**-**	-	**4.37**	nr	**4.70**
extra; bend No.15-7	4.56	5.00	**4.79**	0.15	-	**2.13**	-	**6.92**	nr	**7.44**
extra; three-way intersection No.20-7	6.90	5.00	**7.24**	0.15	-	**2.13**	-	**9.38**	nr	**10.08**
extra; three-way intersection No.25-7	6.62	5.00	**6.95**	0.15	-	**2.13**	-	**9.08**	nr	**9.76**
extra; four-way intersection No.21-7	5.40	5.00	**5.67**	0.15	-	**2.13**	-	**7.80**	nr	**8.38**
extra; four-way intersection No.26-7	5.12	5.00	**5.37**	0.15	-	**2.13**	-	**7.51**	nr	**8.07**
extra; five-way intersection No.35-7	7.67	5.00	**8.05**	0.15	-	**2.13**	-	**10.18**	nr	**10.95**
extra; five-way intersection No.40-7	11.44	5.00	**12.01**	0.15	-	**2.13**	-	**14.14**	nr	**15.20**
extra; floor plate No.61-7	6.08	5.00	**6.38**	0.15	-	**2.13**	-	**8.51**	nr	**9.15**
Infill panel fixed with clips										
50 x 50mm welded mesh	19.60	5.00	**20.58**	0.80	-	**11.37**	0.20	**32.14**	m2	**34.55**
Q40: FENCING: PRECAST CONCRETE POSTS										
Precast concrete fence posts; BS 1722 excavating holes, backfilling around posts in concrete mix C20, disposing of surplus materials, earthwork support										
Fence posts for three wires, housing pattern										
100 x 100mm tapering intermediate post 1570mm long; 600mm into ground	7.43	2.50	**7.57**	0.25	0.25	**6.23**	-	**13.80**	nr	**14.84**
100 x 100mm square strainer post (end or intermediate) 1570mm long; 600mm into ground	8.65	2.50	**8.81**	0.25	0.25	**6.23**	-	**15.04**	nr	**16.17**
Clothes line post										
125 x 125mm tapering, 2670mm long	15.57	2.50	**15.91**	0.33	0.33	**8.22**	-	**24.13**	nr	**25.94**
Close boarded fence post										
94 x 100mm, 2745mm long	12.79	2.50	**13.06**	0.40	0.40	**9.97**	-	**23.03**	nr	**24.76**
Chain link fence post for 1800mm high fencing										
125 x 75mm tapering intermediate post, 2620mm long	11.80	2.50	**12.05**	0.33	0.33	**8.22**	-	**20.27**	nr	**21.79**
125 x 125mm end post, 1 strut, 2620mm long	17.57	2.50	**17.91**	0.66	0.66	**16.45**	-	**34.36**	nr	**36.93**
125 x 125mm angle post, 2 strut, 2620mm long	19.57	2.50	**19.91**	0.99	0.99	**24.67**	-	**44.58**	nr	**47.92**
125 x 125mm gate post, 2620mm long	18.42	2.50	**18.83**	0.33	0.33	**8.22**	-	**27.05**	nr	**29.08**
150 x 150mm gate post, 2620mm long	27.08	2.50	**27.71**	0.33	0.33	**8.22**	-	**35.93**	nr	**38.63**
Q40: FENCING: TIMBER GATES										
Gates; impregnated wrought softwood; featheredge pales; including ring latch and heavy hinges										
Gates (posts included elsewhere)										
900 x 1150mm	82.90	2.50	**84.97**	0.50	0.50	**12.46**	0.17	**97.60**	nr	**104.92**
900 x 1450mm	91.60	2.50	**93.89**	0.50	0.50	**12.46**	0.17	**106.52**	nr	**114.51**
Q40: FENCING: STEEL GATES										
Ornamental steel gates; Ranalah Gates Ltd.; primed at works; fixing in brick openings										
Popular, 1118mm high overall										
single gate reference 50, for 914mm wide opening.	40.34	1.50	**40.95**	1.50	0.75	**29.35**	0.27	**70.56**	nr	**75.86**
Diana, 965mm high overall										
single gate reference 800, for 914mm wide opening	47.70	1.50	**48.41**	1.50	0.75	**29.35**	0.27	**78.03**	nr	**83.88**
single gate reference 801, for 1016mm wide opening	50.69	1.50	**51.45**	1.50	0.75	**29.35**	0.27	**81.06**	nr	**87.14**
double gates reference 810, for 2438mm wide opening	106.90	1.50	**108.50**	2.25	1.13	**44.07**	0.49	**153.07**	nr	**164.55**
Popular arch gate 1568mm high overall										
single gate reference 55, for 914mm wide opening.	74.01	1.50	**75.12**	1.50	0.75	**29.35**	0.27	**104.74**	nr	**112.59**
Thrifty, 914mm high overall										
single gate reference 20, for 914mm wide opening.	33.12	1.50	**33.61**	1.50	0.75	**29.35**	0.27	**63.23**	nr	**67.97**
Elite, 914mm high overall										
single gate reference 40, for 914mm wide opening.	42.84	1.50	**43.48**	1.50	0.75	**29.35**	0.27	**73.10**	nr	**78.58**
double gates reference 41, for 2134mm wide opening	99.23	1.50	**100.72**	2.25	1.13	**44.07**	0.49	**145.29**	nr	**156.18**
double gates reference 42, for 2438mm wide opening	104.73	1.50	**106.31**	2.25	1.13	**44.07**	0.49	**150.87**	nr	**162.19**
Super, 965mm high overall										
single gate reference 60, for 914mm wide opening.	49.91	1.50	**50.66**	1.50	0.75	**29.35**	0.27	**80.28**	nr	**86.30**
double gates reference 62, for 2286mm wide opening	126.84	1.50	**128.74**	2.25	1.13	**44.07**	0.49	**173.30**	nr	**186.30**
double gates reference 63, for 2438mm wide opening	131.47	1.50	**133.44**	2.25	1.13	**44.07**	0.49	**178.00**	nr	**191.35**
Concord, 914mm high overall										
single gate reference 160, for 914mm wide opening	56.48	1.50	**57.32**	1.50	0.75	**29.35**	0.27	**86.94**	nr	**93.46**
double gates reference 162, for 2286mm wide opening	129.77	1.50	**131.71**	2.25	1.13	**44.07**	0.49	**176.28**	nr	**189.50**
double gates reference 163, for 2438mm wide opening	137.66	1.50	**139.72**	2.25	1.13	**44.07**	0.49	**184.29**	nr	**198.11**
Regent, 914mm high overall										
single gate reference 150, for 914mm wide opening	63.38	1.50	**64.33**	1.50	0.75	**29.35**	0.27	**93.95**	nr	**100.99**
double gates reference 151, for 2134mm wide opening	137.89	1.50	**139.96**	2.25	1.13	**44.07**	0.49	**184.52**	nr	**198.36**

Labour hourly rates: (except Specialists) Craft Operatives £14.21 Labourer £10.71 Rates are national average prices. Refer to REGIONAL VARIATIONS for indicative levels of overall pricing in regions	MATERIALS			LABOUR				RATES		
	Del to Site	Waste	Material Cost	Craft Optve	Lab	Labour Cost	Sunds	Nett Rate	Unit	Gross Rate (+7.5%)
	£	%	£	Hrs	Hrs	£	£	£		£
Q40: FENCING: STEEL GATES Cont.										
Ornamental steel gates; Ranalah Gates Ltd.; primed at works; fixing in brick openings Cont.										
Regent, 914mm high overall Cont.										
double gates reference 152, for 2286mm wide opening	141.75	1.50	**143.88**	2.25	1.13	**44.07**	0.49	**188.44**	nr	**202.57**
double gates reference 153, for 2438mm wide opening	158.06	1.50	**160.43**	2.25	1.13	**44.07**	0.49	**205.00**	nr	**220.37**
Brighton, 991mm high overall										
single gate reference 600, for 914mm wide opening	71.72	1.50	**72.80**	1.50	0.75	**29.35**	0.27	**102.42**	nr	**110.10**
double gates reference 602, for 2286mm wide opening	189.83	1.50	**192.68**	2.25	1.13	**44.07**	0.49	**237.24**	nr	**255.03**
double gates reference 603, for 2438mm wide opening	194.51	1.50	**197.42**	2.25	1.13	**44.07**	0.49	**241.99**	nr	**260.14**
Preston, 1067mm high overall										
single gate reference 425, for 914mm wide opening	84.13	1.50	**85.39**	1.50	0.75	**29.35**	0.27	**115.01**	nr	**123.63**
single gate reference 426, for 1016mm wide opening	92.68	1.50	**94.07**	1.50	0.75	**29.35**	0.27	**123.69**	nr	**132.96**
double gates reference 430, for 2134mm wide opening	189.50	1.50	**192.34**	2.25	1.13	**44.07**	0.49	**236.90**	nr	**254.67**
double gates reference 431, for 2438mm wide opening	210.53	1.50	**213.69**	2.25	1.13	**44.07**	0.49	**258.25**	nr	**277.62**
Extra cost of fixing gates to and including pair of steel gate posts; backfilling around posts in concrete mix ST4										
for single or double gates; posts reference 901, size 40 x 40mm	39.67	1.50	**40.30**	0.33	0.17	**6.51**	-	**47.45**	nr	**51.01**
for single or double gates; posts reference 902, size 50 x 50mm	49.47	1.50	**50.25**	0.33	0.17	**6.51**	-	**57.08**	nr	**61.36**
for single or double gates; posts reference 903, size 70 x 70mm	66.06	1.50	**67.09**	0.66	0.33	**12.91**	-	**80.48**	nr	**86.51**
for single or double gates; posts reference 904, size 100 x 100mm	95.99	1.50	**97.49**	0.66	0.33	**12.91**	-	**110.88**	nr	**119.20**
Gates in chain link fencing										
Gate with 40 x 40 x 5mm painted angle iron framing, braces and rails, infilled with 50mm x 10 1/2 G mesh										
1200 x 1800mm; galvanized mesh	205.75	1.50	**208.84**	6.00	6.00	**149.52**	-	**358.36**	nr	**385.23**
1200 x 1800mm; plastic coated mesh	205.75	1.50	**208.84**	6.00	6.00	**149.52**	-	**358.36**	nr	**385.23**
Gate post of 80 x 80 x 6mm painted angle iron; backfilling around posts in concrete mix C30										
to suit 1800mm high gate	49.31	1.50	**50.00**	1.50	1.50	**37.38**	-	**87.38**	nr	**93.94**
Q50: SITE/STREET FURNITURE/EQUIPMENT										
Plain ordinary portland cement precast concrete bollards; setting in ground, excavating hole, removing surplus spoil, filling with concrete mix ST4; working around base										
230mm diameter										
455mm high above ground level; Woodhouse; Smooth Grey	68.00	-	**68.00**	0.50	0.50	**12.46**	3.10	**83.56**	nr	**89.83**
150mm diameter										
915mm high above ground level; Bridgford	76.00	-	**76.00**	0.90	0.90	**22.43**	3.10	**101.53**	nr	**109.14**
Precast concrete plant containers; setting in position										
Plant container (soil filling included elsewhere)										
710mm diameter x 500mm high; Strada	198.00	-	**198.00**	0.80	0.80	**19.94**	0.30	**218.24**	nr	**234.60**
1066mm diameter x 760mm high; Shirley	299.00	-	**299.00**	1.00	1.00	**24.92**	0.32	**324.24**	nr	**348.56**
500 x 500 x 500mm high; Strada	285.00	-	**285.00**	0.60	0.60	**14.95**	0.30	**300.25**	nr	**322.77**
Precast concrete litter bin with galvanized wire plastic coated inner basket Plastic Coated; setting in position										
Plain concrete										
65 litres capacity; Newstead	165.00	-	**165.00**	1.00	1.00	**24.92**	0.32	**190.24**	nr	**204.51**
190 litres capacity; Gransden	189.00	-	**189.00**	1.00	1.00	**24.92**	0.38	**214.30**	nr	**230.37**
Benches and seats with precast concrete supports and hardwood slats to seats and back rests; fixed or free-standing										
Bench; placing in position										
2000mm long; Kelvin; Smooth Grey	250.00	-	**250.00**	0.50	0.50	**12.46**	-	**262.46**	nr	**282.14**
2000mm long; Kelvin; Smooth White	250.00	-	**250.00**	0.50	0.50	**12.46**	-	**262.46**	nr	**282.14**
Steel cycle stands, Mawrob Co. (Engineers) Ltd; galvanized steel construction; free standing										
Single sided										
reference TR6 for 6 cycles	215.00	-	**215.00**	1.00	0.25	**16.89**	-	**231.89**	nr	**249.28**
reference TR8 for 8 cycles	289.50	-	**289.50**	1.25	0.30	**20.98**	-	**310.48**	nr	**333.76**
Double sided										
reference DTR10 for 10 cycles	260.00	-	**260.00**	1.50	0.40	**25.60**	-	**285.60**	nr	**307.02**
reference DTR12 for 12 cycles	388.50	-	**388.50**	1.50	0.40	**25.60**	-	**414.10**	nr	**445.16**

Labour hourly rates: (except Specialists) Craft Operatives £14.21 Labourer £10.71 Rates are national average prices. Refer to REGIONAL VARIATIONS for indicative levels of overall pricing in regions	MATERIALS			LABOUR				RATES		
	Del to Site £	Waste %	Material Cost £	Craft Optve Hrs	Lab Hrs	Labour Cost £	Sunds £	Nett Rate £	Unit	Gross Rate (+7.5%) £
Q50: SITE/STREET FURNITURE/EQUIPMENT Cont.										
Steel cycle stands, Mawrob Co. (Engineers) Ltd; galvanized steel construction; free standing Cont.										
Double sided Cont.										
reference DTR20 for 20 cycles	470.00	-	**470.00**	2.00	0.50	**33.78**	-	**503.78**	nr	541.56
Steel front wheel cycle supports; Mawrob Co. (Engineers) Ltd; galvanized steel construction, pillar mounted; bolting base plate to concrete										
Single sided										
reference MW/MC 6/S for 6 cycles	199.00	-	**199.00**	4.00	1.00	**67.55**	-	**266.55**	nr	286.54
reference MW/MC 9/S for 9 cycles	335.00	-	**335.00**	4.00	1.00	**67.55**	-	**402.55**	nr	432.74
reference MW/MC 12/S for 12 cycles	435.00	-	**435.00**	6.00	1.50	**101.33**	-	**536.33**	nr	576.55
Double sided										
reference MW/MC 6/D for 12 cycles	299.00	-	**299.00**	4.00	1.00	**67.55**	-	**366.55**	nr	394.04
reference MW/MC 9/D for 18 cycles	380.00	-	**380.00**	4.00	1.00	**67.55**	-	**447.55**	nr	481.12
reference MW/MC 12/D for 24 cycles	470.00	-	**470.00**	6.00	1.50	**101.33**	-	**571.33**	nr	614.17
Steel cycle shelters, Mawrob Co. (Engineers) Ltd; steel angle framing; with red oxide priming; corrugated galvanized steel roofing; assembling; bolting to concrete										
Horizontal loading, single sided										
reference MW/AV10 for 10 cycles	689.00	-	**689.00**	8.00	2.00	**135.10**	-	**824.10**	nr	885.91
reference MW/AK10 for 10 cycles	777.00	-	**777.00**	8.00	2.00	**135.10**	-	**912.10**	nr	980.51
reference MW/AW8 for 8 cycles	970.00	-	**970.00**	6.00	1.50	**101.33**	-	**1071.33**	nr	1151.67
reference MW/AW10 for 10 cycles	1162.00	-	**1162.00**	8.00	2.00	**135.10**	-	**1297.10**	nr	1394.38

DISPOSAL SYSTEMS

Labour hourly rates: (except Specialists) Craft Operatives £17.38 Labourer £10.71 Rates are national average prices. Refer to REGIONAL VARIATIONS for indicative levels of overall pricing in regions	MATERIALS			LABOUR				RATES		
	Del to Site £	Waste %	Material Cost £	Craft Optve Hrs	Lab Hrs	Labour Cost £	Sunds £	Nett Rate £	Unit	Gross Rate (+7.5%) £
R10: RAINWATER PIPEWORK/GUTTERS: CAST IRON ROUND RAINWATER PIPES (BS 460)										
Cast iron pipes and fittings, BS 460 Type A sockets; dry joints										
Pipes; straight										
75mm; in or on supports (included elsewhere)	13.99	5.00	**14.69**	0.67	-	**11.64**	-	**26.33**	m	**28.31**
extra; shoes	11.56	2.00	**11.79**	0.84	-	**14.60**	-	**26.39**	nr	**28.37**
extra; bends	9.91	2.00	**10.11**	0.67	-	**11.64**	-	**21.75**	nr	**23.38**
extra; offset bends 75mm projection	12.50	2.00	**12.75**	0.67	-	**11.64**	-	**24.39**	nr	**26.22**
extra; offset bends 150mm projection	12.50	2.00	**12.75**	0.67	-	**11.64**	-	**24.39**	nr	**26.22**
extra; offset bends 225mm projection	14.55	2.00	**14.84**	0.67	-	**11.64**	-	**26.49**	nr	**28.47**
extra; offset bends 300mm projection	17.88	2.00	**18.24**	0.67	-	**11.64**	-	**29.88**	nr	**32.12**
extra; branches	17.35	2.00	**17.70**	0.67	-	**11.64**	-	**29.34**	nr	**31.54**
100mm; in or on supports (included elsewhere)	19.09	5.00	**20.04**	0.80	-	**13.90**	-	**33.95**	m	**36.49**
extra; shoes	15.58	2.00	**15.89**	1.00	-	**17.38**	-	**33.27**	nr	**35.77**
extra; bends	13.99	2.00	**14.27**	0.80	-	**13.90**	-	**28.17**	nr	**30.29**
extra; offset bends 75mm projection	23.57	2.00	**24.04**	0.80	-	**13.90**	-	**37.95**	nr	**40.79**
extra; offset bends 150mm projection	23.57	2.00	**24.04**	0.80	-	**13.90**	-	**37.95**	nr	**40.79**
extra; offset bends 225mm projection	28.55	2.00	**29.12**	0.80	-	**13.90**	-	**43.03**	nr	**46.25**
extra; offset bends 300mm projection	29.10	2.00	**29.68**	0.80	-	**13.90**	-	**43.59**	nr	**46.85**
extra; branches	20.62	2.00	**21.03**	0.80	-	**13.90**	-	**34.94**	nr	**37.56**
Cast iron pipes and fittings, BS 460 Type A sockets; ears cast on; dry joints										
Pipes; straight										
63mm; ears fixing to masonry with galvanized pipe nails and distance pieces	15.29	5.00	**16.06**	0.57	-	**9.91**	-	**25.97**	m	**27.91**
extra; shoes	13.33	2.00	**13.60**	0.71	-	**12.34**	-	**25.94**	nr	**27.88**
extra; bends	8.16	2.00	**8.32**	0.57	-	**9.91**	-	**18.23**	nr	**19.60**
extra; offset bends 75mm projection	12.50	2.00	**12.75**	0.57	-	**9.91**	-	**22.66**	nr	**24.36**
extra; offset bends 150mm projection	12.50	2.00	**12.75**	0.57	-	**9.91**	-	**22.66**	nr	**24.36**
extra; offset bends 225mm projection	14.55	2.00	**14.84**	0.57	-	**9.91**	-	**24.75**	nr	**26.60**
extra; offset bends 300mm projection	17.03	2.00	**17.37**	0.57	-	**9.91**	-	**27.28**	nr	**29.32**
extra; branches	16.04	2.00	**16.36**	0.57	-	**9.91**	-	**26.27**	nr	**28.24**
75mm; ears fixing to masonry with galvanized pipe nails and distance pieces	15.29	5.00	**16.06**	0.67	-	**11.64**	-	**27.70**	m	**29.78**
extra; shoes	13.33	2.00	**13.60**	0.84	-	**14.60**	-	**28.20**	nr	**30.31**
extra; bends	9.91	2.00	**10.11**	0.67	-	**11.64**	-	**21.75**	nr	**23.38**
extra; offset bends 75mm projection	12.50	2.00	**12.75**	0.67	-	**11.64**	-	**24.39**	nr	**26.22**
extra; offset bends 150mm projection	12.50	2.00	**12.75**	0.67	-	**11.64**	-	**24.39**	nr	**26.22**
extra; offset bends 225mm projection	14.55	2.00	**14.84**	0.67	-	**11.64**	-	**26.49**	nr	**28.47**
extra; offset bends 300mm projection	17.88	2.00	**18.24**	0.67	-	**11.64**	-	**29.88**	nr	**32.12**
extra; branches	17.35	2.00	**17.70**	0.67	-	**11.64**	-	**29.34**	nr	**31.54**
100mm; ears fixing to masonry with galvanized pipe nails and distance pieces	20.41	5.00	**21.44**	0.80	-	**13.90**	-	**35.34**	m	**37.99**
extra; shoes	17.70	2.00	**18.05**	1.00	-	**17.38**	-	**35.43**	nr	**38.09**
extra; bends	13.99	2.00	**14.27**	0.80	-	**13.90**	-	**28.17**	nr	**30.29**
extra; offset bends 75mm projection	23.57	2.00	**24.04**	0.80	-	**13.90**	-	**37.95**	nr	**40.79**
extra; offset bends 150mm projection	23.57	2.00	**24.04**	0.80	-	**13.90**	-	**37.95**	nr	**40.79**
extra; offset bends 225mm projection	28.55	2.00	**29.12**	0.80	-	**13.90**	-	**43.03**	nr	**46.25**
extra; offset bends 300mm projection	29.10	2.00	**29.68**	0.80	-	**13.90**	-	**43.59**	nr	**46.85**
extra; branches	20.62	2.00	**21.03**	0.80	-	**13.90**	-	**34.94**	nr	**37.56**
Pipe supports										
Steel holderbats										
for 75mm pipes; for building in (fixing included elsewhere)	6.04	2.00	**6.16**	-	-	**-**	-	**6.16**	nr	**6.62**
for 100mm pipes; for building in (fixing included elsewhere)	6.11	2.00	**6.23**	-	-	**-**	-	**6.23**	nr	**6.70**
for 75mm pipes; fixing to masonry with galvanized screws	4.74	2.00	**4.83**	0.17	-	**2.95**	-	**7.79**	nr	**8.37**
for 100mm pipes; fixing to masonry with galvanized screws	4.80	2.00	**4.90**	0.17	-	**2.95**	-	**7.85**	nr	**8.44**

Labour hourly rates: (except Specialists) Craft Operatives £17.38 Labourer £10.71 Rates are national average prices. Refer to REGIONAL VARIATIONS for indicative levels of overall pricing in regions	MATERIALS			LABOUR				RATES		
	Del to Site £	Waste %	Material Cost £	Craft Optve Hrs	Lab Hrs	Labour Cost £	Sunds £	Nett Rate £	Unit	Gross Rate (+7.5%) £
R10: RAINWATER PIPEWORK/GUTTERS: CAST IRON RECTANGULAR RAINWATER PIPES										
Cast iron rectangular pipes and fittings; ears cast on; dry joints										
Pipes; straight										
100 x 75mm; ears fixing to masonry with galvanized pipe nails and distance pieces	58.32	5.00	**61.24**	1.00	-	**17.38**	-	**78.62**	m	**84.52**
extra; shoes (front)	49.94	2.00	**50.94**	1.00	-	**17.38**	-	**68.32**	nr	**73.44**
extra; bends (front)	40.01	2.00	**40.81**	0.75	-	**13.04**	-	**53.85**	nr	**57.88**
extra; offset bends (front) 150 projection	52.00	2.00	**53.04**	0.75	-	**13.04**	-	**66.08**	nr	**71.03**
extra; offset bends (front) 300 projection	71.57	2.00	**73.00**	0.75	-	**13.04**	-	**86.04**	nr	**92.49**
R10: RAINWATER PIPEWORK/GUTTERS: ALUMINIUM RAINWATER PIPES (BS EN 612)										
Cast aluminium pipes and fittings BS EN 612 Section B, ears cast on; dry joints										
Pipes; straight										
63mm; ears fixing to masonry with galvanized screws	11.04	5.00	**11.59**	0.50	-	**8.69**	-	**20.28**	m	**21.80**
extra; shoes	7.00	2.00	**7.14**	0.63	-	**10.95**	-	**18.09**	nr	**19.45**
extra; bends	8.02	2.00	**8.18**	0.50	-	**8.69**	-	**16.87**	nr	**18.14**
extra; offset bends 75mm projection	15.41	2.00	**15.72**	0.50	-	**8.69**	-	**24.41**	nr	**26.24**
extra; offset bends 150mm projection	19.12	2.00	**19.50**	0.50	-	**8.69**	-	**28.19**	nr	**30.31**
extra; offset bends 225mm projection	21.66	2.00	**22.09**	0.50	-	**8.69**	-	**30.78**	nr	**33.09**
extra; offset bends 300mm projection	24.25	2.00	**24.74**	0.50	-	**8.69**	-	**33.43**	nr	**35.93**
75mm; ears fixing to masonry with galvanized pipe nails	12.83	5.00	**13.47**	0.67	-	**11.64**	-	**25.11**	m	**27.00**
extra; shoes	9.71	2.00	**9.90**	0.84	-	**14.60**	-	**24.50**	nr	**26.34**
extra; bends	10.51	2.00	**10.72**	0.67	-	**11.64**	-	**22.36**	nr	**24.04**
extra; offset bends 75mm projection	18.67	2.00	**19.04**	0.67	-	**11.64**	-	**30.69**	nr	**32.99**
extra; offset bends 150mm projection	21.46	2.00	**21.89**	0.67	-	**11.64**	-	**33.53**	nr	**36.05**
extra; offset bends 225mm projection	23.94	2.00	**24.42**	0.67	-	**11.64**	-	**36.06**	nr	**38.77**
extra; offset bends 300mm projection	26.58	2.00	**27.11**	0.67	-	**11.64**	-	**38.76**	nr	**41.66**
100mm; ears fixing to masonry with galvanized pipe nails	21.85	5.00	**22.94**	0.80	-	**13.90**	-	**36.84**	m	**39.61**
extra; shoes	11.60	2.00	**11.83**	1.00	-	**17.38**	-	**29.21**	nr	**31.40**
extra; bends	15.34	2.00	**15.65**	0.80	-	**13.90**	-	**29.55**	nr	**31.77**
extra; offset bends 75mm projection	21.11	2.00	**21.53**	0.80	-	**13.90**	-	**35.44**	nr	**38.09**
extra; offset bends 150mm projection	24.53	2.00	**25.02**	0.80	-	**13.90**	-	**38.92**	nr	**41.84**
extra; offset bends 225mm projection	27.75	2.00	**28.31**	0.80	-	**13.90**	-	**42.21**	nr	**45.37**
extra; offset bends 300mm projection	30.99	2.00	**31.61**	0.80	-	**13.90**	-	**45.51**	nr	**48.93**
R10: RAINWATER PIPEWORK/GUTTERS: UNPLASTICIZED PVC RAINWATER PIPES (BS EN 607)										
PVC-U pipes and fittings, BS EN 607; push fit joints; pipework and supports self coloured grey										
Pipes; straight										
68mm; in standard holderbats fixing to masonry with galvanized screws	3.44	5.00	**3.59**	0.40	-	**6.95**	-	**10.54**	m	**11.33**
extra; shoes	3.21	2.00	**3.27**	0.31	-	**5.39**	-	**8.66**	nr	**9.31**
extra; bends	3.72	2.00	**3.79**	0.25	-	**4.35**	-	**8.14**	nr	**8.75**
extra; offset bends (spigot)	2.06	2.00	**2.10**	0.25	-	**4.35**	-	**6.45**	nr	**6.93**
extra; offset bends (socket)	2.02	2.00	**2.06**	0.25	-	**4.35**	-	**6.41**	nr	**6.89**
extra; angled branches	7.40	2.00	**7.55**	0.25	-	**4.35**	-	**11.89**	nr	**12.78**
R10: RAINWATER PIPEWORK/GUTTERS: CAST IRON EAVES GUTTERS (BS 460)										
Cast iron half round gutters and fittings, BS 460; bolted and mastic joints										
Gutters; straight										
100mm; in standard fascia brackets fixing to timber with galvanized screws	8.52	5.00	**8.92**	0.50	-	**8.69**	-	**17.61**	m	**18.93**
extra; stopped ends	1.97	2.00	**2.01**	0.50	-	**8.69**	-	**10.70**	nr	**11.50**
extra; running outlets	5.70	2.00	**5.81**	0.50	-	**8.69**	-	**14.50**	nr	**15.59**
extra; stopped ends with outlet	6.69	2.00	**6.82**	0.50	-	**8.69**	-	**15.51**	nr	**16.68**
extra; angles	5.85	2.00	**5.97**	0.50	-	**8.69**	-	**14.66**	nr	**15.76**
114mm; in standard fascia brackets fixing to timber with galvanized screws	8.84	5.00	**9.25**	0.50	-	**8.69**	-	**17.94**	m	**19.29**
extra; stopped ends	2.55	2.00	**2.60**	0.50	-	**8.69**	-	**11.29**	nr	**12.14**
extra; running outlets	6.22	2.00	**6.34**	0.50	-	**8.69**	-	**15.03**	nr	**16.16**
extra; stopped ends with outlet	7.50	2.00	**7.65**	0.50	-	**8.69**	-	**16.34**	nr	**17.57**
extra; angles	6.03	2.00	**6.15**	0.50	-	**8.69**	-	**14.84**	nr	**15.95**
125mm; in standard fascia brackets fixing to timber with galvanized screws	10.21	5.00	**10.69**	0.57	-	**9.91**	-	**20.60**	m	**22.14**
extra; stopped ends	2.55	2.00	**2.60**	0.57	-	**9.91**	-	**12.51**	nr	**13.45**
extra; running outlets	7.11	2.00	**7.25**	0.57	-	**9.91**	-	**17.16**	nr	**18.45**
extra; stopped ends with outlet	8.39	2.00	**8.56**	0.57	-	**9.91**	-	**18.46**	nr	**19.85**
extra; angles	7.11	2.00	**7.25**	0.57	-	**9.91**	-	**17.16**	nr	**18.45**
150mm; in standard fascia brackets fixing to timber with galvanized screws	17.05	5.00	**17.87**	0.75	-	**13.04**	-	**30.90**	m	**33.22**
extra; stopped ends	3.54	2.00	**3.61**	0.75	-	**13.04**	-	**16.65**	nr	**17.89**
extra; running outlets	12.30	2.00	**12.55**	0.75	-	**13.04**	-	**25.58**	nr	**27.50**
extra; stopped ends with outlet	14.07	2.00	**14.35**	0.75	-	**13.04**	-	**27.39**	nr	**29.44**
extra; angles	12.98	2.00	**13.24**	0.75	-	**13.04**	-	**26.27**	nr	**28.25**

Labour hourly rates: (except Specialists)
Craft Operatives £17.38 Labourer £10.71
Rates are national average prices.
Refer to REGIONAL VARIATIONS for indicative levels of overall pricing in regions

	MATERIALS			LABOUR				RATES		
	Del to Site £	Waste %	Material Cost £	Craft Optve Hrs	Lab Hrs	Labour Cost £	Sunds £	Nett Rate £	Unit	Gross Rate (+7.5%) £
R10: RAINWATER PIPEWORK/GUTTERS: CAST IRON EAVES GUTTERS (BS 460) Cont.										
Cast iron OG gutters and fittings, BS 460; bolted and mastic joints										
Gutters; straight										
100mm; fixing to timber with galvanized screws...	8.57	5.00	**9.00**	0.50	-	**8.69**	-	**17.69**	m	**19.02**
extra; stopped ends	2.01	2.00	**2.05**	0.50	-	**8.69**	-	**10.74**	nr	**11.55**
extra; running outlets	6.22	2.00	**6.34**	0.50	-	**8.69**	-	**15.03**	nr	**16.16**
extra; angles	6.11	2.00	**6.23**	0.50	-	**8.69**	-	**14.92**	nr	**16.04**
114mm; fixing to timber with galvanized screws...	9.43	5.00	**9.90**	0.50	-	**8.69**	-	**18.59**	m	**19.99**
extra; stopped ends	2.60	2.00	**2.65**	0.50	-	**8.69**	-	**11.34**	nr	**12.19**
extra; running outlets	6.62	2.00	**6.75**	0.50	-	**8.69**	-	**15.44**	nr	**16.60**
extra; angles	6.62	2.00	**6.75**	0.50	-	**8.69**	-	**15.44**	nr	**16.60**
125mm; fixing to timber with galvanized screws...	9.88	5.00	**10.38**	0.57	-	**9.91**	-	**20.28**	m	**21.80**
extra; stopped ends	2.60	2.00	**2.65**	0.57	-	**9.91**	-	**12.56**	nr	**13.50**
extra; running outlets	7.23	2.00	**7.37**	0.57	-	**9.91**	-	**17.28**	nr	**18.58**
extra; angles	7.23	2.00	**7.37**	0.57	-	**9.91**	-	**17.28**	nr	**18.58**
R10: RAINWATER PIPEWORK/GUTTERS: CAST IRON MOULDED SECTIONS (STOCK SECTIONS)										
Cast iron moulded (stock sections) gutters and fittings; bolted and mastic joints										
Gutters; straight										
100 x 75mm; in standard fascia brackets fixing to timber with galvanized screws....................	16.22	5.00	**16.99**	0.75	-	**13.04**	-	**30.02**	m	**32.27**
extra; stopped ends	5.67	2.00	**5.78**	0.75	-	**13.04**	-	**18.82**	nr	**20.23**
extra; running outlets	15.16	2.00	**15.46**	0.75	-	**13.04**	-	**28.50**	nr	**30.64**
extra; angles	15.16	2.00	**15.46**	0.75	-	**13.04**	-	**28.50**	nr	**30.64**
extra; clips	6.34	2.00	**6.47**	0.75	-	**13.04**	-	**19.50**	nr	**20.96**
125 x 100mm; in standard fascia brackets fixing to timber with galvanized screws..................	23.07	5.00	**24.18**	1.00	-	**17.38**	-	**41.56**	m	**44.68**
extra; stopped ends	7.35	2.00	**7.50**	1.00	-	**17.38**	-	**24.88**	nr	**26.74**
extra; running outlets	21.36	2.00	**21.79**	1.00	-	**17.38**	-	**39.17**	nr	**42.10**
extra; angles	21.36	2.00	**21.79**	1.00	-	**17.38**	-	**39.17**	nr	**42.10**
extra; clips	7.35	2.00	**7.50**	1.00	-	**17.38**	-	**24.88**	nr	**26.74**
R10: RAINWATER PIPEWORK/GUTTERS: CAST IRON BOX GUTTERS										
Cast iron box gutters and fittings; bolted and mastic joints										
Gutters; straight										
100 x 75mm; in or on brackets..................	25.89	5.00	**27.13**	0.75	-	**13.04**	-	**40.16**	m	**43.18**
extra; stopped ends	3.64	2.00	**3.71**	0.75	-	**13.04**	-	**16.75**	nr	**18.00**
extra; running outlets	17.50	2.00	**17.85**	0.75	-	**13.04**	-	**30.89**	nr	**33.20**
extra; angles	17.50	2.00	**17.85**	0.75	-	**13.04**	-	**30.89**	nr	**33.20**
extra; clips	4.61	2.00	**4.70**	0.75	-	**13.04**	-	**17.74**	nr	**19.07**
R10: RAINWATER PIPEWORK/GUTTERS: ALUMINIUM GUTTERS (BS EN 612)										
Cast aluminium half round gutters and fittings, BS EN 612 Section A; bolted and mastic joints										
Gutters; straight										
102mm; in standard fascia brackets fixing to timber with galvanized screws....................	9.49	5.00	**9.94**	0.50	-	**8.69**	-	**18.63**	m	**20.03**
extra; stopped ends	2.58	2.00	**2.63**	0.50	-	**8.69**	-	**11.32**	nr	**12.17**
extra; running outlets	6.11	2.00	**6.23**	0.50	-	**8.69**	-	**14.92**	nr	**16.04**
extra; angles	5.31	2.00	**5.42**	0.50	-	**8.69**	-	**14.11**	nr	**15.16**
127mm; in standard fascia brackets fixing to timber with galvanized screws....................	12.35	5.00	**12.92**	0.57	-	**9.91**	-	**22.83**	m	**24.54**
extra; stopped ends	3.42	2.00	**3.49**	0.57	-	**9.91**	-	**13.40**	nr	**14.40**
extra; running outlets	7.00	2.00	**7.14**	0.57	-	**9.91**	-	**17.05**	nr	**18.33**
extra; angles	6.79	2.00	**6.93**	0.57	-	**9.91**	-	**16.83**	nr	**18.09**
R10: RAINWATER PIPEWORK/GUTTERS: UNPLASTICISED PVC GUTTERS (BS EN 607)										
PVC-U half round gutters and fittings, BS EN 607; push fit connector joints; gutterwork and supports self coloured grey										
Gutters; straight										
114mm; in standard fascia brackets fixing to timber with galvanized screws....................	3.64	2.00	**3.71**	0.33	-	**5.74**	-	**9.44**	m	**10.15**
extra; stopped ends	2.05	2.00	**2.09**	0.33	-	**5.74**	-	**7.83**	nr	**8.41**
extra; running outlets	3.81	2.00	**3.89**	0.33	-	**5.74**	-	**9.62**	nr	**10.34**
extra; angles	4.16	2.00	**4.24**	0.33	-	**5.74**	-	**9.98**	nr	**10.73**
R10: RAINWATER PIPEWORK/GUTTERS: ANCILLARIES										
Aluminium pipework ancillaries										
Rainwater heads; BS EN 612 square type										
250 x 180 x 180mm; 63mm outlet spigot; dry joint to pipe; fixing to masonry with galvanized screws	18.88	1.00	**19.07**	0.50	-	**8.69**	-	**27.76**	nr	**29.84**
250 x 180 x 180mm; 75mm outlet spigot; dry joint to pipe; fixing to masonry with galvanized screws	19.79	1.00	**19.99**	0.67	-	**11.64**	-	**31.63**	nr	**34.00**
250 x 180 x 180mm; 100mm outlet spigot; dry joint to pipe; fixing to masonry with galvanized screws	32.54	1.00	**32.86**	0.80	-	**13.90**	-	**46.77**	nr	**50.28**

Labour hourly rates: (except Specialists) Craft Operatives £17.38 Labourer £10.71 Rates are national average prices. Refer to REGIONAL VARIATIONS for indicative levels of overall pricing in regions	MATERIALS			LABOUR				RATES		
	Del to Site £	Waste %	Material Cost £	Craft Optve Hrs	Lab Hrs	Labour Cost £	Sunds £	Nett Rate £	Unit	Gross Rate (+7.5%) £
R10: RAINWATER PIPEWORK/GUTTERS: RAINWATER HEADS										
Cast iron pipework ancillaries										
Rainwater heads; BS 460 hopper type (flat pattern)										
210 x 160 x 185mm; 65mm outlet spigot; dry joint to pipe; fixing to masonry with galvanized screws	12.44	1.00	**12.56**	0.50	-	**8.69**	-	**21.25**	nr	**22.85**
210 x 160 x 185mm; 75mm outlet spigot; dry joint to pipe; fixing to masonry with galvanized screws	14.12	1.00	**14.26**	0.67	-	**11.64**	-	**25.90**	nr	**27.85**
250 x 180 x 175mm; 100mm outlet spigot; dry joint to pipe; fixing to masonry with galvanized screws	31.19	1.00	**31.50**	0.80	-	**13.90**	-	**45.40**	nr	**48.81**
Rainwater heads; BS 460 square type (flat pattern)										
250 x 180 x 175mm; 65mm outlet spigot; dry joint to pipe; fixing to masonry with galvanized screws	28.53	1.00	**28.81**	0.50	-	**8.69**	-	**37.50**	nr	**40.32**
250 x 180 x 175mm; 75mm outlet spigot; dry joint to pipe; fixing to masonry with galvanized screws	28.53	1.00	**28.81**	0.67	-	**11.64**	-	**40.46**	nr	**43.49**
250 x 180 x 175mm; 100mm outlet spigot; dry joint to pipe; fixing to masonry with galvanized screws	28.53	1.00	**28.81**	0.80	-	**13.90**	-	**42.72**	nr	**45.92**
Plastics pipework ancillaries self coloured black										
Rainwater heads; BS EN 607 square type										
280 x 155 x 230mm; 68mm outlet spigot; push fit joint to plastics pipe; fixing to masonry with galvanized screws	11.95	1.00	**12.07**	0.50	-	**8.69**	-	**20.76**	nr	**22.31**
252 x 195 x 210mm; 110mm outlet spigot; push fit joint to plastics pipe; fixing to masonry with galvanized screws	42.98	1.00	**43.41**	0.80	-	**13.90**	-	**57.31**	nr	**61.61**
R10: RAINWATER PIPEWORK/GUTTERS: ROOF OUTLETS										
Cast iron pipework ancillaries										
Roof outlets; luting flange for asphalt										
flat grating; 100mm outlet spigot; coupling joint to cast iron pipe	76.94	1.00	**77.71**	1.00	-	**17.38**	-	**95.09**	nr	**102.22**
domical grating; 100mm outlet spigot; coupling joint to cast iron pipe	77.89	1.00	**78.67**	1.00	-	**17.38**	-	**96.05**	nr	**103.25**
Plastics pipework ancillaries, self coloured grey										
Roof outlets; BS EN 607; luting flange for asphalt										
domical grating; 75mm outlet spigot; push fit joint to plastics pipe	20.59	1.00	**20.80**	0.37	-	**6.43**	-	**27.23**	nr	**29.27**
domical grating; 100mm outlet spigot; push fit joint to plastics pipe	20.59	1.00	**20.80**	0.37	-	**6.43**	-	**27.23**	nr	**29.27**
R11: FOUL DRAINAGE ABOVE GROUND: CAST IRON SOIL AND WASTE PIPES										
Cast iron pipes and fittings, BS 416; bolted and synthetic rubber gasket couplings										
Pipes; straight										
75mm; in or on supports (included elsewhere)	13.10	5.00	**13.70**	0.67	-	**11.64**	-	**25.34**	m	**27.24**
extra; pipes with inspection door bolted and sealed	33.44	2.00	**34.11**	0.67	-	**11.64**	-	**45.75**	nr	**49.18**
extra; short radius bends	20.60	2.00	**21.01**	0.67	-	**11.64**	-	**32.66**	nr	**35.11**
extra; short radius bends with inspection door bolted and sealed	33.97	2.00	**34.65**	0.67	-	**11.64**	-	**46.29**	nr	**49.77**
extra; offset bends 150mm projection	24.33	2.00	**24.82**	0.67	-	**11.64**	-	**36.46**	nr	**39.20**
extra; single angled branches	30.94	2.00	**31.56**	1.00	-	**17.38**	-	**48.94**	nr	**52.61**
extra; single angled branches with inspection door bolted and sealed	45.62	2.00	**46.53**	1.00	-	**17.38**	-	**63.91**	nr	**68.71**
extra; double angled branches	46.03	2.00	**46.95**	1.33	-	**23.12**	-	**70.07**	nr	**75.32**
boss pipes with one threaded boss	30.38	2.00	**30.99**	0.67	-	**11.64**	-	**42.63**	nr	**45.83**
100mm; in or on supports (included elsewhere)	16.02	5.00	**16.75**	0.80	-	**13.90**	-	**30.65**	m	**32.95**
extra; pipes with inspection door bolted and sealed	38.07	2.00	**38.83**	0.80	-	**13.90**	-	**52.74**	nr	**56.69**
extra; short radius bends	27.61	2.00	**28.16**	0.80	-	**13.90**	-	**42.07**	nr	**45.22**
extra; short radius bends with inspection door bolted and sealed	41.71	2.00	**42.54**	0.80	-	**13.90**	-	**56.45**	nr	**60.68**
extra; offset bends 150mm projection	32.32	2.00	**32.97**	0.80	-	**13.90**	-	**46.87**	nr	**50.39**
extra; offset bends 300mm projection	35.52	2.00	**36.23**	0.80	-	**13.90**	-	**50.13**	nr	**53.89**
extra; single angled branches	41.98	2.00	**42.82**	1.20	-	**20.86**	-	**63.68**	nr	**68.45**
extra; single angled branches with inspection door bolted and sealed	56.10	2.00	**57.22**	1.20	-	**20.86**	-	**78.08**	nr	**83.93**
extra; double angled branches	54.10	2.00	**55.18**	1.60	-	**27.81**	-	**82.99**	nr	**89.21**
extra; double angled branches with inspection door bolted and sealed	68.21	2.00	**69.57**	1.60	-	**27.81**	-	**97.38**	nr	**104.69**
extra; single anti-syphon branches, angled branch	42.93	2.00	**43.79**	1.20	-	**20.86**	-	**64.64**	nr	**69.49**
extra; straight wc connectors	20.25	2.00	**20.66**	0.80	-	**13.90**	-	**34.56**	nr	**37.15**
extra; roof connectors	21.84	2.00	**22.28**	0.80	-	**13.90**	-	**36.18**	nr	**38.89**
boss pipes with one boss	37.56	2.00	**38.31**	0.80	-	**13.90**	-	**52.22**	nr	**56.13**
boss pipes with two bosses	44.47	2.00	**45.36**	0.80	-	**13.90**	-	**59.26**	nr	**63.71**
50mm; in standard wall fixing bracket (fixing included elsewhere)	11.73	5.00	**12.27**	0.55	-	**9.56**	-	**21.83**	m	**23.46**
extra; short radius bends	19.52	2.00	**19.91**	0.55	-	**9.56**	-	**29.47**	nr	**31.68**
extra; short radius bends with inspection door bolted and sealed	32.89	2.00	**33.55**	0.55	-	**9.56**	-	**43.11**	nr	**46.34**
extra; single angled branches	29.32	2.00	**29.91**	0.82	-	**14.25**	-	**44.16**	nr	**47.47**
extra; single angled branches with inspection door bolted and sealed	44.00	2.00	**44.88**	0.82	-	**14.25**	-	**59.13**	nr	**63.57**

Labour hourly rates: (except Specialists) Craft Operatives £17.38 Labourer £10.71 Rates are national average prices. Refer to REGIONAL VARIATIONS for indicative levels of overall pricing in regions	MATERIALS			LABOUR				RATES		
	Del to Site £	Waste %	Material Cost £	Craft Optve Hrs	Lab Hrs	Labour Cost £	Sunds £	Nett Rate £	Unit	Gross Rate (+7.5%) £
R11: FOUL DRAINAGE ABOVE GROUND: CAST IRON SOIL AND WASTE PIPES Cont.										
Cast iron lightweight soil and waste pipes and fittings										
Pipes; straight										
50mm; in or on supports (included elsewhere).....	8.48	5.00	**8.87**	0.55	-	**9.56**	-	**18.43**	m	**19.81**
extra; short pipe with access door..............	21.94	2.00	**22.38**	0.55	-	**9.56**	-	**31.94**	nr	**34.33**
extra; bends	12.39	2.00	**12.64**	0.55	-	**9.56**	-	**22.20**	nr	**23.86**
extra; offset bend 130mm projection.............	16.37	2.00	**16.70**	0.55	-	**9.56**	-	**26.26**	nr	**28.23**
extra; single angled branch.....................	19.17	2.00	**19.55**	0.82	-	**14.25**	-	**33.81**	nr	**36.34**
extra; plug	5.63	2.00	**5.74**	0.27	-	**4.69**	-	**10.44**	nr	**11.22**
extra; universal plug with one inlet............	8.75	2.00	**8.93**	0.14	-	**2.43**	-	**11.36**	nr	**12.21**
extra; step coupling for connection to BS 416 pipe	4.82	2.00	**4.92**	0.40	-	**6.95**	-	**11.87**	nr	**12.76**
75mm; in or on supports (included elsewhere).....	9.74	5.00	**10.20**	0.67	-	**11.64**	-	**21.84**	m	**23.48**
extra; short pipe with access door..............	23.48	2.00	**23.95**	0.67	-	**11.64**	-	**35.59**	nr	**38.26**
extra; bends	13.76	2.00	**14.04**	0.67	-	**11.64**	-	**25.68**	nr	**27.61**
extra; offset bend 130mm projection.............	22.06	2.00	**22.50**	0.67	-	**11.64**	-	**34.15**	nr	**36.71**
extra; single angled branch.....................	20.67	2.00	**21.08**	1.00	-	**17.38**	-	**38.46**	nr	**41.35**
extra; diminishing piece 75/50..................	15.80	2.00	**16.12**	0.67	-	**11.64**	-	**27.76**	nr	**29.84**
extra; plug	6.09	2.00	**6.21**	0.33	-	**5.74**	-	**11.95**	nr	**12.84**
extra; universal plug with one inlet............	9.37	2.00	**9.56**	0.17	-	**2.95**	-	**12.51**	nr	**13.45**
extra; step coupling for connection to BS 416 pipe	5.91	2.00	**6.03**	0.50	-	**8.69**	-	**14.72**	nr	**15.82**
100mm; in or on supports (included elsewhere)....	11.73	5.00	**12.27**	0.80	-	**13.90**	-	**26.17**	m	**28.14**
extra; short pipe with access door..............	27.31	2.00	**27.86**	0.80	-	**13.90**	-	**41.76**	nr	**44.89**
extra; bends	17.17	2.00	**17.51**	0.80	-	**13.90**	-	**31.42**	nr	**33.77**
extra; long radius bend.........................	27.64	2.00	**28.19**	0.80	-	**13.90**	-	**42.10**	nr	**45.25**
extra; offset bend 75mm projection..............	21.29	2.00	**21.72**	0.80	-	**13.90**	-	**35.62**	nr	**38.29**
extra; offset bend 130mm projection.............	28.90	2.00	**29.48**	0.80	-	**13.90**	-	**43.38**	nr	**46.64**
extra; single angled branch.....................	26.67	2.00	**27.20**	1.20	-	**20.86**	-	**48.06**	nr	**51.66**
extra; double angled branch.....................	36.86	2.00	**37.60**	1.60	-	**27.81**	-	**65.41**	nr	**70.31**
extra; diminishing piece 100/50.................	18.45	2.00	**18.82**	0.80	-	**13.90**	-	**32.72**	nr	**35.18**
extra; diminishing piece 100/75.................	18.78	2.00	**19.16**	0.80	-	**13.90**	-	**33.06**	nr	**35.54**
extra; plug	7.60	2.00	**7.75**	0.40	-	**6.95**	-	**14.70**	nr	**15.81**
extra; universal plug with three inlets..........	9.95	2.00	**10.15**	0.40	-	**6.95**	-	**17.10**	nr	**18.38**
extra; step coupling for connection to BS 416 pipe	6.94	2.00	**7.08**	0.60	-	**10.43**	-	**17.51**	nr	**18.82**
extra; traditional joint connector..............	16.19	2.00	**16.51**	0.80	-	**13.90**	-	**30.42**	nr	**32.70**
extra; roof connector (asphalt).................	21.56	2.00	**21.99**	0.80	-	**13.90**	-	**35.90**	nr	**38.59**
Pipe supports										
galvanized steel vertical bracket										
for 50mm pipes; for fixing bolt(fixing included elsewhere)	3.47	2.00	**3.54**	-	-	**-**	-	**3.54**	nr	**3.80**
for 75mm pipes; for fixing bolt(fixing included elsewhere)	3.67	2.00	**3.74**	-	-	**-**	-	**3.74**	nr	**4.02**
for 100mm pipes; for fixing bolt(fixing included elsewhere)	3.90	2.00	**3.98**	-	-	**-**	-	**3.98**	nr	**4.28**
R11: FOUL DRAINAGE ABOVE GROUND: UNPLASTICIZED MPVC WASTE PIPES (SOLVENT WELDED)										
MuPVC pipes and fittings, BS EN 1329-1 Section Three; solvent welded joints										
Pipes; straight										
32mm; in standard plastics pipe brackets fixing to timber with screws	1.81	10.00	**1.95**	0.33	-	**5.74**	-	**7.68**	m	**8.26**
extra; straight expansion couplings.............	1.44	2.00	**1.47**	0.25	-	**4.35**	-	**5.81**	nr	**6.25**
extra; connections to plastics pipe socket; socket adaptor; solvent welded joint...................	1.53	2.00	**1.56**	0.25	-	**4.35**	-	**5.91**	nr	**6.35**
extra; connections to plastics pipe boss; boss adaptor; solvent welded joint...................	1.53	2.00	**1.56**	0.25	-	**4.35**	-	**5.91**	nr	**6.35**
extra; fittings; one end.........................	1.58	2.00	**1.61**	0.13	-	**2.26**	-	**3.87**	nr	**4.16**
extra; fittings; two ends........................	0.88	2.00	**0.90**	0.25	-	**4.35**	-	**5.24**	nr	**5.64**
extra; fittings; three ends......................	1.13	2.00	**1.15**	0.25	-	**4.35**	-	**5.50**	nr	**5.91**
40mm; in standard plastics pipe brackets fixing to timber with screws	2.16	10.00	**2.33**	0.33	-	**5.74**	-	**8.06**	m	**8.67**
extra; straight expansion couplings.............	1.73	2.00	**1.76**	0.25	-	**4.35**	-	**6.11**	nr	**6.57**
extra; connections to plastics pipe socket; socket adaptor; solvent welded joint...................	1.53	2.00	**1.56**	0.25	-	**4.35**	-	**5.91**	nr	**6.35**
extra; connections to plastics pipe boss; boss adaptor; solvent welded joint...................	1.53	2.00	**1.56**	0.25	-	**4.35**	-	**5.91**	nr	**6.35**
extra; fittings; one end.........................	1.57	2.00	**1.60**	0.13	-	**2.26**	-	**3.86**	nr	**4.15**
extra; fittings; two ends........................	1.00	2.00	**1.02**	0.25	-	**4.35**	-	**5.37**	nr	**5.77**
extra; fittings; three ends......................	1.41	2.00	**1.44**	0.25	-	**4.35**	-	**5.78**	nr	**6.22**
50mm; in standard plastics pipe brackets fixing to timber with screws	3.27	10.00	**3.53**	0.40	-	**6.95**	-	**10.48**	m	**11.26**
extra; straight expansion couplings.............	2.35	2.00	**2.40**	0.33	-	**5.74**	-	**8.13**	nr	**8.74**
extra; connections to plastics pipe socket; socket adaptor; solvent welded joint...................	2.19	2.00	**2.23**	0.33	-	**5.74**	-	**7.97**	nr	**8.57**
extra; connections to plastics pipe boss; boss adaptor; solvent welded joint...................	2.19	2.00	**2.23**	0.33	-	**5.74**	-	**7.97**	nr	**8.57**
extra; fittings; one end.........................	2.58	2.00	**2.63**	0.17	-	**2.95**	-	**5.59**	nr	**6.01**
extra; fittings; two ends........................	1.68	2.00	**1.71**	0.33	-	**5.74**	-	**7.45**	nr	**8.01**
extra; fittings; three ends......................	2.76	2.00	**2.82**	0.33	-	**5.74**	-	**8.55**	nr	**9.19**
32mm; in standard plastics pipe brackets fixing to masonry with screws	1.84	10.00	**1.98**	0.39	-	**6.78**	-	**8.76**	m	**9.41**
40mm; in standard plastics pipe brackets fixing to masonry with screws	2.19	10.00	**2.36**	0.39	-	**6.78**	-	**9.13**	m	**9.82**
50mm; in standard plastics pipe brackets fixing to masonry with screws	3.30	10.00	**3.56**	0.46	-	**7.99**	-	**11.55**	m	**12.42**

Labour hourly rates: (except Specialists) Craft Operatives £17.38 Labourer £10.71 Rates are national average prices. Refer to REGIONAL VARIATIONS for indicative levels of overall pricing in regions	MATERIALS			LABOUR				RATES		
	Del to Site £	Waste %	Material Cost £	Craft Optve Hrs	Lab Hrs	Labour Cost £	Sunds £	Nett Rate £	Unit	Gross Rate (+7.5%) £
R11: FOUL DRAINAGE ABOVE GROUND: POLYPROPYLENE WASTE PIPES ('O' RING)										
Polypropylene pipes and fittings; butyl ring joints										
Pipes; straight										
32mm; in standard plastics pipe brackets fixing to timber with screws	1.30	10.00	**1.39**	0.33	-	**5.74**	-	**7.13**	m	**7.66**
extra; connections to plastics pipe socket; socket adaptor; solvent welded joint	1.66	2.00	**1.69**	0.25	-	**4.35**	-	**6.04**	nr	**6.49**
extra; connections to plastics pipe boss; boss adaptor; solvent welded joint	1.66	2.00	**1.69**	0.25	-	**4.35**	-	**6.04**	nr	**6.49**
extra; fittings; one end	1.49	2.00	**1.52**	0.13	-	**2.26**	-	**3.78**	nr	**4.06**
extra; fittings; two ends	0.86	2.00	**0.88**	0.25	-	**4.35**	-	**5.22**	nr	**5.61**
extra; fittings; three ends	1.33	2.00	**1.36**	0.25	-	**4.35**	-	**5.70**	nr	**6.13**
40mm; in standard plastics pipe brackets fixing to timber with screws	1.56	10.00	**1.67**	0.33	-	**5.74**	-	**7.41**	m	**7.96**
extra; connections to plastics pipe socket; socket adaptor; solvent welded joint	1.66	2.00	**1.69**	0.33	-	**5.74**	-	**7.43**	nr	**7.99**
extra; connections to plastics pipe boss; boss adaptor; solvent welded joint	1.66	2.00	**1.69**	0.33	-	**5.74**	-	**7.43**	nr	**7.99**
extra; fittings; one end	1.69	2.00	**1.72**	0.17	-	**2.95**	-	**4.68**	nr	**5.03**
extra; fittings; two ends	0.90	2.00	**0.92**	0.33	-	**5.74**	-	**6.65**	nr	**7.15**
extra; fittings; three ends	1.38	2.00	**1.41**	0.33	-	**5.74**	-	**7.14**	nr	**7.68**
32mm; in standard plastics pipe brackets fixing to masonry with screws	1.33	10.00	**1.42**	0.39	-	**6.78**	-	**8.20**	m	**8.81**
40mm; in standard plastics pipe brackets fixing to masonry with screws	1.59	10.00	**1.70**	0.39	-	**6.78**	-	**8.48**	m	**9.11**
R11: FOUL DRAINAGE ABOVE GROUND: UNPLASTICIZED SOIL AND VENTILATING PIPES										
PVC-U pipes and fittings, BS 4514; rubber ring joints; pipework self coloured grey										
Pipes; straight										
82mm; in or on supports (included elsewhere)	7.69	5.00	**8.03**	0.44	-	**7.65**	-	**15.68**	m	**16.86**
extra; pipes with inspection door	15.00	2.00	**15.30**	0.33	-	**5.74**	-	**21.04**	nr	**22.61**
extra; short radius bends	8.42	2.00	**8.59**	0.33	-	**5.74**	-	**14.32**	nr	**15.40**
extra; single angled branches	12.36	2.00	**12.61**	0.33	-	**5.74**	-	**18.34**	nr	**19.72**
pipes with one boss socket	6.69	2.00	**6.82**	0.33	-	**5.74**	-	**12.56**	nr	**13.50**
110mm; in or on supports (included elsewhere)	6.84	5.00	**7.14**	0.50	-	**8.69**	-	**15.83**	m	**17.02**
extra; pipes with inspection door	14.38	2.00	**14.67**	0.50	-	**8.69**	-	**23.36**	nr	**25.11**
extra; short radius bends	9.85	2.00	**10.05**	0.50	-	**8.69**	-	**18.74**	nr	**20.14**
extra; single angled branches	12.83	2.00	**13.09**	0.50	-	**8.69**	-	**21.78**	nr	**23.41**
extra; straight wc connectors	3.88	2.00	**3.96**	0.50	-	**8.69**	-	**12.65**	nr	**13.60**
pipes with one boss socket	7.21	2.00	**7.35**	0.50	-	**8.69**	-	**16.04**	nr	**17.25**
160mm; in or on supports (included elsewhere)	18.33	5.00	**19.15**	0.75	-	**13.04**	-	**32.19**	m	**34.60**
extra; pipes with inspection door	34.05	2.00	**34.73**	0.75	-	**13.04**	-	**47.77**	nr	**51.35**
extra; short radius bends	27.76	2.00	**28.32**	0.75	-	**13.04**	-	**41.35**	nr	**44.45**
extra; single angled branches	45.72	2.00	**46.63**	0.75	-	**13.04**	-	**59.67**	nr	**64.14**
pipes with one boss socket	13.98	2.00	**14.26**	0.75	-	**13.04**	-	**27.29**	nr	**29.34**
PVC-U pipes and fittings, BS 4514; solvent weld joints; pipework self coloured grey										
Pipes; straight										
110mm; in or on supports (included elsewhere)	6.84	5.00	**7.14**	0.53	-	**9.21**	-	**16.35**	m	**17.58**
extra; pipes with inspection door	14.38	2.00	**14.67**	0.53	-	**9.21**	-	**23.88**	nr	**25.67**
extra; short radius bends	9.85	2.00	**10.05**	0.53	-	**9.21**	-	**19.26**	nr	**20.70**
extra; single angled branches	12.83	2.00	**13.09**	0.53	-	**9.21**	-	**22.30**	nr	**23.97**
extra; straight wc connectors	3.88	2.00	**3.96**	0.53	-	**9.21**	-	**13.17**	nr	**14.16**
pipes with one boss socket	7.21	2.00	**7.35**	0.53	-	**9.21**	-	**16.57**	nr	**17.81**
160mm; in or on supports (included elsewhere)	18.33	5.00	**19.15**	0.80	-	**13.90**	-	**33.06**	m	**35.54**
extra; pipes with inspection door	34.05	2.00	**34.73**	0.80	-	**13.90**	-	**48.64**	nr	**52.28**
extra; short radius bends	27.76	2.00	**28.32**	0.80	-	**13.90**	-	**42.22**	nr	**45.39**
extra; single angled branches	45.72	2.00	**46.63**	0.80	-	**13.90**	-	**60.54**	nr	**65.08**
pipes with one boss socket	13.98	2.00	**14.26**	0.80	-	**13.90**	-	**28.16**	nr	**30.28**
Plastics pipework ancillaries, self coloured grey										
Weathering aprons										
fitted to 82mm pipes	3.07	2.00	**3.13**	0.33	-	**5.74**	-	**8.87**	nr	**9.53**
fitted to 110mm pipes	2.37	2.00	**2.42**	0.50	-	**8.69**	-	**11.11**	nr	**11.94**
Pipe supports for PVC-U soil and vent pipes, rubber ring joints; self coloured grey										
Plastics coated metal holderbats										
for 82mm pipes; for building in (fixing included elsewhere)	2.99	2.00	**3.05**	-	-	**-**	-	**3.05**	nr	**3.28**
for 110mm pipes; for building in (fixing included elsewhere)	4.28	2.00	**4.37**	-	-	**-**	-	**4.37**	nr	**4.69**
for 82mm pipes; fixing to timber with galvanized screws	2.45	2.00	**2.50**	0.17	-	**2.95**	-	**5.46**	nr	**5.87**
for 110mm pipes; fixing to timber with galvanized screws	3.21	2.00	**3.28**	0.17	-	**2.95**	-	**6.23**	nr	**6.70**
for 160mm pipes; fixing to timber with galvanized screws	6.09	2.00	**6.21**	0.17	-	**2.95**	-	**9.17**	nr	**9.86**
Pipe supports; solvent weld joints;										
Plastic brackets										
for 110mm pipes; for building in (fixing included elsewhere)	4.74	2.00	**4.83**	-	-	**-**	-	**4.83**	nr	**5.20**

Labour hourly rates: (except Specialists) Craft Operatives £17.38 Labourer £10.71 Rates are national average prices. Refer to REGIONAL VARIATIONS for indicative levels of overall pricing in regions	MATERIALS			LABOUR				RATES		
	Del to Site £	Waste %	Material Cost £	Craft Optve Hrs	Lab Hrs	Labour Cost £	Sunds £	Nett Rate £	Unit	Gross Rate (+7.5%) £
R11: FOUL DRAINAGE ABOVE GROUND: UNPLASTICIZED SOIL AND VENTILATING PIPES Cont.										
Pipe supports; solvent weld joints; Cont.										
Plastic brackets Cont.										
for 110mm pipes; fixing to timber with galvanized screws	2.32	2.00	**2.37**	0.25	-	**4.35**	-	**6.71**	nr	**7.22**
for 160mm pipes; fixing to timber with galvanized screws	5.52	2.00	**5.63**	0.25	-	**4.35**	-	**9.98**	nr	**10.73**
R11: FOUL DRAINAGE ABOVE GROUND: WIRE BALLOON GRATINGS										
Pipework ancillaries										
Wire balloon gratings										
fitted to 50mm pipes	1.56	3.00	**1.61**	0.10	-	**1.74**	-	**3.34**	nr	**3.60**
fitted to 82mm pipes	2.28	3.00	**2.35**	0.10	-	**1.74**	-	**4.09**	nr	**4.39**
fitted to 110mm pipes	2.37	3.00	**2.44**	0.10	-	**1.74**	-	**4.18**	nr	**4.49**
R11: FOUL DRAINAGE ABOVE GROUND: PLASTIC TRAPS										
Plastics pipework ancillaries self coloured white										
Traps; BS EN 274										
P, two piece, 75mm seal, inlet with coupling nut, outlet with seal ring socket										
32mm outlet	3.34	2.00	**3.41**	0.33	-	**5.74**	-	**9.14**	nr	**9.83**
40mm outlet	3.86	2.00	**3.94**	0.33	-	**5.74**	-	**9.67**	nr	**10.40**
S, two piece, 75mm seal, inlet with coupling nut, outlet with seal ring socket										
32mm outlet	4.23	2.00	**4.31**	0.33	-	**5.74**	-	**10.05**	nr	**10.80**
40mm outlet	4.95	2.00	**5.05**	0.33	-	**5.74**	-	**10.78**	nr	**11.59**
P, two piece, 75mm seal, with flexible polypropylene pipe for overflow connection, inlet with coupling nut, outlet with seal ring socket										
40mm outlet	7.15	2.00	**7.29**	0.50	-	**8.69**	-	**15.98**	nr	**17.18**
R11: FOUL DRAINAGE ABOVE GROUND: COPPER TRAPS										
Copper pipework ancillaries										
Traps; BS 1184 Section Two (Solid Drawn)										
P, two piece, 75mm seal, inlet with coupling nut, compression outlet										
35mm outlet	17.40	2.00	**17.75**	0.33	-	**5.74**	-	**23.48**	nr	**25.24**
42mm outlet	20.50	2.00	**20.91**	0.33	-	**5.74**	-	**26.65**	nr	**28.64**
54mm outlet	73.63	2.00	**75.10**	0.70	-	**12.17**	-	**87.27**	nr	**93.81**
S, two piece, 75mm seal, inlet with coupling nut, compression outlet										
35mm outlet	18.44	2.00	**18.81**	0.33	-	**5.74**	-	**24.54**	nr	**26.39**
42mm outlet	22.18	2.00	**22.62**	0.33	-	**5.74**	-	**28.36**	nr	**30.49**
54mm outlet	77.69	2.00	**79.24**	0.70	-	**12.17**	-	**91.41**	nr	**98.27**
bath, two piece, 75mm seal, overflow connection, cleaning eye and plug, inlet with coupling nut, compression outlet										
42mm outlet	29.10	2.00	**29.68**	0.50	-	**8.69**	-	**38.37**	nr	**41.25**

This section continues on the next page

Labour hourly rates: (except Specialists) Craft Operatives £14.21 Labourer £10.71 Rates are national average prices. Refer to REGIONAL VARIATIONS for indicative levels of overall pricing in regions	PLANT & TRANSPORT			LABOUR				RATES		
	Plant Cost £	Trans Cost £	P & T Cost £	Craft Optve Hrs	Lab Hrs	Labour Cost £	Sunds £	Nett Rate £	Unit	Gross Rate (+7.5%) £
R12: DRAINAGE BELOW GROUND: EXCAVATION FOR DRAINS - BY MACHINE										
Excavating trenches by machine to receive pipes not exceeding 200mm nominal size; disposing of surplus excavated material by removing from site										
Excavations commencing from natural ground level average depth										
500mm	1.32	2.88	4.20	-	0.23	2.46	0.78	7.44	m	8.00
750mm	2.13	2.88	5.01	-	0.34	3.64	1.18	9.83	m	10.57
1000mm	2.88	2.88	5.76	-	0.45	4.82	1.58	12.16	m	13.07
1250mm	3.70	2.88	6.58	-	0.61	6.53	1.97	15.08	m	16.21
1500mm	4.44	2.88	7.32	-	0.73	7.82	2.36	17.51	m	18.82
1750mm	5.26	2.88	8.14	-	0.86	9.21	2.76	20.11	m	21.62
2000mm	5.98	2.88	8.86	-	0.98	10.50	3.15	22.51	m	24.19
2250mm	7.63	2.88	10.51	-	1.69	18.10	3.55	32.16	m	34.58
2500mm	8.47	2.88	11.35	-	1.88	20.13	3.94	35.43	m	38.09
2750mm	9.38	2.88	12.26	-	2.06	22.06	4.34	38.66	m	41.56
3000mm	10.22	2.88	13.10	-	2.25	24.10	4.73	41.93	m	45.07
3250mm	11.13	2.88	14.01	-	2.44	26.13	5.12	45.26	m	48.66
3500mm	11.94	2.88	14.82	-	2.63	28.17	5.51	48.50	m	52.14
3750mm	12.85	2.88	15.73	-	2.82	30.20	5.91	51.85	m	55.74
4000mm	13.72	2.88	16.60	-	3.00	32.13	6.31	55.04	m	59.16
4250mm	14.60	2.88	17.48	-	4.25	45.52	6.70	69.70	m	74.93
4500mm	15.44	2.88	18.32	-	4.50	48.20	7.09	73.61	m	79.13
4750mm	16.35	2.88	19.23	-	4.75	50.87	7.48	77.59	m	83.41
5000mm	17.16	2.88	20.04	-	6.87	73.58	8.67	102.29	m	109.96
5250mm	18.08	2.88	20.96	-	6.56	70.26	8.28	99.49	m	106.95
5500mm	18.91	2.88	21.79	-	6.87	73.58	8.67	104.04	m	111.85
5750mm	19.82	2.88	22.70	-	7.18	76.90	9.06	108.66	m	116.81
6000mm	20.90	2.88	23.78	-	7.50	80.33	9.45	113.56	m	122.07
Excavating trenches by machine to receive pipes 225mm nominal size; disposing of surplus excavated material by removing from site										
Excavations commencing from natural ground level average depth										
500mm	1.42	3.04	4.46	-	0.24	2.57	0.78	7.81	m	8.40
750mm	2.23	3.04	5.27	-	0.36	3.86	1.18	10.31	m	11.08
1000mm	3.05	3.04	6.09	-	0.47	5.03	1.58	12.71	m	13.66
1250mm	3.86	3.04	6.90	-	0.64	6.85	1.97	15.73	m	16.91
1500mm	4.72	3.04	7.76	-	0.77	8.25	2.36	18.37	m	19.74
1750mm	5.53	3.04	8.57	-	0.90	9.64	2.76	20.97	m	22.54
2000mm	6.34	3.04	9.38	-	1.03	11.03	3.15	23.56	m	25.33
2250mm	8.04	3.04	11.08	-	1.77	18.96	3.55	33.59	m	36.11
2500mm	9.00	3.04	12.04	-	1.97	21.10	3.94	37.08	m	39.86
2750mm	9.91	3.04	12.95	-	2.16	23.13	4.34	40.42	m	43.45
3000mm	10.82	3.04	13.86	-	2.36	25.28	4.73	43.86	m	47.15
3250mm	11.73	3.04	14.77	-	2.56	27.42	5.12	47.30	m	50.85
3500mm	12.68	3.04	15.72	-	2.76	29.56	5.51	50.79	m	54.60
3750mm	13.59	3.04	16.63	-	2.96	31.70	5.91	54.25	m	58.32
4000mm	14.43	3.04	17.47	-	3.15	33.74	6.31	57.51	m	61.82
4250mm	15.42	3.04	18.46	-	4.46	47.77	6.70	72.92	m	78.39
4500mm	16.40	3.04	19.44	-	4.70	50.34	7.09	76.87	m	82.63
4750mm	17.28	3.04	20.32	-	4.98	53.34	7.48	81.14	m	87.23
5000mm	18.19	3.04	21.23	-	6.56	70.26	8.67	100.16	m	107.67
5250mm	19.10	3.04	22.14	-	6.89	73.79	8.28	104.21	m	112.03
5500mm	20.09	3.04	23.13	-	7.20	77.11	8.67	108.91	m	117.08
5750mm	20.97	3.04	24.01	-	7.54	80.75	9.06	113.83	m	122.36
6000mm	21.88	3.04	24.92	-	7.85	84.07	9.45	118.45	m	127.33
Excavating trenches by machine to receive pipes 300mm nominal size; disposing of surplus excavated material by removing from site										
Excavations commencing from natural ground level average depth										
500mm	1.56	3.36	4.92	-	0.25	2.68	0.78	8.38	m	9.01
750mm	2.51	3.36	5.87	-	0.37	3.96	1.18	11.00	m	11.83
1000mm	3.42	3.36	6.78	-	0.50	5.36	1.58	13.72	m	14.75
1250mm	4.37	3.36	7.73	-	0.67	7.18	1.97	16.87	m	18.14
1500mm	5.31	3.36	8.67	-	0.80	8.57	2.36	19.60	m	21.07
1750mm	6.23	3.36	9.59	-	0.95	10.17	2.76	22.52	m	24.21
2000mm	7.17	3.36	10.53	-	1.10	11.78	3.15	25.46	m	27.37
2250mm	9.09	3.36	12.45	-	1.86	19.92	3.55	35.92	m	38.61
2500mm	10.14	3.36	13.50	-	2.07	22.17	3.94	39.62	m	42.59
2750mm	11.20	3.36	14.56	-	2.25	24.10	4.34	42.99	m	46.21
3000mm	12.23	3.36	15.59	-	2.48	26.56	4.73	46.87	m	50.39
3250mm	13.28	3.36	16.64	-	2.68	28.70	5.12	50.46	m	54.25
3500mm	14.31	3.36	17.67	-	2.90	31.06	5.51	54.24	m	58.31
3750mm	15.36	3.36	18.72	-	3.10	33.20	5.91	57.84	m	62.18
4000mm	16.42	3.36	19.78	-	3.30	35.34	6.31	61.43	m	66.04
4250mm	17.45	3.36	20.81	-	4.68	50.12	6.70	77.63	m	83.45
4500mm	18.50	3.36	21.86	-	4.95	53.01	7.09	81.97	m	88.11
4750mm	19.53	3.36	22.89	-	5.23	56.01	7.48	86.39	m	92.87
5000mm	20.59	3.36	23.95	-	6.88	73.68	8.67	106.30	m	114.27
5250mm	21.64	3.36	25.00	-	7.22	77.33	8.28	110.60	m	118.90
5500mm	22.67	3.36	26.03	-	7.55	80.86	8.67	115.56	m	124.23
5750mm	23.72	3.36	27.08	-	7.90	84.61	9.06	120.75	m	129.81
6000mm	24.78	3.36	28.14	-	8.25	88.36	9.45	125.95	m	135.40

Labour hourly rates: (except Specialists) Craft Operatives £14.21 Labourer £10.71 Rates are national average prices. Refer to REGIONAL VARIATIONS for indicative levels of overall pricing in regions	PLANT & TRANSPORT			LABOUR				RATES		
	Plant Cost £	Trans Cost £	P & T Cost £	Craft Optve Hrs	Lab Hrs	Labour Cost £	Sunds £	Nett Rate £	Unit	Gross Rate (+7.5%) £
R12: DRAINAGE BELOW GROUND: EXCAVATION FOR DRAINS - BY MACHINE Cont.										
Excavating trenches by machine to receive pipes 400mm nominal size; disposing of surplus excavated material by removing from site										
Excavations commencing from natural ground level average depth										
2250mm	10.62	4.00	**14.62**	-	2.03	**21.74**	3.55	**39.91**	m	**42.90**
2500mm	11.79	4.00	**15.79**	-	2.26	**24.20**	3.94	**43.94**	m	**47.23**
2750mm	13.03	4.00	**17.03**	-	2.47	**26.45**	4.34	**47.82**	m	**51.41**
3000mm	14.23	4.00	**18.23**	-	2.70	**28.92**	4.73	**51.88**	m	**55.77**
3250mm	15.48	4.00	**19.48**	-	2.92	**31.27**	5.12	**55.87**	m	**60.06**
3500mm	16.68	4.00	**20.68**	-	3.16	**33.84**	5.51	**60.03**	m	**64.53**
3750mm	17.92	4.00	**21.92**	-	3.38	**36.20**	5.91	**64.03**	m	**68.84**
4000mm	19.12	4.00	**23.12**	-	3.60	**38.56**	6.31	**67.98**	m	**73.08**
4250mm	20.36	4.00	**24.36**	-	5.10	**54.62**	6.70	**85.68**	m	**92.11**
4500mm	21.54	4.00	**25.54**	-	5.40	**57.83**	7.09	**90.46**	m	**97.25**
4750mm	15.57	4.00	**19.57**	-	5.70	**61.05**	7.83	**88.44**	m	**95.07**
5000mm	23.98	4.00	**27.98**	-	7.50	**80.33**	8.67	**116.98**	m	**125.75**
5250mm	25.22	4.00	**29.22**	-	7.85	**84.07**	8.28	**121.58**	m	**130.69**
5500mm	26.42	4.00	**30.42**	-	8.25	**88.36**	8.67	**127.45**	m	**137.01**
5750mm	27.67	4.00	**31.67**	-	8.60	**92.11**	9.06	**132.84**	m	**142.80**
6000mm	19.58	4.00	**23.58**	-	9.00	**96.39**	9.89	**129.87**	m	**139.61**
Excavating trenches by machine to receive pipes 450mm nominal size; disposing of surplus excavated material by removing from site										
Excavations commencing from natural ground level average depth										
2250mm	11.36	4.16	**15.52**	-	2.11	**22.60**	3.55	**41.67**	m	**44.79**
2500mm	12.67	4.16	**16.83**	-	2.35	**25.17**	3.94	**45.94**	m	**49.39**
2750mm	13.92	4.16	**18.08**	-	2.58	**27.63**	4.34	**50.04**	m	**53.80**
3000mm	15.26	4.16	**19.42**	-	2.81	**30.10**	4.73	**54.24**	m	**58.31**
3250mm	15.70	4.16	**19.86**	-	3.05	**32.67**	5.12	**57.64**	m	**61.97**
3500mm	17.89	4.16	**22.05**	-	3.29	**35.24**	5.51	**62.80**	m	**67.51**
3750mm	19.14	4.16	**23.30**	-	3.52	**37.70**	5.91	**66.91**	m	**71.93**
4000mm	20.48	4.16	**24.64**	-	3.75	**40.16**	6.31	**71.11**	m	**76.44**
4250mm	21.80	4.16	**25.96**	-	5.30	**56.76**	6.70	**89.42**	m	**96.13**
4500mm	23.12	4.16	**27.28**	-	5.63	**60.30**	7.09	**94.66**	m	**101.76**
4750mm	24.36	4.16	**28.52**	-	5.95	**63.72**	7.48	**99.73**	m	**107.21**
5000mm	25.70	4.16	**29.86**	-	7.80	**83.54**	8.67	**122.07**	m	**131.23**
5250mm	27.02	4.16	**31.18**	-	8.20	**87.82**	8.28	**127.28**	m	**136.83**
5500mm	28.34	4.16	**32.50**	-	8.60	**92.11**	8.67	**133.27**	m	**143.27**
5750mm	29.58	4.16	**33.74**	-	8.95	**95.85**	9.06	**138.66**	m	**149.06**
6000mm	30.92	4.16	**35.08**	-	9.35	**100.14**	9.45	**144.68**	m	**155.53**
R12: DRAINAGE BELOW GROUND: EXCAVATION FOR DRAINS - BY HAND										
Excavating trenches by hand to receive pipes not exceeding 200mm nominal size; disposing of surplus excavated material on site										
Excavations commencing from natural ground level average depth										
500mm	-	-	**-**	-	1.15	**12.32**	0.78	**13.10**	m	**14.08**
750mm	-	-	**-**	-	2.05	**21.96**	1.18	**23.13**	m	**24.87**
1000mm	-	-	**-**	-	2.95	**31.59**	1.58	**33.17**	m	**35.66**
1250mm	-	-	**-**	-	3.85	**41.23**	1.97	**43.21**	m	**46.45**
1500mm	-	-	**-**	-	4.75	**50.87**	2.36	**53.24**	m	**57.23**
1750mm	-	-	**-**	-	5.65	**60.51**	2.76	**63.27**	m	**68.01**
2000mm	-	-	**-**	-	6.50	**69.62**	3.15	**72.76**	m	**78.22**
2250mm	-	-	**-**	-	7.65	**81.93**	3.55	**85.48**	m	**91.89**
2500mm	-	-	**-**	-	8.80	**94.25**	3.94	**98.19**	m	**105.56**
2750mm	-	-	**-**	-	9.95	**106.56**	4.34	**110.90**	m	**119.22**
3000mm	-	-	**-**	-	11.10	**118.88**	4.73	**123.61**	m	**132.88**
3250mm	-	-	**-**	-	12.25	**131.20**	5.12	**136.32**	m	**146.54**
3500mm	-	-	**-**	-	13.45	**144.05**	5.51	**149.56**	m	**160.78**
3750mm	-	-	**-**	-	14.60	**156.37**	5.91	**162.28**	m	**174.45**
4000mm	-	-	**-**	-	15.80	**169.22**	6.31	**175.52**	m	**188.69**
Excavating trenches by hand to receive pipes 225mm nominal size; disposing of surplus excavated material on site										
Excavations commencing from natural ground level average depth										
500mm	-	-	**-**	-	1.28	**13.71**	0.78	**14.49**	m	**15.58**
750mm	-	-	**-**	-	2.23	**23.88**	1.18	**25.06**	m	**26.94**
1000mm	-	-	**-**	-	3.18	**34.06**	1.58	**35.64**	m	**38.31**
1250mm	-	-	**-**	-	4.13	**44.23**	1.97	**46.20**	m	**49.67**
1500mm	-	-	**-**	-	5.08	**54.41**	2.36	**56.77**	m	**61.03**
1750mm	-	-	**-**	-	6.03	**64.58**	2.76	**67.34**	m	**72.39**
2000mm	-	-	**-**	-	6.95	**74.43**	3.15	**77.58**	m	**83.40**
2250mm	-	-	**-**	-	8.23	**88.14**	3.55	**91.69**	m	**98.57**
2500mm	-	-	**-**	-	9.45	**101.21**	3.94	**105.15**	m	**113.04**
2750mm	-	-	**-**	-	10.70	**114.60**	4.34	**118.93**	m	**127.85**
3000mm	-	-	**-**	-	11.95	**127.98**	4.73	**132.71**	m	**142.67**
3250mm	-	-	**-**	-	13.20	**141.37**	5.12	**146.49**	m	**157.48**
3500mm	-	-	**-**	-	14.45	**154.76**	5.51	**160.27**	m	**172.29**
3750mm	-	-	**-**	-	15.70	**168.15**	5.91	**174.06**	m	**187.12**
4000mm	-	-	**-**	-	17.00	**182.07**	6.31	**188.38**	m	**202.51**

Labour hourly rates: (except Specialists) Craft Operatives £14.21 Labourer £10.71 Rates are national average prices. Refer to REGIONAL VARIATIONS for indicative levels of overall pricing in regions	PLANT & TRANSPORT			LABOUR				RATES		
	Plant Cost £	Trans Cost £	P & T Cost £	Craft Optve Hrs	Lab Hrs	Labour Cost £	Sunds £	Nett Rate £	Unit	Gross Rate (+7.5%) £
R12: DRAINAGE BELOW GROUND: EXCAVATION FOR DRAINS - BY HAND Cont.										
Excavating trenches by hand to receive pipes 300mm nominal size; disposing of surplus material on site										
Excavations commencing from natural ground level average depth										
500mm	-	-	-	-	1.40	**14.99**	0.78	**15.78**	m	**16.96**
750mm	-	-	-	-	2.40	**25.70**	1.18	**26.88**	m	**28.90**
1000mm	-	-	-	-	3.40	**36.41**	1.58	**37.99**	m	**40.84**
1250mm	-	-	-	-	4.40	**47.12**	1.97	**49.10**	m	**52.78**
1500mm	-	-	-	-	5.40	**57.83**	2.36	**60.20**	m	**64.71**
1750mm	-	-	-	-	6.40	**68.54**	2.76	**71.30**	m	**76.65**
2000mm	-	-	-	-	7.40	**79.25**	3.15	**82.40**	m	**88.58**
2250mm	-	-	-	-	8.80	**94.25**	3.55	**97.80**	m	**105.13**
2500mm	-	-	-	-	10.10	**108.17**	3.94	**112.11**	m	**120.52**
2750mm	-	-	-	-	11.45	**122.63**	4.34	**126.96**	m	**136.49**
3000mm	-	-	-	-	12.75	**136.55**	4.73	**141.28**	m	**151.88**
3250mm	-	-	-	-	14.10	**151.01**	5.12	**156.13**	m	**167.84**
3500mm	-	-	-	-	15.45	**165.47**	5.51	**170.98**	m	**183.80**
3750mm	-	-	-	-	16.80	**179.93**	5.91	**185.84**	m	**199.78**
4000mm	-	-	-	-	18.15	**194.39**	6.31	**200.69**	m	**215.75**

This section continues on the next page

Labour hourly rates: (except Specialists)
Craft Operatives £14.21 Labourer £10.71
Rates are national average prices.
Refer to REGIONAL VARIATIONS for indicative levels of overall pricing in regions

	MATERIALS			LABOUR				RATES		
	Del to Site £	Waste %	Material Cost £	Craft Optve Hrs	Lab Hrs	Labour Cost £	Sunds £	Nett Rate £	Unit	Gross Rate (+7.5%) £
R12: DRAINAGE BELOW GROUND: BEDS, BENCHINGS AND COVERS										
Granular material, 10mm nominal size pea shingle, to be obtained off site										
Beds										
400 x 150mm	1.51	10.00	**1.66**	-	0.15	**1.61**	-	**3.27**	m	**3.51**
450 x 150mm	1.69	10.00	**1.86**	-	0.16	**1.71**	-	**3.57**	m	**3.84**
525 x 150mm	1.97	10.00	**2.16**	-	0.17	**1.82**	-	**3.98**	m	**4.28**
600 x 150mm	2.27	10.00	**2.49**	-	0.22	**2.36**	-	**4.85**	m	**5.21**
700 x 150mm	2.65	10.00	**2.91**	-	0.24	**2.57**	-	**5.48**	m	**5.89**
750 x 150mm	2.85	10.00	**3.13**	-	0.27	**2.89**	-	**6.02**	m	**6.48**
Beds and surrounds										
400 x 150mm bed; 150mm thick surround to 100mm internal diameter pipes -1	4.03	10.00	**4.44**	-	0.46	**4.93**	-	**9.36**	m	**10.06**
450 x 150mm bed; 150mm thick surround to 150mm internal diameter pipes -1	5.09	10.00	**5.60**	-	0.52	**5.57**	-	**11.17**	m	**12.01**
525 x 150mm bed; 150mm thick surround to 225mm internal diameter pipes -1	6.93	10.00	**7.62**	-	0.68	**7.28**	-	**14.91**	m	**16.02**
600 x 150mm bed; 150mm thick surround to 300mm internal diameter pipes -1	9.07	10.00	**9.98**	-	0.84	**9.00**	-	**18.98**	m	**20.40**
700 x 150mm bed; 150mm thick surround to 400mm internal diameter pipes -1	12.35	10.00	**13.58**	-	1.00	**10.71**	-	**24.29**	m	**26.11**
750 x 150mm bed; 150mm thick surround to 450mm internal diameter pipes -1	14.11	10.00	**15.52**	-	1.16	**12.42**	-	**27.95**	m	**30.04**
Beds and filling to half pipe depth										
400 x 150mm overall; to 100mm internal diameter pipes - 1	3.53	10.00	**3.88**	-	0.12	**1.29**	-	**5.17**	m	**5.55**
450 x 225mm overall; to 150mm internal diameter pipes - 1	4.23	10.00	**4.66**	-	0.20	**2.14**	-	**6.80**	m	**7.31**
525 x 263mm overall; to 225mm internal diameter pipes - 1	6.45	10.00	**7.10**	-	0.28	**3.00**	-	**10.10**	m	**10.85**
600 x 300mm overall; to 300mm internal diameter pipes -1	9.07	10.00	**9.98**	-	0.36	**3.86**	-	**13.83**	m	**14.87**
700 x 350mm overall; to 400mm internal diameter pipes -1	13.23	10.00	**14.55**	-	0.49	**5.25**	-	**19.80**	m	**21.29**
750 x 375mm overall; to 450mm internal diameter pipes -1	15.57	10.00	**17.13**	-	0.56	**6.00**	-	**23.13**	m	**24.86**
Sand; to be obtained off site										
Beds										
400 x 50mm	0.29	5.00	**0.31**	-	0.04	**0.43**	-	**0.73**	m	**0.79**
400 x 100mm	0.58	5.00	**0.61**	-	0.08	**0.86**	-	**1.47**	m	**1.58**
450 x 50mm	0.32	5.00	**0.34**	-	0.05	**0.54**	-	**0.87**	m	**0.94**
450 x 100mm	0.66	5.00	**0.69**	-	0.09	**0.96**	-	**1.65**	m	**1.78**
525 x 50mm	0.38	5.00	**0.40**	-	0.06	**0.64**	-	**1.04**	m	**1.12**
525 x 100mm	0.76	5.00	**0.80**	-	0.11	**1.18**	-	**1.97**	m	**2.12**
600 x 50mm	0.44	5.00	**0.46**	-	0.06	**0.64**	-	**1.10**	m	**1.18**
600 x 100mm	0.87	5.00	**0.92**	-	0.12	**1.29**	-	**2.20**	m	**2.37**
700 x 50mm	0.51	5.00	**0.54**	-	0.07	**0.75**	-	**1.29**	m	**1.38**
700 x 100mm	1.02	5.00	**1.07**	-	0.14	**1.50**	-	**2.57**	m	**2.76**
750 x 50mm	0.54	5.00	**0.57**	-	0.08	**0.86**	-	**1.42**	m	**1.53**
750 x 100mm	1.09	5.00	**1.15**	-	0.15	**1.61**	-	**2.75**	m	**2.96**
Plain in-situ concrete; BS 5328, ordinary prescribed mix, ST3, 20mm aggregate										
Beds										
400 x 100mm	2.32	5.00	**2.44**	-	0.18	**1.93**	2.00	**6.36**	m	**6.84**
400 x 150mm	3.48	5.00	**3.66**	-	0.27	**2.89**	3.00	**9.55**	m	**10.26**
450 x 100mm	2.61	5.00	**2.74**	-	0.20	**2.14**	2.00	**6.88**	m	**7.40**
450 x 150mm	3.89	5.00	**4.08**	-	0.30	**3.21**	3.00	**10.29**	m	**11.07**
525 x 100mm	3.02	5.00	**3.17**	-	0.24	**2.57**	2.00	**7.74**	m	**8.32**
525 x 150mm	4.53	5.00	**4.75**	-	0.35	**3.75**	3.00	**11.50**	m	**12.36**
600 x 100mm	3.48	5.00	**3.66**	-	0.27	**2.89**	2.00	**8.55**	m	**9.19**
600 x 150mm	5.22	5.00	**5.48**	-	0.38	**4.07**	3.00	**12.55**	m	**13.49**
700 x 100mm	4.06	5.00	**4.27**	-	0.32	**3.43**	2.00	**9.69**	m	**10.42**
700 x 150mm	6.09	5.00	**6.40**	-	0.47	**5.03**	3.00	**14.43**	m	**15.51**
750 x 100mm	4.35	5.00	**4.57**	-	0.34	**3.64**	2.00	**10.21**	m	**10.98**
750 x 150mm	6.50	5.00	**6.83**	-	0.51	**5.46**	3.00	**15.29**	m	**16.43**
Beds and surrounds										
400 x 100mm bed; 150mm thick surround to 100mm internal diameter pipes -1	8.13	5.00	**8.53**	-	0.72	**7.71**	3.92	**20.16**	m	**21.67**
400 x 150mm bed; 150mm thick surround to 100mm internal diameter pipes -1	9.29	5.00	**9.75**	-	0.81	**8.68**	4.40	**22.82**	m	**24.53**
450 x 100mm bed; 150mm thick surround to 150mm internal diameter pipes -1	10.45	5.00	**10.97**	-	0.91	**9.75**	4.40	**25.11**	m	**27.00**
450 x 150mm bed; 150mm thick surround to 150mm internal diameter pipes -1	11.72	5.00	**12.31**	-	1.01	**10.82**	4.89	**28.01**	m	**30.11**
525 x 100mm bed; 150mm thick surround to 225mm internal diameter pipes -1	14.45	5.00	**15.17**	-	1.24	**13.28**	4.89	**33.34**	m	**35.84**
525 x 150mm bed; 150mm thick surround to 225mm internal diameter pipes -1	15.96	5.00	**16.76**	-	1.36	**14.57**	5.62	**36.94**	m	**39.72**
600 x 100mm bed; 150mm thick surround to 300mm internal diameter pipes -1	19.15	5.00	**20.11**	-	1.62	**17.35**	5.88	**43.34**	m	**46.59**
600 x 150mm bed; 150mm thick surround to 300mm internal diameter pipes -1	20.89	5.00	**21.94**	-	1.76	**18.85**	6.35	**47.14**	m	**50.68**
700 x 100mm bed; 150mm thick surround to 400mm internal diameter pipes -1	26.41	5.00	**27.73**	-	2.21	**23.67**	6.84	**58.24**	m	**62.61**
700 x 150mm bed; 150mm thick surround to 400mm internal diameter pipes -1	28.44	5.00	**29.86**	-	2.36	**25.28**	7.34	**62.48**	m	**67.17**

Labour hourly rates: (except Specialists) Craft Operatives £14.21 Labourer £10.71 Rates are national average prices. Refer to REGIONAL VARIATIONS for indicative levels of overall pricing in regions	MATERIALS			LABOUR				RATES		
	Del to Site £	Waste %	Material Cost £	Craft Optve Hrs	Lab Hrs	Labour Cost £	Sunds £	Nett Rate £	Unit	Gross Rate (+7.5%) £
R12: DRAINAGE BELOW GROUND: BEDS, BENCHINGS AND COVERS Cont.										
Plain in-situ concrete; BS 5328, ordinary prescribed mix, ST3, 20mm aggregate Cont.										
Beds and surrounds Cont.										
750 x 100mm bed; 150mm thick surround to 450mm internal diameter pipes -1	30.47	5.00	**31.99**	-	2.53	**27.10**	7.34	**66.43**	m	**71.42**
750 x 150mm bed; 150mm thick surround to 450mm internal diameter pipes -1	32.62	5.00	**34.25**	-	2.70	**28.92**	7.82	**70.99**	m	**76.31**
Beds and haunchings										
400 x 150mm; to 100mm internal diameter pipes - 1	4.47	5.00	**4.69**	-	0.40	**4.28**	2.94	**11.91**	m	**12.81**
450 x 150mm; to 150mm internal diameter pipes - 1	5.51	5.00	**5.79**	-	0.46	**4.93**	2.94	**13.65**	m	**14.68**
525 x 150mm; to 225mm internal diameter pipes - 1	7.14	5.00	**7.50**	-	0.62	**6.64**	2.94	**17.07**	m	**18.35**
600 x 150mm; to 300mm internal diameter pipes - 1	8.94	5.00	**9.39**	-	0.81	**8.68**	2.94	**21.00**	m	**22.57**
700 x 150mm; to 400mm internal diameter pipes - 1	11.55	5.00	**12.13**	-	1.10	**11.78**	2.94	**26.85**	m	**28.86**
750 x 150mm; to 450mm internal diameter pipes - 1	12.94	5.00	**13.59**	-	1.35	**14.46**	2.94	**30.99**	m	**33.31**
Vertical casings										
400 x 400mm to 100mm internal diameter pipes - 1.	34.89	5.00	**35.35**	-	0.80	**8.57**	-	**43.92**	m	**47.21**
450 x 450mm to 150mm internal diameter pipes - 1.	40.52	5.00	**41.11**	-	1.01	**10.82**	-	**51.93**	m	**55.82**
525 x 525mm to 225mm internal diameter pipes - 1.	49.88	5.00	**50.68**	-	1.38	**14.78**	-	**65.46**	m	**70.37**
600 x 600mm to 300mm internal diameter pipes - 1.	59.29	5.00	**60.34**	-	1.80	**19.28**	-	**79.62**	m	**85.59**
700 x 700mm to 400mm internal diameter pipes - 1.	73.24	5.00	**74.66**	-	2.45	**26.24**	-	**100.90**	m	**108.47**
750 x 750mm to 450mm internal diameter pipes - 1.	80.62	5.00	**82.25**	-	2.81	**30.10**	-	**112.34**	m	**120.77**
R12: DRAINAGE BELOW GROUND: VITRIFIED CLAY FLEXIBLE JOINT DRAINS										
Drains; vitrified clay pipes and fittings, BS EN 295, normal; flexible mechanical joints; Hepworth Hepseal										
Pipework in trenches										
400mm	93.99	5.00	**98.69**	0.55	0.18	**9.74**	-	**108.44**	m	**116.57**
extra; bends	353.19	5.00	**370.85**	0.39	-	**5.54**	-	**376.39**	nr	**404.62**
450mm	124.64	5.00	**130.87**	0.70	0.20	**12.09**	-	**142.96**	m	**153.68**
extra; bends	465.09	5.00	**488.34**	0.70	-	**9.95**	-	**498.29**	nr	**535.66**
Pipework in trenches; vertical										
400mm	93.99	5.00	**98.69**	0.60	0.18	**10.45**	-	**109.15**	m	**117.33**
450mm	124.64	5.00	**130.87**	0.77	0.20	**13.08**	-	**143.95**	m	**154.75**
R12: DRAINAGE BELOW GROUND: VITRIFIED CLAY SLEEVE JOINTS										
Drains; vitrified clay pipes and fittings, BS EN 295; sleeve joints, push-fit polypropylene standard ring flexible couplings; Hepworth Supersleeve										
Pipework in trenches										
100mm	5.13	5.00	**5.39**	0.16	0.06	**2.92**	-	**8.31**	m	**8.93**
extra; bends	6.99	5.00	**7.34**	0.08	-	**1.14**	-	**8.47**	nr	**9.11**
extra; branches	15.26	5.00	**16.02**	0.16	-	**2.27**	-	**18.30**	nr	**19.67**
150mm	9.79	5.00	**10.28**	0.18	0.08	**3.41**	-	**13.70**	m	**14.73**
extra; bends	14.48	5.00	**15.20**	0.09	-	**1.28**	-	**16.48**	nr	**17.71**
extra; branches	22.49	5.00	**23.62**	0.18	-	**2.56**	-	**26.17**	nr	**28.14**
Pipework in trenches; vertical										
100mm	5.13	5.00	**5.39**	0.18	0.06	**3.20**	-	**8.59**	m	**9.23**
150mm	9.79	5.00	**10.28**	0.20	0.08	**3.70**	-	**13.98**	m	**15.03**
Drains; vitrified clay pipes and fittings, BS EN 295; sleeve joints, push-fit polypropylene neoprene ring flexible couplings; Hepworth Supersleeve										
Pipework in trenches										
100mm	5.98	5.00	**6.28**	0.16	0.06	**2.92**	-	**9.19**	m	**9.88**
150mm	11.35	5.00	**11.92**	0.18	0.08	**3.41**	-	**15.33**	m	**16.48**
Pipework in trenches; vertical										
100mm	5.98	5.00	**6.28**	0.18	0.06	**3.20**	-	**9.48**	m	**10.19**
150mm	11.35	5.00	**11.92**	0.20	0.08	**3.70**	-	**15.62**	m	**16.79**
Drains; vitrified clay pipes and fittings, BS EN 295, normal; cement mortar (1:3)										
Pipework in trenches										
100mm	8.04	5.00	**8.45**	0.33	0.03	**5.01**	-	**13.46**	m	**14.47**
extra; double collars	9.68	5.00	**10.17**	0.10	-	**1.42**	-	**11.59**	nr	**12.46**
extra; bends	8.04	5.00	**8.45**	0.20	-	**2.84**	-	**11.29**	nr	**12.14**
extra; rest bends	13.17	5.00	**13.83**	0.20	-	**2.84**	-	**16.67**	nr	**17.92**
extra; branches	14.70	5.00	**15.45**	0.22	-	**3.13**	-	**18.57**	nr	**19.96**
150mm	8.04	5.00	**8.45**	0.44	0.04	**6.68**	-	**15.13**	m	**16.26**
extra; double collars	16.07	5.00	**16.88**	0.15	-	**2.13**	-	**19.01**	nr	**20.44**
extra; bends	12.32	5.00	**12.95**	0.27	-	**3.84**	-	**16.78**	nr	**18.04**
extra; rest bends	22.21	5.00	**23.33**	0.27	-	**3.84**	-	**27.16**	nr	**29.20**
extra; branches	24.29	5.00	**25.51**	0.28	-	**3.98**	-	**29.49**	nr	**31.70**
225mm	24.44	5.00	**25.67**	0.55	0.07	**8.57**	-	**34.24**	m	**36.80**
extra; double collars	37.57	5.00	**39.47**	0.23	-	**3.27**	-	**42.73**	nr	**45.94**
300mm	40.86	5.00	**42.91**	0.80	0.10	**12.44**	-	**55.35**	m	**59.50**
Pipework in trenches; vertical										
100mm	8.04	5.00	**8.45**	0.50	0.03	**7.43**	-	**15.87**	m	**17.07**
150mm	8.04	5.00	**8.45**	0.48	0.04	**7.25**	-	**15.70**	m	**16.88**

Labour hourly rates: (except Specialists) Craft Operatives £14.21 Labourer £10.71 Rates are national average prices. Refer to REGIONAL VARIATIONS for indicative levels of overall pricing in regions	MATERIALS			LABOUR				RATES		
	Del to Site £	Waste %	Material Cost £	Craft Optve Hrs	Lab Hrs	Labour Cost £	Sunds £	Nett Rate £	Unit	Gross Rate (+7.5%) £
R12: DRAINAGE BELOW GROUND: VITRIFIED CLAY SLEEVE JOINTS Cont.										
Drains; vitrified clay pipes and fittings, BS EN 295, normal; cement mortar (1:3) Cont.										
Pipework in trenches; vertical Cont.										
225mm	24.44	5.00	**25.67**	0.60	0.07	**9.28**	-	**34.95**	m	**37.57**
300mm	40.86	5.00	**42.91**	0.88	0.10	**13.58**	-	**56.49**	m	**60.73**
R12: DRAINAGE BELOW GROUND: PVC-U DRAINS										
Drains; Wavin; PVC-U pipes and fittings, BS EN 13598-1; ring seal joints										
Pipework in trenches										
110mm, ref. 4D.076	5.35	5.00	**5.61**	0.14	0.02	**2.20**	-	**7.82**	m	**8.40**
extra; connections to cast iron and clay sockets, ref. 4D.107	10.76	5.00	**11.29**	0.08	-	**1.14**	-	**12.43**	nr	**13.36**
extra; 45 degree bends, ref. 4D.163	9.17	5.00	**9.62**	0.14	0.02	**2.20**	-	**11.83**	nr	**12.71**
extra; branches, ref. 4D.210	13.50	5.00	**14.18**	0.14	0.02	**2.20**	-	**16.38**	nr	**17.61**
extra; connections cast iron and clay spigot, ref. 4D.128	10.76	5.00	**11.29**	0.08	-	**1.14**	-	**12.43**	nr	**13.36**
160mm, ref. 6D.076	11.77	5.00	**12.36**	0.20	0.03	**3.16**	-	**15.53**	m	**16.69**
extra; connections to cast iron and clay sockets, ref. 6D.107	21.60	5.00	**22.68**	0.10	-	**1.42**	-	**24.10**	nr	**25.91**
extra; 45 degree bends, ref. 6D.163	21.62	5.00	**22.70**	0.20	0.03	**3.16**	-	**25.86**	nr	**27.80**
extra; branches, ref. 6D.210	38.60	5.00	**40.52**	0.20	0.03	**3.16**	-	**43.69**	nr	**46.96**
extra; connections to cast iron and clay spigot, ref. 6D.128	21.60	5.00	**22.68**	0.10	-	**1.42**	-	**24.10**	nr	**25.91**
Pipework in trenches; vertical										
110mm, ref. 4D.076	5.35	5.00	**5.61**	0.28	0.02	**4.19**	-	**9.81**	m	**10.54**
160mm, ref. 6D.076	11.77	5.00	**12.36**	0.24	0.03	**3.73**	-	**16.10**	m	**17.30**
Drains; PVC-U solid wall concentric external rib-reinforced Wavin 'Ultra-Rib' pipes and fittings with sealing rings to joints										
Pipework in trenches										
150mm 6UR 046	6.84	5.00	**7.19**	0.20	0.02	**3.06**	-	**10.24**	m	**11.01**
extra; connectors 6UR 205	7.17	5.00	**7.53**	0.05	-	**0.71**	-	**8.24**	nr	**8.86**
extra; bends 6UR 563	9.65	5.00	**10.13**	0.20	-	**2.84**	-	**12.97**	nr	**13.95**
extra; branches 6UR 213	23.34	5.00	**24.50**	0.20	-	**2.84**	-	**27.34**	nr	**29.39**
extra; connections to clayware pipe ends 6UR 129	17.58	5.00	**18.45**	0.10	-	**1.42**	-	**19.87**	nr	**21.37**
225mm 9UR 046	15.14	5.00	**15.89**	0.30	0.03	**4.58**	-	**20.48**	m	**22.01**
extra; connectors 9UR 205	15.00	5.00	**15.75**	0.08	-	**1.14**	-	**16.89**	nr	**18.15**
extra; bends 9UR 563	38.83	5.00	**40.77**	0.30	-	**4.26**	-	**45.03**	nr	**48.41**
extra; branches 9UR 213	73.73	5.00	**77.42**	0.30	-	**4.26**	-	**81.68**	nr	**87.81**
extra; connections to clayware pipe ends 9UR 109	41.27	5.00	**43.33**	0.12	-	**1.71**	-	**45.03**	nr	**48.41**
300mm 12UR 043	25.39	5.00	**26.66**	0.38	0.05	**5.94**	-	**32.60**	m	**35.04**
extra; connectors 12UR 205	30.27	5.00	**31.78**	0.10	-	**1.42**	-	**33.20**	nr	**35.69**
extra; bends 12UR 563	67.15	5.00	**70.50**	0.38	-	**5.40**	-	**75.90**	nr	**81.59**
extra; branches 12UR 213	157.26	5.00	**165.12**	0.38	-	**5.40**	-	**170.52**	nr	**183.31**
extra; connections to clayware pipe ends 12UR 112	108.55	5.00	**113.97**	0.15	-	**2.13**	-	**116.10**	nr	**124.81**
Pipework in trenches; vertical										
150mm	6.84	5.00	**7.19**	0.40	0.02	**5.90**	-	**13.08**	m	**14.07**
225mm	15.14	5.00	**15.89**	0.38	0.03	**5.72**	-	**21.61**	m	**23.23**
300mm	25.39	5.00	**26.66**	0.48	0.05	**7.36**	-	**34.02**	m	**36.57**
R12: DRAINAGE BELOW GROUND: CONCRETE DRAINS										
Drains; unreinforced concrete pipes and fittings BS 5911, Class H, flexible joints										
Pipework in trenches										
300mm	11.50	5.00	**12.08**	0.59	0.24	**10.95**	-	**29.23**	m	**31.43**
extra; bends	115.00	5.00	**120.75**	0.30	0.12	**5.55**	-	**129.40**	nr	**139.11**
extra; 100mm branches	67.50	5.00	**70.88**	0.24	0.06	**4.05**	-	**76.48**	nr	**82.22**
extra; 150mm branches	67.50	5.00	**70.88**	0.40	0.11	**6.86**	-	**80.58**	nr	**86.62**
extra; 225mm branches	137.50	5.00	**144.38**	0.56	0.16	**9.67**	-	**158.18**	nr	**170.05**
extra; 300mm branches	157.50	5.00	**165.38**	0.72	0.22	**12.59**	-	**183.65**	nr	**197.42**
375mm	14.50	5.00	**15.23**	0.72	0.30	**13.44**	-	**36.43**	m	**39.16**
extra; bends	145.00	5.00	**152.25**	0.36	0.15	**6.72**	-	**162.85**	nr	**175.06**
extra; 100mm branches	75.00	5.00	**78.75**	0.24	0.06	**4.05**	-	**84.35**	nr	**90.68**
extra; 150mm branches	75.00	5.00	**78.75**	0.40	0.11	**6.86**	-	**88.46**	nr	**95.09**
extra; 225mm branches	145.00	5.00	**152.25**	0.56	0.16	**9.67**	-	**166.06**	nr	**178.51**
extra; 300mm branches	165.00	5.00	**173.25**	0.72	0.22	**12.59**	-	**191.53**	nr	**205.89**
extra; 375mm branches	185.00	5.00	**194.25**	0.88	0.27	**15.40**	-	**216.63**	nr	**232.87**
450mm	16.75	5.00	**17.59**	0.85	0.37	**16.04**	-	**43.20**	m	**46.44**
extra; bends	167.50	5.00	**175.88**	0.43	0.18	**8.04**	-	**188.57**	nr	**202.71**
extra; 100mm branches	85.00	5.00	**89.25**	0.24	0.06	**4.05**	-	**94.85**	nr	**101.97**
extra; 150mm branches	85.00	5.00	**89.25**	0.40	0.11	**6.86**	-	**98.96**	nr	**106.38**
extra; 225mm branches	155.00	5.00	**162.75**	0.56	0.16	**9.67**	-	**176.56**	nr	**189.80**
extra; 300mm branches	175.00	5.00	**183.75**	0.72	0.22	**12.59**	-	**202.03**	nr	**217.18**
extra; 375mm branches	195.00	5.00	**204.75**	0.88	0.27	**15.40**	-	**227.13**	nr	**244.16**
extra; 450mm branches	230.00	5.00	**241.50**	1.04	0.33	**18.31**	-	**268.35**	nr	**288.47**
525mm	20.25	5.00	**21.26**	0.98	0.43	**18.53**	-	**50.91**	m	**54.73**
extra; bends	202.50	5.00	**212.63**	0.49	0.22	**9.32**	-	**227.63**	nr	**244.71**
extra; 100mm branches	90.00	5.00	**94.50**	0.24	0.06	**4.05**	-	**100.10**	nr	**107.61**
extra; 150mm branches	90.00	5.00	**94.50**	0.40	0.11	**6.86**	-	**104.21**	nr	**112.02**
extra; 225mm branches	160.00	5.00	**168.00**	0.56	0.16	**9.67**	-	**181.81**	nr	**195.44**
extra; 300mm branches	180.00	5.00	**189.00**	0.72	0.22	**12.59**	-	**207.28**	nr	**222.82**
extra; 375mm branches	200.00	5.00	**210.00**	0.88	0.27	**15.40**	-	**232.38**	nr	**249.81**
extra; 450mm branches	235.00	5.00	**246.75**	1.04	0.33	**18.31**	-	**273.60**	nr	**294.12**

Labour hourly rates: (except Specialists) Craft Operatives £14.21 Labourer £10.71 Rates are national average prices. Refer to REGIONAL VARIATIONS for indicative levels of overall pricing in regions	MATERIALS			LABOUR			RATES			
	Del to Site £	Waste %	Material Cost £	Craft Optve Hrs	Lab Hrs	Labour Cost £	Sunds £	Nett Rate £	Unit	Gross Rate (+7.5%) £
R12: DRAINAGE BELOW GROUND: CONCRETE DRAINS Cont.										
Drains; unreinforced concrete pipes and fittings BS 5911, Class H, flexible joints Cont.										
Pipework in trenches Cont.										
600mm	24.75	5.00	**25.99**	1.10	0.50	**20.99**	-	**59.90**	m	**64.39**
extra; bends	247.50	5.00	**259.88**	0.55	0.25	**10.49**	-	**276.83**	nr	**297.59**
extra; 100mm branches	98.00	5.00	**102.90**	0.24	0.06	**4.05**	-	**108.50**	nr	**116.64**
extra; 150mm branches	98.00	5.00	**102.90**	0.40	0.11	**6.86**	-	**112.61**	nr	**121.05**
extra; 225mm branches	168.00	5.00	**176.40**	0.56	0.16	**9.67**	-	**190.21**	nr	**204.47**
extra; 300mm branches	188.00	5.00	**197.40**	0.72	0.22	**12.59**	-	**215.68**	nr	**231.85**
extra; 375mm branches	208.00	5.00	**218.40**	0.88	0.27	**15.40**	-	**240.78**	nr	**258.84**
extra; 450mm branches	243.00	5.00	**255.15**	1.04	0.33	**18.31**	-	**282.00**	nr	**303.15**
Drains; reinforced concrete pipes and fittings, BS 5911, Class H, flexible joints										
Pipework in trenches										
450mm	32.07	5.00	**33.67**	0.85	0.37	**16.04**	-	**59.28**	m	**63.73**
extra; bends	320.72	5.00	**336.76**	0.43	0.18	**8.04**	-	**349.45**	nr	**375.66**
extra; 100mm branches	89.48	5.00	**93.95**	0.24	0.06	**4.05**	-	**99.56**	nr	**107.03**
extra; 150mm branches	91.53	5.00	**96.11**	0.40	0.11	**6.86**	-	**105.81**	nr	**113.75**
extra; 225mm branches	94.60	5.00	**99.33**	0.56	0.16	**9.67**	-	**113.14**	nr	**121.62**
extra; 300mm branches	115.08	5.00	**120.83**	0.72	0.22	**12.59**	-	**139.11**	nr	**149.54**
extra; 375mm branches	130.44	5.00	**136.96**	0.88	0.27	**15.40**	-	**159.34**	nr	**171.29**
extra; 450mm branches	150.92	5.00	**158.47**	1.04	0.33	**18.31**	-	**185.31**	nr	**199.21**
525mm	39.20	5.00	**41.16**	0.98	0.43	**18.53**	-	**70.81**	m	**76.12**
extra; bends	391.99	5.00	**411.59**	0.49	0.22	**9.32**	-	**426.60**	nr	**458.59**
extra; 100mm branches	100.84	5.00	**105.88**	0.24	0.06	**4.05**	-	**111.49**	nr	**119.85**
extra; 150mm branches	102.89	5.00	**108.03**	0.40	0.11	**6.86**	-	**117.74**	nr	**126.57**
extra; 225mm branches	105.96	5.00	**111.26**	0.56	0.16	**9.67**	-	**125.07**	nr	**134.45**
extra; 300mm branches	121.32	5.00	**127.39**	0.72	0.22	**12.59**	-	**145.66**	nr	**156.59**
extra; 375mm branches	141.80	5.00	**148.89**	0.88	0.27	**15.40**	-	**171.27**	nr	**184.11**
extra; 450mm branches	162.28	5.00	**170.39**	1.04	0.33	**18.31**	-	**197.24**	nr	**212.03**
600mm	42.76	5.00	**44.90**	1.10	0.50	**20.99**	-	**78.81**	m	**84.72**
extra; bends	427.62	5.00	**449.00**	0.55	0.25	**10.49**	-	**465.96**	nr	**500.91**
extra; 100mm branches	91.42	5.00	**95.99**	0.24	0.06	**4.05**	-	**101.60**	nr	**109.22**
extra; 150mm branches	113.95	5.00	**119.65**	0.40	0.11	**6.86**	-	**129.35**	nr	**139.06**
extra; 225mm branches	117.02	5.00	**122.87**	0.56	0.16	**9.67**	-	**136.68**	nr	**146.93**
extra; 300mm branches	132.38	5.00	**139.00**	0.72	0.22	**12.59**	-	**157.27**	nr	**169.07**
extra; 375mm branches	152.86	5.00	**160.50**	0.88	0.27	**15.40**	-	**182.88**	nr	**196.60**
extra; 450mm branches	173.34	5.00	**182.01**	1.04	0.33	**18.31**	-	**208.85**	nr	**224.52**

This section continues on the next page

Labour hourly rates: (except Specialists) Craft Operatives £17.38 Labourer £10.71 Rates are national average prices. Refer to REGIONAL VARIATIONS for indicative levels of overall pricing in regions	MATERIALS			LABOUR				RATES		
	Del to Site £	Waste %	Material Cost £	Craft Optve Hrs	Lab Hrs	Labour Cost £	Sunds £	Nett Rate £	Unit	Gross Rate (+7.5%) £
R12: DRAINAGE BELOW GROUND: CAST IRON DRAINS										
Drains; cast iron pipes and cast iron fittings BS 437, coated; bolted cast iron coupling and synthetic rubber gasket joints										
Pipework in trenches										
100mm	21.79	5.00	**22.77**	1.00	-	**17.38**	-	**40.15**	m	**43.16**
extra; connectors, large socket for clayware	25.20	2.00	**25.70**	1.00	-	**17.38**	-	**43.08**	nr	**46.32**
extra; bends	44.62	2.00	**45.51**	1.00	-	**17.38**	-	**62.89**	nr	**67.61**
extra; branches	63.19	2.00	**64.45**	1.50	-	**26.07**	-	**90.52**	nr	**97.31**
extra; branches with access	100.56	2.00	**102.57**	1.50	-	**26.07**	-	**128.64**	nr	**138.29**
150mm	37.90	5.00	**39.65**	1.50	-	**26.07**	-	**65.72**	m	**70.65**
extra; connectors, large socket for clayware	37.62	2.00	**38.37**	1.50	-	**26.07**	-	**64.44**	nr	**69.28**
extra; diminishing pieces, reducing to 100mm	53.59	2.00	**54.66**	1.50	-	**26.07**	-	**80.73**	nr	**86.79**
extra; bends	77.52	2.00	**79.07**	1.50	-	**26.07**	-	**105.14**	nr	**113.03**
extra; branches	112.45	2.00	**114.70**	2.25	-	**39.11**	-	**153.80**	nr	**165.34**
extra; branches with access	163.67	2.00	**166.94**	2.25	-	**39.11**	-	**206.05**	nr	**221.50**
in runs not exceeding 3m										
100mm	29.52	5.00	**30.65**	1.50	-	**26.07**	-	**56.72**	m	**60.97**
150mm	47.25	5.00	**49.19**	2.25	-	**39.11**	-	**88.30**	m	**94.92**
Pipework in trenches; vertical										
100mm	29.52	5.00	**30.65**	1.20	-	**20.86**	-	**51.51**	m	**55.37**
150mm	47.25	5.00	**49.19**	1.80	-	**31.28**	-	**80.48**	m	**86.51**
R12: DRAINAGE BELOW GROUND: STAINLESS STEEL DRAINS										
Drains; stainless steel pipes and fittings, B.M. Stainless Steel Drains Ltd; ring seal push fit joints (AIS 1316)										
Pipework in trenches										
75mm	25.10	5.00	**26.36**	0.60	-	**10.43**	-	**36.78**	m	**39.54**
extra; bends	30.87	2.00	**31.49**	0.60	-	**10.43**	-	**41.92**	nr	**45.06**
extra; branches	35.87	2.00	**36.59**	0.90	-	**15.64**	-	**52.23**	nr	**56.15**
110mm	31.85	5.00	**33.44**	0.75	-	**13.04**	-	**46.48**	m	**49.96**
extra; diminishing pieces, reducing to 75mm	54.22	2.00	**55.30**	0.75	-	**13.04**	-	**68.34**	nr	**73.46**
extra; bends	39.41	2.00	**40.20**	0.75	-	**13.04**	-	**53.23**	nr	**57.23**
extra; branches	41.42	2.00	**42.25**	1.12	-	**19.47**	-	**61.71**	nr	**66.34**
in runs not exceeding 3m										
75mm	25.10	5.00	**26.36**	0.90	-	**15.64**	-	**42.00**	m	**45.15**
110mm	31.85	5.00	**33.44**	1.12	-	**19.47**	-	**52.91**	m	**56.88**
Pipework in trenches; vertical										
75mm	25.10	5.00	**26.36**	0.75	-	**13.04**	-	**39.39**	m	**42.34**
110mm	31.85	5.00	**33.44**	0.94	-	**16.34**	-	**49.78**	m	**53.51**

This section continues on the next page

Labour hourly rates: (except Specialists) Craft Operatives £14.21 Labourer £10.71 Rates are national average prices. Refer to REGIONAL VARIATIONS for indicative levels of overall pricing in regions	MATERIALS			LABOUR				RATES		
	Del to Site £	Waste %	Material Cost £	Craft Optve Hrs	Lab Hrs	Labour Cost £	Sunds £	Nett Rate £	Unit	Gross Rate (+7.5%) £
R12: DRAINAGE BELOW GROUND: VITRIFIED CLAY GULLIES										
Vitrified clay accessories										
Gullies; Supersleve; joint to pipe; bedding and surrounding in concrete to BS 5328, ordinary prescribed mix ST3, 20mm aggregate; Hepworth Building Products										
100mm outlet, reference RG1/1, trapped, round; 255mm diameter top	36.81	5.00	**38.65**	0.60	0.08	**9.38**	-	**48.03**	nr	**51.63**
100mm outlet, reference SG2/1, trapped, square; 150 x 150 top	43.50	5.00	**45.67**	0.60	0.08	**9.38**	-	**55.05**	nr	**59.18**
100mm outlet, reference SG1/1, trapped, reversible, round; 100mm trap; hopper reference SH1, 100mm outlet, 150 x 150mm rebated top	64.97	5.00	**68.21**	1.00	0.10	**15.28**	-	**83.50**	nr	**89.76**
100mm outlet, reference SG1/1, trapped, reversible, round; 100mm trap; hopper reference SH2, 100mm outlet, 100mm horizontal inlet, 150 x 150mm rebated top	65.08	5.00	**68.33**	1.15	0.10	**17.41**	-	**85.74**	nr	**92.18**
100mm outlet, reference SG1/1, trapped, reversible, round; 100mm trap; raising pieces - 1, reference RRP2/2, 100mm diameter, 225mm high; hopper reference SH1, 100mm outlet, 150 x 150mm rebated top	103.20	5.00	**108.36**	1.40	0.15	**21.50**	-	**129.86**	nr	**139.60**
100mm outlet, reference SG1/1, trapped, reversible, round; 100mm trap; raising pieces - 1, reference RRP2/2, 100mm diameter, 225mm high; hopper reference SH2, 100mm outlet, 100mm horizontal inlet, 150 x 150mm rebated top	103.31	5.00	**108.48**	1.45	0.16	**22.32**	-	**130.80**	nr	**140.61**
100mm outlet, reference RGP5, trapped, round; 225mm diameter x 600mm deep; perforated galvanized steel bucket reference IBP3	124.68	5.00	**130.91**	2.50	0.25	**38.20**	-	**169.11**	nr	**181.79**
100mm outlet, reference RGR1, trapped, round; 300mm diameter x 600mm deep	91.15	5.00	**95.70**	2.50	0.33	**39.06**	-	**134.76**	nr	**144.87**
150mm outlet, reference RGR2, trapped, round; 300mm diameter x 600mm deep	92.64	5.00	**97.27**	2.50	0.33	**39.06**	-	**136.33**	nr	**146.55**
150mm outlet, reference RGR3, trapped, round; 400mm diameter x 750mm deep	102.80	5.00	**107.94**	4.00	0.45	**61.66**	-	**169.60**	nr	**182.32**
150mm outlet, reference RGR4, trapped, round; 450mm diameter x 900mm deep	128.85	5.00	**135.29**	5.50	0.50	**83.51**	-	**218.80**	nr	**235.21**
100mm outlet, reference RGU1, trapped; internal size 600 x 450 x 600mm deep; perforated tray and galvanized cover and frame	586.35	5.00	**615.66**	10.00	1.50	**158.17**	-	**773.83**	nr	**831.87**
Rainwater shoes; Supersleve; joint to pipe; bedding and surrounding in concrete to BS 5328, ordinary prescribed mix ST3, 20mm aggregate										
100mm outlet, reference RRW/S3/1 trapless, round; 100mm vertical inlet, 250 x 150mm rectangular access opening	38.19	5.00	**40.10**	0.50	0.05	**7.64**	-	**47.74**	nr	**51.32**
R12: DRAINAGE BELOW GROUND: CONCRETE GULLIES										
Unreinforced concrete accessories										
Gullies; BS 5911; joint to pipe; bedding and surrounding in concrete to BS 5328 ordinary prescribed mix ST3, 20mm aggregate										
150mm outlet, trapped, round; 375mm diameter x 750mm deep internally, stopper	42.65	5.00	**44.79**	3.30	0.50	**52.25**	-	**97.03**	nr	**104.31**
150mm outlet, trapped, round; 375mm diameter x 900mm deep internally, stopper	53.02	5.00	**55.67**	3.50	0.50	**55.09**	-	**110.76**	nr	**119.07**
150mm outlet, trapped, round; 450mm diameter x 750mm deep internally, stopper	43.65	5.00	**45.84**	3.50	0.60	**56.16**	-	**102.00**	nr	**109.65**
150mm outlet, trapped, round; 450mm diameter x 900mm deep internally, stopper	54.02	5.00	**56.72**	3.70	0.60	**59.00**	-	**115.72**	nr	**124.40**
150mm outlet, trapped, round; 450mm diameter x 1050mm deep internally, stopper	60.82	5.00	**63.87**	4.00	0.60	**63.27**	-	**127.13**	nr	**136.67**
150mm outlet, trapped, round; 450mm diameter x 1200mm deep internally, stopper	74.73	5.00	**78.47**	4.50	0.60	**70.37**	-	**148.84**	nr	**160.00**
R12: DRAINAGE BELOW GROUND: PVC-U GULLIES										
PVC-U accessories										
Gullies; Osma; joint to pipe; bedding and surrounding in concrete to BS 5328, ordinary prescribed mix ST3, 20mm aggregate										
110mm outlet trap, base reference 4D.500, outlet bend with access reference 4D.569, inlet raising piece 300mm long, plain hopper reference 4D.503	54.05	5.00	**56.75**	1.00	0.20	**16.35**	-	**73.10**	nr	**78.58**
110mm outlet bottle gulley, reference 4D.900 trapped, round; 200mm diameter top with grating reference 4D.919	45.85	5.00	**48.14**	1.50	0.25	**23.99**	-	**72.13**	nr	**77.54**
Sealed rodding eyes; Osma; joint to pipe; surrounding in concrete to BS 5328; ordinary prescribed mix ST3, 20mm aggregate										
110mm outlet, reference 4D.316, with cover	43.34	5.00	**45.50**	1.80	0.45	**30.40**	-	**75.90**	nr	**81.59**

Labour hourly rates: (except Specialists)
Craft Operatives £14.21 Labourer £10.71
Rates are national average prices.
Refer to REGIONAL VARIATIONS for indicative levels of overall pricing in regions

	MATERIALS			LABOUR				RATES		
	Del to Site £	Waste %	Material Cost £	Craft Optve Hrs	Lab Hrs	Labour Cost £	Sunds £	Nett Rate £	Unit	Gross Rate (+7.5%) £
R12: DRAINAGE BELOW GROUND: GULLY GRATING AND SEALING PLATES										
Cast iron accessories, painted black										
Gratings; Hepworth Building Products										
reference IG6C; 140mm diameter..................	3.09	-	**3.09**	-	0.03	**0.32**	-	**3.41**	nr	**3.66**
reference IG7C; 197mm diameter..................	4.71	-	**4.71**	-	0.03	**0.32**	-	**5.03**	nr	**5.41**
reference IG8C; 284mm diameter..................	10.28	-	**10.28**	-	0.03	**0.32**	-	**10.60**	nr	**11.39**
reference IG2C; 150 x 150mm.....................	3.09	-	**3.09**	-	0.03	**0.32**	-	**3.41**	nr	**3.66**
reference IG3C; 225 x 225mm.....................	9.18	-	**9.18**	-	0.03	**0.32**	-	**9.50**	nr	**10.21**
reference IG4C; 300 x 300mm.....................	20.68	-	**20.68**	-	0.03	**0.32**	-	**21.00**	nr	**22.58**
Gratings and frames; bedding frames in cement mortar (1:3); Hepworth Building Products										
reference IH5C; 100mm diameter..................	11.21	15.00	**12.89**	1.00	-	**14.21**	-	**27.10**	nr	**29.13**
reference IH6C; 150mm diameter..................	19.28	15.00	**22.17**	1.25	-	**17.76**	-	**39.93**	nr	**42.92**
reference IH7C; 225mm diameter..................	38.39	15.00	**44.15**	1.70	-	**24.16**	-	**68.31**	nr	**73.43**
reference IH2C; 150 x 150mm with lock and key....	12.44	15.00	**14.31**	1.25	-	**17.76**	-	**32.07**	nr	**34.48**
reference IH3C; 230 x 230mm with lock and key....	22.65	15.00	**26.05**	1.75	-	**24.87**	-	**50.92**	nr	**54.73**
reference IH4C; 316 x 316mm with lock and key....	59.85	15.00	**68.83**	1.90	-	**27.00**	-	**95.83**	nr	**103.02**
Sealing plates and frames; bedding frames in cement mortar (1:3); Hepworth Building Products										
reference IS5C; 140mm diameter..................	9.79	15.00	**11.26**	1.00	-	**14.21**	-	**25.47**	nr	**27.38**
reference IS6C; 197mm diameter..................	14.02	15.00	**16.13**	1.25	-	**17.76**	-	**33.89**	nr	**36.43**
reference IS7C; 273mm diameter..................	22.37	15.00	**25.73**	1.70	-	**24.16**	-	**49.88**	nr	**53.62**
reference IS2C; 150 x 150mm.....................	12.04	15.00	**13.85**	1.25	-	**17.76**	-	**31.61**	nr	**33.98**
reference IS3C; 225 x 225mm.....................	21.82	15.00	**25.09**	1.75	-	**24.87**	-	**49.96**	nr	**53.70**
Alloy accessories										
Gratings; Hepworth Building Products										
reference IG6; 140mm diameter...................	3.09	15.00	**3.55**	-	0.03	**0.32**	-	**3.87**	nr	**4.16**
reference IG7; 197mm diameter...................	4.71	15.00	**5.42**	-	0.03	**0.32**	-	**5.74**	nr	**6.17**
reference IG8; 284mm diameter...................	10.27	15.00	**11.81**	-	0.03	**0.32**	-	**12.13**	nr	**13.04**
reference IG2; 150 x 150mm......................	3.09	15.00	**3.55**	-	0.03	**0.32**	-	**3.87**	nr	**4.16**
reference IG3; 225 x 225mm......................	9.18	15.00	**10.56**	-	0.03	**0.32**	-	**10.88**	nr	**11.69**
reference IG4; 300 x 300mm......................	20.68	15.00	**23.78**	-	0.03	**0.32**	-	**24.10**	nr	**25.91**
Gratings and frames; bedding frames in cement mortar (1:3); Hepworth Building Products										
reference IH6; 193mm diameter...................	19.28	15.00	**22.17**	1.25	-	**17.76**	-	**39.93**	nr	**42.92**
reference IH2; 150 x 150mm with lock and key.....	12.44	15.00	**14.31**	1.25	-	**17.76**	-	**32.07**	nr	**34.48**
reference IH3; 225 x 225mm with lock and key.....	22.65	15.00	**26.05**	1.75	-	**24.87**	-	**50.92**	nr	**54.73**
Sealing plates and frames; bedding frames in cement mortar (1:3); Hepworth Building Products										
reference IS7; 273mm diameter...................	22.37	15.00	**25.73**	1.70	-	**24.16**	-	**49.88**	nr	**53.62**
reference IS2; 150 x 150mm......................	11.98	15.00	**13.78**	1.25	-	**17.76**	-	**31.54**	nr	**33.91**
Cast iron accessories, painted black										
Gratings and frames; Stanton PLC.; bedding frames in cement mortar (1:3)										
reference HY811 Watergate; 370 x 305mm..........	56.91	15.00	**65.45**	1.00	-	**14.21**	-	**79.66**	nr	**85.64**
reference HY812; 370 x 430mm....................	56.91	15.00	**65.45**	2.50	-	**35.53**	-	**100.98**	nr	**108.55**
reference HY815; 510 x 360mm....................	56.91	15.00	**65.45**	3.33	-	**47.32**	-	**112.77**	nr	**121.23**
reference HP894 Waterflow; 325 x 312mm..........	44.85	15.00	**51.58**	4.25	-	**60.39**	-	**111.97**	nr	**120.37**
reference HP895; 325 x 437mm....................	51.61	15.00	**59.36**	2.75	-	**39.08**	-	**98.43**	nr	**105.82**
R12: DRAINAGE BELOW GROUND: SURFACE DRAINAGE SYSTEMS										
Precast polyester concrete ACO Drain surface drainage systems; Technologies Plc; bedding and haunching in concrete to BS 5328 ordinary prescribed mix ST5, 20mm aggregate										
N100K Channel system with resin composite 'Heelguard' non rusting lockable grating										
constant depth or complete with 0.6% fall........	24.51	5.00	**25.73**	1.50	1.95	**42.20**	-	**67.93**	m	**73.02**
extra; end cap	-	-	**-**	-	0.20	**2.14**	-	**2.14**	nr	**2.30**
ParkDrain Class one piece channel system with integral grating										
constant depth or complete with 0.6% fall........	24.51	5.00	**25.73**	1.70	2.10	**46.65**	-	**72.38**	m	**77.81**
extra; end cap	5.32	5.00	**5.59**	-	0.20	**2.14**	-	**7.73**	nr	**8.31**
Raindrain Class F900 one piece channel system with integral grating										
invert depth 115mm	37.96	5.00	**39.86**	1.40	1.65	**37.57**	-	**77.42**	m	**83.23**
extra; end cap	7.05	5.00	**7.40**	-	0.20	**2.14**	-	**9.55**	nr	**10.26**
extra; rodding access unit.......................	41.95	5.00	**44.05**	-	0.50	**5.36**	-	**49.41**	nr	**53.11**
extra; universal Gully assembly complete with bucket	182.91	5.00	**192.06**	1.00	1.00	**24.92**	-	**216.98**	nr	**233.25**
KerbDrain Class D400 combined kerb drainage system										
305mm x 150mm x 500mm	33.25	5.00	**34.91**	1.20	1.45	**32.58**	-	**67.49**	nr	**72.56**
extra; rodding access	65.33	5.00	**68.60**	-	0.50	**5.36**	-	**73.95**	nr	**79.50**
extra; Class D400 Gully..........................	175.98	5.00	**184.78**	1.00	1.00	**24.92**	-	**209.70**	nr	**225.43**

Labour hourly rates: (except Specialists) Craft Operatives £14.21 Labourer £10.71 Rates are national average prices. Refer to REGIONAL VARIATIONS for indicative levels of overall pricing in regions	PLANT & TRANSPORT			LABOUR				RATES		
	Plant Cost £	Trans Cost £	P & T Cost £	Craft Optve Hrs	Lab Hrs	Labour Cost £	Sunds £	Nett Rate £	Unit	Gross Rate (+7.5%) £
R12: DRAINAGE BELOW GROUND; MANHOLES AND SOAKAWAYS:										
Excavating - by machine										
Excavating pits commencing from natural ground level; maximum depth not exceeding										
0.25m	3.88	-	**3.88**	-	0.84	**9.00**	-	**12.87**	m3	**13.84**
1.00m	5.17	-	**5.17**	-	0.88	**9.42**	-	**14.60**	m3	**15.69**
2.00m	5.69	-	**5.69**	-	1.00	**10.71**	-	**16.40**	m3	**17.63**
4.00m	6.21	-	**6.21**	-	1.27	**13.60**	-	**19.81**	m3	**21.29**
Working space allowance to excavation pits	5.17	-	**5.17**	-	0.60	**6.43**	-	**11.60**	m2	**12.47**
Compacting bottoms of excavations	1.05	-	**1.05**	-	-	**-**	-	**1.05**	m2	**1.13**
Excavated material depositing on site in temporary spoil heaps where directed										
25m	2.07	-	**2.07**	-	-	**-**	-	**2.07**	m3	**2.22**
50m	2.59	-	**2.59**	-	-	**-**	-	**2.59**	m3	**2.78**
100m	3.10	-	**3.10**	-	-	**-**	-	**3.10**	m3	**3.34**
Excavating - by hand										
Excavating pits commencing from natural ground level; maximum depth not exceeding										
0.25m	-	-	**-**	-	3.35	**35.88**	-	**35.88**	m3	**38.57**
1.00m	-	-	**-**	-	3.50	**37.49**	-	**37.49**	m3	**40.30**
2.00m	-	-	**-**	-	4.00	**42.84**	-	**42.84**	m3	**46.05**
4.00m	-	-	**-**	-	5.10	**54.62**	-	**54.62**	m3	**58.72**
Working space allowance to excavation pits	-	-	**-**	-	3.12	**33.42**	-	**33.42**	m2	**35.92**
Compacting bottoms of excavations	-	-	**-**	-	0.12	**1.29**	-	**1.29**	m2	**1.38**
Excavated material depositing on site in temporary spoil heaps where directed										
25m	-	-	**-**	-	1.25	**13.39**	-	**13.39**	m3	**14.39**
50m	-	-	**-**	-	1.50	**16.07**	-	**16.07**	m3	**17.27**
100m	-	-	**-**	-	2.00	**21.42**	-	**21.42**	m3	**23.03**

This section continues on the next page

Labour hourly rates: (except Specialists)
Craft Operatives £14.21 Labourer £10.71
Rates are national average prices.
Refer to REGIONAL VARIATIONS for indicative levels of overall pricing in regions

	MATERIALS			LABOUR				RATES		
	Del to Site £	Waste %	Material Cost £	Craft Optve Hrs	Lab Hrs	Labour Cost £	Sunds £	Nett Rate £	Unit	Gross Rate (+7.5%) £
R12: DRAINAGE BELOW GROUND; MANHOLES AND SOAKAWAYS: Cont.										
Earthwork support										
Earthwork support										
distance between opposing faces not exceeding 2.00m; maximum depth not exceeding										
1.00m	0.74	-	**0.74**	-	0.20	**2.14**	-	**2.88**	m2	**3.10**
2.00m	0.98	-	**0.98**	-	0.25	**2.68**	-	**3.66**	m2	**3.93**
4.00m	1.23	-	**1.23**	-	0.30	**3.21**	-	**4.44**	m2	**4.77**
distance between opposing faces 2.00 - 4.00m; maximum not exceeding										
1.00m	1.47	-	**1.47**	-	0.21	**2.25**	-	**3.72**	m2	**4.00**
2.00m	1.72	-	**1.72**	-	0.26	**2.78**	-	**4.50**	m2	**4.84**
4.00m	1.97	-	**1.97**	-	0.32	**3.43**	-	**5.39**	m2	**5.80**
Earthwork support; in unstable ground										
distance between opposing faces not exceeding 2.00m; maximum depth not exceeding										
1.00m	2.70	-	**2.70**	-	0.33	**3.53**	-	**6.24**	m2	**6.70**
2.00m	2.95	-	**2.95**	-	0.42	**4.50**	-	**7.45**	m2	**8.01**
4.00m	3.19	-	**3.19**	-	0.50	**5.36**	-	**8.55**	m2	**9.19**
distance between opposing faces 2.00 - 4.00m; maximum depth not exceeding										
1.00m	3.19	-	**3.19**	-	0.35	**3.75**	-	**6.94**	m2	**7.46**
2.00m	3.44	-	**3.44**	-	0.43	**4.61**	-	**8.05**	m2	**8.65**
4.00m	3.69	-	**3.69**	-	0.53	**5.68**	-	**9.36**	m2	**10.06**
R12: DRAINAGE BELOW GROUND; MANHOLES AND SOAKAWAYS: FILLING										
Excavated material arising from excavations										
Filling to excavation										
average thickness exceeding 0.25m	-	-	**-**	-	1.33	**14.24**	-	**14.24**	m3	**15.31**
Excavated material obtained from on site spoil heaps 25m distant										
Filling to excavation										
average thickness not exceeding 0.25m	-	-	**-**	-	1.75	**18.74**	-	**18.74**	m3	**20.15**
average thickness exceeding 0.25m	-	-	**-**	-	1.50	**16.07**	-	**16.07**	m3	**17.27**
Excavated material obtained from on site spoil heaps 50m distant										
Filling to excavation										
average thickness not exceeding 0.25m	-	-	**-**	-	2.00	**21.42**	-	**21.42**	m3	**23.03**
average thickness exceeding 0.25m	-	-	**-**	-	1.85	**19.81**	-	**19.81**	m3	**21.30**
Excavated material obtained from on site spoil heaps 100m distant										
Filling to excavation										
average thickness not exceeding 0.25m	-	-	**-**	-	2.50	**26.78**	-	**26.78**	m3	**28.78**
average thickness exceeding 0.25m	-	-	**-**	-	2.25	**24.10**	-	**24.10**	m3	**25.90**
Hard, dry broken brick or stone, 100 - 75mm gauge, to be obtained off site										
Filling to excavation										
average thickness exceeding 0.25m	17.00	25.00	**21.25**	-	1.25	**13.39**	-	**34.64**	m3	**37.24**
R12: DRAINAGE BELOW GROUND; MANHOLES AND SOAKAWAYS: CONCRETE WORK										
Plain in-situ concrete; BS 5328, ordinary prescribed mix ST3, 20mm aggregate										
Beds; poured on or against earth or unblinded hardcore										
thickness 150 - 450mm	58.04	5.00	**60.94**	-	4.00	**42.84**	-	**103.78**	m3	**111.57**
thickness not exceeding 150mm	58.04	5.00	**60.94**	-	4.30	**46.05**	-	**107.00**	m3	**115.02**
Benching in bottoms; rendering with 13mm cement and sand (1:2) before final set										
600 x 450 x average 225mm; trowelling	3.90	5.00	**4.13**	-	0.54	**5.78**	-	**9.92**	nr	**10.66**
675 x 675 x average 225mm; trowelling	6.58	5.00	**6.97**	-	0.92	**9.85**	-	**16.82**	nr	**18.09**
800 x 675 x average 225mm; trowelling	7.80	5.00	**8.27**	-	1.10	**11.78**	-	**20.05**	nr	**21.55**
800 x 800 x average 300mm; trowelling	12.81	5.00	**13.58**	-	1.73	**18.53**	-	**32.10**	nr	**34.51**
1025 x 800 x average 300mm; trowelling	16.02	5.00	**16.97**	-	2.21	**23.67**	-	**40.64**	nr	**43.69**
Reinforced in-situ concrete; BS 5328, designed mix ST4, 20mm aggregate, minimum cement content 240 kg/m3; vibrated										
Slabs										
thickness 150 - 450mm	60.25	5.00	**63.26**	-	3.50	**37.49**	-	**100.75**	m3	**108.30**
thickness not exceeding 150mm	60.25	5.00	**63.26**	-	4.00	**42.84**	-	**106.10**	m3	**114.06**
Formwork and basic finish										
Edges of suspended slabs; plain vertical										
height not exceeding 250mm	1.15	12.50	**1.30**	0.75	0.15	**12.26**	-	**13.56**	m	**14.58**

Labour hourly rates: (except Specialists) Craft Operatives £14.21 Labourer £10.71 Rates are national average prices. Refer to REGIONAL VARIATIONS for indicative levels of overall pricing in regions	MATERIALS			LABOUR				RATES		
	Del to Site £	Waste %	Material Cost £	Craft Optve Hrs	Lab Hrs	Labour Cost £	Sunds £	Nett Rate £	Unit	Gross Rate (+7.5%) £
R12: DRAINAGE BELOW GROUND; MANHOLES AND SOAKAWAYS: CONCRETE WORK Cont.										
Formwork and basic finish Cont.										
Soffits of slabs; horizontal										
slab thickness not exceeding 200mm; height to soffit not exceeding 1.50m......................	15.00	10.00	**16.50**	2.20	0.42	**35.76**	-	**52.26**	m2	**56.18**
slab thickness not exceeding 200mm; height to soffit 1.50 - 3.00m	15.00	10.00	**16.50**	2.00	0.42	**32.92**	-	**49.42**	m2	**53.12**
Holes; rectangular										
girth 1.00 - 2.00m; depth not exceeding 250mm....	2.31	12.50	**2.59**	0.50	0.05	**7.64**	-	**10.23**	nr	**11.00**
girth 2.00 - 3.00m; depth not exceeding 250mm....	3.46	12.50	**3.89**	0.80	0.10	**12.44**	-	**16.33**	nr	**17.55**
Reinforcement fabric; BS 4483; hard drawn plain round steel, welded; including laps, tying wire, all cutting and bending and spacers and chairs which are at the discretion of the Contractor										
Reference A142, 2.22kg/m2; 200mm side laps; 200mm end laps										
generally	1.12	15.00	**1.29**	0.25	0.02	**3.77**	-	**5.05**	m2	**5.43**
R12: DRAINAGE BELOW GROUND; MANHOLES AND SOAKAWAYS: BRICKWORK										
Common bricks, BS EN 772, Category M, 215 x 102.5 x 65mm, compressive strength 20.5 N/mm2; in cement mortar (1:3)										
Walls; vertical										
102mm thick; stretcher bond.....................	13.04	5.00	**13.96**	1.33	1.45	**34.43**	-	**48.39**	m2	**52.02**
215mm thick; English bond.......................	25.92	5.00	**27.75**	2.30	2.54	**59.89**	-	**87.63**	m2	**94.21**
Milton Hall Second hard stock bricks, BS EN 772, Category M, 215 x 102.5 x 65mm; in cement mortar (1:3)										
Walls; vertical										
102mm thick; stretcher bond.....................	22.80	5.00	**24.21**	1.33	1.45	**34.43**	-	**58.64**	m2	**63.04**
215mm thick; English bond.......................	45.28	5.00	**48.08**	2.30	2.54	**59.89**	-	**107.96**	m2	**116.06**
Engineering bricks, Category F, BS EN 772, 215 x 102.5 x 65mm, class B; in cement mortar (1:3)										
Walls; vertical										
102mm thick; stretcher bond.....................	13.65	5.00	**14.60**	1.33	1.45	**34.43**	-	**49.03**	m2	**52.71**
215mm thick; English bond.......................	27.13	5.00	**29.02**	2.30	2.54	**59.89**	-	**88.90**	m2	**95.57**
Common bricks, BS EN 772, Category M, 215 x 102.5 x 65mm, compressive strength 20.5 N/mm2; in cement mortar (1:3); flush smooth pointing as work proceeds										
Walls; vertical										
102mm thick; stretcher bond; facework one side...	13.04	5.00	**13.96**	1.53	1.65	**39.41**	-	**53.38**	m2	**57.38**
215mm thick; English bond; facework one side.....	25.92	5.00	**27.75**	2.50	2.74	**64.87**	-	**92.62**	m2	**99.56**
Milton Hall Second hard stock bricks, BS EN 772, Category M, 215 x 102.5 x 65mm; in cement mortar (1:3); flush smooth pointing as work proceeds										
Walls; vertical										
102mm thick; stretcher bond; facework one side...	22.80	5.00	**24.21**	1.53	1.65	**39.41**	-	**63.62**	m2	**68.40**
215mm thick; English bond; facework one side.....	45.28	5.00	**48.08**	2.50	2.74	**64.87**	-	**112.95**	m2	**121.42**
Engineering bricks, BS EN 772, Category F, 215 x 102.5 x 65mm, class B; in cement mortar (1:3); flush smooth pointing as work proceeds										
Walls; vertical										
102mm thick; stretcher bond; facework one side...	13.65	5.00	**14.60**	1.53	1.65	**39.41**	-	**54.02**	m2	**58.07**
215mm thick; English bond; facework one side.....	27.13	5.00	**29.02**	2.50	2.74	**64.87**	-	**93.89**	m2	**100.93**
Accessories/sundry items for brick/block/stone walling										
Building in ends of pipes; 100mm diameter										
making good fair face one side										
102mm brickwork	0.07	15.00	**0.08**	0.08	0.08	**1.99**	-	**2.07**	nr	**2.23**
215mm brickwork	0.13	15.00	**0.15**	0.13	0.13	**3.24**	-	**3.39**	nr	**3.65**
Building in ends of pipes; 150mm diameter										
making good fair face one side										
102mm brickwork	0.07	15.00	**0.08**	0.13	0.13	**3.24**	-	**3.32**	nr	**3.57**
215mm brickwork	0.13	15.00	**0.15**	0.20	0.20	**4.98**	-	**5.14**	nr	**5.52**
Building in ends of pipes; 225mm diameter										
making good fair face one side										
102mm brickwork	0.13	15.00	**0.15**	0.20	0.20	**4.98**	-	**5.14**	nr	**5.52**
215mm brickwork	0.27	15.00	**0.31**	0.25	0.25	**6.23**	-	**6.54**	nr	**7.03**
Building in ends of pipes; 300mm diameter										
making good fair face one side										
102mm brickwork	0.20	15.00	**0.23**	0.25	0.25	**6.23**	-	**6.46**	nr	**6.95**
215mm brickwork	0.40	15.00	**0.46**	0.33	0.33	**8.22**	-	**8.68**	nr	**9.34**

Labour hourly rates: (except Specialists) Craft Operatives £14.21 Labourer £10.71 Rates are national average prices. Refer to REGIONAL VARIATIONS for indicative levels of overall pricing in regions	MATERIALS			LABOUR				RATES		
	Del to Site £	Waste %	Material Cost £	Craft Optve Hrs	Lab Hrs	Labour Cost £	Sunds £	Nett Rate £	Unit	Gross Rate (+7.5%) £
R12: DRAINAGE BELOW GROUND; MANHOLES AND SOAKAWAYS: BRICKWORK Cont.										
Mortar, cement and sand (1:3)										
13mm work to walls on brickwork base; one coat; trowelled										
width exceeding 300mm	0.87	15.00	**1.00**	0.40	0.40	**9.97**	-	**10.97**	m2	**11.79**
13mm work to floors on concrete base; one coat; trowelled										
level and to falls only not exceeding 15 degrees from horizontal	0.87	15.00	**1.00**	0.60	0.60	**14.95**	-	**15.95**	m2	**17.15**
R12: DRAINAGE BELOW GROUND; MANHOLES AND SOAKAWAYS: CHANNELS, VITRIFIED CLAY										
Channels in bottoms; vitrified clay, BS 65, normal, glazed; cement mortar (1:3) joints; bedding in cement mortar (1:3)										
Half section										
straight; 600mm effective length										
100mm	3.53	5.00	**3.73**	0.30	0.03	**4.58**	-	**8.31**	nr	**8.94**
150mm	5.87	5.00	**6.19**	0.40	0.04	**6.11**	-	**12.30**	nr	**13.22**
straight; 1000mm effective length										
100mm	5.12	5.00	**5.41**	0.50	0.05	**7.64**	-	**13.06**	nr	**14.03**
150mm	8.40	5.00	**8.87**	0.67	0.07	**10.27**	-	**19.14**	nr	**20.58**
225mm	18.33	5.00	**19.31**	0.84	0.08	**12.79**	-	**32.11**	nr	**34.52**
300mm	37.05	5.00	**38.99**	1.25	0.15	**19.37**	-	**58.36**	nr	**62.73**
curved; 500mm effective length										
100mm	5.28	5.00	**5.57**	0.30	0.03	**4.58**	-	**10.15**	nr	**10.91**
150mm	8.66	5.00	**9.12**	0.40	0.25	**8.36**	-	**17.48**	nr	**18.79**
225mm	28.38	5.00	**29.84**	0.50	0.05	**7.64**	-	**37.48**	nr	**40.29**
300mm	57.60	5.00	**60.53**	0.75	0.09	**11.62**	-	**72.15**	nr	**77.56**
curved; 900mm effective length										
100mm	5.48	5.00	**5.80**	0.60	0.06	**9.17**	-	**14.97**	nr	**16.09**
150mm	8.92	5.00	**9.42**	0.80	0.08	**12.22**	-	**21.65**	nr	**23.27**
225mm	28.65	5.00	**30.15**	1.00	0.10	**15.28**	-	**45.43**	nr	**48.84**
300mm	57.86	5.00	**60.84**	1.50	0.18	**23.24**	-	**84.08**	nr	**90.38**
Three quarter section										
branch bends										
100mm	14.40	5.00	**15.52**	0.33	0.03	**5.01**	-	**20.53**	nr	**22.07**
150mm	22.38	5.00	**24.03**	0.50	0.05	**7.64**	-	**31.68**	nr	**34.05**
225mm	57.28	5.00	**60.21**	1.00	0.10	**15.28**	-	**75.49**	nr	**81.15**
Channels in bottoms; P.V.C.; 'O' ring joints; bedding in cement mortar (1:3)										
For 610mm clear opening										
straight										
110mm	25.61	5.00	**27.09**	0.40	0.03	**6.01**	-	**33.09**	nr	**35.57**
160mm	46.99	5.00	**49.61**	0.50	0.04	**7.53**	-	**57.14**	nr	**61.43**
short bend										
110mm	12.90	5.00	**13.74**	0.20	0.01	**2.95**	-	**16.69**	nr	**17.94**
160mm	28.33	5.00	**30.02**	0.30	0.02	**4.48**	-	**34.49**	nr	**37.08**
long radius bend										
110mm	42.71	5.00	**45.24**	0.33	0.02	**4.90**	-	**50.15**	nr	**53.91**
160mm	89.00	5.00	**93.99**	0.50	0.04	**7.53**	-	**101.52**	nr	**109.13**
R12: DRAINAGE BELOW GROUND; MANHOLES AND SOAKAWAYS: PRECAST CONCRETE INSPECTION CHAMBERS										
Precast concrete; standard or stock pattern units; BS 5911; jointing and pointing in cement mortar (1:3)										
Chamber or shaft sections										
900mm internal diameter										
250mm high	20.07	5.00	**21.08**	0.58	1.16	**20.67**	-	**41.74**	nr	**44.87**
500mm high	30.07	5.00	**31.58**	0.67	1.34	**23.87**	-	**55.45**	nr	**59.61**
750mm high	30.07	5.00	**31.58**	0.87	1.74	**31.00**	-	**62.57**	nr	**67.27**
1000mm high	40.07	5.00	**42.08**	0.93	1.86	**33.14**	-	**75.21**	nr	**80.85**
1050mm internal diameter										
305mm high	19.13	5.00	**20.10**	0.81	1.62	**28.86**	-	**48.96**	nr	**52.64**
458mm high	28.63	5.00	**30.08**	0.87	1.74	**31.00**	-	**61.08**	nr	**65.66**
610mm high	28.63	5.00	**30.08**	0.93	1.86	**33.14**	-	**63.21**	nr	**67.96**
762mm high	38.13	5.00	**40.05**	0.99	1.98	**35.27**	-	**75.33**	nr	**80.98**
1200mm internal diameter										
250mm high	24.13	5.00	**25.35**	0.99	1.98	**35.27**	-	**60.63**	nr	**65.17**
500mm high	36.13	5.00	**37.95**	1.04	2.08	**37.06**	-	**75.01**	nr	**80.63**
750mm high	36.13	5.00	**37.95**	1.22	2.44	**43.47**	-	**81.42**	nr	**87.53**
1000mm high	48.13	5.00	**50.55**	1.28	2.56	**45.61**	-	**96.16**	nr	**103.37**
1500mm internal diameter										
500mm high	36.13	5.00	**37.95**	1.04	2.08	**37.06**	-	**75.01**	nr	**80.63**
750mm high	36.13	5.00	**37.95**	1.22	2.44	**43.47**	-	**81.42**	nr	**87.53**
1000mm high	48.13	5.00	**50.55**	1.28	2.56	**45.61**	-	**96.16**	nr	**103.37**
1800mm internal diameter										
500mm high	36.13	5.00	**37.95**	1.04	2.08	**37.06**	-	**75.01**	nr	**80.63**
750mm high	36.13	5.00	**37.95**	1.22	2.44	**43.47**	-	**81.42**	nr	**87.53**
1000mm high	48.13	5.00	**50.55**	1.28	2.56	**45.61**	-	**96.16**	nr	**103.37**
2100mm internal diameter										
500mm high	36.13	5.00	**37.95**	1.04	2.08	**37.06**	-	**75.01**	nr	**80.63**
750mm high	36.13	5.00	**37.95**	1.22	2.44	**43.47**	-	**81.42**	nr	**87.53**
1000mm high	48.13	5.00	**50.55**	1.28	2.56	**45.61**	-	**96.16**	nr	**103.37**

Labour hourly rates: (except Specialists) Craft Operatives £14.21 Labourer £10.71 Rates are national average prices. Refer to REGIONAL VARIATIONS for indicative levels of overall pricing in regions	MATERIALS			LABOUR				RATES		
	Del to Site £	Waste %	Material Cost £	Craft Optve Hrs	Lab Hrs	Labour Cost £	Sunds £	Nett Rate £	Unit	Gross Rate (+7.5%) £
R12: DRAINAGE BELOW GROUND; MANHOLES AND SOAKAWAYS: PRECAST CONCRETE INSPECTION CHAMBERS Cont.										
Precast concrete; standard or stock pattern units; BS 5911; jointing and pointing in cement mortar (1:3) Cont.										
Chamber or shaft sections Cont.										
2400mm internal diameter										
750mm high	36.13	5.00	**37.95**	1.22	2.44	**43.47**	-	**81.42**	nr	**87.53**
1000mm high	48.13	5.00	**50.55**	1.28	2.56	**45.61**	-	**96.16**	nr	**103.37**
2700mm internal diameter										
750mm high	36.13	5.00	**37.95**	1.22	2.44	**43.47**	-	**81.42**	nr	**87.53**
1000mm high	48.13	5.00	**50.55**	1.28	2.56	**45.61**	-	**96.16**	nr	**103.37**
3000mm internal diameter										
750mm high	36.13	5.00	**37.95**	1.22	2.44	**43.47**	-	**81.42**	nr	**87.53**
1000mm high	48.13	5.00	**50.55**	1.28	2.56	**45.61**	-	**96.16**	nr	**103.37**
Cutting holes										
cutting holes for pipes; 100mm diameter; making good; pointing	30.13	-	**30.13**	-	0.08	**0.86**	-	**30.99**	nr	**33.31**
cutting holes for pipes; 150mm diameter; making good; pointing	30.20	-	**30.20**	-	0.10	**1.07**	-	**31.27**	nr	**33.62**
cutting holes for pipes; 225mm diameter; making good; pointing	45.33	-	**45.33**	-	0.12	**1.29**	-	**46.62**	nr	**50.12**
cutting holes for pipes; 300mm diameter; making good; pointing	60.47	-	**60.47**	-	0.15	**1.61**	-	**62.07**	nr	**66.73**
Step irons										
galvanized malleable cast iron step irons BS 1247, cast in	5.00	5.00	**5.25**	-	0.15	**1.61**	-	**6.86**	nr	**7.37**
Cover slabs, heavy duty										
for chamber or shaft sections 900mm internal diameter, 125mm thick; access opening 600mm x 600mm	42.77	5.00	**44.93**	1.50	2.25	**45.41**	-	**90.34**	nr	**97.11**
for chamber or shaft sections 1050mm internal diameter, 125mm thick; access opening 600mm x 600mm	49.08	5.00	**51.55**	2.00	3.00	**60.55**	-	**112.10**	nr	**120.51**
for chamber or shaft sections 1200mm internal diameter, 125mm thick; access opening 600mm x 600mm	64.87	5.00	**68.13**	2.50	3.75	**75.69**	-	**143.82**	nr	**154.60**
for chamber or shaft sections 1350mm internal diameter, 125mm thick; access opening 600mm x 600mm	95.11	5.00	**99.88**	3.00	4.50	**90.83**	-	**190.71**	nr	**205.01**
for chamber or shaft sections 1500mm internal diameter, 150mm thick; access opening 600mm x 600mm	117.49	5.00	**123.39**	3.50	5.25	**105.96**	-	**229.35**	nr	**246.55**
for chamber or shaft sections 1800mm internal diameter, 150mm thick access opening 600mm x 600mm	135.97	5.00	**142.79**	4.00	6.00	**121.10**	-	**263.89**	nr	**283.68**
for chamber or shaft sections 2100mm internal diameter, 150mm thick access opening 600mm x 600mm	281.04	5.00	**295.13**	4.50	6.75	**136.24**	-	**431.37**	nr	**463.72**
for chamber or shaft sections 2400mm internal diameter, 150mm thick access opening 600mm x 600mm	489.57	5.00	**514.09**	5.00	7.50	**151.38**	-	**665.46**	nr	**715.37**
for chamber or shaft sections 2700mm internal diameter, 150mm thick access opening 600mm x 600mm	633.77	5.00	**665.50**	5.50	8.25	**166.51**	-	**832.01**	nr	**894.41**
for chamber or shaft sections 3000mm internal diameter, 150mm thick access opening 600mm x 600mm	746.38	5.00	**783.74**	6.00	9.00	**181.65**	-	**965.39**	nr	**1037.79**
Precast concrete inspection chambers; BS 5911; jointing and pointing in cement mortar (1:3)										
Rectangular inspection chambers, internal size 457 x 610mm										
100mm high	6.39	5.00	**6.73**	0.50	0.35	**10.85**	-	**17.58**	nr	**18.90**
150mm high	7.34	5.00	**7.72**	0.65	0.45	**14.06**	-	**21.78**	nr	**23.41**
250mm high	9.84	5.00	**10.35**	0.75	0.53	**16.33**	-	**26.68**	nr	**28.68**
extra; galvanized malleable cast iron step irons BS 1247, built in	5.00	5.00	**5.25**	-	-	**-**	-	**5.25**	nr	**5.64**
Rectangular inspection chambers, internal size 760 x 610mm										
150mm high	10.36	5.00	**10.90**	1.05	0.75	**22.95**	-	**33.85**	nr	**36.39**
225mm high	13.21	5.00	**13.89**	1.21	0.88	**26.62**	-	**40.51**	nr	**43.55**
extra; galvanized malleable cast iron step irons BS 1247, built in	5.00	5.00	**5.25**	-	-	**-**	-	**5.25**	nr	**5.64**
Rectangular inspection chambers, internal size 990 x 610mm										
150mm high	12.53	5.00	**13.18**	1.05	0.75	**22.95**	-	**36.13**	nr	**38.84**
Rectangular inspection chambers, internal size 1200 x 750mm										
150mm high	17.31	5.00	**18.20**	1.20	0.85	**26.16**	-	**44.35**	nr	**47.68**
200mm high	21.28	5.00	**22.37**	1.38	1.00	**30.32**	-	**52.69**	nr	**56.64**
Base units										
450 x 610mm internal size; 360mm deep	29.50	5.00	**31.00**	1.05	0.75	**22.95**	-	**53.95**	nr	**57.99**
760 x 610mm internal size; 360mm deep	35.73	5.00	**37.54**	1.05	0.75	**22.95**	-	**60.49**	nr	**65.03**
990 x 610mm internal size; 360mm deep	47.57	5.00	**49.99**	1.05	0.75	**22.95**	-	**72.94**	nr	**78.41**
Cover slab with integral cover										
450 x 610mm internal size;	22.98	5.00	**24.15**	1.05	0.75	**22.95**	-	**47.10**	nr	**50.64**
760 x 610mm internal size;	35.06	5.00	**36.84**	1.05	0.75	**22.95**	-	**59.79**	nr	**64.27**
990 x 610mm internal size;	54.20	5.00	**56.95**	1.05	0.75	**22.95**	-	**79.90**	nr	**85.89**

Labour hourly rates: (except Specialists) Craft Operatives £14.21 Labourer £10.71 Rates are national average prices. Refer to REGIONAL VARIATIONS for indicative levels of overall pricing in regions	MATERIALS			LABOUR				RATES		
	Del to Site £	Waste %	Material Cost £	Craft Optve Hrs	Lab Hrs	Labour Cost £	Sunds £	Nett Rate £	Unit	Gross Rate (+7.5%) £
R12: DRAINAGE BELOW GROUND; MANHOLES AND SOAKAWAYS: GRP INSPECTION CHAMBERS										
Inspection chambers; Terrain; GRP with preformed benching; 4 nr pvc adaptors; A15 light duty single seal cast iron cover to BS EN 124 and plastic frame set in position										
Chamber or shaft sections and benching										
470mm internal diameter; depth to invert										
520mm	190.43	5.00	**199.95**	5.00	0.80	**79.62**	-	**279.57**	nr	**300.53**
755mm	227.02	5.00	**238.37**	5.00	0.80	**79.62**	-	**317.99**	nr	**341.84**
990mm	263.61	5.00	**276.79**	5.35	0.85	**85.13**	-	**361.92**	nr	**389.06**
1225mm	300.21	5.00	**315.22**	5.35	0.85	**85.13**	-	**400.34**	nr	**430.37**
R12: DRAINAGE BELOW GROUND; MANHOLES AND SOAKAWAYS: POLYPROPYLENE INSPECTION CHAMBERS										
Osma; Shallow inspection chamber ref. 4D.960 Polypropylene with preformed benching. Light duty single sealed cover and frame ref. 4D.961 set in position, 250mm diameter; for 100mm diameter pipes, depth to invert										
Chamber or shaft sections and benching										
600mm	44.38	5.00	**46.60**	2.00	0.10	**29.49**	-	**76.09**	nr	**81.80**
Osma; universal inspection chamber; Polypropylene; 450mm diameter										
preformed chamber base benching, ref. 4D.922; for 100mm pipes	64.15	5.00	**67.36**	1.10	0.10	**16.70**	-	**84.06**	nr	**90.36**
chamber shaft 230mm effective length, ref. 4D.925	16.36	5.00	**17.18**	-	0.50	**5.36**	-	**22.53**	nr	**24.22**
cast iron cover and frame; single seal light duty to suit, ref. 4D.942	51.54	5.00	**54.11**	0.50	-	**7.11**	-	**61.22**	nr	**65.81**
Polypropylene sealed cover and frame, ref. 4D.943	30.82	5.00	**32.36**	0.50	-	**7.11**	-	**39.47**	nr	**42.43**

This section continues on the next page

Labour hourly rates: (except Specialists) Craft Operatives £17.38 Labourer £10.71 Rates are national average prices. Refer to REGIONAL VARIATIONS for indicative levels of overall pricing in regions	MATERIALS			LABOUR				RATES		
	Del to Site £	Waste %	Material Cost £	Craft Optve Hrs	Lab Hrs	Labour Cost £	Sunds £	Nett Rate £	Unit	Gross Rate (+7.5%) £
R12: DRAINAGE BELOW GROUND; MANHOLES AND SOAKAWAYS: CAST IRON INSPECTION CHAMBERS										
Coated cast iron bolted inspection chambers										
100mm diameter straight chamber; coupling joints; bedding in cement mortar (1:3)										
with one branch one side	70.82	2.00	**72.24**	2.50	-	**43.45**	-	**115.69**	nr	**124.37**
with one branch each side	108.16	2.00	**110.34**	3.25	-	**56.49**	-	**166.82**	nr	**179.33**
with two branches each side	212.85	2.00	**217.12**	4.75	-	**82.56**	-	**299.67**	nr	**322.15**
150 x 100mm diameter straight chamber; coupling joints; bedding in cement mortar (1:3)										
with one branch one side	140.11	2.00	**142.92**	3.25	-	**56.49**	-	**199.40**	nr	**214.36**
150mm diameter straight chamber; coupling joints; bedding in cement mortar (1:3)										
with one branch each side	177.78	2.00	**181.35**	4.50	-	**78.21**	-	**259.56**	nr	**279.02**
with two branches each side	354.94	2.00	**362.05**	6.50	-	**112.97**	-	**475.02**	nr	**510.64**

This section continues on the next page

Labour hourly rates: (except Specialists) Craft Operatives £14.21 Labourer £10.71 Rates are national average prices. Refer to REGIONAL VARIATIONS for indicative levels of overall pricing in regions	MATERIALS			LABOUR				RATES		
	Del to Site £	Waste %	Material Cost £	Craft Optve Hrs	Lab Hrs	Labour Cost £	Sunds £	Nett Rate £	Unit	Gross Rate (+7.5%) £
R12: DRAINAGE BELOW GROUND; MANHOLES AND SOAKAWAYS: MANHOLE ACCESSORIES										
Manhole step irons										
Step irons; BS 1247 malleable cast iron; galvanized; building in to joints										
figure 1 general purpose pattern; 110mm tail.....	3.28	-	**3.28**	0.25	-	**3.55**	-	**6.83**	nr	**7.34**
figure 1 general purpose pattern; 225mm tail.....	4.10	-	**4.10**	0.25	-	**3.55**	-	**7.66**	nr	**8.23**
Manhole covers - A15 light duty										
Access covers; Durey Castings Ltd; BS EN 124; coated; bedding frame in cement mortar (1:3); bedding cover in grease and sand										
reference DU14B; clear opening 450mm dia.........	13.27	2.50	**13.61**	0.65	0.65	**16.20**	-	**29.81**	nr	**32.04**
reference DU10A; clear opening 450 x 450mm.......	17.67	2.50	**18.12**	0.65	0.65	**16.20**	-	**34.32**	nr	**36.89**
reference DU10B; clear opening 600 x 450mm.......	22.07	2.50	**22.63**	0.75	0.75	**18.69**	-	**41.32**	nr	**44.42**
reference DU10C; clear opening 600 x 600mm.......	30.87	2.50	**31.65**	1.25	1.25	**31.15**	-	**62.80**	nr	**67.51**
reference DU15A recessed; clear opening 600 x 450mm	33.51	2.50	**34.36**	1.15	1.15	**28.66**	-	**63.01**	nr	**67.74**
reference DU15A recessed; clear opening 600 x 600mm	41.87	2.50	**42.92**	1.45	1.45	**36.13**	-	**79.06**	nr	**84.99**
reference DU15A recessed; clear opening 750 x 600mm	214.79	2.50	**220.17**	1.60	1.60	**39.87**	-	**260.04**	nr	**279.54**
reference DU30A double seal; clear opening 450 x 450mm	228.87	2.50	**234.60**	1.30	1.30	**32.40**	-	**267.00**	nr	**287.02**
reference DU30A double seal; clear opening 600 x 450mm	241.81	2.50	**247.86**	1.45	1.45	**36.13**	-	**283.99**	nr	**305.29**
reference DU30A double seal; clear opening 600 x 600mm	268.69	2.50	**275.42**	1.60	1.60	**39.87**	-	**315.29**	nr	**338.93**
lifting keys	6.60	2.50	**6.77**	-	-	**-**	-	**6.77**	nr	**7.27**
Manhole covers - B125 medium duty										
Access covers; Stanton PLC; BS EN 124; coated; bedding frame in cement mortar (1:3); bedding cover in grease and sand										
reference H783 Trojan single seal; 600mm diameter	94.84	2.50	**97.38**	1.65	1.65	**41.12**	-	**138.50**	nr	**148.88**
reference H781; 600 x 450mm......................	88.61	2.50	**91.08**	1.43	1.43	**35.64**	-	**126.71**	nr	**136.22**
reference H782; 600 x 600mm......................	109.64	2.50	**112.63**	1.80	1.80	**44.86**	-	**157.49**	nr	**169.30**
reference HA08; 600 x 450mm......................	115.12	2.50	**118.17**	1.69	1.69	**42.11**	-	**160.28**	nr	**172.31**
reference HA09; 600 x 600mm......................	146.30	2.50	**150.30**	2.24	2.24	**55.82**	-	**206.12**	nr	**221.57**
lifting keys	4.71	2.50	**4.83**	-	-	**-**	-	**4.83**	nr	**5.19**
Manhole covers - D400 heavy duty										
Access covers; BS EN 124; coated; bedding frame in cement mortar (1:3); bedding cover in grease and sand										
reference H9330; 600mm diameter...................	119.06	2.50	**122.37**	1.89	1.89	**47.10**	-	**169.47**	nr	**182.18**
reference H9320; 600 x 600mm.....................	102.84	2.50	**105.57**	1.89	1.89	**47.10**	-	**152.67**	nr	**164.12**
reference H9420; 750 x 750mm.....................	248.65	2.50	**255.37**	2.25	2.25	**56.07**	-	**311.44**	nr	**334.80**
lifting keys	5.50	2.50	**5.64**	-	-	**-**	-	**5.64**	nr	**6.06**
R12: DRAINAGE BELOW GROUND; MANHOLES AND SOAKAWAYS: INTERCEPTING TRAPS										
Vitrified clay intercepting traps										
Intercepting traps; BS 65; joint to pipe; building into side of manhole; bedding and surrounding in concrete to BS 5328 ordinary prescribed mix C15P, 20mm aggregate										
cleaning arm and stopper										
100mm outlet, ref. R1 1/1	61.48	5.00	**64.55**	1.25	1.00	**28.47**	-	**93.02**	nr	**100.00**
150mm outlet, ref. R1 1/2	88.12	5.00	**92.53**	1.75	1.50	**40.93**	-	**133.46**	nr	**143.47**
225mm outlet, ref. R1 1/3	272.58	5.00	**286.21**	2.25	2.00	**53.39**	-	**339.60**	nr	**365.07**
R12: DRAINAGE BELOW GROUND; MANHOLES AND SOAKAWAYS: FRESH AIR INLETS										
Aluminium; fresh air inlets										
Air inlet valve, mice flap; set in cement mortar (1:3)										
100mm diameter	28.82	5.00	**30.27**	1.15	-	**16.34**	-	**46.61**	nr	**50.11**
R12: DRAINAGE BELOW GROUND; MANHOLES AND SOAKAWAYS: SETTLEMENT TANKS										
Fibreglass settlement tanks; including cover and frame										
Settlement tank; fibreglass; set in position										
1000mm deep to invert of inlet, standard grade										
2700 litres capacity	486.00	-	**486.00**	0.90	1.80	**32.07**	-	**518.07**	nr	**556.92**
3750 litres capacity	623.00	-	**623.00**	1.00	2.00	**35.63**	-	**658.63**	nr	**708.03**
4500 litres capacity	717.00	-	**717.00**	1.10	2.20	**39.19**	-	**756.19**	nr	**812.91**
7500 litres capacity	1080.00	-	**1080.00**	1.20	2.40	**42.76**	-	**1122.76**	nr	**1206.96**
9000 litres capacity	1155.00	-	**1155.00**	1.30	2.60	**46.32**	-	**1201.32**	nr	**1291.42**
1500mm deep to invert of inlet, heavy grade										
2700 litres capacity	548.00	-	**548.00**	0.90	1.80	**32.07**	-	**580.07**	nr	**623.57**
3750 litres capacity	686.00	-	**686.00**	1.00	2.00	**35.63**	-	**721.63**	nr	**775.75**
4500 litres capacity	780.00	-	**780.00**	1.10	2.20	**39.19**	-	**819.19**	nr	**880.63**
7500 litres capacity	1167.00	-	**1167.00**	1.20	2.40	**42.76**	-	**1209.76**	nr	**1300.49**
9000 litres capacity	1356.00	-	**1356.00**	1.30	2.60	**46.32**	-	**1402.32**	nr	**1507.49**

Labour hourly rates: (except Specialists) Craft Operatives £14.21 Labourer £10.71 Rates are national average prices. Refer to REGIONAL VARIATIONS for indicative levels of overall pricing in regions	MATERIALS			LABOUR				RATES		
	Del to Site £	Waste %	Material Cost £	Craft Optve Hrs	Lab Hrs	Labour Cost £	Sunds £	Nett Rate £	Unit	Gross Rate (+7.5%) £
R12: DRAINAGE BELOW GROUND; WORK TO EXISTING DRAINS										
Work to disused drains										
Fly ash filling to disused drain										
100mm diameter	0.38	-	**0.38**	0.40	-	**5.68**	-	**6.06**	m	**6.52**
150mm diameter	0.70	-	**0.70**	0.60	-	**8.53**	-	**9.22**	m	**9.92**
225mm diameter	1.50	-	**1.50**	0.90	-	**12.79**	-	**14.29**	m	**15.36**
Sealing end of disused drain with plain in-situ concrete										
100mm diameter	0.62	5.00	**0.65**	0.20	-	**2.84**	-	**3.49**	nr	**3.75**
150mm diameter	1.16	5.00	**1.22**	0.30	-	**4.26**	-	**5.48**	nr	**5.89**
225mm diameter	1.86	5.00	**1.95**	0.40	-	**5.68**	-	**7.63**	nr	**8.21**
R13: LAND DRAINAGE										
Drains; clayware pipes, BS 1196; butt joints										
Pipework in trenches										
75mm	3.03	5.00	**3.18**	-	0.22	**2.36**	-	**5.54**	m	**5.95**
100mm	5.21	5.00	**5.47**	-	0.24	**2.57**	-	**8.04**	m	**8.64**
150mm	10.67	5.00	**11.20**	-	0.30	**3.21**	-	**14.42**	m	**15.50**
Drains; porous concrete pipes, BS 5911; butt joints										
Pipework in trenches										
150mm	5.32	5.00	**5.59**	-	0.28	**3.00**	-	**8.58**	m	**9.23**
225mm	6.82	5.00	**7.16**	-	0.36	**3.86**	-	**11.02**	m	**11.84**
300mm	11.21	5.00	**11.77**	-	0.42	**4.50**	-	**16.27**	m	**17.49**
375mm	15.27	5.00	**16.03**	-	0.48	**5.14**	-	**21.17**	m	**22.76**
450mm	21.48	5.00	**22.55**	-	0.54	**5.78**	-	**28.34**	m	**30.46**
525mm	31.82	5.00	**33.41**	-	0.60	**6.43**	-	**39.84**	m	**42.82**
600mm	39.52	5.00	**41.50**	-	0.66	**7.07**	-	**48.56**	m	**52.21**
Drains; OGEE pipes, BS 5911 Part 110; butt joints										
Pipework in trenches										
150mm	4.35	5.00	**4.57**	-	0.28	**3.00**	-	**7.57**	m	**8.13**
225mm	7.03	5.00	**7.38**	-	0.36	**3.86**	-	**11.24**	m	**12.08**
300mm	11.54	5.00	**12.12**	-	0.42	**4.50**	-	**16.62**	m	**17.86**
375mm	15.68	5.00	**16.46**	-	0.48	**5.14**	-	**21.60**	m	**23.23**
450mm	19.91	5.00	**20.91**	-	0.54	**5.78**	-	**26.69**	m	**28.69**
525mm	26.90	5.00	**28.25**	-	0.60	**6.43**	-	**34.67**	m	**37.27**
600mm	36.18	5.00	**37.99**	-	0.66	**7.07**	-	**45.06**	m	**48.44**
750mm	54.01	5.00	**56.71**	-	0.72	**7.71**	-	**64.42**	m	**69.25**
Drains; clayware perforated pipes, butt joints; 150mm hardcore to sides and top										
Pipework in trenches										
100mm	11.19	5.00	**11.95**	-	0.42	**4.50**	-	**16.45**	m	**17.68**
150mm	18.05	5.00	**19.20**	-	0.54	**5.78**	-	**24.98**	m	**26.86**
225mm	30.84	5.00	**32.73**	-	0.60	**6.43**	-	**39.16**	m	**42.09**

PIPED SUPPLY SYSTEMS

Labour hourly rates: (except Specialists) Craft Operatives £17.38 Labourer £10.71 Rates are national average prices. Refer to REGIONAL VARIATIONS for indicative levels of overall pricing in regions	MATERIALS			LABOUR				RATES		
	Del to Site £	Waste %	Material Cost £	Craft Optve Hrs	Lab Hrs	Labour Cost £	Sunds £	Nett Rate £	Unit	Gross Rate (+7.5%) £
S115: HOT AND COLD WATER AND GAS: BLACK AND GALVANIZED STEEL TUBES AND FITTINGS										
Steel pipes, BS 1387 Table 2; steel fittings BS 1740 Part 1; screwed and PTFE tape joints; pipework black										
Pipes, straight in malleable iron pipe brackets fixing to timber with screws										
15mm	2.30	5.00	**2.41**	0.50	-	**8.69**	-	**11.10**	m	**11.94**
15mm: fittings										
extra; made bends	-	-	-	0.17	-	**2.95**	-	**2.95**	nr	**3.18**
extra; connections to copper pipe ends; compression joint	0.59	2.00	**0.60**	0.10	-	**1.74**	-	**2.34**	nr	**2.52**
extra; fittings; one end	0.46	2.00	**0.47**	0.13	-	**2.26**	-	**2.73**	nr	**2.93**
extra; fittings; two ends	0.47	2.00	**0.48**	0.25	-	**4.35**	-	**4.82**	nr	**5.19**
extra; fittings; three ends	0.64	2.00	**0.65**	0.25	-	**4.35**	-	**5.00**	nr	**5.37**
20mm	2.68	5.00	**2.82**	0.57	-	**9.91**	-	**12.72**	m	**13.68**
20mm; fittings										
extra; made bends	-	-	-	0.25	-	**4.35**	-	**4.35**	nr	**4.67**
extra; connections to copper pipe ends; compression joint	0.96	2.00	**0.98**	0.17	-	**2.95**	-	**3.93**	nr	**4.23**
extra; fittings; one end	0.59	2.00	**0.60**	0.17	-	**2.95**	-	**3.56**	nr	**3.82**
extra; fittings; two ends	0.64	2.00	**0.65**	0.33	-	**5.74**	-	**6.39**	nr	**6.87**
extra; fittings; three ends	0.93	2.00	**0.95**	0.33	-	**5.74**	-	**6.68**	nr	**7.19**
25mm	3.72	5.00	**3.91**	0.67	-	**11.64**	-	**15.55**	m	**16.72**
25mm; fittings										
extra; made bends	-	-	-	0.33	-	**5.74**	-	**5.74**	nr	**6.17**
extra; connections to copper pipe ends; compression joint	1.86	2.00	**1.90**	0.25	-	**4.35**	-	**6.24**	nr	**6.71**
extra; fittings; one end	0.66	2.00	**0.67**	0.25	-	**4.35**	-	**5.02**	nr	**5.39**
extra; fittings; two ends	0.99	2.00	**1.01**	0.50	-	**8.69**	-	**9.70**	nr	**10.43**
extra; fittings; three ends	1.34	2.00	**1.37**	0.50	-	**8.69**	-	**10.06**	nr	**10.81**
32mm	4.67	5.00	**4.90**	0.80	-	**13.90**	-	**18.81**	m	**20.22**
32mm; fittings										
extra; made bends	-	-	-	0.50	-	**8.69**	-	**8.69**	nr	**9.34**
extra; connections to copper pipe ends; compression joint	6.32	2.00	**6.45**	0.33	-	**5.74**	-	**12.18**	nr	**13.10**
extra; fittings; one end	1.05	2.00	**1.07**	0.33	-	**5.74**	-	**6.81**	nr	**7.32**
extra; fittings; two ends	1.67	2.00	**1.70**	0.67	-	**11.64**	-	**13.35**	nr	**14.35**
extra; fittings; three ends	2.27	2.00	**2.32**	0.67	-	**11.64**	-	**13.96**	nr	**15.01**
40mm	6.20	5.00	**6.49**	1.00	-	**17.38**	-	**23.87**	m	**25.66**
40mm; fittings										
extra; made bends	-	-	-	0.67	-	**11.64**	-	**11.64**	nr	**12.52**
extra; connections to copper pipe ends; compression joint	10.17	2.00	**10.37**	0.50	-	**8.69**	-	**19.06**	nr	**20.49**
extra; fittings; one end	1.25	2.00	**1.28**	0.42	-	**7.30**	-	**8.57**	nr	**9.22**
extra; fittings; two ends	2.81	2.00	**2.87**	0.83	-	**14.43**	-	**17.29**	nr	**18.59**
extra; fittings; three ends	3.49	2.00	**3.56**	0.83	-	**14.43**	-	**17.99**	nr	**19.33**
50mm	7.71	5.00	**8.09**	1.33	-	**23.12**	-	**31.21**	m	**33.55**
50mm; fittings										
extra; made bends	-	-	-	1.00	-	**17.38**	-	**17.38**	nr	**18.68**
extra; connections to copper pipe ends; compression joint	15.39	2.00	**15.70**	0.67	-	**11.64**	-	**27.34**	nr	**29.39**
extra; fittings; one end	2.61	2.00	**2.66**	0.50	-	**8.69**	-	**11.35**	nr	**12.20**
extra; fittings; two ends	3.29	2.00	**3.36**	1.00	-	**17.38**	-	**20.74**	nr	**22.29**
extra; fittings; three ends	5.02	2.00	**5.12**	1.00	-	**17.38**	-	**22.50**	nr	**24.19**
Pipes straight in malleable iron pipe brackets fixing to masonry with screws										
15mm	2.31	5.00	**2.43**	0.53	-	**9.21**	-	**11.64**	m	**12.51**
20mm	2.70	5.00	**2.83**	0.60	-	**10.43**	-	**13.26**	m	**14.26**
25mm	3.73	5.00	**3.92**	0.70	-	**12.17**	-	**16.09**	m	**17.29**
32mm	4.68	5.00	**4.92**	0.83	-	**14.43**	-	**19.34**	m	**20.80**
40mm	5.54	5.00	**5.82**	1.03	-	**17.90**	-	**23.72**	m	**25.50**
50mm	7.72	5.00	**8.11**	1.36	-	**23.64**	-	**31.75**	m	**34.13**
Pipes straight in malleable iron single rings, 165mm long screwed both ends steel tubes and malleable iron backplates fixing to timber with screws										
15mm	2.77	5.00	**2.91**	0.60	-	**10.43**	-	**13.34**	m	**14.34**
20mm	3.16	5.00	**3.32**	0.67	-	**11.64**	-	**14.96**	m	**16.08**

Labour hourly rates: (except Specialists) Craft Operatives £17.38 Labourer £10.71 Rates are national average prices. Refer to REGIONAL VARIATIONS for indicative levels of overall pricing in regions	MATERIALS			LABOUR				RATES		
	Del to Site £	Waste %	Material Cost £	Craft Optve Hrs	Lab Hrs	Labour Cost £	Sunds £	Nett Rate £	Unit	Gross Rate (+7.5%) £
S115: HOT AND COLD WATER AND GAS: BLACK AND GALVANIZED STEEL TUBES AND FITTINGS Cont.										
Steel pipes, BS 1387 Table 2; steel fittings BS 1740 Part 1; screwed and PTFE tape joints; pipework black Cont.										
Pipes straight in malleable iron single rings, 165mm long screwed both ends steel tubes and malleable iron backplates fixing to timber with screws Cont.										
25mm	4.17	5.00	**4.38**	0.77	-	**13.38**	-	**17.76**	m	**19.09**
32mm	4.96	5.00	**5.21**	0.90	-	**15.64**	-	**20.85**	m	**22.42**
40mm	5.76	5.00	**6.05**	1.10	-	**19.12**	-	**25.17**	m	**27.05**
50mm	7.82	5.00	**8.22**	1.43	-	**24.85**	-	**33.07**	m	**35.55**
Steel pipes, BS 1387 Table 3; steel fittings BS 1740 Part 1; screwed and PTFE tape joints; pipework black										
Pipes, straight in malleable iron pipe brackets fixing to timber with screws										
15mm	2.66	5.00	**2.79**	0.50	-	**8.69**	-	**11.48**	m	**12.34**
15mm; fittings										
extra; made bends	-	-	-	0.17	-	**2.95**	-	**2.95**	nr	**3.18**
extra; connections to copper pipe ends; compression joint	0.59	2.00	**0.60**	0.10	-	**1.74**	-	**2.34**	nr	**2.52**
extra; fittings; one end	0.46	2.00	**0.47**	0.13	-	**2.26**	-	**2.73**	nr	**2.93**
extra; fittings; two ends	0.47	2.00	**0.48**	0.25	-	**4.35**	-	**4.82**	nr	**5.19**
extra; fittings; three ends	0.64	2.00	**0.65**	0.25	-	**4.35**	-	**5.00**	nr	**5.37**
20mm	3.13	5.00	**3.29**	0.57	-	**9.91**	-	**13.20**	m	**14.19**
20mm; fittings										
extra; made bends	-	-	-	0.25	-	**4.35**	-	**4.35**	nr	**4.67**
extra; connections to copper pipe ends; compression joint	0.96	2.00	**0.98**	0.17	-	**2.95**	-	**3.93**	nr	**4.23**
extra; fittings; one end	0.59	2.00	**0.60**	0.17	-	**2.95**	-	**3.56**	nr	**3.82**
extra; fittings; two ends	0.64	2.00	**0.65**	0.33	-	**5.74**	-	**6.39**	nr	**6.87**
extra; fittings; three ends	0.93	2.00	**0.95**	0.33	-	**5.74**	-	**6.68**	nr	**7.19**
25mm	4.44	5.00	**4.66**	0.67	-	**11.64**	-	**16.31**	m	**17.53**
25mm; fittings										
extra; made bends	-	-	-	0.33	-	**5.74**	-	**5.74**	nr	**6.17**
extra; connections to copper pipe ends; compression joint	1.86	2.00	**1.90**	0.25	-	**4.35**	-	**6.24**	nr	**6.71**
extra; fittings; one end	0.66	2.00	**0.67**	0.25	-	**4.35**	-	**5.02**	nr	**5.39**
extra; fittings; two ends	0.99	2.00	**1.01**	0.50	-	**8.69**	-	**9.70**	nr	**10.43**
extra; fittings; three ends	1.34	2.00	**1.37**	0.50	-	**8.69**	-	**10.06**	nr	**10.81**
32mm	5.56	5.00	**5.84**	0.80	-	**13.90**	-	**19.74**	m	**21.22**
32mm; fittings										
extra; made bends	-	-	-	0.50	-	**8.69**	-	**8.69**	nr	**9.34**
extra; connections to copper pipe ends; compression joint	6.32	2.00	**6.45**	0.33	-	**5.74**	-	**12.18**	nr	**13.10**
extra; fittings; one end	1.05	2.00	**1.07**	0.33	-	**5.74**	-	**6.81**	nr	**7.32**
extra; fittings; two ends	1.67	2.00	**1.70**	0.67	-	**11.64**	-	**13.35**	nr	**14.35**
extra; fittings; three ends	2.27	2.00	**2.32**	0.67	-	**11.64**	-	**13.96**	nr	**15.01**
40mm	6.58	5.00	**6.91**	1.00	-	**17.38**	-	**24.29**	m	**26.12**
40mm; fittings										
extra; made bends	-	-	-	0.67	-	**11.64**	-	**11.64**	nr	**12.52**
extra; connections to copper pipe ends; compression joint	10.17	2.00	**10.37**	0.50	-	**8.69**	-	**19.06**	nr	**20.49**
extra; fittings; one end	1.25	2.00	**1.28**	0.42	-	**7.30**	-	**8.57**	nr	**9.22**
extra; fittings; two ends	2.81	2.00	**2.87**	0.83	-	**14.43**	-	**17.29**	nr	**18.59**
extra; fittings; three ends	3.49	2.00	**3.56**	0.83	-	**14.43**	-	**17.99**	nr	**19.33**
50mm	9.09	5.00	**9.54**	1.33	-	**23.12**	-	**32.66**	m	**35.11**
50mm; fittings										
extra; made bends	-	-	-	1.00	-	**17.38**	-	**17.38**	nr	**18.68**
extra; connections to copper pipe ends; compression joint	15.39	2.00	**15.70**	0.67	-	**11.64**	-	**27.34**	nr	**29.39**
extra; fittings; one end	2.61	2.00	**2.66**	0.50	-	**8.69**	-	**11.35**	nr	**12.20**
extra; fittings; two ends	3.29	2.00	**3.36**	1.00	-	**17.38**	-	**20.74**	nr	**22.29**
extra; fittings; three ends	5.02	2.00	**5.12**	1.00	-	**17.38**	-	**22.50**	nr	**24.19**
Pipes straight in malleable iron pipe brackets fixing to masonry with screws										
15mm	2.67	5.00	**2.81**	0.53	-	**9.21**	-	**12.02**	m	**12.92**
20mm	3.15	5.00	**3.31**	0.60	-	**10.43**	-	**13.74**	m	**14.77**
25mm	4.45	5.00	**4.68**	0.70	-	**12.17**	-	**16.84**	m	**18.11**
32mm	5.57	5.00	**5.85**	0.83	-	**14.43**	-	**20.28**	m	**21.80**
40mm	6.60	5.00	**6.93**	1.03	-	**17.90**	-	**24.83**	m	**26.69**
50mm	9.10	5.00	**9.56**	1.36	-	**23.64**	-	**33.20**	m	**35.69**
Pipes straight in malleable iron single rings, 165mm long screwed both ends steel tubes and malleable iron backplates fixing to timber with screws										
15mm	3.13	5.00	**3.29**	0.60	-	**10.43**	-	**13.71**	m	**14.74**
20mm	3.61	5.00	**3.79**	0.67	-	**11.64**	-	**15.43**	m	**16.59**
25mm	4.89	5.00	**5.13**	0.77	-	**13.38**	-	**18.52**	m	**19.91**
32mm	5.85	5.00	**6.15**	0.90	-	**15.64**	-	**21.79**	m	**23.42**
40mm	6.82	5.00	**7.16**	1.10	-	**19.12**	-	**26.28**	m	**28.25**
50mm	9.20	5.00	**9.66**	1.43	-	**24.85**	-	**34.52**	m	**37.11**
Steel pipes, BS 1387 Table 2; steel fittings BS 1740 Part 1; screwed and PTFE tape joints; pipework galvanized										
Pipes, straight in malleable iron pipe brackets fixing to timber with screws										
15mm	3.45	5.00	**3.62**	0.50	-	**8.69**	-	**12.31**	m	**13.23**

Labour hourly rates: (except Specialists) Craft Operatives £17.38 Labourer £10.71 Rates are national average prices. Refer to REGIONAL VARIATIONS for indicative levels of overall pricing in regions	MATERIALS			LABOUR				RATES		
	Del to Site £	Waste %	Material Cost £	Craft Optve Hrs	Lab Hrs	Labour Cost £	Sunds £	Nett Rate £	Unit	Gross Rate (+7.5%) £
S115: HOT AND COLD WATER AND GAS: BLACK AND GALVANIZED STEEL TUBES AND FITTINGS Cont.										
Steel pipes, BS 1387 Table 2; steel fittings BS 1740 Part 1; screwed and PTFE tape joints; pipework galvanized Cont.										
Pipes, straight in malleable iron pipe brackets fixing to timber with screws Cont.										
15mm; fittings										
extra; made bends	-	-	-	0.17	-	2.95	-	2.95	nr	3.18
extra; connections to copper pipe ends; compression joint	0.59	2.00	0.60	0.10	-	1.74	-	2.34	nr	2.52
extra; fittings; one end	0.64	2.00	0.65	0.13	-	2.26	-	2.91	nr	3.13
extra; fittings; two ends	0.64	2.00	0.65	0.25	-	4.35	-	5.00	nr	5.37
extra; fittings; three ends	0.88	2.00	0.90	0.25	-	4.35	-	5.24	nr	5.64
20mm	3.88	5.00	4.08	0.57	-	9.91	-	13.98	m	15.03
20mm; fittings										
extra; made bends	-	-	-	0.25	-	4.35	-	4.35	nr	4.67
extra; connections to copper pipe ends; compression joint	0.96	2.00	0.98	0.17	-	2.95	-	3.93	nr	4.23
extra; fittings; one end	0.73	2.00	0.74	0.17	-	2.95	-	3.70	nr	3.98
extra; fittings; two ends	0.88	2.00	0.90	0.33	-	5.74	-	6.63	nr	7.13
extra; fittings; three ends	1.29	2.00	1.32	0.33	-	5.74	-	7.05	nr	7.58
25mm	5.29	5.00	5.56	0.67	-	11.64	-	17.20	m	18.49
25mm; fittings										
extra; made bends	-	-	-	0.33	-	5.74	-	5.74	nr	6.17
extra; connections to copper pipe ends; compression joint	1.86	2.00	1.90	0.25	-	4.35	-	6.24	nr	6.71
extra; fittings; one end	0.98	2.00	1.00	0.25	-	4.35	-	5.34	nr	5.75
extra; fittings; two ends	1.37	2.00	1.40	0.50	-	8.69	-	10.09	nr	10.84
extra; fittings; three ends	1.86	2.00	1.90	0.50	-	8.69	-	10.59	nr	11.38
32mm	6.63	5.00	6.96	0.80	-	13.90	-	20.86	m	22.43
32mm; fittings										
extra; made bends	-	-	-	0.50	-	8.69	-	8.69	nr	9.34
extra; connections to copper pipe ends; compression joint	6.32	2.00	6.45	0.33	-	5.74	-	12.18	nr	13.10
extra; fittings; one end	1.51	2.00	1.54	0.33	-	5.74	-	7.28	nr	7.82
extra; fittings; two ends	2.57	2.00	2.62	0.67	-	11.64	-	14.27	nr	15.34
extra; fittings; three ends	3.52	2.00	3.59	0.67	-	11.64	-	15.24	nr	16.38
40mm	7.85	5.00	8.25	1.00	-	17.38	-	25.63	m	27.55
40mm; fittings										
extra; made bends	-	-	-	0.67	-	11.64	-	11.64	nr	12.52
extra; connections to copper pipe ends; compression joint	10.17	2.00	10.37	0.50	-	8.69	-	19.06	nr	20.49
extra; fittings; one end	1.92	2.00	1.96	0.42	-	7.30	-	9.26	nr	9.95
extra; fittings; two ends	3.85	2.00	3.93	0.83	-	14.43	-	18.35	nr	19.73
extra; fittings; three ends	4.82	2.00	4.92	0.83	-	14.43	-	19.34	nr	20.79
50mm	10.91	5.00	11.45	1.33	-	23.12	-	34.57	m	37.16
50mm; fittings										
extra; made bends	-	-	-	1.00	-	17.38	-	17.38	nr	18.68
extra; connections to copper pipe ends; compression joint	15.35	2.00	15.66	0.67	-	11.64	-	27.30	nr	29.35
extra; fittings; one end	3.58	2.00	3.65	0.50	-	8.69	-	12.34	nr	13.27
extra; fittings; two ends	4.50	2.00	4.59	1.00	-	17.38	-	21.97	nr	23.62
extra; fittings; three ends	6.95	2.00	7.09	1.00	-	17.38	-	24.47	nr	26.30
Pipes straight in malleable iron pipe brackets fixing to masonry with screws										
15mm	3.67	5.00	3.85	0.53	-	9.21	-	13.06	m	14.04
20mm	4.17	5.00	4.37	0.60	-	10.43	-	14.80	m	15.91
25mm	5.66	5.00	5.93	0.70	-	12.17	-	18.09	m	19.45
32mm	7.33	5.00	7.68	0.83	-	14.43	-	22.10	m	23.76
40mm	8.80	5.00	9.21	1.03	-	17.90	-	27.11	m	29.14
50mm	12.27	5.00	12.84	1.36	-	23.64	-	36.48	m	39.21
Pipes straight in malleable iron single rings, 165mm long screwed both ends steel tubes and malleable iron backplates fixing to timber with screws										
15mm	4.23	5.00	4.44	0.60	-	10.43	-	14.87	m	15.98
20mm	4.74	5.00	4.97	0.67	-	11.64	-	16.61	m	17.86
25mm	6.19	5.00	6.49	0.77	-	13.38	-	19.87	m	21.36
32mm	7.67	5.00	8.03	0.90	-	15.64	-	23.67	m	25.45
40mm	9.04	5.00	9.46	1.10	-	19.12	-	28.58	m	30.72
50mm	12.37	5.00	12.95	1.43	-	24.85	-	37.80	m	40.63
Steel pipes, BS 1387 Table 3; steel fittings BS 1740 Part 1; screwed and PTFE tape joints; pipework galvanized										
Pipes, straight in malleable iron pipe brackets fixing to timber with screws										
15mm	4.19	5.00	4.40	0.50	-	8.69	-	13.09	m	14.07
15mm; fittings										
extra; made bends	-	-	-	0.17	-	2.95	-	2.95	nr	3.18
extra; connections to copper pipe ends; compression joint	0.59	2.00	0.60	0.10	-	1.74	-	2.34	nr	2.52
extra; fittings; one end	0.64	2.00	0.65	0.13	-	2.26	-	2.91	nr	3.13
extra; fittings; two ends	0.64	2.00	0.65	0.25	-	4.35	-	5.00	nr	5.37
extra; fittings; three ends	0.88	2.00	0.90	0.25	-	4.35	-	5.24	nr	5.64
20mm	4.78	5.00	5.02	0.57	-	9.91	-	14.92	m	16.04
20mm; fittings										
extra; made bends	-	-	-	0.25	-	4.35	-	4.35	nr	4.67
extra; connections to copper pipe ends; compression joint	0.96	2.00	0.98	0.17	-	2.95	-	3.93	nr	4.23

Labour hourly rates: (except Specialists) Craft Operatives £17.38 Labourer £10.71 Rates are national average prices. Refer to REGIONAL VARIATIONS for indicative levels of overall pricing in regions	MATERIALS			LABOUR				RATES		
	Del to Site £	Waste %	Material Cost £	Craft Optve Hrs	Lab Hrs	Labour Cost £	Sunds £	Nett Rate £	Unit	Gross Rate (+7.5%) £
S115: HOT AND COLD WATER AND GAS: BLACK AND GALVANIZED STEEL TUBES AND FITTINGS Cont.										
Steel pipes, BS 1387 Table 3; steel fittings BS 1740 Part 1; screwed and PTFE tape joints; pipework galvanized Cont.										
Pipes, straight in malleable iron pipe brackets fixing to timber with screws Cont.										
20mm; fittings Cont.										
extra; fittings; one end	0.73	2.00	**0.74**	0.17	-	**2.95**	-	**3.70**	nr	**3.98**
extra; fittings; two ends	0.88	2.00	**0.90**	0.33	-	**5.74**	-	**6.63**	nr	**7.13**
extra; fittings; three ends	1.29	2.00	**1.32**	0.33	-	**5.74**	-	**7.05**	nr	**7.58**
25mm	6.63	5.00	**6.95**	0.67	-	**11.64**	-	**18.60**	m	**19.99**
25mm; fittings										
extra; made bends	-	-	-	0.33	-	**5.74**	-	**5.74**	nr	**6.17**
extra; connections to copper pipe ends; compression joint	1.86	2.00	**1.90**	0.25	-	**4.35**	-	**6.24**	nr	**6.71**
extra; fittings; one end	0.98	2.00	**1.00**	0.25	-	**4.35**	-	**5.34**	nr	**5.75**
extra; fittings; two ends	1.37	2.00	**1.40**	0.50	-	**8.69**	-	**10.09**	nr	**10.84**
extra; fittings; three ends	1.86	2.00	**1.90**	0.50	-	**8.69**	-	**10.59**	nr	**11.38**
32mm	8.56	5.00	**8.96**	0.80	-	**13.90**	-	**22.87**	m	**24.58**
32mm; fittings										
extra; made bends	-	-	-	0.50	-	**8.69**	-	**8.69**	nr	**9.34**
extra; connections to copper pipe ends; compression joint	6.32	2.00	**6.45**	0.33	-	**5.74**	-	**12.18**	nr	**13.10**
extra; fittings; one end	1.51	2.00	**1.54**	0.33	-	**5.74**	-	**7.28**	nr	**7.82**
extra; fittings; two ends	2.57	2.00	**2.62**	0.67	-	**11.64**	-	**14.27**	nr	**15.34**
extra; fittings; three ends	3.52	2.00	**3.59**	0.67	-	**11.64**	-	**15.24**	nr	**16.38**
40mm	10.26	5.00	**10.75**	1.00	-	**17.38**	-	**28.13**	m	**30.24**
40mm; fittings										
extra; made bends	-	-	-	0.67	-	**11.64**	-	**11.64**	nr	**12.52**
extra; connections to copper pipe ends; compression joint	10.17	2.00	**10.37**	0.50	-	**8.69**	-	**19.06**	nr	**20.49**
extra; fittings; one end	1.92	2.00	**1.96**	0.42	-	**7.30**	-	**9.26**	nr	**9.95**
extra; fittings; two ends	3.85	2.00	**3.93**	0.83	-	**14.43**	-	**18.35**	nr	**19.73**
extra; fittings; three ends	4.82	2.00	**4.92**	0.83	-	**14.43**	-	**19.34**	nr	**20.79**
50mm	14.19	5.00	**14.86**	1.33	-	**23.12**	-	**37.98**	m	**40.83**
50mm; fittings										
extra; made bends	-	-	-	1.00	-	**17.38**	-	**17.38**	nr	**18.68**
extra; connections to copper pipe ends; compression joint	15.35	2.00	**15.66**	0.67	-	**11.64**	-	**27.30**	nr	**29.35**
extra; fittings; one end	3.58	2.00	**3.65**	0.50	-	**8.69**	-	**12.34**	nr	**13.27**
extra; fittings; two ends	4.50	2.00	**4.59**	1.00	-	**17.38**	-	**21.97**	nr	**23.62**
extra; fittings; three ends	6.95	2.00	**7.09**	1.00	-	**17.38**	-	**24.47**	nr	**26.30**
Pipes straight in malleable iron pipe brackets fixing to masonry with screws										
15mm	4.21	5.00	**4.41**	0.53	-	**9.21**	-	**13.62**	m	**14.65**
20mm	4.80	5.00	**5.03**	0.60	-	**10.43**	-	**15.46**	m	**16.62**
25mm	6.65	5.00	**6.97**	0.70	-	**12.17**	-	**19.13**	m	**20.57**
32mm	8.57	5.00	**8.98**	0.83	-	**14.43**	-	**23.40**	m	**25.16**
40mm	10.28	5.00	**10.76**	1.03	-	**17.90**	-	**28.66**	m	**30.81**
50mm	14.21	5.00	**14.88**	1.36	-	**23.64**	-	**38.51**	m	**41.40**
Pipes straight in malleable iron single rings, 165mm long screwed both ends steel tubes and malleable iron backplates fixing to timber with screws										
15mm	4.77	5.00	**5.01**	0.60	-	**10.43**	-	**15.43**	m	**16.59**
20mm	5.37	5.00	**5.63**	0.67	-	**11.64**	-	**17.28**	m	**18.57**
25mm	7.18	5.00	**7.53**	0.77	-	**13.38**	-	**20.91**	m	**22.48**
32mm	8.91	5.00	**9.33**	0.90	-	**15.64**	-	**24.97**	m	**26.85**
40mm	10.52	5.00	**11.01**	1.10	-	**19.12**	-	**30.13**	m	**32.39**
50mm	14.31	5.00	**14.98**	1.43	-	**24.85**	-	**39.84**	m	**42.82**
S115: HOT AND COLD WATER AND GAS: LIGHT GAUGE STAINLESS STEEL TUBES AND FITTINGS										
Stainless steel pipes, BS EN 10312 Part 2; C.P. copper fittings, capillary, BS EN 1254 Part 2										
Pipes, straight in C.P. pipe brackets fixing to timber with screws										
15mm	3.89	5.00	**4.08**	0.33	-	**5.74**	-	**9.81**	m	**10.55**
15mm; fittings										
extra; fittings; one end	3.17	3.00	**3.27**	0.05	-	**0.87**	-	**4.13**	nr	**4.44**
extra; fittings; two ends	0.98	3.00	**1.01**	0.10	-	**1.74**	-	**2.75**	nr	**2.95**
extra; fittings; three ends	1.74	3.00	**1.79**	0.10	-	**1.74**	-	**3.53**	nr	**3.79**
22mm	6.09	5.00	**6.38**	0.36	-	**6.26**	-	**12.64**	m	**13.59**
22mm; fittings										
extra; fittings; one end	3.68	3.00	**3.79**	0.09	-	**1.56**	-	**5.35**	nr	**5.76**
extra; fittings; two ends	2.33	3.00	**2.40**	0.17	-	**2.95**	-	**5.35**	nr	**5.76**
extra; fittings; three ends	5.33	3.00	**5.49**	0.17	-	**2.95**	-	**8.44**	nr	**9.08**
28mm	8.22	5.00	**8.61**	0.40	-	**6.95**	-	**15.56**	m	**16.73**
28mm; fittings										
extra; fittings; one end	6.48	3.00	**6.67**	0.13	-	**2.26**	-	**8.93**	nr	**9.60**
extra; fittings; two ends	5.27	3.00	**5.43**	0.25	-	**4.35**	-	**9.77**	nr	**10.51**
extra; fittings; three ends	8.14	3.00	**8.38**	0.25	-	**4.35**	-	**12.73**	nr	**13.68**
35mm	10.52	5.00	**11.01**	0.50	-	**8.69**	-	**19.70**	m	**21.18**
35mm; fittings										
extra; fittings; one end	14.31	3.00	**14.74**	0.17	-	**2.95**	-	**17.69**	nr	**19.02**
extra; fittings; two ends	12.53	3.00	**12.91**	0.33	-	**5.74**	-	**18.64**	nr	**20.04**
extra; fittings; three ends	20.40	3.00	**21.01**	0.33	-	**5.74**	-	**26.75**	nr	**28.75**
42mm	14.66	5.00	**15.33**	0.67	-	**11.64**	-	**26.97**	m	**29.00**
42mm; fittings										
extra; fittings; one end	24.65	3.00	**25.39**	0.25	-	**4.35**	-	**29.73**	nr	**31.96**

Labour hourly rates: (except Specialists) Craft Operatives £17.38 Labourer £10.71 Rates are national average prices. Refer to REGIONAL VARIATIONS for indicative levels of overall pricing in regions	MATERIALS			LABOUR				RATES		
	Del to Site £	Waste %	Material Cost £	Craft Optve Hrs	Lab Hrs	Labour Cost £	Sunds £	Nett Rate £	Unit	Gross Rate (+7.5%) £
S115: HOT AND COLD WATER AND GAS: LIGHT GAUGE STAINLESS STEEL TUBES AND FITTINGS Cont.										
Stainless steel pipes, BS EN 10312 Part 2; C.P. copper fittings, capillary, BS EN 1254 Part 2 Cont.										
Pipes, straight in C.P. pipe brackets fixing to timber with screws Cont.										
42mm; fittings Cont.										
extra; fittings; two ends	20.71	3.00	21.33	0.50	-	8.69	-	30.02	nr	32.27
extra; fittings; three ends	32.74	3.00	33.72	0.50	-	8.69	-	42.41	nr	45.59
Pipes straight in C.P. pipe brackets fixing to masonry with screws										
15mm	3.90	5.00	4.09	0.36	-	6.26	-	10.35	m	11.13
22mm	6.10	5.00	6.40	0.39	-	6.78	-	13.18	m	14.17
28mm	8.23	5.00	8.63	0.43	-	7.47	-	16.10	m	17.31
35mm	10.54	5.00	11.03	0.53	-	9.21	-	20.24	m	21.75
42mm	14.68	5.00	15.34	0.70	-	12.17	-	27.51	m	29.57
Stainless steel pipes, BS EN 10312 Part 2; C.P. copper alloy fittings, compression, BS EN 1254 Part 2, Type A										
Pipes, straight in C.P. pipe brackets fixing to timber with screws										
15mm	4.12	5.00	4.32	0.33	-	5.74	-	10.06	m	10.81
15mm; fittings										
extra; fittings; one end	3.04	3.00	3.13	0.05	-	0.87	-	4.00	nr	4.30
extra; fittings; two ends	2.79	3.00	2.87	0.10	-	1.74	-	4.61	nr	4.96
extra; fittings; three ends	4.00	3.00	4.12	0.10	-	1.74	-	5.86	nr	6.30
22mm	6.29	5.00	6.59	0.36	-	6.26	-	12.85	m	13.81
22mm; fittings										
extra; fittings; one end	4.08	3.00	4.20	0.09	-	1.56	-	5.77	nr	6.20
extra; fittings; two ends	4.33	3.00	4.46	0.17	-	2.95	-	7.41	nr	7.97
extra; fittings; three ends	6.29	3.00	6.48	0.17	-	2.95	-	9.43	nr	10.14
28mm	8.50	5.00	8.90	0.40	-	6.95	-	15.86	m	17.04
28mm; fittings										
extra; fittings; one end	7.00	3.00	7.21	0.13	-	2.26	-	9.47	nr	10.18
extra; fittings; two ends	10.95	3.00	11.28	0.25	-	4.35	-	15.62	nr	16.80
extra; fittings; three ends	16.95	3.00	17.46	0.25	-	4.35	-	21.80	nr	23.44
35mm	11.34	5.00	11.85	0.50	-	8.69	-	20.54	m	22.08
35mm; fittings										
extra; fittings; one end	14.00	3.00	14.42	0.17	-	2.95	-	17.37	nr	18.68
extra; fittings; two ends	21.27	3.00	21.91	0.33	-	5.74	-	27.64	nr	29.72
extra; fittings; three ends	25.57	3.00	26.34	0.33	-	5.74	-	32.07	nr	34.48
42mm	15.65	5.00	16.34	0.67	-	11.64	-	27.99	m	30.09
42mm; fittings										
extra; fittings; one end	22.90	3.00	23.59	0.25	-	4.35	-	27.93	nr	30.03
extra; fittings; two ends	36.60	3.00	37.70	0.50	-	8.69	-	46.39	nr	49.87
extra; fittings; three ends	47.70	3.00	49.13	0.50	-	8.69	-	57.82	nr	62.16
Pipes straight in C.P. pipe brackets fixing to masonry with screws										
15mm	4.14	5.00	4.34	0.36	-	6.26	-	10.59	m	11.39
22mm	6.31	5.00	6.61	0.39	-	6.78	-	13.39	m	14.39
28mm	8.52	5.00	8.92	0.43	-	7.47	-	16.39	m	17.62
35mm	11.36	5.00	11.87	0.53	-	9.21	-	21.08	m	22.66
42mm	15.66	5.00	16.36	0.70	-	12.17	-	28.53	m	30.67
S115: HOT AND COLD WATER AND GAS: COPPER TUBING AND FITTINGS										
Copper pipes, BS EN 1057 Part 1 Table X; copper fittings, capillary, BS EN 1254 Part 2										
Pipes, straight in two piece copper spacing clips to timber with screws										
15mm	0.86	5.00	0.91	0.33	-	5.74	-	6.64	m	7.14
15mm; fittings										
extra; made bends	-	-	-	0.08	-	1.39	-	1.39	nr	1.49
extra; connections to iron pipe ends; screwed joint	1.61	3.00	1.66	0.10	-	1.74	-	3.40	nr	3.65
extra; fittings; one end	0.82	3.00	0.84	0.05	-	0.87	-	1.71	nr	1.84
extra; fittings; two ends	0.21	3.00	0.22	0.10	-	1.74	-	1.95	nr	2.10
extra; fittings; three ends	0.39	3.00	0.40	0.10	-	1.74	-	2.14	nr	2.30
22mm	1.67	5.00	1.75	0.36	-	6.26	-	8.01	m	8.61
22mm; fittings										
extra; made bends	-	-	-	0.13	-	2.26	-	2.26	nr	2.43
extra; connections to iron pipe ends; screwed joint	2.86	3.00	2.95	0.17	-	2.95	-	5.90	nr	6.34
extra; fittings; one end	1.55	3.00	1.60	0.09	-	1.56	-	3.16	nr	3.40
extra; fittings; two ends	0.53	3.00	0.55	0.17	-	2.95	-	3.50	nr	3.76
extra; fittings; three ends	1.23	3.00	1.27	0.17	-	2.95	-	4.22	nr	4.54
28mm	2.40	5.00	2.52	0.40	-	6.95	-	9.47	m	10.18
28mm; fittings										
extra; made bends	-	-	-	0.17	-	2.95	-	2.95	nr	3.18
extra; connections to iron pipe ends; screwed joint	4.61	3.00	4.75	0.25	-	4.35	-	9.09	nr	9.78
extra; fittings; one end	2.76	3.00	2.84	0.13	-	2.26	-	5.10	nr	5.48
extra; fittings; two ends	1.23	3.00	1.27	0.25	-	4.35	-	5.61	nr	6.03
extra; fittings; three ends	3.40	3.00	3.50	0.25	-	4.35	-	7.85	nr	8.44
35mm	5.26	5.00	5.51	0.50	-	8.69	-	14.20	m	15.27
35mm; fittings										
extra; made bends	-	-	-	0.25	-	4.35	-	4.35	nr	4.67
extra; connections to iron pipe ends; screwed joint	8.12	3.00	8.36	0.33	-	5.74	-	14.10	nr	15.16

Labour hourly rates: (except Specialists) Craft Operatives £17.38 Labourer £10.71 Rates are national average prices. Refer to REGIONAL VARIATIONS for indicative levels of overall pricing in regions	MATERIALS			LABOUR				RATES		
	Del to Site £	Waste %	Material Cost £	Craft Optve Hrs	Lab Hrs	Labour Cost £	Sunds £	Nett Rate £	Unit	Gross Rate (+7.5%) £
S115: HOT AND COLD WATER AND GAS: COPPER TUBING AND FITTINGS Cont.										
Copper pipes, BS EN 1057 Part 1 Table X; copper fittings, capillary, BS EN 1254 Part 2 Cont.										
Pipes, straight in two piece copper spacing clips to timber with screws Cont.										
35mm; fittings Cont.										
extra; fittings; one end	6.09	3.00	**6.27**	0.17	-	**2.95**	-	**9.23**	nr	**9.92**
extra; fittings; two ends	5.33	3.00	**5.49**	0.33	-	**5.74**	-	**11.23**	nr	**12.07**
extra; fittings; three ends	8.68	3.00	**8.94**	0.33	-	**5.74**	-	**14.68**	nr	**15.78**
42mm	6.71	5.00	**7.03**	0.67	-	**11.64**	-	**18.68**	m	**20.08**
42mm; fittings										
extra; made bends	-	-	**-**	0.33	-	**5.74**	-	**5.74**	nr	**6.17**
extra; connections to iron pipe ends; screwed joint	10.46	3.00	**10.77**	0.50	-	**8.69**	-	**19.46**	nr	**20.92**
extra; fittings; one end	10.49	3.00	**10.80**	0.25	-	**4.35**	-	**15.15**	nr	**16.29**
extra; fittings; two ends	8.81	3.00	**9.07**	0.50	-	**8.69**	-	**17.76**	nr	**19.10**
extra; fittings; three ends	13.93	3.00	**14.35**	0.50	-	**8.69**	-	**23.04**	nr	**24.77**
54mm	8.99	5.00	**9.41**	1.00	-	**17.38**	-	**26.79**	m	**28.80**
54mm; fittings										
extra; made bends	-	-	**-**	0.50	-	**8.69**	-	**8.69**	nr	**9.34**
extra; connections to iron pipe ends; screwed joint	15.88	3.00	**16.36**	0.67	-	**11.64**	-	**28.00**	nr	**30.10**
extra; fittings; one end	14.64	3.00	**15.08**	0.33	-	**5.74**	-	**20.81**	nr	**22.38**
extra; fittings; two ends	18.20	3.00	**18.75**	0.67	-	**11.64**	-	**30.39**	nr	**32.67**
extra; fittings; three ends	28.09	3.00	**28.93**	0.67	-	**11.64**	-	**40.58**	nr	**43.62**
Pipes straight in two piece copper spacing clips fixing to masonry with screws										
15mm	0.88	5.00	**0.92**	0.36	-	**6.26**	-	**7.18**	m	**7.72**
22mm	1.69	5.00	**1.77**	0.39	-	**6.78**	-	**8.55**	m	**9.19**
28mm	2.42	5.00	**2.53**	0.43	-	**7.47**	-	**10.01**	m	**10.76**
35mm	5.27	5.00	**5.53**	0.53	-	**9.21**	-	**14.74**	m	**15.84**
42mm	6.73	5.00	**7.05**	0.70	-	**12.17**	-	**19.21**	m	**20.65**
54mm	9.01	5.00	**9.43**	1.03	-	**17.90**	-	**27.33**	m	**29.38**
Pipes straight in pressed brass pipe brackets fixing to timber with screws										
15mm	1.17	5.00	**1.23**	0.33	-	**5.74**	-	**6.97**	m	**7.49**
22mm	2.03	5.00	**2.13**	0.36	-	**6.26**	-	**8.39**	m	**9.02**
28mm	2.83	5.00	**2.96**	0.40	-	**6.95**	-	**9.92**	m	**10.66**
35mm	5.69	5.00	**5.97**	0.50	-	**8.69**	-	**14.66**	m	**15.76**
42mm	7.30	5.00	**7.65**	0.67	-	**11.64**	-	**19.30**	m	**20.74**
54mm	9.78	5.00	**10.24**	1.00	-	**17.38**	-	**27.62**	m	**29.70**
Pipes straight in cast brass pipe brackets fixing to timber with screws										
15mm	1.25	5.00	**1.32**	0.33	-	**5.74**	-	**7.05**	m	**7.58**
22mm	2.19	5.00	**2.29**	0.36	-	**6.26**	-	**8.55**	m	**9.19**
28mm	3.05	5.00	**3.20**	0.40	-	**6.95**	-	**10.15**	m	**10.91**
35mm	6.24	5.00	**6.54**	0.50	-	**8.69**	-	**15.23**	m	**16.38**
42mm	7.95	5.00	**8.34**	0.67	-	**11.64**	-	**19.98**	m	**21.48**
54mm	10.86	5.00	**11.37**	1.00	-	**17.38**	-	**28.75**	m	**30.91**
Pipes straight In cast brass single rings and back plates fixing to timber with screws										
15mm	1.89	5.00	**1.99**	0.36	-	**6.26**	-	**8.24**	m	**8.86**
22mm	2.74	5.00	**2.88**	0.39	-	**6.78**	-	**9.65**	m	**10.38**
28mm	3.57	5.00	**3.75**	0.43	-	**7.47**	-	**11.22**	m	**12.06**
35mm	6.43	5.00	**6.74**	0.53	-	**9.21**	-	**15.95**	m	**17.15**
42mm	7.86	5.00	**8.24**	0.70	-	**12.17**	-	**20.40**	m	**21.93**
54mm	10.24	5.00	**10.72**	1.03	-	**17.90**	-	**28.62**	m	**30.77**
66.7mm	16.75	5.00	**17.51**	1.50	-	**26.07**	-	**43.58**	m	**46.85**
76.1mm	25.77	5.00	**26.89**	2.00	-	**34.76**	-	**61.65**	m	**66.28**
108mm	37.00	5.00	**38.62**	3.00	-	**52.14**	-	**90.76**	m	**97.57**
Copper pipes, BS EN 1057 Part 1 Table X; copper alloy fittings, compression, BS EN 1254 Part 2, Type A										
Pipes, straight in C.P. pipe brackets fixing to timber with screws										
15mm	1.05	5.00	**1.10**	0.33	-	**5.74**	-	**6.84**	m	**7.35**
15mm; fittings										
extra; made bends	-	-	**-**	0.08	-	**1.39**	-	**1.39**	nr	**1.49**
extra; connections to iron pipe ends; screwed joint	0.59	3.00	**0.61**	0.10	-	**1.74**	-	**2.35**	nr	**2.52**
extra; fittings; one end	1.00	3.00	**1.03**	0.05	-	**0.87**	-	**1.90**	nr	**2.04**
extra; fittings; two ends	0.76	3.00	**0.78**	0.10	-	**1.74**	-	**2.52**	nr	**2.71**
extra; fittings; three ends	1.11	3.00	**1.14**	0.10	-	**1.74**	-	**2.88**	nr	**3.10**
22mm	1.98	5.00	**2.07**	0.36	-	**6.26**	-	**8.33**	m	**8.95**
22mm; fittings										
extra; made bends	-	-	**-**	0.13	-	**2.26**	-	**2.26**	nr	**2.43**
extra; connections to iron pipe ends; screwed joint	0.96	3.00	**0.99**	0.17	-	**2.95**	-	**3.94**	nr	**4.24**
extra; fittings; one end	1.19	3.00	**1.23**	0.09	-	**1.56**	-	**2.79**	nr	**3.00**
extra; fittings; two ends	1.30	3.00	**1.34**	0.17	-	**2.95**	-	**4.29**	nr	**4.62**
extra; fittings; three ends	1.82	3.00	**1.87**	0.17	-	**2.95**	-	**4.83**	nr	**5.19**
28mm	3.50	5.00	**3.65**	0.40	-	**6.95**	-	**10.60**	m	**11.39**
28mm; fittings										
extra; made bends	-	-	**-**	0.17	-	**2.95**	-	**2.95**	nr	**3.18**
extra; connections to iron pipe ends; screwed joint	1.86	3.00	**1.92**	0.25	-	**4.35**	-	**6.26**	nr	**6.73**
extra; fittings; one end	3.88	3.00	**4.00**	0.13	-	**2.26**	-	**6.26**	nr	**6.72**
extra; fittings; two ends	4.66	3.00	**4.80**	0.25	-	**4.35**	-	**9.14**	nr	**9.83**

Labour hourly rates: (except Specialists) Craft Operatives £17.38 Labourer £10.71 Rates are national average prices. Refer to REGIONAL VARIATIONS for indicative levels of overall pricing in regions	MATERIALS			LABOUR				RATES		
	Del to Site £	Waste %	Material Cost £	Craft Optve Hrs	Lab Hrs	Labour Cost £	Sunds £	Nett Rate £	Unit	Gross Rate (+7.5%) £
S115: HOT AND COLD WATER AND GAS: COPPER TUBING AND FITTINGS Cont.										
Copper pipes, BS EN 1057 Part 1 Table X; copper alloy fittings, compression, BS EN 1254 Part 2, Type A Cont.										
Pipes, straight in C.P. pipe brackets fixing to timber with screws Cont.										
28mm; fittings Cont.										
extra; fittings; three ends	8.48	3.00	**8.73**	0.25	-	**4.35**	-	**13.08**	nr	14.06
35mm	7.77	5.00	**8.10**	0.50	-	**8.69**	-	**16.79**	m	18.05
35mm; fittings										
extra; made bends	-	-	**-**	0.25	-	**4.35**	-	**4.35**	nr	4.67
extra; connections to iron pipe ends; screwed joint	6.32	3.00	**6.51**	0.33	-	**5.74**	-	**12.25**	nr	13.16
extra; fittings; one end	7.00	3.00	**7.21**	0.17	-	**2.95**	-	**10.16**	nr	10.93
extra; fittings; two ends	11.72	3.00	**12.07**	0.33	-	**5.74**	-	**17.81**	nr	19.14
extra; fittings; three ends	15.50	3.00	**15.97**	0.33	-	**5.74**	-	**21.70**	nr	23.33
42mm	10.09	5.00	**10.51**	0.67	-	**11.64**	-	**22.15**	m	23.82
42mm; fittings										
extra; made bends	-	-	**-**	0.33	-	**5.74**	-	**5.74**	nr	6.17
extra; connections to iron pipe ends; screwed joint	10.17	3.00	**10.48**	0.50	-	**8.69**	-	**19.17**	nr	20.60
extra; fittings; one end	11.45	3.00	**11.79**	0.25	-	**4.35**	-	**16.14**	nr	17.35
extra; fittings; two ends	16.40	3.00	**16.89**	0.50	-	**8.69**	-	**25.58**	nr	27.50
extra; fittings; three ends	23.85	3.00	**24.57**	0.50	-	**8.69**	-	**33.26**	nr	35.75
54mm	13.53	5.00	**14.09**	1.00	-	**17.38**	-	**31.47**	m	33.83
54mm; fittings										
extra; made bends	-	-	**-**	0.50	-	**8.69**	-	**8.69**	nr	9.34
extra; connections to iron pipe ends; screwed joint	15.39	3.00	**15.85**	0.67	-	**11.64**	-	**27.50**	nr	29.56
extra; fittings; one end	15.98	3.00	**16.46**	0.33	-	**5.74**	-	**22.19**	nr	23.86
extra; fittings; two ends	27.88	3.00	**28.72**	0.67	-	**11.64**	-	**40.36**	nr	43.39
extra; fittings; three ends	38.38	3.00	**39.53**	0.67	-	**11.64**	-	**51.18**	nr	55.01
Pipes; straight; in two piece copper spacing clips fixing to masonry with screws										
15mm	1.07	5.00	**1.12**	0.36	-	**6.26**	-	**7.37**	m	7.93
22mm	1.99	5.00	**2.09**	0.39	-	**6.78**	-	**8.87**	m	9.53
28mm	3.51	5.00	**3.66**	0.43	-	**7.47**	-	**11.14**	m	11.97
35mm	7.79	5.00	**8.12**	0.53	-	**9.21**	-	**17.33**	m	18.63
42mm	10.10	5.00	**10.53**	0.70	-	**12.17**	-	**22.69**	m	24.39
54mm	13.54	5.00	**14.10**	1.03	-	**17.90**	-	**32.00**	m	34.40
Pipes straight in pressed brass pipe brackets fixing to timber with screws										
15mm	1.36	5.00	**1.43**	0.33	-	**5.74**	-	**7.16**	m	7.70
22mm	2.34	5.00	**2.45**	0.36	-	**6.26**	-	**8.71**	m	9.36
28mm	3.92	5.00	**4.09**	0.40	-	**6.95**	-	**11.05**	m	11.87
35mm	8.21	5.00	**8.56**	0.50	-	**8.69**	-	**17.25**	m	18.54
42mm	10.68	5.00	**11.13**	0.67	-	**11.64**	-	**22.77**	m	24.48
54mm	14.32	5.00	**14.92**	1.00	-	**17.38**	-	**32.30**	m	34.72
Pipes straight in cast brass pipe brackets fixing to timber with screws										
15mm	1.44	5.00	**1.51**	0.33	-	**5.74**	-	**7.25**	m	7.79
22mm	2.49	5.00	**2.61**	0.36	-	**6.26**	-	**8.87**	m	9.53
28mm	4.15	5.00	**4.33**	0.40	-	**6.95**	-	**11.28**	m	12.13
35mm	8.76	5.00	**9.14**	0.50	-	**8.69**	-	**17.83**	m	19.16
42mm	11.33	5.00	**11.82**	0.67	-	**11.64**	-	**23.46**	m	25.22
54mm	15.39	5.00	**16.04**	1.00	-	**17.38**	-	**33.42**	m	35.93
Pipes straight in cast brass single rings and back plates fixing to timber with screws										
15mm	2.08	5.00	**2.18**	0.33	-	**5.74**	-	**7.92**	m	8.51
22mm	3.05	5.00	**3.19**	0.36	-	**6.26**	-	**9.45**	m	10.16
28mm	4.67	5.00	**4.88**	0.40	-	**6.95**	-	**11.83**	m	12.71
35mm	8.94	5.00	**9.33**	0.50	-	**8.69**	-	**18.02**	m	19.37
42mm	11.24	5.00	**11.72**	0.67	-	**11.64**	-	**23.36**	m	25.11
54mm	14.77	5.00	**15.39**	1.00	-	**17.38**	-	**32.77**	m	35.23
66.7mm	40.07	5.00	**41.53**	1.50	-	**26.07**	-	**67.60**	m	72.67
76.1mm	58.59	5.00	**60.70**	2.00	-	**34.76**	-	**95.46**	m	102.62
108mm	92.23	5.00	**95.51**	3.00	-	**52.14**	-	**147.65**	m	158.73
Copper pipes, BS EN 1057 Part 1 Table X; copper alloy fittings, compression, BS EN 1254 Part 2 Type B										
Pipes, straight in two piece copper spacing clips to timber with screws										
15mm	1.82	5.00	**1.89**	0.33	-	**5.74**	-	**7.62**	m	8.20
15mm; fittings										
extra; made bends	-	-	**-**	0.08	-	**1.39**	-	**1.39**	nr	1.49
extra; connections to iron pipe ends; screwed joint	2.70	3.00	**2.78**	0.13	-	**2.26**	-	**5.04**	nr	5.42
extra; fittings; one end	2.95	3.00	**3.04**	0.07	-	**1.22**	-	**4.26**	nr	4.57
extra; fittings; two ends	3.50	3.00	**3.61**	0.13	-	**2.26**	-	**5.86**	nr	6.30
extra; fittings; three ends	4.90	3.00	**5.05**	0.13	-	**2.26**	-	**7.31**	nr	7.85
22mm	3.20	5.00	**3.33**	0.36	-	**6.26**	-	**9.59**	m	10.31
22mm; fittings										
extra; made bends	-	-	**-**	0.13	-	**2.26**	-	**2.26**	nr	2.43
extra; connections to iron pipe ends; screwed joint	3.85	3.00	**3.97**	0.21	-	**3.65**	-	**7.62**	nr	8.19
extra; fittings; one end	3.60	3.00	**3.71**	0.11	-	**1.91**	-	**5.62**	nr	6.04
extra; fittings; two ends	5.70	3.00	**5.87**	0.21	-	**3.65**	-	**9.52**	nr	10.23
extra; fittings; three ends	8.25	3.00	**8.50**	0.21	-	**3.65**	-	**12.15**	nr	13.06
28mm	4.86	5.00	**5.05**	0.40	-	**6.95**	-	**12.00**	m	12.90

Labour hourly rates: (except Specialists) Craft Operatives £17.38 Labourer £10.71 Rates are national average prices. Refer to REGIONAL VARIATIONS for indicative levels of overall pricing in regions	MATERIALS			LABOUR				RATES		
	Del to Site £	Waste %	Material Cost £	Craft Optve Hrs	Lab Hrs	Labour Cost £	Sunds £	Nett Rate £	Unit	Gross Rate (+7.5%) £
S115: HOT AND COLD WATER AND GAS: COPPER TUBING AND FITTINGS Cont.										
Copper pipes, BS EN 1057 Part 1 Table X; copper alloy fittings, compression, BS EN 1254 Part 2 Type B Cont.										
Pipes, straight in two piece copper spacing clips to timber with screws Cont.										
28mm;fittings										
extra; made bends	-	-	-	0.17	-	2.95	-	2.95	nr	3.18
extra; connections to iron pipe ends; screwed joint	5.90	3.00	6.08	0.31	-	5.39	-	11.46	nr	12.32
extra; fittings; one end	6.75	3.00	6.95	0.16	-	2.78	-	9.73	nr	10.46
extra; fittings; two ends	9.65	3.00	9.94	0.31	-	5.39	-	15.33	nr	16.48
extra; fittings; three ends	13.90	3.00	14.32	0.31	-	5.39	-	19.70	nr	21.18
35mm	9.59	5.00	9.97	0.50	-	8.69	-	18.66	m	20.06
35mm; fittings										
extra; made bends	-	-	-	0.25	-	4.35	-	4.35	nr	4.67
extra; connections to iron pipe ends; screwed joint	11.65	3.00	12.00	0.40	-	6.95	-	18.95	nr	20.37
extra; fittings; one end	11.95	3.00	12.31	0.20	-	3.48	-	15.78	nr	16.97
extra; fittings; two ends	18.55	3.00	19.11	0.40	-	6.95	-	26.06	nr	28.01
extra; fittings; three ends	25.20	3.00	25.96	0.40	-	6.95	-	32.91	nr	35.38
42mm	12.47	5.00	12.97	0.67	-	11.64	-	24.61	m	26.46
42mm; fittings										
extra; made bends	-	-	-	0.33	-	5.74	-	5.74	nr	6.17
extra; connections to iron pipe ends; screwed joint	17.20	3.00	17.72	0.63	-	10.95	-	28.67	nr	30.82
extra; fittings; one end	18.80	3.00	19.36	0.32	-	5.56	-	24.93	nr	26.80
extra; fittings; two ends	27.80	3.00	28.63	0.63	-	10.95	-	39.58	nr	42.55
extra; fittings; three ends	41.00	3.00	42.23	0.63	-	10.95	-	53.18	nr	57.17
54mm	17.26	5.00	17.93	1.00	-	17.38	-	35.31	m	37.96
54mm; fittings										
extra; made bends	-	-	-	0.50	-	8.69	-	8.69	nr	9.34
extra; connections to iron pipe ends; screwed joint	24.90	3.00	25.65	0.83	-	14.43	-	40.07	nr	43.08
extra; fittings; one end	26.75	3.00	27.55	0.42	-	7.30	-	34.85	nr	37.47
extra; fittings; two ends	44.25	3.00	45.58	0.83	-	14.43	-	60.00	nr	64.50
extra; fittings; three ends	65.75	3.00	67.72	0.83	-	14.43	-	82.15	nr	88.31
Pipes straight in two piece copper spacing clips fixing to masonry with screws										
15mm	1.83	5.00	1.91	0.36	-	6.26	-	8.16	m	8.77
22mm	3.22	5.00	3.35	0.39	-	6.78	-	10.13	m	10.89
28mm	4.88	5.00	5.07	0.43	-	7.47	-	12.54	m	13.48
35mm	9.60	5.00	9.99	0.53	-	9.21	-	19.20	m	20.64
42mm	12.49	5.00	12.98	0.70	-	12.17	-	25.15	m	27.03
54mm	17.27	5.00	17.95	1.03	-	17.90	-	35.85	m	38.54
Pipes straight in pressed brass pipe brackets fixing to timber with screws										
15mm	2.13	5.00	2.21	0.36	-	6.26	-	8.47	m	9.11
22mm	3.56	5.00	3.71	0.39	-	6.78	-	10.49	m	11.27
28mm	5.28	5.00	5.50	0.43	-	7.47	-	12.97	m	13.94
35mm	10.02	5.00	10.43	0.53	-	9.21	-	19.64	m	21.11
42mm	13.06	5.00	13.59	0.70	-	12.17	-	25.75	m	27.68
54mm	18.05	5.00	18.76	1.03	-	17.90	-	36.66	m	39.41
Pipes straight in cast brass pipe brackets fixing to timber with screws										
15mm	2.21	5.00	2.30	0.36	-	6.26	-	8.56	m	9.20
22mm	3.72	5.00	3.87	0.39	-	6.78	-	10.65	m	11.45
28mm	5.51	5.00	5.73	0.43	-	7.47	-	13.21	m	14.20
35mm	10.57	5.00	11.00	0.53	-	9.21	-	20.22	m	21.73
42mm	13.72	5.00	14.27	0.70	-	12.17	-	26.44	m	28.42
54mm	19.12	5.00	19.89	1.03	-	17.90	-	37.79	m	40.62
Pipes straight in cast brass single rings and back plates fixing to timber with screws										
15mm	2.85	5.00	2.97	0.36	-	6.26	-	9.23	m	9.92
22mm	4.27	5.00	4.46	0.39	-	6.78	-	11.23	m	12.08
28mm	6.03	5.00	6.28	0.43	-	7.47	-	13.75	m	14.78
35mm	10.76	5.00	11.20	0.53	-	9.21	-	20.41	m	21.94
42mm	13.62	5.00	14.17	0.70	-	12.17	-	26.34	m	28.32
54mm	18.50	5.00	19.24	1.03	-	17.90	-	37.14	m	39.92
Copper pipes, BS EN 1057 Part 1 Table Y; copper fittings, capillary, BS EN 1254 Part 2										
Pipes; straight; in trenches										
15mm	1.77	5.00	1.86	0.17	-	2.95	-	4.81	m	5.17
15mm; fittings										
extra; made bends	-	-	-	0.08	-	1.39	-	1.39	nr	1.49
extra; connections to iron pipe ends; screwed joint	1.61	3.00	1.66	0.13	-	2.26	-	3.92	nr	4.21
extra; fittings; one end	0.82	3.00	0.84	0.07	-	1.22	-	2.06	nr	2.22
extra; fittings; two ends	0.21	3.00	0.22	0.13	-	2.26	-	2.48	nr	2.66
extra; fittings; three ends	0.39	3.00	0.40	0.13	-	2.26	-	2.66	nr	2.86
22mm	3.10	5.00	3.25	0.18	-	3.13	-	6.38	m	6.86
22mm; fittings										
extra; made bends	-	-	-	0.13	-	2.26	-	2.26	nr	2.43
extra; connections to iron pipe ends; screwed joint	2.86	3.00	2.95	0.21	-	3.65	-	6.60	nr	7.09
extra; fittings; one end	1.55	3.00	1.60	0.11	-	1.91	-	3.51	nr	3.77
extra; fittings; two ends	0.53	3.00	0.55	0.21	-	3.65	-	4.20	nr	4.51
extra; fittings; three ends	1.23	3.00	1.27	0.21	-	3.65	-	4.92	nr	5.29

Labour hourly rates: (except Specialists)
Craft Operatives £17.38 Labourer £10.71
Rates are national average prices.
Refer to REGIONAL VARIATIONS for indicative levels of overall pricing in regions

	MATERIALS			LABOUR				RATES		
	Del to Site £	Waste %	Material Cost £	Craft Optve Hrs	Lab Hrs	Labour Cost £	Sunds £	Nett Rate £	Unit	Gross Rate (+7.5%) £
S115: HOT AND COLD WATER AND GAS: COPPER TUBING AND FITTINGS Cont.										
Copper pipes, BS EN 1057 Part 1 Table Y; copper fittings, capillary, BS EN 1254 Part 2 Cont.										
Pipes; straight; in trenches Cont.										
22mm; fittings Cont.										
28mm	4.04	5.00	**4.24**	0.20	-	**3.48**	-	**7.72**	m	**8.30**
28mm; fittings										
extra; made bends	-	-	**-**	0.17	-	**2.95**	-	**2.95**	nr	**3.18**
extra; connections to iron pipe ends; screwed joint	4.61	3.00	**4.75**	0.31	-	**5.39**	-	**10.14**	nr	**10.90**
extra; fittings; one end	2.76	3.00	**2.84**	0.16	-	**2.78**	-	**5.62**	nr	**6.05**
extra; fittings; two ends	1.23	3.00	**1.27**	0.31	-	**5.39**	-	**6.65**	nr	**7.15**
extra; fittings; three ends	3.40	3.00	**3.50**	0.31	-	**5.39**	-	**8.89**	nr	**9.56**
35mm	6.19	5.00	**6.49**	0.22	-	**3.82**	-	**10.32**	m	**11.09**
35mm; fittings										
extra; made bends	-	-	**-**	0.25	-	**4.35**	-	**4.35**	nr	**4.67**
extra; connections to iron pipe ends; screwed joint	8.12	3.00	**8.36**	0.40	-	**6.95**	-	**15.32**	nr	**16.46**
extra; fittings; one end	6.09	3.00	**6.27**	0.20	-	**3.48**	-	**9.75**	nr	**10.48**
extra; fittings; two ends	5.33	3.00	**5.49**	0.40	-	**6.95**	-	**12.44**	nr	**13.38**
extra; fittings; three ends	8.68	3.00	**8.94**	0.40	-	**6.95**	-	**15.89**	nr	**17.08**
42mm	7.58	5.00	**7.95**	0.29	-	**5.04**	-	**12.99**	m	**13.97**
42mm; fittings										
extra; made bends	-	-	**-**	0.33	-	**5.74**	-	**5.74**	nr	**6.17**
extra; connections to iron pipe ends; screwed joint	10.46	3.00	**10.77**	0.63	-	**10.95**	-	**21.72**	nr	**23.35**
extra; fittings; one end	10.49	3.00	**10.80**	0.32	-	**5.56**	-	**16.37**	nr	**17.59**
extra; fittings; two ends	8.81	3.00	**9.07**	0.63	-	**10.95**	-	**20.02**	nr	**21.53**
extra; fittings; three ends	13.93	3.00	**14.35**	0.63	-	**10.95**	-	**25.30**	nr	**27.19**
54mm	12.99	5.00	**13.63**	0.33	-	**5.74**	-	**19.37**	m	**20.82**
54mm; fittings										
extra; made bends	-	-	**-**	0.50	-	**8.69**	-	**8.69**	nr	**9.34**
extra; connections to iron pipe ends; screwed joint	15.88	3.00	**16.36**	0.83	-	**14.43**	-	**30.78**	nr	**33.09**
extra; fittings; one end	14.64	3.00	**15.08**	0.42	-	**7.30**	-	**22.38**	nr	**24.06**
extra; fittings; two ends	18.20	3.00	**18.75**	0.83	-	**14.43**	-	**33.17**	nr	**35.66**
extra; fittings; three ends	28.09	3.00	**28.93**	0.83	-	**14.43**	-	**43.36**	nr	**46.61**
Copper pipes, BS EN 1057 Part 1 Table Y; non-dezincifiable fittings; compression, BS EN 1254 Part 2, Type B										
Pipes; straight; in trenches										
15mm	1.84	5.00	**1.93**	0.17	-	**2.95**	-	**4.88**	m	**5.25**
15mm; fittings										
extra; made bends	-	-	**-**	0.08	-	**1.39**	-	**1.39**	nr	**1.49**
extra; connections to iron pipe ends; screwed joint	0.99	3.00	**1.02**	0.13	-	**2.26**	-	**3.28**	nr	**3.53**
extra; fittings; one end	1.47	3.00	**1.51**	0.07	-	**1.22**	-	**2.73**	nr	**2.94**
extra; fittings; two ends	1.35	3.00	**1.39**	0.13	-	**2.26**	-	**3.65**	nr	**3.92**
extra; fittings; three ends	1.93	3.00	**1.99**	0.13	-	**2.26**	-	**4.25**	nr	**4.57**
22mm	3.20	5.00	**3.36**	0.18	-	**3.13**	-	**6.49**	m	**6.97**
22mm; fittings										
extra; made bends	-	-	**-**	0.13	-	**2.26**	-	**2.26**	nr	**2.43**
extra; connections to iron pipe ends; screwed joint	1.61	3.00	**1.66**	0.21	-	**3.65**	-	**5.31**	nr	**5.71**
extra; fittings; one end	1.97	3.00	**2.03**	0.11	-	**1.91**	-	**3.94**	nr	**4.24**
extra; fittings; two ends	2.09	3.00	**2.15**	0.21	-	**3.65**	-	**5.80**	nr	**6.24**
extra; fittings; three ends	3.04	3.00	**3.13**	0.21	-	**3.65**	-	**6.78**	nr	**7.29**
28mm	4.28	5.00	**4.48**	0.20	-	**3.48**	-	**7.96**	m	**8.56**
28mm; fittings										
extra; made bends	-	-	**-**	0.17	-	**2.95**	-	**2.95**	nr	**3.18**
extra; connections to iron pipe ends; screwed joint	2.60	3.00	**2.68**	0.31	-	**5.39**	-	**8.07**	nr	**8.67**
extra; fittings; one end	4.24	3.00	**4.37**	0.16	-	**2.78**	-	**7.15**	nr	**7.68**
extra; fittings; two ends	5.48	3.00	**5.64**	0.31	-	**5.39**	-	**11.03**	nr	**11.86**
extra; fittings; three ends	8.99	3.00	**9.26**	0.31	-	**5.39**	-	**14.65**	nr	**15.75**
35mm	6.68	5.00	**7.00**	0.22	-	**3.82**	-	**10.82**	m	**11.63**
35mm; fittings										
extra; made bends	-	-	**-**	0.25	-	**4.35**	-	**4.35**	nr	**4.67**
extra; connections to iron pipe ends; screwed joint	6.65	3.00	**6.85**	0.40	-	**6.95**	-	**13.80**	nr	**14.84**
extra; fittings; one end	7.00	3.00	**7.21**	0.20	-	**3.48**	-	**10.69**	nr	**11.49**
extra; fittings; two ends	12.89	3.00	**13.28**	0.40	-	**6.95**	-	**20.23**	nr	**21.75**
extra; fittings; three ends	15.50	3.00	**15.97**	0.40	-	**6.95**	-	**22.92**	nr	**24.64**
42mm	8.13	5.00	**8.52**	0.29	-	**5.04**	-	**13.56**	m	**14.58**
42mm; fittings										
extra; made bends	-	-	**-**	0.33	-	**5.74**	-	**5.74**	nr	**6.17**
extra; connections to iron pipe ends; screwed joint	11.13	3.00	**11.46**	0.63	-	**10.95**	-	**22.41**	nr	**24.09**
extra; fittings; one end	11.45	3.00	**11.79**	0.32	-	**5.56**	-	**17.36**	nr	**18.66**
extra; fittings; two ends	18.30	3.00	**18.85**	0.63	-	**10.95**	-	**29.80**	nr	**32.03**
extra; fittings; three ends	23.85	3.00	**24.57**	0.63	-	**10.95**	-	**35.51**	nr	**38.18**
54mm	13.67	5.00	**14.32**	0.33	-	**5.74**	-	**20.06**	m	**21.56**
54mm; fittings										
extra; made bends	-	-	**-**	0.50	-	**8.69**	-	**8.69**	nr	**9.34**
extra; connections to iron pipe ends; screwed joint	16.23	3.00	**16.72**	0.83	-	**14.43**	-	**31.14**	nr	**33.48**
extra; fittings; one end	15.98	3.00	**16.46**	0.42	-	**7.30**	-	**23.76**	nr	**25.54**
extra; fittings; two ends	30.63	3.00	**31.55**	0.83	-	**14.43**	-	**45.97**	nr	**49.42**
extra; fittings; three ends	38.38	3.00	**39.53**	0.83	-	**14.43**	-	**53.96**	nr	**58.00**

Labour hourly rates: (except Specialists) Craft Operatives £17.38 Labourer £10.71 Rates are national average prices. Refer to REGIONAL VARIATIONS for indicative levels of overall pricing in regions	MATERIALS			LABOUR				RATES		
	Del to Site £	Waste %	Material Cost £	Craft Optve Hrs	Lab Hrs	Labour Cost £	Sunds £	Nett Rate £	Unit	Gross Rate (+7.5%) £
S115: HOT AND COLD WATER AND GAS: POLYBUTYLENE TUBES AND FITTINGS										
Polybutylene pipes; Hepworth Hep20 flexible plumbing system BS 7291 Class H; polybutylene slimline fittings										
Pipes, flexible; in ducts in pipe clips to timber										
15mm	1.18	10.00	**1.27**	0.30	-	**5.21**	-	**6.48**	m	**6.97**
15mm; fittings										
extra; made bend	-	-	-	0.01	-	**0.17**	-	**0.17**	nr	**0.19**
extra; connection to copper pipe ends	0.96	2.00	**0.98**	0.10	-	**1.74**	-	**2.72**	nr	**2.92**
extra; fittings; one end	1.66	2.00	**1.69**	0.05	-	**0.87**	-	**2.56**	nr	**2.75**
extra; fittings; two ends	1.25	2.00	**1.28**	0.10	-	**1.74**	-	**3.01**	nr	**3.24**
extra; fittings; three ends	1.78	2.00	**1.82**	0.10	-	**1.74**	-	**3.55**	nr	**3.82**
22mm	1.63	10.00	**1.77**	0.33	-	**5.74**	-	**7.50**	m	**8.06**
22mm; fittings										
extra; made bend	-	-	-	0.01	-	**0.17**	-	**0.17**	nr	**0.19**
extra; connection to copper pipe ends	1.27	2.00	**1.30**	0.25	-	**4.35**	-	**5.64**	nr	**6.06**
extra; fittings; one end	2.17	2.00	**2.21**	0.13	-	**2.26**	-	**4.47**	nr	**4.81**
extra; fittings; two ends	1.76	2.00	**1.80**	0.25	-	**4.35**	-	**6.14**	nr	**6.60**
extra; fittings; three ends	2.26	2.00	**2.31**	0.25	-	**4.35**	-	**6.65**	nr	**7.15**
Straight pipe fittings in pipe clips fixing to masonry with screws										
15mm	1.27	10.00	**1.37**	0.30	-	**5.21**	-	**6.58**	m	**7.07**
22mm	1.67	10.00	**1.82**	0.33	-	**5.74**	-	**7.55**	m	**8.12**
Polybutylene pipes; Hepworth Hep20 flexible plumbing system BS 7291 Class H; polybutylene demountable fittings										
Pipes, flexible in pipe clips to timber with screws										
15mm	1.18	10.00	**1.27**	0.37	-	**6.43**	-	**7.70**	m	**8.28**
15mm; fittings										
extra; connection to copper pipe ends	0.94	2.00	**0.96**	0.10	-	**1.74**	-	**2.70**	nr	**2.90**
extra; fittings; one end	1.59	2.00	**1.62**	0.05	-	**0.87**	-	**2.49**	nr	**2.68**
extra; fittings; two ends	1.22	2.00	**1.24**	0.10	-	**1.74**	-	**2.98**	nr	**3.21**
extra; fittings; three ends	1.74	2.00	**1.77**	0.10	-	**1.74**	-	**3.51**	nr	**3.78**
22mm	1.62	10.00	**1.76**	0.43	-	**7.47**	-	**9.24**	m	**9.93**
22mm; fittings										
extra; connection to copper pipe ends	1.24	2.00	**1.26**	0.17	-	**2.95**	-	**4.22**	nr	**4.54**
extra; fittings; one end	2.10	2.00	**2.14**	0.09	-	**1.56**	-	**3.71**	nr	**3.98**
extra; fittings; two ends	1.73	2.00	**1.76**	0.17	-	**2.95**	-	**4.72**	nr	**5.07**
extra; fittings; three ends	2.21	2.00	**2.25**	0.17	-	**2.95**	-	**5.21**	nr	**5.60**
28mm	2.25	10.00	**2.44**	0.50	-	**8.69**	-	**11.13**	m	**11.96**
28mm; fittings										
extra; connection to copper pipe ends	2.93	2.00	**2.99**	0.33	-	**5.74**	-	**8.72**	nr	**9.38**
extra; fittings; one end	4.38	2.00	**4.47**	0.17	-	**2.95**	-	**7.42**	nr	**7.98**
extra; fittings; two ends	3.71	2.00	**3.78**	0.33	-	**5.74**	-	**9.52**	nr	**10.23**
extra; fittings; three ends	5.18	2.00	**5.28**	0.33	-	**5.74**	-	**11.02**	nr	**11.85**
Pipes flexible in pipe clips fixing to masonry with screws										
15mm	1.27	10.00	**1.36**	0.33	-	**5.74**	-	**7.10**	m	**7.63**
22mm	1.67	10.00	**1.81**	0.37	-	**6.43**	-	**8.24**	m	**8.86**
28mm	2.28	10.00	**2.47**	0.50	-	**8.69**	-	**11.16**	m	**12.00**
S115: HOT AND COLD WATER AND GAS: POLYETHYLENE TUBES AND FITTINGS										
Polythene pipes, BS 6572 Blue; copper alloy fittings, compression, BS EN 1254 Part 3, Type A										
Pipes; straight; in trenches										
20mm	0.45	10.00	**0.48**	0.14	-	**2.43**	-	**2.91**	m	**3.13**
20mm; fittings										
extra; connections to copper pipe ends; compression joint	4.03	3.00	**4.15**	0.13	-	**2.26**	-	**6.41**	nr	**6.89**
extra; fittings; one end	3.94	3.00	**4.06**	0.07	-	**1.22**	-	**5.27**	nr	**5.67**
extra; fittings; two ends	5.30	3.00	**5.46**	0.13	-	**2.26**	-	**7.72**	nr	**8.30**
extra; fittings; three ends	7.51	3.00	**7.74**	0.13	-	**2.26**	-	**9.99**	nr	**10.74**
25mm	0.62	10.00	**0.66**	0.17	-	**2.95**	-	**3.61**	m	**3.88**
25mm; fittings										
extra; connections to copper pipe ends; compression joint	6.16	3.00	**6.34**	0.21	-	**3.65**	-	**9.99**	nr	**10.74**
extra; fittings; one end	5.93	3.00	**6.11**	0.11	-	**1.91**	-	**8.02**	nr	**8.62**
extra; fittings; two ends	8.07	3.00	**8.31**	0.21	-	**3.65**	-	**11.96**	nr	**12.86**
extra; fittings; three ends	11.71	3.00	**12.06**	0.21	-	**3.65**	-	**15.71**	nr	**16.89**
32mm	0.99	10.00	**1.06**	0.20	-	**3.48**	-	**4.54**	m	**4.88**
32mm; fittings										
extra; connections to copper pipe ends; compression joint	10.54	3.00	**10.86**	0.31	-	**5.39**	-	**16.24**	nr	**17.46**
extra; fittings; one end	13.43	3.00	**13.83**	0.16	-	**2.78**	-	**16.61**	nr	**17.86**
extra; fittings; two ends	14.17	3.00	**14.60**	0.31	-	**5.39**	-	**19.98**	nr	**21.48**
extra; fittings; three ends	17.88	3.00	**18.42**	0.31	-	**5.39**	-	**23.80**	nr	**25.59**
50mm	2.46	10.00	**2.63**	0.33	-	**5.74**	-	**8.37**	m	**8.99**
50mm; fittings										
extra; fittings; two ends	33.05	3.00	**34.04**	0.50	-	**8.69**	-	**42.73**	nr	**45.94**
extra; fittings; three ends	43.93	3.00	**45.25**	0.50	-	**8.69**	-	**53.94**	nr	**57.98**
63mm	5.78	10.00	**6.10**	0.45	-	**7.82**	-	**13.93**	m	**14.97**
63mm; fittings										
extra; fittings; two ends	39.66	3.00	**40.85**	0.83	-	**14.43**	-	**55.28**	nr	**59.42**
extra; fittings; three ends	64.52	3.00	**66.46**	0.83	-	**14.43**	-	**80.88**	nr	**86.95**

Labour hourly rates: (except Specialists) Craft Operatives £17.38 Labourer £10.71 Rates are national average prices. Refer to REGIONAL VARIATIONS for indicative levels of overall pricing in regions	MATERIALS			LABOUR				RATES		
	Del to Site £	Waste %	Material Cost £	Craft Optve Hrs	Lab Hrs	Labour Cost £	Sunds £	Nett Rate £	Unit	Gross Rate (+7.5%) £
S115: HOT AND COLD WATER AND GAS: UNPLASTICISED PVC TUBES AND FITTINGS										
PVC-U pipes, BS EN 1452 Class E; fittings, solvent welded joints, BS 4346										
Pipes, straight in standard plastics pipe brackets fixing to timber with screws										
3/8''	3.07	5.00	**3.18**	0.33	-	**5.74**	-	**8.92**	m	**9.59**
3/8'; fittings										
extra; connections to iron pipe ends; screwed joint	0.96	2.00	**0.98**	0.10	-	**1.74**	-	**2.72**	nr	**2.92**
extra; fittings; one end	0.60	2.00	**0.61**	0.05	-	**0.87**	-	**1.48**	nr	**1.59**
extra; fittings; two ends	0.92	2.00	**0.94**	0.10	-	**1.74**	-	**2.68**	nr	**2.88**
extra; fittings; three ends	1.02	2.00	**1.04**	0.10	-	**1.74**	-	**2.78**	nr	**2.99**
1/2''	3.74	5.00	**3.87**	0.37	-	**6.43**	-	**10.30**	m	**11.08**
1/2'; fittings										
extra; connections to iron pipe ends; screwed joint	1.30	2.00	**1.33**	0.17	-	**2.95**	-	**4.28**	nr	**4.60**
extra; fittings; one end	0.74	2.00	**0.75**	0.09	-	**1.56**	-	**2.32**	nr	**2.49**
extra; fittings; two ends	1.04	2.00	**1.06**	0.17	-	**2.95**	-	**4.02**	nr	**4.32**
extra; fittings; three ends	1.20	2.00	**1.22**	0.17	-	**2.95**	-	**4.18**	nr	**4.49**
3/4''	3.85	5.00	**4.01**	0.43	-	**7.47**	-	**11.49**	m	**12.35**
3/4'; fittings										
extra; connections to iron pipe ends; screwed joint	1.41	2.00	**1.44**	0.25	-	**4.35**	-	**5.78**	nr	**6.22**
extra; fittings; one end	0.87	2.00	**0.89**	0.13	-	**2.26**	-	**3.15**	nr	**3.38**
extra; fittings; two ends	1.24	2.00	**1.26**	0.25	-	**4.35**	-	**5.61**	nr	**6.03**
extra; fittings; three ends	1.52	2.00	**1.55**	0.25	-	**4.35**	-	**5.90**	nr	**6.34**
1''	4.00	5.00	**4.18**	0.50	-	**8.69**	-	**12.87**	m	**13.83**
1'; fittings										
extra; connections to iron pipe ends; screwed joint	2.19	2.00	**2.23**	0.33	-	**5.74**	-	**7.97**	nr	**8.57**
extra; fittings; one end	0.97	2.00	**0.99**	0.17	-	**2.95**	-	**3.94**	nr	**4.24**
extra; fittings; two ends	1.73	2.00	**1.76**	0.33	-	**5.74**	-	**7.50**	nr	**8.06**
extra; fittings; three ends	2.30	2.00	**2.35**	0.33	-	**5.74**	-	**8.08**	nr	**8.69**
1 1/4''	5.72	5.00	**5.97**	0.60	-	**10.43**	-	**16.40**	m	**17.63**
11/4'; fittings										
extra; connections to iron pipe ends; screwed joint	3.07	2.00	**3.13**	0.50	-	**8.69**	-	**11.82**	nr	**12.71**
extra; fittings; one end	1.52	2.00	**1.55**	0.25	-	**4.35**	-	**5.90**	nr	**6.34**
extra; fittings; two ends	3.03	2.00	**3.09**	0.50	-	**8.69**	-	**11.78**	nr	**12.66**
extra; fittings; three ends	3.24	2.00	**3.30**	0.50	-	**8.69**	-	**11.99**	nr	**12.89**
1 1/2''	7.20	5.00	**7.52**	0.75	-	**13.04**	-	**20.56**	m	**22.10**
11/2'; fittings										
extra; connections to iron pipe ends; screwed joint	4.17	2.00	**4.25**	0.67	-	**11.64**	-	**15.90**	nr	**17.09**
extra; fittings; one end	2.56	2.00	**2.61**	0.33	-	**5.74**	-	**8.35**	nr	**8.97**
extra; fittings; two ends	3.91	2.00	**3.99**	0.67	-	**11.64**	-	**15.63**	nr	**16.81**
extra; fittings; three ends	4.69	2.00	**4.78**	0.67	-	**11.64**	-	**16.43**	nr	**17.66**
Pipes straight In standard plastics pipe brackets fixing to masonry with screws										
3/8''	3.19	5.00	**3.31**	0.39	-	**6.78**	-	**10.09**	m	**10.85**
1/2''	3.86	5.00	**4.00**	0.43	-	**7.47**	-	**11.47**	m	**12.33**
3/4''	3.91	5.00	**4.08**	0.49	-	**8.52**	-	**12.59**	m	**13.54**
1''	3.99	5.00	**4.16**	0.56	-	**9.73**	-	**13.89**	m	**14.94**
1 1/4''	5.76	5.00	**6.02**	0.66	-	**11.47**	-	**17.49**	m	**18.80**
1 1/2''	7.28	5.00	**7.60**	0.81	-	**14.08**	-	**21.68**	m	**23.31**
Pipes; straight; in trenches										
3/8''	1.57	5.00	**1.64**	0.17	-	**2.95**	-	**4.60**	m	**4.94**
1/2''	1.99	5.00	**2.09**	0.18	-	**3.13**	-	**5.22**	m	**5.61**
3/4''	2.82	5.00	**2.95**	0.20	-	**3.48**	-	**6.43**	m	**6.91**
1''	3.27	5.00	**3.43**	0.25	-	**4.35**	-	**7.78**	m	**8.36**
1 1/4''	4.85	5.00	**5.08**	0.29	-	**5.04**	-	**10.12**	m	**10.88**
1 1/2''	6.29	5.00	**6.60**	0.33	-	**5.74**	-	**12.33**	m	**13.26**
S115: HOT AND COLD WATER AND GAS: UNPLASTICISED OVERFLOW PIPES										
PVC-U pipes and fittings; solvent welded joints; pipework self coloured white										
Pipes, straight in standard plastics pipe brackets fixing to timber with screws										
19mm	1.17	5.00	**1.21**	0.25	-	**4.35**	-	**5.56**	m	**5.97**
19mm; fittings										
extra; fittings; two ends	0.93	2.00	**0.95**	0.20	-	**3.48**	-	**4.42**	nr	**4.76**
extra; fittings; three ends	1.01	2.00	**1.03**	0.20	-	**3.48**	-	**4.51**	nr	**4.84**
extra; fittings; tank connector	1.22	2.00	**1.24**	0.20	-	**3.48**	-	**4.72**	nr	**5.07**
S115: HOT AND COLD WATER AND GAS: DUCTILE IRON PIPES AND FITTINGS										
Ductile iron pipes and fittings, Tyton socketed flexible joints										
Pipes; straight; in trenches										
80mm	22.58	5.00	**23.71**	0.80	-	**13.90**	-	**37.61**	m	**40.43**
80mm; fittings										
extra; bends, 90 degree	55.25	2.00	**56.36**	0.80	-	**13.90**	-	**70.26**	nr	**75.53**
extra; duckfoot bends, 90 degree	125.94	2.00	**128.46**	0.80	-	**13.90**	-	**142.36**	nr	**153.04**
extra; tees	81.23	2.00	**82.85**	1.20	-	**20.86**	-	**103.71**	nr	**111.49**
extra; hydrant tees	110.98	2.00	**113.20**	1.20	-	**20.86**	-	**134.06**	nr	**144.11**
extra; branches, 45 degree	232.89	2.00	**237.55**	1.20	-	**20.86**	-	**258.40**	nr	**277.78**

Labour hourly rates: (except Specialists) Craft Operatives £17.38 Labourer £10.71 Rates are national average prices. Refer to REGIONAL VARIATIONS for indicative levels of overall pricing in regions	MATERIALS			LABOUR				RATES		
	Del to Site £	Waste %	Material Cost £	Craft Optve Hrs	Lab Hrs	Labour Cost £	Sunds £	Nett Rate £	Unit	Gross Rate (+7.5%) £
S115: HOT AND COLD WATER AND GAS: DUCTILE IRON PIPES AND FITTINGS Cont.										
Ductile iron pipes and fittings, Tyton socketed flexible joints Cont.										
Pipes; straight; in trenches Cont.										
80mm; fittings Cont.										
extra; flanged sockets	43.62	2.00	**44.49**	0.80	-	**13.90**	-	**58.40**	nr	**62.78**
extra; flanged spigots	40.80	2.00	**41.62**	0.80	-	**13.90**	-	**55.52**	nr	**59.68**
100mm	21.45	5.00	**22.52**	1.00	-	**17.38**	-	**39.90**	m	**42.90**
100mm; fittings										
extra; bends, 90 degree	58.17	2.00	**59.33**	1.00	-	**17.38**	-	**76.71**	nr	**82.47**
extra; duckfoot bends, 90 degree	136.48	2.00	**139.21**	1.00	-	**17.38**	-	**156.59**	nr	**168.33**
extra; tees	86.55	2.00	**88.28**	1.50	-	**26.07**	-	**114.35**	nr	**122.93**
extra; hydrant tees	81.84	2.00	**83.48**	1.50	-	**26.07**	-	**109.55**	nr	**117.76**
extra; branches, 45 degree	332.12	2.00	**338.76**	1.50	-	**26.07**	-	**364.83**	nr	**392.19**
extra; flanged sockets	47.19	2.00	**48.13**	1.00	-	**17.38**	-	**65.51**	nr	**70.43**
extra; flanged spigots	43.09	2.00	**43.95**	1.00	-	**17.38**	-	**61.33**	nr	**65.93**
150mm	27.33	5.00	**28.70**	1.50	-	**26.07**	-	**54.77**	m	**58.87**
150mm; fittings										
extra; bends, 90 degree	104.37	2.00	**106.46**	1.50	-	**26.07**	-	**132.53**	nr	**142.47**
extra; duckfoot bends, 90 degree	298.32	2.00	**304.29**	1.50	-	**26.07**	-	**330.36**	nr	**355.13**
extra; tees	161.22	2.00	**164.44**	2.25	-	**39.11**	-	**203.55**	nr	**218.82**
extra; hydrant tees	148.58	2.00	**151.55**	2.25	-	**39.11**	-	**190.66**	nr	**204.96**
extra; branches, 45 degree	420.16	2.00	**428.56**	2.25	-	**39.11**	-	**467.67**	nr	**502.74**
extra; flanged sockets	71.90	2.00	**73.34**	1.50	-	**26.07**	-	**99.41**	nr	**106.86**
extra; flanged spigots	49.97	2.00	**50.97**	1.50	-	**26.07**	-	**77.04**	nr	**82.82**
S115: HOT AND COLD WATER AND GAS: STOPCOCKS										
Brass stopcocks										
Stopcocks; BS 1010, crutch head; DZR; joints to polythene BS 6572										
each end										
20mm	18.22	2.00	**18.58**	0.33	-	**5.74**	-	**24.32**	nr	**26.14**
25mm	29.03	2.00	**29.61**	0.50	-	**8.69**	-	**38.30**	nr	**41.17**
32mm	38.94	2.00	**39.72**	0.67	-	**11.64**	-	**51.36**	nr	**55.22**
50mm	97.34	2.00	**99.29**	0.75	-	**13.04**	-	**112.32**	nr	**120.75**
Stopcocks BS 1010, crutch head; compression joints to copper (Type A)										
each end										
15mm	3.04	2.00	**3.10**	0.25	-	**4.35**	-	**7.45**	nr	**8.00**
22mm	5.29	2.00	**5.40**	0.33	-	**5.74**	-	**11.13**	nr	**11.97**
28mm	12.70	2.00	**12.95**	0.50	-	**8.69**	-	**21.64**	nr	**23.27**
35mm	38.15	2.00	**38.91**	0.67	-	**11.64**	-	**50.56**	nr	**54.35**
42mm	48.15	2.00	**49.11**	0.75	-	**13.04**	-	**62.15**	nr	**66.81**
54mm	72.68	2.00	**74.13**	1.00	-	**17.38**	-	**91.51**	nr	**98.38**
Stopcocks BS 1010, crutch head; DZR; compression joints to copper (Type A)										
each end										
15mm	11.15	2.00	**11.37**	0.25	-	**4.35**	-	**15.72**	nr	**16.90**
22mm	18.31	2.00	**18.68**	0.33	-	**5.74**	-	**24.41**	nr	**26.24**
28mm	30.29	2.00	**30.90**	0.50	-	**8.69**	-	**39.59**	nr	**42.55**
35mm	55.74	2.00	**56.85**	0.67	-	**11.64**	-	**68.50**	nr	**73.64**
42mm	79.79	2.00	**81.39**	0.75	-	**13.04**	-	**94.42**	nr	**101.50**
54mm	108.68	2.00	**110.85**	1.00	-	**17.38**	-	**128.23**	nr	**137.85**
Polybutylene stopcocks										
Stopcocks; fitted with Hep20 ends										
each end										
15mm	3.33	2.00	**3.40**	0.25	-	**4.35**	-	**7.74**	nr	**8.32**
22mm	3.99	2.00	**4.07**	0.33	-	**5.74**	-	**9.81**	nr	**10.54**
S115: HOT AND COLD WATER AND GAS: LEVER BALL VALVES										
Copper alloy ball valves										
Lever ball valves; screwed and PTFE joints										
each end threaded internally										
13mm	0.56	2.00	**0.57**	0.25	-	**4.35**	-	**4.91**	nr	**5.28**
19mm	0.58	2.00	**0.59**	0.33	-	**5.74**	-	**6.33**	nr	**6.81**
25mm	0.54	2.00	**0.55**	0.50	-	**8.69**	-	**9.24**	nr	**9.94**
32mm	0.47	2.00	**0.48**	0.67	-	**11.64**	-	**12.12**	nr	**13.03**
38mm	12.02	2.00	**12.26**	0.75	-	**13.04**	-	**25.30**	nr	**27.19**
51mm	16.35	2.00	**16.68**	1.00	-	**17.38**	-	**34.06**	nr	**36.61**
S115: HOT AND COLD WATER AND GAS: GATE VALVES										
Copper alloy gate valves										
Gate valves; BS EN 12288 series B; compression joints to copper (Type A)										
each end										
15mm	5.48	2.00	**5.59**	0.25	-	**4.35**	-	**9.93**	nr	**10.68**
22mm	6.66	2.00	**6.79**	0.33	-	**5.74**	-	**12.53**	nr	**13.47**
28mm	9.17	2.00	**9.35**	0.50	-	**8.69**	-	**18.04**	nr	**19.40**
35mm	17.26	2.00	**17.61**	0.67	-	**11.64**	-	**29.25**	nr	**31.44**
42mm	28.16	2.00	**28.72**	0.75	-	**13.04**	-	**41.76**	nr	**44.89**
54mm	43.60	2.00	**44.47**	1.00	-	**17.38**	-	**61.85**	nr	**66.49**

Labour hourly rates: (except Specialists) Craft Operatives £17.38 Labourer £10.71 Rates are national average prices. Refer to REGIONAL VARIATIONS for indicative levels of overall pricing in regions	MATERIALS			LABOUR				RATES		
	Del to Site £	Waste %	Material Cost £	Craft Optve Hrs	Lab Hrs	Labour Cost £	Sunds £	Nett Rate £	Unit	Gross Rate (+7.5%) £
S115: HOT AND COLD WATER AND GAS: GATE VALVES Cont.										
Copper alloy gate valves Cont.										
Gate valves; DZR; fitted with Hep20 ends										
each end										
15mm	3.11	2.00	**3.17**	0.25	-	**4.35**	-	**7.52**	nr	**8.08**
22mm	4.39	2.00	**4.48**	0.33	-	**5.74**	-	**10.21**	nr	**10.98**
S115: HOT AND COLD WATER AND GAS: BALL VALVES										
Copper alloy body float valves										
Float operated valves for low pressure; BS 1212; copper float										
inlet threaded externally; fixing to steel										
13mm; part 1	6.29	2.00	**6.42**	0.33	-	**5.74**	-	**12.15**	nr	**13.06**
13mm; part 2	7.43	2.00	**7.58**	0.33	-	**5.74**	-	**13.31**	nr	**14.31**
Float operated valves for low pressure; BS 1212; plastics float										
inlet threaded externally; fixing to steel										
13mm; part 1	4.24	2.00	**4.32**	0.33	-	**5.74**	-	**10.06**	nr	**10.81**
13mm; part 2	5.38	2.00	**5.49**	0.33	-	**5.74**	-	**11.22**	nr	**12.06**
Float operated valves for high pressure; BS 1212; copper float										
inlet threaded externally; fixing to steel										
13mm; part 1	5.04	2.00	**5.14**	0.33	-	**5.74**	-	**10.88**	nr	**11.69**
13mm; part 2	7.27	2.00	**7.42**	0.33	-	**5.74**	-	**13.15**	nr	**14.14**
19mm; part 1	14.76	2.00	**15.06**	0.50	-	**8.69**	-	**23.75**	nr	**25.53**
25mm; part 1	44.29	2.00	**45.18**	1.00	-	**17.38**	-	**62.56**	nr	**67.25**
Float operated valves for high pressure; BS 1212; plastics float										
inlet threaded externally; fixing to steel										
13mm; part 1	2.99	2.00	**3.05**	0.33	-	**5.74**	-	**8.79**	nr	**9.44**
13mm; part 2	5.22	2.00	**5.32**	0.33	-	**5.74**	-	**11.06**	nr	**11.89**
19mm; part 1	9.76	2.00	**9.96**	0.50	-	**8.69**	-	**18.65**	nr	**20.04**
25mm; part 1	37.79	2.00	**38.55**	1.00	-	**17.38**	-	**55.93**	nr	**60.12**
S115: HOT AND COLD WATER AND GAS: DRAINING TAPS										
Brass draining taps										
Drain cocks; BS 2879, square head Type2										
13mm	1.80	2.00	**1.84**	0.10	-	**1.74**	-	**3.57**	nr	**3.84**
19mm	14.52	2.00	**14.81**	0.13	-	**2.26**	-	**17.07**	nr	**18.35**
S115: HOT AND COLD WATER AND GAS: STORAGE TANKS										
GRP water storage tanks										
GRP Sarena mfg Ltd; pre insulated water storage tanks with cover										
reference SC10; 18 litres	76.00	-	**76.00**	1.50	-	**26.07**	-	**102.07**	nr	**109.73**
reference SC20; 55 litres	106.00	-	**106.00**	1.50	-	**26.07**	-	**132.07**	nr	**141.98**
reference SC30; 86 litres	116.00	-	**116.00**	1.75	-	**30.42**	-	**146.42**	nr	**157.40**
reference SC40; 114 litres	142.00	-	**142.00**	1.75	-	**30.42**	-	**172.42**	nr	**185.35**
reference SC50; 159 litres	161.00	-	**161.00**	2.00	-	**34.76**	-	**195.76**	nr	**210.44**
reference SC60; 191 litres	178.00	-	**178.00**	2.00	-	**34.76**	-	**212.76**	nr	**228.72**
reference SC70; 227 litres	198.00	-	**198.00**	2.50	-	**43.45**	-	**241.45**	nr	**259.56**
reference SC100/1; 327 litres	261.00	-	**261.00**	3.00	-	**52.14**	-	**313.14**	nr	**336.63**
reference SC100/2; 336 litres	254.00	-	**254.00**	3.00	-	**52.14**	-	**306.14**	nr	**329.10**
reference SC125; 423 litres	322.00	-	**322.00**	4.00	-	**69.52**	-	**391.52**	nr	**420.88**
reference SC150; 491 litres	383.00	-	**383.00**	4.00	-	**69.52**	-	**452.52**	nr	**486.46**
reference SC200; 709 litres	445.00	-	**445.00**	4.00	-	**69.52**	-	**514.52**	nr	**553.11**
reference SC250; 900 litres	510.00	-	**510.00**	6.00	-	**104.28**	-	**614.28**	nr	**660.35**
reference SC350; 1250 litres	650.00	-	**650.00**	7.00	-	**121.66**	-	**771.66**	nr	**829.53**
reference SC500; 1727 litres	850.00	-	**850.00**	8.00	-	**139.04**	-	**989.04**	nr	**1063.22**
reference SC600; 2137 litres	980.00	-	**980.00**	9.00	-	**156.42**	-	**1136.42**	nr	**1221.65**
reference SC600H; 2350 litres	1100.00	-	**1100.00**	9.50	-	**165.11**	-	**1265.11**	nr	**1359.99**
reference SC1000; 3364 litres	1450.00	-	**1450.00**	12.00	-	**208.56**	-	**1658.56**	nr	**1782.95**
drilled holes for 13mm pipes	-	-	-	0.20	-	**3.48**	-	**3.48**	nr	**3.74**
drilled holes for 19mm pipes	-	-	-	0.20	-	**3.48**	-	**3.48**	nr	**3.74**
drilled holes for 25mm pipes	-	-	-	0.25	-	**4.35**	-	**4.35**	nr	**4.67**
drilled holed for 32mm pipes	-	-	-	0.25	-	**4.35**	-	**4.35**	nr	**4.67**
drilled holes for 38mm pipes	-	-	-	0.33	-	**5.74**	-	**5.74**	nr	**6.17**
drilled holes for 51mm pipes	-	-	-	0.33	-	**5.74**	-	**5.74**	nr	**6.17**
Plastics water storage cisterns										
Rectangular cold water storage cisterns; plastics; BS 4213 with sealed lid and byelaw 30 kit										
18 litres	12.10	2.00	**12.34**	1.25	-	**21.73**	-	**34.07**	nr	**36.62**
114 litres	40.45	2.00	**41.26**	1.75	-	**30.42**	-	**71.67**	nr	**77.05**
182 litres	63.90	2.00	**65.18**	2.00	-	**34.76**	-	**99.94**	nr	**107.43**
227 litres	69.55	2.00	**70.94**	2.50	-	**43.45**	-	**114.39**	nr	**122.97**
drilled holes for 13mm pipes	-	-	-	0.13	-	**2.26**	-	**2.26**	nr	**2.43**
drilled holes for 19mm pipes	-	-	-	0.13	-	**2.26**	-	**2.26**	nr	**2.43**
drilled holes for 25mm pipes	-	-	-	0.17	-	**2.95**	-	**2.95**	nr	**3.18**
drilled holed for 32mm pipes	-	-	-	0.17	-	**2.95**	-	**2.95**	nr	**3.18**
drilled holes for 38mm pipes	-	-	-	0.22	-	**3.82**	-	**3.82**	nr	**4.11**
drilled holes for 51mm pipes	-	-	-	0.22	-	**3.82**	-	**3.82**	nr	**4.11**

Labour hourly rates: (except Specialists) Craft Operatives £17.38 Labourer £10.71 Rates are national average prices. Refer to REGIONAL VARIATIONS for indicative levels of overall pricing in regions	MATERIALS			LABOUR				RATES		
	Del to Site £	Waste %	Material Cost £	Craft Optve Hrs	Lab Hrs	Labour Cost £	Sunds £	Nett Rate £	Unit	Gross Rate (+7.5%) £
S115: HOT AND COLD WATER AND GAS: STORAGE TANKS AND CYLINDERS										
Copper direct hot water cylinders; pre-insulated										
Direct hot water cylinders; copper cylinder, BS 699 Grade 3										
reference 3; 116 litres; four bosses	122.25	2.00	**124.70**	2.00	-	**34.76**	-	**159.46**	nr	**171.41**
reference 7; 120 litres; four bosses	109.25	2.00	**111.44**	2.00	-	**34.76**	-	**146.20**	nr	**157.16**
reference 8; 144 litres; four bosses	136.50	2.00	**139.23**	2.00	-	**34.76**	-	**173.99**	nr	**187.04**
reference 9; 166 litres; four bosses	176.00	2.00	**179.52**	2.00	-	**34.76**	-	**214.28**	nr	**230.35**
Copper indirect hot water cylinders; pre-insulated										
Double feed indirect hot water cylinders; copper cylinder, BS 1566 Part 1 Grade 3										
reference 3; 114 litres; four bosses	130.25	2.00	**132.86**	2.00	-	**34.76**	-	**167.62**	nr	**180.19**
reference 7; 117 litres; four bosses	111.50	2.00	**113.73**	2.00	-	**34.76**	-	**148.49**	nr	**159.63**
reference 8; 140 litres; four bosses	144.00	2.00	**146.88**	2.00	-	**34.76**	-	**181.64**	nr	**195.26**
reference 9; 162 litres; four bosses	180.50	2.00	**184.11**	2.00	-	**34.76**	-	**218.87**	nr	**235.29**
Single feed indirect hot water cylinders; copper cylinder, BS 1566 Part 2 Grade 3										
reference 7; 108 litres; four bosses	232.75	2.00	**237.41**	2.00	-	**34.76**	-	**272.17**	nr	**292.58**
reference 8; 130 litres; four bosses	274.25	2.00	**279.74**	2.00	-	**34.76**	-	**314.50**	nr	**338.08**
Copper combination hot water storage units; pre-insulated										
Combination direct hot water storage units; copper unit, BS 3198 with lid										
450mm diameter x 1050mm high; 115 litres hot water; 45 litres cold water	218.25	2.00	**222.62**	3.00	-	**52.14**	-	**274.76**	nr	**295.36**
450mm diameter x 1200mm high; 115 litres hot water; 115 litres cold water	241.75	2.00	**246.59**	3.00	-	**52.14**	-	**298.73**	nr	**321.13**
Combination double feed indirect hot water storage units; copper unit, BS 3198 with lid										
450mm diameter x 1050mm high; 115 litres hot water; 45 litres cold water	230.50	2.00	**235.11**	3.00	-	**52.14**	-	**287.25**	nr	**308.79**
450mm diameter x 1200mm high; 115 litres hot water; 115 litres cold water	245.00	2.00	**249.90**	3.00	-	**52.14**	-	**302.04**	nr	**324.69**
Combination single feed indirect hot water storage units; copper unit, BS 3198 with lid										
450mm diameter x 1200mm high; 115 litres hot water; 115 litres cold water	280.75	2.00	**286.37**	3.00	-	**52.14**	-	**338.51**	nr	**363.89**
S115: HOT AND COLD WATER AND GAS: PIPE INSULATION										
Thermal insulation; glass fibre preformed lagging, butt joints in the running length; secured with metal bands										
19mm thick insulation to copper pipework around one pipe										
15mm	2.56	3.00	**2.64**	0.20	-	**3.48**	-	**6.11**	m	**6.57**
22mm	2.70	3.00	**2.78**	0.22	-	**3.82**	-	**6.60**	m	**7.10**
28mm	2.86	3.00	**2.95**	0.25	-	**4.35**	-	**7.29**	m	**7.84**
35mm	3.06	3.00	**3.15**	0.30	-	**5.21**	-	**8.37**	m	**8.99**
42mm	3.38	3.00	**3.48**	0.33	-	**5.74**	-	**9.22**	m	**9.91**
54mm	3.84	3.00	**3.96**	0.40	-	**6.95**	-	**10.91**	m	**11.73**
25mm thick insulation to copper pipework around one pipe										
15mm	2.86	3.00	**2.95**	0.20	-	**3.48**	-	**6.42**	m	**6.90**
22mm	2.94	3.00	**3.03**	0.22	-	**3.82**	-	**6.85**	m	**7.37**
28mm	3.15	3.00	**3.24**	0.25	-	**4.35**	-	**7.59**	m	**8.16**
35mm	3.49	3.00	**3.59**	0.30	-	**5.21**	-	**8.81**	m	**9.47**
42mm	3.79	3.00	**3.90**	0.33	-	**5.74**	-	**9.64**	m	**10.36**
54mm	4.45	3.00	**4.58**	0.40	-	**6.95**	-	**11.54**	m	**12.40**
Thermal insulation; foamed polyurethane preformed lagging, butt joints in the running length; secured with adhesive bands										
13mm thick insulation to copper pipework around one pipe										
15mm	1.21	3.00	**1.25**	0.17	-	**2.95**	-	**4.20**	m	**4.52**
22mm	1.42	3.00	**1.46**	0.18	-	**3.13**	-	**4.59**	m	**4.94**
28mm	1.67	3.00	**1.72**	0.20	-	**3.48**	-	**5.20**	m	**5.59**
35mm	1.79	3.00	**1.84**	0.22	-	**3.82**	-	**5.67**	m	**6.09**
42mm	2.08	3.00	**2.14**	0.25	-	**4.35**	-	**6.49**	m	**6.97**
54mm	2.52	3.00	**2.60**	0.33	-	**5.74**	-	**8.33**	m	**8.96**
S115: HOT AND COLD WATER AND GAS: INSULATION JACKETS AND LAGGING UNITS FOR CYLINDERS AND TANKS										
Thermal insulation; glass fibre filled insulating jacket in strips, white pvc covering both sides; secured with metal straps and wire holder ring at top										
75mm thick insulation to equipment sides and tops of cylinders; overall size										
400mm diameter x 1050mm high	9.35	5.00	**9.82**	0.75	-	**13.04**	-	**22.85**	nr	**24.57**
450mm diameter x 900mm high	8.25	5.00	**8.66**	0.75	-	**13.04**	-	**21.70**	nr	**23.32**

Labour hourly rates: (except Specialists) Craft Operatives £17.38 Labourer £10.71 Rates are national average prices. Refer to REGIONAL VARIATIONS for indicative levels of overall pricing in regions	MATERIALS			LABOUR				RATES		
	Del to Site £	Waste %	Material Cost £	Craft Optve Hrs	Lab Hrs	Labour Cost £	Sunds £	Nett Rate £	Unit	Gross Rate (+7.5%) £
S115: HOT AND COLD WATER AND GAS: INSULATION JACKETS AND LAGGING UNITS FOR CYLINDERS AND TANKS Cont.										
Thermal insulation; glass fibre filled insulating jacket in strips, white pvc covering both sides; secured with metal straps and wire holder ring at top Cont.										
75mm thick insulation to equipment Cont.										
sides and tops of cylinders; overall size Cont.										
450mm diameter x 1050mm high	9.35	5.00	**9.82**	0.75	-	**13.04**	-	**22.85**	nr	24.57
450mm diameter x 1200mm high	24.80	5.00	**26.04**	0.75	-	**13.04**	-	**39.08**	nr	42.01
S115: HOT AND COLD WATER AND GAS: FUEL OIL STORAGE TANKS										
Steel fuel oil storage tanks										
Oil storage tanks; carbon steel tank, BS 799 Part 5 Type III, rectangular; 457mm diameter manhole cover, oil tight washer and screwed socket for fill, vent, sludge and draw off; painted one coat black bituminous paint										
14 gauge; 1520 x 610 x 1220mm; 1130 litre capacity	156.00	-	**156.00**	3.50	-	**60.83**	-	**216.83**	nr	233.09
14 gauge; 1830 x 610 x 1220mm; 1360 litre capacity	160.00	-	**160.00**	3.75	-	**65.18**	-	**225.18**	nr	242.06
14 gauge; 1830 x 1220 x 1220mm; 2730 litre capacity	393.50	-	**393.50**	4.00	-	**69.52**	-	**463.02**	nr	497.75

PIPED SUPPLY SYSTEMS

MECHANICAL HEATING/COOLING/REFRIGERATION SYSTEMS

Labour hourly rates: (except Specialists) Craft Operatives £17.38 Labourer £10.71 Rates are national average prices. Refer to REGIONAL VARIATIONS for indicative levels of overall pricing in regions	MATERIALS			LABOUR				RATES		
	Del to Site £	Waste %	Material Cost £	Craft Optve Hrs	Lab Hrs	Labour Cost £	Sunds £	Nett Rate £	Unit	Gross Rate (+7.5%) £
T10: GAS/OIL FIRED BOILERS										
Gas flue pipes comprising galvanised steel outer skin and aluminium inner skin with air space between, BS 715; socketed joints										
Pipes; straight										
100mm	18.57	5.00	**19.50**	0.67	-	**11.64**	-	**31.14**	m	**33.48**
extra; connections to appliance	3.82	2.00	**3.90**	0.50	-	**8.69**	-	**12.59**	nr	**13.53**
extra; adjustable pipes	10.87	2.00	**11.09**	0.50	-	**8.69**	-	**19.78**	nr	**21.26**
extra; terminals	14.17	2.00	**14.45**	0.50	-	**8.69**	-	**23.14**	nr	**24.88**
extra; elbows 0- 90 degrees	12.42	2.00	**12.67**	0.50	-	**8.69**	-	**21.36**	nr	**22.96**
extra; tees	27.55	2.00	**28.10**	0.50	-	**8.69**	-	**36.79**	nr	**39.55**
125mm	22.25	5.00	**23.36**	0.80	-	**13.90**	-	**37.27**	m	**40.06**
extra; connections to appliance	4.32	2.00	**4.41**	0.67	-	**11.64**	-	**16.05**	nr	**17.25**
extra; adjustable pipes	12.22	2.00	**12.46**	0.67	-	**11.64**	-	**24.11**	nr	**25.92**
extra; terminals	15.55	2.00	**15.86**	0.67	-	**11.64**	-	**27.51**	nr	**29.57**
extra; elbows 0- 90 degrees	14.66	2.00	**14.95**	0.67	-	**11.64**	-	**26.60**	nr	**28.59**
extra; tees	29.72	2.00	**30.31**	0.67	-	**11.64**	-	**41.96**	nr	**45.11**
150mm	24.84	5.00	**26.08**	1.00	-	**17.38**	-	**43.46**	m	**46.72**
extra; connections to appliance	4.69	2.00	**4.78**	1.00	-	**17.38**	-	**22.16**	nr	**23.83**
extra; adjustable pipes	15.39	2.00	**15.70**	1.00	-	**17.38**	-	**33.08**	nr	**35.56**
extra; terminals	19.96	2.00	**20.36**	1.00	-	**17.38**	-	**37.74**	nr	**40.57**
extra; elbows 0- 90 degrees	18.38	2.00	**18.75**	1.00	-	**17.38**	-	**36.13**	nr	**38.84**
extra; tees	30.93	2.00	**31.55**	1.00	-	**17.38**	-	**48.93**	nr	**52.60**
Flue supports; galvanised steel wall bands fixing to masonry with screws										
for 100mm pipes	6.08	2.00	**6.20**	0.17	-	**2.95**	-	**9.16**	nr	**9.84**
for 125mm pipes	6.47	2.00	**6.60**	0.17	-	**2.95**	-	**9.55**	nr	**10.27**
for 150mm pipes	8.20	2.00	**8.36**	0.17	-	**2.95**	-	**11.32**	nr	**12.17**
Fire stop spacers										
for 100mm pipes	2.64	2.00	**2.69**	0.25	-	**4.35**	-	**7.04**	nr	**7.57**
for 125mm pipes	2.64	2.00	**2.69**	0.33	-	**5.74**	-	**8.43**	nr	**9.06**
for 150mm pipes	3.01	2.00	**3.07**	0.50	-	**8.69**	-	**11.76**	nr	**12.64**
Gas fired boilers										
Gas fired boilers for central heating and hot water; automatically controlled by thermostat with electrical control, gas governor and flame failure device										
free standing boiler; conventional flue; approximate output rating B Th U per hour										
40000	405.75	1.00	**409.81**	4.00	-	**69.52**	-	**479.33**	nr	**515.28**
50000	424.50	1.00	**428.75**	4.00	-	**69.52**	-	**498.27**	nr	**535.63**
60000	451.75	1.00	**456.27**	4.00	-	**69.52**	-	**525.79**	nr	**565.22**
80000	586.00	1.00	**591.86**	4.00	-	**69.52**	-	**661.38**	nr	**710.98**
100000	751.25	1.00	**758.76**	5.00	-	**86.90**	-	**845.66**	nr	**909.09**
free standing boiler; balanced flue; approximate output rating B Th U per hour										
40000	530.00	1.00	**535.30**	5.00	-	**86.90**	-	**622.20**	nr	**668.87**
50000	549.00	1.00	**554.49**	5.00	-	**86.90**	-	**641.39**	nr	**689.49**
60000	587.00	1.00	**592.87**	5.00	-	**86.90**	-	**679.77**	nr	**730.75**
80000	794.00	1.00	**801.94**	5.00	-	**86.90**	-	**888.84**	nr	**955.50**
100000	1005.00	1.00	**1015.05**	6.00	-	**104.28**	-	**1119.33**	nr	**1203.28**
wall mounted boiler; balanced flue; approximate output rating B Th U per hour										
40000	452.00	1.00	**456.52**	5.50	-	**95.59**	-	**552.11**	nr	**593.52**
50000	490.00	1.00	**494.90**	5.50	-	**95.59**	-	**590.49**	nr	**634.78**
60000	527.00	1.00	**532.27**	5.50	-	**95.59**	-	**627.86**	nr	**674.95**
Oil fired boilers										
Oil fired boilers for central heating and hot water; automatically controlled by thermostat, with electrical control box										
free standing boiler; conventional flue; approximate output rating BThU per hour										
40,000 - 48,000	727.00	1.00	**734.27**	4.00	-	**69.52**	-	**803.79**	nr	**864.07**
50,000 - 65,000	813.00	1.00	**821.13**	4.00	-	**69.52**	-	**890.65**	nr	**957.45**

Labour hourly rates: (except Specialists) Craft Operatives £17.38 Labourer £10.71 Rates are national average prices. Refer to REGIONAL VARIATIONS for indicative levels of overall pricing in regions	MATERIALS			LABOUR				RATES		
	Del to Site £	Waste %	Material Cost £	Craft Optve Hrs	Lab Hrs	Labour Cost £	Sunds £	Nett Rate £	Unit	Gross Rate (+7.5%) £
T10: GAS/OIL FIRED BOILERS Cont.										
Oil fired boilers Cont.										
Oil fired boilers for central heating and hot water; automatically controlled by thermostat, with electrical control box Cont.										
free standing boiler; conventional flue; approximate output rating BThU per hour Cont.										
70,000 - 85,000	898.00	1.00	**906.98**	4.00	-	**69.52**	-	**976.50**	nr	**1049.74**
88,000 - 110,000	1007.00	1.00	**1017.07**	5.00	-	**86.90**	-	**1103.97**	nr	**1186.77**
T11: COAL FIRED BOILERS										
Equipment										
Solid fuel boilers for central heating and hot water; thermostatically controlled; with tools										
free standing boiler, gravity feed; approximate output rating BThU per hour										
45000	1486.00	1.00	**1500.86**	4.00	-	**69.52**	-	**1570.38**	nr	**1688.16**
60000	1731.00	1.00	**1748.31**	4.00	-	**69.52**	-	**1817.83**	nr	**1954.17**
80000	1926.00	1.00	**1945.26**	5.00	-	**86.90**	-	**2032.16**	nr	**2184.57**
T31: LOW TEMPERATURE HOT WATER HEATING										
Accelerator pumps										
Variable head accelerator pumps for forced central heating; small bore indirect systems										
BSP connections; with valves										
20mm	46.75	2.00	**47.69**	0.75	-	**13.04**	-	**60.72**	nr	**65.27**
25mm	47.40	2.00	**48.35**	0.75	-	**13.04**	-	**61.38**	nr	**65.99**
Radiators										
Pressed steel radiators single panel with convector; air cock, plain plug										
450mm high; fixing brackets to masonry with screws										
600mm long	20.60	1.00	**20.81**	3.00	-	**52.14**	-	**72.95**	nr	**78.42**
900mm long	30.88	1.00	**31.19**	3.00	-	**52.14**	-	**83.33**	nr	**89.58**
1200mm long	41.16	1.00	**41.57**	3.00	-	**52.14**	-	**93.71**	nr	**100.74**
1600mm long	62.83	1.00	**63.46**	3.00	-	**52.14**	-	**115.60**	nr	**124.27**
2400mm long	126.76	1.00	**128.03**	3.00	-	**52.14**	-	**180.17**	nr	**193.68**
600mm high; fixing brackets to masonry with screws										
600mm long	26.57	1.00	**26.84**	3.00	-	**52.14**	-	**78.98**	nr	**84.90**
900mm long	39.86	1.00	**40.26**	3.00	-	**52.14**	-	**92.40**	nr	**99.33**
1200mm long	53.16	1.00	**53.69**	3.00	-	**52.14**	-	**105.83**	nr	**113.77**
1600mm long	81.11	1.00	**81.92**	3.00	-	**52.14**	-	**134.06**	nr	**144.12**
2400mm long	163.65	1.00	**165.29**	3.00	-	**52.14**	-	**217.43**	nr	**233.73**
700mm high; fixing brackets to masonry with screws										
600mm long	30.48	1.00	**30.78**	3.00	-	**52.14**	-	**82.92**	nr	**89.14**
900mm long	52.27	1.00	**52.79**	3.00	-	**52.14**	-	**104.93**	nr	**112.80**
1200mm long	93.75	1.00	**94.69**	3.00	-	**52.14**	-	**146.83**	nr	**157.84**
1600mm long	125.02	1.00	**126.27**	3.00	-	**52.14**	-	**178.41**	nr	**191.79**
2400mm long	187.53	1.00	**189.41**	3.00	-	**52.14**	-	**241.55**	nr	**259.66**
Pressed steel radiators; double panel with single convector; air cock, plain plug										
450mm high; fixing brackets to masonry with screws										
600mm long	31.99	1.00	**32.31**	3.00	-	**52.14**	-	**84.45**	nr	**90.78**
900mm long	47.98	1.00	**48.46**	3.00	-	**52.14**	-	**100.60**	nr	**108.14**
1200mm long	73.22	1.00	**73.95**	3.00	-	**52.14**	-	**126.09**	nr	**135.55**
1600mm long	97.64	1.00	**98.62**	3.00	-	**52.14**	-	**150.76**	nr	**162.06**
2400mm long	196.79	1.00	**198.76**	3.00	-	**52.14**	-	**250.90**	nr	**269.72**
600mm high; fixing brackets to masonry with screws										
600mm long	40.31	1.00	**40.71**	3.00	-	**52.14**	-	**92.85**	nr	**99.82**
900mm long	60.46	1.00	**61.06**	3.00	-	**52.14**	-	**113.20**	nr	**121.69**
1200mm long	80.61	1.00	**81.42**	3.00	-	**52.14**	-	**133.56**	nr	**143.57**
1600mm long	123.02	1.00	**124.25**	3.00	-	**52.14**	-	**176.39**	nr	**189.62**
2400mm long	248.18	1.00	**250.66**	3.00	-	**52.14**	-	**302.80**	nr	**325.51**
700mm high; fixing brackets to masonry with screws										
600mm long	51.97	1.00	**52.49**	3.00	-	**52.14**	-	**104.63**	nr	**112.48**
900mm long	104.84	1.00	**105.89**	3.00	-	**52.14**	-	**158.03**	nr	**169.88**
1200mm long	139.76	1.00	**141.16**	3.00	-	**52.14**	-	**193.30**	nr	**207.79**
1600mm long	186.37	1.00	**188.23**	3.00	-	**52.14**	-	**240.37**	nr	**258.40**
2400mm long	279.55	1.00	**282.35**	3.00	-	**52.14**	-	**334.49**	nr	**359.57**
Brass valves for radiators; self colour										
Valves for radiators, BS 2767; wheel head inlet for copper; angle pattern										
15mm	5.92	2.00	**6.04**	0.25	-	**4.35**	-	**10.38**	nr	**11.16**
22mm	8.04	2.00	**8.20**	0.33	-	**5.74**	-	**13.94**	nr	**14.98**
Valves for radiators, BS 2767; lock shield inlet for copper; angle pattern										
15mm	5.92	2.00	**6.04**	0.25	-	**4.35**	-	**10.38**	nr	**11.16**
22mm	8.04	2.00	**8.20**	0.33	-	**5.74**	-	**13.94**	nr	**14.98**
Brass valves for radiators; chromium plated										
Valves for radiators, BS 2767; wheel head inlet for copper; angle pattern										
15mm	8.20	2.00	**8.36**	0.25	-	**4.35**	-	**12.71**	nr	**13.66**
22mm	10.72	2.00	**10.93**	0.33	-	**5.74**	-	**16.67**	nr	**17.92**

Labour hourly rates: (except Specialists) Craft Operatives £17.38 Labourer £10.71 Rates are national average prices. Refer to REGIONAL VARIATIONS for indicative levels of overall pricing in regions	MATERIALS			LABOUR				RATES		
	Del to Site £	Waste %	Material Cost £	Craft Optve Hrs	Lab Hrs	Labour Cost £	Sunds £	Nett Rate £	Unit	Gross Rate (+7.5%) £
T31: LOW TEMPERATURE HOT WATER HEATING Cont.										
Brass valves for radiators; chromium plated Cont.										
Valves for radiators, BS 2767; wheel head Cont.										
inlet for copper; straight pattern										
15mm	9.53	2.00	**9.72**	0.25	-	**4.35**	-	**14.07**	nr	**15.12**
22mm	12.40	2.00	**12.65**	0.33	-	**5.74**	-	**18.38**	nr	**19.76**
Valves for radiators, BS 2767; lock shield										
inlet for copper; angle pattern										
15mm	8.20	2.00	**8.36**	0.25	-	**4.35**	-	**12.71**	nr	**13.66**
22mm	10.72	2.00	**10.93**	0.33	-	**5.74**	-	**16.67**	nr	**17.92**
inlet for copper; straight pattern										
15mm	9.53	2.00	**9.72**	0.25	-	**4.35**	-	**14.07**	nr	**15.12**
22mm	12.40	2.00	**12.65**	0.33	-	**5.74**	-	**18.38**	nr	**19.76**
Brass thermostatic valves for radiators; chromium plated										
Thermostatic valves for radiators, one piece; ABS plastics head										
inlet for copper; angle pattern										
15mm	14.60	2.00	**14.89**	0.25	-	**4.35**	-	**19.24**	nr	**20.68**
inlet for copper; straight pattern										
15mm	14.60	2.00	**14.89**	0.25	-	**4.35**	-	**19.24**	nr	**20.68**
inlet for iron; angle pattern										
19mm	22.50	2.00	**22.95**	0.33	-	**5.74**	-	**28.69**	nr	**30.84**
Skirting heaters										
Pressed metal skirting heaters with fins on copper tube; straight										
900mm long	28.50	2.00	**29.07**	1.25	-	**21.73**	-	**50.80**	nr	**54.60**
1200mm long	37.50	2.00	**38.25**	1.67	-	**29.02**	-	**67.27**	nr	**72.32**
1500mm long	41.25	2.00	**42.08**	2.08	-	**36.15**	-	**78.23**	nr	**84.09**
1800mm long	45.00	2.00	**45.90**	2.50	-	**43.45**	-	**89.35**	nr	**96.05**
extra; internal corners	6.00	2.00	**6.12**	0.33	-	**5.74**	-	**11.86**	nr	**12.74**
extra; external corners	6.00	2.00	**6.12**	0.33	-	**5.74**	-	**11.86**	nr	**12.74**
extra; end stops	6.00	2.00	**6.12**	0.33	-	**5.74**	-	**11.86**	nr	**12.74**
extra; valve boxes	14.25	2.00	**14.54**	0.33	-	**5.74**	-	**20.27**	nr	**21.79**

This section continues on the next page

Labour hourly rates: (except Specialists) Craft Operatives £14.21 Labourer £10.71 Rates are national average prices. Refer to REGIONAL VARIATIONS for indicative levels of overall pricing in regions	MATERIALS			LABOUR				RATES		
	Del to Site £	Waste %	Material Cost £	Craft Optve Hrs	Lab Hrs	Labour Cost £	Sunds £	Nett Rate £	Unit	Gross Rate (+7.5%) £
T32: LOW TEMPERATURE HOT WATER HEATING (SMALL SCALE)										
Central heating installations - indicative prices										
Note the following are indicative prices for installation in two storey three bedroomed dwellings with a floor area of approximately 85m2										
Installation; boiler, copper piping, pressed steel radiators, heated towel rail; providing hot water to sink , bath, lavatory basin; complete with all necessary pumps, controls etc										
solid fuel fired	-	-	-	-	21.00	**224.91**	-	**3174.91**	it	**3413.03**
gas fired	-	-	-	-	19.00	**203.49**	-	**2903.49**	it	**3121.25**
oil fired, including oil storage tank	-	-	-	-	24.00	**257.04**	-	**3607.04**	it	**3877.57**

VENTILATION/AIR CONDITIONING SYSTEMS

Labour hourly rates: (except Specialists) Craft Operatives £20.54 Labourer £10.71 Rates are national average prices. Refer to REGIONAL VARIATIONS for indicative levels of overall pricing in regions	MATERIALS			LABOUR				RATES		
	Del to Site £	Waste %	Material Cost £	Craft Optve Hrs	Lab Hrs	Labour Cost £	Sunds £	Nett Rate £	Unit	Gross Rate (+7.5%) £
U10: GENERAL VENTILATION										
Ventilating fans										
Note: the following prices exclude cutting holes in glass										
Extract fans, window mounted; shutter and incorporated switch unit										
102mm diameter	66.48	2.50	**68.14**	1.25	-	**25.68**	-	**93.82**	nr	**100.85**
102mm diameter - low voltage	95.10	2.50	**97.48**	1.25	-	**25.68**	-	**123.15**	nr	**132.39**
257mm diameter	214.88	2.50	**220.25**	1.50	-	**30.81**	-	**251.06**	nr	**269.89**
324mm diameter	310.28	2.50	**318.04**	1.50	-	**30.81**	-	**348.85**	nr	**375.01**
Extract fans, wall mounted; shutter, wall duct and incorporated switch unit										
Note: the following prices exclude forming hole and making good										
102mm diameter	39.61	2.50	**40.60**	1.50	-	**30.81**	-	**71.41**	nr	**76.77**
102mm diameter - low voltage	68.23	2.50	**69.94**	1.50	-	**30.81**	-	**100.75**	nr	**108.30**
257mm diameter	188.01	2.50	**192.71**	2.00	-	**41.08**	-	**233.79**	nr	**251.32**
324mm diameter	283.41	2.50	**290.50**	2.00	-	**41.08**	-	**331.58**	nr	**356.44**
Extract fans, built in wall; shutter and incorporated switch unit										
290 x 321 mm	168.95	2.50	**173.17**	2.50	0.08	**52.21**	-	**225.38**	nr	**242.28**
409 x 433 mm	306.75	2.50	**314.42**	2.70	0.08	**56.31**	-	**370.73**	nr	**398.54**
473 x 476 mm	370.35	2.50	**379.61**	2.80	0.08	**58.37**	-	**437.98**	nr	**470.83**

ELECTRICAL SUPPLY/POWER/LIGHTING SYSTEMS

Labour hourly rates: (except Specialists) Craft Operatives £20.54 Labourer £10.71 Rates are national average prices. Refer to REGIONAL VARIATIONS for indicative levels of overall pricing in regions	MATERIALS			LABOUR				RATES		
	Del to Site £	Waste %	Material Cost £	Craft Optve Hrs	Lab Hrs	Labour Cost £	Sunds £	Nett Rate £	Unit	Gross Rate (+7.5%) £
V41: STREET/AREA/FLOOD LIGHTING										
Street lighting columns and set in ground; excavate hole, remove surplus spoil, filling with concrete mix ST4, working around base										
Aluminium; BS EN 40-1, nominal mounting height										
4.88m	258.67	-	258.67	4.75	9.50	199.31	21.29	479.27	nr	515.22
8.00m	637.18	-	637.18	5.50	11.00	230.78	33.12	901.08	nr	968.66
10.00m	809.65	-	809.65	6.25	12.50	262.25	72.16	1144.06	nr	1229.87
12.00m	1135.02	-	1135.02	7.00	14.00	293.72	72.16	1500.90	nr	1613.46
Steel; BS EN 40-1, nominal mounting height										
4.00m	134.76	-	134.76	4.75	9.50	199.31	21.29	355.36	nr	382.01
5.00m	138.87	-	138.87	5.50	11.00	230.78	33.12	402.77	nr	432.97
6.00m	171.14	-	171.14	6.25	12.50	262.25	72.16	505.55	nr	543.46
8.00m	253.96	-	253.96	7.00	14.00	293.72	72.16	619.84	nr	666.33
V90: GENERAL LIGHTING AND POWER - COMMERCIAL										
Steel trunking										
Straight lighting trunking; PVC lid										
50 x 50mm	1.81	5.00	1.90	0.40	-	8.22	0.50	10.62	m	11.42
extra; end cap	2.10	-	2.10	0.08	-	1.64	0.25	3.99	nr	4.29
extra; internal angle	10.31	-	10.31	0.25	-	5.14	0.25	15.70	nr	16.87
extra; external angle	10.31	-	10.31	0.25	-	5.14	0.25	15.70	nr	16.87
extra; tee	11.04	-	11.04	0.25	-	5.14	0.25	16.42	nr	17.66
extra; four way intersection	11.43	-	11.43	0.40	-	8.22	0.25	19.89	nr	21.39
Straight dado trunking, two compartment; steel lid										
150 x 30mm	6.50	5.00	6.83	0.45	-	9.24	0.50	16.57	m	17.81
extra; end cap	4.70	-	4.70	0.08	-	1.64	0.25	6.60	nr	7.09
extra; internal angle	21.78	-	21.78	0.35	-	7.19	0.25	29.22	nr	31.41
extra; external angle	21.78	-	21.78	0.35	-	7.19	0.25	29.22	nr	31.41
extra; single socket plate	8.57	-	8.57	0.16	-	3.29	0.25	12.11	nr	13.01
extra; twin socket plate	9.10	-	9.10	0.16	-	3.29	0.25	12.64	nr	13.59
Straight dado trunking, three compartment; steel lid										
200 x 38mm	8.59	5.00	9.02	0.50	-	10.27	0.75	20.04	m	21.54
extra; end cap	5.14	-	5.14	0.08	-	1.64	0.25	7.03	nr	7.56
extra; internal angle	26.88	-	26.88	0.40	-	8.22	0.25	35.34	nr	37.99
extra; external angle	26.88	-	26.88	0.40	-	8.22	0.25	35.34	nr	37.99
extra; single socket plate	8.85	-	8.85	0.16	-	3.29	0.25	12.38	nr	13.31
extra; twin socket plate	9.49	-	9.49	0.16	-	3.29	0.25	13.03	nr	14.00
Straight skirting trunking; two compartments										
200 x 38mm	8.58	5.00	9.00	0.50	-	10.27	0.75	20.02	m	21.53
extra; internal bend	25.41	-	25.41	0.40	-	8.22	0.25	33.88	nr	36.42
extra; external bend	25.41	-	25.41	0.40	-	8.22	0.25	33.88	nr	36.42
Straight underfloor trunking; two compartments										
150 x 25mm	5.83	5.00	6.12	0.70	-	14.38	-	20.50	m	22.04
extra; floor junction boxes; adjustable frame and trap	70.00	-	70.00	1.85	-	38.00	-	108.00	nr	116.10
extra; horizontal bend	17.09	-	17.09	0.50	-	10.27	-	27.36	nr	29.41
PVC trunking; white										
Straight mini-trunking; clip on lid										
16 x 16mm	0.31	5.00	0.32	0.20	-	4.11	0.50	4.93	m	5.30
extra; end cap	0.28	-	0.28	0.04	-	0.82	0.13	1.23	nr	1.32
extra; internal bend	0.28	-	0.28	0.08	-	1.64	0.13	2.05	nr	2.21
extra; external bend	0.28	-	0.28	0.08	-	1.64	0.13	2.05	nr	2.21
extra; flat bend	0.28	-	0.28	0.08	-	1.64	0.13	2.05	nr	2.21
extra; equal tee	0.49	-	0.49	0.17	-	3.49	0.13	4.11	nr	4.42
25 x 16mm	0.34	5.00	0.36	0.20	-	4.11	0.13	4.60	m	4.94
extra; end cap	0.28	-	0.28	0.04	-	0.82	0.13	1.23	nr	1.32
extra; internal bend	0.28	-	0.28	0.08	-	1.64	0.13	2.05	nr	2.21
extra; external bend	0.28	-	0.28	0.08	-	1.64	0.13	2.05	nr	2.21
extra; flat bend	0.28	-	0.28	0.08	-	1.64	0.13	2.05	nr	2.21

Labour hourly rates: (except Specialists) Craft Operatives £20.54 Labourer £10.71 Rates are national average prices. Refer to REGIONAL VARIATIONS for indicative levels of overall pricing in regions	MATERIALS			LABOUR				RATES		
	Del to Site £	Waste %	Material Cost £	Craft Optve Hrs	Lab Hrs	Labour Cost £	Sunds £	Nett Rate £	Unit	Gross Rate (+7.5%) £
V90: GENERAL LIGHTING AND POWER - COMMERCIAL Cont.										
PVC trunking; white Cont.										
Straight mini-trunking; clip on lid Cont.										
extra; equal tee	0.49	-	**0.49**	0.17	-	**3.49**	0.13	**4.11**	nr	**4.42**
40 x 16mm	0.53	5.00	**0.55**	0.25	-	**5.14**	0.13	**5.82**	m	**6.25**
extra; end cap	0.29	-	**0.29**	0.04	-	**0.82**	0.13	**1.24**	nr	**1.33**
extra; internal bend	0.29	-	**0.29**	0.08	-	**1.64**	0.13	**2.06**	nr	**2.22**
extra; external bend	0.29	-	**0.29**	0.08	-	**1.64**	0.13	**2.06**	nr	**2.22**
extra; flat bend	0.29	-	**0.29**	0.08	-	**1.64**	0.13	**2.06**	nr	**2.22**
extra; equal tee	0.51	-	**0.51**	0.17	-	**3.49**	0.13	**4.13**	nr	**4.44**
Straight dado trunking, three compartment; clip on lid										
145 x 40mm	19.99	5.00	**20.99**	0.40	-	**8.22**	0.50	**29.71**	m	**31.93**
extra; end caps	1.40	-	**1.40**	0.08	-	**1.64**	0.25	**3.29**	nr	**3.54**
extra; internal bend	4.81	-	**4.81**	0.30	-	**6.16**	0.25	**11.22**	nr	**12.06**
extra; external bend	4.81	-	**4.81**	0.30	-	**6.16**	0.25	**11.22**	nr	**12.06**
extra; tee	37.80	-	**37.80**	0.40	-	**8.22**	0.25	**46.27**	nr	**49.74**
extra; flat angle	32.40	-	**32.40**	0.30	-	**6.16**	0.25	**38.81**	nr	**41.72**
extra; mini trunking adaptor	1.36	-	**1.36**	0.20	-	**4.11**	0.25	**5.72**	nr	**6.15**
230 x 40mm	33.68	5.00	**35.36**	0.45	-	**9.24**	0.25	**44.86**	m	**48.22**
extra; end caps	4.47	-	**4.47**	0.08	-	**1.64**	0.25	**6.36**	nr	**6.84**
extra; internal bend	9.36	-	**9.36**	0.30	-	**6.16**	0.25	**15.77**	nr	**16.96**
extra; external bend	9.36	-	**9.36**	0.30	-	**6.16**	0.25	**15.77**	nr	**16.96**
extra; mini trunking adaptor	1.39	-	**1.39**	0.20	-	**4.11**	0.25	**5.75**	nr	**6.18**
Straight skirting trunking, three compartment; clip on lid										
100 x 40mm	3.38	5.00	**3.54**	0.35	-	**7.19**	0.50	**11.23**	m	**12.08**
extra; end cap	1.94	-	**1.94**	0.06	-	**1.23**	0.25	**3.42**	nr	**3.67**
extra; internal corner	2.49	-	**2.49**	0.25	-	**5.14**	0.25	**7.87**	nr	**8.47**
extra; external corner	2.49	-	**2.49**	0.25	-	**5.14**	0.25	**7.87**	nr	**8.47**
extra; mini trunking adaptor	1.75	-	**1.75**	0.16	-	**3.29**	0.25	**5.28**	nr	**5.68**
extra; socket adaptor, one gang	3.91	-	**3.91**	0.25	-	**5.14**	0.25	**9.30**	nr	**9.99**
extra; socket adaptor, two gang	5.62	-	**5.62**	0.25	-	**5.14**	0.25	**11.00**	nr	**11.83**
PVC heavy gauge conduit and fittings; push fit joints; spacer bar saddles at 600mm centres										
Conduits; straight										
20mm diameter; plugging to masonry; to surfaces	0.50	10.00	**0.55**	0.17	-	**3.49**	0.42	**4.46**	m	**4.79**
extra; small circular terminal boxes	0.86	2.50	**0.88**	0.10	-	**2.05**	0.27	**3.21**	nr	**3.45**
extra; small circular angle boxes	0.96	2.50	**0.98**	0.11	-	**2.26**	0.27	**3.51**	nr	**3.78**
extra; small circular three way boxes	1.04	2.50	**1.07**	0.13	-	**2.67**	0.27	**4.01**	nr	**4.31**
extra; small circular through way boxes	0.96	2.50	**0.98**	0.11	-	**2.26**	0.27	**3.51**	nr	**3.78**
25mm diameter; plugging to masonry; to surfaces	0.96	10.00	**1.05**	0.18	-	**3.70**	0.42	**5.17**	m	**5.56**
extra; small circular terminal boxes	1.32	2.50	**1.35**	0.10	-	**2.05**	0.27	**3.68**	nr	**3.95**
extra; small circular angle boxes	1.44	2.50	**1.48**	0.11	-	**2.26**	0.27	**4.01**	nr	**4.31**
extra; small circular three way boxes	1.59	2.50	**1.63**	0.13	-	**2.67**	0.27	**4.57**	nr	**4.91**
extra; small circular through way boxes	1.44	2.50	**1.48**	0.11	-	**2.26**	0.27	**4.01**	nr	**4.31**
20mm diameter; plugging to masonry; in chases	0.50	10.00	**0.55**	0.15	-	**3.08**	0.42	**4.05**	m	**4.35**
extra; small circular terminal boxes	0.86	2.50	**0.88**	0.10	-	**2.05**	0.27	**3.21**	nr	**3.45**
extra; small circular angle boxes	0.96	2.50	**0.98**	0.11	-	**2.26**	0.27	**3.51**	nr	**3.78**
extra; small circular three way boxes	1.04	2.50	**1.07**	0.13	-	**2.67**	0.27	**4.01**	nr	**4.31**
extra; small circular through way boxes	0.96	2.50	**0.98**	0.11	-	**2.26**	0.27	**3.51**	nr	**3.78**
25mm diameter; plugging to masonry; in chases	0.96	10.00	**1.05**	0.16	-	**3.29**	0.42	**4.76**	m	**5.12**
extra; small circular terminal boxes	1.32	2.50	**1.35**	0.10	-	**2.05**	0.27	**3.68**	nr	**3.95**
extra; small circular angle boxes	1.44	2.50	**1.48**	0.11	-	**2.26**	0.27	**4.01**	nr	**4.31**
extra; small circular three way boxes	1.59	2.50	**1.63**	0.13	-	**2.67**	0.27	**4.57**	nr	**4.91**
extra; small circular through way boxes	1.44	2.50	**1.48**	0.11	-	**2.26**	0.27	**4.01**	nr	**4.31**
Oval conduits; straight										
20mm (nominal size); plugging to masonry; in chases	0.28	10.00	**0.31**	0.15	-	**3.08**	0.42	**3.81**	m	**4.10**
25mm (nominal size); plugging to masonry in chases	0.30	10.00	**0.33**	0.16	-	**3.29**	0.42	**4.04**	m	**4.34**
20mm (nominal size); fixing to timber	0.28	10.00	**0.31**	0.13	-	**2.67**	0.21	**3.19**	m	**3.43**
25mm (nominal size); fixing to timber	0.30	10.00	**0.33**	0.15	-	**3.08**	0.21	**3.62**	m	**3.89**
Steel welded heavy gauge conduits and fittings, screwed joints; spacer bar saddles at 1000mm centres; enamelled black inside and outside by manufacturer										
Conduits; straight										
20mm diameter; plugging to masonry; to surfaces	0.75	10.00	**0.83**	0.33	-	**6.78**	0.25	**7.86**	m	**8.44**
extra; small circular terminal boxes	0.75	2.50	**0.77**	0.13	-	**2.67**	0.27	**3.71**	nr	**3.99**
extra; small circular angle boxes	0.85	2.50	**0.87**	0.16	-	**3.29**	0.27	**4.43**	nr	**4.76**
extra; small circular three way boxes	0.75	2.50	**0.77**	0.20	-	**4.11**	0.27	**5.15**	nr	**5.53**
extra; small circular through way boxes	0.80	2.50	**0.82**	0.16	-	**3.29**	0.27	**4.38**	nr	**4.70**
25mm diameter; plugging to masonry; to surfaces	0.98	10.00	**1.08**	0.42	-	**8.63**	0.25	**9.95**	m	**10.70**
extra; small circular terminal boxes	0.88	2.50	**0.90**	0.13	-	**2.67**	0.27	**3.84**	nr	**4.13**
extra; small circular angle boxes	0.99	2.50	**1.01**	0.16	-	**3.29**	0.27	**4.57**	nr	**4.91**
extra; small circular three way boxes	0.88	2.50	**0.90**	0.20	-	**4.11**	0.27	**5.28**	nr	**5.68**
extra; small circular through way boxes	0.94	2.50	**0.96**	0.16	-	**3.29**	0.27	**4.52**	nr	**4.86**
32mm diameter; plugging to masonry; to surfaces	2.04	10.00	**2.25**	0.50	-	**10.27**	0.25	**12.77**	m	**13.73**
Steel welded heavy gauge conduits and fittings, screwed joints; spacer bar saddles at 1000mm centres; galvanized inside and outside by manufacturer										
Conduits; straight										
20mm diameter; plugging to masonry; to surfaces	1.53	10.00	**1.69**	0.33	-	**6.78**	0.25	**8.72**	m	**9.37**
extra; small circular terminal boxes	0.89	2.50	**0.91**	0.13	-	**2.67**	0.27	**3.85**	nr	**4.14**
extra; small circular angle boxes	1.01	2.50	**1.04**	0.16	-	**3.29**	0.27	**4.59**	nr	**4.94**
extra; small circular three way boxes	0.89	2.50	**0.91**	0.20	-	**4.11**	0.27	**5.29**	nr	**5.69**
extra; small circular through way boxes	0.95	2.50	**0.97**	0.16	-	**3.29**	0.27	**4.53**	nr	**4.87**

Labour hourly rates: (except Specialists) Craft Operatives £20.54 Labourer £10.71 Rates are national average prices. Refer to REGIONAL VARIATIONS for indicative levels of overall pricing in regions	MATERIALS			LABOUR				RATES		
	Del to Site £	Waste %	Material Cost £	Craft Optve Hrs	Lab Hrs	Labour Cost £	Sunds £	Nett Rate £	Unit	Gross Rate (+7.5%) £
V90: GENERAL LIGHTING AND POWER - COMMERCIAL Cont.										
Steel welded heavy gauge conduits and fittings, screwed joints; spacer bar saddles at 1000mm centres; galvanized inside and outside by manufacturer Cont.										
Conduits; straight Cont.										
25mm diameter; plugging to masonry; to surfaces..	2.03	10.00	**2.23**	0.42	-	**8.63**	0.25	**11.11**	m	**11.94**
extra; small circular terminal boxes............	1.16	2.50	**1.19**	0.13	-	**2.67**	0.27	**4.13**	nr	**4.44**
extra; small circular angle boxes...............	1.31	2.50	**1.34**	0.16	-	**3.29**	0.27	**4.90**	nr	**5.27**
extra; small circular three way boxes...........	1.16	2.50	**1.19**	0.20	-	**4.11**	0.27	**5.57**	nr	**5.98**
extra; small circular through way boxes.........	1.24	2.50	**1.27**	0.16	-	**3.29**	0.27	**4.83**	nr	**5.19**
32mm diameter; plugging to masonry; to surfaces..	2.77	10.00	**3.05**	0.50	-	**10.27**	0.25	**13.57**	m	**14.58**
galvanized standard cable tray and fittings										
Light duty; straight										
50mm tray	1.00	5.00	**1.05**	0.25	-	**5.14**	0.09	**6.28**	m	**6.75**
extra; flat bend	6.57	-	**6.57**	0.20	-	**4.11**	0.25	**10.93**	nr	**11.75**
extra; tee	8.52	-	**8.52**	0.25	-	**5.14**	0.25	**13.91**	nr	**14.95**
extra; cross piece	14.38	-	**14.38**	0.40	-	**8.22**	0.50	**23.10**	nr	**24.83**
extra; riser bend	10.25	-	**10.25**	0.20	-	**4.11**	0.25	**14.61**	nr	**15.70**
100mm tray	1.62	5.00	**1.70**	0.33	-	**6.78**	0.09	**8.57**	m	**9.21**
extra; flat bend	7.09	-	**7.09**	0.23	-	**4.72**	0.25	**12.07**	nr	**12.97**
extra; tee	9.37	-	**9.37**	0.30	-	**6.16**	0.25	**15.78**	nr	**16.97**
extra; cross piece	15.32	-	**15.32**	0.41	-	**8.42**	0.50	**24.25**	nr	**26.06**
extra; riser bend	11.82	-	**11.82**	0.23	-	**4.72**	0.25	**16.79**	nr	**18.05**
extra; reducer 100 to 50mm......................	8.99	-	**8.99**	0.23	-	**4.72**	0.25	**13.96**	nr	**15.01**
150mm tray	2.04	5.00	**2.15**	0.42	-	**8.63**	0.09	**10.86**	m	**11.68**
extra; flat bend	8.28	-	**8.28**	0.25	-	**5.14**	0.25	**13.67**	nr	**14.69**
extra; tee	11.37	-	**11.37**	0.33	-	**6.78**	0.25	**18.40**	nr	**19.78**
extra; cross piece	17.99	-	**17.99**	0.42	-	**8.63**	0.50	**27.12**	nr	**29.15**
extra; riser bend	15.95	-	**15.95**	0.25	-	**5.14**	0.25	**21.34**	nr	**22.94**
extra; reducer 150 to 100mm.....................	11.84	-	**11.84**	0.25	-	**5.14**	0.25	**17.23**	nr	**18.52**
300mm tray	4.00	5.00	**4.20**	0.50	-	**10.27**	0.09	**14.56**	m	**15.65**
extra; flat bend	15.27	-	**15.27**	0.33	-	**6.78**	0.25	**22.30**	nr	**23.97**
extra; tee	20.93	-	**20.93**	0.50	-	**10.27**	0.25	**31.45**	nr	**33.81**
extra; cross piece	32.16	-	**32.16**	0.58	-	**11.91**	0.50	**44.58**	nr	**47.92**
extra; riser bend	28.09	-	**28.09**	0.33	-	**6.78**	0.25	**35.11**	nr	**37.75**
extra; reducer 300 to 150mm.....................	21.54	-	**21.54**	0.33	-	**6.78**	0.25	**28.57**	nr	**30.71**
450mm tray	8.07	5.00	**8.48**	0.58	-	**11.91**	0.09	**20.48**	m	**22.01**
extra; flat bend	24.45	-	**24.45**	0.50	-	**10.27**	0.25	**34.97**	nr	**37.59**
extra; tee	34.14	-	**34.14**	0.75	-	**15.41**	0.25	**49.80**	nr	**53.53**
extra; cross piece	53.08	-	**53.08**	1.00	-	**20.54**	0.50	**74.12**	nr	**79.68**
extra; riser bend	39.78	-	**39.78**	0.50	-	**10.27**	0.25	**50.30**	nr	**54.07**
extra; reducer 450 to 300mm.....................	35.75	-	**35.75**	0.50	-	**10.27**	0.25	**46.27**	nr	**49.74**
Medium duty; straight										
75mm tray	2.04	5.00	**2.15**	0.26	-	**5.34**	0.09	**7.58**	m	**8.14**
extra; flat bend	24.74	-	**24.74**	0.21	-	**4.31**	0.25	**29.30**	nr	**31.50**
extra; tee	32.58	-	**32.58**	0.26	-	**5.34**	0.25	**38.17**	nr	**41.04**
extra; cross piece	51.70	-	**51.70**	0.42	-	**8.63**	0.50	**60.83**	nr	**65.39**
extra; riser bend	13.44	-	**13.44**	0.21	-	**4.31**	0.25	**18.00**	nr	**19.35**
100mm tray	2.83	5.00	**2.97**	0.34	-	**6.98**	0.09	**10.05**	m	**10.80**
extra; flat bend	24.74	-	**24.74**	0.24	-	**4.93**	0.25	**29.92**	nr	**32.16**
extra; tee	32.58	-	**32.58**	0.31	-	**6.37**	0.25	**39.20**	nr	**42.14**
extra; cross piece	50.29	-	**50.29**	0.43	-	**8.83**	0.50	**59.62**	nr	**64.10**
extra; riser bend	12.30	-	**12.30**	0.24	-	**4.93**	0.25	**17.48**	nr	**18.79**
extra; reducer 100 to 75mm......................	11.41	-	**11.41**	0.24	-	**4.93**	0.25	**16.59**	nr	**17.84**
150mm tray	3.33	5.00	**3.50**	0.44	-	**9.04**	0.09	**12.63**	m	**13.57**
extra; flat bend	27.88	-	**27.88**	0.26	-	**5.34**	0.25	**33.47**	nr	**35.98**
extra; tee	37.06	-	**37.06**	0.34	-	**6.98**	0.25	**44.29**	nr	**47.62**
extra; cross piece	55.77	-	**55.77**	0.44	-	**9.04**	0.50	**65.31**	nr	**70.21**
extra; riser bend	16.13	-	**16.13**	0.26	-	**5.34**	0.25	**21.72**	nr	**23.35**
extra; reducer 150 to 100mm.....................	9.91	-	**9.91**	0.26	-	**5.34**	0.25	**15.50**	nr	**16.66**
300mm tray	4.50	5.00	**4.73**	0.52	-	**10.68**	0.09	**15.50**	m	**16.66**
extra; flat bend	39.66	-	**39.66**	0.34	-	**6.98**	0.25	**46.90**	nr	**50.42**
extra; tee	49.18	-	**49.18**	0.52	-	**10.68**	0.25	**60.11**	nr	**64.62**
extra; cross piece	75.71	-	**75.71**	0.61	-	**12.53**	0.50	**88.74**	nr	**95.39**
extra; riser bend	26.59	-	**26.59**	0.34	-	**6.98**	0.25	**33.82**	nr	**36.36**
extra; reducer 300 to 150mm.....................	21.76	-	**21.76**	0.34	-	**6.98**	0.25	**28.99**	nr	**31.16**
450mm tray	6.16	5.00	**6.47**	0.61	-	**12.53**	0.09	**19.09**	m	**20.52**
extra; flat bend	52.48	-	**52.48**	0.52	-	**10.68**	0.25	**63.41**	nr	**68.16**
extra; tee	68.75	-	**68.75**	0.78	-	**16.02**	0.25	**85.02**	nr	**91.40**
extra; cross piece	103.04	-	**103.04**	1.05	-	**21.57**	0.50	**125.11**	nr	**134.49**
extra; riser bend	40.11	-	**40.11**	0.52	-	**10.68**	0.25	**51.04**	nr	**54.87**
extra; reducer 450 to 300mm.....................	35.89	-	**35.89**	0.52	-	**10.68**	0.25	**46.82**	nr	**50.33**
Heavy duty; straight										
75mm tray	3.04	5.00	**3.19**	0.27	-	**5.55**	0.09	**8.83**	m	**9.49**
extra; flat bend	36.36	-	**36.36**	0.22	-	**4.52**	0.25	**41.13**	nr	**44.21**
extra; tee	44.70	-	**44.70**	0.27	-	**5.55**	0.25	**50.49**	nr	**54.28**
extra; cross piece	72.21	-	**72.21**	0.44	-	**9.04**	0.50	**81.74**	nr	**87.87**
extra; riser bend	24.99	-	**24.99**	0.22	-	**4.52**	0.25	**29.76**	nr	**31.99**
100mm tray	3.09	5.00	**3.24**	0.36	-	**7.39**	0.09	**10.73**	m	**11.53**
extra; flat bend	36.84	-	**36.84**	0.25	-	**5.14**	0.25	**42.23**	nr	**45.39**
extra; tee	45.08	-	**45.08**	0.33	-	**6.78**	0.25	**52.11**	nr	**56.02**
extra; cross piece	72.78	-	**72.78**	0.45	-	**9.24**	0.50	**82.52**	nr	**88.71**
extra; riser bend	25.61	-	**25.61**	0.25	-	**5.14**	0.25	**30.99**	nr	**33.32**
extra; reducer 100 to 75mm......................	16.66	-	**16.66**	0.25	-	**5.14**	0.25	**22.05**	nr	**23.70**
150mm tray	3.42	5.00	**3.59**	0.46	-	**9.45**	0.09	**13.13**	m	**14.11**
extra; flat bend	39.98	-	**39.98**	0.27	-	**5.55**	0.25	**45.77**	nr	**49.21**
extra; tee	49.56	-	**49.56**	0.36	-	**7.39**	0.25	**57.20**	nr	**61.49**
extra; cross piece	78.83	-	**78.83**	0.46	-	**9.45**	0.50	**88.77**	nr	**95.43**
extra; riser bend	27.81	-	**27.81**	0.27	-	**5.55**	0.25	**33.61**	nr	**36.13**
extra; reducer 150 to 100mm.....................	16.72	-	**16.72**	0.27	-	**5.55**	0.25	**22.51**	nr	**24.20**

Labour hourly rates: (except Specialists) Craft Operatives £20.54 Labourer £10.71 Rates are national average prices. Refer to REGIONAL VARIATIONS for indicative levels of overall pricing in regions	MATERIALS			LABOUR				RATES		
	Del to Site £	Waste %	Material Cost £	Craft Optve Hrs	Lab Hrs	Labour Cost £	Sunds £	Nett Rate £	Unit	Gross Rate (+7.5%) £
V90: GENERAL LIGHTING AND POWER - COMMERCIAL Cont.										
galvanized standard cable tray and fittings Cont.										
Heavy duty; straight Cont.										
300mm tray	4.98	5.00	**5.23**	0.55	-	**11.30**	0.09	**16.61**	m	**17.86**
extra; flat bend	45.89	-	**45.89**	0.36	-	**7.39**	0.25	**53.53**	nr	**57.55**
extra; tee	59.88	-	**59.88**	0.55	-	**11.30**	0.25	**71.43**	nr	**76.79**
extra; cross piece	94.78	-	**94.78**	0.64	-	**13.15**	0.50	**108.42**	nr	**116.56**
extra; riser bend	30.55	-	**30.55**	0.36	-	**7.39**	0.25	**38.20**	nr	**41.06**
extra; reducer 300 to 150mm	27.42	-	**27.42**	0.36	-	**7.39**	0.25	**35.06**	nr	**37.69**
450mm tray	8.61	5.00	**9.04**	0.64	-	**13.15**	0.09	**22.27**	m	**23.95**
extra; flat bend	77.78	-	**77.78**	0.55	-	**11.30**	0.25	**89.33**	nr	**96.03**
extra; tee	91.16	-	**91.16**	0.83	-	**17.05**	0.50	**108.71**	nr	**116.86**
extra; cross piece	146.34	-	**146.34**	1.10	-	**22.59**	0.25	**169.18**	nr	**181.87**
extra; riser bend	53.67	-	**53.67**	0.55	-	**11.30**	0.25	**65.22**	nr	**70.11**
extra; reducer 450 to 300mm	42.87	-	**42.87**	0.55	-	**11.30**	0.25	**54.41**	nr	**58.49**
Supports for cable tray										
Cantilever arms; mild steel; hot dip galvanized; fixing to masonry										
65mm wide	4.14	2.50	**4.25**	0.33	-	**6.78**	0.25	**11.27**	nr	**12.12**
90mm wide	4.14	2.50	**4.25**	0.33	-	**6.78**	0.25	**11.27**	nr	**12.12**
120mm wide	4.14	2.50	**4.25**	0.33	-	**6.78**	0.25	**11.27**	nr	**12.12**
Stand-off bracket; mild steel; hot dip galvanized; fixing to masonry										
75mm wide	4.14	2.50	**4.25**	0.33	-	**6.78**	0.25	**11.27**	nr	**12.12**
100mm wide	5.33	2.50	**5.47**	0.33	-	**6.78**	0.25	**12.50**	nr	**13.43**
150mm wide	5.33	2.50	**5.47**	0.33	-	**6.78**	0.25	**12.50**	nr	**13.43**
300mm wide	11.76	2.50	**12.05**	0.42	-	**8.63**	0.25	**20.93**	nr	**22.50**
450mm wide	16.10	2.50	**16.50**	0.42	-	**8.63**	0.25	**25.38**	nr	**27.28**
PVC insulated cables; single core; reference 6491X; stranded copper conductors; to BS 6004										
Drawn into conduits or ducts or laid or drawn into trunking										
1.5mm2	0.00	15.00	**0.00**	0.03	-	**0.62**	-	**0.62**	m	**0.66**
2.5mm2	0.00	15.00	**0.00**	0.03	-	**0.62**	-	**0.62**	m	**0.66**
4.0mm2	0.00	15.00	**0.00**	0.04	-	**0.82**	-	**0.82**	m	**0.89**
6.0mm2	0.00	15.00	**0.00**	0.05	-	**1.03**	-	**1.03**	m	**1.11**
10.0mm2	0.00	15.00	**0.00**	0.05	-	**1.03**	-	**1.03**	m	**1.11**
16.0mm2	0.01	15.00	**0.01**	0.08	-	**1.54**	-	**1.55**	m	**1.66**
PVC double insulated cables; single core; reference 6181Y; stranded copper conductors; to BS 6004										
Fixed to timber with clips										
6.0mm2	0.54	15.00	**0.62**	0.07	-	**1.44**	-	**2.06**	m	**2.21**
10.0mm2	0.59	15.00	**0.68**	0.08	-	**1.64**	-	**2.32**	m	**2.50**
16.0mm2	0.78	15.00	**0.90**	0.10	-	**2.05**	-	**2.95**	m	**3.17**
25.0mm2	0.99	15.00	**1.14**	0.15	-	**3.08**	-	**4.22**	m	**4.54**
35.0mm2	2.46	15.00	**2.83**	0.20	-	**4.11**	-	**6.94**	m	**7.46**
PVC insulated and PVC sheathed cables; multicore; copper conductors and bare earth continuity conductor 6242Y; to BS 6004										
Drawn into conduits or ducts or laid or drawn into trunking										
1.0mm2; twin with bare earth	0.00	15.00	**0.00**	0.05	-	**1.03**	-	**1.03**	m	**1.11**
1.5mm2; twin with bare earth	0.00	15.00	**0.00**	0.05	-	**1.03**	-	**1.03**	m	**1.11**
2.5mm2; twin with bare earth	0.00	15.00	**0.00**	0.05	-	**1.03**	-	**1.03**	m	**1.11**
4.0mm2; twin with bare earth	0.00	15.00	**0.01**	0.06	-	**1.23**	-	**1.24**	m	**1.33**
6.0mm2; twin with bare earth	0.00	15.00	**0.01**	0.07	-	**1.44**	-	**1.44**	m	**1.55**
Fixed to timber with clips										
1.0mm2; twin with bare earth	0.00	15.00	**0.00**	0.07	-	**1.44**	0.06	**1.50**	m	**1.61**
1.5mm2; twin with bare earth	0.00	15.00	**0.00**	0.07	-	**1.44**	0.06	**1.50**	m	**1.61**
2.5mm2; twin with bare earth	0.00	15.00	**0.00**	0.07	-	**1.44**	0.07	**1.51**	m	**1.62**
4.0mm2; twin with bare earth	0.00	15.00	**0.01**	0.08	-	**1.64**	0.08	**1.73**	m	**1.86**
6.0mm2; twin with bare earth	0.00	15.00	**0.01**	0.09	-	**1.85**	0.08	**1.93**	m	**2.08**
PVC insulated SWA armoured and PVC sheathed cables; multicore; copper conductors to BS 6346										
Laid and laced on cable tray										
16mm2; three core	2.95	5.00	**3.10**	0.17	-	**3.49**	0.11	**6.70**	m	**7.20**
Fixed to masonry with screwed clips at average 300mm centres; plugging										
16mm2; three core	2.95	5.00	**3.10**	0.25	-	**5.14**	0.94	**9.17**	m	**9.86**
Mineral insulated copper sheathed cables; PVC outer sheath; copper conductors; 600V light duty to BS EN 60702-1										
Fixed to masonry with screwed clips at average 300mm centres; plugging										
1.5mm2; two core	2.17	10.00	**2.39**	0.17	-	**3.49**	0.94	**6.82**	m	**7.33**
extra; termination including gland, seal and shroud	2.77	2.50	**2.84**	0.42	-	**8.63**	-	**11.47**	nr	**12.33**
2.5mm2; two core	3.18	10.00	**3.50**	0.17	-	**3.49**	0.94	**7.93**	m	**8.52**

Labour hourly rates: (except Specialists) Craft Operatives £20.54 Labourer £10.71 Rates are national average prices. Refer to REGIONAL VARIATIONS for indicative levels of overall pricing in regions	MATERIALS			LABOUR				RATES		
	Del to Site £	Waste %	Material Cost £	Craft Optve Hrs	Lab Hrs	Labour Cost £	Sunds £	Nett Rate £	Unit	Gross Rate (+7.5%) £
V90: GENERAL LIGHTING AND POWER - COMMERCIAL Cont.										
Mineral insulated copper sheathed cables; PVC outer sheath; copper conductors; 600V light duty to BS EN 60702-1 Cont.										
Fixed to masonry with screwed clips at average 300mm centres; plugging Cont.										
extra; termination including gland, seal and shroud	2.77	2.50	**2.84**	0.42	-	**8.63**	-	**11.47**	nr	**12.33**
4.0mm2; two core	4.03	10.00	**4.43**	0.20	-	**4.11**	0.94	**9.48**	m	**10.19**
extra; termination including gland, seal and shroud	2.77	2.50	**2.84**	0.50	-	**10.27**	-	**13.11**	nr	**14.09**
HV switchgear										
Switch fuse 500V, metal cased, fixed to masonry with screws; plugging										
32 amp; single pole and neutral; short circuit rating 6.4Ka	74.76	2.50	**76.63**	1.60	-	**32.86**	0.50	**110.00**	nr	**118.25**
Consumer control units, 250V, metal cased, fitted with 100 amp D.P. isolator and MCB's , fixed to masonry with screws; plugging										
eight way, single pole and neutral	85.83	2.50	**87.98**	1.60	-	**32.86**	0.50	**121.34**	nr	**130.44**
Distribution boards 500V, metal cased, MCB pattern, fitted with 100 amp D.P. isolator, fixed to masonry with screws; plugging										
nine way, single pole and neutral, 20 amp	106.68	2.50	**109.35**	1.80	-	**36.97**	0.50	**146.82**	nr	**157.83**
twelve way, triple pole and neutral, 32 amp	473.98	2.50	**485.83**	8.00	-	**164.32**	0.50	**650.65**	nr	**699.45**
Luminaires										
Note										
for lighting fittings and lamps see General Lighting and Power - Domestic Section										
Accessories; white plastics, with boxes, fixing to masonry with screws; plugging										
Flush plate switches										
5 amp; one gang; one way; single pole	1.35	2.50	**1.38**	0.42	-	**8.63**	0.25	**10.26**	nr	**11.03**
5 amp; two gang; one way; single pole	1.85	2.50	**1.90**	0.43	-	**8.83**	0.25	**10.98**	nr	**11.80**
Surface plate switches										
5 amp; one gang; one way; single pole	1.45	2.50	**1.49**	0.42	-	**8.63**	0.25	**10.36**	nr	**11.14**
5 amp; two gang; one way; single pole	1.95	2.50	**2.00**	0.43	-	**8.83**	0.25	**11.08**	nr	**11.91**
Ceiling switches										
5 amp; one gang; one way; single pole	1.85	2.50	**1.90**	0.43	-	**8.83**	0.25	**10.98**	nr	**11.80**
Flush switched socket outlets										
13 amp; one gang	2.60	2.50	**2.67**	0.43	-	**8.83**	0.25	**11.75**	nr	**12.63**
13 amp; two gang	3.25	2.50	**3.33**	0.43	-	**8.83**	0.25	**12.41**	nr	**13.34**
Surface switched socket outlets										
13 amp; one gang	2.70	2.50	**2.77**	0.43	-	**8.83**	0.25	**11.85**	nr	**12.74**
13 amp; two gang	3.45	2.50	**3.54**	0.43	-	**8.83**	0.25	**12.62**	nr	**13.56**
Flush switched socket outlets with RCCB (residual current device) protected at 30 mAmp										
13 amp; one gang	32.50	2.50	**33.31**	0.43	-	**8.83**	0.25	**42.39**	nr	**45.57**
Flush fused connection units										
13 amp; one gang; switched; flexible outlet	3.85	2.50	**3.95**	0.50	-	**10.27**	0.25	**14.47**	nr	**15.55**
13 amp; one gang; switched; flexible outlet; pilot lamp	4.25	2.50	**4.36**	0.50	-	**10.27**	0.25	**14.88**	nr	**15.99**
Accessories, metalclad, with boxes, fixing to masonry with screws; plugging										
Surface plate switches										
5 amp; one gang; two way; single pole	2.85	2.50	**2.92**	0.42	-	**8.63**	0.25	**11.80**	nr	**12.68**
5 amp; two gang; two way; single pole	3.55	2.50	**3.64**	0.43	-	**8.83**	0.25	**12.72**	nr	**13.68**
Surface switched socket outlets										
13 amp; one gang	4.00	2.50	**4.10**	0.43	-	**8.83**	0.25	**13.18**	nr	**14.17**
13 amp; two gang	5.75	2.50	**5.89**	0.43	-	**8.83**	0.25	**14.98**	nr	**16.10**
Weatherproof accessories, with boxes, fixing to masonry with screws; plugging										
Surface plate switches										
5 amp; one gang; two way; single pole	6.04	2.50	**6.19**	0.43	-	**8.83**	0.25	**15.27**	nr	**16.42**

Labour hourly rates: (except Specialists) Craft Operatives £20.54 Labourer £10.71 Rates are national average prices. Refer to REGIONAL VARIATIONS for indicative levels of overall pricing in regions	MATERIALS			LABOUR				RATES		
	Del to Site £	Waste %	Material Cost £	Craft Optve Hrs	Lab Hrs	Labour Cost £	Sunds £	Nett Rate £	Unit	Gross Rate (+7.5%) £
V90: GENERAL LIGHTING AND POWER - DOMESTIC										
Electric wiring										
Note the following approximate prices of various types of installations are dependant on the number and disposition of points; lamps and fittings together with cutting and making good are excluded										
Electric wiring in new building										
Insulated Consumer unit, fixed to masonry withscrews; plugging										
2 usable ways 63a DP main Switch	16.52	2.50	**16.94**	1.00	-	**20.54**	0.25	**37.73**	nr	**40.56**
4 usable ways 100a DP main Switch	18.67	2.50	**19.13**	1.66	-	**34.10**	0.25	**53.48**	nr	**57.49**
6 usable ways 100a DP main Switch	21.75	2.50	**22.29**	2.00	-	**41.08**	0.25	**63.62**	nr	**68.40**
9 usable ways 100a DP main Switch	24.10	2.50	**24.70**	2.30	-	**47.24**	0.25	**72.19**	nr	**77.61**
13 usable ways 100a DP main Switch	30.59	2.50	**31.35**	2.75	-	**56.49**	0.25	**88.09**	nr	**94.69**
2 usable ways 63a 30ma RCD main Switch	53.68	2.50	**55.02**	1.16	-	**23.83**	0.25	**79.10**	nr	**85.03**
4 usable ways 80a 30ma RCD main Switch	59.81	2.50	**61.30**	1.25	-	**25.68**	0.25	**87.23**	nr	**93.77**
6 usable ways 80a 30ma RCD main Switch	66.01	2.50	**67.66**	1.50	-	**30.81**	0.25	**98.72**	nr	**106.12**
9 usable ways 80a 30ma RCD main Switch	71.76	2.50	**73.55**	1.66	-	**34.10**	0.25	**107.90**	nr	**115.99**
Miniature Circuit Breakers fitted in consumer unit; Bsen 60898										
6a Single Pole	5.86	2.50	**6.01**	0.13	-	**2.67**	-	**8.68**	nr	**9.33**
10a Single Pole	5.59	2.50	**5.73**	0.13	-	**2.67**	-	**8.40**	nr	**9.03**
16a Single Pole	5.59	2.50	**5.73**	0.13	-	**2.67**	-	**8.40**	nr	**9.03**
20a Single Pole	5.59	2.50	**5.73**	0.13	-	**2.67**	-	**8.40**	nr	**9.03**
32a Single Pole	5.59	2.50	**5.73**	0.13	-	**2.67**	-	**8.40**	nr	**9.03**
40a Single Pole	5.86	2.50	**6.01**	0.13	-	**2.67**	-	**8.68**	nr	**9.33**
Domestic Metal consumer units and accessories										
2 usable ways 63a DP main Switch	19.83	2.50	**20.33**	1.00	-	**20.54**	0.25	**41.12**	nr	**44.20**
4 usable ways 100a DP main Switch	21.27	2.50	**21.80**	1.66	-	**34.10**	0.25	**56.15**	nr	**60.36**
6 usable ways 100a DP main Switch	24.27	2.50	**24.88**	2.00	-	**41.08**	0.25	**66.21**	nr	**71.17**
9 usable ways 100a DP main Switch	29.24	2.50	**29.97**	2.30	-	**47.24**	0.25	**77.47**	nr	**83.28**
13 usable ways 100a DP main Switch	34.77	2.50	**35.64**	2.75	-	**56.49**	0.25	**92.37**	nr	**99.30**
Miniature Circuit Breakers fitted in consumer unit; Bsen 60898										
6a Single Pole	5.86	2.50	**6.01**	0.13	-	**2.67**	-	**8.68**	nr	**9.33**
10a Single Pole	5.59	2.50	**5.73**	0.13	-	**2.67**	-	**8.40**	nr	**9.03**
16a Single Pole	5.59	2.50	**5.73**	0.13	-	**2.67**	-	**8.40**	nr	**9.03**
20a Single Pole	5.59	2.50	**5.73**	0.13	-	**2.67**	-	**8.40**	nr	**9.03**
32a Single Pole	5.59	2.50	**5.73**	0.13	-	**2.67**	-	**8.40**	nr	**9.03**
40a Single Pole	5.86	2.50	**6.01**	0.13	-	**2.67**	-	**8.68**	nr	**9.33**
Installation with PVC insulated and sheathed cables										
lighting points	6.44	7.50	**6.90**	3.25	-	**66.76**	-	**73.65**	nr	**79.18**
socket outlets; 5A	6.42	7.50	**6.87**	4.40	-	**90.38**	-	**97.25**	nr	**104.54**
socket outlets; 13A ring main	3.76	7.50	**4.02**	3.70	-	**76.00**	-	**80.02**	nr	**86.02**
socket outlets; 13A radial circuit	3.89	7.50	**4.16**	3.85	-	**79.08**	-	**83.23**	nr	**89.48**
cooker points; 45A	12.41	7.50	**13.30**	6.30	-	**129.40**	-	**142.70**	nr	**153.40**
immersion heater points	6.10	7.50	**6.53**	4.60	-	**94.48**	-	**101.02**	nr	**108.59**
shaver sockets (transformer)	15.03	7.50	**16.14**	1.65	-	**33.89**	-	**50.03**	nr	**53.78**
Installation with mineral insulated copper sheathed cables										
lighting points	15.94	7.50	**17.11**	6.20	-	**127.35**	-	**144.46**	nr	**155.29**
socket outlets; 5A	29.54	7.50	**31.71**	4.85	-	**99.62**	-	**131.33**	nr	**141.18**
socket outlets; 13A ring main	20.80	7.50	**22.34**	4.15	-	**85.24**	-	**107.58**	nr	**115.65**
socket outlets; 13A radial circuit	26.58	7.50	**28.55**	5.55	-	**114.00**	-	**142.55**	nr	**153.24**
cooker points; 45A	73.82	7.50	**79.31**	9.20	-	**188.97**	-	**268.28**	nr	**288.40**
immersion heater points	34.56	7.50	**37.13**	6.60	-	**135.56**	-	**172.70**	nr	**185.65**
shaver sockets (transformer)	43.06	7.50	**46.27**	3.00	-	**61.62**	-	**107.89**	nr	**115.98**
Installation with black enamel heavy gauge conduit with PVC cables										
lighting points	28.49	7.50	**30.39**	4.65	-	**95.51**	-	**125.90**	nr	**135.35**
socket outlets; 5A	25.36	7.50	**27.03**	4.70	-	**96.54**	-	**123.57**	nr	**132.84**
socket outlets; 13A ring main	19.82	7.50	**21.14**	4.25	-	**87.30**	-	**108.44**	nr	**116.57**
socket outlets; 13A radial circuit	22.71	7.50	**24.18**	6.00	-	**123.24**	-	**147.42**	nr	**158.48**
cooker points; 45A	38.36	7.50	**40.96**	9.65	-	**198.21**	-	**239.17**	nr	**257.11**
immersion heater points	25.62	7.50	**27.27**	7.15	-	**146.86**	-	**174.13**	nr	**187.19**
shaver sockets (transformer)	40.66	7.50	**43.11**	2.30	-	**47.24**	-	**90.36**	nr	**97.13**
Installation with black enamel heavy gauge conduit with coaxial cable										
T.V. sockets	32.98	7.50	**34.97**	5.75	-	**118.11**	-	**153.08**	nr	**164.56**
Installation with black enamel heavy gauge conduit with draw wire										
telephone points	34.02	7.50	**36.08**	5.45	-	**111.94**	-	**148.03**	nr	**159.13**
Electric wiring in extension to existing building										
Installation with PVC insulated and sheathed cables										
lighting points	6.44	7.50	**6.90**	3.25	-	**66.76**	-	**73.65**	nr	**79.18**
socket outlets; 5A	6.42	7.50	**6.87**	4.40	-	**90.38**	-	**97.25**	nr	**104.54**
socket outlets; 13A ring main	3.76	7.50	**4.02**	3.70	-	**76.00**	-	**80.02**	nr	**86.02**
socket outlets; 13A radial circuit	3.89	7.50	**4.16**	3.85	-	**79.08**	-	**83.23**	nr	**89.48**
cooker points; 45A	12.41	7.50	**13.30**	6.30	-	**129.40**	-	**142.70**	nr	**153.40**
immersion heater points	6.10	7.50	**6.53**	4.60	-	**94.48**	-	**101.02**	nr	**108.59**
shaver sockets (transformer)	15.03	7.50	**16.14**	1.65	-	**33.89**	-	**50.03**	nr	**53.78**

Labour hourly rates: (except Specialists) Craft Operatives £20.54 Labourer £10.71 Rates are national average prices. Refer to REGIONAL VARIATIONS for indicative levels of overall pricing in regions	MATERIALS			LABOUR				RATES		
	Del to Site £	Waste %	Material Cost £	Craft Optve Hrs	Lab Hrs	Labour Cost £	Sunds £	Nett Rate £	Unit	Gross Rate (+7.5%) £
V90: GENERAL LIGHTING AND POWER - DOMESTIC Cont.										
Electric wiring in extension to existing building Cont.										
Installation with mineral insulated copper sheathed cables										
lighting points	15.94	7.50	**17.11**	6.20	-	**127.35**	-	**144.46**	nr	**155.29**
socket outlets; 5A	29.54	7.50	**31.71**	4.85	-	**99.62**	-	**131.33**	nr	**141.18**
socket outlets; 13A ring main	20.80	7.50	**22.34**	4.15	-	**85.24**	-	**107.58**	nr	**115.65**
socket outlets; 13A radial circuit	26.58	7.50	**28.55**	5.55	-	**114.00**	-	**142.55**	nr	**153.24**
cooker points; 45A	73.82	7.50	**79.31**	9.20	-	**188.97**	-	**268.28**	nr	**288.40**
immersion heater points	34.56	7.50	**37.13**	6.60	-	**135.56**	-	**172.70**	nr	**185.65**
shaver sockets (transformer)	43.06	7.50	**46.27**	3.00	-	**61.62**	-	**107.89**	nr	**115.98**
Installation with black enamel heavy gauge conduit with PVC cables										
lighting points	28.49	7.50	**30.39**	4.65	-	**95.51**	-	**125.90**	nr	**135.35**
socket outlets; 5A	25.36	7.50	**27.03**	4.70	-	**96.54**	-	**123.57**	nr	**132.84**
socket outlets; 13A ring main	19.82	7.50	**21.14**	4.25	-	**87.30**	-	**108.44**	nr	**116.57**
socket outlets; 13A radial circuit	22.71	7.50	**24.18**	6.00	-	**123.24**	-	**147.42**	nr	**158.48**
cooker points; 45A	38.36	7.50	**40.96**	9.65	-	**198.21**	-	**239.17**	nr	**257.11**
immersion heater points	25.62	7.50	**27.27**	7.15	-	**146.86**	-	**174.13**	nr	**187.19**
shaver sockets (transformer)	40.66	7.50	**43.11**	2.30	-	**47.24**	-	**90.36**	nr	**97.13**
Installation with black enamel heavy gauge conduit with coaxial cable										
T.V. sockets	32.98	7.50	**34.97**	5.75	-	**118.11**	-	**153.08**	nr	**164.56**
Installation with black enamel heavy gauge conduit with draw wire										
telephone points	34.02	7.50	**36.08**	5.45	-	**111.94**	-	**148.03**	nr	**159.13**
Electric wiring in extending existing installation in existing building										
Installation with PVC insulated and sheathed cables										
lighting points	6.44	7.50	**6.90**	3.75	-	**77.03**	-	**83.92**	nr	**90.22**
socket outlets; 5A	6.42	7.50	**6.87**	5.05	-	**103.73**	-	**110.60**	nr	**118.90**
socket outlets; 13A ring main	3.76	7.50	**4.02**	4.25	-	**87.30**	-	**91.31**	nr	**98.16**
socket outlets; 13A radial circuit	3.89	7.50	**4.16**	4.45	-	**91.40**	-	**95.56**	nr	**102.73**
cooker points; 45A	12.41	7.50	**13.30**	7.25	-	**148.92**	-	**162.21**	nr	**174.38**
immersion heater points	6.10	7.50	**6.53**	5.30	-	**108.86**	-	**115.39**	nr	**124.05**
shaver sockets (transformer)	15.03	7.50	**16.14**	1.90	-	**39.03**	-	**55.16**	nr	**59.30**
Installation with mineral insulated copper sheathed cables										
lighting points	15.94	7.50	**17.11**	7.15	-	**146.86**	-	**163.97**	nr	**176.27**
socket outlets; 5A	29.54	7.50	**31.71**	5.60	-	**115.02**	-	**146.73**	nr	**157.74**
socket outlets; 13A ring main	20.80	7.50	**22.34**	4.80	-	**98.59**	-	**120.93**	nr	**130.00**
socket outlets; 13A radial circuit	26.58	7.50	**28.55**	6.40	-	**131.46**	-	**160.01**	nr	**172.01**
cooker points; 45A	73.82	7.50	**79.31**	10.60	-	**217.72**	-	**297.04**	nr	**319.31**
immersion heater points	34.56	7.50	**37.13**	7.60	-	**156.10**	-	**193.24**	nr	**207.73**
shaver sockets (transformer)	43.06	7.50	**46.27**	3.45	-	**70.86**	-	**117.13**	nr	**125.92**
Installation with black enamel heavy gauge conduit with PVC cables										
lighting points	28.49	7.50	**30.39**	5.35	-	**109.89**	-	**140.28**	nr	**150.80**
socket outlets; 5A	25.36	7.50	**27.03**	5.40	-	**110.92**	-	**137.95**	nr	**148.29**
socket outlets; 13A ring main	19.82	7.50	**21.14**	4.90	-	**100.65**	-	**121.79**	nr	**130.93**
socket outlets; 13A radial circuit	22.71	7.50	**24.18**	6.90	-	**141.73**	-	**165.91**	nr	**178.35**
cooker points; 45A	38.36	7.50	**40.96**	11.10	-	**227.99**	-	**268.96**	nr	**289.13**
immersion heater points	25.62	7.50	**27.27**	8.25	-	**169.46**	-	**196.73**	nr	**211.48**
shaver sockets (transformer)	40.66	7.50	**43.11**	2.65	-	**54.43**	-	**97.54**	nr	**104.86**
Installation with black enamel heavy gauge conduit with coaxial cable										
T.V. sockets	32.98	7.50	**34.97**	6.60	-	**135.56**	-	**170.54**	nr	**183.33**
Installation with black enamel heavy gauge conduit with draw wire										
telephone points	34.02	7.50	**36.08**	6.25	-	**128.38**	-	**164.46**	nr	**176.79**
V90: LAMP FITTINGS, FANS, HEATERS etc., ALL INSTALLATIONS										
Fluorescent lamp fittings, mains voltage operations; switch start; including lamps										
Batten type; single tube										
1200mm; 36W	9.60	2.50	**9.84**	0.75	-	**15.41**	0.50	**25.75**	nr	**27.68**
1500mm; 58W	11.00	2.50	**11.28**	1.00	-	**20.54**	0.50	**32.32**	nr	**34.74**
1800mm; 70W	17.45	2.50	**17.89**	1.00	-	**20.54**	0.50	**38.93**	nr	**41.85**
2400mm; 100W	24.00	2.50	**24.60**	1.25	-	**25.68**	0.50	**50.78**	nr	**54.58**
Batten type; twin tube										
1200mm; 36W	15.55	2.50	**15.94**	0.75	-	**15.41**	0.50	**31.84**	nr	**34.23**
1500mm; 58W	18.56	2.50	**19.02**	1.00	-	**20.54**	0.50	**40.06**	nr	**43.07**
1800mm; 70W	27.55	2.50	**28.24**	1.00	-	**20.54**	0.50	**49.28**	nr	**52.97**
2400mm; 100W	36.50	2.50	**37.41**	1.25	-	**25.68**	0.50	**63.59**	nr	**68.36**
Metal trough reflector; single tube										
1200mm; 36W	32.16	2.50	**32.96**	1.00	-	**20.54**	0.50	**54.00**	nr	**58.05**
1500mm; 58W	35.21	2.50	**36.09**	1.20	-	**24.65**	0.50	**61.24**	nr	**65.83**
1800mm; 70W	41.50	2.50	**42.54**	1.20	-	**24.65**	0.50	**67.69**	nr	**72.76**
2400mm; 100W	59.06	2.50	**60.54**	1.45	-	**29.78**	0.50	**90.82**	nr	**97.63**

Labour hourly rates: (except Specialists) Craft Operatives £20.54 Labourer £10.71 Rates are national average prices. Refer to REGIONAL VARIATIONS for indicative levels of overall pricing in regions	MATERIALS			LABOUR				RATES		
	Del to Site £	Waste %	Material Cost £	Craft Optve Hrs	Lab Hrs	Labour Cost £	Sunds £	Nett Rate £	Unit	Gross Rate (+7.5%) £
V90: LAMP FITTINGS, FANS, HEATERS etc., ALL INSTALLATIONS Cont.										
Fluorescent lamp fittings, mains voltage operations; switch start; including lamps Cont.										
Metal trough reflector; twin tube										
1200mm; 36W	49.73	2.50	**50.97**	1.00	-	**20.54**	0.50	**72.01**	nr	**77.41**
1500mm; 58W	57.17	2.50	**58.60**	1.20	-	**24.65**	0.50	**83.75**	nr	**90.03**
1800mm; 70W	63.16	2.50	**64.74**	1.20	-	**24.65**	0.50	**89.89**	nr	**96.63**
2400mm; 100W	85.92	2.50	**88.07**	1.45	-	**29.78**	0.50	**118.35**	nr	**127.23**
Plastics diffused type; single tube										
1200mm; 36W	33.18	2.50	**34.01**	1.00	-	**20.54**	0.50	**55.05**	nr	**59.18**
1500mm; 58W	37.16	2.50	**38.09**	1.20	-	**24.65**	0.50	**63.24**	nr	**67.98**
1800mm; 70W	45.47	2.50	**46.61**	1.20	-	**24.65**	0.50	**71.75**	nr	**77.14**
2400mm; 100W	60.79	2.50	**62.31**	1.45	-	**29.78**	0.50	**92.59**	nr	**99.54**
Plastics diffused type; twin tube										
1200mm; 36W	61.93	2.50	**63.48**	1.00	-	**20.54**	0.50	**84.52**	nr	**90.86**
1500mm; 58W	75.23	2.50	**77.11**	1.20	-	**24.65**	0.50	**102.26**	nr	**109.93**
1800mm; 70W	84.99	2.50	**87.11**	1.20	-	**24.65**	0.50	**112.26**	nr	**120.68**
2400mm; 100W	108.53	2.50	**111.24**	1.45	-	**29.78**	0.50	**141.53**	nr	**152.14**
Lighting fittings complete with tungsten lamps										
Bulkhead; alloy										
100W G.L.S. lamp	9.00	2.50	**9.23**	0.65	-	**13.35**	0.25	**22.83**	nr	**24.54**
Wall glass; alloy; corner bracket										
100W G.L.S. lamp	26.05	2.50	**26.70**	0.75	-	**15.41**	0.25	**42.36**	nr	**45.53**
Ceiling sphere										
152mm; 60W G.L.S. lamp	14.00	2.50	**14.35**	0.65	-	**13.35**	0.25	**27.95**	nr	**30.05**
203mm; 100W G.L.S. lamp	20.40	2.50	**20.91**	0.65	-	**13.35**	0.25	**34.51**	nr	**37.10**
Down light; semi recessed										
175mm; 150W PAR38 lamp	17.32	2.50	**17.75**	0.65	-	**13.35**	0.25	**31.35**	nr	**33.71**
Lighting track; single circuit; surface mounted										
1250mm; starter pack	17.60	2.50	**18.04**	1.20	-	**24.65**	0.50	**43.19**	nr	**46.43**
plug-in fitting; 100W G.L.S. lamp	13.50	2.50	**13.84**	0.20	-	**4.11**	0.25	**18.20**	nr	**19.56**
Fan heaters										
3kw unit fan heaters										
for commercial application	152.06	2.50	**155.86**	1.50	-	**30.81**	0.50	**187.17**	nr	**201.21**
Thermostatic switch units										
comprising ON/OFF switch and selector switch between winter and summer (fan only) operation	51.86	2.50	**53.16**	1.00	-	**20.54**	0.25	**73.95**	nr	**79.49**
Tubular heaters										
60W per 305mm loading										
600mm long	10.80	2.50	**11.07**	1.00	-	**20.54**	0.75	**32.36**	nr	**34.79**
910mm long	11.89	2.50	**12.19**	1.00	-	**20.54**	0.75	**33.48**	nr	**35.99**
1220mm long	13.66	2.50	**14.00**	1.25	-	**25.68**	1.00	**40.68**	nr	**43.73**
1520mm long	17.85	2.50	**18.30**	1.25	-	**25.68**	1.00	**44.97**	nr	**48.34**
1830mm long	18.24	2.50	**18.70**	1.50	-	**30.81**	1.50	**51.01**	nr	**54.83**
Two way mounting brackets										
pair	5.19	2.50	**5.32**	0.40	-	**8.22**	0.50	**14.04**	nr	**15.09**
Night storage heaters; installed complete in new building including wiring										
Plastics insulated and sheathed cabled										
1.7KW	221.38	2.50	**226.92**	6.50	-	**133.51**	-	**360.43**	nr	**387.46**
2.55KW	291.41	2.50	**298.70**	7.00	-	**143.78**	-	**442.48**	nr	**475.66**
3.4KW	331.67	2.50	**339.97**	7.50	-	**154.05**	-	**494.02**	nr	**531.07**
Mineral insulated copper sheathed cables										
1.7KW	266.57	2.50	**273.23**	10.00	-	**205.40**	-	**478.63**	nr	**514.53**
2.55KW	336.57	2.50	**344.98**	10.50	-	**215.67**	-	**560.65**	nr	**602.70**
3.4KW	376.81	2.50	**386.23**	11.00	-	**225.94**	-	**612.17**	nr	**658.08**
Bell equipment										
Transformers										
3-5-8V	14.47	2.50	**14.83**	0.40	-	**8.22**	0.24	**23.29**	nr	**25.03**
12V	18.52	2.50	**18.98**	0.40	-	**8.22**	0.24	**27.44**	nr	**29.50**
Bells for transformer operation										
chime 2-note	10.40	2.50	**10.66**	0.50	-	**10.27**	0.24	**21.17**	nr	**22.76**
76mm domestic type	5.73	2.50	**5.87**	0.50	-	**10.27**	0.24	**16.38**	nr	**17.61**
150mm round bell type	20.84	2.50	**21.36**	0.40	-	**8.22**	0.24	**29.82**	nr	**32.05**
Bell pushes										
domestic	1.45	2.50	**1.49**	0.25	-	**5.14**	-	**6.62**	nr	**7.12**
industrial	5.72	2.50	**5.86**	0.30	-	**6.16**	-	**12.03**	nr	**12.93**

Labour hourly rates: (except Specialists) Craft Operatives £20.54 Labourer £10.71 Rates are national average prices. Refer to REGIONAL VARIATIONS for indicative levels of overall pricing in regions	MATERIALS			LABOUR				RATES		
	Del to Site £	Waste %	Material Cost £	Craft Optve Hrs	Lab Hrs	Labour Cost £	Sunds £	Nett Rate £	Unit	Gross Rate (+7.5%) £
V90: LAMP FITTINGS, FANS, HEATERS etc., ALL INSTALLATIONS Cont.										
Immersion heaters and thermostats										
Note: the following prices exclude Plumber's work										
2 or 3KW immersion heaters; without flanges										
305mm; non-withdrawable elements	22.91	2.50	**23.48**	0.65	-	**13.35**	4.65	**41.48**	nr	**44.60**
457mm; non-withdrawable elements	23.14	2.50	**23.72**	0.80	-	**16.43**	4.65	**44.80**	nr	**48.16**
686mm; non-withdrawable elements	23.25	2.50	**23.83**	0.95	-	**19.51**	4.65	**47.99**	nr	**51.59**
Thermostats										
immersion heater	10.04	2.50	**10.29**	0.50	-	**10.27**	0.25	**20.81**	nr	**22.37**
15 amp a.c. for air heating; without switch	15.16	2.50	**15.54**	0.40	-	**8.22**	0.25	**24.01**	nr	**25.81**
Water heaters										
Note: the following prices exclude Plumber's work										
Storage type units; free outlet										
7 litre	104.57	-	**104.57**	1.50	-	**30.81**	5.15	**140.53**	nr	**151.07**
23 litre	461.65	-	**461.65**	2.25	-	**46.22**	5.65	**513.52**	nr	**552.03**
55 litre	673.33	-	**673.33**	4.50	-	**92.43**	7.98	**773.74**	nr	**831.77**
Storage type units; multi-point										
50 litre	634.77	-	**634.77**	4.50	-	**92.43**	7.98	**735.18**	nr	**790.32**
75 litre	754.78	-	**754.78**	5.25	-	**107.84**	7.98	**870.60**	nr	**935.89**
125 litre	863.83	-	**863.83**	6.00	-	**123.24**	10.30	**997.37**	nr	**1072.17**
Shower units										
Note: the following prices exclude Plumber's work										
Instantaneous shower units										
complete with fittings	155.24	2.50	**159.12**	4.00	-	**82.16**	7.73	**249.01**	nr	**267.69**
V90: TESTING										
Testing; existing installations										
Point										
from	-	-	**-**	2.00	-	**41.08**	-	**41.08**	nr	**44.16**
to	-	-	**-**	3.50	-	**71.89**	-	**71.89**	nr	**77.28**
Complete installation; three bedroom house										
from	-	-	**-**	10.00	-	**205.40**	-	**205.40**	nr	**220.81**
to	-	-	**-**	16.00	-	**328.64**	-	**328.64**	nr	**353.29**

ELECTRICAL

COMMUNICATIONS/SECURITY/CONTROL SYSTEMS

Labour hourly rates: (except Specialists) Craft Operatives £20.54 Labourer £10.71 Rates are national average prices. Refer to REGIONAL VARIATIONS for indicative levels of overall pricing in regions	MATERIALS			LABOUR				RATES		
	Del to Site £	Waste %	Material Cost £	Craft Optve Hrs	Lab Hrs	Labour Cost £	Sunds £	Nett Rate £	Unit	Gross Rate (+7.5%) £
W50: FIRE DETECTION AND ALARM										
Trunking, conduits, cables etc included in section V										
Fittings and equipment										
fire alarm 1 zone panel with batteries and charger	85.00	2.50	**87.13**	3.00	-	**61.62**	0.50	**149.25**	nr	**160.44**
fire alarm 4 zone panel with batteries and charger	120.00	2.50	**123.00**	4.00	-	**82.16**	0.50	**205.66**	nr	**221.08**
fire alarm 8 zone panel with batteries and charger	170.00	2.50	**174.25**	6.00	-	**123.24**	0.50	**297.99**	nr	**320.34**
Ionisation smoke detector	28.00	2.50	**28.70**	0.25	-	**5.14**	0.50	**34.34**	nr	**36.91**
Optical smoke detector	25.00	2.50	**25.63**	0.25	-	**5.14**	0.50	**31.26**	nr	**33.60**
Smoke detector mounting base with diode..........	6.50	2.50	**6.66**	0.75	-	**15.41**	0.50	**22.57**	nr	**24.26**
Manual surface mounted call point...............	9.75	2.50	**9.99**	0.75	-	**15.41**	0.50	**25.90**	nr	**27.84**
Emergency lighting										
General purpose 8 watt non maintained...........	20.00	2.50	**20.50**	1.00	-	**20.54**	0.50	**41.54**	nr	**44.66**
General purpose 8 watt maintained...............	30.00	2.50	**30.75**	1.00	-	**20.54**	0.50	**51.79**	nr	**55.67**
General purpose 8 watt mains....................	12.00	2.50	**12.30**	1.00	-	**20.54**	0.50	**33.34**	nr	**35.84**

This section continues on the next page

Labour hourly rates: (except Specialists) Craft Operatives £14.21 Labourer £10.71 Rates are national average prices. Refer to REGIONAL VARIATIONS for indicative levels of overall pricing in regions	MATERIALS			LABOUR				RATES		
	Del to Site £	Waste %	Material Cost £	Craft Optve Hrs	Lab Hrs	Labour Cost £	Sunds £	Nett Rate £	Unit	Gross Rate (+7.5%) £
W52: LIGHTNING PROTECTION										
Lightning conductors										
Note: the following are indicative average prices										
Air termination rods 610mm long; in clips, fixing to masonry..........	-	-	**Spclist**	-	-	**Spclist**	-	**39.50**	nr	**42.46**
Air termination tape fixing to masonry	-	-	**Spclist**	-	-	**Spclist**	-	**9.50**	m	**10.21**
Copper tape down conductors										
19 x 3mm; in clips, fixing to masonry...........	-	-	**Spclist**	-	-	**Spclist**	-	**14.35**	m	**15.43**
25 x 3mm; in clips, fixing to masonry...........	-	-	**Spclist**	-	-	**Spclist**	-	**15.35**	m	**16.50**
Test clamps with securing bolts	-	-	**Spclist**	-	-	**Spclist**	-	**22.00**	nr	**23.65**
Copper earth rods; tape connectors 2438 x 16mm; driving into ground.................	-	-	**Spclist**	-	-	**Spclist**	-	**140.00**	nr	**150.50**

TRANSPORT SYSTEMS

Labour hourly rates: (except Specialists) Craft Operatives £14.21 Labourer £10.71 Rates are national average prices. Refer to REGIONAL VARIATIONS for indicative levels of overall pricing in regions	MATERIALS			LABOUR				RATES		
	Del to Site £	Waste %	Material Cost £	Craft Optve Hrs	Lab Hrs	Labour Cost £	Sunds £	Nett Rate £	Unit	Gross Rate (+7.5%) £
X10: LIFTS										
Generally										
Note: the following are average indicative prices										
Light passenger lifts; standard range										
Electro Hydraulic drive; 630 kg, 8 person, 0.63m/s, 3 stop										
basic; primed car and entrances	-	-	Spclist	-	-	Spclist	-	15500.00	nr	16662.50
median; laminate walls, standard carpet floor, painted entrances	-	-	Spclist	-	-	Spclist	-	18200.00	nr	19565.00
ACVF drive; 1000 kg, 13 person, 0.63m/s, 4 stop										
basic; primed car and entrances	-	-	Spclist	-	-	Spclist	-	34600.00	nr	37195.00
median; laminate walls, standard carpet floor, painted entrances	-	-	Spclist	-	-	Spclist	-	35900.00	nr	38592.50
Extras on the above lifts										
800/900mm landing doors in cellulose paint finish	-	-	Spclist	-	-	Spclist	-	150.00	nr	161.25
800/900mm landing doors in brushed stainless steel	-	-	Spclist	-	-	Spclist	-	420.00	nr	451.50
800/900mm landing doors in brushed brass	-	-	Spclist	-	-	Spclist	-	1210.00	nr	1300.75
landing position indicators	-	-	Spclist	-	-	Spclist	-	237.00	nr	254.78
landing direction arrows and gongs	-	-	Spclist	-	-	Spclist	-	140.00	nr	150.50
car preference key switch	-	-	Spclist	-	-	Spclist	-	120.00	nr	129.00
car entrance safety (Progard L)	-	-	Spclist	-	-	Spclist	-	400.00	nr	430.00
fluorescent emergency light	-	-	Spclist	-	-	Spclist	-	372.00	nr	399.90
half height mirror to rear wall	-	-	Spclist	-	-	Spclist	-	420.00	nr	451.50
full height mirror to rear wall	-	-	Spclist	-	-	Spclist	-	610.00	nr	655.75
vandal resistant buttons; per car/entrance	-	-	Spclist	-	-	Spclist	-	136.00	nr	146.20
car bumper rail	-	-	Spclist	-	-	Spclist	-	820.00	nr	881.50
trailing cable for Warden Call System	-	-	Spclist	-	-	Spclist	-	349.00	nr	375.18
car telephone	-	-	Spclist	-	-	Spclist	-	274.00	nr	294.55
General purpose passenger lifts										
ACVF drive; 630 kg, 8 person, 1.0m/s, 4 stop										
median; laminate walls, standard carpet floor, painted entrances	-	-	Spclist	-	-	Spclist	-	35600.00	nr	38270.00
Variable speed a/c drive; 1000 kg, 13 person, 1.6m/s, 5 stop										
median; laminate walls, standard carpet floor, painted entrances	-	-	Spclist	-	-	Spclist	-	38200.00	nr	41065.00
Variable speed a/c drive; 1600 kg, 21 person, 1.6m/s, 6 stop										
median; laminate walls, standard carpet floor, painted entrances	-	-	Spclist	-	-	Spclist	-	92000.00	nr	98900.00
If a high quality architectural finish is required, e.g. veneered panelling walls, marble floors, special ceiling, etc, add to the above prices										
from	-	-	Spclist	-	-	Spclist	-	15000.00	nr	16125.00
to	-	-	Spclist	-	-	Spclist	-	42000.00	nr	45150.00
Bed/Passenger lifts										
Electro Hydraulic drive; 1800 kg, 24 person, 0.63m/s, 3 stop										
standard specification	-	-	Spclist	-	-	Spclist	-	48200.00	nr	51815.00
Variable speed a/c drive; 2000 kg, 26 person, 1.6m/s, 4 stop										
standard specification	-	-	Spclist	-	-	Spclist	-	87500.00	nr	94062.50

Labour hourly rates: (except Specialists) Craft Operatives £14.21 Labourer £10.71 Rates are national average prices. Refer to REGIONAL VARIATIONS for indicative levels of overall pricing in regions	MATERIALS			LABOUR				RATES		
	Del to Site £	Waste %	Material Cost £	Craft Optve Hrs	Lab Hrs	Labour Cost £	Sunds £	Nett Rate £	Unit	Gross Rate (+7.5%) £
X10: LIFTS Cont.										
Goods lifts (direct coupled)										
Electro Hydraulic drive; 1500 kg, 0.4m/s, 3 stop stainless steel car lining with chequer plate floor and galvanised shutters....................	-	-	**Spclist**	-	-	**Spclist**	-	**38200.00**	nr	**41065.00**
Electro Hydraulic drive; 2000 kg, 0.4m/s, 3 stop stainless steel car lining with chequer plate floor and galvanised shutters....................	-	-	**Spclist**	-	-	**Spclist**	-	**39400.00**	nr	**42355.00**
Service hoists										
Single speed a/c drive; 50 kg, 0.63m/s, 2 stop standard specification	-	-	**Spclist**	-	-	**Spclist**	-	**5400.00**	nr	**5805.00**
Single speed a/c drive; 250 kg, 0.4m/s, 2 stop standard specification	-	-	**Spclist**	-	-	**Spclist**	-	**6850.00**	nr	**7363.75**
Extra on the above hoists two hour fire resisting shutters; per landing....	-	-	**Spclist**	-	-	**Spclist**	-	**227.00**	nr	**244.03**
X11: ESCALATORS										
Escalators										
30 degree inclination; 3.00m vertical rise department store specification..................	-	-	**Spclist**	-	-	**Spclist**	-	**58000.00**	nr	**62350.00**

BASIC PRICES OF MATERIALS

This section gives basic prices of materials delivered to site. Prices are exclusive of Value Added Tax

Excavation and Earthwork

Filling		
Broken brick or stone filling	17.00	m3
Clinker ashes filing	9.98	m3
MOT type 1 filling	19.95	m3
MOT type 2 filling	18.48	m3
Imported topsoil	15.00	m3
Sand	14.58	m3
Gravel		
Coarse gravel	17.34	m3
Blinding gravel	24.29	m3
Earthwork support timber	245.70	m3

Concrete Work

Materials for site mixed concrete		
40mm Aggregate	26.25	m3
20mm Aggregate	25.20	m3
10mm Aggregate	25.20	m3
White Portland Cement	350.00	tonne
Sharp sand	23.04	m3
White silver sand	27.00	m3
Ready mixed concrete, prescribed mix		
ST1 - 40mm aggregate	54.96	m3
ST3 - 20mm aggregate	58.80	m3
ST4 - 20mm aggregate	60.02	m3
ST5 - 20mm aggregate	62.08	m3
Ready mixed concrete, design mix		
C 15 - 15 N/mm2 - 20mm	60.24	m3
C 20 - 20 N/mm2 - 20mm	61.25	m3
C 25 - 25 N/mm2 - 20mm	63.45	m3
C 30 - 30 N/mm2 - 20mm	64.91	m3
Site mixed concretes, mortars, various mixes		
1:6, all in aggregate	77.37	m3
1:8, all in aggregate	69.50	m3
1:12, all in aggregate	61.61	m3
Lightweight concrete		
20.5 N/mm2; vibrated	85.44	m3
26.0 N/mm2; vibrated	89.55	m3
41.5 N/mm2; vibrated	97.77	m3
Mild steel rod reinforcement		
32mm diameter	380.00	tonne
25mm diameter	385.00	tonne
20mm diameter	385.00	tonne
16mm diameter	385.00	tonne
12mm diameter	390.00	tonne
10mm diameter	395.00	tonne
8mm diameter	405.00	tonne
6mm diameter	420.00	tonne
High yield rod reinforcement		
32mm diameter	390.00	tonne
25mm diameter	390.00	tonne
20mm diameter	390.00	tonne
16mm diameter	390.00	tonne
12mm diameter	400.00	tonne
10mm diameter	400.00	tonne
8mm diameter	410.00	tonne
6mm diameter	430.00	tonne
Stainless steel rod reinforcement		
25mm diameter	4100.00	tonne
20mm diameter	4100.00	tonne
16mm diameter	4100.00	tonne
12mm diameter	4100.00	tonne
10mm diameter	4100.00	tonne
8mm diameter	4100.00	tonne
Fabric reinforcement, BS Ref 4483		
A393	2.97	m2
A252	2.01	m2
A193	1.47	m2
A142	1.11	m2
A98	0.80	m2
B1131	5.35	m2
B785	3.96	m2
B503	2.88	m2
B385	2.22	m2
B283	1.82	m2
B196	1.50	m2
C785	3.28	m2
C503	2.04	m2
C385	1.65	m2
C283	1.27	m2
D98	0.72	m2
D49	0.77	m2
C636	2.72	m2
Formwork, basic finish		
Sides of foundations; height exceeding 1.00m	3.32	m2
Sides of foundations; height not exceeding 250mm	1.03	m
Sides of ground beams and edges of beds; height exceeding 1.00m	3.32	m2
Sides of ground beams and edges of beds; height not exceeding 250mm	1.03	m
Edges of suspended slabs; height not exceeding 250mm	1.26	m
Edges of suspended slabs; height 250 - 500mm	2.31	m
Sides of upstands; height not exceeding 250mm	1.45	m
Sides of upstands; height 250 - 500mm	1.72	m
Soffits of slabs; horizontal slab thickness not exceeding 200mm	10.02	m2
Soffits of slabs; horizontal slab thickness 200 - 300mm	11.06	m2
Soffits of slabs; sloping not exceeding 15 degrees slab thickness not exceeding 200mm	11.07	m2
Soffits of slabs; sloping exceeding 15 degrees slab thickness not exceeding 200mm	11.53	m2
Walls; plain	6.66	m2
Walls; plain; height exceeding 3.00m above floor level	7.48	m2
Beams; attached to slabs	6.97	m2
Beams; with 50mm wide chamfers	7.11	m2
Beams; isolated	7.20	m2
Beams; isolated with 50mm wide chamfers	7.36	m2
Columns; attached to walls	4.84	m2
Columns; attached to walls, height exceeding 3.00m above floor level	4.53	m2
Wall kickers straight	0.91	m
Wall kickers curved 2m radius	2.35	m
Wall kickers curved 10m radius	1.45	m
Stairflights; 1000mm wide; 155mm thick waist; 178mm risers	14.96	m
Stairflights; 1500mm wide; 180mm thick waist; 178mm risers	18.71	m
Stairflights; 1000mm wide; 155mm thick waist; 178mm undercut risers	14.96	m
Stairflights; 1500mm wide; 180mm thick waist; 178mm undercut risers	18.71	m
Claymaster expanded polystyrene permanent formwork		
2400 x 1200 x 75mm thick Reference 2611, 4.86 kg/m2	35.57	sheet
2400 x 1200 x 100mm thick Reference 2411, 6.34 kg/m2	59.64	sheet
2400 x 1200 x 150mm thick Reference 2811, 3.39 kg/m2	89.45	sheet
Liquid damp proof applications		
Synthaprufe bituminous emulsion	3.52	litre
Bituminous emulsion	3.52	litre
R.I.W. liquid asphaltic composition	4.15	litre
Flexible sheet damp proof membranes		
Polythene sheeting, 125mu	0.33	m2
Polythene sheeting, 250mu	0.38	m2
Polythene sheeting, 300mu	0.50	m2
Waterproof building paper to BS 1521		
British Sisalkraft Ltd 728 damp proof membrane	1.30	m2
Visqueen 1200 super damp proof membrane	0.50	m2
Precast concrete lintels		
75 x 150 x 1200mm lintel; reinf with 1.20 kg of 12mm lintel bars	6.41	nr

Concrete Work (Cont'd)

Precast concrete lintels (Cont'd)		
102 x 150 x 1200mm lintel; reinf with 1.20 kg of 12mm lintel bars	8.72	nr
215 x 150 x 1200mm lintel; reinf with 2.40 kg of 12mm lintel bars	18.37	nr
327 x 150 x 1200mm lintel; reinf with 3.60 kg of 12mm lintel bars	27.94	nr
252 x 150 x 1200mm boot lintel; reinf with 3.60 kg of 12mm boot lintel bars and 0.56 kg of 6mm boot lintel links	26.49	nr
252 x 150 x 1800mm boot lintel; reinf with 5.19 kg of 12mm boot lintel bars and 0.83 kg of 6mm boot lintel links	39.73	nr
Prestressed precast lintels 100 x 70mm		
900mm long	3.23	nr
1050mm long	3.77	nr
1200mm long	4.30	nr
1500mm long	5.38	nr
1800mm long	6.46	nr
2100mm long	7.53	nr
2400mm long	8.61	nr
2700mm long	9.68	nr
3000mm long	10.76	nr
Prestressed precast lintels 140 x 70mm		
1200mm long	5.84	nr
3000mm long	14.62	nr
Prestressed precast lintels 225 x 70mm		
1200mm long	9.01	nr
3000mm long	22.53	nr
Prestressed precast lintels 255 x 70mm		
1200mm long	9.60	nr
3000mm long	23.99	nr
Precast concrete copings ; figure 1;		
75 x 200mm; splayed; rebated joints	10.39	m
100 x 300mm; splayed; rebated joints	20.78	m
75 x 200mm; saddleback; rebated joints	11.08	m
100 x 300mm; saddleback; rebated joints	22.17	m

Brickwork and Blockwork

Bricks		
Common bricks plain	185.00	1000
Common bricks keyed	192.00	1000
Selected regraded bricks	149.00	1000
Second hard stock bricks	365.00	1000
Engineering bricks class A	275.00	1000
Engineering bricks class B	195.00	1000
Staffordshire blue wirecut bricks	355.00	1000
Staffordshire blue pressed bricks	415.00	1000
Staffordshire blue splayed plinth or bullnose	1550.00	1000
Claydon Red Multi	227.00	1000
Tudors	256.00	1000
Milton Buff Ridgefaced	238.00	1000
Heathers	279.00	1000
Brecken Grey	219.00	1000
Leicester Red Stock	273.00	1000
West Hoathly Medium multi-stock	400.00	1000
Himley dark brown rustic	360.00	1000
Tonbridge hand made multi-coloured	580.00	1000
Waingroves smooth red	260.00	1000
Old English Russet	236.00	1000
First hard stocks	350.00	1000
Miscellaneous materials for brickwork and blockwork		
Cement and sand mortar (1:3)	93.27	m3
Cement lime mortar (1:1:6)	100.78	m3
Hydrated Lime	191.08	tonne
Cavity wall insulation		
Glass fibre slabs 50mm	3.50	m2
Glass fibre slabs 75mm	4.79	m2
Butterfly pattern stainless steel wall ties, 203 x 3mm		
Butterfly pattern stainless steel	123.05	1000
Vertical twisted wall ties, 203 x 19 x 3mm		
stainless steel	521.00	1000
Extension Profiles		
Extension profiles 60-250mm ref C2KS	10.88	m
Extension profiles 100-150mm ref 1FUR100S	11.73	m
Extension profiles 190-250mm ref 1FUR215S	21.21	m
Thermabate insulated cavity closers for fixing to masonry with PVCU ties at 225mm centres		
Thermabate 50; vertical	4.21	m3
Thermabate 75; vertical	4.40	m3
Thermalite Shield blocks		
75mm thick blocks	6.56	m2
100mm thick blocks	8.63	m2
150mm thick blocks	12.79	m2
200mm thick blocks	16.93	m2
Thermalite Turbo blocks		
100mm thick blocks	8.77	m2
115mm thick blocks	10.05	m2
130mm thick blocks	11.34	m2
190mm thick blocks	16.47	m2
215mm thick blocks	18.61	m2
Thermalite Party Wall blocks		
215mm thick blocks	18.34	m2
Thermalite Hi-Strength blocks		
100mm thick blocks	10.78	m2
140mm thick blocks	14.94	m2
Thermalite Trench blocks		
255mm thick blocks	21.70	m2
275mm thick blocks	23.37	m2
305mm thick blocks	25.89	m2
355mm thick blocks	30.08	m2
Lignacite blocks solid Fair Face 7.0 N/mm2		
100mm thick blocks	12.13	m2
140mm thick blocks	17.86	m2
190mm thick blocks	24.23	m2
Toplite solid Standard blocks 7.0 N/mm2		
100mm thick blocks	8.47	m2
140mm thick blocks	11.86	m2
190mm thick blocks	16.09	m2
215mm thick blocks	18.21	m2
Hemelite solid Standard blocks 3.5 N/mm2		
75mm thick blocks	5.38	m2
90mm thick blocks	5.98	m2
100mm thick blocks	6.50	m2
140mm thick blocks	9.56	m2
190mm thick blocks	12.97	m2
215mm thick blocks	14.67	m2
Hemelite solid Standard blocks 7.0 N/mm2		
90mm thick blocks	6.28	m2
100mm thick blocks	6.90	m2
140mm thick blocks	10.14	m2
190mm thick blocks	13.77	m2
215mm thick blocks	15.58	m2
Damp proof course		
BS 6398 Class A, bitumen with hessian base	6.91	m2
BS 6398 Class B, bitumen with fibre base	4.13	m2
BS 6398 Class D, bitumen with hessian base laminated with lead	13.70	m2
BS 6398 Class E, bitumen with fibre base laminated with lead	15.22	m2
Hyload Original, pitch polymer	5.52	m2
Synthaprufe bituminous latex emulsion	3.52	litre
Bituminous emulsion	3.52	litre
Slates 350 x 225mm	1335.00	1000
Type G Cavitray without lead		
900mm long cavitray	8.15	nr
220 x 332mm external angle	7.22	nr
230 x 117mm internal angle	7.22	nr
Type X Cavitray with lead, 40 degree pitch		
Intermediate tray	4.43	nr
Ridge tray	9.35	nr
Catchment tray	4.43	nr
Corner catchment angle tray	9.77	nr
Type W		
Cavity weep ventilator	0.45	nr
Extension duct	0.76	nr
Stainless steel mesh reinforcement		
61mm mesh	1.04	m
112mm mesh	1.58	m
178mm mesh	2.73	m
228mm mesh	3.57	m
Air bricks, terra cotta, red or buff, square hole		
225 x 75mm	1.90	nr
225 x 150mm	3.62	nr
225 x 225mm	6.35	nr
Air bricks, iron, light		
225 x 75mm	1.85	nr
225 x 150mm	2.99	nr
225 x 225mm	5.24	nr

Roofing

Profile 3 sheeting, standard grey colour		
coverings	10.92	m2
eaves filler pieces	9.36	m
two piece plain angular adjustable ridge tiles	13.54	m
standard barge boards	7.67	m
apron flashings	9.40	m
standard ridge cap finials	8.47	nr
Profile 6 sheeting, standard coloured		
coverings	10.43	m2
eaves filler pieces	7.76	m
two piece plain angular adjustable ridge tiles	14.28	m
standard barge boards	9.59	m
apron flashings	8.38	m
standard ridge cap finials	9.45	nr
Treated softwood roofing battens		
25 x 19mm	0.22	m
32 x 19mm	0.22	m
32 x 25mm	0.32	m
38 x 19mm	0.27	m
50 x 19mm	0.39	m
50 x 25mm	0.48	m
Roofing underfelts, reinforced bitumen		
type 1F 15kg/10m2	0.53	m2
Plain clay tiles, 268 x 165mm		
Machine made	461.95	1000
hand made	717.15	1000
Redland plain tiles, 268 x 165mm		
Granular faced	300.00	1000
Granular faced, tile and a half	60.00	100
Downland	335.00	1000
Downland, tile and a half	67.00	100

Roofing (Cont'd)

Fibre cement slates		
600 x 300mm	984.00	1000
500 x 250mm	792.00	1000
400 x 200mm	572.00	1000
Ridges or hips	23.66	m
Welsh Blue/grey slates, 6.5mm thick		
600 x 300mm	5124.00	1000
500 x 250mm	2478.00	1000
400 x 200mm	1134.00	1000
Redland Cambrian interlocking riven textured slates		
300 x 336mm	1712.00	1000
slate and a half slates to match 300 x 336mm	3424.00	1000
150 x 6mm fibre cement undercloak	2.25	m
Stainless steel eaves clip to each slate	0.12	nr
Redland third round ridge or hip	3.00	nr
Redland half round ridge or hip	3.00	nr
Redland universal angle ridge or hip	3.26	m
Redland Dry Ridge system with half round ridge tiles	23.51	m
Redland Dry Ridge system with universal angle ridge tiles	24.11	m
Monoridge filler units	34.51	m
Redland Universal valley troughs	8.16	nr
Timloc uPVC ventilators		
Mark 3 eaves ventilators; reference 1126/40	3.68	m
Soffit ventilators; reference 1137	1.97	m
Polypropylene over-fascia ventilators; reference 3011	3.50	m
Redland single lap tiles		
Stonewold 430 x 380mm	1380.00	1000
Regent 418 x 332mm	724.00	1000
Grovebury 418 x 332mm	724.00	1000
Norfolk 381 x 227mm	511.00	1000
Redland 49, 381 x 227mm	499.00	1000
Renown 418 x 330mm	700.00	1000
Redland 50, 418 x 330mm (GRANULAR)	700.00	1000
Felt to BS 747		
Rubervent 3G underlay 2.6 kg/m2	1.55	m2
Ruberfort HP180 underlay 3.4 kg/m2	2.02	m2
Ruberfort HP180 mineral surface top sheet 3.6 kg/m2	2.80	m2
Type 3B -1.8 kg/m2	0.83	m2
Type 3E -2.8 kg/m2	1.67	m2
Hot bitumen bonding compound	0.47	kg
Insulating boards for felt roofing		
25mm Kingspan Thermaroof TR21 (PIR) Board	8.13	m2
30mm Kingspan Thermaroof TR21 (PIR) Board	9.32	m2
40mm Kingspan Thermaroof TR21 (PIR) Board	10.77	m2
50mm Kingspan Thermaroof TR21 (PIR) Board	12.30	m2
Lead sheet		
Code 3	14.64	m2
Code 4	19.92	m2
Code 5	24.79	m2
Aluminium roofing		
0.6 mm thick sheets	9.13	m2
150 mm girth flashings	2.26	m
180 mm girth flashings	3.05	m
240 mm girth flashings	3.50	m
300 mm girth flashings	3.50	m
450 mm girth flashings	5.52	m
180 x 180 mm soakers	0.55	nr
180 x 300 mm soakers	0.92	nr
360mm girth gutter linings	4.62	m
420mm girth gutter linings	4.62	nr
450mm girth gutter linings	4.62	nr
Copper roof sheeting		
0.55 mm thick sheets	24.00	m2
0.70 mm thick sheets	30.00	m2
Zinc roof sheeting		
0.80 mm thick	16.50	m2
Zinc flashings - 0.80 mm thick		
150 mm girth	2.47	m
180 mm girth	2.97	m
240 mm girth	3.96	m
300 mm girth	4.95	m
450 mm girth	7.42	m
Zinc soakers - 0.80 mm thick		
180 x 180 mm	0.54	nr
180 x 300 mm	0.89	nr
450 x 450 mm	3.34	nr

Woodwork

Sawn Softwood General Structural Grade SC3		
32 x 125mm	0.80	m
32 x 150mm	0.96	m
32 x 175mm	1.12	m
32 x 200mm	1.28	m
32 x 225mm	1.44	m
32 x 250mm	1.60	m
32 x 300mm	1.92	m
38 x 125mm	0.95	m
38 x 150mm	1.14	m
38 x 175mm	1.33	m
38 x 200mm	1.52	m
38 x 225mm	1.71	m
38 x 250mm	1.90	m
38 x 300mm	2.28	m
44 x 125mm	1.07	m
44 x 150mm	1.29	m
44 x 175mm	1.50	m
44 x 200mm	1.72	m
44 x 225mm	1.93	m
44 x 250mm	2.15	m
44 x 300mm	2.57	m
50 x 50mm	0.49	m
50 x 63mm	0.61	m
50 x 75mm	0.73	m
50 x 100mm	0.98	m
50 x 125mm	1.22	m
50 x 150mm	1.46	m
50 x 175mm	1.71	m
50 x 200mm	1.95	m
50 x 225mm	2.19	m
50 x 250mm	2.44	m
50 x 300mm	2.92	m
63 x 63mm	0.77	m
63 x 75mm	0.92	m
63 x 100mm	1.23	m
63 x 125mm	1.54	m
63 x 150mm	1.84	m
63 x 175mm	2.15	m
63 x 200mm	2.46	m
63 x 225mm	2.76	m
63 x 250mm	3.07	m
63 x 300mm	3.69	m
75 x 75mm	1.12	m
75 x 100mm	1.50	m
75 x 125mm	1.88	m
75 x 150mm	2.25	m
75 x 175mm	2.62	m
75 x 200mm	3.00	m
75 x 225mm	3.38	m
75 x 250mm	3.75	m
75 x 300mm	4.50	m
100 x 100mm	2.00	m
150 x 150mm	4.50	m
150 x 250mm	7.50	m
150 x 300mm	9.00	m
200 x 200mm	8.00	m
250 x 250mm	12.50	m
300 x 300mm	18.00	m
Sawn Softwood General Structural Grade SC3 Pressure impregnated		
32 x 125mm	0.90	m
32 x 150mm	1.08	m
32 x 175mm	1.26	m
32 x 200mm	1.44	m
32 x 225mm	1.62	m
32 x 250mm	1.80	m
32 x 300mm	2.16	m
38 x 125mm	1.07	m
38 x 150mm	1.28	m
38 x 175mm	1.50	m
38 x 200mm	1.71	m
38 x 225mm	1.92	m
38 x 250mm	2.14	m
38 x 300mm	2.56	m
44 x 125mm	1.21	m
44 x 150mm	1.45	m
44 x 175mm	1.69	m
44 x 200mm	1.94	m
44 x 225mm	2.18	m
44 x 250mm	2.42	m
44 x 300mm	2.90	m
50 x 50mm	0.55	m
50 x 63mm	0.69	m
50 x 75mm	0.82	m
50 x 100mm	1.10	m
50 x 125mm	1.38	m
50 x 150mm	1.65	m
50 x 175mm	1.92	m
50 x 200mm	2.20	m
50 x 225mm	2.48	m
50 x 250mm	2.75	m
50 x 300mm	3.30	m
63 x 63mm	0.87	m
63 x 75mm	1.04	m
63 x 100mm	1.39	m
63 x 125mm	1.73	m
63 x 150mm	2.08	m
63 x 175mm	2.43	m
63 x 200mm	2.77	m
63 x 225mm	3.12	m
63 x 250mm	3.46	m
63 x 300mm	4.16	m
75 x 75mm	1.27	m
75 x 100mm	1.69	m
75 x 125mm	2.11	m
75 x 150mm	2.53	m
75 x 175mm	2.95	m
75 x 200mm	3.38	m
75 x 225mm	3.80	m
75 x 250mm	4.22	m
75 x 300mm	5.06	m
100 x 100mm	2.25	m
150 x 150mm	5.06	m
150 x 225mm	7.59	m
150 x 250mm	8.44	m
150 x 300mm	10.12	m
200 x 200mm	9.00	m
250 x 250mm	14.06	m
300 x 300mm	20.25	m
Softwood boarding, wrot, impregnated		
19mm thick	7.04	m2
25mm thick	9.28	m2
Plywood BB/CC Far Eastern H/W		
18mm thick	10.19	m2
25mm thick	14.16	m2
Pressure Impregnated Softwood Boarding		
25 x 200mm	2.20	m
25 x 225mm	2.47	m
Softwood, wrought; - boarded flooring		
19 x 75mm, square edge	6.72	m2

Woodwork (Cont'd)

Softwood, wrought; - boarded flooring (Cont'd)

25 x 125mm, square edge.	8.40	m2
32 x 150mm, square edge.	10.69	m2
19 x 75mm, tongued and grooved joint	7.68	m2
25 x 125mm, tongued and grooved joint	9.20	m2

Western Red Cedar, wrought; - wall and ceiling boarding, tongued, grooved and veed joints

19mm thick, 100mm wide..	42.70	m2
25mm thick, 100mm wide..	53.60	m2

Knotty Pine, wrought, selected for transparent finish; - wall and ceiling boarding, tongued, grooved and veed joints

19mm thick, 100mm wide..	32.81	m2
25mm thick, 100mm wide..	41.27	m2

Sapele, wrought, selected for transparent finish; - wall and ceiling boarding, tongued, grooved and veed joints

19mm thick, 100mm wide..	55.43	m2
25mm thick, 100mm wide..	67.45	m2

Timber for First and Second Fixings

Iroko	1200.00	m3
Idigbo	1200.00	m3
African Mahogany	1400.00	m3
South American Mahogany	1400.00	m3
African Walnut	2000.00	m3
Beech	1200.00	m3
Afrormosia	1500.00	m3
American Oak	1500.00	m3
Sapele	1200.00	m3
British Columbain Pine	918.00	m3
Teak	4750.00	m3

Blockboard Birch faced, Finnish manufacturer

16mm	9.92	m2
18mm	9.17	m2
25mm	15.08	m2

Blockboard Oak faced

18mm	16.12	m2

Blockboard Sapele faced

18mm	15.20	m2

Blockboard Beech faced

18mm	16.88	m2

Douglas Fir T&G plywood, unsanded select sheathing quality, WBP bonded

15mm thick	6.23	m2
18mm thick	7.89	m2

Birch faced T&G plywood, BB quality, WBP bonded

12mm thick	18.71	m2
15mm thick	22.46	m2
18mm thick	27.24	m2

Wood chipboard, BS EN 634, type II T&G flooring

18mm thick P4 T&G F/G ...	3.85	m2
22mm thick P4 T&G F/G ...	5.28	m2
18mm thick, moisture resisting, Furniture Grade.	2.92	m2

Standard hardboard, BS EN 622, type SHA; butt joints

3.2mm thick linings	1.25	m2
4.8mm thick linings	2.27	m2
6.4mm thick linings	2.84	m2

Low density medium board, BS EN 622, type LME; butt joints

6mm thick linings	7.81	m2
9mm thick linings	9.99	m2
12mm thick linings	12.74	m2

Medium density fibreboard, BS EN 622, type MDF; butt joints, linings to walls

9mm thick	3.28	m2
12mm thick	4.03	m2
18mm thick	5.29	m2
25mm thick	7.31	m2

Tacboard Class O fire resisting boards; butt joints

6mm thick linings	8.85	m2
9mm thick linings	16.31	m2
12mm thick linings	21.23	m2

Softwood standard skirtings, picture rails, architraves and the like; (finished sizes)

13 x 70mm	1.07	m
13 x 95mm	1.37	m
20 x 120mm	2.12	m
13 x 70mm	1.11	m
13 x 95mm	1.43	m
20 x 120mm	2.12	m

MDF skirtings, picture rails, architraves and the like; (finished sizes)

13 x 70mm	1.47	m
20 x 120mm	3.00	m

Structural Steelwork

Universal beams

914 x 419 x 343	675.00	tonne
914 x 419 x 388	675.00	tonne
914 x 305 x 201	670.00	tonne
914 x 305 x 224	670.00	tonne
914 x 305 x 253	670.00	tonne
914 x 305 x 289	670.00	tonne
838 x 292 x 176	665.00	tonne
838 x 292 x 194	665.00	tonne
838 x 292 x 226	665.00	tonne
762 x 267 x 134	665.00	tonne
762 x 267 x 147	665.00	tonne
762 x 267 x 173	665.00	tonne
762 x 267 x 197	665.00	tonne
686 x 254 x 125	665.00	tonne
686 x 254 x 140	665.00	tonne
686 x 254 x 152	665.00	tonne
686 x 254 x 170	665.00	tonne
610 x 305 x 238	655.00	tonne
610 x 305 x 179	655.00	tonne
610 x 305 x 149	655.00	tonne
610 x 229 x 140	655.00	tonne
610 x 229 x 125	655.00	tonne
610 x 229 x 113	655.00	tonne
610 x 229 x 101	655.00	tonne
533 x 210 x 122	640.00	tonne
533 x 210 x 109	640.00	tonne
533 x 210 x 101	640.00	tonne
533 x 210 x 92	640.00	tonne
533 x 210 x 82	640.00	tonne
457 x 191 x 98	630.00	tonne
457 x 191 x 89	630.00	tonne
457 x 191 x 82	630.00	tonne
457 x 191 x 74	630.00	tonne
457 x 191 x 67	630.00	tonne
457 x 152 x 82	630.00	tonne
457 x 152 x 74	630.00	tonne
457 x 152 x 67	630.00	tonne
457 x 152 x 60	630.00	tonne
457 x 152 x 52	630.00	tonne
406 x 178 x 74	635.00	tonne
406 x 178 x 67	635.00	tonne
406 x 178 x 60	635.00	tonne
406 x 178 x 54	635.00	tonne
406 x 140 x 46	635.00	tonne
406 x 140 x 39	635.00	tonne
356 x 171 x 67	635.00	tonne
356 x 171 x 57	635.00	tonne
356 x 171 x 51	635.00	tonne
356 x 171 x 45	635.00	tonne
356 x 127 x 39	635.00	tonne
356 x 127 x 33	635.00	tonne
305 x 165 x 54	630.00	tonne
305 x 165 x 46	630.00	tonne
305 x 165 x 40	630.00	tonne
305 x 127 x 48	630.00	tonne
305 x 127 x 42	630.00	tonne
305 x 127 x 37	630.00	tonne
305 x 102 x 33	630.00	tonne
305 x 102 x 28	630.00	tonne
305 x 102 x 25	630.00	tonne
254 x 102 x 28	625.00	tonne
254 x 102 x 25	625.00	tonne
254 x 102 x 22	625.00	tonne

Universal Columns

356 x 406 x 634	675.00	tonne
356 x 406 x 551	675.00	tonne
356 x 406 x 467	675.00	tonne
356 x 406 x 393	675.00	tonne
356 x 406 x 340	675.00	tonne
356 x 406 x 287	675.00	tonne
356 x 406 x 235	675.00	tonne
356 x 368 x 202	675.00	tonne
356 x 368 x 177	675.00	tonne
356 x 368 x 153	675.00	tonne
356 x 368 x 129	675.00	tonne
305 x 305 x 283	655.00	tonne
305 x 305 x 240	655.00	tonne
305 x 305 x 198	655.00	tonne
305 x 305 x 158	655.00	tonne
305 x 305 x 137	655.00	tonne
305 x 305 x 118	655.00	tonne
305 x 305 x 97	655.00	tonne
254 x 254 x 167	635.00	tonne
254 x 254 x 132	635.00	tonne
254 x 254 x 107	635.00	tonne
254 x 254 x 89	635.00	tonne
254 x 254 x 73	635.00	tonne
203 x 203 x 86	625.00	tonne
203 x 203 x 71	625.00	tonne
203 x 203 x 60	625.00	tonne
203 x 203 x 52	625.00	tonne
203 x 203 x 46	625.00	tonne
152 x 152 x 37	625.00	tonne
152 x 152 x 30	625.00	tonne
152 x 152 x 23	625.00	tonne

Quantity extras

10 tonnes and over	0.00	tonne
Under 10 tonnes to 5 tonnes	10.00	tonne
Under 5 tonnes to 2 tonnes	25.00	tonne
Under 2 tonnes	50.00	tonne

Parallel flange channels

430 x 100 x 64	685.00	tonne
330 x 100 x 54	685.00	tonne
300 x 100 x 46	655.00	tonne
300 x 90 x 41	655.00	tonne
260 x 90 x 35	655.00	tonne
260 x 75 x 28	655.00	tonne
230 x 90 x 32	655.00	tonne
230 x 75 x 26	655.00	tonne
200 x 90 x 30	655.00	tonne
180 x 90 x 26	655.00	tonne
150 x 90 x 24	655.00	tonne
Extra; Parallel flange channels, non-standard sizes	25.00	tonne

Equal and unequal angles

200 x 150 x 18	625.00	tonne
200 x 150 x 15	625.00	tonne
200 x 150 x 12	625.00	tonne
200 x 100 x 15	625.00	tonne
200 x 100 x 12	625.00	tonne
200 x 100 x 10	625.00	tonne
200 x 200 x 24	610.00	tonne
200 x 200 x 20	610.00	tonne
200 x 200 x 18	610.00	tonne
200 x 200 x 16	610.00	tonne
150 x 90 x 15	605.00	tonne
150 x 90 x 12	605.00	tonne
150 x 90 x 10	605.00	tonne
150 x 150 x 18	605.00	tonne
150 x 150 x 15	605.00	tonne
150 x 150 x 12	605.00	tonne
150 x 150 x 10	605.00	tonne
150 x 75 x 15	595.00	tonne
150 x 75 x 12	595.00	tonne
150 x 75 x 10	595.00	tonne
120 x 120 x 15	600.00	tonne
120 x 120 x 12	600.00	tonne
120 x 120 x 10	600.00	tonne
120 x 120 x 8	600.00	tonne
125 x 75 x 12	580.00	tonne

Structural Steelwork (Cont'd)

Equal and unequal angles (Cont'd)

125 x 75 x 10	580.00	tonne
125 x 75 x 8	580.00	tonne
100 x 75 x 12	575.00	tonne
100 x 75 x 10	575.00	tonne
100 x 75 x 8	575.00	tonne
100 x 65 x 10	575.00	tonne
100 x 65 x 8	575.00	tonne
100 x 65 x 7	575.00	tonne
100 x 100 x 15	565.00	tonne
100 x 100 x 12	565.00	tonne
100 x 100 x 10	565.00	tonne
100 x 100 x 8	565.00	tonne
90 x 90 x 12	565.00	tonne
90 x 90 x 10	565.00	tonne
90 x 90 x 8	565.00	tonne
90 x 90 x 7	565.00	tonne
90 x 90 x 6	565.00	tonne
Extra; angles non-standard sizes	25.00	tonne

Cold rolled zed purlins and cladding rails, Metsec purlin sleeved system

ref: 14214; 3.03 kg/m	14.79	m
ref: 14216; 3.47 kg/m	16.70	m
ref: 17214; 3.66 kg/m	17.50	m
ref: 17216; 4.11 kg/m	19.88	m
ref: 20216; 4.49 kg/m	21.71	m
ref: 20218; 5.03 kg/m	24.48	m
ref: 20220; 5.57 kg/m	26.89	m
ref: 23218; 5.73 kg/m	27.79	m
ref: 23220; 6.34 kg/m	30.82	m

Cold rolled zed purlins and cladding rails, Metsec purlin overlap system

ref: 142	9.94	m
ref: 172	10.56	m
ref: 202	11.51	m
ref: 232	12.45	m

Cold rolled zed purlins and cladding rails, Metsec purlin non continuous system

ref: 17215; 3.85 kg/m	18.61	m
ref: 17216; 4.11 kg/m	19.88	m
ref: 20216; 4.49 kg/m	21.71	m
ref: 20218; 5.03 kg/m	24.48	m
ref: 20220; 5.57 kg/m	26.89	m
ref: 23218; 5.73 kg/m	27.79	m
ref: 23223; 7.26 kg/m	35.61	m
ref: 26223; 7.92 kg/m	38.78	m
ref: 26229; 9.88 kg/m	48.09	m

Cold rolled zed purlins and cladding rails, Metsec galvanized purlin cleats

ref: 142 for 142mm deep purlins	5.17	nr
ref: 172 for 172mm deep purlins	5.89	nr
ref: 202 for 202mm deep purlins	6.43	nr
ref: 232 for 232mm deep purlins	7.08	nr
ref: 262 for 262mm deep purlins	7.73	nr

Cold rolled zed purlins and cladding rails, Metsec Round Lok anti-sag rods

1150mm centres	5.20	nr
1350mm centres	6.04	nr
1550mm centres	6.83	nr
1700mm centres	7.69	nr
1950mm centres	8.44	nr

Cold rolled zed purlins and cladding rails, Metsec side rail sleeved/single span system

ref: 14214; 3.03 kg/m	14.79	m
ref: 14216; 3.47 kg/m	16.70	m
ref: 17215; 3.85 kg/m	18.61	m
ref: 17216; 4.11 kg/m	19.88	m
ref: 20216; 4.49 kg/m	21.71	m
ref: 20218; 5.03 kg/m	24.48	m
ref: 20220; 5.57 kg/m	26.89	m
ref: 23218; 5.73 kg/m	27.79	m
ref: 23223; 7.26 kg/m	35.61	m

Cold rolled zed purlins and cladding rails, Metsec side rail cleats

ref: 142 for 142mm rails	5.18	nr
ref: 172 for 172mm rails	7.05	nr
ref: 202 for 202mm rails	8.68	nr
ref: 232 for 232mm rails	10.09	nr

Cold rolled zed purlins and cladding rails, Metsec galvanized side rail suports, 122-262 series

rail 1000mm long	15.20	nr
rail 1400mm long	16.59	nr
rail 1600mm long	18.40	nr
rail 1800mm long	22.14	nr

Cold rolled zed purlins and cladding rails, Metsec diagonal tie wire supports

1700mm long	29.66	nr
2200mm long	33.18	nr
2600mm long	36.70	nr
3600mm long	45.31	nr

Steel short span lattice joists primed at works

200mm deep, 8.2 kg/m	41.52	m
250mm deep, 8.5 kg/m	43.26	m
300mm deep, 10.7 kg/m	54.65	m
350mm deep, 11.6 kg/m	58.75	m
350mm deep, 12.8 kg/m	64.21	m

Steel intermediate span lattice joists primed at works

450mm deep, 12.2 kg/m	56.17	m
500mm deep, 15.8 kg/m	69.45	m
550mm deep, 19.4 kg/m	82.12	m
600mm deep, 22.5 kg/m	93.96	m
650mm deep, 29.7 kg/m	123.79	m

Steel long span lattice joists primed at works

700mm deep, 39.2 kg/m	160.15	m
800mm deep, 44.1 kg/m	171.38	m
900mm deep, 45.3 kg/m	172.37	m
1000mm deep, 46.1 kg/m	176.09	m
1500mm deep, 54.2 kg/m	208.27	m

Metalwork

Steel; galvanised - 30 x 2.5mm standard strapping

600mm long	0.74	nr
800mm long	1.08	nr
1000mm long	1.39	nr
1200mm long	1.68	nr
1600mm long	2.18	nr

Truss clips

for 38mm thick members	0.29	nr
for 50mm thick members	0.31	nr

Joist hangers; BAT Building Products Ltd

SPH type S

for 50 x 100mm joist	1.55	nr
for 50 x 150mm joist	1.59	nr
for 50 x 200mm joist	1.84	nr
for 50 x 250mm joist	2.52	nr
for 63 x 100mm joist	1.98	nr
for 63 x 150mm joist	2.30	nr
for 63 x 200mm joist	2.30	nr
for 63 x 250mm joist	2.63	nr

SPH type R

for 50 x 100mm joist	2.35	nr
for 50 x 150mm joist	2.46	nr
for 50 x 200mm joist	2.72	nr
for 50 x 250mm joist	3.53	nr
for 63 x 100mm joist	2.87	nr
for 63 x 150mm joist	3.23	nr
for 63 x 200mm joist	3.23	nr
for 63 x 250mm joist	3.65	nr

Steel; galvanised - truss plates and framing anchors

51 x 114mm	0.16	nr
76 x 254mm	0.60	nr
114 x 152mm	0.48	nr
114 x 254mm	0.82	nr
152 x 152mm	0.66	nr

Galvanized steel lintels; SUPERGALV (BIRTLEY) lintels reference CB 50

125 x 750mm	20.97	nr
125 x 1200mm	33.07	nr
125 x 1350mm	39.03	nr
125 x 1500mm	43.12	nr
125 x 1650mm	49.16	nr
165 x 1800mm	53.19	nr
165 x 1950mm	58.37	nr
165 x 2100mm	61.10	nr
165 x 2250mm	70.02	nr
165 x 2400mm	74.06	nr
215 x 2550mm	83.48	nr
215 x 2850mm	113.76	nr
215 x 3000mm	121.13	nr
215 x 3300mm	139.65	nr
215 x 3600mm	160.01	nr
215 x 3900mm	195.26	nr

Galvanized steel lintels; SUPERGALV (BIRTLEY) lintels reference CB 50 H.D

165 x 750mm	24.75	nr
165 x 1050mm	32.59	nr
165 x 1200mm	37.13	nr
165 x 1350mm	45.14	nr
165 x 1500mm	49.76	nr
215 x 1650mm	59.54	nr
215 x 1800mm	65.00	nr
215 x 2100mm	76.66	nr
215 x 2250mm	96.39	nr
215 x 2400mm	113.11	nr
215 x 2550mm	122.89	nr
215 x 3150mm	185.39	nr

Galvanized steel lintels; SUPERGALV (BIRTLEY) lintels reference CB 70

120 x 750mm	20.81	nr
120 x 1200mm	33.06	nr
120 x 1500mm	43.13	nr
160 x 1800mm	53.05	nr
215 x 2400mm	74.06	nr
215 x 3000mm	120.78	nr
215 x 3900mm	187.43	nr

Galvanized steel lintels; SUPERGALV (BIRTLEY) lintels reference CB 70 H.D

160 x 750mm	24.77	nr
160 x 1200mm	37.13	nr
160 x 1500mm	49.77	nr
215 x 1800mm	69.51	nr
215 x 2400mm	113.14	nr
215 x 3300mm	185.44	nr

Galvanized steel lintels; SUPERGALV (BIRTLEY) lintels reference CB 50/130

120 x 750mm	21.06	nr
120 x 1200mm	33.47	nr
120 x 1500mm	42.42	nr
165 x 1800mm	53.15	nr
165 x 2400mm	77.47	nr
215 x 3900mm	230.48	nr

Galvanized steel lintels; SUPERGALV (BIRTLEY) lintels reference CB 50/130 H.D

160 x 750mm	30.09	nr
160 x 1200mm	45.04	nr
160 x 1500mm	62.09	nr
215 x 1800mm	76.98	nr
215 x 2400mm	119.19	nr
215 x 3300mm	189.18	nr

Galvanized steel lintels; SUPERGALV (BIRTLEY) lintels reference CB 70/130

115 x 750mm	25.49	nr
115 x 1200mm	41.99	nr
115 x 1500mm	53.04	nr
155 x 1800mm	72.96	nr
155 x 2400mm	121.24	nr
210 x 3300mm	220.09	nr

Galvanized steel lintels; SUPERGALV (BIRTLEY) lintels reference HD 50

225 x 3900mm	229.23	nr
225 x 4200mm	271.10	nr
225 x 4500mm	299.66	nr
225 x 4800mm	319.61	nr

Metalwork (Cont'd)

Galvanized steel lintels; SUPERGALV (BIRTLEY) lintels reference HD 70

225 x 3000mm	186.14	nr
225 x 3300mm	204.74	nr
225 x 3600mm	224.39	nr

Galvanized steel lintels; SUPERGALV (BIRTLEY) lintels reference SB100

75 x 750mm	13.96	nr
75 x 1200mm	21.47	nr
140 x 1800mm	41.12	nr
140 x 2400mm	57.61	nr
215 x 3000mm	105.91	nr

Galvanized steel lintels; SUPERGALV (BIRTLEY) lintels reference SB140

140 x 750mm	17.64	nr
140 x 1200mm	28.70	nr
140 x 1800mm	48.28	nr
140 x 2400mm	68.88	nr
215 x 3000mm	116.62	nr

Galvanized steel lintels; SUPERGALV (BIRTLEY) lintels reference SBL200

142 x 750mm	24.39	nr
142 x 1200mm	38.35	nr
142 x 1800mm	60.61	nr
142 x 2400mm	85.45	nr
218 x 3000mm	144.12	nr

Galvanized steel lintels; SUPERGALV (BIRTLEY) internal door lintels reference INT100

100 x 900mm	11.08	nr
100 x 1050mm	12.94	nr
100 x 1200mm	14.81	nr

Mechanical Installations

Red Primer steel pipes, BS 1387 Table 2; steel fittings BS EN 10241

15mm pipes	1.90	m
15mm fittings; one end	0.46	nr
15mm fittings; two ends	0.47	nr
15mm fittings; three ends	0.64	nr
20mm pipes	2.24	m
20mm fittings; one end	0.59	nr
20mm fittings; two ends	0.64	nr
20mm fittings; three ends	0.93	nr
25mm pipes	3.21	m
25mm fittings; one end	0.66	nr
25mm fittings; two ends	0.99	nr
25mm fittings; three ends	1.34	nr
32mm pipes	3.98	m
32mm fittings; one end	1.05	nr
32mm fittings; two ends	1.67	nr
32mm fittings; three ends	2.27	nr
40mm pipes	4.62	m
40mm fittings; one end	1.25	nr
40mm fittings; two ends	2.81	nr
40mm fittings; three ends	3.49	nr
50mm pipes	6.51	m
50mm fittings; one end	2.61	nr
50mm fittings; two ends	3.29	nr
50mm fittings; three ends	5.02	nr

Red Primer steel pipes, BS 1387 Table 3; steel fittings BS EN 10241

15mm pipes	2.26	m
15mm fittings; one end	0.46	nr
15mm fittings; two ends	0.47	nr
15mm fittings; three ends	0.64	nr
20mm pipes	2.69	m
20mm fittings; one end	0.59	nr
20mm fittings; two ends	0.64	nr
20mm fittings; three ends	0.93	nr
25mm pipes	3.93	m
25mm fittings; one end	0.66	nr
25mm fittings; two ends	0.99	nr
25mm fittings; three ends	1.34	nr
32mm pipes	4.87	m
32mm fittings; one end	1.05	nr
32mm fittings; two ends	1.67	nr
32mm fittings; three ends	2.27	nr
40mm pipes	5.68	m
40mm fittings; one end	1.25	nr
40mm fittings; two ends	2.81	nr
40mm fittings; three ends	3.49	nr
50mm pipes	7.89	m
50mm fittings; one end	2.61	nr
50mm fittings; two ends	3.29	nr
50mm fittings; three ends	5.02	nr

Black steel pipe brackets

15mm malleable iron	0.75	m
20mm malleable iron	0.84	nr
25mm malleable iron	0.97	nr
32mm malleable iron	1.33	nr
40mm malleable iron	1.76	nr
50mm malleable iron	2.35	m

Galvanized steel pipes, BS 1387 Table 2; steel fittings BS EN 10241

15mm pipes	2.93	m
15mm pipe connectors	0.62	nr
15mm fittings; one end	0.64	nr
15mm fittings; two ends	0.64	nr
15mm fittings; three ends	0.88	nr
20mm pipes	3.31	m
20mm pipe connectors	0.82	nr
20mm fittings; one end	0.73	nr
20mm fittings; two ends	0.88	nr
20mm fittings; three ends	1.29	nr
25mm pipes	4.63	m
25mm pipe connectors	1.05	nr
25mm fittings; one end	0.98	nr
25mm fittings; two ends	1.37	nr
25mm fittings; three ends	1.86	nr
32mm pipes	5.73	m
32mm pipe connectors	2.08	nr
32mm fittings; one end	1.51	nr
32mm fittings; two ends	2.57	nr
32mm fittings; three ends	3.52	nr
40mm pipes	6.66	m
40mm pipe connectors	2.81	nr
40mm fittings; one end	1.92	nr
40mm fittings; two ends	3.85	nr
40mm fittings; three ends	4.82	nr
50mm pipes	9.34	m
50mm pipe connectors	4.07	nr
50mm fittings; one end	3.58	nr
50mm fittings; two ends	4.50	nr
50mm fittings; three ends	6.95	nr

Galvanized steel pipes, BS 1387 Table 3; steel fittings BS EN 10241

15mm pipes	3.47	m
15mm pipe connectors	0.62	nr
15mm fittings; one end	0.64	nr
15mm fittings; two ends	0.64	nr
15mm fittings; three ends	0.88	nr
20mm pipes	3.94	m
20mm pipe connectors	0.82	nr
20mm fittings; one end	0.73	nr
20mm fittings; two ends	0.88	nr
20mm fittings; three ends	1.29	nr
25mm pipes	5.62	m
25mm pipe connectors	1.05	nr
25mm fittings; one end	0.98	nr
25mm fittings; two ends	1.37	nr
25mm fittings; three ends	1.86	nr
32mm pipes	6.97	m
32mm pipe connectors	2.08	nr
32mm fittings; one end	1.51	nr
32mm fittings; two ends	2.57	nr
32mm fittings; three ends	3.52	nr
40mm pipes	8.14	m
40mm pipe connectors	2.81	nr
40mm fittings; one end	1.92	nr
40mm fittings; two ends	3.85	nr
40mm fittings; three ends	4.82	nr
50mm pipes	11.28	m
50mm pipe connectors	4.07	nr
50mm fittings; one end	3.58	nr
50mm fittings; two ends	4.50	nr
50mm fittings; three ends	6.95	nr

Galvanized steel pipe brackets

15mm malleable iron	0.99	m
20mm malleable iron	1.10	nr
25mm malleable iron	1.28	nr
32mm malleable iron	1.75	nr
40mm malleable iron	2.34	nr
50mm malleable iron	3.09	m

Stainless steel pipes, BS 4127 Part 2

15mm pipes	3.05	m
22mm pipes	4.90	m
28mm pipes	6.50	m
35mm pipes	7.70	m
42mm pipes	10.35	m

Copper pipes, BS 2871 Part 1 Table X

15mm pipes	0.76	m
22mm pipes	1.53	m
28mm pipes	2.16	m
35mm pipes	4.69	m
42mm pipes	5.75	m
54mm pipes	7.37	m
66.7mm pipes	9.96	m
76.1mm pipes	14.07	m
108mm pipes	20.38	m

Copper pipes, BS 2871 Part 1 Table Y

15mm pipes	1.76	m
22mm pipes	3.08	m
28mm pipes	3.99	m
35mm pipes	6.02	m
42mm pipes	7.30	m
54mm pipes	12.48	m

Capillary fittings for copper pipes

15mm fittings; one end	0.82	nr
15mm fittings; two ends	0.21	nr
15mm fittings; three ends	0.39	nr
22mm fittings; one end	1.55	nr
22mm fittings; two ends	0.53	nr
22mm fittings; three ends	1.23	nr
28mm fittings; one end	2.76	nr
28mm fittings; two ends	1.23	nr
28mm fittings; three ends	3.40	nr
35mm fittings; one end	6.09	nr
35mm fittings; two ends	5.33	nr
35mm fittings; three ends	8.68	nr
42mm fittings; one end	10.49	nr
42mm fittings; two ends	8.81	nr
42mm fittings; three ends	13.93	nr
54mm fittings; one end	14.64	nr
54mm fittings; two ends	18.20	nr
54mm fittings; three ends	28.09	nr

Compression fittings type A for copper pipes

15mm fittings; one end	1.00	nr
15mm fittings; two ends	0.76	nr
15mm fittings; three ends	1.11	nr
22mm fittings; one end	1.19	nr
22mm fittings; two ends	1.30	nr
22mm fittings; three ends	1.82	nr
28mm fittings; one end	3.88	nr
28mm fittings; two ends	4.66	nr
28mm fittings; three ends	8.48	nr
35mm fittings; one end	7.00	nr
35mm fittings; two ends	11.72	nr
35mm fittings; three ends	15.50	nr
42mm fittings; one end	11.45	nr
42mm fittings; two ends	16.40	nr
42mm fittings; three ends	23.85	nr
54mm fittings; one end	15.98	nr
54mm fittings; two ends	27.88	nr
54mm fittings; three ends	38.38	nr

Hepworth Hep20 polybutylene pipes

15mm pipes	0.63	m
22mm pipes	1.24	m
28mm pipes	1.71	m

Blue polythene pipes, BS EN 12201

20mm pipes	0.27	m
25mm pipes	0.36	m
32mm pipes	0.55	m
50mm pipes	1.39	m
63mm pipes	2.13	m

PVC-U pipes, BS 3505 Class E

3/8 pipes	1.45	m
1/2 pipes	1.86	m
3/4 pipes	2.67	m
1 pipes	3.10	m

Mechanical Installations (Cont'd)

PVC-U pipes, BS 3505 Class E (Cont'd)

1 1/4 pipes	4.54	m
1 1/2 pipes	5.93	m

PVC-U pipes self coloured white

19mm pipes	0.74	m

Ductile iron pipes

80mm pipes	22.58	m
100mm pipes	21.45	m
150mm pipes	27.33	m

Brass stopcocks

crutch head; DZR; joints to polythene

20mm	18.22	nr
25mm	29.03	nr
32mm	38.94	nr
50mm	97.34	nr

crutch head; compression joints to copper (Type A)

15mm	3.04	nr
22mm	5.29	nr
28mm	12.70	nr
35mm	38.15	nr
42mm	48.15	nr
54mm	72.68	nr

crutch head; DZR; compression joints to copper (Type A)

15mm	11.15	nr
22mm	18.31	nr
28mm	30.29	nr
35mm	55.74	nr
42mm	79.79	nr
54mm	108.68	nr

Polybutylene stopcocks fitted with Hep2O ends

15mm	3.33	nr
22mm	3.99	nr

Brass gate valves

BS 5154 series B; compression joints to copper (Type A)

15mm	5.48	nr
22mm	6.66	nr
28mm	9.17	nr
35mm	17.26	nr
42mm	28.16	nr
54mm	43.60	nr

DZR; fitted with Hep2O ends

15mm	3.11	nr
22mm	4.39	nr

Lever ball valves, screwed and PTFE joints

13mm	3.14	5.65
19mm	4.24	7.27
25mm	5.45	10.02
32mm	7.52	16.02
38mm	12.02	nr
51mm	16.35	nr

13mm Brass ball valves for low pressure; BS 1212;

part 1, copper float	6.29	nr
part 2, copper float	7.43	nr
part 1, plastics float	4.24	nr
part 2, plastics float	5.38	nr

Brass ball valves for high pressure; BS 1212

13mm; part 1, copper	5.04	nr
13mm; part 2, copper	7.27	nr
19mm; part 1, copper	14.76	nr
25mm; part 1, copper	44.29	nr
13mm; part 1, plastics	2.99	nr
13mm; part 2, plastics	5.22	nr
19mm; part 1, plastics	9.76	nr
25mm; part 1, plastics	37.79	nr

GRP Sarena mfg Ltd water storage tanks with cover

ref. SC10; 18 litres; pre insulated	76.00	nr
ref. SC20; 55 litres; pre insulated	106.00	nr
ref. SC30; 86 litres; pre insulated	116.00	nr
ref. SC40; 114 litres; pre insulated	142.00	nr
ref. SC50; 159 litres; pre insulated	161.00	nr
ref. SC60; 191 litres; pre insulated	178.00	nr
ref. SC70; 227 litres; pre insulated	198.00	nr
ref. SC100/1; 327 litres; pre insulated	261.00	nr
ref. SC100/2; 336 litres; pre insulated	254.00	nr
ref. SC125; 423 litres; pre insulated	322.00	nr
ref. SC150; 491 litres; pre insulated	383.00	nr
ref. SC200; 709 litres; pre insulated	445.00	nr
ref. SC250; 900 litres; pre insulated	510.00	nr
ref. SC350; 1250 litres; pre insulated	650.00	nr
ref. SC500; 1727 litres; pre insulated	850.00	nr
ref. SC600; 2137 litres; pre insulated	980.00	nr

Plastics water storage cisterns with sealed lid / Byelaw 30 thick

18 litres	12.10	nr
114 litres	40.45	nr
182 litres	63.90	nr
227 litres	69.55	nr

Copper direct hot water cylinders; pre-insulated

ref. 3; 116 litres; four bosses	122.25	nr
ref. 7; 120 litres; four bosses	109.25	nr
ref. 8; 144 litres; four bosses	136.50	nr
ref. 9; 166 litres; four bosses	176.00	nr

Copper indirect hot water cylinders; pre-insulated

ref. 7; 108 litres; four bosses	232.75	nr
ref. 8; 130 litres; four bosses	274.25	nr

Cast iron rainwater pipes and fittings, BS 460 Type A sockets; dry joints

75mm pipes	13.99	m
75mm shoes	11.56	nr
75mm bends	9.91	nr
75mm offset bends 150mm projection	12.50	nr
75mm offset bends 225mm projection	14.55	nr
75mm offset bends 300mm projection	17.88	nr
75mm angled branches	17.35	nr
100mm pipes	19.09	m
100mm shoes	15.58	nr
100mm bends	13.99	nr
100mm offset bends 75mm projection	23.57	nr
100mm offset bends 150mm projection	23.57	nr
100mm offset bends 225mm projection	28.55	nr
100mm offset bends 300mm projection	29.10	nr
100mm angled branches	20.62	nr

Cast iron rainwater pipes and fittings, BS 460 Type A sockets; ears cast on; dry joints

65mm pipes	14.95	m
65mm shoes	13.33	nr
65mm bends	8.16	nr
75mm pipes	14.95	m
75mm shoes	13.33	nr
75mm bends	9.91	nr
100mm pipes	20.07	m
100mm shoes	17.70	nr
100mm bends	13.99	nr

PVC-U rainwater pipes and fittings, BS EN 607; push fit joints; self coloured grey

68mm pipes	2.71	m
68mm pipe brackets	1.39	nr
68mm shoes	3.21	nr
68mm bends	3.72	nr

Cast iron half round gutters and fittings, BS 460

100mm gutters	7.68	m
100mm gutter brackets	1.65	nr
100mm stopped ends	1.97	nr
100mm running outlets	5.70	nr
100mm stopped ends with outlet	6.69	nr
100mm angles	5.85	nr
115mm gutters	8.00	m
115mm gutter brackets	1.65	nr
115mm stopped ends	2.55	nr
115mm running outlets	6.22	nr
115mm stopped ends with outlet	7.50	nr
115mm angles	6.03	nr
125mm gutters	9.37	m
125mm gutter brackets	1.65	nr
125mm stopped ends	2.55	nr
125mm running outlets	7.11	nr
125mm stopped ends with outlet	8.39	nr
125mm angles	7.11	nr
150mm gutters	15.99	m
150mm gutter brackets	2.09	nr
150mm stopped ends	3.54	nr
150mm running outlets	12.30	nr
150mm stopped ends with outlet	14.07	nr
150mm angles	12.98	nr

Cast iron box gutters

100 x 75mm gutters	23.95	m
100 x 75mm stopped ends	3.64	nr
100 x 75mm running outlets	17.50	nr
100 x 75mm gutter brackets	3.47	nr

Cast aluminium half round gutters and fittings, BS EN 612 Section A

100mm gutters	8.64	m
100mm gutter brackets	1.52	nr
100mm stopped ends	2.58	nr
100mm running outlets	6.11	nr
100mm angles	5.31	nr
125mm gutters	10.80	m
125mm gutter brackets	2.79	nr
125mm stopped ends	3.42	nr
125mm running outlets	7.00	nr
125mm angles	6.79	nr

PVC-U half round gutters and fittings, BS 4576, Part 1; push fit connector joints; self coloured grey

114mm gutters	2.63	nr
114mm Union	3.20	m
114mm gutter brackets	0.85	nr
114mm stopped ends	2.05	nr
114mm running outlets	3.81	nr
114mm angles	4.16	nr

Cast iron soil and waste pipes and fittings, BS 416

75mm pipes	11.21	m
75mm pipe couplings	5.73	nr
100mm pipes	13.55	m
100mm pipe couplings	7.48	nr
50mm pipes	10.02	m
50mm pipe couplings	5.19	nr

MuPVC waste pipes and fittings, BS EN 1329-1 Section Three; solvent welded joints

32mm pipes	1.24	m
32mm pipe couplings	0.82	nr
32mm pipe brackets	0.26	nr

Mechanical Installations (Cont'd)

MuPVC waste pipes and fittings, BS EN 1329-1 Section Three; solvent welded joints (Cont'd)

32mm fittings; three ends	1.13	nr
40mm pipes	1.54	m
40mm pipe couplings	0.81	nr
40mm pipe brackets	0.31	nr
40mm fittings; three ends	1.41	nr
50mm pipes	2.33	m
50mm pipe couplings	1.49	nr
50mm pipe brackets	0.41	nr
50mm fittings; three ends	2.76	nr

Polypropylene waste pipes and fittings; butyl ring joints

32mm pipes	0.73	m
32mm pipe couplings	0.86	nr
32mm fittings; three ends	1.33	nr
40mm pipes	0.92	m
40mm pipe couplings	0.95	nr
40mm pipe brackets	0.29	nr
40mm fittings; three ends	1.38	nr

PVC-U soil and vent pipes and fittings, BS 4514; rubber ring joints; self coloured grey

82mm pipes	6.36	m
82mm pipe couplings	4.03	nr
110mm pipes	5.44	m
110mm pipe couplings	4.24	nr
160mm pipes	15.13	m
160mm pipe couplings	9.71	nr

Pipe supports for PVC-U soil and vent pipes; for building in

for 82mm pipes	2.99	nr
for 110mm pipes	4.28	nr

Pipe supports for PVC-U soil and vent pipes; for fixing to timber

for 82mm pipes	2.41	nr
for 110mm pipes	3.17	nr
for 160mm pipes	6.05	nr

PVC balloon gratings

50mm pipes	1.56	nr
82mm pipes	2.28	nr
110mm pipes	2.37	nr

Plastic traps, two piece, 75mm seal, inlet with coupling nut, outlet with seal ring socket

32mm outlet, P	3.34	nr
40mm outlet, P	3.86	nr
32mm outlet, S	4.23	nr
40mm outlet, S	4.95	nr
40mm outlet, P, with flexible polypropylene pipe for overflow connection	7.15	nr

Copper traps, two piece, 75mm seal, inlet with coupling nut, compression outlet

35mm outlet, P	17.40	nr
42mm outlet, P	20.50	nr
54mm outlet, P	73.63	nr
35mm outlet, S	18.44	nr
42mm outlet, S	22.18	nr
54mm outlet, S	77.69	nr
42mm outlet, bath, overflow connection, cleaning eye and plug	29.10	nr

Pressed steel radiators single panel with convector; air cock, plain plug

450 x 600mm long	20.60	nr
450 x 900mm long	30.88	nr
450 x 1200mm long	41.16	nr
450 x 1600mm long	62.83	nr
450 x 2400mm long	126.76	nr
600 x 600mm long	26.57	nr
600 x 900mm long	39.86	nr
600 x 1200mm long	53.16	nr
600 x 1600mm long	81.11	nr
600 x 2400mm long	163.65	nr
700 x 600mm long	30.48	nr
700 x 900mm long	52.27	nr
700 x 1200mm long	93.75	nr
700 x 1600mm long	125.02	nr
700 x 2400mm long	187.53	nr

Electrical Installations

Lighting columns, aluminium; BS EN 40-1, nominal mounting height

4.88m	258.67	nr
8.00m	637.18	nr
10.00m	809.65	nr
12.00m	1135.02	nr

Lighting columns, steel; BS EN 40-1, nominal mounting height

4.00m	134.76	nr
5.00m	138.87	nr
6.00m	171.14	nr
8.00m	253.96	nr

Steel straight lighting trunking; PVC lid

50 x 50mm	5.44	3m
end cap	2.10	nr
internal angle	10.31	nr
external angle	10.31	nr
tee	11.04	nr
four way intersection	11.43	nr

Steel straight dado trunking, two compartment; steel lid

150 x 30mm	19.51	3m
end cap	4.70	nr
internal angle	21.78	nr
external angle	21.78	nr
single socket plate	8.57	nr
twin socket plate	9.10	nr

Steel straight dado trunking, three compartment; steel lid

200 x 38mm	25.77	3m
end cap	5.14	nr
internal angle	26.88	nr
external angle	26.88	nr
single socket plate	8.85	nr
twin socket plate	9.49	nr

Steel straight skirting trunking; two compartments

200 x 38mm	25.73	3m
internal bend	25.41	nr
external bend	25.41	nr

Steel straight underfloor trunking; two compartments

150 x 25mm	17.50	3m
floor junction boxes; adjustable frame and trap	70.00	nr
horizontal bend	17.09	nr

PVC trunking; white straight mini-trunking; clip on lid

16 x 16mm	0.92	3m
end cap	0.28	nr
internal bend	0.28	nr
external bend	0.28	nr
flat bend	0.28	nr
equal tee	0.49	nr
25 x 16mm	1.03	3m
end cap	0.28	nr
internal bend	0.28	nr
external bend	0.28	nr
flat bend	0.28	nr
equal tee	0.49	nr
40 x 16mm	1.58	3m
end cap	0.29	nr
internal bend	0.29	nr
external bend	0.29	nr
flat bend	0.29	nr
equal tee	0.51	nr

PVC trunking; white straight dado trunking, three compartment; clip on lid

145 x 40mm (3m)	59.97	3m
end caps	1.40	nr
internal bend	4.81	nr
external bend	4.81	nr
fabricated flat tee	37.80	nr
fabricated flat elbow	32.40	nr
mini trunking adaptor	1.36	nr
230 x 40mm (3m)	101.04	3m
end caps	4.47	nr
internal bend	9.36	nr
external bend	9.36	nr
mini trunking adaptor	1.39	nr

PVC trunking; white straight skirting trunking, three compartment; clip on lid

100 x 40mm	10.13	3m
end cap	1.94	nr
internal corner	2.49	nr
external corner	2.49	nr
mini trunking adaptor	1.75	nr
socket adaptor, one gang	3.91	nr
socket adaptor, two gang	5.62	nr

PVC heavy gauge conduits; straight with small circular boxes

20mm diameter conduit	1.49	3m
terminal boxes	0.86	nr
angle boxes	0.96	nr
three way boxes	1.04	nr
through way boxes	0.96	nr
25mm diameter conduit	2.87	3m
terminal boxes	1.32	nr
angle boxes	1.44	nr
three way boxes	1.59	nr
through way boxes	1.44	nr

PVC light gauge conduits; straight with small circular boxes

20mm diameter conduit	1.05	3m
terminal boxes	0.86	nr
angle boxes	0.96	nr
three way boxes	1.04	nr
through way boxes	0.96	nr
25mm diameter conduit	2.02	3m
terminal boxes	1.32	nr
angle boxes	1.44	nr
three way boxes	1.59	nr
through way boxes	1.44	nr

PVC heavy gauge oval conduits; straight

20mm (nominal size)	0.85	3m
25mm (nominal size)	0.90	3m

Steel welded black enamelled heavy gauge conduits; straight with small circular boxes

20mm diameter conduit	2.82	3.75m
terminal boxes	0.75	nr
angle boxes	0.85	nr
three way boxes	0.75	nr
through way boxes	0.80	nr
25mm diameter conduit	3.67	3.75m
terminal boxes	0.88	nr
angle boxes	0.99	nr
three way boxes	0.88	nr
through way boxes	0.94	nr
32mm diameter conduit	7.67	3.75m

Steel welded galvanised heavy gauge conduits; straight with small circular boxes

20mm diameter conduit	5.75	3.75m
terminal boxes	0.89	nr
angle boxes	1.01	nr
three way boxes	0.89	nr
through way boxes	0.95	nr
25mm diameter conduit	7.60	3.75m
terminal boxes	1.16	nr
angle boxes	1.31	nr
three way boxes	1.16	nr
through way boxes	1.24	nr
32mm diameter conduit	10.39	3.75m

Galvanised light duty standard cable tray and fittings; straight

50mm tray	3.01	3m
flat bend	6.57	nr
tee	8.52	nr
cross piece	14.38	nr
riser bend	10.25	nr
100mm tray	4.85	3m
flat bend	7.09	nr
tee	9.37	nr
cross piece	15.32	nr
riser bend	11.82	nr

Electrical Installations (Cont'd)

Galvanised light duty standard cable tray and fittings; straight (Cont'd)

reducer 100 to 50mm	8.99	nr
150mm tray	6.13	3m
flat bend	8.28	nr
tee	11.37	nr
cross piece	17.99	nr
riser bend	15.95	nr
reducer 150 to 100mm	11.84	nr
300mm tray	11.99	3m
flat bend	15.27	nr
tee	20.93	nr
cross piece	32.16	nr
riser bend	28.09	nr
reducer 300 to 150mm	21.54	nr
450mm tray	24.21	3m
flat bend	24.45	nr
tee	34.14	nr
cross piece	53.08	nr
riser bend	39.78	nr
reducer 450 to 300mm	35.75	nr

Galvanised medium duty standard cable tray and fittings; straight

75mm tray	6.13	3m
flat bend	24.74	nr
tee	32.58	nr
cross piece	51.70	nr
riser bend	13.44	nr
100mm tray	8.49	3m
flat bend	24.74	nr
tee	32.58	nr
cross piece	50.29	nr
riser bend	12.30	nr
reducer 100 to 75mm	11.41	nr
150mm tray	10.00	3m
flat bend	27.88	nr
tee	37.06	nr
cross piece	55.77	nr
riser bend	16.13	nr
reducer 150 to 100mm	9.91	nr
300mm tray	13.51	3m
flat bend	39.66	nr
tee	49.18	nr
cross piece	75.71	nr
riser bend	26.59	nr
reducer 300 to 150mm	21.76	nr
450mm tray	18.48	3m
flat bend	52.48	nr
tee	68.75	nr
cross piece	103.04	nr
riser bend	40.11	nr
reducer 450 to 300mm	35.89	nr

Galvanised heavy duty standard cable tray and fittings; straight

75mm tray	9.12	3m
flat bend	36.36	nr
tee	44.70	nr
cross piece	72.21	nr
riser bend	24.99	nr
100mm tray	9.27	3m
flat bend	36.84	nr
tee	45.08	nr
cross piece	72.78	nr
riser bend	25.61	nr
reducer 100 to 75mm	16.66	nr
150mm tray	10.26	3m
flat bend	39.98	nr
tee	49.56	nr
cross piece	78.83	nr
riser bend	27.81	nr
reducer 150 to 100mm	16.72	nr
300mm tray	14.94	3m
flat bend	45.89	nr
tee	59.88	nr
cross piece	94.78	nr
riser bend	30.55	nr
reducer 300 to 150mm	27.42	nr
450mm tray	25.83	3m
flat bend	77.78	nr
tee	91.16	nr
cross piece	146.34	nr
riser bend	53.67	nr
reducer 450 to 300mm	42.87	nr

Supports for cable tray; ms; galv

65mm cantilever arms	4.14	nr
90mm cantilever arms	4.14	nr
120mm cantilever arms	4.14	nr
75mm stand-off bracket	4.14	nr
100mm stand-off bracket	5.33	nr
150mm stand-off bracket	5.33	nr
300mm stand-off bracket	11.76	nr
450mm stand-off bracket	16.10	nr

PVC insulated cables; single core; reference 6491X; stranded copper conductors; to BS 6004

1.5mm2 (100m)	4.96	100m
2.5mm2 (100m)	7.89	100m
4.0mm2 (100m)	19.20	100m
6.0mm2 (100m)	23.15	100m
10.0mm2 (100m)	33.90	100m
16.0mm2 (100m)	56.15	100m

PVC double insulated cables; single core; reference 6181Y; stranded copper conductors; to BS 6004

16.0mm2 meter tails, 1m long	0.78	m
25.0mm2 meter tails, 1m long	0.99	m
35.0mm2 meter tails, 1m long	2.46	m

PVC insulated and PVC sheathed cables; multicore; copper conductors and bare earth continuity conductor 6242Y; to BS 6004

1.00mm2; twin and earth (100m)	10.95	100m
1.50mm2; twin and earth (100m)	14.45	100m
2.5mm2; twin and earth (100m)	15.99	100m
4.00mm2; twin and earth (100m)	44.60	100m
6.00mm2; twin and earth (100m)	49.50	100m

PVC insulated SWA armoured and LSF sheathed cables; multicore; copper conductors to BS 6724

16mm2; 3 core	2.95	m

PVC Sheathed, polyethylene insulated cables

Co-axial, braid screened, copper conductor 75 ohm (100m)	10.00	100m

Telephone cables

Mineral insulated copper sheathed cables; PVC outer sheath; copper conductors; 600V light duty to BS EN 60702-1

1.5mm2; two core cable	2.17	m
1.5mm2; termination including gland, seal and shroud	2.77	nr
2.5mm2; two core cable	3.18	m
2.5mm2; termination including gland, seal and shroud	2.77	nr
4.0mm2; two core cable	4.03	m
4.0mm2; termination including gland, seal and shroud	2.77	nr

HV switchgear

Switch fuse 500V, metal cased, 32 amp; single pole and neutral; short circuit rating 6.4Ka	74.76	nr
Consumer control units, 250V, metal cased, fitted with 100 amp D.P. isolator and MCB's, eight way, single pole and neutral	85.83	nr
Distribution boards 500V, metal cased, MCB pattern, fitted with 100 amp D.P. isolator, nine way, single pole and neutral, 20 amp	106.68	nr
Distribution boards 500V, metal cased, MCB pattern, fitted with 100 amp D.P. isolator, twelve way, triple pole and neutral, 32 amp	473.98	nr

Accessories; white plastics, with boxes

Flush plate switches 5 amp; one gang; one way; single pole	1.35	nr
Flush plate switches 5 amp; two gang; two way; single pole	1.85	nr
Surface plate switches 5 amp; one gang; one way; single pole	1.45	nr
Surface plate switches 5 amp; two gang; two way; single pole	1.95	nr
Ceiling switches 5 amp; one gang; one way; single pole	1.85	nr
Flush switched socket outlets 13 amp; one gang	2.60	nr
Flush switched socket outlets 13 amp; two gang	3.25	nr
Surface switched socket outlets 13 amp; one gang	2.70	nr
Surface switched socket outlets 13 amp; two gang	3.45	nr
Flush switched socket outlets with RCCB (residual current device) protected at 30 mAmp 13 amp; one gang	32.50	nr
Flush fused connection units 13 amp; one gang; switched; flexible outlet	3.85	nr
Flush fused connection units 13 amp; one gang; switched; flexible outlet; pilot lamp	4.25	nr

Accessories, metalclad, with boxes

Surface plate switches 5 amp; one gang; two way; single pole	2.85	nr
Surface plate switches 5 amp; two gang; two way; single pole	3.55	nr
Surface switched socket outlets 13 amp; one gang	4.00	nr
Surface switched socket outlets 13 amp; two gang	5.75	nr

Weatherproof accessories, with boxes

Surface plate switches 5 amp; one gang; two way; single pole	6.04	nr

Domestic electrical fittings, wiring and ancillaries

Ceiling rose	2.03	nr
Pendant	1.35	nr
Light switch	1.35	nr
Socket outlet 5 Amp	5.25	nr
Cooker point 4 SA	9.75	nr
Immersion heater point	4.25	nr
Shaver socket (transformer)	12.75	nr
Conduit PVC 16mm	0.72	3m
Grommets	0.02	nr
Clips	0.02	nr
MICS cable 1.00mm2	1.89	m
MICS cable 2.50mm2	2.83	m
MICS cable 6.00mm2	6.12	m
Gland	0.55	nr
20mm black enamel conduit	2.82	3.75m

Electrical Installations (Cont'd)

Domestic electrical fittings, wiring and ancillaries (Cont'd)

20mm black enamel conduit bends	1.70	nr
20mm black enamel conduit tees	2.49	nr
20mm black enamel conduit box terminals	3.06	nr
20mm black enamel conduit couplers	0.26	nr
20mm black enamel conduit saddles	0.23	nr
25mm black enamel conduit	3.67	3.75m
25mm black enamel conduit bends	2.45	nr
25mm black enamel conduit tees	3.58	nr
25mm black enamel conduit couplers	0.41	nr
25mm black enamel conduit saddles	0.29	nr
TV Socket and box	5.45	nr
Telephone point	3.50	nr
Draw wire	0.25	m

Fluorescent lamp fittings, mains voltage operations; switch start; including lamps

Batten type; single tube		
1200mm; 36W	9.60	nr
1500mm; 58W	11.00	nr
1800mm; 70W	17.45	nr
2400mm; 100W	24.00	nr
Batten type; twin tube		
1200mm; 36W	15.55	nr
1500mm; 58W	18.56	nr
1800mm; 70W	27.55	nr
2400mm; 100W	36.50	nr
Metal trough reflector; single tube		
1200mm; 36W	32.16	nr
1500mm; 58W	35.21	nr
1800mm; 70W	41.50	nr
2400mm; 100W	59.06	nr
Metal trough reflector; twin tube		
1200mm; 36W	49.73	nr
1500mm; 58W	57.17	nr
1800mm; 70W	63.16	nr
2400mm; 100W	85.92	nr
Plastics diffused type; single tube		
1200mm; 36W	33.18	nr
1500mm; 58W	37.16	nr
1800mm; 70W	45.47	nr
2400mm; 100W	60.79	nr
Plastics diffused type; twin tube		
1200mm; 36W	61.93	nr
1500mm; 58W	75.23	nr
1800mm; 70W	84.99	nr
2400mm; 100W	108.53	nr

Lighting fittings complete with tungsten lamps

Bulkhead; alloy 100W G.L.S. lamp	9.00	nr
Wall glass; alloy; corner bracket 100W G.L.S. lamp	26.05	nr
Ceiling sphere 152mm; 60W G.L.S. lamp	14.00	nr
Ceiling sphere 203mm; 100W G.L.S. lamp	20.40	nr
Down light; semi recessed 175mm; 150W PAR38 lamp	17.32	nr
Lighting track; single circuit; 1250mm; starter pack	17.60	nr
Lighting track; single circuit; plug-in fitting; 100W G.L.S. lamp	13.50	nr

Fan heaters

3kw unit fan heaters for commercial application	152.06	nr
Thermostatic switch units	51.86	nr

Turbular heaters (excluding wire guards), 60W per 305mm loading

600mm long	10.80	nr
910mm long	11.89	nr
1220mm long	13.66	nr
1520mm long	17.85	nr
1830mm long	18.24	nr
Pair two way mounting brackets	5.19	nr

Night storage heaters

1.7KW night storage heater	200.00	nr
2.55KW night storage heater	270.00	nr
3.4KW night storage heater	310.00	nr
MICS 1000v 2 core cable 4.0mm	4.69	m
20 Amp double pole switch	3.25	nr
8 way fuse board	83.74	nr

Bell equipment

Transformers 3-5-8V	14.47	nr
Transformers 12V	18.52	nr
Bells for transformer operation chime 2-note	10.40	nr
Bells for transformer operation 76mm domestic type	5.73	nr
Bells for transformer operation 150mm round bell type	20.84	nr
Bell pushes domestic	1.45	nr
Bell pushes industrial	5.72	nr

2 or 3KW immersion heaters and thermostats; non-withdrawable elements

without flanges 305mm	22.91	nr
without flanges 457mm	23.14	nr
without flanges 686mm	23.25	nr
Thermostats		
immersion heater	10.04	nr
15 amp a.c. for air heating; without switch	15.16	nr

Water heaters, Storage type units

free outlet 7 litre	104.57	nr
free outlet 23 litre	461.65	nr
free outlet 55 litre	673.33	nr
multi-point 50 litre	634.77	nr
multi-point 75 litre	754.78	nr
multi-point 125 litre	863.83	nr

Instantaneous Shower units

complete with fittings	155.24	nr

Extract fans, window mounted; shutter and incorporated switch unit

105mm dia; ref DX100	21.15	nr
257mm dia; ref GX9	169.55	nr
324mm dia; ref GX12	264.95	nr
Louvre shutter	5.29	nr
Window kit	32.16	nr

Extract fans, wall mounted; shutter and incorporated switch unit

105mm dia; ref DX100	21.15	nr
257mm dia; ref GX9	169.55	nr
324mm dia; ref GX12	264.95	nr
473 x 476mm; ref WX12	317.95	nr
Wall duct	5.29	nr
Louvre shutter	5.29	nr
Switch	7.88	nr
Fan controller	44.52	nr

Floor Wall and Ceiling Finishes

Screeds and toppings

Cement and sand (1:3)	112.00	m3
Surface hardener for paving		
epoxy resin based	1.65	litre
polyurethane	6.20	litre
Nitoflor Lithurin	4.85	kg

Latex Flooring

One coat levelling screeds	0.48	kg

Gypsum plasterboard to BS 1230 Part 2

9.5mm handiboard	2.33	m2
9.5mm Duplex handiboard	2.76	m2
12.5mm handiboard	3.43	m2
9.5mm square edge wallboard	1.59	m2
12.5mm square edge wallboard	1.69	m2
15mm square edge wallboard	2.72	m2

Rendering and the like

Cement and sand (1:3)	112.00	m3
Pea shingle for pebbledash	21.35	tonne
Cement for wet slurry	97.90	tonne

Plasters

Board finish	128.40	tonne
Hardwall	186.00	tonne
Multi finish	138.30	tonne
Universal one coat	260.70	tonne
Thistle browning	171.60	tonne
Thistle finish	115.20	tonne
Thistle bonding	171.60	tonne
Thistle tough coat	176.10	tonne
Thistle renovating undercoat	206.40	tonne
Thistle renovating finish	181.50	tonne

Plaster beads and stops

Angle ref: 550	0.54	m
Angle ref: 553	0.54	m
Angle ref: 554	0.70	m
Stop ref: 560	0.74	m
Stop ref: 561	0.74	m
Stop ref: 562	0.69	m
Stop ref: 563	0.69	m
Stop ref: 565	0.88	m
Stop ref: 566	0.88	m
Plasterboard edging ref: 567	1.13	m
Plasterboard edging ref: 568	1.13	m

Bonding fluid

Bonding fluid	3.70	litre

Metal lathing

Expanded metal lath, 9mm mesh x 0.500mm thick x 1.11 kg/m2	4.75	m2
Expanded metal lath 9mm mesh x 0.725mm thick x 1.61 kg/m2	5.55	m2
Expanded metal lath 9mm mesh x 0.950mm thick x 2.50 kg/m2	9.00	m2
Rib-Lath, 0.300mm thick x 1.16 kg/m2	5.90	m2
Rib-Lath, 0.500mm thick x 1.90 kg/m2	6.85	m2
Stainless steel Rib-Lath, 0.300mm thick x 1.16 kg/m2	15.80	m2
Expanded metal lath 9.2mmx 28.6mm mesh x 0.460mm thick x 1.15 kg/m2	8.35	m2
Red-rib lath, 0.500mm thick x 1.91 kg/m2	7.50	m2

Galvanized arch formers, corners

372mm radius	13.80	nr
452mm radius	16.85	nr
602mm radius	21.00	nr
752mm radius	27.60	nr

Plasterboard coving

Gyproc cove 100mm girth	1.01	m
Gyproc cove 127mm girth	0.89	m

Floor Wall and Ceiling Finishes (Cont'd)

Plasterboard coving (Cont'd)

Coving Adhesive	0.74	kg

Clay floor quarry tiles - terracotta - square edge

150 x 150 x 12.5mm	328.00	1000
150 x 150 x 20mm	492.00	1000
225 x 225 x 25mm	1115.00	1000

Ceramic floor tiles - red - square edge

100 x 100 x 9mm	14.85	m2
152 x 152 x 12mm	16.35	m2
200 x 200 x 12mm	17.55	m2

Ceramic floor tiles - cream - square edge

100 x 100 x 9mm	15.75	m2
152 x 152 x 12mm	17.05	m2
200 x 200 x 12mm	18.25	m2

Glazed ceramic wall tiles to BS 6431

108 x 108 x 6.5mm - white	13.50	m2
152 x 152 x 5.5mm - white	9.10	m2
108 x 108 x 6.5mm - light colours	13.50	m2
152 x 152 x 5.5mm - light colours	9.10	m2
108 x 108 x 6.5mm - dark colours	15.70	m2
152 x 152 x 5.5mm - dark colours	11.30	m2

Fixing materials for glazed ceramic wall tiles

Cement and sand (1:3)	112.00	m3
Tile adhesive	1.00	litre
White cement	0.80	kg

Terrazzo tiles

300 x 300 x 28mm units	31.00	m2
305 x 305 x 28mm units rounded nosing	15.00	m
305 x 165 x 28mm units	7.50	m

Terrazzo

Terrazzo mix 1:2	1000.00	m3

Wood block flooring, tongued and grooved, 70 x 230 x 20mm units

Maple wood blocks	26.15	m2
Merbau wood blocks	21.15	m2
Oak wood blocks	30.15	m2

PVC flooring, 300 x 300mm tiles

2mm tiles	7.45	m2
2.5mm tiles	9.10	m2
2mm heavy duty tiles	13.60	m2
2.5mm heavy duty tiles	15.75	m2

Linoleum flooring

500 x 500 x 2mm tiles	14.60	m2
2mm sheet flooring	13.60	m2
3.2mm sheet flooring	17.10	m2
Adhesive for linoleum flooring	2.75	litre

Burmatex carpeting

Adhesive for carpeting	11.50	litre

Edge fixed carpeting

Tredaire underlay	1.30	m2
Tufted wool/nylon	20.00	m2
Axminster wool/nylon	26.00	m2
Wilton wool/nylon	32.50	m2

Wallpapers and wall coverings

Lining paper	1.00	roll
Pulp paper	3.00	roll
Washable	4.00	roll
Vinyl coated	6.50	roll
Embossed	4.50	roll
Textured	5.00	roll
Woodchip	1.50	roll

Plastic faced plasterboard

Thermaline BASIC vapour check grade plasterboard

22mm Gyproc Thermaline	5.87	m2
30mm Gyproc Thermaline	6.03	m2
40mm Gyproc Thermaline	7.17	m2
50mm Gyproc Thermaline	9.08	m2

Cellular partitions

57mm panels	10.20	m2
37 x 19mm softwood battens	0.32	m
37 x 37mm softwood battens	0.41	m

Solidwall partitions plasterboard components

25 x 38mm softwood battens	0.29	m

Metal stud partitions

48mm studs	1.15	m
50mm floor and ceiling channels	1.28	m

Suspended ceiling tiles, bevelled, grooved and rebated

1200 x 600 x 9mm asbestos-free fire resisting	15.80	m2
1200 x 600 x 15.8mm plain mineral fibre	9.60	m2
600 x 600 x 15.8mm plain mineral fibre	9.60	m2

Glazing

Float glass

3mm GG quality	37.63	m2
4mm GG quality	37.63	m2
5mm GG quality	55.01	m2
6mm GG quality	55.01	m2
10mm GG quality	111.43	m2
3mm obscured ground	81.12	m2
4mm obscured ground	89.01	m2

Rough cast glass

6mm	49.46	m2

Pyroshield glass

6mm clear	89.03	m2
6mm clear, aligning panes	89.03	m2
6mm safety clear	104.84	m2
6mm safety clear, aligning panes	104.84	m2
7mm texture	41.19	m2
7mm texture, aligning panes	41.19	m2
7mm safety texture	53.59	m2
7mm safety texture, aligning panes	53.59	m2

Patterned glass

4mm white	34.90	m2
6mm white	52.53	m2
4mm tinted	51.54	m2
6mm tinted	58.32	m2

Antisun float glass

4mm grey	71.87	m2
6mm grey	103.66	m2
10mm grey	185.34	m2
12mm grey	256.78	m2
4mm bronze	71.87	m2
6mm bronze	103.66	m2
10mm bronze	185.34	m2
12mm bronze	255.83	m2
6mm green	132.88	m2

Thick float glass

10mm GG quality	111.43	m2
12mm GG quality	128.47	m2
15mm GG quality	213.74	m2
19mm GG quality	302.76	m2
25mm GG quality	480.61	m2

Toughened safety glass

4mm clear float glasses	52.97	m2
5mm clear float glasses	79.49	m2
6mm clear float glasses	79.49	m2
10mm clear float glasses	126.08	m2
12mm clear float glasses	230.87	m2
4mm white	106.78	m2
6mm white	111.48	m2
4mm tinted	106.79	m2
6mm tinted	111.48	m2

Toughened safety solar control glasses

4mm, antisun grey	97.21	m2
6mm, antisun grey	101.34	m2
10mm, antisun grey	200.58	m2
12mm, antisun grey	263.93	m2
4mm , antisun bronze	97.13	m2
6mm , antisun bronze	101.34	m2
10mm , antisun bronze	201.92	m2
12mm , antisun bronze	263.93	m2

Clear laminated glass - float quality

4.4mm safety glass	70.02	m2
6.4mm safety glass	58.71	m2
8.8mm safety glass	105.66	m2
10.8mm safety glass	123.56	m2
7.5mm anti-bandit glass	122.51	m2
9.5mm anti-bandit glass	143.77	m2
11.5mm anti-bandit glass	164.14	m2
11.5mm anti-bandit glass, large panes	197.04	m2

Georgian wired glass

7mm standard	41.19	m2
7mm safety	53.59	m2
6mm polished	89.03	m2
6mm polishedsafety	104.84	m2

Sundries

linseed oil putty	13.89	25kg
metal casement putty	14.60	25kg
washleather strip, 3mm glass	0.54	m
washleather strip, 6mm glass	0.67	m
rubber glazing strip, 3mm glass	0.50	m
rubber glazing strip, 6mm glass	0.60	m

Painting and Decorating

Timber treatments

Creosote	157.50	200L
Wood preservative	18.75	5L

Emulsion paint

matt brilliant white	6.75	5L
matt colours	7.45	5L
silk brilliant white	7.95	5L
silk colours	8.75	5L

Priming, paint

wood primer	16.55	5L
primer sealer	22.25	5L
aluminium sealer and wood primer	27.50	5L
alkali-resisting primer	24.75	5L
universal primer	24.95	5L
red oxide primer	20.00	5L
calcium plumbate primer	32.75	5L
zinc phosphate primer	25.45	5L
acrylic primer	15.05	5L

Undercoat paint

Brilliant white undercoat	11.70	5L
colours	15.80	5L

Gloss paint

Brilliant white gloss	11.70	5L
colours	15.80	5L

Eggshell paint

Brilliant white eggshell	14.40	5L
colours	15.85	5L

Plastic finish

coating compound	11.75	25kg

Waterproof cement paint

cream or white	43.90	5L
colours	43.90	5L
sealer	20.15	5L

Textured masonry paint

white	17.90	10L
colours	20.35	10L
Stabilising solution	47.10	20L

Mulitcolour painting

primer	13.45	5L

Painting and Decorating (Cont'd)

Mulitcolour painting (Cont'd)

basecoat	26.85	5L
finish	37.75	5L

Metallic paint

aluminium	35.00	5L
gold or bronze	52.70	5L

Bituminous paint

Bituminous paint	10.20	5L

Polyurethane lacquer

clear	25.65	5L

Linseed oil

raw	15.00	5L
boiled	19.75	5L

Fire retardant paints and varnishes

fire retardant paint	46.00	5L
fire retardant varnish	32.50	5L
overcoat varnish	49.15	5L

French polishing

white polish	5.90	litre
button polish	6.15	litre
garnet polish	5.60	litre
methylated spirit	1.80	litre

Road marking materials

solvent based paint	51.50	litre
50mm wide non-reflective self adhesive tape	1.60	m
100mm wide non-reflective self adhesive tape	2.95	m
50mm wide reflective self adhesive tape	2.24	m
100mm wide reflective self adhesive tape	4.13	m
road marking primer	3.60	litre

Surface treatments

silicone based water repellent	15.75	5L
fungicide	2.38	litre

Drainage

Drains; vitrified clay pipes and fittings, BS EN 295, normal; flexible mechanical joints; Hepworth Hepseal

400mm pipes	93.99	m
400mm bends	353.19	nr
450mm pipes	124.64	m
450mm bends	465.09	nr

Drains; vitrified clay pipes and fittings, BS EN 295; sleeve joints, push-fit polypropylene standard ring flexible couplings; Hepworth Supersleeve

100mm pipes	3.51	m
100mm pipe couplings	2.59	nr
100mm bends	6.99	nr
100mm branches	15.26	nr
150mm pipes	7.10	m
150mm pipe couplings	4.71	nr
150mm bends	14.48	nr
150mm branches	22.49	nr

Vitrified clay pipes and fittings, BS EN 295; sleeve joints, push-fit polypropylene neoprene ring flexible couplings; Hepworth Supersleeve

100mm pipes	3.51	m
100mm pipe couplings	3.94	nr
150mm pipes	7.10	m
150mm pipe couplings	7.43	nr

Drains; vitrified clay pipes and fittings, BS EN 295

100mm pipes	7.97	m
100mm double collars	9.61	nr
100mm bends	7.97	nr
100mm rest bends	13.10	nr
100mm branches	14.64	nr
150mm pipes	12.26	m
150mm double collars	16.01	nr
150mm bends	12.26	nr
150mm rest bends	22.14	nr
150mm branches	24.22	nr
225mm pipes	24.30	m
225mm double collars	37.44	nr
300mm pipes	40.72	m

Drains; Wavin; PVC-U pipes and fittings, BS EN 13598-1; ring seal joints

110mm pipes	4.10	m
110mm pipe couplings	3.78	nr
110mm connections to cast iron and clay sockets	10.76	nr
110mm 45 degree bends	9.16	nr
110mm branches	13.50	nr
110mm connections cast iron and clay spigot	10.76	nr
160mm pipes	9.26	m
160mm pipe couplings	7.62	nr
160mm connections to cast iron and clay sockets	21.60	nr
160mm 45 degree bends	21.62	nr
160mm branches	38.60	nr
160mm connections to cast iron and clay spigot	21.60	nr

PVC-U solid wall concentric external rib-reinforced Wavin Ultra-Rib pipes and fittings with sealing rings to joints

150mm 6UR 046 pipes	4.23	m
150mm pipe couplings	7.92	nr
150mm connectors 6UR 205	7.17	nr
150mm bends 6UR 563	9.65	nr
150mm branches 6UR 213	23.34	nr
150mm connections to clayware pipe ends 6UR 129	17.58	nr
225mm 9UR 046 pipes	9.70	m
225mm pipe couplings	16.47	nr
225mm connectors 9UR 205	15.00	nr
225mm bends 9UR 563	38.83	nr
225mm branches 9UR 213	73.73	nr
225mm connections to clayware pipe ends 9UR 109	41.26	nr
300mm 12UR 043 pipes	14.55	m
300mm pipe couplings	32.85	nr
300mm connectors 12UR 205	30.27	nr
300mm bends 12UR 563	67.14	nr
300mm branches 12UR 213	157.26	nr
300mm connections to clayware pipe ends 12UR 112	108.54	nr
Reducers 40/32mm	1.02	nr
32mm Pipe brackets - Metal	1.17	nr
40mm Pipe brackets - Metal	1.40	nr

Unreinforced concrete pipes and fittings BS 5911, Class H, flexible joints

300mm pipes	11.50	m
300mm bends	115.00	nr
300mm, 100mm branches	67.50	nr
300mm, 150mm branches	67.50	nr
300mm, 225mm branches	137.50	nr
300mm, 300mm branches	157.50	nr
375mm pipes	14.50	m
375mm bends	145.00	nr
375mm, 100mm branches	75.00	nr
375mm, 150mm branches	75.00	nr
375mm, 225mm branches	145.00	nr
375mm, 300mm branches	165.00	nr
375mm, 375mm branches	185.00	nr
450mm pipes	16.75	m
450mm bends	167.50	nr
450mm, 100mm branches	85.00	nr
450mm, 150mm branches	85.00	nr
450mm, 225mm branches	155.00	nr
450mm, 300mm branches	175.00	nr
450mm, 375mm branches	195.00	nr
450mm, 450mm branches	230.00	nr
525mm pipes	20.25	m
525mm bends	202.50	nr
525mm, 100mm branches	90.00	nr
525mm, 150mm branches	90.00	nr
525mm, 225mm branches	160.00	nr
525mm, 300mm branches	180.00	nr
525mm, 375mm branches	200.00	nr
525mm, 450mm branches	235.00	nr
600mm pipes	24.75	m
600mm bends	247.50	nr
600mm, 100mm branches	98.00	nr
600mm, 150mm branches	98.00	nr
600mm, 225mm branches	168.00	nr
600mm, 300mm branches	188.00	nr
600mm, 375mm branches	208.00	nr
600mm, 450mm branches	243.00	nr

Reinforced concrete pipes and fittings, BS 5911, Class H, flexible joints

450mm pipes	32.07	m
450mm bends	320.72	nr
450mm, 100mm branches	89.48	nr
450mm, 150mm branches	91.53	nr
450mm, 225mm branches	94.60	nr
450mm, 300mm branches	115.08	nr
450mm, 375mm branches	130.44	nr
450mm, 450mm branches	150.92	nr
525mm pipes	39.20	m
525mm bends	391.99	nr
525mm, 100mm branches	100.84	nr
525mm, 150mm branches	102.89	nr
525mm, 225mm branches	105.96	nr
525mm, 300mm branches	121.32	nr
525mm, 375mm branches	141.80	nr
525mm, 450mm branches	162.28	nr
600mm pipes	42.76	m
600mm bends	427.62	nr
600mm, 100mm branches	91.42	nr
600mm, 150mm branches	113.95	nr
600mm, 225mm branches	117.02	nr
600mm, 300mm branches	132.38	nr
600mm, 375mm branches	152.86	nr
600mm, 450mm branches	173.34	nr

Cast iron pipes and cast iron fittings BS 437, coated; bolted cast iron coupling and synthetic rubber gasket joints

100mm pipes	17.99	m
100mm pipe couplings	11.53	nr
100mm connectors, large socket for clayware	25.20	nr
100mm bends	21.56	nr
100mm branches	28.60	nr
150mm pipes	33.29	m
150mm pipe couplings	13.96	nr
150mm connectors, large socket for clayware	37.62	nr
150mm diminishing pieces, reducing to 100mm	28.10	nr
150mm bends	49.60	nr
150mm branches	70.57	nr

Stainless steel pipes and fittings, B.M. Stainless Steel Drains Ltd; ring seal push fit joints (AIS 1316)

75mm pipes	25.10	m
75mm pipe couplings	20.59	nr
75mm bends	30.87	nr
75mm branches	35.87	nr
110mm pipes	31.85	m
110mm pipe couplings	17.41	nr
110mm diminishing pieces, reducing to 75mm	54.22	nr
110mm bends	39.41	nr
110mm branches	41.42	nr

Precast circular concrete inspection chambers; standard or stock pattern units; BS 5911

250mm high, 900mm internal diameter	20.00	nr
500mm high, 900mm internal diameter	30.00	nr
750mm high, 900mm internal diameter	30.00	nr

Drainage (Cont'd)

Precast circular concrete inspection chambers; standard or stock pattern units; BS 5911 (Cont'd)

1000mm high, 900mm internal diameter	40.00	nr
250mm high, 1050mm internal diameter	19.00	nr
500mm high, 1050mm internal diameter	28.50	nr
750mm high, 1050mm internal diameter	28.50	nr
1000mm high, 1050mm internal diameter	38.00	nr
250mm high, 1200mm internal diameter	24.00	nr
500mm high, 1200mm internal diameter	36.00	nr
750mm high, 1200mm internal diameter	36.00	nr
1000mm high, 1200mm internal diameter	48.00	nr

Galvanized malleable cast iron steps

step irons BS 1247	5.00	nr

Cover slabs, heavy duty, 125mm thick, for chamber sections:

2400mm internal diameter	489.24	nr
2700mm internal diameter	633.37	nr
3000mm internal diameter	745.98	nr

Precast concrete inspection chambers; standard or stock pattern units; rectangular chamber sections

150mm high, internal size 1200 x 750mm	17.04	nr
200mm high, internal size 1200 x 750mm	21.01	nr

GRP inspection chambers with preformed benching; 475mm internal diameter

depth to invert 450mm	75.61	nr
depth to invert 585mm	36.59	nr
depth to invert 750mm	19.06	nr
depth to invert 930mm	78.22	nr

Polypropylene shallow inspection chamber with preformed benching. Light duty single sealed cover and frame ref. 4D.961

300mm diameter; for 100mm diameter pipes ref. 4D.960, depth to invert 600mm	44.38	nr

Polypropylene universal inspection chamber; 475mm diameter

preformed chamber base benching; for 100mm pipes, ref. 4D.922	64.15	nr
chamber shaft 230mm effective length, ref. 4D.925	16.36	nr
cast iron cover and frame; single seal light duty to suit, ref. 4D.942	51.54	nr
Polypropylene sealed cover and frame, ref. 4D.943	30.82	nr

Cast iron inspection chambers

100mm diameter straight with one branch	70.75	nr
100mm diameter straight with one branch	108.10	nr
100mm diameter straight with two branches	212.78	nr
150 x 100mm diameter straight with one branch	140.04	nr
150mm diameter straight with one branch	177.72	nr
150mm diameter straight with two branches	354.87	nr

Manhole covers, cast iron, A15 light duty, Single Seal

Solid top Circular 450mm dia	13.20	nr
Solid top 450mm x 450mm	17.60	nr
Solid top 600mm x 450mm	22.00	nr
Solid top 600mm x 600mm	30.80	nr
Recessed top 600mm x 450mm	33.44	nr
Recessed top 600mm x 600mm	41.80	nr
Recessed top 750mm x 600mm	214.72	nr

Manhole covers, cast iron, A15 light duty, Double Seal

Solid top 450mm x 450mm	228.80	nr
Solid top 600mm x 450mm	241.74	nr
Solid top 600mm x 600mm	268.62	nr
A15 light duty, lifting keys	6.60	nr

Manhole covers, cast iron, B125 medium duty

Ref. H783 Trojan single seal; 600mm diameter	93.50	nr
Ref. H781; 600 x 450mm	86.60	nr
Ref. H782; 600 x 600mm	107.64	nr
Ref. H743 Trojan S; 600 x 450mm	113.78	nr
Ref. H742; 600 x 600mm	143.63	nr
Ref. H741; 750 x 600mm	224.81	nr
B125 medium duty, lifting keys	4.71	nr

Manhole covers, cast iron, D400 heavy duty

Ref. H9330; 600mm diameter	116.38	nr
Ref. H9320; 600 x 600mm	101.50	nr
Ref. H9420; 750 x 750mm	244.64	nr
D400 heavy duty, lifting keys	5.50	nr

Clayware land drain pipes, BS 1196; butt joints

75mm	3.03	m
100mm	5.21	m
150mm	10.67	m

Porous concrete land drain pipes, BS 5911; butt joints

150mm	5.32	m
225mm	6.82	m
300mm	11.21	m
375mm	15.27	m
450mm	21.48	m
525mm	31.82	m
600mm	39.52	m

Fencing

Strained wire fencing - concrete posts

900mm intermediate post	4.88	nr
900mm end/angle post	5.97	nr
900mm strutt	4.75	nr
900mm stretcher bar	1.48	nr
1050mm intermediate post	5.29	nr
1050mm end/angle post	6.58	nr
1050mm strutt	5.29	nr
1050mm stretcher bar	1.98	nr
Ratchet winders	2.60	nr
2.5mm Stirrup wire	35.00	651m
4mm Galvanized line wire	33.95	254m

900mm Chain link fencing - concrete posts - galvanized mesh

intermediate post GLC	4.88	nr
end/angle post GLC	5.97	nr
strutt	4.75	nr
stretcher bar	1.48	nr
chain link	48.25	25m

1200mm Chain link fencing - concrete posts - galvanized mesh

intermediate post GLC	5.78	nr
end/angle post GLC	9.04	nr
strutt	5.48	nr
stretcher bar	1.98	nr
chain link	63.10	25m

1400mm Chain link fencing - concrete posts - galvanized mesh

intermediate post GLC	6.24	nr
end/angle post GLC	7.72	nr
strutt	6.24	nr
stretcher bar	2.34	nr
chain link	70.70	25m

1800mm Chain link fencing - concrete posts - galvanized mesh

intermediate post GLC	8.80	nr
end/angle post GLC	12.19	nr
strutt	7.59	nr
stretcher bar	2.99	nr
chain link	44.78	12.5m

Chain Link fencing sundries

Ratchet winders	2.60	nr
2.5mm Stirrup wire	35.00	651
PPC Line wire	5.20	26m
Ring fasteners	5.22	1000

900mm Chain link fencing - concrete posts - plastic coated mesh

intermediate post PLC	4.88	nr
end/angle post PLC	5.97	nr
strutt	4.75	nr
stretcher bar	1.48	nr
chain link 2.5 / 3.55mm	88.90	25m

1200mm Chain link fencing - concrete posts - plastic coated mesh

intermediate post PLC	5.78	nr
end/angle post PLC	9.04	nr
strutt	5.48	nr
stretcher bar	1.98	nr
chain link	106.85	25m

1400mm Chain link fencing - concrete posts - plastic coated mesh

intermediate post PLC	6.24	nr
end/angle post PLC	7.72	nr
strutt	6.24	nr
stretcher bar	2.34	nr
chain link	131.75	25m

1800mm Chain link fencing - concrete posts - plastic coated mesh

intermediate post PLC	8.80	nr
end/angle post PLC	12.19	nr
strutt	7.59	nr
stretcher bar	2.99	nr
chain link	167.35	25m

900mm Chain link fencing - steel posts - galvanized mesh

intermediate post GLS	9.11	nr
end post with strutts	21.23	nr
angle post with strutts	32.70	nr
chain link 3 x 50mm	48.25	25m

1200mm Chain link fencing - steel posts - galvanized mesh

intermediate post GLS	11.07	nr
end post with strutts	25.44	nr
angle post with strutts	39.67	nr
chain link	63.10	25m

1400mm Chain link fencing - steel posts - galvanized mesh

intermediate post GLS	12.04	nr
end post with strutts	27.64	nr
angle post with strutts	43.10	nr
chain link	70.70	25m

1800mm Chain link fencing - steel posts - galvanized mesh

intermediate post GLS	14.96	nr
end post with strutts	36.87	nr
angle post with strutts	55.19	nr
chain link	44.78	12.5m

900mm Chain link fencing - steel posts - plastic coated mesh

intermediate post PLS	14.14	nr
end post with strutts	28.97	nr
angle post with strutts	43.24	nr
chain link	88.90	25m

Fencing (Cont'd)

1200mm Chain link fencing - steel posts - plastic coated mesh

intermediate post PLS	14.78	nr
end post with strutts	39.84	nr
angle post with strutts	52.90	nr
chain link	106.85	25m

1400mm Chain link fencing - steel posts - plastic coated mesh

intermediate post PLS	16.11	nr
end post with strutts	43.28	nr
angle post with strutts	57.05	nr
chain link	131.75	25m

1800mm Chain link fencing - steel posts - plastic coated mesh

intermediate post PLS	20.11	nr
end post with strutts	59.37	nr
angle post with strutts	74.19	nr
chain link	167.35	25m

Chestnut fencing

900mm high fencing	31.45	9.1m
900mm intermediate post	1.30	nr
900mm end/angle post	4.80	nr
900mm strutts	1.40	nr
1200mm high fencing	39.80	9.1m
1200mm intermediate post	1.55	nr
1200mm end/angle post	5.75	nr
1200mm strutts	1.80	nr
1500mm high fencing	58.60	9.1m
1500mm intermediate post	3.05	nr
1500mm end/angle post	6.60	nr
1500mm strutts	2.20	nr

Close boarded fencing

1200mm posts	10.30	nr
1200mm pales	0.60	nr
1500mm posts	13.70	nr
1500mm pales	0.76	nr
1800mm posts	15.40	nr
1800mm pales	0.92	nr
Rails	1.58	m
Gravel boards	1.71	m
Centre stumps	0.87	nr
50mm nails	46.50	25kg
75mm nails	46.50	25kg

Wooden palisade fencing

1050mm posts	6.75	nr
1050mm pales	0.91	nr
1200mm posts	8.03	nr
1200mm pales	1.07	nr

Galvanized mild steel guard rail of tubing to BS 1387 medium grade with Kee Klamp fittings and intersections

40mm diameter ref. 7-2-G	5.92	m
flanged end	4.16	nr
bend No.15-7	4.56	nr
three-way No.20-7	6.90	nr
three-way No.25-7	6.62	nr
four-way No.21-7	5.40	nr
four-way No.26-7	5.12	nr
five-way No.35-7	7.67	nr
five-way No.40-7	11.44	nr
floor plate No.62-7	6.08	nr

External works and landscaping

Precast concrete kerbs

125 x 255 x 915mm	2.52	nr
125 x 255 x 915mm kerb; curved 10.00m radius	3.99	nr
150 x 305 x 915mm	5.18	nr
150 x 305 x 915mm kerb; curved 10.00m radius	9.86	nr

Precast concrete edgings

50 x 150 x 915mm	1.20	m
50 x 200 x 915mm	1.81	m
50 x 250 x 915mm	2.07	m

Granite edge kerb

150 x 300mm	12.13	m
200 x 300mm	22.44	m

York stone paving

50mm thick	63.69	m2
75mm thick	71.94	m2

Precast concrete flags, BS 7263 Part 1, natural finish

600 x 450 x 50mm	2.32	nr
600 x 600 x 50mm	2.52	nr
600 x 750 x 50mm	3.18	nr
600 x 900 x 50mm	3.36	nr
600 x 450 x 63mm	2.90	nr

Brick paving blocks

25mm thick Brick paviors	300.00	1000

Granite sett paving

125mm thick; new	32.16	m2
150mm thick; new	38.59	m2

COMPOSITE PRICES FOR APPROXIMATE ESTIMATING

The purpose of this section is to provide an easy reference to costs of "composite" items of work, which would usually be measured for Approximate Estimates. Rates for items, which are not included here, may be found by reference to the preceding pages of this volume. Allowance has been made in the rates for the cost of works incidental to and forming part of the described item although not described therein, i.e. the cost of formwork to edges of slabs has been allowed in the rates for slabs; the cost of finishings to reveals has been allowed in the rates for finishings to walls.

The rates are based on rates contained in the preceding Major Works section of the book and include the cost of materials, Labour fixing or laying and allow 7.5% for overheads and profit: Preliminaries are not included within the rates. The rates are related to a housing or simple commercial building contract in the range of £500,00 to £1,500,000. For contracts of a higher or lower value, rates should be adjusted by the percentage adjustments given in Essential Information at the beginning of this Volume. Landfill tax is included for inert waste.

The following cost plans have been included in both the Major Works and Small Works sections, and priced accordingly as examples on how to construct a Cost Plan, for information and to demonstrate how to use the Approximate Estimating Section.

COST PLAN FOR AN OFFICE DEVELOPMENT

The Cost Plan that follows gives an example of an Approximate Estimate for an office building as the attached sketch drawings. The estimate can be amended by using the examples of Composite Prices following or the prices included within the preceding pages. Details of the specification used are indicated within the elements. The cost plan is based on a fully measured bill of quantities priced using the detailed rates within this current edition of Laxton's. Preliminaries are shown separately.

Floor area **932** **m2**

Description	Item	Qty		Rate	Item Total	Cost per m2	Total
1.0 SUBSTRUCTURE							
Note: Rates allow for excavation by machine							
1.1 Substructure (271 m2)							
Clear site vegetation; excavate 150 topsoil and remove spoil.	Site clearance	288	m2	0.77	221.57	0.24	
300 x 600 reinforced concrete perimeter ground beams with attached 180 x 300 toes; damp proof membrane 50 concrete blinding; jablite; 102.5 thick common and facing bricks from toe to top of ground beam; 75 cavity filled with plain concrete; cavity weeps; damp proof courses (excavation included in ground slab excavation)	Ground beam	81	m	196.66	15,929.46	17.09	
600 x 450 reinforced concrete internal ground beams, mix C30; 50 thick concrete blinding; formwork; jablite protection (excavation included in ground slab excavation)	Ground beam	34	m	146.59	4,984.06	5.35	
Excavate to reduced levels; level and compact bottoms; 100 hardcore filling, blinded; damp proof membrane; 50 jablite insulation; 150 reinforced concrete ground slab 2 layers A252 fabric reinforcement	Ground floor slab	271	m2	71.94	19,496.75	20.92	
1900 x 1900 x 1150 lift pit; excavate; level and compact bottoms; working space; earthwork support; 50 concrete blinding; damp proof membrane; 200 reinforced concrete slab, 200 reinforced concrete walls, formwork; Bituthene tanking and protection board	Lift pit	1	nr	1,995.59	1,995.59	2.14	
1500 x 1500 x 1000 stanchion bases; excavate; level and compact bottoms; earthwork support; reinforced concrete foundations; 102.5 bricks from top of foundation to ground slab level	Stanchion base	13	nr	671.55	8,730.21	9.37	51,357.65
					51,357.65	55.10	
(Element unit rate = £ 189.51 m2)							

APPROXIMATE ESTIMATING

COST PLAN FOR AN OFFICE DEVELOPMENT (Cont'd)

Description	Item	Qty		Rate	Item Total	Cost per m2	Total
2.0 SUPERSTRUCTURE							
2.1 Frame (932 m2)							
300 x 300 reinforced concrete columns; formwork two coats RIW to exterior faces	Columns	111	m	91.10	10,111.97	10.85	
29.5m girth x 1.0m high steel frame to support roof over high level louvres	Frame	1	nr	6,573.52	6,573.52	7.05	
300 x 550 reinforced concrete beams; formwork	Beams	61	m	128.82	7,857.79	8.43	24,543.28
					24,543.28	26.33	
(Element unit rate = £ 26.33 m2)							
2.2 Upper floors (614 m2)							
275 reinforced concrete floor slab; formwork	Floor slab	614	m2	108.62	66,689.72	71.56	
300 x 400/675 x 175/300 x 150 reinforced concrete attached beams; formwork	Attached beams	273	m	99.24	27,091.79	29.07	93,781.51
					93,781.51	100.62	
(Element unit rate = £ 152.74 m2)							
2.3 Roof (403 m2 on plan)							
2.3.1 Roof Structure							
Pitched roof trusses, wall plates; 160 thick insulation and Netlon support	Roof structure	403	m2	43.82	17,658.17	18.95	
100 thick blockwork walls in roof space	Blockwork	27	m2	42.20	1,139.50	1.22	
Lift shaft capping internal size 1900 x 1900; 200 thick blockwork walls and 150 reinforced concrete slab	Lift shaft	1	nr	727.19	727.19	0.78	
2.3.2 Roof Coverings							
Concrete interlocking roof tiles, underlay and battens	Roof finish	465	m2	38.84	18,060.60	19.38	
Ridge	Ridge	22	m	15.32	337.04	0.36	
Hip	Hip	50	m	15.91	795.26	0.85	
Valley	Valley	6	m	34.44	206.64	0.22	
Abutments to cavity walls, lead flashings and cavity tray damp proof courses	Abutments	9	m	73.13	658.14	0.71	
Eaves framing, 25 x 225 softwood fascia 50 wide Resoplan soffit and ventilator; eaves tile ventilator and decoration; 164 x 127 pressed Aluminium gutters and fittings	Eaves	102	m	249.42	25,440.46	27.30	
2.3.3 Roof drainage							
100 x 100 Aluminium rainwater pipes and rainwater heads and hopper heads	Rain water pipes	58	m	56.45	3,274.26	3.51	68,297.25
					68,297.25	73.28	
(Element unit rate = £ 169.47 m2)							
2.4 Stairs							
Reinforced in-situ concrete staircase; blockwork support walls screed and vinyl floor finishes; plaster and paint to soffits; to wall							
3n2 1200 wide flights, 1nr half landing and 1nr quarter landing; 3100 rise; handrail one side	Stairs	1	nr	4,366.61	4,366.61	4.69	
4nr 1200 wide flights, 2nr half landings 620 rise; plastics coated steel balustrade one side and handrail one side	Stairs	1	nr	12,830.41	12,830.41	13.77	

COST PLAN FOR AN OFFICE DEVELOPMENT (Cont'd)

Description	Item	Qty		Rate	Item Total	Cost per m2	Total
2.0 SUPERSTRUCTURE (Cont'd)							
2.4 Stairs (Cont'd)							
4nr 1200 wide flights, 2nr half landings mix; 6200 rise; plastics coated steel and glass in fill panels balustrade one side and handrail one side	Stairs	1	nr	12,521.05	12,521.05	13.43	29,718.07
					29,718.07	31.89	
2.5 External Walls (614 m2)							
Cavity walls, 102.5 thick facing bricks (PC £250 per 1000), 75 wide insulated (PC £250 per cavity, 140 thick lightweight concrete blockwork	Facing brick wall	455	m2	114.88	52,270.40	56.08	
Cavity walls, 102.5 thick facing bricks (PC £250 per 1000), built against concrete beams and columns	Facing brick wall	159	m2	88.24	14,030.46	15.05	
880 x 725 x 150 cast stonework pier cappings	Stonework	9	nr	77.97	701.69	0.75	
780 x 1300 x 250 cast stonework spandrels	Stonework	9	nr	155.94	1,403.47	1.51	
6500 x 2500 attached main entrance canopy	Canopy	1	nr	6,836.42	6,836.42	7.34	
Milled lead sheet on impregnated softwood boarding and framing, and building paper	Lead sheet	4	m2	249.52	998.06	1.07	
Expansion joint in facing brickwork	Expansion joint	118	m	16.99	2,004.81	2.15	78,245.31
					78,245.31	83.95	
(Element unit rate = £ 127.44 m2)							
2.6 Windows and External Doors (214 m2)							
2.6.1 Windows (190 m2)							
Polyester powder coated Aluminium vertical pivot double glazed windows	Windows	190	m2	422.42	80,260.69	86.12	
Polyester powder coated Aluminium fixed louvre panels	Louvres	35	m2	176.99	6,194.66	6.65	
Heads 150 x 150 x 10 stainless steel angle bolted to concrete beam, cavity damp proof course, weep vents, plaster and decoration	Heads	98	m	92.48	9,062.86	9.72	
Reveals closing cavity with facing bricks, damp proof course, plaster and decoration	Reveals	99	m	29.25	2,895.71	3.11	
Works to reveals of openings comprising closing cavity with facing bricks, damp proof course, 13 thick cement and sand render with angle beads, and ceramic tile finish	Reveals	27	m	41.46	1,119.53	1.20	
Sills closing cavity with blockwork, damp proof course, and 19 x 150 Ash veneered MDF window boards	Sills	8	m	33.14	265.11	0.28	
Sills 230 x 150 cast stonework sills and damp proof course	Sills	36	m	45.74	1,646.78	1.77	
Sills 230 x 150 cast stonework sills, damp proof course, and 19 x 150 Ash veneered MDF window boards	Sills	36	m	255.87	9,211.18	9.88	
Sills closing cavity with blockwork, damp proof course, and ceramic tile sill finish	Sills	9	m	39.32	353.85	0.38	
Cavity wall lintels	Lintels	15	m	40.36	605.42	0.65	
Surrounds to 900 diameter circular openings; stainless steel radius lintels, 102.5 x 102.5 facing brick on edge surround, 13 thick plaster, and eggshell decoration	Lintels	8	m	153.19	1,225.55	1.31	
300 x 100 cast stonework surrounds to openings	Stonework	20	m	49.37	987.34	1.06	

COST PLAN FOR AN OFFICE DEVELOPMENT (Cont'd)

Description	Item	Qty		Rate	Item Total	Cost per m2	Total
2.0 SUPERSTRUCTURE (Cont'd)							
2.6 Windows and External Doors (214 m2) (Cont'd)							
2.6.2 External Doors (24 m2)							
Polyester powder coated Aluminium double glazed doors and any integral screens	External doors	24	m2	396.92	9,526.17	10.22	
Cavity wall lintels	Lintels	3	m	40.36	121.08	0.13	123,475.92
					123,475.92	132.48	
(Element unit rate = £ 576.99 m2)							
2.7 Internal Walls and Partitions (481 m2)							
100 thick blockwork walls	Block walls	231	m2	33.06	7,636.99	8.19	
200 thick blockwork walls	Block walls	113	m2	57.30	6,475.26	6.95	
215 thick common brickwork walls	Brick walls	84	m2	90.93	7,637.78	8.20	
Demountable partitions	Partitions	42	m2	26.09	1,095.90	1.18	
Extra over demountable partitions for single doors	Partitions	2	nr	25.10	50.20	0.05	
Extra over demountable partitions for double doors	Partitions	1	nr	30.13	30.13	0.03	
WC Cubicles	Cubicles	9	nr	120.88	1,087.96	1.17	
Screens in softwood glazed with 6 thick GWPP glass	Screens	11	m2	283.23	3,115.57	3.34	
Duct panels in Formica Beautyboard and 50 x 75 softwood framework	Ducts	27	m2	55.46	1,497.41	1.61	
Duct panels in veneered MDF board and 50 x 75 softwood framework	Ducts	26	m2	50.20	1,305.28	1.40	29,932.48
					29,932.48	32.12	
(Element unit rate = £ 62.23 m2)							
2.8 Internal Doors (84 m2)							
900 x 2100 half hour fire resistant single doorsets comprising doors, softwood frames, ironmongery and decoration	Doors	45	m2	314.95	14,172.97	15.21	
900 x 2100 one hour fire resistance single doorsets comprising doors, softwood frames, ironmongery and decoration	Doors	9	m2	478.17	4,303.57	4.62	
900 x 2100 half hour fire resistance single doorsets comprising doors with 2nr vision panels, softwood frames, ironmongery and decoration	Doors	11	m2	360.19	3,962.13	4.25	
1500 x 2100 half hour fire resistance double doorsets comprising Ash veneered doors with 2nr vision panels, Ash frames, ironmongery and decoration	Doors	19	m2	340.13	6,462.44	6.93	
100 wide lintels	Lintels	41	m	13.55	555.39	0.60	
200 wide lintels	Lintels	7	m	28.34	198.39	0.21	29,654.89
					29,654.89	31.82	
(Element unit rate = £ 353.03 m2)							
3.0 INTERNAL FINISHES							
3.1 Wall Finishes (1,086 m2)							
13 thick plaster to blockwork walls with eggshell paint finish	Plaster work	703	m2	17.28	12,150.45	13.04	
13 thick plaster to blockwork walls with lining paper and vinyl wall paper finish	Plaster work	124	m2	18.71	2,320.28	2.49	

COST PLAN FOR AN OFFICE DEVELOPMENT (Cont'd)

Description	Item	Qty		Rate	Item Total	Cost per m2	Total
3.0 INTERNAL FINISHES (Cont'd)							
3.1 Wall Finishes (1,086 m2) (Cont'd)							
10 thick plaster to concrete walls with eggshell paint finish	Plaster work	75	m2	24.85	1,863.95	2.00	
13 thick cement and sand render to blockwork walls with 150 x 150 coloured ceramic tiles finish	Tiling	184	m2	44.01	8,097.84	8.69	24,432.52
					24,432.52	26.22	
(Element unit rate = £ 22.50 m2)							
3.2 Floor Finishes (779 m2)							
50 thick sand and cement screed with 300 x 300 x 2.5 vinyl tile finish	Vinyl tiling	19	m2	38.69	735.14	0.79	
50 thick sand and cement screed with 2 thick sheet vinyl finish	Sheet vinyl	108	m2	36.03	3,891.66	4.18	
50 thick sand and cement screed with 250 x 250 x 6 carpet tile finish	Carpet tiling	587	m2	32.27	18,942.26	20.32	
50 thick sand and cement screed with carpet finish	Carpeting	17	m2	43.55	740.34	0.79	
50 thick sand and cement screed with 150 x 150 x 13 ceramic tile finish	Ceramic tiling	46	m2	55.29	2,543.49	2.73	
Entrance mats comprising brass Matt frame, 60 thick cement and sand screed and matting	Matting	2	m2	203.06	406.11	0.44	
150 high x 2.5 thick vinyl skirting	Skirting	45	m	6.02	270.90	0.29	
119 high x 2 thick sheet vinyl skirtings with 20 x 30 lacquered Hardwood cover fillet	Skirting	116	m	19.35	2,244.67	2.41	
25 x 50 Hardwood skirtings with lacquer finish	Skirting	214	m	14.84	3,176.30	3.41	
150 x 150 x 13 ceramic tile skirtings on 13 thick sand and cement render	Skirting	74	m	11.66	862.76	0.93	33,813.63
					33,813.63	36.28	
(Element unit rate = £ 43.41 m2)							
3.3 Ceiling Finishes (809 m2)							
Suspended ceiling systems to offices	Suspended	587	m2	33.25	19,517.75	20.94	
Suspended ceiling systems to toilets	Suspended	46	m2	29.91	1,375.64	1.48	
Suspended ceiling systems to reception area	Suspended	18	m2	29.91	538.29	0.58	
Suspended ceiling plain edge trims and 25 x 38 painted painted softwood shadow battens	Edge trim	300	m	12.14	3,640.80	3.91	
Gyproc MF suspended ceiling system with eggshell paint finish	Suspended	127	m2	25.71	3,264.81	3.50	
12.5 Gyproc Square edge plaster board and skim on softwood supports with eggshell paint finish	Plastering	29	m2	38.39	1,113.18	1.19	
10 thick plaster to concrete with eggshell paint finish	Plastering	2	m2	64.50	128.99	0.14	
Plaster coving	Cove	234	m	6.37	1,490.92	1.60	31,070.39
					31,070.39	33.34	
(Element unit rate = £ 38.41 m2)							
4.0 FITTINGS AND FURNISHINGS							
4.1 Fittings							
2100 x 500 x 900 reception counter in softwood	Counter	1	nr	971.36	971.36	1.04	
2700 x 500 x 900 vanity units in Formica faced chipboard	Units	6	nr	628.83	3,772.98	4.05	

COST PLAN FOR AN OFFICE DEVELOPMENT (Cont'd)

Description	Item	Qty		Rate	Item Total	Cost per m2	Total
4.0 FITTINGS AND FURNISHINGS (Cont'd)							
4.1 Fittings (Cont'd)							
600x 400 x 6 thick mirrors	Mirror	15	nr	73.52	1,102.80	1.18	
Vertical fabric blinds	Blinds	161	m2	47.36	7,624.64	8.18	
900 x 250 x 19 blockboard lipped shelves on spur support system	Shelves	12	nr	43.05	516.60	0.55	
1200 x 600 x 900 sink base units in melamine finished particle board	Sink base unit	3	nr	158.30	474.90	0.51	14,463.28
					14,463.28	15.52	
5.0 SERVICES							
5.1 Sanitary Appliances							
1200 x 600 stainless steel sink units with pillar taps	Sink	3	nr	144.80	434.40	0.47	
White glazed fireclay cleaners sink units with pillar taps	Cleaners sink	3	nr	267.08	801.24	0.86	
Vitreous china vanity basins with pillar taps	Vanity basins	12	nr	216.71	2,600.52	2.79	
Vitreous china wash hand basins and pedestal with pillar taps	Wash hand basin	1	nr	166.53	166.53	0.18	
Vitreous china wash hand basin with pillar taps	Wash hand basin	1	nr	118.61	118.61	0.13	
Vitreous china disabled wash hand basin with pillar taps	Wash hand basin	1	nr	145.30	145.30	0.16	
Close coupled vitreous china WC suites	WC suite	11	nr	196.63	2,162.95	2.32	
Showers comprising shower tray, mixer valve and fittings, and corner shower cubicle	Shower	1	nr	856.71	856.71	0.92	
Vitreous china urinals with automatic cistern	Urinals	3	nr	319.15	957.45	1.03	
Disabled grab rails set	Grab rails	1	nr	376.55	376.55	0.40	
Toilet roll holders	Accessories	11	nr	14.55	160.02	0.17	
Paper towel dispensers	Accessories	8	nr	23.91	191.29	0.21	
Paper towel waste bins	Accessories	8	nr	36.39	291.08	0.31	
Sanitary towel disposal units	Accessories	3	nr	686.15	2,058.44	2.21	
Hand dryers	Accessories	8	nr	239.15	1,913.23	2.05	13,234.32
					13,234.32	14.20	
5.2 Services Equipment		1	nr	2,000.00	2,000.00	2.15	2,000.00
5.3 Disposal Installations							
Upvc waste pipes and fittings to internal drainage	Internal Drainage	37	nr	118.33	4,378.11	4.70	4,378.11
5.4 Water Installations							
Hot and cold water supplies		37	nr	196.69	7,277.52	7.81	7,277.52
5.5 Heat Source							
Gas boiler for radiators and hot water from indirect storage tank		1	nr	76,927.35	76,927.35	82.54	76,927.35
5.6 Space Heating and Air treatment							
Low temperature hot water heating by radiators		1	nr	2,558.52	2,558.52	2.75	2,558.52
5.7 Ventilation Installations							
Extractor fan to toilet and kitchen areas		9	nr	629.66	5,666.95	6.08	5,666.95

COST PLAN FOR AN OFFICE DEVELOPMENT (Cont'd)

Description	Item	Qty		Rate	Item Total	Cost per m2	Total
5.0 SERVICES (Cont'd)							
5.8 Electrical Installations							
Lighting & Power		1	nr	88,124.33	88,124.33	94.55	88,124.33
5.9 Gas Installations							
To boiler room		1	nr	2,002.32	2,002.32	2.15	2,002.32
5.10 Lift and Conveyor Installations							
8 person 3 stop hydraulic lift		1	nr	23,759.66	23,759.66	25.49	23,759.66
5.11 Protective Installations							
Burglar Alarm System		1	nr	5,876.82	5,876.82	6.31	5,876.82
5.12 Communication Installations							
Cabling for telephones		1	nr	7,870.75	7,870.75	8.45	7,870.75
5.13 Special Installations							
Door entry to receptions on all floors		1	nr	1,300.00	1,300.00	1.39	1,300.00
5.14 Builders Work in Connection with Services							
Builders work in connection with plumbing, mechanical and electrical installations		1	nr	6,256.92	6,256.92	6.71	6,256.92
Builders Profit and Attendance on Services		1	nr	1,000.00	1,000.00	1.07	1,000.00
6.0 EXTERNAL WORKS							
6.1 Site Works							
6.1.1 Site Preparation							
Clearing site vegetation, filling root voids with excavated material	Site clearance	1716	m2	0.57	978.12	1.05	
Excavated topsoil 150 deep and preserve on site	Clear topsoil	214	m3	5.32	1,138.89	1.22	
Dispose of surplus topsoil off site	Disposal	70	m3	17.20	1,204.00	1.29	
6.1.2 Surface Treatments							
Excavated to reduce levels 400 deep; dispose of excavated material on site; level and compact bottoms; herbicide; 250 thick granular material filling, levelled, compacted and blinded; 70 thick two coat coated macadam pavings; white road lines and markings	Pavings	719	m2	33.31	23,946.70	25.69	
Excavate to reduce levels 400 deep; dispose of excavated material off site; level and compact bottoms; herbicide; 150 thick granular material filling, levelled compacted, and blinded; 200 x 100 x 65 thick clay paviors	Pavings	40	m2	40.01	1,600.37	1.72	
Excavate to reduce levels 400 deep; dispose of excavated material off site; level and compact bottoms; herbicide; 150 thick granular material filling, levelled, compacted and blinded; 900 x 600 x 50 thick concrete paving flags	Pavings	149	m2	30.78	4,585.86	4.92	
125 x 255 precast concrete road kerbs and foundations	Kerbs	265	m	25.05	6,637.72	7.12	
50 x 150 precast concrete edgings and foundations	Kerbs	48	m	16.53	793.44	0.85	
Excavate to reduce levels 400 deep; dispose of excavated material off site; prepare sub-soil; herbicide; 150 thick preserved topsoil filling; rotavate; turf	Landscaping	55	m2	21.93	1,205.90	1.29	

COST PLAN FOR AN OFFICE DEVELOPMENT (Cont'd)

Description	Item	Qty		Rate	Item Total	Cost per m2	Total
6.0 EXTERNAL WORKS (Cont'd)							
6.1 Site Works (Cont'd)							
6.1.2 Surface Treatments (Cont'd)							
Excavate to reduce levels 400 deep; dispose of excavated material off site; prepare sub-soil; herbicide; 450 thick preserved topsoil filling; cultivate; plant 450 - 600 size herbaceous plants - 3/m2, and 3m high trees - 1nr per 22m2; fertiliser; 25 thick peat mulch	Planting	397	m2	33.55	13,317.81	14.29	
6.1.3 Site Enclosure and Division							
Palisade fencing	Fencing	62	m	40.27	2,496.74	2.68	
1800 high fencing	Fencing	96	m	37.40	3,590.72	3.85	
Extra for 1000 x 2000 gate	Gate	1	nr	385.23	385.23	0.41	61,881.50
					61,881.50	66.40	
6.2 Drainage							
Excavate trench 250 - 500 deep; lay 100 diameter vitrified clay pipe and fittings bedded and surrounded in granular material	Pipes in trenches	44	m	35.27	1,552.06	1.67	
Excavate trench 500 - 750 deep; lay 100 diameter vitrified clay pipe and fittings bedded and surrounded in granular material	Pipes in trenches	59	m	38.03	2,243.95	2.41	
Excavate trench 750 - 1000 deep; lay 100 diameter vitrified clay pipe and fittings bedded and surrounded in granular material	Pipes in trenches	49	m	40.25	1,972.02	2.12	
Excavate trench 1000 - 1250 deep; lay 100 diameter vitrified clay pipe and fittings bedded and surrounded in granular material	Pipes in trenches	4	m	41.71	166.82	0.18	
Excavate trench 250 - 500 deep; lay 100 diameter vitrified clay pipe and fittings bedded and surrounded in in-situ concrete	Pipes in trenches	8	m	43.16	345.24	0.37	
Excavate trench 1000 - 1250 deep; lay 150 diameter vitrified clay pipe and fittings bedded and surrounded in in-situ concrete	Pipes in trenches	23	m	62.08	1,427.75	1.53	
Excavate trench 1500 - 1750 deep; lay 150 diameter vitrified clay pipe and fittings bedded and surrounded in in-situ concrete	Pipes in trenches	10	m	69.71	697.06	0.75	
100 diameter vertical vitrified clay pipe and fittings cased in in-situ concrete	Pipes in trenches	3	m	89.94	269.81	0.29	
100 outlet vitrified clay yard gully with cast iron grating, surrounded in in-situ concrete	Gully	1	nr	93.92	93.92	0.10	
100 outlet vitrified clay road gully with brick kerb and cast iron grating, surrounded in in-situ concrete	Gully	7	nr	167.96	1,175.72	1.26	
Inspection chambers 600 x 900 x 600 deep internally comprising excavation works, in-situ concrete bases, suspended slabs and benchings, engineering brick walls and kerbs, clay channels, galvanised iron step irons and grade A access covers and frames	Insp. chamber	4	nr	724.62	2,898.48	3.11	
Inspection chambers 600 x 900 x 600 deep internally comprising excavation works, in-situ concrete bases, suspended slabs and benchings, engineering brick walls and kerbs, clay channels, galvanised iron step irons and grade B access covers and frames	Insp. chamber	3	nr	711.36	2,134.08	2.29	14,976.91
					14,976.91	16.07	

COST PLAN FOR AN OFFICE DEVELOPMENT (Cont'd)

Description	Item	Qty		Rate	Item Total	Cost per m2	Total
6.0 EXTERNAL WORKS (Cont'd)							
6.3 External Services							
Excavate trench for 4nr ducts; lay 4nr 100 diameter UPVC ducts and fittings; lay warning tape; backfill trench		20	m	43.94	878.88	0.94	
Sewer connection charges		1	nr	5,000.00	5,000.00	5.36	
Gas main and meter charges		1	nr	4,000.00	4,000.00	4.29	
Electricity main and meter charges		1	nr	4,000.00	4,000.00	4.29	
Water main and meter charges		1	nr	3,000.00	3,000.00	3.22	
					16,878.87	18.11	16,878.87
Preliminaries							
Site staff, accommodation, lighting, water safety health and welfare, rubbish disposal, cleaning, drying, protection, security, plant, transport, temporary works and scaffolding with a contract period of 38 weeks		item			82,300.00	88.30	82,300.00
				Total			£ 1,057,057.03
				Cost per m2		**£ 1,134.18**	

COST PLAN FOR AN OFFICE DEVELOPMENT (Cont'd)

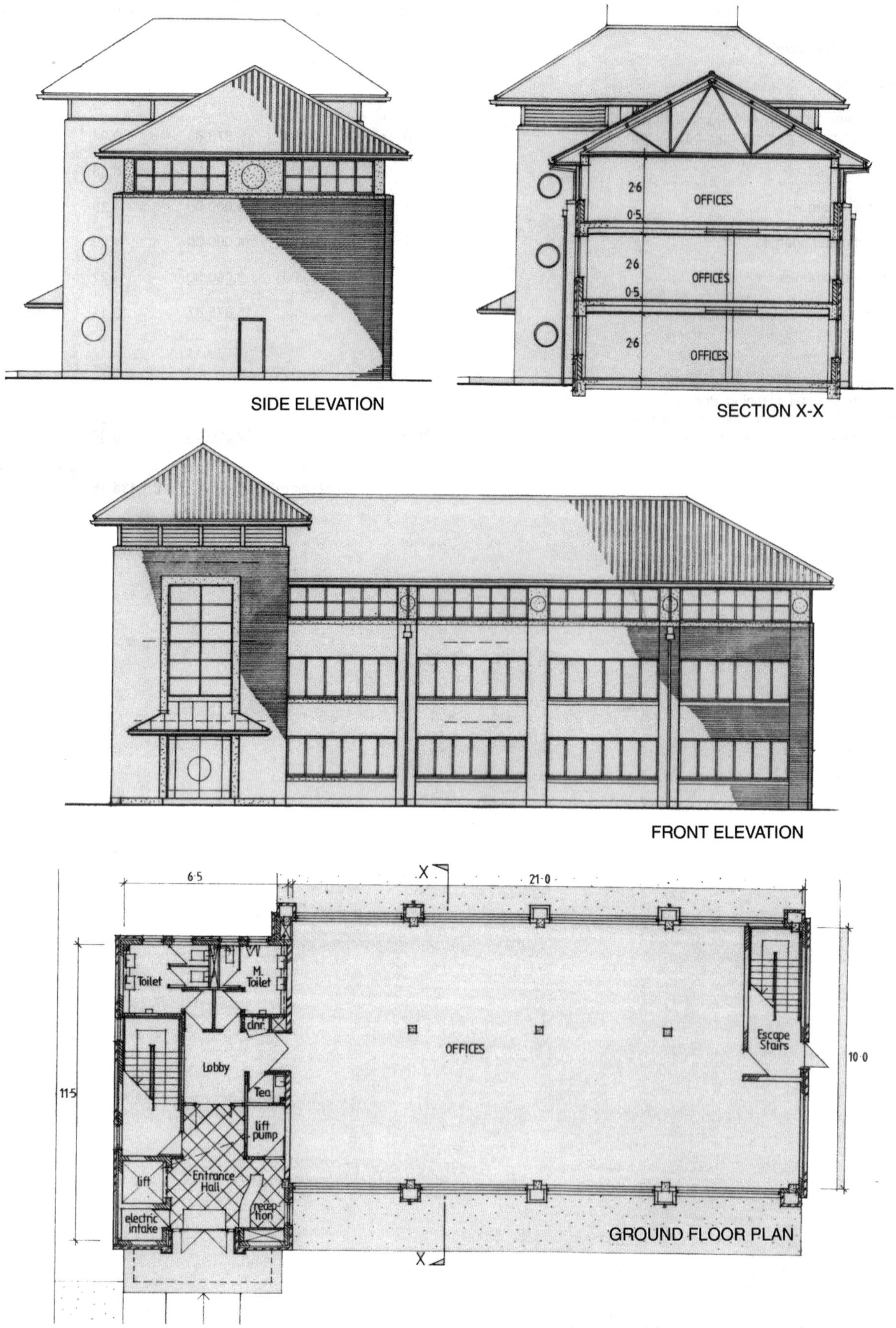

OFFICE DEVELOPMENT

Cotterell Thomas & Thomas Architects

COST PLAN FOR A DETACHED HOUSE

The Cost Plan that follows gives an example of an Approximate Estimate for a detached house with garage and porch as the attached sketch drawings. The estimate can be amended by using the examples of Composite Prices following or the prices included within the preceding pages. Details of the specification used are included within the elements. The cost plan is based on a fully measured bill of quantities priced using the detailed rates within this current edition of Laxton's. Preliminaries are shown separately

COST PLAN FOR A DETACHED HOUSE

Floor area

House (Including porch 5 m2)	120	m2
Garage	22	m2
Total	142	m2

Description	Item	Qty		Rate	Item Total	Cost per m2	Total
1.0 SUBSTRUCTURE							
Note: Rates allow for excavation by machine							
1.1 Substructure (86 m2)							
Clear site vegetation excavate 150 topsoil and remove spoil.	Site clearance	113	m2	3.45	389.85	2.75	
Excavate trench commencing at reduced level, average 800 deep, concrete to within 150 of ground level, construct cavity brick wall 300 high and lay pitch polymer damp proof course	Strip foundation	33	m	131.38	4,335.44	30.53	
Excavate trench commencing at reduced level, average 800mm deep, concrete to within 150 of ground level, construct 103 brick wall 300 high and lay pitch polymer damp proof course	Strip foundation	26	m	102.03	2,652.89	18.68	
Excavate trench commencing at reduced level, average 800mm deep, concrete to within 150 of ground level, construct 215 brick wall 300 high and lay pitch polymer damp proof course.	Strip foundation	2	m	232.19	464.38	3.27	
Excavate to reduce levels average 150 deep, lay 150mm bed of hardcore blinded, dpm and 150 concrete ground slab reinforced with Ref. A193 Fabric reinforcement.	Ground floor slab	85	m2	57.08	4,851.44	34.17	12,694.00
					12,694.00	89.39	
(Element unit rate = £ 147.60 m2)							
2.0 SUPERSTRUCTURE							
2.2 Upper Floors (55 m2)							
175 deep floor joists with 22 chipboard finish	Softwood floor	49	m2	35.09	1,719.35	12.11	
175 deep floor joists with 22 plywood finish	Softwood floor	6	m2	60.37	362.22	2.55	2,081.57
					2,081.57	14.66	
(Element unit rate = £ 37.85 m2)							
2.3 Roof (110 m2 on plan)							
2.3.1 Roof structure							
35 degree pitched roof with 50 x 100 rafters and ceiling joists, 100 glass fibre insulation.	Pitched roof	110	m2	51.23	5,635.25	39.68	
Gable end 275 cavity brick and blockwork facings PC £250/1000	Gable wall	29	m2	101.56	2,945.37	20.74	
Gable end 102 brick wall facings £250/1000	Gable wall	8	m2	74.15	593.19	4.18	

COST PLAN FOR A DETACHED HOUSE (Cont'd)

Description	Item	Qty		Rate	Item Total	Cost per m2	Total
2.0 SUPERSTRUCTURE (Cont'd)							
2.3 Roof (110 m2 on plan) (Cont'd)							
2.3.2 Roof coverings							
Interlocking concrete tiles, felt underlay, battens.	Roof tiling	143	m2	19.42	2,777.06	19.56	
Verge, 150 under cloak.	Verge	38	m	11.37	432.06	3.04	
Ridge, half round	Ridge	14	m	15.32	214.48	1.51	
Valley trough tiles.	Valley	2	m	34.44	68.88	0.49	
Abutments, lead flashings cavity trays.	Abutment	11	m	98.41	1,082.47	7.62	
Sheet lead cladding.	Lead roof	2	m2	129.85	259.70	1.83	
Eaves soffit, fascia and 100 PVC gutter	Eaves	29	m	50.54	1,465.78	10.32	
2.3.3 Roof Drainage							
68 PVCU rainwater pipe.	Rainwater pipe	24	m	21.12	506.94	3.57	15,981.18
					15,981.18	112.54	
(Element unit rate = £ 145.28 m2)							
2.4 Stairs (1nr)							
Softwood staircase 2670 rise, balustrade one side, 910 wide two flights with half landing	Stairs	1	nr	1,339.46	1,339.46	9.43	1,339.46
					1,339.46	9.43	
2.5 External Walls (76 m2)							
Hollow wall, one skin light weight blocks, one skin facings PC £250/1000, ties.	Hollow wall	127	m2	105.69	13,422.62	94.53	
Facing brick wall PC £250/1000 102 thick.	1/2B Facing wall	25	m2	72.61	1,815.22	12.78	
Facing brick wall PC £250/1000 215 thick.	1B Facing wall	3	m2	127.37	382.11	2.69	15,619.95
					15,619.95	110.00	
(Element unit rate = £ 102.76 m2)							
2.6 Windows and External Doors (35 m2)							
2.6.1 Windows (16 m2)							
Hardwood double glazed casement windows.	HW windows	16	m2	402.25	6,436.05	45.32	
Galvanised steel lintel, dpc, brick soldier course.	Lintels	16	m	46.22	739.44	5.21	
Softwood window board, dpc close cavity.	Window board	15	m	9.55	143.25	1.01	
Close cavity at jamb, dpc facings to reveal.	Jamb	32	m	11.86	379.58	2.67	
2.6.2 External Doors (19 m2)							
Hardwood glazed doors, frames, cills, ironmongery	Hardwood doors	6	m2	546.96	3,281.77	23.11	
Hardwood patio doors.	Patio doors	9	m2	378.05	3,402.45	23.96	
Galvanised steel garage door.	Garage door	4	m2	152.03	608.10	4.28	
Galvanised steel lintel, dpc, brick soldier course.	Lintels	12	m	47.44	569.34	4.01	
Close cavity at jamb, dpc, facings to reveal.	Jamb	4	m	11.85	47.40	0.33	15,607.38
					15,607.38	109.91	
(Element unit rate = £ 445.93 m2)							

COST PLAN FOR A DETACHED HOUSE (Cont'd)

Description	Item	Qty		Rate	Item Total	Cost per m2	Total
2.0 SUPERSTRUCTURE (Cont'd)							
2.7 Internal Walls and Partitions (153 m2)							
Lightweight concrete block 100 thick.	Block walls	104	m2	35.17	3,657.96	25.76	
75 timber stud partition plasterboard and skim both sides.	Stud partitions	53	m2	57.19	3,030.96	21.34	6,688.92
					6,688.92	47.11	
(Element unit rate = £ 43.72 m2)							
2.8 Internal doors (40 m2)							
12 hardboard embossed panel door, linings ironmongery PC £20.00	Flush doors	18	m2	250.09	4,501.70	31.70	
3 softwood "GG glazed door, linings ironmongery PC £20.00	Glazed doors	6	m2	262.99	1,577.93	11.11	
6 Plywood faced door, linings, ironmongery PC £20.00	Flush doors	12	m2	228.41	2,740.86	19.30	
1 plywood faced half hour fire check flush door, frame, ironmongery PC £60.00	Fire doors	2	m2	186.55	373.09	2.63	
Plywood ceiling hatch	Hatch	1	nr	77.64	77.64	0.55	
Precast concrete lintels.	Lintels	10	m	18.49	184.94	1.30	9,456.17
					9,456.17	66.59	
(Element unit rate = £ 236.40 m2)							
3.0 INTERNAL FINISHES							
3.1 Wall finishes (264 M2)							
13 light weight plaster.	Plaster work	264	m2	16.21	4,278.25	30.13	
152 x 152 wall tiling coloured PC £20.00 supply	Tiling	13	m2	33.95	441.35	3.11	4,719.60
					4,719.60	33.24	
(Element unit rate = £ 17.88 m2)							
3.2 Floor Finishes (63 M2)							
22 Chipboard, 50 insulation.	Board flooring	63	m2	27.08	1,706.04	12.01	
3 PVC sheet flooring.	Sheet vinyl	18	m2	18.66	335.88	2.37	
25 x 125 softwood skirting.	Skirting	131	m	10.93	1,431.83	10.08	3,473.75
					3,473.75	24.46	
(Element unit rate = £ 55.14 m2)							
3.3 Ceiling Finishes (115 M2)							
9.5 Plasterboard and set.	Plasterboard	109	m2	15.00	1,635.00	11.51	
Plywood finish	Plywood	6	m2	43.84	263.04	1.85	1,898.04
					1,898.04	13.37	
(Element unit rate = £ 16.50 m2)							
4.0 FITTINGS AND FURNITURE							
4.1 Fittings							
Kitchen units :-							
600 base unit		2	nr	207.49	414.98	2.92	
1000 base unit		6	nr	333.13	1,998.78	14.08	
500 broom cupboard		1	nr	287.24	287.24	2.02	
600 wall unit		3	nr	120.89	362.67	2.55	
1000 wall unit		2	nr	181.09	362.18	2.55	
Work tops		10	m	59.14	591.40	4.16	
Gas fire surround		1	nr	591.25	591.25	4.16	4,608.50
					4,608.50	32.45	

COST PLAN FOR A DETACHED HOUSE (Cont'd)

Description	Item	Qty		Rate	Item Total	Cost per m2	Total
5.0 SERVICES							
5.1 Sanitary Appliances							
Lavatory basin with pedestal.		3	nr	153.54	460.62	3.24	
Bath with side panel.		1	nr	249.25	249.25	1.76	
WC suite.		3	nr	234.20	702.60	4.95	
Stainless steel sink and drainer. (500)		1	nr	144.80	144.80	1.02	
Stainless steel sink and drainer. (600)		1	nr	160.48	160.48	1.13	
shower.		1	nr	643.97	643.97	4.54	
5.2 Services Equipment							
Electric oven built in.		1	nr	758.74	758.74	5.34	
Gas hob built in.		1	nr	370.26	370.26	2.61	
Kitchen extract unit		1	nr	302.94	302.94	2.13	
Gas fire with balanced flue.		1	nr	379.37	379.37	2.67	
5.3 Disposal Installations							
Upvc soil and vent pipe.		15	m	75.88	1,138.15	8.02	
Automatic air admittance valve.		3	nr	80.01	240.03	1.69	
5.4 Water Installations							
Cold water, tanks, 9 draw off points		item		1,011.09	1,011.09	7.12	
Hot water cylinder, 7 draw off points		item		1,379.43	1,379.43	9.71	
5.5 Heat Source							
Boiler, pump, and controls.		item		967.50	967.50	6.81	
5.6 Space heating							
Low temperature hot water radiator system.		item		4,192.50	4,192.50	29.52	
5.7 Ventilating Systems							
Extract fan units.		3	nr	111.36	334.09	2.35	
5.8 Electrical Installations							
Electrical Installation comprising: 16 Nr lighting points, 30 Nr outlet points, cooker points, immersion heater point and 9 Nr fittings points to include smoke alarms, bells, telephone and aerial installation.		item		4,256.47	4,256.47	29.98	
5.9 Gas installation							
Installation for boiler, hob and fire		item		561.00	561.00	3.95	
5.10 Builders work							
Building work for service installation.		item		1,442.67	1,442.67	10.16	19,695.96
					19,695.96	138.70	
(Element unit rate = £ 138.70 m2)							
6.0 EXTERNAL WORKS							
6.1 Site work							
Clear site vegetation excavate 150 topsoil and remove spoil.	Site clearance	479	m2	3.45	1,652.55	11.64	
Excavate and lay 50 precast concrete flag paving on hardcore.	Paving	120	m2	53.53	6,423.84	45.24	
Brick wall, excavation 600 deep, concrete foundation, 215 brickwork facing. PC £200/1000 damp proof course brick on edge coping 900 high	Garden wall	13	m	124.09	1,613.22	11.36	
Plastic coated chain link fencing 900 high on angle iron posts	Fencing	92	m	42.58	3,917.20	27.59	
Turfed Areas	Landscaping	103	m2	31.57	3,251.20	22.90	

COST PLAN FOR A DETACHED HOUSE (Cont'd)

Description	Item	Qty		Rate	Item Total	Cost per m2	Total
6.0 EXTERNAL WORKS (Cont'd)							
6.2 Drainage							
100 diameter clay pipes, excavation, 150 bed and surround, pea shingle, backfill 750 deep.		85	m	48.08	4,086.63	28.78	
Polypropylene inspection chambers; universal inspection chamber; polypropylene; 475 diameter; preformed chamber base benching; for 100mm pipe; chamber shaft, 175mm effective length; cast iron cover and frame; single seal A15 light duty to suit.		6	nr	180.39	1,082.34	7.62	
6.3 External Services							
100 diameter clay duct, excavation backfill, polythene warning tape, 500 deep.		38	m	17.27	656.30	4.62	
Provisional sums for External Services Gas, electric, water, telephone.		item		3,999.00	3,999.00	28.16	26,682.27
					26,682.27	187.90	
Preliminaries							
Site staff, accommodation, lighting, water, safety and health and welfare, rubbish disposal cleaning, drying, protection, security, plant, transport, temporary works and scaffolding with a contract period of 28 weeks					11,000.00	77.46	11,000.00
				Total			**£ 151,546.75**
				Cost per m2		**£ 1,067.23**	

COST PLAN FOR A DETACHED HOUSE (cont'd)

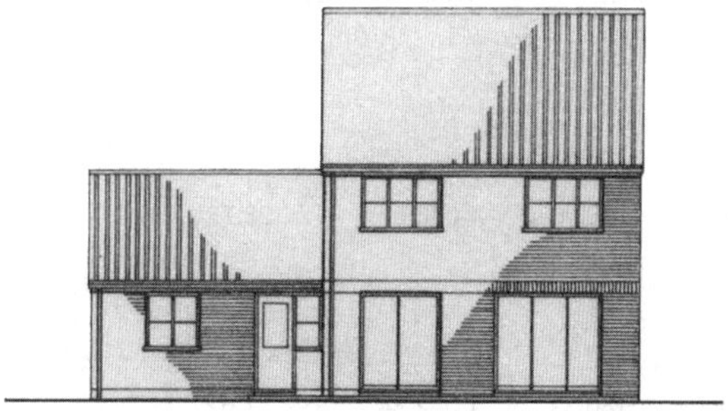

REAR ELEVATION

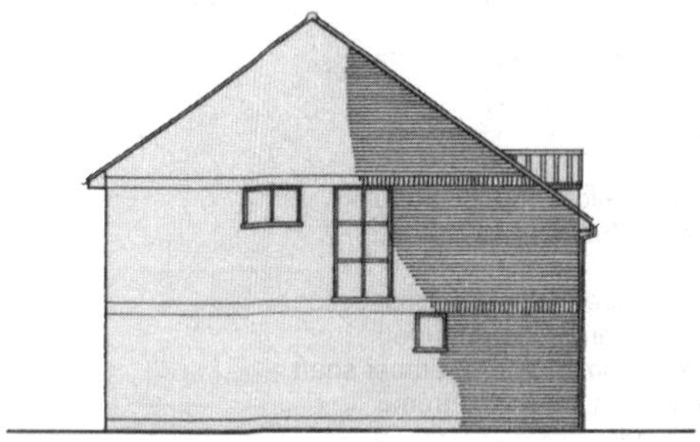

SIDE ELEVATION

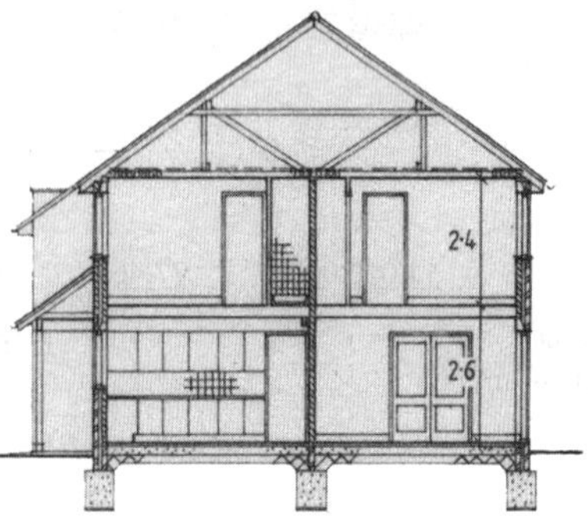

SECTION X-X

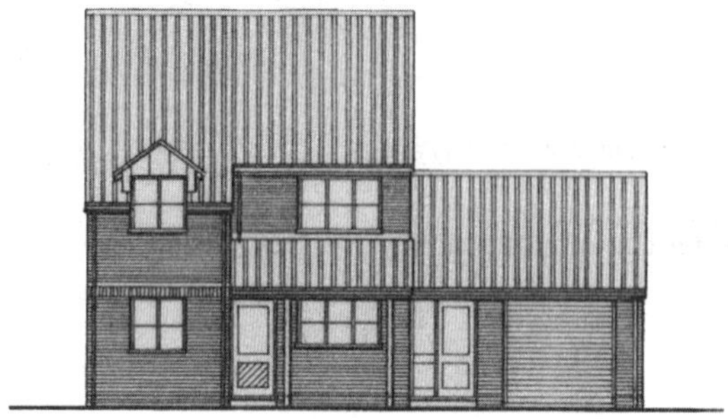

FRONT ELEVATION

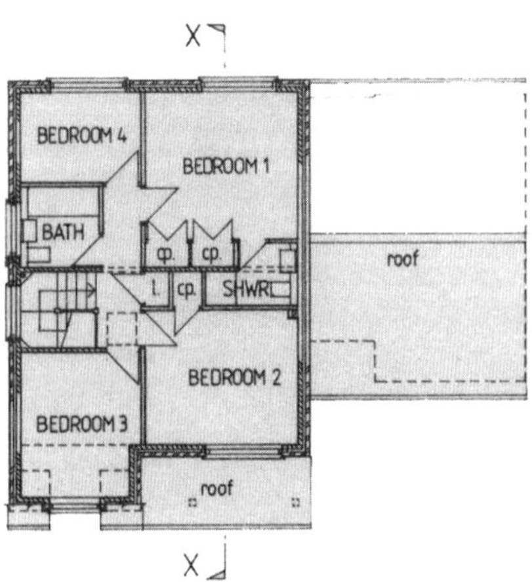

FIRST FLOOR PLAN

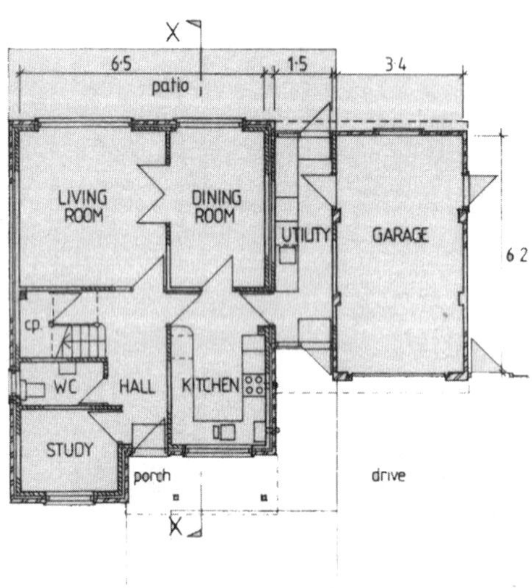

GROUND FLOOR PLAN

4 - BEDROOM DETACHED HOUSE

Cotterell Thomas & Thomas Architects

COMPOSITE PRICES

The following composite prices are based on the detailed rates herein and may be used to amend the previous example cost plans or to create new approximate budget estimates.
The composite prices can be further extended and amended by reference to the various sections
Allowances should be made for location factors, tender values and price changes.

PRELIMINARIES

The following preliminary items need to be added to the composite prices that follow.
A calculation can be made by reference to the preliminaries section.

		£
Typical allowances (major works)		
Management and staff - average each	week	500.00
Site accommodation	m²	26.00
Power and lighting	week	21.00
add setup and remove	item	1,500.00
Heating	week	18.00
Water	item	500.00
add setup and remove	item	1,000.00
Telephone	week	30.00
Safety, health and welfare	week	60.00
Rubbish disposal and cleaning site	week	30.00
Final clean	item	400.00
Drying out building	item	500.00
Security	week	30.00
Small plant and tools	%	1.50
Insurance of works	%	0.20
Travelling time and fares	week	20.00
Miscellaneous		
Rates on temporary buildings	week	20.00
Signboards and notices	item	200.00
Glass breakage	item	100.00
Sample panels	item	300.00
Testing	item	300.00
Attendance; chainman etc (on clerk of works etc)	item	400.00
Setting out equipment, pegs	item	50.00
Provision for winter working	item	1,250.00
Defects	item	1,680.00
Sureties and bonds	item	500.00
Special conditions of contract	item	500.00
Attendance on subcontractors		1,000.00
Mechanical plant		
Cranes - mobile tower	item	80,800.00
Cranes - static tower	item	30,725.00
Cranes - mobile crane	item	6,200.00
Pumping	item	2,550.00
Transport		
On-site transport		720.00
Transport for plant		2,265.00
Temporary roads		1,890.00
Scaffolding		
External access		8,820.00
Support		5,000.00
Internal		6,300.00
Hoardings		2,600.00

DEMOLITIONS AND ALTERATIONS

Demolition

Demolish timber outbuilding size 2.50 x 2.00 x 3.00 high	nr	202.44
Demolish brick lean-to and make good size 1.50 x 1.00 x 2.00 high	nr	256.02

Demolishing parts of structures

Substructures		
incl. Excavation and removal by hand		
break out brickwork	m^3	147.87
break out concrete	m^3	182.41
break out reinforced concrete	m^3	239.98
break up concrete bed		
100 thick	m^2	18.24
150 thick	m^2	27.36
200 thick	m^2	36.48
break up reinforced concrete bed		
100 thick	m^2	24.00
150 thick	m^2	36.00
200 thick	m^2	44.63
Walls		
brick internal		
102 thick	m^2	15.36
215 thick	m^2	31.65
327 thick	m^2	48.47
brick external		
102 thick	m^2	11.67
250 thick cavity wall	m^2	24.28
Floors		
concrete reinforced; screed		
150 thick; 50 screed	m^2	52.41
200 thick; 50 screed	m^2	65.58
timber boarded on joists		
225 x 50 joists; floor boards	m^2	12.14
Roofs		
timber flat roofs		
150 x 50 joists, 25 boards; felt	m^2	12.76
timber pitched roof		
100 x 50 rafters; joists; plain tile	m^2	15.80
100 x 50 rafters; joists; interlocking tiles	m^2	13.74
100 x 50 rafters; joists; slates	m^2	15.72

Alterations

Includes make good finishes		
Cut opening through reinforced concrete floor		
150 thick x 900 x 2134 for staircase	nr	868.48
Cut opening through block wall		
100 thick x 838 x 2032 for doorway	nr	543.45
Cut opening through brick wall		
215 thick x 914 x 2082 for doorway	nr	794.85
250 thick cavity wall x 900 x 900 for window	nr	514.46
Cut opening through stud partition		
100 thick x 838 x 2032 for doorway	nr	382.22
Fill opening in block wall		
100 thick x 838 x 2032; door removed	nr	383.21
Fill opening in brick wall		
102 thick x 838 x 2032; door removed	nr	455.50
215 thick x 914 x 2082; external door removed	nr	732.61
250 thick cavity wall x 914 x 2082; external door removed	nr	683.93
Fill opening in stud partition		
100 thick x 838 x 2032; door removed	nr	344.32

Removal

Remove window		
1486 x 923 metal	nr	21.48
1225 x 1070 wood casement; frame	nr	23.92
915 x 1525 cased frame; sashes	nr	35.68
Remove door and frame		
single internal	nr	17.50
single external	nr	19.49
Remove roof coverings		
3 layer felt	m^2	4.12
asphalt	m^2	5.15
slates	m^2	11.59
tiles	m^2	11.67

COMPOSITE PRICES (Cont'd)

SUBSTRUCTURE

Note: Rates allow for excavation by machine.

Strip Foundations

Excavating, 225x450 concrete, brickwork, dpc, common bricks			
800	deep for 103 wall	m	60.11
1000	deep for 103 wall	m	72.98
1200	deep for 103 wall	m	92.32
800	deep for 215 wall	m	90.66
1000	deep for 215 wall	m	111.69
1200	deep for 215 wall	m	132.57
800	deep for 275 wall	m	107.44
1000	deep for 275 wall	m	134.55
1200	deep for 275 wall	m	160.04
Extra 100mm depth concrete foundation; excavation; concrete		m	6.91
Extra 100mm depth 103 brick wall in foundation; excavation; brickwork		m	6.51
Extra 100mm depth 215 brick wall in foundation; excavation; brickwork		m	10.52
Extra 100mm depth 275 brick wall in foundation; excavation; brickwork		m	11.76
Extra for four courses facing bricks P.C. £250.00 per 1000		m	3.36
Excavating, 225x600 concrete, brickwork, dpc, common bricks			
800	deep for 275 wall	m	112.63
1000	deep for 275 wall	m	140.20
1200	deep for 275 wall	m	163.24
Extra 100mm depth 275 brick wall in foundation; excavation; brickwork		m	11.91
Extra for four courses facing bricks P.C. £250.00 per 1000		m	3.36
Excavating, 225x450 concrete, brickwork, dpc, Engineering bricks Class B			
800	deep for 215 wall	m	96.11
1000	deep for 215 wall	m	118.63
1200	deep for 215 wall	m	141.00
800	deep for 275 wall	m	113.39
1000	deep for 275 wall	m	142.14
1200	deep for 275 wall	m	169.26
Extra 100mm depth concrete foundation; excavation; concrete		m	6.91
Extra 100mm depth 103 brick wall in foundation; excavation; brickwork		m	6.92
Extra 100mm depth 215 brick wall in foundation; excavation; brickwork		m	11.27
Extra 100mm depth 275 brick wall in foundation; excavation; brickwork		m	12.58
Extra for four courses facing bricks P.C. £250.00 per 1000		m	2.13
Excavating, 225x450 concrete, blockwork, dpc, Concrete blocks			
800	deep for 100 wall	m	49.29
1000	deep for 100 wall	m	54.90
1200	deep for 100 wall	m	63.88
Extra 100mm depth concrete foundation; excavation; concrete		m	6.91
Extra 100mm depth 100 block wall in foundation; excavation; blockwork		m	4.57
800	deep for 140 wall	m	51.78
1000	deep for 140 wall	m	62.30
1200	deep for 140 wall	m	72.82
Extra 100mm depth concrete foundation; excavation; concrete		m	6.91
Extra 100mm depth 140 block wall in foundation; excavation; blockwork		m	5.32
800	deep for 190 wall	m	57.60
1000	deep for 190 wall	m	69.56
1200	deep for 190 wall	m	81.53
Extra 100mm depth concrete foundation; excavation; concrete		m	6.91
Extra 100mm depth 190 block wall in foundation; excavation; blockwork		m	6.05
Excavating, 225 x 450 concrete, blockwork, dpc, Thermalite Trenchblocks			
800	deep for 275 wall	m	73.31
1000	deep for 275 wall	m	89.50
1200	deep for 275 wall	m	105.69
Extra 100mm depth concrete foundation; excavation; concrete		m	6.91
Extra 100mm depth 190 block wall in foundation; excavation; blockwork		m	8.21
Excavating, 225x600 concrete, blockwork, dpc, Thermalite Trenchblocks			
800	deep for 275 wall	m	79.12
1000	deep for 275 wall	m	95.77
1200	deep for 275 wall	m	112.27
Extra 100mm depth concrete foundation; excavation; concrete;		m	6.91
Extra 100mm depth 275 brick wall in foundation; excavation; blockwork		m	8.21
Fabric reinforcement ref. A193 (no allowance for working space or formwork)			
350 wide		m	1.13
600 wide		m	1.94
750 wide		m	2.88

Trench Fill Foundations

Excavating 450 wide concrete fill to 150 of ground level, brickwork, dpc			
Common bricks			
800	deep for 102 wall	m	60.72
1000	deep for 102 wall	m	73.28
1200	deep for 102 wall	m	85.85
800	deep for 215 wall	m	72.33
1000	deep for 215 wall	m	84.89
1200	deep for 215 wall	m	97.45
800	deep for 275 wall	m	78.51
1000	deep for 275 wall	m	91.07
1200	deep for 275 wall	m	103.63
Extra 100mm depth concrete foundation; excavation; concrete;		m	6.91
Extra for four courses facing bricks PC £250 per 1000		m	3.36

Column bases

Excavating, 450 concrete		
1000 x 1000 x 1000 deep	nr	77.56
1000 x 1000 x 1200 deep	nr	88.33
1000 x 1000 x 1500 deep	nr	97.91
1200 x 1200 x 1000 deep	nr	108.74
1200 x 1200 x 1200 deep	nr	224.95
1200 x 1200 x 1500 deep	nr	135.41
1500 x 1500 x 1000 deep	nr	165.32
1500 x 1500 x 1200 deep	nr	184.81
1500 x 1500 x 1500 deep	nr	202.87
2000 x 2000 x 1000 deep	nr	285.74
2000 x 2000 x 1200 deep	nr	316.16
2000 x 2000 x 1500 deep	nr	345.18
Excavating, 750 concrete		
1000 x 1000 x 1000 deep	nr	110.77
1000 x 1000 x 1200 deep	nr	121.54
1000 x 1000 x 1500 deep	nr	131.12
1000 x 1000 x 2000 deep	nr	147.08
1200 x 1200 x 1000 deep	nr	156.56
1200 x 1200 x 1200 deep	nr	170.56
1200 x 1200 x 1500 deep	nr	100.00
1200 x 1200 x 2000 deep	nr	204.36
1500 x 1500 x 1000 deep	nr	240.04
1500 x 1500 x 1200 deep	nr	259.53
1500 x 1500 x 1500 deep	nr	277.59
1500 x 1500 x 2000 deep	nr	307.70
2000 x 2000 x 1000 deep	nr	418.58
2000 x 2000 x 1200 deep	nr	449.00
2000 x 2000 x 1500 deep	nr	478.02
2000 x 2000 x 2000 deep	nr	526.38
3000 x 3000 x 1000 deep	nr	923.45
3000 x 3000 x 1200 deep	nr	982.40
3000 x 3000 x 1500 deep	nr	1,040.72

COMPOSITE PRICES (Cont'd)

SUBSTRUCTURE (Cont'd)

Column Bases (Cont'd)

Excavating, 1000 concrete		
1000 x 1000 x 1000 deep	nr	138.44
1000 x 1000 x 1200 deep	nr	149.21
1000 x 1000 x 1500 deep	nr	158.79
1000 x 1000 x 2000 deep	nr	174.75
1200 x 1200 x 1000 deep	nr	196.42
1200 x 1200 x 1200 deep	nr	210.41
1200 x 1200 x 1500 deep	nr	223.08
1200 x 1200 x 2000 deep	nr	244.21
1500 x 1500 x 1000 deep	nr	302.31
1500 x 1500 x 1200 deep	nr	321.80
1500 x 1500 x 1500 deep	nr	339.86
1500 x 1500 x 2000 deep	nr	369.97
2000 x 2000 x 1000 deep	nr	529.28
2000 x 2000 x 1200 deep	nr	559.70
2000 x 2000 x 1500 deep	nr	588.72
2000 x 2000 x 2000 deep	nr	637.08
3000 x 3000 x 1000 deep	nr	1,172.52
3000 x 3000 x 1200 deep	nr	1,231.47
3000 x 3000 x 1500 deep	nr	1,289.79
3000 x 3000 x 2000 deep	nr	1,386.99
Excavating, 1500 concrete		
1000 x 1000 x 1500 deep	nr	214.14
1000 x 1000 x 2000 deep	nr	230.10
1500 x 1500 x 1500 deep	nr	464.40
1500 x 1500 x 2000 deep	nr	494.51
2000 x 2000 x 1500 deep	nr	810.12
2000 x 2000 x 2000 deep	nr	858.48
3000 x 3000 x 1500 deep	nr	1,787.94
3000 x 3000 x 2000 deep	nr	1,885.14
Excavating, 450 reinforced concrete		
1000 x 1000 x 1000 deep	nr	227.65
1000 x 1000 x 1200 deep	nr	247.44
1000 x 1000 x 1500 deep	nr	269.15
1000 x 1000 x 2000 deep	nr	305.33
1200 x 1200 x 1000 deep	nr	308.06
1200 x 1200 x 1200 deep	nr	331.76
1200 x 1200 x 1500 deep	nr	359.00
1200 x 1200 x 2000 deep	nr	404.39
Excavating, 750 reinforced concrete		
1000 x 1000 x 1000 deep	nr	386.37
1000 x 1000 x 1200 deep	nr	405.23
1000 x 1000 x 1500 deep	nr	426.94
1000 x 1000 x 2000 deep	nr	463.12
1200 x 1200 x 1000 deep	nr	517.44
1200 x 1200 x 1200 deep	nr	541.14
1200 x 1200 x 1500 deep	nr	568.37
1200 x 1200 x 2000 deep	nr	613.76

Ground beams

Excavation, blinding, reinforced concrete		
300 x 450	m	75.48
300 x 600	m	118.18
300 x 450	m	102.89
300 x 600	m	153.38

Ground slabs

Excavation, 150 hardcore, blinding, concrete		
150 ground slab	m²	33.99
200 ground slab	m²	38.41
225 ground slab	m²	41.48
Extra: 50 blinding	m²	6.87
Extra: Fabric reinforcement A142	m²	2.77
Extra: Fabric reinforcement A193	m²	3.23
Slab thickening 450 wide x 150 thick	m²	8.29
Damp proof membrane		
250mu Polythene sheeting	m²	1.41
2 coats Bitumen emulsion	m²	12.62
Extra: 50mm Expanded polystyrene insulation	m²	5.99
Extra: 50mm Concrete thickness	m²	5.22
Extra: 100mm Reduced level excavation	m²	1.84
Extra: 100mm hardcore filling	m²	2.67

Beam and block floors

(excludes excavation, blinding etc.)		
155mm beam floor span 3.9m	m²	25.94
Extra: 150mm Reduced level excavation; 50 concrete blinding	m²	8.19

Piling

On/off site charge	nr	4,411.80
450 bored cast piles n/e 6m deep	m	25.84
610 bored cast piles n/e 6m deep	m	39.07
Cutting off tops of piles	nr	8.85

INSIDE EXISTING BUILDING

Note: rates allow for excavation by hand.

Strip foundations

Excavating, 225 concrete, brickwork, dpc		
Common bricks		
800 deep for 103 wall	m	78.69
1000 deep for 103 wall	m	95.54
1200 deep for 103 wall	m	133.91
800 deep for 215 wall	m	110.74
1000 deep for 215 wall	m	136.51
1200 deep for 215 wall	m	168.82
Extra 100mm depth concrete foundation; excavation; concrete	m	10.12

Trench Fill Foundations

Excavating 450 wide concrete fill to 150 of ground level, brickwork, dpc		
Common bricks		
800 deep for 102 wall	m	83.45
1000 deep for 102 wall	m	101.80
800 deep for 215 wall	m	95.06
1000 deep for 215 wall	m	96.76
Extra 100mm depth concrete foundation; excavation; concrete	m	10.12

Ground slabs

Excavation, 150 hardcore, blinding, concrete		
150 ground slab	m²	54.03
200 ground slab	m²	63.56
Slab thickening 450 wide x 150 thick	m	12.87
Damp proof membrane		
250 mu polythene sheeting	m²	1.41
2 coats bitumen emulsion	m²	12.62

FRAME

Reinforced concrete

		£
200 x 200 column	m	51.69
225 x 225 column	m	58.15
300 x 300 column	m	84.85
200 x 200 attached beam	m	53.14
225 x 250 attached beam	m	65.77
300 x 450 attached beam	m	117.93
400 x 650 attached beam	m	186.76
Casing to 152 x 152 steel column	m	43.41
Casing to 203 x 203 steel column	m	56.03
Casing to 146 x 254 steel beam	m	55.68
Casing to 165 x 305 steel beam	m	65.40
Extra: wrought formwork	m²	4.03

Steel

Columns and beams	tonne	1,458.59
Roof trusses	tonne	2,076.72
Purlins and cladding rails	tonne	1,474.72
Small sections	tonne	1,107.25
Two coats primer at works	tonne	176.99
Prime, undercoat and gloss paint	m²	7.08

COMPOSITE PRICES (Cont'd)

FRAME (Cont'd)

Glued laminated softwood beams

65x150x4000 long	nr	69.22
65x250x4000 long	nr	116.91
65x300x6000 long	nr	202.73
65x325x8000 long	nr	286.54
90x200x4000 long	nr	118.47
90x325x8000 long	nr	361.33
90x400x10000 long	nr	547.34
90x450x12000 long	nr	731.13
115x250x4000 long	nr	186.13
115x300x4000 long	nr	224.33
115x350x8000 long	nr	495.13
115x425x12000 long	nr	857.48
115x450x8000 long	nr	619.71
115x500x14000 long	nr	1,178.67
140x350x8000 long	nr	599.65
140x450x12000 long	nr	1,117.03
165x450x12000 long	nr	1,302.66
190x475x14000 long	nr	1,841.62

UPPER FLOORS

Concrete

In-situ reinforced concrete		
150 thick	m^2	67.82
200 thick	m^2	73.24
225 thick	m^2	79.96
In-situ reinforced concrete with permanent Expamet Hy-rib formwork ref 2411 and A142 mesh reinforcement		
100 thick	m^2	67.80
125 thick	m^2	71.25
150 thick	m^2	74.79
Hollow prestressed concrete		
125 thick	m^2	38.81
150 thick	m^2	42.87
200 thick	m^2	46.34
Beam and block		
200 thick	m^2	25.94

Timber

Softwood joist		
100 thick	m^2	15.51
150 thick	m^2	18.28
175 thick	m^2	19.44
200 thick	m^2	21.60
225 thick	m^2	23.43
250 thick	m^2	25.73
Floor boarding		
19 square edge softwood	m^2	19.53
25 square edge softwood	m^2	20.60
19 tongued and grooved softwood	m^2	22.29
25 tongued and grooved softwood	m^2	23.29
18 tongued and grooved chipboard	m^2	7.91
22 tongued and grooved chipboard	m^2	8.44
15 tongued and grooved plywood Douglas Fir unsanded	m^2	12.12
18 tongued and grooved plywood Douglas Fir unsanded	m^2	14.20
15 tongued and grooved plywood Birch Faced BB quality	m^2	30.88
18 tongued and grooved plywood Birch Faced BB quality	m^2	36.56

ROOF

Roof structure

Reinforced in-situ concrete		
100 thick	m^2	57.58
150 thick	m^2	65.78
200 thick	m^2	66.16
50 lightweight screed to falls	m^2	21.22
50 cement sand screed to falls	m^2	14.40
50 vermiculite screed to falls	m^2	19.87
50 flat glass fibre insulation board	m^2	9.43
Extra for vapour barrier	m^2	3.43
Softwood flat roof joists		
50 x 100 joists @ 400 centres	m^2	19.16
50 x 175 joists @ 400 centres	m^2	21.70
50 x 200 joists @ 400 centres	m^2	24.19
50 x 225 joists @ 400 centres	m^2	26.10
50 x 100 joists @ 600 centres	m^2	13.85
50 x 175 joists @ 600 centres	m^2	15.80
50 x 200 joists @ 600 centres	m^2	17.15
50 x 225 joists @ 600 centres	m^2	18.44
Boarding		
18 chipboard	m^2	8.51
25 softwood tongued grooved	m^2	14.20
18 plywood decking	m^2	17.11
24 plywood decking	m^2	22.57
Pitched roof with 50 x 100 rafters and ceiling joists at 400 centres	m^2	34.88
Pitched roof with trussed rafters at 600 centres	m2	20.92

Roof coverings

Three layer fibre based felt	m^2	18.20
Three layer high performance felt	m^2	47.07
20 two coat mastic asphalt	m^2	23.68
50 lightweight screed to falls	m^2	21.22
50 cement sand screed to falls	m^2	14.40
60 Vermiculite screed to falls	m^2	23.68
50 flat glass fibre insulation board	m^2	9.43
Extra for vapour barrier	m^2	3.43
Profile 3 fibre cement sheeting (on slope)	m^2	20.67
Profile 6 fibre cement sheeting (on slope)	m^2	17.92
Colour coated galvanised steel sheeting (on slope)	m^2	25.45
381 x 227 concrete interlocking tiles (on slope)	m^2	24.19
268 x 165 plain concrete tiles (on slope)	m^2	49.67
268 x 165 machine made clay tiles (on slope)	m^2	51.09
265 x 165 hand made tiles (on slope)	m^2	65.67
400 x 200 fibre cement slates (on slope)	m^2	39.43
500 x 250 fibre cement slates (on slope)	m^2	33.97
600 x 300 fibre cement slates (on slope)	m^2	28.83
400 x 200 natural slates (on slope)	m^2	58.30
500 x 250 natural slates (on slope)	m^2	68.11
600 x 300 natural slates (on slope)	m^2	86.15
100 Fibreglass insulation quilt	m^2	6.33
100 Rockwool insulation quilt	m^2	5.15
150 Rockwool insulation quilt	m^2	7.92

Eaves, soffits, fascias and gutters

Eaves soffit, fascia, gutter and decoration		
225 x 150, 114 pvc-u	m	39.12
225 x 150, 100 half round Cast Iron gutter	m	53.28
225 x 150, 114 half round Cast Iron gutter	m	54.28
225 x 150, 125 half round Cast Iron gutter	m	58.44
225 x 150, 150 half round Cast Iron gutter	m	73.96
225 x 150, 100 og Cast Iron gutter	m	53.42
225 x 150, 114 og Cast Iron gutter	m	55.03
225 x 150, 125 og Cast Iron gutter	m	58.15
400 x 200, 114 pvc-u	m	45.75
400 x 200, 100 half round Cast Iron gutter	m	59.91
400 x 200, 114 half round Cast Iron gutter	m	60.91
400 x 200, 125 half round Cast Iron gutter	m	65.07
400 x 200, 150 half round Cast Iron gutter	m	80.59
400 x 200, 100 og Cast Iron gutter	m	60.05
400 x 200, 114 og Cast Iron gutter	m	61.66
400 x 200, 125 og Cast Iron gutter	m	64.78
400 x 200, 102 half round cast aluminium gutter	m	59.39
400 x 200, 127 half round cast aluminium gutter	m	66.13

COMPOSITE PRICES (Cont'd)

ROOF (Cont'd)

Verges, hips, ridges and valleys

Verge undercloak, barge board and decoration		
225 x 25	m	25.46
Hip rafter and capping		
225 x 50, concrete capping tiles	m	23.25
225 x 50, clay machine made capping tiles	m	34.15
225 x 50, clay hand made capping tiles	m	32.30
225 x 50, concrete bonnet hip tiles	m	34.15
225 x 50, clay machine made bonnet hip tiles	m	34.15
225 x 50, clay hand made bonnet hip tiles	m	34.15
Ridge board and capping		
175 x 32, concrete capping tile	m	19.44
175 x 32, machine made clay tile	m	30.34
175 x 32, hand made clay tile	m	28.49
Valley rafter and tile		
225 x 32, concrete angular valley tile	m	22.43
225 x 32, concrete valley trough	m	24.55
225 x 32, machine made clay valley tile	m	34.43
225 x 32, hand made clay valley tile	m	33.41
Valley rafter, sole boards and leadwork		
225 x 32, code 4 lead	m	82.59
225 x 32, code 5 lead	m	88.34

Rain water pipes

Rainwater pipe			
68	pvc-u pipe	m	20.08
63	cast iron pipe and decoration	m	49.80
75	cast iron pipe and decoration	m	55.45
100	cast iron pipe and decoration	m	70.57
63	cast aluminium pipe	m	39.94
75	cast aluminium pipe	m	51.04
100	cast aluminium pipe	m	71.38

STAIRS

Concrete

Reinforced concrete 2670 rise, mild steel balustrade one side, straight flight		
1000 wide granolithic finish	nr	2,468.94
1500 wide granolithic finish	nr	3,219.48
1000 wide pvc sheet finish non slip nosings	nr	2,851.12
1500 wide pvc sheet finish non slip nosings	nr	3,180.69
1000 wide terrazzo finish	nr	3,066.01
1500 wide terrazzo finish	nr	3,411.23
Reinforced concrete 2670 rise, mild steel balustrade one side in two flights with half landing		
1000 wide granolithic finish	nr	2,925.82
1500 wide granolithic finish	nr	3,323.27
1000 wide pvc sheet finish non slip nosings	nr	3,325.27
1500 wide pvc sheet finish non slip nosings	nr	3,630.31
1000 wide terrazzo finish	nr	3,569.05
1500 wide terrazzo finish	nr	4,378.96

Timber

Softwood staircase 2670 rise, balustrade one side.		
910 wide straight flight	nr	939.82
910 wide 1 flight with 3 winders	nr	1,153.89
910 wide 2 flight with half landing	nr	1,153.89

Steel

Mild steel staircase 2670 rise, balustrade, both sides.		
910 wide straight flight	nr	2,769.20
910 wide 2 flights with quarter landing	nr	5,155.16
910 wide 2 flights with half landing	nr	6,243.06
Spiral steel staircase, balustrade, decorated		
1568 diameter 2640 rise, domestic	nr	2,397.75
1568 diameter 3000 rise, shop, office, factory and domestic	nr	2,902.69
2072 diameter 3000 rise, shop, office, factory and domestic	nr	3,232.11
1568 diameter, 3000 rise, shop, office, factory and domestic; galvanised	nr	3,198.03
2072 diameter, 3000 rise, shop, office, factory and domestic; galvanised	nr	3,446.82

EXTERNAL WALLS

Brickwork

Common bricks		
102 wall	m²	49.97
215 wall	m²	88.66
327 wall	m²	117.78
Hollow wall with stainless steel ties		
275 common bricks both skins	m²	102.86
275 facings one side P.C. £150/1000	m²	107.41
275 facings one side P.C. £200/1000	m²	110.74
275 facings one side P.C. £250/1000	m²	114.07
275 facings one side P.C. £300/1000	m²	117.40
Engineering bricks Class B		
102 wall	m²	54.07
215 wall	m²	96.12
327 wall	m²	127.27
Engineering bricks Class A		
102 wall	m²	59.40
215 wall	m²	106.87
327 wall	m²	143.43
Facing bricks		
102 wall P.C. £150 per 1000	m²	54.52
102 wall P.C. £200 per 1000	m²	57.85
102 wall P.C. £250 per 1000	m²	61.18
102 wall P.C. £300 per 1000	m²	64.51
215 wall P.C. £150 per 1000	m²	90.84
215 wall P.C. £200 per 1000	m²	97.56
215 wall P.C. £250 per 1000	m²	104.27
215 wall P.C. £300 per 1000	m²	110.99

Blockwork

Aerated concrete block 4.0n/mm²		
100 wall	m²	30.53
140 wall	m²	37.51
190 wall	m²	44.06
Aerated concrete block 7.0n/mm²		
100 wall	m²	32.96
140 wall	m²	40.88
190 wall	m²	48.62
215 wall	m²	53.42
Dense concrete block 7.0n/mm²		
100 wall	m²	40.49
140 wall	m²	47.54
190 wall	m²	54.95
215 wall	m²	59.99
Fair face concrete block		
100 wall	m²	43.28
140 wall	m²	53.74
190 wall	m²	62.80

Composite

Hollow wall in light weight blocks with galvanised ties.		
275 facings one side P.C. £150/1000	m²	87.97
275 facings one side P.C. £200/1000	m²	91.30
275 facings one side P.C. £250/1000	m²	94.63
275 facings one side P.C. £300/1000	m²	97.96
Extra for:		
50 mm glass fibre slab insulation	m²	7.84
13 mm two coat plain render painted	m²	10.06
15 mm pebbled dash render	m²	17.90

COMPOSITE PRICES (Cont'd)

WINDOWS AND EXTERNAL DOORS

Windows

Softwood single glazed casement window		
type N09V size 488 x 900	nr	148.68
type 110V size 631 x 1050	nr	193.74
type 2N10W size 915 x 1050	nr	239.00
type 212CV size 1220 x 1220	nr	339.08
type 312CVC size 1769 x 1050	nr	506.45
average cost	m²	271.62
purpose made in panes 0.10-0.50m²	m²	239.42
Softwood double glazed casement window		
type N09V size 488 x 900	nr	211.03
type 110V size 631 x 1050	nr	273.19
type 2N10W size 915 x 1050	nr	323.27
type 212CV size 1220 x 1220	nr	458.29
type 310CVC size 1769 x 1050	nr	613.73
type 312CVC size 1769 x 1050	nr	674.90
average cost	m²	366.31
purpose made in panes 0.10-0.50m²	m²	339.65
Hardwood double glazed standard casement windows		
type W107VH size 626 x 750	nr	326.17
type W112VH size 626 x 1200	nr	392.02
type W2N10CVH size 913 x 1050	nr	452.93
type WH212CH size 1220 x 1220	nr	528.52
average cost	m²	499.84
Afrormosia double glazed casement windows		
purpose made in panes 0.l0-0.50m²	m²	502.38
European Oak double glazed casement windows		
purpose made in panes 0.10-0.50m²	m²	596.54
Softwood glazed double hung sash window		
type GS3 size 825 x 1094	nr	610.11
type GS4 size 825 x 1394	nr	668.74
type GSW4 size 1051 x 1394	nr	813.54
average cost	m²	642.41
Softwood glazed double hung sash window 38 thick; cased frames; oak cills; sash chords		
size 900 x 1200	nr	648.62
size 1050 x 1200	nr	782.84
size 1200 x 1500	nr	921.87
size 1200 x 1800	nr	1,121.27
average cost	m²	595.64
Softwood glazed double hung sash window 50 thick; cased frames; oak cills; sash chords		
size 1050 x 1200	nr	829.39
size 1200 x 1500	nr	983.65
size 1200 x 1800	nr	1,193.20
size 1500 x 1800	nr	1,394.73
average cost	m²	600.13
Galvanised steel windows;4mm glass		
type NC6F size 279 x 923	nr	121.02
type NC05F size 508 x 1067	nr	159.66
type NE2 size 997 x 628	nr	182.33
type NC02F size 997 x 1067	nr	339.00
type ND2F size 997 x 1218	nr	330.61
type NC4F size 1486 x 923	nr	424.17
type NC04F size 1486 x 1067	nr	458.44
type ND4 size 1486 x 1218	nr	443.64
type NC11F size 1994 x 923	nr	490.58
type ND11F size 1994 x 1218	nr	563.85
average cost	m²	297.83
Aluminium double glazed; vertical sliders; cill		
size 600 x 800	nr	466.81
size 600 x 1200	nr	620.47
size 800 x 800	nr	762.88
size 800 x 1200	nr	974.87
size 1200 x 800	nr	1,501.92
size 1200 x 1200	nr	1,727.56
Aluminium double glazed; horizontal sliders; cill		
size 1200 x 800	nr	475.02
size 1200 x 1400	nr	556.16
size 1400 x 800	nr	507.28
size 1400 x 1400	nr	591.31
size 1600 x 800	nr	543.19
size 1600 x 1600	nr	662.72
Upvc windows; opening/fixed lights; factory double glazed		
size 600 x 1350 window	nr	373.89
size 750 x 1350 window	nr	404.05
size 900 x 1350 window	nr	436.08
size 1200 x 1350 window	nr	490.34
Upvc windows tilt / turn factory double glazed		
size 600 x 600 window	nr	519.55
size 600 x 1050 window	nr	628.64
size 750 x 1050 window	nr	652.76
size 900 x 1200 window	nr	731.65
size 900 x 1500 window	nr	815.36
size 1200 x 1050 window	nr	776.30
Upvc windows tilt / turn; fixed light; factory double glazed		
size 600 x 1500 window	nr	722.99
size 750 x 1500 window	nr	766.39
size 900 x 1500 window	nr	815.36
size 1200 x 1500 window	nr	921.02
Upvc windows tilt / turn; fixed side light; factory double glazed		
size 1800 x 1050 window	nr	929.53
size 1800 x 1350 window	nr	1,105.36
Upvc windows tilt / turn; fixed centre light; factory double glazed		
size 2400 x 1050 window	nr	1,428.54
size 2400 x 1350 window	nr	1,601.17

Doors

44 External ledged and braced door, 50 x 100 softwood frame, ironmongery P.C. £50.00 per leaf and decorate		
762 x 1981 door	nr	299.57
838 x 1981 door	nr	319.88
1372 x 1981 pair doors	nr	507.04
44 External framed, ledged and braced door, 50 x 100 softwood frame, ironmongery P.C. £50.00 per leaf and decorate		
762 x 1981 door	nr	299.57
838 x 1981 door	nr	334.83
1372 x 1981 pair doors	nr	559.36
44 External plywood flush door, 50 x 100 SW frame, architrave one side, ironmongery P.C. £50.00 per leaf and decorate		
762 x 1981 door	nr	323.33
838 x 1981 door	nr	331.61
1372 x 1981 pair doors	nr	523.77
44 Softwood external panel door glazed panels, 50 x 100 softwood frame, architrave one side, ironmongery P.C. £50.00 decorated.		
2XG size 762 x 1981	nr	375.90
2XG size 838 x 1981	nr	326.63
2XGG size 762 x 1981	nr	320.26
2XGG size 838 x 1981	nr	329.47
KXT size 762 x 1981	nr	407.08
KXT size 838 x 1981	nr	418.16
44 Hardwood Magnet Carolina door, clear glass 50 x 100 hardwood frame architrave one side, ironmongery P.C. £50.00 decorated.		
size 838 x 1981	nr	386.15
size 813 x 2032	nr	449.73
44 Hardwood Magnet Alicante door, bevelled glass 50 x 100 hardwood frame architrave one side, ironmongery P.C. £50.00 decorated.		
size 838 x 1981	nr	521.30
size 813 x 2032	nr	525.54

Garage doors

Galvanised steel garage door		
size 2135 x 1980	nr	726.03
size 4265 x 1980	nr	1,671.26
Western Red Cedar garage door		
size 2135 x 1980 side hung pair	nr	828.64
size 2135 x 1980 up and over	nr	1,032.00

Opening in cavity wall

Galvanised steel lintel, DPC, brick soldier course	m	94.40
Concrete boot lintel		
size 252x150x1200	nr	48.80
size 252x225x1200	nr	67.78
size 252x150x1800	nr	73.17
size 252x225x1800	nr	101.60
Softwood window board, DPC, close cavity	m	22.16
Close cavity at jamb, DPC, facings to reveal	m	12.65

COMPOSITE PRICES (Cont'd)

INTERNAL WALLS

Brickwork

Common bricks		
102 wall	m²	49.97
215 wall	m²	88.66
327 wall	m²	117.78
Engineering bricks		
102 wall	m²	54.07
215 wall	m²	96.12
327 wall	m²	127.27

Blockwork

Light weight concrete block 4.0 n/m²		
75 wall	m²	23.19
90 wall	m²	29.59
100 wall	m²	30.53
140 wall	m²	37.51
190 wall	m²	44.06
Dense concrete block 7.0 n/m²		
90 wall	m²	38.19
100 wall	m²	40.49
140 wall	m²	47.54
190 wall	m²	54.95
215 wall	m²	59.99
Hollow glass blocks		
80 wall - 190 x 190 blocks	m²	314.96
80 wall - 240 x 240 blocks	m²	256.68

Partitions

Timber stud partition with plasterboard both sides		
75 partition, direct decoration	m²	52.94
100 partition direct decoration	m²	56.60
75 partition with skim coat plaster and decorate	m²	64.72
100 partition with skim coat plaster and decorate	m²	68.38
75 partition with plastic faced plasterboard	m²	52.65
100 partition with plastic faced plasterboard	m²	56.31
Extra for 75 insulation	m²	8.53
Cellular core partition		
57 partition direct decoration	m²	48.36
57 partition with skim coat plaster and decorate	m²	60.14
Gyproc laminated partition		
50 partition direct decoration	m²	47.14
65 partition direct decoration	m²	54.37
50 partition with skim coat plaster and decorate	m²	58.92
65 partition with skim coat plaster and decorate	m²	66.15
Metal stud partition with one layer plasterboard each side		
75 partition direct decoration	m²	42.45
75 partition with skim coat plaster and decorate	m²	54.23
Metal stud partition with two layers plasterboard each side		
100 partition direct decoration	m²	49.29
100 partition with skim coat plaster and decorate	m²	61.07

INTERNAL DOORS

Panel doors

34 softwood panelled type SA door; bevelled glass; 38 x 150 lining, stop, architrave both sides, ironmongery PC £20.00 per leaf; decorated		
size 762 x 1981	nr	438.76
34 Magnet Winsoe hardwood six panel door 38 x 150 lining, stop, architrave both sides, ironmongery PC £20.00 per leaf; decorated		
size 686 x 1981	nr	367.14
size 762 x 1981	nr	377.27
34 Magnet Victorian hardwood four panel door; 38 x 150 lining, stop, architrave both sides, ironmongery PC £20.00 per leaf; decorated		
size 686 x 1981	nr	363.78
size 762 x 1981	nr	375.03
34 softwood panelled pattern 2G door; 4mm toughened glass top panel; 38 x 150 lining, stop, architrave both sides, ironmongery PC £20.00 per leaf; decorated		
size 762 x 1981	nr	355.76
34 softwood panelled pattern 2GG door; 4mm toughened glass both panels; 38 x 150 lining, stop, architrave both sides, ironmongery PC £20.00 per leaf; decorated		
size 762 x 1981	nr	373.13
34 softwood panelled pattern 10 door; 4mm toughened glass to one large panel; 38 x 150 lining, stop, architrave both sides, ironmongery PC £20.00 per leaf; decorated		
size 762 x 1981	nr	351.69
Softwood panelled door 38 x 150 lining, stop, architrave both sides, ironmongery PC £20.00 per leaf; decorated		
38 thick x 762 x 1981 two panel	nr	337.16
38 thick x 762 x 1981 four panel	nr	349.45
38 thick x 762 x 1981 six panel	nr	362.17
50 thick x 838 x 1981 six panel	nr	378.97
Afromosia panelled door 38 x 150 swd lining, stop, architrave both sides, ironmongery PC £20.00 per leaf; decorated		
38 thick x 762 x 1981 two panel	nr	553.67
38 thick x 762 x 1981 four panel	nr	574.65
38 thick x 762 x 1981 six panel	nr	595.63
50 thick x 838 x 1981 six panel	nr	633.86

Flush doors

34 flush door 38 x 150 lining, stop, architrave both sides, ironmongery PC £20.00 per leaf, decorated		
762 x 1981 hardboard faced	nr	262.41
838 x 1981 hardboard faced	nr	272.05
1372 x 1981 pair hardboard faced	nr	379.49
762 x 1981 plywood faced	nr	262.41
838 x 1981 plywood faced	nr	280.46
1372 x 1981 pair plywood faced	nr	379.49
762 x 1981 Sapele faced	nr	240.02
838 x 1981 Sapele faced	nr	254.59
1372 x 1981 pair Sapele faced	nr	345.70
44 half hour fire check flush door 50 x 100 frame, architrave both sides, ironmongery PC £60.00 per leaf, decorated.		
762 x 1981 plywood faced	nr	285.58
838 x 1981 plywood faced	nr	313.66
1372 x 1981 pair plywood faced	nr	490.71
762 x 1981 Sapele faced	nr	304.31
838 x 1981 Sapele faced	nr	317.21
Extra 6mm Georgian polished wired glass		
508 x 508 panel	nr	39.48
584 x 584 panel	nr	45.70

Lintels

Precast prestressed concrete lintels		
100 x 70 x 900	nr	8.88
100 x 70 x 1200	nr	11.67
100 x 70 x 1800	nr	16.85
100 x 70 x 2400	nr	21.74
100 x 70 x 3000	nr	26.49
150 x 70 x 1200	nr	13.52
150 x 70 x 3000	nr	31.19
225 x 70 x 1200	nr	17.28
225 x 70 x 3000	nr	40.17
255 x 70 x 1200	nr	17.96
255 x 70 x 3000	nr	41.93
100 x 145 x 1800	nr	24.59
100 x 145 x 3000	nr	39.34
Precast concrete lintels		
100x150x1200	nr	16.05
100x150x1650	nr	21.93
215x150x1200	nr	33.07
215x150x1650	nr	45.51

COMPOSITE PRICES (Cont'd)

INTERNAL DOORS (Cont'd)

Lintels (Cont'd)

Galvanized steel lintels - Birtley		
ref INT100; 100 wide x 900 long	nr	21.05
ref INT100; 100 wide x 1050 long	nr	23.91
ref INT100; 100 wide x 1200 long	nr	26.50
ref SB100; 75 deep x 1200 long	nr	38.12
ref SB100; 140 deep x 1800 long	nr	62.72
ref SB100; 215 deep x 3000 long	nr	142.15
ref SB140; 140 deep x 1200 long	nr	46.09
ref SB140; 140 deep x 1800 long	nr	70.61
ref SB140; 215 deep x 3000 long	nr	154.49
ref SBL200; 142 deep x 1200 long	nr	56.72
ref SBL200; 142 deep x 1800 long	nr	86.07
ref SBL200; 218 deep x 3000 long	nr	184.25
ref HD50; 225 deep x 3900 long	nr	279.91
ref HD50; 225 deep x 4800 long	nr	382.17
ref HD70; 225 deep x 3600 long	nr	271.90

WALL FINISHES

Plaster

Hardwall plaster in two coats		
13 thick	m²	11.78
13 thick, emulsion	m²	15.01
13 thick, oil paint	m²	16.85
13 thick textured plastic coating, stipple	m²	15.23
Lightweight plaster in two coats		
13 thick	m²	10.07
13 thick, emulsion	m²	13.30
13 thick, oil paint	m²	15.14
13 thick textured plastic coating, stipple	m²	13.52

Plasterboard

Plasterboard for direct decoration		
9.5 thick	m²	9.57
9.5 thick, emulsion	m²	12.80
9.5 thick, oil paint	m²	14.64
9.5 thick textured plastic coating, stipple	m²	13.02
12.5 thick	m²	10.73
12.5 thick, emulsion	m²	13.96
12.5 thick, oil paint	m²	15.80
12.5 thick textured plastic coating, stipple	m²	14.18
15 thick	m²	16.26
15 thick, emulsion	m²	19.49
15 thick, oil paint	m²	21.33
15 thick textured plastic coating, stipple	m²	19.71
two layer 9.5 thick total 19mm thick	m²	16.82
two layer 9.5 + 12.5; total 22mm thick	m²	17.98
two layer 12.5; total 25mm thick	m²	19.15
Plasterboard with 3 plaster skim coat finish		
9.5 thick plasterboard; skim	m²	13.39
9.5 thick plasterboard; skim, emulsion	m²	16.62
9.5 thick plasterboard; skim, oil paint	m²	18.46
9.5 thick plasterboard; skim, textured plastic coating, stipple	m²	10.95
12.5 thick plasterboard and skim	m²	14.50
12.5 thick plasterboard and skim, emulsion	m²	17.73
12.5 thick plasterboard and skim, oil paint	m²	19.57
12.5 thick plasterboard and skim, textured plastic coating; stipple	m²	17.95
15 thick plasterboard and skim	m²	17.82
15 thick plasterboard and skim, emulsion	m²	21.05
15 thick plasterboard and skim, oil paint	m²	22.89
15 thick plasterboard and skim, textured plastic coating; stipple	m²	21.27
two layer 9.5 thick total 21mm thick including skim	m²	20.57
two layer 9.5 + 12.5; total 25mm thick including skim	m²	21.79
two layer 12.5; total 28mm thick including skim	m²	22.90
Gyproc plaster cove		
100 girth	m	5.32
127 girth	m	5.18

Extra

lining paper PC £1.00/roll	m²	2.16
wall paper PC £4.00/roll	m²	4.56
vinyl coated wall paper PC £7.00/roll	m²	5.37
embossed paper PC £4.50/roll	m²	8.14
wall tiling 108 x 108 x 6.5 white glazed	m²	45.37
wall tiling 152 x 152 x 5.5 white glazed	m²	33.95
wall tiling 152 x 152 x 5.5 light colour	m²	33.95

Redecorations

Stripping existing paper and lining with paper PC £1.00 per roll	m²	4.93
Two coats emulsion paint on existing	m²	4.11
one undercoat; one gloss paint on existing	m²	5.17
one undercoat; one eggshell paint on existing	m²	5.21
Textured plastic coating on existing painted plaster	m²	5.16

FLOOR FINISHES

Cement; sand bed

25 thick	m²	9.14
32 thick	m²	10.71
38 thick	m²	11.89
50 thick	m²	14.40
25 thick, quarry tile paving	m²	49.59
32 thick, quarry tile paving	m²	51.16
38 thick, quarry tile paving	m²	52.34
50 thick, quarry tile paving	m²	54.85
Cement; sand bed and flooring		
25 thick, 3 pvc sheet	m²	32.87
32 thick, 3 pvc sheet	m²	34.44
38 thick, 3 pvc sheet	m²	35.62
50 thick, 3 pvc sheet	m²	38.13
25 thick, 2.5 pvc tile	m²	32.87
32 thick, 2.5 pvc tile	m²	34.44
38 thick, 2.5 pvc tile	m²	35.62
50 thick, 2.5 pvc tile	m²	38.13
25 thick, 20 oak wood block	m²	84.11
32 thick, 20 oak wood block	m²	85.68
38 thick, 20 oak wood block	m²	86.86
50 thick, 20 oak wood block	m²	89.37
25 thick, 20 maple wood block	m²	79.60
32 thick, 20 maple wood block	m²	81.17
38 thick, 20 maple wood block	m²	82.35
50 thick, 20 maple wood block	m²	84.86
25 thick, 20 Merbau wood block	m²	73.95
32 thick, 20 Merbau wood block	m²	75.52
38 thick, 20 Merbau wood block	m²	76.70
50 thick, 20 Merbau wood block	m²	79.21
Cement; sand bed; Carpet and underlay		
25 thick, carpet PC £28/m²	m²	50.51
32 thick, carpet PC £28/m²	m²	52.08
38 thick, carpet PC £28/m²	m²	53.26
50 thick, carpet PC £28/m²	m²	55.77

Granolithic pavings

25 thick	m²	12.96
32 thick	m²	14.81
38 thick	m²	16.38
50 thick	m²	19.64
Granolithic skirtings		
19 thick x 150 high	m	13.49
25 thick x 150 high	m	14.60
32 thick x 150 high	m	15.54

Levelling screed

Latex levelling screed; one coat		
3 thick average to concrete	m²	9.69

Terrazzo

300x300x28 tiles; 40 bed; polished	m²	67.87
16 thick laid insitu in bays	m²	65.44
6 thick skirtings laid insitu 75 high	m	21.84
16 thick skirtings laid insitu 150 high	m	32.06

COMPOSITE

FLOOR FINISHES (Cont'd)

PVC sheet flooring		
2 thick	m^2	18.66
2.5 thick	m^2	20.52
PVC coved skirting		
100 high set in	m	5.57
100 high sit on	m	6.02
Softwood flooring		
Floor boarding and insulation to concrete		
22 tongued and grooved chipboard and 50 insulation board	m^2	27.08
Softwood skirting decorated		
13 x 95 to BS 1186 ref. 13RS95	m	8.94
20 x 70 to BS 1186 ref. 20CA70	m	8.93
20 x 120 to BS 1186 ref. 20CS120	m	10.38
19 x 100 square	m	9.96
19 x 150 square	m	11.51
25 x 100 square	m	10.93
25 x 150 square	m	12.91
15 x 125 moulded	m	9.68
19 x 125 moulded	m	10.55

CEILING FINISHES

Plaster		
Hardwall plaster in two coats		
13 thick	m^2	13.15
13 thick, emulsion	m^2	15.64
13 thick, oil paint	m^2	18.69
13 thick textured plastic coatings, stipple	m^2	17.88
Lightweight plaster in two coats		
10 thick	m^2	11.29
10 thick, emulsion	m^2	13.78
10 thick, oil paint	m^2	16.83
10 thick textured plastic coatings, stipple	m^2	16.02
Plasterboard		
Plasterboard for direct decoration		
9.5 thick	m^2	10.40
9.5 thick, emulsion	m^2	12.89
9.5 thick, oil paint	m^2	15.94
9.5 thick textured plastic coatings, stipple	m^2	15.13
12.5 thick	m^2	12.32
12.5 thick, emulsion	m^2	14.81
12.5 thick, oil paint	m^2	17.86
12.5 thick textured plastic coating, stipple	m^2	17.05
two layer 9.5 thick total 19mm thick	m^2	18.36
two layer 9.5 + 12.5; total 22mm thick	m^2	19.73
two layer 12.5; total 25mm thick	m^2	20.99
Plasterboard with 3 plaster skim coat finish		
9.5 thick plasterboard and skim	m^2	14.91
9.5 thick plasterboard and skim, emulsion	m^2	17.40
9.5 thick plasterboard and skim, oil paint	m^2	20.45
9.5 thick plasterboard; skim, textured plastic coating, stipple	m^2	19.64
12.5 thick plasterboard and skim	m^2	16.17
12.5 thick plasterboard and skim, emulsion	m^2	18.66
12.5 thick plasterboard and skim, oil paint	m^2	21.71
12.5 thick plasterboard; skim, textured plastic coating, stipple	m^2	20.90
two layer 9.5 thick total 21mm thick including skim	m^2	22.86
two layer 9.5 + 12.5; total 25mm thick including skim	m^2	24.23
two layer 12.5; total 28mm thick including skim	m^2	25.49
Suspended ceilings		
300 x 300 x 15.8 plain mineral fibre tile	m^2	26.03
300 x 300 x 15.8 textured mineral fibre tile	m^2	28.74
300 x 300 patterned tile P.C. £15.00	m^2	32.97
600 x 600 x 15.8 plain mineral fibre tile	m^2	26.30
600 x 600 x 15.8 drilled mineral fibre tile	m^2	29.01
600 x 600 patterned tile P.C. £15.00	m^2	33.25

FITTINGS AND FURNISHINGS

Mirrors		
6mm bevelled edges; silvered; copper backing		
250 x 400	nr	45.08
460 x 600	nr	73.52
600 x 900	nr	110.51
Venetian blinds		
25mm slat; plain colours		
1000 wide x 1200 drop	nr	49.40
2000 wide x 1200 drop	nr	75.15
3000 wide x 1200 drop	nr	100.87
50mm slat; plain colours		
1000 wide x 1200 drop	nr	49.28
2000 wide x 1200 drop	nr	74.81
3000 wide x 1200 drop	nr	100.71
Roller blinds		
self acting roller		
1000 wide x 1200 drop	nr	73.54
2000 wide x 1200 drop	nr	94.37
3000 wide x 1200 drop	nr	115.18
Vertical louvre blinds		
89mm wide louvres		
1000 wide x 1200 drop	nr	80.21
2000 wide x 1200 drop	nr	137.02
3000 wide x 1200 drop	nr	193.84
127mm wide louvres		
1000 wide x 1200 drop	nr	74.59
2000 wide x 1200 drop	nr	125.61
3000 wide x 1200 drop	nr	177.23
Kitchen units		
Pre-assembled domestic kitchen units		
400 floor unit	nr	175.75
500 floor unit	nr	193.16
600 floor unit	nr	207.49
1000 floor unit	nr	333.13
1200 floor unit	nr	362.42
1200 sink unit	nr	255.46
1500 sink unit	nr	371.09
500 floor one drawer unit	nr	231.94
600 floor one drawer unit	nr	246.27
1000 floor one drawer unit	nr	389.39
1200 sink one drawer unit	nr	316.30
500 floor four drawer unit	nr	308.98
600 floor four drawer unit	nr	327.79
500 x 600 high wall unit	nr	112.65
600 x 600 high wall unit	nr	120.89
1000 x 600 high wall unit	nr	181.09
1200 x 600 high wall unit	nr	195.70
500 x 1950 tall store unit	nr	303.06
600 x 1950 tall store unit	nr	333.42
28 work top	m	59.14
40 work top	m	59.14
plinth	m	9.10
cornice	m	12.52
pelmet	m	9.59

SANITARY APPLIANCES

Sanitary appliance with allowance for waste fittings, traps, overflows, taps and builder's work		
Domestic building, with plastic wastes		
lavatory basin	nr	131.44
lavatory basin with pedestal	nr	163.37
bath with panels to front and one end	nr	254.40
low level w.c. suite	nr	168.11
stainless steel sink and drainer	nr	155.20
shower	nr	548.21

COMPOSITE PRICES (Cont'd)

SANITARY APPLIANCES (Cont'd)

Commercial building, with copper wastes		
lavatory basin	nr	163.87
low level w.c. suite	nr	168.11
bowl type urinal	nr	347.79
stainless steel sink	nr	173.44
drinking fountain	nr	414.07
shower	nr	530.57

DISPOSAL INSTALLATIONS

Soil and vent pipe for ground and first floor		
plastic wastes and SVP	nr	250.07
copper wastes and cast iron SVP	nr	588.53

WATER INSTALLATIONS

Hot and cold water services		
domestic building	nr	1,431.46
gas instantaneous sink water heater	nr	679.98
electric instantaneous sink water heater	nr	615.74
undersink water heater	nr	668.30
electric shower unit	nr	409.52

HEATING INSTALLATIONS

Boiler, radiators, pipework, controls and pumps		
100m² domestic building, solid fuel	m²	37.48
100m² domestic building, gas	m²	34.30
100m² domestic building, oil	m²	42.76
150m² domestic building, solid fuel	m²	33.45
150m² domestic building, gas	m²	30.09
150m² domestic building, oil	m²	37.39
500m² commercial building, gas	m²	49.98
500m² commercial building, oil	m²	58.17
1000m² commercial building, gas	m²	59.29
1000m² commercial building, oil	m²	70.32
1500m² commercial building, gas	m²	68.93
1500m² commercial building, oil	m²	80.77
2000m² commercial building, gas	m²	74.89
2000m² commercial building, oil	m²	88.38
Underfloor heating with polymer pipework and insulation - excludes screed.	m²	25.92

ELECTRICAL INSTALLATIONS

Concealed installation with lighting, power, and ancillary cooker, immersion heater etc. circuit.		
100m² domestic building		
PVC insulated		
PVC insulated and sheathed cables	m²	41.88
MICS cables	m²	57.58
steel conduit	m²	67.09
150m² domestic building,		
PVC insulated and sheathed cables	m²	37.58
MICS cables	m²	50.86
steel conduit	m²	58.70
500m² commercial building, light fittings, steel conduit	m²	83.86
1000m² commercial building, light fittings, steel conduit	m²	89.10
1500m² commercial building, light fittings, steel conduit	m²	94.34
2000m² commercial building, light fittings, steel conduit	m²	99.58

LIFT AND ESCALATOR INSTALLATIONS

Light passenger lift, laminate walls, stainless steel doors.		
3 stop, 8 person	nr	22,753.47
4 stop, 13 person	nr	42,487.25
General purpose passenger lift, laminate walls stainless steel doors.		
4 stop, 8 person	nr	42,164.75
5 stop, 13 person	nr	45,666.03
6 stop, 21 person	nr	104,207.31
Service hoist		
2 stop, 50 kg	nr	6,269.40
Escalator 30 degree inclination department store specification		
3.00 m rise	nr	62,350.00

SITE WORK

Kerbs, edgings channels

Precast concrete kerb and foundation		
125 x 255 kerb	m	30.05
150 x 305 kerb	m	35.03
Precast concrete edging and foundation		
50 x 150 edging	m	21.36
50 x 200 edging	m	24.56
50 x 250 edging	m	27.24
Precast concrete channel and foundation		
255 x 125 channel	m	28.84
Granite kerb on edge and foundation		
150 x 300 kerb	m	42.06
200 x 300 kerb	m	57.17
Granite kerb laid flat and foundation		
300 x 150 kerb	m	56.63
300 x 200 kerb	m	68.51

Pavings

Concrete; 250 excavation, 150 hardcore		
150 concrete road tamped	m²	36.92
200 concrete road tamped	m²	41.68
150 concrete and 70 bitumen macadam	m²	47.83
200 concrete and 70 bitumen macadam	m²	52.59
Extra for Ref. A252 fabric reinforcement	m²	4.09
Extra for Ref. C636 fabric reinforcement	m²	5.18
Gravel; 250 excavation; 100 hardcore; 75 clinker; 75 course gravel; 38 blinding gravel	m²	19.78
Gravel; 150 excavation; 100 hardcore; 50 blinding gravel	m²	9.85
Gravel; 150 excavation; 100 hardcore; 50 clinker;		
38 blinding gravel	m²	11.79
38 blinding gravel with 'Colas'; Shingle dressing	m²	14.84
York Stone; 200 excavation, 150 hardcore		
50 paving	m²	90.51
75 paving	m²	102.14
Precast concrete; 200 excavation, 150 hardcore		
600 x 450 x 50 slabs	m²	27.36
600 x 600 x 50 slabs	m²	24.97
600 x 750 x 50 slabs	m²	27.46
600 x 900 x 50 slabs	m²	25.75
600 x 450 x 63 slabs	m²	30.06
600 x 600 x 63 slabs	m²	28.14
600 x 750 x 63 slabs	m²	28.59
600 x 900 x 63 slabs	m²	28.95
50 thick Crazy paving	m²	26.26
Precast concrete; 200 excavation, 150 hardcore		
200 x 100 x 50 Blocks on 50 sand	m²	31.20
Granite Sets; 300 excavation, 100 hardcore		
125 thick on 100 concrete	m²	106.88
150 thick on 100 concrete	m²	118.69
Cobbles; 200 excavation, 150 hardcore		
Cobbles set in 100 concrete	m²	78.05
Westmorland Green slate; 250 excavation, 150 hardcore		
38 slate on 100 concrete	m²	155.55

Seeding/Turfing

Seeding; Cultivating;digging over one spit deep; removing debris; weeding; weedkiller		
Grass seed; fertilizer; maintain	m²	6.87
Seeding; 150 excavation; 150 topsoil; weedkiller		
Grass seed; fertilizer; maintain	m²	10.76
Turfing; Cultivating;digging over one spit deep; removing debris; weeding; weedkiller		
Turf; maintain	m²	9.31
Turfing; 150 excavation; 150 topsoil; weedkiller		
Turf; maintain	m²	13.20

Walls

Brick wall, excavation 600 deep concrete foundation, 215 wall with piers at 3m centres in facings PC £200 / 1000, damp proof course.		
900 high brick on edge coping	m	160.16
1800 high brick on edge coping	m	258.71
900 high 300 x 100 PCC coping	m	181.58
1800 high 300 x 100 PCC coping	m	280.13

COMPOSITE PRICES (Cont'd)

SITE WORK (Cont'd)

Fences

Plastic coated chain link fencing, posts at 2743 centres.		
900 high 38 x 38 angle iron posts, two line wires	m	19.90
1200 high 38 x 38 angle iron posts, two line wires	m	22.21
1400 high 38 x 38 angle iron posts, three line wires	m	24.46
1800 high 38 x 38 angle iron posts, three line wires	m	30.02
900 high concrete posts two line wires	m	19.67
1200 high PCC posts two line wires	m	22.02
1400 high PCC posts three line wires	m	24.49
1800 high PCC posts three line wires	m	30.26
Gates; chain link; posts		
1200 x 1800	nr	573.11
Close boarded fencing, 100 x 100 posts at 2743 centres 25 x 100 pales, 25 x 150 gravel boards, creosote, post holes filled with concrete		
1200 high two arris rails	m	46.28
1500 high three arris rails	m	55.56
1800 high three arris rails	m	60.96
Gates; timber		
900 x 1150	nr	163.08
900 x 1450	nr	172.67

DRAINAGE

Drains

Vitrified clay sleeve joint drain, excavation and backfill.		
100 diameter 500 deep	m	18.02
100 diameter 750 deep	m	21.09
100 diameter 1000 deep	m	24.10
100 diameter 1500 deep	m	30.85
150 diameter 500 deep	m	24.07
150 diameter 750 deep	m	27.20
150 diameter 1000 deep	m	30.27
150 diameter 1500 deep	m	37.15
Vitrified clay sleeve joint drain, excavation 150 bed and surround pea shingle and backfill.		
100 diameter 500 deep	m	29.94
100 diameter 750 deep	m	33.02
100 diameter 1000 deep	m	36.02
100 diameter 1500 deep	m	42.78
150 diameter 500 deep	m	38.34
150 diameter 750 deep	m	41.47
150 diameter 1000 deep	m	44.54
150 diameter 1500 deep	m	51.42
Vitrified clay sleeve joint drain, excavation 150 bed and haunch concrete and backfill.		
100 diameter 500 deep	m	31.63
100 diameter 750 deep	m	34.71
100 diameter 1000 deep	m	37.71
100 diameter 1500 deep	m	44.84
150 diameter 500 deep	m	37.77
150 diameter 750 deep	m	40.90
150 diameter 1000 deep	m	43.97
150 diameter 1500 deep	m	50.85
Vitrified clay sleeve joint drain, excavation 150 bed and surround concrete and back fill.		
100 diameter 500 deep	m	41.55
100 diameter 750 deep	m	44.63
100 diameter 1000 deep	m	47.63
100 diameter 1500 deep	m	54.39
150 diameter 500 deep	m	48.00
150 diameter 750 deep	m	51.13
150 diameter 1000 deep	m	54.20
150 diameter 1500 deep	m	61.08
Upvc drain, excavation 50 mm bed sand and backfill.		
110 diameter 500 deep	m	18.55
110 diameter 750 deep	m	21.63
110 diameter 1000 deep	m	24.63
110 diameter 1500 deep	m	31.38
160 diameter 500 deep	m	27.29
160 diameter 750 deep	m	30.43
160 diameter 1000 deep	m	33.50
160 diameter 1500 deep	m	40.38
Upvc drain, excavation bed, and surround pea shingle and back fill.		
110 diameter 500 deep	m	22.86
110 diameter 750 deep	m	25.94
110 diameter 1000 deep	m	28.94
110 diameter 1500 deep	m	35.70
160 diameter 500 deep	m	32.14
160 diameter 750 deep	m	35.27
160 diameter 1000 deep	m	38.34
160 diameter 1500 deep	m	45.22
Upvc drain, excavation bed and haunch concrete and back fill.		
110 diameter 500 deep	m	40.08
110 diameter 750 deep	m	43.16
110 diameter 1000 deep	m	46.16
110 diameter 1500 deep	m	52.92
160 diameter 500 deep	m	40.07
160 diameter 750 deep	m	43.21
160 diameter 1000 deep	m	46.27
160 diameter 1500 deep	m	53.15
Upvc drain, excavation bed and surround concrete and back fill.		
110 diameter 500 deep	m	41.02
110 diameter 750 deep	m	44.10
110 diameter 1000 deep	m	47.10
110 diameter 1500 deep	m	53.86
160 diameter 500 deep	m	55.30
160 diameter 750 deep	m	58.43
160 diameter 1000 deep	m	61.50
160 diameter 1500 deep	m	68.38
Cast iron drain, excavation and back fill.		
100 diameter 500 deep	m	52.25
100 diameter 750 deep	m	55.32
100 diameter 1000 deep	m	58.33
100 diameter 1500 deep	m	65.08
150 diameter 500 deep	m	79.99
150 diameter 750 deep	m	83.12
150 diameter 1000 deep	m	86.19
150 diameter 1500 deep	m	93.07
Cast iron drain, excavation, bed and haunch concrete and back fill.		
100 diameter 500 deep	m	74.81
100 diameter 750 deep	m	77.88
100 diameter 1000 deep	m	80.89
100 diameter 1500 deep	m	87.64
150 diameter 500 deep	m	93.97
150 diameter 750 deep	m	97.11
150 diameter 1000 deep	m	100.18
150 diameter 1500 deep	m	107.06
Cast iron drain, excavation, bed and surround concrete and back fill.		
100 diameter 500 deep	m	75.78
100 diameter 750 deep	m	78.86
100 diameter 1000 deep	m	81.86
100 diameter 1500 deep	m	88.62
150 diameter 500 deep	m	109.26
150 diameter 750 deep	m	112.39
150 diameter 1000 deep	m	115.46
150 diameter 1500 deep	m	122.34

Brick manhole

Excavation; 150 concrete bed; 215 class B engineering brick walls; 100 clayware main channel; concrete benching; step irons and cast iron manhole cover		
600 x 450, two branches 750 deep	nr	556.04
600 x 450, two branches 1500 deep	nr	895.47
750 x 450, three branches 750 deep	nr	811.75
750 x 450, three branches 1000 deep	nr	932.56
750 x 450, three branches 1500 deep	nr	1,157.38
900 x 600, five branches 750 deep	nr	859.69
900 x 600, five branches 1000 deep	nr	1,011.83
900 x 600, five branches 1500 deep	nr	1,280.51

COMPOSITE PRICES (Cont'd)

DRAINAGE (Cont'd)

Concrete circular manhole		
Excavation; 150 concrete bed; manhole rings; cover slab; 100 clayware main channel; concrete benching; step irons; cast iron manhole cover		
900 dia., two branches 750 deep	nr	631.39
900 dia., two branches 1000 deep	nr	758.65
900 dia., two branches 1500 deep	nr	914.81
1050 dia., three branches 750 deep	nr	724.38
1050 dia., three branches 1000 deep	nr	842.20
1050 dia., three branches 1500 deep	nr	1,054.35
1200 dia., five branches 750 deep	nr	826.71
1200 dia., five branches 1000 deep	nr	954.82
1200 dia., five branches 1500 deep	nr	1,153.94
Concrete rectangular manhole		
Excavation; concrete bed; manhole sections; base unit; cover slab with integral cover		
600 x 450 x 500 deep	nr	228.63
600 x 450 x 750 deep	nr	297.71
600 x 450 x 1000 deep	nr	355.52
GRP Inspection chamber		
Excavation; GRP Inspection chamber cast iron manhole cover		
470 diameter, 520 deep	nr	348.83
470 diameter, 990 deep	nr	471.93
Polypropolene Inspection chamber		
Excavation; Polypropolene Inspection chamber cast iron manhole cover		
250 diameter, 600 deep	nr	112.37
450 diameter, 600 deep	nr	252.64
450 diameter, 1000 deep	nr	327.17
Concrete gullies		
Excavation; concrete bed; gully; bricks;cast iron road grating		
375 diameter x 750 deep	nr	288.17
375 diameter x 900 deep	nr	311.81
450 diameter x 750 deep	nr	310.63
450 diameter x 900 deep	nr	334.36
Vitrified clay gullies		
Excavation; concrete bed; gully; grating trapped round; 100 outlet	nr	81.56
Pvc-u gullies		
Excavation; concrete bed; bottle gully; grating 200 diameter top; trapped round	nr	98.74

SMALL WORKS

The measured rates in this section are intended to apply to Contracts within the price range £50,000 to £100,000 based on National Average prices.

Adjustment to overall contract values for Regional Variations and contract values from £15,000 to £500,000 can be made as noted in the "Essential Information" section.

REGIONAL VARIATIONS

Laxton's Building Price Book is based upon national average prices (=1.00). Indicative levels of pricing in regions at the second quarter 2005 are given on this map and the factors shown may be applied to adjust overall pricing.

MAJOR WORKS - TENDER VALUES

The measured rates are based on contracts valued in the range of £500,000 to £1,500,000.

As a guide to pricing works of larger value and for cost planning or budgetary purposes, the following adjustments may be made to overall project values:

Project value:
£1,500,000 to £2,500,000 ... deduct 2.5%
£2,500,000 to £5,000,000 ... deduct 5%

SMALL WORKS - TENDER VALUES

The measured rates are based on contracts valued in the range of £50,000 to £100,000.

As a guide to pricing works of smaller or larger value and for cost planning or budgetary purposes, the following adjustments may be made to overall project values:

Project value:
£15,000 to £25,000.......... add 10 %
£25,000 to £50,000.......... add 5%
£50,000 to £100,000........ rates as shown
£100,000 to £200,000....... deduct 5%
£200,000 to £500,000....... deduct 7.5%

ALTERATIONS/REPAIRS/MAINTENANCE

The measured rates are based on contracts valued in the range of £5,000 to £15,000.
The rates may be used for orders of lower value however due allowances must be added for disproportionate waste factors, working in occupied premises and preliminary items.

EXISTING SITE/BUILDINGS/SERVICES

Labour hourly rates: (except Specialists) Craft Operatives £14.21 Labourer £10.71 Rates are national average prices. Refer to REGIONAL VARIATIONS for indicative levels of overall pricing in regions	MATERIALS			LABOUR				RATES		
	Del to Site £	Waste %	Material Cost £	Craft Optve Hrs	Lab Hrs	Labour Cost £	Sunds £	Nett Rate £	Unit	Gross Rate (+12.5%) £
C20: DEMOLITION										
Pulling down outbuildings										
Demolishing individual structures										
timber outbuilding 2.50 x 2.00 x 3.00m maximum high	-	-	-	-	11.00	**117.81**	70.51	**188.32**	nr	**211.86**
outbuilding with half brick walls, one brick piers and felt covered timber roof, 2.50 x 2.00 x 3.00m maximum high	-	-	-	-	22.00	**235.62**	141.02	**376.64**	nr	**423.71**
greenhouse with half brick dwarf walls and timber framing to upper walls and roof, 3.00 x 2.00 x 3.00m maximum high	-	-	-	-	18.00	**192.78**	120.87	**313.65**	nr	**352.86**
Demolishing individual structures; setting aside materials for re-use										
metal framed greenhouse, 3.00 x 2.00 x 3.00m maximum high	-	-	-	4.00	7.00	**131.81**	-	**131.81**	nr	**148.29**
prefabricated concrete garage, 5.40 x 2.60 x 2.40m maximum high	-	-	-	6.00	11.00	**203.07**	-	**203.07**	nr	**228.45**
Demolishing individual structures; making good structures										
timber framed lean-to outbuilding and remove flashings and make good facing brick wall at ridge and vertical abutments, 3.00 x 1.90 x 2.40m maximum high	-	-	-	1.50	12.00	**149.84**	74.66	**224.49**	nr	**252.55**
Unit rates for pricing the above and similar work										
pull down building 3.00 x 1.90 x 2.40m maximum high	-	-	-	-	10.00	**107.10**	70.51	**177.61**	nr	**199.81**
remove flashing	-	-	-	-	0.12	**1.29**	0.40	**1.69**	m	**1.90**
make good facing brick wall at ridge	-	-	-	0.20	0.20	**4.98**	0.37	**5.35**	m	**6.02**
make good facing brick wall at vertical abutment	-	-	-	0.17	0.17	**4.24**	0.37	**4.60**	m	**5.18**
make good rendered wall at ridge	-	-	-	0.20	0.20	**4.98**	0.85	**5.84**	m	**6.57**
make good rendered wall at vertical abutment	-	-	-	0.20	0.20	**4.98**	0.85	**5.84**	m	**6.57**
Demolishing individual structures; making good structures										
lean-to outbuilding with half brick walls and slate covered timber roof and remove flashings, hack off plaster to house wall and make good with rendering to match existing, 1.50 x 1.00 x 2.00m maximum high	-	-	-	4.00	12.50	**190.72**	47.45	**238.16**	nr	**267.93**
Unit rates for pricing the above and similar work										
pull down building with half brick walls, 1.50 x 1.00 x 2.00m maximum high	-	-	-	-	5.30	**56.76**	32.23	**89.00**	nr	**100.12**
pull down building with one brick walls, 1.50 x 1.00 x 2.00m maximum high	-	-	-	-	10.00	**107.10**	60.44	**167.54**	nr	**188.48**
hack off plaster or rendering and make good to match existing brick facings	-	-	-	1.00	2.00	**35.63**	1.21	**36.84**	m2	**41.44**
hack off plaster or rendering and make good to match existing rendering	-	-	-	1.30	2.30	**43.11**	4.87	**47.98**	m2	**53.97**
Removal of old work										
Demolishing parts of structures										
in-situ plain concrete bed										
75mm thick	-	-	-	-	0.90	**9.64**	3.02	**12.66**	m2	**14.24**
100mm thick	-	-	-	-	1.20	**12.85**	4.03	**16.88**	m2	**18.99**
150mm thick	-	-	-	-	1.80	**19.28**	6.04	**25.32**	m2	**28.49**
200mm thick	-	-	-	-	2.40	**25.70**	8.06	**33.76**	m2	**37.98**
in-situ reinforced concrete flat roof										
100mm thick	-	-	-	-	1.90	**20.35**	4.03	**24.38**	m2	**27.43**
150mm thick	-	-	-	-	2.85	**30.52**	6.04	**36.57**	m2	**41.14**
200mm thick	-	-	-	-	3.80	**40.70**	8.06	**48.76**	m2	**54.85**
225mm thick	-	-	-	-	4.28	**45.84**	9.07	**54.90**	m2	**61.77**
in-situ reinforced concrete upper floor										
100mm thick	-	-	-	-	1.90	**20.35**	4.03	**24.38**	m2	**27.43**
150mm thick	-	-	-	-	2.85	**30.52**	6.04	**36.57**	m2	**41.14**

Labour hourly rates: (except Specialists) Craft Operatives £14.21 Labourer £10.71 Rates are national average prices. Refer to REGIONAL VARIATIONS for indicative levels of overall pricing in regions	MATERIALS			LABOUR				RATES		
	Del to Site £	Waste %	Material Cost £	Craft Optve Hrs	Lab Hrs	Labour Cost £	Sunds £	Nett Rate £	Unit	Gross Rate (+12.5%) £
C20: DEMOLITION Cont.										
Removal of old work Cont.										
Demolishing parts of structures Cont.										
in-situ reinforced concrete upper floor Cont.										
200mm thick	-	-	-	-	3.80	**40.70**	8.06	**48.76**	m2	**54.85**
225mm thick	-	-	-	-	4.28	**45.84**	9.07	**54.90**	m2	**61.77**
in-situ reinforced concrete beam	-	-	-	-	19.00	**203.49**	40.29	**243.78**	m3	**274.25**
in-situ reinforced concrete column	-	-	-	-	18.00	**192.78**	40.29	**233.07**	m3	**262.20**
in-situ reinforced concrete wall										
100mm thick	-	-	-	-	1.80	**19.28**	4.03	**23.31**	m2	**26.22**
150mm thick	-	-	-	-	2.70	**28.92**	6.04	**34.96**	m2	**39.33**
200mm thick	-	-	-	-	3.60	**38.56**	8.06	**46.61**	m2	**52.44**
225mm thick	-	-	-	-	4.05	**43.38**	9.07	**52.44**	m2	**59.00**
in-situ reinforced concrete casing to beam	-	-	-	-	16.00	**171.36**	40.29	**211.65**	m3	**238.11**
in-situ reinforced concrete casing to column	-	-	-	-	15.25	**163.33**	40.29	**203.62**	m3	**229.07**
brick internal walls in lime mortar										
102mm thick	-	-	-	-	0.63	**6.75**	4.11	**10.86**	m2	**12.22**
215mm thick	-	-	-	-	1.30	**13.92**	8.66	**22.59**	m2	**25.41**
327mm thick	-	-	-	-	2.00	**21.42**	13.18	**34.60**	m2	**38.92**
brick internal walls in cement mortar										
102mm thick	-	-	-	-	0.95	**10.17**	4.11	**14.29**	m2	**16.07**
215mm thick	-	-	-	-	1.94	**20.78**	8.66	**29.44**	m2	**33.12**
327mm thick	-	-	-	-	2.98	**31.92**	13.18	**45.09**	m2	**50.73**
reinforced brick internal walls in cement lime mortar										
102mm thick	-	-	-	-	0.95	**10.17**	4.11	**14.29**	m2	**16.07**
215mm thick	-	-	-	-	1.94	**20.78**	8.66	**29.44**	m2	**33.12**
hollow clay block internal walls in cement-lime mortar										
50mm thick	-	-	-	-	0.30	**3.21**	2.01	**5.23**	m2	**5.88**
75mm thick	-	-	-	-	0.40	**4.28**	3.02	**7.31**	m2	**8.22**
100mm thick	-	-	-	-	0.52	**5.57**	4.03	**9.60**	m2	**10.80**
hollow clay block internal walls in cement mortar										
50mm thick	-	-	-	-	0.32	**3.43**	2.01	**5.44**	m2	**6.12**
75mm thick	-	-	-	-	0.43	**4.61**	3.02	**7.63**	m2	**8.58**
100mm thick	-	-	-	-	0.57	**6.10**	4.03	**10.13**	m2	**11.40**
concrete block internal walls in cement lime mortar										
75mm thick	-	-	-	-	0.35	**3.75**	3.02	**6.77**	m2	**7.62**
100mm thick	-	-	-	-	0.53	**5.68**	4.03	**9.71**	m2	**10.92**
190mm thick	-	-	-	-	0.70	**7.50**	7.66	**15.15**	m2	**17.05**
215mm thick	-	-	-	-	1.41	**15.10**	8.66	**23.76**	m2	**26.73**
concrete block internal walls in cement mortar										
75mm thick	-	-	-	-	0.56	**6.00**	3.02	**9.02**	m2	**10.15**
100mm thick	-	-	-	-	0.73	**7.82**	4.03	**11.85**	m2	**13.33**
190mm thick	-	-	-	-	0.91	**9.75**	7.66	**17.40**	m2	**19.58**
215mm thick	-	-	-	-	1.67	**17.89**	8.66	**26.55**	m2	**29.87**
if internal walls plastered, add per side	-	-	-	-	0.12	**1.29**	0.81	**2.09**	m2	**2.35**
if internal walls rendered, add per side	-	-	-	-	0.18	**1.93**	0.81	**2.73**	m2	**3.08**
brick external walls in lime mortar										
102mm thick	-	-	-	-	0.48	**5.14**	4.11	**9.25**	m2	**10.41**
215mm thick	-	-	-	-	0.98	**10.50**	8.66	**19.16**	m2	**21.55**
327mm thick	-	-	-	-	1.51	**16.17**	13.18	**29.35**	m2	**33.02**
brick external walls in cement mortar										
102mm thick	-	-	-	-	0.63	**6.75**	4.11	**10.86**	m2	**12.22**
215mm thick	-	-	-	-	1.30	**13.92**	8.66	**22.59**	m2	**25.41**
327mm thick	-	-	-	-	2.00	**21.42**	13.18	**34.60**	m2	**38.92**
reinforced brick external walls in cement lime mortar										
102mm thick	-	-	-	-	0.63	**6.75**	4.11	**10.86**	m2	**12.22**
215mm thick	-	-	-	-	1.30	**13.92**	8.66	**22.59**	m2	**25.41**
if external walls plastered, add per side	-	-	-	-	0.09	**0.96**	0.81	**1.77**	m2	**1.99**
if external walls rendered or rough cast, add per side	-	-	-	-	0.12	**1.29**	0.81	**2.09**	m2	**2.35**
clean old bricks in lime mortar and stack for re-use, per thousand	-	-	-	-	15.00	**160.65**	-	**160.65**	nr	**180.73**
clean old bricks in cement mortar and stack for re-use, per thousand	-	-	-	-	25.00	**267.75**	-	**267.75**	nr	**301.22**
rough rubble walling 600mm thick, in lime mortar	-	-	-	-	3.75	**40.16**	24.17	**64.34**	m2	**72.38**
random rubble walling 350mm thick, in lime mortar	-	-	-	-	2.20	**23.56**	14.10	**37.66**	m2	**42.37**
random rubble walling 500mm thick, in lime mortar	-	-	-	-	3.10	**33.20**	20.15	**53.35**	m2	**60.01**
dressed stone walling 100mm thick, in gauged mortar	-	-	-	2.00	2.30	**53.05**	4.03	**57.08**	m2	**64.22**
dressed stone walling 200mm thick, in gauged mortar	-	-	-	3.90	4.50	**103.61**	8.06	**111.67**	m2	**125.63**
stone copings 300 x 50mm	-	-	-	0.20	0.26	**5.63**	0.60	**6.23**	m	**7.01**
stone copings 375 x 100mm	-	-	-	0.40	0.52	**11.25**	1.52	**12.77**	m	**14.37**
stone staircases	-	-	-	-	15.00	**160.65**	40.29	**200.94**	m3	**226.06**
stone steps	-	-	-	-	14.75	**157.97**	40.29	**198.26**	m3	**223.05**
structural timbers										
50 x 75mm	-	-	-	-	0.13	**1.39**	0.15	**1.55**	m	**1.74**
50 x 100mm	-	-	-	-	0.15	**1.61**	0.20	**1.81**	m	**2.03**
50 x 150mm	-	-	-	-	0.18	**1.93**	0.31	**2.24**	m	**2.52**
50 x 225mm	-	-	-	-	0.24	**2.57**	0.45	**3.02**	m	**3.40**
75 x 100mm	-	-	-	-	0.18	**1.93**	0.31	**2.24**	m	**2.52**
75 x 150mm	-	-	-	-	0.20	**2.14**	0.45	**2.59**	m	**2.92**
75 x 225mm	-	-	-	-	0.31	**3.32**	0.68	**4.00**	m	**4.49**
steelwork										
steel beams, joists and lintels not exceeding 10 kg/m	-	-	-	50.00	50.00	**1246.00**	40.29	**1286.29**	t	**1447.08**
steel beams, joists and lintels 10 - 20 kg/m	-	-	-	46.00	46.00	**1146.32**	40.29	**1186.61**	t	**1334.94**

Labour hourly rates: (except Specialists) Craft Operatives £14.21 Labourer £10.71 Rates are national average prices. Refer to REGIONAL VARIATIONS for indicative levels of overall pricing in regions	MATERIALS			LABOUR				RATES		
	Del to Site £	Waste %	Material Cost £	Craft Optve Hrs	Lab Hrs	Labour Cost £	Sunds £	Nett Rate £	Unit	Gross Rate (+12.5%) £
C20: DEMOLITION Cont.										
Removal of old work Cont.										
Demolishing parts of structures Cont.										
steelwork Cont.										
steel beams, joists and lintels 20 - 50 kg/m .	-	-	-	40.00	40.00	**996.80**	40.29	**1037.09**	t	**1166.73**
steel columns and stanchions not exceeding 10 kg/m	-	-	-	60.00	60.00	**1495.20**	40.29	**1535.49**	t	**1727.43**
steel columns and stanchions 10 - 20 kg/m	-	-	-	56.00	56.00	**1395.52**	40.29	**1435.81**	t	**1615.29**
steel columns and stanchions 20 - 50 kg/m	-	-	-	50.00	50.00	**1246.00**	40.29	**1286.29**	t	**1447.08**
steel purlins and rails not exceeding 10 kg/m	-	-	-	50.00	50.00	**1246.00**	40.29	**1286.29**	t	**1447.08**
steel purlins and rails 10 - 20 kg/m	-	-	-	50.00	50.00	**1246.00**	40.29	**1286.29**	t	**1447.08**
C30: SHORING/FACADE RETENTION										
Support of structures not to be demolished										
Provide and erect timber raking shore complete with sole piece, cleats, needles and 25mm brace boarding with										
two 150 x 150mm rakers (total length 8.00m) and 50 x 175mm wall piece	122.00	10.00	**134.20**	4.00	4.00	**99.68**	22.53	**256.41**	nr	**288.46**
weekly cost of maintaining last................	-	-	-	1.00	1.00	**24.92**	1.13	**26.05**	nr	**29.31**
three 225 x 225mm rakers (total length 21.00m) and 50 x 250mm wall piece	461.00	10.00	**507.10**	23.20	23.20	**578.14**	97.94	**1183.18**	nr	**1331.08**
weekly cost of maintaining last................	-	-	-	1.50	1.50	**37.38**	4.84	**42.22**	nr	**47.49**
Provide and erect timber flying shore of 100 x 150mm main member, 50 x 175mm wall pieces and 50 x 100mm straining pieces and 75 x 100mm struts, distance between wall faces										
4.00m	124.00	10.00	**136.40**	15.00	15.00	**373.80**	4.84	**515.04**	nr	**579.41**
weekly cost of maintaining last................	-	-	-	2.50	2.50	**62.30**	1.13	**63.43**	nr	**71.36**
5.00m	144.00	10.00	**158.40**	17.80	17.80	**443.58**	5.63	**607.61**	nr	**683.56**
weekly cost of maintaining last................	-	-	-	3.00	3.00	**74.76**	1.29	**76.05**	nr	**85.56**
6.00m	157.00	10.00	**172.70**	5.10	20.60	**293.10**	6.24	**472.04**	nr	**531.05**
weekly cost of maintaining last................	-	-	-	3.50	3.50	**87.22**	1.48	**88.70**	nr	**99.78**
C40: CLEANING MASONRY/CONCRETE										
Cleaning surfaces										
Thoroughly clean existing concrete surfaces prior to applying damp proof membrane										
floors	-	-	-	-	0.19	**2.03**	0.02	**2.06**	m2	**2.32**
walls	-	-	-	-	0.21	**2.25**	0.02	**2.27**	m2	**2.56**
Thoroughly clean existing concrete surfaces, fill in nail holes and small surface imperfections and leave smooth										
walls	-	-	-	0.35	0.20	**7.12**	0.05	**7.16**	m2	**8.06**
soffits	-	-	-	0.50	0.29	**10.21**	0.05	**10.26**	m2	**11.54**
Thoroughly clean existing brick or block surfaces prior to applying damp proof membrane										
walls	-	-	-	-	0.25	**2.68**	-	**2.68**	m2	**3.01**
Clean out air brick										
225 x 150mm	-	-	-	-	0.20	**2.14**	-	**2.14**	nr	**2.41**
C41: REPAIRING/RENOVATING/CONSERVING MASONRY										
Cut out decayed bricks and replace with new bricks in gauged mortar (1:2:9) and make good surrounding work										
Fair faced common bricks										
singly	0.22	10.00	**0.24**	0.17	0.17	**4.24**	0.15	**4.63**	nr	**5.21**
small patches	13.20	10.00	**14.52**	5.80	5.80	**144.54**	6.43	**165.48**	m2	**186.17**
Picked stock facing bricks										
singly	0.46	10.00	**0.51**	0.17	0.17	**4.24**	0.15	**4.89**	nr	**5.50**
small patches	27.60	10.00	**30.36**	5.80	5.80	**144.54**	6.43	**181.32**	m2	**203.99**
Cut out defective wall and re-build in gauged mortar (1:2:9) and tooth and bond to surrounding work (25% new bricks allowed)										
Half brick wall; stretcher bond										
common bricks	3.30	10.00	**3.63**	4.00	4.00	**99.68**	3.69	**107.00**	m2	**120.37**
One brick wall; English bond										
common bricks	6.60	10.00	**7.26**	7.90	7.90	**196.87**	7.92	**212.05**	m2	**238.56**
common bricks faced one side with picked stock facings	12.00	10.00	**13.20**	8.30	8.30	**206.84**	7.92	**227.96**	m2	**256.45**
picked stock facing bricks faced both sides......	13.80	10.00	**15.18**	8.80	8.80	**219.30**	7.92	**242.40**	m2	**272.70**
One brick wall; Flemish bond										
common bricks	6.60	10.00	**7.26**	7.90	7.90	**196.87**	3.69	**207.82**	m2	**233.79**
common bricks faced one side with picked stock facings	12.00	10.00	**13.20**	8.30	8.30	**206.84**	7.92	**227.96**	m2	**256.45**
picked stock facing bricks faced both sides......	13.80	10.00	**15.18**	8.80	8.80	**219.30**	7.92	**242.40**	m2	**272.70**
275mm hollow wall; skins in stretcher bond										
two half brick skins in common bricks...........	6.60	10.00	**7.26**	8.16	8.16	**203.35**	8.07	**218.68**	m2	**246.02**

Labour hourly rates: (except Specialists) Craft Operatives £14.21 Labourer £10.71 Rates are national average prices. Refer to REGIONAL VARIATIONS for indicative levels of overall pricing in regions	MATERIALS			LABOUR				RATES		
	Del to Site £	Waste %	Material Cost £	Craft Optve Hrs	Lab Hrs	Labour Cost £	Sunds £	Nett Rate £	Unit	Gross Rate (+12.5%) £
C41: REPAIRING/RENOVATING/CONSERVING MASONRY Cont.										
Cut out defective wall and re-build in gauged mortar (1:2:9) and tooth and bond to surrounding work (25% new bricks allowed) Cont.										
275mm hollow wall; skins in stretcher bond Cont.										
one half brick skin in common bricks and one half brick skin in picked stock facings	10.20	10.00	**11.22**	8.60	8.60	**214.31**	8.07	**233.61**	m2	**262.81**
one 100mm skin in concrete blocks and one half brick skin in picked stock facings	9.43	10.00	**10.37**	8.26	8.26	**205.84**	7.55	**223.75**	m2	**251.72**
Cut out crack in brickwork and stitch across with new bricks in gauged mortar(1:2:9); average 450mm wide										
Half brick wall										
common bricks	5.72	10.00	**6.29**	1.90	1.90	**47.35**	3.58	**57.22**	m	**64.37**
common bricks fair faced one side	5.72	10.00	**6.29**	2.30	2.30	**57.32**	3.58	**67.19**	m	**75.59**
picked stock facing bricks faced both sides	11.96	10.00	**13.16**	2.40	2.40	**59.81**	3.58	**76.54**	m	**86.11**
One brick wall										
common bricks	11.44	10.00	**12.58**	3.90	3.90	**97.19**	7.71	**117.48**	m	**132.16**
common bricks fair faced one side	11.44	10.00	**12.58**	4.40	4.40	**109.65**	7.71	**129.94**	m	**146.18**
common bricks faced one side with picked stock facings	20.80	10.00	**22.88**	4.60	4.60	**114.63**	7.71	**145.22**	m	**163.37**
picked stock facing bricks faced both sides	23.92	10.00	**26.31**	4.70	4.70	**117.12**	7.71	**151.14**	m	**170.04**
Cut out defective arch; re-build in picked stock facing bricks in gauged mortar (1:2:9) including centering										
Brick-on-edge flat arch										
102mm on soffit	3.22	10.00	**3.54**	1.00	1.00	**24.92**	0.90	**29.37**	m	**33.04**
215mm on soffit	6.13	10.00	**6.74**	0.95	0.95	**23.67**	1.58	**32.00**	m	**36.00**
Brick-on-end flat arch										
102mm on soffit	6.13	10.00	**6.74**	0.95	0.95	**23.67**	1.58	**32.00**	m	**36.00**
215mm on soffit	12.26	10.00	**13.49**	1.90	1.90	**47.35**	3.16	**64.00**	m	**72.00**
Segmental arch in two half brick rings										
102mm on soffit	6.13	10.00	**6.74**	2.90	2.90	**72.27**	1.58	**80.59**	m	**90.67**
215mm on soffit	12.26	10.00	**13.49**	2.90	2.90	**72.27**	3.16	**88.92**	m	**100.03**
Semi-circular arch in two half brick rings										
102mm on soffit	6.13	10.00	**6.74**	2.90	2.90	**72.27**	1.58	**80.59**	m	**90.67**
215mm on soffit	12.26	10.00	**13.49**	2.90	2.90	**72.27**	3.16	**88.92**	m	**100.03**
Cut out defective arch; replace with precast concrete, BS 5328, designed mix C20, 20mm aggregate lintel; wedge and pin up to brickwork over with slates in cement mortar (1:3) and extend external rendering over										
Remove defective brick-on-end flat arch										
replace with 102 x 215mm precast concrete lintel reinforced with one 12mm mild steel bar	27.27	10.00	**30.00**	1.90	1.90	**47.35**	2.48	**79.83**	m	**89.81**
Cut out defective external sill and re-build										
Shaped brick-on-edge sill in picked stock facing bricks in gauged mortar (1:2:9)										
225mm wide	6.13	10.00	**6.74**	1.00	1.00	**24.92**	2.20	**33.87**	m	**38.10**
Roofing tile sill set weathering and projecting in cement mortar (1:3)										
two courses	6.14	10.00	**6.76**	1.20	1.20	**29.90**	1.14	**37.80**	m	**42.53**
Take down defective brick-on-edge coping and re-build in picked stock facing bricks in gauged mortar (1:2:9)										
Coping										
to one brick wall	6.13	10.00	**6.74**	0.60	0.60	**14.95**	2.20	**23.90**	m	**26.89**
Coping with cement fillets both sides										
to one brick wall; with oversailing course	9.20	10.00	**10.12**	0.67	0.67	**16.70**	3.31	**30.13**	m	**33.89**
to one brick wall; with single course tile creasing	9.20	10.00	**10.12**	1.09	1.09	**27.16**	2.77	**40.06**	m	**45.07**
to one brick wall; with double course tile creasing	12.28	10.00	**13.50**	1.38	1.38	**34.39**	3.91	**51.81**	m	**58.28**
Cut out defective air brick and replace with new; bed and point in gauged mortar (1:2:9)										
Cast iron, light, square hole										
225 x 75mm	3.95	-	**3.95**	0.45	0.45	**11.21**	0.17	**15.34**	nr	**17.25**
225 x 150mm	7.25	-	**7.25**	0.45	0.45	**11.21**	0.28	**18.74**	nr	**21.09**
Terra cotta										
215 x 65mm	1.33	-	**1.33**	0.45	0.45	**11.21**	0.17	**12.72**	nr	**14.31**
215 x 140mm	1.85	-	**1.85**	0.45	0.45	**11.21**	0.28	**13.34**	nr	**15.01**
Rake out joints of old brickwork 20mm deep and re-point in cement mortar (1:3)										
For turned in edge of lead flashing										
flush pointing; horizontal	-	-	**-**	0.50	0.50	**12.46**	0.26	**12.72**	m	**14.31**
flush pointing; stepped	-	-	**-**	0.85	0.85	**21.18**	0.37	**21.55**	m	**24.24**

Labour hourly rates: (except Specialists) Craft Operatives £14.21 Labourer £10.71 Rates are national average prices. Refer to REGIONAL VARIATIONS for indicative levels of overall pricing in regions	MATERIALS			LABOUR			RATES			
	Del to Site £	Waste %	Material Cost £	Craft Optve Hrs	Lab Hrs	Labour Cost £	Sunds £	Nett Rate £	Unit	Gross Rate (+12.5%) £
C41: REPAIRING/RENOVATING/CONSERVING MASONRY Cont.										
Rake out joints of old brickwork 20mm deep and re-point in cement mortar (1:3) Cont.										
For turned in edge of lead flashing Cont.										
weathered pointing; horizontal	-	-	-	0.55	0.55	**13.71**	0.26	**13.96**	m	**15.71**
weathered pointing; stepped	-	-	-	0.90	0.90	**22.43**	0.37	**22.79**	m	**25.64**
ironed in pointing; horizontal	-	-	-	0.55	0.55	**13.71**	0.26	**13.96**	m	**15.71**
ironed in pointing; stepped	-	-	-	0.90	0.90	**22.43**	0.37	**22.79**	m	**25.64**
Rake out joints of old brickwork 20mm deep and re-point in cement-lime mortar(1:1:6)										
Generally; English bond										
flush pointing	-	-	-	1.00	0.66	**21.28**	0.74	**22.02**	m2	**24.77**
weathered pointing	-	-	-	1.00	0.66	**21.28**	0.74	**22.02**	m2	**24.77**
ironed in pointing	-	-	-	1.20	0.80	**25.62**	0.74	**26.36**	m2	**29.66**
Generally; Flemish bond										
flush pointing	-	-	-	1.00	0.66	**21.28**	0.74	**22.02**	m2	**24.77**
weathered pointing	-	-	-	1.00	0.66	**21.28**	0.74	**22.02**	m2	**24.77**
ironed in pointing	-	-	-	1.20	0.80	**25.62**	0.74	**26.36**	m2	**29.66**
Isolated areas not exceeding 1.00m2; English bond										
flush pointing	-	-	-	1.50	1.00	**32.03**	0.74	**32.77**	m2	**36.86**
weathered pointing	-	-	-	1.50	1.00	**32.03**	0.74	**32.77**	m2	**36.86**
ironed in pointing	-	-	-	1.80	1.20	**38.43**	0.74	**39.17**	m2	**44.07**
Isolated areas not exceeding 1.00m2; Flemish bond										
flush pointing	-	-	-	1.50	1.00	**32.03**	0.74	**32.77**	m2	**36.86**
weathered pointing	-	-	-	1.50	1.00	**32.03**	0.74	**32.77**	m2	**36.86**
ironed in pointing	-	-	-	1.80	1.20	**38.43**	0.74	**39.17**	m2	**44.07**
For using coloured mortar										
add	-	-	-	-	-	-	0.54	**0.54**	m2	**0.60**
Rake out joints of old brickwork 20mm deep and re-point in cement mortar (1:3)										
Generally; English bond										
flush pointing	-	-	-	1.20	0.80	**25.62**	0.74	**26.36**	m2	**29.66**
weathered pointing	-	-	-	1.20	0.80	**25.62**	0.74	**26.36**	m2	**29.66**
Generally; Flemish bond										
flush pointing	-	-	-	1.20	0.80	**25.62**	0.74	**26.36**	m2	**29.66**
weathered pointing	-	-	-	1.20	0.80	**25.62**	0.74	**26.36**	m2	**29.66**
Isolated areas not exceeding 1.00m2; English bond										
flush pointing	-	-	-	1.80	1.20	**38.43**	0.74	**39.17**	m2	**44.07**
weathered pointing	-	-	-	1.80	1.20	**38.43**	0.74	**39.17**	m2	**44.07**
Isolated areas not exceeding 1.00m2; Flemish bond										
flush pointing	-	-	-	1.80	1.20	**38.43**	0.74	**39.17**	m2	**44.07**
weathered pointing	-	-	-	1.80	1.20	**38.43**	0.74	**39.17**	m2	**44.07**
Repairing/Renovating stone										
Note restoration and repair of stonework is work for a specialist; prices being obtained for specific projects. Some firms that specialise in this class of work are included within the list at the end of section F-Masonry										
Repairs in Portland stone work and set in gauged mortar (1:2:9) and point to match existing										
Cut out decayed stones in facings of walls, piers or the like, prepare for and set new 75mm thick stone facings										
single stone	357.00	-	**357.00**	15.00	15.00	**373.80**	6.93	**737.73**	m2	**829.94**
areas of two or more adjacent stones	357.00	-	**357.00**	12.00	12.00	**299.04**	6.93	**662.97**	m2	**745.84**
Take out decayed stone sill, prepare for and set new sunk weathered and throated sill										
175 x 100mm	191.00	-	**191.00**	1.80	1.80	**44.86**	0.93	**236.79**	m	**266.39**
250 x 125mm	260.00	-	**260.00**	2.50	2.50	**62.30**	1.58	**323.88**	m	**364.37**
Take out sections of decayed stone coping, prepare for and set new weathered and twice throated coping										
300 x 75mm	232.00	-	**232.00**	1.40	1.40	**34.89**	1.17	**268.06**	m	**301.56**
300 x 100mm	265.00	-	**265.00**	1.70	1.70	**42.36**	1.55	**308.91**	m	**347.53**
Repairs in York stone work and set in gauged mortar (1:2:9) and point to match existing										
Cut out decayed stones in facings of walls, piers or the like, prepare for and set new 75mm thick stone facings										
single stone	268.00	-	**268.00**	17.00	17.00	**423.64**	6.93	**698.57**	m2	**785.89**
areas of two or more adjacent stones	268.00	-	**268.00**	13.75	13.75	**342.65**	6.93	**617.58**	m2	**694.77**
Take out decayed stone sill, prepare for and set new sunk weathered and throated sill										
175 x 100mm	119.00	-	**119.00**	1.80	1.80	**44.86**	1.17	**165.03**	m	**185.65**
250 x 125mm	185.00	-	**185.00**	2.50	2.50	**62.30**	1.55	**248.85**	m	**279.96**

Labour hourly rates: (except Specialists) Craft Operatives £14.21 Labourer £10.71 Rates are national average prices. Refer to REGIONAL VARIATIONS for indicative levels of overall pricing in regions	MATERIALS			LABOUR				RATES		
	Del to Site £	Waste %	Material Cost £	Craft Optve Hrs	Lab Hrs	Labour Cost £	Sunds £	Nett Rate £	Unit	Gross Rate (+12.5%) £
C41: REPAIRING/RENOVATING/CONSERVING MASONRY Cont.										
Repairs in York stone work and set in gauged mortar (1:2:9) and point to match existing Cont.										
Take out sections of decayed stone coping, prepare for and set new weathered and twice throated coping										
300 x 75mm	130.00	-	**130.00**	1.40	1.40	**34.89**	1.17	**166.06**	m	**186.81**
300 x 100mm	154.00	-	**154.00**	1.70	1.70	**42.36**	1.55	**197.91**	m	**222.65**
Rake out decayed mortar joints and re-point in gauged mortar (1:2:9)										
Re-point rubble walling										
coursed	-	-	-	1.25	1.25	**31.15**	1.15	**32.30**	m2	**36.34**
uncoursed	-	-	-	1.32	1.32	**32.89**	1.15	**34.04**	m2	**38.30**
squared	-	-	-	1.25	1.25	**31.15**	1.15	**32.30**	m2	**36.34**
Re-point stonework										
ashlar	-	-	-	1.20	1.20	**29.90**	1.15	**31.05**	m2	**34.93**
coursed block	-	-	-	1.20	1.20	**29.90**	1.15	**31.05**	m2	**34.93**
C42: REPAIRING/ RENOVATING/CONSERVING CONCRETE										
Re-bed loose copings										
Take off loose concrete coping and clean and remove old bedding mortar										
re-bed, joint and point in cement mortar (1:3)	-	-	-	1.00	1.00	**24.92**	1.22	**26.14**	m	**29.41**
Repairing cracks in concrete										
Repair cracks; clean out all dust and debris and fill with mortar mixed with a bonding agent										
up to 5mm wide	-	-	-	0.50	0.50	**12.46**	0.59	**13.05**	m	**14.68**
Repair cracks; cutting out to form groove, treat with bonding agent and fill with fine concrete mixed with a bonding agent										
25 x 25mm deep	-	-	-	0.50	0.50	**12.46**	1.06	**13.52**	m	**15.21**
40 x 40mm deep	-	-	-	0.60	0.60	**14.95**	2.52	**17.47**	m	**19.65**
C45: DAMP PROOF COURSE RENEWAL/INSERTION										
Insert damp proof course in old wall by cutting out one course of brickwork in alternate lengths not exceeding 1.00m and replace old bricks with 50mm bricks in cement mortar (1:3)										
B.S.6398 Class D, bitumen with hessian base laminated with lead in										
one brick wall	3.85	10.00	**4.24**	3.00	3.00	**74.76**	5.94	**84.94**	m	**95.56**
one and a half brick wall	5.79	10.00	**6.37**	4.00	4.00	**99.68**	8.93	**114.98**	m	**129.35**
254mm hollow wall	3.85	10.00	**4.24**	3.50	3.50	**87.22**	5.94	**97.40**	m	**109.58**
Hyload, pitch polymer in										
one brick wall	1.85	10.00	**2.03**	3.00	3.00	**74.76**	5.94	**82.74**	m	**93.08**
one and a half brick wall	2.77	10.00	**3.05**	4.00	4.00	**99.68**	8.93	**111.67**	m	**125.62**
254mm hollow wall	1.85	10.00	**2.03**	3.50	3.50	**87.22**	5.94	**95.20**	m	**107.10**
Two courses slates in cement mortar (1:3) in										
one brick wall	6.04	10.00	**6.64**	3.00	3.00	**74.76**	5.94	**87.35**	m	**98.27**
one and a half brick wall	9.06	10.00	**9.97**	4.00	4.00	**99.68**	8.93	**118.58**	m	**133.40**
254mm hollow wall	6.04	10.00	**6.64**	3.50	3.50	**87.22**	5.94	**99.81**	m	**112.28**
0.6mm bitumen coated copper in										
one brick wall	11.44	10.00	**12.58**	3.00	3.00	**74.76**	5.94	**93.29**	m	**104.95**
one and a half brick wall	17.18	10.00	**18.90**	4.00	4.00	**99.68**	8.93	**127.51**	m	**143.45**
254mm hollow wall	11.44	10.00	**12.58**	3.50	3.50	**87.22**	5.94	**105.75**	m	**118.96**
Insert damp proof course in old wall by hand sawing in 600mm lengths										
B.S.6398 Class D, bitumen with hessian base laminated with lead in										
one brick wall	3.85	10.00	**4.24**	2.50	2.50	**62.30**	8.89	**75.43**	m	**84.86**
one and a half brick wall	5.79	10.00	**6.37**	3.50	3.50	**87.22**	13.34	**106.93**	m	**120.30**
254mm hollow wall	3.85	10.00	**4.24**	3.00	3.00	**74.76**	8.89	**87.89**	m	**98.88**
Hyload, pitch polymer in										
one brick wall	1.85	10.00	**2.03**	2.50	2.50	**62.30**	8.89	**73.22**	m	**82.38**
one and a half brick wall	2.77	10.00	**3.05**	3.50	3.50	**87.22**	13.34	**103.61**	m	**116.56**
254mm hollow wall	1.85	10.00	**2.03**	3.00	3.00	**74.76**	8.89	**85.68**	m	**96.39**
Two courses slates in cement mortar (1:3) in										
one brick wall	6.04	10.00	**6.64**	2.50	2.50	**62.30**	8.89	**77.83**	m	**87.56**
one and a half brick wall	9.06	10.00	**9.97**	3.50	3.50	**87.22**	13.34	**110.53**	m	**124.34**
254mm hollow wall	3.02	10.00	**3.32**	3.00	3.00	**74.76**	8.89	**86.97**	m	**97.84**
0.6mm bitumen coated copper in										
one brick wall	11.44	10.00	**12.58**	2.50	2.50	**62.30**	8.89	**83.77**	m	**94.24**
one and a half brick wall	17.18	10.00	**18.90**	3.50	3.50	**87.22**	13.34	**119.46**	m	**134.39**
254mm hollow wall	11.44	10.00	**12.58**	3.00	3.00	**74.76**	8.89	**96.23**	m	**108.26**

Labour hourly rates: (except Specialists) Craft Operatives £14.21 Labourer £10.71 Rates are national average prices. Refer to REGIONAL VARIATIONS for indicative levels of overall pricing in regions	MATERIALS			LABOUR				RATES		
	Del to Site £	Waste %	Material Cost £	Craft Optve Hrs	Lab Hrs	Labour Cost £	Sunds £	Nett Rate £	Unit	Gross Rate (+12.5%) £
C45: DAMP PROOF COURSE RENEWAL/INSERTION Cont.										
Insert damp proof course in old wall by machine sawing in 600mm lengths										
B.S.6398 Class D, bitumen with hessian base laminated with lead in										
one brick wall	3.85	10.00	**4.24**	2.00	2.00	**49.84**	11.80	**65.88**	m	**74.12**
one and a half brick wall	5.79	10.00	**6.37**	3.00	3.00	**74.76**	17.69	**98.82**	m	**111.18**
254mm hollow wall	3.85	10.00	**4.24**	2.50	2.50	**62.30**	11.80	**78.34**	m	**88.14**
Hyload, pitch polymer in										
one brick wall	1.85	10.00	**2.03**	2.00	2.00	**49.84**	11.80	**63.68**	m	**71.63**
one and a half brick wall	2.77	10.00	**3.05**	3.00	3.00	**74.76**	17.69	**95.51**	m	**107.45**
254mm hollow wall	1.85	10.00	**2.03**	2.50	2.50	**62.30**	11.80	**76.14**	m	**85.65**
Two courses slates in cement mortar (1:3) in										
one brick wall	6.04	10.00	**6.64**	2.00	2.00	**49.84**	11.80	**68.29**	m	**76.82**
one and a half brick wall	9.06	10.00	**9.97**	3.00	3.00	**74.76**	17.69	**102.42**	m	**115.22**
254mm hollow wall	3.02	10.00	**3.32**	2.50	2.50	**62.30**	11.80	**77.43**	m	**87.10**
0.6mm bitumen coated copper in										
one brick wall	11.44	10.00	**12.58**	2.00	2.00	**49.84**	11.80	**74.22**	m	**83.50**
one and a half brick wall	17.18	10.00	**18.90**	3.00	3.00	**74.76**	17.69	**111.35**	m	**125.27**
254mm hollow wall	11.44	10.00	**12.58**	2.50	2.50	**62.30**	11.80	**86.68**	m	**97.52**
Insert cavity tray in old wall by cutting out by hand in short lengths; insert individual trays and make good with picked stock facing bricks in gauged mortar (1:2:9) including pinning up with slates										
Type E Cavitray by Cavity Trays Ltd in										
half brick skin of cavity wall	9.06	10.00	**9.96**	1.75	1.75	**43.61**	7.55	**61.12**	m	**68.76**
external angle	6.25	10.00	**6.88**	1.00	1.00	**24.92**	6.93	**38.73**	nr	**43.57**
internal angle	6.25	10.00	**6.88**	1.25	1.25	**31.15**	6.93	**44.96**	nr	**50.58**
Type X Cavity Trays Ltd to suit 40 degree pitched roof complete with attached code 4 lead flashing and dress over tiles in										
half brick skin of cavity wall	37.66	10.00	**41.42**	4.25	4.25	**105.91**	15.10	**162.43**	m	**182.74**
ridge tray	9.35	10.00	**10.29**	1.00	1.00	**24.92**	3.52	**38.72**	nr	**43.56**
catchment tray	4.43	10.00	**4.87**	0.50	0.50	**12.46**	1.85	**19.18**	nr	**21.58**
corner catchment tray	9.77	10.00	**10.75**	1.00	1.00	**24.92**	3.52	**39.18**	nr	**44.08**
Damp proofing old brick wall by silicone injection method (excluding removal and reinstatement of plaster, etc.)										
Damp proofing										
one brick wall	7.98	-	**7.98**	1.00	1.00	**24.92**	0.81	**33.71**	m	**37.92**
one and a half brick wall	11.92	-	**11.92**	1.50	1.50	**37.38**	1.20	**50.50**	m	**56.82**
254mm hollow wall	7.98	-	**7.98**	1.35	1.35	**33.64**	0.81	**42.43**	m	**47.73**
C50: REPAIRING/RENOVATING/CONSERVING METAL										
Replace defective arch bar										
Take out defective arch bar including cutting away brickwork as necessary, replace with new mild steel arch bar primed all round and make good all work disturbed										
32 x 6mm flat arch bar	1.55	-	**1.55**	1.03	1.03	**25.67**	0.75	**27.96**	m	**31.46**
50 x 6mm flat arch bar	2.40	-	**2.40**	1.12	1.12	**27.91**	1.09	**31.40**	m	**35.33**
50 x 50 x 6mm angle arch bar	4.45	-	**4.45**	1.27	1.27	**31.65**	1.94	**38.04**	m	**42.80**
75 x 50 x 6mm angle arch bar	5.55	-	**5.55**	1.44	1.44	**35.88**	2.45	**43.88**	m	**49.37**
Replace defective baluster										
Cut out defective baluster, clean core and bottom rails, provide new mild steel baluster and weld on										
13mm diameter x 686mm long	0.90	-	**0.90**	0.92	0.92	**22.93**	1.55	**25.38**	nr	**28.55**
13mm diameter x 762mm long	0.90	-	**0.90**	0.94	0.94	**23.42**	1.55	**25.88**	nr	**29.11**
13 x 13 x 686mm long	1.20	-	**1.20**	0.94	0.94	**23.42**	1.84	**26.46**	nr	**29.77**
13 x 13 x 762mm long	1.25	-	**1.25**	0.95	0.95	**23.67**	1.84	**26.76**	nr	**30.11**
Repair damaged weld										
Clean off damaged weld including removing remains of weld and cleaning off surrounding paint back to bright metal and re-weld connection between 10mm rail and										
13mm diameter rod	-	-	**-**	0.50	0.50	**12.46**	2.29	**14.75**	nr	**16.59**
13 x 13mm bar	-	-	**-**	0.53	0.53	**13.21**	2.72	**15.93**	nr	**17.92**
25 x 25mm bar	-	-	**-**	0.60	0.60	**14.95**	3.41	**18.36**	nr	**20.66**
C51: REPAIRING/RENOVATING/CONSERVING TIMBER										
Cut out rotten or infected structural timber members including cutting away wall, etc. and shoring up adjacent work and replace with new pressure impregnated timber										
Plates and bed in cement mortar (1:3)										
50 x 75mm	1.26	10.00	**1.39**	0.30	0.30	**7.48**	0.17	**9.03**	m	**10.16**
75 x 100mm	2.70	10.00	**2.97**	0.50	0.50	**12.46**	0.33	**15.76**	m	**17.73**

Labour hourly rates: (except Specialists) Craft Operatives £14.21 Labourer £10.71 Rates are national average prices. Refer to REGIONAL VARIATIONS for indicative levels of overall pricing in regions	MATERIALS			LABOUR				RATES		
	Del to Site £	Waste %	Material Cost £	Craft Optve Hrs	Lab Hrs	Labour Cost £	Sunds £	Nett Rate £	Unit	Gross Rate (+12.5%) £
C51: REPAIRING/RENOVATING/CONSERVING TIMBER Cont.										
Cut out rotten or infected structural timber members including cutting away wall, etc. and shoring up adjacent work and replace with new pressure impregnated timber Cont.										
Floor or roof joists										
50 x 175mm	2.93	10.00	**3.22**	0.57	0.57	**14.20**	0.38	**17.81**	m	**20.03**
50 x 225mm	3.77	10.00	**4.15**	0.67	0.67	**16.70**	0.49	**21.33**	m	**24.00**
75 x 175mm	4.74	10.00	**5.21**	0.79	0.79	**19.69**	0.58	**25.48**	m	**28.67**
Rafters										
38 x 100mm	1.64	10.00	**1.80**	0.35	0.35	**8.72**	0.17	**10.70**	m	**12.03**
50 x 100mm	1.67	10.00	**1.84**	0.38	0.38	**9.47**	0.22	**11.53**	m	**12.97**
Ceiling joists and collars										
50 x 100mm	1.67	10.00	**1.84**	0.38	0.38	**9.47**	0.22	**11.53**	m	**12.97**
50 x 150mm	2.51	10.00	**2.76**	0.52	0.52	**12.96**	0.33	**16.05**	m	**18.06**
Purlins, ceiling beams and struts										
75 x 100mm	2.70	10.00	**2.97**	0.55	0.55	**13.71**	0.33	**17.01**	m	**19.13**
100 x 225mm	8.52	10.00	**9.37**	1.23	1.23	**30.65**	0.99	**41.01**	m	**46.14**
Roof trusses										
75 x 100mm	2.70	10.00	**2.97**	0.92	0.92	**22.93**	0.33	**26.23**	m	**29.50**
75 x 150mm	4.05	10.00	**4.46**	1.37	1.37	**34.14**	0.49	**39.09**	m	**43.97**
Hangers in roof										
25 x 75mm	0.78	10.00	**0.86**	0.30	0.30	**7.48**	0.09	**8.42**	m	**9.48**
38 x 100mm	1.64	10.00	**1.80**	0.45	0.45	**11.21**	0.17	**13.19**	m	**14.84**
Cut out rotten or infected timber and prepare timber under to receive new										
Roof boarding; provide new 25mm pressure impregnated softwood boarding										
flat boarding; areas not exceeding 0.50m2	11.00	10.00	**12.10**	1.96	1.96	**48.84**	1.09	**62.03**	m2	**69.79**
flat boarding; areas 0.50 - 5.00m2	11.00	10.00	**12.10**	1.63	1.63	**40.62**	1.09	**53.81**	m2	**60.54**
sloping boarding; areas not exceeding 0.50m2	11.00	10.00	**12.10**	2.34	2.34	**58.31**	1.09	**71.50**	m2	**80.44**
sloping boarding; areas 0.50 - 5.00m2	11.00	10.00	**12.10**	1.89	1.89	**47.10**	1.09	**60.29**	m2	**67.82**
Gutter boarding and bearers; provide new 25mm pressure impregnated softwood boarding and 50 x 50mm bearers										
over 300mm wide	14.36	10.00	**15.80**	2.06	2.06	**51.34**	1.31	**68.44**	m2	**77.00**
200mm	3.88	10.00	**4.27**	0.55	0.55	**13.71**	0.44	**18.41**	m	**20.72**
225mm	4.16	10.00	**4.57**	0.62	0.62	**15.45**	0.48	**20.50**	m	**23.06**
Gutter sides; provide new 19mm pressure impregnated softwood boarding										
150mm	1.27	15.00	**1.46**	0.46	0.46	**11.46**	0.12	**13.04**	m	**14.67**
200mm	1.76	15.00	**2.02**	0.50	0.50	**12.46**	0.17	**14.65**	m	**16.49**
225mm	2.15	15.00	**2.47**	0.57	0.57	**14.20**	0.19	**16.87**	m	**18.98**
Eaves or verge soffit and bearers and provide new 25mm pressure impregnated softwood boarding and 25 x 50mm bearers										
over 300mm wide	15.81	15.00	**18.13**	2.10	2.10	**52.33**	1.31	**71.78**	m2	**80.75**
225mm	4.37	15.00	**4.98**	0.67	0.67	**16.70**	0.48	**22.15**	m	**24.92**
Fascia; provide new pressure impregnated softwood										
25 x 150mm	2.28	15.00	**2.62**	0.46	0.46	**11.46**	0.17	**14.26**	m	**16.04**
25 x 200mm	3.00	15.00	**3.45**	0.51	0.51	**12.71**	0.22	**16.38**	m	**18.43**
Barge board; provide new pressure impregnated softwood										
25 x 225mm	3.41	15.00	**3.92**	0.55	0.55	**13.71**	0.25	**17.88**	m	**20.11**
25 x 250mm	3.75	15.00	**4.31**	0.57	0.57	**14.20**	0.28	**18.80**	m	**21.15**
Take up floor boarding, remove all nails and clean joists under, re-lay boarding, clear away and replace damaged boards and sand surface (20% new softwood boards allowed)										
Square edged boarding										
25mm	2.70	15.00	**3.11**	1.73	1.73	**43.11**	0.44	**46.66**	m2	**52.49**
32mm	3.58	15.00	**4.12**	1.76	1.76	**43.86**	0.55	**48.53**	m2	**54.59**
Tongued and grooved boarding										
25mm	2.86	20.00	**3.43**	1.87	1.87	**46.60**	0.44	**50.47**	m2	**56.78**
32mm	3.79	20.00	**4.55**	1.89	1.89	**47.10**	0.55	**52.20**	m2	**58.73**
Take up floor boarding, remove all nails and clean joists under, re-lay boarding, destroy and replace infected boards with new pressure impregnated boards and sand surface (20% new softwood boards allowed)										
Square edged boarding										
25mm	2.97	15.00	**3.42**	1.81	1.81	**45.11**	0.44	**48.96**	m2	**55.08**
32mm	3.94	15.00	**4.53**	1.83	1.83	**45.60**	0.55	**50.68**	m2	**57.02**
Tongued and grooved boarding										
25mm	3.15	20.00	**3.78**	1.94	1.94	**48.34**	0.44	**52.56**	m2	**59.14**
32mm	4.17	20.00	**5.00**	1.95	1.95	**48.59**	0.55	**54.15**	m2	**60.92**

Labour hourly rates: (except Specialists) Craft Operatives £14.21 Labourer £10.71 Rates are national average prices. Refer to REGIONAL VARIATIONS for indicative levels of overall pricing in regions	MATERIALS			LABOUR				RATES		
	Del to Site £	Waste %	Material Cost £	Craft Optve Hrs	Lab Hrs	Labour Cost £	Sunds £	Nett Rate £	Unit	Gross Rate (+12.5%) £
C51: REPAIRING/RENOVATING/CONSERVING TIMBER Cont.										
Take up worn or damaged floor boarding, remove all nails and clean joists under and replace with new softwood boarding										
Square edged boarding										
25mm in areas not exceeding 0.50m2..............	13.50	15.00	**15.53**	1.90	1.90	**47.35**	1.09	**63.96**	m2	**71.96**
32mm in areas not exceeding 0.50m2..............	17.90	15.00	**20.59**	1.94	1.94	**48.34**	1.39	**70.32**	m2	**79.11**
25mm in areas 0.50 - 3.00m2.....................	13.50	15.00	**15.53**	1.57	1.57	**39.12**	1.09	**55.74**	m2	**62.71**
32mm in areas 0.50 - 3.00m2.....................	17.90	15.00	**20.59**	1.61	1.61	**40.12**	1.39	**62.10**	m2	**69.86**
Tongued and grooved boarding										
25mm in areas not exceeding 0.50m2..............	14.31	20.00	**17.17**	2.00	2.00	**49.84**	1.09	**68.10**	m2	**76.61**
32mm in areas not exceeding 0.50m2..............	18.97	20.00	**22.76**	2.06	2.06	**51.34**	1.39	**75.49**	m2	**84.93**
25mm in areas 0.50 - 3.00m2.....................	14.31	20.00	**17.17**	1.82	1.82	**45.35**	1.09	**63.62**	m2	**71.57**
32mm in areas 0.50 - 3.00m2.....................	18.97	20.00	**22.76**	1.87	1.87	**46.60**	1.39	**70.75**	m2	**79.60**
Punch down all nails, remove all tacks, etc. and fill nail holes										
Surface of existing flooring										
generally	-	-	**-**	0.50	0.10	**8.18**	0.03	**8.21**	m2	**9.23**
generally and machine sand in addition..........	-	-	**-**	1.00	0.30	**17.42**	1.18	**18.60**	m2	**20.93**
Note										
removal of old floor coverings included elsewhere										
Easing and overhauling doors										
Ease and adjust door and oil ironmongery										
762 x 1981mm	-	-	**-**	0.33	0.06	**5.33**	0.02	**5.35**	nr	**6.02**
914 x 2134mm	-	-	**-**	0.39	0.06	**6.18**	0.02	**6.20**	nr	**6.98**
Take off door, shave 12mm off bottom edge and re-hang										
762 x 1981mm	-	-	**-**	0.66	0.11	**10.56**	-	**10.56**	nr	**11.88**
914 x 2134mm	-	-	**-**	0.79	0.13	**12.62**	-	**12.62**	nr	**14.20**
Adapting doors										
Take off door and re-hang on opposite hand, piece out as necessary and oil ironmongery										
762 x 1981mm	-	-	**-**	2.50	0.25	**38.20**	0.10	**38.30**	nr	**43.09**
914 x 2134mm	-	-	**-**	2.65	0.44	**42.37**	0.10	**42.47**	nr	**47.78**
Refixing removed doors and frames										
Doors										
762 x 1981mm	-	-	**-**	2.00	0.34	**32.06**	0.14	**32.20**	nr	**36.23**
914 x 2134mm	-	-	**-**	2.25	0.38	**36.04**	0.14	**36.18**	nr	**40.71**
Doors and frames; oil existing ironmongery										
762 x 1981mm	-	-	**-**	4.00	0.66	**63.91**	0.35	**64.26**	nr	**72.29**
914 x 2134mm	-	-	**-**	4.50	0.75	**71.98**	0.35	**72.33**	nr	**81.37**
Repairing doors and frames										
Piece in door where ironmongery removed										
rim lock	-	-	**-**	1.00	0.15	**15.82**	0.16	**15.98**	nr	**17.97**
rim lock or latch and furniture.................	-	-	**-**	1.50	0.25	**23.99**	0.39	**24.38**	nr	**27.43**
mortice lock	-	-	**-**	1.50	0.25	**23.99**	0.39	**24.38**	nr	**27.43**
mortice lock or latch and furniture.............	-	-	**-**	2.00	0.32	**31.85**	0.46	**32.31**	nr	**36.35**
butt hinge	-	-	**-**	0.50	0.08	**7.96**	0.22	**8.18**	nr	**9.20**
Piece in frame where ironmongery removed										
lock or latch keep	-	-	**-**	0.50	0.08	**7.96**	0.22	**8.18**	nr	**9.20**
butt hinge	-	-	**-**	0.50	0.08	**7.96**	0.22	**8.18**	nr	**9.20**
Take off door, dismantle as necessary, cramp and re-wedge with new glued wedges, pin and re-hang										
762 x 1981mm, panelled	-	-	**-**	5.00	0.84	**80.05**	0.16	**80.21**	nr	**90.23**
914 x 2134mm, panelled	-	-	**-**	6.00	1.00	**95.97**	0.16	**96.13**	nr	**108.15**
762 x 1981mm, glazed panelled...................	-	-	**-**	5.50	0.90	**87.79**	0.16	**87.95**	nr	**98.95**
914 x 2134mm, glazed panelled...................	-	-	**-**	6.50	1.08	**103.93**	0.16	**104.09**	nr	**117.10**
Take off softwood panelled door 762 x 1981mm, dismantle as necessary and replace the following damaged members and re-hang										
38 x 125mm top rail	5.39	10.00	**5.93**	2.15	0.36	**34.41**	0.45	**40.78**	nr	**45.88**
38 x 125mm hanging stile........................	14.01	10.00	**15.41**	2.36	0.39	**37.71**	1.14	**54.26**	nr	**61.04**
38 x 125mm locking stile........................	14.01	10.00	**15.41**	2.36	0.39	**37.71**	1.14	**54.26**	nr	**61.04**
38 x 200mm middle rail	8.42	10.00	**9.26**	4.23	0.71	**67.71**	0.74	**77.71**	nr	**87.43**
38 x 200mm bottom rail	8.42	10.00	**9.26**	2.20	0.37	**35.22**	0.74	**45.23**	nr	**50.88**
Take off afromosia panelled door 762 x 1981mm, dismantle as necessary and replace the following damaged members and re-hang										
38 x 125mm top rail	27.75	10.00	**30.53**	5.20	0.87	**83.21**	0.45	**114.19**	nr	**128.46**
38 x 125mm hanging stile........................	72.15	10.00	**79.36**	5.36	0.89	**85.70**	1.14	**166.20**	nr	**186.98**
38 x 125mm locking stile........................	72.15	10.00	**79.36**	5.36	0.89	**85.70**	1.14	**166.20**	nr	**186.98**
38 x 200mm middle rail	43.93	10.00	**48.32**	10.40	1.73	**166.31**	0.74	**215.37**	nr	**242.30**
38 x 200mm bottom rail	43.93	10.00	**48.32**	5.30	0.88	**84.74**	0.74	**133.80**	nr	**150.53**
Take off damaged softwood stop and replace with new										
12 x 38mm	0.46	10.00	**0.51**	0.35	0.06	**5.62**	0.10	**6.22**	m	**7.00**
25 (fin) x 50mm glue and screw on...............	0.69	10.00	**0.76**	0.45	0.08	**7.25**	0.12	**8.13**	m	**9.15**

Labour hourly rates: (except Specialists) Craft Operatives £14.21 Labourer £10.71 Rates are national average prices. Refer to REGIONAL VARIATIONS for indicative levels of overall pricing in regions	MATERIALS			LABOUR				RATES		
	Del to Site	Waste	Material Cost	Craft Optve	Lab	Labour Cost	Sunds	Nett Rate	Unit	Gross Rate (+12.5%)
	£	%	£	Hrs	Hrs	£	£	£		£
C51: REPAIRING/RENOVATING/CONSERVING TIMBER Cont.										
Repairing doors and frames Cont.										
Take off damaged afromosia stop and replace with new										
12 x 38mm	1.85	10.00	**2.04**	0.45	0.08	**7.25**	0.10	**9.39**	m	**10.56**
25 (fin) x 50mm glue and screw on	3.70	10.00	**4.07**	0.69	0.12	**11.09**	0.12	**15.28**	m	**17.19**
Flush facing old doors										
Take off panelled door, remove ironmongery, cover both sides with 3.2mm hardboard; refix ironmongery, adjust stops and re-hang										
762 x 1981mm	7.20	10.00	**7.92**	3.10	0.50	**49.41**	0.25	**57.58**	nr	**64.77**
838 x 1981mm	7.20	10.00	**7.92**	3.15	0.51	**50.22**	0.25	**58.39**	nr	**65.69**
Take off panelled door, remove ironmongery, cover one side with 3.2mm hardboard and other side with 6mm fire resistant sheeting screwed on; refix ironmongery, adjust stops and re-hang										
762 x 1981mm	45.30	10.00	**49.83**	4.00	0.67	**64.02**	0.82	**114.67**	nr	**129.00**
838 x 1981mm	45.30	10.00	**49.83**	4.05	0.68	**64.83**	0.82	**115.48**	nr	**129.92**
Take off panelled door, remove ironmongery, cover one side with 3.2mm hardboard, fill panels on other side with 6mm fire resistant sheeting and cover with 6mm fire resistant sheeting screwed on; refix ironmongery, adjust stops and re-hang										
762 x 1981mm	66.15	10.00	**72.77**	5.00	0.83	**79.94**	1.05	**153.75**	nr	**172.97**
838 x 1981mm	66.15	10.00	**72.77**	5.05	0.84	**80.76**	1.05	**154.57**	nr	**173.89**
Take off flush door, remove ironmongery, cover one side with 6mm fire resistant sheeting screwed on; refix ironmongery and re-hang										
762 x 1981mm	41.70	10.00	**45.87**	2.50	0.42	**40.02**	0.53	**86.42**	nr	**97.23**
838 x 1981mm	41.70	10.00	**45.87**	2.50	0.43	**40.13**	0.53	**86.53**	nr	**97.35**
Easing and overhauling windows										
Ease and adjust and oil ironmongery										
casement opening light	-	-	**-**	0.41	0.07	**6.58**	0.02	**6.60**	nr	**7.42**
double hung sash and renew sash lines with best quality flax cord	1.18	10.00	**1.30**	1.64	0.27	**26.20**	0.05	**27.55**	nr	**30.99**
Refixing removed windows										
Windows										
casement opening light and oil ironmongery	-	-	**-**	1.03	0.17	**16.46**	0.04	**16.50**	nr	**18.56**
double hung sash and provide new best quality flax sash cord	1.18	10.00	**1.30**	1.70	0.28	**27.16**	0.05	**28.51**	nr	**32.07**
Single or multi-light timber casements and frames										
440 x 920mm	-	-	**-**	4.00	0.66	**63.91**	0.04	**63.95**	nr	**71.94**
1225 x 1070mm	-	-	**-**	6.00	1.00	**95.97**	0.05	**96.02**	nr	**108.02**
2395 x 1225mm	-	-	**-**	8.00	1.33	**127.92**	0.06	**127.98**	nr	**143.98**
Repairing windows										
Remove damaged members of cased frame and replace with new softwood										
parting bead	0.58	10.00	**0.64**	1.16	0.19	**18.52**	0.05	**19.21**	m	**21.61**
pulley stile	2.25	10.00	**2.48**	1.68	0.28	**26.87**	0.14	**29.49**	m	**33.17**
stop bead	0.75	10.00	**0.83**	0.62	0.10	**9.88**	0.05	**10.76**	m	**12.10**
Take out sash not exceeding 50mm thick, dismantle as necessary, replace the following damaged members in softwood and re-hang with new best quality flax sash cord										
glazing bar	2.80	10.00	**3.08**	3.14	0.52	**50.19**	0.05	**53.32**	m	**59.99**
stile	3.43	10.00	**3.78**	3.03	0.50	**48.41**	0.09	**52.28**	m	**58.81**
rail	5.22	10.00	**5.75**	2.80	0.45	**44.61**	0.25	**50.60**	m	**56.93**
Take out casement opening light not exceeding 50mm thick, dismantle as necessary, replace the following damaged members in softwood and re-hang										
glazing bar	1.62	10.00	**1.78**	1.97	0.32	**31.42**	0.05	**33.25**	m	**37.41**
hanging stile	2.25	10.00	**2.48**	1.85	0.30	**29.50**	0.10	**32.08**	m	**36.09**
shutting stile	2.25	10.00	**2.48**	1.85	0.30	**29.50**	0.10	**32.08**	m	**36.09**
rail	4.04	10.00	**4.44**	1.61	0.27	**25.77**	0.26	**30.47**	m	**34.28**
Take out sash, cramp and re-wedge with new glued wedges, pin and re-hang										
sliding sash	-	-	**-**	5.65	0.95	**90.46**	0.87	**91.33**	nr	**102.75**
opening casement light	-	-	**-**	5.00	0.85	**80.15**	0.87	**81.02**	nr	**91.15**
Repair corner of sash with 100 x 100mm angle repair plate										
let in flush	0.42	-	**0.42**	1.00	0.17	**16.03**	0.17	**16.62**	nr	**18.70**
Carefully rake out and dry crack in sill and fill with two part wood repair system	-	-	**-**	0.50	0.08	**7.96**	1.50	**9.46**	m	**10.64**

Labour hourly rates: (except Specialists) Craft Operatives £14.21 Labourer £10.71 Rates are national average prices. Refer to REGIONAL VARIATIONS for indicative levels of overall pricing in regions	MATERIALS			LABOUR				RATES		
	Del to Site £	Waste %	Material Cost £	Craft Optve Hrs	Lab Hrs	Labour Cost £	Sunds £	Nett Rate £	Unit	Gross Rate (+12.5%) £
C51: REPAIRING/RENOVATING/CONSERVING TIMBER Cont.										
Repairs to softwood skirtings, picture rails, architraves, etc.										
Take off and refix										
skirting	-	-	-	0.40	0.07	**6.43**	0.04	**6.47**	m	**7.28**
dado or picture rail	-	-	-	0.35	0.06	**5.62**	0.03	**5.65**	m	**6.35**
architrave	-	-	-	0.26	0.05	**4.23**	0.03	**4.26**	m	**4.79**
Cut out damaged section and piece in new										
19 x 100mm square skirting	1.04	10.00	**1.14**	0.69	0.11	**10.98**	0.20	**12.33**	m	**13.87**
25 x 150mm square skirting	2.03	10.00	**2.23**	0.71	0.12	**11.37**	0.24	**13.85**	m	**15.58**
19 x 125mm moulded skirting	4.60	10.00	**5.06**	0.70	0.11	**11.13**	0.23	**16.42**	m	**18.47**
19 x 50mm moulded dado or picture rail	2.47	10.00	**2.72**	0.74	0.12	**11.80**	0.10	**14.62**	m	**16.44**
25 x 50mm moulded or splayed architrave	3.33	10.00	**3.66**	0.51	0.09	**8.21**	0.12	**11.99**	m	**13.49**
38 x 75mm moulded or splayed architrave	4.79	10.00	**5.27**	0.52	0.09	**8.35**	0.23	**13.85**	m	**15.58**
Cut out rotten or infected skirting and grounds and replace with new treated softwood										
19 x 100mm square skirting	1.48	10.00	**1.63**	1.00	0.17	**16.03**	0.27	**17.93**	m	**20.17**
25 x 150mm square skirting	2.47	10.00	**2.72**	1.07	0.18	**17.13**	0.37	**20.22**	m	**22.75**
19 x 125mm moulded skirting	5.04	10.00	**5.54**	1.03	0.17	**16.46**	0.31	**22.31**	m	**25.10**
Repairs to staircase members in softwood										
Cut out damaged members of staircase and piece in new										
32 x 275mm tread	7.89	10.00	**8.68**	4.33	0.72	**69.24**	0.46	**78.38**	m	**88.18**
38 x 275mm tread	9.38	10.00	**10.32**	4.85	0.81	**77.59**	0.53	**88.44**	m	**99.50**
19 x 200mm riser	3.84	10.00	**4.22**	2.25	0.38	**36.04**	0.18	**40.45**	m	**45.50**
25 x 200mm riser	4.95	10.00	**5.45**	3.03	0.51	**48.52**	0.27	**54.23**	m	**61.01**
Cut back old tread for a width of 125mm and provide and fix new nosing with glued and dowelled joints to tread										
32mm with rounded nosing	4.24	10.00	**4.66**	2.16	0.36	**34.55**	0.76	**39.97**	m	**44.97**
38mm with rounded nosing	4.90	10.00	**5.39**	2.55	0.43	**40.84**	0.85	**47.08**	m	**52.97**
32mm with moulded nosing	5.04	10.00	**5.54**	2.54	0.42	**40.59**	0.76	**46.90**	m	**52.76**
38mm with moulded nosing	5.72	10.00	**6.29**	3.06	0.51	**48.94**	0.90	**56.14**	m	**63.15**
Replace missing square bar baluster 914mm high; clean out old mortices										
25 x 25mm	0.75	-	**0.75**	0.56	0.09	**8.92**	-	**9.67**	nr	**10.88**
38 x 38mm	1.36	-	**1.36**	0.75	0.13	**12.05**	-	**13.41**	nr	**15.09**
Cut out damaged square bar baluster 914mm high and piece in new										
25 x 25mm	0.75	-	**0.75**	0.75	0.03	**10.98**	0.03	**11.76**	nr	**13.23**
38 x 38mm	1.36	-	**1.36**	0.91	0.15	**14.54**	0.05	**15.95**	nr	**17.94**
Cut out damaged handrail and piece in new										
50mm mopstick	25.21	10.00	**27.73**	1.95	0.33	**31.24**	3.20	**62.17**	m	**69.95**
75 x 100mm moulded	42.04	10.00	**46.24**	3.97	0.66	**63.48**	3.47	**113.20**	m	**127.35**
Refixing removed handrail										
to new balustrade	-	-	-	0.69	0.12	**11.09**	-	**11.09**	m	**12.48**
Repairs to staircase members in oak										
Cut out damaged members of staircase and piece in new										
32 x 275mm tread	45.40	10.00	**49.94**	6.09	1.02	**97.46**	0.46	**147.86**	m	**166.35**
38 x 275mm tread	54.71	10.00	**60.18**	7.18	1.20	**114.88**	0.53	**175.59**	m	**197.54**
19 x 200mm riser	24.45	10.00	**26.90**	3.34	0.56	**53.46**	0.18	**80.53**	m	**90.60**
25 x 200mm riser	30.27	10.00	**33.30**	4.56	0.76	**72.94**	0.29	**106.52**	m	**119.84**
Cut back old tread for a width of 125mm and provide and fix new nosing with glued and dowelled joints to tread										
32mm with rounded nosing	26.76	10.00	**29.44**	2.23	0.37	**35.65**	0.76	**65.85**	m	**74.08**
38mm with rounded nosing	30.27	10.00	**33.30**	2.92	0.49	**46.74**	0.90	**80.94**	m	**91.06**
32mm with moulded nosing	27.94	10.00	**30.73**	3.20	0.53	**51.15**	0.76	**82.64**	m	**92.97**
38mm with moulded nosing	31.43	10.00	**34.57**	3.95	0.66	**63.20**	0.90	**98.67**	m	**111.00**
Replace missing square bar baluster 914mm high; clean out old mortices										
25 x 25mm	3.95	-	**3.95**	0.94	0.16	**15.07**	-	**19.02**	nr	**21.40**
38 x 38mm	8.03	-	**8.03**	0.99	0.17	**15.89**	-	**23.92**	nr	**26.91**
Cut out damaged square bar baluster 914mm high and piece in new										
25 x 25mm	3.95	-	**3.95**	1.21	0.20	**19.34**	0.03	**23.32**	nr	**26.23**
38 x 38mm	8.03	-	**8.03**	1.32	0.22	**21.11**	0.05	**29.19**	nr	**32.84**
Cut out damaged handrail and piece in new										
50mm mopstick	55.49	10.00	**61.04**	3.64	0.61	**58.26**	3.20	**122.50**	m	**137.81**
75 x 100mm moulded	92.47	10.00	**101.72**	5.50	0.92	**88.01**	3.47	**193.20**	m	**217.34**
Refixing removed handrail										
to new balustrade	-	-	-	1.35	0.23	**21.65**	-	**21.65**	m	**24.35**
Remove broken ironmongery and fix only new to softwood										
Door ironmongery										
75mm butt hinges	-	-	-	0.23	0.04	**3.70**	0.01	**3.71**	nr	**4.17**
100mm butt hinges	-	-	-	0.23	0.04	**3.70**	0.01	**3.71**	nr	**4.17**
tee hinges up to 300mm	-	-	-	0.62	0.10	**9.88**	0.01	**9.89**	nr	**11.13**

Labour hourly rates: (except Specialists) Craft Operatives £14.21 Labourer £10.71 Rates are national average prices. Refer to REGIONAL VARIATIONS for indicative levels of overall pricing in regions	MATERIALS			LABOUR			RATES			
	Del to Site £	Waste %	Material Cost £	Craft Optve Hrs	Lab Hrs	Labour Cost £	Sunds £	Nett Rate £	Unit	Gross Rate (+12.5%) £
C51: REPAIRING/RENOVATING/CONSERVING TIMBER Cont.										
Remove broken ironmongery and fix only new to softwood Cont.										
Door ironmongery Cont.										
single action floor spring hinges	-	-	-	3.00	0.50	**47.99**	1.30	**49.29**	nr	**55.45**
barrel bolt up to 300mm	-	-	-	0.41	0.07	**6.58**	0.01	**6.59**	nr	**7.41**
lock or latch furniture	-	-	-	0.44	0.07	**7.00**	0.01	**7.01**	nr	**7.89**
rim lock	-	-	-	0.80	0.13	**12.76**	0.01	**12.77**	nr	**14.37**
rim latch	-	-	-	0.80	0.13	**12.76**	0.01	**12.77**	nr	**14.37**
mortice lock	-	-	-	1.22	0.20	**19.48**	0.01	**19.49**	nr	**21.92**
mortice latch	-	-	-	1.08	0.18	**17.27**	0.01	**17.28**	nr	**19.45**
pull handle	-	-	-	0.40	0.07	**6.43**	0.03	**6.46**	nr	**7.27**
Window ironmongery										
sash fastener	-	-	-	1.09	0.18	**17.42**	0.01	**17.43**	nr	**19.61**
sash lift	-	-	-	0.23	0.04	**3.70**	0.01	**3.71**	nr	**4.17**
casement stay	-	-	-	0.40	0.07	**6.43**	0.01	**6.44**	nr	**7.25**
casement fastener	-	-	-	0.40	0.07	**6.43**	0.01	**6.44**	nr	**7.25**
Sundry ironmongery										
hat and coat hook	-	-	-	0.29	0.05	**4.66**	0.01	**4.67**	nr	**5.25**
toilet roll holder	-	-	-	0.29	0.05	**4.66**	0.01	**4.67**	nr	**5.25**
Remove broken ironmongery and fix only new to hardwood										
Door ironmongery										
75mm butt hinges	-	-	-	0.30	0.05	**4.80**	0.01	**4.81**	nr	**5.41**
100mm butt hinges	-	-	-	0.30	0.05	**4.80**	0.01	**4.81**	nr	**5.41**
tee hinges up to 300mm	-	-	-	0.86	0.14	**13.72**	0.01	**13.73**	nr	**15.45**
single action floor spring hinges	-	-	-	4.20	0.70	**67.18**	1.30	**68.48**	nr	**77.04**
barrel bolt up to 300mm	-	-	-	0.60	0.10	**9.60**	0.01	**9.61**	nr	**10.81**
lock or latch furniture	-	-	-	0.61	0.10	**9.74**	0.01	**9.75**	nr	**10.97**
rim lock	-	-	-	1.15	0.19	**18.38**	0.01	**18.39**	nr	**20.68**
rim latch	-	-	-	1.15	0.19	**18.38**	0.01	**18.39**	nr	**20.68**
mortice lock	-	-	-	1.77	0.30	**28.36**	0.01	**28.37**	nr	**31.92**
mortice latch	-	-	-	1.58	0.26	**25.24**	0.01	**25.25**	nr	**28.40**
pull handle	-	-	-	0.57	0.10	**9.17**	0.03	**9.20**	nr	**10.35**
Window ironmongery										
sash fastener	-	-	-	1.57	0.26	**25.09**	0.01	**25.10**	nr	**28.24**
sash lift	-	-	-	0.31	0.05	**4.94**	0.01	**4.95**	nr	**5.57**
casement stay	-	-	-	0.56	0.09	**8.92**	0.01	**8.93**	nr	**10.05**
casement fastener	-	-	-	0.56	0.09	**8.92**	0.01	**8.93**	nr	**10.05**
Sundry ironmongery										
hat and coat hook	-	-	-	0.42	0.07	**6.72**	0.01	**6.73**	nr	**7.57**
toilet roll holder	-	-	-	0.42	0.07	**6.72**	0.01	**6.73**	nr	**7.57**
C52: FUNGUS/BEETLE ERADICATION										
Protective treatment of existing timbers										
Treat with two coats of spray applied preservative										
boarding	0.60	10.00	**0.66**	-	0.15	**1.61**	0.04	**2.30**	m2	**2.59**
structural timbers	0.89	10.00	**0.98**	-	0.20	**2.14**	0.05	**3.17**	m2	**3.57**
Treat with two coats of brush applied preservative										
boarding	0.89	10.00	**0.98**	-	0.26	**2.78**	0.03	**3.80**	m2	**4.27**
structural timbers	1.37	10.00	**1.51**	-	0.80	**8.57**	0.05	**10.12**	m2	**11.39**
Insecticide treatment of existing timbers										
Treat worm infected timbers with spray applied proprietary insecticide										
boarding	0.89	10.00	**0.98**	-	0.24	**2.57**	0.05	**3.60**	m2	**4.05**
structural timbers	1.19	10.00	**1.31**	-	0.28	**3.00**	0.06	**4.37**	m2	**4.92**
joinery timbers	1.19	10.00	**1.31**	-	0.32	**3.43**	0.06	**4.80**	m2	**5.40**
Treat worm infected timbers with brush applied proprietary insecticide										
boarding	1.19	10.00	**1.31**	-	0.36	**3.86**	0.03	**5.20**	m2	**5.85**
structural timbers	1.43	10.00	**1.57**	-	0.42	**4.50**	0.05	**6.12**	m2	**6.89**
joinery timbers	1.43	10.00	**1.57**	-	0.48	**5.14**	0.05	**6.76**	m2	**7.61**
Treatment of wall surfaces										
Treat surfaces of concrete and brickwork adjoining areas where infected timbers removed										
with a blow lamp	-	-	-	-	0.25	**2.68**	0.03	**2.71**	m2	**3.05**
C90: ALTERATIONS - SPOT ITEMS										
Removal of fittings and fixtures										
Removing fittings and fixtures										
cupboard with doors, frames, architraves, etc	-	-	-	-	0.59	**6.32**	1.23	**7.55**	m2	**8.50**
worktop with legs and bearers	-	-	-	-	0.56	**6.00**	1.65	**7.64**	m2	**8.60**
shelving exceeding 300mm wide with bearers and brackets	-	-	-	-	0.45	**4.82**	1.04	**5.86**	m2	**6.60**
shelving not exceeding 300mm wide with bearers and brackets	-	-	-	-	0.20	**2.14**	0.33	**2.47**	m	**2.78**
draining board and bearers										
500 x 600mm	-	-	-	-	0.28	**3.00**	0.63	**3.63**	nr	**4.08**
600 x 900mm	-	-	-	-	0.37	**3.96**	1.11	**5.08**	nr	**5.71**

Labour hourly rates: (except Specialists) Craft Operatives £14.21 Labourer £10.71 Rates are national average prices. Refer to REGIONAL VARIATIONS for indicative levels of overall pricing in regions	MATERIALS			LABOUR				RATES		
	Del to Site £	Waste %	Material Cost £	Craft Optve Hrs	Lab Hrs	Labour Cost £	Sunds £	Nett Rate £	Unit	Gross Rate (+12.5%) £
C90: ALTERATIONS - SPOT ITEMS Cont.										
Removal of fittings and fixtures Cont.										
Removing fittings and fixtures Cont.										
wall cupboard unit										
510 x 305 x 305mm	-	-	-	-	0.47	**5.03**	1.00	**6.03**	nr	**6.78**
510 x 305 x 610mm	-	-	-	-	0.65	**6.96**	1.93	**8.89**	nr	**10.00**
1020 x 305 x 610mm	-	-	-	-	0.94	**10.07**	3.92	**13.99**	nr	**15.74**
1220 x 305 x 610mm	-	-	-	-	1.15	**12.32**	4.67	**16.99**	nr	**19.11**
floor cupboard unit										
510 x 510 x 915mm	-	-	-	-	0.92	**9.85**	4.91	**14.76**	nr	**16.60**
610 x 510 x 915mm	-	-	-	-	1.21	**12.96**	5.85	**18.81**	nr	**21.16**
1020 x 510 x 915mm	-	-	-	-	1.65	**17.67**	9.81	**27.48**	nr	**30.92**
1220 x 510 x 915mm	-	-	-	-	1.93	**20.67**	11.76	**32.43**	nr	**36.48**
sink base unit										
585 x 510 x 895mm	-	-	-	-	0.92	**9.85**	5.53	**15.39**	nr	**17.31**
1070 x 510 x 895mm	-	-	-	-	1.65	**17.67**	10.07	**27.74**	nr	**31.21**
tall cupboard unit										
510 x 510 x 1505mm	-	-	-	-	1.56	**16.71**	8.08	**24.79**	nr	**27.89**
610 x 510 x 1505mm	-	-	-	-	1.77	**18.96**	9.66	**28.61**	nr	**32.19**
bath panel and bearers	-	-	-	-	0.47	**5.03**	1.04	**6.08**	nr	**6.84**
steel or concrete clothes line post with concrete base	-	-	-	-	0.50	**5.36**	2.06	**7.42**	nr	**8.34**

This section continues on the next page

Labour hourly rates: (except Specialists) Craft Operatives £17.38 Labourer £10.71 Rates are national average prices. Refer to REGIONAL VARIATIONS for indicative levels of overall pricing in regions	MATERIALS			LABOUR				RATES		
	Del to Site £	Waste %	Material Cost £	Craft Optve Hrs	Lab Hrs	Labour Cost £	Sunds £	Nett Rate £	Unit	Gross Rate (+12.5%) £
C90: ALTERATIONS - SPOT ITEMS										
Removal of plumbing and electrical installations										
Removing plumbing and electrical installations										
100mm cast iron rainwater gutters	-	-	-	0.20	-	3.48	0.20	3.68	m	4.14
75mm cast iron rainwater pipes	-	-	-	0.25	-	4.35	0.23	4.57	m	5.14
100mm cast iron rainwater pipes	-	-	-	0.25	-	4.35	0.38	4.72	m	5.31
cast iron rainwater head	-	-	-	0.30	-	5.21	0.63	5.84	nr	6.57
cast iron soil and vent pipe with caulked joints										
50mm	-	-	-	0.35	-	6.08	0.12	6.20	m	6.98
75mm	-	-	-	0.35	-	6.08	0.23	6.31	m	7.10
100mm	-	-	-	0.35	-	6.08	0.38	6.46	m	7.27
150mm	-	-	-	0.50	-	8.69	0.90	9.59	m	10.79
100mm asbestos cement soil and vent pipe	-	-	-	0.25	-	4.35	0.38	4.72	m	5.31
150mm asbestos cement soil and vent pipe	-	-	-	0.25	-	4.35	0.90	5.25	m	5.90
100mm p.v.c. soil and vent pipe	-	-	-	0.25	-	4.35	0.38	4.72	m	5.31
150mm p.v.c. soil and vent pipe	-	-	-	0.25	-	4.35	0.90	5.25	m	5.90
100mm lead soil and vent pipe	-	-	-	0.35	-	6.08	0.38	6.46	m	7.27
150mm lead soil and vent pipe	-	-	-	0.35	-	6.08	0.90	6.98	m	7.86
w.c. suites	-	-	-	1.55	-	26.94	3.86	30.80	nr	34.65
lavatory basins	-	-	-	1.35	-	23.46	2.32	25.79	nr	29.01
baths	-	-	-	1.55	-	26.94	7.74	34.68	nr	39.01
glazed ware sinks	-	-	-	1.35	-	23.46	3.86	27.33	nr	30.74
stainless steel sinks	-	-	-	1.35	-	23.46	2.32	25.79	nr	29.01
stainless steel sinks with single drainer	-	-	-	1.35	-	23.46	3.10	26.57	nr	29.89
stainless steel sinks with double drainer	-	-	-	1.35	-	23.46	3.86	27.33	nr	30.74
hot and cold water, waste pipes, etc. 8 - 25mm diameter										
copper	-	-	-	0.15	-	2.61	0.05	2.65	m	2.99
lead	-	-	-	0.15	-	2.61	0.05	2.65	m	2.99
polythene or p.v.c.	-	-	-	0.15	-	2.61	0.05	2.65	m	2.99
stainless steel	-	-	-	0.25	-	4.35	0.05	4.39	m	4.94
steel	-	-	-	0.25	-	4.35	0.05	4.39	m	4.94
hot and cold water, waste pipes, etc. 32 - 50mm diameter										
copper	-	-	-	0.20	-	3.48	0.12	3.59	m	4.04
lead	-	-	-	0.20	-	3.48	0.12	3.59	m	4.04
polythene or p.v.c.	-	-	-	0.20	-	3.48	0.12	3.59	m	4.04
stainless steel	-	-	-	0.30	-	5.21	0.12	5.33	m	6.00
steel	-	-	-	0.30	-	5.21	0.12	5.33	m	6.00
cold water cisterns up to 454 litres capacity	-	-	-	2.00	-	34.76	17.56	52.32	nr	58.86
hot water tanks or cylinders up to 227 litre capacity	-	-	-	2.00	-	34.76	8.77	43.53	nr	48.97
solid fuel boilers up to 66000 BTU output	-	-	-	3.75	-	65.18	-	65.18	nr	73.32
gas or oil fired boilers up to 66000 BTU output	-	-	-	3.75	-	65.18	-	65.18	nr	73.32
wall mounted water heaters	-	-	-	2.00	-	34.76	-	34.76	nr	39.11
wall type radiators up to 914mm long, up to 610mm high	-	-	-	1.00	-	17.38	0.78	18.16	nr	20.43
wall type radiators up to 914mm long, 610 - 914mm high	-	-	-	1.35	-	23.46	1.55	25.02	nr	28.14
wall type radiators 914 - 1829mm long, up to 610mm high	-	-	-	1.55	-	26.94	2.37	29.31	nr	32.97
wall type radiators 914 - 1829mm long, 610 - 914mm high	-	-	-	1.95	-	33.89	3.10	37.00	nr	41.62
column type radiators up to 914mm long, up to 610mm high	-	-	-	1.00	-	17.38	4.06	21.44	nr	24.13
column type radiators up to 914mm long, 610 - 914mm high	-	-	-	1.35	-	23.46	8.12	31.58	nr	35.53
column type radiators 914 - 1829mm long, up to 610mm high	-	-	-	1.55	-	26.94	12.18	39.12	nr	44.01
column type radiators 914 - 1829mm long, 610 - 914mm high	-	-	-	1.95	-	33.89	16.23	50.13	nr	56.39
pipe insulation from pipes 8 - 25mm diameter	-	-	-	0.10	-	1.74	0.12	1.86	m	2.09
pipe insulation from pipes 32 - 50mm diameter	-	-	-	0.13	-	2.26	0.23	2.48	m	2.80
water tank or calorifier insulation from tanks up to 227 litres	-	-	-	0.50	-	8.69	8.77	17.46	nr	19.64
water tank or calorifier insulation from tanks 227 - 454 litres	-	-	-	0.75	-	13.04	17.56	30.60	nr	34.42
Removing plumbing and electrical installations; setting aside for re-use										
cast iron soil and vent pipe with caulked joints										
50mm	-	-	-	0.55	-	9.56	-	9.56	m	10.75
75mm	-	-	-	0.60	-	10.43	-	10.43	m	11.73
100mm	-	-	-	0.80	-	13.90	-	13.90	m	15.64
150mm	-	-	-	1.00	-	17.38	-	17.38	m	19.55

This section continues on the next page

Labour hourly rates: (except Specialists) Craft Operatives £20.54 Labourer £10.71 Rates are national average prices. Refer to REGIONAL VARIATIONS for indicative levels of overall pricing in regions	MATERIALS			LABOUR				RATES		
	Del to Site £	Waste %	Material Cost £	Craft Optve Hrs	Lab Hrs	Labour Cost £	Sunds £	Nett Rate £	Unit	Gross Rate (+12.5%) £
C90: ALTERATIONS - SPOT ITEMS Cont.										
Removal of plumbing and electrical installations Cont.										
Removing plumbing and electrical installations										
wall mounted electric fire	-	-	-	2.00	-	**41.08**	2.06	**43.14**	nr	**48.53**
1.5 kW night storage heater	-	-	-	2.50	-	**51.35**	6.19	**57.54**	nr	**64.73**
2 KW night storage heater	-	-	-	2.50	-	**51.35**	8.25	**59.60**	nr	**67.05**
3 KW night storage heater	-	-	-	2.50	-	**51.35**	12.37	**63.72**	nr	**71.69**
Removing plumbing and electrical installations; extending and making good finishings										
lighting points	-	-	-	1.25	-	**25.68**	1.68	**27.36**	nr	**30.78**
flush type switches	-	-	-	1.25	-	**25.68**	1.68	**27.36**	nr	**30.78**
surface mounted type switches	-	-	-	1.00	-	**20.54**	1.35	**21.89**	nr	**24.63**
flush type socket outlets	-	-	-	1.25	-	**25.68**	1.68	**27.36**	nr	**30.78**
surface mounted type socket outlets	-	-	-	1.00	-	**20.54**	1.35	**21.89**	nr	**24.63**
flush type fitting points	-	-	-	1.25	-	**25.68**	1.68	**27.36**	nr	**30.78**
surface mounted type fitting points	-	-	-	1.00	-	**20.54**	1.35	**21.89**	nr	**24.63**
surface mounted p.v.c. insulated and sheathed cables	-	-	-	0.04	-	**0.82**	0.07	**0.89**	m	**1.00**
surface mounted mineral insulated copper sheathed cables	-	-	-	0.10	-	**2.05**	0.14	**2.20**	m	**2.47**
conduits up to 25mm diameter with junction boxes	-	-	-	0.15	-	**3.08**	0.20	**3.28**	m	**3.69**
surface mounted cable trunking up to 100 x 100mm	-	-	-	0.55	-	**11.30**	0.73	**12.03**	m	**13.54**

This section continues on the next page

Labour hourly rates: (except Specialists) Craft Operatives £14.21 Labourer £10.71 Rates are national average prices. Refer to REGIONAL VARIATIONS for indicative levels of overall pricing in regions	MATERIALS			LABOUR				RATES		
	Del to Site £	Waste %	Material Cost £	Craft Optve Hrs	Lab Hrs	Labour Cost £	Sunds £	Nett Rate £	Unit	Gross Rate (+12.5%) £
C90: ALTERATIONS - SPOT ITEMS Cont.										
Removal of floor, wall and ceiling finishings										
Removing finishings										
cement and sand to floors	-	-	-	-	0.86	9.21	2.06	11.27	m2	12.68
granolithic to floors	-	-	-	-	0.92	9.85	2.06	11.92	m2	13.40
granolithic to treads and risers	-	-	-	-	1.43	15.32	2.06	17.38	m2	19.55
plastic or similar tiles to floors	-	-	-	-	0.20	2.14	0.21	2.36	m2	2.65
plastic or similar tiles to floors; cleaning off for new	-	-	-	-	0.47	5.03	0.21	5.25	m2	5.90
plastic or similar tiles to floors and screed under	-	-	-	-	1.05	11.25	2.28	13.52	m2	15.21
linoleum and underlay to floors	-	-	-	-	0.07	0.75	2.06	2.81	m2	3.16
ceramic or quarry tiles to floors	-	-	-	-	1.21	12.96	1.23	14.19	m2	15.97
ceramic or quarry tiles to floors and screed under	-	-	-	-	2.39	25.60	3.29	28.89	m2	32.50
extra; cleaning and keying surface of concrete under	-	-	-	-	0.53	5.68	0.05	5.72	m2	6.44
asphalt to floors on loose underlay	-	-	-	-	0.35	3.75	0.83	4.58	m2	5.15
asphalt to floors keyed to concrete or screed	-	-	-	-	0.76	8.14	0.83	8.97	m2	10.09
granolithic skirtings	-	-	-	-	0.33	3.53	0.21	3.75	m	4.22
plastic or similar skirtings; cleaning off for new	-	-	-	-	0.13	1.39	0.05	1.44	m	1.62
timber skirtings	-	-	-	-	0.12	1.29	0.21	1.50	m	1.69
ceramic or quarry tile skirtings	-	-	-	-	0.43	4.61	0.24	4.84	m	5.45
plaster to walls	-	-	-	-	0.60	6.43	0.83	7.26	m2	8.16
rendering to walls	-	-	-	-	0.87	9.32	0.83	10.15	m2	11.42
rough cast to walls	-	-	-	-	0.90	9.64	1.04	10.68	m2	12.02
match boarding linings to walls with battens	-	-	-	-	0.38	4.07	1.23	5.30	m2	5.96
plywood or similar sheet linings to walls with battens	-	-	-	-	0.31	3.32	1.23	4.55	m2	5.12
insulating board linings to walls with battens	-	-	-	-	0.31	3.32	1.23	4.55	m2	5.12
plasterboard to walls	-	-	-	-	0.31	3.32	0.63	3.95	m2	4.44
plasterboard dry linings to walls	-	-	-	-	0.50	5.36	0.63	5.98	m2	6.73
plasterboard and skim to walls	-	-	-	-	0.60	6.43	0.83	7.26	m2	8.16
lath and plaster to walls	-	-	-	-	0.44	4.71	1.04	5.76	m2	6.47
metal lath and plaster to walls	-	-	-	-	0.33	3.53	1.04	4.58	m2	5.15
ceramic tiles to walls	-	-	-	-	0.75	8.03	0.63	8.66	m2	9.74
ceramic tiles to walls and backing under	-	-	-	-	1.11	11.89	1.04	12.93	m2	14.55
asphalt coverings to walls keyed to concrete or brickwork	-	-	-	-	0.66	7.07	0.83	7.90	m2	8.89
plaster to ceilings	-	-	-	-	0.74	7.93	0.83	8.75	m2	9.85
match boarding linings to ceilings with battens	-	-	-	-	0.54	5.78	1.23	7.02	m2	7.89
plywood or similar sheet linings to ceilings with battens	-	-	-	-	0.38	4.07	1.23	5.30	m2	5.96
insulating board linings to ceilings with battens	-	-	-	-	0.26	2.78	1.23	4.02	m2	4.52
plasterboard to ceilings	-	-	-	-	0.33	3.53	0.63	4.16	m2	4.68
plasterboard and skim to ceilings	-	-	-	-	0.40	4.28	0.83	5.11	m2	5.75
lath and plaster to ceilings	-	-	-	-	0.59	6.32	1.04	7.36	m2	8.28
metal lath and plaster to ceilings	-	-	-	-	0.47	5.03	1.04	6.08	m2	6.84
Hacking surfaces of concrete as key for new finishings										
ceiling	-	-	-	-	0.67	7.18	0.05	7.22	m2	8.13
floor	-	-	-	-	0.53	5.68	0.05	5.72	m2	6.44
wall	-	-	-	-	0.53	5.68	0.05	5.72	m2	6.44
hacking surfaces of brick wall and raking out joints as key for new finishings	-	-	-	-	0.60	6.43	0.05	6.47	m2	7.28
Hacking surfaces of brick wall and raking out joints as key for new finishings										
wall	-	-	-	-	0.60	6.43	0.05	6.47	m2	7.28
Removing finishings; extending and making good finishings										
plaster cornices up to 100mm girth on face	0.35	10.00	0.38	0.78	0.78	19.44	0.08	19.90	m	22.39
plaster cornices 100 - 200mm girth on face	0.57	10.00	0.62	0.88	0.88	21.93	0.21	22.77	m	25.61
plaster cornices 200 - 300mm girth on face	1.00	10.00	1.10	0.97	0.97	24.17	0.63	25.90	m	29.14
plaster ceiling roses up to 300mm diameter	0.17	10.00	0.19	0.41	0.41	10.22	0.17	10.57	nr	11.89
plaster ceiling roses 300 - 450mm diameter	0.35	10.00	0.38	0.64	0.64	15.95	0.33	16.66	nr	18.74
plaster ceiling roses 450 - 600mm diameter	0.68	10.00	0.75	0.90	0.90	22.43	0.65	23.83	nr	26.80
Removing finishings; carefully handling and disposing toxic or other special waste by approved method										
asbestos cement sheet linings to walls	-	-	-	2.00	2.00	49.84	3.29	53.13	m2	59.78
asbestos cement sheet linings to ceilings	-	-	-	3.00	3.00	74.76	3.29	78.05	m2	87.81
Removal of roof coverings										
Removing coverings										
felt to roofs	-	-	-	-	0.28	3.00	0.83	3.83	m2	4.31
felt skirtings to roofs	-	-	-	-	0.33	3.53	0.08	3.62	m	4.07
asphalt to roofs	-	-	-	-	0.35	3.75	1.04	4.79	m2	5.39
asphalt skirtings to roofs; on expanded metal reinforcement	-	-	-	-	0.20	2.14	0.17	2.31	m	2.60
asphalt skirtings to roofs; keyed to concrete, brickwork, etc.	-	-	-	-	0.30	3.21	0.17	3.38	m	3.80
asphalt coverings to roofs; on expanded metal reinforcement; per 100mm of width	-	-	-	-	0.13	1.39	0.13	1.52	m	1.71
asphalt coverings to roofs; keyed to concrete, brickwork, etc.; per 100mm of width	-	-	-	-	0.20	2.14	0.13	2.27	m	2.56
slate to roofs	-	-	-	-	0.34	3.64	0.83	4.47	m2	5.03
tiles to roofs	-	-	-	-	0.31	3.32	1.23	4.55	m2	5.12
extra; removing battens	-	-	-	-	0.28	3.00	0.83	3.83	m2	4.31
extra; removing counter battens	-	-	-	-	0.10	1.07	0.17	1.24	m2	1.39
extra; removing underfelt	-	-	-	-	0.20	2.14	0.33	2.47	m2	2.78
lead to roofs	-	-	-	-	0.48	5.14	0.83	5.97	m2	6.72
lead flashings to roofs; per 25mm of girth	-	-	-	-	0.05	0.54	0.05	0.58	m	0.66

Labour hourly rates: (except Specialists) Craft Operatives £14.21 Labourer £10.71 Rates are national average prices. Refer to REGIONAL VARIATIONS for indicative levels of overall pricing in regions	MATERIALS			LABOUR				RATES		
	Del to Site £	Waste %	Material Cost £	Craft Optve Hrs	Lab Hrs	Labour Cost £	Sunds £	Nett Rate £	Unit	Gross Rate (+12.5%) £
C90: ALTERATIONS - SPOT ITEMS Cont.										
Removal of roof coverings Cont.										
Removing coverings Cont.										
zinc to roofs	-	-	-	-	0.34	**3.64**	0.83	**4.47**	m2	**5.03**
zinc flashings to roofs; per 25mm of girth	-	-	-	-	0.03	**0.32**	0.05	**0.37**	m	**0.41**
copper to roofs	-	-	-	-	0.34	**3.64**	0.83	**4.47**	m2	**5.03**
copper flashings to roofs; per 25mm of girth	-	-	-	-	0.03	**0.32**	0.05	**0.37**	m	**0.41**
corrugated metal sheeting to roofs	-	-	-	-	0.31	**3.32**	1.23	**4.55**	m2	**5.12**
corrugated translucent sheeting to roofs	-	-	-	-	0.31	**3.32**	1.23	**4.55**	m2	**5.12**
board roof decking	-	-	-	-	0.31	**3.32**	1.23	**4.55**	m2	**5.12**
woodwool roof decking	-	-	-	-	0.39	**4.18**	1.23	**5.41**	m2	**6.09**
Removing coverings; setting aside for re-use										
slate to roofs	-	-	-	-	0.60	**6.43**	-	**6.43**	m2	**7.23**
tiles to roofs	-	-	-	-	0.60	**6.43**	-	**6.43**	m2	**7.23**
clean and stack 405 x 205mm slates; per 100	-	-	-	-	1.50	**16.07**	-	**16.07**	nr	**18.07**
clean and stack 510 x 255mm slates; per 100	-	-	-	-	1.89	**20.24**	-	**20.24**	nr	**22.77**
clean and stack 610 x 305mm slates; per 100	-	-	-	-	2.24	**23.99**	-	**23.99**	nr	**26.99**
clean and stack concrete tiles; per 100	-	-	-	-	2.24	**23.99**	-	**23.99**	nr	**26.99**
clean and stack plain tiles; per 100	-	-	-	-	1.50	**16.07**	-	**16.07**	nr	**18.07**
Removing coverings; carefully handling and disposing toxic or other special waste by approved method										
asbestos cement sheeting to roofs	-	-	-	2.00	2.00	**49.84**	3.29	**53.13**	m2	**59.78**
asbestos cement roof decking	-	-	-	2.00	2.00	**49.84**	3.29	**53.13**	m2	**59.78**
Removal of woodwork										
Removing										
stud partitions plastered both sides	-	-	-	-	0.66	**7.07**	3.10	**10.17**	m2	**11.44**
roof boarding	-	-	-	-	0.19	**2.03**	1.04	**3.08**	m2	**3.46**
roof boarding; prepare joists for new	-	-	-	-	0.45	**4.82**	1.04	**5.86**	m2	**6.60**
gutter boarding and the like	-	-	-	-	0.19	**2.03**	1.04	**3.08**	m2	**3.46**
weather boarding and battens	-	-	-	-	0.24	**2.57**	1.23	**3.80**	m2	**4.28**
tilting fillets, angle fillets and the like	-	-	-	-	0.09	**0.96**	0.08	**1.05**	m	**1.18**
fascia boards 150 mm wide	-	-	-	-	0.10	**1.07**	0.17	**1.24**	m	**1.39**
barge boards 200mm wide	-	-	-	-	0.10	**1.07**	0.21	**1.28**	m	**1.44**
soffit boards 300mm wide	-	-	-	-	0.10	**1.07**	0.33	**1.40**	m	**1.58**
floor boarding	-	-	-	-	0.33	**3.53**	1.04	**4.58**	m2	**5.15**
handrails and brackets	-	-	-	-	0.10	**1.07**	0.65	**1.72**	m	**1.94**
balustrades complete down to and with cappings to aprons or strings	-	-	-	0.58	0.10	**9.31**	1.65	**10.96**	m	**12.33**
newel posts; cut off flush with landing or string	-	-	-	0.48	0.08	**7.68**	0.41	**8.09**	nr	**9.10**
ends of treads projecting beyond face of cut outer string including scotia under; make good treads where balusters removed	-	-	-	0.90	0.15	**14.40**	1.23	**15.63**	m	**17.58**
dado or picture rails with grounds	-	-	-	-	0.10	**1.07**	0.17	**1.24**	m	**1.39**
architrave	-	-	-	-	0.06	**0.64**	0.17	**0.81**	m	**0.91**
window boards and bearers	-	-	-	-	0.12	**1.29**	0.33	**1.62**	m	**1.82**
Removing; setting aside for reuse										
handrails and brackets	-	-	-	0.25	0.04	**3.98**	-	**3.98**	m	**4.48**
Removal of windows and doors										
Removing										
metal windows with internal and external sills; in conjunction with demolition										
997 x 923mm	-	-	-	-	0.40	**4.28**	1.98	**6.26**	nr	**7.05**
1486 x 923mm	-	-	-	-	0.50	**5.36**	2.84	**8.20**	nr	**9.22**
1486 x 1513mm	-	-	-	-	0.60	**6.43**	4.67	**11.09**	nr	**12.48**
1994 x 1513mm	-	-	-	-	0.70	**7.50**	6.23	**13.73**	nr	**15.45**
metal windows with internal and external sills; preparatory to filling openings; cut out lugs										
997 x 923mm	-	-	-	-	1.30	**13.92**	1.98	**15.90**	nr	**17.89**
1486 x 923mm	-	-	-	-	1.60	**17.14**	2.84	**19.98**	nr	**22.48**
1486 x 1513mm	-	-	-	-	2.00	**21.42**	4.67	**26.09**	nr	**29.35**
1994 x 1513mm	-	-	-	-	2.30	**24.63**	6.23	**30.87**	nr	**34.72**
wood single or multi-light casements and frames with internal and external sills; in conjunction with demolition										
440 x 920mm	-	-	-	-	0.28	**3.00**	1.28	**4.28**	nr	**4.81**
1225 x 1070mm	-	-	-	-	0.29	**3.11**	4.04	**7.15**	nr	**8.04**
2395 x 1225mm	-	-	-	-	0.30	**3.21**	9.08	**12.29**	nr	**13.83**
wood single or multi-light casements and frames with internal and external sills; preparatory to filling openings; cut out fixing cramps										
440 x 920mm	-	-	-	-	1.25	**13.39**	1.28	**14.67**	nr	**16.50**
1225 x 1070mm	-	-	-	-	1.70	**18.21**	4.04	**22.25**	nr	**25.03**
2395 x 1225mm	-	-	-	-	2.52	**26.99**	9.08	**36.07**	nr	**40.57**
wood cased frames and sashes with internal and external sills complete with accessories and weights; in conjunction with demolition										
610 x 1225mm	-	-	-	-	0.61	**6.53**	3.10	**9.64**	nr	**10.84**
915 x 1525mm	-	-	-	-	0.66	**7.07**	5.77	**12.84**	nr	**14.44**
1370 x 1525mm	-	-	-	-	0.75	**8.03**	8.61	**16.65**	nr	**18.73**
wood cased frames and sashes with internal and external sills complete with accessories and weights; preparatory to filling openings; cut out fixing cramps										
610 x 1225mm	-	-	-	-	1.79	**19.17**	3.10	**22.28**	nr	**25.06**
915 x 1525mm	-	-	-	-	2.56	**27.42**	5.77	**33.19**	nr	**37.34**
1370 x 1525mm	-	-	-	-	4.11	**44.02**	8.61	**52.63**	nr	**59.21**

Labour hourly rates: (except Specialists) Craft Operatives £14.21 Labourer £10.71 Rates are national average prices. Refer to REGIONAL VARIATIONS for indicative levels of overall pricing in regions	MATERIALS			LABOUR			RATES			
	Del to Site	Waste	Material Cost	Craft Optve	Lab	Labour Cost	Sunds	Nett Rate	Unit	Gross Rate (+12.5%)
	£	%	£	Hrs	Hrs	£	£	£		£
C90: ALTERATIONS - SPOT ITEMS Cont.										
Removal of windows and doors Cont.										
Removing Cont.										
wood doors										
single internal doors	-	-	-	-	0.23	**2.46**	2.48	**4.94**	nr	**5.56**
single internal doors and frames; in conjunction with demolition	-	-	-	-	0.18	**1.93**	5.57	**7.50**	nr	**8.43**
single internal doors and frames; preparatory to filling openings; cut out fixing cramps	-	-	-	-	1.00	**10.71**	5.57	**16.28**	nr	**18.31**
double internal doors	-	-	-	-	0.47	**5.03**	4.95	**9.99**	nr	**11.24**
double internal doors and frames; in conjunction with demolition	-	-	-	-	0.19	**2.03**	8.46	**10.50**	nr	**11.81**
double internal doors and frames; preparatory to filling openings; cut out fixing cramps	-	-	-	-	1.20	**12.85**	8.46	**21.31**	nr	**23.98**
single external doors	-	-	-	-	0.25	**2.68**	4.12	**6.80**	nr	**7.65**
single external doors and frames; in conjunction with demolition	-	-	-	-	0.18	**1.93**	7.42	**9.35**	nr	**10.51**
single external doors and frames; preparatory to filling openings; cut out fixing cramps	-	-	-	-	1.00	**10.71**	7.42	**18.13**	nr	**20.39**
double external doors	-	-	-	-	0.47	**5.03**	4.95	**9.99**	nr	**11.24**
double external doors and frames; in conjunction with demolition	-	-	-	-	0.19	**2.03**	8.46	**10.50**	nr	**11.81**
double external doors and frames; preparatory to filling openings; cut out fixing cramps	-	-	-	-	1.20	**12.85**	8.46	**21.31**	nr	**23.98**
single door frames; in conjunction with demolition	-	-	-	-	0.17	**1.82**	3.10	**4.93**	nr	**5.54**
single door frames; preparatory to filling openings; cut out fixing cramps	-	-	-	-	0.77	**8.25**	3.10	**11.35**	nr	**12.77**
double door frames; in conjunction with demolition	-	-	-	-	0.18	**1.93**	3.51	**5.44**	nr	**6.11**
double door frames; preparatory to filling openings; cut out fixing cramps	-	-	-	-	0.82	**8.78**	3.51	**12.29**	nr	**13.83**
Removing; setting aside for reuse										
metal windows with internal and external sills; in conjunction with demolition										
997 x 923mm	-	-	-	0.70	0.50	**15.30**	-	**15.30**	nr	**17.21**
1486 x 923mm	-	-	-	0.85	0.65	**19.04**	-	**19.04**	nr	**21.42**
1486 x 1513mm	-	-	-	0.90	0.75	**20.82**	-	**20.82**	nr	**23.42**
1994 x 1513mm	-	-	-	1.00	0.85	**23.31**	-	**23.31**	nr	**26.23**
metal windows with internal and external sills; preparatory to filling openings; cut out lugs										
997 x 923mm	-	-	-	0.70	1.40	**24.94**	-	**24.94**	nr	**28.06**
1486 x 923mm	-	-	-	0.85	1.75	**30.82**	-	**30.82**	nr	**34.67**
1486 x 1513mm	-	-	-	0.90	2.15	**35.82**	-	**35.82**	nr	**40.29**
1994 x 1513mm	-	-	-	1.00	2.45	**40.45**	-	**40.45**	nr	**45.51**
wood single or multi-light casements and frames with internal and external sills; in conjunction with demolition										
440 x 920mm	-	-	-	0.50	0.36	**10.96**	-	**10.96**	nr	**12.33**
1225 x 1070mm	-	-	-	0.83	0.43	**16.40**	-	**16.40**	nr	**18.45**
2395 x 1225mm	-	-	-	1.00	0.46	**19.14**	-	**19.14**	nr	**21.53**
wood single or multi-light casements and frames with internal and external sills; preparatory to filling openings; cut out fixing cramps										
440 x 920mm	-	-	-	0.50	1.33	**21.35**	-	**21.35**	nr	**24.02**
1225 x 1070mm	-	-	-	0.83	1.84	**31.50**	-	**31.50**	nr	**35.44**
2395 x 1225mm	-	-	-	1.00	2.68	**42.91**	-	**42.91**	nr	**48.28**
wood cased frames and sashes with internal and external sills complete with accessories and weights; in conjunction with demolition										
610 x 1225mm	-	-	-	0.97	0.78	**22.14**	-	**22.14**	nr	**24.90**
915 x 1525mm	-	-	-	1.63	0.93	**33.12**	-	**33.12**	nr	**37.26**
1370 x 1525mm	-	-	-	1.94	1.07	**39.03**	-	**39.03**	nr	**43.91**
wood cased frames and sashes with internal and external sills complete with accessories and weights; preparatory to filling openings; cut out fixing cramps										
610 x 1225mm	-	-	-	0.97	1.96	**34.78**	-	**34.78**	nr	**39.12**
915 x 1525mm	-	-	-	1.63	2.83	**53.47**	-	**53.47**	nr	**60.16**
1370 x 1525mm	-	-	-	1.94	4.43	**75.01**	-	**75.01**	nr	**84.39**
note: the above rates assume that windows and sashes will be re-glazed when re-used										
wood doors										
single internal doors	-	-	-	0.50	0.31	**10.43**	-	**10.43**	nr	**11.73**
single internal doors and frames; in conjunction with demolition	-	-	-	1.00	0.34	**17.85**	-	**17.85**	nr	**20.08**
single internal doors and frames; preparatory to filling openings; cut out fixing cramps	-	-	-	1.00	1.16	**26.63**	-	**26.63**	nr	**29.96**
double internal doors	-	-	-	1.00	0.63	**20.96**	-	**20.96**	nr	**23.58**
double internal doors and frames; in conjunction with demolition	-	-	-	1.47	0.43	**25.49**	-	**25.49**	nr	**28.68**
double internal doors and frames; preparatory to filling openings; cut out fixing cramps	-	-	-	1.47	1.45	**36.42**	-	**36.42**	nr	**40.97**
single external doors	-	-	-	0.50	0.33	**10.64**	-	**10.64**	nr	**11.97**
single external doors and frames; in conjunction with demolition	-	-	-	1.00	0.34	**17.85**	-	**17.85**	nr	**20.08**
single external doors and frames; preparatory to filling openings; cut out fixing cramps	-	-	-	1.00	1.16	**26.63**	-	**26.63**	nr	**29.96**
double external doors	-	-	-	1.00	0.63	**20.96**	-	**20.96**	nr	**23.58**
double external doors and frames; in conjunction with demolition	-	-	-	1.47	0.44	**25.60**	-	**25.60**	nr	**28.80**
double external doors and frames; preparatory to filling openings; cut out fixing cramps	-	-	-	1.47	1.45	**36.42**	-	**36.42**	nr	**40.97**

Labour hourly rates: (except Specialists) Craft Operatives £14.21 Labourer £10.71 Rates are national average prices. Refer to REGIONAL VARIATIONS for indicative levels of overall pricing in regions	MATERIALS			LABOUR				RATES		
	Del to Site £	Waste %	Material Cost £	Craft Optve Hrs	Lab Hrs	Labour Cost £	Sunds £	Nett Rate £	Unit	Gross Rate (+12.5%) £
C90: ALTERATIONS - SPOT ITEMS Cont.										
Removal of windows and doors Cont.										
Removing; setting aside for reuse Cont.										
wood doors Cont.										
single door frames; in conjunction with demolition	-	-	-	0.50	0.25	**9.78**	-	**9.78**	nr	**11.01**
single door frames; preparatory to filling openings; cut out fixing cramps	-	-	-	0.50	0.85	**16.21**	-	**16.21**	nr	**18.23**
double door frames; in conjunction with demolition	-	-	-	0.50	0.26	**9.89**	-	**9.89**	nr	**11.13**
double door frames; preparatory to filling openings; cut out fixing cramps	-	-	-	0.50	0.90	**16.74**	-	**16.74**	nr	**18.84**
Removal of ironmongery										
Removing door ironmongery (piecing in doors and frames included elsewhere)										
butt hinges	-	-	-	-	0.06	**0.64**	0.01	**0.65**	nr	**0.74**
tee hinges up to 300mm	-	-	-	-	0.20	**2.14**	0.01	**2.15**	nr	**2.42**
floor spring hinges and top centres	-	-	-	-	0.92	**9.85**	0.24	**10.09**	nr	**11.35**
barrel bolts up to 200mm	-	-	-	-	0.13	**1.39**	0.01	**1.40**	nr	**1.58**
flush bolts up to 200mm	-	-	-	-	0.17	**1.82**	0.01	**1.83**	nr	**2.06**
indicating bolts	-	-	-	-	0.23	**2.46**	0.01	**2.48**	nr	**2.78**
double panic bolts	-	-	-	-	0.69	**7.39**	0.13	**7.52**	nr	**8.46**
single panic bolts	-	-	-	-	0.58	**6.21**	0.08	**6.29**	nr	**7.08**
Norfolk or Suffolk latches	-	-	-	-	0.23	**2.46**	0.01	**2.48**	nr	**2.78**
cylinder rim night latches	-	-	-	-	0.23	**2.46**	0.01	**2.48**	nr	**2.78**
rim locks or latches and furniture	-	-	-	-	0.23	**2.46**	0.01	**2.48**	nr	**2.78**
mortice dead locks	-	-	-	-	0.17	**1.82**	0.01	**1.83**	nr	**2.06**
mortice locks or latches and furniture	-	-	-	-	0.29	**3.11**	0.01	**3.12**	nr	**3.51**
Bales catches	-	-	-	-	0.12	**1.29**	0.01	**1.30**	nr	**1.46**
overhead door closers, surface fixed	-	-	-	-	0.29	**3.11**	0.13	**3.24**	nr	**3.64**
pull handles	-	-	-	-	0.12	**1.29**	0.01	**1.30**	nr	**1.46**
push plates	-	-	-	-	0.12	**1.29**	0.01	**1.30**	nr	**1.46**
kicking plates	-	-	-	-	0.17	**1.82**	0.05	**1.87**	nr	**2.10**
letter plates	-	-	-	-	0.12	**1.29**	0.01	**1.30**	nr	**1.46**
Removing window ironmongery										
sash centres	-	-	-	-	0.12	**1.29**	0.01	**1.30**	nr	**1.46**
sash fasteners	-	-	-	-	0.17	**1.82**	0.01	**1.83**	nr	**2.06**
sash lifts	-	-	-	-	0.07	**0.75**	0.01	**0.76**	nr	**0.86**
sash screws	-	-	-	-	0.17	**1.82**	0.01	**1.83**	nr	**2.06**
casement fasteners	-	-	-	-	0.12	**1.29**	0.01	**1.30**	nr	**1.46**
casement stays	-	-	-	-	0.12	**1.29**	0.01	**1.30**	nr	**1.46**
quadrant stays	-	-	-	-	0.12	**1.29**	0.01	**1.30**	nr	**1.46**
fanlight catches	-	-	-	-	0.13	**1.39**	0.01	**1.40**	nr	**1.58**
curtain tracks	-	-	-	-	0.15	**1.61**	0.08	**1.69**	m	**1.90**
Removing sundry ironmongery										
hat and coat hooks	-	-	-	-	0.07	**0.75**	0.01	**0.76**	nr	**0.86**
cabin hooks and eyes	-	-	-	-	0.12	**1.29**	0.01	**1.30**	nr	**1.46**
shelf brackets	-	-	-	-	0.07	**0.75**	0.05	**0.80**	nr	**0.90**
toilet roll holders	-	-	-	-	0.09	**0.96**	0.05	**1.01**	nr	**1.14**
towel rollers	-	-	-	-	0.12	**1.29**	0.08	**1.37**	nr	**1.54**
Removal of metalwork										
Removing										
balustrades 1067mm high	-	-	-	-	0.81	**8.68**	1.07	**9.74**	m	**10.96**
wire mesh screens with timber or metal beads screwed on										
generally	-	-	-	-	0.62	**6.64**	1.04	**7.68**	m2	**8.64**
225 x 225mm	-	-	-	-	0.20	**2.14**	0.08	**2.22**	nr	**2.50**
305 x 305mm	-	-	-	-	0.22	**2.36**	0.13	**2.49**	nr	**2.80**
guard bars of vertical bars at 100mm centres welded to horizontal fixing bars at 450mm centres, fixed with bolts	-	-	-	-	0.62	**6.64**	1.04	**7.68**	m2	**8.64**
guard bars of vertical bars at 100mm centres welded to horizontal fixing bars at 450mm centres, fixed with screws	-	-	-	-	0.39	**4.18**	1.04	**5.22**	m2	**5.87**
extra; piecing in softwood after removal of bolts	-	-	-	0.25	0.04	**3.98**	0.11	**4.09**	nr	**4.60**
small bracket	-	-	-	-	0.18	**1.93**	0.08	**2.01**	nr	**2.26**
Removing; making good finishings										
balustrades 1067mm high; making good mortices in treads	-	-	-	0.97	0.97	**24.17**	1.35	**25.52**	m	**28.71**
guard bars of vertical bars at 100mm centres welded to horizontal fixing bars at 450mm centres fixed in mortices; making good concrete or brickwork and plaster	-	-	-	0.60	1.42	**23.73**	1.41	**25.14**	m2	**28.29**
small bracket built or cast in; making good concrete or brickwork and plaster	-	-	-	0.25	0.25	**6.23**	0.56	**6.79**	nr	**7.64**
Removal of manholes										
Removing										
manholes overall size; remove cover and frame; break up brick sides and concrete bottom; fill in void with hardcore										
overall size 914 x 1067mm and 600mm deep to invert	2.00	30.00	**2.60**	-	10.00	**107.10**	4.12	**113.82**	nr	**128.05**
overall size 914 x 1219mm and 900mm deep to invert	4.00	30.00	**5.20**	-	11.50	**123.17**	4.54	**132.90**	nr	**149.52**

Labour hourly rates: (except Specialists) Craft Operatives £14.21 Labourer £10.71 Rates are national average prices. Refer to REGIONAL VARIATIONS for indicative levels of overall pricing in regions	MATERIALS			LABOUR				RATES		
	Del to Site £	Waste %	Material Cost £	Craft Optve Hrs	Lab Hrs	Labour Cost £	Sunds £	Nett Rate £	Unit	Gross Rate (+12.5%) £
C90: ALTERATIONS - SPOT ITEMS Cont.										
Removal of manholes Cont.										
Removing Cont.										
manholes overall size; remove cover and frame; break up brick sides and concrete bottom; fill in void with hardcore Cont.										
manhole covers and frames; clean off brickwork	-	-	-	-	0.50	5.36	0.73	6.09	nr	6.85
fresh air inlets; clean out pipe socket	-	-	-	-	0.50	5.36	0.24	5.59	nr	6.29
Removal of fencing										
Removing										
chestnut pale fencing with posts										
610mm high	-	-	-	-	0.30	3.21	0.65	3.86	m	4.35
914mm high	-	-	-	-	0.34	3.64	1.00	4.64	m	5.22
1219mm high	-	-	-	-	0.38	4.07	1.33	5.40	m	6.07
close boarded fencing with posts										
1219mm high	-	-	-	-	0.68	7.28	1.52	8.80	m	9.90
1524mm high	-	-	-	-	0.81	8.68	1.90	10.57	m	11.89
1829mm high	-	-	-	-	0.91	9.75	2.28	12.02	m	13.52
chain link fencing with posts										
914mm high	-	-	-	-	0.29	3.11	0.73	3.84	m	4.32
1219mm high	-	-	-	-	0.34	3.64	3.06	6.70	m	7.54
1524mm high	-	-	-	-	0.38	4.07	1.23	5.30	m	5.96
1829mm high	-	-	-	-	0.42	4.50	1.52	6.02	m	6.77
timber gate and posts	-	-	-	-	0.98	10.50	3.10	13.60	nr	15.30
Strutting for forming openings										
Strutting generally the cost of strutting in connection with forming openings is to be added to the cost of forming openings; the following are examples of costs for strutting for various sizes and locations of openings. Costs are based on three uses of timber										
Strutting for forming small openings in internal load bearing walls										
opening size 900 x 2000mm	8.91	5.00	9.36	2.00	0.33	31.95	0.78	42.09	nr	47.35
opening size 2000 x 2300mm	19.80	5.00	20.79	5.00	0.83	79.94	1.69	102.42	nr	115.23
Strutting for forming openings in external one brick walls										
opening size 900 x 2000mm	13.37	5.00	14.04	3.00	0.50	47.99	1.35	63.37	nr	71.30
opening size 1500 x 1200mm	13.37	5.00	14.04	3.00	0.50	47.99	1.35	63.37	nr	71.30
Strutting for forming openings in external 280mm brick cavity walls										
opening size 900 x 2000mm	18.32	5.00	19.24	4.00	0.66	63.91	1.52	84.66	nr	95.24
opening size 1500 x 1200mm	18.32	5.00	19.24	4.00	0.66	63.91	1.52	84.66	nr	95.24
Strutting for forming large openings in internal load bearing walls on the ground floor of an average two storey building with timber floors and pitched roof; with load bearing surface 450mm below ground floor; including cutting holes and making good										
opening size 3700 x 2300mm	123.75	5.00	129.94	20.00	20.00	498.40	17.96	646.30	nr	727.09
Unit rates for pricing the above and similar work										
plates, struts, braces and wedges in supports to floor and roof	1.73	5.00	1.82	0.28	0.28	6.98	0.26	9.05	m	10.19
dead shore and needle, sole plates, braces and wedges including cutting holes and making good	62.37	5.00	65.49	10.00	10.00	249.20	8.95	323.64	nr	364.09
Strutting for forming large openings in external one brick walls on the ground floor of an average two storey building as described above										
opening size 6000 x 2500mm	250.47	5.00	262.99	40.00	40.00	996.80	35.63	1295.43	nr	1457.35
Unit rates for pricing the above and similar work										
strutting to window opening over new opening	12.47	5.00	13.09	2.00	2.00	49.84	1.77	64.70	nr	72.79
plates, struts, braces and wedges in supports to floor and roof	1.68	5.00	1.76	0.28	0.28	6.98	0.26	9.00	m	10.13
dead shore and needle, sole plates, braces and wedges including cutting holes and making good	98.11	5.00	103.02	15.70	15.70	391.24	14.04	508.30	nr	571.84
set of two raking shores with 50mm wall piece, wedges and dogs	25.05	5.00	26.30	4.00	4.00	99.68	3.54	129.53	nr	145.72
Strutting for forming large openings in external 280mm brick cavity walls on the ground floor of an average two storey building as above described										
opening size 6000 x 2500mm	274.53	5.00	288.26	43.90	43.90	1093.99	39.18	1421.42	nr	1599.10
Unit rates for pricing the above and similar work										
strutting to window opening over new opening	24.95	5.00	26.20	4.00	4.00	99.68	3.54	129.42	nr	145.60
plates, struts, braces and wedges in supports to floor and roof	1.68	5.00	1.76	0.28	0.28	6.98	0.26	9.00	m	10.13
dead shore and needle, sole plates, braces and wedges including cutting holes and making good	103.95	5.00	109.15	16.70	16.70	416.16	14.97	540.28	nr	607.81
set of two raking shores with 50mm wall piece, wedges and dogs	25.34	5.00	26.61	4.00	4.00	99.68	3.54	129.83	nr	146.06

Labour hourly rates: (except Specialists) Craft Operatives £14.21 Labourer £10.71 Rates are national average prices. Refer to REGIONAL VARIATIONS for indicative levels of overall pricing in regions	MATERIALS			LABOUR				RATES		
	Del to Site £	Waste %	Material Cost £	Craft Optve Hrs	Lab Hrs	Labour Cost £	Sunds £	Nett Rate £	Unit	Gross Rate (+12.5%) £
C90: ALTERATIONS - SPOT ITEMS Cont.										
Spot items										
Spot items generally the following items are usually specified as spot items where all work in all trades is included in one item; to assist in adapting these items for varying circumstances, unit rates have been given for the component operation where possible while these items are not in accordance with SMM7 either by description or measurement, it is felt that the pricing information given will be of value to the reader										
Breaking up and reinstatement of concrete floors for excavations										
Break up concrete bed and 150mm hardcore bed under for a width of 760mm for excavation of trench and reinstate with new hardcore and in-situ concrete, BS 5328, designed mix C20, 20mm aggregate; make good up to new wall and existing concrete bed both side										
with 100mm plain concrete bed	10.83	10.00	**12.20**	-	2.70	**28.92**	7.66	**48.77**	m	**54.86**
with 150mm plain concrete bed	15.53	10.00	**17.37**	-	5.00	**53.55**	9.27	**80.18**	m	**90.21**
with 100mm concrete bed reinforced with steel fabric to BS 4483, Reference A193	12.23	10.00	**13.74**	-	4.05	**43.38**	7.66	**64.77**	m	**72.87**
with 150mm concrete bed reinforced with steel fabric to BS 4483, Reference A193	16.94	10.00	**18.91**	-	7.33	**78.50**	9.27	**106.69**	m	**120.02**
Openings through concrete										
Form opening for door frame through 150mm reinforced concrete wall plastered both sides and with skirting both sides; take off skirtings, cut opening through wall, square up reveals, make good existing plaster up to new frame both sides and form fitted ends on existing skirtings up to new frame; extend cement and sand floor screed and vinyl floor covering through opening and make good up to existing										
838 x 2032mm	18.80	-	**18.80**	22.50	28.00	**619.61**	35.79	**674.19**	nr	**758.47**
Unit rates for pricing the above and similar work										
cut opening through wall										
100mm thick	-	-	-	-	5.30	**56.76**	8.43	**65.19**	m2	**73.34**
150mm thick	-	-	-	-	7.90	**84.61**	12.60	**97.21**	m2	**109.36**
200mm thick	-	-	-	-	10.50	**112.46**	16.77	**129.22**	m2	**145.38**
make good fair face around opening	-	-	-	-	0.80	**8.57**	0.59	**9.16**	m	**10.31**
square up reveals to opening										
100mm wide	-	-	-	0.75	0.75	**18.69**	1.47	**20.16**	m	**22.68**
150mm wide	-	-	-	1.07	1.07	**26.66**	2.07	**28.74**	m	**32.33**
200mm wide	-	-	-	1.38	1.38	**34.39**	3.29	**37.68**	m	**42.39**
make good existing plaster up to new frame	0.32	10.00	**0.35**	0.55	0.55	**13.71**	-	**14.06**	m	**15.82**
13mm two coat hardwall plaster to reveal not exceeding 300mm wide	0.64	10.00	**0.70**	0.64	0.64	**15.95**	-	**16.65**	m	**18.73**
take off old skirting	-	-	-	0.22	-	**3.13**	0.12	**3.24**	m	**3.65**
form fitted end on existing skirting up to new frame	-	-	-	0.38	-	**5.40**	-	**5.40**	nr	**6.07**
short length of old skirting up to 100mm long with mitre with existing one end	-	-	-	0.58	-	**8.24**	0.05	**8.29**	nr	**9.33**
short length of old skirting up to 200mm long with mitres with existing both ends	-	-	-	1.15	-	**16.34**	0.07	**16.41**	nr	**18.46**
38mm cement and sand (1:3) screeded bed in opening	4.93	10.00	**5.43**	1.54	1.54	**38.38**	-	**43.80**	m2	**49.28**
vinyl or similar floor covering to match existing, fixed with adhesive, not exceeding 300mm wide	6.44	10.00	**7.09**	0.15	0.15	**3.74**	0.23	**11.05**	m	**12.43**
iroko threshold; twice oiled; plugged and screwed										
15 x 100mm	5.01	10.00	**5.51**	0.90	-	**12.79**	0.06	**18.36**	m	**20.65**
15 x 150mm	7.33	10.00	**8.06**	1.20	-	**17.05**	0.07	**25.19**	m	**28.33**
15 x 200mm	8.86	10.00	**9.75**	1.55	-	**22.03**	0.11	**31.88**	m	**35.86**
Form opening for staircase through 150mm reinforced concrete suspended floor plastered on soffit and with screed and vinyl floor covering on top; take up floor covering and screed, cut opening through floor, square up edges of slab, make good ceiling plaster up to lining and make good screed and floor covering to lining										
900 x 2134mm	73.99	-	**73.99**	19.00	39.00	**687.68**	46.22	**807.89**	nr	**908.87**
Unit rates for pricing the above and similar work										
take up vinyl or similar floor covering	-	-	-	-	0.55	**5.89**	0.17	**6.06**	m2	**6.81**
hack up screed	-	-	-	-	1.50	**16.07**	1.53	**17.59**	m2	**19.79**
cut opening through 150mm thick slab	-	-	-	-	7.90	**84.61**	12.60	**97.21**	m2	**109.36**
cut opening through 200mm thick slab	-	-	-	-	10.50	**112.46**	16.77	**129.22**	m2	**145.38**
make good fair face around opening	-	-	-	-	0.80	**8.57**	0.56	**9.12**	m	**10.27**
square up edges of 150mm thick slab	-	-	-	1.07	1.07	**26.66**	2.19	**28.86**	m	**32.46**
square up edges of 200mm thick slab	-	-	-	1.38	1.38	**34.39**	2.93	**37.32**	m	**41.98**
make good existing plaster up to new lining	0.32	10.00	**0.35**	0.55	0.55	**13.71**	-	**14.06**	m	**15.82**
make good existing 38mm floor screed up to new lining	0.78	10.00	**0.86**	0.45	0.45	**11.21**	-	**12.07**	m	**13.58**
make good existing vinyl or similar floor covering up to new lining	8.59	10.00	**9.45**	0.67	0.67	**16.70**	0.36	**26.50**	m	**29.81**

EXISTING SITE/BUILDINGS/SERVICES

Labour hourly rates: (except Specialists) Craft Operatives £14.21 Labourer £10.71 Rates are national average prices. Refer to REGIONAL VARIATIONS for indicative levels of overall pricing in regions	MATERIALS			LABOUR				RATES		
	Del to Site £	Waste %	Material Cost £	Craft Optve Hrs	Lab Hrs	Labour Cost £	Sunds £	Nett Rate £	Unit	Gross Rate (+12.5%) £
C90: ALTERATIONS - SPOT ITEMS Cont.										
Filling holes in concrete										
Fill holes in concrete structure where pipes removed with concrete to BS 5328, designed mix C20, 20mm aggregate; formwork										
wall; thickness 100mm; pipe diameter										
50mm	-	-	-	0.60	0.30	11.74	1.81	13.55	nr	15.25
100mm	-	-	-	0.78	0.39	15.26	2.39	17.65	nr	19.86
150mm	-	-	-	0.98	0.49	19.17	2.88	22.05	nr	24.81
wall; thickness 200mm; pipe diameter										
50mm	-	-	-	0.78	0.39	15.26	2.39	17.65	nr	19.86
100mm	-	-	-	0.98	0.49	19.17	2.88	22.05	nr	24.81
150mm	-	-	-	1.00	0.77	22.46	3.70	26.15	nr	29.42
floor; thickness 200mm; pipe diameter										
50mm	-	-	-	0.74	0.37	14.48	2.17	16.65	nr	18.73
100mm	-	-	-	0.93	0.46	18.14	2.77	20.91	nr	23.53
150mm	-	-	-	0.95	0.48	18.64	3.51	22.15	nr	24.92
Make good fair face to filling and surrounding work										
to one side of wall; pipe diameter										
50mm	-	-	-	0.10	0.10	2.49	0.25	2.74	nr	3.08
100mm	-	-	-	0.20	0.20	4.98	0.38	5.36	nr	6.03
150mm	-	-	-	0.25	0.25	6.23	0.46	6.69	nr	7.53
to soffit of floor; pipe diameter										
50mm	-	-	-	0.10	0.10	2.49	0.25	2.74	nr	3.08
100mm	-	-	-	0.20	0.20	4.98	0.38	5.36	nr	6.03
150mm	-	-	-	0.25	0.25	6.23	0.46	6.69	nr	7.53
Make good plaster to filling and surrounding work										
to one side of wall; pipe diameter										
50mm	-	-	-	0.19	0.19	4.73	0.25	4.98	nr	5.61
100mm	-	-	-	0.36	0.36	8.97	0.38	9.35	nr	10.52
150mm	-	-	-	0.45	0.45	11.21	0.46	11.68	nr	13.14
to soffit of floor; pipe diameter										
50mm	-	-	-	0.20	0.20	4.98	0.25	5.23	nr	5.89
100mm	-	-	-	0.40	0.40	9.97	0.38	10.35	nr	11.64
150mm	-	-	-	0.50	0.50	12.46	0.46	12.92	nr	14.54
Make good floor screed and surrounding work										
pipe diameter 50mm	-	-	-	0.19	0.19	4.73	0.11	4.84	nr	5.45
pipe diameter 100mm	-	-	-	0.36	0.36	8.97	0.23	9.20	nr	10.35
pipe diameter 150mm	-	-	-	0.45	0.45	11.21	0.31	11.52	nr	12.96
Make good vinyl or similar floor covering and surrounding work										
pipe diameter 50mm	-	-	-	0.50	0.50	12.46	1.85	14.31	nr	16.10
pipe diameter 100mm	-	-	-	1.00	1.00	24.92	2.48	27.40	nr	30.82
pipe diameter 150mm	-	-	-	1.20	1.20	29.90	3.47	33.38	nr	37.55
Fill holes in concrete structure where metal sections removed with concrete to BS 5328, designed mix C20, 20mm aggregate; formwork										
wall; thickness 100mm; section not exceeding 250mm deep	-	-	-	0.60	0.60	14.95	2.20	17.16	nr	19.30
wall; thickness 200mm; section not exceeding 250mm deep	-	-	-	0.75	0.75	18.69	3.21	21.90	nr	24.64
wall; thickness 100mm; section 250 - 500mm deep	-	-	-	0.75	0.75	18.69	3.21	21.90	nr	24.64
wall; thickness 200mm; section 250 - 500mm deep	-	-	-	0.90	0.90	22.43	3.87	26.30	nr	29.59
Make good fair face to filling and surrounding work										
to one side of wall; section not exceeding 250mm deep	-	-	-	0.22	0.22	5.48	0.21	5.70	nr	6.41
to one side of wall; section 250 - 500mm deep	-	-	-	0.26	0.26	6.48	0.33	6.81	nr	7.66
Make good plaster to filling and surrounding work										
to one side of wall; section not exceeding 250mm deep	-	-	-	0.35	0.35	8.72	0.58	9.30	nr	10.47
to one side of wall; section 250 - 500mm deep	-	-	-	0.43	0.43	10.72	0.88	11.59	nr	13.04
Filling openings in concrete										
Fill opening where door removed in 150mm reinforced concrete wall with concrete reinforced with 12mm mild steel bars at 300mm centres both ways tied to existing structure; hack off plaster to reveals, hack up floor covering and screed in opening, prepare edges of opening to form joint with filling, drill edges of opening and tie in new reinforcement, plaster filling both sides and extended skirting both sides; making good junction of new and existing plaster and skirtings and make good floor screed and vinyl covering up to filling both sides										
838 x 2032mm	124.08	-	124.08	22.50	34.50	689.22	11.41	824.71	nr	927.80
Unit rates for pricing the above and similar work										
hack off plaster to reveal not exceeding 100mm wide	-	-	-	-	0.24	2.57	0.08	2.65	m	2.99
hack off plaster to reveal 100 - 200mm wide	-	-	-	-	0.33	3.53	0.17	3.70	m	4.16
take up vinyl or similar floor covering and screed	-	-	-	-	2.00	21.42	1.69	23.11	m2	26.00
prepare edge of opening 100mm wide to form joint with filling	-	-	-	0.25	0.25	6.23	-	6.23	m	7.01
prepare edge of opening 200mm wide to form joint with filling	-	-	-	0.35	0.35	8.72	-	8.72	m	9.81

Labour hourly rates: (except Specialists) Craft Operatives £14.21 Labourer £10.71 Rates are national average prices. Refer to REGIONAL VARIATIONS for indicative levels of overall pricing in regions	MATERIALS			LABOUR				RATES		
	Del to Site £	Waste %	Material Cost £	Craft Optve Hrs	Lab Hrs	Labour Cost £	Sunds £	Nett Rate £	Unit	Gross Rate (+12.5%) £
C90: ALTERATIONS - SPOT ITEMS Cont.										
Filling openings in concrete Cont.										
Unit rates for pricing the above and similar work Cont.										
concrete to BS 5328, designed mix C20, 20mm aggregate in walls 150 - 450m thick	110.82	5.00	**116.36**	-	12.00	**128.52**	0.49	**245.37**	m3	**276.04**
concrete to BS 5328, designed mix C20, 20mm aggregate in walls not exceeding 150mm thick	110.82	5.00	**116.36**	-	12.00	**128.52**	0.49	**245.37**	m3	**276.04**
12mm mild steel bar reinforcement; straight	968.00	10.00	**1064.80**	85.00	-	**1207.85**	5.34	**2277.99**	t	**2562.74**
drill edge of opening and tie in new reinforcement	-	-	**-**	-	1.00	**10.71**	0.49	**11.20**	m	**12.60**
formwork and basic finish to wall, per side	8.55	10.00	**9.41**	2.50	2.50	**62.30**	1.16	**72.87**	m2	**81.98**
formwork and fine formed finish to wall, per side	8.55	10.00	**9.41**	2.75	2.75	**68.53**	1.30	**79.24**	m2	**89.15**
junction between new and existing fair face	-	-	**-**	-	0.30	**3.21**	-	**3.21**	m	**3.61**
13mm two coat hardwall plaster over 300mm wide	2.14	10.00	**2.36**	1.30	1.30	**32.40**	-	**34.75**	m2	**39.10**
25 x 100mm softwood chamfered skirting, primed all round	2.00	10.00	**2.20**	0.56	0.09	**8.92**	0.13	**11.25**	m	**12.66**
junction with existing	-	-	**-**	0.50	0.08	**7.96**	-	**7.96**	nr	**8.96**
make good floor screed and vinyl or similar floor covering up to filling not exceeding 300mm wide	7.87	10.00	**8.66**	1.12	1.12	**27.91**	0.38	**36.95**	m	**41.56**
Fill opening where staircase removed in 150mm reinforced concrete floor with concrete reinforced with 12mm mild steel bars at 225mm centres both ways tied to existing structure; remove timber lining, prepare edges of opening to form joint with filling, drill edges of opening tie in new reinforcement, plaster filling on soffit and extend cement and sand screed and vinyl floor covering to match existing; make good junction of new and existing plaster and vinyl floor covering										
900 x 2134mm	140.49	-	**140.49**	14.75	22.30	**448.43**	9.48	**598.40**	nr	**673.20**
Unit rates for pricing the above and similar work										
take off timber lining	-	-	**-**	-	0.24	**2.57**	0.12	**2.69**	m	**3.03**
prepare edge of opening 150mm wide to form joint with filling	-	-	**-**	0.30	0.30	**7.48**	-	**7.48**	m	**8.41**
prepare edge of opening 200mm wide to form joint with filling	-	-	**-**	0.35	0.35	**8.72**	-	**8.72**	m	**9.81**
concrete to BS 5328, designed mix C20, 20mm aggregate in slabs 150 - 450mm thick	110.82	5.00	**116.36**	-	12.00	**128.52**	0.49	**245.37**	m3	**276.04**
concrete to BS 5328, designed mix C20, 20mm aggregate in slabs not exceeding 150mm thick	110.82	5.00	**116.36**	-	12.00	**128.52**	0.49	**245.37**	m3	**276.04**
12mm mild steel bar reinforcement; straight	968.00	10.00	**1064.80**	85.00	-	**1207.85**	5.34	**2277.99**	t	**2562.74**
drill edge of opening and tie in new reinforcement	-	-	**-**	-	1.00	**10.71**	0.49	**11.20**	m	**12.60**
formwork and basic finish to soffit, slab thickness not exceeding 200mm; height to soffit 1.50 - 3.00m	9.50	10.00	**10.45**	2.25	2.25	**56.07**	1.15	**67.67**	m2	**76.13**
formwork and basic finish to soffit, slab thickness 200 - 300mm; height to soffit 1.50 - 3.00m	10.50	10.00	**11.55**	2.50	2.50	**62.30**	1.30	**75.15**	m2	**84.55**
formwork and fine formed finish to soffit, slab thickness not exceeding 200mm; height to soffit 1.50 - 3.00m	9.50	10.00	**10.45**	2.50	2.50	**62.30**	1.33	**74.08**	m2	**83.34**
formwork and fine formed finish to soffit, slab thickness 200 - 300mm; height to soffit 1.50 - 3.00m	10.50	10.00	**11.55**	2.75	2.75	**68.53**	1.37	**81.45**	m2	**91.64**
junction between new and existing fair face	-	-	**-**	-	0.30	**3.21**	-	**3.21**	m	**3.61**
10mm two coat lightweight plaster over 300mm wide	2.14	10.00	**2.36**	1.30	1.30	**32.40**	-	**34.75**	m2	**39.10**
38mm cement and sand (1:3) trowelled bed	4.93	10.00	**5.43**	0.63	0.63	**15.70**	-	**21.13**	m2	**23.77**
vinyl or similar floor covering to match existing fixed with adhesive exceeding 300mm wide	21.47	10.00	**23.62**	1.25	0.50	**23.12**	0.66	**47.40**	m2	**53.32**
Openings through brickwork or blockwork										
Form openings for door frames through 100mm block partition plastered both sides and with skirting both sides; take off skirtings, cut opening through partition, insert 100 x 150mm precast concrete lintel and wedge and pin up over, quoin up jambs, extend plaster to faces of lintel and make good junction with existing plaster; make good existing plaster up to new frame both sides and form fitted ends on existing skirtings up to new frame; extend softwood board flooring through opening on and with bearers and make good up to existing										
838 x 2032mm	22.07	-	**22.07**	20.00	17.20	**468.41**	15.05	**505.53**	nr	**568.72**
Unit rates for pricing the above and similar work										
cut opening through 75mm block partition plastered both sides	-	-	**-**	1.10	1.30	**29.55**	4.63	**34.19**	m2	**38.46**
cut opening through 100mm block partition plastered both sides	-	-	**-**	1.30	1.56	**35.18**	5.44	**40.62**	m2	**45.70**
make good fair face around opening	-	-	**-**	0.25	0.25	**6.23**	0.12	**6.35**	m	**7.14**
precast concrete, BS 5328, designed mix C20, 20mm aggregate lintel 75 x 150 x 1200mm, reinforced with 1.01 kg of 12mm mild steel bars	8.70	10.00	**9.57**	0.80	0.80	**19.94**	0.30	**29.80**	nr	**33.52**
precast concrete, BS 5328, designed mix C20, 20mm aggregate lintel 100 x 150 x 1200mm, reinforced with 1.01 kg of 12mm mild steel bars	17.17	10.00	**18.89**	1.00	1.00	**24.92**	0.39	**44.20**	nr	**49.72**
precast concrete, BS 5328 designed mix C20, 20mm aggregate lintel 75 x 150 x 1200mm, reinforced with 1.01 kg of 12mm mild steel bars, fair finish all faces	8.70	10.00	**9.57**	0.80	0.80	**19.94**	0.30	**29.80**	nr	**33.53**

Labour hourly rates: (except Specialists) Craft Operatives £14.21 Labourer £10.71 Rates are national average prices. Refer to REGIONAL VARIATIONS for indicative levels of overall pricing in regions	MATERIALS			LABOUR				RATES		
	Del to Site £	Waste %	Material Cost £	Craft Optve Hrs	Lab Hrs	Labour Cost £	Sunds £	Nett Rate £	Unit	Gross Rate (+12.5%) £
C90: ALTERATIONS - SPOT ITEMS Cont.										
Openings through brickwork or blockwork Cont.										
Unit rates for pricing the above and similar work Cont.										
precast concrete, BS 5328 designed mix C20, 20mm aggregate lintel 100 x 150 x 1200mm, reinforced with 1.01 kg of 12mm mild steel bars, fair finish all faces	17.17	10.00	**18.89**	1.00	1.00	**24.92**	0.39	**44.20**	nr	**49.72**
wedge and pin up over lintel 75mm wide	-	-	-	0.50	0.50	**12.46**	0.73	**13.19**	m	**14.84**
wedge and pin up over lintel 100mm wide	-	-	-	0.66	0.66	**16.45**	0.97	**17.42**	m	**19.60**
quoin up 75mm wide jambs	-	-	-	0.52	0.52	**12.96**	0.73	**13.69**	m	**15.40**
quoin up 100mm wide jambs	-	-	-	0.68	0.68	**16.95**	0.97	**17.92**	m	**20.16**
extend 13mm hardwall plaster to face of lintel	2.14	10.00	**2.36**	2.13	2.13	**53.08**	0.07	**55.51**	m2	**62.45**
make good existing plaster up to new frame	0.32	10.00	**0.35**	0.55	0.55	**13.71**	-	**14.06**	m	**15.82**
take off old skirting	-	-	-	0.22	-	**3.13**	0.12	**3.24**	m	**3.65**
form fitted end on existing skirting up to new frame	-	-	-	0.38	-	**5.40**	-	**5.40**	nr	**6.07**
38mm cement and sand (1:3) screeded bed in opening	4.93	10.00	**5.43**	1.54	1.54	**38.38**	-	**43.80**	m2	**49.28**
extend 25mm softwood board flooring through opening on and with bearers	31.11	20.00	**35.65**	16.20	2.70	**259.12**	0.64	**295.41**	m2	**332.34**
iroko threshold; twice oiled; plugged and screwed										
19 x 75mm	4.43	10.00	**4.87**	0.80	-	**11.37**	0.08	**16.32**	m	**18.36**
19 x 100mm	5.01	10.00	**5.51**	0.90	-	**12.79**	0.09	**18.39**	m	**20.69**
Form openings for door frames through one brick wall plastered one side and with external rendering other side; take off skirting, cut opening through wall, insert 215 x 150mm precast concrete lintel and wedge and pin up over; quoin up jambs; extend plaster to face of lintel and make good junction with existing plaster; make good existing plaster up to new frame one side; extend external rendering to face of lintel and to reveals and make good up to existing rendering and new frame the other side; form fitted ends on existing skirting up to new frame										
914 x 2082mm	36.18	-	**36.18**	28.00	25.40	**669.91**	33.30	**739.39**	nr	**831.82**
Unit rates for pricing the above and similar work										
cut opening through half brick wall	-	-	-	1.95	2.40	**53.41**	5.64	**59.05**	m2	**66.44**
cut opening through one brick wall	-	-	-	3.80	4.70	**104.34**	10.68	**115.01**	m2	**129.39**
make good fair face around opening	-	-	-	0.50	0.50	**12.46**	0.73	**13.19**	m	**14.84**
make good facings to match existing around opening	1.61	10.00	**1.77**	0.50	0.50	**12.46**	0.73	**14.97**	m	**16.84**
precast concrete, BS 5328, designed mix C20, 20mm aggregate lintel 102 x 150 x 1200mm, reinforced with 1.07 kg of 12mm mild steel bars	17.17	10.00	**18.89**	1.00	1.00	**24.92**	0.43	**44.23**	nr	**49.76**
precast concrete, BS 5328, designed mix C20, 20mm aggregate lintel 215 x 150 x 1200mm, reinforced with 2.13 kg of 12mm mild steel bars	35.18	10.00	**38.70**	1.50	1.50	**37.38**	0.88	**76.95**	nr	**86.57**
precast concrete, BS 5328, designed mix C20, 20mm aggregate lintel 102 x 150 x 1200mm, reinforced with 1.07 kg of 12mm mild steel bars, fair finish all faces	17.17	10.00	**18.89**	1.00	1.00	**24.92**	0.43	**44.23**	nr	**49.76**
precast concrete, BS 5328, designed mix C20, 20mm aggregate lintel 215 x 150 x 1200mm, reinforced with 2.13 kg of 12mm mild steel bars, fair finish all faces	37.14	10.00	**40.85**	1.50	1.50	**37.38**	0.85	**79.09**	nr	**88.97**
wedge and pin up over lintel 102mm wide	-	-	-	0.66	0.66	**16.45**	0.97	**17.42**	m	**19.60**
wedge and pin up over lintel 215mm wide	-	-	-	1.00	1.00	**24.92**	2.07	**26.99**	m	**30.37**
quoin up half brick jambs	-	-	-	0.70	0.70	**17.44**	0.97	**18.42**	m	**20.72**
quoin up one brick jambs	-	-	-	1.27	1.27	**31.65**	2.07	**33.72**	m	**37.94**
facings to match existing to margin	1.61	10.00	**1.77**	0.67	0.67	**16.70**	0.97	**19.44**	m	**21.87**
make good existing plaster up to new frame	0.32	10.00	**0.35**	0.55	0.55	**13.71**	-	**14.06**	m	**15.82**
13mm two coat hardwall plaster to reveal not exceeding 300mm wide	0.64	10.00	**0.70**	0.78	0.78	**19.44**	0.02	**20.17**	m	**22.69**
make good existing external rendering up to new frame	0.24	10.00	**0.27**	0.47	0.47	**11.71**	-	**11.98**	m	**13.48**
13mm two coat cement and sand (1:3) external rendering to reveal not exceeding 300mm wide	0.49	10.00	**0.54**	0.74	0.74	**18.44**	0.02	**19.00**	m	**21.38**
take off old skirting	-	-	-	0.23	-	**3.27**	0.11	**3.37**	m	**3.80**
short length of old skirting up to 100mm long with mitre with existing one end	-	-	-	0.58	-	**8.24**	0.05	**8.29**	nr	**9.33**
short length of old skirting up to 300mm long with mitres with existing both ends	-	-	-	1.15	-	**16.34**	0.06	**16.40**	nr	**18.45**
38mm cement and sand (1:3) screeded bed in opening	4.93	10.00	**5.43**	1.54	1.54	**38.38**	-	**43.80**	m2	**49.28**
extend 25mm softwood board flooring through opening on and with bearers	6.06	15.00	**6.80**	3.65	0.61	**58.40**	0.17	**65.37**	nr	**73.54**
iroko threshold, twice oiled, plugged and screwed										
15 x 113mm	5.14	10.00	**5.65**	0.90	-	**12.79**	0.06	**18.50**	m	**20.82**
15 x 225mm	9.50	10.00	**10.45**	1.65	-	**23.45**	0.11	**34.00**	m	**38.25**
Form opening for window through 275mm hollow wall with two half brick skins, plastered one side and faced with picked stock facings the other; cut opening through wall, insert 275 x 225mm precast concrete boot lintel and wedge and pin up over; insert damp proof course and cavity gutter; quoin up jambs, close cavity at jambs with brickwork bonded to existing and with vertical damp proof course and close cavity at sill with one course of slates; face margin externally to match existing and extend plaster to face of lintel and to reveals and make good up to existing plaster and new frame										
900 x 900mm	55.40	-	**55.40**	15.90	16.80	**405.87**	17.30	**478.57**	nr	**538.39**

Labour hourly rates: (except Specialists) Craft Operatives £14.21 Labourer £10.71 Rates are national average prices. Refer to REGIONAL VARIATIONS for indicative levels of overall pricing in regions	MATERIALS			LABOUR				RATES		
	Del to Site £	Waste %	Material Cost £	Craft Optve Hrs	Lab Hrs	Labour Cost £	Sunds £	Nett Rate £	Unit	Gross Rate (+12.5%) £
C90: ALTERATIONS - SPOT ITEMS Cont.										
Openings through brickwork or blockwork Cont.										
Unit rates for pricing the above and similar work										
cut opening through wall with two half brick skins	-	-	-	3.90	4.80	**106.83**	9.67	**116.50**	m2	**131.06**
cut opening through wall with one half brick and one 100mm block skins	-	-	-	3.25	3.96	**88.59**	9.67	**98.26**	m2	**110.55**
precast concrete, BS 5328, designed mix C20, 20mm aggregate boot lintel 275 x 225 x 1200mm long reinforced with 3.60 kg of 12mm mild steel bars and 0.67 kg of 6mm mild steel links	57.19	10.00	**62.91**	2.30	2.30	**57.32**	0.97	**121.20**	nr	**136.35**
close cavity at jambs with blockwork bonded to existing and with 112.5mm wide bituminous hessian based damp proof course, BS 6398, Class A	2.42	10.00	**2.67**	0.50	0.50	**12.46**	0.25	**15.37**	m	**17.30**
close cavity at jambs with brickwork bonded to existing and with 112.5mm wide bituminous hessian based damp proof course, BS 6398, Class A	2.45	10.00	**2.69**	0.70	0.70	**17.44**	0.97	**21.11**	m	**23.75**
close cavity at jambs with slates in cement mortar (1:3) set vertically	3.02	10.00	**3.32**	0.44	0.44	**10.96**	0.25	**14.54**	m	**16.35**
close cavity at sill with one course of slates in cement mortar (1:3)	3.02	10.00	**3.32**	0.44	0.44	**10.96**	0.25	**14.54**	m	**16.35**
lead cored bituminous hessian based damp proof course and cavity tray, BS 6398, Class F, width exceeding 225mm	17.13	10.00	**18.84**	0.80	0.80	**19.94**	0.25	**39.03**	m2	**43.91**
precast concrete, BS 5328, designed mix C20, 20mm aggregate lintel 160 x 225 x 1200mm long reinforced with 2.13 kg of 12mm mild steel bars	39.27	10.00	**43.20**	1.50	1.50	**37.38**	0.73	**81.31**	nr	**91.48**
76 x 76 x 6mm mild steel angle arch bar, primed	6.70	5.00	**7.04**	0.50	0.50	**12.46**	0.73	**20.23**	m	**22.76**
take out three courses of facing bricks and build brick-on-end flat arch in picked stock facing bricks	6.13	10.00	**6.74**	1.00	1.00	**24.92**	2.14	**33.81**	m	**38.04**
For unit rates for finishes, quoining up jambs, etc see previous items										
Form opening for window where door removed, with window head at same level as old door head, through one brick wall plastered one side and faced the other with picked stock facings; remove old lintel and flat arch, cut away for and insert 102 x 225mm precast concrete lintel and wedge and pin up over; build brick-on-end flat arch in facing bricks on 76 x 76 x 6mm mild steel angle arch bar and point to match existing; cut away jambs of old door opening to extend width; quoin up jambs and face margins externally to match existing; fill in lower part of old opening with brickwork in gauged mortar (1:2:9), bond to existing and face externally and point to match existing										
1200 x 900mm; old door opening 914 x 2082mm; extend plaster to face of lintel, reveals and filling and make good up to existing plaster and new frame; extend skirting one side and make good junction with existing	119.28	-	**119.28**	22.00	21.50	**542.89**	29.63	**691.79**	nr	**778.26**
Unit rates for pricing the above and similar work										
take out old lintel	-	-	-	1.00	1.30	**28.13**	4.03	**32.16**	m	**36.18**
cut away one brick wall to increase width of opening	-	-	-	3.80	4.70	**104.34**	10.07	**114.41**	m2	**128.71**
cut away 275mm hollow wall of two half brick skins to increase width of opening	-	-	-	3.90	4.80	**106.83**	10.07	**116.90**	m2	**131.51**
For other unit rates for pricing the above and similar work see previous items and items in 'Filling openings in Brickwork and Blockwork' section										
Form opening for door frame where old window removed, with door head at same level as old window head, through one brick wall plastered one side and faced the other with picked stock facings; remove old lintel and flat arch, cut away for and insert 102 x 225mm precast concrete lintel and wedge and pin up over; build brick-on-end flat arch in facing bricks on 76 x 76 x 6mm mild steel angle arch bar and point, fill old opening at sides of new opening to reduce width with brickwork in gauged mortar (1:2:9), bond to existing, wedge and pin up at soffit, face externally and to margins and point to match existing; take off skirtings, cut away wall below sill of old opening, quion up jambs and face margins externally										
914 x 2082mm, old window opening 1200 x 900mm; extend plaster to face of lintel and filling and make good up to existing plaster and new frame; form fitted ends on existing skirting up to new frame	72.12	-	**72.12**	24.00	23.80	**595.94**	33.77	**701.83**	nr	**789.56**
For unit rates for pricing the above and similar work see previous items and items in 'Filling openings in Brickwork and Blockwork' section										
Air bricks in old walls										
Cut opening through old one brick wall, render all round in cement mortar (1:3); build in clay air brick externally and fibrous plaster ventilator internally and make good facings and plaster										
225 x 150mm	9.17	-	**9.17**	1.75	1.75	**43.61**	0.65	**53.43**	nr	**60.11**
225 x 225mm	7.48	-	**7.48**	2.33	2.33	**58.06**	0.82	**66.36**	nr	**74.66**

Labour hourly rates: (except Specialists) Craft Operatives £14.21 Labourer £10.71 Rates are national average prices. Refer to REGIONAL VARIATIONS for indicative levels of overall pricing in regions	MATERIALS			LABOUR				RATES		
	Del to Site £	Waste %	Material Cost £	Craft Optve Hrs	Lab Hrs	Labour Cost £	Sunds £	Nett Rate £	Unit	Gross Rate (+12.5%) £
C90: ALTERATIONS - SPOT ITEMS Cont.										
Air bricks in old walls Cont.										
Cut opening through old 275mm hollow brick wall and seal cavity with slates; build in clay air brick externally and fibrous plaster ventilator internally and make good facings and plaster										
225 x 150mm	12.19	10.00	12.49	2.14	2.14	53.33	0.65	66.47	nr	74.78
225 x 225mm	10.50	10.00	10.80	2.81	2.81	70.03	0.82	81.64	nr	91.85
Openings through rubble walling										
Form opening for door frame through 600mm rough rubble wall faced both sides; cut opening through wall, insert 600 x 150mm precast concrete lintel finished fair on all exposed faces and wedge and pin up over; quoin and face up jambs and extend softwood board flooring through opening on and with bearers and make good up to existing										
838 x 2032mm	82.62	-	82.62	41.30	43.70	1054.90	72.40	1209.92	nr	1361.16
Unit rates for pricing the above and similar work										
cut opening through 350mm random rubble wall	-	-	-	6.00	7.00	160.23	14.10	174.33	m2	196.12
cut opening through 600mm rough rubble wall	-	-	-	10.00	12.00	270.62	24.17	294.79	m2	331.64
precast concrete, BS 5328, designed mix C20, 20mm aggregate lintel 350 x 150 x 1140mm reinforced with 3.04 kg of 12mm mild steel bars; fair finish all faces	57.43	10.00	63.17	2.25	2.25	56.07	0.85	120.10	nr	135.11
precast concrete, BS 5328, designed mix C20, 20mm aggregate lintel 600 x 150 x 1140mm reinforced with 5.00 kg of 12mm mild steel bars; fair finish all faces	98.45	10.00	108.30	3.65	3.65	90.96	1.34	200.59	nr	225.67
wedge and pin up over lintel 350mm wide	-	-	-	2.00	2.00	49.84	2.44	52.28	m	58.82
wedge and pin up over lintel 600mm wide	-	-	-	3.00	3.00	74.76	4.28	79.04	m	88.92
quoin up 350mm wide jambs	-	-	-	2.00	2.00	49.84	2.44	52.28	m	58.82
quoin up 600mm wide jambs	-	-	-	3.00	3.00	74.76	4.28	79.04	m	88.92
extend 25mm softwood board flooring through opening on and with bearers	15.15	15.00	17.00	9.70	1.60	154.97	0.46	172.44	nr	193.99
Openings through stone faced walls										
Form opening for window through wall of half brick backing, plastered, and 100mm stone facing; cut opening through wall, insert 102 x 150mm precast concrete lintel and wedge and pin up over and build in 76 x 76 x 6mm mild steel angle arch bar to support stone facing, clean soffit of stone and point; quoin up jambs and face up and point externally; extend plaster to face of lintel and to reveals and make good up to existing plaster and new frame										
900 x 900mm	27.20	-	27.20	19.20	19.80	484.89	22.40	534.49	nr	601.30
Unit rates for pricing the above and similar work										
cut opening through wall	-	-	-	5.15	5.85	135.84	9.67	145.50	m2	163.69
precast concrete, BS 5328, designed mix C20, 20mm aggregate lintel 102 x 150 x 1200mm reinforced with 1.01 kg of 12mm mild steel bars	17.17	10.00	18.89	1.05	1.05	26.17	0.37	45.42	nr	51.10
76 x 76 x 6mm mild steel angle arch bar, primed	6.70	5.00	7.04	0.50	0.50	12.46	0.73	20.23	m	22.76
quoin up jambs	-	-	-	0.70	0.70	17.44	1.10	18.55	m	20.86
face stone jambs and point	-	-	-	2.10	2.10	52.33	3.06	55.39	m	62.31
clean stone head and point	-	-	-	2.10	2.10	52.33	0.37	52.70	m	59.29
For unit rates for finishes, etc. see previous items										
Filling openings in brickwork or blockwork										
Fill opening where door removed in 100mm block wall with concrete blocks in gauged mortar (1:2:9); provide 50mm preservative treated softwood sole plate on timber floor in opening, bond to blockwork at jambs and wedge and pin up at head; plaster filling both sides and extend skirting both sides; make good junction of new and existing plaster and skirtings										
838 x 2032mm	58.78	-	58.78	14.10	8.15	287.65	10.05	356.48	nr	401.04
Unit rates for pricing the above and similar work										
75mm blockwork in filling	8.42	10.00	9.26	1.00	1.25	27.60	1.34	38.20	m2	42.97
100mm blockwork in filling	10.10	10.00	11.11	1.20	1.50	33.12	1.84	46.06	m2	51.82
75mm blockwork in filling and fair face and flush smooth pointing one side	8.42	10.00	9.26	1.25	1.50	33.83	1.59	44.68	m2	50.26
100mm blockwork in filling and fair face and flush smooth pointing one side	10.10	10.00	11.11	1.45	1.75	39.35	1.96	52.41	m2	58.96
preservative treated softwood sole plate										
50 x 75mm	1.26	10.00	1.39	0.30	0.05	4.80	0.25	6.43	m	7.24
50 x 100mm	1.67	10.00	1.84	0.33	0.06	5.33	0.33	7.50	m	8.44
lead cored bituminous hessian based damp proof course, BS 6398, Class F, width not exceeding 225mm	17.13	10.00	18.84	0.80	0.80	19.94	0.49	39.26	m2	44.17
wedge and pin up at head 75mm wide	-	-	-	0.10	0.10	2.49	0.49	2.98	m	3.35
wedge and pin up at head 100mm wide	-	-	-	0.12	0.12	2.99	0.73	3.73	m	4.19
cut pockets and bond 75mm filling to existing blockwork at jambs	-	-	-	0.40	0.40	9.97	0.85	10.82	m	12.17
cut pockets and bond 100mm filling to existing blockwork at jambs	-	-	-	0.50	0.50	12.46	1.21	13.67	m	15.38

Labour hourly rates: (except Specialists) Craft Operatives £14.21 Labourer £10.71 Rates are national average prices. Refer to REGIONAL VARIATIONS for indicative levels of overall pricing in regions	MATERIALS			LABOUR				RATES		
	Del to Site £	Waste %	Material Cost £	Craft Optve Hrs	Lab Hrs	Labour Cost £	Sunds £	Nett Rate £	Unit	Gross Rate (+12.5%) £
C90: ALTERATIONS - SPOT ITEMS Cont.										
Filling openings in brickwork or blockwork Cont.										
Unit rates for pricing the above and similar work Cont.										
preservative treated softwood sole plate Cont.										
12mm (average) cement and sand dubbing over 300mm wide on filling	1.56	10.00	**1.71**	0.54	0.54	**13.46**	-	**15.17**	m2	**17.07**
13mm two coat lightweight plaster over 300mm wide on filling	2.14	10.00	**2.36**	1.27	1.27	**31.65**	0.02	**34.03**	m2	**38.28**
19 x 100mm softwood chamfered skirting, primed all round	1.63	10.00	**1.79**	0.50	0.08	**7.96**	0.13	**9.89**	m	**11.12**
junction with existing	-	-	**-**	0.50	0.08	**7.96**	-	**7.96**	nr	**8.96**
make good vinyl or similar floor covering up to filling	7.15	10.00	**7.86**	0.50	0.50	**12.46**	0.17	**20.49**	m	**23.05**
Fill opening where door removed in half brick wall with brickwork in gauged mortar (1:2:9); provide lead cored damp proof course in opening, lapped with existing, bond to existing brickwork at jambs and wedge and pin up at head; plaster filling both sides and extend skirting both sides; make good junction of new and existing plaster and skirtings										
838 x 2032mm	51.75	-	**51.75**	15.70	12.75	**359.65**	12.32	**423.72**	nr	**476.69**
Unit rates for pricing the above and similar work										
half brick filling in common bricks	13.20	10.00	**14.52**	2.00	2.00	**49.84**	3.06	**67.42**	m2	**75.84**
half brick filling in common bricks fair faced and flush pointed one side	13.20	10.00	**14.52**	2.33	2.33	**58.06**	3.29	**75.88**	m2	**85.36**
50 x 100mm softwood sole plate	1.67	10.00	**1.84**	0.33	0.06	**5.33**	0.30	**7.47**	m	**8.40**
lead cored bituminous hessian based damp proof course, BS 6398, Class F, width not exceeding 225mm	17.13	10.00	**18.84**	0.80	0.80	**19.94**	0.49	**39.26**	m2	**44.17**
wedge and pin up at head 102mm wide	-	-	**-**	0.12	0.12	**2.99**	0.73	**3.73**	m	**4.19**
cut pockets and bond half brick filling to existing brickwork at jambs	-	-	**-**	0.50	0.50	**12.46**	1.22	**13.68**	m	**15.39**
For unit rates for finishes, etc see previous items										
Fill opening where door removed in one brick external wall with brickwork in gauged mortar (1:2:9) faced externally with picked stock facings pointed to match existing; remove old lintel and arch, provide lead cored damp proof course in opening lapped with existing, bond to existing brickwork at jambs and wedge and pin up at head; plaster filling one side and extend skirting; make good junction of new and existing plaster and skirting										
914 x 2082mm	169.93	-	**169.93**	19.65	18.85	**481.11**	30.45	**681.49**	nr	**766.68**
Unit rates for pricing the above and similar work										
remove old lintel and arch	-	-	**-**	1.46	1.64	**38.31**	2.42	**40.73**	m	**45.82**
one brick filling in common bricks	26.40	10.00	**29.04**	4.00	4.00	**99.68**	7.32	**136.04**	m2	**153.05**
one brick filling in common bricks faced one side with picked stock facings and point to match existing	48.00	10.00	**52.80**	5.00	5.00	**124.60**	7.32	**184.72**	m2	**207.81**
one brick filling in common bricks fair faced and flush smooth pointing one side	26.40	10.00	**29.04**	4.33	4.33	**107.90**	7.32	**144.27**	m2	**162.30**
two course slate damp proof course, width not exceeding 225mm	24.16	10.00	**26.58**	1.02	1.02	**25.42**	0.85	**52.85**	m2	**59.45**
lead cored bituminous hessian based damp proof course, BS 6398, Class F, width not exceeding 225mm	17.13	10.00	**18.84**	0.80	0.80	**19.94**	0.37	**39.15**	m2	**44.04**
wedge and pin up at head 215mm wide	-	-	**-**	0.20	0.20	**4.98**	1.71	**6.69**	m	**7.53**
cut pockets and bond one brick filling to existing brickwork at jambs	-	-	**-**	1.00	1.00	**24.92**	2.81	**27.73**	m	**31.19**
12mm cement and sand (1:3) two coat external rendering over 300mm wide on filling	1.56	10.00	**1.71**	1.25	1.25	**31.15**	0.05	**32.91**	m2	**37.03**
For unit rates for internal finishes, etc. see previous items										
Fill opening where door removed in 275mm external hollow wall with inner skin in common bricks and outer skin in picked stock facings pointed to match existing, in gauged mortar (1:2:9); remove old lintel and arch, provide lead cored combined damp proof course and cavity gutter lapped with existing; form cavity with ties; bond to existing brickwork at jambs and wedge and pin up at head; plaster filling one side and extend skirting; make good junction of new and existing plaster and skirting										
914 x 2082mm	117.77	-	**117.77**	19.90	19.10	**487.34**	31.11	**636.22**	nr	**715.74**
Unit rates for pricing the above and similar work										
half brick filling in common bricks	13.20	10.00	**14.52**	2.00	2.00	**49.84**	3.06	**67.42**	m2	**75.84**
half brick filling in common bricks and fair face and flush smooth pointing one side	13.20	10.00	**14.52**	2.33	2.33	**58.06**	3.29	**75.88**	m2	**85.36**
half brick filling in picked stock facings and point to match existing	27.60	10.00	**30.36**	2.95	2.95	**73.51**	3.29	**107.17**	m2	**120.56**
form 50mm wide cavity with BS 1243 Fig. 3 galvanised steel twisted wall ties built in	1.27	10.00	**1.40**	0.20	0.20	**4.98**	-	**6.39**	m2	**7.18**
lead cored bituminous hessian based damp proof course and cavity tray, BS 6398, Class F, width exceeding 225mm	17.13	10.00	**18.84**	0.80	0.80	**19.94**	0.37	**39.15**	m2	**44.04**
wedge and pin up at head 102mm wide	-	-	**-**	0.12	0.12	**2.99**	0.73	**3.73**	m	**4.19**

EXISTING SITE/BUILDINGS/SERVICES

Labour hourly rates: (except Specialists) Craft Operatives £14.21 Labourer £10.71 Rates are national average prices. Refer to REGIONAL VARIATIONS for indicative levels of overall pricing in regions	MATERIALS			LABOUR			RATES			
	Del to Site £	Waste %	Material Cost £	Craft Optve Hrs	Lab Hrs	Labour Cost £	Sunds £	Nett Rate £	Unit	Gross Rate (+12.5%) £
C90: ALTERATIONS - SPOT ITEMS Cont.										
Filling openings in brickwork or blockwork Cont.										
Unit rates for pricing the above and similar work Cont.										
cut pockets and bond half brick wall in common bricks to existing brickwork at jambs	-	-	-	0.50	0.50	**12.46**	1.22	**13.68**	m	**15.39**
cut pockets and bond half brick wall in facing bricks to existing brickwork at jambs and make good facings	-	-	-	0.60	0.60	**14.95**	2.56	**17.51**	m	**19.70**
For unit rates for internal finishes, etc. see previous items										
Fillings openings in rubble walling										
Fill opening where window removed in 600mm rough rubble wall with rubble walling in lime mortar (1:3) pointed to match existing; bond to existing walling at jambs and wedge and pin up at head; plaster filling one side and make good junction of new and existing plaster										
900 x 900mm	117.26	-	**117.26**	12.00	12.00	**299.04**	18.84	**435.14**	nr	**489.53**
Unit rates for pricing the above and similar work										
rough rubble walling in filling										
600mm thick	121.00	10.00	**133.10**	6.00	6.00	**149.52**	14.65	**297.27**	m2	**334.42**
random rubble walling in filling										
300mm thick	88.00	10.00	**96.80**	5.80	5.80	**144.54**	7.32	**248.66**	m2	**279.74**
500mm thick	130.00	10.00	**143.00**	10.90	10.90	**271.63**	12.21	**426.83**	m2	**480.19**
wedge and pin up at head										
300mm wide	-	-	-	0.28	0.28	**6.98**	2.44	**9.42**	m	**10.60**
500mm wide	-	-	-	0.47	0.47	**11.71**	4.15	**15.86**	m	**17.84**
600mm wide	-	-	-	0.56	0.56	**13.96**	4.88	**18.84**	m	**21.19**
cut pockets and bond walling to existing at jambs										
300mm	6.60	10.00	**7.26**	0.83	0.83	**20.68**	0.73	**28.68**	m	**32.26**
500mm	9.75	10.00	**10.73**	1.38	1.38	**34.39**	1.10	**46.22**	m	**51.99**
600mm	9.08	10.00	**9.98**	1.75	1.75	**43.61**	1.34	**54.93**	m	**61.80**
For unit rates for internal finishes, etc. see previous items										
Work to chimney stacks										
SEAL AND VENTILATE TOP OF FLUE - remove chimney pot and flaunching and provide 50mm precast concrete, BS 5328, designed mix C20, 20mm aggregate, weathered and throated capping and bed in cement mortar (1:3) to seal top of stack; cut opening through brick side of stack and build in 225 x 225mm clay air brick to ventilate flue and make good facings										
stack size 600 x 600mm with one flue	21.26	10.00	**22.88**	4.90	4.90	**122.11**	5.12	**150.10**	nr	**168.87**
stack size 825 x 600mm with two flues	32.42	10.00	**34.64**	9.00	9.00	**224.28**	9.43	**268.35**	nr	**301.90**
stack size 1050 x 600mm with three flues	43.58	10.00	**46.41**	13.10	13.10	**326.45**	12.93	**385.79**	nr	**434.01**
RENEW DEFECTIVE CHIMNEY POT - remove chimney pot and provide new clay chimney pot set and flaunched in cement mortar (1:3)										
150mm diameter x 600mm high	25.90	-	**25.90**	2.50	2.50	**62.30**	4.63	**92.83**	nr	**104.44**
REBUILD DEFECTIVE CHIMNEY STACK - pull down defective stack to below roof level for a height of 2.00m; prepare for raising and rebuild in common brickwork faced with picked stock facings in gauged mortar (1:2:9) pointed to match existing; parge and core flues; provide No. 4 lead flashings and soakers; provide 150mm diameter clay chimney pots 600mm high set and flaunched in cement mortar (1:3); make good roof tiling and all other work disturbed										
600 x 600mm with one flue	280.77	10.00	**306.26**	35.30	37.55	**903.77**	69.92	**1279.94**	nr	**1439.94**
825 x 600mm with two flues	367.90	10.00	**399.51**	47.15	47.90	**1183.01**	97.17	**1679.69**	nr	**1889.65**
REMOVE DEFECTIVE CHIMNEY STACK - pull down stack to below roof level; remove all flashings and soakers; piece in 50 x 100mm rafters; extend roof tiling with machine made plain tiles to 100mm gauge nailed every fourth course with galvanised nails to and with 25 x 19mm battens; bond new tiling to existing										
850 x 600mm for a height of 2.00m	36.02	10.00	**39.62**	26.00	22.30	**608.29**	42.66	**690.58**	nr	**776.90**
Work to chimney breasts and fireplaces										
TAKE OUT FIREPLACE AND FILL IN OPENING - remove fire surround, fire back and hearth; hack up screed to hearth and extend 25mm tongued and grooved softwood floor boarding over hearth on and with bearers; fill in opening where fireplace removed with 75mm concrete blocks in gauged mortar (1:2:9) bonded to existing at jambs and wedged and pinned up to soffits; form 225 x 225mm opening and provide and build in 225 x 225mm fibrous plaster louvered air vent; plaster filling with 12mm two coat lightweight plaster on 12mm (average) dubbing plaster; extend 19 x 100mm softwood chamfered skirting over filling and join with existing; make good all new finishings to existing										
fireplace opening 570 x 685mm, hearth 1030 x 405mm	39.79	-	**39.79**	8.00	7.45	**193.47**	12.21	**245.47**	nr	**276.15**
fireplace opening 800 x 760mm, hearth 1260 x 405mm	48.64	-	**48.64**	9.65	8.95	**232.98**	14.22	**295.84**	nr	**332.82**

Labour hourly rates: (except Specialists) Craft Operatives £14.21 Labourer £10.71 Rates are national average prices. Refer to REGIONAL VARIATIONS for indicative levels of overall pricing in regions	MATERIALS			LABOUR				RATES		
	Del to Site £	Waste %	Material Cost £	Craft Optve Hrs	Lab Hrs	Labour Cost £	Sunds £	Nett Rate £	Unit	Gross Rate (+12.5%) £
C90: ALTERATIONS - SPOT ITEMS Cont.										
Work to chimney breasts and fireplaces Cont.										
REMOVE CHIMNEY BREAST - pull down chimney breast for full height from ground to roof level (two storeys) including removing two fire surrounds and hearths complete; make out brickwork where flues removed and make out brickwork where breasts removed; extend 50 x 175mm floor joists and 25mm tongued and grooved softwood floor boarding (ground and first floors); extend 50 x 100mm ceiling joists (first floor); plaster walls with 12mm two coat lightweight plaster on 12mm (average) plaster dubbing; extend ceiling plaster with expanded metal lathing and 19mm three coat lightweight plaster; run 19 x 100mm softwood chamfered skirting to walls to join with existing; make good all new finishings up to existing										
chimney breast 2.00m wide and 7.50m high including gathering in roof space	163.90	-	**163.90**	62.00	84.00	**1780.66**	170.64	**2115.20**	nr	**2379.60**
Form trap door in ceiling										
Form new trap door in ceiling with 100mm deep joists and lath and plaster finish; cut away and trim ceiling joists around opening and insert new trimming and trimmer joists; provide 32x 127mm softwood lining with 15 x 25mm stop and 20 x 70mm architrave; provide and place in position 18mm blockboard trap door, lipped all round; make good existing ceiling plaster around openings										
600 x 900mm	43.60	-	**43.60**	21.85	5.60	**370.46**	10.78	**424.85**	nr	**477.95**
Unit rates for pricing the above and similar work										
cut away plastered plasterboard ceiling	-	-	-	1.50	0.33	**24.85**	1.01	**25.86**	m2	**29.09**
cut away lath and plaster ceiling	-	-	-	2.00	0.41	**32.81**	1.21	**34.02**	m2	**38.27**
cut away and trim existing 100mm deep joists around opening										
664 x 964mm	-	-	-	0.95	0.16	**15.21**	0.84	**16.05**	nr	**18.06**
964 x 1444mm	-	-	-	1.90	0.33	**30.53**	1.69	**32.23**	nr	**36.26**
cut away and trim existing 150mm deep joists around opening										
664 x 964mm	-	-	-	1.43	0.24	**22.89**	1.29	**24.18**	nr	**27.21**
964 x 1444mm	-	-	-	2.86	0.48	**45.78**	2.54	**48.32**	nr	**54.36**
softwood trimming or trimmer joists										
75 x 100mm	2.70	10.00	**2.97**	0.95	0.16	**15.21**	0.43	**18.61**	m	**20.94**
75 x 150mm	4.05	10.00	**4.46**	1.00	0.17	**16.03**	0.46	**20.95**	m	**23.57**
softwood lining										
32 x 125mm	2.10	10.00	**2.31**	1.50	0.25	**23.99**	0.13	**26.43**	m	**29.74**
32 x 150mm	2.18	10.00	**2.40**	1.60	0.27	**25.63**	0.09	**28.12**	m	**31.64**
15 x 25mm softwood stop	0.45	10.00	**0.50**	0.33	0.05	**5.22**	0.06	**5.78**	m	**6.50**
19 x 50mm softwood twice rounded architrave to Patt. 18	0.82	10.00	**0.90**	0.33	0.05	**5.22**	0.06	**6.19**	m	**6.96**
18mm blockboard trap door lipped all round										
600 x 900mm	12.94	10.00	**14.23**	3.00	0.50	**47.99**	0.20	**62.42**	nr	**70.22**
900 x 1280mm	29.60	10.00	**32.56**	6.00	1.00	**95.97**	0.36	**128.89**	nr	**145.00**
75mm steel butt hinge to softwood	0.35	-	**0.35**	0.66	0.11	**10.56**	0.20	**11.11**	nr	**12.50**
make good plastered plasterboard ceiling around opening	-	-	-	0.75	0.75	**18.69**	1.46	**20.15**	m	**22.67**
make good lath and plaster ceiling around opening	-	-	-	0.83	0.83	**20.68**	1.79	**22.47**	m	**25.28**
Openings through stud partitions										
Form opening for door frame through lath and plaster finished 50 x 100mm stud partition; take off skirtings, cut away plaster, cut away and trim studding around opening and insert new studding at head and jambs of opening; make good existing lath and plaster up to new frame both sides and form fitted ends on existing skirtings up to new frame; extend softwood board flooring through opening on and with bearers and make good up to existing										
838 x 2032mm	13.94	-	**13.94**	18.10	6.10	**322.53**	19.08	**355.55**	nr	**399.99**
Unit rates for pricing the above and similar work										
cut away plastered plasterboard	-	-	-	1.00	0.25	**16.89**	1.01	**17.89**	m2	**20.13**
cut away lath and plaster	-	-	-	1.10	0.26	**18.42**	1.21	**19.62**	m2	**22.08**
cut away and trim existing 75mm studding around opening										
838 x 2032mm	-	-	-	2.05	0.36	**32.99**	2.58	**35.57**	nr	**40.02**
914 x 2032mm	-	-	-	2.05	0.36	**32.99**	2.58	**35.57**	nr	**40.02**
cut away and trim existing 100mm studding around opening										
838 x 2032mm	-	-	-	2.55	0.45	**41.06**	3.38	**44.43**	nr	**49.99**
914 x 2032mm	-	-	-	2.55	0.45	**41.06**	3.38	**44.43**	nr	**49.99**
50 x 75mm softwood studding	1.26	10.00	**1.39**	0.50	0.08	**7.96**	0.13	**9.48**	m	**10.66**
50 x 100mm softwood studding	1.67	10.00	**1.84**	0.60	0.10	**9.60**	0.14	**11.58**	m	**13.02**
make good plastered plasterboard around opening	-	-	-	0.75	0.75	**18.69**	1.46	**20.15**	m	**22.67**
make good lath and plaster around opening	-	-	-	0.83	0.83	**20.68**	1.79	**22.47**	m	**25.28**
take off skirting	-	-	-	0.21	-	**2.98**	0.21	**3.20**	m	**3.60**
form fitted end on existing skirting	-	-	-	0.57	-	**8.10**	-	**8.10**	nr	**9.11**
extend 25mm softwood tongued and grooved board flooring through opening on and with bearers	3.11	20.00	**3.57**	1.50	0.25	**23.99**	0.07	**27.63**	nr	**31.08**

Labour hourly rates: (except Specialists) Craft Operatives £14.21 Labourer £10.71 Rates are national average prices. Refer to REGIONAL VARIATIONS for indicative levels of overall pricing in regions	MATERIALS			LABOUR				RATES		
	Del to Site £	Waste %	Material Cost £	Craft Optve Hrs	Lab Hrs	Labour Cost £	Sunds £	Nett Rate £	Unit	Gross Rate (+12.5%) £
C90: ALTERATIONS - SPOT ITEMS Cont.										
Filling openings in stud partitions										
Fill opening where door removed in stud partition with 50 x 100mm studding and sole plate, covered on both sides with 9.5mm plasterboard baseboard with 5mm one coat board finish plaster; extend skirting on both sides; make good junction of new and existing plaster and skirtings										
838 x 2032mm	31.98	-	**31.98**	16.00	5.36	**284.77**	3.56	**320.30**	nr	**360.34**
Unit rates for pricing the above and similar work										
50 x 75mm softwood sole plate	1.26	10.00	**1.39**	0.50	0.08	**7.96**	0.13	**9.48**	m	**10.66**
50 x 100mm softwood sole plate	1.67	10.00	**1.84**	0.60	0.10	**9.60**	0.17	**11.60**	m	**13.05**
50 x 75mm softwood studding	1.26	10.00	**1.39**	0.50	0.08	**7.96**	0.13	**9.48**	m	**10.66**
50 x 100mm softwood studding	1.67	10.00	**1.84**	0.60	0.10	**9.60**	0.17	**11.60**	m	**13.05**
9.5mm plasterboard baseboard to filling	2.17	10.00	**2.39**	0.50	0.10	**8.18**	0.33	**10.89**	m2	**12.26**
9.5mm plasterboard baseboard to filling including packing out not exceeding 10mm	2.17	10.00	**2.39**	1.00	0.17	**16.03**	0.44	**18.86**	m2	**21.21**
5mm one coat board finish plaster on plasterboard to filling	1.14	10.00	**1.25**	0.91	0.91	**22.68**	0.04	**23.97**	m2	**26.96**
10mm two coat hardwall plaster on plasterboard to filling	1.76	10.00	**1.93**	1.27	1.27	**31.65**	0.04	**33.61**	m2	**37.82**
19 x 100mm softwood chamfered skirting, primed all round	1.63	10.00	**1.79**	0.54	0.09	**8.64**	0.21	**10.64**	m	**11.97**
junction with existing	-	-	-	0.57	-	**8.10**	-	**8.10**	nr	**9.11**
25 x 150mm softwood chamfered skirting, primed all round	2.92	10.00	**3.21**	0.55	0.09	**8.78**	0.28	**12.28**	m	**13.81**
junction with existing	-	-	-	0.57	-	**8.10**	-	**8.10**	nr	**9.11**
Work to existing drains										
BREAK UP CONCRETE PAVING FOR EXCAVATING DRAIN TRENCH AND REINSTATE - break up 100mm concrete paving and 150mm hardcore bed under for excavation of drain trench and reinstate with new hardcore and concrete, BS 5328, designed mix C20, 20mm aggregate and make good up to existing paving both sides										
400mm wide	5.70	30.00	**6.42**	-	3.27	**35.02**	4.03	**45.47**	m	**51.15**
TAKE UP FLAG PAVING FOR EXCAVATING DRAIN TRENCH AND REINSTATE - carefully take up 50mm precast concrete flag paving and set aside for re-use and break up 150mm hardcore bed under for excavation of drain trench and reinstate with new hardcore and re-lay salvaged flag paving on and with 25mm sand bed and grout in lime sand (1:3) and make good up to existing paving on both sides										
400mm wide	1.98	30.00	**2.55**	1.47	1.47	**36.63**	3.71	**42.89**	m	**48.25**
INSERT NEW JUNCTION IN EXISTING DRAIN -excavate for and trace and expose existing drain, break into glazed vitrified clay drain and insert junction with 100mm branch, short length of new pipe and double collar and joint to existing drain; support earthwork, make good concrete bed and haunching and backfill										
existing 100mm diameter drain, invert depth										
600mm	29.75	10.00	**32.73**	3.00	12.00	**171.15**	10.25	**214.13**	nr	**240.89**
750mm	29.75	10.00	**32.73**	3.00	14.25	**195.25**	11.08	**239.05**	nr	**268.93**
900mm	29.75	10.00	**32.73**	3.00	16.50	**219.35**	12.15	**264.22**	nr	**297.24**
existing 150mm diameter drain invert depth										
750mm	54.13	10.00	**59.54**	4.50	15.75	**232.63**	12.80	**304.97**	nr	**343.09**
900mm	54.13	10.00	**59.54**	4.50	18.00	**256.73**	14.52	**330.78**	nr	**372.13**
1050mm	54.13	10.00	**59.54**	4.50	20.25	**280.82**	15.88	**356.24**	nr	**400.78**
1200mm	54.13	10.00	**59.54**	4.50	22.50	**304.92**	17.36	**381.82**	nr	**429.55**
REPAIR DEFECTIVE DRAIN - excavate for and trace and expose existing drain, break out fractured glazed vitrified clay drain pipe, replace with new pipe and double collars and joint to existing drain; support earthwork, make good concrete bed and haunching and backfill										
existing 100mm diameter drain; single pipe length; invert depth										
450mm	25.11	10.00	**27.62**	3.00	11.50	**165.80**	7.76	**201.18**	nr	**226.32**
600mm	25.11	10.00	**27.62**	3.00	13.50	**187.22**	9.36	**224.20**	nr	**252.22**
750mm	25.11	10.00	**27.62**	3.00	16.00	**213.99**	11.44	**253.05**	nr	**284.68**
add for each additional pipe length; invert depth										
450mm	16.23	10.00	**17.85**	2.00	6.75	**100.71**	7.41	**125.97**	nr	**141.72**
600mm	16.23	10.00	**17.85**	2.00	7.60	**109.82**	8.95	**136.62**	nr	**153.69**
750mm	16.23	10.00	**17.85**	2.00	8.80	**122.67**	10.72	**151.25**	nr	**170.15**
Work to existing manholes										
RAISE TOP OF EXISTING MANHOLE - take off cover and frame and set aside for re-use, prepare level bed on existing one brick sides of manhole internal size 610 x 457mm and raise with Class B engineering brickwork in cement mortar (1:3) finished with a fair face and flush pointed; refix salvaged cover and frame, bed frame in cement mortar and cover in grease and sand										
raising 150mm high	16.74	10.00	**18.41**	5.30	5.30	**132.08**	3.76	**154.25**	nr	**173.53**
raising 225mm high	25.11	10.00	**27.62**	6.75	6.75	**168.21**	5.57	**201.40**	nr	**226.58**
raising 300mm high	33.48	10.00	**36.83**	8.50	8.50	**211.82**	7.13	**255.78**	nr	**287.75**

Labour hourly rates: (except Specialists) Craft Operatives £14.21 Labourer £10.71 Rates are national average prices. Refer to REGIONAL VARIATIONS for indicative levels of overall pricing in regions	MATERIALS			LABOUR				RATES		
	Del to Site £	Waste %	Material Cost £	Craft Optve Hrs	Lab Hrs	Labour Cost £	Sunds £	Nett Rate £	Unit	Gross Rate (+12.5%) £
C90: ALTERATIONS - SPOT ITEMS Cont.										
Work to existing manholes Cont.										
INSERT NEW BRANCH BEND IN BOTTOM OF EXISTING MANHOLE - break into bottom and one brick side of manhole, insert new glazed vitrified clay three quarter section branch bend to discharge over existing main channel, build in end of new drain and make good benching and side of manhole to match existing										
100mm diameter branch bend	10.58	10.00	**11.64**	5.75	5.75	**143.29**	2.69	**157.62**	nr	**177.32**
EXTEND EXISTING MANHOLE - take off cover and frame and set aside; break up one brick end wall of 457mm wide (internal) manhole and excavate for and extend manhole 225mm; support earthwork, level and compact bottom, part backfill and remove surplus spoil from site; extend bottom with 150mm concrete, BS 5328, ordinary prescribed mix C15P, 20mmaggregate bed; extend one brick sides and end with Class B engineering brickwork in cement mortar (1:3) fair faced and flush pointed and bond to existing; extend 100mm main channel and insert 100mm three quarter section branch channel bend; extend benching and build in end of new 100mm drain to one brick side, corbel over end of manhole manholes of the following invert depths; refix salvaged cover and frame, bed frame in cement mortar and cover in grease and sand										
450mm	51.06	10.00	**56.17**	17.60	17.90	**441.81**	10.38	**508.35**	nr	**571.90**
600mm	56.21	10.00	**61.83**	19.80	20.20	**497.70**	13.34	**572.87**	nr	**644.48**
750mm	61.53	10.00	**67.68**	22.00	22.50	**553.60**	16.31	**637.58**	nr	**717.28**
900mm	66.75	10.00	**73.43**	24.20	24.80	**609.49**	19.27	**702.18**	nr	**789.96**
Temporary weatherproof coverings										
Providing and erecting and clearing away on completion; for scaffold tube framework see preliminaries										
corrugated iron sheeting	6.32	10.00	**6.95**	0.15	0.50	**7.49**	0.94	**15.37**	m2	**17.30**
flexible reinforced plastic sheeting	2.73	10.00	**3.00**	0.10	0.25	**4.10**	0.76	**7.86**	m2	**8.84**
tarpaulins	5.80	10.00	**6.38**	0.10	0.25	**4.10**	0.76	**11.24**	m2	**12.64**
Temporary internal screens										
Providing and erecting and clearing away on completion; temporary screen of 50 x 50mm softwood framing and cover with										
reinforced building paper	2.60	10.00	**2.86**	0.43	0.30	**9.32**	0.26	**12.44**	m2	**14.00**
heavy duty polythene sheeting	1.78	10.00	**1.96**	0.43	0.30	**9.32**	0.24	**11.52**	m2	**12.96**
3mm hardboard	2.68	10.00	**2.95**	0.53	0.32	**10.96**	0.39	**14.30**	m2	**16.08**
Providing and erecting and clearing away on completion; temporary dustproof screen of 50 x 75mm softwood framing securely fixed to walls, floor and ceiling, the joints and edges of lining sealed with masking tape and cover with										
3mm hardboard lining one side	3.45	10.00	**3.80**	1.10	0.18	**17.56**	0.52	**21.88**	m2	**24.61**
3mm hardboard lining both sides	4.76	10.00	**5.24**	1.30	0.22	**20.83**	0.65	**26.72**	m2	**30.06**
6mm plywood lining one side	6.74	10.00	**7.41**	1.15	0.19	**18.38**	0.58	**26.37**	m2	**29.67**
6mm plywood lining both sides	11.28	10.00	**12.41**	1.40	0.23	**22.36**	0.91	**35.68**	m2	**40.14**
providing 35mm hardboard faced flush door size 838 x 1981mm in dustproof screen with 25 x 87mm softwood frame, 18 x 25mm softwood stop, pair of 100mm butts, pull handle and ball catch	137.36	10.00	**151.10**	1.50	0.25	**23.99**	3.29	**178.38**	nr	**200.68**
Temporary dustproof corridors										
Note for prices of walls and doors, see dustproof screens above										
Providing and erecting and clearing away on completion; ceiling of 50 x 75mm softwood joists at 450mm centres, the joints and edges sealed with masking tape and cover with										
3mm hardboard lining to soffit	3.72	10.00	**4.09**	1.10	0.18	**17.56**	0.52	**22.17**	m2	**24.94**
6mm plywood lining to soffit	6.98	10.00	**7.68**	1.15	0.19	**18.38**	0.65	**26.71**	m2	**30.04**
Temporary timber balustrades										
Providing and erecting and clearing away on completion; temporary softwood balustrade consisting of 50 x 75mm plate, 50 x 50mm standards at 900mm centres, four 25 x 150mm intermediate rails and 50 x 50mm handrail										
1150mm high	6.83	10.00	**7.51**	1.00	0.26	**16.99**	1.49	**26.00**	m	**29.25**
Temporary steel balustrades										
Providing and erecting and clearing away on completion; temporary steel balustrade constructed of 50mm diameter galvanised scaffold tubing and fittings with standards at 900mm centres intermediate rail, handrail, plastic mesh infill and with plates on ends of standards fixed to floor										
1150mm high	6.80	-	**6.80**	0.18	0.90	**12.20**	0.36	**19.35**	m	**21.77**
weekly cost of hire and maintenance	0.67	-	**0.67**	0.02	0.10	**1.36**	0.05	**2.07**	m	**2.33**

Labour hourly rates: (except Specialists) Craft Operatives £14.21 Labourer £10.71 Rates are national average prices. Refer to REGIONAL VARIATIONS for indicative levels of overall pricing in regions	MATERIALS			LABOUR				RATES		
	Del to Site £	Waste %	Material Cost £	Craft Optve Hrs	Lab Hrs	Labour Cost £	Sunds £	Nett Rate £	Unit	Gross Rate (+12.5%) £
C90: ALTERATIONS - SPOT ITEMS Cont.										
Temporary fillings to openings in external walls										
Providing and erecting and clearing away on completion; temporary filling to window or door opening of 50 x 100mm softwood framing and cover with										
corrugated iron sheeting	13.92	10.00	**15.31**	0.45	0.45	**11.21**	1.55	**28.08**	m2	**31.59**
25mm softwood boarding	12.42	10.00	**13.66**	0.60	0.60	**14.95**	1.81	**30.43**	m2	**34.23**
12mm external quality plywood	11.32	10.00	**12.45**	0.50	0.50	**12.46**	1.17	**26.09**	m2	**29.35**
C90: REPAIRING/RENOVATING ASPHALT WORK										
Cut out crack in old covering and make good with new material to match existing										
Floor or tanking										
20mm thick in two coats, horizontal	-	-	-	0.66	0.66	**16.45**	0.59	**17.04**	m	**19.17**
30mm thick in three coats, horizontal	-	-	-	0.90	0.90	**22.43**	0.89	**23.32**	m	**26.23**
30mm thick in three coats, vertical	-	-	-	1.19	1.19	**29.65**	0.89	**30.54**	m	**34.36**
Cut out detached blister in old covering and make good with new material to match existing										
Floor or tanking										
20mm thick in two coats, horizontal	-	-	-	0.30	0.30	**7.48**	0.30	**7.77**	nr	**8.74**
30mm thick in three coats, horizontal	-	-	-	0.35	0.35	**8.72**	0.41	**9.14**	nr	**10.28**
30mm thick in three coats, vertical	-	-	-	0.48	0.48	**11.96**	0.41	**12.38**	nr	**13.92**
Cut out crack in old covering and make good with new material to match existing										
Roof covering										
20mm thick in two coats	-	-	-	0.66	0.66	**16.45**	0.59	**17.04**	m	**19.17**
Cut out detached blister in old covering and make good with new material to match existing										
20mm thick in two coats	-	-	-	0.30	0.30	**7.48**	0.30	**7.77**	nr	**8.74**
C90: REPAIRING/RENOVATING ROOF COVERINGS										
Slate roofing repairs to roofs covered with 610 x 300mm slates										
Remove damaged slates and replace with new; sloping or vertical										
one slate	5.73	5.00	**6.02**	0.32	0.32	**7.97**	0.11	**14.10**	nr	**15.86**
patch of 10 slates	57.30	5.00	**60.17**	1.10	1.10	**27.41**	0.83	**88.41**	nr	**99.46**
patch of slates 1.00 - 3.00m2	70.19	5.00	**73.70**	1.25	1.25	**31.15**	0.90	**105.75**	m2	**118.97**
Examine battens, remove defective and provide 20% new										
19 x 38mm	0.21	10.00	**0.23**	0.09	0.09	**2.24**	0.06	**2.53**	m2	**2.85**
25 x 50mm	0.36	10.00	**0.40**	0.09	0.09	**2.24**	0.08	**2.72**	m2	**3.06**
Re-cover roof with slates previously removed and stacked and fix with slate nails										
with 75mm lap	-	-	-	0.77	0.77	**19.19**	0.11	**19.30**	m2	**21.71**
extra for providing 20% new slates	14.04	5.00	**14.74**	-	-	-	0.21	**14.95**	m2	**16.82**
Remove double course at eaves and										
refix	-	-	-	0.24	0.24	**5.98**	0.37	**6.35**	m	**7.14**
extra for providing 20% new slates	4.87	5.00	**5.11**	-	-	-	0.11	**5.22**	m	**5.87**
Remove double course at verge and replace with new slates and bed and point in mortar	21.43	5.00	**22.50**	0.29	0.29	**7.23**	0.69	**30.42**	m	**34.22**
Slate roofing repairs for roofs covered with 510 x 250mm slates										
Remove damaged slates and replace with new; sloping or vertical										
one slate	2.76	5.00	**2.90**	0.32	0.32	**7.97**	0.11	**10.98**	nr	**12.35**
patch of 10 slates	27.60	5.00	**28.98**	1.00	1.00	**24.92**	0.70	**54.60**	nr	**61.42**
patch of slates 1.00 - 3.00m2	49.68	5.00	**52.16**	1.49	1.49	**37.13**	0.89	**90.18**	m2	**101.46**
Examine battens, remove defective and provide 20% new										
19 x 38mm	0.26	10.00	**0.28**	0.11	0.11	**2.74**	0.08	**3.11**	m2	**3.50**
25 x 50mm	0.44	10.00	**0.49**	0.11	0.11	**2.74**	0.11	**3.33**	m2	**3.75**
Re-cover roof with slates previously removed and stacked and fix with slate nails										
with 75mm lap	-	-	-	1.00	1.00	**24.92**	0.15	**25.07**	m2	**28.21**
extra for providing 20% new slates	9.94	5.00	**10.43**	-	-	-	0.21	**10.65**	m2	**11.98**
Remove double course at eaves and										
refix	-	-	-	0.25	0.25	**6.23**	0.63	**6.86**	m	**7.72**
extra for providing 20% new slates	3.09	5.00	**3.25**	-	-	-	0.09	**3.34**	m	**3.76**
Remove double course at verge and replace with new slates and bed and point in mortar	12.70	5.00	**13.33**	0.29	0.29	**7.23**	0.68	**21.23**	m	**23.89**

Labour hourly rates: (except Specialists) Craft Operatives £14.21 Labourer £10.71 Rates are national average prices. Refer to REGIONAL VARIATIONS for indicative levels of overall pricing in regions	MATERIALS			LABOUR				RATES		
	Del to Site	Waste	Material Cost	Craft Optve	Lab	Labour Cost	Sunds	Nett Rate	Unit	Gross Rate (+12.5%)
	£	%	£	Hrs	Hrs	£	£	£		£
C90: REPAIRING/RENOVATING ROOF COVERINGS Cont.										
Slate roofing repairs for roofs covered with 405 x 200mm slates										
Remove damaged slates and replace with new; sloping or vertical										
one slate	1.36	5.00	**1.43**	0.32	0.32	**7.97**	0.09	**9.50**	nr	**10.68**
patch of 10 slates	13.60	5.00	**14.28**	0.90	0.90	**22.43**	0.60	**37.31**	nr	**41.98**
patch of slates 1.00 - 3.00m2	40.80	5.00	**42.84**	2.07	2.07	**51.58**	0.98	**95.41**	m2	**107.33**
Examine battens, remove defective and provide 20% new										
19 x 38mm	0.34	10.00	**0.38**	0.13	0.13	**3.24**	0.09	**3.71**	m2	**4.18**
25 x 50mm	0.59	10.00	**0.65**	0.13	0.13	**3.24**	0.11	**4.00**	m2	**4.50**
Re-cover roof with slates previously removed and stacked and fix with slate nails										
with 75mm lap	-	-	-	1.14	1.14	**28.41**	0.23	**28.63**	m2	**32.21**
extra for providing 20% new slates	8.16	5.00	**8.57**	-	-	-	0.21	**8.78**	m2	**9.88**
Remove double course at eaves and										
refix	-	-	-	0.26	0.26	**6.48**	0.62	**7.10**	m	**7.98**
extra for providing 20% new slates	1.85	5.00	**1.94**	-	-	-	0.09	**2.04**	m	**2.29**
Remove double course at verge and replace with new slates and bed and point in mortar	7.93	5.00	**8.33**	0.30	0.30	**7.48**	0.75	**16.55**	m	**18.62**
Tile roofing repairs for roofs covered with machine made clay plain tiles laid to 100mm gauge										
Remove damaged tiles and replace with new, sloping or vertical										
one tile	0.51	10.00	**0.56**	0.32	0.32	**7.97**	0.04	**8.57**	nr	**9.64**
patch of 10 tiles	5.12	10.00	**5.63**	0.90	0.90	**22.43**	0.30	**28.36**	nr	**31.90**
patch of tiles 1.00 - 3.00m2	31.23	10.00	**34.36**	1.72	1.72	**42.86**	1.90	**79.11**	m2	**89.00**
Examine battens, remove defective and provide 20% new										
19 x 25mm	0.46	10.00	**0.51**	0.13	0.13	**3.24**	0.09	**3.84**	m2	**4.32**
19 x 38mm	0.56	10.00	**0.62**	0.14	0.14	**3.49**	0.11	**4.21**	m2	**4.74**
19 x 50mm	0.84	10.00	**0.92**	0.14	0.14	**3.49**	0.13	**4.54**	m2	**5.11**
Re-cover roof with tiles previously removed and stacked										
nail every fourth course with galvanised nails	-	-	-	1.00	1.00	**24.92**	0.20	**25.12**	m2	**28.26**
extra for nailing every course	-	-	-	0.20	0.20	**4.98**	0.53	**5.52**	m2	**6.21**
extra for providing 20% new tiles	6.25	10.00	**6.87**	-	-	-	0.21	**7.08**	m2	**7.97**
Tile roofing repairs to roofs covered with machine made clay plain tiles laid to 90mm gauge										
Remove damaged tiles and replace with new, sloping or vertical										
one tile	0.51	10.00	**0.56**	0.32	0.32	**7.97**	0.04	**8.57**	nr	**9.64**
patch of 10 tiles	5.12	10.00	**5.63**	0.90	0.90	**22.43**	0.30	**28.36**	nr	**31.90**
patch of tiles 1.00 - 3.00m2	34.82	10.00	**38.30**	2.05	2.05	**51.09**	2.11	**91.49**	m2	**102.93**
Examine battens, remove defective and provide 20% new										
19 x 25mm	0.51	10.00	**0.56**	0.14	0.14	**3.49**	0.11	**4.16**	m2	**4.68**
19 x 38mm	0.62	10.00	**0.68**	0.15	0.15	**3.74**	0.13	**4.55**	m2	**5.12**
19 x 50mm	0.93	10.00	**1.03**	0.15	0.15	**3.74**	0.14	**4.91**	m2	**5.52**
Re-cover roof with tiles previously removed and stacked										
nail every fourth course with galvanised nails	-	-	-	1.15	1.15	**28.66**	0.23	**28.88**	m2	**32.49**
extra for nailing every course	-	-	-	0.23	0.23	**5.73**	0.56	**6.29**	m2	**7.07**
extra for providing 20% new tiles	6.96	10.00	**7.66**	-	-	-	0.21	**7.87**	m2	**8.86**
Tile roofing repairs for roofs covered with 381 x 227mm concrete interlocking tiles laid to 75mm laps										
Remove damaged tiles and replace with new, sloping or vertical										
one tile	0.62	10.00	**0.69**	0.32	0.32	**7.97**	0.08	**8.74**	nr	**9.84**
patch of 10 tiles	6.24	10.00	**6.86**	0.90	0.90	**22.43**	0.87	**30.16**	nr	**33.93**
patch of tiles 1.00 - 3.00m2	10.61	10.00	**11.67**	0.85	0.85	**21.18**	1.37	**34.23**	m2	**38.50**
Examine battens, remove defective and provide 20% new										
19 x 25mm	0.17	10.00	**0.18**	0.05	0.05	**1.25**	0.08	**1.51**	m2	**1.70**
19 x 38mm	0.20	10.00	**0.22**	0.06	0.06	**1.50**	0.08	**1.80**	m2	**2.02**
19 x 50mm	0.30	10.00	**0.33**	0.07	0.07	**1.74**	0.09	**2.17**	m2	**2.44**
Re-cover roof with tiles previously removed and stacked										
nail every fourth course with galvanised nails	-	-	-	0.50	0.50	**12.46**	0.02	**12.48**	m2	**14.04**
extra for nailing every course	-	-	-	0.05	0.05	**1.25**	0.07	**1.32**	m2	**1.48**
extra for providing 20% new tiles	2.50	10.00	**2.75**	-	-	-	0.26	**3.01**	m2	**3.38**
Tile roofing repairs generally										
Remove damaged clay tiles and provide new										
half round ridge or hip and bed in mortar	19.73	15.00	**22.69**	0.60	0.60	**14.95**	1.29	**38.94**	m	**43.80**
bonnet hip	27.60	5.00	**28.98**	0.75	0.75	**18.69**	0.36	**48.03**	m	**54.03**
trough valley	27.60	5.00	**28.98**	0.75	0.75	**18.69**	0.36	**48.03**	m	**54.03**
vertical angle	27.60	10.00	**30.36**	0.75	0.75	**18.69**	0.36	**49.41**	m	**55.58**

Labour hourly rates: (except Specialists) Craft Operatives £14.21 Labourer £10.71 Rates are national average prices. Refer to REGIONAL VARIATIONS for indicative levels of overall pricing in regions	MATERIALS			LABOUR				RATES		
	Del to Site £	Waste %	Material Cost £	Craft Optve Hrs	Lab Hrs	Labour Cost £	Sunds £	Nett Rate £	Unit	Gross Rate (+12.5%) £
C90: REPAIRING/RENOVATING ROOF COVERINGS Cont.										
Tile roofing repairs generally Cont.										
Remove double course and replace with new clay tiles and bed and point in mortar										
at eaves	6.14	10.00	**6.76**	0.23	0.23	**5.73**	0.53	**13.02**	m	**14.65**
at verge	12.80	10.00	**14.08**	0.38	0.38	**9.47**	1.65	**25.20**	m	**28.35**
Rake out defective pointing and re-point in cement mortar (1:3)										
ridge or hip tiles	-	-	**-**	0.34	0.34	**8.47**	0.20	**8.67**	m	**9.76**
Hack off defective cement mortar fillet and renew in cement mortar (1:3)	-	-	**-**	0.34	0.34	**8.47**	0.40	**8.88**	m	**9.99**
Remove defective hip hook and replace with new	2.07	5.00	**2.17**	0.34	0.34	**8.47**	0.08	**10.73**	nr	**12.07**
Corrugated fibre cement roofing repairs										
Remove damaged sheets and provide and fix new with screws and washers to timber purlins and ease and adjust edges of adjoining sheets as necessary										
one sheet; 1825mm long Profile 3 natural grey	16.51	10.00	**18.16**	1.23	1.23	**30.65**	2.36	**51.17**	nr	**57.57**
one sheet; 1825mm long Profile 3 standard colour	19.81	10.00	**21.79**	1.23	1.23	**30.65**	2.36	**54.80**	nr	**61.65**
one sheet; 1825mm long Profile 6 natural grey	20.72	10.00	**22.79**	1.78	1.78	**44.36**	3.48	**70.63**	nr	**79.46**
one sheet; 1825mm long Profile 6 standard colour	24.87	10.00	**27.36**	1.78	1.78	**44.36**	3.48	**75.20**	nr	**84.60**
over one sheet; Profile 3 natural grey	15.01	10.00	**16.51**	0.89	0.89	**22.18**	2.23	**40.92**	m2	**46.03**
over one sheet; Profile 3 standard colour	18.01	10.00	**19.81**	0.89	0.89	**22.18**	2.23	**44.22**	m2	**49.74**
over one sheet; Profile 6 natural grey	12.19	10.00	**13.41**	0.89	0.89	**22.18**	3.28	**38.87**	m2	**43.73**
over one sheet; Profile 6 standard colour	14.63	10.00	**16.09**	0.89	0.89	**22.18**	3.28	**41.55**	m2	**46.75**
Remove damaged sheets and provide and fix new with hook bolts and washers to steel purlins and ease and adjust edges of adjoining sheets as necessary										
one sheet; 1825mm long Profile 3 natural grey	16.51	10.00	**18.16**	1.37	1.37	**34.14**	2.36	**54.66**	nr	**61.49**
one sheet; 1825mm long Profile 3 standard colour	19.81	10.00	**21.79**	1.37	1.37	**34.14**	2.36	**58.29**	nr	**65.58**
one sheet; 1825mm long Profile 6 natural grey	20.72	10.00	**22.79**	1.98	1.98	**49.34**	3.48	**75.62**	nr	**85.07**
one sheet; 1825mm long Profile 6 standard colour	24.87	10.00	**27.36**	1.98	1.98	**49.34**	3.48	**80.18**	nr	**90.21**
over one sheet; Profile 3 natural grey	15.01	10.00	**16.51**	1.03	1.03	**25.67**	2.23	**44.41**	m2	**49.96**
over one sheet; Profile 3 standard colour	18.01	10.00	**19.81**	1.03	1.03	**25.67**	2.23	**47.71**	m2	**53.67**
over one sheet; Profile 6 natural grey	12.19	10.00	**13.41**	1.03	1.03	**25.67**	3.28	**42.36**	m2	**47.65**
over one sheet; Profile 6 standard colour	14.63	10.00	**16.09**	1.03	1.03	**25.67**	3.28	**45.04**	m2	**50.67**
Felt roofing repairs for roofs covered with bituminous felt roofing, BS 747										
Sweep clean, cut out defective layer, re-bond adjacent felt and cover with one layer of felt bonded in hot bitumen; patches not exceeding 1.00m2										
fibre based roofing felt type 1B	1.42	15.00	**1.63**	0.88	0.88	**21.93**	2.82	**26.38**	m2	**29.68**
for each defective underlayer cut out and replaced, add	1.42	15.00	**1.63**	0.88	0.88	**21.93**	2.82	**26.38**	m2	**29.68**
fibre based mineral surfaced roofing felt type 1E	2.70	15.00	**3.10**	1.05	1.05	**26.17**	2.82	**32.09**	m2	**36.10**
high performance polyester based roofing felt type 5B	5.50	15.00	**6.33**	1.02	1.02	**25.42**	2.82	**34.56**	m2	**38.88**
for each defective underlayer cut out and replaced, add	5.50	15.00	**6.33**	1.02	1.02	**25.42**	2.82	**34.56**	m2	**38.88**
high performance polyester based mineral surfaced roofing felt type 5E	3.90	15.00	**4.49**	1.13	1.13	**28.16**	2.82	**35.46**	m2	**39.90**
Sweep clean, cut out defective layer, re-bond adjacent felt and cover with one layer of felt bonded in hot bitumen; patches 1.00 - 3.00m2										
fibre based roofing felt type 1B	1.42	15.00	**1.63**	0.43	0.43	**10.72**	2.82	**15.17**	m2	**17.07**
for each defective underlayer cut out and replaced, add	1.42	15.00	**1.63**	0.43	0.43	**10.72**	2.82	**15.17**	m2	**17.07**
fibre based mineral surfaced roofing felt type 1E	2.70	15.00	**3.10**	0.56	0.56	**13.96**	2.82	**19.88**	m2	**22.36**
high performance polyester based roofing felt type 5B	5.50	15.00	**6.33**	0.53	0.53	**13.21**	2.82	**22.35**	m2	**25.15**
for each defective underlayer cut out and replaced, add	5.50	15.00	**6.33**	0.53	0.53	**13.21**	2.82	**22.35**	m2	**25.15**
high performance polyester based mineral surfaced roofing felt type 5E	3.90	15.00	**4.49**	0.65	0.65	**16.20**	2.82	**23.50**	m2	**26.44**
Sweep clean surface of existing roof, prime with hot bitumen and dress with										
13mm layer of limestone or granite chippings	1.80	10.00	**1.98**	-	0.23	**2.46**	1.58	**6.02**	m2	**6.77**
13mm layer of pea shingle	1.25	10.00	**1.38**	-	0.36	**3.86**	1.58	**6.81**	m2	**7.66**
extra; removing existing chippings or shingle	-	-	**-**	-	0.24	**2.57**	0.47	**3.04**	m2	**3.42**
Liquid bitumen proofing (black) brush applied on old roof covering including cleaning old covering										
Corrugated asbestos roofing										
one coat	2.37	10.00	**2.61**	-	0.21	**2.25**	-	**4.86**	m2	**5.46**
two coats	4.42	10.00	**4.86**	-	0.25	**2.68**	-	**7.54**	m2	**8.48**
Felt roofing										
one coat	2.05	10.00	**2.26**	-	0.17	**1.82**	-	**4.08**	m2	**4.59**
two coats	3.29	10.00	**3.62**	-	0.21	**2.25**	-	**5.87**	m2	**6.60**
For top coat in green in lieu black add	0.63	10.00	**0.69**	-	-	**-**	-	**0.69**	m2	**0.78**

Labour hourly rates: (except Specialists) Craft Operatives £17.38 Labourer £10.71 Rates are national average prices. Refer to REGIONAL VARIATIONS for indicative levels of overall pricing in regions	MATERIALS			LABOUR				RATES		
	Del to Site £	Waste %	Material Cost £	Craft Optve Hrs	Lab Hrs	Labour Cost £	Sunds £	Nett Rate £	Unit	Gross Rate (+12.5%) £
C90: REPAIRING/RENOVATING ROOF COVERINGS Cont.										
Repairs to lead roofing										
Repair crack										
clean out and fill with copper bit solder........	-	-	-	0.80	-	13.90	0.36	14.26	m	16.04
Turn back flashing										
and re-dress	-	-	-	0.48	-	8.34	-	8.34	m	9.39
Repairs to lead flashings, etc.; replace with new lead										
Remove flashing and provide new code Nr4 lead; wedge into groove										
150mm girth	3.44	5.00	3.61	1.06	-	18.42	0.31	22.34	m	25.14
240mm girth	5.50	5.00	5.78	1.49	-	25.90	0.31	31.98	m	35.98
300mm girth	6.88	5.00	7.22	1.82	-	31.63	0.31	39.16	m	44.06
Remove stepped flashing and provide new code Nr 4 lead; wedge into groove										
180mm girth	4.13	5.00	4.34	1.54	-	26.77	0.31	31.41	m	35.34
240mm girth	5.50	5.00	5.78	1.94	-	33.72	0.31	39.80	m	44.78
300mm girth	6.88	5.00	7.22	2.34	-	40.67	0.31	48.20	m	54.23
Remove apron flashing and provide new code Nr 5 lead; wedge into groove										
150mm girth	4.28	5.00	4.49	1.06	-	18.42	0.31	23.22	m	26.13
300mm girth	8.56	5.00	8.99	1.82	-	31.63	0.31	40.93	m	46.04
450mm girth	12.83	5.00	13.47	2.58	-	44.84	0.47	58.79	m	66.13
Remove lining to valley gutter and provide new code Nr 5 lead										
240mm girth	6.84	5.00	7.18	1.49	-	25.90	-	33.08	m	37.21
300mm girth	8.56	5.00	8.99	1.82	-	31.63	-	40.62	m	45.70
450mm girth	12.83	5.00	13.47	2.58	-	44.84	-	58.31	m	65.60
Remove gutter lining and provide new code Nr 6 lead										
360mm girth	12.17	5.00	12.78	2.12	-	36.85	-	49.62	m	55.83
420mm girth	14.19	5.00	14.90	2.43	-	42.23	-	57.13	m	64.27
450mm girth	15.20	5.00	15.96	2.58	-	44.84	-	60.80	m	68.40
Repairs to zinc flashings, etc.; replace with 0.8mm zinc										
Remove flashing and provide new; wedge into groove										
150mm girth	2.72	5.00	2.86	1.12	-	19.47	0.21	22.53	m	25.35
240mm girth	4.36	5.00	4.58	1.60	-	27.81	0.21	32.60	m	36.67
300mm girth	5.45	5.00	5.72	1.94	-	33.72	0.31	39.75	m	44.72
Remove stepped flashing and provide new; wedge into groove										
180mm girth	3.27	5.00	3.43	1.65	-	28.68	0.21	32.32	m	36.36
240mm girth	4.36	5.00	4.58	2.08	-	36.15	0.21	40.94	m	46.06
300mm girth	5.45	5.00	5.72	2.52	-	43.80	0.31	49.83	m	56.06
Remove apron flashing and provide new; wedge into groove										
150mm girth	2.72	5.00	2.86	1.12	-	19.47	0.21	22.53	m	25.35
300mm girth	5.45	5.00	5.72	1.94	-	33.72	0.31	39.75	m	44.72
450mm girth	8.17	5.00	8.58	2.76	-	47.97	0.38	56.93	m	64.04
Remove lining to valley gutter and provide new										
240mm girth	4.36	5.00	4.58	1.60	-	27.81	-	32.39	m	36.43
300mm girth	5.45	5.00	5.72	1.94	-	33.72	-	39.44	m	44.37
450mm girth	8.17	5.00	8.58	2.76	-	47.97	-	56.55	m	63.62
Remove gutter lining and provide new										
360mm girth	6.53	5.00	6.86	2.27	-	39.45	-	46.31	m	52.10
420mm girth	7.62	5.00	8.00	2.61	-	45.36	-	53.36	m	60.03
450mm girth	8.17	5.00	8.58	2.76	-	47.97	-	56.55	m	63.62
Repairs to copper flashings, etc.; replace with 0.55mm copper										
Remove flashing and provide new; wedge into groove										
150mm girth	3.87	5.00	4.06	1.06	-	18.42	0.43	22.91	m	25.78
240mm girth	6.19	5.00	6.50	1.49	-	25.90	0.43	32.82	m	36.93
300mm girth	7.74	5.00	8.13	1.82	-	31.63	0.43	40.19	m	45.21
Remove stepped flashing and provide new; wedge into groove										
180mm girth	4.64	5.00	4.87	1.54	-	26.77	0.43	32.06	m	36.07
240mm girth	6.19	5.00	6.50	1.94	-	33.72	0.43	40.64	m	45.72
300mm girth	7.74	5.00	8.13	2.34	-	40.67	0.43	49.22	m	55.38
Remove apron flashing and provide new; wedge into groove										
150mm girth	3.87	5.00	4.06	1.06	-	18.42	0.43	22.91	m	25.78
300mm girth	7.74	5.00	8.13	1.82	-	31.63	0.43	40.19	m	45.21
450mm girth	11.61	5.00	12.19	2.58	-	44.84	0.64	57.67	m	64.88
Remove lining to valley gutter and provide new										
240mm girth	6.19	5.00	6.50	1.49	-	25.90	-	32.40	m	36.45
300mm girth	7.74	5.00	8.13	1.82	-	31.63	-	39.76	m	44.73
450mm girth	11.61	5.00	12.19	2.58	-	44.84	-	57.03	m	64.16

Labour hourly rates: (except Specialists)
Craft Operatives £17.38 Labourer £10.71
Rates are national average prices.
Refer to REGIONAL VARIATIONS for indicative levels of overall pricing in regions

	MATERIALS			LABOUR				RATES		
	Del to Site £	Waste %	Material Cost £	Craft Optve Hrs	Lab Hrs	Labour Cost £	Sunds £	Nett Rate £	Unit	Gross Rate (+12.5%) £
C90: REPAIRING/RENOVATING ROOF COVERINGS Cont.										
Repairs to copper flashings, etc.; replace with 0.55mm copper Cont.										
Remove gutter lining and provide new										
360mm girth	9.29	5.00	9.75	2.12	-	36.85	-	46.60	m	52.43
420mm girth	10.84	5.00	11.38	2.43	-	42.23	-	53.62	m	60.32
450mm girth	11.61	5.00	12.19	2.58	-	44.84	-	57.03	m	64.16
Repairs to aluminium flashings etc; replace with 0.60mm aluminium										
Remove flashings and provide new; wedge into groove										
150mm girth	2.75	5.00	2.89	1.06	-	18.42	0.19	21.50	m	24.19
240mm girth	4.13	5.00	4.34	1.49	-	25.90	0.19	30.42	m	34.23
300mm girth	4.13	5.00	4.34	1.82	-	31.63	0.19	36.16	m	40.68
Remove stepped flashing and provide new; wedge into groove										
180mm girth	3.63	5.00	3.81	1.54	-	26.77	0.19	30.77	m	34.61
240mm girth	4.13	5.00	4.34	1.94	-	33.72	0.19	38.24	m	43.02
300mm girth	4.13	5.00	4.34	2.34	-	40.67	0.19	45.20	m	50.84
Remove apron flashing and provide new; wedge into groove										
150mm girth	2.75	5.00	2.89	1.06	-	18.42	0.19	21.50	m	24.19
300mm girth	4.13	5.00	4.34	1.82	-	31.63	0.19	36.16	m	40.68
450mm girth	6.38	5.00	6.70	2.56	-	44.49	3.56	54.75	m	61.59
Remove lining to valley gutter and provide new										
240mm girth	4.13	5.00	4.34	1.49	-	25.90	-	30.23	m	34.01
300mm girth	4.13	5.00	4.34	1.82	-	31.63	-	35.97	m	40.46
450mm girth	6.38	5.00	6.70	2.58	-	44.84	-	51.54	m	57.98
Remove gutter lining and provide new										
360mm girth	5.38	5.00	5.65	2.12	-	36.85	-	42.49	m	47.81
420mm girth	5.38	5.00	5.65	2.43	-	42.23	-	47.88	m	53.87
450mm girth	5.38	5.00	5.65	2.58	-	44.84	-	50.49	m	56.80
C90: REPAIRING/RENOVATING PLUMBING										
Repairs to rainwater goods										
Clean out and make good loose fixings										
eaves gutter	-	-	-	0.40	-	6.95	-	6.95	m	7.82
rainwater pipe	-	-	-	0.40	-	6.95	-	6.95	m	7.82
rainwater head	-	-	-	0.75	-	13.04	-	13.04	nr	14.66
Clean out defective joint to existing eaves gutter and re-make joint										
with jointing compound and new bolt	1.00	-	1.00	0.75	-	13.04	-	14.04	nr	15.79
Remove remains of old balloon grating and provide new plastic grating in outlet or end of pipe										
50mm diameter	2.86	2.00	2.92	0.50	-	8.69	-	11.61	nr	13.06
82mm diameter	2.90	2.00	2.96	0.50	-	8.69	-	11.65	nr	13.10
110mm diameter	2.92	2.00	2.98	0.50	-	8.69	-	11.67	nr	13.13
Repair or replacement of stopcocks, valves, etc.										
Turn off water supply drain down as necessary and re-new washer to the following up to 19mm										
main stopcock	0.50	-	0.50	2.00	-	34.76	-	35.26	nr	39.67
service stopcock	0.50	-	0.50	2.00	-	34.76	-	35.26	nr	39.67
bib or pillar cock	0.50	-	0.50	2.00	-	34.76	-	35.26	nr	39.67
supatap	0.50	-	0.50	2.00	-	34.76	-	35.26	nr	39.67
draining tap	0.50	-	0.50	2.00	-	34.76	-	35.26	nr	39.67
Turn off water supply, drain down as necessary, take out old valves and re-new with joints to copper										
13mm brass stopcock (BS 1010)	4.06	2.00	4.14	2.00	-	34.76	-	38.90	nr	43.76
13mm chromium plated sink pillar cock (BS 1010)	14.65	2.00	14.94	2.00	-	34.76	-	49.70	nr	55.92
13mm brass bib cock (BS 1010)	8.80	2.00	8.98	2.00	-	34.76	-	43.74	nr	49.20
13mm drain tap (BS 2879) type 2	2.41	2.00	2.46	2.00	-	34.76	-	37.22	nr	41.87
19mm drain tap (BS 2879) type 2	18.88	2.00	19.26	2.00	-	34.76	-	54.02	nr	60.77
13mm high pressure ball valve (BS 1212) piston type with copper ball	6.72	2.00	6.85	2.00	-	34.76	-	41.61	nr	46.82
13mm high pressure ball valve (BS 1212) piston type with plastic ball	3.94	2.00	4.02	2.00	-	34.76	-	38.78	nr	43.63
13mm low pressure ball valve (BS 1212) piston type with copper ball	8.38	2.00	8.55	2.00	-	34.76	-	43.31	nr	48.72
13mm low pressure ball valve (BS 1212) piston type with plastic ball	5.60	2.00	5.71	2.00	-	34.76	-	40.47	nr	45.53
19mm high pressure ball valve (BS 1212) piston type with copper ball	19.67	2.00	20.06	2.50	-	43.45	-	63.51	nr	71.45
19mm high pressure ball valve (BS 1212) piston type with plastic ball	19.05	2.00	19.43	2.50	-	43.45	-	62.88	nr	70.74
Replacement of traps										
Take out old plastic trap and replace with new 76mm seal trap (BS EN 274) with 'O'' ring joint outlet										
36mm diameter P trap	4.86	2.00	4.96	1.25	-	21.73	-	26.68	nr	30.02
42mm diameter P trap	5.61	2.00	5.72	1.50	-	26.07	-	31.79	nr	35.77
36mm diameter S trap	6.16	2.00	6.28	1.25	-	21.73	-	28.01	nr	31.51

Labour hourly rates: (except Specialists) Craft Operatives £17.38 Labourer £10.71 Rates are national average prices. Refer to REGIONAL VARIATIONS for indicative levels of overall pricing in regions	MATERIALS			LABOUR				RATES		
	Del to Site £	Waste %	Material Cost £	Craft Optve Hrs	Lab Hrs	Labour Cost £	Sunds £	Nett Rate £	Unit	Gross Rate (+12.5%) £
C90: REPAIRING/RENOVATING PLUMBING Cont.										
Replacement of traps Cont.										
Take out old plastic trap and replace with new 76mm seal trap (BS EN 274) with 'O'' ring joint outlet Cont.										
42mm diameter S trap	7.20	2.00	**7.34**	1.50	-	**26.07**	-	**33.41**	nr	**37.59**
42mm diameter bath trap with overflow connection	10.40	2.00	**10.61**	1.75	-	**30.42**	-	**41.02**	nr	**46.15**
Take out old copper trap and replace with new 76mm seal two piece trap (BS 1184) with compression joint										
35mm diameter P trap	23.20	2.00	**23.66**	1.25	-	**21.73**	-	**45.39**	nr	**51.06**
42mm diameter P trap	27.33	2.00	**27.88**	1.50	-	**26.07**	-	**53.95**	nr	**60.69**
35mm diameter S trap	24.58	2.00	**25.07**	1.25	-	**21.73**	-	**46.80**	nr	**52.65**
42mm diameter S trap	29.57	2.00	**30.16**	1.50	-	**26.07**	-	**56.23**	nr	**63.26**
42mm diameter bath trap with male iron overflow connection	38.80	2.00	**39.58**	1.75	-	**30.42**	-	**69.99**	nr	**78.74**
Preparation of pipe for insertion of new pipe or fitting										
Cut into cast iron pipe and take out a length up to 1.00m for insertion of new pipe or fitting										
50mm diameter	-	-	-	3.70	-	**64.31**	-	**64.31**	nr	**72.34**
64 and 76mm diameter	-	-	-	4.30	-	**74.73**	-	**74.73**	nr	**84.08**
89 and 102mm diameter	-	-	-	4.90	-	**85.16**	-	**85.16**	nr	**95.81**
Cut into cast iron pipe with caulked joint and take out a length up to 1.00m for insertion of new pipe or fitting										
50mm diameter	-	-	-	3.70	-	**64.31**	-	**64.31**	nr	**72.34**
64 and 76mm diameter	-	-	-	4.30	-	**74.73**	-	**74.73**	nr	**84.08**
89 and 102mm diameter	-	-	-	4.90	-	**85.16**	-	**85.16**	nr	**95.81**
Cut into asbestos cement pipe and take out a length up to 1.00m for insertion of new pipe or fitting										
50mm diameter	-	-	-	3.15	-	**54.75**	-	**54.75**	nr	**61.59**
64 and 76mm diameter	-	-	-	3.75	-	**65.18**	-	**65.18**	nr	**73.32**
89 and 102mm diameter	-	-	-	4.35	-	**75.60**	-	**75.60**	nr	**85.05**
Cut into polythene or p.v.c. pipe and take out a length up to 1.00m for insertion of new pipe or fitting										
up to 25mm diameter	-	-	-	2.50	-	**43.45**	-	**43.45**	nr	**48.88**
32 - 50mm diameter	-	-	-	3.15	-	**54.75**	-	**54.75**	nr	**61.59**
64 and 76mm diameter	-	-	-	3.75	-	**65.18**	-	**65.18**	nr	**73.32**
89 and 102mm diameter	-	-	-	4.40	-	**76.47**	-	**76.47**	nr	**86.03**
Cut into steel pipe and take out a length up to 1.00m for insertion of new pipe or fitting										
up to 25mm diameter	-	-	-	3.05	-	**53.01**	-	**53.01**	nr	**59.64**
32 - 50mm diameter	-	-	-	3.70	-	**64.31**	-	**64.31**	nr	**72.34**
64 and 76mm diameter	-	-	-	4.30	-	**74.73**	-	**74.73**	nr	**84.08**
89 and 102mm diameter	-	-	-	4.90	-	**85.16**	-	**85.16**	nr	**95.81**
Cut into stainless steel pipe and take out a length up to 1.00m for insertion of new pipe or fitting										
up to 25mm diameter	-	-	-	2.90	-	**50.40**	-	**50.40**	nr	**56.70**
32 - 50mm diameter	-	-	-	3.50	-	**60.83**	-	**60.83**	nr	**68.43**
64 and 76mm diameter	-	-	-	4.10	-	**71.26**	-	**71.26**	nr	**80.17**
89 and 102mm diameter	-	-	-	4.75	-	**82.56**	-	**82.56**	nr	**92.87**
Cut into copper pipe and take out a length up to 1.00m for insertion of new pipe or fitting										
up to 25mm diameter	-	-	-	2.90	-	**50.40**	-	**50.40**	nr	**56.70**
32 - 50mm diameter	-	-	-	3.50	-	**60.83**	-	**60.83**	nr	**68.43**
64 and 76mm diameter	-	-	-	4.10	-	**71.26**	-	**71.26**	nr	**80.17**
89 and 102mm diameter	-	-	-	4.75	-	**82.56**	-	**82.56**	nr	**92.87**
Cut into lead pipe and take out a length up to 1.00m for insertion of new pipe or fitting										
up to 25mm diameter	-	-	-	3.05	-	**53.01**	-	**53.01**	nr	**59.64**
32 - 50mm diameter	-	-	-	3.70	-	**64.31**	-	**64.31**	nr	**72.34**
64 and 76mm diameter	-	-	-	4.30	-	**74.73**	-	**74.73**	nr	**84.08**
89 and 102mm diameter	-	-	-	4.90	-	**85.16**	-	**85.16**	nr	**95.81**
Removal of traps, valves, etc., for re-use										
Take out the following, including unscrewing or uncoupling, for re-use and including draining down as necessary										
trap up to 25mm diameter	-	-	-	0.70	-	**12.17**	-	**12.17**	nr	**13.69**
trap 32 - 50mm diameter	-	-	-	0.85	-	**14.77**	-	**14.77**	nr	**16.62**
stop valve up to 25mm diameter	-	-	-	2.00	-	**34.76**	-	**34.76**	nr	**39.11**
stop valve 32 - 50mm diameter	-	-	-	2.25	-	**39.11**	-	**39.11**	nr	**43.99**
radiator valve up to 25mm diameter	-	-	-	2.00	-	**34.76**	-	**34.76**	nr	**39.11**
radiator valve 32 - 50mm diameter	-	-	-	2.25	-	**39.11**	-	**39.11**	nr	**43.99**
tap up to 25mm diameter	-	-	-	2.00	-	**34.76**	-	**34.76**	nr	**39.11**
tap 32 - 50mm diameter	-	-	-	2.25	-	**39.11**	-	**39.11**	nr	**43.99**
extra; burning out one soldered joint up to 25mm diameter	-	-	-	0.50	-	**8.69**	-	**8.69**	nr	**9.78**
extra; burning out one soldered joint 32 - 50mm diameter	-	-	-	0.80	-	**13.90**	-	**13.90**	nr	**15.64**

Labour hourly rates: (except Specialists) Craft Operatives £17.38 Labourer £10.71 Rates are national average prices. Refer to REGIONAL VARIATIONS for indicative levels of overall pricing in regions	MATERIALS			LABOUR				RATES		
	Del to Site £	Waste %	Material Cost £	Craft Optve Hrs	Lab Hrs	Labour Cost £	Sunds £	Nett Rate £	Unit	Gross Rate (+12.5%) £
C90: REPAIRING/RENOVATING PLUMBING Cont.										
Replacement of old sanitary fittings										
Disconnect trap, valves and services pipes, remove old fitting and provide and fix new including re-connecting trap, valves and services pipes										
560 x 405mm vitreous china lavatory basin with 32mm chromium plated waste outlet, plug, chain and stay and pair of cast iron brackets	54.00	5.00	**56.70**	6.00	-	**104.28**	-	**160.98**	nr	**181.10**
600 x 1000mm stainless steel sink with single drainer with 38mm chromium plated waste outlet, plug, chain and stay	91.75	5.00	**96.34**	6.00	-	**104.28**	-	**200.62**	nr	**225.69**
600 x 1500mm stainless steel sink with double drainer with 38mm chromium plated waste outlet, overflow plug, chain and stay	112.25	5.00	**117.86**	6.50	-	**112.97**	-	**230.83**	nr	**259.69**
1500mm pressed steel vitreous enamelled rectangular top bath with cradles and with 38mm chromium plated waste outlet, plug, chain and stay and 32mm chromium plated overflow and including removing old cast iron bath	97.25	5.00	**102.11**	8.00	-	**139.04**	-	**241.15**	nr	**271.30**
1700mm pressed steel vitreous enamelled rectangular top bath with cradles and with 38mm chromium plated waste outlet, plug, chain and stay and 32mm chromium plated overflow and including removing old cast iron bath	97.25	5.00	**102.11**	8.00	-	**139.04**	-	**241.15**	nr	**271.30**
Take off old W.C. seat and provide and fix new ring pattern black plastic seat and cover	12.50	5.00	**13.13**	1.50	-	**26.07**	-	**39.20**	nr	**44.09**
Disconnect supply pipe, overflow and flush pipe and take out old W.C. cistern and provide and fix new 9 litre black plastic W.C. cistern with cover and connect pipe and ball valve										
low level	47.75	5.00	**50.14**	5.00	-	**86.90**	-	**137.04**	nr	**154.17**
high level	58.25	5.00	**61.16**	5.00	-	**86.90**	-	**148.06**	nr	**166.57**
Disconnect flush pipe, take off seat and remove old W.C. pan, provide and fix new white china pan, connect flush pipe, form cement joint with drain and refix old seat										
S or P trap pan	28.00	5.00	**29.40**	5.00	-	**86.90**	-	**116.30**	nr	**130.84**
Disconnect and remove old high level W.C. suite complete and provide and fix new suite with P or S trap pan, plastic ring seat and cover, plastic flush pipe or bend, 9 litre plastic W.C. cistern with cover, chain and pull or lever handle and 13mm low pressure ball valve; adapt supply pipe and overflow as necessary and connect to new suite and cement joint to drain										
new high level suite	102.00	5.00	**107.10**	8.50	-	**147.73**	-	**254.83**	nr	**286.68**
new low level suite (vitreous china cistern)	95.50	5.00	**100.28**	9.00	-	**156.42**	-	**256.70**	nr	**288.78**
Disconnect and remove old low level W.C. suite complete and provide and fix new suite with P or S trap pan, plastic ring seat and cover, plastic flush bend, 9 litre vitreous china W.C. cistern with cover, lever handle and 13mm low pressure ball valve with plastic ball; re-connect supply pipe and overflow and cement joint to drain										
new low level suite	95.50	5.00	**100.28**	8.50	-	**147.73**	-	**248.01**	nr	**279.01**
Replacement of old water tanks										
Disconnect all pipework, set aside ball valve and take out cold water storage cistern in roof space; replace with new GRP water storage tank (BS EN 13280); reconnect all pipework and ball valve; remove metal filings and clean inside of tank										
type SC70 size 915 x 610 x 580mm (227 litre)	198.00	2.00	**201.96**	9.00	-	**156.42**	0.51	**358.89**	nr	**403.75**

This section continues on the next page

Labour hourly rates: (except Specialists) Craft Operatives £20.54 Labourer £10.71 Rates are national average prices. Refer to REGIONAL VARIATIONS for indicative levels of overall pricing in regions	MATERIALS			LABOUR				RATES		
	Del to Site £	Waste %	Material Cost £	Craft Optve Hrs	Lab Hrs	Labour Cost £	Sunds £	Nett Rate £	Unit	Gross Rate (+12.5%) £
C90: REPAIRING/RENOVATING ELECTRICAL INSTALLATIONS										
Generally										
Note										
the following approximate estimates of costs are dependent on the number and disposition of the points. Lamps and fittings together with cutting and making good are excluded										
Strip out and re-wire complete										
Provide new lamp holders, flush switches, flush socket outlets, fitting outlets and consumer unit with circuit breakers; a three bedroom house with ten lighting points and associated switches, sixteen 13 amp socket outlets, three 13 amp fitting outlets, one 30 amp cooker control panel, three earth bonding points and one eight way consumer unit										
PVC cables in existing conduit	1001.63	7.50	**1076.75**	52.85	-	**1085.54**	30.00	**2192.29**	nr	**2466.33**
PVC cables and conduit	1110.62	7.50	**1193.92**	99.00	-	**2033.46**	30.00	**3257.38**	nr	**3664.55**
Strip out and re-wire with PVC insulated and sheathed cable in existing conduit										
Point to junction box, re-using existing lamp holder, socket outlet etc										
5 amp socket outlet	9.00	7.50	**9.68**	0.85	-	**17.46**	0.45	**27.58**	nr	**31.03**
13 amp socket or fitting outlet	12.00	7.50	**12.90**	0.85	-	**17.46**	0.45	**30.81**	nr	**34.66**
lighting point or switch	5.28	7.50	**5.68**	0.98	-	**20.13**	0.45	**26.26**	nr	**29.54**
Strip out and re-wire with PVC insulated and sheathed cable and new conduit										
Point to junction box, re-using existing lamp holder, socket outlet etc										
5 amp socket outlet	9.92	7.50	**10.66**	1.20	-	**24.65**	0.50	**35.81**	nr	**40.29**
13 amp socket outlet	12.92	7.50	**13.89**	1.20	-	**24.65**	0.50	**39.04**	nr	**43.92**
lighting point or switch	7.38	7.50	**7.93**	1.30	-	**26.70**	0.30	**34.94**	nr	**39.30**
Luminaires and accessories										
Take out damaged and renew										
flourescent lamp fitting 1500mm, batten type, single tube	12.40	2.50	**12.71**	0.56	-	**11.50**	0.25	**24.46**	nr	**27.52**
flourescent lamp fitting 1500mm, batten type, twin tube	20.00	2.50	**20.50**	0.56	-	**11.50**	0.25	**32.25**	nr	**36.28**
flourescent lamp fitting 1500mm, plastic diffused type, twin tube	39.77	2.50	**40.76**	0.56	-	**11.50**	0.25	**52.52**	nr	**59.08**
flourescent lamp fitting 1500mm, plastic diffused type, twin tube	80.52	2.50	**82.53**	0.56	-	**11.50**	0.25	**94.29**	nr	**106.07**
lamp holder, batten type	2.65	7.50	**2.85**	0.56	-	**11.50**	0.25	**14.60**	nr	**16.43**
lamp holder, pendant type with rose	1.30	7.50	**1.40**	1.06	-	**21.77**	0.25	**23.42**	nr	**26.35**
switch, flush type	1.15	7.50	**1.24**	0.75	-	**15.41**	0.45	**17.09**	nr	**19.23**
switch, surface mounted type	2.00	7.50	**2.15**	0.75	-	**15.41**	0.45	**18.01**	nr	**20.26**
13 amp socket outlet, flush type	2.25	7.50	**2.42**	0.75	-	**15.41**	0.45	**18.27**	nr	**20.56**
13 amp socket outlet, surface mounted type	4.00	7.50	**4.30**	0.88	-	**18.08**	0.45	**22.83**	nr	**25.68**
5 amp socket outlet, flush type	5.40	7.50	**5.81**	0.70	-	**14.38**	0.45	**20.63**	nr	**23.21**
5 amp socket outlet, surface mounted type	6.10	7.50	**6.56**	0.88	-	**18.08**	0.45	**25.08**	nr	**28.22**

This section continues on the next page

Labour hourly rates: (except Specialists) Craft Operatives £14.21 Labourer £10.71 Rates are national average prices. Refer to REGIONAL VARIATIONS for indicative levels of overall pricing in regions	MATERIALS			LABOUR				RATES		
	Del to Site £	Waste %	Material Cost £	Craft Optve Hrs	Lab Hrs	Labour Cost £	Sunds £	Nett Rate £	Unit	Gross Rate (+12.5%) £
C90: REPAIRING/RENOVATING FINISHINGS										
Repairs to old wall and ceiling finishings										
Cut out damaged two coat plaster to wall in patches and make out in new plaster to match existing										
under 1.00m2	2.70	10.00	**2.98**	2.13	2.13	**53.08**	0.29	**56.35**	m2	**63.39**
1.00 - 2.00m2	2.70	10.00	**2.98**	1.76	1.76	**43.86**	0.29	**47.13**	m2	**53.02**
2.00 - 4.00m2	2.70	10.00	**2.98**	1.39	1.39	**34.64**	0.29	**37.91**	m2	**42.64**
Cut out damaged three coat plaster to wall in patches and make out in new plaster to match existing										
under 1.00m2	3.59	10.00	**3.95**	2.50	2.50	**62.30**	0.45	**66.70**	m2	**75.04**
1.00 - 2.00m2	3.59	10.00	**3.95**	2.06	2.06	**51.34**	0.45	**55.74**	m2	**62.70**
2.00 - 4.00m2	3.59	10.00	**3.95**	1.62	1.62	**40.37**	0.45	**44.77**	m2	**50.37**
Cut out damaged two coat rendering to wall in patches and make out in new rendering to match existing										
under 1.00m2	1.58	10.00	**1.74**	1.93	1.93	**48.10**	0.29	**50.13**	m2	**56.40**
1.00 - 2.00m2	1.58	10.00	**1.74**	1.60	1.60	**39.87**	0.29	**41.91**	m2	**47.14**
2.00 - 4.00m2	1.58	10.00	**1.74**	1.26	1.26	**31.40**	0.29	**33.43**	m2	**37.61**
Cut out damaged three coat rendering to wall in patches and make out in new rendering to match existing										
under 1.00m2	2.44	10.00	**2.68**	2.28	2.28	**56.82**	0.45	**59.95**	m2	**67.44**
1.00 - 2.00m2	2.44	10.00	**2.68**	1.87	1.87	**46.60**	0.45	**49.73**	m2	**55.95**
2.00 - 4.00m2	2.44	10.00	**2.68**	1.47	1.47	**36.63**	0.45	**39.76**	m2	**44.73**
Cut out damaged two coat plaster to ceiling in patches and make out in new plaster to match existing										
under 1.00m2	2.70	10.00	**2.98**	2.38	2.38	**59.31**	0.29	**62.58**	m2	**70.40**
1.00 - 2.00m2	2.70	10.00	**2.98**	1.96	1.96	**48.84**	0.29	**52.11**	m2	**58.62**
2.00 - 4.00m2	2.70	10.00	**2.98**	1.54	1.54	**38.38**	0.29	**41.64**	m2	**46.85**
Cut out damaged three coat plaster to ceiling in patches and make out in new plaster to match existing										
under 1.00m2	3.59	10.00	**3.95**	2.79	2.79	**69.53**	0.45	**73.93**	m2	**83.17**
1.00 - 2.00m2	3.59	10.00	**3.95**	2.30	2.30	**57.32**	0.45	**61.72**	m2	**69.43**
2.00 - 4.00m2	3.59	10.00	**3.95**	1.81	1.81	**45.11**	0.45	**49.51**	m2	**55.70**
Cut out damaged 9.5mm plasterboard and two coat plaster to wall in patches and make out in new plasterboard and plaster to match existing										
under 1.00m2	4.51	10.00	**4.96**	2.23	2.23	**55.57**	0.50	**61.03**	m2	**68.66**
1.00 - 2.00m2	4.51	10.00	**4.96**	1.78	1.78	**44.36**	0.50	**49.82**	m2	**56.04**
2.00 - 4.00m2	4.51	10.00	**4.96**	1.36	1.36	**33.89**	0.50	**39.35**	m2	**44.27**
Cut out damaged 12.5mm plasterboard and two coat plaster to wall in patches and make out in new plasterboard and plaster to match existing										
under 1.00m2	4.70	10.00	**5.17**	2.28	2.28	**56.82**	0.55	**62.54**	m2	**70.36**
1.00 - 2.00m2	4.70	10.00	**5.17**	1.83	1.83	**45.60**	0.55	**51.33**	m2	**57.74**
2.00 - 4.00m2	4.70	10.00	**5.17**	1.38	1.38	**34.39**	0.55	**40.11**	m2	**45.13**
Cut out damaged 9.5mm plasterboard and two coat plaster to ceiling in patches and make out in new plasterboard and plaster to match existing										
under 1.00m2	4.51	10.00	**4.96**	2.54	2.54	**63.30**	0.50	**68.76**	m2	**77.35**
1.00 - 2.00m2	4.51	10.00	**4.96**	2.03	2.03	**50.59**	0.50	**56.05**	m2	**63.05**
2.00 - 4.00m2	4.51	10.00	**4.96**	1.55	1.55	**38.63**	0.50	**44.09**	m2	**49.60**
Cut out damaged 12.5mm plasterboard and two coat plaster to ceiling in patches and make out in new plasterboard and plaster to match existing										
under 1.00m2	4.70	10.00	**5.17**	2.56	2.56	**63.80**	0.55	**69.52**	m2	**78.21**
1.00 - 2.00m2	4.70	10.00	**5.17**	2.06	2.06	**51.34**	0.55	**57.06**	m2	**64.19**
2.00 - 4.00m2	4.70	10.00	**5.17**	1.57	1.57	**39.12**	0.55	**44.85**	m2	**50.45**
Cut out damaged lath and plaster to wall in patches and make out with new metal lathing and three coat plaster to match existing										
under 1.00m2	6.99	10.00	**7.69**	2.53	2.53	**63.05**	0.55	**71.30**	m2	**80.21**
1.00 - 2.00m2	6.99	10.00	**7.69**	2.14	2.14	**53.33**	0.55	**61.58**	m2	**69.27**
2.00 - 4.00m2	6.99	10.00	**7.69**	1.73	1.73	**43.11**	0.55	**51.36**	m2	**57.78**
Cut out damaged lath and plaster to ceiling in patches and make out with new metal lathing and three coat plaster to match existing										
under 1.00m2	6.99	10.00	**7.69**	2.87	2.87	**71.52**	0.55	**79.77**	m2	**89.74**
1.00 - 2.00m2	6.99	10.00	**7.69**	2.43	2.43	**60.56**	0.55	**68.80**	m2	**77.40**
2.00 - 4.00m2	6.99	10.00	**7.69**	1.95	1.95	**48.59**	0.55	**56.84**	m2	**63.95**
Cut out crack in plaster, form dovetailed key and make good with plaster										
not exceeding 50mm wide	0.13	10.00	**0.14**	0.20	0.20	**4.98**	0.02	**5.14**	m	**5.78**
Make good plaster where the following removed										
small pipe	-	-	-	0.16	0.16	**3.99**	0.02	**4.00**	nr	**4.50**
small steel section	-	-	-	0.21	0.21	**5.23**	0.02	**5.26**	nr	**5.91**
Cut out damaged plaster moulding or cornice and make out with new to match existing										
up to 100mm girth on face	0.84	10.00	**0.92**	1.25	0.63	**24.51**	0.12	**25.55**	m	**28.75**
100 - 150mm girth on face	1.40	10.00	**1.54**	1.86	0.93	**36.39**	0.23	**38.16**	m	**42.93**
150 - 200mm girth on face	2.80	10.00	**3.08**	2.25	1.13	**44.07**	0.41	**47.57**	m	**53.52**

Labour hourly rates: (except Specialists) Craft Operatives £14.21 Labourer £10.71 Rates are national average prices. Refer to REGIONAL VARIATIONS for indicative levels of overall pricing in regions	MATERIALS			LABOUR				RATES		
	Del to Site £	Waste %	Material Cost £	Craft Optve Hrs	Lab Hrs	Labour Cost £	Sunds £	Nett Rate £	Unit	Gross Rate (+12.5%) £
C90: REPAIRING/RENOVATING FINISHINGS Cont.										
Repairs to old wall and ceiling finishings Cont.										
Remove damaged fibrous plaster louvered ventilator, replace with new and make good										
229 x 75mm	1.60	5.00	**1.68**	0.50	0.50	**12.46**	0.06	**14.20**	nr	**15.97**
229 x 150mm	1.92	-	**1.92**	0.60	0.60	**14.95**	0.10	**16.97**	nr	**19.09**
Take out damaged 152 x 152 x 5.5mm white glazed wall tiles and renew to match existing										
isolated tile	0.45	5.00	**0.47**	0.32	0.32	**7.97**	0.03	**8.48**	nr	**9.54**
patch of 5 tiles	2.24	5.00	**2.35**	0.60	0.60	**14.95**	0.17	**17.47**	nr	**19.66**
patch 0.50 - 1.00m2	19.71	5.00	**20.70**	2.32	2.32	**57.81**	1.32	**79.83**	m2	**89.81**
patch 1.00 - 2.00m2	19.71	5.00	**20.70**	1.66	1.66	**41.37**	1.32	**63.38**	m2	**71.30**
Repairs to old floor finishings										
Take out damaged 150 x 150 x 12.5mm clay floor tiles and renew to match existing										
isolated tile	0.32	5.00	**0.34**	0.32	0.21	**6.80**	0.03	**7.17**	nr	**8.06**
patch of 5 tiles	1.60	5.00	**1.68**	0.65	0.45	**14.06**	0.17	**15.90**	nr	**17.89**
patch 0.50 - 1.00m2	14.40	5.00	**15.12**	2.40	1.61	**51.35**	1.32	**67.78**	m2	**76.26**
patch 1.00 - 2.00m2	14.40	5.00	**15.12**	1.90	1.27	**40.60**	1.32	**57.04**	m2	**64.17**
Cut out damaged 25mm granolithic paving in patches and make out in new granolithic to match existing										
under 1.00m2	7.18	5.00	**7.53**	1.50	1.50	**37.38**	0.62	**45.53**	m2	**51.22**
1.00 - 2.00m2	7.18	5.00	**7.53**	1.25	1.25	**31.15**	0.62	**39.30**	m2	**44.21**
2.00 - 4.00m2	7.18	5.00	**7.53**	1.07	1.07	**26.66**	0.62	**34.81**	m2	**39.17**
Cut out damaged 32mm granolithic paving in patches and make out in new granolithic to match existing										
under 1.00m2	9.18	5.00	**9.64**	1.68	1.68	**41.87**	0.72	**52.23**	m2	**58.76**
1.00 - 2.00m2	9.18	5.00	**9.64**	1.38	1.38	**34.39**	0.72	**44.76**	m2	**50.35**
2.00 - 4.00m2	9.18	5.00	**9.64**	1.15	1.15	**28.66**	0.72	**39.02**	m2	**43.90**
Cut out damaged 38mm granolithic paving in patches and make out in new granolithic to match existing										
under 1.00m2	10.91	5.00	**11.45**	1.80	1.80	**44.86**	0.85	**57.16**	m2	**64.30**
1.00 - 2.00m2	10.91	5.00	**11.45**	1.50	1.50	**37.38**	0.85	**49.68**	m2	**55.89**
2.00 - 4.00m2	10.91	5.00	**11.45**	1.25	1.25	**31.15**	0.85	**43.45**	m2	**48.88**
Cut out damaged 50mm granolithic paving in patches and make out in new granolithic to match existing										
under 1.00m2	14.35	5.00	**15.07**	2.12	2.12	**52.83**	1.04	**68.93**	m2	**77.55**
1.00 - 2.00m2	14.35	5.00	**15.07**	1.73	1.73	**43.11**	1.04	**59.22**	m2	**66.62**
2.00 - 4.00m2	14.35	5.00	**15.07**	1.45	1.45	**36.13**	1.04	**52.24**	m2	**58.77**
Cut out damaged granolithic paving to tread 275mm wide and make out in new granolithic to match existing										
25mm thick	2.01	5.00	**2.11**	0.44	0.44	**10.96**	0.17	**13.25**	m	**14.90**
32mm thick	2.58	5.00	**2.71**	0.48	0.48	**11.96**	0.20	**14.87**	m	**16.73**
38mm thick	2.87	5.00	**3.01**	0.50	0.50	**12.46**	0.23	**15.70**	m	**17.67**
50mm thick	4.02	5.00	**4.22**	0.59	0.59	**14.70**	0.29	**19.21**	m	**21.61**
Cut out damaged granolithic covering to plain riser 175mm wide and make out in new granolithic to match existing										
13mm thick	0.57	5.00	**0.60**	0.80	0.80	**19.94**	1.58	**22.12**	m	**24.88**
19mm thick	0.86	5.00	**0.90**	0.84	0.84	**20.93**	1.59	**23.43**	m	**26.36**
25mm thick	1.15	5.00	**1.21**	0.88	0.88	**21.93**	1.61	**24.74**	m	**27.83**
32mm thick	1.72	5.00	**1.81**	0.93	0.93	**23.18**	1.62	**26.60**	m	**29.93**
Cut out damaged granolithic covering to undercut riser 175mm wide and make out in new granolithic to match existing										
13mm thick	0.57	5.00	**0.60**	0.87	0.87	**21.68**	1.58	**23.86**	m	**26.85**
19mm thick	0.86	5.00	**0.90**	0.93	0.93	**23.18**	1.59	**25.67**	m	**28.88**
25mm thick	1.15	5.00	**1.21**	0.98	0.98	**24.42**	1.61	**27.23**	m	**30.64**
32mm thick	1.72	5.00	**1.81**	1.04	1.04	**25.92**	1.62	**29.34**	m	**33.01**
Cut out damaged granolithic skirting 150mm wide and make out in new granolithic to match existing										
13mm thick	0.57	5.00	**0.60**	0.72	0.72	**17.94**	1.35	**19.89**	m	**22.38**
19mm thick	0.86	5.00	**0.90**	0.76	0.76	**18.94**	1.37	**21.21**	m	**23.86**
25mm thick	1.15	5.00	**1.21**	0.80	0.80	**19.94**	1.38	**22.52**	m	**25.33**
32mm thick	1.44	5.00	**1.51**	0.82	0.82	**20.43**	1.39	**23.34**	m	**26.25**
Cut out crack in granolithic paving not exceeding 50mm wide, form dovetailed key and make good to match existing										
25mm thick	0.29	5.00	**0.30**	0.60	0.60	**14.95**	0.03	**15.28**	m	**17.19**
32mm thick	0.57	5.00	**0.60**	0.66	0.66	**16.45**	0.03	**17.08**	m	**19.22**
38mm thick	0.57	5.00	**0.60**	0.69	0.69	**17.19**	0.04	**17.84**	m	**20.07**
50mm thick	0.86	5.00	**0.90**	0.79	0.79	**19.69**	0.05	**20.64**	m	**23.22**
C90: REPAIRING GLAZING										
Resecure glass										
Remove decayed putties, paint one coat on edge of rebate and re-putty										
wood window	0.13	10.00	**0.15**	0.69	-	**9.80**	0.02	**9.97**	m	**11.22**
metal window	0.13	10.00	**0.15**	0.76	-	**10.80**	0.02	**10.97**	m	**12.34**

Labour hourly rates: (except Specialists) Craft Operatives £14.21 Labourer £10.71 Rates are national average prices. Refer to REGIONAL VARIATIONS for indicative levels of overall pricing in regions	MATERIALS			LABOUR				RATES		
	Del to Site £	Waste %	Material Cost £	Craft Optve Hrs	Lab Hrs	Labour Cost £	Sunds £	Nett Rate £	Unit	Gross Rate (+12.5%) £
C90: REPAIRING GLAZING Cont.										
Resecure glass Cont.										
Remove beads, remove decayed bedding materials, paint one coat on edge of rebate, re-bed glass and re-fix beads										
wood window	0.13	10.00	**0.15**	1.22	-	**17.34**	0.02	**17.51**	m	**19.69**
metal window	0.13	10.00	**0.15**	1.32	-	**18.76**	0.02	**18.93**	m	**21.29**
C90: REPAIRING/RENOVATING FENCING										
Repairs to softwood fencing										
Remove defective timber fence post and replace with new 127 x 102mm post; letting post 450mm into ground; backfilling around in concrete mix C20; securing ends of existing arris rails and gravel boards to new post										
to suit fencing 914mm high	13.50	10.00	**14.05**	2.04	2.04	**50.84**	0.41	**65.30**	nr	**73.46**
to suit fencing 1219mm high	14.36	10.00	**14.91**	2.06	2.06	**51.34**	0.47	**66.72**	nr	**75.06**
to suit fencing 1524mm high	15.26	10.00	**15.81**	2.08	2.08	**51.83**	0.55	**68.20**	nr	**76.72**
to suit fencing 1829mm high	16.62	10.00	**17.17**	2.10	2.10	**52.33**	0.64	**70.15**	nr	**78.91**
Remove defective pale from palisade fence and replace with new 19 x 75mm pointed pale; fixing with galvanised nails										
764mm long	0.72	-	**0.72**	0.55	0.09	**8.78**	0.18	**9.68**	nr	**10.89**
1069mm long	1.01	-	**1.01**	0.56	0.09	**8.92**	0.22	**10.15**	nr	**11.42**
1374mm long	1.30	-	**1.30**	0.58	0.10	**9.31**	0.25	**10.86**	nr	**12.22**
Remove defective pales from close boarded fence and replace with new 100mm wide feather edged pales; fixing with galvanised nails										
singly	1.63	-	**1.63**	0.25	0.04	**3.98**	0.21	**5.82**	m	**6.55**
areas up to 1.00m2	9.94	20.00	**11.93**	1.32	0.22	**21.11**	1.75	**34.79**	m2	**39.14**
areas over 1.00m2	9.94	20.00	**11.93**	0.98	0.16	**15.64**	1.75	**29.32**	m2	**32.98**
Remove defective gravel board and centre stump and replace with new 25 x 150mm gravel board and 50 x 50mm centre stump; fixing with galvanised nails										
securing to fence posts	2.44	5.00	**2.56**	0.53	0.09	**8.50**	0.21	**11.27**	m	**12.68**
Repairs to oak fencing										
Remove defective timber fence post and replace with new 127 x 102mm post; letting post 450mm into ground; backfilling around in concrete mix C20; securing ends of existing arris rails and gravel boards to new post										
to suit fencing 914mm high	23.28	10.00	**23.83**	2.17	2.17	**54.08**	0.41	**78.32**	nr	**88.11**
to suit fencing 1219mm high	27.09	10.00	**27.64**	2.21	2.21	**55.07**	0.47	**83.19**	nr	**93.59**
to suit fencing 1524mm high	30.89	10.00	**31.44**	2.25	2.25	**56.07**	0.55	**88.06**	nr	**99.07**
to suit fencing 1829mm high	34.70	10.00	**35.25**	2.29	2.29	**57.07**	0.64	**92.96**	nr	**104.58**
Remove defective pale from palisade fence and replace with new 19 x 75mm pointed pale; fixing with galvanised nails										
764mm long	1.90	-	**1.90**	0.70	0.12	**11.23**	0.18	**13.31**	nr	**14.98**
1069mm long	2.64	-	**2.64**	0.76	0.13	**12.19**	0.22	**15.05**	nr	**16.93**
1374mm long	3.41	-	**3.41**	0.83	0.14	**13.29**	0.25	**16.95**	nr	**19.07**
Remove defective pales from close boarded fence and replace with new 100mm wide feather edged pales; fixing with galvanised nails										
singly	4.27	-	**4.27**	0.27	0.05	**4.37**	0.21	**8.85**	m	**9.96**
areas up to 1.00m2	26.05	20.00	**31.26**	2.56	0.43	**40.98**	1.75	**73.99**	m2	**83.24**
areas over 1.00m2	26.05	20.00	**31.26**	2.07	0.35	**33.16**	1.75	**66.17**	m2	**74.44**
Remove defective gravel board and centre stump and replace with new 25 x 150mm gravel board and 50 x 50mm centre stump; fixing with galvanised nails										
securing to fence posts	6.36	5.00	**6.68**	1.10	0.18	**17.56**	0.21	**24.45**	m	**27.50**
Repairs to posts										
100 x 100mm precast concrete spur 1219mm long; setting into ground; backfilling around in concrete mix C20										
bolting to existing timber post	14.02	10.00	**14.57**	0.76	0.76	**18.94**	1.66	**35.17**	nr	**39.57**
C90: REPAIRING/RENOVATING EXTERNAL PAVINGS										
Repairs to precast concrete flag paving										
Take up uneven paving and set aside for re-use; level up and consolidate existing hardcore and relay salvaged paving										
bedding on 25mm sand bed; grouting in lime mortar (1:3)	0.94	33.00	**1.22**	0.67	0.93	**19.48**	0.12	**20.82**	m2	**23.43**
Repairs to York stone paving										
Take up old paving; re-square; relay random in random sizes										
bedding, jointing and pointing in cement mortar (1:3)	4.39	10.00	**4.83**	1.10	1.10	**27.41**	0.90	**33.14**	m2	**37.28**

Labour hourly rates: (except Specialists) Craft Operatives £14.21 Labourer £10.71 Rates are national average prices. Refer to REGIONAL VARIATIONS for indicative levels of overall pricing in regions	MATERIALS			LABOUR				RATES		
	Del to Site £	Waste %	Material Cost £	Craft Optve Hrs	Lab Hrs	Labour Cost £	Sunds £	Nett Rate £	Unit	Gross Rate (+12.5%) £
C90: REPAIRING/RENOVATING EXTERNAL PAVINGS Cont.										
Repairs to granite sett paving										
Take up, clean and stack old paving	-	-	-	-	0.90	**9.64**	-	**9.64**	m2	**10.84**
Take up old paving, clean and re-lay bedding and grouting in cement mortar (1:3)......	-	-	-	1.65	2.55	**50.76**	1.20	**51.96**	m2	**58.45**
Clean out joints of old paving and re-grout in cement mortar (1:3)	-	-	-	-	0.55	**5.89**	0.52	**6.41**	m2	**7.21**
Repairs to bitumen paving										
Cut out damaged 50mm paving to footpath and fill with bitumen macadam under 1.00m2.......	36.00	10.00	**39.60**	1.65	2.55	**50.76**	1.20	**91.56**	m2	**103.00**
Cut out damaged 75mm paving to road fill with macadam and seal all around with hot bitumen under 1.00m2	54.00	10.00	**59.40**	1.65	2.55	**50.76**	1.20	**111.36**	m2	**125.28**

GROUNDWORK

Labour hourly rates: (except Specialists) Craft Operatives £14.21 Labourer £10.71 Rates are national average prices. Refer to REGIONAL VARIATIONS for indicative levels of overall pricing in regions	PLANT & TRANSPORT			LABOUR			RATES			
	Plant Cost £	Trans Cost £	P & T Cost £	Craft Optve Hrs	Lab Hrs	Labour Cost £	Sunds £	Nett Rate £	Unit	Gross Rate (+12.5%) £
D20: EXCAVATING AND FILLING										
Site preparation										
Removing trees; filling voids left by removal of roots with selected material arising from excavation										
girth 0.60 - 1.50m	76.05	-	**76.05**	-	11.00	**117.81**	-	**193.86**	nr	**218.09**
girth 1.50 - 3.00m	140.40	-	**140.40**	-	23.00	**246.33**	-	**386.73**	nr	**435.07**
girth exceeding 3.00m	204.75	-	**204.75**	-	35.00	**374.85**	-	**579.60**	nr	**652.05**
Clearing site vegetation; filling voids left by removal of roots with selected material arising from excavation										
bushes, scrub, undergrowth, hedges, trees and tree stumps not exceeding 600mm girth	0.62	-	**0.62**	-	-	**-**	-	**0.62**	m2	**0.70**
Lifting turf for preservation										
stacking on site average 100m distant for future use; watering	-	-	**-**	-	0.60	**6.43**	-	**6.43**	m2	**7.23**
Excavating - by machine										
Top soil for preservation										
average 150mm deep	0.33	-	**0.33**	-	-	**-**	-	**0.33**	m2	**0.37**
To reduce levels										
maximum depth not exceeding										
0.25m	2.54	-	**2.54**	-	-	**-**	-	**2.54**	m3	**2.86**
1.00m	1.33	-	**1.33**	-	-	**-**	-	**1.33**	m3	**1.50**
2.00m	1.53	-	**1.53**	-	-	**-**	-	**1.53**	m3	**1.72**
4.00m	1.49	-	**1.49**	-	-	**-**	-	**1.49**	m3	**1.67**
6.00m	1.83	-	**1.83**	-	-	**-**	-	**1.83**	m3	**2.06**
Basements and the like										
maximum depth not exceeding										
0.25m	2.54	-	**2.54**	-	0.45	**4.82**	-	**7.36**	m3	**8.28**
1.00m	1.60	-	**1.60**	-	0.50	**5.36**	-	**6.96**	m3	**7.83**
2.00m	1.87	-	**1.87**	-	0.55	**5.89**	-	**7.76**	m3	**8.73**
4.00m	2.29	-	**2.29**	-	0.70	**7.50**	-	**9.79**	m3	**11.01**
6.00m	3.22	-	**3.22**	-	0.85	**9.10**	-	**12.32**	m3	**13.86**
Trenches width not exceeding 0.30m										
maximum depth not exceeding										
0.25m	6.55	-	**6.55**	-	2.00	**21.42**	-	**27.97**	m3	**31.47**
1.00m	6.79	-	**6.79**	-	2.10	**22.49**	-	**29.28**	m3	**32.94**
Trenches width exceeding 0.30m										
maximum depth not exceeding										
0.25m	6.54	-	**6.54**	-	0.65	**6.96**	-	**13.50**	m3	**15.19**
1.00m	5.60	-	**5.60**	-	0.70	**7.50**	-	**13.10**	m3	**14.74**
2.00m	5.60	-	**5.60**	-	0.80	**8.57**	-	**14.17**	m3	**15.94**
4.00m	5.74	-	**5.74**	-	1.00	**10.71**	-	**16.45**	m3	**18.50**
6.00m	5.74	-	**5.74**	-	1.20	**12.85**	-	**18.59**	m3	**20.91**
commencing 1.50m below existing ground level, maximum depth not exceeding										
0.25m	5.62	-	**5.62**	-	1.00	**10.71**	-	**16.33**	m3	**18.37**
1.00m	5.62	-	**5.62**	-	1.05	**11.25**	-	**16.86**	m3	**18.97**
2.00m	4.83	-	**4.83**	-	1.30	**13.92**	-	**18.75**	m3	**21.10**
commencing 3.00m below existing ground level, maximum depth not exceeding										
0.25m	4.83	-	**4.83**	-	1.45	**15.53**	-	**20.36**	m3	**22.90**
1.00m	5.74	-	**5.74**	-	1.55	**16.60**	-	**22.34**	m3	**25.13**
2.00m	5.74	-	**5.74**	-	1.75	**18.74**	-	**24.48**	m3	**27.54**
For pile caps and ground beams between piles										
maximum depth not exceeding										
0.25m	6.80	-	**6.80**	-	0.95	**10.17**	-	**16.97**	m3	**19.09**
1.00m	6.67	-	**6.67**	-	1.00	**10.71**	-	**17.38**	m3	**19.55**
2.00m	6.67	-	**6.67**	-	1.15	**12.32**	-	**18.99**	m3	**21.36**

Labour hourly rates: (except Specialists)
Craft Operatives £14.21 Labourer £10.71
Rates are national average prices.
Refer to REGIONAL VARIATIONS for indicative levels of overall pricing in regions

	PLANT & TRANSPORT			LABOUR				RATES		
	Plant Cost £	Trans Cost £	P & T Cost £	Craft Optve Hrs	Lab Hrs	Labour Cost £	Sunds £	Nett Rate £	Unit	Gross Rate (+12.5%) £
D20: EXCAVATING AND FILLING Cont.										
Excavating - by machine Cont.										
Pits										
maximum depth not exceeding										
0.25m	5.73	-	5.73	-	0.95	10.17	-	15.91	m3	17.90
1.00m	7.07	-	7.07	-	1.00	10.71	-	17.78	m3	20.00
2.00m	7.07	-	7.07	-	1.15	12.32	-	19.38	m3	21.81
4.00m	6.88	-	6.88	-	1.45	15.53	-	22.41	m3	25.21
Extra over any types of excavating irrespective of depth										
excavating below ground water level	5.80	-	5.80	-	-	-	-	5.80	m3	6.53
excavating in running silt, running sand or liquid mud	10.07	-	10.07	-	-	-	-	10.07	m3	11.33
excavating below ground water level in running silt, running sand or liquid mud	13.85	-	13.85	-	-	-	-	13.85	m3	15.58
excavating in heavy soil or clay	1.29	-	1.29	-	-	-	-	1.29	m3	1.45
excavating in gravel	2.01	-	2.01	-	-	-	-	2.01	m3	2.26
excavating in brash, loose rock or chalk	2.41	-	2.41	-	-	-	-	2.41	m3	2.71
Breaking out existing materials; extra over any types of excavating irrespective of depth										
sandstone	6.56	-	6.56	-	-	-	-	6.56	m3	7.38
hard rock	9.87	-	9.87	-	-	-	-	9.87	m3	11.10
concrete	8.19	-	8.19	-	-	-	-	8.19	m3	9.22
reinforced concrete	12.30	-	12.30	-	-	-	-	12.30	m3	13.84
brickwork, blockwork or stonework	5.49	-	5.49	-	-	-	-	5.49	m3	6.18
drain and concrete bed under										
100mm diameter	0.77	-	0.77	-	-	-	-	0.77	m	0.86
150mm diameter	0.93	-	0.93	-	-	-	-	0.93	m	1.04
225mm diameter	1.07	-	1.07	-	-	-	-	1.07	m	1.21
Breaking out existing hard pavings; extra over any types of excavating irrespective of depth										
concrete 150mm thick	1.29	-	1.29	-	-	-	-	1.29	m2	1.45
reinforced concrete 200mm thick	2.59	-	2.59	-	-	-	-	2.59	m2	2.92
coated macadam or asphalt 75mm thick	0.22	-	0.22	-	-	-	-	0.22	m2	0.24
Working space allowance to excavation; including additional earthwork support, disposal (excluding landfill tax) and backfilling										
reduce levels, basements and the like	5.87	-	5.87	-	0.30	3.21	-	9.09	m2	10.22
pits	6.67	-	6.67	-	0.70	7.50	-	14.17	m2	15.94
trenches	5.87	-	5.87	-	0.50	5.36	-	11.23	m2	12.63
Compacting										
bottoms of excavations	0.21	-	0.21	-	-	-	-	0.21	m2	0.24
Compacting with 680 kg vibratory roller										
ground	0.22	-	0.22	-	-	-	-	0.22	m2	0.25
Compacting with 6-8 tonnes smooth wheeled roller										
ground	0.36	-	0.36	-	-	-	-	0.36	m2	0.41
Trimming										
sides of cuttings; vertical or battered	0.56	-	0.56	-	-	-	-	0.56	m2	0.63
sides of embankments; vertical or battered	0.56	-	0.56	-	-	-	-	0.56	m2	0.63
Disposal of excavated material										
depositing on site in temporary spoil heaps where directed										
25m	-	2.04	2.04	-	0.12	1.29	-	3.33	m3	3.74
50m	-	2.13	2.13	-	0.12	1.29	-	3.42	m3	3.84
100m	-	2.33	2.33	-	0.12	1.29	-	3.61	m3	4.06
400m	-	2.76	2.76	-	0.12	1.29	-	4.04	m3	4.55
800m	-	3.58	3.58	-	0.12	1.29	-	4.86	m3	5.47
1200m	-	3.82	3.82	-	0.12	1.29	-	5.11	m3	5.74
1600m	-	4.05	4.05	-	0.12	1.29	-	5.34	m3	6.00
extra for each additional 1600m	-	0.63	0.63	-	-	-	-	0.63	m3	0.71
removing from site to tip a) Inert	-	17.00	17.00	-	-	-	-	17.00	m3	19.13
removing from site to tip b) Active	-	40.00	40.00	-	-	-	-	40.00	m3	45.00
removing from site to tip c) Contaminated (Guide price - always seek a quotation for specialist disposal costs.)	-	80.00	80.00	-	-	-	-	80.00	m3	90.00
Disposal of preserved top soil										
depositing on site in temporary spoil heaps where directed										
25m	-	2.04	2.04	-	0.12	1.29	-	3.33	m3	3.74
50m	-	2.13	2.13	-	0.12	1.29	-	3.42	m3	3.84
100m	-	2.33	2.33	-	0.12	1.29	-	3.61	m3	4.06
Excavating - by hand										
Top soil for preservation										
average 150mm deep	-	-	-	-	0.33	3.53	-	3.53	m2	3.98
To reduce levels										
maximum depth not exceeding										
0.25m	-	-	-	-	2.75	29.45	-	29.45	m3	33.13
1.00m	-	-	-	-	2.90	31.06	-	31.06	m3	34.94
2.00m	-	-	-	-	3.30	35.34	-	35.34	m3	39.76
4.00m	-	-	-	-	4.30	46.05	-	46.05	m3	51.81
6.00m	-	-	-	-	5.35	57.30	-	57.30	m3	64.46

Labour hourly rates: (except Specialists) Craft Operatives £14.21 Labourer £10.71 Rates are national average prices. Refer to REGIONAL VARIATIONS for indicative levels of overall pricing in regions	PLANT & TRANSPORT			LABOUR				RATES		
	Plant Cost £	Trans Cost £	P & T Cost £	Craft Optve Hrs	Lab Hrs	Labour Cost £	Sunds £	Nett Rate £	Unit	Gross Rate (+12.5%) £
D20: EXCAVATING AND FILLING Cont.										
Excavating - by hand Cont.										
Basements and the like										
maximum depth not exceeding										
0.25m	-	-	-	-	2.85	30.52	-	30.52	m3	34.34
1.00m	-	-	-	-	3.05	32.67	-	32.67	m3	36.75
2.00m	-	-	-	-	3.40	36.41	-	36.41	m3	40.97
4.00m	-	-	-	-	4.40	47.12	-	47.12	m3	53.01
6.00m	-	-	-	-	5.35	57.30	-	57.30	m3	64.46
Trenches width not exceeding 0.30m										
maximum depth not exceeding										
0.25m	-	-	-	-	4.70	50.34	-	50.34	m3	56.63
1.00m	-	-	-	-	4.95	53.01	-	53.01	m3	59.64
Trenches width exceeding 0.30m										
maximum depth not exceeding										
0.25m	-	-	-	-	3.14	33.63	-	33.63	m3	37.83
1.00m	-	-	-	-	3.30	35.34	-	35.34	m3	39.76
2.00m	-	-	-	-	3.85	41.23	-	41.23	m3	46.39
4.00m	-	-	-	-	4.80	51.41	-	51.41	m3	57.83
6.00m	-	-	-	-	6.05	64.80	-	64.80	m3	72.89
commencing 1.50m below existing ground level, maximum depth not exceeding										
0.25m	-	-	-	-	4.70	50.34	-	50.34	m3	56.63
1.00m	-	-	-	-	4.95	53.01	-	53.01	m3	59.64
2.00m	-	-	-	-	6.20	66.40	-	66.40	m3	74.70
commencing 3.00m below existing ground level, maximum depth not exceeding										
0.25m	-	-	-	-	7.00	74.97	-	74.97	m3	84.34
1.00m	-	-	-	-	7.30	78.18	-	78.18	m3	87.96
2.00m	-	-	-	-	8.40	89.96	-	89.96	m3	101.21
For pile caps and ground beams between piles										
maximum depth not exceeding										
0.25m	-	-	-	-	3.70	39.63	-	39.63	m3	44.58
1.00m	-	-	-	-	3.85	41.23	-	41.23	m3	46.39
2.00m	-	-	-	-	4.40	47.12	-	47.12	m3	53.01
Pits										
maximum depth not exceeding										
0.25m	-	-	-	-	3.70	39.63	-	39.63	m3	44.58
1.00m	-	-	-	-	3.85	41.23	-	41.23	m3	46.39
2.00m	-	-	-	-	4.40	47.12	-	47.12	m3	53.01
4.00m	-	-	-	-	5.60	59.98	-	59.98	m3	67.47
Extra over any types of excavating irrespective of depth										
excavating below ground water level	-	-	-	-	1.65	17.67	-	17.67	m3	19.88
excavating in running silt, running sand or liquid mud	-	-	-	-	4.95	53.01	-	53.01	m3	59.64
excavating below ground water level in running silt, running sand or liquid mud	-	-	-	-	6.60	70.69	-	70.69	m3	79.52
excavating in heavy soil or clay	-	-	-	-	0.70	7.50	-	7.50	m3	8.43
excavating in gravel	-	-	-	-	1.80	19.28	-	19.28	m3	21.69
excavating in brash, loose rock or chalk	-	-	-	-	3.70	39.63	-	39.63	m3	44.58
Breaking out existing materials; extra over any types of excavating irrespective of depth										
sandstone	-	-	-	-	7.30	78.18	-	78.18	m3	87.96
hard rock	-	-	-	-	14.70	157.44	-	157.44	m3	177.12
concrete	-	-	-	-	11.00	117.81	-	117.81	m3	132.54
reinforced concrete	-	-	-	-	16.50	176.72	-	176.72	m3	198.80
brickwork, blockwork or stonework	-	-	-	-	7.70	82.47	-	82.47	m3	92.78
drain and concrete bed under										
100mm diameter	-	-	-	-	0.90	9.64	-	9.64	m	10.84
150mm diameter	-	-	-	-	1.10	11.78	-	11.78	m	13.25
225mm diameter	-	-	-	-	1.25	13.39	-	13.39	m	15.06
Breaking out existing hard pavings; extra over any types of excavating irrespective of depth										
concrete 150mm thick	-	-	-	-	1.65	17.67	-	17.67	m2	19.88
reinforced concrete 200mm thick	-	-	-	-	3.30	35.34	-	35.34	m2	39.76
coated macadam or asphalt 75mm thick	-	-	-	-	0.30	3.21	-	3.21	m2	3.61
Working space allowance to excavation; including additional earthwork support, disposal (excluding landfill tax) and backfilling										
reduce levels, basements and the like	-	-	-	-	2.85	30.52	-	30.52	m2	34.34
pits	-	-	-	-	3.45	36.95	-	36.95	m2	41.57
trenches	-	-	-	-	3.10	33.20	-	33.20	m2	37.35
Compacting										
bottoms of excavations	-	-	-	-	0.15	1.61	-	1.61	m2	1.81
Trimming										
sides of cuttings; vertical or battered	-	-	-	-	0.22	2.36	-	2.36	m2	2.65
sides of embankments; vertical or battered	-	-	-	-	0.22	2.36	-	2.36	m2	2.65
Disposal of excavated material										
depositing on site in temporary spoil heaps where directed										
25m	-	-	-	-	1.40	14.99	-	14.99	m3	16.87
50m	-	-	-	-	1.65	17.67	-	17.67	m3	19.88

Labour hourly rates: (except Specialists) Craft Operatives £14.21 Labourer £10.71 Rates are national average prices. Refer to REGIONAL VARIATIONS for indicative levels of overall pricing in regions	PLANT & TRANSPORT			LABOUR				RATES		
	Plant Cost £	Trans Cost £	P & T Cost £	Craft Optve Hrs	Lab Hrs	Labour Cost £	Sunds £	Nett Rate £	Unit	Gross Rate (+12.5%) £
D20: EXCAVATING AND FILLING Cont.										
Excavating - by hand Cont.										
Disposal of excavated material Cont.										
depositing on site in temporary spoil heaps where directed Cont.										
100m	-	-	-	-	2.20	**23.56**	-	**23.56**	m3	**26.51**
400m	-	2.76	**2.76**	-	1.72	**18.42**	-	**21.18**	m3	**23.82**
800m	-	3.58	**3.58**	-	1.72	**18.42**	-	**22.00**	m3	**24.75**
1200m	-	3.82	**3.82**	-	1.72	**18.42**	-	**22.24**	m3	**25.02**
1600m	-	4.05	**4.05**	-	1.72	**18.42**	-	**22.47**	m3	**25.28**
extra for each additional 1600m	-	0.63	**0.63**	-	-	**-**	-	**0.63**	m3	**0.70**
removing from site to tip a) Inert	-	17.00	**17.00**	-	1.50	**16.07**	-	**33.07**	m3	**37.20**
removing from site to tip b) Active	-	40.00	**40.00**	-	1.50	**16.07**	-	**56.07**	m3	**63.07**
removing from site to tip c) Contaminated (Guide price - always seek a quotation for specialist disposal costs.)	-	80.00	**80.00**	-	1.50	**16.07**	-	**96.07**	m3	**108.07**
Disposal of preserved top soil										
depositing on site in temporary spoil heaps where directed										
25m	-	-	-	-	1.40	**14.99**	-	**14.99**	m3	**16.87**
50m	-	-	-	-	1.65	**17.67**	-	**17.67**	m3	**19.88**
100m	-	-	-	-	2.20	**23.56**	-	**23.56**	m3	**26.51**
Excavating inside existing building - by hand										
To reduce levels										
maximum depth not exceeding										
0.25m	-	-	-	-	3.85	**41.23**	-	**41.23**	m3	**46.39**
1.00m	-	-	-	-	4.10	**43.91**	-	**43.91**	m3	**49.40**
2.00m	-	-	-	-	4.65	**49.80**	-	**49.80**	m3	**56.03**
4.00m	-	-	-	-	6.00	**64.26**	-	**64.26**	m3	**72.29**
Trenches width not exceeding 0.30m										
maximum depth not exceeding										
0.25m	-	-	-	-	6.65	**71.22**	-	**71.22**	m3	**80.12**
1.00m	-	-	-	-	7.00	**74.97**	-	**74.97**	m3	**84.34**
Trenches width exceeding 0.30m										
maximum depth not exceeding										
0.25m	-	-	-	-	4.45	**47.66**	-	**47.66**	m3	**53.62**
1.00m	-	-	-	-	4.70	**50.34**	-	**50.34**	m3	**56.63**
2.00m	-	-	-	-	5.50	**58.91**	-	**58.91**	m3	**66.27**
4.00m	-	-	-	-	6.80	**72.83**	-	**72.83**	m3	**81.93**
Pits										
maximum depth not exceeding										
0.25m	-	-	-	-	5.20	**55.69**	-	**55.69**	m3	**62.65**
1.00m	-	-	-	-	5.40	**57.83**	-	**57.83**	m3	**65.06**
2.00m	-	-	-	-	6.20	**66.40**	-	**66.40**	m3	**74.70**
4.00m	-	-	-	-	7.85	**84.07**	-	**84.07**	m3	**94.58**
Extra over any types of excavating irrespective of depth										
excavating below ground water level	-	-	-	-	2.35	**25.17**	-	**25.17**	m3	**28.31**
Breaking out existing materials; extra over any types of excavating irrespective of depth										
concrete	-	-	-	-	12.00	**128.52**	-	**128.52**	m3	**144.59**
reinforced concrete	-	-	-	-	17.30	**185.28**	-	**185.28**	m3	**208.44**
brickwork, blockwork or stonework	-	-	-	-	7.80	**83.54**	-	**83.54**	m3	**93.98**
drain and concrete bed under										
100mm diameter	-	-	-	-	1.25	**13.39**	-	**13.39**	m	**15.06**
150mm diameter	-	-	-	-	1.50	**16.07**	-	**16.07**	m	**18.07**
Breaking out existing hard pavings; extra over any types of excavating irrespective of depth										
concrete 150mm thick	-	-	-	-	1.65	**17.67**	-	**17.67**	m2	**19.88**
reinforced concrete 200mm thick	-	-	-	-	3.30	**35.34**	-	**35.34**	m2	**39.76**
Working space allowance to excavation; including additional earthwork support, disposal (excluding landfill tax) and backfilling										
pits	-	-	-	-	4.40	**47.12**	-	**47.12**	m2	**53.01**
trenches	-	-	-	-	4.10	**43.91**	-	**43.91**	m2	**49.40**
Compacting										
bottoms of excavations	-	-	-	-	0.15	**1.61**	-	**1.61**	m2	**1.81**
Disposal of excavated material										
depositing on site in temporary spoil heaps where directed										
25m	-	-	-	-	1.95	**20.88**	-	**20.88**	m3	**23.50**
50m	-	-	-	-	2.15	**23.03**	-	**23.03**	m3	**25.90**
removing from site to tip a) Inert	-	17.00	**17.00**	-	2.50	**26.78**	-	**43.78**	m3	**49.25**
removing from site to tip b) Active	-	40.00	**40.00**	-	2.50	**26.78**	-	**66.78**	m3	**75.12**
removing from site to tip c) Contaminated (Guide price - always seek a quotation for specialist disposal costs.)	-	80.00	**80.00**	-	2.50	**26.78**	-	**106.78**	m3	**120.12**

Labour hourly rates: (except Specialists) Craft Operatives £14.21 Labourer £10.71 Rates are national average prices. Refer to REGIONAL VARIATIONS for indicative levels of overall pricing in regions	MATERIALS			LABOUR			RATES			
	Del to Site	Waste	Material Cost	Craft Optve	Lab	Labour Cost	Sunds	Nett Rate	Unit	Gross Rate (+12.5%)
	£	%	£	Hrs	Hrs	£	£	£		£
D20: EXCAVATING AND FILLING Cont.										
Earthwork support										
Earthwork support										
distance between opposing faces not exceeding 2.00m; maximum depth not exceeding										
1.00m	1.10	-	**1.10**	-	0.21	**2.25**	0.03	**3.38**	m2	**3.81**
2.00m	1.38	-	**1.38**	-	0.26	**2.78**	0.04	**4.21**	m2	**4.74**
4.00m	1.57	-	**1.57**	-	0.32	**3.43**	0.05	**5.05**	m2	**5.68**
6.00m	1.78	-	**1.78**	-	0.37	**3.96**	0.05	**5.80**	m2	**6.52**
distance between opposing faces 2.00 - 4.00m; maximum depth not exceeding										
1.00m	1.20	-	**1.20**	-	0.22	**2.36**	0.04	**3.60**	m2	**4.04**
2.00m	1.46	-	**1.46**	-	0.27	**2.89**	0.04	**4.40**	m2	**4.95**
4.00m	1.73	-	**1.73**	-	0.34	**3.64**	0.05	**5.43**	m2	**6.11**
6.00m	1.96	-	**1.96**	-	0.39	**4.18**	0.07	**6.20**	m2	**6.97**
distance between opposing faces exceeding 4.00m; maximum depth not exceeding										
1.00m	1.36	-	**1.36**	-	0.23	**2.46**	0.04	**3.87**	m2	**4.35**
2.00m	1.58	-	**1.58**	-	0.29	**3.11**	0.05	**4.74**	m2	**5.34**
4.00m	1.87	-	**1.87**	-	0.35	**3.75**	0.07	**5.69**	m2	**6.40**
6.00m	2.11	-	**2.11**	-	0.41	**4.39**	0.07	**6.57**	m2	**7.39**
Earthwork support; unstable ground										
distance between opposing faces not exceeding 2.00m; maximum depth not exceeding										
1.00m	1.87	-	**1.87**	-	0.35	**3.75**	0.07	**5.69**	m2	**6.40**
2.00m	2.25	-	**2.25**	-	0.44	**4.71**	0.08	**7.04**	m2	**7.92**
4.00m	2.60	-	**2.60**	-	0.53	**5.68**	0.09	**8.37**	m2	**9.41**
6.00m	2.96	-	**2.96**	-	0.62	**6.64**	0.10	**9.70**	m2	**10.92**
distance between opposing faces 2.00 - 4.00m; maximum depth not exceeding										
1.00m	2.06	-	**2.06**	-	0.37	**3.96**	0.07	**6.09**	m2	**6.85**
2.00m	2.46	-	**2.46**	-	0.45	**4.82**	0.08	**7.36**	m2	**8.28**
4.00m	2.87	-	**2.87**	-	0.56	**6.00**	0.09	**8.96**	m2	**10.08**
6.00m	3.29	-	**3.29**	-	0.65	**6.96**	0.10	**10.35**	m2	**11.65**
distance between opposing faces exceeding 4.00m; maximum depth not exceeding										
1.00m	2.25	-	**2.25**	-	0.37	**3.96**	0.08	**6.29**	m2	**7.08**
2.00m	2.65	-	**2.65**	-	0.49	**5.25**	0.09	**7.98**	m2	**8.98**
4.00m	3.15	-	**3.15**	-	0.58	**6.21**	0.10	**9.46**	m2	**10.65**
6.00m	3.57	-	**3.57**	-	0.68	**7.28**	0.11	**10.97**	m2	**12.34**
Earthwork support; next to roadways										
distance between opposing faces not exceeding 2.00m; maximum depth not exceeding										
1.00m	2.18	-	**2.18**	-	0.42	**4.50**	0.07	**6.74**	m2	**7.59**
2.00m	2.65	-	**2.65**	-	0.53	**5.68**	0.09	**8.41**	m2	**9.46**
4.00m	3.15	-	**3.15**	-	0.63	**6.75**	0.10	**10.00**	m2	**11.25**
6.00m	3.57	-	**3.57**	-	0.74	**7.93**	0.11	**11.61**	m2	**13.06**
distance between opposing faces 2.00 - 4.00m; maximum depth not exceeding										
1.00m	2.46	-	**2.46**	-	0.44	**4.71**	0.08	**7.25**	m2	**8.16**
2.00m	2.94	-	**2.94**	-	0.55	**5.89**	0.10	**8.93**	m2	**10.05**
4.00m	3.47	-	**3.47**	-	0.67	**7.18**	0.11	**10.75**	m2	**12.10**
6.00m	3.91	-	**3.91**	-	0.78	**8.35**	0.13	**12.40**	m2	**13.95**
distance between opposing faces exceeding 4.00m; maximum depth not exceeding										
1.00m	2.65	-	**2.65**	-	0.46	**4.93**	0.09	**7.66**	m2	**8.62**
2.00m	3.21	-	**3.21**	-	0.59	**6.32**	0.10	**9.63**	m2	**10.83**
4.00m	3.75	-	**3.75**	-	0.69	**7.39**	0.12	**11.26**	m2	**12.67**
6.00m	4.29	-	**4.29**	-	0.82	**8.78**	0.14	**13.21**	m2	**14.87**
Earthwork support; left in										
distance between opposing faces not exceeding 2.00m; maximum depth not exceeding										
1.00m	3.96	-	**3.96**	-	0.17	**1.82**	0.07	**5.85**	m2	**6.58**
2.00m	3.96	-	**3.96**	-	0.21	**2.25**	0.08	**6.29**	m2	**7.07**
4.00m	3.96	-	**3.96**	-	0.25	**2.68**	0.09	**6.73**	m2	**7.57**
6.00m	3.96	-	**3.96**	-	0.29	**3.11**	0.10	**7.17**	m2	**8.06**
distance between opposing faces 2.00 - 4.00m; maximum depth not exceeding										
1.00m	3.96	-	**3.96**	-	0.18	**1.93**	0.07	**5.95**	m2	**6.70**
2.00m	3.96	-	**3.96**	-	0.22	**2.36**	0.09	**6.40**	m2	**7.21**
4.00m	3.96	-	**3.96**	-	0.27	**2.89**	0.10	**6.95**	m2	**7.82**
6.00m	3.96	-	**3.96**	-	0.32	**3.43**	0.12	**7.51**	m2	**8.45**
distance between opposing faces exceeding 4.00m; maximum depth not exceeding										
1.00m	5.27	-	**5.27**	-	0.19	**2.03**	0.08	**7.38**	m2	**8.31**
2.00m	5.27	-	**5.27**	-	0.23	**2.46**	0.09	**7.82**	m2	**8.80**
4.00m	5.27	-	**5.27**	-	0.27	**2.89**	0.11	**8.27**	m2	**9.31**
6.00m	5.27	-	**5.27**	-	0.33	**3.53**	0.12	**8.93**	m2	**10.04**
Earthwork support; next to roadways; left in										
distance between opposing faces not exceeding 2.00m; maximum depth not exceeding										
1.00m	13.18	-	**13.18**	-	0.34	**3.64**	0.12	**16.94**	m2	**19.06**
2.00m	13.18	-	**13.18**	-	0.42	**4.50**	0.15	**17.83**	m2	**20.06**
4.00m	13.18	-	**13.18**	-	0.51	**5.46**	0.18	**18.82**	m2	**21.17**
6.00m	13.18	-	**13.18**	-	0.59	**6.32**	0.20	**19.70**	m2	**22.16**
distance between opposing faces 2.00 - 4.00m; maximum depth not exceeding										
1.00m	15.82	-	**15.82**	-	0.36	**3.86**	0.14	**19.82**	m2	**22.29**
2.00m	15.82	-	**15.82**	-	0.44	**4.71**	0.18	**20.71**	m2	**23.30**
4.00m	15.82	-	**15.82**	-	0.54	**5.78**	0.22	**21.82**	m2	**24.55**

Labour hourly rates: (except Specialists) Craft Operatives £14.21 Labourer £10.71 Rates are national average prices. Refer to REGIONAL VARIATIONS for indicative levels of overall pricing in regions	MATERIALS			LABOUR				RATES		
	Del to Site	Waste	Material Cost	Craft Optve	Lab	Labour Cost	Sunds	Nett Rate	Unit	Gross Rate (+12.5%)
	£	%	£	Hrs	Hrs	£	£	£		£
D20: EXCAVATING AND FILLING Cont.										
Earthwork support Cont.										
Earthwork support; next to roadways; left in Cont.										
distance between opposing faces 2.00 - 4.00m; maximum depth not exceeding Cont.										
6.00m	15.82	-	**15.82**	-	0.62	**6.64**	0.24	**22.70**	m2	**25.54**
distance between opposing faces exceeding 4.00m; maximum depth not exceeding										
1.00m	19.78	-	**19.78**	-	0.37	**3.96**	0.16	**23.91**	m2	**26.90**
2.00m	19.78	-	**19.78**	-	0.47	**5.03**	0.19	**25.00**	m2	**28.12**
4.00m	19.78	-	**19.78**	-	0.56	**6.00**	0.23	**26.01**	m2	**29.26**
6.00m	19.78	-	**19.78**	-	0.65	**6.96**	0.27	**27.02**	m2	**30.39**
Earthwork support; unstable ground; left in										
distance between opposing faces not exceeding 2.00m; maximum depth not exceeding										
1.00m	9.23	-	**9.23**	-	0.27	**2.89**	0.11	**12.23**	m2	**13.76**
2.00m	9.23	-	**9.23**	-	0.36	**3.86**	0.13	**13.22**	m2	**14.87**
4.00m	9.23	-	**9.23**	-	0.42	**4.50**	0.14	**13.87**	m2	**15.61**
6.00m	9.23	-	**9.23**	-	0.49	**5.25**	0.16	**14.65**	m2	**16.48**
distance between opposing faces 2.00 - 4.00m; maximum depth not exceeding										
1.00m	10.55	-	**10.55**	-	0.29	**3.11**	0.12	**13.77**	m2	**15.49**
2.00m	10.55	-	**10.55**	-	0.36	**3.86**	0.14	**14.54**	m2	**16.36**
4.00m	10.55	-	**10.55**	-	0.44	**4.71**	0.16	**15.42**	m2	**17.35**
6.00m	10.55	-	**10.55**	-	0.53	**5.68**	0.19	**16.41**	m2	**18.46**
distance between opposing faces exceeding 4.00m; maximum depth not exceeding										
1.00m	13.18	-	**13.18**	-	0.32	**3.43**	0.13	**16.74**	m2	**18.83**
2.00m	13.18	-	**13.18**	-	0.40	**4.28**	0.15	**17.62**	m2	**19.82**
4.00m	13.18	-	**13.18**	-	0.46	**4.93**	0.18	**18.28**	m2	**20.57**
6.00m	13.18	-	**13.18**	-	0.55	**5.89**	0.20	**19.27**	m2	**21.68**
Earthwork support; inside existing building										
distance between opposing faces not exceeding 2.00m; maximum depth not exceeding										
1.00m	1.15	-	**1.15**	-	0.23	**2.46**	0.03	**3.64**	m2	**4.10**
2.00m	1.43	-	**1.43**	-	0.29	**3.11**	0.04	**4.58**	m2	**5.15**
4.00m	1.76	-	**1.76**	-	0.32	**3.43**	0.05	**5.24**	m2	**5.89**
distance between opposing faces 2.00 - 4.00m; maximum depth not exceeding										
1.00m	1.27	-	**1.27**	-	0.24	**2.57**	0.04	**3.88**	m2	**4.36**
2.00m	1.57	-	**1.57**	-	0.30	**3.21**	0.05	**4.84**	m2	**5.44**
4.00m	1.78	-	**1.78**	-	0.37	**3.96**	0.05	**5.80**	m2	**6.52**
distance between opposing faces exceeding 4.00m; maximum depth not exceeding										
1.00m	1.42	-	**1.42**	-	0.25	**2.68**	0.04	**4.14**	m2	**4.66**
2.00m	1.70	-	**1.70**	-	0.33	**3.53**	0.05	**5.29**	m2	**5.95**
4.00m	1.96	-	**1.96**	-	0.38	**4.07**	0.07	**6.09**	m2	**6.85**
Earthwork support; unstable ground; inside existing building										
distance between opposing faces not exceeding 2.00m; maximum depth not exceeding										
1.00m	1.96	-	**1.96**	-	0.38	**4.07**	0.07	**6.09**	m2	**6.85**
2.00m	2.37	-	**2.37**	-	0.48	**5.14**	0.08	**7.58**	m2	**8.53**
4.00m	2.74	-	**2.74**	-	0.58	**6.21**	0.09	**9.04**	m2	**10.17**
distance between opposing faces 2.00 - 4.00m; maximum depth not exceeding										
1.00m	2.16	-	**2.16**	-	0.41	**4.39**	0.07	**6.61**	m2	**7.44**
2.00m	2.58	-	**2.58**	-	0.49	**5.25**	0.08	**7.90**	m2	**8.89**
4.00m	3.02	-	**3.02**	-	0.61	**6.53**	0.10	**9.65**	m2	**10.86**
distance between opposing faces exceeding 4.00m; maximum depth not exceeding										
1.00m	2.37	-	**2.37**	-	0.43	**4.61**	0.07	**7.04**	m2	**7.92**
2.00m	2.79	-	**2.79**	-	0.55	**5.89**	0.09	**8.77**	m2	**9.86**
4.00m	3.29	-	**3.29**	-	0.64	**6.85**	0.10	**10.25**	m2	**11.53**
Excavated material arising from excavation - by machine										
Filling to excavation										
average thickness exceeding 0.25m	-	-	**-**	-	-	**-**	3.36	**3.36**	m3	**3.78**
Compacting										
filling	-	-	**-**	-	-	**-**	0.36	**0.36**	m2	**0.40**
Selected excavated material obtained from on site spoil heaps 25m distant - by machine										
Filling to make up levels; depositing in layers 150mm maximum thickness										
average thickness not exceeding 0.25m	-	-	**-**	-	-	**-**	4.85	**4.85**	m3	**5.46**
average thickness exceeding 0.25m	-	-	**-**	-	-	**-**	4.18	**4.18**	m3	**4.70**
Compacting with 680 kg vibratory roller										
filling	-	-	**-**	-	-	**-**	0.26	**0.26**	m2	**0.29**
Compacting with 6 - 8 tonnes smooth wheeled roller										
filling	-	-	**-**	-	-	**-**	0.50	**0.50**	m2	**0.57**
Trimming										
sides of cuttings; vertical or battered	-	-	**-**	-	-	**-**	0.70	**0.70**	m2	**0.79**
sides of embankments; vertical or battered	-	-	**-**	-	-	**-**	0.70	**0.70**	m2	**0.79**

Labour hourly rates: (except Specialists) Craft Operatives £14.21 Labourer £10.71 Rates are national average prices. Refer to REGIONAL VARIATIONS for indicative levels of overall pricing in regions	MATERIALS			LABOUR				RATES		
	Del to Site £	Waste %	Material Cost £	Craft Optve Hrs	Lab Hrs	Labour Cost £	Sunds £	Nett Rate £	Unit	Gross Rate (+12.5%) £
D20: EXCAVATING AND FILLING Cont.										
Selected excavated material obtained from on site spoil heaps 50m distant - by machine										
Filling to make up levels; depositing in layers 150mm maximum thickness										
average thickness not exceeding 0.25m............	-	-	-	-	-	-	5.53	**5.53**	m3	**6.22**
average thickness exceeding 0.25m................	-	-	-	-	-	-	5.14	**5.14**	m3	**5.78**
Compacting with 680 kg vibratory roller										
filling	-	-	-	-	-	-	0.26	**0.26**	m2	**0.29**
Compacting with 6 - 8 tonnes smooth wheeled roller										
filling	-	-	-	-	-	-	0.50	**0.50**	m2	**0.57**
Trimming										
sides of cuttings; vertical or battered..........	-	-	-	-	-	-	0.70	**0.70**	m2	**0.79**
sides of embankments; vertical or battered.......	-	-	-	-	-	-	0.70	**0.70**	m2	**0.79**
Preserved topsoil obtained from on site spoil heaps not exceeding 100m distant - by machine										
Filling to make up levels										
average thickness not exceeding 0.25m............	-	-	-	-	-	-	8.04	**8.04**	m3	**9.05**
Imported topsoil; wheeling not exceeding 50m - by machine										
Filling to make up levels										
average thickness not exceeding 0.25m............	15.00	10.00	**16.50**	-	-	-	9.71	**26.21**	m3	**29.49**
Hard, dry, broken brick or stone to be obtained off site; wheeling not exceeding 25m - by machine										
Filling to make up levels; depositing in layers 150mm maximum thickness										
average thickness not exceeding 0.25m............	17.85	25.00	**22.31**	-	-	-	4.18	**26.49**	m3	**29.80**
average thickness exceeding 0.25m................	17.85	25.00	**22.31**	-	-	-	3.47	**25.78**	m3	**29.00**
Surface packing to filling										
to vertical or battered faces...................	-	-	-	-	-	-	0.70	**0.70**	m2	**0.79**
Compacting with 680 kg vibratory roller										
filling; blinding with sand, ashes or similar fine material	0.18	40.00	**0.25**	-	0.05	**0.54**	0.26	**1.04**	m2	**1.18**
Compacting with 6 - 8 tonnes smooth wheeled roller										
filling; blinding with sand, ashes or similar fine material	0.18	40.00	**0.25**	-	0.05	**0.54**	0.50	**1.29**	m2	**1.45**
Hard, dry, broken brick or stone to be obtained off site; wheeling not exceeding 50m - by machine										
Filling to make up levels; depositing in layers 150mm maximum thickness										
average thickness not exceeding 0.25m............	17.85	25.00	**22.31**	-	-	-	4.85	**27.17**	m3	**30.56**
average thickness exceeding 0.25m................	17.85	25.00	**22.31**	-	-	-	4.18	**26.49**	m3	**29.80**
Surface packing to filling										
to vertical or battered faces...................	-	-	-	-	-	-	0.70	**0.70**	m2	**0.79**
Compacting with 680 kg vibratory roller										
filling; blinding with sand, ashes or similar fine material	0.18	40.00	**0.25**	-	0.05	**0.54**	0.26	**1.04**	m2	**1.18**
Compacting with 6 - 8 tonnes smooth wheeled roller										
filling; blinding with sand, ashes or similar fine material	0.18	40.00	**0.25**	-	0.05	**0.54**	0.50	**1.29**	m2	**1.45**
MOT Type 1 to be obtained off site; wheeling not exceeding 25m - by machine										
Filling to make up levels; depositing in layers 150mm maximum thickness										
average thickness not exceeding 0.25m............	20.95	25.00	**26.19**	-	-	-	4.18	**30.37**	m3	**34.16**
average thickness exceeding 0.25m................	20.95	25.00	**26.19**	-	-	-	3.47	**29.65**	m3	**33.36**
Surface packing to filling										
to vertical or battered faces...................	-	-	-	-	-	-	0.70	**0.70**	m2	**0.79**
Compacting with 680 kg vibratory roller										
filling; blinding with sand, ashes or similar fine material	0.18	40.00	**0.25**	-	0.05	**0.54**	0.26	**1.04**	m2	**1.18**
Compacting with 6 - 8 tonnes smooth wheeled roller										
filling; blinding with sand, ashes or similar fine material	0.18	40.00	**0.25**	-	0.05	**0.54**	0.50	**1.29**	m2	**1.45**
MOT Type 1 to be obtained off site; wheeling not exceeding 50m - by machine										
Filling to make up levels; depositing in layers 150mm maximum thickness										
average thickness not exceeding 0.25m............	20.95	25.00	**26.19**	-	-	-	4.85	**31.04**	m3	**34.92**
average thickness exceeding 0.25m................	20.95	25.00	**26.19**	-	-	-	4.18	**30.37**	m3	**34.16**

Labour hourly rates: (except Specialists) Craft Operatives £14.21 Labourer £10.71 Rates are national average prices. Refer to REGIONAL VARIATIONS for indicative levels of overall pricing in regions	MATERIALS			LABOUR				RATES		
	Del to Site £	Waste %	Material Cost £	Craft Optve Hrs	Lab Hrs	Labour Cost £	Sunds £	Nett Rate £	Unit	Gross Rate (+12.5%) £
D20: EXCAVATING AND FILLING Cont.										
MOT Type 1 to be obtained off site; wheeling not exceeding 50m - by machine Cont.										
Surface packing to filling										
to vertical or battered faces....................	-	-	**-**	-	-	**-**	0.70	**0.70**	m2	**0.79**
Compacting with 680 kg vibratory roller										
filling; blinding with sand, ashes or similar fine material	0.18	40.00	**0.25**	-	0.05	**0.54**	0.26	**1.04**	m2	**1.18**
Compacting with 6 - 8 tonnes smooth wheeled roller										
filling; blinding with sand, ashes or similar fine material	0.18	40.00	**0.25**	-	0.05	**0.54**	0.50	**1.29**	m2	**1.45**
MOT Type 2 to be obtained off site; wheeling not exceeding 25m - by machine										
Filling to make up levels; depositing in layers 150mm maximum thickness										
average thickness not exceeding 0.25m............	19.40	25.00	**24.26**	-	-	**-**	4.18	**28.44**	m3	**31.99**
average thickness exceeding 0.25m................	19.40	25.00	**24.26**	-	-	**-**	3.47	**27.72**	m3	**31.19**
Surface packing to filling										
to vertical or battered faces....................	-	-	**-**	-	-	**-**	0.70	**0.70**	m2	**0.79**
Compacting with 680 kg vibratory roller										
filling; blinding with sand, ashes or similar fine material	0.18	40.00	**0.25**	-	0.05	**0.54**	0.26	**1.04**	m2	**1.18**
Compacting with 6 - 8 tonnes smooth wheeled roller										
filling; blinding with sand, ashes or similar fine material	0.18	40.00	**0.25**	-	0.05	**0.54**	0.50	**1.29**	m2	**1.45**
MOT Type 2 to be obtained off site; wheeling not exceeding 50m - by machine										
Filling to make up levels; depositing in layers 150mm maximum thickness										
average thickness not exceeding 0.25m............	19.40	25.00	**24.26**	-	-	**-**	4.85	**29.11**	m3	**32.75**
average thickness exceeding 0.25m................	19.40	25.00	**24.26**	-	-	**-**	4.18	**28.44**	m3	**31.99**
Surface packing to filling										
to vertical or battered faces....................	-	-	**-**	-	-	**-**	0.70	**0.70**	m2	**0.79**
Compacting with 680 kg vibratory roller										
filling; blinding with sand, ashes or similar fine material	0.18	40.00	**0.25**	-	0.05	**0.54**	0.26	**1.04**	m2	**1.18**
Compacting with 6 - 8 tonnes smooth wheeled roller										
filling; blinding with sand, ashes or similar fine material	0.18	40.00	**0.25**	-	0.05	**0.54**	0.50	**1.29**	m2	**1.45**
Ashes to be obtained off site; wheeling not exceeding 25m - by machine										
Filling to make up levels; depositing in layers 150mm maximum thickness										
average thickness not exceeding 0.25m............	12.00	40.00	**16.80**	-	-	**-**	4.43	**21.23**	m3	**23.88**
Compacting with 680 kg vibratory roller										
filling	-	-	**-**	-	-	**-**	0.26	**0.26**	m2	**0.29**
Compacting with 6 - 8 tonnes smooth wheeled roller										
filling	-	-	**-**	-	-	**-**	0.50	**0.50**	m2	**0.57**
Ashes to be obtained off site; wheeling not exceeding 50m - by machine										
Filling to make up levels; depositing in layers 150mm maximum thickness										
average thickness not exceeding 0.25m............	12.00	40.00	**16.80**	-	-	**-**	5.14	**21.94**	m3	**24.68**
Compacting with 680 kg vibratory roller										
filling	-	-	**-**	-	-	**-**	0.26	**0.26**	m2	**0.29**
Compacting with 6 - 8 tonnes smooth wheeled roller										
filling	-	-	**-**	-	-	**-**	0.50	**0.50**	m2	**0.57**
Sand to be obtained off site; wheeling not exceeding 25m - by machine										
Filling to make up levels; depositing in layers 150mm maximum thickness										
average thickness not exceeding 0.25m............	15.31	33.00	**20.36**	-	-	**-**	4.85	**25.22**	m3	**28.37**
average thickness exceeding 0.25m................	15.31	33.00	**20.36**	-	-	**-**	4.17	**24.53**	m3	**27.60**
Compacting with 680 kg vibratory roller										
filling	-	-	**-**	-	-	**-**	0.26	**0.26**	m2	**0.29**
Compacting with 6 - 8 tonnes smooth wheeled roller										
filling	-	-	**-**	-	-	**-**	0.50	**0.50**	m2	**0.57**

Labour hourly rates: (except Specialists) Craft Operatives £14.21 Labourer £10.71 Rates are national average prices. Refer to REGIONAL VARIATIONS for indicative levels of overall pricing in regions	MATERIALS			LABOUR			RATES			
	Del to Site £	Waste %	Material Cost £	Craft Optve Hrs	Lab Hrs	Labour Cost £	Sunds £	Nett Rate £	Unit	Gross Rate (+12.5%) £
D20: EXCAVATING AND FILLING Cont.										
Sand to be obtained off site; wheeling not exceeding 50m - by machine										
Filling to make up levels; depositing in layers 150mm maximum thickness										
average thickness not exceeding 0.25m............	15.31	33.00	**20.36**	-	-	-	5.52	**25.88**	m3	**29.11**
average thickness exceeding 0.25m................	15.31	33.00	**20.36**	-	-	-	4.85	**25.22**	m3	**28.37**
Compacting with 680 kg vibratory roller filling	-	-	-	-	-	-	0.26	**0.26**	m2	**0.29**
Compacting with 6 - 8 tonnes smooth wheeled roller filling	-	-	-	-	-	-	0.50	**0.50**	m2	**0.57**
Hoggin to be obtained off site; wheeling not exceeding 25m - by machine										
Filling to make up levels; depositing in layers 150mm maximum thickness										
average thickness not exceeding 0.25m............	16.80	33.00	**22.34**	-	-	-	4.17	**26.51**	m3	**29.83**
average thickness exceeding 0.25m................	16.80	33.00	**22.34**	-	-	-	3.47	**25.81**	m3	**29.04**
Compacting with 680 kg vibratory roller filling	-	-	-	-	-	-	0.26	**0.26**	m2	**0.29**
Compacting with 6 - 8 tonnes smooth wheeled roller filling	-	-	-	-	-	-	0.50	**0.50**	m2	**0.57**
Hoggin to be obtained off site; wheeling not exceeding 50m - by machine										
Filling to make up levels; depositing in layers 150mm maximum thickness										
average thickness not exceeding 0.25m............	16.80	33.00	**22.34**	-	-	-	4.85	**27.20**	m3	**30.60**
average thickness exceeding 0.25m................	16.80	33.00	**22.34**	-	-	-	4.17	**26.51**	m3	**29.83**
Compacting with 680 kg vibratory roller filling	-	-	-	-	-	-	0.26	**0.26**	m2	**0.29**
Compacting with 6 - 8 tonnes smooth wheeled roller filling	-	-	-	-	-	-	0.50	**0.50**	m2	**0.57**
Excavated material arising from excavation - by hand										
Filling to excavation average thickness exceeding 0.25m................	-	-	-	-	1.44	**15.42**	-	**15.42**	m3	**17.35**
Compacting filling	-	-	-	-	-	-	0.50	**0.50**	m2	**0.57**
Selected excavated material obtained from on site spoil heaps 25m distant - by hand										
Filling to make up levels; depositing in layers 150mm maximum thickness										
average thickness not exceeding 0.25m............	-	-	-	-	1.85	**19.81**	-	**19.81**	m3	**22.29**
average thickness exceeding 0.25m................	-	-	-	-	1.60	**17.14**	-	**17.14**	m3	**19.28**
Compacting with 680 kg vibratory roller filling	-	-	-	-	-	-	0.26	**0.26**	m2	**0.29**
Compacting with 6 - 8 tonnes smooth wheeled roller filling	-	-	-	-	-	-	0.50	**0.50**	m2	**0.57**
Trimming										
sides of cuttings; vertical or battered..........	-	-	-	-	0.22	**2.36**	-	**2.36**	m2	**2.65**
sides of embankments; vertical or battered.......	-	-	-	-	0.22	**2.36**	-	**2.36**	m2	**2.65**
Selected excavated material obtained from on site spoil heaps 50m distant - by hand										
Filling to make up levels; depositing in layers 150mm maximum thickness										
average thickness not exceeding 0.25m............	-	-	-	-	2.10	**22.49**	-	**22.49**	m3	**25.30**
average thickness exceeding 0.25m................	-	-	-	-	1.95	**20.88**	-	**20.88**	m3	**23.50**
Compacting with 680 kg vibratory roller filling	-	-	-	-	-	-	0.26	**0.26**	m2	**0.29**
Compacting with 6 - 8 tonnes smooth wheeled roller filling	-	-	-	-	-	-	0.50	**0.50**	m2	**0.57**
Trimming										
sides of cuttings; vertical or battered..........	-	-	-	-	0.22	**2.36**	-	**2.36**	m2	**2.65**
sides of embankments; vertical or battered.......	-	-	-	-	0.22	**2.36**	-	**2.36**	m2	**2.65**
Preserved topsoil obtained from on site spoil heaps not exceeding 100m distant - by hand										
Filling to make up levels average thickness not exceeding 0.25m............	-	-	-	-	4.40	**47.12**	-	**47.12**	m3	**53.01**
Filling to external planters average thickness not exceeding 0.25m............	-	-	-	-	5.50	**58.91**	-	**58.91**	m3	**66.27**

Labour hourly rates: (except Specialists) Craft Operatives £14.21 Labourer £10.71 Rates are national average prices. Refer to REGIONAL VARIATIONS for indicative levels of overall pricing in regions	MATERIALS			LABOUR				RATES		
	Del to Site £	Waste %	Material Cost £	Craft Optve Hrs	Lab Hrs	Labour Cost £	Sunds £	Nett Rate £	Unit	Gross Rate (+12.5%) £
D20: EXCAVATING AND FILLING Cont.										
Imported topsoil; wheeling not exceeding 50m - by hand										
Filling to make up levels										
average thickness not exceeding 0.25m...........	15.00	10.00	**16.50**	-	3.85	**41.23**	-	**57.73**	m3	**64.95**
Filling to external planters										
average thickness exceeding 0.25m...............	15.00	10.00	**16.50**	-	4.95	**53.01**	-	**69.51**	m3	**78.20**
Hard, dry, broken brick or stone to be obtained off site; wheeling not exceeding 25m - by hand										
Filling to make up levels; depositing in layers 150mm maximum thickness										
average thickness not exceeding 0.25m...........	17.85	25.00	**22.31**	-	1.58	**16.92**	-	**39.23**	m3	**44.14**
average thickness exceeding 0.25m...............	17.85	25.00	**22.31**	-	1.31	**14.03**	-	**36.34**	m3	**40.89**
Surface packing to filling										
to vertical or battered faces...................	-	-	-	-	0.22	**2.36**	-	**2.36**	m2	**2.65**
Compacting with 680 kg vibratory roller										
filling; blinding with sand, ashes or similar fine material	0.18	40.00	**0.25**	-	0.05	**0.54**	0.26	**1.04**	m2	**1.18**
Compacting with 6 - 8 tonnes smooth wheeled roller										
filling; blinding with sand, ashes or similar fine material	0.18	40.00	**0.25**	-	0.05	**0.54**	0.50	**1.29**	m2	**1.45**
Hard, dry, broken brick or stone to be obtained off site; wheeling not exceeding 50m - by hand										
Filling to make up levels; depositing in layers 150mm maximum thickness										
average thickness not exceeding 0.25m...........	17.85	25.00	**22.31**	-	1.85	**19.81**	-	**42.13**	m3	**47.39**
average thickness exceeding 0.25m...............	17.85	25.00	**22.31**	-	1.58	**16.92**	-	**39.23**	m3	**44.14**
Surface packing to filling										
to vertical or battered faces...................	-	-	-	-	0.22	**2.36**	-	**2.36**	m2	**2.65**
Compacting with 680 kg vibratory roller										
filling; blinding with sand, ashes or similar fine material	0.18	40.00	**0.25**	-	0.05	**0.54**	0.26	**1.04**	m2	**1.18**
Compacting with 6 - 8 tonnes smooth wheeled roller										
filling; blinding with sand, ashes or similar fine material	0.18	40.00	**0.25**	-	0.05	**0.54**	0.50	**1.29**	m2	**1.45**
MOT Type 1 to be obtained off site; wheeling not exceeding 25m - by hand										
Filling to make up levels; depositing in layers 150mm maximum thickness										
average thickness not exceeding 0.25m...........	20.95	25.00	**26.19**	-	1.58	**16.92**	-	**43.11**	m3	**48.50**
average thickness exceeding 0.25m...............	20.95	25.00	**26.19**	-	1.31	**14.03**	-	**40.22**	m3	**45.24**
Surface packing to filling										
to vertical or battered faces...................	-	-	-	-	0.22	**2.36**	-	**2.36**	m2	**2.65**
Compacting with 680 kg vibratory roller										
filling; blinding with sand, ashes or similar fine material	0.18	40.00	**0.25**	-	0.05	**0.54**	0.26	**1.04**	m2	**1.18**
Compacting with 6 - 8 tonnes smooth wheeled roller										
filling; blinding with sand, ashes or similar fine material	0.18	40.00	**0.25**	-	0.05	**0.54**	0.50	**1.29**	m2	**1.45**
MOT Type 1 to be obtained off site; wheeling not exceeding 50m - by hand										
Filling to make up levels; depositing in layers 150mm maximum thickness										
average thickness not exceeding 0.25m...........	20.95	25.00	**26.19**	-	1.85	**19.81**	-	**46.00**	m3	**51.75**
average thickness exceeding 0.25m...............	20.95	25.00	**26.19**	-	1.58	**16.92**	-	**43.11**	m3	**48.50**
Surface packing to filling										
to vertical or battered faces...................	-	-	-	-	0.22	**2.36**	-	**2.36**	m2	**2.65**
Compacting with 680 kg vibratory roller										
filling; blinding with sand, ashes or similar fine material	0.18	40.00	**0.25**	-	0.05	**0.54**	0.26	**1.04**	m2	**1.18**
Compacting with 6 - 8 tonnes smooth wheeled roller										
filling; blinding with sand, ashes or similar fine material	0.18	40.00	**0.25**	-	0.05	**0.54**	0.50	**1.29**	m2	**1.45**
MOT Type 2 to be obtained off site; wheeling not exceeding 25m - by hand										
Filling to make up levels; depositing in layers 150mm maximum thickness										
average thickness not exceeding 0.25m...........	19.40	25.00	**24.26**	-	1.58	**16.92**	-	**41.18**	m3	**46.32**
average thickness exceeding 0.25m...............	19.40	25.00	**24.26**	-	1.31	**14.03**	-	**38.29**	m3	**43.07**
Surface packing to filling										
to vertical or battered faces...................	-	-	-	-	0.22	**2.36**	-	**2.36**	m2	**2.65**

Labour hourly rates: (except Specialists) Craft Operatives £14.21 Labourer £10.71 Rates are national average prices. Refer to REGIONAL VARIATIONS for indicative levels of overall pricing in regions	MATERIALS			LABOUR				RATES		
	Del to Site £	Waste %	Material Cost £	Craft Optve Hrs	Lab Hrs	Labour Cost £	Sunds £	Nett Rate £	Unit	Gross Rate (+12.5%) £
D20: EXCAVATING AND FILLING Cont.										
MOT Type 2 to be obtained off site; wheeling not exceeding 25m - by hand Cont.										
Compacting with 680 kg vibratory roller filling; blinding with sand, ashes or similar fine material	0.18	40.00	**0.25**	-	0.05	**0.54**	0.26	**1.04**	m2	**1.18**
Compacting with 6 - 8 tonnes smooth wheeled roller filling; blinding with sand, ashes or similar fine material	0.18	40.00	**0.25**	-	0.05	**0.54**	0.50	**1.29**	m2	**1.45**
MOT Type 2 to be obtained off site; wheeling not exceeding 50m - by hand										
Filling to make up levels; depositing in layers 150mm maximum thickness										
average thickness not exceeding 0.25m	19.40	25.00	**24.26**	-	1.85	**19.81**	-	**44.07**	m3	**49.58**
average thickness exceeding 0.25m	19.40	25.00	**24.26**	-	1.58	**16.92**	-	**41.18**	m3	**46.32**
Surface packing to filling to vertical or battered faces	-	-	**-**	-	0.22	**2.36**	-	**2.36**	m2	**2.65**
Compacting with 680 kg vibratory roller filling; blinding with sand, ashes or similar fine material	0.18	40.00	**0.25**	-	0.05	**0.54**	0.26	**1.04**	m2	**1.18**
Compacting with 6 - 8 tonnes smooth wheeled roller filling; blinding with sand, ashes or similar fine material	0.18	40.00	**0.25**	-	0.05	**0.54**	0.50	**1.29**	m2	**1.45**
Ashes to be obtained off site; wheeling not exceeding 25m - by hand										
Filling to make up levels; depositing in layers 150mm maximum thickness										
average thickness not exceeding 0.25m	12.00	40.00	**16.80**	-	1.70	**18.21**	-	**35.01**	m3	**39.38**
Compacting with 680 kg vibratory roller filling	-	-	**-**	-	-	**-**	0.26	**0.26**	m2	**0.29**
Compacting with 6 - 8 tonnes smooth wheeled roller filling	-	-	**-**	-	-	**-**	0.50	**0.50**	m2	**0.57**
Ashes to be obtained off site; wheeling not exceeding 50m - by hand										
Filling to make up levels; depositing in layers 150mm maximum thickness										
average thickness not exceeding 0.25m	12.00	40.00	**16.80**	-	1.94	**20.78**	-	**37.58**	m3	**42.27**
Compacting with 680 kg vibratory roller filling	-	-	**-**	-	-	**-**	0.26	**0.26**	m2	**0.29**
Compacting with 6 - 8 tonnes smooth wheeled roller filling	-	-	**-**	-	-	**-**	0.50	**0.50**	m2	**0.57**
Sand to be obtained off site; wheeling not exceeding 25m - by hand										
Filling to make up levels; depositing in layers 150mm maximum thickness										
average thickness not exceeding 0.25m	15.31	33.00	**20.36**	-	1.84	**19.71**	-	**40.07**	m3	**45.08**
average thickness exceeding 0.25m	15.31	33.00	**20.36**	-	1.58	**16.92**	-	**37.28**	m3	**41.94**
Compacting with 680 kg vibratory roller filling	-	-	**-**	-	-	**-**	0.26	**0.26**	m2	**0.29**
Compacting with 6 - 8 tonnes smooth wheeled roller filling	-	-	**-**	-	-	**-**	0.50	**0.50**	m2	**0.57**
Sand to be obtained off site; wheeling not exceeding 50m - by hand										
Filling to make up levels; depositing in layers 150mm maximum thickness										
average thickness not exceeding 0.25m	15.31	33.00	**20.36**	-	2.10	**22.49**	-	**42.85**	m3	**48.21**
average thickness exceeding 0.25m	15.31	33.00	**20.36**	-	1.84	**19.71**	-	**40.07**	m3	**45.08**
Compacting with 680 kg vibratory roller filling	-	-	**-**	-	-	**-**	0.26	**0.26**	m2	**0.29**
Compacting with 6 - 8 tonnes smooth wheeled roller filling	-	-	**-**	-	-	**-**	0.50	**0.50**	m2	**0.57**
Hoggin to be obtained off site; wheeling not exceeding 25m - by hand										
Filling to make up levels; depositing in layers 150mm maximum thickness										
average thickness not exceeding 0.25m	16.80	33.00	**22.34**	-	1.55	**16.60**	-	**38.94**	m3	**43.81**
average thickness exceeding 0.25m	16.80	33.00	**22.34**	-	1.30	**13.92**	-	**36.27**	m3	**40.80**
Compacting with 680 kg vibratory roller filling	-	-	**-**	-	-	**-**	0.26	**0.26**	m2	**0.29**

Labour hourly rates: (except Specialists) Craft Operatives £14.21 Labourer £10.71 Rates are national average prices. Refer to REGIONAL VARIATIONS for indicative levels of overall pricing in regions	MATERIALS			LABOUR				RATES		
	Del to Site £	Waste %	Material Cost £	Craft Optve Hrs	Lab Hrs	Labour Cost £	Sunds £	Nett Rate £	Unit	Gross Rate (+12.5%) £
D20: EXCAVATING AND FILLING Cont.										
Hoggin to be obtained off site; wheeling not exceeding 25m - by hand Cont.										
Compacting with 6 - 8 tonnes smooth wheeled roller										
filling	-	-	-	-	-	-	0.50	**0.50**	m2	**0.57**
Hoggin to be obtained off site; wheeling not exceeding 50m - by hand										
Filling to make up levels; depositing in layers 150mm maximum thickness										
average thickness not exceeding 0.25m...........	16.80	33.00	**22.34**	-	1.80	**19.28**	-	**41.62**	m3	**46.82**
average thickness exceeding 0.25m...............	16.80	33.00	**22.34**	-	1.55	**16.60**	-	**38.94**	m3	**43.81**
Compacting with 680 kg vibratory roller										
filling	-	-	-	-	-	-	0.26	**0.26**	m2	**0.29**
Compacting with 6 - 8 tonnes smooth wheeled roller										
filling	-	-	-	-	-	-	0.50	**0.50**	m2	**0.57**
Excavated material arising from excavation; work inside existing building - by hand										
Filling to excavation										
average thickness exceeding 0.25m...............	-	-	-	-	2.80	**29.99**	-	**29.99**	m3	**33.74**
Compacting										
filling	-	-	-	-	-	-	0.26	**0.26**	m2	**0.29**
Selected excavated material obtained from on site spoil heaps 25m distant; work inside existing building - by hand										
Filling to make up levels; depositing in layers 150mm maximum thickness										
average thickness not exceeding 0.25m...........	-	-	-	-	3.26	**34.91**	-	**34.91**	m3	**39.28**
average thickness exceeding 0.25m...............	-	-	-	-	3.08	**32.99**	-	**32.99**	m3	**37.11**
Compacting										
filling	-	-	-	-	-	-	0.26	**0.26**	m2	**0.29**
Selected excavated material obtained from on site spoil heaps 50m distant; work inside existing building - by hand										
Filling to make up levels; depositing in layers 150mm maximum thickness										
average thickness not exceeding 0.25m...........	-	-	-	-	3.52	**37.70**	-	**37.70**	m3	**42.41**
average thickness exceeding 0.25m...............	-	-	-	-	3.36	**35.99**	-	**35.99**	m3	**40.48**
Compacting										
filling	-	-	-	-	-	-	0.26	**0.26**	m2	**0.29**
Hard, dry, broken brick or stone to be obtained off site; wheeling not exceeding 25m; work inside existing building - by hand										
Filling to make up levels; depositing in layers 150mm maximum thickness										
average thickness not exceeding 0.25m...........	17.85	25.00	**22.31**	-	2.99	**32.02**	-	**54.34**	m3	**61.13**
average thickness exceeding 0.25m...............	17.85	25.00	**22.31**	-	2.81	**30.10**	-	**52.41**	m3	**58.96**
Surface packing to filling										
to vertical or battered faces...................	-	-	-	-	0.22	**2.36**	-	**2.36**	m2	**2.65**
Compacting										
filling; blinding with sand, ashes or similar fine material	0.18	40.00	**0.25**	-	0.05	**0.54**	0.26	**1.04**	m2	**1.18**
Hard, dry, broken brick or stone to be obtained off site; wheeling not exceeding 50m; work inside existing building - by hand										
Filling to make up levels; depositing in layers 150mm maximum thickness										
average thickness not exceeding 0.25m...........	17.85	25.00	**22.31**	-	3.26	**34.91**	-	**57.23**	m3	**64.38**
average thickness exceeding 0.25m...............	17.85	25.00	**22.31**	-	2.99	**32.02**	-	**54.34**	m3	**61.13**
Surface packing to filling										
to vertical or battered faces...................	-	-	-	-	0.22	**2.36**	-	**2.36**	m2	**2.65**
Compacting										
filling; blinding with sand, ashes or similar fine material	0.18	40.00	**0.25**	-	0.05	**0.54**	0.26	**1.04**	m2	**1.18**
Ashes to be obtained off site; wheeling not exceeding 25m; work inside existing building - by hand										
Filling to make up levels; depositing in layers 150mm maximum thickness										
average thickness not exceeding 0.25m...........	12.00	40.00	**16.80**	-	3.10	**33.20**	-	**50.00**	m3	**56.25**
Compacting										
filling	-	-	-	-	-	-	0.26	**0.26**	m2	**0.29**

Labour hourly rates: (except Specialists) Craft Operatives £14.21 Labourer £10.71 Rates are national average prices. Refer to REGIONAL VARIATIONS for indicative levels of overall pricing in regions	MATERIALS			LABOUR			RATES			
	Del to Site £	Waste %	Material Cost £	Craft Optve Hrs	Lab Hrs	Labour Cost £	Sunds £	Nett Rate £	Unit	Gross Rate (+12.5%) £
D20: EXCAVATING AND FILLING Cont.										
Ashes to be obtained off site; wheeling not exceeding 50m; work inside existing building - by hand										
Filling to make up levels; depositing in layers 150mm maximum thickness										
average thickness not exceeding 0.25m	12.00	40.00	**16.80**	-	3.36	**35.99**	-	**52.79**	m3	**59.38**
Compacting										
filling	-	-	**-**	-	-	**-**	0.26	**0.26**	m2	**0.29**
Sand to be obtained off site; wheeling not exceeding 25m; work inside existing building - by hand										
Filling to make up levels; depositing in layers 150mm maximum thickness										
average thickness not exceeding 0.25m	15.31	33.00	**20.36**	-	3.26	**34.91**	-	**55.28**	m3	**62.19**
average thickness exceeding 0.25m	15.31	33.00	**20.36**	-	2.99	**32.02**	-	**52.38**	m3	**58.93**
Compacting										
filling	-	-	**-**	-	-	**-**	0.26	**0.26**	m2	**0.29**
Sand to be obtained off site; wheeling not exceeding 50m; work inside existing building - by hand										
Filling to make up levels; depositing in layers 150mm maximum thickness										
average thickness not exceeding 0.25m	15.31	33.00	**20.36**	-	3.52	**37.70**	-	**58.06**	m3	**65.32**
average thickness exceeding 0.25m	15.31	33.00	**20.36**	-	3.26	**34.91**	-	**55.28**	m3	**62.19**
Compacting										
filling	-	-	**-**	-	-	**-**	0.26	**0.26**	m2	**0.29**
D30: CAST IN PLACE PILING										
Prices include for a 21 N/mm2 concrete mix, nominal reinforcement and a minimum number of 50 piles on any one contract. The working loads sizes and lengths given below will depend on the nature of the soils in which the piles will be founded as well as structure to be supported (KELLER GROUND ENGINEERING)										
On/Off site charge										
in addition to the following prices add approximately	-	-	**Spclist**	-	-	**Spclist**	-	**5183.94**	Sm	**5831.94**
Short auger piles										
up to 6m long; 450mm nominal diameter; 8 tonnes normal working load	-	-	**Spclist**	-	-	**Spclist**	-	**30.36**	m	**34.15**
up to 6m long; 610mm nominal diameter; 30 tonnes normal working load	-	-	**Spclist**	-	-	**Spclist**	-	**45.92**	m	**51.66**
Bored cast in-situ piles										
up to 15m long; 450mm nominal diameter; 40 tonnes normal working load	-	-	**Spclist**	-	-	**Spclist**	-	**28.01**	m	**31.52**
up to 15m long; 610mm nominal diameter; 120 tonnes normal working load	-	-	**Spclist**	-	-	**Spclist**	-	**43.59**	m	**49.04**
Auger piles										
up to 20m long; 450mm nominal diameter	-	-	**Spclist**	-	-	**Spclist**	-	**28.41**	m	**31.96**
up to 30m long; 1200mm nominal diameter	-	-	**Spclist**	-	-	**Spclist**	-	**140.89**	m	**158.50**
Large diameter auger piles (plain shaft)										
up to 20m long; 610mm nominal diameter; 150 tonnes normal working load	-	-	**Spclist**	-	-	**Spclist**	-	**43.23**	m	**48.63**
up to 30m long; 1525mm nominal diameter; 600 tonnes normal working load	-	-	**Spclist**	-	-	**Spclist**	-	**219.18**	m	**246.58**
Large diameter auger piles (belled base)										
up to 20m long; 610mm nominal diameter; 150 tonnes normal working load	-	-	**Spclist**	-	-	**Spclist**	-	**59.17**	m	**66.56**
up to 30m long; 1525mm nominal diameter; 1000 tonnes normal working load	-	-	**Spclist**	-	-	**Spclist**	-	**281.86**	m	**317.10**
Boring through obstructions, rock like formations, etc undertaken on a time basis at a rate per piling rig per hour										
Cutting off tops of piles; including preparation and integration of reinforcement into pile cap or ground beam and disposal										
450-610mm nominal diameter	-	-	**Spclist**	-	-	**Spclist**	-	**10.40**	m	**11.70**
610-1525mm nominal diameter	-	-	**Spclist**	-	-	**Spclist**	-	**40.05**	m	**45.06**

Labour hourly rates: (except Specialists) Craft Operatives £14.21 Labourer £10.71 Rates are national average prices. Refer to REGIONAL VARIATIONS for indicative levels of overall pricing in regions	MATERIALS			LABOUR				RATES		
	Del to Site £	Waste %	Material Cost £	Craft Optve Hrs	Lab Hrs	Labour Cost £	Sunds £	Nett Rate £	Unit	Gross Rate (+12.5%) £
D32: STEEL PILING										
Mild steel Universal Bearing Piles 'H' section, to BS EN 10025 Grade S275 and high yield steel Grade S275JR in lengths 9 - 15m supplied, handled, pitched and driven vertically with landbased plant (KELLER GROUND ENGINEERING)										
Note; the following prices are based on quantities of 25 - 150 tonnes										
On/Off site charge										
in addition to the following prices, add approximately	-	-	Spclist	-	-	Spclist	-	4467.97	Sm	5026.46
Mild steel piles										
203 x 203 x 45 kg/m, SWL 40 tonnes	-	-	Spclist	-	-	Spclist	-	43.21	m	48.61
203 x 203 x 54 kg/m, SWL 50 tonnes	-	-	Spclist	-	-	Spclist	-	48.73	m	54.83
254 x 254 x 63 kg/m, SWL 60 tonnes	-	-	Spclist	-	-	Spclist	-	55.09	m	61.97
254 x 254 x 71 kg/m, SWL 70 tonnes	-	-	Spclist	-	-	Spclist	-	59.32	m	66.74
305 x 305 x 79 kg/m, SWL 75 tonnes	-	-	Spclist	-	-	Spclist	-	69.86	m	78.59
254 x 254 x 85 kg/m, SWL 80 tonnes	-	-	Spclist	-	-	Spclist	-	71.90	m	80.89
305 x 305 x 88 kg/m, SWL 85 tonnes	-	-	Spclist	-	-	Spclist	-	75.75	m	85.22
305 x 305 x 95 kg/m, SWL 90 tonnes	-	-	Spclist	-	-	Spclist	-	80.54	m	90.61
356 x 368 x 109 kg/m, SWL 105 tonnes	-	-	Spclist	-	-	Spclist	-	96.50	m	108.56
305 x 305 x 110 kg/m, SWL 105 tonnes	-	-	Spclist	-	-	Spclist	-	90.78	m	102.13
305 x 305 x 126 kg/m, SWL 120 tonnes	-	-	Spclist	-	-	Spclist	-	103.56	m	116.51
305 x 305 x 149 kg/m, SWL 140 tonnes	-	-	Spclist	-	-	Spclist	-	114.29	m	128.58
356 x 368 x 133 kg/m, SWL 130 tonnes	-	-	Spclist	-	-	Spclist	-	119.63	m	134.58
356 x 368 x 152 kg/m, SWL 140 tonnes	-	-	Spclist	-	-	Spclist	-	126.96	m	142.83
356 x 368 x 174 kg/m, SWL 165 tonnes	-	-	Spclist	-	-	Spclist	-	146.01	m	164.26
305 x 305 x 186 kg/m, SWL 175 tonnes	-	-	Spclist	-	-	Spclist	-	158.76	m	178.60
305 x 305 x 223 kg/m, SWL 210 tonnes	-	-	Spclist	-	-	Spclist	-	173.75	m	195.47
Extra for high yield steel										
add	-	-	Spclist	-	-	Spclist	-	47.56	T	53.50
For quiet piling, add										
to terminal charge, lump sum	-	-	Spclist	-	-	Spclist	-	948.77	Sm	1067.37
plus for all piles	-	-	Spclist	-	-	Spclist	-	2.30	m	2.59
Mild steel sheet piling to BS EN 10025 Grade S275 or high yield steel Grade S275JR in lengths 4.5-15m supplied, handled, pitched and driven by land based plant in one visit (KELLER GROUND ENGINEERING)										
On/Off site charge										
in addition to the following prices, add approximately	-	-	Spclist	-	-	Spclist	-	4261.36	Sm	4794.03
Mild steel sheet piling										
Larssen Section 6W	-	-	Spclist	-	-	Spclist	-	93.66	m2	105.37
extra for high yield steel	-	-	Spclist	-	-	Spclist	-	5.13	m2	5.77
extra for one coat L.B.V.(Lowca Varnish to BS 1070 Type 2) before driving	-	-	Spclist	-	-	Spclist	-	3.75	m2	4.21
extra for corners	-	-	Spclist	-	-	Spclist	-	37.43	m	42.11
extra for junctions	-	-	Spclist	-	-	Spclist	-	51.05	m	57.43
Frodingham Section 1N and Larssen Section 9W	-	-	Spclist	-	-	Spclist	-	103.45	m2	116.38
extra for high yield steel	-	-	Spclist	-	-	Spclist	-	4.26	m2	4.79
extra for one coat L.B.V.(Lowca Varnish to BS 1070 Type 2) before driving	-	-	Spclist	-	-	Spclist	-	3.75	m2	4.21
extra for corners	-	-	Spclist	-	-	Spclist	-	37.43	m	42.11
extra for junctions	-	-	Spclist	-	-	Spclist	-	51.05	m	57.43
Frodingham Section 2N and Larssen Section 12W	-	-	Spclist	-	-	Spclist	-	115.15	m2	129.54
extra for high yield steel	-	-	Spclist	-	-	Spclist	-	5.45	m2	6.13
extra for one coat L.B.V.(Lowca Varnish to BS 1070 Type 2) before driving	-	-	Spclist	-	-	Spclist	-	3.75	m2	4.21
extra for corners	-	-	Spclist	-	-	Spclist	-	37.43	m	42.11
extra for junctions	-	-	Spclist	-	-	Spclist	-	51.05	m	57.43
Frodingham Section 1BXN	-	-	Spclist	-	-	Spclist	-	127.62	m2	143.57
extra for high yield steel	-	-	Spclist	-	-	Spclist	-	8.45	m2	9.50
extra for one coat L.B.V.(Lowca Varnish to BS 1070 Type 2) before driving	-	-	Spclist	-	-	Spclist	-	3.41	m2	3.83
extra for corners	-	-	Spclist	-	-	Spclist	-	40.84	m	45.95
extra for junctions	-	-	Spclist	-	-	Spclist	-	51.05	m	57.43
Frodingham Section 3N and Larssen Section 16W	-	-	Spclist	-	-	Spclist	-	128.76	m2	144.85
extra for high yield steel	-	-	Spclist	-	-	Spclist	-	6.99	m2	7.86
extra for one coat L.B.V.(Lowca Varnish to BS 1070 Type 2) before driving	-	-	Spclist	-	-	Spclist	-	3.41	m2	3.83
extra for corners	-	-	Spclist	-	-	Spclist	-	40.84	m	45.95
extra for junctions	-	-	Spclist	-	-	Spclist	-	51.05	m	57.43
Larssen Section 20W	-	-	Spclist	-	-	Spclist	-	145.69	m2	163.90
extra for high yield steel	-	-	Spclist	-	-	Spclist	-	7.66	m2	8.61
extra for one coat L.B.V.(Lowca Varnish to BS 1070 Type 2) before driving	-	-	Spclist	-	-	Spclist	-	3.83	m2	4.31
extra for corners	-	-	Spclist	-	-	Spclist	-	44.24	m	49.77
extra for junctions	-	-	Spclist	-	-	Spclist	-	59.54	m	66.99
Frodingham Section 4N and Larssen Section 25W	-	-	Spclist	-	-	Spclist	-	161.32	m2	181.49
extra for high yield steel	-	-	Spclist	-	-	Spclist	-	8.34	m2	9.38
extra for one coat L.B.V.(Lowca Varnish to BS 1070 Type 2) before driving	-	-	Spclist	-	-	Spclist	-	4.26	m2	4.79
extra for corners	-	-	Spclist	-	-	Spclist	-	47.64	m	53.59
extra for junctions	-	-	Spclist	-	-	Spclist	-	59.54	m	66.99
Larssen Section 32W	-	-	Spclist	-	-	Spclist	-	190.47	m2	214.28
extra for high yield steel	-	-	Spclist	-	-	Spclist	-	9.27	m2	10.43
extra for one coat L.B.V.(Lowca Varnish to BS 1070 Type 2) before driving	-	-	Spclist	-	-	Spclist	-	4.60	m2	5.17

Labour hourly rates: (except Specialists) Craft Operatives £14.21 Labourer £10.71 Rates are national average prices. Refer to REGIONAL VARIATIONS for indicative levels of overall pricing in regions	MATERIALS			LABOUR				RATES		
	Del to Site £	Waste %	Material Cost £	Craft Optve Hrs	Lab Hrs	Labour Cost £	Sunds £	Nett Rate £	Unit	Gross Rate (+12.5%) £
D32: STEEL PILING Cont.										
Mild steel sheet piling to BS EN 10025 Grade S275 or high yield steel Grade S275JR in lengths 4.5-15m supplied, handled, pitched and driven by land based plant in one visit (KELLER GROUND ENGINEERING) Cont.										
Mild steel sheet piling Cont.										
extra for corners	-	-	Spclist	-	-	Spclist	-	51.05	m	57.43
extra for junctions	-	-	Spclist	-	-	Spclist	-	63.02	m	70.89
Frodingham Section 5	-	-	Spclist	-	-	Spclist	-	214.94	m2	241.81
extra for high yield steel	-	-	Spclist	-	-	Spclist	-	12.25	m2	13.79
extra for one coat L.B.V.(Lowca Varnish to BS 1070 Type 2) before driving	-	-	Spclist	-	-	Spclist	-	7.99	m2	8.99
extra for corners	-	-	Spclist	-	-	Spclist	-	51.05	m	57.43
extra for junctions	-	-	Spclist	-	-	Spclist	-	68.07	m	76.57
For quiet piling, add										
lump sum	-	-	Spclist	-	-	Spclist	-	865.18	Sm	973.33
plus for all sections, from	-	-	Spclist	-	-	Spclist	-	3.95	m2	4.44
to	-	-	Spclist	-	-	Spclist	-	6.60	m2	7.42
D40: EMBEDDED RETAINING WALLS										
Embedded retaining walls; contiguous panel construction; panel lengths not exceeding 5m. Note:- the following prices are indicative only: firm quotations should always be obtained (CEMENTATION FOUNDATIONS SKANSKA LIMITED)										
On/Off site charge										
in addition to the following prices, add for bringing plant to site, erecting and dismantling, maintaining and removing from site, approximately	-	-	Spclist	-	-	Spclist	-	85668.64	Sm	96377.22
Excavation and Bentonite slurry and disposal										
600mm thick wall; maximum depth										
5m	-	-	Spclist	-	-	Spclist	-	257.64	m3	289.85
10m	-	-	Spclist	-	-	Spclist	-	257.64	m3	289.85
15m	-	-	Spclist	-	-	Spclist	-	257.64	m3	289.85
20m	-	-	Spclist	-	-	Spclist	-	257.64	m3	289.85
800mm thick wall; maximum depth										
5m	-	-	Spclist	-	-	Spclist	-	214.49	m3	241.30
10m	-	-	Spclist	-	-	Spclist	-	214.49	m3	241.30
15m	-	-	Spclist	-	-	Spclist	-	214.49	m3	241.30
20m	-	-	Spclist	-	-	Spclist	-	214.49	m3	241.30
1000mm thick wall; maximum depth										
5m	-	-	Spclist	-	-	Spclist	-	171.34	m3	192.75
10m	-	-	Spclist	-	-	Spclist	-	171.34	m3	192.75
15m	-	-	Spclist	-	-	Spclist	-	171.34	m3	192.75
20m	-	-	Spclist	-	-	Spclist	-	171.34	m3	192.75
Excavating through obstructions, rock like formations, etc										
undertaken on a time basis at a rate per rig per hour	-	-	Spclist	-	-	Spclist	-	500.05	hr	562.56
Reinforced concrete; BS 5328, designed mix C25, 20mm aggregate, minimum cement content 400 kg/m3										
600mm thick wall	-	-	Spclist	-	-	Spclist	-	106.61	m3	119.94
800mm thick wall	-	-	Spclist	-	-	Spclist	-	106.61	m3	119.94
1000mm thick wall	-	-	Spclist	-	-	Spclist	-	106.61	m3	119.94
Reinforced in-situ concrete; sulphate resisting; BS 5328, designed mix C25, 20mm aggregate, minimum cement content 400 kg/m3										
600mm thick wall	-	-	Spclist	-	-	Spclist	-	121.84	m3	137.07
800mm thick wall	-	-	Spclist	-	-	Spclist	-	121.84	m3	137.07
1000mm thick wall	-	-	Spclist	-	-	Spclist	-	121.84	m3	137.07
Reinforcement bars; BS 4449 hot rolled plain round mild steel; including hooks tying wire, and spacers and chairs which are at the discretion of the Contractor										
16mm; straight	-	-	Spclist	-	-	Spclist	-	764.04	t	859.54
20mm; straight	-	-	Spclist	-	-	Spclist	-	764.04	t	859.54
25mm; straight	-	-	Spclist	-	-	Spclist	-	764.04	t	859.54
32mm; straight	-	-	Spclist	-	-	Spclist	-	764.04	t	859.54
40mm; straight	-	-	Spclist	-	-	Spclist	-	764.04	t	859.54
16mm; bent	-	-	Spclist	-	-	Spclist	-	764.04	t	859.54
20mm; bent	-	-	Spclist	-	-	Spclist	-	764.04	t	859.54
25mm; bent	-	-	Spclist	-	-	Spclist	-	764.04	t	859.54
32mm; bent	-	-	Spclist	-	-	Spclist	-	762.83	t	858.19
40mm; bent	-	-	Spclist	-	-	Spclist	-	762.83	t	858.19
Reinforcement bars; BS 4449 hot rolled deformed high yield steel; including hooks tying wire, and spacers and chairs which are at the discretion of the Contractor										
16mm; straight	-	-	Spclist	-	-	Spclist	-	762.83	t	858.19
20mm; straight	-	-	Spclist	-	-	Spclist	-	762.83	t	858.19
25mm; straight	-	-	Spclist	-	-	Spclist	-	762.83	t	858.19
32mm; straight	-	-	Spclist	-	-	Spclist	-	762.83	t	858.19
40mm; straight	-	-	Spclist	-	-	Spclist	-	762.83	t	858.19
16mm; bent	-	-	Spclist	-	-	Spclist	-	762.83	t	858.19
20mm; bent	-	-	Spclist	-	-	Spclist	-	762.83	t	858.19

Labour hourly rates: (except Specialists) Craft Operatives £14.21 Labourer £10.71 Rates are national average prices. Refer to REGIONAL VARIATIONS for indicative levels of overall pricing in regions	MATERIALS			LABOUR				RATES		
	Del to Site £	Waste %	Material Cost £	Craft Optve Hrs	Lab Hrs	Labour Cost £	Sunds £	Nett Rate £	Unit	Gross Rate (+12.5%) £
D40: EMBEDDED RETAINING WALLS Cont.										
Embedded retaining walls; contiguous panel construction; panel lengths not exceeding 5m. Note:- the following prices are indicative only: firm quotations should always be obtained (CEMENTATION FOUNDATIONS SKANSKA LIMITED) Cont.										
Reinforcement bars; BS 4449 hot rolled deformed high yield steel; including hooks tying wire, and spacers and chairs which are at the discretion of the Contractor Cont.										
25mm; bent	-	-	**Spclist**	-	-	**Spclist**	-	**762.83**	t	**858.19**
32mm; bent	-	-	**Spclist**	-	-	**Spclist**	-	**762.83**	t	**858.19**
40mm; bent	-	-	**Spclist**	-	-	**Spclist**	-	**762.83**	t	**858.19**
Guide walls; excavation, disposal and support; reinforced in-situ concrete; BS 5328, designed mix C25, 20mm aggregate, minimum cement content 290 kg/m3; reinforced with one layer fabric BS 4483 reference A252, 3.95 kg/m2 including laps, tying wire, all cutting and bending, and spacers and chairs which are at the discretion of the Contractor; formwork both sides										
both sides; 1000mm high, 600mm apart; propped top and bottom at 2000mm centres	-	-	**Spclist**	-	-	**Spclist**	-	**263.99**	m	**296.98**
both sides; 1000mm high, 800mm apart; propped top and bottom at 2000mm centres	-	-	**Spclist**	-	-	**Spclist**	-	**263.99**	m	**296.98**
both sides; 1000mm high, 1000mm apart; propped top and bottom at 2000mm centres	-	-	**Spclist**	-	-	**Spclist**	-	**263.99**	m	**296.98**
both sides; 1500mm high, 600mm apart; propped top and bottom at 2000mm centres	-	-	**Spclist**	-	-	**Spclist**	-	**406.13**	m	**456.90**
both sides; 1500mm high, 800mm apart; propped top and bottom at 2000mm centres	-	-	**Spclist**	-	-	**Spclist**	-	**406.13**	m	**456.90**
both sides; 1500mm high, 1000mm apart; propped top and bottom at 2000mm centres	-	-	**Spclist**	-	-	**Spclist**	-	**406.13**	m	**456.90**
D41: CRIB WALLS/GABIONS/REINFORCED EARTH										
Retaining Walls										
Betoflor precast concrete landscape retaining walls including soil filling to pockets but excluding excavation, concrete foundations, stone backfill to rear of wall and planting which are all deemed measured separately										
Betoflor interlocking units 500mm long x 250mm wide x 200mm modular deep in wall 250mm wide	44.11	5.00	**46.32**	-	5.00	**53.55**	-	**99.87**	m2	**112.35**
Extra over for colours	4.22	5.00	**4.43**	-	0.50	**5.36**	-	**9.79**	m2	**11.01**
Betoatlas interlocking units 250mm long x 500mm wide x 200mm modular deep in wall 500mm wide	61.02	5.00	**64.07**	-	7.00	**74.97**	-	**139.04**	m2	**156.42**
Extra over for colours	5.51	5.00	**5.79**	-	0.10	**1.07**	-	**6.86**	m2	**7.72**
Betonap 150/50 woven mesh geotextile as reinforcement to backfill	3.82	5.00	**4.01**	-	0.30	**3.21**	-	**7.22**	m2	**8.12**
graded stone filling behind Betoflor or Betoatlas	26.46	5.00	**27.78**	-	4.00	**42.84**	-	**70.62**	m3	**79.45**
concrete haunching to base of blocks	4.64	5.00	**4.87**	-	0.45	**4.82**	-	**9.69**	m	**10.90**
Betoflor range; Betojard precast concrete vertical acoustic wall including soil filling to pockets but excluding excavation, concrete foundations, stone backfill to rear of wall and planting which are all deemed measured separately										
Betojard standard 500mm long x 250mm wide x 200mm deep units in wall 250mm wide	54.89	5.00	**57.63**	-	5.75	**61.58**	-	**119.22**	m2	**134.12**
Extra over for colours	1.83	5.00	**1.92**	-	0.05	**0.54**	-	**2.45**	m2	**2.76**
Betojard pivot 290mm x 250mm wide x 200mm deep	54.89	5.00	**57.63**	-	7.00	**74.97**	-	**132.60**	m2	**149.18**
Extra over for colours	1.83	5.00	**1.92**	-	0.05	**0.54**	-	**2.45**	m2	**2.76**
Extra over fill core to pivot block with reinforced concrete with 4 No 10mm diameter vertical mild steel bars and 6mm links	5.76	7.50	**6.19**	-	2.00	**21.42**	-	**27.61**	m	**31.06**
Extra over dowelled connection between pivot and standard block with concrete infill to standard block and 1 No 10mm diameter x 200mm long galvanized dowel debonded one end at 300mm centres vertically	8.30	5.00	**8.84**	-	1.00	**10.71**	-	**19.55**	m	**21.99**
concrete haunching to base of blocks	4.64	5.00	**4.87**	-	0.45	**4.82**	-	**9.69**	m	**10.90**
D50: UNDERPINNING										
Information										
The work of underpinning in this section comprises work to be carried out in short lengths; prices are exclusive of shoring and other temporary supports										
Excavating										
Preliminary trenches										
maximum depth not exceeding										
1.00m	-	-	**-**	-	5.00	**53.55**	-	**53.55**	m3	**60.24**
2.00m	-	-	**-**	-	6.00	**64.26**	-	**64.26**	m3	**72.29**
4.00m	-	-	**-**	-	7.00	**74.97**	-	**74.97**	m3	**84.34**
Underpinning pits										
maximum depth not exceeding										
1.00m	-	-	**-**	-	5.85	**62.65**	-	**62.65**	m3	**70.49**

Labour hourly rates: (except Specialists) Craft Operatives £14.21 Labourer £10.71 Rates are national average prices. Refer to REGIONAL VARIATIONS for indicative levels of overall pricing in regions	MATERIALS			LABOUR				RATES		
	Del to Site £	Waste %	Material Cost £	Craft Optve Hrs	Lab Hrs	Labour Cost £	Sunds £	Nett Rate £	Unit	Gross Rate (+12.5%) £
D50: UNDERPINNING Cont.										
Excavating Cont.										
Underpinning pits Cont.										
maximum depth not exceeding Cont.										
2.00m	-	-	**-**	-	7.30	**78.18**	-	**78.18**	m3	**87.96**
4.00m	-	-	**-**	-	8.80	**94.25**	-	**94.25**	m3	**106.03**
Earthwork support										
preliminary trenches; distance between opposing faces not exceeding 2.00m; maximum depth not exceeding										
1.00m	1.58	-	**1.58**	-	0.53	**5.68**	0.02	**7.28**	m2	**8.19**
2.00m	1.99	-	**1.99**	-	0.67	**7.18**	0.03	**9.20**	m2	**10.35**
4.00m	2.27	-	**2.27**	-	0.72	**7.71**	0.03	**10.02**	m2	**11.27**
preliminary trenches; distance between opposing faces 2.00 - 4.00m; maximum depth not exceeding										
1.00m	1.72	-	**1.72**	-	0.55	**5.89**	0.02	**7.63**	m2	**8.59**
2.00m	2.18	-	**2.18**	-	0.70	**7.50**	0.03	**9.71**	m2	**10.92**
4.00m	2.39	-	**2.39**	-	0.84	**9.00**	0.04	**11.43**	m2	**12.86**
underpinning pits; distance between opposing faces not exceeding 2.00m; maximum depth not exceeding										
1.00m	1.86	-	**1.86**	-	0.61	**6.53**	0.02	**8.42**	m2	**9.47**
2.00m	2.31	-	**2.31**	-	0.77	**8.25**	0.03	**10.59**	m2	**11.91**
4.00m	2.59	-	**2.59**	-	0.83	**8.89**	0.04	**11.52**	m2	**12.96**
underpinning pits; distance between opposing faces 2.00 - 4.00m; maximum depth not exceeding										
1.00m	1.97	-	**1.97**	-	0.64	**6.85**	0.02	**8.84**	m2	**9.95**
2.00m	2.50	-	**2.50**	-	0.80	**8.57**	0.03	**11.10**	m2	**12.48**
4.00m	2.87	-	**2.87**	-	0.95	**10.17**	0.04	**13.09**	m2	**14.73**
Cutting away existing projecting foundations										
masonry; maximum width 103mm; maximum depth 150mm	-	-	**-**	-	0.90	**9.64**	-	**9.64**	m	**10.84**
masonry; maximum width 154mm; maximum depth 225mm	-	-	**-**	-	1.07	**11.46**	-	**11.46**	m	**12.89**
concrete; maximum width 253mm; maximum depth 190mm	-	-	**-**	-	1.29	**13.82**	-	**13.82**	m	**15.54**
concrete; maximum width 304mm; maximum depth 300mm	-	-	**-**	-	1.52	**16.28**	-	**16.28**	m	**18.31**
Preparing the underside of the existing work to receive the pinning up of the new work										
350mm wide	-	-	**-**	-	0.45	**4.82**	-	**4.82**	m	**5.42**
500mm wide	-	-	**-**	-	0.56	**6.00**	-	**6.00**	m	**6.75**
1000mm wide	-	-	**-**	-	1.13	**12.10**	-	**12.10**	m	**13.62**
Compacting										
bottoms of excavations	-	-	**-**	-	0.45	**4.82**	-	**4.82**	m2	**5.42**
Disposal of excavated material										
removing from site to tip (including tipping charges but excluding landfill tax)	-	-	**-**	-	3.38	**36.20**	12.81	**49.01**	m3	**55.14**
Excavated material arising from excavations										
Filling to excavations										
average thickness exceeding 0.25m	-	-	**-**	-	2.25	**24.10**	-	**24.10**	m3	**27.11**
Compacting										
filling	-	-	**-**	-	0.23	**2.46**	-	**2.46**	m2	**2.77**
Plain in-situ concrete; BS 5328, ordinary prescribed mix ST3, 20mm aggregate										
Foundations; poured on or against earth or unblinded hardcore										
generally	62.20	7.50	**66.86**	-	5.34	**57.19**	-	**124.06**	m3	**139.56**
Plain in-situ concrete; BS 5328, ordinary prescribed mix ST4, 20mm aggregate										
Foundations; poured on or against earth or unblinded hardcore										
generally	63.93	7.50	**68.73**	-	5.34	**57.19**	-	**125.92**	m3	**141.66**
Formwork and basic finish										
Sides of foundations; plain vertical										
height exceeding 1.00m	7.73	10.00	**8.50**	3.94	0.79	**64.45**	1.80	**74.75**	m2	**84.09**
height not exceeding 250mm	2.31	10.00	**2.54**	1.19	0.25	**19.59**	0.54	**22.66**	m	**25.49**
height 250 - 500mm	4.24	10.00	**4.67**	2.16	0.43	**35.30**	0.99	**40.96**	m	**46.08**
height 0.50m - 1.00m	7.80	10.00	**8.58**	4.03	0.81	**65.94**	1.83	**76.35**	m	**85.89**
Common bricks, BS EN 772, Category M, 215 x 102.5 x 65mm, compressive strength not less than 5.2 N/mm2; in cement mortar (1:3)										
Walls; vertical										
215mm thick; English bond	24.28	5.00	**25.49**	4.70	5.00	**120.34**	7.18	**153.01**	m2	**172.13**
327mm thick; English bond	36.52	5.00	**38.34**	5.70	6.15	**146.86**	11.29	**196.49**	m2	**221.05**
440mm thick; English bond	48.55	5.00	**50.98**	6.30	6.90	**163.42**	15.38	**229.78**	m2	**258.51**
Bonding to existing including extra material										
thickness of new work 215mm	2.65	5.00	**2.78**	1.32	1.32	**32.89**	0.77	**36.45**	m	**41.01**
thickness of new work 327mm	4.08	5.00	**4.28**	1.90	1.90	**47.35**	1.16	**52.79**	m	**59.39**
thickness of new work 440mm	5.30	5.00	**5.57**	2.50	2.50	**62.30**	1.55	**69.42**	m	**78.09**

Labour hourly rates: (except Specialists) Craft Operatives £14.21 Labourer £10.71 Rates are national average prices. Refer to REGIONAL VARIATIONS for indicative levels of overall pricing in regions	MATERIALS			LABOUR				RATES		
	Del to Site £	Waste %	Material Cost £	Craft Optve Hrs	Lab Hrs	Labour Cost £	Sunds £	Nett Rate £	Unit	Gross Rate (+12.5%) £
D50: UNDERPINNING Cont.										
Second Hard Stock bricks, BS EN 772, Category M, 215 x 102.5 x 65mm, in cement mortar (1:3)										
Walls; vertical										
215mm thick; English bond	47.84	5.00	**50.23**	4.70	5.00	**120.34**	7.18	**177.75**	m2	**199.97**
327mm thick; English bond	71.96	5.00	**75.56**	5.70	6.15	**146.86**	11.29	**233.71**	m2	**262.92**
440mm thick; English bond	95.68	5.00	**100.46**	6.30	6.90	**163.42**	15.38	**279.27**	m2	**314.17**
Bonding to existing including extra material										
thickness of new work 215mm	5.23	5.00	**5.49**	1.90	1.90	**47.35**	0.77	**53.61**	m	**60.31**
thickness of new work 327mm	8.04	5.00	**8.44**	1.90	1.90	**47.35**	1.16	**56.95**	m	**64.07**
thickness of new work 440mm	10.45	5.00	**10.97**	2.50	2.50	**62.30**	1.55	**74.82**	m	**84.18**
Engineering bricks, BS EN 772, Category F, 215 x 102.5 x 65mm, class A; in cement mortar (1:3)										
Walls; vertical										
215mm thick; English bond	35.94	5.00	**37.73**	5.20	5.50	**132.80**	7.18	**177.71**	m2	**199.93**
327mm thick; English bond	54.06	5.00	**56.76**	6.30	6.75	**161.82**	11.29	**229.86**	m2	**258.60**
440mm thick; English bond	71.88	5.00	**75.47**	6.95	7.55	**179.62**	15.38	**270.47**	m2	**304.28**
Bonding to existing including extra material										
thickness of new work 215mm	3.93	5.00	**4.12**	1.45	1.45	**36.13**	0.77	**41.03**	m	**46.16**
thickness of new work 327mm	6.04	5.00	**6.34**	2.10	2.10	**52.33**	1.16	**59.84**	m	**67.31**
thickness of new work 440mm	7.85	5.00	**8.24**	2.75	2.75	**68.53**	1.55	**78.32**	m	**88.11**
Engineering bricks, BS EN 772, Category F, 215 x 102.5 x 65mm, Class B; in cement mortar (1:3)										
Walls; vertical										
215mm thick; English bond	25.47	5.00	**26.74**	5.20	5.50	**132.80**	7.18	**166.72**	m2	**187.56**
327mm thick; English bond	38.31	5.00	**40.22**	6.30	6.75	**161.82**	11.29	**213.32**	m2	**239.99**
440mm thick; English bond	50.93	5.00	**53.48**	6.95	7.55	**179.62**	15.38	**248.48**	m2	**279.54**
Bonding to existing including extra material										
thickness of new work 215mm	2.78	5.00	**2.92**	1.45	1.45	**36.13**	0.77	**39.83**	m	**44.81**
thickness of new work 327mm	4.28	5.00	**4.49**	2.10	2.10	**52.33**	1.16	**57.99**	m	**65.24**
thickness of new work 440mm	5.56	5.00	**5.84**	2.75	2.75	**68.53**	1.55	**75.92**	m	**85.41**
Sundry items										
Wedging and pinning up to underside of existing construction with two courses slates in cement mortar (1:3)										
215mm walls	8.58	15.00	**9.87**	0.60	0.60	**14.95**	0.20	**25.02**	m	**28.15**
327mm walls	12.87	15.00	**14.80**	0.87	0.87	**21.68**	0.25	**36.73**	m	**41.32**
440mm walls	17.28	15.00	**19.87**	1.14	1.14	**28.41**	0.27	**48.54**	m	**54.61**

IN-SITU CONCRETE/LARGE PRECAST CONCRETE

Labour hourly rates: (except Specialists) Craft Operatives £14.21 Labourer £10.71 Rates are national average prices. Refer to REGIONAL VARIATIONS for indicative levels of overall pricing in regions	MATERIALS			LABOUR				RATES		
	Del to Site £	Waste %	Material Cost £	Craft Optve Hrs	Lab Hrs	Labour Cost £	Sunds £	Nett Rate £	Unit	Gross Rate (+12.5%) £
E10: MIXING/CASTING/CURING IN-SITU CONCRETE - (READY MIXED)										
Plain in-situ concrete; BS 5328, ordinary prescribed mix ST3, 20mm aggregate										
Foundations poured on or against earth or unblinded hardcore										
generally	62.20	7.50	**66.86**	-	2.30	**24.63**	-	**91.50**	m3	**102.93**
Isolated foundations; poured on or against earth or unblinded hardcore										
generally	62.20	7.50	**66.86**	-	2.75	**29.45**	-	**96.32**	m3	**108.36**
Beds; poured on or against earth or unblinded hardcore										
thickness exceeding 450mm	62.20	7.50	**66.86**	-	2.20	**23.56**	-	**90.43**	m3	**101.73**
thickness 150 - 450mm	62.20	7.50	**66.86**	-	2.50	**26.78**	-	**93.64**	m3	**105.34**
thickness not exceeding 150mm	62.20	7.50	**66.86**	-	3.10	**33.20**	-	**100.06**	m3	**112.57**
Filling hollow walls										
thickness not exceeding 150mm	62.20	5.00	**65.31**	-	5.50	**58.91**	-	**124.21**	m3	**139.74**
Plain in-situ concrete; BS 5328, ordinary prescribed mix ST4, 20mm aggregate										
Foundations poured on or against earth or unblinded hardcore										
generally	63.93	7.50	**68.73**	-	2.30	**24.63**	-	**93.36**	m3	**105.03**
Isolated foundations; poured on or against earth or unblinded hardcore										
generally	63.93	7.50	**68.73**	-	2.75	**29.45**	-	**98.18**	m3	**110.45**
Beds; poured on or against earth or unblinded hardcore										
thickness exceeding 450mm	63.93	7.50	**68.73**	-	2.20	**23.56**	-	**92.29**	m3	**103.83**
thickness 150 - 450mm	63.93	7.50	**68.73**	-	2.50	**26.78**	-	**95.50**	m3	**107.44**
thickness not exceeding 150mm	63.93	7.50	**68.73**	-	3.10	**33.20**	-	**101.93**	m3	**114.67**
Filling hollow walls										
thickness not exceeding 150mm	63.93	5.00	**67.13**	-	5.50	**58.91**	-	**126.03**	m3	**141.79**
Plain in-situ concrete; BS 5328, ordinary prescribed mix ST5, 20mm aggregate										
Foundations poured on or against earth or unblinded hardcore										
generally	66.12	7.50	**71.08**	-	2.30	**24.63**	-	**95.71**	m3	**107.68**
Isolated foundations; poured on or against earth or unblinded hardcore										
generally	66.12	7.50	**71.08**	-	2.75	**29.45**	-	**100.53**	m3	**113.10**
Beds; poured on or against earth or unblinded hardcore										
thickness exceeding 450mm	66.12	7.50	**71.08**	-	2.20	**23.56**	-	**94.64**	m3	**106.47**
thickness 150 - 450mm	66.12	7.50	**71.08**	-	2.50	**26.78**	-	**97.85**	m3	**110.09**
thickness not exceeding 150mm	66.12	7.50	**71.08**	-	3.10	**33.20**	-	**104.28**	m3	**117.32**
Filling hollow walls										
thickness not exceeding 150mm	66.12	5.00	**69.43**	-	5.50	**58.91**	-	**128.33**	m3	**144.37**
Reinforced in-situ concrete; BS 5328, designed mix C15, 20mm aggregate, minimum cement content 220 kg/m3; vibrated										
Foundations										
generally	64.17	5.00	**67.37**	-	2.60	**27.85**	0.61	**95.83**	m3	**107.81**
Ground beams										
generally	64.17	5.00	**67.37**	-	3.60	**38.56**	0.61	**106.54**	m3	**119.86**

Labour hourly rates: (except Specialists) Craft Operatives £14.21 Labourer £10.71 Rates are national average prices. Refer to REGIONAL VARIATIONS for indicative levels of overall pricing in regions	MATERIALS			LABOUR				RATES		
	Del to Site £	Waste %	Material Cost £	Craft Optve Hrs	Lab Hrs	Labour Cost £	Sunds £	Nett Rate £	Unit	Gross Rate (+12.5%) £
E10: MIXING/CASTING/CURING IN-SITU CONCRETE - (READY MIXED) Cont.										
Reinforced in-situ concrete; BS 5328, designed mix C15, 20mm aggregate, minimum cement content 220 kg/m3; vibrated Cont.										
Isolated foundations										
generally	64.17	5.00	**67.37**	-	3.00	**32.13**	0.61	**100.11**	m3	**112.63**
Beds										
thickness exceeding 450mm	64.17	5.00	**67.37**	-	2.50	**26.78**	0.61	**94.76**	m3	**106.60**
thickness 150 - 450mm	64.17	5.00	**67.37**	-	2.75	**29.45**	0.61	**97.44**	m3	**109.62**
thickness not exceeding 150mm	64.17	5.00	**67.37**	-	3.50	**37.49**	0.61	**105.47**	m3	**118.65**
Slabs										
thickness exceeding 450mm	64.17	2.50	**65.77**	-	3.20	**34.27**	0.61	**100.65**	m3	**113.23**
thickness 150 - 450mm	64.17	2.50	**65.77**	-	3.45	**36.95**	0.61	**103.33**	m3	**116.25**
thickness not exceeding 150mm	64.17	2.50	**65.77**	-	4.55	**48.73**	0.61	**115.11**	m3	**129.50**
Walls										
thickness exceeding 450mm	64.17	2.50	**65.77**	-	3.50	**37.49**	0.61	**103.87**	m3	**116.85**
thickness 150 - 450mm	64.17	2.50	**65.77**	-	3.90	**41.77**	0.61	**108.15**	m3	**121.67**
thickness not exceeding 150mm	64.17	2.50	**65.77**	-	5.00	**53.55**	0.61	**119.93**	m3	**134.92**
Beams										
isolated	64.17	2.50	**65.77**	-	4.70	**50.34**	0.61	**116.72**	m3	**131.31**
isolated deep	64.17	2.50	**65.77**	-	4.80	**51.41**	0.61	**117.79**	m3	**132.51**
attached deep	64.17	2.50	**65.77**	-	4.70	**50.34**	0.61	**116.72**	m3	**131.31**
Beam casings										
isolated	64.17	2.50	**65.77**	-	5.00	**53.55**	0.61	**119.93**	m3	**134.92**
isolated deep	64.17	2.50	**65.77**	-	5.10	**54.62**	0.61	**121.00**	m3	**136.13**
attached deep	64.17	2.50	**65.77**	-	5.00	**53.55**	0.61	**119.93**	m3	**134.92**
Columns										
generally	64.17	2.50	**65.77**	-	6.30	**67.47**	0.61	**133.85**	m3	**150.58**
Column casings										
generally	64.17	2.50	**65.77**	-	6.60	**70.69**	0.61	**137.07**	m3	**154.20**
Staircases										
generally	64.17	2.50	**65.77**	-	5.50	**58.91**	0.61	**125.29**	m3	**140.95**
Upstands										
generally	64.17	2.50	**65.77**	-	7.70	**82.47**	0.61	**148.85**	m3	**167.45**
Reinforced in-situ concrete; BS 5328, designed mix C20, 20mm aggregate, minimum cement content 240 kg/m3; vibrated										
Foundations										
generally	65.24	5.00	**68.51**	-	2.60	**27.85**	0.61	**96.96**	m3	**109.08**
Ground beams										
generally	65.24	5.00	**68.51**	-	3.60	**38.56**	0.61	**107.67**	m3	**121.13**
Isolated foundations										
generally	65.24	5.00	**68.51**	-	3.00	**32.13**	0.61	**101.25**	m3	**113.90**
Beds										
thickness exceeding 450mm	65.24	5.00	**68.51**	-	2.50	**26.78**	0.61	**95.89**	m3	**107.88**
thickness 150 - 450mm	65.24	5.00	**68.51**	-	2.75	**29.45**	0.61	**98.57**	m3	**110.89**
thickness not exceeding 150mm	65.24	5.00	**68.51**	-	3.50	**37.49**	0.61	**106.60**	m3	**119.93**
Slabs										
thickness exceeding 450mm	65.24	2.50	**66.87**	-	3.20	**34.27**	0.61	**101.76**	m3	**114.48**
thickness 150 - 450mm	65.24	2.50	**66.87**	-	3.45	**36.95**	0.61	**104.43**	m3	**117.49**
thickness not exceeding 150mm	65.24	2.50	**66.87**	-	4.55	**48.73**	0.61	**116.22**	m3	**130.74**
Walls										
thickness exceeding 450mm	65.24	2.50	**66.87**	-	3.50	**37.49**	0.61	**104.97**	m3	**118.09**
thickness 150 - 450mm	65.24	2.50	**66.87**	-	3.90	**41.77**	0.61	**109.25**	m3	**122.91**
thickness not exceeding 150mm	65.24	2.50	**66.87**	-	5.00	**53.55**	0.61	**121.04**	m3	**136.16**
Beams										
isolated	65.24	2.50	**66.87**	-	4.70	**50.34**	0.61	**117.82**	m3	**132.55**
isolated deep	65.24	2.50	**66.87**	-	4.80	**51.41**	0.61	**118.89**	m3	**133.75**
attached deep	65.24	2.50	**66.87**	-	4.70	**50.34**	0.61	**117.82**	m3	**132.55**
Beam casings										
isolated	65.24	2.50	**66.87**	-	5.00	**53.55**	0.61	**121.04**	m3	**136.16**
isolated deep	65.24	2.50	**66.87**	-	5.10	**54.62**	0.61	**122.11**	m3	**137.37**
attached deep	65.24	2.50	**66.87**	-	5.00	**53.55**	0.61	**121.04**	m3	**136.16**
Columns										
generally	65.24	2.50	**66.87**	-	6.30	**67.47**	0.61	**134.96**	m3	**151.83**
Column casings										
generally	65.24	2.50	**66.87**	-	6.60	**70.69**	0.61	**138.17**	m3	**155.44**
Staircases										
generally	65.24	2.50	**66.87**	-	5.50	**58.91**	0.61	**126.39**	m3	**142.19**
Upstands										
generally	65.24	2.50	**66.87**	-	7.70	**82.47**	0.61	**149.95**	m3	**168.70**

Labour hourly rates: (except Specialists) Craft Operatives £14.21 Labourer £10.71 Rates are national average prices. Refer to REGIONAL VARIATIONS for indicative levels of overall pricing in regions	MATERIALS			LABOUR				RATES		
	Del to Site £	Waste %	Material Cost £	Craft Optve Hrs	Lab Hrs	Labour Cost £	Sunds £	Nett Rate £	Unit	Gross Rate (+12.5%) £
E10: MIXING/CASTING/CURING IN-SITU CONCRETE - (READY MIXED) Cont.										
Reinforced in-situ concrete; BS 5328, designed mix C25, 20mm aggregate, minimum cement content 290 kg/m3; vibrated										
Foundations										
generally	67.58	5.00	**70.96**	-	2.60	**27.85**	0.61	**99.42**	m3	**111.85**
Ground beams										
generally	67.58	5.00	**70.96**	-	3.60	**38.56**	0.61	**110.13**	m3	**123.90**
Isolated foundations										
generally	67.58	5.00	**70.96**	-	3.00	**32.13**	0.61	**103.70**	m3	**116.67**
Beds										
thickness exceeding 450mm	67.58	5.00	**70.96**	-	2.50	**26.78**	0.61	**98.35**	m3	**110.64**
thickness 150 - 450mm	67.58	5.00	**70.96**	-	2.75	**29.45**	0.61	**101.03**	m3	**113.65**
thickness not exceeding 150mm	67.58	5.00	**70.96**	-	3.50	**37.49**	0.61	**109.06**	m3	**122.69**
Slabs										
thickness exceeding 450mm	67.58	2.50	**69.27**	-	3.20	**34.27**	0.61	**104.16**	m3	**117.18**
thickness 150 - 450mm	67.58	2.50	**69.27**	-	3.45	**36.95**	0.61	**106.83**	m3	**120.19**
thickness not exceeding 150mm	67.58	2.50	**69.27**	-	4.55	**48.73**	0.61	**118.61**	m3	**133.44**
Walls										
thickness exceeding 450mm	67.58	2.50	**69.27**	-	3.50	**37.49**	0.61	**107.37**	m3	**120.79**
thickness 150 - 450mm	67.58	2.50	**69.27**	-	3.90	**41.77**	0.61	**111.65**	m3	**125.61**
thickness not exceeding 150mm	67.58	2.50	**69.27**	-	5.00	**53.55**	0.61	**123.43**	m3	**138.86**
Beams										
isolated	67.58	2.50	**69.27**	-	4.70	**50.34**	0.61	**120.22**	m3	**135.25**
isolated deep	67.58	2.50	**69.27**	-	4.80	**51.41**	0.61	**121.29**	m3	**136.45**
attached deep	67.58	2.50	**69.27**	-	4.70	**50.34**	0.61	**120.22**	m3	**135.25**
Beam casings										
isolated	67.58	2.50	**69.27**	-	5.00	**53.55**	0.61	**123.43**	m3	**138.86**
isolated deep	67.58	2.50	**69.27**	-	5.10	**54.62**	0.61	**124.50**	m3	**140.07**
attached deep	67.58	2.50	**69.27**	-	5.00	**53.55**	0.61	**123.43**	m3	**138.86**
Columns										
generally	67.58	2.50	**69.27**	-	6.30	**67.47**	0.61	**137.36**	m3	**154.53**
Column casings										
generally	67.58	2.50	**69.27**	-	6.60	**70.69**	0.61	**140.57**	m3	**158.14**
Staircases										
generally	67.58	2.50	**69.27**	-	5.50	**58.91**	0.61	**128.79**	m3	**144.89**
Upstands										
generally	67.58	2.50	**69.27**	-	7.70	**82.47**	0.61	**152.35**	m3	**171.39**
Reinforced in-situ concrete; BS 5328, designed mix C30, 20mm aggregate, minimum cement content 290 kg/m3; vibrated										
Foundations										
generally	69.13	5.00	**72.58**	-	2.60	**27.85**	0.61	**101.04**	m3	**113.67**
Ground beams										
generally	69.13	5.00	**72.58**	-	3.60	**38.56**	0.61	**111.75**	m3	**125.72**
Isolated foundations										
generally	69.13	5.00	**72.58**	-	3.00	**32.13**	0.61	**105.32**	m3	**118.49**
Beds										
thickness exceeding 450mm	69.13	5.00	**72.58**	-	2.50	**26.78**	0.61	**99.97**	m3	**112.46**
thickness 150 - 450mm	69.13	5.00	**72.58**	-	2.75	**29.45**	0.61	**102.65**	m3	**115.48**
thickness not exceeding 150mm	69.13	5.00	**72.58**	-	3.50	**37.49**	0.61	**110.68**	m3	**124.51**
Slabs										
thickness exceeding 450mm	69.13	2.50	**70.85**	-	3.20	**34.27**	0.61	**105.74**	m3	**118.95**
thickness 150 - 450mm	69.13	2.50	**70.85**	-	3.45	**36.95**	0.61	**108.41**	m3	**121.97**
thickness not exceeding 150mm	69.13	2.50	**70.85**	-	4.55	**48.73**	0.61	**120.20**	m3	**135.22**
Walls										
thickness exceeding 450mm	69.13	2.50	**70.85**	-	3.50	**37.49**	0.61	**108.95**	m3	**122.57**
thickness 150 - 450mm	69.13	2.50	**70.85**	-	3.90	**41.77**	0.61	**113.23**	m3	**127.39**
thickness not exceeding 150mm	69.13	2.50	**70.85**	-	5.00	**53.55**	0.61	**125.02**	m3	**140.64**
Beams										
isolated	69.13	2.50	**70.85**	-	4.70	**50.34**	0.61	**121.80**	m3	**137.03**
isolated deep	69.13	2.50	**70.85**	-	4.80	**51.41**	0.61	**122.87**	m3	**138.23**
attached deep	69.13	2.50	**70.85**	-	4.70	**50.34**	0.61	**121.80**	m3	**137.03**
Beam casings										
isolated	69.13	2.50	**70.85**	-	5.00	**53.55**	0.61	**125.02**	m3	**140.64**
isolated deep	69.13	2.50	**70.85**	-	5.10	**54.62**	0.61	**126.09**	m3	**141.85**
attached deep	69.13	2.50	**70.85**	-	5.00	**53.55**	0.61	**125.02**	m3	**140.64**
Columns										
generally	69.13	2.50	**70.85**	-	6.30	**67.47**	0.61	**138.94**	m3	**156.31**
Column casings										
generally	69.13	2.50	**70.85**	-	6.60	**70.69**	0.61	**142.15**	m3	**159.92**

Labour hourly rates: (except Specialists) Craft Operatives £14.21 Labourer £10.71 Rates are national average prices. Refer to REGIONAL VARIATIONS for indicative levels of overall pricing in regions	MATERIALS			LABOUR				RATES		
	Del to Site £	Waste %	Material Cost £	Craft Optve Hrs	Lab Hrs	Labour Cost £	Sunds £	Nett Rate £	Unit	Gross Rate (+12.5%) £
E10: MIXING/CASTING/CURING IN-SITU CONCRETE - (READY MIXED) Cont.										
Reinforced in-situ concrete; BS 5328, designed mix C30, 20mm aggregate, minimum cement content 290 kg/m3; vibrated Cont.										
Upstands										
generally	69.13	2.50	**70.85**	-	7.70	**82.47**	0.61	**153.93**	m3	**173.17**
E10: MIXING/CASTING/CURING IN-SITU CONCRETE - (SITE MIXED)										
Plain in-situ concrete; mix 1:6, all in aggregate										
Foundations poured on or against earth or unblinded hardcore										
generally	79.06	7.50	**84.99**	-	3.40	**36.41**	2.20	**123.61**	m3	**139.06**
Isolated foundations; poured on or against earth or unblinded hardcore										
generally	79.06	7.50	**84.99**	-	3.90	**41.77**	2.20	**128.96**	m3	**145.08**
Beds; poured on or against earth or unblinded hardcore										
thickness 150 - 450mm	79.06	7.50	**84.99**	-	3.60	**38.56**	2.20	**125.75**	m3	**141.47**
thickness not exceeding 150mm	79.06	7.50	**84.99**	-	4.25	**45.52**	2.20	**132.71**	m3	**149.30**
Beds; sloping not exceeding 15 degrees; poured on or against earth or unblinded hardcore										
thickness 150 - 450mm	79.06	7.50	**84.99**	-	3.80	**40.70**	2.20	**127.89**	m3	**143.88**
thickness not exceeding 150mm	79.06	7.50	**84.99**	-	4.45	**47.66**	2.20	**134.85**	m3	**151.71**
Beds; sloping exceeding 15 degrees; poured on or against earth or unblinded hardcore										
thickness 150 - 450mm	79.06	7.50	**84.99**	-	4.00	**42.84**	2.20	**130.03**	m3	**146.29**
thickness not exceeding 150mm	79.06	7.50	**84.99**	-	4.70	**50.34**	2.20	**137.53**	m3	**154.72**
Plain in-situ concrete; mix 1:8, all in aggregate										
Foundations poured on or against earth or unblinded hardcore										
generally	70.66	7.50	**75.96**	-	3.40	**36.41**	2.20	**114.58**	m3	**128.90**
Isolated foundations; poured on or against earth or unblinded hardcore										
generally	70.66	7.50	**75.96**	-	3.90	**41.77**	2.20	**119.93**	m3	**134.92**
Beds; poured on or against earth or unblinded hardcore										
thickness 150 - 450mm	70.66	7.50	**75.96**	-	3.60	**38.56**	2.20	**116.72**	m3	**131.31**
thickness not exceeding 150mm	70.66	7.50	**75.96**	-	4.25	**45.52**	2.20	**123.68**	m3	**139.14**
Beds; sloping not exceeding 15 degrees; poured on or against earth or unblinded hardcore										
thickness 150 - 450mm	70.66	7.50	**75.96**	-	3.80	**40.70**	2.20	**118.86**	m3	**133.72**
thickness not exceeding 150mm	70.66	7.50	**75.96**	-	4.45	**47.66**	2.20	**125.82**	m3	**141.55**
Beds; sloping exceeding 15 degrees; poured on or against earth or unblinded hardcore										
thickness 150 - 450mm	70.66	7.50	**75.96**	-	4.00	**42.84**	2.20	**121.00**	m3	**136.13**
thickness not exceeding 150mm	70.66	7.50	**75.96**	-	4.70	**50.34**	2.20	**128.50**	m3	**144.56**
Plain in-situ concrete; mix 1:12, all in aggregate										
Foundations poured on or against earth or unblinded hardcore										
generally	62.26	7.50	**66.93**	-	3.40	**36.41**	2.20	**105.54**	m3	**118.73**
Isolated foundations; poured on or against earth or unblinded hardcore										
generally	62.26	7.50	**66.93**	-	3.90	**41.77**	2.20	**110.90**	m3	**124.76**
Beds; poured on or against earth or unblinded hardcore										
thickness 150 - 450mm	62.26	7.50	**66.93**	-	3.60	**38.56**	2.20	**107.68**	m3	**121.14**
thickness not exceeding 150mm	62.26	7.50	**66.93**	-	4.25	**45.52**	2.20	**114.65**	m3	**128.98**
Beds; sloping not exceeding 15 degrees; poured on or against earth or unblinded hardcore										
thickness 150 - 450mm	62.26	7.50	**66.93**	-	3.80	**40.70**	2.20	**109.83**	m3	**123.55**
thickness not exceeding 150mm	62.26	7.50	**66.93**	-	4.45	**47.66**	2.20	**116.79**	m3	**131.39**
Beds; sloping exceeding 15 degrees; poured on or against earth or unblinded hardcore										
thickness 150 - 450mm	62.26	7.50	**66.93**	-	4.00	**42.84**	2.20	**111.97**	m3	**125.96**
thickness not exceeding 150mm	62.26	7.50	**66.93**	-	4.70	**50.34**	2.20	**119.47**	m3	**134.40**
Plain in-situ concrete; BS 5328, ordinary prescribed mix C15P, 20mm aggregate										
Foundations poured on or against earth or unblinded hardcore										
generally	80.16	7.50	**86.18**	-	3.40	**36.41**	2.20	**124.79**	m3	**140.39**
Isolated foundations; poured on or against earth or unblinded hardcore										
generally	80.16	7.50	**86.18**	-	3.90	**41.77**	2.20	**130.14**	m3	**146.41**

Labour hourly rates: (except Specialists) Craft Operatives £14.21 Labourer £10.71 Rates are national average prices. Refer to REGIONAL VARIATIONS for indicative levels of overall pricing in regions	MATERIALS			LABOUR				RATES		
	Del to Site £	Waste %	Material Cost £	Craft Optve Hrs	Lab Hrs	Labour Cost £	Sunds £	Nett Rate £	Unit	Gross Rate (+12.5%) £
E10: MIXING/CASTING/CURING IN-SITU CONCRETE - (SITE MIXED) Cont.										
Plain in-situ concrete; BS 5328, ordinary prescribed mix C15P, 20mm aggregate Cont.										
Beds; poured on or against earth or unblinded hardcore										
thickness exceeding 450mm	80.16	7.50	**86.18**	-	3.30	**35.34**	2.20	**123.72**	m3	**139.18**
thickness 150 - 450mm	80.16	7.50	**86.18**	-	3.60	**38.56**	2.20	**126.93**	m3	**142.80**
thickness not exceeding 150mm	80.16	7.50	**86.18**	-	4.25	**45.52**	2.20	**133.89**	m3	**150.63**
Filling hollow walls										
thickness not exceeding 150mm	80.16	7.50	**86.18**	-	6.60	**70.69**	2.20	**159.06**	m3	**178.94**
Plain in-situ concrete; BS 5328, ordinary prescribed mix C20P, 20mm aggregate										
Foundations poured on or against earth or unblinded hardcore										
generally	82.72	7.50	**88.92**	-	3.40	**36.41**	2.20	**127.53**	m3	**143.47**
Isolated foundations; poured on or against earth or unblinded hardcore										
generally	82.72	7.50	**88.92**	-	3.90	**41.77**	2.20	**132.89**	m3	**149.50**
Beds; poured on or against earth or unblinded hardcore										
thickness exceeding 450mm	82.72	7.50	**88.92**	-	3.30	**35.34**	2.20	**126.46**	m3	**142.27**
thickness 150 - 450mm	82.72	7.50	**88.92**	-	3.60	**38.56**	2.20	**129.68**	m3	**145.88**
thickness not exceeding 150mm	82.72	7.50	**88.92**	-	4.25	**45.52**	2.20	**136.64**	m3	**153.72**
Filling hollow walls										
thickness not exceeding 150mm	82.72	7.50	**88.92**	-	6.60	**70.69**	2.20	**161.81**	m3	**182.03**
Plain in-situ concrete; BS 5328, ordinary prescribed mix C25P, 20mm aggregate										
Foundations poured on or against earth or unblinded hardcore										
generally	86.33	7.50	**92.80**	-	3.40	**36.41**	2.20	**131.42**	m3	**147.85**
Isolated foundations; poured on or against earth or unblinded hardcore										
generally	86.33	7.50	**92.80**	-	3.90	**41.77**	2.20	**136.77**	m3	**153.87**
Beds; poured on or against earth or unblinded hardcore										
thickness exceeding 450mm	86.33	7.50	**92.80**	-	3.30	**35.34**	2.20	**130.35**	m3	**146.64**
thickness 150 - 450mm	86.33	7.50	**92.80**	-	3.60	**38.56**	2.20	**133.56**	m3	**150.26**
thickness not exceeding 150mm	86.33	7.50	**92.80**	-	4.25	**45.52**	2.20	**140.52**	m3	**158.09**
Filling hollow walls										
thickness not exceeding 150mm	86.33	7.50	**92.80**	-	6.60	**70.69**	2.20	**165.69**	m3	**186.40**
Reinforced in-situ concrete; BS 5328, designed mix C15, 20mm aggregate, minimum cement content 220 kg/m3; vibrated										
Foundations										
generally	74.69	5.00	**78.43**	-	3.70	**39.63**	2.81	**120.86**	m3	**135.97**
Ground beams										
generally	74.69	5.00	**78.43**	-	4.70	**50.34**	2.81	**131.57**	m3	**148.02**
Isolated foundations										
generally	74.69	5.00	**78.43**	-	4.10	**43.91**	2.81	**125.15**	m3	**140.79**
Beds										
thickness exceeding 450mm	74.69	5.00	**78.43**	-	3.60	**38.56**	2.81	**119.79**	m3	**134.77**
thickness 150 - 450mm	74.69	5.00	**78.43**	-	3.90	**41.77**	2.81	**123.01**	m3	**138.38**
thickness not exceeding 150mm	74.69	5.00	**78.43**	-	4.50	**48.20**	2.81	**129.43**	m3	**145.61**
Reinforced in-situ concrete; BS 5328, designed mix C20, 20mm aggregate, minimum cement content 240 kg/m3; vibrated										
Foundations										
generally	76.12	5.00	**79.93**	-	3.70	**39.63**	2.81	**122.37**	m3	**137.66**
Ground beams										
generally	76.12	5.00	**79.93**	-	4.70	**50.34**	2.81	**133.08**	m3	**149.71**
Isolated foundations										
generally	76.12	5.00	**79.93**	-	4.10	**43.91**	2.81	**126.65**	m3	**142.48**
Beds										
thickness exceeding 450mm	76.12	5.00	**79.93**	-	3.60	**38.56**	2.81	**121.29**	m3	**136.46**
thickness 150 - 450mm	76.12	5.00	**79.93**	-	3.90	**41.77**	2.81	**124.51**	m3	**140.07**
thickness not exceeding 150mm	76.12	5.00	**79.93**	-	4.50	**48.20**	2.81	**130.93**	m3	**147.30**
Reinforced in-situ concrete; BS 5328, designed mix C25, 20mm aggregate, minimum cement content 290 kg/m3; vibrated										
Foundations										
generally	80.86	5.00	**84.90**	-	3.70	**39.63**	2.81	**127.34**	m3	**143.26**

Labour hourly rates: (except Specialists) Craft Operatives £14.21 Labourer £10.71 Rates are national average prices. Refer to REGIONAL VARIATIONS for indicative levels of overall pricing in regions	MATERIALS			LABOUR				RATES		
	Del to Site £	Waste %	Material Cost £	Craft Optve Hrs	Lab Hrs	Labour Cost £	Sunds £	Nett Rate £	Unit	Gross Rate (+12.5%) £
E10: MIXING/CASTING/CURING IN-SITU CONCRETE - (SITE MIXED) Cont.										
Reinforced in-situ concrete; BS 5328, designed mix C25, 20mm aggregate, minimum cement content 290 kg/m3; vibrated Cont.										
Ground beams										
generally	80.86	5.00	**84.90**	-	4.70	**50.34**	2.81	**138.05**	m3	**155.31**
Isolated foundations										
generally	80.86	5.00	**84.90**	-	4.10	**43.91**	2.81	**131.62**	m3	**148.08**
Beds										
thickness exceeding 450mm	80.86	5.00	**84.90**	-	3.60	**38.56**	2.81	**126.27**	m3	**142.05**
thickness 150 - 450mm	80.86	5.00	**84.90**	-	3.90	**41.77**	2.81	**129.48**	m3	**145.67**
thickness not exceeding 150mm	80.86	5.00	**84.90**	-	4.50	**48.20**	2.81	**135.91**	m3	**152.90**
Cement and sand (1:3); for grouting and the like										
Grouting.										
stanchion bases.	2.49	10.00	**2.74**	-	0.45	**4.82**	1.28	**8.84**	nr	**9.94**
E10: MIXING/CASTING/CURING IN-SITU CONCRETE - (SUNDRIES)										
The foregoing concrete is based on the use of Portland cement. For other cements and waterproofers ADD as follows										
Concrete 1:6										
rapid hardening cement	3.96	-	**3.96**	-	-	-	-	**3.96**	m3	**4.46**
sulphate resisting cement	4.97	-	**4.97**	-	-	-	-	**4.97**	m3	**5.59**
waterproofing powder	11.45	-	**11.45**	-	-	-	-	**11.45**	m3	**12.89**
waterproofing liquid	5.84	-	**5.84**	-	-	-	-	**5.84**	m3	**6.57**
Concrete 1:8										
rapid hardening cement	2.98	-	**2.98**	-	-	-	-	**2.98**	m3	**3.35**
sulphate resisting cement	3.73	-	**3.73**	-	-	-	-	**3.73**	m3	**4.20**
waterproofing powder	8.60	-	**8.60**	-	-	-	-	**8.60**	m3	**9.67**
waterproofing liquid	5.84	-	**5.84**	-	-	-	-	**5.84**	m3	**6.57**
Concrete 1:12										
rapid hardening cement	1.99	-	**1.99**	-	-	-	-	**1.99**	m3	**2.24**
sulphate resisting cement	2.48	-	**2.48**	-	-	-	-	**2.48**	m3	**2.79**
waterproofing powder	5.72	-	**5.72**	-	-	-	-	**5.72**	m3	**6.44**
waterproofing liquid	5.84	-	**5.84**	-	-	-	-	**5.84**	m3	**6.57**
Concrete ST3 and C15										
rapid hardening cement	2.65	-	**2.65**	-	-	-	-	**2.65**	m3	**2.98**
sulphate resisting cement	3.30	-	**3.30**	-	-	-	-	**3.30**	m3	**3.71**
waterproofing powder	7.64	-	**7.64**	-	-	-	-	**7.64**	m3	**8.59**
waterproofing liquid	5.84	-	**5.84**	-	-	-	-	**5.84**	m3	**6.57**
Concrete ST4 and C20										
rapid hardening cement	3.92	-	**3.92**	-	-	-	-	**3.92**	m3	**4.41**
sulphate resisting cement	4.91	-	**4.91**	-	-	-	-	**4.91**	m3	**5.52**
waterproofing powder	11.29	-	**11.29**	-	-	-	-	**11.29**	m3	**12.71**
waterproofing liquid	5.84	-	**5.84**	-	-	-	-	**5.84**	m3	**6.57**
Concrete ST5 and C25										
rapid hardening cement	5.29	-	**5.29**	-	-	-	-	**5.29**	m3	**5.95**
sulphate resisting cement	6.61	-	**6.61**	-	-	-	-	**6.61**	m3	**7.44**
waterproofing powder	15.26	-	**15.26**	-	-	-	-	**15.26**	m3	**17.16**
waterproofing liquid	5.84	-	**5.84**	-	-	-	-	**5.84**	m3	**6.57**
Reinforced in-situ lightweight concrete; 20.5 N/mm2; vibrated										
Walls										
thickness exceeding 450mm	88.00	2.50	**90.20**	-	4.85	**51.94**	2.81	**144.96**	m3	**163.08**
thickness 150 - 450mm	88.00	2.50	**90.20**	-	5.10	**54.62**	2.81	**147.63**	m3	**166.09**
thickness not exceeding 150mm	88.00	2.50	**90.20**	-	5.60	**59.98**	2.81	**152.99**	m3	**172.11**
Beam casings										
isolated	88.00	2.50	**90.20**	-	6.05	**64.80**	2.81	**157.81**	m3	**177.53**
Reinforced in-situ lightweight concrete; 26.0 N/mm2; vibrated										
Slabs										
thickness exceeding 450mm	92.23	2.50	**94.54**	-	4.40	**47.12**	2.81	**144.47**	m3	**162.53**
thickness 150 - 450mm	92.23	2.50	**94.54**	-	4.50	**48.20**	2.81	**145.54**	m3	**163.73**
thickness not exceeding 150mm	92.23	2.50	**94.54**	-	5.10	**54.62**	2.81	**151.97**	m3	**170.96**
Reinforced in-situ lightweight concrete; 41.5 N/mm2; vibrated										
Beams										
isolated	100.70	2.50	**103.22**	-	5.50	**58.91**	2.81	**164.93**	m3	**185.55**
Polythene sheeting as temporary protection to surface of concrete (use and waste)										
Polythene sheeting and laying										
125mu	0.35	10.00	**0.38**	-	0.03	**0.32**	-	**0.70**	m2	**0.79**

Labour hourly rates: (except Specialists) Craft Operatives £14.21 Labourer £10.71 Rates are national average prices. Refer to REGIONAL VARIATIONS for indicative levels of overall pricing in regions	MATERIALS			LABOUR				RATES		
	Del to Site £	Waste %	Material Cost £	Craft Optve Hrs	Lab Hrs	Labour Cost £	Sunds £	Nett Rate £	Unit	Gross Rate (+12.5%) £
E10: MIXING/CASTING/CURING IN-SITU CONCRETE - (SUNDRIES) Cont.										
Waterproof building paper, BS 1521; laying on hardcore to receive concrete										
Waterproof building paper and laying										
Grade B1F	0.34	25.00	**0.43**	-	0.06	**0.64**	-	**1.07**	m2	**1.21**
Grade B2	1.33	25.00	**1.67**	-	0.06	**0.64**	-	**2.31**	m2	**2.60**
Polythene sheeting, laying on hardcore to receive concrete										
Polythene sheeting and laying										
300mu	0.56	20.00	**0.67**	-	0.06	**0.64**	-	**1.31**	m2	**1.48**
E20: FORMWORK FOR IN-SITU CONCRETE										
Formwork and basic finish										
Sides of foundations; plain vertical										
height exceeding 1.00m	3.48	12.50	**3.92**	1.93	0.39	**31.60**	0.91	**36.43**	m2	**40.99**
height not exceeding 250mm	1.08	12.50	**1.22**	0.58	0.12	**9.53**	0.28	**11.03**	m	**12.41**
height 250 - 500mm	1.90	12.50	**2.14**	1.06	0.21	**17.31**	0.53	**19.98**	m	**22.48**
height 0.50m - 1.00m	3.56	12.50	**4.01**	1.97	0.40	**32.28**	0.91	**37.20**	m	**41.85**
Sides of foundations; plain vertical; left in										
height exceeding 1.00m	15.71	-	**15.71**	1.10	0.22	**17.99**	0.55	**34.25**	m2	**38.53**
height not exceeding 250mm	4.68	-	**4.68**	0.33	0.07	**5.44**	0.20	**10.32**	m	**11.61**
height 250 - 500mm	8.65	-	**8.65**	0.61	0.12	**9.95**	0.29	**18.90**	m	**21.26**
height 0.50m - 1.00m	16.09	-	**16.09**	1.13	0.23	**18.52**	0.57	**35.18**	m	**39.58**
Sides of ground beams and edges of beds; plain vertical										
height exceeding 1.00m	3.48	12.50	**3.92**	1.93	0.39	**31.60**	0.91	**36.43**	m2	**40.99**
height not exceeding 250mm	1.08	12.50	**1.22**	0.58	0.12	**9.53**	0.28	**11.03**	m	**12.41**
height 250 - 500mm	1.90	12.50	**2.14**	1.06	0.21	**17.31**	0.53	**19.98**	m	**22.48**
height 0.50m - 1.00m	3.56	12.50	**4.01**	1.97	0.40	**32.28**	0.91	**37.20**	m	**41.85**
Sides of ground beams and edges of beds; plain vertical; left in										
height exceeding 1.00m	15.71	-	**15.71**	1.10	0.22	**17.99**	0.53	**34.22**	m2	**38.50**
height not exceeding 250mm	4.68	-	**4.68**	0.33	0.07	**5.44**	0.20	**10.32**	m	**11.61**
height 250 - 500mm	8.65	-	**8.65**	0.61	0.12	**9.95**	0.29	**18.90**	m	**21.26**
height 0.50m - 1.00m	16.09	-	**16.09**	1.13	0.23	**18.52**	0.57	**35.18**	m	**39.58**
Edges of suspended slabs; plain vertical										
height not exceeding 250mm	1.30	12.50	**1.47**	0.73	0.14	**11.87**	0.32	**13.66**	m	**15.36**
height 250 - 500mm	2.43	12.50	**2.74**	1.32	0.26	**21.54**	0.63	**24.91**	m	**28.03**
extra; recesses; plain rectangular										
12 x 25mm	0.02	-	**0.02**	0.05	-	**0.71**	-	**0.73**	m	**0.82**
25 x 50mm	0.05	-	**0.05**	0.07	-	**0.99**	-	**1.04**	m	**1.17**
50 x 75mm	0.17	-	**0.17**	0.09	-	**1.28**	-	**1.45**	m	**1.63**
50 x 100mm	0.28	-	**0.28**	0.12	-	**1.71**	-	**1.98**	m	**2.23**
extra; rebates; plain rectangular										
12 x 25mm	0.02	-	**0.02**	0.07	-	**0.99**	-	**1.02**	m	**1.14**
25 x 50mm	0.05	-	**0.05**	0.08	-	**1.14**	-	**1.18**	m	**1.33**
50 x 75mm	0.18	-	**0.18**	0.11	-	**1.56**	-	**1.75**	m	**1.97**
50 x 100mm	0.29	-	**0.29**	0.14	-	**1.99**	-	**2.28**	m	**2.56**
extra; chamfers										
60mm wide	0.08	-	**0.08**	0.05	-	**0.71**	-	**0.79**	m	**0.89**
75mm wide	0.10	-	**0.10**	0.07	-	**0.99**	-	**1.10**	m	**1.23**
100mm wide	0.21	-	**0.21**	0.12	-	**1.71**	-	**1.91**	m	**2.15**
Edges of suspended slabs; plain vertical; curved 10m radius										
height not exceeding 250mm	1.96	12.50	**2.21**	1.05	0.22	**17.28**	0.55	**20.03**	m	**22.54**
Edges of suspended slabs; plain vertical; curved 1m radius										
height not exceeding 250mm	3.31	12.50	**3.72**	1.73	0.36	**28.44**	0.88	**33.04**	m	**37.17**
Edges of suspended slabs; plain vertical; curved 20m radius										
height not exceeding 250mm	1.48	12.50	**1.66**	0.78	0.17	**12.90**	0.35	**14.92**	m	**16.78**
Sides of upstands; plain vertical										
height not exceeding 250mm	1.55	12.50	**1.74**	0.79	0.17	**13.05**	0.35	**15.14**	m	**17.03**
height 250 - 500mm	1.81	12.50	**2.04**	0.92	0.20	**15.22**	0.47	**17.72**	m	**19.94**
extra; recesses; plain rectangular										
12 x 25mm	0.02	-	**0.02**	0.05	-	**0.71**	-	**0.73**	m	**0.82**
25 x 50mm	0.03	-	**0.03**	0.08	-	**1.14**	-	**1.17**	m	**1.32**
50 x 75mm	0.17	-	**0.17**	0.10	-	**1.42**	-	**1.59**	m	**1.79**
50 x 100mm	0.24	-	**0.24**	0.12	-	**1.71**	-	**1.95**	m	**2.19**
extra; rebates; plain rectangular										
12 x 25mm	0.02	-	**0.02**	0.08	-	**1.14**	-	**1.16**	m	**1.30**
25 x 50mm	0.03	-	**0.03**	0.09	-	**1.28**	-	**1.31**	m	**1.48**
50 x 75mm	0.18	-	**0.18**	0.11	-	**1.56**	-	**1.75**	m	**1.97**
50 x 100mm	0.28	-	**0.28**	0.14	-	**1.99**	-	**2.27**	m	**2.55**
extra; chamfers										
60mm wide	0.08	-	**0.08**	0.06	-	**0.85**	-	**0.93**	m	**1.05**
75mm wide	0.10	-	**0.10**	0.08	-	**1.14**	-	**1.24**	m	**1.39**
100mm wide	0.21	-	**0.21**	0.12	-	**1.71**	-	**1.91**	m	**2.15**
Machine bases and plinths; plain vertical										
height exceeding 1.00m	3.84	12.50	**4.32**	2.12	0.42	**34.62**	1.02	**39.96**	m2	**44.96**

Labour hourly rates: (except Specialists) Craft Operatives £14.21 Labourer £10.71 Rates are national average prices. Refer to REGIONAL VARIATIONS for indicative levels of overall pricing in regions	MATERIALS			LABOUR				RATES		
	Del to Site £	Waste %	Material Cost £	Craft Optve Hrs	Lab Hrs	Labour Cost £	Sunds £	Nett Rate £	Unit	Gross Rate (+12.5%) £
E20: FORMWORK FOR IN-SITU CONCRETE Cont.										
Formwork and basic finish Cont.										
Machine bases and plinths; plain vertical Cont.										
height not exceeding 250mm	1.13	12.50	**1.27**	0.62	0.12	**10.10**	0.29	**11.66**	m	**13.12**
height 250 - 500mm	2.10	12.50	**2.36**	1.17	0.22	**18.98**	0.55	**21.90**	m	**24.63**
height 0.50m - 1.00m	6.01	12.50	**6.76**	2.18	0.43	**35.58**	1.06	**43.41**	m	**48.83**
Soffits of slabs; horizontal										
slab thickness not exceeding 200mm; height to soffit										
1.50 - 3.00m	10.53	10.00	**11.59**	1.65	0.33	**26.98**	1.28	**39.84**	m2	**44.82**
3.00 - 4.50m	10.53	10.00	**11.59**	1.89	0.39	**31.03**	1.74	**44.36**	m2	**49.91**
4.50 - 6.00m	10.53	10.00	**11.59**	2.13	0.44	**34.98**	2.22	**48.79**	m2	**54.89**
slab thickness 200 - 300mm; height to soffit										
1.50 - 3.00m	11.59	10.00	**12.75**	1.82	0.36	**29.72**	1.44	**43.91**	m2	**49.40**
3.00 - 4.50m	11.59	10.00	**12.75**	2.06	0.42	**33.77**	1.88	**48.41**	m2	**54.46**
4.50 - 6.00m	11.59	10.00	**12.75**	2.30	0.47	**37.72**	2.35	**52.82**	m2	**59.42**
Soffits of slabs; horizontal; left in										
slab thickness not exceeding 200mm; height to soffit										
not exceeding 1.50m	28.25	-	**28.25**	1.10	0.22	**17.99**	0.82	**47.05**	m2	**52.93**
1.50 - 3.00m	28.25	-	**28.25**	1.10	0.22	**17.99**	0.91	**47.15**	m2	**53.04**
Soffits of slabs; horizontal; with frequent uses of prefabricated panels										
slab thickness not exceeding 200mm; height to soffit										
1.50 - 3.00m	3.13	15.00	**3.59**	1.65	0.33	**26.98**	1.28	**31.85**	m2	**35.83**
3.00 - 4.50m	3.13	15.00	**3.59**	1.89	0.39	**31.03**	1.78	**36.41**	m2	**40.96**
4.50 - 6.00m	3.13	15.00	**3.59**	2.13	0.44	**34.98**	2.22	**40.80**	m2	**45.90**
slab thickness 200 - 300mm; height to soffit										
1.50 - 3.00m	3.43	15.00	**3.94**	1.82	0.36	**29.72**	1.44	**35.10**	m2	**39.49**
3.00 - 4.50m	3.43	15.00	**3.94**	2.06	0.42	**33.77**	1.88	**39.60**	m2	**44.54**
4.50 - 6.00m	3.43	15.00	**3.94**	2.30	0.47	**37.72**	2.35	**44.01**	m2	**49.51**
Soffits of slabs; sloping not exceeding 15 degrees										
slab thickness not exceeding 200mm; height to soffit										
1.50 - 3.00m	11.60	10.00	**12.76**	1.82	0.36	**29.72**	1.43	**43.91**	m2	**49.40**
3.00 - 4.50m	11.60	10.00	**12.76**	2.06	0.42	**33.77**	1.86	**48.39**	m2	**54.44**
4.50 - 6.00m	11.60	10.00	**12.76**	2.36	0.47	**38.57**	2.32	**53.65**	m2	**60.36**
Soffits of slabs; sloping exceeding 15 degrees										
slab thickness not exceeding 200mm; height to soffit										
1.50 - 3.00m	12.13	10.00	**13.35**	1.90	0.39	**31.18**	1.64	**46.16**	m2	**51.93**
3.00 - 4.50m	12.13	10.00	**13.35**	2.15	0.44	**35.26**	2.09	**50.71**	m2	**57.04**
4.50 - 6.00m	12.13	10.00	**13.35**	2.39	0.50	**39.32**	2.54	**55.20**	m2	**62.10**
Soffits of landings; horizontal										
slab thickness not exceeding 200mm; height to soffit										
not exceeding 1.50m	10.53	10.00	**11.59**	1.98	0.40	**32.42**	1.28	**45.28**	m2	**50.94**
1.50 - 3.00m	10.53	10.00	**11.59**	2.22	0.45	**36.37**	1.74	**49.69**	m2	**55.91**
3.00 - 4.50m	10.53	10.00	**11.59**	2.46	0.51	**40.42**	2.22	**54.23**	m2	**61.01**
Top formwork										
sloping exceeding 15 degrees	4.08	10.00	**4.49**	1.65	0.33	**26.98**	0.73	**32.20**	m2	**36.22**
Walls; vertical										
plain	8.53	15.00	**9.81**	1.76	0.35	**28.76**	1.18	**39.75**	m2	**44.72**
plain; height exceeding 3.00m above floor level	7.87	15.00	**9.05**	1.76	0.35	**28.76**	1.33	**39.14**	m2	**44.03**
interrupted	7.87	15.00	**9.05**	1.93	0.39	**31.60**	1.31	**41.96**	m2	**47.20**
interrupted; height exceeding 3.00m above floor level	8.40	15.00	**9.66**	1.93	0.39	**31.60**	1.45	**42.71**	m2	**48.05**
extra; recesses; plain rectangular										
12 x 25mm	0.03	-	**0.03**	0.06	-	**0.85**	-	**0.88**	m	**0.99**
25 x 50mm	0.05	-	**0.05**	0.08	-	**1.14**	-	**1.19**	m	**1.34**
50 x 75mm	0.20	-	**0.20**	0.10	-	**1.42**	-	**1.62**	m	**1.82**
50 x 100mm	0.28	-	**0.28**	0.12	-	**1.71**	-	**1.98**	m	**2.23**
Walls; vertical; curved 1m radius										
plain	17.45	15.00	**20.07**	4.40	0.88	**71.95**	2.95	**94.96**	m2	**106.83**
Walls; vertical; curved 10m radius										
plain	10.50	15.00	**12.08**	2.64	0.53	**43.19**	1.80	**57.07**	m2	**64.20**
Walls; vertical; curved 20m radius										
plain	7.85	15.00	**9.03**	1.98	0.46	**33.06**	1.33	**43.43**	m2	**48.86**
Walls; battered										
plain	8.36	15.00	**9.61**	2.11	0.42	**34.48**	1.45	**45.54**	m2	**51.24**
Beams; attached to slabs										
regular shaped; rectangular; height to soffit										
1.50 - 3.00m	7.36	10.00	**8.10**	2.48	0.45	**40.06**	1.13	**49.29**	m2	**55.45**
3.00 - 4.50m	7.36	10.00	**8.10**	2.48	0.45	**40.06**	1.39	**49.55**	m2	**55.74**
regular shaped; rectangular; with 50mm wide chamfers; height to soffit										
1.50 - 3.00m	7.49	10.00	**8.24**	2.54	0.45	**40.91**	1.13	**50.29**	m2	**56.58**
3.00 - 4.50m	7.49	10.00	**8.24**	2.54	0.45	**40.91**	1.39	**50.55**	m2	**56.86**

Labour hourly rates: (except Specialists) Craft Operatives £14.21 Labourer £10.71 Rates are national average prices. Refer to REGIONAL VARIATIONS for indicative levels of overall pricing in regions	MATERIALS			LABOUR				RATES		
	Del to Site £	Waste %	Material Cost £	Craft Optve Hrs	Lab Hrs	Labour Cost £	Sunds £	Nett Rate £	Unit	Gross Rate (+12.5%) £
E20: FORMWORK FOR IN-SITU CONCRETE Cont.										
Formwork and basic finish Cont.										
Beams; isolated										
regular shaped; rectangular; height to soffit										
1.50 - 3.00m	7.72	10.00	**8.49**	2.59	0.47	**41.84**	1.13	**51.46**	m2	**57.90**
3.00 - 4.50m	7.72	10.00	**8.49**	2.59	0.47	**41.84**	1.39	**51.72**	m2	**58.18**
regular shaped; rectangular; with 50mm wide chamfers; height to soffit										
1.50 - 3.00m	7.87	10.00	**8.65**	2.65	0.47	**42.69**	1.13	**52.48**	m2	**59.04**
3.00 - 4.50m	7.87	10.00	**8.65**	2.65	0.47	**42.69**	1.39	**52.74**	m2	**59.33**
Columns; attached to walls										
regular shaped; rectangular	5.11	10.00	**5.62**	2.53	0.46	**40.88**	1.06	**47.56**	m2	**53.51**
regular shaped; rectangular, height exceeding 3.00m above floor level	4.75	10.00	**5.23**	2.53	0.46	**40.88**	1.29	**47.39**	m2	**53.31**
Columns; isolated										
regular shaped; rectangular	5.08	10.00	**5.59**	2.48	0.45	**40.06**	1.06	**46.71**	m2	**52.55**
regular shaped; rectangular, height exceeding 3.00m above floor level	5.08	10.00	**5.59**	2.48	0.45	**40.06**	1.29	**46.94**	m2	**52.80**
regular shaped; rectangular; with 50mm wide chamfers	5.49	10.00	**6.04**	2.63	0.45	**42.19**	1.13	**49.37**	m2	**55.54**
regular shaped; rectangular; with 50mm wide chamfers; height exceeding 3.00m above floor level	5.49	10.00	**6.04**	2.63	0.45	**42.19**	1.38	**49.61**	m2	**55.81**
Column casings; attached to walls										
regular shaped; rectangular	5.08	10.00	**5.59**	2.53	0.46	**40.88**	1.06	**47.53**	m2	**53.47**
regular shaped; rectangular, height exceeding 3.00m above floor level	5.08	10.00	**5.59**	2.53	0.46	**40.88**	1.29	**47.75**	m2	**53.72**
Column casings; isolated										
regular shaped; rectangular	5.47	10.00	**6.02**	2.48	0.45	**40.06**	1.13	**47.21**	m2	**53.11**
regular shaped; rectangular, height exceeding 3.00m above floor level	5.47	10.00	**6.02**	2.48	0.45	**40.06**	1.29	**47.36**	m2	**53.28**
regular shaped; rectangular; with 50mm wide chamfers	5.49	10.00	**6.04**	2.63	0.45	**42.19**	1.13	**49.37**	m2	**55.54**
regular shaped; rectangular; with 50mm wide chamfers; height exceeding 3.00m above floor level	5.49	10.00	**6.04**	2.63	0.45	**42.19**	1.29	**49.52**	m2	**55.71**
Extra over a basic finish for a fine formed finish										
slabs	-	-	-	-	0.39	**4.18**	-	**4.18**	m2	**4.70**
walls	-	-	-	-	0.39	**4.18**	-	**4.18**	m2	**4.70**
beams	-	-	-	-	0.39	**4.18**	-	**4.18**	m2	**4.70**
columns	-	-	-	-	0.39	**4.18**	-	**4.18**	m2	**4.70**
Wall kickers										
straight	1.02	15.00	**1.17**	0.29	0.08	**4.98**	0.21	**6.36**	m	**7.15**
curved 2m radius	2.49	15.00	**2.86**	0.73	0.16	**12.09**	0.48	**15.43**	m	**17.36**
curved 10m radius	1.51	15.00	**1.74**	0.44	0.09	**7.22**	0.29	**9.25**	m	**10.40**
curved 20m radius	1.11	15.00	**1.27**	0.33	0.08	**5.55**	0.25	**7.06**	m	**7.95**
Suspended wall kickers										
straight	1.10	15.00	**1.26**	0.32	0.07	**5.30**	0.23	**6.79**	m	**7.64**
curved 2m radius	2.76	15.00	**3.17**	0.80	0.18	**13.30**	0.54	**17.00**	m	**19.13**
curved 10m radius	1.63	15.00	**1.87**	0.44	0.10	**7.32**	0.32	**9.51**	m	**10.70**
curved 20m radius	1.26	15.00	**1.45**	0.36	0.09	**6.08**	0.27	**7.79**	m	**8.77**
Wall ends, soffits and steps in walls; plain										
width not exceeding 250mm	2.06	15.00	**2.37**	0.58	0.12	**9.53**	0.37	**12.28**	m	**13.81**
Openings in walls; plain										
width not exceeding 250mm	2.06	15.00	**2.37**	0.58	0.12	**9.53**	0.37	**12.28**	m	**13.81**
Stairflights; 1000mm wide; 155mm thick waist; 178mm risers; includes formwork to soffits, risers and strings										
strings 300mm wide	15.72	10.00	**17.29**	6.66	1.33	**108.88**	2.69	**128.87**	m	**144.98**
string 300mm wide; junction with wall	15.72	10.00	**17.29**	6.66	1.33	**108.88**	2.69	**128.87**	m	**144.98**
Stairflights; 1500mm wide; 180mm thick waist; 178mm risers; includes formwork to soffits, risers and strings										
strings 325mm wide	19.68	10.00	**21.65**	8.32	1.66	**136.01**	3.37	**161.02**	m	**181.15**
string 325mm wide; junction with wall	19.68	10.00	**21.65**	8.32	1.66	**136.01**	3.37	**161.02**	m	**181.15**
Stairflights; 1000mm wide; 155mm thick waist; 178mm undercut risers; includes formwork to soffits, risers and strings										
strings 300mm wide	15.72	10.00	**17.29**	6.66	1.33	**108.88**	2.69	**128.87**	m	**144.98**
string 300mm wide; junction with wall	15.72	10.00	**17.29**	6.66	1.33	**108.88**	2.69	**128.87**	m	**144.98**
Stairflights; 1500mm wide; 180mm thick waist; 178mm undercut risers; includes formwork to soffits, risers and strings										
strings 325mm wide	19.67	10.00	**21.63**	8.32	1.66	**136.01**	3.37	**161.01**	m	**181.13**
string 325mm wide; junction with wall	19.67	10.00	**21.63**	8.32	1.66	**136.01**	3.37	**161.01**	m	**181.13**
Mortices; rectangular										
girth not exceeding 500mm; depth not exceeding 250mm	0.93	15.00	**1.07**	0.53	0.11	**8.71**	0.26	**10.04**	nr	**11.30**
girth 0.50 - 1.00m; depth 250 - 500mm	3.03	15.00	**3.49**	1.69	0.34	**27.66**	0.80	**31.94**	nr	**35.93**
girth 0.50 - 1.00m; depth 0.50 - 1.00m	5.99	15.00	**6.89**	3.39	0.68	**55.45**	1.61	**63.95**	nr	**71.95**

Labour hourly rates: (except Specialists) Craft Operatives £14.21 Labourer £10.71 Rates are national average prices. Refer to REGIONAL VARIATIONS for indicative levels of overall pricing in regions	MATERIALS Del to Site £	 Waste %	 Material Cost £	LABOUR Craft Optve Hrs	 Lab Hrs	 Labour Cost £	 Sunds £	RATES Nett Rate £	 Unit	 Gross Rate (+12.5%) £
E20: FORMWORK FOR IN-SITU CONCRETE Cont.										
Formwork and basic finish Cont.										
Mortices; circular										
girth not exceeding 500mm; depth not exceeding 250mm	1.88	15.00	**2.16**	1.06	0.22	**17.42**	0.50	**20.08**	nr	**22.59**
Holes; rectangular										
girth not exceeding 500mm; depth not exceeding 250mm	0.93	15.00	**1.07**	0.53	0.11	**8.71**	0.26	**10.04**	nr	**11.30**
girth not exceeding 500mm; depth 250 - 500mm	1.90	15.00	**2.19**	0.97	0.20	**15.93**	0.48	**18.59**	nr	**20.92**
girth 0.50 - 1.00m; depth not exceeding 250mm	1.64	15.00	**1.88**	0.94	0.19	**15.39**	0.46	**17.73**	nr	**19.95**
girth 0.50 - 1.00m; depth 250 - 500mm	3.03	15.00	**3.49**	1.69	0.34	**27.66**	0.80	**31.94**	nr	**35.93**
girth 1.00m - 2.00m; depth not exceeding 250mm	2.35	15.00	**2.71**	1.33	0.26	**21.68**	0.63	**25.02**	nr	**28.15**
girth 2.00m - 3.00m; depth not exceeding 250mm	3.07	15.00	**3.53**	1.73	0.34	**28.22**	0.83	**32.58**	nr	**36.66**
girth 3.00m - 4.00m; depth not exceeding 250mm	3.85	15.00	**4.43**	2.12	0.43	**34.73**	0.99	**40.15**	nr	**45.17**
girth 1.00m - 2.00m; depth 250 - 500mm	4.30	15.00	**4.95**	2.42	0.48	**39.53**	1.13	**45.61**	nr	**51.31**
girth 2.00m - 3.00m; depth 250 - 500mm	5.59	15.00	**6.43**	3.14	0.63	**51.37**	1.49	**59.28**	nr	**66.69**
girth 3.00m - 4.00m; depth 250 - 500mm	6.94	15.00	**7.98**	3.85	0.77	**62.96**	1.80	**72.74**	nr	**81.83**
Holes; circular										
girth not exceeding 500mm; depth not exceeding 250mm	1.88	15.00	**2.16**	1.06	0.22	**17.42**	0.50	**20.08**	nr	**22.59**
girth not exceeding 500mm; depth 250 - 500mm	3.43	15.00	**3.94**	1.94	0.40	**31.85**	0.91	**36.71**	nr	**41.29**
girth 0.50 - 1.00m; depth not exceeding 250mm	3.32	15.00	**3.82**	1.87	0.37	**30.54**	0.88	**35.23**	nr	**39.64**
girth 0.50 - 1.00m; depth 250 - 500mm	5.99	15.00	**6.89**	3.39	0.68	**55.45**	1.61	**63.95**	nr	**71.95**
girth 1.00m - 2.00m; depth not exceeding 250mm	4.73	15.00	**5.44**	2.66	0.53	**43.47**	1.32	**50.24**	nr	**56.51**
girth 2.00m - 3.00m; depth not exceeding 250mm	6.14	15.00	**7.06**	3.45	0.68	**56.31**	1.64	**65.00**	nr	**73.13**
girth 3.00m - 4.00m; depth not exceeding 250mm	7.72	15.00	**8.87**	4.25	0.86	**69.60**	2.01	**80.49**	nr	**90.55**
girth 1.00m - 2.00m; depth 250 - 500mm	8.59	15.00	**9.88**	4.84	0.97	**79.17**	2.20	**91.25**	nr	**102.65**
girth 2.00m - 3.00m; depth 250 - 500mm	11.15	15.00	**12.83**	6.27	1.25	**102.48**	2.95	**118.26**	nr	**133.04**
girth 3.00m - 4.00m; depth 250 - 500mm	19.74	15.00	**22.70**	7.70	1.54	**125.91**	3.65	**152.26**	nr	**171.29**
Formwork and basic finish; coating with retarding agent										
Sides of foundations; plain vertical										
height exceeding 1.00m	4.21	12.50	**4.74**	1.93	0.65	**34.39**	0.91	**40.03**	m2	**45.04**
height not exceeding 250mm	1.15	12.50	**1.30**	0.58	0.15	**9.85**	0.28	**11.43**	m	**12.85**
height 250 - 500mm	2.18	12.50	**2.45**	1.06	0.31	**18.38**	0.51	**21.35**	m	**24.02**
height 0.50m - 1.00m	4.13	12.50	**4.65**	1.97	0.59	**34.31**	0.91	**39.87**	m	**44.85**
Sides of ground beams and edges of beds; plain vertical										
height exceeding 1.00m	4.23	12.50	**4.76**	1.93	0.65	**34.39**	0.91	**40.06**	m2	**45.07**
height not exceeding 250mm	1.15	12.50	**1.30**	0.58	0.15	**9.85**	0.28	**11.43**	m	**12.85**
height 250 - 500mm	2.18	12.50	**2.45**	1.06	0.31	**18.38**	0.51	**21.35**	m	**24.02**
height 0.50m - 1.00m	4.13	12.50	**4.65**	1.97	0.59	**34.31**	0.91	**39.87**	m	**44.85**
Edges of suspended slabs; plain vertical										
height not exceeding 250mm	1.37	12.50	**1.54**	0.73	0.18	**12.30**	0.32	**14.16**	m	**15.93**
height 250 - 500mm	2.66	12.50	**3.00**	1.32	0.39	**22.93**	0.63	**26.56**	m	**29.88**
Edges of suspended slabs; plain vertical; curved 10m radius										
height not exceeding 250mm	3.39	12.50	**3.81**	1.82	0.40	**30.15**	1.37	**35.33**	m	**39.74**
Edges of suspended slabs; plain vertical; curved 1m radius										
height not exceeding 250mm	2.05	12.50	**2.31**	1.10	0.25	**18.31**	0.55	**21.17**	m	**23.81**
Edges of suspended slabs; plain vertical; curved 20m radius										
height not exceeding 250mm	1.56	12.50	**1.75**	0.81	0.20	**13.65**	0.35	**15.75**	m	**17.72**
Sides of upstands; plain vertical										
height not exceeding 250mm	1.70	12.50	**1.91**	0.83	0.20	**13.94**	0.35	**16.19**	m	**18.22**
height 250 - 500mm	2.09	12.50	**2.35**	0.97	0.30	**17.00**	0.48	**19.82**	m	**22.30**
Machine bases and plinths; plain vertical										
height exceeding 1.00m	4.56	12.50	**5.13**	2.12	0.68	**37.41**	1.02	**43.55**	m2	**49.00**
height not exceeding 250mm	1.35	12.50	**1.52**	0.64	0.15	**10.70**	0.29	**12.51**	m	**14.07**
height 250 - 500mm	2.37	12.50	**2.66**	1.17	0.32	**20.05**	0.55	**23.26**	m	**26.17**
height 0.50m - 1.00m	4.44	12.50	**5.00**	2.18	0.63	**37.73**	1.06	**43.79**	m	**49.26**
Soffits of slabs; horizontal										
slab thickness not exceeding 200mm; height to soffit										
1.50 - 3.00m	11.27	10.00	**12.39**	1.65	0.59	**29.77**	1.46	**43.62**	m2	**49.08**
3.00 - 4.50m	11.27	10.00	**12.39**	1.89	0.65	**33.82**	1.74	**47.96**	m2	**53.95**
4.50 - 6.00m	11.27	10.00	**12.39**	2.13	0.76	**38.41**	2.22	**53.02**	m2	**59.65**
slab thickness 200 - 300mm; height to soffit										
1.50 - 3.00m	12.33	10.00	**13.56**	1.82	0.63	**32.61**	1.44	**47.61**	m2	**53.56**
3.00 - 4.50m	12.33	10.00	**13.56**	2.06	0.68	**36.56**	1.88	**52.00**	m2	**58.50**
4.50 - 6.00m	12.33	10.00	**13.56**	2.30	0.74	**40.61**	2.35	**56.52**	m2	**63.59**
Soffits of slabs; horizontal; with frequent uses of prefabricated panels										
slab thickness not exceeding 200mm; height to soffit										
1.50 - 3.00m	3.87	15.00	**4.44**	1.65	0.59	**29.77**	1.46	**35.67**	m2	**40.13**
3.00 - 4.50m	3.87	15.00	**4.44**	1.89	0.65	**33.82**	1.74	**40.01**	m2	**45.01**
4.50 - 6.00m	3.87	15.00	**4.44**	2.13	0.70	**37.76**	2.22	**44.43**	m2	**49.99**
slab thickness 200 - 300mm; height to soffit										
1.50 - 3.00m	4.60	15.00	**5.29**	1.82	0.86	**35.07**	1.44	**41.81**	m2	**47.03**

Labour hourly rates: (except Specialists) Craft Operatives £14.21 Labourer £10.71 Rates are national average prices. Refer to REGIONAL VARIATIONS for indicative levels of overall pricing in regions	MATERIALS			LABOUR				RATES		
	Del to Site £	Waste %	Material Cost £	Craft Optve Hrs	Lab Hrs	Labour Cost £	Sunds £	Nett Rate £	Unit	Gross Rate (+12.5%) £
E20: FORMWORK FOR IN-SITU CONCRETE Cont.										
Formwork and basic finish; coating with retarding agent Cont.										
Soffits of slabs; horizontal; with frequent uses of prefabricated panels Cont.										
slab thickness 200 - 300mm; height to soffit Cont.										
3.00 - 4.50m	4.60	15.00	**5.29**	2.06	0.91	**39.02**	1.88	**46.20**	m2	**51.97**
4.50 - 6.00m	4.60	15.00	**5.29**	2.30	0.97	**43.07**	2.35	**50.72**	m2	**57.06**
Soffits of slabs; sloping not exceeding 15 degrees										
slab thickness not exceeding 200mm; height to soffit										
1.50 - 3.00m	12.37	15.00	**14.22**	1.82	0.63	**32.61**	1.43	**48.26**	m2	**54.29**
3.00 - 4.50m	12.37	15.00	**14.22**	2.06	0.68	**36.56**	1.86	**52.64**	m2	**59.21**
4.50 - 6.00m	12.37	15.00	**14.22**	2.30	0.74	**40.61**	2.32	**57.14**	m2	**64.29**
Soffits of slabs; sloping exceeding 15 degrees										
slab thickness not exceeding 200mm; height to soffit										
1.50 - 3.00m	12.88	15.00	**14.82**	1.90	0.65	**33.96**	1.64	**50.42**	m2	**56.72**
3.00 - 4.50m	12.88	15.00	**14.82**	2.15	0.70	**38.05**	2.09	**54.96**	m2	**61.83**
4.50 - 6.00m	12.88	15.00	**14.82**	2.33	0.76	**41.25**	2.54	**58.60**	m2	**65.93**
Soffits of landings; horizontal										
slab thickness not exceeding 200mm; height to soffit										
not exceeding 1.50m	11.27	15.00	**12.96**	1.98	0.66	**35.20**	1.46	**49.63**	m2	**55.83**
1.50 - 3.00m	11.27	15.00	**12.96**	2.22	0.72	**39.26**	1.74	**53.96**	m2	**60.70**
3.00 - 4.50m	11.27	15.00	**12.96**	2.46	0.77	**43.20**	2.22	**58.38**	m2	**65.68**
Top formwork										
sloping exceeding 15 degrees	4.68	15.00	**5.39**	1.65	0.59	**29.77**	0.75	**35.90**	m2	**40.39**
Walls; vertical										
plain	7.91	15.00	**9.09**	1.76	0.62	**31.65**	1.18	**41.92**	m2	**47.16**
plain; height exceeding 3.00m above floor level	8.72	15.00	**10.02**	1.76	0.62	**31.65**	1.33	**43.01**	m2	**48.38**
interrupted	8.53	15.00	**9.81**	1.93	0.65	**34.39**	1.31	**45.51**	m2	**51.20**
interrupted; height exceeding 3.00m above floor level	9.13	15.00	**10.50**	1.93	0.65	**34.39**	1.45	**46.34**	m2	**52.13**
Walls; vertical; curved 1m radius										
plain	18.26	15.00	**21.00**	4.40	1.14	**74.73**	2.95	**98.68**	m2	**111.01**
Walls; vertical; curved 10m radius										
plain	11.37	15.00	**13.07**	2.64	0.79	**45.98**	1.80	**60.85**	m2	**68.45**
Walls; vertical; curved 20m radius										
plain	8.70	15.00	**10.01**	1.98	0.66	**35.20**	1.21	**46.42**	m2	**52.22**
Walls; battered										
plain	9.21	15.00	**10.59**	2.11	0.68	**37.27**	1.45	**49.31**	m2	**55.47**
Beams; attached to slabs										
regular shaped; rectangular; height to soffit										
1.50 - 3.00m	8.19	10.00	**9.00**	2.48	0.76	**43.38**	1.13	**53.52**	m2	**60.21**
3.00 - 4.50m	8.19	10.00	**9.00**	2.48	0.76	**43.38**	1.39	**53.78**	m2	**60.50**
regular shaped; rectangular; with 50mm wide chamfers; height to soffit										
1.50 - 3.00m	8.32	10.00	**9.15**	2.54	0.76	**44.23**	1.13	**54.52**	m2	**61.33**
3.00 - 4.50m	8.32	10.00	**9.15**	2.54	0.76	**44.23**	1.39	**54.77**	m2	**61.62**
Beams; isolated										
regular shaped; rectangular; height to soffit										
1.50 - 3.00m	8.50	10.00	**9.35**	2.59	0.78	**45.16**	1.13	**55.65**	m2	**62.60**
3.00 - 4.50m	8.50	10.00	**9.35**	2.59	0.78	**45.16**	1.39	**55.90**	m2	**62.89**
regular shaped; rectangular; with 50mm wide chamfers; height to soffit										
1.50 - 3.00m	9.42	10.00	**10.36**	2.65	0.78	**46.01**	1.13	**57.51**	m2	**64.69**
3.00 - 4.50m	9.42	10.00	**10.36**	2.65	0.78	**46.01**	1.39	**57.76**	m2	**64.98**
Columns; attached to walls										
regular shaped; rectangular	5.98	10.00	**6.57**	2.53	0.77	**44.20**	1.06	**51.84**	m2	**58.32**
regular shaped; rectangular, height exceeding 3.00m above floor level	5.98	10.00	**6.57**	2.68	0.78	**46.44**	1.29	**54.30**	m2	**61.08**
Columns; isolated										
regular shaped; rectangular	5.98	10.00	**6.57**	2.48	0.76	**43.38**	1.13	**51.09**	m2	**57.48**
regular shaped; rectangular, height exceeding 3.00m above floor level	5.98	10.00	**6.57**	2.63	0.78	**45.73**	1.29	**53.59**	m2	**60.29**
regular shaped; rectangular; with 50mm wide chamfers	6.31	10.00	**6.94**	2.63	0.76	**45.51**	1.08	**53.53**	m2	**60.22**
regular shaped; rectangular; with 50mm wide chamfers; height exceeding 3.00m above floor level	6.31	10.00	**6.94**	2.78	0.80	**48.07**	1.38	**56.39**	m2	**63.44**
Column casings; attached to walls										
regular shaped; rectangular	5.98	10.00	**6.57**	2.53	0.77	**44.20**	1.29	**52.06**	m2	**58.57**
regular shaped; rectangular, height exceeding 3.00m above floor level	5.98	10.00	**6.57**	2.68	0.78	**46.44**	1.06	**54.07**	m2	**60.83**
Column casings; isolated										
regular shaped; rectangular	5.98	10.00	**6.57**	2.48	0.76	**43.38**	1.07	**51.02**	m2	**57.40**
regular shaped; rectangular, height exceeding 3.00m above floor level	5.98	10.00	**6.57**	2.63	0.78	**45.73**	1.29	**53.59**	m2	**60.29**

Labour hourly rates: (except Specialists) Craft Operatives £14.21 Labourer £10.71 Rates are national average prices. Refer to REGIONAL VARIATIONS for indicative levels of overall pricing in regions	MATERIALS			LABOUR			RATES			
	Del to Site £	Waste %	Material Cost £	Craft Optve Hrs	Lab Hrs	Labour Cost £	Sunds £	Nett Rate £	Unit	Gross Rate (+12.5%) £
E20: FORMWORK FOR IN-SITU CONCRETE Cont.										
Formwork and basic finish; coating with retarding agent Cont.										
Column casings; isolated Cont.										
regular shaped; rectangular; with 50mm wide chamfers	6.31	10.00	**6.94**	2.63	0.76	**45.51**	1.13	**53.59**	m2	**60.28**
regular shaped; rectangular; with 50mm wide chamfers; height exceeding 3.00m above floor level	6.31	10.00	**6.94**	2.78	0.80	**48.07**	2.11	**57.12**	m2	**64.26**
Extra over a basic finish for a fine formed finish										
slabs	-	-	**-**	-	0.39	**4.18**	-	**4.18**	m2	**4.70**
walls	-	-	**-**	-	0.39	**4.18**	-	**4.18**	m2	**4.70**
beams	-	-	**-**	-	0.39	**4.18**	-	**4.18**	m2	**4.70**
columns	-	-	**-**	-	0.39	**4.18**	-	**4.18**	m2	**4.70**
Wall kickers										
straight	1.11	15.00	**1.27**	0.29	0.10	**5.19**	0.21	**6.68**	m	**7.51**
curved 2m radius	2.61	15.00	**3.00**	0.73	0.19	**12.41**	0.49	**15.90**	m	**17.88**
curved 10m radius	1.61	15.00	**1.86**	0.44	0.13	**7.64**	0.29	**9.79**	m	**11.02**
curved 20m radius	1.26	15.00	**1.45**	0.33	0.11	**5.87**	0.25	**7.56**	m	**8.50**
Suspended wall kickers										
straight	1.25	15.00	**1.43**	0.32	0.11	**5.73**	0.23	**7.39**	m	**8.32**
curved 2m radius	2.86	15.00	**3.29**	0.80	0.21	**13.62**	0.54	**17.45**	m	**19.63**
curved 10m radius	1.79	15.00	**2.06**	0.44	0.14	**7.75**	0.32	**10.12**	m	**11.39**
curved 20m radius	1.36	15.00	**1.57**	0.36	0.12	**6.40**	0.27	**8.24**	m	**9.26**
Wall ends, soffits and steps in walls; plain width not exceeding 250mm	2.35	15.00	**2.71**	0.58	0.19	**10.28**	0.37	**13.36**	m	**15.03**
Openings in walls; plain width not exceeding 250mm	2.35	15.00	**2.71**	0.58	0.19	**10.28**	0.37	**13.36**	m	**15.03**
Stairflights; 1000mm wide; 155mm thick waist; 178mm risers; includes formwork to soffits, risers and strings										
strings 300mm wide	17.35	10.00	**19.08**	6.66	1.91	**115.09**	2.69	**136.87**	m	**153.98**
string 300mm wide; junction with wall	17.35	10.00	**19.08**	6.66	1.91	**115.09**	2.69	**136.87**	m	**153.98**
Stairflights; 1500mm wide; 180mm thick waist; 178mm risers; includes formwork to soffits, risers and strings										
strings 325mm wide	21.68	10.00	**23.85**	8.32	2.39	**143.82**	3.37	**171.05**	m	**192.43**
string 325mm wide; junction with wall	21.68	10.00	**23.85**	8.32	2.39	**143.82**	3.37	**171.05**	m	**192.43**
Stairflights; 1000mm wide; 155mm thick waist; 178mm undercut risers; includes formwork to soffits, risers and strings										
strings 300mm wide	17.35	10.00	**19.08**	6.66	1.91	**115.09**	2.69	**136.87**	m	**153.98**
string 300mm wide; junction with wall	17.35	10.00	**19.08**	6.66	1.91	**115.09**	2.69	**136.87**	m	**153.98**
Stairflights; 1500mm wide; 180mm thick waist; 178mm undercut risers; includes formwork to soffits, risers and strings										
strings 325mm wide	21.68	10.00	**23.85**	8.32	2.39	**143.82**	3.37	**171.05**	m	**192.43**
string 325mm wide; junction with wall	21.68	10.00	**23.85**	8.32	2.39	**143.82**	3.37	**171.05**	m	**192.43**
Claymaster low density expanded polystyrene permanent formwork; fixing with Clayfix hooks - 3/m2										
Sides of foundations; plain vertical 75mm thick										
height exceeding 1.00m	14.54	5.00	**15.27**	-	0.07	**0.75**	0.74	**16.76**	m2	**18.85**
height not exceeding 250mm	3.64	5.00	**3.82**	-	0.02	**0.21**	0.25	**4.28**	m	**4.81**
height 250-500mm	7.27	5.00	**7.63**	-	0.08	**0.86**	0.25	**8.74**	m	**9.83**
height 0.50m-1.00m	10.91	5.00	**11.45**	-	0.08	**0.86**	0.49	**12.80**	m	**14.40**
Sides of foundations; plain vertical 100mm thick										
height exceeding 1.00m	21.75	5.00	**22.83**	-	0.09	**0.96**	0.74	**24.53**	m2	**27.60**
height not exceeding 250mm	5.44	5.00	**5.71**	-	0.03	**0.32**	0.25	**6.28**	m	**7.06**
height 250-500mm	10.87	5.00	**11.42**	-	0.05	**0.54**	0.25	**12.20**	m	**13.72**
height 0.50m-1.00m	16.31	5.00	**17.12**	-	0.10	**1.07**	0.49	**18.69**	m	**21.02**
Sides of foundations; plain vertical 150mm thick										
height exceeding 1.00m	32.61	5.00	**34.24**	-	0.11	**1.18**	0.74	**36.16**	m2	**40.68**
height not exceeding 250mm	8.15	5.00	**8.56**	-	0.03	**0.32**	0.25	**9.13**	m	**10.27**
height 250-500mm	16.31	5.00	**17.12**	-	0.07	**0.75**	0.25	**18.12**	m	**20.38**
height 0.50m-1.00m	24.46	5.00	**25.68**	-	0.12	**1.29**	0.49	**27.46**	m	**30.89**
To underside of foundations; laid on earth or hardcore; 75mm thick										
horizontal	14.54	5.00	**15.27**	-	0.05	**0.54**	-	**15.80**	m2	**17.78**
To underside of foundations; laid on earth or hardcore; 100mm thick										
horizontal	21.75	5.00	**22.83**	-	0.07	**0.75**	-	**23.58**	m2	**26.53**
To underside of foundations; laid on earth or hardcore; 150mm thick										
horizontal	32.61	5.00	**34.24**	-	0.08	**0.86**	-	**35.10**	m2	**39.49**

Labour hourly rates: (except Specialists) Craft Operatives £14.21 Labourer £10.71 Rates are national average prices. Refer to REGIONAL VARIATIONS for indicative levels of overall pricing in regions	MATERIALS			LABOUR				RATES		
	Del to Site £	Waste %	Material Cost £	Craft Optve Hrs	Lab Hrs	Labour Cost £	Sunds £	Nett Rate £	Unit	Gross Rate (+12.5%) £
E20: FORMWORK FOR IN-SITU CONCRETE Cont.										
Expamet Hy-rib permanent shuttering and reinforcement										
Reference 2611, 4.86 kg/m2 to soffits of slabs; horizontal; one rib side laps; 150mm end laps										
slab thickness 75mm; strutting and supports at 750mm centres; height to soffit										
1.50 - 3.00m	15.51	7.50	**16.67**	1.31	0.33	**22.15**	1.31	**40.13**	m2	**45.14**
3.00 - 4.50m	15.51	7.50	**16.67**	1.51	0.37	**25.42**	1.66	**43.75**	m2	**49.22**
slab thickness 100mm; strutting and supports at 650mm centres; height to soffit										
1.50 - 3.00m	15.51	7.50	**16.67**	1.33	0.33	**22.43**	1.52	**40.62**	m2	**45.70**
3.00 - 4.50m	15.51	7.50	**16.67**	1.53	0.39	**25.92**	2.06	**44.65**	m2	**50.23**
slab thickness 125mm; strutting and supports at 550mm centres; height to soffit										
1.50 - 3.00m	15.51	7.50	**16.67**	1.35	0.34	**22.82**	1.73	**41.23**	m2	**46.38**
3.00 - 4.50m	15.51	7.50	**16.67**	1.55	0.39	**26.20**	2.47	**45.34**	m2	**51.01**
slab thickness 150mm; strutting and supports at 450mm centres; height to soffit										
1.50 - 3.00m	15.51	7.50	**16.67**	1.38	0.34	**23.25**	1.92	**41.84**	m2	**47.07**
3.00 - 4.50m	15.51	7.50	**16.67**	1.58	0.40	**26.74**	2.88	**46.28**	m2	**52.07**
Reference 2411, 6.34 kg/m2 to soffits of slabs; horizontal; one rib side laps; 150mm end laps										
slab thickness 75mm; strutting and supports at 850mm centres; height to soffit										
1.50 - 3.00m	20.79	7.50	**22.35**	1.34	0.34	**22.68**	1.31	**46.34**	m2	**52.14**
3.00 - 4.50m	20.79	7.50	**22.35**	1.54	0.39	**26.06**	2.88	**51.29**	m2	**57.70**
slab thickness 100mm; strutting and supports at 750mm centres; height to soffit										
1.50 - 3.00m	20.79	7.50	**22.35**	1.36	0.34	**22.97**	1.52	**46.84**	m2	**52.69**
3.00 - 4.50m	20.79	7.50	**22.35**	1.57	0.40	**26.59**	2.06	**51.00**	m2	**57.38**
slab thickness 125mm; strutting and supports at 650mm centres; height to soffit										
1.50 - 3.00m	20.79	7.50	**22.35**	1.38	0.35	**23.36**	1.73	**47.44**	m2	**53.37**
3.00 - 4.50m	20.79	7.50	**22.35**	1.60	0.40	**27.02**	2.47	**51.84**	m2	**58.32**
slab thickness 150mm; strutting and supports at 550mm centres; height to soffit										
1.50 - 3.00m	20.79	7.50	**22.35**	1.41	0.35	**23.78**	1.92	**48.05**	m2	**54.06**
3.00 - 4.50m	20.79	7.50	**22.35**	1.62	0.41	**27.41**	2.88	**52.64**	m2	**59.22**
slab thickness 175mm; strutting and supports at 450mm centres; height to soffit										
1.50 - 3.00m	20.79	7.50	**22.35**	1.44	0.36	**24.32**	2.21	**48.88**	m2	**54.99**
3.00 - 4.50m	20.79	7.50	**22.35**	1.66	0.42	**28.09**	3.44	**53.88**	m2	**60.61**
slab thickness 200mm; strutting and supports at 350mm centres; height to soffit										
1.50 - 3.00m	20.79	7.50	**22.35**	1.47	0.37	**24.85**	2.50	**49.71**	m2	**55.92**
3.00 - 4.50m	20.79	7.50	**22.35**	1.69	0.43	**28.62**	4.00	**54.97**	m2	**61.84**
Reference 2611, 4.86 kg/m2 to soffits of arched slabs; one rib side laps										
900mm span; 75mm rise; height to soffit										
1.50 - 3.00m	15.51	7.50	**16.67**	1.44	0.36	**24.32**	0.97	**41.96**	m2	**47.20**
1200mm span; 75mm rise; one row strutting and supports per span; height to soffit										
1.50 - 3.00m	15.51	7.50	**16.67**	1.44	0.36	**24.32**	1.31	**42.30**	m2	**47.58**
3.00 - 4.50m	15.51	7.50	**16.67**	1.64	0.41	**27.70**	1.66	**46.03**	m2	**51.78**
Reference 2411, 6.34 kg/m2 to soffits of arched slabs; one rib side laps										
1500mm span; 100mm rise; two rows strutting and supports per span; height to soffit										
1.50 - 3.00m	20.79	7.50	**22.35**	1.78	0.44	**30.01**	2.88	**55.24**	m2	**62.14**
3.00 - 4.50m	20.79	7.50	**22.35**	2.02	0.51	**34.17**	3.81	**60.33**	m2	**67.87**
1800mm span; 150mm rise; two rows strutting and supports per span; height to soffit										
1.50 - 3.00m	20.79	7.50	**22.35**	1.78	0.44	**30.01**	2.88	**55.24**	m2	**62.14**
3.00 - 4.50m	20.79	7.50	**22.35**	2.02	0.51	**34.17**	3.81	**60.33**	m2	**67.87**
E30: REINFORCEMENT FOR IN-SITU CONCRETE										
Reinforcement bars; BS 4449, hot rolled plain round mild steel including hooks and tying wire, and spacers and chairs which are at the discretion of the Contractor										
Straight										
6mm	441.00	7.50	**474.08**	51.00	5.00	**778.26**	28.41	**1280.74**	t	**1440.83**
8mm	425.25	7.50	**457.14**	44.00	5.00	**678.79**	25.15	**1161.08**	t	**1306.22**
10mm	414.75	7.50	**445.86**	36.00	5.00	**565.11**	20.42	**1031.38**	t	**1160.31**
12mm	409.50	7.50	**440.21**	31.00	5.00	**494.06**	17.46	**951.73**	t	**1070.70**
16mm	404.25	7.50	**434.57**	24.00	5.00	**394.59**	14.36	**843.52**	t	**948.96**
20mm	404.25	7.50	**434.57**	20.00	5.00	**337.75**	7.85	**780.17**	t	**877.69**
25mm	404.25	7.50	**434.57**	20.00	5.00	**337.75**	7.85	**780.17**	t	**877.69**
32mm	399.00	7.50	**428.93**	20.00	5.00	**337.75**	7.85	**774.52**	t	**871.34**
Bent										
6mm	441.00	7.50	**474.08**	51.00	5.00	**778.26**	28.41	**1280.74**	t	**1440.83**
8mm	425.25	7.50	**457.14**	44.00	5.00	**678.79**	25.15	**1161.08**	t	**1306.22**
10mm	414.75	7.50	**445.86**	36.00	5.00	**565.11**	20.42	**1031.38**	t	**1160.31**
12mm	409.50	7.50	**440.21**	31.00	5.00	**494.06**	17.46	**951.73**	t	**1070.70**
16mm	404.25	7.50	**434.57**	24.00	5.00	**394.59**	14.36	**843.52**	t	**948.96**
20mm	404.25	7.50	**434.57**	20.00	5.00	**337.75**	7.85	**780.17**	t	**877.69**
25mm	404.25	7.50	**434.57**	20.00	5.00	**337.75**	7.85	**780.17**	t	**877.69**
32mm	399.00	7.50	**428.93**	20.00	5.00	**337.75**	7.85	**774.52**	t	**871.34**

Labour hourly rates: (except Specialists) Craft Operatives £14.21 Labourer £10.71 Rates are national average prices. Refer to REGIONAL VARIATIONS for indicative levels of overall pricing in regions	MATERIALS			LABOUR				RATES		
	Del to Site £	Waste %	Material Cost £	Craft Optve Hrs	Lab Hrs	Labour Cost £	Sunds £	Nett Rate £	Unit	Gross Rate (+12.5%) £
E30: REINFORCEMENT FOR IN-SITU CONCRETE Cont.										
Reinforcement bars; BS 4449, hot rolled plain round mild steel including hooks and tying wire, and spacers and chairs which are at the discretion of the Contractor Cont.										
Links										
6mm	441.00	7.50	474.08	66.00	5.00	991.41	28.41	1493.89	t	1680.63
8mm	425.25	7.50	457.14	66.00	5.00	991.41	25.15	1473.70	t	1657.92
Reinforcement bars; BS 4449, hot rolled deformed high yield steel including hooks and tying wire, and spacers and chairs which are at the discretion of the Contractor										
Straight										
6mm	451.50	7.50	485.36	51.00	5.00	778.26	28.41	1292.03	t	1453.53
8mm	430.50	7.50	462.79	44.00	5.00	678.79	25.15	1166.73	t	1312.57
10mm	420.00	7.50	451.50	36.00	5.00	565.11	20.42	1037.03	t	1166.65
12mm	420.00	7.50	451.50	31.00	5.00	494.06	17.46	963.02	t	1083.39
16mm	409.50	7.50	440.21	24.00	5.00	394.59	14.36	849.16	t	955.31
20mm	409.50	7.50	440.21	20.00	5.00	337.75	7.85	785.81	t	884.04
25mm	409.50	7.50	440.21	20.00	5.00	337.75	7.85	785.81	t	884.04
32mm	409.50	7.50	440.21	20.00	5.00	337.75	7.85	785.81	t	884.04
Bent										
6mm	451.50	7.50	485.36	51.00	5.00	778.26	28.41	1292.03	t	1453.53
8mm	430.50	7.50	462.79	44.00	5.00	678.79	25.15	1166.73	t	1312.57
10mm	420.00	7.50	451.50	36.00	5.00	565.11	20.42	1037.03	t	1166.65
12mm	420.00	7.50	451.50	31.00	5.00	494.06	17.46	963.02	t	1083.39
16mm	409.50	7.50	440.21	24.00	5.00	394.59	14.36	849.16	t	955.31
20mm	409.50	7.50	440.21	20.00	5.00	337.75	7.85	785.81	t	884.04
25mm	409.50	7.50	440.21	20.00	5.00	337.75	7.85	785.81	t	884.04
32mm	409.50	7.50	440.21	20.00	5.00	337.75	7.85	785.81	t	884.04
Links										
6mm	451.50	7.50	485.36	66.00	5.00	991.41	28.41	1505.18	t	1693.33
8mm	430.50	7.50	462.79	66.00	5.00	991.41	25.15	1479.35	t	1664.27
Take delivery, cut, bend and fix reinforcing rods; including hooks and tying wire, and spacers and chairs which are at the discretion of the Contractor										
Straight										
6mm	-	-	-	55.00	5.00	835.10	28.41	863.51	t	971.44
8mm	-	-	-	48.00	5.00	735.63	25.15	760.78	t	855.88
10mm	-	-	-	40.00	5.00	621.95	20.42	642.37	t	722.66
12mm	-	-	-	34.00	5.00	536.69	17.46	554.15	t	623.42
16mm	-	-	-	26.00	5.00	423.01	14.36	437.37	t	492.04
20mm	-	-	-	22.00	5.00	366.17	7.85	374.02	t	420.77
25mm	-	-	-	22.00	5.00	366.17	7.85	374.02	t	420.77
32mm	-	-	-	22.00	5.00	366.17	7.85	374.02	t	420.77
Bent										
6mm	-	-	-	75.00	5.00	1119.30	28.41	1147.71	t	1291.17
8mm	-	-	-	66.00	5.00	991.41	25.15	1016.56	t	1143.63
10mm	-	-	-	55.00	5.00	835.10	20.42	855.52	t	962.46
12mm	-	-	-	47.00	5.00	721.42	17.46	738.88	t	831.24
16mm	-	-	-	37.00	5.00	579.32	14.36	593.68	t	667.89
20mm	-	-	-	31.00	5.00	494.06	7.85	501.91	t	564.65
25mm	-	-	-	29.00	5.00	465.64	7.85	473.49	t	532.67
32mm	-	-	-	29.00	5.00	465.64	7.85	473.49	t	532.67
Links										
6mm	-	-	-	99.00	5.00	1460.34	28.41	1488.75	t	1674.84
8mm	-	-	-	99.00	5.00	1460.34	25.15	1485.49	t	1671.18
Take delivery and fix reinforcing rods supplied cut to length and bent; including tying wire, and spacers and chairs which are at the discretion of the Contractor										
Straight										
6mm	-	-	-	51.00	5.00	778.26	28.41	806.67	t	907.50
8mm	-	-	-	44.00	5.00	678.79	25.15	703.94	t	791.93
10mm	-	-	-	36.00	5.00	565.11	20.42	585.53	t	658.72
12mm	-	-	-	31.00	5.00	494.06	17.46	511.52	t	575.46
16mm	-	-	-	24.00	5.00	394.59	14.36	408.95	t	460.07
20mm	-	-	-	20.00	5.00	337.75	7.85	345.60	t	388.80
25mm	-	-	-	20.00	5.00	337.75	7.85	345.60	t	388.80
32mm	-	-	-	20.00	5.00	337.75	7.85	345.60	t	388.80
Bent										
6mm	-	-	-	51.00	5.00	778.26	28.41	806.67	t	907.50
8mm	-	-	-	44.00	5.00	678.79	25.15	703.94	t	791.93
10mm	-	-	-	36.00	5.00	565.11	20.42	585.53	t	658.72
12mm	-	-	-	31.00	5.00	494.06	17.46	511.52	t	575.46
16mm	-	-	-	24.00	5.00	394.59	14.36	408.95	t	460.07
20mm	-	-	-	20.00	5.00	337.75	7.85	345.60	t	388.80
25mm	-	-	-	20.00	5.00	337.75	7.85	345.60	t	388.80
Links										
6mm	-	-	-	66.00	5.00	991.41	28.41	1019.82	t	1147.29
8mm	-	-	-	66.00	5.00	991.41	25.15	1016.56	t	1143.63

Labour hourly rates: (except Specialists) Craft Operatives £14.21 Labourer £10.71 Rates are national average prices. Refer to REGIONAL VARIATIONS for indicative levels of overall pricing in regions	MATERIALS			LABOUR			RATES			
	Del to Site £	Waste %	Material Cost £	Craft Optve Hrs	Lab Hrs	Labour Cost £	Sunds £	Nett Rate £	Unit	Gross Rate (+12.5%) £
E30: REINFORCEMENT FOR IN-SITU CONCRETE Cont.										
Reinforcement bars; high yield stainless steel BS 6744, Type 2 (minimum yield stress 460 N/mm2); including hooks and tying wire, and spacers and chairs which are at the discretion of the Contractor										
Straight										
8mm	4250.00	7.50	**4568.75**	83.00	10.00	**1286.53**	97.64	**5952.92**	t	**6697.04**
10mm	4250.00	7.50	**4568.75**	69.00	10.00	**1087.59**	70.45	**5726.79**	t	**6442.64**
12mm	4250.00	7.50	**4568.75**	60.00	10.00	**959.70**	66.65	**5595.10**	t	**6294.48**
16mm	4250.00	7.50	**4568.75**	50.00	10.00	**817.60**	57.35	**5443.70**	t	**6124.16**
20mm	4250.00	7.50	**4568.75**	41.00	10.00	**689.71**	42.62	**5301.08**	t	**5963.72**
25mm	4250.00	7.50	**4568.75**	33.00	10.00	**576.03**	31.00	**5175.78**	t	**5822.75**
Bent										
8mm	4250.00	7.50	**4568.75**	83.00	10.00	**1286.53**	97.64	**5952.92**	t	**6697.04**
10mm	4250.00	7.50	**4568.75**	69.00	10.00	**1087.59**	70.45	**5726.79**	t	**6442.64**
12mm	4250.00	7.50	**4568.75**	60.00	10.00	**959.70**	66.65	**5595.10**	t	**6294.48**
16mm	4250.00	7.50	**4568.75**	50.00	10.00	**817.60**	57.35	**5443.70**	t	**6124.16**
20mm	4250.00	7.50	**4568.75**	41.00	10.00	**689.71**	42.62	**5301.08**	t	**5963.72**
25mm	4250.00	7.50	**4568.75**	33.00	10.00	**576.03**	31.00	**5175.78**	t	**5822.75**
Links										
6mm	4250.00	7.50	**4568.75**	130.00	10.00	**1954.40**	97.64	**6620.79**	t	**7448.39**
8mm	4250.00	7.50	**4568.75**	115.00	10.00	**1741.25**	70.45	**6380.45**	t	**7178.01**
Reinforcement fabric; BS 4483, hard drawn plain round steel; welded; including laps, tying wire, all cutting and bending, and spacers and chairs which are at the discretion of the Contractor										
Reference A98, 1.54 kg/m2; 200mm side laps; 200mm end laps										
generally	0.80	17.50	**0.94**	0.07	0.01	**1.10**	0.06	**2.09**	m2	**2.36**
strips in one width										
750mm wide	0.80	17.50	**0.94**	0.11	0.01	**1.67**	0.06	**2.66**	m2	**2.99**
900mm wide	0.80	17.50	**0.94**	0.10	0.01	**1.53**	0.06	**2.52**	m2	**2.83**
1050mm wide	0.80	17.50	**0.94**	0.09	0.01	**1.39**	0.06	**2.38**	m2	**2.67**
1200mm wide	0.80	17.50	**0.94**	0.08	0.01	**1.24**	0.06	**2.24**	m2	**2.52**
Reference A142, 2.22kg/m2; 200mm side laps; 200mm end laps										
generally	1.11	17.50	**1.31**	0.08	0.02	**1.35**	0.08	**2.74**	m2	**3.08**
strips in one width										
750mm wide	1.11	17.50	**1.31**	0.12	0.02	**1.92**	0.08	**3.31**	m2	**3.72**
900mm wide	1.11	17.50	**1.31**	0.11	0.02	**1.78**	0.08	**3.17**	m2	**3.56**
1050mm wide	1.11	17.50	**1.31**	0.10	0.02	**1.64**	0.08	**3.03**	m2	**3.40**
1200mm wide	1.11	17.50	**1.31**	0.10	0.02	**1.64**	0.08	**3.03**	m2	**3.40**
Reference A193, 3.02kg/m2; 200mm side laps; 200mm end laps										
generally	1.47	17.50	**1.72**	0.08	0.02	**1.35**	0.11	**3.19**	m2	**3.58**
strips in one width										
750mm wide	1.47	17.50	**1.72**	0.12	0.02	**1.92**	0.11	**3.75**	m2	**4.22**
900mm wide	1.47	17.50	**1.72**	0.11	0.02	**1.78**	0.11	**3.61**	m2	**4.06**
1050mm wide	1.47	17.50	**1.72**	0.10	0.02	**1.64**	0.11	**3.47**	m2	**3.90**
1200mm wide	1.47	17.50	**1.72**	0.10	0.02	**1.64**	0.11	**3.47**	m2	**3.90**
Reference A252, 3.95kg/m2; 200mm side laps; 200mm end laps										
generally	2.01	17.50	**2.36**	0.09	0.02	**1.49**	0.14	**4.00**	m2	**4.50**
strips in one width										
750mm wide	2.01	17.50	**2.36**	0.14	0.02	**2.20**	0.14	**4.71**	m2	**5.30**
900mm wide	2.01	17.50	**2.36**	0.13	0.02	**2.06**	0.14	**4.57**	m2	**5.14**
1050mm wide	2.01	17.50	**2.36**	0.12	0.02	**1.92**	0.14	**4.42**	m2	**4.98**
1200mm wide	2.01	17.50	**2.36**	0.11	0.02	**1.78**	0.14	**4.28**	m2	**4.82**
Reference A393, 6.16 kg/m2; 200mm side laps; 200mm end laps										
generally	2.97	17.50	**3.49**	0.10	0.02	**1.64**	0.21	**5.33**	m2	**6.00**
strips in one width										
750mm wide	2.97	17.50	**3.49**	0.15	0.02	**2.35**	0.21	**6.04**	m2	**6.80**
900mm wide	2.97	17.50	**3.49**	0.14	0.02	**2.20**	0.21	**5.90**	m2	**6.64**
1050mm wide	2.97	17.50	**3.49**	0.13	0.02	**2.06**	0.21	**5.76**	m2	**6.48**
1200mm wide	2.97	17.50	**3.49**	0.12	0.02	**1.92**	0.21	**5.62**	m2	**6.32**
Reference B1131, 10.90 kg/m2; 100mm side laps; 200mm end laps										
generally	5.35	17.50	**6.29**	0.14	0.03	**2.31**	0.38	**8.98**	m2	**10.10**
strips in one width										
750mm wide	5.35	17.50	**6.29**	0.21	0.03	**3.31**	0.38	**9.98**	m2	**11.22**
900mm wide	5.35	17.50	**6.29**	0.20	0.03	**3.16**	0.38	**9.83**	m2	**11.06**
1050mm wide	5.35	17.50	**6.29**	0.18	0.03	**2.88**	0.38	**9.55**	m2	**10.74**
1200mm wide	5.35	17.50	**6.29**	0.17	0.03	**2.74**	0.38	**9.41**	m2	**10.58**
Reference B196, 3.05 kg/m2; 100mm side laps; 200mm end laps										
generally	1.50	17.50	**1.77**	0.08	0.02	**1.35**	0.21	**3.33**	m2	**3.75**
strips in one width										
750mm wide	1.50	17.50	**1.77**	0.12	0.02	**1.92**	0.21	**3.90**	m2	**4.39**
900mm wide	1.50	17.50	**1.77**	0.11	0.02	**1.78**	0.21	**3.76**	m2	**4.23**
1050mm wide	1.50	17.50	**1.77**	0.10	0.02	**1.64**	0.21	**3.61**	m2	**4.07**
1200mm wide	1.50	17.50	**1.77**	0.10	0.02	**1.64**	0.21	**3.61**	m2	**4.07**

Labour hourly rates: (except Specialists) Craft Operatives £14.21 Labourer £10.71 Rates are national average prices. Refer to REGIONAL VARIATIONS for indicative levels of overall pricing in regions	MATERIALS			LABOUR				RATES		
	Del to Site £	Waste %	Material Cost £	Craft Optve Hrs	Lab Hrs	Labour Cost £	Sunds £	Nett Rate £	Unit	Gross Rate (+12.5%) £
E30: REINFORCEMENT FOR IN-SITU CONCRETE Cont.										
Reinforcement fabric; BS 4483, hard drawn plain round steel; welded; including laps, tying wire, all cutting and bending, and spacers and chairs which are at the discretion of the Contractor Cont.										
Reference B283, 3.73 kg/m2; 100mm side laps; 200mm end laps										
generally	1.83	17.50	**2.14**	0.08	0.02	**1.35**	0.13	**3.62**	m2	**4.07**
strips in one width										
750mm wide	1.83	17.50	**2.14**	0.12	0.02	**1.92**	0.13	**4.19**	m2	**4.71**
900mm wide	1.83	17.50	**2.14**	0.11	0.02	**1.78**	0.13	**4.05**	m2	**4.55**
1050mm wide	1.83	17.50	**2.14**	0.10	0.02	**1.64**	0.13	**3.91**	m2	**4.39**
1200mm wide	1.83	17.50	**2.14**	0.10	0.02	**1.64**	0.13	**3.91**	m2	**4.39**
Reference B385, 4.53 kg/m2; 100mm side laps; 200mm end laps										
generally	2.22	17.50	**2.61**	0.09	0.02	**1.49**	0.15	**4.26**	m2	**4.79**
strips in one width										
750mm wide	2.22	17.50	**2.61**	0.14	0.02	**2.20**	0.15	**4.97**	m2	**5.59**
900mm wide	2.22	17.50	**2.61**	0.13	0.02	**2.06**	0.15	**4.83**	m2	**5.43**
1050mm wide	2.22	17.50	**2.61**	0.12	0.02	**1.92**	0.15	**4.69**	m2	**5.27**
1200mm wide	2.22	17.50	**2.61**	0.11	0.02	**1.78**	0.15	**4.54**	m2	**5.11**
Reference B503, 5.93 kg/m2; 100mm side laps; 200mm end laps										
generally	2.88	17.50	**3.39**	0.10	0.02	**1.64**	0.21	**5.23**	m2	**5.89**
strips in one width										
750mm wide	2.88	17.50	**3.39**	0.15	0.02	**2.35**	0.21	**5.94**	m2	**6.69**
900mm wide	2.88	17.50	**3.39**	0.14	0.02	**2.20**	0.21	**5.80**	m2	**6.53**
1050mm wide	2.88	17.50	**3.39**	0.13	0.02	**2.06**	0.21	**5.66**	m2	**6.37**
1200mm wide	2.88	17.50	**3.39**	0.12	0.02	**1.92**	0.21	**5.52**	m2	**6.21**
Reference B785, 8.14 kg/m2; 100mm side laps; 200mm end laps										
generally	3.96	17.50	**4.66**	0.11	0.03	**1.88**	0.28	**6.82**	m2	**7.67**
strips in one width										
750mm wide	3.96	17.50	**4.66**	0.16	0.03	**2.59**	0.28	**7.53**	m2	**8.47**
900mm wide	3.96	17.50	**4.66**	0.15	0.03	**2.45**	0.28	**7.39**	m2	**8.31**
1050mm wide	3.96	17.50	**4.66**	0.14	0.03	**2.31**	0.28	**7.25**	m2	**8.15**
1200mm wide	3.96	17.50	**4.66**	0.13	0.03	**2.17**	0.28	**7.11**	m2	**7.99**
Reference C283, 2.61 kg/m2; 100mm side laps; 400mm end laps										
generally	1.27	17.50	**1.49**	0.08	0.02	**1.35**	0.10	**2.94**	m2	**3.31**
strips in one width										
750mm wide	1.27	17.50	**1.49**	0.12	0.02	**1.92**	0.10	**3.51**	m2	**3.95**
900mm wide	1.27	17.50	**1.49**	0.11	0.02	**1.78**	0.10	**3.36**	m2	**3.79**
1050mm wide	1.27	17.50	**1.49**	0.10	0.02	**1.64**	0.10	**3.22**	m2	**3.63**
1200mm wide	1.27	17.50	**1.49**	0.10	0.02	**1.64**	0.10	**3.22**	m2	**3.63**
Reference C385, 3.41 kg/m2; 100mm side laps; 400mm end laps										
generally	1.65	17.50	**1.94**	0.08	0.02	**1.35**	0.13	**3.42**	m2	**3.85**
strips in one width										
750mm wide	1.65	17.50	**1.94**	0.12	0.02	**1.92**	0.13	**3.99**	m2	**4.49**
900mm wide	1.65	17.50	**1.94**	0.11	0.02	**1.78**	0.13	**3.85**	m2	**4.33**
1050mm wide	1.65	17.50	**1.94**	0.10	0.02	**1.64**	0.13	**3.70**	m2	**4.17**
1200mm wide	1.65	17.50	**1.94**	0.10	0.02	**1.64**	0.13	**3.70**	m2	**4.17**
Reference C503, 4.34 kg/m2; 100mm side laps; 400mm end laps										
generally	2.04	17.50	**2.39**	0.09	0.02	**1.49**	0.15	**4.04**	m2	**4.55**
strips in one width										
750mm wide	2.04	17.50	**2.39**	0.14	0.02	**2.20**	0.15	**4.75**	m2	**5.34**
900mm wide	2.04	17.50	**2.39**	0.13	0.02	**2.06**	0.15	**4.61**	m2	**5.18**
1050mm wide	2.04	17.50	**2.39**	0.12	0.02	**1.92**	0.15	**4.47**	m2	**5.03**
1200mm wide	2.04	17.50	**2.39**	0.11	0.02	**1.78**	0.15	**4.32**	m2	**4.87**
Reference C636, 5.55 kg/m2; 100mm side laps; 400mm end laps										
generally	2.72	17.50	**3.19**	0.10	0.02	**1.64**	0.20	**5.03**	m2	**5.66**
strips in one width										
750mm wide	2.72	17.50	**3.19**	0.15	0.02	**2.35**	0.20	**5.74**	m2	**6.46**
900mm wide	2.72	17.50	**3.19**	0.14	0.02	**2.20**	0.20	**5.60**	m2	**6.30**
1050mm wide	2.72	17.50	**3.19**	0.13	0.02	**2.06**	0.20	**5.45**	m2	**6.14**
1200mm wide	2.72	17.50	**3.19**	0.12	0.02	**1.92**	0.20	**5.31**	m2	**5.98**
Reference C785, 6.72 kg/m2; 100mm side laps; 400mm end laps										
generally	3.28	17.50	**3.85**	0.10	0.02	**1.64**	0.24	**5.73**	m2	**6.44**
strips in one width										
750mm wide	3.28	17.50	**3.85**	0.15	0.02	**2.35**	0.24	**6.44**	m2	**7.24**
900mm wide	3.28	17.50	**3.85**	0.14	0.02	**2.20**	0.24	**6.30**	m2	**7.08**
1050mm wide	3.28	17.50	**3.85**	0.13	0.02	**2.06**	0.24	**6.15**	m2	**6.92**
1200mm wide	3.28	17.50	**3.85**	0.12	0.02	**1.92**	0.24	**6.01**	m2	**6.76**
Reference D49, 0.77 kg/m2; 100mm side laps; 100mm end laps										
bent	0.77	17.50	**0.90**	0.04	0.01	**0.68**	0.08	**1.66**	m2	**1.87**
Reference D98, 1.54 kg/m2; 200mm side laps; 200mm end laps										
bent	0.80	17.50	**0.94**	0.04	0.01	**0.68**	0.08	**1.70**	m2	**1.91**

Labour hourly rates: (except Specialists) Craft Operatives £14.21 Labourer £10.71 Rates are national average prices. Refer to REGIONAL VARIATIONS for indicative levels of overall pricing in regions	MATERIALS			LABOUR			RATES			
	Del to Site £	Waste %	Material Cost £	Craft Optve Hrs	Lab Hrs	Labour Cost £	Sunds £	Nett Rate £	Unit	Gross Rate (+12.5%) £
E30: REINFORCEMENT FOR IN-SITU CONCRETE Cont.										
Expamet 76mm mesh expanded steel reinforcement (uncoated); including laps, tying wire, all cutting and bending, and spacers and chairs which are at the discretion of the Contractor										
Reference 8, 3.80 kg/m2; 150mm side laps; 150mm end laps										
generally	11.59	17.50	**13.62**	0.08	0.02	**1.35**	0.59	**15.56**	m2	**17.51**
strips in one width										
750mm wide	11.59	17.50	**13.62**	0.12	0.02	**1.92**	0.59	**16.13**	m2	**18.15**
900mm wide	11.59	17.50	**13.62**	0.11	0.02	**1.78**	0.59	**15.99**	m2	**17.99**
1050mm wide	11.59	17.50	**13.62**	0.10	0.02	**1.64**	0.59	**15.85**	m2	**17.83**
1200mm wide	11.59	17.50	**13.62**	0.10	0.02	**1.64**	0.59	**15.85**	m2	**17.83**
Expamet 'Securilath' flattened security mesh expanded steel reinforcement (uncoated); including laps										
Reference 2073F; 150mm side laps; 150mm end laps										
generally	25.73	17.50	**30.24**	0.12	0.02	**1.92**	0.28	**32.44**	m2	**36.49**
E40: DESIGNED JOINTS IN IN-SITU CONCRETE										
Formed joints										
Incorporating 10mm thick Korkpak, Grace Construction Products Limited - Servicised; formwork; reinforcement laid continuously across joint										
in concrete, depth not exceeding 150mm; horizontal	3.78	5.00	**3.97**	-	0.08	**0.86**	5.73	**10.55**	m	**11.87**
in concrete, depth 150 - 300mm; horizontal	6.61	5.00	**6.94**	-	0.09	**0.96**	10.03	**17.94**	m	**20.18**
in concrete, depth 300 - 450mm; horizontal	9.44	5.00	**9.92**	-	0.10	**1.07**	14.59	**25.58**	m	**28.78**
Incorporating 13mm thick Korkpak, Grace Construction Products Limited - Servicised; formwork; reinforcement laid continuously across joint										
in concrete, depth not exceeding 150mm; horizontal	3.86	5.00	**4.05**	-	0.08	**0.86**	5.73	**10.64**	m	**11.97**
in concrete, depth 150 - 300mm; horizontal	6.75	5.00	**7.09**	-	0.09	**0.96**	10.03	**18.09**	m	**20.35**
in concrete, depth 300 - 450mm; horizontal	9.65	5.00	**10.13**	-	0.10	**1.07**	14.59	**25.79**	m	**29.02**
Incorporating 19mm thick Korkpak, Grace Construction Products Limited - Servicised; formwork; reinforcement laid continuously across joint										
in concrete, depth not exceeding 150mm; horizontal	5.43	5.00	**5.71**	-	0.08	**0.86**	5.73	**12.29**	m	**13.83**
in concrete, depth 150 - 300mm; horizontal	9.51	5.00	**9.99**	-	0.09	**0.96**	10.03	**20.98**	m	**23.61**
in concrete, depth 300 - 450mm; horizontal	13.59	5.00	**14.27**	-	0.10	**1.07**	14.59	**29.93**	m	**33.67**
Incorporating 195mm wide 'Serviseal 195', external face type p.v.c waterstop, Grace Construction Products Limited - Servicised; heat welded joints; formwork; reinforcement laid continuously across joint										
in concrete, depth not exceeding 150mm; horizontal	6.12	10.00	**6.73**	-	0.44	**4.71**	5.73	**17.18**	m	**19.33**
in concrete, depth 150 - 300mm; horizontal	6.12	10.00	**6.73**	-	0.44	**4.71**	10.03	**21.48**	m	**24.16**
in concrete, depth 300 - 450mm; horizontal	6.12	10.00	**6.73**	-	0.44	**4.71**	14.59	**26.04**	m	**29.29**
vertical L piece	16.37	10.00	**18.01**	-	0.22	**2.36**	-	**20.36**	nr	**22.91**
flat L piece	9.11	10.00	**10.03**	-	0.22	**2.36**	-	**12.38**	nr	**13.93**
flat T piece	15.64	10.00	**17.20**	-	0.28	**3.00**	-	**20.20**	nr	**22.72**
flat X piece	23.58	10.00	**25.94**	-	0.33	**3.53**	-	**29.48**	nr	**33.16**
in concrete, width not exceeding 150mm; vertical	6.12	10.00	**6.73**	-	0.55	**5.89**	5.73	**18.36**	m	**20.65**
in concrete, width 150 - 300mm; vertical	6.12	10.00	**6.73**	-	0.55	**5.89**	10.03	**22.66**	m	**25.49**
in concrete, width 300 - 450mm; vertical	6.12	10.00	**6.73**	-	0.55	**5.89**	14.59	**27.22**	m	**30.62**
flat L piece	9.11	10.00	**10.03**	-	0.22	**2.36**	-	**12.38**	nr	**13.93**
flat T piece	15.64	10.00	**17.20**	-	0.28	**3.00**	-	**20.20**	nr	**22.72**
flat X piece	23.58	10.00	**25.94**	-	0.33	**3.53**	-	**29.48**	nr	**33.16**
Incorporating 240mm wide 'Serviseal 240', heavy duty section external face type p.v.c. waterstop, Grace Construction Products Limited - Servicised; heat welded joints; formwork; reinforcement laid continuously across joint										
in concrete, depth not exceeding 150mm; horizontal	7.48	10.00	**8.22**	-	0.50	**5.36**	5.73	**19.31**	m	**21.72**
in concrete, depth 150 - 300mm; horizontal	7.48	10.00	**8.22**	-	0.50	**5.36**	10.03	**23.61**	m	**26.56**
in concrete, depth 300 - 450mm; horizontal	7.48	10.00	**8.22**	-	0.50	**5.36**	14.59	**28.17**	m	**31.69**
vertical L piece	17.12	10.00	**18.83**	-	0.22	**2.36**	-	**21.18**	nr	**23.83**
flat L piece	10.36	10.00	**11.40**	-	0.22	**2.36**	-	**13.76**	nr	**15.48**
flat T piece	17.53	10.00	**19.28**	-	0.28	**3.00**	-	**22.28**	nr	**25.06**
flat X piece	26.09	10.00	**28.70**	-	0.33	**3.53**	-	**32.24**	nr	**36.27**
in concrete, width not exceeding 150mm; vertical	7.48	10.00	**8.22**	-	0.61	**6.53**	5.73	**20.49**	m	**23.05**
in concrete, width 150 - 300mm; vertical	7.48	10.00	**8.22**	-	0.61	**6.53**	10.03	**24.79**	m	**27.89**
in concrete, width 300 - 450mm; vertical	7.48	10.00	**8.22**	-	0.61	**6.53**	14.59	**29.35**	m	**33.02**
flat L piece	10.36	10.00	**11.40**	-	0.22	**2.36**	-	**13.76**	nr	**15.48**
flat T piece	17.53	10.00	**19.28**	-	0.28	**3.00**	-	**22.28**	nr	**25.06**
flat X piece	26.09	10.00	**28.70**	-	0.33	**3.53**	-	**32.24**	nr	**36.27**
Incorporating 320mm wide 'Serviseal Kicker 320', external face type p.v.c. kickerjoint waterstop, Grace Construction Products Limited - Servicised; heat welded joints; formwork; reinforcement laid continuously across joint										
in concrete, depth not exceeding 150mm; vertical	10.46	10.00	**11.50**	-	0.66	**7.07**	5.73	**24.30**	m	**27.34**
in concrete, depth 150 - 300mm; vertical	10.46	10.00	**11.50**	-	0.66	**7.07**	10.03	**28.61**	m	**32.18**
in concrete, depth 300 - 450mm; vertical	10.46	10.00	**11.50**	-	0.66	**7.07**	14.59	**33.16**	m	**37.31**
vertical L piece	19.87	10.00	**21.85**	-	0.28	**3.00**	-	**24.85**	nr	**27.96**
flat L piece	23.34	10.00	**25.68**	-	0.28	**3.00**	-	**28.68**	nr	**32.26**
flat T piece	34.18	10.00	**37.60**	-	0.33	**3.53**	-	**41.13**	nr	**46.27**
flat X piece	42.57	10.00	**46.82**	-	0.39	**4.18**	-	**51.00**	nr	**57.38**

Labour hourly rates: (except Specialists) Craft Operatives £14.21 Labourer £10.71 Rates are national average prices. Refer to REGIONAL VARIATIONS for indicative levels of overall pricing in regions	MATERIALS			LABOUR				RATES		
	Del to Site	Waste	Material Cost	Craft Optve	Lab	Labour Cost	Sunds	Nett Rate	Unit	Gross Rate (+12.5%)
	£	%	£	Hrs	Hrs	£	£	£		£
E40: DESIGNED JOINTS IN IN-SITU CONCRETE Cont.										
Formed joints Cont.										
Incorporating 210mm wide 'PVC Edgetie 210' centre bulb type internally placed p.v.c. waterstop, Grace Construction Products Limited - Servicised; heat welded joints; formwork; reinforcement laid continuously across joint										
in concrete, depth not exceeding 150mm; horizontal	5.36	10.00	**5.89**	-	0.44	**4.71**	5.73	**16.33**	m	**18.38**
in concrete, depth 150 - 300mm; horizontal	5.36	10.00	**5.89**	-	0.44	**4.71**	10.03	**20.64**	m	**23.21**
in concrete, depth 300 - 450mm; horizontal	5.36	10.00	**5.89**	-	0.44	**4.71**	14.59	**25.19**	m	**28.34**
flat L piece	11.53	10.00	**12.68**	-	0.22	**2.36**	-	**15.04**	nr	**16.92**
flat T piece	15.24	10.00	**16.76**	-	0.28	**3.00**	-	**19.76**	nr	**22.23**
flat X piece	23.05	10.00	**25.35**	-	0.33	**3.53**	-	**28.89**	nr	**32.50**
in concrete, width not exceeding 150mm; vertical	5.36	10.00	**5.89**	-	0.55	**5.89**	5.73	**17.51**	m	**19.70**
in concrete, width 150 - 300mm; vertical	5.36	10.00	**5.89**	-	0.55	**5.89**	10.03	**21.81**	m	**24.54**
in concrete, width 300 - 450mm; vertical	5.36	10.00	**5.89**	-	0.55	**5.89**	14.59	**26.37**	m	**29.67**
vertical L piece	14.86	10.00	**16.34**	-	0.22	**2.36**	-	**18.70**	nr	**21.04**
flat L piece	11.53	10.00	**12.68**	-	0.22	**2.36**	-	**15.04**	nr	**16.92**
vertical T piece	23.24	10.00	**25.56**	-	0.28	**3.00**	-	**28.56**	nr	**32.13**
flat T piece	15.24	10.00	**16.76**	-	0.28	**3.00**	-	**19.76**	nr	**22.23**
flat X piece	23.05	10.00	**25.35**	-	0.33	**3.53**	-	**28.89**	nr	**32.50**
Incorporating 260mm wide 'PVC Edgetie 260' centre bulb type internally placed p.v.c. waterstop, Grace Construction Products Limited - Servicised; heat welded joints; formwork; reinforcement laid continuously across joint										
in concrete, depth not exceeding 150mm; horizontal	6.13	10.00	**6.75**	-	0.50	**5.36**	5.73	**17.83**	m	**20.06**
in concrete, depth 150 - 300mm; horizontal	6.13	10.00	**6.75**	-	0.50	**5.36**	10.03	**22.13**	m	**24.90**
in concrete, depth 300 - 450mm; horizontal	6.13	10.00	**6.75**	-	0.50	**5.36**	14.59	**26.69**	m	**30.03**
flat L piece	12.98	10.00	**14.28**	-	0.22	**2.36**	-	**16.63**	nr	**18.71**
flat T piece	19.95	10.00	**21.95**	-	0.28	**3.00**	-	**24.94**	nr	**28.06**
flat X piece	28.01	10.00	**30.82**	-	0.35	**3.75**	-	**34.56**	nr	**38.88**
in concrete, width not exceeding 150mm; vertical	6.13	10.00	**6.75**	-	0.61	**6.53**	5.73	**19.01**	m	**21.39**
in concrete, width 150 - 300mm; vertical	6.13	10.00	**6.75**	-	0.61	**6.53**	10.03	**23.31**	m	**26.22**
in concrete, width 300 - 450mm; vertical	6.13	10.00	**6.75**	-	0.61	**6.53**	14.59	**27.87**	m	**31.35**
vertical L piece	20.62	10.00	**22.68**	-	0.22	**2.36**	-	**25.04**	nr	**28.17**
flat L piece	12.98	10.00	**14.28**	-	0.22	**2.36**	-	**16.63**	nr	**18.71**
vertical T piece	27.83	10.00	**30.61**	-	0.28	**3.00**	-	**33.61**	nr	**37.81**
flat T piece	19.95	10.00	**21.95**	-	0.28	**3.00**	-	**24.94**	nr	**28.06**
flat X piece	28.01	10.00	**30.82**	-	0.33	**3.53**	-	**34.35**	nr	**38.64**
Incorporating 445mm wide Reference 2411, 6.34 kg/m2 Expamet Hy-rib permanent shuttering and reinforcement; 150mm end laps; temporary supports; formwork laid continuously across joint										
in concrete, depth not exceeding 150mm; horizontal	3.26	5.00	**3.42**	0.25	0.07	**4.30**	6.17	**13.90**	m	**15.64**
in concrete, depth 150 - 300mm; horizontal	6.52	5.00	**6.85**	0.25	0.07	**4.30**	10.79	**21.94**	m	**24.69**
in concrete, width not exceeding 150mm; vertical	3.26	5.00	**3.42**	0.39	0.10	**6.61**	6.17	**16.21**	m	**18.24**
in concrete, width 150 - 300mm; vertical	6.52	5.00	**6.85**	0.39	0.10	**6.61**	10.79	**24.25**	m	**27.28**
Incorporating 445mm wide Reference 2611, 4.86 kg/m2 Expamet Hy-rib permanent shuttering and reinforcement; 150mm end laps; temporary supports; formwork laid continuously across joint										
in concrete, depth not exceeding 150mm; horizontal	2.18	5.00	**2.29**	0.25	0.07	**4.30**	6.17	**12.76**	m	**14.36**
in concrete, depth 150 - 300mm; horizontal	4.36	5.00	**4.58**	0.25	0.07	**4.30**	10.79	**19.67**	m	**22.13**
in concrete, width not exceeding 150mm; vertical	2.18	5.00	**2.29**	0.39	0.10	**6.61**	6.17	**15.07**	m	**16.96**
in concrete, width 150 - 300mm; vertical	4.36	5.00	**4.58**	0.39	0.10	**6.61**	10.79	**21.98**	m	**24.73**
Incorporating 445mm wide Reference 2811, 3.39 kg/m2 Expamet Hy-rib permanent shuttering and reinforcement; 150mm end laps; temporary supports; formwork laid continuously across joint										
in concrete, depth not exceeding 150mm; horizontal	4.89	5.00	**5.14**	0.25	0.07	**4.30**	6.17	**15.61**	m	**17.56**
in concrete, depth 150 - 300mm; horizontal	9.78	5.00	**10.27**	0.25	0.07	**4.30**	10.79	**25.37**	m	**28.54**
in concrete, width not exceeding 150mm; vertical	4.89	5.00	**5.14**	0.39	0.10	**6.61**	6.17	**17.92**	m	**20.16**
in concrete, width 150 - 300mm; vertical	9.78	5.00	**10.27**	0.39	0.10	**6.61**	10.79	**27.68**	m	**31.14**
Sealant to joint; Grace Construction Products Limited - Servicised. 'Servigard DW' moisture tolerant 2 part dark grey compound; including preparation, cleaning and priming with water, mixing sealant with cement										
10 x 25mm; horizontal	3.62	5.00	**3.80**	-	0.15	**1.61**	-	**5.40**	m	**6.08**
13 x 25mm; horizontal	4.63	5.00	**4.86**	-	0.20	**2.14**	-	**7.00**	m	**7.87**
19 x 25mm; horizontal	6.64	5.00	**6.98**	-	0.25	**2.68**	-	**9.65**	m	**10.86**
Sealant to joint; Grace Construction Products Limited - Servicised. 'Paraseal' pouring grade 2 part grey polysulphide sealant ; including preparation, cleaners, primers and sealant										
10 x 25mm; horizontal	2.47	5.00	**2.60**	-	0.11	**1.18**	-	**3.77**	m	**4.25**
13 x 25mm; horizontal	3.21	5.00	**3.37**	-	0.12	**1.29**	-	**4.66**	m	**5.24**
19 x 25mm; horizontal	4.70	5.00	**4.93**	-	0.13	**1.39**	-	**6.32**	m	**7.11**
E41: WORKED FINISHES/ CUTTING ON IN-SITU CONCRETE										
Worked finishes on in-situ concrete										
Vacuum dewatering; power floating and power trowelling surfaces	-	-	-	-	0.22	**2.36**	0.27	**2.63**	m2	**2.95**
Tamping unset concrete surfaces	-	-	-	-	0.22	**2.36**	0.35	**2.71**	m2	**3.05**

Labour hourly rates: (except Specialists) Craft Operatives £14.21 Labourer £10.71 Rates are national average prices. Refer to REGIONAL VARIATIONS for indicative levels of overall pricing in regions	MATERIALS			LABOUR				RATES		
	Del to Site £	Waste %	Material Cost £	Craft Optve Hrs	Lab Hrs	Labour Cost £	Sunds £	Nett Rate £	Unit	Gross Rate (+12.5%) £
E41: WORKED FINISHES/ CUTTING ON IN-SITU CONCRETE Cont.										
Worked finishes on in-situ concrete Cont.										
Tamping unset concrete Cont.										
surfaces to falls	-	-	-	-	0.45	4.82	0.46	5.28	m2	5.94
Trowelling										
surfaces	-	-	-	-	0.36	3.86	-	3.86	m2	4.34
surfaces to falls	-	-	-	-	0.39	4.18	-	4.18	m2	4.70
Hacking; by hand										
surfaces	-	-	-	-	0.55	5.89	-	5.89	m2	6.63
to soffits	-	-	-	-	0.72	7.71	-	7.71	m2	8.68
Hacking; by machine										
surfaces	-	-	-	-	0.22	2.36	0.28	2.64	m2	2.97
to soffits	-	-	-	-	0.28	3.00	0.28	3.28	m2	3.69
Bush hammering										
surfaces	-	-	-	1.43	-	20.32	0.79	21.11	m2	23.75
to soffits	-	-	-	2.20	-	31.26	1.05	32.31	m2	36.35
Cutting on in-situ concrete										
Cutting chases										
depth not exceeding 50mm	-	-	-	-	0.19	2.03	-	2.03	m	2.29
Cutting chases; making good										
depth not exceeding 50mm	-	-	-	-	0.22	2.36	0.31	2.66	m	2.99
Cutting mortices										
50 x 50mm; depth not exceeding 100mm	-	-	-	-	0.44	4.71	-	4.71	nr	5.30
75 x 75mm; depth not exceeding 100mm	-	-	-	-	0.55	5.89	-	5.89	nr	6.63
100 x 100mm; depth 100 - 200mm	-	-	-	-	0.66	7.07	-	7.07	nr	7.95
38mm diameter; depth not exceeding 100mm	-	-	-	-	0.39	4.18	-	4.18	nr	4.70
Cutting holes										
225 x 150mm; depth not exceeding 100mm	-	-	-	-	1.25	13.39	0.42	13.80	nr	15.53
300 x 150mm; depth not exceeding 100mm	-	-	-	-	1.45	15.53	0.46	15.99	nr	17.99
300 x 300mm; depth not exceeding 100mm	-	-	-	-	1.60	17.14	0.46	17.60	nr	19.80
225 x 150mm; depth 100 - 200mm	-	-	-	-	1.89	20.24	0.59	20.83	nr	23.43
300 x 150mm; depth 100 - 200mm	-	-	-	-	2.11	22.60	0.66	23.26	nr	26.17
300 x 300mm; depth 100 - 200mm	-	-	-	-	2.32	24.85	0.66	25.51	nr	28.70
225 x 150mm; depth 200 - 300mm	-	-	-	-	2.20	23.56	0.68	24.25	nr	27.28
300 x 150mm; depth 200 - 300mm	-	-	-	-	2.43	26.03	0.75	26.77	nr	30.12
300 x 300mm; depth 200 - 300mm	-	-	-	-	2.65	28.38	0.75	29.13	nr	32.77
50mm diameter; depth not exceeding 100mm	-	-	-	-	0.85	9.10	0.24	9.35	nr	10.52
100mm diameter; depth not exceeding 100mm	-	-	-	-	0.94	10.07	0.32	10.39	nr	11.68
150mm diameter; depth not exceeding 100mm	-	-	-	-	1.03	11.03	0.32	11.35	nr	12.77
50mm diameter; depth 100 - 200mm	-	-	-	-	1.37	14.67	0.40	15.08	nr	16.96
100mm diameter; depth 100 - 200mm	-	-	-	-	1.51	16.17	0.46	16.64	nr	18.72
150mm diameter; depth 100 - 200mm	-	-	-	-	1.62	17.35	0.46	17.81	nr	20.04
50mm diameter; depth 200 - 300mm	-	-	-	-	1.63	17.46	0.53	17.98	nr	20.23
100mm diameter; depth 200 - 300mm	-	-	-	-	1.79	19.17	0.56	19.73	nr	22.20
150mm diameter; depth 200 - 300mm	-	-	-	-	1.97	21.10	0.56	21.66	nr	24.37
Grouting into mortices with cement mortar (1:1); around steel										
50 x 50mm; depth not exceeding 100mm	-	-	-	-	0.12	1.29	0.62	1.91	nr	2.15
75 x 75mm; depth not exceeding 100mm	-	-	-	-	0.13	1.39	0.62	2.02	nr	2.27
100 x 100mm; depth 100 - 200mm	-	-	-	-	0.17	1.82	0.79	2.62	nr	2.94
38mm diameter; depth not exceeding 100mm	-	-	-	-	0.09	0.96	0.45	1.42	nr	1.59
Cutting on reinforced in-situ concrete										
Cutting holes										
225 x 150mm; depth not exceeding 100mm	-	-	-	-	1.64	17.56	0.53	18.09	nr	20.35
300 x 150mm; depth not exceeding 100mm	-	-	-	-	1.81	19.39	0.56	19.95	nr	22.44
300 x 300mm; depth not exceeding 100mm	-	-	-	-	1.99	21.31	0.56	21.88	nr	24.61
225 x 150mm; depth 100 - 200mm	-	-	-	-	2.39	25.60	0.77	26.37	nr	29.66
300 x 150mm; depth 100 - 200mm	-	-	-	-	2.63	28.17	0.81	28.97	nr	32.60
300 x 300mm; depth 100 - 200mm	-	-	-	-	2.89	30.95	0.81	31.76	nr	35.73
225 x 150mm; depth 200 - 300mm	-	-	-	-	2.78	29.77	0.87	30.64	nr	34.47
300 x 150mm; depth 200 - 300mm	-	-	-	-	3.06	32.77	0.95	33.73	nr	37.94
300 x 300mm; depth 200 - 300mm	-	-	-	-	3.33	35.66	0.95	36.62	nr	41.19
50mm diameter; depth not exceeding 100mm	-	-	-	-	1.05	11.25	0.33	11.58	nr	13.02
100mm diameter; depth not exceeding 100mm	-	-	-	-	1.18	12.64	0.38	13.02	nr	14.64
150mm diameter; depth not exceeding 100mm	-	-	-	-	1.30	13.92	0.38	14.30	nr	16.09
50mm diameter; depth 100 - 200mm	-	-	-	-	1.72	18.42	0.54	18.96	nr	21.33
100mm diameter; depth 100 - 200mm	-	-	-	-	1.89	20.24	0.57	20.82	nr	23.42
150mm diameter; depth 100 - 200mm	-	-	-	-	2.02	21.63	0.57	22.21	nr	24.98
50mm diameter; depth 200 - 300mm	-	-	-	-	2.04	21.85	0.64	22.48	nr	25.29
100mm diameter; depth 200 - 300mm	-	-	-	-	2.24	23.99	0.71	24.70	nr	27.79
150mm diameter; depth 200 - 300mm	-	-	-	-	2.46	26.35	0.71	27.06	nr	30.44
E42: ACCESSORIES CAST INTO IN-SITU CONCRETE										
Cast in accessories										
Galvanized steel dowels										
12mm diameter x 150mm long	0.70	10.00	0.77	-	0.09	0.96	-	1.74	nr	1.96
12mm diameter x 300mm long	1.03	10.00	1.14	-	0.08	0.86	-	1.99	nr	2.24
12mm diameter x 600mm long	1.56	10.00	1.72	-	0.12	1.29	-	3.01	nr	3.38

Labour hourly rates: (except Specialists) Craft Operatives £14.21 Labourer £10.71 Rates are national average prices. Refer to REGIONAL VARIATIONS for indicative levels of overall pricing in regions	MATERIALS			LABOUR				RATES		
	Del to Site £	Waste %	Material Cost £	Craft Optve Hrs	Lab Hrs	Labour Cost £	Sunds £	Nett Rate £	Unit	Gross Rate (+12.5%) £
E42: ACCESSORIES CAST INTO IN-SITU CONCRETE Cont.										
Cast in accessories Cont.										
Galvanized expanded metal tie										
300 x 50mm; bending and temporarily fixing to formwork	0.84	10.00	**0.92**	-	0.13	**1.39**	-	**2.31**	nr	**2.60**
Galvanized steel dovetailed masonry slots with 3mm thick twisted tie to suit 50mm cavity										
100mm long; temporarily fixing to formwork	0.85	10.00	**0.94**	0.11	-	**1.56**	-	**2.50**	nr	**2.81**
Stainless steel channel with anchors at 250mm centres										
28 x 15mm; temporarily fixing to formwork	12.58	5.00	**13.21**	0.28	-	**3.98**	-	**17.19**	m	**19.34**
28 x 15 x 150mm long; temporarily fixing to formwork	2.18	5.00	**2.28**	0.12	-	**1.71**	-	**3.99**	nr	**4.49**
M10 bolt 50mm long with 'T' head nut and washers	4.30	5.00	**4.51**	0.05	-	**0.71**	-	**5.22**	nr	**5.87**
fishtailed tie to suit 50mm cavity	1.81	5.00	**1.90**	0.06	-	**0.85**	-	**2.76**	nr	**3.10**
Stainless steel angle drilled at 450mm centres										
70 x 90 x 5mm	30.84	5.00	**32.38**	0.80	-	**11.37**	-	**43.75**	m	**49.22**
Copper dovetailed masonry slots with 3mm thick twisted tie to suit 50mm cavity										
100mm long; temporarily fixing to formwork	9.00	10.00	**9.89**	0.11	-	**1.56**	-	**11.46**	nr	**12.89**
Bolt boxes; The Expanded Metal Co. Ltd										
reference 220 for use with poured concrete; 75mm diameter										
150mm long	1.45	10.00	**1.59**	0.28	-	**3.98**	0.70	**6.27**	nr	**7.05**
225mm long	1.78	10.00	**1.96**	0.33	-	**4.69**	0.70	**7.34**	nr	**8.26**
300mm long	1.99	10.00	**2.19**	0.39	-	**5.54**	0.70	**8.43**	nr	**9.49**
reference 220 for use with poured concrete; 100mm diameter										
375mm long	3.23	10.00	**3.55**	0.44	-	**6.25**	0.70	**10.50**	nr	**11.81**
450mm long	3.63	10.00	**4.00**	0.50	-	**7.11**	0.70	**11.80**	nr	**13.27**
600mm long	4.03	10.00	**4.43**	0.55	-	**7.82**	0.70	**12.94**	nr	**14.56**
reference 220 for use with vibrated concrete; wrap with single layer of thin polythene sheet; 75mm diameter										
150mm long	1.58	10.00	**1.74**	0.28	-	**3.98**	0.84	**6.56**	nr	**7.38**
225mm long	1.90	10.00	**2.09**	0.33	-	**4.69**	0.84	**7.62**	nr	**8.57**
300mm long	2.20	10.00	**2.42**	0.39	-	**5.54**	0.84	**8.80**	nr	**9.90**
reference 220 for use with vibrated concrete; wrap with single layer of thin polythene sheet; 100mm diameter										
375mm long	3.42	10.00	**3.76**	0.44	-	**6.25**	0.84	**10.86**	nr	**12.21**
450mm long	3.86	10.00	**4.24**	0.50	-	**7.11**	0.84	**12.19**	nr	**13.71**
600mm long	4.25	10.00	**4.68**	0.55	-	**7.82**	0.84	**13.33**	nr	**15.00**
Steel rag bolts										
M 10 x 100mm long	1.40	5.00	**1.47**	0.10	-	**1.42**	-	**2.89**	nr	**3.26**
M 10 x 160mm long	1.51	5.00	**1.59**	0.12	-	**1.71**	-	**3.29**	nr	**3.71**
M 12 x 100mm long	1.55	5.00	**1.63**	0.10	-	**1.42**	-	**3.05**	nr	**3.43**
M 12 x 160mm long	1.80	5.00	**1.89**	0.12	-	**1.71**	-	**3.60**	nr	**4.05**
M 12 x 200mm long	1.94	5.00	**2.04**	0.14	-	**1.99**	-	**4.03**	nr	**4.53**
M 16 x 120mm long	2.42	5.00	**2.54**	0.10	-	**1.42**	-	**3.96**	nr	**4.45**
M 16 x 160mm long	2.78	5.00	**2.92**	0.12	-	**1.71**	-	**4.62**	nr	**5.20**
M 16 x 200mm long	3.08	5.00	**3.24**	0.14	-	**1.99**	-	**5.22**	nr	**5.88**
E60: PRECAST/COMPOSITE CONCRETE DECKING										
Prestressed concrete beams and 100mm building blocks; hoist bed and grout										
155mm thick floors (Bison 'House Beam')										
beams at 510mm centres	12.70	2.50	**13.02**	-	1.15	**12.32**	0.55	**25.88**	m2	**29.12**
beams at 510mm and 285mm centres	14.77	2.50	**15.13**	-	1.05	**11.25**	0.55	**26.93**	m2	**30.30**
double beams at 624mm centres	17.36	2.50	**17.80**	-	1.00	**10.71**	0.55	**29.06**	m2	**32.69**
beams at 285mm centres	18.49	2.50	**18.95**	-	1.15	**12.32**	0.55	**31.81**	m2	**35.79**
double beams at 399mm centres	24.45	2.50	**25.06**	-	1.05	**11.25**	0.55	**36.86**	m2	**41.47**
beams at 114mm centres with concrete between	41.10	2.50	**42.13**	-	1.00	**10.71**	0.55	**53.39**	m2	**60.06**
225mm thick floors (Bison 'Deep Beam')										
beams at 530mm centres	17.34	2.50	**17.77**	-	1.15	**12.32**	0.55	**30.64**	m2	**34.47**
beams at 305mm centres	26.17	2.50	**26.83**	-	1.05	**11.25**	0.55	**38.62**	m2	**43.45**
beams at 140mm centres with concrete between	53.23	2.50	**54.56**	-	1.00	**10.71**	0.55	**65.82**	m2	**74.05**
Precast concrete hollowcore floors and hoist bed and grout (HANSON CONCRETE PRODUCTS LTD)										
150 mm thick planks 750 wide										
span between supports 7.6m; UDL not exceeding 2KN/m2	38.74	2.50	**39.71**	-	-	**-**	-	**39.71**	m2	**44.67**
span between supports 6.6m; UDL not exceeding 4KN/m2	38.74	2.50	**39.71**	-	-	**-**	-	**39.71**	m2	**44.67**
span between supports 5.9m; UDL not exceeding 6KN/m2	38.74	2.50	**39.71**	-	-	**-**	-	**39.71**	m2	**44.67**
span between supports 5.4m; UDL not exceeding 8KN/m2	38.74	2.50	**39.71**	-	-	**-**	-	**39.71**	m2	**44.67**
span between supports 5.0m; UDL not exceeding 10KN/m2	38.74	2.50	**39.71**	-	-	**-**	-	**39.71**	m2	**44.67**
200 mm thick planks 750 wide										
span between supports 9.3m; UDL not exceeding 2KN/m2	42.40	2.50	**43.46**	-	-	**-**	-	**43.46**	m2	**48.90**

Labour hourly rates: (except Specialists) Craft Operatives £14.21 Labourer £10.71 Rates are national average prices. Refer to REGIONAL VARIATIONS for indicative levels of overall pricing in regions	MATERIALS			LABOUR				RATES		
	Del to Site £	Waste %	Material Cost £	Craft Optve Hrs	Lab Hrs	Labour Cost £	Sunds £	Nett Rate £	Unit	Gross Rate (+12.5%) £
E60: PRECAST/COMPOSITE CONCRETE DECKING Cont.										
Precast concrete hollowcore floors and hoist bed and grout (HANSON CONCRETE PRODUCTS LTD) Cont.										
200 mm thick planks 750 wide Cont.										
span between supports 8.1m; UDL not exceeding 4KN/m2	42.40	2.50	**43.46**	-	-	-	-	**43.46**	m2	**48.90**
span between supports 7.3m; UDL not exceeding 6KN/m2	42.40	2.50	**43.46**	-	-	-	-	**43.46**	m2	**48.90**
span between supports 6.7m; UDL not exceeding 8KN/m2	42.40	2.50	**43.46**	-	-	-	-	**43.46**	m2	**48.90**
span between supports 6.2m; UDL not exceeding 10KN/m2	42.40	2.50	**43.46**	-	-	-	-	**43.46**	m2	**48.90**
250 mm thick planks 750 wide										
span between supports 10.6m; UDL not exceeding 2KN/m2	46.27	2.50	**47.42**	-	-	-	-	**47.42**	m2	**53.35**
span between supports 9.3m; UDL not exceeding 4KN/m2	46.27	2.50	**47.42**	-	-	-	-	**47.42**	m2	**53.35**
span between supports 8.5m; UDL not exceeding 6KN/m2	46.27	2.50	**47.42**	-	-	-	-	**47.42**	m2	**53.35**
span between supports 7.8m; UDL not exceeding 8KN/m2	46.27	2.50	**47.42**	-	-	-	-	**47.42**	m2	**53.35**
span between supports 7.2m; UDL not exceeding 10KN/m2	46.27	2.50	**47.42**	-	-	-	-	**47.42**	m2	**53.35**
300 mm thick planks 750 wide										
span between supports 11.5m; UDL not exceeding 2KN/m2	50.11	2.50	**51.36**	-	-	-	-	**51.36**	m2	**57.78**
span between supports 10.3m; UDL not exceeding 4KN/m2	50.11	2.50	**51.36**	-	-	-	-	**51.36**	m2	**57.78**
span between supports 9.4m; UDL not exceeding 6KN/m2	50.11	2.50	**51.36**	-	-	-	-	**51.36**	m2	**57.78**
span between supports 8.7m; UDL not exceeding 8KN/m2	50.11	2.50	**51.36**	-	-	-	-	**51.36**	m2	**57.78**
span between supports 8.1m; UDL not exceeding 10KN/m2	50.11	2.50	**51.36**	-	-	-	-	**51.36**	m2	**57.78**
Passive fall arrest provision										
Add to the foregoing rates for providing safety decking, airbags or netting as appropriate to the works										
total area less than 100 m2	5.01	2.50	**5.13**	-	-	-	-	**5.13**	m2	**5.77**
total area over 100m2	3.80	2.50	**3.90**	-	-	-	-	**3.90**	m2	**4.39**

MASONRY

Labour hourly rates: (except Specialists) Craft Operatives £14.21 Labourer £10.71 Rates are national average prices. Refer to REGIONAL VARIATIONS for indicative levels of overall pricing in regions	MATERIALS			LABOUR				RATES		
	Del to Site £	Waste %	Material Cost £	Craft Optve Hrs	Lab Hrs	Labour Cost £	Sunds £	Nett Rate £	Unit	Gross Rate (+12.5%) £
F10: BRICK/BLOCK WALLING										
Common bricks, BS EN 772, Category M, 215 x 102.5 x 65mm, compressive strength 20.5 N/mm2; in cement-lime mortar (1:2:9)										
Walls; vertical										
102mm thick; stretcher bond	12.04	5.00	**12.64**	1.45	1.60	**37.74**	2.90	**53.28**	m2	**59.94**
215mm thick; English bond	24.28	5.00	**25.49**	2.35	2.65	**61.78**	6.76	**94.03**	m2	**105.78**
327mm thick; English bond	36.52	5.00	**38.34**	2.85	3.30	**75.84**	10.63	**124.82**	m2	**140.42**
extra; grooved bricks; 50% wall surface keyed	0.18	5.00	**0.18**	-	-	**-**	-	**0.18**	m2	**0.21**
extra; grooved bricks; 100% wall surface keyed	0.35	5.00	**0.36**	-	-	**-**	-	**0.36**	m2	**0.41**
Walls; building against concrete (ties measured separately); vertical										
102mm thick; stretcher bond	12.04	5.00	**12.64**	1.60	1.75	**41.48**	2.90	**57.02**	m2	**64.15**
Walls; building against old brickwork; tie new to old with 3mm diameter galvanized twisted wire ties - 6/m2; vertical										
102mm thick; stretcher bond	12.85	10.00	**13.53**	1.60	1.75	**41.48**	2.90	**57.92**	m2	**65.16**
215mm thick; English bond	25.09	10.00	**26.39**	2.65	2.95	**69.25**	6.76	**102.40**	m2	**115.20**
Walls; bonding to stonework; including extra material; vertical										
102mm thick; stretcher bond	12.72	5.00	**13.35**	1.60	1.75	**41.48**	2.90	**57.73**	m2	**64.95**
215mm thick; English bond	25.64	5.00	**26.92**	2.65	2.95	**69.25**	6.76	**102.93**	m2	**115.80**
327mm thick; English bond	37.88	5.00	**39.77**	3.15	3.60	**83.32**	10.63	**133.72**	m2	**150.43**
440mm thick; English bond	49.91	5.00	**52.41**	3.70	4.30	**98.63**	14.49	**165.53**	m2	**186.22**
Walls; bonding to old brickwork; cutting pockets; including extra material; vertical										
102mm thick; stretcher bond	12.72	5.00	**13.35**	3.65	3.80	**92.56**	2.90	**108.82**	m2	**122.42**
215mm thick; English bond	25.64	5.00	**26.92**	4.60	4.90	**117.85**	6.76	**151.53**	m2	**170.47**
Walls; curved 2m radius; including extra material; vertical										
102mm thick; stretcher bond	12.04	5.00	**12.64**	2.85	3.00	**72.63**	2.90	**88.17**	m2	**99.19**
215mm thick; English bond	24.28	5.00	**25.49**	5.20	5.50	**132.80**	6.76	**165.05**	m2	**185.68**
327mm thick; English bond	36.52	5.00	**38.34**	7.10	6.81	**173.83**	10.63	**222.80**	m2	**250.65**
Walls; curved 6m radius; including extra material; vertical										
102mm thick; stretcher bond	12.04	5.00	**12.64**	2.15	2.30	**55.18**	2.90	**70.72**	m2	**79.57**
215mm thick; English bond	24.28	5.00	**25.49**	3.75	4.05	**96.66**	6.76	**128.91**	m2	**145.03**
327mm thick; English bond	36.52	5.00	**38.34**	4.95	5.40	**128.17**	10.63	**177.15**	m2	**199.29**
Isolated piers; vertical										
215mm thick; English bond	24.28	5.00	**25.49**	4.70	5.00	**120.34**	6.76	**152.59**	m2	**171.66**
327mm thick; English bond	36.52	5.00	**38.34**	5.70	6.15	**146.86**	10.63	**195.84**	m2	**220.32**
440mm thick; English bond	48.55	5.00	**50.98**	6.60	7.20	**170.90**	14.49	**236.37**	m2	**265.91**
Chimney stacks; vertical										
440mm thick; English bond	48.55	5.00	**50.98**	6.60	7.20	**170.90**	14.49	**236.37**	m2	**265.91**
890mm thick; English bond	97.10	5.00	**101.96**	10.40	11.60	**272.02**	28.98	**402.96**	m2	**453.33**
Projections; vertical										
215mm wide x 112mm projection	2.72	5.00	**2.86**	0.30	0.35	**8.01**	0.77	**11.64**	m	**13.10**
215mm wide x 215mm projection	5.44	5.00	**5.71**	0.50	0.60	**13.53**	1.55	**20.79**	m	**23.39**
327mm wide x 112mm projection	4.08	5.00	**4.28**	0.45	0.55	**12.29**	1.16	**17.73**	m	**19.95**
327mm wide x 215mm projection	8.16	5.00	**8.57**	0.75	0.95	**20.83**	2.32	**31.72**	m	**35.69**
Projections; horizontal										
215mm wide x 112mm projection	2.72	5.00	**2.86**	0.30	0.35	**8.01**	0.77	**11.64**	m	**13.10**
215mm wide x 215mm projection	5.44	5.00	**5.71**	0.50	0.60	**13.53**	1.55	**20.79**	m	**23.39**
327mm wide x 112mm projection	4.08	5.00	**4.28**	0.45	0.55	**12.29**	1.16	**17.73**	m	**19.95**
327mm wide x 215mm projection	8.16	5.00	**8.57**	0.75	0.95	**20.83**	2.32	**31.72**	m	**35.69**
Projections; bonding to old brickwork; cutting pockets; including extra material; vertical										
215mm wide x 112mm projection	4.53	5.00	**4.76**	0.93	0.98	**23.71**	1.29	**29.76**	m	**33.48**

Labour hourly rates: (except Specialists) Craft Operatives £14.21 Labourer £10.71 Rates are national average prices. Refer to REGIONAL VARIATIONS for indicative levels of overall pricing in regions	MATERIALS			LABOUR				RATES		
	Del to Site £	Waste %	Material Cost £	Craft Optve Hrs	Lab Hrs	Labour Cost £	Sunds £	Nett Rate £	Unit	Gross Rate (+12.5%) £
F10: BRICK/BLOCK WALLING Cont.										
Common bricks, BS EN 772, Category M, 215 x 102.5 x 65mm, compressive strength 20.5 N/mm2; in cement-lime mortar (1:2:9) Cont.										
Projections; bonding to old brickwork; cutting pockets; including extra material; vertical Cont.										
215mm wide x 215mm projection	7.25	5.00	**7.61**	1.13	1.23	**29.23**	2.06	**38.91**	m	**43.77**
327mm wide x 112mm projection	6.80	5.00	**7.14**	1.40	1.50	**35.96**	1.94	**45.03**	m	**50.66**
327mm wide x 215mm projection	9.97	5.00	**10.47**	1.70	1.90	**44.51**	2.84	**57.81**	m	**65.04**
Projections; bonding to old brickwork; cutting pockets; including extra material; horizontal										
215mm wide x 112mm projection	4.53	5.00	**4.76**	0.93	0.93	**23.18**	1.29	**29.22**	m	**32.88**
215mm wide x 215mm projection	7.25	5.00	**7.61**	1.13	1.23	**29.23**	2.06	**38.91**	m	**43.77**
327mm wide x 112mm projection	6.80	5.00	**7.14**	1.40	1.50	**35.96**	1.94	**45.03**	m	**50.66**
327mm wide x 215mm projection	9.97	5.00	**10.47**	1.70	1.90	**44.51**	2.84	**57.81**	m	**65.04**
Arches including centering; flat										
112mm high on face; 215mm thick; width of exposed soffit 215mm; bricks-on-edge	2.72	5.00	**2.86**	1.50	1.25	**34.70**	0.77	**38.33**	m	**43.12**
215mm high on face; 112mm thick; width of exposed soffit 112mm; bricks-on-end	2.72	5.00	**2.86**	1.45	1.25	**33.99**	0.77	**37.62**	m	**42.32**
Arches including centering; semi-circular										
112mm high on face; 215mm thick; width of exposed soffit 215mm; bricks-on-edge	4.27	5.00	**4.48**	2.15	1.60	**47.69**	1.21	**53.39**	m	**60.06**
215mm high on face; 215mm thick; width of exposed soffit 215mm; bricks-on-edge	8.54	5.00	**8.97**	2.95	2.45	**68.16**	2.43	**79.56**	m	**89.50**
Closing cavities; vertical										
50mm wide with brickwork 102mm thick	1.33	5.00	**1.39**	0.40	0.40	**9.97**	0.38	**11.74**	m	**13.20**
75mm wide with brickwork 102mm thick	1.33	5.00	**1.39**	0.40	0.40	**9.97**	0.38	**11.74**	m	**13.20**
Closing cavities; horizontal										
50mm wide with slates	2.11	5.00	**2.22**	0.66	0.66	**16.45**	0.38	**19.04**	m	**21.42**
75mm wide with slates	2.11	5.00	**2.22**	0.66	0.66	**16.45**	0.38	**19.04**	m	**21.42**
50mm wide with brickwork 102mm thick	1.33	5.00	**1.39**	0.40	0.40	**9.97**	0.38	**11.74**	m	**13.20**
75mm wide with brickwork 102mm thick	1.33	5.00	**1.39**	0.40	0.40	**9.97**	0.38	**11.74**	m	**13.20**
Bonding ends to existing common brickwork; cutting pockets; extra material										
walls; bonding every third course										
102mm thick	0.91	5.00	**0.95**	0.37	0.37	**9.22**	0.26	**10.43**	m	**11.73**
215mm thick	1.81	5.00	**1.90**	0.66	0.66	**16.45**	0.52	**18.87**	m	**21.22**
327mm thick	2.72	5.00	**2.86**	0.95	0.95	**23.67**	0.77	**27.30**	m	**30.72**
Second hard stock bricks, BS EN 772, Category M, 215 x 102.5 x 65mm; in cement-lime mortar (1:2:9)										
Walls; vertical										
102mm thick; stretcher bond	23.72	5.00	**24.90**	1.45	1.60	**37.74**	2.90	**65.55**	m2	**73.74**
215mm thick; English bond	47.84	5.00	**50.23**	2.35	2.65	**61.78**	6.76	**118.77**	m2	**133.61**
327mm thick; English bond	71.96	5.00	**75.56**	2.85	3.30	**75.84**	10.63	**162.03**	m2	**182.28**
Walls; building against concrete (ties measured separately); vertical										
102mm thick; stretcher bond	23.72	5.00	**24.90**	1.60	1.75	**41.48**	2.90	**69.28**	m2	**77.95**
Walls; building against old brickwork; tie new to old with 3mm diameter galvanized twisted wire ties - 6/m2; vertical										
102mm thick; stretcher bond	24.53	10.00	**25.80**	1.60	1.75	**41.48**	2.90	**70.18**	m2	**78.95**
215mm thick; English bond	48.65	10.00	**51.13**	2.65	2.95	**69.25**	6.76	**127.14**	m2	**143.03**
Walls; bonding to stonework; including extra material; vertical										
102mm thick; stretcher bond	25.06	5.00	**26.31**	1.60	1.75	**41.48**	3.06	**70.85**	m2	**79.71**
215mm thick; English bond	50.52	5.00	**53.05**	2.65	2.95	**69.25**	7.15	**129.45**	m2	**145.63**
327mm thick; English bond	74.64	5.00	**78.37**	3.15	3.60	**83.32**	11.03	**172.72**	m2	**194.31**
440mm thick; English bond	98.36	5.00	**103.28**	3.70	4.30	**98.63**	14.90	**216.80**	m2	**243.91**
Walls; bonding to old brickwork; cutting pockets; including extra material; vertical										
102mm thick; stretcher bond	25.06	5.00	**26.31**	3.65	3.80	**92.56**	3.06	**121.94**	m2	**137.18**
215mm thick; English bond	50.52	5.00	**53.05**	4.60	4.90	**117.85**	7.15	**178.04**	m2	**200.29**
Walls; curved 2m radius; including extra material; vertical										
102mm thick; stretcher bond	23.72	5.00	**24.90**	2.85	3.00	**72.63**	2.90	**100.43**	m2	**112.99**
215mm thick; English bond	47.84	5.00	**50.23**	5.20	5.50	**132.80**	6.76	**189.79**	m2	**213.51**
327mm thick; English bond	71.96	5.00	**75.56**	7.10	6.81	**173.83**	10.63	**260.01**	m2	**292.52**
Walls; curved 6m radius; including extra material; vertical										
102mm thick; stretcher bond	23.72	5.00	**24.90**	2.15	2.30	**55.18**	2.90	**82.99**	m2	**93.36**
215mm thick; English bond	47.84	5.00	**50.23**	3.75	4.05	**96.66**	6.76	**153.65**	m2	**172.86**
327mm thick; English bond	71.96	5.00	**75.56**	4.95	5.40	**128.17**	10.63	**214.36**	m2	**241.16**
Isolated piers; vertical										
215mm thick; English bond	47.84	5.00	**50.23**	4.70	5.00	**120.34**	6.76	**177.33**	m2	**199.49**
327mm thick; English bond	71.96	5.00	**75.56**	5.70	6.15	**146.86**	10.63	**233.05**	m2	**262.18**
440mm thick; English bond	95.68	5.00	**100.46**	6.60	7.20	**170.90**	14.49	**285.85**	m2	**321.58**

Labour hourly rates: (except Specialists) Craft Operatives £14.21 Labourer £10.71 Rates are national average prices. Refer to REGIONAL VARIATIONS for indicative levels of overall pricing in regions	MATERIALS			LABOUR				RATES		
	Del to Site £	Waste %	Material Cost £	Craft Optve Hrs	Lab Hrs	Labour Cost £	Sunds £	Nett Rate £	Unit	Gross Rate (+12.5%) £
F10: BRICK/BLOCK WALLING Cont.										
Second hard stock bricks, BS EN 772, Category M, 215 x 102.5 x 65mm; in cement-lime mortar (1:2:9) Cont.										
Chimney stacks; vertical										
440mm thick; English bond	95.68	5.00	**100.46**	6.60	7.20	**170.90**	14.49	**285.85**	m2	**321.58**
890mm thick; English bond	191.35	5.00	**200.92**	10.40	11.60	**272.02**	28.98	**501.92**	m2	**564.66**
Projections; vertical										
215mm wide x 112mm projection	5.36	5.00	**5.63**	0.30	0.35	**8.01**	0.77	**14.41**	m	**16.21**
215mm wide x 215mm projection	10.72	5.00	**11.25**	0.50	0.60	**13.53**	1.55	**26.33**	m	**29.62**
327mm wide x 112mm projection	8.04	5.00	**8.44**	0.45	0.55	**12.29**	1.16	**21.89**	m	**24.62**
327mm wide x 215mm projection	16.08	5.00	**16.88**	0.75	0.95	**20.83**	2.32	**40.04**	m	**45.04**
Projections; horizontal										
215mm wide x 112mm projection	5.36	5.00	**5.63**	0.30	0.35	**8.01**	0.77	**14.41**	m	**16.21**
215mm wide x 215mm projection	10.72	5.00	**11.25**	0.50	0.60	**13.53**	1.55	**26.33**	m	**29.62**
327mm wide x 112mm projection	8.04	5.00	**8.44**	0.45	0.55	**12.29**	1.16	**21.89**	m	**24.62**
327mm wide x 215mm projection	16.08	5.00	**16.88**	0.75	0.95	**20.83**	2.32	**40.04**	m	**45.04**
Projections; bonding to old brickwork; cutting pockets; including extra material; vertical										
215mm wide x 112mm projection	8.93	5.00	**9.37**	0.93	0.98	**23.71**	1.29	**34.38**	m	**38.67**
215mm wide x 215mm projection	14.29	5.00	**15.00**	1.13	1.23	**29.23**	2.06	**46.30**	m	**52.08**
327mm wide x 112mm projection	13.40	5.00	**14.07**	1.40	1.50	**35.96**	1.94	**51.96**	m	**58.46**
327mm wide x 215mm projection	19.65	5.00	**20.63**	1.70	1.90	**44.51**	2.84	**67.98**	m	**76.47**
Projections; bonding to old brickwork; cutting pockets; including extra material; horizontal										
215mm wide x 112mm projection	8.93	5.00	**9.37**	0.93	0.93	**23.18**	1.29	**33.84**	m	**38.07**
215mm wide x 215mm projection	14.29	5.00	**15.00**	1.13	1.23	**29.23**	2.06	**46.30**	m	**52.08**
327mm wide x 112mm projection	13.40	5.00	**14.07**	1.40	1.50	**35.96**	1.94	**51.96**	m	**58.46**
327mm wide x 215mm projection	19.65	5.00	**20.63**	1.70	1.90	**44.51**	2.84	**67.98**	m	**76.47**
Arches including centering; flat										
112mm high on face; 215mm thick; width of exposed soffit 215mm; bricks-on-edge	5.36	5.00	**5.63**	1.50	1.25	**34.70**	0.77	**41.10**	m	**46.24**
215mm high on face; 112mm thick; width of exposed soffit 112mm; bricks-on-end	5.36	5.00	**5.63**	1.45	1.25	**33.99**	0.77	**40.39**	m	**45.44**
Arches including centering; semi-circular										
112mm high on face; 215mm thick; width of exposed soffit 215mm; bricks-on-edge	8.41	5.00	**8.83**	2.15	1.60	**47.69**	1.21	**57.74**	m	**64.95**
215mm high on face; 215mm thick; width of exposed soffit 215mm; bricks-on-edge	16.83	5.00	**17.67**	2.95	2.45	**68.16**	2.43	**88.26**	m	**99.29**
Closing cavities; vertical										
50mm wide with brickwork 102mm thick	2.61	5.00	**2.74**	0.40	0.40	**9.97**	0.38	**13.09**	m	**14.72**
75mm wide with brickwork 102mm thick	2.61	5.00	**2.74**	0.40	0.40	**9.97**	0.38	**13.09**	m	**14.72**
Closing cavities; horizontal										
50mm wide with slates	2.11	5.00	**2.22**	0.66	0.66	**16.45**	0.38	**19.04**	m	**21.42**
75mm wide with slates	2.11	5.00	**2.22**	0.66	0.66	**16.45**	0.38	**19.04**	m	**21.42**
50mm wide with brickwork 102mm thick	2.61	5.00	**2.74**	0.40	0.40	**9.97**	0.38	**13.09**	m	**14.72**
75mm wide with brickwork 102mm thick	2.61	5.00	**2.74**	0.40	0.40	**9.97**	0.38	**13.09**	m	**14.72**
Bonding ends to existing common brickwork; cutting pockets; extra material										
walls; bonding every third course										
102mm thick	1.78	5.00	**1.87**	0.37	0.37	**9.22**	0.26	**11.35**	m	**12.77**
215mm thick	3.57	5.00	**3.75**	0.66	0.66	**16.45**	0.52	**20.71**	m	**23.30**
327mm thick	5.36	5.00	**5.63**	0.95	0.95	**23.67**	0.77	**30.07**	m	**33.83**
Engineering bricks, BS EN 772, Category F, 215 x 102.5 x 65, Class B; in cement-lime mortar (1:2:9)										
Walls; vertical										
102mm thick; stretcher bond	12.63	5.00	**13.26**	1.60	1.75	**41.48**	2.90	**57.64**	m2	**64.84**
215mm thick; English bond	25.47	5.00	**26.74**	2.60	2.90	**68.01**	6.76	**101.51**	m2	**114.19**
327mm thick; English bond	38.31	5.00	**40.22**	3.15	3.60	**83.32**	10.63	**134.17**	m2	**150.94**
Walls; building against concrete (ties measured separately); vertical										
102mm thick; stretcher bond	12.63	5.00	**13.26**	1.75	1.90	**45.22**	2.90	**61.38**	m2	**69.05**
Walls; building against old brickwork; tie new to old with 3mm diameter galvanized twisted wire ties - 6/m2; vertical										
102mm thick; stretcher bond	13.44	10.00	**14.15**	1.75	1.90	**45.22**	2.90	**62.27**	m2	**70.06**
215mm thick; English bond	26.28	10.00	**27.64**	2.90	3.20	**75.48**	6.76	**109.88**	m2	**123.61**
Walls; bonding to stonework; including extra material; vertical										
102mm thick; stretcher bond	13.34	5.00	**14.01**	1.75	1.90	**45.22**	3.06	**62.29**	m2	**70.07**
215mm thick; English bond	26.89	5.00	**28.24**	2.90	3.20	**75.48**	7.15	**110.87**	m2	**124.73**
327mm thick; English bond	39.73	5.00	**41.72**	3.45	3.90	**90.79**	11.03	**143.54**	m2	**161.49**
440mm thick; English bond	52.36	5.00	**54.98**	4.05	4.65	**107.35**	14.90	**177.23**	m2	**199.38**
Walls; bonding to old brickwork; cutting pockets; including extra material; vertical										
102mm thick; stretcher bond	13.34	5.00	**14.01**	4.00	4.15	**101.29**	3.06	**118.36**	m2	**133.15**
215mm thick; English bond	26.89	5.00	**28.24**	5.05	5.35	**129.06**	7.15	**164.45**	m2	**185.00**

Labour hourly rates: (except Specialists) Craft Operatives £14.21 Labourer £10.71 Rates are national average prices. Refer to REGIONAL VARIATIONS for indicative levels of overall pricing in regions	MATERIALS			LABOUR				RATES		
	Del to Site £	Waste %	Material Cost £	Craft Optve Hrs	Lab Hrs	Labour Cost £	Sunds £	Nett Rate £	Unit	Gross Rate (+12.5%) £
F10: BRICK/BLOCK WALLING Cont.										
Engineering bricks, BS EN 772, Category F, 215 x 102.5 x 65, Class B; in cement-lime mortar (1:2:9) Cont.										
Walls; curved 2m radius; including extra material; vertical										
102mm thick; stretcher bond	12.63	5.00	**13.26**	3.15	3.30	**80.10**	2.90	**96.26**	m2	**108.30**
215mm thick; English bond	25.47	5.00	**26.74**	5.70	6.00	**145.26**	6.76	**178.76**	m2	**201.10**
327mm thick; English bond	38.31	5.00	**40.22**	7.80	8.25	**199.20**	10.63	**250.05**	m2	**281.30**
Walls; curved 6m radius; including extra material; vertical										
102mm thick; stretcher bond	12.63	5.00	**13.26**	2.35	2.50	**60.17**	2.90	**76.33**	m2	**85.87**
215mm thick; English bond	25.47	5.00	**26.74**	4.10	4.40	**105.39**	6.76	**138.89**	m2	**156.25**
327mm thick; English bond	38.31	5.00	**40.22**	5.45	5.90	**140.63**	10.63	**191.49**	m2	**215.42**
Isolated piers; vertical										
215mm thick; English bond	25.47	5.00	**26.74**	5.15	5.45	**131.55**	6.76	**165.05**	m2	**185.68**
327mm thick; English bond	38.31	5.00	**40.22**	6.25	6.70	**160.57**	10.63	**211.42**	m2	**237.85**
440mm thick; English bond	50.93	5.00	**53.48**	7.25	7.85	**187.10**	14.49	**255.07**	m2	**286.95**
Chimney stacks; vertical										
440mm thick; English bond	50.93	5.00	**53.48**	7.25	7.85	**187.10**	14.49	**255.07**	m2	**286.95**
890mm thick; English bond	101.86	5.00	**106.96**	11.45	12.65	**298.19**	28.98	**434.13**	m2	**488.39**
Projections; vertical										
215mm wide x 112mm projection	2.85	5.00	**3.00**	0.33	0.38	**8.76**	0.77	**12.53**	m	**14.09**
215mm wide x 215mm projection	5.71	5.00	**5.99**	0.55	0.65	**14.78**	1.55	**22.32**	m	**25.10**
327mm wide x 112mm projection	4.28	5.00	**4.49**	0.50	0.60	**13.53**	1.16	**19.19**	m	**21.58**
327mm wide x 215mm projection	8.56	5.00	**8.99**	0.83	1.03	**22.83**	2.32	**34.14**	m	**38.40**
Projections; horizontal										
215mm wide x 112mm projection	2.85	5.00	**3.00**	0.33	0.38	**8.76**	0.77	**12.53**	m	**14.09**
215mm wide x 215mm projection	5.71	5.00	**5.99**	0.55	0.65	**14.78**	1.55	**22.32**	m	**25.10**
327mm wide x 112mm projection	4.28	5.00	**4.49**	0.50	0.60	**13.53**	1.16	**19.19**	m	**21.58**
327mm wide x 215mm projection	8.56	5.00	**8.99**	0.83	1.03	**22.83**	2.32	**34.14**	m	**38.40**
Projections; bonding to old brickwork; cutting pockets; including extra material; vertical										
215mm wide x 112mm projection	4.75	5.00	**4.99**	1.02	1.07	**25.95**	1.29	**32.23**	m	**36.26**
215mm wide x 215mm projection	7.61	5.00	**7.99**	1.24	1.34	**31.97**	2.06	**42.02**	m	**47.27**
327mm wide x 112mm projection	7.13	5.00	**7.49**	1.54	1.64	**39.45**	1.94	**48.87**	m	**54.98**
327mm wide x 215mm projection	10.46	5.00	**10.98**	1.87	2.07	**48.74**	2.84	**62.56**	m	**70.38**
Projections; bonding to old brickwork; cutting pockets; including extra material; horizontal										
215mm wide x 112mm projection	4.75	5.00	**4.99**	1.02	1.07	**25.95**	1.29	**32.23**	m	**36.26**
215mm wide x 215mm projection	7.61	5.00	**7.99**	1.24	1.34	**31.97**	2.06	**42.02**	m	**47.27**
327mm wide x 112mm projection	7.13	5.00	**7.49**	1.54	1.64	**39.45**	1.94	**48.87**	m	**54.98**
327mm wide x 215mm projection	10.46	5.00	**10.98**	1.87	2.07	**48.74**	2.84	**62.56**	m	**70.38**
Arches including centering; flat										
112mm high on face; 215mm thick; width of exposed soffit 215mm; bricks-on-edge	2.85	5.00	**3.00**	1.60	1.30	**36.66**	0.77	**40.43**	m	**45.48**
215mm high on face; 112mm thick; width of exposed soffit 112mm; bricks-on-end	2.85	5.00	**3.00**	1.55	1.35	**36.48**	0.77	**40.25**	m	**45.28**
Arches including centering; semi-circular										
112mm high on face; 215mm thick; width of exposed soffit 215mm; bricks-on-edge	4.48	5.00	**4.70**	2.25	1.65	**49.64**	1.21	**55.56**	m	**62.51**
215mm high on face; 215mm thick; width of exposed soffit 215mm; bricks-on-edge	8.96	5.00	**9.41**	3.10	2.55	**71.36**	2.43	**83.20**	m	**93.60**
Closing cavities; vertical										
50mm wide with brickwork 102mm thick	1.39	5.00	**1.46**	0.44	0.44	**10.96**	0.38	**12.80**	m	**14.40**
75mm wide with brickwork 102mm thick	1.39	5.00	**1.46**	0.44	0.44	**10.96**	0.38	**12.80**	m	**14.40**
Closing cavities; horizontal										
50mm wide with slates	2.11	5.00	**2.22**	0.66	0.66	**16.45**	0.38	**19.04**	m	**21.42**
75mm wide with slates	2.11	5.00	**2.22**	0.66	0.66	**16.45**	0.38	**19.04**	m	**21.42**
50mm wide with brickwork 102mm thick	1.39	5.00	**1.46**	0.44	0.44	**10.96**	0.38	**12.80**	m	**14.40**
75mm wide with brickwork 102mm thick	1.39	5.00	**1.46**	0.44	0.44	**10.96**	0.38	**12.80**	m	**14.40**
Bonding ends to existing common brickwork; cutting pockets; extra material										
walls; bonding every third course										
102mm thick	0.95	5.00	**1.00**	0.41	0.41	**10.22**	0.26	**11.47**	m	**12.91**
215mm thick	1.90	5.00	**2.00**	0.73	0.73	**18.19**	0.52	**20.70**	m	**23.29**
327mm thick	2.85	5.00	**3.00**	0.95	0.95	**23.67**	0.77	**27.44**	m	**30.87**
Engineering bricks, BS EN 772, Category F, 215 x 102.5 x 65mm, Class A; in cement-lime mortar (1:2:9)										
Walls; vertical										
102mm thick; stretcher bond	17.82	5.00	**18.71**	1.60	1.75	**41.48**	2.90	**63.09**	m2	**70.98**
215mm thick; English bond	35.94	5.00	**37.73**	2.60	2.90	**68.01**	6.76	**112.50**	m2	**126.56**
327mm thick; English bond	54.06	5.00	**56.76**	3.15	3.60	**83.32**	10.63	**150.71**	m2	**169.55**
Walls; building against concrete (ties measured separately); vertical										
102mm thick; stretcher bond	17.82	5.00	**18.71**	1.75	1.90	**45.22**	2.90	**66.83**	m2	**75.18**

Labour hourly rates: (except Specialists) Craft Operatives £14.21 Labourer £10.71 Rates are national average prices. Refer to REGIONAL VARIATIONS for indicative levels of overall pricing in regions	MATERIALS			LABOUR				RATES		
	Del to Site £	Waste %	Material Cost £	Craft Optve Hrs	Lab Hrs	Labour Cost £	Sunds £	Nett Rate £	Unit	Gross Rate (+12.5%) £
F10: BRICK/BLOCK WALLING Cont.										
Engineering bricks, BS EN 772, Category F, 215 x 102.5 x 65mm, Class A; in cement-lime mortar (1:2:9) Cont.										
Walls; building against old brickwork; tie new to old with 3mm diameter galvanized twisted wire ties - 6/m2; vertical										
102mm thick; stretcher bond	18.63	10.00	**19.61**	1.75	1.90	**45.22**	2.90	**67.72**	m2	**76.19**
215mm thick; English bond	36.75	10.00	**38.63**	2.90	3.20	**75.48**	6.76	**120.87**	m2	**135.98**
Walls; bonding to stonework; including extra material; vertical										
102mm thick; stretcher bond	18.82	5.00	**19.76**	1.75	1.90	**45.22**	3.06	**68.05**	m2	**76.55**
215mm thick; English bond	37.95	5.00	**39.85**	2.90	3.20	**75.48**	7.15	**122.48**	m2	**137.79**
327mm thick; English bond	56.07	5.00	**58.88**	3.45	3.90	**90.79**	11.03	**160.70**	m2	**180.79**
440mm thick; English bond	73.89	5.00	**77.58**	4.05	4.65	**107.35**	14.90	**199.84**	m2	**224.82**
Walls; bonding to old brickwork; cutting pockets; including extra material; vertical										
102mm thick; stretcher bond	18.82	5.00	**19.76**	4.00	4.15	**101.29**	3.06	**124.12**	m2	**139.63**
215mm thick; English bond	37.95	5.00	**39.85**	5.05	5.35	**129.06**	7.15	**176.06**	m2	**198.06**
Walls; curved 2m radius; including extra material; vertical										
102mm thick; stretcher bond	17.82	5.00	**18.71**	3.15	3.30	**80.10**	2.90	**101.72**	m2	**114.43**
215mm thick; English bond	35.94	5.00	**37.73**	5.70	6.00	**145.26**	6.76	**189.75**	m2	**213.47**
327mm thick; English bond	54.06	5.00	**56.76**	7.80	8.25	**199.20**	10.63	**266.59**	m2	**299.91**
Walls; curved 6m radius; including extra material; vertical										
102mm thick; stretcher bond	17.82	5.00	**18.71**	2.35	2.50	**60.17**	2.90	**81.78**	m2	**92.00**
215mm thick; English bond	35.94	5.00	**37.73**	4.10	4.40	**105.39**	6.76	**149.88**	m2	**168.62**
327mm thick; English bond	54.06	5.00	**56.76**	5.45	5.90	**140.63**	10.63	**208.03**	m2	**234.03**
Isolated piers; vertical										
215mm thick; English bond	35.94	5.00	**37.73**	5.15	5.45	**131.55**	6.76	**176.05**	m2	**198.05**
327mm thick; English bond	54.06	5.00	**56.76**	6.25	6.70	**160.57**	10.63	**227.96**	m2	**256.46**
440mm thick; English bond	71.88	5.00	**75.47**	7.25	7.85	**187.10**	14.49	**277.06**	m2	**311.69**
Chimney stacks; vertical										
440mm thick; English bond	71.88	5.00	**75.47**	7.25	7.85	**187.10**	14.49	**277.06**	m2	**311.69**
890mm thick; English bond	143.75	5.00	**150.94**	11.45	12.65	**298.19**	28.98	**478.11**	m2	**537.87**
Projections; vertical										
215mm wide x 112mm projection	4.03	5.00	**4.23**	0.33	0.38	**8.76**	0.77	**13.76**	m	**15.48**
215mm wide x 215mm projection	8.05	5.00	**8.45**	0.55	0.65	**14.78**	1.55	**24.78**	m	**27.88**
327mm wide x 112mm projection	6.04	5.00	**6.34**	0.50	0.60	**13.53**	1.16	**21.03**	m	**23.66**
327mm wide x 215mm projection	12.08	5.00	**12.68**	0.83	1.03	**22.83**	2.32	**37.83**	m	**42.56**
Projections; horizontal										
215mm wide x 112mm projection	4.03	5.00	**4.23**	0.33	0.38	**8.76**	0.77	**13.76**	m	**15.48**
215mm wide x 215mm projection	8.05	5.00	**8.45**	0.55	0.65	**14.78**	1.55	**24.78**	m	**27.88**
327mm wide x 112mm projection	6.04	5.00	**6.34**	0.50	0.60	**13.53**	1.16	**21.03**	m	**23.66**
327mm wide x 215mm projection	12.08	5.00	**12.68**	0.83	1.03	**22.83**	2.32	**37.83**	m	**42.56**
Projections; bonding to old brickwork; cutting pockets; including extra material; vertical										
215mm wide x 112mm projection	6.71	5.00	**7.04**	1.02	1.07	**25.95**	1.29	**34.29**	m	**38.57**
215mm wide x 215mm projection	10.73	5.00	**11.27**	1.24	1.34	**31.97**	2.06	**45.31**	m	**50.97**
327mm wide x 112mm projection	10.07	5.00	**10.57**	1.54	1.64	**39.45**	1.94	**51.95**	m	**58.45**
327mm wide x 215mm projection	14.76	5.00	**15.50**	1.87	2.07	**48.74**	2.84	**67.08**	m	**75.47**
Projections; bonding to old brickwork; cutting pockets; including extra material; horizontal										
215mm wide x 112mm projection	6.71	5.00	**7.04**	1.02	1.07	**25.95**	1.29	**34.29**	m	**38.57**
215mm wide x 215mm projection	10.73	5.00	**11.27**	1.24	1.34	**31.97**	2.06	**45.31**	m	**50.97**
327mm wide x 112mm projection	10.07	5.00	**10.57**	1.54	1.64	**39.45**	1.94	**51.95**	m	**58.45**
327mm wide x 215mm projection	14.76	5.00	**15.50**	1.87	2.07	**48.74**	2.84	**67.08**	m	**75.47**
Arches including centering; flat										
112mm high on face; 215mm thick; width of exposed soffit 215mm; bricks-on-edge	4.03	5.00	**4.23**	1.60	1.30	**36.66**	0.77	**41.66**	m	**46.87**
215mm high on face; 112mm thick; width of exposed soffit 112mm; bricks-on-end	4.03	5.00	**4.23**	1.55	1.35	**36.48**	0.77	**41.48**	m	**46.67**
Arches including centering; semi-circular										
112mm high on face; 215mm thick; width of exposed soffit 215mm; bricks-on-edge	6.32	5.00	**6.64**	2.25	1.65	**49.64**	1.21	**57.50**	m	**64.68**
215mm high on face; 215mm thick; width of exposed soffit 215mm; bricks-on-edge	12.64	5.00	**13.28**	3.10	2.55	**71.36**	2.43	**87.07**	m	**97.95**
Closing cavities; vertical										
50mm wide with brickwork 102mm thick	1.96	5.00	**2.06**	0.44	0.44	**10.96**	0.38	**13.40**	m	**15.08**
75mm wide with brickwork 102mm thick	1.96	5.00	**2.06**	0.44	0.44	**10.96**	0.38	**13.40**	m	**15.08**
Closing cavities; horizontal										
50mm wide with slates	2.11	5.00	**2.22**	0.66	0.66	**16.45**	0.38	**19.04**	m	**21.42**
75mm wide with slates	2.11	5.00	**2.22**	0.66	0.66	**16.45**	0.38	**19.04**	m	**21.42**
50mm wide with brickwork 102mm thick	1.96	5.00	**2.06**	0.44	0.44	**10.96**	0.38	**13.40**	m	**15.08**
75mm wide with brickwork 102mm thick	1.96	5.00	**2.06**	0.44	0.44	**10.96**	0.38	**13.40**	m	**15.08**
Bonding ends to existing common brickwork; cutting pockets; extra material										
walls; bonding every third course										
102mm thick	1.34	5.00	**1.41**	0.41	0.41	**10.22**	0.26	**11.88**	m	**13.37**

Labour hourly rates: (except Specialists) Craft Operatives £14.21 Labourer £10.71 Rates are national average prices. Refer to REGIONAL VARIATIONS for indicative levels of overall pricing in regions	MATERIALS			LABOUR				RATES		
	Del to Site	Waste	Material Cost	Craft Optve	Lab	Labour Cost	Sunds	Nett Rate	Unit	Gross Rate (+12.5%)
	£	%	£	Hrs	Hrs	£	£	£		£
F10: BRICK/BLOCK WALLING Cont.										
Engineering bricks, BS EN 772, Category F, 215 x 102.5 x 65mm, Class A; in cement-lime mortar (1:2:9) Cont.										
Bonding ends to existing common brickwork; cutting pockets; extra material Cont.										
walls; bonding every third course Cont.										
215mm thick	2.68	5.00	**2.82**	0.73	0.73	**18.19**	0.52	**21.52**	m	**24.21**
327mm thick	4.03	5.00	**4.23**	0.95	0.95	**23.67**	0.77	**28.67**	m	**32.26**
For other mortar mixes, ADD or DEDUCT as follows:										
For each half brick thickness										
1:1:6 cement-lime mortar	-	-	**-**	-	-	**-**	0.42	**0.42**	m2	**0.47**
1:4 cement mortar	-	-	**-**	-	-	**-**	0.13	**0.13**	m2	**0.15**
1:3 cement mortar	-	-	**-**	-	-	**-**	0.18	**0.18**	m2	**0.21**
For using Sulphate Resisting cement in lieu of Portland cement, ADD as follows										
For each half brick thickness										
1:2:9 cement-lime mortar	-	-	**-**	-	-	**-**	0.12	**0.12**	m2	**0.13**
1:1:6 cement-lime mortar	-	-	**-**	-	-	**-**	0.17	**0.17**	m2	**0.19**
1:4 cement mortar	-	-	**-**	-	-	**-**	0.27	**0.27**	m2	**0.30**
1:3 cement mortar	-	-	**-**	-	-	**-**	0.34	**0.34**	m2	**0.39**
Staffordshire wirecut bricks, BS EN 772, Category F, 215 x 102.5 x 65mm; in cement mortar (1:3)										
Walls; vertical										
215mm thick; English bond	46.41	5.00	**48.73**	2.60	2.90	**68.01**	7.18	**123.92**	m2	**139.41**
Staffordshire pressed bricks, BS EN 772, Category F, 215 x 102.5 x 65mm; in cement mortar (1:3)										
Walls; vertical										
215mm thick; English bond	54.26	5.00	**56.98**	2.60	2.90	**68.01**	7.18	**132.16**	m2	**148.68**
Composite walling of Staffordshire bricks, BS EN 772, Category F, 215 x 102.5 x 65mm; in cement mortar (1:3); wirecut bricks backing; pressed bricks facing; weather struck pointing as work proceeds										
Walls; vertical										
English bond; facework one side	51.89	5.00	**54.49**	2.90	3.20	**75.48**	7.18	**137.15**	m2	**154.29**
Flemish bond; facework one side	51.23	5.00	**53.80**	2.90	3.20	**75.48**	7.18	**136.46**	m2	**153.51**
Common bricks, BS EN 772, Category M, 215 x 102.5 x 65mm, compressive strength 20.5 N/mm2; in cement-lime mortar (1:2:9); flush smooth pointing as work proceeds										
Walls; vertical										
stretcher bond; facework one side										
102mm thick	12.04	5.00	**12.64**	1.75	1.90	**45.22**	2.90	**60.76**	m2	**68.35**
stretcher bond; facework both sides										
102mm thick	12.04	5.00	**12.64**	2.05	2.20	**52.69**	2.90	**68.23**	m2	**76.76**
English bond; facework one side										
215mm thick	24.28	5.00	**25.49**	2.65	2.95	**69.25**	6.76	**101.50**	m2	**114.19**
English bond; facework both sides										
215mm thick	24.28	5.00	**25.49**	2.95	3.25	**76.73**	6.76	**108.98**	m2	**122.60**
Engineering bricks, BS EN 772, Category F, 215 x 102.5 x 65mm, Class B; in cement-lime mortar (1:2:9); flush smooth pointing as work proceeds										
Walls; vertical										
stretcher bond; facework one side										
102mm thick	12.63	5.00	**13.26**	1.90	2.05	**48.95**	2.90	**65.11**	m2	**73.25**
stretcher bond; facework both sides										
102mm thick	12.63	5.00	**13.26**	2.20	2.35	**56.43**	2.90	**72.59**	m2	**81.66**
English bond; facework one side										
215mm thick	25.47	5.00	**26.74**	2.90	3.20	**75.48**	6.76	**108.98**	m2	**122.60**
English bond; facework both sides										
215mm thick	25.47	5.00	**26.74**	3.20	3.50	**82.96**	6.76	**116.46**	m2	**131.02**
Engineering bricks, BS EN 772, Category F, 215 x 102.5 x 65mm, Class A; in cement-lime mortar (1:2:9); flush smooth pointing as work proceeds										
Walls; vertical										
stretcher bond; facework one side										
102mm thick	17.82	5.00	**18.71**	1.90	2.05	**48.95**	2.90	**70.57**	m2	**79.39**
stretcher bond; facework both sides										
102mm thick	17.82	5.00	**18.71**	2.20	2.35	**56.43**	2.90	**78.04**	m2	**87.80**
English bond; facework one side										
215mm thick	35.94	5.00	**37.73**	2.90	3.20	**75.48**	6.76	**119.98**	m2	**134.97**
English bond; facework both sides										
215mm thick	35.94	5.00	**37.73**	3.20	3.50	**82.96**	6.76	**127.45**	m2	**143.39**

Labour hourly rates: (except Specialists) Craft Operatives £14.21 Labourer £10.71 Rates are national average prices. Refer to REGIONAL VARIATIONS for indicative levels of overall pricing in regions	MATERIALS			LABOUR				RATES		
	Del to Site £	Waste %	Material Cost £	Craft Optve Hrs	Lab Hrs	Labour Cost £	Sunds £	Nett Rate £	Unit	Gross Rate (+12.5%) £
F10: BRICK/BLOCK WALLING Cont.										
Facing bricks, second hard stocks, BS EN 772, Category M, 215 x 102.5 x 65mm; in cement-lime mortar (1:2:9); flush smooth pointing as work proceeds										
Walls; vertical										
stretcher bond; facework one side										
102mm thick	23.72	5.00	**24.90**	1.75	1.90	**45.22**	2.90	**73.02**	m2	**82.15**
stretcher bond; facework both sides										
102mm thick	23.72	5.00	**24.90**	2.05	2.20	**52.69**	2.90	**80.50**	m2	**90.56**
English bond; facework one side										
215mm thick	47.84	5.00	**50.23**	2.65	2.95	**69.25**	6.76	**126.24**	m2	**142.02**
English bond; facework both sides										
215mm thick	47.84	5.00	**50.23**	2.95	3.25	**76.73**	6.76	**133.72**	m2	**150.43**
Facing bricks p.c. £150.00 per 1000, 215 x 102.5 x 65mm; in cement-lime mortar (1:2:9); flush smooth pointing as work proceeds										
Walls; vertical										
stretcher bond; facework one side										
102mm thick	8.85	5.00	**9.29**	1.75	1.90	**45.22**	2.90	**57.41**	m2	**64.59**
stretcher bond; facework both sides										
102mm thick	8.85	5.00	**9.29**	2.05	2.20	**52.69**	2.90	**64.89**	m2	**73.00**
English bond; facework both sides										
215mm thick	17.85	5.00	**18.74**	2.65	2.95	**69.25**	6.76	**94.76**	m2	**106.60**
Flemish bond; facework both sides										
215mm thick	17.85	5.00	**18.74**	2.95	3.25	**76.73**	6.76	**102.23**	m2	**115.01**
extra; special bricks; vertical angles; squint; BS 4729, type AN.1.1.	25.73	5.00	**27.01**	0.25	0.25	**6.23**	-	**33.24**	m	**37.40**
extra; special bricks; intersections; birdsmouth; BS 4729, type AN.4.1.	77.05	5.00	**80.90**	0.25	0.25	**6.23**	-	**87.13**	m	**98.02**
Walls; building overhand; vertical										
stretcher bond; facework one side										
102mm thick	8.85	5.00	**9.29**	2.15	2.30	**55.18**	2.90	**67.38**	m2	**75.80**
stretcher bond; facework both sides										
102mm thick	8.85	5.00	**9.29**	2.45	2.60	**62.66**	2.90	**74.86**	m2	**84.21**
English bond; facework both sides										
215mm thick	17.85	5.00	**18.74**	3.05	3.35	**79.22**	6.76	**104.72**	m2	**117.81**
Flemish bond; facework both sides										
215mm thick	17.85	5.00	**18.74**	3.35	3.65	**86.70**	6.76	**112.20**	m2	**126.22**
Arches including centering; flat										
112mm high on face; 215mm thick; width of exposed soffit 112mm; bricks-on-edge	2.00	5.00	**2.10**	1.05	1.05	**26.17**	0.77	**29.04**	m	**32.67**
112mm high on face; 215mm thick; width of exposed soffit 215mm; bricks-on-edge	2.00	5.00	**2.10**	1.50	1.25	**34.70**	0.77	**37.58**	m	**42.27**
215mm high on face; 112mm thick; width of exposed soffit 112mm; bricks-on-end	2.00	5.00	**2.10**	1.45	1.25	**33.99**	0.77	**36.87**	m	**41.47**
Arches including centering; segmental										
215mm high on face; 215mm thick; width of exposed soffit 112mm; bricks-on-edge	4.00	5.00	**4.20**	2.40	2.15	**57.13**	1.55	**62.88**	m	**70.74**
215mm high on face; 215mm thick; width of exposed soffit 215mm; bricks-on-edge	4.00	5.00	**4.20**	2.95	2.45	**68.16**	1.55	**73.91**	m	**83.14**
Arches including centering; semi-circular										
215mm high on face; 215mm thick; width of exposed soffit 112mm; bricks-on-edge	4.00	5.00	**4.20**	2.40	2.15	**57.13**	1.55	**62.88**	m	**70.74**
215mm high on face; 215mm thick; width of exposed soffit 215mm; bricks-on-edge	4.00	5.00	**4.20**	2.95	2.45	**68.16**	1.55	**73.91**	m	**83.14**
Facework copings; bricks-on-edge; pointing top and each side										
215 x 102.5mm; horizontal	2.00	5.00	**2.10**	0.70	0.75	**17.98**	0.77	**20.85**	m	**23.46**
215 x 102.5mm; with oversailing course and cement fillets both sides; horizontal	2.00	5.00	**2.10**	1.10	1.15	**27.95**	0.77	**30.82**	m	**34.67**
327 x 102.5mm; horizontal	3.00	5.00	**3.15**	1.05	1.10	**26.70**	1.16	**31.01**	m	**34.89**
327 x 102.5mm; with oversailing course and cement fillets both sides; horizontal	3.00	5.00	**3.15**	1.50	1.55	**37.92**	1.16	**42.23**	m	**47.50**
extra; galvanized iron coping cramp at angle or end	2.34	5.00	**2.46**	0.04	0.04	**1.00**	-	**3.45**	nr	**3.89**
Facing bricks p.c. £200.00 per 1000, 215 x 102.5 x 65mm; in cement-lime mortar (1:2:9); flush smooth pointing as work proceeds										
Walls; vertical										
stretcher bond; facework one side										
102mm thick	11.80	5.00	**12.39**	1.75	1.90	**45.22**	2.90	**60.51**	m2	**68.07**
stretcher bond; facework both sides										
102mm thick	11.80	5.00	**12.39**	2.05	2.20	**52.69**	2.90	**67.99**	m2	**76.48**
English bond; facework both sides										
215mm thick	23.80	5.00	**24.99**	2.65	2.95	**69.25**	6.76	**101.00**	m2	**113.63**
Flemish bond; facework both sides										
215mm thick	23.80	5.00	**24.99**	2.95	3.25	**76.73**	6.76	**108.48**	m2	**122.04**
extra; special bricks; vertical angles; squint; BS 4729, type AN.1.1.	25.73	5.00	**27.01**	0.25	0.25	**6.23**	-	**33.24**	m	**37.40**
extra; special bricks; intersections; birdsmouth; BS 4729, type AN.4.1.	77.05	5.00	**80.90**	0.25	0.25	**6.23**	-	**87.13**	m	**98.02**
Walls; building overhand; vertical										
stretcher bond; facework one side										
102mm thick	11.80	5.00	**12.39**	2.15	2.30	**55.18**	2.90	**70.48**	m2	**79.29**

Labour hourly rates: (except Specialists) Craft Operatives £14.21 Labourer £10.71 Rates are national average prices. Refer to REGIONAL VARIATIONS for indicative levels of overall pricing in regions	MATERIALS			LABOUR				RATES		
	Del to Site £	Waste %	Material Cost £	Craft Optve Hrs	Lab Hrs	Labour Cost £	Sunds £	Nett Rate £	Unit	Gross Rate (+12.5%) £
F10: BRICK/BLOCK WALLING Cont.										
Facing bricks p.c. £200.00 per 1000, 215 x 102.5 x 65mm; in cement-lime mortar (1:2:9); flush smooth pointing as work proceeds Cont.										
Walls; building overhand; vertical Cont.										
stretcher bond; facework both sides										
102mm thick	11.80	5.00	**12.39**	2.45	2.60	**62.66**	2.90	**77.95**	m2	**87.70**
English bond; facework both sides										
215mm thick	23.80	5.00	**24.99**	3.05	3.35	**79.22**	6.76	**110.97**	m2	**124.84**
Flemish bond; facework both sides										
215mm thick	23.80	5.00	**24.99**	3.35	3.65	**86.70**	6.76	**118.45**	m2	**133.25**
Arches including centering; flat										
112mm high on face; 215mm thick; width of exposed soffit 112mm; bricks-on-edge	2.67	5.00	**2.80**	1.05	1.05	**26.17**	0.77	**29.74**	m	**33.46**
112mm high on face; 215mm thick; width of exposed soffit 215mm; bricks-on-edge	2.67	5.00	**2.80**	1.50	1.25	**34.70**	0.77	**38.28**	m	**43.06**
215mm high on face; 112mm thick; width of exposed soffit 112mm; bricks-on-end	2.67	5.00	**2.80**	1.45	1.25	**33.99**	0.77	**37.57**	m	**42.26**
Arches including centering; segmental										
215mm high on face; 215mm thick; width of exposed soffit 112mm; bricks-on-edge	5.33	5.00	**5.60**	2.40	2.15	**57.13**	1.55	**64.28**	m	**72.31**
215mm high on face; 215mm thick; width of exposed soffit 215mm; bricks-on-edge	5.33	5.00	**5.60**	2.95	2.45	**68.16**	1.55	**75.31**	m	**84.72**
Arches including centering; semi-circular										
215mm high on face; 215mm thick; width of exposed soffit 112mm; bricks-on-edge	5.33	5.00	**5.60**	2.40	2.15	**57.13**	1.55	**64.28**	m	**72.31**
215mm high on face; 215mm thick; width of exposed soffit 215mm; bricks-on-edge	5.33	5.00	**5.60**	2.95	2.45	**68.16**	1.55	**75.31**	m	**84.72**
Facework copings; bricks-on-edge; pointing top and each side										
215 x 102.5mm; horizontal	2.67	5.00	**2.80**	0.70	0.75	**17.98**	0.77	**21.55**	m	**24.25**
215 x 102.5mm; with oversailing course and cement fillets both sides; horizontal	2.67	5.00	**2.80**	1.10	1.15	**27.95**	0.77	**31.52**	m	**35.46**
327 x 102.5mm; horizontal	4.00	5.00	**4.20**	1.05	1.10	**26.70**	1.16	**32.06**	m	**36.07**
327 x 102.5mm; with oversailing course and cement fillets both sides; horizontal	4.00	5.00	**4.20**	1.50	1.55	**37.92**	1.16	**43.28**	m	**48.69**
extra; galvanized iron coping cramp at angle or end	2.34	5.00	**2.46**	0.04	0.04	**1.00**	-	**3.45**	nr	**3.89**
Facing bricks p.c. £250.00 per 1000, 215 x 102.5 x 65mm; in cement-lime mortar (1:2:9); flush smooth pointing as work proceeds										
Walls; vertical										
stretcher bond; facework one side										
102mm thick	14.75	5.00	**15.49**	1.75	1.90	**45.22**	2.90	**63.61**	m2	**71.56**
stretcher bond; facework both sides										
102mm thick	14.75	5.00	**15.49**	2.05	2.20	**52.69**	2.90	**71.08**	m2	**79.97**
English bond; facework both sides										
215mm thick	29.75	5.00	**31.24**	2.65	2.95	**69.25**	6.76	**107.25**	m2	**120.66**
Flemish bond; facework both sides										
215mm thick	29.75	5.00	**31.24**	2.95	3.25	**76.73**	6.76	**114.73**	m2	**129.07**
extra; special bricks; vertical angles; squint; BS 4729, type AN.1.1.	25.73	5.00	**27.01**	0.25	0.25	**6.23**	-	**33.24**	m	**37.40**
extra; special bricks; intersections; birdsmouth; BS 4729, type AN.4.1.	77.05	5.00	**80.90**	0.25	0.25	**6.23**	-	**87.13**	m	**98.02**
Walls; building overhand; vertical										
stretcher bond; facework one side										
102mm thick	14.75	5.00	**15.49**	2.15	2.30	**55.18**	2.90	**73.57**	m2	**82.77**
stretcher bond; facework both sides										
102mm thick	14.75	5.00	**15.49**	2.45	2.60	**62.66**	2.90	**81.05**	m2	**91.18**
English bond; facework both sides										
215mm thick	29.75	5.00	**31.24**	3.05	3.35	**79.22**	6.76	**117.22**	m2	**131.87**
Flemish bond; facework both sides										
215mm thick	29.75	5.00	**31.24**	3.35	3.65	**86.70**	6.76	**124.69**	m2	**140.28**
Arches including centering; flat										
112mm high on face; 215mm thick; width of exposed soffit 112mm; bricks-on-edge	3.33	5.00	**3.50**	1.05	1.05	**26.17**	0.77	**30.44**	m	**34.24**
112mm high on face; 215mm thick; width of exposed soffit 215mm; bricks-on-edge	3.33	5.00	**3.50**	1.50	1.25	**34.70**	0.77	**38.98**	m	**43.85**
215mm high on face; 112mm thick; width of exposed soffit 112mm; bricks-on-end	3.33	5.00	**3.50**	1.45	1.25	**33.99**	0.77	**38.27**	m	**43.05**
Arches including centering; segmental										
215mm high on face; 215mm thick; width of exposed soffit 112mm; bricks-on-edge	6.67	5.00	**7.00**	2.40	2.15	**57.13**	1.55	**65.68**	m	**73.89**
215mm high on face; 215mm thick; width of exposed soffit 215mm; bricks-on-edge	6.67	5.00	**7.00**	2.95	2.45	**68.16**	1.55	**76.71**	m	**86.29**
Arches including centering; semi-circular										
215mm high on face; 215mm thick; width of exposed soffit 112mm; bricks-on-edge	6.67	5.00	**7.00**	2.40	2.15	**57.13**	1.55	**65.68**	m	**73.89**
215mm high on face; 215mm thick; width of exposed soffit 215mm; bricks-on-edge	6.67	5.00	**7.00**	2.95	2.45	**68.16**	1.55	**76.71**	m	**86.29**
Facework copings; bricks-on-edge; pointing top and each side										
215 x 102.5mm; horizontal	3.33	5.00	**3.50**	0.70	0.75	**17.98**	0.77	**22.25**	m	**25.03**

Labour hourly rates: (except Specialists) Craft Operatives £14.21 Labourer £10.71 Rates are national average prices. Refer to REGIONAL VARIATIONS for indicative levels of overall pricing in regions	MATERIALS			LABOUR			RATES			
	Del to Site £	Waste %	Material Cost £	Craft Optve Hrs	Lab Hrs	Labour Cost £	Sunds £	Nett Rate £	Unit	Gross Rate (+12.5%) £
F10: BRICK/BLOCK WALLING Cont.										
Facing bricks p.c. £250.00 per 1000, 215 x 102.5 x 65mm; in cement-lime mortar (1:2:9); flush smooth pointing as work proceeds Cont.										
Facework copings; bricks-on-edge; pointing top and each side Cont.										
215 x 102.5mm; with oversailing course and cement fillets both sides; horizontal	3.33	5.00	**3.50**	1.10	1.15	**27.95**	0.77	**32.22**	m	**36.25**
327 x 102.5mm; horizontal	5.00	5.00	**5.25**	1.05	1.10	**26.70**	1.16	**33.11**	m	**37.25**
327 x 102.5mm; with oversailing course and cement fillets both sides; horizontal	5.00	5.00	**5.25**	1.50	1.55	**37.92**	1.16	**44.33**	m	**49.87**
extra; galvanized iron coping cramp at angle or end	2.34	5.00	**2.46**	0.04	0.04	**1.00**	-	**3.45**	nr	**3.89**
Facing bricks p.c. £300.00 per 1000, 215 x 102.5 x 65mm; in cement-lime mortar (1:2:9); flush smooth pointing as work proceeds										
Walls; vertical										
stretcher bond; facework one side										
102mm thick	17.70	5.00	**18.59**	1.75	1.90	**45.22**	2.90	**66.70**	m2	**75.04**
stretcher bond; facework both sides										
102mm thick	17.70	5.00	**18.59**	2.05	2.20	**52.69**	2.90	**74.18**	m2	**83.45**
English bond; facework both sides										
215mm thick	35.70	5.00	**37.49**	2.65	2.95	**69.25**	6.76	**113.50**	m2	**127.68**
Flemish bond; facework both sides										
215mm thick	35.70	5.00	**37.49**	2.95	3.25	**76.73**	6.76	**120.97**	m2	**136.10**
extra; special bricks; vertical angles; squint; BS 4729, type AN.1.1.	25.73	5.00	**27.01**	0.25	0.25	**6.23**	-	**33.24**	m	**37.40**
extra; special bricks; intersections; birdsmouth; BS 4729, type AN.4.1.	77.05	5.00	**80.90**	0.25	0.25	**6.23**	-	**87.13**	m	**98.02**
Walls; building overhand; vertical										
stretcher bond; facework one side										
102mm thick	17.70	5.00	**18.59**	2.15	2.30	**55.18**	2.90	**76.67**	m2	**86.26**
stretcher bond; facework both sides										
102mm thick	17.70	5.00	**18.59**	2.45	2.60	**62.66**	2.90	**84.15**	m2	**94.67**
English bond; facework both sides										
215mm thick	35.70	5.00	**37.49**	3.05	3.35	**79.22**	6.76	**123.47**	m2	**138.90**
Flemish bond; facework both sides										
215mm thick	35.70	5.00	**37.49**	3.35	3.65	**86.70**	6.76	**130.94**	m2	**147.31**
Arches including centering; flat										
112mm high on face; 215mm thick; width of exposed soffit 112mm; bricks-on-edge	4.00	5.00	**4.20**	1.05	1.05	**26.17**	0.77	**31.14**	m	**35.03**
112mm high on face; 215mm thick; width of exposed soffit 215mm; bricks-on-edge	4.00	5.00	**4.20**	1.50	1.25	**34.70**	0.77	**39.68**	m	**44.63**
215mm high on face; 112mm thick; width of exposed soffit 112mm; bricks-on-end	4.00	5.00	**4.20**	1.45	1.25	**33.99**	0.77	**38.96**	m	**43.84**
Arches including centering; segmental										
215mm high on face; 215mm thick; width of exposed soffit 112mm; bricks-on-edge	8.00	5.00	**8.40**	2.40	2.15	**57.13**	1.55	**67.08**	m	**75.46**
215mm high on face; 215mm thick; width of exposed soffit 215mm; bricks-on-edge	8.00	5.00	**8.40**	2.95	2.45	**68.16**	1.55	**78.10**	m	**87.87**
Arches including centering; semi-circular										
215mm high on face; 215mm thick; width of exposed soffit 112mm; bricks-on-edge	8.00	5.00	**8.40**	2.40	2.15	**57.13**	1.55	**67.08**	m	**75.46**
215mm high on face; 215mm thick; width of exposed soffit 215mm; bricks-on-edge	8.00	5.00	**8.40**	2.95	2.45	**68.16**	1.55	**78.10**	m	**87.87**
Facework copings; bricks-on-edge; pointing top and each side										
215 x 102.5mm; horizontal	4.00	5.00	**4.20**	0.70	0.75	**17.98**	0.77	**22.95**	m	**25.82**
215 x 102.5mm; with oversailing course and cement fillets both sides; horizontal	4.00	5.00	**4.20**	1.10	1.15	**27.95**	0.77	**32.92**	m	**37.04**
327 x 102.5mm; horizontal	6.00	5.00	**6.30**	1.05	1.10	**26.70**	1.16	**34.16**	m	**38.43**
327 x 102.5mm; with oversailing course and cement fillets both sides; horizontal	6.00	5.00	**6.30**	1.50	1.55	**37.92**	1.16	**45.38**	m	**51.05**
extra; galvanized iron coping cramp at angle or end	2.34	5.00	**2.46**	0.04	0.04	**1.00**	-	**3.45**	nr	**3.89**
Facing bricks p.c. £350.00 per 1000, 215 x 102.5 x 65mm; in cement-lime mortar (1:2:9); flush smooth pointing as work proceeds										
Walls; vertical										
stretcher bond; facework one side										
102mm thick	20.65	5.00	**21.68**	1.75	1.90	**45.22**	2.90	**69.80**	m2	**78.53**
stretcher bond; facework both sides										
102mm thick	20.65	5.00	**21.68**	2.05	2.20	**52.69**	2.90	**77.28**	m2	**86.94**
English bond; facework both sides										
215mm thick	41.65	5.00	**43.73**	2.65	2.95	**69.25**	6.76	**119.75**	m2	**134.71**
Flemish bond; facework both sides										
215mm thick	41.65	5.00	**43.73**	2.95	3.25	**76.73**	6.76	**127.22**	m2	**143.12**
extra; special bricks; vertical angles; squint; BS 4729, type AN.1.1.	25.73	5.00	**27.01**	0.25	0.25	**6.23**	-	**33.24**	m	**37.40**
extra; special bricks; intersections; birdsmouth; BS 4729, type AN.4.1.	77.05	5.00	**80.90**	0.25	0.25	**6.23**	-	**87.13**	m	**98.02**
Walls; building overhand; vertical										
stretcher bond; facework one side										
102mm thick	20.65	5.00	**21.68**	2.15	2.30	**55.18**	2.90	**79.77**	m2	**89.74**

Labour hourly rates: (except Specialists) Craft Operatives £14.21 Labourer £10.71 Rates are national average prices. Refer to REGIONAL VARIATIONS for indicative levels of overall pricing in regions	MATERIALS			LABOUR				RATES		
	Del to Site £	Waste %	Material Cost £	Craft Optve Hrs	Lab Hrs	Labour Cost £	Sunds £	Nett Rate £	Unit	Gross Rate (+12.5%) £
F10: BRICK/BLOCK WALLING Cont.										
Facing bricks p.c. £350.00 per 1000, 215 x 102.5 x 65mm; in cement-lime mortar (1:2:9); flush smooth pointing as work proceeds Cont.										
Walls; building overhand; vertical Cont.										
stretcher bond; facework both sides										
102mm thick	20.65	5.00	**21.68**	2.45	2.60	**62.66**	2.90	**87.25**	m2	**98.15**
English bond; facework both sides										
215mm thick	41.65	5.00	**43.73**	3.05	3.35	**79.22**	6.76	**129.71**	m2	**145.93**
Flemish bond; facework both sides										
215mm thick	41.65	5.00	**43.73**	3.35	3.65	**86.70**	6.76	**137.19**	m2	**154.34**
Arches including centering; flat										
112mm high on face; 215mm thick; width of exposed soffit 112mm; bricks-on-edge	4.67	5.00	**4.90**	1.05	1.05	**26.17**	0.77	**31.84**	m	**35.82**
112mm high on face; 215mm thick; width of exposed soffit 215mm; bricks-on-edge	4.67	5.00	**4.90**	1.50	1.25	**34.70**	0.77	**40.38**	m	**45.42**
215mm high on face; 112mm thick; width of exposed soffit 112mm; bricks-on-end	4.67	5.00	**4.90**	1.45	1.25	**33.99**	0.77	**39.66**	m	**44.62**
Arches including centering; segmental										
215mm high on face; 215mm thick; width of exposed soffit 112mm; bricks-on-edge	9.33	5.00	**9.80**	2.40	2.15	**57.13**	1.55	**68.48**	m	**77.04**
215mm high on face; 215mm thick; width of exposed soffit 215mm; bricks-on-edge	9.33	5.00	**9.80**	2.95	2.45	**68.16**	1.55	**79.50**	m	**89.44**
Arches including centering; semi-circular										
215mm high on face; 215mm thick; width of exposed soffit 112mm; bricks-on-edge	9.33	5.00	**9.80**	2.40	2.15	**57.13**	1.55	**68.48**	m	**77.04**
215mm high on face; 215mm thick; width of exposed soffit 215mm; bricks-on-edge	9.33	5.00	**9.80**	2.95	2.45	**68.16**	1.55	**79.50**	m	**89.44**
Facework copings; bricks-on-edge; pointing top and each side										
215 x 102.5mm; horizontal	4.67	5.00	**4.90**	0.70	0.75	**17.98**	0.77	**23.65**	m	**26.61**
215 x 102.5mm; with oversailing course and cement fillets both sides; horizontal	4.67	5.00	**4.90**	1.10	1.15	**27.95**	0.77	**33.62**	m	**37.82**
327 x 102.5mm; horizontal	7.00	5.00	**7.35**	1.05	1.10	**26.70**	1.16	**35.21**	m	**39.61**
327 x 102.5mm; with oversailing course and cement fillets both sides; horizontal	7.00	5.00	**7.35**	1.50	1.55	**37.92**	1.16	**46.43**	m	**52.23**
extra; galvanized iron coping cramp at angle or end	2.34	5.00	**2.46**	0.04	0.04	**1.00**	-	**3.45**	nr	**3.89**
Composite walling of bricks 215 x 102.5x 65mm; in cement-lime mortar (1:2:9); common bricks BS EN 772 Category M backing, compressive strength 20.5 N/mm2; facing bricks p.c. £150.00 per 1000; flush smooth pointing as work proceeds										
Walls; vertical										
English bond; facework one side										
215mm thick	19.27	5.00	**20.23**	2.65	2.95	**69.25**	6.76	**96.24**	m2	**108.27**
327mm thick	31.30	5.00	**32.87**	3.15	3.60	**83.32**	10.63	**126.82**	m2	**142.67**
Flemish bond; facework one side										
215mm thick	19.81	5.00	**20.80**	2.65	2.95	**69.25**	6.76	**96.81**	m2	**108.91**
327mm thick	31.84	5.00	**33.43**	3.15	3.60	**83.32**	10.63	**127.38**	m2	**143.31**
Walls; building overhand; vertical										
English bond; facework one side										
215mm thick	19.27	5.00	**20.23**	3.05	3.35	**79.22**	6.76	**106.21**	m2	**119.49**
327mm thick	31.30	5.00	**32.87**	3.55	4.00	**93.29**	10.63	**136.78**	m2	**153.88**
Flemish bond; facework one side										
215mm thick	19.81	5.00	**20.80**	3.05	3.35	**79.22**	6.76	**106.78**	m2	**120.12**
327mm thick	31.84	5.00	**33.43**	3.55	4.00	**93.29**	10.63	**137.35**	m2	**154.52**
Composite walling of bricks 215 x 102.5x 65mm; in cement-lime mortar (1:2:9); common bricks BS EN 772 Category M backing, compressive strength 20.5 N/mm2; facing bricks p.c. £200.00 per 1000; flush smooth pointing as work proceeds										
Walls; vertical										
English bond; facework one side										
215mm thick	23.72	5.00	**24.90**	2.65	2.95	**69.25**	6.76	**100.91**	m2	**113.53**
327mm thick	35.75	5.00	**37.54**	3.15	3.60	**83.32**	10.63	**131.49**	m2	**147.92**
Flemish bond; facework one side										
215mm thick	23.72	5.00	**24.90**	2.65	2.95	**69.25**	6.76	**100.91**	m2	**113.53**
327mm thick	35.79	5.00	**37.58**	3.15	3.60	**83.32**	10.63	**131.53**	m2	**147.97**
Walls; building overhand; vertical										
English bond; facework one side										
215mm thick	23.72	5.00	**24.90**	3.05	3.35	**79.22**	6.76	**110.88**	m2	**124.74**
327mm thick	35.75	5.00	**37.54**	3.55	4.00	**93.29**	10.63	**141.46**	m2	**159.14**
Flemish bond; facework one side										
215mm thick	23.72	5.00	**24.90**	3.05	3.35	**79.22**	6.76	**110.88**	m2	**124.74**
327mm thick	35.79	5.00	**37.58**	3.55	4.00	**93.29**	10.63	**141.50**	m2	**159.19**

Labour hourly rates: (except Specialists) Craft Operatives £14.21 Labourer £10.71 Rates are national average prices. Refer to REGIONAL VARIATIONS for indicative levels of overall pricing in regions	MATERIALS			LABOUR			RATES			
	Del to Site	Waste	Material Cost	Craft Optve	Lab	Labour Cost	Sunds	Nett Rate	Unit	Gross Rate (+12.5%)
	£	%	£	Hrs	Hrs	£	£	£		£
F10: BRICK/BLOCK WALLING Cont.										
Composite walling of bricks 215 x 102.5 x 65mm; in cement-lime mortar (1:2:9); common bricks BS EN 772 Category M backing, compressive strength 20.5 N/mm2; facing bricks p.c. £250.00 per 1000; flush smooth pointing as work proceeds										
Walls; vertical										
English bond; facework one side										
215mm thick	28.17	5.00	**29.57**	2.65	2.95	**69.25**	6.76	**105.59**	m2	**118.79**
327mm thick	40.20	5.00	**42.21**	3.15	3.60	**83.32**	10.63	**136.16**	m2	**153.18**
Flemish bond; facework one side										
215mm thick	28.17	5.00	**29.57**	2.65	2.95	**69.25**	6.76	**105.59**	m2	**118.79**
327mm thick	39.74	5.00	**41.73**	3.15	3.60	**83.32**	10.63	**135.68**	m2	**152.64**
Walls; building overhand; vertical										
English bond; facework one side										
215mm thick	28.17	5.00	**29.57**	3.05	3.35	**79.22**	6.76	**115.56**	m2	**130.00**
327mm thick	40.20	5.00	**42.21**	3.55	4.00	**93.29**	10.63	**146.13**	m2	**164.40**
Flemish bond; facework one side										
215mm thick	28.17	5.00	**29.57**	3.05	3.35	**79.22**	6.76	**115.56**	m2	**130.00**
327mm thick	39.74	5.00	**41.73**	3.55	4.00	**93.29**	10.63	**145.65**	m2	**163.85**
Composite walling of bricks 215 x 102.5 x 65mm; in cement-lime mortar (1:2:9); common bricks BS EN 772 Category M backing, compressive strength 20.5 N/mm2; facing bricks p.c. £300.00 per 1000; flush smooth pointing as work proceeds										
Walls; vertical										
English bond; facework one side										
215mm thick	32.62	5.00	**34.25**	2.65	2.95	**69.25**	6.76	**110.26**	m2	**124.04**
327mm thick	44.65	5.00	**46.88**	3.15	3.60	**83.32**	10.63	**140.83**	m2	**158.44**
Flemish bond; facework one side										
215mm thick	32.62	5.00	**34.25**	2.65	2.95	**69.25**	6.76	**110.26**	m2	**124.04**
327mm thick	43.69	5.00	**45.88**	3.15	3.60	**83.32**	10.63	**139.83**	m2	**157.30**
Walls; building overhand; vertical										
English bond; facework one side										
215mm thick	32.62	5.00	**34.25**	3.05	3.35	**79.22**	6.76	**120.23**	m2	**135.26**
327mm thick	44.65	5.00	**46.88**	3.55	4.00	**93.29**	10.63	**150.80**	m2	**169.65**
Flemish bond; facework one side										
215mm thick	32.62	5.00	**34.25**	3.05	3.35	**79.22**	6.76	**120.23**	m2	**135.26**
327mm thick	43.69	5.00	**45.88**	3.55	4.00	**93.29**	10.63	**149.79**	m2	**168.52**
Composite walling of bricks 215 x 102.5 x 65mm; in cement-lime mortar (1:2:9); common bricks BS EN 772 Category M backing, compressive strength 20.5 N/mm2; facing bricks p.c. £350.00 per 1000; flush smooth pointing as work proceeds										
Walls; vertical										
English bond; facework one side										
215mm thick	37.07	5.00	**38.92**	2.65	2.95	**69.25**	6.76	**114.93**	m2	**129.30**
327mm thick	49.10	5.00	**51.56**	3.15	3.60	**83.32**	10.63	**145.51**	m2	**163.69**
Flemish bond; facework one side										
215mm thick	37.07	5.00	**38.92**	2.65	2.95	**69.25**	6.76	**114.93**	m2	**129.30**
327mm thick	47.64	5.00	**50.02**	3.15	3.60	**83.32**	10.63	**143.97**	m2	**161.97**
Walls; building overhand; vertical										
English bond; facework one side										
215mm thick	37.07	5.00	**38.92**	3.05	3.35	**79.22**	6.76	**124.90**	m2	**140.51**
327mm thick	49.10	5.00	**51.56**	3.55	4.00	**93.29**	10.63	**155.47**	m2	**174.91**
Flemish bond; facework one side										
215mm thick	37.07	5.00	**38.92**	3.05	3.35	**79.22**	6.76	**124.90**	m2	**140.51**
327mm thick	47.64	5.00	**50.02**	3.55	4.00	**93.29**	10.63	**153.94**	m2	**173.18**
For each £10.00 difference in cost of 1000 facing bricks, add or deduct the following										
Wall in facing bricks										
102mm thick; English bond; facework one side	0.89	5.00	**0.93**	-	-	**-**	-	**0.93**	m2	**1.05**
215mm thick; facework both sides	1.19	5.00	**1.25**	-	-	**-**	-	**1.25**	m2	**1.41**
Composite wall; common brick backing; facing bricks facework										
215mm thick; English bond; facework one side	0.89	5.00	**0.93**	-	-	**-**	-	**0.93**	m2	**1.05**
215mm thick; Flemish bond; facework one side	0.79	5.00	**0.83**	-	-	**-**	-	**0.83**	m2	**0.93**
Extra over flush smooth pointing for the following types of pointing										
Pointing as work proceeds										
ironing in joints	-	-	**-**	0.03	0.03	**0.75**	-	**0.75**	m2	**0.84**
weathered pointing	-	-	**-**	0.03	0.03	**0.75**	-	**0.75**	m2	**0.84**
recessed pointing	-	-	**-**	0.04	0.04	**1.00**	-	**1.00**	m2	**1.12**
Raking out joints and pointing on completion										
flush smooth pointing with cement-lime mortar (1:2:9)	-	-	**-**	0.75	0.75	**18.69**	0.19	**18.88**	m2	**21.24**
flush smooth pointing with coloured cement-lime mortar (1:2:9)	-	-	**-**	0.75	0.75	**18.69**	0.26	**18.95**	m2	**21.32**
weathered pointing with cement-lime mortar (1:2:9)	-	-	**-**	0.83	0.83	**20.68**	0.19	**20.88**	m2	**23.49**
weathered pointing with coloured cement-lime mortar (1:2:9)	-	-	**-**	0.83	0.83	**20.68**	0.26	**20.94**	m2	**23.56**

MASONRY

Labour hourly rates: (except Specialists) Craft Operatives £14.21 Labourer £10.71 Rates are national average prices. Refer to REGIONAL VARIATIONS for indicative levels of overall pricing in regions	MATERIALS			LABOUR			RATES			
	Del to Site £	Waste %	Material Cost £	Craft Optve Hrs	Lab Hrs	Labour Cost £	Sunds £	Nett Rate £	Unit	Gross Rate (+12.5%) £
F10: BRICK/BLOCK WALLING Cont.										
Extra over cement-lime mortar (1:2:9) for bedding and pointing for using coloured cement-lime mortar (1:2:9) for the following										
Wall										
102mm thick; pointing both sides	-	-	-	-	-	-	0.98	**0.98**	m2	**1.10**
215mm thick; pointing both sides	-	-	-	-	-	-	2.28	**2.28**	m2	**2.56**
Aerated concrete blocks, Thermalite, 440 x 215mm, Shield blocks (4.0 N/mm2);in cement-lime mortar (1:1:6)										
Walls; vertical										
75mm thick; stretcher bond	7.22	5.00	**7.58**	0.70	0.80	**18.52**	0.88	**26.98**	m2	**30.35**
90mm thick; stretcher bond	8.58	5.00	**9.01**	0.80	0.90	**21.01**	1.11	**31.12**	m2	**35.01**
100mm thick; stretcher bond	9.49	5.00	**9.96**	0.80	0.90	**21.01**	1.11	**32.08**	m2	**36.09**
140mm thick; stretcher bond	13.16	5.00	**13.82**	0.90	1.05	**24.03**	1.67	**39.52**	m2	**44.46**
190mm thick; stretcher bond	17.71	5.00	**18.60**	0.95	1.10	**25.28**	2.21	**46.09**	m2	**51.85**
Closing cavities; vertical										
50mm wide with blockwork 100mm thick	2.37	5.00	**2.49**	0.25	0.25	**6.23**	0.24	**8.96**	m	**10.08**
75mm wide with blockwork 100mm thick	2.37	5.00	**2.49**	0.25	0.25	**6.23**	0.24	**8.96**	m	**10.08**
Closing cavities; horizontal										
50mm wide with blockwork 100mm thick	2.37	5.00	**2.49**	0.25	0.25	**6.23**	0.24	**8.96**	m	**10.08**
75mm wide with blockwork 100mm thick	2.37	5.00	**2.49**	0.25	0.25	**6.23**	0.24	**8.96**	m	**10.08**
Bonding ends to common brickwork; forming pockets; extra material										
walls; bonding every third course										
75mm	1.81	5.00	**1.90**	0.20	0.20	**4.98**	-	**6.88**	m	**7.74**
90mm	2.15	5.00	**2.25**	0.25	0.25	**6.23**	-	**8.48**	m	**9.54**
100mm	2.37	5.00	**2.49**	0.25	0.25	**6.23**	-	**8.72**	m	**9.81**
140mm	3.29	5.00	**3.45**	0.35	0.35	**8.72**	-	**12.18**	m	**13.70**
190mm	4.43	5.00	**4.65**	0.40	0.40	**9.97**	-	**14.62**	m	**16.44**
Bonding ends to existing common brickwork; cutting pockets; extra material										
walls; bonding every third course										
75mm	1.81	5.00	**1.90**	0.40	0.40	**9.97**	-	**11.86**	m	**13.35**
90mm	2.15	5.00	**2.25**	0.50	0.50	**12.46**	-	**14.71**	m	**16.55**
100mm	2.37	5.00	**2.49**	0.50	0.50	**12.46**	-	**14.95**	m	**16.82**
140mm	3.29	5.00	**3.45**	0.70	0.70	**17.44**	-	**20.90**	m	**23.51**
190mm	4.43	5.00	**4.65**	0.80	0.80	**19.94**	-	**24.58**	m	**27.66**
Bonding ends to existing concrete blockwork; cutting pockets; extra material										
walls; bonding every third course										
75mm	1.81	5.00	**1.90**	0.30	0.30	**7.48**	-	**9.37**	m	**10.54**
90mm	2.15	5.00	**2.25**	0.35	0.35	**8.72**	-	**10.97**	m	**12.35**
100mm	2.37	5.00	**2.49**	0.40	0.40	**9.97**	-	**12.46**	m	**14.02**
140mm	3.29	5.00	**3.45**	0.55	0.55	**13.71**	-	**17.16**	m	**19.31**
190mm	4.43	5.00	**4.65**	0.65	0.65	**16.20**	-	**20.85**	m	**23.45**
Aerated concrete blocks, Thermalite, 440 x 215mm, Turbo blocks (2.8 N/mm2); in cement-lime mortar (1:1:6)										
Walls; vertical										
100mm thick; stretcher bond	9.65	5.00	**10.13**	0.80	0.80	**19.94**	1.11	**31.18**	m2	**35.07**
115mm thick; stretcher bond	11.05	5.00	**11.60**	0.80	0.90	**21.01**	1.11	**33.72**	m2	**37.93**
125mm thick; stretcher bond	12.00	5.00	**12.60**	0.85	1.00	**22.79**	1.67	**37.05**	m2	**41.69**
130mm thick; stretcher bond	12.47	5.00	**13.09**	0.85	1.00	**22.79**	1.67	**37.55**	m2	**42.24**
150mm thick; stretcher bond	14.36	5.00	**15.08**	0.90	1.05	**24.03**	1.67	**40.78**	m2	**45.88**
190mm thick; stretcher bond	18.12	5.00	**19.03**	0.95	1.10	**25.28**	2.21	**46.52**	m2	**52.34**
200mm thick; stretcher bond	19.05	5.00	**20.00**	1.00	1.20	**27.06**	2.21	**49.28**	m2	**55.44**
215mm thick; stretcher bond	20.47	5.00	**21.49**	1.00	1.20	**27.06**	2.77	**51.33**	m2	**57.75**
Closing cavities; vertical										
50mm wide with blockwork 100mm thick	2.41	5.00	**2.53**	0.25	0.25	**6.23**	0.24	**9.00**	m	**10.12**
75mm wide with blockwork 100mm thick	2.41	5.00	**2.53**	0.25	0.25	**6.23**	0.24	**9.00**	m	**10.12**
Closing cavities; horizontal										
50mm wide with blockwork 100mm thick	2.41	5.00	**2.53**	0.25	0.25	**6.23**	0.24	**9.00**	m	**10.12**
75mm wide with blockwork 100mm thick	2.41	5.00	**2.53**	0.25	0.25	**6.23**	0.24	**9.00**	m	**10.12**
Aerated concrete blocks, Thermalite, 440 x 215mm, Turbo blocks (2.8 N/mm2); in thin joint mortar										
Walls; vertical										
115mm thick; stretcher bond	12.45	5.00	**13.07**	0.64	0.72	**16.81**	1.09	**30.96**	m2	**34.83**
Aerated concrete blocks, Thermalite, (Large Format) Turbo blocks (2.8 N/mm2); in thin joint mortar										
Walls; vertical										
115mm thick; stretcher bond	12.45	5.00	**13.07**	0.52	0.59	**13.65**	1.09	**27.81**	m2	**31.29**
Aerated concrete blocks, Thermalite, 440 x 215mm, Party wall blocks (4.0 N/mm2); in cement-lime mortar (1:1:6)										
Walls; vertical										
215mm thick; stretcher bond	20.18	5.00	**21.19**	1.00	1.20	**27.06**	2.77	**51.02**	m2	**57.40**

Labour hourly rates: (except Specialists) Craft Operatives £14.21 Labourer £10.71 Rates are national average prices. Refer to REGIONAL VARIATIONS for indicative levels of overall pricing in regions	MATERIALS			LABOUR				RATES		
	Del to Site £	Waste %	Material Cost £	Craft Optve Hrs	Lab Hrs	Labour Cost £	Sunds £	Nett Rate £	Unit	Gross Rate (+12.5%) £
F10: BRICK/BLOCK WALLING Cont.										
Aerated concrete blocks, Thermalite, 440 x 215mm, Hi-Strength 7 blocks (7.0 N/mm2); in cement-lime mortar (1:1:6)										
Walls; vertical										
100mm thick; stretcher bond	11.86	5.00	**12.45**	0.80	0.90	**21.01**	1.11	**34.57**	m2	**38.89**
140mm thick; stretcher bond	16.43	5.00	**17.25**	0.90	1.05	**24.03**	1.67	**42.95**	m2	**48.32**
150mm thick; stretcher bond	17.58	5.00	**18.46**	0.90	1.05	**24.03**	1.67	**44.16**	m2	**49.68**
190mm thick; stretcher bond	22.16	5.00	**23.27**	0.95	1.10	**25.28**	2.21	**50.76**	m2	**57.11**
200mm thick; stretcher bond	23.30	5.00	**24.47**	1.00	1.20	**27.06**	2.21	**53.74**	m2	**60.46**
215mm thick; stretcher bond	25.01	5.00	**26.26**	1.00	1.20	**27.06**	2.77	**56.10**	m2	**63.11**
Closing cavities; vertical										
50mm wide with blockwork 100mm thick	2.97	5.00	**3.11**	0.25	0.25	**6.23**	0.24	**9.58**	m	**10.78**
75mm wide with blockwork 100mm thick	2.97	5.00	**3.11**	0.25	0.25	**6.23**	0.24	**9.58**	m	**10.78**
Closing cavities; horizontal										
50mm wide with blockwork 100mm thick	2.97	5.00	**3.11**	0.25	0.25	**6.23**	0.24	**9.58**	m	**10.78**
75mm wide with blockwork 100mm thick	2.97	5.00	**3.11**	0.25	0.25	**6.23**	0.24	**9.58**	m	**10.78**
Aerated concrete blocks, Thermalite, 440 x 215mm, Trenchblocks (4.0 N/mm2); in cement mortar (1:4)										
Walls; vertical										
255mm thick; stretcher bond	23.87	5.00	**25.06**	1.30	1.50	**34.54**	2.30	**61.90**	m2	**69.64**
275mm thick; stretcher bond	25.70	5.00	**26.99**	1.45	1.65	**38.28**	2.76	**68.02**	m2	**76.53**
305mm thick; stretcher bond	28.48	5.00	**29.90**	1.65	1.85	**43.26**	2.76	**75.93**	m2	**85.42**
355mm thick; stretcher bond	33.09	5.00	**34.74**	1.95	2.15	**50.74**	3.23	**88.71**	m2	**99.79**
Aerated concrete blocks, Tarmac Toplite foundation blocks; 440 x 215mm, (3.5 N/mm2); in cement mortar (1:4)										
Walls; vertical										
260mm thick; stretcher bond	20.16	5.00	**21.17**	1.40	1.40	**34.89**	2.76	**58.82**	m2	**66.17**
275mm thick; stretcher bond	21.33	5.00	**22.40**	1.60	1.60	**39.87**	2.76	**65.03**	m2	**73.16**
300mm thick; stretcher bond	23.27	5.00	**24.43**	1.70	1.70	**42.36**	2.76	**69.56**	m2	**78.26**
Tarmac Topblock, medium density concrete block, Hemelite, 440 x 215mm, solid Standard blocks (3.5 N/mm2); in cement-lime mortar (1:1:6)										
Walls; vertical										
75mm thick; stretcher bond	5.91	5.00	**6.21**	0.75	0.90	**20.30**	0.88	**27.38**	m2	**30.81**
90mm thick; stretcher bond	6.58	5.00	**6.91**	0.90	1.05	**24.03**	1.11	**32.05**	m2	**36.06**
100mm thick; stretcher bond	7.16	5.00	**7.52**	0.90	1.05	**24.03**	1.11	**32.66**	m2	**36.74**
140mm thick; stretcher bond	10.52	5.00	**11.05**	1.00	1.20	**27.06**	1.67	**39.77**	m2	**44.75**
190mm thick; stretcher bond	14.27	5.00	**14.98**	1.05	1.25	**28.31**	2.21	**45.51**	m2	**51.19**
215mm thick; stretcher bond	16.14	5.00	**16.95**	1.10	1.30	**29.55**	2.77	**49.27**	m2	**55.43**
Closing cavities; vertical										
50mm wide with blockwork 100mm thick	1.79	5.00	**1.88**	0.25	0.25	**6.23**	0.24	**8.35**	m	**9.39**
75mm wide with blockwork 100mm thick	1.79	5.00	**1.88**	0.25	0.25	**6.23**	0.24	**8.35**	m	**9.39**
Closing cavities; horizontal										
50mm wide with blockwork 100mm thick	1.79	5.00	**1.88**	0.25	0.25	**6.23**	0.24	**8.35**	m	**9.39**
75mm wide with blockwork 100mm thick	1.79	5.00	**1.88**	0.25	0.25	**6.23**	0.24	**8.35**	m	**9.39**
Bonding ends to common brickwork; forming pockets; extra material										
walls; bonding every third course										
75mm	1.48	5.00	**1.55**	0.25	0.25	**6.23**	-	**7.78**	m	**8.75**
90mm	1.65	5.00	**1.73**	0.30	0.30	**7.48**	-	**9.20**	m	**10.35**
100mm	1.79	5.00	**1.88**	0.30	0.30	**7.48**	-	**9.36**	m	**10.52**
140mm	2.63	5.00	**2.76**	0.40	0.40	**9.97**	-	**12.73**	m	**14.32**
190mm	3.57	5.00	**3.75**	0.45	0.45	**11.21**	-	**14.96**	m	**16.83**
215mm	4.04	5.00	**4.24**	0.55	0.55	**13.71**	-	**17.94**	m	**20.19**
Bonding ends to existing common brickwork; cutting pockets; extra material										
walls; bonding every third course										
75mm	1.48	5.00	**1.55**	0.50	0.50	**12.46**	-	**14.01**	m	**15.76**
90mm	1.65	5.00	**1.73**	0.60	0.60	**14.95**	-	**16.68**	m	**18.76**
100mm	1.79	5.00	**1.88**	0.60	0.60	**14.95**	-	**16.83**	m	**18.94**
140mm	2.63	5.00	**2.76**	0.80	0.80	**19.94**	-	**22.70**	m	**25.53**
190mm	3.57	5.00	**3.75**	0.90	0.90	**22.43**	-	**26.17**	m	**29.45**
215mm	4.04	5.00	**4.24**	1.10	1.10	**27.41**	-	**31.65**	m	**35.60**
Bonding ends to existing concrete blockwork; cutting pockets; extra material										
walls; bonding every third course										
75mm	1.48	5.00	**1.55**	0.40	0.40	**9.97**	-	**11.52**	m	**12.96**
90mm	1.65	5.00	**1.73**	0.45	0.45	**11.21**	-	**12.94**	m	**14.56**
100mm	1.79	5.00	**1.88**	0.45	0.45	**11.21**	-	**13.09**	m	**14.73**
140mm	2.63	5.00	**2.76**	0.60	0.60	**14.95**	-	**17.71**	m	**19.93**
190mm	3.57	5.00	**3.75**	0.70	0.70	**17.44**	-	**21.19**	m	**23.84**
215mm	4.04	5.00	**4.24**	0.85	0.85	**21.18**	-	**25.42**	m	**28.60**
Tarmac Topblock, medium density concrete block, Hemelite, 440 x 215mm, solid Standard blocks (7.0 N/mm2); in cement-lime mortar (1:1:6)										
Walls; vertical										
90mm thick; stretcher bond	6.91	5.00	**7.26**	0.90	1.05	**24.03**	1.11	**32.40**	m2	**36.45**

Labour hourly rates: (except Specialists) Craft Operatives £14.21 Labourer £10.71 Rates are national average prices. Refer to REGIONAL VARIATIONS for indicative levels of overall pricing in regions	MATERIALS			LABOUR				RATES		
	Del to Site £	Waste %	Material Cost £	Craft Optve Hrs	Lab Hrs	Labour Cost £	Sunds £	Nett Rate £	Unit	Gross Rate (+12.5%) £
F10: BRICK/BLOCK WALLING Cont.										
Tarmac Topblock, medium density concrete block, Hemelite, 440 x 215mm, solid Standard blocks (7.0 N/mm2); in cement-lime mortar (1:1:6) Cont.										
Walls; vertical Cont.										
100mm thick; stretcher bond	7.59	5.00	7.97	0.90	1.05	24.03	1.11	33.11	m2	37.25
140mm thick; stretcher bond	11.16	5.00	11.72	1.00	1.20	27.06	1.67	40.45	m2	45.50
190mm thick; stretcher bond	15.15	5.00	15.91	1.05	1.25	28.31	2.21	46.43	m2	52.23
215mm thick; stretcher bond	17.13	5.00	17.99	1.10	1.30	29.55	2.77	50.31	m2	56.60
Closing cavities; vertical										
50mm wide with blockwork 100mm thick	1.90	5.00	1.99	0.25	0.25	6.23	0.24	8.46	m	9.52
75mm wide with blockwork 100mm thick	1.90	5.00	1.99	0.25	0.25	6.23	0.24	8.46	m	9.52
Closing cavities; horizontal										
50mm wide with blockwork 100mm thick	1.90	5.00	1.99	0.25	0.25	6.23	0.24	8.46	m	9.52
75mm wide with blockwork 100mm thick	1.90	5.00	1.99	0.25	0.25	6.23	0.24	8.46	m	9.52
Tarmac Topblock, fair face concrete blocks, Lignacite, 440 x 215mm, solid Standard blocks (7.0 N/mm2); in cement-lime mortar (1:1:6)										
Walls; vertical										
100mm thick; stretcher bond	13.35	5.00	14.02	0.95	1.10	25.28	1.11	40.41	m2	45.46
140mm thick; stretcher bond	19.64	5.00	20.62	1.10	1.25	29.02	1.67	51.31	m2	57.72
190mm thick; stretcher bond	26.66	5.00	27.99	1.15	1.35	30.80	2.21	61.01	m2	68.63
Closing cavities; vertical										
50mm wide with blockwork 100mm thick	3.34	5.00	3.50	0.25	0.25	6.23	0.24	9.97	m	11.22
75mm wide with blockwork 100mm thick	3.34	5.00	3.50	0.25	0.25	6.23	0.24	9.97	m	11.22
Closing cavities; horizontal										
50mm wide with blockwork 100mm thick	3.34	5.00	3.50	0.25	0.25	6.23	0.24	9.97	m	11.22
75mm wide with blockwork 100mm thick	3.34	5.00	3.50	0.25	0.25	6.23	0.24	9.97	m	11.22
Tarmac Topblock, dense concrete blocks, Topcrete, 440 x 215mm, solid Standard blocks (7.0 N/mm2); in cement-lime mortar (1:1:6)										
Walls; vertical										
90mm thick; stretcher bond	8.38	5.00	8.80	1.15	1.35	30.80	1.11	40.71	m2	45.79
100mm thick; stretcher bond	9.32	5.00	9.79	1.20	1.40	32.05	1.11	42.94	m2	48.31
140mm thick; stretcher bond	13.05	5.00	13.70	1.30	1.50	34.54	1.67	49.91	m2	56.15
190mm thick; stretcher bond	17.70	5.00	18.59	1.40	1.60	37.03	2.21	57.83	m2	65.06
215mm thick; stretcher bond	20.04	5.00	21.04	1.50	1.70	39.52	2.77	63.34	m2	71.25
Closing cavities; vertical										
50mm wide with blockwork 100mm thick	2.33	5.00	2.45	0.25	0.25	6.23	0.24	8.91	m	10.03
75mm wide with blockwork 100mm thick	2.33	5.00	2.45	0.25	0.25	6.23	0.24	8.91	m	10.03
Closing cavities; horizontal										
50mm wide with blockwork 100mm thick	2.33	5.00	2.45	0.25	0.25	6.23	0.24	8.91	m	10.03
75mm wide with blockwork 100mm thick	2.33	5.00	2.45	0.25	0.25	6.23	0.24	8.91	m	10.03
Tarmac Topblock, dense concrete blocks, Topcrete, 440 x 215mm hollow standard blocks (7.0 N/mm2); in cement-lime mortar (1:1:6)										
Walls; vertical										
215mm thick; stretcher bond	13.95	5.00	14.65	1.50	1.56	38.02	2.77	55.44	m2	62.37
Aerated concrete blocks, Thermalite, 440 x 215mm, Smooth Face blocks (4.0 N/mm2); in cement-lime mortar (1:1:6); flush smooth pointing as work proceeds										
Walls; vertical										
stretcher bond; facework one side										
100mm thick	15.88	5.00	16.67	1.50	1.56	38.02	1.11	55.80	m2	62.78
140mm thick	22.08	5.00	23.18	1.10	1.25	29.02	1.67	53.87	m2	60.60
190mm thick	29.83	5.00	31.32	1.15	1.35	30.80	2.21	64.34	m2	72.38
215mm thick	33.72	5.00	35.41	1.20	1.40	32.05	2.77	70.23	m2	79.00
stretcher bond; facework both sides										
100mm thick	15.88	5.00	16.67	1.20	1.35	31.51	1.11	49.29	m2	55.45
140mm thick	22.08	5.00	23.18	1.30	1.45	34.00	1.67	58.85	m2	66.21
190mm thick	29.83	5.00	31.32	1.35	1.55	35.78	2.21	69.32	m2	77.99
215mm thick	33.72	5.00	35.41	1.40	1.60	37.03	2.77	75.21	m2	84.61
Tarmac Topblock, fair face concrete blocks, Lignacite, 440 x 215mm, solid Standard blocks (7.0 N/mm2); in cement-lime mortar (1:1:6); flush smooth pointing as work proceeds										
Walls; vertical										
stretcher bond; facework one side										
100mm thick	13.35	5.00	14.02	1.15	1.35	30.80	1.11	45.92	m2	51.67
140mm thick	19.64	5.00	20.62	1.30	1.50	34.54	1.67	56.83	m2	63.93
190mm thick	26.66	5.00	27.99	1.35	1.55	35.78	2.21	65.99	m2	74.24
stretcher bond; facework both sides										
100mm thick	13.35	5.00	14.02	1.35	1.55	35.78	1.11	50.91	m2	57.27
140mm thick	19.64	5.00	20.62	1.50	1.70	39.52	1.67	61.81	m2	69.54
190mm thick	26.66	5.00	27.99	1.55	1.75	40.77	2.21	70.98	m2	79.85

Labour hourly rates: (except Specialists) Craft Operatives £14.21 Labourer £10.71 Rates are national average prices. Refer to REGIONAL VARIATIONS for indicative levels of overall pricing in regions	MATERIALS			LABOUR				RATES		
	Del to Site £	Waste %	Material Cost £	Craft Optve Hrs	Lab Hrs	Labour Cost £	Sunds £	Nett Rate £	Unit	Gross Rate (+12.5%) £
F10: BRICK/BLOCK WALLING Cont.										
Reconstructed stone blocks, Marshalls Mono Ltd. Cromwell coursed random length Split Face buff walling blocks in cement-lime mortar (1:1:6); flat recessed pointing as work proceeds										
Walls; vertical										
100mm thick; alternate courses 102mm and 140mm high blocks; stretcher bond; facework one side...	19.15	5.00	**20.11**	1.90	2.05	**48.95**	1.67	**70.73**	m2	**79.57**
100mm thick; one course 102mm high blocks and two courses 140mm high blocks; stretcher bond; facework one side	19.15	5.00	**20.11**	1.90	2.05	**48.95**	1.67	**70.73**	m2	**79.57**
Reconstructed stone blocks, Marshalls Mono Ltd. Cromwell coursed random length Pitched Face buff walling blocks in cement-lime mortar (1:1:6); flat recessed pointing as work proceeds										
Walls; vertical										
90mm thick; alternate courses 102mm and 140mm high blocks; stretcher bond; facework one side........	21.71	5.00	**22.80**	1.90	2.05	**48.95**	1.67	**73.42**	m2	**82.59**
90mm thick; one course 102mm high blocks and two courses 140mm high blocks; stretcher bond; facework one side	21.71	5.00	**22.80**	1.90	2.05	**48.95**	1.67	**73.42**	m2	**82.59**
Firebricks, 215 x 102.5 x 65mm; in fire cement										
Flue linings; bonding to surrounding brickwork with headers -4/m2										
112mm thick; stretcher bond......................	110.93	5.00	**116.48**	1.90	2.05	**48.95**	7.92	**173.36**	m2	**195.03**
Flue linings; built clear of main brickwork but with one header in each course set projecting to contact main work										
112mm thick; stretcher bond......................	110.93	5.00	**116.48**	1.90	2.05	**48.95**	7.92	**173.36**	m2	**195.03**
112mm thick; stretcher bond; in segmental top to flue	110.93	5.00	**116.48**	2.50	2.65	**63.91**	7.92	**188.31**	m2	**211.85**
HR Supra Flue lining bricks, 230 x 114 x 76mm, Hepworth Refractories, Sheffield, MPK 21 mortar										
Flue linings; bonding to surrounding brickwork with headers -4/m2										
114mm thick; stretcher bond......................	44.81	5.00	**47.05**	1.90	2.05	**48.95**	4.57	**100.58**	m2	**113.15**
230mm thick; English bond.........................	89.63	5.00	**94.11**	3.15	3.40	**81.18**	8.56	**183.84**	m2	**206.82**
F11: GLASS BLOCK WALLING - SCREENS AND PANELS										
Hollow glass blocks, white, cross ribbed, in cement mortar (1:3); continuous joints; flat recessed pointing as work proceeds										
Screens or panels; facework both sides										
190 x 190mm blocks; vertical; 80mm thick.........	164.75	10.00	**181.23**	3.30	3.75	**87.06**	58.86	**327.14**	m2	**368.03**
240 x 240mm blocks; vertical; 80mm thick.........	133.44	10.00	**146.78**	2.75	3.25	**73.89**	47.38	**268.04**	m2	**301.55**
F20: NATURAL STONE RUBBLE WALLING										
Stone rubble work; random stones of Yorkshire limestone; bedding and jointing in lime mortar (1:3); uncoursed										
Walls; tapering both sides; including extra material										
500mm thick	55.00	15.00	**63.25**	3.75	4.30	**99.34**	13.34	**175.93**	m2	**197.92**
600mm thick	60.50	15.00	**69.58**	4.50	5.15	**119.10**	15.39	**204.07**	m2	**229.57**
Stone rubble work; squared rubble face stones of Yorkshire limestone; bedding and jointing in lime mortar (1:3); irregular coursed; courses average 150mm high										
Walls; vertical; face stones 100 - 150mm on bed; bonding to brickwork; including extra material; scappled or axed face; weather struck pointing as work proceeds										
150mm thick; faced one side......................	44.00	15.00	**50.60**	3.00	3.45	**79.58**	4.10	**134.28**	m2	**151.07**
Walls; vertical; face stones 100 - 150mm on bed; bonding to brickwork; including extra material; hammer dressed face; weather struck pointing as work proceeds										
150mm thick; faced one side......................	44.00	15.00	**50.60**	3.30	3.80	**87.59**	4.10	**142.30**	m2	**160.08**
Walls; vertical; face stones 100 - 150mm on bed; bonding to brickwork; including extra material; rock worked face; pointing with a parallel joint as work proceeds										
150mm thick; faced one side......................	46.00	15.00	**52.90**	3.60	4.15	**95.60**	4.10	**152.61**	m2	**171.68**

MASONRY

Labour hourly rates: (except Specialists) Craft Operatives £14.21 Labourer £10.71 Rates are national average prices. Refer to REGIONAL VARIATIONS for indicative levels of overall pricing in regions	MATERIALS			LABOUR				RATES		
	Del to Site £	Waste %	Material Cost £	Craft Optve Hrs	Lab Hrs	Labour Cost £	Sunds £	Nett Rate £	Unit	Gross Rate (+12.5%) £
F20: NATURAL STONE RUBBLE WALLING Cont.										
Stone rubble work; squared rubble face stones of Yorkshire limestone; bedding and jointing in lime mortar (1:3); regular coursed; courses average 150mm high										
Walls; vertical; face stones 100 - 150mm on bed; bonding to brickwork; including extra material; scappled or axed face; weather struck pointing as work proceeds										
150mm thick; faced one side.....................	60.50	15.00	**69.58**	3.20	3.70	**85.10**	4.10	**158.78**	m2	**178.63**
Walls; vertical; face stones 100 - 150mm on bed; bonding to brickwork; including extra material; hammer dressed face; weather struck pointing as work proceeds										
150mm thick; faced one side.....................	60.50	15.00	**69.58**	3.50	4.00	**92.58**	4.10	**166.25**	m2	**187.04**
Walls; vertical; face stones 100 - 150mm on bed; bonding to brickwork; including extra material; rock worked face; pointing with a parallel joint as work proceeds										
150mm thick; faced one side.....................	63.00	15.00	**72.45**	3.80	4.40	**101.12**	4.10	**177.68**	m2	**199.89**
Stone rubble work; random rubble backing and squared rubble face stones of Yorkshire limestone; bedding and jointing in lime mortar (1:3); irregular coursed; courses average 150mm high										
Walls; vertical; face stones 100 - 150mm on bed; scappled or axed face; weather struck pointing as work proceeds										
350mm thick; faced one side.....................	70.50	15.00	**81.08**	4.00	4.60	**106.11**	9.23	**196.42**	m2	**220.97**
500mm thick; faced one side.....................	89.00	15.00	**102.35**	4.75	5.45	**125.87**	13.34	**241.56**	m2	**271.75**
Walls; vertical; face stones 100 - 150mm on bed; hammer dressed face; weather struck pointing as work proceeds										
350mm thick; faced one side.....................	70.50	15.00	**81.08**	4.30	4.95	**114.12**	9.23	**204.43**	m2	**229.98**
500mm thick; faced one side.....................	89.00	15.00	**102.35**	5.05	5.80	**133.88**	13.34	**249.57**	m2	**280.76**
Walls; vertical; face stones 100 - 150mm on bed; rock worked face; pointing with a parallel joint as work proceeds										
350mm thick; faced one side.....................	70.50	15.00	**81.08**	4.60	5.30	**122.13**	9.23	**212.44**	m2	**238.99**
500mm thick; faced one side.....................	89.00	15.00	**102.35**	5.35	6.15	**141.89**	13.34	**257.58**	m2	**289.78**
Quoin stones; scappled or axed face; attached faced two adjacent faces										
250 x 200 x 350mm...........................	28.50	5.00	**29.93**	0.60	0.30	**11.74**	0.51	**42.18**	nr	**47.45**
250 x 200 x 500mm...........................	41.00	5.00	**43.05**	0.90	0.45	**17.61**	0.67	**61.33**	nr	**68.99**
250 x 250 x 350mm...........................	37.00	5.00	**38.85**	0.75	0.40	**14.94**	0.56	**54.36**	nr	**61.15**
250 x 250 x 500mm...........................	51.00	5.00	**53.55**	1.00	0.50	**19.57**	0.87	**73.99**	nr	**83.24**
380 x 200 x 350mm...........................	43.00	5.00	**45.15**	0.90	0.45	**17.61**	0.82	**63.58**	nr	**71.53**
380 x 200 x 500mm...........................	60.50	5.00	**63.53**	1.15	0.55	**22.23**	0.92	**86.68**	nr	**97.52**
Quoin stones; rock worked face; attached faced two adjacent faces										
250 x 200 x 350mm...........................	27.00	5.00	**28.35**	0.65	0.30	**12.45**	0.51	**41.31**	nr	**46.48**
250 x 200 x 500mm...........................	36.50	5.00	**38.33**	0.95	0.45	**18.32**	0.67	**57.31**	nr	**64.47**
250 x 250 x 350mm...........................	31.50	5.00	**33.08**	0.80	0.40	**15.65**	0.56	**49.29**	nr	**55.45**
250 x 250 x 500mm...........................	44.50	5.00	**46.73**	1.10	0.55	**21.52**	0.87	**69.12**	nr	**77.76**
380 x 200 x 350mm...........................	38.50	5.00	**40.43**	0.95	0.50	**18.85**	0.82	**60.10**	nr	**67.61**
380 x 200 x 500mm...........................	54.00	5.00	**56.70**	1.25	0.60	**24.19**	0.92	**81.81**	nr	**92.04**
Arches; relieving										
225mm high on face, 180mm wide on soffit.........	104.50	5.00	**109.73**	1.75	0.85	**33.97**	1.13	**144.82**	m	**162.93**
225mm high on face, 250mm wide on soffit.........	143.00	5.00	**150.15**	2.30	1.15	**45.00**	1.59	**196.74**	m	**221.33**
Grooves										
12 x 25mm	-	-	-	0.30	-	**4.26**	-	**4.26**	m	**4.80**
12 x 38mm	-	-	-	0.35	-	**4.97**	-	**4.97**	m	**5.60**
Stone rubble work; random rubble backing and squared rubble face stones of Yorkshire limestone; bedding and jointing in lime mortar (1:3); regular coursed; courses average 150mm high										
Walls; vertical; face stones 100 - 150mm on bed; scappled or axed face; weather struck pointing as work proceeds										
350mm thick; faced one side.....................	89.00	15.00	**102.35**	4.85	4.85	**120.86**	9.23	**232.45**	m2	**261.50**
Walls; vertical; face stones 125 - 200mm on bed; scappled or axed face; weather struck pointing as work proceeds										
500mm thick; faced one side.....................	109.00	15.00	**125.35**	5.70	5.70	**142.04**	13.34	**280.73**	m2	**315.82**
Walls; vertical; face stones 100 - 150mm on bed; hammer dressed face; weather struck pointing as work proceeds										
350mm thick; faced one side.....................	89.00	15.00	**102.35**	5.15	5.15	**128.34**	9.23	**239.92**	m2	**269.91**
500mm thick; faced one side.....................	109.00	15.00	**125.35**	6.00	6.00	**149.52**	13.34	**288.21**	m2	**324.23**

Labour hourly rates: (except Specialists) Craft Operatives £14.21 Labourer £10.71 Rates are national average prices. Refer to REGIONAL VARIATIONS for indicative levels of overall pricing in regions	MATERIALS			LABOUR			RATES			
	Del to Site £	Waste %	Material Cost £	Craft Optve Hrs	Lab Hrs	Labour Cost £	Sunds £	Nett Rate £	Unit	Gross Rate (+12.5%) £
F20: NATURAL STONE RUBBLE WALLING Cont.										
Stone rubble work; random rubble backing and squared rubble face stones of Yorkshire limestone; bedding and jointing in lime mortar (1:3); regular coursed; courses average 150mm high Cont.										
Walls; vertical; face stones 100 - 150mm on bed; rock worked face; pointing with a parallel joint as work proceeds										
350mm thick; faced one side	89.00	15.00	**102.35**	5.50	5.50	**137.06**	9.23	**248.64**	m2	**279.72**
500mm thick; faced one side	109.00	15.00	**125.35**	6.40	6.40	**159.49**	13.34	**298.18**	m2	**335.45**
F21: NATURAL STONE ASHLAR WALLING/DRESSINGS										
Natural stonework										
Note stonework is work for a specialist; prices being obtained for specific projects. Some firms that specialise in this class of work are included within the list at the end of this section										
Natural stonework dressings; natural Dorset limestone; Portland Whitbed; bedding and jointing in mason's mortar (1:3:12); flush smooth pointing as work proceeds; slurrying with weak lime mortar and cleaning down on completion										
Walls; vertical; building against brickwork; BS 1243 Fig 1 specification 3.5 wall ties - 4/m2 built in										
50mm thick; plain and rubbed one side	149.50	2.50	**153.24**	4.60	3.00	**97.50**	6.21	**256.94**	m2	**289.06**
75mm thick; plain and rubbed one side	206.25	2.50	**211.41**	5.20	3.60	**112.45**	6.77	**330.63**	m2	**371.95**
100mm thick; plain and rubbed one side	269.50	2.50	**276.24**	5.75	4.25	**127.23**	7.34	**410.80**	m2	**462.15**
Natural stonework; natural Dorset limestone, Portland Whitbed; bedding and jointing in cement-lime mortar (1:2:9); flush smooth pointing as work proceeds										
Lintels plain and rubbed faces -3; splayed and rubbed faces -1										
200 x 100mm	66.00	2.50	**67.65**	1.45	1.00	**31.31**	1.44	**100.40**	m	**112.95**
225 x 125mm	84.50	2.50	**86.61**	1.70	1.25	**37.54**	1.85	**126.00**	m	**141.75**
Sills plain and rubbed faces -3; sunk weathered and rubbed faces -1; grooves -1; throats -1										
200 x 75mm	88.00	2.50	**90.20**	1.15	0.75	**24.37**	2.15	**116.73**	m	**131.32**
250 x 75mm	106.50	2.50	**109.16**	1.40	0.90	**29.53**	2.57	**141.26**	m	**158.92**
300 x 75mm	123.00	2.50	**126.08**	1.60	1.15	**35.05**	2.77	**163.90**	m	**184.38**
Jamb stones; attached plain and rubbed faces -3; splayed and rubbed faces -1; rebates -1; grooves -1										
175 x 75mm	68.00	2.50	**69.70**	1.15	0.75	**24.37**	1.90	**95.97**	m	**107.97**
200 x 100mm	74.00	2.50	**75.85**	1.45	1.00	**31.31**	2.46	**109.63**	m	**123.33**
Band courses; moulded; horizontal										
225 x 125mm	154.00	2.50	**157.85**	1.85	1.30	**40.21**	3.39	**201.45**	m	**226.63**
250 x 150mm	185.00	2.50	**189.63**	2.05	1.55	**45.73**	3.95	**239.31**	m	**269.22**
300 x 150mm	218.00	2.50	**223.45**	2.30	1.80	**51.96**	4.62	**280.03**	m	**315.03**
Copings; horizontal plain and rubbed faces -2; weathered and rubbed faces -1; throats -2; cramped joints with stainless steel cramps										
300 x 50mm	106.50	2.50	**109.16**	0.85	0.60	**18.50**	4.77	**132.44**	m	**148.99**
300 x 75mm	124.50	2.50	**127.61**	1.00	0.80	**22.78**	5.34	**155.73**	m	**175.19**
375 x 100mm	171.50	2.50	**175.79**	1.55	1.15	**34.34**	6.41	**216.54**	m	**243.61**
Natural stonework, natural Yorkshire sandstone, Bolton Wood; bedding and jointing in cement-lime mortar (1:2:9); flush smooth pointing as work proceeds										
Sills plain and rubbed faces -3; sunk weathered and rubbed faces -1; throats -1										
175 x 100mm	64.00	2.50	**65.60**	1.45	1.00	**31.31**	1.13	**98.04**	m	**110.30**
250 x 125mm	99.00	2.50	**101.48**	1.60	1.15	**35.05**	1.90	**138.43**	m	**155.73**
Copings; horizontal plain and rubbed faces -2; weathered and rubbed faces -2; throats -2; cramped joints with stainless steel cramps										
300 x 75mm	70.00	2.50	**71.75**	1.05	0.80	**23.49**	4.36	**99.60**	m	**112.05**
300 x 100mm	82.50	2.50	**84.56**	1.25	0.90	**27.40**	4.87	**116.84**	m	**131.44**
Kerbs; horizontal plain and sawn faces - 4										
150 x 150mm	43.00	2.50	**44.08**	1.00	0.75	**22.24**	1.28	**67.60**	m	**76.05**
225 x 150mm	65.00	2.50	**66.63**	1.40	1.15	**32.21**	1.90	**100.73**	m	**113.33**
Cover stones										
75 x 300; rough edges -2; plain and sawn face -1.	37.00	2.50	**37.93**	1.00	0.75	**22.24**	2.51	**62.68**	m	**70.52**

Labour hourly rates: (except Specialists) Craft Operatives £14.21 Labourer £10.71 Rates are national average prices. Refer to REGIONAL VARIATIONS for indicative levels of overall pricing in regions	MATERIALS			LABOUR			RATES			
	Del to Site	Waste	Material Cost	Craft Optve	Lab	Labour Cost	Sunds	Nett Rate	Unit	Gross Rate (+12.5%)
	£	%	£	Hrs	Hrs	£	£	£		£
F21: NATURAL STONE ASHLAR WALLING/DRESSINGS Cont.										
Natural stonework, natural Yorkshire sandstone, Bolton Wood; bedding and jointing in cement-lime mortar (1:2:9); flush smooth pointing as work proceeds Cont.										
Templates										
150 x 300; rough edges -2; plain and sawn face -1	74.25	2.50	**76.11**	1.70	1.40	**39.15**	3.08	**118.34**	m	**133.13**
Steps; plain										
plain and rubbed top and front										
225 x 75mm	29.75	2.50	**30.49**	0.85	0.65	**19.04**	2.51	**52.05**	m	**58.55**
225 x 150mm	59.50	2.50	**60.99**	1.50	1.15	**33.63**	3.08	**97.70**	m	**109.91**
300 x 150mm	79.00	2.50	**80.98**	1.85	1.50	**42.35**	3.74	**127.07**	m	**142.96**
Landings										
75 x 900 x 900mm; sawn edges -4; plain and rubbed face -1	109.00	2.50	**111.73**	2.30	2.00	**54.10**	3.74	**169.57**	nr	**190.77**
F22: CAST STONE ASHLAR WALLING/DRESSINGS										
Cast stonework dressings; simulated Dorset limestone; Portland Whitbed; bedding and jointing in cement-lime mortar (1:2:9); flush smooth pointing as work proceeds										
Walls; vertical building against brickwork; BS 1243 Fig 1 specification 3.5 wall ties -4/m2 built in										
100mm thick; plain and rubbed one side	88.00	5.00	**92.40**	4.60	2.90	**96.43**	6.26	**195.08**	m2	**219.47**
Fair raking cutting										
100 thick	17.60	5.00	**18.48**	1.15	0.60	**22.77**	-	**41.25**	m	**46.40**
Cast stonework; simulated Dorset limestone; Portland Whitbed; bedding and jointing in cement-lime mortar (1:2:9); flush smooth pointing as work proceeds										
Lintels										
plain and rubbed faces -3; splayed and rubbed faces -1										
200 x 100mm	25.30	5.00	**26.57**	1.15	0.70	**23.84**	1.03	**51.43**	m	**57.86**
225 x 125mm	37.40	5.00	**39.27**	1.45	0.85	**29.71**	1.28	**70.26**	m	**79.04**
Sills										
plain and rubbed faces -3; sunk weathered and rubbed faces -1; grooves -1; throats -1										
200 x 75mm	23.10	5.00	**24.26**	1.05	0.60	**21.35**	1.28	**46.88**	m	**52.74**
extra; stoolings	5.50	5.00	**5.78**	-	-	**-**	-	**5.78**	nr	**6.50**
250 x 75mm	25.30	5.00	**26.57**	1.15	0.70	**23.84**	1.64	**52.05**	m	**58.55**
extra; stoolings	5.50	5.00	**5.78**	-	-	**-**	-	**5.78**	nr	**6.50**
300 x 75mm	29.70	5.00	**31.19**	1.45	0.85	**29.71**	1.95	**62.84**	m	**70.70**
extra; stoolings	5.50	5.00	**5.78**	-	-	**-**	-	**5.78**	nr	**6.50**
Jamb stones; attached										
plain and rubbed faces -3; splayed and rubbed faces -1; rebates -1; grooves -1										
175 x 75mm	24.75	5.00	**25.99**	1.05	0.60	**21.35**	0.92	**48.26**	m	**54.29**
200 x 100mm	32.00	5.00	**33.60**	1.15	0.70	**23.84**	1.95	**59.39**	m	**66.81**
Band courses; plain; horizontal										
225 x 125mm	24.75	5.00	**25.99**	1.60	1.15	**35.05**	1.64	**62.68**	m	**70.52**
extra; external return	17.60	5.00	**18.48**	-	-	**-**	-	**18.48**	nr	**20.79**
250 x 150mm	33.00	5.00	**34.65**	1.75	1.25	**38.26**	1.95	**74.85**	m	**84.21**
extra; external return	17.60	5.00	**18.48**	-	-	**-**	-	**18.48**	nr	**20.79**
300 x 150mm	36.50	5.00	**38.33**	1.85	1.40	**41.28**	2.26	**81.86**	m	**92.10**
extra; external return	17.60	5.00	**18.48**	-	-	**-**	-	**18.48**	nr	**20.79**
Copings; horizontal										
plain and rubbed faces -2; weathered and rubbed faces -2; throats -2; cramped joints with stainless steel cramps										
300 x 50mm	19.25	5.00	**20.21**	0.60	0.35	**12.27**	3.85	**36.33**	m	**40.88**
extra; internal angles	17.60	5.00	**18.48**	-	-	**-**	-	**18.48**	nr	**20.79**
extra; external angles	17.60	5.00	**18.48**	-	-	**-**	-	**18.48**	nr	**20.79**
300 x 75mm	24.75	5.00	**25.99**	0.70	0.40	**14.23**	4.10	**44.32**	m	**49.86**
extra; internal angles	17.60	5.00	**18.48**	-	-	**-**	-	**18.48**	nr	**20.79**
extra; external angles	17.60	5.00	**18.48**	-	-	**-**	-	**18.48**	nr	**20.79**
375 x 100mm	41.25	5.00	**43.31**	0.85	0.60	**18.50**	4.87	**66.69**	m	**75.03**
extra; internal angles	17.60	5.00	**18.48**	-	-	**-**	-	**18.48**	nr	**20.79**
extra; external angles	17.60	5.00	**18.48**	-	-	**-**	-	**18.48**	nr	**20.79**
Steps; plain										
300 x 150mm; plain and rubbed tread and riser	39.60	5.00	**41.58**	1.45	1.05	**31.85**	2.57	**76.00**	m	**85.49**
Steps; spandril										
250mm wide tread; 180mm high riser; plain and rubbed tread and riser; carborundum finish to tread	49.50	5.00	**51.98**	2.30	1.45	**48.21**	1.95	**102.14**	m	**114.90**
Landings										
150 x 900 x 900mm; plain and rubbed top surface	88.00	5.00	**92.40**	1.75	1.15	**37.18**	2.57	**132.15**	nr	**148.67**

Labour hourly rates: (except Specialists) Craft Operatives £14.21 Labourer £10.71 Rates are national average prices. Refer to REGIONAL VARIATIONS for indicative levels of overall pricing in regions	MATERIALS			LABOUR				RATES		
	Del to Site £	Waste %	Material Cost £	Craft Optve Hrs	Lab Hrs	Labour Cost £	Sunds £	Nett Rate £	Unit	Gross Rate (+12.5%) £
F30: ACCESSORIES/SUNDRY ITEMS FOR BRICK/BLOCK/STONE WALLING										
Forming cavities in hollow walls										
Cavity with BS 1243 Fig 1 stainless steel wire butterfly wall ties, -4/m2 built in										
width of cavity 50mm	0.54	10.00	**0.60**	0.10	0.10	**2.49**	-	**3.09**	m2	**3.48**
Cavity with Staifix HRT4; 200mm Housing stainless steel wall ties -4/m2 built in; 50 fibreglass resin bonded slab cavity insulation										
width of cavity 50mm	4.10	10.00	**4.51**	0.35	0.35	**8.72**	1.16	**14.39**	m2	**16.19**
Cavity with BS 1243 Fig 3 stainless steel twisted wall ties -4/m2 built in; 25 Celotex CW035 tuff-R Zero cavity insulation boards										
width of cavity 50mm	8.24	10.00	**9.06**	0.35	0.35	**8.72**	1.16	**18.95**	m2	**21.32**
Cavity with 200mm stainless steel vertical twisted wall ties -4/m2 built in; 50 fibreglass resin bonded slab cavity insulation										
width of cavity 50mm	6.15	10.00	**6.76**	0.35	0.35	**8.72**	1.16	**16.64**	m2	**18.72**
Cavity with Staifix HRT4; 200mm Housing stainless steel wall ties -4/m2 built in; 75 fibreglass resin bonded slab cavity insulation										
width of cavity 75mm	5.53	10.00	**6.08**	0.40	0.40	**9.97**	1.16	**17.21**	m2	**19.36**
Cavity with stainless steel vertical twisted wall ties -4/m2 built in; 75 fibreglass resin bonded slab cavity insulation										
width of cavity 75mm	7.58	10.00	**8.33**	0.40	0.40	**9.97**	1.16	**19.46**	m2	**21.90**
Damp proof courses										
BS 6398 Class A, bitumen with hessian base; 100mm laps; bedding in cement mortar (1:3); no allowance made for laps										
width not exceeding 225mm; vertical	7.60	5.00	**7.98**	0.50	-	**7.11**	-	**15.09**	m2	**16.97**
width exceeding 225mm; vertical	7.60	5.00	**7.98**	0.45	-	**6.39**	-	**14.37**	m2	**16.17**
width not exceeding 225mm; horizontal	7.60	5.00	**7.98**	0.35	-	**4.97**	-	**12.95**	m2	**14.57**
width exceeding 225mm; horizontal	7.60	5.00	**7.98**	0.30	-	**4.26**	-	**12.24**	m2	**13.77**
cavity trays; width not exceeding 225mm; horizontal	7.60	5.00	**7.98**	0.55	-	**7.82**	-	**15.80**	m2	**17.77**
cavity trays; width exceeding 225mm; horizontal	7.60	5.00	**7.98**	0.50	-	**7.11**	-	**15.09**	m2	**16.97**
BS 6398 Class B, bitumen with fibre base; 100mm laps; bedding in cement mortar (1:3); no allowance made for laps										
width not exceeding 225mm; vertical	4.55	5.00	**4.78**	0.50	-	**7.11**	-	**11.88**	m2	**13.37**
width exceeding 225mm; vertical	4.55	5.00	**4.78**	0.45	-	**6.39**	-	**11.17**	m2	**12.57**
width not exceeding 225mm; horizontal	4.55	5.00	**4.78**	0.35	-	**4.97**	-	**9.75**	m2	**10.97**
width exceeding 225mm; horizontal	4.55	5.00	**4.78**	0.30	-	**4.26**	-	**9.04**	m2	**10.17**
BS 6398 Class D, bitumen with hessian base laminated with lead; 100mm laps; bedding in cement mortar (1:3); no allowance made for laps										
width not exceeding 225mm; vertical	15.07	5.00	**15.82**	0.50	-	**7.11**	-	**22.93**	m2	**25.79**
width exceeding 225mm; vertical	15.07	5.00	**15.82**	0.45	-	**6.39**	-	**22.22**	m2	**25.00**
width not exceeding 225mm; horizontal	15.07	5.00	**15.82**	0.35	-	**4.97**	-	**20.80**	m2	**23.40**
width exceeding 225mm; horizontal	15.07	5.00	**15.82**	0.30	-	**4.26**	-	**20.09**	m2	**22.60**
BS 6398 Class E, bitumen with fibre base laminated with lead; 100mm laps; bedding in cement mortar (1:3); no allowance made for laps										
width not exceeding 225mm; vertical	16.74	5.00	**17.58**	0.50	-	**7.11**	-	**24.68**	m2	**27.77**
width exceeding 225mm; vertical	16.74	5.00	**17.58**	0.45	-	**6.39**	-	**23.97**	m2	**26.97**
width not exceeding 225mm; horizontal	16.74	5.00	**17.58**	0.35	-	**4.97**	-	**22.55**	m2	**25.37**
width exceeding 225mm; horizontal	16.74	5.00	**17.58**	0.30	-	**4.26**	-	**21.84**	m2	**24.57**
Hyload, pitch polymer; 100mm laps; bedding in cement lime mortar (1:1:6); no allowance made for laps										
width not exceeding 225mm; vertical	6.08	5.00	**6.38**	0.50	-	**7.11**	-	**13.49**	m2	**15.18**
width exceeding 225mm; vertical	6.08	5.00	**6.38**	0.45	-	**6.39**	-	**12.78**	m2	**14.38**
width not exceeding 225mm; horizontal	6.08	5.00	**6.38**	0.35	-	**4.97**	-	**11.36**	m2	**12.78**
width exceeding 225mm; horizontal	6.08	5.00	**6.38**	0.30	-	**4.26**	-	**10.65**	m2	**11.98**
Synthaprufe bituminous latex emulsion; two coats brushed on; blinded with sand										
width not exceeding 225mm; vertical	5.39	5.00	**5.66**	-	0.60	**6.43**	0.17	**12.26**	m2	**13.79**
width exceeding 225mm; vertical	5.39	5.00	**5.66**	-	0.35	**3.75**	0.17	**9.58**	m2	**10.78**
Synthaprufe bituminous latex emulsion; three coats brushed on; blinded with sand										
width not exceeding 225mm; vertical	7.76	5.00	**8.15**	-	0.90	**9.64**	0.19	**17.98**	m2	**20.23**
width exceeding 225mm; vertical	7.76	5.00	**8.15**	-	0.50	**5.36**	0.19	**13.70**	m2	**15.41**
Bituminous emulsion; two coats brushed on										
width not exceeding 225mm; vertical	3.88	5.00	**4.07**	-	0.55	**5.89**	0.17	**10.14**	m2	**11.40**
width exceeding 225mm; vertical	3.88	5.00	**4.07**	-	0.36	**3.86**	0.17	**8.10**	m2	**9.11**
Bituminous emulsion; three coats brushed on										
width not exceeding 225mm; vertical	5.39	5.00	**5.66**	-	0.75	**8.03**	0.19	**13.89**	m2	**15.62**
width exceeding 225mm; vertical	5.39	5.00	**5.66**	-	0.40	**4.28**	0.19	**10.14**	m2	**11.41**

Labour hourly rates: (except Specialists) Craft Operatives £14.21 Labourer £10.71 Rates are national average prices. Refer to REGIONAL VARIATIONS for indicative levels of overall pricing in regions	MATERIALS			LABOUR				RATES		
	Del to Site £	Waste %	Material Cost £	Craft Optve Hrs	Lab Hrs	Labour Cost £	Sunds £	Nett Rate £	Unit	Gross Rate (+12.5%) £
F30: ACCESSORIES/SUNDRY ITEMS FOR BRICK/BLOCK/STONE WALLING Cont.										
Damp proof courses Cont.										
One course slates in cement mortar (1:3)										
width not exceeding 225mm; vertical	17.69	15.00	**20.34**	1.30	1.30	**32.40**	-	**52.74**	m2	**59.33**
width exceeding 225mm; vertical	17.69	15.00	**20.34**	1.20	1.20	**29.90**	-	**50.25**	m2	**56.53**
width not exceeding 225mm; horizontal	17.69	15.00	**20.34**	0.80	0.80	**19.94**	-	**40.28**	m2	**45.31**
width exceeding 225mm; horizontal	17.69	15.00	**20.34**	0.70	0.70	**17.44**	-	**37.79**	m2	**42.51**
Two courses slates in cement mortar (1:3)										
width not exceeding 225mm; vertical	35.38	15.00	**40.68**	2.20	2.20	**54.82**	-	**95.51**	m2	**107.44**
width exceeding 225mm; vertical	35.38	15.00	**40.68**	2.00	2.00	**49.84**	-	**90.52**	m2	**101.84**
width not exceeding 225mm; horizontal	35.38	15.00	**40.68**	1.30	1.30	**32.40**	-	**73.08**	m2	**82.21**
width exceeding 225mm; horizontal	35.38	15.00	**40.68**	1.20	1.20	**29.90**	-	**70.59**	m2	**79.41**
Cavity trays										
Type G Cavitray by Cavity Trays Ltd in										
half brick skin of cavity wall	9.95	5.00	**10.44**	0.20	-	**2.84**	-	**13.28**	m	**14.95**
external angle	7.94	5.00	**8.34**	0.10	-	**1.42**	-	**9.76**	nr	**10.98**
internal angle	7.94	5.00	**8.34**	0.10	-	**1.42**	-	**9.76**	nr	**10.98**
Type X Cavitray by Cavity Trays Ltd to suit 40 degree pitched roof complete with attached code 4 lead flashing and dress over tiles in										
half brick skin of cavity wall	41.48	5.00	**43.55**	0.20	-	**2.84**	-	**46.40**	m	**52.20**
ridge tray	10.29	5.00	**10.80**	0.10	-	**1.42**	-	**12.23**	nr	**13.75**
catchment tray	4.88	5.00	**5.12**	0.10	-	**1.42**	-	**6.55**	nr	**7.36**
corner catchment angle tray	10.75	5.00	**11.29**	0.10	-	**1.42**	-	**12.71**	nr	**14.30**
Type W Cavity weep/ventilator by Cavity Trays Ltd in										
half brick skin of cavity wall	0.54	5.00	**0.57**	0.10	-	**1.42**	-	**1.99**	nr	**2.24**
extension duct	0.82	5.00	**0.86**	0.10	-	**1.42**	-	**2.28**	nr	**2.57**
Joint reinforcement										
Expamet grade 304/S15 Exmet reinforcement, stainless steel; 150mm laps;										
65mm wide	1.14	5.00	**1.20**	0.04	0.04	**1.00**	0.03	**2.23**	m	**2.50**
115mm wide	1.74	5.00	**1.83**	0.05	0.05	**1.25**	0.03	**3.11**	m	**3.49**
175mm wide	3.01	5.00	**3.16**	0.06	0.06	**1.50**	0.05	**4.71**	m	**5.30**
225mm wide	3.93	5.00	**4.13**	0.07	0.07	**1.74**	0.06	**5.94**	m	**6.68**
Bed Joint Reinforcement (UK) Ltd Netspan reinforcement, stainless steel grade 304/S15; 150mm laps;										
61mm wide	1.14	5.00	**1.20**	0.04	0.04	**1.00**	0.03	**2.23**	m	**2.50**
112mm wide	1.74	5.00	**1.83**	0.05	0.05	**1.25**	0.03	**3.11**	m	**3.49**
178mm wide	3.01	5.00	**3.16**	0.05	0.05	**1.25**	0.03	**4.44**	m	**4.99**
228mm wide	3.93	5.00	**4.13**	0.06	0.06	**1.50**	0.05	**5.68**	m	**6.38**
Bed Joint Reinforcement (UK) Ltd grade 304/S15 Wallspan reinforcement, stainless steel; 150mm laps;										
210mm wide	6.78	5.00	**7.12**	0.06	0.06	**1.50**	0.05	**8.67**	m	**9.75**
222.5mm wide	6.78	5.00	**7.12**	0.06	0.06	**1.50**	0.05	**8.67**	m	**9.75**
235mm wide	6.78	5.00	**7.12**	0.07	0.07	**1.74**	0.06	**8.93**	m	**10.04**
250mm wide	6.78	5.00	**7.12**	0.07	0.07	**1.74**	0.06	**8.93**	m	**10.04**
275mm wide	6.78	5.00	**7.12**	0.07	0.07	**1.74**	0.06	**8.93**	m	**10.04**
Brickforce , galvanised steel; 150mm laps;										
60mm wide ref GBF30W60	0.42	5.00	**0.45**	0.04	0.04	**1.00**	0.03	**1.47**	m	**1.66**
60mm wide ref GBF35W60	0.57	5.00	**0.60**	0.04	0.04	**1.00**	0.03	**1.63**	m	**1.83**
60mm wide ref GBF40W60	0.70	5.00	**0.73**	0.04	0.04	**1.00**	0.03	**1.76**	m	**1.98**
60mm wide ref GBF45W60	0.92	5.00	**0.97**	0.04	0.04	**1.00**	0.03	**2.00**	m	**2.25**
60mm wide ref GBF50W60	1.14	5.00	**1.20**	0.04	0.04	**1.00**	0.03	**2.23**	m	**2.51**
160mm wide ref GBF30W160	0.60	5.00	**0.63**	0.05	0.05	**1.25**	0.03	**1.91**	m	**2.15**
160mm wide ref GBF35W160	0.83	5.00	**0.87**	0.05	0.05	**1.25**	0.03	**2.15**	m	**2.41**
160mm wide ref GBF40W160	1.00	5.00	**1.05**	0.05	0.05	**1.25**	0.03	**2.32**	m	**2.62**
160mm wide ref GBF45W160	1.34	5.00	**1.40**	0.05	0.05	**1.25**	0.03	**2.68**	m	**3.02**
160mm wide ref GBF50W160	1.66	5.00	**1.75**	0.05	0.05	**1.25**	0.03	**3.03**	m	**3.40**
Brickforce , stainless steel; 150mm laps;										
60mm wide ref SBF30W60	1.11	5.00	**1.17**	0.04	0.04	**1.00**	0.03	**2.20**	m	**2.47**
60mm wide ref SBF35W60	1.52	5.00	**1.59**	0.04	0.04	**1.00**	0.03	**2.62**	m	**2.95**
60mm wide ref SBF40W60	1.99	5.00	**2.09**	0.04	0.04	**1.00**	0.03	**3.12**	m	**3.51**
60mm wide ref SBF45W60	2.46	5.00	**2.58**	0.04	0.04	**1.00**	0.03	**3.61**	m	**4.06**
60mm wide ref SBF50W60	3.02	5.00	**3.17**	0.04	0.04	**1.00**	0.03	**4.20**	m	**4.73**
160mm wide ref SBF30W160	1.17	5.00	**1.22**	0.05	0.05	**1.25**	0.03	**2.50**	m	**2.82**
160mm wide ref SBF35W160	1.59	5.00	**1.67**	0.05	0.05	**1.25**	0.03	**2.95**	m	**3.32**
160mm wide ref SBF40W160	2.11	5.00	**2.21**	0.05	0.05	**1.25**	0.03	**3.49**	m	**3.93**
160mm wide ref SBF45W160	2.58	5.00	**2.70**	0.05	0.05	**1.25**	0.03	**3.98**	m	**4.48**
160mm wide ref SBF50W160	3.17	5.00	**3.33**	0.05	0.05	**1.25**	0.03	**4.61**	m	**5.18**
Pointing in flashings										
Cement mortar (1:3)										
horizontal	-	-	-	0.45	-	**6.39**	0.26	**6.65**	m	**7.48**
horizontal; in old wall	-	-	-	0.50	-	**7.11**	0.26	**7.36**	m	**8.28**
stepped	-	-	-	0.80	-	**11.37**	0.41	**11.78**	m	**13.25**
stepped; in old wall	-	-	-	0.85	-	**12.08**	0.41	**12.49**	m	**14.05**

Labour hourly rates: (except Specialists) Craft Operatives £14.21 Labourer £10.71 Rates are national average prices. Refer to REGIONAL VARIATIONS for indicative levels of overall pricing in regions	MATERIALS			LABOUR				RATES		
	Del to Site £	Waste %	Material Cost £	Craft Optve Hrs	Lab Hrs	Labour Cost £	Sunds £	Nett Rate £	Unit	Gross Rate (+12.5%) £
F30: ACCESSORIES/SUNDRY ITEMS FOR BRICK/BLOCK/STONE WALLING Cont.										
Wedging and pinning										
Two courses slates in cement mortar (1:3)										
width of wall 215mm	8.84	15.00	**10.17**	0.50	0.50	**12.46**	0.15	**22.78**	m	**25.63**
Joints										
Expansion joints in facing brickwork										
13mm wide; vertical; filling with Servicised Ltd, Fibrepack filler, Vertiseal compound pointing one side; including preparation, cleaners, primers and sealers										
102mm thick wall	3.44	7.50	**3.70**	0.10	0.20	**3.56**	-	**7.26**	m	**8.17**
215mm thick wall	5.46	7.50	**5.87**	0.11	0.22	**3.92**	-	**9.79**	m	**11.01**
Expansion joints in blockwork										
13mm wide; vertical; filling with Servicised Ltd, Fibrepack filler, Vertiseal compound pointing one side; including preparation, cleaners, primers and sealers										
100mm thick wall	3.44	7.50	**3.70**	0.10	0.20	**3.56**	-	**7.26**	m	**8.17**
Expansion joints in glass blockwork										
10mm wide; vertical; in filling with compressible material, polysulphide sealant both sides including preparation, cleaners, primers and sealers										
80mm thick wall	4.58	7.50	**4.92**	0.10	0.20	**3.56**	-	**8.49**	m	**9.55**
Slates and tiles for creasing										
Nibless creasing tiles, red, machine made, 265 x 165 x 10mm; in cement-lime mortar (1:1:6)										
one course 253mm wide	2.66	5.00	**2.80**	0.50	0.50	**12.46**	0.28	**15.54**	m	**17.48**
one course 365mm wide	4.00	5.00	**4.20**	0.70	0.70	**17.44**	0.39	**22.03**	m	**24.78**
two courses 253mm wide	5.33	5.00	**5.59**	0.85	0.85	**21.18**	0.56	**27.34**	m	**30.75**
two courses 365mm wide	7.99	5.00	**8.39**	1.15	1.15	**28.66**	0.77	**37.82**	m	**42.55**
Slate and tile sills										
Clay plain roofing tiles, red, BS EN 1304, machine made, 265 x 165 x 13mm; in cement-lime mortar										
one course 150mm wide; set weathering	1.78	5.00	**1.86**	0.45	0.45	**11.21**	0.26	**13.34**	m	**15.00**
two courses 150mm wide; set weathering	3.55	5.00	**3.73**	0.75	0.75	**18.69**	0.52	**22.94**	m	**25.80**
Fires and fire parts										
Solid one piece or two piece firebacks; BS 1251; bed and joint in fire cement; concrete filling at back										
fire size 400mm	38.22	2.50	**39.18**	1.00	1.00	**24.92**	1.28	**65.37**	nr	**73.55**
fire size 450mm	41.06	2.50	**42.09**	1.10	1.10	**27.41**	1.42	**70.92**	nr	**79.78**
Solid one piece or two piece firebacks with cut out for boiler; BS 1251; bed and joint in fire cement; concrete filling at back										
fire size 400mm	46.29	2.50	**47.45**	1.00	1.00	**24.92**	1.28	**73.65**	nr	**82.85**
fire size 450mm	53.93	2.50	**55.28**	1.10	1.10	**27.41**	1.42	**84.11**	nr	**94.62**
Frets and stools; black vitreous enamelled; place in position										
fire size 400mm	32.24	-	**32.24**	0.33	0.33	**8.22**	-	**40.46**	nr	**45.52**
fire size 450mm	50.39	-	**50.39**	0.36	0.36	**8.97**	-	**59.36**	nr	**66.78**
Frets and stools; lustre finish; place in position										
fire size 400mm	52.49	-	**52.49**	0.33	0.33	**8.22**	-	**60.71**	nr	**68.30**
fire size 450mm	34.39	-	**34.39**	0.36	0.36	**8.97**	-	**43.36**	nr	**48.78**
Continuous burning open fire; vitreous finish; self contained open fire; plugging and screwing; sealing to opening										
fire size 400mm	161.39	-	**161.39**	3.50	3.50	**87.22**	2.61	**251.22**	nr	**282.63**
fire size 450mm	184.34	-	**184.34**	3.75	3.75	**93.45**	5.06	**282.85**	nr	**318.21**
Gas ignited smokeless fuel fires; assembling; plugging and screwing; bedding and jointing in fire cement; sealing to opening										
grate only; black										
fire size 400mm	104.76	-	**104.76**	2.50	2.50	**62.30**	-	**167.06**	nr	**187.94**
fire size 450mm	106.18	-	**106.18**	2.50	2.50	**62.30**	-	**168.48**	nr	**189.54**
grate only; colours										
fire size 400mm	106.18	-	**106.18**	2.50	2.50	**62.30**	-	**168.48**	nr	**189.54**
fire size 450mm	107.57	-	**107.57**	2.50	2.50	**62.30**	-	**169.87**	nr	**191.10**
boiler and self contained flue										
fire size 400mm	243.02	-	**243.02**	4.50	4.50	**112.14**	2.61	**357.77**	nr	**402.49**
fire size 450mm	258.44	-	**258.44**	4.50	4.50	**112.14**	5.06	**375.64**	nr	**422.60**
threefold brick sets										
fire size 400mm	32.27	-	**32.27**	1.00	1.00	**24.92**	1.28	**58.47**	nr	**65.78**
fire size 450mm	32.27	-	**32.27**	1.00	1.00	**24.92**	1.42	**58.61**	nr	**65.94**
extra; chrome finish to fire front; fire size 400 or 450mm	32.27	-	**32.27**	-	-	**-**	-	**32.27**	nr	**36.30**
extra; back boiler unit with Bower Barffed rustless boiler; fire size 400 or 450mm	251.42	-	**251.42**	-	-	**-**	-	**251.42**	nr	**282.85**

MASONRY

Labour hourly rates: (except Specialists) Craft Operatives £14.21 Labourer £10.71 Rates are national average prices. Refer to REGIONAL VARIATIONS for indicative levels of overall pricing in regions	MATERIALS			LABOUR				RATES		
	Del to Site £	Waste %	Material Cost £	Craft Optve Hrs	Lab Hrs	Labour Cost £	Sunds £	Nett Rate £	Unit	Gross Rate (+12.5%) £
F30: ACCESSORIES/SUNDRY ITEMS FOR BRICK/BLOCK/STONE WALLING Cont.										
Fires and fire parts Cont.										
Gas ignited smokeless fuel fires; assembling; plugging and screwing; bedding and jointing in fire cement; sealing to opening Cont.										
threefold brick sets Cont.										
extra; back boiler unit with copper boiler; 400 or 450mm	384.10	-	**384.10**	-	-	-	-	**384.10**	nr	**432.11**
Flue linings										
Clay flue linings, BS 1181; rebated joints; jointed in cement mortar (1:3)										
150mm diameter, Type 2	20.32	5.00	**21.34**	0.40	0.45	**10.50**	0.34	**32.18**	m	**36.21**
terminal Type 6F	19.43	5.00	**20.40**	0.45	0.40	**10.68**	0.18	**31.26**	nr	**35.17**
185mm diameter, Type 2	26.04	5.00	**27.34**	0.50	0.40	**11.39**	0.45	**39.18**	m	**44.08**
terminal Type 6F	21.13	5.00	**22.19**	0.50	0.40	**11.39**	0.20	**33.78**	nr	**38.00**
225mm diameter, Type 2	43.41	5.00	**45.58**	0.60	0.80	**17.09**	0.48	**63.16**	m	**71.05**
terminal Type 6F	22.74	5.00	**23.88**	0.60	0.50	**13.88**	0.26	**38.02**	nr	**42.77**
185 x 185mm, Type 1	27.96	5.00	**29.36**	0.60	0.80	**17.09**	0.45	**46.90**	m	**52.77**
terminal Type 4D	31.63	5.00	**33.21**	0.45	0.40	**10.68**	0.20	**44.09**	nr	**49.61**
Chimney pots										
Clay chimney pots; set and flaunched in cement mortar (1:3)										
tapered roll top; 600mm high	38.94	5.00	**40.89**	0.80	0.95	**21.54**	0.45	**62.88**	nr	**70.74**
tapered roll top; 750mm high	53.02	5.00	**55.67**	0.95	1.15	**25.82**	0.57	**82.06**	nr	**92.31**
tapered roll top; 900mm high	69.93	5.00	**73.43**	1.10	1.40	**30.63**	0.68	**104.73**	nr	**117.82**
Air bricks										
Clay air bricks, BS 493, square hole pattern; opening with slate lintel over										
common brick wall 102mm thick; opening size										
225 x 75mm	2.09	-	**2.09**	0.20	0.20	**4.98**	0.68	**7.75**	nr	**8.72**
225 x 150mm	3.99	-	**3.99**	0.25	0.25	**6.23**	0.72	**10.94**	nr	**12.31**
225 x 225mm	6.99	-	**6.99**	0.30	0.30	**7.48**	0.77	**15.24**	nr	**17.15**
common brick wall 215mm thick; opening size										
225 x 75mm	2.09	-	**2.09**	0.30	0.30	**7.48**	1.27	**10.83**	nr	**12.19**
225 x 150mm	3.99	-	**3.99**	0.38	0.38	**9.47**	1.33	**14.79**	nr	**16.64**
225 x 225mm	6.99	-	**6.99**	0.45	0.45	**11.21**	1.43	**19.63**	nr	**22.09**
common brick wall 215mm thick; facework one side; opening size										
225 x 75mm	2.09	-	**2.09**	0.40	0.40	**9.97**	1.27	**13.33**	nr	**14.99**
225 x 150mm	3.99	-	**3.99**	0.50	0.50	**12.46**	1.33	**17.78**	nr	**20.01**
225 x 225mm	6.99	-	**6.99**	0.60	0.60	**14.95**	1.43	**23.37**	nr	**26.29**
common brick wall 327mm thick; facework one side; opening size										
225 x 75mm	2.09	-	**2.09**	0.60	0.60	**14.95**	2.06	**19.11**	nr	**21.49**
225 x 150mm	3.99	-	**3.99**	0.70	0.70	**17.44**	2.17	**23.61**	nr	**26.56**
225 x 225mm	6.99	-	**6.99**	0.80	0.80	**19.94**	2.23	**29.15**	nr	**32.80**
common brick wall 440mm thick; facework one side; opening size										
225 x 75mm	2.09	-	**2.09**	0.80	0.80	**19.94**	2.74	**24.77**	nr	**27.86**
225 x 150mm	3.99	-	**3.99**	0.90	0.90	**22.43**	2.84	**29.26**	nr	**32.91**
225 x 225mm	6.99	-	**6.99**	1.00	1.00	**24.92**	2.93	**34.84**	nr	**39.20**
cavity wall 252mm thick with 102mm facing brick outer skin, 100mm block inner skin and 50mm cavity; sealing cavity with slates in cement mortar (1:3); rendering all round with cement mortar (1:3); opening size										
225 x 75mm	2.09	-	**2.09**	0.60	0.60	**14.95**	2.47	**19.51**	nr	**21.95**
225 x 150mm	3.99	-	**3.99**	0.70	0.70	**17.44**	3.04	**24.48**	nr	**27.54**
225 x 225mm	6.99	-	**6.99**	0.80	0.80	**19.94**	3.56	**30.48**	nr	**34.29**
Gas flue blocks										
Marflex HP system precast refractory concrete gas flue blocks; bedding and jointing in Flue joint refractory mortar										
recess block reference HP1, 405 x 140 x 222mm	2.09	5.00	**2.19**	0.42	-	**5.97**	0.68	**8.84**	nr	**9.94**
cover block reference HP2, 385 x 140 x 222mm	5.30	5.00	**5.57**	0.51	-	**7.25**	0.68	**13.49**	nr	**15.18**
Standard block reference HP3, 355 x 140 x 72mm	2.64	5.00	**2.77**	0.34	-	**4.83**	0.68	**8.28**	nr	**9.32**
Standard block reference HP3, 355 x 140 x 112mm	2.75	5.00	**2.89**	0.34	-	**4.83**	0.68	**8.40**	nr	**9.45**
Standard block reference HP3, 355 x 140 x 222mm	3.62	5.00	**3.80**	0.34	-	**4.83**	0.68	**9.31**	nr	**10.47**
Vent unit for bathroom/kitchen extract fans reference HP1 and HP2	5.90	5.00	**6.20**	0.34	-	**4.83**	0.68	**11.70**	nr	**13.17**
standard block reference HP4, 280 x 140 x 72mm	2.75	5.00	**2.89**	0.28	-	**3.98**	0.59	**7.46**	nr	**8.39**
standard block reference HP4, 280 x 140 x 112mm	2.86	5.00	**3.00**	0.28	-	**3.98**	0.59	**7.57**	nr	**8.52**
standard block reference HP4, 280 x 140 x 222mm	3.63	5.00	**3.81**	0.28	-	**3.98**	0.59	**8.38**	nr	**9.43**
120mm side offset block reference HP5, 400 x 140 x 222mm	3.97	5.00	**4.17**	0.42	-	**5.97**	0.86	**11.00**	nr	**12.37**
70mm back offset block reference HP6, 280 x 210 x 222mm	13.75	5.00	**14.44**	0.39	-	**5.54**	0.86	**20.84**	nr	**23.44**
vertical exit block reference HP7, 280 x 181 x 222mm	7.98	5.00	**8.38**	0.39	-	**5.54**	0.68	**14.60**	nr	**16.42**
angled entry/exit block reference HP8, 280 x 140 x 230mm	7.98	5.00	**8.38**	0.45	-	**6.39**	0.92	**15.70**	nr	**17.66**
double rebate block used when in conjunction with a boiler reference HP9, 280 x 140 x 222mm	5.73	5.00	**6.02**	0.45	-	**6.39**	0.59	**13.00**	nr	**14.63**
corbel block reference HP10	7.87	5.00	**8.26**	0.45	-	**6.39**	0.68	**15.34**	nr	**17.25**

Labour hourly rates: (except Specialists) Craft Operatives £14.21 Labourer £10.71 Rates are national average prices. Refer to REGIONAL VARIATIONS for indicative levels of overall pricing in regions	MATERIALS			LABOUR			RATES			
	Del to Site £	Waste %	Material Cost £	Craft Optve Hrs	Lab Hrs	Labour Cost £	Sunds £	Nett Rate £	Unit	Gross Rate (+12.5%) £
F30: ACCESSORIES/SUNDRY ITEMS FOR BRICK/BLOCK/STONE WALLING Cont.										
Gas flue blocks Cont.										
Marflex HP system precast refractory concrete gas flue blocks; bedding and jointing in Flue joint refractory mortar Cont.										
conversion unit for back boiler recess units reference HP25, 280 x 262 x 222mm	11.79	5.00	**12.38**	0.45	-	**6.39**	0.59	**19.37**	nr	**21.79**
Arch bars										
Steel flat arch bar										
30 x 6mm	2.26	-	**2.26**	0.80	-	**11.37**	-	**13.63**	m	**15.33**
50 x 6mm	3.49	-	**3.49**	0.80	-	**11.37**	-	**14.86**	m	**16.72**
Steel angle arch bar										
50 x 50 x 6mm	6.54	-	**6.54**	0.80	-	**11.37**	-	**17.91**	m	**20.15**
75 x 50 x 6mm	8.28	-	**8.28**	0.90	-	**12.79**	-	**21.07**	m	**23.70**
Building in										
Building in metal windows; building in lugs; bedding in cement mortar (1:3), pointing with Secomastic standard mastic one side										
200mm high; lugs to brick jambs; plugging and screwing to brick sill and concrete head										
500mm wide	-	-	-	0.56	0.35	**11.71**	1.11	**12.81**	nr	**14.42**
600mm wide	-	-	-	0.61	0.38	**12.74**	1.37	**14.10**	nr	**15.87**
900mm wide	-	-	-	0.78	0.46	**16.01**	1.88	**17.89**	nr	**20.13**
1200mm wide	-	-	-	0.95	0.54	**19.28**	2.43	**21.71**	nr	**24.43**
1500mm wide	-	-	-	1.11	0.61	**22.31**	2.93	**25.24**	nr	**28.40**
1800mm wide	-	-	-	1.28	0.69	**25.58**	3.44	**29.02**	nr	**32.65**
500mm high; lugs to brick jambs; plugging and screwing to brick sill and concrete head										
500mm wide	-	-	-	0.67	0.39	**13.70**	1.70	**15.40**	nr	**17.32**
600mm wide	-	-	-	0.72	0.43	**14.84**	1.87	**16.71**	nr	**18.80**
900mm wide	-	-	-	0.89	0.50	**18.00**	2.38	**20.38**	nr	**22.92**
1200mm wide	-	-	-	1.07	0.58	**21.42**	2.89	**24.31**	nr	**27.35**
1500mm wide	-	-	-	1.24	0.68	**24.90**	3.44	**28.34**	nr	**31.89**
1800mm wide	-	-	-	1.41	0.76	**28.18**	3.91	**32.09**	nr	**36.10**
700mm high; lugs to brick jambs; plugging and screwing to brick sill and concrete head										
500mm wide	-	-	-	0.79	0.44	**15.94**	2.03	**17.97**	nr	**20.22**
600mm wide	-	-	-	0.84	0.47	**16.97**	2.20	**19.17**	nr	**21.57**
900mm wide	-	-	-	1.02	0.55	**20.38**	2.74	**23.13**	nr	**26.02**
1200mm wide	-	-	-	1.19	0.65	**23.87**	3.25	**27.12**	nr	**30.51**
1500mm wide	-	-	-	1.37	0.73	**27.29**	3.76	**31.05**	nr	**34.93**
1800mm wide	-	-	-	1.53	0.82	**30.52**	4.26	**34.78**	nr	**39.13**
900mm high; lugs to brick jambs; plugging and screwing to brick sill and concrete head										
500mm wide	-	-	-	0.89	0.47	**17.68**	2.35	**20.03**	nr	**22.54**
600mm wide	-	-	-	0.96	0.50	**19.00**	2.48	**21.48**	nr	**24.16**
900mm wide	-	-	-	1.13	0.60	**22.48**	3.05	**25.54**	nr	**28.73**
1200mm wide	-	-	-	1.31	0.69	**26.01**	3.56	**29.56**	nr	**33.26**
1500mm wide	-	-	-	1.49	0.79	**29.63**	4.09	**33.72**	nr	**37.93**
1800mm wide	-	-	-	1.66	0.87	**32.91**	4.57	**37.48**	nr	**42.16**
1100mm high; lugs to brick jambs; plugging and screwing to brick sill and concrete head										
500mm wide	-	-	-	1.01	0.52	**19.92**	2.73	**22.65**	nr	**25.48**
600mm wide	-	-	-	1.07	0.55	**21.10**	2.85	**23.94**	nr	**26.94**
900mm wide	-	-	-	1.25	0.65	**24.72**	3.39	**28.11**	nr	**31.62**
1200mm wide	-	-	-	1.43	0.74	**28.25**	3.90	**32.15**	nr	**36.17**
1500mm wide	-	-	-	1.62	0.84	**32.02**	4.43	**36.45**	nr	**41.00**
1800mm wide	-	-	-	1.80	0.93	**35.54**	4.95	**40.48**	nr	**45.54**
1300mm high; lugs to brick jambs; plugging and screwing to brick sill and concrete head										
500mm wide	-	-	-	1.12	0.57	**22.02**	3.02	**25.04**	nr	**28.17**
600mm wide	-	-	-	1.18	0.60	**23.19**	3.19	**26.39**	nr	**29.68**
900mm wide	-	-	-	1.37	0.69	**26.86**	3.71	**30.57**	nr	**34.39**
1200mm wide	-	-	-	1.55	0.79	**30.49**	4.22	**34.71**	nr	**39.05**
1500mm wide	-	-	-	1.73	0.90	**34.22**	4.76	**38.98**	nr	**43.86**
1800mm wide	-	-	-	1.92	0.99	**37.89**	5.27	**43.15**	nr	**48.55**
1500mm high; lugs to brick jambs; plugging and screwing to brick sill and concrete head										
500mm wide	-	-	-	1.24	0.61	**24.15**	3.36	**27.52**	nr	**30.96**
600mm wide	-	-	-	1.29	0.65	**25.29**	3.52	**28.81**	nr	**32.41**
900mm wide	-	-	-	1.49	0.74	**29.10**	4.05	**33.15**	nr	**37.29**
1200mm wide	-	-	-	1.67	0.85	**32.83**	4.56	**37.39**	nr	**42.07**
1500mm wide	-	-	-	1.86	0.95	**36.61**	5.05	**41.66**	nr	**46.86**
1800mm wide	-	-	-	2.05	1.06	**40.48**	5.60	**46.08**	nr	**51.84**
2100mm high; lugs to brick jambs; plugging and screwing to brick sill and concrete head										
500mm wide	-	-	-	1.46	0.69	**28.14**	4.36	**32.50**	nr	**36.56**
600mm wide	-	-	-	1.52	0.73	**29.42**	4.54	**33.95**	nr	**38.20**
900mm wide	-	-	-	1.72	0.84	**33.44**	5.04	**38.48**	nr	**43.29**
1200mm wide	-	-	-	1.91	0.95	**37.32**	5.60	**42.92**	nr	**48.28**
1500mm wide	-	-	-	2.11	1.07	**41.44**	6.07	**47.52**	nr	**53.46**
1800mm wide	-	-	-	2.30	1.18	**45.32**	6.62	**51.94**	nr	**58.44**

Labour hourly rates: (except Specialists) Craft Operatives £14.21 Labourer £10.71 Rates are national average prices. Refer to REGIONAL VARIATIONS for indicative levels of overall pricing in regions	MATERIALS			LABOUR			RATES			
	Del to Site £	Waste %	Material Cost £	Craft Optve Hrs	Lab Hrs	Labour Cost £	Sunds £	Nett Rate £	Unit	Gross Rate (+12.5%) £
F30: ACCESSORIES/SUNDRY ITEMS FOR BRICK/BLOCK/STONE WALLING Cont.										
Building in Cont.										
Building in factory glazed metal windows; screwing with galvanized screws; bedding in cement mortar (1:3), pointing with Secomastic standard mastic one side										
300mm high; plugging and screwing lugs to brick jambs and sill and concrete head										
600mm wide	-	-	-	0.74	0.48	15.66	1.53	17.18	nr	19.33
900mm wide	-	-	-	0.88	0.55	18.40	2.06	20.46	nr	23.02
1200mm wide	-	-	-	1.03	0.60	21.06	2.57	23.63	nr	26.59
1500mm wide	-	-	-	1.18	0.65	23.73	3.10	26.83	nr	30.18
1800mm wide	-	-	-	1.32	0.71	26.36	3.59	29.95	nr	33.70
2400mm wide	-	-	-	1.62	0.84	32.02	4.63	36.65	nr	41.23
3000mm wide	-	-	-	1.91	0.95	37.32	5.64	42.96	nr	48.33
700mm high; plugging and screwing lugs to brick jambs and sill and concrete head										
600mm wide	-	-	-	1.01	0.63	21.10	2.20	23.30	nr	26.22
900mm wide	-	-	-	1.23	0.76	25.62	2.74	28.36	nr	31.90
1200mm wide	-	-	-	1.44	0.90	30.10	3.25	33.35	nr	37.52
1500mm wide	-	-	-	1.66	1.01	34.41	3.76	38.17	nr	42.94
1800mm wide	-	-	-	1.88	1.15	39.03	4.26	43.29	nr	48.70
2400mm wide	-	-	-	2.32	1.42	48.18	5.31	53.49	nr	60.17
3000mm wide	-	-	-	2.75	1.67	56.96	6.36	63.33	nr	71.24
900mm high; plugging and screwing lugs to brick jambs and sill and concrete head										
600mm wide	-	-	-	1.14	0.71	23.80	2.55	26.35	nr	29.65
900mm wide	-	-	-	1.40	0.87	29.21	3.05	32.26	nr	36.30
1200mm wide	-	-	-	1.65	1.04	34.58	3.56	38.14	nr	42.91
1500mm wide	-	-	-	1.90	1.20	39.85	4.09	43.94	nr	49.43
1800mm wide	-	-	-	2.16	1.37	45.37	4.57	49.94	nr	56.18
2400mm wide	-	-	-	2.67	1.70	56.15	5.62	61.77	nr	69.49
3000mm wide	-	-	-	3.17	2.03	66.79	6.67	73.45	nr	82.63
1100mm high; plugging and screwing lugs to brick jambs and sill and concrete head										
600mm wide	-	-	-	1.28	0.79	26.65	2.86	29.51	nr	33.20
900mm wide	-	-	-	1.56	0.98	32.66	3.39	36.05	nr	40.56
1200mm wide	-	-	-	1.86	1.18	39.07	3.89	42.96	nr	48.33
1500mm wide	-	-	-	2.14	1.39	45.30	4.43	49.73	nr	55.94
1800mm wide	-	-	-	2.44	1.59	51.70	4.95	56.65	nr	63.73
2400mm wide	-	-	-	3.01	2.00	64.19	5.97	70.16	nr	78.93
3000mm wide	-	-	-	3.59	2.39	76.61	7.02	83.63	nr	94.08
1300mm high; plugging and screwing lugs to brick jambs and sill and concrete head										
600mm wide	-	-	-	1.41	0.85	29.14	3.39	32.53	nr	36.59
900mm wide	-	-	-	1.74	1.10	36.51	3.71	40.22	nr	45.24
1200mm wide	-	-	-	2.06	1.34	43.62	4.25	47.87	nr	53.85
1500mm wide	-	-	-	2.39	1.56	50.67	4.76	55.43	nr	62.36
1800mm wide	-	-	-	2.72	1.81	58.04	5.27	63.30	nr	71.22
2400mm wide	-	-	-	3.37	2.28	72.31	6.27	78.57	nr	88.40
3000mm wide	-	-	-	4.02	2.76	86.68	7.30	93.98	nr	105.73
1500mm high; plugging and screwing lugs to brick jambs and sill and concrete head										
600mm wide	-	-	-	1.54	0.93	31.84	3.52	35.36	nr	39.78
900mm wide	-	-	-	1.91	1.21	40.10	4.05	44.15	nr	49.67
1200mm wide	-	-	-	2.27	1.48	48.11	4.56	52.67	nr	59.25
1500mm wide	-	-	-	2.64	1.75	56.26	5.10	61.35	nr	69.02
1800mm wide	-	-	-	2.99	2.03	64.23	5.60	69.83	nr	78.56
2400mm wide	-	-	-	3.72	2.58	80.49	6.63	87.13	nr	98.02
3000mm wide	-	-	-	4.44	3.12	96.51	7.64	104.15	nr	117.17
2100mm high; plugging and screwing lugs to brick jambs and sill and concrete head										
600mm wide	-	-	-	1.95	1.17	40.24	4.54	44.78	nr	50.37
900mm wide	-	-	-	2.43	1.54	51.02	5.02	56.04	nr	63.05
1200mm wide	-	-	-	2.89	1.92	61.63	5.59	67.22	nr	75.62
1500mm wide	-	-	-	3.36	2.30	72.38	6.07	78.45	nr	88.26
1800mm wide	-	-	-	3.83	2.69	83.23	6.58	89.81	nr	101.04
2400mm wide	-	-	-	4.77	3.45	104.73	7.62	112.35	nr	126.40
3000mm wide	-	-	-	5.70	4.21	126.09	8.64	134.73	nr	151.57
Building in wood windows; building in lugs; bedding in cement mortar (1:3), pointing with Secomastic standard mastic one side										
768mm high; lugs to brick jambs										
438mm wide	-	-	-	1.52	0.25	24.28	2.06	26.34	nr	29.63
641mm wide	-	-	-	1.75	0.28	27.87	2.43	30.30	nr	34.08
1225mm wide	-	-	-	2.39	0.32	37.39	3.45	40.84	nr	45.94
1809mm wide	-	-	-	3.05	0.35	47.09	4.50	51.59	nr	58.04
2394mm wide	-	-	-	3.68	0.38	56.36	5.57	61.93	nr	69.67
920mm high; lugs to brick jambs										
438mm wide	-	-	-	1.72	0.28	27.44	2.29	29.73	nr	33.45
641mm wide	-	-	-	1.94	0.32	30.99	2.64	33.64	nr	37.84
1225mm wide	-	-	-	2.61	0.35	40.84	3.71	44.55	nr	50.11
1809mm wide	-	-	-	3.27	0.41	50.86	4.76	55.62	nr	62.57
2394mm wide	-	-	-	3.95	0.44	60.84	5.79	66.64	nr	74.97
1073mm high; lugs to brick jambs										
438mm wide	-	-	-	1.92	0.32	30.71	2.57	33.28	nr	37.44
641mm wide	-	-	-	2.14	0.35	34.16	2.93	37.09	nr	41.73
1225mm wide	-	-	-	2.84	0.38	44.43	3.99	48.41	nr	54.47
1809mm wide	-	-	-	3.52	0.44	54.73	5.01	59.74	nr	67.21
2394mm wide	-	-	-	4.21	0.50	65.18	6.06	71.24	nr	80.15
1225mm high; lugs to brick jambs										
438mm wide	-	-	-	2.10	0.32	33.27	2.84	36.11	nr	40.62

Labour hourly rates: (except Specialists) Craft Operatives £14.21 Labourer £10.71 Rates are national average prices. Refer to REGIONAL VARIATIONS for indicative levels of overall pricing in regions	MATERIALS			LABOUR				RATES		
	Del to Site £	Waste %	Material Cost £	Craft Optve Hrs	Lab Hrs	Labour Cost £	Sunds £	Nett Rate £	Unit	Gross Rate (+12.5%) £
F30: ACCESSORIES/SUNDRY ITEMS FOR BRICK/BLOCK/STONE WALLING Cont.										
Building in Cont.										
Building in wood windows; building in lugs; bedding in cement mortar (1:3), pointing with Secomastic standard mastic one side Cont.										
1225mm high; lugs to brick jambs Cont.										
641mm wide	-	-	-	2.35	0.35	37.14	3.16	40.30	nr	45.34
1225mm wide	-	-	-	3.06	0.41	47.87	4.22	52.10	nr	58.61
1809mm wide	-	-	-	3.75	0.50	58.64	5.22	63.87	nr	71.85
2394mm wide	-	-	-	4.46	0.57	69.48	6.29	75.77	nr	85.24
1378mm high; lugs to brick jambs										
438mm wide	-	-	-	2.30	0.32	36.11	3.07	39.18	nr	44.08
641mm wide	-	-	-	2.54	0.35	39.84	3.44	43.28	nr	48.69
1225mm wide	-	-	-	3.27	0.44	51.18	4.49	55.67	nr	62.63
1809mm wide	-	-	-	3.99	0.54	62.48	5.53	68.01	nr	76.51
2394mm wide	-	-	-	4.72	0.63	73.82	6.57	80.39	nr	90.44
Building in fireplace interior with slabbed tile surround and loose hearth tiles for 400mm opening; bed and joint interior in fire cement and fill with concrete at back; plug and screw on surround lugs; bed hearth tiles in cement mortar (1:3) and point in										
with firebrick back	-	-	-	5.00	5.00	124.60	5.22	129.82	nr	146.05
with back boiler unit and self contained flue	-	-	-	6.00	6.00	149.52	6.29	155.81	nr	175.28
Proprietary items										
Precast prestressed concrete lintels; Stressline Ltd; Roughcast; bedding in cement-lime mortar (1:1:6)										
100 x 70mm										
900mm long	3.56	2.50	3.65	0.28	0.14	5.48	0.28	9.41	nr	10.58
1050mm long	4.15	2.50	4.25	0.31	0.16	6.12	0.30	10.67	nr	12.01
1200mm long	4.73	2.50	4.85	0.36	0.18	7.04	0.37	12.26	nr	13.79
1500mm long	5.92	2.50	6.07	0.44	0.22	8.61	0.44	15.12	nr	17.01
1800mm long	7.10	2.50	7.28	0.51	0.26	10.03	0.48	17.79	nr	20.02
2100mm long	8.28	2.50	8.49	0.58	0.29	11.35	0.56	20.39	nr	22.94
2400mm long	9.48	2.50	9.72	0.63	0.32	12.38	0.65	22.74	nr	25.58
2700mm long	10.65	2.50	10.92	0.69	0.35	13.55	0.69	25.16	nr	28.30
3000mm long	11.84	2.50	12.14	0.76	0.38	14.87	0.76	27.77	nr	31.24
150 x 70mm										
900mm long	4.83	2.50	4.95	0.29	0.15	5.73	0.28	10.96	nr	12.33
1050mm long	5.64	2.50	5.78	0.32	0.16	6.26	0.32	12.36	nr	13.91
1200mm long	6.42	2.50	6.58	0.37	0.19	7.29	0.37	14.24	nr	16.02
1500mm long	8.04	2.50	8.24	0.45	0.23	8.86	0.45	17.55	nr	19.74
1800mm long	9.66	2.50	9.90	0.52	0.26	10.17	0.49	20.57	nr	23.14
2100mm long	11.25	2.50	11.53	0.59	0.30	11.60	0.59	23.72	nr	26.68
2400mm long	12.87	2.50	13.19	0.66	0.33	12.91	0.65	26.75	nr	30.09
2700mm long	14.48	2.50	14.84	0.71	0.36	13.94	0.68	29.46	nr	33.15
3000mm long	16.09	2.50	16.49	0.78	0.39	15.26	0.78	32.54	nr	36.60
225 x 70mm										
900mm long	7.43	2.50	7.62	0.30	0.15	5.87	0.28	13.76	nr	15.49
1050mm long	8.68	2.50	8.90	0.33	0.17	6.51	0.32	15.73	nr	17.70
1200mm long	9.91	2.50	10.16	0.38	0.19	7.43	0.37	17.96	nr	20.20
1500mm long	12.40	2.50	12.71	0.46	0.23	9.00	0.45	22.16	nr	24.93
1800mm long	14.88	2.50	15.25	0.53	0.27	10.42	0.49	26.17	nr	29.44
2100mm long	17.35	2.50	17.78	0.60	0.30	11.74	0.59	30.11	nr	33.88
2400mm long	19.83	2.50	20.33	0.67	0.34	13.16	0.68	34.17	nr	38.44
2700mm long	22.31	2.50	22.87	0.72	0.36	14.09	0.75	37.71	nr	42.42
3000mm long	24.79	2.50	25.41	0.79	0.40	15.51	0.78	41.70	nr	46.92
255 x 70mm										
900mm long	7.93	2.50	8.13	0.30	0.15	5.87	0.31	14.31	nr	16.10
1050mm long	9.24	2.50	9.47	0.35	0.18	6.90	0.34	16.72	nr	18.81
1200mm long	10.56	2.50	10.82	0.38	0.19	7.43	0.39	18.65	nr	20.98
1500mm long	13.20	2.50	13.53	0.46	0.23	9.00	0.45	22.98	nr	25.85
1800mm long	15.82	2.50	16.22	0.54	0.27	10.57	0.54	27.32	nr	30.73
2100mm long	18.48	2.50	18.94	0.61	0.32	12.10	0.62	31.66	nr	35.62
2400mm long	21.11	2.50	21.64	0.69	0.35	13.55	0.67	35.86	nr	40.34
2700mm long	23.76	2.50	24.35	0.74	0.37	14.48	0.75	39.58	nr	44.53
3000mm long	26.38	2.50	27.04	0.81	0.41	15.90	0.78	43.73	nr	49.19
100 x 145mm										
900mm long	7.18	2.50	7.36	0.30	0.15	5.87	0.28	13.51	nr	15.20
1050mm long	8.38	2.50	8.59	0.33	0.17	6.51	0.32	15.42	nr	17.35
1200mm long	9.59	2.50	9.83	0.38	0.19	7.43	0.39	17.65	nr	19.86
1500mm long	11.98	2.50	12.28	0.46	0.23	9.00	0.45	21.73	nr	24.45
1800mm long	14.37	2.50	14.73	0.53	0.27	10.42	0.54	25.69	nr	28.90
2100mm long	16.77	2.50	17.19	0.60	0.30	11.74	0.59	29.52	nr	33.21
2400mm long	19.17	2.50	19.65	0.67	0.34	13.16	0.68	33.49	nr	37.67
2700mm long	21.56	2.50	22.10	0.72	0.36	14.09	0.75	36.94	nr	41.56
3000mm long	23.95	2.50	24.55	0.79	0.40	15.51	0.78	40.84	nr	45.95
Galvanized steel lintels; SUPERGALV (BIRTLEY) lintels reference CB 50; bedding in cement-lime mortar (1:1:6)										
125mm deep x 750mm long	23.07	2.50	23.65	0.32	0.32	7.97	-	31.62	nr	35.57
125mm deep x 1200mm long	36.38	2.50	37.29	0.44	0.44	10.96	-	48.25	nr	54.29
125mm deep x 1350mm long	42.93	2.50	44.00	0.48	0.48	11.96	-	55.96	nr	62.96
125mm deep x 1500mm long	47.43	2.50	48.62	0.51	0.51	12.71	-	61.32	nr	68.99
125mm deep x 1650mm long	54.08	2.50	55.43	0.54	0.54	13.46	-	68.89	nr	77.50
165mm deep x 1800mm long	58.51	2.50	59.97	0.58	0.58	14.45	-	74.43	nr	83.73
165mm deep x 1950mm long	64.21	2.50	65.82	0.61	0.61	15.20	-	81.02	nr	91.14
165mm deep x 2100mm long	67.20	2.50	68.88	0.65	0.65	16.20	-	85.08	nr	95.71
165mm deep x 2250mm long	77.02	2.50	78.95	0.69	0.69	17.19	-	96.14	nr	108.16
165mm deep x 2400mm long	81.47	2.50	83.51	0.73	0.73	18.19	-	101.70	nr	114.41
215mm deep x 2550mm long	91.84	2.50	94.14	0.77	0.77	19.19	-	113.32	nr	127.49

Labour hourly rates: (except Specialists) Craft Operatives £14.21 Labourer £10.71 Rates are national average prices. Refer to REGIONAL VARIATIONS for indicative levels of overall pricing in regions	MATERIALS			LABOUR				RATES		
	Del to Site £	Waste %	Material Cost £	Craft Optve Hrs	Lab Hrs	Labour Cost £	Sunds £	Nett Rate £	Unit	Gross Rate (+12.5%) £
F30: ACCESSORIES/SUNDRY ITEMS FOR BRICK/BLOCK/STONE WALLING Cont.										
Proprietary items Cont.										
Galvanized steel lintels; SUPERGALV (BIRTLEY) lintels reference CB 50; bedding in cement-lime mortar (1:1:6) Cont.										
215mm deep x 2850mm long	125.14	2.50	**128.27**	0.82	0.82	**20.43**	-	**148.70**	nr	**167.29**
215mm deep x 3000mm long	133.24	2.50	**136.57**	0.87	0.87	**21.68**	-	**158.25**	nr	**178.03**
215mm deep x 3300mm long	153.62	2.50	**157.46**	0.94	0.94	**23.42**	-	**180.89**	nr	**203.50**
215mm deep x 3600mm long	176.01	2.50	**180.41**	1.02	1.02	**25.42**	-	**205.83**	nr	**231.56**
215mm deep x 3900mm long	214.79	2.50	**220.16**	1.09	1.09	**27.16**	-	**247.32**	nr	**278.24**
Galvanized steel lintels; SUPERGALV (BIRTLEY) lintels reference CB 50 H.D.; bedding in cement-lime mortar (1:1:6)										
165mm deep x 750mm long	27.23	2.50	**27.91**	0.33	0.33	**8.22**	-	**36.13**	nr	**40.65**
165mm deep x 1050mm long	35.85	2.50	**36.75**	0.40	0.40	**9.97**	-	**46.71**	nr	**52.55**
165mm deep x 1200mm long	40.84	2.50	**41.86**	0.44	0.44	**10.96**	-	**52.83**	nr	**59.43**
165mm deep x 1350mm long	49.66	2.50	**50.90**	0.48	0.48	**11.96**	-	**62.86**	nr	**70.72**
165mm deep x 1500mm long	54.74	2.50	**56.11**	0.51	0.51	**12.71**	-	**68.82**	nr	**77.42**
215mm deep x 1650mm long	65.49	2.50	**67.13**	0.54	0.54	**13.46**	-	**80.58**	nr	**90.66**
215mm deep x 1800mm long	71.49	2.50	**73.28**	0.58	0.58	**14.45**	-	**87.73**	nr	**98.70**
215mm deep x 2100mm long	84.33	2.50	**86.44**	0.65	0.65	**16.20**	-	**102.64**	nr	**115.47**
215mm deep x 2250mm long	106.03	2.50	**108.68**	0.69	0.69	**17.19**	-	**125.88**	nr	**141.61**
215mm deep x 2400mm long	124.42	2.50	**127.53**	0.73	0.73	**18.19**	-	**145.72**	nr	**163.94**
215mm deep x 2550mm long	135.18	2.50	**138.56**	0.77	0.77	**19.19**	-	**157.75**	nr	**177.47**
215mm deep x 3000mm long	203.93	2.50	**209.03**	0.90	0.90	**22.43**	-	**231.46**	nr	**260.39**
Galvanized steel lintels; SUPERGALV (BIRTLEY) lintels reference CB 70; bedding in cement-lime mortar (1:1:6)										
120mm deep x 750mm long	22.90	2.50	**23.47**	0.34	0.34	**8.47**	-	**31.95**	nr	**35.94**
120mm deep x 1200mm long	36.37	2.50	**37.28**	0.46	0.46	**11.46**	-	**48.74**	nr	**54.84**
120mm deep x 1350mm long	42.72	2.50	**43.79**	0.50	0.50	**12.46**	-	**56.25**	nr	**63.28**
120mm deep x 1500mm long	47.44	2.50	**48.63**	0.53	0.53	**13.21**	-	**61.83**	nr	**69.56**
120mm deep x 1650mm long	53.75	2.50	**55.09**	0.57	0.57	**14.20**	-	**69.30**	nr	**77.96**
160mm deep x 1800mm long	58.36	2.50	**59.82**	0.61	0.61	**15.20**	-	**75.02**	nr	**84.40**
160mm deep x 1950mm long	65.16	2.50	**66.79**	0.64	0.64	**15.95**	-	**82.74**	nr	**93.08**
160mm deep x 2100mm long	68.07	2.50	**69.77**	0.68	0.68	**16.95**	-	**86.72**	nr	**97.56**
215mm deep x 2250mm long	76.38	2.50	**78.29**	0.72	0.72	**17.94**	-	**96.23**	nr	**108.26**
215mm deep x 2400mm long	81.47	2.50	**83.51**	0.76	0.76	**18.94**	-	**102.45**	nr	**115.25**
215mm deep x 2550mm long	93.62	2.50	**95.96**	0.81	0.81	**20.19**	-	**116.15**	nr	**130.66**
215mm deep x 2850mm long	123.68	2.50	**126.77**	0.86	0.86	**21.43**	-	**148.20**	nr	**166.73**
215mm deep x 3000mm long	132.86	2.50	**136.18**	0.91	0.91	**22.68**	-	**158.86**	nr	**178.72**
215mm deep x 3300mm long	153.64	2.50	**157.48**	0.98	0.98	**24.42**	-	**181.90**	nr	**204.64**
215mm deep x 3600mm long	168.88	2.50	**173.10**	1.06	1.06	**26.42**	-	**199.52**	nr	**224.46**
215mm deep x 3900mm long	206.17	2.50	**211.32**	1.14	1.14	**28.41**	-	**239.73**	nr	**269.70**
Galvanized steel lintels; SUPERGALV (BIRTLEY) lintels reference CB 70 H.D.; bedding in cement-lime mortar (1:1:6)										
160mm deep x 750mm long	27.25	2.50	**27.93**	0.33	0.33	**8.22**	-	**36.15**	nr	**40.67**
160mm deep x 1050mm long	36.45	2.50	**37.36**	0.41	0.41	**10.22**	-	**47.58**	nr	**53.53**
160mm deep x 1200mm long	40.84	2.50	**41.86**	0.46	0.46	**11.46**	-	**53.32**	nr	**59.99**
160mm deep x 1350mm long	50.34	2.50	**51.60**	0.49	0.49	**12.21**	-	**63.81**	nr	**71.79**
160mm deep x 1500mm long	54.75	2.50	**56.12**	0.53	0.53	**13.21**	-	**69.33**	nr	**77.99**
215mm deep x 1650mm long	71.86	2.50	**73.66**	0.56	0.56	**13.96**	-	**87.61**	nr	**98.56**
215mm deep x 1800mm long	76.47	2.50	**78.38**	0.61	0.61	**15.20**	-	**93.58**	nr	**105.28**
215mm deep x 2100mm long	87.93	2.50	**90.13**	0.68	0.68	**16.95**	-	**107.07**	nr	**120.46**
215mm deep x 2250mm long	108.46	2.50	**111.17**	0.72	0.72	**17.94**	-	**129.11**	nr	**145.25**
215mm deep x 2400mm long	124.45	2.50	**127.56**	0.76	0.76	**18.94**	-	**146.50**	nr	**164.81**
215mm deep x 2550mm long	137.01	2.50	**140.44**	0.81	0.81	**20.19**	-	**160.62**	nr	**180.70**
215mm deep x 3300mm long	203.98	2.50	**209.08**	0.94	0.94	**23.42**	-	**232.50**	nr	**261.57**
Galvanized steel lintels; SUPERGALV (BIRTLEY) lintels reference CB 50/130; bedding in cement-lime mortar (1:1:6)										
120mm deep x 750mm long	23.17	2.50	**23.75**	0.36	0.36	**8.97**	-	**32.72**	nr	**36.81**
120mm deep x 1200mm long	36.82	2.50	**37.74**	0.48	0.48	**11.96**	-	**49.70**	nr	**55.91**
120mm deep x 1350mm long	42.20	2.50	**43.26**	0.52	0.52	**12.96**	-	**56.21**	nr	**63.24**
120mm deep x 1500mm long	46.67	2.50	**47.84**	0.55	0.55	**13.71**	-	**61.54**	nr	**69.24**
120mm deep x 1650mm long	54.28	2.50	**55.64**	0.60	0.60	**14.95**	-	**70.59**	nr	**79.41**
165mm deep x 1800mm long	58.47	2.50	**59.93**	0.63	0.63	**15.70**	-	**75.63**	nr	**85.09**
165mm deep x 1950mm long	64.50	2.50	**66.11**	0.67	0.67	**16.70**	-	**82.81**	nr	**93.16**
165mm deep x 2100mm long	69.68	2.50	**71.42**	0.71	0.71	**17.69**	-	**89.12**	nr	**100.25**
165mm deep x 2250mm long	79.88	2.50	**81.88**	0.76	0.76	**18.94**	-	**100.82**	nr	**113.42**
165mm deep x 2400mm long	85.22	2.50	**87.35**	0.79	0.79	**19.69**	-	**107.04**	nr	**120.42**
215mm deep x 2550mm long	96.57	2.50	**98.98**	0.85	0.85	**21.18**	-	**120.17**	nr	**135.19**
215mm deep x 2850mm long	136.08	2.50	**139.48**	0.90	0.90	**22.43**	-	**161.91**	nr	**182.15**
215mm deep x 3000mm long	140.67	2.50	**144.19**	0.95	0.95	**23.67**	-	**167.86**	nr	**188.84**
215mm deep x 3300mm long	157.10	2.50	**161.03**	1.02	1.02	**25.42**	-	**186.45**	nr	**209.75**
215mm deep x 3600mm long	173.01	2.50	**177.34**	1.12	1.12	**27.91**	-	**205.25**	nr	**230.90**
215mm deep x 3900mm long	253.53	2.50	**259.87**	1.18	1.18	**29.41**	-	**289.27**	nr	**325.43**
Galvanized steel lintels; SUPERGALV (BIRTLEY) lintels reference CB 50/130 H.D.; bedding in cement-lime mortar (1:1:6)										
160mm deep x 750mm long	33.10	2.50	**33.93**	0.36	0.36	**8.97**	-	**42.90**	nr	**48.26**
160mm deep x 1200mm long	49.54	2.50	**50.78**	0.48	0.48	**11.96**	-	**62.74**	nr	**70.58**
160mm deep x 1350mm long	62.00	2.50	**63.55**	0.52	0.52	**12.96**	-	**76.51**	nr	**86.07**
160mm deep x 1500mm long	68.30	2.50	**70.01**	0.55	0.55	**13.71**	-	**83.71**	nr	**94.18**
215mm deep x 1650mm long	78.23	2.50	**80.19**	0.60	0.60	**14.95**	-	**95.14**	nr	**107.03**
215mm deep x 1800mm long	84.70	2.50	**86.82**	0.63	0.63	**15.70**	-	**102.52**	nr	**115.33**
215mm deep x 1950mm long	105.42	2.50	**108.06**	0.67	0.67	**16.70**	-	**124.75**	nr	**140.35**
215mm deep x 2250mm long	121.30	2.50	**124.33**	0.76	0.76	**18.94**	-	**143.27**	nr	**161.18**

Labour hourly rates: (except Specialists) Craft Operatives £14.21 Labourer £10.71 Rates are national average prices. Refer to REGIONAL VARIATIONS for indicative levels of overall pricing in regions	MATERIALS			LABOUR				RATES		
	Del to Site £	Waste %	Material Cost £	Craft Optve Hrs	Lab Hrs	Labour Cost £	Sunds £	Nett Rate £	Unit	Gross Rate (+12.5%) £
F30: ACCESSORIES/SUNDRY ITEMS FOR BRICK/BLOCK/STONE WALLING Cont.										
Proprietary items Cont.										
Galvanized steel lintels; SUPERGALV (BIRTLEY) lintels reference CB 50/130 H.D.; bedding in cement-lime mortar (1:1:6) Cont.										
215mm deep x 2400mm long	131.11	2.50	**134.39**	0.79	0.79	**19.69**	-	**154.07**	nr	**173.33**
215mm deep x 2550mm long	143.74	2.50	**147.33**	0.85	0.85	**21.18**	-	**168.52**	nr	**189.58**
215mm deep x 2700mm long	151.83	2.50	**155.63**	0.92	0.92	**22.93**	-	**178.55**	nr	**200.87**
215mm deep x 3300mm long	208.10	2.50	**213.30**	0.99	0.99	**24.67**	-	**237.97**	nr	**267.72**
Galvanized steel lintels; SUPERGALV (BIRTLEY) lintels reference CB 70/130; bedding in cement-lime mortar (1:1:6)										
115mm deep x 750mm long	28.04	2.50	**28.74**	0.36	0.36	**8.97**	-	**37.71**	nr	**42.43**
115mm deep x 1200mm long	46.18	2.50	**47.33**	0.48	0.48	**11.96**	-	**59.30**	nr	**66.71**
115mm deep x 1350mm long	53.18	2.50	**54.51**	0.52	0.52	**12.96**	-	**67.47**	nr	**75.90**
115mm deep x 1500mm long	58.35	2.50	**59.81**	0.55	0.55	**13.71**	-	**73.51**	nr	**82.70**
115mm deep x 1650mm long	71.96	2.50	**73.76**	0.60	0.60	**14.95**	-	**88.71**	nr	**99.80**
155mm deep x 1800mm long	80.25	2.50	**82.26**	0.63	0.63	**15.70**	-	**97.96**	nr	**110.20**
155mm deep x 1950mm long	86.94	2.50	**89.11**	0.67	0.67	**16.70**	-	**105.81**	nr	**119.04**
155mm deep x 2100mm long	92.42	2.50	**94.73**	0.71	0.71	**17.69**	-	**112.42**	nr	**126.48**
155mm deep x 2250mm long	125.03	2.50	**128.16**	0.76	0.76	**18.94**	-	**147.09**	nr	**165.48**
155mm deep x 2400mm long	133.36	2.50	**136.69**	0.79	0.79	**19.69**	-	**156.38**	nr	**175.93**
210mm deep x 2550mm long	147.81	2.50	**151.51**	0.85	0.85	**21.18**	-	**172.69**	nr	**194.27**
210mm deep x 2850mm long	197.10	2.50	**202.03**	0.90	0.90	**22.43**	-	**224.46**	nr	**252.51**
210mm deep x 3000mm long	220.10	2.50	**225.60**	0.95	0.95	**23.67**	-	**249.28**	nr	**280.44**
210mm deep x 3300mm long	242.10	2.50	**248.15**	1.02	1.02	**25.42**	-	**273.57**	nr	**307.77**
Galvanized steel lintels; SUPERGALV (BIRTLEY) lintels reference HD50; bedding in cement-lime mortar (1:1:6)										
225mm deep x 3900mm long	252.15	2.50	**258.45**	1.17	1.17	**29.16**	-	**287.61**	nr	**323.56**
225mm deep x 4200mm long	298.21	2.50	**305.67**	1.21	1.21	**30.15**	-	**335.82**	nr	**377.80**
225mm deep x 4500mm long	329.63	2.50	**337.87**	1.25	1.25	**31.15**	-	**369.02**	nr	**415.15**
225mm deep x 4800mm long	351.57	2.50	**360.36**	1.29	1.29	**32.15**	-	**392.51**	nr	**441.57**
Galvanized steel lintels; SUPERGALV (BIRTLEY) lintels reference HD 70; bedding in cement-lime mortar (1:1:6)										
225mm deep x 3000mm long	204.75	2.50	**209.87**	0.98	0.98	**24.42**	-	**234.29**	nr	**263.58**
225mm deep x 3300mm long	225.21	2.50	**230.84**	1.02	1.02	**25.42**	-	**256.26**	nr	**288.29**
225mm deep x 3600mm long	246.83	2.50	**253.00**	1.06	1.06	**26.42**	-	**279.42**	nr	**314.34**
Galvanized steel lintels; SUPERGALV (BIRTLEY) lintels reference SB100; bedding in cement-lime mortar (1:1:6)										
75mm deep x 750mm long	15.36	2.50	**15.74**	0.52	0.52	**12.96**	-	**28.70**	nr	**32.29**
75mm deep x 900mm long	17.93	2.50	**18.38**	0.55	0.55	**13.71**	-	**32.08**	nr	**36.09**
75mm deep x 1050mm long	21.24	2.50	**21.77**	0.59	0.59	**14.70**	-	**36.47**	nr	**41.03**
75mm deep x 1200mm long	23.62	2.50	**24.21**	0.62	0.62	**15.45**	-	**39.66**	nr	**44.62**
75mm deep x 1500mm long	29.56	2.50	**30.30**	0.69	0.69	**17.19**	-	**47.49**	nr	**53.43**
140mm deep x 1800mm long	45.23	2.50	**46.36**	0.75	0.75	**18.69**	-	**65.05**	nr	**73.18**
140mm deep x 1950mm long	51.47	2.50	**52.76**	0.79	0.79	**19.69**	-	**72.44**	nr	**81.50**
140mm deep x 2100mm long	54.66	2.50	**56.03**	0.84	0.84	**20.93**	-	**76.96**	nr	**86.58**
140mm deep x 2250mm long	58.79	2.50	**60.26**	0.89	0.89	**22.18**	-	**82.44**	nr	**92.74**
140mm deep x 2400mm long	63.37	2.50	**64.95**	0.94	0.94	**23.42**	-	**88.38**	nr	**99.43**
140mm deep x 2550mm long	68.37	2.50	**70.08**	1.00	1.00	**24.92**	-	**95.00**	nr	**106.87**
140mm deep x 2700mm long	72.66	2.50	**74.48**	1.12	1.12	**27.91**	-	**102.39**	nr	**115.19**
215mm deep x 2850mm long	109.91	2.50	**112.66**	1.07	1.07	**26.66**	-	**139.32**	nr	**156.74**
215mm deep x 3000mm long	116.50	2.50	**119.41**	1.09	1.09	**27.16**	-	**146.58**	nr	**164.90**
Galvanized steel lintels; SUPERGALV (BIRTLEY) lintels reference SB140; bedding in cement-lime mortar (1:1:6)										
140mm deep x 750mm long	19.41	2.50	**19.90**	0.52	0.52	**12.96**	-	**32.85**	nr	**36.96**
140mm deep x 900mm long	23.43	2.50	**24.02**	0.55	0.55	**13.71**	-	**37.72**	nr	**42.44**
140mm deep x 1050mm long	27.61	2.50	**28.30**	0.59	0.59	**14.70**	-	**43.00**	nr	**48.37**
140mm deep x 1200mm long	31.57	2.50	**32.36**	0.62	0.62	**15.45**	-	**47.81**	nr	**53.79**
140mm deep x 1350mm long	35.81	2.50	**36.71**	0.66	0.66	**16.45**	-	**53.16**	nr	**59.80**
140mm deep x 1500mm long	40.68	2.50	**41.70**	0.69	0.69	**17.19**	-	**58.89**	nr	**66.25**
140mm deep x 1650mm long	48.86	2.50	**50.08**	0.71	0.71	**17.69**	-	**67.77**	nr	**76.24**
140mm deep x 1800mm long	53.11	2.50	**54.43**	0.75	0.75	**18.69**	-	**73.12**	nr	**82.26**
140mm deep x 1950mm long	60.02	2.50	**61.52**	0.79	0.79	**19.69**	-	**81.21**	nr	**91.36**
140mm deep x 2100mm long	65.44	2.50	**67.07**	0.84	0.84	**20.93**	-	**88.01**	nr	**99.01**
140mm deep x 2250mm long	72.16	2.50	**73.96**	0.89	0.89	**22.18**	-	**96.14**	nr	**108.15**
140mm deep x 2400mm long	75.77	2.50	**77.66**	0.94	0.94	**23.42**	-	**101.09**	nr	**113.72**
140mm deep x 2550mm long	80.30	2.50	**82.30**	1.00	1.00	**24.92**	-	**107.22**	nr	**120.63**
140mm deep x 2700mm long	85.01	2.50	**87.13**	1.06	1.06	**26.42**	-	**113.55**	nr	**127.74**
215mm deep x 2850mm long	123.23	2.50	**126.31**	1.09	1.09	**27.16**	-	**153.47**	nr	**172.66**
215mm deep x 3000mm long	128.29	2.50	**131.50**	1.12	1.12	**27.91**	-	**159.41**	nr	**179.33**
Galvanized steel lintels; SUPERGALV (BIRTLEY) lintels reference SBL200; bedding in cement-lime mortar (1:1:6)										
142mm deep x 750mm long	26.83	2.50	**27.50**	0.52	0.52	**12.96**	-	**40.46**	nr	**45.52**
142mm deep x 900mm long	32.16	2.50	**32.96**	0.55	0.55	**13.71**	-	**46.67**	nr	**52.50**
142mm deep x 1050mm long	37.29	2.50	**38.22**	0.59	0.59	**14.70**	-	**52.93**	nr	**59.54**
142mm deep x 1200mm long	42.18	2.50	**43.24**	0.62	0.62	**15.45**	-	**58.69**	nr	**66.02**
142mm deep x 1350mm long	47.76	2.50	**48.95**	0.66	0.66	**16.45**	-	**65.40**	nr	**73.58**
142mm deep x 1500mm long	53.86	2.50	**55.21**	0.69	0.69	**17.19**	-	**72.41**	nr	**81.46**
142mm deep x 1650mm long	61.08	2.50	**62.61**	0.77	0.77	**19.19**	-	**81.80**	nr	**92.02**
142mm deep x 1800mm long	66.67	2.50	**68.33**	0.83	0.83	**20.68**	-	**89.02**	nr	**100.15**
142mm deep x 1950mm long	72.89	2.50	**74.71**	0.87	0.87	**21.68**	-	**96.39**	nr	**108.44**
142mm deep x 2100mm long	78.04	2.50	**79.99**	0.92	0.89	**22.61**	-	**102.59**	nr	**115.42**
142mm deep x 2250mm long	89.46	2.50	**91.70**	0.89	0.92	**22.50**	-	**114.20**	nr	**128.47**
142mm deep x 2400mm long	93.99	2.50	**96.34**	0.94	0.94	**23.42**	-	**119.77**	nr	**134.74**
142mm deep x 2550mm long	101.76	2.50	**104.30**	1.00	1.00	**24.92**	-	**129.22**	nr	**145.37**

Labour hourly rates: (except Specialists) Craft Operatives £14.21 Labourer £10.71 Rates are national average prices. Refer to REGIONAL VARIATIONS for indicative levels of overall pricing in regions	MATERIALS			LABOUR				RATES		
	Del to Site £	Waste %	Material Cost £	Craft Optve Hrs	Lab Hrs	Labour Cost £	Sunds £	Nett Rate £	Unit	Gross Rate (+12.5%) £
F30: ACCESSORIES/SUNDRY ITEMS FOR BRICK/BLOCK/STONE WALLING Cont.										
Proprietary items Cont.										
Galvanized steel lintels; SUPERGALV (BIRTLEY) lintels reference SBL200; bedding in cement-lime mortar (1:1:6) Cont.										
142mm deep x 2700mm long	105.39	2.50	108.02	1.06	1.06	26.42	-	134.44	nr	151.24
218mm deep x 2850mm long	149.46	2.50	153.19	1.07	1.07	26.66	-	179.86	nr	202.34
218mm deep x 3000mm long	158.53	2.50	162.50	1.09	1.09	27.16	-	189.66	nr	213.37
Galvanized steel lintels; SUPERGALV (BIRTLEY) internal door lintels reference INT100; bedding in cement-lime mortar (1:1:6)										
100mm wide, 900mm long	12.19	2.50	12.49	0.38	0.38	9.47	-	21.96	nr	24.70
100mm wide, 1050mm long	14.24	2.50	14.60	0.41	0.41	10.22	-	24.81	nr	27.91
100mm wide, 1200mm long	16.29	2.50	16.70	0.44	0.44	10.96	-	27.67	nr	31.12
Stainless steel Furfix wall extension profiles; single flange; plugging and screwing to brickwork and building ties in to joints of new walls										
100mm	11.97	5.00	12.57	0.30	-	4.26	-	16.83	m	18.94
150mm	12.90	5.00	13.55	0.35	-	4.97	-	18.52	m	20.83
215mm	23.34	5.00	24.51	0.40	-	5.68	-	30.19	m	33.96
Harris and Edgar Ltd; Hemax restraint channel ref. 36/8; fixing to steelwork with Hemax M8 screws complete with plate washer at 450mm centres; incorporating ties at 450mm centres										
100mm long ties reference BP36 fishtail	10.74	5.00	11.28	0.60	0.60	14.95	0.74	26.97	m	30.34
125mm long ties reference BP36 fishtail	10.79	5.00	11.33	0.60	0.60	14.95	0.87	27.15	m	30.55
150mm long ties reference BP36 fishtail	10.85	5.00	11.39	0.60	0.60	14.95	0.99	27.33	m	30.75
Thermabate insulated cavity closers; fixing to timber with nails										
Thermabate 50; vertical	4.63	5.00	4.86	0.40	0.40	9.97	-	14.83	m	16.68
Thermabate 75; vertical	4.84	5.00	5.08	0.40	0.40	9.97	-	15.05	m	16.93
Thermabate 50; horizontal	4.63	5.00	4.86	0.40	0.40	9.97	-	14.83	m	16.68
Thermabate 75; horizontal	4.84	5.00	5.08	0.40	0.40	9.97	-	15.05	m	16.93
Thermabate insulated cavity closers; fixing to masonry with PVC-U ties at 225mm centres										
Thermabate 50; vertical	4.63	5.00	4.86	0.40	0.40	9.97	-	14.83	m	16.68
Thermabate 75; vertical	4.84	5.00	5.08	0.40	0.40	9.97	-	15.05	m	16.93
Thermabate 50; horizontal	4.63	5.00	4.86	0.40	0.40	9.97	-	14.83	m	16.68
Thermabate 75; horizontal	4.84	5.00	5.08	0.40	0.40	9.97	-	15.05	m	16.93
F31: PRECAST CONCRETE SILLS/LINTELS/COPINGS/FEATURES										
Precast concrete sills; BS 5642 Part 1; bedding in cement lime mortar (1:1:6)										
Sills; figure 2 or figure 4										
50 x 150 x 300mm splayed and grooved	2.29	2.50	2.34	0.17	0.17	4.24	0.13	6.71	nr	7.55
50 x 150 x 400mm splayed and grooved	3.05	2.50	3.13	0.22	0.22	5.48	0.15	8.76	nr	9.85
50 x 150 x 700mm splayed and grooved	5.33	2.50	5.47	0.33	0.33	8.22	0.23	13.92	nr	15.66
50 x 150 x 1300mm splayed and grooved	9.91	2.50	10.15	0.55	0.55	13.71	0.34	24.20	nr	27.23
extra for stooled end	2.59	-	2.59	-	-	-	-	2.59	nr	2.91
75 x 150 x 300mm splayed and grooved	3.43	2.50	3.51	0.19	0.19	4.73	0.13	8.38	nr	9.43
75 x 150 x 400mm splayed and grooved	4.57	2.50	4.69	0.24	0.24	5.98	0.15	10.82	nr	12.17
75 x 150 x 700mm splayed and grooved	8.00	2.50	8.20	0.38	0.38	9.47	0.23	17.90	nr	20.13
75 x 150 x 1300mm splayed and grooved	14.86	2.50	15.23	0.60	0.60	14.95	0.34	30.53	nr	34.34
extra for stooled end	3.24	-	3.24	-	-	-	-	3.24	nr	3.64
100 x 150 x 300mm splayed and grooved	4.57	2.50	4.69	0.22	0.22	5.48	0.13	10.30	nr	11.58
100 x 150 x 400mm splayed and grooved	6.10	2.50	6.25	0.28	0.28	6.98	0.15	13.38	nr	15.05
100 x 150 x 700mm splayed and grooved	10.67	2.50	10.93	0.44	0.44	10.96	0.23	22.13	nr	24.89
100 x 150 x 1300mm splayed and grooved	19.81	2.50	20.31	0.66	0.66	16.45	0.34	37.10	nr	41.73
extra for stooled end	3.88	-	3.88	-	-	-	-	3.88	nr	4.37
Precast concrete copings BS 5642 Part 2; bedding in cement-lime mortar (1:1:6)										
Copings; figure 1; horizontal										
75 x 200mm; splayed; rebated joints	11.43	2.50	11.72	0.28	0.14	5.48	0.19	17.39	m	19.56
extra; stopped ends	1.29	-	1.29	-	-	-	-	1.29	nr	1.46
extra; internal angles	2.59	-	2.59	-	-	-	-	2.59	nr	2.91
100 x 300mm; splayed; rebated joints	22.86	2.50	23.43	0.55	0.28	10.81	0.31	34.56	m	38.88
extra; stopped ends	1.94	-	1.94	-	-	-	-	1.94	nr	2.18
extra; internal angles	3.24	-	3.24	-	-	-	-	3.24	nr	3.64
75 x 200mm; saddleback; rebated joints	12.19	2.50	12.50	0.28	0.14	5.48	0.19	18.17	m	20.44
extra; hipped ends	1.29	-	1.29	-	-	-	-	1.29	nr	1.46
extra; internal angles	2.59	-	2.59	-	-	-	-	2.59	nr	2.91
100 x 300mm; saddleback; rebated joints	24.38	2.50	24.99	0.55	0.28	10.81	0.31	36.12	m	40.63
extra; hipped ends	1.94	-	1.94	-	-	-	-	1.94	nr	2.18
extra; internal angles	3.24	-	3.24	-	-	-	-	3.24	nr	3.64
Precast concrete lintels; BS 5328, designed mix C25, 20mm aggregate minimum cement content 360 kg/m3; vibrated; reinforcement bars BS 4449 hot rolled plain round mild steel; bedding in cement-lime mortar (1:1:6)										
Lintels										
75 x 150 x 1200mm; reinforced with 1.20 kg of 12mm bars	7.05	2.50	7.23	0.20	0.10	3.91	0.13	11.27	nr	12.68

Labour hourly rates: (except Specialists) Craft Operatives £14.21 Labourer £10.71 Rates are national average prices. Refer to REGIONAL VARIATIONS for indicative levels of overall pricing in regions	MATERIALS			LABOUR				RATES		
	Del to Site £	Waste %	Material Cost £	Craft Optve Hrs	Lab Hrs	Labour Cost £	Sunds £	Nett Rate £	Unit	Gross Rate (+12.5%) £
F31: PRECAST CONCRETE SILLS/LINTELS/COPINGS/FEATURES Cont.										
Precast concrete lintels; BS 5328, designed mix C25, 20mm aggregate minimum cement content 360 kg/m3; vibrated; reinforcement bars BS 4449 hot rolled plain round mild steel; bedding in cement-lime mortar (1:1:6) Cont.										
Lintels Cont.										
75 x 150 x 1650mm; reinforced with 1.60 kg of 12mm bars	9.69	2.50	**9.94**	0.28	0.14	**5.48**	0.17	**15.59**	nr	**17.53**
102 x 150 x 1200mm; reinforced with 1.20 kg of 12mm bars	9.59	2.50	**9.83**	0.33	0.17	**6.51**	0.13	**16.47**	nr	**18.53**
102 x 150 x 1650mm; reinforced with 1.60 kg of 12mm bars	13.19	2.50	**13.51**	0.45	0.22	**8.75**	0.17	**22.44**	nr	**25.24**
215 x 150 x 1200mm; reinforced with 2.40 kg of 12mm bars	20.21	2.50	**20.72**	0.66	0.33	**12.91**	0.22	**33.84**	nr	**38.07**
215 x 150 x 1650mm; reinforced with 3.20 kg of 12mm bars	27.79	2.50	**28.48**	0.91	0.53	**18.61**	0.28	**47.37**	nr	**53.29**
215 x 225 x 1200mm; reinforced with 2.40 kg of 12mm bars	30.32	2.50	**31.07**	0.86	0.43	**16.83**	0.23	**48.12**	nr	**54.14**
215 x 225 x 1650mm; reinforced with 3.20 kg of 12mm bars	41.68	2.50	**42.73**	1.18	0.58	**22.98**	0.30	**66.01**	nr	**74.26**
327 x 150 x 1200mm; reinforced with 3.60 kg of 12mm bars	30.74	2.50	**31.51**	0.86	0.43	**16.83**	0.25	**48.58**	nr	**54.65**
327 x 150 x 1650mm; reinforced with 4.80 kg of 12mm bars	42.27	2.50	**43.32**	1.18	0.59	**23.09**	0.30	**66.71**	nr	**75.05**
327 x 225 x 1200mm; reinforced with 3.60 kg of 12mm bars	46.11	2.50	**47.26**	1.32	0.66	**25.83**	0.27	**73.36**	nr	**82.52**
327 x 225 x 1650mm; reinforced with 4.80 kg of 12mm bars	63.40	2.50	**64.98**	1.82	0.91	**35.61**	0.37	**100.96**	nr	**113.58**
75 x 150 x 1200mm; reinforced with 1.20 kg of 12mm bars; fair finish two faces	8.67	2.50	**8.89**	0.20	0.10	**3.91**	0.13	**12.93**	nr	**14.55**
75 x 150 x 1650mm; reinforced with 1.60 kg of 12mm bars; fair finish two faces	11.92	2.50	**12.22**	0.28	0.14	**5.48**	0.17	**17.87**	nr	**20.11**
102 x 150 x 1200mm; reinforced with 1.20 kg of 12mm bars; fair finish two faces	11.79	2.50	**12.09**	0.33	0.17	**6.51**	0.13	**18.73**	nr	**21.07**
102 x 150 x 1650mm; reinforced with 1.60 kg of 12mm bars; fair finish two faces	16.22	2.50	**16.62**	0.45	0.22	**8.75**	0.17	**25.54**	nr	**28.74**
215 x 150 x 1200mm; reinforced with 2.40 kg of 12mm bars; fair finish two faces	24.86	2.50	**25.48**	0.66	0.33	**12.91**	0.22	**38.61**	nr	**43.43**
215 x 150 x 1650mm; reinforced with 3.20 kg of 12mm bars; fair finish two faces	34.18	2.50	**35.03**	0.91	0.45	**17.75**	0.28	**53.06**	nr	**59.70**
215 x 225 x 1200mm; reinforced with 2.40 kg of 12mm bars; fair finish two faces	37.28	2.50	**38.22**	0.86	0.43	**16.83**	0.23	**55.27**	nr	**62.18**
215 x 225 x 1650mm; reinforced with 3.20 kg of 12mm bars; fair finish two faces	51.27	2.50	**52.55**	1.18	0.58	**22.98**	0.30	**75.83**	nr	**85.31**
327 x 150 x 1200mm; reinforced with 3.60 kg of 12mm bars; fair finish two faces	37.81	2.50	**38.75**	0.86	0.43	**16.83**	0.23	**55.80**	nr	**62.78**
327 x 150 x 1650mm; reinforced with 4.80 kg of 12mm bars; fair finish two faces	51.98	2.50	**53.28**	1.18	0.59	**23.09**	0.30	**76.67**	nr	**86.25**
327 x 225 x 1200mm; reinforced with 3.60 kg of 12mm bars; fair finish two faces	56.71	2.50	**58.13**	1.32	0.66	**25.83**	0.26	**84.21**	nr	**94.74**
327 x 225 x 1650mm; reinforced with 4.80 kg of 12mm bars; fair finish two faces	77.97	2.50	**79.92**	1.82	0.91	**35.61**	0.35	**115.89**	nr	**130.37**
Boot lintels										
215 x 150 x 1200mm; reinforced with 3.60 kg of 12mm bars and 0.51 kg of 6mm links	24.86	2.50	**25.48**	0.66	0.33	**12.91**	0.20	**38.59**	nr	**43.42**
215 x 150 x 1800mm; reinforced with 5.19 kg of 12mm bars and 0.77 kg of 6mm links	37.28	2.50	**38.22**	1.00	0.33	**17.74**	0.30	**56.26**	nr	**63.29**
215 x 225 x 1200mm; reinforced with 3.60 kg of 12mm bars and 0.62 kg of 6mm links	37.28	2.50	**38.22**	0.86	0.43	**16.83**	0.22	**55.26**	nr	**62.16**
215 x 225 x 1800mm; reinforced with 5.19 kg of 12mm bars and 0.93 kg of 6mm links	55.93	2.50	**57.33**	1.30	0.65	**25.43**	0.31	**83.07**	nr	**93.46**
252 x 150 x 1200mm; reinforced with 3.60 kg of 12mm bars and 0.56 kg of 6mm links	29.13	2.50	**29.86**	0.79	0.40	**15.51**	0.24	**45.61**	nr	**51.31**
252 x 150 x 1800mm; reinforced with 5.19 kg of 12mm bars and 0.83 kg of 6mm links	43.70	2.50	**44.79**	1.19	0.59	**23.23**	0.35	**68.38**	nr	**76.93**
252 x 225 x 1200mm; reinforced with 3.60 kg of 12mm bars and 0.67 kg of 6mm links	43.70	2.50	**44.79**	0.86	0.43	**16.83**	0.26	**61.88**	nr	**69.61**
252 x 225 x 1800mm; reinforced with 5.19 kg of 12mm bars and 1.00 kg of 6mm links	65.55	2.50	**67.19**	1.30	0.65	**25.43**	0.35	**92.98**	nr	**104.60**
Precast concrete padstones; BS 5328, designed mix C25, 20mm aggregate, minimum cement content 360 kg/m3; vibrated; reinforced at contractor's discretion; bedding in cement-lime mortar (1:1:6)										
Padstones										
215 x 215 x 75mm	2.29	2.50	**2.35**	0.07	0.03	**1.32**	0.12	**3.78**	nr	**4.25**
215 x 215 x 150mm	4.58	2.50	**4.69**	0.09	0.04	**1.71**	0.12	**6.52**	nr	**7.33**
327 x 215 x 150mm	6.96	2.50	**7.14**	0.12	0.06	**2.35**	0.13	**9.61**	nr	**10.81**
327 x 327 x 150mm	10.59	2.50	**10.85**	0.19	0.09	**3.66**	0.14	**14.66**	nr	**16.49**
440 x 215 x 150mm	9.37	2.50	**9.60**	0.19	0.09	**3.66**	0.15	**13.42**	nr	**15.09**
440 x 327 x 150mm	14.25	2.50	**14.60**	0.22	0.11	**4.30**	0.16	**19.07**	nr	**21.45**
440 x 440 x 150mm	19.17	2.50	**19.65**	0.22	0.11	**4.30**	0.17	**24.12**	nr	**27.14**

NATURAL STONEWORK

Natural stonework including restoration and repair is work for the specialist and the cost depends on many factors, prices should be obtained for specific projects.

There are a number of reputable firms who specialize in this class of work and who will submit quotations. Among these are the following:-

STONEWEST LTD.
Lamberts Place, St James Road, Croydon, Surrey, CR9 2HX.
Tel: 020 8684 6646
Fax: 020 8684 9323

A. F. JONES (STONEMASONS).
33 Bedford Road, Reading, Berkshire, RG1 7EX.
Tel: 0118 957 3537
Fax: 0118 957 4334

DIESPEKER MARBLE & TERRAZZO LTD.
132-136 Ormside Street, Peckham, London, SE15 1TF
Tel: 020 7358 0160
Fax: 020 7369 1695

C.W.O. LIMITED.
Terminus Road, Chichester, West Sussex, PO19 2TX.
Tel: 01243 784225
Fax: 01243 813700

LAING STONEMASONRY
Dalston Road, Carlisle, Cumbria, CA2 5NR.
Tel: 01228 521404
Fax: 01228 511306

MACC STONE CO.
196-198 New Street, Biddulph Moor, Staffordshire, Moorlands, ST8 7NW.
Tel: 01782 514353
Fax: 01782 516783

STONEGUARD (LONDON) LTD.
St Martins House, The Runway, South Ruislip, Middlesex, H04 6SG.
Tel: 020 8839 9977
Fax: 020 8839 9988

APS MASONRY LTD.
Osney Mead, Oxford, OX2 0EQ.
Tel: 01865 246677
Fax: 01865 254617

DUNHOUSE QUARRY CO. LTD.
Staine Drop, Darlington, Co. Durham DL2 3QU
Tel: 01833 660208
Fax: 01833 660748

STRUCTURAL/CARCASSING METAL/TIMBER

Labour hourly rates: (except Specialists) Craft Operatives £14.21 Labourer £10.71 Rates are national average prices. Refer to REGIONAL VARIATIONS for indicative levels of overall pricing in regions	MATERIALS			LABOUR				RATES		
	Del to Site £	Waste %	Material Cost £	Craft Optve Hrs	Lab Hrs	Labour Cost £	Sunds £	Nett Rate £	Unit	Gross Rate (+12.5%) £
G10: STRUCTURAL STEEL FRAMING										
Prices generally for fabricated steelwork										
Note										
prices can vary considerably dependent upon the character of the work; the following are average prices for work fabricated and delivered to site										
Framing, fabrication; including shop and site black bolts, nuts and washers for structural framing to structural framing connections										
Weldable steel, BS EN 10025 : 1993 Grade S275JR (formerly BS 7668 Grade 43B), hot rolled sections BS 4-1 (Euronorm 54); welded fabrication in accordance with BS 5950-1										
Columns										
weight less than 40 Kg/m	1030.00	-	**1030.00**	-	-	-	-	**1030.00**	t	**1158.75**
weight 40-100 Kg/m	1030.00	-	**1030.00**	-	-	-	-	**1030.00**	t	**1158.75**
weight exceeding 100 Kg/m	1030.00	-	**1030.00**	-	-	-	-	**1030.00**	t	**1158.75**
Beams										
weight less than 40 Kg/m	1030.00	-	**1030.00**	-	-	-	-	**1030.00**	t	**1158.75**
weight 40-100 Kg/m	1030.00	-	**1030.00**	-	-	-	-	**1030.00**	t	**1158.75**
weight exceeding 100 Kg/m	1030.00	-	**1030.00**	-	-	-	-	**1030.00**	t	**1158.75**
Beams, castellated										
weight less than 40 Kg/m	1235.00	-	**1235.00**	-	-	-	-	**1235.00**	t	**1389.38**
weight 40-100 Kg/m	1235.00	-	**1235.00**	-	-	-	-	**1235.00**	t	**1389.38**
weight exceeding 100 Kg/m	1235.00	-	**1235.00**	-	-	-	-	**1235.00**	t	**1389.38**
Beams, curved										
weight less than 40 Kg/m	1500.00	-	**1500.00**	-	-	-	-	**1500.00**	t	**1687.50**
weight 40-100 Kg/m	1500.00	-	**1500.00**	-	-	-	-	**1500.00**	t	**1687.50**
weight exceeding 100 Kg/m	1500.00	-	**1500.00**	-	-	-	-	**1500.00**	t	**1687.50**
Bracings, tubular										
weight less than 40 Kg/m	1080.00	-	**1080.00**	-	-	-	-	**1080.00**	t	**1215.00**
weight 40-100 Kg/m	1080.00	-	**1080.00**	-	-	-	-	**1080.00**	t	**1215.00**
weight exceeding 100 Kg/m	1080.00	-	**1080.00**	-	-	-	-	**1080.00**	t	**1215.00**
Purlins and cladding rails										
weight less than 40 Kg/m	1045.00	-	**1045.00**	-	-	-	-	**1045.00**	t	**1175.63**
weight 40-100 Kg/m	1045.00	-	**1045.00**	-	-	-	-	**1045.00**	t	**1175.63**
weight exceeding 100 Kg/m	1045.00	-	**1045.00**	-	-	-	-	**1045.00**	t	**1175.63**
portal frames	975.00	-	**975.00**	-	-	-	-	**975.00**	t	**1096.88**
trusses 12m to 18m span	1550.00	-	**1550.00**	-	-	-	-	**1550.00**	t	**1743.75**
trusses 6m to 12m span	1605.00	-	**1605.00**	-	-	-	-	**1605.00**	t	**1805.63**
trusses up to 6m span	1710.00	-	**1710.00**	-	-	-	-	**1710.00**	t	**1923.75**
Weldable steel BS EN 10210 : 1994 Grade S275JOH (formerly BS 7668 Grade 43C), hot rolled sections BS EN 10210-2 (Euronorm 57); welded fabrication in accordance with BS 5950-1										
column, square hollow section										
weight less than 40 Kg/m	1175.00	-	**1175.00**	-	-	-	-	**1175.00**	t	**1321.88**
weight 40-100 Kg/m	1175.00	-	**1175.00**	-	-	-	-	**1175.00**	t	**1321.88**
weight exceeding 100 Kg/m	1175.00	-	**1175.00**	-	-	-	-	**1175.00**	t	**1321.88**
column, rectangular hollow section										
weight less than 40 Kg/m	1190.00	-	**1190.00**	-	-	-	-	**1190.00**	t	**1338.75**
weight 40-100 Kg/m	1190.00	-	**1190.00**	-	-	-	-	**1190.00**	t	**1338.75**
weight exceeding 100 Kg/m	1190.00	-	**1190.00**	-	-	-	-	**1190.00**	t	**1338.75**
Holding down bolts or assemblies; mild steel										
rag or indented bolts; M 10 with nuts and washers										
100mm long	1.33	5.00	**1.40**	0.09	-	**1.28**	-	**2.68**	nr	**3.01**
160mm long	1.38	5.00	**1.44**	0.13	-	**1.85**	-	**3.29**	nr	**3.70**
rag or indented bolts; M 12 with nuts and washers										
100mm long	1.43	5.00	**1.50**	0.12	-	**1.71**	-	**3.21**	nr	**3.61**
160mm long	1.67	5.00	**1.75**	0.15	-	**2.13**	-	**3.88**	nr	**4.36**
200mm long	1.87	5.00	**1.96**	0.16	-	**2.27**	-	**4.23**	nr	**4.76**
rag or indented bolts; M 16 with nuts and washers										
120mm long	2.15	5.00	**2.26**	0.13	-	**1.85**	-	**4.11**	nr	**4.62**
160mm long	2.48	5.00	**2.60**	0.16	-	**2.27**	-	**4.87**	nr	**5.48**
200mm long	2.89	5.00	**3.03**	0.18	-	**2.56**	-	**5.59**	nr	**6.29**

Labour hourly rates: (except Specialists) Craft Operatives £14.21 Labourer £10.71 Rates are national average prices. Refer to REGIONAL VARIATIONS for indicative levels of overall pricing in regions	MATERIALS			LABOUR				RATES		
	Del to Site	Waste	Material Cost	Craft Optve	Lab	Labour Cost	Sunds	Nett Rate	Unit	Gross Rate (+12.5%)
	£	%	£	Hrs	Hrs	£	£	£		£
G10: STRUCTURAL STEEL FRAMING Cont.										
Framing, fabrication; including shop and site black bolts, nuts and washers for structural framing to structural framing connections Cont.										
Holding down bolts or assemblies; mild steel Cont.										
holding down bolt assembly; M 20 bolt with 100 x 100 x 10mm plate washer tack welded to head; with nuts and washers										
300mm long	4.98	5.00	5.23	0.28	-	3.98	-	9.21	nr	10.36
350mm long	5.68	5.00	5.97	0.32	-	4.55	-	10.51	nr	11.83
400mm long	6.45	5.00	6.77	0.35	-	4.97	-	11.74	nr	13.21
450mm long	7.38	5.00	7.75	0.55	-	7.82	-	15.57	nr	17.51
High strength friction grip bolts; BS 4395 Part 1 - general grade										
M 16; with nuts and washers										
50mm long	0.81	5.00	0.85	0.09	-	1.28	-	2.13	nr	2.40
60mm long	0.88	5.00	0.92	0.09	-	1.28	-	2.20	nr	2.47
70mm long	0.94	5.00	0.99	0.10	-	1.42	-	2.41	nr	2.71
80mm long	1.02	5.00	1.07	0.10	-	1.42	-	2.49	nr	2.80
90mm long	1.09	5.00	1.14	0.12	-	1.71	-	2.85	nr	3.20
100mm long	1.12	5.00	1.18	0.12	-	1.71	-	2.88	nr	3.24
120mm long	1.24	5.00	1.31	0.13	-	1.85	-	3.15	nr	3.55
140mm long	1.41	5.00	1.48	0.14	-	1.99	-	3.47	nr	3.90
150mm long	1.42	5.00	1.49	0.14	-	1.99	-	3.48	nr	3.92
160mm long	1.57	5.00	1.64	0.15	-	2.13	-	3.77	nr	4.25
180mm long	1.78	5.00	1.86	0.16	-	2.27	-	4.14	nr	4.66
200mm long	4.51	5.00	4.73	0.16	-	2.27	-	7.01	nr	7.88
220mm long	4.78	5.00	5.02	0.16	-	2.27	-	7.30	nr	8.21
M 20; with nuts and washers										
60mm long	1.53	5.00	1.61	0.09	-	1.28	-	2.89	nr	3.25
70mm long	1.63	5.00	1.71	0.10	-	1.42	-	3.13	nr	3.53
80mm long	1.70	5.00	1.78	0.10	-	1.42	-	3.20	nr	3.60
90mm long	1.79	5.00	1.88	0.12	-	1.71	-	3.58	nr	4.03
100mm long	1.85	5.00	1.95	0.12	-	1.71	-	3.65	nr	4.11
120mm long	2.07	5.00	2.17	0.13	-	1.85	-	4.02	nr	4.52
140mm long	2.32	5.00	2.44	0.15	-	2.13	-	4.57	nr	5.14
150mm long	2.58	5.00	2.70	0.15	-	2.13	-	4.84	nr	5.44
160mm long	2.64	5.00	2.77	0.16	-	2.27	-	5.05	nr	5.68
180mm long	3.02	5.00	3.17	0.17	-	2.42	-	5.59	nr	6.28
200mm long	5.24	5.00	5.50	0.18	-	2.56	-	8.06	nr	9.07
220mm long	5.65	5.00	5.93	0.20	-	2.84	-	8.77	nr	9.87
240mm long	6.11	5.00	6.41	0.21	-	2.98	-	9.39	nr	10.57
260mm long	6.53	5.00	6.85	0.23	-	3.27	-	10.12	nr	11.39
280mm long	6.97	5.00	7.32	0.24	-	3.41	-	10.73	nr	12.07
300mm long	7.42	5.00	7.79	0.25	-	3.55	-	11.34	nr	12.76
M 24; with nuts and washers										
70mm long	2.73	5.00	2.87	0.13	-	1.85	-	4.71	nr	5.30
80mm long	3.09	5.00	3.24	0.13	-	1.85	-	5.09	nr	5.72
90mm long	3.14	5.00	3.30	0.14	-	1.99	-	5.29	nr	5.95
100mm long	3.31	5.00	3.47	0.14	-	1.99	-	5.46	nr	6.15
120mm long	3.72	5.00	3.90	0.16	-	2.27	-	6.18	nr	6.95
140mm long	4.22	5.00	4.43	0.17	-	2.42	-	6.84	nr	7.70
150mm long	4.34	5.00	4.56	0.17	-	2.42	-	6.97	nr	7.84
160mm long	4.66	5.00	4.90	0.18	-	2.56	-	7.45	nr	8.38
180mm long	5.30	5.00	5.56	0.20	-	2.84	-	8.40	nr	9.45
200mm long	7.59	5.00	7.97	0.21	-	2.98	-	10.96	nr	12.33
220mm long	8.27	5.00	8.68	0.23	-	3.27	-	11.95	nr	13.45
240mm long	8.90	5.00	9.35	0.24	-	3.41	-	12.76	nr	14.35
260mm long	9.58	5.00	10.06	0.25	-	3.55	-	13.61	nr	15.31
280mm long	10.21	5.00	10.72	0.28	-	3.98	-	14.70	nr	16.54
300mm long	10.90	5.00	11.45	0.29	-	4.12	-	15.57	nr	17.51
Framing, erection; in accordance with BS 5950-1										
Note										
the following prices for erection include for the site to be reasonably clear, ease of access and the erection carried out during normal working hours										
Permanent erection of fabricated steelwork on site with bolted connections										
framing	-	-	-	25.00	-	355.25	-	355.25	t	399.66
Cold rolled zed purlins and cladding rails; Metsec, galvanized steel										
Purlin sleeved system (section only); fixing to cleats on frame members at 6000mm centres; hoist and fix 3.00m above ground level; fixing with bolts										
reference 14214; 3.03 kg/m	14.79	2.50	15.16	0.23	0.03	3.59	-	18.75	m	21.09
reference 14216; 3.47 kg/m	16.70	2.50	17.12	0.23	0.03	3.59	-	20.71	m	23.30
reference 17214; 3.66 kg/m	17.50	2.50	17.94	0.29	0.03	4.44	-	22.38	m	25.18
reference 17216; 4.11 kg/m	19.88	2.50	20.38	0.29	0.03	4.44	-	24.82	m	27.92
reference 20216; 4.49 kg/m	21.71	2.50	22.25	0.35	0.05	5.51	-	27.76	m	31.23
reference 20218; 5.03 kg/m	24.48	2.50	25.09	0.35	0.05	5.51	-	30.60	m	34.43
reference 20220; 5.57 kg/m	26.89	2.50	27.56	0.35	0.05	5.51	-	33.07	m	37.21
reference 23218; 5.73 kg/m	27.79	2.50	28.48	0.38	0.05	5.94	-	34.42	m	38.72
reference 23220; 6.34 kg/m	30.82	2.50	31.59	0.38	0.05	5.94	-	37.53	m	42.22
Purlin sleeved system (section only); fixing to cleats on frame members at 6000mm centres; hoist and fix 6.00m above ground level; fixing with bolts										
reference 14214; 3.03 kg/m	14.79	2.50	15.16	0.25	0.03	3.87	-	19.03	m	21.41

Labour hourly rates: (except Specialists) Craft Operatives £14.21 Labourer £10.71 Rates are national average prices. Refer to REGIONAL VARIATIONS for indicative levels of overall pricing in regions	MATERIALS			LABOUR				RATES		
	Del to Site £	Waste %	Material Cost £	Craft Optve Hrs	Lab Hrs	Labour Cost £	Sunds £	Nett Rate £	Unit	Gross Rate (+12.5%) £
G10: STRUCTURAL STEEL FRAMING Cont.										
Cold rolled zed purlins and cladding rails; Metsec, galvanized steel Cont.										
Purlin sleeved system (section only); fixing to cleats on frame members at 6000mm centres; hoist and fix 6.00m above ground level; fixing with bolts Cont.										
reference 14216; 3.47 kg/m	16.70	2.50	**17.12**	0.25	0.03	**3.87**	-	**20.99**	m	**23.62**
reference 17214; 3.66 kg/m	17.50	2.50	**17.94**	0.32	0.05	**5.08**	-	**23.02**	m	**25.90**
reference 17216; 4.11 kg/m	19.88	2.50	**20.38**	0.32	0.05	**5.08**	-	**25.46**	m	**28.64**
reference 20216; 4.49 kg/m	21.71	2.50	**22.25**	0.38	0.05	**5.94**	-	**28.19**	m	**31.71**
reference 20218; 5.03 kg/m	24.48	2.50	**25.09**	0.38	0.05	**5.94**	-	**31.03**	m	**34.91**
reference 20220; 5.57 kg/m	26.89	2.50	**27.56**	0.38	0.05	**5.94**	-	**33.50**	m	**37.68**
reference 23218; 5.73 kg/m	27.79	2.50	**28.48**	0.41	0.06	**6.47**	-	**34.95**	m	**39.32**
reference 23220; 6.34 kg/m	30.82	2.50	**31.59**	0.41	0.06	**6.47**	-	**38.06**	m	**42.82**
Purlin overlap system (cleat only); fixing to cleats on frame members at 6000mm centres; hoist and fix 3.00m above ground level; fixing with bolts galvanized										
reference 142	9.94	2.50	**10.19**	0.23	0.03	**3.59**	-	**13.78**	m	**15.50**
reference 172	10.57	2.50	**10.83**	0.23	0.03	**3.59**	-	**14.42**	m	**16.22**
reference 202	11.51	2.50	**11.80**	0.35	0.04	**5.40**	-	**17.20**	m	**19.35**
reference 232	12.45	2.50	**12.76**	0.38	0.05	**5.94**	-	**18.70**	m	**21.03**
Purlin overlap system (cleat only); fixing to cleats on frame members at 6000mm centres; hoist and fix 6.00m above ground level; fixing with bolts galvanized										
reference 142	9.94	2.50	**10.19**	0.25	0.03	**3.87**	-	**14.06**	m	**15.82**
reference 172	10.57	2.50	**10.83**	0.32	0.05	**5.08**	-	**15.91**	m	**17.90**
reference 202	11.51	2.50	**11.80**	0.38	0.05	**5.94**	-	**17.73**	m	**19.95**
reference 232	12.45	2.50	**12.76**	0.41	0.06	**6.47**	-	**19.23**	m	**21.63**
Purlin Non-Continuous system (section only) ; fixing to cleats on frame members at 6000mm centres; hoist and fix 3.00m above ground level; fixing with bolts										
reference 17215; 3.85 kg/m	18.61	2.50	**19.08**	0.29	0.03	**4.44**	-	**23.52**	m	**26.46**
reference 17216; 4.11 kg/m	19.88	2.50	**20.38**	0.29	0.03	**4.44**	-	**24.82**	m	**27.92**
reference 20216; 4.49 kg/m	21.71	2.50	**22.25**	0.35	0.05	**5.51**	-	**27.76**	m	**31.23**
reference 20218; 5.03 kg/m	24.48	2.50	**25.09**	0.35	0.05	**5.51**	-	**30.60**	m	**34.43**
reference 20220; 5.57 kg/m	26.89	2.50	**27.56**	0.35	0.05	**5.51**	-	**33.07**	m	**37.21**
reference 23218; 5.73 kg/m	27.79	2.50	**28.48**	0.38	0.05	**5.94**	-	**34.42**	m	**38.72**
reference 23223; 7.26 kg/m	35.61	2.50	**36.50**	0.38	0.05	**5.94**	-	**42.44**	m	**47.74**
reference 26223; 7.92 kg/m	38.78	2.50	**39.75**	0.46	0.06	**7.18**	-	**46.93**	m	**52.79**
reference 26229; 9.88 kg/m	48.09	2.50	**49.29**	0.46	0.06	**7.18**	-	**56.47**	m	**63.53**
Purlin Non-Continuous system (section only) ; fixing to cleats on frame members at 6000mm centres; hoist and fix 6.00m above ground level; fixing with bolts										
reference 17215; 3.85 kg/m	18.61	2.50	**19.08**	0.32	0.05	**5.08**	-	**24.16**	m	**27.18**
reference 17216; 4.11 kg/m	19.88	2.50	**20.38**	0.32	0.05	**5.08**	-	**25.46**	m	**28.64**
reference 20216; 4.49 kg/m	21.71	2.50	**22.25**	0.38	0.05	**5.94**	-	**28.19**	m	**31.71**
reference 20218; 5.03 kg/m	24.48	2.50	**25.09**	0.38	0.05	**5.94**	-	**31.03**	m	**34.91**
reference 20220; 5.57 kg/m	26.89	2.50	**27.56**	0.38	0.05	**5.94**	-	**33.50**	m	**37.68**
reference 23218; 5.73 kg/m	27.79	2.50	**28.48**	0.41	0.06	**6.47**	-	**34.95**	m	**39.32**
reference 23223; 7.26 kg/m	35.61	2.50	**36.50**	0.41	0.06	**6.47**	-	**42.97**	m	**48.34**
reference 26223; 7.92 kg/m	38.78	2.50	**39.75**	0.51	0.07	**8.00**	-	**47.75**	m	**53.71**
reference 26229; 9.88 kg/m	48.09	2.50	**49.29**	0.51	0.07	**8.00**	-	**57.29**	m	**64.45**
Galvanized purlin cleats; weld on										
Reference 142 for 142mm deep purlins	5.17	2.50	**5.30**	-	-	**-**	-	**5.30**	nr	**5.96**
Reference 172 for 172mm deep purlins	5.89	2.50	**6.04**	-	-	**-**	-	**6.04**	nr	**6.79**
Reference 202 for 202mm deep purlins	6.43	2.50	**6.59**	-	-	**-**	-	**6.59**	nr	**7.41**
Reference 232 for 232mm deep purlins	7.08	2.50	**7.26**	-	-	**-**	-	**7.26**	nr	**8.16**
Reference 262 for 262mm deep purlins	7.73	2.50	**7.92**	-	-	**-**	-	**7.92**	nr	**8.91**
Round Lok anti-sag rods; push fit to purlins										
purlins at 1150mm centres	5.20	2.50	**5.33**	0.29	0.03	**4.44**	-	**9.77**	nr	**10.99**
purlins at 1350mm centres	6.04	2.50	**6.19**	0.29	0.03	**4.44**	-	**10.63**	nr	**11.96**
purlins at 1550mm centres	6.83	2.50	**7.00**	0.38	0.05	**5.94**	-	**12.94**	nr	**14.55**
purlins at 1700mm centres	7.69	2.50	**7.88**	0.38	0.05	**5.94**	-	**13.82**	nr	**15.54**
purlins at 1950mm centres	8.44	2.50	**8.65**	0.38	0.05	**5.94**	-	**14.59**	nr	**16.41**
Side rail sleeved system (section only); fixing to cleats on stanchions at 6000mm centres; hoist and fix 3.00m above ground level; fixing with bolts										
reference 14214; 3.03 kg/m	14.79	2.50	**15.16**	0.23	0.03	**3.59**	-	**18.75**	m	**21.09**
reference 14216; 3.47 kg/m	16.70	2.50	**17.12**	0.23	0.03	**3.59**	-	**20.71**	m	**23.30**
reference 17215; 3.85 kg/m	18.61	2.50	**19.08**	0.29	0.03	**4.44**	-	**23.52**	m	**26.46**
reference 17216; 4.11 kg/m	19.88	2.50	**20.38**	0.29	0.03	**4.44**	-	**24.82**	m	**27.92**
reference 20216; 4.49 kg/m	21.71	2.50	**22.25**	0.35	0.05	**5.51**	-	**27.76**	m	**31.23**
reference 20218; 5.03 kg/m	24.48	2.50	**25.09**	0.35	0.05	**5.51**	-	**30.60**	m	**34.43**
reference 20220; 5.57 kg/m	26.89	2.50	**27.56**	0.35	0.05	**5.51**	-	**33.07**	m	**37.21**
reference 23218; 5.73 kg/m	27.79	2.50	**28.48**	0.38	0.05	**5.94**	-	**34.42**	m	**38.72**
reference 23223; 7.26 kg/m	35.61	2.50	**36.50**	0.38	0.05	**5.94**	-	**42.44**	m	**47.74**
Side rail sleeved system (section only); fixing to cleats on stanchions at 6000mm centres; hoist and fix 6.00m above ground level; fixing with bolts										
reference 14214; 3.03 kg/m	14.79	2.50	**15.16**	0.25	0.03	**3.87**	-	**19.03**	m	**21.41**
reference 14216; 3.47 kg/m	16.70	2.50	**17.12**	0.25	0.03	**3.87**	-	**20.99**	m	**23.62**
reference 17215; 3.85 kg/m	18.61	2.50	**19.08**	0.32	0.05	**5.08**	-	**24.16**	m	**27.18**
reference 17216; 4.11 kg/m	19.88	2.50	**20.38**	0.32	0.05	**5.08**	-	**25.46**	m	**28.64**
reference 20216; 4.49 kg/m	21.71	2.50	**22.25**	0.38	0.05	**5.94**	-	**28.19**	m	**31.71**
reference 20218; 5.03 kg/m	24.48	2.50	**25.09**	0.38	0.05	**5.94**	-	**31.03**	m	**34.91**
reference 20220; 5.57 kg/m	26.89	2.50	**27.56**	0.38	0.05	**5.94**	-	**33.50**	m	**37.68**

Labour hourly rates: (except Specialists) Craft Operatives £14.21 Labourer £10.71 Rates are national average prices. Refer to REGIONAL VARIATIONS for indicative levels of overall pricing in regions	MATERIALS			LABOUR				RATES		
	Del to Site £	Waste %	Material Cost £	Craft Optve Hrs	Lab Hrs	Labour Cost £	Sunds £	Nett Rate £	Unit	Gross Rate (+12.5%) £
G10: STRUCTURAL STEEL FRAMING Cont.										
Cold rolled zed purlins and cladding rails; Metsec, galvanized steel Cont.										
Side rail sleeved system (section only); fixing to cleats on stanchions at 6000mm centres; hoist and fix 6.00m above ground level; fixing with bolts Cont.										
reference 23218; 5.73 kg/m	27.79	2.50	**28.48**	0.41	0.06	**6.47**	-	**34.95**	m	**39.32**
reference 23223; 7.26 kg/m	35.61	2.50	**36.50**	0.41	0.06	**6.47**	-	**42.97**	m	**48.34**
Side rail single span system (section only) ; fixing to cleats on stanchions at 6000mm centres; hoist and fix 3.00m above ground level; fixing with bolts										
reference 14214; 3.03 kg/m	14.79	2.50	**15.16**	0.23	0.03	**3.59**	-	**18.75**	m	**21.09**
reference 14216; 3.47 kg/m	16.70	2.50	**17.12**	0.23	0.03	**3.59**	-	**20.71**	m	**23.30**
reference 17215; 3.85 kg/m	18.61	2.50	**19.08**	0.29	0.03	**4.44**	-	**23.52**	m	**26.46**
reference 17216; 4.11 kg/m	19.88	2.50	**20.38**	0.29	0.03	**4.44**	-	**24.82**	m	**27.92**
reference 20216; 4.49 kg/m	21.71	2.50	**22.25**	0.35	0.05	**5.51**	-	**27.76**	m	**31.23**
reference 20218; 5.03 kg/m	24.48	2.50	**25.09**	0.35	0.05	**5.51**	-	**30.60**	m	**34.43**
reference 20220; 5.57 kg/m	26.89	2.50	**27.56**	0.35	0.05	**5.51**	-	**33.07**	m	**37.21**
reference 23218; 5.73 kg/m	27.79	2.50	**28.48**	0.38	0.05	**5.94**	-	**34.42**	m	**38.72**
reference 23223; 7.26 kg/m	35.61	2.50	**36.50**	0.38	0.05	**5.94**	-	**42.44**	m	**47.74**
Side rail single span system (section only) ; fixing to cleats on stanchions at 6000mm centres; hoist and fix 6.00m above ground level; fixing with bolts										
reference 14214; 3.03 kg/m	14.79	2.50	**15.16**	0.25	0.03	**3.87**	-	**19.03**	m	**21.41**
reference 14216; 3.47 kg/m	16.70	2.50	**17.12**	0.25	0.03	**3.87**	-	**20.99**	m	**23.62**
reference 17215; 3.85 kg/m	18.61	2.50	**19.08**	0.32	0.05	**5.08**	-	**24.16**	m	**27.18**
reference 17216; 4.11 kg/m	19.88	2.50	**20.38**	0.32	0.05	**5.08**	-	**25.46**	m	**28.64**
reference 20216; 4.49 kg/m	21.71	2.50	**22.25**	0.38	0.05	**5.94**	-	**28.19**	m	**31.71**
reference 20218; 5.03 kg/m	24.48	2.50	**25.09**	0.38	0.05	**5.94**	-	**31.03**	m	**34.91**
reference 20220; 5.57 kg/m	26.89	2.50	**27.56**	0.38	0.05	**5.94**	-	**33.50**	m	**37.68**
reference 23218; 5.73 kg/m	27.79	2.50	**28.48**	0.41	0.06	**6.47**	-	**34.95**	m	**39.32**
reference 23223; 7.26 kg/m	35.61	2.50	**36.50**	0.41	0.06	**6.47**	-	**42.97**	m	**48.34**
Galvanized side rail cleats; weld on										
Reference 142 for 142mm wide rails	5.18	2.50	**5.31**	-	-	-	-	**5.31**	nr	**5.97**
Reference 172 for 172mm wide rails	7.05	2.50	**7.23**	-	-	-	-	**7.23**	nr	**8.13**
Reference 202 for 202mm wide rails	8.68	2.50	**8.90**	-	-	-	-	**8.90**	nr	**10.01**
Reference 232 for 232mm wide rails	10.09	2.50	**10.34**	-	-	-	-	**10.34**	nr	**11.64**
Galvanized side rail supports; 122-262 series; weld on										
rail 1000mm long	15.20	2.50	**15.58**	0.40	0.05	**6.22**	-	**21.80**	nr	**24.52**
rail 1400mm long	16.59	2.50	**17.00**	0.40	0.05	**6.22**	-	**23.22**	nr	**26.13**
rail 1600mm long	18.40	2.50	**18.86**	0.52	0.07	**8.14**	-	**27.00**	nr	**30.37**
rail 1800mm long	22.14	2.50	**22.69**	0.52	0.07	**8.14**	-	**30.83**	nr	**34.69**
Diagonal tie wire ropes (assembled with end brackets); bolt on										
1700mm long	29.66	2.50	**30.40**	0.35	0.05	**5.51**	-	**35.91**	nr	**40.40**
2200mm long	33.18	2.50	**34.01**	0.35	0.05	**5.51**	-	**39.52**	nr	**44.46**
2600mm long	36.70	2.50	**37.62**	0.46	0.06	**7.18**	-	**44.80**	nr	**50.40**
3600mm long	45.31	2.50	**46.44**	0.46	0.06	**7.18**	-	**53.62**	nr	**60.32**
Surface preparation at works										
Note notwithstanding the requirement of SMM7 to measure painting on structural steelwork in m2, painting off site has been given in tonnes of structural steelwork in accordance with normal steelwork contractors practice										
Blast cleaning										
surfaces of steelwork	-	-	**Spclist**	-	-	**Spclist**	-	**90.55**	t	**101.87**
Surface treatment at works										
One coat micaceous oxide primer, 75 microns										
surfaces of steelwork	-	-	**Spclist**	-	-	**Spclist**	-	**101.14**	t	**113.78**
Two coats micaceous oxide primer, 150 microns										
surfaces of steelwork	-	-	**Spclist**	-	-	**Spclist**	-	**188.16**	t	**211.68**
G12: ISOLATED STRUCTURAL METAL MEMBERS										
Isolated unfabricated structural members										
Weldable steel, BS EN 10025 : 1993 Grade S275 (formerly BS 7668 Grade 43A), hot rolled sections										
plain member; beam										
weight less than 40 Kg/m	950.00	-	**950.00**	50.00	-	**710.50**	-	**1660.50**	t	**1868.06**
weight 40-100 Kg/m	950.00	-	**950.00**	47.05	-	**668.58**	-	**1618.58**	t	**1820.90**
weight exceeding 100 Kg/m	950.00	-	**950.00**	41.20	-	**585.45**	-	**1535.45**	t	**1727.38**
Steel short span lattice joists primed at works; treated timber inserts in top and bottom chords; full depth end seatings for bolting to supporting structure; hoist and fix 3.00m above ground level										
200mm deep, 8.2 kg/m	46.11	2.50	**47.26**	0.76	0.09	**11.76**	-	**59.02**	m	**66.40**
250mm deep, 8.5 kg/m	48.02	2.50	**49.22**	0.78	0.10	**12.15**	-	**61.38**	m	**69.05**
300mm deep, 10.7 kg/m	60.66	2.50	**62.18**	0.86	0.10	**13.29**	-	**75.47**	m	**84.90**
350mm deep, 11.6 kg/m	65.24	2.50	**66.87**	0.93	0.12	**14.50**	-	**81.37**	m	**91.54**
350mm deep, 12.8 kg/m	71.27	2.50	**73.06**	1.04	0.13	**16.17**	-	**89.23**	m	**100.38**

Labour hourly rates: (except Specialists) Craft Operatives £14.21 Labourer £10.71 Rates are national average prices. Refer to REGIONAL VARIATIONS for indicative levels of overall pricing in regions	MATERIALS			LABOUR				RATES		
	Del to Site £	Waste %	Material Cost £	Craft Optve Hrs	Lab Hrs	Labour Cost £	Sunds £	Nett Rate £	Unit	Gross Rate (+12.5%) £
G12: ISOLATED STRUCTURAL METAL MEMBERS Cont.										
Isolated unfabricated structural members Cont.										
Steel short span lattice joists primed at works; treated timber inserts in top and bottom chords; full depth end seatings for bolting to supporting structure; hoist and fix 6.00m above ground level										
200mm deep, 8.2 kg/m	46.11	2.50	47.26	0.83	0.10	12.87	-	60.12	m	67.64
250mm deep, 8.5 kg/m	48.02	2.50	49.22	0.86	0.10	13.29	-	62.52	m	70.33
300mm deep, 10.7 kg/m	60.66	2.50	62.18	0.94	0.12	14.64	-	76.82	m	86.42
350mm deep, 11.6 kg/m	65.24	2.50	66.87	1.02	0.13	15.89	-	82.76	m	93.10
350mm deep, 12.8 kg/m	71.27	2.50	73.06	1.13	0.14	17.56	-	90.61	m	101.94
Steel intermediate span lattice joists primed at works; treated timber inserts in top and bottom chords; full depth end seatings for bolting to supporting structure; hoist and fix 3.00m above ground level										
450mm deep, 12.2 kg/m	62.34	2.50	63.90	0.98	0.13	15.32	-	79.22	m	89.12
500mm deep, 15.8 kg/m	77.10	2.50	79.02	1.18	0.15	18.37	-	97.40	m	109.57
550mm deep, 19.4 kg/m	91.14	2.50	93.42	1.45	0.18	22.53	-	115.95	m	130.45
600mm deep, 22.5 kg/m	104.32	2.50	106.92	1.55	0.20	24.17	-	131.09	m	147.48
650mm deep, 29.7 kg/m	137.40	2.50	140.84	2.05	0.25	31.81	-	172.65	m	194.23
Steel intermediate span lattice joists primed at works; treated timber inserts in top and bottom chords; full depth end seatings for bolting to supporting structure; hoist and fix 6.00m above ground level										
450mm deep, 12.2 kg/m	62.34	2.50	63.90	1.08	0.14	16.85	-	80.75	m	90.84
500mm deep, 15.8 kg/m	77.10	2.50	79.02	1.30	0.16	20.19	-	99.21	m	111.61
550mm deep, 19.4 kg/m	91.14	2.50	93.42	1.60	0.20	24.88	-	118.30	m	133.09
600mm deep, 22.5 kg/m	104.32	2.50	106.92	1.71	0.22	26.66	-	133.58	m	150.28
650mm deep, 29.7 kg/m	137.40	2.50	140.84	2.25	0.29	35.08	-	175.92	m	197.90
Steel long span lattice joists primed at works; treated timber inserts in top and bottom chords; full depth end seatings for bolting to supporting structure; hoist and fix 3.00m above ground level										
700mm deep, 39.2 kg/m	177.69	2.50	182.13	2.25	0.29	35.08	-	217.21	m	244.36
800mm deep, 44.1 kg/m	190.25	2.50	195.01	2.54	0.32	39.52	-	234.53	m	263.84
900mm deep, 45.3 kg/m	191.32	2.50	196.11	2.61	0.32	40.52	-	236.62	m	266.20
1000mm deep, 46.1 kg/m	195.46	2.50	200.34	2.66	0.33	41.33	-	241.68	m	271.89
1500mm deep, 54.2 kg/m	231.15	2.50	236.93	2.76	0.35	42.97	-	279.90	m	314.88
Steel long span lattice joists primed at works; treated timber inserts in top and bottom chords; full depth end seatings for bolting to supporting structure; hoist and fix 6.00m above ground level										
700mm deep, 39.2 kg/m	177.69	2.50	182.13	2.48	0.31	38.56	-	220.69	m	248.28
800mm deep, 44.1 kg/m	190.25	2.50	195.01	2.79	0.35	43.39	-	238.40	m	268.20
900mm deep, 45.3 kg/m	191.32	2.50	196.11	2.88	0.36	44.78	-	240.89	m	271.00
1000mm deep, 46.1 kg/m	195.46	2.50	200.34	2.92	0.37	45.46	-	245.80	m	276.53
1500mm deep, 54.2 kg/m	231.15	2.50	236.93	3.04	0.38	47.27	-	284.20	m	319.72
Fixing bolts; steel										
Black bolts, BS 4190 grade 4.6										
M 8; with nuts and washers										
30mm long	0.14	5.00	0.15	0.07	-	0.99	-	1.15	nr	1.29
40mm long	0.16	5.00	0.16	0.07	-	0.99	-	1.16	nr	1.30
50mm long	0.17	5.00	0.18	0.07	-	0.99	-	1.17	nr	1.32
60mm long	0.19	5.00	0.20	0.07	-	0.99	-	1.19	nr	1.34
70mm long	0.20	5.00	0.21	0.08	-	1.14	-	1.35	nr	1.52
80mm long	0.22	5.00	0.23	0.08	-	1.14	-	1.37	nr	1.54
90mm long	0.26	5.00	0.27	0.08	-	1.14	-	1.40	nr	1.58
100mm long	0.28	5.00	0.29	0.08	-	1.14	-	1.43	nr	1.61
120mm long	0.33	5.00	0.35	0.09	-	1.28	-	1.63	nr	1.83
M 10; with nuts and washers										
40mm long	0.22	5.00	0.23	0.07	-	0.99	-	1.23	nr	1.38
50mm long	0.23	5.00	0.24	0.07	-	0.99	-	1.24	nr	1.39
60mm long	0.27	5.00	0.28	0.07	-	0.99	-	1.27	nr	1.43
70mm long	0.28	5.00	0.29	0.08	-	1.14	-	1.43	nr	1.61
80mm long	0.34	5.00	0.36	0.08	-	1.14	-	1.50	nr	1.69
90mm long	0.37	5.00	0.38	0.08	-	1.14	-	1.52	nr	1.71
100mm long	0.42	5.00	0.44	0.08	-	1.14	-	1.58	nr	1.78
120mm long	0.46	5.00	0.48	0.09	-	1.28	-	1.76	nr	1.98
140mm long	0.59	5.00	0.62	0.10	-	1.42	-	2.04	nr	2.29
150mm long	0.67	5.00	0.70	0.12	-	1.71	-	2.40	nr	2.71
M 12; with nuts and washers										
40mm long	0.32	5.00	0.34	0.08	-	1.14	-	1.47	nr	1.66
50mm long	0.33	5.00	0.35	0.08	-	1.14	-	1.49	nr	1.67
60mm long	0.36	5.00	0.37	0.08	-	1.14	-	1.51	nr	1.70
70mm long	0.41	5.00	0.43	0.08	-	1.14	-	1.57	nr	1.76
80mm long	0.46	5.00	0.48	0.08	-	1.14	-	1.61	nr	1.82
90mm long	0.47	5.00	0.49	0.08	-	1.14	-	1.63	nr	1.83
100mm long	0.51	5.00	0.54	0.09	-	1.28	-	1.82	nr	2.04
120mm long	0.61	5.00	0.64	0.09	-	1.28	-	1.92	nr	2.16
140mm long	0.76	5.00	0.79	0.12	-	1.71	-	2.50	nr	2.81
150mm long	0.96	5.00	1.00	0.12	-	1.71	-	2.71	nr	3.05
160mm long	1.38	5.00	1.44	0.12	-	1.71	-	3.15	nr	3.54
180mm long	1.44	5.00	1.52	0.13	-	1.85	-	3.36	nr	3.78
200mm long	1.51	5.00	1.59	0.14	-	1.99	-	3.57	nr	4.02
220mm long	1.59	5.00	1.67	0.15	-	2.13	-	3.80	nr	4.27
240mm long	1.65	5.00	1.74	0.16	-	2.27	-	4.01	nr	4.51

Labour hourly rates: (except Specialists) Craft Operatives £14.21 Labourer £10.71 Rates are national average prices. Refer to REGIONAL VARIATIONS for indicative levels of overall pricing in regions	MATERIALS			LABOUR				RATES		
	Del to Site £	Waste %	Material Cost £	Craft Optve Hrs	Lab Hrs	Labour Cost £	Sunds £	Nett Rate £	Unit	Gross Rate (+12.5%) £
G12: ISOLATED STRUCTURAL METAL MEMBERS Cont.										
Fixing bolts; steel Cont.										
Black bolts, BS 4190 grade 4.6 Cont.										
M 12; with nuts and washers Cont.										
260mm long	1.73	5.00	**1.82**	0.17	-	**2.42**	-	**4.23**	nr	**4.76**
280mm long	1.82	5.00	**1.91**	0.18	-	**2.56**	-	**4.47**	nr	**5.03**
300mm long	1.89	5.00	**1.98**	0.20	-	**2.84**	-	**4.82**	nr	**5.43**
M 16; with nuts and washers										
50mm long	0.59	5.00	**0.62**	0.08	-	**1.14**	-	**1.75**	nr	**1.97**
60mm long	0.66	5.00	**0.69**	0.08	-	**1.14**	-	**1.82**	nr	**2.05**
70mm long	0.72	5.00	**0.76**	0.08	-	**1.14**	-	**1.89**	nr	**2.13**
80mm long	0.79	5.00	**0.83**	0.09	-	**1.28**	-	**2.11**	nr	**2.37**
90mm long	0.84	5.00	**0.89**	0.09	-	**1.28**	-	**2.17**	nr	**2.44**
100mm long	0.99	5.00	**1.04**	0.10	-	**1.42**	-	**2.46**	nr	**2.77**
120mm long	1.14	5.00	**1.20**	0.10	-	**1.42**	-	**2.62**	nr	**2.95**
140mm long	1.28	5.00	**1.34**	0.12	-	**1.71**	-	**3.05**	nr	**3.43**
150mm long	1.34	5.00	**1.41**	0.13	-	**1.85**	-	**3.26**	nr	**3.66**
160mm long	1.87	5.00	**1.96**	0.14	-	**1.99**	-	**3.95**	nr	**4.44**
180mm long	1.94	5.00	**2.04**	0.14	-	**1.99**	-	**4.03**	nr	**4.53**
200mm long	2.07	5.00	**2.17**	0.15	-	**2.13**	-	**4.30**	nr	**4.84**
220mm long	2.19	5.00	**2.30**	0.15	-	**2.13**	-	**4.43**	nr	**4.98**
M 20; with nuts and washers										
60mm long	1.11	5.00	**1.17**	0.08	-	**1.14**	-	**2.30**	nr	**2.59**
70mm long	1.22	5.00	**1.28**	0.08	-	**1.14**	-	**2.42**	nr	**2.72**
80mm long	1.39	5.00	**1.46**	0.09	-	**1.28**	-	**2.74**	nr	**3.08**
90mm long	1.40	5.00	**1.47**	0.10	-	**1.42**	-	**2.89**	nr	**3.25**
100mm long	1.60	5.00	**1.68**	0.10	-	**1.42**	-	**3.10**	nr	**3.49**
120mm long	1.88	5.00	**1.97**	0.12	-	**1.71**	-	**3.68**	nr	**4.13**
140mm long	2.19	5.00	**2.30**	0.13	-	**1.85**	-	**4.14**	nr	**4.66**
150mm long	2.24	5.00	**2.35**	0.14	-	**1.99**	-	**4.34**	nr	**4.89**
160mm long	2.75	5.00	**2.89**	0.15	-	**2.13**	-	**5.02**	nr	**5.65**
180mm long	2.91	5.00	**3.05**	0.15	-	**2.13**	-	**5.18**	nr	**5.83**
200mm long	3.11	5.00	**3.26**	0.16	-	**2.27**	-	**5.54**	nr	**6.23**
220mm long	3.22	5.00	**3.38**	0.17	-	**2.42**	-	**5.80**	nr	**6.52**
240mm long	3.41	5.00	**3.58**	0.18	-	**2.56**	-	**6.14**	nr	**6.90**
260mm long	3.51	5.00	**3.68**	0.20	-	**2.84**	-	**6.53**	nr	**7.34**
280mm long	3.60	5.00	**3.78**	0.21	-	**2.98**	-	**6.76**	nr	**7.60**
300mm long	3.81	5.00	**4.00**	0.23	-	**3.27**	-	**7.27**	nr	**8.17**
M 24; with nuts and washers										
70mm long	2.79	5.00	**2.93**	0.10	-	**1.42**	-	**4.35**	nr	**4.89**
80mm long	2.91	5.00	**3.05**	0.10	-	**1.42**	-	**4.47**	nr	**5.03**
90mm long	3.00	5.00	**3.15**	0.12	-	**1.71**	-	**4.85**	nr	**5.46**
100mm long	3.11	5.00	**3.26**	0.12	-	**1.71**	-	**4.97**	nr	**5.59**
120mm long	3.32	5.00	**3.48**	0.12	-	**1.71**	-	**5.19**	nr	**5.84**
140mm long	3.73	5.00	**3.92**	0.13	-	**1.85**	-	**5.76**	nr	**6.48**
150mm long	3.95	5.00	**4.15**	0.15	-	**2.13**	-	**6.28**	nr	**7.07**
160mm long	4.09	5.00	**4.29**	0.13	-	**1.85**	-	**6.14**	nr	**6.90**
180mm long	4.20	5.00	**4.41**	0.15	-	**2.13**	-	**6.54**	nr	**7.35**
200mm long	4.40	5.00	**4.62**	0.16	-	**2.27**	-	**6.89**	nr	**7.75**
220mm long	4.85	5.00	**5.09**	0.17	-	**2.42**	-	**7.51**	nr	**8.45**
240mm long	4.98	5.00	**5.23**	0.18	-	**2.56**	-	**7.79**	nr	**8.76**
260mm long	5.32	5.00	**5.58**	0.21	-	**2.98**	-	**8.57**	nr	**9.64**
280mm long	5.48	5.00	**5.76**	0.23	-	**3.27**	-	**9.03**	nr	**10.15**
300mm long	5.65	5.00	**5.93**	0.25	-	**3.55**	-	**9.49**	nr	**10.67**
Black bolts, BS 4190 grade 4.6; galvanized										
M 8; with nuts and washers										
30mm long	0.20	5.00	**0.21**	0.07	-	**0.99**	-	**1.20**	nr	**1.36**
40mm long	0.22	5.00	**0.23**	0.07	-	**0.99**	-	**1.23**	nr	**1.38**
50mm long	0.24	5.00	**0.26**	0.07	-	**0.99**	-	**1.25**	nr	**1.41**
60mm long	0.27	5.00	**0.28**	0.07	-	**0.99**	-	**1.27**	nr	**1.43**
70mm long	0.29	5.00	**0.30**	0.08	-	**1.14**	-	**1.44**	nr	**1.62**
80mm long	0.31	5.00	**0.33**	0.08	-	**1.14**	-	**1.46**	nr	**1.65**
90mm long	0.34	5.00	**0.36**	0.08	-	**1.14**	-	**1.50**	nr	**1.69**
100mm long	0.42	5.00	**0.44**	0.08	-	**1.14**	-	**1.58**	nr	**1.78**
120mm long	0.51	5.00	**0.54**	0.09	-	**1.28**	-	**1.82**	nr	**2.04**
M 10; with nuts and washers										
40mm long	0.31	5.00	**0.33**	0.07	-	**0.99**	-	**1.32**	nr	**1.49**
50mm long	0.34	5.00	**0.36**	0.07	-	**0.99**	-	**1.36**	nr	**1.53**
60mm long	0.39	5.00	**0.41**	0.07	-	**0.99**	-	**1.40**	nr	**1.58**
70mm long	0.42	5.00	**0.44**	0.08	-	**1.14**	-	**1.58**	nr	**1.78**
80mm long	0.52	5.00	**0.55**	0.08	-	**1.14**	-	**1.68**	nr	**1.90**
90mm long	0.58	5.00	**0.61**	0.08	-	**1.14**	-	**1.74**	nr	**1.96**
100mm long	0.62	5.00	**0.65**	0.08	-	**1.14**	-	**1.79**	nr	**2.01**
120mm long	0.67	5.00	**0.70**	0.09	-	**1.28**	-	**1.98**	nr	**2.23**
140mm long	0.87	5.00	**0.91**	0.10	-	**1.42**	-	**2.33**	nr	**2.62**
150mm long	1.00	5.00	**1.05**	0.12	-	**1.71**	-	**2.75**	nr	**3.10**
M 12; with nuts and washers										
40mm long	0.52	5.00	**0.55**	0.08	-	**1.14**	-	**1.68**	nr	**1.90**
50mm long	0.54	5.00	**0.57**	0.08	-	**1.14**	-	**1.71**	nr	**1.92**
60mm long	0.59	5.00	**0.62**	0.08	-	**1.14**	-	**1.75**	nr	**1.97**
70mm long	0.66	5.00	**0.69**	0.08	-	**1.14**	-	**1.82**	nr	**2.05**
80mm long	0.71	5.00	**0.75**	0.08	-	**1.14**	-	**1.88**	nr	**2.12**
90mm long	0.73	5.00	**0.77**	0.08	-	**1.14**	-	**1.91**	nr	**2.14**
100mm long	0.80	5.00	**0.84**	0.09	-	**1.28**	-	**2.12**	nr	**2.38**
120mm long	0.99	5.00	**1.04**	0.09	-	**1.28**	-	**2.32**	nr	**2.61**
140mm long	1.15	5.00	**1.21**	0.12	-	**1.71**	-	**2.92**	nr	**3.28**
150mm long	1.47	5.00	**1.54**	0.12	-	**1.71**	-	**3.24**	nr	**3.65**
160mm long	2.07	5.00	**2.17**	0.12	-	**1.71**	-	**3.87**	nr	**4.36**
180mm long	2.18	5.00	**2.28**	0.13	-	**1.85**	-	**4.13**	nr	**4.65**
200mm long	2.26	5.00	**2.38**	0.14	-	**1.99**	-	**4.37**	nr	**4.91**
220mm long	2.39	5.00	**2.51**	0.15	-	**2.13**	-	**4.64**	nr	**5.22**
240mm long	2.48	5.00	**2.60**	0.16	-	**2.27**	-	**4.87**	nr	**5.48**
260mm long	2.62	5.00	**2.75**	0.17	-	**2.42**	-	**5.17**	nr	**5.81**

Labour hourly rates: (except Specialists) Craft Operatives £14.21 Labourer £10.71 Rates are national average prices. Refer to REGIONAL VARIATIONS for indicative levels of overall pricing in regions	MATERIALS			LABOUR				RATES		
	Del to Site £	Waste %	Material Cost £	Craft Optve Hrs	Lab Hrs	Labour Cost £	Sunds £	Nett Rate £	Unit	Gross Rate (+12.5%) £
G12: ISOLATED STRUCTURAL METAL MEMBERS Cont.										
Fixing bolts; steel Cont.										
Black bolts, BS 4190 grade 4.6; galvanized Cont.										
M 12; with nuts and washers Cont.										
280mm long	2.71	5.00	**2.84**	0.18	-	**2.56**	-	**5.40**	nr	**6.08**
300mm long	2.82	5.00	**2.96**	0.20	-	**2.84**	-	**5.80**	nr	**6.53**
M 16; with nuts and washers										
50mm long	0.94	5.00	**0.99**	0.08	-	**1.14**	-	**2.13**	nr	**2.39**
60mm long	1.02	5.00	**1.07**	0.08	-	**1.14**	-	**2.21**	nr	**2.48**
70mm long	1.11	5.00	**1.17**	0.08	-	**1.14**	-	**2.30**	nr	**2.59**
80mm long	1.20	5.00	**1.26**	0.09	-	**1.28**	-	**2.54**	nr	**2.86**
90mm long	1.27	5.00	**1.33**	0.09	-	**1.28**	-	**2.61**	nr	**2.93**
100mm long	1.33	5.00	**1.40**	0.10	-	**1.42**	-	**2.82**	nr	**3.17**
120mm long	1.73	5.00	**1.82**	0.10	-	**1.42**	-	**3.24**	nr	**3.64**
140mm long	1.93	5.00	**2.03**	0.12	-	**1.71**	-	**3.73**	nr	**4.20**
150mm long	2.05	5.00	**2.16**	0.13	-	**1.85**	-	**4.00**	nr	**4.50**
160mm long	2.78	5.00	**2.91**	0.14	-	**1.99**	-	**4.90**	nr	**5.52**
180mm long	2.94	5.00	**3.09**	0.14	-	**1.99**	-	**5.08**	nr	**5.71**
200mm long	3.13	5.00	**3.29**	0.15	-	**2.13**	-	**5.42**	nr	**6.10**
220mm long	3.26	5.00	**3.43**	0.15	-	**2.13**	-	**5.56**	nr	**6.25**
M 20; with nuts and washers										
60mm long	1.71	5.00	**1.79**	0.08	-	**1.14**	-	**2.93**	nr	**3.30**
70mm long	1.88	5.00	**1.97**	0.08	-	**1.14**	-	**3.11**	nr	**3.49**
80mm long	2.12	5.00	**2.23**	0.09	-	**1.28**	-	**3.50**	nr	**3.94**
90mm long	2.13	5.00	**2.24**	0.10	-	**1.42**	-	**3.66**	nr	**4.12**
100mm long	2.44	5.00	**2.56**	0.10	-	**1.42**	-	**3.99**	nr	**4.48**
120mm long	2.82	5.00	**2.96**	0.12	-	**1.71**	-	**4.67**	nr	**5.25**
140mm long	3.30	5.00	**3.46**	0.13	-	**1.85**	-	**5.31**	nr	**5.97**
150mm long	3.37	5.00	**3.54**	0.14	-	**1.99**	-	**5.53**	nr	**6.22**
160mm long	4.17	5.00	**4.38**	0.15	-	**2.13**	-	**6.51**	nr	**7.33**
180mm long	4.31	5.00	**4.52**	0.15	-	**2.13**	-	**6.65**	nr	**7.49**
200mm long	4.58	5.00	**4.81**	0.16	-	**2.27**	-	**7.09**	nr	**7.97**
220mm long	4.77	5.00	**5.01**	0.17	-	**2.42**	-	**7.43**	nr	**8.36**
240mm long	5.06	5.00	**5.32**	0.18	-	**2.56**	-	**7.87**	nr	**8.86**
260mm long	5.18	5.00	**5.44**	0.20	-	**2.84**	-	**8.29**	nr	**9.32**
280mm long	5.32	5.00	**5.58**	0.21	-	**2.98**	-	**8.57**	nr	**9.64**
300mm long	5.64	5.00	**5.92**	0.23	-	**3.27**	-	**9.19**	nr	**10.34**
M 24; with nuts and washers										
70mm long	4.12	5.00	**4.32**	0.10	-	**1.42**	-	**5.74**	nr	**6.46**
80mm long	4.31	5.00	**4.52**	0.10	-	**1.42**	-	**5.94**	nr	**6.69**
90mm long	4.44	5.00	**4.66**	0.12	-	**1.71**	-	**6.37**	nr	**7.16**
100mm long	4.61	5.00	**4.84**	0.12	-	**1.71**	-	**6.54**	nr	**7.36**
120mm long	4.86	5.00	**5.11**	0.12	-	**1.71**	-	**6.81**	nr	**7.66**
140mm long	5.52	5.00	**5.79**	0.13	-	**1.85**	-	**7.64**	nr	**8.60**
150mm long	5.86	5.00	**6.15**	0.15	-	**2.13**	-	**8.29**	nr	**9.32**
160mm long	6.05	5.00	**6.35**	0.13	-	**1.85**	-	**8.20**	nr	**9.22**
180mm long	6.19	5.00	**6.50**	0.15	-	**2.13**	-	**8.64**	nr	**9.71**
200mm long	6.26	5.00	**6.57**	0.16	-	**2.27**	-	**8.85**	nr	**9.95**
220mm long	6.84	5.00	**7.18**	0.17	-	**2.42**	-	**9.60**	nr	**10.80**
240mm long	6.98	5.00	**7.33**	0.18	-	**2.56**	-	**9.89**	nr	**11.13**
260mm long	7.34	5.00	**7.70**	0.21	-	**2.98**	-	**10.69**	nr	**12.02**
280mm long	7.53	5.00	**7.90**	0.25	-	**3.55**	-	**11.45**	nr	**12.89**
300mm long	7.77	5.00	**8.16**	0.27	-	**3.84**	-	**12.00**	nr	**13.49**
G20: CARPENTRY/TIMBER FRAMING/FIRST FIXING										
Sasco Trussed Rafters in softwood, sawn, BS 4978, GS grade, impregnated; note:- prices are guide prices only, based on contracts within 30 mile radius of Widnes and are subject to adjustment for quantity: firm quotations should always be obtained										
Trussed rafters; Hydro-Nail plated joints										
22.5 degree Standard duo pitch; 450mm overhangs; fixing with clips (included elsewhere); span over wall plates										
5000mm	40.01	2.50	**41.01**	1.30	0.16	**20.19**	2.18	**63.38**	nr	**71.30**
6000mm	46.04	2.50	**47.20**	1.40	0.17	**21.71**	2.38	**71.29**	nr	**80.20**
7000mm	51.52	2.50	**52.81**	1.50	0.19	**23.35**	2.57	**78.74**	nr	**88.58**
8000mm	61.00	2.50	**62.52**	1.60	0.20	**24.88**	2.79	**90.19**	nr	**101.46**
9000mm	70.61	2.50	**72.37**	1.70	0.21	**26.41**	2.97	**101.75**	nr	**114.47**
10000mm	85.49	2.50	**87.63**	1.80	0.22	**27.93**	3.17	**118.73**	nr	**133.57**
35 degree Standard duo pitch; 450mm overhangs; fixing with clips (included elsewhere); span over wall plates										
5000mm	41.59	2.50	**42.63**	1.30	0.16	**20.19**	2.18	**65.00**	nr	**73.12**
6000mm	48.07	2.50	**49.27**	1.40	0.17	**21.71**	2.38	**73.36**	nr	**82.53**
7000mm	53.56	2.50	**54.90**	1.50	0.19	**23.35**	2.57	**80.83**	nr	**90.93**
8000mm	64.30	2.50	**65.90**	1.60	0.20	**24.88**	2.79	**93.57**	nr	**105.27**
9000mm	71.57	2.50	**73.36**	1.70	0.21	**26.41**	2.97	**102.73**	nr	**115.57**
45 degree Standard duo pitch; 450mm overhangs; fixing with clips (included elsewhere); span over wall plates										
5000mm	65.29	2.50	**66.92**	1.30	0.16	**20.19**	2.18	**89.29**	nr	**100.45**
6000mm	78.13	2.50	**80.08**	1.40	0.17	**21.71**	2.38	**104.17**	nr	**117.20**
7000mm	91.13	2.50	**93.41**	1.50	0.19	**23.35**	2.57	**119.33**	nr	**134.25**
8000mm	104.20	2.50	**106.81**	1.60	0.20	**24.88**	2.79	**134.47**	nr	**151.28**
9000mm	117.42	2.50	**120.35**	1.70	0.21	**26.41**	2.97	**149.73**	nr	**168.44**
22.5 degree Bobtail duo pitch; 450mm overhangs; fixing with clips (included elsewhere); span over wall plates										
4000mm	43.99	2.50	**45.09**	1.20	0.15	**18.66**	1.98	**65.72**	nr	**73.94**
5000mm	54.82	2.50	**56.19**	1.30	0.16	**20.19**	2.18	**78.56**	nr	**88.38**
6000mm	65.86	2.50	**67.50**	1.40	0.17	**21.71**	2.38	**91.59**	nr	**103.04**

Labour hourly rates: (except Specialists) Craft Operatives £14.21 Labourer £10.71 Rates are national average prices. Refer to REGIONAL VARIATIONS for indicative levels of overall pricing in regions	MATERIALS			LABOUR				RATES		
	Del to Site £	Waste %	Material Cost £	Craft Optve Hrs	Lab Hrs	Labour Cost £	Sunds £	Nett Rate £	Unit	Gross Rate (+12.5%) £
G20: CARPENTRY/TIMBER FRAMING/FIRST FIXING Cont.										
Sasco Trussed Rafters in softwood, sawn, BS 4978, GS grade, impregnated; note:- prices are guide prices only, based on contracts within 30 mile radius of Widnes and are subject to adjustment for quantity: firm quotations should always be obtained Cont.										
Trussed rafters; Hydro-Nail plated joints Cont.										
22.5 degree Bobtail duo pitch; 450mm overhangs; fixing with clips (included elsewhere); span over wall plates Cont.										
7000mm	76.92	2.50	**78.85**	1.50	0.19	**23.35**	2.57	**104.77**	nr	**117.87**
8000mm	87.97	2.50	**90.17**	1.60	0.20	**24.88**	2.79	**117.84**	nr	**132.57**
9000mm	99.13	2.50	**101.61**	1.70	0.21	**26.41**	2.97	**130.98**	nr	**147.36**
35 degree Bobtail duo pitch; 450mm overhangs; fixing with clips (included elsewhere); span over wall plates										
4000mm	45.88	2.50	**47.03**	1.20	0.15	**18.66**	1.98	**67.67**	nr	**76.13**
5000mm	57.18	2.50	**58.61**	1.30	0.16	**20.19**	2.18	**80.98**	nr	**91.10**
6000mm	68.80	2.50	**70.52**	1.40	0.17	**21.71**	2.38	**94.61**	nr	**106.44**
7000mm	79.96	2.50	**81.96**	1.50	0.19	**23.35**	2.57	**107.88**	nr	**121.36**
8000mm	91.45	2.50	**93.74**	1.60	0.20	**24.88**	2.79	**121.40**	nr	**136.58**
9000mm	102.77	2.50	**105.34**	1.70	0.21	**26.41**	2.97	**134.71**	nr	**151.55**
45 degree Bobtail duo pitch; 450mm overhangs; fixing with clips (included elsewhere); span over wall plates										
4000mm	71.18	2.50	**72.96**	1.20	0.15	**18.66**	1.98	**93.59**	nr	**105.29**
5000mm	88.90	2.50	**91.12**	1.30	0.16	**20.19**	2.18	**113.48**	nr	**127.67**
6000mm	106.81	2.50	**109.48**	1.40	0.17	**21.71**	2.38	**133.57**	nr	**150.27**
7000mm	124.62	2.50	**127.73**	1.50	0.19	**23.35**	2.57	**153.66**	nr	**172.86**
8000mm	153.71	2.50	**157.55**	1.60	0.20	**24.88**	2.79	**185.21**	nr	**208.37**
9000mm	160.27	2.50	**164.27**	1.70	0.21	**26.41**	2.97	**193.65**	nr	**217.86**
22.5 degree Monopitch; 450mm overhangs; fixing with clips (included elsewhere); span over wall plates										
2000mm	21.53	2.50	**22.07**	1.00	0.13	**15.60**	1.58	**39.26**	nr	**44.16**
3000mm	28.52	2.50	**29.23**	1.10	0.14	**17.13**	1.78	**48.15**	nr	**54.16**
4000mm	35.51	2.50	**36.40**	1.20	0.15	**18.66**	1.98	**57.03**	nr	**64.16**
5000mm	43.40	2.50	**44.49**	1.30	0.16	**20.19**	2.18	**66.85**	nr	**75.21**
6000mm	56.05	2.50	**57.45**	1.40	0.17	**21.71**	2.38	**81.54**	nr	**91.73**
35 degree Monopitch; 450mm overhangs; fixing with clips (included elsewhere); span over wall plates										
2000mm	23.02	2.50	**23.60**	1.00	0.13	**15.60**	1.58	**40.78**	nr	**45.88**
3000mm	30.33	2.50	**31.09**	1.10	0.14	**17.13**	1.78	**50.00**	nr	**56.25**
4000mm	37.78	2.50	**38.72**	1.20	0.15	**18.66**	1.98	**59.36**	nr	**66.78**
5000mm	46.59	2.50	**47.76**	1.30	0.16	**20.19**	2.18	**70.12**	nr	**78.89**
6000mm	60.02	2.50	**61.52**	1.40	0.17	**21.71**	2.38	**85.61**	nr	**96.31**
45 degree Monopitch; 450mm overhangs; fixing with clips (included elsewhere); span over wall plates										
2000mm	24.58	2.50	**25.20**	1.00	0.13	**15.60**	1.58	**42.38**	nr	**47.68**
3000mm	46.82	2.50	**48.00**	1.10	0.14	**17.13**	1.78	**66.91**	nr	**75.27**
4000mm	55.37	2.50	**56.76**	1.20	0.15	**18.66**	1.98	**77.40**	nr	**87.07**
22.5 degree duo pitch Girder Truss; fixing with clips (included elsewhere); span over wall plates										
5000mm	161.60	2.50	**165.64**	1.30	0.16	**20.19**	2.18	**188.00**	nr	**211.50**
6000mm	182.92	2.50	**187.49**	1.40	0.17	**21.71**	2.38	**211.58**	nr	**238.03**
7000mm	198.26	2.50	**203.22**	1.50	0.19	**23.35**	2.57	**229.14**	nr	**257.78**
8000mm	229.96	2.50	**235.70**	1.60	0.20	**24.88**	2.79	**263.37**	nr	**296.29**
9000mm	255.66	2.50	**262.05**	1.70	0.21	**26.41**	2.97	**291.42**	nr	**327.85**
10000mm	301.26	2.50	**308.79**	1.80	0.22	**27.93**	3.17	**339.89**	nr	**382.38**
35 degree duo pitch Girder Truss; fixing with clips (included elsewhere); span over wall plates										
5000mm	169.28	2.50	**173.51**	1.30	0.16	**20.19**	2.18	**195.87**	nr	**220.36**
6000mm	188.56	2.50	**193.27**	1.40	0.17	**21.71**	2.38	**217.36**	nr	**244.53**
7000mm	206.17	2.50	**211.32**	1.50	0.19	**23.35**	2.57	**237.25**	nr	**266.90**
8000mm	237.87	2.50	**243.81**	1.60	0.20	**24.88**	2.79	**271.48**	nr	**305.41**
9000mm	263.55	2.50	**270.14**	1.70	0.21	**26.41**	2.97	**299.51**	nr	**336.95**
10000mm	308.99	2.50	**316.71**	1.80	0.22	**27.93**	3.17	**347.82**	nr	**391.29**
45 degree duo pitch Girder Truss; fixing with clips (included elsewhere); span over wall plates										
5000mm	229.16	2.50	**234.89**	1.30	0.16	**20.19**	2.18	**257.25**	nr	**289.41**
6000mm	278.57	2.50	**285.54**	1.40	0.17	**21.71**	2.38	**309.63**	nr	**348.33**
7000mm	319.15	2.50	**327.13**	1.50	0.19	**23.35**	2.57	**353.06**	nr	**397.19**
8000mm	357.50	2.50	**366.44**	1.60	0.20	**24.88**	2.79	**394.10**	nr	**443.37**
9000mm	396.98	2.50	**406.91**	1.70	0.21	**26.41**	2.97	**436.28**	nr	**490.82**
22.5 degree pitch gable ladder; 450mm overhang; span over wall plate										
5000mm	21.32	2.50	**21.85**	1.30	0.16	**20.19**	2.18	**44.22**	nr	**49.74**
6000mm	24.35	2.50	**24.96**	1.40	0.17	**21.71**	2.38	**49.05**	nr	**55.18**
7000mm	27.30	2.50	**27.98**	1.50	0.19	**23.35**	2.57	**53.90**	nr	**60.64**
8000mm	32.72	2.50	**33.54**	1.60	0.20	**24.88**	2.79	**61.21**	nr	**68.86**
9000mm	37.48	2.50	**38.41**	1.70	0.21	**26.41**	2.97	**67.79**	nr	**76.26**
10000mm	45.46	2.50	**46.60**	1.80	0.22	**27.93**	3.17	**77.70**	nr	**87.41**
35 degree pitch gable ladder; 450mm overhang; span over wall plate										
5000mm	22.12	2.50	**22.67**	1.30	0.16	**20.19**	2.18	**45.04**	nr	**50.66**
6000mm	25.59	2.50	**26.23**	1.40	0.17	**21.71**	2.38	**50.32**	nr	**56.62**
7000mm	28.06	2.50	**28.76**	1.50	0.19	**23.35**	2.57	**54.68**	nr	**61.52**
8000mm	34.05	2.50	**34.91**	1.60	0.20	**24.88**	2.79	**62.57**	nr	**70.39**
9000mm	38.90	2.50	**39.87**	1.70	0.21	**26.41**	2.97	**69.24**	nr	**77.90**
10000mm	47.14	2.50	**48.32**	1.80	0.22	**27.93**	3.17	**79.42**	nr	**89.35**
45 degree pitch gable ladder; 450mm overhang; span over wall plate										
5000mm	34.75	2.50	**35.61**	1.30	0.16	**20.19**	2.18	**57.98**	nr	**65.23**
6000mm	41.59	2.50	**42.63**	1.40	0.17	**21.71**	2.38	**66.72**	nr	**75.06**

Labour hourly rates: (except Specialists) Craft Operatives £14.21 Labourer £10.71 Rates are national average prices. Refer to REGIONAL VARIATIONS for indicative levels of overall pricing in regions	MATERIALS			LABOUR				RATES		
	Del to Site £	Waste %	Material Cost £	Craft Optve Hrs	Lab Hrs	Labour Cost £	Sunds £	Nett Rate £	Unit	Gross Rate (+12.5%) £
G20: CARPENTRY/TIMBER FRAMING/FIRST FIXING Cont.										
Sasco Trussed Rafters in softwood, sawn, BS 4978, GS grade, impregnated; note:- prices are guide prices only, based on contracts within 30 mile radius of Widnes and are subject to adjustment for quantity: firm quotations should always be obtained Cont.										
Trussed rafters; Hydro-Nail plated joints Cont.										
45 degree pitch gable ladder; 450mm overhang; span over wall plate Cont.										
7000mm	48.37	2.50	**49.58**	1.50	0.19	**23.35**	2.57	**75.50**	nr	**84.94**
8000mm	54.38	2.50	**55.74**	1.60	0.20	**24.88**	2.79	**83.41**	nr	**93.83**
9000mm	61.83	2.50	**63.38**	1.70	0.21	**26.41**	2.97	**92.75**	nr	**104.35**
10000mm	68.14	2.50	**69.85**	1.80	0.22	**27.93**	3.17	**100.95**	nr	**113.57**
Softwood, wrot, BS 4978, GS grade - laminated beams										
Glued laminated beams; BS EN 386										
65 x 150 x 4000mm	65.07	2.50	**66.70**	0.60	0.07	**9.28**	1.25	**77.23**	nr	**86.88**
65 x 175 x 4000mm	75.80	2.50	**77.69**	0.80	0.10	**12.44**	1.40	**91.53**	nr	**102.97**
65 x 200 x 4000mm	86.49	2.50	**88.65**	0.90	0.11	**13.97**	1.57	**104.19**	nr	**117.21**
65 x 225 x 4000mm	97.70	2.50	**100.14**	1.10	0.14	**17.13**	1.95	**119.22**	nr	**134.12**
65 x 250 x 4000mm	108.92	2.50	**111.65**	1.20	0.15	**18.66**	2.11	**132.42**	nr	**148.97**
65 x 250 x 6000mm	163.30	2.50	**167.38**	1.30	0.16	**20.19**	2.29	**189.86**	nr	**213.60**
65 x 275 x 4000mm	119.56	2.50	**122.54**	1.30	0.16	**20.19**	2.29	**145.02**	nr	**163.15**
65 x 275 x 6000mm	181.65	2.50	**186.19**	1.40	0.18	**21.82**	2.46	**210.47**	nr	**236.78**
65 x 300 x 4000mm	128.91	2.50	**132.13**	1.40	0.18	**21.82**	2.46	**156.41**	nr	**175.96**
65 x 300 x 6000mm	194.48	2.50	**199.34**	1.50	0.19	**23.35**	2.66	**225.35**	nr	**253.52**
65 x 325 x 4000mm	143.19	2.50	**146.77**	1.60	0.20	**24.88**	2.82	**174.47**	nr	**196.27**
65 x 325 x 6000mm	210.54	2.50	**215.80**	1.70	0.21	**26.41**	3.00	**245.21**	nr	**275.86**
65 x 325 x 8000mm	281.63	2.50	**288.67**	1.80	0.22	**27.93**	3.17	**319.77**	nr	**359.75**
90 x 150 x 4000mm	82.51	2.50	**84.58**	0.90	0.11	**13.97**	1.57	**100.11**	nr	**112.62**
90 x 175 x 4000mm	97.24	2.50	**99.67**	1.10	0.14	**17.13**	1.95	**118.75**	nr	**133.59**
90 x 200 x 4000mm	110.62	2.50	**113.39**	1.20	0.15	**18.66**	2.11	**134.16**	nr	**150.93**
90 x 225 x 4000mm	123.55	2.50	**126.64**	1.30	0.16	**20.19**	2.29	**149.12**	nr	**167.75**
90 x 250 x 4000mm	139.51	2.50	**143.00**	1.40	0.18	**21.82**	2.46	**167.28**	nr	**188.19**
90 x 250 x 6000mm	206.10	2.50	**211.25**	1.50	0.19	**23.35**	2.66	**237.26**	nr	**266.92**
90 x 275 x 4000mm	152.55	2.50	**156.37**	1.50	0.19	**23.35**	2.66	**182.37**	nr	**205.17**
90 x 275 x 6000mm	229.73	2.50	**235.47**	1.60	0.20	**24.88**	2.82	**263.17**	nr	**296.07**
90 x 300 x 4000mm	165.04	2.50	**169.17**	1.70	0.21	**26.41**	3.00	**198.58**	nr	**223.40**
90 x 300 x 6000mm	251.57	2.50	**257.86**	1.80	0.22	**27.93**	3.17	**288.96**	nr	**325.08**
90 x 325 x 4000mm	182.43	2.50	**186.99**	1.80	0.22	**27.93**	3.17	**218.09**	nr	**245.35**
90 x 325 x 6000mm	271.20	2.50	**277.98**	1.90	0.24	**29.57**	3.37	**310.91**	nr	**349.78**
90 x 325 x 8000mm	359.07	2.50	**368.05**	2.00	0.25	**31.10**	3.51	**402.66**	nr	**453.00**
90 x 350 x 4000mm	192.70	2.50	**197.52**	1.90	0.24	**29.57**	3.37	**230.46**	nr	**259.26**
90 x 350 x 6000mm	292.62	2.50	**299.93**	2.00	0.25	**31.10**	3.51	**334.55**	nr	**376.36**
90 x 350 x 8000mm	390.75	2.50	**400.52**	2.10	0.26	**32.63**	3.70	**436.84**	nr	**491.45**
90 x 375 x 4000mm	206.10	2.50	**211.25**	2.00	0.25	**31.10**	3.51	**245.86**	nr	**276.60**
90 x 375 x 6000mm	310.45	2.50	**318.21**	2.10	0.26	**32.63**	3.70	**354.54**	nr	**398.85**
90 x 375 x 8000mm	416.18	2.50	**426.58**	2.20	0.28	**34.26**	3.86	**464.70**	nr	**522.79**
90 x 375 x 10000mm	520.12	2.50	**533.13**	2.30	0.29	**35.79**	4.04	**572.96**	nr	**644.58**
90 x 400 x 4000mm	223.92	2.50	**229.52**	2.20	0.28	**34.26**	3.86	**267.64**	nr	**301.10**
90 x 400 x 6000mm	331.87	2.50	**340.17**	2.30	0.29	**35.79**	4.04	**380.00**	nr	**427.50**
90 x 400 x 8000mm	443.39	2.50	**454.48**	2.40	0.30	**37.32**	4.22	**496.02**	nr	**558.02**
90 x 400 x 10000mm	553.56	2.50	**567.39**	2.50	0.31	**38.85**	4.41	**610.64**	nr	**686.98**
90 x 425 x 6000mm	371.13	2.50	**380.40**	2.40	0.30	**37.32**	4.22	**421.95**	nr	**474.69**
90 x 425 x 8000mm	467.47	2.50	**479.16**	2.50	0.31	**38.85**	4.41	**522.41**	nr	**587.71**
90 x 425 x 10000mm	582.56	2.50	**597.13**	2.60	0.33	**40.48**	4.57	**642.18**	nr	**722.45**
90 x 425 x 12000mm	699.42	2.50	**716.90**	2.70	0.34	**42.01**	4.75	**763.66**	nr	**859.12**
90 x 450 x 6000mm	374.70	2.50	**384.06**	2.50	0.31	**38.85**	4.41	**427.31**	nr	**480.73**
90 x 450 x 8000mm	498.70	2.50	**511.17**	2.60	0.33	**40.48**	4.57	**556.22**	nr	**625.75**
90 x 450 x 10000mm	621.80	2.50	**637.35**	2.70	0.34	**42.01**	4.75	**684.11**	nr	**769.62**
90 x 450 x 12000mm	747.60	2.50	**766.29**	2.80	0.35	**43.54**	4.93	**814.76**	nr	**916.61**
115 x 250 x 4000mm	176.20	2.50	**180.60**	1.70	0.21	**26.41**	3.00	**210.01**	nr	**236.26**
115 x 250 x 6000mm	260.59	2.50	**267.11**	1.80	0.22	**27.93**	3.17	**298.21**	nr	**335.49**
115 x 275 x 4000mm	191.36	2.50	**196.14**	1.80	0.22	**27.93**	3.17	**227.25**	nr	**255.65**
115 x 275 x 6000mm	292.62	2.50	**299.93**	1.90	0.24	**29.57**	3.37	**332.87**	nr	**374.48**
115 x 300 x 4000mm	211.88	2.50	**217.18**	1.90	0.24	**29.57**	3.37	**250.11**	nr	**281.38**
115 x 300 x 6000mm	318.04	2.50	**325.99**	2.00	0.25	**31.10**	3.51	**360.60**	nr	**405.68**
115 x 325 x 4000mm	229.73	2.50	**235.47**	2.00	0.25	**31.10**	3.51	**270.09**	nr	**303.85**
115 x 325 x 6000mm	345.26	2.50	**353.89**	2.10	0.26	**32.63**	3.70	**390.21**	nr	**438.98**
115 x 325 x 8000mm	461.20	2.50	**472.73**	2.20	0.28	**34.26**	3.86	**510.85**	nr	**574.71**
115 x 350 x 4000mm	251.12	2.50	**257.40**	2.20	0.28	**34.26**	3.86	**295.52**	nr	**332.46**
115 x 350 x 6000mm	374.70	2.50	**384.06**	2.30	0.29	**35.79**	4.04	**423.90**	nr	**476.88**
115 x 350 x 8000mm	498.70	2.50	**511.17**	2.40	0.30	**37.32**	4.22	**552.71**	nr	**621.80**
115 x 375 x 4000mm	268.97	2.50	**275.70**	2.30	0.29	**35.79**	4.04	**315.53**	nr	**354.97**
115 x 375 x 6000mm	400.57	2.50	**410.58**	2.40	0.30	**37.32**	4.22	**452.12**	nr	**508.64**
115 x 375 x 8000mm	533.90	2.50	**547.25**	2.50	0.31	**38.85**	4.41	**590.50**	nr	**664.31**
115 x 400 x 4000mm	284.60	2.50	**291.71**	2.40	0.30	**37.32**	4.22	**333.25**	nr	**374.91**
115 x 400 x 6000mm	428.22	2.50	**438.92**	2.50	0.31	**38.85**	4.41	**482.17**	nr	**542.44**
115 x 400 x 8000mm	570.96	2.50	**585.23**	2.60	0.33	**40.48**	4.57	**630.28**	nr	**709.07**
115 x 400 x 10000mm	714.59	2.50	**732.46**	2.70	0.34	**42.01**	4.75	**779.22**	nr	**876.62**
115 x 425 x 6000mm	443.39	2.50	**454.48**	2.60	0.33	**40.48**	4.57	**499.53**	nr	**561.97**
115 x 425 x 8000mm	592.80	2.50	**607.62**	2.70	0.34	**42.01**	4.75	**654.38**	nr	**736.17**
115 x 425 x 10000mm	737.79	2.50	**756.23**	2.80	0.35	**43.54**	4.93	**804.70**	nr	**905.29**
115 x 425 x 12000mm	883.20	2.50	**905.28**	2.90	0.36	**45.06**	5.12	**955.46**	nr	**1074.89**
115 x 450 x 6000mm	469.25	2.50	**480.98**	2.80	0.35	**43.54**	4.93	**529.45**	nr	**595.63**
115 x 450 x 8000mm	626.27	2.50	**641.92**	2.90	0.36	**45.06**	5.12	**692.10**	nr	**778.62**
115 x 450 x 10000mm	785.06	2.50	**804.69**	3.00	0.38	**46.70**	5.28	**856.67**	nr	**963.76**
115 x 450 x 12000mm	942.08	2.50	**965.64**	3.10	0.39	**48.23**	5.46	**1019.32**	nr	**1146.74**
115 x 475 x 6000mm	498.70	2.50	**511.17**	2.90	0.36	**45.06**	5.12	**561.35**	nr	**631.52**
115 x 475 x 8000mm	665.51	2.50	**682.15**	3.00	0.38	**46.70**	5.28	**734.13**	nr	**825.89**
115 x 475 x 10000mm	832.51	2.50	**853.33**	3.10	0.39	**48.23**	5.46	**907.01**	nr	**1020.39**
115 x 475 x 12000mm	994.72	2.50	**1019.59**	3.20	0.40	**49.76**	5.64	**1074.99**	nr	**1209.36**
115 x 475 x 14000mm	1159.76	2.50	**1188.76**	3.30	0.41	**51.28**	5.81	**1245.85**	nr	**1401.58**

Labour hourly rates: (except Specialists)
Craft Operatives £14.21 Labourer £10.71
Rates are national average prices.
Refer to REGIONAL VARIATIONS for indicative levels of overall pricing in regions

	MATERIALS			LABOUR				RATES		
	Del to Site £	Waste %	Material Cost £	Craft Optve Hrs	Lab Hrs	Labour Cost £	Sunds £	Nett Rate £	Unit	Gross Rate (+12.5%) £
G20: CARPENTRY/TIMBER FRAMING/FIRST FIXING Cont.										
Softwood, wrot, BS 4978, GS grade - laminated beams Cont.										
Glued laminated beams; BS EN 386 Cont.										
115 x 500 x 6000mm	526.35	2.50	**539.51**	3.00	0.38	**46.70**	5.28	**591.49**	nr	**665.43**
115 x 500 x 8000mm	698.53	2.50	**716.00**	3.10	0.39	**48.23**	5.46	**769.69**	nr	**865.90**
115 x 500 x 10000mm	877.41	2.50	**899.34**	3.20	0.40	**49.76**	5.64	**954.74**	nr	**1074.08**
115 x 500 x 12000mm	1051.76	2.50	**1078.05**	3.30	0.41	**51.28**	5.81	**1135.14**	nr	**1277.04**
115 x 500 x 14000mm	1224.89	2.50	**1255.51**	3.40	0.42	**52.81**	5.97	**1314.29**	nr	**1478.58**
140 x 350 x 4000mm	301.98	2.50	**309.53**	2.40	0.30	**37.32**	4.22	**351.07**	nr	**394.96**
140 x 350 x 6000mm	457.21	2.50	**468.64**	2.50	0.31	**38.85**	4.41	**511.89**	nr	**575.88**
140 x 350 x 8000mm	608.42	2.50	**623.63**	2.60	0.33	**40.48**	4.57	**668.68**	nr	**752.27**
140 x 375 x 4000mm	321.60	2.50	**329.64**	2.50	0.31	**38.85**	4.41	**372.89**	nr	**419.50**
140 x 375 x 6000mm	504.50	2.50	**517.11**	2.60	0.33	**40.48**	4.57	**562.16**	nr	**632.43**
140 x 375 x 8000mm	645.90	2.50	**662.04**	2.70	0.34	**42.01**	4.75	**708.80**	nr	**797.40**
140 x 375 x 10000mm	806.48	2.50	**826.65**	2.80	0.35	**43.54**	4.93	**875.12**	nr	**984.51**
140 x 400 x 6000mm	510.28	2.50	**523.04**	2.80	0.35	**43.54**	4.93	**571.51**	nr	**642.95**
140 x 400 x 8000mm	685.15	2.50	**702.28**	2.90	0.36	**45.06**	5.12	**752.46**	nr	**846.52**
140 x 400 x 10000mm	853.76	2.50	**875.10**	3.00	0.38	**46.70**	5.28	**927.08**	nr	**1042.97**
140 x 425 x 6000mm	551.33	2.50	**565.11**	2.90	0.36	**45.06**	5.12	**615.29**	nr	**692.20**
140 x 425 x 8000mm	737.79	2.50	**756.23**	3.00	0.38	**46.70**	5.28	**808.21**	nr	**909.24**
140 x 425 x 10000mm	922.44	2.50	**945.50**	3.10	0.39	**48.23**	5.46	**999.19**	nr	**1124.09**
140 x 425 x 12000mm	1107.12	2.50	**1134.80**	3.20	0.40	**49.76**	5.64	**1190.20**	nr	**1338.98**
140 x 450 x 6000mm	580.77	2.50	**595.29**	3.00	0.38	**46.70**	5.28	**647.27**	nr	**728.18**
140 x 450 x 8000mm	775.26	2.50	**794.65**	3.10	0.39	**48.23**	5.46	**848.34**	nr	**954.38**
140 x 450 x 10000mm	967.06	2.50	**991.23**	3.20	0.40	**49.76**	5.64	**1046.63**	nr	**1177.46**
140 x 450 x 12000mm	1159.76	2.50	**1188.76**	3.30	0.41	**51.28**	5.81	**1245.85**	nr	**1401.58**
140 x 475 x 6000mm	608.42	2.50	**623.63**	3.10	0.39	**48.23**	5.46	**677.32**	nr	**761.98**
140 x 475 x 8000mm	810.49	2.50	**830.75**	3.20	0.40	**49.76**	5.64	**886.15**	nr	**996.92**
140 x 475 x 10000mm	1014.80	2.50	**1040.17**	3.30	0.41	**51.28**	5.81	**1097.26**	nr	**1234.42**
140 x 475 x 12000mm	1218.64	2.50	**1249.11**	3.40	0.42	**52.81**	5.97	**1307.89**	nr	**1471.38**
140 x 475 x 14000mm	1414.90	2.50	**1450.28**	3.50	0.44	**54.45**	6.15	**1510.88**	nr	**1699.74**
165 x 425 x 6000mm	649.47	2.50	**665.70**	3.10	0.39	**48.23**	5.46	**719.39**	nr	**809.32**
165 x 425 x 8000mm	867.13	2.50	**888.81**	3.20	0.40	**49.76**	5.64	**944.21**	nr	**1062.23**
165 x 425 x 10000mm	1083.72	2.50	**1110.81**	3.30	0.41	**51.28**	5.81	**1167.90**	nr	**1313.89**
165 x 425 x 12000mm	1298.25	2.50	**1330.70**	3.40	0.42	**52.81**	5.97	**1389.49**	nr	**1563.17**
165 x 450 x 6000mm	677.13	2.50	**694.06**	3.20	0.40	**49.76**	5.64	**749.46**	nr	**843.14**
165 x 450 x 8000mm	904.62	2.50	**927.24**	3.30	0.41	**51.28**	5.81	**984.33**	nr	**1107.37**
165 x 450 x 10000mm	1130.32	2.50	**1158.58**	3.40	0.42	**52.81**	5.97	**1217.36**	nr	**1369.53**
165 x 450 x 12000mm	1357.81	2.50	**1391.76**	3.50	0.44	**54.45**	6.15	**1452.36**	nr	**1633.90**
165 x 475 x 6000mm	718.17	2.50	**736.13**	3.40	0.42	**52.81**	5.97	**794.91**	nr	**894.28**
165 x 475 x 8000mm	959.92	2.50	**983.92**	3.50	0.44	**54.45**	6.15	**1044.52**	nr	**1175.08**
165 x 475 x 10000mm	1197.22	2.50	**1227.16**	3.60	0.45	**55.98**	6.34	**1289.47**	nr	**1450.65**
165 x 475 x 12000mm	1440.33	2.50	**1476.33**	3.70	0.46	**57.50**	6.52	**1540.36**	nr	**1732.90**
165 x 475 x 14000mm	1678.08	2.50	**1720.03**	3.80	0.48	**59.14**	6.68	**1785.85**	nr	**2009.08**
190 x 475 x 6000mm	831.91	2.50	**852.70**	3.60	0.45	**55.98**	6.34	**915.02**	nr	**1029.39**
190 x 475 x 8000mm	1106.66	2.50	**1134.33**	3.70	0.46	**57.50**	6.52	**1198.35**	nr	**1348.14**
190 x 475 x 10000mm	1383.68	2.50	**1418.28**	3.80	0.48	**59.14**	6.68	**1484.10**	nr	**1669.61**
190 x 475 x 12000mm	1660.24	2.50	**1701.75**	3.90	0.49	**60.67**	6.86	**1769.28**	nr	**1990.44**
190 x 475 x 14000mm	1936.80	2.50	**1985.22**	4.00	0.50	**62.20**	7.05	**2054.46**	nr	**2311.27**
Softwood, sawn, BS 4978, GS grade - carcassing										
Floor members										
38 x 75mm	0.63	15.00	**0.72**	0.14	0.02	**2.20**	0.01	**2.93**	m	**3.30**
38 x 100mm	0.84	15.00	**0.96**	0.17	0.02	**2.63**	0.01	**3.60**	m	**4.05**
50 x 75mm	0.80	15.00	**0.92**	0.17	0.02	**2.63**	0.01	**3.56**	m	**4.01**
50 x 100mm	1.07	15.00	**1.23**	0.19	0.02	**2.91**	0.01	**4.16**	m	**4.68**
50 x 150mm	1.61	15.00	**1.85**	0.24	0.03	**3.73**	0.01	**5.60**	m	**6.30**
50 x 175mm	1.88	15.00	**2.16**	0.29	0.04	**4.55**	0.02	**6.72**	m	**7.56**
50 x 200mm	2.15	15.00	**2.47**	0.32	0.04	**4.98**	0.02	**7.46**	m	**8.39**
50 x 225mm	2.41	15.00	**2.78**	0.35	0.04	**5.40**	0.02	**8.20**	m	**9.22**
50 x 250mm	2.68	15.00	**3.08**	0.38	0.05	**5.94**	0.02	**9.04**	m	**10.17**
75 x 150mm	2.48	15.00	**2.85**	0.35	0.04	**5.40**	0.02	**8.27**	m	**9.30**
75 x 175mm	2.89	15.00	**3.32**	0.40	0.05	**6.22**	0.02	**9.56**	m	**10.75**
75 x 200mm	3.30	15.00	**3.80**	0.42	0.05	**6.50**	0.02	**10.32**	m	**11.61**
75 x 225mm	3.71	15.00	**4.27**	0.48	0.06	**7.46**	0.03	**11.76**	m	**13.23**
75 x 250mm	4.13	15.00	**4.74**	0.54	0.07	**8.42**	0.03	**13.20**	m	**14.85**
Wall or partition members										
38 x 75mm	0.63	15.00	**0.72**	0.22	0.03	**3.45**	0.01	**4.18**	m	**4.70**
38 x 100mm	0.84	15.00	**0.96**	0.25	0.03	**3.87**	0.01	**4.85**	m	**5.45**
50 x 75mm	0.80	15.00	**0.92**	0.25	0.03	**3.87**	0.01	**4.81**	m	**5.41**
50 x 100mm	1.07	15.00	**1.23**	0.29	0.04	**4.55**	0.02	**5.80**	m	**6.52**
75 x 100mm	1.65	15.00	**1.90**	0.36	0.04	**5.54**	0.02	**7.46**	m	**8.39**
38 x 75mm; fixing to masonry	0.63	15.00	**0.72**	0.43	0.05	**6.65**	0.02	**7.39**	m	**8.31**
38 x 100mm; fixing to masonry	0.84	15.00	**0.96**	0.47	0.06	**7.32**	0.03	**8.31**	m	**9.35**
50 x 75mm; fixing to masonry	0.80	15.00	**0.92**	0.47	0.06	**7.32**	0.03	**8.27**	m	**9.31**
50 x 100mm; fixing to masonry	1.07	15.00	**1.23**	0.52	0.07	**8.14**	0.03	**9.40**	m	**10.58**
75 x 100mm; fixing to masonry	1.65	15.00	**1.90**	0.50	0.07	**7.85**	0.03	**9.78**	m	**11.00**
Plates										
38 x 75mm	0.63	15.00	**0.72**	0.22	0.03	**3.45**	0.01	**4.18**	m	**4.70**
38 x 100mm	0.84	15.00	**0.96**	0.25	0.03	**3.87**	0.01	**4.85**	m	**5.45**
50 x 75mm	0.80	15.00	**0.92**	0.25	0.03	**3.87**	0.01	**4.81**	m	**5.41**
50 x 100mm	1.07	15.00	**1.23**	0.29	0.04	**4.55**	0.02	**5.80**	m	**6.52**
75 x 150mm	1.65	15.00	**1.90**	0.48	0.06	**7.46**	0.03	**9.39**	m	**10.56**
38 x 75mm; fixing by bolting	0.63	15.00	**0.72**	0.26	0.03	**4.02**	-	**4.74**	m	**5.33**
38 x 100mm; fixing by bolting	0.84	15.00	**0.96**	0.30	0.04	**4.69**	-	**5.65**	m	**6.36**
50 x 75mm; fixing by bolting	0.80	15.00	**0.92**	0.30	0.04	**4.69**	-	**5.62**	m	**6.32**
50 x 100mm; fixing by bolting	1.07	15.00	**1.23**	0.35	0.04	**5.40**	-	**6.64**	m	**7.46**
75 x 150mm; fixing by bolting	1.65	15.00	**1.90**	0.58	0.07	**8.99**	-	**10.89**	m	**12.25**

Labour hourly rates: (except Specialists) Craft Operatives £14.21 Labourer £10.71 Rates are national average prices. Refer to REGIONAL VARIATIONS for indicative levels of overall pricing in regions	MATERIALS			LABOUR			RATES			
	Del to Site £	Waste %	Material Cost £	Craft Optve Hrs	Lab Hrs	Labour Cost £	Sunds £	Nett Rate £	Unit	Gross Rate (+12.5%) £
G20: CARPENTRY/TIMBER FRAMING/FIRST FIXING Cont.										
Softwood, sawn, BS 4978, GS grade - carcassing Cont.										
Roof members; flat										
50 x 175mm	1.88	15.00	**2.16**	0.29	0.04	**4.55**	0.02	**6.72**	m	**7.56**
50 x 200mm	2.15	15.00	**2.47**	0.32	0.04	**4.98**	0.02	**7.46**	m	**8.39**
50 x 225mm	2.41	15.00	**2.78**	0.35	0.04	**5.40**	0.02	**8.20**	m	**9.22**
50 x 250mm	2.68	15.00	**3.08**	0.38	0.05	**5.94**	0.02	**9.04**	m	**10.17**
75 x 175mm	2.89	15.00	**3.32**	0.40	0.05	**6.22**	0.02	**9.56**	m	**10.75**
75 x 200mm	3.30	15.00	**3.80**	0.42	0.05	**6.50**	0.02	**10.32**	m	**11.61**
75 x 225mm	3.71	15.00	**4.27**	0.48	0.06	**7.46**	0.03	**11.76**	m	**13.23**
75 x 250mm	4.13	15.00	**4.74**	0.54	0.07	**8.42**	0.03	**13.20**	m	**14.85**
Roof members; pitched										
25 x 75mm	0.41	15.00	**0.47**	0.17	0.02	**2.63**	0.01	**3.11**	m	**3.50**
25 x 125mm	0.69	15.00	**0.79**	0.19	0.02	**2.91**	0.01	**3.71**	m	**4.18**
32 x 175mm	1.23	15.00	**1.42**	0.26	0.03	**4.02**	0.01	**5.45**	m	**6.13**
32 x 225mm	1.58	15.00	**1.82**	0.30	0.04	**4.69**	0.02	**6.53**	m	**7.35**
38 x 75mm	0.63	15.00	**0.72**	0.18	0.02	**2.77**	0.01	**3.50**	m	**3.94**
38 x 100mm	0.84	15.00	**0.96**	0.22	0.03	**3.45**	0.01	**4.42**	m	**4.97**
50 x 75mm	0.80	15.00	**0.92**	0.22	0.03	**3.45**	0.01	**4.39**	m	**4.93**
50 x 100mm	1.07	15.00	**1.23**	0.24	0.03	**3.73**	0.01	**4.98**	m	**5.60**
50 x 125mm	1.34	15.00	**1.54**	0.28	0.04	**4.41**	0.02	**5.97**	m	**6.71**
50 x 150mm	1.61	15.00	**1.85**	0.31	0.04	**4.83**	0.02	**6.70**	m	**7.54**
50 x 225mm	2.41	15.00	**2.78**	0.43	0.05	**6.65**	0.02	**9.44**	m	**10.62**
50 x 250mm	2.68	15.00	**3.08**	0.48	0.06	**7.46**	0.03	**10.57**	m	**11.90**
75 x 100mm	1.65	15.00	**1.90**	0.31	0.04	**4.83**	0.02	**6.75**	m	**7.59**
75 x 150mm	2.48	15.00	**2.85**	0.42	0.05	**6.50**	0.02	**9.37**	m	**10.54**
100 x 150mm	3.30	15.00	**3.80**	0.52	0.07	**8.14**	0.03	**11.96**	m	**13.46**
100 x 225mm	4.95	15.00	**5.69**	0.72	0.09	**11.20**	0.04	**16.93**	m	**19.04**
Joist strutting; herringbone; depth of joist 175mm										
50 x 50mm	0.97	15.00	**1.11**	0.78	0.10	**12.15**	0.04	**13.31**	m	**14.97**
Joist strutting; herringbone; depth of joist 200mm										
50 x 50mm	1.13	15.00	**1.30**	0.78	0.10	**12.15**	0.04	**13.49**	m	**15.18**
Joist strutting; herringbone; depth of joist 225mm										
50 x 50mm	1.29	15.00	**1.48**	0.78	0.10	**12.15**	0.04	**13.68**	m	**15.39**
Joist strutting; herringbone; depth of joist 250mm										
50 x 50mm	1.45	15.00	**1.67**	0.78	0.10	**12.15**	0.04	**13.86**	m	**15.60**
Joist strutting; block; depth of joist 100mm										
50 x 100mm	1.07	15.00	**1.23**	0.78	0.10	**12.15**	0.04	**13.43**	m	**15.11**
Joist strutting; block; depth of joist 150mm										
50 x 150mm	1.61	15.00	**1.85**	0.78	0.10	**12.15**	0.04	**14.05**	m	**15.80**
Joist strutting; block; depth of joist 175mm										
50 x 175mm	1.88	15.00	**2.16**	0.84	0.11	**13.11**	0.05	**15.32**	m	**17.23**
Joist strutting; block; depth of joist 200mm										
50 x 200mm	2.15	15.00	**2.47**	0.84	0.11	**13.11**	0.05	**15.63**	m	**17.58**
Joist strutting; block; depth of joist 225mm										
50 x 225mm	2.41	15.00	**2.78**	0.90	0.11	**13.97**	0.05	**16.79**	m	**18.89**
Joist strutting; block; depth of joist 250mm										
50 x 250mm	2.68	15.00	**3.08**	0.90	0.11	**13.97**	0.05	**17.10**	m	**19.24**
Noggings to joists										
50 x 50mm	0.54	15.00	**0.62**	0.34	0.04	**5.26**	0.02	**5.89**	m	**6.63**
50 x 75mm	0.80	15.00	**0.92**	0.40	0.05	**6.22**	0.02	**7.16**	m	**8.06**
Wrot surfaces										
plain; 50mm wide	-	-	**-**	0.18	-	**2.56**	-	**2.56**	m	**2.88**
plain; 75mm wide	-	-	**-**	0.24	-	**3.41**	-	**3.41**	m	**3.84**
plain; 100mm wide	-	-	**-**	0.30	-	**4.26**	-	**4.26**	m	**4.80**
plain; 150mm wide	-	-	**-**	0.36	-	**5.12**	-	**5.12**	m	**5.76**
plain; 200mm wide	-	-	**-**	0.42	-	**5.97**	-	**5.97**	m	**6.71**
Softwood, sawn, BS 4978, GS grade, impregnated - carcassing										
Floor members										
38 x 75mm	0.71	15.00	**0.81**	0.14	0.02	**2.20**	0.01	**3.02**	m	**3.40**
38 x 100mm	0.94	15.00	**1.08**	0.17	0.02	**2.63**	0.01	**3.72**	m	**4.19**
50 x 75mm	0.91	15.00	**1.04**	0.17	0.02	**2.63**	0.01	**3.68**	m	**4.14**
50 x 100mm	1.21	15.00	**1.39**	0.19	0.02	**2.91**	0.01	**4.32**	m	**4.85**
50 x 150mm	1.82	15.00	**2.09**	0.24	0.03	**3.73**	0.01	**5.83**	m	**6.56**
50 x 175mm	2.12	15.00	**2.44**	0.29	0.04	**4.55**	0.02	**7.00**	m	**7.88**
50 x 200mm	2.42	15.00	**2.78**	0.32	0.04	**4.98**	0.02	**7.78**	m	**8.75**
50 x 225mm	2.72	15.00	**3.13**	0.35	0.04	**5.40**	0.02	**8.55**	m	**9.62**
50 x 250mm	3.03	15.00	**3.48**	0.38	0.05	**5.94**	0.02	**9.43**	m	**10.61**
75 x 150mm	2.78	15.00	**3.20**	0.35	0.04	**5.40**	0.02	**8.62**	m	**9.70**
75 x 175mm	3.25	15.00	**3.74**	0.40	0.05	**6.22**	0.02	**9.97**	m	**11.22**
75 x 200mm	3.71	15.00	**4.27**	0.42	0.05	**6.50**	0.02	**10.80**	m	**12.15**
75 x 225mm	4.18	15.00	**4.80**	0.48	0.06	**7.46**	0.03	**12.29**	m	**13.83**
75 x 250mm	4.64	15.00	**5.34**	0.54	0.07	**8.42**	0.03	**13.79**	m	**15.51**
Wall or partition members										
38 x 75mm	0.71	15.00	**0.81**	0.22	0.03	**3.45**	0.01	**4.27**	m	**4.81**
38 x 100mm	0.94	15.00	**1.08**	0.25	0.03	**3.87**	0.01	**4.97**	m	**5.59**
50 x 75mm	0.91	15.00	**1.04**	0.25	0.03	**3.87**	0.01	**4.93**	m	**5.55**

Labour hourly rates: (except Specialists) Craft Operatives £14.21 Labourer £10.71 Rates are national average prices. Refer to REGIONAL VARIATIONS for indicative levels of overall pricing in regions	MATERIALS			LABOUR				RATES		
	Del to Site £	Waste %	Material Cost £	Craft Optve Hrs	Lab Hrs	Labour Cost £	Sunds £	Nett Rate £	Unit	Gross Rate (+12.5%) £
G20: CARPENTRY/TIMBER FRAMING/FIRST FIXING Cont.										
Softwood, sawn, BS 4978, GS grade, impregnated - carcassing Cont.										
Wall or partition members Cont.										
50 x 100mm	1.21	15.00	**1.39**	0.29	0.04	**4.55**	0.02	**5.96**	m	**6.70**
75 x 100mm	1.86	15.00	**2.14**	0.36	0.04	**5.54**	0.02	**7.70**	m	**8.66**
38 x 75mm; fixing to masonry	0.71	15.00	**0.81**	0.43	0.05	**6.65**	0.02	**7.48**	m	**8.41**
38 x 100mm; fixing to masonry	0.94	15.00	**1.08**	0.47	0.06	**7.32**	0.03	**8.43**	m	**9.48**
50 x 75mm; fixing to masonry	0.91	15.00	**1.04**	0.47	0.06	**7.32**	0.03	**8.39**	m	**9.44**
50 x 100mm; fixing to masonry	1.21	15.00	**1.39**	0.52	0.07	**8.14**	0.03	**9.56**	m	**10.76**
75 x 100mm; fixing to masonry	1.86	15.00	**2.14**	0.50	0.07	**7.85**	0.03	**10.02**	m	**11.27**
Plates										
38 x 75mm	0.71	15.00	**0.81**	0.22	0.03	**3.45**	0.01	**4.27**	m	**4.81**
38 x 100mm	0.94	15.00	**1.08**	0.25	0.03	**3.87**	0.01	**4.97**	m	**5.59**
50 x 75mm	0.91	15.00	**1.04**	0.25	0.03	**3.87**	0.01	**4.93**	m	**5.55**
50 x 100mm	1.21	15.00	**1.39**	0.29	0.04	**4.55**	0.02	**5.96**	m	**6.70**
75 x 150mm	1.86	15.00	**2.14**	0.48	0.06	**7.46**	0.03	**9.63**	m	**10.83**
38 x 75mm; fixing by bolting	0.71	15.00	**0.81**	0.26	0.03	**4.02**	-	**4.83**	m	**5.43**
38 x 100mm; fixing by bolting	0.94	15.00	**1.08**	0.30	0.04	**4.69**	-	**5.77**	m	**6.49**
50 x 75mm; fixing by bolting	0.91	15.00	**1.04**	0.30	0.04	**4.69**	-	**5.74**	m	**6.45**
50 x 100mm; fixing by bolting	1.21	15.00	**1.39**	0.35	0.04	**5.40**	-	**6.79**	m	**7.64**
75 x 150mm; fixing by bolting	1.86	15.00	**2.14**	0.58	0.07	**8.99**	-	**11.13**	m	**12.52**
Roof members; flat										
50 x 100mm	1.21	15.00	**1.39**	0.25	0.03	**3.87**	0.02	**5.28**	m	**5.94**
50 x 175mm	2.12	15.00	**2.44**	0.29	0.04	**4.55**	0.02	**7.00**	m	**7.88**
50 x 200mm	2.42	15.00	**2.78**	0.32	0.04	**4.98**	0.02	**7.78**	m	**8.75**
50 x 225mm	2.72	15.00	**3.13**	0.35	0.04	**5.40**	0.02	**8.55**	m	**9.62**
50 x 250mm	3.03	15.00	**3.48**	0.38	0.05	**5.94**	0.02	**9.43**	m	**10.61**
75 x 175mm	3.25	15.00	**3.74**	0.40	0.05	**6.22**	0.02	**9.97**	m	**11.22**
75 x 200mm	3.71	15.00	**4.27**	0.42	0.05	**6.50**	0.02	**10.80**	m	**12.15**
75 x 225mm	4.18	15.00	**4.80**	0.48	0.06	**7.46**	0.03	**12.29**	m	**13.83**
75 x 250mm	4.64	15.00	**5.34**	0.54	0.07	**8.42**	0.03	**13.79**	m	**15.51**
Roof members; pitched										
25 x 75mm	0.46	15.00	**0.53**	0.17	0.02	**2.63**	0.01	**3.17**	m	**3.57**
25 x 125mm	0.77	15.00	**0.89**	0.19	0.02	**2.91**	0.01	**3.81**	m	**4.29**
32 x 175mm	1.39	15.00	**1.59**	0.26	0.03	**4.02**	0.01	**5.62**	m	**6.33**
32 x 225mm	1.78	15.00	**2.05**	0.30	0.04	**4.69**	0.02	**6.76**	m	**7.60**
38 x 75mm	0.71	15.00	**0.81**	0.18	0.02	**2.77**	0.01	**3.59**	m	**4.04**
38 x 100mm	0.94	15.00	**1.08**	0.22	0.03	**3.45**	0.01	**4.54**	m	**5.11**
50 x 75mm	0.91	15.00	**1.04**	0.22	0.03	**3.45**	0.01	**4.50**	m	**5.07**
50 x 100mm	1.21	15.00	**1.39**	0.24	0.03	**3.73**	0.01	**5.14**	m	**5.78**
50 x 125mm	1.51	15.00	**1.74**	0.28	0.04	**4.41**	0.02	**6.16**	m	**6.93**
50 x 150mm	1.82	15.00	**2.09**	0.31	0.04	**4.83**	0.02	**6.94**	m	**7.80**
50 x 225mm	2.72	15.00	**3.13**	0.43	0.05	**6.65**	0.02	**9.80**	m	**11.02**
50 x 250mm	3.03	15.00	**3.48**	0.48	0.06	**7.46**	0.03	**10.97**	m	**12.34**
75 x 100mm	1.86	15.00	**2.14**	0.31	0.04	**4.83**	0.02	**6.99**	m	**7.86**
75 x 150mm	2.78	15.00	**3.20**	0.42	0.05	**6.50**	0.02	**9.73**	m	**10.94**
100 x 150mm	3.71	15.00	**4.27**	0.52	0.07	**8.14**	0.03	**12.44**	m	**13.99**
100 x 225mm	5.57	15.00	**6.40**	0.72	0.09	**11.20**	0.04	**17.64**	m	**19.84**
Joist strutting; herringbone; depth of joist 175mm										
50 x 50mm	1.09	15.00	**1.25**	0.78	0.10	**12.15**	0.04	**13.45**	m	**15.13**
Joist strutting; herringbone; depth of joist 200mm										
50 x 50mm	1.27	15.00	**1.46**	0.78	0.10	**12.15**	0.04	**13.66**	m	**15.37**
Joist strutting; herringbone; depth of joist 225mm										
50 x 50mm	1.45	15.00	**1.67**	0.78	0.10	**12.15**	0.04	**13.87**	m	**15.60**
Joist strutting; herringbone; depth of joist 250mm										
50 x 50mm	1.63	15.00	**1.88**	0.78	0.10	**12.15**	0.04	**14.08**	m	**15.84**
Joist strutting; block; depth of joist 150mm										
50 x 150mm	1.82	15.00	**2.09**	0.78	0.10	**12.15**	0.04	**14.28**	m	**16.07**
Joist strutting; block; depth of joist 175mm										
50 x 175mm	2.12	15.00	**2.44**	0.84	0.11	**13.11**	0.05	**15.60**	m	**17.55**
Joist strutting; block; depth of joist 200mm										
50 x 200mm	2.42	15.00	**2.78**	0.84	0.11	**13.11**	0.05	**15.94**	m	**17.94**
Joist strutting; block; depth of joist 225mm										
50 x 225mm	2.72	15.00	**3.13**	0.90	0.11	**13.97**	0.05	**17.15**	m	**19.29**
Joist strutting; block; depth of joist 250mm										
50 x 250mm	3.03	15.00	**3.48**	0.90	0.11	**13.97**	0.05	**17.50**	m	**19.68**
Noggings to joists										
50 x 50mm	0.61	15.00	**0.70**	0.34	0.04	**5.26**	0.02	**5.97**	m	**6.72**
50 x 75mm	0.91	15.00	**1.04**	0.40	0.05	**6.22**	0.02	**7.28**	m	**8.19**
Wrot surfaces										
plain; 50mm wide	-	-	-	0.18	-	**2.56**	-	**2.56**	m	**2.88**
plain; 75mm wide	-	-	-	0.24	-	**3.41**	-	**3.41**	m	**3.84**
plain; 100mm wide	-	-	-	0.30	-	**4.26**	-	**4.26**	m	**4.80**
plain; 150mm wide	-	-	-	0.36	-	**5.12**	-	**5.12**	m	**5.76**
plain; 200mm wide	-	-	-	0.42	-	**5.97**	-	**5.97**	m	**6.71**

Labour hourly rates: (except Specialists) Craft Operatives £14.21 Labourer £10.71 Rates are national average prices. Refer to REGIONAL VARIATIONS for indicative levels of overall pricing in regions	MATERIALS			LABOUR				RATES		
	Del to Site £	Waste %	Material Cost £	Craft Optve Hrs	Lab Hrs	Labour Cost £	Sunds £	Nett Rate £	Unit	Gross Rate (+12.5%) £
G20: CARPENTRY/TIMBER FRAMING/FIRST FIXING Cont.										
Oak, sawn - carcassing										
Floor members										
38 x 75mm	4.70	10.00	**5.17**	0.24	0.03	**3.73**	0.01	**8.91**	m	**10.02**
38 x 100mm	6.27	10.00	**6.90**	0.30	0.04	**4.69**	0.02	**11.60**	m	**13.06**
50 x 75mm	6.18	10.00	**6.80**	0.30	0.04	**4.69**	0.02	**11.51**	m	**12.95**
50 x 100mm	8.25	10.00	**9.07**	0.33	0.04	**5.12**	0.02	**14.21**	m	**15.99**
50 x 175mm	14.43	10.00	**15.87**	0.51	0.06	**7.89**	0.03	**23.79**	m	**26.76**
50 x 200mm	16.50	10.00	**18.15**	0.56	0.07	**8.71**	0.03	**26.89**	m	**30.25**
50 x 225mm	18.56	10.00	**20.41**	0.61	0.08	**9.52**	0.03	**29.97**	m	**33.72**
50 x 250mm	20.62	10.00	**22.69**	0.66	0.08	**10.24**	0.04	**32.96**	m	**37.08**
75 x 150mm	18.56	10.00	**20.41**	0.61	0.08	**9.52**	0.03	**29.97**	m	**33.72**
75 x 175mm	21.67	10.00	**23.83**	0.70	0.09	**10.91**	0.04	**34.78**	m	**39.13**
75 x 200mm	24.75	10.00	**27.22**	0.73	0.09	**11.34**	0.04	**38.60**	m	**43.43**
75 x 225mm	27.86	10.00	**30.65**	0.84	0.10	**13.01**	0.05	**43.70**	m	**49.16**
75 x 250mm	30.93	10.00	**34.02**	0.94	0.12	**14.64**	0.05	**48.72**	m	**54.81**
Wall or partition members										
38 x 75mm	4.70	10.00	**5.17**	0.38	0.05	**5.94**	0.02	**11.12**	m	**12.51**
38 x 100mm	6.27	10.00	**6.90**	0.44	0.06	**6.90**	0.02	**13.82**	m	**15.54**
50 x 75mm	6.18	10.00	**6.80**	0.44	0.06	**6.90**	0.02	**13.72**	m	**15.43**
50 x 100mm	8.25	10.00	**9.07**	0.51	0.06	**7.89**	0.03	**16.99**	m	**19.11**
75 x 100mm	12.37	10.00	**13.61**	0.63	0.08	**9.81**	0.03	**23.45**	m	**26.39**
38 x 75mm; fixing to masonry	4.70	10.00	**5.17**	0.64	0.08	**9.95**	0.03	**15.15**	m	**17.04**
38 x 100mm; fixing to masonry	6.27	10.00	**6.90**	0.71	0.09	**11.05**	0.04	**17.99**	m	**20.24**
50 x 75mm; fixing to masonry	6.18	10.00	**6.80**	0.71	0.09	**11.05**	0.04	**17.89**	m	**20.13**
50 x 100mm; fixing to masonry	8.25	10.00	**9.07**	0.78	0.10	**12.15**	0.04	**21.27**	m	**23.93**
75 x 100mm; fixing to masonry	12.37	10.00	**13.61**	0.94	0.12	**14.64**	0.05	**28.30**	m	**31.84**
Plates										
38 x 75mm	4.70	10.00	**5.17**	0.31	0.04	**4.83**	0.02	**10.02**	m	**11.27**
38 x 100mm	6.27	10.00	**6.90**	0.38	0.05	**5.94**	0.02	**12.85**	m	**14.46**
50 x 75mm	6.18	10.00	**6.80**	0.38	0.05	**5.94**	0.02	**12.75**	m	**14.35**
50 x 100mm	8.25	10.00	**9.07**	0.42	0.05	**6.50**	0.02	**15.60**	m	**17.55**
75 x 150mm	12.37	10.00	**13.61**	0.73	0.09	**11.34**	0.04	**24.99**	m	**28.11**
38 x 75mm; fixing by bolting	4.70	10.00	**5.17**	0.46	0.06	**7.18**	-	**12.34**	m	**13.89**
38 x 100mm; fixing by bolting	6.27	10.00	**6.90**	0.53	0.07	**8.28**	-	**15.18**	m	**17.08**
50 x 75mm; fixing by bolting	6.18	10.00	**6.80**	0.53	0.07	**8.28**	-	**15.08**	m	**16.97**
50 x 100mm; fixing by bolting	8.25	10.00	**9.07**	0.60	0.08	**9.38**	-	**18.46**	m	**20.77**
75 x 150mm; fixing by bolting	12.37	10.00	**13.61**	0.74	0.09	**11.48**	-	**25.09**	m	**28.23**
Roof members; flat										
50 x 175mm	14.43	10.00	**15.87**	0.51	0.06	**7.89**	0.03	**23.79**	m	**26.76**
50 x 200mm	16.50	10.00	**18.15**	0.56	0.07	**8.71**	0.03	**26.89**	m	**30.25**
50 x 225mm	18.56	10.00	**20.41**	0.61	0.08	**9.52**	0.03	**29.97**	m	**33.72**
50 x 250mm	20.62	10.00	**22.69**	0.66	0.08	**10.24**	0.04	**32.96**	m	**37.08**
75 x 175mm	21.67	10.00	**23.83**	0.70	0.09	**10.91**	0.04	**34.78**	m	**39.13**
75 x 200mm	24.75	10.00	**27.22**	0.73	0.09	**11.34**	0.04	**38.60**	m	**43.43**
75 x 225mm	27.86	10.00	**30.65**	0.84	0.10	**13.01**	0.05	**43.70**	m	**49.16**
75 x 250mm	30.93	10.00	**34.02**	0.94	0.12	**14.64**	0.05	**48.72**	m	**54.81**
Roof members; pitched										
25 x 75mm	3.11	10.00	**3.42**	0.30	0.04	**4.69**	0.02	**8.13**	m	**9.14**
25 x 125mm	5.17	10.00	**5.68**	0.33	0.04	**5.12**	0.02	**10.82**	m	**12.17**
32 x 175mm	9.24	10.00	**10.16**	0.45	0.06	**7.04**	0.02	**17.22**	m	**19.38**
32 x 225mm	11.88	10.00	**13.07**	0.52	0.07	**8.14**	0.03	**21.24**	m	**23.89**
38 x 75mm	4.70	10.00	**5.17**	0.31	0.04	**4.83**	0.02	**10.02**	m	**11.27**
38 x 100mm	6.27	10.00	**6.90**	0.38	0.05	**5.94**	0.02	**12.85**	m	**14.46**
50 x 75mm	6.18	10.00	**6.80**	0.38	0.05	**5.94**	0.02	**12.75**	m	**14.35**
50 x 100mm	8.25	10.00	**9.07**	0.42	0.05	**6.50**	0.02	**15.60**	m	**17.55**
50 x 125mm	10.31	10.00	**11.34**	0.49	0.06	**7.61**	0.03	**18.97**	m	**21.34**
50 x 150mm	12.37	10.00	**13.61**	0.54	0.07	**8.42**	0.03	**22.07**	m	**24.82**
50 x 225mm	18.56	10.00	**20.41**	0.75	0.09	**11.62**	0.04	**32.07**	m	**36.08**
50 x 250mm	20.62	10.00	**22.69**	0.84	0.10	**13.01**	0.05	**35.74**	m	**40.21**
75 x 100mm	12.37	10.00	**13.61**	0.54	0.07	**8.42**	0.03	**22.07**	m	**24.82**
75 x 150mm	18.56	10.00	**20.41**	0.73	0.09	**11.34**	0.04	**31.79**	m	**35.76**
100 x 150mm	24.75	10.00	**27.22**	0.94	0.12	**14.64**	0.05	**41.92**	m	**47.16**
100 x 225mm	37.12	10.00	**40.84**	1.26	0.16	**19.62**	0.07	**60.52**	m	**68.09**
Joist strutting; herringbone; depth of joist 175mm 50 x 50mm	7.42	10.00	**8.17**	1.36	0.17	**21.15**	0.07	**29.39**	m	**33.06**
Joist strutting; herringbone; depth of joist 200mm 50 x 50mm	8.66	10.00	**9.53**	1.36	0.17	**21.15**	0.07	**30.75**	m	**34.59**
Joist strutting; herringbone; depth of joist 225mm 50 x 50mm	9.90	10.00	**10.89**	1.36	0.17	**21.15**	0.07	**32.11**	m	**36.12**
Joist strutting; herringbone; depth of joist 250mm 50 x 50mm	11.14	10.00	**12.25**	1.36	0.17	**21.15**	0.07	**33.47**	m	**37.65**
Joist strutting; block; depth of joist 150mm 50 x 150mm	12.37	10.00	**13.61**	1.36	0.17	**21.15**	0.07	**34.83**	m	**39.19**
Joist strutting; block; depth of joist 175mm 50 x 175mm	14.43	10.00	**15.87**	1.47	0.18	**22.82**	0.08	**38.77**	m	**43.62**
Joist strutting; block; depth of joist 200mm 50 x 200mm	16.50	10.00	**18.15**	1.47	0.18	**22.82**	0.08	**41.05**	m	**46.18**
Joist strutting; block; depth of joist 225mm 50 x 225mm	18.56	10.00	**20.41**	1.57	0.20	**24.45**	0.09	**44.95**	m	**50.57**

Labour hourly rates: (except Specialists) Craft Operatives £14.21 Labourer £10.71 Rates are national average prices. Refer to REGIONAL VARIATIONS for indicative levels of overall pricing in regions	MATERIALS			LABOUR				RATES		
	Del to Site £	Waste %	Material Cost £	Craft Optve Hrs	Lab Hrs	Labour Cost £	Sunds £	Nett Rate £	Unit	Gross Rate (+12.5%) £
G20: CARPENTRY/TIMBER FRAMING/FIRST FIXING Cont.										
Oak, sawn - carcassing Cont.										
Joist strutting; block; depth of joist 250mm										
50 x 250mm	20.62	10.00	22.69	1.57	0.20	24.45	0.09	47.22	m	53.13
Noggings to joists										
50 x 50mm	4.12	10.00	4.54	0.60	0.08	9.38	0.03	13.95	m	15.70
50 x 75mm	6.18	10.00	6.80	0.70	0.09	10.91	0.04	17.75	m	19.96
Wrot surfaces										
plain; 50mm wide	-	-	-	0.36	-	5.12	-	5.12	m	5.76
plain; 75mm wide	-	-	-	0.48	-	6.82	-	6.82	m	7.67
plain; 100mm wide	-	-	-	0.60	-	8.53	-	8.53	m	9.59
plain; 150mm wide	-	-	-	0.72	-	10.23	-	10.23	m	11.51
plain; 200mm wide	-	-	-	0.84	-	11.94	-	11.94	m	13.43
Stress grading softwood										
General Structural (GS) grade included in rates										
Special Structural (SS) grade add 10% to materials prices										
Flame proofing treatment to softwood										
For timbers treated with proofing process to Class 1 add 100% to materials prices										
Softwood, sawn - supports										
Butt jointed supports										
width exceeding 300mm; 19 x 38mm at 300mm centres; fixing to masonry	0.53	10.00	0.58	1.20	0.15	18.66	0.07	19.30	m2	21.72
width exceeding 300mm; 25 x 50mm at 300mm centres; fixing to masonry	0.92	10.00	1.01	1.30	0.16	20.19	0.07	21.26	m2	23.92
Framed supports										
width exceeding 300mm; 38 x 50mm members at 400mm centres one way; 38 x 50mm subsidiary members at 400mm centres one way	2.09	10.00	2.30	1.20	0.15	18.66	0.07	21.03	m2	23.66
width exceeding 300mm; 38 x 50mm members at 500mm centres one way; 38 x 50mm subsidiary members at 400mm centres one way	1.88	10.00	2.07	0.95	0.12	14.78	0.05	16.90	m2	19.02
width exceeding 300mm; 50 x 50mm members at 400mm centres one way; 50 x 50mm subsidiary members at 400mm centres one way	2.75	10.00	3.03	1.30	0.16	20.19	0.07	23.28	m2	26.19
width exceeding 300mm; 50 x 50mm members at 500mm centres one way; 50 x 50mm subsidiary members at 400mm centres one way	2.48	10.00	2.72	1.10	0.14	17.13	0.06	19.91	m2	22.40
width exceeding 300mm; 19 x 38mm members at 300mm centres one way; 19 x 38mm subsidiary members at 300mm centres one way; fixing to masonry	1.05	10.00	1.16	2.70	0.34	42.01	0.15	43.31	m2	48.72
width exceeding 300mm; 25 x 50mm members at 300mm centres one way; 25 x 50mm subsidiary members at 300mm centres one way; fixing to masonry	1.83	10.00	2.01	2.80	0.35	43.54	0.15	45.70	m2	51.42
width not exceeding 300mm; 38 x 50mm members longitudinally; 38 x 50mm subsidiary members at 400mm centres laterally	2.09	10.00	2.30	0.45	0.06	7.04	0.02	9.36	m	10.53
width not exceeding 300mm; 50 x 50mm members longitudinally; 50 x 50mm subsidiary members at 400mm centres laterally	2.75	10.00	3.03	0.50	0.06	7.75	0.03	10.80	m	12.15
Individual supports										
6 x 38mm	0.09	10.00	0.10	0.10	0.01	1.53	0.01	1.63	m	1.84
12 x 38mm	0.13	10.00	0.15	0.10	0.01	1.53	0.01	1.68	m	1.89
19 x 38mm	0.16	10.00	0.17	0.11	0.01	1.67	0.01	1.85	m	2.08
25 x 50mm	0.28	10.00	0.30	0.12	0.02	1.92	0.01	2.23	m	2.51
38 x 50mm	0.42	10.00	0.46	0.13	0.02	2.06	0.01	2.53	m	2.84
50 x 50mm	0.55	10.00	0.61	0.14	0.02	2.20	0.01	2.82	m	3.17
50 x 150mm	1.65	10.00	1.82	0.20	0.03	3.16	0.01	4.99	m	5.61
50 x 300mm	3.30	10.00	3.63	0.28	0.04	4.41	0.02	8.05	m	9.06
6 x 38mm; fixing to masonry	0.09	10.00	0.10	0.28	0.04	4.41	0.02	4.52	m	5.09
12 x 38mm; fixing to masonry	0.13	10.00	0.15	0.28	0.04	4.41	0.02	4.57	m	5.14
19 x 38mm; fixing to masonry	0.16	10.00	0.17	0.29	0.04	4.55	0.02	4.74	m	5.33
25 x 50mm; fixing to masonry	0.28	10.00	0.30	0.30	0.04	4.69	0.02	5.01	m	5.64
38 x 50mm; fixing to masonry	0.42	10.00	0.46	0.32	0.04	4.98	0.02	5.45	m	6.13
50 x 50mm; fixing to masonry	0.55	10.00	0.61	0.34	0.04	5.26	0.02	5.88	m	6.62
Softwood, sawn, impregnated - supports										
Butt jointed supports										
width exceeding 300mm; 50mm wide x 25mm average depth at 450mm centres	1.37	10.00	1.51	0.36	0.05	5.65	0.02	7.18	m2	8.08
width exceeding 300mm; 50mm wide x 25mm average depth at 600mm centres	1.03	10.00	1.13	0.28	0.04	4.41	0.02	5.55	m2	6.25
width exceeding 300mm; 50mm wide x 50mm average depth at 450mm centres	2.75	10.00	3.02	0.43	0.05	6.65	0.02	9.69	m2	10.90
width exceeding 300mm; 50mm wide x 50mm average depth at 600mm centres	2.05	10.00	2.26	0.32	0.04	4.98	0.02	7.25	m2	8.16
width exceeding 300mm; 50mm wide x 63mm average depth at 450mm centres	3.46	10.00	3.81	0.48	0.06	7.46	0.03	11.30	m2	12.71
width exceeding 300mm; 50mm wide x 63mm average depth at 600mm centres	2.59	10.00	2.85	0.36	0.05	5.65	0.02	8.52	m2	9.58
width exceeding 300mm; 50mm wide x 75mm average depth at 450mm centres	4.12	10.00	4.53	0.52	0.07	8.14	0.03	12.70	m2	14.29

Labour hourly rates: (except Specialists) Craft Operatives £14.21 Labourer £10.71 Rates are national average prices. Refer to REGIONAL VARIATIONS for indicative levels of overall pricing in regions	MATERIALS			LABOUR				RATES		
	Del to Site £	Waste %	Material Cost £	Craft Optve Hrs	Lab Hrs	Labour Cost £	Sunds £	Nett Rate £	Unit	Gross Rate (+12.5%) £
G20: CARPENTRY/TIMBER FRAMING/FIRST FIXING Cont.										
Softwood, sawn, impregnated - supports Cont.										
Butt jointed supports Cont.										
width exceeding 300mm; 50mm wide x 75mm average depth at 600mm centres	3.08	10.00	**3.39**	0.40	0.05	**6.22**	0.02	**9.63**	m2	**10.83**
Individual supports										
25 x 25mm; 1 labours	0.19	10.00	**0.21**	0.13	0.02	**2.06**	0.01	**2.28**	m	**2.57**
25 x 25mm; 2 labours	0.23	10.00	**0.26**	0.13	0.02	**2.06**	0.01	**2.32**	m	**2.61**
25 x 38mm; 1 labours	0.29	10.00	**0.32**	0.13	0.02	**2.06**	0.01	**2.39**	m	**2.69**
25 x 38mm; 2 labours	0.35	10.00	**0.39**	0.13	0.02	**2.06**	0.01	**2.46**	m	**2.76**
triangular; extreme dimensions										
25 x 25mm	0.19	10.00	**0.21**	0.13	0.02	**2.06**	0.01	**2.28**	m	**2.57**
38 x 75mm	0.88	10.00	**0.97**	0.16	0.02	**2.49**	0.01	**3.47**	m	**3.90**
50 x 75mm	1.16	10.00	**1.28**	0.18	0.02	**2.77**	0.01	**4.06**	m	**4.57**
50 x 100mm	1.55	10.00	**1.70**	0.20	0.03	**3.16**	0.01	**4.87**	m	**5.48**
75 x 100mm	2.32	10.00	**2.55**	0.23	0.03	**3.59**	0.01	**6.15**	m	**6.92**
rounded roll for lead										
50 x 50mm	1.86	10.00	**2.04**	0.26	0.03	**4.02**	0.01	**6.07**	m	**6.83**
50 x 75mm	2.78	10.00	**3.06**	0.30	0.04	**4.69**	0.02	**7.77**	m	**8.74**
rounded roll for lead; birdsmouthed on to ridge or hip										
50 x 50mm	1.86	10.00	**2.04**	0.46	0.06	**7.18**	0.03	**9.25**	m	**10.40**
50 x 75mm	2.78	10.00	**3.06**	0.49	0.06	**7.61**	0.03	**10.69**	m	**12.03**
roll for zinc; 32 x 44mm	1.04	10.00	**1.14**	0.19	0.02	**2.91**	0.01	**4.07**	m	**4.58**
Softwood, wrot, impregnated - gutter boarding										
Gutter boards including sides, tongued and grooved joints										
width exceeding 300mm; 19mm thick	7.74	10.00	**8.52**	2.40	0.30	**37.32**	0.13	**45.96**	m2	**51.71**
width exceeding 300mm; 25mm thick	10.21	10.00	**11.23**	2.70	0.34	**42.01**	0.15	**53.38**	m2	**60.05**
width not exceeding 300mm; 19mm thick										
150mm wide	1.16	10.00	**1.28**	0.36	0.05	**5.65**	0.02	**6.95**	m	**7.82**
225mm wide	1.74	10.00	**1.92**	0.48	0.06	**7.46**	0.03	**9.41**	m	**10.58**
width not exceeding 300mm; 25mm thick										
150mm wide	1.53	10.00	**1.68**	0.40	0.05	**6.22**	0.02	**7.92**	m	**8.91**
225mm wide	2.30	10.00	**2.53**	0.53	0.07	**8.28**	0.03	**10.84**	m	**12.19**
cesspool 225 x 225 x 150mm deep	1.89	10.00	**2.08**	1.80	0.23	**28.04**	0.10	**30.21**	nr	**33.99**
Chimney gutter boards, butt joints										
width not exceeding 300mm; 25mm thick x 100mm average wide	1.02	10.00	**1.12**	0.60	0.08	**9.38**	0.03	**10.54**	m	**11.86**
gusset end	0.35	10.00	**0.39**	0.42	0.05	**6.50**	0.02	**6.91**	nr	**7.78**
width not exceeding 300mm; 25mm thick x 175mm average wide	1.79	10.00	**1.96**	0.78	0.10	**12.15**	0.04	**14.16**	m	**15.93**
gusset end	0.49	10.00	**0.54**	0.48	0.06	**7.46**	0.03	**8.03**	nr	**9.04**
Lier boards; tongued and grooved joints										
width exceeding 300mm; 19mm thick	7.74	10.00	**8.52**	1.10	0.14	**17.13**	0.06	**25.71**	m2	**28.92**
width exceeding 300mm; 25mm thick	10.21	10.00	**11.23**	1.20	0.15	**18.66**	0.07	**29.95**	m2	**33.70**
Valley sole boards; butt joints										
width not exceeding 300mm; 25mm thick										
100mm wide	1.02	10.00	**1.12**	0.55	0.07	**8.57**	0.03	**9.72**	m	**10.93**
150mm wide	1.53	10.00	**1.68**	0.60	0.08	**9.38**	0.03	**11.10**	m	**12.49**
Plywood BS EN 636, II/III grade, WBP bonded, butt joints - gutter boarding										
Gutter boards including sides, butt joints										
width exceeding 300mm; 18mm thick	11.21	10.00	**12.33**	1.80	0.23	**28.04**	0.10	**40.47**	m2	**45.53**
width exceeding 300mm; 25mm thick	15.58	10.00	**17.14**	2.00	0.25	**31.10**	0.11	**48.34**	m2	**54.39**
width not exceeding 300mm; 18mm thick										
150mm wide	1.68	10.00	**1.85**	0.24	0.03	**3.73**	0.01	**5.59**	m	**6.29**
225mm wide	2.52	10.00	**2.77**	0.26	0.03	**4.02**	0.01	**6.80**	m	**7.65**
width not exceeding 300mm; 25mm thick										
150mm wide	2.34	10.00	**2.57**	0.28	0.04	**4.41**	0.02	**6.99**	m	**7.87**
225mm wide	3.51	10.00	**3.86**	0.30	0.04	**4.69**	0.02	**8.56**	m	**9.63**
cesspool 225 x 225 x 150mm deep	2.88	10.00	**3.17**	1.80	0.23	**28.04**	0.10	**31.31**	nr	**35.22**
Softwood, wrot, impregnated - eaves and verge boarding										
Fascia and barge boards										
width not exceeding 300mm; 25mm thick										
150mm wide	1.53	10.00	**1.68**	0.26	0.03	**4.02**	0.01	**5.71**	m	**6.43**
200mm wide	2.04	10.00	**2.25**	0.27	0.03	**4.16**	0.01	**6.42**	m	**7.22**
Fascia and barge boards, tongued, grooved and veed joints										
width not exceeding 300mm; 25mm thick										
225mm wide	2.30	10.00	**2.53**	0.36	0.05	**5.65**	0.02	**8.20**	m	**9.22**
250mm wide	2.55	10.00	**2.81**	0.38	0.05	**5.94**	0.02	**8.76**	m	**9.86**
Eaves or verge soffit boards										
width exceeding 300mm; 25mm thick	10.21	10.00	**11.23**	1.20	0.15	**18.66**	0.07	**29.95**	m2	**33.70**
width not exceeding 300mm; 25mm thick										
225mm wide	2.30	10.00	**2.53**	0.36	0.05	**5.65**	0.02	**8.20**	m	**9.22**

Labour hourly rates: (except Specialists) Craft Operatives £14.21 Labourer £10.71 Rates are national average prices. Refer to REGIONAL VARIATIONS for indicative levels of overall pricing in regions	MATERIALS			LABOUR				RATES		
	Del to Site £	Waste %	Material Cost £	Craft Optve Hrs	Lab Hrs	Labour Cost £	Sunds £	Nett Rate £	Unit	Gross Rate (+12.5%) £
G20: CARPENTRY/TIMBER FRAMING/FIRST FIXING Cont.										
Plywood BS EN 636, II/III grade, WBP bonded, butt joints - eaves and verge boarding										
Fascia and barge boards										
width not exceeding 300mm; 18mm thick										
150mm wide	1.68	10.00	**1.85**	0.26	0.03	**4.02**	0.01	**5.88**	m	**6.61**
225mm wide	2.52	10.00	**2.77**	0.27	0.03	**4.16**	0.01	**6.95**	m	**7.81**
250mm wide	2.80	10.00	**3.08**	0.28	0.04	**4.41**	0.02	**7.51**	m	**8.44**
Eaves or verge soffit boards										
width exceeding 300mm; 18mm thick	11.21	10.00	**12.33**	0.90	0.11	**13.97**	0.05	**26.35**	m2	**29.64**
width not exceeding 300mm; 18mm thick										
225mm wide	2.52	10.00	**2.77**	0.30	0.04	**4.69**	0.02	**7.48**	m	**8.42**
Fascia and bargeboards; Caradon Celuform Ltd										
Fascia reference 4247; butt joint with joint trims reference 4560; fixing to timber with Polytop screws										
150mm wide	12.24	5.00	**12.85**	0.24	0.03	**3.73**	0.17	**16.76**	m	**18.86**
200mm wide	14.73	5.00	**15.47**	0.24	0.03	**3.73**	0.17	**19.37**	m	**21.80**
250mm wide	19.36	5.00	**20.33**	0.24	0.03	**3.73**	0.35	**24.41**	m	**27.46**
Elite solid fascia board reference 4245; butt joint with joint trims reference 4560; fixing to timber with Polytop screws										
150mm wide	12.31	5.00	**12.92**	0.24	0.03	**3.73**	0.17	**16.83**	m	**18.94**
200mm wide	15.15	5.00	**15.91**	0.24	0.03	**3.73**	0.17	**19.81**	m	**22.29**
250mm wide	20.44	5.00	**21.46**	0.24	0.03	**3.73**	0.35	**25.54**	m	**28.73**
Supaliner fascia board reference 4270; butt joint with joint trims reference 4586; fixing to timber with Polytop screws										
150mm wide	9.91	5.00	**10.41**	0.35	0.04	**5.40**	0.17	**15.99**	m	**17.99**
170mm wide	11.42	5.00	**11.99**	0.35	0.04	**5.40**	0.17	**17.57**	m	**19.77**
225mm wide	14.49	5.00	**15.21**	0.35	0.04	**5.40**	0.17	**20.79**	m	**23.39**
250mm wide	16.01	5.00	**16.81**	0.40	0.06	**6.33**	0.35	**23.49**	m	**26.42**
Celuvent multi purpose/soffite board reference 4258; butt joint with panel joint trim reference 4570; fixing to timber with stainless steel nails; fixing soffite board channel reference 4572 to masonry with Polytop screws										
100mm wide	12.03	5.00	**12.63**	0.45	0.07	**7.14**	0.17	**19.95**	m	**22.44**
225mm wide	20.15	5.00	**21.16**	0.45	0.07	**7.14**	0.35	**28.65**	m	**32.24**
325mm wide	26.58	5.00	**27.91**	0.70	0.09	**10.91**	0.52	**39.34**	m	**44.26**
Multi purpose/soffite board reference 4558; butt joint with panel joint trim reference 4570; fixing to timber with stainless steel nails; fixing soffite board channel reference 4572 to masonry with Polytop screws										
125mm wide	13.48	5.00	**14.15**	0.45	0.07	**7.14**	0.17	**21.47**	m	**24.16**
175mm wide	16.41	5.00	**17.23**	0.45	0.07	**7.14**	0.17	**24.55**	m	**27.62**
200mm wide	17.10	5.00	**17.95**	0.45	0.07	**7.14**	0.35	**25.45**	m	**28.63**
300mm wide	24.89	5.00	**26.13**	0.70	0.09	**10.91**	0.52	**37.56**	m	**42.26**
Steel - tie rods and straps										
Tie rods										
19mm diameter; threaded	6.29	2.50	**6.45**	0.30	0.04	**4.69**	-	**11.14**	m	**12.54**
extra; with nut and washer	0.31	2.50	**0.32**	0.25	0.03	**3.87**	-	**4.19**	nr	**4.71**
25mm diameter; threaded	8.51	2.50	**8.72**	0.35	0.04	**5.40**	-	**14.12**	m	**15.89**
extra; with nut and washer	0.55	2.50	**0.57**	0.35	0.04	**5.40**	-	**5.97**	nr	**6.72**
Straps										
3 x 38; holes -4; fixing with bolts (bolts included elsewhere)										
500mm long	1.31	2.50	**1.34**	0.16	0.02	**2.49**	-	**3.83**	nr	**4.31**
750mm long	1.94	2.50	**1.99**	0.23	0.03	**3.59**	-	**5.58**	nr	**6.27**
1000mm long	2.62	2.50	**2.69**	0.30	0.04	**4.69**	-	**7.38**	nr	**8.30**
6 x 50; holes -4; fixing with bolts (bolts included elsewhere)										
500mm long	2.87	2.50	**2.94**	0.19	0.02	**2.91**	-	**5.86**	nr	**6.59**
750mm long	4.24	2.50	**4.35**	0.28	0.03	**4.30**	-	**8.65**	nr	**9.73**
1000mm long	5.70	2.50	**5.84**	0.36	0.04	**5.54**	-	**11.39**	nr	**12.81**
Steel; galvanized - straps										
Straps and clips; BAT Building Products Ltd.										
30 x 2.5mm standard strapping; fixing with nails										
600mm long	0.88	5.00	**0.93**	0.18	0.02	**2.77**	0.01	**3.71**	nr	**4.17**
800mm long	1.29	5.00	**1.36**	0.20	0.03	**3.16**	0.01	**4.53**	nr	**5.10**
1000mm long	1.67	5.00	**1.75**	0.22	0.03	**3.45**	0.01	**5.21**	nr	**5.87**
1200mm long	2.01	5.00	**2.11**	0.23	0.03	**3.59**	0.01	**5.71**	nr	**6.43**
1600mm long	2.61	5.00	**2.74**	0.24	0.03	**3.73**	0.01	**6.49**	nr	**7.30**
30 x 2.5mm standard strapping; twists -1; fixing with nails										
600mm long	0.97	5.00	**1.02**	0.18	0.02	**2.77**	0.01	**3.80**	nr	**4.28**
800mm long	1.39	5.00	**1.46**	0.20	0.03	**3.16**	0.01	**4.64**	nr	**5.22**
1000mm long	1.75	5.00	**1.84**	0.22	0.03	**3.45**	0.01	**5.30**	nr	**5.96**
1200mm long	2.09	5.00	**2.20**	0.23	0.03	**3.59**	0.01	**5.80**	nr	**6.52**
1600mm long	2.71	5.00	**2.84**	0.24	0.03	**3.73**	0.01	**6.59**	nr	**7.41**
30 x 5mm M 305 strapping; bends -1; nailing one end (building in included elsewhere)										
600mm long	1.88	5.00	**1.97**	0.10	0.01	**1.53**	0.01	**3.50**	nr	**3.94**

Labour hourly rates: (except Specialists) Craft Operatives £14.21 Labourer £10.71 Rates are national average prices. Refer to REGIONAL VARIATIONS for indicative levels of overall pricing in regions	MATERIALS			LABOUR				RATES		
	Del to Site £	Waste %	Material Cost £	Craft Optve Hrs	Lab Hrs	Labour Cost £	Sunds £	Nett Rate £	Unit	Gross Rate (+12.5%) £
G20: CARPENTRY/TIMBER FRAMING/FIRST FIXING Cont.										
Steel; galvanized - straps Cont.										
Straps and clips; BAT Building Products Ltd. Cont.										
30 x 5mm M 305 strapping; bends -1; nailing one end (building in included elsewhere) Cont.										
800mm long	2.40	5.00	**2.52**	0.10	0.01	**1.53**	0.01	**4.05**	nr	**4.56**
1000mm long	3.05	5.00	**3.21**	0.11	0.01	**1.67**	0.01	**4.88**	nr	**5.49**
1200mm long	3.67	5.00	**3.85**	0.11	0.01	**1.67**	0.01	**5.53**	nr	**6.22**
1600mm long	4.83	5.00	**5.07**	0.12	0.02	**1.92**	0.01	**7.00**	nr	**7.87**
30 x 5mm M305 strapping; bends -1; twists -1; nailing one end (building in included elsewhere)										
600mm long	1.95	5.00	**2.04**	0.10	0.01	**1.53**	0.01	**3.58**	nr	**4.03**
800mm long	2.45	5.00	**2.57**	0.10	0.01	**1.53**	0.01	**4.11**	nr	**4.62**
1000mm long	3.13	5.00	**3.28**	0.11	0.01	**1.67**	0.01	**4.96**	nr	**5.58**
1200mm long	3.75	5.00	**3.94**	0.11	0.01	**1.67**	0.01	**5.62**	nr	**6.32**
1600mm long	4.92	5.00	**5.17**	0.12	0.02	**1.92**	0.01	**7.09**	nr	**7.98**
truss clips; fixing with nails										
for 38mm thick members	0.35	5.00	**0.37**	0.24	0.03	**3.73**	0.01	**4.11**	nr	**4.62**
for 50mm thick members	0.37	5.00	**0.39**	0.24	0.03	**3.73**	0.01	**4.13**	nr	**4.65**
Steel; galvanized - joist hangers										
Joist hangers; BAT Building Products Ltd; (building in where required included elsewhere)										
SPH type S, for 50 x 100mm joist	1.75	2.50	**1.80**	-	-	**-**	-	**1.80**	nr	**2.02**
SPH type S, for 50 x 125mm joist	1.78	2.50	**1.83**	-	-	**-**	-	**1.83**	nr	**2.06**
SPH type S, for 50 x 150mm joist	1.78	2.50	**1.83**	-	-	**-**	-	**1.83**	nr	**2.06**
SPH type S, for 50 x 175mm joist	1.87	2.50	**1.91**	-	-	**-**	-	**1.91**	nr	**2.15**
SPH type S, for 50 x 200mm joist	2.07	2.50	**2.12**	-	-	**-**	-	**2.12**	nr	**2.39**
SPH type S, for 50 x 225mm joist	2.21	2.50	**2.27**	-	-	**-**	-	**2.27**	nr	**2.55**
SPH type S, for 50 x 250mm joist	2.83	2.50	**2.90**	-	-	**-**	-	**2.90**	nr	**3.26**
SPH type S, for 63 x 100mm joist	2.23	2.50	**2.29**	-	-	**-**	-	**2.29**	nr	**2.58**
SPH type S, for 63 x 125mm joist	2.27	2.50	**2.32**	-	-	**-**	-	**2.32**	nr	**2.61**
SPH type S, for 63 x 150mm joist	2.59	2.50	**2.66**	-	-	**-**	-	**2.66**	nr	**2.99**
SPH type S, for 63 x 175mm joist	2.44	2.50	**2.50**	-	-	**-**	-	**2.50**	nr	**2.81**
SPH type S, for 63 x 200mm joist	2.59	2.50	**2.66**	-	-	**-**	-	**2.66**	nr	**2.99**
SPH type S, for 63 x 225mm joist	2.78	2.50	**2.85**	-	-	**-**	-	**2.85**	nr	**3.20**
SPH type S, for 63 x 250mm joist	2.96	2.50	**3.04**	-	-	**-**	-	**3.04**	nr	**3.42**
SPH type R, for 50 x 100mm joist	2.64	2.50	**2.71**	-	-	**-**	-	**2.71**	nr	**3.05**
SPH type R, for 50 x 125mm joist	2.69	2.50	**2.75**	-	-	**-**	-	**2.75**	nr	**3.10**
SPH type R, for 50 x 150mm joist	2.78	2.50	**2.85**	-	-	**-**	-	**2.85**	nr	**3.20**
SPH type R, for 50 x 175mm joist	2.88	2.50	**2.95**	-	-	**-**	-	**2.95**	nr	**3.32**
SPH type R, for 50 x 200mm joist	3.06	2.50	**3.14**	-	-	**-**	-	**3.14**	nr	**3.53**
SPH type R, for 50 x 225mm joist	3.35	2.50	**3.44**	-	-	**-**	-	**3.44**	nr	**3.86**
SPH type R, for 50 x 250mm joist	3.98	2.50	**4.08**	-	-	**-**	-	**4.08**	nr	**4.59**
SPH type R, for 63 x 100mm joist	3.23	2.50	**3.31**	-	-	**-**	-	**3.31**	nr	**3.72**
SPH type R, for 63 x 125mm joist	3.28	2.50	**3.36**	-	-	**-**	-	**3.36**	nr	**3.78**
SPH type R, for 63 x 150mm joist	3.64	2.50	**3.73**	-	-	**-**	-	**3.73**	nr	**4.20**
SPH type R, for 63 x 175mm joist	3.42	2.50	**3.51**	-	-	**-**	-	**3.51**	nr	**3.95**
SPH type R, for 63 x 200mm joist	3.64	2.50	**3.73**	-	-	**-**	-	**3.73**	nr	**4.20**
SPH type R, for 63 x 225mm joist	3.90	2.50	**3.99**	-	-	**-**	-	**3.99**	nr	**4.49**
SPH type R, for 63 x 250mm joist	4.11	2.50	**4.21**	-	-	**-**	-	**4.21**	nr	**4.74**
SPH type ST for 50 x 100mm joist	4.63	2.50	**4.75**	-	-	**-**	-	**4.75**	nr	**5.34**
SPH type ST for 50 x 125mm joist	4.77	2.50	**4.89**	-	-	**-**	-	**4.89**	nr	**5.50**
SPH type ST for 50 x 150mm joist	4.53	2.50	**4.64**	-	-	**-**	-	**4.64**	nr	**5.22**
SPH type ST for 50 x 175mm joist	5.18	2.50	**5.31**	-	-	**-**	-	**5.31**	nr	**5.97**
SPH type ST for 50 x 200mm joist	5.42	2.50	**5.56**	-	-	**-**	-	**5.56**	nr	**6.25**
SPH type ST for 50 x 225mm joist	5.93	2.50	**6.08**	-	-	**-**	-	**6.08**	nr	**6.84**
SPH type ST for 50 x 250mm joist	7.20	2.50	**7.38**	-	-	**-**	-	**7.38**	nr	**8.30**
Speedy Minor type for the following size joists; fixing with nails										
38 x 100mm	0.49	2.50	**0.50**	0.18	0.02	**2.77**	0.01	**3.29**	nr	**3.70**
50 x 100mm	0.49	2.50	**0.50**	0.18	0.02	**2.77**	0.01	**3.29**	nr	**3.70**
Speedy short leg type for the following size joists fixing with nails										
38 x 100mm	0.90	2.50	**0.92**	0.18	0.02	**2.77**	0.01	**3.71**	nr	**4.17**
50 x 175mm	0.90	2.50	**0.92**	0.18	0.02	**2.77**	0.01	**3.71**	nr	**4.17**
Speedy Standard Leg type for the following size joists; fixing with nails										
38 x 100mm	0.98	2.50	**1.01**	0.18	0.02	**2.77**	0.01	**3.79**	nr	**4.26**
50 x 175mm	0.98	2.50	**1.01**	0.18	0.02	**2.77**	0.01	**3.79**	nr	**4.26**
63 x 225mm	1.03	2.50	**1.05**	0.18	0.02	**2.77**	0.01	**3.83**	nr	**4.31**
75 x 225mm	1.03	2.50	**1.05**	0.20	0.03	**3.16**	0.01	**4.22**	nr	**4.75**
100 x 225mm	1.14	2.50	**1.17**	0.20	0.03	**3.16**	0.01	**4.34**	nr	**4.88**
Steel; galvanized - joist struts										
Herringbone joist struts; BAT Building Products Ltd. to suit joists at the following centres; fixing with nails										
400mm	0.35	5.00	**0.37**	0.12	0.02	**1.92**	0.01	**2.29**	m	**2.58**
450mm	0.39	5.00	**0.41**	0.12	0.02	**1.92**	0.01	**2.33**	m	**2.63**
600mm	0.45	5.00	**0.47**	0.12	0.02	**1.92**	0.01	**2.40**	m	**2.70**
Steel; galvanized - truss plates and framing anchors										
Truss plates; BAT Building Products Ltd. fixing with nails										
51 x 114mm	0.18	5.00	**0.19**	0.40	0.05	**6.22**	0.02	**6.43**	nr	**7.24**
76 x 254mm	0.68	5.00	**0.71**	1.20	0.15	**18.66**	0.07	**19.43**	nr	**21.86**
114 x 152mm	0.54	5.00	**0.57**	1.20	0.15	**18.66**	0.07	**19.29**	nr	**21.71**

Labour hourly rates: (except Specialists) Craft Operatives £14.21 Labourer £10.71 Rates are national average prices. Refer to REGIONAL VARIATIONS for indicative levels of overall pricing in regions	MATERIALS			LABOUR				RATES		
	Del to Site £	Waste %	Material Cost £	Craft Optve Hrs	Lab Hrs	Labour Cost £	Sunds £	Nett Rate £	Unit	Gross Rate (+12.5%) £
G20: CARPENTRY/TIMBER FRAMING/FIRST FIXING Cont.										
Steel; galvanized - truss plates and framing anchors Cont.										
Truss plates; BAT Building Products Ltd. Cont. fixing with nails Cont.										
114 x 254mm	0.93	5.00	**0.98**	1.80	0.23	**28.04**	0.10	**29.12**	nr	**32.76**
152 x 152mm	0.75	5.00	**0.79**	1.90	0.24	**29.57**	0.10	**30.46**	nr	**34.26**
Framing anchors; BAT Building Products Ltd.										
type A; fixing with nails	0.46	5.00	**0.48**	0.25	0.03	**3.87**	0.01	**4.37**	nr	**4.92**
type B; fixing with nails	0.46	5.00	**0.48**	0.25	0.03	**3.87**	0.01	**4.37**	nr	**4.92**
type C; fixing with nails	0.46	5.00	**0.48**	0.25	0.03	**3.87**	0.01	**4.37**	nr	**4.92**
Steel - timber connectors										
Connectors BS EN 912										
split ring connectors table 1; 64mm diameter	1.21	5.00	**1.27**	0.10	0.01	**1.53**	-	**2.80**	nr	**3.15**
split ring connectors table 1; 100 mm diameter	3.32	5.00	**3.49**	0.12	0.02	**1.92**	-	**5.41**	nr	**6.08**
shear plate connectors, table 2; 67mm diameter	1.98	5.00	**2.08**	0.10	0.01	**1.53**	-	**3.61**	nr	**4.06**
single sided round toothed-plate connectors, table 4										
38mm diameter	0.21	5.00	**0.22**	0.10	0.01	**1.53**	-	**1.74**	nr	**1.96**
51mm diameter	0.28	5.00	**0.29**	0.10	0.01	**1.53**	-	**1.82**	nr	**2.05**
64mm diameter	0.37	5.00	**0.39**	0.10	0.01	**1.53**	-	**1.92**	nr	**2.15**
76mm diameter	0.53	5.00	**0.56**	0.10	0.01	**1.53**	-	**2.09**	nr	**2.35**
double sided round toothed-plate connector, table 4										
38mm diameter	0.28	5.00	**0.29**	0.10	0.01	**1.53**	-	**1.82**	nr	**2.05**
51mm diameter	0.32	5.00	**0.33**	0.10	0.01	**1.53**	-	**1.86**	nr	**2.09**
64mm diameter	0.40	5.00	**0.42**	0.10	0.01	**1.53**	-	**1.95**	nr	**2.19**
76mm diameter	0.57	5.00	**0.60**	0.10	0.01	**1.53**	-	**2.13**	nr	**2.40**
Cast iron - connectors										
Connectors BS EN 912										
shear plate connectors, table 2; 102mm diameter	7.36	5.00	**7.73**	0.14	0.02	**2.20**	-	**9.93**	nr	**11.17**
Steel - bolts and nuts										
Black bolts, BS 4190 grade 4.6, hexagon head										
M 12; with nuts and washers										
100mm long	0.54	5.00	**0.57**	0.11	0.01	**1.67**	-	**2.24**	nr	**2.52**
120mm long	0.67	5.00	**0.70**	0.13	0.02	**2.06**	-	**2.76**	nr	**3.11**
160mm long	1.43	5.00	**1.50**	0.13	0.02	**2.06**	-	**3.57**	nr	**4.01**
180mm long	1.48	5.00	**1.55**	0.14	0.02	**2.20**	-	**3.75**	nr	**4.22**
200mm long	1.54	5.00	**1.62**	0.17	0.02	**2.63**	-	**4.25**	nr	**4.78**
M 16; with nuts and washers										
100mm long	0.76	5.00	**0.79**	0.11	0.01	**1.67**	-	**2.46**	nr	**2.77**
120mm long	1.18	5.00	**1.24**	0.13	0.02	**2.06**	-	**3.30**	nr	**3.71**
160mm long	1.91	5.00	**2.00**	0.14	0.02	**2.20**	-	**4.21**	nr	**4.73**
180mm long	2.01	5.00	**2.11**	0.16	0.02	**2.49**	-	**4.60**	nr	**5.17**
200mm long	2.13	5.00	**2.24**	0.17	0.02	**2.63**	-	**4.87**	nr	**5.48**
M 20; with nuts and washers										
100mm long	1.65	5.00	**1.74**	0.12	0.02	**1.92**	-	**3.66**	nr	**4.11**
120mm long	2.25	5.00	**2.37**	0.11	0.01	**1.67**	-	**4.04**	nr	**4.54**
160mm long	2.84	5.00	**2.98**	0.17	0.02	**2.63**	-	**5.61**	nr	**6.32**
180mm long	2.86	5.00	**3.01**	0.17	0.02	**2.63**	-	**5.64**	nr	**6.34**
200mm long	3.16	5.00	**3.32**	0.17	0.02	**2.63**	-	**5.95**	nr	**6.70**
M 12; with nuts and 38 x 38 x 3mm square plate washers										
100mm long	0.56	5.00	**0.58**	0.11	0.01	**1.67**	-	**2.25**	nr	**2.53**
120mm long	0.69	5.00	**0.72**	0.13	0.02	**2.06**	-	**2.78**	nr	**3.13**
160mm long	1.42	5.00	**1.49**	0.13	0.02	**2.06**	-	**3.55**	nr	**4.00**
180mm long	1.53	5.00	**1.61**	0.14	0.02	**2.20**	-	**3.81**	nr	**4.29**
200mm long	1.58	5.00	**1.65**	0.17	0.02	**2.63**	-	**4.28**	nr	**4.82**
M 12; with nuts and 50 x 50 x 3mm square plate washers										
100mm long	0.60	5.00	**0.63**	0.11	0.01	**1.67**	-	**2.30**	nr	**2.59**
120mm long	0.71	5.00	**0.75**	0.13	0.02	**2.06**	-	**2.81**	nr	**3.16**
160mm long	1.47	5.00	**1.54**	0.13	0.02	**2.06**	-	**3.60**	nr	**4.05**
180mm long	1.53	5.00	**1.61**	0.14	0.02	**2.20**	-	**3.81**	nr	**4.29**
200mm long	1.61	5.00	**1.69**	0.17	0.02	**2.63**	-	**4.32**	nr	**4.86**
Black bolts, BS 4933 Grade 4.6, cup head, square neck										
M 6; with nuts and washers										
25mm long	0.08	5.00	**0.09**	0.08	0.01	**1.24**	-	**1.33**	nr	**1.50**
50mm long	0.09	5.00	**0.10**	0.08	0.01	**1.24**	-	**1.34**	nr	**1.51**
75mm long	0.13	5.00	**0.14**	0.10	0.01	**1.53**	-	**1.67**	nr	**1.88**
100mm long	0.19	5.00	**0.20**	0.10	0.01	**1.53**	-	**1.73**	nr	**1.94**
150mm long	0.27	5.00	**0.28**	0.11	0.01	**1.67**	-	**1.95**	nr	**2.19**
M 8; with nuts and washers										
25mm long	0.13	5.00	**0.14**	0.08	0.01	**1.24**	-	**1.38**	nr	**1.56**
50mm long	0.14	5.00	**0.15**	0.08	0.01	**1.24**	-	**1.40**	nr	**1.57**
75mm long	0.19	5.00	**0.20**	0.10	0.01	**1.53**	-	**1.73**	nr	**1.94**
100mm long	0.28	5.00	**0.29**	0.10	0.01	**1.53**	-	**1.82**	nr	**2.05**
150mm long	0.40	5.00	**0.42**	0.11	0.01	**1.67**	-	**2.09**	nr	**2.35**
M 10; with nuts and washers										
50mm long	0.23	5.00	**0.24**	0.08	0.01	**1.24**	-	**1.49**	nr	**1.67**
75mm long	0.27	5.00	**0.28**	0.10	0.01	**1.53**	-	**1.81**	nr	**2.03**
100mm long	0.40	5.00	**0.42**	0.10	0.01	**1.53**	-	**1.95**	nr	**2.19**
150mm long	0.58	5.00	**0.61**	0.12	0.02	**1.92**	-	**2.53**	nr	**2.84**
M 12; with nuts and washers										
50mm long	0.34	5.00	**0.36**	0.10	0.01	**1.53**	-	**1.89**	nr	**2.13**
75mm long	0.46	5.00	**0.48**	0.10	0.01	**1.53**	-	**2.01**	nr	**2.26**

Labour hourly rates: (except Specialists) Craft Operatives £14.21 Labourer £10.71 Rates are national average prices. Refer to REGIONAL VARIATIONS for indicative levels of overall pricing in regions	MATERIALS			LABOUR				RATES		
	Del to Site £	Waste %	Material Cost £	Craft Optve Hrs	Lab Hrs	Labour Cost £	Sunds £	Nett Rate £	Unit	Gross Rate (+12.5%) £
G20: CARPENTRY/TIMBER FRAMING/FIRST FIXING Cont.										
Steel - bolts and nuts Cont.										
Black bolts, BS 4933 Grade 4.6, cup head, square neck Cont.										
M 12; with nuts and washers Cont.										
100mm long	0.56	5.00	**0.58**	0.11	0.01	**1.67**	-	**2.25**	nr	**2.53**
150mm long	0.78	5.00	**0.82**	0.13	0.02	**2.06**	-	**2.88**	nr	**3.24**
Steel; galvanized - bolts and nuts										
Black bolts, BS 4190 grade 4.6, hexagon head										
M 12; with nuts and washers										
100mm long	0.84	5.00	**0.88**	0.11	0.01	**1.67**	-	**2.55**	nr	**2.87**
120mm long	1.04	5.00	**1.09**	0.13	0.02	**2.06**	-	**3.15**	nr	**3.55**
160mm long	2.17	5.00	**2.28**	0.13	0.02	**2.06**	-	**4.34**	nr	**4.88**
180mm long	2.28	5.00	**2.40**	0.14	0.02	**2.20**	-	**4.60**	nr	**5.18**
200mm long	2.38	5.00	**2.50**	0.17	0.02	**2.63**	-	**5.13**	nr	**5.77**
M 16; with nuts and washers										
100mm long	1.40	5.00	**1.47**	0.11	0.01	**1.67**	-	**3.14**	nr	**3.53**
120mm long	1.82	5.00	**1.91**	0.13	0.02	**2.06**	-	**3.97**	nr	**4.47**
160mm long	2.91	5.00	**3.06**	0.14	0.02	**2.20**	-	**5.26**	nr	**5.92**
180mm long	3.09	5.00	**3.24**	0.16	0.02	**2.49**	-	**5.73**	nr	**6.45**
200mm long	3.29	5.00	**3.45**	0.17	0.02	**2.63**	-	**6.08**	nr	**6.84**
M 20; with nuts and washers										
100mm long	2.56	5.00	**2.69**	0.12	0.02	**1.92**	-	**4.61**	nr	**5.19**
120mm long	3.45	5.00	**3.62**	0.11	0.01	**1.67**	-	**5.29**	nr	**5.95**
160mm long	4.36	5.00	**4.58**	0.17	0.02	**2.63**	-	**7.21**	nr	**8.11**
180mm long	4.52	5.00	**4.75**	0.17	0.02	**2.63**	-	**7.38**	nr	**8.30**
200mm long	4.81	5.00	**5.05**	0.17	0.02	**2.63**	-	**7.68**	nr	**8.65**
M 12; with nuts and 38 x 38 x 3mm square plate washers										
100mm long	0.92	5.00	**0.97**	0.11	0.01	**1.67**	-	**2.64**	nr	**2.97**
120mm long	1.11	5.00	**1.16**	0.13	0.02	**2.06**	-	**3.22**	nr	**3.63**
160mm long	2.25	5.00	**2.36**	0.13	0.02	**2.06**	-	**4.42**	nr	**4.98**
180mm long	2.37	5.00	**2.49**	0.14	0.02	**2.20**	-	**4.69**	nr	**5.28**
200mm long	2.47	5.00	**2.59**	0.17	0.02	**2.63**	-	**5.22**	nr	**5.88**
M 12; with nuts and 50 x 50 x 3mm square plate washers										
100mm long	0.98	5.00	**1.03**	0.11	0.01	**1.67**	-	**2.70**	nr	**3.04**
120mm long	1.15	5.00	**1.21**	0.13	0.02	**2.06**	-	**3.27**	nr	**3.68**
160mm long	2.31	5.00	**2.42**	0.13	0.02	**2.06**	-	**4.48**	nr	**5.05**
180mm long	2.44	5.00	**2.56**	0.14	0.02	**2.20**	-	**4.76**	nr	**5.36**
200mm long	2.60	5.00	**2.73**	0.17	0.02	**2.63**	-	**5.36**	nr	**6.03**
Steel - expanding bolts										
Expanding bolts; bolt projecting Rawlbolts; drilling masonry										
with nuts and washers; reference										
44 - 505 (M6 10P)	0.94	5.00	**0.98**	0.24	0.03	**3.73**	-	**4.72**	nr	**5.31**
44 - 510 (M6 25P)	1.05	5.00	**1.10**	0.24	0.03	**3.73**	-	**4.84**	nr	**5.44**
44 - 515 (M6 60P)	1.10	5.00	**1.16**	0.24	0.03	**3.73**	-	**4.89**	nr	**5.50**
44 - 555 (M8 10P)	1.18	5.00	**1.24**	0.26	0.03	**4.02**	-	**5.26**	nr	**5.91**
44 - 560 (M8 25P)	1.25	5.00	**1.31**	0.26	0.03	**4.02**	-	**5.32**	nr	**5.99**
44 - 565 (M8 60P)	1.33	5.00	**1.40**	0.26	0.03	**4.02**	-	**5.41**	nr	**6.09**
44 - 565 (M8 60P)	1.62	5.00	**1.70**	0.30	0.04	**4.69**	-	**6.39**	nr	**7.19**
44 - 610 (M10 30P)	1.70	5.00	**1.78**	0.30	0.04	**4.69**	-	**6.47**	nr	**7.28**
44 - 615 (M10 60P)	1.77	5.00	**1.86**	0.30	0.04	**4.69**	-	**6.55**	nr	**7.37**
44 - 655 (M12 15P)	2.58	5.00	**2.71**	0.32	0.04	**4.98**	-	**7.68**	nr	**8.64**
44 - 660 (M12 30P)	2.77	5.00	**2.91**	0.32	0.04	**4.98**	-	**7.89**	nr	**8.87**
44 - 665 (M12 75P)	3.45	5.00	**3.63**	0.32	0.04	**4.98**	-	**8.60**	nr	**9.68**
44 - 705 (M16 15P)	6.10	5.00	**6.41**	0.36	0.05	**5.65**	-	**12.06**	nr	**13.56**
44 - 710 (M16 35P)	6.60	5.00	**6.93**	0.36	0.05	**5.65**	-	**12.58**	nr	**14.15**
44 - 715 (M16 75P)	6.92	5.00	**7.26**	0.36	0.05	**5.65**	-	**12.91**	nr	**14.53**
44 - 755 (M20 15P)	9.19	5.00	**9.65**	0.40	0.05	**6.22**	-	**15.87**	nr	**17.85**
44 - 760 (M20 30P)	9.93	5.00	**10.42**	0.40	0.05	**6.22**	-	**16.64**	nr	**18.72**
44 - 765 (M20 100P)	11.29	5.00	**11.85**	0.40	0.05	**6.22**	-	**18.07**	nr	**20.33**
Expanding bolts; Sleeve Anchor (Rawlok); bolt projecting type; drilling masonry										
with nuts and washers; reference										
69 - 504 (M4.5 9P)	0.27	5.00	**0.28**	0.19	0.02	**2.91**	-	**3.19**	nr	**3.59**
69 - 506 (M4.5 27P)	0.33	5.00	**0.34**	0.19	0.02	**2.91**	-	**3.26**	nr	**3.66**
69 - 508 (M6 9P)	0.34	5.00	**0.36**	0.22	0.03	**3.45**	-	**3.81**	nr	**4.28**
69 - 510 (M6 35P)	0.43	5.00	**0.45**	0.22	0.03	**3.45**	-	**3.90**	nr	**4.38**
69 - 512 (M6 60P)	0.57	5.00	**0.60**	0.22	0.03	**3.45**	-	**4.05**	nr	**4.55**
69 - 514 (M8 9P)	0.49	5.00	**0.52**	0.24	0.03	**3.73**	-	**4.25**	nr	**4.78**
69 - 516 (M8 36P)	0.63	5.00	**0.66**	0.24	0.03	**3.73**	-	**4.40**	nr	**4.95**
69 - 518 (M8 60P)	0.82	5.00	**0.86**	0.24	0.03	**3.73**	-	**4.60**	nr	**5.17**
69 - 520 (M10 9P)	0.75	5.00	**0.79**	0.26	0.03	**4.02**	-	**4.81**	nr	**5.41**
69 - 522 (M10 22P)	0.84	5.00	**0.88**	0.26	0.03	**4.02**	-	**4.90**	nr	**5.51**
69 - 524 (M10 50P)	1.16	5.00	**1.22**	0.26	0.03	**4.02**	-	**5.24**	nr	**5.89**
69 - 525 (M10 80P)	1.44	5.00	**1.52**	0.26	0.03	**4.02**	-	**5.53**	nr	**6.22**
69 - 526 (M12 13P)	1.53	5.00	**1.61**	0.30	0.04	**4.69**	-	**6.30**	nr	**7.08**
69 - 528 (M12 55P)	1.80	5.00	**1.89**	0.30	0.04	**4.69**	-	**6.58**	nr	**7.40**
69 - 530 (M12 90P)	2.58	5.00	**2.71**	0.30	0.04	**4.69**	-	**7.40**	nr	**8.33**
69 - 533 (M16 25P)	2.60	5.00	**2.73**	0.32	0.04	**4.98**	-	**7.70**	nr	**8.67**
69 - 534 (M16 57P)	3.35	5.00	**3.51**	0.32	0.04	**4.98**	-	**8.49**	nr	**9.55**
69 - 536 (M16 100P)	4.03	5.00	**4.24**	0.32	0.04	**4.98**	-	**9.21**	nr	**10.36**
Expanding bolts; loose bolt Rawlbolts; drilling masonry										
with washers; reference										
44 - 015 (M6 10L)	0.94	5.00	**0.99**	0.24	0.03	**3.73**	-	**4.72**	nr	**5.31**

Labour hourly rates: (except Specialists) Craft Operatives £14.21 Labourer £10.71 Rates are national average prices. Refer to REGIONAL VARIATIONS for indicative levels of overall pricing in regions	MATERIALS			LABOUR				RATES		
	Del to Site £	Waste %	Material Cost £	Craft Optve Hrs	Lab Hrs	Labour Cost £	Sunds £	Nett Rate £	Unit	Gross Rate (+12.5%) £
G20: CARPENTRY/TIMBER FRAMING/FIRST FIXING Cont.										
Steel - expanding bolts Cont.										
Expanding bolts; loose bolt Rawlbolts; drilling masonry Cont.										
with washers; reference Cont.										
44 - 020 (M6 25L)	1.00	5.00	1.04	0.24	0.03	3.73	-	4.78	nr	5.37
44 - 025 (M6 40L)	1.00	5.00	1.05	0.24	0.03	3.73	-	4.78	nr	5.38
44 - 055 (M8 10L)	1.17	5.00	1.23	0.26	0.03	4.02	-	5.25	nr	5.90
44 - 060 (M8 25L)	1.22	5.00	1.28	0.26	0.03	4.02	-	5.30	nr	5.96
44 - 065 (M8 40L)	1.30	5.00	1.36	0.26	0.03	4.02	-	5.38	nr	6.05
44 - 105 (M10 10L)	1.56	5.00	1.64	0.30	0.04	4.69	-	6.33	nr	7.13
44 - 110 (M10 25L)	1.61	5.00	1.69	0.30	0.04	4.69	-	6.38	nr	7.18
44 - 115 (M10 50L)	1.70	5.00	1.79	0.30	0.04	4.69	-	6.48	nr	7.29
44 - 120 (M10 75L)	1.76	5.00	1.85	0.30	0.04	4.69	-	6.54	nr	7.36
44 - 155 (M12 10L)	2.34	5.00	2.45	0.32	0.04	4.98	-	7.43	nr	8.36
44 - 160 (M12 25L)	2.58	5.00	2.71	0.32	0.04	4.98	-	7.68	nr	8.64
44 - 165 (M12 40L)	2.69	5.00	2.83	0.32	0.04	4.98	-	7.80	nr	8.78
44 - 170 (M12 60L)	2.84	5.00	2.98	0.32	0.04	4.98	-	7.95	nr	8.95
44 - 205 (M16 15L)	5.44	5.00	5.71	0.36	0.05	5.65	-	11.36	nr	12.78
44 - 210 (M16 30L)	6.27	5.00	6.59	0.36	0.05	5.65	-	12.24	nr	13.77
44 - 215 (M16 60L)	6.79	5.00	7.13	0.36	0.05	5.65	-	12.78	nr	14.37
44 - 255 (M20 60L)	10.65	5.00	11.18	0.40	0.05	6.22	-	17.40	nr	19.58
44 - 260 (M20 100L)	11.05	5.00	11.60	0.40	0.05	6.22	-	17.82	nr	20.05
Spit Mega high performance safety anchors; loose nut version; drilling masonry										
with nuts and washers, reference										
E10 - 15/20	2.37	5.00	2.49	0.30	0.03	4.58	-	7.08	nr	7.96
E10 - 15/45	2.63	5.00	2.76	0.30	0.03	4.58	-	7.34	nr	8.26
E10 - 15/65	2.79	5.00	2.93	0.30	0.03	4.58	-	7.52	nr	8.46
E12 - 18/25	3.35	5.00	3.51	0.32	0.03	4.87	-	8.38	nr	9.43
E12 - 18/45	3.48	5.00	3.65	0.32	0.03	4.87	-	8.52	nr	9.59
E12 - 18/65	3.72	5.00	3.91	0.32	0.03	4.87	-	8.78	nr	9.87
E16 - 24/25	9.43	5.00	9.90	0.36	0.03	5.44	-	15.34	nr	17.25
E16 - 24/45	10.00	5.00	10.49	0.36	0.04	5.54	-	16.04	nr	18.04
E16 - 24/95	12.28	5.00	12.90	0.36	0.04	5.54	-	18.44	nr	20.75
E20 - 28/25	13.78	5.00	14.46	0.40	0.04	6.11	-	20.58	nr	23.15
E20 - 28/45	16.49	5.00	17.31	0.40	0.04	6.11	-	23.42	nr	26.35
E20 - 28/95	16.58	5.00	17.41	0.40	0.04	6.11	-	23.52	nr	26.46
Spit Mega high performance safety anchors; bolt head version; drilling masonry										
with washers, reference										
V10 - 15/20	2.37	5.00	2.49	0.30	0.03	4.58	-	7.08	nr	7.96
V12 - 18/25	3.36	5.00	3.53	0.32	0.03	4.87	-	8.40	nr	9.44
V16 - 24/25	7.26	5.00	7.62	0.36	0.04	5.54	-	13.17	nr	14.81
V20 - 28/25	13.76	5.00	14.45	0.40	0.04	6.11	-	20.56	nr	23.14
Spit Fix high performance through bolt BZP anchors; drilling masonry										
with nuts and washers, reference										
6/10/45	0.34	5.00	0.35	0.22	0.03	3.45	-	3.80	nr	4.28
6/20/55	0.35	5.00	0.37	0.22	0.03	3.45	-	3.81	nr	4.29
6/50/85	0.39	5.00	0.41	0.22	0.03	3.45	-	3.85	nr	4.33
8/10/50	0.41	5.00	0.43	0.24	0.03	3.73	-	4.16	nr	4.68
8/50/90	0.53	5.00	0.56	0.24	0.03	3.73	-	4.29	nr	4.82
8/90/130	0.61	5.00	0.64	0.24	0.03	3.73	-	4.38	nr	4.92
10/10/60	0.63	5.00	0.66	0.26	0.03	4.02	-	4.67	nr	5.26
10/25/75	0.64	5.00	0.67	0.26	0.03	4.02	-	4.69	nr	5.27
10/45/95	0.69	5.00	0.72	0.26	0.03	4.02	-	4.74	nr	5.33
12/10/70	0.83	5.00	0.87	0.30	0.03	4.58	-	5.46	nr	6.14
12/40/100	1.13	5.00	1.19	0.30	0.03	4.58	-	5.77	nr	6.49
12/80/140	1.31	5.00	1.38	0.30	0.03	4.58	-	5.96	nr	6.71
12/120/180	1.66	5.00	1.74	0.30	0.03	4.58	-	6.33	nr	7.12
12/160/220	2.84	5.00	2.98	0.30	0.03	4.58	-	7.57	nr	8.51
16/10/90	2.23	5.00	2.34	0.32	0.03	4.87	-	7.21	nr	8.11
16/45/125	2.26	5.00	2.38	0.32	0.03	4.87	-	7.24	nr	8.15
16/95/175	2.31	5.00	2.43	0.32	0.03	4.87	-	7.30	nr	8.21
20/20/120	2.42	5.00	2.54	0.36	0.03	5.44	-	7.98	nr	8.98
20/60/160	2.66	5.00	2.80	0.36	0.04	5.54	-	8.34	nr	9.38
20/115/215	2.95	5.00	3.10	0.36	0.04	5.54	-	8.64	nr	9.72
Steel; plated - expanding bolts										
Expanding bolts; hook Rawlbolts; drilling masonry										
with nuts and washers; reference										
44 - 401 (M6)	1.26	5.00	1.32	0.24	0.03	3.73	-	5.06	nr	5.69
44 - 406 (M8)	1.54	5.00	1.62	0.26	0.03	4.02	-	5.63	nr	6.34
44 - 411 (M10)	2.48	5.00	2.61	0.30	0.04	4.69	-	7.30	nr	8.21
44 - 416 (M12)	3.92	5.00	4.12	0.32	0.04	4.98	-	9.09	nr	10.23
Expanding bolts; eye Rawlbolts; drilling masonry										
with nuts and washers; reference										
44 - 432 (M6)	2.79	5.00	2.93	0.24	0.03	3.73	-	6.66	nr	7.49
44 - 437 (M8)	3.18	5.00	3.34	0.26	0.03	4.02	-	7.36	nr	8.28
44 - 442 (M10)	4.26	5.00	4.47	0.30	0.04	4.69	-	9.16	nr	10.31
44 - 447 (M12)	6.37	5.00	6.68	0.32	0.04	4.98	-	11.66	nr	13.12
Expanding bolts; bolt projecting Through Bolts; drilling masonry										
with nuts and washers; reference										
56 - 104 (M6 x 55mm)	0.41	5.00	0.43	0.22	0.03	3.45	-	3.87	nr	4.36
56 - 110 (M6 x 85mm)	0.53	5.00	0.55	0.22	0.03	3.45	-	4.00	nr	4.50
56 - 114 (M8 x 50mm)	0.57	5.00	0.60	0.24	0.03	3.73	-	4.33	nr	4.87

Labour hourly rates: (except Specialists) Craft Operatives £14.21 Labourer £10.71 Rates are national average prices. Refer to REGIONAL VARIATIONS for indicative levels of overall pricing in regions	MATERIALS			LABOUR				RATES		
	Del to Site £	Waste %	Material Cost £	Craft Optve Hrs	Lab Hrs	Labour Cost £	Sunds £	Nett Rate £	Unit	Gross Rate (+12.5%) £
G20: CARPENTRY/TIMBER FRAMING/FIRST FIXING Cont.										
Steel; plated - expanding bolts Cont.										
Expanding bolts; bolt projecting Through Bolts; drilling masonry Cont.										
with nuts and washers; reference Cont.										
56 - 116 (M8 x 65mm)	0.59	5.00	**0.62**	0.24	0.03	**3.73**	-	**4.35**	nr	**4.90**
56 - 120 (M8 x 80mm)	0.62	5.00	**0.65**	0.24	0.03	**3.73**	-	**4.38**	nr	**4.93**
56 - 124 (M8 x 105mm)	0.66	5.00	**0.70**	0.24	0.03	**3.73**	-	**4.43**	nr	**4.98**
56 - 125 (M8 x 115mm)	0.73	5.00	**0.76**	0.24	0.03	**3.73**	-	**4.50**	nr	**5.06**
56 - 128 (M10 x 80mm)	0.76	5.00	**0.80**	0.26	0.03	**4.02**	-	**4.81**	nr	**5.41**
56 - 129 (M10 x 65mm)	0.72	5.00	**0.76**	0.26	0.03	**4.02**	-	**4.77**	nr	**5.37**
56 - 132 (M10 x 95mm)	0.80	5.00	**0.84**	0.26	0.03	**4.02**	-	**4.85**	nr	**5.46**
56 - 136 (M10 x 115mm)	0.86	5.00	**0.90**	0.26	0.03	**4.02**	-	**4.91**	nr	**5.53**
56 - 138 (M10 x 130mm)	0.99	5.00	**1.04**	0.26	0.03	**4.02**	-	**5.05**	nr	**5.69**
56 - 139 (M12 x 80mm)	1.14	5.00	**1.19**	0.30	0.04	**4.69**	-	**5.89**	nr	**6.62**
56 - 140 (M12 x 100mm)	1.15	5.00	**1.21**	0.30	0.04	**4.69**	-	**5.90**	nr	**6.63**
56 - 144 (M12 x 120mm)	1.24	5.00	**1.30**	0.30	0.04	**4.69**	-	**5.99**	nr	**6.74**
56 - 148 (M12 x 135mm)	1.32	5.00	**1.39**	0.30	0.04	**4.69**	-	**6.08**	nr	**6.84**
56 - 150 (M12 x 150mm)	1.39	5.00	**1.46**	0.30	0.04	**4.69**	-	**6.15**	nr	**6.92**
56 - 152 (M16 x 140mm)	2.07	5.00	**2.17**	0.32	0.04	**4.98**	-	**7.15**	nr	**8.04**
56 - 153 (M16 x 105mm)	1.85	5.00	**1.94**	0.32	0.04	**4.98**	-	**6.91**	nr	**7.78**
56 - 156 (M16 x 180mm)	2.51	5.00	**2.64**	0.32	0.04	**4.98**	-	**7.61**	nr	**8.56**
56 - 158 (M16 x 220mm)	3.25	5.00	**3.41**	0.32	0.04	**4.98**	-	**8.39**	nr	**9.43**
56 - 159 (M20 x 125mm)	3.26	5.00	**3.42**	0.36	0.05	**5.65**	-	**9.07**	nr	**10.20**
56 - 160 (M20 x 160mm)	3.55	5.00	**3.73**	0.36	0.05	**5.65**	-	**9.38**	nr	**10.55**
56 - 164 (M20 x 200mm)	4.53	5.00	**4.75**	0.36	0.05	**5.65**	-	**10.40**	nr	**11.71**
56 - 166 (M20 x 300mm)	5.65	5.00	**5.94**	0.36	0.05	**5.65**	-	**11.59**	nr	**13.04**
56 - 168 (M24 x 180mm)	6.49	5.00	**6.82**	0.40	0.05	**6.22**	-	**13.04**	nr	**14.67**
56 - 172 (M24 x 260mm)	7.97	5.00	**8.37**	0.40	0.05	**6.22**	-	**14.59**	nr	**16.42**
Stainless steel grade 316 - expanding bolts										
Expanding bolts; bolt projecting Through Bolts; drilling masonry										
with nuts and washers; reference										
56 - 604 (M6 x 55mm)	1.92	5.00	**2.01**	0.22	0.03	**3.45**	-	**5.46**	nr	**6.14**
56 - 616 (M8 x 65mm)	2.34	5.00	**2.46**	0.24	0.03	**3.73**	-	**6.19**	nr	**6.97**
56 - 624 (M8 x 105mm)	2.90	5.00	**3.04**	0.24	0.03	**3.73**	-	**6.77**	nr	**7.62**
56 - 628 (M10 x 80mm)	3.02	5.00	**3.17**	0.26	0.03	**4.02**	-	**7.19**	nr	**8.08**
56 - 636 (M10 x 115mm)	3.57	5.00	**3.74**	0.26	0.03	**4.02**	-	**7.76**	nr	**8.73**
56 - 638 (M10 x 130mm)	3.99	5.00	**4.19**	0.26	0.03	**4.02**	-	**8.20**	nr	**9.23**
56 - 640 (M12 x 100mm)	5.48	5.00	**5.75**	0.30	0.04	**4.69**	-	**10.45**	nr	**11.75**
56 - 648 (M12 x 135mm)	6.60	5.00	**6.93**	0.30	0.04	**4.69**	-	**11.62**	nr	**13.07**
56 - 650 (M12 x 150mm)	7.34	5.00	**7.71**	0.30	0.04	**4.69**	-	**12.40**	nr	**13.95**
56 - 652 (M16 x 140mm)	9.44	5.00	**9.91**	0.32	0.04	**4.98**	-	**14.89**	nr	**16.75**
56 - 658 (M16 x 220mm)	13.64	5.00	**14.32**	0.32	0.04	**4.98**	-	**19.29**	nr	**21.71**
56 - 660 (M20 x 160mm)	18.49	5.00	**19.42**	0.36	0.05	**5.65**	-	**25.07**	nr	**28.20**
56 - 666 (M20 x 300mm)	27.42	5.00	**28.79**	0.36	0.05	**5.65**	-	**34.45**	nr	**38.75**
56 - 672 (M24 x 260mm)	37.84	5.00	**39.73**	0.40	0.05	**6.22**	-	**45.95**	nr	**51.69**
Chemical anchors										
Chemical anchors; Kemfix capsules and standard studs; drilling masonry										
capsule reference 60-428; stud reference 60-649; with nuts and washers (M8 x 110mm)	1.91	5.00	**2.01**	0.29	0.04	**4.55**	-	**6.56**	nr	**7.38**
capsule reference 60-430; stud reference 60-653; with nuts and washers (M10 x 130mm)	2.14	5.00	**2.25**	0.32	0.04	**4.98**	-	**7.23**	nr	**8.13**
capsule reference 60-432; stud reference 60-661; with nuts and washers (M12 x 160mm)	2.63	5.00	**2.76**	0.36	0.05	**5.65**	-	**8.41**	nr	**9.46**
Chemical anchors; Kemfix capsules and stainless steel studs; drilling masonry										
capsule reference 60-428; stud reference 60-906; with nuts and washers (M8 x 110mm)	3.29	5.00	**3.46**	0.29	0.04	**4.55**	-	**8.01**	nr	**9.01**
capsule reference 60-430; stud reference 60-911; with nuts and washers (M10 x 130mm)	4.40	5.00	**4.61**	0.32	0.04	**4.98**	-	**9.59**	nr	**10.79**
capsule reference 60-432; stud reference 60-916; with nuts and washers (M12 x 160mm)	6.04	5.00	**6.34**	0.36	0.05	**5.65**	-	**11.99**	nr	**13.49**
capsule reference 60-436; stud reference 60-921; with nuts and washers (M16 x 190mm)	10.29	5.00	**10.80**	0.40	0.05	**6.22**	-	**17.02**	nr	**19.15**
capsule reference 60-442; stud reference 60-926; with nuts and washers (M20 x 240mm)	16.18	5.00	**16.99**	0.42	0.05	**6.50**	-	**23.49**	nr	**26.43**
capsule reference 60-446; stud reference 60-931; with nuts and washers (M24 x 295mm)	26.37	5.00	**27.68**	0.46	0.06	**7.18**	-	**34.86**	nr	**39.22**
Chemical anchors; Kemfix capsules and standard internal threaded sockets; drilling masonry										
capsule reference 60-428; socket reference 60-622 (M8 x 75mm)	2.57	5.00	**2.70**	0.36	0.05	**5.65**	-	**8.35**	nr	**9.39**
capsule reference 60-430; socket reference 60-624 (M10 x 75mm)	2.96	5.00	**3.10**	0.40	0.05	**6.22**	-	**9.32**	nr	**10.49**
capsule reference 60-432; socket reference 60-628 (M12 x 100mm)	4.01	5.00	**4.21**	0.42	0.05	**6.50**	-	**10.71**	nr	**12.05**
capsule reference 60-436; socket reference 60-630 (M16 x 125mm)	7.40	5.00	**7.77**	0.46	0.06	**7.18**	-	**14.95**	nr	**16.82**
Chemcial anchors; Kemfix capsules and stainless steel internal threaded sockets; drilling masonry										
capsule reference 60-428; socket reference 60-987	4.42	5.00	**4.64**	0.36	0.05	**5.65**	-	**10.29**	nr	**11.58**
capsule reference 60-430; socket reference 60-989	5.72	5.00	**6.01**	0.40	0.05	**6.22**	-	**12.23**	nr	**13.76**
capsule reference 60-432; socket reference 60-993	7.35	5.00	**7.72**	0.42	0.05	**6.50**	-	**14.22**	nr	**16.00**
capsule reference 60-436; socket reference 60-995	14.52	5.00	**15.24**	0.46	0.06	**7.18**	-	**22.42**	nr	**25.23**

Labour hourly rates: (except Specialists) Craft Operatives £14.21 Labourer £10.71 Rates are national average prices. Refer to REGIONAL VARIATIONS for indicative levels of overall pricing in regions	MATERIALS			LABOUR				RATES		
	Del to Site £	Waste %	Material Cost £	Craft Optve Hrs	Lab Hrs	Labour Cost £	Sunds £	Nett Rate £	Unit	Gross Rate (+12.5%) £
G20: CARPENTRY/TIMBER FRAMING/FIRST FIXING Cont.										
Chemical anchors Cont.										
Chemical anchors in low density material; Kemfix capsules, mesh sleeves and standard studs; drilling masonry										
capsule reference 60-428; sleeve reference 60-805; stud reference 60-805; with nuts and washers.....	2.69	5.00	**2.82**	0.25	0.03	**3.87**	-	**6.70**	nr	**7.53**
capsule reference 60-430; sleeve reference 60-807; stud reference 60-807; with nuts and washers.....	3.21	5.00	**3.37**	0.32	0.04	**4.98**	-	**8.34**	nr	**9.38**
capsule reference 60-432; sleeve reference 60-550; stud reference 60-113; with nuts and washers.....	3.78	5.00	**3.97**	0.36	0.05	**5.65**	-	**9.62**	nr	**10.82**
capsule reference 60-436; sleeve reference 60-562; stud reference 60-117; with nuts and washers.....	5.29	5.00	**5.56**	0.42	0.05	**6.50**	-	**12.06**	nr	**13.57**
Chemical anchors in low density material; Kemfix capsules, mesh sleeves and stainless steel studs; drilling masonry										
capsule reference 60-428; sleeve reference 60-805; stud reference 60-905; with nuts and washers.....	4.12	5.00	**4.32**	0.29	0.04	**4.55**	-	**8.87**	nr	**9.98**
capsule reference 60-430; sleeve reference 60-807; stud reference 60-910; with nuts and washers.....	5.51	5.00	**5.78**	0.32	0.04	**4.98**	-	**10.76**	nr	**12.10**
capsule reference 60-432; sleeve reference 60-113; stud reference 60-915; with nuts and washers.....	7.27	5.00	**7.63**	0.36	0.05	**5.65**	-	**13.28**	nr	**14.94**
capsule reference 60-436; sleeve reference 60-117; stud reference 60-920; with nuts and washers.....	11.57	5.00	**12.15**	0.42	0.05	**6.50**	-	**18.65**	nr	**20.98**
Chemical anchors in low density material; Kemfix capsules, mesh sleeves and standard internal threaded sockets; drilling masonry										
capsule reference 60-428; sleeve reference 60-805; socket reference 60-662..........................	3.39	5.00	**3.56**	0.29	0.04	**4.55**	-	**8.11**	nr	**9.13**
capsule reference 60-430; sleeve reference 60-807; socket reference 60-668..........................	4.07	5.00	**4.27**	0.32	0.04	**4.98**	-	**9.25**	nr	**10.40**
capsule reference 60-432; sleeve reference 60-113; socket reference 60-674..........................	5.24	5.00	**5.50**	0.36	0.05	**5.65**	-	**11.15**	nr	**12.54**
capsule reference 60-436; sleeve reference 60-117; socket reference 60-676..........................	8.68	5.00	**9.11**	0.40	0.05	**6.22**	-	**15.33**	nr	**17.25**
Chemical anchors in low density material; Kemfix capsules, mesh sleeves and stainless steel internal threaded sockets; drilling masonry										
capsule reference 60-428; sleeve reference 60-805; socket reference 60-947..........................	5.25	5.00	**5.51**	0.29	0.04	**4.55**	-	**10.06**	nr	**11.31**
capsule reference 60-430; sleeve reference 60-807; socket reference 60-949..........................	6.83	5.00	**7.17**	0.32	0.04	**4.98**	-	**12.15**	nr	**13.67**
capsule reference 60-436; sleeve reference 60-117; socket reference 60-955..........................	15.80	5.00	**16.59**	0.40	0.05	**6.22**	-	**22.81**	nr	**25.66**
Spit Maxi high performance chemical anchors; capsules and zinc coated steelstuds; drilling masonry										
capsule reference M8 stud reference SM8; with nuts and washers	1.57	5.00	**1.64**	0.28	0.03	**4.30**	-	**5.94**	nr	**6.69**
capsule reference M10; stud reference SM10; with nuts and washers	1.90	5.00	**2.00**	0.31	0.04	**4.83**	-	**6.83**	nr	**7.68**
capsule reference M12; stud reference SM12; with nuts and washers	2.43	5.00	**2.55**	0.35	0.03	**5.29**	-	**7.85**	nr	**8.83**
capsule reference M16; stud reference SM16; with nuts and washers	3.37	5.00	**3.54**	0.38	0.03	**5.72**	-	**9.26**	nr	**10.42**
capsule reference M20; stud reference SM20; with nuts and washers	5.76	5.00	**6.04**	0.40	0.04	**6.11**	-	**12.16**	nr	**13.68**
capsule reference M24; stud reference SM24; with nuts and washers	10.12	5.00	**10.62**	0.44	0.04	**6.68**	-	**17.30**	nr	**19.46**
Spit Maxi high performance chemical anchors; capsules and stainless steel studs Grade 316 (A4); drilling masonry										
capsule reference M8 stud reference SM8i; with nuts and washers	1.78	5.00	**1.87**	0.28	0.03	**4.30**	-	**6.17**	nr	**6.94**
capsule reference M10; stud reference SM10i; with nuts and washers	2.37	5.00	**2.49**	0.31	0.03	**4.73**	-	**7.22**	nr	**8.12**
capsule reference M12; stud reference SM12i; with nuts and washers	3.40	5.00	**3.57**	0.35	0.03	**5.29**	-	**8.86**	nr	**9.97**
capsule reference M16; stud reference SM16i; with nuts and washers	5.39	5.00	**5.66**	0.38	0.03	**5.72**	-	**11.38**	nr	**12.81**
capsule reference M20; stud reference SM20i; with nuts and washers	9.71	5.00	**10.19**	0.40	0.04	**6.11**	-	**16.30**	nr	**18.34**
capsule reference M24; stud reference SM24i; with nuts and washers	14.17	5.00	**14.88**	0.44	0.04	**6.68**	-	**21.56**	nr	**24.26**
G30: METAL PROFILED SHEET DECKING										
Galvanized steel troughed decking; 0.7mm thick metal, 35mm overall depth; natural soffit and top surface; 150mm end laps and one corrugation side laps										
Decking; fixed to steel rails at 1200mm centres with self tapping screws; drilling holes										
sloping; 10 degrees pitch......................	-	-	**Spclist**	-	-	**Spclist**	-	**31.41**	m2	**35.34**
Extra over decking for										
raking cutting	-	-	**Spclist**	-	-	**Spclist**	-	**5.20**	m	**5.85**
holes 50mm diameter; formed on site.............	-	-	**Spclist**	-	-	**Spclist**	-	**14.26**	nr	**16.04**

Labour hourly rates: (except Specialists) Craft Operatives £14.21 Labourer £10.71 Rates are national average prices. Refer to REGIONAL VARIATIONS for indicative levels of overall pricing in regions	MATERIALS			LABOUR				RATES		
	Del to Site £	Waste %	Material Cost £	Craft Optve Hrs	Lab Hrs	Labour Cost £	Sunds £	Nett Rate £	Unit	Gross Rate (+12.5%) £
30: METAL PROFILED SHEET DECKING Cont.										
alvanized steel troughed decking; 0.7mm thick metal, 8mm overall depth; natural soffit and top surface; 50mm end laps and one corrugation side laps										
ecking; fixed to steel rails at 1500mm centres with elf tapping screws; drilling holes										
sloping; 10 degrees pitch	-	-	**Spclist**	-	-	**Spclist**	-	**31.72**	m2	**35.69**
xtra over decking for										
raking cutting	-	-	**Spclist**	-	-	**Spclist**	-	**5.20**	m	**5.85**
holes 50mm diameter; formed on site	-	-	**Spclist**	-	-	**Spclist**	-	**14.26**	nr	**16.04**
alvanized steel troughed decking; 0.7mm thick metal, 3mm overall depth; natural soffit and top surface; 50mm end laps and one corrugation side laps										
ecking; fixed to steel rails at 2000mm centres with elf tapping screws; drilling holes										
sloping; 10 degrees pitch	-	-	**Spclist**	-	-	**Spclist**	-	**33.32**	m2	**37.49**
xtra over decking for										
raking cutting	-	-	**Spclist**	-	-	**Spclist**	-	**5.20**	m	**5.85**
holes 50mm diameter; formed on site	-	-	**Spclist**	-	-	**Spclist**	-	**14.26**	nr	**16.04**
alvanized steel troughed decking; 0.7mm thick metal, 00mm overall depth; natural soffit and top surface; 50mm end laps and one corrugation side laps										
ecking; fixed to steel rails at 3500mm centres with elf tapping screws; drilling holes										
sloping; 10 degrees pitch	-	-	**Spclist**	-	-	**Spclist**	-	**35.10**	m2	**39.49**
xtra over decking for										
raking cutting	-	-	**Spclist**	-	-	**Spclist**	-	**5.20**	m	**5.85**
holes 50mm diameter; formed on site	-	-	**Spclist**	-	-	**Spclist**	-	**14.26**	nr	**16.04**
alvanized steel troughed decking; 0.9mm thick metal, 5mm overall depth; natural soffit and top surface; 50mm end laps and one corrugation side laps										
ecking; fixed to steel rails at 1500mm centres with elf tapping screws; drilling holes										
sloping; 10 degrees pitch	-	-	**Spclist**	-	-	**Spclist**	-	**34.73**	m2	**39.07**
xtra over decking for										
raking cutting	-	-	**Spclist**	-	-	**Spclist**	-	**5.20**	m	**5.85**
holes 50mm diameter; formed on site	-	-	**Spclist**	-	-	**Spclist**	-	**14.26**	nr	**16.04**
alvanized steel troughed decking; 0.9mm thick metal, 8mm overall depth; natural soffit and top surface; 50mm end laps and one corrugation side laps										
ecking; fixing to steel rails at 2000mm centres with elf tapping screws; drilling holes										
sloping; 10 degrees pitch	-	-	**Spclist**	-	-	**Spclist**	-	**35.17**	m2	**39.57**
xtra over decking for										
raking cutting	-	-	**Spclist**	-	-	**Spclist**	-	**5.20**	m	**5.85**
holes 50mm diameter; formed on site	-	-	**Spclist**	-	-	**Spclist**	-	**14.26**	nr	**16.04**
alvanized steel troughed decking; 0.9mm thick metal, 3mm overall depth; natural soffit and top surface; 50mm end laps and one corrugation side laps										
ecking; fixing to steel rails at 2500mm centres with elf tapping screws; drilling holes										
sloping; 10 degrees pitch	-	-	**Spclist**	-	-	**Spclist**	-	**36.51**	m2	**41.07**
xtra over decking for										
raking cutting	-	-	**Spclist**	-	-	**Spclist**	-	**5.20**	m	**5.85**
holes 50mm diameter; formed on site	-	-	**Spclist**	-	-	**Spclist**	-	**14.26**	nr	**16.04**
alvanized steel troughed decking; 0.9mm thick metal, 00mm overall depth; natural soffit and top surface; 50mm end laps and one corrugation side laps										
ecking; fixing to steel rails at 4000mm centres with elf tapping screws; drilling holes										
sloping; 10 degrees pitch	-	-	**Spclist**	-	-	**Spclist**	-	**38.76**	m2	**43.61**
xtra over decking for										
raking cutting	-	-	**Spclist**	-	-	**Spclist**	-	**5.20**	m	**5.85**
holes 50mm diameter; formed on site	-	-	**Spclist**	-	-	**Spclist**	-	**14.26**	nr	**16.04**
alvanized steel troughed decking; 1.2mm thick metal, 5mm overall depth; natural soffit and top surface; 50mm end laps and one corrugation side laps										
ecking; fixing to steel rails at 2000mm centres with elf tapping screws; drilling holes										
sloping; 10 degrees pitch	-	-	**Spclist**	-	-	**Spclist**	-	**38.92**	m2	**43.79**
xtra over decking for										
raking cutting	-	-	**Spclist**	-	-	**Spclist**	-	**5.20**	m	**5.85**

Labour hourly rates: (except Specialists) Craft Operatives £14.21 Labourer £10.71 Rates are national average prices. Refer to REGIONAL VARIATIONS for indicative levels of overall pricing in regions	MATERIALS			LABOUR				RATES		
	Del to Site £	Waste %	Material Cost £	Craft Optve Hrs	Lab Hrs	Labour Cost £	Sunds £	Nett Rate £	Unit	Gross Rate (+12.5%) £
G30: METAL PROFILED SHEET DECKING Cont.										
Galvanized steel troughed decking; 1.2mm thick metal, 35mm overall depth; natural soffit and top surface; 150mm end laps and one corrugation side laps Cont.										
Extra over decking for Cont.										
holes 50mm diameter; formed on site	-	-	Spclist	-	-	Spclist	-	14.26	nr	16.04
Galvanized steel troughed decking; 1.2mm thick metal, 48mm overall depth; natural soffit and top surface; 150mm end laps and one corrugation side laps										
Decking; fixing to steel rails at 2500mm centres with self tapping screws; drilling holes										
sloping; 10 degrees pitch	-	-	Spclist	-	-	Spclist	-	39.45	m2	44.38
Extra over decking for										
raking cutting	-	-	Spclist	-	-	Spclist	-	5.20	m	5.85
holes 50mm diameter; formed on site	-	-	Spclist	-	-	Spclist	-	14.26	nr	16.04
Galvanized steel troughed decking; 1.2mm thick metal, 63mm overall depth; natural soffit and top surface; 150mm end laps and one corrugation side laps										
Decking; fixing to steel rails at 3000mm centres with self tapping screws; drilling holes										
sloping; 10 degrees pitch	-	-	Spclist	-	-	Spclist	-	41.19	m2	46.34
Extra over decking for										
raking cutting	-	-	Spclist	-	-	Spclist	-	5.20	m	5.85
holes 50mm diameter; formed on site	-	-	Spclist	-	-	Spclist	-	14.26	nr	16.04
Galvanized steel troughed decking; 1.2mm thick metal, 100mm overall depth; natural soffit and top surface; 150mm end laps and one corrugation side laps										
Decking; fixing to steel rails at 4500mm centres with self tapping screws; drilling holes										
sloping; 10 degrees pitch	-	-	Spclist	-	-	Spclist	-	44.12	m2	49.64
Extra over decking for										
raking cutting	-	-	Spclist	-	-	Spclist	-	5.20	m	5.85
holes 50mm diameter; formed on site	-	-	Spclist	-	-	Spclist	-	14.26	nr	16.04
Aluminium troughed decking; 0.9mm thick metal 35mm overall depth; natural soffit and top surface; 150mm end laps and one corrugation side laps										
Decking; fixing to steel rails at 900mm centres with self tapping screws; drilling holes										
sloping; 10 degrees pitch	-	-	Spclist	-	-	Spclist	-	37.44	m2	42.12
Extra over decking for										
raking cutting	-	-	Spclist	-	-	Spclist	-	5.20	m	5.85
holes 50mm diameter; formed on site	-	-	Spclist	-	-	Spclist	-	14.26	nr	16.04
Aluminium troughed decking; 0.9mm thick metal, 48mm overall depth; natural soffit and top surface; 150mm end laps and one corrugation side laps										
Decking; fixing to steel rails at 1200mm centres with self tapping screws; drilling holes										
sloping; 10 degrees pitch	-	-	Spclist	-	-	Spclist	-	38.03	m2	42.78
Extra over decking for										
raking cutting	-	-	Spclist	-	-	Spclist	-	5.20	m	5.85
holes 50mm diameter; formed on site	-	-	Spclist	-	-	Spclist	-	14.26	nr	16.04
Aluminium troughed decking; 0.9mm thick metal, 63mm overall depth; natural soffit and top surface; 150mm end laps and one corrugation side laps										
Decking; fixing to steel rails at 1500mm centres with self tapping screws; drilling holes										
sloping; 10 degrees pitch	-	-	Spclist	-	-	Spclist	-	40.08	m2	45.09
Extra over decking for										
raking cutting	-	-	Spclist	-	-	Spclist	-	5.20	m	5.85
holes 50mm diameter; formed on site	-	-	Spclist	-	-	Spclist	-	14.26	nr	16.04
Aluminium troughed decking; 0.9mm thick metal, 100mm overall depth; natural soffit and top surface; 150mm end laps and one corrugation side laps										
Decking; fixing to steel rails at 2900mm centres with self tapping screws; drilling holes										
sloping; 10 degrees pitch	-	-	Spclist	-	-	Spclist	-	43.26	m2	48.67
Extra over decking for										
raking cutting	-	-	Spclist	-	-	Spclist	-	5.20	m	5.85
holes 50mm diameter; formed on site	-	-	Spclist	-	-	Spclist	-	14.26	nr	16.04

Labour hourly rates: (except Specialists) Craft Operatives £14.21 Labourer £10.71 Rates are national average prices. Refer to REGIONAL VARIATIONS for indicative levels of overall pricing in regions	MATERIALS			LABOUR				RATES		
	Del to Site £	Waste %	Material Cost £	Craft Optve Hrs	Lab Hrs	Labour Cost £	Sunds £	Nett Rate £	Unit	Gross Rate (+12.5%) £
G30: METAL PROFILED SHEET DECKING Cont.										
Aluminium troughed decking; 1.2mm thick metal, 35mm overall depth; natural soffit and top surface; 150mm end laps and one corrugation side laps										
Decking; fixing to steel rails at 1200mm centres with self tapping screws; drilling holes										
sloping; 10 degrees pitch	-	-	**Spclist**	-	-	**Spclist**	-	**41.71**	m2	**46.92**
Extra over decking for										
raking cutting	-	-	**Spclist**	-	-	**Spclist**	-	**5.20**	m	**5.85**
holes 50mm diameter; formed on site	-	-	**Spclist**	-	-	**Spclist**	-	**14.26**	nr	**16.04**
Aluminium troughed decking; 1.2mm thick metal, 48mm overall depth; natural soffit and top surface; 150mm end laps and one corrugation side laps										
Decking; fixing to steel rails at 1700mm centres with self tapping screws; drilling holes										
sloping; 10 degrees pitch	-	-	**Spclist**	-	-	**Spclist**	-	**42.28**	m2	**47.57**
Extra over decking for										
raking cutting	-	-	**Spclist**	-	-	**Spclist**	-	**5.20**	m	**5.85**
holes 50mm diameter; formed on site	-	-	**Spclist**	-	-	**Spclist**	-	**14.26**	nr	**16.04**
Aluminium troughed decking; 1.2mm thick metal, 63mm overall depth; natural soffit and top surface; 150mm end laps and one corrugation side laps										
Decking; fixing to steel rails at 2000mm centres with self tapping screws; drilling holes										
sloping; 10 degrees pitch	-	-	**Spclist**	-	-	**Spclist**	-	**44.40**	m2	**49.95**
Extra over decking for										
raking cutting	-	-	**Spclist**	-	-	**Spclist**	-	**5.20**	m	**5.85**
holes 50mm diameter; formed on site	-	-	**Spclist**	-	-	**Spclist**	-	**14.26**	nr	**16.04**
Aluminium troughed decking; 1.2mm thick metal, 100mm overall depth; natural soffit and top surface; 150mm end laps and one corrugation side laps										
Decking; fixing to steel rails at 3000mm centres with self tapping screws; drilling holes										
sloping; 10 degrees pitch	-	-	**Spclist**	-	-	**Spclist**	-	**48.02**	m2	**54.02**
Extra over decking for										
raking cutting	-	-	**Spclist**	-	-	**Spclist**	-	**5.20**	m	**5.85**
holes 50mm diameter; formed on site	-	-	**Spclist**	-	-	**Spclist**	-	**14.26**	nr	**16.04**

Labour hourly rates: (except Specialists) Craft Operatives £14.21 Labourer £10.71 Rates are national average prices. Refer to REGIONAL VARIATIONS for indicative levels of overall pricing in regions	MATERIALS			LABOUR				RATES		
	Del to Site £	Waste %	Material Cost £	Craft Optve Hrs	Lab Hrs	Labour Cost £	Sunds £	Nett Rate £	Unit	Gross Rate (+12.5%) £
H10: PATENT GLAZING										
Patent glazing with aluminium alloy bars 2000mm long with aluminium wings and seatings for glass, spaced at approximately 600mm centres, glazed with BS 952 Georgian wired cast, 7mm thick										
Roof areas										
single tier	116.97	5.00	**122.82**	1.68	1.68	**41.87**	2.58	**167.26**	m2	**188.17**
multi-tier	125.48	5.00	**131.75**	1.95	1.95	**48.59**	2.58	**182.93**	m2	**205.79**
Vertical surfaces										
single tier	130.20	5.00	**136.71**	1.70	1.70	**42.36**	2.58	**181.65**	m2	**204.36**
multi-tier	138.81	5.00	**145.75**	1.95	1.95	**48.59**	2.58	**196.92**	m2	**221.54**
Patent glazing with aluminium alloy bars 2000mm long with aluminium wings and seatings for glass, spaced at approximately 600mm centres, glazed with BS 952 Georgian wired polished, 6mm thick										
Roof areas										
single tier	182.81	5.00	**191.95**	1.90	1.90	**47.35**	2.58	**241.88**	m2	**272.11**
multi-tier	196.46	5.00	**206.28**	2.20	2.20	**54.82**	2.58	**263.69**	m2	**296.65**
Vertical surfaces										
single tier	196.25	5.00	**206.06**	1.90	1.90	**47.35**	2.58	**255.99**	m2	**287.99**
multi-tier	208.64	5.00	**219.07**	2.20	2.20	**54.82**	2.58	**276.48**	m2	**311.04**
Patent glazing with aluminium alloy bars 3000mm long with aluminium wings and seatings for glass, spaced at approximately 600mm centres, glazed with BS 952 Georgian wired cast, 7mm thick										
Roof areas										
single tier	139.02	5.00	**145.97**	1.70	1.70	**42.36**	2.58	**190.92**	m2	**214.78**
multi-tier	147.95	5.00	**155.35**	1.95	1.95	**48.59**	2.58	**206.52**	m2	**232.34**
Vertical surfaces										
single tier	151.10	5.00	**158.66**	1.70	1.70	**42.36**	2.58	**203.60**	m2	**229.05**
multi-tier	161.28	5.00	**169.34**	1.95	1.95	**48.59**	2.58	**220.52**	m2	**248.08**
Patent glazing with aluminium alloy bars 3000mm long with aluminium wings and seatings for glass, spaced at approximately 600mm centres, glazed with BS 952 Georgian wired polished, 6mm thick										
Roof areas										
single tier	203.81	5.00	**214.00**	1.90	1.90	**47.35**	2.58	**263.93**	m2	**296.92**
multi-tier	219.03	5.00	**229.98**	2.20	2.20	**54.82**	2.58	**287.39**	m2	**323.31**
Vertical surfaces										
single tier	216.93	5.00	**227.78**	1.90	1.90	**47.35**	2.58	**277.70**	m2	**312.42**
multi-tier	232.37	5.00	**243.99**	2.20	2.20	**54.82**	2.58	**301.39**	m2	**339.07**
Extra over patent glazing in roof areas with aluminium alloy bars 2000mm long with aluminium wings and seatings for glass, spaced at approximately 600mm centres, glazed with BS 952 Georgian wired polished, 6mm thick										
Opening lights including opening gear										
600 x 600mm	247.72	5.00	**260.11**	5.00	5.00	**124.60**	-	**384.71**	nr	**432.79**
600 x 900mm	300.86	5.00	**315.90**	6.30	6.30	**157.00**	-	**472.90**	nr	**532.01**

Labour hourly rates: (except Specialists) Craft Operatives £14.21 Labourer £10.71 Rates are national average prices. Refer to REGIONAL VARIATIONS for indicative levels of overall pricing in regions	MATERIALS			LABOUR				RATES		
	Del to Site £	Waste %	Material Cost £	Craft Optve Hrs	Lab Hrs	Labour Cost £	Sunds £	Nett Rate £	Unit	Gross Rate (+12.5%) £
H10: PATENT GLAZING Cont.										
Extra over patent glazing in roof areas with aluminium alloy bars 2000mm long with aluminium wings and seatings for glass, spaced at approximately 600mm centres, glazed with BS 952 Georgian wired cast, 7mm thick										
Opening lights including opening gear										
600 x 600mm	247.72	5.00	**260.11**	4.60	4.60	**114.63**	-	**374.74**	nr	**421.58**
600 x 900mm	300.86	5.00	**315.90**	5.75	5.75	**143.29**	-	**459.19**	nr	**516.59**
Extra over patent glazing in vertical surfaces with aluminium alloy bars 3000mm long with aluminium wings and seatings for glass, spaced at approximately 600mm centres, glazed with BS 952 Georgian wired cast, 7mm thick										
Opening lights including opening gear										
600 x 600mm	361.53	5.00	**379.61**	2.30	2.30	**57.32**	-	**436.92**	nr	**491.54**
600 x 900mm	439.50	5.00	**461.48**	3.50	3.50	**87.22**	-	**548.70**	nr	**617.28**
Extra over patent glazing to vertical surfaces with aluminium alloy bars 3000mm long with aluminium wings and seatings for glass, spaced at approximately 600mm centres, glazed with BS 952 Georgian wired polished, 6mm thick										
Opening lights including opening gear										
600 x 600mm	361.53	5.00	**379.61**	2.55	2.55	**63.55**	-	**443.15**	nr	**498.55**
600 x 900mm	439.50	5.00	**461.48**	3.75	3.75	**93.45**	-	**554.93**	nr	**624.29**
H21: TIMBER WEATHERBOARDING										
Softwood, wrought, impregnated										
Boarding to walls, shiplapped joints; 19mm thick, 150mm wide boards width exceeding 300mm; fixing to timber	8.03	10.00	**8.83**	0.85	0.10	**13.15**	0.14	**22.12**	m2	**24.89**
Boarding to walls, shiplapped joints; 25mm thick, 150mm wide boards width exceeding 300mm; fixing to timber	10.56	10.00	**11.62**	0.90	0.11	**13.97**	0.15	**25.73**	m2	**28.95**
Western Red Cedar										
Boarding to walls, shiplapped joints; 19mm thick, 150mm wide boards width exceeding 300mm; fixing to timber	27.59	10.00	**30.35**	0.90	0.12	**14.07**	0.15	**44.57**	m2	**50.14**
Boarding to walls, shiplapped joints; 25mm thick, 150mm wide boards width exceeding 300mm; fixing to timber	36.30	10.00	**39.93**	0.95	0.12	**14.78**	0.15	**54.86**	m2	**61.72**
Abutments										
19mm thick softwood boarding										
raking cutting	0.80	10.00	**0.88**	0.07	-	**0.99**	-	**1.88**	m	**2.11**
curved cutting	1.20	10.00	**1.32**	0.10	-	**1.42**	-	**2.75**	m	**3.09**
25mm thick softwood boarding										
raking cutting	1.06	10.00	**1.16**	0.08	-	**1.14**	-	**2.30**	m	**2.59**
curved cutting	1.58	10.00	**1.74**	0.12	-	**1.71**	-	**3.45**	m	**3.88**
19mm thick Western Red Cedar boarding										
raking cutting	2.76	10.00	**3.03**	0.09	-	**1.28**	-	**4.31**	m	**4.85**
curved cutting	4.14	10.00	**4.55**	0.13	-	**1.85**	-	**6.40**	m	**7.20**
25mm thick Western Red Cedar boarding										
raking cutting	3.63	10.00	**3.99**	0.10	-	**1.42**	-	**5.41**	m	**6.09**
curved cutting	5.45	10.00	**5.99**	0.15	-	**2.13**	-	**8.12**	m	**9.14**
H30: FIBRE CEMENT PROFILED SHEET CLADDING/COVERING/SIDING										
Roof coverings; corrugated reinforced cement Profile 3 sheeting, standard grey colour; lapped one and a half corrugations at sides and 150mm at ends										
Coverings; fixing to timber joists at 900mm general spacing with galvanized mild steel drive screws and washers; drilling holes pitch 30 degrees from horizontal	12.28	15.00	**14.12**	0.28	0.28	**6.98**	0.48	**21.58**	m2	**24.28**
Coverings; fixing to steel purlins at 900mm general spacing with galvanized hook bolts and washers; drilling holes pitch 30 degrees from horizontal	12.28	15.00	**14.12**	0.38	0.38	**9.47**	0.98	**24.57**	m2	**27.64**
Eaves										
eaves filler pieces	10.53	10.00	**11.58**	0.12	0.12	**2.99**	-	**14.57**	m	**16.40**
Ridges										
two piece plain angular adjustable ridge tiles	15.23	10.00	**16.75**	0.23	0.23	**5.73**	-	**22.48**	m	**25.30**

Labour hourly rates: (except Specialists) Craft Operatives £14.21 Labourer £10.71 Rates are national average prices. Refer to REGIONAL VARIATIONS for indicative levels of overall pricing in regions	MATERIALS			LABOUR			RATES			
	Del to Site £	Waste %	Material Cost £	Craft Optve Hrs	Lab Hrs	Labour Cost £	Sunds £	Nett Rate £	Unit	Gross Rate (+12.5%) £
H30: FIBRE CEMENT PROFILED SHEET CLADDING/COVERING/SIDING Cont.										
Roof coverings; corrugated reinforced cement Profile 3 sheeting, standard grey colour; lapped one and a half corrugations at sides and 150mm at ends Cont.										
Barge boards										
standard barge boards	8.63	10.00	**9.49**	0.12	0.12	**2.99**	-	**12.48**	m	**14.04**
Aprons/sills										
apron flashings	10.58	10.00	**11.64**	0.15	0.15	**3.74**	-	**15.38**	m	**17.30**
Finials										
standard ridge cap finials	9.53	10.00	**10.48**	0.17	0.17	**4.24**	-	**14.72**	nr	**16.56**
Cutting										
raking	2.46	15.00	**2.82**	0.28	0.28	**6.98**	-	**9.80**	m	**11.03**
Holes										
for pipes, standards or the like	-	-	**-**	0.58	-	**8.24**	-	**8.24**	nr	**9.27**
Roof coverings; corrugated reinforced cement Profile 3 sheeting, standard coloured; lapped one and a half corrugations at sides and 150mm at ends										
Coverings; fixing to timber joists at 900mm general spacing with galvanized mild steel drive screws and washers; drilling holes										
pitch 30 degrees from horizontal	14.74	15.00	**16.95**	0.28	0.28	**6.98**	0.64	**24.57**	m2	**27.64**
Coverings; fixing to steel purlins at 900mm general spacing with galvanized hook bolts and washers; drilling holes										
pitch 30 degrees from horizontal	14.74	15.00	**16.95**	0.38	0.38	**9.47**	1.26	**27.68**	m2	**31.14**
Eaves										
eaves filler pieces	13.17	10.00	**14.49**	0.12	0.12	**2.99**	-	**17.48**	m	**19.66**
Ridges										
two piece plain angular adjustable ridge tiles	18.37	10.00	**20.21**	0.23	0.23	**5.73**	-	**25.94**	m	**29.18**
Barge boards										
standard barge boards	10.79	10.00	**11.87**	0.12	0.12	**2.99**	-	**14.86**	m	**16.72**
Aprons/sills										
apron flashings	13.21	10.00	**14.53**	0.15	0.15	**3.74**	-	**18.27**	m	**20.55**
Finials										
standard ridge cap finials	10.83	10.00	**11.91**	0.17	0.17	**4.24**	-	**16.15**	nr	**18.17**
Cutting										
raking	2.95	15.00	**3.39**	0.28	0.28	**6.98**	-	**10.37**	m	**11.66**
Holes										
for pipes, standards or the like	-	-	**-**	0.58	-	**8.24**	-	**8.24**	nr	**9.27**
Roof coverings; corrugated reinforced cement Profile 6 sheeting, standard grey colour; lapped half a corrugation at sides and 150mm at ends										
Coverings; fixing to timber joists at 900mm general spacing with galvanized mild steel drive screws and washers; drilling holes										
pitch 30 degrees from horizontal	9.77	15.00	**11.24**	0.28	0.28	**6.98**	0.52	**18.73**	m2	**21.07**
Coverings; fixing to steel purlins at 900mm general spacing with galvanized hook bolts and washers; drilling holes										
pitch 30 degrees from horizontal	9.77	15.00	**11.24**	0.38	0.38	**9.47**	1.02	**21.73**	m2	**24.44**
Eaves										
eaves filler pieces	6.99	10.00	**7.69**	0.12	0.12	**2.99**	-	**10.68**	m	**12.01**
Ridges										
two piece plain angular adjustable ridge tiles	12.84	10.00	**14.12**	0.23	0.23	**5.73**	-	**19.86**	m	**22.34**
Barge boards										
standard barge boards	8.63	10.00	**9.49**	0.12	0.12	**2.99**	-	**12.48**	m	**14.04**
Aprons/sills										
apron flashings	7.53	10.00	**8.28**	0.15	0.15	**3.74**	-	**12.02**	m	**13.52**
Finials										
standard ridge cap finials	8.51	10.00	**9.36**	0.17	0.17	**4.24**	-	**13.60**	nr	**15.30**
Cutting										
raking	1.95	15.00	**2.25**	0.28	0.28	**6.98**	-	**9.22**	m	**10.38**
Holes										
for pipes, standards or the like	-	-	**-**	0.58	-	**8.24**	-	**8.24**	nr	**9.27**

Labour hourly rates: (except Specialists) Craft Operatives £14.21 Labourer £10.71 Rates are national average prices. Refer to REGIONAL VARIATIONS for indicative levels of overall pricing in regions	MATERIALS			LABOUR				RATES		
	Del to Site £	Waste %	Material Cost £	Craft Optve Hrs	Lab Hrs	Labour Cost £	Sunds £	Nett Rate £	Unit	Gross Rate (+12.5%) £
H30: FIBRE CEMENT PROFILED SHEET CLADDING/COVERING/SIDING Cont.										
Roof coverings; corrugated reinforced cement Profile 6 sheeting, standard coloured; lapped half a corrugation at sides and 150mm at ends										
Coverings; fixing to timber joists at 900mm general spacing with galvanized mild steel drive screws and washers; drilling holes										
pitch 30 degrees from horizontal	11.73	15.00	**13.49**	0.28	0.28	**6.98**	0.63	**21.10**	m2	**23.73**
Coverings; fixing to steel purlins at 900mm general spacing with galvanized hook bolts and washers; drilling holes										
pitch 30 degrees from horizontal	11.73	15.00	**13.49**	0.38	0.38	**9.47**	1.26	**24.22**	m2	**27.25**
Eaves										
eaves filler pieces	8.73	10.00	**9.60**	0.12	0.12	**2.99**	-	**12.59**	m	**14.17**
Ridges										
two piece plain angular adjustable ridge tiles	16.06	10.00	**17.67**	0.23	0.23	**5.73**	-	**23.40**	m	**26.32**
Barge boards										
standard barge boards	10.79	10.00	**11.87**	0.12	0.12	**2.99**	-	**14.86**	m	**16.72**
Aprons/sills										
apron flashings	9.42	10.00	**10.36**	0.15	0.15	**3.74**	-	**14.10**	m	**15.86**
Finials										
standard ridge cap finials	10.63	10.00	**11.69**	0.17	0.17	**4.24**	-	**15.93**	nr	**17.92**
Cutting										
raking	2.35	15.00	**2.70**	0.28	0.28	**6.98**	-	**9.68**	m	**10.88**
Holes										
for pipes, standards or the like	-	-	**-**	0.58	-	**8.24**	-	**8.24**	nr	**9.27**
H31: METAL PROFILED/FLAT SHEET CLADDING/COVERING/SIDING										
Wall cladding; PVC colour coated both sides galvanized steel profiled sheeting 0.70mm thick and with sheets secured at seams and laps										
Coverings; fixing to steel rails at 900mm general spacing with galvanized hook bolts and washers; drilling holes										
pitch 90 degrees from horizontal	16.56	5.00	**17.39**	0.40	0.40	**9.97**	1.18	**28.54**	m2	**32.10**
Vertical angles										
vertical corner flashings	6.19	5.00	**6.50**	0.17	0.17	**4.24**	-	**10.74**	m	**12.08**
Filler blocks										
polyethylene	0.70	5.00	**0.74**	0.12	0.12	**2.99**	-	**3.73**	m	**4.19**
PVC	1.97	5.00	**2.07**	0.12	0.12	**2.99**	-	**5.06**	m	**5.69**
black synthetic rubber	2.19	5.00	**2.30**	0.12	0.12	**2.99**	-	**5.29**	m	**5.95**
Cutting										
raking	3.31	5.00	**3.48**	0.28	0.28	**6.98**	-	**10.46**	m	**11.76**
Roof coverings; PVC colour coated both sides galvanized steel profiled sheeting 0.70mm thick and with sheets secured at seams and laps										
Coverings; fixing to steel purlins at 900mm general spacing with galvanized hook bolts and washers; drilling holes										
pitch 30 degrees from horizontal	16.56	5.00	**17.39**	0.35	0.35	**8.72**	1.22	**27.33**	m2	**30.75**
Ridges										
ridge cappings	6.19	5.00	**6.50**	0.17	0.17	**4.24**	-	**10.74**	m	**12.08**
Flashings										
gable flashings	6.19	5.00	**6.50**	0.17	0.17	**4.24**	-	**10.74**	m	**12.08**
eaves flashings	4.12	5.00	**4.33**	0.17	0.17	**4.24**	-	**8.56**	m	**9.63**
Filler blocks										
polyethylene	0.70	5.00	**0.74**	0.12	0.12	**2.99**	-	**3.73**	m	**4.19**
PVC	1.97	5.00	**2.07**	0.12	0.12	**2.99**	-	**5.06**	m	**5.69**
black synthetic rubber	2.19	5.00	**2.30**	0.12	0.12	**2.99**	-	**5.29**	m	**5.95**
Cutting										
raking	3.31	5.00	**3.48**	0.28	0.28	**6.98**	-	**10.46**	m	**11.76**
H32: PLASTICS PROFILED SHEET CLADDING/ COVERING/SIDING										
Roof coverings; standard corrugated glass fibre reinforced translucent sheeting; lapped one and a half corrugations at sides and 150mm at ends										
Coverings; fixing to timber joists at 900mm general spacing with galvanized mild steel drive screws and washers; drilling holes										
pitch 30 degrees from horizontal	14.93	5.00	**15.68**	0.28	0.28	**6.98**	0.33	**22.98**	m2	**25.86**

Labour hourly rates: (except Specialists) Craft Operatives £14.21 Labourer £10.71 Rates are national average prices. Refer to REGIONAL VARIATIONS for indicative levels of overall pricing in regions	MATERIALS			LABOUR			RATES			
	Del to Site £	Waste %	Material Cost £	Craft Optve Hrs	Lab Hrs	Labour Cost £	Sunds £	Nett Rate £	Unit	Gross Rate (+12.5%) £
32: PLASTICS PROFILED SHEET CLADDING/ COVERING/SIDING Cont.										
Roof coverings; standard corrugated glass fibre reinforced translucent sheeting; lapped one and a half corrugations at sides and 150mm at ends Cont.										
Coverings; fixing to steel purlins at 900mm general spacing with galvanized hook bolts and washers; drilling holes										
pitch 30 degrees from horizontal	14.93	5.00	**15.68**	0.35	0.35	**8.72**	0.41	**24.81**	m2	**27.91**
Cutting										
raking	2.99	5.00	**3.14**	0.28	0.28	**6.98**	-	**10.11**	m	**11.38**
Roof coverings; fire resisting corrugated glass fibre reinforced translucent sheeting; lapped one and a half corrugations at sides and 150mm at ends										
Coverings; fixing to timber joists at 900mm general spacing with galvanized mild steel drive screws and washers; drilling holes										
pitch 30 degrees from horizontal	17.54	5.00	**18.42**	0.28	0.28	**6.98**	0.33	**25.72**	m2	**28.94**
Coverings; fixing to steel purlins at 900mm general spacing with galvanized hook bolts and washers; drilling holes										
pitch 30 degrees from horizontal	17.54	5.00	**18.42**	0.35	0.35	**8.72**	0.53	**27.67**	m2	**31.13**
Cutting										
raking	3.51	5.00	**3.68**	0.28	0.28	**6.98**	-	**10.66**	m	**11.99**
Roof coverings; standard vinyl corrugated sheeting; lapped one and a half corrugations at sides and 150mm at ends										
Coverings; fixing to timber joists at 900mm general spacing with galvanized mild steel drive screws and washers; drilling holes										
pitch 30 degrees from horizontal	19.84	5.00	**20.83**	0.28	0.28	**6.98**	0.53	**28.34**	m2	**31.88**
Coverings; fixing to steel purlins at 900mm general spacing with galvanized hook bolts and washers; drilling holes										
pitch 30 degrees from horizontal	19.84	5.00	**20.83**	0.35	0.35	**8.72**	0.53	**30.08**	m2	**33.84**
Cutting										
raking	3.97	5.00	**4.17**	0.28	0.28	**6.98**	-	**11.14**	m	**12.54**
Wall cladding; Swish Products high impact rigid uPVC profiled sections; colour white; secured with starter sections and clips										
Coverings with shiplap profile code C002 giving 150mm cover; fixing to timber vertical cladding										
sections applied horizontally	40.22	5.00	**42.23**	0.42	0.42	**10.47**	1.97	**54.67**	m2	**61.50**
sections applied vertically	40.22	5.00	**42.23**	0.42	0.42	**10.47**	1.97	**54.67**	m2	**61.50**
Coverings with open V profile code C003 giving 150mm cover; fixing to timber vertical cladding										
sections applied horizontally	40.22	5.00	**42.23**	0.45	0.45	**11.21**	2.81	**56.26**	m2	**63.29**
sections applied vertically	40.22	5.00	**42.23**	0.45	0.45	**11.21**	2.81	**56.26**	m2	**63.29**
Coverings with open V profile code C269 giving 100mm cover; fixing to timber vertical cladding										
sections applied horizontally	45.51	5.00	**47.79**	0.38	0.38	**9.47**	1.97	**59.23**	m2	**66.63**
sections applied vertically	45.51	5.00	**47.79**	0.38	0.38	**9.47**	1.97	**59.23**	m2	**66.63**
Vertical angles										
section C030 for vertically applied section	4.60	5.00	**4.83**	0.11	0.11	**2.74**	0.16	**7.73**	m	**8.70**
H51: NATURAL STONE SLAB CLADDING/FEATURES										
English blue/grey slate facings (Best Quality); natural riven finish; bedding, jointing and pointing in gauged mortar (1:2:9)										
450 x 600 x 20mm units to walls on brickwork or blockwork base										
plain, width exceeding 300mm	95.50	2.50	**97.89**	2.00	2.00	**49.84**	0.51	**148.24**	m2	**166.77**
450 x 600 x 30mm units to walls on brickwork or blockwork base										
plain, width exceeding 300mm	125.27	2.50	**128.40**	2.60	2.60	**64.79**	0.57	**193.76**	m2	**217.98**
450 x 600 x 40mm units to walls on brickwork or blockwork base										
plain, width exceeding 300mm	137.68	2.50	**141.12**	2.90	2.90	**72.27**	0.79	**214.18**	m2	**240.95**
450 x 600 x 50mm units to walls on brickwork or blockwork base										
plain, width exceeding 300mm	169.92	2.50	**174.17**	3.45	3.45	**85.97**	0.97	**261.11**	m2	**293.75**

Labour hourly rates: (except Specialists) Craft Operatives £14.21 Labourer £10.71 Rates are national average prices. Refer to REGIONAL VARIATIONS for indicative levels of overall pricing in regions	MATERIALS			LABOUR				RATES		
	Del to Site £	Waste %	Material Cost £	Craft Optve Hrs	Lab Hrs	Labour Cost £	Sunds £	Nett Rate £	Unit	Gross Rate (+12.5%) £
H51: NATURAL STONE SLAB CLADDING/FEATURES Cont.										
English blue/grey slate facings (Best quality); fine rubbed finish, one face; bedding, jointing and pointing in gauged mortar (1:2:9)										
750 x 1200 x 20mm units to walls on brickwork or blockwork base										
plain, width exceeding 300mm	238.50	2.50	**244.46**	2.30	2.30	**57.32**	1.47	**303.25**	m2	**341.15**
extra; rubbed square edges	19.13	-	**19.13**	-	-	**-**	-	**19.13**	m	**21.52**
extra; half rounded edges	58.62	-	**58.62**	-	-	**-**	-	**58.62**	m	**65.95**
extra; full rounded edges	58.62	-	**58.62**	-	-	**-**	-	**58.62**	m	**65.95**
extra; rebated joints	27.70	-	**27.70**	-	-	**-**	-	**27.70**	m	**31.16**
750 x 1200 x 30mm units to walls on brickwork or blockwork base										
plain, width exceeding 300mm	285.01	2.50	**292.14**	2.60	2.60	**64.79**	1.56	**358.49**	m2	**403.30**
extra; rubbed square edges	24.13	-	**24.13**	-	-	**-**	-	**24.13**	m	**27.15**
extra; half rounded edges	66.94	-	**66.94**	-	-	**-**	-	**66.94**	m	**75.31**
extra; full rounded edges	66.94	-	**66.94**	-	-	**-**	-	**66.94**	m	**75.31**
extra; rebated joints	32.92	-	**32.92**	-	-	**-**	-	**32.92**	m	**37.04**
750 x 1200 x 40mm units to walls on brickwork or blockwork base										
plain, width exceeding 300mm	292.16	2.50	**299.46**	2.90	2.90	**72.27**	1.76	**373.49**	m2	**420.18**
extra; rubbed square edges	24.13	-	**24.13**	-	-	**-**	-	**24.13**	m	**27.15**
extra; half rounded edges	66.94	-	**66.94**	-	-	**-**	-	**66.94**	m	**75.31**
extra; full rounded edges	66.94	-	**66.94**	-	-	**-**	-	**66.94**	m	**75.31**
extra; rebated joints	32.92	-	**32.92**	-	-	**-**	-	**32.92**	m	**37.04**
750 x 1200 x 50mm units to walls on brickwork or blockwork base										
plain, width exceeding 300mm	310.05	2.50	**317.80**	3.45	3.45	**85.97**	1.95	**405.73**	m2	**456.44**
extra; rubbed square edges	24.13	-	**24.13**	-	-	**-**	-	**24.13**	m	**27.15**
extra; half rounded edges	66.94	-	**66.94**	-	-	**-**	-	**66.94**	m	**75.31**
extra; full rounded edges	66.94	-	**66.94**	-	-	**-**	-	**66.94**	m	**75.31**
extra; rebated joints	32.20	-	**32.20**	-	-	**-**	-	**32.20**	m	**36.23**
English blue/grey slate facings (Best quality); fine rubbed finish, both faces; bedding, jointing and pointing in gauged mortar (1:2:9)										
750 x 1200 x 20mm units to walls on brickwork or blockwork base										
plain, width exceeding 300mm	277.88	2.50	**284.83**	2.90	2.90	**72.27**	2.44	**359.54**	m2	**404.48**
750 x 1200 x 30mm units to walls on brickwork or blockwork base										
plain, width exceeding 300mm	324.38	2.50	**332.49**	3.15	3.15	**78.50**	2.57	**413.56**	m2	**465.25**
750 x 1200 x 40mm units to walls on brickwork or blockwork base										
plain, width exceeding 300mm	331.54	2.50	**339.83**	3.45	3.45	**85.97**	2.77	**428.57**	m2	**482.14**
750 x 1200 x 50mm units to walls on brickwork or blockwork base										
plain, width exceeding 300mm	349.43	2.50	**358.17**	4.00	4.00	**99.68**	2.90	**460.75**	m2	**518.34**
Kirkstone green slate; bedding, jointing and pointing in gauged mortar (1:2:9)										
30mm thick units to sills on brickwork or blockwork base										
width 125mm; weathered, throated and grooved	67.61	2.50	**69.30**	0.65	0.65	**16.20**	0.64	**86.14**	m	**96.91**
width 190mm; weathered, throated and grooved	86.49	2.50	**88.65**	0.70	0.70	**17.44**	0.89	**106.99**	m	**120.36**
40mm thick units to sills on brickwork or blockwork base										
width 125mm	75.88	2.50	**77.78**	0.70	0.70	**17.44**	0.64	**95.86**	m	**107.84**
extra; stooling for jambs	25.04	-	**25.04**	-	-	**-**	-	**25.04**	nr	**28.17**
extra; notching for jambs and mullions	25.04	-	**25.04**	-	-	**-**	-	**25.04**	nr	**28.17**
width 190mm	99.58	2.50	**102.07**	0.85	0.85	**21.18**	0.83	**124.08**	m	**139.59**
extra; stooling for jambs	25.04	-	**25.04**	-	-	**-**	-	**25.04**	nr	**28.17**
extra; notching for jambs and mullions	25.04	-	**25.04**	-	-	**-**	-	**25.04**	nr	**28.17**
50mm thick units to sills on brickwork or blockwork base										
width 125mm	89.84	2.50	**92.09**	0.85	0.85	**21.18**	0.64	**113.91**	m	**128.15**
extra; stooling for jambs	25.04	-	**25.04**	-	-	**-**	-	**25.04**	nr	**28.17**
extra; notching for jambs and mullions	25.04	-	**25.04**	-	-	**-**	-	**25.04**	nr	**28.17**
width 190mm	120.27	2.50	**123.28**	1.20	1.20	**29.90**	0.83	**154.01**	m	**173.26**
extra; stooling for jambs	25.04	-	**25.04**	-	-	**-**	-	**25.04**	nr	**28.17**
extra; notching for jambs and mullions	25.04	-	**25.04**	-	-	**-**	-	**25.04**	nr	**28.17**
30mm units to window boards on brickwork or blockwork base										
width 230mm	85.61	2.50	**87.75**	0.85	0.85	**21.18**	1.05	**109.98**	m	**123.73**
25mm units to combined sills and window boards on brickwork or blockwork base										
width 360mm	123.37	2.50	**126.45**	1.15	1.15	**28.66**	1.47	**156.58**	m	**176.16**

Labour hourly rates: (except Specialists) Craft Operatives £14.21 Labourer £10.71 Rates are national average prices. Refer to REGIONAL VARIATIONS for indicative levels of overall pricing in regions	MATERIALS			LABOUR			RATES			
	Del to Site £	Waste %	Material Cost £	Craft Optve Hrs	Lab Hrs	Labour Cost £	Sunds £	Nett Rate £	Unit	Gross Rate (+12.5%) £
H60: PLAIN ROOF TILING										
Quantities required for 1m2 of tiling										
Plain tiles to 65mm lap										
tiles, 60nr										
battens, 10m										
Plain tiles to 85mm lap										
tiles, 70nr										
battens, 11m										
Roof coverings; clayware Machine made plain tiles BS EN 1304, red, 265 x 165mm; fixing every fourth course with two galvanized nails per tile to 65mm lap; 38 x 19mm pressure impregnated softwood battens fixed with galvanized nails; underlay BS 747 type 1F reinforced bitumen felt; 150mm laps; fixing with galvanized steel clout nails										
Pitched 50 degrees from horizontal										
generally	34.51	5.00	**36.43**	0.72	0.36	**14.09**	1.56	**52.08**	m2	**58.59**
holes	-	-	-	0.58	0.29	**11.35**	-	**11.35**	nr	**12.77**
abutments; square	3.12	5.00	**3.27**	0.28	0.14	**5.48**	-	**8.75**	m	**9.85**
abutments; raking	4.42	5.00	**4.64**	0.42	0.21	**8.22**	-	**12.86**	m	**14.46**
double course at eaves; Machine made eaves tile	3.18	5.00	**3.34**	0.35	0.18	**6.90**	0.55	**10.79**	m	**12.14**
verges; bed and point in coloured cement-lime mortar (1:1:6)	3.12	5.00	**3.27**	0.40	0.20	**7.83**	0.42	**11.52**	m	**12.96**
verges; single extra undercloak course plain tiles; bed and point in coloured cement-lime mortar (1:1:6)	4.16	5.00	**4.37**	0.58	0.29	**11.35**	0.42	**16.13**	m	**18.15**
Machine made ridge tiles, half round; in 300mm effective lengths; butt jointed; bedding and pointing in coloured cement-lime mortar (1:1:6)	18.38	5.00	**19.30**	0.35	0.18	**6.90**	0.42	**26.62**	m	**29.95**
Machine made hip tiles, half round; in 300mm effective lengths; butt jointed; bedding and pointing in coloured cement-lime mortar (1:1:6)	18.38	5.00	**19.30**	0.35	0.18	**6.90**	0.42	**26.62**	m	**29.95**
Machine made hip tiles; angular	18.38	5.00	**19.30**	0.35	0.18	**6.90**	0.42	**26.62**	m	**29.95**
Machine made hip tiles; bonnet pattern; bedding and pointing in coloured cement-lime mortar (1:1:6)	18.38	5.00	**19.30**	0.35	0.18	**6.90**	0.42	**26.62**	m	**29.95**
Machine made valley tiles; angular	18.38	5.00	**19.30**	0.50	0.25	**9.78**	0.42	**29.50**	m	**33.19**
Roof coverings; clayware Machine made plain tiles BS EN 1304, red, 265 x 165mm; fixing every fourth course with two galvanized nails per tile to 65mm lap; 50 x 19mm pressure impregnated softwood battens fixed with galvanized nails; underlay BS 747 type 1F reinforced bitumen felt; 150mm laps; fixing with galvanized steel clout nails										
Pitched 50 degrees from horizontal										
generally	35.91	5.00	**37.97**	0.70	0.35	**13.70**	1.56	**53.23**	m2	**59.88**
Roof coverings; clayware Machine made plain tiles BS EN 1304, red, 265 x 165mm; fixing every course with two galvanized nails per tile to 65mm lap; 38 x 19mm pressure impregnated softwood battens fixed with galvanized nails; underlay BS 747 type 1F reinforced bitumen felt; 150mm laps; fixing with galvanized steel clout nails										
Pitched 50 degrees from horizontal										
generally	34.51	5.00	**36.43**	0.85	0.43	**16.68**	3.00	**56.11**	m2	**63.13**
Roof coverings; clayware Machine made plain tiles BS EN 1304, red, 265 x 165mm; fixing every fourth course with two copper nails per tile to 65mm lap; 38 x19mm pressure impregnated softwood battens fixed with galvanized nails; underlay BS 747 type 1F reinforced bitumen felt; 150mm laps; fixing with galvanized steel clout nails										
Pitched 50 degrees from horizontal										
generally	34.51	5.00	**36.43**	0.70	0.35	**13.70**	2.40	**52.53**	m2	**59.09**
Roof coverings; clayware Machine made plain tiles BS EN 1304, red, 265 x 165mm; fixing every course with two copper nails per tile to 65mm lap; 38 x 19mm pressure impregnated softwood battens fixed with galvanized nails; underlay BS 747 type 1F reinforced bitumen felt; 150mm laps; fixing with galvanized steel clout nails										
Pitched 50 degrees from horizontal										
generally	34.51	5.00	**36.43**	0.85	0.43	**16.68**	5.50	**58.61**	m2	**65.94**
Roof coverings; clayware Machine made plain tiles BS EN 1304, red, 265 x 165mm; fixing every fourth course with two galvanized nails per tile to 85mm lap; 38 x 19mm pressure impregnated softwood battens fixed with galvanized nails; underlay BS 747 type 1F reinforced bitumen felt; 150mm laps; fixing with galvanized steel clout nails										
Pitched 40 degrees from horizontal										
generally	39.99	5.00	**42.19**	0.80	0.40	**15.65**	1.56	**59.41**	m2	**66.83**

Labour hourly rates: (except Specialists) Craft Operatives £14.21 Labourer £10.71 Rates are national average prices. Refer to REGIONAL VARIATIONS for indicative levels of overall pricing in regions	MATERIALS			LABOUR			RATES			
	Del to Site £	Waste %	Material Cost £	Craft Optve Hrs	Lab Hrs	Labour Cost £	Sunds £	Nett Rate £	Unit	Gross Rate (+12.5%) £
H60: PLAIN ROOF TILING Cont.										
Wall coverings; clayware Machine made plain tiles BS EN 1304, red, 265 x 165mm;fixing every course with two galvanized nails per tile to 38mm lap; 38 x 19mm pressure impregnated softwood battens fixed with galvanized nails; underlay BS 747 type 1F reinforced bitumen felt; 150mm laps; fixing with galvanized steel clout nails										
Vertical										
generally	30.72	5.00	**32.45**	0.85	0.43	**16.68**	2.28	**51.41**	m2	**57.84**
Roof coverings; clayware Hand made plain tiles BS EN 1304, red, 265 x 165mm; fixing every fourth course with two galvanized nails per tile to 65mm lap; 38 x 19mm pressure impregnated softwood battens fixed with galvanized nails; underlay BS 747 type 1F reinforced bitumen felt; 150mm laps; fixing with galvanized steel clout nails										
Pitched 50 degrees from horizontal										
generally	50.45	5.00	**53.17**	0.72	0.36	**14.09**	1.56	**68.82**	m2	**77.42**
holes	-	-	**-**	0.58	0.29	**11.35**	-	**11.35**	nr	**12.77**
abutments; square	4.71	5.00	**4.95**	0.28	0.14	**5.48**	-	**10.43**	m	**11.73**
abutments; raking	6.68	5.00	**7.01**	0.42	0.21	**8.22**	-	**15.23**	m	**17.13**
double course at eaves; Hand made eaves tile	5.40	5.00	**5.67**	0.35	0.18	**6.90**	0.55	**13.12**	m	**14.76**
verges; bed and point in coloured cement-lime mortar (1:1:6)	4.71	5.00	**4.95**	0.40	0.20	**7.83**	0.42	**13.19**	m	**14.84**
verges; single extra undercloak course plain tiles; bed and point in coloured cement-lime mortar (1:1:6)	6.28	5.00	**6.60**	0.58	0.29	**11.35**	0.42	**18.36**	m	**20.66**
Hand made ridge tiles, half round; in 300mm effective lengths; butt jointed; bedding and pointing in coloured cement-lime mortar (1:1:6)	16.52	5.00	**17.34**	0.35	0.18	**6.90**	0.42	**24.66**	m	**27.75**
Hand made hip tiles, half round; in 300mm effective lengths; butt jointed; bedding and pointing in coloured cement-lime mortar (1:1:6)	16.52	5.00	**17.34**	0.35	0.18	**6.90**	0.42	**24.66**	m	**27.75**
Hand made hip tiles; angular	23.48	5.00	**24.65**	0.35	0.18	**6.90**	0.42	**31.97**	m	**35.97**
Hand made hip tiles; bonnet pattern; bedding and pointing in coloured cement-lime mortar (1:1:6)	18.38	5.00	**19.30**	0.35	0.18	**6.90**	0.42	**26.62**	m	**29.95**
Hand made valley tiles; angular	17.35	5.00	**18.22**	0.50	0.25	**9.78**	0.42	**28.42**	m	**31.97**
Roof coverings; clayware Hand made plain tiles BS EN 1304, red, 265 x 165mm; fixing every fourth course with two galvanized nails per tile to 85mm lap; 38 x 19mm pressure impregnated softwood battens fixed with galvanized nails; underlay BS 747 type 1F reinforced bitumen felt; 150mm laps; fixing with galvanized steel clout nails										
Pitched 40 degrees from horizontal										
generally	58.59	5.00	**61.72**	0.80	0.40	**15.65**	1.53	**78.91**	m2	**88.77**
Fittings										
Hip irons										
galvanized mild steel; 32 x 3 x 380mm girth; scrolled; fixing with galvanized steel screws to timber	2.07	5.00	**2.17**	0.12	0.06	**2.35**	-	**4.52**	nr	**5.09**
Lead soakers										
fixing only	-	-	**-**	0.35	0.18	**6.90**	-	**6.90**	nr	**7.76**
Roof coverings; Redland Plain granular faced tiles, 268 x 165mm; fixing every fifth course with two aluminium nails to each tile to 65mm lap; 38 x 19mm pressure impregnated softwood battens fixed with galvanized nails; underlay BS 747 type 1F bitumen reinforced felt; 150mm laps; fixing with galvanized steel clout nails										
Pitched 40 degrees from horizontal										
generally	23.30	5.00	**24.64**	0.70	0.35	**13.70**	1.15	**39.48**	m2	**44.42**
double course at eaves and nailing each tile with two aluminium nails	2.03	5.00	**2.13**	0.35	0.18	**6.90**	0.51	**9.54**	m	**10.73**
verges; Redland Dry Verge system with plain tiles and tile-and-a-half tiles in alternate courses, clips and aluminium nails	11.75	5.00	**12.34**	0.28	0.14	**5.48**	1.08	**18.90**	m	**21.26**
verges; plain tile undercloak; tiles and undercloak bedded and pointed in tinted cement mortar (1:3) and with plain tiles and tile-and-a-half tiles in alternate courses and aluminium nails	1.01	5.00	**1.06**	0.28	0.14	**5.48**	0.45	**6.99**	m	**7.87**
double course at top edges with clips and nailing each tile with two aluminium nails	2.05	5.00	**2.15**	0.34	0.17	**6.65**	3.81	**12.61**	m	**14.19**
ridge or hip tiles, Redland Plain angle; butt jointed; bedding and pointing in tinted cement mortar (1:3)	18.05	5.00	**18.95**	0.35	0.18	**6.90**	0.42	**26.27**	m	**29.56**
ridge or hip tiles, Redland half round; butt jointed; bedding and pointing in tinted cement mortar (1:3)	7.51	5.00	**7.89**	0.35	0.18	**6.90**	0.42	**15.21**	m	**17.11**
valley tiles , Redland Plain valley tiles	9.60	5.00	**10.08**	0.35	0.18	**6.90**	0.42	**17.40**	m	**19.58**

Labour hourly rates: (except Specialists) Craft Operatives £14.21 Labourer £10.71 Rates are national average prices. Refer to REGIONAL VARIATIONS for indicative levels of overall pricing in regions	MATERIALS			LABOUR				RATES		
	Del to Site £	Waste %	Material Cost £	Craft Optve Hrs	Lab Hrs	Labour Cost £	Sunds £	Nett Rate £	Unit	Gross Rate (+12.5%) £
H60: PLAIN ROOF TILING Cont.										
Roof coverings; Redland Plain granular faced tiles, 268 x 165mm; fixing every fifth course with two aluminium nails to each tile to 65mm lap; 50 x 25mm pressure impregnated softwood battens fixed with galvanized nails; underlay; BS 747 type 1F bitumen reinforced felt; 150mm laps; fixing with galvanized steel clout nails										
Pitched 40 degrees from horizontal										
generally	23.30	5.00	**24.64**	0.70	0.35	**13.70**	1.15	**39.48**	m2	**44.42**
abutments; square	2.03	5.00	**2.13**	0.28	0.15	**5.59**	-	**7.71**	m	**8.68**
abutments; raking	2.90	5.00	**3.05**	0.43	0.21	**8.36**	-	**11.41**	m	**12.84**
abutments; curved to 3000mm radius	4.05	5.00	**4.26**	0.56	0.28	**10.96**	-	**15.21**	m	**17.11**
Roof coverings; Redland Plain granular faced tiles, 268 x 165mm; fixing each tile with two aluminium nails to 65mm lap; 38 x 19mm pressure impregnated softwood battens fixed with galvanized nails; underlay; BS 747 type 1F bitumen reinforced felt; 150mm laps; fixing with galvanized steel clout nails										
Pitched 40 degrees from horizontal										
generally	23.30	5.00	**24.64**	0.83	0.42	**16.29**	1.44	**42.37**	m2	**47.67**
Roof coverings; Plain through coloured tiles, 268 x 165mm; fixing every fifth course with two aluminium nails to each tile to 65mm lap; 38 x 19mm pressure impregnated softwood battens fixed with galvanized nails; underlay; BS 747 type 1F bitumen reinforced felt; 150mm laps; fixing with galvanized steel clout nails										
Pitched 40 degrees from horizontal										
generally	23.60	5.00	**24.97**	0.70	0.35	**13.70**	1.15	**39.81**	m2	**44.79**
double course at eaves and nailing each tile with two aluminium nails	2.03	5.00	**2.13**	0.34	0.17	**6.65**	0.51	**9.29**	m	**10.45**
verges; Redland Dry Verge system with plain tiles and tile-and-a-half tiles in alternate courses, clips and aluminium nails	11.75	5.00	**12.34**	0.28	0.14	**5.48**	1.08	**18.90**	m	**21.26**
verges; plain tile undercloak; tiles and undercloak bedded and pointed in tinted cement mortar (1:3) and with plain tiles and tile-and-a-half tiles in alternate courses and aluminium nails	2.03	5.00	**2.13**	0.28	0.14	**5.48**	0.45	**8.06**	m	**9.06**
double course at top edges with clips and nailing each tile with two aluminium nails	2.03	5.00	**2.13**	0.34	0.17	**6.65**	3.81	**12.59**	m	**14.16**
ridge or hip tiles, Redland Plain angle; butt jointed; bedding and pointing in tinted cement mortar (1:3)	18.05	5.00	**18.95**	0.35	0.18	**6.90**	0.42	**26.27**	m	**29.56**
ridge or hip tiles, Redland half round; butt jointed; bedding and pointing in tinted cement mortar (1:3)	7.50	5.00	**7.88**	0.35	0.18	**6.90**	0.42	**15.20**	m	**17.10**
valley tiles, Redland Plain valley tiles	9.59	5.00	**10.07**	0.35	0.18	**6.90**	0.42	**17.39**	m	**19.57**
Roof coverings; Redland Downland plain granular faced tiles, 268 x 165mm; fixing every fifth course with two aluminium nails to each tile to 65mm lap; 38 x 19mm pressure impregnated softwood battens fixed with galvanized nails; underlay; BS 747 type 1F bitumen reinforced felt; 150mm laps; fixing with galvanized steel clout nails										
Pitched 40 degrees from horizontal										
generally	25.94	5.00	**27.43**	0.70	0.35	**13.70**	1.15	**42.28**	m2	**47.56**
double course at eaves and nailing each tile with two aluminium nails	2.26	5.00	**2.37**	0.34	0.17	**6.65**	0.51	**9.54**	m	**10.73**
verges; Redland Dry Verge system with plain tiles and tile-and-a-half tiles in alternate courses, clips and aluminium nails	11.87	5.00	**12.46**	0.28	0.14	**5.48**	1.08	**19.02**	m	**21.40**
verges; plain tile undercloak; tiles and undercloak bedded and pointed in tinted cement mortar (1:3) and with plain tiles and tile-and-a-half tiles in alternate courses and aluminium nails	2.26	5.00	**2.37**	0.28	0.14	**5.48**	0.45	**8.30**	m	**9.34**
double course at top edges with clips and nailing each tile with two aluminium nails	2.26	5.00	**2.37**	0.34	0.17	**6.65**	3.81	**12.84**	m	**14.44**
ridge or hip tiles, Redland Plain angle; butt jointed; bedding and pointing in tinted cement mortar (1:3)	18.05	5.00	**18.95**	0.35	0.18	**6.90**	0.42	**26.27**	m	**29.56**
ridge or hip tiles, Redland half round; butt jointed; bedding and pointing in tinted cement mortar (1:3)	7.50	5.00	**7.88**	0.35	0.18	**6.90**	0.42	**15.20**	m	**17.10**
valley tiles, Redland Plain valley tiles	11.06	5.00	**11.61**	0.35	0.18	**6.90**	0.42	**18.93**	m	**21.30**
Roof coverings; Redland Downland plain through coloured tiles, 268 x 165mm; fixing every fifth course with two aluminium nails to each tile to 65mm lap; 38 x 19mm pressure impregnated softwood battens fixed with galvanized nails; underlay; BS 747 type 1F bitumen reinforced felt; 150mm laps; fixing with galvanized steel clout nails										
Pitched 40 degrees from horizontal										
generally	25.94	5.00	**27.43**	0.70	0.35	**13.70**	1.15	**42.28**	m2	**47.56**

Labour hourly rates: (except Specialists) Craft Operatives £14.21 Labourer £10.71 Rates are national average prices. Refer to REGIONAL VARIATIONS for indicative levels of overall pricing in regions	MATERIALS			LABOUR				RATES		
	Del to Site £	Waste %	Material Cost £	Craft Optve Hrs	Lab Hrs	Labour Cost £	Sunds £	Nett Rate £	Unit	Gross Rate (+12.5%) £
H60: PLAIN ROOF TILING Cont.										
Roof coverings; Redland Downland plain through coloured tiles, 268 x 165mm; fixing every fifth course with two aluminium nails to each tile to 65mm lap; 38 x 19mm pressure impregnated softwood battens fixed with galvanized nails; underlay; BS 747 type 1F bitumen reinforced felt; 150mm laps; fixing with galvanized steel clout nails Cont.										
Pitched 40 degrees from horizontal Cont.										
double course at eaves and nailing each tile with two aluminium nails	2.26	5.00	**2.37**	0.34	0.17	**6.65**	0.51	**9.54**	m	**10.73**
verges; Redland Dry Verge system with plain tiles and tile-and-a-half tiles in alternate courses, clips and aluminium nails	11.87	5.00	**12.46**	0.28	0.14	**5.48**	1.08	**19.02**	m	**21.40**
verges; plain tile undercloak; tiles and undercloak bedded and pointed in tinted cement mortar (1:3) and with plain tiles and tile-and-a-half tiles in alternate courses and aluminium nails	2.26	5.00	**2.37**	0.28	0.14	**5.48**	0.45	**8.30**	m	**9.34**
double course at top edges with clips and nailing each tile with two aluminium nails	2.26	5.00	**2.37**	0.34	0.17	**6.65**	3.81	**12.84**	m	**14.44**
ridge or hip tiles, Redland Plain angle; butt jointed; bedding and pointing in tinted cement mortar (1:3)	18.05	5.00	**18.95**	0.35	0.18	**6.90**	0.42	**26.27**	m	**29.56**
ridge or hip tiles, Redland half round; butt jointed; bedding and pointing in tinted cement mortar (1:3)	11.26	5.00	**11.82**	0.35	0.18	**6.90**	0.42	**19.14**	m	**21.53**
valley tiles, Redland Plain valley tiles	11.06	5.00	**11.61**	0.35	0.18	**6.90**	0.42	**18.93**	m	**21.30**
Fittings										
Ventilator tiles										
Red Vent ridge ventilation terminal 450mm long for										
half round ridge	50.85	5.00	**53.39**	1.45	0.73	**28.42**	0.66	**82.48**	nr	**92.78**
universal angle ridge	50.85	5.00	**53.39**	1.45	0.73	**28.42**	0.66	**82.48**	nr	**92.78**
Redvent eaves ventilators fixing with aluminium nails	9.18	5.00	**9.64**	0.56	0.28	**10.96**	0.03	**20.63**	m	**23.20**
Red Vent Thruvent tiles for										
Stonewold slates	38.81	5.00	**40.75**	0.56	0.28	**10.96**	0.25	**51.96**	nr	**58.45**
Regent tiles	38.81	5.00	**40.75**	0.56	0.28	**10.96**	0.25	**51.96**	nr	**58.45**
Grovebury double pantiles	38.81	5.00	**40.75**	0.56	0.28	**10.96**	0.25	**51.96**	nr	**58.45**
Norfolk pantiles	38.81	5.00	**40.75**	0.56	0.28	**10.96**	0.25	**51.96**	nr	**58.45**
Redland 49 tiles	38.81	5.00	**40.75**	0.56	0.28	**10.96**	0.25	**51.96**	nr	**58.45**
Renown tiles	38.81	5.00	**40.75**	0.56	0.28	**10.96**	0.25	**51.96**	nr	**58.45**
Redland 50 double roman tiles	38.81	5.00	**40.75**	0.56	0.28	**10.96**	0.25	**51.96**	nr	**58.45**
Redland plain tiles	45.11	5.00	**47.37**	0.56	0.28	**10.96**	0.25	**58.57**	nr	**65.89**
Redland Downland plain tiles	47.70	5.00	**50.09**	0.56	0.28	**10.96**	0.25	**61.29**	nr	**68.95**
Red line ventilation tile complete with underlay seal and fixing clips for										
Stonewold slates	48.38	5.00	**50.80**	1.12	0.56	**21.91**	0.25	**72.96**	nr	**82.08**
Regent tiles	48.38	5.00	**50.80**	1.12	0.56	**21.91**	0.25	**72.96**	nr	**82.08**
Grovebury double pantiles	48.38	5.00	**50.80**	1.12	0.56	**21.91**	0.25	**72.96**	nr	**82.08**
Renown tiles	48.38	5.00	**50.80**	1.12	0.56	**21.91**	0.25	**72.96**	nr	**82.08**
Redland 50 double roman tiles	48.38	5.00	**50.80**	1.12	0.56	**21.91**	0.25	**72.96**	nr	**82.08**
Gas terminals										
Gas Flue ridge terminal Mark III, 450mm long with sealing gasket and fixing brackets										
half round ridge	49.84	5.00	**52.33**	1.68	0.84	**32.87**	0.66	**85.86**	nr	**96.59**
universal angle ridge	54.78	5.00	**57.52**	1.68	0.84	**32.87**	0.66	**91.05**	nr	**102.43**
Gas Flue ridge terminal Mark III, 450mm long with sealing gasket and fixing brackets with 150mm R type adaptor for										
half round ridge	68.63	5.00	**72.06**	1.68	0.84	**32.87**	0.66	**105.59**	nr	**118.79**
universal angle ridge	73.87	5.00	**77.56**	1.68	2.84	**54.29**	0.66	**132.51**	nr	**149.08**
extra for extension adaptor and gasket	18.79	5.00	**19.73**	0.35	0.17	**6.79**	0.25	**26.77**	nr	**30.12**
Hip irons										
galvanized mild steel; 32 x 3 x 380mm girth; scrolled; fixing with galvanized steel screws to timber	2.07	5.00	**2.17**	0.11	0.06	**2.21**	-	**4.38**	nr	**4.93**
Lead soakers										
fixing only	-	-	-	0.34	0.17	**6.65**	-	**6.65**	nr	**7.48**
H61: FIBRE CEMENT SLATING										
Quantities required for 1m2 of slating										
Slates to 70mm lap										
400 x 240mm, 25.3nr										
Slates to 102mm lap										
600 x 300mm, 13nr										
500 x 250mm, 19.5nr										
400 x 200mm, 32nr										
Roof coverings; asbestos-free cement slates, 600 x 300mm; centre fixing with copper nails and copper disc rivets to 102mm lap; 38 x 19mm pressure impregnated softwood battens fixed with galvanized nails; underlay; BS 747 type 1F bitumen reinforced felt; 150mm laps; fixing with galvanized steel clout nails										
Pitched 30 degrees from horizontal										
generally	16.04	5.00	**16.95**	0.60	0.30	**11.74**	1.60	**30.29**	m2	**34.08**

Labour hourly rates: (except Specialists) Craft Operatives £14.21 Labourer £10.71 Rates are national average prices. Refer to REGIONAL VARIATIONS for indicative levels of overall pricing in regions	MATERIALS			LABOUR			RATES			
	Del to Site £	Waste %	Material Cost £	Craft Optve Hrs	Lab Hrs	Labour Cost £	Sunds £	Nett Rate £	Unit	Gross Rate (+12.5%) £
H61: FIBRE CEMENT SLATING Cont.										
Roof coverings; asbestos-free cement slates, 600 x 300mm; centre fixing with copper nails and copper disc rivets to 102mm lap; 38 x 19mm pressure impregnated softwood battens fixed with galvanized nails; underlay; BS 747 type 1F bitumen reinforced felt; 150mm laps; fixing with galvanized steel clout nails Cont.										
Pitched 30 degrees from horizontal Cont.										
holes	-	-	-	0.58	0.29	11.35	-	11.35	nr	12.77
abutments; square	1.44	5.00	1.51	0.23	0.12	4.55	-	6.06	m	6.82
abutments; raking	2.16	5.00	2.27	0.28	0.14	5.48	-	7.74	m	8.71
abutments; curved to 3000mm radius	2.88	5.00	3.02	0.35	0.18	6.90	-	9.92	m	11.16
double course at eaves	3.87	5.00	4.07	0.17	0.09	3.38	0.73	8.18	m	9.20
verges; slate undercloak and point in cement mortar (1:3)	2.21	5.00	2.32	0.09	0.05	1.81	0.60	4.74	m	5.33
ridges or hips; asbestos free cement; fixing with nails	26.65	5.00	27.98	0.23	0.12	4.55	0.42	32.96	m	37.08
Roof coverings; asbestos-free cement slates, 600 x 300mm; centre fixing with copper nails and copper disc rivets to 102mm lap; 50 x 25mm pressure impregnated softwood battens fixed with galvanized nails; underlay; BS 747 type 1F bitumen reinforced felt; 150mm laps; fixing with galvanized steel clout nails										
Pitched 30 degrees from horizontal										
generally	17.00	5.00	18.01	0.64	0.32	12.52	1.60	32.13	m2	36.15
Roof coverings; asbestos-free cement slates, 600 x 300mm; centre fixing with copper nails and copper disc rivets to 102mm lap (close boarding on rafters included elsewhere); underlay; BS 747 type 1B underslating felt; 150mm laps; fixing with galvanized steel clout nails										
Pitched 30 degrees from horizontal										
generally	14.92	5.00	15.72	0.48	0.24	9.39	1.17	26.28	m2	29.57
Roof coverings; asbestos-free cement slates, 500 x 250mm; centre fixing with copper nails and copper disc rivets to 102mm lap; 38 x 19mm pressure impregnated softwood battens fixed with galvanized nails; underlay; BS 747 type 1F bitumen reinforced felt; 150mm laps; fixing with galvanized steel clout nails										
Pitched 30 degrees from horizontal										
generally	19.32	5.00	20.41	0.67	0.34	13.16	2.18	35.76	m2	40.22
holes	-	-	-	0.58	0.29	11.35	-	11.35	nr	12.77
abutments; square	1.74	5.00	1.83	0.23	0.12	4.55	-	6.38	m	7.18
abutments; raking	2.61	5.00	2.74	0.35	0.18	6.90	-	9.65	m	10.85
abutments; curved to 3000mm radius	3.48	5.00	3.65	0.46	0.23	9.00	-	12.65	m	14.23
double course at eaves	3.57	5.00	3.75	0.21	0.11	4.16	0.78	8.69	m	9.77
verges; slate undercloak and point in cement mortar (1:3)	2.23	5.00	2.34	0.12	0.06	2.35	0.64	5.33	m	6.00
ridges or hips; asbestos free cement; fixing with nails	26.65	5.00	27.98	0.23	0.12	4.55	0.42	32.96	m	37.08
Roof coverings; asbestos-free cement slates, 500 x 250mm; centre fixing with copper nails and copper disc rivets to 102mm lap; 50 x 25mm pressure impregnated softwood battens fixed with galvanized nails; underlay; BS 747 type 1F bitumen reinforced felt; 150mm laps; fixing with galvanized steel clout nails										
Pitched 30 degrees from horizontal										
generally	20.52	5.00	21.73	0.72	0.36	14.09	2.18	38.00	m2	42.75
Roof coverings; asbestos-free cement slates, 400 x 200mm; centre fixing with copper nails and copper disc rivets to 102mm lap; 38 x 19mm pressure impregnated softwood battens fixed with galvanized nails; underlay; BS 747 type 1F bitumen reinforced felt; 150mm laps; fixing with galvanized steel clout nails										
Pitched 30 degrees from horizontal										
generally	22.82	5.00	24.10	0.71	0.36	13.94	3.31	41.35	m2	46.52
holes	-	-	-	0.58	0.29	11.35	-	11.35	nr	12.77
abutments; square	2.06	5.00	2.16	0.29	0.15	5.73	-	7.89	m	8.88
abutments; raking	3.09	5.00	3.25	0.44	0.22	8.61	-	11.85	m	13.34
abutments; curved to 3000mm radius	4.12	5.00	4.33	0.58	0.29	11.35	-	15.68	m	17.63
double course at eaves	3.22	5.00	3.38	0.23	0.97	13.66	0.87	17.91	m	20.15
verges; slate undercloak and point in cement mortar (1:3)	1.93	5.00	2.03	0.14	0.07	2.74	0.68	5.45	m	6.13
ridges or hips; asbestos free cement; fixing with nails	26.65	5.00	27.98	0.23	0.12	4.55	0.42	32.96	m	37.08

Labour hourly rates: (except Specialists) Craft Operatives £14.21 Labourer £10.71 Rates are national average prices. Refer to REGIONAL VARIATIONS for indicative levels of overall pricing in regions	MATERIALS			LABOUR			RATES			
	Del to Site £	Waste %	Material Cost £	Craft Optve Hrs	Lab Hrs	Labour Cost £	Sunds £	Nett Rate £	Unit	Gross Rate (+12.5%) £
H61: FIBRE CEMENT SLATING Cont.										
Roof coverings; asbestos-free cement slates, 400 x 200mm; centre fixing with copper nails and copper disc rivets to 102mm lap; 50 x 25mm pressure impregnated softwood battens fixed with galvanized nails; underlay; BS 747 type 1F bitumen reinforced felt; 150mm laps; fixing with galvanized steel clout nails										
Pitched 30 degrees from horizontal										
generally	24.26	5.00	**25.68**	0.80	0.40	**15.65**	3.31	**44.64**	m2	**50.22**
Fittings										
Hip irons										
galvanized mild steel; 32 x 3 x 380mm girth; scrolled; fixing with galvanized steel screws to timber	2.07	5.00	**2.17**	0.11	0.06	**2.21**	-	**4.38**	nr	**4.93**
Lead soakers										
fixing only	-	-	**-**	0.34	0.17	**6.65**	-	**6.65**	nr	**7.48**
H62: NATURAL SLATING										
Quantities required for 1m2 of slating										
Slates to 75mm lap										
600 x 300mm, 12.65nr										
500 x 250mm, 18.9nr										
400 x 200mm, 31nr										
Roof coverings; blue/grey slates, 600 x 300mm, 6.5mm thick; fixing with slate nails to 75mm lap; 38 x 19mm pressure impregnated softwood battens fixed with galvanized nails; underlay; BS 747 type 1F bitumen reinforced felt; 150mm laps; fixing with galvanized steel clout nails										
Pitched 30 degrees from horizontal										
generally	74.58	5.00	**78.42**	0.58	0.29	**11.35**	0.56	**90.32**	m2	**101.61**
holes	-	-	**-**	0.58	0.29	**11.35**	-	**11.35**	nr	**12.77**
abutments; square	7.09	5.00	**7.45**	0.17	0.09	**3.38**	-	**10.83**	m	**12.18**
abutments; raking	10.67	5.00	**11.20**	0.26	0.13	**5.09**	-	**16.29**	m	**18.32**
double course at eaves	20.18	5.00	**21.19**	0.16	0.08	**3.13**	0.50	**24.82**	m	**27.92**
verges; slate undercloak and point in cement mortar (1:3)	11.53	5.00	**12.11**	0.09	0.05	**1.81**	0.46	**14.38**	m	**16.18**
Roof coverings; blue/grey slates, 600 x 300mm, 6.5mm thick; fixing with aluminium nails to 75mm lap; 38 x 19mm pressure impregnated softwood battens fixed with galvanized nails; underlay; BS 747 type 1F bitumen reinforced felt; 150mm laps; fixing with galvanized steel clout nails										
Pitched 30 degrees from horizontal										
generally	74.58	5.00	**78.42**	0.58	0.29	**11.35**	1.11	**90.87**	m2	**102.23**
Roof coverings; blue/grey slates, 600 x 300mm, 6.5mm thick; fixing with copper nails to 75mm laps; 38 x 19mm pressure impregnated softwood battens fixed with galvanized nails; underlay; BS 747 type 1F bitumen reinforced felt; 150mm laps; fixing with galvanized steel clout nails										
Pitched 30 degrees from horizontal										
generally	74.58	5.00	**78.42**	0.58	0.29	**11.35**	1.38	**91.14**	m2	**102.54**
Roof coverings; blue/grey slates, 600 x 300mm, 6.5mm thick; fixing with slate nails to 75mm lap; 50 x 25mm pressure impregnated softwood battens fixed with galvanized nails; underlay; BS 747 type 1F bitumen reinforced felt; 150mm laps; fixing with galvanized steel clout nails										
Pitched 30 degrees from horizontal										
generally	75.54	5.00	**79.47**	0.62	0.31	**12.13**	0.56	**92.16**	m2	**103.68**
Roof coverings; blue/grey slates, 600 x 300mm, 6.5mm thick; fixing with slate nails to 75mm lap; 50 x 25mm pressure impregnated softwood counterbattens at 1067mm centres and 38 x 19mm pressure impregnated softwood battens fixed with galvanized nails; underlay; BS 747 type 1B underslating felt; 150mm laps; fixing with galvanized steel clout nails										
Pitched 30 degrees from horizontal										
generally	75.54	5.00	**79.47**	0.62	0.31	**12.13**	1.14	**92.74**	m2	**104.33**
Roof coverings; blue/grey slates, 500 x 250mm, 6.5mm thick; fixing with slate nails to 75mm lap; 38 x 19mm pressure impregnated softwood battens fixed with galvanized nails; underlay; BS 747 type 1F bitumen reinforced felt; 150mm laps; fixing with galvanized steel clout nails										
Pitched 30 degrees from horizontal										
generally	54.62	5.00	**57.48**	0.67	0.34	**13.16**	0.83	**71.47**	m2	**80.40**

Labour hourly rates: (except Specialists) Craft Operatives £14.21 Labourer £10.71 Rates are national average prices. Refer to REGIONAL VARIATIONS for indicative levels of overall pricing in regions	MATERIALS			LABOUR				RATES		
	Del to Site £	Waste %	Material Cost £	Craft Optve Hrs	Lab Hrs	Labour Cost £	Sunds £	Nett Rate £	Unit	Gross Rate (+12.5%) £
H62: NATURAL SLATING Cont.										
Roof coverings; blue/grey slates, 500 x 250mm, 6.5mm thick; fixing with slate nails to 75mm lap; 38 x 19mm pressure impregnated softwood battens fixed with galvanized nails; underlay; BS 747 type 1F bitumen reinforced felt; 150mm laps; fixing with galvanized steel clout nails Cont.										
Pitched 30 degrees from horizontal Cont.										
holes	-	-	**-**	0.58	0.29	**11.35**	-	**11.35**	nr	**12.77**
abutments; square	5.07	5.00	**5.33**	0.23	0.12	**4.55**	-	**9.88**	m	**11.12**
abutments; raking	7.61	5.00	**7.99**	0.35	0.18	**6.90**	-	**14.89**	m	**16.75**
double course at eaves	11.15	5.00	**11.71**	0.20	0.10	**3.91**	0.80	**16.42**	m	**18.48**
verges; slate undercloak and point in cement mortar (1:3)	6.97	5.00	**7.32**	0.12	0.06	**2.35**	0.47	**10.14**	m	**11.40**
Roof coverings; blue/grey slates, 500 x 250mm, 6.5mm thick; fixing with aluminium nails to 75mm lap; 38 x 19mm pressure impregnated softwood battens fixed with galvanized nails; underlay; BS 747 type 1F bitumen reinforced felt; 150mm laps; fixing with galvanized steel clout nails										
Pitched 30 degrees from horizontal										
generally	54.62	5.00	**57.48**	0.67	0.34	**13.16**	1.65	**72.29**	m2	**81.33**
Roof coverings; blue/grey slates, 500 x 250mm, 6.5mm thick; fixing with copper nails to 75mm lap; 38 x 19mm pressure impregnated softwood battens fixed with galvanized nails; underlay; BS 747 type 1F bitumen reinforced felt; 150mm laps; fixing with galvanized steel clout nails										
Pitched 30 degrees from horizontal										
generally	54.62	5.00	**57.48**	0.67	0.34	**13.16**	2.06	**72.70**	m2	**81.79**
Roof coverings; blue/grey slates, 500 x 250mm, 6.5mm thick; fixing with slate nails to 75mm lap; 50 x 25mm pressure impregnated softwood battens fixed with galvanized nails; underlay; BS 747 type 1F bitumen reinforced felt; 150mm laps; fixing with galvanized steel clout nails										
Pitched 30 degrees from horizontal										
generally	55.82	5.00	**58.80**	0.72	0.36	**14.09**	0.83	**73.71**	m2	**82.93**
Roof coverings; blue/grey slates, 400 x 200mm, 6.5mm thick; fixing with slate nails to 75mm lap; 38 x 19mm pressure impregnated softwood battens fixed with galvanized nails; underlay; BS 747 type 1F bitumen reinforced felt; 150mm laps; fixing with galvanized steel clout nails										
Pitched 30 degrees from horizontal										
generally	41.77	5.00	**43.99**	0.80	0.40	**15.65**	1.38	**61.02**	m2	**68.65**
holes	-	-	**-**	0.58	0.29	**11.35**	-	**11.35**	nr	**12.77**
abutments; square	3.83	5.00	**4.02**	0.29	0.15	**5.73**	-	**9.75**	m	**10.97**
abutments; raking	5.74	5.00	**6.03**	0.44	0.22	**8.61**	-	**14.64**	m	**16.47**
double course at eaves	7.66	5.00	**8.04**	0.23	0.12	**4.55**	0.41	**13.00**	m	**14.63**
verges; slate undercloak and point in cement mortar (1:3)	3.83	5.00	**4.02**	0.14	0.07	**2.74**	0.48	**7.24**	m	**8.14**
Roof coverings; blue/grey slates, 400 x 200mm, 6.5mm thick; fixing with aluminium nails to 75mm lap; 38 x 19mm pressure impregnated softwood battens fixed with galvanized nails; underlay; BS 747 type 1F bitumen reinforced felt; 150mm laps; fixing with galvanized steel clout nails										
Pitched 30 degrees from horizontal										
generally	41.77	5.00	**43.99**	0.80	0.40	**15.65**	2.73	**62.37**	m2	**70.17**
Roof coverings; blue/grey slates, 400 x200mm, 6.5mm thick; fixing with copper nails to 75mmlap; 38 x 19mm pressure impregnated softwood battens fixed with galvanized nails; underlay; BS 747 type 1F bitumen reinforced felt; 150mm laps; fixing with galvanized steel clout nails										
Pitched 30 degrees from horizontal										
generally	41.77	5.00	**43.99**	0.80	0.40	**15.65**	3.41	**63.05**	m2	**70.93**
Roof coverings; blue/grey slates, 400 x 200mm, 6.5mm thick; fixing with slate nails to 75mm lap; 50 x 25mm pressure impregnated softwood battens fixed with galvanized nails; underlay; BS 747 type 1F bitumen reinforced felt; 150mm laps; fixing with galvanized steel clout nails										
Pitched 30 degrees from horizontal										
generally	43.21	5.00	**45.58**	0.87	0.44	**17.08**	1.38	**64.03**	m2	**72.03**

Labour hourly rates: (except Specialists) Craft Operatives £14.21 Labourer £10.71 Rates are national average prices. Refer to REGIONAL VARIATIONS for indicative levels of overall pricing in regions	MATERIALS			LABOUR				RATES		
	Del to Site £	Waste %	Material Cost £	Craft Optve Hrs	Lab Hrs	Labour Cost £	Sunds £	Nett Rate £	Unit	Gross Rate (+12.5%) £
H62: NATURAL SLATING Cont.										
Roof coverings; Burlington green slates, best random fixed in diminishing courses with alloy nails to 75mm lap; 38 x 19mm pressure impregnated softwood battens fixed with galvanized nails; underlay; BS 747 type 1F bitumen reinforced felt; 150mm laps; fixing with galvanized steel clout nails										
Pitched 30 degrees from horizontal										
generally	123.76	5.00	**130.09**	0.86	0.43	**16.83**	2.03	**148.94**	m2	**167.56**
abutments; square	12.16	5.00	**12.76**	0.23	0.12	**4.55**	-	**17.32**	m	**19.48**
abutments; raking	18.23	5.00	**19.14**	0.35	0.18	**6.90**	-	**26.05**	m	**29.30**
double course at eaves	26.52	5.00	**27.85**	0.23	0.12	**4.55**	1.01	**33.41**	m	**37.59**
verges; slate undercloak and point in cement mortar (1:3)	17.68	5.00	**18.56**	0.23	0.12	**4.55**	0.62	**23.74**	m	**26.70**
hips, valleys and angles; close mitred	19.89	5.00	**20.88**	0.40	0.20	**7.83**	0.65	**29.36**	m	**33.03**
Fittings										
Hip irons										
galvanized mild steel; 32 x 3 x 380mm girth; scrolled; fixing with galvanized steel screws to timber	2.07	5.00	**2.17**	0.11	0.06	**2.21**	-	**4.38**	nr	**4.93**
Lead soakers										
fixing only	-	-	**-**	0.34	0.17	**6.65**	-	**6.65**	nr	**7.48**
H63: RECONSTRUCTED STONE SLATING/ TILING										
Roof coverings; Redland Cambrian interlocking riven textured slates, 300 x 336mm; fixing every slate with two stainless steel nails and one stainless steel clip to 50mm laps; 38 x 19mm pressure impregnated softwood battens fixed with galvanized nails; underlay; BS 747 type 1F bitumen reinforced felt; 150mm laps; fixing with galvanized steel clout nails										
Pitched 40 degrees from horizontal										
generally	27.32	5.00	**28.80**	0.54	0.27	**10.57**	1.96	**41.32**	m2	**46.49**
abutments; square	2.56	5.00	**2.69**	0.11	0.06	**2.21**	-	**4.90**	m	**5.51**
abutments; raking	3.85	5.00	**4.04**	0.17	0.09	**3.38**	-	**7.42**	m	**8.35**
abutments; curved to 3000mm radius	5.12	5.00	**5.38**	0.23	0.12	**4.55**	-	**9.93**	m	**11.17**
supplementary fixing at eaves with stainless steel eaves clip to each slate	0.52	5.00	**0.55**	0.05	0.03	**1.03**	-	**1.58**	m	**1.78**
verges; 150 x 6mm fibre cement undercloak; slates and undercloak bedded and pointed in tinted cement mortar (1:3) with full slates and slate and a half slates in alternate courses and stainless steel verge clips	8.31	5.00	**8.72**	0.23	0.12	**4.55**	2.27	**15.55**	m	**17.49**
verges; 150 x 6mm fibre cement undercloak; slates and undercloak bedded and pointed in tinted cement mortar (1:3) with verge slates and slate and a half verge slates in alternate courses and stainless steel verge clips	8.31	5.00	**8.72**	0.23	0.12	**4.55**	2.27	**15.55**	m	**17.49**
ridge or hip tiles, Redland third round; butt jointed; bedding and pointing in tinted cement mortar (1:3)	7.77	5.00	**8.16**	0.23	0.12	**4.55**	0.42	**13.14**	m	**14.78**
ridge or hip tiles, Redland half round; butt jointed; bedding and pointing in tinted cement mortar (1:3)	7.77	5.00	**8.16**	0.23	0.12	**4.55**	0.42	**13.14**	m	**14.78**
ridge or hip tiles, Redland universal angle; butt jointed; bedding and pointing in tinted cement mortar (1:3)	8.44	5.00	**8.86**	0.23	0.12	**4.55**	0.42	**13.84**	m	**15.57**
ridge tiles, Redland Dry Ridge system with half round ridge tiles; fixing with stainless steel batten straps and ring shanked fixing nails with neoprene washers and sleeves; polypropylene ridge seals and uPVC profile filler units	26.44	5.00	**27.76**	0.45	0.23	**8.86**	-	**36.62**	m	**41.20**
ridge tiles, Redland Dry Ridge system with universal angle ridge tiles; fixing with stainless steel batten straps and ring shanked fixing nails with neoprene washers and sleeves; polypropylene ridge seals and uPVC profile filler units	27.12	5.00	**28.48**	0.45	0.23	**8.86**	-	**37.33**	m	**42.00**
ridge tiles, Redland Dry Vent Ridge system with half round ridge tiles; fixing with stainless steel batten straps and ring shanked nails with neoprene washers and sleeves; polypropylene ridge seals, PVC air flow control units and uPVC ventilated profile filler units	26.44	5.00	**27.76**	0.45	0.23	**8.86**	-	**36.62**	m	**41.20**
ridge tiles, Redland Dry Vent Ridge system with Universal angle ridge tiles; fixing with stainless steel batten straps and ring shanked nails with neoprene washers and sleeves; polypropylene ridge seals, PVC air flow control units and uPVC ventilated profile filler units	27.12	5.00	**28.48**	0.45	0.23	**8.86**	-	**37.33**	m	**42.00**
Monoridge filler units in conjunction with top tile, bed tile in nonsetting mastic sealant and screw on through filler unit with screws, washers and caps	38.83	5.00	**40.77**	0.20	0.12	**4.13**	-	**44.90**	m	**50.51**
valley tiles, Redland Universal valley troughs; laid with 100mm laps	27.54	5.00	**28.92**	0.50	0.28	**10.10**	0.42	**39.44**	m	**44.37**

Labour hourly rates: (except Specialists) Craft Operatives £14.21 Labourer £10.71 Rates are national average prices. Refer to REGIONAL VARIATIONS for indicative levels of overall pricing in regions	MATERIALS			LABOUR				RATES		
	Del to Site £	Waste %	Material Cost £	Craft Optve Hrs	Lab Hrs	Labour Cost £	Sunds £	Nett Rate £	Unit	Gross Rate (+12.5%) £
H63: RECONSTRUCTED STONE SLATING/ TILING Cont.										
Roof coverings; Redland Cambrian interlocking riven textured slates, 300 x 336mm; fixing every slate with two stainless steel nails and one stainless steel clip to 90mm lap; 38 x 25mm pressure impregnated softwood battens fixed with galvanized nails; underlay; BS 747 type 1F bitumen reinforced felt; 150mm laps; fixing with galvanized steel clout nails										
Pitched 40 degrees from horizontal										
generally	32.55	5.00	**34.30**	0.62	0.31	**12.13**	2.53	**48.96**	m2	**55.08**
abutments; square	3.06	5.00	**3.22**	0.11	0.06	**2.21**	-	**5.42**	m	**6.10**
abutments; raking	4.58	5.00	**4.81**	0.17	0.09	**3.38**	-	**8.19**	m	**9.22**
abutments; curved to 3000mm radius	6.12	5.00	**6.43**	0.23	0.12	**4.55**	-	**10.98**	m	**12.36**
Fittings										
Ventilator tiles										
Red Vent ridge ventilation terminal 450mm long for										
half round ridge	49.84	5.00	**52.33**	1.45	0.73	**28.42**	-	**80.75**	nr	**90.85**
universal angle ridge	49.84	5.00	**52.33**	1.45	0.73	**28.42**	-	**80.75**	nr	**90.85**
Red Vent eaves ventilators; fixing with stainless steel nails	9.19	5.00	**9.65**	0.45	0.23	**8.86**	-	**18.51**	m	**20.82**
Redland Cambrian Thruvent interlocking slate complete with weather cap, underlay seal and fixing clips	40.72	5.00	**42.76**	0.56	0.28	**10.96**	-	**53.71**	nr	**60.43**
Gas terminals										
Gas Flue ridge terminal Mark III, 450mm long with sealing gasket and fixing brackets										
half round ridge	49.84	5.00	**52.33**	1.50	0.75	**29.35**	-	**81.68**	nr	**91.89**
Gas Flue ridge terminal Mark III, 450mm long with sealing gasket and fixing brackets with 150mm R type adaptor for										
half round ridge	49.84	5.00	**52.33**	1.50	0.75	**29.35**	32.06	**113.74**	nr	**127.96**
Hip irons										
galvanized mild steel; 32 x 3 x 380mm girth; scrolled; fixing with galvanized steel screws to timber	2.07	5.00	**2.17**	0.11	0.06	**2.21**	-	**4.38**	nr	**4.93**
Lead soakers										
fixing only	-	-	**-**	0.34	0.17	**6.65**	-	**6.65**	nr	**7.48**
Fittings										
Timloc uPVC Mark 3 eaves ventilators; reference 1126/40 fixing with nails to timber at 400mm centres										
330mm wide	4.14	5.00	**4.35**	0.28	0.14	**5.48**	0.30	**10.13**	m	**11.39**
Timloc uPVC Mark 3 eaves ventilators; reference 1123 fixing with nails to timber at 600mm centres										
330mm wide	2.76	5.00	**2.90**	0.22	0.11	**4.30**	0.30	**7.50**	m	**8.44**
Timlock uPVC soffit ventilators; reference 1137, 10mm airflow; fixing with screws to timber										
type C	2.22	5.00	**2.33**	0.17	0.06	**3.06**	0.12	**5.51**	m	**6.20**
Timlock uPVC Mark 2 eaves ventilators; reference 1122; fitting between trusses at 600mm centres										
300mm girth	2.76	5.00	**2.90**	0.22	0.11	**4.30**	0.30	**7.50**	m	**8.44**
Timlock polypropylene over-fascia ventilators; reference 3011; fixing with screws to timber										
to top of fascia	3.94	5.00	**4.14**	0.22	0.11	**4.30**	0.18	**8.62**	m	**9.70**
H65: SINGLE LAP ROOF TILING										
Roof coverings; Redland Stonewold through coloured slates, 430 x 380mm; fixing every fourth course with galvanized nails to 75mm lap; 38 x 19mm pressure impregnated softwood battens fixed with galvanized nails; underlay; BS 747 type 1F reinforced bitumen felt; 150mm laps; fixing with galvanized steel clout nails										
Pitched 40 degrees from horizontal										
generally	14.10	5.00	**14.90**	0.48	0.24	**9.39**	0.81	**25.10**	m2	**28.23**
abutments; square	1.27	5.00	**1.34**	0.12	0.06	**2.35**	-	**3.68**	m	**4.14**
abutments; raking	1.91	5.00	**2.00**	0.17	0.09	**3.38**	-	**5.38**	m	**6.06**
abutments; curved to 3000mm radius	2.55	5.00	**2.67**	0.23	0.12	**4.55**	-	**7.23**	m	**8.13**
supplementary fixing eaves course with one clip per slate	0.36	5.00	**0.38**	0.03	0.02	**0.64**	-	**1.02**	m	**1.15**
verges; Redland Dry Verge system with half slates, full slates and clips	15.40	5.00	**16.17**	0.28	0.14	**5.48**	1.92	**23.56**	m	**26.51**
verges; Redland Dry Verge system with half slates, verge slates and clips	15.86	5.00	**16.65**	0.28	0.14	**5.48**	1.92	**24.05**	m	**27.06**
verges; 150 x 6mm fibre cement undercloak, slates and undercloak bedded and pointed in tinted cement mortar (1:3) and with half slates, full slates and clips	17.77	5.00	**18.66**	0.23	0.12	**4.55**	2.27	**25.48**	m	**28.67**
verges; 150 x 6mm fibre cement undercloak, slates and undercloak bedded and pointed in tinted cement mortar (1:3) and with half slates, verge slates and clips	7.19	5.00	**7.55**	0.23	0.12	**4.55**	2.27	**14.37**	m	**16.16**

Labour hourly rates: (except Specialists) Craft Operatives £14.21 Labourer £10.71 Rates are national average prices. Refer to REGIONAL VARIATIONS for indicative levels of overall pricing in regions	MATERIALS			LABOUR				RATES		
	Del to Site £	Waste %	Material Cost £	Craft Optve Hrs	Lab Hrs	Labour Cost £	Sunds £	Nett Rate £	Unit	Gross Rate (+12.5%) £
H65: SINGLE LAP ROOF TILING Cont.										
Roof coverings; Redland Stonewold through coloured slates, 430 x 380mm; fixing every fourth course with galvanized nails to 75mm lap; 38 x 19mm pressure impregnated softwood battens fixed with galvanized nails; underlay; BS 747 type 1F reinforced bitumen felt; 150mm laps; fixing with galvanized steel clout nails Cont.										
Pitched 40 degrees from horizontal Cont.										
ridge or hip tiles, Redland third round; butt jointed; bedding and pointing in tinted cement mortar (1:3)	7.77	5.00	**8.16**	0.23	0.12	**4.55**	0.42	**13.14**	m	**14.78**
ridge or hip tiles, Redland half round; butt jointed; bedding and pointing in tinted cement mortar (1:3)	7.77	5.00	**8.16**	0.23	0.12	**4.55**	0.42	**13.14**	m	**14.78**
ridge tiles, Redland universal angle; butt jointed; bedding and pointing in tinted cement mortar (1:3)	8.44	5.00	**8.86**	0.23	0.12	**4.55**	0.42	**13.84**	m	**15.57**
ridge tiles, Redland universal Stonewold type Monopitch; butt jointed; fixing with aluminium nails and bedding and pointing in tinted cement mortar (1:3)	20.28	5.00	**21.29**	0.23	0.12	**4.55**	0.42	**26.27**	m	**29.55**
ridge tiles, Redland universal half round type Monopitch; butt jointed; fixing with aluminium nails and bedding and pointing in tinted cement mortar (1:3)	20.28	5.00	**21.29**	0.23	0.12	**4.55**	0.42	**26.27**	m	**29.55**
ridge tiles, Redland Dry Ridge system with half round ridge tiles; fixing with stainless steel batten straps and ring shanked fixing nails with neoprene washers and sleeves; polypropylene ridge seals and uPVC profile filler units	27.89	5.00	**29.28**	0.45	0.23	**8.86**	-	**38.14**	m	**42.91**
ridge tiles, Redland Dry Ridge system with universal angle ridge tiles; fixing with stainless steel batten straps and ring shanked fixing nails with neoprene washers and sleeves; polypropylene ridge seals and uPVC profile filler units	27.34	5.00	**28.71**	0.45	0.23	**8.86**	-	**37.56**	m	**42.26**
ridge tiles, Redland Dry Vent Ridge system with half round ridge tiles; fixing with stainless steel batten straps and ring shanked nails with neoprene washers and sleeves; polypropylene ridge seals, PVC air flow control units and uPVC ventilated profile filler units	28.97	5.00	**30.42**	0.45	0.23	**8.86**	-	**39.28**	m	**44.19**
ridge tiles, Redland Dry Vent Ridge system with Universal angle ridge tiles; fixing with stainless steel batten straps and ring shanked nails with neoprene washers and sleeves; polypropylene ridge seals, PVC air flow control units and uPVC ventilated profile filler units	27.34	5.00	**28.71**	0.45	0.23	**8.86**	-	**37.56**	m	**42.26**
Monoridge filler units in conjunction with top tile, bed tile in nonsetting mastic sealant and screw on through filler unit with screws, washers and caps	35.43	5.00	**37.20**	0.23	0.12	**4.55**	0.42	**42.18**	m	**47.45**
valley tiles, Redland Universal valley troughs; laid with 100mm laps	27.63	5.00	**29.01**	0.56	0.28	**10.96**	0.42	**40.39**	m	**45.44**
Roof coverings; Redland Stonewold through coloured slates, 430 x 380mm; fixing every fourth course with galvanized nails to 75mm lap; 38 x 25mm pressure impregnated softwood battens fixed with galvanized nails; underlay; BS 747 type 1F reinforced bitumen felt; 150mm laps; fixing with galvanized steel clout nails										
Pitched 40 degrees from horizontal										
generally	14.25	5.00	**15.06**	0.47	0.24	**9.25**	0.81	**25.12**	m2	**28.26**
Roof coverings; Redland Stonewold through coloured slates, 430 x 380mm; fixing each course with galvanized nails to 75mm; 38 x 25mm pressure impregnated softwood battens fixed with galvanized nails; underlay; B.S 747 type 1F reinforced bitumen felt; 150mm laps; fixing with galvanized steel clout nails										
Pitched 40 degrees from horizontal										
generally	14.25	5.00	**15.06**	0.52	0.26	**10.17**	2.34	**27.58**	m2	**31.02**
Roof coverings; Redland Stonewold through coloured slates, 430 x 380mm; fixing every fourth course with galvanized nails to 75mm lap; 38 x 25mm pressure impregnated softwood battens fixed with galvanized nails; underlay, BS 747 type 1F aluminium foil surfaced reinforced bitumen felt; 150mm laps; fixing with galvanized steel clout nails										
Pitched 40 degrees from horizontal										
generally	14.25	5.00	**15.06**	0.48	0.24	**9.39**	0.81	**25.26**	m2	**28.42**

Labour hourly rates: (except Specialists) Craft Operatives £14.21 Labourer £10.71 Rates are national average prices. Refer to REGIONAL VARIATIONS for indicative levels of overall pricing in regions	MATERIALS			LABOUR				RATES		
	Del to Site £	Waste %	Material Cost £	Craft Optve Hrs	Lab Hrs	Labour Cost £	Sunds £	Nett Rate £	Unit	Gross Rate (+12.5%) £
H65: SINGLE LAP ROOF TILING Cont.										
Roof coverings; Redland Stonewold through coloured slates, 430 x 380mm; fixing every fourth course with galvanized nails to 75mm lap; 38 x 25mm pressure impregnated softwood battens fixed with galvanized nails; underlay, BS 747 type 1F reinforced bitumen felt with 50mm glass fibre insulation bonded on; 150mm laps; fixing with galvanized steel clout nails										
Pitched 40 degrees from horizontal										
generally	14.25	5.00	**15.06**	0.58	0.29	**11.35**	0.81	**27.22**	m2	**30.62**
Roof coverings; Redland Regent granular faced tiles, 418 x 332mm; fixing every fourth course with galvanized nails to 75mm lap; 38 x 25mm pressure impregnated softwood battens fixed with galvanized nails; underlay; BS 747 type 1F bitumen reinforced felt; 150mm laps; fixing with galvanized steel clout nails										
Pitched 40 degrees from horizontal										
generally	9.43	5.00	**10.00**	0.47	0.24	**9.25**	0.91	**20.16**	m2	**22.68**
reform eaves filler unit and eaves clip to each eaves tile	0.36	5.00	**0.38**	0.07	0.04	**1.42**	0.71	**2.51**	m	**2.82**
verges; cloaked verge tiles and aluminium nails	9.96	5.00	**10.46**	0.17	0.09	**3.38**	0.07	**13.91**	m	**15.65**
verges; half tiles, cloaked verge tiles and aluminium nails	11.59	5.00	**12.17**	0.23	0.12	**4.55**	0.12	**16.84**	m	**18.95**
verges; 150 x 6mm fibre cement undercloak; tiles and undercloak bedded and pointed in tinted cement mortar (1:3) and with standard tiles and clips	4.98	5.00	**5.22**	0.23	0.12	**4.55**	2.27	**12.05**	m	**13.55**
verges; 150 x 6mm fibre cement undercloak; tiles and undercloak bedded and pointed in tinted cement mortar (1:3) and with verge tiles and clips	4.98	5.00	**5.22**	0.23	0.12	**4.55**	2.27	**12.05**	m	**13.55**
verges; 150 x 6mm fibre cement undercloak; tiles and undercloak bedded and pointed in tinted cement mortar (1:3) and with half tiles, standard tiles and clips	8.88	5.00	**9.32**	0.23	0.12	**4.55**	2.27	**16.14**	m	**18.16**
verges; 150 x 6mm fibre cement undercloak; tiles and undercloak bedded and pointed in tinted cement mortar (1:3) and with half tiles, verge tiles and clips	8.88	5.00	**9.32**	0.23	0.12	**4.55**	2.27	**16.14**	m	**18.16**
ridge or hip tiles, Redland third round; butt jointed; bedding and pointing in tinted cement mortar (1:3); dentil slips in pan of each tile set in bedding	7.44	5.00	**7.81**	0.34	0.17	**6.65**	1.63	**16.09**	m	**18.10**
ridge or hip tiles, Redland half round; butt jointed; bedding and pointing in tinted cement mortar (1:3); dentil slips in pan of each tile set in bedding	7.44	5.00	**7.81**	0.34	0.17	**6.65**	1.63	**16.09**	m	**18.10**
ridge tiles, Redland universal half round type Monopitch; butt jointed; fixing with aluminium nails and bedding and pointing in tinted cement mortar (1:3); dentil slips in pan of each tile set in bedding	20.28	5.00	**21.29**	0.34	0.17	**6.65**	1.63	**29.58**	m	**33.27**
valley tiles, Redland universal valley troughs; laid with 100mm laps	27.63	5.00	**29.01**	0.34	0.17	**6.65**	0.42	**36.08**	m	**40.59**
Roof coverings; Redland Regent granular faced tiles, 418 x 332mm; fixing every fourth course with aluminium nails to 75mm lap; 38 x 25mm pressure impregnated softwood battens fixed with galvanized nails; underlay; BS 747 type 1F bitumen reinforced felt; 150mm laps; fixing with galvanized steel clout nails										
Pitched 40 degrees from horizontal										
generally	9.43	5.00	**10.00**	0.47	0.24	**9.25**	1.07	**20.32**	m2	**22.86**
Roof coverings; Redland Regent granular faced tiles, 418 x 332mm; fixing every fourth course with galvanized nails to 100mm lap; 38 x 25mm pressure impregnated softwood battens fixed with galvanized nails; underlay; BS 747 type 1F bitumen reinforced felt; 150mm laps; fixing with galvanized steel clout nails										
Pitched 40 degrees from horizontal										
generally	10.08	5.00	**10.68**	0.47	0.24	**9.25**	0.65	**20.58**	m2	**23.16**
abutments; square	0.86	5.00	**0.90**	0.11	0.06	**2.21**	-	**3.10**	m	**3.49**
abutments; raking	1.28	5.00	**1.34**	0.17	0.09	**3.38**	-	**4.72**	m	**5.31**
abutments; curved to 3000mm radius	1.71	5.00	**1.80**	0.23	0.12	**4.55**	-	**6.35**	m	**7.14**
Roof coverings; Redland Regent through coloured tiles, 418 x 332mm; fixing every fourth course with galvanized nails to 75mm lap; 38 x 25mm pressure impregnated softwood battens fixed with galvanized nails; underlay; BS 747 type 1F bitumen reinforced felt; 150mm laps; fixing with galvanized steel clout nails										
Pitched 40 degrees from horizontal										
generally	9.43	5.00	**10.00**	0.47	0.24	**9.25**	0.91	**20.16**	m2	**22.68**
reform eaves filler unit and eaves clip to each eaves tile	0.36	5.00	**0.38**	0.07	0.04	**1.42**	0.71	**2.51**	m	**2.82**
verges; cloaked verge tiles and aluminium nails	9.96	5.00	**10.46**	0.17	0.09	**3.38**	0.07	**13.91**	m	**15.65**

Labour hourly rates: (except Specialists) Craft Operatives £14.21 Labourer £10.71 Rates are national average prices. Refer to REGIONAL VARIATIONS for indicative levels of overall pricing in regions	MATERIALS			LABOUR				RATES		
	Del to Site £	Waste %	Material Cost £	Craft Optve Hrs	Lab Hrs	Labour Cost £	Sunds £	Nett Rate £	Unit	Gross Rate (+12.5%) £
H65: SINGLE LAP ROOF TILING Cont.										
Roof coverings; Redland Regent through coloured tiles, 418 x 332mm; fixing every fourth course with galvanized nails to 75mm lap; 38 x 25mm pressure impregnated softwood battens fixed with galvanized nails; underlay; BS 747 type 1F bitumen reinforced felt; 150mm laps; fixing with galvanized steel clout nails Cont.										
Pitched 40 degrees from horizontal Cont.										
verges; 150 x 6mm fibre cement undercloak; tiles and undercloak bedded and pointed in tinted cement mortar (1:3) and with standard tiles and clips...	4.98	5.00	**5.22**	0.23	0.12	**4.55**	2.27	**12.05**	m	**13.55**
verges; 150 x 6mm fibre cement undercloak; tiles and undercloak bedded and pointed in tinted cement mortar (1:3) and with verge tiles and clips......	4.98	5.00	**5.22**	0.23	0.12	**4.55**	2.27	**12.05**	m	**13.55**
verges; 150 x 6mm fibre cement undercloak; tiles and undercloak bedded and pointed in tinted cement mortar (1:3) and with half tiles, standard tiles and clips	8.88	5.00	**9.32**	0.23	0.12	**4.55**	2.27	**16.14**	m	**18.16**
verges; 150 x 6mm fibre cement undercloak; tiles and undercloak bedded and pointed in tinted cement mortar (1:3) and with half tiles, verge tiles and clips	8.88	5.00	**9.32**	0.23	0.12	**4.55**	2.27	**16.14**	m	**18.16**
valley tiles, Redland universal valley troughs; laid with 100mm laps	27.63	5.00	**29.01**	0.34	0.17	**6.65**	0.42	**36.08**	m	**40.59**
Roof coverings; Redland Regent through coloured tiles, 418 x 332mm; fixing every fourth course with galvanized nails to 100mm lap; 38 x 25mm pressure impregnated softwood battens fixed with galvanized nails; underlay; BS 747 type 1F bitumen reinforced felt; 150mm laps; fixing with galvanized steel clout nails										
Pitched 40 degrees from horizontal										
generally	10.08	5.00	**10.68**	0.47	0.24	**9.25**	0.91	**20.84**	m2	**23.45**
abutments; square	0.86	5.00	**0.90**	0.11	0.06	**2.21**	-	**3.10**	m	**3.49**
abutments; raking	1.28	5.00	**1.34**	0.17	0.09	**3.38**	-	**4.72**	m	**5.31**
abutments; curved to 3000mm radius..............	1.71	5.00	**1.80**	0.23	0.12	**4.55**	-	**6.35**	m	**7.14**
Roof coverings; Redland Grovebury granular faced double pantiles, 418 x 332mm; fixing every fourth course with galvanized nails to 75mm lap; 38 x 22mm pressure impregnated softwood battens fixed with galvanized nails; underlay; BS 747 type 1F bitumen reinforced felt; 150mm laps; fixing with galvanized steel clout nails										
Pitched 40 degrees from horizontal										
generally	9.43	5.00	**10.00**	0.47	0.24	**9.25**	0.91	**20.16**	m2	**22.68**
reform eaves filler unit and eaves clip to each eaves tile	0.36	5.00	**0.38**	0.07	0.04	**1.42**	0.71	**2.51**	m	**2.82**
verges; cloaked verge tiles and aluminium nails..	9.96	5.00	**10.46**	0.17	0.09	**3.38**	0.07	**13.91**	m	**15.65**
verges; half tiles, cloaked verge tiles and aluminium nails	11.59	5.00	**12.17**	0.23	0.12	**4.55**	0.11	**16.83**	m	**18.94**
verges; 150 x 6mm fibre cement undercloak; tiles and undercloak bedded and pointed in tinted cement mortar (1:3) and with standard tiles and clips...	4.98	5.00	**5.22**	0.23	0.12	**4.55**	2.27	**12.05**	m	**13.55**
verges; 150 x 6mm fibre cement undercloak; tiles and undercloak bedded and pointed in tinted cement mortar (1:3) and with treble roll verge tiles and clips	4.98	5.00	**5.22**	0.23	0.12	**4.55**	2.27	**12.05**	m	**13.55**
ridge or hip tiles, Redland third round; butt jointed; bedding and pointing in tinted cement mortar (1:3)...................................	7.44	5.00	**7.81**	0.34	0.17	**6.65**	2.27	**16.73**	m	**18.82**
ridge or hip tiles, Redland half round; butt jointed; bedding and pointing in tinted cement mortar (1:3)	7.44	5.00	**7.81**	0.34	0.17	**6.65**	2.27	**16.73**	m	**18.82**
valley tiles, Redland Universal valley troughs; laid with 100mm laps	27.63	5.00	**29.01**	0.34	0.17	**6.65**	0.42	**36.08**	m	**40.59**
Roof coverings; Redland Grovebury through coloured double pantiles, 418 x 332mm; fixing every fourth course with galvanized nails to 75mm lap; 38 x 22mm pressure impregnated softwood battens fixed with galvanized nails; underlay; BS 747 type 1F bitumen reinforced felt; 150mm laps; fixing with galvanized steel clout nails										
Pitched 40 degrees from horizontal										
generally	9.43	5.00	**10.00**	0.47	0.24	**9.25**	0.91	**20.16**	m2	**22.68**
reform eaves filler unit and eaves clip to each eaves tile	0.36	5.00	**0.38**	0.07	0.04	**1.42**	0.71	**2.51**	m	**2.82**
verges; cloaked verge tiles and aluminium nails..	9.96	5.00	**10.46**	0.17	0.09	**3.38**	0.07	**13.91**	m	**15.65**
verges; half tiles, cloaked verge tiles and aluminium nails	11.59	5.00	**12.17**	0.23	0.12	**4.55**	0.11	**16.83**	m	**18.94**
verges; 150 x 6mm fibre cement undercloak; tiles and undercloak bedded and pointed in tinted cement mortar (1:3) and with standard tiles and clips...	4.98	5.00	**5.22**	0.23	0.12	**4.55**	2.27	**12.05**	m	**13.55**
verges; 150 x 6mm fibre cement undercloak; tiles and undercloak bedded and pointed in tinted cement mortar (1:3) and with treble roll verge tiles and clips	4.98	5.00	**5.22**	0.23	0.12	**4.55**	2.27	**12.05**	m	**13.55**

Labour hourly rates: (except Specialists) Craft Operatives £14.21 Labourer £10.71 Rates are national average prices. Refer to REGIONAL VARIATIONS for indicative levels of overall pricing in regions	MATERIALS			LABOUR				RATES		
	Del to Site £	Waste %	Material Cost £	Craft Optve Hrs	Lab Hrs	Labour Cost £	Sunds £	Nett Rate £	Unit	Gross Rate (+12.5%) £
H65: SINGLE LAP ROOF TILING Cont.										
Roof coverings; Redland Grovebury through coloured double pantiles, 418 x 332mm; fixing every fourth course with galvanized nails to 75mm lap; 38 x 22mm pressure impregnated softwood battens fixed with galvanized nails; underlay; BS 747 type 1F bitumen reinforced felt; 150mm laps; fixing with galvanized steel clout nails Cont.										
Pitched 40 degrees from horizontal Cont.										
ridge or hip tiles, Redland third round; butt jointed; bedding and pointing in tinted cement mortar (1:3)	7.44	5.00	**7.81**	0.34	0.17	**6.65**	0.42	**14.88**	m	**16.74**
ridge or hip tiles, Redland half round; butt jointed ; bedding and pointing in tinted cement mortar (1:3)	7.44	5.00	**7.81**	0.34	0.17	**6.65**	0.42	**14.88**	m	**16.74**
valley tiles, Redland Universal valley troughs; laid with 100mm laps	27.63	5.00	**29.01**	0.34	0.17	**6.65**	0.42	**36.08**	m	**40.59**
Roof coverings; Redland Norfolk through coloured pantiles, 381 x 227mm; fixing every fourth course with galvanized nails to 75mm lap; 38 x 25mm pressure impregnated softwood battens fixed with galvanized nails; underlay; BS 747 type 1F bitumen reinforced felt; 150mm laps; fixing with galvanized steel clout nails										
Pitched 40 degrees from horizontal										
generally	11.22	5.00	**11.90**	0.61	0.31	**11.99**	1.32	**25.21**	m2	**28.36**
abutments; square	0.94	5.00	**0.98**	0.11	0.06	**2.21**	-	**3.19**	m	**3.59**
abutments; raking	1.40	5.00	**1.47**	0.17	0.09	**3.38**	-	**4.85**	m	**5.46**
abutments; curved to 3000mm radius	1.87	5.00	**1.97**	0.23	0.12	**4.55**	-	**6.52**	m	**7.34**
reform eaves filler unit and aluminium nails to each eaves tile	0.36	5.00	**0.38**	0.07	0.04	**1.42**	0.71	**2.51**	m	**2.82**
verges; plain tile undercloak; tiles and undercloak bedded and pointed in tinted cement mortar (1:3) and with standard tiles and aluminium nails	1.73	5.00	**1.81**	0.11	0.06	**2.21**	0.45	**4.47**	m	**5.03**
verges; plain tile undercloak; tiles and undercloak bedded and pointed in tinted cement mortar (1:3) and with standard tiles and clips...	1.73	5.00	**1.81**	0.11	0.06	**2.21**	3.45	**7.47**	m	**8.40**
verges; 150 x 6mm fibre cement undercloak; tiles and undercloak bedded and pointed in tinted cement mortar (1:3) and with standard tiles and aluminium nails	4.26	5.00	**4.47**	0.17	0.09	**3.38**	0.48	**8.33**	m	**9.37**
verges; 150 x 6mm fibre cement undercloak; tiles and undercloak bedded and pointed in tinted cement mortar (1:3) and with standard tiles and clips...	4.26	5.00	**4.47**	0.17	0.09	**3.38**	3.45	**11.30**	m	**12.71**
ridge or hip tiles, Redland third round; butt jointed; bedding and pointing in tinted cement mortar (1:3)	7.77	5.00	**8.16**	0.34	0.17	**6.65**	0.42	**15.23**	m	**17.14**
ridge or hip tiles, Redland half round; butt jointed; bedding and pointing in tinted cement mortar (1:3)	7.77	5.00	**8.16**	0.34	0.17	**6.65**	0.42	**15.23**	m	**17.14**
valley tiles, Redland Universal valley troughs; laid with 100mm laps	27.63	5.00	**29.01**	0.34	0.17	**6.65**	0.42	**36.08**	m	**40.59**
Roof coverings; Redland Norfolk through coloured pantiles, 381 x 227mm; fixing every fourth course with aluminium nails to 75mm lap; 38 x 25mm pressure impregnated softwood battens fixed with galvanized nails; underlay; BS 747 type 1F bitumen reinforced felt; 150mm laps; fixing with galvanized steel clout nails										
Pitched 40 degrees from horizontal										
generally	11.22	5.00	**11.90**	0.61	0.31	**11.99**	1.32	**25.21**	m2	**28.36**
Roof coverings; Redland Norfolk through coloured pantiles, 381 x 227mm; fixing every fourth course with galvanized nails to 100mm lap; 38 x 25mm pressure impregnated softwood battens fixed with galvanized nails; underlay; BS 747 type 1F bitumen reinforced felt; 150mm laps; fixing with galvanized steel clout nails										
Pitched 40 degrees from horizontal										
generally	12.09	5.00	**12.81**	0.66	0.33	**12.91**	1.41	**27.13**	m2	**30.52**
Roof coverings; Redland 49 granular faced tiles, 381 x 227mm; fixing every fourth course with galvanized nails to 75mm lap; 38 x 25mm pressure impregnated softwood battens fixed with galvanized nails; underlay BS 747 type 1F bitumen reinforced felt; 150mm laps; fixing with galvanized steel clout nails										
Pitched 40 degrees from horizontal										
generally	10.68	5.00	**11.32**	0.61	0.31	**11.99**	1.32	**24.63**	m2	**27.70**
supplementary fixing eaves course with one aluminium nail per tile	-	-	-	0.02	0.01	**0.39**	0.03	**0.42**	m	**0.47**
verges; Redland Dry Verge system with standard tiles and clips	12.43	5.00	**13.05**	0.28	0.14	**5.48**	1.92	**20.45**	m	**23.00**
verges; Redland Dry Verge system with verge tiles and clips	12.43	5.00	**13.05**	0.28	0.14	**5.48**	1.92	**20.45**	m	**23.00**

Labour hourly rates: (except Specialists) Craft Operatives £14.21 Labourer £10.71 Rates are national average prices. Refer to REGIONAL VARIATIONS for indicative levels of overall pricing in regions	MATERIALS			LABOUR				RATES		
	Del to Site	Waste	Material Cost	Craft Optve	Lab	Labour Cost	Sunds	Nett Rate	Unit	Gross Rate (+12.5%)
	£	%	£	Hrs	Hrs	£	£	£		£
H65: SINGLE LAP ROOF TILING Cont.										
Roof coverings; Redland 49 granular faced tiles, 381 x 227mm; fixing every fourth course with galvanized nails to 75mm lap; 38 x 25mm pressure impregnated softwood battens fixed with galvanized nails; underlay BS 747 type 1F bitumen reinforced felt; 150mm laps; fixing with galvanized steel clout nails Cont.										
Pitched 40 degrees from horizontal Cont.										
verges; plain tile undercloak; tiles and undercloak bedded and pointed in tinted cement mortar (1:3) and with standard tiles and aluminium nails	1.69	5.00	1.77	0.11	0.06	2.21	0.48	4.46	m	5.01
verges; plain tile undercloak; tiles and undercloak bedded and pointed in tinted cement mortar (1:3) and with verge tiles and aluminium nails	1.69	5.00	1.77	0.11	0.06	2.21	0.48	4.46	m	5.01
verges; 150 x 6mm fibre cement undercloak; tiles and undercloak bedded and pointed in tinted cement mortar (1:3) and with standard tiles and aluminium nails	4.22	5.00	4.43	0.17	0.09	3.38	0.48	8.29	m	9.32
verges; 150 x 6mm fibre cement undercloak; tiles and undercloak bedded and pointed in tinted cement mortar (1:3) and with verge tiles and aluminium nails	4.22	5.00	4.43	0.17	0.09	3.38	0.48	8.29	m	9.32
ridge or hip tiles, Redland third round; butt jointed; bedding and pointing in tinted cement mortar (1:3)	7.77	5.00	8.16	0.34	0.17	6.65	0.42	15.23	m	17.14
ridge or hip tiles, Redland half round; butt jointed; bedding and pointing in tinted cement mortar (1:3)	7.77	5.00	8.16	0.34	0.17	6.65	0.42	15.23	m	17.14
valley tiles, Redland Universal valley troughs; laid with 100mm laps	27.63	5.00	29.01	0.34	0.17	6.65	0.42	36.08	m	40.59
Roof coverings; Redland 49 granular faced tiles, 381 x 227mm; fixing every fourth course with galvanized nails to 100mm lap; 38 x 25mm pressure impregnated softwood battens fixed with galvanized nails; underlay; BS 747 type 1F bitumen reinforced felt; 150mm laps; fixing with galvanized steel clout nails										
Pitched 40 degrees from horizontal										
generally	11.52	5.00	12.20	0.66	0.33	12.91	1.41	26.53	m2	29.84
abutments; square	1.00	5.00	1.05	0.11	0.06	2.21	-	3.26	m	3.66
abutments; raking	1.50	5.00	1.58	0.17	0.09	3.38	-	4.96	m	5.57
abutments; curved to 3000mm radius	1.58	5.00	1.66	0.23	0.12	4.55	-	6.22	m	6.99
Roof coverings; Redland 49 through coloured tiles, 381 x 227mm; fixing every fourth course with galvanized nails to 75mm lap; 38 x 25mm pressure impregnated softwood battens fixed with galvanized nails; underlay; BS 747 type 1F bitumen reinforced felt; 150mm laps; fixing with galvanized steel clout nails										
Pitched 40 degrees from horizontal										
generally	10.68	5.00	11.32	0.61	0.31	11.99	1.32	24.63	m2	27.70
abutments; square	0.92	5.00	0.96	0.11	0.06	2.21	-	3.17	m	3.56
abutments; raking	1.38	5.00	1.45	0.17	0.09	3.38	-	4.83	m	5.43
abutments; curved to 3000mm radius	1.85	5.00	1.95	0.23	0.12	4.55	-	6.50	m	7.31
supplementary fixing eaves course with one aluminium nail per tile	-	-	-	0.02	0.01	0.39	0.03	0.42	m	0.47
verges; Redland Dry Verge system with standard tiles and clips	12.43	5.00	13.05	0.28	0.14	5.48	1.92	20.45	m	23.00
verges; Redland Dry Verge system with verge tiles and clips	12.43	5.00	13.05	0.28	0.14	5.48	1.92	20.45	m	23.00
verges; plain tile undercloak; tiles and undercloak bedded and pointed in tinted cement mortar (1:3) and with standard tiles and aluminium nails	1.69	5.00	1.77	0.11	0.06	2.21	0.48	4.46	m	5.01
verges; plain tile undercloak; tiles and undercloak bedded and pointed in tinted cement mortar (1:3) and with verge tiles and aluminium nails	1.69	5.00	1.77	0.11	0.06	2.21	0.48	4.46	m	5.01
verges; 150 x 6mm fibre cement undercloak; tiles and undercloak bedded and pointed in tinted cement mortar (1:3) and with standard tiles and aluminium nails	4.22	5.00	4.43	0.17	0.09	3.38	0.48	8.29	m	9.32
verges; 150 x 6mm fibre cement undercloak; tiles and undercloak bedded and pointed in tinted cement mortar (1:3) and with verge tiles and aluminium nails	4.22	5.00	4.43	0.17	0.09	3.38	0.48	8.29	m	9.32
ridge or hip tiles, Redland third round; butt jointed; bedding and pointing in tinted cement mortar (1:3)	7.77	5.00	8.16	0.34	0.17	6.65	0.42	15.23	m	17.14
ridge or hip tiles, Redland half round; butt jointed; bedding and pointing in tinted cement mortar (1:3)	7.77	5.00	8.16	0.34	0.17	6.65	0.42	15.23	m	17.14
valley tiles, Redland Universal valley troughs; laid with 100mm laps	27.63	5.00	29.01	0.34	0.17	6.65	0.42	36.08	m	40.59

Labour hourly rates: (except Specialists) Craft Operatives £14.21 Labourer £10.71 Rates are national average prices. Refer to REGIONAL VARIATIONS for indicative levels of overall pricing in regions	MATERIALS			LABOUR				RATES		
	Del to Site £	Waste %	Material Cost £	Craft Optve Hrs	Lab Hrs	Labour Cost £	Sunds £	Nett Rate £	Unit	Gross Rate (+12.5%) £
H65: SINGLE LAP ROOF TILING Cont.										
Roof coverings; Redland 49 through coloured tiles, 381 x 227mm; fixing every fourth course with galvanized nails to 100mm lap; 38 x 25mm pressure impregnated softwood battens fixed with galvanized nails; underlay; BS 747 type 1F bitumen reinforced felt; 150mm laps; fixing with galvanized steel clout nails										
Pitched 40 degrees from horizontal										
generally	11.52	5.00	**12.20**	0.66	0.33	**12.91**	1.41	**26.53**	m2	**29.84**
Roof coverings; Redland Renown granular faced tiles, 418 x 330mm; fixing every fourth course with galvanized nails to 75mm lap; 38 x 25mm pressure impregnated softwood battens fixed with galvanized nails; underlay; BS 747 type 1F bitumen reinforced felt; 150mm laps; fixing with galvanized steel clout nails										
Pitched 40 degrees from horizontal										
generally	9.16	5.00	**9.72**	0.47	0.24	**9.25**	0.57	**19.54**	m2	**21.99**
abutments; square	0.76	5.00	**0.80**	0.11	0.06	**2.21**	-	**3.01**	m	**3.38**
abutments; raking	1.15	5.00	**1.21**	0.17	0.09	**3.38**	-	**4.59**	m	**5.16**
abutments; curved to 3000mm radius	1.53	5.00	**1.61**	0.23	0.12	**4.55**	-	**6.16**	m	**6.93**
reform eaves filler unit and aluminium nails to each eaves tile	0.36	5.00	**0.38**	0.07	0.04	**1.42**	0.71	**2.51**	m	**2.82**
verges; cloaked verge tiles and aluminium nails	2.60	5.00	**2.73**	0.17	0.09	**3.38**	0.11	**6.22**	m	**7.00**
verges; half tiles, cloaked verge tiles and aluminium nails	12.32	5.00	**12.94**	0.17	0.09	**3.38**	0.48	**16.80**	m	**18.90**
verges; plain tile undercloak; tiles and undercloak bedded and pointed in tinted cement mortar (1:3) and with standard tiles and aluminium nails	2.36	5.00	**2.48**	0.11	0.06	**2.21**	0.48	**5.17**	m	**5.81**
verges; plain tile undercloak; tiles and undercloak bedded and pointed in tinted cement mortar (1:3) and with verge tiles and aluminium nails	2.36	5.00	**2.48**	0.11	0.06	**2.21**	0.48	**5.17**	m	**5.81**
verges; 150 x 6mm fibre cement undercloak; tiles and undercloak bedded and pointed in tinted cement mortar (1:3) and with standard tiles and aluminium nails	4.89	5.00	**5.14**	0.17	0.09	**3.38**	0.48	**9.00**	m	**10.12**
verges; 150 x 6mm fibre cement undercloak; tiles and undercloak bedded and pointed in tinted cement mortar (1:3) and with verge tiles and aluminium nails	4.89	5.00	**5.14**	0.17	0.09	**3.38**	0.48	**9.00**	m	**10.12**
ridge or hip tiles, Redland third round; butt jointed; bedding and pointing in tinted cement mortar (1:3)	7.77	5.00	**8.16**	0.34	0.17	**6.65**	0.42	**15.23**	m	**17.14**
ridge or hip tiles, Redland half round; butt jointed; bedding and pointing in tinted cement mortar (1:3)	7.77	5.00	**8.16**	0.34	0.17	**6.65**	0.42	**15.23**	m	**17.14**
valley tiles, Redland Universal valley troughs; laid with 100mm laps	21.18	5.00	**22.24**	0.34	0.17	**6.65**	0.42	**29.31**	m	**32.98**
Roof coverings; Redland 50 granular faced double roman tiles, 418 x 330mm; fixing every fourth course with galvanized nails to 75mm lap; 38 x 25mm pressure impregnated softwood battens fixed with galvanized nails; underlay; BS 747 type 1F bitumen reinforced felt; 150mm laps; fixing with galvanized steel clout nails										
Pitched 40 degrees from horizontal										
generally	9.16	5.00	**9.72**	0.47	0.24	**9.25**	0.69	**19.66**	m2	**22.12**
reform eaves filler unit and aluminium nails to each eaves tile	0.36	5.00	**0.38**	0.07	0.04	**1.42**	0.71	**2.51**	m	**2.82**
verges; cloaked verge tiles and aluminium nails	12.32	5.00	**12.94**	0.17	0.09	**3.38**	0.07	**16.39**	m	**18.44**
verges; half tiles, cloaked verge tiles and aluminium nails	11.54	5.00	**12.11**	0.23	0.12	**4.55**	0.07	**16.74**	m	**18.83**
verges; plain tile undercloak; tiles and undercloak bedded and pointed in tinted cement mortar (1:3) and with standard tiles and aluminium nails	2.36	5.00	**2.48**	0.11	0.06	**2.21**	0.48	**5.17**	m	**5.81**
verges; plain tile undercloak; tiles and undercloak bedded and pointed in tinted cement mortar (1:3) and with treble roll verge tiles and aluminium nails	2.36	5.00	**2.48**	0.11	0.06	**2.21**	0.48	**5.17**	m	**5.81**
verges; 150 x 6mm fibre cement undercloak; tiles and undercloak bedded and pointed in tinted cement mortar (1:3) and with standard tiles and aluminium nails	4.89	5.00	**5.14**	0.17	0.09	**3.38**	0.48	**9.00**	m	**10.12**
verges; 150 x 6mm fibre cement undercloak; tiles and undercloak bedded and pointed in tinted cement mortar (1:3) and with treble roll verge tiles and aluminium nails	4.89	5.00	**5.14**	0.17	0.09	**3.38**	0.48	**9.00**	m	**10.12**
ridge or hip tiles, Redland third round; butt jointed; bedding and pointing in tinted cement mortar (1:3)	7.77	5.00	**8.16**	0.34	0.17	**6.65**	0.42	**15.23**	m	**17.14**
ridge or hip tiles, Redland half round; butt jointed; bedding and pointing in tinted cement mortar (1:3)	7.77	5.00	**8.16**	0.34	0.17	**6.65**	0.42	**15.23**	m	**17.14**
valley tiles, Redland Universal valley troughs; laid with 100mm laps	27.63	5.00	**29.01**	0.34	0.17	**6.65**	0.42	**36.08**	m	**40.59**

Labour hourly rates: (except Specialists)
Craft Operatives £14.21 Labourer £10.71
Rates are national average prices.
Refer to REGIONAL VARIATIONS for indicative levels of overall pricing in regions

	MATERIALS			LABOUR				RATES		
	Del to Site £	Waste %	Material Cost £	Craft Optve Hrs	Lab Hrs	Labour Cost £	Sunds £	Nett Rate £	Unit	Gross Rate (+12.5%) £
H65: SINGLE LAP ROOF TILING Cont.										
Roof coverings; Redland 50 through coloured double roman tiles, 418 x 330mm; fixing every fourth course with galvanized nails to 75mm lap; 38 x 25mm pressure impregnated softwood battens fixed with galvanized nails; underlay; BS 747 type 1F bitumen reinforced felt; 150mm laps; fixing with galvanized steel clout nails										
Pitched 40 degrees from horizontal										
generally	9.16	5.00	**9.72**	0.47	0.24	**9.25**	0.69	**19.66**	m2	**22.12**
reform eaves filler unit and aluminium nails to each eaves tile	0.36	5.00	**0.38**	0.07	0.04	**1.42**	0.71	**2.51**	m	**2.82**
verges; cloaked verge tiles and aluminium nails	12.32	5.00	**12.94**	0.17	0.09	**3.38**	0.07	**16.39**	m	**18.44**
verges; half tiles, cloaked verge tiles and aluminium nails	12.32	5.00	**12.94**	0.17	0.09	**3.38**	0.07	**16.39**	m	**18.44**
verges; plain tile undercloak; tiles and undercloak bedded and pointed in tinted cement mortar (1:3) and with standard tiles and aluminium nails	2.36	5.00	**2.48**	0.11	0.06	**2.21**	0.48	**5.17**	m	**5.81**
verges; plain tile undercloak; tiles and undercloak bedded and pointed in tinted cement mortar (1:3) and with treble roll verge tiles and aluminium nails	2.36	5.00	**2.48**	0.11	0.06	**2.21**	0.48	**5.17**	m	**5.81**
verges; 150 x 6mm fibre cement undercloak; tiles and undercloak bedded and pointed in tinted cement mortar (1:3) and with standard tiles and aluminium nails	4.89	5.00	**5.14**	0.17	0.09	**3.38**	0.48	**9.00**	m	**10.12**
verges; 150 x 6mm fibre cement undercloak; tiles and undercloak bedded and pointed in tinted cement mortar (1:3) and with treble roll verge tiles and aluminium nails	4.89	5.00	**5.14**	0.17	0.09	**3.38**	0.48	**9.00**	m	**10.12**
ridge or hip tiles, Redland third round; butt jointed; bedding and pointing in tinted cement mortar (1:3)	7.77	5.00	**8.16**	0.34	0.17	**6.65**	0.42	**15.23**	m	**17.14**
ridge or hip tiles, Redland half round; butt jointed; bedding and pointing in tinted cement mortar (1:3)	7.77	5.00	**8.16**	0.34	0.17	**6.65**	0.42	**15.23**	m	**17.14**
valley tiles, Redland Universal valley troughs; laid with 100mm laps	27.63	5.00	**29.01**	0.34	0.17	**6.65**	0.42	**36.08**	m	**40.59**
Fittings										
Ventilator tiles										
Red Vent ridge ventilation terminal 450mm long for										
half round ridge	50.85	5.00	**53.39**	1.45	0.73	**28.42**	0.66	**82.48**	nr	**92.78**
universal angle ridge	50.85	5.00	**53.39**	1.45	0.73	**28.42**	0.66	**82.48**	nr	**92.78**
Redvent eaves ventilators fixing with aluminium nails	9.19	5.00	**9.65**	0.56	0.28	**10.96**	0.03	**20.64**	m	**23.22**
Red Vent Thruvent tiles for										
Stonewold slates	38.81	5.00	**40.75**	0.56	0.28	**10.96**	0.25	**51.96**	nr	**58.45**
Regent tiles	38.81	5.00	**40.75**	0.56	0.28	**10.96**	0.25	**51.96**	nr	**58.45**
Grovebury double pantiles	38.81	5.00	**40.75**	0.56	0.28	**10.96**	0.25	**51.96**	nr	**58.45**
Norfolk pantiles	38.81	5.00	**40.75**	0.56	0.28	**10.96**	0.25	**51.96**	nr	**58.45**
Redland 49 tiles	38.81	5.00	**40.75**	0.56	0.28	**10.96**	0.25	**51.96**	nr	**58.45**
Renown tiles	38.81	5.00	**40.75**	0.56	0.28	**10.96**	0.25	**51.96**	nr	**58.45**
Redland 50 double roman tiles	38.81	5.00	**40.75**	0.56	0.28	**10.96**	0.25	**51.96**	nr	**58.45**
Red line ventilation tile complete with underlay seal and fixing clips for										
Stonewold slates	48.09	5.00	**50.49**	1.12	0.56	**21.91**	0.25	**72.66**	nr	**81.74**
Regent tiles	48.09	5.00	**50.49**	1.12	0.56	**21.91**	0.25	**72.66**	nr	**81.74**
Grovebury double pantiles	48.09	5.00	**50.49**	1.12	0.56	**21.91**	0.25	**72.66**	nr	**81.74**
Renown tiles	48.09	5.00	**50.49**	1.12	0.56	**21.91**	0.25	**72.66**	nr	**81.74**
Redland 50 double roman tiles	48.09	5.00	**50.49**	1.12	0.56	**21.91**	0.25	**72.66**	nr	**81.74**
Gas terminals										
Gas Flue ridge terminal Mark III, 450mm long with sealing gasket and fixing brackets										
half round ridge	49.84	5.00	**52.33**	1.69	0.84	**33.01**	0.66	**86.00**	nr	**96.75**
extra for 150mm extension adaptor	18.79	5.00	**19.73**	0.34	0.17	**6.65**	0.25	**26.63**	nr	**29.96**
Hip irons										
galvanized mild steel; 32 x 3 x 380mm girth; scrolled; fixing with galvanized steel screws to timber	2.07	5.00	**2.17**	0.11	0.06	**2.21**	-	**4.38**	nr	**4.93**
Lead soakers										
fixing only	-	-	**-**	0.34	0.17	**6.65**	-	**6.65**	nr	**7.48**

This section continues on the next page

Labour hourly rates: (except Specialists) Craft Operatives £17.38 Labourer £10.71 Rates are national average prices. Refer to REGIONAL VARIATIONS for indicative levels of overall pricing in regions	MATERIALS			LABOUR				RATES		
	Del to Site	Waste	Material Cost	Craft Optve	Lab	Labour Cost	Sunds	Nett Rate	Unit	Gross Rate (+12.5%)
	£	%	£	Hrs	Hrs	£	£	£		£
H71: LEAD SHEET COVERINGS/FLASHINGS										
Sheet lead										
Technical data										
Code 3, 1.32mm thick, 14.97 kg/m2, colour code green										
Code 4, 1.80mm thick, 20.41 kg/m2, colour code blue										
Code 5, 2.24mm thick, 25.40 kg/m2, colour code red										
Code 6, 2.65mm thick, 30.05 kg/m2, colour code black										
Code 7, 3.15mm thick, 35.72 kg/m2, colour code white										
Code 8, 3.55mm thick, 40.26 kg/m2, colour code orange										
Milled lead sheet, BS EN 12588										
Nr 3 roof coverings; fixing to timber with milled lead cleats and galvanized screws										
pitch 7.5 degrees from horizontal	16.84	5.00	**17.68**	4.40	-	**76.47**	0.26	**94.41**	m2	**106.21**
pitch 40 degrees from horizontal	16.84	5.00	**17.68**	5.05	-	**87.77**	0.26	**105.71**	m2	**118.92**
pitch 75 degrees from horizontal	16.84	5.00	**17.68**	5.70	-	**99.07**	0.26	**117.00**	m2	**131.63**
Nr 4 roof coverings; fixing to timber with milled lead cleats and galvanized screws										
pitch 7.5 degrees from horizontal	22.91	5.00	**24.06**	5.05	-	**87.77**	0.31	**112.13**	m2	**126.15**
pitch 40 degrees from horizontal	22.91	5.00	**24.06**	5.70	-	**99.07**	0.31	**123.43**	m2	**138.86**
pitch 75 degrees from horizontal	22.91	5.00	**24.06**	6.35	-	**110.36**	0.31	**134.72**	m2	**151.57**
Nr 5 roof coverings; fixing to timber with milled lead cleats and galvanized screws										
pitch 7.5 degrees from horizontal	28.51	5.00	**29.94**	5.70	-	**99.07**	0.37	**129.37**	m2	**145.54**
pitch 40 degrees from horizontal	28.51	5.00	**29.94**	6.35	-	**110.36**	0.37	**140.67**	m2	**158.25**
pitch 75 degrees from horizontal	28.51	5.00	**29.94**	6.95	-	**120.79**	0.37	**151.09**	m2	**169.98**
extra; oil patination to surfaces	1.85	10.00	**2.04**	0.23	-	**4.00**	-	**6.03**	m2	**6.79**
Nr 3 wall coverings; fixing to timber with milled lead cleats and galvanized screws										
vertical	16.84	5.00	**17.68**	8.00	-	**139.04**	0.26	**156.98**	m2	**176.60**
Nr 4 wall coverings; fixing to timber with milled lead cleats and galvanized screws										
vertical	22.91	5.00	**24.06**	8.65	-	**150.34**	0.31	**174.70**	m2	**196.54**
Nr 5 wall coverings; fixing to timber with milled lead cleats and galvanized screws										
vertical	28.51	5.00	**29.94**	9.25	-	**160.77**	0.37	**191.07**	m2	**214.95**
Flashings; horizontal										
Nr 3; 150mm lapped joints; fixing to masonry with milled lead clips and lead wedges										
150mm girth	2.53	5.00	**2.66**	0.46	-	**7.99**	0.26	**10.91**	m	**12.27**
240mm girth	4.04	5.00	**4.24**	0.74	-	**12.86**	0.26	**17.36**	m	**19.53**
300mm girth	5.05	5.00	**5.30**	0.92	-	**15.99**	0.26	**21.55**	m	**24.24**
Nr 4; 150mm lapped joints; fixing to masonry with milled lead clips and lead wedges										
150mm girth	3.44	5.00	**3.61**	0.50	-	**8.69**	0.31	**12.61**	m	**14.18**
240mm girth	5.50	5.00	**5.78**	0.82	-	**14.25**	0.31	**20.33**	m	**22.87**
300mm girth	6.88	5.00	**7.22**	1.02	-	**17.73**	0.31	**25.26**	m	**28.41**
Nr 5; 150mm lapped joints; fixing to masonry with milled lead clips and lead wedges										
150mm girth	4.28	5.00	**4.49**	0.58	-	**10.08**	0.37	**14.94**	m	**16.81**
240mm girth	6.84	5.00	**7.18**	0.92	-	**15.99**	0.37	**23.54**	m	**26.48**
300mm girth	8.56	5.00	**8.99**	1.15	-	**19.99**	0.37	**29.34**	m	**33.01**
Nr 3; 150mm lapped joints; fixing to timber with copper nails										
150mm girth	2.53	5.00	**2.66**	0.46	-	**7.99**	0.30	**10.95**	m	**12.32**
240mm girth	4.04	5.00	**4.24**	0.74	-	**12.86**	0.30	**17.40**	m	**19.57**
300mm girth	5.05	5.00	**5.30**	0.92	-	**15.99**	0.30	**21.59**	m	**24.29**
Nr 4; 150mm lapped joints; fixing to timber with copper nails										
150mm girth	3.44	5.00	**3.61**	0.50	-	**8.69**	0.30	**12.60**	m	**14.17**
240mm girth	5.50	5.00	**5.78**	0.82	-	**14.25**	0.30	**20.32**	m	**22.86**
300mm girth	6.88	5.00	**7.22**	1.02	-	**17.73**	0.30	**25.25**	m	**28.40**
Nr 5; 150mm lapped joints; fixing to timber with copper nails										
150mm girth	4.28	5.00	**4.49**	0.58	-	**10.08**	0.30	**14.87**	m	**16.73**
240mm girth	6.84	5.00	**7.18**	0.92	-	**15.99**	0.30	**23.47**	m	**26.40**
300mm girth	8.56	5.00	**8.99**	1.15	-	**19.99**	0.30	**29.27**	m	**32.93**
Flashings; stepped										
Nr 3; 150mm lapped joints; fixing to masonry with milled lead clips and lead wedges										
180mm girth	3.04	5.00	**3.19**	0.74	-	**12.86**	0.36	**16.41**	m	**18.46**
240mm girth	4.04	5.00	**4.24**	0.98	-	**17.03**	0.36	**21.63**	m	**24.34**
300mm girth	5.05	5.00	**5.30**	1.24	-	**21.55**	0.36	**27.21**	m	**30.61**

Labour hourly rates: (except Specialists) Craft Operatives £17.38 Labourer £10.71 Rates are national average prices. Refer to REGIONAL VARIATIONS for indicative levels of overall pricing in regions	MATERIALS			LABOUR				RATES		
	Del to Site £	Waste %	Material Cost £	Craft Optve Hrs	Lab Hrs	Labour Cost £	Sunds £	Nett Rate £	Unit	Gross Rate (+12.5%) £
H71: LEAD SHEET COVERINGS/FLASHINGS Cont.										
Flashings; stepped Cont.										
Nr 4; 150mm lapped joints; fixing to masonry with milled lead clips and lead wedges										
180mm girth	4.13	5.00	**4.34**	0.82	-	**14.25**	0.47	**19.06**	m	**21.44**
240mm girth	5.50	5.00	**5.78**	1.10	-	**19.12**	0.47	**25.36**	m	**28.53**
300mm girth	6.88	5.00	**7.22**	1.36	-	**23.64**	0.47	**31.33**	m	**35.25**
Nr 5; 150mm lapped joints; fixing to masonry with milled lead clips and lead wedges										
180mm girth	5.13	5.00	**5.39**	0.92	-	**15.99**	0.56	**21.94**	m	**24.68**
240mm girth	6.84	5.00	**7.18**	1.22	-	**21.20**	0.56	**28.95**	m	**32.56**
300mm girth	8.56	5.00	**8.99**	1.52	-	**26.42**	0.56	**35.97**	m	**40.46**
Aprons; horizontal										
Nr 3; 150mm lapped joints; fixing to masonry with milled lead clips and lead wedges										
150mm girth	2.53	5.00	**2.66**	0.46	-	**7.99**	0.26	**10.91**	m	**12.27**
240mm girth	4.04	5.00	**4.24**	0.74	-	**12.86**	0.26	**17.36**	m	**19.53**
300mm girth	5.05	5.00	**5.30**	0.92	-	**15.99**	0.26	**21.55**	m	**24.24**
450mm girth	7.58	5.00	**7.96**	1.38	-	**23.98**	0.36	**32.30**	m	**36.34**
Nr 4; 150mm lapped joints; fixing to masonry with milled lead clips and lead wedges										
150mm girth	3.44	5.00	**3.61**	0.50	-	**8.69**	0.31	**12.61**	m	**14.18**
240mm girth	5.50	5.00	**5.78**	0.82	-	**14.25**	0.31	**20.33**	m	**22.87**
300mm girth	6.88	5.00	**7.22**	1.02	-	**17.73**	0.31	**25.26**	m	**28.41**
450mm girth	10.30	5.00	**10.82**	1.52	-	**26.42**	0.47	**37.70**	m	**42.41**
Nr 5; 150mm lapped joints; fixing to masonry with milled lead clips and lead wedges										
150mm girth	4.28	5.00	**4.49**	0.58	-	**10.08**	0.37	**14.94**	m	**16.81**
240mm girth	6.84	5.00	**7.18**	0.92	-	**15.99**	0.37	**23.54**	m	**26.48**
300mm girth	8.56	5.00	**8.99**	1.15	-	**19.99**	0.37	**29.34**	m	**33.01**
450mm girth	12.83	5.00	**13.47**	1.72	-	**29.89**	0.56	**43.93**	m	**49.42**
Hips; sloping; dressing over slating and tiling										
Nr 3; 150mm lapped joints; fixing to timber with milled lead clips										
240mm girth	4.04	5.00	**4.24**	0.74	-	**12.86**	0.26	**17.36**	m	**19.53**
300mm girth	5.05	5.00	**5.30**	0.92	-	**15.99**	0.26	**21.55**	m	**24.24**
450mm girth	7.58	5.00	**7.96**	1.38	-	**23.98**	0.36	**32.30**	m	**36.34**
Nr 4; 150mm lapped joints; fixing to timber with milled lead clips										
240mm girth	5.50	5.00	**5.78**	0.82	-	**14.25**	0.31	**20.33**	m	**22.87**
300mm girth	6.88	5.00	**7.22**	1.02	-	**17.73**	0.31	**25.26**	m	**28.41**
450mm girth	10.30	5.00	**10.82**	1.52	-	**26.42**	0.47	**37.70**	m	**42.41**
Nr 5; 150mm lapped joints; fixing to timber with milled lead clips										
240mm girth	6.84	5.00	**7.18**	0.92	-	**15.99**	0.37	**23.54**	m	**26.48**
300mm girth	8.56	5.00	**8.99**	1.15	-	**19.99**	0.37	**29.34**	m	**33.01**
450mm girth	12.83	5.00	**13.47**	1.72	-	**29.89**	0.56	**43.93**	m	**49.42**
Kerbs; horizontal										
Nr 3; 150mm lapped joints; fixing to timber with copper nails										
240mm girth	4.04	5.00	**4.24**	0.74	-	**12.86**	0.26	**17.36**	m	**19.53**
300mm girth	5.05	5.00	**5.30**	0.92	-	**15.99**	0.26	**21.55**	m	**24.24**
450mm girth	7.58	5.00	**7.96**	1.38	-	**23.98**	0.36	**32.30**	m	**36.34**
Nr 4; 150mm lapped joints; fixing to timber with copper nails										
240mm girth	5.50	5.00	**5.78**	0.82	-	**14.25**	0.31	**20.33**	m	**22.87**
300mm girth	6.88	5.00	**7.22**	1.02	-	**17.73**	0.31	**25.26**	m	**28.41**
450mm girth	10.30	5.00	**10.82**	1.52	-	**26.42**	0.47	**37.70**	m	**42.41**
Nr 5; 150mm lapped joints; fixing to timber with copper nails										
240mm girth	6.84	5.00	**7.18**	0.92	-	**15.99**	0.37	**23.54**	m	**26.48**
300mm girth	8.56	5.00	**8.99**	1.15	-	**19.99**	0.37	**29.34**	m	**33.01**
450mm girth	12.83	5.00	**13.47**	1.72	-	**29.89**	0.56	**43.93**	m	**49.42**
Ridges; horizontal; dressing over slating and tiling										
Nr 3; 150mm lapped joints; fixing to timber with milled lead clips										
240mm girth	4.04	5.00	**4.24**	0.74	-	**12.86**	0.26	**17.36**	m	**19.53**
300mm girth	5.05	5.00	**5.30**	0.92	-	**15.99**	0.26	**21.55**	m	**24.24**
450mm girth	7.58	5.00	**7.96**	1.38	-	**23.98**	0.36	**32.30**	m	**36.34**
Nr 4; 150mm lapped joints; fixing to timber with milled lead clips										
240mm girth	5.50	5.00	**5.78**	0.82	-	**14.25**	0.31	**20.33**	m	**22.87**
300mm girth	6.88	5.00	**7.22**	1.02	-	**17.73**	0.31	**25.26**	m	**28.41**
450mm girth	10.30	5.00	**10.82**	1.52	-	**26.42**	0.47	**37.70**	m	**42.41**
Nr 5; 150mm lapped joints; fixing to timber with milled lead clips										
240mm girth	6.84	5.00	**7.18**	0.92	-	**15.99**	0.37	**23.54**	m	**26.48**

Labour hourly rates: (except Specialists) Craft Operatives £17.38 Labourer £10.71 Rates are national average prices. Refer to REGIONAL VARIATIONS for indicative levels of overall pricing in regions	MATERIALS			LABOUR				RATES		
	Del to Site £	Waste %	Material Cost £	Craft Optve Hrs	Lab Hrs	Labour Cost £	Sunds £	Nett Rate £	Unit	Gross Rate (+12.5%) £
H71: LEAD SHEET COVERINGS/FLASHINGS Cont.										
Ridges; horizontal; dressing over slating and tiling Cont.										
Nr 5; 150mm lapped joints; fixing to timber with milled lead clips Cont.										
300mm girth	8.56	5.00	**8.99**	1.15	-	**19.99**	0.37	**29.34**	m	**33.01**
450mm girth	12.83	5.00	**13.47**	1.72	-	**29.89**	0.56	**43.93**	m	**49.42**
Valleys; sloping										
Nr 3; dressing over tilting fillets -1; 150mm lapped joints; fixing to timber with copper nails										
240mm girth	4.04	5.00	**4.24**	0.74	-	**12.86**	0.26	**17.36**	m	**19.53**
300mm girth	5.05	5.00	**5.30**	0.92	-	**15.99**	0.26	**21.55**	m	**24.24**
450mm girth	7.58	5.00	**7.96**	1.38	-	**23.98**	0.36	**32.30**	m	**36.34**
Nr 4; dressing over tilting fillets -1; 150mm lapped joints; fixing to timber with copper nails										
240mm girth	5.50	5.00	**5.78**	0.82	-	**14.25**	0.31	**20.33**	m	**22.87**
300mm girth	6.88	5.00	**7.22**	1.02	-	**17.73**	0.31	**25.26**	m	**28.41**
450mm girth	10.30	5.00	**10.82**	1.52	-	**26.42**	0.47	**37.70**	m	**42.41**
Nr 5; dressing over tilting fillets -1; 150mm lapped joints; fixing to timber with copper nails										
240mm girth	6.84	5.00	**7.18**	0.92	-	**15.99**	0.37	**23.54**	m	**26.48**
300mm girth	8.56	5.00	**8.99**	1.15	-	**19.99**	0.37	**29.34**	m	**33.01**
450mm girth	12.83	5.00	**13.47**	1.72	-	**29.89**	0.56	**43.93**	m	**49.42**
Cavity gutters										
Nr 3; 150mm lapped joints; bedding in cement mortar (1:3)										
225mm girth	4.04	5.00	**4.24**	0.74	-	**12.86**	0.47	**17.57**	m	**19.77**
345mm girth	6.06	5.00	**6.36**	1.10	-	**19.12**	0.70	**26.18**	m	**29.46**
Nr 3 edges										
welted	-	-	**-**	0.20	-	**3.48**	-	**3.48**	m	**3.91**
beaded	-	-	**-**	0.20	-	**3.48**	-	**3.48**	m	**3.91**
Nr 4 edges										
welted	-	-	**-**	0.20	-	**3.48**	-	**3.48**	m	**3.91**
beaded	-	-	**-**	0.20	-	**3.48**	-	**3.48**	m	**3.91**
Nr 5 edges										
welted	-	-	**-**	0.20	-	**3.48**	-	**3.48**	m	**3.91**
beaded	-	-	**-**	0.20	-	**3.48**	-	**3.48**	m	**3.91**
Nr 3 seams										
leadburned	0.95	-	**0.95**	0.23	-	**4.00**	-	**4.95**	m	**5.57**
Nr 4 seams										
leadburned	1.20	-	**1.20**	0.23	-	**4.00**	-	**5.20**	m	**5.85**
Nr 5 seams										
leadburned	1.55	-	**1.55**	0.23	-	**4.00**	-	**5.55**	m	**6.24**
Nr 3 dressings										
corrugated roofing; fibre cement; down corrugations	-	-	**-**	0.29	-	**5.04**	-	**5.04**	m	**5.67**
corrugated roofing; fibre cement; across corrugations	-	-	**-**	0.29	-	**5.04**	-	**5.04**	m	**5.67**
glass and glazing bars; timber	-	-	**-**	0.29	-	**5.04**	-	**5.04**	m	**5.67**
Nr 4 dressings										
corrugated roofing; fibre cement; down corrugations	-	-	**-**	0.29	-	**5.04**	-	**5.04**	m	**5.67**
corrugated roofing; fibre cement; across corrugations	-	-	**-**	0.29	-	**5.04**	-	**5.04**	m	**5.67**
glass and glazing bars; timber	-	-	**-**	0.29	-	**5.04**	-	**5.04**	m	**5.67**
Nr 5 dressings										
corrugated roofing; fibre cement; down corrugations	-	-	**-**	0.29	-	**5.04**	-	**5.04**	m	**5.67**
corrugated roofing; fibre cement; across corrugations	-	-	**-**	0.29	-	**5.04**	-	**5.04**	m	**5.67**
glass and glazing bars; timber	-	-	**-**	0.29	-	**5.04**	-	**5.04**	m	**5.67**
Nr 3 soakers and slates handed to others for fixing										
180 x 180mm	0.55	5.00	**0.58**	0.23	-	**4.00**	-	**4.57**	nr	**5.15**
180 x 300mm	0.91	5.00	**0.96**	0.23	-	**4.00**	-	**4.95**	nr	**5.57**
450 x 450mm	3.42	5.00	**3.59**	0.23	-	**4.00**	-	**7.59**	nr	**8.54**
Nr 4 soakers and slates handed to others for fixing										
180 x 180mm	0.75	5.00	**0.79**	0.23	-	**4.00**	-	**4.78**	nr	**5.38**
180 x 300mm	1.24	5.00	**1.30**	0.23	-	**4.00**	-	**5.30**	nr	**5.96**
450 x 450mm	4.63	5.00	**4.86**	0.23	-	**4.00**	-	**8.86**	nr	**9.97**
Nr 5 soakers and slates handed to others for fixing										
180 x 180mm	0.92	5.00	**0.97**	0.23	-	**4.00**	-	**4.96**	nr	**5.58**
180 x 300mm	1.54	5.00	**1.62**	0.23	-	**4.00**	-	**5.61**	nr	**6.32**
450 x 450mm	5.77	5.00	**6.06**	0.23	-	**4.00**	-	**10.06**	nr	**11.31**

Labour hourly rates: (except Specialists) Craft Operatives £17.38 Labourer £10.71 Rates are national average prices. Refer to REGIONAL VARIATIONS for indicative levels of overall pricing in regions	MATERIALS			LABOUR				RATES		
	Del to Site £	Waste %	Material Cost £	Craft Optve Hrs	Lab Hrs	Labour Cost £	Sunds £	Nett Rate £	Unit	Gross Rate (+12.5%) £
H71: LEAD SHEET COVERINGS/FLASHINGS Cont.										
Cavity gutters Cont.										
Nr 3 collars around pipes, standards and the like 150mm long; soldered joints to metal covering										
50mm diameter	4.37	5.00	**4.59**	1.00	-	**17.38**	-	**21.97**	nr	**24.71**
100mm diameter	4.92	5.00	**5.17**	1.15	-	**19.99**	-	**25.15**	nr	**28.30**
Nr 4 collars around pipes, standards and the like 150mm long; soldered joints to metal covering										
50mm diameter	5.61	5.00	**5.89**	1.15	-	**19.99**	-	**25.88**	nr	**29.11**
100mm diameter	6.26	5.00	**6.57**	1.15	-	**19.99**	-	**26.56**	nr	**29.88**
Nr 5 collars around pipes, standards and the like 150mm long; soldered joints to metal covering										
50mm diameter	6.99	5.00	**7.34**	1.15	-	**19.99**	-	**27.33**	nr	**30.74**
100mm diameter	7.81	5.00	**8.20**	1.15	-	**19.99**	-	**28.19**	nr	**31.71**
Nr 3 dots										
cast lead	0.75	5.00	**0.79**	0.77	-	**13.38**	-	**14.17**	nr	**15.94**
soldered	7.15	5.00	**7.51**	0.86	-	**14.95**	-	**22.45**	nr	**25.26**
Nr 4 dots										
cast lead	1.03	5.00	**1.08**	0.77	-	**13.38**	-	**14.46**	nr	**16.27**
soldered	7.15	5.00	**7.51**	0.86	-	**14.95**	-	**22.45**	nr	**25.26**
Nr 5 dots										
cast lead	1.21	5.00	**1.27**	0.77	-	**13.38**	-	**14.65**	nr	**16.48**
soldered	7.15	5.00	**7.51**	0.86	-	**14.95**	-	**22.45**	nr	**25.26**
H72: ALUMINIUM SHEET STRIP COVERINGS/FLASHINGS										
Aluminium sheet, BS EN 485 grade S1BO, commercial purity										
0.60mm thick roof coverings; fixing to timber with aluminium cleats and aluminium alloy screws										
pitch 7.5 degrees from horizontal	10.42	5.00	**10.94**	5.05	-	**87.77**	0.14	**98.85**	m2	**111.21**
pitch 40 degrees from horizontal	10.42	5.00	**10.94**	5.70	-	**99.07**	0.14	**110.15**	m2	**123.92**
pitch 75 degrees from horizontal	10.42	5.00	**10.94**	6.35	-	**110.36**	0.14	**121.45**	m2	**136.63**
Flashings; horizontal										
0.60mm thick; 150mm lapped joints; fixing to masonry with aluminium clips and wedges										
150mm girth	2.75	5.00	**2.89**	0.51	-	**8.86**	0.14	**11.89**	m	**13.38**
240mm girth	4.13	5.00	**4.34**	0.82	-	**14.25**	0.14	**18.73**	m	**21.07**
300mm girth	4.13	5.00	**4.34**	1.02	-	**17.73**	0.14	**22.21**	m	**24.98**
Flashings; stepped										
0.60mm thick; 150mm lapped joints; fixing to masonry with aluminium clips and wedges										
180mm girth	3.63	5.00	**3.81**	0.82	-	**14.25**	0.21	**18.28**	m	**20.56**
240mm girth	4.13	5.00	**4.34**	1.09	-	**18.94**	0.21	**23.49**	m	**26.43**
300mm girth	4.13	5.00	**4.34**	1.37	-	**23.81**	0.21	**28.36**	m	**31.91**
Aprons; horizontal										
0.60mm thick; 150mm lapped joints; fixing to masonry with aluminium clips and wedges										
150mm girth	2.75	5.00	**2.89**	0.51	-	**8.86**	0.14	**11.89**	m	**13.38**
180mm girth	3.63	5.00	**3.81**	0.82	-	**14.25**	0.14	**18.21**	m	**20.48**
240mm girth	4.13	5.00	**4.34**	1.02	-	**17.73**	0.14	**22.21**	m	**24.98**
450mm girth	6.38	5.00	**6.70**	1.53	-	**26.59**	0.21	**33.50**	m	**37.69**
Hips; sloping										
0.60mm thick; 150mm lapped joints; fixing to timber with aluminium clips										
240mm girth	4.13	5.00	**4.34**	0.82	-	**14.25**	0.14	**18.73**	m	**21.07**
300mm girth	4.13	5.00	**4.34**	1.02	-	**17.73**	0.14	**22.21**	m	**24.98**
450mm girth	6.38	5.00	**6.70**	1.53	-	**26.59**	0.21	**33.50**	m	**37.69**
Kerbs; horizontal										
0.60mm thick; 150mm lapped joints; fixing to timber with aluminium clips										
240mm girth	4.13	5.00	**4.34**	0.82	-	**14.25**	0.14	**18.73**	m	**21.07**
300mm girth	4.13	5.00	**4.34**	1.02	-	**17.73**	0.14	**22.21**	m	**24.98**
450mm girth	6.38	5.00	**6.70**	1.53	-	**26.59**	0.21	**33.50**	m	**37.69**
Ridges; horizontal										
0.60mm thick; 150mm lapped joints; fixing to timber with aluminium clips										
240mm girth	4.13	5.00	**4.34**	0.82	-	**14.25**	0.14	**18.73**	m	**21.07**
300mm girth	4.13	5.00	**4.34**	1.02	-	**17.73**	0.14	**22.21**	m	**24.98**
450mm girth	6.38	5.00	**6.70**	1.53	-	**26.59**	0.21	**33.50**	m	**37.69**
Valleys; sloping										
0.60mm thick; 150mm lapped joints; fixing to timber with aluminium clips										
240mm girth	4.13	5.00	**4.34**	0.82	-	**14.25**	0.14	**18.73**	m	**21.07**
300mm girth	4.13	5.00	**4.34**	1.02	-	**17.73**	0.14	**22.21**	m	**24.98**
450mm girth	6.38	5.00	**6.70**	1.53	-	**26.59**	0.21	**33.50**	m	**37.69**
0.60mm thick edges										
welted	-	-	**-**	0.20	-	**3.48**	-	**3.48**	m	**3.91**
beaded	-	-	**-**	0.20	-	**3.48**	-	**3.48**	m	**3.91**

Labour hourly rates: (except Specialists) Craft Operatives £17.38 Labourer £10.71 Rates are national average prices. Refer to REGIONAL VARIATIONS for indicative levels of overall pricing in regions	MATERIALS			LABOUR				RATES		
	Del to Site £	Waste %	Material Cost £	Craft Optve Hrs	Lab Hrs	Labour Cost £	Sunds £	Nett Rate £	Unit	Gross Rate (+12.5%) £
H72: ALUMINIUM SHEET STRIP COVERINGS/FLASHINGS Cont.										
Aluminium sheet, BS EN 485 grade S1BO, commercial purity Cont.										
0.60mm thick soakers and slates handed to others for fixing										
180 x 180mm	0.65	5.00	0.68	0.23	-	4.00	-	4.68	nr	5.26
180 x 300mm	1.09	5.00	1.14	0.23	-	4.00	-	5.14	nr	5.78
H73: COPPER STRIP SHEET COVERINGS/FLASHINGS										
Copper sheet; BS EN 1172										
0.60mm thick roof coverings; fixing to timber with copper cleats and copper nails										
pitch 7.5 degrees from horizontal	25.80	5.00	27.09	5.05	-	87.77	0.42	115.28	m2	129.69
pitch 40 degrees from horizontal	25.80	5.00	27.09	5.70	-	99.07	0.42	126.57	m2	142.40
pitch 75 degrees from horizontal	25.80	5.00	27.09	6.35	-	110.36	0.42	137.87	m2	155.11
0.70mm thick roof coverings; fixing to timber with copper cleats and copper nails										
pitch 7.5 degrees from horizontal	32.25	5.00	33.86	5.70	-	99.07	0.53	133.46	m2	150.14
pitch 40 degrees from horizontal	32.25	5.00	33.86	6.35	-	110.36	0.53	144.76	m2	162.85
pitch 75 degrees from horizontal	32.25	5.00	33.86	6.95	-	120.79	0.53	155.18	m2	174.58
Flashings; horizontal										
0.60mm thick; 150mm lapped joints; fixing to masonry with copper clips and wedges										
150mm girth	3.87	5.00	4.06	0.51	-	8.86	0.42	13.35	m	15.01
240mm girth	6.19	5.00	6.50	0.82	-	14.25	0.42	21.17	m	23.82
300mm girth	7.74	5.00	8.13	1.02	-	17.73	0.42	26.27	m	29.56
0.70mm thick; 150mm lapped joints; fixing to masonry with copper clips and wedges										
150mm girth	4.84	5.00	5.08	0.58	-	10.08	0.53	15.69	m	17.65
240mm girth	7.74	5.00	8.13	0.92	-	15.99	0.53	24.65	m	27.73
300mm girth	9.67	5.00	10.15	1.15	-	19.99	0.53	30.67	m	34.50
Flashings; stepped										
0.60mm thick; 150mm lapped joints; fixing to masonry with copper clips and wedges										
180mm girth	4.64	5.00	4.87	0.82	-	14.25	0.63	19.76	m	22.23
240mm girth	6.19	5.00	6.50	1.09	-	18.94	0.63	26.08	m	29.34
300mm girth	7.74	5.00	8.13	1.37	-	23.81	0.63	32.57	m	36.64
0.70mm thick; 150mm lapped joints; fixing to masonry with copper clips and wedges										
180mm girth	5.81	5.00	6.10	0.92	-	15.99	0.80	22.89	m	25.75
240mm girth	7.74	5.00	8.13	1.23	-	21.38	0.80	30.30	m	34.09
300mm girth	9.67	5.00	10.15	1.53	-	26.59	0.80	37.54	m	42.23
Aprons; horizontal										
0.60mm thick; 150mm lapped joints; fixing to timber with copper clips										
150mm girth	3.87	5.00	4.06	0.51	-	8.86	0.42	13.35	m	15.01
240mm girth	6.19	5.00	6.50	0.82	-	14.25	0.42	21.17	m	23.82
300mm girth	7.74	5.00	8.13	1.02	-	17.73	0.42	26.27	m	29.56
450mm girth	11.61	5.00	12.19	1.53	-	26.59	0.63	39.41	m	44.34
0.70mm thick; 150mm lapped joints; fixing to timber with copper clips										
150mm girth	4.84	5.00	5.08	0.58	-	10.08	0.53	15.69	m	17.65
240mm girth	7.74	5.00	8.13	0.92	-	15.99	0.53	24.65	m	27.73
300mm girth	9.67	5.00	10.15	1.23	-	21.38	0.53	32.06	m	36.07
450mm girth	14.51	5.00	15.24	1.53	-	26.59	0.80	42.62	m	47.95
Hips; sloping										
0.60mm thick; 150mm lapped joints; fixing to timber with copper clips										
240mm girth	6.19	5.00	6.50	0.71	-	12.34	0.37	19.21	m	21.61
300mm girth	7.74	5.00	8.13	1.02	-	17.73	0.42	26.27	m	29.56
450mm girth	11.61	5.00	12.19	1.53	-	26.59	0.63	39.41	m	44.34
0.70mm thick; 150mm lapped joints; fixing to timber with copper clips										
240mm girth	7.74	5.00	8.13	0.92	-	15.99	0.53	24.65	m	27.73
300mm girth	9.67	5.00	10.15	1.23	-	21.38	0.53	32.06	m	36.07
450mm girth	14.51	5.00	15.24	1.53	-	26.59	0.80	42.62	m	47.95
Kerbs; horizontal										
0.60mm thick; 150mm lapped joints; fixing to timber with copper clips										
240mm girth	6.19	5.00	6.50	0.82	-	14.25	0.42	21.17	m	23.82
300mm girth	7.74	5.00	8.13	1.02	-	17.73	0.42	26.27	m	29.56
450mm girth	11.61	5.00	12.19	1.53	-	26.59	0.63	39.41	m	44.34
0.70mm thick; 150mm lapped joints; fixing to timber with copper clips										
240mm girth	7.74	5.00	8.13	0.92	-	15.99	0.53	24.65	m	27.73
300mm girth	9.67	5.00	10.15	1.23	-	21.38	0.53	32.06	m	36.07
450mm girth	14.51	5.00	15.24	1.53	-	26.59	0.80	42.62	m	47.95
Ridges; horizontal										
0.60mm thick; 150mm lapped joints; fixing to timber with copper clips										
240mm girth	6.19	5.00	6.50	0.82	-	14.25	0.42	21.17	m	23.82
300mm girth	7.74	5.00	8.13	1.02	-	17.73	0.42	26.27	m	29.56
450mm girth	11.61	5.00	12.19	1.53	-	26.59	0.63	39.41	m	44.34

Labour hourly rates: (except Specialists) Craft Operatives £17.38 Labourer £10.71 Rates are national average prices. Refer to REGIONAL VARIATIONS for indicative levels of overall pricing in regions	MATERIALS			LABOUR				RATES		
	Del to Site £	Waste %	Material Cost £	Craft Optve Hrs	Lab Hrs	Labour Cost £	Sunds £	Nett Rate £	Unit	Gross Rate (+12.5%) £
H73: COPPER STRIP SHEET COVERINGS/FLASHINGS Cont.										
Copper sheet; BS EN 1172 Cont.										
Ridges; horizontal Cont.										
0.70mm thick; 150mm lapped joints; fixing to timber with copper clips										
240mm girth	7.74	5.00	**8.13**	0.92	-	**15.99**	0.53	**24.65**	m	**27.73**
300mm girth	9.67	5.00	**10.15**	1.23	-	**21.38**	0.53	**32.06**	m	**36.07**
450mm girth	14.51	5.00	**15.24**	1.53	-	**26.59**	0.80	**42.62**	m	**47.95**
Valleys; sloping										
0.60mm thick; 150mm lapped joints; fixing to timber with copper clips										
240mm girth	6.19	5.00	**6.50**	0.82	-	**14.25**	0.42	**21.17**	m	**23.82**
300mm girth	7.74	5.00	**8.13**	1.02	-	**17.73**	0.42	**26.27**	m	**29.56**
450mm girth	11.61	5.00	**12.19**	1.53	-	**26.59**	0.63	**39.41**	m	**44.34**
0.70mm thick; 150mm lapped joints; fixing to timber with copper clips										
240mm girth	7.74	5.00	**8.13**	0.80	-	**13.90**	0.46	**22.49**	m	**25.30**
300mm girth	9.67	5.00	**10.15**	1.07	-	**18.60**	0.46	**29.21**	m	**32.86**
450mm girth	14.51	5.00	**15.24**	1.33	-	**23.12**	0.69	**39.04**	m	**43.93**
0.60mm thick edges										
welted	-	-	**-**	0.17	-	**2.95**	-	**2.95**	m	**3.32**
beaded	-	-	**-**	0.17	-	**2.95**	-	**2.95**	m	**3.32**
0.70mm thick edges										
welted	-	-	**-**	0.17	-	**2.95**	-	**2.95**	m	**3.32**
beaded	-	-	**-**	0.17	-	**2.95**	-	**2.95**	m	**3.32**
H74: ZINC STRIP SHEET COVERINGS/FLASHINGS										
Zinc alloy sheet; BS EN 988										
0.8mm thick roof coverings; roll cap; fixing to timber with zinc clips										
pitch 7.5 degrees from horizontal	18.15	5.00	**19.06**	5.70	-	**99.07**	0.20	**118.33**	m2	**133.12**
pitch 40 degrees from horizontal	18.15	5.00	**19.06**	6.35	-	**110.36**	0.20	**129.62**	m2	**145.83**
pitch 75 degrees from horizontal	18.15	5.00	**19.06**	6.95	-	**120.79**	0.20	**140.05**	m2	**157.56**
Flashings; horizontal										
0.8mm thick; 150mm lapped joints; fixing to masonry with zinc clips and wedges										
150mm girth	2.72	5.00	**2.86**	0.51	-	**8.86**	0.15	**11.87**	m	**13.36**
240mm girth	4.36	5.00	**4.58**	0.82	-	**14.25**	0.15	**18.98**	m	**21.36**
300mm girth	5.45	5.00	**5.72**	1.02	-	**17.73**	0.15	**23.60**	m	**26.55**
Flashings; stepped										
0.8mm thick; 150mm lapped joints; fixing to masonry with zinc clips and wedges										
180mm girth	3.27	5.00	**3.43**	0.71	-	**12.34**	0.20	**15.98**	m	**17.97**
240mm girth	4.36	5.00	**4.58**	1.09	-	**18.94**	0.23	**23.76**	m	**26.73**
300mm girth	5.45	5.00	**5.72**	1.37	-	**23.81**	0.23	**29.77**	m	**33.49**
Aprons; horizontal										
0.8mm thick; 150mm lapped joints; fixing to masonry with zinc clips and wedges										
150mm girth	2.72	5.00	**2.86**	0.92	-	**15.99**	0.32	**19.16**	m	**21.56**
240mm girth	4.36	5.00	**4.58**	0.82	-	**14.25**	0.15	**18.98**	m	**21.36**
300mm girth	5.45	5.00	**5.72**	1.02	-	**17.73**	0.15	**23.60**	m	**26.55**
450mm girth	8.17	5.00	**8.58**	1.53	-	**26.59**	0.23	**35.40**	m	**39.83**
Hips; sloping										
0.8mm thick; 150mm lapped joints; fixing to timber with zinc clips										
240mm girth	4.36	5.00	**4.58**	0.82	-	**14.25**	0.15	**18.98**	m	**21.36**
300mm girth	5.45	5.00	**5.72**	1.02	-	**17.73**	0.15	**23.60**	m	**26.55**
450mm girth	8.17	5.00	**8.58**	1.53	-	**26.59**	0.23	**35.40**	m	**39.83**
Kerbs; horizontal										
0.8mm thick; 150mm lapped joints; fixing to timber with zinc clips										
240mm girth	4.36	5.00	**4.58**	0.82	-	**14.25**	0.15	**18.98**	m	**21.36**
300mm girth	5.45	5.00	**5.72**	1.02	-	**17.73**	0.15	**23.60**	m	**26.55**
450mm girth	8.17	5.00	**8.58**	1.53	-	**26.59**	0.23	**35.40**	m	**39.83**
Ridges; horizontal										
0.8mm thick; 150mm lapped joints; fixing to timber with zinc clips										
240mm girth	4.36	5.00	**4.58**	0.82	-	**14.25**	0.15	**18.98**	m	**21.36**
300mm girth	5.45	5.00	**5.72**	1.02	-	**17.73**	0.15	**23.60**	m	**26.55**
450mm girth	8.17	5.00	**8.58**	1.53	-	**26.59**	0.23	**35.40**	m	**39.83**
Valleys; sloping										
0.8mm thick; 150mm lapped joints; fixing to timber with zinc clips										
240mm girth	4.36	5.00	**4.58**	0.82	-	**14.25**	0.15	**18.98**	m	**21.36**
300mm girth	5.45	5.00	**5.72**	1.02	-	**17.73**	0.15	**23.60**	m	**26.55**
450mm girth	8.17	5.00	**8.58**	1.53	-	**26.59**	0.23	**35.40**	m	**39.83**
0.65mm thick edges										
welted	-	-	**-**	0.20	-	**3.48**	-	**3.48**	m	**3.91**
beaded	-	-	**-**	0.20	-	**3.48**	-	**3.48**	m	**3.91**

Labour hourly rates: (except Specialists) Craft Operatives £17.38 Labourer £10.71 Rates are national average prices. Refer to REGIONAL VARIATIONS for indicative levels of overall pricing in regions	MATERIALS			LABOUR				RATES		
	Del to Site £	Waste %	Material Cost £	Craft Optve Hrs	Lab Hrs	Labour Cost £	Sunds £	Nett Rate £	Unit	Gross Rate (+12.5%) £
H74: ZINC STRIP SHEET COVERINGS/FLASHINGS Cont.										
Zinc alloy sheet; BS EN 988 Cont.										
0.8mm thick edges										
welted	-	-	-	0.20	-	**3.48**	-	**3.48**	m	**3.91**
beaded	-	-	-	0.20	-	**3.48**	-	**3.48**	m	**3.91**
0.8mm thick soakers and slates - handed to others for fixing										
180 x 180mm	0.59	5.00	**0.62**	0.23	-	**4.00**	-	**4.62**	nr	**5.19**
180 x 300mm	0.98	5.00	**1.03**	0.23	-	**4.00**	-	**5.03**	nr	**5.65**
450 x 450mm	3.68	5.00	**3.86**	0.23	-	**4.00**	-	**7.86**	nr	**8.84**
0.8mm thick collars around pipes, standards and the like										
150mm long; soldered joints to metal covering										
50mm diameter	3.52	5.00	**3.70**	1.00	-	**17.38**	-	**21.08**	nr	**23.71**
100mm diameter	3.93	5.00	**4.13**	1.00	-	**17.38**	-	**21.51**	nr	**24.19**

WATERPROOFING

Labour hourly rates: (except Specialists) Craft Operatives £14.21 Labourer £10.71 Rates are national average prices. Refer to REGIONAL VARIATIONS for indicative levels of overall pricing in regions	MATERIALS			LABOUR				RATES		
	Del to Site £	Waste %	Material Cost £	Craft Optve Hrs	Lab Hrs	Labour Cost £	Sunds £	Nett Rate £	Unit	Gross Rate (+12.5%) £
J20: MASTIC ASPHALT TANKING/ DAMP PROOF MEMBRANES										
Tanking and damp proofing BS 6925 (limestone aggregate)										
Tanking and damp proofing 13mm thick; coats of asphalt -1; to concrete base; horizontal										
work subsequently covered										
width not exceeding 150mm	-	-	Spclist	-	-	Spclist	-	6.43	lm	7.23
width 150 - 225mm	-	-	Spclist	-	-	Spclist	-	7.67	lm	8.63
width 225 - 300mm	-	-	Spclist	-	-	Spclist	-	9.35	lm	10.52
width exceeding 300mm	-	-	Spclist	-	-	Spclist	-	15.78	m2	17.75
Tanking and damp proofing 20mm thick; coats of asphalt -2; to concrete base; horizontal										
work subsequently covered										
width not exceeding 150mm	-	-	Spclist	-	-	Spclist	-	7.05	lm	7.93
width 150 - 225mm	-	-	Spclist	-	-	Spclist	-	8.85	lm	9.96
width 225 - 300mm	-	-	Spclist	-	-	Spclist	-	10.93	lm	12.30
width exceeding 300mm	-	-	Spclist	-	-	Spclist	-	23.01	m2	25.89
Tanking and damp proofing 30mm thick; coats of asphalt -3; to concrete base; horizontal										
work subsequently covered										
width not exceeding 150mm	-	-	Spclist	-	-	Spclist	-	10.59	lm	11.91
width 150 - 225mm	-	-	Spclist	-	-	Spclist	-	13.25	lm	14.91
width 225 - 300mm	-	-	Spclist	-	-	Spclist	-	16.39	lm	18.44
width exceeding 300mm	-	-	Spclist	-	-	Spclist	-	32.77	m2	36.87
Tanking and damp proofing 20mm thick; coats of asphalt -2; to concrete base; vertical										
work subsequently covered										
width not exceeding 150mm	-	-	Spclist	-	-	Spclist	-	13.88	lm	15.62
width 150 - 225mm	-	-	Spclist	-	-	Spclist	-	16.23	lm,	18.26
width 225 - 300mm	-	-	Spclist	-	-	Spclist	-	18.51	lm	20.82
width exceeding 300mm	-	-	Spclist	-	-	Spclist	-	47.82	m2	53.80
Tanking and damp proofing 20mm thick; coats of asphalt -3; to concrete base; vertical										
work subsequently covered										
width not exceeding 150mm	-	-	Spclist	-	-	Spclist	-	20.87	lm	23.48
width 150 - 225mm	-	-	Spclist	-	-	Spclist	-	24.44	lm	27.50
width 225 - 300mm	-	-	Spclist	-	-	Spclist	-	27.87	lm	31.35
width exceeding 300mm	-	-	Spclist	-	-	Spclist	-	66.40	m2	74.70
Internal angle fillets to concrete base; priming base with bitumen										
two coats; work subsequently covered	-	-	Spclist	-	-	Spclist	-	6.21	m	6.99
J21: MASTIC ASPHALT ROOFING/INSULATION/FINISHES										
Roofing; BS 6925 (limestone aggregate); sand rubbing; BS 747 type 4A sheathing felt isolating membrane, butt joints										
Roofing 20mm thick; coats of asphalt -2; to concrete base; pitch not exceeding 6 degrees from horizontal										
width exceeding 300mm	-	-	Spclist	-	-	Spclist	-	24.78	m2	27.88
extra; solid water check roll	-	-	Spclist	-	-	Spclist	-	12.43	m	13.98
Roofing 20mm thick; coats of asphalt -2; to concrete base; pitch 30 degrees from horizontal										
width exceeding 300mm	-	-	Spclist	-	-	Spclist	-	37.90	m2	42.64
Roofing 20mm thick; coats of asphalt -2; to concrete base; pitch 45 degrees from horizontal										
width exceeding 300mm	-	-	Spclist	-	-	Spclist	-	48.50	m2	54.56
Paint two coats of Solar reflective roof paint on surfaces of asphalt										
width exceeding 300mm	-	-	Spclist	-	-	Spclist	-	6.75	m2	7.59

Labour hourly rates: (except Specialists) Craft Operatives £14.21 Labourer £10.71 Rates are national average prices. Refer to REGIONAL VARIATIONS for indicative levels of overall pricing in regions	MATERIALS			LABOUR				RATES		
	Del to Site	Waste	Material Cost	Craft Optve	Lab	Labour Cost	Sunds	Nett Rate	Unit	Gross Rate (+12.5%)
	£	%	£	Hrs	Hrs	£	£	£		£
J21: MASTIC ASPHALT ROOFING/INSULATION/FINISHES Cont.										
Roofing; BS 6925 (limestone aggregate); sand rubbing; BS 747 type 4A sheathing felt isolating membrane, butt joints Cont.										
Skirtings 13mm thick; coats of asphalt -2; to brickwork base										
girth not exceeding 150mm	-	-	Spclist	-	-	Spclist	-	14.72	m	16.56
Coverings to kerbs 20mm thick; coats of asphalt -2; to concrete base										
girth 600mm	-	-	Spclist	-	-	Spclist	-	31.21	m	35.11
Jointing new roofing to existing										
20mm thick	-	-	Spclist	-	-	Spclist	-	18.65	m	20.98
Edge trim; aluminium; silver anodized										
priming with bituminous primer; butt joints with internal jointing sleeves; fixing with aluminium alloy screws to timber; working two coat asphalt into grooves -1										
63.5mm wide x 44.4mm face depth	7.85	5.00	8.24	0.40	-	5.68	1.38	15.31	m	17.22
63.5mm wide x 76.2mm face depth	9.20	5.00	9.66	0.40	-	5.68	1.44	16.78	m	18.88
104.8mm wide x 44.4mm face depth	8.80	5.00	9.24	0.40	-	5.68	1.44	16.36	m	18.41
104.8mm wide x 76.2mm face depth	9.22	5.00	9.68	0.50	-	7.11	1.64	18.43	m	20.73
76.2mm fixing arm at 10 degrees x 38.1mm face depth	9.30	5.00	9.77	0.40	-	5.68	1.44	16.89	m	19.00
Edge trim; glass fibre reinforced										
butt joints; fixing with stainless steel screws to timber; working two coat asphalt into grooves -1.	10.83	5.00	11.37	0.35	-	4.97	1.64	17.99	m	20.23
Roof Ventilators										
plastic; setting in position	8.32	2.50	8.53	0.85	-	12.08	0.54	21.15	nr	23.79
aluminium; setting in position	9.24	2.50	9.47	0.85	-	12.08	0.54	22.09	nr	24.85
Roofing; BS 6925 (limestone aggregate); covering with 13mm white spar chippings in hot bitumen; BS 747 type 4A sheathing felt isolating membrane; butt joints										
Roofing 20mm thick; coats of asphalt -2; to concrete base; pitch not exceeding 6 degrees from horizontal										
width exceeding 300mm	-	-	Spclist	-	-	Spclist	-	29.79	m2	33.51
Roofing 20mm thick; coats of asphalt -2; to concrete base; pitch 30 degrees from horizontal										
width exceeding 300mm	-	-	Spclist	-	-	Spclist	-	42.89	m2	48.25
J30: LIQUID APPLIED TANKING/DAMP PROOF MEMBRANES										
Synthaprufe bituminous emulsion; blinding with sand										
Coverings, coats -2; to concrete base; horizontal										
width not exceeding 150mm	5.20	10.00	5.72	-	0.46	4.93	0.14	10.79	m2	12.13
width 150 - 225mm	5.20	10.00	5.72	-	0.38	4.07	0.14	9.93	m2	11.17
width 225 - 300mm	5.20	10.00	5.72	-	0.30	3.21	0.14	9.07	m2	10.21
width exceeding 300mm	5.20	10.00	5.72	-	0.23	2.46	0.14	8.32	m2	9.36
Coverings, coats -3; to concrete base; horizontal										
width not exceeding 150mm	7.76	10.00	8.54	-	0.64	6.85	0.14	15.53	m2	17.47
width 150 - 225mm	7.76	10.00	8.54	-	0.53	5.68	0.14	14.35	m2	16.15
width 225 - 300mm	7.76	10.00	8.54	-	0.42	4.50	0.14	13.17	m2	14.82
width exceeding 300mm	7.76	10.00	8.54	-	0.32	3.43	0.14	12.10	m2	13.62
Coverings; coats -2; to concrete base; vertical										
width not exceeding 150mm	5.20	10.00	5.72	-	0.69	7.39	0.14	13.25	m2	14.91
width 150 - 225mm	5.20	10.00	5.72	-	0.57	6.10	0.14	11.96	m2	13.46
width 225 - 300mm	5.20	10.00	5.72	-	0.45	4.82	0.14	10.68	m2	12.01
width exceeding 300mm	5.20	10.00	5.72	-	0.35	3.75	0.14	9.61	m2	10.81
Coverings; coats -3; to concrete base; vertical										
width not exceeding 150mm	7.76	10.00	8.54	-	0.96	10.28	0.14	18.96	m2	21.33
width 150 - 225mm	7.76	10.00	8.54	-	0.80	8.57	0.14	17.24	m2	19.40
width 225 - 300mm	7.76	10.00	8.54	-	0.63	6.75	0.14	15.42	m2	17.35
width exceeding 300mm	7.76	10.00	8.54	-	0.48	5.14	0.14	13.82	m2	15.54
Bituminous emulsion										
Coverings, coats -2; to concrete base; horizontal										
width not exceeding 150mm	3.88	10.00	4.27	-	0.39	4.18	0.14	8.58	m2	9.66
width 150 - 225mm	3.88	10.00	4.27	-	0.32	3.43	0.14	7.84	m2	8.81
width 225 - 300mm	3.88	10.00	4.27	-	0.25	2.68	0.14	7.09	m2	7.97
width exceeding 300mm	3.88	10.00	4.27	-	0.20	2.14	0.14	6.55	m2	7.37
Coverings, coats -3; to concrete base; horizontal										
width not exceeding 150mm	5.43	10.00	5.98	-	0.55	5.89	0.14	12.01	m2	13.51
width 150 - 225mm	5.43	10.00	5.98	-	0.45	4.82	0.14	10.93	m2	12.30
width 225 - 300mm	5.43	10.00	5.98	-	0.36	3.86	0.14	9.97	m2	11.22
width exceeding 300mm	5.43	10.00	5.98	-	0.28	3.00	0.14	9.11	m2	10.25
Coverings; coats -2; to concrete base; vertical										
width not exceeding 150mm	3.88	10.00	4.27	-	0.58	6.21	0.14	10.62	m2	11.95
width 150 - 225mm	3.88	10.00	4.27	-	0.48	5.14	0.14	9.55	m2	10.74
width 225 - 300mm	3.88	10.00	4.27	-	0.38	4.07	0.14	8.48	m2	9.54
width exceeding 300mm	3.88	10.00	4.27	-	0.30	3.21	0.14	7.62	m2	8.57

Labour hourly rates: (except Specialists) Craft Operatives £14.21 Labourer £10.71 Rates are national average prices. Refer to REGIONAL VARIATIONS for indicative levels of overall pricing in regions	MATERIALS			LABOUR				RATES		
	Del to Site £	Waste %	Material Cost £	Craft Optve Hrs	Lab Hrs	Labour Cost £	Sunds £	Nett Rate £	Unit	Gross Rate (+12.5%) £
J30: LIQUID APPLIED TANKING/DAMP PROOF MEMBRANES Cont.										
Bituminous emulsion Cont.										
Coverings; coats -3; to concrete base; vertical										
width not exceeding 150mm	5.43	10.00	**5.98**	-	0.82	**8.78**	0.14	**14.90**	m2	**16.76**
width 150 - 225mm	5.43	10.00	**5.98**	-	0.68	**7.28**	0.14	**13.40**	m2	**15.07**
width 225 - 300mm	5.43	10.00	**5.98**	-	0.54	**5.78**	0.14	**11.90**	m2	**13.39**
width exceeding 300mm	5.43	10.00	**5.98**	-	0.42	**4.50**	0.14	**10.61**	m2	**11.94**
R.I.W. liquid asphaltic composition										
Coverings, coats -2; to concrete base; horizontal										
width not exceeding 150mm	8.04	10.00	**8.85**	-	0.46	**4.93**	0.14	**13.91**	m2	**15.65**
width 150 - 225mm	8.04	10.00	**8.85**	-	0.38	**4.07**	0.14	**13.06**	m2	**14.69**
width 225 - 300mm	8.04	10.00	**8.85**	-	0.30	**3.21**	0.14	**12.20**	m2	**13.73**
width exceeding 300mm	8.04	10.00	**8.85**	-	0.23	**2.46**	0.14	**11.45**	m2	**12.88**
Coverings, coats -3; to concrete base; horizontal										
width not exceeding 150mm	12.11	10.00	**13.32**	-	0.64	**6.85**	0.14	**20.32**	m2	**22.86**
width 150 - 225mm	12.11	10.00	**13.32**	-	0.53	**5.68**	0.14	**19.14**	m2	**21.53**
width 225 - 300mm	12.11	10.00	**13.32**	-	0.42	**4.50**	0.14	**17.96**	m2	**20.20**
width exceeding 300mm	12.11	10.00	**13.32**	-	0.32	**3.43**	0.14	**16.89**	m2	**19.00**
Coverings; coats -2; to concrete base; vertical										
width not exceeding 150mm	8.04	10.00	**8.85**	-	0.69	**7.39**	0.14	**16.38**	m2	**18.42**
width 150 - 225mm	8.04	10.00	**8.85**	-	0.57	**6.10**	0.14	**15.09**	m2	**16.98**
width 225 - 300mm	8.04	10.00	**8.85**	-	0.45	**4.82**	0.14	**13.81**	m2	**15.53**
width exceeding 300mm	8.04	10.00	**8.85**	-	0.35	**3.75**	0.14	**12.74**	m2	**14.33**
Coverings; coats -3; to concrete base; vertical										
width not exceeding 150mm	12.11	10.00	**13.32**	-	0.96	**10.28**	0.14	**23.74**	m2	**26.71**
width 150 - 225mm	12.11	10.00	**13.32**	-	0.80	**8.57**	0.14	**22.03**	m2	**24.78**
width 225 - 300mm	12.11	10.00	**13.32**	-	0.63	**6.75**	0.14	**20.21**	m2	**22.73**
width exceeding 300mm	12.11	10.00	**13.32**	-	0.48	**5.14**	0.14	**18.60**	m2	**20.93**
J40: FLEXIBLE SHEET TANKING/DAMP PROOFING MEMBRANES										
Polythene sheeting; 100mm welted laps										
Tanking and damp proofing; 125mu										
horizontal; on concrete base	0.36	20.00	**0.43**	-	0.06	**0.64**	-	**1.07**	m2	**1.21**
vertical; on concrete base	0.36	20.00	**0.43**	-	0.09	**0.96**	-	**1.40**	m2	**1.57**
Tanking and damp proofing; 250mu										
horizontal; on concrete base	0.42	20.00	**0.50**	-	0.09	**0.96**	-	**1.47**	m2	**1.65**
vertical; on concrete base	0.42	20.00	**0.50**	-	0.14	**1.50**	-	**2.00**	m2	**2.25**
Tanking and damp proofing; 300mu										
horizontal; on concrete base	0.56	20.00	**0.67**	-	0.11	**1.18**	-	**1.85**	m2	**2.08**
vertical; on concrete base	0.56	20.00	**0.67**	-	0.17	**1.82**	-	**2.49**	m2	**2.80**
Bituthene 500X self adhesive damp proof membrane; 25mm lapped joints										
Tanking and damp proofing										
horizontal; on concrete base	4.63	20.00	**5.56**	-	0.18	**1.93**	-	**7.48**	m2	**8.42**
vertical; on concrete base	4.63	20.00	**5.56**	-	0.27	**2.89**	-	**8.45**	m2	**9.50**
Bituthene 500X self adhesive damp proof membrane; 75mm lapped joints										
Tanking and damp proofing										
horizontal; on concrete base	5.07	20.00	**6.09**	-	0.23	**2.46**	-	**8.55**	m2	**9.62**
vertical; on concrete base; priming with Servicised B primer	5.07	20.00	**6.09**	-	0.35	**3.75**	0.26	**10.09**	m2	**11.36**
extra; Bituthene internal angle fillet; 40 x 40mm	5.70	10.00	**6.27**	-	0.12	**1.29**	-	**7.56**	m	**8.50**
British Sisalkraft Ltd 728 damp proof membrane; 150mm laps sealed with tape										
Tanking and damp proofing										
horizontal; on concrete base	1.44	20.00	**1.73**	-	0.08	**0.86**	0.11	**2.69**	m2	**3.03**
Visqueen 1200 super damp proof membrane; 150mm laps sealed with tape and mastic										
Tanking and damp proofing										
horizontal; on concrete base	0.56	20.00	**0.67**	-	0.08	**0.86**	0.11	**1.64**	m2	**1.84**
J41: BUILT UP FELT ROOF COVERINGS										
Roofing; felt BS 747, comprising 3G Rubervent underlay 2.6 kg/m2, 1 layer Ruberfort HP180 underlay 3.4 kg/m2, 1 layer Ruberfort HP180 mineral surface top sheet 3.6 kg/m2										
Roof coverings; layers of felt -3; bonding with hot bitumen compound to timber base										
pitch 7.5 degrees from horizontal	10.34	15.00	**12.05**	0.46	0.06	**7.18**	-	**19.23**	m2	**21.63**
pitch 40 degrees from horizontal	10.34	15.00	**12.05**	0.76	0.09	**11.76**	-	**23.81**	m2	**26.79**
pitch 75 degrees from horizontal	10.34	15.00	**12.05**	0.92	0.12	**14.36**	-	**26.41**	m2	**29.71**
Roof coverings; layers of felt -3; bonding with hot bitumen compound to cement and sand or concrete base										
pitch 7.5 degrees from horizontal	10.34	15.00	**12.05**	0.46	0.06	**7.18**	-	**19.23**	m2	**21.63**

WATERPROOFING

Labour hourly rates: (except Specialists) Craft Operatives £14.21 Labourer £10.71 Rates are national average prices. Refer to REGIONAL VARIATIONS for indicative levels of overall pricing in regions	MATERIALS			LABOUR				RATES		
	Del to Site £	Waste %	Material Cost £	Craft Optve Hrs	Lab Hrs	Labour Cost £	Sunds £	Nett Rate £	Unit	Gross Rate (+12.5%) £
J41: BUILT UP FELT ROOF COVERINGS Cont.										
Roofing; felt BS 747, comprising 3G Rubervent underlay 2.6 kg/m2, 1 layer Ruberfort HP180 underlay 3.4 kg/m2, 1 layer Ruberfort HP180 mineral surface top sheet 3.6 kg/m2 Cont.										
Roof coverings; layers of felt -3; bonding with hot bitumen compound to cement and sand or concrete base Cont.										
pitch 40 degrees from horizontal	10.34	15.00	12.05	0.76	0.09	11.76	-	23.81	m2	26.79
pitch 75 degrees from horizontal	10.34	15.00	12.05	0.92	0.12	14.36	-	26.41	m2	29.71
Skirtings; layers of felt - 3; bonding with hot bitumen compound to brickwork base										
girth 0 - 200mm	2.07	15.00	2.41	0.28	0.03	4.30	-	6.71	m	7.55
girth 200 - 400mm	4.14	15.00	4.82	0.46	0.06	7.18	-	12.00	m	13.50
Collars around pipes; standard and the like; layers of felt -1; bonding with hot bitumen compound to metal base										
50mm diameter x 150mm long; hole in 3 layer covering	0.87	15.00	1.00	0.17	0.02	2.63	-	3.63	nr	4.08
100mm diameter x 150mm long; hole in 3 layer covering	1.10	15.00	1.28	0.22	0.03	3.45	-	4.73	nr	5.32
Eaves trim; aluminium; silver anodized butt joints over matching sleeve pieces 200mm long; fixing to timber base with aluminium alloy screws; bedding in mastic										
63.5mm wide x 44.4mm face depth	7.85	5.00	8.24	0.40	-	5.68	1.50	15.43	m	17.35
63.5mm wide x 76.2mm face depth	9.20	5.00	9.66	0.40	-	5.68	1.58	16.92	m	19.04
104.8mm wide x 44.4mm face depth	8.80	5.00	9.24	0.40	-	5.68	1.58	16.50	m	18.57
104.8mm wide x 76.2mm face depth	9.22	5.00	9.68	0.50	-	7.11	1.77	18.56	m	20.88
Eaves trim; glass fibre reinforced butt joints; fixing to timber base with stainless steel screws; bedding in mastic	10.83	5.00	11.37	0.35	-	4.97	1.77	18.12	m	20.38
Roof ventilators										
plastic; setting in position	8.32	2.50	8.53	0.85	-	12.08	0.55	21.16	nr	23.80
aluminium; setting in position	9.24	2.50	9.47	0.85	-	12.08	0.55	22.10	nr	24.86
Roofing; BS 747, all layers type 3B -1.8 kg/m2, 75mm laps; fully bonding layers with hot bitumen bonding compound										
Roof coverings; layers of felt -2; bonding with hot bitumen compound to timber base										
pitch 7.5 degrees from horizontal	4.00	15.00	4.71	0.38	0.05	5.94	-	10.64	m2	11.97
pitch 40 degrees from horizontal	4.00	15.00	4.71	0.63	0.08	9.81	-	14.52	m2	16.33
pitch 75 degrees from horizontal	4.00	15.00	4.71	0.76	0.09	11.76	-	16.47	m2	18.53
Roof coverings; layers of felt -2; bonding with hot bitumen compound to cement and sand or concrete base										
pitch 7.5 degrees from horizontal	4.00	15.00	4.71	0.38	0.05	5.94	-	10.64	m2	11.97
pitch 40 degrees from horizontal	4.00	15.00	4.71	0.63	0.08	9.81	-	14.52	m2	16.33
pitch 75 degrees from horizontal	4.00	15.00	4.71	0.76	0.09	11.76	-	16.47	m2	18.53
Roof coverings; layers of felt -3; bonding with hot bitumen compound to timber base										
pitch 7.5 degrees from horizontal	6.00	15.00	7.06	0.55	0.07	8.57	-	15.62	m2	17.58
pitch 40 degrees from horizontal	6.00	15.00	7.06	0.92	0.12	14.36	-	21.42	m2	24.09
pitch 75 degrees from horizontal	6.00	15.00	7.06	1.10	0.14	17.13	-	24.19	m2	27.21
Roof coverings; layers of felt -3; bonding with hot bitumen compound to cement and sand or concrete base										
pitch 7.5 degrees from horizontal	6.00	15.00	7.06	0.55	0.07	8.57	-	15.62	m2	17.58
pitch 40 degrees from horizontal	6.00	15.00	7.06	0.92	0.12	14.36	-	21.42	m2	24.09
pitch 75 degrees from horizontal	6.00	15.00	7.06	1.10	0.14	17.13	-	24.19	m2	27.21
Roofing; felt BS 747, bottom and intermediate layers type 3B -1.8 kg/m2, top layer type 3E -2.8 kg/m2, 75mm laps; fully bonding layers with hot bitumen bonding compound										
Roof coverings; layers of felt -2; bonding with hot bitumen compound to timber base										
pitch 7.5 degrees from horizontal	4.94	15.00	5.79	0.38	0.05	5.94	-	11.72	m2	13.19
pitch 40 degrees from horizontal	4.94	15.00	5.79	0.63	0.08	9.81	-	15.60	m2	17.55
pitch 75 degrees from horizontal	4.94	15.00	5.79	0.76	0.09	11.76	-	17.55	m2	19.74
Roof coverings; layers of felt -2; bonding with hot bitumen compound to cement and sand or concrete base										
pitch 7.5 degrees from horizontal	4.94	15.00	5.79	0.38	0.05	5.94	-	11.72	m2	13.19
pitch 40 degrees from horizontal	4.94	15.00	5.79	0.63	0.08	9.81	-	15.60	m2	17.55
pitch 75 degrees from horizontal	4.94	15.00	5.79	0.76	0.09	11.76	-	17.55	m2	19.74
Roof coverings; layers of felt -3; bonding with hot bitumen compound to timber base										
pitch 7.5 degrees from horizontal	6.94	15.00	8.14	0.55	0.07	8.57	-	16.71	m2	18.79
pitch 40 degrees from horizontal	6.94	15.00	8.14	0.92	0.12	14.36	-	22.50	m2	25.31
pitch 75 degrees from horizontal	6.94	15.00	8.14	1.10	0.14	17.13	-	25.27	m2	28.43

Labour hourly rates: (except Specialists) Craft Operatives £14.21 Labourer £10.71 Rates are national average prices. Refer to REGIONAL VARIATIONS for indicative levels of overall pricing in regions	MATERIALS			LABOUR				RATES		
	Del to Site £	Waste %	Material Cost £	Craft Optve Hrs	Lab Hrs	Labour Cost £	Sunds £	Nett Rate £	Unit	Gross Rate (+12.5%) £
J41: BUILT UP FELT ROOF COVERINGS Cont.										
Roofing; felt BS 747, bottom and intermediate layers type 3B -1.8 kg/m2, top layer type 3E -2.8 kg/m2, 75mm laps; fully bonding layers with hot bitumen bonding compound Cont.										
Roof coverings; layers of felt -3; bonding with hot bitumen compound to cement and sand or concrete base										
pitch 7.5 degrees from horizontal................	6.94	15.00	**8.14**	0.55	0.07	**8.57**	-	**16.71**	m2	**18.79**
pitch 40 degrees from horizontal.................	6.94	15.00	**8.14**	0.92	0.12	**14.36**	-	**22.50**	m2	**25.31**
pitch 75 degrees from horizontal.................	6.94	15.00	**8.14**	1.10	0.14	**17.13**	-	**25.27**	m2	**28.43**
Skirtings; layers of felt - 3; bonding with hot bitumen compound to brickwork base										
girth 0 - 200mm	1.39	15.00	**1.63**	0.28	0.03	**4.30**	-	**5.93**	m	**6.67**
girth 200 - 400mm	2.78	15.00	**3.26**	0.46	0.06	**7.18**	-	**10.44**	m	**11.74**
Collars around pipes; standard and the like; layers of felt -1; bonding with hot bitumen compound to metal base										
50mm diameter x 150mm long; hole in 3 layer covering	0.55	15.00	**0.64**	0.17	0.02	**2.63**	-	**3.27**	nr	**3.67**
100mm diameter x 150mm long; hole in 3 layer covering	0.72	15.00	**0.84**	0.22	0.03	**3.45**	-	**4.29**	nr	**4.82**
Roofing; sheet, Ruberoid Building Products Superflex Ultrabond system, bottom layer Rubervent BS 747 type 3G; intermediate layer Superbase Ultrabond; top layer Superflex Ultrabond slate surfaced cap sheet, 50mm side laps, 75mm end laps; partially bonding bottom layer, fully bonding other layers in Ruberoid 95/25 hot bonding bitumen										
Roof coverings; layers of sheet -3; bonding with hot bitumen compound to timber base										
pitch not exceeding 5.0 degrees from horizontal..	15.98	15.00	**18.54**	0.55	0.28	**10.81**	-	**29.35**	m2	**33.02**
Roof coverings; layers of sheet -3; bonding with hot bitumen compound to cement and sand or concrete base										
pitch not exceeding 5.0 degrees from horizontal..	15.98	15.00	**18.54**	0.55	0.28	**10.81**	-	**29.35**	m2	**33.02**
Roof coverings; layers of sheet -3; bonding with hot bitumen compound to metal base										
pitch not exceeding 5.0 degrees from horizontal..	15.98	15.00	**18.54**	0.55	0.28	**10.81**	-	**29.35**	m2	**33.02**
Skirtings; layers of sheet -2; bonding with hot bitumen compound to brickwork base										
girth 0 - 200mm	1.96	15.00	**2.27**	0.28	0.05	**4.51**	-	**6.78**	m	**7.63**
girth 200 - 400mm	3.91	15.00	**4.54**	0.46	0.06	**7.18**	-	**11.72**	m	**13.19**
Roofing; sheet, Ruberoid Building Products Superflex Ultrabond system, bottom layer Rubervent BS 747 type 3G; intermediate layer Superbase Ultrabond; top layer Superflex Ultrabond slate surfaced cap sheet, 50mm side laps, 75mm end laps; partially bonding bottom layer, fully bonding other layers in Ruberoid 95/25 hot bonding bitumen; 25mm Kingspan Thermaroof TR21 (PIR) Board										
Roof coverings; layers of sheet -3; bonding with hot bitumen compound to timber base										
pitch not exceeding 5.0 degrees from horizontal..	24.52	15.00	**28.36**	1.10	0.36	**19.49**	1.11	**48.95**	m2	**55.07**
Roof coverings; layers of sheet -3; bonding with hot bitumen compound to cement and sand or concrete base										
pitch not exceeding 5.0 degrees from horizontal..	24.52	15.00	**28.36**	1.16	0.36	**20.34**	1.11	**49.81**	m2	**56.03**
Roof coverings; layers of sheet -3; bonding with hot bitumen compound to metal base										
pitch not exceeding 5.0 degrees from horizontal..	24.52	15.00	**28.36**	1.14	0.36	**20.06**	1.11	**49.52**	m2	**55.71**
Roofing; sheet, Ruberoid Building Products Superflex Ultrabond system, bottom layer Rubervent BS 747 type 3G; intermediate layer Superbase Ultrabond; top layer Superflex Ultrabond slate surfaced cap sheet, 50mm side laps, 75mm end laps; partially bonding bottom layer, fully bonding other layers in Ruberoid 95/25 hot bonding bitumen; 30mm Kingspan Thermaroof TR21 (PIR) Board										
Roof coverings; layers of sheet -3; bonding with hot bitumen compound to timber base										
pitch not exceeding 5.0 degrees from horizontal..	25.77	15.00	**29.79**	1.12	0.36	**19.77**	-	**49.57**	m2	**55.76**

Labour hourly rates: (except Specialists) Craft Operatives £14.21 Labourer £10.71 Rates are national average prices. Refer to REGIONAL VARIATIONS for indicative levels of overall pricing in regions	MATERIALS			LABOUR				RATES		
	Del to Site £	Waste %	Material Cost £	Craft Optve Hrs	Lab Hrs	Labour Cost £	Sunds £	Nett Rate £	Unit	Gross Rate (+12.5%) £
J41: BUILT UP FELT ROOF COVERINGS Cont.										
Roofing; sheet, Ruberoid Building Products Superflex Ultrabond system, bottom layer Rubervent BS 747 type 3G; intermediate layer Superbase Ultrabond; top layer Superflex Ultrabond slate surfaced cap sheet, 50mm side laps, 75mm end laps; partially bonding bottom layer, fully bonding other layers in Ruberoid 95/25 hot bonding bitumen; 40mm Kingspan Thermaroof TR21 (PIR) Board										
Roof coverings; layers of sheet -3; bonding with hot bitumen compound to timber base										
pitch not exceeding 5.0 degrees from horizontal ..	27.29	15.00	**31.54**	1.18	0.36	**20.62**	-	**52.17**	m2	**58.69**
Roofing; sheet, Ruberoid Building Products Superflex Ultrabond system, bottom layer Rubervent BS 747 type 3G; intermediate layer Superbase Ultrabond; top layer Superflex Ultrabond slate surfaced cap sheet, 50mm side laps, 75mm end laps; partially bonding bottom layer, fully bonding other layers in Ruberoid 95/25 hot bonding bitumen; 50mm Kingspan Thermaboard TR21 (PIR) Board										
Roof coverings; layers of sheet -3; bonding with hot bitumen compound to timber base										
pitch not exceeding 5.0 degrees from horizontal ..	28.90	15.00	**33.39**	1.20	0.36	**20.91**	-	**54.30**	m2	**61.09**
Roofing; sheet, Ruberoid Building Products Superflex Ultratorch system, bottom layer Superbase Ultratorch; top layer Superflex Ultratorch slate surfaced cap sheet, 75mm side and end laps; fully bonding both layers (by torching); SuperRock Torch, mineral wool insulation with bitumen primed surface; Superbar vapour control layer										
Roof coverings; layers of sheet -3; bonding to timber base										
pitch not exceeding 5.0 degrees from horizontal ..	40.35	15.00	**46.40**	0.55	0.28	**10.81**	2.87	**60.09**	m2	**67.60**
Roof coverings; layers of sheet -3; bonding to cement and sand or concrete base										
pitch not exceeding 5.0 degrees from horizontal ..	40.35	15.00	**46.40**	0.55	0.28	**10.81**	2.87	**60.09**	m2	**67.60**
Roof coverings; layers of sheet -3; bonding to metal base										
pitch not exceeding 5.0 degrees from horizontal ..	40.35	15.00	**46.40**	0.59	0.28	**11.38**	2.87	**60.66**	m2	**68.24**
Skirtings; layers of sheet -2; bonding with hot bitumen compound to brickwork base										
girth 0 - 200mm	3.03	15.00	**3.49**	0.28	0.05	**4.51**	0.58	**8.58**	m	**9.65**
girth 200 - 400mm	6.06	15.00	**6.97**	0.46	0.06	**7.18**	1.16	**15.31**	m	**17.23**
Roofing; sheet, Ruberoid Building Products Superflex Ultratorch system, bottom layer Superbase Ultratorch; top layer Superflex Ultratorch slate surfaced cap sheet, 75mm side and end laps; fully bonding both layers (by torching); SuperRock Torch, 25mm Kingspan Thermaroof TR21 (PIR Board) insulation with bitumen primed surface; Superbar vapour control layer										
Roof coverings; layers of sheet -3; bonding to timber base										
pitch not exceeding 5.0 degrees from horizontal ..	48.89	15.00	**56.22**	1.10	0.36	**19.49**	2.87	**78.58**	m2	**88.40**
Roof coverings; layers of sheet -3; bonding to cement and sand or concrete base										
pitch not exceeding 5.0 degrees from horizontal ..	48.89	15.00	**56.22**	1.10	0.36	**19.49**	2.87	**78.58**	m2	**88.40**
Roof coverings; layers of sheet -3; bonding to metal base										
pitch not exceeding 5.0 degrees from horizontal ..	48.89	15.00	**56.22**	1.15	0.36	**20.20**	2.87	**79.29**	m2	**89.20**
Roofing; sheet, Ruberoid Building Products Superflex Ultratorch system, bottom layer Superbase Ultratorch; top layer Superflex Ultratorch slate surfaced cap sheet, 75mm side and end laps; fully bonding both layers (by torching); SuperRock Torch, 30mm Kingspan Thermaroof TR21 (PIR Board) insulation with bitumen primed surface; Superbar vapour control layer										
Roof coverings; layers of sheet -3; bonding to timber base										
pitch not exceeding 5.0 degrees from horizontal ..	50.14	15.00	**57.66**	1.12	0.36	**19.77**	2.87	**80.30**	m2	**90.34**
Roofing; sheet, Ruberoid Building Products Superflex Ultratorch system, bottom layer Superbase Ultratorch; top layer Superflex Ultratorch slate surfaced cap sheet, 75mm side and end laps; fully bonding both layers (by torching); SuperRock Torch, 40mm Kingspan Thermaroof TR21 (PIR Board) insulation with bitumen primed surface; Superbar vapour control layer										
Roof coverings; layers of sheet -3; bonding to timber base										
pitch not exceeding 5.0 degrees from horizontal ..	51.66	15.00	**59.41**	1.18	0.36	**20.62**	2.87	**82.90**	m2	**93.27**

LININGS/SHEATHING/DRY PARTITIONING

Labour hourly rates: (except Specialists) Craft Operatives £14.21 Labourer £10.71 Rates are national average prices. Refer to REGIONAL VARIATIONS for indicative levels of overall pricing in regions	MATERIALS			LABOUR				RATES		
	Del to Site £	Waste %	Material Cost £	Craft Optve Hrs	Lab Hrs	Labour Cost £	Sunds £	Nett Rate £	Unit	Gross Rate (+12.5%) £
K10: PLASTERBOARD DRY LINING/PARTITIONS/CEILINGS										
Linings; tapered edge sheeting of one layer 9.5mm thick Gyproc wall board for direct decoration, BS 1230, butt joints; fixing with galvanized nails to timber base; butt joints filled with joint filler tape and joint finish, spot filling										
Walls										
height 2100 - 2400mm	5.28	5.00	**5.55**	0.96	0.48	**18.78**	-	**24.33**	m	**27.37**
height 2400 - 2700mm	5.94	5.00	**6.24**	1.08	0.54	**21.13**	-	**27.37**	m	**30.79**
height 2700 - 3000mm	6.52	5.00	**6.85**	1.20	0.60	**23.48**	-	**30.32**	m	**34.11**
Beams; 3nr faces										
total girth 600 - 1200mm	2.68	10.00	**2.92**	1.20	0.24	**19.62**	-	**22.54**	m	**25.36**
Columns; 4nr faces										
total girth 600 - 1200mm	2.74	10.00	**2.99**	1.20	0.24	**19.62**	-	**22.61**	m	**25.44**
total girth 1200 - 1800mm	4.06	10.00	**4.42**	1.80	0.36	**29.43**	-	**33.85**	m	**38.09**
Reveals and soffits of openings and recesses										
width not exceeding 300mm	0.72	10.00	**0.78**	0.24	0.06	**4.05**	-	**4.83**	m	**5.44**
Ceilings										
generally	2.21	5.00	**2.33**	0.48	0.20	**8.96**	-	**11.29**	m2	**12.70**
Linings; tapered edge sheeting of one layer 12.5mm thick Gyproc wallboard for direct decoration, BS 1230 butt joints; fixing with galvanized nails to timber base; butt joints filled with joint filler tape and joint finish, spot filling										
Walls										
height 2100 - 2400mm	5.55	5.00	**5.82**	1.13	0.56	**22.05**	-	**27.88**	m	**31.36**
height 2400 - 2700mm	6.24	5.00	**6.55**	1.27	0.63	**24.79**	-	**31.35**	m	**35.26**
height 2700 - 3000mm	6.93	5.00	**7.28**	1.41	0.71	**27.64**	-	**34.92**	m	**39.28**
Beams; 3nr faces										
total girth 600 - 1200mm	2.81	10.00	**3.07**	1.42	0.28	**23.18**	-	**26.24**	m	**29.53**
Columns; 4nr faces										
total girth 600 - 1200mm	2.87	10.00	**3.13**	1.42	0.28	**23.18**	-	**26.31**	m	**29.60**
total girth 1200 - 1800mm	4.25	10.00	**4.64**	2.12	0.42	**34.62**	-	**39.26**	m	**44.17**
Reveals and soffits of openings and recesses										
width not exceeding 300mm	0.75	10.00	**0.82**	0.28	0.07	**4.73**	-	**5.55**	m	**6.24**
Ceilings										
generally	2.32	5.00	**2.44**	0.56	0.24	**10.53**	-	**12.97**	m2	**14.59**
Linings; tapered edge sheeting of one layer 15mm thick Gyproc wall board for direct decoration, BS 1230, butt joints; fixing with galvanized nails to timber base; butt joints filled with joint filler tape and joint finish, spot filling										
Walls										
height 2100 - 2400mm	8.33	5.00	**8.75**	1.30	0.65	**25.43**	-	**34.18**	m	**38.45**
height 2400 - 2700mm	9.37	5.00	**9.84**	1.46	0.73	**28.56**	-	**38.40**	m	**43.21**
height 2700 - 3000mm	10.33	5.00	**10.84**	1.62	0.81	**31.70**	-	**42.54**	m	**47.86**
Beams; 3nr faces										
total girth 600 - 1200mm	4.20	10.00	**4.60**	1.64	0.33	**26.84**	-	**31.44**	m	**35.37**
Columns; 4nr faces										
total girth 600 - 1200mm	4.27	10.00	**4.66**	1.64	0.33	**26.84**	-	**31.50**	m	**35.44**
total girth 1200 - 1800mm	6.34	10.00	**6.93**	2.46	0.49	**40.20**	-	**47.14**	m	**53.03**
Reveals and soffits of openings and recesses										
width not exceeding 300mm	1.10	10.00	**1.20**	0.32	0.08	**5.40**	-	**6.60**	m	**7.43**

Labour hourly rates: (except Specialists) Craft Operatives £14.21 Labourer £10.71 Rates are national average prices. Refer to REGIONAL VARIATIONS for indicative levels of overall pricing in regions	MATERIALS			LABOUR				RATES		
	Del to Site £	Waste %	Material Cost £	Craft Optve Hrs	Lab Hrs	Labour Cost £	Sunds £	Nett Rate £	Unit	Gross Rate (+12.5%) £
K10: PLASTERBOARD DRY LINING/PARTITIONS/CEILINGS Cont.										
Linings; tapered edge sheeting of one layer 15mm thick Gyproc wall board for direct decoration, BS 1230, butt joints; fixing with galvanized nails to timber base; butt joints filled with joint filler tape and joint finish, spot filling Cont.										
Ceilings										
generally	3.48	5.00	**3.66**	0.64	0.28	**12.09**	-	**15.75**	m2	**17.72**
Linings; tapered edge sheeting of one layer 9.5mm thick Gyproc Duplex wallboard for direct decoration, BS 1230 butt joints; fixing with galvanized nails to timber base; butt joints filled with joint filler tape and joint finish, spot filling										
Walls										
height 2100 - 2400mm	6.75	5.00	**7.08**	0.96	0.48	**18.78**	-	**25.87**	m	**29.10**
height 2400 - 2700mm	7.59	5.00	**7.97**	1.08	0.54	**21.13**	-	**29.10**	m	**32.74**
height 2700 - 3000mm	8.35	5.00	**8.77**	1.20	0.60	**23.48**	-	**32.24**	m	**36.27**
Beams; 3nr faces										
total girth 600 - 1200mm	3.41	10.00	**3.73**	1.20	0.24	**19.62**	-	**23.35**	m	**26.27**
Columns; 4nr faces										
total girth 600 - 1200mm	3.47	10.00	**3.79**	1.20	0.24	**19.62**	-	**23.42**	m	**26.34**
total girth 1200 - 1800mm	5.15	10.00	**5.63**	1.80	0.36	**29.43**	-	**35.06**	m	**39.44**
Reveals and soffits of openings and recesses										
width not exceeding 300mm	0.90	10.00	**0.98**	0.24	0.06	**4.05**	-	**5.04**	m	**5.67**
Ceilings										
generally	2.82	5.00	**2.97**	0.48	0.20	**8.96**	-	**11.93**	m2	**13.42**
Linings; tapered edge sheeting of one layer 12.5mm thick Gyproc Duplex wallboard for direct decoration, BS 1230, butt joints; fixing with galvanized nails to timber base; butt joints filled with joint filler tape and joint finish, spot filling										
Walls										
height 2100 - 2400mm	7.13	5.00	**7.49**	1.13	0.56	**22.05**	-	**29.54**	m	**33.23**
height 2400 - 2700mm	8.02	5.00	**8.42**	1.27	0.63	**24.79**	-	**33.22**	m	**37.37**
height 2700 - 3000mm	8.91	5.00	**9.36**	1.41	0.71	**27.64**	-	**37.00**	m	**41.62**
Beams; 3nr faces										
total girth 600 - 1200mm	3.60	10.00	**3.94**	1.42	0.28	**23.18**	-	**27.12**	m	**30.51**
Columns; 4nr faces										
total girth 600 - 1200mm	3.67	10.00	**4.00**	1.42	0.28	**23.18**	-	**27.18**	m	**30.58**
total girth 1200 - 1800mm	5.44	10.00	**5.94**	2.12	0.42	**34.62**	-	**40.57**	m	**45.64**
Reveals and soffits of openings and recesses										
width not exceeding 300mm	0.95	10.00	**1.04**	0.28	0.07	**4.73**	-	**5.76**	m	**6.48**
Ceilings										
generally	2.98	5.00	**3.13**	0.56	0.24	**10.53**	-	**13.66**	m2	**15.37**
Linings; tapered edge sheeting of one layer 15mm thick Gyproc Duplex wallboard for direct decoration, BS 1230, butt joints; fixing with galvanized nails to timber base; butt joints filled with joint filler tape and joint finish, spot filling										
Walls										
height 2100 - 2400mm	10.85	5.00	**11.39**	1.30	0.65	**25.43**	-	**36.83**	m	**41.43**
height 2400 - 2700mm	12.21	5.00	**12.82**	1.46	0.73	**28.56**	-	**41.38**	m	**46.55**
height 2700 - 3000mm	13.44	5.00	**14.11**	1.62	0.81	**31.70**	-	**45.81**	m	**51.53**
Beams; 3nr faces										
total girth 600 - 1200mm	5.46	10.00	**5.98**	1.64	0.33	**26.84**	-	**32.82**	m	**36.93**
Columns; 4nr faces										
total girth 600 - 1200mm	5.53	10.00	**6.05**	1.64	0.33	**26.84**	-	**32.89**	m	**37.00**
total girth 1200 - 1800mm	8.23	10.00	**9.01**	2.46	0.49	**40.20**	-	**49.22**	m	**55.37**
Reveals and soffits of openings and recesses										
width not exceeding 300mm	1.41	10.00	**1.55**	0.32	0.08	**5.40**	-	**6.95**	m	**7.82**
Ceilings										
generally	4.53	5.00	**4.76**	0.64	0.28	**12.09**	-	**16.85**	m2	**18.96**
Linings; tapered edge sheeting of one layer 19mm thick Gyproc plank for direct decoration, BS 1230, butt joints; fixing with galvanized nails to timber base; butt joints filled with joint filler tape and joint finish, spot filling										
Walls										
height 2100 - 2400mm	11.76	5.00	**12.35**	1.44	0.72	**28.17**	-	**40.52**	m	**45.59**
height 2400 - 2700mm	13.23	5.00	**13.89**	1.62	0.81	**31.70**	-	**45.58**	m	**51.28**
height 2700 - 3000mm	14.62	5.00	**15.35**	1.80	0.90	**35.22**	-	**50.57**	m	**56.89**
Beams; 3nr faces										
total girth 600 - 1200mm	5.89	10.00	**6.44**	1.80	0.36	**29.43**	-	**35.88**	m	**40.36**

Labour hourly rates: (except Specialists)
Craft Operatives £14.21 Labourer £10.71
Rates are national average prices.
Refer to REGIONAL VARIATIONS for indicative levels of overall pricing in regions

	MATERIALS			LABOUR			RATES			
	Del to Site £	Waste %	Material Cost £	Craft Optve Hrs	Lab Hrs	Labour Cost £	Sunds £	Nett Rate £	Unit	Gross Rate (+12.5%) £
K10: PLASTERBOARD DRY LINING/PARTITIONS/CEILINGS Cont.										
Linings; tapered edge sheeting of one layer 19mm thick Gyproc plank for direct decoration, BS 1230, butt joints; fixing with galvanized nails to timber base; butt joints filled with joint filler tape and joint finish, spot filling Cont.										
Columns; 4nr faces										
total girth 600 - 1200mm	5.95	10.00	**6.51**	1.80	0.36	**29.43**	-	**35.94**	m	**40.43**
total girth 1200 - 1800mm	8.86	10.00	**9.70**	2.70	0.54	**44.15**	-	**53.85**	m	**60.58**
Reveals and soffits of openings and recesses										
width not exceeding 300mm	1.52	10.00	**1.66**	0.36	0.09	**6.08**	-	**7.74**	m	**8.71**
Ceilings										
generally	4.91	5.00	**5.16**	0.72	0.30	**13.44**	-	**18.60**	m2	**20.93**
Linings; two layers of Gypsum wallboard, first layer square edge sheeting 9.5mm thick, second layer tapered edge sheeting 9.5mm thick for direct decoration, BS 1230, butt joints; fixing with galvanized nails to timber base; butt joints of second layer filled with joint filler tape and joint finish, spot filling										
Walls										
height 2100 - 2400mm	10.04	5.00	**10.54**	1.68	0.84	**32.87**	-	**43.41**	m	**48.84**
height 2400 - 2700mm	11.30	5.00	**11.86**	1.89	0.95	**37.03**	-	**48.89**	m	**55.00**
height 2700 - 3000mm	12.47	5.00	**13.09**	2.10	1.05	**41.09**	-	**54.18**	m	**60.95**
Beams; 3nr faces										
total girth 600 - 1200mm	2.91	10.00	**3.17**	2.10	0.42	**34.34**	-	**37.50**	m	**42.19**
Columns; 4nr faces										
total girth 600 - 1200mm	5.12	10.00	**5.59**	2.10	0.42	**34.34**	-	**39.93**	m	**44.92**
total girth 1200 - 1800mm	7.63	10.00	**8.33**	3.15	0.63	**51.51**	-	**59.84**	m	**67.32**
Reveals and soffits of openings and recesses										
width not exceeding 300mm	1.31	10.00	**1.43**	0.42	0.11	**7.15**	-	**8.58**	m	**9.65**
Ceilings										
generally	4.20	5.00	**4.41**	0.84	0.35	**15.68**	-	**20.09**	m2	**22.60**
Linings; two layers of Gypsum wallboard, first layer square edge sheeting 9.5mm thick, second layer tapered edge sheeting 12.5mm thick for direct decoration, BS 1230, butt joints; fixing with galvanized nails to timber base; butt joints of second layer filled with joint filler tape and joint finish, spot filling										
Walls										
height 2100 - 2400mm	10.30	5.00	**10.82**	1.92	0.96	**37.56**	-	**48.38**	m	**54.43**
height 2400 - 2700mm	11.59	5.00	**12.17**	2.16	1.08	**42.26**	-	**54.43**	m	**61.24**
height 2700 - 3000mm	12.80	5.00	**13.44**	2.40	1.20	**46.96**	-	**60.39**	m	**67.94**
Beams; 3nr faces										
total girth 600 - 1200mm	5.19	10.00	**5.57**	2.40	0.48	**39.24**	-	**44.81**	m	**50.41**
Columns; 4nr faces										
total girth 600 - 1200mm	5.25	10.00	**5.63**	2.40	0.48	**39.24**	-	**44.88**	m	**50.49**
total girth 1200 - 1800mm	7.82	10.00	**8.39**	3.60	0.72	**58.87**	-	**67.25**	m	**75.66**
Reveals and soffits of openings and recesses										
width not exceeding 300mm	1.35	10.00	**1.44**	0.48	0.12	**8.11**	-	**9.55**	m	**10.74**
Ceilings										
generally	4.31	5.00	**4.52**	0.96	0.40	**17.93**	-	**22.45**	m2	**25.25**
Linings; two layers of Gypsum wallboard, first layer square edge sheeting 12.5mm thick, second layer tapered edge sheeting 12.5mm thick for direct decoration, BS 1230, butt joints; fixing with galvanized nails to timber base; butt joints of second layer filled with joint filler tape and joint finish, spot filling										
Walls										
height 2100 - 2400mm	10.57	5.00	**11.10**	2.04	1.02	**39.91**	-	**51.01**	m	**57.39**
height 2400 - 2700mm	11.89	5.00	**12.48**	2.30	1.15	**45.00**	-	**57.48**	m	**64.67**
height 2700 - 3000mm	13.13	5.00	**13.78**	2.55	1.28	**49.94**	-	**63.73**	m	**71.69**
Beams; 3nr faces										
total girth 600 - 1200mm	5.32	10.00	**5.82**	2.56	0.51	**41.84**	-	**47.66**	m	**53.62**
Columns; 4nr faces										
total girth 600 - 1200mm	5.39	10.00	**5.88**	2.56	0.51	**41.84**	-	**47.72**	m	**53.69**
total girth 1200 - 1800mm	8.02	10.00	**8.76**	3.83	0.77	**62.67**	-	**71.44**	m	**80.37**
Reveals and soffits of openings and recesses										
width not exceeding 300mm	1.38	10.00	**1.51**	0.51	0.13	**8.64**	-	**10.14**	m	**11.41**
Ceilings										
generally	4.42	5.00	**4.64**	1.02	0.43	**19.10**	-	**23.74**	m2	**26.70**

Labour hourly rates: (except Specialists) Craft Operatives £14.21 Labourer £10.71 Rates are national average prices. Refer to REGIONAL VARIATIONS for indicative levels of overall pricing in regions	MATERIALS			LABOUR				RATES		
	Del to Site £	Waste %	Material Cost £	Craft Optve Hrs	Lab Hrs	Labour Cost £	Sunds £	Nett Rate £	Unit	Gross Rate (+12.5%) £
K10: PLASTERBOARD DRY LINING/PARTITIONS/CEILINGS Cont.										
Linings; two layers of Gypsum wallboard, first layer square edge sheeting 15mm thick, second layer tapered edge sheeting 15mm thick for direct decoration, BS 1230, butt joints; fixing with galvanized nails to timber base; butt joints of second layer filled with joint filler tape and joint finish, spot filling										
Walls										
height 2100 - 2400mm	15.31	5.00	**16.07**	2.16	1.08	**42.26**	-	**58.33**	m	**65.62**
height 2400 - 2700mm	18.15	5.00	**19.06**	2.43	1.22	**47.60**	-	**66.66**	m	**74.99**
height 2700 - 3000mm	20.09	5.00	**21.09**	2.70	1.35	**52.83**	-	**73.92**	m	**83.16**
Beams; 3nr faces										
total girth 600 - 1200mm	8.11	10.00	**8.88**	2.70	0.54	**44.15**	-	**53.03**	m	**59.66**
Columns; 4nr faces										
total girth 600 - 1200mm	8.17	10.00	**8.95**	2.70	0.54	**44.15**	-	**53.10**	m	**59.73**
total girth 1200 - 1800mm	12.20	10.00	**13.36**	4.05	0.81	**66.23**	-	**79.58**	m	**89.53**
Reveals and soffits of openings and recesses										
width not exceeding 300mm	2.08	10.00	**2.27**	0.54	0.14	**9.17**	-	**11.44**	m	**12.87**
Ceilings										
generally	6.74	5.00	**7.07**	1.08	0.46	**20.27**	-	**27.35**	m2	**30.77**
Linings; two layers of Gypsum wallboard, first layer Duplex square edge sheeting 9.5mm thick, second layer tapered edge sheeting 9.5mm thick for direct decoration, BS 1230, butt joints; fixing with galvanized nails to timber base; butt joints of second layer filled with joint filler tape and joint finish, spot filling										
Walls										
height 2100 - 2400mm	11.50	5.00	**12.08**	1.68	0.84	**32.87**	-	**44.95**	m	**50.57**
height 2400 - 2700mm	12.94	5.00	**13.59**	1.89	0.95	**37.03**	-	**50.62**	m	**56.95**
height 2700 - 3000mm	14.30	5.00	**15.01**	2.10	1.05	**41.09**	-	**56.10**	m	**63.11**
Beams; 3nr faces										
total girth 600 - 1200mm	5.79	10.00	**6.19**	2.10	0.42	**34.34**	-	**40.53**	m	**45.59**
Columns; 4nr faces										
total girth 600 - 1200mm	5.85	10.00	**6.25**	2.10	0.42	**34.34**	-	**40.59**	m	**45.67**
total girth 1200 - 1800mm	8.72	10.00	**9.32**	3.15	0.63	**51.51**	-	**60.83**	m	**68.43**
Reveals and soffits of openings and recesses										
width not exceeding 300mm	1.50	10.00	**1.60**	0.42	0.11	**7.15**	-	**8.74**	m	**9.84**
Ceilings										
generally	6.60	5.00	**6.93**	0.84	0.35	**15.68**	-	**22.61**	m2	**25.44**
Linings; two layers of Gypsum wallboard, first layer Duplex square edge sheeting 9.5mm thick, second layer tapered edge sheeting 12.5mm thick for direct decoration, BS 1230, butt joints; fixing with galvanized nails to timber base; butt joints of second layer filled with joint filler tape and joint finish, spot filling										
Walls										
height 2100 - 2400mm	11.77	5.00	**12.36**	1.92	0.96	**37.56**	-	**49.92**	m	**56.16**
height 2400 - 2700mm	13.24	5.00	**13.90**	2.16	1.08	**42.26**	-	**56.16**	m	**63.18**
height 2700 - 3000mm	14.63	5.00	**15.36**	2.40	1.20	**46.96**	-	**62.31**	m	**70.10**
Beams; 3nr faces										
total girth 600 - 1200mm	5.92	10.00	**6.33**	2.40	0.48	**39.24**	-	**45.58**	m	**51.28**
Columns; 4nr faces										
total girth 600 - 1200mm	5.99	10.00	**6.40**	2.40	0.48	**39.24**	-	**45.64**	m	**51.35**
total girth 1200 - 1800mm	8.92	10.00	**9.54**	3.60	0.72	**58.87**	-	**68.41**	m	**76.96**
Reveals and soffits of openings and recesses										
width not exceeding 300mm	1.53	10.00	**1.63**	0.48	0.12	**8.11**	-	**9.74**	m	**10.96**
Ceilings										
generally	4.92	5.00	**5.16**	0.96	0.40	**17.93**	-	**23.09**	m2	**25.97**
Linings; two layers of Gypsum wallboard, first layer Duplex square edge sheeting 12.5mm thick, second layer tapered edge sheeting 12.5mm thick for direct decoration, BS 1230, butt joints; fixing with galvanized nails to timber base; butt joints of second layer filled with joint filler tape and joint finish, spot filling										
Walls										
height 2100 - 2400mm	12.15	5.00	**12.76**	2.04	1.02	**39.91**	-	**52.67**	m	**59.26**
height 2400 - 2700mm	13.67	5.00	**14.36**	2.30	1.15	**45.00**	-	**59.36**	m	**66.77**
height 2700 - 3000mm	15.11	5.00	**15.86**	2.55	1.28	**49.94**	-	**65.81**	m	**74.03**
Beams; 3nr faces										
total girth 600 - 1200mm	6.12	10.00	**6.54**	2.56	0.51	**41.84**	-	**48.38**	m	**54.42**

Labour hourly rates: (except Specialists) Craft Operatives £14.21 Labourer £10.71 Rates are national average prices. Refer to REGIONAL VARIATIONS for indicative levels of overall pricing in regions	MATERIALS			LABOUR				RATES		
	Del to Site £	Waste %	Material Cost £	Craft Optve Hrs	Lab Hrs	Labour Cost £	Sunds £	Nett Rate £	Unit	Gross Rate (+12.5%) £
K10: PLASTERBOARD DRY LINING/PARTITIONS/CEILINGS Cont.										
Linings; two layers of Gypsum wallboard, first layer Duplex square edge sheeting 12.5mm thick, second layer tapered edge sheeting 12.5mm thick for direct decoration, BS 1230, butt joints; fixing with galvanized nails to timber base; butt joints of second layer filled with joint filler tape and joint finish, spot filling Cont.										
Columns; 4nr faces										
total girth 600 - 1200mm	6.18	10.00	**6.60**	2.56	0.51	**41.84**	-	**48.44**	m	**54.50**
total girth 1200 - 1800mm	9.21	10.00	**9.84**	3.83	0.77	**62.67**	-	**72.51**	m	**81.58**
Reveals and soffits of openings and recesses										
width not exceeding 300mm	1.58	10.00	**1.68**	0.51	0.13	**8.64**	-	**10.32**	m	**11.61**
Ceilings										
generally	5.08	5.00	**5.33**	1.02	0.43	**19.10**	-	**24.43**	m2	**27.48**
Linings; two layers of Gypsum wallboard, first layer Duplex square edge sheeting 15mm thick, second layer tapered edge sheeting 15mm thick for direct decoration, BS 1230, butt joints; fixing with galvanized nails to timber base; butt joints of second layer filled with joint filler tape and joint finish, spot filling										
Walls										
height 2100 - 2400mm	17.83	5.00	**18.72**	2.16	1.08	**42.26**	-	**60.98**	m	**68.60**
height 2400 - 2700mm	20.99	5.00	**22.04**	2.43	1.22	**47.60**	-	**69.63**	m	**78.34**
height 2700 - 3000mm	23.24	5.00	**24.40**	2.70	1.35	**52.83**	-	**77.22**	m	**86.88**
Beams; 3nr faces										
total girth 600 - 1200mm	9.37	10.00	**10.02**	2.70	0.54	**44.15**	-	**54.17**	m	**60.94**
Columns; 4nr faces										
total girth 600 - 1200mm	9.43	10.00	**10.09**	2.70	0.54	**44.15**	-	**54.24**	m	**61.02**
total girth 1200 - 1800mm	14.09	10.00	**15.07**	4.05	0.81	**66.23**	-	**81.29**	m	**91.45**
Reveals and soffits of openings and recesses										
width not exceeding 300mm	2.39	10.00	**2.56**	0.54	0.14	**9.17**	-	**11.73**	m	**13.19**
Ceilings										
generally	7.79	5.00	**8.18**	1.08	0.46	**20.27**	-	**28.45**	m2	**32.01**
K10: PLASTERBOARD DRY LINING/PARTITIONS/CEILINGS: PLASTIC FACED PLASTERBOARD										
Linings; square edge sheeting of one layer 12.5mm thick Gyproc Industrial Grade plastic faced plasterboard, BS 1230, butt joints; fixing with galvanized nails to timber base; butt joints sealed with self adhesive white tape										
Walls										
height 2100 - 2400mm	10.10	5.00	**10.65**	0.84	0.42	**16.43**	-	**27.09**	m	**30.47**
height 2400 - 2700mm	11.37	5.00	**11.98**	0.95	0.47	**18.53**	-	**30.52**	m	**34.33**
height 2700 - 3000mm	12.63	5.00	**13.31**	1.05	0.53	**20.60**	-	**33.91**	m	**38.15**
Beams; 3nr faces										
total girth 600 - 1200mm	5.09	10.00	**5.59**	1.06	0.21	**17.31**	-	**22.90**	m	**25.77**
Columns; 4nr faces										
total girth 600 - 1200mm	5.49	10.00	**6.03**	1.06	0.21	**17.31**	-	**23.34**	m	**26.26**
total girth 1200 - 1800mm	7.84	10.00	**8.61**	1.58	0.32	**25.88**	-	**34.49**	m	**38.80**
Reveals and soffits of openings and recesses										
width not exceeding 300mm	1.79	10.00	**1.97**	0.21	0.05	**3.52**	-	**5.49**	m	**6.17**
Ceilings										
generally	4.28	5.00	**4.52**	0.42	0.18	**7.90**	-	**12.41**	m2	**13.97**
3.50 - 5.00m above floor	4.28	5.00	**4.52**	0.46	0.18	**8.46**	-	**12.98**	m2	**14.61**
K10: PLASTERBOARD DRY LINING/PARTITIONS/CEILING - THERMAL BOARD										
Linings; tapered edged sheeting of one layer 22mm thick Gyproc Thermaline BASIC board vapour check grade for direct decoration, butt joints; fixing with galvanized nails to timber base; butt joints filled with joint filler tape and joint finish, spot filling										
Walls										
height 2100 - 2400mm	18.05	5.00	**18.95**	1.72	0.86	**33.65**	-	**52.60**	m	**59.18**
height 2400 - 2700mm	20.31	5.00	**21.32**	1.94	0.97	**37.96**	-	**59.28**	m	**66.69**
height 2700 - 3000mm	22.48	5.00	**23.60**	2.15	1.08	**42.12**	-	**65.72**	m	**73.94**
Reveals and soffits of openings and recesses										
width not exceeding 300mm	2.31	10.00	**2.53**	0.43	0.11	**7.29**	-	**9.82**	m	**11.05**
Ceilings										
generally	7.53	5.00	**7.91**	0.86	0.38	**16.29**	-	**24.20**	m2	**27.23**

Labour hourly rates: (except Specialists) Craft Operatives £14.21 Labourer £10.71 Rates are national average prices. Refer to REGIONAL VARIATIONS for indicative levels of overall pricing in regions	MATERIALS			LABOUR				RATES		
	Del to Site £	Waste %	Material Cost £	Craft Optve Hrs	Lab Hrs	Labour Cost £	Sunds £	Nett Rate £	Unit	Gross Rate (+12.5%) £
K10: PLASTERBOARD DRY LINING/PARTITIONS/CEILING - THERMAL BOARD Cont.										
Linings; tapered edged sheeting of one layer 30mm thick Gyproc thermaline BASIC board vapour check grade for direct decoration, butt joints; fixing with galvanized nails to timber base; butt joints filled with joint filler tape and joint finish, spot filling										
Walls										
height 2100 - 2400mm	18.51	5.00	**19.43**	2.02	1.01	**39.52**	-	**58.95**	m	**66.32**
height 2400 - 2700mm	20.82	5.00	**21.86**	2.27	1.14	**44.47**	-	**66.33**	m	**74.62**
height 2700 - 3000mm	23.05	5.00	**24.20**	2.52	1.26	**49.30**	-	**73.50**	m	**82.69**
Reveals and soffits of openings and recesses										
width not exceeding 300mm	2.37	10.00	**2.60**	0.51	0.13	**8.64**	-	**11.24**	m	**12.64**
Ceilings										
generally	7.72	5.00	**8.11**	1.01	0.43	**18.96**	-	**27.07**	m2	**30.45**
Linings; tapered edged sheeting of one layer 40mm thick Gyproc Thermaline BASIC board vapour check grade for direct decoration, butt joints; fixing with galvanized nails to timber base; butt joints filled with joint filler tape and joint finish, spot filling										
Walls										
height 2100 - 2400mm	20.47	5.00	**21.50**	2.30	1.15	**45.00**	-	**66.50**	m	**74.81**
height 2400 - 2700mm	23.03	5.00	**24.18**	2.59	1.30	**50.73**	-	**74.91**	m	**84.28**
height 2700 - 3000mm	25.51	5.00	**26.78**	2.87	1.44	**56.21**	-	**82.99**	m	**93.36**
Reveals and soffits of openings and recesses										
width not exceeding 300mm	2.62	10.00	**2.87**	0.58	0.14	**9.74**	-	**12.61**	m	**14.19**
Ceilings										
generally	8.54	5.00	**8.97**	1.15	0.49	**21.59**	-	**30.56**	m2	**34.38**
Linings; tapered edged sheeting of one layer 50mm thick Gyproc Thermaline BASIC board vapour check grade for direct decoration, butt joints; fixing with galvanized nails to timber base; butt joints filled with joint filler tape and joint finish, spot filling										
Walls										
height 2100 - 2400mm	25.63	5.00	**26.92**	2.59	1.30	**50.73**	-	**77.64**	m	**87.35**
height 2400 - 2700mm	28.84	5.00	**30.28**	2.91	1.46	**56.99**	-	**87.27**	m	**98.18**
height 2700 - 3000mm	31.96	5.00	**33.56**	3.24	1.62	**63.39**	-	**96.95**	m	**109.07**
Reveals and soffits of openings and recesses										
width not exceeding 300mm	3.26	10.00	**3.58**	0.65	0.16	**10.95**	-	**14.53**	m	**16.34**
Ceilings										
generally	10.69	5.00	**11.23**	1.30	0.56	**24.47**	-	**35.70**	m2	**40.16**
K10: PLASTERBOARD DRY LINING/PARTITIONS/CEILING - CELLULAR PARTITIONS										
Proprietary partitions; Panelwall PDP single leaf, tapered edge panels for direct decoration, butt joints; fixing with galvanized nails to 19 x 37mm (f sizes) impregnated wrot softwood sole plate with 19 x 37 x 300mm (f sizes) locating blocks at 900mm centres and 19x 37mm (f sizes) impregnated wrought softwood wall and ceiling battens; applying continuous beads of Gyproc acoustical sealant to perimeters of construction both sides; butt joints filled with joint filler tape and joint finish, spot filling; fixing to masonry with cartridge fired nails										
Height 2100 - 2400mm, 57mm thick										
boarded both sides	36.32	5.00	**38.14**	2.85	1.43	**55.81**	-	**93.95**	m	**105.70**
Height 2400 - 2700mm, 57mm thick										
boarded both sides	40.13	5.00	**42.14**	3.95	1.98	**77.34**	-	**119.48**	m	**134.41**
Angles to partitions; 37 x 19mm softwood batten, 37 x 37mm softwood batten, fixing together through partitions; one panel rebated by cutting back one face by full panel thickness; joints filled with joint filler tape and joint finish										
Plain										
57mm thick partitions	0.73	5.00	**0.77**	0.30	0.15	**5.87**	-	**6.64**	m	**7.47**
Tee junctions to partitions; 37 x 19mm softwood batten, 37 x 37mm softwood batten, fixing together through partitions; joints filled with joint filler tape and joint finish										
Plain										
57mm thick partitions	0.73	5.00	**0.77**	0.20	0.10	**3.91**	-	**4.68**	m	**5.26**

Labour hourly rates: (except Specialists) Craft Operatives £14.21 Labourer £10.71 Rates are national average prices. Refer to REGIONAL VARIATIONS for indicative levels of overall pricing in regions	MATERIALS			LABOUR				RATES		
	Del to Site £	Waste %	Material Cost £	Craft Optve Hrs	Lab Hrs	Labour Cost £	Sunds £	Nett Rate £	Unit	Gross Rate (+12.5%) £
K10: PLASTERBOARD DRY LINING/PARTITIONS/CEILING - CELLULAR PARTITIONS Cont.										
Fair ends to partitions										
57 mm thick partitions										
facing with 9.5mm Gyproc wallboard; fixing to 37 x 37mm softwood batten with galvanized nails.......	0.51	10.00	**0.54**	0.15	0.08	**2.99**	-	**3.53**	m	**3.97**
Fixings for heavy fittings										
Sinks										
insert 37 x 37mm softwood fixing block into core prior to erection of panels; not exceeding 150mm long	0.06	5.00	**0.06**	0.13	-	**1.85**	-	**1.91**	nr	**2.15**
insert 37 x 37mm softwood fixing block into core prior to erection panels; 150 - 300mm long.......	0.12	5.00	**0.13**	0.13	-	**1.85**	-	**1.98**	nr	**2.22**
Radiators										
insert 37 x 37mm softwood fixing block into core prior to erection of panels; not exceeding 150mm long	0.06	5.00	**0.06**	0.13	-	**1.85**	-	**1.91**	nr	**2.15**
insert 37 x 37mm softwood fixing block into core prior to erection panels; 150 - 300mm long.......	0.12	5.00	**0.13**	0.13	-	**1.85**	-	**1.98**	nr	**2.22**
K10: PLASTERBOARD DRY LINING/PARTITIONS/CEILINGS: LAMINATED PARTITIONS										
Proprietary partitions; Gyproc SolidWall laminated partition; tapered edge sheeting of one layer 12.5mm thick Gyproc wall board each side for direct decoration, BS 1230, butt joints; applying continuous beads of Gyproc acoustical sealant to perimeters of construction on both sides, fixing with nails to timber frame comprising 25 x 38mm (f sizes) perimeter members with one layer 19mm thick Gyproc plank infill; butt joints filled with joint filler tape and joint finish, spot filling; fixing to masonry with cartridge fired nails										
Height 2100 - 2400mm, 50mm thick										
boarded both sides	27.57	5.00	**28.95**	3.60	1.80	**70.43**	-	**99.39**	m	**111.81**
Height 2400 - 2700mm, 50mm thick										
boarded both sides	30.35	5.00	**31.87**	3.95	1.98	**77.34**	-	**109.20**	m	**122.85**
Proprietary partitions; Gyproc SolidWall laminated partition; tapered edge sheeting of one layer 19mm thick Gyproc plank for direct decoration, B.S 1230, butt joints; applying continuous beads of Gyproc acoustical sealant to perimeters of construction on both sides, fixing with nails to timber frame comprising 25 x 38mm (f sizes) perimeter members with one layer 19mm thick Gyproc plank infill; butt joints filled with joint filler tape and joint finish, spot filling; fixing to masonry with cartridge fired nails										
Height 2100 - 2400mm, 65mm thick										
boarded both sides	39.57	5.00	**41.55**	3.80	1.90	**74.35**	-	**115.90**	m	**130.39**
Height 2400 - 2700mm, 65mm thick										
boarded both sides	43.85	5.00	**46.04**	4.20	2.10	**82.17**	-	**128.22**	m	**144.24**
Height 2700 - 3000mm, 65mm thick										
boarded both sides	48.13	5.00	**50.53**	4.60	2.30	**90.00**	-	**140.53**	m	**158.10**
Angles to partitions; 38 x 25mm softwood batten, fixing together through partitions; each panel rebated by cutting back core and inner layers; joints filled with joint filler tape and joint finish										
Plain										
50mm thick partitions	0.29	5.00	**0.30**	0.30	-	**4.26**	-	**4.57**	m	**5.14**
65mm thick partitions	0.29	5.00	**0.30**	0.30	-	**4.26**	-	**4.57**	m	**5.14**
Tee junctions to partitions; forming groove by cutting back one face layer; bedding end in groove in bonding compound										
Plain										
50mm thick partitions	-	-	-	0.25	-	**3.55**	-	**3.55**	m	**4.00**
65mm thick partitions	-	-	-	0.25	-	**3.55**	-	**3.55**	m	**4.00**

Labour hourly rates: (except Specialists) Craft Operatives £14.21 Labourer £10.71 Rates are national average prices. Refer to REGIONAL VARIATIONS for indicative levels of overall pricing in regions	MATERIALS			LABOUR				RATES		
	Del to Site £	Waste %	Material Cost £	Craft Optve Hrs	Lab Hrs	Labour Cost £	Sunds £	Nett Rate £	Unit	Gross Rate (+12.5%) £
K10: PLASTERBOARD DRY LINING/PARTITIONS/CEILINGS: METAL STUD PARTITIONS										
Proprietary partitions; Gyproc Gypwall metal stud; tapered edge sheeting of one layer 12.5mm thick Gyproc wallboard for direct decoration, BS 1230, butt joints; applying continuous beads of Gyproc acoustical sealant to perimeters of construction both sides, fixing with Pozidriv head screws to steel frame comprising 50mm head and floor channels, 48mm vertical studs at 600mm centres; butt joints filled with joint filler tape and joint finish, spot filling; fixing to masonry with cartridge fired nails										
Height 2100 - 2400mm, 75mm thick boarded both sides	24.83	5.00	**26.08**	2.85	1.43	**55.81**	-	**81.89**	m	**92.13**
Proprietary partitions; Gyproc Gypwall metal stud; tapered edge sheeting of two layers 12.5mm thick Gyproc wallboard for direct decoration, BS 1230, butt joints; applying continuous beads of Gyproc acoustical sealant to perimeters of construction both sides, fixing with Pozidriv head screws to steel frame comprising 50mm head and floor channels, 48mm vertical studs at 600mm centres; butt joints filled with joint filler tape and joint finish, spot filling; fixing to masonry with cartridge fired nails										
Height 2100 - 2400mm, 100mm thick boarded both sides	34.87	5.00	**36.61**	3.25	1.63	**63.64**	-	**100.25**	m	**112.79**
Height 2400 - 2700mm, 100mm thick boarded both sides	38.23	5.00	**40.14**	3.65	1.83	**71.47**	-	**111.60**	m	**125.55**
Height 2700 - 3000mm, 100mm thick boarded both sides	41.58	5.00	**43.66**	4.05	2.03	**79.29**	-	**122.95**	m	**138.32**
Angles to partitions; fixing end studs together through partition; extending sheeting to one side across end of partition; joints filled with joint filler tape and joint finish										
Plain										
75mm thick partitions	1.43	10.00	**1.51**	0.13	-	**1.85**	-	**3.36**	m	**3.78**
100mm thick partitions	1.48	10.00	**1.56**	0.13	-	**1.85**	-	**3.41**	m	**3.84**
K10: PLASTERBOARD DRY LINING/PARTITIONS/CEILINGS: STAIRCASE AREAS										
Linings; tapered edge sheeting of one layer 9.5mm thick Gyproc wall board for direct decoration, BS 1230, butt joints; fixing with galvanized nails to timber base; butt joints filled with joint filler tape and joint finish, spot filling										
Walls										
height 2100 - 2400mm	5.28	5.00	**5.55**	1.20	0.48	**22.19**	-	**27.74**	m	**31.21**
height 2400 - 2700mm	5.94	5.00	**6.24**	1.35	0.54	**24.97**	-	**31.21**	m	**35.11**
height 2700 - 3000mm	6.52	5.00	**6.85**	1.50	0.60	**27.74**	-	**34.59**	m	**38.91**
Reveals and soffits of openings and recesses width not exceeding 300mm	0.72	10.00	**0.78**	0.27	0.06	**4.48**	-	**5.26**	m	**5.92**
Ceilings generally	2.21	5.00	**2.33**	0.54	0.20	**9.82**	-	**12.14**	m2	**13.66**
Linings; tapered edge sheeting of one layer 12.5mm thick Gyproc wallboard for direct decoration, BS 1230 butt joints; fixing with galvanized nails to timber base; butt joints filled with joint filler tape and joint finish, spot filling										
Walls										
height 2100 - 2400mm	5.55	5.00	**5.82**	1.42	0.56	**26.18**	-	**32.00**	m	**36.00**
height 2400 - 2700mm	6.24	5.00	**6.55**	1.59	0.63	**29.34**	-	**35.89**	m	**40.38**
height 2700 - 3000mm	6.93	5.00	**7.28**	1.77	0.71	**32.76**	-	**40.03**	m	**45.04**
Reveals and soffits of openings and recesses width not exceeding 300mm	0.75	10.00	**0.82**	0.32	0.07	**5.30**	-	**6.11**	m	**6.88**
Ceilings generally	2.32	5.00	**2.44**	0.63	0.24	**11.52**	-	**13.96**	m2	**15.71**
Linings; tapered edge sheeting of one layer 15mm thick Gyproc wallboard for direct decoration, BS 1230 butt joints; fixing with galvanized nails to timber base; butt joints filled with joint filler tape and joint finish, spot filling										
Walls										
height 2100 - 2400mm	8.33	5.00	**8.75**	1.64	0.65	**30.27**	-	**39.01**	m	**43.89**
height 2400 - 2700mm	9.37	5.00	**9.84**	1.84	0.73	**33.96**	-	**43.80**	m	**49.28**
height 2700 - 3000mm	10.33	5.00	**10.84**	1.98	0.81	**36.81**	-	**47.66**	m	**53.61**
Reveals and soffits of openings and recesses width not exceeding 300mm	1.10	10.00	**1.20**	0.37	0.08	**6.11**	-	**7.31**	m	**8.23**

Labour hourly rates: (except Specialists) Craft Operatives £14.21 Labourer £10.71 Rates are national average prices. Refer to REGIONAL VARIATIONS for indicative levels of overall pricing in regions	MATERIALS			LABOUR				RATES		
	Del to Site £	Waste %	Material Cost £	Craft Optve Hrs	Lab Hrs	Labour Cost £	Sunds £	Nett Rate £	Unit	Gross Rate (+12.5%) £
K10: PLASTERBOARD DRY LINING/PARTITIONS/CEILINGS: STAIRCASE AREAS Cont.										
Linings; tapered edge sheeting of one layer 15mm thick Gyproc wallboard for direct decoration, BS 1230 butt joints; fixing with galvanized nails to timber base; butt joints filled with joint filler tape and joint finish, spot filling Cont.										
Ceilings										
generally	3.48	5.00	**3.66**	0.72	0.28	**13.23**	-	**16.89**	m2	**19.00**
Linings; tapered edge sheeting of one layer 9.5mm thick Gyproc Duplex wallboard for direct decoration, BS 1230 butt joints; fixing with galvanized nails to timber base; butt joints filled with joint filler tape and joint finish, spot filling										
Walls										
height 2100 - 2400mm	6.75	10.00	**7.37**	1.20	0.48	**22.19**	-	**29.56**	m	**33.26**
height 2400 - 2700mm	7.59	5.00	**7.97**	1.35	0.54	**24.97**	-	**32.94**	m	**37.05**
height 2700 - 3000mm	8.35	5.00	**8.77**	1.50	0.60	**27.74**	-	**36.51**	m	**41.07**
Reveals and soffits of openings and recesses										
width not exceeding 300mm	0.90	10.00	**0.98**	0.27	0.06	**4.48**	-	**5.46**	m	**6.14**
Ceilings										
generally	2.82	5.00	**2.97**	0.54	0.20	**9.82**	-	**12.78**	m2	**14.38**
Linings; tapered edge sheeting of one layer 12.5mm thick Gyproc Duplex wallboard for direct decoration, BS 1230, butt joints; fixing with galvanized nails to timber base; butt joints filled with joint filler tape and joint finish, spot filling										
Walls										
height 2100 - 2400mm	7.13	5.00	**7.49**	1.42	0.56	**26.18**	-	**33.66**	m	**37.87**
height 2400 - 2700mm	8.02	5.00	**8.42**	1.59	0.63	**29.34**	-	**37.76**	m	**42.48**
height 2700 - 3000mm	8.91	5.00	**9.36**	1.77	0.71	**32.76**	-	**42.11**	m	**47.38**
Reveals and soffits of openings and recesses										
width not exceeding 300mm	0.95	10.00	**1.04**	0.32	0.07	**5.30**	-	**6.33**	m	**7.12**
Ceilings										
generally	2.98	5.00	**3.13**	0.63	0.24	**11.52**	-	**14.66**	m2	**16.49**
Linings; tapered edge sheeting of one layer 15mm thick Gyproc Duplex wallboard for direct decoration, BS 1230 butt joints; fixing with galvanized nails to timber base; butt joints filled with joint filler tape and joint finish, spot filling										
Walls										
height 2100 - 2400mm	10.85	5.00	**11.39**	1.64	0.65	**30.27**	-	**41.66**	m	**46.87**
height 2400 - 2700mm	12.21	5.00	**12.82**	1.84	0.73	**33.96**	-	**46.78**	m	**52.63**
height 2700 - 3000mm	13.44	5.00	**14.11**	1.98	0.81	**36.81**	-	**50.92**	m	**57.29**
Reveals and soffits of openings and recesses										
width not exceeding 300mm	1.41	10.00	**1.55**	0.37	0.08	**6.11**	-	**7.66**	m	**8.62**
Ceilings										
generally	4.53	5.00	**4.76**	0.72	0.28	**13.23**	-	**17.99**	m2	**20.24**
Linings; tapered edge sheeting of one layer 19mm thick Gyproc plank for direct decoration, BS 1230, butt joints; fixing with galvanized nails to timber base; butt joints filled with joint filler tape and joint finish, spot filling										
Walls										
height 2100 - 2400mm	11.76	5.00	**12.35**	1.80	0.72	**33.29**	-	**45.64**	m	**51.34**
height 2400 - 2700mm	13.23	5.00	**13.89**	2.03	0.81	**37.52**	-	**51.41**	m	**57.84**
height 2700 - 3000mm	14.62	5.00	**15.35**	2.25	0.90	**41.61**	-	**56.97**	m	**64.09**
Reveals and soffits of openings and recesses										
width not exceeding 300mm	1.52	10.00	**1.66**	0.41	0.09	**6.79**	-	**8.45**	m	**9.51**
Ceilings										
generally	4.91	5.00	**5.16**	0.81	0.30	**14.72**	-	**19.88**	m2	**22.37**
Linings; two layers of Gypsum wallboard, first layer square edge sheeting 9.5mm thick, second layer tapered edge sheeting 9.5mm thick for direct decoration, BS 1230, butt joints; fixing with galvanized nails to timber base; butt joints of second layer filled with joint filler tape and joint finish, spot filling										
Walls										
height 2100 - 2400mm	10.04	5.00	**10.54**	2.11	0.84	**38.98**	-	**49.52**	m	**55.71**
height 2400 - 2700mm	11.30	5.00	**11.86**	2.38	0.95	**43.99**	-	**55.86**	m	**62.84**
height 2700 - 3000mm	12.47	5.00	**13.09**	2.64	1.05	**48.76**	-	**61.85**	m	**69.58**
Reveals and soffits of openings and recesses										
width not exceeding 300mm	1.31	10.00	**1.43**	0.47	0.11	**7.86**	-	**9.29**	m	**10.45**

Labour hourly rates: (except Specialists) Craft Operatives £14.21 Labourer £10.71 Rates are national average prices. Refer to REGIONAL VARIATIONS for indicative levels of overall pricing in regions	MATERIALS			LABOUR				RATES		
	Del to Site £	Waste %	Material Cost £	Craft Optve Hrs	Lab Hrs	Labour Cost £	Sunds £	Nett Rate £	Unit	Gross Rate (+12.5%) £
K10: PLASTERBOARD DRY LINING/PARTITIONS/CEILINGS: STAIRCASE AREAS Cont.										
Linings; two layers of Gypsum wallboard, first layer square edge sheeting 9.5mm thick, second layer tapered edge sheeting 9.5mm thick for direct decoration, BS 1230, butt joints; fixing with galvanized nails to timber base; butt joints of second layer filled with joint filler tape and joint finish, spot filling Cont.										
Ceilings										
generally	4.20	5.00	**4.41**	0.95	0.35	**17.25**	-	**21.66**	m2	**24.36**
Linings; two layers of Gypsum wallboard, first layer square edge sheeting 9.5mm thick, second layer tapered edge sheeting 12.5mm thick for direct decoration, BS 1230, butt joints; fixing with galvanized nails to timber base; butt joints of second layer filled with joint filler tape and joint finish, spot filling										
Walls										
height 2100 - 2400mm	10.30	5.00	**10.82**	2.40	0.96	**44.39**	-	**55.21**	m	**62.11**
height 2400 - 2700mm	11.59	5.00	**12.17**	2.70	1.08	**49.93**	-	**62.11**	m	**69.87**
height 2700 - 3000mm	12.80	5.00	**13.44**	3.00	1.20	**55.48**	-	**68.92**	m	**77.53**
Reveals and soffits of openings and recesses										
width not exceeding 300mm	1.35	10.00	**1.44**	0.48	0.12	**8.11**	-	**9.55**	m	**10.74**
Ceilings										
generally	4.31	5.00	**4.52**	1.08	0.40	**19.63**	-	**24.15**	m2	**27.17**
Linings; two layers of Gypsum wallboard, first layer square edge sheeting 12.5mm thick, second layer tapered edge sheeting 12.5mm thick for direct decoration, BS 1230, butt joints; fixing with galvanized nails to timber base; butt joints of second layer filled with joint filler tape and joint finish, spot filling										
Walls										
height 2100 - 2400mm	10.57	5.00	**11.10**	2.54	1.02	**47.02**	-	**58.11**	m	**65.38**
height 2400 - 2700mm	11.89	5.00	**12.48**	2.86	1.15	**52.96**	-	**65.44**	m	**73.62**
height 2700 - 3000mm	13.13	5.00	**13.78**	3.18	1.28	**58.90**	-	**72.68**	m	**81.76**
Reveals and soffits of openings and recesses										
width not exceeding 300mm	1.38	10.00	**1.51**	0.57	0.13	**9.49**	-	**11.00**	m	**12.37**
Ceilings										
generally	4.42	5.00	**4.64**	1.15	0.43	**20.95**	-	**25.58**	m2	**28.78**
Linings; two layers of Gypsum wallboard, first layer square edge sheeting 15mm thick, second layer tapered edge sheeting 15mm thick for direct decoration, BS 1230, butt joints; fixing with galvanized nails to timber base; butt joints of second layer filled with joint filler tape and joint finish, spot filling										
Walls										
height 2100 - 2400mm	15.31	5.00	**16.07**	2.68	1.08	**49.65**	-	**65.72**	m	**73.94**
height 2400 - 2700mm	18.15	5.00	**19.06**	3.01	1.22	**55.84**	-	**74.90**	m	**84.26**
height 2700 - 3000mm	20.09	5.00	**21.09**	3.35	1.35	**62.06**	-	**83.15**	m	**93.55**
Reveals and soffits of openings and recesses										
width not exceeding 300mm	2.08	10.00	**2.27**	0.66	0.14	**10.88**	-	**13.15**	m	**14.79**
Ceilings										
generally	6.74	5.00	**7.07**	1.22	0.46	**22.26**	-	**29.34**	m2	**33.00**
Linings; two layers of Gypsum wallboard, first layer Duplex square edge sheeting 9.5mm thick, second layer tapered edge sheeting 9.5mm thick for direct decoration, BS 1230, butt joints; finish with galvanized nails to timber base; butt joint on second layer filled with joint filler tape and joint finish, spot filling										
Walls										
height 2100 - 2400mm	11.50	5.00	**12.08**	2.11	0.84	**38.98**	-	**51.06**	m	**57.44**
height 2400 - 2700mm	12.94	5.00	**13.59**	2.38	0.95	**43.99**	-	**57.58**	m	**64.78**
height 2700 - 3000mm	14.30	5.00	**15.01**	2.64	1.05	**48.76**	-	**63.77**	m	**71.74**
Reveals and soffits of openings and recesses										
width not exceeding 300mm	1.50	10.00	**1.60**	0.47	0.11	**7.86**	-	**9.45**	m	**10.64**
Ceilings										
generally	4.81	5.00	**5.05**	0.95	0.35	**17.25**	-	**22.30**	m2	**25.08**

Labour hourly rates: (except Specialists) Craft Operatives £14.21 Labourer £10.71 Rates are national average prices. Refer to REGIONAL VARIATIONS for indicative levels of overall pricing in regions	MATERIALS			LABOUR				RATES		
	Del to Site £	Waste %	Material Cost £	Craft Optve Hrs	Lab Hrs	Labour Cost £	Sunds £	Nett Rate £	Unit	Gross Rate (+12.5%) £
K10: PLASTERBOARD DRY LINING/PARTITIONS/CEILINGS: STAIRCASE AREAS Cont.										
Linings; two layers of Gypsum wallboard, first layer Duplex square edge sheeting 9.5mm thick, second layer tapered edge sheeting 12.5mm thick for direct decoration, BS 1230, butt joints; fixing with galvanized nails to timber base; butt joints of second layer filled with joint filler tape and joint finish, spot filling										
Walls										
height 2100 - 2400mm	11.77	5.00	12.36	2.40	0.96	44.39	-	56.74	m	63.84
height 2400 - 2700mm	13.24	5.00	13.90	2.70	1.08	49.93	-	63.84	m	71.82
height 2700 - 3000mm	14.63	5.00	15.36	3.00	1.20	55.48	-	70.84	m	79.69
Reveals and soffits of openings and recesses										
width not exceeding 300mm	1.53	10.00	1.63	0.48	0.12	8.11	-	9.74	m	10.96
Ceilings										
generally	4.92	5.00	5.16	1.08	0.40	19.63	-	24.79	m2	27.89
Linings; two layers of Gypsum wallboard, first layer Duplex square edge sheeting 12.5mm thick, second layer tapered edge sheeting 12.5mm thick for direct decoration, BS 1230, butt joints; fixing with galvanized nails to timber base; butt joints of second layer filled with joint filler tape and joint finish, spot filling										
Walls										
height 2100 - 2400mm	12.15	5.00	12.76	2.54	1.02	47.02	-	59.78	m	67.25
height 2400 - 2700mm	13.67	5.00	14.36	2.86	1.15	52.96	-	67.31	m	75.73
height 2700 - 3000mm	15.11	5.00	15.86	3.18	1.28	58.90	-	74.76	m	84.10
Reveals and soffits of openings and recesses										
width not exceeding 300mm	1.58	10.00	1.68	0.57	0.13	9.49	-	11.18	m	12.57
Ceilings										
generally	5.08	5.00	5.33	1.15	0.43	20.95	-	26.28	m2	29.56
Linings; two layers of Gypsum wallboard, first layer Duplex square edge sheeting 15mm thick, second layer tapered edge sheeting 15mm thick for direct decoration, BS 1230, butt joints; fixing with galvanized nails to timber base; butt joints of second layer filled with joint filler tape and joint finish, spot filling										
Walls										
height 2100 - 2400mm	17.83	5.00	18.72	2.68	1.08	49.65	-	68.37	m	76.91
height 2400 - 2700mm	20.99	5.00	22.04	3.01	1.22	55.84	-	77.88	m	87.61
height 2700 - 3000mm	23.24	5.00	24.40	3.35	1.35	62.06	-	86.46	m	97.27
Reveals and soffits of openings and recesses										
width not exceeding 300mm	2.39	10.00	2.56	0.66	0.14	10.88	-	13.43	m	15.11
Ceilings										
generally	7.79	5.00	8.18	1.22	0.46	22.26	-	30.44	m2	34.24
K10: PLASTERBOARD DRY LINING/PARTITIONS/CEILING: STAIRCASE AREAS: PLASTIC FACED										
Linings; square edge sheeting of one layer 12.5mm thick Gyproc Industrial grade plastic faced plasterboard, BS 1230, butt joints; fixing with galvanized nails to timber base; butt joints sealed with self adhesive white tape										
Walls										
height 2100 - 2400mm	10.10	5.00	10.65	1.06	0.42	19.56	-	30.21	m	33.99
height 2400 - 2700mm	11.37	5.00	11.98	1.19	0.47	21.94	-	33.93	m	38.17
height 2700 - 3000mm	12.63	5.00	13.32	1.32	0.53	24.43	-	37.75	m	42.47
Reveals and soffits of openings and recesses										
width not exceeding 300mm	1.79	10.00	1.97	0.24	0.05	3.95	-	5.91	m	6.65
Ceilings										
generally	4.28	5.00	4.52	0.47	0.18	8.61	-	13.13	m2	14.77
K11: RIGID SHEET FLOORING/SHEATHING/LININGS/CASINGS										
Douglas Fir plywood, unsanded select sheathing quality, WBP bonded; tongued and grooved joints long edges										
15mm thick sheeting to floors										
width exceeding 300mm	6.85	7.50	7.37	0.35	0.04	5.40	0.04	12.80	m2	14.41
18mm thick sheeting to floors										
width exceeding 300mm	8.68	7.50	9.34	0.36	0.04	5.54	0.04	14.92	m2	16.78

Labour hourly rates: (except Specialists) Craft Operatives £14.21 Labourer £10.71 Rates are national average prices. Refer to REGIONAL VARIATIONS for indicative levels of overall pricing in regions	MATERIALS			LABOUR				RATES		
	Del to Site £	Waste %	Material Cost £	Craft Optve Hrs	Lab Hrs	Labour Cost £	Sunds £	Nett Rate £	Unit	Gross Rate (+12.5%) £
K11: RIGID SHEET FLOORING/SHEATHING/LININGS/CASINGS Cont.										
Birch faced plywood, BB quality, WBP bonded; tongued and grooved joints all edges										
12mm thick sheeting to floors										
width exceeding 300mm	20.58	7.50	**22.12**	0.32	0.04	**4.98**	0.04	**27.14**	m2	**30.53**
15mm thick sheeting to floors										
width exceeding 300mm	24.71	7.50	**26.56**	0.35	0.04	**5.40**	0.04	**32.00**	m2	**36.00**
18mm thick sheeting to floors										
width exceeding 300mm	29.97	7.50	**32.21**	0.36	0.04	**5.54**	0.04	**37.80**	m2	**42.52**
Wood chipboard, BS EN 312, type II flooring, square edges; butt joints										
18mm thick sheeting to floors										
width exceeding 300mm	2.70	7.50	**2.90**	0.35	0.04	**5.40**	0.04	**8.34**	m2	**9.38**
22mm thick sheeting to floors										
width exceeding 300mm	3.20	7.50	**3.44**	0.36	0.04	**5.54**	0.04	**9.02**	m2	**10.15**
Wood chipboard, BS EN 312, type II flooring; tongued and grooved joints										
18mm thick sheeting to floors										
width exceeding 300mm	4.24	7.50	**4.56**	0.35	0.04	**5.40**	0.04	**10.00**	m2	**11.25**
22mm thick sheeting to floors										
width exceeding 300mm	5.81	7.50	**6.25**	0.36	0.04	**5.54**	0.04	**11.83**	m2	**13.30**
Abutments										
12mm thick plywood linings										
raking cutting	2.06	7.50	**2.21**	0.04	-	**0.57**	-	**2.78**	m	**3.13**
curved cutting	3.09	7.50	**3.32**	0.06	-	**0.85**	-	**4.17**	m	**4.69**
15mm thick plywood linings										
raking cutting	2.47	7.50	**2.66**	0.05	-	**0.71**	-	**3.37**	m	**3.79**
curved cutting	3.71	7.50	**3.98**	0.07	-	**0.99**	-	**4.98**	m	**5.60**
18mm thick plywood linings										
raking cutting	3.00	7.50	**3.22**	0.05	-	**0.71**	-	**3.93**	m	**4.42**
curved cutting	4.50	7.50	**4.83**	0.07	-	**0.99**	-	**5.83**	m	**6.56**
18mm thick wood chipboard linings										
raking cutting	0.27	7.50	**0.29**	0.04	-	**0.57**	-	**0.86**	m	**0.97**
curved cutting	0.41	7.50	**0.44**	0.06	-	**0.85**	-	**1.29**	m	**1.45**
22mm thick wood chipboard linings										
raking cutting	0.32	7.50	**0.34**	0.05	-	**0.71**	-	**1.05**	m	**1.19**
curved cutting	0.48	7.50	**0.52**	0.07	-	**0.99**	-	**1.51**	m	**1.70**
K11: RIGID SHEET FLOORING/SHEATHING/LININGS/CASINGS: SOUND INSULATED FLOORS										
Foam backed softwood battens covered with tongued and grooved chipboard or plywood panels										
19mm chipboard panels to level sub-floors										
width exceeding 300mm; battens at 450mm centres; elevation										
50mm	8.38	10.00	**9.12**	0.35	0.04	**5.40**	0.10	**14.62**	m2	**16.45**
75mm	9.02	10.00	**9.80**	0.35	0.04	**5.40**	0.11	**15.30**	m2	**17.22**
width exceeding 300mm; battens at 600mm centres; elevation										
50mm	7.89	10.00	**8.61**	0.35	0.04	**5.40**	0.08	**14.09**	m2	**15.85**
75mm	8.37	10.00	**9.11**	0.35	0.04	**5.40**	0.09	**14.60**	m2	**16.43**
19mm chipboard panels with improved moisture resistance to level sub-floors										
width exceeding 300mm; battens at 450mm centres; elevation										
50mm	5.14	10.00	**5.56**	0.35	0.04	**5.40**	0.10	**11.06**	m2	**12.44**
75mm	5.79	10.00	**6.24**	0.35	0.04	**5.40**	0.11	**11.75**	m2	**13.21**
width exceeding 300mm; battens at 600mm centres; elevation										
50mm	4.66	10.00	**5.05**	0.35	0.04	**5.40**	0.08	**10.53**	m2	**11.85**
75mm	5.14	10.00	**5.56**	0.35	0.04	**5.40**	0.09	**11.04**	m2	**12.42**
19mm plywood panels to level sub-floors										
width exceeding 300mm; battens at 450mm centres; elevation										
50mm	10.62	10.00	**11.58**	0.35	0.04	**5.40**	0.10	**17.08**	m2	**19.22**
75mm	11.26	10.00	**12.26**	0.35	0.04	**5.40**	0.11	**17.77**	m2	**19.99**
K11: RIGID SHEET FLOORING/SHEATHING/LININGS/CASINGS: FIBREBOARDS AND HARDBOARDS										
Standard hardboard, BS EN 622; butt joints										
3.2mm thick linings to walls										
width exceeding 300mm	1.37	5.00	**1.44**	0.38	0.05	**5.94**	0.13	**7.50**	m2	**8.44**
width not exceeding 300mm	0.41	10.00	**0.45**	0.23	0.02	**3.48**	0.08	**4.01**	m	**4.51**

Labour hourly rates: (except Specialists) Craft Operatives £14.21 Labourer £10.71 Rates are national average prices. Refer to REGIONAL VARIATIONS for indicative levels of overall pricing in regions	MATERIALS			LABOUR				RATES		
	Del to Site £	Waste %	Material Cost £	Craft Optve Hrs	Lab Hrs	Labour Cost £	Sunds £	Nett Rate £	Unit	Gross Rate (+12.5%) £
K11: RIGID SHEET FLOORING/SHEATHING/LININGS/CASINGS: FIBREBOARDS AND HARDBOARDS Cont.										
Standard hardboard, BS EN 622; butt joints Cont.										
4.8mm thick linings to walls										
width exceeding 300mm	2.50	5.00	**2.62**	0.40	0.05	**6.22**	0.13	**8.97**	m2	**10.10**
width not exceeding 300mm	0.75	10.00	**0.82**	0.24	0.02	**3.62**	0.08	**4.53**	m	**5.09**
6.4mm thick linings to walls										
width exceeding 300mm	3.12	5.00	**3.28**	0.41	0.05	**6.36**	0.14	**9.78**	m2	**11.00**
width not exceeding 300mm	0.94	10.00	**1.03**	0.25	0.02	**3.77**	0.08	**4.88**	m	**5.49**
3.2mm thick sheeting to floors										
width exceeding 300mm	1.37	5.00	**1.44**	0.19	0.02	**2.91**	0.06	**4.42**	m2	**4.97**
3.2mm thick linings to ceilings										
width exceeding 300mm	1.37	5.00	**1.44**	0.48	0.06	**7.46**	0.16	**9.06**	m2	**10.20**
width not exceeding 300mm	0.41	10.00	**0.45**	0.29	0.02	**4.34**	0.10	**4.88**	m	**5.49**
4.8mm thick linings to ceilings										
width exceeding 300mm	2.50	5.00	**2.62**	0.50	0.06	**7.75**	0.17	**10.54**	m2	**11.85**
width not exceeding 300mm	0.75	10.00	**0.82**	0.30	0.02	**4.48**	0.10	**5.40**	m	**6.08**
6.4mm thick linings to ceilings										
width exceeding 300mm	3.12	5.00	**3.28**	0.51	0.06	**7.89**	0.17	**11.34**	m2	**12.75**
width not exceeding 300mm	0.94	10.00	**1.03**	0.31	0.02	**4.62**	0.10	**5.75**	m	**6.47**
Tempered hardboard, BS EN 622; butt joints										
3.2mm thick linings to walls										
width exceeding 300mm	1.97	5.00	**2.07**	0.38	0.05	**5.94**	0.13	**8.13**	m2	**9.15**
width not exceeding 300mm	0.59	10.00	**0.65**	0.23	0.02	**3.48**	0.08	**4.21**	m	**4.74**
4.8mm thick linings to walls										
width exceeding 300mm	2.87	5.00	**3.02**	0.40	0.05	**6.22**	0.13	**9.37**	m2	**10.54**
width not exceeding 300mm	0.86	10.00	**0.95**	0.24	0.02	**3.62**	0.08	**4.65**	m	**5.23**
6.4mm thick linings to walls										
width exceeding 300mm	3.63	5.00	**3.82**	0.41	0.05	**6.36**	0.14	**10.31**	m2	**11.60**
width not exceeding 300mm	1.09	10.00	**1.20**	0.25	0.02	**3.77**	0.08	**5.05**	m	**5.68**
3.2mm thick linings to ceilings										
width exceeding 300mm	1.97	5.00	**2.07**	0.48	0.06	**7.46**	0.16	**9.69**	m2	**10.91**
width not exceeding 300mm	0.59	10.00	**0.65**	0.29	0.02	**4.34**	0.10	**5.08**	m	**5.72**
4.8mm thick linings to ceilings										
width exceeding 300mm	2.87	5.00	**3.02**	0.50	0.06	**7.75**	0.17	**10.93**	m2	**12.30**
width not exceeding 300mm	0.86	10.00	**0.95**	0.30	0.02	**4.48**	0.10	**5.52**	m	**6.21**
6.4mm thick linings to ceilings										
width exceeding 300mm	3.63	5.00	**3.82**	0.51	0.06	**7.89**	0.17	**11.87**	m2	**13.36**
width not exceeding 300mm	1.09	10.00	**1.20**	0.31	0.02	**4.62**	0.10	**5.92**	m	**6.66**
Flame retardant hardboard, BS EN 622 tested to BS 476 Part 7 Class 1; butt joints										
3.2mm thick linings to walls										
width exceeding 300mm	3.12	5.00	**3.28**	0.38	0.05	**5.94**	0.13	**9.34**	m2	**10.51**
width not exceeding 300mm	0.94	10.00	**1.03**	0.23	0.02	**3.48**	0.08	**4.59**	m	**5.16**
4.8mm thick linings to walls										
width exceeding 300mm	6.25	5.00	**6.56**	0.41	0.05	**6.36**	0.14	**13.05**	m2	**14.69**
width not exceeding 300mm	1.87	10.00	**2.06**	0.25	0.02	**3.77**	0.08	**5.91**	m	**6.65**
6.4mm thick linings to walls										
width exceeding 300mm	6.87	5.00	**7.21**	0.42	0.05	**6.50**	0.14	**13.86**	m2	**15.59**
width not exceeding 300mm	2.06	10.00	**2.27**	0.25	0.02	**3.77**	0.08	**6.12**	m	**6.88**
3.2mm thick linings to ceilings										
width exceeding 300mm	3.12	5.00	**3.28**	0.48	0.06	**7.46**	0.16	**10.90**	m2	**12.26**
width not exceeding 300mm	0.94	10.00	**1.03**	0.29	0.02	**4.34**	0.10	**5.46**	m	**6.14**
4.8mm thick linings to ceilings										
width exceeding 300mm	6.25	5.00	**6.56**	0.51	0.06	**7.89**	0.17	**14.62**	m2	**16.44**
width not exceeding 300mm	1.87	10.00	**2.06**	0.31	0.02	**4.62**	0.10	**6.78**	m	**7.63**
6.4mm thick linings to ceilings										
width exceeding 300mm	6.87	5.00	**7.21**	0.53	0.07	**8.28**	0.17	**15.67**	m2	**17.63**
width not exceeding 300mm	2.06	10.00	**2.27**	0.32	0.02	**4.76**	0.11	**7.13**	m	**8.03**
Low density medium board, BS EN 622, type LME; butt joints										
6mm thick linings to walls										
width exceeding 300mm	8.59	5.00	**9.02**	0.41	0.05	**6.36**	0.14	**15.51**	m2	**17.45**
width not exceeding 300mm	2.58	10.00	**2.83**	0.25	0.02	**3.77**	0.08	**6.68**	m	**7.52**
9mm thick linings to walls										
width exceeding 300mm	10.99	5.00	**11.54**	0.42	0.05	**6.50**	0.14	**18.18**	m2	**20.46**
width not exceeding 300mm	3.30	10.00	**3.63**	0.25	0.02	**3.77**	0.08	**7.48**	m	**8.41**
12mm thick linings to walls										
width exceeding 300mm	14.02	5.00	**14.72**	0.44	0.06	**6.90**	0.15	**21.76**	m2	**24.48**
width not exceeding 300mm	4.20	10.00	**4.63**	0.26	0.02	**3.91**	0.09	**8.62**	m	**9.70**

Labour hourly rates: (except Specialists) Craft Operatives £14.21 Labourer £10.71 Rates are national average prices. Refer to REGIONAL VARIATIONS for indicative levels of overall pricing in regions	MATERIALS			LABOUR				RATES		
	Del to Site £	Waste %	Material Cost £	Craft Optve Hrs	Lab Hrs	Labour Cost £	Sunds £	Nett Rate £	Unit	Gross Rate (+12.5%) £
K11: RIGID SHEET FLOORING/SHEATHING/LININGS/CASINGS: FIBREBOARDS AND HARDBOARDS Cont.										
Low density medium board, BS EN 622, type LME; butt joints Cont.										
6mm thick linings to ceilings										
width exceeding 300mm	8.59	5.00	**9.02**	0.51	0.06	**7.89**	0.17	**17.08**	m2	**19.21**
width not exceeding 300mm	2.58	10.00	**2.83**	0.31	0.02	**4.62**	0.02	**7.48**	m	**8.41**
9mm thick linings to ceilings										
width exceeding 300mm	10.99	5.00	**11.54**	0.53	0.07	**8.28**	0.17	**20.00**	m2	**22.50**
width not exceeding 300mm	3.30	10.00	**3.63**	0.32	0.02	**4.76**	0.11	**8.49**	m	**9.56**
12mm thick linings to ceilings										
width exceeding 300mm	14.02	5.00	**14.72**	0.55	0.07	**8.57**	0.18	**23.46**	m2	**26.40**
width not exceeding 300mm	4.20	10.00	**4.63**	0.33	0.02	**4.90**	0.11	**9.64**	m	**10.84**
Low density medium board, BS EN 622, type LMN; butt joints										
6mm thick linings to walls										
width exceeding 300mm	7.35	5.00	**7.72**	0.41	0.05	**6.36**	0.14	**14.22**	m2	**15.99**
width not exceeding 300mm	2.21	10.00	**2.43**	0.25	0.02	**3.77**	0.08	**6.28**	m	**7.06**
12mm thick linings to walls										
width exceeding 300mm	9.14	5.00	**9.59**	0.42	0.05	**6.50**	0.14	**16.24**	m2	**18.27**
width not exceeding 300mm	2.74	10.00	**3.02**	0.29	0.02	**4.34**	0.10	**7.45**	m	**8.38**
6mm thick linings to ceilings										
width exceeding 300mm	7.35	5.00	**7.72**	0.51	0.06	**7.89**	0.17	**15.78**	m2	**17.75**
width not exceeding 300mm	2.21	10.00	**2.43**	0.31	0.02	**4.62**	0.10	**7.15**	m	**8.04**
12mm thick linings to ceilings										
width exceeding 300mm	9.14	5.00	**9.59**	0.53	0.07	**8.28**	0.17	**18.05**	m2	**20.31**
width not exceeding 300mm	2.74	10.00	**3.02**	0.32	0.02	**4.76**	0.11	**7.88**	m	**8.87**
Medium density fibreboard, BS EN 622, type MDF; butt joints										
9mm thick linings to walls										
width exceeding 300mm	3.61	5.00	**3.79**	0.41	0.05	**6.36**	0.14	**10.29**	m2	**11.57**
width not exceeding 300mm	1.08	10.00	**1.19**	0.25	0.02	**3.77**	0.08	**5.04**	m	**5.67**
12mm thick linings to walls										
width exceeding 300mm	4.43	5.00	**4.65**	0.44	0.06	**6.90**	0.15	**11.69**	m2	**13.15**
width not exceeding 300mm	1.33	10.00	**1.46**	0.26	0.02	**3.91**	0.09	**5.46**	m	**6.14**
18mm thick linings to walls										
width exceeding 300mm	5.82	5.00	**6.11**	0.47	0.06	**7.32**	0.16	**13.59**	m2	**15.29**
width not exceeding 300mm	1.75	10.00	**1.92**	0.28	0.02	**4.19**	0.09	**6.20**	m	**6.97**
25mm thick linings to walls										
width exceeding 300mm	8.04	5.00	**8.44**	0.50	0.06	**7.75**	0.17	**16.35**	m2	**18.40**
width not exceeding 300mm	2.41	10.00	**2.65**	0.30	0.02	**4.48**	0.10	**7.23**	m	**8.13**
9mm thick linings to ceilings										
width exceeding 300mm	3.61	5.00	**3.79**	0.51	0.06	**7.89**	0.17	**11.85**	m2	**13.33**
width not exceeding 300mm	1.08	10.00	**1.19**	0.31	0.02	**4.62**	0.10	**5.91**	m	**6.65**
12mm thick linings to ceilings										
width exceeding 300mm	4.43	5.00	**4.65**	0.55	0.07	**8.57**	0.18	**13.40**	m2	**15.07**
width not exceeding 300mm	1.33	10.00	**1.46**	0.33	0.02	**4.90**	0.11	**6.47**	m	**7.28**
18mm thick linings to ceilings										
width exceeding 300mm	5.82	5.00	**6.11**	0.59	0.07	**9.13**	0.19	**15.44**	m2	**17.37**
width not exceeding 300mm	1.75	10.00	**1.92**	0.35	0.02	**5.19**	0.12	**7.22**	m	**8.13**
25mm thick linings to ceilings										
width exceeding 300mm	8.04	5.00	**8.44**	0.63	0.08	**9.81**	0.21	**18.46**	m2	**20.77**
width not exceeding 300mm	2.41	10.00	**2.65**	0.38	0.02	**5.61**	0.13	**8.39**	m	**9.44**
Insulating softboard, BS EN 622; butt joints										
13mm thick linings to walls										
width exceeding 300mm	2.14	5.00	**2.25**	0.44	0.06	**6.90**	0.15	**9.29**	m2	**10.45**
width not exceeding 300mm	0.64	10.00	**0.71**	0.26	0.02	**3.91**	0.09	**4.70**	m	**5.29**
13mm thick linings to ceilings										
width exceeding 300mm	2.14	5.00	**2.25**	0.55	0.07	**8.57**	0.18	**11.00**	m2	**12.37**
width not exceeding 300mm	0.64	10.00	**0.71**	0.33	0.02	**4.90**	0.11	**5.72**	m	**6.43**
Bitumen impregnated insulating board, BS EN 622 ; butt joints										
15mm thick linings to walls										
width exceeding 300mm	2.47	5.00	**2.60**	0.44	0.06	**6.90**	0.15	**9.64**	m2	**10.84**
width not exceeding 300mm	0.74	10.00	**0.82**	0.26	0.02	**3.91**	0.09	**4.81**	m	**5.41**
15mm thick linings to ceilings										
width exceeding 300mm	2.47	5.00	**2.60**	0.55	0.07	**8.57**	0.18	**11.34**	m2	**12.76**
width not exceeding 300mm	0.74	10.00	**0.82**	0.33	0.02	**4.90**	0.11	**5.83**	m	**6.56**

Labour hourly rates: (except Specialists) Craft Operatives £14.21 Labourer £10.71 Rates are national average prices. Refer to REGIONAL VARIATIONS for indicative levels of overall pricing in regions	MATERIALS			LABOUR			RATES			
	Del to Site £	Waste %	Material Cost £	Craft Optve Hrs	Lab Hrs	Labour Cost £	Sunds £	Nett Rate £	Unit	Gross Rate (+12.5%) £
K11: RIGID SHEET FLOORING/SHEATHING/LININGS/CASINGS: ASBESTOS FREE BOARD										
Masterboard Class 0 fire resisting boards; butt joints										
6mm thick linings to walls										
width exceeding 300mm	11.31	5.00	**11.88**	0.72	0.09	**11.20**	0.24	**23.31**	m2	**26.22**
width not exceeding 300mm	3.40	10.00	**3.74**	0.43	0.03	**6.43**	0.14	**10.31**	m	**11.60**
9mm thick linings to walls										
width exceeding 300mm	23.89	5.00	**25.08**	0.80	0.10	**12.44**	0.26	**37.79**	m2	**42.51**
width not exceeding 300mm	7.17	10.00	**7.88**	0.48	0.03	**7.14**	0.16	**15.18**	m	**17.08**
12mm thick linings to walls										
width exceeding 300mm	31.50	5.00	**33.08**	0.92	0.12	**14.36**	0.30	**47.74**	m2	**53.70**
width not exceeding 300mm	9.45	10.00	**10.40**	0.55	0.04	**8.24**	0.18	**18.82**	m	**21.17**
6mm thick linings to ceilings										
width exceeding 300mm	11.31	5.00	**11.88**	0.77	0.10	**12.01**	0.25	**24.14**	m2	**27.16**
width not exceeding 300mm	3.40	10.00	**3.74**	0.46	0.03	**6.86**	0.15	**10.75**	m	**12.09**
9mm thick linings to ceilings										
width exceeding 300mm	23.89	5.00	**25.08**	0.90	0.11	**13.97**	0.30	**39.35**	m2	**44.27**
width not exceeding 300mm	7.17	10.00	**7.88**	0.54	0.03	**7.99**	0.18	**16.06**	m	**18.06**
12mm thick linings to ceilings										
width exceeding 300mm	31.50	5.00	**33.08**	0.99	0.12	**15.35**	0.33	**48.75**	m2	**54.85**
width not exceeding 300mm	9.45	10.00	**10.40**	0.54	0.03	**7.99**	0.18	**18.57**	m	**20.89**
Supalux fire resisting boards, Sanded; butt joints; fixing to timber with screws, countersinking										
6mm thick linings to walls										
width exceeding 300mm	15.66	5.00	**16.44**	0.72	0.09	**11.20**	0.24	**27.88**	m2	**31.36**
width not exceeding 300mm	4.70	10.00	**5.17**	0.43	0.03	**6.43**	0.14	**11.74**	m	**13.21**
9mm thick linings to walls										
width exceeding 300mm	26.66	5.00	**27.99**	0.80	0.10	**12.44**	0.26	**40.70**	m2	**45.78**
width not exceeding 300mm	8.00	10.00	**8.80**	0.48	0.03	**7.14**	0.16	**16.10**	m	**18.11**
12mm thick linings to walls										
width exceeding 300mm	35.29	5.00	**37.05**	0.92	0.12	**14.36**	0.30	**51.72**	m2	**58.18**
width not exceeding 300mm	10.59	10.00	**11.65**	0.55	0.04	**8.24**	0.18	**20.07**	m	**22.58**
6mm thick linings to ceilings										
width exceeding 300mm	15.66	5.00	**16.44**	0.77	0.10	**12.01**	0.25	**28.71**	m2	**32.30**
width not exceeding 300mm	4.70	10.00	**5.17**	0.46	0.03	**6.86**	0.15	**12.18**	m	**13.70**
9mm thick linings to ceilings										
width exceeding 300mm	26.66	5.00	**27.99**	0.90	0.11	**13.97**	0.30	**42.26**	m2	**47.54**
width not exceeding 300mm	8.00	10.00	**8.80**	0.54	0.03	**7.99**	0.18	**16.97**	m	**19.09**
12mm thick linings to ceilings										
width exceeding 300mm	35.29	5.00	**37.05**	0.99	0.12	**15.35**	0.33	**52.73**	m2	**59.33**
width not exceeding 300mm	10.59	10.00	**11.65**	0.59	0.04	**8.81**	0.19	**20.65**	m	**23.23**
6mm thick casings to isolated beams or the like										
total girth not exceeding 600mm	15.66	10.00	**17.23**	1.44	0.18	**22.39**	0.48	**40.09**	m2	**45.10**
total girth 600 - 1200mm	15.66	5.00	**16.44**	0.86	0.11	**13.40**	0.28	**30.13**	m2	**33.89**
9mm thick casings to isolated beams or the like										
total girth not exceeding 600mm	26.66	10.00	**29.33**	1.66	0.21	**25.84**	0.55	**55.71**	m2	**62.68**
total girth 600 - 1200mm	26.66	5.00	**27.99**	1.00	0.13	**15.60**	0.33	**43.93**	m2	**49.42**
12mm thick casings to isolated beams or the like										
total girth not exceeding 600mm	35.29	10.00	**38.82**	1.84	0.23	**28.61**	0.61	**68.04**	m2	**76.54**
total girth 600 - 1200mm	35.29	5.00	**37.05**	1.09	0.14	**16.99**	0.36	**54.40**	m2	**61.20**
6mm thick casings to isolated columns or the like										
total girth 600 - 1200mm	15.66	10.00	**17.23**	1.44	0.18	**22.39**	0.48	**40.09**	m2	**45.10**
total girth 1200 - 1800mm	15.66	5.00	**16.44**	0.86	0.11	**13.40**	0.28	**30.13**	m2	**33.89**
9mm thick casings to isolated columns or the like										
total girth 600 - 1200mm	26.66	10.00	**29.33**	1.66	0.21	**25.84**	0.55	**55.71**	m2	**62.68**
total girth 1200 - 1800mm	26.66	5.00	**27.99**	1.00	0.13	**15.60**	0.33	**43.93**	m2	**49.42**
12mm thick casings to isolated columns or the like										
total girth 600 - 1200mm	35.29	10.00	**38.82**	1.84	0.23	**28.61**	0.61	**68.04**	m2	**76.54**
total girth 1200 - 1800mm	35.29	5.00	**37.05**	1.09	0.14	**16.99**	0.36	**54.40**	m2	**61.20**
Vermiculux fire resisting boards; butt joints; fixing to steel with self tapping screws, countersinking										
20mm thick casings to isolated beams or the like										
total girth not exceeding 600mm	24.15	10.00	**26.57**	1.88	0.23	**29.18**	0.62	**56.37**	m2	**63.41**
total girth 600 - 1200mm	24.15	5.00	**25.36**	1.13	0.14	**17.56**	0.37	**43.29**	m2	**48.70**
30mm thick casings to isolated beams or the like										
total girth not exceeding 600mm	39.50	10.00	**43.45**	2.06	0.26	**32.06**	0.68	**76.19**	m2	**85.72**
total girth 600 - 1200mm	39.52	5.00	**41.49**	1.25	0.16	**19.48**	0.41	**61.38**	m2	**69.06**
40mm thick casings to isolated beams or the like										
total girth not exceeding 600mm	59.90	10.00	**65.88**	2.24	0.28	**34.83**	0.74	**101.45**	m2	**114.13**
total girth 600 - 1200mm	59.90	5.00	**62.89**	1.37	0.17	**21.29**	0.45	**84.63**	m2	**95.21**

Labour hourly rates: (except Specialists) Craft Operatives £14.21 Labourer £10.71 Rates are national average prices. Refer to REGIONAL VARIATIONS for indicative levels of overall pricing in regions	MATERIALS			LABOUR			RATES			
	Del to Site £	Waste %	Material Cost £	Craft Optve Hrs	Lab Hrs	Labour Cost £	Sunds £	Nett Rate £	Unit	Gross Rate (+12.5%) £
K11: RIGID SHEET FLOORING/SHEATHING/LININGS/CASINGS: ASBESTOS FREE BOARD Cont.										
Vermiculux fire resisting boards; butt joints; fixing to steel with self tapping screws, countersinking Cont.										
50mm thick casings to isolated beams or the like										
total girth not exceeding 600mm.................	79.76	10.00	**87.74**	2.42	0.30	**37.60**	0.80	**126.14**	m2	**141.90**
total girth 600 - 1200mm........................	79.76	5.00	**83.75**	1.49	0.19	**23.21**	0.49	**107.45**	m2	**120.88**
60mm thick casings to isolated beams or the like										
total girth not exceeding 600mm.................	95.11	10.00	**104.62**	2.60	0.33	**40.48**	0.86	**145.96**	m2	**164.20**
total girth 600 - 1200mm........................	95.11	5.00	**99.87**	1.61	0.20	**25.02**	0.53	**125.42**	m2	**141.09**
20mm thick casings to isolated columns or the like										
total girth 600 - 1200mm........................	24.15	10.00	**26.57**	1.88	0.23	**29.18**	0.62	**56.37**	m2	**63.41**
total girth 1200 - 1800mm.......................	24.15	5.00	**25.36**	1.13	0.14	**17.56**	0.37	**43.29**	m2	**48.70**
30mm thick casings to isolated columns or the like										
total girth 600 - 1200mm........................	39.50	10.00	**43.45**	2.06	0.26	**32.06**	0.68	**76.19**	m2	**85.72**
total girth 1200 - 1800mm.......................	39.52	5.00	**41.49**	1.25	0.16	**19.48**	0.41	**61.38**	m2	**69.06**
40mm thick casings to isolated columns or the like										
total girth 600 - 1200mm........................	59.90	10.00	**65.88**	2.24	0.28	**34.83**	0.74	**101.45**	m2	**114.13**
total girth 1200 - 1800mm.......................	59.90	5.00	**62.89**	1.37	0.17	**21.29**	0.45	**84.63**	m2	**95.21**
50mm thick casings to isolated columns or the like										
total girth 600 - 1200mm........................	79.76	10.00	**87.74**	2.42	0.30	**37.60**	0.80	**126.14**	m2	**141.90**
total girth 1200 - 1800mm.......................	79.76	5.00	**83.75**	1.49	0.19	**23.21**	0.49	**107.45**	m2	**120.88**
60mm thick casings to isolated columns or the like										
total girth 600 - 1200mm........................	95.11	10.00	**104.62**	2.60	0.33	**40.48**	0.86	**145.96**	m2	**164.20**
total girth 1200 - 1800mm.......................	95.11	5.00	**99.87**	1.61	0.20	**25.02**	0.53	**125.42**	m2	**141.09**
50 x 25mm x 20g mild steel fixing angle										
fixing to masonry	1.88	5.00	**1.98**	0.48	0.08	**7.68**	0.16	**9.81**	m	**11.04**
fixing to steel	1.88	5.00	**1.98**	0.48	0.08	**7.68**	0.16	**9.81**	m	**11.04**
Tacboard Class 0 fire resisting boards; butt joints										
6mm thick linings to walls										
width exceeding 300mm	9.74	5.00	**10.22**	0.57	0.07	**8.85**	0.19	**19.26**	m2	**21.67**
width not exceeding 300mm.......................	2.92	10.00	**3.21**	0.34	0.02	**5.05**	0.11	**8.37**	m	**9.42**
9mm thick linings to walls										
width exceeding 300mm	17.94	5.00	**18.84**	0.67	0.08	**10.38**	0.22	**29.43**	m2	**33.11**
width not exceeding 300mm.......................	5.38	10.00	**5.92**	0.40	0.02	**5.90**	0.13	**11.95**	m	**13.44**
12mm thick linings to walls										
width exceeding 300mm	23.36	5.00	**24.52**	0.73	0.09	**11.34**	0.24	**36.10**	m2	**40.61**
width not exceeding 300mm.......................	7.01	10.00	**7.71**	0.44	0.03	**6.57**	0.15	**14.43**	m	**16.23**
6mm thick linings to ceilings										
width exceeding 300mm	9.74	5.00	**10.22**	0.72	0.09	**11.20**	0.24	**21.66**	m2	**24.36**
width not exceeding 300mm.......................	2.92	10.00	**3.21**	0.43	0.03	**6.43**	0.14	**9.79**	m	**11.01**
New Tacfire Class 1 fire resisting boards; butt joints										
6mm thick linings to walls										
width exceeding 300mm	13.02	5.00	**13.67**	0.57	0.07	**8.85**	0.19	**22.71**	m2	**25.54**
width not exceeding 300mm.......................	3.91	10.00	**4.30**	0.34	0.02	**5.05**	0.11	**9.45**	m	**10.64**
9mm thick linings to walls										
width exceeding 300mm	19.94	5.00	**20.94**	0.67	0.08	**10.38**	0.22	**31.54**	m2	**35.48**
width not exceeding 300mm.......................	5.98	10.00	**6.58**	0.40	0.02	**5.90**	0.13	**12.61**	m	**14.19**
12mm thick linings to walls										
width exceeding 300mm	26.29	5.00	**27.61**	0.73	0.09	**11.34**	0.24	**39.18**	m2	**44.08**
width not exceeding 300mm.......................	7.89	10.00	**8.68**	0.44	0.03	**6.57**	0.15	**15.39**	m	**17.32**
6mm thick linings to ceilings; 6mm cover fillets										
width exceeding 300mm	14.39	5.00	**15.04**	0.72	0.09	**11.20**	0.24	**26.48**	m2	**29.79**
width not exceeding 300mm.......................	4.32	10.00	**4.71**	0.43	0.03	**6.43**	0.14	**11.28**	m	**12.69**
9mm thick linings to ceilings; 9mm cover fillets										
width exceeding 300mm	22.50	5.00	**23.50**	0.80	0.10	**12.44**	0.26	**36.20**	m2	**40.73**
width not exceeding 300mm.......................	6.75	10.00	**7.35**	0.48	0.03	**7.14**	0.16	**14.65**	m	**16.48**
12mm thick linings to ceilings; 9mm cover fillets										
width exceeding 300mm	28.85	5.00	**30.17**	0.92	0.11	**14.25**	0.30	**44.72**	m2	**50.31**
width not exceeding 300mm.......................	8.66	10.00	**9.44**	0.55	0.03	**8.14**	0.18	**17.76**	m	**19.98**
Viculad fire resisting boards; butt joints; fixing with screws; countersinking										
18mm thick casings to isolated beams or the like; including noggins										
total girth not exceeding 600mm.................	21.30	10.00	**23.35**	1.78	0.22	**27.65**	0.59	**51.59**	m2	**58.04**
total girth 600 - 1200mm........................	21.30	5.00	**22.37**	1.07	0.07	**15.95**	0.35	**38.68**	m2	**43.51**
25mm thick casings to isolated beams or the like; including noggins										
total girth not exceeding 600mm.................	24.22	10.00	**26.55**	1.98	0.25	**30.81**	0.65	**58.01**	m2	**65.26**
total girth 600 - 1200mm........................	24.22	5.00	**25.43**	1.19	0.15	**18.52**	0.39	**44.34**	m2	**49.88**

Labour hourly rates: (except Specialists) Craft Operatives £14.21 Labourer £10.71 Rates are national average prices. Refer to REGIONAL VARIATIONS for indicative levels of overall pricing in regions	MATERIALS			LABOUR			RATES			
	Del to Site £	Waste %	Material Cost £	Craft Optve Hrs	Lab Hrs	Labour Cost £	Sunds £	Nett Rate £	Unit	Gross Rate (+12.5%) £
K11: RIGID SHEET FLOORING/SHEATHING/LININGS/CASINGS: ASBESTOS FREE BOARD Cont.										
Viculad fire resisting boards; butt joints; fixing with screws; countersinking Cont.										
35mm thick casings to isolated beams or the like; including noggins										
total girth not exceeding 600mm	29.33	10.00	**32.14**	2.06	0.26	**32.06**	0.68	**64.88**	m2	**72.99**
total girth 600 - 1200mm	29.32	5.00	**30.79**	1.25	0.16	**19.48**	0.41	**50.67**	m2	**57.01**
18mm thick casings to isolated columns or the like										
total girth 600 - 1200mm	20.33	10.00	**22.33**	1.78	0.22	**27.65**	0.59	**50.56**	m2	**56.88**
total girth 1200 - 1800mm	20.33	5.00	**21.34**	1.07	0.07	**15.95**	0.35	**37.65**	m2	**42.36**
25mm thick casings to isolated columns or the like										
total girth 600 - 1200mm	23.24	10.00	**25.52**	1.98	0.25	**30.81**	0.65	**56.99**	m2	**64.11**
total girth 1200 - 1800mm	23.24	5.00	**24.40**	1.19	0.15	**18.52**	0.39	**43.31**	m2	**48.73**
35mm thick casings to isolated columns or the like										
total girth 600 - 1200mm	28.35	10.00	**31.12**	2.06	0.26	**32.06**	0.68	**63.85**	m2	**71.84**
total girth 1200 - 1800mm	28.35	5.00	**29.77**	1.25	0.16	**19.48**	0.41	**49.65**	m2	**55.86**
K11: RIGID SHEET FLOORING/SHEATHING/LININGS/CASINGS: DUCT FRONTS										
Plywood BS EN 636, II/III grade, MR bonded; butt joints; fixing to timber with screws, countersinking										
12mm thick linings to walls										
width exceeding 300mm	13.84	20.00	**16.61**	1.15	0.14	**17.84**	0.06	**34.51**	m2	**38.83**
width exceeding 300mm; fixing to timber with brass screws and cups	13.84	20.00	**16.61**	1.70	0.21	**26.41**	0.21	**43.23**	m2	**48.63**
Blockboard, BS 3444, 2/2 grade, MR bonded; butt joints										
18mm thick linings to walls										
width exceeding 300mm	10.09	20.00	**12.11**	1.40	0.17	**21.71**	0.08	**33.90**	m2	**38.14**
width exceeding 300mm; fixing to timber with brass screws and cups	10.09	20.00	**12.11**	2.00	0.25	**31.10**	0.21	**43.42**	m2	**48.85**
Extra over plywood BS EN 636, II/III grade, MR bonded; butt joints; 12mm thick linings to walls; width exceeding 300mm										
Access panels										
300 x 600mm	-	-	**-**	0.60	0.08	**9.38**	0.03	**9.42**	nr	**10.59**
600 x 600mm	-	-	**-**	0.75	0.09	**11.62**	0.04	**11.66**	nr	**13.12**
600 x 900mm	-	-	**-**	0.90	0.11	**13.97**	0.05	**14.02**	nr	**15.77**
Extra over blockboard, BS 3444, 2/2 grade, MR bonded, butt joints; 18mm thick linings to walls; width exceeding 300mm										
Access panels										
300 x 600mm	-	-	**-**	0.70	0.09	**10.91**	0.04	**10.95**	nr	**12.32**
600 x 600mm	-	-	**-**	0.90	0.11	**13.97**	0.05	**14.02**	nr	**15.77**
600 x 900mm	-	-	**-**	1.25	0.16	**19.48**	0.07	**19.55**	nr	**21.99**
K11: RIGID SHEET FLOORING/SHEATHING/LININGS/CASINGS: PIPE CASINGS										
Standard hardboard, BS EN 622, type SHA; butt joints										
3mm thick linings to pipe ducts and casings										
width exceeding 300mm	1.31	20.00	**1.57**	0.90	0.11	**13.97**	0.05	**15.59**	m2	**17.54**
width not exceeding 300mm	0.39	20.00	**0.47**	0.30	0.04	**4.69**	0.02	**5.18**	m	**5.83**
Plywood BS EN 636, II/III grade, MR bonded; butt joints; fixing to timber with screws, countersinking										
12mm thick linings to pipe ducts and casings										
width exceeding 300mm	13.84	20.00	**16.61**	1.70	0.21	**26.41**	0.09	**43.11**	m2	**48.49**
width not exceeding 300mm	4.15	20.00	**4.98**	0.60	0.08	**9.38**	0.03	**14.40**	m	**16.20**
Supalux Class 0 fire resisting boards, sanded; butt joints; fixing to timber with screws, countersinking										
9mm thick linings to pipe ducts and casings										
width exceeding 300mm	26.66	20.00	**31.99**	1.70	0.21	**26.41**	0.09	**58.49**	m2	**65.80**
width not exceeding 300mm	8.00	20.00	**9.60**	0.60	0.08	**9.38**	0.03	**19.01**	m	**21.39**
Pendock Profiles Ltd preformed plywood casings; white melamine finish; butt joints; fixing to timber with polytop white screws, countersinking										
5mm thick casings; horizontal										
reference TK45/110, 45 x 110mm	7.73	20.00	**9.27**	0.46	0.06	**7.18**	0.08	**16.53**	m	**18.60**
extra; internal corner	7.44	2.50	**7.63**	0.30	0.04	**4.69**	0.08	**12.40**	nr	**13.94**
extra; external corner	9.70	2.50	**9.94**	0.30	0.04	**4.69**	0.08	**14.71**	nr	**16.55**
extra; stop end	6.72	2.50	**6.89**	0.20	0.02	**3.06**	0.04	**9.99**	nr	**11.24**
reference TK45/150, 45 x 150mm	9.77	20.00	**11.72**	0.48	0.06	**7.46**	0.08	**19.26**	m	**21.67**
extra; internal corner	8.79	2.50	**9.01**	0.30	0.04	**4.69**	0.08	**13.78**	nr	**15.50**
extra; external corner	11.15	2.50	**11.43**	0.30	0.04	**4.69**	0.08	**16.19**	nr	**18.22**
extra; stop end	9.41	2.50	**9.65**	0.20	0.02	**3.06**	0.04	**12.74**	nr	**14.33**
reference TK45/190, 45 x 190mm	12.83	20.00	**15.39**	0.50	0.06	**7.75**	0.08	**23.22**	m	**26.12**

Labour hourly rates: (except Specialists) Craft Operatives £14.21 Labourer £10.71 Rates are national average prices. Refer to REGIONAL VARIATIONS for indicative levels of overall pricing in regions	MATERIALS			LABOUR				RATES		
	Del to Site £	Waste %	Material Cost £	Craft Optve Hrs	Lab Hrs	Labour Cost £	Sunds £	Nett Rate £	Unit	Gross Rate (+12.5%) £
K11: RIGID SHEET FLOORING/SHEATHING/LININGS/CASINGS: PIPE CASINGS Cont.										
Pendock Profiles Ltd preformed plywood casings; white melamine finish; butt joints; fixing to timber with polytop white screws, countersinking Cont.										
5mm thick casings; horizontal Cont.										
extra; internal corner	8.85	2.50	**9.07**	0.30	0.04	**4.69**	0.08	**13.84**	nr	**15.57**
extra; external corner	11.21	2.50	**11.49**	0.30	0.04	**4.69**	0.08	**16.25**	nr	**18.28**
extra; stop end	11.21	2.50	**11.49**	0.20	0.02	**3.06**	0.04	**14.58**	nr	**16.40**
reference TK25/125, 25 x 125mm	7.71	20.00	**9.26**	0.46	0.06	**7.18**	0.08	**16.51**	m	**18.58**
extra; internal corner	7.46	2.50	**7.64**	0.30	0.04	**4.69**	0.08	**12.41**	nr	**13.96**
extra; external corner	9.70	2.50	**9.94**	0.30	0.04	**4.69**	0.08	**14.71**	nr	**16.55**
extra; stop end	6.72	2.50	**6.89**	0.20	0.02	**3.06**	0.04	**9.99**	nr	**11.24**
5mm thick casings; vertical										
reference TK45/110, 45 x 110mm	7.04	20.00	**8.45**	0.46	0.06	**7.18**	0.08	**15.70**	m	**17.67**
reference TK45/135, 45 x 135mm	10.27	20.00	**12.33**	0.47	0.06	**7.32**	0.08	**19.72**	m	**22.19**
reference TK45/150, 45 x 150mm	9.01	20.00	**10.81**	0.48	0.06	**7.46**	0.08	**18.35**	m	**20.64**
reference TK45/190, 45 x 190mm	12.05	20.00	**14.46**	0.50	0.06	**7.75**	0.08	**22.29**	m	**25.07**
5mm thick casings; horizontal										
reference MX 100/100, 100 x 100mm	12.89	20.00	**15.46**	0.44	0.06	**6.90**	0.08	**22.43**	m	**25.24**
extra; stop end	11.18	2.50	**11.46**	0.20	0.02	**3.06**	0.04	**14.55**	nr	**16.37**
8mm thick casings; horizontal										
reference MX 150/150, 150 x 150mm	19.51	20.00	**23.41**	0.52	0.06	**8.03**	0.08	**31.52**	m	**35.46**
extra; internal corner	13.89	2.50	**14.24**	0.30	0.04	**4.69**	0.08	**19.01**	nr	**21.38**
extra; external corner	20.95	2.50	**21.47**	0.30	0.04	**4.69**	0.08	**26.24**	nr	**29.52**
extra; stop end	11.18	2.50	**11.46**	0.20	0.02	**3.06**	0.04	**14.55**	nr	**16.37**
reference MX 150/200, 200 x 150mm	21.62	20.00	**25.95**	0.53	0.07	**8.28**	0.08	**34.30**	m	**38.59**
extra; stop end	11.18	2.50	**11.46**	0.20	0.02	**3.06**	0.04	**14.55**	nr	**16.37**
reference MX 200/200, 200 x 200mm	23.58	20.00	**28.29**	0.54	0.07	**8.42**	0.08	**36.79**	m	**41.39**
extra; internal corner	14.01	2.50	**14.36**	0.30	0.04	**4.69**	0.08	**19.12**	nr	**21.52**
extra; external corner	21.18	2.50	**21.71**	0.30	0.04	**4.69**	0.08	**26.47**	nr	**29.78**
extra; stop end	11.18	2.50	**11.46**	0.20	0.02	**3.06**	0.04	**14.55**	nr	**16.37**
reference MX 200/300, 200 x 300mm	27.67	20.00	**33.20**	0.55	0.07	**8.57**	0.08	**41.84**	m	**47.08**
extra; stop end	12.27	2.50	**12.58**	0.20	0.02	**3.06**	0.04	**15.67**	nr	**17.63**
reference MX 150/300, 300 x 150mm	26.16	20.00	**31.39**	0.55	0.07	**8.57**	0.08	**40.03**	m	**45.04**
extra; internal corner	14.87	-	**14.87**	0.30	0.04	**4.69**	0.08	**19.64**	nr	**22.09**
extra; external corner	22.10	-	**22.10**	0.30	0.04	**4.69**	0.08	**26.86**	nr	**30.22**
extra; stop end	12.27	2.50	**12.58**	0.20	0.02	**3.06**	0.04	**15.67**	nr	**17.63**
reference MX 300/300, 300 x 300mm	31.55	20.00	**37.86**	0.57	0.07	**8.85**	0.08	**46.78**	m	**52.63**
extra; internal corner	15.86	2.50	**16.26**	0.30	0.04	**4.69**	0.08	**21.02**	nr	**23.65**
extra; external corner	23.10	2.50	**23.68**	0.30	0.04	**4.69**	0.08	**28.45**	nr	**32.00**
extra; stop end	15.24	2.50	**15.62**	0.20	0.02	**3.06**	0.04	**18.72**	nr	**21.06**
5mm thick casings; vertical										
reference MX 100/100, 100 x 100mm	10.79	20.00	**12.95**	0.48	0.06	**7.46**	0.08	**20.49**	m	**23.05**
8mm thick casings; vertical										
reference MX 150/150, 150 x 150mm	17.38	20.00	**20.86**	0.52	0.06	**8.03**	0.08	**28.97**	m	**32.59**
reference MX 150/200, 200 x 150mm	19.52	20.00	**23.43**	0.53	0.07	**8.28**	0.08	**31.79**	m	**35.76**
reference MX 200/200, 200 x 200mm	21.49	20.00	**25.79**	0.54	0.07	**8.42**	0.08	**34.29**	m	**38.58**
reference MX 200/300, 200 x 300mm	25.57	20.00	**30.69**	0.55	0.07	**8.57**	0.08	**39.33**	m	**44.24**
reference MX 150/300, 300 x 150mm	24.06	20.00	**28.88**	0.56	0.07	**8.71**	0.08	**37.66**	m	**42.37**
reference MX 300/300, 300 x 300mm	29.44	20.00	**35.32**	0.57	0.07	**8.85**	0.08	**44.25**	m	**49.78**
K11: RIGID SHEET FLOORING/SHEATHING/LININGS/CASINGS: ROOF BOARDING										
Plywood BS EN 636, II/III grade, WBP bonded; butt joints										
18mm thick sheeting to roofs; external										
width exceeding 300mm	11.21	10.00	**12.33**	0.40	0.05	**6.22**	0.02	**18.57**	m2	**20.89**
extra; cesspool 225 x 225 x 150mm	2.24	10.00	**2.47**	1.70	0.21	**26.41**	0.09	**28.96**	nr	**32.59**
extra; cesspool 300 x 300 x 150mm	3.08	10.00	**3.39**	1.80	0.22	**27.93**	0.10	**31.42**	nr	**35.35**
width exceeding 300mm; sloping	11.21	10.00	**12.33**	0.50	0.06	**7.75**	0.03	**20.11**	m2	**22.62**
24mm thick sheeting to roofs; external										
width exceeding 300mm	15.58	10.00	**17.14**	0.45	0.06	**7.04**	0.02	**24.20**	m2	**27.22**
extra; cesspool 225 x 225 x 150mm	3.12	10.00	**3.43**	1.80	0.22	**27.93**	0.10	**31.46**	nr	**35.39**
extra; cesspool 300 x 300 x 150mm	4.28	10.00	**4.71**	1.90	0.24	**29.57**	0.10	**34.38**	nr	**38.68**
width exceeding 300mm; sloping	15.58	10.00	**17.14**	0.55	0.07	**8.57**	0.03	**25.73**	m2	**28.95**
Wood chipboard, BS EN 312, type 1 standard; butt joints										
12mm thick sheeting to roofs; external										
width exceeding 300mm	2.37	10.00	**2.61**	0.30	0.04	**4.69**	0.02	**7.31**	m2	**8.23**
extra; cesspool 225 x 225 x 150mm	0.47	10.00	**0.52**	1.55	0.19	**24.06**	0.08	**24.66**	nr	**27.75**
extra; cesspool 300 x 300 x 150mm	0.65	10.00	**0.72**	1.65	0.21	**25.70**	0.09	**26.50**	nr	**29.81**
width exceeding 300mm; sloping	2.37	10.00	**2.61**	0.40	0.05	**6.22**	0.02	**8.85**	m2	**9.95**
18mm thick sheeting to roofs; external										
width exceeding 300mm	3.21	10.00	**3.53**	0.35	0.04	**5.40**	0.02	**8.95**	m2	**10.07**
extra; cesspool 225 x 225 x 150mm	0.64	10.00	**0.71**	1.65	0.21	**25.70**	0.09	**26.49**	nr	**29.80**
extra; cesspool 300 x 300 x 150mm	0.88	10.00	**0.97**	1.75	0.22	**27.22**	0.09	**28.29**	nr	**31.82**
width exceeding 300mm; sloping	3.21	10.00	**3.53**	0.45	0.06	**7.04**	0.02	**10.59**	m2	**11.92**
Abutments										
18mm thick plywood linings										
raking cutting	1.12	10.00	**1.23**	0.35	-	**4.97**	-	**6.21**	m	**6.98**

Labour hourly rates: (except Specialists) Craft Operatives £14.21 Labourer £10.71 Rates are national average prices. Refer to REGIONAL VARIATIONS for indicative levels of overall pricing in regions	MATERIALS			LABOUR				RATES		
	Del to Site £	Waste %	Material Cost £	Craft Optve Hrs	Lab Hrs	Labour Cost £	Sunds £	Nett Rate £	Unit	Gross Rate (+12.5%) £
K11: RIGID SHEET FLOORING/SHEATHING/LININGS/CASINGS: ROOF BOARDING Cont.										
Abutments Cont.										
18mm thick plywood linings Cont.										
curved cutting	1.68	10.00	**1.85**	0.55	-	**7.82**	-	**9.67**	m	**10.87**
rebates	-	-	-	0.22	-	**3.13**	-	**3.13**	m	**3.52**
grooves	-	-	-	0.22	-	**3.13**	-	**3.13**	m	**3.52**
chamfers	-	-	-	0.16	-	**2.27**	-	**2.27**	m	**2.56**
24mm thick plywood linings										
raking cutting	1.56	10.00	**1.71**	0.50	-	**7.11**	-	**8.82**	m	**9.92**
curved cutting	2.34	10.00	**2.57**	0.70	-	**9.95**	-	**12.52**	m	**14.08**
rebates	-	-	-	0.22	-	**3.13**	-	**3.13**	m	**3.52**
grooves	-	-	-	0.22	-	**3.13**	-	**3.13**	m	**3.52**
chamfers	-	-	-	0.16	-	**2.27**	-	**2.27**	m	**2.56**
12mm thick wood chipboard linings										
raking cutting	0.24	10.00	**0.26**	0.30	-	**4.26**	-	**4.52**	m	**5.09**
curved cutting	0.36	10.00	**0.39**	0.40	-	**5.68**	-	**6.08**	m	**6.83**
rebates	-	-	-	0.22	-	**3.13**	-	**3.13**	m	**3.52**
grooves	-	-	-	0.22	-	**3.13**	-	**3.13**	m	**3.52**
chamfers	-	-	-	0.16	-	**2.27**	-	**2.27**	m	**2.56**
18mm thick wood chipboard linings										
raking cutting	0.32	10.00	**0.35**	0.35	-	**4.97**	-	**5.33**	m	**5.99**
curved cutting	0.48	10.00	**0.53**	0.55	-	**7.82**	-	**8.35**	m	**9.39**
rebates	-	-	-	0.22	-	**3.13**	-	**3.13**	m	**3.52**
grooves	-	-	-	0.22	-	**3.13**	-	**3.13**	m	**3.52**
chamfers	-	-	-	0.16	-	**2.27**	-	**2.27**	m	**2.56**
K11: RIGID SHEET FLOORING/SHEATHING/ LININGS/CASINGS; STAIRCASE AREAS - ASBESTOS FREE BOARD										
Masterboard Class 0 fire resisting boards; butt joints										
6mm thick linings to walls										
width exceeding 300mm	11.31	5.00	**11.88**	0.72	0.09	**11.20**	0.24	**23.31**	m2	**26.22**
width not exceeding 300mm	3.40	10.00	**3.74**	0.43	0.03	**6.43**	0.14	**10.31**	m	**11.60**
9mm thick linings to walls										
width exceeding 300mm	23.89	5.00	**25.08**	0.80	0.10	**12.44**	0.26	**37.79**	m2	**42.51**
width not exceeding 300mm	7.17	10.00	**7.88**	0.48	0.03	**7.14**	0.16	**15.18**	m	**17.08**
12mm thick linings to walls										
width exceeding 300mm	31.50	5.00	**33.08**	0.92	0.12	**14.36**	0.30	**47.74**	m2	**53.70**
width not exceeding 300mm	9.45	10.00	**10.40**	0.55	0.04	**8.24**	0.18	**18.82**	m	**21.17**
6mm thick linings to ceilings										
width exceeding 300mm	11.31	5.00	**11.88**	0.77	0.10	**12.01**	0.25	**24.14**	m2	**27.16**
width not exceeding 300mm	3.40	10.00	**3.74**	0.46	0.03	**6.86**	0.15	**10.75**	m	**12.09**
9mm thick linings to ceilings										
width exceeding 300mm	23.89	5.00	**25.08**	0.90	0.11	**13.97**	0.30	**39.35**	m2	**44.27**
width not exceeding 300mm	7.17	10.00	**7.88**	0.54	0.03	**7.99**	0.18	**16.06**	m	**18.06**
12mm thick linings to ceilings										
width exceeding 300mm	31.50	5.00	**33.08**	0.99	0.12	**15.35**	0.33	**48.75**	m2	**54.85**
width not exceeding 300mm	9.45	10.00	**10.40**	0.59	0.04	**8.81**	0.19	**19.40**	m	**21.83**
Supalux fire resisting boards, sanded; butt joints; fixing to timber with screws, countersinking										
6mm thick linings to walls										
width exceeding 300mm	15.66	5.00	**16.44**	0.72	0.09	**11.20**	0.24	**27.88**	m2	**31.36**
width not exceeding 300mm	4.70	10.00	**5.17**	0.43	0.03	**6.43**	0.14	**11.74**	m	**13.21**
9mm thick linings to walls										
width exceeding 300mm	26.66	5.00	**27.99**	0.80	0.10	**12.44**	0.26	**40.70**	m2	**45.78**
width not exceeding 300mm	8.00	10.00	**8.80**	0.48	0.03	**7.14**	0.16	**16.10**	m	**18.11**
12mm thick linings to walls										
width exceeding 300mm	35.29	5.00	**37.05**	0.92	0.12	**14.36**	0.30	**51.72**	m2	**58.18**
width not exceeding 300mm	10.59	10.00	**11.65**	0.55	0.04	**8.24**	0.18	**20.07**	m	**22.58**
6mm thick linings to ceilings										
width exceeding 300mm	15.66	5.00	**16.44**	0.77	0.10	**12.01**	0.25	**28.71**	m2	**32.30**
width not exceeding 300mm	4.70	10.00	**5.17**	0.46	0.03	**6.86**	0.15	**12.18**	m	**13.70**
9mm thick linings to ceilings										
width exceeding 300mm	26.66	5.00	**27.99**	0.90	0.11	**13.97**	0.30	**42.26**	m2	**47.54**
width not exceeding 300mm	8.00	10.00	**8.80**	0.54	0.03	**7.99**	0.18	**16.97**	m	**19.09**
12mm thick linings to ceilings										
width exceeding 300mm	35.29	5.00	**37.05**	0.99	0.12	**15.35**	0.33	**52.73**	m2	**59.33**
width not exceeding 300mm	10.59	10.00	**11.65**	0.59	0.04	**8.81**	0.19	**20.65**	m	**23.23**

LININGS/SHEATHING/DRY PARTITIONING

Labour hourly rates: (except Specialists) Craft Operatives £14.21 Labourer £10.71 Rates are national average prices. Refer to REGIONAL VARIATIONS for indicative levels of overall pricing in regions	MATERIALS			LABOUR			RATES			
	Del to Site £	Waste %	Material Cost £	Craft Optve Hrs	Lab Hrs	Labour Cost £	Sunds £	Nett Rate £	Unit	Gross Rate (+12.5%) £
K12: UNDER PURLIN/INSIDE RAIL PANEL LININGS: PLASTERBOARD										
Gypliner metal grid fixing system; square edge sheeting of one layer 12.5mm thick Gyproc wallboard, BS 1230, butt joints;screwing to mild steel grid.										
12.5mm thick linings to walls										
width exceeding 300mm	6.42	5.00	**6.77**	1.00	0.50	**19.57**	0.13	**26.47**	m2	**29.78**
width not exceeding 300mm	6.52	5.00	**6.85**	0.60	0.15	**10.13**	0.13	**17.11**	m	**19.25**
12.5mm thick linings to ceilings										
width exceeding 300mm	6.35	5.00	**6.69**	1.25	0.63	**24.51**	0.13	**31.34**	m2	**35.25**
width not exceeding 300mm	6.52	5.00	**6.85**	0.75	0.19	**12.69**	0.13	**19.67**	m	**22.13**
K13: RIGID SHEET FINE LININGS/PANELLING										
Softwood, wrot - wall panelling										
19mm thick panelled linings to walls; square framed; including grounds										
width exceeding 300mm	78.50	2.50	**80.47**	1.90	0.24	**29.57**	0.11	**110.14**	m2	**123.91**
width exceeding 300mm; fixing to masonry with screws	78.50	2.50	**80.47**	2.90	0.36	**45.06**	0.17	**125.71**	m2	**141.42**
25mm thick panelled linings to walls; square framed; including grounds										
width exceeding 300mm	79.57	2.50	**81.55**	2.05	0.26	**31.92**	0.12	**113.59**	m2	**127.78**
width exceeding 300mm; fixing to masonry with screws	79.57	2.50	**81.55**	3.05	0.38	**47.41**	0.18	**129.15**	m2	**145.29**
19mm thick panelled linings to walls; square framed; obstructed by integral services; including grounds										
width exceeding 300mm	78.50	2.50	**80.47**	2.40	0.30	**37.32**	0.14	**117.93**	m2	**132.67**
width exceeding 300mm; fixing to masonry with screws	78.50	2.50	**80.47**	3.40	0.42	**52.81**	0.21	**133.49**	m2	**150.18**
25mm thick panelled linings to walls; square framed; obstructed by integral services; including grounds										
width exceeding 300mm	79.57	2.50	**81.55**	2.55	0.32	**39.66**	0.15	**121.37**	m2	**136.54**
width exceeding 300mm; fixing to masonry with screws	79.57	2.50	**81.55**	3.55	0.44	**55.16**	0.22	**136.93**	m2	**154.05**
19mm thick panelled linings to walls; moulded; including grounds										
width exceeding 300mm	85.93	2.50	**88.08**	2.05	0.26	**31.92**	0.12	**120.11**	m2	**135.12**
width exceeding 300mm; fixing to masonry with screws	85.93	2.50	**88.08**	3.05	0.38	**47.41**	0.18	**135.67**	m2	**152.63**
25mm thick panelled linings to walls; moulded; including grounds										
width exceeding 300mm	89.11	2.50	**91.34**	2.15	0.27	**33.44**	0.15	**124.94**	m2	**140.55**
width exceeding 300mm; fixing to masonry with screws	89.11	2.50	**91.34**	3.15	0.39	**48.94**	0.19	**140.47**	m2	**158.03**
19mm thick panelled linings to walls; moulded; obstructed by integral services; including grounds										
width exceeding 300mm	85.93	2.50	**88.08**	2.55	0.32	**39.66**	0.15	**127.89**	m2	**143.88**
width exceeding 300mm; fixing to masonry with screws	85.93	2.50	**88.08**	3.55	0.44	**55.16**	0.22	**143.45**	m2	**161.39**
25mm thick panelled linings to walls; moulded; obstructed by integral services; including grounds										
width exceeding 300mm	89.11	2.50	**91.34**	2.65	0.33	**41.19**	0.16	**132.69**	m2	**149.27**
width exceeding 300mm; fixing to masonry with screws	89.11	2.50	**91.34**	3.65	0.46	**56.79**	0.22	**148.36**	m2	**166.90**
Afromosia, wrot, selected for transparent finish - wall panelling										
19mm thick panelled linings to walls; square framed; including grounds										
width exceeding 300mm	223.56	2.50	**229.15**	2.70	0.34	**42.01**	0.16	**271.31**	m2	**305.23**
width exceeding 300mm; fixing to masonry with screws	223.56	2.50	**229.15**	3.70	0.46	**57.50**	0.22	**286.87**	m2	**322.73**
25mm thick panelled linings to walls; square framed; including grounds										
width exceeding 300mm	229.77	2.50	**235.51**	2.80	0.35	**43.54**	0.16	**279.21**	m2	**314.11**
width exceeding 300mm; fixing to masonry with screws	229.77	2.50	**235.51**	3.80	0.47	**59.03**	0.23	**294.78**	m2	**331.62**
19mm thick panelled linings to walls; square framed; obstructed by integral services; including grounds										
width exceeding 300mm	223.56	2.50	**229.15**	3.20	0.40	**49.76**	0.18	**279.09**	m2	**313.98**
width exceeding 300mm; fixing to masonry with screws	223.56	2.50	**229.15**	4.20	0.52	**65.25**	0.25	**294.65**	m2	**331.49**
25mm thick panelled linings to walls; square framed; obstructed by integral services; including grounds										
width exceeding 300mm	229.77	2.50	**235.51**	3.30	0.41	**51.28**	0.18	**286.98**	m2	**322.85**
width exceeding 300mm; fixing to masonry with screws	229.77	2.50	**235.51**	4.30	0.54	**66.89**	0.25	**302.65**	m2	**340.48**

Labour hourly rates: (except Specialists) Craft Operatives £14.21 Labourer £10.71 Rates are national average prices. Refer to REGIONAL VARIATIONS for indicative levels of overall pricing in regions	MATERIALS			LABOUR				RATES		
	Del to Site £	Waste %	Material Cost £	Craft Optve Hrs	Lab Hrs	Labour Cost £	Sunds £	Nett Rate £	Unit	Gross Rate (+12.5%) £
K13: RIGID SHEET FINE LININGS/PANELLING Cont.										
Afromosia, wrot, selected for transparent finish - wall panelling Cont.										
19mm thick panelled linings to walls; moulded; including grounds										
width exceeding 300mm	240.12	2.50	**246.12**	2.90	0.36	**45.06**	0.16	**291.35**	m2	**327.76**
width exceeding 300mm; fixing to masonry with screws	240.12	2.50	**246.12**	3.90	0.49	**60.67**	0.22	**307.01**	m2	**345.39**
25mm thick panelled linings to walls; moulded; including grounds										
width exceeding 300mm	255.65	2.50	**262.04**	3.00	0.37	**46.59**	0.17	**308.79**	m2	**347.39**
width exceeding 300mm; fixing to masonry with screws	255.65	2.50	**262.04**	4.00	0.50	**62.20**	0.23	**324.46**	m2	**365.02**
19mm thick panelled linings to walls; moulded; obstructed by integral services; including grounds										
width exceeding 300mm	240.12	2.50	**246.12**	3.40	0.42	**52.81**	0.18	**299.12**	m2	**336.51**
width exceeding 300mm; fixing to masonry with screws	240.12	2.50	**246.12**	4.40	0.55	**68.41**	0.25	**314.79**	m2	**354.14**
25mm thick panelled linings to walls; moulded; obstructed by integral services; including grounds										
width exceeding 300mm	255.65	2.50	**262.04**	3.50	0.44	**54.45**	0.19	**316.68**	m2	**356.26**
width exceeding 300mm; fixing to masonry with screws	255.65	2.50	**262.04**	4.50	0.56	**69.94**	0.26	**332.24**	m2	**373.77**
Sapele, wrot, selected for transparent finish - wall panelling										
19mm thick panelled linings to walls; square framed; including grounds										
width exceeding 300mm	180.00	2.50	**184.50**	2.45	0.31	**38.13**	0.14	**222.77**	m2	**250.62**
width exceeding 300mm; fixing to masonry with screws	180.00	2.50	**184.50**	3.45	0.43	**53.63**	0.20	**238.33**	m2	**268.12**
25mm thick panelled linings to walls; square framed; including grounds										
width exceeding 300mm	188.80	2.50	**193.52**	2.60	0.33	**40.48**	0.14	**234.14**	m2	**263.41**
width exceeding 300mm; fixing to masonry with screws	188.80	2.50	**193.52**	3.60	0.45	**55.98**	0.20	**249.70**	m2	**280.91**
19mm thick panelled linings to walls; square framed; obstructed by integral services; including grounds										
width exceeding 300mm	180.00	2.50	**184.50**	2.95	0.37	**45.88**	0.17	**230.55**	m2	**259.37**
width exceeding 300mm; fixing to masonry with screws	180.00	2.50	**184.50**	3.95	0.49	**61.38**	0.23	**246.11**	m2	**276.87**
25mm thick panelled linings to walls; square framed; obstructed by integral services; including grounds										
width exceeding 300mm	188.80	2.50	**193.52**	3.10	0.39	**48.23**	0.17	**241.92**	m2	**272.16**
width exceeding 300mm; fixing to masonry with screws	188.80	2.50	**193.52**	4.10	0.51	**63.72**	0.23	**257.48**	m2	**289.66**
19mm thick panelled linings to walls; moulded; including grounds										
width exceeding 300mm	198.70	2.50	**203.67**	2.60	0.33	**40.48**	0.14	**244.29**	m2	**274.83**
width exceeding 300mm; fixing to masonry with screws	198.70	2.50	**203.67**	3.60	0.45	**55.98**	0.20	**259.85**	m2	**292.33**
25mm thick panelled linings to walls; moulded; including grounds										
width exceeding 300mm	206.20	2.50	**211.36**	2.75	0.34	**42.72**	0.15	**254.23**	m2	**286.00**
width exceeding 300mm; fixing to masonry with screws	206.20	2.50	**211.36**	3.75	0.47	**58.32**	0.22	**269.89**	m2	**303.63**
19mm thick panelled linings to walls; moulded; obstructed by integral services; including grounds										
width exceeding 300mm	198.70	2.50	**203.67**	3.10	0.39	**48.23**	0.17	**252.07**	m2	**283.58**
width exceeding 300mm; fixing to masonry with screws	198.70	2.50	**203.67**	4.10	0.51	**63.72**	0.23	**267.62**	m2	**301.08**
25mm thick panelled linings to walls; moulded; obstructed by integral services; including grounds										
width exceeding 300mm	206.20	2.50	**211.36**	3.25	0.41	**50.57**	0.18	**262.11**	m2	**294.87**
width exceeding 300mm; fixing to masonry with screws	206.20	2.50	**211.36**	4.25	0.53	**66.07**	0.25	**277.67**	m2	**312.38**
K13: RIGID SHEET FINE LININGS/PANELLING: VENEERED PLYWOOD PANELLING										
Pre-finished, decorative veneers; butt joints - wall panelling. Note. prices are for panelling with Afromosia, Ash, Beech, Cherry, Elm, Oak, Knotted Pine, Sapele or Teak faced veneers										
4mm thick linings to walls										
width exceeding 300mm	10.99	15.00	**12.64**	0.95	0.12	**14.78**	0.05	**27.48**	m2	**30.91**
width exceeding 300 mm; fixing to timber with adhesive	10.99	15.00	**12.64**	1.10	0.14	**17.13**	0.73	**30.50**	m2	**34.31**
width exceeding 300mm; fixing to plaster with adhesive	10.99	15.00	**12.64**	1.05	0.13	**16.31**	0.73	**29.68**	m2	**33.39**

Labour hourly rates: (except Specialists) Craft Operatives £14.21 Labourer £10.71 Rates are national average prices. Refer to REGIONAL VARIATIONS for indicative levels of overall pricing in regions	MATERIALS			LABOUR				RATES		
	Del to Site	Waste	Material Cost	Craft Optve	Lab	Labour Cost	Sunds	Nett Rate	Unit	Gross Rate (+12.5%)
	£	%	£	Hrs	Hrs	£	£	£		£
K13: RIGID SHEET FINE LININGS/PANELLING: VENEERED PLYWOOD PANELLING Cont.										
Flame retardant, pre-finished, decorative, veneers not matched; random V-grooves on face; butt joints - wall panelling. Note. prices are for panelling with Afromosia, Ash, Beech, Cherry, Elm, Oak, Knotted Pine, Sapele or Teak faced veneers										
4mm thick linings to walls										
width exceeding 300mm	29.50	15.00	**33.92**	0.95	0.12	**14.78**	0.05	**48.76**	m2	**54.85**
width exceeding 300 mm; fixing to timber with adhesive	29.50	15.00	**33.92**	1.10	0.14	**17.13**	0.73	**51.78**	m2	**58.25**
width exceeding 300mm; fixing to plaster with adhesive	29.50	15.00	**33.92**	1.05	0.13	**16.31**	0.73	**50.96**	m2	**57.33**
K20: TIMBER BOARD FLOORING/SHEATHING/LININGS/CASINGS										
Softwood, sawn; fixing to timber - boarded flooring										
Boarding to floors, square edges; 19mm thick, 75mm wide boards										
width exceeding 300mm	4.43	7.50	**4.77**	0.70	0.09	**10.91**	0.04	**15.71**	m2	**17.68**
Boarding to floors, square edges; 25mm thick, 125mm wide boards										
width exceeding 300mm	5.55	7.50	**5.97**	0.65	0.08	**10.09**	0.04	**16.10**	m2	**18.11**
Boarding to floors, square edges; 32mm thick, 150mm wide boards										
width exceeding 300mm	7.05	7.50	**7.58**	0.65	0.08	**10.09**	0.04	**17.71**	m2	**19.92**
Softwood, wrought; fixing to timber - boarded flooring										
Boarding to floors, square edges; 19mm thick, 75mm wide boards										
width exceeding 300mm	7.39	7.50	**7.95**	0.85	0.10	**13.15**	0.05	**21.14**	m2	**23.78**
Boarding to floors, square edges; 25mm thick, 125mm wide boards										
width exceeding 300mm	9.25	7.50	**9.94**	0.80	0.10	**12.44**	0.04	**22.42**	m2	**25.22**
Boarding to floors, square edges; 32mm thick, 150mm wide boards										
width exceeding 300mm	11.76	7.50	**12.64**	0.80	0.10	**12.44**	0.04	**25.12**	m2	**28.26**
Boarding to floors, tongued and grooved joints; 19mm thick, 75mm wide boards										
width exceeding 300mm	8.45	7.50	**9.08**	0.95	0.12	**14.78**	0.05	**23.92**	m2	**26.90**
Boarding to floors, tongued and grooved joints; 25mm thick, 125mm wide boards										
width exceeding 300mm	10.13	7.50	**10.88**	0.90	0.11	**13.97**	0.05	**24.90**	m2	**28.01**
Boarding to floors, tongued and grooved joints; 32mm thick, 150mm wide boards										
width exceeding 300mm	12.67	7.50	**13.62**	0.90	0.11	**13.97**	0.05	**27.64**	m2	**31.09**
Abutments										
19mm thick softwood boarding										
raking cutting	0.74	7.50	**0.79**	0.07	-	**0.99**	-	**1.79**	m	**2.01**
curved cutting	1.11	7.50	**1.19**	0.10	-	**1.42**	-	**2.61**	m	**2.94**
25mm thick softwood boarding										
raking cutting	0.92	7.50	**0.99**	0.07	-	**0.99**	-	**1.99**	m	**2.24**
curved cutting	1.39	7.50	**1.49**	0.10	-	**1.42**	-	**2.91**	m	**3.28**
32mm thick softwood boarding										
raking cutting	1.18	7.50	**1.26**	0.07	-	**0.99**	-	**2.26**	m	**2.54**
curved cutting	1.76	7.50	**1.90**	0.10	-	**1.42**	-	**3.32**	m	**3.73**
K20: TIMBER BOARD FLOORING/SHEATHING/LININGS/CASINGS: SURFACE TREATMENT OF EXISTING WOOD FLOORING										
Sanding and sealing existing wood flooring										
Machine sanding										
width exceeding 300mm	-	-	**-**	0.55	-	**7.82**	0.73	**8.54**	m2	**9.61**
Prepare, one priming coat and one finish coat of seal										
width exceeding 300mm	-	-	**-**	0.18	-	**2.56**	0.36	**2.92**	m2	**3.29**
K20: TIMBER BOARD FLOORING/SHEATHING/LININGS/CASINGS: WALL AND CEILING BOARDING										
Softwood, wrought; fixing to timber										
Boarding to walls, tongued, grooved and veed joints; 19mm thick, 100mm wide boards										
width exceeding 300mm	12.14	10.00	**13.36**	1.20	0.15	**18.66**	0.07	**32.08**	m2	**36.09**
Boarding to walls, tongued, grooved and veed joints; 19mm thick, 150mm wide boards										
width exceeding 300mm	11.09	10.00	**12.20**	0.85	0.10	**13.15**	0.05	**25.39**	m2	**28.57**

Labour hourly rates: (except Specialists) Craft Operatives £14.21 Labourer £10.71 Rates are national average prices. Refer to REGIONAL VARIATIONS for indicative levels of overall pricing in regions	MATERIALS			LABOUR			RATES			
	Del to Site £	Waste %	Material Cost £	Craft Optve Hrs	Lab Hrs	Labour Cost £	Sunds £	Nett Rate £	Unit	Gross Rate (+12.5%) £
K20: TIMBER BOARD FLOORING/SHEATHING/LININGS/CASINGS: WALL AND CEILING BOARDING Cont.										
Softwood, wrought; fixing to timber Cont.										
Boarding to walls, tongued, grooved and veed joints; 25mm thick, 100mm wide boards width exceeding 300mm	15.84	10.00	**17.42**	1.25	0.16	**19.48**	0.07	**36.97**	m2	**41.59**
Boarding to walls, tongued, grooved and veed joints; 25mm thick, 150mm wide boards width exceeding 300mm	14.78	10.00	**16.26**	0.90	0.11	**13.97**	0.05	**30.28**	m2	**34.06**
Boarding to ceilings, tongued, grooved and veed joints; 19mm thick, 100mm wide boards width exceeding 300mm	12.14	10.00	**13.36**	1.80	0.22	**27.93**	0.10	**41.39**	m2	**46.56**
Boarding to ceilings, tongued, grooved and veed joints; 19mm thick, 150mm wide boards width exceeding 300mm	11.09	10.00	**12.20**	1.20	0.15	**18.66**	0.07	**30.92**	m2	**34.79**
Boarding to ceilings, tongued, grooved and veed joints; 25mm thick, 100mm wide boards width exceeding 300mm	15.84	10.00	**17.42**	1.85	0.23	**28.75**	0.10	**46.27**	m2	**52.06**
Boarding to ceilings, tongued, grooved and veed joints; 25mm thick, 150mm wide boards width exceeding 300mm	14.78	10.00	**16.26**	1.25	0.16	**19.48**	0.07	**35.80**	m2	**40.28**
Western Red Cedar, wrought; fixing to timber										
Boarding to walls, tongued, grooved and veed joints; 19mm thick, 100mm wide boards width exceeding 300mm	46.97	10.00	**51.67**	1.25	0.16	**19.48**	0.07	**71.21**	m2	**80.11**
Boarding to walls, tongued, grooved and veed joints; 19mm thick, 150mm wide boards width exceeding 300mm	40.06	10.00	**44.07**	0.90	0.12	**14.07**	0.05	**58.19**	m2	**65.47**
Boarding to walls, tongued, grooved and veed joints; 25mm thick, 100mm wide boards width exceeding 300mm	58.96	10.00	**64.86**	1.30	0.17	**20.29**	0.07	**85.22**	m2	**95.88**
Boarding to walls, tongued, grooved and veed joints; 25mm thick, 150mm wide boards width exceeding 300mm	53.13	10.00	**58.44**	0.95	0.12	**14.78**	0.05	**73.28**	m2	**82.44**
Knotty Pine, wrought, selected for transparent finish; fixing to timber										
Boarding to walls, tongued, grooved and veed joints; 19mm thick, 100mm wide boards width exceeding 300mm	36.09	10.00	**39.70**	1.55	0.19	**24.06**	0.08	**63.84**	m2	**71.82**
Boarding to walls, tongued, grooved and veed joints; 19mm thick, 150mm wide boards width exceeding 300mm	30.82	10.00	**33.90**	1.05	0.13	**16.31**	0.06	**50.27**	m2	**56.56**
Boarding to walls, tongued, grooved and veed joints; 25mm thick, 100mm wide boards width exceeding 300mm	45.40	10.00	**49.94**	1.60	0.20	**24.88**	0.09	**74.90**	m2	**84.26**
Boarding to walls, tongued, grooved and veed joints; 25mm thick, 150mm wide boards width exceeding 300mm	40.89	10.00	**44.98**	1.10	0.14	**17.13**	0.06	**62.17**	m2	**69.94**
Boarding to ceilings, tongued, grooved and veed joints; 19mm thick, 100mm wide boards width exceeding 300mm	36.09	10.00	**39.70**	2.20	0.28	**34.26**	0.12	**74.08**	m2	**83.34**
Boarding to ceilings, tongued, grooved and veed joints; 19mm thick, 150mm wide boards width exceeding 300mm	30.82	10.00	**33.90**	1.60	0.20	**24.88**	0.09	**58.87**	m2	**66.23**
Boarding to ceilings, tongued, grooved and veed joints; 25mm thick, 100mm wide boards width exceeding 300mm	45.40	10.00	**49.94**	2.25	0.29	**35.08**	0.12	**85.14**	m2	**95.78**
Boarding to ceilings, tongued, grooved and veed joints; 25mm thick, 150mm wide boards width exceeding 300mm	40.89	10.00	**44.98**	1.65	0.21	**25.70**	0.09	**70.76**	m2	**79.61**
Sapele, wrought, selected for transparent finish; fixing to timber										
Boarding to walls, tongued, grooved and veed joints; 19mm thick, 100mm wide boards width exceeding 300mm	58.29	10.00	**64.12**	1.75	0.22	**27.22**	0.09	**91.44**	m2	**102.86**
Boarding to walls, tongued, grooved and veed joints; 19mm thick, 150mm wide boards width exceeding 300mm	50.99	10.00	**56.09**	1.20	0.15	**18.66**	0.07	**74.81**	m2	**84.17**
Boarding to walls, tongued, grooved and veed joints; 25mm thick, 100mm wide boards width exceeding 300mm	74.20	10.00	**81.62**	1.80	0.22	**27.93**	0.10	**109.65**	m2	**123.36**

Labour hourly rates: (except Specialists) Craft Operatives £14.21 Labourer £10.71 Rates are national average prices. Refer to REGIONAL VARIATIONS for indicative levels of overall pricing in regions	MATERIALS			LABOUR				RATES		
	Del to Site £	Waste %	Material Cost £	Craft Optve Hrs	Lab Hrs	Labour Cost £	Sunds £	Nett Rate £	Unit	Gross Rate (+12.5%) £
K20: TIMBER BOARD FLOORING/SHEATHING/LININGS/CASINGS: WALL AND CEILING BOARDING Cont.										
Sapele, wrought, selected for transparent finish; fixing to timber Cont.										
Boarding to walls, tongued, grooved and veed joints; 25mm thick, 150mm wide boards										
width exceeding 300mm	66.73	10.00	**73.40**	1.25	0.16	**19.48**	0.07	**92.95**	m2	**104.56**
Boarding to ceilings, tongued, grooved and veed joints; 19mm thick, 100mm wide boards										
width exceeding 300mm	58.29	10.00	**64.12**	2.40	0.30	**37.32**	0.13	**101.56**	m2	**114.26**
Boarding to ceilings, tongued, grooved and veed joints; 19mm thick, 150mm wide boards										
width exceeding 300mm	50.99	10.00	**56.09**	1.80	0.22	**27.93**	0.10	**84.12**	m2	**94.63**
Boarding to ceilings, tongued, grooved and veed joints; 25mm thick, 100mm wide boards										
width exceeding 300mm	74.20	10.00	**81.62**	2.45	0.31	**38.13**	0.13	**119.89**	m2	**134.87**
Boarding to ceilings, tongued, grooved and veed joints; 25mm thick, 150mm wide boards										
width exceeding 300mm	66.73	10.00	**73.40**	1.85	0.23	**28.75**	0.10	**102.25**	m2	**115.04**
Abutments										
19mm thick softwood boarding										
raking cutting	1.21	10.00	**1.34**	0.07	-	**0.99**	-	**2.33**	m	**2.62**
curved cutting	1.82	10.00	**2.00**	0.11	-	**1.56**	-	**3.57**	m	**4.01**
25mm thick softwood boarding										
raking cutting	1.58	10.00	**1.74**	0.08	-	**1.14**	-	**2.88**	m	**3.24**
curved cutting	2.38	10.00	**2.61**	0.12	-	**1.71**	-	**4.32**	m	**4.86**
19mm thick Western Red Cedar boarding										
raking cutting	4.70	10.00	**5.17**	0.08	-	**1.14**	-	**6.30**	m	**7.09**
curved cutting	7.05	10.00	**7.75**	0.12	-	**1.71**	-	**9.46**	m	**10.64**
25mm thick Western Red Cedar boarding										
raking cutting	5.90	10.00	**6.49**	0.10	-	**1.42**	-	**7.91**	m	**8.90**
curved cutting	8.84	10.00	**9.73**	0.13	-	**1.85**	-	**11.58**	m	**13.02**
19 mm thick hardwood boarding										
raking cutting	5.83	10.00	**6.41**	0.17	-	**2.42**	-	**8.83**	m	**9.93**
curved cutting	8.74	10.00	**9.62**	0.25	-	**3.55**	-	**13.17**	m	**14.82**
25 mm thick hardwood boarding										
raking cutting	7.42	10.00	**8.16**	0.19	-	**2.70**	-	**10.86**	m	**12.22**
curved cutting	11.13	10.00	**12.24**	0.29	-	**4.12**	-	**16.36**	m	**18.41**
Finished angles										
External; 19mm thick softwood boarding										
tongued and mitred	-	-	-	0.80	-	**11.37**	-	**11.37**	m	**12.79**
External; 25mm thick softwood boarding										
tongued and mitred	-	-	-	0.85	-	**12.08**	-	**12.08**	m	**13.59**
External; 19mm thick hardwood boarding										
tongued and mitred	-	-	-	1.15	-	**16.34**	-	**16.34**	m	**18.38**
External; 25mm thick hardwood boarding										
tongued and mitred	-	-	-	1.20	-	**17.05**	-	**17.05**	m	**19.18**
K20: TIMBER BOARD FLOORING/SHEATHING/LININGS/CASINGS: ROOF BOARDING										
Softwood, sawn, impregnated; fixing to timber										
Boarding to roofs, butt joints; 19mm thick, 75mm wide boards; external										
width exceeding 300mm	4.58	7.50	**4.92**	0.75	0.09	**11.62**	0.04	**16.58**	m2	**18.65**
width exceeding 300mm; sloping	4.58	7.50	**4.92**	0.85	0.11	**13.26**	0.05	**18.22**	m2	**20.50**
Boarding to roofs, butt joints; 25mm thick, 125mm wide boards; external										
width exceeding 300mm	6.02	7.50	**6.47**	0.70	0.09	**10.91**	0.04	**17.42**	m2	**19.60**
width exceeding 300mm; sloping	6.02	7.50	**6.47**	0.80	0.10	**12.44**	0.04	**18.95**	m2	**21.32**
Softwood, wrought, impregnated; fixing to timber										
Boarding to roofs, tongued and grooved joints; 19mm thick, 75mm wide boards external										
width exceeding 300mm	9.15	7.50	**9.84**	0.95	0.12	**14.78**	0.05	**24.67**	m2	**27.76**
extra; cross rebated and rounded drip; 50mm wide	0.95	10.00	**1.05**	0.24	0.03	**3.73**	0.01	**4.79**	m	**5.39**
extra; dovetailed cesspool 225 x 225 x 150mm	1.53	10.00	**1.68**	1.70	0.21	**26.41**	0.09	**28.18**	nr	**31.70**
extra; dovetailed cesspool 300 x 300 x 150mm	1.91	10.00	**2.10**	1.90	0.24	**29.57**	0.10	**31.78**	nr	**35.75**
width exceeding 300mm; sloping	9.15	7.50	**9.84**	1.05	0.13	**16.31**	0.06	**26.21**	m2	**29.48**
Boarding to roofs, tongued and grooved joints; 25mm thick, 125mm wide boards external										
width exceeding 300mm	10.96	7.50	**11.78**	0.90	0.11	**13.97**	0.05	**25.80**	m2	**29.02**
extra; cross rebated and rounded drip; 50mm wide	0.95	10.00	**1.05**	0.24	0.03	**3.73**	0.01	**4.79**	m	**5.39**
extra; dovetailed cesspool 225 x 225 x 150mm	1.91	10.00	**2.10**	1.80	0.23	**28.04**	0.10	**30.24**	nr	**34.02**

Labour hourly rates: (except Specialists) Craft Operatives £14.21 Labourer £10.71 Rates are national average prices. Refer to REGIONAL VARIATIONS for indicative levels of overall pricing in regions	MATERIALS			LABOUR				RATES		
	Del to Site £	Waste %	Material Cost £	Craft Optve Hrs	Lab Hrs	Labour Cost £	Sunds £	Nett Rate £	Unit	Gross Rate (+12.5%) £
K20: TIMBER BOARD FLOORING/SHEATHING/LININGS/CASINGS: ROOF BOARDING Cont.										
Softwood, wrought, impregnated; fixing to timber Cont.										
Boarding to roofs, tongued and grooved joints; 25mm thick, 125mm wide boards external Cont.										
extra; dovetailed cesspool 300 x 300 x 150mm.....	2.57	10.00	**2.83**	2.00	0.25	**31.10**	0.11	**34.03**	nr	**38.29**
width exceeding 300mm; sloping..................	10.96	7.50	**11.78**	1.00	0.12	**15.50**	0.05	**27.33**	m2	**30.74**
Abutments										
19mm thick softwood boarding										
raking cutting	0.92	7.50	**0.98**	0.40	-	**5.68**	-	**6.67**	m	**7.50**
curved cutting	1.37	7.50	**1.48**	0.65	-	**9.24**	-	**10.71**	m	**12.05**
rebates	-	-	-	0.22	-	**3.13**	-	**3.13**	m	**3.52**
grooves	-	-	-	0.22	-	**3.13**	-	**3.13**	m	**3.52**
chamfers	-	-	-	0.16	-	**2.27**	-	**2.27**	m	**2.56**
25mm thick softwood boarding										
raking cutting	1.08	7.50	**1.16**	0.55	-	**7.82**	-	**8.98**	m	**10.10**
curved cutting	1.63	7.50	**1.75**	0.80	-	**11.37**	-	**13.12**	m	**14.75**
rebates	-	-	-	0.22	-	**3.13**	-	**3.13**	m	**3.52**
grooves	-	-	-	0.22	-	**3.13**	-	**3.13**	m	**3.52**
chamfers	-	-	-	0.16	-	**2.27**	-	**2.27**	m	**2.56**
K21: TIMBER STRIP/BOARD FINE FLOORING/LININGS: STRIP FLOORING										
Hardwood, wrought, selected for transparent finish; sanded, two coats sealer finish										
Boarding to floors, tongued and grooved joints; 22mm thick, 75mm wide boards; fixing on and with 25 x 50mm impregnated softwood battens to masonry										
width exceeding 300 mm; secret fixing to timber..	40.38	7.50	**43.39**	1.50	0.19	**23.35**	0.08	**66.82**	m2	**75.17**
Boarding to floors, tongued and grooved joints; 12mm thick, 75mm wide boards; overlay										
width exceeding 300 mm; secret fixing to timber..	32.92	7.50	**35.39**	1.20	0.15	**18.66**	0.07	**54.12**	m2	**60.88**
K21: TIMBER STRIP/BOARD FINE FLOORING/LININGS: SEMI-SPRUNG FLOORS										
Foam backed softwood battens covered with hardwood tongued and grooved strip flooring										
22mm Standard Beech strip flooring to level sub-floors										
width exceeding 300mm; battens at approximately 400mm centres; elevation										
75mm...	-	-	**Spclist**	-	-	**Spclist**	-	**56.47**	m2	**63.53**
22mm 'Sylva Squash' Beech strip flooring to level sub-floors										
width exceeding 300mm; battens at approximately 430mm centres; elevation										
75mm...	-	-	**Spclist**	-	-	**Spclist**	-	**60.38**	m2	**67.93**
20mm Prime Maple strip flooring to level sub-floors										
width exceeding 300mm; battens at approximately 300mm centres; elevation										
75mm...	-	-	**Spclist**	-	-	**Spclist**	-	**45.94**	m2	**51.68**
20mm First Grade Maple strip flooring to level sub-floors										
width exceeding 300mm; battens at approximately 300mm centres; elevation										
75mm...	-	-	**Spclist**	-	-	**Spclist**	-	**49.88**	m2	**56.12**
K40: DEMOUNTABLE SUSPENDED CEILINGS										
Suspended ceilings; 1200 x 600mm bevelled, grooved and rebated asbestos-free fire resisting tiles; laying in position in metal suspension system of main channel members on wire or rod hangers										
Depth of suspension 150 - 500mm										
9mm thick linings; fixing hangers to masonry.....	17.45	5.00	**18.32**	0.60	0.30	**11.74**	4.80	**34.86**	m2	**39.22**
9mm thick linings not exceeding 300mm wide; fixing hangers to masonry	5.24	10.00	**5.76**	0.30	0.08	**5.07**	2.90	**13.72**	m	**15.44**
Suspended ceilings; 600 x 600mm bevelled, grooved and rebated asbestos-free fire resisting tiles; laying in position in metal suspension system of main channel members on wire or rod hangers										
Depth of suspension 150 - 500mm										
9mm thick linings; fixing hangers to masonry.....	17.10	5.00	**17.96**	0.72	0.36	**14.09**	2.95	**34.99**	m2	**39.37**
9mm thick linings not exceeding 300mm wide; fixing hangers to masonry	5.13	10.00	**5.64**	0.36	0.11	**6.29**	2.90	**14.84**	m	**16.69**

Labour hourly rates: (except Specialists) Craft Operatives £14.21 Labourer £10.71 Rates are national average prices. Refer to REGIONAL VARIATIONS for indicative levels of overall pricing in regions	MATERIALS			LABOUR				RATES		
	Del to Site £	Waste %	Material Cost £	Craft Optve Hrs	Lab Hrs	Labour Cost £	Sunds £	Nett Rate £	Unit	Gross Rate (+12.5%) £
K40: DEMOUNTABLE SUSPENDED CEILINGS Cont.										
Suspended ceilings; 1200 x 600mm bevelled, grooved and rebated plain mineral fibre tiles; laying in position in metal suspension system of main channel members on wire or rod hangers										
Depth of suspension 150 - 500mm										
15.8mm thick linings; fixing hangers to masonry..	10.60	5.00	**11.13**	0.60	0.30	**11.74**	4.85	**27.72**	m2	**31.18**
15.8mm thick linings not exceeding 300mm wide; fixing hangers to masonry......................	3.18	10.00	**3.50**	0.30	0.08	**5.07**	2.90	**11.46**	m	**12.90**
Suspended ceilings; 600 x 600mm bevelled, grooved and rebated plain mineral fibre tiles; laying in position in metal suspension system of main channel members on wire or rod hangers										
Depth of suspension 150 - 500mm										
15.8mm thick linings; fixing hangers to masonry..	10.60	5.00	**11.13**	0.72	0.36	**14.09**	2.95	**28.17**	m2	**31.69**
15.8mm thick linings not exceeding 300mm wide; fixing hangers to masonry......................	3.98	10.00	**4.37**	0.43	0.11	**7.29**	1.75	**13.41**	m	**15.09**
Suspended ceilings; 1200 x 600mm bevelled, grooved and rebated textured mineral fibre tiles; laying in position in metal suspension system of main channel members on wire or rod hangers										
Depth of suspension 150 - 500mm										
15.8mm thick linings; fixing hangers to masonry..	13.25	5.00	**13.91**	0.60	0.30	**11.74**	4.85	**30.50**	m2	**34.31**
15.8mm thick linings not exceeding 300mm wide; fixing hangers to masonry......................	3.98	10.00	**4.37**	0.30	0.08	**5.07**	2.90	**12.34**	m	**13.88**
Suspended ceilings; 600 x 600mm bevelled, grooved and rebated regular drilled mineral fibre tiles; laying in position in metal suspension system of main channel members on wire or rod hangers										
Depth of suspension 150 - 500mm										
15.8mm thick linings; fixing hangers to masonry..	13.25	5.00	**13.91**	0.72	0.36	**14.09**	4.10	**32.10**	m2	**36.11**
15.8mm thick linings not exceeding 300mm wide; fixing hangers to masonry......................	3.98	10.00	**4.37**	0.43	0.11	**7.29**	2.45	**14.11**	m	**15.87**
Suspended ceilings; 1200 x 600mm bevelled, grooved and rebated patterned tiles p.c. £15.00/m2; laying in position in metal suspension system of main channel members on wire or rod hangers										
Depth of suspension 150 - 500mm										
15.8mm thick linings; fixing hangers to masonry..	15.75	5.00	**16.54**	0.60	0.30	**11.74**	4.85	**33.13**	m2	**37.27**
15.8mm thick linings not exceeding 300mm wide; fixing hangers to masonry......................	4.95	10.00	**5.45**	0.30	0.08	**5.07**	2.90	**13.41**	m	**15.09**
Suspended ceilings; 600 x 600mm bevelled, grooved and rebated patterned tiles p.c. £15.00/m2; laying in position in metal suspension system of main channel members on wire or rod hangers										
Depth of suspension 150 - 500mm										
15.8mm thick linings; fixing hangers to masonry..	15.75	5.00	**16.54**	0.72	0.36	**14.09**	4.10	**34.72**	m2	**39.06**
15.8mm thick linings not exceeding 300mm wide; fixing hangers to masonry......................	4.95	10.00	**5.45**	0.43	0.11	**7.29**	2.45	**15.18**	m	**17.08**
Suspended ceiling; British Gypsum Ltd, 1200 x 600mm Glasroc GRG tiles; fixing into Casoline MF grid system										
Depth of suspension 150-500mm										
10mm thick linings; fixing hangers to masonry....	5.75	5.00	**6.04**	0.43	0.11	**7.29**	4.10	**17.43**	m2	**19.60**
10mm thick linings not exceeding 300mm wide; fixing hangers to masonry......................	1.73	10.00	**1.90**	0.43	0.11	**7.29**	4.10	**13.29**	m	**14.95**
Edge trims										
Plain										
angle section; white stove enamelled aluminium; fixing to masonry with screws at 450mm centres...	1.25	5.00	**1.31**	0.50	0.25	**9.78**	-	**11.10**	m	**12.48**
K41: RAISED ACCESS FLOORS										
Flooring; Microfloor 'Bonded 600' light grade full access system; 600 x 600mm high density particle board panels, BS EN 312 and DIN 68761; 100 x 100mm precast lightweight concrete pedestals at 600mm centres fixed to sub-floor with epoxy resin adhesive										
Thickness of panel 30mm										
finished floor height										
50mm..............................	-	-	**Spclist**	-	-	**Spclist**	-	**19.24**	m2	**21.65**
75mm..............................	-	-	**Spclist**	-	-	**Spclist**	-	**19.38**	m2	**21.80**
100mm.............................	-	-	**Spclist**	-	-	**Spclist**	-	**19.62**	m2	**22.07**
125mm.............................	-	-	**Spclist**	-	-	**Spclist**	-	**19.83**	m2	**22.31**
150mm.............................	-	-	**Spclist**	-	-	**Spclist**	-	**20.23**	m2	**22.76**
175mm.............................	-	-	**Spclist**	-	-	**Spclist**	-	**20.55**	m2	**23.12**
200mm.............................	-	-	**Spclist**	-	-	**Spclist**	-	**20.73**	m2	**23.32**

Labour hourly rates: (except Specialists) Craft Operatives £14.21 Labourer £10.71 Rates are national average prices. Refer to REGIONAL VARIATIONS for indicative levels of overall pricing in regions	MATERIALS			LABOUR				RATES		
	Del to Site £	Waste %	Material Cost £	Craft Optve Hrs	Lab Hrs	Labour Cost £	Sunds £	Nett Rate £	Unit	Gross Rate (+12.5%) £
K41: RAISED ACCESS FLOORS Cont.										
Flooring; Microfloor 'Bonded 600' medium grade full access system; 600 x 600mm high density particle board panels, BS EN 312 and DIN 68761; 100 x 100mm precast lightweight concrete pedestals at 600mm centres fixed to sub-floor with epoxy resin adhesive										
Thickness of panel 38mm finished floor height										
50mm	-	-	Spclist	-	-	Spclist	-	21.88	m2	24.62
75mm	-	-	Spclist	-	-	Spclist	-	21.88	m2	24.62
100mm	-	-	Spclist	-	-	Spclist	-	22.30	m2	25.09
125mm	-	-	Spclist	-	-	Spclist	-	22.49	m2	25.30
150mm	-	-	Spclist	-	-	Spclist	-	23.22	m2	26.12
175mm	-	-	Spclist	-	-	Spclist	-	23.41	m2	26.34
200mm	-	-	Spclist	-	-	Spclist	-	23.59	m2	26.54
Flooring; Microfloor 'Bonded 600' office loadings grade full access system; 600 x 600mm high density particle board panels, BS EN 634 and DIN 68761; 100 x 100mm precast lightweight concrete pedestals at 600mm centres fixed to sub-floor with epoxy resin adhesive										
Thickness of panel 30mm finished floor height										
50mm	-	-	Spclist	-	-	Spclist	-	14.98	m2	16.85
75mm	-	-	Spclist	-	-	Spclist	-	15.07	m2	16.95
100mm	-	-	Spclist	-	-	Spclist	-	15.40	m2	17.33
125mm	-	-	Spclist	-	-	Spclist	-	15.59	m2	17.54
150mm	-	-	Spclist	-	-	Spclist	-	15.97	m2	17.97
175mm	-	-	Spclist	-	-	Spclist	-	16.33	m2	18.37
200mm	-	-	Spclist	-	-	Spclist	-	16.50	m2	18.56

WINDOWS/DOORS/STAIRS

Labour hourly rates: (except Specialists) Craft Operatives £14.21 Labourer £10.71 Rates are national average prices. Refer to REGIONAL VARIATIONS for indicative levels of overall pricing in regions	MATERIALS			LABOUR				RATES		
	Del to Site £	Waste %	Material Cost £	Craft Optve Hrs	Lab Hrs	Labour Cost £	Sunds £	Nett Rate £	Unit	Gross Rate (+12.5%) £
L10: WINDOWS/ROOFLIGHTS/SCREENS/LOUVRES: TIMBER WINDOWS										
Casements in softwood, base coat stain by manufacturer										
Magnet Trade; softwood Statesman Plus casement windows without bars; softwood sub-sills; hinges; fasteners										
488 x 900mm overall; N09V	82.16	2.50	**84.22**	1.55	0.77	**30.27**	1.62	**116.11**	nr	**130.62**
631 x 750mm overall; 107V	95.60	2.50	**97.99**	1.55	0.77	**30.27**	1.62	**129.88**	nr	**146.12**
631 x 900mm overall; 109V	99.70	2.50	**102.19**	1.70	0.85	**33.26**	1.62	**137.07**	nr	**154.20**
631 x 1050mm overall; 110V	103.48	2.50	**106.06**	1.80	0.90	**35.22**	1.62	**142.90**	nr	**160.76**
631 x 1200mm overall; 112v	107.47	2.50	**110.15**	1.90	0.95	**37.17**	1.62	**148.95**	nr	**167.57**
915 x 900mm overall; 2N09W	124.37	2.50	**127.48**	1.90	0.95	**37.17**	1.62	**166.28**	nr	**187.06**
915 x 1050mm overall; 2N10W	121.96	2.50	**125.01**	2.05	1.02	**40.05**	1.62	**166.68**	nr	**187.52**
915 x 1200mm overall; 2N12W	132.67	2.50	**135.98**	2.15	1.07	**42.01**	1.62	**179.62**	nr	**202.07**
1200 x 900mm overall; 209W	145.48	2.50	**149.11**	2.15	1.07	**42.01**	1.62	**192.75**	nr	**216.84**
1200 x 1050mm overall; 210CV	176.35	2.50	**180.76**	2.30	1.15	**45.00**	1.62	**227.38**	nr	**255.80**
1200 x 1050mm overall; 210W	148.21	2.50	**151.91**	2.30	1.15	**45.00**	1.62	**198.53**	nr	**223.35**
1200 x 1200mm overall; 212CV	176.14	2.50	**180.54**	2.40	1.20	**46.96**	1.62	**229.12**	nr	**257.76**
1200 x 1200mm overall; 212W	156.92	2.50	**160.85**	2.40	1.20	**46.96**	1.62	**209.42**	nr	**235.60**
1524 x 1050mm overall; 3NN10WW	190.31	2.50	**195.07**	2.50	1.25	**48.91**	1.62	**245.60**	nr	**276.30**
1769 x 1050mm overall; 310CVC	255.31	2.50	**261.69**	2.90	1.45	**56.74**	1.62	**320.05**	nr	**360.06**
1769 x 1200mm overall; 312CVC	286.07	2.50	**293.22**	3.00	1.50	**58.70**	1.62	**353.54**	nr	**397.73**
Fully reversible windows in softwood, wrought; base coat stain by manufacturer										
Magnet Trade Hi-Profile; softwood horizontal tilt windows; locking catch; high security espagnolette bolt; fully weatherstripped and Factory Double Glazed; clear low E glass										
600 x 900mm overall; HP0906	294.07	2.50	**301.42**	1.60	0.80	**31.30**	1.62	**334.34**	nr	**376.13**
900 x 900mm overall; HP0909	325.25	2.50	**333.38**	1.90	0.95	**37.17**	1.62	**372.18**	nr	**418.70**
900 x 1050mm overall; HP0910	338.54	2.50	**347.00**	2.00	1.00	**39.13**	1.62	**387.75**	nr	**436.22**
900 x 1200mm overall; HP0912	353.71	2.50	**362.56**	2.10	1.05	**41.09**	1.62	**405.26**	nr	**455.92**
1200 x 900mm overall; HP1209	365.40	2.50	**374.54**	2.10	1.05	**41.09**	1.62	**417.24**	nr	**469.40**
1200 x 1050mm overall; HP1210	386.64	2.50	**396.31**	2.25	1.12	**43.97**	1.62	**441.90**	nr	**497.13**
1200 x 1200mm overall; HP1212	405.32	2.50	**415.45**	2.40	1.20	**46.96**	1.62	**464.02**	nr	**522.03**
Sash windows in softwood, wrought; base coat stain by manufacturer										
Magnet Victorian Statesman plus non bar sash softwood windows; glazing beads; fasteners; base coat stain by manufacturer										
1200 x 900mm overall; VBS209B	198.59	2.50	**203.55**	2.90	1.45	**56.74**	1.62	**261.91**	nr	**294.65**
1200 x 1050mm overall; VBS210B	207.95	2.50	**213.15**	3.05	1.52	**59.62**	1.62	**274.39**	nr	**308.69**
1200 x 1200mm overall; VBS212B	207.95	2.50	**213.15**	3.30	1.65	**64.56**	1.62	**279.33**	nr	**314.25**
1200 x 1350mm overall; VBS213B	215.23	2.50	**220.61**	3.60	1.80	**70.43**	1.62	**292.66**	nr	**329.25**
1769 x 1050mm overall; VBS310BB	375.39	2.50	**384.77**	3.70	1.85	**72.39**	1.62	**458.79**	nr	**516.13**
1769 x 1200mm overall; VBS312BB	308.07	2.50	**315.77**	4.10	2.05	**80.22**	1.62	**397.61**	nr	**447.31**
1769 x 1350mm overall; VBS313BB	387.30	2.50	**396.98**	4.30	2.15	**84.13**	1.62	**482.73**	nr	**543.07**
1769 x 1050mm overall; VBN310BBMOE	614.36	2.50	**629.72**	3.70	1.85	**72.39**	1.62	**703.73**	nr	**791.70**
1769 x 1200mm overall; VBN312BBMOE	632.26	2.50	**648.07**	4.10	2.05	**80.22**	1.62	**729.90**	nr	**821.14**
1769 x 1350mm overall; VBN313BBMOE	645.04	2.50	**661.17**	4.30	2.15	**84.13**	1.62	**746.92**	nr	**840.28**
Windows in softwood, base coat stained by manufacturer										
Jeld-Wen Ltd. Softwood Sovereign windows plain casement non bar type side hung; glazing beads; weatherstripping, hinges; fasteners; base coat stain by manufacturer										
630 x 750mm overall; W107C	104.85	2.50	**107.47**	1.90	0.95	**37.17**	1.62	**146.26**	nr	**164.55**
630 x 900mm overall; W109C	111.25	2.50	**114.03**	2.20	1.10	**43.04**	1.62	**158.69**	nr	**178.53**
630 x 1050mm overall; W110C	111.71	2.50	**114.50**	2.45	1.22	**47.88**	1.62	**164.00**	nr	**184.50**
630 x 1200mm overall; W112C	115.87	2.50	**118.77**	2.65	1.32	**51.79**	1.62	**172.18**	nr	**193.70**
630 x 1350mm overall; W113C	129.91	2.50	**133.16**	2.90	1.45	**56.74**	1.62	**191.52**	nr	**215.46**
1200 x 750mm overall; W207C	136.89	2.50	**140.31**	2.60	1.30	**50.87**	1.62	**192.80**	nr	**216.90**
1200 x 900mm overall; W209C	140.05	2.50	**143.55**	2.90	1.45	**56.74**	1.62	**201.91**	nr	**227.15**
1200 x 1050mm overall; W210C	142.84	2.50	**146.41**	3.05	1.52	**59.62**	1.62	**207.65**	nr	**233.61**
1200 x 1200mm overall; W212C	145.62	2.50	**149.26**	3.30	1.65	**64.56**	1.62	**215.45**	nr	**242.38**

Labour hourly rates: (except Specialists) Craft Operatives £14.21 Labourer £10.71 Rates are national average prices. Refer to REGIONAL VARIATIONS for indicative levels of overall pricing in regions	MATERIALS			LABOUR				RATES		
	Del to Site £	Waste %	Material Cost £	Craft Optve Hrs	Lab Hrs	Labour Cost £	Sunds £	Nett Rate £	Unit	Gross Rate (+12.5%) £
L10: WINDOWS/ROOFLIGHTS/SCREENS/LOUVRES: TIMBER WINDOWS Cont.										
Windows in softwood, base coat stained by manufacturer Cont.										
Jeld-Wen Ltd. Softwood Sovereign windows plain casement non bar type side hung; glazing beads; weatherstripping, hinges; fasteners; base coat stain by manufacturer Cont.										
1200 x 1350mm overall; W213C	163.22	2.50	167.30	3.55	1.77	69.40	1.62	238.32	nr	268.11
1770 x 750mm overall; W307CC	201.51	2.50	206.55	3.30	1.65	64.56	1.62	272.73	nr	306.82
1770 x 900mm overall; W309CC	199.67	2.50	204.66	3.55	1.77	69.40	1.62	275.68	nr	310.14
1770 x 1050mm overall; W310C; Landscape	168.07	2.50	172.28	3.70	1.85	72.39	1.62	246.29	nr	277.07
1770 x 1050mm overall; W310CC	207.14	2.50	212.32	3.70	1.85	72.39	1.62	286.33	nr	322.12
1770 x 1200mm overall; W312C; Landscape	214.67	2.50	220.04	4.00	2.00	78.26	1.62	299.92	nr	337.41
1770 x 1200mm overall; W312CC	207.71	2.50	212.90	4.00	2.00	78.26	1.62	292.78	nr	329.38
1770 x 1350mm overall; W313C; Landscape	181.74	2.50	186.29	4.30	2.15	84.13	1.62	272.04	nr	306.04
1770 x 1350mm overall; W313CC	242.13	2.50	248.18	4.30	2.15	84.13	1.62	333.93	nr	375.67
2339 x 900mm overall; W409CMC	257.92	2.50	264.37	4.30	2.15	84.13	1.62	350.12	nr	393.88
2339 x 1050mm overall; W410CMC	268.86	2.50	275.58	4.60	2.30	90.00	1.62	367.20	nr	413.10
2339 x 1200mm overall; W412CMC	280.21	2.50	287.22	4.85	2.42	94.84	1.62	383.67	nr	431.63
2339 x 1350mm overall; W413CMC	298.07	2.50	305.52	5.10	2.55	99.78	1.62	406.92	nr	457.79
Jeld-Wen Ltd. Softwood Alpha Sovereign windows casement with vents top hung type; glazing beads; weatherstripping, hinges; fasteners; base coat stain by manufacturer										
630 x 750mm overall; W107V	102.77	2.50	105.34	1.80	0.90	35.22	1.62	142.18	nr	159.95
630 x 900mm overall; W109V	103.11	2.50	105.69	1.90	0.95	37.17	1.62	144.48	nr	162.54
630 x 1050mm overall; W110V	105.54	2.50	108.18	2.20	1.10	43.04	1.62	152.84	nr	171.95
630 x 1200mm overall; W112V	113.85	2.50	116.70	2.45	1.22	47.88	1.62	166.20	nr	186.97
630 x 1350mm overall; W113V	117.08	2.50	120.01	2.65	1.32	51.79	1.62	173.42	nr	195.10
915 x 750mm overall; W2N07CV	118.68	2.50	121.65	2.15	1.07	42.01	1.62	165.28	nr	185.94
915 x 900mm overall; W2N09CV	121.77	2.50	124.81	2.35	1.17	45.92	1.62	172.36	nr	193.90
915 x 1050mm overall; W2N10CV	123.43	2.50	126.51	2.60	1.30	50.87	1.62	179.00	nr	201.38
915 x 1200mm overall; W2N12CV	133.18	2.50	136.51	2.80	1.40	54.78	1.62	192.91	nr	217.03
1200 x 900mm overall; W209T	172.82	2.50	177.14	2.60	1.30	50.87	1.62	229.63	nr	258.33
1200 x 1050mm overall; W210T	173.04	2.50	177.37	2.90	1.45	56.74	1.62	235.72	nr	265.19
1200 x 1200mm overall; W212T	178.52	2.50	182.98	3.05	1.52	59.62	1.62	244.22	nr	274.75
1200 x 1350mm overall; W213T	197.67	2.50	202.61	3.30	1.65	64.56	1.62	268.80	nr	302.40
1200 x 1500mm overall; W215T	245.03	2.50	251.16	3.20	1.60	62.61	1.62	315.38	nr	354.81
1770 x 1050mm overall; W310TXT	253.02	2.50	259.35	3.30	1.65	64.56	1.62	325.53	nr	366.22
1770 x 1350mm overall; W313TXT	277.04	2.50	283.97	3.70	1.85	72.39	1.62	357.98	nr	402.72
1770 x 1500mm overall; W315TXT	246.52	2.50	252.68	4.00	2.00	78.26	1.62	332.56	nr	374.13
Windows in hardwood, base coat stained by manufacturer										
Jeld-Wen Ltd. Hardwood Sovereign windows plain casement non bar type side hung; glazing beads; weatherstripping, hinges; fasteners; base coat stain by manufacturer										
630 x 750mm overall; W107CH	196.26	2.50	201.17	2.45	1.22	47.88	1.62	250.67	nr	282.00
630 x 900mm overall; W109CH	204.72	2.50	209.84	2.65	1.32	51.79	1.62	263.25	nr	296.16
630 x 1050mm overall; W110CH	213.59	2.50	218.93	2.65	1.32	51.79	1.62	272.34	nr	306.39
630 x 1200mm overall; W112CH	222.39	2.50	227.95	2.75	1.37	53.75	1.62	283.32	nr	318.73
630 x 1350mm overall; W113CH	238.57	2.50	244.53	2.75	1.37	53.75	1.62	299.90	nr	337.39
1200 x 750mm overall; W207CH	252.47	2.50	258.78	2.60	1.30	50.87	1.62	311.27	nr	350.18
1200 x 900mm overall; W209CH	265.03	2.50	271.66	2.80	1.40	54.78	1.62	328.06	nr	369.06
1200 x 1050mm overall; W210CH	274.63	2.50	281.50	3.05	1.52	59.62	1.62	342.74	nr	385.58
1200 x 1200mm overall; W212CH	286.25	2.50	293.41	3.30	1.65	64.56	1.62	359.59	nr	404.54
1200 x 1350mm overall; W213CH	305.74	2.50	313.38	3.55	1.77	69.40	1.62	384.41	nr	432.46
1770 x 750mm overall; W307CCH	383.09	2.50	392.67	3.30	1.65	64.56	1.62	458.85	nr	516.21
1770 x 900mm overall; W309CCH	388.69	2.50	398.41	3.55	1.77	69.40	1.62	469.43	nr	528.11
1770 x 1050mm overall; W310CH; Landscape	312.62	2.50	320.43	3.70	1.85	72.39	1.62	394.44	nr	443.75
1770 x 1050mm overall; W310CCH	417.81	2.50	428.26	3.70	1.85	72.39	1.62	502.27	nr	565.05
1770 x 1200mm overall; W312CH; Landscape	434.27	2.50	445.13	4.00	2.00	78.26	1.62	525.01	nr	590.63
1770 x 1200mm overall; W312CCH	420.17	2.50	430.67	4.00	2.00	78.26	1.62	510.55	nr	574.37
1770 x 1350mm overall; W313CH; Landscape	329.87	2.50	338.11	4.30	2.15	84.13	1.62	423.86	nr	476.85
1770 x 1350mm overall; W313CCH	465.35	2.50	476.98	4.30	2.15	84.13	1.62	562.73	nr	633.07
2339 x 900mm overall; W409CMCH	506.99	2.50	519.66	4.50	2.25	88.04	1.62	609.33	nr	685.49
2339 x 1050mm overall; W410CMCH	519.07	2.50	532.05	4.60	2.30	90.00	1.62	623.67	nr	701.62
2339 x 1200mm overall; W412CMCH	542.12	2.50	555.67	4.85	2.42	94.84	1.62	652.13	nr	733.65
2339 x 1350mm overall; W413CMCH	587.05	2.50	601.73	5.05	2.52	98.75	1.62	702.10	nr	789.86
Jeld-Wen Ltd. Hardwood Sovereign windows casement with vents; glazing beads; weatherstripping, hinges; fasteners; base coat stain by manufacturer										
630 x 750mm overall; W107VH	188.48	2.50	193.19	1.95	0.95	37.88	1.62	232.70	nr	261.78
630 x 900mm overall; W109VH	195.11	2.50	199.99	2.65	1.08	49.22	1.62	250.83	nr	282.18
630 x 1050mm overall; W110VH	200.10	2.50	205.10	2.45	1.22	47.88	1.62	254.60	nr	286.43
630 x 1200mm overall; W112VH	205.74	2.50	210.88	2.60	1.30	50.87	1.62	263.37	nr	296.29
630 x 1350mm overall; W113VH	212.52	2.50	217.83	3.00	1.50	58.70	1.62	278.15	nr	312.92
915 x 750mm overall; W2N07CVH	227.50	2.50	233.19	2.45	1.22	47.88	1.62	282.69	nr	318.03
915 x 900mm overall; W2N09CVH	235.94	2.50	241.83	2.60	1.30	50.87	1.62	294.32	nr	331.11
915 x 1050mm overall; W2N10CVH	239.74	2.50	245.73	2.85	1.42	55.71	1.62	303.06	nr	340.94
915 x 1200mm overall; W2N12CVH	245.75	2.50	251.90	2.95	1.47	57.66	1.62	311.18	nr	350.08
1200 x 900mm overall; W209TH	334.59	2.50	342.95	2.95	1.47	57.66	1.62	402.24	nr	452.52
1200 x 1050mm overall; W210TH	344.88	2.50	353.50	3.05	1.52	59.62	1.62	414.74	nr	466.58
1200 x 1200mm overall; W212TH	357.04	2.50	365.97	3.30	1.65	64.56	1.62	432.15	nr	486.17
1200 x 1350mm overall; W213TH	375.50	2.50	384.89	3.60	1.80	70.43	1.62	456.94	nr	514.06
1200 x 1500mm overall; W215TH	470.99	2.50	482.76	3.60	1.80	70.43	1.62	554.82	nr	624.17
1770 x 1050mm overall; W310TXTH	487.11	2.50	499.29	3.80	1.90	74.35	1.62	575.25	nr	647.16
1770 x 1350mm overall; W313TXTH	535.88	2.50	549.28	4.30	2.15	84.13	1.62	635.03	nr	714.40
1770 x 1500mm overall; W315TXTH	460.85	2.50	472.37	4.50	2.25	88.04	1.62	562.03	nr	632.28

Labour hourly rates: (except Specialists) Craft Operatives £14.21 Labourer £10.71 Rates are national average prices. Refer to REGIONAL VARIATIONS for indicative levels of overall pricing in regions	MATERIALS			LABOUR				RATES		
	Del to Site £	Waste %	Material Cost £	Craft Optve Hrs	Lab Hrs	Labour Cost £	Sunds £	Nett Rate £	Unit	Gross Rate (+12.5%) £
L10: WINDOWS/ROOFLIGHTS/SCREENS/LOUVRES: TIMBER WINDOWS Cont.										
Windows in hardwood, base coat stained by manufacturer Cont.										
Jeld-Wen Ltd. Hardwood Sovereign windows, horizontal glazing bar type; glazing beads; weatherstripping, hinges; fasteners; base coat stain by manufacturer										
630 x 900mm overall; WH109CH	123.33	2.50	**126.41**	2.20	1.10	**43.04**	1.62	**171.08**	nr	**192.46**
630 x 1050mm overall; WH110CH	127.23	2.50	**130.41**	2.45	1.22	**47.88**	1.62	**179.91**	nr	**202.40**
630 x 1200mm overall; WH112CH	131.40	2.50	**134.69**	2.65	1.32	**51.79**	1.62	**188.10**	nr	**211.61**
630 x 1350mm overall; WH113CH	139.50	2.50	**142.99**	2.90	1.45	**56.74**	1.62	**201.35**	nr	**226.51**
1200 x 900mm overall; WH209CH	216.07	2.50	**221.47**	2.90	1.45	**56.74**	1.62	**279.83**	nr	**314.81**
1200 x 1050mm overall; WH210CH	223.66	2.50	**229.25**	3.05	1.52	**59.62**	1.62	**290.49**	nr	**326.80**
1200 x 1200mm overall; WH212CH	231.86	2.50	**237.66**	3.30	1.65	**64.56**	1.62	**303.84**	nr	**341.82**
1200 x 1350mm overall; WH213CH	248.58	2.50	**254.79**	3.60	1.80	**70.43**	1.62	**326.85**	nr	**367.70**
1770 x 1050mm overall; WH310CCH	248.78	2.50	**255.00**	3.70	1.85	**72.39**	1.62	**329.01**	nr	**370.14**
1770 x 1200mm overall; WH312CCH	257.11	2.50	**263.54**	4.10	2.05	**80.22**	1.62	**345.37**	nr	**388.55**
1770 x 1350mm overall; WH313CCH	272.62	2.50	**279.44**	4.30	2.15	**84.13**	1.62	**365.19**	nr	**410.83**
2339 x 1050mm overall; WH410CMCH	310.16	2.50	**317.91**	4.60	2.30	**90.00**	1.62	**409.53**	nr	**460.72**
2339 x 1200mm overall; WH412CMCH	320.97	2.50	**328.99**	4.85	2.42	**94.84**	1.62	**425.45**	nr	**478.63**
2339 x 1350mm overall; WH413CMCH	339.43	2.50	**347.92**	5.15	2.57	**100.71**	1.62	**450.24**	nr	**506.52**
Purpose made windows in softwood, wrought										
38mm moulded casements or fanlights										
in one pane	106.15	2.50	**108.80**	1.75	0.22	**27.22**	0.28	**136.30**	m2	**153.34**
divided into panes 0.10 - 0.50m2	122.07	2.50	**125.12**	1.75	0.22	**27.22**	0.28	**152.62**	m2	**171.70**
divided into panes not exceeding 0.10m2	137.99	2.50	**141.44**	1.75	0.22	**27.22**	0.28	**168.94**	m2	**190.06**
50mm moulded casements or fanlights										
in one pane	122.07	2.50	**125.12**	1.85	0.23	**28.75**	0.30	**154.17**	m2	**173.44**
divided into panes 0.10 - 0.50m2	137.99	2.50	**141.44**	1.85	0.23	**28.75**	0.30	**170.49**	m2	**191.80**
divided into panes not exceeding 0.10m2	153.91	2.50	**157.76**	1.85	0.23	**28.75**	0.30	**186.81**	m2	**210.16**
38mm moulded casements with semi-circular heads (measured square)										
in one pane	212.29	2.50	**217.60**	1.90	0.24	**29.57**	0.31	**247.48**	m2	**278.41**
divided into panes 0.10 - 0.50m2	228.21	2.50	**233.92**	1.90	0.24	**29.57**	0.31	**263.80**	m2	**296.77**
divided into panes not exceeding 0.10m2	244.13	2.50	**250.24**	1.90	0.24	**29.57**	0.31	**280.12**	m2	**315.13**
50mm moulded casements with semi-circular heads (measured square)										
in one pane	244.13	2.50	**250.24**	2.00	0.25	**31.10**	0.33	**281.66**	m2	**316.87**
divided into panes 0.10 - 0.50m2	260.06	2.50	**266.56**	2.00	0.25	**31.10**	0.33	**297.98**	m2	**335.23**
divided into panes not exceeding 0.10m2	275.98	2.50	**282.88**	2.00	0.25	**31.10**	0.33	**314.30**	m2	**353.59**
38mm bullseye casements										
457mm diameter in one pane	159.22	2.50	**163.20**	1.15	0.14	**17.84**	0.19	**181.23**	nr	**203.88**
762mm diameter in one pane	212.29	2.50	**217.60**	1.75	0.22	**27.22**	0.28	**245.10**	nr	**275.74**
50mm bullseye casements										
457mm diameter in one pane	169.83	2.50	**174.08**	1.25	0.16	**19.48**	0.21	**193.76**	nr	**217.98**
762mm diameter in one pane	228.21	2.50	**233.92**	1.90	0.24	**29.57**	0.31	**263.80**	nr	**296.77**
Labours										
check throated edge	0.54	2.50	**0.55**	-	-	**-**	-	**0.55**	m	**0.62**
rebated and splayed bottom rail	0.54	2.50	**0.55**	-	-	**-**	-	**0.55**	m	**0.62**
rebated and beaded meeting stile	0.75	2.50	**0.77**	-	-	**-**	-	**0.77**	m	**0.86**
fitting and hanging casement or fanlight on butts (included elsewhere)										
38mm	-	-	**-**	0.75	0.09	**11.62**	-	**11.62**	nr	**13.07**
50mm	-	-	**-**	0.80	0.10	**12.44**	-	**12.44**	nr	**13.99**
fitting and hanging casement or fanlight on sash centres (included elsewhere)										
38mm	-	-	**-**	1.65	0.21	**25.70**	-	**25.70**	nr	**28.91**
50mm	-	-	**-**	1.80	0.22	**27.93**	-	**27.93**	nr	**31.43**
Purpose made windows in Afrormosia, wrought										
38mm moulded casements or fanlights										
in one pane	204.70	2.50	**209.82**	2.60	0.32	**40.37**	0.42	**250.61**	m2	**281.94**
divided into panes 0.10 - 0.50m2	235.41	2.50	**241.29**	2.60	0.32	**40.37**	0.42	**282.08**	m2	**317.34**
divided into panes not exceeding 0.10m2	266.11	2.50	**272.76**	2.60	0.32	**40.37**	0.42	**313.56**	m2	**352.75**
50mm moulded casements or fanlights										
in one pane	235.41	2.50	**241.29**	2.75	0.34	**42.72**	0.45	**284.46**	m2	**320.02**
divided into panes 0.10 - 0.50m2	266.11	2.50	**272.76**	2.75	0.34	**42.72**	0.45	**315.93**	m2	**355.42**
divided into panes not exceeding 0.10m2	296.82	2.50	**304.24**	2.75	0.34	**42.72**	0.45	**347.40**	m2	**390.83**
38mm moulded casements with semi-circular heads (measured square)										
in one pane	409.40	2.50	**419.64**	2.80	0.35	**43.54**	0.46	**463.63**	m2	**521.59**
divided into panes 0.10 - 0.50m2	440.11	2.50	**451.11**	2.80	0.35	**43.54**	0.46	**495.10**	m2	**556.99**
divided into panes not exceeding 0.10m2	470.81	2.50	**482.58**	2.80	0.35	**43.54**	0.46	**526.58**	m2	**592.40**
50mm moulded casements with semi-circular heads (measured square)										
in one pane	470.81	2.50	**482.58**	3.00	0.37	**46.59**	0.49	**529.66**	m2	**595.87**
divided into panes 0.10 - 0.50m2	501.52	2.50	**514.05**	3.00	0.37	**46.59**	0.49	**561.14**	m2	**631.28**
divided into panes not exceeding 0.10m2	532.22	2.50	**545.53**	3.00	0.37	**46.59**	0.49	**592.61**	m2	**666.68**
38mm bullseye casements										
457mm diameter in one pane	307.05	2.50	**314.73**	1.75	0.22	**27.22**	0.28	**342.23**	nr	**385.01**
762mm diameter in one pane	409.40	2.50	**419.64**	2.60	0.32	**40.37**	0.42	**460.43**	nr	**517.98**

Labour hourly rates: (except Specialists) Craft Operatives £14.21 Labourer £10.71 Rates are national average prices. Refer to REGIONAL VARIATIONS for indicative levels of overall pricing in regions	MATERIALS			LABOUR				RATES		
	Del to Site £	Waste %	Material Cost £	Craft Optve Hrs	Lab Hrs	Labour Cost £	Sunds £	Nett Rate £	Unit	Gross Rate (+12.5%) £
L10: WINDOWS/ROOFLIGHTS/SCREENS/LOUVRES: TIMBER WINDOWS Cont.										
Purpose made windows in Afrormosia, wrought Cont.										
50mm bullseye casements										
457mm diameter in one pane	327.52	2.50	**335.71**	1.90	0.24	**29.57**	0.31	**365.59**	nr	**411.29**
762mm diameter in one pane	440.11	2.50	**451.11**	2.80	0.35	**43.54**	0.46	**495.10**	nr	**556.99**
Labours										
check throated edge	1.02	2.50	**1.05**	-	-	-	-	**1.05**	m	**1.18**
rebated and splayed bottom rail	1.02	2.50	**1.05**	-	-	-	-	**1.05**	m	**1.18**
rebated and beaded meeting stile	1.44	2.50	**1.47**	-	-	-	-	**1.47**	m	**1.66**
fitting and hanging casement or fanlight on butts (included elsewhere)										
38mm	-	-	-	1.30	0.16	**20.19**	-	**20.19**	nr	**22.71**
50mm	-	-	-	1.40	0.17	**21.71**	-	**21.71**	nr	**24.43**
fitting and hanging casement or fanlight on sash centres (included elsewhere)										
38mm	-	-	-	2.85	0.36	**44.35**	-	**44.35**	nr	**49.90**
50mm	-	-	-	3.20	0.40	**49.76**	-	**49.76**	nr	**55.98**
Purpose made windows in European Oak, wrought										
38mm moulded casements or fanlights										
in one pane	269.45	2.50	**276.18**	3.45	0.43	**53.63**	0.56	**330.37**	m2	**371.67**
divided into panes 0.10 - 0.50m2	309.86	2.50	**317.61**	3.45	0.43	**53.63**	0.56	**371.80**	m2	**418.27**
divided into panes not exceeding 0.10m2	350.28	2.50	**359.04**	3.45	0.43	**53.63**	0.56	**413.23**	m2	**464.88**
50mm moulded casements or fanlights										
in one pane	309.86	2.50	**317.61**	3.70	0.46	**57.50**	0.60	**375.71**	m2	**422.68**
divided into panes 0.10 - 0.50m2	350.28	2.50	**359.04**	3.70	0.46	**57.50**	0.60	**417.14**	m2	**469.28**
divided into panes not exceeding 0.10m2	390.70	2.50	**400.46**	3.70	0.46	**57.50**	0.60	**458.57**	m2	**515.89**
38mm moulded casements with semi-circular heads (measured square)										
in one pane	538.89	2.50	**552.36**	3.80	0.48	**59.14**	0.62	**612.12**	m2	**688.64**
divided into panes 0.10 - 0.50m2	579.31	2.50	**593.79**	3.80	0.48	**59.14**	0.62	**653.55**	m2	**735.24**
divided into panes not exceeding 0.10m2	619.72	2.50	**635.22**	3.80	0.48	**59.14**	0.62	**694.98**	m2	**781.85**
50mm moulded casements with semi-circular heads (measured square)										
in one pane	619.72	2.50	**635.22**	4.00	0.50	**62.20**	0.65	**698.06**	m2	**785.32**
divided into panes 0.10 - 0.50m2	660.14	2.50	**676.64**	4.00	0.50	**62.20**	0.65	**739.49**	m2	**831.92**
divided into panes not exceeding 0.10m2	700.56	2.50	**718.07**	4.00	0.50	**62.20**	0.65	**780.92**	m2	**878.53**
38mm bullseye casements										
457mm diameter in one pane	404.17	2.50	**414.27**	2.30	0.29	**35.79**	0.37	**450.43**	nr	**506.73**
762mm diameter in one pane	538.89	2.50	**552.36**	3.45	0.43	**53.63**	0.56	**606.55**	nr	**682.37**
50mm bullseye casements										
457mm diameter in one pane	431.11	2.50	**441.89**	2.50	0.31	**38.85**	0.41	**481.14**	nr	**541.29**
762mm diameter in one pane	579.31	2.50	**593.79**	3.80	0.47	**59.03**	0.62	**653.44**	nr	**735.12**
Labours										
check throated edge	1.35	2.50	**1.39**	-	-	-	-	**1.39**	m	**1.56**
rebated and splayed bottom rail	1.35	2.50	**1.39**	-	-	-	-	**1.39**	m	**1.56**
rebated and beaded meeting stile	1.88	2.50	**1.93**	-	-	-	-	**1.93**	m	**2.17**
fitting and hanging casement or fanlight on butts (included elsewhere)										
38mm	-	-	-	1.45	0.18	**22.53**	-	**22.53**	nr	**25.35**
50mm	-	-	-	1.60	0.20	**24.88**	-	**24.88**	nr	**27.99**
fitting and hanging casement or fanlight on sash centres (included elsewhere)										
38mm	-	-	-	3.30	0.41	**51.28**	-	**51.28**	nr	**57.69**
50mm	-	-	-	3.60	0.45	**55.98**	-	**55.98**	nr	**62.97**
Sash windows in softwood, wrought, preservative treated; priming or staining by manufacturer										
Magnet Trade; weather stripping; spiral balances; brass plated fittings; fixing to masonry with galvanized steel cramps -4 nr, 25 x 3 x 150mm girth, flat section holes -2										
825 x 1094mm overall; SS3	327.78	2.50	**335.97**	3.80	1.90	**74.35**	1.55	**411.87**	nr	**463.35**
825 x 1394mm overall; SS4	323.55	2.50	**331.64**	4.15	2.07	**81.14**	1.55	**414.33**	nr	**466.12**
920 x 1225mm overall; SS2	330.19	2.50	**338.45**	4.15	2.07	**81.14**	1.55	**421.14**	nr	**473.78**
1051 x 1094mm overall; SSW3	325.36	2.50	**333.49**	4.15	2.07	**81.14**	1.55	**416.19**	nr	**468.21**
1051 x 1394mm overall; SSW4	338.04	2.50	**346.49**	4.50	2.25	**88.04**	1.55	**436.08**	nr	**490.59**
1051 x 1694mm overall; SSW5	401.43	2.50	**411.47**	4.80	2.40	**93.91**	1.55	**506.93**	nr	**570.30**
Magnet Trade; Georgian, open rebated for glass; weatherstripping; spiral balances; brass plated fittings; fixing to masonry with galvanized steel cramps -4 nr, 25 x 3 x 150mm girth, flat section, holes -2										
825 x 1094mm overall; GS3	373.53	2.50	**382.87**	2.60	1.30	**50.87**	1.55	**435.28**	nr	**489.69**
825 x 1394mm overall; GS4	382.35	2.50	**391.91**	3.00	1.50	**58.70**	1.55	**452.15**	nr	**508.67**
1051 x 1394mm overall; GSW4	423.93	2.50	**434.53**	3.40	1.70	**66.52**	1.55	**502.60**	nr	**565.42**

Labour hourly rates: (except Specialists) Craft Operatives £14.21 Labourer £10.71 Rates are national average prices. Refer to REGIONAL VARIATIONS for indicative levels of overall pricing in regions	MATERIALS			LABOUR				RATES		
	Del to Site £	Waste %	Material Cost £	Craft Optve Hrs	Lab Hrs	Labour Cost £	Sunds £	Nett Rate £	Unit	Gross Rate (+12.5%) £
L10: WINDOWS/ROOFLIGHTS/SCREENS/LOUVRES: TIMBER WINDOWS Cont.										
Sash windows mainly in softwood, wrought										
Cased frames; 25mm inside and outside linings; 32mm pulley stiles and head; 10mm beads, back linings, etc; 76mm oak sunk weathered and throated sills; 38mm moulded sashes										
in one pane	165.55	2.50	**169.69**	2.50	0.31	**38.85**	0.41	**208.94**	m2	**235.06**
divided into panes 0.50 - 1.00m2	173.83	2.50	**178.17**	2.50	0.31	**38.85**	0.41	**217.43**	m2	**244.61**
divided into panes 0.10 - 0.50m2	190.38	2.50	**195.14**	2.50	0.31	**38.85**	0.41	**234.40**	m2	**263.70**
divided into panes not exceeding 0.10m2	215.22	2.50	**220.60**	2.50	0.31	**38.85**	0.41	**259.85**	m2	**292.33**
Cased frames; 25mm inside and outside linings; 32mm pulley stiles and head; 10mm beads, back linings, etc; 76mm oak sunk weathered and throated sills; 50mm moulded sashes										
in one pane	190.38	2.50	**195.14**	2.75	0.34	**42.72**	0.45	**238.31**	m2	**268.10**
divided into panes 0.50 - 1.00m2	198.66	2.50	**203.63**	2.75	0.34	**42.72**	0.45	**246.80**	m2	**277.64**
divided into panes 0.10 - 0.50m2	215.22	2.50	**220.60**	2.75	0.34	**42.72**	0.45	**263.76**	m2	**296.73**
divided into panes not exceeding 0.10m2	240.05	2.50	**246.05**	2.75	0.34	**42.72**	0.45	**289.22**	m2	**325.37**
extra; windows in three lights with boxed mullions	33.11	2.50	**33.94**	0.75	0.09	**11.62**	0.12	**45.68**	nr	**51.39**
extra; windows with moulded horns	41.39	2.50	**42.42**	-	-	**-**	-	**42.42**	nr	**47.72**
extra; deep bottom rails and draught beads	4.14	2.50	**4.24**	-	-	**-**	-	**4.24**	m	**4.77**
Labours										
throats	0.43	2.50	**0.44**	-	-	**-**	-	**0.44**	m	**0.49**
grooves for jamb linings	0.50	2.50	**0.51**	-	-	**-**	-	**0.51**	m	**0.57**
fitting and hanging sashes in double hung windows; providing brass faced iron pulleys, best flax cords and iron weights										
38mm thick weighing 14 lbs per sash	44.00	2.50	**45.10**	1.55	0.19	**24.06**	-	**69.16**	nr	**77.81**
50mm thick weighing 20 lbs per sash	49.50	2.50	**50.74**	1.55	0.19	**24.06**	-	**74.80**	nr	**84.15**
Note for small frames and sashes, i.e. 1.25m2 and under, add 20% to the foregoing prices										
Sash windows in Afrormosia, wrought										
Cased frames; 25mm inside and outside linings; 32mm pulley stiles and head; 10mm beads, back linings, etc; 76mm oak sunk weathered and throated sills; 38mm moulded sashes										
in one pane	384.99	2.50	**394.61**	4.40	0.55	**68.41**	0.72	**463.75**	m2	**521.72**
divided into panes 0.50 - 1.00m2	404.24	2.50	**414.34**	4.40	0.55	**68.41**	0.72	**483.48**	m2	**543.91**
divided into panes 0.10 - 0.50m2	442.74	2.50	**453.81**	4.40	0.55	**68.41**	0.72	**522.94**	m2	**588.31**
divided into panes not exceeding 0.10m2	500.49	2.50	**513.00**	4.40	0.55	**68.41**	0.72	**582.14**	m2	**654.90**
Cased frames; 25mm inside and outside linings; 32mm pulley stiles and head; 10mm beads, back linings, etc; 76mm oak sunk weathered and throated sills; 50mm moulded sashes										
in one pane	442.74	2.50	**453.81**	4.80	0.60	**74.63**	0.78	**529.22**	m2	**595.37**
divided into panes 0.50 - 1.00m2	461.99	2.50	**473.54**	4.80	0.60	**74.63**	0.78	**548.95**	m2	**617.57**
divided into panes 0.10 - 0.50m2	500.49	2.50	**513.00**	4.80	0.60	**74.63**	0.78	**588.42**	m2	**661.97**
divided into panes not exceeding 0.10m2	558.24	2.50	**572.19**	4.80	0.60	**74.63**	0.78	**647.61**	m2	**728.56**
extra; windows in three lights with boxed mullions	77.00	2.50	**78.93**	1.30	0.16	**20.19**	0.21	**99.32**	nr	**111.74**
extra; windows with moulded horns	96.25	2.50	**98.66**	-	-	**-**	-	**98.66**	nr	**110.99**
extra; deep bottom rails and draught beads	9.63	2.50	**9.87**	-	-	**-**	-	**9.87**	m	**11.10**
Labours										
throats	0.97	2.50	**0.99**	-	-	**-**	-	**0.99**	m	**1.12**
grooves for jamb linings	1.16	2.50	**1.18**	-	-	**-**	-	**1.18**	m	**1.33**
fitting and hanging sashes in double hung windows; providing brass faced iron pulleys, best flax cords and iron weights										
38mm thick weighing 14 lbs per sash	44.00	2.50	**45.10**	2.75	0.34	**42.72**	-	**87.82**	nr	**98.80**
50mm thick weighing 20 lbs per sash	49.50	2.50	**50.74**	2.75	0.34	**42.72**	-	**93.46**	nr	**105.14**
Note for small frames and sashes, i.e. 1.25m2 and under, add 20% to the foregoing prices										
Sash windows in European Oak, wrought										
Cased frames; 25mm inside and outside linings; 32mm pulley stiles and head; 10mm beads, back linings, etc; 76mm oak sunk weathered and throated sills; 38mm moulded sashes										
in one pane	508.25	2.50	**520.95**	5.00	0.62	**77.69**	0.81	**599.45**	m2	**674.39**
divided into panes 0.50 - 1.00m2	533.66	2.50	**547.00**	5.00	0.62	**77.69**	0.81	**625.50**	m2	**703.69**
divided into panes 0.10 - 0.50m2	584.48	2.50	**599.09**	5.00	0.62	**77.69**	0.81	**677.59**	m2	**762.29**
divided into panes not exceeding 0.10m2	660.72	2.50	**677.24**	5.00	0.62	**77.69**	0.81	**755.74**	m2	**850.21**
Cased frames; 25mm inside and outside linings; 32mm pulley stiles and head; 10mm beads, back linings, etc; 76mm oak sunk weathered and throated sills; 50mm moulded sashes										
in one pane	584.48	2.50	**599.09**	5.50	0.69	**85.54**	0.90	**685.54**	m2	**771.23**
divided into panes 0.50 - 1.00m2	609.89	2.50	**625.14**	5.50	0.69	**85.54**	0.90	**711.58**	m2	**800.53**
divided into panes 0.10 - 0.50m2	660.72	2.50	**677.24**	5.50	0.69	**85.54**	0.90	**763.68**	m2	**859.14**
divided into panes not exceeding 0.10m2	736.95	2.50	**755.37**	5.50	0.69	**85.54**	0.90	**841.82**	m2	**947.05**
extra; windows in three lights with boxed mullions	130.69	2.50	**133.96**	1.55	0.19	**24.06**	0.25	**158.27**	nr	**178.05**
extra; windows with moulded horns	127.06	2.50	**130.24**	-	-	**-**	-	**130.24**	nr	**146.52**
extra; deep bottom rails and draught beads	12.72	2.50	**13.04**	-	-	**-**	-	**13.04**	m	**14.66**

Labour hourly rates: (except Specialists) Craft Operatives £14.21 Labourer £10.71 Rates are national average prices. Refer to REGIONAL VARIATIONS for indicative levels of overall pricing in regions	MATERIALS			LABOUR				RATES		
	Del to Site £	Waste %	Material Cost £	Craft Optve Hrs	Lab Hrs	Labour Cost £	Sunds £	Nett Rate £	Unit	Gross Rate (+12.5%) £
L10: WINDOWS/ROOFLIGHTS/SCREENS/LOUVRES: TIMBER WINDOWS Cont.										
Sash windows in European Oak, wrought Cont.										
Labours										
throats	1.27	2.50	1.30	-	-	-	-	1.30	m	1.47
grooves for jamb linings	1.27	2.50	1.30	-	-	-	-	1.30	m	1.47
fitting and hanging sashes in double hung windows; providing brass faced iron pulleys, best flax cords and iron weights										
38mm thick weighing 14 lbs per sash	44.00	2.50	45.10	3.10	0.39	48.23	-	93.33	nr	104.99
50mm thick weighing 20 lbs per sash	49.50	2.50	50.74	3.10	0.39	48.23	-	98.97	nr	111.34
Note										
for small frames and sashes, i.e. 1.25m2 and under, add 20% to the foregoing prices										
L10: WINDOWS/ROOFLIGHTS/SCREENS/LOUVRES: TIMBER ROOFLIGHTS										
Skylights in softwood, wrought										
Chamfered, straight bar										
38mm	79.76	2.50	81.75	1.50	0.19	23.35	0.24	105.34	m2	118.51
50mm	92.35	2.50	94.66	1.75	0.22	27.22	0.28	122.17	m2	137.44
63mm	104.95	2.50	107.57	2.00	0.25	31.10	0.33	139.00	m2	156.37
Moulded, straight bar										
38mm	83.96	2.50	86.06	1.50	0.19	23.35	0.24	109.65	m2	123.35
50mm	96.55	2.50	98.96	1.75	0.22	27.22	0.28	126.47	m2	142.28
63mm	109.14	2.50	111.87	2.00	0.25	31.10	0.33	143.30	m2	161.21
Skylights in Oak, wrought										
Chamfered, straight bar										
38mm	244.84	2.50	250.96	2.30	0.29	35.79	0.37	287.12	m2	323.01
50mm	283.50	2.50	290.59	2.70	0.34	42.01	0.44	333.04	m2	374.67
63mm	322.16	2.50	330.22	3.10	0.39	48.23	0.50	378.94	m2	426.31
Moulded, straight bar										
38mm	257.73	2.50	264.17	2.30	0.29	35.79	0.37	300.33	m2	337.87
50mm	296.39	2.50	303.80	2.70	0.34	42.01	0.44	346.25	m2	389.53
63mm	335.05	2.50	343.43	3.10	0.39	48.23	0.50	392.15	m2	441.17
Skylight kerbs in softwood, wrought										
Kerbs; dovetailed at angles										
38 x 225mm	10.63	2.50	10.89	0.31	0.04	4.83	0.05	15.78	m	17.75
50 x 225mm	12.74	2.50	13.06	0.35	0.04	5.40	0.05	18.51	m	20.82
38 x 225mm; chamfers -1 nr	11.15	2.50	11.42	0.31	0.04	4.83	0.05	16.31	m	18.35
50 x 225mm; chamfers -1 nr	13.27	2.50	13.60	0.35	0.04	5.40	0.05	19.05	m	21.44
Kerbs; in two thicknesses to circular skylights										
38 x 225mm	31.83	2.50	32.63	0.47	0.06	7.32	0.08	40.03	m	45.03
50 x 225mm	38.16	2.50	39.12	0.53	0.07	8.28	0.08	47.48	m	53.41
38 x 225mm; chamfers -1 nr	32.91	2.50	33.73	0.47	0.06	7.32	0.08	41.13	m	46.27
50 x 225mm; chamfers -1 nr	39.27	2.50	40.25	0.53	0.07	8.28	0.08	48.61	m	54.69
Skylight kerbs in Oak, wrought										
Kerbs; dovetailed at angles										
38 x 225mm	52.03	2.50	53.33	0.62	0.08	9.67	0.10	63.10	m	70.98
50 x 225mm	62.44	2.50	64.00	0.69	0.08	10.66	0.11	74.77	m	84.11
38 x 225mm; chamfers -1 nr	54.63	2.50	56.00	0.62	0.08	9.67	0.10	65.76	m	73.98
50 x 225mm; chamfers -1 nr	65.04	2.50	66.66	0.69	0.08	10.66	0.11	77.44	m	87.11
Kerbs; in two thicknesses to circular skylights										
38 x 225mm	156.09	2.50	159.99	0.94	0.12	14.64	0.15	174.78	m	196.63
50 x 225mm	187.31	2.50	191.99	1.06	0.13	16.45	0.17	208.62	m	234.69
38 x 225mm; chamfers -1 nr	161.29	2.50	165.33	0.94	0.12	14.64	0.15	180.12	m	202.63
50 x 225mm; chamfers -1 nr	192.51	2.50	197.32	1.06	0.13	16.45	0.17	213.95	m	240.69
Roof windows in Nordic red pine, wrought, treated										
Velux roof windows GGL/GHL 3000 Range, factory glazed clear float double glazed sealed unit; type EDL 0000; fixing to timber with screws; dressing flashings										
778 x 550mm overall, C02 GGL Range	151.01	2.50	154.79	5.50	2.75	107.61	-	262.39	nr	295.19
978 x 550mm overall, C04 GGL Range	168.12	2.50	172.32	6.00	3.00	117.39	-	289.71	nr	325.93
1178 x 660mm overall, F06 GGL Range	198.74	2.50	203.71	6.00	3.00	117.39	-	321.10	nr	361.24
978 x 780mm overall, M04 GGL Range	190.21	2.50	194.97	6.00	3.00	117.39	-	312.36	nr	351.40
1178 x 780mm overall, M06 GGL Range	218.41	2.50	223.87	6.00	3.00	117.39	-	341.26	nr	383.92
1398 x 780mm overall, M08 GGL Range	227.44	2.50	233.13	7.00	3.50	136.96	-	370.08	nr	416.34
1600 x 942mm overall, P10 GGL Range	284.64	2.50	291.76	7.00	3.50	136.96	-	428.71	nr	482.30
1178 x 1140mm overall, S06 GGL Range	253.31	2.50	259.65	8.00	4.00	156.52	-	416.17	nr	468.19
778 x 550mm overall, C02 GHL Range	151.01	2.50	154.79	7.50	3.75	146.74	-	301.52	nr	339.21
978 x 550mm overall, C04 GHL Range	231.27	2.50	237.05	7.50	3.75	146.74	-	383.79	nr	431.76
1178 x 660mm overall, F06 GHL Range	264.86	2.50	271.48	7.50	3.75	146.74	-	418.22	nr	470.50
978 x 780mm overall, M04 GHL Range	276.79	2.50	283.71	7.50	3.75	146.74	-	430.44	nr	484.25
1178 x 780mm overall, M06 GHL Range	279.69	2.50	286.68	8.00	4.00	156.52	-	443.20	nr	498.61
1398 x 780mm overall, M08 GHL Range	298.46	2.50	305.92	8.00	4.00	156.52	-	462.44	nr	520.25
1600 x 942mm overall, P10 GHL Range	284.64	2.50	291.76	8.00	4.00	156.52	-	448.28	nr	504.31
1178 x 1140mm overall, S06 GHL Range	321.21	2.50	329.24	8.00	4.00	156.52	-	485.76	nr	546.48

Labour hourly rates: (except Specialists) Craft Operatives £14.21 Labourer £10.71 Rates are national average prices. Refer to REGIONAL VARIATIONS for indicative levels of overall pricing in regions	MATERIALS			LABOUR				RATES		
	Del to Site £	Waste %	Material Cost £	Craft Optve Hrs	Lab Hrs	Labour Cost £	Sunds £	Nett Rate £	Unit	Gross Rate (+12.5%) £
L10: WINDOWS/ROOFLIGHTS/SCREENS/LOUVRES: TIMBER ROOFLIGHTS Cont.										
Roof windows in Nordic red pine, wrought, treated Cont.										
Velux roof windows GGL/GHL 3000 Range, factory glazed clear float double glazed sealed unit; type EDZ 0000; fixing to timber with screws; dressing flashings										
778 x 550mm overall, C02 GGL Range	154.71	2.50	**158.58**	5.50	2.75	**107.61**	-	**266.19**	nr	**299.46**
978 x 550mm overall, C04 GGL Range	171.07	2.50	**175.35**	6.00	3.00	**117.39**	-	**292.74**	nr	**329.33**
1178 x 660mm overall, F06 GGL Range	201.57	2.50	**206.61**	6.00	3.00	**117.39**	-	**324.00**	nr	**364.50**
978 x 780mm overall, M04 GGL Range	194.04	2.50	**198.89**	6.00	3.00	**117.39**	-	**316.28**	nr	**355.81**
1178 x 780mm overall, M06 GGL Range	223.36	2.50	**228.94**	6.00	3.00	**117.39**	-	**346.33**	nr	**389.62**
1398 x 780mm overall, M08 GGL Range	233.12	2.50	**238.95**	7.00	3.50	**136.96**	-	**375.90**	nr	**422.89**
1600 x 942mm overall, P10 GGL Range	289.59	2.50	**296.83**	7.00	3.50	**136.96**	-	**433.78**	nr	**488.01**
1178 x 1140mm overall, S06 GGL Range	257.93	2.50	**264.37**	8.00	4.00	**156.52**	-	**420.89**	nr	**473.51**
778 x 550mm overall, C02 GHL Range	154.71	2.50	**158.58**	7.50	3.75	**146.74**	-	**305.32**	nr	**343.48**
978 x 550mm overall, C04 GHL Range	234.22	2.50	**240.07**	7.50	3.75	**146.74**	-	**386.81**	nr	**435.16**
1178 x 660mm overall, F06 GHL Range	267.72	2.50	**274.41**	7.50	3.75	**146.74**	-	**421.15**	nr	**473.79**
978 x 780mm overall, M04 GHL Range	257.87	2.50	**264.32**	7.50	3.75	**146.74**	-	**411.05**	nr	**462.44**
1178 x 780mm overall, M06 GHL Range	284.64	2.50	**291.76**	8.00	4.00	**156.52**	-	**448.28**	nr	**504.31**
1398 x 780mm overall, M08 GHL Range	304.14	2.50	**311.74**	8.00	4.00	**156.52**	-	**468.26**	nr	**526.80**
1600 x 942mm overall, P10 GHL Range	289.59	2.50	**296.83**	8.00	4.00	**156.52**	-	**453.35**	nr	**510.02**
1178 x 1140mm overall, S06 GHL Range	325.71	2.50	**333.85**	8.00	4.00	**156.52**	-	**490.37**	nr	**551.66**
Velux roof windows GGL/GHL 3073 Range, factory glazed clear float/heat absorbing glass double glazed sealed unit; type EDZ 0000; fixing to timber with screws; dressing flashings										
778 x 550mm overall, C02 GGL Range	195.92	2.50	**200.82**	5.50	2.75	**107.61**	-	**308.43**	nr	**346.98**
978 x 550mm overall, C04 GGL Range	216.43	2.50	**221.84**	6.00	3.00	**117.39**	-	**339.23**	nr	**381.64**
1178 x 660mm overall, F06 GGL Range	254.97	2.50	**261.35**	6.00	3.00	**117.39**	-	**378.74**	nr	**426.08**
978 x 780mm overall, M04 GGL Range	245.15	2.50	**251.27**	6.00	3.00	**117.39**	-	**368.66**	nr	**414.75**
1178 x 780mm overall, M06 GGL Range	269.81	2.50	**276.55**	6.00	3.00	**117.39**	-	**393.94**	nr	**443.19**
1398 x 780mm overall, M08 GGL Range	294.50	2.50	**301.87**	7.00	3.50	**136.96**	-	**438.82**	nr	**493.67**
1600 x 942mm overall, P10 GGL Range	358.76	2.50	**367.73**	7.00	3.50	**136.96**	-	**504.68**	nr	**567.77**
1178 x 1140mm overall, S06 GGL Range	329.12	2.50	**337.34**	8.00	4.00	**156.52**	-	**493.86**	nr	**555.60**
778 x 550mm overall, C02 GHL Range	195.92	2.50	**200.82**	7.50	3.75	**146.74**	-	**347.56**	nr	**391.00**
978 x 550mm overall, C04 GHL Range	271.78	2.50	**278.58**	7.50	3.75	**146.74**	-	**425.31**	nr	**478.48**
1178 x 660mm overall, F06 GHL Range	314.28	2.50	**322.14**	7.50	3.75	**146.74**	-	**468.88**	nr	**527.49**
978 x 780mm overall, M04 GHL Range	301.48	2.50	**309.02**	7.50	3.75	**146.74**	-	**455.76**	nr	**512.73**
1178 x 780mm overall, M06 GHL Range	328.59	2.50	**336.80**	8.00	4.00	**156.52**	-	**493.32**	nr	**554.99**
1398 x 780mm overall, M08 GHL Range	364.68	2.50	**373.80**	8.00	4.00	**156.52**	-	**530.32**	nr	**596.61**
1600 x 942mm overall, P10 GHL Range	358.76	2.50	**367.73**	8.00	4.00	**156.52**	-	**524.25**	nr	**589.78**
1178 x 1140mm overall, S06 GHL Range	395.33	2.50	**405.21**	8.00	4.00	**156.52**	-	**561.73**	nr	**631.94**
Velux roof windows GGL/GHL 3073 Range, factory glazed clear float/heat absorbing glass double glazed sealed unit; type EDL 0000; fixing to timber with screws; dressing flashings										
778 x 550mm overall, C02 GGL Range	199.63	2.50	**204.62**	5.50	2.75	**107.61**	-	**312.22**	nr	**351.25**
978 x 550mm overall, C04 GGL Range	219.38	2.50	**224.87**	6.00	3.00	**117.39**	-	**342.26**	nr	**385.04**
1178 x 660mm overall, F06 GGL Range	257.80	2.50	**264.25**	6.00	3.00	**117.39**	-	**381.64**	nr	**429.34**
978 x 780mm overall, M04 GGL Range	248.97	2.50	**255.20**	6.00	3.00	**117.39**	-	**372.59**	nr	**419.16**
1178 x 780mm overall, M06 GGL Range	274.76	2.50	**281.62**	6.00	3.00	**117.39**	-	**399.01**	nr	**448.89**
1398 x 780mm overall, M08 GGL Range	300.18	2.50	**307.68**	7.00	3.50	**136.96**	-	**444.64**	nr	**500.22**
1600 x 942mm overall, P10 GGL Range	363.71	2.50	**372.80**	7.00	3.50	**136.96**	-	**509.76**	nr	**573.47**
1178 x 1140mm overall, S06 GGL Range	333.73	2.50	**342.07**	8.00	4.00	**156.52**	-	**498.59**	nr	**560.91**
778 x 550mm overall, C02 GHL Range	199.63	2.50	**204.62**	7.50	3.75	**146.74**	-	**351.35**	nr	**395.27**
978 x 550mm overall, C04 GHL Range	274.73	2.50	**281.60**	7.50	3.75	**146.74**	-	**428.34**	nr	**481.88**
1178 x 660mm overall, F06 GHL Range	317.11	2.50	**325.04**	7.50	3.75	**146.74**	-	**471.78**	nr	**530.75**
978 x 780mm overall, M04 GHL Range	305.31	2.50	**312.94**	7.50	3.75	**146.74**	-	**459.68**	nr	**517.14**
1178 x 780mm overall, M06 GHL Range	333.31	2.50	**341.65**	8.00	4.00	**156.52**	-	**498.17**	nr	**560.44**
1398 x 780mm overall, M08 GHL Range	364.68	2.50	**373.80**	8.00	4.00	**156.52**	-	**530.32**	nr	**596.61**
1600 x 942mm overall, P10 GHL Range	358.76	2.50	**367.73**	8.00	4.00	**156.52**	-	**524.25**	nr	**589.78**
1178 x 1140mm overall, S06 GHL Range	395.33	2.50	**405.21**	8.00	4.00	**156.52**	-	**561.73**	nr	**631.94**
Bedding and pointing frames										
Bedding in cement mortar (1:3); pointing with Secomastic standard mastic one side										
wood frames	0.75	5.00	**0.81**	0.20	-	**2.84**	-	**3.65**	m	**4.11**
Bedding in cement mortar (1:3); pointing with coloured two part polysulphide mastic one side										
wood frames	0.78	5.00	**0.84**	0.20	-	**2.84**	-	**3.68**	m	**4.14**
L10: WINDOWS/ROOFLIGHTS/SCREENS/LOUVRES: METAL WINDOWS										
Windows in galvanized steel										
Crittall Windows Ltd. Duralife Homelight; weatherstripping; fixing to masonry with lugs										
fixed lights										
508 x 292mm, NG5	20.08	2.50	**20.58**	0.61	0.25	**11.35**	-	**31.92**	nr	**35.91**
508 x 923mm, NC5	27.53	2.50	**28.21**	0.98	0.35	**17.67**	-	**45.89**	nr	**51.62**
508 x 1218mm, ND5	31.43	2.50	**32.21**	1.23	0.41	**21.87**	-	**54.08**	nr	**60.84**
997 x 628mm, NE13	36.34	2.50	**37.25**	1.12	0.40	**20.20**	-	**57.45**	nr	**64.63**
997 x 923mm, NC13	42.39	2.50	**43.45**	1.24	0.44	**22.33**	-	**65.78**	nr	**74.01**
997 x 1218mm, ND13	46.94	2.50	**48.11**	1.50	0.51	**26.78**	-	**74.89**	nr	**84.25**
1486 x 628mm, NE14	43.49	2.50	**44.58**	1.50	0.53	**26.99**	-	**71.57**	nr	**80.52**
1486 x 923mm, NC14	48.89	2.50	**50.12**	1.63	0.58	**29.37**	-	**79.49**	nr	**89.43**
1486 x 1218mm, ND14	55.33	2.50	**56.71**	1.90	0.66	**34.07**	-	**90.78**	nr	**102.12**
top hung lights										
508 x 292mm, NG1	67.37	2.50	**69.05**	0.61	0.25	**11.35**	-	**80.39**	nr	**90.44**

Labour hourly rates: (except Specialists) Craft Operatives £14.21 Labourer £10.71 Rates are national average prices. Refer to REGIONAL VARIATIONS for indicative levels of overall pricing in regions	MATERIALS			LABOUR				RATES		
	Del to Site £	Waste %	Material Cost £	Craft Optve Hrs	Lab Hrs	Labour Cost £	Sunds £	Nett Rate £	Unit	Gross Rate (+12.5%) £
L10: WINDOWS/ROOFLIGHTS/SCREENS/LOUVRES: METAL WINDOWS Cont.										
Windows in galvanized steel Cont.										
Crittall Windows Ltd. Duralife Homelight; weatherstripping; fixing to masonry with lugs Cont.										
top hung lights Cont.										
508 x 457mm, NH1	73.44	2.50	75.28	0.74	0.25	13.19	-	88.47	nr	99.53
508 x 628mm, NE1	79.01	2.50	80.98	0.86	0.29	15.33	-	96.31	nr	108.34
997 x 628mm, NE13E	110.85	2.50	113.62	1.12	0.32	19.34	-	132.96	nr	149.58
997 x 923mm, NC13C	119.30	2.50	122.29	1.24	0.40	21.90	-	144.19	nr	162.21
bottom hung lights										
508 x 628mm, NL1	102.99	2.50	105.57	0.86	0.32	15.65	-	121.22	nr	136.37
side hung lights										
508 x 628mm, NES1	86.03	2.50	88.18	0.86	0.32	15.65	-	103.83	nr	116.81
508 x 923mm, NC1	92.44	2.50	94.75	0.98	0.35	17.67	-	112.43	nr	126.48
508 x 1067mm, NCO1	97.99	2.50	100.44	1.10	0.38	19.70	-	120.14	nr	135.16
508 x 1218mm, ND1	104.30	2.50	106.91	1.23	0.41	21.87	-	128.78	nr	144.87
mixed lights										
279 x 923mm, NC6F	70.87	2.50	72.64	0.98	0.35	17.67	-	90.31	nr	101.60
508 x 923mm, NC5F	81.79	2.50	83.83	0.98	0.35	17.67	-	101.50	nr	114.19
508 x 1067mm, NCO5F	83.73	2.50	85.83	1.10	0.38	19.70	-	105.53	nr	118.72
997 x 292mm, NG2	82.81	2.50	84.88	0.85	0.33	15.61	-	100.49	nr	113.05
997 x 457mm, NH2	90.52	2.50	92.78	0.98	0.37	17.89	-	110.67	nr	124.50
997 x 628mm, NE2	98.34	2.50	100.79	1.12	0.40	20.20	-	120.99	nr	136.12
997 x 628mm, NES2	116.68	2.50	119.60	1.12	0.40	20.20	-	139.80	nr	157.27
997 x 923mm, NC2	127.43	2.50	130.61	1.24	0.44	22.33	-	152.95	nr	172.07
997 x 923mm, NC2F	169.15	2.50	173.37	1.24	0.44	22.33	-	195.71	nr	220.17
997 x 1067mm, NCO2	138.59	2.50	142.05	1.37	0.47	24.50	-	166.55	nr	187.37
997 x 1067mm, NCO2F	184.46	2.50	189.07	1.37	0.47	24.50	-	213.57	nr	240.27
997 x 1218mm, ND2	147.65	2.50	151.34	1.50	0.51	26.78	-	178.12	nr	200.39
997 x 1218mm, ND2F	186.50	2.50	191.16	1.50	0.51	26.78	-	217.94	nr	245.18
997 x 1513mm, NDV2FSB	214.94	2.50	220.31	1.63	0.58	29.37	-	249.68	nr	280.90
1486 x 628mm, NE3	123.87	2.50	126.97	1.50	0.51	26.78	-	153.74	nr	172.96
1486 x 923mm, NC4	210.73	2.50	215.99	1.63	0.58	29.37	-	245.37	nr	276.04
1486 x 923mm, NC4F	257.95	2.50	264.40	1.63	0.58	29.37	-	293.77	nr	330.49
1486 x 1067mm, NCO4	226.97	2.50	232.64	1.63	0.58	29.37	-	262.01	nr	294.77
1486 x 1067mm, NCO4F	272.66	2.50	279.47	1.77	0.61	31.68	-	311.16	nr	350.05
1486 x 1218mm, ND4	242.17	2.50	248.22	1.90	0.66	34.07	-	282.29	nr	317.57
1486 x 1218mm, ND4F	287.90	2.50	295.10	1.90	0.66	34.07	-	329.17	nr	370.31
1486 x 1513mm, NDV4FSB	332.89	2.50	341.21	2.04	0.69	36.38	-	377.59	nr	424.79
1994 x 923mm, NC11F	279.75	2.50	286.74	2.01	0.69	35.95	-	322.70	nr	363.03
1994 x 1218mm, ND11F	302.76	2.50	310.32	2.30	0.81	41.36	-	351.68	nr	395.64
1994 x 1513mm, NDV11FSB	353.85	2.50	362.69	2.47	0.86	44.31	-	407.00	nr	457.88
Extra for										
mullions										
292mm high	11.69	2.50	11.98	0.23	0.12	4.55	-	16.53	nr	18.60
457mm high	16.34	2.50	16.75	0.35	0.17	6.79	-	23.55	nr	26.49
628mm high	21.11	2.50	21.63	0.46	0.23	9.00	-	30.63	nr	34.46
923mm high	29.35	2.50	30.08	0.58	0.29	11.35	-	41.43	nr	46.60
1067mm high	33.98	2.50	34.83	0.69	0.35	13.55	-	48.38	nr	54.43
1218mm high	38.24	2.50	39.19	0.92	0.46	18.00	-	57.19	nr	64.34
1513mm high	46.47	2.50	47.63	1.15	0.58	22.55	-	70.18	nr	78.95
transoms										
279mm wide	12.12	2.50	12.42	0.23	0.12	4.55	-	16.98	nr	19.10
508mm wide	18.68	2.50	19.14	0.35	0.17	6.79	-	25.94	nr	29.18
997mm wide	48.71	2.50	49.93	0.69	0.35	13.55	-	63.48	nr	71.42
1143mm wide	54.10	2.50	55.45	1.15	0.58	22.55	-	78.01	nr	87.76
Controlair Ventilators, permanent, mill finish										
508mm wide	31.20	2.50	31.98	-	-	-	-	31.98	nr	35.98
997mm wide	41.90	2.50	42.95	-	-	-	-	42.95	nr	48.32
1486mm wide	94.39	2.50	96.75	-	-	-	-	96.75	nr	108.84
Controlair Ventilators, permanent, mill finish, flyscreen										
508mm wide	37.08	2.50	38.01	-	-	-	-	38.01	nr	42.76
997mm wide	52.83	2.50	54.15	-	-	-	-	54.15	nr	60.92
1486mm wide	109.25	2.50	111.99	-	-	-	-	111.99	nr	125.98
Controlair Ventilators, adjustable										
508mm wide 1-LT	37.19	2.50	38.11	-	-	-	-	38.11	nr	42.88
997mm wide 2-LT	47.02	2.50	48.19	-	-	-	-	48.19	nr	54.22
Controlair Ventilators, adjustable, flyscreen										
508mm wide 1-LT	42.84	2.50	43.91	-	-	-	-	43.91	nr	49.40
997mm wide 2-LT	57.95	2.50	59.40	-	-	-	-	59.40	nr	66.82
locks with Parkes locking handles for										
side hung lights	38.58	2.50	39.55	-	-	-	-	39.55	nr	44.49
horizontally pivoted lights	38.58	2.50	39.55	-	-	-	-	39.55	nr	44.49
Windows in galvanized steel; polyester powder coated; white matt										
Crittall Windows Ltd. Duralife Homelight; weatherstripping; fixing to masonry with lugs										
fixed lights										
508 x 292mm, NG5	23.57	2.50	24.16	0.61	0.25	11.35	-	35.50	nr	39.94
508 x 923mm, NC5	32.29	2.50	33.09	0.98	0.35	17.67	-	50.77	nr	57.11
508 x 1218mm, ND5	36.88	2.50	37.80	1.23	0.41	21.87	-	59.67	nr	67.13
997 x 628mm, NE13	42.66	2.50	43.72	1.12	0.40	20.20	-	63.92	nr	71.91
997 x 923mm, NC13	49.94	2.50	51.18	1.24	0.44	22.33	-	73.52	nr	82.71
997 x 1218mm, ND13	57.89	2.50	59.34	1.50	0.51	26.78	-	86.11	nr	96.88
1486 x 628mm, NE14	51.19	2.50	52.47	1.50	0.53	26.99	-	79.47	nr	89.40
1486 x 923mm, NC14	57.36	2.50	58.80	1.63	0.58	29.37	-	88.17	nr	99.19
1486 x 1218mm, ND14	64.92	2.50	66.54	1.90	0.66	34.07	-	100.61	nr	113.18
top hung lights										
508 x 292mm, NG1	75.05	2.50	76.92	0.61	0.25	11.35	-	88.27	nr	99.30

Labour hourly rates: (except Specialists) Craft Operatives £14.21 Labourer £10.71 Rates are national average prices. Refer to REGIONAL VARIATIONS for indicative levels of overall pricing in regions	MATERIALS			LABOUR				RATES		
	Del to Site £	Waste %	Material Cost £	Craft Optve Hrs	Lab Hrs	Labour Cost £	Sunds £	Nett Rate £	Unit	Gross Rate (+12.5%) £
L10: WINDOWS/ROOFLIGHTS/SCREENS/LOUVRES: METAL WINDOWS Cont.										
Windows in galvanized steel; polyester powder coated; white matt Cont.										
Crittall Windows Ltd. Duralife Homelight; weatherstripping; fixing to masonry with lugs Cont.										
top hung lights Cont.										
508 x 457mm, NH1	83.92	2.50	86.01	0.74	0.25	13.19	-	99.21	nr	111.61
508 x 628mm, NE1	90.04	2.50	92.29	0.86	0.29	15.33	-	107.62	nr	121.07
997 x 628mm, NE13E	124.90	2.50	128.02	1.12	0.32	19.34	-	147.36	nr	165.79
997 x 923mm, NC13C	135.54	2.50	138.93	1.24	0.40	21.90	-	160.84	nr	180.94
bottom hung lights										
508 x 628mm, NL1	127.72	2.50	130.91	0.86	0.32	15.65	-	146.56	nr	164.88
side hung lights										
508 x 628mm, NES1	96.78	2.50	99.20	0.86	0.32	15.65	-	114.85	nr	129.20
508 x 923mm, NC1	104.09	2.50	106.70	0.98	0.35	17.67	-	124.37	nr	139.92
508 x 1067mm, NC01	110.34	2.50	113.10	1.10	0.38	19.70	-	132.80	nr	149.40
508 x 1218mm, ND1	117.62	2.50	120.56	1.23	0.41	21.87	-	142.43	nr	160.23
mixed lights										
279 x 923mm, NC6F	81.25	2.50	83.28	0.98	0.35	17.67	-	100.95	nr	113.57
508 x 923mm, NC5F	91.92	2.50	94.21	0.98	0.35	17.67	-	111.89	nr	125.87
508 x 1067mm, NC05F	94.19	2.50	96.55	1.10	0.38	19.70	-	116.25	nr	130.78
997 x 292mm, NG2	97.60	2.50	100.04	0.85	0.33	15.61	-	115.66	nr	130.11
997 x 457mm, NH2	102.77	2.50	105.33	0.98	0.37	17.89	-	123.22	nr	138.63
997 x 628mm, NE2	113.25	2.50	116.08	1.12	0.40	20.20	-	136.28	nr	153.31
997 x 628mm, NES2	131.92	2.50	135.21	1.12	0.40	20.20	-	155.41	nr	174.84
997 x 923mm, NC2	143.88	2.50	147.47	1.24	0.44	22.33	-	169.81	nr	191.03
997 x 923mm, NC2F	189.14	2.50	193.87	1.24	0.44	22.33	-	216.20	nr	243.23
997 x 1067mm, NC02	155.70	2.50	159.59	1.37	0.47	24.50	-	184.09	nr	207.10
997 x 1067mm, NC02F	206.41	2.50	211.57	1.37	0.47	24.50	-	236.07	nr	265.58
997 x 1218mm, ND2	168.25	2.50	172.46	1.50	0.51	26.78	-	199.24	nr	224.14
997 x 1218mm, ND2F	208.67	2.50	213.88	1.50	0.51	26.78	-	240.66	nr	270.74
997 x 1513mm, NDV2FSB	242.03	2.50	248.08	1.63	0.58	29.37	-	277.45	nr	312.13
1486 x 628mm, NE3	147.20	2.50	150.87	1.50	0.51	26.78	-	177.65	nr	199.86
1486 x 923mm, NC4	235.60	2.50	241.49	1.63	0.58	29.37	-	270.86	nr	304.72
1486 x 923mm, NC4F	288.01	2.50	295.21	1.63	0.58	29.37	-	324.59	nr	365.16
1486 x 1067mm, NC04	265.88	2.50	272.53	1.63	0.58	29.37	-	301.90	nr	339.64
1486 x 1067mm, NC04F	307.72	2.50	315.42	1.77	0.61	31.68	-	347.10	nr	390.49
1486 x 1218mm, ND4	286.73	2.50	293.90	1.90	0.66	34.07	-	327.97	nr	368.96
1486 x 1218mm, ND4F	324.43	2.50	332.54	1.90	0.66	34.07	-	366.61	nr	412.44
1486 x 1513mm, NDV4FSB	395.04	2.50	404.91	2.04	0.69	36.38	-	441.29	nr	496.45
1994 x 923mm, NC11F	315.00	2.50	322.88	2.01	0.69	35.95	-	358.83	nr	403.68
1994 x 1218mm, ND11F	342.13	2.50	350.68	2.30	0.81	41.36	-	392.04	nr	441.04
1994 x 1513mm, NDV11FSB	398.05	2.50	408.00	2.47	0.86	44.31	-	452.31	nr	508.84
Extra for										
mullions										
292mm high	14.24	2.50	14.59	0.23	0.12	4.55	-	19.15	nr	21.54
457mm high	18.91	2.50	19.38	0.35	0.17	6.79	-	26.17	nr	29.45
628mm high	23.84	2.50	24.44	0.46	0.23	9.00	-	33.44	nr	37.62
923mm high	32.16	2.50	32.96	0.58	0.29	11.35	-	44.31	nr	49.85
1067mm high	36.98	2.50	37.90	0.69	0.35	13.55	-	51.46	nr	57.89
1218mm high	41.23	2.50	42.26	0.92	0.46	18.00	-	60.26	nr	67.79
1513mm high	49.62	2.50	50.86	1.15	0.58	22.55	-	73.41	nr	82.58
transoms										
279mm wide	15.12	2.50	15.50	0.23	0.12	4.55	-	20.05	nr	22.56
508mm wide	21.88	2.50	22.43	0.35	0.17	6.79	-	29.22	nr	32.88
997mm wide	59.64	2.50	61.13	0.69	0.35	13.55	-	74.68	nr	84.02
1143mm wide	65.32	2.50	66.95	1.15	0.58	22.55	-	89.50	nr	100.69
Controlair Ventilators, permanent, mill finish										
508mm wide	31.20	2.50	31.98	-	-	-	-	31.98	nr	35.98
997mm wide	41.90	2.50	42.95	-	-	-	-	42.95	nr	48.32
1486mm wide	94.39	2.50	96.75	-	-	-	-	96.75	nr	108.84
Controlair Ventilators, permanent, mill finish, flyscreen										
508mm wide	37.08	2.50	38.01	-	-	-	-	38.01	nr	42.76
997mm wide	52.83	2.50	54.15	-	-	-	-	54.15	nr	60.92
1486mm wide	109.25	2.50	111.99	-	-	-	-	111.99	nr	125.98
Controlair Ventilators, adjustable										
508mm wide 1-LT	37.19	2.50	38.11	-	-	-	-	38.11	nr	42.88
997mm wide 2-LT	47.02	2.50	48.19	-	-	-	-	48.19	nr	54.22
Controlair Ventilators, adjustable, flyscreen										
508mm wide 1-LT	42.84	2.50	43.91	-	-	-	-	43.91	nr	49.40
997mm wide 2-LT	57.95	2.50	59.40	-	-	-	-	59.40	nr	66.82
locks with Parkes locking handles for										
side hung lights	38.58	2.50	39.55	-	-	-	-	39.55	nr	44.49
horizontally pivoted lights	38.58	2.50	39.55	-	-	-	-	39.55	nr	44.49
Windows in aluminium; polyester powder finish; white matt										
Kawneer Products 102 Casement range; factory glazed, 4mm, one pane clear glass; fixing to masonry with lugs										
fixed lights										
600 x 400mm	82.89	2.50	84.96	0.81	0.35	15.26	-	100.22	nr	112.75
600 x 800mm	101.18	2.50	103.71	1.25	0.46	22.69	-	126.40	nr	142.19
600 x 1000mm	113.37	2.50	116.20	1.40	0.58	26.11	-	142.31	nr	160.10
600 x 1200mm	121.90	2.50	124.95	1.69	0.68	31.30	-	156.25	nr	175.78
600 x 1600mm	145.06	2.50	148.69	1.69	0.68	31.30	-	179.99	nr	202.48
800 x 400mm	90.21	2.50	92.46	0.97	0.39	17.96	-	110.42	nr	124.22
800 x 800mm	113.37	2.50	116.20	1.53	0.63	28.49	-	144.69	nr	162.78
800 x 1000mm	124.34	2.50	127.45	1.71	0.71	31.90	-	159.35	nr	179.27
800 x 1200mm	137.75	2.50	141.19	2.09	0.89	39.23	-	180.42	nr	202.97
800 x 1600mm	163.35	2.50	167.43	2.09	0.89	39.23	-	206.66	nr	232.49

Labour hourly rates: (except Specialists) Craft Operatives £14.21 Labourer £10.71 Rates are national average prices. Refer to REGIONAL VARIATIONS for indicative levels of overall pricing in regions	MATERIALS			LABOUR				RATES		
	Del to Site £	Waste %	Material Cost £	Craft Optve Hrs	Lab Hrs	Labour Cost £	Sunds £	Nett Rate £	Unit	Gross Rate (+12.5%) £
L10: WINDOWS/ROOFLIGHTS/SCREENS/LOUVRES: METAL WINDOWS Cont.										
Windows in aluminium; polyester powder finish; white matt Cont.										
Kawneer Products 102 Casement range; factory glazed, 4mm, one pane clear glass; fixing to masonry with lugs Cont.										
fixed lights Cont.										
1200 x 400mm	108.49	2.50	111.20	1.13	0.44	20.77	-	131.97	nr	148.47
1200 x 800mm	128.00	2.50	131.19	1.81	0.76	33.86	-	165.05	nr	185.69
1200 x 1000mm	152.38	2.50	156.18	2.04	0.86	38.20	-	194.38	nr	218.68
1200 x 1200mm	167.00	2.50	171.18	2.48	1.91	55.70	-	226.87	nr	255.23
1200 x 1600mm	196.26	2.50	201.17	2.48	1.91	55.70	-	256.86	nr	288.97
1400 x 800mm	148.72	2.50	152.44	2.08	1.71	47.87	-	200.31	nr	225.35
1400 x 1000mm	165.78	2.50	169.93	2.35	1.85	53.21	-	223.14	nr	251.03
1400 x 1400mm	197.48	2.50	202.41	2.89	2.12	63.77	-	266.19	nr	299.46
1400 x 1600mm	236.49	2.50	242.40	2.89	2.12	63.77	-	306.17	nr	344.44
top hung casement										
600 x 800mm	216.98	2.50	222.41	1.25	0.52	23.33	-	245.74	nr	276.46
600 x 1000mm	235.27	2.50	241.15	1.40	0.58	26.11	-	267.25	nr	300.66
600 x 1200mm	249.90	2.50	256.14	1.69	0.68	31.30	-	287.44	nr	323.37
800 x 800mm	235.27	2.50	241.15	1.53	0.63	28.49	-	269.64	nr	303.34
800 x 1000mm	253.55	2.50	259.89	1.71	0.71	31.90	-	291.79	nr	328.27
800 x 1200mm	270.62	2.50	277.38	2.09	0.89	39.23	-	316.61	nr	356.19
top hung fanlights										
600 x400mm	215.76	2.50	221.16	0.81	0.35	15.26	-	236.42	nr	265.97
800 x 600mm	240.14	2.50	246.15	0.97	0.39	17.96	-	264.11	nr	297.12
1200 x 600mm	260.87	2.50	267.39	1.13	0.44	20.77	-	288.16	nr	324.18
Kawneer Products 102 Casement range; factory glazed, 4mm, one pane obscure glass; fixing to masonry with lugs										
fixed lights										
600 x 400mm	82.89	2.50	84.96	0.81	0.35	15.26	-	100.22	nr	112.75
600 x 800mm	102.40	2.50	104.96	1.25	0.46	22.69	-	127.65	nr	143.60
600 x 1000mm	113.37	2.50	116.20	1.40	0.58	26.11	-	142.31	nr	160.10
600 x 1200mm	124.34	2.50	127.45	1.69	0.68	31.30	-	158.74	nr	178.59
600 x 1600mm	148.72	2.50	152.44	1.69	0.68	31.30	-	183.73	nr	206.70
800 x 400mm	90.21	2.50	92.46	0.97	0.39	17.96	-	110.42	nr	124.22
800 x 800mm	114.59	2.50	117.45	1.53	0.63	28.49	-	145.94	nr	164.18
800 x 1000mm	126.78	2.50	129.95	1.71	0.71	31.90	-	161.85	nr	182.08
1200 x 400mm	109.71	2.50	112.45	1.13	0.44	20.77	-	133.22	nr	149.88
1200 x 800mm	140.19	2.50	143.69	1.81	0.76	33.86	-	177.55	nr	199.74
top hung casement										
600 x 800mm	218.20	2.50	223.66	1.25	0.52	23.33	-	246.99	nr	277.86
600 x 1000mm	236.49	2.50	242.40	1.40	0.58	26.11	-	268.50	nr	302.07
600 x 1200mm	252.33	2.50	258.64	1.69	0.68	31.30	-	289.94	nr	326.18
800 x 800mm	236.49	2.50	242.40	1.53	0.63	28.49	-	270.89	nr	304.75
800 x 1000mm	254.77	2.50	261.14	1.71	0.71	31.90	-	293.04	nr	329.67
800 x 1200mm	271.84	2.50	278.63	2.09	0.89	39.23	-	317.86	nr	357.60
top hung fanlights										
600 x400mm	218.20	2.50	223.66	0.81	0.35	15.26	-	238.91	nr	268.78
800 x 600mm	242.58	2.50	248.65	0.97	0.39	17.96	-	266.61	nr	299.93
1200 x 600mm	266.96	2.50	273.64	1.13	0.44	20.77	-	294.40	nr	331.21
Sliding windows in aluminium; polyester powder finish; white matt										
Kawneer Products Kingsley Equal series; factory double glazed, 24mm - 4/16/4 units with clear 'Pilkington K Glass' glass; fixing to masonry with lugs										
vertical sliders										
600 x 800mm	355.05	2.50	363.92	1.25	0.52	23.33	-	387.25	nr	435.66
600 x 1000mm	416.26	2.50	426.67	1.40	0.58	26.11	-	452.77	nr	509.37
600 x 1200mm	487.49	2.50	499.68	1.54	0.62	28.52	-	528.20	nr	594.23
800 x 800mm	603.25	2.50	618.33	1.53	0.63	28.49	-	646.82	nr	727.67
800 x 1000mm	703.42	2.50	721.00	1.71	0.71	31.90	-	752.90	nr	847.02
800 x 1200mm	786.89	2.50	806.56	1.91	0.81	35.82	-	842.38	nr	947.68
1200 x 800mm	1259.92	2.50	1291.41	2.04	0.86	38.20	-	1329.61	nr	1495.81
1200 x 1000mm	1341.17	2.50	1374.69	2.25	0.98	42.47	-	1417.16	nr	1594.31
1200 x 1200mm	1454.69	2.50	1491.06	2.48	1.08	46.81	-	1537.87	nr	1730.10
1400 x 800mm	1594.93	2.50	1634.80	2.62	1.14	49.44	-	1684.24	nr	1894.77
1400 x 1000mm	1699.55	2.50	1742.04	2.89	1.28	54.78	-	1796.82	nr	2021.42
1400 x 1200mm	1777.46	2.50	1821.90	2.89	1.11	52.96	-	1874.85	nr	2109.21
horizontal sliders										
1200 x 800mm	331.67	2.50	339.97	1.81	0.76	33.86	-	373.83	nr	420.55
1200 x 1000mm	349.48	2.50	358.22	2.04	0.86	38.20	-	396.42	nr	445.97
1200 x 1400mm	394.00	2.50	403.85	2.25	0.98	42.47	-	446.32	nr	502.11
1400 x 800mm	348.37	2.50	357.08	2.08	0.87	38.87	-	395.95	nr	445.45
1400 x 1000mm	371.74	2.50	381.04	2.35	1.01	44.21	-	425.25	nr	478.40
1400 x 1400mm	411.81	2.50	422.11	2.62	1.14	49.44	-	471.54	nr	530.49
1400 x 1600mm	432.96	2.50	443.78	2.89	1.28	54.78	-	498.56	nr	560.88
1600 x 800mm	369.52	2.50	378.75	2.37	1.00	44.39	-	423.14	nr	476.03
1600 x 1000mm	392.89	2.50	402.71	2.67	1.16	50.36	-	453.08	nr	509.71
1600 x 1600mm	457.44	2.50	468.88	3.28	1.48	62.46	-	531.34	nr	597.76
Window sills in galvanized pressed steel BS 6510; fixing to metal										
for opening										
508mm wide; AWA	23.36	2.50	23.95	0.58	0.29	11.35	-	35.29	nr	39.71
508mm wide; AWB	25.85	2.50	26.50	0.58	0.29	11.35	-	37.85	nr	42.58
508mm wide; PS	26.79	2.50	27.46	0.58	0.29	11.35	-	38.80	nr	43.65
508mm wide; RPS	26.79	2.50	27.46	0.58	0.29	11.35	-	38.80	nr	43.65
628mm wide; AWA	25.95	2.50	26.60	0.69	0.35	13.55	-	40.15	nr	45.17

Labour hourly rates: (except Specialists) Craft Operatives £14.21 Labourer £10.71 Rates are national average prices. Refer to REGIONAL VARIATIONS for indicative levels of overall pricing in regions	MATERIALS			LABOUR				RATES		
	Del to Site £	Waste %	Material Cost £	Craft Optve Hrs	Lab Hrs	Labour Cost £	Sunds £	Nett Rate £	Unit	Gross Rate (+12.5%) £
L10: WINDOWS/ROOFLIGHTS/SCREENS/LOUVRES: METAL WINDOWS Cont.										
Sliding windows in aluminium; polyester powder finish; white matt Cont.										
Window sills in galvanized pressed steel BS 6510; fixing to metal Cont.										
for opening Cont.										
628mm wide; AWB	28.22	2.50	**28.92**	0.69	0.35	**13.55**	-	**42.48**	nr	**47.79**
628mm wide; PS	31.84	2.50	**32.64**	0.69	0.35	**13.55**	-	**46.19**	nr	**51.97**
628mm wide; RPS	31.84	2.50	**32.64**	0.69	0.35	**13.55**	-	**46.19**	nr	**51.97**
997mm wide; AWA	36.46	2.50	**37.38**	0.81	0.41	**15.90**	-	**53.28**	nr	**59.94**
997mm wide; AWB	41.13	2.50	**42.16**	0.81	0.41	**15.90**	-	**58.06**	nr	**65.31**
997mm wide; PS	36.79	2.50	**37.71**	0.81	0.41	**15.90**	-	**53.61**	nr	**60.32**
997mm wide; RPS	37.90	2.50	**38.84**	0.81	0.41	**15.90**	-	**54.74**	nr	**61.59**
1237mm wide; AWA	41.13	2.50	**42.16**	0.92	0.46	**18.00**	-	**60.16**	nr	**67.68**
1237mm wide; AWB	42.46	2.50	**43.53**	0.92	0.46	**18.00**	-	**61.53**	nr	**69.22**
1237mm wide; PS	38.37	2.50	**39.33**	0.92	0.46	**18.00**	-	**57.33**	nr	**64.50**
1237mm wide; RPS	39.77	2.50	**40.77**	0.92	0.46	**18.00**	-	**58.77**	nr	**66.11**
1486mm wide; AWA	46.00	2.50	**47.15**	1.04	0.52	**20.35**	-	**67.50**	nr	**75.94**
1486mm wide; AWB	47.46	2.50	**48.64**	1.04	0.52	**20.35**	-	**68.99**	nr	**77.61**
1486mm wide; PS	42.52	2.50	**43.58**	1.04	0.52	**20.35**	-	**63.93**	nr	**71.92**
1486mm wide; RPS	44.18	2.50	**45.29**	1.04	0.52	**20.35**	-	**65.63**	nr	**73.84**
1846mm wide; AWA	49.27	2.50	**50.50**	1.15	0.58	**22.55**	-	**73.05**	nr	**82.19**
1846mm wide; AWB	51.07	2.50	**52.35**	1.15	0.58	**22.55**	-	**74.90**	nr	**84.26**
1846mm wide; PS	44.90	2.50	**46.02**	1.15	0.58	**22.55**	-	**68.58**	nr	**77.15**
1846mm wide; RPS	46.96	2.50	**48.13**	1.15	0.58	**22.55**	-	**70.69**	nr	**79.52**
L10: WINDOWS/ROOFLIGHTS/SCREENS/LOUVRES: METAL ROOFLIGHTS										
Rooflights, mainly in mill finish aluminium										
Rooflights; factory glazed with one layer 6mm thick Georgian wired cast glass										
non-ventilating base frame; fixing to masonry with screws										
600 x 600mm	159.50	2.50	**163.49**	2.00	1.00	**39.13**	-	**202.62**	nr	**227.94**
600 x 900mm	195.50	2.50	**200.39**	2.30	1.15	**45.00**	-	**245.39**	nr	**276.06**
900 x 900mm	262.50	2.50	**269.06**	2.30	1.15	**45.00**	-	**314.06**	nr	**353.32**
900 x 1200mm	311.50	2.50	**319.29**	2.50	1.25	**48.91**	-	**368.20**	nr	**414.23**
1200 x 1200mm	378.50	2.50	**387.96**	2.80	1.40	**54.78**	-	**442.74**	nr	**498.09**
ventilating base frame; fixing to masonry with screws										
600 x 600mm	235.46	2.50	**241.35**	2.00	1.00	**39.13**	-	**280.48**	nr	**315.54**
600 x 900mm	296.50	2.50	**303.91**	2.30	1.15	**45.00**	-	**348.91**	nr	**392.53**
900 x 900mm	357.54	2.50	**366.48**	2.30	1.15	**45.00**	-	**411.48**	nr	**462.91**
900 x 1200mm	438.66	2.50	**449.63**	2.50	1.25	**48.91**	-	**498.54**	nr	**560.86**
1200 x 1200mm	519.78	2.50	**532.77**	2.80	1.40	**54.78**	-	**587.56**	nr	**661.00**
ventilating base frame and kerb; fixing to masonry with screws										
600 x 600mm	279.16	2.50	**286.14**	2.00	1.00	**39.13**	-	**325.27**	nr	**365.93**
600 x 900mm	340.50	2.50	**349.01**	2.30	1.15	**45.00**	-	**394.01**	nr	**443.26**
900 x 900mm	401.54	2.50	**411.58**	2.30	1.15	**45.00**	-	**456.58**	nr	**513.65**
900 x 1200mm	467.17	2.50	**478.85**	2.50	1.25	**48.91**	-	**527.76**	nr	**593.73**
1200 x 1200mm	532.80	2.50	**546.12**	2.80	1.40	**54.78**	-	**600.90**	nr	**676.01**
L10: WINDOWS/ROOFLIGHTS/SCREENS/LOUVRES: METAL SCREENS, BORROWED LIGHTS, FRAMES AND GRILLES										
Screens in mild steel										
Screens; 6 x 50 x 50mm angle framing to perimeter; 75 x 75 x 3mm mesh infill; welded connections										
2000 x 2000mm overall; fixing to masonry with screws	380.19	5.00	**399.20**	6.50	3.25	**127.17**	-	**526.37**	nr	**592.17**
3000 x 2000mm overall; fixing to masonry with screws	470.33	5.00	**493.85**	9.00	4.50	**176.09**	-	**669.93**	nr	**753.67**
Screens; 6 x 50 x 50mm angle framing to perimeter; 75 x 75 x 5mm mesh infill; welded connections										
2000 x 2000mm overall; fixing to masonry with screws	394.06	5.00	**413.76**	6.50	3.25	**127.17**	-	**540.94**	nr	**608.55**
Screens; 6 x 50 x 75mm angle framing to perimeter; 75 x 75 x 3mm mesh infill; welded connections										
3000 x 2000mm overall; fixing to masonry with screws	578.96	5.00	**607.90**	9.00	4.50	**176.09**	-	**783.99**	nr	**881.99**
Screens; 6 x 50 x 50mm angle framing to perimeter; 6 x 51 x 102mm tee mullion; 75 x 75 x 3mm mesh infill; welded connections										
2000 x 4000mm overall; mullions -1 nr; fixing to masonry with screws	991.51	5.00	**1041.08**	9.00	4.50	**176.09**	-	**1217.17**	nr	**1369.31**
Screens; 38.1 x 38.1 x 3.2mm hollow section framing to perimeter; 75 x 75 x 3mm mesh infill; welded connections										
2000 x 2000mm overall; fixing to masonry with screws	508.46	5.00	**533.89**	6.50	3.25	**127.17**	-	**661.06**	nr	**743.69**

Labour hourly rates: (except Specialists) Craft Operatives £14.21 Labourer £10.71 Rates are national average prices. Refer to REGIONAL VARIATIONS for indicative levels of overall pricing in regions	MATERIALS			LABOUR			RATES			
	Del to Site £	Waste %	Material Cost £	Craft Optve Hrs	Lab Hrs	Labour Cost £	Sunds £	Nett Rate £	Unit	Gross Rate (+12.5%) £
L10: WINDOWS/ROOFLIGHTS/SCREENS/LOUVRES: METAL SCREENS, BORROWED LIGHTS, FRAMES AND GRILLES Cont.										
Screens in mild steel Cont.										
Screens; 38.1 x 38.1 x 3.2mm hollow section framing to perimeter; 75 x 75 x 5mm mesh infill; welded connections										
2000 x 2000mm overall; fixing to masonry with screws	546.60	5.00	573.93	6.50	3.25	127.17	-	701.10	nr	788.74
Screens; 76.2 x 38.1 x 4mm hollow section framing to perimeter; 75 x 75 x 3mm mesh infill; welded connections										
2000 x 2000mm overall; fixing to masonry with screws	554.69	5.00	582.42	6.50	3.25	127.17	-	709.59	nr	798.29
Grilles in mild steel										
Grilles; 13 x 51mm flat bar framing to perimeter; 13mm diameter vertical infill bars at 150mm centres; welded connections										
1000 x 1000mm overall; fixing to masonry with screws	387.13	5.00	406.48	4.50	2.25	88.04	-	494.52	nr	556.34
2000 x 1000mm overall; fixing to masonry with screws	520.02	5.00	546.02	6.50	3.25	127.17	-	673.19	nr	757.34
Grilles; 13 x 51mm flat bar framing to perimeter; 13mm diameter vertical infill bars at 300mm centres; welded connections										
1000 x 1000mm overall; fixing to masonry with screws	369.79	5.00	388.28	4.50	2.25	88.04	-	476.32	nr	535.86
Grilles; 13 x 51mm flat bar framing to perimeter; 18mm diameter vertical infill bars at 150mm centres; welded connections										
1000 x 1000mm overall; fixing to masonry with screws	375.57	5.00	394.35	4.50	2.25	88.04	-	482.39	nr	542.69
Glazing frames in mild steel										
Glazing frames; 13 x 38 x 3mm angle framing to perimeter; welded connections; 13 x 13mm glazing beads, fixed with screws										
500 x 500mm overall; fixing to timber with screws	228.81	5.00	240.25	3.50	1.75	68.48	-	308.73	nr	347.32
Glazing frames; 15 x 21 x 3mm angle framing to perimeter; welded connections; 13 x 13mm glazing beads, fixed with screws										
500 x 1000mm overall; fixing to timber with screws	369.79	5.00	388.28	3.50	1.75	68.48	-	456.76	nr	513.85
Glazing frames; 18 x 25 x 3mm angle framing to perimeter; welded connections; 9 x 25mm glazing beads, fixed with screws										
1000 x 1000mm overall; fixing to timber with screws	444.91	5.00	467.15	4.50	2.25	88.04	-	555.19	nr	624.59
Glazing frames; 18 x 25 x 3mm angle framing to perimeter; welded connections; 13 x 13mm glazing beads, fixed with screws										
1000 x 1000mm overall; fixing to timber with screws	479.57	5.00	503.55	4.50	2.25	88.04	-	591.60	nr	665.54
Glazing frames; 18 x 38 x 3mm angle framing to perimeter; welded connections; 9 x 25mm glazing beads, fixed with screws										
1000 x 1000mm overall; fixing to timber with screws	597.45	5.00	627.32	3.50	1.75	68.48	-	695.79	nr	782.77
Glazing frames; 18 x 25 x 3mm angle framing to perimeter; welded connections; 15 x 15 x 3mm channel glazing beads; fixed with screws										
1000 x 1000mm overall; fixing to timber with screws	597.45	5.00	627.32	3.50	1.75	68.48	-	695.79	nr	782.77
Glazing frames; 18 x 38 x 3mm angle framing to perimeter; welded connections; 15 x 15 x 3mm channel glazing beads; fixed with screws										
1000 x 1000mm overall; fixing to timber with screws	597.45	5.00	627.32	3.50	1.75	68.48	-	695.79	nr	782.77
Glazing frames in aluminium										
Glazing frames; 13 x 38 x 3mm angle framing to perimeter; welded connections; 13 x 13mm glazing beads, fixed with screws										
500 x 500mm overall; fixing to timber with screws	390.59	5.00	410.12	3.50	1.75	68.48	-	478.60	nr	538.43
Glazing frames; 15 x 21 x 3mm angle framing to perimeter; welded connections; 13 x 13mm glazing beads, fixed with screws										
500 x 1000mm overall; fixing to timber with screws	390.59	5.00	410.12	3.50	1.75	68.48	-	478.60	nr	538.43

Labour hourly rates: (except Specialists) Craft Operatives £14.21 Labourer £10.71 Rates are national average prices. Refer to REGIONAL VARIATIONS for indicative levels of overall pricing in regions	MATERIALS			LABOUR				RATES		
	Del to Site £	Waste %	Material Cost £	Craft Optve Hrs	Lab Hrs	Labour Cost £	Sunds £	Nett Rate £	Unit	Gross Rate (+12.5%) £
L10: WINDOWS/ROOFLIGHTS/SCREENS/LOUVRES: METAL SCREENS, BORROWED LIGHTS, FRAMES AND GRILLES Cont.										
Glazing frames in aluminium Cont.										
Glazing frames; 18 x 25 x 3mm angle framing to perimeter; welded connections; 9 x 25mm glazing beads, fixed with screws										
1000 x 1000mm overall; fixing to timber with screws	419.48	5.00	**440.46**	4.50	2.25	**88.04**	-	**528.50**	nr	**594.56**
Glazing frames; 18 x 25 x 3mm angle framing to perimeter; welded connections; 13 x 13mm glazing beads, fixed with screws										
1000 x 1000mm overall; fixing to timber with screws	432.19	5.00	**453.80**	4.50	2.25	**88.04**	-	**541.85**	nr	**609.58**
Glazing frames; 18 x 38 x 3mm angle framing to perimeter; welded connections; 9 x 25mm glazing beads, fixed with screws										
1000 x 1000mm overall; fixing to timber with screws	489.97	5.00	**514.47**	4.50	2.25	**88.04**	-	**602.52**	nr	**677.83**
Glazing frames; 18 x 25 x 3mm angle framing to perimeter; welded connections; 15 x 15 x 3mm channel glazing beads; fixed with screws										
1000 x 1000mm overall; fixing to timber with screws	489.97	5.00	**514.47**	4.50	2.25	**88.04**	-	**602.52**	nr	**677.83**
Glazing frames; 18 x 38 x 3mm angle framing to perimeter; welded connections; 15 x 15 x 3mm channel glazing beads; fixed with screws										
1000 x 1000mm overall; fixing to timber with screws	553.53	5.00	**581.21**	4.50	2.25	**88.04**	-	**669.25**	nr	**752.91**
Bedding and pointing frames										
Pointing with Secomastic standard metal frames one side	0.73	10.00	**0.80**	0.14	-	**1.99**	-	**2.79**	m	**3.14**
Bedding in cement mortar (1:3); pointing with Secomastic standard mastic one side metal frames	0.90	5.00	**0.97**	0.20	-	**2.84**	-	**3.82**	m	**4.29**
L10 - WINDOWS/ROOFLIGHTS/SCREENS/LOUVRES: PLASTICS WINDOWS										
Windows in uPVC; white										
Windows; factory glazed 20mm double glazed units; hinges; fastenings										
fixed light; fixing to masonry with cleats and screws; overall size										
600 x 600mm	159.18	2.50	**163.16**	2.60	1.30	**50.87**	1.55	**215.58**	nr	**242.53**
600 x 900mm	170.59	2.50	**174.85**	3.25	1.62	**63.53**	1.55	**239.93**	nr	**269.93**
600 x 1050mm	181.95	2.50	**186.49**	3.55	1.77	**69.40**	1.55	**257.45**	nr	**289.63**
600 x 1200mm	196.28	2.50	**201.19**	3.90	1.95	**76.30**	1.55	**279.04**	nr	**313.92**
600 x 1500mm	230.37	2.50	**236.13**	4.55	2.27	**88.97**	1.55	**326.65**	nr	**367.48**
750 x 600mm	164.86	2.50	**168.99**	2.95	1.47	**57.66**	1.55	**228.20**	nr	**256.72**
750 x 900mm	193.32	2.50	**198.15**	3.60	1.80	**70.43**	1.55	**270.14**	nr	**303.90**
750 x 1050mm	204.68	2.50	**209.80**	3.90	1.95	**76.30**	1.55	**287.65**	nr	**323.61**
750 x 1200mm	216.06	2.50	**221.46**	4.20	2.10	**82.17**	1.55	**305.18**	nr	**343.33**
750 x 1500mm	253.00	2.50	**259.32**	4.85	2.42	**94.84**	1.55	**355.71**	nr	**400.17**
900 x 600mm	179.10	2.50	**183.58**	3.25	1.62	**63.53**	1.55	**248.66**	nr	**279.74**
900 x 900mm	235.94	2.50	**241.84**	3.90	1.95	**76.30**	1.55	**319.69**	nr	**359.66**
900 x 1050mm	227.42	2.50	**233.10**	4.10	2.05	**80.22**	1.55	**314.87**	nr	**354.23**
900 x 1200mm	244.47	2.50	**250.58**	4.55	2.27	**88.97**	1.55	**341.10**	nr	**383.74**
900 x 1500mm	287.25	2.50	**294.43**	5.15	2.57	**100.71**	1.55	**396.68**	nr	**446.27**
1200 x 600mm	196.28	2.50	**201.19**	3.65	1.82	**71.36**	1.55	**274.10**	nr	**308.36**
1200 x 900mm	236.05	2.50	**241.96**	4.30	2.15	**84.13**	1.55	**327.64**	nr	**368.59**
1200 x 1050mm	258.80	2.50	**265.27**	4.60	2.30	**90.00**	1.55	**356.82**	nr	**401.43**
1200 x 1200mm	281.55	2.50	**288.59**	5.00	2.50	**97.83**	1.55	**387.97**	nr	**436.46**
1200 x 1500mm	335.45	2.50	**343.84**	5.70	2.85	**111.52**	1.55	**456.91**	nr	**514.02**
fixed light with fixed light; fixing to masonry with cleats and screws; overall size										
600 x 1350mm	284.29	2.50	**291.39**	4.20	2.10	**82.17**	1.55	**375.12**	nr	**422.01**
600 x 1500mm	307.02	2.50	**314.70**	4.55	2.27	**88.97**	1.55	**405.21**	nr	**455.87**
600 x 1800mm	338.41	2.50	**346.87**	4.85	2.42	**94.84**	1.55	**443.26**	nr	**498.66**
600 x 2100mm	369.54	2.50	**378.78**	5.15	2.57	**100.71**	1.55	**481.04**	nr	**541.17**
750 x 1350mm	307.02	2.50	**314.70**	4.55	2.27	**88.97**	1.55	**405.21**	nr	**455.87**
750 x 1500mm	327.03	2.50	**335.21**	4.85	2.42	**94.84**	1.55	**431.60**	nr	**485.55**
750 x 1800mm	369.54	2.50	**378.78**	5.15	2.57	**100.71**	1.55	**481.04**	nr	**541.17**
750 x 2100mm	415.04	2.50	**425.42**	5.50	2.75	**107.61**	1.55	**534.58**	nr	**601.40**
900 x 1350mm	332.73	2.50	**341.05**	4.85	2.42	**94.84**	1.55	**437.43**	nr	**492.11**
900 x 1500mm	352.50	2.50	**361.32**	5.15	2.57	**100.71**	1.55	**463.57**	nr	**521.52**
900 x 1800mm	395.27	2.50	**405.15**	5.40	2.70	**105.65**	1.55	**512.35**	nr	**576.39**
900 x 2100mm	437.55	2.50	**448.49**	5.80	2.90	**113.48**	1.55	**563.52**	nr	**633.96**
1200 x 1350mm	375.24	2.50	**384.62**	5.40	2.70	**105.65**	1.55	**491.82**	nr	**553.30**
1200 x 1500mm	400.96	2.50	**410.98**	5.70	2.85	**111.52**	1.55	**524.06**	nr	**589.56**
1200 x 1800mm	452.11	2.50	**463.41**	6.05	3.02	**118.31**	1.55	**583.28**	nr	**656.19**
1200 x 2100mm	500.31	2.50	**512.82**	6.35	3.17	**124.18**	1.55	**638.56**	nr	**718.38**
overall, tilt/turn; fixing to masonry with cleats and screws; overall size										
600 x 600mm	463.47	2.50	**475.06**	2.60	1.30	**50.87**	1.55	**527.48**	nr	**593.41**
600 x 900mm	511.70	2.50	**524.49**	3.25	1.62	**63.53**	1.55	**589.58**	nr	**663.27**
600 x 1050mm	551.25	2.50	**565.03**	3.60	1.80	**70.43**	1.55	**637.02**	nr	**716.65**

Labour hourly rates: (except Specialists) Craft Operatives £14.21 Labourer £10.71 Rates are national average prices. Refer to REGIONAL VARIATIONS for indicative levels of overall pricing in regions	MATERIALS			LABOUR			RATES			
	Del to Site £	Waste %	Material Cost £	Craft Optve Hrs	Lab Hrs	Labour Cost £	Sunds £	Nett Rate £	Unit	Gross Rate (+12.5%) £
L10 - WINDOWS/ROOFLIGHTS/SCREENS/LOUVRES: PLASTICS WINDOWS Cont.										
Windows in uPVC; white Cont.										
Windows; factory glazed 20mm double glazed units; hinges; fastenings Cont.										
overall, tilt/turn; fixing to masonry with cleats and screws; overall size Cont.										
600 x 1200mm	568.55	2.50	**582.76**	3.90	1.95	**76.30**	1.55	**660.61**	nr	**743.19**
600 x 1350mm	608.10	2.50	**623.30**	4.20	2.10	**82.17**	1.55	**707.02**	nr	**795.40**
600 x 1500mm	639.74	2.50	**655.73**	4.55	2.27	**88.97**	1.55	**746.25**	nr	**839.53**
750 x 600mm	481.76	2.50	**493.80**	2.95	1.47	**57.66**	1.55	**553.02**	nr	**622.14**
750 x 900mm	537.41	2.50	**550.85**	3.60	1.80	**70.43**	1.55	**622.83**	nr	**700.68**
750 x 1050mm	569.06	2.50	**583.29**	3.90	1.95	**76.30**	1.55	**661.14**	nr	**743.79**
750 x 1200mm	599.92	2.50	**614.92**	4.20	2.10	**82.17**	1.55	**698.64**	nr	**785.97**
750 x 1350mm	631.10	2.50	**646.87**	4.55	2.27	**88.97**	1.55	**737.39**	nr	**829.57**
750 x 1500mm	659.51	2.50	**676.00**	4.85	2.42	**94.84**	1.55	**772.39**	nr	**868.93**
900 x 900mm	568.55	2.50	**582.76**	3.90	1.95	**76.30**	1.55	**660.61**	nr	**743.19**
900 x 1050mm	602.65	2.50	**617.72**	4.10	2.05	**80.22**	1.55	**699.49**	nr	**786.92**
900 x 1200mm	634.06	2.50	**649.91**	4.55	2.27	**88.97**	1.55	**740.42**	nr	**832.98**
900 x 1350mm	665.19	2.50	**681.82**	4.85	2.42	**94.84**	1.55	**778.21**	nr	**875.48**
900 x 1500mm	704.98	2.50	**722.61**	5.15	2.57	**100.71**	1.55	**824.86**	nr	**927.97**
1200 x 1050mm	676.57	2.50	**693.48**	4.60	2.30	**90.00**	1.55	**785.03**	nr	**883.16**
1200 x 1200mm	716.37	2.50	**734.28**	5.00	2.50	**97.83**	1.55	**833.65**	nr	**937.86**
1200 x 1350mm	747.76	2.50	**766.45**	5.40	2.70	**105.65**	1.55	**873.65**	nr	**982.86**
1200 x 1500mm	787.56	2.50	**807.25**	5.70	2.85	**111.52**	1.55	**920.32**	nr	**1035.36**
tilt/turn with fixed lights; fixing to masonry with cleats and screws; overall size										
600 x 1350mm	594.24	2.50	**609.10**	4.20	2.10	**82.17**	1.55	**692.82**	nr	**779.42**
600 x 1500mm	625.42	2.50	**641.05**	4.55	2.27	**88.97**	1.55	**731.57**	nr	**823.02**
600 x 2100mm	722.05	2.50	**740.10**	5.15	2.57	**100.71**	1.55	**842.36**	nr	**947.65**
750 x 1350mm	634.06	2.50	**649.91**	4.55	2.27	**88.97**	1.55	**740.42**	nr	**832.98**
750 x 1500mm	662.47	2.50	**679.03**	4.85	2.42	**94.84**	1.55	**775.42**	nr	**872.35**
750 x 2100mm	756.41	2.50	**775.32**	5.50	2.75	**107.61**	1.55	**884.48**	nr	**995.04**
900 x 1350mm	679.79	2.50	**696.79**	4.85	2.42	**94.84**	1.55	**793.17**	nr	**892.32**
900 x 1500mm	704.98	2.50	**722.61**	5.15	2.57	**100.71**	1.55	**824.86**	nr	**927.97**
900 x 2100mm	807.34	2.50	**827.52**	5.80	2.90	**113.48**	1.55	**942.55**	nr	**1060.36**
1200 x 1350mm	767.53	2.50	**786.72**	5.40	2.70	**105.65**	1.55	**893.92**	nr	**1005.66**
1200 x 1500mm	798.92	2.50	**818.89**	5.70	2.85	**111.52**	1.55	**931.96**	nr	**1048.46**
1200 x 2100mm	883.94	2.50	**906.04**	6.35	3.17	**124.18**	1.55	**1031.77**	nr	**1160.74**
tilt/turn sash with fixed side light; fixing to masonry with cleats and screws; overall size										
1800 x 900mm	756.41	2.50	**775.32**	5.30	2.65	**103.69**	1.55	**880.57**	nr	**990.64**
1800 x 1050mm	807.34	2.50	**827.52**	5.65	2.82	**110.49**	1.55	**939.56**	nr	**1057.00**
1800 x 1200mm	858.51	2.50	**879.98**	6.05	3.02	**118.31**	1.55	**999.84**	nr	**1124.82**
1800 x 1350mm	909.66	2.50	**932.40**	6.50	3.25	**127.17**	1.55	**1061.13**	nr	**1193.77**
1800 x 1500mm	960.82	2.50	**984.85**	6.80	3.40	**133.04**	1.55	**1119.44**	nr	**1259.37**
tilt/turn sash with centre fixed light; fixing to masonry with cleats and screws; overall size										
2400 x 900mm	1193.95	2.50	**1223.80**	5.30	2.65	**103.69**	1.55	**1329.04**	nr	**1495.17**
2400 x 1050mm	1279.21	2.50	**1311.19**	6.90	3.45	**135.00**	1.55	**1447.73**	nr	**1628.70**
2400 x 1200mm	1364.49	2.50	**1398.60**	7.30	3.65	**142.82**	1.55	**1542.98**	nr	**1735.85**
2400 x 1350mm	1435.68	2.50	**1471.58**	7.70	3.85	**150.65**	1.55	**1623.78**	nr	**1826.75**
2400 x 1500mm	746.27	2.50	**764.93**	8.00	4.00	**156.52**	1.55	**923.00**	nr	**1038.37**
L10 - WINDOWS/ROOFLIGHTS/SCREENS/LOUVRES: PLASTIC ROOFLIGHTS										
Roof domelights in glass reinforced plastics										
Domelights, translucent										
plain; fixing to masonry with screws										
610 x 610mm	93.13	2.50	**95.46**	1.15	1.15	**28.66**	0.31	**124.43**	nr	**139.98**
914 x 914mm	119.67	2.50	**122.66**	2.30	2.30	**57.32**	0.63	**180.60**	nr	**203.18**
1219 x 1219mm	159.11	2.50	**163.09**	3.50	3.50	**87.22**	0.96	**251.27**	nr	**282.68**
fixed ventilator; fixing to masonry with screws										
610 x 610mm	165.59	2.50	**169.73**	1.15	1.15	**28.66**	0.31	**198.70**	nr	**223.54**
914 x 914mm	199.21	2.50	**204.19**	2.30	2.30	**57.32**	0.63	**262.14**	nr	**294.91**
1219 x 1219mm	245.81	2.50	**251.95**	3.50	3.50	**87.22**	0.96	**340.13**	nr	**382.65**
controlled ventilator; fixing to masonry with screws										
610 x 610mm	185.65	2.50	**190.29**	1.15	1.15	**28.66**	0.31	**219.26**	nr	**246.66**
914 x 914mm	252.25	2.50	**258.55**	2.30	2.30	**57.32**	0.63	**316.50**	nr	**356.06**
1219 x 1219mm	338.36	2.50	**346.81**	3.50	3.50	**87.22**	0.96	**434.99**	nr	**489.37**
Bedding and pointing frames										
Pointing with Secomastic standard mastic plastics frames one side	1.31	10.00	**1.45**	0.14	-	**1.99**	-	**3.43**	m	**3.86**
Bedding in cement mortar (1:3); pointing with Secomastic standard mastic one side plastics frames	1.85	5.00	**2.02**	0.20	-	**2.84**	-	**4.86**	m	**5.47**
L20: DOORS/SHUTTERS/HATCHES: TIMBER DOORS										
Doors in softwood, wrought										
Internal panelled doors; SA; Magnet Trade 762 x 1981 x 34mm (f sizes)	128.94	2.50	**132.16**	1.40	0.18	**21.82**	-	**153.99**	nr	**173.23**
Internal panelled doors; SA; glazed with bevelled glass; Magnet Trade 762 x 1981 x 34mm (f sizes)	163.77	2.50	**167.86**	4.85	0.61	**75.45**	-	**243.31**	nr	**273.73**

Labour hourly rates: (except Specialists) Craft Operatives £14.21 Labourer £10.71 Rates are national average prices. Refer to REGIONAL VARIATIONS for indicative levels of overall pricing in regions	MATERIALS			LABOUR				RATES		
	Del to Site £	Waste %	Material Cost £	Craft Optve Hrs	Lab Hrs	Labour Cost £	Sunds £	Nett Rate £	Unit	Gross Rate (+12.5%) £
L20: DOORS/SHUTTERS/HATCHES: TIMBER DOORS Cont.										
Doors in softwood, wrought, preservative treated										
External ledged and braced doors; L & B Magnet Trade										
686 x 1981 x 44mm (f sizes)	80.89	2.50	82.91	1.25	0.16	19.48	-	102.39	nr	115.19
762 x 1981 x 44mm (f sizes)	75.24	2.50	77.12	1.50	0.16	23.03	-	100.15	nr	112.67
838 x 1981 x 44mm (f sizes)	88.43	2.50	90.64	1.75	0.22	27.22	-	117.86	nr	132.59
External framed, ledged and braced doors; YX; Magnet Trade										
686 x 1981 x 44mm (f sizes)	107.27	2.50	109.95	1.25	0.16	19.48	-	129.42	nr	145.60
762 x 1981 x 44mm (f sizes)	75.24	2.50	77.12	1.50	0.19	23.35	-	100.47	nr	113.03
813 x 1981 x 44mm (f sizes)	124.82	2.50	127.94	1.75	0.22	27.22	-	155.16	nr	174.55
838 x 1981 x 44mm (f sizes)	103.50	2.50	106.09	1.75	0.22	27.22	-	133.31	nr	149.98
External panelled doors; Richmond; Magnet Trade										
762 x 1981 x 44mm (f sizes)	178.66	2.50	183.13	1.50	0.19	23.35	-	206.48	nr	232.29
838 x 1981 x 44mm (f sizes)	182.39	2.50	186.95	1.75	0.22	27.22	-	214.18	nr	240.95
External panelled doors; KXT; Magnet Trade										
762 x 1981 x 44mm (f sizes)	107.86	2.50	110.56	1.50	0.19	23.35	-	133.91	nr	150.65
External panelled doors; Stable; softwood; Magnet Trade										
762 x 1981 x 44mm (f sizes)	160.01	2.50	164.01	2.75	0.34	42.72	-	206.72	nr	232.57
838 x 1981 x 44mm (f sizes)	170.58	2.50	174.85	3.00	0.38	46.70	-	221.55	nr	249.24
External panelled doors; Stable upper door glazed with Geneva glazing panel, Magnet Trade										
762 x 1981 x 44mm (f sizes)	225.49	2.50	231.13	3.10	0.39	48.23	-	279.36	nr	314.28
838 x 1981 x 44mm (f sizes)	237.77	2.50	243.71	3.35	0.42	52.10	-	295.81	nr	332.79
External panelled doors; 2XG; Magnet Trade										
762 x 1981 x 44mm (f sizes)	65.83	2.50	67.47	1.50	0.19	23.35	-	90.82	nr	102.17
813 x 2032 x 44mm (f sizes)	69.59	2.50	71.33	1.75	0.22	27.22	-	98.55	nr	110.87
838 x 1981 x 44mm (f sizes)	67.71	2.50	69.41	1.75	0.22	27.22	-	96.63	nr	108.71
External panelled doors; 2XGG; Magnet Trade										
762 x 1981 x 44mm (f sizes)	70.57	2.50	72.34	1.50	0.19	23.35	-	95.69	nr	107.65
838 x 1981 x 44mm (f sizes)	70.57	2.50	72.34	1.75	0.22	27.22	-	99.56	nr	112.01
External panelled doors; 2XG, glazed with 4mm clear toughend glass; Magnet Trade										
762 x 1981 x 44mm (f sizes)	90.07	2.50	92.32	1.95	0.24	30.28	-	122.60	nr	137.92
813 x 2032 x 44mm (f sizes)	103.43	2.50	106.02	2.20	0.28	34.26	-	140.28	nr	157.82
838 x 1981 x 44mm (f sizes)	97.60	2.50	100.04	2.20	0.28	34.26	-	134.30	nr	151.09
External panelled doors; 2XG, glazed with 4mm obscure toughend glass; Magnet Trade										
762 x 1981 x 44mm (f sizes)	94.10	2.50	96.45	1.95	0.24	30.28	-	126.73	nr	142.57
813 x 2032 x 44mm (f sizes)	100.79	2.50	103.31	2.20	0.28	34.26	-	137.57	nr	154.77
838 x 1981 x 44mm (f sizes)	98.91	2.50	101.39	2.20	0.28	34.26	-	135.65	nr	152.60
External panelled doors; 2XGG, glazed with 4mm obscure toughend glass; Magnet Trade										
762 x 1981 x 44mm (f sizes)	116.26	2.50	119.17	2.30	0.29	35.79	-	154.96	nr	174.33
838 x 1981 x 44mm (f sizes)	120.59	2.50	123.60	2.55	0.32	39.66	-	163.27	nr	183.67
Garage doors; MFL, pair, side hung; Magnet Trade										
1981 x 2134 x 44mm (f sizes) overall	316.30	2.50	324.21	4.00	0.50	62.20	-	386.41	nr	434.71
2134 x 2134 x 44mm (f sizes) overall	325.34	2.50	333.48	4.20	0.53	65.36	-	398.84	nr	448.69
Garage doors; 301, pair, side hung; Magnet Trade										
1981 x 2134 x 44mm (f sizes) overall	372.69	2.50	382.01	4.00	0.50	62.20	-	444.21	nr	499.73
2134 x 2134 x 44mm (f sizes) overall	381.74	2.50	391.28	4.20	0.53	65.36	-	456.64	nr	513.72
Doors in Hemlock, wrought										
Internal panelled doors; 2G; Magnet Trade										
762 x 1981 x 34mm (f sizes)	96.56	2.50	98.98	1.40	0.18	21.82	-	120.80	nr	135.90
Internal panelled doors; 2GG; Magnet Trade										
762 x 1981 x 34mm (f sizes)	90.91	2.50	93.19	1.40	0.18	21.82	-	115.01	nr	129.38
Internal panelled doors; SA; Magnet Trade										
762 x 1981 x 34mm (f sizes)	127.65	2.50	130.84	1.40	0.18	21.82	-	152.67	nr	171.75
Internal panelled doors; 10; Magnet Trade										
762 x 1981 x 34mm (f sizes)	71.48	2.50	73.26	1.40	0.18	21.82	-	95.09	nr	106.97
Internal panelled doors; 2G, glazed with 4mm clear toughend glass; Magnet Trade										
762 x 1981 x 34mm (f sizes)	123.96	2.50	127.06	1.85	0.23	28.75	-	155.81	nr	175.28
Internal panelled doors; 2GG, glazed with 4mm clear toughend glass; Magnet Trade										
762 x 1981 x 34mm (f sizes)	136.33	2.50	139.74	2.20	0.28	34.26	-	174.00	nr	195.75
Internal panelled doors; 10, glazed with 4mm clear toughend glass; Magnet Trade										
762 x 1981 x 34mm (f sizes)	118.08	2.50	121.03	1.95	0.24	30.28	-	151.31	nr	170.23
Internal panelled doors; 2G, glazed with 4mm obscure toughend glass; Magnet Trade										
762 x 1981 x 34mm (f sizes)	127.99	2.50	131.19	1.85	0.23	28.75	-	159.94	nr	179.94

Labour hourly rates: (except Specialists) Craft Operatives £14.21 Labourer £10.71 Rates are national average prices. Refer to REGIONAL VARIATIONS for indicative levels of overall pricing in regions	MATERIALS			LABOUR				RATES		
	Del to Site £	Waste %	Material Cost £	Craft Optve Hrs	Lab Hrs	Labour Cost £	Sunds £	Nett Rate £	Unit	Gross Rate (+12.5%) £
L20: DOORS/SHUTTERS/HATCHES: TIMBER DOORS Cont.										
Doors in Hemlock, wrought Cont.										
Internal panelled doors; 2GG, glazed with 4mm obscure toughend glass; Magnet Trade										
762 x 1981 x 34mm (f sizes)	139.75	2.50	143.25	2.20	0.28	34.26		177.51	nr	199.70
Internal panelled doors; 10, glazed with 4mm obscure toughend glass; Magnet Trade										
762 x 1981 x 34mm (f sizes)	119.56	2.50	122.55	1.95	0.24	30.28	-	152.83	nr	171.93
Internal panelled doors; SA, glazed with bevelled glass; Magnet Trade										
762 x 1981 x 34mm (f sizes)	163.77	2.50	167.86	4.85	0.61	75.45	-	243.31	nr	273.73
External panelled doors; 2XG; Magnet Trade										
762 x 1981 x 44mm (f sizes)	75.24	2.50	77.12	1.50	0.19	23.35		100.47	nr	113.03
813 x 2032 x 44mm (f sizes)	92.19	2.50	94.50	1.75	0.22	27.22		121.72	nr	136.93
838 x 1981 x 44mm (f sizes)	84.66	2.50	86.78	1.75	0.22	27.22		114.00	nr	128.26
External panelled doors; 2XGG; Magnet Trade										
762 x 1981 x 44mm (f sizes)	88.43	2.50	90.64	1.50	0.19	23.35	-	113.99	nr	128.24
813 x 2032 x 44mm (f sizes)	80.18	2.50	82.18	1.75	0.22	27.22	-	109.41	nr	123.08
838 x 1981 x 44mm (f sizes)	95.96	2.50	98.36	1.75	0.22	27.22	-	125.59	nr	141.29
External panelled doors; KXT; Magnet Trade										
762 x 1981 x 44mm (f sizes)	158.12	2.50	162.07	1.50	0.19	23.35	-	185.42	nr	208.60
838 x 1981 x 44mm (f sizes)	160.01	2.50	164.01	1.75	0.22	27.22	-	191.23	nr	215.13
External panelled doors; SA; Magnet Trade										
762 x 1981 x 44mm (f sizes)	122.33	2.50	125.39	1.50	0.19	23.35	-	148.74	nr	167.33
838 x 1981 x 44mm (f sizes)	181.88	2.50	186.43	1.75	0.22	27.22	-	213.65	nr	240.36
External panelled doors; 10; Magnet Trade										
762 x 1981 x 44mm (f sizes)	82.78	2.50	84.85	1.50	0.19	23.35	-	108.20	nr	121.72
External panelled doors; 2XG, glazed with 4mm clear toughend glass; Magnet Trade										
762 x 1981 x 44mm (f sizes)	99.48	2.50	101.97	1.95	0.24	30.28	-	132.25	nr	148.78
813 x 2032 x 44mm (f sizes)	126.04	2.50	129.19	2.20	0.28	34.26	-	163.45	nr	183.88
838 x 1981 x 44mm (f sizes)	114.55	2.50	117.42	2.20	0.28	34.26	-	151.68	nr	170.64
External panelled doors; 2XGG, glazed with 4mm clear toughend glass; Magnet Trade										
762 x 1981 x 44mm (f sizes)	130.69	2.50	133.96	2.30	0.29	35.79	-	169.75	nr	190.97
813 x 2032 x 44mm (f sizes)	132.61	2.50	135.93	2.55	0.32	39.66	-	175.59	nr	197.54
838 x 1981 x 44mm (f sizes)	144.44	2.50	148.06	2.55	0.32	39.66	-	187.72	nr	211.18
External panelled doors; 10, glazed with 4mm clear toughend glass; Magnet Trade										
762 x 1981 x 44mm (f sizes)	126.79	2.50	129.96	2.10	0.26	32.63	-	162.59	nr	182.91
External panelled doors; 2XG, glazed with 4mm obscure toughend glass; Magnet Trade										
762 x 1981 x 44mm (f sizes)	103.51	2.50	106.10	1.95	0.24	30.28	-	136.38	nr	153.43
813 x 2032 x 44mm (f sizes)	123.39	2.50	126.48	2.20	0.28	34.26	-	160.74	nr	180.83
838 x 1981 x 44mm (f sizes)	115.87	2.50	118.76	2.20	0.28	34.26	-	153.02	nr	172.15
External panelled doors; 2XGG, glazed with 4mm obscure toughend glass; Magnet Trade										
762 x 1981 x 44mm (f sizes)	134.12	2.50	137.47	2.30	0.29	35.79	-	173.26	nr	194.91
813 x 2032 x 44mm (f sizes)	130.15	2.50	133.40	2.55	0.32	39.66	-	173.07	nr	194.70
838 x 1981 x 44mm (f sizes)	145.98	2.50	149.63	2.55	0.32	39.66	-	189.29	nr	212.96
External panelled doors; SA, glazed with 6mm bevelled glass; Magnet Trade										
762 x 1981 x 44mm (f sizes)	246.58	2.50	252.74	4.95	0.62	76.98	-	329.72	nr	370.94
838 x 1981 x 44mm (f sizes)	314.04	2.50	321.90	5.20	0.65	80.85	-	402.75	nr	453.09
Doors in Hardwood, wrought										
External panelled doors; Alicante; Magnet Trade										
813 x 2032 x 44mm (f sizes)	144.03	2.50	147.63	2.65	0.33	41.19	-	188.82	nr	212.42
838 x 1981 x 44mm (f sizes)	144.03	2.50	147.63	2.65	0.33	41.19	-	188.82	nr	212.42
External panelled doors; Carolina, glazed in clear glass; Magnet Trade										
813 x 2032 x 44mm (f sizes)	174.07	2.50	178.43	2.65	0.33	41.19	-	219.62	nr	247.07
838 x 1981 x 44mm (f sizes)	174.07	2.50	178.43	2.65	0.33	41.19	-	219.62	nr	247.07
External panelled doors; Alicante, double glazed with clear glass; Magnet Trade										
813 x 2032 x 44mm (f sizes)	220.25	2.50	225.76	4.70	0.59	73.11	-	298.86	nr	336.22
838 x 1981 x 44mm (f sizes)	220.25	2.50	225.76	4.70	0.59	73.11	-	298.86	nr	336.22
Doors in Hardwood (solid, laminated or veneered), wrought										
External panelled doors, fire resisting; FD30 2XGG; Magnet Trade										
838 x 1981 x 44mm (f sizes)	176.23	2.50	180.64	2.55	0.32	39.66	-	220.30	nr	247.84
External panelled doors; Kendall; Magnet Trade										
762 x 1981 x 44mm (f sizes)	146.85	2.50	150.52	2.40	0.30	37.32	-	187.84	nr	211.32
838 x 1981 x 44mm (f sizes)	146.85	2.50	150.52	2.65	0.33	41.19	-	191.71	nr	215.68

Labour hourly rates: (except Specialists) Craft Operatives £14.21 Labourer £10.71 Rates are national average prices. Refer to REGIONAL VARIATIONS for indicative levels of overall pricing in regions	MATERIALS			LABOUR				RATES		
	Del to Site £	Waste %	Material Cost £	Craft Optve Hrs	Lab Hrs	Labour Cost £	Sunds £	Nett Rate £	Unit	Gross Rate (+12.5%) £
L20: DOORS/SHUTTERS/HATCHES: TIMBER DOORS Cont.										
Doors in Hardwood (solid, laminated or veneered), wrought Cont.										
External panelled doors; Steeton; Magnet Trade										
838 x 1981 x 44mm (f sizes)	198.83	2.50	**203.81**	2.65	0.33	**41.19**	-	**245.00**	nr	**275.62**
External panelled doors; Conway; Magnet Trade										
762 x 1981 x 44mm (f sizes)	187.53	2.50	**192.22**	2.40	0.30	**37.32**	-	**229.54**	nr	**258.23**
813 x 2032 x 44mm (f sizes)	187.53	2.50	**192.22**	2.65	0.33	**41.19**	-	**233.41**	nr	**262.59**
838 x 1981 x 44mm (f sizes)	187.53	2.50	**192.22**	2.65	0.33	**41.19**	-	**233.41**	nr	**262.59**
External panelled doors; Elizabethan; Magnet Trade										
813 x 2032 x 44mm (f sizes)	177.36	2.50	**181.80**	2.65	0.33	**41.19**	-	**222.99**	nr	**250.86**
838 x 1981 x 44mm (f sizes)	158.12	2.50	**162.07**	2.65	0.33	**41.19**	-	**203.26**	nr	**228.67**
External panelled doors; Manilla; Magnet Trade										
762 x 1981 x 44mm (f sizes)	225.92	2.50	**231.57**	2.40	0.30	**37.32**	-	**268.89**	nr	**302.50**
813 x 2032 x 44mm (f sizes)	225.92	2.50	**231.57**	2.65	0.33	**41.19**	-	**272.76**	nr	**306.86**
838 x 1981 x 44mm (f sizes)	182.39	2.50	**186.95**	2.65	0.33	**41.19**	-	**228.14**	nr	**256.66**
External panelled doors; Richmond; Magnet Trade										
838 x 1981 x 44mm (f sizes)	209.57	2.50	**214.81**	2.65	0.33	**41.19**	-	**256.00**	nr	**288.00**
External panelled doors; Rutland; Magnet Trade										
762 x 1981 x 44mm (f sizes)	142.33	2.50	**145.89**	2.40	0.30	**37.32**	-	**183.21**	nr	**206.11**
838 x 1981 x 44mm (f sizes)	142.33	2.50	**145.89**	2.65	0.33	**41.19**	-	**187.08**	nr	**210.47**
External panelled doors; Stable door; Magnet Trade										
762 x 1981 x 44mm (f sizes)	222.00	2.50	**227.55**	4.15	0.52	**64.54**	-	**292.09**	nr	**328.60**
838 x 1981 x 44mm (f sizes)	232.17	2.50	**237.98**	4.45	0.56	**69.23**	-	**307.21**	nr	**345.61**
External panelled doors; Stourbridge; Magnet Trade										
838 x 1981 x 44mm (f sizes)	357.04	2.50	**365.97**	2.65	0.33	**41.19**	-	**407.16**	nr	**458.06**
External panelled doors; Stuart; Magnet Trade										
762 x 1981 x 44mm (f sizes)	146.85	2.50	**150.52**	2.40	0.30	**37.32**	-	**187.84**	nr	**211.32**
838 x 1981 x 44mm (f sizes)	146.85	2.50	**150.52**	2.65	0.33	**41.19**	-	**191.71**	nr	**215.68**
External panelled doors; Kendall, Double glazed with clear glass; Magnet Trade										
762 x 1981 x 44mm (f sizes)	232.68	2.50	**238.50**	2.90	0.36	**45.06**	-	**283.56**	nr	**319.01**
838 x 1981 x 44mm (f sizes)	231.55	2.50	**237.34**	3.10	0.39	**48.23**	-	**285.57**	nr	**321.26**
External panelled doors; Kendall, Double glazed with obscure glass; Magnet Trade										
762 x 1981 x 44mm (f sizes)	240.03	2.50	**246.03**	2.90	0.36	**45.06**	-	**291.09**	nr	**327.48**
838 x 1981 x 44mm (f sizes)	241.72	2.50	**247.76**	3.10	0.39	**48.23**	-	**295.99**	nr	**332.99**
External panelled doors; Rutland, Double glazed with clear glass; Magnet Trade										
762 x 1981 x 44mm (f sizes)	275.57	2.50	**282.45**	3.20	0.40	**49.76**	-	**332.21**	nr	**373.74**
838 x 1981 x 44mm (f sizes)	274.44	2.50	**281.30**	3.45	0.43	**53.63**	-	**334.93**	nr	**376.79**
External panelled doors; Rutland, Double glazed with obscure glass; Magnet Trade										
838 x 1981 x 44mm (f sizes)	286.87	2.50	**294.04**	3.45	0.43	**53.63**	-	**347.67**	nr	**391.13**
External panelled doors; Stourbridge, Double glazed with clear glass; Magnet Trade										
838 x 1981 x 44mm (f sizes)	595.43	2.50	**610.32**	6.10	0.76	**94.82**	-	**705.14**	nr	**793.28**
External panelled doors; Stuart, glazed with 6mm bevelled glass; Magnet Trade										
762 x 1981 x 44mm (f sizes)	265.45	2.50	**272.09**	5.85	0.73	**90.95**	-	**363.03**	nr	**408.41**
838 x 1981 x 44mm (f sizes)	276.75	2.50	**283.67**	6.10	0.76	**94.82**	-	**378.49**	nr	**425.80**
External panelled doors; Stable door, upper door glazed with decorative leaded obscured glazing; Magnet Trade										
762 x 1981 x 44mm (f sizes)	287.49	2.50	**294.68**	5.75	0.72	**89.42**	-	**384.09**	nr	**432.11**
838 x 1981 x 44mm (f sizes)	299.35	2.50	**306.84**	6.05	0.76	**94.11**	-	**400.95**	nr	**451.07**
External panelled doors; Steeton, glazed with clear glass double glazing unit; Magnet Trade										
838 x 1981 x 44mm (f sizes)	266.58	2.50	**273.25**	3.00	0.38	**46.70**	-	**319.95**	nr	**359.94**
External panelled doors; Richmond, glazed with decorative leaded obscured double glazing; Magnet Trade										
838 x 1981 x 44mm (f sizes)	407.84	2.50	**418.04**	3.35	0.42	**52.10**	-	**470.14**	nr	**528.90**
External panelled doors; Rutland, glazed with clear glass double glazing unit; Magnet Trade										
762 x 1981 x 44mm (f sizes)	275.57	2.50	**282.45**	3.20	0.40	**49.76**	-	**332.21**	nr	**373.74**
838 x 1981 x 44mm (f sizes)	274.44	2.50	**281.30**	3.45	0.45	**53.84**	-	**335.14**	nr	**377.03**
Doors in Hardwood (solid, laminated or fibreboard core and/or veneered), wrought										
Internal panelled doors; SA; Magnet Trade										
686 x 1981 x 34mm (f sizes)	84.64	2.50	**86.76**	2.20	0.28	**34.26**	-	**121.02**	nr	**136.15**
762 x 1981 x 34mm (f sizes)	84.64	2.50	**86.76**	2.40	0.30	**37.32**	-	**124.07**	nr	**139.58**

Labour hourly rates: (except Specialists) Craft Operatives £14.21 Labourer £10.71 Rates are national average prices. Refer to REGIONAL VARIATIONS for indicative levels of overall pricing in regions	MATERIALS			LABOUR				RATES		
	Del to Site £	Waste %	Material Cost £	Craft Optve Hrs	Lab Hrs	Labour Cost £	Sunds £	Nett Rate £	Unit	Gross Rate (+12.5%) £
L20: DOORS/SHUTTERS/HATCHES: TIMBER DOORS Cont.										
Doors in Hardwood (solid, laminated or fibreboard core and/or veneered), wrought Cont.										
Internal panelled doors; Victorian; Magnet Trade										
686 x 1981 x 34mm (f sizes)	132.73	2.50	**136.04**	2.20	0.28	**34.26**	-	**170.30**	nr	**191.59**
762 x 1981 x 34mm (f sizes)	134.99	2.50	**138.36**	2.40	0.30	**37.32**	-	**175.68**	nr	**197.64**
Internal panelled doors; Louis; Magnet Trade										
686 x 1981 x 34mm (f sizes)	335.12	2.50	**343.50**	2.20	0.28	**34.26**	-	**377.76**	nr	**424.98**
762 x 1981 x 34mm (f sizes)	335.12	2.50	**343.50**	2.40	0.30	**37.32**	-	**380.82**	nr	**428.42**
Internal panelled doors; Windsor; Magnet Trade										
686 x 1981 x 34mm (f sizes)	136.12	2.50	**139.52**	2.20	0.28	**34.26**	-	**173.78**	nr	**195.50**
762 x 1981 x 34mm (f sizes)	137.25	2.50	**140.68**	2.40	0.30	**37.32**	-	**177.99**	nr	**200.24**
Internal panelled doors; 10; Magnet Trade										
762 x 1981 x 34mm (f sizes)	80.74	2.50	**82.76**	2.40	0.30	**37.32**	-	**120.08**	nr	**135.09**
Internal panelled doors; 10, glazed with 4mm clear toughend glass; Magnet Trade										
762 x 1981 x 34mm (f sizes)	124.19	2.50	**127.30**	3.00	0.38	**46.70**	-	**174.00**	nr	**195.75**
Internal panelled doors; 10, glazed with 4mm obscure toughend glass; Magnet Trade										
762 x 1981 x 34mm (f sizes)	125.67	2.50	**128.82**	3.00	0.38	**46.70**	-	**175.52**	nr	**197.46**
Internal panelled doors; Malaga; Magnet Trade										
762 x 1981 x 34mm (f sizes)	158.15	2.50	**162.11**	2.40	0.30	**37.32**	-	**199.42**	nr	**224.35**
Internal panelled doors; SA, glazed with bevelled glass; Magnet Trade										
762 x 1981 x 34mm (f sizes)	150.58	2.50	**154.35**	5.85	0.73	**90.95**	-	**245.29**	nr	**275.95**
Doors in solid or laminated construction, Oak facings										
Internal panelled doors; Louis; Magnet Trade										
762 x 1981 x 34mm (f sizes)	150.58	2.50	**154.34**	2.40	0.30	**37.32**	-	**191.66**	nr	**215.62**
Doors in hardboard, embossed										
Internal panelled doors; Colonist; Magnet Trade										
610 x 1981 x 34mm (f sizes)	28.16	2.50	**28.87**	1.85	0.23	**28.75**	-	**57.62**	nr	**64.82**
686 x 1981 x 34mm (f sizes)	28.16	2.50	**28.87**	1.95	0.24	**30.28**	-	**59.15**	nr	**66.54**
762 x 1981 x 34mm (f sizes)	28.16	2.50	**28.87**	2.10	0.26	**32.63**	-	**61.49**	nr	**69.18**
Internal panelled doors; Sentinel; Magnet Trade										
686 x 1981 x 34mm (f sizes)	52.64	2.50	**53.95**	1.95	0.24	**30.28**	-	**84.23**	nr	**94.76**
762 x 1981 x 34mm (f sizes)	41.34	2.50	**42.37**	2.10	0.26	**32.63**	-	**75.00**	nr	**84.37**
Flush doors										
Internal; hardboard facings; Magnet Trade										
305 x 1981 x 34mm	31.02	2.50	**31.80**	1.15	0.14	**17.84**	-	**49.64**	nr	**55.84**
381 x 1981 x 34mm	29.33	2.50	**30.06**	1.25	0.16	**19.48**	-	**49.53**	nr	**55.73**
457 x 1981 x 34mm	29.33	2.50	**30.06**	1.40	0.18	**21.82**	-	**51.88**	nr	**58.37**
533 x 1981 x 34mm	28.16	2.50	**28.87**	1.50	0.19	**23.35**	-	**52.22**	nr	**58.74**
610 x 1981 x 34mm	28.16	2.50	**28.87**	1.60	0.20	**24.88**	-	**53.74**	nr	**60.46**
686 x 1981 x 34mm	28.16	2.50	**28.87**	1.75	0.22	**27.22**	-	**56.09**	nr	**63.10**
711 x 1981 x 34mm	30.46	2.50	**31.22**	1.85	0.23	**28.75**	-	**59.97**	nr	**67.46**
762 x 1981 x 34mm	28.16	2.50	**28.87**	1.95	0.24	**30.28**	-	**59.15**	nr	**66.54**
813 x 2032 x 34mm	33.85	2.50	**34.69**	2.20	0.28	**34.26**	-	**68.95**	nr	**77.57**
838 x 1981 x 34mm	29.10	2.50	**29.83**	2.20	0.28	**34.26**	-	**64.09**	nr	**72.10**
Internal; Magnaseal; Magnet Trade										
610 x 1981 x 34mm	23.74	2.50	**24.33**	1.95	0.24	**30.28**	-	**54.61**	nr	**61.44**
686 x 1981 x 34mm	23.74	2.50	**24.33**	2.20	0.28	**34.26**	-	**58.59**	nr	**65.92**
762 x 1981 x 34mm	21.91	2.50	**22.46**	2.20	0.28	**34.26**	-	**56.72**	nr	**63.81**
838 x 1981 x 34mm	31.10	2.50	**31.88**	2.55	0.32	**39.66**	-	**71.54**	nr	**80.48**
Internal; plywood facings; Magnet Trade										
381 x 1981 x 34mm	35.69	2.50	**36.58**	1.25	0.16	**19.48**	-	**56.06**	nr	**63.06**
457 x 1981 x 34mm	35.69	2.50	**36.58**	1.40	0.18	**21.82**	-	**58.40**	nr	**65.70**
533 x 1981 x 34mm	36.64	2.50	**37.55**	1.50	0.19	**23.35**	-	**60.90**	nr	**68.52**
610 x 1981 x 34mm	35.69	2.50	**36.58**	1.60	0.20	**24.88**	-	**61.46**	nr	**69.14**
686 x 1981 x 34mm	28.16	2.50	**28.87**	1.75	0.22	**27.22**	-	**56.09**	nr	**63.10**
711 x 1981 x 34mm	35.69	2.50	**36.58**	1.85	0.23	**28.75**	-	**65.33**	nr	**73.50**
762 x 1981 x 34mm	28.16	2.50	**28.87**	1.95	0.24	**30.28**	-	**59.15**	nr	**66.54**
813 x 2032 x 34mm	39.46	2.50	**40.45**	2.20	0.28	**34.26**	-	**74.71**	nr	**84.05**
838 x 1981 x 34mm	37.57	2.50	**38.51**	2.20	0.28	**34.26**	-	**72.77**	nr	**81.87**
Internal; Sapele Showpiece; Magnet Trade										
457 x 1981 x 34mm	51.93	2.50	**53.22**	1.60	0.20	**24.88**	-	**78.10**	nr	**87.87**
533 x 1981 x 34mm	51.93	2.50	**53.22**	1.75	0.22	**27.22**	-	**80.45**	nr	**90.50**
610 x 1981 x 34mm	45.11	2.50	**46.24**	1.85	0.23	**28.75**	-	**74.99**	nr	**84.37**
686 x 1981 x 34mm	37.57	2.50	**38.51**	1.95	0.24	**30.28**	-	**68.79**	nr	**77.39**
762 x 1981 x 34mm	31.92	2.50	**32.72**	2.40	0.30	**37.32**	-	**70.04**	nr	**78.79**
838 x 1981 x 34mm	44.16	2.50	**45.27**	2.40	0.30	**37.32**	-	**82.58**	nr	**92.91**
External; MF1X, plywood facings; Magnet Trade										
762 x 1981 x 44mm	67.71	2.50	**69.41**	2.10	0.26	**32.63**	-	**102.03**	nr	**114.79**
838 x 1981 x 44mm	69.59	2.50	**71.33**	2.30	0.29	**35.79**	-	**107.12**	nr	**120.51**
External; MF2X, plywood facings; Magnet Trade										
762 x 1981 x 44mm; glazing aperture 457 x 457mm	92.61	2.50	**94.92**	2.10	0.26	**32.63**	-	**127.55**	nr	**143.49**

Labour hourly rates: (except Specialists) Craft Operatives £14.21 Labourer £10.71 Rates are national average prices. Refer to REGIONAL VARIATIONS for indicative levels of overall pricing in regions	MATERIALS			LABOUR				RATES		
	Del to Site £	Waste %	Material Cost £	Craft Optve Hrs	Lab Hrs	Labour Cost £	Sunds £	Nett Rate £	Unit	Gross Rate (+12.5%) £
L20: DOORS/SHUTTERS/HATCHES: TIMBER DOORS Cont.										
Flush doors Cont.										
External; MF2X, plywood facings; Magnet Trade Cont.										
838 x 1981 x 44mm; glazing aperture 457 x 457mm..	92.61	2.50	**94.92**	2.30	0.29	**35.79**	-	**130.71**	nr	**147.05**
External; MF4X, plywood facings; Magnet Trade										
762 x 1981 x 44mm; glazing aperture 559 x 864mm..	91.48	2.50	**93.77**	2.10	0.26	**32.63**	-	**126.39**	nr	**142.19**
838 x 1981 x 44mm; glazing aperture 635 x 864mm..	95.43	2.50	**97.82**	2.30	0.29	**35.79**	-	**133.61**	nr	**150.31**
Internal; fire resisting; Magnaseal; FD30; Magnet Trade										
813 x 2032 x 44mm	66.39	2.50	**68.05**	2.55	0.32	**39.66**	-	**107.71**	nr	**121.18**
Internal; fire resisting; Plywood facings; FD30; Magnet Trade										
686 x 1981 x 44mm	66.39	2.50	**68.05**	2.20	0.28	**34.26**	-	**102.31**	nr	**115.10**
762 x 1981 x 44mm	66.39	2.50	**68.05**	2.20	0.28	**34.26**	-	**102.31**	nr	**115.10**
813 x 2032 x 44mm	69.33	2.50	**71.06**	2.55	0.32	**39.66**	-	**110.73**	nr	**124.57**
838 x 1981 x 44mm	69.33	2.50	**71.06**	2.55	0.32	**39.66**	-	**110.73**	nr	**124.57**
Internal; fire resisting; plywood facings; FDG30; Magnet Trade										
762 x 1981 x 44mm; glazing aperture 508 x 508mm..	82.78	2.50	**84.85**	2.20	0.28	**34.26**	-	**119.11**	nr	**134.00**
813 x 2032 x 44mm; glazing aperture 559 x 559mm..	98.26	2.50	**100.72**	2.55	0.32	**39.66**	-	**140.38**	nr	**157.93**
838 x 1981 x 44mm; glazing aperture 584 x 584mm..	84.66	2.50	**86.78**	2.55	0.32	**39.66**	-	**126.44**	nr	**142.25**
External; fire resisting; plywood facings; FD30; Magnet Trade										
762 x 1981 x 44mm	30.04	2.50	**30.79**	2.20	0.28	**34.26**	-	**65.05**	nr	**73.18**
838 x 1981 x 44mm	30.04	2.50	**30.79**	2.65	0.33	**41.19**	-	**71.98**	nr	**80.98**
Internal; fire resisting; Sapele facings; FD30; Magnet Trade										
762 x 1981 x 44mm	94.87	2.50	**97.24**	2.55	0.32	**39.66**	-	**136.90**	nr	**154.02**
838 x 1981 x 44mm	102.21	2.50	**104.77**	2.75	0.34	**42.72**	-	**147.49**	nr	**165.92**
Trap doors in softwood, wrought										
19mm matchboarding on 25 x 75mm ledges										
457 x 610mm	25.62	2.50	**26.26**	0.21	0.03	**3.31**	-	**29.56**	nr	**33.26**
610 x 610mm	31.68	2.50	**32.47**	0.22	0.03	**3.45**	-	**35.92**	nr	**40.41**
762 x 610mm	38.41	2.50	**39.37**	0.23	0.03	**3.59**	-	**42.96**	nr	**48.33**
Trap doors in B.C. Pine, wrought										
19mm matchboarding on 25 x 75mm ledges										
457 x 610mm	31.72	2.50	**32.52**	0.21	0.03	**3.31**	-	**35.82**	nr	**40.30**
610 x 610mm	44.86	2.50	**45.98**	0.22	0.03	**3.45**	-	**49.42**	nr	**55.60**
762 x 610mm	57.34	2.50	**58.77**	0.23	0.03	**3.59**	-	**62.36**	nr	**70.16**
Panelled doors										
Note the following prices for panelled doors are for doors to detail in moderate numbers: doors in large numbers to one pattern would cost considerably less										
Panelled doors in softwood, wrought										
38mm square framed (or chamfered or moulded one or both sides)										
two panel	72.46	2.50	**74.27**	1.00	0.13	**15.60**	-	**89.87**	m2	**101.10**
four panel	80.58	2.50	**82.59**	1.00	0.13	**15.60**	-	**98.20**	m2	**110.47**
six panel	89.00	2.50	**91.22**	1.00	0.13	**15.60**	-	**106.82**	m2	**120.18**
add if upper panels open moulded in small squares for glass	17.49	2.50	**17.93**	-	-	**-**	-	**17.93**	m2	**20.17**
50mm square framed (or chamfered or moulded one or both sides)										
two panel	80.58	2.50	**82.59**	1.15	0.14	**17.84**	-	**100.43**	m2	**112.99**
four panel	89.00	2.50	**91.22**	1.15	0.14	**17.84**	-	**109.06**	m2	**122.70**
six panel	97.44	2.50	**99.87**	1.15	0.14	**17.84**	-	**117.71**	m2	**132.43**
add if upper panels open moulded in small squares for glass	17.49	2.50	**17.93**	-	-	**-**	-	**17.93**	m2	**20.17**
Panelled doors in Afromosia, wrought										
38mm square framed (or chamfered or moulded one or both sides)										
two panel	208.01	2.50	**213.21**	1.75	0.22	**27.22**	-	**240.43**	m2	**270.49**
four panel	221.89	2.50	**227.44**	1.75	0.22	**27.22**	-	**254.66**	m2	**286.50**
six panel	235.76	2.50	**241.66**	1.75	0.22	**27.22**	-	**268.88**	m2	**302.49**
add if upper panels open moulded in small squares for glass	39.52	2.50	**40.51**	-	-	**-**	-	**40.51**	m2	**45.57**
50mm square framed (or chamfered or moulded one or both sides)										
two panel	228.83	2.50	**234.55**	2.00	0.25	**31.10**	-	**265.65**	m2	**298.86**
four panel	242.69	2.50	**248.76**	2.00	0.25	**31.10**	-	**279.86**	m2	**314.84**
six panel	256.56	2.50	**262.98**	2.00	0.25	**31.10**	-	**294.08**	m2	**330.84**
add if upper panels open moulded in small squares for glass	39.52	2.50	**40.51**	-	-	**-**	-	**40.51**	m2	**45.57**

Labour hourly rates: (except Specialists) Craft Operatives £14.21 Labourer £10.71 Rates are national average prices. Refer to REGIONAL VARIATIONS for indicative levels of overall pricing in regions	MATERIALS			LABOUR				RATES		
	Del to Site	Waste	Material Cost	Craft Optve	Lab	Labour Cost	Sunds	Nett Rate	Unit	Gross Rate (+12.5%)
	£	%	£	Hrs	Hrs	£	£	£		£
L20: DOORS/SHUTTERS/HATCHES: TIMBER DOORS Cont.										
Panelled doors in Sapele, wrought										
38mm square framed (or chamfered or moulded one or both sides)										
two panel	181.84	2.50	**186.39**	1.45	0.18	**22.53**	-	**208.92**	m2	**235.04**
four panel	196.05	2.50	**200.95**	1.45	0.18	**22.53**	-	**223.48**	m2	**251.41**
six panel	210.26	2.50	**215.52**	1.45	0.18	**22.53**	-	**238.05**	m2	**267.81**
add if upper panels open moulded in small squares for glass	34.10	2.50	**34.95**	-	-	**-**	-	**34.95**	m2	**39.32**
50mm square framed (or chamfered or moulded one or both sides)										
two panel	196.05	2.50	**200.95**	1.75	0.22	**27.22**	-	**228.17**	m2	**256.69**
four panel	210.26	2.50	**215.52**	1.75	0.22	**27.22**	-	**242.74**	m2	**273.09**
six panel	224.47	2.50	**230.08**	1.75	0.22	**27.22**	-	**257.30**	m2	**289.46**
add if upper panels open moulded in small squares for glass	34.10	2.50	**34.95**	-	-	**-**	-	**34.95**	m2	**39.32**
Garage doors in softwood, wrought										
Side hung; framed, tongued and grooved boarded										
2134 x 1981 x 44mm overall, pair	319.72	2.50	**327.71**	4.50	0.56	**69.94**	-	**397.66**	nr	**447.36**
2134 x 2134 x 44mm overall, pair	332.51	2.50	**340.83**	4.70	0.59	**73.11**	-	**413.93**	nr	**465.67**
Up and over; framed, tongued and grooved boarded; up and over door gear										
2134 x 1981 x 44mm	447.60	2.50	**458.79**	9.00	1.10	**139.67**	-	**598.46**	nr	**673.27**
Garage doors in Western Red Cedar, wrought										
Side hung; framed, tongued and grooved boarded										
2134 x 1981 x 44mm overall, pair	448.72	2.50	**459.94**	5.00	0.62	**77.69**	-	**537.63**	nr	**604.84**
2134 x 2134 x 44mm overall, pair	460.85	2.50	**472.37**	5.25	0.66	**81.67**	-	**554.04**	nr	**623.29**
Up and over; framed, tongued and grooved boarded; up and over door gear										
2134 x 1981 x 44mm	570.00	2.50	**584.25**	9.25	1.15	**143.76**	-	**728.01**	nr	**819.01**
Garage doors in Oak, wrought										
Side hung; framed, tongued and grooved boarded										
2134 x 1981 x 44mm overall, pair	1199.94	2.50	**1229.93**	7.80	0.98	**121.33**	-	**1351.27**	nr	**1520.18**
2134 x 2134 x 44mm overall, pair	1266.61	2.50	**1298.27**	8.10	1.01	**125.92**	-	**1424.19**	nr	**1602.21**
Up and over; framed, tongued and grooved boarded; up and over door gear										
2134 x 1981 x 44mm	1399.94	2.50	**1434.94**	13.00	1.62	**202.08**	-	**1637.02**	nr	**1841.64**
Patio doors in hardwood, wrought										
Double glazed clear toughened glass, Magnet Trade, one opening leaf, doors and frames treated with base coat stains, fixing frame to masonry with screws										
2073 x 1805mm overall, HP 6	1399.94	2.50	**1434.94**	12.00	1.50	**186.59**	-	**1621.52**	nr	**1824.21**
2073 X 2387mm overall, HP 8	1399.94	2.50	**1434.94**	12.00	1.50	**186.59**	-	**1621.52**	nr	**1824.21**
extra; ventilator head, for HP 6	22.00	2.50	**22.55**	-	-	**-**	-	**22.55**	nr	**25.37**
extra; ventilator head, for HP 8	22.00	2.50	**22.55**	-	-	**-**	-	**22.55**	nr	**25.37**
Doorsets mainly in softwood, wrought										
Doorsets; 28mm thick jambs, head and transome with 12mm thick stop to suit 100mm thick wall; 14mm thick hardwood threshold; honeycomb core plywood faced flush door lipped two long edges; 6mm plywood transome panel fixed with pinned beads; 65mm snap in hinges; 57mm backset mortice latch; fixing frame to masonry with screws										
526 x 2040 x 40mm flush door -1 nr; basic dimensions										
600 x 2100mm	132.77	2.50	**136.09**	1.25	0.16	**19.48**	-	**155.57**	nr	**175.01**
600 x 2400mm	162.34	2.50	**166.40**	1.35	0.17	**21.00**	-	**187.40**	nr	**210.83**
626 x 2040 x 40mm flush door -1 nr; basic dimensions										
700 x 2100mm	132.77	2.50	**136.09**	1.25	0.16	**19.48**	-	**155.57**	nr	**175.01**
700 x 2400mm	162.34	2.50	**166.40**	1.35	0.17	**21.00**	-	**187.40**	nr	**210.83**
726 x 2040 x 40mm flush door -1 nr; basic dimensions										
800 x 2100mm	132.77	2.50	**136.09**	1.25	0.16	**19.48**	-	**155.57**	nr	**175.01**
800 x 2400mm	162.34	2.50	**166.40**	1.35	0.17	**21.00**	-	**187.40**	nr	**210.83**
826 x 2040 x 40mm flush door -1 nr; basic dimensions										
900 x 2100mm	138.55	2.50	**142.01**	1.25	0.16	**19.48**	-	**161.49**	nr	**181.68**
900 x 2400mm	165.80	2.50	**169.95**	1.35	0.17	**21.00**	-	**190.95**	nr	**214.82**

Labour hourly rates: (except Specialists) Craft Operatives £14.21 Labourer £10.71 Rates are national average prices. Refer to REGIONAL VARIATIONS for indicative levels of overall pricing in regions	MATERIALS			LABOUR				RATES		
	Del to Site £	Waste %	Material Cost £	Craft Optve Hrs	Lab Hrs	Labour Cost £	Sunds £	Nett Rate £	Unit	Gross Rate (+12.5%) £
L20: DOORS/SHUTTERS/HATCHES: TIMBER DOORS Cont.										
Doorsets mainly in softwood, wrought Cont.										
Note										
actual frame size is 14mm narrower and 14mm shorter than basic size dimensions										
L20: DOORS/SHUTTERS/HATCHES: ROLLER SHUTTERS										
Roller shutters in wood										
Note										
roller shutters are always purpose made to order and the following prices are indicative only. Firm quotations should always be obtained										
Pole and hook operation										
2134 x 2134mm	1500.00	-	**1500.00**	12.00	12.00	**299.04**	25.00	**1824.04**	nr	**2052.05**
L20: DOORS/SHUTTERS/HATCHES: LININGS										
Door frames and door lining sets in softwood, wrought										
Sets										
linings; fixing to masonry with screws										
32 x 113mm	5.48	2.50	**5.62**	0.45	0.06	**7.04**	0.09	**12.75**	m	**14.34**
32 x 150mm	6.58	2.50	**6.75**	0.45	0.06	**7.04**	0.09	**13.87**	m	**15.61**
32 x 225mm	9.85	2.50	**10.09**	0.70	0.08	**10.80**	0.13	**21.03**	m	**23.65**
32 x 330mm	15.47	2.50	**15.85**	0.70	0.08	**10.80**	0.13	**26.79**	m	**30.13**
38 x 113mm	6.36	2.50	**6.52**	0.45	0.06	**7.04**	0.09	**13.64**	m	**15.35**
38 x 150mm	7.69	2.50	**7.89**	0.45	0.06	**7.04**	0.09	**15.01**	m	**16.89**
38 x 225mm	11.32	2.50	**11.60**	0.70	0.08	**10.80**	0.13	**22.54**	m	**25.35**
38 x 330mm	18.29	2.50	**18.74**	0.70	0.08	**10.80**	0.13	**29.68**	m	**33.39**
linings										
32 x 113mm; labours -1	5.63	2.50	**5.77**	0.23	0.03	**3.59**	0.04	**9.40**	m	**10.58**
32 x 113mm; labours -2	5.71	2.50	**5.85**	0.23	0.03	**3.59**	0.04	**9.48**	m	**10.67**
32 x 113mm; labours -3	5.77	2.50	**5.91**	0.23	0.03	**3.59**	0.04	**9.54**	m	**10.73**
32 x 113mm; labours -4	5.85	2.50	**5.99**	0.23	0.03	**3.59**	0.04	**9.62**	m	**10.82**
38 x 113mm; labours -1	6.51	2.50	**6.68**	0.23	0.03	**3.59**	0.04	**10.31**	m	**11.60**
38 x 113mm; labours -2	6.58	2.50	**6.75**	0.23	0.03	**3.59**	0.04	**10.38**	m	**11.67**
38 x 113mm; labours -3	6.66	2.50	**6.83**	0.23	0.03	**3.59**	0.04	**10.46**	m	**11.77**
38 x 113mm; labours -4	6.73	2.50	**6.90**	0.23	0.03	**3.59**	0.04	**10.53**	m	**11.84**
Internal door lining sets; supplied unassembled; fixing to masonry with screws										
32 x 115mm rebated linings; assembling										
for 610 x 1981mm doors	24.44	2.50	**25.05**	2.30	0.29	**35.79**	0.37	**61.21**	nr	**68.86**
for 686 x 1981mm doors	24.44	2.50	**25.05**	2.30	0.29	**35.79**	0.37	**61.21**	nr	**68.86**
for 762 x 1981mm doors	24.44	2.50	**25.05**	2.30	0.29	**35.79**	0.37	**61.21**	nr	**68.86**
32 x 140mm rebated linings; assembling										
for 610 x 1981mm doors	26.66	2.50	**27.33**	2.30	0.29	**35.79**	0.37	**63.48**	nr	**71.42**
for 686 x 1981mm doors	26.66	2.50	**27.33**	2.30	0.29	**35.79**	0.37	**63.48**	nr	**71.42**
for 762 x 1981mm doors	26.66	2.50	**27.33**	2.30	0.29	**35.79**	0.37	**63.48**	nr	**71.42**
Door frames and door lining sets in Afromosia, wrought, selected for transparent finish										
Sets										
linings; fixing to masonry with screws										
32 x 113mm	13.23	2.50	**13.56**	0.70	0.09	**10.91**	0.13	**24.60**	m	**27.68**
32 x 150mm	16.02	2.50	**16.42**	0.70	0.09	**10.91**	0.13	**27.46**	m	**30.89**
32 x 225mm	23.73	2.50	**24.32**	1.00	0.12	**15.50**	0.18	**40.00**	m	**44.99**
32 x 330mm	36.95	2.50	**37.87**	1.00	0.12	**15.50**	0.18	**53.55**	m	**60.24**
38 x 113mm	15.43	2.50	**15.82**	0.70	0.09	**10.91**	0.13	**26.86**	m	**30.22**
38 x 150mm	18.48	2.50	**18.94**	0.70	0.09	**10.91**	0.13	**29.98**	m	**33.73**
38 x 225mm	27.09	2.50	**27.77**	1.00	0.12	**15.50**	0.18	**43.45**	m	**48.88**
38 x 330mm	43.11	2.50	**44.19**	1.00	0.12	**15.50**	0.18	**59.86**	m	**67.34**
linings										
32 x 113mm; labours -1	13.42	2.50	**13.76**	0.35	0.04	**5.40**	0.07	**19.23**	m	**21.63**
32 x 113mm; labours -2	13.55	2.50	**13.89**	0.35	0.04	**5.40**	0.07	**19.36**	m	**21.78**
32 x 113mm; labours -3	13.67	2.50	**14.01**	0.35	0.04	**5.40**	0.07	**19.49**	m	**21.92**
32 x 113mm; labours -4	13.82	2.50	**14.16**	0.35	0.04	**5.40**	0.07	**19.63**	m	**22.09**
38 x 113mm; labours -1	15.62	2.50	**16.01**	0.35	0.04	**5.40**	0.07	**21.48**	m	**24.17**
38 x 113mm; labours -2	15.75	2.50	**16.15**	0.35	0.04	**5.40**	0.07	**21.62**	m	**24.32**
38 x 113mm; labours -3	15.88	2.50	**16.28**	0.35	0.04	**5.40**	0.07	**21.75**	m	**24.47**
38 x 113mm; labours -4	16.02	2.50	**16.42**	0.35	0.04	**5.40**	0.07	**21.89**	m	**24.62**
Door frames and door lining sets in Sapele, wrought, selected for transparent finish										
Sets										
linings; fixing to masonry with screws										
32 x 113mm	11.52	2.50	**11.81**	0.65	0.08	**10.09**	0.12	**22.03**	m	**24.78**
32 x 150mm	13.83	2.50	**14.18**	0.65	0.08	**10.09**	0.12	**24.39**	m	**27.44**
32 x 225mm	20.74	2.50	**21.26**	0.92	0.11	**14.25**	0.17	**35.68**	m	**40.14**
32 x 330mm	32.27	2.50	**33.07**	0.92	0.11	**14.25**	0.17	**47.49**	m	**53.43**
38 x 113mm	13.49	2.50	**13.83**	0.65	0.08	**10.09**	0.12	**24.04**	m	**27.05**
38 x 150mm	16.13	2.50	**16.53**	0.65	0.08	**10.09**	0.12	**26.74**	m	**30.09**
38 x 225mm	23.70	2.50	**24.29**	0.92	0.11	**14.25**	0.17	**38.71**	m	**43.55**
38 x 330mm	38.18	2.50	**39.14**	0.92	0.11	**14.25**	0.17	**53.56**	m	**60.26**
linings										
32 x 113mm; labours -1	11.72	2.50	**12.02**	0.31	0.04	**4.83**	0.07	**16.92**	m	**19.04**
32 x 113mm; labours -2	11.85	2.50	**12.15**	0.31	0.04	**4.83**	0.07	**17.05**	m	**19.18**
32 x 113mm; labours -3	11.99	2.50	**12.29**	0.31	0.04	**4.83**	0.07	**17.19**	m	**19.34**
32 x 113mm; labours -4	12.11	2.50	**12.42**	0.31	0.04	**4.83**	0.07	**17.32**	m	**19.48**
38 x 113mm; labours -1	13.69	2.50	**14.04**	0.31	0.04	**4.83**	0.07	**18.94**	m	**21.31**
38 x 113mm; labours -2	13.83	2.50	**14.18**	0.31	0.04	**4.83**	0.07	**19.08**	m	**21.47**

Labour hourly rates: (except Specialists) Craft Operatives £14.21 Labourer £10.71 Rates are national average prices. Refer to REGIONAL VARIATIONS for indicative levels of overall pricing in regions	MATERIALS			LABOUR				RATES		
	Del to Site £	Waste %	Material Cost £	Craft Optve Hrs	Lab Hrs	Labour Cost £	Sunds £	Nett Rate £	Unit	Gross Rate (+12.5%) £
L20: DOORS/SHUTTERS/HATCHES: LININGS Cont.										
Door frames and door lining sets in Sapele, wrought, selected for transparent finish Cont.										
Sets Cont.										
linings Cont.										
38 x 113mm; labours -3	13.96	2.50	**14.31**	0.31	0.04	**4.83**	0.07	**19.21**	m	**21.61**
38 x 113mm; labours -4	14.09	2.50	**14.45**	0.31	0.04	**4.83**	0.07	**19.35**	m	**21.77**
L20: DOORS/SHUTTERS/HATCHES: FRAMES										
Door frames and door lining sets in softwood, wrought - frames										
Sets										
38 x 75mm jambs	5.13	2.50	**5.26**	0.21	0.03	**3.31**	0.04	**8.61**	m	**9.68**
38 x 100mm jambs	5.94	2.50	**6.09**	0.21	0.03	**3.31**	0.04	**9.43**	m	**10.61**
50 x 75mm jambs	5.94	2.50	**6.09**	0.23	0.03	**3.59**	0.04	**9.71**	m	**10.93**
50 x 100mm jambs	6.89	2.50	**7.06**	0.23	0.03	**3.59**	0.04	**10.69**	m	**12.03**
50 x 125mm jambs	7.69	2.50	**7.89**	0.23	0.03	**3.59**	0.04	**11.51**	m	**12.95**
63 x 100mm jambs	7.69	2.50	**7.89**	0.23	0.03	**3.59**	0.04	**11.51**	m	**12.95**
75 x 100mm jambs	9.85	2.50	**10.09**	0.25	0.03	**3.87**	0.04	**14.01**	m	**15.76**
75 x 113mm jambs	11.25	2.50	**11.53**	0.25	0.03	**3.87**	0.04	**15.45**	m	**17.38**
100 x 100mm jambs	12.43	2.50	**12.74**	0.29	0.04	**4.55**	0.05	**17.34**	m	**19.51**
100 x 113mm jambs	14.29	2.50	**14.64**	0.31	0.04	**4.83**	0.05	**19.53**	m	**21.97**
100 x 125mm jambs	16.15	2.50	**16.55**	0.31	0.04	**4.83**	0.05	**21.43**	m	**24.11**
113 x 113mm jambs	16.15	2.50	**16.55**	0.31	0.04	**4.83**	0.05	**21.43**	m	**24.11**
113 x 125mm jambs	17.90	2.50	**18.35**	0.35	0.04	**5.40**	0.06	**23.81**	m	**26.79**
113 x 150mm jambs	19.70	2.50	**20.20**	0.35	0.04	**5.40**	0.06	**25.66**	m	**28.86**
38 x 75mm jambs; labours -1	5.26	2.50	**5.39**	0.21	0.03	**3.31**	0.04	**8.73**	m	**9.83**
38 x 75mm jambs; labours -2	5.31	2.50	**5.45**	0.21	0.03	**3.31**	0.04	**8.79**	m	**9.89**
38 x 75mm jambs; labours -3	5.42	2.50	**5.55**	0.21	0.03	**3.31**	0.04	**8.90**	m	**10.01**
38 x 75mm jambs; labours -4	5.48	2.50	**5.62**	0.21	0.03	**3.31**	0.04	**8.97**	m	**10.09**
38 x 75mm heads	5.13	2.50	**5.26**	0.21	0.03	**3.31**	0.04	**8.61**	m	**9.68**
38 x 100mm heads	5.94	2.50	**6.09**	0.21	0.03	**3.31**	0.04	**9.43**	m	**10.61**
50 x 75mm heads	5.94	2.50	**6.09**	0.23	0.03	**3.59**	0.04	**9.71**	m	**10.93**
50 x 100mm heads	6.89	2.50	**7.06**	0.23	0.03	**3.59**	0.04	**10.69**	m	**12.03**
50 x 125mm heads	7.69	2.50	**7.89**	0.23	0.03	**3.59**	0.04	**11.51**	m	**12.95**
63 x 100mm heads	7.69	2.50	**7.89**	0.23	0.03	**3.59**	0.04	**11.51**	m	**12.95**
75 x 100mm heads	9.85	2.50	**10.09**	0.25	0.03	**3.87**	0.04	**14.01**	m	**15.76**
75 x 113mm heads	11.25	2.50	**11.53**	0.25	0.03	**3.87**	0.04	**15.45**	m	**17.38**
100 x 100mm heads	12.43	2.50	**12.74**	0.29	0.04	**4.55**	0.05	**17.34**	m	**19.51**
100 x 113mm heads	14.29	2.50	**14.64**	0.31	0.04	**4.83**	0.05	**19.53**	m	**21.97**
100 x 125mm heads	16.16	2.50	**16.56**	0.31	0.04	**4.83**	0.05	**21.44**	m	**24.12**
113 x 113mm heads	16.16	2.50	**16.56**	0.31	0.04	**4.83**	0.05	**21.44**	m	**24.12**
113 x 125mm heads	17.90	2.50	**18.35**	0.35	0.04	**5.40**	0.05	**23.80**	m	**26.78**
113 x 150mm heads	19.70	2.50	**20.20**	0.35	0.04	**5.40**	0.05	**25.65**	m	**28.85**
38 x 75mm heads; labours -1	5.26	2.50	**5.39**	0.21	0.03	**3.31**	0.03	**8.72**	m	**9.81**
38 x 75mm heads; labours -2	5.31	2.50	**5.45**	0.21	0.03	**3.31**	0.03	**8.78**	m	**9.88**
38 x 75mm heads; labours -3	5.42	2.50	**5.55**	0.21	0.03	**3.31**	0.03	**8.89**	m	**10.00**
38 x 75mm heads; labours -4	5.48	2.50	**5.62**	0.21	0.03	**3.31**	0.03	**8.96**	m	**10.08**
63 x 125mm sills	11.69	2.50	**11.98**	0.26	0.03	**4.02**	0.04	**16.04**	m	**18.05**
75 x 125mm sills	13.32	2.50	**13.66**	0.28	0.04	**4.41**	0.05	**18.11**	m	**20.38**
75 x 150mm sills	14.96	2.50	**15.33**	0.29	0.04	**4.55**	0.05	**19.93**	m	**22.42**
63 x 125mm sills; labours -1	11.84	2.50	**12.14**	0.26	0.03	**4.02**	0.04	**16.19**	m	**18.22**
63 x 125mm sills; labours -2	11.90	2.50	**12.19**	0.26	0.03	**4.02**	0.04	**16.25**	m	**18.28**
63 x 125mm sills; labours -3	12.00	2.50	**12.30**	0.26	0.03	**4.02**	0.04	**16.35**	m	**18.40**
63 x 125mm sills; labours -4	12.06	2.50	**12.36**	0.26	0.03	**4.02**	0.04	**16.41**	m	**18.46**
63 x 100mm mullions	7.69	2.50	**7.89**	0.06	0.01	**0.96**	-	**8.85**	m	**9.95**
75 x 100mm mullions	9.85	2.50	**10.09**	0.06	0.01	**0.96**	-	**11.05**	m	**12.43**
75 x 113mm mullions	11.25	2.50	**11.53**	0.06	0.01	**0.96**	-	**12.49**	m	**14.05**
100 x 100mm mullions	12.43	2.50	**12.74**	0.06	0.01	**0.96**	-	**13.70**	m	**15.41**
100 x 125mm mullions	16.16	2.50	**16.56**	0.06	0.01	**0.96**	-	**17.52**	m	**19.71**
63 x 100mm mullions; labours -1	7.85	2.50	**8.05**	0.06	0.01	**0.96**	-	**9.01**	m	**10.13**
63 x 100mm mullions; labours -2	7.92	2.50	**8.12**	0.06	0.01	**0.96**	-	**9.08**	m	**10.21**
63 x 100mm mullions; labours -3	8.01	2.50	**8.21**	0.06	0.01	**0.96**	-	**9.17**	m	**10.32**
63 x 100mm mullions; labours -4	8.07	2.50	**8.27**	0.06	0.01	**0.96**	-	**9.23**	m	**10.38**
63 x 100mm transoms	7.69	2.50	**7.89**	0.06	0.01	**0.96**	-	**8.85**	m	**9.95**
75 x 100mm transoms	9.85	2.50	**10.09**	0.06	0.01	**0.96**	-	**11.05**	m	**12.43**
75 x 113mm transoms	11.25	2.50	**11.53**	0.06	0.01	**0.96**	-	**12.49**	m	**14.05**
100 x 100mm transoms	12.43	2.50	**12.74**	0.06	0.01	**0.96**	-	**13.70**	m	**15.41**
100 x 125mm transoms	16.16	2.50	**16.56**	0.06	0.01	**0.96**	-	**17.52**	m	**19.71**
63 x 100mm transoms; labours -1	7.85	2.50	**8.05**	0.06	0.01	**0.96**	-	**9.01**	m	**10.13**
63 x 100mm transoms; labours -2	7.92	2.50	**8.12**	0.06	0.01	**0.96**	-	**9.08**	m	**10.21**
63 x 100mm transoms; labours -3	8.01	2.50	**8.21**	0.06	0.01	**0.96**	-	**9.17**	m	**10.32**
63 x 100mm transoms; labours -4	8.07	2.50	**8.27**	0.06	0.01	**0.96**	-	**9.23**	m	**10.38**
Internal door frame sets; supplied unassembled										
38 x 63mm jambs and head; stops (supplied loose); assembling										
for 686 x 1981mm doors	22.22	2.50	**22.77**	1.15	0.14	**17.84**	0.19	**40.80**	nr	**45.91**
for 762 x 1981mm doors	22.22	2.50	**22.77**	1.15	0.14	**17.84**	0.19	**40.80**	nr	**45.91**
38 x 75mm jambs and head; stops (supplied loose); assembling										
for 686 x 1981mm doors	23.69	2.50	**24.28**	1.15	0.14	**17.84**	0.19	**42.31**	nr	**47.60**
for 762 x 1981mm doors	23.69	2.50	**24.28**	1.15	0.14	**17.84**	0.19	**42.31**	nr	**47.60**
50 x 100mm jambs and head; stops (supplied loose); assembling										
for 686 x 1981mm doors	37.02	2.50	**37.94**	1.15	0.14	**17.84**	0.19	**55.97**	nr	**62.97**
for 762 x 1981mm doors	37.02	2.50	**37.94**	1.15	0.14	**17.84**	0.19	**55.97**	nr	**62.97**
50 x 113mm jambs and head; stops (supplied loose); assembling										
for 686 x 1981mm doors	45.91	2.50	**47.06**	1.15	0.14	**17.84**	0.19	**65.09**	nr	**73.22**
for 762 x 1981mm doors	45.91	2.50	**47.06**	1.15	0.14	**17.84**	0.19	**65.09**	nr	**73.22**

Labour hourly rates: (except Specialists) Craft Operatives £14.21 Labourer £10.71 Rates are national average prices. Refer to REGIONAL VARIATIONS for indicative levels of overall pricing in regions	MATERIALS			LABOUR				RATES		
	Del to Site £	Waste %	Material Cost £	Craft Optve Hrs	Lab Hrs	Labour Cost £	Sunds £	Nett Rate £	Unit	Gross Rate (+12.5%) £
L20: DOORS/SHUTTERS/HATCHES: FRAMES Cont.										
Door frames and door lining sets in softwood, wrought - frames Cont.		-								
External door frame sets; one coat external primer before delivery to site										
63 x 75mm jambs and head; rebates -1										
for 762 x 1981 x 50mm doors	39.97	2.50	**40.97**	1.15	0.14	**17.84**	0.19	**59.00**	nr	**66.38**
for 838 x 1981 x 50mm doors	41.46	2.50	**42.49**	1.15	0.14	**17.84**	0.19	**60.52**	nr	**68.09**
63 x 88mm jambs and head; rebates -1										
for 762 x 1981 x 50mm doors	41.46	2.50	**42.49**	1.15	0.14	**17.84**	0.19	**60.52**	nr	**68.09**
for 838 x 1981 x 50mm doors	42.95	2.50	**44.03**	1.15	0.14	**17.84**	0.19	**62.06**	nr	**69.81**
63 x 75mm jambs and head, rebates -1 nr; hardwood sill										
for 762 x 1981 x 50mm doors	88.84	2.50	**91.06**	1.45	0.18	**22.53**	0.24	**113.83**	nr	**128.06**
for 838 x 1981 x 50mm doors	90.32	2.50	**92.58**	1.45	0.18	**22.53**	0.24	**115.35**	nr	**129.77**
63 x 88mm jambs and head, rebates -1 nr; hardwood sill										
for 762 x 1981 x 50mm doors	91.78	2.50	**94.08**	1.45	0.18	**22.53**	0.24	**116.85**	nr	**131.46**
for 838 x 1981 x 50mm doors	93.28	2.50	**95.61**	1.45	0.18	**22.53**	0.24	**118.38**	nr	**133.18**
for 1168 x 1981 x 50mm doors	94.74	2.50	**97.11**	1.75	0.22	**27.22**	0.28	**124.61**	nr	**140.19**
External garage door frame sets; one coat external primer before delivery to site; supplied unassembled										
75 x 100mm jambs and head; assembling										
for 2134 x 1981mm side hung doors	65.35	2.50	**66.99**	2.50	0.31	**38.85**	0.41	**106.24**	nr	**119.52**
for 2134 x 2134mm side hung doors	69.69	2.50	**71.43**	2.50	0.31	**38.85**	0.41	**110.69**	nr	**124.52**
75 x 75mm jambs and head; assembling										
for 2134 x 1981mm up and over doors	60.69	2.50	**62.21**	2.50	0.31	**38.85**	0.41	**101.47**	nr	**114.15**
Door frames and door lining sets in Afromosia, wrought, selected for transparent finish - frames										
Sets										
38 x 75mm jambs	12.32	2.50	**12.63**	0.36	0.05	**5.65**	0.06	**18.34**	m	**20.63**
38 x 100mm jambs	14.19	2.50	**14.54**	0.36	0.05	**5.65**	0.06	**20.26**	m	**22.79**
50 x 75mm jambs	14.19	2.50	**14.54**	0.40	0.05	**6.22**	0.06	**20.82**	m	**23.43**
50 x 100mm jambs	16.67	2.50	**17.08**	0.40	0.05	**6.22**	0.06	**23.36**	m	**26.28**
50 x 125mm jambs	18.48	2.50	**18.94**	0.40	0.05	**6.22**	0.06	**25.22**	m	**28.37**
63 x 100mm jambs	18.48	2.50	**18.94**	0.40	0.05	**6.22**	0.06	**25.22**	m	**28.37**
75 x 100mm jambs	23.73	2.50	**24.32**	0.44	0.06	**6.90**	0.07	**31.29**	m	**35.20**
75 x 113mm jambs	27.09	2.50	**27.77**	0.44	0.06	**6.90**	0.07	**34.74**	m	**39.08**
100 x 100mm jambs	30.21	2.50	**30.96**	0.51	0.06	**7.89**	0.08	**38.93**	m	**43.80**
100 x 113mm jambs	34.50	2.50	**35.36**	0.54	0.07	**8.42**	0.09	**43.87**	m	**49.36**
100 x 125mm jambs	38.84	2.50	**39.81**	0.60	0.08	**9.38**	0.10	**49.29**	m	**55.46**
113 x 113mm jambs	38.84	2.50	**39.81**	0.54	0.07	**8.42**	0.09	**48.33**	m	**54.37**
113 x 125mm jambs	43.11	2.50	**44.19**	0.60	0.08	**9.38**	0.10	**53.67**	m	**60.38**
113 x 150mm jambs	47.47	2.50	**48.65**	0.60	0.08	**9.38**	0.10	**58.13**	m	**65.40**
38 x 75mm jambs; labours -1	12.51	2.50	**12.82**	0.36	0.05	**5.65**	0.06	**18.53**	m	**20.85**
38 x 75mm jambs; labours -2	12.64	2.50	**12.95**	0.36	0.05	**5.65**	0.06	**18.67**	m	**21.00**
38 x 75mm jambs; labours -3	12.84	2.50	**13.16**	0.36	0.05	**5.65**	0.06	**18.87**	m	**21.23**
38 x 75mm jambs; labours -4	13.04	2.50	**13.36**	0.36	0.05	**5.65**	0.06	**19.07**	m	**21.46**
38 x 75mm heads	12.32	2.50	**12.63**	0.36	0.05	**5.65**	0.06	**18.34**	m	**20.63**
38 x 100mm heads	14.19	2.50	**14.54**	0.36	0.05	**5.65**	0.06	**20.26**	m	**22.79**
50 x 75mm heads	14.19	2.50	**14.54**	0.40	0.05	**6.22**	0.06	**20.82**	m	**23.43**
50 x 100mm heads	16.67	2.50	**17.08**	0.40	0.05	**6.22**	0.06	**23.36**	m	**26.28**
50 x 125mm heads	18.48	2.50	**18.94**	0.40	0.05	**6.22**	0.06	**25.22**	m	**28.37**
63 x 100mm heads	18.48	2.50	**18.94**	0.40	0.05	**6.22**	0.06	**25.22**	m	**28.37**
75 x 100mm heads	23.73	2.50	**24.32**	0.44	0.06	**6.90**	0.07	**31.29**	m	**35.20**
75 x 113mm heads	27.09	2.50	**27.77**	0.44	0.06	**6.90**	0.07	**34.74**	m	**39.08**
100 x 100mm heads	30.21	2.50	**30.96**	0.51	0.06	**7.89**	0.08	**38.93**	m	**43.80**
100 x 113mm heads	34.50	2.50	**35.36**	0.54	0.07	**8.42**	0.09	**43.87**	m	**49.36**
100 x 125mm heads	38.84	2.50	**39.81**	0.60	0.08	**9.38**	0.10	**49.29**	m	**55.46**
113 x 113mm heads	38.84	2.50	**39.81**	0.54	0.07	**8.42**	0.09	**48.33**	m	**54.37**
113 x 125mm heads	43.11	2.50	**44.19**	0.60	0.08	**9.38**	0.10	**53.67**	m	**60.38**
113 x 150mm heads	47.47	2.50	**48.65**	0.60	0.08	**9.38**	0.10	**58.13**	m	**65.40**
38 x 75mm heads; labours -1	12.51	2.50	**12.82**	0.36	0.05	**5.65**	0.06	**18.53**	m	**20.85**
38 x 75mm heads; labours -2	12.64	2.50	**12.95**	0.36	0.05	**5.65**	0.06	**18.67**	m	**21.00**
38 x 75mm heads; labours -3	12.84	2.50	**13.16**	0.36	0.05	**5.65**	0.06	**18.87**	m	**21.23**
38 x 75mm heads; labours -4	13.04	2.50	**13.36**	0.36	0.05	**5.65**	0.06	**19.07**	m	**21.46**
63 x 125mm sills	28.01	2.50	**28.71**	0.46	0.06	**7.18**	0.08	**35.97**	m	**40.46**
75 x 125mm sills	32.03	2.50	**32.83**	0.51	0.06	**7.89**	0.08	**40.80**	m	**45.90**
75 x 150mm sills	36.06	2.50	**36.96**	0.52	0.07	**8.14**	0.09	**45.19**	m	**50.84**
63 x 125mm sills; labours -1	28.34	2.50	**29.04**	0.46	0.06	**7.18**	0.08	**36.30**	m	**40.84**
63 x 125mm sills; labours -2	28.47	2.50	**29.18**	0.46	0.06	**7.18**	0.08	**36.44**	m	**40.99**
63 x 125mm sills; labours -3	28.59	2.50	**29.30**	0.46	0.06	**7.18**	0.08	**36.56**	m	**41.13**
63 x 125mm sills; labours -4	28.72	2.50	**29.44**	0.46	0.06	**7.18**	0.08	**36.70**	m	**41.29**
63 x 100mm mullions	18.48	2.50	**18.94**	0.06	0.01	**0.96**	-	**19.90**	m	**22.39**
75 x 100mm mullions	23.73	2.50	**24.32**	0.06	0.01	**0.96**	-	**25.28**	m	**28.44**
75 x 113mm mullions	27.09	2.50	**27.77**	0.06	0.01	**0.96**	-	**28.73**	m	**32.32**
100 x 100mm mullions	30.21	2.50	**30.96**	0.06	0.01	**0.96**	-	**31.92**	m	**35.91**
100 x 125mm mullions	38.84	2.50	**39.81**	0.06	0.01	**0.96**	-	**40.77**	m	**45.87**
63 x 100mm mullions; labours -1	18.67	2.50	**19.13**	0.06	0.01	**0.96**	-	**20.09**	m	**22.61**
63 x 100mm mullions; labours -2	18.80	2.50	**19.27**	0.06	0.01	**0.96**	-	**20.23**	m	**22.76**
63 x 100mm mullions; labours -3	18.93	2.50	**19.40**	0.06	0.01	**0.96**	-	**20.36**	m	**22.91**
63 x 100mm mullions; labours -4	19.06	2.50	**19.54**	0.06	0.01	**0.96**	-	**20.50**	m	**23.06**
63 x 100mm transoms	18.48	2.50	**18.94**	0.06	0.01	**0.96**	-	**19.90**	m	**22.39**
75 x 100mm transoms	23.73	2.50	**24.32**	0.06	0.01	**0.96**	-	**25.28**	m	**28.44**
75 x 113mm transoms	27.09	2.50	**27.77**	0.06	0.01	**0.96**	-	**28.73**	m	**32.32**
100 x 100mm transoms	30.21	2.50	**30.96**	0.06	0.01	**0.96**	-	**31.92**	m	**35.91**
100 x 125mm transoms	38.84	2.50	**39.81**	0.06	0.01	**0.96**	-	**40.77**	m	**45.87**
63 x 100mm transoms; labours -1	18.67	2.50	**19.13**	0.06	0.01	**0.96**	-	**20.09**	m	**22.61**
63 x 100mm transoms; labours -2	18.80	2.50	**19.27**	0.06	0.01	**0.96**	-	**20.23**	m	**22.76**
63 x 100mm transoms; labours -3	18.93	2.50	**19.40**	0.06	0.01	**0.96**	-	**20.36**	m	**22.91**
63 x 100mm transoms; labours -4	19.06	2.50	**19.54**	0.06	0.01	**0.96**	-	**20.50**	m	**23.06**

Labour hourly rates: (except Specialists) Craft Operatives £14.21 Labourer £10.71 Rates are national average prices. Refer to REGIONAL VARIATIONS for indicative levels of overall pricing in regions	MATERIALS			LABOUR				RATES		
	Del to Site £	Waste %	Material Cost £	Craft Optve Hrs	Lab Hrs	Labour Cost £	Sunds £	Nett Rate £	Unit	Gross Rate (+12.5%) £
L20: DOORS/SHUTTERS/HATCHES: FRAMES Cont.										
Door frames and door lining sets in Sapele, wrought, selected for transparent finish - frames										
Sets										
38 x 75mm jambs	10.80	2.50	11.07	0.31	0.04	4.83	0.05	15.95	m	17.94
38 x 100mm jambs	12.44	2.50	12.75	0.31	0.04	4.83	0.05	17.63	m	19.84
50 x 75mm jambs	12.44	2.50	12.75	0.35	0.04	5.40	0.06	18.21	m	20.49
50 x 100mm jambs	14.48	2.50	14.85	0.35	0.04	5.40	0.06	20.31	m	22.85
50 x 125mm jambs	16.13	2.50	16.53	0.35	0.04	5.40	0.06	21.99	m	24.74
63 x 100mm jambs	16.13	2.50	16.53	0.35	0.04	5.40	0.06	21.99	m	24.74
75 x 100mm jambs	20.74	2.50	21.26	0.38	0.05	5.94	0.06	27.25	m	30.66
75 x 113mm jambs	23.70	2.50	24.29	0.38	0.05	5.94	0.06	30.29	m	34.07
100 x 100mm jambs	26.21	2.50	26.86	0.43	0.05	6.65	0.07	33.58	m	37.78
100 x 113mm jambs	30.02	2.50	30.77	0.46	0.06	7.18	0.08	38.03	m	42.78
100 x 125mm jambs	33.91	2.50	34.76	0.46	0.06	7.18	0.08	42.02	m	47.27
113 x 113mm jambs	33.91	2.50	34.76	0.46	0.06	7.18	0.08	42.02	m	47.27
113 x 125mm jambs	37.80	2.50	38.74	0.52	0.07	8.14	0.09	46.97	m	52.84
113 x 150mm jambs	41.48	2.50	42.52	0.52	0.07	8.14	0.09	50.75	m	57.09
38 x 75mm jambs; labours -1	11.00	2.50	11.27	0.31	0.04	4.83	0.05	16.16	m	18.18
38 x 75mm jambs; labours -2	11.12	2.50	11.40	0.31	0.04	4.83	0.05	16.28	m	18.32
38 x 75mm jambs; labours -3	11.26	2.50	11.54	0.31	0.04	4.83	0.05	16.43	m	18.48
38 x 75mm jambs; labours -4	11.39	2.50	11.67	0.31	0.04	4.83	0.05	16.55	m	18.62
38 x 75mm heads	10.80	2.50	11.07	0.31	0.04	4.83	0.05	15.95	m	17.94
38 x 100mm heads	12.44	2.50	12.75	0.31	0.04	4.83	0.05	17.63	m	19.84
50 x 75mm heads	12.44	2.50	12.75	0.35	0.04	5.40	0.06	18.21	m	20.49
50 x 100mm heads	14.48	2.50	14.85	0.35	0.04	5.40	0.06	20.31	m	22.85
50 x 125mm heads	16.13	2.50	16.53	0.35	0.04	5.40	0.06	21.99	m	24.74
63 x 100mm heads	16.13	2.50	16.53	0.35	0.04	5.40	0.06	21.99	m	24.74
75 x 100mm heads	16.79	2.50	17.21	0.38	0.05	5.94	0.06	23.21	m	26.11
75 x 113mm heads	23.70	2.50	24.29	0.38	0.05	5.94	0.06	30.29	m	34.07
100 x 100mm heads	26.21	2.50	26.86	0.43	0.05	6.65	0.07	33.58	m	37.78
100 x 113mm heads	30.02	2.50	30.77	0.46	0.06	7.18	0.08	38.03	m	42.78
100 x 125mm heads	33.91	2.50	34.76	0.46	0.06	7.18	0.08	42.02	m	47.27
113 x 113mm heads	33.91	2.50	34.76	0.46	0.06	7.18	0.08	42.02	m	47.27
113 x 125mm heads	37.80	2.50	38.74	0.52	0.07	8.14	0.09	46.97	m	52.84
113 x 150mm heads	41.48	2.50	42.52	0.52	0.07	8.14	0.09	50.75	m	57.09
38 x 75mm heads; labours -1	11.00	2.50	11.27	0.31	0.04	4.83	0.05	16.16	m	18.18
38 x 75mm heads; labours -2	11.00	2.50	11.27	0.31	0.04	4.83	0.05	16.16	m	18.18
38 x 75mm heads; labours -3	11.26	2.50	11.54	0.31	0.04	4.83	0.05	16.43	m	18.48
38 x 75mm heads; labours -4	11.39	2.50	11.67	0.31	0.04	4.83	0.05	16.55	m	18.62
63 x 125mm sills	24.63	2.50	25.24	0.40	0.05	6.22	0.06	31.52	m	35.46
75 x 125mm sills	27.98	2.50	28.68	0.42	0.05	6.50	0.07	35.25	m	39.66
75 x 150mm sills	31.60	2.50	32.39	0.43	0.05	6.65	0.07	39.11	m	43.99
63 x 125mm sills; labours -1	24.82	2.50	25.44	0.40	0.05	6.22	0.06	31.72	m	35.68
63 x 125mm sills; labours -2	24.95	2.50	25.58	0.40	0.05	6.22	0.06	31.86	m	35.84
63 x 125mm sills; labours -3	25.08	2.50	25.71	0.40	0.05	6.22	0.06	31.99	m	35.98
63 x 125mm sills; labours -4	25.22	2.50	25.85	0.40	0.05	6.22	0.06	32.13	m	36.14
63 x 100mm mullions	16.13	2.50	16.53	0.06	0.01	0.96	-	17.49	m	19.68
75 x 100mm mullions	16.79	2.50	17.21	0.06	0.01	0.96	-	18.17	m	20.44
75 x 113mm mullions	23.70	2.50	24.29	0.06	0.01	0.96	-	25.25	m	28.41
100 x 100mm mullions	26.21	2.50	26.86	0.06	0.01	0.96	-	27.82	m	31.30
100 x 125mm mullions	33.91	2.50	34.76	0.06	0.01	0.96	-	35.72	m	40.18
63 x 100mm mullions; labours -1	16.33	2.50	16.74	0.06	0.01	0.96	-	17.69	m	19.91
63 x 100mm mullions; labours -2	16.47	2.50	16.88	0.06	0.01	0.96	-	17.84	m	20.07
63 x 100mm mullions; labours -3	16.59	2.50	17.01	0.06	0.01	0.96	-	17.96	m	20.21
63 x 100mm mullions; labours -4	16.73	2.50	17.15	0.06	0.01	0.96	-	18.11	m	20.37
63 x 100mm transoms	16.13	2.50	16.53	0.06	0.01	0.96	-	17.49	m	19.68
75 x 100mm transoms	16.79	2.50	17.21	0.06	0.01	0.96	-	18.17	m	20.44
75 x 113mm transoms	23.70	2.50	24.29	0.06	0.01	0.96	-	25.25	m	28.41
100 x 100mm transoms	26.21	2.50	26.86	0.06	0.01	0.96	-	27.82	m	31.30
100 x 125mm transoms	33.91	2.50	34.76	0.06	0.01	0.96	-	35.72	m	40.18
63 x 100mm transoms; labours -1	16.33	2.50	16.74	0.06	0.01	0.96	-	17.69	m	19.91
63 x 100mm transoms; labours -2	16.47	2.50	16.88	0.06	0.01	0.96	-	17.84	m	20.07
63 x 100mm transoms; labours -3	16.59	2.50	17.01	0.06	0.01	0.96	-	17.96	m	20.21
63 x 100mm transoms; labours -4	16.73	2.50	17.15	0.06	0.01	0.96	-	18.11	m	20.37
Intumescent strips and smoke seals										
Albi-Flex, white PVC sleeved self adhesive; Rentokil Ltd.										
12 x 4mm intumescent strip, half hour application; setting into groove in timber frame or door	4.04	10.00	4.45	0.17	-	2.42	-	6.86	m	7.72
18 x 4mm intumescent strip, one hour application; setting into groove in timber frame or door	5.67	10.00	6.24	0.17	-	2.42	-	8.66	m	9.74
22 x 4mm intumescent strip with integral cold smoke seal, half hour application; setting into groove in timber frame or door	8.59	10.00	9.45	0.17	-	2.42	-	11.87	m	13.35
28 x 4mm intumescent strip with integral cold smoke seal, one hour application; setting into groove in timber frame or door	9.96	10.00	10.95	0.17	-	2.42	-	13.37	m	15.04
12 x 4mm intumescent strip, half hour application; fixing to both sides of glass behind glazing beads	8.11	10.00	8.92	0.35	-	4.97	-	13.89	m	15.63
25 x 4mm intumescent strip, one hour application; fixing to both sides of glass behind glazing beads	14.56	10.00	16.01	0.35	-	4.97	-	20.99	m	23.61
10 x 4mm cold smoke seal; setting into groove in timber frame or door	4.44	10.00	4.88	0.17	-	2.42	-	7.30	m	8.21
Lorient Polyproducts Ltd; System 36										
15 x 12mm glazing channel reference LG1512; fitting over the edge of the pane	1.89	10.00	2.08	0.17	-	2.42	-	4.49	m	5.05
Lorient Polyproducts Ltd; System 90										
intumescent lining reference LX4402 to suit 44mm thick doors; fixing to timber with adhesive	4.11	10.00	4.53	0.35	-	4.97	-	9.50	m	10.69

Labour hourly rates: (except Specialists) Craft Operatives £14.21 Labourer £10.71 Rates are national average prices. Refer to REGIONAL VARIATIONS for indicative levels of overall pricing in regions	MATERIALS			LABOUR				RATES		
	Del to Site £	Waste %	Material Cost £	Craft Optve Hrs	Lab Hrs	Labour Cost £	Sunds £	Nett Rate £	Unit	Gross Rate (+12.5%) £
L20: DOORS/SHUTTERS/HATCHES: FRAMES Cont.										
Intumescent strips and smoke seals Cont.										
Lorient Polyproducts Ltd; System 90 Cont.										
intumescent lining reference LX5402 to suit 54mm thick doors; fixing to timber with adhesive	4.61	10.00	**5.07**	0.35	-	**4.97**	-	**10.04**	m	**11.30**
Mann McGowan Fabrications Ltd										
Pyroglaze 30 half hour application; fixing to both sides of glass behind glazing beads	3.33	10.00	**3.66**	0.40	-	**5.68**	-	**9.34**	m	**10.51**
Pyroglaze 60 one hour application; fixing to both sides of glass behind glazing beads	5.22	10.00	**5.74**	0.40	-	**5.68**	-	**11.42**	m	**12.85**
Bedding and pointing frames										
Bedding in cement mortar (1:3); pointing one side wood frames	0.32	5.00	**0.33**	0.14	-	**1.99**	-	**2.32**	m	**2.61**
Bedding in cement mortar (1:3); pointing each side wood frames	0.42	5.00	**0.44**	0.20	-	**2.84**	-	**3.28**	m	**3.69**
Bedding in cement mortar (1:3); pointing with Secomastic standard mastic one side wood frames one side	1.19	5.00	**1.29**	0.20	-	**2.84**	-	**4.14**	m	**4.65**
L20: DOORS/SHUTTERS/HATCHES: METAL DOORS										
Doors and sidelights in galvanized steel										
Crittall Windows Ltd. Duralife Homelight; weatherstripping; fixing to masonry with lugs										
761 x 2056mm doors NA15	695.03	2.50	**712.41**	3.20	1.60	**62.61**	-	**775.02**	nr	**871.89**
997 x 2056mm doors NA2	1046.15	2.50	**1072.31**	3.45	1.70	**67.23**	-	**1139.54**	nr	**1281.98**
1143 x 2056mm doors NA25	1065.83	2.50	**1092.47**	3.65	1.85	**71.68**	-	**1164.15**	nr	**1309.67**
279 x 2056mm sidelights, NA6	165.42	2.50	**169.55**	1.40	0.70	**27.39**	-	**196.94**	nr	**221.56**
508 x 2056mm sidelights, NA5	201.04	2.50	**206.07**	1.70	0.85	**33.26**	-	**239.33**	nr	**269.25**
997 x 2056mm sidelights, NA13F	382.48	2.50	**392.04**	2.00	1.05	**39.67**	-	**431.71**	nr	**485.67**
Doors and sidelights in galvanized steel; polyester powder coated; white matt										
Crittall Windows Ltd. Duralife Homelight; weatherstripping; fixing to masonry with lugs										
761 x 2056mm doors NA15	875.13	2.50	**897.01**	3.20	1.60	**62.61**	-	**959.62**	nr	**1079.57**
997 x 2056mm doors NA2	1316.67	2.50	**1349.58**	3.45	1.70	**67.23**	-	**1416.82**	nr	**1593.92**
1143 x 2056mm doors NA25	1341.81	2.50	**1375.36**	3.65	1.85	**71.68**	-	**1447.04**	nr	**1627.92**
279 x 2056mm sidelights, NA6	205.14	2.50	**210.27**	1.40	0.70	**27.39**	-	**237.66**	nr	**267.37**
508 x 2056mm sidelights, NA5	249.33	2.50	**255.56**	1.70	0.85	**33.26**	-	**288.82**	nr	**324.93**
997 x 2056mm sidelights, NA13F	465.07	2.50	**476.69**	2.00	1.05	**39.67**	-	**516.36**	nr	**580.91**
Garage doors; Henderson Durham Castle in galvanized steel; one coat primer by manufacturer										
Overhead garage doors; tensioning device Tracked spring counterbalanced										
fixing to timber with screws; for opening size										
2134 x 1981mm	278.25	2.50	**285.21**	6.00	6.00	**149.52**	-	**434.73**	nr	**489.07**
2134 x 2134mm	317.10	2.50	**325.03**	6.00	6.00	**149.52**	-	**474.55**	nr	**533.87**
Overhead garage doors; counterbalanced by springs										
4267 x 1981mm	970.20	2.50	**994.46**	8.00	8.00	**199.36**	-	**1193.82**	nr	**1343.04**
4267 x 2134mm	970.20	2.50	**994.46**	8.00	8.00	**199.36**	-	**1193.82**	nr	**1343.04**
L20: DOORS/SHUTTERS/HATCHES: ROLLER SHUTTERS										
Roller shutters in steel, galvanized										
Note										
roller shutters are always purpose made to order and the following prices are indicative only. Firm quotations should always be obtained										
Crank handle operation										
4572 x 3658mm overall; fixing to masonry with screws	1306.47	2.50	**1339.13**	24.00	24.00	**598.08**	-	**1937.21**	nr	**2179.36**
Endless hand chain operation										
5486 x 4267mm overall; fixing to masonry with screws	1371.79	2.50	**1406.09**	24.00	24.00	**598.08**	-	**2004.17**	nr	**2254.69**
Electric motor operation										
6096 x 6096mm overall; fixing to masonry with screws	2553.56	2.50	**2617.39**	30.00	30.00	**747.60**	-	**3364.99**	nr	**3785.62**
Roller shutters in aluminium										
Endless hand chain operation										
3553 x 3048mm overall; fixing to masonry with screws	1190.03	2.50	**1219.78**	24.00	24.00	**598.08**	-	**1817.86**	nr	**2045.09**
Bedding and pointing frames										
Pointing with Secomastic standard mastic metal frames one side	0.88	10.00	**0.96**	0.12	0.06	**2.35**	-	**3.31**	m	**3.73**

Labour hourly rates: (except Specialists) Craft Operatives £14.21 Labourer £10.71 Rates are national average prices. Refer to REGIONAL VARIATIONS for indicative levels of overall pricing in regions	MATERIALS			LABOUR				RATES		
	Del to Site £	Waste %	Material Cost £	Craft Optve Hrs	Lab Hrs	Labour Cost £	Sunds £	Nett Rate £	Unit	Gross Rate (+12.5%) £
L20: DOORS/SHUTTERS/HATCHES: ROLLER SHUTTERS Cont.										
Bedding and pointing frames Cont.										
Bedding in cement mortar (1:3); pointing with Secomastic standard mastic one side										
metal frames one side	1.05	5.00	**1.13**	0.20	0.10	**3.91**	-	**5.05**	m	**5.68**
L20: DOORS/SHUTTERS/HATCHES: PLASTICS/RUBBER DOORS										
Flexible doors mainly in plastics and rubber										
Mancuna standard doors; Mandor Engineering Ltd; 43mm diameter steel tube frame; top and bottom plates with pivoting arrangements and spring unit; 8mm rubber panel doors with triangular perspex windows in each leaf										
fixing to masonry; for opening size										
1800 x 2100mm	937.13	2.50	**960.55**	4.40	4.40	**109.65**	-	**1070.20**	nr	**1203.98**
2440 x 2400mm	1047.38	2.50	**1073.56**	5.50	5.50	**137.06**	-	**1210.62**	nr	**1361.95**
Mancuna standard doors; Mandor Engineering Ltd; 43mm diameter steel tube frame; top and bottom plates with pivoting arrangements and spring unit; 12mm rubber panel doors with triangular perspex windows in each leaf										
fixing to masonry; for opening size										
1800 x 2100mm	1047.38	2.50	**1073.56**	4.95	4.95	**123.35**	-	**1196.91**	nr	**1346.53**
2440 x 2400mm	1196.21	2.50	**1226.12**	6.05	6.05	**150.77**	-	**1376.88**	nr	**1548.99**
Mancuna standard doors; Mandor Engineering Ltd; 43mm diameter steel tube frame; top and bottom plates with pivoting arrangements and spring unit; 6mm clear flexible panel doors										
fixing to masonry; for opening size										
1800 x 2100mm	793.80	2.50	**813.65**	3.85	3.85	**95.94**	-	**909.59**	nr	**1023.29**
2440 x 2400mm	843.41	2.50	**864.50**	4.40	4.40	**109.65**	-	**974.15**	nr	**1095.91**
Mancuna standard doors; Mandor Engineering Ltd; 43mm diameter steel tube frame; top and bottom plates with pivoting arrangements and spring unit; 9mm clear flexible panel doors										
fixing to masonry; for opening size										
1800 x 2100mm	882.00	2.50	**904.05**	4.13	4.13	**102.92**	-	**1006.97**	nr	**1132.84**
2440 x 2400mm	948.15	2.50	**971.85**	4.68	4.68	**116.63**	-	**1088.48**	nr	**1224.54**
Mandor heavy duty doors; Mandor Engineering Ltd; 63mm diameter steel tube frame; top and bottom plates with pivoting arrangements and spring unit; 12mm rubber panel doors with oval panel windows in each leaf										
fixing to masonry; for opening size										
2440 x 3050mm	1598.63	2.50	**1638.59**	8.80	8.80	**219.30**	-	**1857.89**	nr	**2090.12**
2740 x 3350mm	1874.25	2.50	**1921.11**	9.90	9.90	**246.71**	-	**2167.81**	nr	**2438.79**
3050 x 3660mm	1973.48	2.50	**2022.81**	11.00	11.00	**274.12**	-	**2296.93**	nr	**2584.05**
Mandor heavy duty doors; Mandor Engineering Ltd; 63mm diameter steel tube frame; top and bottom plates with pivoting arrangements and spring unit; 9mm clear flexible panel doors										
fixing to masonry; for opening size										
2440 x 3050mm	1488.38	2.50	**1525.58**	6.60	6.60	**164.47**	-	**1690.06**	nr	**1901.31**
2740 x 3350mm	1730.93	2.50	**1774.20**	7.70	7.70	**191.88**	-	**1966.08**	nr	**2211.84**
3050 x 3660mm	2028.60	2.50	**2079.32**	8.80	8.80	**219.30**	-	**2298.61**	nr	**2585.94**
Flexible strip curtains mainly in pvc										
Mandor strip curtains; Mandor Engineering Ltd; 200mm wide x 2.0mm thick clear pvc strip curtains, suspended from 80 x 80 x 6mm steel angle; fixing angle to masonry with bolts										
double overlap; for opening size										
1800 x 2100mm	121.80	2.50	**124.85**	2.20	2.20	**54.82**	-	**179.67**	nr	**202.13**
2100 x 2440mm	171.15	2.50	**175.43**	2.53	2.53	**63.05**	-	**238.48**	nr	**268.29**
single overlap; for opening size										
1800 x 2100mm	97.65	2.50	**100.09**	1.98	1.98	**49.34**	-	**149.43**	nr	**168.11**
2100 x 2440mm	135.45	2.50	**138.84**	2.31	2.31	**57.57**	-	**196.40**	nr	**220.95**
Mandor strip curtains; Mandor Engineering Ltd; 300mm wide x 2.5mm thick clear pvc strip curtains, suspended from 80 x 80 x 6mm steel angle; fixing angle to masonry with bolts										
double overlap; for opening size										
2440 x 2740mm	238.35	2.50	**244.31**	2.90	2.90	**72.27**	-	**316.58**	nr	**356.15**
2740 x 2900mm	281.40	2.50	**288.44**	3.23	3.23	**80.49**	-	**368.93**	nr	**415.04**
single overlap; for opening size										
2440 x 2740mm	183.75	2.50	**188.34**	2.68	2.68	**66.79**	-	**255.13**	nr	**287.02**
2740 x 2900mm	218.40	2.50	**223.86**	3.01	3.01	**75.01**	-	**298.87**	nr	**336.23**

Labour hourly rates: (except Specialists) Craft Operatives £14.21 Labourer £10.71 Rates are national average prices. Refer to REGIONAL VARIATIONS for indicative levels of overall pricing in regions	MATERIALS			LABOUR				RATES		
	Del to Site £	Waste %	Material Cost £	Craft Optve Hrs	Lab Hrs	Labour Cost £	Sunds £	Nett Rate £	Unit	Gross Rate (+12.5%) £
L30 STAIRS/WALKWAYS/BALUSTRADES: TIMBER STAIRS										
Stairs in softwood, wrought										
Straight flight staircase and balustrade; 25mm treads; 19mm risers, glued; wedged and blocked; 25mm wall string; 38mm outer string; 75 x 75mm newels; 25 x 25mm balusters; 38 x 75mm hardwood handrail										
864mm wide x 2600mm rise overall; balustrade to one side; fixing to masonry with screws	700.09	2.50	**717.59**	15.00	1.90	**233.50**	2.40	**953.49**	nr	**1072.67**
864mm wide x 2600mm rise overall with 3 nr winders at bottom; balustrade to one side; fixing to masonry with screws	840.11	2.50	**861.11**	20.00	2.50	**310.98**	3.25	**1175.33**	nr	**1322.25**
864mm wide x 2600mm rise overall with 3 nr winders at top; balustrade to one side; fixing to masonry with screws	840.11	2.50	**861.11**	20.00	2.50	**310.98**	3.25	**1175.33**	nr	**1322.25**
Balustrades in softwood, wrought										
Isolated balustrades; 25 x 25mm balusters at 150mm centres, housed construction (handrail included elsewhere)										
914mm high; fixing to timber with screws	36.77	2.50	**37.69**	0.80	0.10	**12.44**	0.13	**50.26**	m	**56.54**
Isolated balustrades; 38 x 38mm balusters at 150mm centres, housed construction (handrail included elsewhere)										
914mm high; fixing to timber with screws	53.89	2.50	**55.24**	1.15	0.14	**17.84**	0.19	**73.27**	m	**82.43**
Isolated balustrades; 50 x 50mm balusters at 150mm centres, housed construction (handrail included elsewhere)										
914mm high; fixing to timber with screws	70.80	2.50	**72.57**	1.30	0.16	**20.19**	0.21	**92.97**	m	**104.59**
Isolated balustrades; 25 x 25mm balusters at 150mm centres, 50mm mopstick handrail, housed construction										
914mm high; fixing to timber with screws	61.63	2.50	**63.17**	1.20	0.15	**18.66**	0.20	**82.03**	m	**92.28**
extra; ramps	37.83	2.50	**38.77**	1.10	0.14	**17.13**	0.18	**56.08**	nr	**63.09**
extra; wreaths	75.64	2.50	**77.53**	2.15	0.27	**33.44**	0.35	**111.33**	nr	**125.24**
extra; bends	37.83	2.50	**38.77**	0.70	0.09	**10.91**	0.11	**49.79**	nr	**56.02**
Isolated balustrades; 38 x 38mm balusters at 150mm centres, 50mm mopstick handrail, housed construction										
914mm high; fixing to timber with screws	78.44	2.50	**80.40**	1.55	0.19	**24.06**	0.25	**104.71**	m	**117.80**
extra; ramps	37.83	2.50	**38.77**	1.10	0.14	**17.13**	0.18	**56.08**	nr	**63.09**
extra; wreaths	75.64	2.50	**77.53**	2.15	0.27	**33.44**	0.35	**111.33**	nr	**125.24**
extra; bends	37.83	2.50	**38.77**	0.70	0.09	**10.91**	0.11	**49.79**	nr	**56.02**
Isolated balustrades; 50 x 50mm balusters at 150mm centres, 50mm mopstick handrail, housed construction										
914mm high; fixing to timber with screws	95.24	2.50	**97.63**	1.75	0.22	**27.22**	0.28	**125.13**	m	**140.77**
extra; ramps	37.83	2.50	**38.77**	1.10	0.14	**17.13**	0.18	**56.08**	nr	**63.09**
extra; wreaths	75.64	2.50	**77.53**	2.15	0.27	**33.44**	0.35	**111.33**	nr	**125.24**
extra; bends	37.83	2.50	**38.77**	0.70	0.09	**10.91**	0.11	**49.79**	nr	**56.02**
Isolated balustrades; 25 x 25mm balusters at 150mm centres, 75 x 100mm moulded handrail; housed construction										
914mm high; fixing to timber with screws	75.64	2.50	**77.53**	1.25	0.16	**19.48**	0.20	**97.21**	m	**109.36**
extra; ramps	56.03	2.50	**57.43**	2.40	0.30	**37.32**	0.39	**95.14**	nr	**107.03**
extra; wreaths	112.06	2.50	**114.86**	4.80	0.60	**74.63**	0.78	**190.27**	nr	**214.06**
extra; bends	56.03	2.50	**57.43**	1.45	0.18	**22.53**	0.24	**80.20**	nr	**90.23**
Isolated balustrades; 38 x 38mm balusters at 150mm centres, 75 x 100mm moulded handrail; housed construction										
914mm high; fixing to timber with screws	92.44	2.50	**94.76**	1.60	0.20	**24.88**	0.26	**119.89**	m	**134.88**
extra; ramps	56.03	2.50	**57.43**	2.40	0.30	**37.32**	0.39	**95.14**	nr	**107.03**
extra; wreaths	112.06	2.50	**114.86**	4.80	0.60	**74.63**	0.78	**190.27**	nr	**214.06**
extra; bends	56.03	2.50	**57.43**	1.45	0.18	**22.53**	0.24	**80.20**	nr	**90.23**
Isolated balustrades; 50 x 50mm balusters at 150mm centres, 75 x 100mm moulded handrail; housed construction										
914mm high; fixing to timber with screws	109.26	2.50	**111.99**	1.80	0.23	**28.04**	0.29	**140.32**	m	**157.86**
extra; ramps	56.03	2.50	**57.43**	2.40	0.30	**37.32**	0.39	**95.14**	nr	**107.03**
extra; wreaths	112.06	2.50	**114.86**	4.80	0.60	**74.63**	0.78	**190.27**	nr	**214.06**
extra; bends	56.03	2.50	**57.43**	1.45	0.18	**22.53**	0.24	**80.20**	nr	**90.23**
Balustrades in European Oak, wrought, selected for transparent finish										
Isolated balustrades; 25 x 25mm balusters at 150mm centres, 50mm mopstick handrail, housed construction										
914mm high; fixing to timber with screws	135.59	2.50	**138.98**	2.05	0.26	**31.92**	0.33	**171.22**	m	**192.62**
extra; ramps	83.22	2.50	**85.30**	1.80	0.23	**28.04**	0.29	**113.63**	nr	**127.83**
extra; wreaths	166.40	2.50	**170.56**	3.60	0.45	**55.98**	0.59	**227.13**	nr	**255.52**
extra; bends	83.22	2.50	**85.30**	1.30	0.16	**20.19**	0.21	**105.70**	nr	**118.91**
Isolated balustrades; 38 x 38mm balusters at 150mm centres, 50mm mopstick handrail, housed construction										
914mm high; fixing to timber with screws	172.57	2.50	**176.89**	2.60	0.26	**39.73**	0.41	**217.03**	m	**244.16**
extra; ramps	83.22	2.50	**85.30**	1.80	0.23	**28.04**	0.29	**113.63**	nr	**127.83**
extra; wreaths	166.40	2.50	**170.56**	3.60	0.45	**55.98**	0.59	**227.13**	nr	**255.52**
extra; bends	83.22	2.50	**85.30**	1.30	0.16	**20.19**	0.21	**105.70**	nr	**118.91**

Labour hourly rates: (except Specialists) Craft Operatives £14.21 Labourer £10.71 Rates are national average prices. Refer to REGIONAL VARIATIONS for indicative levels of overall pricing in regions	MATERIALS			LABOUR				RATES		
	Del to Site £	Waste %	Material Cost £	Craft Optve Hrs	Lab Hrs	Labour Cost £	Sunds £	Nett Rate £	Unit	Gross Rate (+12.5%) £
L30 STAIRS/WALKWAYS/BALUSTRADES: TIMBER STAIRS Cont.										
Balustrades in European Oak, wrought, selected for transparent finish Cont.										
Isolated balustrades; 50 x 50mm balusters at 150mm centres, 50mm mopstick handrail, housed construction										
914mm high; fixing to timber with screws	209.55	2.50	**214.79**	3.00	0.38	**46.70**	0.49	**261.98**	m	**294.73**
extra; ramps	83.22	2.50	**85.30**	1.80	0.23	**28.04**	0.29	**113.63**	nr	**127.83**
extra; wreaths	166.40	2.50	**170.56**	3.60	0.45	**55.98**	0.59	**227.13**	nr	**255.52**
extra; bends	83.22	2.50	**85.30**	1.30	0.16	**20.19**	0.21	**105.70**	nr	**118.91**
Isolated balustrades; 25 x 25mm balusters at 150mm centres, 75 x 100mm moulded handrail; housed construction										
914mm high; fixing to timber with screws	166.40	2.50	**170.56**	2.15	0.27	**33.44**	0.35	**204.35**	m	**229.90**
extra; ramps	123.26	2.50	**126.35**	3.70	0.46	**57.50**	0.60	**184.45**	nr	**207.51**
extra; wreaths	246.53	2.50	**252.69**	7.45	0.93	**115.82**	1.21	**369.73**	nr	**415.94**
extra; bends	123.26	2.50	**126.35**	2.60	0.33	**40.48**	0.42	**167.25**	nr	**188.15**
Isolated balustrades; 38 x 38mm balusters at 150mm centres, 75 x 100mm moulded handrail; housed construction										
914mm high; fixing to timber with screws	203.39	2.50	**208.48**	2.70	0.34	**42.01**	0.44	**250.92**	m	**282.29**
extra; ramps	123.26	2.50	**126.35**	3.70	0.46	**57.50**	0.60	**184.45**	nr	**207.51**
extra; wreaths	246.53	2.50	**252.69**	7.45	0.93	**115.82**	1.21	**369.73**	nr	**415.94**
extra; bends	123.26	2.50	**126.35**	2.60	0.33	**40.48**	0.42	**167.25**	nr	**188.15**
Isolated balustrades; 50 x 50mm balusters at 150mm centres, 75 x 100mm moulded handrail; housed construction										
914mm high; fixing to timber with screws	240.37	2.50	**246.38**	3.10	0.39	**48.23**	0.50	**295.10**	m	**331.99**
extra; ramps	123.26	2.50	**126.35**	3.70	0.46	**57.50**	0.60	**184.45**	nr	**207.51**
extra; wreaths	246.53	2.50	**252.69**	7.45	0.93	**115.82**	1.21	**369.73**	nr	**415.94**
extra; bends	123.26	2.50	**126.35**	2.60	0.33	**40.48**	0.42	**167.25**	nr	**188.15**
Handrails in softwood, wrought										
Associated handrails										
50mm mopstick; fixing through metal backgrounds with screws	21.01	5.00	**22.06**	0.40	0.05	**6.22**	0.06	**28.34**	m	**31.89**
extra; ramps	37.83	5.00	**39.72**	1.20	0.15	**18.66**	0.20	**58.58**	nr	**65.90**
extra; wreaths	75.64	5.00	**79.42**	2.40	0.30	**37.32**	0.39	**117.13**	nr	**131.77**
extra; bends	37.83	5.00	**39.72**	1.20	0.15	**18.66**	0.20	**58.58**	nr	**65.90**
75 x 100mm; moulded; fixing through metal backgrounds with screws	35.03	5.00	**36.78**	0.50	0.06	**7.75**	0.08	**44.61**	m	**50.18**
extra; ramps	56.03	5.00	**58.83**	1.45	0.18	**22.53**	0.24	**81.60**	nr	**91.80**
extra; wreaths	112.06	5.00	**117.66**	2.90	0.36	**45.06**	0.47	**163.20**	nr	**183.60**
extra; bends	56.03	5.00	**58.83**	1.45	0.18	**22.53**	0.24	**81.60**	nr	**91.80**
Handrails in African Mahogany, wrought, selected for transparent finish										
Associated handrails										
50mm mopstick; fixing through metal backgrounds with screws	26.27	5.00	**27.59**	0.60	0.08	**9.38**	0.10	**37.07**	m	**41.70**
extra; ramps	47.29	5.00	**49.65**	1.80	0.23	**28.04**	0.29	**77.98**	nr	**87.73**
extra; wreaths	94.55	5.00	**99.28**	3.60	0.45	**55.98**	0.59	**155.84**	nr	**175.32**
extra; bends	47.29	5.00	**49.65**	1.80	0.23	**28.04**	0.29	**77.98**	nr	**87.73**
75 x 100mm; moulded; fixing through metal backgrounds with screws	43.78	5.00	**45.97**	0.70	0.09	**10.91**	0.11	**56.99**	m	**64.11**
extra; ramps	70.04	5.00	**73.54**	2.15	0.27	**33.44**	0.35	**107.34**	nr	**120.75**
extra; wreaths	140.07	5.00	**147.08**	4.30	0.54	**66.89**	0.70	**214.66**	nr	**241.50**
extra; bends	70.04	5.00	**73.54**	2.15	0.27	**33.44**	0.35	**107.34**	nr	**120.75**
Handrails in European Oak, wrought, selected for transparent finish										
Associated handrails										
50mm mopstick; fixing through metal backgrounds with screws	46.24	5.00	**48.55**	0.60	0.08	**9.38**	0.10	**58.04**	m	**65.29**
extra; ramps	83.22	5.00	**87.38**	1.80	0.23	**28.04**	0.29	**115.71**	nr	**130.18**
extra; wreaths	166.40	5.00	**174.72**	3.60	0.45	**55.98**	0.59	**231.29**	nr	**260.20**
extra; bends	83.22	5.00	**87.38**	1.80	0.23	**28.04**	0.29	**115.71**	nr	**130.18**
75 x 100mm; moulded; fixing through metal backgrounds with screws	77.06	5.00	**80.91**	0.70	0.09	**10.91**	0.11	**91.93**	m	**103.42**
extra; ramps	123.26	5.00	**129.43**	2.15	0.27	**33.44**	0.35	**163.22**	nr	**183.62**
extra; wreaths	246.53	5.00	**258.85**	4.30	0.54	**66.89**	0.70	**326.44**	nr	**367.25**
extra; bends	123.26	5.00	**129.43**	2.15	0.26	**33.34**	0.35	**163.11**	nr	**183.50**
L30: STAIRS/WALKWAYS/BALUSTRADES: METAL STAIRS AND BALUSTRADES										
Staircases in steel										
Straight flight staircases; 180 x 10mm flat stringers, shaped ends to top and bottom; 6 x 250mm on plain raised pattern plate treads, with 50 x 50 x 6mm shelf angles and 40 x 40 x 6mm stiffening bars, bolted to stringers; welded, cleated and bolted connections										
770mm wide x 3000mm going x 2600mm rise overall; fixing to masonry with 4 nr Rawlbolts	-	-	**Spclist**	-	-	**Spclist**	-	**2242.50**	nr	**2522.81**
920mm wide x 3000mm going x 2600mm rise overall; fixing to masonry with 4 nr Rawlbolts	-	-	**Spclist**	-	-	**Spclist**	-	**2558.75**	nr	**2878.59**

Labour hourly rates: (except Specialists) Craft Operatives £14.21 Labourer £10.71 Rates are national average prices. Refer to REGIONAL VARIATIONS for indicative levels of overall pricing in regions	MATERIALS			LABOUR				RATES		
	Del to Site £	Waste %	Material Cost £	Craft Optve Hrs	Lab Hrs	Labour Cost £	Sunds £	Nett Rate £	Unit	Gross Rate (+12.5%) £
L30: STAIRS/WALKWAYS/BALUSTRADES: METAL STAIRS AND BALUSTRADES Cont.										
Staircases in steel Cont.										
Straight flight staircases, 180 x 10mm flat stringers, shaped ends to top and bottom; 6 x 250mm tray treads, with three 6mm diameter reinforcing bars welded to inside, and 50 x 50 x 6mm shelf angles, bolted to stringers; welded, cleated and bolted connections										
770mm wide x 3000mm going x 2600mm rise overall; fixing to masonry with 4 nr Rawlbolts	-	-	Spclist	-	-	Spclist	-	2127.50	nr	2393.44
920mm wide x 3000mm going x 2600mm rise overall; fixing to masonry with 4 nr Rawlbolts	-	-	Spclist	-	-	Spclist	-	2202.25	nr	2477.53
Straight flight staircases; 178 x 76mm channel stringers, shaped ends to top and bottom; 6 x 250mm on plain raised pattern plate treads, with 50 x 50 x 6mm shelf angles and 40 x 40 x 6mm stiffening bars, bolted to stringers; welded, cleated and bolted connections										
770mm wide x 3000mm going x 2600mm rise overall; fixing to masonry with 4 nr Rawlbolts	-	-	Spclist	-	-	Spclist	-	2426.50	nr	2729.81
920mm wide x 3000mm going x 2600mm rise overall; fixing to masonry with 4 nr Rawlbolts	-	-	Spclist	-	-	Spclist	-	2771.50	nr	3117.94
Straight flight staircases and balustrades; 180 x 10mm flat stringers, shaped ends to top and bottom; 6 x 250mm on plain raised pattern plate treads with 50 x 50 x 6mm shelf angles and 40 x 40 x 6mm stiffening bars, bolted to stringers; 915mm high balustrade to both sides consisting of 25mm diameter solid bar handrail and 32mm diameter solid bar standards at 250mm centres with base plate welded on and bolted to face of stringer, and ball type joints at intersections; welded, cleated and bolted connections										
770mm wide x 3000mm going x 2600mm rise overall; fixing to masonry with 4 nr Rawlbolts	-	-	Spclist	-	-	Spclist	-	3312.00	nr	3726.00
Straight flight staircases and balustrades; 180 x 10mm flat stringers, shaped ends to top and bottom; 6 x 250mm on plain raised pattern plate treads with 50 x 50 x 6mm shelf angles and 40 x 40 x 6mm stiffening bars, bolted to stringers; 915mm high balustrade to both sides consisting of 25mm diameter solid bar handrail and 32mm diameter solid bar standards at 250mm centres with base plate welded on and bolted to top of stringer, and ball type joints at intersections; welded, cleated and bolted connections										
770mm wide x 3000mm going x 2600mm rise overall; fixing to masonry with 4 nr Rawlbolts	-	-	Spclist	-	-	Spclist	-	3266.00	nr	3674.25
Straight flight staircases and balustrades; 180 x 10mm flat stringers, shaped ends to top and bottom; 6 x 250mm on plain raised pattern plate treads with 50 x 50 x 6mm shelf angles and 40 x 40 x 6mm stiffening bars, bolted to stringers; 915mm high balustrade to one side consisting of 25mm diameter solid bar handrail and 32mm diameter solid bar standards at 250mm centres with base plate welded on and bolted to face of stringer, and ball type joints at intersections; welded, cleated and bolted connections										
770mm wide x 3000mm going x 2600mm rise overall; fixing to masonry with 4 nr Rawlbolts	-	-	Spclist	-	-	Spclist	-	3266.00	nr	3674.25
Straight flight staircases and balustrades; 180 x 10mm flat stringers, shaped ends to top and bottom; 6 x 250mm on plain raised pattern plate treads with 50 x 50 x 6mm shelf angles and 40 x 40 x 6mm stiffening bars, bolted to stringers; 915mm high balustrade to one side consisting of 25mm diameter solid bar handrail and 32mm diameter solid bar standards at 250mm centres with base plate welded on and bolted to top of stringer, and ball type joints at intersections; welded, cleated and bolted connections										
770mm wide x 3000mm going x 2600mm rise overall; fixing to masonry with 4 nr Rawlbolts	-	-	Spclist	-	-	Spclist	-	3266.00	nr	3674.25
Straight flight staircases and balustrades; 180 x 10mm flat stringers, shaped ends to top and bottom; 6 x 250mm on plain raised pattern plate treads with 50 x 50 x 6mm shelf angles and 40 x 40 x 6mm stiffening bars, bolted to stringers; 1070mm high balustrade to both sides consisting of 25mm diameter solid bar handrail and 32mm diameter solid bar standards at 250mm centres with base plate welded on and bolted to top of stringer, and ball type joints at intersections; welded, cleated and bolted connections										
770mm wide x 3000mm going x 2600mm rise overall; fixing to masonry with 4 nr Rawlbolts	-	-	Spclist	-	-	Spclist	-	3294.75	nr	3706.59

Labour hourly rates: (except Specialists) Craft Operatives £14.21 Labourer £10.71 Rates are national average prices. Refer to REGIONAL VARIATIONS for indicative levels of overall pricing in regions	MATERIALS			LABOUR				RATES		
	Del to Site £	Waste %	Material Cost £	Craft Optve Hrs	Lab Hrs	Labour Cost £	Sunds £	Nett Rate £	Unit	Gross Rate (+12.5%) £
L30: STAIRS/WALKWAYS/BALUSTRADES: METAL STAIRS AND BALUSTRADES Cont.										
Staircases in steel Cont.										
Straight flight staircases and balustrades; 180 x 10mm flat stringers, shaped ends to top and bottom; 6 x 250mm on plain raised pattern plate treads with 50 x 50 x 6mm shelf angles and 40 x 40 x 6mm stiffening bars, bolted to stringers; 1070mm high balustrade to one side consisting of 25mm diameter solid bar handrail and 32mm diameter solid bar standards at 250mm centres with base plate welded on and bolted to top of stringer, and ball type joints at intersections; welded, cleated and bolted connections										
770mm wide x 3000mm going x 2600mm rise overall; fixing to masonry with 4 nr Rawlbolts............	-	-	**Spclist**	-	-	**Spclist**	-	**3070.50**	nr	**3454.31**
Straight flight staircases and balustrades; 180 x 10mm flat stringers, shaped ends to top and bottom; 6 x 250mm on plain raised pattern plate treads with 50 x 50 x 6mm shelf angles and 40 x 40 x 6mm stiffening bars, bolted to stringers; 915mm high balustrade to one side consisting of 25mm diameter solid bar handrail and intermediate rail, and 32mm diameter solid bar standards at 250mm centres with base plate welded on and bolted to top of stringer, and ball type joints at intersections										
770mm wide x 3000mm going x 2600mm rise overall; fixing to masonry with 4 nr Rawlbolts............	-	-	**Spclist**	-	-	**Spclist**	-	**3047.50**	nr	**3428.44**
Straight flight staircases and balustrades; 180 x 10mm flat stringers, shaped ends to top and bottom; 6 x 250mm on plain raised pattern plate treads with 50 x 50 x 6mm shelf angles and 40 x 40 x 6mm stiffening bars, bolted to stringers; 1070mm high balustrade to one side consisting of 25mm diameter solid bar handrail and intermediate rail, and 32mm diameter solid bar standards at 250mm centres with base plate welded on and bolted to top of stringer, and ball type joints at intersections										
770mm wide x 3000mm going x 2600mm rise overall; fixing to masonry with 4 nr Rawlbolts............	-	-	**Spclist**	-	-	**Spclist**	-	**3070.50**	nr	**3454.31**
Quarter landing staircases, in two flights; 180 x 10mm flat stringers, shaped ends to top and bottom; 6 x 250mm on plain raised pattern plate treads, with 50 x 50 x 6mm shelf angles and 40 x 40 x 6mm stiffening bars, bolted to stringers; 6mm on plain raised pattern plate landing, welded on; 100 x 8mm flat kicking plates welded on; welded, cleated and bolted connections										
770mm wide x 2000mm going first flight excluding landing x 1000mm going second flight excluding landing x 2600mm rise overall; 770 x 770mm landing overall; fixing to masonry with 4 nr Rawlbolts...	-	-	**Spclist**	-	-	**Spclist**	-	**4904.75**	nr	**5517.84**
920mm wide x 2000mm going first flight excluding landing x 1000mm going second flight excluding landing x 2600mm rise overall; 920 x 920mm landing overall; fixing to masonry with 4 nr Rawlbolts...	-	-	**Spclist**	-	-	**Spclist**	-	**5157.75**	nr	**5802.47**
Half landing staircases, in two flights; 180 x 10mm flat stringers, shaped ends to top and bottom; 6 x 250mm on plain raised pattern plate treads, with 50 x 50 x 6mm shelf angles and 40 x 40 x 6mm stiffening bars, bolted to stringers; 6mm on plain raised pattern plate landing, welded on; 100 x 8mm flat kicking plates welded on; welded, cleated and bolted connections										
770mm wide x 2000mm going first flight excluding landing x 1000mm going second flight excluding landing x 2600mm rise overall; 770 x 1640mm landing overall; fixing to masonry with 4 nr Rawlbolts	-	-	**Spclist**	-	-	**Spclist**	-	**5502.75**	nr	**6190.59**
920mm wide x 2000mm going first flight excluding landing x 1000mm going second flight excluding landing x 2600mm rise overall; 920 x 1940mm landing overall; fixing to masonry with 4 nr Rawlbolts	-	-	**Spclist**	-	-	**Spclist**	-	**6244.50**	nr	**7025.06**
Balustrades in steel										
Isolated balustrades; 6 x 38mm flat core rail; 13mm diameter balusters at 250mm centres; welded fabrication ground to smooth finish; casting into mortices in concrete; wedging in position; temporary wedges										
838mm high; level	154.70	2.50	**158.57**	1.10	0.55	**21.52**	-	**180.09**	m	**202.60**
extra; ramps	26.46	2.50	**27.12**	1.10	0.55	**21.52**	-	**48.65**	nr	**54.73**
extra; wreaths	44.14	2.50	**45.25**	2.20	1.10	**43.04**	-	**88.29**	nr	**99.32**
extra; bends	24.08	2.50	**24.68**	1.10	0.55	**21.52**	-	**46.20**	nr	**51.98**
838mm high; raking	159.94	2.50	**163.93**	1.40	0.70	**27.39**	-	**191.32**	m	**215.24**
914mm high; level	170.00	2.50	**174.25**	1.10	0.55	**21.52**	-	**195.77**	m	**220.24**
extra; ramps	26.46	2.50	**27.12**	1.10	0.55	**21.52**	-	**48.65**	nr	**54.73**
extra; wreaths	44.14	2.50	**45.25**	2.20	1.10	**43.04**	-	**88.29**	nr	**99.32**
extra; bends	24.08	2.50	**24.68**	1.10	0.55	**21.52**	-	**46.20**	nr	**51.98**

Labour hourly rates: (except Specialists) Craft Operatives £14.21 Labourer £10.71 Rates are national average prices. Refer to REGIONAL VARIATIONS for indicative levels of overall pricing in regions	MATERIALS			LABOUR				RATES		
	Del to Site £	Waste %	Material Cost £	Craft Optve Hrs	Lab Hrs	Labour Cost £	Sunds £	Nett Rate £	Unit	Gross Rate (+12.5%) £
L30: STAIRS/WALKWAYS/BALUSTRADES: METAL STAIRS AND BALUSTRADES Cont.										
Balustrades in steel Cont.										
Isolated balustrades; 6 x 38mm flat core rail; 13mm diameter balusters at 250mm centres; welded fabrication ground to smooth finish; casting into mortices in concrete; wedging in position; temporary wedges Cont.										
914mm high; raking	179.71	2.50	**184.20**	1.40	0.70	**27.39**	-	**211.59**	m	**238.04**
Isolated balustrades; 6 x 38mm flat core rail; 13 x 13mm balusters at 250mm centres; welded fabrication ground to smooth finish; casting into mortices in concrete; wedging in position; temporary wedges										
838mm high; level	161.68	2.50	**165.72**	1.10	0.55	**21.52**	-	**187.24**	m	**210.65**
838mm high; raking	167.21	2.50	**171.39**	1.10	0.55	**21.52**	-	**192.91**	m	**217.02**
914mm high; level	175.64	2.50	**180.03**	1.10	0.55	**21.52**	-	**201.55**	m	**226.74**
914mm high; raking	187.27	2.50	**191.95**	1.10	0.55	**21.52**	-	**213.47**	m	**240.16**
Isolated balustrades; 13 x 51mm rounded handrail; 13mm diameter balusters at 250mm centres; welded fabrication ground to smooth finish; casting into mortices in concrete; wedging in position; temporary wedges										
838mm high; level	187.27	2.50	**191.95**	1.10	0.55	**21.52**	-	**213.47**	m	**240.16**
extra; ramps	29.08	2.50	**29.81**	1.10	0.55	**21.52**	-	**51.33**	nr	**57.74**
extra; wreaths	48.39	2.50	**49.60**	2.20	1.10	**43.04**	-	**92.64**	nr	**104.22**
extra; bends	26.29	2.50	**26.94**	1.10	0.55	**21.52**	-	**48.47**	nr	**54.52**
838mm high; raking	190.76	2.50	**195.53**	1.10	0.55	**21.52**	-	**217.05**	m	**244.18**
914mm high; level	198.90	2.50	**203.87**	1.10	0.55	**21.52**	-	**225.39**	m	**253.57**
extra; ramps	29.08	2.50	**29.81**	1.10	0.55	**21.52**	-	**51.33**	nr	**57.74**
extra; wreaths	48.39	2.50	**49.60**	2.20	1.10	**43.04**	-	**92.64**	nr	**104.22**
extra; bends	26.29	2.50	**26.94**	1.10	0.55	**21.52**	-	**48.47**	nr	**54.52**
914mm high; raking	207.04	2.50	**212.22**	1.10	0.55	**21.52**	-	**233.74**	m	**262.96**
Isolated balustrades; 6 x 38mm flat core rail; 10 x 51mm flat bottom rail; 13mm diameter infill balusters at 250mm centres; 25 x 25mm standards at 3000mm centres; welded fabrication ground to smooth finish; casting into mortices in concrete; wedging in position; temporary wedges										
838mm high; level	207.04	2.50	**212.22**	1.10	0.55	**21.52**	-	**233.74**	m	**262.96**
extra; ramps	49.78	2.50	**51.03**	1.10	0.55	**21.52**	-	**72.55**	nr	**81.62**
extra; wreaths	62.23	2.50	**63.79**	2.20	1.10	**43.04**	-	**106.83**	nr	**120.18**
extra; bends	44.14	2.50	**45.25**	1.10	0.55	**21.52**	-	**66.77**	nr	**75.11**
838mm high; raking	214.02	2.50	**219.37**	1.10	0.55	**21.52**	-	**240.89**	m	**271.00**
914mm high; level	227.98	2.50	**233.68**	1.10	0.55	**21.52**	-	**255.20**	m	**287.10**
extra; ramps	49.78	2.50	**51.03**	1.10	0.55	**21.52**	-	**72.55**	nr	**81.62**
extra; wreaths	62.23	2.50	**63.79**	2.20	1.10	**43.04**	-	**106.83**	nr	**120.18**
extra; bends	44.14	2.50	**45.25**	1.10	0.55	**21.52**	-	**66.77**	nr	**75.11**
Balustrades in steel tubing, BS 1387, medium grade; galvanized after fabrication										
Isolated balustrades; 40mm diameter handrails; 40mm diameter standards at 1500mm centres; welded fabrication ground to smooth finish; casting into mortices in concrete; wedging in position; temporary wedges										
838mm high; level	96.72	2.50	**99.13**	0.85	0.40	**16.36**	-	**115.50**	m	**129.93**
838mm high; raking	103.64	2.50	**106.23**	0.85	0.40	**16.36**	-	**122.59**	m	**137.92**
914mm high; level	103.64	2.50	**106.23**	0.85	0.40	**16.36**	-	**122.59**	m	**137.92**
914mm high; raking	110.50	2.50	**113.26**	0.85	0.40	**16.36**	-	**129.63**	m	**145.83**
Isolated balustrades; 40mm diameter handrails; 40mm diameter standards at 1500mm centres, with 6 x 75mm diameter fixing plates welded on to end, holes -3 nr; welded fabrication ground to smooth finish; fixing to masonry with Rawlbolts										
838mm high; level	110.50	2.50	**113.26**	1.10	0.55	**21.52**	-	**134.78**	m	**151.63**
838mm high; raking	117.48	2.50	**120.42**	1.10	0.55	**21.52**	-	**141.94**	m	**159.68**
914mm high; level	131.44	2.50	**134.72**	1.10	0.55	**21.52**	-	**156.24**	m	**175.77**
914mm high; raking	145.10	2.50	**148.73**	1.10	0.55	**21.52**	-	**170.25**	m	**191.53**
Spiral staircases in steel; one coat red oxide primer before delivery to site										
Crescent 'H' range domestic spiral staircases; model 15 'H'; Crescent of Cambridge Ltd.; comprising tread modules complete with centre column, tread support, wooden tread, tread baluster, handrail section, PVC handrail cover, baluster wicket infill panel and tread riser bar, and with attachment brackets for fixing to floor and newel extending from centre column; assembling and bolting together										
1568mm diameter; for floor to floor height units between 2160mm and 2640mm; plywood treads with hardwood edgings; fixing base and ground plates to timber	1179.78	2.50	**1209.27**	31.50	16.00	**618.98**	-	**1828.25**	nr	**2056.78**
extra; additional tread modules, 180 - 220mm per rise	86.81	2.50	**88.98**	1.65	0.85	**32.55**	-	**121.53**	nr	**136.73**
extra; mini landing to match treads, at upper floor level	217.04	2.50	**222.46**	2.10	1.05	**41.09**	-	**263.55**	nr	**296.49**

Labour hourly rates: (except Specialists) Craft Operatives £14.21 Labourer £10.71 Rates are national average prices. Refer to REGIONAL VARIATIONS for indicative levels of overall pricing in regions	MATERIALS			LABOUR				RATES		
	Del to Site £	Waste %	Material Cost £	Craft Optve Hrs	Lab Hrs	Labour Cost £	Sunds £	Nett Rate £	Unit	Gross Rate (+12.5%) £
L30: STAIRS/WALKWAYS/BALUSTRADES: METAL STAIRS AND BALUSTRADES Cont.										
Spiral staircases in steel; one coat red oxide primer before delivery to site Cont.										
Crescent 'H' range domestic spiral staircases; model 15 'H'; Crescent of Cambridge Ltd.; comprising tread modules complete with centre column, tread support, wooden tread, tread baluster, handrail section, PVC handrail cover, baluster wicket infill panel and tread riser bar, and with attachment brackets for fixing to floor and newel extending from centre column; assembling and bolting together Cont.										
1568mm diameter; for floor to floor height units between 2160mm and 2640mm; plywood treads with carpet recess; fixing base and ground plates to timber	1346.73	2.50	**1380.40**	31.50	16.00	**618.98**	-	**1999.37**	nr	**2249.29**
extra; additional tread modules, 180 - 220mm per rise	103.51	2.50	**106.10**	2.10	1.05	**41.09**	-	**147.18**	nr	**165.58**
1568mm diameter; for floor to floor height units between 2160mm and 2640mm; hardwood treads, plain or with hardwood surround providing carpet recess; fixing base and ground plates to timber	1530.38	2.50	**1568.63**	31.50	16.00	**618.98**	-	**2187.61**	nr	**2461.06**
extra; additional tread modules, 180 - 220mm per rise	117.98	2.50	**120.93**	2.10	1.05	**41.09**	-	**162.01**	nr	**182.27**
extra; mini landing to match treads, at upper floor level	260.44	2.50	**266.95**	2.10	1.05	**41.09**	-	**308.04**	nr	**346.54**
stairwell balustrading with straight bar infill and PVC handrail cover	96.83	2.50	**99.25**	31.50	16.00	**618.98**	-	**718.23**	m	**808.01**
stairwell balustrading to match staircase including PVC handrail cover	158.05	2.50	**162.00**	1.95	1.00	**38.42**	-	**200.42**	m	**225.47**
Crescent 'H' range domestic spiral staircases; model 18 'H'; Crescent of Cambridge Ltd.; comprising tread modules complete with centre column, tread support, wooden tread, tread baluster, handrail section, PVC handrail cover, baluster wicket infill panel and tread riser bar, and with attachment brackets for fixing to floor and newel extending from centre column; assembling and bolting together										
1956 diameter; for floor to floor height units between 2160mm and 2640mm; plywood treads with hardwood edgings; fixing base and ground plates to timber	1218.74	2.50	**1249.20**	31.50	16.00	**618.98**	-	**1868.18**	nr	**2101.70**
extra; additional tread modules. 180-220mm per rise	91.27	2.50	**93.55**	1.65	0.85	**32.55**	-	**126.10**	nr	**141.86**
extra; mini landing to match treads, at upper floor level	224.83	2.50	**230.45**	2.10	1.05	**41.09**	-	**271.53**	nr	**305.47**
1956 diameter; for floor to floor height units between 2160mm and 2640mm; plywood treads with carpet recess fixing base and ground plates to timber	1391.25	2.50	**1426.03**	34.00	17.00	**665.21**	-	**2091.24**	nr	**2352.65**
extra; additional tread modules, 180-220mm per rise	104.62	2.50	**107.24**	1.65	0.85	**32.55**	-	**139.79**	nr	**157.26**
1956 diameter; for floor to floor height units between 2160mm and 2640mm; hardwood treads, plain or with hardwood surround providing carpet recess; fixing base and ground plates to timber	1558.20	2.50	**1597.16**	31.50	16.50	**624.33**	-	**2221.49**	nr	**2499.17**
extra; additional tread modules, 180-220mm per rise	119.09	2.50	**122.07**	1.65	0.85	**32.55**	-	**154.62**	nr	**173.95**
extra; mini landing to match treads, at upper floor level	267.12	2.50	**273.80**	2.10	1.05	**41.09**	-	**314.88**	nr	**354.25**
stairwell balustrading with straight bar infill and PVC handrail cover	96.83	2.50	**99.25**	1.95	1.00	**38.42**	-	**137.67**	m	**154.88**
stairwell balustrading to match staircase including PVC handrail cover	158.05	2.50	**162.00**	1.95	1.00	**38.42**	-	**200.42**	nr	**225.47**
Crescent 'H' range domestic spiral staircases; model 20 'H'; Crescent of Cambridge Ltd.; comprising tread modules complete with centre column, tread support, wooden tread, tread baluster, handrail section, PVC handrail cover, baluster wicket infill panel and tread riser bar, and with attachment brackets for fixing to floor and newel extending from centre column; assembling and bolting together										
2092mm diameter; for floor to floor height units between 2160mm and 2640mm; plywood treads with hardwood edgings; fixing base and ground plates to timber	1268.82	2.50	**1300.54**	31.50	16.00	**618.98**	-	**1919.52**	nr	**2159.45**
extra; additional tread modules, 180 - 220mm per rise	93.49	2.50	**95.83**	1.65	0.85	**32.55**	-	**128.38**	nr	**144.43**
extra; mini landing to match treads, at upper floor level	233.73	2.50	**239.57**	2.10	1.05	**41.09**	-	**280.66**	nr	**315.74**
2092mm diameter; for floor to floor height units between 2160mm and 2640mm; plywood treads with carpet recess; fixing base and ground plates to timber	1435.77	2.50	**1471.66**	31.50	16.00	**618.98**	-	**2090.64**	nr	**2351.97**
extra; additional tread modules, 180 - 220mm per rise	105.74	2.50	**108.38**	1.65	0.85	**32.55**	-	**140.93**	nr	**158.54**
2092mm diameter; for floor to floor height units between 2160mm and 2640mm; hardwood treads, plain or with hardwood surround providing carpet recess; fixing base and ground plates to timber	1586.03	2.50	**1625.68**	31.50	16.00	**618.98**	-	**2244.65**	nr	**2525.23**
extra; additional tread modules, 180 - 220mm per rise	122.43	2.50	**125.49**	1.65	0.85	**32.55**	-	**158.04**	nr	**177.80**

Labour hourly rates: (except Specialists) Craft Operatives £14.21 Labourer £10.71 Rates are national average prices. Refer to REGIONAL VARIATIONS for indicative levels of overall pricing in regions	MATERIALS			LABOUR				RATES		
	Del to Site £	Waste %	Material Cost £	Craft Optve Hrs	Lab Hrs	Labour Cost £	Sunds £	Nett Rate £	Unit	Gross Rate (+12.5%) £
L30: STAIRS/WALKWAYS/BALUSTRADES: METAL STAIRS AND BALUSTRADES Cont.										
Spiral staircases in steel; one coat red oxide primer before delivery to site Cont.										
Crescent 'H' range domestic spiral staircases; model 20 'H'; Crescent of Cambridge Ltd.; comprising tread modules complete with centre column, tread support, wooden tread, tread baluster, handrail section, PVC handrail cover, baluster wicket infill panel and tread riser bar, and with attachment brackets for fixing to floor and newel extending from centre column; assembling and bolting together Cont.										
extra; mini landing to match treads, at upper floor level	274.91	2.50	**281.78**	2.10	1.05	**41.09**	-	**322.87**	nr	**363.23**
stairwell balustrading with straight bar infill and PVC handrail cover	96.83	2.50	**99.25**	1.95	1.00	**38.42**	-	**137.67**	m	**154.88**
stairwell balustrading to match staircase including PVC handrail cover	158.05	2.50	**162.00**	1.95	1.00	**38.42**	-	**200.42**	m	**225.47**
Crescent 'S' range shop, office, factory and domestic spiral staircases; model 15 'S'; Crescent of Cambridge Ltd.; comprising tread modules complete with tread/collar section, baluster bar and section of handrail, and with upper floor attachment details with flanged column tube section, newel and top handrail baluster; assembling and bolting together										
1548mm diameter; for floor to floor height units between 2520mm and 3080mm; plain or ribbed treads; fixing base and ground plates and upper floor attachment to masonry with expanding bolts	1303.32	2.50	**1335.91**	37.00	18.50	**723.91**	-	**2059.81**	nr	**2317.29**
extra; additional tread modules, 180 - 220mm per rise	86.81	2.50	**88.98**	1.65	0.85	**32.55**	-	**121.53**	nr	**136.73**
1548mm diameter; for floor to floor height units between 2520mm and 3080mm; recessed treads; fixing base and ground plates and upper floor attachment to masonry with expanding bolts	1391.25	2.50	**1426.03**	39.00	19.50	**763.04**	-	**2189.07**	nr	**2462.70**
extra; additional tread modules, 180 - 220mm per rise	92.38	2.50	**94.69**	1.65	0.85	**32.55**	-	**127.24**	nr	**143.14**
1548mm diameter; for floor to floor height units between 2520mm and 3080mm; recessed treads with studded rubber tread profile covers; fixing base and ground plates and upper floor attachment to masonry with expanding bolts	178.08	2.50	**182.53**	39.00	19.50	**763.04**	-	**945.57**	nr	**1063.76**
extra; additional tread modules, 180 - 220mm per rise	119.09	2.50	**122.07**	1.65	0.85	**32.55**	-	**154.62**	nr	**173.95**
extra; 825 x 825mm plain or Durbar plate landing at upper floor level, complete with column attachment section, column newel, top handrail, baluster and handrail section, and balustrading complete to one side	395.12	2.50	**404.99**	3.20	1.60	**62.61**	-	**467.60**	nr	**526.05**
extra; 825 x 825mm recessed landing at upper floor level complete with column attachment section, column newel, top handrail, baluster and handrail section, and balustrading complete to one side	430.73	2.50	**441.50**	3.20	1.60	**62.61**	-	**504.11**	nr	**567.12**
extra; stairwell balustrade including handrail balusters and ground rail	82.36	2.50	**84.42**	1.95	1.00	**38.42**	-	**122.84**	m	**138.20**
extra; additional tread balusters	8.90	2.50	**9.13**	0.85	0.45	**16.90**	-	**26.02**	nr	**29.28**
extra; riser bars	7.35	2.50	**7.53**	0.86	0.45	**17.04**	-	**24.57**	nr	**27.64**
extra; PVC handrail cover	4.45	2.50	**4.56**	0.32	0.15	**6.15**	-	**10.72**	m	**12.06**
Crescent 'S' range shop, office, factory and domestic spiral staircases; model 20 'S'; Crescent of Cambridge Ltd.; comprising tread modules complete with tread/collar section, baluster bar and section of handrail, and with upper floor attachment details with flanged column tube section, newel and top handrail baluster; assembling and bolting together										
2072mm diameter; for floor to floor height units between 2520mm and 3080mm; plain or ribbed treads; fixing base and ground plates and upper floor attachment to masonry with expanding bolts	1446.90	2.50	**1483.07**	39.00	19.50	**763.04**	-	**2246.11**	nr	**2526.87**
extra; additional tread modules, 180 - 220mm per rise	96.83	2.50	**99.25**	1.65	0.85	**32.55**	-	**131.80**	nr	**148.28**
2072mm diameter; for floor to floor height units between 2520mm and 3080mm; recessed treads; fixing base and ground plates and upper floor attachment to masonry with expanding bolts	1524.81	2.50	**1562.93**	39.00	19.50	**763.04**	-	**2325.97**	nr	**2616.71**
extra; additional tread modules, 180 - 220mm per rise	102.40	2.50	**104.96**	1.65	0.85	**32.55**	-	**137.51**	nr	**154.69**
2072mm diameter; for floor to floor height units between 2520mm and 3080mm; recessed treads with studded rubber tread profile covers; fixing base and ground plates and upper floor attachment to masonry with expanding bolts	2029.00	2.50	**2079.72**	39.00	19.50	**763.04**	-	**2842.76**	nr	**3198.10**
extra; additional tread modules, 180 - 220mm per rise	138.01	2.50	**141.46**	1.65	0.85	**32.55**	-	**174.01**	nr	**195.76**
extra; 1100 x 1100mm plain or Durbar plate landing at upper floor level, complete with column attachment section, column newel, top handrail, baluster and handrail section, and balustrading complete to one side	471.91	2.50	**483.71**	3.20	1.60	**62.61**	-	**546.32**	nr	**614.61**

Labour hourly rates: (except Specialists) Craft Operatives £14.21 Labourer £10.71 Rates are national average prices. Refer to REGIONAL VARIATIONS for indicative levels of overall pricing in regions	MATERIALS			LABOUR				RATES		
	Del to Site £	Waste %	Material Cost £	Craft Optve Hrs	Lab Hrs	Labour Cost £	Sunds £	Nett Rate £	Unit	Gross Rate (+12.5%) £
L30: STAIRS/WALKWAYS/BALUSTRADES: METAL STAIRS AND BALUSTRADES Cont.										
Spiral staircases in steel; one coat red oxide primer before delivery to site Cont.										
Crescent 'S' range shop, office, factory and domestic spiral staircases; model 20 'S'; Crescent of Cambridge Ltd.; comprising tread modules complete with tread/collar section, baluster bar and section of handrail, and with upper floor attachment details with flanged column tube section, newel and top handrail baluster; assembling and bolting together Cont.										
extra; 1100 x 1100mm recessed landing at upper floor level complete with column attachment section, column newel, top handrail, baluster and handrail section, and balustrading complete to one side	511.98	2.50	**524.78**	3.70	1.85	**72.39**	-	**597.17**	nr	**671.82**
extra; stairwell balustrade including handrail balusters and ground rail	80.14	2.50	**82.14**	1.95	1.00	**38.42**	-	**120.56**	m	**135.63**
extra; additional tread balusters	8.90	2.50	**9.13**	0.85	0.45	**16.90**	-	**26.02**	nr	**29.28**
extra; riser bars	7.35	2.50	**7.53**	0.85	0.45	**16.90**	-	**24.43**	nr	**27.48**
extra; PVC handrail cover	4.45	2.50	**4.56**	0.32	0.16	**6.26**	-	**10.82**	m	**12.18**
Spiral staircases in steel; hot dip galvanized finish										
Crescent 'S' range shop, office, factory and domestic spiral staircases; model 15 'S'; Crescent of Cambridge Ltd.; comprising tread modules complete with tread/collar section, baluster bar and section of handrail, and with upper floor attachment details with flanged column tube section, newel and top handrail baluster; assembling and bolting together										
1548mm diameter; for floor to floor height units between 2520mm and 3080mm; plain or ribbed treads; fixing base and ground plates and upper floor attachment to masonry with expanding bolts	1391.25	2.50	**1426.03**	39.00	19.50	**763.04**	-	**2189.07**	nr	**2462.70**
extra; additional tread modules, 180 - 220mm per rise	93.49	2.50	**95.83**	1.65	0.85	**32.55**	-	**128.38**	nr	**144.43**
1548mm diameter; for floor to floor height units between 2520mm and 3080mm; recessed treads; fixing base and ground plates and upper floor attachment to masonry with expanding bolts	1480.29	2.50	**1517.30**	39.00	19.50	**763.04**	-	**2280.33**	nr	**2565.37**
extra; additional tread modules, 180 - 220mm per rise	100.17	2.50	**102.67**	1.65	0.85	**32.55**	-	**135.22**	nr	**152.13**
extra; 825 x 825mm plain or Durbar plate landing at upper floor level, complete with column attachment section, column newel, top handrail, baluster and handrail section, and balustrading complete to one side	440.75	2.50	**451.77**	3.15	1.60	**61.90**	-	**513.66**	nr	**577.87**
extra; 825 x 825mm recessed landing at upper floor level complete with column attachment section, column newel, top handrail, baluster and handrail section, and balustrading complete to one side	474.14	2.50	**485.99**	3.70	1.85	**72.39**	-	**558.38**	nr	**628.18**
extra; stairwell balustrade including handrail balusters and ground rail	89.71	2.50	**91.95**	1.95	1.00	**38.42**	-	**130.37**	m	**146.67**
extra; additional tread balusters	10.24	2.50	**10.50**	0.85	0.45	**16.90**	-	**27.39**	nr	**30.82**
extra; riser bars	8.96	2.50	**9.18**	0.85	0.45	**16.90**	-	**26.08**	nr	**29.34**
Crescent 'S' range shop, office, factory and domestic spiral staircases; model 20 'S'; Crescent of Cambridge Ltd.; comprising tread modules complete with tread/collar section, baluster bar and section of handrail, and with upper floor attachment details with flanged column tube section, newel and top handrail baluster; assembling and bolting together										
2072mm diameter; for floor to floor height units between 2520mm and 3080mm; plain or ribbed treads; fixing base and ground plates and upper floor attachment to masonry with expanding bolts	1530.38	2.50	**1568.63**	39.00	19.50	**763.04**	-	**2331.67**	nr	**2623.13**
extra; additional tread modules, 180 - 220mm per rise	104.62	2.50	**107.24**	1.65	0.85	**32.55**	-	**139.79**	nr	**157.26**
2072mm diameter; for floor to floor height units between 2520mm and 3080mm; recessed treads; fixing base and ground plates and upper floor attachment to masonry with expanding bolts	1604.95	2.50	**1645.07**	39.00	19.50	**763.04**	-	**2408.10**	nr	**2709.12**
extra; additional tread modules, 180 - 220mm per rise	110.19	2.50	**112.94**	1.65	0.85	**32.55**	-	**145.49**	nr	**163.68**
extra; 1100 x 1100mm plain or Durbar plate landing at upper floor level, complete with column attachment section, column newel, top handrail, baluster and handrail section, and balustrading complete to one side	536.47	2.50	**549.88**	3.20	1.60	**62.61**	-	**612.49**	nr	**689.05**
extra; 1100 x 1100mm recessed landing at upper floor level complete with column attachment section, column newel, top handrail, baluster and handrail section, and balustrading complete to one side	584.33	2.50	**598.93**	3.70	1.85	**72.39**	-	**671.32**	nr	**755.24**
extra; stairwell balustrade including handrail balusters and ground rail	89.71	2.50	**91.95**	1.95	1.00	**38.42**	-	**130.37**	m	**146.67**
extra; additional tread balusters	10.24	2.50	**10.50**	0.85	0.45	**16.90**	-	**27.39**	nr	**30.82**
extra; riser bars	8.96	2.50	**9.18**	0.85	0.45	**16.90**	-	**26.08**	nr	**29.34**

Labour hourly rates: (except Specialists) Craft Operatives £14.21 Labourer £10.71 Rates are national average prices. Refer to REGIONAL VARIATIONS for indicative levels of overall pricing in regions	MATERIALS			LABOUR				RATES		
	Del to Site £	Waste %	Material Cost £	Craft Optve Hrs	Lab Hrs	Labour Cost £	Sunds £	Nett Rate £	Unit	Gross Rate (+12.5%) £
L30: STAIRS/WALKWAYS/BALUSTRADES: METAL STAIRS AND BALUSTRADES Cont.										
Spiral staircases in steel; hot dip galvanized finish Cont.										
Crescent modular spiral fire escape staircases; model 15; Crescent of Cambridge Ltd.; comprising tread modules with plain or ribbed treads, and complete with centre core, spacers, baluster bar and section of handrail, tread riser bars and standard landing with balustrade one side, bars at 115mm centres; assembling and bolting together										
1548mm diameter; ground to first floor assembly, with base and ground plates; for floor to floor heights between 2520mm and 3080mm; fixing base plates and landing to masonry with expanding bolts	2270.52	2.50	2327.28	42.00	21.00	821.73	-	3149.01	nr	3542.64
extra; base extension	146.92	2.50	150.59	2.10	1.05	41.09	-	191.68	nr	215.63
1548mm diameter; intermediate floor assembly; for floor to floor heights between 2520mm and 3080mm; fixing landing to masonry with expanding bolts...	2183.71	2.50	2238.30	42.00	21.00	821.73	-	3060.03	nr	3442.53
1548mm diameter; top floor assembly with two additional lengths of balustrade to landing; for floor to floor heights between 2520mm and 3080mm; fixing landing to masonry with expanding bolts...	2534.30	2.50	2597.66	31.50	16.00	618.98	-	3216.63	nr	3618.71
extra; additional tread modules 180 - 220mm per rise	117.98	2.50	120.93	1.65	0.85	32.55	-	153.48	nr	172.66
extra; shield type infill panel in lieu of additional balusters to tread module	173.63	2.50	177.97	1.05	0.55	20.81	-	198.78	nr	223.63
extra; 'wicket bar' type infill panel in lieu of additional balusters to tread module	267.12	2.50	273.80	1.05	0.55	20.81	-	294.61	nr	331.44
extra; PVC handrail cover	4.50	2.50	4.61	0.32	0.16	6.26	-	10.87	m	12.23
Crescent modular spiral fire escape staircases; model 20; Crescent of Cambridge Ltd.; comprising tread modules with plain or ribbed treads, and complete with centre core, spacers, baluster bar and section of handrail, tread riser bars and standard landing with balustrade one side, bars at 115mm centres; assembling and bolting together										
2072mm diameter; ground to first floor assembly, with base and ground plates; for floor to floor heights between 2520mm and 3080mm; fixing base plates and landing to masonry with expanding bolts	2699.03	2.50	2766.50	42.00	21.00	821.73	-	3588.23	nr	4036.76
extra; base extension	162.50	2.50	166.56	2.10	1.05	41.09	-	207.65	nr	233.60
2072mm diameter; intermediate floor assembly; for floor to floor heights between 2520mm and 3080mm; fixing landing to masonry with expanding bolts...	2609.99	2.50	2675.23	31.50	16.00	618.98	-	3294.21	nr	3705.99
2072mm diameter; top floor assembly with two additional lengths of balustrade to landing; for floor to floor heights between 2520mm and 3080mm; fixing landing to masonry with expanding bolts...	2960.58	2.50	3034.59	31.50	16.00	618.98	-	3653.57	nr	4110.27
extra; additional tread modules 180 - 220mm per rise	116.87	2.50	119.79	1.65	0.85	32.55	-	152.34	nr	171.38
extra; 'wicket bar' type infill panel in lieu of additional balusters to tread module	146.92	2.50	150.59	1.05	0.55	20.81	-	171.40	nr	192.82
extra; PVC handrail cover	4.50	2.50	4.61	0.32	0.16	6.26	-	10.87	m	12.23
Ladders in steel										
Vertical ladders; 65 x 10mm flat stringers; 65 x 10 x 250mm girth stringer brackets, bent once, bolted to stringers; 20mm diameter solid bar rungs welded to stringers; fixing to masonry with expanding bolts										
395mm wide x 3000mm long overall; brackets -6 nr; rungs -7 nr	555.99	2.50	569.89	11.00	11.00	274.12	-	844.01	nr	949.51
395mm wide x 5000mm long overall; brackets -8 nr; rungs -14 nr	555.99	2.50	569.89	13.50	13.50	336.42	-	906.31	nr	1019.60
470mm wide x 3000mm long overall; brackets -6 nr; rungs -7 nr	471.08	2.50	482.86	11.00	11.00	274.12	-	756.98	nr	851.60
470mm wide x 5000mm long overall; brackets -8 nr; rungs -14 nr	825.84	2.50	846.49	13.50	13.50	336.42	-	1182.91	nr	1330.77
395mm wide x 3000mm long overall, stringers rising and returning 900mm above top fixing points; brackets -6 nr; rungs -7 nr	471.08	2.50	482.86	11.00	11.00	274.12	-	756.98	nr	851.60
395mm wide x 5000mm long overall, stringers rising and returning 900mm above top fixing points; brackets -8 nr; rungs -14 nr	825.84	2.50	846.49	13.50	13.50	336.42	-	1182.91	nr	1330.77
470mm wide x 3000mm long overall, stringers rising and returning 900mm above top fixing points; brackets -6 nr; rungs -7 nr	825.84	2.50	846.49	11.00	11.00	274.12	-	1120.61	nr	1260.69
470mm wide x 5000mm long overall, stringers rising and returning 900mm above top fixing points; brackets -8 nr; rungs -14 nr	1000.32	2.50	1025.33	13.50	13.52	336.63	-	1361.96	nr	1532.21
ertical ladders; 65 x 10mm flat stringers; 65 x 10 x 50mm girth stringer brackets, bent once, bolted to tringers; 20mm diameter solid bar rungs welded to tringers; 65 x 10mm flat back hoops 900mm diameter, elded to stringers										
395mm wide x 3000mm long overall; brackets -6 nr; rungs -7 nr; hoops -7 nr	1000.32	2.50	1025.33	16.50	16.50	411.18	-	1436.51	nr	1616.07
395mm wide x 5000mm long overall; brackets -8 nr; rungs -14 nr; hoops -14 nr	1209.69	2.50	1239.93	21.00	21.00	523.32	-	1763.25	nr	1983.65
470mm wide x 3000mm long overall; brackets -6 nr; rungs -7 nr; hoops -7 nr	1139.90	2.50	1168.39	16.50	16.50	411.18	-	1579.57	nr	1777.02

Labour hourly rates: (except Specialists) Craft Operatives £14.21 Labourer £10.71 Rates are national average prices. Refer to REGIONAL VARIATIONS for indicative levels of overall pricing in regions	MATERIALS			LABOUR				RATES		
	Del to Site	Waste	Material Cost	Craft Optve	Lab	Labour Cost	Sunds	Nett Rate	Unit	Gross Rate (+12.5%)
	£	%	£	Hrs	Hrs	£	£	£		£
L30: STAIRS/WALKWAYS/BALUSTRADES: METAL STAIRS AND BALUSTRADES Cont.										
Ladders in steel Cont.										
Vertical ladders; 65 x 10mm flat stringers; 65 x 10 x 250mm girth stringer brackets, bent once, bolted to stringers; 20mm diameter solid bar rungs welded to stringers; 65 x 10mm flat back hoops 900mm diameter, welded to stringers Cont.										
470mm wide x 5000mm long overall; brackets -8 nr; rungs -14 nr; hoops -14 nr	1343.45	2.50	**1377.04**	21.00	21.00	**523.32**	-	**1900.36**	nr	**2137.90**
395mm wide x 3000mm long overall, stringers rising and returning 900mm above top fixing points; brackets -6 nr; rungs -7 nr; hoops -7 nr	959.61	2.50	**983.60**	16.50	16.50	**411.18**	-	**1394.78**	nr	**1569.12**
395mm wide x 5000mm long overall, stringers rising and returning 900mm above top fixing points; brackets -8 nr; rungs -14 nr; hoops -14 nr	1192.24	2.50	**1222.04**	21.00	21.00	**523.32**	-	**1745.36**	nr	**1963.54**
470mm wide x 3000mm long overall, stringers rising and returning 900mm above top fixing points; brackets -6 nr; rungs -7 nr; hoops -7 nr	1087.56	2.50	**1114.74**	16.50	16.50	**411.18**	-	**1525.92**	nr	**1716.66**
470mm wide x 5000mm long overall, stringers rising and returning 900mm above top fixing points; brackets -8 nr; rungs -14 nr; hoops -14 nr	1419.06	2.50	**1454.53**	21.00	21.00	**523.32**	-	**1977.85**	nr	**2225.08**
Ship type ladders; 75 x 10mm flat stringers, shaped and cleated ends to tops and bottoms; 65 x 10 x 350mm girth stringer brackets, bent once, bolted to stringers; 8 x 75mm on plain raised pattern plate treads with 50 x 50 x 6mm shelf angles, bolted to stringers; 25mm diameter solid bar handrails to both sides, with flattened ends bolted to stringers; 32mm diameter x 175mm long solid bar standards with three way ball type joint at intersection with handrail, and flattened ends bolted to stringers; fixing to masonry with expanding bolts										
420mm wide x 3000mm long overall, handrails rising and returning 900mm above top of stringers; brackets -4 nr; treads -12 nr; standards -6 nr	796.77	2.50	**816.68**	11.00	11.00	**274.12**	-	**1090.80**	nr	**1227.15**
420mm wide x 5000mm long overall, handrails rising and returning 900mm above top of stringers; brackets -4 nr; treads -21 nr; standards -10 nr	1122.45	2.50	**1150.51**	13.00	13.00	**323.96**	-	**1474.47**	nr	**1658.78**
470mm wide x 3000mm long overall, handrails rising and returning 900mm above top of stringers; brackets -4 nr; treads -12 nr; standards -6 nr	924.71	2.50	**947.83**	11.00	11.00	**274.12**	-	**1221.95**	nr	**1374.69**
470mm wide x 5000mm long overall, handrails rising and returning 900mm above top of stringers; brackets -4 nr; treads -21 nr; standards -10 nr	1244.58	2.50	**1275.70**	13.00	13.00	**323.96**	-	**1599.66**	nr	**1799.61**
Ship type ladders; 75 x 10mm flat stringers, shaped and cleated ends to tops and bottoms; 65 x 10 x 350mm girth stringer brackets, bent once, bolted to stringers; 8 x 75mm on plain raised pattern plate treads with 50 x 50 x 6mm shelf angles, bolted to stringers; 25mm diameter solid bar handrails to both sides, with flattened ends bolted to stringers; 32mm diameter x 330mm long solid bar standards with three way ball type joint at intersection with handrail, and flattened ends bolted to stringers; fixing to masonry with expanding bolts										
420mm wide x 3000mm long overall, handrails rising and returning 900mm above top of stringers; brackets -4 nr; treads -12 nr; standards -6 nr	796.77	2.50	**816.68**	11.00	11.00	**274.12**	-	**1090.80**	nr	**1227.15**
420mm wide x 5000mm long overall, handrails rising and returning 900mm above top of stringers; brackets -4 nr; treads -21 nr; standards -10 nr	1139.90	2.50	**1168.39**	13.00	13.00	**323.96**	-	**1492.35**	nr	**1678.90**
470mm wide x 3000mm long overall, handrails rising and returning 900mm above top of stringers; brackets -4 nr; treads -12 nr; standards -6 nr	988.69	2.50	**1013.40**	11.00	11.00	**274.12**	-	**1287.52**	nr	**1448.46**
470mm wide x 5000mm long overall, handrails rising and returning 900mm above top of stringers; brackets -4 nr; treads -21 nr; standards -10 nr	1262.03	2.50	**1293.58**	13.00	13.00	**323.96**	-	**1617.54**	nr	**1819.73**
L40: GENERAL GLAZING										
Glass; BS 952, clear float										
3mm thick to wood rebates with BS 544 putty										
not exceeding 0.15m2	40.45	5.00	**42.47**	0.96	0.10	**14.71**	0.35	**57.54**	m2	**64.73**
0.15 - 4.00m2	40.45	5.00	**42.47**	0.48	0.05	**7.36**	0.18	**50.00**	m2	**56.26**
4mm thick to wood rebates with BS 544 putty										
not exceeding 0.15m2	40.45	5.00	**42.47**	0.96	0.10	**14.71**	0.35	**57.54**	m2	**64.73**
0.15 - 4.00m2	40.45	5.00	**42.47**	0.48	0.05	**7.36**	0.18	**50.00**	m2	**56.26**
5mm thick to wood rebates with BS 544 putty										
not exceeding 0.15m2	59.14	5.00	**62.10**	1.08	0.10	**16.42**	0.35	**78.87**	m2	**88.73**
0.15 - 4.00m2	59.14	5.00	**62.10**	0.54	0.05	**8.21**	0.18	**70.48**	m2	**79.29**
6mm thick to wood rebates with BS 544 putty										
not exceeding 0.15m2	59.14	5.00	**62.10**	1.08	0.10	**16.42**	0.35	**78.87**	m2	**88.73**
0.15 - 4.00m2	59.14	5.00	**62.10**	0.54	0.05	**8.21**	0.18	**70.48**	m2	**79.29**
3mm thick to wood rebates with bradded wood beads (included elsewhere) and BS 544 putty										
not exceeding 0.15m2	40.45	5.00	**42.47**	1.20	0.10	**18.12**	0.35	**60.95**	m2	**68.57**

Labour hourly rates: (except Specialists) Craft Operatives £14.21 Labourer £10.71 Rates are national average prices. Refer to REGIONAL VARIATIONS for indicative levels of overall pricing in regions	MATERIALS			LABOUR				RATES		
	Del to Site £	Waste %	Material Cost £	Craft Optve Hrs	Lab Hrs	Labour Cost £	Sunds £	Nett Rate £	Unit	Gross Rate (+12.5%) £
L40: GENERAL GLAZING Cont.										
Glass; BS 952, clear float Cont.										
3mm thick to wood rebates with bradded wood beads (included elsewhere) and BS 544 putty Cont.										
0.15 - 4.00m2	40.45	5.00	**42.47**	0.60	0.05	**9.06**	0.18	**51.71**	m2	**58.17**
4mm thick to wood rebates with bradded wood beads (included elsewhere) and BS 544 putty										
not exceeding 0.15m2	40.45	5.00	**42.47**	1.20	0.10	**18.12**	0.35	**60.95**	m2	**68.57**
0.15 - 4.00m2	40.45	5.00	**42.47**	0.60	0.05	**9.06**	0.18	**51.71**	m2	**58.17**
5mm thick to wood rebates with bradded wood beads (included elsewhere) and BS 544 putty										
not exceeding 0.15m2	59.14	5.00	**62.10**	1.32	0.10	**19.83**	0.35	**82.28**	m2	**92.56**
0.15 - 4.00m2	59.14	5.00	**62.10**	0.66	0.05	**9.91**	0.18	**72.19**	m2	**81.21**
6mm thick to wood rebates with bradded wood beads (included elsewhere) and BS 544 putty										
not exceeding 0.15m2	59.14	5.00	**62.10**	1.32	0.10	**19.83**	0.35	**82.28**	m2	**92.56**
0.15 - 4.00m2	59.14	5.00	**62.10**	0.66	0.05	**9.91**	0.18	**72.19**	m2	**81.21**
3mm thick to wood rebates with screwed wood beads and glazing strip (included elsewhere)										
not exceeding 0.15m2	40.45	5.00	**42.47**	1.68	0.10	**24.94**	0.35	**67.77**	m2	**76.24**
0.15 - 4.00m2	40.45	5.00	**42.47**	0.84	0.05	**12.47**	0.18	**55.12**	m2	**62.01**
4mm thick to wood rebates with screwed wood beads and glazing strip (included elsewhere)										
not exceeding 0.15m2	40.45	5.00	**42.47**	1.68	0.10	**24.94**	0.35	**67.77**	m2	**76.24**
0.15 - 4.00m2	40.45	5.00	**42.47**	0.84	0.05	**12.47**	0.18	**55.12**	m2	**62.01**
5mm thick to wood rebates with screwed wood beads and glazing strip (included elsewhere)										
not exceeding 0.15m2	59.14	5.00	**62.10**	1.80	0.10	**26.65**	0.35	**89.10**	m2	**100.24**
0.15 - 4.00m2	59.14	5.00	**62.10**	0.90	0.05	**13.32**	0.18	**75.60**	m2	**85.05**
6mm thick to wood rebates with screwed wood beads and glazing strip (included elsewhere)										
not exceeding 0.15m2	59.14	5.00	**62.10**	1.80	0.10	**26.65**	0.35	**89.10**	m2	**100.24**
0.15 - 4.00m2	59.14	5.00	**62.10**	0.90	0.05	**13.32**	0.18	**75.60**	m2	**85.05**
3mm thick to metal rebates with metal casement glazing compound										
not exceeding 0.15m2	40.45	5.00	**42.47**	1.08	0.10	**16.42**	0.41	**59.30**	m2	**66.71**
0.15 - 4.00m2	40.45	5.00	**42.47**	0.54	0.05	**8.21**	0.20	**50.88**	m2	**57.24**
4mm thick to metal rebates with metal casement glazing compound										
not exceeding 0.15m2	40.45	5.00	**42.47**	1.08	0.10	**16.42**	0.41	**59.30**	m2	**66.71**
0.15 - 4.00m2	40.45	5.00	**42.47**	0.54	0.05	**8.21**	0.20	**50.88**	m2	**57.24**
5mm thick to metal rebates with metal casement glazing compound										
not exceeding 0.15m2	59.14	5.00	**62.10**	1.20	0.10	**18.12**	0.41	**80.63**	m2	**90.71**
0.15 - 4.00m2	59.14	5.00	**62.10**	0.60	0.05	**9.06**	0.20	**71.36**	m2	**80.28**
6mm thick to metal rebates with metal casement glazing compound										
not exceeding 0.15m2	59.14	5.00	**62.10**	1.20	0.10	**18.12**	0.41	**80.63**	m2	**90.71**
0.15 - 4.00m2	59.14	5.00	**62.10**	0.60	0.05	**9.06**	0.20	**71.36**	m2	**80.28**
3mm thick to metal rebates with clipped metal beads and gaskets (included elsewhere)										
not exceeding 0.15m2	40.45	5.00	**42.47**	1.44	0.10	**21.53**	0.35	**64.36**	m2	**72.40**
0.15 - 4.00m2	40.45	5.00	**42.47**	0.72	0.05	**10.77**	0.18	**53.42**	m2	**60.09**
4mm thick to metal rebates with clipped metal beads and gaskets (included elsewhere)										
not exceeding 0.15m2	40.45	5.00	**42.47**	1.44	0.10	**21.53**	0.35	**64.36**	m2	**72.40**
0.15 - 4.00m2	40.45	5.00	**42.47**	0.72	0.05	**10.77**	0.18	**53.42**	m2	**60.09**
5mm thick to metal rebates with clipped metal beads and gaskets (included elsewhere)										
not exceeding 0.15m2	59.14	5.00	**62.10**	1.56	0.10	**23.24**	0.35	**85.69**	m2	**96.40**
0.15 - 4.00m2	59.14	5.00	**62.10**	0.78	0.05	**11.62**	0.18	**73.89**	m2	**83.13**
6mm thick to metal rebates with clipped metal beads and gaskets (included elsewhere)										
not exceeding 0.15m2	59.14	5.00	**62.10**	1.56	0.10	**23.24**	0.35	**85.69**	m2	**96.40**
0.15 - 4.00m2	59.14	5.00	**62.10**	0.78	0.05	**11.62**	0.18	**73.89**	m2	**83.13**
Glass; BS 952, rough cast										
6mm thick to wood rebates with BS 544 putty										
not exceeding 0.15m2	53.28	5.00	**55.94**	1.08	0.10	**16.42**	0.35	**72.71**	m2	**81.80**
0.15 - 4.00m2	53.28	5.00	**55.94**	0.54	0.05	**8.21**	0.18	**64.33**	m2	**72.37**
6mm thick to wood rebates with bradded wood beads (included elsewhere) and BS 544 putty										
not exceeding 0.15m2	53.28	5.00	**55.94**	1.32	0.10	**19.83**	0.35	**76.12**	m2	**85.64**
0.15 - 4.00m2	53.28	5.00	**55.94**	0.66	0.05	**9.91**	0.18	**66.03**	m2	**74.29**

Labour hourly rates: (except Specialists) Craft Operatives £14.21 Labourer £10.71 Rates are national average prices. Refer to REGIONAL VARIATIONS for indicative levels of overall pricing in regions	MATERIALS			LABOUR				RATES		
	Del to Site £	Waste %	Material Cost £	Craft Optve Hrs	Lab Hrs	Labour Cost £	Sunds £	Nett Rate £	Unit	Gross Rate (+12.5%) £
L40: GENERAL GLAZING Cont.										
Glass; BS 952, rough cast Cont.										
6mm thick to wood rebates with screwed wood beads and glazing strip (included elsewhere)										
not exceeding 0.15m2	53.28	5.00	55.94	1.80	0.10	26.65	0.35	82.95	m2	93.31
0.15 - 4.00m2	53.28	5.00	55.94	0.90	0.05	13.32	0.18	69.44	m2	78.13
6mm thick to metal rebates with metal casement glazing compound										
not exceeding 0.15m2	53.28	5.00	55.94	1.20	0.10	18.12	0.41	74.47	m2	83.78
0.15 - 4.00m2	53.28	5.00	55.94	0.60	0.05	9.06	0.20	65.20	m2	73.35
6mm thick to metal rebates with clipped metal beads and gaskets (included elsewhere)										
not exceeding 0.15m2	53.28	5.00	55.94	1.56	0.10	23.24	0.41	79.59	m2	89.54
0.15 - 4.00m2	53.28	5.00	55.94	0.78	0.05	11.62	0.20	67.76	m2	76.23
Glass; BS 952, Pyroshield clear										
6mm thick to wood rebates with screwed beads and intumescent glazing strip (included elsewhere)										
not exceeding 0.15m2	95.71	5.00	100.50	2.16	0.10	31.76	-	132.26	m2	148.79
not exceeding 0.15m2; aligning panes with adjacent panes	95.71	8.50	103.85	2.40	0.10	35.18	-	139.02	m2	156.40
0.15 - 4.00m2	95.71	5.00	100.50	1.08	0.05	15.88	-	116.38	m2	130.93
0.15 - 4.00m2; aligning panes with adjacent panes	95.71	8.50	103.85	1.20	0.05	17.59	-	121.43	m2	136.61
6mm thick to wood rebates with screwed beads and Pyroglazing strip (included elsewhere)										
not exceeding 0.15m2	95.71	5.00	100.50	2.16	0.10	31.76	-	132.26	m2	148.79
not exceeding 0.15m2; aligning panes with adjacent panes	95.71	8.50	103.85	2.40	0.10	35.18	-	139.02	m2	156.40
0.15 - 4.00m2	95.71	5.00	100.50	1.08	0.05	15.88	-	116.38	m2	130.93
0.15 - 4.00m2; aligning panes with adjacent panes	95.71	8.50	103.85	1.20	0.05	17.59	-	121.43	m2	136.61
6mm thick to wood rebates with screwed beads and Lorient System 90 (included elsewhere)										
not exceeding 0.15m2	95.71	5.00	100.50	2.40	0.10	35.18	-	135.67	m2	152.63
not exceeding 0.15m2; aligning panes with adjacent panes	95.71	8.50	103.85	2.64	0.10	38.59	-	142.43	m2	160.23
0.15 - 4.00m2	95.71	5.00	100.50	1.20	0.05	17.59	-	118.08	m2	132.84
0.15 - 4.00m2; aligning panes with adjacent panes	95.71	8.50	103.85	1.32	0.05	19.29	-	123.14	m2	138.53
6mm thick to metal rebates with screwed metal beads and mild steel flat strips (included elsewhere)										
not exceeding 0.15m2	95.71	5.00	100.50	2.16	0.10	31.76	-	132.26	m2	148.79
not exceeding 0.15m2; aligning panes with adjacent panes	95.71	8.50	103.85	2.40	0.10	35.18	-	139.02	m2	156.40
0.15 - 4.00m2	95.71	5.00	100.50	1.08	0.05	15.88	-	116.38	m2	130.93
0.15 - 4.00m2; aligning panes with adjacent panes	95.71	8.50	103.85	1.20	0.05	17.59	-	121.43	m2	136.61
Glass; BS 952, Pyroshield safety clear										
6mm thick to wood rebates with screwed beads and intumescent glazing strip (included elsewhere)										
not exceeding 0.15m2	112.70	5.00	118.34	2.16	0.10	31.76	-	150.10	m2	168.86
not exceeding 0.15m2; aligning panes with adjacent panes	112.70	8.50	122.28	2.40	0.10	35.18	-	157.45	m2	177.14
0.15 - 4.00m2	112.70	5.00	118.34	1.08	0.05	15.88	-	134.22	m2	150.99
0.15 - 4.00m2; aligning panes with adjacent panes	112.70	8.50	122.28	1.20	0.05	17.59	-	139.87	m2	157.35
6mm thick to wood rebates with screwed beads and Pyroglazing strip (included elsewhere)										
not exceeding 0.15m2	112.70	5.00	118.34	2.16	0.10	31.76	-	150.10	m2	168.86
not exceeding 0.15m2; aligning panes with adjacent panes	112.70	8.50	122.28	2.40	0.10	35.18	-	157.45	m2	177.14
0.15 - 4.00m2	112.70	5.00	118.34	1.08	0.05	15.88	-	134.22	m2	150.99
0.15 - 4.00m2; aligning panes with adjacent panes	112.70	8.50	122.28	1.20	0.05	17.59	-	139.87	m2	157.35
6mm thick to wood rebates with screwed beads and Lorient System 90 (included elsewhere)										
not exceeding 0.15m2	112.70	5.00	118.34	2.40	0.10	35.18	-	153.51	m2	172.70
not exceeding 0.15m2; aligning panes with adjacent panes	112.70	8.50	122.28	2.64	0.10	38.59	-	160.86	m2	180.97
0.15 - 4.00m2	112.70	5.00	118.34	1.20	0.05	17.59	-	135.92	m2	152.91
0.15 - 4.00m2; aligning panes with adjacent panes	112.70	8.50	122.28	1.32	0.05	19.29	-	141.57	m2	159.27
6mm thick to metal rebates with screwed metal beads and mild steel flat strips (included elsewhere)										
not exceeding 0.15m2	112.70	5.00	118.34	2.16	0.10	31.76	-	150.10	m2	168.86
not exceeding 0.15m2; aligning panes with adjacent panes	112.70	8.50	122.28	2.40	0.10	35.18	-	157.45	m2	177.14
0.15 - 4.00m2	112.70	5.00	118.34	1.08	0.05	15.88	-	134.22	m2	150.99
0.15 - 4.00m2; aligning panes with adjacent panes	112.70	8.50	122.28	1.20	0.05	17.59	-	139.87	m2	157.35
Glass; BS 952, Pyroshield texture										
7mm thick to wood rebates with screwed beads and intumescent glazing strip (included elsewhere)										
not exceeding 0.15m2	44.28	5.00	46.49	2.16	0.10	31.76	-	78.26	m2	88.04
not exceeding 0.15m2; aligning panes with adjacent panes	44.28	8.50	48.04	2.40	0.10	35.18	-	83.22	m2	93.62
0.15 - 4.00m2	44.28	5.00	46.49	1.08	0.05	15.88	-	62.38	m2	70.17
0.15 - 4.00m2; aligning panes with adjacent panes	44.28	8.50	48.04	1.20	0.05	17.59	-	65.63	m2	73.84

Labour hourly rates: (except Specialists) Craft Operatives £14.21 Labourer £10.71 Rates are national average prices. Refer to REGIONAL VARIATIONS for indicative levels of overall pricing in regions	MATERIALS			LABOUR				RATES		
	Del to Site £	Waste %	Material Cost £	Craft Optve Hrs	Lab Hrs	Labour Cost £	Sunds £	Nett Rate £	Unit	Gross Rate (+12.5%) £
L40: GENERAL GLAZING Cont.										
Glass; BS 952, Pyroshield texture Cont.										
7mm thick to wood rebates with screwed beads and Pyroglazing strip (included elsewhere)										
not exceeding 0.15m2	44.28	5.00	**46.49**	2.16	0.10	**31.76**	-	**78.26**	m2	**88.04**
not exceeding 0.15m2; aligning panes with adjacent panes	44.28	8.50	**48.04**	2.40	0.10	**35.18**	-	**83.22**	m2	**93.62**
0.15 - 4.00m2	44.28	5.00	**46.49**	1.08	0.05	**15.88**	-	**62.38**	m2	**70.17**
0.15 - 4.00m2; aligning panes with adjacent panes	44.28	8.50	**48.04**	1.20	0.05	**17.59**	-	**65.63**	m2	**73.84**
7mm thick to wood rebates with screwed beads and Lorient System 90 (included elsewhere)										
not exceeding 0.15m2	44.28	5.00	**46.49**	2.40	0.10	**35.18**	-	**81.67**	m2	**91.88**
not exceeding 0.15m2; aligning panes with adjacent panes	44.28	8.50	**48.04**	2.64	0.10	**38.59**	-	**86.63**	m2	**97.46**
0.15 - 4.00m2	44.28	5.00	**46.49**	1.20	0.05	**17.59**	-	**64.08**	m2	**72.09**
0.15 - 4.00m2; aligning panes with adjacent panes	44.28	8.50	**48.04**	1.32	0.05	**19.29**	-	**67.34**	m2	**75.75**
7mm thick to metal rebates with screwed metal beads and mild steel flat strips (included elsewhere)										
not exceeding 0.15m2	44.28	5.00	**46.49**	2.16	0.10	**31.76**	-	**78.26**	m2	**88.04**
not exceeding 0.15m2; aligning panes with adjacent panes	44.28	8.50	**48.04**	2.40	0.10	**35.18**	-	**83.22**	m2	**93.62**
0.15 - 4.00m2	44.28	5.00	**46.49**	1.08	0.05	**15.88**	-	**62.38**	m2	**70.17**
0.15 - 4.00m2; aligning panes with adjacent panes	44.28	8.50	**48.04**	1.20	0.05	**17.59**	-	**65.63**	m2	**73.84**
Glass; BS 952, Pyroshield safety texture										
7mm thick to wood rebates with screwed beads and intumescent glazing strip (included elsewhere)										
not exceeding 0.15m2	57.61	5.00	**60.49**	2.16	0.10	**31.76**	-	**92.26**	m2	**103.79**
not exceeding 0.15m2; aligning panes with adjacent panes	57.61	8.50	**62.51**	2.40	0.10	**35.18**	-	**97.68**	m2	**109.89**
0.15 - 4.00m2	57.61	5.00	**60.49**	1.08	0.05	**15.88**	-	**76.37**	m2	**85.92**
0.15 - 4.00m2; aligning panes with adjacent panes	57.61	8.50	**62.51**	1.20	0.05	**17.59**	-	**80.09**	m2	**90.11**
7mm thick to wood rebates with screwed beads and Pyroglazing strip (included elsewhere)										
not exceeding 0.15m2	57.61	5.00	**60.49**	2.16	0.10	**31.76**	-	**92.26**	m2	**103.79**
not exceeding 0.15m2; aligning panes with adjacent panes	57.61	8.50	**62.51**	2.40	0.10	**35.18**	-	**97.68**	m2	**109.89**
0.15 - 4.00m2	57.61	5.00	**60.49**	1.08	0.05	**15.88**	-	**76.37**	m2	**85.92**
0.15 - 4.00m2; aligning panes with adjacent panes	57.61	8.50	**62.51**	1.20	0.05	**17.59**	-	**80.09**	m2	**90.11**
7mm thick to wood rebates with screwed beads and Lorient System 90 (included elsewhere)										
not exceeding 0.15m2	57.61	5.00	**60.49**	2.40	0.10	**35.18**	-	**95.67**	m2	**107.62**
not exceeding 0.15m2; aligning panes with adjacent panes	57.61	8.50	**62.51**	2.64	0.10	**38.59**	-	**101.09**	m2	**113.73**
0.15 - 4.00m2	57.61	5.00	**60.49**	1.20	0.05	**17.59**	-	**78.08**	m2	**87.84**
0.15 - 4.00m2; aligning panes with adjacent panes	57.61	8.50	**62.51**	1.32	0.05	**19.29**	-	**81.80**	m2	**92.02**
7mm thick to metal rebates with screwed metal beads and mild steel flat strips (included elsewhere)										
not exceeding 0.15m2	57.61	5.00	**60.49**	2.16	0.10	**31.76**	-	**92.26**	m2	**103.79**
not exceeding 0.15m2; aligning panes with adjacent panes	57.61	8.50	**62.51**	2.40	0.10	**35.18**	-	**97.68**	m2	**109.89**
0.15 - 4.00m2	57.61	5.00	**60.49**	1.08	0.05	**15.88**	-	**76.37**	m2	**85.92**
0.15 - 4.00m2; aligning panes with adjacent panes	57.61	8.50	**62.51**	1.20	0.05	**17.59**	-	**80.09**	m2	**90.11**
Glass; BS 952, white patterned										
4mm thick to wood rebates with BS 544 putty										
not exceeding 0.15m2	37.52	5.00	**39.40**	0.96	0.10	**14.71**	0.35	**54.46**	m2	**61.27**
0.15 - 4.00m2	37.52	5.00	**39.40**	0.48	0.05	**7.36**	0.18	**46.93**	m2	**52.79**
6mm thick to wood rebates with BS 544 putty										
not exceeding 0.15m2	56.47	5.00	**59.29**	1.08	0.10	**16.42**	0.35	**76.06**	m2	**85.57**
0.15 - 4.00m2	56.47	5.00	**59.29**	0.54	0.05	**8.21**	0.18	**67.68**	m2	**76.14**
4mm thick to wood rebates with bradded wood beads (included elsewhere) and BS 544 putty										
not exceeding 0.15m2	37.52	5.00	**39.40**	1.20	0.10	**18.12**	0.35	**57.87**	m2	**65.10**
0.15 - 4.00m2	37.52	5.00	**39.40**	0.60	0.05	**9.06**	0.18	**48.63**	m2	**54.71**
6mm thick to wood rebates with bradded wood beads (included elsewhere) and BS 544 putty										
not exceeding 0.15m2	56.47	5.00	**59.29**	1.32	0.10	**19.83**	0.35	**79.47**	m2	**89.41**
0.15 - 4.00m2	56.47	5.00	**59.29**	0.66	0.05	**9.91**	0.18	**69.38**	m2	**78.06**
4mm thick to wood rebates with screwed wood beads and glazing strip (included elsewhere)										
not exceeding 0.15m2	37.52	5.00	**39.40**	1.68	0.10	**24.94**	0.35	**64.69**	m2	**72.78**
0.15 - 4.00m2	37.52	5.00	**39.40**	0.84	0.05	**12.47**	0.18	**52.04**	m2	**58.55**
6mm thick to wood rebates with screwed wood beads and glazing strip (included elsewhere)										
not exceeding 0.15m2	56.47	5.00	**59.29**	1.80	0.10	**26.65**	0.35	**86.29**	m2	**97.08**
0.15 - 4.00m2	56.47	5.00	**59.29**	0.90	0.05	**13.32**	0.18	**72.79**	m2	**81.89**
4mm thick to metal rebates with metal casement glazing compound										
not exceeding 0.15m2	37.52	5.00	**39.40**	1.08	0.10	**16.42**	0.41	**56.22**	m2	**63.25**
0.15 - 4.00m2	37.52	5.00	**39.40**	0.54	0.05	**8.21**	0.20	**47.80**	m2	**53.78**

Labour hourly rates: (except Specialists) Craft Operatives £14.21 Labourer £10.71 Rates are national average prices. Refer to REGIONAL VARIATIONS for indicative levels of overall pricing in regions	MATERIALS			LABOUR			RATES			
	Del to Site £	Waste %	Material Cost £	Craft Optve Hrs	Lab Hrs	Labour Cost £	Sunds £	Nett Rate £	Unit	Gross Rate (+12.5%) £
L40: GENERAL GLAZING Cont.										
Glass; BS 952, white patterned Cont.										
6mm thick to metal rebates with metal casement glazing compound										
not exceeding 0.15m2	56.47	5.00	**59.29**	1.20	0.10	**18.12**	0.41	**77.82**	m2	**87.55**
0.15 - 4.00m2	56.47	5.00	**59.29**	0.60	0.05	**9.06**	0.20	**68.55**	m2	**77.12**
4mm thick to metal rebates with clipped metal beads and gaskets (included elsewhere)										
not exceeding 0.15m2	37.52	5.00	**39.40**	1.44	0.10	**21.53**	0.35	**61.28**	m2	**68.94**
0.15 - 4.00m2	37.52	5.00	**39.40**	0.72	0.05	**10.77**	0.18	**50.34**	m2	**56.63**
6mm thick to metal rebates with clipped metal beads and gaskets (included elsewhere)										
not exceeding 0.15m2	56.47	5.00	**59.29**	1.56	0.10	**23.24**	0.35	**82.88**	m2	**93.24**
0.15 - 4.00m2	56.47	5.00	**59.29**	0.78	0.05	**11.62**	0.18	**71.09**	m2	**79.97**
Glass; BS 952, tinted patterned										
4mm thick to wood rebates with BS 544 putty										
not exceeding 0.15m2	55.40	5.00	**58.17**	0.96	0.10	**14.71**	0.35	**73.23**	m2	**82.39**
0.15 - 4.00m2	55.40	5.00	**58.17**	0.48	0.05	**7.36**	0.18	**65.70**	m2	**73.92**
6mm thick to wood rebates with BS 544 putty										
not exceeding 0.15m2	62.69	5.00	**65.82**	1.08	0.10	**16.42**	0.35	**82.59**	m2	**92.92**
0.15 - 4.00m2	62.69	5.00	**65.82**	0.54	0.05	**8.21**	0.18	**74.21**	m2	**83.49**
4mm thick to wood rebates with bradded wood beads (included elsewhere) and BS 544 putty										
not exceeding 0.15m2	55.40	5.00	**58.17**	1.20	0.10	**18.12**	0.35	**76.65**	m2	**86.23**
0.15 - 4.00m2	55.40	5.00	**58.17**	0.60	0.05	**9.06**	0.18	**67.41**	m2	**75.83**
6mm thick to wood rebates with bradded wood beads (included elsewhere) and BS 544 putty										
not exceeding 0.15m2	62.69	5.00	**65.82**	1.32	0.10	**19.83**	0.35	**86.00**	m2	**96.76**
0.15 - 4.00m2	62.69	5.00	**65.82**	0.66	0.05	**9.91**	0.18	**75.91**	m2	**85.40**
4mm thick to wood rebates with screwed wood beads and glazing strip (included elsewhere)										
not exceeding 0.15m2	55.40	5.00	**58.17**	1.68	0.10	**24.94**	0.35	**83.47**	m2	**93.90**
0.15 - 4.00m2	55.40	5.00	**58.17**	0.84	0.05	**12.47**	0.18	**70.82**	m2	**79.67**
6mm thick to wood rebates with screwed wood beads and glazing strip (included elsewhere)										
not exceeding 0.15m2	62.69	5.00	**65.82**	1.80	0.10	**26.65**	0.35	**92.83**	m2	**104.43**
0.15 - 4.00m2	62.69	5.00	**65.82**	0.90	0.05	**13.32**	0.18	**79.33**	m2	**89.24**
4mm thick to metal rebates with metal casement glazing compound										
not exceeding 0.15m2	55.40	5.00	**58.17**	1.08	0.10	**16.42**	0.40	**74.98**	m2	**84.36**
0.15 - 4.00m2	55.40	5.00	**58.17**	0.54	0.05	**8.21**	0.20	**66.58**	m2	**74.90**
6mm thick to metal rebates with metal casement glazing compound										
not exceeding 0.15m2	62.69	5.00	**65.82**	1.20	0.10	**18.12**	0.40	**84.34**	m2	**94.89**
0.15 - 4.00m2	62.69	5.00	**65.82**	0.60	0.05	**9.06**	0.20	**75.08**	m2	**84.47**
4mm thick to metal rebates with clipped metal beads and gaskets (included elsewhere)										
not exceeding 0.15m2	55.40	5.00	**58.17**	1.44	0.10	**21.53**	0.35	**80.06**	m2	**90.06**
0.15 - 4.00m2	55.40	5.00	**58.17**	0.72	0.05	**10.77**	0.18	**69.11**	m2	**77.75**
6mm thick to metal rebates with clipped metal beads and gaskets (included elsewhere)										
not exceeding 0.15m2	62.69	5.00	**65.82**	1.56	0.10	**23.24**	0.35	**89.42**	m2	**100.59**
0.15 - 4.00m2	62.69	5.00	**65.82**	0.78	0.05	**11.62**	0.18	**77.62**	m2	**87.32**
Patterns available										
4mm Bronze tint										
Autumn; Cotswold; Everglade; Sycamore										
4mm white										
Arctic										
6mm white										
Deep Flemish										
4 and 6mm white										
Autumn; Cotswold; Driftwood; Everglade; Flemish; Linkon; Mayflower; Reeded; Stippolyte										
Glass; BS 952, antisun float; grey										
4mm thick to wood rebates with BS 544 putty										
not exceeding 2400 x 1200mm........	77.23	5.00	**81.09**	0.72	0.12	**11.52**	0.25	**92.86**	m2	**104.47**
6mm thick to wood rebates with BS 544 putty										
not exceeding 5950 x 3150mm........	111.45	5.00	**117.02**	0.78	0.12	**12.37**	0.25	**129.64**	m2	**145.85**
10mm thick to wood rebates with BS 544 putty										
not exceeding 5950 x 3150mm........	199.24	5.00	**209.20**	1.80	0.12	**26.86**	0.32	**236.38**	m2	**265.93**
12mm thick to wood rebates with BS 544 putty										
not exceeding 5950 x 3150mm........	275.01	5.00	**288.76**	2.40	0.12	**35.39**	0.40	**324.55**	m2	**365.11**

Labour hourly rates: (except Specialists) Craft Operatives £14.21 Labourer £10.71 Rates are national average prices. Refer to REGIONAL VARIATIONS for indicative levels of overall pricing in regions	MATERIALS			LABOUR				RATES		
	Del to Site £	Waste %	Material Cost £	Craft Optve Hrs	Lab Hrs	Labour Cost £	Sunds £	Nett Rate £	Unit	Gross Rate (+12.5%) £
L40: GENERAL GLAZING Cont.										
Glass; BS 952, antisun float; grey Cont.										
4mm thick to wood rebates with bradded wood beads (included elsewhere) and BS 544 putty not exceeding 2400 x 1200mm	77.23	5.00	81.09	0.84	0.12	13.22	0.25	94.57	m2	106.39
6mm thick to wood rebates with bradded wood beads (included elsewhere) and BS 544 putty not exceeding 5950 x 3150mm	111.45	5.00	117.02	0.90	0.12	14.07	0.25	131.35	m2	147.77
10mm thick to wood rebates with bradded wood beads (included elsewhere) and BS 544 putty not exceeding 5950 x 3150mm	199.24	5.00	209.20	1.92	0.12	28.57	0.32	238.09	m2	267.85
12mm thick to wood rebates with bradded wood beads (included elsewhere) and BS 544 putty not exceeding 5950 x 3150mm	275.01	5.00	288.76	2.52	0.12	37.09	0.40	326.25	m2	367.03
4mm thick to wood rebates with screwed wood beads and glazing strip (included elsewhere) not exceeding 2400 x 1200mm	77.23	5.00	81.09	1.08	0.12	16.63	-	97.72	m2	109.94
6mm thick to wood rebates with screwed wood beads and glazing strip (included elsewhere) not exceeding 5950 x 3150mm	111.45	5.00	117.02	1.14	0.12	17.48	-	134.51	m2	151.32
10mm thick to wood rebates with screwed wood beads and glazing strip (included elsewhere) not exceeding 5950 x 3150mm	199.24	5.00	209.20	2.04	0.12	30.27	-	239.48	m2	269.41
12mm thick to wood rebates with screwed wood beads and glazing strip (included elsewhere) not exceeding 5950 x 3150mm	275.01	5.00	288.76	2.64	0.12	38.80	-	327.56	m2	368.51
Glass; BS 952, antisun float; bronze										
4mm thick to wood rebates with BS 544 putty not exceeding 2400 x 1200mm	77.23	5.00	81.09	0.72	0.12	11.52	0.25	92.86	m2	104.47
6mm thick to wood rebates with BS 544 putty not exceeding 5950 x 3150mm	111.45	5.00	117.02	0.78	0.12	12.37	0.25	129.64	m2	145.85
10mm thick to wood rebates with BS 544 putty not exceeding 5950 x 3150mm	199.24	5.00	209.20	1.80	0.12	26.86	0.32	236.38	m2	265.93
12mm thick to wood rebates with BS 544 putty not exceeding 5950 x 3150mm	275.01	5.00	288.76	2.40	0.12	35.39	0.40	324.55	m2	365.11
4mm thick to wood rebates with bradded wood beads (included elsewhere) and BS 544 putty not exceeding 2400 x 1200mm	77.23	5.00	81.09	0.84	0.12	13.22	0.25	94.57	m2	106.39
6mm thick to wood rebates with bradded wood beads (included elsewhere) and BS 544 putty not exceeding 5950 x 3150mm	111.45	5.00	117.02	0.90	0.12	14.07	0.25	131.35	m2	147.77
10mm thick to wood rebates with bradded wood beads (included elsewhere) and BS 544 putty not exceeding 5950 x 3150mm	199.24	5.00	209.20	1.92	0.12	28.57	0.32	238.09	m2	267.85
12mm thick to wood rebates with bradded wood beads (included elsewhere) and BS 544 putty not exceeding 5950 x 3150mm	275.01	5.00	288.76	2.52	0.12	37.09	0.40	326.25	m2	367.03
4mm thick to wood rebates with screwed wood beads and glazing strip (included elsewhere) not exceeding 2400 x 1200mm	77.23	5.00	81.09	1.08	0.12	16.63	-	97.72	m2	109.94
6mm thick to wood rebates with screwed wood beads and glazing strip (included elsewhere) not exceeding 5950 x 3150mm	111.45	5.00	117.02	1.14	0.12	17.48	-	134.51	m2	151.32
10mm thick to wood rebates with screwed wood beads and glazing strip (included elsewhere) not exceeding 5950 x 3150mm	199.24	5.00	209.20	2.04	0.12	30.27	-	239.48	m2	269.41
12mm thick to wood rebates with screwed wood beads and glazing strip (included elsewhere) not exceeding 5950 x 3150mm	275.01	5.00	288.76	2.64	0.12	38.80	-	327.56	m2	368.51
Glass; BS 952, antisun float; green										
6mm thick to wood rebates with BS 544 putty not exceeding 3150 x 2050mm	142.84	5.00	149.98	0.78	0.12	12.37	0.25	162.60	m2	182.93
6mm thick to wood rebates with bradded wood beads (included elsewhere) and BS 544 putty not exceeding 3150 x 2050mm	142.84	5.00	149.98	0.90	0.12	14.07	0.25	164.31	m2	184.85
6mm thick to wood rebates with screwed wood beads and glazing strip (included elsewhere) not exceeding 3150 x 2050mm	142.84	5.00	149.98	1.14	0.12	17.48	-	167.47	m2	188.40

Labour hourly rates: (except Specialists) Craft Operatives £14.21 Labourer £10.71 Rates are national average prices. Refer to REGIONAL VARIATIONS for indicative levels of overall pricing in regions	MATERIALS			LABOUR				RATES		
	Del to Site £	Waste %	Material Cost £	Craft Optve Hrs	Lab Hrs	Labour Cost £	Sunds £	Nett Rate £	Unit	Gross Rate (+12.5%) £
L40: GENERAL GLAZING Cont.										
Glass; BS 952, clear float										
10mm thick to wood rebates with screwed wood beads and glazing strip (included elsewhere) not exceeding 5950 x 3150mm	119.79	5.00	**125.78**	2.70	0.12	**39.65**	-	**165.43**	m2	**186.11**
12mm thick to wood rebates with screwed wood beads and glazing strip (included elsewhere) not exceeding 5950 x 3150mm	138.10	5.00	**145.01**	3.30	0.12	**48.18**	-	**193.18**	m2	**217.33**
15mm thick to wood rebates with screwed wood beads and glazing strip (included elsewhere) not exceeding 2950 x 2000mm	229.77	5.00	**241.26**	4.20	0.12	**60.97**	-	**302.23**	m2	**340.00**
19mm thick to wood rebates with screwed wood beads and glazing strip (included elsewhere) not exceeding 2950 x 2000mm	325.47	5.00	**341.74**	4.80	0.12	**69.49**	-	**411.24**	m2	**462.64**
25mm thick to wood rebates with screwed wood beads and glazing strip (included elsewhere) not exceeding 2950 x 2000mm	516.65	5.00	**542.48**	5.40	0.12	**78.02**	-	**620.50**	m2	**698.06**
Glass; BS 952, toughened clear float										
4mm thick to metal rebates with screwed metal beads and gaskets (included elsewhere) not exceeding 2400 x 1300mm	56.94	5.00	**59.79**	0.72	0.12	**11.52**	-	**71.30**	m2	**80.22**
5mm thick to metal rebates with screwed metal beads and gaskets (included elsewhere) not exceeding 2500 x 1520mm	85.45	5.00	**89.72**	0.72	0.12	**11.52**	-	**101.24**	m2	**113.89**
6mm thick to metal rebates with screwed metal beads and gaskets (included elsewhere) not exceeding 2500 x 1520mm	85.45	5.00	**89.72**	0.78	0.12	**12.37**	-	**102.09**	m2	**114.85**
10mm thick to metal rebates with screwed metal beads and gaskets (included elsewhere) not exceeding 2500 x 1520mm	135.53	5.00	**142.31**	1.80	0.12	**26.86**	-	**169.17**	m2	**190.32**
12mm thick to metal rebates with screwed metal beads and gaskets (included elsewhere) not exceeding 2500 x 1520mm	248.19	5.00	**260.60**	2.40	0.12	**35.39**	-	**295.99**	m2	**332.99**
Glass; BS 952, toughened white patterned										
4mm thick to metal rebates with screwed metal beads and gaskets (included elsewhere) not exceeding 2100 x 1300mm	114.79	5.00	**120.53**	0.72	0.12	**11.52**	-	**132.05**	m2	**148.55**
6mm thick to metal rebates with screwed metal beads and gaskets (included elsewhere) not exceeding 2100 x 1300mm	119.86	5.00	**125.85**	0.78	0.12	**12.37**	-	**138.22**	m2	**155.50**
Glass; BS 952, toughened tinted patterned										
4mm thick to metal rebates with screwed metal beads and gaskets (included elsewhere) not exceeding 2100 x 1300mm	114.77	5.00	**120.51**	0.72	0.12	**11.52**	-	**132.02**	m2	**148.53**
6mm thick to metal rebates with screwed metal beads and gaskets (included elsewhere) not exceeding 2100 x 1300mm	119.86	5.00	**125.85**	0.78	0.12	**12.37**	-	**138.22**	m2	**155.50**
Patterns available										
4mm tinted Everglade										
4mm white Reeded										
4mm tinted Autumn; Cotswold										
4 and 6mm white Autumn; Cotswold; Driftwood; Everglade; Flemish; Linkon; Mayflower; Stippolyte; Sycamore										
Glass; BS 952, toughened antisun float; grey										
4mm thick to metal rebates with screwed metal beads and gaskets (included elsewhere) not exceeding 2100 x 1250mm	104.41	5.00	**109.63**	0.72	0.12	**11.52**	-	**121.15**	m2	**136.29**
6mm thick to metal rebates with screwed metal beads and gaskets (included elsewhere) not exceeding 2500 x 1520mm	108.96	5.00	**114.41**	0.78	0.12	**12.37**	-	**126.78**	m2	**142.62**
10mm thick to metal rebates with screwed metal beads and gaskets (included elsewhere) not exceeding 2500 x 1520mm	215.55	5.00	**226.33**	1.80	0.12	**26.86**	-	**253.19**	m2	**284.84**
12mm thick to metal rebates with screwed metal beads and gaskets (included elsewhere) not exceeding 2500 x 1520mm	283.71	5.00	**297.90**	2.40	0.12	**35.39**	-	**333.28**	m2	**374.95**

Labour hourly rates: (except Specialists) Craft Operatives £14.21 Labourer £10.71 Rates are national average prices. Refer to REGIONAL VARIATIONS for indicative levels of overall pricing in regions	MATERIALS			LABOUR				RATES		
	Del to Site £	Waste %	Material Cost £	Craft Optve Hrs	Lab Hrs	Labour Cost £	Sunds £	Nett Rate £	Unit	Gross Rate (+12.5%) £
L40: GENERAL GLAZING Cont.										
Glass; BS 952, toughened antisun float; bronze										
4mm thick to metal rebates with screwed metal beads and gaskets (included elsewhere)										
not exceeding 2100 x 1250mm	104.41	5.00	109.63	0.72	0.12	11.52	-	121.15	m2	136.29
6mm thick to metal rebates with screwed metal beads and gaskets (included elsewhere)										
not exceeding 2500 x 1520mm	108.96	5.00	114.41	0.78	0.12	12.37	-	126.78	m2	142.62
10mm thick to metal rebates with screwed metal beads and gaskets (included elsewhere)										
not exceeding 2500 x 1520mm	215.62	5.00	226.40	1.80	0.12	26.86	-	253.26	m2	284.92
12mm thick to metal rebates with screwed metal beads and gaskets (included elsewhere)										
not exceeding 2500 x 1520mm	283.77	5.00	297.96	2.40	0.12	35.39	-	333.35	m2	375.02
Glass; BS 952, laminated safety, clear float										
4.4mm thick to metal rebates with screwed metal beads and gaskets (included elsewhere)										
not exceeding 2100 x 1200mm	75.27	5.00	79.03	1.92	0.12	28.57	-	107.60	m2	121.05
6.4mm thick to metal rebates with screwed metal beads and gaskets (included elsewhere)										
not exceeding 3210 x 2000mm	63.11	5.00	66.27	2.40	0.12	35.39	-	101.65	m2	114.36
Glass; BS 952, laminated anti-bandit, clear float										
7.5mm thick to metal rebates with screwed metal beads and gaskets (included elsewhere)										
not exceeding 3210 x 2000mm	123.56	5.00	129.74	3.00	0.12	43.92	-	173.65	m2	195.36
9.5mm thick to metal rebates with screwed metal beads and gaskets (included elsewhere)										
not exceeding 3600 x 2500mm	154.54	5.00	162.27	3.30	0.12	48.18	-	210.45	m2	236.75
11.5mm thick to metal rebates with screwed metal beads and gaskets (included elsewhere)										
not exceeding 3180 x 2000mm	176.44	5.00	185.26	3.90	0.12	56.70	-	241.97	m2	272.21
not exceeding 4500 x 2500mm	211.84	5.00	222.43	3.90	0.12	56.70	-	279.14	m2	314.03
Materials resembling glass; UVA stabilised polycarbonate sheet latex paper masked both sides										
3mm thick to metal rebates with screwed metal beads and gaskets (included elsewhere)										
standard grade	40.93	5.00	42.98	1.20	0.12	18.34	-	61.31	m2	68.98
4mm thick to metal rebates with screwed metal beads and gaskets (included elsewhere)										
standard grade	54.58	5.00	57.31	1.20	0.12	18.34	-	75.65	m2	85.10
5mm thick to metal rebates with screwed metal beads and gaskets (included elsewhere)										
standard grade	68.22	5.00	71.63	1.20	0.12	18.34	-	89.97	m2	101.21
6mm thick to metal rebates with screwed metal beads and gaskets (included elsewhere)										
standard grade	81.90	5.00	86.00	1.38	0.12	20.90	-	106.89	m2	120.25
8mm thick to metal rebates with screwed metal beads and gaskets (included elsewhere)										
standard grade	109.16	5.00	114.62	1.62	0.12	24.31	-	138.92	m2	156.29
9.5mm thick to metal rebates with screwed metal beads and gaskets (included elsewhere)										
standard grade	129.62	5.00	136.10	1.80	0.12	26.86	-	162.96	m2	183.33
3mm thick to metal rebates with screwed metal beads and gaskets (included elsewhere)										
abrasion resistant hard coated grade	105.49	5.00	110.76	1.20	0.12	18.34	-	129.10	m2	145.24
4mm thick to metal rebates with screwed metal beads and gaskets (included elsewhere)										
abrasion resistant hard coated grade	123.64	5.00	129.82	1.20	0.12	18.34	-	148.16	m2	166.68
mm thick to metal rebates with screwed metal beads and gaskets (included elsewhere)										
abrasion resistant hard coated grade	142.53	5.00	149.66	1.20	0.12	18.34	-	167.99	m2	188.99
mm thick to metal rebates with screwed metal beads nd gaskets (included elsewhere)										
abrasion resistant hard coated grade	167.01	5.00	175.36	1.38	0.12	20.90	-	196.26	m2	220.79
mm thick to metal rebates with screwed metal beads nd gaskets (included elsewhere)										
abrasion resistant hard coated grade	215.44	5.00	226.21	1.62	0.12	24.31	-	250.52	m2	281.83
.5mm thick to metal rebates with screwed metal beads nd gaskets (included elsewhere)										
abrasion resistant hard coated grade	252.13	5.00	264.74	1.80	0.12	26.86	-	291.60	m2	328.05

Labour hourly rates: (except Specialists) Craft Operatives £14.21 Labourer £10.71 Rates are national average prices. Refer to REGIONAL VARIATIONS for indicative levels of overall pricing in regions	MATERIALS			LABOUR				RATES		
	Del to Site	Waste	Material Cost	Craft Optve	Lab	Labour Cost	Sunds	Nett Rate	Unit	Gross Rate (+12.5%)
	£	%	£	Hrs	Hrs	£	£	£		£
L40: GENERAL GLAZING Cont.										
Materials resembling glass; Meshlite vandal resistant glazing; Georgian, clear, smooth both faces or crinkle one face										
3mm thick to wood rebates with screwed wood beads (included elsewhere) and BS 544 putty										
general purpose grade	30.66	5.00	**32.19**	1.15	0.12	**17.63**	0.17	**49.98**	m2	**56.23**
fire retardant grade, class 2	36.98	5.00	**38.83**	1.15	0.12	**17.63**	0.17	**56.62**	m2	**63.70**
fire retardant grade, class 0	43.43	5.00	**45.60**	1.15	0.12	**17.63**	0.17	**63.39**	m2	**71.32**
4mm thick to wood rebates with screwed wood beads (included elsewhere) and BS 544 putty										
general purpose grade	39.61	5.00	**41.59**	1.15	0.12	**17.63**	0.17	**59.38**	m2	**66.80**
fire retardant grade, class 2	47.83	5.00	**50.22**	1.15	0.12	**17.63**	0.17	**68.01**	m2	**76.51**
fire retardant grade, class 0	56.31	5.00	**59.13**	1.15	0.12	**17.63**	0.17	**76.92**	m2	**86.53**
6mm thick to wood rebates with screwed wood beads (included elsewhere) and BS 544 putty										
general purpose grade	52.95	5.00	**55.60**	1.30	0.12	**19.76**	0.17	**75.52**	m2	**84.96**
fire retardant grade, class 2	67.10	5.00	**70.46**	1.30	0.12	**19.76**	0.17	**90.38**	m2	**101.68**
fire retardant grade, class 0	81.49	5.00	**85.56**	1.30	0.12	**19.76**	0.17	**105.49**	m2	**118.67**
3mm thick to metal rebates with screwed metal beads and gaskets (included elsewhere)										
general purpose grade	30.66	5.00	**32.19**	1.15	0.12	**17.63**	-	**49.82**	m2	**56.05**
fire retardant grade, class 2	36.98	5.00	**38.83**	1.15	0.12	**17.63**	-	**56.46**	m2	**63.51**
fire retardant grade, class 0	43.43	5.00	**45.60**	1.15	0.12	**17.63**	-	**63.23**	m2	**71.13**
4mm thick to metal rebates with screwed metal beads and gaskets (included elsewhere)										
general purpose grade	39.61	5.00	**41.59**	1.15	0.12	**17.63**	-	**59.22**	m2	**66.62**
fire retardant grade, class 2	47.83	5.00	**50.22**	1.15	0.12	**17.63**	-	**67.85**	m2	**76.33**
fire retardant grade, class 0	56.31	5.00	**59.13**	1.15	0.12	**17.63**	-	**76.75**	m2	**86.35**
6mm thick to metal rebates with screwed metal beads and gaskets (included elsewhere)										
general purpose grade	52.95	5.00	**55.60**	1.30	0.12	**19.76**	-	**75.36**	m2	**84.78**
fire retardant grade, class 2	67.10	5.00	**70.46**	1.30	0.12	**19.76**	-	**90.21**	m2	**101.49**
fire retardant grade, class 0	81.49	5.00	**85.56**	1.30	0.12	**19.76**	-	**105.32**	m2	**118.49**
Materials resembling glass; Meshlite Vandal resistant glazing; plain, opaque colours or clear, smooth both faces or crinkle one face										
2mm thick to wood rebates with screwed wood beads (included elsewhere) and BS 544 putty										
general purpose grade	23.22	5.00	**24.38**	1.15	0.12	**17.63**	0.17	**42.17**	m2	**47.44**
3mm thick to wood rebates with screwed wood beads (included elsewhere) and BS 544 putty										
general purpose grade	27.79	5.00	**29.18**	1.15	0.12	**17.63**	0.17	**46.97**	m2	**52.84**
fire retardant grade, class 2	34.95	5.00	**36.70**	1.15	0.12	**17.63**	0.17	**54.49**	m2	**61.30**
fire retardant grade, class 0	42.10	5.00	**44.21**	1.15	0.12	**17.63**	0.17	**62.00**	m2	**69.75**
4mm thick to wood rebates with screwed wood beads (included elsewhere) and BS 544 putty										
general purpose grade	36.98	5.00	**38.83**	1.15	0.12	**17.63**	0.17	**56.62**	m2	**63.70**
fire retardant grade, class 2	45.42	5.00	**47.69**	1.15	0.12	**17.63**	0.17	**65.48**	m2	**73.67**
fire retardant grade, class 0	54.93	5.00	**57.68**	1.15	0.12	**17.63**	0.17	**75.47**	m2	**84.90**
6mm thick to wood rebates with screwed wood beads (included elsewhere) and BS 544 putty										
general purpose grade	48.83	5.00	**51.27**	1.30	0.12	**19.76**	0.17	**71.19**	m2	**80.09**
fire retardant grade, class 2	64.38	5.00	**67.60**	1.30	0.12	**19.76**	0.17	**87.52**	m2	**98.46**
fire retardant grade, class 0	79.94	5.00	**83.94**	1.30	0.12	**19.76**	0.17	**103.86**	m2	**116.84**
2mm thick to metal rebates with screwed metal beads and gaskets (included elsewhere)										
general purpose grade; plain	23.22	5.00	**24.38**	1.15	0.12	**17.63**	-	**42.01**	m2	**47.26**
3mm thick to metal rebates with screwed metal beads and gaskets (included elsewhere)										
general purpose grade	27.79	5.00	**29.18**	1.15	0.12	**17.63**	-	**46.81**	m2	**52.66**
fire retardant grade, class 2	34.95	5.00	**36.70**	1.15	0.12	**17.63**	-	**54.32**	m2	**61.11**
fire retardant grade, class 0	42.10	5.00	**44.21**	1.15	0.12	**17.63**	-	**61.83**	m2	**69.56**
4mm thick to metal rebates with screwed metal beads and gaskets (included elsewhere)										
general purpose grade	36.98	5.00	**38.83**	1.15	0.12	**17.63**	-	**56.46**	m2	**63.51**
fire retardant grade, class 2	45.42	5.00	**47.69**	1.15	0.12	**17.63**	-	**65.32**	m2	**73.48**
fire retardant grade, class 0	54.93	5.00	**57.68**	1.15	0.12	**17.63**	-	**75.30**	m2	**84.72**
6mm thick to metal rebates with screwed metal beads and gaskets (included elsewhere)										
general purpose grade	48.83	5.00	**51.27**	1.30	0.12	**19.76**	-	**71.03**	m2	**79.91**
fire retardant grade, class 2	64.38	5.00	**67.60**	1.30	0.12	**19.76**	-	**87.36**	m2	**98.28**
fire retardant grade, class 0	79.94	5.00	**83.94**	1.30	0.12	**19.76**	-	**103.70**	m2	**116.66**

Labour hourly rates: (except Specialists) Craft Operatives £14.21 Labourer £10.71 Rates are national average prices. Refer to REGIONAL VARIATIONS for indicative levels of overall pricing in regions	MATERIALS			LABOUR				RATES		
	Del to Site £	Waste %	Material Cost £	Craft Optve Hrs	Lab Hrs	Labour Cost £	Sunds £	Nett Rate £	Unit	Gross Rate (+12.5%) £
L40: GENERAL GLAZING Cont.										
Materials resembling glass; Meshlite Vandal resistant glazing; diamond, clear, smooth both faces or crinkle one face										
3mm thick to wood rebates with screwed wood beads (included elsewhere) and BS 544 putty										
general purpose grade	31.24	5.00	**32.80**	1.15	0.12	**17.63**	0.17	**50.59**	m2	**56.92**
fire retardant grade, class 2	36.95	5.00	**38.80**	1.15	0.12	**17.63**	0.17	**56.59**	m2	**63.66**
fire retardant grade, class 0	42.79	5.00	**44.93**	1.15	0.12	**17.63**	0.17	**62.72**	m2	**70.56**
4mm thick to wood rebates with screwed wood beads (included elsewhere) and BS 544 putty										
general purpose grade	40.89	5.00	**42.93**	1.15	0.12	**17.63**	0.17	**60.73**	m2	**68.32**
fire retardant grade, class 2	48.39	5.00	**50.81**	1.15	0.12	**17.63**	0.17	**68.60**	m2	**77.18**
fire retardant grade, class 0	55.88	5.00	**58.67**	1.15	0.12	**17.63**	0.17	**76.47**	m2	**86.02**
6mm thick to wood rebates with screwed wood beads (included elsewhere) and BS 544 putty										
general purpose grade	53.77	5.00	**56.46**	1.30	0.12	**19.76**	0.17	**76.38**	m2	**85.93**
fire retardant grade, class 2	67.50	5.00	**70.88**	1.30	0.12	**19.76**	0.17	**90.80**	m2	**102.15**
fire retardant grade, class 0	80.93	5.00	**84.98**	1.30	0.12	**19.76**	0.17	**104.90**	m2	**118.01**
3mm thick to metal rebates with screwed metal beads and gaskets (included elsewhere)										
general purpose grade	31.24	5.00	**32.80**	1.15	0.12	**17.63**	-	**50.43**	m2	**56.73**
fire retardant grade, class 2	36.95	5.00	**38.80**	1.15	0.12	**17.63**	-	**56.42**	m2	**63.48**
fire retardant grade, class 0	42.79	5.00	**44.93**	1.15	0.12	**17.63**	-	**62.56**	m2	**70.38**
4mm thick to metal rebates with screwed metal beads and gaskets (included elsewhere)										
general purpose grade	40.89	5.00	**42.93**	1.15	0.12	**17.63**	-	**60.56**	m2	**68.13**
fire retardant grade, class 2	48.39	5.00	**50.81**	1.15	0.12	**17.63**	-	**68.44**	m2	**76.99**
fire retardant grade, class 0	55.88	5.00	**58.67**	1.15	0.12	**17.63**	-	**76.30**	m2	**85.84**
6mm thick to metal rebates with screwed metal beads and gaskets (included elsewhere)										
general purpose grade	53.77	5.00	**56.46**	1.30	0.12	**19.76**	-	**76.22**	m2	**85.74**
fire retardant grade, class 2	67.50	5.00	**70.88**	1.30	0.12	**19.76**	-	**90.63**	m2	**101.96**
fire retardant grade, class 0	80.93	5.00	**84.98**	1.30	0.12	**19.76**	-	**104.73**	m2	**117.83**
Factory made double glazed hermetically sealed units										
Two panes BS 952, clear float 3 or 4mm thick; to metal rebates with screwed metal beads and gaskets (included elsewhere)										
521mm wide x 421mm high	19.08	5.00	**20.03**	1.15	-	**16.34**	-	**36.38**	nr	**40.92**
521mm wide x 621mm high	28.15	5.00	**29.56**	1.45	-	**20.60**	-	**50.16**	nr	**56.43**
740mm wide x 740mm high	47.62	5.00	**50.00**	2.00	-	**28.42**	-	**78.42**	nr	**88.22**
848mm wide x 848mm high	62.55	5.00	**65.68**	2.30	-	**32.68**	-	**98.36**	nr	**110.66**
1048mm wide x 1048mm high	95.53	5.00	**100.31**	2.90	-	**41.21**	-	**141.52**	nr	**159.20**
1148mm wide x 1248mm high	124.61	5.00	**130.84**	3.45	-	**49.02**	-	**179.87**	nr	**202.35**
Two panes BS 952, clear float 5 or 6mm thick; to metal rebates with screwed metal beads and gaskets (included elsewhere)										
521mm wide x 421mm high	28.09	5.00	**29.49**	1.25	-	**17.76**	-	**47.26**	nr	**53.16**
521mm wide x 621mm high	41.45	5.00	**43.52**	1.60	-	**22.74**	-	**66.26**	nr	**74.54**
740mm wide x 740mm high	70.17	5.00	**73.68**	2.20	-	**31.26**	-	**104.94**	nr	**118.06**
848mm wide x 848mm high	92.14	5.00	**96.75**	2.50	-	**35.53**	-	**132.27**	nr	**148.81**
1048mm wide x 1048mm high	113.87	5.00	**119.56**	3.20	-	**45.47**	-	**165.04**	nr	**185.66**
1148mm wide x 1248mm high	168.86	5.00	**177.30**	3.80	-	**54.00**	-	**231.30**	nr	**260.21**
Inner pane BS 952, clear float 4mm thick; outer pane BS 952, white patterned 4mm thick; to metal rebates with screwed metal beads and gaskets (included elsewhere)										
521mm wide x 421mm high	23.43	5.00	**24.60**	1.15	-	**16.34**	-	**40.94**	nr	**46.06**
521mm wide x 621mm high	34.40	5.00	**36.12**	1.45	-	**20.60**	-	**56.72**	nr	**63.82**
740mm wide x 740mm high	58.22	5.00	**61.13**	2.00	-	**28.42**	-	**89.55**	nr	**100.74**
848mm wide x 848mm high	76.44	5.00	**80.26**	2.30	-	**32.68**	-	**112.95**	nr	**127.06**
1048mm wide x 1048mm high	94.48	5.00	**99.20**	2.90	-	**41.21**	-	**140.41**	nr	**157.96**
1148mm wide x 1248mm high	152.32	5.00	**159.94**	3.45	-	**49.02**	-	**208.96**	nr	**235.08**
Inner pane BS 952, clear float 4mm thick; outer pane BS 952, white patterned 6mm thick; to metal rebates with screwed metal beads and gaskets (included elsewhere)										
521mm wide x 421mm high	27.92	5.00	**29.32**	1.25	-	**17.76**	-	**47.08**	nr	**52.96**
521mm wide x 621mm high	41.19	5.00	**43.25**	1.60	-	**22.74**	-	**65.99**	nr	**74.23**
740mm wide x 740mm high	69.71	5.00	**73.20**	2.20	-	**31.26**	-	**104.46**	nr	**117.51**
848mm wide x 848mm high	91.54	5.00	**96.12**	2.50	-	**35.53**	-	**131.64**	nr	**148.10**
1048mm wide x 1048mm high	113.13	5.00	**118.79**	3.20	-	**45.47**	-	**164.26**	nr	**184.79**
1148mm wide x 1248mm high	182.38	5.00	**191.50**	3.80	-	**54.00**	-	**245.50**	nr	**276.18**
Polyester window films; Durable Berkeley Company Ltd.,										
3M Scotchshield Safety films; applying to glass										
type SH4CLL, optically clear	-	-	**Spclist**	-	-	**Spclist**	-	**28.92**	m2	**32.54**
type SH4S1L, combination solar/safety	-	-	**Spclist**	-	-	**Spclist**	-	**34.70**	m2	**39.04**
3M Scotchtint Solar Control films; applying to glass										
type P18, silver	-	-	**Spclist**	-	-	**Spclist**	-	**31.81**	m2	**35.79**
type RE15S1X, external	-	-	**Spclist**	-	-	**Spclist**	-	**40.48**	m2	**45.54**
type RE35NEARL, neutral	-	-	**Spclist**	-	-	**Spclist**	-	**31.81**	m2	**35.79**

Labour hourly rates: (except Specialists) Craft Operatives £14.21 Labourer £10.71 Rates are national average prices. Refer to REGIONAL VARIATIONS for indicative levels of overall pricing in regions	MATERIALS			LABOUR				RATES		
	Del to Site £	Waste %	Material Cost £	Craft Optve Hrs	Lab Hrs	Labour Cost £	Sunds £	Nett Rate £	Unit	Gross Rate (+12.5%) £
L40: GENERAL GLAZING Cont.										
Polyester window films; Durable Berkeley Company Ltd., Cont.										
3M Scotchtint Plus All Seasons insulating films; applying to glass										
type LE20 S1AR (silver)............................	-	-	**Spclist**	-	-	**Spclist**	-	**40.48**	m2	**45.54**
type LE35 AMARL (bronze)...........................	-	-	**Spclist**	-	-	**Spclist**	-	**40.48**	m2	**45.54**
type LE50 AMARL (bronze)...........................	-	-	**Spclist**	-	-	**Spclist**	-	**40.48**	m2	**45.54**
Glass doors; glass, BS 952, toughened clear, polished plate; fittings finish BMA, satin chrome or polished chrome; prices include the provision of floor springs but exclude handles										
12mm thick										
750mm wide x 2150mm high...........................	1595.00	-	**1595.00**	9.00	9.00	**224.28**	-	**1819.28**	nr	**2046.69**
762mm wide x 2134mm high...........................	1507.00	-	**1507.00**	9.00	9.00	**224.28**	-	**1731.28**	nr	**1947.69**
800mm wide x 2150mm high...........................	1650.00	-	**1650.00**	9.00	9.00	**224.28**	-	**1874.28**	nr	**2108.57**
838mm wide x 2134mm high...........................	1583.00	-	**1583.00**	9.00	9.00	**224.28**	-	**1807.28**	nr	**2033.19**
850mm wide x 2150mm high...........................	1698.00	-	**1698.00**	9.00	9.00	**224.28**	-	**1922.28**	nr	**2162.57**
900mm wide x 2150mm high...........................	1755.00	-	**1755.00**	9.00	9.00	**224.28**	-	**1979.28**	nr	**2226.69**
914mm wide x 2134mm high...........................	1649.00	-	**1649.00**	9.00	9.00	**224.28**	-	**1873.28**	nr	**2107.44**
950mm wide x 2150mm high...........................	1849.00	-	**1849.00**	10.00	10.00	**249.20**	-	**2098.20**	nr	**2360.48**
1000mm wide x 2150mm high..........................	1977.00	-	**1977.00**	10.00	10.00	**249.20**	-	**2226.20**	nr	**2504.48**
1100mm wide x 2150mm high..........................	2208.00	-	**2208.00**	11.00	11.00	**274.12**	-	**2482.12**	nr	**2792.39**
1200mm wide x 2150mm high..........................	2313.00	-	**2313.00**	11.00	11.00	**274.12**	-	**2587.12**	nr	**2910.51**
Drilling										
Hole through sheet or float glass; not exceeding 6mm thick										
6 - 15mm diameter	3.98	-	**3.98**	-	-	-	-	**3.98**	nr	**4.48**
16 - 38mm diameter	5.62	-	**5.62**	-	-	-	-	**5.62**	nr	**6.32**
exceeding 38mm diameter............................	11.34	-	**11.34**	-	-	-	-	**11.34**	nr	**12.76**
Hole through sheet or float glass; not exceeding 10mm thick										
6 - 15mm diameter	5.15	-	**5.15**	-	-	-	-	**5.15**	nr	**5.80**
16 - 38mm diameter	7.95	-	**7.95**	-	-	-	-	**7.95**	nr	**8.94**
exceeding 38mm diameter............................	13.77	-	**13.77**	-	-	-	-	**13.77**	nr	**15.49**
Hole through sheet or float glass; not exceeding 12mm thick										
6 - 15mm diameter	6.38	-	**6.38**	-	-	-	-	**6.38**	nr	**7.18**
16 - 38mm diameter	9.03	-	**9.03**	-	-	-	-	**9.03**	nr	**10.16**
exceeding 38mm diameter............................	16.34	-	**16.34**	-	-	-	-	**16.34**	nr	**18.38**
Hole through sheet or float glass; not exceeding 19mm thick										
6 - 15mm diameter	7.98	-	**7.98**	-	-	-	-	**7.98**	nr	**8.98**
16 - 38mm diameter	11.34	-	**11.34**	-	-	-	-	**11.34**	nr	**12.76**
exceeding 38mm diameter............................	20.17	-	**20.17**	-	-	-	-	**20.17**	nr	**22.69**
Hole through sheet or float glass; not exceeding 25mm thick										
6 - 15mm diameter	10.03	-	**10.03**	-	-	-	-	**10.03**	nr	**11.29**
16 - 38mm diameter	14.22	-	**14.22**	-	-	-	-	**14.22**	nr	**15.99**
exceeding 38mm diameter............................	25.06	-	**25.06**	-	-	-	-	**25.06**	nr	**28.19**
For wired and laminated glass add 50%										
For countersunk holes add 33 1/3%										
Bedding edges of panes										
Wash leather strips										
to edges of 3mm thick glass or the like..........	0.58	5.00	**0.61**	0.06	-	**0.85**	-	**1.46**	m	**1.64**
to edges of 6mm thick glass or the like..........	0.70	5.00	**0.74**	0.07	-	**0.99**	-	**1.73**	m	**1.95**
Rubber glazing strips										
to edges of 3mm thick glass or the like..........	0.54	5.00	**0.57**	0.06	-	**0.85**	-	**1.42**	m	**1.60**
to edges of 6mm thick glass or the like..........	0.65	5.00	**0.68**	0.07	-	**0.99**	-	**1.68**	m	**1.89**
Hacking out existing glass; preparing for re-glazing										
Float glass										
wood rebates	-	-	-	0.44	0.44	**10.96**	0.19	**11.15**	m	**12.55**
wood rebates and screwed wood beads; storing beads for re-use	-	-	-	0.52	0.52	**12.96**	0.19	**13.15**	m	**14.79**
metal rebates	-	-	-	0.46	0.46	**11.46**	0.19	**11.65**	m	**13.11**
metal rebates and screwed metal beads; storing beads for re-use	-	-	-	0.55	0.55	**13.71**	0.19	**13.89**	m	**15.63**
Float glass behind guard bars in position										
wood rebates and screwed wood beads; storing beads for re-use	-	-	-	0.85	0.85	**21.18**	0.19	**21.37**	m	**24.04**
metal rebates and screwed metal beads; storing beads for re-use	-	-	-	0.90	0.90	**22.43**	0.19	**22.62**	m	**25.44**

SURFACE FINISHES

Labour hourly rates: (except Specialists) Craft Operatives £14.21 Labourer £10.71 Rates are national average prices. Refer to REGIONAL VARIATIONS for indicative levels of overall pricing in regions	MATERIALS			LABOUR			RATES			
	Del to Site £	Waste %	Material Cost £	Craft Optve Hrs	Lab Hrs	Labour Cost £	Sunds £	Nett Rate £	Unit	Gross Rate (+12.5%) £
M10: CEMENT: SAND/CONCRETE SCREEDS/TOPPINGS										
Mortar, cement and sand (1:3) - screeds										
13mm work to walls on brickwork or blockwork base; one coat; screeded										
width exceeding 300mm	1.57	5.00	1.64	0.46	0.23	9.00	-	10.64	m2	11.98
width not exceeding 300mm	0.48	10.00	0.53	0.28	0.07	4.73	-	5.26	m	5.92
13mm work to walls on brickwork or blockwork base; one coat; troweled										
width exceeding 300mm	1.57	5.00	1.64	0.61	0.23	11.13	-	12.78	m2	14.37
width not exceeding 300mm	0.48	10.00	0.53	0.37	0.07	6.01	-	6.54	m	7.35
13mm work to walls on brickwork or blockwork base; one coat; floated										
width exceeding 300mm	1.57	5.00	1.64	0.58	0.23	10.71	-	12.35	m2	13.89
width not exceeding 300mm	0.48	10.00	0.53	0.35	0.07	5.72	-	6.25	m	7.04
19mm work to floors on concrete base; one coat; screeded										
level and to falls only not exceeding 15 degrees from horizontal	2.46	5.00	2.57	0.26	0.17	5.52	-	8.09	m2	9.10
to falls and crossfalls and to slopes not exceeding 15 degrees from horizontal	2.46	5.00	2.57	0.38	0.17	7.22	-	9.79	m2	11.02
25mm work to floors on concrete base; one coat; screeded										
level and to falls only not exceeding 15 degrees from horizontal	3.18	5.00	3.33	0.29	0.18	6.05	-	9.38	m2	10.55
to falls and crossfalls and to slopes not exceeding 15 degrees from horizontal	3.18	5.00	3.33	0.41	0.18	7.75	-	11.09	m2	12.47
32mm work to floors on concrete base; one coat; screeded										
level and to falls only not exceeding 15 degrees from horizontal	4.02	5.00	4.22	0.32	0.29	7.65	-	11.87	m2	13.35
to falls and crossfalls and to slopes not exceeding 15 degrees from horizontal	4.02	5.00	4.22	0.44	0.20	8.39	-	12.61	m2	14.19
38mm work to floors on concrete base; one coat; screeded										
level and to falls only not exceeding 15 degrees from horizontal	4.75	5.00	4.98	0.34	0.21	7.08	-	12.06	m2	13.56
to falls and crossfalls and to slopes not exceeding 15 degrees from horizontal	4.75	5.00	4.98	0.46	0.21	8.79	-	13.76	m2	15.48
50mm work to floors on concrete base; one coat; screeded										
level and to falls only not exceeding 15 degrees from horizontal	6.19	5.00	6.49	0.40	0.24	8.25	-	14.75	m2	16.59
to falls and crossfalls and to slopes not exceeding 15 degrees from horizontal	6.19	5.00	6.49	0.52	0.24	9.96	-	16.45	m2	18.51
19mm work to floors on concrete base; one coat; troweled										
level and to falls only not exceeding 15 degrees from horizontal	2.46	5.00	2.57	0.41	0.17	7.65	-	10.22	m2	11.50
to falls and crossfalls and to slopes not exceeding 15 degrees from horizontal	2.46	5.00	2.57	0.53	0.17	9.35	-	11.92	m2	13.42
25mm work to floors on concrete base; one coat; troweled										
level and to falls only not exceeding 15 degrees from horizontal	3.18	5.00	3.33	0.44	0.18	8.18	-	11.51	m2	12.95
to falls and crossfalls and to slopes not exceeding 15 degrees from horizontal	3.18	5.00	3.33	0.56	0.18	9.89	-	13.22	m2	14.87

Labour hourly rates: (except Specialists) Craft Operatives £14.21 Labourer £10.71 Rates are national average prices. Refer to REGIONAL VARIATIONS for indicative levels of overall pricing in regions	MATERIALS			LABOUR				RATES		
	Del to Site £	Waste %	Material Cost £	Craft Optve Hrs	Lab Hrs	Labour Cost £	Sunds £	Nett Rate £	Unit	Gross Rate (+12.5%) £
M10: CEMENT: SAND/CONCRETE SCREEDS/TOPPINGS Cont.										
Mortar, cement and sand (1:3) - screeds Cont.										
32mm work to floors on concrete base; one coat; troweled										
level and to falls only not exceeding 15 degrees from horizontal	4.02	5.00	4.22	0.47	0.20	8.82	-	13.04	m2	14.67
to falls and crossfalls and to slopes not exceeding 15 degrees from horizontal	4.02	5.00	4.22	0.59	0.20	10.53	-	14.74	m2	16.59
38mm work to floors on concrete base; one coat; troweled										
level and to falls only not exceeding 15 degrees from horizontal	4.75	5.00	4.98	0.49	0.21	9.21	-	14.19	m2	15.96
to falls and crossfalls and to slopes not exceeding 15 degrees from horizontal	4.75	5.00	4.98	0.61	0.21	10.92	-	15.89	m2	17.88
50mm work to floors on concrete base; one coat; troweled										
level and to falls only not exceeding 15 degrees from horizontal	6.19	5.00	6.49	0.55	0.24	10.39	-	16.88	m2	18.99
to falls and crossfalls and to slopes not exceeding 15 degrees from horizontal	6.19	5.00	6.49	0.67	0.24	12.09	-	18.59	m2	20.91
19mm work to floors on concrete base; one coat; floated										
level and to falls only not exceeding 15 degrees from horizontal	2.46	5.00	2.57	0.38	0.17	7.22	-	9.79	m2	11.02
to falls and crossfalls and to slopes not exceeding 15 degrees from horizontal	2.46	5.00	2.57	0.50	0.17	8.93	-	11.50	m2	12.94
25mm work to floors on concrete base; one coat; floated										
level and to falls only not exceeding 15 degrees from horizontal	3.18	5.00	3.33	0.41	0.18	7.75	-	11.09	m2	12.47
to falls and crossfalls and to slopes not exceeding 15 degrees from horizontal	3.18	5.00	3.33	0.53	0.18	9.46	-	12.79	m2	14.39
32mm work to floors on concrete base; one coat; floated										
level and to falls only not exceeding 15 degrees from horizontal	4.02	5.00	4.22	0.44	0.20	8.39	-	12.61	m2	14.19
to falls and crossfalls and to slopes not exceeding 15 degrees from horizontal	4.02	5.00	4.22	0.56	0.20	10.10	-	14.32	m2	16.11
38mm work to floors on concrete base; one coat; floated										
level and to falls only not exceeding 15 degrees from horizontal	4.75	5.00	4.98	0.46	0.21	8.79	-	13.76	m2	15.48
to falls and crossfalls and to slopes not exceeding 15 degrees from horizontal	4.75	5.00	4.98	0.58	0.21	10.49	-	15.47	m2	17.40
50mm work to floors on concrete base; one coat; floated										
level and to falls only not exceeding 15 degrees from horizontal	6.19	5.00	6.49	0.52	0.24	9.96	-	16.45	m2	18.51
to falls and crossfalls and to slopes not exceeding 15 degrees from horizontal	6.19	5.00	6.49	0.64	0.24	11.66	-	18.16	m2	20.43
Mortar, cement and sand (1:3) - paving										
25mm work to floors on concrete base; one coat; troweled										
level and to falls only not exceeding 15 degrees from horizontal	3.18	5.00	3.33	0.44	0.18	8.18	-	11.51	m2	12.95
to falls and crossfalls and to slopes not exceeding 15 degrees from horizontal	3.18	5.00	3.33	0.56	0.18	9.89	-	13.22	m2	14.87
to slopes exceeding 15 degrees from horizontal	3.18	5.00	3.33	0.62	0.18	10.74	-	14.07	m2	15.83
32mm work to floors on concrete base; one coat; troweled										
level and to falls only not exceeding 15 degrees from horizontal	4.02	5.00	4.22	0.47	0.20	8.82	-	13.04	m2	14.67
to falls and crossfalls and to slopes not exceeding 15 degrees from horizontal	4.02	5.00	4.22	0.59	0.20	10.53	-	14.74	m2	16.59
to slopes exceeding 15 degrees from horizontal	4.02	5.00	4.22	0.65	0.20	11.38	-	15.60	m2	17.55
38mm work to floors on concrete base; one coat; troweled										
level and to falls only not exceeding 15 degrees from horizontal	4.75	5.00	4.98	0.49	0.21	9.21	-	14.19	m2	15.96
to falls and crossfalls and to slopes not exceeding 15 degrees from horizontal	4.75	5.00	4.98	0.61	0.21	10.92	-	15.89	m2	17.88
to slopes exceeding 15 degrees from horizontal	4.75	5.00	4.98	0.67	0.21	11.77	-	16.75	m2	18.84
50mm work to floors on concrete base; one coat; troweled										
level and to falls only not exceeding 15 degrees from horizontal	6.19	5.00	6.49	0.55	0.24	10.39	-	16.88	m2	18.99
to falls and crossfalls and to slopes not exceeding 15 degrees from horizontal	6.19	5.00	6.49	0.67	0.24	12.09	-	18.59	m2	20.91
to slopes exceeding 15 degrees from horizontal	6.19	5.00	6.49	0.73	0.24	12.94	-	19.44	m2	21.87
If paving oil proofed, add										
25mm work	0.19	10.00	0.21	-	0.03	0.32	-	0.53	m2	0.60

Labour hourly rates: (except Specialists) Craft Operatives £14.21 Labourer £10.71 Rates are national average prices. Refer to REGIONAL VARIATIONS for indicative levels of overall pricing in regions	MATERIALS			LABOUR				RATES		
	Del to Site £	Waste %	Material Cost £	Craft Optve Hrs	Lab Hrs	Labour Cost £	Sunds £	Nett Rate £	Unit	Gross Rate (+12.5%) £
M10: CEMENT: SAND/CONCRETE SCREEDS/TOPPINGS Cont.										
Mortar, cement and sand (1:3) - paving Cont.										
If paving oil proofed, add Cont.										
32mm work	0.25	10.00	**0.27**	-	0.03	**0.32**	-	**0.59**	m2	**0.67**
38mm work	0.29	10.00	**0.32**	-	0.04	**0.43**	-	**0.75**	m2	**0.84**
50mm work	0.38	10.00	**0.42**	-	0.05	**0.54**	-	**0.96**	m2	**1.08**
19mm work to skirtings on brickwork or blockwork base										
height 100mm	0.36	10.00	**0.40**	0.80	0.05	**11.90**	-	**12.30**	m	**13.84**
height 150mm	0.48	10.00	**0.53**	0.90	0.05	**13.32**	-	**13.85**	m	**15.59**
25mm work to skirtings on brickwork or blockwork base										
height 100mm	0.48	10.00	**0.53**	0.85	0.05	**12.61**	-	**13.14**	m	**14.79**
height 150mm	0.72	10.00	**0.80**	0.95	0.05	**14.04**	-	**14.83**	m	**16.68**
Granolithic screed; steel troweled										
25mm work to floors on concrete base; one coat										
level and to falls only not exceeding 15 degrees from horizontal	4.19	5.00	**4.38**	0.48	0.21	**9.07**	-	**13.45**	m2	**15.13**
to falls and crossfalls and to slopes not exceeding 15 degrees from horizontal	4.19	5.00	**4.38**	0.60	0.21	**10.78**	-	**15.15**	m2	**17.05**
to slopes exceeding 15 degrees from horizontal	4.19	5.00	**4.38**	0.66	0.21	**11.63**	-	**16.01**	m2	**18.01**
32mm work to floors on concrete base; one coat										
level and to falls only not exceeding 15 degrees from horizontal	5.27	5.00	**5.51**	0.51	0.22	**9.60**	-	**15.11**	m2	**17.00**
to falls and crossfalls and to slopes not exceeding 15 degrees from horizontal	5.27	5.00	**5.51**	0.63	0.22	**11.31**	-	**16.82**	m2	**18.92**
to slopes exceeding 15 degrees from horizontal	5.27	5.00	**5.51**	0.69	0.22	**12.16**	-	**17.67**	m2	**19.88**
38mm work to floors on concrete base; one coat										
level and to falls only not exceeding 15 degrees from horizontal	6.19	5.00	**6.48**	0.55	0.24	**10.39**	-	**16.87**	m2	**18.98**
to falls and crossfalls and to slopes not exceeding 15 degrees from horizontal	6.19	5.00	**6.48**	0.67	0.24	**12.09**	-	**18.57**	m2	**20.89**
to slopes exceeding 15 degrees from horizontal	6.19	5.00	**6.48**	0.73	0.24	**12.94**	-	**19.43**	m2	**21.85**
50mm work to floors on concrete base; one coat										
level and to falls only not exceeding 15 degrees from horizontal	8.04	5.00	**8.42**	0.61	0.27	**11.56**	-	**19.98**	m2	**22.48**
to falls and crossfalls and to slopes not exceeding 15 degrees from horizontal	8.04	5.00	**8.42**	0.73	0.27	**13.27**	-	**21.69**	m2	**24.40**
to slopes exceeding 15 degrees from horizontal	8.04	5.00	**8.42**	0.79	0.27	**14.12**	-	**22.54**	m2	**25.36**
If paving tinted, add according to tint										
25mm work	1.95	10.00	**2.15**	-	0.03	**0.32**	-	**2.47**	m2	**2.77**
32mm work	2.50	10.00	**2.75**	-	0.03	**0.32**	-	**3.07**	m2	**3.45**
38mm work	2.96	10.00	**3.26**	-	0.04	**0.43**	-	**3.69**	m2	**4.15**
50mm work	3.90	10.00	**4.29**	-	0.05	**0.54**	-	**4.83**	m2	**5.43**
If carborundum troweled into surface add	3.85	10.00	**4.24**	0.10	0.05	**1.96**	-	**6.19**	m2	**6.97**
25mm linings to channels on concrete base										
150mm girth on face; to falls	0.97	10.00	**1.06**	0.25	0.05	**4.09**	-	**5.15**	m	**5.79**
225mm girth on face; to falls	1.32	10.00	**1.44**	0.30	0.05	**4.80**	-	**6.24**	m	**7.02**
32mm linings to channels on concrete base										
150mm girth on face; to falls	1.12	10.00	**1.23**	0.28	0.05	**4.51**	-	**5.74**	m	**6.46**
225mm girth on face; to falls	1.78	10.00	**1.95**	0.33	0.05	**5.22**	-	**7.17**	m	**8.07**
38mm linings to channels on concrete base										
150mm girth on face; to falls	1.43	10.00	**1.57**	0.30	0.05	**4.80**	-	**6.37**	m	**7.16**
225mm girth on face; to falls	2.09	10.00	**2.29**	0.35	0.05	**5.51**	-	**7.80**	m	**8.77**
50mm linings to channels on concrete base										
150mm girth on face; to falls	1.74	10.00	**1.91**	0.35	0.05	**5.51**	-	**7.41**	m	**8.34**
225mm girth on face; to falls	2.70	10.00	**2.96**	0.40	0.05	**6.22**	-	**9.18**	m	**10.33**
13mm work to skirtings on concrete base										
height 150mm	0.46	10.00	**0.51**	0.85	0.05	**12.61**	-	**13.12**	m	**14.76**
19mm work to skirtings on concrete base										
height 150mm	0.62	10.00	**0.68**	0.90	0.05	**13.32**	-	**14.00**	m	**15.75**
25mm work to skirtings on concrete base										
height 150mm	0.92	10.00	**1.02**	0.95	0.05	**14.04**	-	**15.05**	m	**16.93**
32mm work to skirtings on concrete base										
height 150mm	1.08	10.00	**1.19**	1.00	0.05	**14.75**	-	**15.93**	m	**17.92**
Rounded angles and intersections										
10 - 100mm radius	-	-	**-**	0.20	-	**2.84**	-	**2.84**	m	**3.20**
Surface hardeners on paving										
Proprietary surface hardener (Epoxy Resin Based)										
two coats	1.80	10.00	**1.98**	-	0.23	**2.46**	-	**4.44**	m2	**5.00**
three coats	2.47	10.00	**2.71**	-	0.29	**3.11**	-	**5.82**	m2	**6.55**
Polyurethane floor sealer										
two coats	2.08	10.00	**2.28**	-	0.23	**2.46**	-	**4.75**	m2	**5.34**
three coats	2.68	10.00	**2.95**	-	0.29	**3.11**	-	**6.05**	m2	**6.81**

Labour hourly rates: (except Specialists) Craft Operatives £14.21 Labourer £10.71 Rates are national average prices. Refer to REGIONAL VARIATIONS for indicative levels of overall pricing in regions	MATERIALS			LABOUR				RATES		
	Del to Site £	Waste %	Material Cost £	Craft Optve Hrs	Lab Hrs	Labour Cost £	Sunds £	Nett Rate £	Unit	Gross Rate (+12.5%) £
M10: CEMENT: SAND/CONCRETE SCREEDS/TOPPINGS Cont.										
Surface hardeners on paving Cont.										
Nitoflor Lithurin concrete surface dressing										
one coat	0.41	10.00	**0.45**	-	0.21	**2.25**	-	**2.70**	m2	**3.04**
two coats	0.82	10.00	**0.90**	-	0.30	**3.21**	-	**4.11**	m2	**4.63**
If on old floors										
add for cleaning and degreasing	0.35	5.00	**0.37**	-	0.40	**4.28**	-	**4.66**	m2	**5.24**
Vermiculite screed consisting of cement and Vermiculite aggregate finished with 20mm cement and sand (1:4) screeded bed										
45mm work to roofs on concrete base; two coats to falls and crossfalls and to slopes not exceeding 15 degrees from horizontal	7.19	5.00	**7.54**	0.55	0.60	**14.24**	-	**21.79**	m2	**24.51**
60mm work to roofs on concrete base; two coats to falls and crossfalls and to slopes not exceeding 15 degrees from horizontal	10.09	5.00	**10.58**	0.60	0.65	**15.49**	-	**26.07**	m2	**29.33**
70mm work to roofs on concrete base; two coats to falls and crossfalls and to slopes not exceeding 15 degrees from horizontal	12.02	5.00	**12.61**	0.65	0.70	**16.73**	-	**29.34**	m2	**33.01**
80mm work to roofs on concrete base; two coats to falls and crossfalls and to slopes not exceeding 15 degrees from horizontal	13.95	5.00	**14.64**	0.70	0.75	**17.98**	-	**32.62**	m2	**36.69**
Lightweight concrete screed consisting of cement and lightweight aggregate, medium grade, 10 - 5 gauge, 800 kg/m3 (1:10) finished with 15mm cement and sand (1:4) troweled bed										
50mm work to roofs on concrete base; two coats to falls and crossfalls and to slopes not exceeding 15 degrees from horizontal	6.26	5.00	**6.57**	0.73	0.62	**17.01**	-	**23.58**	m2	**26.53**
75mm work to roofs on concrete base; two coats to falls and crossfalls and to slopes not exceeding 15 degrees from horizontal	9.44	5.00	**9.90**	0.78	0.67	**18.26**	-	**28.16**	m2	**31.68**
100mm work to roofs on concrete base; two coats to falls and crossfalls and to slopes not exceeding 15 degrees from horizontal	12.61	5.00	**13.24**	0.83	0.72	**19.51**	-	**32.74**	m2	**36.83**
Screed reinforcement										
Reinforcement; galvanized wire netting, BS EN 10223-2, 13mm mesh, 22 gauge wire; 150mm laps; placing in position										
floors	2.44	5.00	**2.56**	0.05	-	**0.71**	-	**3.27**	m2	**3.68**
Reinforcement; galvanized wire netting, BS EN 10223-2, 25mm mesh, 20 gauge wire; 150mm laps; placing in position										
floors	1.71	5.00	**1.80**	0.05	-	**0.71**	-	**2.51**	m2	**2.82**
Reinforcement; galvanized wire netting, BS EN 10223-2, 38mm mesh, 19 gauge wire; 150mm laps; placing in position										
floors	1.42	5.00	**1.49**	0.05	-	**0.71**	-	**2.20**	m2	**2.48**
Reinforcement; galvanized wire netting, BS EN 10223-2, 50mm mesh, 19 gauge wire; 150mm laps; placing in position										
floors	1.05	5.00	**1.10**	0.05	-	**0.71**	-	**1.81**	m2	**2.04**
Division strips										
Aluminium dividing strips; setting in bed										
5 x 25mm; flat section	4.85	5.00	**5.09**	0.21	-	**2.98**	-	**8.08**	m	**9.09**
Brass dividing strips; setting in bed										
5 x 25mm; flat section	6.10	5.00	**6.41**	0.21	-	**2.98**	-	**9.39**	m	**10.56**
M10: CEMENT: SAND/CONCRETE SCREEDS/TOPPINGS; STAIRCASE AREAS										
Mortar, cement and sand (1:3) - paving										
13mm work to walls on brickwork or blockwork base; one coat; screeded										
width exceeding 300mm	1.57	5.00	**1.64**	0.69	0.23	**12.27**	-	**13.91**	m2	**15.65**
width not exceeding 300mm	0.48	10.00	**0.53**	0.41	0.07	**6.58**	-	**7.11**	m	**7.99**
13mm work to walls on brickwork or blockwork base; one coat; troweled										
width exceeding 300mm	1.57	5.00	**1.64**	0.92	0.23	**15.54**	-	**17.18**	m2	**19.33**
width not exceeding 300mm	0.48	10.00	**0.53**	0.55	0.07	**8.57**	-	**9.10**	m	**10.23**
13mm work to walls on brickwork or blockwork base; one coat; floated										
width exceeding 300mm	1.57	5.00	**1.64**	0.87	0.23	**14.83**	-	**16.47**	m2	**18.53**

Labour hourly rates: (except Specialists) Craft Operatives £14.21 Labourer £10.71 Rates are national average prices. Refer to REGIONAL VARIATIONS for indicative levels of overall pricing in regions	MATERIALS			LABOUR			RATES			
	Del to Site £	Waste %	Material Cost £	Craft Optve Hrs	Lab Hrs	Labour Cost £	Sunds £	Nett Rate £	Unit	Gross Rate (+12.5%) £
M10: CEMENT: SAND/CONCRETE SCREEDS/TOPPINGS; STAIRCASE AREAS Cont.										
Mortar, cement and sand (1:3) - paving Cont.										
13mm work to walls on brickwork or blockwork base; one coat; floated Cont.										
width not exceeding 300mm	0.48	10.00	0.53	0.52	0.07	8.14	-	8.67	m	9.75
19mm work to floors on concrete base; one coat; screeded										
level and to falls only not exceeding 15 degrees from horizontal	2.46	5.00	2.57	0.39	0.17	7.36	-	9.94	m2	11.18
25mm work to floors on concrete base; one coat; screeded										
level and to falls only not exceeding 15 degrees from horizontal	3.18	5.00	3.33	0.44	0.18	8.18	-	11.51	m2	12.95
32mm work to floors on concrete base; one coat; screeded										
level and to falls only not exceeding 15 degrees from horizontal	4.02	5.00	4.22	0.48	0.29	9.93	-	14.14	m2	15.91
38mm work to floors on concrete base; one coat; screeded										
level and to falls only not exceeding 15 degrees from horizontal	4.75	5.00	4.98	0.51	0.21	9.50	-	14.47	m2	16.28
50mm work to floors on concrete base; one coat; screeded										
level and to falls only not exceeding 15 degrees from horizontal	6.19	5.00	6.49	0.60	0.24	11.10	-	17.59	m2	19.79
19mm work to floors on concrete base; one coat; troweled										
level and to falls only not exceeding 15 degrees from horizontal	2.46	5.00	2.57	0.62	0.17	10.63	-	13.20	m2	14.85
25mm work to floors on concrete base; one coat; troweled										
level and to falls only not exceeding 15 degrees from horizontal	3.18	5.00	3.33	0.66	0.18	11.31	-	14.64	m2	16.47
32mm work to floors on concrete base; one coat; troweled										
level and to falls only not exceeding 15 degrees from horizontal	4.02	5.00	4.22	0.71	0.20	12.23	-	16.45	m2	18.50
38mm work to floors on concrete base; one coat; troweled										
level and to falls only not exceeding 15 degrees from horizontal	4.75	5.00	4.98	0.74	0.21	12.76	-	17.74	m2	19.96
50mm work to floors on concrete base; one coat; troweled										
level and to falls only not exceeding 15 degrees from horizontal	6.19	5.00	6.49	0.83	0.24	14.36	-	20.86	m2	23.47
19mm work to floors on concrete base; one coat; floated										
level and to falls only not exceeding 15 degrees from horizontal	2.46	5.00	2.57	0.57	0.17	9.92	-	12.49	m2	14.05
25mm work to floors on concrete base; one coat; floated										
level and to falls only not exceeding 15 degrees from horizontal	3.18	5.00	3.33	0.62	0.18	10.74	-	14.07	m2	15.83
32mm work to floors on concrete base; one coat; floated										
level and to falls only not exceeding 15 degrees from horizontal	4.02	5.00	4.22	0.66	0.20	11.52	-	15.74	m2	17.71
38mm work to floors on concrete base; one coat; floated										
level and to falls only not exceeding 15 degrees from horizontal	4.75	5.00	4.98	0.69	0.21	12.05	-	17.03	m2	19.16
50mm work to floors on concrete base; one coat; floated										
level and to falls only not exceeding 15 degrees from horizontal	6.19	5.00	6.49	0.78	0.24	13.65	-	20.15	m2	22.67
19mm work to treads on concrete base; screeded										
width 275mm	1.01	10.00	1.10	0.20	0.04	3.27	-	4.37	m	4.92
25mm work to treads on concrete base; screeded										
width 275mm	1.25	10.00	1.37	0.21	0.04	3.41	-	4.78	m	5.38
32mm work to treads on concrete base; screeded										
width 275mm	1.61	10.00	1.77	0.23	0.05	3.80	-	5.57	m	6.27
38mm work to treads on concrete base; screeded										
width 275mm	1.97	10.00	2.16	0.25	0.05	4.09	-	6.25	m	7.03

Labour hourly rates: (except Specialists) Craft Operatives £14.21 Labourer £10.71 Rates are national average prices. Refer to REGIONAL VARIATIONS for indicative levels of overall pricing in regions	MATERIALS			LABOUR				RATES		
	Del to Site £	Waste %	Material Cost £	Craft Optve Hrs	Lab Hrs	Labour Cost £	Sunds £	Nett Rate £	Unit	Gross Rate (+12.5%) £
M10: CEMENT: SAND/CONCRETE SCREEDS/TOPPINGS; STAIRCASE AREAS Cont.										
Mortar, cement and sand (1:3) - paving Cont.										
19mm work to treads on concrete base; floated width 275mm	1.01	10.00	1.10	0.25	0.04	3.98	-	5.08	m	5.72
25mm work to treads on concrete base; floated width 275mm	1.25	10.00	1.37	0.26	0.04	4.12	-	5.49	m	6.18
32mm work to treads on concrete base; floated width 275mm	1.61	10.00	1.77	0.28	0.05	4.51	-	6.28	m	7.06
38mm work to treads on concrete base; floated width 275mm	1.97	10.00	2.16	0.30	0.05	4.80	-	6.96	m	7.83
19mm work to treads on concrete base; troweled width 275mm	1.01	10.00	1.10	0.30	0.04	4.69	-	5.79	m	6.52
25mm work to treads on concrete base; troweled width 275mm	1.25	10.00	1.37	0.31	0.04	4.83	-	6.20	m	6.98
32mm work to treads on concrete base; troweled width 275mm	1.61	10.00	1.77	0.33	0.05	5.22	-	6.99	m	7.86
38mm work to treads on concrete base; troweled width 275mm	1.97	10.00	2.16	0.35	0.05	5.51	-	7.67	m	8.63
19mm work to plain risers on concrete base; keyed height 175mm	0.60	10.00	0.66	0.35	0.03	5.29	-	5.96	m	6.70
25mm work to plain risers on concrete base; keyed height 175mm	0.84	10.00	0.93	0.40	0.03	6.01	-	6.93	m	7.80
32mm work to plain risers on concrete base; keyed height 175mm	0.96	10.00	1.06	0.45	0.03	6.72	-	7.78	m	8.75
19mm work to plain risers on concrete base; troweled height 175mm	0.60	10.00	0.66	0.55	0.03	8.14	-	8.80	m	9.90
25mm work to plain risers on concrete base; troweled height 175mm	0.84	10.00	0.93	0.60	0.03	8.85	-	9.78	m	11.00
32mm work to plain risers on concrete base; troweled height 175mm	0.96	10.00	1.06	0.65	0.03	9.56	-	10.62	m	11.95
19mm work to undercut risers on concrete base; keyed height 175mm	0.60	10.00	0.66	0.40	0.03	6.01	-	6.67	m	7.50
25mm work to undercut risers on concrete base; keyed height 175mm	0.84	10.00	0.93	0.45	0.03	6.72	-	7.64	m	8.60
32mm work to undercut risers on concrete base; keyed height 175mm	0.96	10.00	1.06	0.50	0.03	7.43	-	8.49	m	9.55
19mm work to undercut risers on concrete base; troweled height 175mm	0.60	10.00	0.66	0.60	0.03	8.85	-	9.51	m	10.70
25mm work to undercut risers on concrete base; troweled height 175mm	0.84	10.00	0.93	0.65	0.03	9.56	-	10.49	m	11.80
32mm work to undercut risers on concrete base; troweled height 175mm	0.96	10.00	1.06	0.70	0.03	10.27	-	11.33	m	12.74
Granolithic screed; steel troweled										
25mm work to floors on concrete base; one coat level and to falls only not exceeding 15 degrees from horizontal	4.19	5.00	4.38	0.72	0.21	12.48	-	16.86	m2	18.97
32mm work to floors on concrete base; one coat level and to falls only not exceeding 15 degrees from horizontal	5.27	5.00	5.51	0.77	0.22	13.30	-	18.81	m2	21.16
38mm work to floors on concrete base; one coat level and to falls only not exceeding 15 degrees from horizontal	6.19	5.00	6.48	0.83	0.24	14.36	-	20.85	m2	23.45
50mm work to floors on concrete base; one coat level and to falls only not exceeding 15 degrees from horizontal	8.04	5.00	8.42	0.92	0.27	15.96	-	24.39	m2	27.44
25mm work to treads on concrete base width 275mm	1.62	10.00	1.78	0.30	0.06	4.91	-	6.68	m	7.52
32mm work to treads on concrete base width 275mm	2.09	10.00	2.29	0.35	0.06	5.62	-	7.90	m	8.89
38mm work to treads on concrete base width 275mm	2.55	10.00	2.79	0.40	0.07	6.43	-	9.23	m	10.38

Labour hourly rates: (except Specialists) Craft Operatives £14.21 Labourer £10.71 Rates are national average prices. Refer to REGIONAL VARIATIONS for indicative levels of overall pricing in regions	MATERIALS Del to Site £	Waste %	Material Cost £	LABOUR Craft Optve Hrs	Lab Hrs	Labour Cost £	Sunds £	RATES Nett Rate £	Unit	Gross Rate (+12.5%) £
M10: CEMENT: SAND/CONCRETE SCREEDS/TOPPINGS; STAIRCASE AREAS Cont.										
Granolithic screed; steel troweled Cont.										
50mm work to treads on concrete base width 275mm	3.32	10.00	3.64	0.45	0.07	7.14	-	10.79	m	12.13
13mm work to plain risers on concrete base height 175mm	0.46	10.00	0.51	0.55	0.05	8.35	-	8.86	m	9.97
19mm work to plain risers on concrete base height 175mm	0.77	10.00	0.85	0.60	0.05	9.06	-	9.91	m	11.15
25mm work to plain risers on concrete base height 175mm	1.08	10.00	1.19	0.65	0.05	9.77	-	10.96	m	12.33
32mm work to plain risers on concrete base height 175mm	1.23	10.00	1.36	0.70	0.05	10.48	-	11.84	m	13.32
13mm work to undercut risers on concrete base height 175mm	0.46	10.00	0.51	0.60	0.05	9.06	-	9.57	m	10.77
19mm work to undercut risers on concrete base height 175mm	0.77	10.00	0.85	0.65	0.05	9.77	-	10.62	m	11.95
25mm work to undercut risers on concrete base height 175mm	1.08	10.00	1.19	0.70	0.05	10.48	-	11.67	m	13.13
32mm work to undercut risers on concrete base height 175mm	1.23	10.00	1.36	0.75	0.05	11.19	-	12.55	m	14.12
13mm work to strings on concrete base height 300mm	0.92	10.00	1.02	0.75	0.05	11.19	-	12.21	m	13.74
19mm work to strings on concrete base height 300mm	1.39	10.00	1.52	0.80	0.05	11.90	-	13.43	m	15.11
25mm work to strings on concrete base height 300mm	1.69	10.00	1.86	0.85	0.06	12.72	-	14.58	m	16.41
32mm work to strings on concrete base height 300mm	2.16	10.00	2.37	0.90	0.07	13.54	-	15.91	m	17.90
13mm work to strings on brickwork or blockwork base height 275mm	0.77	10.00	0.85	0.70	0.05	10.48	-	11.33	m	12.75
19mm work to strings on brickwork or blockwork base height 275mm	1.23	10.00	1.36	0.75	0.05	11.19	-	12.55	m	14.12
25mm work to strings on brickwork or blockwork base height 275mm	1.54	10.00	1.69	0.80	0.06	12.01	-	13.70	m	15.42
32mm work to strings on brickwork or blockwork base height 275mm	2.00	10.00	2.20	0.85	0.07	12.83	-	15.03	m	16.91
13mm work to aprons on concrete base height 150mm	0.46	10.00	0.51	0.55	0.05	8.35	-	8.86	m	9.97
19mm work to aprons on concrete base height 150mm	0.62	10.00	0.68	0.60	0.05	9.06	-	9.74	m	10.96
25mm work to aprons on concrete base height 150mm	0.92	10.00	1.02	0.65	0.05	9.77	-	10.79	m	12.14
32mm work to aprons on concrete base height 150mm	1.08	10.00	1.19	0.70	0.05	10.48	-	11.67	m	13.13
M11: MASTIC ASPHALT FLOORING/FLOOR UNDERLAYS										
Paving; BS 6925 (limestone aggregate) , brown, smooth floated finish										
Flooring and underlay 15mm thick; coats of asphalt - 1; to concrete base; flat										
width not exceeding 150mm	-	-	-	-	-	-	-	26.25	m2	29.53
width 150 - 225mm	-	-	-	-	-	-	-	22.00	m2	24.75
width 225 - 300mm	-	-	-	-	-	-	-	17.40	m2	19.58
width exceeding 300mm	-	-	-	-	-	-	-	13.15	m2	14.79
Skirtings 15mm thick; coats of asphalt - 1; no underlay; to brickwork base										
girth not exceeding 150mm	-	-	-	-	-	-	-	8.45	m	9.51
Paving; BS 6925 (limestone aggregate) , red, smooth floated finish										
Flooring and underlay 15mm thick; coats of asphalt - 1; to concrete base; flat										
width not exceeding 150mm	-	-	-	-	-	-	-	25.25	m2	28.41
width 150 - 225mm	-	-	-	-	-	-	-	23.70	m2	26.66
width 225 - 300mm	-	-	-	-	-	-	-	18.95	m2	21.32
width exceeding 300mm	-	-	-	-	-	-	-	14.05	m2	15.81

Labour hourly rates: (except Specialists) Craft Operatives £14.21 Labourer £10.71 Rates are national average prices. Refer to REGIONAL VARIATIONS for indicative levels of overall pricing in regions	MATERIALS			LABOUR				RATES		
	Del to Site	Waste	Material Cost	Craft Optve	Lab	Labour Cost	Sunds	Nett Rate	Unit	Gross Rate (+12.5%)
	£	%	£	Hrs	Hrs	£	£	£		£
M11: MASTIC ASPHALT FLOORING/FLOOR UNDERLAYS Cont.										
Paving; BS 6925 (limestone aggregate) , red, smooth floated finish Cont.										
Skirtings 15mm thick; coats of asphalt - 1; no underlay; to brickwork base										
girth not exceeding 150mm	-	-	-	-	-	-	-	9.40	m	10.58
M12: TROWELED BITUMEN/RESIN/RUBBER LATEX FLOORING										
One coat levelling screeds										
3mm work to floors on concrete base; one coat level and to falls only not exceeding 15 degrees from horizontal	3.50	5.00	3.67	0.46	0.05	7.07	-	10.74	m2	12.09
3mm work to floors on existing timber boarded base; one coat level and to falls only not exceeding 15 degrees from horizontal	3.50	5.00	3.67	0.58	0.06	8.88	-	12.56	m2	14.13
M20: PLASTERED/RENDERED/ROUGHCAST COATINGS - PLASTERBOARD										
Baseboarding; Gyproc Handiboard; 5mm joints, filled with plaster and scrimmed; fixing with Galvanized nails, 30mm long at 400mm ccs; internal										
9.5mm work to walls on timber base										
width exceeding 300mm	3.00	5.00	3.16	0.29	0.15	5.73	-	8.89	m2	10.00
width not exceeding 300mm	0.98	10.00	1.07	0.17	0.05	2.95	-	4.02	m	4.52
9.5mm work to ceilings on timber base										
width exceeding 300mm	3.00	5.00	3.16	0.35	0.15	6.58	-	9.74	m2	10.95
width not exceeding 300mm	1.03	10.00	1.12	0.21	0.05	3.52	-	4.64	m	5.22
9.5mm work to isolated beams on timber base										
width exceeding 300mm	3.00	5.00	3.16	0.44	0.15	7.86	-	11.02	m2	12.39
width not exceeding 300mm	1.03	10.00	1.12	0.26	0.05	4.23	-	5.35	m	6.02
9.5mm work to isolated columns on timber base										
width exceeding 300mm	3.00	5.00	3.16	0.44	0.15	7.86	-	11.02	m2	12.39
width not exceeding 300mm	1.03	10.00	1.12	0.26	0.05	4.23	-	5.35	m	6.02
Baseboarding; Gyproc Handiboard; 5mm joints, filled with plaster and scrimmed; fixing with adhesive dabs; internal										
9.5mm work to walls on masonry base										
width exceeding 300mm	2.97	5.00	3.13	0.29	0.15	5.73	-	8.86	m2	9.96
width not exceeding 300mm	0.96	10.00	1.05	0.17	0.05	2.95	-	4.00	m	4.50
Baseboarding; Gyproc Handiboard Duplex; 5mm joints filled with plaster and scrimmed; fixing with Galvanized nails, 30mm long at 400mm ccs; internal										
9.5mm work to walls on timber base										
width exceeding 300mm	3.48	5.00	3.66	0.29	0.15	5.73	-	9.39	m2	10.56
width not exceeding 300mm	1.12	10.00	1.23	0.17	0.05	2.95	-	4.18	m	4.70
9.5mm work to ceilings on timber base										
width exceeding 300mm	3.48	5.00	3.66	0.35	0.15	6.58	-	10.24	m2	11.52
width not exceeding 300mm	1.18	10.00	1.28	0.21	0.05	3.52	-	4.80	m	5.40
9.5mm work to isolated beams on timber base										
width exceeding 300mm	3.48	5.00	3.66	0.44	0.15	7.86	-	11.52	m2	12.96
width not exceeding 300mm	1.18	10.00	1.28	0.26	0.05	4.23	-	5.51	m	6.20
9.5mm work to isolated columns on timber base										
width exceeding 300mm	3.48	5.00	3.66	0.44	0.15	7.86	-	11.52	m2	12.96
width not exceeding 300mm	1.18	10.00	1.28	0.26	0.05	4.23	-	5.51	m	6.20
Baseboarding; Gyproc Handiboard; 3mm joints, filled with plaster; fixing with Galvanized nails, 30mm long at 400mm ccs; internal										
12.5mm work to walls on timber base										
width exceeding 300mm	4.24	5.00	4.46	0.35	0.18	6.90	-	11.36	m2	12.78
width not exceeding 300mm	1.35	10.00	1.48	0.21	0.05	3.52	-	5.00	m	5.62
12.5mm work to ceilings on timber base										
width exceeding 300mm	4.05	5.00	4.26	0.42	0.18	7.90	-	12.15	m2	13.67
width not exceeding 300mm	1.35	10.00	1.48	0.25	0.05	4.09	-	5.57	m	6.26
12.5mm work to isolated beams on timber base										
width exceeding 300mm	4.24	5.00	4.46	0.53	0.18	9.46	-	13.92	m2	15.66
width not exceeding 300mm	1.35	10.00	1.48	0.32	0.05	5.08	-	6.56	m	7.38
12.5mm work to isolated columns on timber base										
width exceeding 300mm	4.24	5.00	4.46	0.53	0.18	9.46	-	13.92	m2	15.66
width not exceeding 300mm	1.35	10.00	1.48	0.32	0.05	5.08	-	6.56	m	7.38

Labour hourly rates: (except Specialists) Craft Operatives £14.21 Labourer £10.71 Rates are national average prices. Refer to REGIONAL VARIATIONS for indicative levels of overall pricing in regions	MATERIALS			LABOUR				RATES		
	Del to Site	Waste	Material Cost	Craft Optve	Lab	Labour Cost	Sunds	Nett Rate	Unit	Gross Rate (+12.5%)
	£	%	£	Hrs	Hrs	£	£	£		£
M20: PLASTERED/RENDERED/ROUGHCAST COATINGS - PLASTERBOARD Cont.										
Baseboarding; Gyproc plank, square edge; 5mm joints, filled with plaster and scrimmed; internal										
19mm work to walls on timber base										
width exceeding 300mm	4.96	5.00	**5.21**	0.47	0.24	**9.25**	-	**14.46**	m2	**16.27**
width not exceeding 300mm...........................	1.58	10.00	**1.73**	0.28	0.07	**4.73**	-	**6.46**	m	**7.27**
19mm work to ceilings on timber base										
width exceeding 300mm	4.96	5.00	**5.21**	0.56	0.24	**10.53**	-	**15.74**	m2	**17.71**
width not exceeding 300mm...........................	1.58	10.00	**1.73**	0.34	0.07	**5.58**	-	**7.31**	m	**8.23**
19mm work to isolated beams on timber base										
width exceeding 300mm	4.96	5.00	**5.21**	0.71	0.24	**12.66**	-	**17.87**	m2	**20.11**
width not exceeding 300mm...........................	1.58	10.00	**1.73**	0.43	0.07	**6.86**	-	**8.59**	m	**9.66**
19mm work to isolated columns on timber base										
width exceeding 300mm	4.96	5.00	**5.21**	0.71	0.24	**12.66**	-	**17.87**	m2	**20.11**
width not exceeding 300mm...........................	1.58	10.00	**1.73**	0.43	0.07	**6.86**	-	**8.59**	m	**9.66**
Baseboarding; Gyproc square edge wallboard; 3mm joints filled with plaster and scrimmed; fixing with Galvanized nails, 30mm long at 400mm ccs; internal										
9.5mm work to walls on timber base										
width exceeding 300mm	2.17	5.00	**2.29**	0.29	0.15	**5.73**	-	**8.01**	m2	**9.02**
width not exceeding 300mm...........................	0.73	10.00	**0.79**	0.17	0.05	**2.95**	-	**3.75**	m	**4.21**
12.5mm work to walls on timber base										
width exceeding 300mm	2.28	5.00	**2.40**	0.35	0.18	**6.90**	-	**9.30**	m2	**10.47**
width not exceeding 300mm...........................	0.76	10.00	**0.83**	0.21	0.05	**3.52**	-	**4.35**	m	**4.89**
15mm work to walls on timber base										
width exceeding 300mm	3.44	5.00	**3.62**	0.41	0.21	**8.08**	-	**11.69**	m2	**13.16**
width not exceeding 300mm...........................	1.11	10.00	**1.21**	0.25	0.06	**4.20**	-	**5.41**	m	**6.08**
9.5mm work to ceilings on timber base										
width exceeding 300mm	2.17	5.00	**2.29**	0.35	0.15	**6.58**	-	**8.87**	m2	**9.97**
width not exceeding 300mm...........................	0.73	10.00	**0.79**	0.21	0.05	**3.52**	-	**4.31**	m	**4.85**
12.5mm work to ceilings on timber base										
width exceeding 300mm	2.28	5.00	**2.40**	0.42	0.18	**7.90**	-	**10.30**	m2	**11.58**
width not exceeding 300mm...........................	0.76	10.00	**0.83**	0.25	0.05	**4.09**	-	**4.92**	m	**5.53**
15mm work to ceilings on timber base										
width exceeding 300mm	3.44	5.00	**3.62**	0.49	0.25	**9.64**	-	**13.26**	m2	**14.92**
width not exceeding 300mm...........................	1.11	10.00	**1.21**	0.29	0.08	**4.98**	-	**6.19**	m	**6.96**
9.5mm work to isolated beams on timber base										
width exceeding 300mm	2.17	5.00	**2.29**	0.44	0.15	**7.86**	-	**10.15**	m2	**11.41**
width not exceeding 300mm...........................	0.73	10.00	**0.79**	0.26	0.05	**4.23**	-	**5.02**	m	**5.65**
12.5mm work to isolated beams on timber base										
width exceeding 300mm	2.28	5.00	**2.40**	0.53	0.18	**9.46**	-	**11.86**	m2	**13.34**
width not exceeding 300mm...........................	0.76	10.00	**0.83**	0.32	0.05	**5.08**	-	**5.91**	m	**6.65**
15mm work to isolated beams on timber base										
width exceeding 300mm	3.44	5.00	**3.62**	0.62	0.21	**11.06**	-	**14.68**	m2	**16.51**
width not exceeding 300mm...........................	1.11	10.00	**1.21**	0.37	0.06	**5.90**	-	**7.11**	m	**8.00**
9.5mm work to isolated columns on timber base										
width exceeding 300mm	2.17	5.00	**2.29**	0.44	0.15	**7.86**	-	**10.15**	m2	**11.41**
width not exceeding 300mm...........................	0.73	10.00	**0.79**	0.26	0.05	**4.23**	-	**5.02**	m	**5.65**
12.5mm work to isolated columns on timber base										
width exceeding 300mm	2.28	5.00	**2.40**	0.53	0.18	**9.46**	-	**11.86**	m2	**13.34**
width not exceeding 300mm...........................	0.76	10.00	**0.83**	0.32	0.05	**5.08**	-	**5.91**	m	**6.65**
15mm work to isolated columns on timber base										
width exceeding 300mm	3.44	5.00	**3.62**	0.62	0.21	**11.06**	-	**14.68**	m2	**16.51**
width not exceeding 300mm...........................	1.11	10.00	**1.21**	0.37	0.06	**5.90**	-	**7.11**	m	**8.00**
Baseboarding; Gyproc square edge Duplex wallboard; 3mm joints filled with plaster and scrimmed; fixing with Galvanized nails, 30mm long at 400mm ccs; internal										
9.5mm work to walls on timber base										
width exceeding 300mm	2.78	5.00	**2.93**	0.29	0.15	**5.73**	-	**8.65**	m2	**9.74**
width not exceeding 300mm...........................	0.91	10.00	**1.00**	0.17	0.05	**2.95**	-	**3.95**	m	**4.44**
12.5mm work to walls on timber base										
width exceeding 300mm	2.94	5.00	**3.09**	0.35	0.18	**6.90**	-	**10.00**	m2	**11.25**
width not exceeding 300mm...........................	0.96	10.00	**1.05**	0.21	0.05	**3.52**	-	**4.57**	m	**5.14**
15mm work to walls on timber base										
width exceeding 300mm	4.49	5.00	**4.72**	0.41	0.21	**8.08**	-	**12.80**	m2	**14.40**
width not exceeding 300mm...........................	1.43	10.00	**1.56**	0.25	0.06	**4.20**	-	**5.75**	m	**6.47**
9.5mm work to ceilings on timber base										
width exceeding 300mm	2.78	5.00	**2.93**	0.35	0.15	**6.58**	-	**9.51**	m2	**10.70**
width not exceeding 300mm...........................	0.91	10.00	**1.00**	0.21	0.05	**3.52**	-	**4.51**	m	**5.08**
12.5mm work to ceilings on timber base										
width exceeding 300mm	2.94	5.00	**3.09**	0.42	0.18	**7.90**	-	**10.99**	m2	**12.36**

Labour hourly rates: (except Specialists) Craft Operatives £14.21 Labourer £10.71 Rates are national average prices. Refer to REGIONAL VARIATIONS for indicative levels of overall pricing in regions	MATERIALS			LABOUR				RATES		
	Del to Site £	Waste %	Material Cost £	Craft Optve Hrs	Lab Hrs	Labour Cost £	Sunds £	Nett Rate £	Unit	Gross Rate (+12.5%) £
M20: PLASTERED/RENDERED/ROUGHCAST COATINGS - PLASTERBOARD Cont.										
Baseboarding; Gyproc square edge Duplex wallboard; 3mm joints filled with plaster and scrimmed; fixing with Galvanized nails, 30mm long at 400mm ccs; internal Cont.										
12.5mm work to ceilings on timber base Cont.										
width not exceeding 300mm	0.96	10.00	1.05	0.25	0.05	4.09	-	5.14	m	5.78
15mm work to ceilings on timber base										
width exceeding 300mm	4.49	5.00	4.72	0.49	0.25	9.64	-	14.36	m2	16.16
width not exceeding 300mm	1.43	10.00	1.56	0.29	0.08	4.98	-	6.54	m	7.35
9.5mm work to isolated beams on timber base										
width exceeding 300mm	2.78	5.00	2.93	0.44	0.15	7.86	-	10.79	m2	12.13
width not exceeding 300mm	0.91	10.00	1.00	0.26	0.05	4.23	-	5.23	m	5.88
12.5mm work to isolated beams on timber base										
width exceeding 300mm	2.94	5.00	3.09	0.53	0.18	9.46	-	12.55	m2	14.12
width not exceeding 300mm	0.96	10.00	1.05	0.32	0.05	5.08	-	6.13	m	6.90
15mm work to isolated beams on timber base										
width exceeding 300mm	4.49	5.00	4.72	0.62	0.21	11.06	-	15.78	m2	17.75
width not exceeding 300mm	1.43	10.00	1.56	0.37	0.06	5.90	-	7.46	m	8.39
9.5mm work to isolated columns on timber base										
width exceeding 300mm	2.78	5.00	2.93	0.44	0.15	7.86	-	10.79	m2	12.13
width not exceeding 300mm	0.91	10.00	1.00	0.26	0.05	4.23	-	5.23	m	5.88
12.5mm work to isolated columns on timber base										
width exceeding 300mm	2.94	5.00	3.09	0.53	0.18	9.46	-	12.55	m2	14.12
width not exceeding 300mm	0.96	10.00	1.05	0.32	0.05	5.08	-	6.13	m	6.90
15mm work to isolated columns on timber base										
width exceeding 300mm	4.49	5.00	4.72	0.62	0.21	11.06	-	15.78	m2	17.75
width not exceeding 300mm	1.43	10.00	1.56	0.37	0.06	5.90	-	7.46	m	8.39
Baseboarding; two layers 9.5mm Gyproc square edge wallboard; second layer with 3mm joints filled with plaster and scrimmed; fixing with Galvanized nails, 30mm long at 400mm ccs; internal										
19mm work to walls on timber base										
width exceeding 300mm	4.15	5.00	4.36	0.60	0.30	11.74	-	16.10	m2	18.11
width not exceeding 300mm	1.38	10.00	1.50	0.36	0.09	6.08	-	7.58	m	8.53
19mm work to ceilings on timber base										
width exceeding 300mm	4.15	5.00	4.36	0.72	0.30	13.44	-	17.80	m2	20.03
width not exceeding 300mm	1.38	10.00	1.50	0.43	0.09	7.07	-	8.57	m	9.65
19mm work to isolated beams on timber base										
width exceeding 300mm	4.15	5.00	4.36	0.80	0.30	14.58	-	18.94	m2	21.31
width not exceeding 300mm	1.38	10.00	1.50	0.48	0.09	7.78	-	9.29	m	10.45
19mm work to isolated columns on timber base										
width exceeding 300mm	4.15	5.00	4.36	0.80	0.30	14.58	-	18.94	m2	21.31
width not exceeding 300mm	1.38	10.00	1.50	0.48	0.09	7.78	-	9.29	m	10.45
Baseboarding; one layer 9.5mm and one layer 12.5mm Gyproc square edge wallboard; second layer with 3mm joints filled with plaster and scrimmed; fixing with Galvanized nails, 30mm long at 400mm ccs, internal										
22mm work to walls on timber base										
width exceeding 300mm	4.26	5.00	4.47	0.65	0.33	12.77	-	17.24	m2	19.40
width not exceeding 300mm	1.41	10.00	1.54	0.39	0.10	6.61	-	8.15	m	9.17
22mm work to ceilings on timber base										
width exceeding 300mm	4.26	5.00	4.47	0.78	0.33	14.62	-	19.09	m2	21.48
width not exceeding 300mm	1.41	10.00	1.54	0.47	0.10	7.75	-	9.29	m	10.45
22mm work to isolated beams on timber base										
width exceeding 300mm	4.26	5.00	4.47	0.98	0.33	17.46	-	21.93	m2	24.68
width not exceeding 300mm	1.41	10.00	1.54	0.59	0.10	9.45	-	10.99	m	12.37
22mm work to isolated columns on timber base										
width exceeding 300mm	4.26	5.00	4.47	0.98	0.33	17.46	-	21.93	m2	24.68
width not exceeding 300mm	1.41	10.00	1.54	0.59	0.10	9.45	-	10.99	m	12.37
Baseboarding; two layers 12.5mm Gyproc square edge wallboard; second layer with 3mm joints filled with plaster and scrimmed; fixing with Galvanized nails, 30mm long at 400mm ccs; internal										
25mm work to walls on timber base										
width exceeding 300mm	4.37	5.00	4.59	0.70	0.35	13.70	-	18.29	m2	20.57
width not exceeding 300mm	1.45	10.00	1.57	0.42	0.11	7.15	-	8.72	m	9.81
25mm work to ceilings on timber base										
width exceeding 300mm	4.37	5.00	4.59	0.84	0.35	15.68	-	20.27	m2	22.81
width not exceeding 300mm	1.45	10.00	1.57	0.50	0.11	8.28	-	9.86	m	11.09
25mm work to isolated beams on timber base										
width exceeding 300mm	4.37	5.00	4.59	1.05	0.35	18.67	-	23.26	m2	26.17

Labour hourly rates: (except Specialists) Craft Operatives £14.21 Labourer £10.71 Rates are national average prices. Refer to REGIONAL VARIATIONS for indicative levels of overall pricing in regions	MATERIALS			LABOUR				RATES		
	Del to Site £	Waste %	Material Cost £	Craft Optve Hrs	Lab Hrs	Labour Cost £	Sunds £	Nett Rate £	Unit	Gross Rate (+12.5%) £
M20: PLASTERED/RENDERED/ROUGHCAST COATINGS - PLASTERBOARD Cont.										
Baseboarding; two layers 12.5mm Gyproc square edge wallboard; second layer with 3mm joints filled with plaster and scrimmed; fixing with Galvanized nails, 30mm long at 400mm ccs; internal Cont.										
25mm work to isolated beams on timber base Cont.										
width not exceeding 300mm	1.45	10.00	**1.57**	0.64	0.11	**10.27**	-	**11.85**	m	**13.33**
25mm work to isolated columns on timber base										
width exceeding 300mm	4.37	5.00	**4.59**	1.05	0.35	**18.67**	-	**23.26**	m2	**26.17**
width not exceeding 300mm	1.45	10.00	**1.57**	0.64	0.11	**10.27**	-	**11.85**	m	**13.33**
Baseboarding; two layers 15mm Gyproc square edge wallboard; second layer with 3mm joints filled with plaster and scrimmed; fixing with Galvanized nails, 30mm long at 400mm ccs; internal										
30mm work to walls on timber base										
width exceeding 300mm	6.69	5.00	**7.03**	0.75	0.38	**14.73**	-	**21.75**	m2	**24.47**
width not exceeding 300mm	2.14	10.00	**2.34**	0.45	0.11	**7.57**	-	**9.91**	m	**11.15**
30mm work to ceilings on timber base										
width exceeding 300mm	6.69	5.00	**7.03**	0.90	0.38	**16.86**	-	**23.88**	m2	**26.87**
width not exceeding 300mm	2.14	10.00	**2.34**	0.54	0.11	**8.85**	-	**11.19**	m	**12.59**
30mm work to isolated beams on timber base										
width exceeding 300mm	6.69	5.00	**7.03**	1.13	0.38	**20.13**	-	**27.15**	m2	**30.55**
width not exceeding 300mm	2.14	10.00	**2.34**	0.68	0.11	**10.84**	-	**13.18**	m	**14.83**
30mm work to isolated columns on timber base										
width exceeding 300mm	6.69	5.00	**7.03**	1.13	0.38	**20.13**	-	**27.15**	m2	**30.55**
width not exceeding 300mm	2.14	10.00	**2.34**	0.68	0.11	**10.84**	-	**13.18**	m	**14.83**
Baseboarding; two layers; first layer 9.5mm Gyproc Duplex wallboard, second layer 9.5mm Gyproc wallboard both layers with square edge boards; second layer with 3mm joints filled with plaster and scrimmed; fixing with Galvanized nails, 30mm long at 400mm ccs; internal										
19mm work to walls on timber base										
width exceeding 300mm	4.76	5.00	**5.00**	0.60	0.30	**11.74**	-	**16.74**	m2	**18.83**
width not exceeding 300mm	1.56	10.00	**1.70**	0.36	0.09	**6.08**	-	**7.78**	m	**8.75**
19mm work to ceilings on timber base										
width exceeding 300mm	4.76	5.00	**5.00**	0.72	0.30	**13.44**	-	**18.44**	m2	**20.75**
width not exceeding 300mm	1.56	10.00	**1.70**	0.43	0.09	**7.07**	-	**8.78**	m	**9.87**
19mm work to isolated beams on timber base										
width exceeding 300mm	4.76	5.00	**5.00**	0.80	0.30	**14.58**	-	**19.58**	m2	**22.03**
width not exceeding 300mm	1.56	10.00	**1.70**	0.48	0.09	**7.78**	-	**9.49**	m	**10.67**
19mm work to isolated columns on timber base										
width exceeding 300mm	4.76	5.00	**5.00**	0.80	0.30	**14.58**	-	**19.58**	m2	**22.03**
width not exceeding 300mm	1.56	10.00	**1.70**	0.48	0.09	**7.78**	-	**9.49**	m	**10.67**
Baseboarding; two layers; first layer 12.5mm Gyproc Duplex wallboard, second layer 9.5mm Gyproc wallboard, both layers with square edge boards; second layer with 3mm joints filled with plaster and scrimmed; fixing with Galvanized nails, 30mm long at 400mm ccs; internal										
22mm work to walls on timber base										
width exceeding 300mm	4.92	5.00	**5.17**	0.65	0.33	**12.77**	-	**17.94**	m2	**20.18**
width not exceeding 300mm	1.61	10.00	**1.75**	0.39	0.10	**6.61**	-	**8.37**	m	**9.41**
22mm work to ceilings on timber base										
width exceeding 300mm	4.92	5.00	**5.17**	0.78	0.33	**14.62**	-	**19.79**	m2	**22.26**
width not exceeding 300mm	1.61	10.00	**1.75**	0.47	0.10	**7.75**	-	**9.50**	m	**10.69**
22mm work to isolated beams on timber base										
width exceeding 300mm	4.92	5.00	**5.17**	0.98	0.33	**17.46**	-	**22.63**	m2	**25.46**
width not exceeding 300mm	1.61	10.00	**1.75**	0.59	0.10	**9.45**	-	**11.21**	m	**12.61**
22mm work to isolated columns on timber base										
width exceeding 300mm	4.92	5.00	**5.17**	0.98	0.33	**17.46**	-	**22.63**	m2	**25.46**
width not exceeding 300mm	1.61	10.00	**1.75**	0.59	0.10	**9.45**	-	**11.21**	m	**12.61**
Baseboarding; two layers; first layer 12.5mm Gyproc Duplex wallboard, second layer 12.5mm Gyproc wallboard, both layers with square edge boards; second layer with 3mm joints filled with plaster and scrimmed; fixing with Galvanized nails, 30mm long at 400mm ccs; internal										
25mm work to walls on timber base										
width exceeding 300mm	5.03	5.00	**5.28**	0.70	0.35	**13.70**	-	**18.98**	m2	**21.35**
width not exceeding 300mm	1.65	10.00	**1.79**	0.42	0.11	**7.15**	-	**8.94**	m	**10.05**
25mm work to ceilings on timber base										
width exceeding 300mm	5.03	5.00	**5.28**	0.84	0.35	**15.68**	-	**20.97**	m2	**23.59**
width not exceeding 300mm	1.65	10.00	**1.79**	0.50	0.11	**8.28**	-	**10.07**	m	**11.33**

Labour hourly rates: (except Specialists) Craft Operatives £14.21 Labourer £10.71 Rates are national average prices. Refer to REGIONAL VARIATIONS for indicative levels of overall pricing in regions	MATERIALS			LABOUR				RATES		
	Del to Site	Waste	Material Cost	Craft Optve	Lab	Labour Cost	Sunds	Nett Rate	Unit	Gross Rate (+12.5%)
	£	%	£	Hrs	Hrs	£	£	£		£
M20: PLASTERED/RENDERED/ROUGHCAST COATINGS - PLASTERBOARD Cont.										
Baseboarding; two layers; first layer 12.5mm Gyproc Duplex wallboard, second layer 12.5mm Gyproc wallboard, both layers with square edge boards; second layer with 3mm joints filled with plaster and scrimmed; fixing with Galvanized nails, 30mm long at 400mm ccs; internal Cont.										
25mm work to isolated beams on timber base										
width exceeding 300mm	5.03	5.00	5.28	1.05	0.35	18.67	-	23.95	m2	26.95
width not exceeding 300mm	1.65	10.00	1.79	0.64	0.11	10.27	-	12.06	m	13.57
25mm work to isolated columns on timber base										
width exceeding 300mm	5.03	5.00	5.28	1.05	0.35	18.67	-	23.95	m2	26.95
width not exceeding 300mm	1.65	10.00	1.79	0.64	0.11	10.27	-	12.06	m	13.57
Baseboarding; two layers; first layer 15mm Gyproc Duplex wallboard, second layer 15mm Gyproc wallboard, both layers with square edge boards; second layer with 3mm joints filled with plaster and scrimmed; fixing with Galvanized nails, 30mm long at 400mm ccs; internal										
30mm work to walls on timber base										
width exceeding 300mm	7.74	5.00	8.13	0.75	0.38	14.73	-	22.86	m2	25.71
width not exceeding 300mm	2.46	10.00	2.69	0.45	0.11	7.57	-	10.26	m	11.54
30mm work to ceilings on timber base										
width exceeding 300mm	7.74	5.00	8.13	0.90	0.38	16.86	-	24.99	m2	28.11
width not exceeding 300mm	2.46	10.00	2.69	0.54	0.11	8.85	-	11.54	m	12.98
30mm work to isolated beams on timber base										
width exceeding 300mm	7.74	5.00	8.13	1.13	0.38	20.13	-	28.26	m2	31.79
width not exceeding 300mm	2.46	10.00	2.69	0.68	0.11	10.84	-	13.53	m	15.22
30mm work to isolated columns on timber base										
width exceeding 300mm	7.74	5.00	8.13	1.13	0.38	20.13	-	28.26	m2	31.79
width not exceeding 300mm	2.46	10.00	2.69	0.68	0.11	10.84	-	13.53	m	15.22
M20: PLASTERED/RENDERED/ROUGHCAST COATINGS: DAMP WALL TREATMENT										
Newlath; fixing with masonry nails; internal										
Work to walls on brickwork or blockwork base										
width exceeding 300mm	12.35	5.00	12.84	0.40	0.20	7.83	-	20.67	m2	23.25
width not exceeding 300mm	3.71	5.00	3.85	0.24	0.06	4.05	-	7.91	m	8.89
M20: PLASTERED/RENDERED/ROUGHCAST COATINGS: PORTLAND CEMENT WORK										
Plain face, first and finishing coats cement and sand (1:3), total 13mm thick; wood floated; internal										
13mm work to walls on brickwork or blockwork base										
width exceeding 300mm	1.57	5.00	1.64	0.60	0.30	11.74	-	13.38	m2	15.06
width exceeding 300mm; dubbing average 6mm thick	2.29	5.00	2.40	0.84	0.44	16.65	-	19.05	m2	21.43
width exceeding 300mm; dubbing average 12mm thick	3.01	5.00	3.16	0.88	0.58	18.72	-	21.88	m2	24.61
width exceeding 300mm; dubbing average 19mm thick	3.86	5.00	4.05	0.92	0.74	21.00	-	25.05	m2	28.18
width exceeding 300mm; curved to 3000mm radius	1.57	5.00	1.64	0.80	0.30	14.58	-	16.23	m2	18.25
width not exceeding 300mm	0.48	10.00	0.53	0.36	0.09	6.08	-	6.61	m	7.44
width not exceeding 300mm; dubbing average 6mm thick	0.72	10.00	0.80	0.53	0.17	9.35	-	10.15	m	11.42
width not exceeding 300mm; dubbing average 12mm thick	0.96	10.00	1.06	0.53	0.17	9.35	-	10.41	m	11.71
width not exceeding 300mm; dubbing average 19mm thick	1.21	10.00	1.33	0.55	0.22	10.17	-	11.50	m	12.93
width not exceeding 300mm; curved to 3000mm radius	0.48	10.00	0.53	0.48	0.09	7.78	-	8.31	m	9.35
If plain face waterproofed										
add	0.32	10.00	0.36	-	0.02	0.21	-	0.57	m2	0.64
Rounded angles										
radius 10 - 100mm	-	-	-	0.30	-	4.26	-	4.26	m	4.80
Work forming flush skirting; 13mm thick										
height 150mm	0.24	50.00	0.36	0.60	0.05	9.06	-	9.42	m	10.60
height 225mm	0.36	50.00	0.54	0.65	0.07	9.99	-	10.53	m	11.84
height 150mm; curved to 3000mm radius	0.24	50.00	0.36	0.80	0.05	11.90	-	12.27	m	13.80
height 225mm; curved to 3000mm radius	0.36	50.00	0.54	0.86	0.07	12.97	-	13.51	m	15.20
Work forming projecting skirting; 13mm projection										
height 150mm	0.24	50.00	0.36	0.90	0.05	13.32	-	13.69	m	15.40
height 225mm	0.36	50.00	0.54	0.98	0.07	14.68	-	15.22	m	17.12
height 150mm; curved to 3000mm radius	0.24	50.00	0.36	1.20	0.05	17.59	-	17.95	m	20.19
height 225mm; curved to 3000mm radius	0.36	50.00	0.54	1.29	0.07	19.08	-	19.62	m	22.08

Labour hourly rates: (except Specialists) Craft Operatives £14.21 Labourer £10.71 Rates are national average prices. Refer to REGIONAL VARIATIONS for indicative levels of overall pricing in regions	MATERIALS			LABOUR			RATES			
	Del to Site £	Waste %	Material Cost £	Craft Optve Hrs	Lab Hrs	Labour Cost £	Sunds £	Nett Rate £	Unit	Gross Rate (+12.5%) £
M20: PLASTERED/RENDERED/ROUGHCAST COATINGS: ROUGH CAST										
Render and dry dash; first and finishing coats cement and sand (1:3), total 15mm thick; wood floated; dry dash of pea shingle; external										
15mm work to walls on brickwork or blockwork base										
width exceeding 300mm	1.98	5.00	**2.08**	0.85	0.43	**16.68**	-	**18.77**	m2	**21.11**
extra; spatterdash coat	0.85	10.00	**0.94**	0.23	0.12	**4.55**	-	**5.49**	m2	**6.18**
width exceeding 300mm; curved to 3000mm radius	1.98	5.00	**2.08**	1.15	0.43	**20.95**	-	**23.03**	m2	**25.91**
width not exceeding 300mm	0.43	10.00	**0.48**	0.51	0.13	**8.64**	-	**9.12**	m	**10.26**
extra; spatterdash coat	0.02	10.00	**0.03**	0.14	0.04	**2.42**	-	**2.44**	m	**2.75**
width not exceeding 300mm; curved to 3000mm radius	0.43	10.00	**0.48**	0.69	0.13	**11.20**	-	**11.68**	m	**13.13**
Render and wet dash; first and finishing coats cement and sand (1:3), total 15mm thick; wood floated; wet dash of crushed stone or shingle and cement slurry										
15mm work to walls on brickwork or blockwork base										
width exceeding 300mm	2.74	5.00	**2.92**	0.95	0.48	**18.64**	-	**21.56**	m2	**24.26**
extra; spatterdash coat	0.51	10.00	**0.56**	0.28	0.14	**5.48**	-	**6.04**	m2	**6.79**
width exceeding 300mm; curved to 3000mm radius	2.74	5.00	**2.92**	1.24	0.48	**22.76**	-	**25.68**	m2	**28.89**
width not exceeding 300mm	0.87	10.00	**0.96**	0.57	0.14	**9.60**	-	**10.56**	m	**11.87**
extra; spatterdash coat	0.24	10.00	**0.27**	0.17	0.04	**2.84**	-	**3.11**	m	**3.50**
width not exceeding 300mm; curved to 3000mm radius	1.53	10.00	**1.69**	0.74	0.14	**12.01**	-	**13.70**	m	**15.41**
M20: PLASTERED/RENDERED/ROUGHCAST COATINGS: TYROLEAN FINISH										
Render and Tyrolean finish; first and finishing coats cement and sand (1:3), total 15mm thick; wood floated; Tyrolean finish of 'Cullamix' mixture applied by machine										
15mm work to walls on brickwork or blockwork base										
width exceeding 300mm	2.45	5.00	**2.57**	1.05	0.30	**18.13**	-	**21.36**	m2	**24.03**
width not exceeding 300mm	0.82	10.00	**0.90**	0.63	0.09	**9.92**	-	**11.21**	m	**12.61**
M20: PLASTERED/RENDERED/ROUGHCAST COATINGS: HARDWALL PLASTERING										
Plaster, BS 1191, Part 1, Class B; finishing coat of board finish, 3mm thick; steel troweled; internal										
3mm work to walls on concrete or plasterboard base										
width exceeding 300mm	0.54	5.00	**0.56**	0.33	0.17	**6.51**	0.01	**7.09**	m2	**7.97**
width exceeding 300mm; curved to 3000mm radius	0.54	5.00	**0.56**	0.44	0.17	**8.07**	0.01	**8.65**	m2	**9.73**
width not exceeding 300mm	0.28	10.00	**0.31**	0.20	0.05	**3.38**	0.00	**3.69**	m	**4.15**
width not exceeding 300mm; curved to 3000mm radius	0.28	10.00	**0.31**	0.27	0.05	**4.37**	0.00	**4.68**	m	**5.27**
3mm work to ceilings on concrete or plasterboard base										
width exceeding 300mm	0.54	5.00	**0.56**	0.40	0.17	**7.50**	0.01	**8.08**	m2	**9.09**
width exceeding 300mm; 3.50 - 5.00m above floor	0.54	5.00	**0.56**	0.43	0.17	**7.93**	0.01	**8.51**	m2	**9.57**
width not exceeding 300mm	0.28	10.00	**0.31**	0.24	0.05	**3.95**	0.00	**4.26**	m	**4.79**
width not exceeding 300mm; 3.50 - 5.00m above floor	0.28	10.00	**0.31**	0.26	0.05	**4.23**	0.00	**4.54**	m	**5.11**
3mm work to isolated beams on concrete or plasterboard base										
width exceeding 300mm	0.54	5.00	**0.56**	0.50	0.17	**8.93**	0.01	**9.50**	m2	**10.69**
width not exceeding 300mm	0.28	10.00	**0.31**	0.30	0.05	**4.80**	0.00	**5.11**	m	**5.75**
3mm work to isolated columns on concrete or plasterboard base										
width exceeding 300mm	0.54	5.00	**0.56**	0.50	0.17	**8.93**	0.01	**9.50**	m2	**10.69**
width not exceeding 300mm	0.28	10.00	**0.31**	0.30	0.05	**4.80**	0.00	**5.11**	m	**5.75**
Plaster; first coat of hardwall, 11mm thick; finishing coat of multi-finish 2mm thick; steel troweled; internal										
13mm work to walls on brickwork or blockwork base										
width exceeding 300mm	2.15	5.00	**2.25**	0.60	0.30	**11.74**	0.05	**14.05**	m2	**15.80**
width exceeding 300mm; curved to 3000mm radius	2.15	5.00	**2.25**	0.80	0.30	**14.58**	0.05	**16.89**	m2	**19.00**
width not exceeding 300mm	0.97	10.00	**1.07**	0.36	0.09	**6.08**	0.02	**7.16**	m	**8.06**
width not exceeding 300mm; curved to 3000mm radius	0.97	10.00	**1.07**	0.48	0.09	**7.78**	0.02	**8.87**	m	**9.98**
13mm work to ceilings on concrete base										
width exceeding 300mm	2.15	5.00	**2.25**	0.72	0.30	**13.44**	0.05	**15.75**	m2	**17.72**
width exceeding 300mm; 3.50 - 5.00m above floor	2.15	5.00	**2.25**	0.78	0.30	**14.30**	0.05	**16.60**	m2	**18.68**
width not exceeding 300mm	0.97	10.00	**1.07**	0.43	0.09	**7.07**	0.02	**8.16**	m	**9.18**
width not exceeding 300mm; 3.50 - 5.00m above floor	0.97	10.00	**1.07**	0.47	0.09	**7.64**	0.02	**8.73**	m	**9.82**
13mm work to isolated beams on concrete base										
width exceeding 300mm	2.15	5.00	**2.25**	0.90	0.30	**16.00**	0.05	**18.31**	m2	**20.60**
width not exceeding 300mm	0.97	10.00	**1.07**	0.54	0.09	**8.64**	0.02	**9.72**	m	**10.94**
13mm work to isolated columns on concrete base										
width exceeding 300mm	2.15	5.00	**2.25**	0.90	0.30	**16.00**	0.05	**18.31**	m2	**20.60**
width not exceeding 300mm	0.97	10.00	**1.07**	0.54	0.09	**8.64**	0.02	**9.72**	m	**10.94**
Rounded angles										
radius 10 - 100mm	-	-	-	0.30	-	**4.26**	-	**4.26**	m	**4.80**

Labour hourly rates: (except Specialists) Craft Operatives £14.21 Labourer £10.71 Rates are national average prices. Refer to REGIONAL VARIATIONS for indicative levels of overall pricing in regions	MATERIALS			LABOUR				RATES		
	Del to Site	Waste	Material Cost	Craft Optve	Lab	Labour Cost	Sunds	Nett Rate	Unit	Gross Rate (+12.5%)
	£	%	£	Hrs	Hrs	£	£	£		£
M20: PLASTERED/RENDERED/ROUGHCAST COATINGS: HARDWALL PLASTERING Cont.										
Plaster; Thistle Universal one coat plaster, 13mm thick; steel troweled; internal										
Note: the thickness is from the face of the metal lathing										
13mm work to walls on metal lathing base										
width exceeding 300mm	3.44	5.00	3.61	0.50	0.25	9.78	0.09	13.48	m2	15.17
width exceeding 300mm; curved to 3000mm radius	3.44	5.00	3.61	0.67	0.25	12.20	0.09	15.90	m2	17.88
width not exceeding 300mm	1.43	10.00	1.58	0.30	0.08	5.12	0.03	6.72	m	7.57
width not exceeding 300mm; curved to 3000mm radius	1.43	10.00	1.58	0.40	0.08	6.54	0.03	8.15	m	9.16
13mm work to ceilings on metal lathing base										
width exceeding 300mm	3.44	5.00	3.61	0.60	0.25	11.20	0.09	14.90	m2	16.77
width exceeding 300mm; 3.50 - 5.00m above floor	3.44	5.00	3.61	0.65	0.25	11.91	0.09	15.61	m2	17.56
width not exceeding 300mm	1.43	10.00	1.58	0.36	0.08	5.97	0.03	7.58	m	8.52
width not exceeding 300mm; 3.50 - 5.00m above floor	1.43	10.00	1.58	0.39	0.08	6.40	0.03	8.00	m	9.00
13mm work to isolated beams on metal lathing base										
width exceeding 300mm	3.44	5.00	3.61	0.75	0.25	13.34	0.09	17.03	m2	19.16
width not exceeding 300mm	1.43	10.00	1.58	0.45	0.08	7.25	0.03	8.86	m	9.96
13mm work to isolated columns on metal lathing base										
width exceeding 300mm	3.44	5.00	3.61	0.75	0.25	13.34	0.09	17.03	m2	19.16
width not exceeding 300mm	1.43	10.00	1.58	0.45	0.08	7.25	0.03	8.86	m	9.96
Rounded angles										
radius 10 - 100mm	-	-	-	0.22	-	3.13	-	3.13	m	3.52
M20: PLASTERED/RENDERED/ROUGHCAST COATINGS: LIGHTWEIGHT PLASTERING										
Plaster, Thistle; pre-mixed; floating coat of browning, 11mm thick; finishing coat of finish, 2mm thick; steel troweled; internal										
13mm work to walls on brickwork or blockwork base										
width exceeding 300mm	1.57	5.00	1.65	0.50	0.25	9.78	0.04	11.48	m2	12.91
width exceeding 300mm; curved to 3000mm radius	1.57	5.00	1.65	0.67	0.25	12.20	0.04	13.89	m2	15.63
width exceeding 300mm; dubbing average 6mm thick	2.33	5.00	2.45	0.78	0.48	16.22	0.06	18.73	m2	21.07
width exceeding 300mm; dubbing average 6mm thick; curved to 3000mm radius	2.33	5.00	2.45	0.94	0.48	18.50	0.06	21.00	m2	23.63
width exceeding 300mm; dubbing average 12mm thick	2.90	5.00	3.04	0.78	0.48	16.22	0.08	19.34	m2	21.76
width exceeding 300mm; dubbing average 12mm thick; curved to 3000mm radius	2.90	5.00	3.04	0.94	0.48	18.50	0.08	21.61	m2	24.32
width exceeding 300mm; dubbing average 19mm thick	3.65	5.00	3.83	0.80	0.62	18.01	0.08	21.92	m2	24.66
width exceeding 300mm; dubbing average 19mm thick; curved to 3000mm radius	3.84	5.00	4.03	0.97	0.62	20.42	0.08	24.53	m2	27.60
width not exceeding 300mm	0.69	10.00	0.76	0.30	0.08	5.12	0.01	5.89	m	6.63
width not exceeding 300mm; curved to 3000mm radius	0.69	10.00	0.76	0.40	0.08	6.54	0.01	7.32	m	8.23
width not exceeding 300mm; dubbing average 6mm thick	1.07	10.00	1.18	0.47	0.15	8.29	0.02	9.48	m	10.67
width not exceeding 300mm; dubbing average 6mm thick; curved to 3000mm radius	1.07	10.00	1.18	0.65	0.15	10.84	0.02	12.04	m	13.54
width not exceeding 300mm; dubbing average 12mm thick	1.26	10.00	1.39	0.47	0.15	8.29	0.02	9.69	m	10.90
width not exceeding 300mm; dubbing average 12mm thick; curved to 3000mm radius	1.26	10.00	1.39	0.65	0.15	10.84	0.02	12.25	m	13.78
width not exceeding 300mm; dubbing average 19mm thick	1.64	10.00	1.80	0.48	0.19	8.86	0.02	10.68	m	12.01
width not exceeding 300mm; dubbing average 19mm thick; curved to 3000mm radius	1.64	10.00	1.80	0.64	0.19	11.13	0.02	12.95	m	14.57
Rounded angles										
radius 10 - 100mm	-	-	-	0.30	-	4.26	-	4.26	m	4.80
Plaster, Thistle; pre-mixed; floating coat of bonding 8mm thick; finishing coat of finish, 2mm thick; steel troweled; internal										
10mm work to walls on concrete or plasterboard base										
width exceeding 300mm	1.57	5.00	1.65	0.50	0.25	9.78	0.04	11.48	m2	12.91
width exceeding 300mm; curved to 3000mm radius	1.57	5.00	1.65	0.67	0.25	12.20	0.04	13.89	m2	15.63
width not exceeding 300mm	0.69	10.00	0.76	0.30	0.08	5.12	0.01	5.89	m	6.63
width not exceeding 300mm; curved to 3000mm radius	0.69	10.00	0.76	0.40	0.08	6.54	0.01	7.32	m	8.23
10mm work to ceilings on concrete or plasterboard base										
width exceeding 300mm	1.57	5.00	1.65	0.60	0.25	11.20	0.04	12.90	m2	14.51
width exceeding 300mm; 3.50 - 5.00m above floor	1.57	5.00	1.65	0.65	0.25	11.91	0.04	13.61	m2	15.31
width not exceeding 300mm	0.69	10.00	0.76	0.36	0.08	5.97	0.01	6.75	m	7.59
width not exceeding 300mm; 3.50 - 5.00m above floor	0.69	10.00	0.76	0.39	0.08	6.40	0.01	7.17	m	8.07
10mm work to isolated beams on concrete or plasterboard base										
width exceeding 300mm	1.57	5.00	1.65	0.75	0.25	13.34	0.04	15.03	m2	16.91
width not exceeding 300mm	0.69	10.00	0.76	0.45	0.08	7.25	0.01	8.03	m	9.03
10mm work to isolated columns on concrete or plasterboard base										
width exceeding 300mm	1.57	5.00	1.65	0.75	0.25	13.34	0.04	15.03	m2	16.91
width not exceeding 300mm	0.69	10.00	0.76	0.45	0.08	7.25	0.01	8.03	m	9.03

Labour hourly rates: (except Specialists) Craft Operatives £14.21 Labourer £10.71 Rates are national average prices. Refer to REGIONAL VARIATIONS for indicative levels of overall pricing in regions	MATERIALS			LABOUR				RATES		
	Del to Site £	Waste %	Material Cost £	Craft Optve Hrs	Lab Hrs	Labour Cost £	Sunds £	Nett Rate £	Unit	Gross Rate (+12.5%) £
M20: PLASTERED/RENDERED/ROUGHCAST COATINGS: LIGHTWEIGHT PLASTERING Cont.										
Plaster, Thistle; pre-mixed; floating coat of bonding 8mm thick; finishing coat of finish, 2mm thick; steel troweled; internal Cont.										
Rounded angles										
radius 10 - 100mm	-	-	-	0.30	-	4.26	-	4.26	m	4.80
Plaster, Thistle; pre-mixed; floating coat of bonding 11mm thick; finishing coat of finish, 2mm thick; steel troweled; internal										
13mm work to ceilings on precast concrete beam and infill blocks base										
width exceeding 300mm	2.14	5.00	2.25	0.60	0.25	11.20	0.05	13.51	m2	15.19
width not exceeding 300mm	1.07	10.00	1.18	0.36	0.08	5.97	0.02	7.17	m	8.06
Plaster, Thistle; tough coat, 11mm thick;finish 2mm thick steel troweled; internal										
Note: the thickness is from the face of the metal lathing										
13mm work to walls on metal lathing base										
width exceeding 300mm	1.61	5.00	1.69	0.70	0.35	13.70	0.04	15.43	m2	17.35
width exceeding 300mm; curved to 3000mm radius	1.61	5.00	1.69	0.93	0.35	16.96	0.04	18.69	m2	21.03
width not exceeding 300mm	0.71	10.00	0.78	0.42	0.11	7.15	0.01	7.94	m	8.93
width not exceeding 300mm; curved to 3000mm radius	0.71	10.00	0.78	0.56	0.11	9.14	0.01	9.93	m	11.17
13mm work to ceilings on metal lathing base										
width exceeding 300mm	1.61	5.00	1.69	0.84	0.35	15.68	0.04	17.42	m2	19.59
width exceeding 300mm; 3.50 - 5.00m above floor	1.61	5.00	1.69	0.91	0.35	16.68	0.04	18.41	m2	20.71
width not exceeding 300mm	0.71	10.00	0.78	0.50	0.11	8.28	0.01	9.07	m	10.21
width not exceeding 300mm; 3.50 - 5.00m above floor	0.71	10.00	0.78	0.55	0.11	8.99	0.01	9.79	m	11.01
13mm work to isolated beams on metal lathing base										
width exceeding 300mm	1.61	5.00	1.69	1.05	0.35	18.67	0.04	20.40	m2	22.95
width not exceeding 300mm	0.71	10.00	0.78	0.63	0.11	10.13	0.01	10.92	m	12.29
13mm work to isolated columns on metal lathing base										
width exceeding 300mm	1.61	5.00	1.69	1.05	0.35	18.67	0.04	20.40	m2	22.95
width not exceeding 300mm	0.71	10.00	0.78	0.63	0.11	10.13	0.01	10.92	m	12.29
Rounded angles										
radius 10 - 100mm	-	-	-	0.40	-	5.68	-	5.68	m	6.39
M20: PLASTERED/RENDERED/ROUGHCAST COATINGS: RENOVATING PLASTER										
Renovating plaster, Thistle; undercoat 11mm thick; finishing coat 2mm thick; steel troweled; internal										
13mm work to walls on existing brickwork or blockwork base										
width exceeding 300mm	2.44	5.00	2.56	0.65	0.33	12.77	0.06	15.40	m2	17.32
width exceeding 300mm; curved to 3000mm radius	2.44	5.00	2.56	0.87	0.33	15.90	0.06	18.52	m2	20.84
width exceeding 300mm; dubbing average 6mm thick	3.35	5.00	3.52	0.90	0.51	18.25	0.09	21.86	m2	24.59
width exceeding 300mm; dubbing average 6mm thick; curved to 3000mm radius	3.35	5.00	3.52	1.20	0.51	22.51	0.09	26.12	m2	29.39
width exceeding 300mm; dubbing average 12mm thick	4.49	5.00	4.71	0.93	0.69	20.61	0.12	25.43	m2	28.61
width exceeding 300mm; dubbing average 12mm thick; curved to 3000mm radius	4.49	5.00	4.71	1.24	0.69	25.01	0.12	29.84	m2	33.57
width not exceeding 300mm	1.11	10.00	1.22	0.39	0.10	6.61	0.02	7.85	m	8.83
width not exceeding 300mm; curved to 3000mm radius	1.11	10.00	1.22	0.52	0.10	8.46	0.02	9.70	m	10.91
width not exceeding 300mm; dubbing average 6mm thick	1.56	10.00	1.72	0.54	0.15	9.28	0.03	11.02	m	12.40
width not exceeding 300mm; dubbing average 6mm thick; curved to 3000mm radius	1.56	10.00	1.72	0.72	0.15	11.84	0.03	13.58	m	15.28
width not exceeding 300mm; dubbing average 12mm thick	2.02	10.00	2.22	0.56	0.21	10.21	0.03	12.46	m	14.02
width not exceeding 300mm; dubbing average 12mm thick; curved to 3000mm radius	2.02	10.00	2.22	0.75	0.21	12.91	0.03	15.16	m	17.05
Rounded angles										
radius 10 - 100mm	-	-	-	0.40	-	5.68	-	5.68	m	6.39
M20: PLASTERED/RENDERED/ROUGHCAST COATINGS: PLASTER BEADS AND STOPS										
Accessories										
galvanized steel angle beads										
Expamet fixing to brickwork or blockwork with masonry nails										
Reference 550	0.59	10.00	0.65	0.10	-	1.42	-	2.07	m	2.33
Reference 553 for 3mm	0.59	10.00	0.65	0.10	-	1.42	-	2.07	m	2.33
Reference 554 for 6mm	0.76	10.00	0.83	0.10	-	1.42	-	2.25	m	2.54
galvanized steel stop beads										
Expamet fixing to brickwork or blockwork with masonry nails										
Reference 560 for 3mm	0.79	10.00	0.86	0.10	-	1.42	-	2.28	m	2.57
Reference 561for 6mm	0.79	10.00	0.86	0.10	-	1.42	-	2.28	m	2.57
Reference 562 for 10mm	0.75	10.00	0.82	0.10	-	1.42	-	2.24	m	2.52

Labour hourly rates: (except Specialists) Craft Operatives £14.21 Labourer £10.71 Rates are national average prices. Refer to REGIONAL VARIATIONS for indicative levels of overall pricing in regions	MATERIALS			LABOUR				RATES		
	Del to Site £	Waste %	Material Cost £	Craft Optve Hrs	Lab Hrs	Labour Cost £	Sunds £	Nett Rate £	Unit	Gross Rate (+12.5%) £
M20: PLASTERED/RENDERED/ROUGHCAST COATINGS: PLASTER BEADS AND STOPS Cont.										
Accessories Cont.										
galvanized steel stop beads Cont.										
Expamet fixing to brickwork or blockwork with masonry nails Cont.										
Reference 563 for 13mm	0.75	10.00	**0.82**	0.10	-	**1.42**	-	**2.24**	m	**2.52**
Reference 565 for 16mm	0.95	10.00	**1.04**	0.10	-	**1.42**	-	**2.46**	m	**2.77**
Reference 566 for 19mm	0.95	10.00	**1.04**	0.10	-	**1.42**	-	**2.46**	m	**2.77**
galvanized steel plasterboard edging beads										
Expamet fixing to timber with galvanized nails, 30mm long at 400mm ccs										
Reference 567 for 10mm	8.41	10.00	**8.53**	0.10	-	**1.42**	-	**9.95**	m	**11.19**
Reference 568 for 13mm	8.41	10.00	**8.53**	0.10	-	**1.42**	-	**9.95**	m	**11.19**
Reference 576 for 15mm	8.41	10.00	**8.53**	0.10	-	**1.42**	-	**9.95**	m	**11.19**
Accessories; external										
galvanized steel stop beads										
Expamet fixing to brickwork or blockwork with masonry nails										
Reference 570	0.73	10.00	**0.80**	0.10	-	**1.42**	-	**2.22**	m	**2.49**
Stainless steel angle beads										
Expamet fixing to brickwork or blockwork with masonry nails										
Reference 545	1.97	10.00	**2.16**	0.10	-	**1.42**	-	**3.59**	m	**4.03**
Stainless steel stop beads										
Expamet fixing to brickwork or blockwork with masonry nails										
Reference 547	1.91	10.00	**2.09**	0.10	-	**1.42**	-	**3.52**	m	**3.95**
M20: PLASTERED/RENDERED/ROUGHCAST COATINGS: ANTI-CRACK STRIPS										
Accessories										
Lathing; BB galvanized expanded metal lath, 9mm mesh X 0.500mm thick x 1.11 kg/m2; butt joints fixing with nails										
100mm wide to walls; one edge to timber; one edge to brickwork or blockwork	0.50	5.00	**0.52**	0.10	-	**1.42**	0.03	**1.97**	m	**2.22**
100mm wide to walls; one edge to timber; one edge to concrete	0.50	5.00	**0.52**	0.12	-	**1.71**	0.04	**2.26**	m	**2.55**
100mm wide to walls; one edge to brickwork or blockwork; one edge to concrete	0.50	5.00	**0.52**	0.14	-	**1.99**	0.05	**2.56**	m	**2.87**
Lathing; BB galvanized expanded metal lath, 9mm mesh x 0.725mm thick x 1.61 kg/m2; butt joints; fixing with nails										
100mm wide to walls; one edge to timber; one edge to brickwork or blockwork	0.58	5.00	**0.61**	0.10	-	**1.42**	0.03	**2.06**	m	**2.32**
100mm wide to walls; one edge to timber; one edge to concrete	0.58	5.00	**0.61**	0.12	-	**1.71**	0.04	**2.35**	m	**2.65**
100mm wide to walls; one edge to brickwork or blockwork; one edge to concrete	0.58	5.00	**0.61**	0.14	-	**1.99**	0.05	**2.64**	m	**2.98**
M20: PLASTERED/RENDERED/ROUGHCAST COATINGS: BONDING FLUID										
Prepare and apply bonding fluid to receive plaster or cement rendering										
Work to walls on existing cement and sand base width exceeding 300mm	0.57	10.00	**0.62**	0.25	-	**3.55**	-	**4.17**	m2	**4.70**
Work to walls on existing glazed tiling base width exceeding 300mm	0.43	10.00	**0.47**	0.20	-	**2.84**	-	**3.31**	m2	**3.72**
Work to walls on existing painted surface width exceeding 300mm	0.51	10.00	**0.56**	0.23	-	**3.27**	-	**3.83**	m2	**4.31**
Work to walls on existing concrete base width exceeding 300mm	0.57	10.00	**0.62**	0.25	-	**3.55**	-	**4.17**	m2	**4.70**
Work to ceilings on existing cement and sand base width exceeding 300mm	0.57	10.00	**0.62**	0.31	-	**4.41**	-	**5.03**	m2	**5.66**
Work to ceilings on existing painted surface width exceeding 300mm	0.51	10.00	**0.56**	0.28	-	**3.98**	-	**4.54**	m2	**5.11**
Work to ceilings on existing concrete base width exceeding 300mm	0.57	10.00	**0.62**	0.31	-	**4.41**	-	**5.03**	m2	**5.66**

Labour hourly rates: (except Specialists) Craft Operatives £14.21 Labourer £10.71 Rates are national average prices. Refer to REGIONAL VARIATIONS for indicative levels of overall pricing in regions	MATERIALS			LABOUR				RATES		
	Del to Site £	Waste %	Material Cost £	Craft Optve Hrs	Lab Hrs	Labour Cost £	Sunds £	Nett Rate £	Unit	Gross Rate (+12.5%) £
M20: PLASTERED/RENDERED/ROUGHCAST COATINGS; STAIRCASE AREAS - PLASTERBOARD										
Baseboarding; Gyproc Handiboard; 5mm joints, filled with plaster and scrimmed; fixing with galvanized nails, 30mm long at 400mm ccs; internal										
9.5mm work to walls on timber base										
width exceeding 300mm	3.00	5.00	**3.16**	0.36	0.15	**6.72**	-	**9.88**	m2	**11.11**
width not exceeding 300mm	0.98	10.00	**1.07**	0.22	0.05	**3.66**	-	**4.73**	m	**5.32**
9.5mm work to ceilings on timber base										
width exceeding 300mm	3.00	5.00	**3.16**	0.44	0.15	**7.86**	-	**11.02**	m2	**12.39**
width not exceeding 300mm	1.03	10.00	**1.12**	0.26	0.05	**4.23**	-	**5.35**	m	**6.02**
Baseboarding; Gyproc Handiboard Duplex; 5mm joints filled with plaster and scrimmed; fixing with galvanized nails, 30mm long at 400mm ccs; internal										
9.5mm work to walls on timber base										
width exceeding 300mm	3.48	5.00	**3.66**	0.36	0.15	**6.72**	-	**10.38**	m2	**11.68**
width not exceeding 300mm	1.12	10.00	**1.23**	0.22	0.05	**3.66**	-	**4.89**	m	**5.50**
9.5mm work to ceilings on timber base										
width exceeding 300mm	3.48	5.00	**3.66**	0.44	0.15	**7.86**	-	**11.52**	m2	**12.96**
width not exceeding 300mm	1.18	10.00	**1.28**	0.26	0.05	**4.23**	-	**5.51**	m	**6.20**
Baseboarding; Gyproc Handiboard; 3mm joints, filled with plaster; fixing with galvanized nails, 30mm long at 400mm ccs; internal										
12.5mm work to walls on timber base										
width exceeding 300mm	4.24	5.00	**4.46**	0.44	0.18	**8.18**	-	**12.64**	m2	**14.22**
width not exceeding 300mm	1.35	10.00	**1.48**	0.26	0.05	**4.23**	-	**5.71**	m	**6.42**
12.5mm work to ceilings on timber base										
width exceeding 300mm	4.05	5.00	**4.26**	0.47	0.18	**8.61**	-	**12.86**	m2	**14.47**
width not exceeding 300mm	1.35	10.00	**1.48**	0.28	0.05	**4.51**	-	**5.99**	m	**6.74**
Baseboarding; Gyproc plank, square edge; 5mm joints, filled with plaster and scrimmed; internal										
19mm work to walls on timber base										
width exceeding 300mm	4.96	5.00	**5.21**	0.59	0.24	**10.95**	-	**16.17**	m2	**18.19**
width not exceeding 300mm	1.58	10.00	**1.73**	0.35	0.07	**5.72**	-	**7.45**	m	**8.39**
19mm work to ceilings on timber base										
width exceeding 300mm	4.96	5.00	**5.21**	0.63	0.24	**11.52**	-	**16.74**	m2	**18.83**
width not exceeding 300mm	1.58	10.00	**1.73**	0.38	0.07	**6.15**	-	**7.88**	m	**8.87**
Baseboarding; Gyproc square edge wallboard; 3mm joints filled with plaster and scrimmed; fixing with galvanized nails, 30mm long at 400mm ccs; internal										
9.5mm work to walls on timber base										
width exceeding 300mm	2.17	5.00	**2.29**	0.36	0.15	**6.72**	-	**9.01**	m2	**10.13**
width not exceeding 300mm	0.73	10.00	**0.79**	0.22	0.05	**3.66**	-	**4.46**	m	**5.01**
12.5mm work to walls on timber base										
width exceeding 300mm	2.28	5.00	**2.40**	0.44	0.18	**8.18**	-	**10.58**	m2	**11.90**
width not exceeding 300mm	0.76	10.00	**0.83**	0.26	0.05	**4.23**	-	**5.06**	m	**5.69**
15mm work to walls on timber base										
width exceeding 300mm	3.44	5.00	**3.62**	0.52	0.21	**9.64**	-	**13.26**	m2	**14.92**
width not exceeding 300mm	1.11	10.00	**1.21**	0.31	0.06	**5.05**	-	**6.26**	m	**7.04**
9.5mm work to ceilings on timber base										
width exceeding 300mm	2.17	5.00	**2.29**	0.39	0.15	**7.15**	-	**9.43**	m2	**10.61**
width not exceeding 300mm	0.73	10.00	**0.79**	0.23	0.05	**3.80**	-	**4.60**	m	**5.17**
12.5mm work to ceilings on timber base										
width exceeding 300mm	2.28	5.00	**2.40**	0.47	0.18	**8.61**	-	**11.01**	m2	**12.38**
width not exceeding 300mm	0.76	10.00	**0.83**	0.28	0.05	**4.51**	-	**5.34**	m	**6.01**
15mm work to ceilings on timber base										
width exceeding 300mm	3.44	5.00	**3.62**	0.55	0.25	**10.49**	-	**14.11**	m2	**15.88**
width not exceeding 300mm	1.11	10.00	**1.21**	0.33	0.08	**5.55**	-	**6.76**	m	**7.60**
Baseboarding; Gyproc square edge Duplex wallboard; 3mm joints filled with plaster and scrimmed; fixing with galvanized nails, 30mm long at 400mm ccs; internal										
9.5mm work to walls on timber base										
width exceeding 300mm	2.78	5.00	**2.93**	0.36	0.15	**6.72**	-	**9.65**	m2	**10.85**
width not exceeding 300mm	0.91	10.00	**1.00**	0.22	0.05	**3.66**	-	**4.66**	m	**5.24**
12.5mm work to walls on timber base										
width exceeding 300mm	2.94	5.00	**3.09**	0.44	0.18	**8.18**	-	**11.27**	m2	**12.68**
width not exceeding 300mm	0.96	10.00	**1.05**	0.26	0.05	**4.23**	-	**5.28**	m	**5.94**
15mm work to walls on timber base										
width exceeding 300mm	4.49	5.00	**4.72**	0.52	0.21	**9.64**	-	**14.36**	m2	**16.16**
width not exceeding 300mm	1.43	10.00	**1.56**	0.31	0.06	**5.05**	-	**6.61**	m	**7.43**
9.5mm work to ceilings on timber base										
width exceeding 300mm	2.78	5.00	**2.93**	0.39	0.15	**7.15**	-	**10.08**	m2	**11.33**

Labour hourly rates: (except Specialists) Craft Operatives £14.21 Labourer £10.71 Rates are national average prices. Refer to REGIONAL VARIATIONS for indicative levels of overall pricing in regions	MATERIALS			LABOUR				RATES		
	Del to Site £	Waste %	Material Cost £	Craft Optve Hrs	Lab Hrs	Labour Cost £	Sunds £	Nett Rate £	Unit	Gross Rate (+12.5%) £
M20: PLASTERED/RENDERED/ROUGHCAST COATINGS; STAIRCASE AREAS - PLASTERBOARD Cont.										
Baseboarding; Gyproc square edge Duplex wallboard; 3mm joints filled with plaster and scrimmed; fixing with galvanized nails, 30mm long at 400mm ccs; internal Cont.										
9.5mm work to ceilings on timber base Cont.										
width not exceeding 300mm	0.91	10.00	1.00	0.23	0.05	3.80	-	4.80	m	5.40
12.5mm work to ceilings on timber base										
width exceeding 300mm	2.94	5.00	3.09	0.47	0.18	8.61	-	11.70	m2	13.16
width not exceeding 300mm	0.96	10.00	1.05	0.28	0.05	4.51	-	5.56	m	6.26
15mm work to ceilings on timber base										
width exceeding 300mm	4.49	5.00	4.72	0.55	0.25	10.49	-	15.22	m2	17.12
width not exceeding 300mm	1.43	10.00	1.56	0.33	0.08	5.55	-	7.11	m	7.99
Baseboarding; two layers 9.5mm Gyproc square edge wallboard; second layer with 3mm joints filled with plaster and scrimmed; fixing with galvanized nails, 30mm long at 400mm ccs; internal										
19mm work to walls on timber base										
width exceeding 300mm	4.15	5.00	4.36	0.75	0.30	13.87	-	18.23	m2	20.51
width not exceeding 300mm	1.38	10.00	1.50	0.45	0.09	7.36	-	8.86	m	9.97
19mm work to ceilings on timber base										
width exceeding 300mm	4.15	5.00	4.36	0.81	0.30	14.72	-	19.08	m2	21.47
width not exceeding 300mm	1.38	10.00	1.50	0.49	0.09	7.93	-	9.43	m	10.61
Baseboarding; one layer 9.5mm and one layer 12.5mm Gyproc square edge wallboard; second layer with 3mm joints filled with plaster and scrimmed; fixing with Galvanized nails, 30mm long at 400mm ccs, internal										
22mm work to walls on timber base										
width exceeding 300mm	4.26	5.00	4.47	0.81	0.33	15.04	-	19.52	m2	21.96
width not exceeding 300mm	1.41	10.00	1.54	0.49	0.10	8.03	-	9.57	m	10.77
22mm work to ceilings on timber base										
width exceeding 300mm	4.26	5.00	4.47	0.88	0.33	16.04	-	20.51	m2	23.08
width not exceeding 300mm	1.41	10.00	1.54	0.53	0.10	8.60	-	10.14	m	11.41
Baseboarding; two layers 12.5mm Gyproc square edge wallboard; second layer with 3mm joints filled with plaster and scrimmed; fixing with galvanized nails, 30mm long at 400mm ccs; internal										
25mm work to walls on timber base										
width exceeding 300mm	4.37	5.00	4.59	0.88	0.35	16.25	-	20.84	m2	23.45
width not exceeding 300mm	1.45	10.00	1.57	0.53	0.11	8.71	-	10.28	m	11.57
25mm work to ceilings on timber base										
width exceeding 300mm	4.37	5.00	4.59	0.95	0.35	17.25	-	21.84	m2	24.57
width not exceeding 300mm	1.45	10.00	1.57	0.57	0.11	9.28	-	10.85	m	12.21
Baseboarding; two layers 15mm Gyproc square edge wallboard; second layer with 3mm joints filled with plaster and scrimmed; fixing with galvanized nails, 30mm long at 400mm ccs; internal										
30mm work to walls on timber base										
width exceeding 300mm	6.69	5.00	7.03	0.95	0.38	17.57	-	24.59	m2	27.67
width not exceeding 300mm	2.14	10.00	2.34	0.57	0.11	9.28	-	11.62	m	13.07
30mm work to ceilings on timber base										
width exceeding 300mm	6.69	5.00	7.03	1.02	0.38	18.56	-	25.59	m2	28.79
width not exceeding 300mm	2.14	10.00	2.34	0.61	0.11	9.85	-	12.18	m	13.71
Baseboarding; two layers; first layer 9.5mm Gyproc Duplex wallboard, second layer 9.5mm Gyproc wallboard both layers with square edge boards; second layer with 3mm joints filled with plaster and scrimmed; fixing with galvanized nails, 30mm long at 400mm ccs; internal										
19mm work to walls on timber base										
width exceeding 300mm	4.76	5.00	5.00	0.75	0.30	13.87	-	18.87	m2	21.23
width not exceeding 300mm	1.56	10.00	1.70	0.45	0.09	7.36	-	9.06	m	10.19
19mm work to ceilings on timber base										
width exceeding 300mm	4.76	5.00	5.00	0.81	0.30	14.72	-	19.72	m2	22.19
width not exceeding 300mm	1.56	10.00	1.70	0.49	0.09	7.93	-	9.63	m	10.83
Baseboarding; two layers; first layer 12.5mm Gyproc Duplex wallboard, second layer 9.5mm Gyproc wallboard, both layers with square edge boards; second layer with 3mm joints filled with plaster and scrimmed; fixing with galvanized nails, 30mm long at 400mm ccs; internal										
22mm work to walls on timber base										
width exceeding 300mm	4.92	5.00	5.17	0.81	0.33	15.04	-	20.21	m2	22.74
width not exceeding 300mm	1.61	10.00	1.75	0.49	0.10	8.03	-	9.79	m	11.01

Labour hourly rates: (except Specialists) Craft Operatives £14.21 Labourer £10.71 Rates are national average prices. Refer to REGIONAL VARIATIONS for indicative levels of overall pricing in regions	MATERIALS			LABOUR				RATES		
	Del to Site	Waste	Material Cost	Craft Optve	Lab	Labour Cost	Sunds	Nett Rate	Unit	Gross Rate (+12.5%)
	£	%	£	Hrs	Hrs	£	£	£		£
M20: PLASTERED/RENDERED/ROUGHCAST COATINGS; STAIRCASE AREAS - PLASTERBOARD Cont.										
Baseboarding; two layers; first layer 12.5mm Gyproc Duplex wallboard, second layer 9.5mm Gyproc wallboard, both layers with square edge boards; second layer with 3mm joints filled with plaster and scrimmed; fixing with galvanized nails, 30mm long at 400mm ccs; internal Cont.										
22mm work to ceilings on timber base										
width exceeding 300mm	4.92	5.00	**5.17**	0.88	0.33	**16.04**	-	**21.21**	m2	**23.86**
width not exceeding 300mm........................	1.61	10.00	**1.75**	0.53	0.10	**8.60**	-	**10.36**	m	**11.65**
Baseboarding; two layers; first layer 12.5mm Gyproc Duplex wallboard, second layer 12.5mm Gyproc wallboard, both layers with square edge boards; second layer with 3mm joints filled with plaster and scrimmed; fixing with galvanized nails, 30mm long at 400mm ccs; internal										
25mm work to walls on timber base										
width exceeding 300mm	5.03	5.00	**5.28**	0.88	0.35	**16.25**	-	**21.54**	m2	**24.23**
width not exceeding 300mm........................	1.65	10.00	**1.79**	0.53	0.11	**8.71**	-	**10.50**	m	**11.81**
25mm work to ceilings on timber base										
width exceeding 300mm	5.03	5.00	**5.28**	0.95	0.35	**17.25**	-	**22.53**	m2	**25.35**
width not exceeding 300mm........................	1.65	10.00	**1.79**	0.57	0.11	**9.28**	-	**11.07**	m	**12.45**
Baseboarding; two layers; first layer 15mm Gyproc Duplex wallboard, second layer 15mm Gyproc wallboard, both layers with square edge boards; second layer with 3mm joints filled with plaster and scrimmed; fixing with galvanized nails, 30mm long at 400mm ccs; internal										
30mm work to walls on timber base										
width exceeding 300mm	7.74	5.00	**8.13**	0.95	0.38	**17.57**	-	**25.70**	m2	**28.91**
width not exceeding 300mm........................	2.46	10.00	**2.69**	0.57	0.11	**9.28**	-	**11.96**	m	**13.46**
30mm work to ceilings on timber base										
width exceeding 300mm	7.74	5.00	**8.13**	1.02	0.38	**18.56**	-	**26.69**	m2	**30.03**
width not exceeding 300mm........................	2.46	10.00	**2.69**	0.61	0.11	**9.85**	-	**12.53**	m	**14.10**
M20: PLASTERED/RENDERED/ROUGHCAST COATINGS: STAIRCASE AREAS - HARDWALL PLASTERING										
Plaster, BS 1191, Part 1, Class B; finishing coat of board finish, 3mm thick; steel troweled; internal										
3mm work to walls on concrete or plasterboard base										
width exceeding 300mm	0.54	5.00	**0.56**	0.41	0.17	**7.65**	0.01	**8.22**	m2	**9.25**
width not exceeding 300mm........................	0.28	10.00	**0.31**	0.25	0.05	**4.09**	0.00	**4.40**	m	**4.95**
3mm work to ceilings on concrete or plasterboard base										
width exceeding 300mm	0.54	5.00	**0.56**	0.45	0.17	**8.22**	0.01	**8.79**	m2	**9.89**
width not exceeding 300mm........................	0.28	10.00	**0.31**	0.27	0.05	**4.37**	0.00	**4.68**	m	**5.27**
Plaster; first coat of hardwall, 11mm thick; finishing coat of multi-finish 2mm thick; steel troweled; internal										
13mm work to walls on brickwork or blockwork base										
width exceeding 300mm	2.15	5.00	**2.25**	0.75	0.30	**13.87**	0.05	**16.18**	m2	**18.20**
width not exceeding 300mm........................	0.97	10.00	**1.07**	0.45	0.09	**7.36**	0.02	**8.44**	m	**9.50**
13mm work to ceilings on concrete base										
width exceeding 300mm	2.15	5.00	**2.25**	0.81	0.30	**14.72**	0.05	**17.03**	m2	**19.16**
width not exceeding 300mm........................	0.97	10.00	**1.07**	0.49	0.09	**7.93**	0.02	**9.01**	m	**10.14**
Plaster; Thistle Universal one coat plaster, 13mm thick; steel troweled; internal										
Note: the thickness is from the face of the metal lathing										
13mm work to walls on metal lathing base										
width exceeding 300mm	3.44	5.00	**3.61**	0.63	0.25	**11.63**	0.09	**15.33**	m2	**17.25**
width not exceeding 300mm........................	1.43	10.00	**1.58**	0.38	0.08	**6.26**	0.03	**7.86**	m	**8.84**
13mm work to ceilings on metal lathing base										
width exceeding 300mm	3.44	5.00	**3.61**	0.68	0.25	**12.34**	0.09	**16.04**	m2	**18.04**
width not exceeding 300mm........................	1.43	10.00	**1.58**	0.41	0.08	**6.68**	0.03	**8.29**	m	**9.32**
M20: PLASTERED/RENDERED/ROUGHCAST COATINGS: STAIRCASE AREAS - LIGHTWEIGHT PLASTERING										
Plaster, Thistle; pre-mixed; floating coat of browning, 11mm thick; finishing coat of finish, 2mm thick; steel troweled; internal										
13mm work to walls on brickwork or blockwork base										
width exceeding 300mm	1.57	5.00	**1.65**	0.63	0.25	**11.63**	0.04	**13.32**	m2	**14.99**
width not exceeding 300mm........................	0.50	10.00	**0.55**	0.38	0.08	**6.26**	0.01	**6.82**	m	**7.68**

Labour hourly rates: (except Specialists) Craft Operatives £14.21 Labourer £10.71 Rates are national average prices. Refer to REGIONAL VARIATIONS for indicative levels of overall pricing in regions	MATERIALS			LABOUR				RATES		
	Del to Site	Waste	Material Cost	Craft Optve	Lab	Labour Cost	Sunds	Nett Rate	Unit	Gross Rate (+12.5%)
	£	%	£	Hrs	Hrs	£	£	£		£
M20: PLASTERED/RENDERED/ROUGHCAST COATINGS: STAIRCASE AREAS - LIGHTWEIGHT PLASTERING Cont.										
Plaster, Thistle; pre-mixed; floating coat of bonding 8mm thick; finishing coat of finish, 2mm thick; steel troweled; internal										
10mm work to walls on concrete or plasterboard base										
width exceeding 300mm	1.57	5.00	**1.65**	0.63	0.25	**11.63**	0.04	**13.32**	m2	**14.99**
width not exceeding 300mm	0.69	10.00	**0.76**	0.38	0.08	**6.26**	0.01	**7.03**	m	**7.91**
10mm work to ceilings on concrete or plasterboard base										
width exceeding 300mm	1.57	5.00	**1.65**	0.68	0.25	**12.34**	0.04	**14.03**	m2	**15.79**
width not exceeding 300mm	0.69	10.00	**0.76**	0.41	0.08	**6.68**	0.01	**7.46**	m	**8.39**
Plaster, Thistle; tough coat, 11mm thick;finish 2mm thick steel troweled; internal										
Note: the thickness is from the face of the metal lathing										
13mm work to walls on metal lathing base										
width exceeding 300mm	1.61	5.00	**1.69**	0.88	0.35	**16.25**	0.04	**17.98**	m2	**20.23**
width not exceeding 300mm	0.71	10.00	**0.78**	0.53	0.11	**8.71**	0.01	**9.50**	m	**10.69**
13mm work to ceilings on metal lathing base										
width exceeding 300mm	1.61	5.00	**1.69**	0.95	0.35	**17.25**	0.04	**18.98**	m2	**21.35**
width not exceeding 300mm	0.71	10.00	**0.78**	0.57	0.11	**9.28**	0.01	**10.07**	m	**11.33**
M30: METAL MESH LATHING/ANCHORED REINFORCEMENT FOR PLASTERED COATINGS										
Lathing; BB galvanized expanded metal lath, 9mm mesh x 0.500mm thick x 1.11 kg/m2; butt joints; fixing with galvanized staples										
Work to walls										
width exceeding 300mm	4.95	5.00	**5.20**	0.18	0.09	**3.52**	0.06	**8.78**	m2	**9.88**
width not exceeding 300mm	1.73	10.00	**1.91**	0.11	0.03	**1.88**	0.04	**3.83**	m	**4.30**
Work to ceilings										
width exceeding 300mm	4.95	5.00	**5.20**	0.23	0.09	**4.23**	0.08	**9.51**	m2	**10.69**
width not exceeding 300mm	1.73	10.00	**1.91**	0.14	0.03	**2.31**	0.05	**4.26**	m	**4.80**
Work to isolated beams										
width exceeding 300mm	4.95	5.00	**5.20**	0.27	0.09	**4.80**	0.09	**10.09**	m2	**11.35**
width not exceeding 300mm	1.73	10.00	**1.91**	0.16	0.03	**2.59**	0.05	**4.55**	m	**5.12**
Work to isolated columns										
width exceeding 300mm	4.95	5.00	**5.20**	0.27	0.09	**4.80**	0.09	**10.09**	m2	**11.35**
width not exceeding 300mm	1.73	10.00	**1.91**	0.16	0.03	**2.59**	0.05	**4.55**	m	**5.12**
Lathing; BB galvanized expanded metal lath 9mm mesh x 0.725mm thick x 1.61 kg/m2; butt joints; fixing with galvanized staples										
Work to walls										
width exceeding 300mm	5.80	5.00	**6.09**	0.18	0.09	**3.52**	0.06	**9.67**	m2	**10.88**
width not exceeding 300mm	2.03	10.00	**2.23**	0.11	0.03	**1.88**	0.04	**4.15**	m	**4.67**
Work to ceilings										
width exceeding 300mm	5.80	5.00	**6.09**	0.23	0.09	**4.23**	0.08	**10.40**	m2	**11.70**
width not exceeding 300mm	2.03	10.00	**2.23**	0.14	0.03	**2.31**	0.05	**4.59**	m	**5.16**
Work to isolated beams										
width exceeding 300mm	5.80	5.00	**6.09**	0.27	0.09	**4.80**	0.09	**10.98**	m2	**12.35**
width not exceeding 300mm	2.03	10.00	**2.23**	0.16	0.03	**2.59**	0.05	**4.88**	m	**5.49**
Work to isolated columns										
width exceeding 300mm	5.80	5.00	**6.09**	0.27	0.09	**4.80**	0.09	**10.98**	m2	**12.35**
width not exceeding 300mm	2.03	10.00	**2.23**	0.16	0.03	**2.59**	0.05	**4.88**	m	**5.49**
Lathing; galvanized expanded metal lath 9mm mesh x 0.950mm thick x 2.50 kg/m2; butt joints; fixing with galvanized staples										
Work to walls										
width exceeding 300mm	9.45	5.00	**9.92**	0.18	0.09	**3.52**	0.06	**13.50**	m2	**15.19**
width not exceeding 300mm	3.31	10.00	**3.64**	0.11	0.03	**1.88**	0.04	**5.56**	m	**6.25**
Work to ceilings										
width exceeding 300mm	9.45	5.00	**9.92**	0.23	0.09	**4.23**	0.08	**14.23**	m2	**16.01**
width not exceeding 300mm	3.31	10.00	**3.64**	0.14	0.03	**2.31**	0.05	**6.00**	m	**6.74**
Work to isolated beams										
width exceeding 300mm	9.45	5.00	**9.92**	0.27	0.09	**4.80**	0.09	**14.81**	m2	**16.66**
width not exceeding 300mm	3.31	10.00	**3.64**	0.16	0.03	**2.59**	0.05	**6.29**	m	**7.07**
Work to isolated columns										
width exceeding 300mm	9.45	5.00	**9.92**	0.27	0.09	**4.80**	0.09	**14.81**	m2	**16.66**
width not exceeding 300mm	3.31	10.00	**3.64**	0.16	0.03	**2.59**	0.05	**6.29**	m	**7.07**

Labour hourly rates: (except Specialists) Craft Operatives £14.21 Labourer £10.71 Rates are national average prices. Refer to REGIONAL VARIATIONS for indicative levels of overall pricing in regions	MATERIALS			LABOUR				RATES		
	Del to Site £	Waste %	Material Cost £	Craft Optve Hrs	Lab Hrs	Labour Cost £	Sunds £	Nett Rate £	Unit	Gross Rate (+12.5%) £
M30: METAL MESH LATHING/ANCHORED REINFORCEMENT FOR PLASTERED COATINGS Cont.										
Lathing; Expamet galvanized Rib-Lath, 0.300mm thick x 1.16 kg/m2; butt joints; fixing with galvanized staples										
Work to walls										
width exceeding 300mm	6.20	5.00	**6.51**	0.18	0.09	**3.52**	0.06	**10.09**	m2	**11.35**
width not exceeding 300mm	2.17	10.00	**2.39**	0.11	0.03	**1.88**	0.04	**4.31**	m	**4.85**
Work to ceilings										
width exceeding 300mm	6.20	5.00	**6.51**	0.23	0.09	**4.23**	0.08	**10.82**	m2	**12.17**
width not exceeding 300mm	2.17	10.00	**2.39**	0.14	0.03	**2.31**	0.05	**4.74**	m	**5.34**
Work to isolated beams										
width exceeding 300mm	6.20	5.00	**6.51**	0.27	0.09	**4.80**	0.09	**11.40**	m2	**12.82**
width not exceeding 300mm	2.17	10.00	**2.39**	0.16	0.03	**2.59**	0.05	**5.03**	m	**5.66**
Work to isolated columns										
width exceeding 300mm	6.20	5.00	**6.51**	0.27	0.09	**4.80**	0.09	**11.40**	m2	**12.82**
width not exceeding 300mm	2.17	10.00	**2.39**	0.16	0.03	**2.59**	0.05	**5.03**	m	**5.66**
Lathing; Expamet galvanized Rib-Lath, 0.500mm thick x 1.90 kg/m2; butt joints; fixing with galvanized staples										
Work to walls										
width exceeding 300mm	7.20	5.00	**7.56**	0.18	0.09	**3.52**	0.06	**11.14**	m2	**12.53**
width not exceeding 300mm	2.52	10.00	**2.77**	0.11	0.03	**1.88**	0.04	**4.69**	m	**5.28**
Work to ceilings										
width exceeding 300mm	7.20	5.00	**7.56**	0.23	0.09	**4.23**	0.08	**11.87**	m2	**13.35**
width not exceeding 300mm	2.52	10.00	**2.77**	0.14	0.03	**2.31**	0.05	**5.13**	m	**5.77**
Work to isolated beams										
width exceeding 300mm	7.20	5.00	**7.56**	0.27	0.09	**4.80**	0.09	**12.45**	m2	**14.01**
width not exceeding 300mm	2.52	10.00	**2.77**	0.16	0.03	**2.59**	0.05	**5.42**	m	**6.10**
Work to isolated columns										
width exceeding 300mm	7.20	5.00	**7.56**	0.27	0.09	**4.80**	0.09	**12.45**	m2	**14.01**
width not exceeding 300mm	2.52	10.00	**2.77**	0.16	0.03	**2.59**	0.05	**5.42**	m	**6.10**
Lathing; Expamet stainless steel Rib-Lath, 0.300mm thick x 1.16 kg/m2; butt joints; fixing with stainless steel staples										
Work to walls										
width exceeding 300mm	16.55	5.00	**17.38**	0.18	0.09	**3.52**	0.06	**20.96**	m2	**23.58**
width not exceeding 300mm	5.79	10.00	**6.37**	0.11	0.03	**1.88**	0.04	**8.29**	m	**9.33**
Work to ceilings										
width exceeding 300mm	16.55	5.00	**17.38**	0.23	0.09	**4.23**	0.08	**21.69**	m2	**24.40**
width not exceeding 300mm	5.79	10.00	**6.37**	0.14	0.03	**2.31**	0.05	**8.73**	m	**9.82**
Work to isolated beams										
width exceeding 300mm	16.55	5.00	**17.38**	0.27	0.09	**4.80**	0.09	**22.27**	m2	**25.05**
width not exceeding 300mm	5.79	10.00	**6.37**	0.16	0.03	**2.59**	0.05	**9.02**	m	**10.15**
Work to isolated columns										
width exceeding 300mm	16.55	5.00	**17.38**	0.27	0.09	**4.80**	0.09	**22.27**	m2	**25.05**
width not exceeding 300mm	5.79	10.00	**6.37**	0.16	0.03	**2.59**	0.05	**9.02**	m	**10.15**
Lathing; galvanized expanded metal lath 9.2mm x 28.6mm mesh x 0.460mm thick x 1.15 kg/m2 for render; butt joints; fixing with galvanized staples										
Work to walls										
width exceeding 300mm	8.75	5.00	**9.19**	0.18	0.09	**3.52**	0.06	**12.77**	m2	**14.36**
width not exceeding 300mm	3.06	10.00	**3.37**	0.11	0.03	**1.88**	0.04	**5.29**	m	**5.95**
Work to ceilings										
width exceeding 300mm	8.75	5.00	**9.19**	0.23	0.09	**4.23**	0.08	**13.50**	m2	**15.18**
width not exceeding 300mm	3.06	10.00	**3.37**	0.14	0.03	**2.31**	0.05	**5.73**	m	**6.44**
Work to isolated beams										
width exceeding 300mm	8.75	5.00	**9.19**	0.27	0.09	**4.80**	0.09	**14.08**	m2	**15.84**
width not exceeding 300mm	3.06	10.00	**3.37**	0.16	0.03	**2.59**	0.05	**6.02**	m	**6.77**
Work to isolated columns										
width exceeding 300mm	8.75	5.00	**9.19**	0.27	0.09	**4.80**	0.09	**14.08**	m2	**15.84**
width not exceeding 300mm	3.06	10.00	**3.37**	0.16	0.03	**2.59**	0.05	**6.02**	m	**6.77**
Lathing; galvanized Red-rib lath, 0.500mm thick x 1.91kg/m2, butt joints; fixing with galvanized staples										
Work to walls										
width exceeding 300mm	7.90	5.00	**8.30**	0.18	0.09	**3.52**	0.06	**11.88**	m2	**13.36**
width not exceeding 300mm	2.77	10.00	**3.04**	0.11	0.03	**1.88**	0.04	**4.96**	m	**5.58**
Work to ceilings										
width exceeding 300mm	7.90	5.00	**8.30**	0.23	0.09	**4.23**	0.08	**12.60**	m2	**14.18**
width not exceeding 300mm	2.77	10.00	**3.04**	0.14	0.03	**2.31**	0.05	**5.40**	m	**6.07**
Work to isolated beams										
width exceeding 300mm	7.90	5.00	**8.30**	0.27	0.09	**4.80**	0.09	**13.18**	m2	**14.83**

Labour hourly rates: (except Specialists) Craft Operatives £14.21 Labourer £10.71 Rates are national average prices. Refer to REGIONAL VARIATIONS for indicative levels of overall pricing in regions	MATERIALS			LABOUR				RATES		
	Del to Site	Waste	Material Cost	Craft Optve	Lab	Labour Cost	Sunds	Nett Rate	Unit	Gross Rate (+12.5%)
	£	%	£	Hrs	Hrs	£	£	£		£
M30: METAL MESH LATHING/ANCHORED REINFORCEMENT FOR PLASTERED COATINGS Cont.										
Lathing; galvanized Red-rib lath, 0.500mm thick x 1.91kg/m2, butt joints; fixing with galvanized staples Cont.										
Work to isolated beams Cont.										
width not exceeding 300mm	2.77	10.00	3.04	0.16	0.03	2.59	0.05	5.69	m	6.40
Work to isolated columns										
width exceeding 300mm	7.90	5.00	8.30	0.27	0.09	4.80	0.09	13.18	m2	14.83
width not exceeding 300mm	2.77	10.00	3.04	0.16	0.03	2.59	0.05	5.69	m	6.40
Extra cost of fixing lathing to brickwork, blockwork or concrete with cartridge fired nails in lieu of to timber with staples										
Work to walls										
width exceeding 300mm	-	-	-	0.12	-	1.71	0.16	1.87	m2	2.10
width not exceeding 300mm	-	-	-	0.07	-	0.99	0.09	1.09	m	1.22
Extra cost of fixing lathing to steel with tying wire in lieu of to timber with staples										
Work to walls										
width exceeding 300mm	-	-	-	0.18	-	2.56	0.18	2.74	m2	3.08
width not exceeding 300mm	-	-	-	0.11	-	1.56	0.11	1.67	m	1.88
Work to ceilings										
width exceeding 300mm	-	-	-	0.18	-	2.56	0.18	2.74	m2	3.08
width not exceeding 300mm	-	-	-	0.11	-	1.56	0.11	1.67	m	1.88
Work to isolated beams										
width exceeding 300mm	-	-	-	0.18	-	2.56	0.18	2.74	m2	3.08
width not exceeding 300mm	-	-	-	0.11	-	1.56	0.11	1.67	m	1.88
Work to isolated columns										
width exceeding 300mm	-	-	-	0.18	-	2.56	0.18	2.74	m2	3.08
width not exceeding 300mm	-	-	-	0.11	-	1.56	0.11	1.67	m	1.88
Expamet galvanized arch formers and fix to brickwork or blockwork with galvanized nails, 30mm long at 400mm ccs										
Arch corners										
372mm radius	15.35	5.00	16.12	0.40	0.20	7.83	0.15	24.10	nr	27.11
452mm radius	18.70	5.00	19.64	0.40	0.20	7.83	0.19	27.65	nr	31.10
602mm radius	23.30	5.00	24.47	0.40	0.20	7.83	0.23	32.52	nr	36.59
752mm radius	30.65	5.00	32.18	0.40	0.20	7.83	0.31	40.32	nr	45.35
Semi-circular arches										
372mm radius	30.55	5.00	32.08	0.80	0.40	15.65	0.31	48.04	nr	54.04
397mm radius	31.40	5.00	32.97	0.80	0.40	15.65	0.31	48.94	nr	55.05
412mm radius	32.45	5.00	34.07	0.80	0.40	15.65	0.32	50.05	nr	56.31
452mm radius	37.55	5.00	39.43	0.80	0.40	15.65	0.38	55.46	nr	62.39
602mm radius	47.30	5.00	49.67	0.80	0.40	15.65	0.47	65.79	nr	74.01
752mm radius	65.10	5.00	68.36	0.80	0.40	15.65	0.65	84.66	nr	95.24
Elliptical arches										
1220mm wide x 340mm rise	58.85	5.00	61.79	0.80	0.40	15.65	0.59	78.03	nr	87.79
1370mm wide x 360mm rise	60.00	5.00	63.00	0.80	0.40	15.65	0.60	79.25	nr	89.16
1520mm wide x 380mm rise	66.20	5.00	69.51	0.80	0.40	15.65	0.66	85.82	nr	96.55
1830mm wide x 410mm rise	70.30	5.00	73.82	1.20	0.60	23.48	0.70	98.00	nr	110.25
2130mm wide x 430mm rise	75.00	5.00	78.75	1.20	0.60	23.48	0.75	102.98	nr	115.85
2440mm wide x 440mm rise	77.20	5.00	81.06	1.20	0.60	23.48	0.77	105.31	nr	118.47
3050mm wide x 520mm rise	79.60	5.00	83.58	1.20	0.60	23.48	0.80	107.85	nr	121.34
Spandrel arches										
760mm wide x 180mm radius x 220mm rise	39.80	5.00	41.79	0.80	0.40	15.65	0.40	57.84	nr	65.07
910mm wide x 180mm radius x 240mm rise	42.15	5.00	44.26	0.80	0.40	15.65	0.42	60.33	nr	67.87
1220mm wide x 230mm radius x 290mm rise	54.25	5.00	56.96	0.80	0.40	15.65	0.54	73.16	nr	82.30
1520mm wide x 230mm radius x 330mm rise	60.25	5.00	63.26	0.80	0.40	15.65	0.60	79.52	nr	89.46
1830mm wide x 230mm radius x 360mm rise	60.90	5.00	63.95	0.80	0.40	15.65	0.61	80.21	nr	90.23
2130mm wide x 230mm radius x 370mm rise	72.35	5.00	75.97	0.80	0.40	15.65	0.72	92.34	nr	103.89
2440mm wide x 230mm radius x 390mm rise	76.65	5.00	80.48	1.60	0.80	31.30	0.77	112.55	nr	126.62
3050mm wide x 230mm radius x 440mm rise	80.95	5.00	85.00	1.60	0.80	31.30	0.81	117.11	nr	131.75
Bulls-eyes										
222mm radius	38.25	5.00	40.16	0.20	0.10	3.91	0.38	44.46	nr	50.02
Soffit strips										
155mm wide	3.85	5.00	4.04	0.10	0.05	1.96	0.04	6.04	m	6.79
Make-up pieces										
600mm long	9.30	5.00	9.77	0.20	0.10	3.91	0.09	13.77	nr	15.49
Expamet galvanized circular window formers and fix to brickwork or blockwork with galvanized nails, 30mm long at 400mm ccs										
Circular windows										
594mm diameter	33.35	5.00	35.02	1.60	0.80	31.30	0.33	66.66	nr	74.99

Labour hourly rates: (except Specialists) Craft Operatives £14.21 Labourer £10.71 Rates are national average prices. Refer to REGIONAL VARIATIONS for indicative levels of overall pricing in regions	MATERIALS			LABOUR			RATES			
	Del to Site £	Waste %	Material Cost £	Craft Optve Hrs	Lab Hrs	Labour Cost £	Sunds £	Nett Rate £	Unit	Gross Rate (+12.5%) £
M30: METAL MESH LATHING/ANCHORED REINFORCEMENT FOR PLASTERED COATINGS; STAIRCASE AREAS										
Lathing; BB galvanized expanded metal lath, 9mm mesh x 0.500mm thick x 1.11 kg/m2; butt joints; fixing with galvanized staples										
Work to walls										
width exceeding 300mm	4.95	5.00	5.20	0.23	0.09	4.23	0.08	9.51	m2	10.69
width not exceeding 300mm	1.73	10.00	1.91	0.14	0.03	2.31	0.05	4.26	m	4.80
Work to ceilings										
width exceeding 300mm	4.95	5.00	5.20	0.29	0.09	5.08	0.10	10.38	m2	11.68
width not exceeding 300mm	1.73	10.00	1.91	0.17	0.03	2.74	0.06	4.70	m	5.29
Lathing; BB galvanized expanded metal lath 9mm mesh x 0.725mm thick x 1.61 kg/m2; butt joints; fixing with galvanized staples										
Work to walls										
width exceeding 300mm	5.80	5.00	6.09	0.23	0.09	4.23	0.08	10.40	m2	11.70
width not exceeding 300mm	2.03	10.00	2.23	0.14	0.03	2.31	0.05	4.59	m	5.16
Work to ceilings										
width exceeding 300mm	5.80	5.00	6.09	0.29	0.09	5.08	0.10	11.27	m2	12.68
width not exceeding 300mm	2.03	10.00	2.23	0.17	0.03	2.74	0.06	5.03	m	5.65
Lathing; galvanized expanded metal lath 9mm mesh x 0.950mm thick x 2.50 kg/m2; butt joints; fixing with galvanized staples										
Work to walls										
width exceeding 300mm	9.45	5.00	9.92	0.23	0.09	4.23	0.08	14.23	m2	16.01
width not exceeding 300mm	3.31	10.00	3.64	0.14	0.03	2.31	0.05	6.00	m	6.74
Work to ceilings										
width exceeding 300mm	9.45	5.00	9.92	0.29	0.09	5.08	0.10	15.10	m2	16.99
width not exceeding 300mm	3.31	10.00	3.64	0.17	0.03	2.74	0.06	6.43	m	7.24
Lathing; Expamet galvanized Rib-Lath, 0.300mm thick x 1.16 kg/m2; butt joints; fixing with galvanized staples										
Work to walls										
width exceeding 300mm	6.20	5.00	6.51	0.23	0.09	4.23	0.08	10.82	m2	12.17
width not exceeding 300mm	2.17	10.00	2.39	0.14	0.03	2.31	0.05	4.74	m	5.34
Work to ceilings										
width exceeding 300mm	6.20	5.00	6.51	0.29	0.09	5.08	0.10	11.69	m2	13.15
width not exceeding 300mm	2.17	10.00	2.39	0.17	0.03	2.74	0.06	5.18	m	5.83
Lathing; Expamet galvanized Rib-Lath, 0.500mm thick x 1.90 kg/m2; butt joints; fixing with galvanized staples										
Work to walls										
width exceeding 300mm	7.20	5.00	7.56	0.23	0.09	4.23	0.08	11.87	m2	13.35
width not exceeding 300mm	2.52	10.00	2.77	0.14	0.03	2.31	0.05	5.13	m	5.77
Work to ceilings										
width exceeding 300mm	7.20	5.00	7.56	0.29	0.09	5.08	0.10	12.74	m2	14.33
width not exceeding 300mm	2.52	10.00	2.77	0.17	0.03	2.74	0.06	5.57	m	6.26
Lathing; Expamet stainless steel Rib-Lath, 0.300mm thick x 1.16 kg/m2; butt joints; fixing with stainless steel staples										
Work to walls										
width exceeding 300mm	16.55	5.00	17.38	0.23	0.09	4.23	0.08	21.69	m2	24.40
width not exceeding 300mm	5.79	10.00	6.37	0.14	0.03	2.31	0.05	8.73	m	9.82
Work to ceilings										
width exceeding 300mm	16.55	5.00	17.38	0.29	0.09	5.08	0.10	22.56	m2	25.38
width not exceeding 300mm	5.79	10.00	6.37	0.17	0.03	2.74	0.06	9.16	m	10.31
Lathing; galvanized Red-rib lath, 0.500mm thick x 1.91kg/m2, butt joints; fixing with galvanized staples										
Work to walls										
width exceeding 300mm	7.90	5.00	8.30	0.23	0.09	4.23	0.08	12.60	m2	14.18
width not exceeding 300mm	2.77	10.00	3.04	0.14	0.03	2.31	0.05	5.40	m	6.07
Work to ceilings										
width exceeding 300mm	7.90	5.00	8.30	0.29	0.09	5.08	0.10	13.48	m2	15.16
width not exceeding 300mm	2.77	10.00	3.04	0.17	0.03	2.74	0.06	5.83	m	6.56
M31: FIBROUS PLASTER: PLASTERBOARD COVE										
Coves										
Gyproc cove; fixing with adhesive										
100mm girth	1.28	10.00	1.41	0.18	0.09	3.52	-	4.93	m	5.55
extra; ends	-	-	-	0.06	-	0.85	-	0.85	nr	0.96
extra; internal angles	-	-	-	0.30	-	4.26	-	4.26	nr	4.80
extra; external angles	-	-	-	0.30	-	4.26	-	4.26	nr	4.80
127mm girth	1.15	10.00	1.26	0.18	0.09	3.52	-	4.79	m	5.38

Labour hourly rates: (except Specialists) Craft Operatives £14.21 Labourer £10.71 Rates are national average prices. Refer to REGIONAL VARIATIONS for indicative levels of overall pricing in regions	MATERIALS			LABOUR				RATES		
	Del to Site	Waste	Material Cost	Craft Optve	Lab	Labour Cost	Sunds	Nett Rate	Unit	Gross Rate (+12.5%)
	£	%	£	Hrs	Hrs	£	£	£		£
M31: FIBROUS PLASTER: PLASTERBOARD COVE Cont.										
Coves Cont.										
Gyproc cove; fixing with adhesive Cont.										
extra; ends	-	-	-	0.06	-	0.85	-	0.85	nr	0.96
extra; internal angles	-	-	-	0.30	-	4.26	-	4.26	nr	4.80
extra; external angles	-	-	-	0.30	-	4.26	-	4.26	nr	4.80
Gyproc cove; fixing with nails to timber										
100mm girth	1.13	10.00	1.24	0.12	0.06	2.35	0.04	3.63	m	4.08
extra; ends	-	-	-	0.06	-	0.85	-	0.85	nr	0.96
extra; internal angles	-	-	-	0.30	-	4.26	-	4.26	nr	4.80
extra; external angles	-	-	-	0.30	-	4.26	-	4.26	nr	4.80
127mm girth	1.00	10.00	1.10	0.12	0.06	2.35	0.04	3.49	m	3.92
extra; ends	-	-	-	0.06	-	0.85	-	0.85	nr	0.96
extra; internal angles	-	-	-	0.30	-	4.26	-	4.26	nr	4.80
extra; external angles	-	-	-	0.30	-	4.26	-	4.26	nr	4.80
M31: FIBROUS PLASTER: PLASTER VENTILATORS										
Fibrous plaster ventilator; fixing in plastered wall										
Plain										
229 x 79mm	1.07	5.00	1.12	0.15	-	2.13	-	3.26	nr	3.66
229 x 152mm	1.28	5.00	1.34	0.18	-	2.56	-	3.90	nr	4.39
229 x 229mm	1.58	5.00	1.66	0.20	-	2.84	-	4.50	nr	5.06
Flyproof										
229 x 79mm	1.21	5.00	1.27	0.20	-	2.84	-	4.11	nr	4.63
229 x 152mm	1.53	5.00	1.61	0.23	-	3.27	-	4.87	nr	5.48
229 x 229mm	1.98	5.00	2.08	0.25	-	3.55	-	5.63	nr	6.34
M40: STONE/CONCRETE/QUARRY/CERAMIC TILING/MOSAIC: CLAY TILE PAVING										
Clay floor quarries, BS 6431, terracotta; 3mm joints, symmetrical layout; bedding in 10mm cement mortar (1:3); pointing in cement mortar (1:3); on cement and sand base										
150 x 150 x 12.5mm units to floors level or to falls only not exceeding 15 degrees from horizontal										
plain	16.60	5.00	17.43	0.95	0.60	19.93	-	37.35	m2	42.02
150 x 150 x 20mm units to floors level or to falls only not exceeding 15 degrees from horizontal										
plain	24.00	5.00	25.20	1.20	0.73	24.87	-	50.07	m2	56.32
225 x 225 x 25mm units to floors level or to falls only not exceeding 15 degrees from horizontal										
plain	25.21	5.00	26.47	0.85	0.55	17.97	-	44.44	m2	49.99
150 x 150 x 12.5mm units to floors to falls and crossfalls and to slopes not exceeding 15 degrees from horizontal										
plain	16.60	5.00	17.43	1.05	0.60	21.35	-	38.78	m2	43.62
150 x 150 x 20mm units to floors to falls and crossfalls and to slopes not exceeding 15 degrees from horizontal										
plain	24.00	5.00	25.20	1.30	0.73	26.29	-	51.49	m2	57.92
225 x 225 x 25mm units to floors to falls and crossfalls and to slopes not exceeding 15 degrees from horizontal										
plain	25.21	5.00	26.47	0.65	0.43	13.84	-	40.31	m2	45.35
Extra for pointing with tinted mortar										
150 x 150 x 12.5mm tiles	0.93	10.00	1.03	0.11	-	1.56	-	2.59	m2	2.91
150 x 150 x 20mm tiles	0.93	10.00	1.03	0.11	-	1.56	-	2.59	m2	2.91
225 x 225 x 25mm tiles	0.93	10.00	1.03	0.08	-	1.14	-	2.16	m2	2.43
150 x 150 x 12.5mm units to skirtings on brickwork or blockwork base										
height 150mm; square top edge	2.77	10.00	3.03	0.36	0.09	6.08	-	9.11	m	10.25
height 150mm; rounded top edge	3.00	10.00	3.28	0.36	0.09	6.08	-	9.36	m	10.53
height 150mm; rounded top edge and cove at bottom	4.72	10.00	5.17	0.36	0.09	6.08	-	11.25	m	12.65
150 x 150 x 20mm units to skirtings on brickwork or blockwork base										
height 150mm; square top edge	3.98	10.00	4.36	0.46	0.11	7.71	-	12.08	m	13.58
height 150mm; rounded top edge	4.58	10.00	5.01	0.46	0.11	7.71	-	12.73	m	14.32
Clay floor quarries, BS 6431, blended; 3mm joints symmetrical layout; bedding in 10mm cement mortar (1:3); pointing in cement mortar (1:3);on cement and sand base										
150 x 150 x 12.5mm units to floors level or to falls only not exceeding 15 degrees from horizontal										
plain	20.94	5.00	21.99	0.95	0.60	19.93	-	41.92	m2	47.15
150 x 150 x 20mm units to floors level or to falls only not exceeding 15 degrees from horizontal										
plain	23.61	5.00	24.79	1.20	0.73	24.87	-	49.66	m2	55.87

Labour hourly rates: (except Specialists) Craft Operatives £14.21 Labourer £10.71 Rates are national average prices. Refer to REGIONAL VARIATIONS for indicative levels of overall pricing in regions	MATERIALS			LABOUR				RATES		
	Del to Site	Waste	Material Cost	Craft Optve	Lab	Labour Cost	Sunds	Nett Rate	Unit	Gross Rate (+12.5%)
	£	%	£	Hrs	Hrs	£	£	£		£
M40: STONE/CONCRETE/QUARRY/CERAMIC TILING/MOSAIC: CLAY TILE PAVING Cont.										
Clay floor quarries, BS 6431, blended; 3mm joints symmetrical layout; bedding in 10mm cement mortar (1:3); pointing in cement mortar (1:3);on cement and sand base Cont.										
225 x 225 x 25mm units to floors level or to falls only not exceeding 15 degrees from horizontal										
plain	26.01	5.00	**27.31**	0.85	0.55	**17.97**	-	**45.28**	m2	**50.94**
150 x 150 x 12.5mm units to floors to falls and crossfalls and to slopes not exceeding 15 degrees from horizontal										
plain	20.94	5.00	**21.99**	1.05	0.60	**21.35**	-	**43.34**	m2	**48.75**
150 x 150 x 20mm units to floors to falls and crossfalls and to slopes not exceeding 15 degrees from horizontal										
plain	23.61	5.00	**24.79**	1.30	0.73	**26.29**	-	**51.08**	m2	**57.47**
225 x 225 x 25mm units to floors to falls and crossfalls and to slopes not exceeding 15 degrees from horizontal										
plain	26.01	5.00	**27.31**	0.65	0.43	**13.84**	-	**41.15**	m2	**46.29**
Extra for pointing with tinted mortar										
150 x 150 x 12.5mm tiles	0.93	10.00	**1.03**	0.11	-	**1.56**	-	**2.59**	m2	**2.91**
150 x 150 x 20mm tiles	0.93	10.00	**1.03**	0.11	-	**1.56**	-	**2.59**	m2	**2.91**
225 x 225 x 25mm tiles	0.93	10.00	**1.03**	0.08	-	**1.14**	-	**2.16**	m2	**2.43**
150 x 150 x 12.5mm units to skirtings on brickwork or blockwork base										
height 150mm; square top edge	3.48	10.00	**3.81**	0.36	0.09	**6.08**	-	**9.89**	m	**11.12**
height 150mm; rounded top edge	3.91	10.00	**4.28**	0.36	0.09	**6.08**	-	**10.36**	m	**11.66**
height 150mm; rounded top edge and cove at bottom	4.88	10.00	**5.35**	0.36	0.09	**6.08**	-	**11.43**	m	**12.86**
150 x 150 x 20mm units to skirtings on brickwork or blockwork base										
height 150mm; square top edge	3.93	10.00	**4.31**	0.46	0.11	**7.71**	-	**12.02**	m	**13.52**
height 150mm; rounded top edge	4.22	10.00	**4.62**	0.46	0.11	**7.71**	-	**12.34**	m	**13.88**
Clay floor quarries, BS 6431, dark; 3mm joints, symmetrical layout; bedding in 10mm cement mortar (1:3); pointing in cement mortar (1:3); on cement and sand base										
150 x 150 x 12.5mm units to floors level or to falls only not exceeding 15degrees from horizontal										
plain	23.78	5.00	**24.97**	0.95	0.60	**19.93**	-	**44.90**	m2	**50.51**
150 x 150 x 12.5mm units to floors to falls and crossfalls and to slopes not exceeding 15 degrees from horizontal										
plain	23.78	5.00	**24.97**	1.05	0.60	**21.35**	-	**46.32**	m2	**52.11**
150 x 150 x 12.5mm units to skirtings on brickwork or blockwork base										
height 150mm; square top edge	3.94	10.00	**4.31**	0.36	0.09	**6.08**	-	**10.39**	m	**11.69**
height 150mm; rounded top edge	4.18	10.00	**4.58**	0.36	0.09	**6.08**	-	**10.66**	m	**11.99**
height 150mm; rounded top edge and cove at bottom	5.64	10.00	**6.19**	0.36	0.09	**6.08**	-	**12.26**	m	**13.80**
Ceramic floor tiles, BS 6431, fully vitrified, red; 3mm joints, symmetrical layout; bedding in 10mm cement mortar (1:3); pointing in cement mortar (1:3);on cement and sand base										
100 x 100 x 9mm units to floors level or to falls only not exceeding 15 degrees from horizontal										
plain	17.36	5.00	**18.23**	1.95	1.10	**39.49**	-	**57.72**	m2	**64.93**
152 x 152 x 12mm units to floors level or to falls only not exceeding 15 degrees from horizontal										
plain	18.86	5.00	**19.80**	1.10	0.68	**22.91**	-	**42.71**	m2	**48.05**
200 x 200 x 12mm units to floors level or to falls only not exceeding 15 degrees from horizontal										
plain	20.16	5.00	**21.17**	0.75	0.50	**16.01**	-	**37.18**	m2	**41.83**
100 x 100 x 9mm units to floors to falls and crossfalls and to slopes not exceeding 15 degrees from horizontal										
plain	17.36	5.00	**18.23**	2.15	1.10	**42.33**	-	**60.56**	m2	**68.13**
152 x 152 x 12mm units to floors to falls and crossfalls and to slopes not exceeding 15 degrees from horizontal										
plain	18.86	5.00	**19.80**	1.21	0.68	**24.48**	-	**44.28**	m2	**49.81**
200 x 200 x 12mm units to floors to falls and crossfalls and to slopes not exceeding 15 degrees from horizontal										
plain	20.16	5.00	**21.17**	0.83	0.50	**17.15**	-	**38.31**	m2	**43.10**

Labour hourly rates: (except Specialists) Craft Operatives £14.21 Labourer £10.71 Rates are national average prices. Refer to REGIONAL VARIATIONS for indicative levels of overall pricing in regions	MATERIALS			LABOUR				RATES		
	Del to Site	Waste	Material Cost	Craft Optve	Lab	Labour Cost	Sunds	Nett Rate	Unit	Gross Rate (+12.5%)
	£	%	£	Hrs	Hrs	£	£	£		£
M40: STONE/CONCRETE/QUARRY/CERAMIC TILING/MOSAIC: CLAY TILE PAVING Cont.										
Ceramic floor tiles, BS 6431, fully vitrified, red; 3mm joints, symmetrical layout; bedding in 10mm cement mortar (1:3); pointing in cement mortar (1:3);on cement and sand base Cont.										
Extra for pointing with tinted mortar										
100 x 100 x 9mm tiles	0.93	10.00	1.03	0.18	-	2.56	-	3.58	m2	4.03
152 x 152 x 12mm tiles	0.93	10.00	1.03	0.12	-	1.71	-	2.73	m2	3.07
200 x 200 x 12mm tiles	0.93	10.00	1.03	0.10	-	1.42	-	2.45	m2	2.75
100 x 100 x 9mm units to skirtings on brickwork or blockwork base										
height 100mm; square top edge	1.80	10.00	1.96	0.40	0.11	6.86	-	8.83	m	9.93
height 100mm; rounded top edge	2.31	10.00	2.53	0.40	0.11	6.86	-	9.39	m	10.57
152 x 152 x 12mm units to skirtings on brickwork or blockwork base										
height 152mm; square top edge	3.15	10.00	3.45	0.36	0.10	6.19	-	9.63	m	10.84
height 152mm; rounded top edge	3.75	10.00	4.11	0.36	0.10	6.19	-	10.29	m	11.58
height 152mm; rounded top edge and cove at bottom	5.26	10.00	5.77	0.36	0.10	6.19	-	11.96	m	13.45
200 x 200 x 12mm units to skirtings on brickwork or blockwork base										
height 200mm; square top edge	4.28	10.00	4.68	0.33	0.10	5.76	-	10.44	m	11.75
Ceramic floor tiles, BS 6431, fully vitrified, cream; 3mm joints, symmetrical layout; bedding in 10mm cement mortar (1:3); pointing in cement mortar (1:3); on cement and sand base										
100 x 100 x 9mm units to floors level or to falls only not exceeding 15 degrees from horizontal										
plain	18.31	5.00	19.22	1.95	1.10	39.49	-	58.71	m2	66.05
152 x 152 x 12mm units to floors level or to falls only not exceeding 15 degrees from horizontal										
plain	19.66	5.00	20.64	1.10	0.68	22.91	-	43.55	m2	49.00
200 x 200 x 12mm units to floors level or to falls only not exceeding 15 degrees from horizontal										
plain	20.91	5.00	21.95	0.75	0.50	16.01	-	37.97	m2	42.71
100 x 100 x 9mm units to floors to falls and crossfalls and to slopes not exceeding 15 degrees from horizontal										
plain	18.31	5.00	19.22	2.15	1.10	42.33	-	61.56	m2	69.25
152 x 152 x 12mm units to floors to falls and crossfalls and to slopes not exceeding 15 degrees from horizontal										
plain	19.66	5.00	20.64	1.21	0.68	24.48	-	45.12	m2	50.76
200 x 200 x 12mm units to floors to falls and crossfalls and to slopes not exceeding 15 degrees from horizontal										
plain	20.91	5.00	21.95	0.83	0.50	17.15	-	39.10	m2	43.99
Extra for pointing with tinted mortar										
100 x 100 x 9mm tiles	0.93	10.00	1.03	0.18	-	2.56	-	3.58	m2	4.03
152 x 152 x 12mm tiles	0.93	10.00	1.03	0.12	-	1.71	-	2.73	m2	3.07
200 x 200 x 12mm tiles	0.93	10.00	1.03	0.10	-	1.42	-	2.45	m2	2.75
100 x 100 x 9mm units to skirtings on brickwork or blockwork base										
height 100mm; square top edge	1.89	10.00	2.07	0.40	0.11	6.86	-	8.93	m	10.05
height 100mm; rounded top edge	2.41	10.00	2.63	0.40	0.11	6.86	-	9.50	m	10.68
152 x 152 x 12mm units to skirtings on brickwork or blockwork base										
height 152mm; square top edge	3.07	10.00	3.36	0.36	0.10	6.19	-	9.55	m	10.74
height 152mm; rounded top edge	3.80	10.00	4.17	0.36	0.10	6.19	-	10.35	m	11.65
height 152mm; rounded top edge and cove at bottom	5.28	10.00	5.79	0.36	0.10	6.19	-	11.98	m	13.47
200 x 200 x 12mm units to skirtings on brickwork or blockwork base										
height 200mm; square top edge	4.42	10.00	4.83	0.33	0.10	5.76	-	10.59	m	11.92
Ceramic floor tiles, BS 6431, fully vitrified, black; 3mm joints, symmetrical layout; bedding in 10mm cement mortar (1:3); pointing in cement mortar (1:3); on cement and sand base										
100 x 100 x 9mm units to floors level or to falls only not exceeding 15 degrees from horizontal										
plain	22.11	5.00	23.21	1.95	1.10	39.49	-	62.70	m2	70.54
152 x 152 x 12mm units to floors level or to falls only not exceeding 15 degrees from horizontal										
plain	23.11	5.00	24.26	1.10	0.68	22.91	-	47.18	m2	53.07
200 x 200 x 12mm units to floors level or to falls only not exceeding 15 degrees from horizontal										
plain	24.01	5.00	25.21	0.75	0.50	16.01	-	41.22	m2	46.37

Labour hourly rates: (except Specialists) Craft Operatives £14.21 Labourer £10.71 Rates are national average prices. Refer to REGIONAL VARIATIONS for indicative levels of overall pricing in regions	MATERIALS			LABOUR				RATES		
	Del to Site £	Waste %	Material Cost £	Craft Optve Hrs	Lab Hrs	Labour Cost £	Sunds £	Nett Rate £	Unit	Gross Rate (+12.5%) £
M40: STONE/CONCRETE/QUARRY/CERAMIC TILING/MOSAIC: CLAY TILE PAVING Cont.										
Ceramic floor tiles, BS 6431, fully vitrified, black; 3mm joints, symmetrical layout; bedding in 10mm cement mortar (1:3); pointing in cement mortar (1:3); on cement and sand base Cont.										
100 x 100 x 9mm units to floors to falls and crossfalls and to slopes not exceeding 15 degrees from horizontal										
plain	22.11	5.00	**23.21**	2.15	1.10	**42.33**	-	**65.55**	m2	**73.74**
152 x 152 x 12mm units to floors to falls and crossfalls and to slopes not exceeding 15 degrees from horizontal										
plain	23.11	5.00	**24.26**	1.21	0.68	**24.48**	-	**48.74**	m2	**54.83**
200 x 200 x 12mm units to floors to falls and crossfalls and to slopes not exceeding 15 degrees from horizontal										
plain	24.01	5.00	**25.21**	0.83	0.50	**17.15**	-	**42.36**	m2	**47.65**
Extra for pointing with tinted mortar										
100 x 100 x 9mm tiles	0.93	10.00	**1.03**	0.18	-	**2.56**	-	**3.58**	m2	**4.03**
152 x 152 x 12mm tiles	0.93	10.00	**1.03**	0.12	-	**1.71**	-	**2.73**	m2	**3.07**
200 x 200 x 12mm tiles	0.93	10.00	**1.03**	0.10	-	**1.42**	-	**2.45**	m2	**2.75**
100 x 100 x 9mm units to skirtings on brickwork or blockwork base										
height 100mm; square top edge	2.27	10.00	**2.49**	0.40	0.11	**6.86**	-	**9.35**	m	**10.52**
height 100mm; rounded top edge	2.79	10.00	**3.05**	0.40	0.11	**6.86**	-	**9.91**	m	**11.15**
152 x 152 x 12mm units to skirtings on brickwork or blockwork base										
height 152mm; square top edge	3.60	10.00	**3.94**	0.36	0.10	**6.19**	-	**10.13**	m	**11.39**
height 152mm; rounded top edge	4.06	10.00	**4.45**	0.36	0.10	**6.19**	-	**10.64**	m	**11.97**
height 152mm; rounded top edge and cove at bottom	5.26	10.00	**5.77**	0.36	0.10	**6.19**	-	**11.96**	m	**13.45**
200 x 200 x 12mm units to skirtings on brickwork or blockwork base										
height 200mm; square top edge	5.04	10.00	**5.52**	0.33	0.10	**5.76**	-	**11.28**	m	**12.69**
M40: STONE/CONCRETE/QUARRY/CERAMIC TILING/MOSAIC: TILE WALL LININGS										
Ceramic tiles, BS 6431, glazed white; 2mm joints, symmetrical layout; bedding in 10mm cement mortar (1:3); pointing with neat white cement; on brickwork or blockwork base										
108 x 108 x 6.5mm units to walls										
plain, width exceeding 300mm	18.38	5.00	**19.31**	1.60	0.90	**32.38**	-	**51.69**	m2	**58.15**
plain, width not exceeding 300mm	6.08	10.00	**6.67**	0.95	0.57	**19.60**	-	**26.27**	m	**29.56**
152 x 152 x 5.5mm units to walls										
plain, width exceeding 300mm	12.66	5.00	**13.30**	1.25	0.73	**25.58**	-	**38.88**	m2	**43.74**
plain, width not exceeding 300mm	3.93	10.00	**4.30**	0.75	0.45	**15.48**	-	**19.78**	m	**22.25**
Ceramic tiles, BS 6431, glazed white; 2mm joints; symmetrical layout; fixing with thin bed adhesive; pointing with neat white cement; on plaster base										
108 x 108 x 6.5mm units to walls										
plain, width exceeding 300mm	19.39	5.00	**20.49**	1.60	0.90	**32.38**	-	**52.87**	m2	**59.47**
plain, width not exceeding 300mm	6.72	10.00	**7.39**	0.95	0.57	**19.60**	-	**26.99**	m	**30.37**
152 x 152 x 5.5mm units to walls										
plain, width exceeding 300mm	13.67	5.00	**14.48**	1.25	0.73	**25.58**	-	**40.06**	m2	**45.07**
plain, width not exceeding 300mm	4.56	10.00	**5.02**	0.75	0.45	**15.48**	-	**20.50**	m	**23.06**
Ceramic tiles, BS 6431, glazed white; 2mm joints, symmetrical layout; fixing with thick bed adhesive; pointing with neat white cement; on cement and sand base										
108 x 108 x 6.5mm units to walls										
plain, width exceeding 300mm	20.51	5.00	**21.72**	1.60	0.90	**32.38**	-	**54.10**	m2	**60.86**
plain, width not exceeding 300mm	7.22	10.00	**7.94**	0.95	0.57	**19.60**	-	**27.54**	m	**30.98**
152 x 152 x 5.5mm units to walls										
plain, width exceeding 300mm	14.79	5.00	**15.71**	1.25	0.73	**25.58**	-	**41.29**	m2	**46.45**
plain, width not exceeding 300mm	5.06	10.00	**5.57**	0.75	0.45	**15.48**	-	**21.05**	m	**23.68**
Ceramic tiles, BS 6431, glazed light colour; 2mm joints symmetrical layout; bedding in 10mm cement mortar (1:3); pointing with neat white cement; on cement and sand base										
108 x 108 x 6.5mm units to walls										
plain, width exceeding 300mm	18.38	5.00	**19.31**	1.60	0.90	**32.38**	-	**51.69**	m2	**58.15**
plain, width not exceeding 300mm	6.08	10.00	**6.67**	0.95	0.57	**19.60**	-	**26.27**	m	**29.56**
152 x 152 x 5.5mm units to walls										
plain, width exceeding 300mm	12.66	5.00	**13.30**	1.25	0.73	**25.58**	-	**38.88**	m2	**43.74**
plain, width not exceeding 300mm	3.93	10.00	**4.30**	0.75	0.45	**15.48**	-	**19.78**	m	**22.25**

Labour hourly rates: (except Specialists) Craft Operatives £14.21 Labourer £10.71 Rates are national average prices. Refer to REGIONAL VARIATIONS for indicative levels of overall pricing in regions	MATERIALS			LABOUR				RATES		
	Del to Site £	Waste %	Material Cost £	Craft Optve Hrs	Lab Hrs	Labour Cost £	Sunds £	Nett Rate £	Unit	Gross Rate (+12.5%) £
M40: STONE/CONCRETE/QUARRY/CERAMIC TILING/MOSAIC: TILE WALL LININGS Cont.										
Ceramic tiles, BS 6431, glazed light colour; 2mm joints, symmetrical layout; fixing with thin bed adhesive; pointing with neat white cement; on plaster base										
108 x 108 x 6.5mm units to walls										
plain, width exceeding 300mm	19.39	5.00	**20.49**	1.60	0.90	**32.38**	-	**52.87**	m2	**59.47**
plain, width not exceeding 300mm	6.72	10.00	**7.39**	0.95	0.57	**19.60**	-	**26.99**	m	**30.37**
152 x 152 x 5.5mm units to walls										
plain, width exceeding 300mm	13.67	5.00	**14.48**	1.25	0.73	**25.58**	-	**40.06**	m2	**45.07**
plain, width not exceeding 300mm	4.56	10.00	**5.02**	0.75	0.45	**15.48**	-	**20.50**	m	**23.06**
Ceramic tiles, BS 6431, glazed light colour; 2mm joints, symmetrical layout; fixing with thick bed adhesive; pointing with neat white cement; on cement and sand base										
108 x 108 x 6.5mm units to walls										
plain, width exceeding 300mm	20.51	5.00	**21.72**	1.60	0.90	**32.38**	-	**54.10**	m2	**60.86**
plain, width not exceeding 300mm	7.22	10.00	**7.94**	0.95	0.57	**19.60**	-	**27.54**	m	**30.98**
152 x 152 x 5.5mm units to walls										
plain, width exceeding 300mm	14.79	5.00	**15.71**	1.25	0.73	**25.58**	-	**41.29**	m2	**46.45**
plain, width not exceeding 300mm	5.06	10.00	**5.57**	0.75	0.45	**15.48**	-	**21.05**	m	**23.68**
Ceramic tiles, BS 6431, glazed dark colour; 2mm joints, symmetrical layout; bedding in 10mm cement mortar (1:3); pointing with neat white cement; on brickwork or blockwork base										
108 x 108 x 6.5mm units to walls										
plain, width exceeding 300mm	21.03	5.00	**22.10**	1.60	0.90	**32.38**	-	**54.47**	m2	**61.28**
plain, width not exceeding 300mm	6.95	10.00	**7.63**	0.95	0.57	**19.60**	-	**27.23**	m	**30.64**
152 x 152 x 5.5mm units to walls										
plain, width exceeding 300mm	15.51	5.00	**16.29**	1.25	0.73	**25.58**	-	**41.88**	m2	**47.11**
plain, width not exceeding 300mm	4.80	10.00	**5.27**	0.75	0.45	**15.48**	-	**20.74**	m	**23.34**
Ceramic tiles, BS 6431, glazed dark colour; 2mm joints, symmetrical layout; fixing with thin bed adhesive; pointing with neat white cement; on plaster base										
108 x 108 x 6.5mm units to walls										
plain, width exceeding 300mm	22.04	5.00	**23.27**	1.60	0.90	**32.38**	-	**55.65**	m2	**62.60**
plain, width not exceeding 300mm	7.59	10.00	**8.35**	0.95	0.57	**19.60**	-	**27.95**	m	**31.45**
152 x 152 x 5.5mm units to walls										
plain, width exceeding 300mm	16.52	5.00	**17.47**	1.25	0.73	**25.58**	-	**43.05**	m2	**48.43**
plain, width not exceeding 300mm	5.44	10.00	**5.99**	0.75	0.45	**15.48**	-	**21.46**	m	**24.15**
Ceramic tiles, BS 6431, glazed dark colour; 2mm joints, symmetrical layout; fixing with thick bed adhesive; pointing with neat white cement; on cement and sand base										
108 x 108 x 6.5mm units to walls										
plain, width exceeding 300mm	23.16	5.00	**24.50**	1.60	0.90	**32.38**	-	**56.88**	m2	**63.99**
plain, width not exceeding 300mm	8.09	10.00	**8.90**	0.95	0.57	**19.60**	-	**28.50**	m	**32.07**
152 x 152 x 5.5mm units to walls										
plain, width exceeding 300mm	17.64	5.00	**18.70**	1.25	0.73	**25.58**	-	**44.28**	m2	**49.82**
plain, width not exceeding 300mm	5.94	10.00	**6.54**	0.75	0.45	**15.48**	-	**22.01**	m	**24.77**
Extra over ceramic tiles, BS 6431, glazed white; 2mm joints, symmetrical layout; bedding or fixing in any material to general surfaces on any base										
Special tiles										
rounded edge	3.69	5.00	**3.87**	-	-	**-**	-	**3.87**	m	**4.35**
external angle	7.37	5.00	**7.74**	-	-	**-**	-	**7.74**	m	**8.71**
internal angle to skirting	-	-	**-**	0.10	-	**1.42**	-	**1.42**	nr	**1.60**
external angle to skirting	-	-	**-**	0.10	-	**1.42**	-	**1.42**	nr	**1.60**
internal angle	-	-	**-**	0.14	-	**1.99**	-	**1.99**	m	**2.24**
Extra over ceramic tiles, BS 6431, glazed light colour; 2mm joints, symmetrical layout; bedding or fixing in any material to general surfaces on any base										
Special tiles										
rounded edge	3.69	5.00	**3.87**	-	-	**-**	-	**3.87**	m	**4.35**
external angle	7.37	5.00	**7.74**	-	-	**-**	-	**7.74**	m	**8.71**
internal angle to skirting	-	-	**-**	0.10	-	**1.42**	-	**1.42**	nr	**1.60**
external angle to skirting	-	-	**-**	0.10	-	**1.42**	-	**1.42**	nr	**1.60**
internal angle	-	-	**-**	0.14	-	**1.99**	-	**1.99**	m	**2.24**
Extra over ceramic tiles, BS 6431, glazed dark colour; 2mm joints, symmetrical layout; bedding or fixing in any material to general surfaces on any base										
Special tiles										
rounded edge	4.27	5.00	**4.48**	-	-	**-**	-	**4.48**	m	**5.04**

Labour hourly rates: (except Specialists) Craft Operatives £14.21 Labourer £10.71 Rates are national average prices. Refer to REGIONAL VARIATIONS for indicative levels of overall pricing in regions	MATERIALS			LABOUR				RATES		
	Del to Site £	Waste %	Material Cost £	Craft Optve Hrs	Lab Hrs	Labour Cost £	Sunds £	Nett Rate £	Unit	Gross Rate (+12.5%) £
M40: STONE/CONCRETE/QUARRY/CERAMIC TILING/MOSAIC: TILE WALL LININGS Cont.										
Extra over ceramic tiles, BS 6431, glazed dark colour; 2mm joints, symmetrical layout; bedding or fixing in any material to general surfaces on any base Cont.										
Special tiles Cont.										
external angle bead	4.27	5.00	4.48	0.22	-	3.13	-	7.61	m	8.56
internal angle to skirting	-	-	-	0.10	-	1.42	-	1.42	nr	1.60
external angle to skirting	-	-	-	0.10	-	1.42	-	1.42	nr	1.60
internal angle	-	-	-	0.14	-	1.99	-	1.99	m	2.24
M40: STONE/CONCRETE/QUARRY/CERAMIC TILING/MOSAIC: TILE CILLS										
Ceramic tiles, BS 6431, glazed white; 2mm joints, symmetrical layout; bedding in 10mm cement mortar (1:3); pointing with neat white cement; on brickwork or blockwork base - cills										
108 x 108 x 4mm units to cills on brickwork or blockwork base										
width 150mm; rounded angle	3.10	5.00	3.26	0.44	0.14	7.75	-	11.01	m	12.39
width 225mm; rounded angle	5.11	5.00	5.37	0.60	0.20	10.67	-	16.04	m	18.05
width 300mm; rounded angle	6.32	5.00	6.64	0.75	0.27	13.55	-	20.19	m	22.72
152 x 152 x 5.5mm units to cills on brickwork or blockwork base										
width 150mm; rounded angle	2.84	5.00	2.99	0.36	0.11	6.29	-	9.28	m	10.44
width 225mm; rounded angle	3.84	5.00	4.04	0.50	0.17	8.93	-	12.97	m	14.59
width 300mm; rounded angle	4.96	5.00	5.22	0.60	0.22	10.88	-	16.10	m	18.11
Clay floor quarries, BS 6431, terracotta; 3mm joints, symmetrical layout; bedding in 10mm cement mortar (1:3); pointing in cement mortar (1:3); on cement and sand base - cills										
152 x 152 x 12.5mm units to cills on brickwork or blockwork base										
width 152mm; rounded angle	3.00	10.00	3.28	0.33	0.10	5.76	-	9.04	m	10.17
width 225mm; rounded angle	4.45	5.00	4.80	0.44	0.15	7.86	-	12.66	m	14.24
width 300mm; rounded angle	5.89	5.00	6.32	0.56	0.20	10.10	-	16.42	m	18.47
M40: STONE/CONCRETE/QUARRY/CERAMIC TILING/MOSAIC; PLANT ROOMS										
Clay floor quarries, BS 6431, terracotta; 3mm joints, symmetrical layout; bedding in 10mm cement mortar (1:3); pointing in cement mortar (1:3); on cement and sand base										
150 x 150 x 12.5mm units to floors level or to falls only not exceeding 15 degrees from horizontal										
plain	16.60	5.00	17.43	1.05	0.60	21.35	-	38.78	m2	43.62
150 x 150 x 20mm units to floors level or to falls only not exceeding 15 degrees from horizontal										
plain	24.00	5.00	25.20	1.32	0.73	26.58	-	51.77	m2	58.24
225 x 225 x 25mm units to floors level or to falls only not exceeding 15 degrees from horizontal										
plain	25.21	5.00	26.47	0.75	0.55	16.55	-	43.02	m2	48.39
150 x 150 x 12.5mm units to floors to falls and crossfalls and to slopes not exceeding 15 degrees from horizontal										
plain	16.60	5.00	17.43	1.14	0.60	22.63	-	40.05	m2	45.06
150 x 150 x 20mm units to floors to falls and crossfalls and to slopes not exceeding 15 degrees from horizontal										
plain	24.00	5.00	25.20	1.44	0.73	28.28	-	53.48	m2	60.16
225 x 225 x 25mm units to floors to falls and crossfalls and to slopes not exceeding 15 degrees from horizontal										
plain	25.21	5.00	26.47	0.72	0.43	14.84	-	41.30	m2	46.47
Clay floor quarries, BS 6431, blended; 3mm joints symmetrical layout; bedding in 10mm cement mortar (1:3); pointing in cement mortar (1:3);on cement and sand base										
150 x 150 x 12.5mm units to floors level or to falls only not exceeding 15 degrees from horizontal										
plain	20.94	5.00	21.99	1.05	0.60	21.35	-	43.34	m2	48.75
150 x 150 x 20mm units to floors level or to falls only not exceeding 15 degrees from horizontal										
plain	23.61	5.00	24.79	1.32	0.73	26.58	-	51.36	m2	57.78
225 x 225 x 25mm units to floors level or to falls only not exceeding 15 degrees from horizontal										
plain	26.01	5.00	27.31	0.66	0.55	15.27	-	42.58	m2	47.90

Labour hourly rates: (except Specialists) Craft Operatives £14.21 Labourer £10.71 Rates are national average prices. Refer to REGIONAL VARIATIONS for indicative levels of overall pricing in regions	MATERIALS			LABOUR				RATES		
	Del to Site £	Waste %	Material Cost £	Craft Optve Hrs	Lab Hrs	Labour Cost £	Sunds £	Nett Rate £	Unit	Gross Rate (+12.5%) £
M40: STONE/CONCRETE/QUARRY/CERAMIC TILING/MOSAIC; PLANT ROOMS Cont.										
Clay floor quarries, BS 6431, blended; 3mm joints symmetrical layout; bedding in 10mm cement mortar (1:3); pointing in cement mortar (1:3);on cement and sand base Cont.										
150 x 150 x 12.5mm units to floors to falls and crossfalls and to slopes not exceeding 15 degrees from horizontal										
plain	20.94	5.00	21.99	1.14	0.60	22.63	-	44.62	m2	50.19
150 x 150 x 20mm units to floors to falls and crossfalls and to slopes not exceeding 15 degrees from horizontal										
plain	23.61	5.00	24.79	1.44	0.73	28.28	-	53.07	m2	59.70
225 x 225 x 25mm units to floors to falls and crossfalls and to slopes not exceeding 15 degrees from horizontal										
plain	26.01	5.00	27.31	0.96	0.55	19.53	-	46.84	m2	52.69
Clay floor quarries, BS 6431, dark; 3mm joints, symmetrical layout; bedding in 10mm cement mortar (1:3); pointing in cement mortar (1:3); on cement and sand base										
150 x 150 x 12.5mm units to floors level or to falls only not exceeding 15degrees from horizontal										
plain	23.78	5.00	24.97	1.05	0.60	21.35	-	46.32	m2	52.11
150 x 150 x 12.5mm units to floors to falls and crossfalls and to slopes not exceeding 15 degrees from horizontal										
plain	23.78	5.00	24.97	1.14	0.60	22.63	-	47.59	m2	53.54
M41: TERRAZZO TILING/IN SITU TERRAZZO										
Terrazzo tiles, BS 4131, aggregate size random, ground, grouted and polished, standard colour range; 3mm joints, symmetrical layout; bedding in 40mm cement mortar (1:3); grouting with white cement; in-situ margins										
300 x 300 x 28mm units to floors on concrete base; level or to falls only not exceeding 15 degrees from horizontal										
plain	38.81	5.00	40.90	1.10	0.65	22.59	-	63.50	m2	71.43
Terrazzo, white cement and white marble chippings (2:5); polished; on cement and sand base										
6mm work to walls										
width exceeding 300mm	6.00	5.00	6.30	4.30	2.15	84.13	-	90.43	m2	101.73
width not exceeding 300mm	3.00	10.00	3.30	2.58	0.65	43.62	-	46.92	m	52.79
16mm work to floors; one coat; floated level and to falls only not exceeding 15 degrees from horizontal; laid in bays, average size 610 x 610mm	16.00	5.00	16.80	2.70	1.35	52.83	-	69.63	m2	78.33
6mm work to skirtings										
height 75mm	1.00	5.00	1.05	1.50	0.16	23.03	-	24.08	m	27.09
height 75mm; curved to 3000mm radius	1.00	5.00	1.05	2.00	0.16	30.13	-	31.18	m	35.08
16mm work to skirtings on brickwork or blockwork base										
height 150mm	4.00	5.00	4.20	1.90	0.32	30.43	-	34.63	m	38.95
height 150mm; curved to 3000mm radius	4.00	5.00	4.20	2.53	0.32	39.38	-	43.58	m	49.03
Rounded angles and intersections										
10 - 100mm radius	-	-	-	0.21	-	2.98	-	2.98	m	3.36
Coves										
25mm girth	-	-	-	0.48	-	6.82	-	6.82	m	7.67
40mm girth	-	-	-	0.72	-	10.23	-	10.23	m	11.51
Accessories										
Plastic dividing strips; setting in bed and finishing										
6 x 16mm; flat section	2.00	-	2.00	0.30	-	4.26	-	6.26	m	7.05
M41: TERRAZZO TILING/IN SITU TERRAZZO; STAIRCASE AREAS										
Terrazzo tiles, BS 4131, aggregate size random, ground, grouted and polished, standard colour range; 3mm joints, symmetrical layout; bedding in 40mm cement mortar (1:3); grouting with white cement; in-situ margins										
300 x 300 x 28mm units to floors on concrete base; level or to falls only not exceeding 15 degrees from horizontal										
plain	38.81	5.00	40.90	2.05	0.65	36.09	-	76.99	m2	86.62

Labour hourly rates: (except Specialists) Craft Operatives £14.21 Labourer £10.71 Rates are national average prices. Refer to REGIONAL VARIATIONS for indicative levels of overall pricing in regions	MATERIALS			LABOUR				RATES		
	Del to Site	Waste	Material Cost	Craft Optve	Lab	Labour Cost	Sunds	Nett Rate	Unit	Gross Rate (+12.5%)
	£	%	£	Hrs	Hrs	£	£	£		£
M41: TERRAZZO TILING/IN SITU TERRAZZO; STAIRCASE AREAS Cont.										
Terrazzo tiles, BS 4131, aggregate size random, ground, grouted and polished, standard colour range; 3mm joints, symmetrical layout; bedding in 40mm cement mortar (1:3); grouting with white cement; in-situ margins Cont.										
305 x 305 x 28mm units to treads on concrete base										
width 292mm; rounded nosing	17.76	5.00	**18.71**	0.95	0.19	**15.53**	-	**34.24**	m	**38.53**
305 x 305 x 28mm units to plain risers on concrete base										
height 165mm	5.70	5.00	**6.02**	0.65	0.11	**10.41**	-	**16.43**	m	**18.49**
Terrazzo, white cement and white marble chippings (2:5); polished; on cement and sand base										
16mm work to treads; one coat										
width 279mm	7.00	5.00	**7.35**	3.60	0.38	**55.23**	-	**62.58**	m	**70.40**
extra; two line carborundum non-slip inlay	8.80	5.00	**9.24**	0.30	-	**4.26**	-	**13.50**	m	**15.19**
16mm work to undercut risers; one coat										
height 178mm	4.00	5.00	**4.20**	2.40	0.24	**36.67**	-	**40.87**	m	**45.98**
6mm work to strings; one coat										
height 150mm	1.00	5.00	**1.05**	2.15	0.32	**33.98**	-	**35.03**	m	**39.41**
height 200mm	2.00	5.00	**2.10**	3.00	0.43	**47.24**	-	**49.34**	m	**55.50**
M42: WOOD BLOCK/COMPOSITION BLOCK/ PARQUET FLOORING										
Maple wood blocks, tongued and grooved joints; symmetrical herringbone pattern layout, two block plain borders; fixing with adhesive; sanding, one coat sealer										
70 x 230 x 20mm units to floors on cement and sand base; level or to falls only not exceeding 15 degrees from horizontal										
plain	32.29	5.00	**33.90**	1.80	1.45	**41.11**	-	**75.01**	m2	**84.39**
Merbau wood blocks, tongued and grooved joints; symmetrical herringbone pattern layout, two block plain borders; fixing with adhesive; sanding, one coat sealer										
70 x 230 x 20mm units to floors on cement and sand base; level or to falls only not exceeding 15 degrees from horizontal										
plain	26.79	5.00	**28.13**	1.80	1.45	**41.11**	-	**69.23**	m2	**77.89**
Oak wood blocks, tongued and grooved joints; symmetrical herringbone pattern layout, two block plain borders; fixing with adhesive; sanding, one coat sealer										
70 x 230 x 20mm units to floors on cement and sand base; level or to falls only not exceeding 15 degrees from horizontal										
plain	36.69	5.00	**38.52**	1.80	1.45	**41.11**	-	**79.63**	m2	**89.58**
M50: RUBBER/PLASTICS/CORK/LINO/CARPET TILING/SHEETING: RUBBER FLOORING										
Rubber floor tiles, BS 1711; butt joints, symmetrical layout; fixing with adhesive; on cement and sand base										
610 x 610 x 2mm units to floors level or to falls only not exceeding 15 degrees from horizontal										
width exceeding 300mm	26.70	5.00	**28.16**	0.40	0.25	**8.36**	-	**36.52**	m2	**41.08**
width not exceeding 300mm	8.39	10.00	**9.22**	0.24	0.08	**4.27**	-	**13.49**	m	**15.18**
610 x 610 x 3.0mm units to floors level or to falls only not exceeding 15 degrees from horizontal										
width exceeding 300mm	30.90	5.00	**32.57**	0.40	0.25	**8.36**	-	**40.93**	m2	**46.04**
width not exceeding 300mm	9.65	10.00	**10.61**	0.24	0.08	**4.27**	-	**14.88**	m	**16.74**
Smooth finish rubber matting, BS 1711; butt joints; fixing with adhesive; on cement and sand base										
2mm work to floors level or to falls only not exceeding 15 degrees from horizontal										
width exceeding 300mm	24.10	5.00	**25.43**	0.20	0.15	**4.45**	-	**29.88**	m2	**33.61**
width not exceeding 300mm	7.61	10.00	**8.37**	0.12	0.05	**2.24**	-	**10.61**	m	**11.93**
3mm work to floors level or to falls only not exceeding 15 degrees from horizontal										
width exceeding 300mm	27.65	5.00	**29.15**	0.20	0.15	**4.45**	-	**33.60**	m2	**37.80**
width not exceeding 300mm	8.67	10.00	**9.54**	0.12	0.05	**2.24**	-	**11.78**	m	**13.25**
5mm work to floors level or to falls only not exceeding 15 degrees from horizontal										
width exceeding 300mm	31.10	5.00	**32.78**	0.20	0.15	**4.45**	-	**37.23**	m2	**41.88**
width not exceeding 300mm	9.71	10.00	**10.68**	0.12	0.05	**2.24**	-	**12.92**	m	**14.53**

Labour hourly rates: (except Specialists) Craft Operatives £14.21 Labourer £10.71 Rates are national average prices. Refer to REGIONAL VARIATIONS for indicative levels of overall pricing in regions	MATERIALS			LABOUR				RATES		
	Del to Site £	Waste %	Material Cost £	Craft Optve Hrs	Lab Hrs	Labour Cost £	Sunds £	Nett Rate £	Unit	Gross Rate (+12.5%) £
M50: RUBBER/PLASTICS/CORK/LINO/CARPET TILING/SHEETING: PVC FLOORING										
P.V.C. floor tiles, BS EN 654; butt joints, symmetrical layout; fixing with adhesive; two coats sealer; on cement and sand base										
300 x 300 x 2mm units to floors level or to falls only not exceeding 15 degrees from horizontal										
width exceeding 300mm	11.60	5.00	**12.36**	0.60	0.30	**11.74**	-	**24.10**	m2	**27.12**
width not exceeding 300mm	4.04	10.00	**4.45**	0.36	0.09	**6.08**	-	**10.53**	m	**11.84**
300 x 300 x 2.5mm units to floors level or to falls only not exceeding 15 degrees from horizontal										
width exceeding 300mm	13.25	5.00	**14.10**	0.60	0.30	**11.74**	-	**25.83**	m2	**29.06**
width not exceeding 300mm	4.54	10.00	**4.99**	0.36	0.09	**6.08**	-	**11.07**	m	**12.45**
Fully flexible PVC heavy duty floor tiles, BS EN 649 type A; butt joints, symmetrical layout; fixing with adhesive; two coats sealer; on cement and sand base										
610 x 610 x 2mm units to floors level or to falls only not exceeding 15 degrees from horizontal										
width exceeding 300mm	18.00	5.00	**19.08**	0.60	0.30	**11.74**	-	**30.82**	m2	**34.68**
width not exceeding 300mm	5.96	10.00	**6.56**	0.36	0.09	**6.08**	-	**12.64**	m	**14.22**
610 x 610 x 2.5mm units to floors level or to falls only not exceeding 15 degrees from horizontal										
width exceeding 300mm	19.50	5.00	**20.66**	0.60	0.30	**11.74**	-	**32.40**	m2	**36.45**
width not exceeding 300mm	6.41	10.00	**7.05**	0.36	0.09	**6.08**	-	**13.13**	m	**14.77**
Fully flexible P.V.C. heavy duty sheet, BS EN 649 Type A; butt joints; welded; fixing with adhesive; two coats sealer; on cement and sand base										
2mm work to floors level or to falls only not exceeding 15 degrees from horizontal										
width exceeding 300mm	11.60	5.00	**12.36**	0.40	0.20	**7.83**	-	**20.19**	m2	**22.71**
width not exceeding 300mm	4.04	10.00	**4.45**	0.24	0.06	**4.05**	-	**8.50**	m	**9.56**
2.5mm work to floors level or to falls only not exceeding 15 degrees from horizontal										
width exceeding 300mm	13.25	5.00	**14.10**	0.40	0.20	**7.83**	-	**21.92**	m2	**24.66**
width not exceeding 300mm	4.54	10.00	**4.99**	0.24	0.06	**4.05**	-	**9.04**	m	**10.17**
P.V.C. coved skirting; fixing with adhesive										
Skirtings on plaster base										
height 100mm - Set in	2.31	5.00	**2.46**	0.25	0.02	**3.77**	-	**6.22**	m	**7.00**
height 100mm - Sit on	2.66	5.00	**2.82**	0.25	0.02	**3.77**	-	**6.59**	m	**7.41**
M50: RUBBER/PLASTICS/CORK/LINO/CARPET TILING/SHEETING: CORK FLOORING										
Cork tiles, BS 8203; butt joints; symmetrical layout; fixing with adhesive; two coats seal; on cement and sand base										
300 x 300 x 3.2mm units to floors level or to falls only not exceeding 15 degrees from horizontal										
width exceeding 300mm	15.87	5.00	**16.87**	0.90	0.28	**15.79**	-	**32.66**	m2	**36.74**
width not exceeding 300mm	5.18	10.00	**5.69**	0.54	0.08	**8.53**	-	**14.22**	m	**16.00**
300 x 300 x 4.75mm units to floors level or to falls only not exceeding 15 degrees from horizontal										
width exceeding 300mm	17.82	5.00	**18.92**	0.90	0.28	**15.79**	-	**34.70**	m2	**39.04**
width not exceeding 300mm	5.77	10.00	**6.35**	0.54	0.08	**8.53**	-	**14.88**	m	**16.74**
M50: RUBBER/PLASTICS/CORK/LINO/CARPET TILING/SHEETING: LINOLEUM FLOORING										
Linoleum tiles, BS EN 548 & 670 butt joints, symmetrical layout; fixing with adhesive; on cement and sand base										
500 x 500 x 2mm units to floors level or to falls only not exceeding 15 degrees from horizontal										
width exceeding 300mm	16.26	5.00	**17.12**	0.45	0.28	**9.39**	-	**26.51**	m2	**29.82**
width not exceeding 300mm	5.03	10.00	**5.53**	0.27	0.08	**4.69**	-	**10.22**	m	**11.50**
500 x 500 x 3.2mm units to floors level or to falls only not exceeding 15 degrees from horizontal										
width exceeding 300mm	18.46	5.00	**19.43**	0.45	0.28	**9.39**	-	**28.82**	m2	**32.42**
width not exceeding 300mm	5.69	10.00	**6.25**	0.27	0.08	**4.69**	-	**10.95**	m	**12.32**
Linoleum sheet, BS EN 548 & 670 butt joints; laying loose; on cement and sand base										
2mm work to floors level or to falls only not exceeding 15 degrees from horizontal										
width exceeding 300mm	14.25	5.00	**14.96**	0.20	0.15	**4.45**	-	**19.41**	m2	**21.84**
width not exceeding 300mm	4.28	10.00	**4.70**	0.12	0.05	**2.24**	-	**6.94**	m	**7.81**
3.2mm work to floors level or to falls only not exceeding 15 degrees from horizontal										
width exceeding 300mm	17.95	5.00	**18.85**	0.20	0.15	**4.45**	-	**23.30**	m2	**26.21**

Labour hourly rates: (except Specialists) Craft Operatives £14.21 Labourer £10.71 Rates are national average prices. Refer to REGIONAL VARIATIONS for indicative levels of overall pricing in regions	MATERIALS			LABOUR				RATES		
	Del to Site £	Waste %	Material Cost £	Craft Optve Hrs	Lab Hrs	Labour Cost £	Sunds £	Nett Rate £	Unit	Gross Rate (+12.5%) £
M50: RUBBER/PLASTICS/CORK/LINO/CARPET TILING/SHEETING: LINOLEUM FLOORING Cont.										
Linoleum sheet, BS EN 548 & 670 butt joints; laying loose; on cement and sand base Cont.										
3.2mm work to floors level or to falls only not exceeding 15 degrees from horizontal Cont.										
width not exceeding 300mm	5.39	10.00	**5.92**	0.12	0.05	**2.24**	-	**8.16**	m	**9.18**
M50: RUBBER/PLASTICS/CORK/LINO/CARPET TILING/SHEETING; STAIRCASE AREAS - STAIR NOSINGS AND TREADS										
Accessories										
Heavy duty aluminium alloy stair nosings with anti-slip inserts; 46mm wide with single line insert; fixing to timber with screws										
11.5mm drop	8.59	5.00	**9.02**	0.46	-	**6.54**	0.01	**15.56**	m	**17.50**
22mm drop	9.99	5.00	**10.49**	0.46	-	**6.54**	0.01	**17.03**	m	**19.16**
25mm drop	10.74	5.00	**11.28**	0.46	-	**6.54**	0.01	**17.82**	m	**20.05**
32mm drop	12.16	5.00	**12.76**	0.46	-	**6.54**	0.01	**19.31**	m	**21.72**
33mm drop	12.16	5.00	**12.76**	0.46	-	**6.54**	0.01	**19.31**	m	**21.72**
38mm drop	13.27	5.00	**13.93**	0.46	-	**6.54**	0.01	**20.47**	m	**23.03**
46mm drop	13.97	5.00	**14.67**	0.46	-	**6.54**	0.01	**21.21**	m	**23.86**
Heavy duty aluminium alloy stair nosings with anti-slip inserts; 80mm wide with two line inserts; fixing to timber with screws										
22mm drop	14.31	5.00	**15.03**	0.62	-	**8.81**	0.08	**23.92**	m	**26.91**
25mm drop	17.87	5.00	**18.76**	0.62	-	**8.81**	0.08	**27.65**	m	**31.11**
32mm drop	19.34	5.00	**20.31**	0.62	-	**8.81**	0.08	**29.20**	m	**32.85**
33mm drop	19.34	5.00	**20.31**	0.62	-	**8.81**	0.08	**29.20**	m	**32.85**
51mm drop	20.03	5.00	**21.03**	0.62	-	**8.81**	0.08	**29.93**	m	**33.67**
63mm drop	21.83	5.00	**22.92**	0.62	-	**8.81**	0.08	**31.82**	m	**35.80**
Heavy duty aluminium alloy stair nosings with anti-slip inserts; 46mm wide with single line insert; fixing to masonry with screws										
11.5mm drop	8.59	5.00	**9.02**	0.75	-	**10.66**	0.08	**19.76**	m	**22.23**
22mm drop	9.99	5.00	**10.49**	0.75	-	**10.66**	0.08	**21.23**	m	**23.89**
25mm drop	10.74	5.00	**11.28**	0.75	-	**10.66**	0.08	**22.02**	m	**24.77**
32mm drop	12.16	5.00	**12.76**	0.75	-	**10.66**	0.08	**23.51**	m	**26.44**
33mm drop	12.16	5.00	**12.76**	0.75	-	**10.66**	0.08	**23.51**	m	**26.44**
38mm drop	13.27	5.00	**13.93**	0.75	-	**10.66**	0.08	**24.67**	m	**27.76**
46mm drop	13.97	5.00	**14.67**	0.75	-	**10.66**	0.08	**25.41**	m	**28.59**
Heavy duty aluminium alloy stair nosings with anti-slip inserts; 80mm wide with two line inserts; fixing to masonry with screws										
22mm drop	14.31	5.00	**15.03**	0.90	-	**12.79**	0.12	**27.93**	m	**31.42**
25mm drop	17.87	5.00	**18.76**	0.90	-	**12.79**	0.12	**31.67**	m	**35.63**
32mm drop	19.34	5.00	**20.31**	0.90	-	**12.79**	0.12	**33.21**	m	**37.37**
33mm drop	19.34	5.00	**20.31**	0.90	-	**12.79**	0.12	**33.21**	m	**37.37**
51mm drop	20.03	5.00	**21.03**	0.90	-	**12.79**	0.12	**33.94**	m	**38.18**
63mm drop	21.83	5.00	**22.92**	0.90	-	**12.79**	0.12	**35.83**	m	**40.31**
Aluminium stair nosings with anti-slip inserts; 46mm wide with single line insert; fixing to timber with screws										
22mm drop	7.13	5.00	**7.48**	0.46	-	**6.54**	0.06	**14.08**	m	**15.83**
28mm drop	7.13	5.00	**7.48**	0.46	-	**6.54**	0.06	**14.08**	m	**15.83**
33mm drop	9.28	5.00	**9.74**	0.46	-	**6.54**	0.06	**16.34**	m	**18.38**
Aluminium stair nosings with anti-slip inserts; 66mm wide with single line insert; fixing to timber with screws										
22mm drop	9.99	5.00	**10.49**	0.52	-	**7.39**	0.08	**17.97**	m	**20.21**
25mm drop	10.74	5.00	**11.28**	0.52	-	**7.39**	0.08	**18.75**	m	**21.09**
32mm drop	12.16	5.00	**12.76**	0.52	-	**7.39**	0.08	**20.24**	m	**22.77**
33mm drop	12.84	5.00	**13.48**	0.52	-	**7.39**	0.08	**20.95**	m	**23.57**
Aluminium stair nosings with anti-slip inserts; 80mm wide with two line inserts; fixing to timber with screws										
22mm drop	11.47	5.00	**12.04**	0.62	-	**8.81**	0.08	**20.93**	m	**23.55**
28mm drop	12.84	5.00	**13.48**	0.62	-	**8.81**	0.08	**22.37**	m	**25.17**
33mm drop	13.62	5.00	**14.30**	0.62	-	**8.81**	0.08	**23.19**	m	**26.09**
46mm drop	14.31	5.00	**15.03**	0.62	-	**8.81**	0.08	**23.92**	m	**26.91**
51mm drop	15.02	5.00	**15.77**	0.62	-	**8.81**	0.08	**24.67**	m	**27.75**
Aluminium alloy stair nosings with anti-slip inserts; 46mm wide with single line insert; fixing to masonry with screws										
22mm drop	7.13	5.00	**7.48**	0.75	-	**10.66**	0.08	**18.22**	m	**20.50**
28mm drop	7.13	5.00	**7.48**	0.75	-	**10.66**	0.08	**18.22**	m	**20.50**
33mm drop	9.28	5.00	**9.74**	0.75	-	**10.66**	0.08	**20.48**	m	**23.05**
Aluminium alloy stair nosings with anti-slip inserts; 66mm wide with single line insert; fixing to masonry with screws										
22mm drop	9.99	5.00	**10.49**	0.80	-	**11.37**	0.12	**21.98**	m	**24.73**
25mm drop	10.74	5.00	**11.28**	0.80	-	**11.37**	0.12	**22.76**	m	**25.61**
32mm drop	12.16	5.00	**12.76**	0.80	-	**11.37**	0.12	**24.25**	m	**27.28**
33mm drop	12.84	5.00	**13.48**	0.80	-	**11.37**	0.12	**24.96**	m	**28.08**

Labour hourly rates: (except Specialists) Craft Operatives £14.21 Labourer £10.71 Rates are national average prices. Refer to REGIONAL VARIATIONS for indicative levels of overall pricing in regions	MATERIALS			LABOUR				RATES		
	Del to Site	Waste	Material Cost	Craft Optve	Lab	Labour Cost	Sunds	Nett Rate	Unit	Gross Rate (+12.5%)
	£	%	£	Hrs	Hrs	£	£	£		£
M50: RUBBER/PLASTICS/CORK/LINO/CARPET TILING/SHEETING; STAIRCASE AREAS - STAIR NOSINGS AND TREADS Cont.										
Accessories Cont.										
Aluminium alloy stair nosings with anti-slip inserts; 80mm wide with two line inserts; fixing to masonry with screws										
22mm drop	11.47	5.00	12.04	0.90	-	12.79	0.12	24.95	m	28.06
28mm drop	12.84	5.00	13.48	0.90	-	12.79	0.12	26.39	m	29.68
33mm drop	13.62	5.00	14.30	0.90	-	12.79	0.12	27.21	m	30.61
46mm drop	14.31	5.00	15.03	0.90	-	12.79	0.12	27.93	m	31.42
51mm drop	15.02	5.00	15.77	0.90	-	12.79	0.12	28.68	m	32.27
M51: EDGE FIXED CARPETING										
Underlay to carpeting; Tredaire; fixing with tacks; on timber base										
Work to floors; level or to falls only not exceeding 15 degrees from horizontal										
width exceeding 300mm	1.75	5.00	1.84	0.15	0.08	2.99	0.17	4.99	m2	5.62
Fitted carpeting; Tufted wool/nylon, p.c. £20.00 m2; fixing with tackless grippers; on timber base										
Work to floors; level or to falls only not exceeding 15 degrees from horizontal										
width exceeding 300mm	20.00	5.00	21.00	0.40	0.20	7.83	0.17	28.99	m2	32.62
Fitted carpeting; Axminster wool/nylon, P.C. £26.00m2; fixing with tackless grippers; on timber base										
Work to floors; level or to falls only not exceeding 15 degrees from horizontal										
width exceeding 300mm	26.00	5.00	27.30	0.40	0.20	7.83	0.17	35.29	m2	39.71
Fitted carpeting; Wilton wool/nylon, p.c. £32.5000 m2; fixing with tackless grippers; on timber base										
Work to floors; level or to falls only not exceeding 15 degrees from horizontal										
width exceeding 300mm	32.50	5.00	34.13	0.40	0.20	7.83	0.17	42.12	m2	47.38
M51: EDGE FIXED CARPETING; STAIRCASE AREAS - STAIR NOSINGS AND TREADS										
Accessories										
Aluminium alloy carpet nosings with anti-slip inserts; with single line insert; fixing to timber with screws										
65mm wide x 24mm drop	16.26	5.00	17.07	0.52	-	7.39	0.08	24.54	m	27.61
116mm wide overall with carpet gripper x 42mm drop	23.23	5.00	24.39	0.58	-	8.24	0.12	32.75	m	36.84
Aluminium alloy carpet nosings with anti-slip inserts; with two line inserts; fixing to timber with screws										
80mm wide x 31mm drop	18.58	5.00	19.51	0.62	-	8.81	0.08	28.40	m	31.96
Aluminium alloy carpet nosings with anti-slip inserts; with single line insert; fixing to masonry with screws										
65mm wide x 24mm drop	16.26	5.00	17.07	0.80	-	11.37	0.12	28.56	m	32.13
116mm wide overall with carpet gripper x 42mm drop	23.23	5.00	24.39	0.85	-	12.08	0.16	36.62	m	41.20
Aluminium alloy carpet nosings with anti-slip inserts; with two line inserts; fixing to masonry with screws										
80mm wide x 31mm drop	18.58	5.00	19.51	0.90	-	12.79	0.12	32.42	m	36.47
M52: DECORATIVE PAPERS/FABRICS										
Lining paper (prime cost sum for supply and allowance for waste included elsewhere); sizing; applying adhesive; hanging; butt joints										
Plaster walls and columns exceeding 0.50m2	0.05	-	0.05	0.13	-	1.85	0.04	1.93	m2	2.18
Plaster ceilings and beams exceeding 0.50m2	0.05	-	0.05	0.14	-	1.99	0.04	2.08	m2	2.34
Pulp paper (prime cost sum for supply and allowance for waste included elsewhere); sizing; applying adhesive; hanging; butt joints										
Plaster walls and columns exceeding 0.50m2	0.05	-	0.05	0.31	-	4.41	0.08	4.54	m2	5.11
Plaster ceilings and beams exceeding 0.50m2	0.05	-	0.05	0.34	-	4.83	0.09	4.98	m2	5.60
Washable paper (prime cost sum for supply and allowance for waste included elsewhere); sizing; applying adhesive; hanging; butt joints										
Plaster walls and columns exceeding 0.50m2	0.07	-	0.07	0.24	-	3.41	0.06	3.54	m2	3.99

Labour hourly rates: (except Specialists) Craft Operatives £14.21 Labourer £10.71 Rates are national average prices. Refer to REGIONAL VARIATIONS for indicative levels of overall pricing in regions	MATERIALS			LABOUR				RATES		
	Del to Site £	Waste %	Material Cost £	Craft Optve Hrs	Lab Hrs	Labour Cost £	Sunds £	Nett Rate £	Unit	Gross Rate (+12.5%) £
M52: DECORATIVE PAPERS/FABRICS Cont.										
Washable paper (prime cost sum for supply and allowance for waste included elsewhere); sizing; applying adhesive; hanging; butt joints Cont.										
Plaster ceilings and beams exceeding 0.50m2	0.07	-	**0.07**	0.26	-	**3.69**	0.07	**3.83**	m2	**4.31**
Vinyl coated paper (prime cost sum for supply and allowance for waste included elsewhere); sizing; applying adhesive; hanging; butt joints										
Plaster walls and columns exceeding 0.50m2	0.07	-	**0.07**	0.24	-	**3.41**	0.06	**3.54**	m2	**3.99**
Plaster ceilings and beams exceeding 0.50m2	0.07	-	**0.07**	0.26	-	**3.69**	0.07	**3.83**	m2	**4.31**
Embossed or textured paper (prime cost sum for supply and allowance for waste included elsewhere); sizing; applying adhesive; hanging; butt joints										
Plaster walls and columns exceeding 0.50m2	0.07	-	**0.07**	0.24	-	**3.41**	0.06	**3.54**	m2	**3.99**
Plaster ceilings and beams exceeding 0.50m2	0.14	-	**0.14**	0.26	-	**3.69**	0.07	**3.91**	m2	**4.39**
Woodchip paper (prime cost sum for supply and allowance for waste included elsewhere); sizing; applying adhesive; hanging; butt joints										
Plaster walls and columns exceeding 0.50m2	0.05	-	**0.05**	0.24	-	**3.41**	0.06	**3.53**	m2	**3.97**
Plaster ceilings and beams exceeding 0.50m2	0.05	-	**0.05**	0.26	-	**3.69**	0.07	**3.82**	m2	**4.29**
Paper border strips (prime cost sum for supply and allowance for waste included elsewhere); sizing; applying adhesive; hanging; butt joints										
Border strips										
25mm wide to papered walls and columns	0.01	-	**0.01**	0.09	-	**1.28**	0.02	**1.31**	m	**1.47**
75mm wide to papered walls and columns	0.02	-	**0.02**	0.10	-	**1.42**	0.03	**1.47**	m	**1.65**
Lining paper p.c. £1.00 roll; sizing; applying adhesive; hanging; butt joints										
Plaster walls and columns exceeding 0.50m2	0.25	12.00	**0.28**	0.13	-	**1.85**	0.04	**2.16**	m2	**2.43**
Plaster ceilings and beams exceeding 0.50m2	0.25	12.00	**0.28**	0.14	-	**1.99**	0.04	**2.30**	m2	**2.59**
Pulp paper p.c. £3.00 roll; sizing; applying adhesive; hanging; butt joints										
Plaster walls and columns exceeding 0.50m2	0.65	30.00	**0.83**	0.24	-	**3.41**	0.06	**4.31**	m2	**4.85**
Plaster ceilings and beams exceeding 0.50m2	0.65	25.00	**0.80**	0.26	-	**3.69**	0.07	**4.57**	m2	**5.14**
Pulp paper (24' drop pattern match) p.c. £4.00 roll; sizing; applying adhesive; hanging; butt joints										
Plaster walls and columns exceeding 0.50m2	0.85	40.00	**1.17**	0.31	-	**4.41**	0.08	**5.66**	m2	**6.37**
Plaster ceilings and beams exceeding 0.50m2	0.85	35.00	**1.13**	0.33	-	**4.69**	0.09	**5.91**	m2	**6.65**
Washable paper p.c. £4.00 roll; sizing; applying adhesive; hanging; butt joints										
Plaster walls and columns exceeding 0.50m2	0.87	25.00	**1.07**	0.24	-	**3.41**	0.06	**4.54**	m2	**5.11**
Plaster ceilings and beams exceeding 0.50m2	0.87	20.00	**1.03**	0.26	-	**3.69**	0.07	**4.79**	m2	**5.39**
Vinyl coated paper p.c. £6.50 roll; sizing; applying adhesive; hanging; butt joints										
Plaster walls and columns exceeding 0.50m2	1.37	25.00	**1.69**	0.24	-	**3.41**	0.06	**5.17**	m2	**5.82**
Plaster ceilings and beams exceeding 0.50m2	1.37	20.00	**1.63**	0.26	-	**3.69**	0.07	**5.39**	m2	**6.07**

Labour hourly rates: (except Specialists) Craft Operatives £14.21 Labourer £10.71 Rates are national average prices. Refer to REGIONAL VARIATIONS for indicative levels of overall pricing in regions	MATERIALS			LABOUR				RATES		
	Del to Site	Waste	Material Cost	Craft Optve	Lab	Labour Cost	Sunds	Nett Rate	Unit	Gross Rate (+12.5%)
	£	%	£	Hrs	Hrs	£	£	£		£
M52: DECORATIVE PAPERS/FABRICS Cont.										
Vinyl coated paper p.c. £7.00 roll; sizing; applying adhesive; hanging; butt joints										
Plaster walls and columns exceeding 0.50m2	1.47	25.00	1.82	0.24	-	3.41	0.06	5.29	m2	5.96
Plaster ceilings and beams exceeding 0.50m2	1.47	20.00	1.75	0.26	-	3.69	0.07	5.51	m2	6.20
Embossed paper p.c. £4.50 roll; sizing; applying adhesive; hanging; butt joints										
Plaster walls and columns exceeding 0.50m2	1.04	25.00	1.27	0.24	-	3.41	0.06	4.74	m2	5.33
Plaster ceilings and beams exceeding 0.50m2	1.04	20.00	1.22	0.26	-	3.69	0.07	4.99	m2	5.61
Textured paper p.c. £5.00 roll; sizing; applying adhesive; hanging; butt joints										
Plaster walls and columns exceeding 0.50m2	1.11	25.00	1.36	0.24	-	3.41	0.06	4.82	m2	5.43
Plaster ceilings and beams exceeding 0.50m2	1.11	20.00	1.31	0.26	-	3.69	0.06	5.06	m2	5.70
Woodchip paper p.c. £1.50 roll; sizing; applying adhesive; hanging; butt joints										
Plaster walls and columns exceeding 0.50m2	0.35	25.00	0.43	0.24	-	3.41	0.06	3.90	m2	4.38
Plaster ceilings and beams exceeding 0.50m2	0.35	20.00	0.41	0.26	-	3.69	0.06	4.17	m2	4.69
Paper border strips p.c. £0.25m; applying adhesive; hanging; butt joints										
Border strips 25mm wide to papered walls and columns	0.26	10.00	0.28	0.09	-	1.28	0.02	1.58	m	1.78
Paper border strips p.c. £0.50m; applying adhesive; hanging; butt joints										
Border strips 75mm wide to papered walls and columns	0.52	10.00	0.57	0.10	-	1.42	0.03	2.02	m	2.27
Hessian wall covering (prime cost sum for supply and allowance for waste included elsewhere); sizing; applying adhesive; hanging; butt joints										
Plaster walls and columns exceeding 0.50m2	0.21	-	0.21	0.44	-	6.25	0.12	6.58	m2	7.40
Textile hessian paper backed wall covering (prime cost sum for supply and allowance for waste included elsewhere); sizing; applying adhesive; hanging; butt joints										
Plaster walls and columns exceeding 0.50m2	0.21	-	0.21	0.36	-	5.12	0.10	5.42	m2	6.10
Hessian wall covering p.c. £4.00m2; sizing; applying adhesive; hanging; butt joints										
Plaster walls and columns exceeding 0.50m2	4.21	30.00	5.41	0.24	-	3.41	0.06	8.89	m2	10.00
Textile hessian paper backed wall covering p.c. £7.50m2; sizing; applying adhesive; hanging; butt joints										
Plaster walls and columns exceeding 0.50m2	7.71	25.00	9.59	0.36	-	5.12	0.10	14.80	m2	16.65
M52: DECORATIVE PAPERS/FABRICS - REDECORATIONS										
Stripping existing paper; lining paper (prime cost sum for supply and allowance for waste included elsewhere); sizing; applying adhesive hanging; butt joints										
Hard building board walls and columns exceeding 0.50m2	0.05	-	0.05	0.32	-	4.55	0.19	4.79	m2	5.39
Plaster walls and columns exceeding 0.50m2	0.05	-	0.05	0.28	-	3.98	0.19	4.22	m2	4.75
Hard building board ceilings and beams exceeding 0.50m2	0.05	-	0.05	0.35	-	4.97	0.23	5.26	m2	5.92
Plaster ceilings and beams exceeding 0.50m2	0.05	-	0.05	0.30	-	4.26	0.20	4.52	m2	5.08

Labour hourly rates: (except Specialists) Craft Operatives £14.21 Labourer £10.71 Rates are national average prices. Refer to REGIONAL VARIATIONS for indicative levels of overall pricing in regions	MATERIALS			LABOUR				RATES		
	Del to Site £	Waste %	Material Cost £	Craft Optve Hrs	Lab Hrs	Labour Cost £	Sunds £	Nett Rate £	Unit	Gross Rate (+12.5%) £
M52: DECORATIVE PAPERS/FABRICS - REDECORATIONS Cont.										
Stripping existing paper; lining paper and cross lining (prime cost sum for supply and allowance for waste included elsewhere); sizing; applying adhesive; hanging; butt joints										
Hard building board walls and columns exceeding 0.50m2	0.11	-	**0.11**	0.45	-	**6.39**	0.30	**6.80**	m2	**7.65**
Plaster walls and columns exceeding 0.50m2	0.11	-	**0.11**	0.41	-	**5.83**	0.27	**6.21**	m2	**6.98**
Hard building board ceilings and beams exceeding 0.50m2	0.11	-	**0.11**	0.50	-	**7.11**	0.34	**7.55**	m2	**8.49**
Plaster ceilings and beams exceeding 0.50m2	0.11	-	**0.11**	0.44	-	**6.25**	0.29	**6.65**	m2	**7.48**
Stripping existing paper; pulp paper (prime cost sum for supply and allowance for waste included elsewhere); applying adhesive; hanging; butt joints										
Hard building board walls and columns exceeding 0.50m2	0.05	-	**0.05**	0.43	-	**6.11**	0.29	**6.45**	m2	**7.26**
Plaster walls and columns exceeding 0.50m2	0.05	-	**0.05**	0.39	-	**5.54**	0.26	**5.86**	m2	**6.59**
Hard building board ceilings and beams exceeding 0.50m2	0.05	-	**0.05**	0.47	-	**6.68**	0.31	**7.05**	m2	**7.93**
Plaster ceilings and beams exceeding 0.50m2	0.05	-	**0.05**	0.42	-	**5.97**	0.28	**6.30**	m2	**7.09**
Stripping existing washable or vinyl coated paper; washable paper (prime cost sum for supply and allowance for waste included elsewhere); applying adhesive; hanging; butt joints										
Hard building board walls and columns exceeding 0.50m2	0.07	-	**0.07**	0.44	-	**6.25**	0.29	**6.62**	m2	**7.44**
Plaster walls and columns exceeding 0.50m2	0.07	-	**0.07**	0.40	-	**5.68**	0.27	**6.02**	m2	**6.77**
Hard building board ceilings and beams exceeding 0.50m2	0.07	-	**0.07**	0.48	-	**6.82**	0.32	**7.21**	m2	**8.11**
Plaster ceilings and beams exceeding 0.50m2	0.07	-	**0.07**	0.43	-	**6.11**	0.29	**6.47**	m2	**7.28**
Stripping existing washable or vinyl coated paper; vinyl coated paper (prime cost sum for supply and allowance for waste included elsewhere); sizing; applying adhesive; hanging; butt joints										
Hard building board walls and columns exceeding 0.50m2	0.07	-	**0.07**	0.44	-	**6.25**	0.29	**6.62**	m2	**7.44**
Plaster walls and columns exceeding 0.50m2	0.07	-	**0.07**	0.40	-	**5.68**	0.27	**6.02**	m2	**6.77**
Hard building board ceilings and beams exceeding 0.50m2	0.07	-	**0.07**	0.48	-	**6.82**	0.32	**7.21**	m2	**8.11**
Plaster ceilings and beams exceeding 0.50m2	0.07	-	**0.07**	0.43	-	**6.11**	0.29	**6.47**	m2	**7.28**
Stripping existing paper; embossed or textured paper; prime cost sum for supply and allowance for waste included elsewhere); sizing; applying adhesive; hanging; butt joints										
Hard building board walls and columns exceeding 0.50m2	0.14	-	**0.14**	0.43	-	**6.11**	0.29	**6.54**	m2	**7.36**
Plaster walls and columns exceeding 0.50m2	0.14	-	**0.14**	0.39	-	**5.54**	0.26	**5.94**	m2	**6.69**
Hard building board ceilings and beams exceeding 0.50m2	0.14	-	**0.14**	0.47	-	**6.68**	0.31	**7.13**	m2	**8.03**
Plaster ceilings and beams exceeding 0.50m2	0.14	-	**0.14**	0.42	-	**5.97**	0.28	**6.39**	m2	**7.19**
Stripping existing paper; woodchip paper (prime cost sum for supply and allowance for waste included elsewhere); sizing; applying adhesive; hanging; butt joints										
Hard building board walls and columns exceeding 0.50m2	0.05	-	**0.05**	0.43	-	**6.11**	0.29	**6.45**	m2	**7.26**
Plaster walls and columns exceeding 0.50m2	0.05	-	**0.05**	0.39	-	**5.54**	0.26	**5.86**	m2	**6.59**

SURFACE FINISHES

Labour hourly rates: (except Specialists) Craft Operatives £14.21 Labourer £10.71 Rates are national average prices. Refer to REGIONAL VARIATIONS for indicative levels of overall pricing in regions	MATERIALS			LABOUR			RATES			
	Del to Site £	Waste %	Material Cost £	Craft Optve Hrs	Lab Hrs	Labour Cost £	Sunds £	Nett Rate £	Unit	Gross Rate (+12.5%) £
M52: DECORATIVE PAPERS/FABRICS - REDECORATIONS Cont.										
Stripping existing paper; woodchip paper (prime cost sum for supply and allowance for waste included elsewhere); sizing; applying adhesive; hanging; butt joints Cont.										
Hard building board ceilings and beams exceeding 0.50m2	0.05	-	**0.05**	0.47	-	**6.68**	0.31	**7.05**	m2	**7.93**
Plaster ceilings and beams exceeding 0.50m2	0.05	-	**0.05**	0.42	-	**5.97**	0.28	**6.30**	m2	**7.09**
Stripping existing varnished or painted paper; woodchip paper (prime cost sum for supply and allowance for waste included elsewhere); sizing; applying adhesive; hanging; butt joints										
Hard building board walls and columns exceeding 0.50m2	0.05	-	**0.05**	0.57	-	**8.10**	0.38	**8.53**	m2	**9.60**
Plaster walls and columns exceeding 0.50m2	0.05	-	**0.05**	0.53	-	**7.53**	0.36	**7.94**	m2	**8.93**
Hard building board ceilings and beams exceeding 0.50m2	0.05	-	**0.05**	0.63	-	**8.95**	0.42	**9.43**	m2	**10.61**
Plaster ceilings and beams exceeding 0.50m2	0.05	-	**0.05**	0.58	-	**8.24**	0.39	**8.68**	m2	**9.77**
Stripping existing paper; lining paper p.c. £1.00 roll; sizing; applying adhesive; hanging; butt joints										
Hard building board walls and columns exceeding 0.50m2	0.25	12.00	**0.28**	0.32	-	**4.55**	0.40	**5.22**	m2	**5.87**
Plaster walls and columns exceeding 0.50m2	0.25	12.00	**0.28**	0.28	-	**3.98**	0.19	**4.44**	m2	**5.00**
Hard building board ceilings and beams exceeding 0.50m2	0.25	12.00	**0.28**	0.35	-	**4.97**	0.23	**5.48**	m2	**6.17**
Plaster ceilings and beams exceeding 0.50m2	0.25	12.00	**0.28**	0.30	-	**4.26**	0.20	**4.74**	m2	**5.33**
Stripping existing paper; lining paper p.c. £1.00 roll and cross lining P.C. £1.00 roll; sizing; applying adhesive; hanging; butt joints										
Hard building board walls and columns exceeding 0.50m2	0.51	12.00	**0.55**	0.45	-	**6.39**	0.30	**7.25**	m2	**8.16**
Plaster walls and columns exceeding 0.50m2	0.51	12.00	**0.55**	0.41	-	**5.83**	0.25	**6.63**	m2	**7.46**
Hard building board ceilings and beams exceeding 0.50m2	0.51	12.00	**0.55**	0.50	-	**7.11**	0.34	**7.99**	m2	**8.99**
Plaster ceilings and beams exceeding 0.50m2	0.51	12.00	**0.55**	0.44	-	**6.25**	0.29	**7.10**	m2	**7.99**
Stripping existing paper; pulp paper p.c. £3.00 roll; sizing; applying adhesive; hanging; butt joints										
Hard building board walls and columns exceeding 0.50m2	0.65	30.00	**0.83**	0.43	-	**6.11**	0.29	**7.23**	m2	**8.13**
Plaster walls and columns exceeding 0.50m2	0.65	30.00	**0.83**	0.39	-	**5.54**	0.26	**6.64**	m2	**7.47**
Hard building board ceilings and beams exceeding 0.50m2	0.65	25.00	**0.80**	0.47	-	**6.68**	0.31	**7.80**	m2	**8.77**
Plaster ceilings and beams exceeding 0.50m2	0.65	25.00	**0.80**	0.42	-	**5.97**	0.28	**7.05**	m2	**7.93**
Stripping existing paper; pulp paper with 24' drop pattern; p.c. £4.00 roll; sizing; applying adhesive; hanging; butt joints										
Hard building board walls and columns exceeding 0.50m2	0.85	40.00	**1.17**	0.50	-	**7.11**	0.34	**8.61**	m2	**9.6**
Plaster walls and columns exceeding 0.50m2	0.85	40.00	**1.17**	0.45	-	**6.39**	0.30	**7.87**	m2	**8.8**
Hard building board ceilings and beams exceeding 0.50m2	0.85	35.00	**1.13**	0.54	-	**7.67**	0.36	**9.17**	m2	**10.3**
Plaster ceilings and beams exceeding 0.50m2	0.85	35.00	**1.13**	0.48	-	**6.82**	0.32	**8.27**	m2	**9.3**

Labour hourly rates: (except Specialists) Craft Operatives £14.21 Labourer £10.71 Rates are national average prices. Refer to REGIONAL VARIATIONS for indicative levels of overall pricing in regions	MATERIALS			LABOUR				RATES		
	Del to Site £	Waste %	Material Cost £	Craft Optve Hrs	Lab Hrs	Labour Cost £	Sunds £	Nett Rate £	Unit	Gross Rate (+12.5%) £
M52: DECORATIVE PAPERS/FABRICS - REDECORATIONS Cont.										
Stripping existing washable or vinyl coated paper; washable paper p.c. £4.00 roll; sizing; applying adhesive; hanging; butt joints										
Hard building board walls and columns exceeding 0.50m2	0.87	25.00	**1.07**	0.44	-	**6.25**	0.29	**7.62**	m2	**8.57**
Plaster walls and columns exceeding 0.50m2	0.87	25.00	**1.07**	0.40	-	**5.68**	0.27	**7.02**	m2	**7.90**
Hard building board ceilings and beams exceeding 0.50m2	0.87	20.00	**1.03**	0.48	-	**6.82**	0.32	**8.17**	m2	**9.19**
Plaster ceilings and beams exceeding 0.50m2	0.87	20.00	**1.03**	0.43	-	**6.11**	0.29	**7.43**	m2	**8.36**
Stripping existing washable or vinyl coated paper; vinyl coated paper P.C. £6.50 roll; sizing; applying adhesive; hanging; butt joints										
Hard building board walls and columns exceeding 0.50m2	1.37	25.00	**1.69**	0.44	-	**6.25**	0.29	**8.24**	m2	**9.27**
Plaster walls and columns exceeding 0.50m2	1.37	25.00	**1.69**	0.40	-	**5.68**	0.27	**7.65**	m2	**8.60**
Hard building board ceilings and beams exceeding 0.50m2	1.37	20.00	**1.63**	0.48	-	**6.82**	0.32	**8.77**	m2	**9.87**
Plaster ceilings and beams exceeding 0.50m2	1.37	20.00	**1.63**	0.43	-	**6.11**	0.29	**8.03**	m2	**9.03**
Stripping existing washable or vinyl coated paper; vinyl coated paper P.C. £8.00 roll; sizing; applying adhesive; hanging; butt joints										
Hard building board walls and columns exceeding 0.50m2	1.67	25.00	**2.07**	0.44	-	**6.25**	0.29	**8.62**	m2	**9.69**
Plaster walls and columns exceeding 0.50m2	1.67	25.00	**2.07**	0.40	-	**5.68**	0.27	**8.02**	m2	**9.02**
Hard building board ceilings and beams exceeding 0.50m2	1.67	20.00	**1.99**	0.48	-	**6.82**	0.32	**9.13**	m2	**10.27**
Plaster ceilings and beams exceeding 0.50m2	1.67	20.00	**1.99**	0.43	-	**6.11**	0.29	**8.39**	m2	**9.44**
Stripping existing paper; embossed paper p.c. £4.50 roll; sizing; applying adhesive; hanging; butt joints										
Hard building board walls and columns exceeding 0.50m2	1.04	25.00	**1.27**	0.43	-	**6.11**	0.29	**7.66**	m2	**8.62**
Plaster walls and columns exceeding 0.50m2	1.04	25.00	**1.27**	0.39	-	**5.54**	0.26	**7.07**	m2	**7.95**
Hard building board ceilings and beams exceeding 0.50m2	1.04	20.00	**1.22**	0.47	-	**6.68**	0.31	**8.21**	m2	**9.24**
Plaster ceilings and beams exceeding 0.50m2	1.04	20.00	**1.22**	0.42	-	**5.97**	0.28	**7.47**	m2	**8.40**
Stripping existing paper; textured paper p.c. £5.00 roll; sizing; applying adhesive; hanging; butt joints										
Hard building board walls and columns exceeding 0.50m2	1.11	25.00	**1.36**	0.43	-	**6.11**	0.29	**7.75**	m2	**8.72**
Plaster walls and columns exceeding 0.50m2	1.11	25.00	**1.36**	0.39	-	**5.54**	0.26	**7.16**	m2	**8.05**
Hard building board ceilings and beams exceeding 0.50m2	1.11	20.00	**1.31**	0.47	-	**6.68**	0.31	**8.30**	m2	**9.34**
Plaster ceilings and beams exceeding 0.50m2	1.11	20.00	**1.31**	0.42	-	**5.97**	0.28	**7.55**	m2	**8.50**
Stripping existing paper; woodchip paper p.c. £1.50 roll; sizing; applying adhesive; hanging; butt joints										
Hard building board walls and columns exceeding 0.50m2	0.35	25.00	**0.43**	0.43	-	**6.11**	0.29	**6.83**	m2	**7.68**
Plaster walls and columns exceeding 0.50m2	0.35	25.00	**0.43**	0.39	-	**5.54**	0.26	**6.23**	m2	**7.01**
Hard building board ceilings and beams exceeding 0.50m2	0.35	20.00	**0.41**	0.47	-	**6.68**	0.31	**7.41**	m2	**8.33**
Plaster ceilings and beams exceeding 0.50m2	0.35	20.00	**0.41**	0.42	-	**5.97**	0.28	**6.66**	m2	**7.49**

Labour hourly rates: (except Specialists) Craft Operatives £14.21 Labourer £10.71 Rates are national average prices. Refer to REGIONAL VARIATIONS for indicative levels of overall pricing in regions	MATERIALS			LABOUR				RATES		
	Del to Site £	Waste %	Material Cost £	Craft Optve Hrs	Lab Hrs	Labour Cost £	Sunds £	Nett Rate £	Unit	Gross Rate (+12.5%) £
M52: DECORATIVE PAPERS/FABRICS - REDECORATIONS Cont.										
Washing down old distempered or painted surfaces; lining paper p.c. £1.00 roll; sizing; applying adhesive; hanging; butt joints										
Hard building board walls and columns exceeding 0.50m2	0.25	12.00	**0.28**	0.22	-	**3.13**	0.06	**3.46**	m2	**3.89**
Plaster walls and columns exceeding 0.50m2	0.25	12.00	**0.28**	0.22	-	**3.13**	0.06	**3.46**	m2	**3.89**
Hard building board ceilings and beams exceeding 0.50m2	0.25	12.00	**0.28**	0.23	-	**3.27**	0.06	**3.61**	m2	**4.06**
Plaster ceilings and beams exceeding 0.50m2	0.25	12.00	**0.28**	0.23	-	**3.27**	0.06	**3.61**	m2	**4.06**
Lining papered walls and columns exceeding 0.50m2	0.25	12.00	**0.28**	0.22	-	**3.13**	0.06	**3.46**	m2	**3.89**
Lining papered ceilings and beams exceeding 0.50m2	0.25	12.00	**0.28**	0.23	-	**3.27**	0.06	**3.61**	m2	**4.06**
Washing down old distempered or painted surfaces; lining paper p.c. £1.00 roll; and cross lining p.c. £1.00 roll; sizing; applying adhesive; hanging; butt joints										
Hard building board walls and columns exceeding 0.50m2	0.51	12.00	**0.55**	0.35	-	**4.97**	0.09	**5.62**	m2	**6.32**
Plaster walls and columns exceeding 0.50m2	0.51	12.00	**0.55**	0.35	-	**4.97**	0.09	**5.62**	m2	**6.32**
Hard building board ceilings and beams exceeding 0.50m2	0.51	12.00	**0.55**	0.37	-	**5.26**	0.10	**5.91**	m2	**6.65**
Plaster ceilings and beams exceeding 0.50m2	0.51	12.00	**0.55**	0.37	-	**5.26**	0.10	**5.91**	m2	**6.65**
Washing down old distempered or painted surfaces; pulp paper p.c. £3.00 roll; sizing; applying adhesive; hanging; butt joints										
Hard building board walls and columns exceeding 0.50m2	0.65	30.00	**0.83**	0.33	-	**4.69**	0.09	**5.61**	m2	**6.31**
Plaster walls and columns exceeding 0.50m2	0.65	30.00	**0.83**	0.33	-	**4.69**	0.09	**5.61**	m2	**6.31**
Hard building board ceilings and beams exceeding 0.50m2	0.65	25.00	**0.80**	0.35	-	**4.97**	0.09	**5.87**	m2	**6.60**
Plaster ceilings and beams exceeding 0.50m2	0.65	25.00	**0.80**	0.35	-	**4.97**	0.09	**5.87**	m2	**6.60**
Washing down old distempered or painted surfaces; pulp paper with 24' drop pattern; p.c. £4.00 roll; sizing; applying adhesive; hanging; butt joints										
Hard building board walls and columns exceeding 0.50m2	0.85	40.00	**1.17**	0.40	-	**5.68**	0.11	**6.96**	m2	**7.84**
Plaster walls and columns exceeding 0.50m2	0.85	40.00	**1.17**	0.40	-	**5.68**	0.11	**6.96**	m2	**7.84**
Hard building board ceilings and beams exceeding 0.50m2	0.85	35.00	**1.13**	0.42	-	**5.97**	0.11	**7.21**	m2	**8.12**
Plaster ceilings and beams exceeding 0.50m2	0.85	35.00	**1.13**	0.42	-	**5.97**	0.11	**7.21**	m2	**8.12**
Washing down old distempered or painted surfaces; washable paper p.c. £4.00 roll; sizing; applying adhesive; hanging; butt joints										
Hard building board walls and columns exceeding 0.50m2	0.87	25.00	**1.07**	0.33	-	**4.69**	0.09	**5.85**	m2	**6.58**
Plaster walls and columns exceeding 0.50m2	0.87	25.00	**1.07**	0.33	-	**4.69**	0.09	**5.85**	m2	**6.58**
Hard building board ceilings and beams exceeding 0.50m2	0.87	20.00	**1.03**	0.35	-	**4.97**	0.09	**6.10**	m2	**6.86**
Plaster ceilings and beams exceeding 0.50m2	0.87	20.00	**1.03**	0.35	-	**4.97**	0.09	**6.09**	m2	**6.85**
Washing down old distempered or painted surfaces; vinyl coated paper p.c. £6.50 roll; sizing; applying adhesive; hanging; butt joints										
Hard building board walls and columns exceeding 0.50m2	1.37	25.00	**1.69**	0.33	-	**4.69**	0.09	**6.47**	m2	**7.28**

Labour hourly rates: (except Specialists) Craft Operatives £14.21 Labourer £10.71 Rates are national average prices. Refer to REGIONAL VARIATIONS for indicative levels of overall pricing in regions	MATERIALS			LABOUR				RATES		
	Del to Site £	Waste %	Material Cost £	Craft Optve Hrs	Lab Hrs	Labour Cost £	Sunds £	Nett Rate £	Unit	Gross Rate (+12.5%) £
M52: DECORATIVE PAPERS/FABRICS - REDECORATIONS Cont.										
Washing down old distempered or painted surfaces; vinyl coated paper p.c. £6.50 roll; sizing; applying adhesive; hanging; butt joints Cont.										
Plaster walls and columns exceeding 0.50m2	1.37	25.00	**1.69**	0.33	-	**4.69**	0.09	**6.47**	m2	**7.28**
Hard building board ceilings and beams exceeding 0.50m2	1.37	20.00	**1.63**	0.35	-	**4.97**	0.09	**6.70**	m2	**7.53**
Plaster ceilings and beams exceeding 0.50m2	1.37	20.00	**1.63**	0.35	-	**4.97**	0.09	**6.70**	m2	**7.53**
Washing down old distempered or painted surfaces; vinyl coated paper p.c. £7.00 roll; sizing; applying adhesive; hanging; butt joints										
Hard building board walls and columns exceeding 0.50m2	1.47	25.00	**1.82**	0.33	-	**4.69**	0.09	**6.60**	m2	**7.42**
Plaster walls and columns exceeding 0.50m2	1.47	25.00	**1.82**	0.33	-	**4.69**	0.09	**6.60**	m2	**7.42**
Hard building board ceilings and beams exceeding 0.50m2	1.47	20.00	**1.75**	0.35	-	**4.97**	0.09	**6.82**	m2	**7.67**
Plaster ceilings and beams exceeding 0.50m2	1.47	20.00	**1.75**	0.35	-	**4.97**	0.09	**6.82**	m2	**7.67**
Washing down old distempered or painted surfaces; embossed paper p.c. £4.50 roll; sizing; applying adhesive; hanging; butt joints										
Hard building board walls and columns exceeding 0.50m2	1.04	25.00	**1.27**	0.33	-	**4.69**	0.09	**6.04**	m2	**6.80**
Plaster walls and columns exceeding 0.50m2	1.04	25.00	**1.27**	0.33	-	**4.69**	0.09	**6.04**	m2	**6.80**
Hard building board ceilings and beams exceeding 0.50m2	1.04	20.00	**1.22**	0.35	-	**4.97**	0.09	**6.29**	m2	**7.07**
Plaster ceilings and beams exceeding 0.50m2	1.04	20.00	**1.22**	0.35	-	**4.97**	0.09	**6.29**	m2	**7.07**
Washing down old distempered or painted surfaces; textured paper p.c. £5.00 roll; sizing; applying adhesive; hanging; butt joints										
Hard building board walls and columns exceeding 0.50m2	1.11	25.00	**1.36**	0.33	-	**4.69**	0.09	**6.13**	m2	**6.90**
Plaster walls and columns exceeding 0.50m2	1.11	25.00	**1.36**	0.33	-	**4.69**	0.09	**6.13**	m2	**6.90**
Hard building board ceilings and beams exceeding 0.50m2	1.11	20.00	**1.31**	0.35	-	**4.97**	0.09	**6.37**	m2	**7.17**
Plaster ceilings and beams exceeding 0.50m2	1.11	20.00	**1.31**	0.35	-	**4.97**	0.09	**6.37**	m2	**7.17**
Washing down old distempered or painted surfaces; woodchip paper p.c. £1.50 roll; sizing; applying adhesive; hanging; butt joints										
Hard building board walls and columns exceeding 0.50m2	0.35	25.00	**0.43**	0.33	-	**4.69**	0.09	**5.21**	m2	**5.86**
Plaster walls and columns exceeding 0.50m2	0.35	25.00	**0.43**	0.33	-	**4.69**	0.09	**5.21**	m2	**5.86**
Hard building board ceilings and beams exceeding 0.50m2	0.35	20.00	**0.41**	0.35	-	**4.97**	0.09	**5.48**	m2	**6.17**
Plaster ceilings and beams exceeding 0.50m2	0.35	20.00	**0.41**	0.35	-	**4.97**	0.09	**5.48**	m2	**6.17**
Stripping existing paper; hessian wall covering (prime cost sum for supply and allowance for waste included elsewhere); sizing; applying adhesive; hanging; butt joints										
Hard building board walls and columns exceeding 0.50m2	0.21	-	**0.21**	0.63	-	**8.95**	0.42	**9.58**	m2	**10.78**
Plaster walls and columns exceeding 0.50m2	0.21	-	**0.21**	0.58	-	**8.24**	0.39	**8.84**	m2	**9.95**

Labour hourly rates: (except Specialists) Craft Operatives £14.21 Labourer £10.71 Rates are national average prices. Refer to REGIONAL VARIATIONS for indicative levels of overall pricing in regions	MATERIALS			LABOUR				RATES		
	Del to Site £	Waste %	Material Cost £	Craft Optve Hrs	Lab Hrs	Labour Cost £	Sunds £	Nett Rate £	Unit	Gross Rate (+12.5%) £
M52: DECORATIVE PAPERS/FABRICS - REDECORATIONS Cont.										
Stripping existing paper; textile hessian paper backed wall covering (prime cost sum for supply and allowance for waste included elsewhere); sizing; applying adhesive; hanging; butt joints										
Hard building board walls and columns exceeding 0.50m2	0.21	-	**0.21**	0.55	-	**7.82**	0.37	**8.39**	m2	**9.44**
Plaster walls and columns exceeding 0.50m2	0.21	-	**0.21**	0.51	-	**7.25**	0.34	**7.80**	m2	**8.77**
Stripping existing hessian; hessian surfaced wall paper (prime cost sum for supply and allowance for waste included elsewhere); sizing; applying adhesive; hanging; butt joints										
Hard building board walls and columns exceeding 0.50m2	0.21	-	**0.21**	1.03	-	**14.64**	0.69	**15.54**	m2	**17.48**
Plaster walls and columns exceeding 0.50m2	0.21	-	**0.21**	0.99	-	**14.07**	0.66	**14.94**	m2	**16.81**
Stripping existing hessian; textile hessian paper backed wall covering (prime cost sum for supply and allowance for waste included elsewhere); sizing; applying adhesive; hanging; butt joints										
Hard building board walls and columns exceeding 0.50m2	0.21	-	**0.21**	0.96	-	**13.64**	0.64	**14.49**	m2	**16.31**
Plaster walls and columns exceeding 0.50m2	0.21	-	**0.21**	0.91	-	**12.93**	0.61	**13.75**	m2	**15.47**
Stripping existing hessian, paper backed; hessian surfaced wall paper (prime cost sum for supply and allowance for waste included elsewhere); sizing; applying adhesive; hanging; butt joints										
Hard building board walls and columns exceeding 0.50m2	0.21	-	**0.21**	0.63	-	**8.95**	0.39	**9.55**	m2	**10.74**
Plaster walls and columns exceeding 0.50m2	0.21	-	**0.21**	0.58	-	**8.24**	0.39	**8.84**	m2	**9.95**
Stripping existing hessian, paper backed; textile hessian paper backed wall covering (prime cost sum for supply and allowance for waste included elsewhere); sizing; applying adhesive; hanging; butt joints										
Hard building board walls and columns exceeding 0.50m2	0.21	-	**0.21**	0.55	-	**7.82**	0.37	**8.39**	m2	**9.44**
Plaster walls and columns exceeding 0.50m2	0.21	-	**0.21**	0.51	-	**7.25**	0.34	**7.80**	m2	**8.77**
Stripping existing paper; hessian surfaced wall paper p.c. £4.00m2; sizing; applying adhesive; hanging; butt joints										
Hard building board walls and columns exceeding 0.50m2	4.21	30.00	**5.41**	0.63	-	**8.95**	0.42	**14.78**	m2	**16.63**
Plaster walls and columns exceeding 0.50m2	4.21	30.00	**5.41**	0.58	-	**8.24**	0.39	**14.04**	m2	**15.80**
Stripping existing paper; textile hessian paper backed wall covering P.C. £7.50m2; sizing; applying adhesive; hanging; butt joints										
Hard building board walls and columns exceeding 0.50m2	4.21	30.00	**5.41**	0.55	-	**7.82**	0.37	**13.59**	m2	**15.29**
Plaster walls and columns exceeding 0.50m2	4.21	30.00	**5.41**	0.51	-	**7.25**	0.34	**13.00**	m2	**14.62**
Stripping existing hessian; hessian surfaced wall paper p.c. £4.00m2; sizing; applying adhesive; hanging ; butt joints										
Hard building board walls and columns exceeding 0.50m2	4.21	30.00	**5.41**	1.03	-	**14.64**	0.69	**20.74**	m2	**23.33**
Plaster walls and columns exceeding 0.50m2	4.21	30.00	**5.41**	0.99	-	**14.07**	0.66	**20.14**	m2	**22.66**
Stripping existing hessian; textile hessian paper backed wall covering P.C. £7.5m2; sizing; applying adhesive; hanging; butt joints										
Hard building board walls and columns exceeding 0.50m2	4.21	30.00	**5.41**	0.96	-	**13.64**	0.64	**19.69**	m2	**22.16**
Plaster walls and columns exceeding 0.50m2	4.21	30.00	**5.41**	0.91	-	**12.93**	0.61	**18.95**	m2	**21.32**

Labour hourly rates: (except Specialists) Craft Operatives £14.21 Labourer £10.71 Rates are national average prices. Refer to REGIONAL VARIATIONS for indicative levels of overall pricing in regions	MATERIALS			LABOUR			RATES			
	Del to Site £	Waste %	Material Cost £	Craft Optve Hrs	Lab Hrs	Labour Cost £	Sunds £	Nett Rate £	Unit	Gross Rate (+12.5%) £
M52: DECORATIVE PAPERS/FABRICS - REDECORATIONS Cont.										
Stripping existing hessian, paper backed; hessian surfaced wall paper p.c. £4.00m2; sizing; applying adhesive; hanging; butt joints										
Hard building board walls and columns exceeding 0.50m2	4.21	30.00	**5.41**	0.63	-	**8.95**	0.42	**14.78**	m2	**16.63**
Plaster walls and columns exceeding 0.50m2	4.21	30.00	**5.41**	0.58	-	**8.24**	0.36	**14.01**	m2	**15.76**
Stripping existing hessian, paper backed; textile hessian paper backed wall covering p.c. £7.50m2; sizing; applying adhesive; hanging; butt joints										
Hard building board walls and columns exceeding 0.50m2	7.71	25.00	**9.59**	0.55	-	**7.82**	0.34	**17.74**	m2	**19.95**
Plaster walls and columns exceeding 0.50m2	0.87	25.00	**1.05**	0.51	-	**7.25**	-	**8.30**	m2	**9.34**
Expanded polystyrene sheet 2mm thick; sizing; applying adhesive; hanging; butt joints										
Plaster walls and columns exceeding 0.50m2	0.40	25.00	**0.50**	0.36	-	**5.12**	0.22	**5.83**	m2	**6.56**
Plaster ceilings and beams exceeding 0.50m2	0.40	25.00	**0.50**	0.43	-	**6.11**	0.22	**6.83**	m2	**7.68**
M60: PAINTING/CLEAR FINISHING - EMULSION PAINTING										
Mist coat, one full coat emulsion paint										
Concrete general surfaces girth exceeding 300mm	0.19	10.00	**0.20**	0.15	-	**2.13**	0.04	**2.38**	m2	**2.67**
Concrete general surfaces 3.50 - 5.00m above floor girth exceeding 300mm	0.19	10.00	**0.20**	0.17	-	**2.42**	0.05	**2.67**	m2	**3.00**
Plaster general surfaces girth exceeding 300mm	0.15	10.00	**0.16**	0.13	-	**1.85**	0.04	**2.05**	m2	**2.30**
isolated surfaces, girth not exceeding 300mm	0.44	10.00	**0.49**	0.07	-	**0.99**	0.02	**1.50**	m	**1.69**
Plaster general surfaces 3.50 - 5.00m above floor girth exceeding 300mm	0.15	10.00	**0.16**	0.15	-	**2.13**	0.04	**2.33**	m2	**2.63**
Plasterboard general surfaces girth exceeding 300mm	0.15	10.00	**0.16**	0.13	-	**1.85**	0.04	**2.05**	m2	**2.30**
Plasterboard general surfaces 3.50 - 5.00m above floor girth exceeding 300mm	0.15	10.00	**0.16**	0.15	-	**2.13**	0.04	**2.33**	m2	**2.63**
Brickwork general surfaces girth exceeding 300mm	0.19	10.00	**0.20**	0.17	-	**2.42**	0.05	**2.67**	m2	**3.00**
Paper covered general surfaces girth exceeding 300mm	0.15	10.00	**0.16**	0.14	-	**1.99**	0.04	**2.19**	m2	**2.46**
Paper covered general surfaces 3.50 - 5.00m above floor girth exceeding 300mm	0.15	10.00	**0.16**	0.16	-	**2.27**	0.04	**2.48**	m2	**2.79**
Mist coat, two full coats emulsion paint										
Concrete general surfaces girth exceeding 300mm	0.31	10.00	**0.34**	0.21	-	**2.98**	0.06	**3.38**	m2	**3.81**
Concrete general surfaces 3.50 - 5.00m above floor girth exceeding 300mm	0.31	10.00	**0.34**	0.24	-	**3.41**	0.06	**3.82**	m2	**4.29**
Plaster general surfaces girth exceeding 300mm	0.25	10.00	**0.28**	0.19	-	**2.70**	0.05	**3.03**	m2	**3.41**
isolated surfaces, girth not exceeding 300mm	0.07	10.00	**0.08**	0.10	-	**1.42**	0.03	**1.53**	m	**1.72**
Plaster general surfaces 3.50 - 5.00m above floor girth exceeding 300mm	0.25	10.00	**0.28**	0.22	-	**3.13**	0.06	**3.46**	m2	**3.90**
Plasterboard general surfaces girth exceeding 300mm	0.25	10.00	**0.28**	0.19	-	**2.70**	0.05	**3.03**	m2	**3.41**
Plasterboard general surfaces 3.50 - 5.00m above floor girth exceeding 300mm	0.25	10.00	**0.28**	0.22	-	**3.13**	0.06	**3.46**	m2	**3.90**
Brickwork general surfaces girth exceeding 300mm	0.31	10.00	**0.34**	0.24	-	**3.41**	0.06	**3.82**	m2	**4.29**
Paper covered general surfaces girth exceeding 300mm	0.25	10.00	**0.28**	0.20	-	**2.84**	0.05	**3.17**	m2	**3.57**
Paper covered general surfaces 3.50 - 5.00m above floor girth exceeding 300mm	0.25	10.00	**0.28**	0.30	-	**4.26**	0.08	**4.62**	m2	**5.20**

Labour hourly rates: (except Specialists) Craft Operatives £14.21 Labourer £10.71 Rates are national average prices. Refer to REGIONAL VARIATIONS for indicative levels of overall pricing in regions	MATERIALS			LABOUR			RATES			
	Del to Site £	Waste %	Material Cost £	Craft Optve Hrs	Lab Hrs	Labour Cost £	Sunds £	Nett Rate £	Unit	Gross Rate (+12.5%) £
M60: PAINTING/CLEAR FINISHING - CEMENT PAINTING										
One coat waterproof cement paint; external work										
Cement rendered general surfaces										
girth exceeding 300mm	1.15	10.00	**1.27**	0.11	-	**1.56**	0.03	**2.86**	m2	**3.22**
isolated surfaces, girth not exceeding 300mm	1.15	10.00	**1.27**	0.04	-	**0.57**	0.01	**1.85**	m	**2.08**
Concrete general surfaces										
girth exceeding 300mm	1.15	10.00	**1.27**	0.11	-	**1.56**	0.03	**2.86**	m2	**3.22**
Brickwork general surfaces										
girth exceeding 300mm	1.34	10.00	**1.47**	0.13	-	**1.85**	0.04	**3.35**	m2	**3.77**
Rough cast general surfaces										
girth exceeding 300mm	1.52	10.00	**1.67**	0.17	-	**2.42**	0.05	**4.14**	m2	**4.65**
Two coats waterproof cement paint; external work										
Cement rendered general surfaces										
girth exceeding 300mm	3.04	10.00	**3.35**	0.22	-	**3.13**	0.06	**6.53**	m2	**7.35**
isolated surfaces, girth not exceeding 300mm	0.92	10.00	**1.01**	0.08	-	**1.14**	0.02	**2.17**	m	**2.44**
Concrete general surfaces										
girth exceeding 300mm	2.31	10.00	**2.54**	0.22	-	**3.13**	0.06	**5.72**	m2	**6.44**
Brickwork general surfaces										
girth exceeding 300mm	2.67	10.00	**2.94**	0.24	-	**3.41**	0.06	**6.42**	m2	**7.22**
Rough cast general surfaces										
girth exceeding 300mm	3.04	10.00	**3.35**	0.33	-	**4.69**	0.09	**8.13**	m2	**9.14**
One coat sealer, one coat waterproof cement paint; external work										
Cement rendered general surfaces										
girth exceeding 300mm	2.45	10.00	**2.70**	0.18	-	**2.56**	0.05	**5.30**	m2	**5.97**
isolated surfaces, girth not exceeding 300mm	0.73	10.00	**0.80**	0.06	-	**0.85**	0.02	**1.67**	m	**1.88**
Concrete general surfaces										
girth exceeding 300mm	2.08	10.00	**2.29**	0.18	-	**2.56**	0.05	**4.90**	m2	**5.51**
Brickwork general surfaces										
girth exceeding 300mm	2.27	10.00	**2.49**	0.21	-	**2.98**	0.06	**5.53**	m2	**6.23**
Rough cast general surfaces										
girth exceeding 300mm	2.45	10.00	**2.70**	0.27	-	**3.84**	0.07	**6.61**	m2	**7.43**
Two coats sealer, one coat waterproof cement paint; external work										
Cement rendered general surfaces										
girth exceeding 300mm	3.09	10.00	**3.40**	0.34	-	**4.83**	0.09	**8.33**	m2	**9.37**
isolated surfaces, girth not exceeding 300mm	0.93	10.00	**1.02**	0.11	-	**1.56**	0.03	**2.61**	m	**2.94**
Concrete general surfaces										
girth exceeding 300mm	2.53	10.00	**2.78**	0.34	-	**4.83**	0.09	**7.70**	m2	**8.66**
Brickwork general surfaces										
girth exceeding 300mm	2.91	10.00	**3.20**	0.39	-	**5.54**	0.11	**8.85**	m2	**9.95**
Rough cast general surfaces										
girth exceeding 300mm	3.09	10.00	**3.40**	0.51	-	**7.25**	0.14	**10.79**	m2	**12.14**
One coat textured masonry paint; external work										
Cement rendered general surfaces										
girth exceeding 300mm	0.20	10.00	**0.22**	0.11	-	**1.56**	0.03	**1.81**	m2	**2.04**
isolated surfaces, girth not exceeding 300mm	0.08	10.00	**0.09**	0.04	-	**0.57**	0.01	**0.67**	m	**0.75**
Concrete general surfaces										
girth exceeding 300mm	0.28	10.00	**0.30**	0.11	-	**1.56**	0.03	**1.90**	m2	**2.13**
Brickwork general surfaces										
girth exceeding 300mm	0.49	10.00	**0.54**	0.13	-	**1.85**	0.04	**2.42**	m2	**2.73**
Rough cast general surfaces										
girth exceeding 300mm	0.65	10.00	**0.71**	0.19	-	**2.70**	0.05	**3.46**	m2	**3.90**
Two coats textured masonry paint; external work										
Cement rendered general surfaces										
girth exceeding 300mm	0.55	10.00	**0.61**	0.24	-	**3.41**	0.06	**4.08**	m2	**4.59**
isolated surfaces, girth not exceeding 300mm	0.16	10.00	**0.17**	0.09	-	**1.28**	0.02	**1.48**	m	**1.66**
Concrete general surfaces										
girth exceeding 300mm	0.55	10.00	**0.61**	0.24	-	**3.41**	0.06	**4.08**	m2	**4.59**
Brickwork general surfaces										
girth exceeding 300mm	0.98	10.00	**1.08**	0.26	-	**3.69**	0.07	**4.85**	m2	**5.45**
Rough cast general surfaces										
girth exceeding 300mm	1.30	10.00	**1.43**	0.36	-	**5.12**	0.10	**6.64**	m2	**7.47**

Labour hourly rates: (except Specialists) Craft Operatives £14.21 Labourer £10.71 Rates are national average prices. Refer to REGIONAL VARIATIONS for indicative levels of overall pricing in regions	MATERIALS			LABOUR				RATES		
	Del to Site £	Waste %	Material Cost £	Craft Optve Hrs	Lab Hrs	Labour Cost £	Sunds £	Nett Rate £	Unit	Gross Rate (+12.5%) £
M60: PAINTING/CLEAR FINISHING - CEMENT PAINTING Cont.										
One coat stabilizing solution, one coat textured masonry paint; external work										
Cement rendered general surfaces										
girth exceeding 300mm	0.54	10.00	**0.59**	0.20	-	**2.84**	0.05	**3.49**	m2	**3.92**
isolated surfaces, girth not exceeding 300mm.....	0.16	10.00	**0.17**	0.07	-	**0.99**	0.02	**1.19**	m	**1.33**
Concrete general surfaces										
girth exceeding 300mm	0.54	10.00	**0.59**	0.20	-	**2.84**	0.05	**3.49**	m2	**3.92**
Brickwork general surfaces										
girth exceeding 300mm	0.75	10.00	**0.83**	0.23	-	**3.27**	0.06	**4.16**	m2	**4.68**
Rough cast general surfaces										
girth exceeding 300mm	0.91	10.00	**1.00**	0.30	-	**4.26**	0.08	**5.35**	m2	**6.01**
One coat stabilizing solution, two coats textured masonry paint; external work										
Cement rendered general surfaces										
girth exceeding 300mm	0.81	10.00	**0.89**	0.37	-	**5.26**	0.10	**6.25**	m2	**7.03**
isolated surfaces, girth not exceeding 300mm.....	0.24	10.00	**0.26**	0.12	-	**1.71**	0.03	**2.00**	m	**2.25**
Concrete general surfaces										
girth exceeding 300mm	0.81	10.00	**0.89**	0.37	-	**5.26**	0.10	**6.25**	m2	**7.03**
Brickwork general surfaces										
girth exceeding 300mm	1.24	10.00	**1.37**	0.43	-	**6.11**	0.12	**7.60**	m2	**8.54**
Rough cast general surfaces										
girth exceeding 300mm	1.56	10.00	**1.71**	0.56	-	**7.96**	0.15	**9.82**	m2	**11.05**
M60: PAINTING/CLEAR FINISHING - PRESERVATIVE TREATMENT										
One coat creosote BS 144; external work (sawn timber)										
Wood general surfaces										
girth exceeding 300mm	0.27	10.00	**0.30**	0.11	-	**1.56**	0.03	**1.89**	m2	**2.13**
isolated surfaces, girth not exceeding 300mm.....	0.08	10.00	**0.09**	0.04	-	**0.57**	0.01	**0.67**	m	**0.75**
Two coats creosote BS 144; external work (sawn timber)										
Wood general surfaces										
girth exceeding 300mm	0.55	10.00	**0.61**	0.22	-	**3.13**	0.06	**3.80**	m2	**4.27**
isolated surfaces, girth not exceeding 300mm.....	0.17	10.00	**0.18**	0.08	-	**1.14**	0.02	**1.34**	m	**1.51**
One coat wood preservative, internal work (sawn timber)										
Wood general surfaces										
girth exceeding 300mm	0.31	10.00	**0.35**	0.12	-	**1.71**	0.03	**2.08**	m2	**2.34**
isolated surfaces, girth not exceeding 300mm.....	0.10	10.00	**0.11**	0.04	-	**0.57**	0.01	**0.69**	m	**0.77**
One coat wood preservative, internal work (wrought timber)										
Wood general surfaces										
girth exceeding 300mm	0.26	10.00	**0.29**	0.10	-	**1.42**	0.03	**1.74**	m2	**1.95**
isolated surfaces, girth not exceeding 300mm.....	0.08	10.00	**0.09**	0.03	-	**0.43**	0.01	**0.52**	m	**0.59**
One coat wood preservative, external work (sawn timber)										
Wood general surfaces										
girth exceeding 300mm	0.31	10.00	**0.35**	0.11	-	**1.56**	0.03	**1.94**	m2	**2.18**
isolated surfaces, girth not exceeding 300mm.....	0.10	10.00	**0.11**	0.04	-	**0.57**	0.01	**0.69**	m	**0.77**
One coat wood preservative, external work (wrought timber)										
Wood general surfaces										
girth exceeding 300mm	0.26	10.00	**0.29**	0.09	-	**1.28**	0.02	**1.59**	m2	**1.79**
isolated surfaces, girth not exceeding 300mm.....	0.08	10.00	**0.09**	0.02	-	**0.28**	0.01	**0.38**	m	**0.42**
Two coats wood preservative, internal work (sawn timber)										
Wood general surfaces										
girth exceeding 300mm	0.63	10.00	**0.69**	0.24	-	**3.41**	0.06	**4.17**	m2	**4.69**
isolated surfaces, girth not exceeding 300mm.....	0.20	10.00	**0.22**	0.08	-	**1.14**	0.02	**1.37**	m	**1.55**
Two coats wood preservative, internal work (wrought timber)										
Wood general surfaces										
girth exceeding 300mm	0.52	10.00	**0.57**	0.22	-	**3.13**	0.06	**3.76**	m2	**4.23**
isolated surfaces, girth not exceeding 300mm.....	0.16	10.00	**0.17**	0.06	-	**0.85**	0.02	**1.04**	m	**1.17**
Two coats wood preservative, external work (sawn timber)										
Wood general surfaces										
girth exceeding 300mm	0.63	10.00	**0.69**	0.22	-	**3.13**	0.06	**3.88**	m2	**4.36**
isolated surfaces, girth not exceeding 300mm.....	0.20	10.00	**0.22**	0.08	-	**1.14**	0.02	**1.37**	m	**1.55**

Labour hourly rates: (except Specialists) Craft Operatives £14.21 Labourer £10.71 Rates are national average prices. Refer to REGIONAL VARIATIONS for indicative levels of overall pricing in regions	MATERIALS			LABOUR				RATES		
	Del to Site £	Waste %	Material Cost £	Craft Optve Hrs	Lab Hrs	Labour Cost £	Sunds £	Nett Rate £	Unit	Gross Rate (+12.5%) £
M60: PAINTING/CLEAR FINISHING - PRESERVATIVE TREATMENT Cont.										
Two coats wood preservative, external work (wrought timber)										
Wood general surfaces										
girth exceeding 300mm	0.52	10.00	0.57	0.18	-	2.56	0.05	3.18	m2	3.58
isolated surfaces, girth not exceeding 300mm	0.16	10.00	0.17	0.05	-	0.71	0.01	0.90	m	1.01
Two coats Sadolins High Performance Varnish										
Wood general surfaces										
girth exceeding 300mm	1.20	10.00	1.32	0.30	-	4.26	0.08	5.67	m2	6.37
isolated surfaces, girth not exceeding 300mm	0.36	10.00	0.40	0.11	-	1.56	0.03	1.99	m	2.24
Wood glazed doors										
girth exceeding 300mm; panes, area not exceeding 0.10m2	0.69	10.00	0.76	0.98	-	13.93	0.26	14.95	m2	16.81
girth exceeding 300mm; panes, area 0.10 - 0.50m2	0.49	10.00	0.53	0.71	-	10.09	0.19	10.82	m2	12.17
girth exceeding 300mm; panes, area 0.50 - 1.00m2	0.43	10.00	0.47	0.60	-	8.53	0.16	9.16	m2	10.30
girth exceeding 300mm; panes, area exceeding 1.00m2	0.43	10.00	0.47	0.52	-	7.39	0.14	8.00	m2	9.00
Wood partially glazed doors										
girth exceeding 300mm; panes, area 0.50 - 1.00m2	0.66	10.00	0.72	0.60	-	8.53	0.16	9.41	m2	10.59
girth exceeding 300mm; panes, area exceeding 1.00m2	0.66	10.00	0.72	0.58	-	8.24	0.16	9.12	m2	10.26
Wood windows and screens										
girth exceeding 300mm; panes, area not exceeding 0.10m2	0.69	10.00	0.76	1.09	-	15.49	0.29	16.54	m2	18.61
girth exceeding 300mm; panes, area 0.10 - 0.50m2	0.49	10.00	0.53	0.78	-	11.08	0.21	11.83	m2	13.31
girth exceeding 300mm; panes, area 0.50 - 1.00m2	0.43	10.00	0.47	0.66	-	9.38	0.18	10.03	m2	11.29
girth exceeding 300mm; panes, area exceeding 1.00m2	0.43	10.00	0.47	0.58	-	8.24	0.16	8.87	m2	9.98
One coat Sadolins Classic; two coats Sadolins Extra; external work										
Wood general surfaces										
girth exceeding 300mm	1.95	10.00	2.15	0.55	-	7.82	0.15	10.11	m2	11.38
isolated surfaces, girth not exceeding 300mm	0.58	10.00	0.63	0.19	-	2.70	0.05	3.38	m	3.81
Wood glazed doors										
girth exceeding 300mm; panes, area not exceeding 0.10m2	1.07	10.00	1.18	0.86	-	12.22	0.23	13.63	m2	15.33
girth exceeding 300mm; panes, area 0.10 - 0.50m2	0.77	10.00	0.84	0.62	-	8.81	0.17	9.82	m2	11.05
girth exceeding 300mm; panes, area 0.50 - 1.00m2	0.69	10.00	0.76	0.52	-	7.39	0.14	8.29	m2	9.33
girth exceeding 300mm; panes, area exceeding 1.00m2	0.69	10.00	0.76	0.46	-	6.54	0.12	7.42	m2	8.35
Wood partially glazed doors										
girth exceeding 300mm; panes, area 0.50 - 1.00m2	1.07	10.00	1.18	0.52	-	7.39	0.14	8.71	m2	9.79
girth exceeding 300mm; panes, area exceeding 1.00m2	1.07	10.00	1.18	0.50	-	7.11	0.14	8.42	m2	9.47
Wood windows and screens										
girth exceeding 300mm; panes, area not exceeding 0.10m2	1.07	10.00	1.18	0.98	-	13.93	0.26	15.37	m2	17.29
girth exceeding 300mm; panes, area 0.10 - 0.50m2	0.77	10.00	0.84	0.72	-	10.23	0.19	11.27	m2	12.68
girth exceeding 300mm; panes, area 0.50 - 1.00m2	0.69	10.00	0.76	0.61	-	8.67	0.16	9.59	m2	10.79
girth exceeding 300mm; panes, area exceeding 1.00m2	0.69	10.00	0.76	0.54	-	7.67	0.15	8.58	m2	9.65
Wood railings fences and gates; open type										
girth exceeding 300mm	1.49	10.00	1.64	0.55	-	7.82	0.15	9.61	m2	10.81
isolated surfaces, girth not exceeding 300mm	0.58	10.00	0.63	0.19	-	2.70	0.05	3.38	m	3.81
Wood railings fences and gates; close type										
girth exceeding 300mm	1.95	10.00	2.15	0.55	-	7.82	0.15	10.11	m2	11.38
M60: PAINTING/CLEAR FINISHING - OIL PAINTING WALLS AND CEILINGS										
One coat primer, one undercoat, one coat full gloss finish										
Concrete general surfaces										
girth exceeding 300mm	0.87	10.00	0.96	0.33	-	4.69	0.09	5.73	m2	6.45
Concrete general surfaces 3.50 - 5.00m above floor										
girth exceeding 300mm	0.87	10.00	0.96	0.36	-	5.12	0.10	6.17	m2	6.94
Brickwork general surfaces										
girth exceeding 300mm	1.02	10.00	1.12	0.37	-	5.26	0.10	6.47	m2	7.28
Plasterboard general surfaces										
girth exceeding 300mm	0.80	10.00	0.87	0.30	-	4.26	0.08	5.22	m2	5.87
Plasterboard general surfaces 3.50 - 5.00m above floor										
girth exceeding 300mm	0.80	10.00	0.87	0.33	-	4.69	0.09	5.65	m2	6.36
Plaster general surfaces										
girth exceeding 300mm	0.80	10.00	0.87	0.30	-	4.26	0.08	5.22	m2	5.87
isolated surfaces, girth not exceeding 300mm	1.57	10.00	1.72	0.13	-	1.85	0.04	3.60	m	4.05

Labour hourly rates: (except Specialists) Craft Operatives £14.21 Labourer £10.71 Rates are national average prices. Refer to REGIONAL VARIATIONS for indicative levels of overall pricing in regions	MATERIALS			LABOUR				RATES		
	Del to Site £	Waste %	Material Cost £	Craft Optve Hrs	Lab Hrs	Labour Cost £	Sunds £	Nett Rate £	Unit	Gross Rate (+12.5%) £
M60: PAINTING/CLEAR FINISHING - OIL PAINTING WALLS AND CEILINGS Cont.										
One coat primer, one undercoat, one coat full gloss finish Cont.										
Plaster general surfaces 3.50 - 5.00m above floor girth exceeding 300mm	0.80	10.00	**0.87**	0.33	-	**4.69**	0.09	**5.65**	m2	**6.36**
One coat primer, two undercoats, one coat full gloss finish										
Concrete general surfaces girth exceeding 300mm	1.03	10.00	**1.13**	0.44	-	**6.25**	0.12	**7.50**	m2	**8.44**
Concrete general surfaces 3.50 - 5.00m above floor girth exceeding 300mm	1.03	10.00	**1.13**	0.48	-	**6.82**	0.13	**8.08**	m2	**9.09**
Brickwork general surfaces girth exceeding 300mm	1.25	10.00	**1.37**	0.50	-	**7.11**	0.14	**8.61**	m2	**9.69**
Plasterboard general surfaces girth exceeding 300mm	1.05	10.00	**1.16**	0.40	-	**5.68**	0.11	**6.95**	m2	**7.82**
Plasterboard general surfaces 3.50 - 5.00m above floor girth exceeding 300mm	0.95	10.00	**1.05**	0.44	-	**6.25**	0.12	**7.42**	m2	**8.35**
Plaster general surfaces girth exceeding 300mm	0.95	10.00	**1.05**	0.40	-	**5.68**	0.11	**6.84**	m2	**7.70**
isolated surfaces, girth not exceeding 300mm.....	0.29	10.00	**0.32**	0.18	-	**2.56**	0.05	**2.93**	m	**3.30**
Plaster general surfaces 3.50 - 5.00m above floor girth exceeding 300mm	0.95	10.00	**1.05**	0.44	-	**6.25**	0.12	**7.42**	m2	**8.35**
One coat primer, one undercoat, one coat eggshell finish										
Concrete general surfaces girth exceeding 300mm	0.91	10.00	**1.01**	0.33	-	**4.69**	0.09	**5.78**	m2	**6.51**
Concrete general surfaces 3.50 - 5.00m above floor girth exceeding 300mm	0.91	10.00	**1.01**	0.36	-	**5.12**	0.10	**6.22**	m2	**7.00**
Brickwork general surfaces girth exceeding 300mm	1.13	10.00	**1.25**	0.37	-	**5.26**	0.10	**6.60**	m2	**7.43**
Plasterboard general surfaces girth exceeding 300mm	0.84	10.00	**0.92**	0.30	-	**4.26**	0.08	**5.27**	m2	**5.92**
Plasterboard general surfaces 3.50 - 5.00m above floor girth exceeding 300mm	0.84	10.00	**0.92**	0.33	-	**4.69**	0.09	**5.70**	m2	**6.41**
Plaster general surfaces girth exceeding 300mm	0.84	10.00	**0.92**	0.30	-	**4.26**	0.08	**5.27**	m2	**5.92**
isolated surfaces, girth not exceeding 300mm.....	1.58	10.00	**1.74**	0.13	-	**1.85**	0.04	**3.62**	m	**4.07**
Plaster general surfaces 3.50 - 5.00m above floor girth exceeding 300mm	0.84	10.00	**0.92**	0.33	-	**4.69**	0.09	**5.70**	m2	**6.41**
One coat primer, two undercoats, one coat eggshell finish										
Concrete general surfaces girth exceeding 300mm	1.07	10.00	**1.18**	0.44	-	**6.25**	0.12	**7.55**	m2	**8.50**
Concrete general surfaces 3.50 - 5.00m above floor girth exceeding 300mm	1.07	10.00	**1.18**	0.48	-	**6.82**	0.13	**8.13**	m2	**9.15**
Brickwork general surfaces girth exceeding 300mm	1.30	10.00	**1.44**	0.50	-	**7.11**	0.14	**8.68**	m2	**9.76**
Plasterboard general surfaces girth exceeding 300mm	1.00	10.00	**1.10**	0.40	-	**5.68**	0.11	**6.89**	m2	**7.75**
Plasterboard general surfaces 3.50 - 5.00m above floor girth exceeding 300mm	1.00	10.00	**1.10**	0.44	-	**6.25**	0.12	**7.47**	m2	**8.40**
Plaster general surfaces girth exceeding 300mm	1.00	10.00	**1.10**	0.40	-	**5.68**	0.11	**6.89**	m2	**7.75**
isolated surfaces, girth not exceeding 300mm.....	0.31	10.00	**0.34**	0.18	-	**2.56**	0.05	**2.95**	m	**3.31**
Plaster general surfaces 3.50 - 5.00m above floor girth exceeding 300mm	1.00	10.00	**1.10**	0.44	-	**6.25**	0.02	**7.37**	m2	**8.29**
M60: PAINTING/CLEAR FINISHING - SPRAY PAINTING										
Spray one coat primer, one undercoat, one coat full gloss finish										
Concrete general surfaces girth exceeding 300mm	0.94	10.00	**1.04**	0.29	-	**4.12**	0.08	**5.24**	m2	**5.89**
Brickwork general surfaces girth exceeding 300mm	1.15	10.00	**1.27**	0.33	-	**4.69**	0.09	**6.04**	m2	**6.80**

Labour hourly rates: (except Specialists) Craft Operatives £14.21 Labourer £10.71 Rates are national average prices. Refer to REGIONAL VARIATIONS for indicative levels of overall pricing in regions	MATERIALS			LABOUR				RATES		
	Del to Site £	Waste %	Material Cost £	Craft Optve Hrs	Lab Hrs	Labour Cost £	Sunds £	Nett Rate £	Unit	Gross Rate (+12.5%) £
M60: PAINTING/CLEAR FINISHING - SPRAY PAINTING Cont.										
Spray one coat primer, one undercoat, one coat full gloss finish Cont.										
Plaster general surfaces girth exceeding 300mm	0.87	10.00	**0.96**	0.26	-	**3.69**	0.07	**4.72**	m2	**5.31**
Spray one coat primer, two undercoats, one coat full gloss finish										
Concrete general surfaces girth exceeding 300mm	1.13	10.00	**1.24**	0.34	-	**4.83**	0.09	**6.16**	m2	**6.93**
Brickwork general surfaces girth exceeding 300mm	1.40	10.00	**1.53**	0.39	-	**5.54**	0.11	**7.18**	m2	**8.08**
Plaster general surfaces girth exceeding 300mm	1.05	10.00	**1.16**	0.31	-	**4.41**	0.08	**5.65**	m2	**6.35**
Spray one coat primer, one basecoat, one coat multicolour finish										
Concrete general surfaces girth exceeding 300mm	1.44	10.00	**1.58**	0.43	-	**6.11**	0.12	**7.81**	m2	**8.78**
Brickwork general surfaces girth exceeding 300mm	1.72	10.00	**1.90**	0.45	-	**6.39**	0.12	**8.41**	m2	**9.46**
Plaster general surfaces girth exceeding 300mm	1.30	10.00	**1.43**	0.41	-	**5.83**	0.11	**7.37**	m2	**8.29**
M60: PAINTING/CLEAR FINISHING - ACRYLATED RUBBER PAINTING										
One coat primer, two coats acrylated rubber paint										
Concrete general surfaces girth exceeding 300mm	3.31	10.00	**3.64**	0.33	-	**4.69**	0.09	**8.42**	m2	**9.47**
Brickwork general surfaces girth exceeding 300mm	3.95	10.00	**4.34**	0.35	-	**4.97**	0.09	**9.41**	m2	**10.58**
Plasterboard general surfaces girth exceeding 300mm	2.97	10.00	**3.27**	0.32	-	**4.55**	0.09	**7.90**	m2	**8.89**
Plaster general surfaces girth exceeding 300mm	2.97	10.00	**3.27**	0.32	-	**4.55**	0.09	**7.90**	m2	**8.89**
isolated surfaces, girth not exceeding 300mm	0.86	10.00	**0.95**	0.14	-	**1.99**	0.04	**2.98**	m	**3.35**
M60: PAINTING/CLEAR FINISHING - PLASTIC FINISH										
Textured plastic coating - Stippled finish										
Concrete general surfaces girth exceeding 300mm	0.35	10.00	**0.38**	0.28	-	**3.98**	0.08	**4.44**	m2	**4.99**
Concrete general surfaces 3.50 - 5.00m above floor girth exceeding 300mm	0.35	10.00	**0.38**	0.31	-	**4.41**	0.08	**4.87**	m2	**5.48**
Brickwork general surfaces girth exceeding 300mm	0.41	10.00	**0.46**	0.33	-	**4.69**	0.09	**5.23**	m2	**5.89**
Plasterboard general surfaces girth exceeding 300mm	0.31	10.00	**0.34**	0.22	-	**3.13**	0.06	**3.53**	m2	**3.97**
Plasterboard general surfaces 3.50 - 5.00m above floor girth exceeding 300mm	0.31	10.00	**0.34**	0.24	-	**3.41**	0.06	**3.82**	m2	**4.29**
Plaster general surfaces girth exceeding 300mm	0.31	10.00	**0.34**	0.22	-	**3.13**	0.06	**3.53**	m2	**3.97**
Plaster general surfaces 3.50 - 5.00m above floor girth exceeding 300mm	0.31	10.00	**0.34**	0.24	-	**3.41**	0.06	**3.82**	m2	**4.29**
Textured plastic coating - Combed Finish										
Concrete general surfaces girth exceeding 300mm	0.41	10.00	**0.46**	0.33	-	**4.69**	0.09	**5.23**	m2	**5.89**
Concrete general surfaces 3.50 - 5.00m above floor girth exceeding 300mm	0.41	10.00	**0.46**	0.36	-	**5.12**	0.10	**5.67**	m2	**6.38**
Brickwork general surfaces girth exceeding 300mm	0.50	10.00	**0.55**	0.39	-	**5.54**	0.11	**6.19**	m2	**6.97**
Plasterboard general surfaces girth exceeding 300mm	0.37	10.00	**0.41**	0.28	-	**3.98**	0.08	**4.46**	m2	**5.02**
Plasterboard general surfaces 3.50 - 5.00m above floor girth exceeding 300mm	0.37	10.00	**0.41**	0.30	-	**4.26**	0.08	**4.75**	m2	**5.35**
Plaster general surfaces girth exceeding 300mm	0.37	10.00	**0.41**	0.28	-	**3.98**	0.08	**4.46**	m2	**5.02**

Labour hourly rates: (except Specialists) Craft Operatives £14.21 Labourer £10.71 Rates are national average prices. Refer to REGIONAL VARIATIONS for indicative levels of overall pricing in regions	MATERIALS			LABOUR				RATES		
	Del to Site £	Waste %	Material Cost £	Craft Optve Hrs	Lab Hrs	Labour Cost £	Sunds £	Nett Rate £	Unit	Gross Rate (+12.5%) £
M60: PAINTING/CLEAR FINISHING - PLASTIC FINISH Cont.										
Textured plastic coating - Combed Finish Cont.										
Plaster general surfaces 3.50 - 5.00m above floor										
girth exceeding 300mm	0.37	10.00	**0.41**	0.30	-	**4.26**	0.08	**4.75**	m2	**5.35**
M60: PAINTING/CLEAR FINISHING - OIL PAINTING METALWORK										
One undercoat, one coat full gloss finish on ready primed metal surfaces										
Iron or steel structural work										
girth exceeding 300mm	0.32	10.00	**0.35**	0.31	-	**4.41**	0.08	**4.84**	m2	**5.44**
isolated surfaces, girth not exceeding 300mm	0.10	10.00	**0.11**	0.13	-	**1.85**	0.04	**1.99**	m	**2.24**
Iron or steel structural members of roof trusses, lattice girders, purlins and the like										
girth exceeding 300mm	0.32	10.00	**0.35**	0.40	-	**5.68**	0.11	**6.14**	m2	**6.91**
isolated surfaces, girth not exceeding 300mm	0.10	10.00	**0.11**	0.13	-	**1.85**	0.04	**1.99**	m	**2.24**
Two undercoats, one coat full gloss finish on ready primed metal surfaces										
Iron or steel structural work										
girth exceeding 300mm	0.48	10.00	**0.53**	0.46	-	**6.54**	0.12	**7.19**	m2	**8.08**
isolated surfaces, girth not exceeding 300mm	0.15	10.00	**0.16**	0.15	-	**2.13**	0.04	**2.33**	m	**2.63**
Iron or steel structural members of roof trusses, lattice girders, purlins and the like										
girth exceeding 300mm	0.48	10.00	**0.53**	0.59	-	**8.38**	0.16	**9.07**	m2	**10.20**
isolated surfaces, girth not exceeding 300mm	0.15	10.00	**0.16**	0.20	-	**2.84**	0.05	**3.06**	m	**3.44**
One coat primer, one undercoat, one coat full gloss finish on metal surfaces										
Iron or steel general surfaces										
girth exceeding 300mm	0.92	10.00	**1.01**	0.43	-	**6.11**	0.12	**7.24**	m2	**8.15**
isolated surfaces, girth not exceeding 300mm	0.29	10.00	**0.32**	0.14	-	**1.99**	0.04	**2.35**	m	**2.64**
Galvanized glazed doors, windows or screens										
girth exceeding 300mm; panes, area not exceeding 0.10m2	0.50	10.00	**0.55**	0.79	-	**11.23**	0.21	**11.99**	m2	**13.49**
girth exceeding 300mm; panes, area 0.10 - 0.50m2	0.37	10.00	**0.41**	0.56	-	**7.96**	0.15	**8.52**	m2	**9.58**
girth exceeding 300mm; panes, area 0.50 - 1.00m2	0.32	10.00	**0.35**	0.47	-	**6.68**	0.13	**7.15**	m2	**8.05**
girth exceeding 300mm; panes, area exceeding 1.00m2	0.32	10.00	**0.35**	0.43	-	**6.11**	0.12	**6.58**	m2	**7.40**
Iron or steel structural work										
girth exceeding 300mm	0.92	10.00	**1.01**	0.46	-	**6.54**	0.12	**7.68**	m2	**8.63**
isolated surfaces, girth not exceeding 300mm	0.29	10.00	**0.32**	0.20	-	**2.84**	0.05	**3.22**	m	**3.62**
Iron or steel structural members of roof trusses, lattice girders, purlins and the like										
girth exceeding 300mm	0.92	10.00	**1.01**	0.59	-	**8.38**	0.16	**9.56**	m2	**10.75**
isolated surfaces, girth not exceeding 300mm	0.29	10.00	**0.32**	0.20	-	**2.84**	0.05	**3.22**	m	**3.62**
Iron or steel services										
girth exceeding 300mm	0.92	10.00	**1.01**	0.59	-	**8.38**	0.16	**9.56**	m2	**10.75**
isolated surfaces, girth not exceeding 300mm	0.29	10.00	**0.32**	0.20	-	**2.84**	0.05	**3.22**	m	**3.62**
isolated areas not exceeding 0.50m2 irrespective of girth	0.47	10.00	**0.52**	0.59	-	**8.38**	0.16	**9.06**	nr	**10.20**
Copper services										
girth exceeding 300mm	0.92	10.00	**1.01**	0.59	-	**8.38**	0.16	**9.56**	m2	**10.75**
isolated surfaces, girth not exceeding 300mm	0.29	10.00	**0.32**	0.20	-	**2.84**	0.05	**3.22**	m	**3.62**
Galvanized services										
girth exceeding 300mm	0.92	10.00	**1.01**	0.59	-	**8.38**	0.16	**9.56**	m2	**10.75**
isolated surfaces, girth not exceeding 300mm	0.29	10.00	**0.32**	0.20	-	**2.84**	0.05	**3.22**	m	**3.62**
One coat primer, one undercoat, one coat full gloss finish on metal surfaces ; external work										
Iron or steel general surfaces										
girth exceeding 300mm	0.92	10.00	**1.01**	0.46	-	**6.54**	0.12	**7.68**	m2	**8.63**
isolated surfaces, girth not exceeding 300mm	0.29	10.00	**0.32**	0.18	-	**2.56**	0.05	**2.93**	m	**3.29**
Galvanized glazed doors, windows or screens										
girth exceeding 300mm; panes, area not exceeding 0.10m2	0.50	10.00	**0.55**	0.83	-	**11.79**	0.22	**12.57**	m2	**14.14**
girth exceeding 300mm; panes, area 0.10 - 0.50m2	0.37	10.00	**0.41**	0.59	-	**8.38**	0.16	**8.95**	m2	**10.07**
girth exceeding 300mm; panes, area 0.50 - 1.00m2	0.32	10.00	**0.35**	0.51	-	**7.25**	0.14	**7.73**	m2	**8.70**
girth exceeding 300mm; panes, area exceeding 1.00m2	0.32	10.00	**0.35**	0.46	-	**6.54**	0.12	**7.01**	m2	**7.89**
Iron or steel structural work										
girth exceeding 300mm	0.92	10.00	**1.01**	0.46	-	**6.54**	0.12	**7.68**	m2	**8.63**
isolated surfaces, girth not exceeding 300mm	0.29	10.00	**0.32**	0.20	-	**2.84**	0.05	**3.22**	m	**3.62**
Iron or steel structural members of roof trusses, lattice girders, purlins and the like										
girth exceeding 300mm	0.92	10.00	**1.01**	0.59	-	**8.38**	0.16	**9.56**	m2	**10.75**
isolated surfaces, girth not exceeding 300mm	0.29	10.00	**0.32**	0.20	-	**2.84**	0.05	**3.22**	m	**3.62**

Labour hourly rates: (except Specialists) Craft Operatives £14.21 Labourer £10.71 Rates are national average prices. Refer to REGIONAL VARIATIONS for indicative levels of overall pricing in regions	MATERIALS			LABOUR			RATES			
	Del to Site £	Waste %	Material Cost £	Craft Optve Hrs	Lab Hrs	Labour Cost £	Sunds £	Nett Rate £	Unit	Gross Rate (+12.5%) £
M60: PAINTING/CLEAR FINISHING - OIL PAINTING METALWORK Cont.										
One coat primer, one undercoat, one coat full gloss finish on metal surfaces ; external work Cont.										
Iron or steel railings, fences and gates; plain open type										
girth exceeding 300mm	0.71	10.00	**0.78**	0.43	-	**6.11**	0.12	**7.01**	m2	**7.89**
isolated surfaces, girth not exceeding 300mm	0.29	10.00	**0.32**	0.14	-	**1.99**	0.04	**2.35**	m	**2.64**
Iron or steel railings, fences and gates; close type										
girth exceeding 300mm	0.92	10.00	**1.01**	0.36	-	**5.12**	0.10	**6.23**	m2	**7.01**
Iron or steel railings, fences and gates; ornamental type										
girth exceeding 300mm	0.71	10.00	**0.78**	0.73	-	**10.37**	0.20	**11.35**	m2	**12.77**
Iron or steel eaves gutters										
girth exceeding 300mm	0.92	10.00	**1.01**	0.50	-	**7.11**	0.14	**8.25**	m2	**9.29**
isolated surfaces, girth not exceeding 300mm	0.29	10.00	**0.32**	0.17	-	**2.42**	0.05	**2.78**	m	**3.13**
Galvanized eaves gutters										
girth exceeding 300mm	0.92	10.00	**1.01**	0.50	-	**7.11**	0.14	**8.25**	m2	**9.29**
isolated surfaces, girth not exceeding 300mm	0.29	10.00	**0.32**	0.17	-	**2.42**	0.05	**2.78**	m	**3.13**
Iron or steel services										
girth exceeding 300mm	0.92	10.00	**1.01**	0.59	-	**8.38**	0.16	**9.56**	m2	**10.75**
isolated surfaces, girth not exceeding 300mm	0.29	10.00	**0.32**	0.20	-	**2.84**	0.05	**3.22**	m	**3.62**
isolated areas not exceeding 0.50m2 irrespective of girth	0.47	10.00	**0.52**	0.59	-	**8.38**	0.16	**9.06**	nr	**10.20**
Copper services										
girth exceeding 300mm	0.92	10.00	**1.01**	0.59	-	**8.38**	0.16	**9.56**	m2	**10.75**
isolated surfaces, girth not exceeding 300mm	0.29	10.00	**0.32**	0.20	-	**2.84**	0.05	**3.22**	m	**3.62**
Galvanized services										
girth exceeding 300mm	0.92	10.00	**1.01**	0.59	-	**8.38**	0.16	**9.56**	m2	**10.75**
isolated surfaces, girth not exceeding 300mm	0.29	10.00	**0.32**	0.20	-	**2.84**	0.05	**3.22**	m	**3.62**
One coat primer, two undercoats, one coat full gloss finish on metal surfaces										
Iron or steel general surfaces										
girth exceeding 300mm	1.08	10.00	**1.19**	0.57	-	**8.10**	0.15	**9.44**	m2	**10.62**
isolated surfaces, girth not exceeding 300mm	0.34	10.00	**0.37**	0.19	-	**2.70**	0.05	**3.12**	m	**3.51**
Galvanized glazed doors, windows or screens										
girth exceeding 300mm; panes, area not exceeding 0.10m2	0.59	10.00	**0.65**	1.06	-	**15.06**	0.29	**15.99**	m2	**17.99**
girth exceeding 300mm; panes, area 0.10 - 0.50m2	0.43	10.00	**0.47**	0.75	-	**10.66**	0.20	**11.33**	m2	**12.75**
girth exceeding 300mm; panes, area 0.50 - 1.00m2	0.37	10.00	**0.40**	0.63	-	**8.95**	0.17	**9.53**	m2	**10.72**
girth exceeding 300mm; panes, area exceeding 1.00m2	0.37	10.00	**0.40**	0.57	-	**8.10**	0.15	**8.66**	m2	**9.74**
Iron or steel structural work										
girth exceeding 300mm	1.08	10.00	**1.19**	0.62	-	**8.81**	0.17	**10.17**	m2	**11.44**
isolated surfaces, girth not exceeding 300mm	0.34	10.00	**0.37**	0.24	-	**3.41**	0.06	**3.85**	m	**4.33**
Iron or steel structural members of roof trusses, lattice girders, purlins and the like										
girth exceeding 300mm	1.08	10.00	**1.19**	0.79	-	**11.23**	0.21	**12.63**	m2	**14.21**
isolated surfaces, girth not exceeding 300mm	0.34	10.00	**0.37**	0.24	-	**3.41**	0.06	**3.85**	m	**4.33**
Iron or steel services										
girth exceeding 300mm	1.08	10.00	**1.19**	0.69	-	**9.80**	0.19	**11.18**	m2	**12.58**
isolated surfaces, girth not exceeding 300mm	0.34	10.00	**0.37**	0.23	-	**3.27**	0.06	**3.70**	m	**4.17**
isolated areas not exceeding 0.50m2 irrespective of girth	0.55	10.00	**0.60**	0.69	-	**9.80**	0.19	**10.59**	nr	**11.92**
Copper services										
girth exceeding 300mm	1.08	10.00	**1.19**	0.69	-	**9.80**	0.19	**11.18**	m2	**12.58**
isolated surfaces, girth not exceeding 300mm	0.34	10.00	**0.37**	0.23	-	**3.27**	0.06	**3.70**	m	**4.17**
Galvanized services										
girth exceeding 300mm	1.08	10.00	**1.19**	0.69	-	**9.80**	0.19	**11.18**	m2	**12.58**
isolated surfaces, girth not exceeding 300mm	0.34	10.00	**0.37**	0.23	-	**3.27**	0.06	**3.70**	m	**4.17**
One coat primer, two undercoats, one coat full gloss finish on metal surfaces; external work										
Iron or steel general surfaces										
girth exceeding 300mm	1.08	10.00	**1.19**	0.62	-	**8.81**	0.17	**10.17**	m2	**11.44**
isolated surfaces, girth not exceeding 300mm	0.34	10.00	**0.37**	0.21	-	**2.98**	0.06	**3.41**	m	**3.84**
Galvanized glazed doors, windows or screens										
girth exceeding 300mm; panes, area not exceeding 0.10m2	0.59	10.00	**0.65**	1.10	-	**15.63**	0.30	**16.57**	m2	**18.64**
girth exceeding 300mm; panes, area 0.10 - 0.50m2	0.43	10.00	**0.47**	0.79	-	**11.23**	0.21	**11.91**	m2	**13.40**
girth exceeding 300mm; panes, area 0.50 - 1.00m2	0.37	10.00	**0.40**	0.67	-	**9.52**	0.18	**10.10**	m2	**11.37**
girth exceeding 300mm; panes, area exceeding 1.00m2	0.37	10.00	**0.40**	0.62	-	**8.81**	0.17	**9.38**	m2	**10.55**
Iron or steel structural work										
girth exceeding 300mm	1.08	10.00	**1.19**	0.62	-	**8.81**	0.17	**10.17**	m2	**11.44**
isolated surfaces, girth not exceeding 300mm	0.34	10.00	**0.37**	0.24	-	**3.41**	0.06	**3.85**	m	**4.33**

Labour hourly rates: (except Specialists) Craft Operatives £14.21 Labourer £10.71 Rates are national average prices. Refer to REGIONAL VARIATIONS for indicative levels of overall pricing in regions	MATERIALS			LABOUR				RATES		
	Del to Site £	Waste %	Material Cost £	Craft Optve Hrs	Lab Hrs	Labour Cost £	Sunds £	Nett Rate £	Unit	Gross Rate (+12.5%) £
M60: PAINTING/CLEAR FINISHING - OIL PAINTING METALWORK Cont.										
One coat primer, two undercoats, one coat full gloss finish on metal surfaces; external work Cont.										
Iron or steel structural members of roof trusses, lattice girders, purlins and the like										
girth exceeding 300mm	1.08	10.00	**1.19**	0.79	-	**11.23**	0.21	**12.63**	m2	**14.21**
isolated surfaces, girth not exceeding 300mm	0.34	10.00	**0.37**	0.24	-	**3.41**	0.06	**3.85**	m	**4.33**
Iron or steel railings, fences and gates; plain open type										
girth exceeding 300mm	0.83	10.00	**0.92**	0.57	-	**8.10**	0.15	**9.17**	m2	**10.32**
isolated surfaces, girth not exceeding 300mm	0.34	10.00	**0.37**	0.19	-	**2.70**	0.05	**3.12**	m	**3.51**
Iron or steel railings, fences and gates; close type										
girth exceeding 300mm	1.08	10.00	**1.19**	0.48	-	**6.82**	0.13	**8.14**	m2	**9.16**
Iron or steel railings, fences and gates; ornamental type										
girth exceeding 300mm	0.83	10.00	**0.92**	0.97	-	**13.78**	0.26	**14.96**	m2	**16.83**
Iron or steel eaves gutters										
girth exceeding 300mm	1.08	10.00	**1.19**	0.66	-	**9.38**	0.18	**10.75**	m2	**12.09**
isolated surfaces, girth not exceeding 300mm	0.34	10.00	**0.37**	0.22	-	**3.13**	0.06	**3.56**	m	**4.00**
Galvanized eaves gutters										
girth exceeding 300mm	1.08	10.00	**1.19**	0.66	-	**9.38**	0.18	**10.75**	m2	**12.09**
isolated surfaces, girth not exceeding 300mm	0.34	10.00	**0.37**	0.22	-	**3.13**	0.06	**3.56**	m	**4.00**
Iron or steel services										
girth exceeding 300mm	1.08	10.00	**1.19**	0.69	-	**9.80**	0.19	**11.18**	m2	**12.58**
isolated surfaces, girth not exceeding 300mm	0.34	10.00	**0.37**	0.23	-	**3.27**	0.06	**3.70**	m	**4.17**
isolated areas not exceeding 0.50m2 irrespective of girth	0.56	10.00	**0.62**	0.69	-	**9.80**	0.19	**10.61**	nr	**11.93**
Copper services										
girth exceeding 300mm	1.08	10.00	**1.19**	0.69	-	**9.80**	0.19	**11.18**	m2	**12.58**
isolated surfaces, girth not exceeding 300mm	0.34	10.00	**0.37**	0.23	-	**3.27**	0.06	**3.70**	m	**4.17**
Galvanized services										
girth exceeding 300mm	1.08	10.00	**1.19**	0.69	-	**9.80**	0.19	**11.18**	m2	**12.58**
isolated surfaces, girth not exceeding 300mm	0.34	10.00	**0.37**	0.23	-	**3.27**	0.06	**3.70**	m	**4.17**
One coat primer, one undercoat, two coats full gloss finish on metal surfaces										
Iron or steel general surfaces										
girth exceeding 300mm	1.08	10.00	**1.19**	0.57	-	**8.10**	0.15	**9.44**	m2	**10.62**
isolated surfaces, girth not exceeding 300mm	0.34	10.00	**0.37**	0.19	-	**2.70**	0.05	**3.12**	m	**3.51**
Galvanized glazed doors, windows or screens										
girth exceeding 300mm; panes, area not exceeding 0.10m2	0.59	10.00	**0.65**	1.06	-	**15.06**	0.29	**15.99**	m2	**17.99**
girth exceeding 300mm; panes, area 0.10 - 0.50m2	0.43	10.00	**0.47**	0.75	-	**10.66**	0.20	**11.33**	m2	**12.75**
girth exceeding 300mm; panes, area 0.50 - 1.00m2	0.37	10.00	**0.40**	0.63	-	**8.95**	0.17	**9.53**	m2	**10.72**
girth exceeding 300mm; panes, area exceeding 1.00m2	0.37	10.00	**0.40**	0.57	-	**8.10**	0.15	**8.66**	m2	**9.74**
Iron or steel structural work										
girth exceeding 300mm	1.08	10.00	**1.19**	0.62	-	**8.81**	0.17	**10.17**	m2	**11.44**
isolated surfaces, girth not exceeding 300mm	0.34	10.00	**0.37**	0.24	-	**3.41**	0.06	**3.85**	m	**4.33**
Iron or steel structural members of roof trusses, lattice girders, purlins and the like										
girth exceeding 300mm	1.08	10.00	**1.19**	0.79	-	**11.23**	0.21	**12.63**	m2	**14.21**
isolated surfaces, girth not exceeding 300mm	0.34	10.00	**0.37**	0.24	-	**3.41**	0.06	**3.85**	m	**4.33**
Iron or steel services										
girth exceeding 300mm	1.08	10.00	**1.19**	0.69	-	**9.80**	0.19	**11.18**	m2	**12.58**
isolated surfaces, girth not exceeding 300mm	0.34	10.00	**0.37**	0.23	-	**3.27**	0.06	**3.70**	m	**4.17**
isolated areas not exceeding 0.50m2 irrespective of girth	0.56	10.00	**0.62**	0.69	-	**9.80**	0.19	**10.61**	nr	**11.93**
Copper services										
girth exceeding 300mm	1.08	10.00	**1.19**	0.69	-	**9.80**	0.19	**11.18**	m2	**12.58**
isolated surfaces, girth not exceeding 300mm	0.34	10.00	**0.37**	0.23	-	**3.27**	0.06	**3.70**	m	**4.17**
Galvanized services										
girth exceeding 300mm	1.08	10.00	**1.19**	0.69	-	**9.80**	0.19	**11.18**	m2	**12.58**
isolated surfaces, girth not exceeding 300mm	0.34	10.00	**0.37**	0.23	-	**3.27**	0.06	**3.70**	m	**4.17**
One coat primer, one undercoat, two coats full gloss finish on metal surfaces; external work										
Iron or steel general surfaces										
girth exceeding 300mm	1.08	10.00	**1.19**	0.62	-	**8.81**	0.17	**10.17**	m2	**11.44**
isolated surfaces, girth not exceeding 300mm	0.34	10.00	**0.37**	0.21	-	**2.98**	0.06	**3.41**	m	**3.84**
Galvanized glazed doors, windows or screens										
girth exceeding 300mm; panes, area not exceeding 0.10m2	0.59	10.00	**0.65**	1.10	-	**15.63**	0.30	**16.57**	m2	**18.64**
girth exceeding 300mm; panes, area 0.10 - 0.50m2	0.43	10.00	**0.47**	0.79	-	**11.23**	0.21	**11.91**	m2	**13.40**
girth exceeding 300mm; panes, area 0.50 - 1.00m2	0.37	10.00	**0.40**	0.67	-	**9.52**	0.18	**10.10**	m2	**11.37**

Labour hourly rates: (except Specialists) Craft Operatives £14.21 Labourer £10.71 Rates are national average prices. Refer to REGIONAL VARIATIONS for indicative levels of overall pricing in regions	MATERIALS			LABOUR			RATES			
	Del to Site	Waste	Material Cost	Craft Optve	Lab	Labour Cost	Sunds	Nett Rate	Unit	Gross Rate (+12.5%)
	£	%	£	Hrs	Hrs	£	£	£		£
M60: PAINTING/CLEAR FINISHING - OIL PAINTING METALWORK Cont.										
One coat primer, one undercoat, two coats full gloss finish on metal surfaces; external work Cont.										
Galvanized glazed doors, windows or screens Cont.										
girth exceeding 300mm; panes, area exceeding 1.00m2	0.37	10.00	**0.40**	0.62	-	**8.81**	0.17	**9.38**	m2	**10.55**
Iron or steel structural work										
girth exceeding 300mm	1.08	10.00	**1.19**	0.62	-	**8.81**	0.17	**10.17**	m2	**11.44**
isolated surfaces, girth not exceeding 300mm	0.34	10.00	**0.37**	0.62	-	**8.81**	0.17	**9.35**	m2	**10.52**
Iron or steel structural members of roof trusses, lattice girders, purlins and the like										
girth exceeding 300mm	1.08	10.00	**1.19**	0.62	-	**8.81**	0.17	**10.17**	m2	**11.44**
isolated surfaces, girth not exceeding 300mm	0.34	10.00	**0.37**	0.24	-	**3.41**	0.06	**3.85**	m	**4.33**
Iron or steel railings, fences and gates; plain open type										
girth exceeding 300mm	0.83	10.00	**0.92**	0.57	-	**8.10**	0.15	**9.17**	m2	**10.32**
isolated surfaces, girth not exceeding 300mm	0.34	10.00	**0.37**	0.19	-	**2.70**	0.05	**3.12**	m	**3.51**
Iron or steel railings, fences and gates; close type										
girth exceeding 300mm	1.08	10.00	**1.19**	0.48	-	**6.82**	0.13	**8.14**	m2	**9.16**
Iron or steel railings, fences and gates; ornamental type										
girth exceeding 300mm	0.83	10.00	**0.92**	0.97	-	**13.78**	0.26	**14.96**	m2	**16.83**
Iron or steel eaves gutters										
girth exceeding 300mm	1.08	10.00	**1.19**	0.66	-	**9.38**	0.18	**10.75**	m2	**12.09**
isolated surfaces, girth not exceeding 300mm	0.34	10.00	**0.37**	0.22	-	**3.13**	0.06	**3.56**	m	**4.00**
Galvanized eaves gutters										
girth exceeding 300mm	1.08	10.00	**1.19**	0.66	-	**9.38**	0.18	**10.75**	m2	**12.09**
isolated surfaces, girth not exceeding 300mm	0.34	10.00	**0.37**	0.22	-	**3.13**	0.06	**3.56**	m	**4.00**
Iron or steel services										
girth exceeding 300mm	1.08	10.00	**1.19**	0.69	-	**9.80**	0.19	**11.18**	m2	**12.58**
isolated surfaces, girth not exceeding 300mm	0.34	10.00	**0.37**	0.23	-	**3.27**	0.06	**3.70**	m	**4.17**
isolated areas not exceeding 0.50m2 irrespective of girth	0.56	10.00	**0.62**	0.69	-	**9.80**	0.19	**10.61**	nr	**11.93**
Copper services										
girth exceeding 300mm	1.08	10.00	**1.19**	0.69	-	**9.80**	0.19	**11.18**	m2	**12.58**
isolated surfaces, girth not exceeding 300mm	0.34	10.00	**0.37**	0.23	-	**3.27**	0.06	**3.70**	m	**4.17**
Galvanized services										
girth exceeding 300mm	1.08	10.00	**1.19**	0.69	-	**9.80**	0.19	**11.18**	m2	**12.58**
isolated surfaces, girth not exceeding 300mm	0.34	10.00	**0.37**	0.23	-	**3.27**	0.06	**3.70**	m	**4.17**
M60: PAINTING/CLEAR FINISHING - METALLIC PAINTING										
One coat aluminium metallic paint										
Iron or steel radiators; panel type										
girth exceeding 300mm	0.51	10.00	**0.56**	0.15	-	**2.13**	0.04	**2.73**	m2	**3.07**
Iron or steel radiators; column type										
girth exceeding 300mm	0.51	10.00	**0.56**	0.22	-	**3.13**	0.06	**3.74**	m2	**4.21**
Iron or steel services										
isolated surfaces, girth not exceeding 300mm	0.16	10.00	**0.17**	0.06	-	**0.85**	0.02	**1.04**	m	**1.17**
Copper services										
isolated surfaces, girth not exceeding 300mm	0.16	10.00	**0.17**	0.06	-	**0.85**	0.02	**1.04**	m	**1.17**
Galvanized services										
isolated surfaces, girth not exceeding 300mm	0.16	10.00	**0.17**	0.06	-	**0.85**	0.02	**1.04**	m	**1.17**
Two coats aluminium metallic paint										
Iron or steel radiators; panel type										
girth exceeding 300mm	1.01	10.00	**1.12**	0.31	-	**4.41**	0.08	**5.60**	m2	**6.30**
Iron or steel radiators; column type										
girth exceeding 300mm	1.01	10.00	**1.12**	0.44	-	**6.25**	0.12	**7.49**	m2	**8.42**
Iron or steel services										
isolated surfaces, girth not exceeding 300mm	0.31	10.00	**0.34**	0.11	-	**1.56**	0.03	**1.94**	m	**2.18**
Copper services										
isolated surfaces, girth not exceeding 300mm	0.31	10.00	**0.34**	0.11	-	**1.56**	0.03	**1.94**	m	**2.18**
Galvanized services										
isolated surfaces, girth not exceeding 300mm	0.31	10.00	**0.34**	0.11	-	**1.56**	0.03	**1.94**	m	**2.18**
One coat gold or bronze metallic paint										
Iron or steel radiators; panel type										
girth exceeding 300mm	0.98	10.00	**1.08**	0.15	-	**2.13**	0.04	**3.25**	m2	**3.66**
Iron or steel radiators; column type										
girth exceeding 300mm	0.98	10.00	**1.08**	0.22	-	**3.13**	0.06	**4.27**	m2	**4.80**

Labour hourly rates: (except Specialists) Craft Operatives £14.21 Labourer £10.71 Rates are national average prices. Refer to REGIONAL VARIATIONS for indicative levels of overall pricing in regions	MATERIALS			LABOUR			RATES			
	Del to Site £	Waste %	Material Cost £	Craft Optve Hrs	Lab Hrs	Labour Cost £	Sunds £	Nett Rate £	Unit	Gross Rate (+12.5%) £
M60: PAINTING/CLEAR FINISHING - METALLIC PAINTING Cont.										
One coat gold or bronze metallic paint Cont.										
Iron or steel services										
isolated surfaces, girth not exceeding 300mm.....	0.29	10.00	**0.32**	0.06	-	**0.85**	0.02	**1.19**	m	**1.34**
Copper services										
isolated surfaces, girth not exceeding 300mm.....	0.29	10.00	**0.32**	0.06	-	**0.85**	0.02	**1.19**	m	**1.34**
Galvanized services										
isolated surfaces, girth not exceeding 300mm.....	0.29	10.00	**0.32**	0.06	-	**0.85**	0.02	**1.19**	m	**1.34**
Two coats gold or bronze metallic paint										
Iron or steel radiators; panel type										
girth exceeding 300mm	1.97	10.00	**2.17**	0.31	-	**4.41**	0.08	**6.65**	m2	**7.49**
Iron or steel radiators; column type										
girth exceeding 300mm	1.97	10.00	**2.17**	0.44	-	**6.25**	0.12	**8.54**	m2	**9.60**
Iron or steel services										
isolated surfaces, girth not exceeding 300mm.....	0.58	10.00	**0.64**	0.11	-	**1.56**	0.03	**2.23**	m	**2.51**
Copper services										
isolated surfaces, girth not exceeding 300mm.....	0.58	10.00	**0.64**	0.11	-	**1.56**	0.03	**2.23**	m	**2.51**
Galvanized services										
isolated surfaces, girth not exceeding 300mm.....	0.58	10.00	**0.64**	0.11	-	**1.56**	0.03	**2.23**	m	**2.51**
M60: PAINTING/CLEAR FINISHING - BITUMINOUS PAINT										
One coat black bitumen paint; external work										
Iron or steel general surfaces										
girth exceeding 300mm	0.25	10.00	**0.27**	0.14	-	**1.99**	0.04	**2.30**	m2	**2.59**
Iron or steel eaves gutters										
girth exceeding 300mm	0.25	10.00	**0.27**	0.15	-	**2.13**	0.04	**2.44**	m2	**2.75**
Iron or steel services										
girth exceeding 300mm	0.25	10.00	**0.27**	0.17	-	**2.42**	0.05	**2.73**	m2	**3.08**
isolated surfaces, girth not exceeding 300mm.....	0.08	10.00	**0.09**	0.06	-	**0.85**	0.02	**0.96**	m	**1.07**
Two coats black bitumen paint; external work										
Iron or steel general surfaces										
girth exceeding 300mm	0.50	10.00	**0.54**	0.29	-	**4.12**	0.08	**4.74**	m2	**5.34**
Iron or steel eaves gutters										
girth exceeding 300mm	0.50	10.00	**0.54**	0.31	-	**4.41**	0.08	**5.03**	m2	**5.66**
isolated surfaces, girth not exceeding 300mm.....	0.15	10.00	**0.16**	0.10	-	**1.42**	0.03	**1.61**	m	**1.81**
Iron or steel services										
girth exceeding 300mm	0.50	10.00	**0.54**	0.33	-	**4.69**	0.09	**5.32**	m2	**5.99**
isolated surfaces, girth not exceeding 300mm.....	0.15	10.00	**0.16**	0.10	-	**1.42**	0.03	**1.61**	m	**1.81**
M60: PAINTING/CLEAR FINISHING - OIL PAINTING WOODWORK										
One coat primer; carried out on site before fixing members										
Wood general surfaces										
girth exceeding 300mm	0.24	10.00	**0.27**	0.18	-	**2.56**	0.05	**2.87**	m2	**3.23**
isolated surfaces, girth not exceeding 300mm.....	0.07	10.00	**0.08**	0.07	-	**0.99**	0.02	**1.09**	m	**1.23**
One undercoat, one coat full gloss finish on ready primed wood surfaces										
Wood general surfaces										
girth exceeding 300mm	0.32	10.00	**0.35**	0.35	-	**4.97**	0.09	**5.42**	m2	**6.10**
isolated surfaces, girth not exceeding 300mm.....	0.10	10.00	**0.11**	0.12	-	**1.71**	0.03	**1.85**	m	**2.08**
Wood glazed doors										
girth exceeding 300mm; panes, area not exceeding 0.10m2	0.17	10.00	**0.19**	0.59	-	**8.38**	0.16	**8.73**	m2	**9.82**
girth exceeding 300mm; panes, area 0.10 - 0.50m2.	0.12	10.00	**0.13**	0.42	-	**5.97**	0.11	**6.22**	m2	**6.99**
girth exceeding 300mm; panes, area 0.50 - 1.00m2.	0.12	10.00	**0.13**	0.35	-	**4.97**	0.09	**5.20**	m2	**5.85**
girth exceeding 300mm; panes, area exceeding 1.00m2	0.12	10.00	**0.13**	0.32	-	**4.55**	0.09	**4.77**	m2	**5.36**
Wood partially glazed doors										
girth exceeding 300mm; panes, area not exceeding 0.10m2	0.25	10.00	**0.27**	0.47	-	**6.68**	0.13	**7.08**	m2	**7.96**
girth exceeding 300mm; panes, area 0.10 - 0.50m2.	0.20	10.00	**0.22**	0.39	-	**5.54**	0.11	**5.86**	m2	**6.60**
girth exceeding 300mm; panes, area 0.50 - 1.00m2.	0.17	10.00	**0.19**	0.35	-	**4.97**	0.09	**5.26**	m2	**5.91**
girth exceeding 300mm; panes, area exceeding 1.00m2	0.17	10.00	**0.19**	0.34	-	**4.83**	0.09	**5.11**	m2	**5.75**
Wood windows and screens										
girth exceeding 300mm; panes, area not exceeding 0.10m2	0.25	10.00	**0.27**	0.65	-	**9.24**	0.18	**9.68**	m2	**10.89**
girth exceeding 300mm; panes, area 0.10 - 0.50m2.	0.20	10.00	**0.22**	0.46	-	**6.54**	0.12	**6.88**	m2	**7.74**
girth exceeding 300mm; panes, area 0.50 - 1.00m2.	0.17	10.00	**0.19**	0.39	-	**5.54**	0.11	**5.84**	m2	**6.57**

Labour hourly rates: (except Specialists) Craft Operatives £14.21 Labourer £10.71 Rates are national average prices. Refer to REGIONAL VARIATIONS for indicative levels of overall pricing in regions	MATERIALS			LABOUR			RATES			
	Del to Site £	Waste %	Material Cost £	Craft Optve Hrs	Lab Hrs	Labour Cost £	Sunds £	Nett Rate £	Unit	Gross Rate (+12.5%) £
M60: PAINTING/CLEAR FINISHING - OIL PAINTING WOODWORK Cont.										
One undercoat, one coat full gloss finish on ready primed wood surfaces Cont.										
Wood windows and screens Cont.										
girth exceeding 300mm; panes, area exceeding 1.00m2	0.17	10.00	**0.19**	0.35	-	**4.97**	0.09	**5.26**	m2	**5.91**
One undercoat, one coat full gloss finish on ready primed wood surfaces; external work										
Wood general surfaces										
girth exceeding 300mm	0.32	10.00	**0.35**	0.37	-	**5.26**	0.10	**5.71**	m2	**6.42**
isolated surfaces, girth not exceeding 300mm	0.10	10.00	**0.11**	0.13	-	**1.85**	0.04	**1.99**	m	**2.24**
Wood glazed doors										
girth exceeding 300mm; panes, area not exceeding 0.10m2	0.17	10.00	**0.19**	0.59	-	**8.38**	0.16	**8.73**	m2	**9.82**
girth exceeding 300mm; panes, area 0.10 - 0.50m2	0.12	10.00	**0.13**	0.42	-	**5.97**	0.11	**6.22**	m2	**6.99**
girth exceeding 300mm; panes, area 0.50 - 1.00m2	0.12	10.00	**0.13**	0.35	-	**4.97**	0.09	**5.20**	m2	**5.85**
girth exceeding 300mm; panes, area exceeding 1.00m2	0.12	10.00	**0.13**	0.32	-	**4.55**	0.09	**4.77**	m2	**5.36**
Wood partially glazed doors										
girth exceeding 300mm; panes, area not exceeding 0.10m2	0.25	10.00	**0.27**	0.47	-	**6.68**	0.13	**7.08**	m2	**7.96**
girth exceeding 300mm; panes, area 0.10 - 0.50m2	0.20	10.00	**0.22**	0.39	-	**5.54**	0.11	**5.86**	m2	**6.60**
girth exceeding 300mm; panes, area 0.50 - 1.00m2	0.17	10.00	**0.19**	0.35	-	**4.97**	0.09	**5.26**	m2	**5.91**
girth exceeding 300mm; panes, area exceeding 1.00m2	0.17	10.00	**0.19**	0.34	-	**4.83**	0.09	**5.11**	m2	**5.75**
Wood windows and screens										
girth exceeding 300mm; panes, area not exceeding 0.10m2	0.25	10.00	**0.27**	0.67	-	**9.52**	0.18	**9.97**	m2	**11.22**
girth exceeding 300mm; panes, area 0.10 - 0.50m2	0.20	10.00	**0.22**	0.48	-	**6.82**	0.13	**7.17**	m2	**8.06**
girth exceeding 300mm; panes, area 0.50 - 1.00m2	0.17	10.00	**0.19**	0.41	-	**5.83**	0.11	**6.13**	m2	**6.89**
girth exceeding 300mm; panes, area exceeding 1.00m2	0.17	10.00	**0.19**	0.37	-	**5.26**	0.10	**5.55**	m2	**6.24**
Wood railings fences and gates; open type										
girth exceeding 300mm	0.25	10.00	**0.27**	0.28	-	**3.98**	0.08	**4.32**	m2	**4.86**
isolated surfaces, girth not exceeding 300mm	0.10	10.00	**0.11**	0.09	-	**1.28**	0.02	**1.41**	m	**1.59**
Wood railings fences and gates; close type										
girth exceeding 300mm	0.32	10.00	**0.35**	0.24	-	**3.41**	0.06	**3.83**	m2	**4.30**
Two undercoats, one coat full gloss finish on ready primed wood surfaces										
Wood general surfaces										
girth exceeding 300mm	0.48	10.00	**0.53**	0.51	-	**7.25**	0.14	**7.91**	m2	**8.90**
isolated surfaces, girth not exceeding 300mm	0.15	10.00	**0.16**	0.17	-	**2.42**	0.05	**2.62**	m	**2.95**
Wood glazed doors										
girth exceeding 300mm; panes, area not exceeding 0.10m2	0.26	10.00	**0.28**	0.84	-	**11.94**	0.23	**12.45**	m2	**14.00**
girth exceeding 300mm; panes, area 0.10 - 0.50m2	0.18	10.00	**0.20**	0.61	-	**8.67**	0.16	**9.03**	m2	**10.16**
girth exceeding 300mm; panes, area 0.50 - 1.00m2	0.18	10.00	**0.20**	0.51	-	**7.25**	0.14	**7.59**	m2	**8.54**
girth exceeding 300mm; panes, area exceeding 1.00m2	0.18	10.00	**0.20**	0.45	-	**6.39**	0.12	**6.72**	m2	**7.56**
Wood partially glazed doors										
girth exceeding 300mm; panes, area not exceeding 0.10m2	0.37	10.00	**0.40**	0.67	-	**9.52**	0.18	**10.11**	m2	**11.37**
girth exceeding 300mm; panes, area 0.10 - 0.50m2	0.29	10.00	**0.32**	0.56	-	**7.96**	0.15	**8.43**	m2	**9.49**
girth exceeding 300mm; panes, area 0.50 - 1.00m2	0.26	10.00	**0.28**	0.51	-	**7.25**	0.14	**7.67**	m2	**8.63**
girth exceeding 300mm; panes, area exceeding 1.00m2	0.26	10.00	**0.28**	0.48	-	**6.82**	0.13	**7.23**	m2	**8.14**
Wood windows and screens										
girth exceeding 300mm; panes, area not exceeding 0.10m2	0.37	10.00	**0.40**	0.92	-	**13.07**	0.25	**13.73**	m2	**15.44**
girth exceeding 300mm; panes, area 0.10 - 0.50m2	0.29	10.00	**0.32**	0.67	-	**9.52**	0.18	**10.03**	m2	**11.28**
girth exceeding 300mm; panes, area 0.50 - 1.00m2	0.26	10.00	**0.28**	0.56	-	**7.96**	0.15	**8.39**	m2	**9.44**
girth exceeding 300mm; panes, area exceeding 1.00m2	0.26	10.00	**0.28**	0.50	-	**7.11**	0.14	**7.52**	m2	**8.46**
Two undercoats, one coat full gloss finish on ready primed wood surfaces; external work										
Wood general surfaces										
girth exceeding 300mm	0.48	10.00	**0.53**	0.54	-	**7.67**	0.15	**8.34**	m2	**9.39**
isolated surfaces, girth not exceeding 300mm	0.15	10.00	**0.16**	0.19	-	**2.70**	0.05	**2.91**	m	**3.28**
Wood glazed doors										
girth exceeding 300mm; panes, area not exceeding 0.10m2	0.26	10.00	**0.28**	0.84	-	**11.94**	0.23	**12.45**	m2	**14.00**
girth exceeding 300mm; panes, area 0.10 - 0.50m2	0.18	10.00	**0.20**	0.61	-	**8.67**	0.16	**9.03**	m2	**10.16**
girth exceeding 300mm; panes, area 0.50 - 1.00m2	0.18	10.00	**0.20**	0.51	-	**7.25**	0.14	**7.59**	m2	**8.54**
girth exceeding 300mm; panes, area exceeding 1.00m2	0.18	10.00	**0.20**	0.45	-	**6.39**	0.12	**6.72**	m2	**7.56**

Labour hourly rates: (except Specialists) Craft Operatives £14.21 Labourer £10.71 Rates are national average prices. Refer to REGIONAL VARIATIONS for indicative levels of overall pricing in regions	MATERIALS			LABOUR				RATES		
	Del to Site £	Waste %	Material Cost £	Craft Optve Hrs	Lab Hrs	Labour Cost £	Sunds £	Nett Rate £	Unit	Gross Rate (+12.5%) £
M60: PAINTING/CLEAR FINISHING - OIL PAINTING WOODWORK Cont.										
Two undercoats, one coat full gloss finish on ready primed wood surfaces; external work Cont.										
Wood partially glazed doors										
girth exceeding 300mm; panes, area not exceeding 0.10m2	0.37	10.00	**0.40**	0.67	-	**9.52**	0.18	**10.11**	m2	**11.37**
girth exceeding 300mm; panes, area 0.10 - 0.50m2	0.29	10.00	**0.32**	0.56	-	**7.96**	0.15	**8.43**	m2	**9.49**
girth exceeding 300mm; panes, area 0.50 - 1.00m2	0.26	10.00	**0.28**	0.51	-	**7.25**	0.14	**7.67**	m2	**8.63**
girth exceeding 300mm; panes, area exceeding 1.00m2	0.26	10.00	**0.28**	0.48	-	**6.82**	0.13	**7.23**	m2	**8.14**
Wood windows and screens										
girth exceeding 300mm; panes, area not exceeding 0.10m2	0.37	10.00	**0.40**	0.96	-	**13.64**	0.26	**14.31**	m2	**16.09**
girth exceeding 300mm; panes, area 0.10 - 0.50m2	0.29	10.00	**0.32**	0.70	-	**9.95**	0.19	**10.46**	m2	**11.77**
girth exceeding 300mm; panes, area 0.50 - 1.00m2	0.26	10.00	**0.28**	0.59	-	**8.38**	0.16	**8.83**	m2	**9.93**
girth exceeding 300mm; panes, area exceeding 1.00m2	0.26	10.00	**0.28**	0.53	-	**7.53**	0.14	**7.96**	m2	**8.95**
Wood railings fences and gates; open type										
girth exceeding 300mm	0.37	10.00	**0.40**	0.42	-	**5.97**	0.11	**6.49**	m2	**7.30**
isolated surfaces, girth not exceeding 300mm	0.15	10.00	**0.16**	0.14	-	**1.99**	0.04	**2.19**	m	**2.46**
Wood railings fences and gates; close type										
girth exceeding 300mm	0.48	10.00	**0.53**	0.36	-	**5.12**	0.10	**5.74**	m2	**6.46**
One coat primer, one undercoat, one coat full gloss finish on wood surfaces										
Wood general surfaces										
girth exceeding 300mm	0.56	10.00	**0.62**	0.51	-	**7.25**	0.14	**8.00**	m2	**9.00**
isolated surfaces, girth not exceeding 300mm	0.17	10.00	**0.18**	0.17	-	**2.42**	0.05	**2.65**	m	**2.98**
Wood glazed doors										
girth exceeding 300mm; panes, area not exceeding 0.10m2	0.31	10.00	**0.34**	0.84	-	**11.94**	0.23	**12.50**	m2	**14.07**
girth exceeding 300mm; panes, area 0.10 - 0.50m2	0.23	10.00	**0.25**	0.61	-	**8.67**	0.16	**9.08**	m2	**10.22**
girth exceeding 300mm; panes, area 0.50 - 1.00m2	0.21	10.00	**0.23**	0.51	-	**7.25**	0.14	**7.61**	m2	**8.57**
girth exceeding 300mm; panes, area exceeding 1.00m2	0.21	10.00	**0.23**	0.45	-	**6.39**	0.12	**6.75**	m2	**7.59**
Wood partially glazed doors										
girth exceeding 300mm; panes, area not exceeding 0.10m2	0.44	10.00	**0.48**	0.67	-	**9.52**	0.18	**10.18**	m2	**11.45**
girth exceeding 300mm; panes, area 0.10 - 0.50m2	0.33	10.00	**0.37**	0.56	-	**7.96**	0.15	**8.48**	m2	**9.54**
girth exceeding 300mm; panes, area 0.50 - 1.00m2	0.31	10.00	**0.34**	0.51	-	**7.25**	0.14	**7.73**	m2	**8.69**
girth exceeding 300mm; panes, area exceeding 1.00m2	0.31	10.00	**0.34**	0.48	-	**6.82**	0.13	**7.29**	m2	**8.20**
Wood windows and screens										
girth exceeding 300mm; panes, area not exceeding 0.10m2	0.44	10.00	**0.48**	0.92	-	**13.07**	0.25	**13.80**	m2	**15.53**
girth exceeding 300mm; panes, area 0.10 - 0.50m2	0.33	10.00	**0.37**	0.67	-	**9.52**	0.18	**10.07**	m2	**11.33**
girth exceeding 300mm; panes, area 0.50 - 1.00m2	0.31	10.00	**0.34**	0.56	-	**7.96**	0.15	**8.45**	m2	**9.51**
girth exceeding 300mm; panes, area exceeding 1.00m2	0.31	10.00	**0.34**	0.50	-	**7.11**	0.14	**7.58**	m2	**8.53**
One coat primer, one undercoat, one coat full gloss finish on wood surfaces; external work										
Wood general surfaces										
girth exceeding 300mm	0.56	10.00	**0.62**	0.54	-	**7.67**	0.15	**8.44**	m2	**9.49**
isolated surfaces, girth not exceeding 300mm	0.17	10.00	**0.18**	0.19	-	**2.70**	0.05	**2.94**	m	**3.30**
Wood glazed doors										
girth exceeding 300mm; panes, area not exceeding 0.10m2	0.31	10.00	**0.34**	0.84	-	**11.94**	0.23	**12.50**	m2	**14.07**
girth exceeding 300mm; panes, area 0.10 - 0.50m2	0.23	10.00	**0.25**	0.61	-	**8.67**	0.16	**9.08**	m2	**10.22**
girth exceeding 300mm; panes, area 0.50 - 1.00m2	0.21	10.00	**0.23**	0.51	-	**7.25**	0.02	**7.50**	m2	**8.44**
girth exceeding 300mm; panes, area exceeding 1.00m2	0.21	10.00	**0.23**	0.45	-	**6.39**	0.12	**6.75**	m2	**7.59**
Wood partially glazed doors										
girth exceeding 300mm; panes, area not exceeding 0.10m2	0.44	10.00	**0.48**	0.67	-	**9.52**	0.18	**10.18**	m2	**11.45**
girth exceeding 300mm; panes, area 0.10 - 0.50m2	0.33	10.00	**0.37**	0.56	-	**7.96**	0.15	**8.48**	m2	**9.54**
girth exceeding 300mm; panes, area 0.50 - 1.00m2	0.31	10.00	**0.34**	0.51	-	**7.25**	0.14	**7.73**	m2	**8.69**
girth exceeding 300mm; panes, area exceeding 1.00m2	0.31	10.00	**0.34**	0.48	-	**6.82**	0.13	**7.29**	m2	**8.20**
Wood windows and screens										
girth exceeding 300mm; panes, area not exceeding 0.10m2	0.44	10.00	**0.48**	0.96	-	**13.64**	0.26	**14.38**	m2	**16.18**
girth exceeding 300mm; panes, area 0.10 - 0.50m2	0.33	10.00	**0.37**	0.70	-	**9.95**	0.19	**10.50**	m2	**11.82**
girth exceeding 300mm; panes, area 0.50 - 1.00m2	0.31	10.00	**0.34**	0.59	-	**8.38**	0.16	**8.88**	m2	**10.00**
girth exceeding 300mm; panes, area exceeding 1.00m2	0.31	10.00	**0.34**	0.53	-	**7.53**	0.14	**8.02**	m2	**9.02**
Wood railings fences and gates; open type										
girth exceeding 300mm	0.44	10.00	**0.48**	0.42	-	**5.97**	0.11	**6.56**	m2	**7.38**
isolated surfaces, girth not exceeding 300mm	0.17	10.00	**0.18**	0.14	-	**1.99**	0.04	**2.21**	m	**2.49**

Labour hourly rates: (except Specialists) Craft Operatives £14.21 Labourer £10.71 Rates are national average prices. Refer to REGIONAL VARIATIONS for indicative levels of overall pricing in regions	MATERIALS			LABOUR				RATES		
	Del to Site	Waste	Material Cost	Craft Optve	Lab	Labour Cost	Sunds	Nett Rate	Unit	Gross Rate (+12.5%)
	£	%	£	Hrs	Hrs	£	£	£		£
M60: PAINTING/CLEAR FINISHING - OIL PAINTING WOODWORK Cont.										
One coat primer, one undercoat, one coat full gloss finish on wood surfaces; external work Cont.										
Wood railings fences and gates; close type										
girth exceeding 300mm	0.56	10.00	**0.62**	0.36	-	**5.12**	0.10	**5.83**	m2	**6.56**
One coat primer, two undercoats, one coat full gloss finish on wood surfaces										
Wood general surfaces										
girth exceeding 300mm	0.72	10.00	**0.79**	0.66	-	**9.38**	0.18	**10.35**	m2	**11.64**
isolated surfaces, girth not exceeding 300mm	0.22	10.00	**0.24**	0.22	-	**3.13**	0.06	**3.42**	m	**3.85**
Wood glazed doors										
girth exceeding 300mm; panes, area not exceeding 0.10m2	0.40	10.00	**0.44**	1.10	-	**15.63**	0.30	**16.36**	m2	**18.41**
girth exceeding 300mm; panes, area 0.10 - 0.50m2	0.29	10.00	**0.32**	0.79	-	**11.23**	0.21	**11.76**	m2	**13.23**
girth exceeding 300mm; panes, area 0.50 - 1.00m2	0.27	10.00	**0.30**	0.66	-	**9.38**	0.18	**9.85**	m2	**11.09**
girth exceeding 300mm; panes, area exceeding 1.00m2	0.27	10.00	**0.30**	0.59	-	**8.38**	0.16	**8.84**	m2	**9.95**
Wood partially glazed doors										
girth exceeding 300mm; panes, area not exceeding 0.10m2	0.56	10.00	**0.61**	0.88	-	**12.50**	0.24	**13.36**	m2	**15.03**
girth exceeding 300mm; panes, area 0.10 - 0.50m2	0.43	10.00	**0.48**	0.73	-	**10.37**	0.20	**11.05**	m2	**12.43**
girth exceeding 300mm; panes, area 0.50 - 1.00m2	0.40	10.00	**0.44**	0.66	-	**9.38**	0.18	**9.99**	m2	**11.24**
girth exceeding 300mm; panes, area exceeding 1.00m2	0.40	10.00	**0.44**	0.63	-	**8.95**	0.17	**9.56**	m2	**10.75**
Wood windows and screens										
girth exceeding 300mm; panes, area not exceeding 0.10m2	0.56	10.00	**0.61**	1.21	-	**17.19**	0.33	**18.13**	m2	**20.40**
girth exceeding 300mm; panes, area 0.10 - 0.50m2	0.43	10.00	**0.48**	0.87	-	**12.36**	0.23	**13.07**	m2	**14.71**
girth exceeding 300mm; panes, area 0.50 - 1.00m2	0.40	10.00	**0.44**	0.73	-	**10.37**	0.20	**11.01**	m2	**12.38**
girth exceeding 300mm; panes, area exceeding 1.00m2	0.40	10.00	**0.44**	0.65	-	**9.24**	0.18	**9.85**	m2	**11.08**
One coat primer, two undercoats, one coat full gloss finish on wood surfaces; external work										
Wood general surfaces										
girth exceeding 300mm	0.72	10.00	**0.79**	0.70	-	**9.95**	0.19	**10.93**	m2	**12.29**
isolated surfaces, girth not exceeding 300mm	0.22	10.00	**0.24**	0.24	-	**3.41**	0.06	**3.71**	m2	**4.18**
Wood glazed doors										
girth exceeding 300mm; panes, area not exceeding 0.10m2	0.40	10.00	**0.44**	1.10	-	**15.63**	0.30	**16.36**	m2	**18.41**
girth exceeding 300mm; panes, area 0.10 - 0.50m2	0.29	10.00	**0.32**	0.79	-	**11.23**	0.21	**11.76**	m2	**13.23**
girth exceeding 300mm; panes, area 0.50 - 1.00m2	0.27	10.00	**0.30**	0.66	-	**9.38**	0.18	**9.85**	m2	**11.09**
girth exceeding 300mm; panes, area exceeding 1.00m2	0.27	10.00	**0.30**	0.59	-	**8.38**	0.16	**8.84**	m2	**9.95**
Wood partially glazed doors										
girth exceeding 300mm; panes, area not exceeding 0.10m2	0.56	10.00	**0.61**	0.88	-	**12.50**	0.24	**13.36**	m2	**15.03**
girth exceeding 300mm; panes, area 0.10 - 0.50m2	0.43	10.00	**0.48**	0.73	-	**10.37**	0.20	**11.05**	m2	**12.43**
girth exceeding 300mm; panes, area 0.50 - 1.00m2	0.40	10.00	**0.44**	0.66	-	**9.38**	0.18	**9.99**	m2	**11.24**
girth exceeding 300mm; panes, area exceeding 1.00m2	0.40	10.00	**0.44**	0.63	-	**8.95**	0.17	**9.56**	m2	**10.75**
Wood windows and screens										
girth exceeding 300mm; panes, area not exceeding 0.10m2	0.56	10.00	**0.61**	1.25	-	**17.76**	0.34	**18.71**	m2	**21.05**
girth exceeding 300mm; panes, area 0.10 - 0.50m2	0.43	10.00	**0.48**	0.91	-	**12.93**	0.25	**13.65**	m2	**15.36**
girth exceeding 300mm; panes, area 0.50 - 1.00m2	0.40	10.00	**0.44**	0.77	-	**10.94**	0.21	**11.59**	m2	**13.03**
girth exceeding 300mm; panes, area exceeding 1.00m2	0.40	10.00	**0.44**	0.69	-	**9.80**	0.19	**10.43**	m2	**11.73**
Wood railings fences and gates; open type										
girth exceeding 300mm	0.56	10.00	**0.61**	0.55	-	**7.82**	0.15	**8.58**	m2	**9.65**
isolated surfaces, girth not exceeding 300mm	0.22	10.00	**0.24**	0.19	-	**2.70**	0.05	**2.99**	m	**3.36**
Wood railings fences and gates; close type										
girth exceeding 300mm	0.72	10.00	**0.79**	0.48	-	**6.82**	0.13	**7.74**	m2	**8.71**
One coat primer, one undercoat, two coats full gloss finish on wood surfaces										
Wood general surfaces										
girth exceeding 300mm	0.72	10.00	**0.79**	0.66	-	**9.38**	0.18	**10.35**	m2	**11.64**
isolated surfaces, girth not exceeding 300mm	0.22	10.00	**0.24**	0.22	-	**3.13**	0.06	**3.42**	m	**3.85**
Wood glazed doors										
girth exceeding 300mm; panes, area not exceeding 0.10m2	0.40	10.00	**0.44**	1.10	-	**15.63**	0.30	**16.36**	m2	**18.41**
girth exceeding 300mm; panes, area 0.10 - 0.50m2	0.29	10.00	**0.32**	0.79	-	**11.23**	0.21	**11.76**	m2	**13.23**
girth exceeding 300mm; panes, area 0.50 - 1.00m2	0.27	10.00	**0.30**	0.66	-	**9.38**	0.18	**9.85**	m2	**11.09**
girth exceeding 300mm; panes, area exceeding 1.00m2	0.27	10.00	**0.30**	0.59	-	**8.38**	0.16	**8.84**	m2	**9.95**
Wood partially glazed doors										
girth exceeding 300mm; panes, area not exceeding 0.10m2	0.56	10.00	**0.61**	0.88	-	**12.50**	0.24	**13.36**	m2	**15.03**

Labour hourly rates: (except Specialists) Craft Operatives £14.21 Labourer £10.71 Rates are national average prices. Refer to REGIONAL VARIATIONS for indicative levels of overall pricing in regions	MATERIALS			LABOUR				RATES		
	Del to Site £	Waste %	Material Cost £	Craft Optve Hrs	Lab Hrs	Labour Cost £	Sunds £	Nett Rate £	Unit	Gross Rate (+12.5%) £
M60: PAINTING/CLEAR FINISHING - OIL PAINTING WOODWORK Cont.										
One coat primer, one undercoat, two coats full gloss finish on wood surfaces Cont.										
Wood partially glazed doors Cont.										
girth exceeding 300mm; panes, area 0.10 - 0.50m2 .	0.43	10.00	**0.48**	0.73	-	**10.37**	0.20	**11.05**	m2	**12.43**
girth exceeding 300mm; panes, area 0.50 - 1.00m2 .	0.40	10.00	**0.44**	0.66	-	**9.38**	0.18	**9.99**	m2	**11.24**
girth exceeding 300mm; panes, area exceeding 1.00m2	0.40	10.00	**0.44**	0.63	-	**8.95**	0.17	**9.56**	m2	**10.75**
Wood windows and screens										
girth exceeding 300mm; panes, area not exceeding 0.10m2	0.56	10.00	**0.61**	1.21	-	**17.19**	0.33	**18.13**	m2	**20.40**
girth exceeding 300mm; panes, area 0.10 - 0.50m2 .	0.43	10.00	**0.48**	0.87	-	**12.36**	0.23	**13.07**	m2	**14.71**
girth exceeding 300mm; panes, area 0.50 - 1.00m2 .	0.40	10.00	**0.44**	0.73	-	**10.37**	0.20	**11.01**	m2	**12.38**
girth exceeding 300mm; panes, area exceeding 1.00m2	0.40	10.00	**0.44**	0.65	-	**9.24**	0.18	**9.85**	m2	**11.08**
One coat primer, one undercoat, two coats full gloss finish wood surfaces; external work										
Wood general surfaces										
girth exceeding 300mm	0.72	10.00	**0.79**	0.70	-	**9.95**	0.19	**10.93**	m2	**12.29**
isolated surfaces, girth not exceeding 300mm.....	0.22	10.00	**0.24**	0.24	-	**3.41**	0.06	**3.71**	m	**4.18**
Wood glazed doors										
girth exceeding 300mm; panes, area not exceeding 0.10m2	0.40	10.00	**0.44**	1.10	-	**15.63**	0.30	**16.36**	m2	**18.41**
girth exceeding 300mm; panes, area 0.10 - 0.50m2 .	0.29	10.00	**0.32**	0.79	-	**11.23**	0.21	**11.76**	m2	**13.23**
girth exceeding 300mm; panes, area 0.50 - 1.00m2 .	0.27	10.00	**0.30**	0.66	-	**9.38**	0.18	**9.85**	m2	**11.09**
girth exceeding 300mm; panes, area exceeding 1.00m2	0.27	10.00	**0.30**	0.59	-	**8.38**	0.16	**8.84**	m2	**9.95**
Wood partially glazed doors										
girth exceeding 300mm; panes, area not exceeding 0.10m2	0.56	10.00	**0.61**	0.88	-	**12.50**	0.24	**13.36**	m2	**15.03**
girth exceeding 300mm; panes, area 0.10 - 0.50m2 .	0.43	10.00	**0.48**	0.73	-	**10.37**	0.20	**11.05**	m2	**12.43**
girth exceeding 300mm; panes, area 0.50 - 1.00m2 .	0.40	10.00	**0.44**	0.66	-	**9.38**	0.18	**9.99**	m2	**11.24**
girth exceeding 300mm; panes, area exceeding 1.00m2	0.40	10.00	**0.44**	0.63	-	**8.95**	0.17	**9.56**	m2	**10.75**
Wood windows and screens										
girth exceeding 300mm; panes, area not exceeding 0.10m2	0.56	10.00	**0.61**	1.25	-	**17.76**	0.34	**18.71**	m2	**21.05**
girth exceeding 300mm; panes, area 0.10 - 0.50m2 .	0.43	10.00	**0.48**	0.91	-	**12.93**	0.25	**13.65**	m2	**15.36**
girth exceeding 300mm; panes, area 0.50 - 1.00m2 .	0.40	10.00	**0.44**	0.77	-	**10.94**	0.21	**11.59**	m2	**13.03**
girth exceeding 300mm; panes, area exceeding 1.00m2	0.40	10.00	**0.44**	0.69	-	**9.80**	0.19	**10.43**	m2	**11.73**
Wood railings fences and gates; open type										
girth exceeding 300mm	0.56	10.00	**0.61**	0.55	-	**7.82**	0.15	**8.58**	m2	**9.65**
isolated surfaces, girth not exceeding 300mm.....	0.22	10.00	**0.24**	0.19	-	**2.70**	0.05	**2.99**	m	**3.36**
Wood railings fences and gates; close type										
girth exceeding 300mm	0.72	10.00	**0.79**	0.48	-	**6.82**	0.13	**7.74**	m2	**8.71**
M60: PAINTING/CLEAR FINISHING - POLYURETHANE LACQUER										
Two coats polyurethane lacquer										
Wood general surfaces										
girth exceeding 300mm	0.85	10.00	**0.93**	0.44	-	**6.25**	0.12	**7.30**	m2	**8.22**
isolated surfaces, girth not exceeding 300mm.....	0.25	10.00	**0.28**	0.14	-	**1.99**	0.04	**2.31**	m	**2.60**
Wood glazed doors										
girth exceeding 300mm; panes, area not exceeding 0.10m2	0.48	10.00	**0.53**	0.74	-	**10.52**	0.20	**11.24**	m2	**12.65**
girth exceeding 300mm; panes, area 0.10 - 0.50m2 .	0.34	10.00	**0.37**	0.53	-	**7.53**	0.14	**8.05**	m2	**9.05**
girth exceeding 300mm; panes, area 0.50 - 1.00m2 .	0.31	10.00	**0.34**	0.44	-	**6.25**	0.12	**6.71**	m2	**7.55**
girth exceeding 300mm; panes, area exceeding 1.00m2	0.31	10.00	**0.34**	0.40	-	**5.68**	0.11	**6.13**	m2	**6.90**
Wood partially glazed doors										
girth exceeding 300mm; panes, area not exceeding 0.10m2	0.68	10.00	**0.75**	0.59	-	**8.38**	0.16	**9.29**	m2	**10.45**
girth exceeding 300mm; panes, area 0.10 - 0.50m2 .	0.51	10.00	**0.56**	0.48	-	**6.82**	0.13	**7.51**	m2	**8.45**
girth exceeding 300mm; panes, area 0.50 - 1.00m2 .	0.48	10.00	**0.53**	0.44	-	**6.25**	0.12	**6.90**	m2	**7.76**
girth exceeding 300mm; panes, area exceeding 1.00m2	0.48	10.00	**0.53**	0.42	-	**5.97**	0.11	**6.61**	m2	**7.44**
Wood windows and screens										
girth exceeding 300mm; panes, area not exceeding 0.10m2	0.68	10.00	**0.75**	0.81	-	**11.51**	0.22	**12.47**	m2	**14.03**
girth exceeding 300mm; panes, area 0.10 - 0.50m2 .	0.51	10.00	**0.56**	0.58	-	**8.24**	0.16	**8.96**	m2	**10.08**
girth exceeding 300mm; panes, area 0.50 - 1.00m2 .	0.48	10.00	**0.53**	0.48	-	**6.82**	0.13	**7.48**	m2	**8.41**
girth exceeding 300mm; panes, area exceeding 1.00m2	0.48	10.00	**0.53**	0.44	-	**6.25**	0.12	**6.90**	m2	**7.76**
Two coats polyurethane lacquer; external work										
Wood general surfaces										
girth exceeding 300mm	0.85	10.00	**0.93**	0.46	-	**6.54**	0.12	**7.59**	m2	**8.54**
isolated surfaces, girth not exceeding 300mm.....	0.25	10.00	**0.28**	0.15	-	**2.13**	0.04	**2.45**	m	**2.76**

Labour hourly rates: (except Specialists) Craft Operatives £14.21 Labourer £10.71 Rates are national average prices. Refer to REGIONAL VARIATIONS for indicative levels of overall pricing in regions	MATERIALS			LABOUR			RATES			
	Del to Site £	Waste %	Material Cost £	Craft Optve Hrs	Lab Hrs	Labour Cost £	Sunds £	Nett Rate £	Unit	Gross Rate (+12.5%) £
M60: PAINTING/CLEAR FINISHING - POLYURETHANE LACQUER Cont.										
Two coats polyurethane lacquer; external work Cont.										
Wood glazed doors										
girth exceeding 300mm; panes, area not exceeding 0.10m2	0.48	10.00	**0.53**	0.74	-	**10.52**	0.20	**11.24**	m2	**12.65**
girth exceeding 300mm; panes, area 0.10 - 0.50m2	0.34	10.00	**0.37**	0.53	-	**7.53**	0.14	**8.05**	m2	**9.05**
girth exceeding 300mm; panes, area 0.50 - 1.00m2	0.31	10.00	**0.34**	0.44	-	**6.25**	0.12	**6.71**	m2	**7.55**
girth exceeding 300mm; panes, area exceeding 1.00m2	0.31	10.00	**0.34**	0.40	-	**5.68**	0.11	**6.13**	m2	**6.90**
Wood partially glazed doors										
girth exceeding 300mm; panes, area not exceeding 0.10m2	0.68	10.00	**0.75**	0.59	-	**8.38**	0.16	**9.29**	m2	**10.45**
girth exceeding 300mm; panes, area 0.10 - 0.50m2	0.51	10.00	**0.56**	0.48	-	**6.82**	0.13	**7.51**	m2	**8.45**
girth exceeding 300mm; panes, area 0.50 - 1.00m2	0.48	10.00	**0.53**	0.44	-	**6.25**	0.12	**6.90**	m2	**7.76**
girth exceeding 300mm; panes, area exceeding 1.00m2	0.48	10.00	**0.53**	0.42	-	**5.97**	0.11	**6.61**	m2	**7.44**
Wood windows and screens										
girth exceeding 300mm; panes, area not exceeding 0.10m2	0.68	10.00	**0.75**	0.84	-	**11.94**	0.23	**12.91**	m2	**14.52**
girth exceeding 300mm; panes, area 0.10 - 0.50m2	0.51	10.00	**0.56**	0.61	-	**8.67**	0.16	**9.39**	m2	**10.57**
girth exceeding 300mm; panes, area 0.50 - 1.00m2	0.48	10.00	**0.53**	0.51	-	**7.25**	0.14	**7.91**	m2	**8.90**
girth exceeding 300mm; panes, area exceeding 1.00m2	0.48	10.00	**0.53**	0.46	-	**6.54**	0.12	**7.19**	m2	**8.09**
Three coats polyurethane lacquer										
Wood general surfaces										
girth exceeding 300mm	1.24	10.00	**1.37**	0.52	-	**7.39**	0.14	**8.90**	m2	**10.01**
isolated surfaces, girth not exceeding 300mm	0.37	10.00	**0.40**	0.18	-	**2.56**	0.05	**3.01**	m	**3.39**
Wood glazed doors										
girth exceeding 300mm; panes, area not exceeding 0.10m2	0.71	10.00	**0.78**	0.86	-	**12.22**	0.23	**13.23**	m2	**14.88**
girth exceeding 300mm; panes, area 0.10 - 0.50m2	0.51	10.00	**0.56**	0.62	-	**8.81**	0.17	**9.54**	m2	**10.73**
girth exceeding 300mm; panes, area 0.50 - 1.00m2	0.45	10.00	**0.50**	0.52	-	**7.39**	0.14	**8.03**	m2	**9.03**
girth exceeding 300mm; panes, area exceeding 1.00m2	0.45	10.00	**0.50**	0.46	-	**6.54**	0.12	**7.16**	m2	**8.05**
Wood partially glazed doors										
girth exceeding 300mm; panes, area not exceeding 0.10m2	0.99	10.00	**1.09**	0.69	-	**9.80**	0.19	**11.08**	m2	**12.46**
girth exceeding 300mm; panes, area 0.10 - 0.50m2	0.73	10.00	**0.81**	0.63	-	**8.95**	0.17	**9.93**	m2	**11.17**
girth exceeding 300mm; panes, area 0.50 - 1.00m2	0.68	10.00	**0.75**	0.52	-	**7.39**	0.14	**8.28**	m2	**9.31**
girth exceeding 300mm; panes, area exceeding 1.00m2	0.68	10.00	**0.75**	0.50	-	**7.11**	0.14	**7.99**	m2	**8.98**
Wood windows and screens										
girth exceeding 300mm; panes, area not exceeding 0.10m2	0.71	10.00	**0.78**	0.95	-	**13.50**	0.26	**14.53**	m2	**16.35**
girth exceeding 300mm; panes, area 0.10 - 0.50m2	0.51	10.00	**0.56**	0.68	-	**9.66**	0.18	**10.41**	m2	**11.71**
girth exceeding 300mm; panes, area 0.50 - 1.00m2	0.45	10.00	**0.50**	0.57	-	**8.10**	0.15	**8.75**	m2	**9.84**
girth exceeding 300mm; panes, area exceeding 1.00m2	0.45	10.00	**0.50**	0.51	-	**7.25**	0.14	**7.88**	m2	**8.87**
Three coats polyurethane lacquer; external work										
Wood general surfaces										
girth exceeding 300mm	1.24	10.00	**1.37**	0.55	-	**7.82**	0.15	**9.33**	m2	**10.50**
isolated surfaces, girth not exceeding 300mm	0.37	10.00	**0.40**	0.19	-	**2.70**	0.05	**3.16**	m	**3.55**
Wood glazed doors										
girth exceeding 300mm; panes, area not exceeding 0.10m2	0.71	10.00	**0.78**	0.86	-	**12.22**	0.23	**13.23**	m2	**14.88**
girth exceeding 300mm; panes, area 0.10 - 0.50m2	0.51	10.00	**0.56**	0.62	-	**8.81**	0.17	**9.54**	m2	**10.73**
girth exceeding 300mm; panes, area 0.50 - 1.00m2	0.45	10.00	**0.50**	0.52	-	**7.39**	0.14	**8.03**	m2	**9.03**
girth exceeding 300mm; panes, area exceeding 1.00m2	0.45	10.00	**0.50**	0.46	-	**6.54**	0.12	**7.16**	m2	**8.05**
Wood partially glazed doors										
girth exceeding 300mm; panes, area not exceeding 0.10m2	0.99	10.00	**1.09**	0.69	-	**9.80**	0.19	**11.08**	m2	**12.46**
girth exceeding 300mm; panes, area 0.10 - 0.50m2	0.73	10.00	**0.81**	0.63	-	**8.95**	0.17	**9.93**	m2	**11.17**
girth exceeding 300mm; panes, area 0.50 - 1.00m2	0.68	10.00	**0.75**	0.52	-	**7.39**	0.14	**8.28**	m2	**9.31**
girth exceeding 300mm; panes, area exceeding 1.00m2	0.68	10.00	**0.75**	0.50	-	**7.11**	0.14	**7.99**	m2	**8.98**
Wood windows and screens										
girth exceeding 300mm; panes, area not exceeding 0.10m2	0.71	10.00	**0.78**	0.98	-	**13.93**	0.26	**14.97**	m2	**16.84**
girth exceeding 300mm; panes, area 0.10 - 0.50m2	0.51	10.00	**0.56**	0.72	-	**10.23**	0.19	**10.98**	m2	**12.36**
girth exceeding 300mm; panes, area 0.50 - 1.00m2	0.45	10.00	**0.50**	0.61	-	**8.67**	0.16	**9.33**	m2	**10.50**
girth exceeding 300mm; panes, area exceeding 1.00m2	0.45	10.00	**0.50**	0.54	-	**7.67**	0.15	**8.32**	m2	**9.36**
M60: PAINTING/CLEAR FINISHING - FIRE RETARDANT PAINTS AND VARNISHES										
Two coats fire retardant paint										
Wood general surfaces										
girth exceeding 300mm	3.18	10.00	**3.50**	0.44	-	**6.25**	0.12	**9.87**	m2	**11.11**

Labour hourly rates: (except Specialists) Craft Operatives £14.21 Labourer £10.71 Rates are national average prices. Refer to REGIONAL VARIATIONS for indicative levels of overall pricing in regions	MATERIALS			LABOUR				RATES		
	Del to Site £	Waste %	Material Cost £	Craft Optve Hrs	Lab Hrs	Labour Cost £	Sunds £	Nett Rate £	Unit	Gross Rate (+12.5%) £
M60: PAINTING/CLEAR FINISHING - FIRE RETARDANT PAINTS AND VARNISHES Cont.										
Two coats fire retardant paint Cont.										
Wood general surfaces Cont.										
isolated surfaces, girth not exceeding 300mm	0.97	10.00	**1.06**	0.14	-	**1.99**	0.04	**3.09**	m	**3.47**
Two coats fire retardant varnish, one overcoat varnish										
Wood general surfaces										
girth exceeding 300mm	3.44	10.00	**3.78**	0.59	-	**8.38**	0.16	**12.33**	m2	**13.87**
isolated surfaces, girth not exceeding 300mm	1.06	10.00	**1.17**	0.20	-	**2.84**	0.05	**4.06**	m	**4.57**
M60: PAINTING/CLEAR FINISHING - OILING HARDWOOD										
Two coats raw linseed oil										
Wood general surfaces										
girth exceeding 300mm	1.26	10.00	**1.38**	0.48	-	**6.82**	0.13	**8.33**	m2	**9.37**
isolated surfaces, girth not exceeding 300mm	0.37	10.00	**0.41**	0.17	-	**2.42**	0.05	**2.87**	m	**3.23**
M60: PAINTING/CLEAR FINISHING - FRENCH AND WAX POLISHING										
Stain; two coats white polish										
Wood general surfaces										
girth exceeding 300mm	2.77	10.00	**3.04**	0.56	-	**7.96**	0.15	**11.15**	m2	**12.55**
isolated surfaces, girth not exceeding 300mm	0.87	10.00	**0.96**	0.19	-	**2.70**	0.05	**3.71**	m	**4.17**
Open grain French polish										
Wood general surfaces										
girth exceeding 300mm	2.79	10.00	**3.07**	2.64	-	**37.51**	0.71	**41.30**	m2	**46.46**
isolated surfaces, girth not exceeding 300mm	0.88	10.00	**0.97**	0.88	-	**12.50**	0.24	**13.71**	m	**15.42**
Stain; body in; fully French polish										
Wood general surfaces										
girth exceeding 300mm	4.69	10.00	**5.16**	3.96	-	**56.27**	1.07	**62.50**	m2	**70.31**
isolated surfaces, girth not exceeding 300mm	1.48	10.00	**1.63**	1.32	-	**18.76**	0.36	**20.74**	m	**23.33**
M60: PAINTING/CLEAR FINISHING - ROAD MARKINGS										
One coat solvent based road marking paint; external work										
Concrete general surfaces										
isolated surfaces, 50mm wide	1.35	5.00	**1.42**	0.12	-	**1.71**	0.03	**3.16**	m	**3.55**
isolated surfaces, 100mm wide	2.70	5.00	**2.84**	0.14	-	**1.99**	0.04	**4.86**	m	**5.47**
Two coats solvent based road marking paint; external work										
Concrete general surfaces										
isolated surfaces, 50mm wide	2.70	5.00	**2.84**	0.18	-	**2.56**	0.05	**5.44**	m	**6.12**
isolated surfaces, 100mm wide	5.40	5.00	**5.67**	0.21	-	**2.98**	0.06	**8.71**	m	**9.80**
Prime and apply non-reflective self adhesive road marking tape; external work										
Concrete general surfaces										
isolated surfaces, 50mm wide	1.64	5.00	**1.72**	0.12	-	**1.71**	0.03	**3.46**	m	**3.89**
isolated surfaces, 100mm wide	3.02	5.00	**3.17**	0.15	-	**2.13**	0.04	**5.35**	m	**6.01**
Prime and apply reflective self adhesive road marking tape; external work										
Concrete general surfaces										
isolated surfaces, 50mm wide	2.28	5.00	**2.39**	0.12	-	**1.71**	0.03	**4.13**	m	**4.64**
isolated surfaces, 100mm wide	4.20	5.00	**4.41**	0.13	-	**1.85**	0.04	**6.29**	m	**7.08**
M60: PAINTING/CLEAR FINISHING (REDECORATIONS) - EMULSION PAINTING										
Generally										
Note: the following rates include for the cost of all preparatory work, e.g. washing down, etc.										
Two coats emulsion paint, existing emulsion painted surfaces										
Concrete general surfaces										
girth exceeding 300mm	0.25	10.00	**0.28**	0.32	-	**4.55**	0.09	**4.91**	m2	**5.52**
Plaster general surfaces										
girth exceeding 300mm	0.20	10.00	**0.22**	0.28	-	**3.98**	0.08	**4.27**	m2	**4.81**
Plasterboard general surfaces										
girth exceeding 300mm	0.20	10.00	**0.22**	0.29	-	**4.12**	0.08	**4.42**	m2	**4.97**
Brickwork general surfaces										
girth exceeding 300mm	0.25	10.00	**0.28**	0.32	-	**4.55**	0.09	**4.91**	m2	**5.52**

SURFACE FINISHES

Labour hourly rates: (except Specialists) Craft Operatives £14.21 Labourer £10.71 Rates are national average prices. Refer to REGIONAL VARIATIONS for indicative levels of overall pricing in regions	MATERIALS			LABOUR			RATES			
	Del to Site £	Waste %	Material Cost £	Craft Optve Hrs	Lab Hrs	Labour Cost £	Sunds £	Nett Rate £	Unit	Gross Rate (+12.5%) £
M60: PAINTING/CLEAR FINISHING (REDECORATIONS) - EMULSION PAINTING Cont.										
Two coats emulsion paint, existing emulsion painted surfaces Cont.										
Paper covered general surfaces girth exceeding 300mm	0.20	10.00	**0.22**	0.30	-	**4.26**	0.08	**4.56**	m2	**5.13**
Two full coats emulsion paint, existing washable distempered surfaces										
Concrete general surfaces girth exceeding 300mm	0.25	10.00	**0.28**	0.33	-	**4.69**	0.09	**5.06**	m2	**5.69**
Plaster general surfaces girth exceeding 300mm	0.20	10.00	**0.22**	0.29	-	**4.12**	0.08	**4.42**	m2	**4.97**
Plasterboard general surfaces girth exceeding 300mm	0.20	10.00	**0.22**	0.30	-	**4.26**	0.08	**4.56**	m2	**5.13**
Brickwork general surfaces girth exceeding 300mm	0.25	10.00	**0.28**	0.33	-	**4.69**	0.09	**5.06**	m2	**5.69**
Paper covered general surfaces girth exceeding 300mm	0.20	10.00	**0.22**	0.30	-	**4.26**	0.08	**4.56**	m2	**5.13**
Two full coats emulsion paint, existing textured plastic coating surfaces										
Concrete general surfaces girth exceeding 300mm	0.25	10.00	**0.28**	0.36	-	**5.12**	0.10	**5.49**	m2	**6.18**
Plaster general surfaces girth exceeding 300mm	0.20	10.00	**0.22**	0.32	-	**4.55**	0.09	**4.85**	m2	**5.46**
Plasterboard general surfaces girth exceeding 300mm	0.20	10.00	**0.22**	0.33	-	**4.69**	0.09	**5.00**	m2	**5.62**
Brickwork general surfaces girth exceeding 300mm	0.26	10.00	**0.28**	0.36	-	**5.12**	0.10	**5.50**	m2	**6.18**
Paper covered general surfaces girth exceeding 300mm	0.20	10.00	**0.22**	0.33	-	**4.69**	0.09	**5.00**	m2	**5.62**
Mist coat, two full coats emulsion paint, existing non-washable distempered surfaces										
Concrete general surfaces girth exceeding 300mm	0.31	10.00	**0.34**	0.38	-	**5.40**	0.10	**5.84**	m2	**6.57**
Plaster general surfaces girth exceeding 300mm	0.25	10.00	**0.28**	0.34	-	**4.83**	0.09	**5.20**	m2	**5.85**
Plasterboard general surfaces girth exceeding 300mm	0.25	10.00	**0.28**	0.35	-	**4.97**	0.09	**5.34**	m2	**6.01**
Brickwork general surfaces girth exceeding 300mm	0.31	10.00	**0.34**	0.38	-	**5.40**	0.10	**5.84**	m2	**6.57**
Paper covered general surfaces girth exceeding 300mm	0.25	10.00	**0.28**	0.35	-	**4.97**	0.09	**5.34**	m2	**6.01**
Two full coats emulsion paint, existing gloss painted surfaces										
Concrete general surfaces girth exceeding 300mm	0.25	10.00	**0.28**	0.33	-	**4.69**	0.09	**5.06**	m2	**5.69**
Plaster general surfaces girth exceeding 300mm	0.20	10.00	**0.22**	0.29	-	**4.12**	0.08	**4.42**	m2	**4.97**
Plasterboard general surfaces girth exceeding 300mm	0.20	10.00	**0.22**	0.30	-	**4.26**	0.08	**4.56**	m2	**5.13**
Brickwork general surfaces girth exceeding 300mm	0.26	10.00	**0.28**	0.33	-	**4.69**	0.09	**5.06**	m2	**5.70**
Paper covered general surfaces girth exceeding 300mm	0.20	10.00	**0.22**	0.30	-	**4.26**	0.08	**4.56**	m2	**5.13**
M60: PAINTING/CLEAR FINISHING (REDECORATIONS) - CEMENT PAINTING										
Generally										
Note: the following rates include for the cost of all preparatory work, e.g. washing down, etc.										
One coat sealer, one coat waterproof cement paint; external work										
Cement rendered general surfaces girth exceeding 300mm	2.45	10.00	**2.70**	0.36	-	**5.12**	0.10	**7.91**	m2	**8.90**
Concrete general surfaces girth exceeding 300mm	2.08	10.00	**2.29**	0.43	-	**6.11**	0.12	**8.52**	m2	**9.58**

Labour hourly rates: (except Specialists) Craft Operatives £14.21 Labourer £10.71 Rates are national average prices. Refer to REGIONAL VARIATIONS for indicative levels of overall pricing in regions	MATERIALS			LABOUR			RATES			
	Del to Site £	Waste %	Material Cost £	Craft Optve Hrs	Lab Hrs	Labour Cost £	Sunds £	Nett Rate £	Unit	Gross Rate (+12.5%) £
M60: PAINTING/CLEAR FINISHING (REDECORATIONS) - CEMENT PAINTING Cont.										
One coat sealer, one coat waterproof cement paint; external work Cont.										
Brickwork general surfaces girth exceeding 300mm	2.27	10.00	**2.49**	0.43	-	**6.11**	0.12	**8.72**	m2	**9.81**
Rough cast general surfaces girth exceeding 300mm	2.45	10.00	**2.70**	0.52	-	**7.39**	0.14	**10.23**	m2	**11.50**
Two coats sealer, one coat waterproof cement paint; external work										
Cement rendered general surfaces girth exceeding 300mm	3.09	10.00	**3.40**	0.47	-	**6.68**	0.13	**10.21**	m2	**11.49**
Concrete general surfaces girth exceeding 300mm	2.53	10.00	**2.78**	0.58	-	**8.24**	0.16	**11.18**	m2	**12.57**
Brickwork general surfaces girth exceeding 300mm	2.91	10.00	**3.20**	0.58	-	**8.24**	0.16	**11.60**	m2	**13.05**
Rough cast general surfaces girth exceeding 300mm	3.09	10.00	**3.40**	0.70	-	**9.95**	0.19	**13.54**	m2	**15.23**
One coat waterproof cement paint existing cement painted surfaces ; external work										
Cement rendered general surfaces girth exceeding 300mm	1.15	10.00	**1.27**	0.27	-	**3.84**	0.07	**5.18**	m2	**5.82**
Concrete general surfaces girth exceeding 300mm	1.15	10.00	**1.27**	0.34	-	**4.83**	0.09	**6.19**	m2	**6.96**
Brickwork general surfaces girth exceeding 300mm	1.34	10.00	**1.47**	0.34	-	**4.83**	0.09	**6.39**	m2	**7.19**
Rough cast general surfaces girth exceeding 300mm	1.52	10.00	**1.67**	0.44	-	**6.25**	0.12	**8.04**	m2	**9.05**
Two coats waterproof cement paint existing cement painted surfaces ; external work										
Cement rendered general surfaces girth exceeding 300mm	3.04	10.00	**3.35**	0.38	-	**5.40**	0.10	**8.85**	m2	**9.96**
Concrete general surfaces girth exceeding 300mm	2.31	10.00	**2.54**	0.49	-	**6.96**	0.13	**9.63**	m2	**10.83**
Brickwork general surfaces girth exceeding 300mm	2.67	10.00	**2.94**	0.49	-	**6.96**	0.13	**10.04**	m2	**11.29**
Rough cast general surfaces girth exceeding 300mm	3.04	10.00	**3.35**	0.60	-	**8.53**	0.16	**12.03**	m2	**13.54**
One coat stabilizing solution, one coat textured masonry paint, existing undecorated surfaces; external work										
Cement rendered general surfaces girth exceeding 300mm	0.46	10.00	**0.50**	0.35	-	**4.97**	0.09	**5.57**	m2	**6.27**
Concrete general surfaces girth exceeding 300mm	0.54	10.00	**0.59**	0.42	-	**5.97**	0.11	**6.67**	m2	**7.51**
Brickwork general surfaces girth exceeding 300mm	0.75	10.00	**0.83**	0.42	-	**5.97**	0.11	**6.91**	m2	**7.77**
Rough cast general surfaces girth exceeding 300mm	0.91	10.00	**1.00**	0.52	-	**7.39**	0.14	**8.53**	m2	**9.60**
One coat stabilizing solution, two coats textured masonry paint, existing undecorated surfaces; external work										
Cement rendered general surfaces girth exceeding 300mm	0.81	10.00	**0.89**	0.47	-	**6.68**	0.13	**7.70**	m2	**8.66**
Concrete general surfaces girth exceeding 300mm	0.81	10.00	**0.89**	0.57	-	**8.10**	0.15	**9.15**	m2	**10.29**
Brickwork general surfaces girth exceeding 300mm	1.24	10.00	**1.37**	0.57	-	**8.10**	0.15	**9.62**	m2	**10.83**
Rough cast general surfaces girth exceeding 300mm	1.56	10.00	**1.71**	0.70	-	**9.95**	0.19	**11.85**	m2	**13.33**
One coat textured masonry paint, existing cement painted surfaces; external work										
Cement rendered general surfaces girth exceeding 300mm	0.20	10.00	**0.22**	0.26	-	**3.69**	0.07	**3.98**	m2	**4.48**

Labour hourly rates: (except Specialists) Craft Operatives £14.21 Labourer £10.71 Rates are national average prices. Refer to REGIONAL VARIATIONS for indicative levels of overall pricing in regions	MATERIALS			LABOUR			RATES			
	Del to Site £	Waste %	Material Cost £	Craft Optve Hrs	Lab Hrs	Labour Cost £	Sunds £	Nett Rate £	Unit	Gross Rate (+12.5%) £
M60: PAINTING/CLEAR FINISHING (REDECORATIONS) - CEMENT PAINTING Cont.										
One coat textured masonry paint, existing cement painted surfaces; external work Cont.										
Concrete general surfaces girth exceeding 300mm	0.28	10.00	**0.30**	0.33	-	**4.69**	0.09	**5.08**	m2	**5.72**
Brickwork general surfaces girth exceeding 300mm	0.49	10.00	**0.54**	0.33	-	**4.69**	0.09	**5.32**	m2	**5.98**
Rough cast general surfaces girth exceeding 300mm	0.65	10.00	**0.71**	0.43	-	**6.11**	0.12	**6.94**	m2	**7.81**
Two coats textured masonry paint, existing cement painted surfaces; external work										
Cement rendered general surfaces girth exceeding 300mm	0.55	10.00	**0.61**	0.39	-	**5.54**	0.11	**6.25**	m2	**7.03**
Concrete general surfaces girth exceeding 300mm	0.55	10.00	**0.61**	0.48	-	**6.82**	0.13	**7.56**	m2	**8.50**
Brickwork general surfaces girth exceeding 300mm	0.98	10.00	**1.08**	0.48	-	**6.82**	0.13	**8.03**	m2	**9.04**
Rough cast general surfaces girth exceeding 300mm	1.30	10.00	**1.43**	0.62	-	**8.81**	0.17	**10.40**	m2	**11.70**
M60: PAINTING/CLEAR FINISHING (REDECORATIONS) - OIL PAINTING WALLS AND CEILINGS										
Generally										
Note: the following rates include for the cost of all preparatory work, e.g. washing down, etc.										
One undercoat, one coat full gloss finish, existing gloss painted surfaces										
Concrete general surfaces girth exceeding 300mm	0.33	10.00	**0.36**	0.38	-	**5.40**	0.10	**5.87**	m2	**6.60**
Brickwork general surfaces girth exceeding 300mm	0.36	10.00	**0.39**	0.39	-	**5.54**	0.11	**6.04**	m2	**6.79**
Plasterboard general surfaces girth exceeding 300mm	0.31	10.00	**0.34**	0.35	-	**4.97**	0.09	**5.40**	m2	**6.08**
Plaster general surfaces girth exceeding 300mm	0.31	10.00	**0.34**	0.34	-	**4.83**	0.09	**5.26**	m2	**5.92**
One undercoat, one coat eggshell finish, existing gloss painted surfaces										
Concrete general surfaces girth exceeding 300mm	0.38	10.00	**0.41**	0.38	-	**5.40**	0.10	**5.92**	m2	**6.66**
Brickwork general surfaces girth exceeding 300mm	0.47	10.00	**0.52**	0.39	-	**5.54**	0.11	**6.17**	m2	**6.94**
Plasterboard general surfaces girth exceeding 300mm	0.35	10.00	**0.38**	0.35	-	**4.97**	0.09	**5.45**	m2	**6.13**
Plaster general surfaces girth exceeding 300mm	0.35	10.00	**0.38**	0.34	-	**4.83**	0.09	**5.31**	m2	**5.97**
Two undercoats, one coat full gloss finish, existing gloss painted surfaces										
Concrete general surfaces girth exceeding 300mm	0.49	10.00	**0.54**	0.48	-	**6.82**	0.13	**7.49**	m2	**8.43**
Brickwork general surfaces girth exceeding 300mm	0.59	10.00	**0.65**	0.48	-	**6.82**	0.13	**7.60**	m2	**8.55**
Plasterboard general surfaces girth exceeding 300mm	0.47	10.00	**0.51**	0.45	-	**6.39**	0.12	**7.03**	m2	**7.91**
Plaster general surfaces girth exceeding 300mm	0.47	10.00	**0.51**	0.44	-	**6.25**	0.12	**6.88**	m2	**7.74**
Two undercoats, one coat eggshell finish, existing gloss painted surfaces										
Concrete general surfaces girth exceeding 300mm	0.54	10.00	**0.59**	0.48	-	**6.82**	0.13	**7.54**	m2	**8.48**
Brickwork general surfaces girth exceeding 300mm	0.64	10.00	**0.71**	0.48	-	**6.82**	0.13	**7.66**	m2	**8.62**
Plasterboard general surfaces girth exceeding 300mm	0.51	10.00	**0.56**	0.45	-	**6.39**	0.12	**7.07**	m2	**7.96**
Plaster general surfaces girth exceeding 300mm	0.51	10.00	**0.56**	0.44	-	**6.25**	0.12	**6.93**	m2	**7.80**

Labour hourly rates: (except Specialists) Craft Operatives £14.21 Labourer £10.71 Rates are national average prices. Refer to REGIONAL VARIATIONS for indicative levels of overall pricing in regions	MATERIALS			LABOUR				RATES		
	Del to Site £	Waste %	Material Cost £	Craft Optve Hrs	Lab Hrs	Labour Cost £	Sunds £	Nett Rate £	Unit	Gross Rate (+12.5%) £
M60: PAINTING/CLEAR FINISHING (REDECORATIONS) - OIL PAINTING WALLS AND CEILINGS Cont.										
One coat primer, one undercoat, one coat full gloss finish, existing washable distempered or emulsion painted surfaces										
Concrete general surfaces girth exceeding 300mm	0.93	10.00	**1.03**	0.48	-	**6.82**	0.13	**7.98**	m2	**8.98**
Brickwork general surfaces girth exceeding 300mm	1.10	10.00	**1.21**	0.48	-	**6.82**	0.13	**8.16**	m2	**9.18**
Plasterboard general surfaces girth exceeding 300mm	0.86	10.00	**0.94**	0.45	-	**6.39**	0.12	**7.46**	m2	**8.39**
Plaster general surfaces girth exceeding 300mm	0.86	10.00	**0.94**	0.44	-	**6.25**	0.12	**7.31**	m2	**8.23**
One coat primer, one undercoat, one coat eggshell finish, existing washable distempered or emulsion painted surfaces										
Concrete general surfaces girth exceeding 300mm	0.98	10.00	**1.08**	0.48	-	**6.82**	0.13	**8.03**	m2	**9.03**
Brickwork general surfaces girth exceeding 300mm	1.21	10.00	**1.34**	0.48	-	**6.82**	0.13	**8.29**	m2	**9.32**
Plasterboard general surfaces girth exceeding 300mm	0.90	10.00	**0.99**	0.45	-	**6.39**	0.12	**7.50**	m2	**8.44**
Plaster general surfaces girth exceeding 300mm	0.90	10.00	**0.99**	0.44	-	**6.25**	0.12	**7.36**	m2	**8.28**
One coat primer, two undercoats, one coat full gloss finish, existing washable distempered or emulsion painted surfaces										
Concrete general surfaces girth exceeding 300mm	1.09	10.00	**1.20**	0.58	-	**8.24**	0.16	**9.60**	m2	**10.80**
Brickwork general surfaces girth exceeding 300mm	1.33	10.00	**1.46**	0.59	-	**8.38**	0.16	**10.01**	m2	**11.26**
Plasterboard general surfaces girth exceeding 300mm	1.01	10.00	**1.12**	0.55	-	**7.82**	0.15	**9.08**	m2	**10.21**
Plaster general surfaces girth exceeding 300mm	1.01	10.00	**1.12**	0.54	-	**7.67**	0.15	**8.94**	m2	**10.05**
One coat primer, two undercoats, one coat eggshell finish, existing washable distempered or emulsion painted surfaces										
Concrete general surfaces girth exceeding 300mm	1.14	10.00	**1.25**	0.58	-	**8.24**	0.16	**9.65**	m2	**10.86**
Brickwork general surfaces girth exceeding 300mm	1.39	10.00	**1.52**	0.59	-	**8.38**	0.16	**10.07**	m2	**11.33**
Plasterboard general surfaces girth exceeding 300mm	1.06	10.00	**1.16**	0.55	-	**7.82**	0.15	**9.13**	m2	**10.27**
Plaster general surfaces girth exceeding 300mm	1.06	10.00	**1.16**	0.54	-	**7.67**	0.15	**8.98**	m2	**10.10**
One coat primer, one undercoat, one coat full gloss finish, existing non-washable distempered surfaces										
Concrete general surfaces girth exceeding 300mm	0.93	10.00	**1.03**	0.48	-	**6.82**	0.13	**7.98**	m2	**8.98**
Brickwork general surfaces girth exceeding 300mm	1.10	10.00	**1.21**	0.48	-	**6.82**	0.13	**8.16**	m2	**9.18**
Plasterboard general surfaces girth exceeding 300mm	0.86	10.00	**0.94**	0.45	-	**6.39**	0.12	**7.46**	m2	**8.39**
Plaster general surfaces girth exceeding 300mm	0.86	10.00	**0.94**	0.44	-	**6.25**	0.12	**7.31**	m2	**8.23**
One coat primer, one undercoat, one coat eggshell finish, existing non-washable distempered surfaces										
Concrete general surfaces girth exceeding 300mm	0.98	10.00	**1.08**	0.48	-	**6.82**	0.13	**8.03**	m2	**9.03**
Brickwork general surfaces girth exceeding 300mm	1.21	10.00	**1.34**	0.48	-	**6.82**	0.13	**8.29**	m2	**9.32**
Plasterboard general surfaces girth exceeding 300mm	0.90	10.00	**0.99**	0.45	-	**6.39**	0.12	**7.50**	m2	**8.44**
Plaster general surfaces girth exceeding 300mm	0.90	10.00	**0.99**	0.44	-	**6.25**	0.12	**7.36**	m2	**8.28**

Labour hourly rates: (except Specialists) Craft Operatives £14.21 Labourer £10.71 Rates are national average prices. Refer to REGIONAL VARIATIONS for indicative levels of overall pricing in regions	MATERIALS			LABOUR				RATES		
	Del to Site £	Waste %	Material Cost £	Craft Optve Hrs	Lab Hrs	Labour Cost £	Sunds £	Nett Rate £	Unit	Gross Rate (+12.5%) £
M60: PAINTING/CLEAR FINISHING (REDECORATIONS) - OIL PAINTING WALLS AND CEILINGS Cont.										
One coat primer, two undercoats, one coat full gloss finish, existing non-washable distempered surfaces										
Concrete general surfaces girth exceeding 300mm	1.09	10.00	**1.20**	0.58	-	**8.24**	0.16	**9.60**	m2	**10.80**
Brickwork general surfaces girth exceeding 300mm	1.33	10.00	**1.46**	0.59	-	**8.38**	0.16	**10.01**	m2	**11.26**
Plasterboard general surfaces girth exceeding 300mm	1.01	10.00	**1.12**	0.55	-	**7.82**	0.15	**9.08**	m2	**10.21**
Plaster general surfaces girth exceeding 300mm	1.01	10.00	**1.12**	0.54	-	**7.67**	0.15	**8.94**	m2	**10.05**
One coat primer, two undercoats, one coat eggshell finish, existing non-washable distempered surfaces										
Concrete general surfaces girth exceeding 300mm	1.14	10.00	**1.25**	0.58	-	**8.24**	0.16	**9.65**	m2	**10.86**
Brickwork general surfaces girth exceeding 300mm	1.39	10.00	**1.52**	0.59	-	**8.38**	0.16	**10.07**	m2	**11.33**
Plasterboard general surfaces girth exceeding 300mm	1.06	10.00	**1.16**	0.55	-	**7.82**	0.15	**9.13**	m2	**10.27**
Plaster general surfaces girth exceeding 300mm	1.06	10.00	**1.16**	0.54	-	**7.67**	0.15	**8.98**	m2	**10.10**
M60: PAINTING/CLEAR FINISHING (REDECORATIONS) - PLASTIC FINISH										
Generally										
Note: the following rates include for the cost of all preparatory work, e.g. washing down, etc.										
Textured plastic coating - stippled finish, existing washable distempered or emulsion painted surfaces										
Concrete general surfaces girth exceeding 300mm	0.35	10.00	**0.38**	0.40	-	**5.68**	0.11	**6.17**	m2	**6.95**
Brickwork general surfaces girth exceeding 300mm	0.41	10.00	**0.46**	0.45	-	**6.39**	0.12	**6.97**	m2	**7.84**
Plasterboard general surfaces girth exceeding 300mm	0.31	10.00	**0.34**	0.34	-	**4.83**	0.09	**5.27**	m2	**5.92**
Plaster general surfaces girth exceeding 300mm	0.31	10.00	**0.34**	0.34	-	**4.83**	0.09	**5.27**	m2	**5.92**
Textured plastic coating - combed finish, existing washable distempered or emulsion painted surfaces										
Concrete general surfaces girth exceeding 300mm	0.41	10.00	**0.46**	0.45	-	**6.39**	0.12	**6.97**	m2	**7.84**
Brickwork general surfaces girth exceeding 300mm	0.50	10.00	**0.55**	0.50	-	**7.11**	0.14	**7.79**	m2	**8.76**
Plasterboard general surfaces girth exceeding 300mm	0.37	10.00	**0.41**	0.40	-	**5.68**	0.11	**6.20**	m2	**6.98**
Plaster general surfaces girth exceeding 300mm	0.37	10.00	**0.41**	0.40	-	**5.68**	0.11	**6.20**	m2	**6.98**
Textured plastic coating - stippled finish, existing non-washable distempered surfaces										
Concrete general surfaces girth exceeding 300mm	0.35	10.00	**0.38**	0.45	-	**6.39**	0.12	**6.90**	m2	**7.76**
Brickwork general surfaces girth exceeding 300mm	0.41	10.00	**0.46**	0.50	-	**7.11**	0.14	**7.70**	m2	**8.66**
Plasterboard general surfaces girth exceeding 300mm	0.31	10.00	**0.34**	0.40	-	**5.68**	0.11	**6.13**	m2	**6.90**
Plaster general surfaces girth exceeding 300mm	0.31	10.00	**0.34**	0.40	-	**5.68**	0.11	**6.13**	m2	**6.90**
Textured plastic coating - combed finish, existing non-washable distempered surfaces										
Concrete general surfaces girth exceeding 300mm	0.41	10.00	**0.46**	0.50	-	**7.11**	0.14	**7.70**	m2	**8.66**
Brickwork general surfaces girth exceeding 300mm	0.50	10.00	**0.55**	0.56	-	**7.96**	0.15	**8.66**	m2	**9.74**
Plasterboard general surfaces girth exceeding 300mm	0.37	10.00	**0.41**	0.45	-	**6.39**	0.12	**6.93**	m2	**7.79**

Labour hourly rates: (except Specialists) Craft Operatives £14.21 Labourer £10.71 Rates are national average prices. Refer to REGIONAL VARIATIONS for indicative levels of overall pricing in regions	MATERIALS			LABOUR				RATES		
	Del to Site £	Waste %	Material Cost £	Craft Optve Hrs	Lab Hrs	Labour Cost £	Sunds £	Nett Rate £	Unit	Gross Rate (+12.5%) £
M60: PAINTING/CLEAR FINISHING (REDECORATIONS) - PLASTIC FINISH Cont.										
Textured plastic coating - combed finish, existing non-washable distempered surfaces Cont.										
Plaster general surfaces										
girth exceeding 300mm	0.37	10.00	**0.41**	0.45	-	**6.39**	0.12	**6.93**	m2	**7.79**
M60: PAINTING/CLEAR FINISHING (REDECORATIONS) - CLEAN OUT GUTTERS										
Generally										
Note: the following rates include for the cost of all preparatory work, e.g. washing down, etc.										
Clean out gutters prior to repainting, staunch joints with red lead, bituminous compound or mastic										
Iron or steel eaves gutters										
generally	-	-	-	0.06	-	**0.85**	0.06	**0.91**	m	**1.03**
M60: PAINTING/CLEAR FINISHING (REDECORATIONS) - OIL PAINTING METALWORK										
Generally										
Note: the following rates include for the cost of all preparatory work, e.g. washing down, etc.										
One undercoat, one coat full gloss finish, existing gloss painted metal surfaces										
Iron or steel general surfaces										
girth exceeding 300mm	0.32	10.00	**0.35**	0.47	-	**6.68**	0.13	**7.16**	m2	**8.05**
isolated surfaces, girth not exceeding 300mm	0.10	10.00	**0.11**	0.16	-	**2.27**	0.04	**2.42**	m	**2.73**
Galvanized glazed doors, windows or screens										
girth exceeding 300mm; panes, area not exceeding 0.10m2	0.17	10.00	**0.19**	0.88	-	**12.50**	0.24	**12.93**	m2	**14.55**
girth exceeding 300mm; panes, area 0.10 - 0.50m2	0.12	10.00	**0.13**	0.64	-	**9.09**	0.17	**9.40**	m2	**10.58**
girth exceeding 300mm; panes, area 0.50 - 1.00m2	0.10	10.00	**0.11**	0.53	-	**7.53**	0.14	**7.78**	m2	**8.75**
girth exceeding 300mm; panes, area exceeding 1.00m2	0.10	10.00	**0.11**	0.46	-	**6.54**	0.12	**6.77**	m2	**7.61**
Iron or steel structural work										
girth exceeding 300mm	0.32	10.00	**0.35**	0.52	-	**7.39**	0.14	**7.88**	m2	**8.86**
isolated surfaces, girth not exceeding 300mm	0.10	10.00	**0.11**	0.21	-	**2.98**	0.06	**3.15**	m	**3.54**
Iron or steel structural members of roof trusses, lattice girders, purlins and the like										
girth exceeding 300mm	0.32	10.00	**0.35**	0.65	-	**9.24**	0.18	**9.76**	m2	**10.98**
isolated surfaces, girth not exceeding 300mm	0.10	10.00	**0.11**	0.22	-	**3.13**	0.06	**3.29**	m	**3.71**
Iron or steel services										
girth exceeding 300mm	0.32	10.00	**0.35**	0.65	-	**9.24**	0.18	**9.76**	m2	**10.98**
isolated surfaces, girth not exceeding 300mm	0.10	10.00	**0.11**	0.22	-	**3.13**	0.06	**3.29**	m	**3.71**
isolated areas not exceeding 0.50m2 irrespective of girth	0.17	10.00	**0.19**	0.65	-	**9.24**	0.18	**9.60**	nr	**10.80**
Copper services										
girth exceeding 300mm	0.32	10.00	**0.35**	0.65	-	**9.24**	0.18	**9.76**	m2	**10.98**
isolated surfaces, girth not exceeding 300mm	0.10	10.00	**0.11**	0.22	-	**3.13**	0.06	**3.29**	m	**3.71**
Galvanized services										
girth exceeding 300mm	0.32	10.00	**0.35**	0.65	-	**9.24**	0.18	**9.76**	m2	**10.98**
isolated surfaces, girth not exceeding 300mm	0.10	10.00	**0.11**	0.22	-	**3.13**	0.06	**3.29**	m	**3.71**
One undercoat, one coat full gloss finish, existing gloss painted metal surfaces; external work										
Iron or steel general surfaces										
girth exceeding 300mm	0.32	10.00	**0.35**	0.50	-	**7.11**	0.14	**7.59**	m2	**8.54**
isolated surfaces, girth not exceeding 300mm	0.10	10.00	**0.11**	0.17	-	**2.42**	0.05	**2.57**	m	**2.89**
Galvanized glazed doors, windows or screens										
girth exceeding 300mm; panes, area not exceeding 0.10m2	0.17	10.00	**0.19**	0.90	-	**12.79**	0.24	**13.22**	m2	**14.87**
girth exceeding 300mm; panes, area 0.10 - 0.50m2	0.12	10.00	**0.13**	0.67	-	**9.52**	0.18	**9.84**	m2	**11.07**
girth exceeding 300mm; panes, area 0.50 - 1.00m2	0.10	10.00	**0.11**	0.55	-	**7.82**	0.15	**8.07**	m2	**9.08**
girth exceeding 300mm; panes, area exceeding 1.00m2	0.10	10.00	**0.11**	0.49	-	**6.96**	0.13	**7.20**	m2	**8.10**
Iron or steel structural work										
girth exceeding 300mm	0.32	10.00	**0.35**	0.52	-	**7.39**	0.14	**7.88**	m2	**8.86**
isolated surfaces, girth not exceeding 300mm	0.10	10.00	**0.11**	0.21	-	**2.98**	0.06	**3.15**	m	**3.54**
Iron or steel structural members of roof trusses, lattice girders, purlins and the like										
girth exceeding 300mm	0.32	10.00	**0.35**	0.65	-	**9.24**	0.18	**9.76**	m2	**10.98**
isolated surfaces, girth not exceeding 300mm	0.10	10.00	**0.11**	0.22	-	**3.13**	0.06	**3.29**	m	**3.71**
Iron or steel railings, fences and gates; plain open type										
girth exceeding 300mm	0.25	10.00	**0.27**	0.49	-	**6.96**	0.13	**7.36**	m2	**8.29**
isolated surfaces, girth not exceeding 300mm	0.10	10.00	**0.11**	0.16	-	**2.27**	0.04	**2.42**	m	**2.73**

Labour hourly rates: (except Specialists) Craft Operatives £14.21 Labourer £10.71 Rates are national average prices. Refer to REGIONAL VARIATIONS for indicative levels of overall pricing in regions	MATERIALS Del to Site £	Waste %	Material Cost £	LABOUR Craft Optve Hrs	Lab Hrs	Labour Cost £	Sunds £	RATES Nett Rate £	Unit	Gross Rate (+12.5%) £
M60: PAINTING/CLEAR FINISHING (REDECORATIONS) - OIL PAINTING METALWORK Cont.										
One undercoat, one coat full gloss finish, existing gloss painted metal surfaces; external work Cont.										
Iron or steel railings, fences and gates; close type										
girth exceeding 300mm	0.32	10.00	**0.35**	0.45	-	**6.39**	0.12	**6.87**	m2	**7.72**
Iron or steel railings, fences and gates; ornamental type										
girth exceeding 300mm	0.25	10.00	**0.27**	0.90	-	**12.79**	0.24	**13.30**	m2	**14.96**
Iron or steel eaves gutters										
girth exceeding 300mm	0.32	10.00	**0.35**	0.53	-	**7.53**	0.14	**8.02**	m2	**9.03**
isolated surfaces, girth not exceeding 300mm	0.10	10.00	**0.11**	0.17	-	**2.42**	0.05	**2.57**	m	**2.89**
Galvanized eaves gutters										
girth exceeding 300mm	0.32	10.00	**0.35**	0.53	-	**7.53**	0.14	**8.02**	m2	**9.03**
isolated surfaces, girth not exceeding 300mm	0.10	10.00	**0.11**	0.17	-	**2.42**	0.05	**2.57**	m	**2.89**
Iron or steel services										
girth exceeding 300mm	0.32	10.00	**0.35**	0.65	-	**9.24**	0.18	**9.76**	m2	**10.98**
isolated surfaces, girth not exceeding 300mm	0.10	10.00	**0.11**	0.22	-	**3.13**	0.06	**3.29**	m	**3.71**
isolated areas not exceeding 0.50m2 irrespective of girth	0.17	10.00	**0.19**	0.65	-	**9.24**	0.18	**9.60**	nr	**10.80**
Copper services										
girth exceeding 300mm	0.32	10.00	**0.35**	0.65	-	**9.24**	0.18	**9.76**	m2	**10.98**
isolated surfaces, girth not exceeding 300mm	0.10	10.00	**0.11**	0.22	-	**3.13**	0.06	**3.29**	m	**3.71**
Galvanized services										
girth exceeding 300mm	0.32	10.00	**0.35**	0.65	-	**9.24**	0.18	**9.76**	m2	**10.98**
isolated surfaces, girth not exceeding 300mm	0.10	10.00	**0.11**	0.22	-	**3.13**	0.06	**3.29**	m	**3.71**
Two coats full gloss finish, existing gloss painted metal surfaces										
Iron or steel general surfaces										
girth exceeding 300mm	0.32	10.00	**0.35**	0.47	-	**6.68**	0.13	**7.16**	m2	**8.05**
isolated surfaces, girth not exceeding 300mm	0.10	10.00	**0.11**	0.16	-	**2.27**	0.04	**2.42**	m	**2.73**
Galvanized glazed doors, windows or screens										
girth exceeding 300mm; panes, area not exceeding 0.10m2	0.17	10.00	**0.19**	0.88	-	**12.50**	0.24	**12.93**	m2	**14.55**
girth exceeding 300mm; panes, area 0.10 - 0.50m2	0.12	10.00	**0.13**	0.64	-	**9.09**	0.17	**9.40**	m2	**10.58**
girth exceeding 300mm; panes, area 0.50 - 1.00m2	0.10	10.00	**0.11**	0.53	-	**7.53**	0.14	**7.78**	m2	**8.75**
girth exceeding 300mm; panes, area exceeding 1.00m2	0.10	10.00	**0.11**	0.46	-	**6.54**	0.12	**6.77**	m2	**7.61**
Iron or steel structural work										
girth exceeding 300mm	0.32	10.00	**0.35**	0.52	-	**7.39**	0.14	**7.88**	m2	**8.86**
isolated surfaces, girth not exceeding 300mm	0.10	10.00	**0.11**	0.21	-	**2.98**	0.06	**3.15**	m	**3.54**
Iron or steel structural members of roof trusses, lattice girders, purlins and the like										
girth exceeding 300mm	0.32	10.00	**0.35**	0.65	-	**9.24**	0.18	**9.76**	m2	**10.98**
isolated surfaces, girth not exceeding 300mm	0.10	10.00	**0.11**	0.22	-	**3.13**	0.06	**3.29**	m	**3.71**
Iron or steel services										
girth exceeding 300mm	0.32	10.00	**0.35**	0.65	-	**9.24**	0.18	**9.76**	m2	**10.98**
isolated surfaces, girth not exceeding 300mm	0.10	10.00	**0.11**	0.22	-	**3.13**	0.06	**3.29**	m	**3.71**
isolated areas not exceeding 0.50m2 irrespective of girth	0.16	10.00	**0.18**	0.65	-	**9.24**	0.18	**9.59**	nr	**10.79**
Copper services										
girth exceeding 300mm	0.32	10.00	**0.35**	0.65	-	**9.24**	0.18	**9.76**	m2	**10.98**
isolated surfaces, girth not exceeding 300mm	0.10	10.00	**0.11**	0.22	-	**3.13**	0.06	**3.29**	m	**3.71**
Galvanized services										
girth exceeding 300mm	0.32	10.00	**0.35**	0.65	-	**9.24**	0.18	**9.76**	m2	**10.98**
isolated surfaces, girth not exceeding 300mm	0.10	10.00	**0.11**	0.22	-	**3.13**	0.06	**3.29**	m	**3.71**
Two coats full gloss finish, existing gloss painted metal surfaces; external work										
Iron or steel general surfaces										
girth exceeding 300mm	0.32	10.00	**0.35**	0.50	-	**7.11**	0.14	**7.59**	m2	**8.54**
isolated surfaces, girth not exceeding 300mm	0.10	10.00	**0.11**	0.17	-	**2.42**	0.05	**2.57**	m	**2.89**
Galvanized glazed doors, windows or screens										
girth exceeding 300mm; panes, area not exceeding 0.10m2	0.17	10.00	**0.19**	0.90	-	**12.79**	0.24	**13.22**	m2	**14.87**
girth exceeding 300mm; panes, area 0.10 - 0.50m2	0.12	10.00	**0.13**	0.67	-	**9.52**	0.18	**9.84**	m2	**11.07**
girth exceeding 300mm; panes, area 0.50 - 1.00m2	0.10	10.00	**0.11**	0.55	-	**7.82**	0.15	**8.07**	m	**9.08**
girth exceeding 300mm; panes, area exceeding 1.00m2	0.10	10.00	**0.11**	0.49	-	**6.96**	0.13	**7.20**	m2	**8.10**
Iron or steel structural work										
girth exceeding 300mm	0.32	10.00	**0.35**	0.52	-	**7.39**	0.14	**7.88**	m2	**8.86**
isolated surfaces, girth not exceeding 300mm	0.10	10.00	**0.11**	0.21	-	**2.98**	0.06	**3.15**	m	**3.54**
Iron or steel structural members of roof trusses, lattice girders, purlins and the like										
girth exceeding 300mm	0.32	10.00	**0.35**	0.65	-	**9.24**	0.18	**9.76**	m2	**10.98**
isolated surfaces, girth not exceeding 300mm	0.10	10.00	**0.11**	0.22	-	**3.13**	0.06	**3.29**	m	**3.71**

Labour hourly rates: (except Specialists) Craft Operatives £14.21 Labourer £10.71 Rates are national average prices. Refer to REGIONAL VARIATIONS for indicative levels of overall pricing in regions	MATERIALS			LABOUR				RATES		
	Del to Site £	Waste %	Material Cost £	Craft Optve Hrs	Lab Hrs	Labour Cost £	Sunds £	Nett Rate £	Unit	Gross Rate (+12.5%) £
M60: PAINTING/CLEAR FINISHING (REDECORATIONS) - OIL PAINTING METALWORK Cont.										
Two coats full gloss finish, existing gloss painted metal surfaces; external work Cont.										
Iron or steel railings, fences and gates; plain open type										
girth exceeding 300mm	0.25	10.00	**0.27**	0.49	-	**6.96**	0.13	**7.36**	m2	**8.29**
isolated surfaces, girth not exceeding 300mm	0.10	10.00	**0.11**	0.16	-	**2.27**	0.04	**2.42**	m	**2.73**
Iron or steel railings, fences and gates; close type										
girth exceeding 300mm	0.32	10.00	**0.35**	0.45	-	**6.39**	0.12	**6.87**	m2	**7.72**
Iron or steel railings, fences and gates; ornamental type										
girth exceeding 300mm	0.25	10.00	**0.27**	0.82	-	**11.65**	0.22	**12.14**	m2	**13.66**
Iron or steel eaves gutters										
girth exceeding 300mm	0.32	10.00	**0.35**	0.53	-	**7.53**	0.14	**8.02**	m2	**9.03**
isolated surfaces, girth not exceeding 300mm	0.10	10.00	**0.11**	0.17	-	**2.42**	0.05	**2.57**	m	**2.89**
Galvanized eaves gutters										
girth exceeding 300mm	0.32	10.00	**0.35**	0.53	-	**7.53**	0.14	**8.02**	m2	**9.03**
isolated surfaces, girth not exceeding 300mm	0.10	10.00	**0.11**	0.17	-	**2.42**	0.05	**2.57**	m	**2.89**
Iron or steel services										
girth exceeding 300mm	0.32	10.00	**0.35**	0.65	-	**9.24**	0.18	**9.76**	m2	**10.98**
isolated surfaces, girth not exceeding 300mm	0.10	10.00	**0.11**	0.22	-	**3.13**	0.06	**3.29**	m	**3.71**
isolated areas not exceeding 0.50m2 irrespective of girth	0.16	10.00	**0.18**	0.65	-	**9.24**	0.18	**9.59**	nr	**10.79**
Copper services										
girth exceeding 300mm	0.32	10.00	**0.35**	0.65	-	**9.24**	0.18	**9.76**	m2	**10.98**
isolated surfaces, girth not exceeding 300mm	0.10	10.00	**0.11**	0.22	-	**3.13**	0.06	**3.29**	m	**3.71**
Galvanized services										
girth exceeding 300mm	0.32	10.00	**0.35**	0.65	-	**9.24**	0.18	**9.76**	m2	**10.98**
isolated surfaces, girth not exceeding 300mm	0.10	10.00	**0.11**	0.22	-	**3.13**	0.06	**3.29**	m	**3.71**
Two undercoats, one coat full gloss finish, existing gloss painted metal surfaces										
Iron or steel general surfaces										
girth exceeding 300mm	0.48	10.00	**0.53**	0.62	-	**8.81**	0.17	**9.50**	m2	**10.69**
isolated surfaces, girth not exceeding 300mm	0.15	10.00	**0.16**	0.21	-	**2.98**	0.06	**3.20**	m	**3.60**
Galvanized glazed doors, windows or screens										
girth exceeding 300mm; panes, area not exceeding 0.10m2	0.26	10.00	**0.28**	1.15	-	**16.34**	0.31	**16.93**	m2	**19.05**
girth exceeding 300mm; panes, area 0.10 - 0.50m2	0.18	10.00	**0.20**	0.83	-	**11.79**	0.22	**12.22**	m2	**13.75**
girth exceeding 300mm; panes, area 0.50 - 1.00m2	0.15	10.00	**0.16**	0.69	-	**9.80**	0.19	**10.15**	m2	**11.42**
girth exceeding 300mm; panes, area exceeding 1.00m2	0.15	10.00	**0.16**	0.61	-	**8.67**	0.16	**8.99**	m2	**10.12**
Iron or steel structural work										
girth exceeding 300mm	0.48	10.00	**0.53**	0.67	-	**9.52**	0.18	**10.23**	m2	**11.51**
isolated surfaces, girth not exceeding 300mm	0.15	10.00	**0.16**	0.27	-	**3.84**	0.07	**4.07**	m	**4.58**
Iron or steel structural members of roof trusses, lattice girders, purlins and the like										
girth exceeding 300mm	0.48	10.00	**0.53**	0.84	-	**11.94**	0.23	**12.69**	m2	**14.27**
isolated surfaces, girth not exceeding 300mm	0.15	10.00	**0.16**	0.28	-	**3.98**	0.08	**4.22**	m	**4.74**
Iron or steel services										
girth exceeding 300mm	0.48	10.00	**0.53**	0.84	-	**11.94**	0.23	**12.69**	m2	**14.27**
isolated surfaces, girth not exceeding 300mm	0.15	10.00	**0.16**	0.28	-	**3.98**	0.08	**4.22**	m	**4.74**
isolated areas not exceeding 0.50m2 irrespective of girth	0.25	10.00	**0.27**	0.84	-	**11.94**	0.23	**12.43**	nr	**13.99**
Copper services										
girth exceeding 300mm	0.48	10.00	**0.53**	0.84	-	**11.94**	0.23	**12.69**	m2	**14.27**
isolated surfaces, girth not exceeding 300mm	0.15	10.00	**0.16**	0.28	-	**3.98**	0.08	**4.22**	m	**4.74**
Galvanized services										
girth exceeding 300mm	0.48	10.00	**0.53**	0.84	-	**11.94**	0.23	**12.69**	m2	**14.27**
isolated surfaces, girth not exceeding 300mm	0.15	10.00	**0.16**	0.28	-	**3.98**	0.08	**4.22**	m	**4.74**
Two undercoats, one coat full gloss finish, existing gloss painted metal surfaces; external work										
Iron or steel general surfaces										
girth exceeding 300mm	0.48	10.00	**0.53**	0.65	-	**9.24**	0.18	**9.94**	m2	**11.18**
isolated surfaces, girth not exceeding 300mm	0.15	10.00	**0.16**	0.22	-	**3.13**	0.06	**3.35**	m	**3.77**
Galvanized glazed doors, windows or screens										
girth exceeding 300mm; panes, area not exceeding 0.10m2	0.26	10.00	**0.28**	1.18	-	**16.77**	0.32	**17.37**	m2	**19.54**
girth exceeding 300mm; panes, area 0.10 - 0.50m2	0.18	10.00	**0.20**	0.86	-	**12.22**	0.23	**12.65**	m2	**14.24**
girth exceeding 300mm; panes, area 0.50 - 1.00m2	0.15	10.00	**0.16**	0.72	-	**10.23**	0.19	**10.59**	m2	**11.91**
girth exceeding 300mm; panes, area exceeding 1.00m2	0.15	10.00	**0.16**	0.64	-	**9.09**	0.17	**9.43**	m2	**10.61**
Iron or steel structural work										
girth exceeding 300mm	0.48	10.00	**0.53**	0.67	-	**9.52**	0.18	**10.23**	m2	**11.51**
isolated surfaces, girth not exceeding 300mm	0.15	10.00	**0.16**	0.27	-	**3.84**	0.07	**4.07**	m	**4.58**

Labour hourly rates: (except Specialists) Craft Operatives £14.21 Labourer £10.71 Rates are national average prices. Refer to REGIONAL VARIATIONS for indicative levels of overall pricing in regions	MATERIALS Del to Site £	 Waste %	 Material Cost £	LABOUR Craft Optve Hrs	 Lab Hrs	 Labour Cost £	 Sunds £	RATES Nett Rate £	 Unit	 Gross Rate (+12.5%) £
M60: PAINTING/CLEAR FINISHING (REDECORATIONS) - OIL PAINTING METALWORK Cont.										
Two undercoats, one coat full gloss finish, existing gloss painted metal surfaces; external work Cont.										
Iron or steel structural members of roof trusses, lattice girders, purlins and the like										
girth exceeding 300mm	0.48	10.00	**0.53**	0.84	-	**11.94**	0.23	**12.69**	m2	**14.27**
isolated surfaces, girth not exceeding 300mm	0.15	10.00	**0.16**	0.28	-	**3.98**	0.08	**4.22**	m	**4.74**
Iron or steel railings, fences and gates; plain open type										
girth exceeding 300mm	0.37	10.00	**0.40**	0.63	-	**8.95**	0.17	**9.53**	m2	**10.72**
isolated surfaces, girth not exceeding 300mm	0.15	10.00	**0.16**	0.21	-	**2.98**	0.06	**3.20**	m	**3.60**
Iron or steel railings, fences and gates; close type										
girth exceeding 300mm	0.48	10.00	**0.53**	0.57	-	**8.10**	0.15	**8.78**	m2	**9.88**
Iron or steel railings, fences and gates; ornamental type										
girth exceeding 300mm	0.37	10.00	**0.40**	1.06	-	**15.06**	0.29	**15.75**	m2	**17.72**
Iron or steel eaves gutters										
girth exceeding 300mm	0.48	10.00	**0.53**	0.68	-	**9.66**	0.18	**10.37**	m2	**11.67**
isolated surfaces, girth not exceeding 300mm	0.15	10.00	**0.16**	0.23	-	**3.27**	0.06	**3.49**	m	**3.93**
Galvanized eaves gutters										
girth exceeding 300mm	0.48	10.00	**0.53**	0.68	-	**9.66**	0.18	**10.37**	m2	**11.67**
isolated surfaces, girth not exceeding 300mm	0.15	10.00	**0.16**	0.23	-	**3.27**	0.06	**3.49**	m	**3.93**
Iron or steel services										
girth exceeding 300mm	0.48	10.00	**0.53**	0.84	-	**11.94**	0.23	**12.69**	m2	**14.27**
isolated surfaces, girth not exceeding 300mm	0.15	10.00	**0.16**	0.28	-	**3.98**	0.08	**4.22**	m	**4.74**
isolated areas not exceeding 0.50m2 irrespective of girth	0.25	10.00	**0.27**	0.84	-	**11.94**	0.23	**12.43**	nr	**13.99**
Copper services										
girth exceeding 300mm	0.48	10.00	**0.53**	0.84	-	**11.94**	0.23	**12.69**	m2	**14.27**
isolated surfaces, girth not exceeding 300mm	0.15	10.00	**0.16**	0.28	-	**3.98**	0.08	**4.22**	m	**4.74**
Galvanized services										
girth exceeding 300mm	0.48	10.00	**0.53**	0.84	-	**11.94**	0.23	**12.69**	m2	**14.27**
isolated surfaces, girth not exceeding 300mm	0.15	10.00	**0.16**	0.28	-	**3.98**	0.08	**4.22**	m	**4.74**
One undercoat, two coats full gloss finish, existing gloss painted metal surfaces										
Iron or steel general surfaces										
girth exceeding 300mm	0.48	10.00	**0.53**	0.62	-	**8.81**	0.17	**9.50**	m2	**10.69**
isolated surfaces, girth not exceeding 300mm	0.15	10.00	**0.16**	0.21	-	**2.98**	0.06	**3.20**	m	**3.60**
Galvanized glazed doors, windows or screens										
girth exceeding 300mm; panes, area not exceeding 0.10m2	0.26	10.00	**0.28**	1.16	-	**16.48**	0.31	**17.08**	m2	**19.21**
girth exceeding 300mm; panes, area 0.10 - 0.50m2	0.18	10.00	**0.20**	0.83	-	**11.79**	0.22	**12.22**	m2	**13.75**
girth exceeding 300mm; panes, area 0.50 - 1.00m2	0.15	10.00	**0.16**	0.69	-	**9.80**	0.19	**10.15**	m2	**11.42**
girth exceeding 300mm; panes, area exceeding 1.00m2	0.15	10.00	**0.16**	0.61	-	**8.67**	0.16	**8.99**	m2	**10.12**
Iron or steel structural work										
girth exceeding 300mm	0.48	10.00	**0.53**	0.67	-	**9.52**	0.18	**10.23**	m2	**11.51**
isolated surfaces, girth not exceeding 300mm	0.15	10.00	**0.16**	0.27	-	**3.84**	0.07	**4.07**	m	**4.58**
Iron or steel structural members of roof trusses, lattice girders, purlins and the like										
girth exceeding 300mm	0.48	10.00	**0.53**	0.84	-	**11.94**	0.23	**12.69**	m2	**14.27**
isolated surfaces, girth not exceeding 300mm	0.15	10.00	**0.16**	0.28	-	**3.98**	0.08	**4.22**	m	**4.74**
Iron or steel services										
girth exceeding 300mm	0.48	10.00	**0.53**	0.84	-	**11.94**	0.23	**12.69**	m2	**14.27**
isolated surfaces, girth not exceeding 300mm	0.15	10.00	**0.16**	0.28	-	**3.98**	0.08	**4.22**	m	**4.74**
isolated areas not exceeding 0.50m2 irrespective of girth	0.26	10.00	**0.28**	0.84	-	**11.94**	0.23	**12.45**	nr	**14.00**
Copper services										
girth exceeding 300mm	0.48	10.00	**0.53**	0.84	-	**11.94**	0.23	**12.69**	m2	**14.27**
isolated surfaces, girth not exceeding 300mm	0.15	10.00	**0.16**	0.28	-	**3.98**	0.08	**4.22**	m	**4.74**
Galvanized services										
girth exceeding 300mm	0.48	10.00	**0.53**	0.84	-	**11.94**	0.23	**12.69**	m2	**14.27**
isolated surfaces, girth not exceeding 300mm	0.15	10.00	**0.16**	0.28	-	**3.98**	0.08	**4.22**	m	**4.74**
One undercoat, two coats full gloss finish, existing gloss painted metal surfaces; external work										
Iron or steel general surfaces										
girth exceeding 300mm	0.48	10.00	**0.53**	0.65	-	**9.24**	0.18	**9.94**	m2	**11.18**
isolated surfaces, girth not exceeding 300mm	0.15	10.00	**0.16**	0.22	-	**3.13**	0.06	**3.35**	m	**3.77**
Galvanized glazed doors, windows or screens										
girth exceeding 300mm; panes, area not exceeding 0.10m2	0.26	10.00	**0.28**	1.18	-	**16.77**	0.32	**17.37**	m2	**19.54**
girth exceeding 300mm; panes, area 0.10 - 0.50m2	0.18	10.00	**0.20**	0.86	-	**12.22**	0.23	**12.65**	m2	**14.24**
girth exceeding 300mm; panes, area 0.50 - 1.00m2	0.15	10.00	**0.16**	0.72	-	**10.23**	0.19	**10.59**	m2	**11.91**

Labour hourly rates: (except Specialists) Craft Operatives £14.21 Labourer £10.71 Rates are national average prices. Refer to REGIONAL VARIATIONS for indicative levels of overall pricing in regions	MATERIALS			LABOUR				RATES		
	Del to Site £	Waste %	Material Cost £	Craft Optve Hrs	Lab Hrs	Labour Cost £	Sunds £	Nett Rate £	Unit	Gross Rate (+12.5%) £
M60: PAINTING/CLEAR FINISHING (REDECORATIONS) - OIL PAINTING METALWORK Cont.										
One undercoat, two coats full gloss finish, existing gloss painted metal surfaces; external work Cont.										
Galvanized glazed doors, windows or screens Cont.										
girth exceeding 300mm; panes, area exceeding 1.00m2	0.15	10.00	**0.16**	0.64	-	**9.09**	0.17	**9.43**	m2	**10.61**
Iron or steel structural work										
girth exceeding 300mm	0.48	10.00	**0.53**	0.67	-	**9.52**	0.18	**10.23**	m2	**11.51**
isolated surfaces, girth not exceeding 300mm	0.15	10.00	**0.16**	0.27	-	**3.84**	0.07	**4.07**	m	**4.58**
Iron or steel structural members of roof trusses, lattice girders, purlins and the like										
girth exceeding 300mm	0.48	10.00	**0.53**	0.84	-	**11.94**	0.23	**12.69**	m2	**14.27**
isolated surfaces, girth not exceeding 300mm	0.15	10.00	**0.16**	0.28	-	**3.98**	0.08	**4.22**	m	**4.74**
Iron or steel railings, fences and gates; plain open type										
girth exceeding 300mm	0.37	10.00	**0.40**	0.64	-	**9.09**	0.17	**9.67**	m2	**10.88**
isolated surfaces, girth not exceeding 300mm	0.15	10.00	**0.16**	0.21	-	**2.98**	0.06	**3.20**	m	**3.60**
Iron or steel railings, fences and gates; close type										
girth exceeding 300mm	0.48	10.00	**0.53**	0.57	-	**8.10**	0.15	**8.78**	m2	**9.88**
Iron or steel railings, fences and gates; ornamental type										
girth exceeding 300mm	0.37	10.00	**0.40**	1.06	-	**15.06**	0.29	**15.75**	m2	**17.72**
Iron or steel eaves gutters										
girth exceeding 300mm	0.48	10.00	**0.53**	0.68	-	**9.66**	0.18	**10.37**	m2	**11.67**
isolated surfaces, girth not exceeding 300mm	0.15	10.00	**0.16**	0.23	-	**3.27**	0.06	**3.49**	m	**3.93**
Galvanized eaves gutters										
girth exceeding 300mm	0.48	10.00	**0.53**	0.68	-	**9.66**	0.18	**10.37**	m2	**11.67**
isolated surfaces, girth not exceeding 300mm	0.15	10.00	**0.16**	0.23	-	**3.27**	0.06	**3.49**	m	**3.93**
Iron or steel services										
girth exceeding 300mm	0.48	10.00	**0.53**	0.84	-	**11.94**	0.23	**12.69**	m2	**14.27**
isolated surfaces, girth not exceeding 300mm	0.15	10.00	**0.16**	0.28	-	**3.98**	0.08	**4.22**	m	**4.74**
isolated areas not exceeding 0.50m2 irrespective of girth	0.26	10.00	**0.28**	0.84	-	**11.94**	0.23	**12.45**	nr	**14.00**
Copper services										
girth exceeding 300mm	0.48	10.00	**0.53**	0.84	-	**11.94**	0.23	**12.69**	m2	**14.27**
isolated surfaces, girth not exceeding 300mm	0.15	10.00	**0.16**	0.28	-	**3.98**	0.08	**4.22**	m	**4.74**
Galvanized services										
girth exceeding 300mm	0.48	10.00	**0.53**	0.84	-	**11.94**	0.23	**12.69**	m2	**14.27**
isolated surfaces, girth not exceeding 300mm	0.15	10.00	**0.16**	0.28	-	**3.98**	0.08	**4.22**	m	**4.74**
M60: PAINTING/CLEAR FINISHING (REDECORATIONS) - OIL PAINTING WOODWORK										
Generally										
Note: the following rates include for the cost of all preparatory work, e.g. washing down, etc.										
One undercoat, one coat full gloss finish, existing gloss painted wood surfaces										
Wood general surfaces										
girth exceeding 300mm	0.32	10.00	**0.35**	0.45	-	**6.39**	0.12	**6.87**	m2	**7.72**
isolated surfaces, girth not exceeding 300mm	0.10	10.00	**0.11**	0.15	-	**2.13**	0.04	**2.28**	m	**2.56**
Wood glazed doors										
girth exceeding 300mm; panes, area not exceeding 0.10m2	0.17	10.00	**0.19**	0.79	-	**11.23**	0.21	**11.63**	m2	**13.08**
girth exceeding 300mm; panes, area 0.10 - 0.50m2	0.12	10.00	**0.13**	0.57	-	**8.10**	0.15	**8.39**	m2	**9.44**
girth exceeding 300mm; panes, area 0.50 - 1.00m2	0.12	10.00	**0.13**	0.48	-	**6.82**	0.13	**7.09**	m2	**7.97**
girth exceeding 300mm; panes, area exceeding 1.00m2	0.12	10.00	**0.13**	0.41	-	**5.83**	0.11	**6.07**	m2	**6.83**
Wood partially glazed doors										
girth exceeding 300mm; panes, area not exceeding 0.10m2	0.25	10.00	**0.27**	0.67	-	**9.52**	0.18	**9.97**	m2	**11.22**
girth exceeding 300mm; panes, area 0.10 - 0.50m2	0.20	10.00	**0.22**	0.54	-	**7.67**	0.15	**8.03**	m2	**9.04**
girth exceeding 300mm; panes, area 0.50 - 1.00m2	0.17	10.00	**0.19**	0.48	-	**6.82**	0.13	**7.14**	m2	**8.03**
girth exceeding 300mm; panes, area exceeding 1.00m2	0.17	10.00	**0.19**	0.43	-	**6.11**	0.12	**6.42**	m2	**7.22**
Wood windows and screens										
girth exceeding 300mm; panes, area not exceeding 0.10m2	0.25	10.00	**0.27**	0.85	-	**12.08**	0.23	**12.58**	m2	**14.15**
girth exceeding 300mm; panes, area 0.10 - 0.50m2	0.20	10.00	**0.22**	0.61	-	**8.67**	0.16	**9.05**	m2	**10.18**
girth exceeding 300mm; panes, area 0.50 - 1.00m2	0.17	10.00	**0.19**	0.52	-	**7.39**	0.14	**7.72**	m2	**8.68**
girth exceeding 300mm; panes, area exceeding 1.00m2	0.17	10.00	**0.19**	0.44	-	**6.25**	0.12	**6.56**	m2	**7.38**
One undercoat, one coat full gloss finish, existing gloss painted wood surfaces; external work										
Wood general surfaces										
girth exceeding 300mm	0.32	10.00	**0.35**	0.48	-	**6.82**	0.13	**7.30**	m2	**8.21**

Labour hourly rates: (except Specialists) Craft Operatives £14.21 Labourer £10.71 Rates are national average prices. Refer to REGIONAL VARIATIONS for indicative levels of overall pricing in regions	MATERIALS			LABOUR				RATES		
	Del to Site £	Waste %	Material Cost £	Craft Optve Hrs	Lab Hrs	Labour Cost £	Sunds £	Nett Rate £	Unit	Gross Rate (+12.5%) £
M60: PAINTING/CLEAR FINISHING (REDECORATIONS) - OIL PAINTING WOODWORK Cont.										
One undercoat, one coat full gloss finish, existing gloss painted wood surfaces; external work Cont.										
Wood general surfaces Cont.										
isolated surfaces, girth not exceeding 300mm	0.10	10.00	**0.11**	0.16	-	**2.27**	0.04	**2.42**	m	**2.73**
Wood glazed doors										
girth exceeding 300mm; panes, area not exceeding 0.10m2	0.17	10.00	**0.19**	0.81	-	**11.51**	0.22	**11.92**	m2	**13.41**
girth exceeding 300mm; panes, area 0.10 - 0.50m2	0.12	10.00	**0.13**	0.59	-	**8.38**	0.16	**8.68**	m2	**9.76**
girth exceeding 300mm; panes, area 0.50 - 1.00m2	0.12	10.00	**0.13**	0.49	-	**6.96**	0.13	**7.23**	m2	**8.13**
girth exceeding 300mm; panes, area exceeding 1.00m2	0.12	10.00	**0.13**	0.42	-	**5.97**	0.11	**6.22**	m2	**6.99**
Wood partially glazed doors										
girth exceeding 300mm; panes, area not exceeding 0.10m2	0.25	10.00	**0.27**	0.69	-	**9.80**	0.19	**10.26**	m2	**11.54**
girth exceeding 300mm; panes, area 0.10 - 0.50m2	0.20	10.00	**0.22**	0.56	-	**7.96**	0.15	**8.32**	m2	**9.36**
girth exceeding 300mm; panes, area 0.50 - 1.00m2	0.17	10.00	**0.19**	0.49	-	**6.96**	0.13	**7.28**	m2	**8.19**
girth exceeding 300mm; panes, area exceeding 1.00m2	0.17	10.00	**0.19**	0.44	-	**6.25**	0.12	**6.56**	m2	**7.38**
Wood windows and screens										
girth exceeding 300mm; panes, area not exceeding 0.10m2	0.25	10.00	**0.27**	0.89	-	**12.65**	0.24	**13.16**	m2	**14.80**
girth exceeding 300mm; panes, area 0.10 - 0.50m2	0.20	10.00	**0.22**	0.65	-	**9.24**	0.18	**9.63**	m2	**10.83**
girth exceeding 300mm; panes, area 0.50 - 1.00m2	0.17	10.00	**0.19**	0.55	-	**7.82**	0.15	**8.15**	m2	**9.17**
girth exceeding 300mm; panes, area exceeding 1.00m2	0.17	10.00	**0.19**	0.47	-	**6.68**	0.13	**6.99**	m2	**7.87**
Wood railings fences and gates; open type										
girth exceeding 300mm	0.25	10.00	**0.27**	0.41	-	**5.83**	0.11	**6.21**	m2	**6.98**
isolated surfaces, girth not exceeding 300mm	0.10	10.00	**0.11**	0.13	-	**1.85**	0.04	**1.99**	m	**2.24**
Wood railings fences and gates; close type										
girth exceeding 300mm	0.32	10.00	**0.35**	0.37	-	**5.26**	0.10	**5.71**	m2	**6.42**
Two coats full gloss finish, existing gloss painted wood surfaces										
Wood general surfaces										
girth exceeding 300mm	0.32	10.00	**0.35**	0.45	-	**6.39**	0.12	**6.87**	m2	**7.72**
isolated surfaces, girth not exceeding 300mm	0.10	10.00	**0.11**	0.15	-	**2.13**	0.04	**2.28**	m	**2.56**
Wood glazed doors										
girth exceeding 300mm; panes, area not exceeding 0.10m2	0.17	10.00	**0.19**	0.79	-	**11.23**	0.21	**11.63**	m2	**13.08**
girth exceeding 300mm; panes, area 0.10 - 0.50m2	0.12	10.00	**0.13**	0.57	-	**8.10**	0.15	**8.39**	m2	**9.44**
girth exceeding 300mm; panes, area 0.50 - 1.00m2	0.12	10.00	**0.13**	0.48	-	**6.82**	0.13	**7.09**	m2	**7.97**
girth exceeding 300mm; panes, area exceeding 1.00m2	0.12	10.00	**0.13**	0.41	-	**5.83**	0.11	**6.07**	m2	**6.83**
Wood partially glazed doors										
girth exceeding 300mm; panes, area not exceeding 0.10m2	0.25	10.00	**0.27**	0.67	-	**9.52**	0.18	**9.97**	m2	**11.22**
girth exceeding 300mm; panes, area 0.10 - 0.50m2	0.20	10.00	**0.22**	0.54	-	**7.67**	0.15	**8.03**	m2	**9.04**
girth exceeding 300mm; panes, area 0.50 - 1.00m2	0.17	10.00	**0.19**	0.48	-	**6.82**	0.13	**7.14**	m2	**8.03**
girth exceeding 300mm; panes, area exceeding 1.00m2	0.17	10.00	**0.19**	0.43	-	**6.11**	0.12	**6.42**	m2	**7.22**
Wood windows and screens										
girth exceeding 300mm; panes, area not exceeding 0.10m2	0.25	10.00	**0.27**	0.85	-	**12.08**	0.23	**12.58**	m2	**14.15**
girth exceeding 300mm; panes, area 0.10 - 0.50m2	0.20	10.00	**0.22**	0.61	-	**8.67**	0.16	**9.05**	m2	**10.18**
girth exceeding 300mm; panes, area 0.50 - 1.00m2	0.17	10.00	**0.19**	0.52	-	**7.39**	0.14	**7.72**	m2	**8.68**
girth exceeding 300mm; panes, area exceeding 1.00m2	0.17	10.00	**0.19**	0.44	-	**6.25**	0.12	**6.56**	m2	**7.38**
Two coats full gloss finish, existing gloss painted wood surfaces; external work										
Wood general surfaces										
girth exceeding 300mm	0.32	10.00	**0.35**	0.48	-	**6.82**	0.13	**7.30**	m2	**8.21**
isolated surfaces, girth not exceeding 300mm	0.10	10.00	**0.11**	0.16	-	**2.27**	0.04	**2.42**	m	**2.73**
Wood glazed doors										
girth exceeding 300mm; panes, area not exceeding 0.10m2	0.17	10.00	**0.19**	0.81	-	**11.51**	0.22	**11.92**	m2	**13.41**
girth exceeding 300mm; panes, area 0.10 - 0.50m2	0.12	10.00	**0.13**	0.59	-	**8.38**	0.16	**8.68**	m2	**9.76**
girth exceeding 300mm; panes, area 0.50 - 1.00m2	0.12	10.00	**0.13**	0.49	-	**6.96**	0.13	**7.23**	m2	**8.13**
girth exceeding 300mm; panes, area exceeding 1.00m2	0.12	10.00	**0.13**	0.42	-	**5.97**	0.11	**6.22**	m2	**6.99**
Wood partially glazed doors										
girth exceeding 300mm; panes, area not exceeding 0.10m2	0.25	10.00	**0.27**	0.69	-	**9.80**	0.19	**10.26**	m2	**11.54**
girth exceeding 300mm; panes, area 0.10 - 0.50m2	0.20	10.00	**0.22**	0.56	-	**7.96**	0.15	**8.32**	m2	**9.36**
girth exceeding 300mm; panes, area 0.50 - 1.00m2	0.17	10.00	**0.19**	0.49	-	**6.96**	0.13	**7.28**	m2	**8.19**
girth exceeding 300mm; panes, area exceeding 1.00m2	0.17	10.00	**0.19**	0.44	-	**6.25**	0.12	**6.56**	m2	**7.38**

Labour hourly rates: (except Specialists) Craft Operatives £14.21 Labourer £10.71 Rates are national average prices. Refer to REGIONAL VARIATIONS for indicative levels of overall pricing in regions	MATERIALS			LABOUR			RATES			
	Del to Site £	Waste %	Material Cost £	Craft Optve Hrs	Lab Hrs	Labour Cost £	Sunds £	Nett Rate £	Unit	Gross Rate (+12.5%) £
M60: PAINTING/CLEAR FINISHING (REDECORATIONS) - OIL PAINTING WOODWORK Cont.										
Two coats full gloss finish, existing gloss painted wood surfaces; external work Cont.										
Wood windows and screens										
girth exceeding 300mm; panes, area not exceeding 0.10m2	0.25	10.00	**0.27**	0.89	-	**12.65**	0.24	**13.16**	m2	**14.80**
girth exceeding 300mm; panes, area 0.10 - 0.50m2	0.20	10.00	**0.22**	0.65	-	**9.24**	0.18	**9.63**	m2	**10.83**
girth exceeding 300mm; panes, area 0.50 - 1.00m2	0.17	10.00	**0.19**	0.55	-	**7.82**	0.15	**8.15**	m2	**9.17**
girth exceeding 300mm; panes, area exceeding 1.00m2	0.17	10.00	**0.19**	0.47	-	**6.68**	0.13	**6.99**	m2	**7.87**
Wood railings fences and gates; open type										
girth exceeding 300mm	0.25	10.00	**0.27**	0.41	-	**5.83**	0.11	**6.21**	m2	**6.98**
isolated surfaces, girth not exceeding 300mm	0.10	10.00	**0.11**	0.13	-	**1.85**	0.04	**1.99**	m	**2.24**
Wood railings fences and gates; close type										
girth exceeding 300mm	0.32	10.00	**0.35**	0.37	-	**5.26**	0.10	**5.71**	m2	**6.42**
Two undercoats, one coat full gloss finish, existing gloss painted wood surfaces										
Wood general surfaces										
girth exceeding 300mm	0.48	10.00	**0.53**	0.61	-	**8.67**	0.16	**9.36**	m2	**10.53**
isolated surfaces, girth not exceeding 300mm	0.15	10.00	**0.16**	0.20	-	**2.84**	0.05	**3.06**	m	**3.44**
Wood glazed doors										
girth exceeding 300mm; panes, area not exceeding 0.10m2	0.26	10.00	**0.28**	1.04	-	**14.78**	0.28	**15.34**	m2	**17.26**
girth exceeding 300mm; panes, area 0.10 - 0.50m2	0.18	10.00	**0.20**	0.76	-	**10.80**	0.21	**11.21**	m2	**12.61**
girth exceeding 300mm; panes, area 0.50 - 1.00m2	0.18	10.00	**0.20**	0.64	-	**9.09**	0.17	**9.47**	m2	**10.65**
girth exceeding 300mm; panes, area exceeding 1.00m2	0.18	10.00	**0.20**	0.54	-	**7.67**	0.15	**8.02**	m2	**9.02**
Wood partially glazed doors										
girth exceeding 300mm; panes, area not exceeding 0.10m2	0.37	10.00	**0.40**	0.87	-	**12.36**	0.23	**13.00**	m2	**14.63**
girth exceeding 300mm; panes, area 0.10 - 0.50m2	0.29	10.00	**0.32**	0.71	-	**10.09**	0.19	**10.60**	m2	**11.93**
girth exceeding 300mm; panes, area 0.50 - 1.00m2	0.26	10.00	**0.28**	0.64	-	**9.09**	0.17	**9.55**	m2	**10.74**
girth exceeding 300mm; panes, area exceeding 1.00m2	0.26	10.00	**0.28**	0.57	-	**8.10**	0.15	**8.54**	m2	**9.60**
Wood windows and screens										
girth exceeding 300mm; panes, area not exceeding 0.10m2	0.37	10.00	**0.40**	1.12	-	**15.92**	0.30	**16.62**	m2	**18.70**
girth exceeding 300mm; panes, area 0.10 - 0.50m2	0.29	10.00	**0.32**	0.82	-	**11.65**	0.22	**12.20**	m2	**13.72**
girth exceeding 300mm; panes, area 0.50 - 1.00m2	0.26	10.00	**0.28**	0.69	-	**9.80**	0.19	**10.27**	m2	**11.56**
girth exceeding 300mm; panes, area exceeding 1.00m2	0.26	10.00	**0.28**	0.59	-	**8.38**	0.16	**8.83**	m2	**9.93**
Two undercoats, one coat full gloss finish, existing gloss painted wood surfaces; external work										
Wood general surfaces										
girth exceeding 300mm	0.48	10.00	**0.53**	0.65	-	**9.24**	0.18	**9.94**	m2	**11.18**
isolated surfaces, girth not exceeding 300mm	0.15	10.00	**0.16**	0.22	-	**3.13**	0.06	**3.35**	m	**3.77**
Wood glazed doors										
girth exceeding 300mm; panes, area not exceeding 0.10m2	0.26	10.00	**0.28**	1.26	-	**17.90**	0.34	**18.53**	m2	**20.84**
girth exceeding 300mm; panes, area 0.10 - 0.50m2	0.18	10.00	**0.20**	0.78	-	**11.08**	0.21	**11.50**	m2	**12.93**
girth exceeding 300mm; panes, area 0.50 - 1.00m2	0.18	10.00	**0.20**	0.65	-	**9.24**	0.18	**9.61**	m2	**10.82**
girth exceeding 300mm; panes, area exceeding 1.00m2	0.18	10.00	**0.20**	0.55	-	**7.82**	0.15	**8.17**	m2	**9.19**
Wood partially glazed doors										
girth exceeding 300mm; panes, area not exceeding 0.10m2	0.37	10.00	**0.40**	0.89	-	**12.65**	0.24	**13.29**	m2	**14.95**
girth exceeding 300mm; panes, area 0.10 - 0.50m2	0.29	10.00	**0.32**	0.73	-	**10.37**	0.20	**10.89**	m2	**12.26**
girth exceeding 300mm; panes, area 0.50 - 1.00m2	0.26	10.00	**0.28**	0.65	-	**9.24**	0.18	**9.69**	m2	**10.91**
girth exceeding 300mm; panes, area exceeding 1.00m2	0.26	10.00	**0.28**	0.58	-	**8.24**	0.16	**8.68**	m2	**9.77**
Wood windows and screens										
girth exceeding 300mm; panes, area not exceeding 0.10m2	0.37	10.00	**0.40**	1.18	-	**16.77**	0.32	**17.49**	m2	**19.68**
girth exceeding 300mm; panes, area 0.10 - 0.50m2	0.29	10.00	**0.32**	0.87	-	**12.36**	0.23	**12.92**	m2	**14.54**
girth exceeding 300mm; panes, area 0.50 - 1.00m2	0.26	10.00	**0.28**	0.73	-	**10.37**	0.20	**10.85**	m2	**12.21**
girth exceeding 300mm; panes, area exceeding 1.00m2	0.26	10.00	**0.28**	0.63	-	**8.95**	0.17	**9.41**	m2	**10.58**
Wood railings fences and gates; open type										
girth exceeding 300mm	0.37	10.00	**0.40**	0.55	-	**7.82**	0.15	**8.37**	m2	**9.41**
isolated surfaces, girth not exceeding 300mm	0.15	10.00	**0.16**	0.18	-	**2.56**	0.05	**2.77**	m	**3.11**
Wood railings fences and gates; close type										
girth exceeding 300mm	0.48	10.00	**0.53**	0.49	-	**6.96**	0.13	**7.62**	m2	**8.57**
One undercoat, two coats full gloss finish, existing gloss painted wood surfaces										
Wood general surfaces										
girth exceeding 300mm	0.48	10.00	**0.53**	0.61	-	**8.67**	0.16	**9.36**	m2	**10.53**
isolated surfaces, girth not exceeding 300mm	0.15	10.00	**0.16**	0.20	-	**2.84**	0.05	**3.06**	m	**3.44**

Labour hourly rates: (except Specialists) Craft Operatives £14.21 Labourer £10.71 Rates are national average prices. Refer to REGIONAL VARIATIONS for indicative levels of overall pricing in regions	MATERIALS			LABOUR				RATES		
	Del to Site £	Waste %	Material Cost £	Craft Optve Hrs	Lab Hrs	Labour Cost £	Sunds £	Nett Rate £	Unit	Gross Rate (+12.5%) £
M60: PAINTING/CLEAR FINISHING (REDECORATIONS) - OIL PAINTING WOODWORK Cont.										
One undercoat, two coats full gloss finish, existing gloss painted wood surfaces Cont.										
Wood glazed doors										
girth exceeding 300mm; panes, area not exceeding 0.10m2	0.26	10.00	**0.28**	1.04	-	**14.78**	0.28	**15.34**	m2	**17.26**
girth exceeding 300mm; panes, area 0.10 - 0.50m2	0.18	10.00	**0.20**	0.76	-	**10.80**	0.21	**11.21**	m2	**12.61**
girth exceeding 300mm; panes, area 0.50 - 1.00m2	0.18	10.00	**0.20**	0.64	-	**9.09**	0.17	**9.47**	m2	**10.65**
girth exceeding 300mm; panes, area exceeding 1.00m2	0.18	10.00	**0.20**	0.54	-	**7.67**	0.15	**8.02**	m2	**9.02**
Wood partially glazed doors										
girth exceeding 300mm; panes, area not exceeding 0.10m2	0.37	10.00	**0.40**	0.87	-	**12.36**	0.23	**13.00**	m2	**14.63**
girth exceeding 300mm; panes, area 0.10 - 0.50m2	0.29	10.00	**0.32**	0.71	-	**10.09**	0.19	**10.60**	m2	**11.93**
girth exceeding 300mm; panes, area 0.50 - 1.00m2	0.26	10.00	**0.28**	0.64	-	**9.09**	0.17	**9.55**	m2	**10.74**
girth exceeding 300mm; panes, area exceeding 1.00m2	0.26	10.00	**0.28**	0.57	-	**8.10**	0.15	**8.54**	m2	**9.60**
Wood windows and screens										
girth exceeding 300mm; panes, area not exceeding 0.10m2	0.37	10.00	**0.40**	1.12	-	**15.92**	0.30	**16.62**	m2	**18.70**
girth exceeding 300mm; panes, area 0.10 - 0.50m2	0.29	10.00	**0.32**	0.82	-	**11.65**	0.22	**12.20**	m2	**13.72**
girth exceeding 300mm; panes, area 0.50 - 1.00m2	0.26	10.00	**0.28**	0.69	-	**9.80**	0.19	**10.27**	m2	**11.56**
girth exceeding 300mm; panes, area exceeding 1.00m2	0.26	10.00	**0.28**	0.59	-	**8.38**	0.16	**8.83**	m2	**9.93**
One undercoat, two coats full gloss finish, existing gloss painted wood surfaces; external work										
Wood general surfaces										
girth exceeding 300mm	0.48	10.00	**0.53**	0.65	-	**9.24**	0.18	**9.94**	m2	**11.18**
isolated surfaces, girth not exceeding 300mm	0.15	10.00	**0.16**	0.22	-	**3.13**	0.06	**3.35**	m	**3.77**
Wood glazed doors										
girth exceeding 300mm; panes, area not exceeding 0.10m2	0.26	10.00	**0.28**	1.06	-	**15.06**	0.29	**15.63**	m2	**17.59**
girth exceeding 300mm; panes, area 0.10 - 0.50m2	0.18	10.00	**0.20**	0.78	-	**11.08**	0.21	**11.50**	m2	**12.93**
girth exceeding 300mm; panes, area 0.50 - 1.00m2	0.18	10.00	**0.20**	0.65	-	**9.24**	0.18	**9.61**	m2	**10.82**
girth exceeding 300mm; panes, area exceeding 1.00m2	0.18	10.00	**0.20**	0.55	-	**7.82**	0.15	**8.17**	m2	**9.19**
Wood partially glazed doors										
girth exceeding 300mm; panes, area not exceeding 0.10m2	0.37	10.00	**0.40**	0.89	-	**12.65**	0.24	**13.29**	m2	**14.95**
girth exceeding 300mm; panes, area 0.10 - 0.50m2	0.29	10.00	**0.32**	0.73	-	**10.37**	0.20	**10.89**	m2	**12.26**
girth exceeding 300mm; panes, area 0.50 - 1.00m2	0.26	10.00	**0.28**	0.65	-	**9.24**	0.18	**9.69**	m2	**10.91**
girth exceeding 300mm; panes, area exceeding 1.00m2	0.26	10.00	**0.28**	0.58	-	**8.24**	0.16	**8.68**	m2	**9.77**
Wood windows and screens										
girth exceeding 300mm; panes, area not exceeding 0.10m2	0.37	10.00	**0.40**	1.18	-	**16.77**	0.32	**17.49**	m2	**19.68**
girth exceeding 300mm; panes, area 0.10 - 0.50m2	0.29	10.00	**0.32**	0.87	-	**12.36**	0.23	**12.92**	m2	**14.54**
girth exceeding 300mm; panes, area 0.50 - 1.00m2	0.26	10.00	**0.28**	0.73	-	**10.37**	0.20	**10.85**	m2	**12.21**
girth exceeding 300mm; panes, area exceeding 1.00m2	0.26	10.00	**0.28**	0.63	-	**8.95**	0.17	**9.41**	m2	**10.58**
Wood railings fences and gates; open type										
girth exceeding 300mm	0.37	10.00	**0.40**	0.55	-	**7.82**	0.15	**8.37**	m2	**9.41**
isolated surfaces, girth not exceeding 300mm	0.15	10.00	**0.16**	0.18	-	**2.56**	0.05	**2.77**	m	**3.11**
Wood railings fences and gates; close type										
girth exceeding 300mm	0.48	10.00	**0.53**	0.49	-	**6.96**	0.13	**7.62**	m2	**8.57**
Burn off, one coat primer, one undercoat, one coat full gloss finish, existing painted wood surfaces										
Wood general surfaces										
girth exceeding 300mm	0.56	10.00	**0.62**	0.90	-	**12.79**	0.60	**14.01**	m2	**15.76**
isolated surfaces, girth not exceeding 300mm	0.17	10.00	**0.18**	0.26	-	**3.69**	0.17	**4.05**	m	**4.56**
Wood glazed doors										
girth exceeding 300mm; panes, area not exceeding 0.10m2	0.31	10.00	**0.34**	1.50	-	**21.32**	1.01	**22.66**	m2	**25.49**
girth exceeding 300mm; panes, area 0.10 - 0.50m2	0.23	10.00	**0.25**	1.08	-	**15.35**	0.72	**16.32**	m2	**18.36**
girth exceeding 300mm; panes, area 0.50 - 1.00m2	0.21	10.00	**0.23**	0.90	-	**12.79**	0.60	**13.62**	m2	**15.32**
girth exceeding 300mm; panes, area exceeding 1.00m2	0.21	10.00	**0.23**	0.80	-	**11.37**	0.54	**12.13**	m2	**13.65**
Wood partially glazed doors										
girth exceeding 300mm; panes, area not exceeding 0.10m2	0.44	10.00	**0.48**	1.20	-	**17.05**	0.80	**18.34**	m2	**20.63**
girth exceeding 300mm; panes, area 0.10 - 0.50m2	0.33	10.00	**0.37**	0.99	-	**14.07**	0.66	**15.10**	m2	**16.99**
girth exceeding 300mm; panes, area 0.50 - 1.00m2	0.31	10.00	**0.34**	0.90	-	**12.79**	0.60	**13.73**	m2	**15.45**
girth exceeding 300mm; panes, area exceeding 1.00m2	0.31	10.00	**0.34**	0.86	-	**12.22**	0.58	**13.14**	m2	**14.78**
Wood windows and screens										
girth exceeding 300mm; panes, area not exceeding 0.10m2	0.44	10.00	**0.48**	1.65	-	**23.45**	1.11	**25.03**	m2	**28.16**
girth exceeding 300mm; panes, area 0.10 - 0.50m2	0.33	10.00	**0.37**	1.18	-	**16.77**	0.79	**17.93**	m2	**20.17**
girth exceeding 300mm; panes, area 0.50 - 1.00m2	0.31	10.00	**0.34**	0.99	-	**14.07**	0.66	**15.07**	m2	**16.96**

Labour hourly rates: (except Specialists) Craft Operatives £14.21 Labourer £10.71 Rates are national average prices. Refer to REGIONAL VARIATIONS for indicative levels of overall pricing in regions	MATERIALS			LABOUR				RATES		
	Del to Site £	Waste %	Material Cost £	Craft Optve Hrs	Lab Hrs	Labour Cost £	Sunds £	Nett Rate £	Unit	Gross Rate (+12.5%) £
M60: PAINTING/CLEAR FINISHING (REDECORATIONS) - OIL PAINTING WOODWORK Cont.										
Burn off, one coat primer, one undercoat, one coat full gloss finish, existing painted wood surfaces Cont.										
Wood windows and screens Cont.										
girth exceeding 300mm; panes, area exceeding 1.00m2	0.31	10.00	**0.34**	0.88	-	**12.50**	0.59	**13.44**	m2	**15.12**
Burn off, one coat primer, one undercoat, one coat full gloss finish, existing painted wood surfaces; external work										
Wood general surfaces										
girth exceeding 300mm	0.56	10.00	**0.62**	0.95	-	**13.50**	0.64	**14.75**	m2	**16.60**
isolated surfaces, girth not exceeding 300mm	0.17	10.00	**0.18**	0.29	-	**4.12**	0.19	**4.50**	m	**5.06**
Wood glazed doors										
girth exceeding 300mm; panes, area not exceeding 0.10m2	0.31	10.00	**0.34**	1.50	-	**21.32**	1.01	**22.66**	m2	**25.49**
girth exceeding 300mm; panes, area 0.10 - 0.50m2	0.23	10.00	**0.25**	1.08	-	**15.35**	0.72	**16.32**	m2	**18.36**
girth exceeding 300mm; panes, area 0.50 - 1.00m2	0.21	10.00	**0.23**	0.90	-	**12.79**	0.60	**13.62**	m2	**15.32**
girth exceeding 300mm; panes, area exceeding 1.00m2	0.21	10.00	**0.23**	0.80	-	**11.37**	0.54	**12.13**	m2	**13.65**
Wood partially glazed doors										
girth exceeding 300mm; panes, area not exceeding 0.10m2	0.44	10.00	**0.48**	1.20	-	**17.05**	0.80	**18.34**	m2	**20.63**
girth exceeding 300mm; panes, area 0.10 - 0.50m2	0.33	10.00	**0.37**	0.99	-	**14.07**	0.66	**15.10**	m2	**16.99**
girth exceeding 300mm; panes, area 0.50 - 1.00m2	0.31	10.00	**0.34**	0.90	-	**12.79**	0.60	**13.73**	m2	**15.45**
girth exceeding 300mm; panes, area exceeding 1.00m2	0.31	10.00	**0.34**	0.86	-	**12.22**	0.58	**13.14**	m2	**14.78**
Wood windows and screens										
girth exceeding 300mm; panes, area not exceeding 0.10m2	0.44	10.00	**0.48**	1.69	-	**24.01**	1.13	**25.63**	m2	**28.83**
girth exceeding 300mm; panes, area 0.10 - 0.50m2	0.33	10.00	**0.37**	1.22	-	**17.34**	0.82	**18.52**	m2	**20.84**
girth exceeding 300mm; panes, area 0.50 - 1.00m2	0.31	10.00	**0.34**	1.03	-	**14.64**	0.69	**15.67**	m2	**17.63**
girth exceeding 300mm; panes, area exceeding 1.00m2	0.31	10.00	**0.34**	0.92	-	**13.07**	0.62	**14.03**	m2	**15.78**
Wood railings fences and gates; open type										
girth exceeding 300mm	0.44	10.00	**0.48**	0.81	-	**11.51**	0.54	**12.53**	m2	**14.10**
isolated surfaces, girth not exceeding 300mm	0.17	10.00	**0.18**	0.24	-	**3.41**	0.16	**3.76**	m	**4.22**
Wood railings fences and gates; close type										
girth exceeding 300mm	0.56	10.00	**0.62**	0.76	-	**10.80**	0.51	**11.93**	m2	**13.42**
Burn off, one coat primer, two undercoats, one coat full gloss finish, existing painted wood surfaces										
Wood general surfaces										
girth exceeding 300mm	0.72	10.00	**0.79**	1.06	-	**15.06**	0.71	**16.57**	m2	**18.64**
isolated surfaces, girth not exceeding 300mm	0.22	10.00	**0.24**	0.32	-	**4.55**	0.21	**5.00**	m	**5.62**
Wood glazed doors										
girth exceeding 300mm; panes, area not exceeding 0.10m2	0.40	10.00	**0.44**	1.76	-	**25.01**	1.18	**26.62**	m2	**29.95**
girth exceeding 300mm; panes, area 0.10 - 0.50m2	0.29	10.00	**0.32**	1.27	-	**18.05**	0.85	**19.21**	m2	**21.62**
girth exceeding 300mm; panes, area 0.50 - 1.00m2	0.27	10.00	**0.30**	1.06	-	**15.06**	0.71	**16.07**	m2	**18.08**
girth exceeding 300mm; panes, area exceeding 1.00m2	0.27	10.00	**0.30**	0.95	-	**13.50**	0.64	**14.43**	m2	**16.24**
Wood partially glazed doors										
girth exceeding 300mm; panes, area not exceeding 0.10m2	0.56	10.00	**0.61**	1.41	-	**20.04**	0.94	**21.59**	m2	**24.29**
girth exceeding 300mm; panes, area 0.10 - 0.50m2	0.43	10.00	**0.48**	1.17	-	**16.63**	0.78	**17.89**	m2	**20.12**
girth exceeding 300mm; panes, area 0.50 - 1.00m2	0.40	10.00	**0.44**	1.06	-	**15.06**	0.71	**16.21**	m2	**18.23**
girth exceeding 300mm; panes, area exceeding 1.00m2	0.40	10.00	**0.44**	1.00	-	**14.21**	0.67	**15.32**	m2	**17.23**
Wood windows and screens										
girth exceeding 300mm; panes, area not exceeding 0.10m2	0.56	10.00	**0.61**	1.94	-	**27.57**	1.30	**29.48**	m2	**33.17**
girth exceeding 300mm; panes, area 0.10 - 0.50m2	0.43	10.00	**0.48**	1.40	-	**19.89**	0.94	**21.31**	m2	**23.97**
girth exceeding 300mm; panes, area 0.50 - 1.00m2	0.40	10.00	**0.44**	1.17	-	**16.63**	0.78	**17.85**	m2	**20.08**
girth exceeding 300mm; panes, area exceeding 1.00m2	0.40	10.00	**0.44**	1.03	-	**14.64**	0.69	**15.76**	m2	**17.73**
Burn off, one coat primer, two undercoats, one coat full gloss finish, existing painted wood surfaces; external work										
Wood general surfaces										
girth exceeding 300mm	0.72	10.00	**0.79**	1.11	-	**15.77**	0.74	**17.31**	m2	**19.47**
isolated surfaces, girth not exceeding 300mm	0.22	10.00	**0.24**	0.35	-	**4.97**	0.23	**5.45**	m	**6.13**
Wood glazed doors										
girth exceeding 300mm; panes, area not exceeding 0.10m2	0.40	10.00	**0.44**	1.76	-	**25.01**	1.18	**26.62**	m2	**29.95**
girth exceeding 300mm; panes, area 0.10 - 0.50m2	0.29	10.00	**0.32**	1.27	-	**18.05**	0.85	**19.21**	m2	**21.62**
girth exceeding 300mm; panes, area 0.50 - 1.00m2	0.27	10.00	**0.30**	1.06	-	**15.06**	0.71	**16.07**	m2	**18.08**
girth exceeding 300mm; panes, area exceeding 1.00m2	0.27	10.00	**0.30**	0.95	-	**13.50**	0.64	**14.43**	m2	**16.24**

Labour hourly rates: (except Specialists) Craft Operatives £14.21 Labourer £10.71 Rates are national average prices. Refer to REGIONAL VARIATIONS for indicative levels of overall pricing in regions	MATERIALS			LABOUR				RATES		
	Del to Site £	Waste %	Material Cost £	Craft Optve Hrs	Lab Hrs	Labour Cost £	Sunds £	Nett Rate £	Unit	Gross Rate (+12.5%) £
M60: PAINTING/CLEAR FINISHING (REDECORATIONS) - OIL PAINTING WOODWORK Cont.										
Burn off, one coat primer, two undercoats, one coat full gloss finish, existing painted wood surfaces; external work Cont.										
Wood partially glazed doors										
girth exceeding 300mm; panes, area not exceeding 0.10m2	0.56	10.00	**0.61**	1.41	-	**20.04**	0.94	**21.59**	m2	**24.29**
girth exceeding 300mm; panes, area 0.10 - 0.50m2	0.43	10.00	**0.48**	1.17	-	**16.63**	0.78	**17.89**	m2	**20.12**
girth exceeding 300mm; panes, area 0.50 - 1.00m2	0.40	10.00	**0.44**	1.06	-	**15.06**	0.71	**16.21**	m2	**18.23**
girth exceeding 300mm; panes, area exceeding 1.00m2	0.40	10.00	**0.44**	1.00	-	**14.21**	0.67	**15.32**	m2	**17.23**
Wood windows and screens										
girth exceeding 300mm; panes, area not exceeding 0.10m2	0.56	10.00	**0.61**	1.99	-	**28.28**	1.33	**30.23**	m2	**34.00**
girth exceeding 300mm; panes, area 0.10 - 0.50m2	0.43	10.00	**0.48**	1.45	-	**20.60**	0.97	**22.05**	m2	**24.81**
girth exceeding 300mm; panes, area 0.50 - 1.00m2	0.40	10.00	**0.44**	1.22	-	**17.34**	0.82	**18.59**	m2	**20.91**
girth exceeding 300mm; panes, area exceeding 1.00m2	0.40	10.00	**0.44**	1.09	-	**15.49**	0.73	**16.65**	m2	**18.74**
Wood railings fences and gates; open type										
girth exceeding 300mm	0.56	10.00	**0.61**	0.95	-	**13.50**	0.64	**14.75**	m2	**16.59**
isolated surfaces, girth not exceeding 300mm	0.22	10.00	**0.24**	0.29	-	**4.12**	0.19	**4.55**	m	**5.12**
Wood railings fences and gates; close type										
girth exceeding 300mm	0.72	10.00	**0.79**	0.88	-	**12.50**	0.59	**13.89**	m2	**15.62**
Burn off, one coat primer, one undercoat, two coats full gloss finish, existing painted wood surfaces										
Wood general surfaces										
girth exceeding 300mm	0.72	10.00	**0.79**	1.06	-	**15.06**	0.71	**16.57**	m2	**18.64**
isolated surfaces, girth not exceeding 300mm	0.22	10.00	**0.24**	0.32	-	**4.55**	0.21	**5.00**	m	**5.62**
Wood glazed doors										
girth exceeding 300mm; panes, area not exceeding 0.10m2	0.40	10.00	**0.44**	1.76	-	**25.01**	1.18	**26.62**	m2	**29.95**
girth exceeding 300mm; panes, area 0.10 - 0.50m2	0.29	10.00	**0.32**	1.27	-	**18.05**	0.85	**19.21**	m2	**21.62**
girth exceeding 300mm; panes, area 0.50 - 1.00m2	0.27	10.00	**0.30**	1.06	-	**15.06**	0.71	**16.07**	m2	**18.08**
girth exceeding 300mm; panes, area exceeding 1.00m2	0.27	10.00	**0.30**	0.95	-	**13.50**	0.64	**14.43**	m2	**16.24**
Wood partially glazed doors										
girth exceeding 300mm; panes, area not exceeding 0.10m2	0.56	10.00	**0.61**	1.41	-	**20.04**	0.94	**21.59**	m2	**24.29**
girth exceeding 300mm; panes, area 0.10 - 0.50m2	0.43	10.00	**0.48**	1.17	-	**16.63**	0.78	**17.89**	m2	**20.12**
girth exceeding 300mm; panes, area 0.50 - 1.00m2	0.40	10.00	**0.44**	1.06	-	**15.06**	0.71	**16.21**	m2	**18.23**
girth exceeding 300mm; panes, area exceeding 1.00m2	0.40	10.00	**0.44**	1.00	-	**14.21**	0.67	**15.32**	m2	**17.23**
Wood windows and screens										
girth exceeding 300mm; panes, area not exceeding 0.10m2	0.56	10.00	**0.61**	1.94	-	**27.57**	1.30	**29.48**	m2	**33.17**
girth exceeding 300mm; panes, area 0.10 - 0.50m2	0.43	10.00	**0.48**	1.40	-	**19.89**	0.94	**21.31**	m2	**23.97**
girth exceeding 300mm; panes, area 0.50 - 1.00m2	0.40	10.00	**0.44**	1.17	-	**16.63**	0.78	**17.85**	m2	**20.08**
girth exceeding 300mm; panes, area exceeding 1.00m2	0.40	10.00	**0.44**	1.03	-	**14.64**	0.69	**15.76**	m2	**17.73**
Burn off, one coat primer, one undercoat, two coats full gloss finish, existing painted wood surfaces; external work										
Wood general surfaces										
girth exceeding 300mm	0.72	10.00	**0.79**	1.11	-	**15.77**	0.74	**17.31**	m2	**19.47**
isolated surfaces, girth not exceeding 300mm	0.22	10.00	**0.24**	0.35	-	**4.97**	0.23	**5.45**	m	**6.13**
Wood glazed doors										
girth exceeding 300mm; panes, area not exceeding 0.10m2	0.40	10.00	**0.44**	1.76	-	**25.01**	1.18	**26.62**	m2	**29.95**
girth exceeding 300mm; panes, area 0.10 - 0.50m2	0.29	10.00	**0.32**	1.27	-	**18.05**	0.85	**19.21**	m2	**21.62**
girth exceeding 300mm; panes, area 0.50 - 1.00m2	0.27	10.00	**0.30**	1.06	-	**15.06**	0.71	**16.07**	m2	**18.08**
girth exceeding 300mm; panes, area exceeding 1.00m2	0.27	10.00	**0.30**	0.95	-	**13.50**	0.64	**14.43**	m2	**16.24**
Wood partially glazed doors										
girth exceeding 300mm; panes, area not exceeding 0.10m2	0.56	10.00	**0.61**	1.41	-	**20.04**	0.94	**21.59**	m2	**24.29**
girth exceeding 300mm; panes, area 0.10 - 0.50m2	0.43	10.00	**0.48**	1.17	-	**16.63**	0.78	**17.89**	m2	**20.12**
girth exceeding 300mm; panes, area 0.50 - 1.00m2	0.40	10.00	**0.44**	1.06	-	**15.06**	0.71	**16.21**	m2	**18.23**
girth exceeding 300mm; panes, area exceeding 1.00m2	0.40	10.00	**0.44**	1.00	-	**14.21**	0.67	**15.32**	m2	**17.23**
Wood windows and screens										
girth exceeding 300mm; panes, area not exceeding 0.10m2	0.56	10.00	**0.61**	1.99	-	**28.28**	1.33	**30.23**	m2	**34.00**
girth exceeding 300mm; panes, area 0.10 - 0.50m2	0.43	10.00	**0.48**	1.45	-	**20.60**	0.97	**22.05**	m2	**24.81**
girth exceeding 300mm; panes, area 0.50 - 1.00m2	0.40	10.00	**0.44**	1.22	-	**17.34**	0.82	**18.59**	m2	**20.91**
girth exceeding 300mm; panes, area exceeding 1.00m2	0.40	10.00	**0.44**	1.09	-	**15.49**	0.73	**16.65**	m2	**18.74**
Wood railings fences and gates; open type										
girth exceeding 300mm	0.56	10.00	**0.61**	0.95	-	**13.50**	0.64	**14.75**	m2	**16.59**
isolated surfaces, girth not exceeding 300mm	0.22	10.00	**0.24**	0.29	-	**4.12**	0.19	**4.55**	m	**5.12**

Labour hourly rates: (except Specialists) Craft Operatives £14.21 Labourer £10.71 Rates are national average prices. Refer to REGIONAL VARIATIONS for indicative levels of overall pricing in regions	MATERIALS			LABOUR				RATES		
	Del to Site £	Waste %	Material Cost £	Craft Optve Hrs	Lab Hrs	Labour Cost £	Sunds £	Nett Rate £	Unit	Gross Rate (+12.5%) £
M60: PAINTING/CLEAR FINISHING (REDECORATIONS) - OIL PAINTING WOODWORK Cont.										
Burn off, one coat primer, one undercoat, two coats full gloss finish, existing painted wood surfaces; external work Cont.										
Wood railings fences and gates; close type										
girth exceeding 300mm	0.72	10.00	**0.79**	0.88	-	**12.50**	0.59	**13.89**	m2	**15.62**
M60: PAINTING/CLEAR FINISHING (REDECORATIONS) - POLYURETHANE LACQUER										
Generally										
Note: the following rates include for the cost of all preparatory work, e.g. washing down, etc.										
Two coats polyurethane lacquer, existing lacquered surfaces										
Wood general surfaces										
girth exceeding 300mm	0.85	10.00	**0.93**	0.54	-	**7.67**	0.15	**8.75**	m2	**9.85**
isolated surfaces, girth not exceeding 300mm	0.25	10.00	**0.28**	0.18	-	**2.56**	0.05	**2.89**	m	**3.25**
Wood glazed doors										
girth exceeding 300mm; panes, area not exceeding 0.10m2	0.48	10.00	**0.53**	0.92	-	**13.07**	0.25	**13.85**	m2	**15.58**
girth exceeding 300mm; panes, area 0.10 - 0.50m2	0.34	10.00	**0.37**	0.67	-	**9.52**	0.18	**10.07**	m2	**11.33**
girth exceeding 300mm; panes, area 0.50 - 1.00m2	0.31	10.00	**0.34**	0.55	-	**7.82**	0.15	**8.31**	m2	**9.34**
girth exceeding 300mm; panes, area exceeding 1.00m2	0.31	10.00	**0.34**	0.48	-	**6.82**	0.13	**7.29**	m2	**8.20**
Wood partially glazed doors										
girth exceeding 300mm; panes, area not exceeding 0.10m2	0.68	10.00	**0.75**	0.77	-	**10.94**	0.21	**11.90**	m2	**13.38**
girth exceeding 300mm; panes, area 0.10 - 0.50m2	0.51	10.00	**0.56**	0.62	-	**8.81**	0.17	**9.54**	m2	**10.73**
girth exceeding 300mm; panes, area 0.50 - 1.00m2	0.48	10.00	**0.53**	0.55	-	**7.82**	0.15	**8.49**	m2	**9.55**
girth exceeding 300mm; panes, area exceeding 1.00m2	0.48	10.00	**0.53**	0.50	-	**7.11**	0.14	**7.77**	m2	**8.74**
Wood windows and screens										
girth exceeding 300mm; panes, area not exceeding 0.10m2	0.68	10.00	**0.75**	0.99	-	**14.07**	0.27	**15.08**	m2	**16.97**
girth exceeding 300mm; panes, area 0.10 - 0.50m2	0.51	10.00	**0.56**	0.65	-	**9.24**	0.18	**9.97**	m2	**11.22**
girth exceeding 300mm; panes, area 0.50 - 1.00m2	0.48	10.00	**0.53**	0.59	-	**8.38**	0.16	**9.07**	m2	**10.21**
girth exceeding 300mm; panes, area exceeding 1.00m2	0.48	10.00	**0.53**	0.52	-	**7.39**	0.14	**8.06**	m2	**9.07**
Two coats polyurethane lacquer, existing lacquered surfaces; external work										
Wood general surfaces										
girth exceeding 300mm	0.85	10.00	**0.93**	0.57	-	**8.10**	0.15	**9.19**	m2	**10.33**
isolated surfaces, girth not exceeding 300mm	0.25	10.00	**0.28**	0.19	-	**2.70**	0.05	**3.03**	m	**3.41**
Wood glazed doors										
girth exceeding 300mm; panes, area not exceeding 0.10m2	0.48	10.00	**0.53**	0.94	-	**13.36**	0.25	**14.14**	m2	**15.91**
girth exceeding 300mm; panes, area 0.10 - 0.50m2	0.34	10.00	**0.37**	0.68	-	**9.66**	0.18	**10.22**	m2	**11.50**
girth exceeding 300mm; panes, area 0.50 - 1.00m2	0.31	10.00	**0.34**	0.56	-	**7.96**	0.15	**8.45**	m2	**9.51**
girth exceeding 300mm; panes, area exceeding 1.00m2	0.31	10.00	**0.34**	0.49	-	**6.96**	0.13	**7.44**	m2	**8.37**
Wood partially glazed doors										
girth exceeding 300mm; panes, area not exceeding 0.10m2	0.68	10.00	**0.75**	0.79	-	**11.23**	0.21	**12.19**	m2	**13.71**
girth exceeding 300mm; panes, area 0.10 - 0.50m2	0.51	10.00	**0.56**	0.63	-	**8.95**	0.17	**9.68**	m2	**10.89**
girth exceeding 300mm; panes, area 0.50 - 1.00m2	0.48	10.00	**0.53**	0.56	-	**7.96**	0.15	**8.64**	m2	**9.72**
girth exceeding 300mm; panes, area exceeding 1.00m2	0.48	10.00	**0.53**	0.51	-	**7.25**	0.14	**7.91**	m2	**8.90**
Wood windows and screens										
girth exceeding 300mm; panes, area not exceeding 0.10m2	0.68	10.00	**0.75**	1.04	-	**14.78**	0.28	**15.81**	m2	**17.78**
girth exceeding 300mm; panes, area 0.10 - 0.50m2	0.51	10.00	**0.56**	0.76	-	**10.80**	0.21	**11.56**	m2	**13.01**
girth exceeding 300mm; panes, area 0.50 - 1.00m2	0.48	10.00	**0.53**	0.63	-	**8.95**	0.17	**9.65**	m2	**10.86**
girth exceeding 300mm; panes, area exceeding 1.00m2	0.48	10.00	**0.53**	0.55	-	**7.82**	0.15	**8.49**	m2	**9.55**
Three coats polyurethane lacquer, existing lacquered surfaces										
Wood general surfaces										
girth exceeding 300mm	1.24	10.00	**1.37**	0.62	-	**8.81**	0.17	**10.34**	m2	**11.64**
isolated surfaces, girth not exceeding 300mm	0.37	10.00	**0.40**	0.21	-	**2.98**	0.06	**3.44**	m	**3.88**
Wood glazed doors										
girth exceeding 300mm; panes, area not exceeding 0.10m2	0.71	10.00	**0.78**	1.04	-	**14.78**	0.28	**15.84**	m2	**17.82**
girth exceeding 300mm; panes, area 0.10 - 0.50m2	0.51	10.00	**0.56**	0.76	-	**10.80**	0.21	**11.56**	m2	**13.01**
girth exceeding 300mm; panes, area 0.50 - 1.00m2	0.45	10.00	**0.50**	0.63	-	**8.95**	0.17	**9.62**	m2	**10.82**
girth exceeding 300mm; panes, area exceeding 1.00m2	0.45	10.00	**0.50**	0.54	-	**7.67**	0.15	**8.32**	m2	**9.36**

Labour hourly rates: (except Specialists) Craft Operatives £14.21 Labourer £10.71 Rates are national average prices. Refer to REGIONAL VARIATIONS for indicative levels of overall pricing in regions	MATERIALS			LABOUR				RATES		
	Del to Site £	Waste %	Material Cost £	Craft Optve Hrs	Lab Hrs	Labour Cost £	Sunds £	Nett Rate £	Unit	Gross Rate (+12.5%) £
M60: PAINTING/CLEAR FINISHING (REDECORATIONS) - POLYURETHANE LACQUER Cont.										
Three coats polyurethane lacquer, existing lacquered surfaces Cont.										
Wood partially glazed doors										
girth exceeding 300mm; panes, area not exceeding 0.10m2	0.99	10.00	**1.09**	0.87	-	**12.36**	0.23	**13.69**	m2	**15.40**
girth exceeding 300mm; panes, area 0.10 - 0.50m2	0.73	10.00	**0.81**	0.77	-	**10.94**	0.21	**11.96**	m2	**13.45**
girth exceeding 300mm; panes, area 0.50 - 1.00m2	0.68	10.00	**0.75**	0.63	-	**8.95**	0.17	**9.87**	m2	**11.10**
girth exceeding 300mm; panes, area exceeding 1.00m2	0.68	10.00	**0.75**	0.58	-	**8.24**	0.16	**9.14**	m2	**10.29**
Wood windows and screens										
girth exceeding 300mm; panes, area not exceeding 0.10m2	0.71	10.00	**0.78**	1.13	-	**16.06**	0.31	**17.14**	m2	**19.28**
girth exceeding 300mm; panes, area 0.10 - 0.50m2	0.51	10.00	**0.56**	0.82	-	**11.65**	0.22	**12.43**	m2	**13.99**
girth exceeding 300mm; panes, area 0.50 - 1.00m2	0.45	10.00	**0.50**	0.68	-	**9.66**	0.18	**10.34**	m2	**11.64**
girth exceeding 300mm; panes, area exceeding 1.00m2	0.45	10.00	**0.50**	0.59	-	**8.38**	0.16	**9.04**	m2	**10.17**
Three coats polyurethane lacquer, existing lacquered surfaces; external work										
Wood general surfaces										
girth exceeding 300mm	1.24	10.00	**1.37**	0.66	-	**9.38**	0.18	**10.92**	m2	**12.29**
isolated surfaces, girth not exceeding 300mm	0.37	10.00	**0.40**	0.22	-	**3.13**	0.06	**3.59**	m	**4.04**
Wood glazed doors										
girth exceeding 300mm; panes, area not exceeding 0.10m2	0.71	10.00	**0.78**	1.06	-	**15.06**	0.29	**16.13**	m2	**18.14**
girth exceeding 300mm; panes, area 0.10 - 0.50m2	0.51	10.00	**0.56**	0.77	-	**10.94**	0.21	**11.71**	m2	**13.17**
girth exceeding 300mm; panes, area 0.50 - 1.00m2	0.45	10.00	**0.50**	0.64	-	**9.09**	0.17	**9.76**	m2	**10.98**
girth exceeding 300mm; panes, area exceeding 1.00m2	0.45	10.00	**0.50**	0.54	-	**7.67**	0.15	**8.32**	m2	**9.36**
Wood partially glazed doors										
girth exceeding 300mm; panes, area not exceeding 0.10m2	0.99	10.00	**1.09**	0.89	-	**12.65**	0.24	**13.97**	m2	**15.72**
girth exceeding 300mm; panes, area 0.10 - 0.50m2	0.73	10.00	**0.81**	0.78	-	**11.08**	0.21	**12.10**	m2	**13.62**
girth exceeding 300mm; panes, area 0.50 - 1.00m2	0.68	10.00	**0.75**	0.64	-	**9.09**	0.17	**10.01**	m2	**11.26**
girth exceeding 300mm; panes, area exceeding 1.00m2	0.68	10.00	**0.75**	0.59	-	**8.38**	0.16	**9.29**	m2	**10.45**
Wood windows and screens										
girth exceeding 300mm; panes, area not exceeding 0.10m2	0.71	10.00	**0.78**	1.18	-	**16.77**	0.32	**17.86**	m2	**20.10**
girth exceeding 300mm; panes, area 0.10 - 0.50m2	0.51	10.00	**0.56**	0.87	-	**12.36**	0.23	**13.16**	m2	**14.80**
girth exceeding 300mm; panes, area 0.50 - 1.00m2	0.45	10.00	**0.50**	0.73	-	**10.37**	0.20	**11.07**	m2	**12.45**
girth exceeding 300mm; panes, area exceeding 1.00m2	0.45	10.00	**0.50**	0.63	-	**8.95**	0.17	**9.62**	m2	**10.82**
M60: PAINTING/CLEAR FINISHING (REDECORATIONS) - FRENCH AND WAX POLISHING										
Generally										
Note: the following rates include for the cost of all preparatory work, e.g. washing down, etc.										
Stain; two coats white polish										
Wood general surfaces										
girth exceeding 300mm	2.76	10.00	**3.03**	0.40	-	**5.68**	0.11	**8.82**	m2	**9.93**
isolated surfaces, girth not exceeding 300mm	0.87	10.00	**0.95**	0.14	-	**1.99**	0.04	**2.98**	m	**3.35**
Strip old polish, oil, two coats wax polish, existing polished surfaces										
Wood general surfaces										
girth exceeding 300mm	2.76	10.00	**3.03**	1.23	-	**17.48**	0.33	**20.84**	m2	**23.45**
isolated surfaces, girth not exceeding 300mm	0.87	10.00	**0.95**	0.37	-	**5.26**	0.10	**6.31**	m	**7.10**
Strip old polish, oil, stain, two coats wax polish, existing polished surfaces										
Wood general surfaces										
girth exceeding 300mm	2.76	10.00	**3.03**	1.35	-	**19.18**	0.36	**22.58**	m2	**25.40**
isolated surfaces, girth not exceeding 300mm	0.87	10.00	**0.95**	0.42	-	**5.97**	0.11	**7.04**	m	**7.91**
Open grain French polish existing polished surfaces										
Wood general surfaces										
girth exceeding 300mm	2.79	10.00	**3.07**	2.64	-	**37.51**	0.71	**41.30**	m2	**46.46**
isolated surfaces, girth not exceeding 300mm	0.88	10.00	**0.97**	0.88	-	**12.50**	0.24	**13.71**	m	**15.42**
Fully French polish existing polished surfaces										
Wood general surfaces										
girth exceeding 300mm	4.69	10.00	**5.16**	1.54	-	**21.88**	0.42	**27.46**	m2	**30.89**
isolated surfaces, girth not exceeding 300mm	1.48	10.00	**1.63**	0.52	-	**7.39**	0.14	**9.16**	m	**10.30**

Labour hourly rates: (except Specialists) Craft Operatives £14.21 Labourer £10.71 Rates are national average prices. Refer to REGIONAL VARIATIONS for indicative levels of overall pricing in regions	MATERIALS			LABOUR				RATES		
	Del to Site £	Waste %	Material Cost £	Craft Optve Hrs	Lab Hrs	Labour Cost £	Sunds £	Nett Rate £	Unit	Gross Rate (+12.5%) £
M60: PAINTING/CLEAR FINISHING (REDECORATIONS) - FRENCH AND WAX POLISHING Cont.										
Strip old polish, oil, open grain French polish, existing polished surfaces										
Wood general surfaces										
girth exceeding 300mm	2.79	10.00	**3.07**	3.43	-	**48.74**	0.93	**52.74**	m2	**59.33**
isolated surfaces, girth not exceeding 300mm	0.88	10.00	**0.97**	1.14	-	**16.20**	0.31	**17.47**	m	**19.66**
Strip old polish, oil, stain, open grain French polish, existing polished surfaces										
Wood general surfaces										
girth exceeding 300mm	2.79	10.00	**3.07**	3.55	-	**50.45**	0.96	**54.47**	m2	**61.28**
isolated surfaces, girth not exceeding 300mm	0.88	10.00	**0.97**	1.19	-	**16.91**	0.32	**18.20**	m	**20.47**
Strip old polish, oil, body in, fully French polish existing polished surfaces										
Wood general surfaces										
girth exceeding 300mm	4.69	10.00	**5.16**	4.63	-	**65.79**	1.25	**72.20**	m2	**81.23**
isolated surfaces, girth not exceeding 300mm	1.48	10.00	**1.63**	1.54	-	**21.88**	0.42	**23.93**	m	**26.92**
Strip old polish, oil, stain, body in, fully French polish, existing polished surfaces										
Wood general surfaces										
girth exceeding 300mm	4.69	10.00	**5.16**	4.74	-	**67.36**	1.28	**73.80**	m2	**83.02**
isolated surfaces, girth not exceeding 300mm	1.48	10.00	**1.63**	1.58	-	**22.45**	0.43	**24.51**	m	**27.57**
M60: PAINTING/CLEAR FINISHING (REDECORATIONS) - WATER REPELLENT										
One coat solvent based water repellent, existing surfaces										
Cement rendered general surfaces										
girth exceeding 300mm	0.16	5.00	**0.16**	0.12	-	**1.71**	0.03	**1.90**	m2	**2.14**
Stone general surfaces										
girth exceeding 300mm	0.19	5.00	**0.20**	0.13	-	**1.85**	0.04	**2.08**	m2	**2.34**
Brickwork general surfaces										
girth exceeding 300mm	0.19	5.00	**0.20**	0.13	-	**1.85**	0.04	**2.08**	m2	**2.34**
M60: PAINTING/CLEAR FINISHING (REDECORATIONS) - REMOVAL OF MOULD GROWTH										
Apply fungicide to existing decorated and infected surfaces										
Cement rendered general surfaces										
girth exceeding 300mm	0.12	5.00	**0.12**	0.22	-	**3.13**	0.06	**3.31**	m2	**3.72**
Concrete general surfaces										
girth exceeding 300mm	0.14	5.00	**0.15**	0.22	-	**3.13**	0.06	**3.34**	m2	**3.75**
Plaster general surfaces										
girth exceeding 300mm	0.10	5.00	**0.11**	0.22	-	**3.13**	0.06	**3.29**	m2	**3.71**
Brickwork general surfaces										
girth exceeding 300mm	0.14	5.00	**0.15**	0.24	-	**3.41**	0.06	**3.63**	m2	**4.08**
Apply fungicide to existing decorated and infected surfaces; external work										
Cement rendered general surfaces										
girth exceeding 300mm	0.12	5.00	**0.12**	0.24	-	**3.41**	0.06	**3.60**	m2	**4.05**
Concrete general surfaces										
girth exceeding 300mm	0.14	5.00	**0.15**	0.24	-	**3.41**	0.06	**3.63**	m2	**4.08**
Plaster general surfaces										
girth exceeding 300mm	0.10	5.00	**0.11**	0.24	-	**3.41**	0.06	**3.58**	m2	**4.03**
Brickwork general surfaces										
girth exceeding 300mm	0.14	5.00	**0.15**	0.26	-	**3.69**	0.07	**3.91**	m2	**4.40**
M60: PAINTING/CLEAR FINISHING; STAIRCASE AREAS: EMULSION PAINTING										
Mist coat, one full coat emulsion paint										
Concrete general surfaces										
girth exceeding 300mm	0.15	10.00	**0.16**	0.23	-	**3.27**	0.06	**3.49**	m2	**3.93**
Plaster general surfaces										
girth exceeding 300mm	0.13	10.00	**0.14**	0.21	-	**2.98**	0.06	**3.18**	m2	**3.58**
Plasterboard general surfaces										
girth exceeding 300mm	0.13	10.00	**0.14**	0.21	-	**2.98**	0.06	**3.18**	m2	**3.58**
Brickwork general surfaces										
girth exceeding 300mm	0.19	10.00	**0.20**	0.25	-	**3.55**	0.07	**3.82**	m2	**4.30**

Labour hourly rates: (except Specialists) Craft Operatives £14.21 Labourer £10.71 Rates are national average prices. Refer to REGIONAL VARIATIONS for indicative levels of overall pricing in regions	MATERIALS			LABOUR				RATES		
	Del to Site £	Waste %	Material Cost £	Craft Optve Hrs	Lab Hrs	Labour Cost £	Sunds £	Nett Rate £	Unit	Gross Rate (+12.5%) £
M60: PAINTING/CLEAR FINISHING; STAIRCASE AREAS: EMULSION PAINTING Cont.										
Mist coat, one full coat emulsion paint Cont.										
Paper covered general surfaces girth exceeding 300mm	0.13	10.00	**0.15**	0.22	-	**3.13**	0.06	**3.33**	m2	**3.75**
Mist coat, two full coats emulsion paint										
Concrete general surfaces girth exceeding 300mm	0.27	10.00	**0.29**	0.33	-	**4.69**	0.09	**5.07**	m2	**5.71**
Plaster general surfaces girth exceeding 300mm	0.22	10.00	**0.24**	0.31	-	**4.41**	0.08	**4.73**	m2	**5.32**
Plasterboard general surfaces girth exceeding 300mm	0.22	10.00	**0.24**	0.31	-	**4.41**	0.08	**4.73**	m2	**5.32**
Brickwork general surfaces girth exceeding 300mm	0.33	10.00	**0.37**	0.36	-	**5.12**	0.10	**5.58**	m2	**6.28**
Paper covered general surfaces girth exceeding 300mm	0.24	10.00	**0.27**	0.32	-	**4.55**	0.09	**4.90**	m2	**5.51**
M60: PAINTING/CLEAR FINISHING; STAIRCASE AREAS: CEMENT PAINTING										
One coat textured masonry paint										
Cement rendered general surfaces girth exceeding 300mm	0.32	10.00	**0.36**	0.15	-	**2.13**	0.04	**2.53**	m2	**2.84**
Concrete general surfaces girth exceeding 300mm	0.25	10.00	**0.27**	0.15	-	**2.13**	0.04	**2.44**	m2	**2.75**
Brickwork general surfaces girth exceeding 300mm	0.28	10.00	**0.31**	0.18	-	**2.56**	0.05	**2.92**	m2	**3.28**
Rough cast general surfaces girth exceeding 300mm	0.65	10.00	**0.71**	0.22	-	**3.13**	0.06	**3.90**	m2	**4.39**
Two coats textured masonry paint										
Cement rendered general surfaces girth exceeding 300mm	0.65	10.00	**0.71**	0.31	-	**4.41**	0.08	**5.20**	m2	**5.85**
Concrete general surfaces girth exceeding 300mm	0.49	10.00	**0.54**	0.31	-	**4.41**	0.08	**5.03**	m2	**5.66**
Brickwork general surfaces girth exceeding 300mm	0.57	10.00	**0.63**	0.33	-	**4.69**	0.09	**5.41**	m2	**6.08**
Rough cast general surfaces girth exceeding 300mm	1.30	10.00	**1.43**	0.40	-	**5.68**	0.11	**7.22**	m2	**8.12**
One coat stabilizing solution, one coat textured masonry paint										
Cement rendered general surfaces girth exceeding 300mm	0.65	10.00	**0.72**	0.26	-	**3.69**	0.07	**4.48**	m2	**5.04**
Concrete general surfaces girth exceeding 300mm	0.53	10.00	**0.59**	0.26	-	**3.69**	0.07	**4.35**	m2	**4.90**
Brickwork general surfaces girth exceeding 300mm	0.61	10.00	**0.67**	0.30	-	**4.26**	0.08	**5.02**	m2	**5.64**
Rough cast general surfaces girth exceeding 300mm	0.98	10.00	**1.07**	0.36	-	**5.12**	0.10	**6.29**	m2	**7.07**
One coat sealer, two coats waterproof cement paint; external work										
Cement rendered general surfaces girth exceeding 300mm	7.02	10.00	**7.72**	0.42	-	**5.97**	0.11	**13.80**	m2	**15.52**
Concrete general surfaces girth exceeding 300mm	5.54	10.00	**6.09**	0.42	-	**5.97**	0.11	**12.18**	m2	**13.70**
Brickwork general surfaces girth exceeding 300mm	3.66	10.00	**4.03**	0.45	-	**6.39**	0.12	**10.54**	m2	**11.86**
Rough cast general surfaces girth exceeding 300mm	7.02	10.00	**7.72**	0.54	-	**7.67**	0.15	**15.54**	m2	**17.48**
M60: PAINTING/CLEAR FINISHING; STAIRCASE AREAS: OIL PAINTING										
One coat primer, one undercoat, one coat full gloss finish										
Concrete general surfaces girth exceeding 300mm	0.93	10.00	**1.02**	0.43	-	**6.11**	0.12	**7.25**	m2	**8.16**

Labour hourly rates: (except Specialists) Craft Operatives £14.21 Labourer £10.71 Rates are national average prices. Refer to REGIONAL VARIATIONS for indicative levels of overall pricing in regions	MATERIALS			LABOUR			RATES			
	Del to Site £	Waste %	Material Cost £	Craft Optve Hrs	Lab Hrs	Labour Cost £	Sunds £	Nett Rate £	Unit	Gross Rate (+12.5%) £
60: PAINTING/CLEAR FINISHING; STAIRCASE AREAS: OIL AINTING Cont.										
ne coat primer, one undercoat, one coat full gloss inish Cont.										
rickwork general surfaces girth exceeding 300mm	1.15	10.00	**1.27**	0.47	-	**6.68**	0.13	**8.07**	m2	**9.08**
lasterboard general surfaces girth exceeding 300mm	0.76	10.00	**0.83**	0.40	-	**5.68**	0.11	**6.63**	m2	**7.45**
laster general surfaces girth exceeding 300mm	0.76	10.00	**0.83**	0.40	-	**5.68**	0.11	**6.63**	m2	**7.45**
ne coat primer, two undercoats, one coat full gloss inish										
oncrete general surfaces girth exceeding 300mm	1.13	10.00	**1.24**	0.57	-	**8.10**	0.15	**9.49**	m2	**10.68**
rickwork general surfaces girth exceeding 300mm	1.40	10.00	**1.53**	0.63	-	**8.95**	0.17	**10.66**	m2	**11.99**
lasterboard general surfaces girth exceeding 300mm	0.92	10.00	**1.01**	0.53	-	**7.53**	0.14	**8.68**	m2	**9.77**
laster general surfaces girth exceeding 300mm	0.92	10.00	**1.01**	0.53	-	**7.53**	0.14	**8.68**	m2	**9.77**
ne coat primer, one undercoat, one coat eggshell inish										
oncrete general surfaces girth exceeding 300mm	0.99	10.00	**1.09**	0.43	-	**6.11**	0.12	**7.31**	m2	**8.23**
rickwork general surfaces girth exceeding 300mm	1.22	10.00	**1.34**	0.47	-	**6.68**	0.13	**8.15**	m2	**9.17**
lasterboard general surfaces girth exceeding 300mm	0.80	10.00	**0.89**	0.40	-	**5.68**	0.11	**6.68**	m2	**7.51**
laster general surfaces girth exceeding 300mm	0.80	10.00	**0.89**	0.40	-	**5.68**	0.11	**6.68**	m2	**7.51**
ne coat primer, two undercoats, one coat eggshell inish										
oncrete general surfaces girth exceeding 300mm	1.18	10.00	**1.30**	0.57	-	**8.10**	0.15	**9.55**	m2	**10.75**
rickwork general surfaces girth exceeding 300mm	1.47	10.00	**1.61**	0.63	-	**8.95**	0.17	**10.74**	m2	**12.08**
lasterboard general surfaces girth exceeding 300mm	0.96	10.00	**1.06**	0.53	-	**7.53**	0.14	**8.73**	m2	**9.83**
laster general surfaces girth exceeding 300mm	0.96	10.00	**1.06**	0.53	-	**7.53**	0.14	**8.73**	m2	**9.83**
60: PAINTING/CLEAR FINISHING; STAIRCASE AREAS: LASTIC FINISH										
extured plastic coating - Stippled finish										
oncrete general surfaces girth exceeding 300mm	0.41	10.00	**0.45**	0.37	-	**5.26**	0.10	**5.81**	m2	**6.53**
rickwork general surfaces girth exceeding 300mm	0.49	10.00	**0.54**	0.48	-	**6.82**	0.13	**7.49**	m2	**8.43**
lasterboard general surfaces girth exceeding 300mm	0.33	10.00	**0.37**	0.29	-	**4.12**	0.08	**4.57**	m2	**5.14**
laster general surfaces girth exceeding 300mm	0.33	10.00	**0.37**	0.29	-	**4.12**	0.08	**4.57**	m2	**5.14**
extured plastic coating - Combed Finish										
oncrete general surfaces girth exceeding 300mm	0.41	10.00	**0.45**	0.43	-	**6.11**	0.12	**6.67**	m2	**7.51**
rickwork general surfaces girth exceeding 300mm	0.49	10.00	**0.54**	0.59	-	**8.38**	0.16	**9.08**	m2	**10.22**
lasterboard general surfaces girth exceeding 300mm	0.33	10.00	**0.37**	0.34	-	**4.83**	0.09	**5.29**	m2	**5.95**
laster general surfaces girth exceeding 300mm	0.41	10.00	**0.45**	0.34	-	**4.83**	0.09	**5.37**	m2	**6.04**

Labour hourly rates: (except Specialists)
Craft Operatives £14.21 Labourer £10.71
Rates are national average prices.
Refer to REGIONAL VARIATIONS for indicative levels of overall pricing in regions

	MATERIALS			LABOUR				RATES		
	Del to Site £	Waste %	Material Cost £	Craft Optve Hrs	Lab Hrs	Labour Cost £	Sunds £	Nett Rate £	Unit	Gross Rate (+12.5%) £
M60: PAINTING/CLEAR FINISHING; STAIRCASE AREAS - REDECORATIONS: EMULSION PAINTING										
Generally										
Note: the following rates include for the cost of all preparatory work, e.g. washing down, etc.										
Two coats emulsion paint, existing emulsion painted surfaces										
Concrete general surfaces girth exceeding 300mm	0.24	10.00	**0.26**	0.39	-	**5.54**	0.11	**5.91**	m2	**6.65**
Plaster general surfaces girth exceeding 300mm	0.20	10.00	**0.22**	0.35	-	**4.97**	0.09	**5.29**	m2	**5.95**
Plasterboard general surfaces girth exceeding 300mm	0.20	10.00	**0.22**	0.36	-	**5.12**	0.10	**5.43**	m2	**6.11**
Brickwork general surfaces girth exceeding 300mm	0.30	10.00	**0.33**	0.39	-	**5.54**	0.11	**5.97**	m2	**6.72**
Paper covered general surfaces girth exceeding 300mm	0.22	10.00	**0.24**	0.36	-	**5.12**	0.10	**5.46**	m2	**6.14**
Two full coats emulsion paint, existing washable distempered surfaces										
Concrete general surfaces girth exceeding 300mm	0.24	10.00	**0.26**	0.40	-	**5.68**	0.11	**6.05**	m2	**6.81**
Plaster general surfaces girth exceeding 300mm	0.20	10.00	**0.22**	0.36	-	**5.12**	0.10	**5.43**	m2	**6.11**
Plasterboard general surfaces girth exceeding 300mm	0.20	10.00	**0.22**	0.37	-	**5.26**	0.10	**5.58**	m2	**6.27**
Brickwork general surfaces girth exceeding 300mm	0.30	10.00	**0.33**	0.40	-	**5.68**	0.11	**6.12**	m2	**6.88**
Paper covered general surfaces girth exceeding 300mm	0.22	10.00	**0.24**	0.37	-	**5.26**	0.10	**5.60**	m2	**6.30**
Two full coats emulsion paint, existing textured plastic coated surfaces										
Concrete general surfaces girth exceeding 300mm	0.24	10.00	**0.26**	0.44	-	**6.25**	0.12	**6.63**	m2	**7.46**
Plaster general surfaces girth exceeding 300mm	0.20	10.00	**0.22**	0.39	-	**5.54**	0.11	**5.87**	m2	**6.60**
Plasterboard general surfaces girth exceeding 300mm	0.20	10.00	**0.22**	0.40	-	**5.68**	0.11	**6.01**	m2	**6.76**
Brickwork general surfaces girth exceeding 300mm	0.30	10.00	**0.33**	0.43	-	**6.11**	0.12	**6.55**	m2	**7.37**
Paper covered general surfaces girth exceeding 300mm	0.22	10.00	**0.24**	0.40	-	**5.68**	0.11	**6.04**	m2	**6.79**
Mist coat, two full coats emulsion paint, existing non-washable distempered surfaces										
Concrete general surfaces girth exceeding 300mm	0.27	10.00	**0.29**	0.49	-	**6.96**	0.13	**7.39**	m2	**8.31**
Plaster general surfaces girth exceeding 300mm	0.22	10.00	**0.24**	0.45	-	**6.39**	0.12	**6.76**	m2	**7.61**
Plasterboard general surfaces girth exceeding 300mm	0.22	10.00	**0.24**	0.46	-	**6.54**	0.12	**6.91**	m2	**7.77**
Brickwork general surfaces girth exceeding 300mm	0.33	10.00	**0.37**	0.49	-	**6.96**	0.13	**7.46**	m2	**8.39**
Paper covered general surfaces girth exceeding 300mm	0.24	10.00	**0.27**	0.46	-	**6.54**	0.12	**6.93**	m2	**7.80**
Two full coats emulsion paint, existing gloss painted surfaces										
Concrete general surfaces girth exceeding 300mm	0.24	10.00	**0.26**	0.48	-	**6.82**	0.13	**7.21**	m2	**8.11**
Plaster general surfaces girth exceeding 300mm	0.20	10.00	**0.22**	0.34	-	**4.83**	0.09	**5.14**	m2	**5.79**
Plasterboard general surfaces girth exceeding 300mm	0.20	10.00	**0.22**	0.35	-	**4.97**	0.09	**5.29**	m2	**5.95**
Brickwork general surfaces girth exceeding 300mm	0.30	10.00	**0.33**	0.38	-	**5.40**	0.10	**5.83**	m2	**6.56**
Paper covered general surfaces girth exceeding 300mm	0.22	10.00	**0.24**	0.35	-	**4.97**	0.09	**5.31**	m2	**5.98**

Labour hourly rates: (except Specialists) Craft Operatives £14.21 Labourer £10.71 Rates are national average prices. Refer to REGIONAL VARIATIONS for indicative levels of overall pricing in regions	MATERIALS			LABOUR				RATES		
	Del to Site £	Waste %	Material Cost £	Craft Optve Hrs	Lab Hrs	Labour Cost £	Sunds £	Nett Rate £	Unit	Gross Rate (+12.5%) £
M60: PAINTING/CLEAR FINISHING; STAIRCASE AREAS - REDECORATIONS: OIL PAINTING										
Generally										
Note: the following rates include for the cost of all preparatory work, e.g. washing down, etc.										
One undercoat, one coat full gloss finish, existing gloss painted surfaces										
Concrete general surfaces girth exceeding 300mm	0.33	10.00	**0.36**	0.44	-	**6.25**	0.12	**6.74**	m2	**7.58**
Brickwork general surfaces girth exceeding 300mm	0.36	10.00	**0.39**	0.45	-	**6.39**	0.12	**6.91**	m2	**7.77**
Plasterboard general surfaces girth exceeding 300mm	0.31	10.00	**0.34**	0.41	-	**5.83**	0.11	**6.27**	m2	**7.06**
Plaster general surfaces girth exceeding 300mm	0.31	10.00	**0.34**	0.40	-	**5.68**	0.11	**6.13**	m2	**6.89**
One undercoat, one coat eggshell finish, existing gloss painted surfaces										
Concrete general surfaces girth exceeding 300mm	0.38	10.00	**0.41**	0.44	-	**6.25**	0.12	**6.79**	m2	**7.63**
Brickwork general surfaces girth exceeding 300mm	0.47	10.00	**0.52**	0.45	-	**6.39**	0.12	**7.04**	m2	**7.92**
Plasterboard general surfaces girth exceeding 300mm	0.35	10.00	**0.38**	0.41	-	**5.83**	0.11	**6.32**	m2	**7.11**
Plaster general surfaces girth exceeding 300mm	0.35	10.00	**0.38**	0.40	-	**5.68**	0.11	**6.18**	m2	**6.95**
Two undercoats, one coat full gloss finish, existing gloss painted surfaces										
Concrete general surfaces girth exceeding 300mm	0.49	10.00	**0.54**	0.57	-	**8.10**	0.15	**8.79**	m2	**9.89**
Brickwork general surfaces girth exceeding 300mm	0.59	10.00	**0.65**	0.57	-	**8.10**	0.15	**8.90**	m2	**10.01**
Plasterboard general surfaces girth exceeding 300mm	0.47	10.00	**0.51**	0.54	-	**7.67**	0.15	**8.33**	m2	**9.37**
Plaster general surfaces girth exceeding 300mm	0.47	10.00	**0.51**	0.53	-	**7.53**	0.14	**8.19**	m2	**9.21**
Two undercoats, one coat eggshell finish, existing gloss painted surfaces										
Concrete general surfaces girth exceeding 300mm	0.54	10.00	**0.59**	0.57	-	**8.10**	0.15	**8.84**	m2	**9.95**
Brickwork general surfaces girth exceeding 300mm	0.64	10.00	**0.71**	0.57	-	**8.10**	0.15	**8.96**	m2	**10.08**
Plasterboard general surfaces girth exceeding 300mm	0.51	10.00	**0.56**	0.54	-	**7.67**	0.15	**8.38**	m2	**9.43**
Plaster general surfaces girth exceeding 300mm	0.51	10.00	**0.56**	0.53	-	**7.53**	0.14	**8.23**	m2	**9.26**
One coat primer, one undercoat, one coat full gloss finish, existing washable distempered or emulsion painted surfaces										
Concrete general surfaces girth exceeding 300mm	0.93	10.00	**1.03**	0.57	-	**8.10**	0.15	**9.28**	m2	**10.44**
Brickwork general surfaces girth exceeding 300mm	1.10	10.00	**1.21**	0.57	-	**8.10**	0.15	**9.46**	m2	**10.64**
Plasterboard general surfaces girth exceeding 300mm	0.69	10.00	**0.76**	0.54	-	**7.67**	0.18	**8.61**	m2	**9.68**
Plaster general surfaces girth exceeding 300mm	0.80	10.00	**0.88**	0.53	-	**7.53**	0.14	**8.55**	m2	**9.62**
One coat primer, one undercoat, one coat eggshell finish, existing washable distempered or emulsion painted surfaces										
Concrete general surfaces girth exceeding 300mm	0.98	10.00	**1.08**	0.57	-	**8.10**	0.15	**9.33**	m2	**10.50**
Brickwork general surfaces girth exceeding 300mm	1.21	10.00	**1.34**	0.57	-	**8.10**	0.15	**9.59**	m2	**10.79**
Plasterboard general surfaces girth exceeding 300mm	0.90	10.00	**0.99**	0.54	-	**7.67**	0.15	**8.81**	m2	**9.91**

Labour hourly rates: (except Specialists) Craft Operatives £14.21 Labourer £10.71 Rates are national average prices. Refer to REGIONAL VARIATIONS for indicative levels of overall pricing in regions	MATERIALS			LABOUR				RATES		
	Del to Site	Waste	Material Cost	Craft Optve	Lab	Labour Cost	Sunds	Nett Rate	Unit	Gross Rate (+12.5%)
	£	%	£	Hrs	Hrs	£	£	£		£
M60: PAINTING/CLEAR FINISHING; STAIRCASE AREAS - REDECORATIONS: OIL PAINTING Cont.										
One coat primer, one undercoat, one coat eggshell finish, existing washable distempered or emulsion painted surfaces Cont.										
Plaster general surfaces girth exceeding 300mm	0.90	10.00	0.99	0.53	-	7.53	0.14	8.66	m2	9.74
One coat primer, two undercoats, one coat full gloss finish, existing washable distempered or emulsion painted surfaces										
Concrete general surfaces girth exceeding 300mm	1.09	10.00	1.20	0.70	-	9.95	0.19	11.34	m2	12.76
Brickwork general surfaces girth exceeding 300mm	1.33	10.00	1.46	0.71	-	10.09	0.19	11.74	m2	13.21
Plasterboard general surfaces girth exceeding 300mm	1.11	10.00	1.22	0.67	-	9.52	0.18	10.93	m2	12.29
Plaster general surfaces girth exceeding 300mm	1.01	10.00	1.12	0.66	-	9.38	0.18	10.67	m2	12.01
One coat primer, two undercoats, one coat eggshell finish, existing washable distempered or emulsion painted surfaces										
Concrete general surfaces girth exceeding 300mm	1.14	10.00	1.25	0.70	-	9.95	0.19	11.39	m2	12.81
Brickwork general surfaces girth exceeding 300mm	1.39	10.00	1.52	0.71	-	10.09	0.19	11.81	m2	13.28
Plasterboard general surfaces girth exceeding 300mm	1.06	10.00	1.16	0.67	-	9.52	0.18	10.86	m2	12.22
Plaster general surfaces girth exceeding 300mm	1.06	10.00	1.16	0.66	-	9.38	0.18	10.72	m2	12.06
One coat primer, one undercoat, one coat full gloss finish, existing non-washable distempered surfaces										
Concrete general surfaces girth exceeding 300mm	0.93	10.00	1.03	0.57	-	8.10	0.15	9.28	m2	10.44
Brickwork general surfaces girth exceeding 300mm	1.10	10.00	1.21	0.57	-	8.10	0.15	9.46	m2	10.64
Plasterboard general surfaces girth exceeding 300mm	0.86	10.00	0.94	0.54	-	7.67	0.15	8.76	m2	9.85
Plaster general surfaces girth exceeding 300mm	0.86	10.00	0.94	0.53	-	7.53	0.14	8.62	m2	9.69
One coat primer, one undercoat, one coat eggshell finish, existing non-washable distempered surfaces										
Concrete general surfaces girth exceeding 300mm	0.98	10.00	1.08	0.57	-	8.10	0.15	9.33	m2	10.50
Brickwork general surfaces girth exceeding 300mm	1.21	10.00	1.34	0.57	-	8.10	0.15	9.59	m2	10.79
Plasterboard general surfaces girth exceeding 300mm	0.90	10.00	0.99	0.54	-	7.67	0.15	8.81	m2	9.91
Plaster general surfaces girth exceeding 300mm	0.90	10.00	0.99	0.53	-	7.53	0.14	8.66	m2	9.74
One coat primer, two undercoats, one coat full gloss finish, existing non-washable distempered surfaces										
Concrete general surfaces girth exceeding 300mm	1.09	10.00	1.20	0.70	-	9.95	0.19	11.34	m2	12.76
Brickwork general surfaces girth exceeding 300mm	1.33	10.00	1.46	0.71	-	10.09	0.19	11.74	m2	13.21
Plasterboard general surfaces girth exceeding 300mm	1.11	10.00	1.22	0.67	-	9.52	0.18	10.93	m2	12.29
Plaster general surfaces girth exceeding 300mm	1.01	10.00	1.12	0.66	-	9.38	0.18	10.67	m2	12.01
One coat primer, two undercoats, one coat eggshell finish, existing non-washable distempered surfaces										
Concrete general surfaces girth exceeding 300mm	1.14	10.00	1.25	0.70	-	9.95	0.19	11.39	m2	12.81
Brickwork general surfaces girth exceeding 300mm	1.39	10.00	1.52	0.71	-	10.09	0.19	11.81	m2	13.28

Labour hourly rates: (except Specialists) Craft Operatives £14.21 Labourer £10.71 Rates are national average prices. Refer to REGIONAL VARIATIONS for indicative levels of overall pricing in regions	MATERIALS			LABOUR				RATES		
	Del to Site £	Waste %	Material Cost £	Craft Optve Hrs	Lab Hrs	Labour Cost £	Sunds £	Nett Rate £	Unit	Gross Rate (+12.5%) £
M60: PAINTING/CLEAR FINISHING; STAIRCASE AREAS - REDECORATIONS: OIL PAINTING Cont.										
One coat primer, two undercoats, one coat eggshell finish, existing non-washable distempered surfaces Cont.										
Plasterboard general surfaces girth exceeding 300mm	1.06	10.00	**1.16**	0.67	-	**9.52**	0.18	**10.86**	m2	**12.22**
Plaster general surfaces girth exceeding 300mm	1.06	10.00	**1.16**	0.66	-	**9.38**	0.18	**10.72**	m2	**12.06**
M60: PAINTING/CLEAR FINISHING; STAIRCASE AREAS - REDECORATIONS: PLASTIC FINISH										
Generally										
Note: the following rates include for the cost of all preparatory work, e.g. washing down, etc.										
Textured plastic coating - stippled finish, existing washable distempered or emulsion painted surfaces										
Concrete general surfaces girth exceeding 300mm	0.35	10.00	**0.38**	0.50	-	**7.11**	0.14	**7.62**	m2	**8.57**
Brickwork general surfaces girth exceeding 300mm	0.41	10.00	**0.46**	0.55	-	**7.82**	0.15	**8.42**	m2	**9.47**
Plasterboard general surfaces girth exceeding 300mm	0.31	10.00	**0.34**	0.43	-	**6.11**	0.12	**6.57**	m2	**7.39**
Plaster general surfaces girth exceeding 300mm	0.31	10.00	**0.34**	0.42	-	**5.97**	0.11	**6.42**	m2	**7.23**
Textured plastic coating - combed finish, existing washable distempered or emulsion painted surfaces										
Concrete general surfaces girth exceeding 300mm	0.41	10.00	**0.46**	0.55	-	**7.82**	0.15	**8.42**	m2	**9.47**
Brickwork general surfaces girth exceeding 300mm	0.50	10.00	**0.55**	0.60	-	**8.53**	0.16	**9.24**	m2	**10.39**
Plasterboard general surfaces girth exceeding 300mm	0.37	10.00	**0.41**	0.48	-	**6.82**	0.13	**7.36**	m2	**8.28**
Plaster general surfaces girth exceeding 300mm	0.37	10.00	**0.41**	0.47	-	**6.68**	0.13	**7.22**	m2	**8.12**
Textured plastic coating - stippled finish, existing non-washable distempered surfaces										
Concrete general surfaces girth exceeding 300mm	0.35	10.00	**0.38**	0.55	-	**7.82**	0.15	**8.35**	m2	**9.39**
Brickwork general surfaces girth exceeding 300mm	0.41	10.00	**0.46**	0.60	-	**8.53**	0.16	**9.14**	m2	**10.29**
Plasterboard general surfaces girth exceeding 300mm	0.31	10.00	**0.34**	0.48	-	**6.82**	0.13	**7.29**	m2	**8.20**
Plaster general surfaces girth exceeding 300mm	0.31	10.00	**0.34**	0.47	-	**6.68**	0.13	**7.15**	m2	**8.04**
Textured plastic coating - combed finish, existing non-washable distempered surfaces										
Concrete general surfaces girth exceeding 300mm	0.41	10.00	**0.46**	0.60	-	**8.53**	0.16	**9.14**	m2	**10.29**
Brickwork general surfaces girth exceeding 300mm	0.50	10.00	**0.55**	0.65	-	**9.24**	0.18	**9.96**	m2	**11.20**
Plasterboard general surfaces girth exceeding 300mm	0.37	10.00	**0.41**	0.53	-	**7.53**	0.14	**8.08**	m2	**9.10**
Plaster general surfaces girth exceeding 300mm	0.37	10.00	**0.41**	0.52	-	**7.39**	0.14	**7.94**	m2	**8.93**

FURNITURE/EQUIPMENT

Labour hourly rates: (except Specialists) Craft Operatives £14.21 Labourer £10.71 Rates are national average prices. Refer to REGIONAL VARIATIONS for indicative levels of overall pricing in regions	MATERIALS			LABOUR				RATES		
	Del to Site £	Waste %	Material Cost £	Craft Optve Hrs	Lab Hrs	Labour Cost £	Sunds £	Nett Rate £	Unit	Gross Rate (+12.5%) £
N10: GENERAL FIXTURES/FURNISHING/EQUIPMENT: MIRRORS										
BS 952, clear float, SG, silvered and protected with copper backing										
6mm thick; fixing to masonry with brass screws, chromium plated dome covers, rubber sleeves and washers										
holes 6mm diameter -4; edges polished										
254 x 400mm	27.85	5.00	**29.24**	0.50	-	**7.11**	0.45	**36.80**	nr	**41.40**
300 x 460mm	32.33	5.00	**33.95**	0.50	-	**7.11**	0.45	**41.50**	nr	**46.69**
360 x 500mm	37.24	5.00	**39.10**	0.50	-	**7.11**	0.45	**46.66**	nr	**52.49**
460 x 560mm	46.06	5.00	**48.36**	0.65	-	**9.24**	0.45	**58.05**	nr	**65.31**
460 x 600mm	48.17	5.00	**50.58**	0.65	-	**9.24**	0.45	**60.27**	nr	**67.80**
460 x 900mm	64.02	5.00	**67.22**	0.70	-	**9.95**	0.45	**77.62**	nr	**87.32**
500 x 680mm	55.30	5.00	**58.07**	0.65	-	**9.24**	0.45	**67.75**	nr	**76.22**
600 x 900mm	76.98	5.00	**80.83**	0.75	-	**10.66**	0.45	**91.94**	nr	**103.43**
holes 6mm diameter -4; edges bevelled										
254 x 400mm	35.23	5.00	**36.99**	0.50	-	**7.11**	0.45	**44.55**	nr	**50.12**
300 x 460mm	40.91	5.00	**42.96**	0.50	-	**7.11**	0.45	**50.51**	nr	**56.83**
360 x 500mm	46.96	5.00	**49.31**	0.50	-	**7.11**	0.45	**56.86**	nr	**63.97**
460 x 560mm	57.57	5.00	**60.45**	0.65	-	**9.24**	0.45	**70.14**	nr	**78.90**
460 x 600mm	60.13	5.00	**63.14**	0.65	-	**9.24**	0.45	**72.82**	nr	**81.93**
460 x 900mm	79.37	5.00	**83.34**	0.70	-	**9.95**	0.45	**93.74**	nr	**105.45**
500 x 680mm	68.63	5.00	**72.06**	0.65	-	**9.24**	0.45	**81.75**	nr	**91.97**
600 x 900mm	93.91	5.00	**98.61**	0.75	-	**10.66**	0.45	**109.71**	nr	**123.43**
N10: GENERAL FIXTURES/FURNISHING/EQUIPMENT: CURTAIN TRACKS										
Fixing only curtain track										
Metal or plastic track with fittings										
fixing with screws to softwood	-	-	**-**	0.50	-	**7.11**	-	**7.11**	m	**7.99**
fixing with screws to hardwood	-	-	**-**	0.70	-	**9.95**	-	**9.95**	m	**11.19**
N10: GENERAL FIXTURES/FURNISHING/EQUIPMENT: BLINDS										
Internal blinds										
Venetian blinds; stove enamelled aluminium alloy slats 25mm wide; plain colours										
1200mm drop; fixing to timber with screws										
1000mm wide	32.51	2.50	**33.33**	1.10	0.14	**17.13**	0.06	**50.52**	nr	**56.83**
2000mm wide	49.88	2.50	**51.12**	1.65	0.21	**25.70**	0.09	**76.91**	nr	**86.52**
3000mm wide	67.21	2.50	**68.89**	2.20	0.28	**34.26**	0.12	**103.27**	nr	**116.18**
Venetian blinds; stove enamelled aluminium alloy slats 50mm wide; plain colours										
1200mm drop; fixing to timber with screws										
1000mm wide	29.98	2.50	**30.73**	1.27	0.16	**19.76**	0.07	**50.56**	nr	**56.88**
2000mm wide	46.25	2.50	**47.40**	1.90	0.24	**29.57**	0.10	**77.08**	nr	**86.71**
3000mm wide	62.88	2.50	**64.46**	2.50	0.31	**38.85**	0.14	**103.44**	nr	**116.37**
Venetian blinds; stove enamelled aluminium alloy slats 25mm wide; plain colours with single cord control										
1200mm drop; fixing to timber with screws										
1000mm wide	46.25	2.50	**47.40**	1.10	0.14	**17.13**	0.06	**64.59**	nr	**72.67**
2000mm wide	72.26	2.50	**74.07**	1.65	0.21	**25.70**	0.09	**99.85**	nr	**112.34**
3000mm wide	97.93	2.50	**100.38**	2.20	0.28	**34.26**	0.12	**134.76**	nr	**151.60**
Venetian blinds; stove enamelled aluminium alloy slats 50mm wide; plain colours with single cord control										
1200mm drop; fixing to timber with screws										
1000mm wide	41.91	2.50	**42.96**	1.27	0.16	**19.76**	0.07	**62.79**	nr	**70.64**
2000mm wide	63.59	2.50	**65.18**	1.90	0.24	**29.57**	0.10	**94.85**	nr	**106.71**
3000mm wide	87.44	2.50	**89.63**	2.50	0.31	**38.85**	0.14	**128.61**	nr	**144.69**

Labour hourly rates: (except Specialists) Craft Operatives £14.21 Labourer £10.71 Rates are national average prices. Refer to REGIONAL VARIATIONS for indicative levels of overall pricing in regions	MATERIALS			LABOUR				RATES		
	Del to Site £	Waste %	Material Cost £	Craft Optve Hrs	Lab Hrs	Labour Cost £	Sunds £	Nett Rate £	Unit	Gross Rate (+12.5%) £
N10: GENERAL FIXTURES/FURNISHING/EQUIPMENT: BLINDS Cont.										
Internal blinds Cont.										
Venetian blinds; stove enamelled aluminium alloy slats 25mm wide; plain colours with channels for dimout										
1200mm drop; fixing to timber with screws										
1000mm wide	78.77	2.50	**80.74**	2.20	0.28	**34.26**	0.12	**115.12**	nr	**129.51**
2000mm wide	96.83	2.50	**99.25**	2.75	0.34	**42.72**	0.15	**142.12**	nr	**159.88**
3000mm wide	114.17	2.50	**117.03**	3.30	0.41	**51.28**	0.18	**168.49**	nr	**189.55**
Venetian blinds; stove enamelled aluminium alloy slats 50mm wide; plain colours with channels for dimout										
1200mm drop; fixing to timber with screws										
1000mm wide	75.87	2.50	**77.77**	2.50	0.31	**38.85**	0.14	**116.75**	nr	**131.34**
2000mm wide	93.22	2.50	**95.55**	3.10	0.39	**48.23**	0.17	**143.95**	nr	**161.94**
3000mm wide	109.84	2.50	**112.58**	3.70	0.46	**57.50**	0.20	**170.29**	nr	**191.57**
Roller blinds; automatic ratchet action; fire resistant material										
1200mm drop; fixing to timber with screws										
1000mm wide	60.70	2.50	**62.22**	1.10	0.14	**17.13**	0.06	**79.41**	nr	**89.33**
2000mm wide	74.44	2.50	**76.30**	1.65	0.21	**25.70**	0.09	**102.08**	nr	**114.84**
3000mm wide	88.15	2.50	**90.35**	2.20	0.28	**34.26**	0.12	**124.73**	nr	**140.33**
Roller blinds; automatic ratchet action; holland type material										
1200mm drop; fixing to timber with screws										
1000mm wide	58.54	2.50	**60.00**	1.10	0.14	**17.13**	0.06	**77.19**	nr	**86.84**
2000mm wide	70.10	2.50	**71.85**	1.65	0.21	**25.70**	0.09	**97.63**	nr	**109.84**
3000mm wide	81.62	2.50	**83.66**	2.20	0.28	**34.26**	0.12	**118.04**	nr	**132.80**
Roller blinds; self acting roller; blackout material										
1200mm drop; fixing to timber with screws										
1000mm wide	54.20	2.50	**55.56**	1.27	0.16	**19.76**	0.07	**75.39**	nr	**84.81**
2000mm wide	65.76	2.50	**67.41**	1.90	0.24	**29.57**	0.10	**97.08**	nr	**109.21**
3000mm wide	77.33	2.50	**79.26**	2.50	0.31	**38.85**	0.14	**118.24**	nr	**133.02**
Roller blinds; 100% blackout; natural anodised box and channels										
1200mm drop; fixing to timber with screws										
1000mm wide	179.22	2.50	**183.70**	2.20	0.28	**34.26**	0.12	**218.08**	nr	**245.34**
2000mm wide	223.25	2.50	**228.83**	3.05	0.31	**46.66**	0.16	**275.65**	nr	**310.10**
3000mm wide	273.15	2.50	**279.98**	3.85	0.48	**59.85**	0.21	**340.04**	nr	**382.54**
Vertical louvre blinds; 89mm wide louvres in standard material										
1200mm drop; fixing to timber with screws										
1000mm wide	65.04	2.50	**66.66**	1.00	0.13	**15.60**	0.05	**82.32**	nr	**92.61**
2000mm wide	114.17	2.50	**117.03**	1.50	0.19	**23.35**	0.08	**140.46**	nr	**158.02**
3000mm wide	163.32	2.50	**167.41**	2.00	0.25	**31.10**	0.11	**198.61**	nr	**223.44**
Vertical louvre blinds; 127mm wide louvres in standard material										
1200mm drop; fixing to timber with screws										
1000mm wide	58.54	2.50	**60.00**	1.05	0.13	**16.31**	0.06	**76.37**	nr	**85.92**
2000mm wide	101.90	2.50	**104.44**	1.55	0.19	**24.06**	0.08	**128.59**	nr	**144.66**
3000mm wide	145.97	2.50	**149.62**	2.03	0.25	**31.52**	0.11	**181.25**	nr	**203.91**
External blinds; manually operated										
Venetian blinds; stove enamelled aluminium slats 80mm wide; natural anodised side guides; excluding boxing										
1200mm drop; fixing to timber with screws										
1000mm wide	-	-	**-**	-	-	**-**	-	**266.96**	nr	**300.33**
2000mm wide	-	-	**-**	-	-	**-**	-	**397.32**	nr	**446.99**
3000mm wide	-	-	**-**	-	-	**-**	-	**535.15**	nr	**602.04**
Queensland awnings; acrylic material; natural anodised arms and boxing										
1000mm projection										
2000mm long	-	-	**-**	-	-	**-**	-	**625.78**	nr	**704.00**
3000mm long	-	-	**-**	-	-	**-**	-	**938.09**	nr	**1055.35**
4000mm long	-	-	**-**	-	-	**-**	-	**1251.56**	nr	**1408.01**
Rollscreen vertical drop roller blinds; mesh material; natural anodised side guides and boxing										
1200mm drop; fixing to timber with screws										
1000mm wide	-	-	**-**	-	-	**-**	-	**260.71**	nr	**293.30**
2000mm wide	-	-	**-**	-	-	**-**	-	**391.12**	nr	**440.01**
3000mm wide	-	-	**-**	-	-	**-**	-	**521.26**	nr	**586.42**
Quadrant canopy; acrylic material; natural anodised frames										
1000mm projection										
2000mm long	-	-	**-**	-	-	**-**	-	**345.17**	nr	**388.32**
3000mm long	-	-	**-**	-	-	**-**	-	**527.69**	nr	**593.65**
4000mm long	-	-	**-**	-	-	**-**	-	**717.66**	nr	**807.37**
Foldaway awning; acrylic material; natural anodised arms and front rail										
2000mm projection										
2000mm long	-	-	**-**	-	-	**-**	-	**559.98**	nr	**629.98**
3000mm long	-	-	**-**	-	-	**-**	-	**821.95**	nr	**924.69**
4000mm long	-	-	**-**	-	-	**-**	-	**1069.05**	nr	**1202.68**

Labour hourly rates: (except Specialists) Craft Operatives £14.21 Labourer £10.71 Rates are national average prices. Refer to REGIONAL VARIATIONS for indicative levels of overall pricing in regions	MATERIALS			LABOUR				RATES		
	Del to Site £	Waste %	Material Cost £	Craft Optve Hrs	Lab Hrs	Labour Cost £	Sunds £	Nett Rate £	Unit	Gross Rate (+12.5%) £
N10: GENERAL FIXTURES/FURNISHING/EQUIPMENT: BLINDS Cont.										
External blinds; electrically operated										
Venetian blinds; stove enamelled aluminium slats 80mm wide; natural anodised side guides; excluding boxing										
1200mm drop; fixing to timber with screws										
1000mm wide	-	-	-	-	-	-	-	613.36	nr	690.03
2000mm wide	-	-	-	-	-	-	-	749.95	nr	843.69
3000mm wide	-	-	-	-	-	-	-	879.08	nr	988.97
Queensland awnings; acrylic material; natural anodised arms and boxing										
1000mm projection										
2000mm long	-	-	-	-	-	-	-	717.66	nr	807.37
3000mm long	-	-	-	-	-	-	-	1024.34	nr	1152.38
4000mm long	-	-	-	-	-	-	-	1303.71	nr	1466.67
Rollscreen vertical drop roller blinds; mesh material; natural anodised side guides and boxing										
1200mm drop; fixing to timber with screws										
1000mm wide	-	-	-	-	-	-	-	496.65	nr	558.73
2000mm wide	-	-	-	-	-	-	-	625.78	nr	704.00
3000mm wide	-	-	-	-	-	-	-	756.15	nr	850.67
Quadrant canopy; acrylic material; natural anodised frames										
1000mm projection										
2000mm long	-	-	-	-	-	-	-	449.47	nr	505.65
3000mm long	-	-	-	-	-	-	-	625.78	nr	704.00
4000mm long	-	-	-	-	-	-	-	834.37	nr	938.67
Foldaway awning; acrylic material; natural anodised arms and front rail										
2000mm projection										
2000mm long	-	-	-	-	-	-	-	820.72	nr	923.31
3000mm long	-	-	-	-	-	-	-	1075.25	nr	1209.66
4000mm long	-	-	-	-	-	-	-	1365.79	nr	1536.51
N10: GENERAL FIXTURES/FURNISHING/EQUIPMENT: STORAGE SYSTEMS										
Vista two tone coloured steel boltless storage systems; Welconstruct Co. Ltd.; units 900mm wide x 1800mm high; assembling										
Open shelving with top, bottom, three shelves and braces; placing in position										
300mm deep code L106-101	132.62	2.50	135.93	1.20	0.60	23.48	-	159.41	nr	179.33
450mm deep code L106-102	166.95	2.50	171.12	1.40	0.70	27.39	-	198.51	nr	223.33
600mm deep code L106-103	192.68	2.50	197.49	1.60	0.80	31.30	-	228.80	nr	257.40
Closed shelving with top, bottom, and three shelves; placing in position										
300mm deep code L106-111	186.90	2.50	191.57	1.30	0.65	25.43	-	217.01	nr	244.13
extra for additional shelf code L106-301	13.02	2.50	13.35	0.22	0.11	4.30	-	17.65	nr	19.86
450mm deep code L106-112	224.70	2.50	230.32	1.50	0.75	29.35	-	259.67	nr	292.12
extra for additional shelf code L106-302	19.85	2.50	20.34	0.22	0.11	4.30	-	24.65	nr	27.73
600mm deep code L106-113	253.05	2.50	259.38	1.70	0.85	33.26	-	292.64	nr	329.22
extra for additional shelf code L106-303	24.78	2.50	25.40	0.22	0.11	4.30	-	29.70	nr	33.42
Extra over unit for										
pair of lockable doors code L106-401	318.15	2.50	326.10	0.22	0.11	4.30	-	330.41	nr	371.71
N10: GENERAL FIXTURES/FURNISHING/EQUIPMENT: SHELVING SYSTEMS										
Spanwel wide access boltless steel shelving systems; Welconstruct Co Ltd.; upright frames connected with shelf beams and with shelves; assembling										
Unit 1800mm high with 2 upright frames and three standard duty steel shelves including top and bottom; placing in position										
1200mm long x 650mm deep code L108-201	283.50	2.50	290.59	1.60	0.80	31.30	-	321.89	nr	362.13
extra; additional standard duty steel shelf and shelf beam code L108-225	61.85	2.50	63.39	0.24	0.12	4.70	-	68.09	nr	76.60
1200mm long x 800mm deep code L108-202	299.25	2.50	306.73	1.70	0.85	33.26	-	339.99	nr	382.49
extra; additional standard duty steel shelf and shelf beam L108-226	65.42	2.50	67.05	0.24	0.12	4.70	-	71.75	nr	80.71
1200mm long x 950mm deep code L108-203	313.95	2.50	321.80	1.80	0.90	35.22	-	357.02	nr	401.64
extra; additional standard duty steel shelf and shelf beam code L108-227	69.41	2.50	71.14	0.24	0.12	4.70	-	75.84	nr	85.32
2400mm long x 650mm deep code L108-204	405.30	2.50	415.43	1.70	0.85	33.26	-	448.69	nr	504.78
extra; additional standard duty steel shelf and shelf beam code L108-228	102.27	2.50	104.83	0.24	0.12	4.70	-	109.52	nr	123.21
2400mm long x 800mm deep code L108-205	430.50	2.50	441.26	1.80	0.90	35.22	-	476.48	nr	536.04
extra; additional standard duty steel shelf and shelf beam code L108-229	109.10	2.50	111.82	0.24	0.12	4.70	-	116.52	nr	131.08
2400mm long x 950mm deep code L108-206	456.75	2.50	468.17	1.90	0.95	37.17	-	505.34	nr	568.51
extra; additional standard duty steel shelf and shelf beam code L108-230	116.87	2.50	119.79	0.24	0.12	4.70	-	124.48	nr	140.04

Labour hourly rates: (except Specialists) Craft Operatives £14.21 Labourer £10.71 Rates are national average prices. Refer to REGIONAL VARIATIONS for indicative levels of overall pricing in regions	MATERIALS			LABOUR				RATES		
	Del to Site £	Waste %	Material Cost £	Craft Optve Hrs	Lab Hrs	Labour Cost £	Sunds £	Nett Rate £	Unit	Gross Rate (+12.5%) £
N10: GENERAL FIXTURES/FURNISHING/EQUIPMENT: SHELVING SYSTEMS Cont.										
Spanwel wide access boltless steel shelving systems; Welconstruct Co Ltd.; upright frames connected with shelf beams and with shelves; assembling Cont.										
Unit 2400mm high with 3 upright frames and five standard duty steel shelves including top and bottom; placing in position										
2400mm long x 650mm deep	703.50	2.50	**721.09**	3.20	1.60	**62.61**	-	**783.70**	nr	**881.66**
2400mm long x 800mm deep	742.35	2.50	**760.91**	3.45	1.75	**67.77**	-	**828.68**	nr	**932.26**
2400mm long x 950mm deep	799.05	2.50	**819.03**	3.70	1.85	**72.39**	-	**891.42**	nr	**1002.84**
N10: GENERAL FIXTURES/FURNISHING/EQUIPMENT: SHELVING SYSTEMS - SHELVING SUPPORT SYSTEMS										
Spur patent shelf supports in steel with white enamelled finish; Spur Shelving Ltd; Spur Steel-Lok components										
Wall uprights										
type 9011, 430mm long; fixing to masonry with screws	1.65	2.50	**1.69**	0.25	-	**3.55**	-	**5.24**	nr	**5.90**
type 9013, 1000mm long; fixing to masonry with screws	3.25	2.50	**3.33**	0.45	-	**6.39**	-	**9.73**	nr	**10.94**
type 9016, 1600mm long; fixing to masonry with screws	4.95	2.50	**5.07**	0.55	-	**7.82**	-	**12.89**	nr	**14.50**
type 9020, 2400mm long; fixing to masonry with screws	6.95	2.50	**7.12**	0.70	-	**9.95**	-	**17.07**	nr	**19.20**
Straight brackets										
type 9001, 120mm long	0.95	2.50	**0.97**	0.05	-	**0.71**	-	**1.68**	nr	**1.89**
type 9003, 270mm long	1.40	2.50	**1.44**	0.05	-	**0.71**	-	**2.15**	nr	**2.41**
type 9004, 360mm long	1.95	2.50	**2.00**	0.07	-	**0.99**	-	**2.99**	nr	**3.37**
type 9006, 470mm long	3.25	2.50	**3.33**	0.07	-	**0.99**	-	**4.33**	nr	**4.87**
N10: GENERAL FIXTURES/FURNISHING/EQUIPMENT: MAT RIMS										
Matwells in Aluminium										
Mat frames; welded fabrication										
34 x 26 x 6mm angle section; angles mitred; plain lugs -4, welded on; mat space										
610 x 457mm	38.39	2.50	**39.35**	1.10	1.10	**27.41**	-	**66.77**	nr	**75.11**
762 x 457mm	44.01	2.50	**45.11**	1.40	1.40	**34.89**	-	**80.00**	nr	**90.00**
914 x 610mm	55.08	2.50	**56.46**	1.70	1.70	**42.36**	-	**98.82**	nr	**111.17**
Matwells in polished brass										
Mat frames; brazed fabrication										
38 x 38 x 6mm angle section; angles mitred; plain lugs -4, welded on; mat space										
610 x 457mm	118.80	2.50	**121.77**	1.40	1.40	**34.89**	-	**156.66**	nr	**176.24**
762 x 457mm	140.40	2.50	**143.91**	1.70	1.70	**42.36**	-	**186.27**	nr	**209.56**
914 x 610mm	183.60	2.50	**188.19**	1.90	1.90	**47.35**	-	**235.54**	nr	**264.98**
N10: GENERAL FIXTURES/FURNISHING/EQUIPMENT: CLOTHES LOCKERS										
Individual compartment clothes lockers; Welconstruct Co. Ltd.; steel, stove enamelled finish; standard colour solid doors and with cam locks 442 series. 12-19 ordered										
2 compartment; placing in position										
305 x 305 x 1830mm code L442-021	74.03	2.50	**75.88**	-	0.60	**6.43**	-	**82.30**	nr	**92.59**
305 x 460 x 1830mm code L442-022	80.54	2.50	**82.55**	-	0.65	**6.96**	-	**89.51**	nr	**100.70**
380 x 380 x 1830mm code L442-024	82.74	2.50	**84.81**	-	0.70	**7.50**	-	**92.31**	nr	**103.84**
460 x 460 x 1830mm code L442-025	102.27	2.50	**104.83**	-	0.80	**8.57**	-	**113.39**	nr	**127.57**
3 compartment; placing in position										
305 x 305 x 1830mm code L442-026	79.49	2.50	**81.47**	-	0.60	**6.43**	-	**87.90**	nr	**98.89**
305 x 460 x 1830mm code L442-027	89.25	2.50	**91.48**	-	0.65	**6.96**	-	**98.44**	nr	**110.75**
380 x 380 x 1830mm code L442-029	89.25	2.50	**91.48**	-	0.70	**7.50**	-	**98.98**	nr	**111.35**
460 x 460 x 1830mm code L442-030	107.73	2.50	**110.42**	-	0.80	**8.57**	-	**118.99**	nr	**133.87**
4 compartment; placing in position										
305 x 305 x 1830mm code L442-031	86.00	2.50	**88.14**	-	0.60	**6.43**	-	**94.57**	nr	**106.39**
305 x 460 x 1830mm code L442-032	94.71	2.50	**97.08**	-	0.65	**6.96**	-	**104.04**	nr	**117.04**
380 x 380 x 1830mm code L442-034	96.92	2.50	**99.34**	-	0.70	**7.50**	-	**106.83**	nr	**120.19**
460 x 460 x 1830mm code L442-035	118.65	2.50	**121.62**	-	0.80	**8.57**	-	**130.18**	nr	**146.46**
6 compartment; placing in position										
305 x 305 x 1830mm code L442-036	106.68	2.50	**109.35**	-	0.60	**6.43**	-	**115.77**	nr	**130.24**
305 x 460 x 1830mm code L442-037	115.40	2.50	**118.28**	-	0.65	**6.96**	-	**125.24**	nr	**140.90**
380 x 380 x 1830mm code L442-039	115.40	2.50	**118.28**	-	0.70	**7.50**	-	**125.78**	nr	**141.50**
460 x 460 x 1830mm code L442-040	137.13	2.50	**140.56**	-	0.80	**8.57**	-	**149.13**	nr	**167.77**
Swimming bath lockers; Welconstruct Co Ltd.; galvanized steel, stove enamelled finish; standard colour doors with mastered series cam locks 452 series. 20-49 ordered										
1 compartment; placing in position										
305 x 305 x 1830mm code L452-001	105.63	2.50	**108.27**	-	0.60	**6.43**	-	**114.70**	nr	**129.03**

Labour hourly rates: (except Specialists) Craft Operatives £14.21 Labourer £10.71 Rates are national average prices. Refer to REGIONAL VARIATIONS for indicative levels of overall pricing in regions	MATERIALS			LABOUR				RATES		
	Del to Site £	Waste %	Material Cost £	Craft Optve Hrs	Lab Hrs	Labour Cost £	Sunds £	Nett Rate £	Unit	Gross Rate (+12.5%) £
N10: GENERAL FIXTURES/FURNISHING/EQUIPMENT: CLOTHES LOCKERS Cont.										
Swimming bath lockers; Welconstruct Co Ltd.; galvanized steel, stove enamelled finish; standard colour doors with mastered series cam locks 452 series. 20-49 ordered Cont.										
2 compartment; placing in position										
305 x 305 x 1830mm code L452-002	136.92	2.50	**140.34**	-	0.60	**6.43**	-	**146.77**	nr	**165.12**
305 x 460 x 1830mm code L452-003	143.43	2.50	**147.02**	-	0.65	**6.96**	-	**153.98**	nr	**173.22**
3 compartment; placing in position										
305 x 305 x 1830mm code L452-004	161.18	2.50	**165.20**	-	0.50	**5.36**	-	**170.56**	nr	**191.88**
305 x 460 x 1830mm code L452-005	169.58	2.50	**173.81**	-	0.55	**5.89**	-	**179.70**	nr	**202.17**
4 compartment; placing in position										
305 x 305 x 1830mm code L452-006	173.78	2.50	**178.12**	-	0.60	**6.43**	-	**184.55**	nr	**207.61**
305 x 460 x 1830mm code L452-007	185.85	2.50	**190.50**	-	0.65	**6.96**	-	**197.46**	nr	**222.14**
N10: GENERAL FIXTURES/FURNISHING/EQUIPMENT: CLOAK ROOM EQUIPMENT										
Static square tube double sided coat racks; Welconstruct Co. Ltd.; hardwood seats; in assembled units										
Racks 1500mm long x 1675mm high x 610mm deep; placing in position; 40mm square tube										
10 hooks and angle framed steel mesh single shoe tray, reference L456-102 AND 109; placing in position	491.72	2.50	**504.01**	-	0.60	**6.43**	-	**510.43**	nr	**574.24**
10 hooks, reference L456-102; placing in position	425.25	2.50	**435.88**	-	0.60	**6.43**	-	**442.31**	nr	**497.60**
10 hooks and 5 baskets L456-102 and 106	495.29	2.50	**507.67**	-	0.60	**6.43**	-	**514.09**	nr	**578.35**
10 hooks and 10 baskets L456-102 and 107	544.11	2.50	**557.71**	-	0.60	**6.43**	-	**564.14**	nr	**634.66**
Static square tube single sided coat racks; Welconstruct Co. Ltd.; mobile; in assembled units										
Racks 1500mm long x 1825mm high x 600mm deep; placing in position; 40mm square tube										
15 hangers and top tray, reference L456-103	410.55	2.50	**420.81**	-	0.60	**6.43**	-	**427.24**	nr	**480.64**
Free-standing bench seats; Welconstruct Co. Ltd.; steel square tube framing; hardwood seats; placing in position										
Bench seat with shelf 1500mm long x 300mm deep and 450mm high; placing in position; 40mm square tube										
single sided 300mm deep, reference L456-115	198.98	2.50	**203.95**	-	0.70	**7.50**	-	**211.45**	nr	**237.88**
double sided 600mm deep L456-116	320.25	2.50	**328.26**	-	0.70	**7.50**	-	**335.75**	nr	**377.72**
Wall rack Welconstruct Co. Ltd; hardwood										
1500mm long; fixing to masonry										
5 hooks, reference L456-029	47.72	2.50	**48.92**	-	0.70	**7.50**	-	**56.41**	nr	**63.46**
N10: GENERAL FIXTURES/FURNISHING/EQUIPMENT: HAT AND COAT RAILS										
Hat and coat rails										
Rails in softwood, wrought										
25 x 75mm; chamfers -2	1.42	10.00	**1.56**	0.29	0.04	**4.55**	0.02	**6.12**	m	**6.89**
25 x 100mm; chamfers -2	1.90	10.00	**2.09**	0.29	0.04	**4.55**	0.02	**6.66**	m	**7.49**
25 x 75mm; chamfers -2; fixing to masonry with screws	1.42	10.00	**1.56**	0.57	0.07	**8.85**	0.03	**10.44**	m	**11.74**
25 x 100mm; chamfers -2; fixing to masonry with screws	1.90	10.00	**2.09**	0.57	0.07	**8.85**	0.03	**10.97**	m	**12.34**
Rails in Afromosia, wrought, selected for transparent finish										
25 x 75mm; chamfers -2	4.25	10.00	**4.67**	0.57	0.07	**8.85**	0.03	**13.55**	m	**15.24**
25 x 100mm; chamfers -2	5.68	10.00	**6.24**	0.57	0.07	**8.85**	0.03	**15.12**	m	**17.01**
25 x 75mm; chamfers -2; fixing to masonry with screws, countersinking and pellating	4.25	10.00	**4.67**	1.15	0.14	**17.84**	0.06	**22.57**	m	**25.40**
25 x 100mm; chamfers -2; fixing to masonry with screws, countersinking and pellating	5.68	10.00	**6.24**	1.15	0.14	**17.84**	0.06	**24.15**	m	**27.17**
Hat and coat hooks										
B.M.A. finish	2.20	2.50	**2.26**	0.40	-	**5.68**	-	**7.94**	nr	**8.93**
chromium plated	2.20	2.50	**2.26**	0.40	-	**5.68**	-	**7.94**	nr	**8.93**
SAA finish	1.10	2.50	**1.13**	0.40	-	**5.68**	-	**6.81**	nr	**7.66**
N11: DOMESTIC KITCHEN FITTINGS										
Standard melamine finish on chipboard units, with backs										
Wall units										
fixing to masonry with screws										
400 x 300 x 600mm	71.49	2.50	**73.28**	2.00	0.25	**31.10**	-	**104.38**	nr	**117.42**
400 x 300 x 900mm	93.00	2.50	**95.33**	2.10	0.26	**32.63**	-	**127.95**	nr	**143.95**
500 x 300 x 600mm	76.78	2.50	**78.70**	2.20	0.28	**34.26**	-	**112.96**	nr	**127.08**
500 x 300 x 900mm	97.35	2.50	**99.78**	2.25	0.28	**34.97**	-	**134.75**	nr	**151.59**
600 x 300 x 600mm	83.88	2.50	**85.98**	2.25	0.28	**34.97**	-	**120.95**	nr	**136.07**

Labour hourly rates: (except Specialists) Craft Operatives £14.21 Labourer £10.71 Rates are national average prices. Refer to REGIONAL VARIATIONS for indicative levels of overall pricing in regions	MATERIALS			LABOUR				RATES		
	Del to Site £	Waste %	Material Cost £	Craft Optve Hrs	Lab Hrs	Labour Cost £	Sunds £	Nett Rate £	Unit	Gross Rate (+12.5%) £
N11: DOMESTIC KITCHEN FITTINGS Cont.										
Standard melamine finish on chipboard units, with backs Cont.										
Wall units Cont.										
fixing to masonry with screws Cont.										
600 x 300 x 900mm	101.47	2.50	**104.01**	2.35	0.29	**36.50**	-	**140.51**	nr	**158.07**
1000 x 300 x 600mm	134.76	2.50	**138.13**	2.70	0.34	**42.01**	-	**180.14**	nr	**202.66**
1000 x 300 x 900mm	164.25	2.50	**168.36**	2.90	0.36	**45.06**	-	**213.42**	nr	**240.10**
1200 x 300 x 600mm	146.36	2.50	**150.02**	2.90	0.36	**45.06**	-	**195.09**	nr	**219.47**
1200 x 300 x 900mm	166.65	2.50	**170.82**	3.10	0.39	**48.23**	-	**219.05**	nr	**246.43**
Floor units on plinths and with plastic faced worktops; without drawers										
fixing to masonry with screws										
400 x 600 x 900mm	135.31	2.50	**138.69**	2.25	0.28	**34.97**	-	**173.66**	nr	**195.37**
500 x 600 x 900mm	148.64	2.50	**152.35**	2.55	0.32	**39.66**	-	**192.01**	nr	**216.02**
600 x 600 x 900mm	159.13	2.50	**163.11**	2.75	0.34	**42.72**	-	**205.83**	nr	**231.56**
1000 x 600 x 900mm	265.88	2.50	**272.52**	3.70	0.46	**57.50**	-	**330.03**	nr	**371.28**
1200 x 600 x 900mm	288.92	2.50	**296.15**	4.00	0.50	**62.20**	-	**358.34**	nr	**403.14**
Floor units on plinths and with plastic faced worktops; with one drawer										
fixing to masonry with screws										
500 x 600 x 900mm	185.51	2.50	**190.14**	2.55	0.32	**39.66**	-	**229.81**	nr	**258.53**
600 x 600 x 900mm	196.00	2.50	**200.90**	2.75	0.34	**42.72**	-	**243.62**	nr	**274.07**
1000 x 600 x 900mm	319.37	2.50	**327.35**	3.70	0.46	**57.50**	-	**384.85**	nr	**432.96**
1200 x 600 x 900mm	345.12	2.50	**353.75**	4.00	0.50	**62.20**	-	**415.94**	nr	**467.93**
Floor units on plinths and with plastic faced worktops; with four drawers										
fixing to masonry with screws										
500 x 600 x 900mm	258.75	2.50	**265.22**	2.55	0.32	**39.66**	-	**304.88**	nr	**342.99**
600 x 600 x 900mm	273.51	2.50	**280.35**	2.75	0.34	**42.72**	-	**323.07**	nr	**363.45**
Sink units on plinths										
1200 x 600 x 900mm, with one drawer; fixing to masonry with screws	245.08	2.50	**251.21**	4.00	0.50	**62.20**	-	**313.40**	nr	**352.58**
1200 x 600 x 900mm, without drawer; fixing to masonry with screws	187.23	2.50	**191.91**	4.00	0.50	**62.20**	-	**254.11**	nr	**285.87**
1500 x 600 x 900mm, without drawer; fixing to masonry with screws	293.21	2.50	**300.54**	4.30	0.55	**66.99**	-	**367.53**	nr	**413.47**
Store cupboards on plinths										
fixing to masonry with screws										
500 x 600 x 1950mm without shelves	215.88	2.50	**221.28**	4.00	0.50	**62.20**	-	**283.48**	nr	**318.91**
500 x 600 x 1950mm with shelves	230.93	2.50	**236.70**	4.00	0.50	**62.20**	-	**298.89**	nr	**336.26**
600 x 600 x 1950mm without shelves	235.00	2.50	**240.88**	4.50	0.56	**69.94**	-	**310.82**	nr	**349.67**
600 x 600 x 1950mm with shelves	253.44	2.50	**259.78**	4.50	0.56	**69.94**	-	**329.72**	nr	**370.93**
Laminated plastic faced units complete with ironmongery and with backs										
Wall units										
400 x 300 x 600mm, with one door; fixing to masonry with screws	142.92	2.50	**146.49**	2.00	0.25	**31.10**	-	**177.59**	nr	**199.79**
500 x 300 x 600mm, with one door; fixing to masonry with screws	153.53	2.50	**157.37**	2.10	0.26	**32.63**	-	**189.99**	nr	**213.74**
1000 x 300 x 600mm, with two doors; fixing to masonry with screws	269.51	2.50	**276.25**	2.70	0.34	**42.01**	-	**318.26**	nr	**358.04**
1200 x 300 x 600mm, with two doors; fixing to masonry with screws	292.71	2.50	**300.03**	2.85	0.36	**44.35**	-	**344.38**	nr	**387.43**
Floor units on plinths and with worktops										
500 x 600 x 900mm, with one drawer and one cupboard; fixing to masonry with screws	371.02	2.50	**380.30**	2.20	0.28	**34.26**	-	**414.56**	nr	**466.38**
500 x 600 x 900mm, with four drawers; fixing to masonry with screws	517.50	2.50	**530.44**	2.50	0.31	**38.85**	-	**569.28**	nr	**640.44**
1000 x 600 x 900mm, with two drawers and two cupboards; fixing to masonry with screws	638.72	2.50	**654.69**	3.50	0.45	**54.55**	-	**709.24**	nr	**797.90**
1500 x 600 x 900mm, with three drawers and three cupboards; fixing to masonry with screws	825.22	2.50	**845.85**	4.20	0.53	**65.36**	-	**911.20**	nr	**1025.10**
Sink units on plinths										
1000 x 600 x 900mm, with one drawer and two cupboards, fixing to masonry with screws	489.94	2.50	**502.19**	4.20	0.52	**65.25**	-	**567.44**	nr	**638.37**
1500 x 600 x 900mm, with two drawers and three cupboards; fixing to masonry with screws	732.85	2.50	**751.17**	4.30	0.54	**66.89**	-	**818.06**	nr	**920.31**
Tall units on plinths										
500 x 600 x 1950mm, broom cupboard; fixing to masonry with screws	450.63	2.50	**461.90**	4.30	0.54	**66.89**	-	**528.79**	nr	**594.88**
600 x 600 x 1950mm, larder unit; fixing to masonry with screws	470.20	2.50	**481.95**	4.30	0.54	**66.89**	-	**548.84**	nr	**617.45**
Hardwood veneered units complete with ironmongery but without backs										
Wardrobe units										
fixing to masonry with screws										
1000 x 600 x 2175mm	521.40	2.50	**534.44**	4.75	0.60	**73.92**	-	**608.36**	nr	**684.40**
1500 x 600 x 2175mm	655.05	2.50	**671.43**	5.50	0.69	**85.54**	-	**756.97**	nr	**851.59**

Labour hourly rates: (except Specialists) Craft Operatives £14.21 Labourer £10.71 Rates are national average prices. Refer to REGIONAL VARIATIONS for indicative levels of overall pricing in regions	MATERIALS			LABOUR				RATES		
	Del to Site £	Waste %	Material Cost £	Craft Optve Hrs	Lab Hrs	Labour Cost £	Sunds £	Nett Rate £	Unit	Gross Rate (+12.5%) £
N11: DOMESTIC KITCHEN FITTINGS Cont.										
Pre-Assembled kitchen units, white melamine finish; Module 500 range; Magnet PLC										
Drawerline units; fixing to masonry with screws										
floor units										
300mm F300	87.53	2.50	**89.72**	1.50	0.34	**24.96**	-	**114.67**	nr	**129.01**
500mm F1	100.04	2.50	**102.55**	1.50	0.38	**25.38**	-	**127.93**	nr	**143.92**
600mm F6	104.51	2.50	**107.12**	1.50	0.38	**25.38**	-	**132.50**	nr	**149.07**
1000mm F2	164.43	2.50	**168.54**	2.00	0.50	**33.78**	-	**202.32**	nr	**227.61**
hob floor units										
1000mm H2	127.70	2.50	**130.89**	2.00	0.50	**33.78**	-	**164.67**	nr	**185.25**
sink units										
1000mm S2	121.49	2.50	**124.52**	2.00	0.50	**33.78**	-	**158.30**	nr	**178.08**
Tall units; fixing to masonry with screws										
floor units										
500mm DF1	78.55	2.50	**80.51**	3.00	0.31	**45.95**	-	**126.46**	nr	**142.27**
Drawer units; fixing to masonry with screws										
four drawer floor units										
500mm FD1	194.90	2.50	**199.77**	1.50	0.51	**26.78**	-	**226.55**	nr	**254.87**
Wall units; fixing to masonry with screws										
standard units 600mm high										
300mm W306	43.70	2.50	**44.79**	2.50	0.35	**39.27**	-	**84.07**	nr	**94.58**
500mm W7	80.34	2.50	**82.34**	2.50	0.40	**39.81**	-	**122.15**	nr	**137.42**
600mm W5	43.70	2.50	**44.79**	2.50	0.41	**39.92**	-	**84.71**	nr	**95.30**
1000mm W2	93.74	2.50	**96.09**	3.00	0.52	**48.20**	-	**144.29**	nr	**162.32**
standard corner units 900mm high										
600 x 600mm CW6	128.59	2.50	**131.81**	3.50	0.51	**55.20**	-	**187.01**	nr	**210.38**
Worktops; fixing with screws										
round front edge										
38mm worktops, round front edge ADEN Range	26.81	25.00	**33.51**	1.50	0.19	**23.35**	-	**56.86**	m	**63.97**
Plinth										
Minster 500	3.89	25.00	**4.86**	0.25	0.03	**3.87**	-	**8.73**	m	**9.82**
Cornice										
Minster 500	6.58	25.00	**8.23**	0.25	0.03	**3.87**	-	**12.10**	m	**13.62**
Pelmet										
Minster 500	4.27	25.00	**5.34**	0.25	0.03	**3.87**	-	**9.22**	m	**10.37**
Worktop trim										
coloured	7.77	25.00	**9.71**	0.25	0.03	**3.87**	-	**13.59**	m	**15.28**
N13: SANITARY APPLIANCES/FITTINGS										
Attendance on sanitary fittings										
Note: Sanitary fittings are usually included as p.c. items or provisional sums, the following are the allowances for attendance by the main contractor i.e. unloading, storing and distributing fittings for fixing by the plumber and returning empty cases and packings										
Sinks, fireclay										
610 x 457 x 254mm	-	-	**-**	-	1.15	**12.32**	0.66	**12.98**	nr	**14.60**
762 x 508 x 254mm	-	-	**-**	-	1.40	**14.99**	0.84	**15.83**	nr	**17.81**
add if including tubular stands	-	-	**-**	-	0.85	**9.10**	0.55	**9.65**	nr	**10.86**
Wash basins, 559 x 406mm; earthenware or fireclay										
single	-	-	**-**	-	0.90	**9.64**	0.55	**10.19**	nr	**11.46**
range of 4 with cover overlaps between basins	-	-	**-**	-	2.75	**29.45**	1.49	**30.94**	nr	**34.81**
Combined sinks and drainers; stainless steel										
1067 x 533mm	-	-	**-**	-	1.60	**17.14**	0.98	**18.12**	nr	**20.38**
1600 x 610mm	-	-	**-**	-	1.90	**20.35**	1.16	**21.51**	nr	**24.20**
add if including tubular stands	-	-	**-**	-	0.85	**9.10**	0.55	**9.65**	nr	**10.86**
Combined sinks and drainers; porcelain enamel										
1067 x 533mm	-	-	**-**	-	1.60	**17.14**	0.98	**18.12**	nr	**20.38**
1600 x 610mm	-	-	**-**	-	1.90	**20.35**	1.16	**21.51**	nr	**24.20**
Baths excluding panels; pressed steel										
694 x 1688 x 570mm overall	-	-	**-**	-	2.75	**29.45**	1.45	**30.90**	nr	**34.76**
W.C. suites; china										
complete with WWP, flush pipe etc.,	-	-	**-**	-	1.60	**17.14**	0.98	**18.12**	nr	**20.38**
Block pattern urinals; with glazed ends, back and channel in one piece with separate tread including flushing cistern etc.,										
single stall	-	-	**-**	-	3.00	**32.13**	1.65	**33.78**	nr	**38.01**
range of four with loose overlaps	-	-	**-**	-	7.00	**74.97**	3.86	**78.83**	nr	**88.68**
Wall urinals, bowl type, with flushing cistern, etc.,										
single	-	-	**-**	-	1.60	**17.14**	0.98	**18.12**	nr	**20.38**
range of 3 with two divisions	-	-	**-**	-	3.75	**40.16**	1.98	**42.14**	nr	**47.41**

Labour hourly rates: (except Specialists) Craft Operatives £14.21 Labourer £10.71 Rates are national average prices. Refer to REGIONAL VARIATIONS for indicative levels of overall pricing in regions	MATERIALS			LABOUR			RATES			
	Del to Site £	Waste %	Material Cost £	Craft Optve Hrs	Lab Hrs	Labour Cost £	Sunds £	Nett Rate £	Unit	Gross Rate (+12.5%) £
N13: SANITARY APPLIANCES/FITTINGS Cont.										
Attendance on sanitary fittings Cont.										
Slab urinals with one return end, flushing cistern, sparge pipe, etc.,										
1219mm long x 1067mm high........................	-	-	-	-	3.15	**33.74**	2.20	**35.94**	nr	**40.43**
1829mm long x 1067mm high........................	-	-	-	-	4.25	**45.52**	3.30	**48.82**	nr	**54.93**

This section continues on the next page

Labour hourly rates: (except Specialists) Craft Operatives £17.38 Labourer £10.71 Rates are national average prices. Refer to REGIONAL VARIATIONS for indicative levels of overall pricing in regions	MATERIALS			LABOUR				RATES		
	Del to Site £	Waste %	Material Cost £	Craft Optve Hrs	Lab Hrs	Labour Cost £	Sunds £	Nett Rate £	Unit	Gross Rate (+12.5%) £
N13: SANITARY APPLIANCES/FITTINGS Cont.										
Fix only appliances (prime cost sum for supply included elsewhere)										
Sink units, stainless steel; combined overflow and waste outlet, plug and chain; pair pillar taps fixing on base unit with metal clips										
1000 x 600mm	-	-	-	3.60	-	**62.57**	1.02	**63.59**	nr	**71.54**
1600 x 600mm	-	-	-	4.20	-	**73.00**	1.02	**74.02**	nr	**83.27**
Sinks; white glazed fireclay; waste outlet, plug, chain and stay; cantilever brackets; pair bib taps fixing brackets to masonry with screws; sealing at back with white sealant										
610 x 457 x 254mm	-	-	-	4.80	-	**83.42**	1.63	**85.06**	nr	**95.69**
762 x 508 x 254mm	-	-	-	4.80	-	**83.42**	1.63	**85.06**	nr	**95.69**
Sinks; white glazed fireclay; waste outlet, plug, chain and stay; legs and screw-to-wall bearers; pair bib taps fixing legs to masonry with screws; sealing at back with mastic sealant										
610 x 457 x 254mm	-	-	-	5.40	-	**93.85**	1.63	**95.48**	nr	**107.42**
762 x 508 x 254mm	-	-	-	5.40	-	**93.85**	1.63	**95.48**	nr	**107.42**
Wash basins; vitreous china; waste outlet, plug, chain and stay; screw to wall brackets; pair pillar taps										
560 x 406mm; fixing brackets to masonry with screws; sealing at back with mastic sealant	-	-	-	3.60	-	**62.57**	1.63	**64.20**	nr	**72.23**
range of four, each 559 x 406mm with overlap strips; fixing brackets to masonry with screws; sealing overlap strips and at back with mastic sealant	-	-	-	15.30	-	**265.91**	6.12	**272.03**	nr	**306.04**
Wash basins; vitreous china; waste outlet, plug, chain and stay; legs and screw-to-wall bearers										
560 x 406mm; fixing legs to masonry with screws; sealing at back with mastic sealant	-	-	-	4.20	-	**73.00**	1.63	**74.63**	nr	**83.96**
range of 4, each 559 x 406mm with overlap strips; fixing legs to masonry with screws; sealing overlap strips and at back with mastic sealant	-	-	-	16.50	-	**286.77**	6.12	**292.89**	nr	**329.50**
Baths; enamelled pressed steel; combined overflow and waste outlet, plug and chain; metal cradles; pair pillar taps										
700 x 1700 x 570mm overall; sealing at walls with mastic sealant	-	-	-	6.60	-	**114.71**	3.57	**118.28**	nr	**133.06**
High level W.C. suites; vitreous china pan; plastics cistern with valveless fittings, ball valve, chain and pull handle; flush pipe; plastics seat and cover										
fixing pan and cistern brackets to masonry with screws; bedding pan in mastic; jointing pan to drain with cement mortar (1:2) and gaskin joint	-	-	-	5.40	-	**93.85**	2.14	**95.99**	nr	**107.99**
fixing pan and cistern brackets to masonry with screws; bedding pan in mastic; jointing pan to soil pipe with Multikwik connector	-	-	-	5.40	-	**93.85**	2.14	**95.99**	nr	**107.99**
Low level W.C. suites; vitreous china pan; plastics cistern with valveless fittings and ball valve; flush pipe; plastics seat and cover										
fixing pan and cistern brackets to masonry with screws; bedding pan in mastic; jointing pan to drain with cement mortar (1:2) and gaskin joint	-	-	-	5.40	-	**93.85**	2.14	**95.99**	nr	**107.99**
fixing pan and cistern brackets to masonry with screws; bedding pan in mastic; jointing pan to soil pipe with Multikwik connector	-	-	-	5.40	-	**93.85**	2.14	**95.99**	nr	**107.99**
Wall urinals; vitreous china bowl; automatic cistern; spreads and flush pipe; waste outlet										
single bowl; fixing bowl, cistern and pipe brackets to masonry with screws	-	-	-	5.40	-	**93.85**	1.68	**95.54**	nr	**107.48**
range of three bowls; fixing bowls, divisions, cistern and pipe brackets to masonry with screws	-	-	-	16.20	-	**281.56**	3.37	**284.92**	nr	**320.54**
Single stall urinals; white glazed fireclay in one piece; white glazed fireclay tread; plastics automatic cistern; spreader and flush pipe; waste outlet and domed grating										
bedding stall and tread in cement mortar (1:4) and jointing tread with waterproof jointing compound; fixing cistern and pipe brackets to masonry with screws	-	-	-	8.40	-	**145.99**	4.28	**150.28**	nr	**169.06**
Slab urinals; range of four; white glazed fireclay back, ends, divisions, channel and tread; automatic cistern; flush and sparge pipes; waste outlet and domed grating										
bedding back, ends, channel and tread in cement mortar (1:4) and jointing with waterproof jointing compound; fixing divisions, cistern and pipe brackets to masonry with screws	-	-	-	22.80	-	**396.26**	13.26	**409.52**	nr	**460.71**

FURNITURE/EQUIPMENT

Labour hourly rates: (except Specialists) Craft Operatives £17.38 Labourer £10.71 Rates are national average prices. Refer to REGIONAL VARIATIONS for indicative levels of overall pricing in regions	MATERIALS			LABOUR				RATES		
	Del to Site £	Waste %	Material Cost £	Craft Optve Hrs	Lab Hrs	Labour Cost £	Sunds £	Nett Rate £	Unit	Gross Rate (+12.5%) £
N13: SANITARY APPLIANCES/FITTINGS Cont.										
Fix only appliances (prime cost sum for supply included elsewhere) Cont.										
Showers; white glazed fireclay tray 760 x 760 x 180mm; waste outlet; recessed valve and spray head										
bedding tray in cement mortar (1:4); fixing valve head to masonry with screws; sealing at walls with mastic sealant	-	-	-	5.70	-	**99.07**	5.10	**104.17**	nr	**117.19**
Shower curtains with rails, hooks, supports and fittings										
straight, 914mm long x 1829mm drop; fixing supports to masonry with screws	-	-	-	1.20	-	**20.86**	0.51	**21.37**	nr	**24.04**
angled, 1676mm girth x 1829mm drop; fixing supports to masonry with screws	-	-	-	1.80	-	**31.28**	0.66	**31.95**	nr	**35.94**
Supply and fix appliances - sinks										
Note: the following prices include joints to copper services and wastes but do not include traps										
Sink units; single bowl with single drainer and back ledge; stainless steel; BS EN 13310 Type A; 38mm waste, plug and chain to BS EN 274 Part 1 with combined overflow; pair 13mm pillar taps to BS 1010 Part 2										
fixing on base unit with metal clips										
1000 x 600mm	91.75	2.50	**94.04**	3.00	-	**52.14**	0.82	**147.00**	nr	**165.37**
1200 x 600mm	98.25	2.50	**100.71**	3.50	-	**60.83**	0.82	**162.35**	nr	**182.65**
Sink units; single bowl with double drainer and back ledge; stainless steel; BS EN 13310 Type B; 38mm waste, plug and chain to BS EN 274 Part 1 with combined overflow; pair 13mm pillar taps to BS 1010 Part 2										
fixing on base unit with metal clips										
1500 x 600mm	112.25	2.50	**115.06**	3.00	-	**52.14**	1.07	**168.27**	nr	**189.30**
Sinks; white glazed fireclay; BS 1206 with weir overflow; 38mm slotted waste, chain, stay and plug to BS EN 274 Part 1; aluminium alloy cantilever brackets										
screwing chain stay; fixing brackets to masonry with screws; sealing at back with mastic sealant										
455 x 380 x 205mm	96.25	2.50	**98.66**	3.50	-	**60.83**	1.33	**160.81**	nr	**180.91**
610 x 455 x 255mm	134.25	2.50	**137.61**	3.50	-	**60.83**	1.33	**199.76**	nr	**224.73**
760 x 455 x 255mm	193.25	2.50	**198.08**	3.50	-	**60.83**	1.33	**260.24**	nr	**292.77**
Sinks; white glazed fireclay; BS 1206 with wier overflow; 38mm slotted waste, chain, stay and plug to BS EN 274 Part 1; painted legs and screw-to-wall bearers										
screwing chain stay; fixing legs to masonry with screws; sealing at back with mastic sealant										
455 x 380 x 205mm	115.50	2.50	**118.39**	3.50	-	**60.83**	1.33	**180.54**	nr	**203.11**
610 x 455 x 255mm	150.00	2.50	**153.75**	3.50	-	**60.83**	1.33	**215.91**	nr	**242.89**
760 x 455 x 255mm	209.00	2.50	**214.23**	3.50	-	**60.83**	1.33	**276.38**	nr	**310.93**
Supply and fix appliances - lavatory basins										
Wash basins; white vitreous china; BS 5506 Part 3; 32mm slotted waste, chain, stay and plug to BS EN 274 Part 1; painted cantilever brackets; pair 13 pillar taps to BS 1010 Part 2										
sealing at back with mastic sealant										
560 x 406mm	54.00	2.50	**55.35**	3.50	-	**60.83**	1.33	**117.51**	nr	**132.19**
Wash basins; white vitreous china; BS 5506 Part 3; 32mm slotted waste, chain, stay and plug to BS EN 274 Part 1; painted legs and screw-to-wall bearers; pair 13mm taps to BS 1010 Part 2										
fixing legs to masonry with screws; sealing at back with mastic sealant										
560 x 406mm	63.25	2.50	**64.83**	4.90	-	**85.16**	1.63	**151.63**	nr	**170.58**
Wash basins; white vitreous china; BS 5506 Part 3; 32mm slotted waste, chain, stay and plug to BS EN 274 Part 1; painted towel rail brackets; pair 13mm pillar taps to BS 1010 Part 2										
fixing brackets to masonry with screws; sealing at back with mastic sealant										
560 x 406mm	47.50	2.50	**48.69**	4.80	-	**83.42**	1.63	**133.74**	nr	**150.46**
Wash basins; coloured vitreous china; BS 5506 Part 3; 32mm slotted waste, chain, stay and plug to BS EN 274 Part 1; painted build-in brackets; pair 13mm pillar taps to BS 1010 Part 2										
building in brackets to masonry; sealing at back with mastic sealant										
560 x 406mm	79.00	2.50	**80.98**	4.20	-	**73.00**	1.63	**155.60**	nr	**175.05**

Labour hourly rates: (except Specialists) Craft Operatives £17.38 Labourer £10.71 Rates are national average prices. Refer to REGIONAL VARIATIONS for indicative levels of overall pricing in regions	MATERIALS			LABOUR				RATES		
	Del to Site £	Waste %	Material Cost £	Craft Optve Hrs	Lab Hrs	Labour Cost £	Sunds £	Nett Rate £	Unit	Gross Rate (+12.5%) £
N13: SANITARY APPLIANCES/FITTINGS Cont.										
Supply and fix appliances - lavatory basins Cont.										
Wash basins; coloured vitreous china; BS 5506 Part 3; 32mm slotted waste, chain, stay and plug to BS EN 274 Part 1; painted legs and screw-to-wall bearers; pair 13mm pillar taps to BS 1010 Part 2										
fixing legs to masonry with screws; sealing at back with mastic sealant										
560 x 406mm	88.25	2.50	**90.46**	4.90	-	**85.16**	1.63	**177.25**	nr	**199.41**
Wash basins; coloured vitreous china; BS 5506 Part 3; 32mm slotted waste, chain, stay and plug to BS EN 274 Part 1; painted towel rail brackets; pair 13mm pillar taps to BS 1010 Part 2										
fixing brackets to masonry with screws; sealing at back with mastic sealant										
560 x 406mm	72.50	2.50	**74.31**	4.80	-	**83.42**	1.63	**159.37**	nr	**179.29**
Wash basins, angle type; white vitreous china; BS 5506 Part 3; 32mm slotted waste, chain, stay and plug to BS EN 274 Part 1; concealed brackets; pair 13mm pillar taps to BS 1010 Part 2										
fixing brackets to masonry with screws; sealing at back with mastic sealant										
457 x 431mm	72.25	2.50	**74.06**	4.80	-	**83.42**	1.63	**159.11**	nr	**179.00**
Wash basins, angle type; coloured vitreous china; BS 5506 Part 3; 32mm slotted waste, chain, stay and plug to BS EN 274 Part 1; concealed brackets; pair 13mm pillar taps to BS 1010 Part 2										
fixing brackets to masonry with screws; sealing at back with mastic sealant										
457 x 431mm	84.50	2.50	**86.61**	4.80	-	**83.42**	1.63	**171.67**	nr	**193.13**
Wash basins with pedestal; white vitreous china; BS 5506 Part 3; 32mm slotted waste, chain, stay and plug to BS EN 274 Part 1; pair 13mm pillar taps to BS 1010 Part 2										
fixing basin to masonry with screws; sealing at back with mastic sealant										
560 x 406mm	71.00	2.50	**72.78**	4.50	-	**78.21**	1.63	**152.62**	nr	**171.69**
Wash basins with pedestal; coloured vitreous china; BS 5506 Part 3; 32mm slotted waste, chain, stay and plug to BS EN 274 Part 1; pair 13mm pillar taps to BS 1010 Part 2										
fixing basin to masonry with screws; sealing at back with mastic sealant										
560 x 406mm	99.25	2.50	**101.73**	4.50	-	**78.21**	1.63	**181.57**	nr	**204.27**
Supply and fix appliances - baths										
Baths; white enamelled pressed steel; BS EN 232; 32mm overflow with front grid and 38mm waste, chain and plug to BS EN 274 Part 1 with combined overflow; metal cradles with adjustable feet; pair 19mm pillar taps to BS 1010 Part 2										
sealing at walls with mastic sealant										
700 x 1700 x 570mm overall	97.25	2.50	**99.68**	6.60	-	**114.71**	3.57	**217.96**	nr	**245.20**
724 x 1500 x 570mm overall	97.25	2.50	**99.68**	6.60	-	**114.71**	3.57	**217.96**	nr	**245.20**
Baths; coloured enamelled pressed steel; BS EN 232; 32mm overflow with front grid and 38mm waste, chain and plug to BS EN 274 Part 1 with combined overflow; metal cradles with adjustable feet; pair 19mm pillar taps to BS 1010 Part 2										
sealing at walls with mastic sealant										
700 x 1700 x 570mm overall	97.25	2.50	**99.68**	6.60	-	**114.71**	3.57	**217.96**	nr	**245.20**
Baths; white acrylic; BS 4305; 32mm overflow with front grid and 38mm waste, chain and plug to BS EN 274 Part 1 with combined overflow; metal cradles with adjustable feet and wall fixing brackets; pair 19mm pillar taps to BS 1010 Part 2										
700 x 1700 x 570mm overall; fixing brackets to masonry with screws; sealing at walls with mastic sealant	124.00	2.50	**127.10**	6.00	-	**104.28**	3.57	**234.95**	nr	**264.32**
Baths; coloured acrylic; BS 4305; 32mm overflow with front grid and 38mm waste, chain and plug to BS EN 274 Part 1 with combined overflow; metal cradles with adjustable feet and wall fixing brackets; pair 19mm pillar taps to BS 1010 Part 2										
700 x 1700 x 570mm overall; fixing brackets to masonry with screws; sealing at walls with mastic sealant	124.00	2.50	**127.10**	6.00	-	**104.28**	3.57	**234.95**	nr	**264.32**
Supply and fix appliances - W.C. pans										
White vitreous china pans to BS 5503										
pan with horizontal outlet; fixing pan to timber with screws; bedding pan in mastic; jointing pan to soil pipe with Multikwik connector	28.00	2.50	**28.70**	3.00	-	**52.14**	0.92	**81.76**	nr	**91.98**

Labour hourly rates: (except Specialists) Craft Operatives £17.38 Labourer £10.71 Rates are national average prices. Refer to REGIONAL VARIATIONS for indicative levels of overall pricing in regions	MATERIALS			LABOUR				RATES		
	Del to Site £	Waste %	Material Cost £	Craft Optve Hrs	Lab Hrs	Labour Cost £	Sunds £	Nett Rate £	Unit	Gross Rate (+12.5%) £
N13: SANITARY APPLIANCES/FITTINGS Cont.										
Supply and fix appliances - W.C. pans Cont.										
White vitreous china pans to BS 5503 Cont.										
pan with horizontal outlet; fixing pan to masonry with screws; bedding pan in mastic; jointing pan to soil pipe with Multikwik connector...........	28.00	2.50	**28.70**	3.60	-	**62.57**	1.22	**92.49**	nr	**104.05**
Coloured vitreous china pans to BS 5503										
pan with horizontal outlet; fixing pan to timber with screws; bedding pan in mastic; jointing pan to soil pipe with Multikwik connector...........	54.25	2.50	**55.61**	3.00	-	**52.14**	0.92	**108.66**	nr	**122.25**
pan with horizontal outlet; fixing pan to masonry with screws; bedding pan in mastic; jointing pan to soil pipe with Multikwik connector...........	54.25	2.50	**55.61**	3.60	-	**62.57**	1.22	**119.40**	nr	**134.32**
Plastics W.C. seats; BS 1254, type 1										
black; fixing to pan	8.90	2.50	**9.12**	0.30	-	**5.21**	-	**14.34**	nr	**16.13**
Plastics W.C. seats and covers; BS 1254, type 1										
black; fixing to pan	12.50	2.50	**12.81**	0.30	-	**5.21**	-	**18.03**	nr	**20.28**
coloured; fixing to pan.....................	22.50	2.50	**23.06**	0.30	-	**5.21**	-	**28.28**	nr	**31.81**
white; fixing to pan	20.25	2.50	**20.76**	0.30	-	**5.21**	-	**25.97**	nr	**29.22**
Supply and fix appliances - cisterns										
9 litre black plastics cistern with valveless fittings, chain and pull handle to BS 1125; 13mm piston type high pressure ball valve to BS 1212 Part 1 with 127mm plastics float to BS 2456										
fixing cistern brackets to masonry with screws...	58.25	2.50	**59.71**	1.20	-	**20.86**	0.46	**81.02**	nr	**91.15**
Plastics flush pipe to BS 1125; adjustable										
for back wall fixing	2.75	2.50	**2.82**	0.90	-	**15.64**	0.46	**18.92**	nr	**21.28**
for side wall fixing	3.35	2.50	**3.43**	0.90	-	**15.64**	0.46	**19.53**	nr	**21.98**
Supply and fix appliances - W.C. suites										
High level W.C. suites; white vitreous china pan to BS 5503; 9 litre black plastics cistern with valveless fittings, chain and pull handle and plastics flush pipe to BS 1125; 13mm piston type high pressure ball valve to BS 1212 Part 1 with 127mm plastics float to BS 2456; black plastics seat and cover to BS 1254 type 1										
pan with S or P trap conversion bend to BS 5627; fixing pan to timber with screws; fixing cistern brackets to masonry with screws; bedding pan in mastic; jointing pan to soil pipe with Multikwik connector	102.00	2.50	**104.55**	4.80	-	**83.42**	2.14	**190.12**	nr	**213.88**
pan with S or P trap conversion bend to BS 5627; fixing pan and cistern brackets to masonry with screws; bedding pan in mastic; jointing pan to soil pipe with Multikwik connector..............	102.00	2.50	**104.55**	5.40	-	**93.85**	2.14	**200.54**	nr	**225.61**
Low level W.C. suites; white vitreous china pan to BS 5503; 9 litre white vitreous china cistern with valveless fittings and plastics flush bend to BS 1125; 13mm piston type high pressure ball valve to BS 1212 Part 1 with 127mm plastics float to BS 2456; white plastics seat and cover to BS 1254 type 1										
pan with S or P trap conversion bend to BS 5627; fixing pan to timber with screws; fixing cistern brackets to masonry with screws; bedding pan in mastic; jointing pan to soil pipe with Multikwik connector	95.50	2.50	**97.89**	4.80	-	**83.42**	2.14	**183.45**	nr	**206.39**
pan with S or P trap conversion bend to BS 5627; fixing pan and cistern brackets to masonry with screws; bedding pan in mastic; jointing pan to soil pipe with Multikwik connector..............	95.50	2.50	**97.89**	5.40	-	**93.85**	2.14	**193.88**	nr	**218.12**
Low level W.C. suites; coloured vitreous china pan to BS 5503; 9 litre coloured vitreous china cistern with valveless fittings and plastics flush bend to BS 1125; 13mm piston type high pressure ball valve to BS 1212 Part 1 with 127mm plastics float to BS 2456; coloured plastics seat and cover to BS 1254 type 1										
pan with S or P trap conversion bend to BS 5627; fixing pan to timber with screws; fixing cistern brackets to masonry with screws; bedding pan in mastic; jointing pan to soil pipe with Multikwik connector	154.75	2.50	**158.62**	4.80	-	**83.42**	2.14	**244.18**	nr	**274.71**
pan with S or P trap conversion bend to BS 5627; fixing pan and cistern brackets to masonry with screws; bedding pan in mastic; jointing pan to soil pipe with Multikwik connector..............	154.75	2.50	**158.62**	5.40	-	**93.85**	2.14	**254.61**	nr	**286.44**
Supply and fix appliances - urinals										
Wall urinals; white vitreous china bowl to BS 5520; 4.5 litre white vitreous china automatic cistern to BS 1876; spreader and stainless steel flush pipe; 32mm plastics waste outlet										
fixing bowl, cistern and pipe brackets to masonry with screws	243.75	2.50	**249.84**	5.40	-	**93.85**	1.68	**345.38**	nr	**388.55**

Labour hourly rates: (except Specialists) Craft Operatives £17.38 Labourer £10.71 Rates are national average prices. Refer to REGIONAL VARIATIONS for indicative levels of overall pricing in regions	MATERIALS			LABOUR				RATES		
	Del to Site £	Waste %	Material Cost £	Craft Optve Hrs	Lab Hrs	Labour Cost £	Sunds £	Nett Rate £	Unit	Gross Rate (+12.5%) £
N13: SANITARY APPLIANCES/FITTINGS Cont.										
Supply and fix appliances - urinals Cont.										
Wall urinals; range of 2; white vitreous china bowls to BS 5520; white vitreous china divisions; 9 litre white plastics automatic cistern to BS 1876; stainless steel flush pipes and spreaders; 32mm plastics waste outlets										
fixing bowls, divisions, cistern and pipe brackets to masonry with screws	458.25	2.50	**469.71**	10.80	-	**187.70**	2.50	**659.91**	nr	**742.40**
Wall urinals; range of 3; white vitreous china bowls to BS 5520; white vitreous china divisions; 14 litre white plastics automatic cistern to BS 1876; stainless steel flush pipes and spreaders; 32mm plastics waste outlets										
fixing bowls, divisions, cistern and pipe brackets to masonry with screws	704.50	2.50	**722.11**	16.20	-	**281.56**	3.32	**1006.98**	nr	**1132.86**
1200mm long stainless steel wall hung trough urinal with plastic; white; automatic cistern to BS 1876; spreader and stainless steel flush pipe; waste outlet and domed grating										
fixing urinal cistern and pipe brackets to masonry with screws	662.00	2.50	**678.55**	7.00	-	**121.66**	4.23	**804.44**	nr	**905.00**
1800mm long stainless steel wall hung trough urinal with plastic; white; automatic cistern to BS 1876; spreader and stainless steel flush pipe; waste outlet and domed grating										
fixing urinal cistern and pipe brackets to masonry with screws	765.00	2.50	**784.13**	9.50	-	**165.11**	7.14	**956.38**	nr	**1075.92**
2400mm long stainless steel wall hung trough urinal with plastic; white; automatic cistern to BS 1876; spreader and stainless steel flush pipe; waste outlet and domed grating										
fixing urinal cistern and pipe brackets to masonry with screws	788.00	2.50	**807.70**	12.00	-	**208.56**	10.20	**1026.46**	nr	**1154.77**
Supply and fix appliances - showers										
Showers; white glazed Armastone tray, 760 x 760 x 180mm; 38mm waste outlet; surface mechanical valve										
bedding tray in cement mortar (1:4); fixing valve and head to masonry with screws; sealing at walls with mastic sealant	277.00	2.50	**283.93**	5.70	-	**99.07**	5.10	**388.09**	nr	**436.60**
Showers; white glazed Armastone tray, 760 x 760 x 180mm; 38mm waste outlet; recessed thermostatic valve and swivel spray head										
bedding tray in cement mortar (1:4); fixing valve and head to masonry with screws; sealing at walls with mastic sealant	426.50	2.50	**437.16**	5.70	-	**99.07**	5.10	**541.33**	nr	**608.99**
Showers; coloured glazed acrylic tray, 760 x 760 x 180mm; 38mm waste outlet; recessed thermostatic valve and swivel spray head										
bedding tray in cement mortar (1:4); fixing valve and head to masonry with screws; sealing at walls with mastic sealant	474.75	2.50	**486.62**	5.70	-	**99.07**	5.10	**590.78**	nr	**664.63**
Showers; white moulded acrylic tray with removable front and side panels, 760 x 760 x 260mm with adjustable metal cradle; 38mm waste outlet; surface fixing mechanical valve, flexible tube hand spray and slide bar										
fixing cradle, valve and slide bar to masonry with screws; sealing at walls with mastic sealant	325.25	2.50	**333.38**	5.70	-	**99.07**	3.57	**436.02**	nr	**490.52**
Showers; white moulded acrylic tray with removable front and side panels, 760 x 760 x 260mm with adjustable metal cradle; 38mm waste outlet; surface fixing thermostatic valve, flexible tube hand spray and slide bar										
fixing cradle, valve and slide bar to masonry with screws; sealing at walls with mastic sealant	474.75	2.50	**486.62**	5.70	-	**99.07**	3.57	**589.25**	nr	**662.91**
Shower curtains; nylon; anodised aluminium rail, glider hooks, end and suspension fittings										
straight, 914mm long x 1829mm drop	37.25	2.50	**38.18**	1.20	-	**20.86**	0.51	**59.55**	nr	**66.99**
angled, 1676mm girth x 1829mm drop	52.50	2.50	**53.81**	1.80	-	**31.28**	0.66	**85.76**	nr	**96.48**
Shower curtains; heavy duty plastic; anodised aluminium rail, glider hooks, end and suspension fitting; fixing supports to masonry with screws										
straight, 914mm long x 1829mm drop	42.00	2.50	**43.05**	1.20	-	**20.86**	0.51	**64.42**	nr	**72.47**
angled, 1676mm girth x 1829mm drop	56.75	2.50	**58.17**	1.80	-	**31.28**	0.66	**90.12**	nr	**101.38**

Labour hourly rates: (except Specialists) Craft Operatives £14.21 Labourer £10.71 Rates are national average prices. Refer to REGIONAL VARIATIONS for indicative levels of overall pricing in regions	MATERIALS			LABOUR				RATES		
	Del to Site £	Waste %	Material Cost £	Craft Optve Hrs	Lab Hrs	Labour Cost £	Sunds £	Nett Rate £	Unit	Gross Rate (+12.5%) £
N15: SIGNS/NOTICES										
Signwriting in gloss paint; one coat										
Letters or numerals, Helvetica medium style, on painted or varnished surfaces										
50mm high	0.01	10.00	**0.01**	0.15	-	**2.13**	0.00	**2.15**	nr	**2.42**
extra; shading	0.01	10.00	**0.01**	0.09	-	**1.28**	0.00	**1.30**	nr	**1.46**
extra; outline	0.01	10.00	**0.01**	0.13	-	**1.85**	0.00	**1.86**	nr	**2.10**
100mm high	0.01	10.00	**0.01**	0.31	-	**4.41**	0.01	**4.42**	nr	**4.98**
extra; shading	0.01	10.00	**0.01**	0.17	-	**2.42**	0.00	**2.43**	nr	**2.74**
extra; outline	0.01	10.00	**0.01**	0.26	-	**3.69**	0.01	**3.71**	nr	**4.18**
150mm high	0.02	10.00	**0.03**	0.45	-	**6.39**	0.01	**6.43**	nr	**7.23**
extra; shading	0.01	10.00	**0.01**	0.26	-	**3.69**	0.01	**3.71**	nr	**4.18**
extra; outline	0.01	10.00	**0.01**	0.39	-	**5.54**	0.01	**5.56**	nr	**6.26**
200mm high	0.02	10.00	**0.03**	0.60	-	**8.53**	0.01	**8.56**	nr	**9.63**
extra; shading	0.01	10.00	**0.01**	0.34	-	**4.83**	0.01	**4.85**	nr	**5.46**
extra; outline	0.01	10.00	**0.01**	0.52	-	**7.39**	0.01	**7.41**	nr	**8.34**
300mm high	0.04	10.00	**0.04**	0.90	-	**12.79**	0.02	**12.85**	nr	**14.45**
extra; shading	0.02	10.00	**0.03**	0.52	-	**7.39**	0.01	**7.43**	nr	**8.36**
extra; outline	0.02	10.00	**0.03**	0.77	-	**10.94**	0.02	**10.98**	nr	**12.36**
stops	-	-	**-**	0.04	-	**0.57**	-	**0.57**	nr	**0.64**
Signwriting in gloss paint; two coats										
Letters or numerals, Helvetica medium style, on painted or varnished surfaces										
50mm high	0.01	10.00	**0.01**	0.27	-	**3.84**	0.01	**3.86**	nr	**4.34**
100mm high	0.02	10.00	**0.03**	0.54	-	**7.67**	0.01	**7.71**	nr	**8.68**
150mm high	0.02	10.00	**0.03**	0.81	-	**11.51**	0.02	**11.55**	nr	**13.00**
200mm high	0.04	10.00	**0.04**	1.08	-	**15.35**	0.02	**15.41**	nr	**17.33**
250mm high	0.04	10.00	**0.04**	1.34	-	**19.04**	0.03	**19.11**	nr	**21.50**
300mm high	0.05	10.00	**0.05**	1.61	-	**22.88**	0.03	**22.96**	nr	**25.83**
stops	-	-	**-**	0.08	-	**1.14**	0.00	**1.14**	nr	**1.28**

BUILDING FABRIC SUNDRIES

Labour hourly rates: (except Specialists) Craft Operatives £14.21 Labourer £10.71 Rates are national average prices. Refer to REGIONAL VARIATIONS for indicative levels of overall pricing in regions	MATERIALS			LABOUR				RATES		
	Del to Site £	Waste %	Material Cost £	Craft Optve Hrs	Lab Hrs	Labour Cost £	Sunds £	Nett Rate £	Unit	Gross Rate (+12.5%) £
P10: SUNDRY INSULATION/PROOFING WORK/ FIRE STOPS										
Waterproof reinforced building paper BS 1521; grade A1F; 150mm lapped joints										
Across members at 450mm centres vertical	0.83	5.00	**0.87**	0.20	0.03	**3.16**	0.03	**4.06**	m2	**4.57**
Waterproof reinforced building paper; reflection (thermal) grade, single sided; 150mm lapped joints										
Across members at 450mm centres vertical	1.23	5.00	**1.30**	0.20	0.03	**3.16**	0.03	**4.49**	m2	**5.05**
Waterproof reinforced building paper; reflection (thermal) grade, double sided; 150mm lapped joints										
Across members at 450mm centres vertical	1.83	5.00	**1.92**	0.20	0.03	**3.16**	0.03	**5.11**	m2	**5.75**
British Sisalkraft Ltd Insulex 714 vapour control layer; 150mm laps sealed with tape										
Across members at 450mm centres horizontal; fixing to timber with stainless steel staples	1.84	10.00	**2.03**	0.10	0.01	**1.53**	0.06	**3.61**	m2	**4.07**
Plain areas horizontal; laid loose	1.84	10.00	**2.03**	0.06	0.01	**0.96**	-	**2.98**	m2	**3.36**
British Sisalkraft Ltd SK860 vapour control layer; 100mm laps sealed with tape										
Plain areas horizontal; laid loose	1.42	10.00	**1.56**	0.08	0.01	**1.24**	0.04	**2.85**	m2	**3.20**
Mineral wool insulation quilt; butt joints										
60mm thick; across members at 450mm centres horizontal; laid loose	2.38	15.00	**2.74**	0.12	0.01	**1.81**	-	**4.55**	m2	**5.12**
Paper faced mineral wool insulation quilt; butt joints										
60mm thick; between members at 450mm centres vertical	4.61	15.00	**5.30**	0.25	0.03	**3.87**	0.03	**9.20**	m2	**10.35**
Glass fibre insulation quilt; butt joints										
60mm thick; across members at 450mm centres horizontal; laid loose	2.47	15.00	**2.84**	0.12	0.02	**1.92**	-	**4.76**	m2	**5.35**
80mm thick; across members at 450mm centres horizontal; laid loose	3.44	15.00	**3.96**	0.12	0.02	**1.92**	-	**5.88**	m2	**6.61**
100mm thick; across members at 450mm centres horizontal; laid loose	4.28	15.00	**4.92**	0.12	0.02	**1.92**	-	**6.84**	m2	**7.69**
60mm thick; between members at 450mm centres horizontal; laid loose	2.47	10.00	**2.72**	0.18	0.02	**2.77**	-	**5.49**	m2	**6.17**
80mm thick; between members at 450mm centres horizontal; laid loose	3.44	10.00	**3.78**	0.18	0.02	**2.77**	-	**6.56**	m2	**7.38**
100mm thick; between members at 450mm centres horizontal; laid loose	4.28	10.00	**4.71**	0.18	0.02	**2.77**	-	**7.48**	m2	**8.41**
Paper faced glass fibre insulation quilt; butt joints										
60mm thick; across members at 450mm centres horizontal; laid loose	4.61	15.00	**5.30**	0.12	0.02	**1.92**	-	**7.22**	m2	**8.12**

Labour hourly rates: (except Specialists) Craft Operatives £14.21 Labourer £10.71 Rates are national average prices. Refer to REGIONAL VARIATIONS for indicative levels of overall pricing in regions	MATERIALS			LABOUR				RATES		
	Del to Site	Waste	Material Cost	Craft Optve	Lab	Labour Cost	Sunds	Nett Rate	Unit	Gross Rate (+12.5%)
	£	%	£	Hrs	Hrs	£	£	£		£
P10: SUNDRY INSULATION/PROOFING WORK/ FIRE STOPS Cont.										
Paper faced glass fibre insulation quilt; butt joints Cont.										
80mm thick; across members at 450mm centres horizontal; laid loose	5.58	15.00	**6.42**	0.12	0.02	**1.92**	-	**8.33**	m2	**9.38**
Owens Corning Building Products (UK) Ltd; Crown wool insulation quilt; butt joints										
100mm thick; between members at 450mm centres horizontal; laid loose	3.11	10.00	**3.42**	0.11	0.02	**1.78**	-	**5.20**	m2	**5.85**
150mm thick; between members at 450mm centres horizontal; laid loose	4.72	10.00	**5.19**	0.11	0.02	**1.78**	-	**6.97**	m2	**7.84**
200mm thick; between members at 450mm centres horizontal; laid loose	6.62	10.00	**7.28**	0.22	0.04	**3.55**	-	**10.83**	m2	**12.18**
100mm thick; between members at 450mm centres; 50mm thick across members at 450mm centres horizontal; laid loose	4.70	10.00	**5.17**	0.30	0.06	**4.91**	-	**10.07**	m2	**11.33**
100mm thick; between members at 450mm centres; 100mm thick across members at 450mm centres horizontal; laid loose	6.22	10.00	**6.84**	0.40	0.06	**6.33**	-	**13.17**	m2	**14.81**
Rockwool Rollbatts; insulation quilt; butt joints										
100mm thick; between members at 450mm centres horizontal; laid loose	3.24	10.00	**3.57**	0.11	0.02	**1.78**	-	**5.34**	m2	**6.01**
150mm thick; between members at 450mm centres horizontal; laid loose	5.01	10.00	**5.51**	0.17	0.03	**2.74**	-	**8.24**	m2	**9.27**
100mm thick; between members at 450mm centres; 100mm thick across members at 450mm centres horizontal; laid loose	6.48	10.00	**7.13**	0.22	0.04	**3.55**	-	**10.69**	m2	**12.02**
Glass fibre medium density insulation board; butt joints										
30mm thick; plain areas horizontal; bedding in bitumen	2.13	10.00	**2.34**	0.23	0.03	**3.59**	1.65	**7.58**	m2	**8.53**
50mm thick; plain areas horizontal; bedding in bitumen	3.80	10.00	**4.18**	0.25	0.03	**3.87**	1.65	**9.71**	m2	**10.92**
75mm thick; plain areas horizontal; bedding in bitumen	5.56	10.00	**6.11**	0.28	0.03	**4.30**	1.65	**12.06**	m2	**13.57**
30mm thick; across members at 450mm centres vertical	2.13	10.00	**2.34**	0.23	0.03	**3.59**	0.03	**5.96**	m2	**6.71**
50mm thick; across members at 450mm centres vertical	3.80	10.00	**4.18**	0.25	0.03	**3.87**	0.03	**8.09**	m2	**9.10**
75mm thick; across members at 450mm centres vertical	5.56	10.00	**6.11**	0.28	0.03	**4.30**	0.03	**10.44**	m2	**11.75**
Dow Construction Products; Styrofoam Floormate 200; butt joints										
25mm thick; plain areas horizontal; laid loose	6.03	10.00	**6.63**	0.12	0.02	**1.92**	-	**8.55**	m2	**9.62**
35mm thick; plain areas horizontal; laid loose	8.26	10.00	**9.08**	0.17	0.03	**2.74**	-	**11.82**	m2	**13.30**
50mm thick; plain areas horizontal; laid loose	11.58	10.00	**12.73**	0.22	0.04	**3.55**	-	**16.29**	m2	**18.32**
Sempatap latex foam sheeting; butt joints										
5mm thick; plain areas soffit; fixing with adhesive	10.15	15.00	**11.68**	1.50	1.50	**37.38**	3.00	**52.06**	m2	**58.56**
vertical; fixing with adhesive	10.15	15.00	**11.68**	1.20	1.20	**29.90**	3.00	**44.58**	m2	**50.15**
Sempafloor SBR latex foam sheeting with coated non woven polyester surface; butt joints										
4.5mm thick; plain areas horizontal; fixing with adhesive	11.33	15.00	**13.03**	0.90	0.90	**22.43**	2.55	**38.01**	m2	**42.76**
Expanded polystyrene sheeting; butt joints										
13mm thick; plain areas horizontal; laid loose	0.86	5.00	**0.90**	0.09	0.05	**1.81**	-	**2.72**	m2	**3.06**
vertical; fixing with adhesive	0.86	5.00	**0.90**	0.27	0.14	**5.34**	1.10	**7.34**	m2	**8.26**
19mm thick; plain areas horizontal; laid loose	1.35	5.00	**1.41**	0.10	0.05	**1.96**	-	**3.37**	m2	**3.79**
vertical; fixing with adhesive	1.35	5.00	**1.41**	0.28	1.06	**15.33**	1.10	**17.84**	m2	**20.07**

Labour hourly rates: (except Specialists) Craft Operatives £14.21 Labourer £10.71 Rates are national average prices. Refer to REGIONAL VARIATIONS for indicative levels of overall pricing in regions	MATERIALS			LABOUR				RATES		
	Del to Site £	Waste %	Material Cost £	Craft Optve Hrs	Lab Hrs	Labour Cost £	Sunds £	Nett Rate £	Unit	Gross Rate (+12.5%) £
P10: SUNDRY INSULATION/PROOFING WORK/ FIRE STOPS Cont.										
Expanded polystyrene sheeting; butt joints Cont.										
25mm thick; plain areas										
horizontal; laid loose	1.70	5.00	**1.78**	0.12	0.06	**2.35**	-	**4.13**	m2	**4.65**
vertical; fixing with adhesive	1.70	5.00	**1.78**	0.30	0.15	**5.87**	1.10	**8.75**	m2	**9.85**
50mm thick; plain areas										
horizontal; laid loose	3.36	5.00	**3.53**	0.14	0.07	**2.74**	-	**6.27**	m2	**7.05**
vertical; fixing with adhesive	3.36	5.00	**3.53**	0.32	0.16	**6.26**	1.10	**10.89**	m2	**12.25**
75mm thick; plain areas										
horizontal; laid loose	5.06	5.00	**5.31**	0.18	0.09	**3.52**	-	**8.84**	m2	**9.94**
vertical; fixing with adhesive	5.06	5.00	**5.31**	0.36	0.18	**7.04**	1.10	**13.46**	m2	**15.14**
Expanded polystyrene sheeting, non-flammable; butt joints										
13mm thick; plain areas										
horizontal; laid loose	1.36	5.00	**1.42**	0.09	0.05	**1.81**	-	**3.24**	m2	**3.64**
vertical; fixing with adhesive	1.36	5.00	**1.42**	0.27	0.14	**5.34**	1.10	**7.86**	m2	**8.84**
19mm thick; plain areas										
horizontal; laid loose	1.61	5.00	**1.69**	0.10	0.05	**1.96**	-	**3.65**	m2	**4.10**
vertical; fixing with adhesive	1.61	5.00	**1.69**	0.28	0.14	**5.48**	1.10	**8.27**	m2	**9.30**
25mm thick; plain areas										
horizontal; laid loose	1.94	5.00	**2.04**	0.12	0.06	**2.35**	-	**4.39**	m2	**4.93**
vertical; fixing with adhesive	1.94	5.00	**2.04**	0.30	0.15	**5.87**	1.10	**9.01**	m2	**10.13**
50mm thick; plain areas										
horizontal; laid loose	4.65	5.00	**4.89**	0.14	0.07	**2.74**	-	**7.62**	m2	**8.58**
vertical; fixing with adhesive	4.65	5.00	**4.89**	0.32	0.16	**6.26**	1.10	**12.25**	m2	**13.78**
75mm thick; plain areas										
horizontal; laid loose	7.00	5.00	**7.35**	0.18	0.09	**3.52**	-	**10.87**	m2	**12.23**
vertical; fixing with adhesive	7.00	5.00	**7.35**	0.36	0.18	**7.04**	1.10	**15.49**	m2	**17.43**
Perforated zinc sheeting; butt joints										
Nr 7 gauge; across members at 450mm centres										
soffit	24.98	20.00	**29.98**	1.50	0.19	**23.35**	0.24	**53.57**	m2	**60.27**
vertical	24.98	20.00	**29.98**	1.00	0.13	**15.60**	0.16	**45.74**	m2	**51.46**
Nr 8 gauge; across members at 450mm centres										
soffit	27.39	20.00	**32.86**	1.50	0.19	**23.35**	0.24	**56.45**	m2	**63.51**
vertical	27.39	20.00	**32.86**	1.00	0.13	**15.60**	0.16	**48.63**	m2	**54.70**
Nr 9 gauge; across members at 450mm centres										
soffit	30.62	20.00	**36.74**	1.50	0.19	**23.35**	0.24	**60.33**	m2	**67.87**
vertical	30.62	20.00	**36.74**	1.00	0.13	**15.60**	0.16	**52.50**	m2	**59.06**
galvanized wire netting, BS EN 10223-2; butt joints										
13mm mesh x Nr 22 gauge; across members at 450mm centres										
soffit	2.54	10.00	**2.79**	-	0.14	**1.50**	0.03	**4.32**	m2	**4.86**
vertical	2.54	10.00	**2.79**	-	0.10	**1.07**	0.03	**3.90**	m2	**4.38**
25mm mesh x Nr 19 gauge; across members at 450mm centres										
soffit	2.04	10.00	**2.24**	-	0.14	**1.50**	0.03	**3.77**	m2	**4.25**
vertical	2.04	10.00	**2.24**	-	0.10	**1.07**	0.03	**3.35**	m2	**3.76**
38mm mesh x Nr 19 gauge; across members at 450mm centres										
soffit	1.46	10.00	**1.61**	-	0.14	**1.50**	0.03	**3.14**	m2	**3.53**
vertical	1.46	10.00	**1.61**	-	0.10	**1.07**	0.03	**2.71**	m2	**3.05**
50mm mesh x Nr 19 gauge; across members at 450mm centres										
soffit	1.16	10.00	**1.28**	-	0.14	**1.50**	0.03	**2.81**	m2	**3.16**
vertical	1.16	10.00	**1.28**	-	0.10	**1.07**	0.03	**2.38**	m2	**2.67**
P20: UNFRAMED ISOLATED TRIMS/SKIRTINGS/SUNDRY ITEMS										
Softwood, wrought - skirtings, architraves, picture rails and cover fillets to BS EN 942										
Skirtings, picture rails, architraves and the like; (finished sizes)										
13 x 45mm; reference 13CA45	0.91	10.00	**1.00**	0.25	0.03	**3.87**	0.03	**4.90**	m	**5.52**
20 x 70mm; reference 20CA70	1.42	10.00	**1.56**	0.27	0.03	**4.16**	0.03	**5.75**	m	**6.47**
13 x 45mm; reference 13CP45	0.91	10.00	**1.00**	0.26	0.03	**4.02**	0.03	**5.05**	m	**5.68**
13 x 70mm; reference 13CP70	1.17	10.00	**1.29**	0.26	0.03	**4.02**	0.03	**5.34**	m	**6.00**
13 x 70mm; reference 13CS70	1.17	10.00	**1.29**	0.28	0.03	**4.30**	0.03	**5.62**	m	**6.32**
13 x 95mm; reference 13CS95	1.51	10.00	**1.66**	0.26	0.03	**4.02**	0.03	**5.71**	m	**6.42**
20 x 120mm; reference 20CS120	2.33	10.00	**2.56**	0.31	0.03	**4.73**	0.04	**7.33**	m	**8.25**
9 x 33mm; reference 9RA33	0.67	10.00	**0.74**	0.23	0.02	**3.48**	0.03	**4.25**	m	**4.79**
13 x 45mm; reference 13RA45	0.91	10.00	**1.00**	0.26	0.03	**4.02**	0.03	**5.05**	m	**5.68**
20 x 70mm; reference 20RA70	1.51	10.00	**1.66**	0.27	0.03	**4.16**	0.03	**5.85**	m	**6.58**
13 x 70mm; reference 13RS70	1.22	10.00	**1.34**	0.28	0.03	**4.30**	0.03	**5.67**	m	**6.38**
13 x 95mm; reference 13RS95	1.57	10.00	**1.73**	0.26	0.03	**4.02**	0.03	**5.77**	m	**6.49**
20 x 120mm; reference 20RS120	2.33	10.00	**2.56**	0.31	0.03	**4.73**	0.04	**7.33**	m	**8.25**
13 x 45mm; reference 13RP45	0.95	10.00	**1.05**	0.26	0.03	**4.02**	0.03	**5.09**	m	**5.73**

Labour hourly rates: (except Specialists) Craft Operatives £14.21 Labourer £10.71 Rates are national average prices. Refer to REGIONAL VARIATIONS for indicative levels of overall pricing in regions	MATERIALS			LABOUR				RATES		
	Del to Site £	Waste %	Material Cost £	Craft Optve Hrs	Lab Hrs	Labour Cost £	Sunds £	Nett Rate £	Unit	Gross Rate (+12.5%) £
P20: UNFRAMED ISOLATED TRIMS/SKIRTINGS/SUNDRY ITEMS Cont.										
Softwood, wrought - skirtings, architraves, picture rails and cover fillets to BS EN 942 Cont.										
Skirtings, picture rails, architraves and the like; (finished sizes) Cont.										
13 x 70mm; reference 13RP70	1.28	10.00	1.40	0.28	0.03	4.30	0.03	5.73	m	6.45
Cover fillets, stops, trims, beads, nosings and the like; (finished sizes)										
11 x 33mm; reference C33	0.67	10.00	0.74	0.22	0.02	3.34	0.02	4.10	m	4.61
11 x 45mm; reference C45	0.70	10.00	0.77	0.22	0.02	3.34	0.02	4.13	m	4.65
13 x 33mm; reference HR33	0.79	10.00	0.87	0.22	0.02	3.34	0.02	4.23	m	4.76
20 x 45mm; reference HR45	1.06	10.00	1.16	0.24	0.03	3.73	0.03	4.92	m	5.54
11 x 11mm; reference Q11	0.35	10.00	0.39	0.22	0.02	3.34	0.02	3.75	m	4.22
13 x 13mm; reference Q13	0.56	10.00	0.61	0.22	0.02	3.34	0.02	3.97	m	4.47
20 x 20mm; reference Q20	0.70	10.00	0.77	0.22	0.02	3.34	0.02	4.13	m	4.65
13 x 13mm; reference S13	0.56	10.00	0.61	0.22	0.02	3.34	0.02	3.97	m	4.47
20 x 20mm; reference S20	0.70	10.00	0.77	0.22	0.02	3.34	0.02	4.13	m	4.65
27 x 27mm; reference S27	0.91	10.00	1.00	0.24	0.03	3.73	0.03	4.76	m	5.36
12 x 95mm; reference 12HRCS95	1.67	10.00	1.84	0.28	0.03	4.30	0.04	6.18	m	6.95
15 x 95mm; reference 15HRCS95	1.77	10.00	1.95	0.31	0.03	4.73	0.04	6.72	m	7.56
Softwood, wrought - skirtings, architraves, picture rails and cover fillets										
Skirtings, picture rails, architraves and the like; (finished sizes)										
19 x 100mm	2.23	10.00	2.45	0.29	0.04	4.55	0.05	7.05	m	7.93
19 x 150mm	3.12	10.00	3.43	0.32	0.04	4.98	0.05	8.46	m	9.52
25 x 100mm	2.84	10.00	3.13	0.31	0.04	4.83	0.05	8.01	m	9.01
25 x 150mm	4.00	10.00	4.40	0.36	0.05	5.65	0.06	10.11	m	11.38
25 x 50mm; splays -1	1.61	10.00	1.77	0.25	0.03	3.87	0.04	5.69	m	6.40
25 x 63mm; splays -1	1.88	10.00	2.06	0.28	0.04	4.41	0.05	6.52	m	7.34
25 x 75mm; splays -1	2.40	10.00	2.64	0.29	0.04	4.55	0.05	7.24	m	8.15
32 x 100mm; splays -1	3.47	10.00	3.82	0.34	0.04	5.26	0.05	9.13	m	10.27
38 x 75mm; splays -1	3.20	10.00	3.51	0.32	0.04	4.98	0.06	8.55	m	9.62
15 x 50mm; mouldings -1	1.06	10.00	1.16	0.24	0.03	3.73	0.04	4.93	m	5.55
15 x 125mm; mouldings -1	1.96	10.00	2.16	0.29	0.04	4.55	0.05	6.76	m	7.60
19 x 50mm; mouldings -1	1.33	10.00	1.47	0.25	0.03	3.87	0.04	5.38	m	6.05
19 x 125mm; mouldings -1	2.49	10.00	2.74	0.31	0.04	4.83	0.05	7.62	m	8.58
25 x 50mm; mouldings -1	1.61	10.00	1.77	0.25	0.03	3.87	0.04	5.69	m	6.40
25 x 63mm; mouldings -1	1.88	10.00	2.06	0.28	0.04	4.41	0.05	6.52	m	7.34
25 x 75mm; mouldings -1	2.40	10.00	2.64	0.29	0.04	4.55	0.05	7.24	m	8.15
32 x 100mm; mouldings -1	3.47	10.00	3.82	0.34	0.04	5.26	0.05	9.13	m	10.27
38 x 75mm; mouldings -1	3.20	10.00	3.51	0.32	0.04	4.98	0.06	8.55	m	9.62
15 x 125mm; chamfers -1	1.96	10.00	2.16	0.29	0.04	4.55	0.05	6.76	m	7.60
19 x 125mm; chamfers -1	2.49	10.00	2.74	0.30	0.04	4.69	0.05	7.48	m	8.42
25 x 75mm; chamfers -2	2.49	10.00	2.74	0.29	0.04	4.55	0.05	7.34	m	8.26
25 x 100mm; chamfers -2	3.03	10.00	3.34	0.30	0.04	4.69	0.05	8.08	m	9.09
Cover fillets, stops, trims, beads, nosings and the like; (finished sizes)										
12 x 50mm; mouldings -1	0.91	10.00	1.00	0.24	0.03	3.73	0.04	4.77	m	5.37
19 x 19mm; mouldings -1	0.70	10.00	0.77	0.22	0.03	3.45	0.04	4.26	m	4.79
19 x 50mm; mouldings -1	1.33	10.00	1.47	0.25	0.03	3.87	0.04	5.38	m	6.05
25 x 25mm; mouldings -1	0.91	10.00	1.00	0.24	0.03	3.73	0.04	4.77	m	5.37
25 x 38mm; mouldings -1	1.26	10.00	1.39	0.25	0.03	3.87	0.04	5.30	m	5.96
19 x 19mm; chamfers -1	0.70	10.00	0.77	0.22	0.03	3.45	0.04	4.26	m	4.79
25 x 25mm; chamfers -1	0.91	10.00	1.00	0.24	0.03	3.73	0.04	4.77	m	5.37
25 x 38mm; chamfers -1	1.26	10.00	1.39	0.25	0.03	3.87	0.04	5.30	m	5.96
Afrormosia, wrought - skirtings, architraves, picture rails and cover fillets										
Skirtings, picture rails, architraves and the like; (finished sizes)										
19 x 100mm	6.63	10.00	7.30	0.48	0.06	7.46	0.08	14.84	m	16.69
19 x 150mm	8.59	10.00	9.45	0.55	0.07	8.57	0.09	18.11	m	20.37
25 x 100mm	8.18	10.00	9.00	0.53	0.07	8.28	0.09	17.37	m	19.55
25 x 150mm	11.07	10.00	12.17	0.55	0.07	8.57	0.09	20.83	m	23.43
extra; ends	1.69	10.00	1.86	0.06	0.01	0.96	0.01	2.83	nr	3.19
extra; angles	2.75	-	2.75	0.11	0.02	1.78	0.02	4.55	nr	5.12
extra; mitres	2.75	-	2.75	0.11	0.02	1.78	0.02	4.55	nr	5.12
extra; intersections	2.75	-	2.75	0.11	0.02	1.78	0.02	4.55	nr	5.12
25 x 50mm; splays -1	4.80	10.00	5.28	0.43	0.05	6.65	0.07	11.99	m	13.49
25 x 63mm; splays -1	5.64	10.00	6.21	0.47	0.06	7.32	0.08	13.61	m	15.31
25 x 75mm; splays -1	6.97	10.00	7.67	0.48	0.06	7.46	0.08	15.21	m	17.12
32 x 100mm; splays -1	9.58	10.00	10.54	0.55	0.07	8.57	0.09	19.19	m	21.59
38 x 75mm; splays -1	8.89	10.00	9.78	0.55	0.07	8.57	0.09	18.43	m	20.74
extra; ends	1.33	-	1.33	0.06	0.01	0.96	0.01	2.30	nr	2.59
extra; angles	2.26	-	2.26	0.11	0.02	1.78	0.02	4.05	nr	4.56
extra; mitres	2.26	-	2.26	0.11	0.02	1.78	0.02	4.05	nr	4.56
extra; intersections	2.26	-	2.26	0.11	0.02	1.78	0.02	4.05	nr	4.56
15 x 50mm; mouldings -1	3.11	10.00	3.42	0.41	0.05	6.36	0.07	9.86	m	11.09
15 x 125mm; mouldings -1	5.93	10.00	6.52	0.48	0.06	7.46	0.08	14.07	m	15.82
19 x 50mm; mouldings -1	3.88	10.00	4.27	0.43	0.05	6.65	0.07	10.99	m	12.36
19 x 125mm; mouldings -1	7.47	10.00	8.22	0.53	0.07	8.28	0.09	16.59	m	18.66
25 x 50mm; mouldings -1	4.80	10.00	5.28	0.43	0.05	6.65	0.07	11.99	m	13.49
25 x 63mm; mouldings -1	5.64	10.00	6.21	0.47	0.06	7.32	0.08	13.61	m	15.31
25 x 75mm; mouldings -1	6.97	10.00	7.67	0.48	0.06	7.46	0.08	15.21	m	17.12
32 x 100mm; mouldings -1	9.58	10.00	10.54	0.55	0.07	8.57	0.09	19.19	m	21.59
38 x 75mm; mouldings -1	8.89	10.00	9.78	0.55	0.07	8.57	0.09	18.43	m	20.74
extra; ends	1.33	-	1.33	0.06	0.01	0.96	0.01	2.30	nr	2.59

Labour hourly rates: (except Specialists) Craft Operatives £14.21 Labourer £10.71 Rates are national average prices. Refer to REGIONAL VARIATIONS for indicative levels of overall pricing in regions	MATERIALS			LABOUR			RATES			
	Del to Site £	Waste %	Material Cost £	Craft Optve Hrs	Lab Hrs	Labour Cost £	Sunds £	Nett Rate £	Unit	Gross Rate (+12.5%) £
P20: UNFRAMED ISOLATED TRIMS/SKIRTINGS/SUNDRY ITEMS Cont.										
Afrormosia, wrought - skirtings, architraves, picture rails and cover fillets Cont.										
Skirtings, picture rails, architraves and the like; (finished sizes) Cont.										
extra; angles	2.26	-	**2.26**	0.11	0.02	**1.78**	0.02	**4.05**	nr	**4.56**
extra; mitres	2.26	-	**2.26**	0.11	0.02	**1.78**	0.02	**4.05**	nr	**4.56**
extra; intersections	2.26	-	**2.26**	0.11	0.02	**1.78**	0.02	**4.05**	nr	**4.56**
15 x 125mm; chamfers -1	5.93	10.00	**6.52**	0.48	0.06	**7.46**	0.08	**14.07**	m	**15.82**
19 x 125mm; chamfers -1	7.47	10.00	**8.22**	0.53	0.07	**8.28**	0.09	**16.59**	m	**18.66**
25 x 75mm; chamfers -2	7.19	10.00	**7.91**	0.48	0.06	**7.46**	0.08	**15.46**	m	**17.39**
25 x 100mm; chamfers -2	8.59	10.00	**9.45**	0.51	0.06	**7.89**	0.08	**17.42**	m	**19.60**
Cover fillets, stops, trims, beads, nosings and the like; (finished sizes)										
12 x 50mm; mouldings -1	2.62	10.00	**2.88**	0.41	0.05	**6.36**	0.07	**9.31**	m	**10.48**
19 x 19mm; mouldings -1	1.76	10.00	**1.94**	0.37	0.05	**5.79**	0.06	**7.79**	m	**8.76**
19 x 50mm; mouldings -1	3.88	10.00	**4.27**	0.43	0.05	**6.65**	0.07	**10.99**	m	**12.36**
25 x 25mm; mouldings -1	2.68	10.00	**2.95**	0.41	0.05	**6.36**	0.07	**9.38**	m	**10.56**
25 x 38mm; mouldings -1	3.59	10.00	**3.94**	0.43	0.05	**6.65**	0.07	**10.66**	m	**11.99**
19 x 19mm; chamfers -1	1.76	10.00	**1.94**	0.37	0.05	**5.79**	0.06	**7.79**	m	**8.76**
25 x 25mm; chamfers -1	2.68	10.00	**2.95**	0.41	0.05	**6.36**	0.07	**9.38**	m	**10.56**
25 x 38mm; chamfers -1	3.59	10.00	**3.94**	0.43	0.05	**6.65**	0.07	**10.66**	m	**11.99**
Medium density fibreboard, BS EN 622, skirtings, architraves, picture rails and cover fillets to BS 1186 Part 3										
Skirtings, picture rails, architraves and the like; (finished sizes)										
13 x 45mm; reference 13CA45	1.28	10.00	**1.41**	0.32	0.05	**5.08**	0.07	**6.57**	m	**7.39**
20 x 70mm; reference 20CA70	2.01	10.00	**2.21**	0.34	0.05	**5.37**	0.07	**7.65**	m	**8.60**
13 x 45mm; reference 13CP45	1.28	10.00	**1.41**	0.33	0.05	**5.22**	0.07	**6.71**	m	**7.55**
13 x 70mm; reference 13CP70	1.62	10.00	**1.78**	0.33	0.05	**5.22**	0.07	**7.07**	m	**7.96**
13 x 70mm; reference 13CS70	1.62	10.00	**1.78**	0.35	0.05	**5.51**	0.07	**7.36**	m	**8.28**
13 x 95mm; reference 13CS95	2.13	10.00	**2.34**	0.33	0.05	**5.22**	0.07	**7.64**	m	**8.59**
20 x 120mm; reference 20CS120	3.30	10.00	**3.62**	0.38	0.05	**5.94**	0.07	**9.63**	m	**10.83**
9 x 33mm; reference 9RA33	0.96	10.00	**1.06**	0.28	0.03	**4.30**	0.07	**5.43**	m	**6.11**
13 x 45mm; reference 13RA45	1.28	10.00	**1.41**	0.33	0.05	**5.22**	0.07	**6.71**	m	**7.55**
20 x 70mm; reference 20RA70	2.13	10.00	**2.34**	0.34	0.05	**5.37**	0.07	**7.78**	m	**8.75**
13 x 70mm; reference 13RS70	1.71	10.00	**1.88**	0.35	0.05	**5.51**	0.07	**7.46**	m	**8.40**
13 x 95mm; reference 13RS95	2.22	10.00	**2.45**	0.33	0.05	**5.22**	0.07	**7.74**	m	**8.71**
20 x 120mm; reference 20RS120	3.30	10.00	**3.62**	0.38	0.05	**5.94**	0.07	**9.63**	m	**10.83**
13 x 45mm; reference 13RP45	1.34	10.00	**1.48**	0.33	0.05	**5.22**	0.07	**6.77**	m	**7.62**
13 x 70mm; reference 13RP70	1.81	10.00	**1.99**	0.35	0.05	**5.51**	0.07	**7.57**	m	**8.51**
Cover fillets, stops, trims, beads, nosings and the like; (finished sizes)										
11 x 33mm; reference C33	0.96	10.00	**1.06**	0.27	0.02	**4.05**	0.06	**5.17**	m	**5.82**
11 x 45mm; reference C45	1.01	10.00	**1.11**	0.27	0.02	**4.05**	0.06	**5.22**	m	**5.88**
13 x 33mm; reference HR33	1.13	10.00	**1.24**	0.27	0.02	**4.05**	0.06	**5.35**	m	**6.02**
20 x 45mm; reference HR45	1.52	10.00	**1.68**	0.29	0.04	**4.55**	0.06	**6.28**	m	**7.07**
11 x 11mm; reference Q11	0.51	10.00	**0.56**	0.27	0.02	**4.05**	0.06	**4.67**	m	**5.26**
13 x 13mm; reference Q13	0.79	10.00	**0.86**	0.27	0.02	**4.05**	0.06	**4.97**	m	**5.60**
20 x 20mm; reference Q20	1.01	10.00	**1.11**	0.27	0.02	**4.05**	0.06	**5.22**	m	**5.88**
13 x 13mm; reference S13	0.79	10.00	**0.86**	0.27	0.02	**4.05**	0.06	**4.97**	m	**5.60**
20 x 20mm; reference S20	1.01	10.00	**1.11**	0.27	0.02	**4.05**	0.06	**5.22**	m	**5.88**
27 x 27mm; reference S27	1.28	10.00	**1.41**	0.29	0.04	**4.55**	0.06	**6.02**	m	**6.78**
12 x 95mm; reference 12HRCS95	2.38	10.00	**2.62**	0.35	0.05	**5.51**	0.06	**8.19**	m	**9.21**
15 x 95mm; reference 15HRCS95	2.50	10.00	**2.75**	0.38	0.05	**5.94**	0.06	**8.74**	m	**9.84**
Softwood, wrought - cappings										
Cover fillets, stops, trims, beads, nosings and the like; (finished sizes)										
25 x 50mm; level; rebates -1; mouldings -1	1.61	10.00	**1.77**	0.25	0.03	**3.87**	0.04	**5.69**	m	**6.40**
50 x 75mm; level; rebates -1; mouldings -1	4.63	10.00	**5.09**	0.36	0.05	**5.65**	0.06	**10.81**	m	**12.16**
50 x 100mm; level; rebates -1; mouldings -1	5.79	10.00	**6.37**	0.39	0.05	**6.08**	0.06	**12.51**	m	**14.07**
25 x 50mm; ramped; rebates -1; mouldings -1	1.61	10.00	**1.77**	0.38	0.05	**5.94**	0.06	**7.77**	m	**8.74**
50 x 75mm; ramped; rebates -1; mouldings -1	4.63	10.00	**5.09**	0.53	0.07	**8.28**	0.09	**13.47**	m	**15.15**
50 x 100mm; ramped; rebates -1; mouldings -1	5.79	10.00	**6.37**	0.59	0.07	**9.13**	0.10	**15.60**	m	**17.55**
Afrormosia, wrought - cappings										
Cover fillets, stops, trims, beads, nosings and the like; (finished sizes)										
25 x 50mm; level; rebates -1; mouldings -1	4.94	10.00	**5.43**	0.43	0.05	**6.65**	0.07	**12.15**	m	**13.67**
50 x 75mm; level; rebates -1; mouldings -1	12.69	10.00	**13.96**	0.61	0.08	**9.52**	0.10	**23.59**	m	**26.54**
extra; ends	1.90	-	**1.90**	0.09	0.01	**1.39**	0.01	**3.30**	nr	**3.71**
extra; angles	3.18	-	**3.18**	0.15	0.02	**2.35**	0.02	**5.54**	nr	**6.24**
extra; mitres	3.18	-	**3.18**	0.15	0.02	**2.35**	0.02	**5.54**	nr	**6.24**
extra; rounded corners not exceeding 300mm girth	12.69	-	**12.69**	0.61	0.08	**9.52**	0.10	**22.32**	nr	**25.11**
50 x 100mm; level; rebates -1; mouldings -1	15.79	10.00	**17.36**	0.67	0.08	**10.38**	0.11	**27.85**	m	**31.33**
extra; ends	2.40	-	**2.40**	0.10	0.01	**1.53**	0.02	**3.95**	nr	**4.44**
extra; angles	3.95	-	**3.95**	0.16	0.02	**2.49**	0.03	**6.47**	nr	**7.28**
extra; mitres	3.95	-	**3.95**	0.16	0.02	**2.49**	0.03	**6.47**	nr	**7.28**
extra; rounded corners not exceeding 300mm girth	15.79	-	**15.79**	0.67	0.08	**10.38**	0.11	**26.27**	nr	**29.56**
25 x 50mm; ramped; rebates -1; mouldings -1	4.94	10.00	**5.43**	0.63	0.08	**9.81**	0.10	**15.34**	m	**17.26**
50 x 75mm; ramped; rebates -1; mouldings -1	12.69	10.00	**13.96**	0.92	0.12	**14.36**	0.15	**28.47**	m	**32.03**
extra; ends	1.90	-	**1.90**	0.14	0.02	**2.20**	0.02	**4.13**	nr	**4.64**
extra; angles	3.18	-	**3.18**	0.23	0.03	**3.59**	0.04	**6.81**	nr	**7.66**
extra; mitres	3.18	-	**3.18**	0.23	0.03	**3.59**	0.04	**6.81**	nr	**7.66**
extra; rounded corners not exceeding 300mm girth	12.69	-	**12.69**	0.92	0.12	**14.36**	0.15	**27.20**	nr	**30.60**

Labour hourly rates: (except Specialists) Craft Operatives £14.21 Labourer £10.71 Rates are national average prices. Refer to REGIONAL VARIATIONS for indicative levels of overall pricing in regions	MATERIALS			LABOUR				RATES		
	Del to Site £	Waste %	Material Cost £	Craft Optve Hrs	Lab Hrs	Labour Cost £	Sunds £	Nett Rate £	Unit	Gross Rate (+12.5%) £
P20: UNFRAMED ISOLATED TRIMS/SKIRTINGS/SUNDRY ITEMS Cont.										
Afrormosia, wrought - cappings Cont.										
Cover fillets, stops, trims, beads, nosings and the like; (finished sizes) Cont.										
50 x 100mm; ramped; rebates -1; mouldings -1	15.79	10.00	**17.36**	1.05	0.13	**16.31**	0.17	**33.85**	m	**38.08**
extra; ends	2.40	-	**2.40**	0.15	0.02	**2.35**	0.02	**4.76**	nr	**5.36**
extra; angles	3.95	-	**3.95**	0.25	0.03	**3.87**	0.04	**7.86**	nr	**8.85**
extra; mitres	3.95	-	**3.95**	0.25	0.03	**3.87**	0.04	**7.86**	nr	**8.85**
extra; rounded corners not exceeding 300mm girth	15.79	-	**15.79**	1.05	0.13	**16.31**	0.17	**32.27**	nr	**36.30**
Sapele; wrought - cappings										
Cover fillets, stops, trims, beads, nosings and the like; (finished sizes)										
25 x 50mm; level; rebates -1; mouldings -1	4.09	10.00	**4.50**	0.38	0.05	**5.94**	0.06	**10.49**	m	**11.80**
50 x 75mm; level; rebates -1; mouldings -1	10.21	10.00	**11.23**	0.53	0.07	**8.28**	0.08	**19.59**	m	**22.04**
extra; ends	1.52	-	**1.52**	0.08	0.01	**1.24**	0.01	**2.77**	nr	**3.12**
extra; angles	2.57	-	**2.57**	0.14	0.02	**2.20**	0.02	**4.79**	nr	**5.39**
extra; mitres	2.57	-	**2.57**	0.14	0.02	**2.20**	0.02	**4.79**	nr	**5.39**
extra; rounded corners not exceeding 300mm girth	10.21	-	**10.21**	0.53	0.07	**8.28**	0.09	**18.58**	nr	**20.90**
50 x 100mm; level; rebates -1; mouldings -1	12.51	10.00	**13.77**	0.57	0.07	**8.85**	0.09	**22.71**	m	**25.54**
extra; ends	1.91	-	**1.91**	0.09	0.01	**1.39**	0.01	**3.30**	nr	**3.71**
extra; angles	3.16	-	**3.16**	0.15	0.02	**2.35**	0.02	**5.53**	nr	**6.22**
extra; mitres	3.16	-	**3.16**	0.15	0.02	**2.35**	0.02	**5.53**	nr	**6.22**
extra; rounded corners not exceeding 300mm girth	12.51	-	**12.51**	0.57	0.07	**8.85**	0.09	**21.45**	nr	**24.14**
25 x 50mm; ramped; rebates -1; mouldings -1	4.09	10.00	**4.50**	0.57	0.07	**8.85**	0.09	**13.44**	m	**15.12**
50 x 75mm; ramped; rebates -1; mouldings -1	10.21	10.00	**11.23**	0.80	0.10	**12.44**	0.13	**23.80**	m	**26.77**
extra; ends	1.52	10.00	**1.67**	0.11	0.01	**1.67**	0.02	**3.36**	nr	**3.78**
extra; angles	2.57	-	**2.57**	0.20	0.03	**3.16**	0.03	**5.76**	nr	**6.48**
extra; mitres	2.57	-	**2.57**	0.20	0.03	**3.16**	0.03	**5.76**	nr	**6.48**
extra; rounded corners not exceeding 300mm girth	10.21	-	**10.21**	0.80	0.10	**12.44**	0.13	**22.78**	nr	**25.62**
50 x 100mm; ramped; rebates -1; mouldings -1	12.51	10.00	**13.77**	0.85	0.11	**13.26**	0.14	**27.16**	m	**30.56**
extra; ends	1.91	-	**1.91**	0.13	0.02	**2.06**	0.02	**3.99**	nr	**4.49**
extra; angles	3.16	-	**3.16**	0.22	0.03	**3.45**	0.04	**6.65**	nr	**7.48**
extra; mitres	3.16	-	**3.16**	0.22	0.03	**3.45**	0.04	**6.65**	nr	**7.48**
extra; rounded corners not exceeding 300mm girth	12.51	-	**12.51**	0.85	0.11	**13.26**	0.14	**25.91**	nr	**29.15**
Softwood, wrought - window boards										
Cover fillets, stops, trims, beads, nosings and the like; (finished sizes)										
mouldings -1; tongued on										
25 x 50mm	1.88	10.00	**2.06**	0.29	0.04	**4.55**	0.05	**6.66**	m	**7.50**
25 x 75mm	3.03	10.00	**3.34**	0.30	0.04	**4.69**	0.05	**8.08**	m	**9.09**
32 x 50mm	2.33	10.00	**2.56**	0.32	0.04	**4.98**	0.05	**7.59**	m	**8.54**
32 x 75mm	3.58	10.00	**3.93**	0.33	0.04	**5.12**	0.05	**9.10**	m	**10.24**
rounded edges -1; tongued on										
25 x 50mm	1.88	10.00	**2.06**	0.29	0.04	**4.55**	0.05	**6.66**	m	**7.50**
25 x 75mm	3.03	10.00	**3.34**	0.30	0.04	**4.69**	0.05	**8.08**	m	**9.09**
32 x 50mm	2.33	10.00	**2.56**	0.32	0.04	**4.98**	0.05	**7.59**	m	**8.54**
32 x 75mm	3.58	10.00	**3.93**	0.33	0.04	**5.12**	0.05	**9.10**	m	**10.24**
Window boards										
25 x 150mm; mouldings -1	4.10	10.00	**4.51**	0.34	0.04	**5.26**	0.05	**9.82**	m	**11.05**
32 x 150mm; mouldings -1	4.82	10.00	**5.30**	0.38	0.05	**5.94**	0.06	**11.30**	m	**12.71**
25 x 150mm; rounded edges -1	4.10	10.00	**4.51**	0.34	0.04	**5.26**	0.05	**9.82**	m	**11.05**
32 x 150mm; rounded edges -1	4.82	10.00	**5.30**	0.38	0.05	**5.94**	0.06	**11.30**	m	**12.71**
Afrormosia, wrought - window boards										
Cover fillets, stops, trims, beads, nosings and the like; (finished sizes)										
mouldings -1; tongued on										
25 x 50mm	4.58	10.00	**5.03**	0.44	0.06	**6.90**	0.07	**12.00**	m	**13.50**
25 x 75mm	7.11	10.00	**7.82**	0.46	0.06	**7.18**	0.08	**15.08**	m	**16.96**
32 x 50mm	5.51	10.00	**6.06**	0.48	0.06	**7.46**	0.08	**13.61**	m	**15.31**
32 x 75mm	8.44	10.00	**9.28**	0.50	0.06	**7.75**	0.08	**17.11**	m	**19.25**
rounded edges -1; tongued on										
25 x 50mm	4.58	10.00	**5.03**	0.44	0.06	**6.90**	0.07	**12.00**	m	**13.50**
25 x 75mm	7.11	10.00	**7.82**	0.46	0.06	**7.18**	0.08	**15.08**	m	**16.96**
32 x 50mm	5.51	10.00	**6.06**	0.48	0.06	**7.46**	0.08	**13.61**	m	**15.31**
32 x 75mm	8.44	10.00	**9.28**	0.50	0.06	**7.75**	0.08	**17.11**	m	**19.25**
Window boards										
25 x 150mm; mouldings -1	11.30	10.00	**12.43**	0.60	0.08	**9.38**	0.10	**21.91**	m	**24.65**
32 x 150mm; mouldings -1	13.42	10.00	**14.76**	0.67	0.08	**10.38**	0.11	**25.25**	m	**28.41**
25 x 150mm; rounded edges -1	11.30	10.00	**12.43**	0.60	0.08	**9.38**	0.10	**21.91**	m	**24.65**
32 x 150mm; rounded edges -1	13.42	10.00	**14.76**	0.67	0.08	**10.38**	0.11	**25.25**	m	**28.41**
Sapele; wrought - window boards										
Cover fillets, stops, trims, beads, nosings and the like; (finished sizes)										
mouldings -1; tongued on										
25 x 50mm	3.82	10.00	**4.21**	0.43	0.05	**6.65**	0.07	**10.92**	m	**12.29**
25 x 75mm	5.86	10.00	**6.44**	0.45	0.06	**7.04**	0.07	**13.55**	m	**15.24**
32 x 50mm	4.48	10.00	**4.92**	0.47	0.06	**7.32**	0.08	**12.33**	m	**13.87**
32 x 75mm	6.98	10.00	**7.68**	0.49	0.06	**7.61**	0.08	**15.37**	m	**17.29**
rounded edges -1; tongued on										
25 x 50mm	3.82	10.00	**4.21**	0.43	0.05	**6.65**	0.07	**10.92**	m	**12.29**
25 x 75mm	5.86	10.00	**6.44**	0.45	0.06	**7.04**	0.07	**13.55**	m	**15.24**
32 x 50mm	4.48	10.00	**4.92**	0.47	0.06	**7.32**	0.08	**12.33**	m	**13.87**
32 x 75mm	6.98	10.00	**7.68**	0.49	0.06	**7.61**	0.08	**15.37**	m	**17.29**

Labour hourly rates: (except Specialists) Craft Operatives £14.21 Labourer £10.71 Rates are national average prices. Refer to REGIONAL VARIATIONS for indicative levels of overall pricing in regions	MATERIALS			LABOUR				RATES		
	Del to Site £	Waste %	Material Cost £	Craft Optve Hrs	Lab Hrs	Labour Cost £	Sunds £	Nett Rate £	Unit	Gross Rate (+12.5%) £
P20: UNFRAMED ISOLATED TRIMS/SKIRTINGS/SUNDRY ITEMS Cont.										
Sapele; wrought - window boards Cont.										
Window boards										
25 x 150mm; mouldings -1	9.28	10.00	**10.21**	0.52	0.07	**8.14**	0.09	**18.44**	m	**20.74**
32 x 150mm; mouldings -1	11.00	10.00	**12.10**	0.57	0.07	**8.85**	0.09	**21.04**	m	**23.67**
25 x 150mm; rounded edges -1	9.28	10.00	**10.21**	0.52	0.07	**8.14**	0.09	**18.44**	m	**20.74**
32 x 150mm; rounded edges -1	11.00	10.00	**12.10**	0.57	0.07	**8.85**	0.09	**21.04**	m	**23.67**
Moisture resistant medium density fibreboard, BS EN 622										
Window boards										
25 x 225mm; nosed and tongued	6.27	10.00	**6.90**	0.34	0.05	**5.37**	0.06	**12.32**	m	**13.86**
25 x 250mm; nosed and tongued	6.97	10.00	**7.66**	0.36	0.05	**5.65**	0.06	**13.38**	m	**15.05**
Softwood, wrought - isolated shelves and worktops										
Cover fillets, stops, trims, beads, nosings and the like; (finished sizes)										
19 x 12mm	0.37	10.00	**0.41**	0.21	0.03	**3.31**	0.04	**3.76**	m	**4.23**
19 x 16mm	0.45	10.00	**0.50**	0.22	0.03	**3.45**	0.04	**3.99**	m	**4.49**
19 x 18mm	0.52	10.00	**0.57**	0.22	0.03	**3.45**	0.04	**4.06**	m	**4.57**
19 x 22mm	0.57	10.00	**0.63**	0.23	0.03	**3.59**	0.04	**4.26**	m	**4.79**
19 x 25mm	0.64	10.00	**0.71**	0.23	0.03	**3.59**	0.04	**4.33**	m	**4.88**
Isolated shelves and worktops										
25 x 150mm	2.44	10.00	**2.69**	0.23	0.03	**3.59**	0.04	**6.32**	m	**7.11**
25 x 225mm	3.65	10.00	**4.01**	0.29	0.04	**4.55**	0.05	**8.61**	m	**9.69**
25 x 300mm	5.14	10.00	**5.66**	0.34	0.04	**5.26**	0.05	**10.97**	m	**12.34**
32 x 150mm	3.10	10.00	**3.41**	0.25	0.03	**3.87**	0.04	**7.32**	m	**8.24**
32 x 225mm	4.71	10.00	**5.19**	0.31	0.04	**4.83**	0.05	**10.07**	m	**11.33**
32 x 300mm	6.24	10.00	**6.86**	0.38	0.05	**5.94**	0.06	**12.86**	m	**14.46**
25 x 450mm; cross tongued	10.96	2.50	**11.24**	0.40	0.05	**6.22**	0.06	**17.52**	m	**19.71**
25 x 600mm; cross tongued	14.61	2.50	**14.98**	0.52	0.07	**8.14**	0.09	**23.20**	m	**26.11**
32 x 450mm; cross tongued	14.04	2.50	**14.39**	0.46	0.06	**7.18**	0.08	**21.65**	m	**24.35**
32 x 600mm; cross tongued	18.67	2.50	**19.14**	0.57	0.07	**8.85**	0.09	**28.08**	m	**31.59**
38 x 450mm; cross tongued	16.64	2.50	**17.06**	0.52	0.07	**8.14**	0.09	**25.29**	m	**28.45**
38 x 600mm; cross tongued	22.52	2.50	**23.08**	0.63	0.08	**9.81**	0.10	**32.99**	m	**37.11**
25 x 450mm overall; 25 x 50mm slats spaced 25mm apart	5.69	10.00	**6.26**	1.15	0.14	**17.84**	0.19	**24.29**	m	**27.33**
25 x 450mm overall; 25 x 50mm slats spaced 32mm apart	4.87	10.00	**5.36**	1.05	0.13	**16.31**	0.17	**21.84**	m	**24.58**
25 x 600mm overall; 25 x 50mm slats spaced 25mm apart	7.32	10.00	**8.05**	1.45	0.18	**22.53**	0.24	**30.82**	m	**34.68**
25 x 600mm overall; 25 x 50mm slats spaced 32mm apart	6.49	10.00	**7.14**	1.35	0.17	**21.00**	0.22	**28.36**	m	**31.91**
25 x 900mm overall; 25 x 50mm slats spaced 25mm apart	10.56	10.00	**11.62**	2.20	0.27	**34.15**	0.36	**46.13**	m	**51.90**
25 x 900mm overall; 25 x 50mm slats spaced 32mm apart	9.75	10.00	**10.72**	2.05	0.26	**31.92**	0.33	**42.97**	m	**48.34**
Afrormosia, wrought - isolated shelves and worktops										
Cover fillets, stops, trims, beads, nosings and the like; (finished sizes)										
19 x 12mm	1.10	10.00	**1.21**	0.35	0.04	**5.40**	0.06	**6.67**	m	**7.51**
19 x 16mm	1.30	10.00	**1.43**	0.38	0.05	**5.94**	0.06	**7.42**	m	**8.35**
19 x 18mm	1.50	10.00	**1.65**	0.38	0.05	**5.94**	0.06	**7.64**	m	**8.60**
19 x 22mm	1.62	10.00	**1.78**	0.40	0.05	**6.22**	0.06	**8.06**	m	**9.07**
19 x 25mm	1.88	10.00	**2.07**	0.40	0.05	**6.22**	0.06	**8.35**	m	**9.39**
Isolated shelves and worktops										
25 x 150mm	8.43	10.00	**9.27**	0.40	0.05	**6.22**	0.06	**15.55**	m	**17.49**
25 x 225mm	12.65	10.00	**13.92**	0.51	0.06	**7.89**	0.08	**21.88**	m	**24.62**
25 x 300mm	16.86	10.00	**18.55**	0.61	0.08	**9.52**	0.10	**28.17**	m	**31.70**
32 x 150mm	10.70	10.00	**11.77**	0.44	0.06	**6.90**	0.07	**18.74**	m	**21.08**
32 x 225mm	15.88	10.00	**17.47**	0.54	0.07	**8.42**	0.09	**25.99**	m	**29.23**
32 x 300mm	21.40	10.00	**23.53**	0.64	0.08	**9.95**	0.10	**33.59**	m	**37.78**
25 x 450mm; cross tongued	35.01	2.50	**35.89**	0.70	0.09	**10.91**	0.11	**46.91**	m	**52.77**
25 x 600mm; cross tongued	46.68	2.50	**47.85**	0.90	0.11	**13.97**	0.15	**61.97**	m	**69.71**
32 x 450mm; cross tongued	44.73	2.50	**45.84**	0.80	0.10	**12.44**	0.13	**58.41**	m	**65.71**
32 x 600mm; cross tongued	59.64	2.50	**61.13**	1.00	0.13	**15.60**	0.16	**76.90**	m	**86.51**
38 x 450mm; cross tongued	53.16	2.50	**54.49**	0.90	0.11	**13.97**	0.15	**68.61**	m	**77.19**
38 x 600mm; cross tongued	71.32	2.50	**73.11**	1.10	0.14	**17.13**	0.18	**90.42**	m	**101.72**
Plywood BS EN 636, II/III grade, INT bonded; butt joints - isolated shelves and worktops										
Cover fillets, stops, trims, beads, nosings and the like										
50 x 6.5mm	0.27	20.00	**0.32**	0.17	0.02	**2.63**	0.06	**3.01**	m	**3.39**
50 x 12mm	0.37	20.00	**0.45**	0.18	0.02	**2.77**	0.06	**3.28**	m	**3.69**
50 x 15mm	0.46	20.00	**0.55**	0.20	0.03	**3.16**	0.06	**3.78**	m	**4.25**
50 x 19mm	0.54	20.00	**0.65**	0.21	0.03	**3.31**	0.06	**4.01**	m	**4.51**
100 x 6.5mm	0.41	20.00	**0.49**	0.18	0.02	**2.77**	0.06	**3.32**	m	**3.74**
100 x 12mm	0.73	20.00	**0.88**	0.20	0.03	**3.16**	0.06	**4.10**	m	**4.61**
100 x 15mm	0.95	20.00	**1.14**	0.21	0.03	**3.31**	0.06	**4.50**	m	**5.07**
100 x 19mm	1.06	20.00	**1.28**	0.22	0.03	**3.45**	0.06	**4.78**	m	**5.38**
Isolated shelves and worktops										
6.5 x 150mm	0.88	20.00	**1.06**	0.21	0.03	**3.31**	0.06	**4.43**	m	**4.98**
6.5 x 225mm	1.36	20.00	**1.63**	0.26	0.03	**4.02**	0.06	**5.70**	m	**6.42**
6.5 x 300mm	1.81	20.00	**2.17**	0.32	0.04	**4.98**	0.06	**7.20**	m	**8.10**

Labour hourly rates: (except Specialists) Craft Operatives £14.21 Labourer £10.71 Rates are national average prices. Refer to REGIONAL VARIATIONS for indicative levels of overall pricing in regions	MATERIALS			LABOUR			RATES			
	Del to Site	Waste	Material Cost	Craft Optve	Lab	Labour Cost	Sunds	Nett Rate	Unit	Gross Rate (+12.5%)
	£	%	£	Hrs	Hrs	£	£	£		£
P20: UNFRAMED ISOLATED TRIMS/SKIRTINGS/SUNDRY ITEMS Cont.										
Plywood BS EN 636, II/III grade, INT bonded; butt joints - isolated shelves and worktops Cont.										
Isolated shelves and worktops Cont.										
6.5 x 450mm	2.69	20.00	3.23	0.38	0.05	5.94	0.06	9.22	m	10.37
6.5 x 600mm	3.57	20.00	4.29	0.44	0.06	6.90	0.06	11.24	m	12.65
12 x 150mm	1.66	20.00	2.00	0.22	0.03	3.45	0.06	5.50	m	6.19
12 x 225mm	2.48	20.00	2.98	0.28	0.04	4.41	0.06	7.45	m	8.38
12 x 300mm	3.34	20.00	4.01	0.33	0.04	5.12	0.06	9.19	m	10.34
12 x 450mm	4.98	20.00	5.98	0.39	0.05	6.08	0.06	12.11	m	13.63
12 x 600mm	6.63	20.00	7.96	0.45	0.06	7.04	0.06	15.06	m	16.94
15 x 150mm	2.00	20.00	2.40	0.23	0.03	3.59	0.06	6.05	m	6.80
15 x 225mm	3.01	20.00	3.61	0.29	0.04	4.55	0.06	8.22	m	9.25
15 x 300mm	4.02	20.00	4.82	0.34	0.04	5.26	0.06	10.14	m	11.41
15 x 450mm	6.02	20.00	7.22	0.40	0.05	6.22	0.06	13.50	m	15.19
15 x 600mm	8.03	20.00	9.63	0.46	0.06	7.18	0.06	16.87	m	18.98
19 x 150mm	2.22	20.00	2.66	0.23	0.03	3.59	0.06	6.31	m	7.10
19 x 225mm	3.33	20.00	3.99	0.29	0.04	4.55	0.06	8.60	m	9.68
19 x 300mm	4.40	20.00	5.29	0.34	0.04	5.26	0.06	10.61	m	11.93
19 x 450mm	6.61	20.00	7.93	0.40	0.05	6.22	0.06	14.21	m	15.98
19 x 600mm	8.81	20.00	10.57	0.46	0.06	7.18	0.06	17.81	m	20.04
Blockboard, BS 3444, 2/2 grade, INT bonded, butt joints - isolated shelves and worktops										
Isolated shelves and worktops										
16 x 150mm	5.00	20.00	6.00	0.23	0.03	3.59	0.09	9.68	m	10.88
16 x 225mm	7.49	20.00	8.99	0.29	0.04	4.55	0.09	13.63	m	15.34
16 x 300mm	9.99	20.00	11.99	0.34	0.04	5.26	0.09	17.34	m	19.51
16 x 450mm	14.99	20.00	17.99	0.40	0.05	6.22	0.09	24.30	m	27.33
16 x 600mm	19.91	20.00	23.89	0.46	0.06	7.18	0.09	31.16	m	35.06
18 x 150mm	5.15	20.00	6.18	0.23	0.03	3.59	0.09	9.86	m	11.09
18 x 225mm	7.72	20.00	9.26	0.29	0.04	4.55	0.09	13.90	m	15.64
18 x 300mm	10.27	20.00	12.32	0.34	0.04	5.26	0.09	17.67	m	19.88
18 x 450mm	15.39	20.00	18.47	0.40	0.05	6.22	0.09	24.78	m	27.87
18 x 600mm	20.56	20.00	24.67	0.46	0.06	7.18	0.09	31.94	m	35.93
22 x 150mm	6.95	20.00	8.33	0.23	0.03	3.59	0.09	12.01	m	13.52
22 x 225mm	10.42	20.00	12.50	0.29	0.04	4.55	0.09	17.14	m	19.28
22 x 300mm	13.42	20.00	16.10	0.34	0.04	5.26	0.09	21.45	m	24.13
22 x 450mm	20.84	20.00	25.00	0.40	0.05	6.22	0.09	31.31	m	35.23
22 x 600mm	27.78	20.00	33.34	0.46	0.06	7.18	0.09	40.61	m	45.68
25 x 150mm	7.52	20.00	9.02	0.23	0.03	3.59	0.09	12.70	m	14.29
25 x 225mm	11.22	20.00	13.46	0.29	0.04	4.55	0.09	18.10	m	20.36
25 x 300mm	14.99	20.00	17.99	0.34	0.04	5.26	0.09	23.34	m	26.25
25 x 450mm	22.48	20.00	26.98	0.40	0.05	6.22	0.09	33.29	m	37.45
25 x 600mm	29.98	20.00	35.98	0.46	0.06	7.18	0.09	43.25	m	48.65
Wood chipboard BS EN 312, type I standard; butt joints - isolated shelves and worktops										
Isolated shelves and worktops										
12 x 150mm	0.55	20.00	0.66	0.22	0.03	3.45	0.06	4.17	m	4.69
12 x 225mm	0.79	20.00	0.95	0.28	0.04	4.41	0.06	5.42	m	6.10
12 x 300mm	1.08	20.00	1.29	0.33	0.04	5.12	0.06	6.47	m	7.28
12 x 450mm	1.60	20.00	1.92	0.39	0.05	6.08	0.06	8.06	m	9.07
12 x 600mm	2.15	20.00	2.58	0.45	0.06	7.04	0.06	9.68	m	10.89
18 x 150mm	0.72	20.00	0.86	0.23	0.03	3.59	0.06	4.51	m	5.07
18 x 225mm	1.06	20.00	1.28	0.29	0.04	4.55	0.06	5.88	m	6.62
18 x 300mm	1.40	20.00	1.67	0.34	0.04	5.26	0.06	6.99	m	7.87
18 x 450mm	2.10	20.00	2.52	0.40	0.05	6.22	0.06	8.80	m	9.90
18 x 600mm	2.79	20.00	3.35	0.46	0.06	7.18	0.06	10.59	m	11.91
25 x 150mm	1.32	20.00	1.58	0.23	0.03	3.59	0.06	5.23	m	5.89
25 x 225mm	1.97	20.00	2.37	0.29	0.04	4.55	0.06	6.98	m	7.85
25 x 300mm	2.62	20.00	3.15	0.34	0.04	5.26	0.06	8.47	m	9.53
25 x 450mm	3.93	20.00	4.72	0.40	0.05	6.22	0.06	11.00	m	12.37
25 x 600mm	5.24	20.00	6.28	0.46	0.06	7.18	0.06	13.52	m	15.21
Blockboard, BS 3444, 2/2 grade, INT bonded, faced with 1.5mm laminated plastic sheet, BS EN 438, classified HGS, with 1.2mm laminated plastic sheet balance veneer; butt joints										
Isolated shelves and worktops										
16 x 150mm	31.60	20.00	37.92	0.46	0.06	7.18	0.12	45.22	m	50.88
16 x 225mm	42.16	20.00	50.59	0.57	0.07	8.85	0.12	59.56	m	67.01
16 x 300mm	52.72	20.00	63.26	0.69	0.09	10.77	0.12	74.15	m	83.42
16 x 450mm	71.68	20.00	86.02	0.80	0.10	12.44	0.12	98.58	m	110.90
16 x 600mm	84.32	20.00	101.19	0.92	0.12	14.36	0.12	115.67	m	130.12
18 x 150mm	31.60	20.00	37.92	0.46	0.06	7.18	0.12	45.22	m	50.88
18 x 225mm	42.16	20.00	50.59	0.57	0.07	8.85	0.12	59.56	m	67.01
18 x 300mm	52.72	20.00	63.26	0.69	0.09	10.77	0.12	74.15	m	83.42
18 x 450mm	71.68	20.00	86.02	0.80	0.10	12.44	0.12	98.58	m	110.90
18 x 600mm	83.60	20.00	100.33	0.92	0.12	14.36	0.12	114.80	m	129.15
22 x 150mm	33.73	20.00	40.47	0.46	0.06	7.18	0.12	47.77	m	53.75
22 x 225mm	45.97	20.00	55.16	0.57	0.07	8.85	0.12	64.13	m	72.15
22 x 300mm	54.80	20.00	65.76	0.69	0.09	10.77	0.12	76.65	m	86.23
22 x 450mm	73.79	20.00	88.55	0.80	0.10	12.44	0.12	101.11	m	113.75
22 x 600mm	86.32	20.00	103.58	0.92	0.11	14.25	0.12	117.95	m	132.70
25 x 150mm	35.91	20.00	43.10	0.46	0.06	7.18	0.12	50.39	m	56.69
25 x 225mm	46.38	20.00	55.66	0.57	0.07	8.85	0.12	64.63	m	72.71
25 x 300mm	57.01	20.00	68.42	0.69	0.09	10.77	0.12	79.31	m	89.22
25 x 450mm	75.93	20.00	91.12	0.80	0.10	12.44	0.12	103.68	m	116.64
25 x 600mm	88.55	20.00	106.26	0.92	0.11	14.25	0.12	120.63	m	135.70

Labour hourly rates: (except Specialists) Craft Operatives £14.21 Labourer £10.71 Rates are national average prices. Refer to REGIONAL VARIATIONS for indicative levels of overall pricing in regions	MATERIALS			LABOUR				RATES		
	Del to Site £	Waste %	Material Cost £	Craft Optve Hrs	Lab Hrs	Labour Cost £	Sunds £	Nett Rate £	Unit	Gross Rate (+12.5%) £
P20: UNFRAMED ISOLATED TRIMS/SKIRTINGS/SUNDRY ITEMS Cont.										
Laminated plastic sheet, BS EN 438 classification HGS										
Cover fillets, stops, trims, beads, nosings and the like; fixing with adhesive										
1.5 x 16mm	0.68	20.00	**0.81**	0.23	0.03	**3.59**	0.07	**4.47**	m	**5.03**
1.5 x 18mm	0.76	20.00	**0.91**	0.24	0.03	**3.73**	0.08	**4.72**	m	**5.31**
1.5 x 22mm	0.82	20.00	**0.99**	0.26	0.03	**4.02**	0.09	**5.09**	m	**5.73**
1.5 x 25mm	0.90	20.00	**1.08**	0.29	0.04	**4.55**	0.10	**5.73**	m	**6.44**
1.5 x 75mm	1.72	20.00	**2.06**	0.34	0.04	**5.26**	0.30	**7.62**	m	**8.58**
Isolated shelves and worktops; fixing with adhesive										
1.5 x 150mm	3.41	20.00	**4.10**	0.40	0.05	**6.22**	0.60	**10.92**	m	**12.28**
1.5 x 225mm	5.11	20.00	**6.13**	0.46	0.06	**7.18**	0.90	**14.21**	m	**15.98**
1.5 x 300mm	6.92	20.00	**8.30**	0.52	0.07	**8.14**	1.20	**17.64**	m	**19.85**
1.5 x 450mm	10.33	20.00	**12.40**	0.61	0.08	**9.52**	1.80	**23.72**	m	**26.69**
1.5 x 600mm	13.75	20.00	**16.50**	0.69	0.09	**10.77**	2.40	**29.66**	m	**33.37**
P21: IRONMONGERY										
Fix only ironmongery (prime cost sum for supply included elsewhere)										
To softwood										
butt hinges; 50mm	-	-	-	0.17	-	**2.42**	-	**2.42**	nr	**2.72**
butt hinges; 100mm	-	-	-	0.17	-	**2.42**	-	**2.42**	nr	**2.72**
rising hinges; 100mm	-	-	-	0.30	-	**4.26**	-	**4.26**	nr	**4.80**
tee hinges; 150mm	-	-	-	0.45	-	**6.39**	-	**6.39**	nr	**7.19**
tee hinges; 300mm	-	-	-	0.55	-	**7.82**	-	**7.82**	nr	**8.79**
tee hinges; 450mm	-	-	-	0.65	-	**9.24**	-	**9.24**	nr	**10.39**
hook and band hinges; 300mm	-	-	-	0.90	-	**12.79**	-	**12.79**	nr	**14.39**
hook and band hinges; 450mm	-	-	-	1.10	-	**15.63**	-	**15.63**	nr	**17.58**
hook and band hinges; 900mm	-	-	-	1.65	-	**23.45**	-	**23.45**	nr	**26.38**
collinge hinges; 600mm	-	-	-	1.00	-	**14.21**	-	**14.21**	nr	**15.99**
collinge hinges; 750mm	-	-	-	1.25	-	**17.76**	-	**17.76**	nr	**19.98**
collinge hinges; 900mm	-	-	-	1.55	-	**22.03**	-	**22.03**	nr	**24.78**
single action floor springs and top centres	-	-	-	2.75	-	**39.08**	-	**39.08**	nr	**43.96**
double action floor springs and top centres	-	-	-	3.30	-	**46.89**	-	**46.89**	nr	**52.75**
coil springs	-	-	-	0.30	-	**4.26**	-	**4.26**	nr	**4.80**
overhead door closers	-	-	-	1.65	-	**23.45**	-	**23.45**	nr	**26.38**
concealed overhead door closers	-	-	-	2.75	-	**39.08**	-	**39.08**	nr	**43.96**
Perko door closers	-	-	-	1.65	-	**23.45**	-	**23.45**	nr	**26.38**
door selectors	-	-	-	2.20	0.28	**34.26**	-	**34.26**	nr	**38.54**
cabin hooks	-	-	-	0.25	-	**3.55**	-	**3.55**	nr	**4.00**
fanlight catches	-	-	-	0.40	-	**5.68**	-	**5.68**	nr	**6.39**
roller catches	-	-	-	0.40	-	**5.68**	-	**5.68**	nr	**6.39**
casement fasteners	-	-	-	0.40	-	**5.68**	-	**5.68**	nr	**6.39**
sash fasteners	-	-	-	1.10	-	**15.63**	-	**15.63**	nr	**17.58**
sash screws	-	-	-	0.55	-	**7.82**	-	**7.82**	nr	**8.79**
mortice latches	-	-	-	0.70	-	**9.95**	-	**9.95**	nr	**11.19**
night latches	-	-	-	1.10	-	**15.63**	-	**15.63**	nr	**17.58**
Norfolk latches	-	-	-	0.90	-	**12.79**	-	**12.79**	nr	**14.39**
rim latches	-	-	-	0.95	-	**13.50**	-	**13.50**	nr	**15.19**
Suffolk latches	-	-	-	0.90	-	**12.79**	-	**12.79**	nr	**14.39**
budget locks	-	-	-	0.75	-	**10.66**	-	**10.66**	nr	**11.99**
cylinder locks	-	-	-	1.10	-	**15.63**	-	**15.63**	nr	**17.58**
dead locks	-	-	-	0.90	-	**12.79**	-	**12.79**	nr	**14.39**
mortice locks	-	-	-	1.10	-	**15.63**	-	**15.63**	nr	**17.58**
rim locks	-	-	-	0.90	-	**12.79**	-	**12.79**	nr	**14.39**
automatic coin collecting locks	-	-	-	2.20	-	**31.26**	-	**31.26**	nr	**35.17**
casement stays	-	-	-	0.30	-	**4.26**	-	**4.26**	nr	**4.80**
quadrant stays	-	-	-	0.30	-	**4.26**	-	**4.26**	nr	**4.80**
barrel bolts; 150mm	-	-	-	0.35	-	**4.97**	-	**4.97**	nr	**5.60**
barrel bolts; 250mm	-	-	-	0.45	-	**6.39**	-	**6.39**	nr	**7.19**
door bolts; 150mm	-	-	-	0.35	-	**4.97**	-	**4.97**	nr	**5.60**
door bolts; 250mm	-	-	-	0.45	-	**6.39**	-	**6.39**	nr	**7.19**
monkey tail bolts										
300mm	-	-	-	0.40	-	**5.68**	-	**5.68**	nr	**6.39**
450mm	-	-	-	0.45	-	**6.39**	-	**6.39**	nr	**7.19**
600mm	-	-	-	0.55	-	**7.82**	-	**7.82**	nr	**8.79**
flush bolts; 200mm	-	-	-	0.75	-	**10.66**	-	**10.66**	nr	**11.99**
flush bolts; 450mm	-	-	-	1.10	-	**15.63**	-	**15.63**	nr	**17.58**
indicating bolts	-	-	-	1.10	-	**15.63**	-	**15.63**	nr	**17.58**
panic bolts; to single door	-	-	-	1.65	-	**23.45**	-	**23.45**	nr	**26.38**
panic bolts; to double doors	-	-	-	2.20	-	**31.26**	-	**31.26**	nr	**35.17**
knobs	-	-	-	0.25	-	**3.55**	-	**3.55**	nr	**4.00**
lever handles	-	-	-	0.25	-	**3.55**	-	**3.55**	nr	**4.00**
sash lifts	-	-	-	0.20	-	**2.84**	-	**2.84**	nr	**3.20**
pull handles; 150mm	-	-	-	0.20	-	**2.84**	-	**2.84**	nr	**3.20**
pull handles; 225mm	-	-	-	0.25	-	**3.55**	-	**3.55**	nr	**4.00**
pull handles; 300mm	-	-	-	0.30	-	**4.26**	-	**4.26**	nr	**4.80**
back plates	-	-	-	0.30	-	**4.26**	-	**4.26**	nr	**4.80**
escutcheon plates	-	-	-	0.25	-	**3.55**	-	**3.55**	nr	**4.00**
kicking plates	-	-	-	0.55	-	**7.82**	-	**7.82**	nr	**8.79**
letter plates	-	-	-	1.65	-	**23.45**	-	**23.45**	nr	**26.38**
push plates; 225mm	-	-	-	0.25	-	**3.55**	-	**3.55**	nr	**4.00**
push plates; 300mm	-	-	-	0.30	-	**4.26**	-	**4.26**	nr	**4.80**
shelf brackets	-	-	-	0.30	-	**4.26**	-	**4.26**	nr	**4.80**
sash cleats	-	-	-	0.25	-	**3.55**	-	**3.55**	nr	**4.00**
To softwood and brickwork										
cabin hooks	-	-	-	0.30	-	**4.26**	-	**4.26**	nr	**4.80**

Labour hourly rates: (except Specialists) Craft Operatives £14.21 Labourer £10.71 Rates are national average prices. Refer to REGIONAL VARIATIONS for indicative levels of overall pricing in regions	MATERIALS			LABOUR				RATES		
	Del to Site £	Waste %	Material Cost £	Craft Optve Hrs	Lab Hrs	Labour Cost £	Sunds £	Nett Rate £	Unit	Gross Rate (+12.5%) £
P21: IRONMONGERY Cont.										
Fix only ironmongery (prime cost sum for supply included elsewhere)										
To hardwood or the like										
butt hinges; 50mm	-	-	-	0.25	-	3.55	-	3.55	nr	4.00
butt hinges; 100mm	-	-	-	0.25	-	3.55	-	3.55	nr	4.00
rising hinges; 100mm	-	-	-	0.40	-	5.68	-	5.68	nr	6.39
tee hinges; 150mm	-	-	-	0.60	-	8.53	-	8.53	nr	9.59
tee hinges; 300mm	-	-	-	0.75	-	10.66	-	10.66	nr	11.99
tee hinges; 450mm	-	-	-	0.90	-	12.79	-	12.79	nr	14.39
hook and band hinges; 300mm	-	-	-	1.20	-	17.05	-	17.05	nr	19.18
hook and band hinges; 450mm	-	-	-	1.55	-	22.03	-	22.03	nr	24.78
hook and band hinges; 900mm	-	-	-	2.20	-	31.26	-	31.26	nr	35.17
collinge hinges; 600mm	-	-	-	1.45	-	20.60	-	20.60	nr	23.18
collinge hinges; 750mm	-	-	-	1.85	-	26.29	-	26.29	nr	29.57
collinge hinges; 900mm	-	-	-	2.25	-	31.97	-	31.97	nr	35.97
single action floor springs and top centres	-	-	-	4.10	-	58.26	-	58.26	nr	65.54
double action floor springs and top centres	-	-	-	4.90	-	69.63	-	69.63	nr	78.33
coil springs	-	-	-	0.45	-	6.39	-	6.39	nr	7.19
overhead door closers	-	-	-	2.45	-	34.81	-	34.81	nr	39.17
concealed overhead door closers	-	-	-	4.10	-	58.26	-	58.26	nr	65.54
Perko door closers	-	-	-	2.45	-	34.81	-	34.81	nr	39.17
door selectors	-	-	-	3.30	0.40	51.18	-	51.18	nr	57.57
cabin hooks	-	-	-	0.35	-	4.97	-	4.97	nr	5.60
fanlight catches	-	-	-	0.55	-	7.82	-	7.82	nr	8.79
roller catches	-	-	-	0.55	-	7.82	-	7.82	nr	8.79
casement fasteners	-	-	-	0.55	-	7.82	-	7.82	nr	8.79
sash fasteners	-	-	-	1.65	-	23.45	-	23.45	nr	26.38
sash screws	-	-	-	0.80	-	11.37	-	11.37	nr	12.79
mortice latches	-	-	-	1.10	-	15.63	-	15.63	nr	17.58
night latches	-	-	-	1.65	-	23.45	-	23.45	nr	26.38
Norfolk latches	-	-	-	1.30	-	18.47	-	18.47	nr	20.78
rim latches	-	-	-	0.80	-	11.37	-	11.37	nr	12.79
Suffolk latches	-	-	-	1.30	-	18.47	-	18.47	nr	20.78
budget locks	-	-	-	1.10	-	15.63	-	15.63	nr	17.58
cylinder locks	-	-	-	1.65	-	23.45	-	23.45	nr	26.38
dead locks	-	-	-	1.30	-	18.47	-	18.47	nr	20.78
mortice locks	-	-	-	1.65	-	23.45	-	23.45	nr	26.38
rim locks	-	-	-	1.30	-	18.47	-	18.47	nr	20.78
automatic coin collecting locks	-	-	-	3.30	-	46.89	-	46.89	nr	52.75
casement stays	-	-	-	0.40	-	5.68	-	5.68	nr	6.39
quadrant stays	-	-	-	0.40	-	5.68	-	5.68	nr	6.39
barrel bolts; 150mm	-	-	-	0.45	-	6.39	-	6.39	nr	7.19
barrel bolts; 250mm	-	-	-	0.60	-	8.53	-	8.53	nr	9.59
door bolts; 150mm	-	-	-	0.45	-	6.39	-	6.39	nr	7.19
door bolts; 250mm	-	-	-	0.60	-	8.53	-	8.53	nr	9.59
monkey tail bolts										
300mm	-	-	-	0.55	-	7.82	-	7.82	nr	8.79
450mm	-	-	-	0.65	-	9.24	-	9.24	nr	10.39
600mm	-	-	-	0.75	-	10.66	-	10.66	nr	11.99
flush bolts; 200mm	-	-	-	1.10	-	15.63	-	15.63	nr	17.58
flush bolts; 450mm	-	-	-	1.65	-	23.45	-	23.45	nr	26.38
indicating bolts	-	-	-	1.65	-	23.45	-	23.45	nr	26.38
panic bolts; to single door	-	-	-	2.45	-	34.81	-	34.81	nr	39.17
panic bolts; to double doors	-	-	-	3.30	-	46.89	-	46.89	nr	52.75
knobs	-	-	-	0.35	-	4.97	-	4.97	nr	5.60
lever handles	-	-	-	0.35	-	4.97	-	4.97	nr	5.60
sash lifts	-	-	-	0.35	-	4.97	-	4.97	nr	5.60
pull handles; 150mm	-	-	-	0.30	-	4.26	-	4.26	nr	4.80
pull handles; 225mm	-	-	-	0.35	-	4.97	-	4.97	nr	5.60
pull handles; 300mm	-	-	-	0.40	-	5.68	-	5.68	nr	6.39
back plates	-	-	-	0.40	-	5.68	-	5.68	nr	6.39
escutcheon plates	-	-	-	0.35	-	4.97	-	4.97	nr	5.60
kicking plates	-	-	-	0.80	-	11.37	-	11.37	nr	12.79
letter plates	-	-	-	2.45	-	34.81	-	34.81	nr	39.17
push plates; 225mm	-	-	-	0.35	-	4.97	-	4.97	nr	5.60
push plates; 300mm	-	-	-	0.40	-	5.68	-	5.68	nr	6.39
shelf brackets	-	-	-	0.45	-	6.39	-	6.39	nr	7.19
sash cleats	-	-	-	0.35	-	4.97	-	4.97	nr	5.60
To hardwood and brickwork										
cabin hooks	-	-	-	0.40	-	5.68	-	5.68	nr	6.39
Water bars; steel; galvanized										
Water bars; to concrete										
flat section; setting in groove in mastic										
25 x 3 x 900mm long	2.32	5.00	2.43	0.75	0.09	11.62	0.45	14.50	nr	16.32
40 x 3 x 900mm long	3.68	5.00	3.87	0.75	0.09	11.62	0.55	16.04	nr	18.04
40 x 6 x 900mm long	5.29	5.00	5.56	0.75	0.09	11.62	0.55	17.73	nr	19.94
50 x 6 x 900mm long	7.13	5.00	7.49	0.75	0.09	11.62	0.55	19.66	nr	22.12
Water bars; to hardwood										
flat section; setting in groove in mastic										
25 x 3 x 900mm long	2.32	5.00	2.43	0.65	0.08	10.09	0.45	12.97	nr	14.60
40 x 3 x 900mm long	3.68	5.00	3.87	0.65	0.08	10.09	0.55	14.51	nr	16.32
40 x 6 x 900mm long	5.29	5.00	5.56	0.65	0.08	10.09	0.55	16.20	nr	18.22
50 x 6 x 900mm long	7.13	5.00	7.49	0.65	0.08	10.09	0.55	18.13	nr	20.40

Labour hourly rates: (except Specialists) Craft Operatives £14.21 Labourer £10.71 Rates are national average prices. Refer to REGIONAL VARIATIONS for indicative levels of overall pricing in regions	MATERIALS			LABOUR				RATES		
	Del to Site £	Waste %	Material Cost £	Craft Optve Hrs	Lab Hrs	Labour Cost £	Sunds £	Nett Rate £	Unit	Gross Rate (+12.5%) £
P21: IRONMONGERY Cont.										
Sliding door gear, P.C Henderson Ltd.										
To softwood										
interior straight sliding door gear sets for commercial and domestic doors; Senator single door set comprising track, hangers, end stops and bottom guide; for doors 20 - 35mm thick, maximum weight 25kg maximum 900mm wide	25.52	2.50	**26.16**	1.65	-	**23.45**	-	**49.61**	nr	**55.81**
extra for pelmet 1855mm long	20.28	2.50	**20.79**	0.70	-	**9.95**	-	**30.74**	nr	**34.58**
extra for pelmet end cap	2.22	2.50	**2.28**	0.11	-	**1.56**	-	**3.84**	nr	**4.32**
extra for Doorseal deflector guide	9.49	2.50	**9.73**	0.40	-	**5.68**	-	**15.41**	nr	**17.34**
interior straight sliding door gear sets for commercial and domestic doors; Phantom single door set comprising top assembly, hangers, adjustable nylon guide and door stops; for doors 30 - 50mm thick, maximum weight 45kg, 610 - 915mm wide	41.16	2.50	**42.19**	1.95	-	**27.71**	-	**69.90**	nr	**78.63**
extra for soffit fixing bracket and bolt	1.67	2.50	**1.71**	0.22	-	**3.13**	-	**4.84**	nr	**5.44**
extra for pelmet 1550mm long	16.96	2.50	**17.38**	0.60	-	**8.53**	-	**25.91**	nr	**29.15**
extra for pelmet 1855mm long	20.28	2.50	**20.79**	0.70	-	**9.95**	-	**30.74**	nr	**34.58**
extra for pelmet end cap	2.22	2.50	**2.28**	0.11	-	**1.56**	-	**3.84**	nr	**4.32**
extra for Doorseal deflector guide	9.49	2.50	**9.73**	0.40	-	**5.68**	-	**15.41**	nr	**17.34**
interior straight sliding door gear sets for commercial and domestic doors; Marathon Junior Nr J2 single door set comprising top assembly, hangers, end stops, inverted guide channel and nylon guide; for doors 32 - 50mm thick, maximum weight, 55kg, 400 - 750mm wide; forming groove for guide channel	38.27	2.50	**39.22**	2.65	-	**37.66**	-	**76.88**	nr	**86.49**
interior straight sliding door gear sets for commercial and domestic doors; Marathon Junior Nr J3 single door set comprising top assembly, hangers, end stops, inverted guide channel and nylon guide; for doors 32 - 50mm thick, maximum weight, 55kg, 750 - 900mm wide; forming groove for guide channel	40.38	2.50	**41.39**	2.85	-	**40.50**	-	**81.89**	nr	**92.12**
interior straight sliding door gear sets for commercial and domestic doors; Marathon Junior Nr J4 single door set comprising top assembly, hangers, end stops, inverted guide channel and nylon guide; for doors 32 - 50mm thick, maximum weight, 55kg, 900 - 1050mm wide; forming groove for guide channel	43.70	2.50	**44.80**	3.10	-	**44.05**	-	**88.85**	nr	**99.95**
interior straight sliding door gear sets for commercial and domestic doors; Marathon Junior Nr J5 single door set comprising top assembly, hangers, end stops, inverted guide channel and nylon guide; for doors 32 - 50mm thick, maximum weight, 55kg, 1050 - 1200mm wide; forming groove for guide channel	49.39	2.50	**50.63**	3.30	-	**46.89**	-	**97.52**	nr	**109.71**
interior straight sliding door gear sets for commercial and domestic doors; Marathon Junior Nr J6 single door set comprising top assembly, hangers, end stops, inverted guide channel and nylon guide; for doors 32 - 50mm thick, maximum weight, 55kg, 1200 - 1500mm wide; forming groove for guide channel	81.98	2.50	**84.03**	3.50	-	**49.74**	-	**133.77**	nr	**150.49**
extra for soffit fixing bracket and bolt	1.67	2.50	**1.71**	0.22	-	**3.13**	-	**4.84**	nr	**5.44**
extra for pelmet 1550mm long	16.96	2.50	**17.38**	0.60	-	**8.53**	-	**25.91**	nr	**29.15**
extra for pelmet 1855mm long	20.28	2.50	**20.79**	0.70	-	**9.95**	-	**30.74**	nr	**34.58**
extra for pelmet end cap	2.22	2.50	**2.28**	0.11	-	**1.56**	-	**3.84**	nr	**4.32**
extra for Doorseal deflector guide	9.49	2.50	**9.73**	0.40	-	**5.68**	-	**15.41**	nr	**17.34**
interior straight sliding door gear sets for commercial and domestic doors; Marathon Senior Nr S3 single door set comprising top assembly, hangers, end stops, inverted guide channel and nylon guide; for doors 32 - 50mm thick maximum weight 90kg, 750 - 900mm wide; forming groove for guide channel	54.17	2.50	**55.53**	3.10	-	**44.05**	-	**99.58**	nr	**112.03**
interior straight sliding door gear sets for commercial and domestic doors; Marathon Senior Nr S4 single door set comprising top assembly, hangers, end stops, inverted guide channel and nylon guide; for doors 32 - 50mm thick maximum weight 90kg, 900 - 1050mm wide; forming groove for guide channel	60.16	2.50	**61.66**	3.30	-	**46.89**	-	**108.55**	nr	**122.12**
interior straight sliding door gear sets for commercial and domestic doors; Marathon Senior Nr S5 single door set comprising top assembly, hangers, end stops, inverted guide channel and nylon guide; for doors 32 - 50mm thick maximum weight 90kg, 1050 - 1200mm wide; forming groove for guide channel	66.37	2.50	**68.03**	3.50	-	**49.74**	-	**117.77**	nr	**132.49**
interior straight sliding door gear sets for commercial and domestic doors; Marathon Senior Nr S6 single door set comprising top assembly, hangers, end stops, inverted guide channel and nylon guide; for doors 32 - 50mm thick maximum weight 90kg, 1200 - 1500mm wide; forming groove for guide channel	106.43	2.50	**109.09**	3.65	-	**51.87**	-	**160.96**	nr	**181.08**
extra for soffit fixing bracket and bolt	1.67	2.50	**1.71**	0.22	-	**3.13**	-	**4.84**	nr	**5.44**
extra for pelmet 1550mm long	16.96	2.50	**17.38**	0.60	-	**8.53**	-	**25.91**	nr	**29.15**
extra for pelmet 1855mm long	20.28	2.50	**20.79**	0.70	-	**9.95**	-	**30.74**	nr	**34.58**
extra for pelmet end cap	2.22	2.50	**2.28**	0.11	-	**1.56**	-	**3.84**	nr	**4.32**
extra for Doorseal deflector guide	9.49	2.50	**9.73**	0.35	-	**4.97**	-	**14.70**	nr	**16.54**

Labour hourly rates: (except Specialists) Craft Operatives £14.21 Labourer £10.71 Rates are national average prices. Refer to REGIONAL VARIATIONS for indicative levels of overall pricing in regions	MATERIALS			LABOUR				RATES		
	Del to Site £	Waste %	Material Cost £	Craft Optve Hrs	Lab Hrs	Labour Cost £	Sunds £	Nett Rate £	Unit	Gross Rate (+12.5%) £
P21: IRONMONGERY Cont.										
Sliding door gear, P.C Henderson Ltd. Cont.										
To softwood Cont.										
interior straight sliding door gear sets for wardrobe and cupboard doors; Single Top Nr ST12 single door set comprising track, hangers, guides and safety stop; for door 16 - 35mm thick, maximum weight 25kg, maximum 900mm wide; one door to 600mm wide opening	15.57	2.50	**15.96**	1.40	-	**19.89**	-	**35.86**	nr	**40.34**
interior straight sliding door gear sets for wardrobe and cupboard doors; Single Top Nr ST15 single door set comprising track, hangers, guides and safety stop; for door 16 - 35mm thick, maximum weight 25kg, maximum 900mm wide; one door to 750mm wide opening	17.55	2.50	**17.99**	1.50	-	**21.32**	-	**39.30**	nr	**44.22**
interior straight sliding door gear sets for wardrobe and cupboard doors; Single Top Nr ST18 single door set comprising track, hangers, guides and safety stop; for door 16 - 35mm thick, maximum weight 25kg, maximum 900mm wide; one door to 900mm wide opening	19.86	2.50	**20.36**	1.60	-	**22.74**	-	**43.10**	nr	**48.48**
interior straight sliding door gear sets for wardrobe and cupboard doors; Double Top Nr W12 bi-passing door set comprising double track section, hangers, guides and safety stop; for doors 16 - 35mm thick, maximum weight 25kg, maximum 900mm wide; two doors in 1200mm wide opening	26.71	2.50	**27.38**	2.20	-	**31.26**	-	**58.64**	nr	**65.97**
interior straight sliding door gear sets for wardrobe and cupboard doors; Double Top Nr W15 bi-passing door set comprising double track section, hangers, guides and safety stop; for doors 16 - 35mm thick, maximum weight 25kg, maximum 900mm wide; two doors in 1500mm wide opening	29.78	2.50	**30.52**	2.30	-	**32.68**	-	**63.20**	nr	**71.10**
interior straight sliding door gear sets for wardrobe and cupboard doors; Double Top Nr W18 bi-passing door set comprising double track section, hangers, guides and safety stop; for doors 16 - 35mm thick, maximum weight 25kg, maximum 900mm wide; two doors in 1800mm wide opening	33.04	2.50	**33.86**	2.40	-	**34.10**	-	**67.97**	nr	**76.46**
interior straight sliding door gear sets for wardrobe and cupboard doors; Double Top Nr W24 bi-passing door set comprising double track section, hangers, guides and safety stop; for doors 16 - 35mm thick, maximum weight 25kg, maximum 900mm wide; three doors in 2400mm wide opening	43.30	2.50	**44.38**	2.65	-	**37.66**	-	**82.04**	nr	**92.29**
interior straight sliding door gear sets for wardrobe and cupboard doors; Bi-Fold Nr B10-2 folding door set comprising top guide track, top and bottom pivots; top guide and hinges; for doors 20 - 35mm thick; maximum weight 14kg each leaf, maximum 530mm wide; two doors in 1065mm wide opening	24.56	2.50	**25.17**	3.30	-	**46.89**	-	**72.07**	nr	**81.07**
interior straight sliding door gear sets for wardrobes and cupboard doors; Bi-Fold Nr B15-4 folding door set comprising top guide track, top and bottom pivots, top guide and hinges; for doors 20 - 35mm thick, maximum weight 14kg each leaf, maximum 530mm wide; four doors in 1525mm wide opening with aligner	41.68	2.50	**42.72**	6.60	-	**93.79**	-	**136.51**	nr	**153.57**
interior straight sliding door gear sets for wardrobes and cupboard doors; Bi-Fold Nr B20-4 folding door set comprising top guide track, top and bottom pivots, top guide and hinges; for doors 20 - 35mm thick, maximum weight 14kg each leaf, maximum 530mm wide; four doors in 2135mm wide opening with aligner	45.62	2.50	**46.76**	7.15	-	**101.60**	-	**148.36**	nr	**166.91**
interior straight sliding door gear sets for built in cupboard doors; Slipper Nr SS4 double passing door set comprising two top tracks, sliders, safety stop, flush pulls and guides; for doors 16 - 30mm thick, maximum weight 9kg, maximum 900mm wide; two doors in 1200mm wide opening	18.82	2.50	**19.29**	2.05	-	**29.13**	-	**48.43**	nr	**54.48**
interior straight sliding door gear sets for built in cupboard doors; Slipper Nr SS5 double passing door set comprising two top tracks, sliders, safety stop, flush pulls and guides; for doors 16 - 30mm thick, maximum weight 9kg, maximum 900mm wide; two doors in 1500mm wide opening	22.16	2.50	**22.72**	2.20	-	**31.26**	-	**53.98**	nr	**60.72**
interior straight sliding door gear sets for built in cupboard doors; Slipper Nr SS6 double passing door set comprising two top tracks, sliders, safety stop, flush pulls and guides; for doors 16 - 30mm thick, maximum weight 9kg, maximum 900mm wide; two doors in 1800mm wide opening	24.74	2.50	**25.36**	2.35	-	**33.39**	-	**58.76**	nr	**66.10**
interior straight sliding door gear sets for cupboards, book cases and cabinet work; Loretto Nr D4 bi-passing door set comprising two top guide channels, two bottom rails, nylon guides and bottom rollers; for doors 20 - 45mm thick; maximum weight 23kg, maximum 900mm wide; two doors in 1200mm wide opening; forming grooves for top guide channels and bottom rails; forming sinkings for bottom rollers	31.91	2.50	**32.71**	2.75	-	**39.08**	-	**71.79**	nr	**80.76**

Labour hourly rates: (except Specialists) Craft Operatives £14.21 Labourer £10.71 Rates are national average prices. Refer to REGIONAL VARIATIONS for indicative levels of overall pricing in regions	MATERIALS			LABOUR				RATES		
	Del to Site £	Waste %	Material Cost £	Craft Optve Hrs	Lab Hrs	Labour Cost £	Sunds £	Nett Rate £	Unit	Gross Rate (+12.5%) £
P21: IRONMONGERY Cont.										
Sliding door gear, P.C Henderson Ltd. Cont.										
To softwood Cont.										
interior straight sliding door gear sets for cupboards, book cases and cabinet work; Loretto Nr D5 bi-passing door set comprising two top guide channels, two bottom rails, nylon guides and bottom rollers; for doors 20 - 45mm thick; maximum weight 23kg, maximum 900mm wide; two doors in 1500mm wide opening; forming grooves for top guide channels and bottom rails; forming sinkings for bottom rollers	35.51	2.50	**36.40**	2.85	-	**40.50**	-	**76.90**	nr	**86.51**
interior straight sliding door gear sets for cupboards, book cases and cabinet work; Loretto Nr D6 bi-passing door set comprising two top guide channels, two bottom rails, nylon guides and bottom rollers; for doors 20 - 45mm thick; maximum weight 23kg, maximum 900mm wide; two doors in 1800mm wide opening; forming grooves for top guide channels and bottom rails; forming sinkings for bottom rollers	40.12	2.50	**41.12**	3.00	-	**42.63**	-	**83.75**	nr	**94.22**
extra for retractable roller guide bolt	7.53	2.50	**7.72**	0.40	-	**5.68**	-	**13.40**	nr	**15.07**
extra for flush pull	3.26	2.50	**3.34**	0.30	-	**4.26**	-	**7.61**	nr	**8.56**
extra for safety stop	1.03	2.50	**1.05**	0.11	-	**1.56**	-	**2.61**	nr	**2.94**
extra for cylinder lock	15.76	2.50	**16.15**	0.65	-	**9.24**	-	**25.39**	nr	**28.56**
interior straight sliding/folding room divider gear sets; Husky Folding Nr HF-2 folding door set comprising top track, brackets, top and bottom pivots, hangers and hinges, bottom channel and roller guides; for doors 20 - 40mm thick, maximum weight 25kg, maximum 600mm wide, maximum 2400mm high; two doors in 1200mm wide opening; forming groove for bottom channel	67.76	2.50	**69.45**	6.40	-	**90.94**	-	**160.40**	nr	**180.44**
interior straight sliding/folding room divider gear sets; Husky Folding Nr HF-4 folding door set comprising top track, brackets, top and bottom pivots, hangers and hinges, bottom channel and roller guides; for doors 20 - 40mm thick, maximum weight 25kg, maximum 600mm wide, maximum 2400mm high; four doors in 2400mm wide opening; forming groove for bottom channel	128.64	2.50	**131.86**	12.65	-	**179.76**	-	**311.62**	nr	**350.57**
interior straight sliding door gear sets for glass panels; Zenith 2 Nr 12 double passing door set comprising double top guide, bottom rail, glass rail with rubber glazing strip, end caps and bottom rollers; for panels 6mm thick, maximum weight 16kg or 1m2 per panel; two panels nr 1200mm wide opening	48.44	2.50	**49.65**	1.65	-	**23.45**	-	**73.10**	nr	**82.23**
extra for dust seal	4.56	2.50	**4.67**	0.35	-	**4.97**	-	**9.65**	m	**10.85**
interior straight sliding door gear sets for glass panels; Zenith 2 Nr 15 double passing door set comprising double top guide, bottom rail, glass rail with rubber glazing strip, end caps and bottom rollers; for panels 6mm thick, maximum weight 16kg or 1m2 per panel; two panels nr 1500mm wide opening	55.79	2.50	**57.19**	1.80	-	**25.58**	-	**82.77**	nr	**93.11**
extra for dust seal	4.56	2.50	**4.67**	0.35	-	**4.97**	-	**9.65**	m	**10.85**
interior straight sliding door gear sets for glass panels; Zenith 2 Nr 18 double passing door set comprising double top guide, bottom rail, glass rail with rubber glazing strip, end caps and bottom rollers; for panels 6mm thick, maximum weight 16kg or 1m2 per panel; two panels nr 1800mm wide opening	62.47	2.50	**64.03**	2.00	-	**28.42**	-	**92.45**	nr	**104.01**
extra for finger pull	3.26	2.50	**3.34**	0.45	-	**6.39**	-	**9.74**	nr	**10.96**
extra for cylinder lock	15.76	2.50	**16.15**	1.10	-	**15.63**	-	**31.78**	nr	**35.76**
extra for dust seal	4.56	2.50	**4.67**	0.35	-	**4.97**	-	**9.65**	m	**10.85**
To softwood and concrete										
exterior straight sliding bottom roller gear; Sterling 225 components comprising top guide reference 900, top guide brackets for face fixing for single run reference 1/900, top guide rollers for face fixing reference 54/900, bottom rollers reference 5 and bottom rail reference 299; for timber doors 44 - 50mm thick, maximum weight 225kg, maximum 3300mm high; single door in 1200mm wide opening; forming notches for bottom rollers; casting bottom rail into concrete	320.59	2.50	**328.61**	6.75	-	**95.92**	-	**424.53**	nr	**477.59**
exterior straight sliding bottom roller gear; Sterling 225 components comprising top guide reference 900, top guide brackets for soffit fixing for single run reference 3/900, top guide rollers for edge fixing reference 203/900, bottom rollers reference 5 and bottom rail reference 299; for timber doors 44 - 50mm thick, maximum weight 225kg, maximum 3300mm high; single door in 1200mm wide opening; forming notches for bottom rollers; casting bottom rail into concrete	346.78	2.50	**355.45**	7.00	-	**99.47**	-	**454.92**	nr	**511.79**

BUILDING FABRIC SUNDRIES

Labour hourly rates: (except Specialists) Craft Operatives £14.21 Labourer £10.71 Rates are national average prices. Refer to REGIONAL VARIATIONS for indicative levels of overall pricing in regions	MATERIALS			LABOUR				RATES		
	Del to Site	Waste	Material Cost	Craft Optve	Lab	Labour Cost	Sunds	Nett Rate	Unit	Gross Rate (+12.5%)
	£	%	£	Hrs	Hrs	£	£	£		£
P21: IRONMONGERY Cont.										
Sliding door gear, P.C Henderson Ltd. Cont.										
To softwood and concrete Cont.										
exterior straight sliding bottom roller gear; Sterling 225 components comprising top guide reference 900, top guide brackets for face fixing for single run reference 1/900, top guide rollers for face fixing reference 54/900, bottom rollers reference 5 and bottom rail reference 299; for timber doors 44 - 50mm thick, maximum weight 225kg, maximum 3300mm high; two doors in 2400mm wide opening; forming notches for bottom rollers; casting bottom rails into concrete	631.71	2.50	**647.50**	13.20	-	**187.57**	-	**835.07**	nr	**939.46**
exterior straight sliding bottom roller gear; Sterling 225 components comprising top guides reference 900, top guide brackets for face fixing for double run reference 5/900, top guide rollers for face fixing reference 54/900, bottom rollers reference 5 and bottom rails reference 299; for timber doors 44 - 50mm thick, maximum weight 225kg, maximum 3300 high; two doors in 2400mm wide opening; forming notches for bottom rollers; casting bottom rails into concrete	718.51	2.50	**736.47**	13.60	-	**193.26**	-	**929.73**	nr	**1045.95**
exterior straight sliding bottom roller gear; Sterling 350 components comprising top guide reference 99, top guide brackets for face fixing for single run reference 31 and 31x, top guide rollers reference 53/99, bottom rollers reference 2 and bottom rail reference 299; for timber doors 44 - 50mm thick, maximum weight 350kg, maximum 4000mm high; single door in 2400mm wide opening; forming notches for bottom rollers; casting bottom rail into concrete	710.68	2.50	**728.44**	14.20	-	**201.78**	-	**930.22**	nr	**1046.50**
exterior straight sliding bottom roller gear; Sterling 350 components comprising top guide reference 99, top guide brackets for face fixing for single run reference 31 and 31X, top guide rollers reference 53/99, bottom rollers reference 2 and bottom rail reference 299; for timber doors 44 - 50mm thick, maximum weight 350kg, maximum 4000mm high; two doors in 3600mm wide opening; forming notches for bottom rollers; casting bottom rail into concrete	1193.04	2.50	**1222.86**	18.00	-	**255.78**	-	**1478.64**	nr	**1663.47**
exterior straight sliding bottom roller gear; Sterling 800 components comprising top guide reference 99, top guide brackets for face fixing for single run reference 31 and 31X, top guide rollers reference 53/99, bottom rollers reference 3 and bottom rail reference 298; for timber doors 54 - 63mm thick, maximum weight 800kg, maximum 5200mm high; single door in 3600mm wide opening; forming notches for bottom rollers; casting bottom rail into concrete	1028.43	2.50	**1054.14**	16.10	-	**228.78**	-	**1282.92**	nr	**1443.28**
exterior straight sliding bottom roller gear, Sterling 800 components comprising top guide, reference 99, top guide brackets for face fixing for single run reference 31 and 31X, top guide rollers reference 53/99, bottom rollers reference 3 and bottom rail reference 298; for timber doors 54 - 63mm thick, maximum weight 800kg, maximum 5200mm high; two doors in 5400mm wide opening; forming notches for bottom rollers; casting bottom rail into concrete	1693.23	2.50	**1735.56**	25.90	-	**368.04**	-	**2103.60**	nr	**2366.55**
Hinges; standard quality										
To softwood										
backflap hinges; steel										
25mm	0.25	2.50	**0.26**	0.15	-	**2.13**	-	**2.39**	nr	**2.69**
38mm	0.30	2.50	**0.31**	0.17	-	**2.42**	-	**2.72**	nr	**3.06**
50mm	0.39	2.50	**0.40**	0.17	-	**2.42**	-	**2.82**	nr	**3.17**
63mm	1.14	2.50	**1.17**	0.17	-	**2.42**	-	**3.58**	nr	**4.03**
75mm	1.82	2.50	**1.87**	0.17	-	**2.42**	-	**4.28**	nr	**4.82**
butt hinges; steel; light medium pattern										
38mm	0.15	2.50	**0.15**	0.17	-	**2.42**	-	**2.57**	nr	**2.89**
50mm	0.15	2.50	**0.15**	0.17	-	**2.42**	-	**2.57**	nr	**2.89**
63mm	0.18	2.50	**0.18**	0.17	-	**2.42**	-	**2.60**	nr	**2.93**
75mm	0.18	2.50	**0.18**	0.17	-	**2.42**	-	**2.60**	nr	**2.93**
100mm	0.37	2.50	**0.38**	0.17	-	**2.42**	-	**2.79**	nr	**3.14**
butt hinges; steel; strong pattern										
75mm	0.70	2.50	**0.72**	0.17	-	**2.42**	-	**3.13**	nr	**3.52**
100mm	0.77	2.50	**0.79**	0.17	-	**2.42**	-	**3.20**	nr	**3.61**
butt hinges; cast iron; light										
50mm	0.92	2.50	**0.94**	0.17	-	**2.42**	-	**3.36**	nr	**3.78**
63mm	1.03	2.50	**1.06**	0.17	-	**2.42**	-	**3.47**	nr	**3.91**
75mm	1.14	2.50	**1.17**	0.17	-	**2.42**	-	**3.58**	nr	**4.03**
100mm	1.24	2.50	**1.27**	0.17	-	**2.42**	-	**3.69**	nr	**4.15**
butt hinges; brass, brass pin										
38mm	0.87	2.50	**0.89**	0.17	-	**2.42**	-	**3.31**	nr	**3.72**
50mm	1.00	2.50	**1.03**	0.17	-	**2.42**	-	**3.44**	nr	**3.87**
63mm	1.18	2.50	**1.21**	0.17	-	**2.42**	-	**3.63**	nr	**4.08**
75mm	1.28	2.50	**1.31**	0.17	-	**2.42**	-	**3.73**	nr	**4.19**
100mm	2.95	2.50	**3.02**	0.17	-	**2.42**	-	**5.44**	nr	**6.12**

Labour hourly rates: (except Specialists) Craft Operatives £14.21 Labourer £10.71 Rates are national average prices. Refer to REGIONAL VARIATIONS for indicative levels of overall pricing in regions	MATERIALS			LABOUR				RATES		
	Del to Site £	Waste %	Material Cost £	Craft Optve Hrs	Lab Hrs	Labour Cost £	Sunds £	Nett Rate £	Unit	Gross Rate (+12.5%) £
P21: IRONMONGERY Cont.										
Hinges; standard quality Cont.										
To softwood Cont.										
butt hinges; brass, steel washers, steel pin										
75mm	2.04	2.50	2.09	0.17	-	2.42	-	4.51	nr	5.07
100mm	3.49	2.50	3.58	0.17	-	2.42	-	5.99	nr	6.74
butt hinges; brass, B.M.A., steel washers, steel pin										
75mm	3.85	2.50	3.95	0.17	-	2.42	-	6.36	nr	7.16
100mm	6.26	2.50	6.42	0.17	-	2.42	-	8.83	nr	9.94
butt hinges; brass, chromium plated, steel washers, steel pin										
75mm	3.85	2.50	3.95	0.17	-	2.42	-	6.36	nr	7.16
100mm	6.73	2.50	6.90	0.17	-	2.42	-	9.31	nr	10.48
butt hinges; aluminium, stainless steel pin and washers										
75mm	3.01	2.50	3.09	0.17	-	2.42	-	5.50	nr	6.19
100mm	3.73	2.50	3.82	0.17	-	2.42	-	6.24	nr	7.02
rising hinges; steel										
75mm	1.12	2.50	1.15	0.30	-	4.26	-	5.41	nr	6.09
100mm	1.51	2.50	1.55	0.30	-	4.26	-	5.81	nr	6.54
rising hinges; cast iron										
75mm	2.17	2.50	2.22	0.30	-	4.26	-	6.49	nr	7.30
100mm	3.18	2.50	3.26	0.30	-	4.26	-	7.52	nr	8.46
spring hinges; steel, japanned; single action										
100mm	14.06	2.50	14.41	0.57	-	8.10	-	22.51	nr	25.33
125mm	15.81	2.50	16.21	0.62	-	8.81	-	25.02	nr	28.14
150mm	17.44	2.50	17.88	0.67	-	9.52	-	27.40	nr	30.82
spring hinges; steel, japanned; double action										
75mm	15.20	2.50	15.58	0.67	-	9.52	-	25.10	nr	28.24
100mm	17.44	2.50	17.88	0.67	-	9.52	-	27.40	nr	30.82
125mm	20.44	2.50	20.95	0.77	-	10.94	-	31.89	nr	35.88
150mm	23.44	2.50	24.03	0.77	-	10.94	-	34.97	nr	39.34
tee hinges; japanned; light										
150mm	0.39	2.50	0.40	0.45	-	6.39	-	6.79	nr	7.64
230mm	0.57	2.50	0.58	0.50	-	7.11	-	7.69	nr	8.65
300mm	0.75	2.50	0.77	0.55	-	7.82	-	8.58	nr	9.66
375mm	1.07	2.50	1.10	0.60	-	8.53	-	9.62	nr	10.83
450mm	1.47	2.50	1.51	0.65	-	9.24	-	10.74	nr	12.09
tee hinges; self colour; heavy										
150mm	0.68	2.50	0.70	0.50	-	7.11	-	7.80	nr	8.78
230mm	0.92	2.50	0.94	0.55	-	7.82	-	8.76	nr	9.85
300mm	1.47	2.50	1.51	0.60	-	8.53	-	10.03	nr	11.29
375mm	2.23	2.50	2.29	0.65	-	9.24	-	11.52	nr	12.96
450mm	2.82	2.50	2.89	0.70	-	9.95	-	12.84	nr	14.44
600mm	5.29	2.50	5.42	0.75	-	10.66	-	16.08	nr	18.09
hook and band hinges; on plate; heavy										
300mm	2.71	2.50	2.78	0.90	-	12.79	-	15.57	nr	17.51
450mm	4.44	2.50	4.55	1.10	-	15.63	-	20.18	nr	22.70
610mm	7.04	2.50	7.22	1.30	-	18.47	-	25.69	nr	28.90
914mm	12.33	2.50	12.64	1.65	-	23.45	-	36.08	nr	40.60
collinge hinges; cup for wood; best quality										
610mm	33.41	2.50	34.25	1.00	-	14.21	-	48.46	nr	54.51
762mm	41.17	2.50	42.20	1.25	-	17.76	-	59.96	nr	67.46
914mm	52.58	2.50	53.89	1.55	-	22.03	-	75.92	nr	85.41
parliament hinges; steel; 100mm	3.78	2.50	3.87	0.30	-	4.26	-	8.14	nr	9.15
cellar flap hinges; wrought, welded, plain joint; 450mm	24.64	2.50	25.26	0.55	-	7.82	-	33.07	nr	37.21
To hardwood or the like										
backflap hinges; steel										
25mm	0.25	2.50	0.26	0.25	-	3.55	-	3.81	nr	4.28
38mm	0.30	2.50	0.31	0.25	-	3.55	-	3.86	nr	4.34
50mm	0.39	2.50	0.40	0.25	-	3.55	-	3.95	nr	4.45
63mm	1.14	2.50	1.17	0.25	-	3.55	-	4.72	nr	5.31
75mm	1.82	2.50	1.87	0.25	-	3.55	-	5.42	nr	6.10
butt hinges; steel; light medium pattern										
38mm	0.15	2.50	0.15	0.25	-	3.55	-	3.71	nr	4.17
50mm	0.15	2.50	0.15	0.25	-	3.55	-	3.71	nr	4.17
63mm	0.18	2.50	0.18	0.25	-	3.55	-	3.74	nr	4.20
75mm	0.18	2.50	0.18	0.25	-	3.55	-	3.74	nr	4.20
100mm	0.37	2.50	0.38	0.25	-	3.55	-	3.93	nr	4.42
butt hinges; steel; strong pattern										
75mm	0.70	2.50	0.72	0.25	-	3.55	-	4.27	nr	4.80
100mm	0.77	2.50	0.79	0.25	-	3.55	-	4.34	nr	4.88
butt hinges; cast iron; light										
50mm	0.92	2.50	0.94	0.25	-	3.55	-	4.50	nr	5.06
63mm	1.03	2.50	1.06	0.25	-	3.55	-	4.61	nr	5.18
75mm	1.14	2.50	1.17	0.25	-	3.55	-	4.72	nr	5.31
100mm	1.24	2.50	1.27	0.25	-	3.55	-	4.82	nr	5.43
butt hinges; brass, brass pin										
38mm	0.87	2.50	0.89	0.25	-	3.55	-	4.44	nr	5.00
50mm	1.00	2.50	1.03	0.25	-	3.55	-	4.58	nr	5.15
63mm	1.18	2.50	1.21	0.25	-	3.55	-	4.76	nr	5.36
75mm	1.28	2.50	1.31	0.25	-	3.55	-	4.86	nr	5.47
100mm	2.95	2.50	3.02	0.25	-	3.55	-	6.58	nr	7.40
butt hinges; brass, steel washers, steel pin										
75mm	2.04	2.50	2.09	0.25	-	3.55	-	5.64	nr	6.35
100mm	3.49	2.50	3.58	0.25	-	3.55	-	7.13	nr	8.02
butt hinges; brass, B.M.A., steel washers, steel pin										
75mm	3.85	2.50	3.95	0.25	-	3.55	-	7.50	nr	8.44
100mm	6.26	2.50	6.42	0.25	-	3.55	-	9.97	nr	11.22

Labour hourly rates: (except Specialists) Craft Operatives £14.21 Labourer £10.71 Rates are national average prices. Refer to REGIONAL VARIATIONS for indicative levels of overall pricing in regions	MATERIALS			LABOUR				RATES		
	Del to Site	Waste	Material Cost	Craft Optve	Lab	Labour Cost	Sunds	Nett Rate	Unit	Gross Rate (+12.5%)
	£	%	£	Hrs	Hrs	£	£	£		£
P21: IRONMONGERY Cont.										
Hinges; standard quality Cont.										
To hardwood or the like Cont.										
butt hinges; brass, chromium plated, steel washers, steel pin										
75mm	3.85	2.50	3.95	0.25	-	3.55	-	7.50	nr	8.44
100mm	6.73	2.50	6.90	0.25	-	3.55	-	10.45	nr	11.76
butt hinges; aluminium, stainless steel pin and washers										
75mm	3.01	2.50	3.09	0.25	-	3.55	-	6.64	nr	7.47
100mm	3.73	2.50	3.82	0.25	-	3.55	-	7.38	nr	8.30
rising hinges; steel										
75mm	1.12	2.50	1.15	0.40	-	5.68	-	6.83	nr	7.69
100mm	1.51	2.50	1.55	0.40	-	5.68	-	7.23	nr	8.14
rising hinges; cast iron										
75mm	2.17	2.50	2.22	0.40	-	5.68	-	7.91	nr	8.90
100mm	3.18	2.50	3.26	0.40	-	5.68	-	8.94	nr	10.06
spring hinges; steel, japanned; single action										
100mm	14.06	2.50	14.41	0.85	-	12.08	-	26.49	nr	29.80
125mm	15.81	2.50	16.21	0.90	-	12.79	-	28.99	nr	32.62
150mm	17.44	2.50	17.88	0.95	-	13.50	-	31.38	nr	35.30
spring hinges; steel, japanned; double action										
75mm	15.20	2.50	15.58	0.90	-	12.79	-	28.37	nr	31.92
100mm	17.44	2.50	17.88	0.95	-	13.50	-	31.38	nr	35.30
125mm	20.44	2.50	20.95	1.05	-	14.92	-	35.87	nr	40.36
150mm	23.44	2.50	24.03	1.10	-	15.63	-	39.66	nr	44.61
tee hinges; japanned; light										
150mm	0.39	2.50	0.40	0.60	-	8.53	-	8.93	nr	10.04
230mm	0.57	2.50	0.58	0.65	-	9.24	-	9.82	nr	11.05
300mm	0.75	2.50	0.77	0.75	-	10.66	-	11.43	nr	12.85
375mm	1.07	2.50	1.10	0.80	-	11.37	-	12.46	nr	14.02
450mm	1.47	2.50	1.51	0.90	-	12.79	-	14.30	nr	16.08
tee hinges; self colour; heavy										
150mm	0.68	2.50	0.70	0.65	-	9.24	-	9.93	nr	11.18
230mm	0.92	2.50	0.94	0.70	-	9.95	-	10.89	nr	12.25
300mm	1.47	2.50	1.51	0.80	-	11.37	-	12.87	nr	14.48
375mm	2.23	2.50	2.29	0.85	-	12.08	-	14.36	nr	16.16
450mm	2.82	2.50	2.89	0.90	-	12.79	-	15.68	nr	17.64
600mm	5.29	2.50	5.42	1.00	-	14.21	-	19.63	nr	22.09
hook and band hinges; on plate; heavy										
300mm	2.71	2.50	2.78	1.20	-	17.05	-	19.83	nr	22.31
450mm	4.44	2.50	4.55	1.55	-	22.03	-	26.58	nr	29.90
610mm	7.04	2.50	7.22	1.90	-	27.00	-	34.22	nr	38.49
914mm	12.33	2.50	12.64	2.20	-	31.26	-	43.90	nr	49.39
collinge hinges; cup for wood; best quality										
610mm	33.41	2.50	34.25	1.45	-	20.60	-	54.85	nr	61.71
762mm	41.17	2.50	42.20	1.85	-	26.29	-	68.49	nr	77.05
914mm	52.58	2.50	53.89	2.25	-	31.97	-	85.87	nr	96.60
parliament hinges; steel; 100mm	3.78	2.50	3.87	0.40	-	5.68	-	9.56	nr	10.75
cellar flap hinges; wrought, welded, plain joint; 450mm	24.64	2.50	25.26	0.85	-	12.08	-	37.33	nr	42.00
Floor springs; standard quality										
To softwood										
single action floor springs and top centres; B.M.A.	204.34	2.50	209.45	2.75	-	39.08	-	248.53	nr	279.59
single action floor springs and top centres; chromium plated	204.34	2.50	209.45	2.75	-	39.08	-	248.53	nr	279.59
double action floor springs and top centres; B.M.A.	246.40	2.50	252.56	3.30	-	46.89	-	299.45	nr	336.88
double action floor springs and top centres; chromium plated	246.40	2.50	252.56	3.30	-	46.89	-	299.45	nr	336.88
To hardwood or the like										
single action floor springs and top centres; B.M.A.	204.34	2.50	209.45	4.10	-	58.26	-	267.71	nr	301.17
single action floor springs and top centres; chromium plated	204.34	2.50	209.45	4.10	-	58.26	-	267.71	nr	301.17
double action floor springs and top centres; B.M.A.	246.40	2.50	252.56	4.90	-	69.63	-	322.19	nr	362.46
double action floor springs and top centres; chromium plated	246.40	2.50	252.56	4.90	-	69.63	-	322.19	nr	362.46
Door closers; standard quality										
To softwood										
coil door springs; japanned	4.32	2.50	4.43	0.30	-	4.26	-	8.69	nr	9.78
overhead door closers; liquid check and spring; 'Briton' 2000 series; silver; light doors	70.28	2.50	72.04	1.65	-	23.45	-	95.48	nr	107.42
overhead door closers; liquid check and spring; 'Briton' 2000 series; silver; medium doors	81.73	2.50	83.77	1.65	-	23.45	-	107.22	nr	120.62
overhead door closers; liquid check and spring; 'Briton' 2000 series; silver; heavy doors	105.17	2.50	107.80	1.65	-	23.45	-	131.25	nr	147.65
concealed overhead door closers; liquid check and spring; medium doors	117.19	2.50	120.12	2.75	-	39.08	-	159.20	nr	179.10
Perko door closers; brass plate	7.33	2.50	7.51	1.65	-	23.45	-	30.96	nr	34.83
To hardwood or the like										
coil door springs; japanned	4.32	2.50	4.43	0.45	-	6.39	-	10.82	nr	12.18
overhead door closers; liquid check and spring; 'Briton' 2000 series; silver; light doors	70.28	2.50	72.04	2.45	-	34.81	-	106.85	nr	120.21
overhead door closers; liquid check and spring; 'Briton' 2000 series; silver; medium doors	81.73	2.50	83.77	2.45	-	34.81	-	118.59	nr	133.41

Labour hourly rates: (except Specialists) Craft Operatives £14.21 Labourer £10.71 Rates are national average prices. Refer to REGIONAL VARIATIONS for indicative levels of overall pricing in regions	MATERIALS			LABOUR				RATES		
	Del to Site £	Waste %	Material Cost £	Craft Optve Hrs	Lab Hrs	Labour Cost £	Sunds £	Nett Rate £	Unit	Gross Rate (+12.5%) £
P21: IRONMONGERY Cont.										
Door closers; standard quality Cont.										
To hardwood or the like Cont.										
overhead door closers; liquid check and spring; 'Briton' 2000 series; silver; heavy doors	105.17	2.50	**107.80**	2.45	-	**34.81**	-	**142.61**	nr	**160.44**
concealed overhead door closers; liquid check and spring; medium doors	117.19	2.50	**120.12**	4.10	-	**58.26**	-	**178.38**	nr	**200.68**
Perko door closers; brass plate	7.33	2.50	**7.51**	2.45	-	**34.81**	-	**42.33**	nr	**47.62**
Door selectors; standard quality										
To softwood or the like										
Union 8815 door selector; aluminium anodised silver finish	52.66	2.50	**53.98**	2.20	0.28	**34.26**	-	**88.24**	nr	**99.27**
Close Rite door selector; satin nickel plated	39.31	2.50	**40.29**	2.50	0.31	**38.85**	-	**79.14**	nr	**89.03**
To hardwood or the like										
Union 8815 door selector; aluminium anodised silver finish	52.66	2.50	**53.98**	3.30	0.41	**51.28**	-	**105.26**	nr	**118.42**
Close Rite door selector; satin nickel plated	39.31	2.50	**40.29**	3.90	0.49	**60.67**	-	**100.96**	nr	**113.58**
Locks and latches; standard quality										
To softwood										
magnetic catches; 6lb pull	0.57	2.50	**0.58**	0.25	-	**3.55**	-	**4.14**	nr	**4.65**
magnetic catches; 9lb pull	0.70	2.50	**0.72**	0.25	-	**3.55**	-	**4.27**	nr	**4.80**
magnetic catches; 13lb pull	1.17	2.50	**1.20**	0.25	-	**3.55**	-	**4.75**	nr	**5.35**
roller catches; mortice; nylon; 18mm	1.17	2.50	**1.20**	0.40	-	**5.68**	-	**6.88**	nr	**7.74**
roller catches; mortice; nylon; 27mm	1.17	2.50	**1.20**	0.40	-	**5.68**	-	**6.88**	nr	**7.74**
roller catches; surface, adjustable; nylon; 16mm	2.04	2.50	**2.09**	0.30	-	**4.26**	-	**6.35**	nr	**7.15**
roller catches; mortice, adjustable; satin chrome plated; 25 x 22 x 10mm	4.98	2.50	**5.10**	0.45	-	**6.39**	-	**11.50**	nr	**12.94**
roller catches; double ball; brass; 43mm	0.76	2.50	**0.78**	0.35	-	**4.97**	-	**5.75**	nr	**6.47**
mortice latches; stamped steel case; 75mm	1.12	2.50	**1.15**	0.70	-	**9.95**	-	**11.10**	nr	**12.48**
mortice latches; locking; stamped steel case; 75mm	2.23	2.50	**2.29**	0.70	-	**9.95**	-	**12.23**	nr	**13.76**
cylinder mortice latches; 'Union' key operated outside, knob inside, bolt held by slide; B.M.A.; for end set to suit 13mm rebate	43.93	2.50	**45.03**	0.80	-	**11.37**	-	**56.40**	nr	**63.45**
cylinder mortice latches; 'Union' key operated outside, knob inside, bolt held by slide; chromium plated; for end set to suit 13mm rebate	43.93	2.50	**45.03**	0.80	-	**11.37**	-	**56.40**	nr	**63.45**
cylinder rim night latches; 'Legge' 707; chromium plated	14.31	2.50	**14.67**	1.10	-	**15.63**	-	**30.30**	nr	**34.09**
cylinder rim night latches; 'Union' 1022; chromium plated	16.71	2.50	**17.13**	1.10	-	**15.63**	-	**32.76**	nr	**36.85**
cylinder rim night latches; 'Yale' 88; chromium plated	18.57	2.50	**19.03**	1.10	-	**15.63**	-	**34.67**	nr	**39.00**
Suffolk latches; japanned; medium; size Nr 2	2.23	2.50	**2.29**	0.90	-	**12.79**	-	**15.07**	nr	**16.96**
Suffolk latches; japanned; medium; size Nr 3	2.88	2.50	**2.95**	0.90	-	**12.79**	-	**15.74**	nr	**17.71**
Suffolk latches; japanned; heavy; size Nr 4	4.69	2.50	**4.81**	0.90	-	**12.79**	-	**17.60**	nr	**19.80**
Suffolk latches; galvanized; size Nr 4	4.98	2.50	**5.10**	0.90	-	**12.79**	-	**17.89**	nr	**20.13**
cupboard locks; 1 lever; japanned; 63mm	4.39	2.50	**4.50**	0.80	-	**11.37**	-	**15.87**	nr	**17.85**
cupboard locks; 2 lever; brass; 50mm	5.95	2.50	**6.10**	0.80	-	**11.37**	-	**17.47**	nr	**19.65**
cupboard locks; 2 lever; brass; 63mm	5.95	2.50	**6.10**	0.80	-	**11.37**	-	**17.47**	nr	**19.65**
cupboard locks; 4 lever; brass; 63mm	5.98	2.50	**6.13**	0.80	-	**11.37**	-	**17.50**	nr	**19.68**
rim dead lock; japanned case; 100mm	3.78	2.50	**3.87**	0.90	-	**12.79**	-	**16.66**	nr	**18.75**
rim dead lock; japanned case; 125mm	4.39	2.50	**4.50**	0.90	-	**12.79**	-	**17.29**	nr	**19.45**
rim dead lock; japanned case; 150mm	5.29	2.50	**5.42**	0.90	-	**12.79**	-	**18.21**	nr	**20.49**
mortice dead locks; japanned case; 75mm	5.53	2.50	**5.67**	0.90	-	**12.79**	-	**18.46**	nr	**20.76**
mortice locks; three levers	4.32	2.50	**4.43**	1.10	-	**15.63**	-	**20.06**	nr	**22.57**
mortice locks; five levers	11.42	2.50	**11.71**	1.10	-	**15.63**	-	**27.34**	nr	**30.75**
rim locks; japanned case; 150 x 100mm	5.29	2.50	**5.42**	0.90	-	**12.79**	-	**18.21**	nr	**20.49**
rim locks; japanned case; 150 x 75mm	5.90	2.50	**6.05**	0.90	-	**12.79**	-	**18.84**	nr	**21.19**
rim locks; japanned case; strong pattern; 150 x 100mm	7.93	2.50	**8.13**	0.90	-	**12.79**	-	**20.92**	nr	**23.53**
padlocks, 'Squire', anti-pilfer bolt, warded spring; zinc plated; 32mm	1.51	2.50	**1.55**	-	-	-	-	**1.55**	nr	**1.74**
padlocks, 'Squire', anti-pilfer bolt, warded spring; zinc plated; 38mm	2.34	2.50	**2.40**	-	-	-	-	**2.40**	nr	**2.70**
padlocks, 'Squire', anti-pilfer bolt, warded spring; zinc plated; 44mm	2.59	2.50	**2.65**	-	-	-	-	**2.65**	nr	**2.99**
padlocks; 'Squire', anti-pilfer bolt, 4 pin tumblers; zinc plated; 32mm	4.75	2.50	**4.87**	-	-	-	-	**4.87**	nr	**5.48**
padlocks; 'Squire', anti-pilfer bolt, 4 pin tumblers; zinc plated; 38mm	6.26	2.50	**6.42**	-	-	-	-	**6.42**	nr	**7.22**
padlocks; 'Squire', anti-pilfer bolt, 4 pin tumblers; zinc plated; 44mm	7.39	2.50	**7.57**	-	-	-	-	**7.57**	nr	**8.52**
padlocks; 'Squire', 4 levers; galvanized case	3.73	2.50	**3.82**	-	-	-	-	**3.82**	nr	**4.30**
padlocks, 'Squire', 4 levers; brass case	4.81	2.50	**4.93**	-	-	-	-	**4.93**	nr	**5.55**
padlocks, 'Squire', strong pattern; chromium plated	32.21	2.50	**33.02**	-	-	-	-	**33.02**	nr	**37.14**
hasps and staples; japanned wire; light; 75mm	0.62	2.50	**0.64**	0.35	-	**4.97**	-	**5.61**	nr	**6.31**
hasps and staples; japanned wire; light; 100mm	0.73	2.50	**0.75**	0.35	-	**4.97**	-	**5.72**	nr	**6.44**
hasps and staples; galvanized wire; heavy; 75mm	1.34	2.50	**1.37**	0.35	-	**4.97**	-	**6.35**	nr	**7.14**
hasps and staples; galvanized wire; heavy; 100mm	1.45	2.50	**1.49**	0.35	-	**4.97**	-	**6.46**	nr	**7.27**
hasps and staples; galvanized wire; heavy; 150mm	1.51	2.50	**1.55**	0.35	-	**4.97**	-	**6.52**	nr	**7.34**
locking bars; japanned; 200mm	5.65	2.50	**5.79**	0.50	-	**7.11**	-	**12.90**	nr	**14.51**
locking bars; japanned; 250mm	5.90	2.50	**6.05**	0.50	-	**7.11**	-	**13.15**	nr	**14.80**
locking bars; japanned; 300mm	7.46	2.50	**7.65**	0.50	-	**7.11**	-	**14.75**	nr	**16.60**
locking bars; japanned; 350mm	8.78	2.50	**9.00**	0.50	-	**7.11**	-	**16.10**	nr	**18.12**
To hardwood or the like										
magnetic catches; 6lb pull	0.57	2.50	**0.58**	0.35	-	**4.97**	-	**5.56**	nr	**6.25**
magnetic catches; 9lb pull	0.70	2.50	**0.72**	0.35	-	**4.97**	-	**5.69**	nr	**6.40**

Labour hourly rates: (except Specialists) Craft Operatives £14.21 Labourer £10.71 Rates are national average prices. Refer to REGIONAL VARIATIONS for indicative levels of overall pricing in regions	MATERIALS			LABOUR				RATES		
	Del to Site £	Waste %	Material Cost £	Craft Optve Hrs	Lab Hrs	Labour Cost £	Sunds £	Nett Rate £	Unit	Gross Rate (+12.5%) £
P21: IRONMONGERY Cont.										
Locks and latches; standard quality Cont.										
To hardwood or the like Cont.										
magnetic catches; 13lb pull	1.17	2.50	**1.20**	0.35	-	**4.97**	-	**6.17**	nr	**6.94**
roller catches; mortice; nylon; 18mm	1.17	2.50	**1.20**	0.55	-	**7.82**	-	**9.01**	nr	**10.14**
roller catches; mortice; nylon; 27mm	1.17	2.50	**1.20**	0.55	-	**7.82**	-	**9.01**	nr	**10.14**
roller catches; surface, adjustable; nylon; 16mm	2.04	2.50	**2.09**	0.45	-	**6.39**	-	**8.49**	nr	**9.55**
roller catches; mortice, adjustable; satin chrome plated; 25 x 22 x 10mm	4.98	2.50	**5.10**	0.65	-	**9.24**	-	**14.34**	nr	**16.13**
roller catches; double ball; brass; 43mm	0.76	2.50	**0.78**	0.50	-	**7.11**	-	**7.88**	nr	**8.87**
mortice latches; stamped steel case; 75mm	1.12	2.50	**1.15**	1.10	-	**15.63**	-	**16.78**	nr	**18.88**
mortice latches; locking; stamped steel case; 75mm	2.23	2.50	**2.29**	1.10	-	**15.63**	-	**17.92**	nr	**20.16**
cylinder mortice latches; 'Union' key operated outside, knob inside, bolt held by slide; B.M.A.; for end set to suit 13mm rebate	43.93	2.50	**45.03**	1.20	-	**17.05**	-	**62.08**	nr	**69.84**
cylinder mortice latches; 'Union' key operated outside, knob inside, bolt held by slide; chromium plated; for end set to suit 13mm rebate	43.93	2.50	**45.03**	1.20	-	**17.05**	-	**62.08**	nr	**69.84**
cylinder rim night latches; 'Legge' 707; chromium plated	14.31	2.50	**14.67**	1.65	-	**23.45**	-	**38.11**	nr	**42.88**
cylinder rim night latches; 'Union' 1022; chromium plated	16.71	2.50	**17.13**	1.65	-	**23.45**	-	**40.57**	nr	**45.65**
cylinder rim night latches; 'Yale' 88; chromium plated	18.57	2.50	**19.03**	1.65	-	**23.45**	-	**42.48**	nr	**47.79**
Suffolk latches; japanned; medium; size Nr 2	2.23	2.50	**2.29**	1.30	-	**18.47**	-	**20.76**	nr	**23.35**
Suffolk latches; japanned; medium; size Nr 3	2.88	2.50	**2.95**	1.30	-	**18.47**	-	**21.43**	nr	**24.10**
Suffolk latches; japanned; heavy; size Nr 4	4.69	2.50	**4.81**	1.30	-	**18.47**	-	**23.28**	nr	**26.19**
Suffolk latches; galvanized; size Nr 4	4.98	2.50	**5.10**	1.30	-	**18.47**	-	**23.58**	nr	**26.52**
cupboard locks; 1 lever; japanned; 63mm	4.39	2.50	**4.50**	1.20	-	**17.05**	-	**21.55**	nr	**24.25**
cupboard locks; 2 lever; brass; 50mm	5.95	2.50	**6.10**	1.20	-	**17.05**	-	**23.15**	nr	**26.04**
cupboard locks; 2 lever; brass; 63mm	5.95	2.50	**6.10**	1.20	-	**17.05**	-	**23.15**	nr	**26.04**
cupboard locks; 4 lever; brass; 63mm	5.98	2.50	**6.13**	1.20	-	**17.05**	-	**23.18**	nr	**26.08**
rim dead lock; japanned case; 100mm	3.78	2.50	**3.87**	1.30	-	**18.47**	-	**22.35**	nr	**25.14**
rim dead lock; japanned case; 125mm	4.39	2.50	**4.50**	1.30	-	**18.47**	-	**22.97**	nr	**25.84**
rim dead lock; japanned case; 150mm	5.29	2.50	**5.42**	1.30	-	**18.47**	-	**23.90**	nr	**26.88**
mortice dead locks; japanned case; 75mm	5.53	2.50	**5.67**	1.30	-	**18.47**	-	**24.14**	nr	**27.16**
mortice locks; three levers	4.32	2.50	**4.43**	1.65	-	**23.45**	-	**27.87**	nr	**31.36**
mortice locks; five levers	11.42	2.50	**11.71**	1.65	-	**23.45**	-	**35.15**	nr	**39.55**
rim locks; japanned case; 150 x 100mm	5.29	2.50	**5.42**	1.30	-	**18.47**	-	**23.90**	nr	**26.88**
rim locks; japanned case; 150 x 75mm	5.90	2.50	**6.05**	1.30	-	**18.47**	-	**24.52**	nr	**27.59**
rim locks; japanned case; strong pattern; 150 x 100mm	7.93	2.50	**8.13**	1.30	-	**18.47**	-	**26.60**	nr	**29.93**
padlocks, 'Squire', anti-pilfer bolt, warded spring; zinc plated; 32mm	1.51	2.50	**1.55**	-	-	**-**	-	**1.55**	nr	**1.74**
padlocks, 'Squire', anti-pilfer bolt, warded spring; zinc plated; 38mm	2.34	2.50	**2.40**	-	-	**-**	-	**2.40**	nr	**2.70**
padlocks, 'Squire', anti-pilfer bolt, warded spring; zinc plated; 44mm	2.59	2.50	**2.65**	-	-	**-**	-	**2.65**	nr	**2.99**
padlocks; 'Squire', anti-pilfer bolt, 4 pin tumblers; zinc plated; 32mm	4.75	2.50	**4.87**	-	-	**-**	-	**4.87**	nr	**5.48**
padlocks; 'Squire', anti-pilfer bolt, 4 pin tumblers; zinc plated; 38mm	6.26	2.50	**6.42**	-	-	**-**	-	**6.42**	nr	**7.22**
padlocks; 'Squire', anti-pilfer bolt, 4 pin tumblers; zinc plated; 44mm	7.39	2.50	**7.57**	-	-	**-**	-	**7.57**	nr	**8.52**
padlocks; 'Squire', 4 levers; galvanized case	3.73	2.50	**3.82**	-	-	**-**	-	**3.82**	nr	**4.30**
padlocks, 'Squire', 4 levers; brass case	4.81	2.50	**4.93**	-	-	**-**	-	**4.93**	nr	**5.55**
padlocks; 'Squire', strong pattern; chromium plated	32.21	2.50	**33.02**	-	-	**-**	-	**33.02**	nr	**37.14**
hasps and staples; japanned wire; light; 75mm	0.62	2.50	**0.64**	0.50	-	**7.11**	-	**7.74**	nr	**8.71**
hasps and staples; japanned wire; light; 100mm	0.73	2.50	**0.75**	0.55	-	**7.82**	-	**8.56**	nr	**9.63**
hasps and staples; galvanized wire; heavy; 75mm	1.34	2.50	**1.37**	0.55	-	**7.82**	-	**9.19**	nr	**10.34**
hasps and staples; galvanized wire; heavy; 100mm	1.45	2.50	**1.49**	0.55	-	**7.82**	-	**9.30**	nr	**10.46**
hasps and staples; galvanized wire; heavy; 150mm	1.51	2.50	**1.55**	0.55	-	**7.82**	-	**9.36**	nr	**10.53**
locking bars; japanned; 200mm	5.65	2.50	**5.79**	0.65	-	**9.24**	-	**15.03**	nr	**16.91**
locking bars; japanned; 250mm	5.90	2.50	**6.05**	0.65	-	**9.24**	-	**15.28**	nr	**17.19**
locking bars; japanned; 300mm	7.46	2.50	**7.65**	0.65	-	**9.24**	-	**16.88**	nr	**18.99**
locking bars; japanned; 350mm	8.78	2.50	**9.00**	0.65	-	**9.24**	-	**18.24**	nr	**20.52**
Bolts; standard quality										
To softwood										
barrel bolts; japanned, steel barrel; medium										
100mm	1.14	2.50	**1.17**	0.30	-	**4.26**	-	**5.43**	nr	**6.11**
150mm	1.45	2.50	**1.49**	0.35	-	**4.97**	-	**6.46**	nr	**7.27**
200mm	1.87	2.50	**1.92**	0.40	-	**5.68**	-	**7.60**	nr	**8.55**
barrel bolts; japanned, steel barrel; heavy										
100mm	2.88	2.50	**2.95**	0.30	-	**4.26**	-	**7.22**	nr	**8.12**
150mm	3.49	2.50	**3.58**	0.35	-	**4.97**	-	**8.55**	nr	**9.62**
200mm	4.21	2.50	**4.32**	0.40	-	**5.68**	-	**10.00**	nr	**11.25**
250mm	4.98	2.50	**5.10**	0.45	-	**6.39**	-	**11.50**	nr	**12.94**
300mm	5.90	2.50	**6.05**	0.50	-	**7.11**	-	**13.15**	nr	**14.80**
barrel bolts; extruded brass, round brass shoot; 25mm wide										
75mm	1.14	2.50	**1.17**	0.30	-	**4.26**	-	**5.43**	nr	**6.11**
100mm	1.45	2.50	**1.49**	0.30	-	**4.26**	-	**5.75**	nr	**6.47**
150mm	1.74	2.50	**1.78**	0.35	-	**4.97**	-	**6.76**	nr	**7.60**
barrel bolts; extruded brass; B.M.A., round brass shoot; 25mm wide										
75mm	1.56	2.50	**1.60**	0.30	-	**4.26**	-	**5.86**	nr	**6.59**
100mm	1.87	2.50	**1.92**	0.30	-	**4.26**	-	**6.18**	nr	**6.95**
150mm	2.34	2.50	**2.40**	0.35	-	**4.97**	-	**7.37**	nr	**8.29**
barrel bolts; extruded brass, chromium plated, round brass shoot; 25mm wide										
75mm	1.56	2.50	**1.60**	0.30	-	**4.26**	-	**5.86**	nr	**6.59**

Labour hourly rates: (except Specialists) Craft Operatives £14.21 Labourer £10.71 Rates are national average prices. Refer to REGIONAL VARIATIONS for indicative levels of overall pricing in regions	MATERIALS			LABOUR				RATES		
	Del to Site £	Waste %	Material Cost £	Craft Optve Hrs	Lab Hrs	Labour Cost £	Sunds £	Nett Rate £	Unit	Gross Rate (+12.5%) £
P21: IRONMONGERY Cont.										
Bolts; standard quality Cont.										
To softwood Cont.										
barrel bolts; extruded brass, chromium plated, round brass shoot; 25mm wide Cont.										
100mm	1.87	2.50	**1.92**	0.30	-	**4.26**	-	**6.18**	nr	**6.95**
150mm	2.34	2.50	**2.40**	0.35	-	**4.97**	-	**7.37**	nr	**8.29**
barrel bolts; extruded aluminium, S.A.A., round aluminium shoot; 25mm wide										
75mm	1.14	2.50	**1.17**	0.30	-	**4.26**	-	**5.43**	nr	**6.11**
100mm	1.25	2.50	**1.28**	0.30	-	**4.26**	-	**5.54**	nr	**6.24**
150mm	1.45	2.50	**1.49**	0.35	-	**4.97**	-	**6.46**	nr	**7.27**
monkey tail bolts										
japanned; 300mm	8.78	2.50	**9.00**	0.40	-	**5.68**	-	**14.68**	nr	**16.52**
japanned; 450mm	11.72	2.50	**12.01**	0.45	-	**6.39**	-	**18.41**	nr	**20.71**
japanned; 610mm	14.06	2.50	**14.41**	0.55	-	**7.82**	-	**22.23**	nr	**25.01**
flush bolts; B.M.A.										
100mm	7.57	2.50	**7.76**	0.55	-	**7.82**	-	**15.57**	nr	**17.52**
150mm	10.93	2.50	**11.20**	0.65	-	**9.24**	-	**20.44**	nr	**22.99**
200mm	18.76	2.50	**19.23**	0.75	-	**10.66**	-	**29.89**	nr	**33.62**
flush bolts; S.C.P.										
100mm	8.17	2.50	**8.37**	0.55	-	**7.82**	-	**16.19**	nr	**18.21**
150mm	11.12	2.50	**11.40**	0.65	-	**9.24**	-	**20.63**	nr	**23.21**
200mm	17.55	2.50	**17.99**	0.75	-	**10.66**	-	**28.65**	nr	**32.23**
lever action flush bolts; B.M.A.										
150mm	9.37	2.50	**9.60**	0.65	-	**9.24**	-	**18.84**	nr	**21.20**
200 x 25mm	9.98	2.50	**10.23**	0.75	-	**10.66**	-	**20.89**	nr	**23.50**
lever action flush bolts; S.A.A.										
150mm	7.24	2.50	**7.42**	0.65	-	**9.24**	-	**16.66**	nr	**18.74**
200 x 25mm	7.82	2.50	**8.02**	0.75	-	**10.66**	-	**18.67**	nr	**21.01**
necked bolts; extruded brass, round brass shoot; 25mm wide										
75mm	1.45	2.50	**1.49**	0.30	-	**4.26**	-	**5.75**	nr	**6.47**
100mm	1.57	2.50	**1.61**	0.30	-	**4.26**	-	**5.87**	nr	**6.61**
150mm	2.10	2.50	**2.15**	0.35	-	**4.97**	-	**7.13**	nr	**8.02**
necked bolts; extruded aluminium, S.A.A., round aluminium shoot; 25mm wide										
75mm	1.03	2.50	**1.06**	0.30	-	**4.26**	-	**5.32**	nr	**5.98**
100mm	1.14	2.50	**1.17**	0.30	-	**4.26**	-	**5.43**	nr	**6.11**
150mm	1.45	2.50	**1.49**	0.35	-	**4.97**	-	**6.46**	nr	**7.27**
indicating bolts; S.A.A.	6.26	2.50	**6.42**	1.10	-	**15.63**	-	**22.05**	nr	**24.80**
indicating bolts; B.M.A.	6.03	2.50	**6.18**	1.10	-	**15.63**	-	**21.81**	nr	**24.54**
panic bolts; to single door; iron, bronzed or silver	58.59	2.50	**60.05**	1.65	-	**23.45**	-	**83.50**	nr	**93.94**
panic bolts; to single door; aluminium box, steel shoots and cross rail, anodised silver	58.29	2.50	**59.75**	1.65	-	**23.45**	-	**83.19**	nr	**93.59**
panic bolts; to double doors; iron bronzed or silver	67.72	2.50	**69.41**	2.20	-	**31.26**	-	**100.68**	nr	**113.26**
panic bolts; to double doors; aluminium box, steel shoots and cross rail, anodised silver	67.72	2.50	**69.41**	2.20	-	**31.26**	-	**100.68**	nr	**113.26**
padlock bolts; galvanized; heavy										
150mm	3.43	2.50	**3.52**	0.35	-	**4.97**	-	**8.49**	nr	**9.55**
200mm	3.66	2.50	**3.75**	0.35	-	**4.97**	-	**8.73**	nr	**9.82**
250mm	4.39	2.50	**4.50**	0.50	-	**7.11**	-	**11.60**	nr	**13.06**
300mm	5.41	2.50	**5.55**	0.50	-	**7.11**	-	**12.65**	nr	**14.23**
To hardwood or the like										
barrel bolts; japanned, steel barrel; medium										
100mm	1.14	2.50	**1.17**	0.40	-	**5.68**	-	**6.85**	nr	**7.71**
150mm	1.45	2.50	**1.49**	0.45	-	**6.39**	-	**7.88**	nr	**8.87**
200mm	1.87	2.50	**1.92**	0.55	-	**7.82**	-	**9.73**	nr	**10.95**
barrel bolts; japanned, steel barrel; heavy										
100mm	2.88	2.50	**2.95**	0.40	-	**5.68**	-	**8.64**	nr	**9.72**
150mm	3.49	2.50	**3.58**	0.45	-	**6.39**	-	**9.97**	nr	**11.22**
200mm	4.21	2.50	**4.32**	0.55	-	**7.82**	-	**12.13**	nr	**13.65**
250mm	4.98	2.50	**5.10**	0.60	-	**8.53**	-	**13.63**	nr	**15.33**
300mm	5.90	2.50	**6.05**	0.70	-	**9.95**	-	**15.99**	nr	**17.99**
barrel bolts; extruded brass, round brass shoot; 25mm wide										
75mm	1.14	2.50	**1.17**	0.40	-	**5.68**	-	**6.85**	nr	**7.71**
100mm	1.45	2.50	**1.49**	0.40	-	**5.68**	-	**7.17**	nr	**8.07**
150mm	1.74	2.50	**1.78**	0.45	-	**6.39**	-	**8.18**	nr	**9.20**
barrel bolts; extruded brass; B.M.A., round brass shoot; 25mm wide										
75mm	1.56	2.50	**1.60**	0.40	-	**5.68**	-	**7.28**	nr	**8.19**
100mm	1.87	2.50	**1.92**	0.40	-	**5.68**	-	**7.60**	nr	**8.55**
150mm	2.34	2.50	**2.40**	0.45	-	**6.39**	-	**8.79**	nr	**9.89**
barrel bolts; extruded brass, chromium plated, round brass shoot; 25mm wide										
75mm	1.56	2.50	**1.60**	0.40	-	**5.68**	-	**7.28**	nr	**8.19**
100mm	1.87	2.50	**1.92**	0.40	-	**5.68**	-	**7.60**	nr	**8.55**
150mm	2.34	2.50	**2.40**	0.45	-	**6.39**	-	**8.79**	nr	**9.89**
barrel bolts; extruded aluminium, S.A.A., round aluminium shoot; 25mm wide										
75mm	1.14	2.50	**1.17**	0.40	-	**5.68**	-	**6.85**	nr	**7.71**
100mm	1.25	2.50	**1.28**	0.40	-	**5.68**	-	**6.97**	nr	**7.84**
150mm	1.45	2.50	**1.49**	0.45	-	**6.39**	-	**7.88**	nr	**8.87**
monkey tail bolts										
japanned; 300mm	8.78	2.50	**9.00**	0.55	-	**7.82**	-	**16.82**	nr	**18.92**
japanned; 450mm	11.72	2.50	**12.01**	0.65	-	**9.24**	-	**21.25**	nr	**23.91**
japanned; 610mm	14.06	2.50	**14.41**	0.75	-	**10.66**	-	**25.07**	nr	**28.20**
flush bolts; B.M.A.										
100mm	7.57	2.50	**7.76**	0.90	-	**12.79**	-	**20.55**	nr	**23.12**
150mm	10.93	2.50	**11.20**	1.00	-	**14.21**	-	**25.41**	nr	**28.59**

Labour hourly rates: (except Specialists) Craft Operatives £14.21 Labourer £10.71 Rates are national average prices. Refer to REGIONAL VARIATIONS for indicative levels of overall pricing in regions	MATERIALS			LABOUR				RATES		
	Del to Site £	Waste %	Material Cost £	Craft Optve Hrs	Lab Hrs	Labour Cost £	Sunds £	Nett Rate £	Unit	Gross Rate (+12.5%) £
P21: IRONMONGERY Cont.										
Bolts; standard quality Cont.										
To hardwood or the like Cont.										
flush bolts; B.M.A. Cont.										
200mm	18.76	2.50	19.23	1.10	-	15.63	-	34.86	nr	39.22
flush bolts; S.C.P.										
100mm	8.17	2.50	8.37	0.90	-	12.79	-	21.16	nr	23.81
150mm	11.12	2.50	11.40	1.00	-	14.21	-	25.61	nr	28.81
200mm	17.55	2.50	17.99	1.10	-	15.63	-	33.62	nr	37.82
lever action flush bolts; B.M.A.										
150mm	9.37	2.50	9.60	1.00	-	14.21	-	23.81	nr	26.79
200 x 25mm	9.98	2.50	10.23	1.10	-	15.63	-	25.86	nr	29.09
lever action flush bolts; S.A.A.										
150mm	7.24	2.50	7.42	1.00	-	14.21	-	21.63	nr	24.33
200 x 25mm	7.82	2.50	8.02	1.10	-	15.63	-	23.65	nr	26.60
necked bolts; extruded brass, round brass shoot; 25mm wide										
75mm	1.45	2.50	1.49	0.40	-	5.68	-	7.17	nr	8.07
100mm	1.57	2.50	1.61	0.40	-	5.68	-	7.29	nr	8.20
150mm	2.10	2.50	2.15	0.45	-	6.39	-	8.55	nr	9.62
necked bolts; extruded aluminium, S.A.A., round aluminium shoot; 25mm wide										
75mm	1.03	2.50	1.06	0.40	-	5.68	-	6.74	nr	7.58
100mm	1.14	2.50	1.17	0.40	-	5.68	-	6.85	nr	7.71
150mm	1.45	2.50	1.49	0.45	-	6.39	-	7.88	nr	8.87
indicating bolts; S.A.A.	6.26	2.50	6.42	1.65	-	23.45	-	29.86	nr	33.60
indicating bolts; B.M.A.	6.03	2.50	6.18	1.65	-	23.45	-	29.63	nr	33.33
panic bolts; to single door; iron, bronzed or silver	58.59	2.50	60.05	2.45	-	34.81	-	94.87	nr	106.73
panic bolts; to single door; aluminium box, steel shoots and cross rail, anodised silver	58.29	2.50	59.75	2.45	-	34.81	-	94.56	nr	106.38
panic bolts; to double doors; iron bronzed or silver	67.72	2.50	69.41	3.30	-	46.89	-	116.31	nr	130.84
panic bolts; to double doors; aluminium box, steel shoots and cross rail, anodised silver	67.72	2.50	69.41	3.30	-	46.89	-	116.31	nr	130.84
padlock bolts; galvanized; heavy										
150mm	3.43	2.50	3.52	0.50	-	7.11	-	10.62	nr	11.95
200mm	3.66	2.50	3.75	0.50	-	7.11	-	10.86	nr	12.21
250mm	4.39	2.50	4.50	0.75	-	10.66	-	15.16	nr	17.05
300mm	5.41	2.50	5.55	0.75	-	10.66	-	16.20	nr	18.23
Door handles; standard quality										
To softwood										
lever handles; spring action; B.M.A., best quality; 41 x 150mm	27.06	2.50	27.74	0.25	-	3.55	-	31.29	nr	35.20
lever handles; spring action; chromium plated, housing quality; 41 x 150mm	7.04	2.50	7.22	0.25	-	3.55	-	10.77	nr	12.11
lever handles; spring action; chromium plated, best quality; 41 x 150mm	24.64	2.50	25.26	0.25	-	3.55	-	28.81	nr	32.41
lever handles; spring action; S.A.A. housing quality; 41 x 150mm	4.27	2.50	4.38	0.25	-	3.55	-	7.93	nr	8.92
lever handles; spring action; S.A.A., best quality; 41 x 150mm	16.27	2.50	16.68	0.25	-	3.55	-	20.23	nr	22.76
lever handles; spring action; plastic/nylon, housing quality; 41 x 150mm	3.43	2.50	3.52	0.25	-	3.55	-	7.07	nr	7.95
pull handles; B.M.A.; 150mm	9.86	2.50	10.11	0.20	-	2.84	-	12.95	nr	14.57
pull handles; B.M.A.; 225mm	18.47	2.50	18.93	0.25	-	3.55	-	22.48	nr	25.29
pull handles; B.M.A.; 300mm	20.79	2.50	21.31	0.30	-	4.26	-	25.57	nr	28.77
pull handles; S.A.A.; 150mm	3.78	2.50	3.87	0.20	-	2.84	-	6.72	nr	7.56
pull handles; S.A.A.; 225mm	5.90	2.50	6.04	0.25	-	3.55	-	9.60	nr	10.80
pull handles; S.A.A.; 300mm	7.57	2.50	7.76	0.30	-	4.26	-	12.02	nr	13.53
pull handles on 50 x 250mm plate, lettered; B.M.A.; 225mm	42.68	2.50	43.75	0.45	-	6.39	-	50.14	nr	56.41
pull handles on 50 x 250mm plate, lettered; S.A.A.; 225mm	29.88	2.50	30.63	0.45	-	6.39	-	37.02	nr	41.65
To hardwood or the like										
lever handles; spring action; B.M.A., best quality; 41 x 150mm	27.06	2.50	27.74	0.35	-	4.97	-	32.71	nr	36.80
lever handles; spring action; chromium plated, housing quality; 41 x 150mm	7.04	2.50	7.22	0.35	-	4.97	-	12.19	nr	13.71
lever handles; spring action; chromium plated, best quality; 41 x 150mm	24.64	2.50	25.26	0.35	-	4.97	-	30.23	nr	34.01
lever handles; spring action; S.A.A. housing quality; 41 x 150mm	4.27	2.50	4.38	0.35	-	4.97	-	9.35	nr	10.52
lever handles; spring action; S.A.A., best quality; 41 x 150mm	16.27	2.50	16.68	0.35	-	4.97	-	21.65	nr	24.36
lever handles; spring action; plastic/nylon, housing quality; 41 x 150mm	3.43	2.50	3.52	0.35	-	4.97	-	8.49	nr	9.55
pull handles; B.M.A.; 150mm	9.86	2.50	10.11	0.30	-	4.26	-	14.37	nr	16.17
pull handles; B.M.A.; 225mm	18.47	2.50	18.93	0.35	-	4.97	-	23.91	nr	26.89
pull handles; B.M.A.; 300mm	20.79	2.50	21.31	0.40	-	5.68	-	26.99	nr	30.37
pull handles; S.A.A.; 150mm	3.78	2.50	3.87	0.30	-	4.26	-	8.14	nr	9.15
pull handles; S.A.A.; 225mm	5.90	2.50	6.04	0.35	-	4.97	-	11.02	nr	12.39
pull handles; S.A.A.; 300mm	7.57	2.50	7.76	0.40	-	5.68	-	13.44	nr	15.12
pull handles on 50 x 250mm plate, lettered; B.M.A.; 225mm	42.68	2.50	43.75	0.65	-	9.24	-	52.98	nr	59.61
pull handles on 50 x 250mm plate, lettered; S.A.A.; 225mm	29.88	2.50	30.63	0.65	-	9.24	-	39.86	nr	44.85

Labour hourly rates: (except Specialists) Craft Operatives £14.21 Labourer £10.71 Rates are national average prices. Refer to REGIONAL VARIATIONS for indicative levels of overall pricing in regions	MATERIALS			LABOUR				RATES		
	Del to Site £	Waste %	Material Cost £	Craft Optve Hrs	Lab Hrs	Labour Cost £	Sunds £	Nett Rate £	Unit	Gross Rate (+12.5%) £
P21: IRONMONGERY Cont.										
Door furniture; standard quality										
To softwood										
knobs; real B.M.A.; surface fixing	9.08	2.50	**9.31**	0.25	-	**3.55**	-	**12.86**	nr	**14.47**
knobs; S.A.A.; surface fixing	5.29	2.50	**5.42**	0.25	-	**3.55**	-	**8.97**	nr	**10.10**
knobs; real B.M.A.; secret fixing	14.67	2.50	**15.04**	0.35	-	**4.97**	-	**20.01**	nr	**22.51**
knobs; S.A.A.; secret fixing	7.63	2.50	**7.82**	0.35	-	**4.97**	-	**12.79**	nr	**14.39**
To hardwood or the like										
knobs; real B.M.A.; surface fixing	9.08	2.50	**9.31**	0.35	-	**4.97**	-	**14.28**	nr	**16.07**
knobs; S.A.A.; surface fixing	5.29	2.50	**5.42**	0.35	-	**4.97**	-	**10.40**	nr	**11.70**
knobs; real B.M.A.; secret fixing	14.67	2.50	**15.04**	0.50	-	**7.11**	-	**22.14**	nr	**24.91**
knobs; S.A.A.; secret fixing	7.63	2.50	**7.82**	0.50	-	**7.11**	-	**14.93**	nr	**16.79**
Window furniture; standard quality										
To softwood										
fanlight catches; brass	5.41	2.50	**5.55**	0.40	-	**5.68**	-	**11.23**	nr	**12.63**
fanlight catches; B.M.A.	5.90	2.50	**6.05**	0.40	-	**5.68**	-	**11.73**	nr	**13.20**
fanlight catches; chromium plated	6.01	2.50	**6.16**	0.40	-	**5.68**	-	**11.84**	nr	**13.32**
casement fasteners; wedge plate; black malleable iron	1.14	2.50	**1.17**	0.40	-	**5.68**	-	**6.85**	nr	**7.71**
casement fasteners; wedge plate; B.M.A.	2.65	2.50	**2.72**	0.40	-	**5.68**	-	**8.40**	nr	**9.45**
casement fasteners; wedge plate; chromium plated	2.65	2.50	**2.72**	0.40	-	**5.68**	-	**8.40**	nr	**9.45**
casement fasteners; wedge plate; S.A.A.	2.10	2.50	**2.15**	0.40	-	**5.68**	-	**7.84**	nr	**8.82**
sash fasteners; brass; 70mm	6.07	2.50	**6.22**	1.10	-	**15.63**	-	**21.85**	nr	**24.58**
sash fasteners; B.M.A.; 70mm	6.49	2.50	**6.65**	1.10	-	**15.63**	-	**22.28**	nr	**25.07**
sash fasteners; chromium plated; 70mm	7.15	2.50	**7.33**	1.10	-	**15.63**	-	**22.96**	nr	**25.83**
casement stays; two pins; grey malleable iron										
200mm	1.51	2.50	**1.55**	0.40	-	**5.68**	-	**7.23**	nr	**8.14**
250mm	1.62	2.50	**1.66**	0.40	-	**5.68**	-	**7.34**	nr	**8.26**
300mm	1.74	2.50	**1.78**	0.40	-	**5.68**	-	**7.47**	nr	**8.40**
casement stays; two pins; B.M.A.										
200mm	3.49	2.50	**3.58**	0.40	-	**5.68**	-	**9.26**	nr	**10.42**
250mm	3.78	2.50	**3.87**	0.40	-	**5.68**	-	**9.56**	nr	**10.75**
300mm	4.09	2.50	**4.19**	0.40	-	**5.68**	-	**9.88**	nr	**11.11**
casement stays; two pins; chromium plated										
200mm	3.49	2.50	**3.58**	0.40	-	**5.68**	-	**9.26**	nr	**10.42**
250mm	3.78	2.50	**3.87**	0.40	-	**5.68**	-	**9.56**	nr	**10.75**
300mm	4.09	2.50	**4.19**	0.40	-	**5.68**	-	**9.88**	nr	**11.11**
casement stays; two pins; S.A.A.										
200mm	1.93	2.50	**1.98**	0.40	-	**5.68**	-	**7.66**	nr	**8.62**
250mm	2.14	2.50	**2.19**	0.40	-	**5.68**	-	**7.88**	nr	**8.86**
300mm	2.23	2.50	**2.29**	0.40	-	**5.68**	-	**7.97**	nr	**8.97**
sash lifts; polished brass; 50mm	1.20	2.50	**1.23**	0.25	-	**3.55**	-	**4.78**	nr	**5.38**
sash lifts; B.M.A.; 50mm	1.45	2.50	**1.49**	0.25	-	**3.55**	-	**5.04**	nr	**5.67**
sash lifts; chromium plated; 50mm	1.62	2.50	**1.66**	0.25	-	**3.55**	-	**5.21**	nr	**5.86**
flush lifts; brass; 75mm	1.56	2.50	**1.60**	0.55	-	**7.82**	-	**9.41**	nr	**10.59**
flush lifts; B.M.A.; 75mm	1.87	2.50	**1.92**	0.55	-	**7.82**	-	**9.73**	nr	**10.95**
flush lifts; chromium plated; 75mm	1.87	2.50	**1.92**	0.55	-	**7.82**	-	**9.73**	nr	**10.95**
sash cleats; polished brass; 76mm	1.87	2.50	**1.92**	0.25	-	**3.55**	-	**5.47**	nr	**6.15**
sash cleats; B.M.A.; 76mm	1.93	2.50	**1.98**	0.25	-	**3.55**	-	**5.53**	nr	**6.22**
sash cleats; chromium plated; 76mm	2.34	2.50	**2.40**	0.25	-	**3.55**	-	**5.95**	nr	**6.69**
sash pulleys; frame and wheel	2.40	2.50	**2.46**	0.55	-	**7.82**	-	**10.28**	nr	**11.56**
To hardwood or the like										
fanlight catches; brass	5.41	2.50	**5.55**	0.55	-	**7.82**	-	**13.36**	nr	**15.03**
fanlight catches; B.M.A.	5.90	2.50	**6.05**	0.55	-	**7.82**	-	**13.86**	nr	**15.60**
fanlight catches; chromium plated	6.01	2.50	**6.16**	0.55	-	**7.82**	-	**13.98**	nr	**15.72**
casement fasteners; wedge plate; black malleable iron	1.14	2.50	**1.17**	0.55	-	**7.82**	-	**8.98**	nr	**10.11**
casement fasteners; wedge plate; B.M.A.	2.65	2.50	**2.72**	0.55	-	**7.82**	-	**10.53**	nr	**11.85**
casement fasteners; wedge plate; chromium plated	2.65	2.50	**2.72**	0.55	-	**7.82**	-	**10.53**	nr	**11.85**
casement fasteners; wedge plate; S.A.A.	2.10	2.50	**2.15**	0.55	-	**7.82**	-	**9.97**	nr	**11.21**
sash fasteners; brass; 70mm	6.07	2.50	**6.22**	1.65	-	**23.45**	-	**29.67**	nr	**33.38**
sash fasteners; B.M.A.; 70mm	6.49	2.50	**6.65**	1.65	-	**23.45**	-	**30.10**	nr	**33.86**
sash fasteners; chromium plated; 70mm	7.15	2.50	**7.33**	1.65	-	**23.45**	-	**30.78**	nr	**34.62**
casement stays; two pins; grey malleable iron										
200mm	1.51	2.50	**1.55**	0.55	-	**7.82**	-	**9.36**	nr	**10.53**
250mm	1.62	2.50	**1.66**	0.55	-	**7.82**	-	**9.48**	nr	**10.66**
300mm	1.74	2.50	**1.78**	0.55	-	**7.82**	-	**9.60**	nr	**10.80**
casement stays; two pins; B.M.A.										
200mm	3.49	2.50	**3.58**	0.55	-	**7.82**	-	**11.39**	nr	**12.82**
250mm	3.78	2.50	**3.87**	0.55	-	**7.82**	-	**11.69**	nr	**13.15**
300mm	4.09	2.50	**4.19**	0.55	-	**7.82**	-	**12.01**	nr	**13.51**
casement stays; two pins; chromium plated										
200mm	3.49	2.50	**3.58**	0.55	-	**7.82**	-	**11.39**	nr	**12.82**
250mm	3.78	2.50	**3.87**	0.55	-	**7.82**	-	**11.69**	nr	**13.15**
300mm	4.09	2.50	**4.19**	0.55	-	**7.82**	-	**12.01**	nr	**13.51**
casement stays; two pins; S.A.A.										
200mm	1.93	2.50	**1.98**	0.55	-	**7.82**	-	**9.79**	nr	**11.02**
250mm	2.14	2.50	**2.19**	0.55	-	**7.82**	-	**10.01**	nr	**11.26**
300mm	2.23	2.50	**2.29**	0.55	-	**7.82**	-	**10.10**	nr	**11.36**
sash lifts; polished brass; 50mm	1.20	2.50	**1.23**	0.35	-	**4.97**	-	**6.20**	nr	**6.98**
sash lifts; B.M.A.; 50mm	1.45	2.50	**1.49**	0.35	-	**4.97**	-	**6.46**	nr	**7.27**
sash lifts; chromium plated; 50mm	1.62	2.50	**1.66**	0.35	-	**4.97**	-	**6.63**	nr	**7.46**
flush lifts; brass; 75mm	1.56	2.50	**1.60**	0.80	-	**11.37**	-	**12.97**	nr	**14.59**
flush lifts; B.M.A.; 75mm	1.87	2.50	**1.92**	0.80	-	**11.37**	-	**13.28**	nr	**14.95**
flush lifts; chromium plated; 75mm	1.87	2.50	**1.92**	0.80	-	**11.37**	-	**13.28**	nr	**14.95**
sash cleats; polished brass; 76mm	1.87	2.50	**1.92**	0.35	-	**4.97**	-	**6.89**	nr	**7.75**
sash cleats; B.M.A.; 76mm	1.93	2.50	**1.98**	0.35	-	**4.97**	-	**6.95**	nr	**7.82**
sash cleats; chromium plated; 76mm	2.34	2.50	**2.40**	0.35	-	**4.97**	-	**7.37**	nr	**8.29**
sash pulleys; frame and wheel	2.40	2.50	**2.46**	0.80	-	**11.37**	-	**13.83**	nr	**15.56**

Labour hourly rates: (except Specialists) Craft Operatives £14.21 Labourer £10.71 Rates are national average prices. Refer to REGIONAL VARIATIONS for indicative levels of overall pricing in regions	MATERIALS			LABOUR				RATES		
	Del to Site £	Waste %	Material Cost £	Craft Optve Hrs	Lab Hrs	Labour Cost £	Sunds £	Nett Rate £	Unit	Gross Rate (+12.5%) £
P21: IRONMONGERY Cont.										
Window furniture; Willan Building Services Ltd - to softwood or the like										
Glidevale frame vent; aluminium with epoxy paint finish										
reference TV FV2; 198mm long; fixing with screws and clip on covers	8.11	2.50	**8.31**	0.22	0.03	**3.45**	0.17	**11.93**	nr	**13.42**
reference TV FV4; 358mm long; fixing with screws and clip on covers	9.78	2.50	**10.02**	0.33	0.04	**5.12**	0.20	**15.34**	nr	**17.26**
Glidevale canopy grille; aluminium with epoxy paint finish										
reference TV CG2; 155 x 15mm slot size; fixing with screws and clip on covers	4.06	2.50	**4.16**	0.22	0.03	**3.45**	0.08	**7.68**	nr	**8.65**
reference TV CG4; 308 x 15mm slot size; fixing with screws and clip on covers	5.74	2.50	**5.88**	0.33	0.03	**5.01**	0.12	**11.01**	nr	**12.39**
Glidevale condensation drainage channel; aluminium with powder coated finish										
reference CDN2500; to suit 4 or 6mm glazing; fixing with screws	25.54	2.50	**26.17**	0.45	0.06	**7.04**	0.53	**33.74**	m	**37.96**
Window furniture; Willan Building Services Ltd - to hardwood or the like										
Glidevale frame vent; aluminium with epoxy paint finish										
reference TV FV2; 198mm long; fixing with screws and clip on covers	8.11	2.50	**8.31**	0.33	0.05	**5.22**	0.17	**13.71**	nr	**15.42**
reference TV FV4; 358mm long; fixing with screws and clip on covers	9.78	2.50	**10.02**	0.50	0.07	**7.85**	0.20	**18.08**	nr	**20.34**
Glidevale canopy grille; aluminium with epoxy paint finish										
reference TV CG2; 155 x 15mm slot size; fixing with screws and clip on covers	4.06	2.50	**4.16**	0.33	0.05	**5.22**	0.08	**9.46**	nr	**10.64**
reference TV CG4; 308 x 15mm slot size; fixing with screws and clip on covers	5.74	2.50	**5.88**	0.50	0.07	**7.85**	0.12	**13.85**	nr	**15.59**
Glidevale condensation drainage channel; aluminium with powder coated finish										
reference CDN2500; to suit 4 or 6mm glazing; fixing with screws	25.54	2.50	**26.17**	0.70	0.10	**11.02**	0.55	**37.74**	m	**42.46**
Letter plates; standard quality										
To softwood										
letter plates; plain; real B.M.A.; 356mm wide	48.08	2.50	**49.28**	1.65	-	**23.45**	-	**72.73**	nr	**81.82**
letter plates; plain; anodised silver; 356mm wide	7.57	2.50	**7.76**	1.65	-	**23.45**	-	**31.21**	nr	**35.11**
letter plates; gravity flap; B.M.A.; 254mm wide	9.13	2.50	**9.36**	1.65	-	**23.45**	-	**32.80**	nr	**36.91**
letter plates; gravity flap; chromium plated; 254mm wide	9.13	2.50	**9.36**	1.65	-	**23.45**	-	**32.80**	nr	**36.91**
letter plates; gravity flap, two numerals; anodised silver; 254mm wide	7.93	2.50	**8.13**	1.75	-	**24.87**	-	**33.00**	nr	**37.12**
postal knockers; S.A.A.; frame 260 x 83mm; opening 203 x 44m	10.52	2.50	**10.78**	1.75	-	**24.87**	-	**35.65**	nr	**40.11**
To hardwood or the like										
letter plates; plain; real B.M.A.; 356mm wide	48.08	2.50	**49.28**	2.50	-	**35.53**	-	**84.81**	nr	**95.41**
letter plates; plain; anodised silver; 356mm wide	7.57	2.50	**7.76**	2.50	-	**35.53**	-	**43.28**	nr	**48.69**
letter plates; gravity flap; B.M.A.; 254mm wide	9.13	2.50	**9.36**	2.50	-	**35.53**	-	**44.88**	nr	**50.49**
letter plates; gravity flap; chromium plated; 254mm wide	9.13	2.50	**9.36**	2.50	-	**35.53**	-	**44.88**	nr	**50.49**
letter plates; gravity flap, two numerals; anodised silver; 254mm wide	7.93	2.50	**8.13**	2.65	-	**37.66**	-	**45.78**	nr	**51.51**
postal knockers; S.A.A.; frame 260 x 83mm; opening 203 x 44m	10.52	2.50	**10.78**	2.65	-	**37.66**	-	**48.44**	nr	**54.49**
Security locks; standard quality										
To hardwood or the like										
window catches; locking; polished aluminium	4.98	2.50	**5.10**	1.20	-	**17.05**	-	**22.16**	nr	**24.93**
dual screws; sanded brass	2.29	2.50	**2.35**	1.00	-	**14.21**	-	**16.56**	nr	**18.63**
window stops; locking; brass	5.29	2.50	**5.42**	1.00	-	**14.21**	-	**19.63**	nr	**22.09**
mortice latches; locking	58.30	2.50	**59.76**	3.50	-	**49.74**	-	**109.49**	nr	**123.18**
double cylinder automatic deadlatches; B.M.A.	34.86	2.50	**35.73**	3.50	-	**49.74**	-	**85.47**	nr	**96.15**
double cylinder automatic deadlatches; chromium plated	33.66	2.50	**34.50**	3.50	-	**49.74**	-	**84.24**	nr	**94.77**
mortice deadlocks; Chubb	27.06	2.50	**27.74**	3.50	-	**49.74**	-	**77.47**	nr	**87.16**
mortice locks; two bolt	33.66	2.50	**34.50**	3.50	-	**49.74**	-	**84.24**	nr	**94.77**
metal window locks; white	6.01	2.50	**6.16**	1.50	-	**21.32**	-	**27.48**	nr	**30.91**
security door chains; steel chain; brass	4.87	2.50	**4.99**	0.60	-	**8.53**	-	**13.52**	nr	**15.21**
security door chains; steel chain; chromium plated	5.41	2.50	**5.55**	0.60	-	**8.53**	-	**14.07**	nr	**15.83**
security mortice bolts; loose keys; chromium plated; 16mm diameter	4.57	2.50	**4.68**	1.20	-	**17.05**	-	**21.74**	nr	**24.45**
security mortice bolts; loose keys; chromium plated; 32mm diameter	6.07	2.50	**6.22**	1.30	-	**18.47**	-	**24.69**	nr	**27.78**
security hinge bolts; silver anodised	4.93	2.50	**5.05**	1.80	-	**25.58**	-	**30.63**	nr	**34.46**
door viewers; chromium plated	4.57	2.50	**4.68**	1.50	-	**21.32**	-	**26.00**	nr	**29.25**

Labour hourly rates: (except Specialists) Craft Operatives £14.21 Labourer £10.71 Rates are national average prices. Refer to REGIONAL VARIATIONS for indicative levels of overall pricing in regions	MATERIALS			LABOUR				RATES		
	Del to Site £	Waste %	Material Cost £	Craft Optve Hrs	Lab Hrs	Labour Cost £	Sunds £	Nett Rate £	Unit	Gross Rate (+12.5%) £
P21: IRONMONGERY Cont.										
Shelf brackets; standard quality										
To softwood										
shelf brackets; grey finished										
100 x 75mm	0.18	2.50	**0.18**	0.25	-	**3.55**	-	**3.74**	nr	**4.20**
150 x 125mm	0.24	2.50	**0.25**	0.25	-	**3.55**	-	**3.80**	nr	**4.27**
225 x 175mm	0.39	2.50	**0.40**	0.25	-	**3.55**	-	**3.95**	nr	**4.45**
300 x 250mm	0.57	2.50	**0.58**	0.25	-	**3.55**	-	**4.14**	nr	**4.65**
350 x 300mm	0.75	2.50	**0.77**	0.25	-	**3.55**	-	**4.32**	nr	**4.86**
Cabin hooks; standard quality										
To softwood										
cabin hooks and eyes; black japanned										
100mm	1.14	2.50	**1.17**	0.25	-	**3.55**	-	**4.72**	nr	**5.31**
150mm	1.45	2.50	**1.49**	0.25	-	**3.55**	-	**5.04**	nr	**5.67**
200mm	2.10	2.50	**2.15**	0.25	-	**3.55**	-	**5.71**	nr	**6.42**
250mm	2.34	2.50	**2.40**	0.25	-	**3.55**	-	**5.95**	nr	**6.69**
cabin hooks and eyes; polished brass										
100mm	4.81	2.50	**4.93**	0.25	-	**3.55**	-	**8.48**	nr	**9.54**
150mm	5.41	2.50	**5.55**	0.25	-	**3.55**	-	**9.10**	nr	**10.23**
To hardwood or the like										
cabin hooks and eyes; black japanned										
100mm	1.14	2.50	**1.17**	0.35	-	**4.97**	-	**6.14**	nr	**6.91**
150mm	1.45	2.50	**1.49**	0.35	-	**4.97**	-	**6.46**	nr	**7.27**
200mm	2.10	2.50	**2.15**	0.35	-	**4.97**	-	**7.13**	nr	**8.02**
250mm	2.34	2.50	**2.40**	0.35	-	**4.97**	-	**7.37**	nr	**8.29**
cabin hooks and eyes; polished brass										
100mm	4.81	2.50	**4.93**	0.35	-	**4.97**	-	**9.90**	nr	**11.14**
150mm	5.41	2.50	**5.55**	0.35	-	**4.97**	-	**10.52**	nr	**11.83**
To softwood and brickwork										
cabin hooks and eyes; black japanned										
100mm	1.14	2.50	**1.17**	0.30	-	**4.26**	-	**5.43**	nr	**6.11**
150mm	1.45	2.50	**1.49**	0.30	-	**4.26**	-	**5.75**	nr	**6.47**
200mm	2.10	2.50	**2.15**	0.30	-	**4.26**	-	**6.42**	nr	**7.22**
250mm	2.34	2.50	**2.40**	0.30	-	**4.26**	-	**6.66**	nr	**7.49**
cabin hooks and eyes; polished brass										
100mm	4.81	2.50	**4.93**	0.30	-	**4.26**	-	**9.19**	nr	**10.34**
150mm	5.41	2.50	**5.55**	0.30	-	**4.26**	-	**9.81**	nr	**11.03**
To hardwood and brickwork										
cabin hooks and eyes; black japanned										
100mm	1.14	2.50	**1.17**	0.40	-	**5.68**	-	**6.85**	nr	**7.71**
150mm	1.45	2.50	**1.49**	0.40	-	**5.68**	-	**7.17**	nr	**8.07**
200mm	2.10	2.50	**2.15**	0.40	-	**5.68**	-	**7.84**	nr	**8.82**
250mm	2.34	2.50	**2.40**	0.40	-	**5.68**	-	**8.08**	nr	**9.09**
cabin hooks and eyes; polished brass										
100mm	4.81	2.50	**4.93**	0.40	-	**5.68**	-	**10.61**	nr	**11.94**
150mm	5.41	2.50	**5.55**	0.40	-	**5.68**	-	**11.23**	nr	**12.63**
Draught seals and strips; standard quality										
To softwood										
draught excluders; plastic foam, self-adhesive										
900mm long	0.18	10.00	**0.20**	0.15	-	**2.13**	-	**2.33**	nr	**2.62**
2000mm long	0.36	10.00	**0.40**	0.30	-	**4.26**	-	**4.66**	nr	**5.24**
draught excluders; aluminium section, rubber tubing										
900mm long	7.51	10.00	**8.26**	0.17	0.02	**2.63**	-	**10.89**	nr	**12.25**
2000mm long	17.02	10.00	**18.72**	0.35	0.04	**5.40**	-	**24.12**	nr	**27.14**
draught excluders; aluminium section, vinyl seal										
900mm long	7.51	10.00	**8.26**	0.17	0.02	**2.63**	-	**10.89**	nr	**12.25**
2000mm long	17.02	10.00	**18.72**	0.35	0.04	**5.40**	-	**24.12**	nr	**27.14**
draught excluders; plastic moulding, nylon brush; 900mm long	6.44	10.00	**7.08**	0.17	0.02	**2.63**	-	**9.71**	nr	**10.93**
draught excluders; aluminium base, flexible arch; 900mm long	17.13	10.00	**18.84**	0.25	0.03	**3.87**	-	**22.72**	nr	**25.56**
draught excluders; aluminium threshold, flexible arch; 900mm long	17.13	10.00	**18.84**	0.30	0.04	**4.69**	-	**23.53**	nr	**26.48**
Draught seals and strips; Sealmaster Ltd										
Threshold seals										
reference BDA; fixing to masonry with screws	10.74	2.50	**11.00**	0.45	0.06	**7.04**	0.22	**18.26**	m	**20.54**
reference BDB; fixing to masonry with screws	13.76	2.50	**14.11**	0.45	0.06	**7.04**	0.30	**21.44**	m	**24.12**
reference BDWB (weather board); fixing to timber with screws	16.72	2.50	**17.14**	0.70	0.10	**11.02**	0.35	**28.51**	m	**32.07**
reference TTM (stop seal); fixing to timber with screws	10.15	2.50	**10.41**	0.33	0.05	**5.22**	0.22	**15.85**	m	**17.83**
Drawer pulls; standard quality										
To softwood										
drawer pulls; brass; 100mm	1.56	2.50	**1.60**	0.20	-	**2.84**	-	**4.44**	nr	**5.00**
drawer pulls; B.M.A.; 100mm	2.04	2.50	**2.09**	0.20	-	**2.84**	-	**4.93**	nr	**5.55**
drawer pulls; chromium plated; 100mm	1.98	2.50	**2.03**	0.20	-	**2.84**	-	**4.87**	nr	**5.48**
drawer pulls; S.A.A.; 100mm	1.20	2.50	**1.23**	0.20	-	**2.84**	-	**4.07**	nr	**4.58**
To hardwood or the like										
drawer pulls; brass; 100mm	1.56	2.50	**1.60**	0.30	-	**4.26**	-	**5.86**	nr	**6.59**
drawer pulls; B.M.A.; 100mm	2.04	2.50	**2.09**	0.30	-	**4.26**	-	**6.35**	nr	**7.15**
drawer pulls; chromium plated; 100mm	1.98	2.50	**2.03**	0.30	-	**4.26**	-	**6.29**	nr	**7.08**

Labour hourly rates: (except Specialists) Craft Operatives £14.21 Labourer £10.71 Rates are national average prices. Refer to REGIONAL VARIATIONS for indicative levels of overall pricing in regions	MATERIALS			LABOUR				RATES		
	Del to Site £	Waste %	Material Cost £	Craft Optve Hrs	Lab Hrs	Labour Cost £	Sunds £	Nett Rate £	Unit	Gross Rate (+12.5%) £
P21: IRONMONGERY Cont.										
Drawer pulls; standard quality Cont.										
To hardwood or the like Cont.										
drawer pulls; S.A.A.; 100mm	1.20	2.50	**1.23**	0.30	-	**4.26**	-	**5.49**	nr	**6.18**
Hooks; standard quality										
To softwood										
cup hooks; polished brass; 25mm	0.18	2.50	**0.18**	0.07	-	**0.99**	-	**1.18**	nr	**1.33**
hat and coat hooks; B.M.A.	1.91	2.50	**1.96**	0.30	-	**4.26**	-	**6.22**	nr	**7.00**
hat and coat hooks; chromium plated	1.91	2.50	**1.96**	0.30	-	**4.26**	-	**6.22**	nr	**7.00**
hat and coat hooks; S.A.A.	0.84	2.50	**0.86**	0.30	-	**4.26**	-	**5.12**	nr	**5.76**
To hardwood or the like										
cup hooks; polished brass; 25mm	0.18	2.50	**0.18**	0.10	-	**1.42**	-	**1.61**	nr	**1.81**
hat and coat hooks; B.M.A.	1.91	2.50	**1.96**	0.45	-	**6.39**	-	**8.35**	nr	**9.40**
hat and coat hooks; chromium plated	1.91	2.50	**1.96**	0.45	-	**6.39**	-	**8.35**	nr	**9.40**
hat and coat hooks; S.A.A.	0.84	2.50	**0.86**	0.45	-	**6.39**	-	**7.26**	nr	**8.16**

This section continues on the next page

Labour hourly rates: (except Specialists) Craft Operatives £14.21 Labourer £10.71 Rates are national average prices. Refer to REGIONAL VARIATIONS for indicative levels of overall pricing in regions	PLANT & TRANSPORT			LABOUR				RATES		
	Plant Cost £	Trans Cost £	P & T Cost £	Craft Optve Hrs	Lab Hrs	Labour Cost £	Sunds £	Nett Rate £	Unit	Gross Rate (+12.5%) £
P30: TRENCHES/PIPEWAYS/PITS FOR BURIED ENGINEERING SERVICES										
Site preparation										
Lifting turf for preservation										
stacking on site average 50 metres distance for immediate use; watering	0.00	-	**0.00**	-	0.60	**6.43**	-	**6.43**	m2	**7.23**
Re-laying turf										
taking from stack average 50 metres distance, laying on prepared bed, watering, maintaining	0.00	-	**0.00**	-	0.60	**6.43**	-	**6.43**	m2	**7.23**
Excavating trenches to receive services not exceeding 200mm nominal size - by machine										
Excavations commencing from natural ground level; compacting; backfilling with excavated material										
not exceeding 1m deep; average 750mm deep	1.76	0.51	**2.27**	-	0.25	**2.68**	2.47	**7.43**	m	**8.35**
not exceeding 1m deep; average 750mm deep; next to roadways	1.90	0.51	**2.41**	-	0.25	**2.68**	3.88	**8.97**	m	**10.09**
Excavations commencing from existing ground level; levelling and grading backfilling to receive turf										
not exceeding 1m deep; average 750mm deep	1.90	0.51	**2.41**	-	0.25	**2.68**	3.72	**8.81**	m	**9.91**
Extra over excavating trenches irrespective of depth; breaking out existing materials										
hard rock	17.29	-	**17.29**	-	-	-	-	**17.29**	m3	**19.45**
concrete	14.14	-	**14.14**	-	-	-	-	**14.14**	m3	**15.91**
reinforced concrete	21.22	-	**21.22**	-	-	-	-	**21.22**	m3	**23.87**
brickwork, blockwork or stonework	9.43	-	**9.43**	-	-	-	-	**9.43**	m3	**10.61**
Extra over excavating trenches irrespective of depth; breaking out existing hard pavings										
concrete 150mm thick	2.20	-	**2.20**	-	-	-	-	**2.20**	m2	**2.48**
reinforced concrete 200mm thick	4.56	-	**4.56**	-	-	-	-	**4.56**	m2	**5.13**
concrete 150mm thick; reinstating	2.67	-	**2.67**	-	-	-	13.71	**16.38**	m2	**18.43**
macadam paving 75mm thick	0.39	-	**0.39**	-	-	-	-	**0.39**	m2	**0.44**
macadam paving 75mm thick; reinstating	0.39	-	**0.39**	-	-	-	17.83	**18.22**	m2	**20.50**
concrete flag paving 50mm thick	-	-	-	-	0.30	**3.21**	-	**3.21**	m2	**3.61**
concrete flag paving 50mm thick; re-instating	-	-	-	0.60	0.90	**18.17**	11.52	**29.68**	m2	**33.39**
Extra over excavating trenches irrespective of depth; excavating next to existing services										
electricity services -1	-	-	-	-	5.75	**61.58**	-	**61.58**	m	**69.28**
gas services -1	-	-	-	-	5.75	**61.58**	-	**61.58**	m	**69.28**
water services -1	-	-	-	-	5.75	**61.58**	-	**61.58**	m	**69.28**
Extra over excavating trenches irrespective of depth; excavating around existing services crossing trench										
electricity services; cables crossing -1	-	-	-	-	8.50	**91.04**	-	**91.04**	nr	**102.41**
gas services services crossing -1	-	-	-	-	8.50	**91.04**	-	**91.04**	nr	**102.41**
water services services crossing -1	-	-	-	-	8.50	**91.04**	-	**91.04**	nr	**102.41**
Site preparation										
Lifting turf for preservation										
stacking on site average 50 metres distance for immediate use; watering	-	-	-	-	0.60	**6.43**	-	**6.43**	m2	**7.23**
Re-laying turf										
taking from stack average 50 metres distance, laying on prepared bed, watering, maintaining	-	-	-	-	0.60	**6.43**	-	**6.43**	m2	**7.23**
Excavating trenches to receive services not exceeding 200mm nominal size - by hand										
Excavations commencing from natural ground level; compacting; backfilling with excavated material										
not exceeding 1m deep; average 750mm deep	-	-	-	-	1.20	**12.85**	8.06	**20.91**	m	**23.52**
not exceeding 1m deep; average 750mm deep; next to roadways	-	-	-	-	1.20	**12.85**	9.08	**21.93**	m	**24.67**
Excavations commencing from existing ground level; levelling and grading backfilling to receive turf										
not exceeding 1m deep; average 750mm deep	-	-	-	-	1.20	**12.85**	9.44	**22.29**	m	**25.08**
Extra over excavating trenches irrespective of depth; breaking out existing materials										
hard rock	-	-	-	-	14.70	**157.44**	-	**157.44**	m3	**177.12**
concrete	-	-	-	-	11.00	**117.81**	-	**117.81**	m3	**132.54**
reinforced concrete	-	-	-	-	16.50	**176.72**	-	**176.72**	m3	**198.80**
brickwork, blockwork or stonework	-	-	-	-	7.70	**82.47**	-	**82.47**	m3	**92.78**
Extra over excavating trenches irrespective of depth; breaking out existing hard pavings										
concrete 150mm thick	-	-	-	-	1.65	**17.67**	-	**17.67**	m2	**19.88**
reinforced concrete 150mm thick	-	-	-	-	3.30	**35.34**	-	**35.34**	m2	**39.76**
concrete 150mm thick; reinstating	-	-	-	-	3.30	**35.34**	16.01	**51.35**	m2	**57.77**
macadam paving 75mm thick	-	-	-	-	0.30	**3.21**	-	**3.21**	m2	**3.61**
macadam paving 75mm thick; reinstating	-	-	-	-	0.30	**3.21**	20.82	**24.03**	m2	**27.03**
concrete flag paving 50mm thick	-	-	-	-	0.32	**3.43**	-	**3.43**	m2	**3.86**
concrete flag paving 50mm thick; re-instating	-	-	-	-	0.88	**9.42**	14.12	**23.55**	m2	**26.49**

Labour hourly rates: (except Specialists) Craft Operatives £14.21 Labourer £10.71 Rates are national average prices. Refer to REGIONAL VARIATIONS for indicative levels of overall pricing in regions	PLANT & TRANSPORT			LABOUR				RATES		
	Plant Cost	Trans Cost	P & T Cost	Craft Optve	Lab	Labour Cost	Sunds	Nett Rate	Unit	Gross Rate (+12.5%)
	£	£	£	Hrs	Hrs	£	£	£		£
P30: TRENCHES/PIPEWAYS/PITS FOR BURIED ENGINEERING SERVICES Cont.										
Excavating trenches to receive services not exceeding 200mm nominal size - by hand Cont.										
Extra over excavating trenches irrespective of depth; excavating next existing services										
electricity services -1	-	-	-	-	5.75	**61.58**	-	**61.58**	m	**69.28**
gas services -1	-	-	-	-	5.75	**61.58**	-	**61.58**	m	**69.28**
water services -1	-	-	-	-	5.75	**61.58**	-	**61.58**	m	**69.28**
Extra over excavating trenches irrespective of depth; excavating around existing services crossing trench										
electricity services; cables crossing -1	-	-	-	-	8.50	**91.04**	-	**91.04**	nr	**102.41**
gas services services crossing -1	-	-	-	-	8.50	**91.04**	-	**91.04**	nr	**102.41**
water services services crossing -1	-	-	-	-	8.50	**91.04**	-	**91.04**	nr	**102.41**

This section continues on the next page

Labour hourly rates: (except Specialists) Craft Operatives £14.21 Labourer £10.71 Rates are national average prices. Refer to REGIONAL VARIATIONS for indicative levels of overall pricing in regions	MATERIALS			LABOUR				RATES		
	Del to Site £	Waste %	Material Cost £	Craft Optve Hrs	Lab Hrs	Labour Cost £	Sunds £	Nett Rate £	Unit	Gross Rate (+12.5%) £
P30: TRENCHES/PIPEWAYS/PITS FOR BURIED ENGINEERING SERVICES Cont.										
Underground ducts; vitrified clayware, BS 65 extra strength; flexible joints										
Straight										
100mm nominal size single way duct; laid in position in trench	3.84	5.00	**4.03**	0.20	0.03	**3.16**	-	**7.20**	m	**8.09**
100mm nominal size bonded to form 2 way duct; laid in position in trench	7.68	5.00	**8.06**	0.30	0.04	**4.69**	-	**12.76**	m	**14.35**
100mm nominal size bonded to form 4 way duct; laid in position in trench	15.36	5.00	**16.13**	0.45	0.06	**7.04**	-	**23.17**	m	**26.06**
100mm nominal size bonded to form 6 way duct; laid in position in trench	23.04	5.00	**24.19**	0.55	0.07	**8.57**	-	**32.76**	m	**36.85**
extra; providing and laying in position nylon draw wire	0.15	10.00	**0.17**	0.02	-	**0.28**	-	**0.45**	m	**0.51**
Stop cock pits, valve chambers and the like										
Stop cock pits; half brick thick sides of common bricks, BS EN 772, Category M, 215 x 102.5 x 65mm, compressive strength 20.5 N/mm2, in cement mortar (1:3); 100mm thick base and top of plain in-situ concrete, BS 5328, ordinary prescribed mix ST4, 20mm aggregate, formwork; including all excavation, backfilling, disposal of surplus excavated material, earthwork support and compaction of ground										
600mm deep in clear; (surface boxes included elsewhere)										
internal size 225 x 225mm	18.07	5.00	**19.23**	1.75	1.75	**43.61**	0.06	**65.45**	nr	**73.64**
internal size 338 x 338mm	30.82	5.00	**33.30**	2.55	2.55	**63.55**	0.10	**100.86**	nr	**113.46**
aggregate, formwork										
750mm deep in clear; (surface boxes included elsewhere)										
150mm diameter	33.28	5.00	**34.12**	1.30	1.30	**32.40**	0.07	**69.14**	nr	**77.78**
Cast iron surface boxes and covers, coated to BS 4147; bedding in cement mortar (1:3)										
Surface boxes, BS 5834 Part 2 marked S.V.										
light grade, hinged lid; 150 x 150mm overall top size, 75mm deep	11.47	15.00	**11.58**	0.85	0.43	**16.68**	0.25	**28.51**	nr	**32.07**
Surface boxes, BS 750, marked 'FIRE HYDRANT'										
heavy grade; minimum clear opening 230 x 380mm, minimum depth 125mm	50.66	15.00	**50.89**	1.40	0.70	**27.39**	1.21	**79.49**	nr	**89.43**
Surface boxes, BS 5834 Part 2 marked 'W' or WATER										
medium grade; double triangular cover; minimum clear opening 300 x 300mm, minimum depth 100mm	49.66	15.00	**49.89**	1.70	0.85	**33.26**	1.21	**84.37**	nr	**94.91**
Accessories/Sundry items										
Stop cock keys										
tee	18.73	-	**18.73**	-	-	**-**	-	**18.73**	nr	**21.07**
P31: HOLES/CHASES/COVERS/SUPPORTS FOR SERVICES; BUILDERS WORK IN CONNECTION WITH MECHANICAL SERVICES										
Cutting holes for services										
For ducts through concrete; making good										
rectangular ducts not exceeding 1.00m girth										
150mm thick	-	-	**-**	1.84	-	**26.15**	0.57	**26.71**	nr	**30.05**
200mm thick	-	-	**-**	2.18	-	**30.98**	0.68	**31.66**	nr	**35.62**
300mm thick	-	-	**-**	2.52	-	**35.81**	0.80	**36.61**	nr	**41.18**
rectangular ducts 1.00-2.00m girth										
150mm thick	-	-	**-**	3.68	-	**52.29**	1.16	**53.45**	nr	**60.13**
200mm thick	-	-	**-**	4.35	-	**61.81**	1.34	**63.16**	nr	**71.05**
300mm thick	-	-	**-**	5.06	-	**71.90**	1.56	**73.47**	nr	**82.65**
rectangular ducts 2.00-3.00m girth										
150mm thick	-	-	**-**	4.60	-	**65.37**	1.44	**66.80**	nr	**75.16**
200mm thick	-	-	**-**	5.46	-	**77.59**	1.69	**79.28**	nr	**89.19**
300mm thick	-	-	**-**	6.33	-	**89.95**	1.94	**91.89**	nr	**103.38**
For pipes through concrete; making good										
pipes not exceeding 55mm nominal size										
150mm thick	-	-	**-**	1.04	-	**14.78**	0.36	**15.14**	nr	**17.03**
200mm thick	-	-	**-**	1.32	-	**18.76**	0.45	**19.21**	nr	**21.61**
300mm thick	-	-	**-**	1.67	-	**23.73**	0.58	**24.31**	nr	**27.35**
pipes 55 - 110mm nominal size										
150mm thick	-	-	**-**	1.28	-	**18.19**	0.44	**18.63**	nr	**20.96**
200mm thick	-	-	**-**	1.56	-	**22.17**	0.51	**22.68**	nr	**25.52**
300mm thick	-	-	**-**	2.08	-	**29.56**	0.70	**30.26**	nr	**34.04**
pipes exceeding 110mm nominal size										
150mm thick	-	-	**-**	1.55	-	**22.03**	0.51	**22.54**	nr	**25.36**
200mm thick	-	-	**-**	1.98	-	**28.14**	0.67	**28.81**	nr	**32.41**
300mm thick	-	-	**-**	2.50	-	**35.53**	0.85	**36.38**	nr	**40.92**
For ducts through reinforced concrete; making good										
rectangular ducts not exceeding 1.00m girth										
150mm thick	-	-	**-**	2.76	-	**39.22**	0.85	**40.07**	nr	**45.08**
200mm thick	-	-	**-**	3.28	-	**46.61**	1.04	**47.65**	nr	**53.60**
300mm thick	-	-	**-**	3.80	-	**54.00**	1.22	**55.22**	nr	**62.12**

Labour hourly rates: (except Specialists) Craft Operatives £14.21 Labourer £10.71 Rates are national average prices. Refer to REGIONAL VARIATIONS for indicative levels of overall pricing in regions	MATERIALS			LABOUR				RATES		
	Del to Site £	Waste %	Material Cost £	Craft Optve Hrs	Lab Hrs	Labour Cost £	Sunds £	Nett Rate £	Unit	Gross Rate (+12.5%) £
P31: HOLES/CHASES/COVERS/SUPPORTS FOR SERVICES; BUILDERS WORK IN CONNECTION WITH MECHANICAL SERVICES Cont.										
Cutting holes for services Cont.										
For ducts through reinforced concrete; making good Cont.										
rectangular ducts 1.00-2.00m girth										
150mm thick	-	-	-	5.52	-	78.44	1.70	80.14	nr	90.16
200mm thick	-	-	-	6.56	-	93.22	2.06	95.28	nr	107.19
300mm thick	-	-	-	7.60	-	108.00	2.37	110.37	nr	124.17
rectangular ducts 2.00-3.00m girth										
150mm thick	-	-	-	6.90	-	98.05	2.05	100.10	nr	112.61
200mm thick	-	-	-	8.05	-	114.39	2.42	116.81	nr	131.41
300mm thick	-	-	-	9.48	-	134.71	2.77	137.48	nr	154.67
For pipes through reinforced concrete; making good										
pipes not exceeding 55mm nominal size										
150mm thick	-	-	-	1.55	-	22.03	0.50	22.53	nr	25.35
200mm thick	-	-	-	2.00	-	28.42	0.65	29.07	nr	32.70
300mm thick	-	-	-	2.50	-	35.53	0.82	36.34	nr	40.89
pipes 55 - 110mm nominal size										
150mm thick	-	-	-	1.96	-	27.85	0.62	28.47	nr	32.03
200mm thick	-	-	-	2.36	-	33.54	0.76	34.29	nr	38.58
300mm thick	-	-	-	3.10	-	44.05	1.01	45.06	nr	50.69
pipes exceeding 110mm nominal size										
150mm thick	-	-	-	2.30	-	32.68	0.76	33.44	nr	37.62
200mm thick	-	-	-	3.00	-	42.63	0.96	43.59	nr	49.03
300mm thick	-	-	-	3.70	-	52.58	1.22	53.80	nr	60.52
For ducts through brickwork; making good										
rectangular ducts not exceeding 1.00m girth										
102.5mm thick	-	-	-	1.38	-	19.61	0.40	20.01	nr	22.51
215mm thick	-	-	-	1.60	-	22.74	0.48	23.22	nr	26.12
327.5mm thick	-	-	-	1.90	-	27.00	0.58	27.58	nr	31.02
rectangular ducts 1.00-2.00m girth										
102.5mm thick	-	-	-	2.76	-	39.22	0.81	40.03	nr	45.03
215mm thick	-	-	-	3.22	-	45.76	0.97	46.72	nr	52.56
327.5mm thick	-	-	-	3.80	-	54.00	1.16	55.15	nr	62.05
rectangular ducts 2.00-3.00m girth										
102.5mm thick	-	-	-	3.45	-	49.02	1.05	50.07	nr	56.33
215mm thick	-	-	-	4.02	-	57.12	1.21	58.33	nr	65.62
327.5mm thick	-	-	-	4.72	-	67.07	1.45	68.52	nr	77.09
For pipes through brickwork; making good										
pipes not exceeding 55mm nominal size										
102.5mm thick	-	-	-	0.46	-	6.54	0.16	6.69	nr	7.53
215mm thick	-	-	-	0.76	-	10.80	0.25	11.05	nr	12.43
327.5mm thick	-	-	-	1.15	-	16.34	0.39	16.73	nr	18.82
pipes 55 - 110mm nominal size										
102.5mm thick	-	-	-	0.60	-	8.53	0.21	8.74	nr	9.83
215mm thick	-	-	-	1.02	-	14.49	0.35	14.84	nr	16.70
327.5mm thick	-	-	-	1.53	-	21.74	0.51	22.26	nr	25.04
pipes exceeding 110mm nominal size										
102.5mm thick	-	-	-	0.77	-	10.94	0.26	11.20	nr	12.60
215mm thick	-	-	-	1.26	-	17.90	0.43	18.34	nr	20.63
327.5mm thick	-	-	-	1.92	-	27.28	0.65	27.93	nr	31.43
For ducts through brickwork; making good fair face or facings one side										
rectangular ducts not exceeding 1.00m girth										
102.5mm thick	-	-	-	1.67	-	23.73	0.64	24.37	nr	27.42
215mm thick	-	-	-	1.95	-	27.71	0.76	28.47	nr	32.02
327.5mm thick	-	-	-	2.18	-	30.98	0.81	31.79	nr	35.76
rectangular ducts 1.00-2.00m girth										
102.5mm thick	-	-	-	3.10	-	44.05	1.27	45.32	nr	50.99
215mm thick	-	-	-	3.55	-	50.45	1.50	51.95	nr	58.44
327.5mm thick	-	-	-	4.14	-	58.83	1.62	60.45	nr	68.00
rectangular ducts 2.00-3.00m girth										
102.5mm thick	-	-	-	3.90	-	55.42	1.62	57.04	nr	64.17
215mm thick	-	-	-	4.48	-	63.66	1.91	65.57	nr	73.77
327.5mm thick	-	-	-	5.18	-	73.61	2.03	75.63	nr	85.09
For pipes through brickwork; making good fair face or facings one side										
pipes not exceeding 55mm nominal size										
102.5mm thick	-	-	-	0.58	-	8.24	0.21	8.45	nr	9.51
215mm thick	-	-	-	0.87	-	12.36	0.30	12.67	nr	14.25
327.5mm thick	-	-	-	1.27	-	18.05	0.43	18.48	nr	20.79
pipes 55 - 110mm nominal size										
102.5mm thick	-	-	-	0.78	-	11.08	0.25	11.34	nr	12.75
215mm thick	-	-	-	1.18	-	16.77	0.38	17.15	nr	19.29
327.5mm thick	-	-	-	1.70	-	24.16	0.57	24.72	nr	27.81
pipes exceeding 110mm nominal size										
102.5mm thick	-	-	-	1.00	-	14.21	0.32	14.53	nr	16.34
215mm thick	-	-	-	1.50	-	21.32	0.46	21.78	nr	24.50
327.5mm thick	-	-	-	2.15	-	30.55	0.68	31.23	nr	35.14
For ducts through blockwork; making good										
rectangular ducts not exceeding 1.00m girth										
100mm thick	-	-	-	1.04	-	14.78	0.41	15.19	nr	17.09
140mm thick	-	-	-	1.15	-	16.34	0.48	16.82	nr	18.93
190mm thick	-	-	-	1.26	-	17.90	0.58	18.48	nr	20.79
rectangular ducts 1.00-2.00m girth										
100mm thick	-	-	-	2.08	-	29.56	0.81	30.37	nr	34.16

Labour hourly rates: (except Specialists) Craft Operatives £14.21 Labourer £10.71 Rates are national average prices. Refer to REGIONAL VARIATIONS for indicative levels of overall pricing in regions	MATERIALS			LABOUR				RATES		
	Del to Site £	Waste %	Material Cost £	Craft Optve Hrs	Lab Hrs	Labour Cost £	Sunds £	Nett Rate £	Unit	Gross Rate (+12.5%) £
P31: HOLES/CHASES/COVERS/SUPPORTS FOR SERVICES; BUILDERS WORK IN CONNECTION WITH MECHANICAL SERVICES Cont.										
Cutting holes for services Cont.										
For ducts through blockwork; making good Cont.										
rectangular ducts 1.00-2.00m girth Cont.										
140mm thick	-	-	-	2.24	-	31.83	0.97	32.80	nr	36.90
190mm thick	-	-	-	2.42	-	34.39	1.16	35.54	nr	39.99
rectangular ducts 2.00-3.00m girth										
100mm thick	-	-	-	2.58	-	36.66	1.04	37.70	nr	42.41
140mm thick	-	-	-	2.82	-	40.07	1.22	41.29	nr	46.45
190mm thick	-	-	-	3.05	-	43.34	1.45	44.79	nr	50.39
For pipes through blockwork; making good										
pipes not exceeding 55mm nominal size										
100mm thick	-	-	-	0.35	-	4.97	0.15	5.12	nr	5.76
140mm thick	-	-	-	0.46	-	6.54	0.24	6.78	nr	7.63
190mm thick	-	-	-	0.58	-	8.24	0.37	8.61	nr	9.69
pipes 55 - 110mm nominal size										
100mm thick	-	-	-	0.46	-	6.54	0.20	6.74	nr	7.58
140mm thick	-	-	-	0.61	-	8.67	0.33	8.99	nr	10.12
190mm thick	-	-	-	0.76	-	10.80	0.49	11.29	nr	12.70
pipes exceeding 110mm nominal size										
100mm thick	-	-	-	0.58	-	8.24	0.25	8.49	nr	9.56
140mm thick	-	-	-	0.78	-	11.08	0.41	11.49	nr	12.93
190mm thick	-	-	-	0.95	-	13.50	0.62	14.12	nr	15.88
For ducts through existing brickwork with plaster finish; making good each side										
rectangular ducts not exceeding 1.00m girth										
102.5mm thick	-	-	-	2.24	-	31.83	0.68	32.51	nr	36.58
215mm thick	-	-	-	2.58	-	36.66	0.86	37.52	nr	42.21
327.5mm thick	-	-	-	2.93	-	41.64	0.99	42.62	nr	47.95
rectangular ducts 1.00-2.00m girth										
102.5mm thick	-	-	-	4.14	-	58.83	1.39	60.22	nr	67.74
215mm thick	-	-	-	4.84	-	68.78	1.85	70.62	nr	79.45
327.5mm thick	-	-	-	5.52	-	78.44	1.97	80.41	nr	90.46
rectangular ducts 2.00-3.00m girth										
102.5mm thick	-	-	-	5.18	-	73.61	2.08	75.69	nr	85.15
215mm thick	-	-	-	6.04	-	85.83	2.60	88.43	nr	99.49
327.5mm thick	-	-	-	6.95	-	98.76	2.95	101.71	nr	114.42
For ducts through existing brickwork with plaster finish; making good one side										
rectangular ducts not exceeding 1.00m girth										
102.5mm thick	-	-	-	2.00	-	28.42	0.63	29.05	nr	32.68
215mm thick	-	-	-	2.35	-	33.39	0.79	34.18	nr	38.45
327.5mm thick	-	-	-	2.65	-	37.66	0.89	38.55	nr	43.37
rectangular ducts 1.00-2.00m girth										
102.5mm thick	-	-	-	3.74	-	53.15	1.24	54.38	nr	61.18
215mm thick	-	-	-	4.38	-	62.24	1.56	63.80	nr	71.78
327.5mm thick	-	-	-	5.00	-	71.05	1.79	72.84	nr	81.94
rectangular ducts 2.00-3.00m girth										
102.5mm thick	-	-	-	4.65	-	66.08	1.85	67.92	nr	76.42
215mm thick	-	-	-	5.46	-	77.59	2.31	79.90	nr	89.88
327.5mm thick	-	-	-	6.26	-	88.95	2.66	91.61	nr	103.06
For pipes through existing brickwork with plaster finish; making good each side										
pipes not exceeding 55mm nominal size										
102.5mm thick	-	-	-	0.75	-	10.66	0.25	10.91	nr	12.27
215mm thick	-	-	-	1.15	-	16.34	0.36	16.70	nr	18.79
327.5mm thick	-	-	-	1.66	-	23.59	0.50	24.09	nr	27.10
pipes 55 - 110mm nominal size										
102.5mm thick	-	-	-	1.10	-	15.63	0.30	15.94	nr	17.93
215mm thick	-	-	-	1.55	-	22.03	0.46	22.49	nr	25.30
327.5mm thick	-	-	-	2.24	-	31.83	0.68	32.51	nr	36.58
pipes exceeding 110mm nominal size										
102.5mm thick	-	-	-	1.38	-	19.61	0.37	19.98	nr	22.47
215mm thick	-	-	-	2.00	-	28.42	0.56	28.98	nr	32.60
327.5mm thick	-	-	-	2.88	-	40.92	0.82	41.74	nr	46.96
For pipes through existing brickwork with plaster finish; making good one side										
pipes not exceeding 55mm nominal size										
102.5mm thick	-	-	-	0.70	-	9.95	0.21	10.16	nr	11.43
215mm thick	-	-	-	1.04	-	14.78	0.30	15.08	nr	16.97
327.5mm thick	-	-	-	1.50	-	21.32	0.43	21.75	nr	24.46
pipes 55 - 110mm nominal size										
102.5mm thick	-	-	-	0.98	-	13.93	0.25	14.18	nr	15.95
215mm thick	-	-	-	1.38	-	19.61	0.38	19.99	nr	22.49
327.5mm thick	-	-	-	2.00	-	28.42	0.57	28.99	nr	32.61
pipes exceeding 110mm nominal size										
102.5mm thick	-	-	-	1.26	-	17.90	0.32	18.22	nr	20.50
215mm thick	-	-	-	1.84	-	26.15	0.46	26.61	nr	29.93
327.5mm thick	-	-	-	2.58	-	36.66	0.68	37.34	nr	42.01
For pipes through softwood										
pipes not exceeding 55mm nominal size										
19mm thick	-	-	-	0.12	-	1.71	-	1.71	nr	1.92
50mm thick	-	-	-	0.15	-	2.13	-	2.13	nr	2.40
pipes 55 - 110mm nominal size										
19mm thick	-	-	-	0.17	-	2.42	-	2.42	nr	2.72
50mm thick	-	-	-	0.23	-	3.27	-	3.27	nr	3.68

Labour hourly rates: (except Specialists) Craft Operatives £14.21 Labourer £10.71 Rates are national average prices. Refer to REGIONAL VARIATIONS for indicative levels of overall pricing in regions	MATERIALS			LABOUR				RATES		
	Del to Site £	Waste %	Material Cost £	Craft Optve Hrs	Lab Hrs	Labour Cost £	Sunds £	Nett Rate £	Unit	Gross Rate (+12.5%) £
P31: HOLES/CHASES/COVERS/SUPPORTS FOR SERVICES; BUILDERS WORK IN CONNECTION WITH MECHANICAL SERVICES Cont.										
Cutting holes for services Cont.										
For pipes through plywood										
pipes not exceeding 55mm nominal size										
13mm thick	-	-	-	0.12	-	**1.71**	-	**1.71**	nr	**1.92**
19mm thick	-	-	-	0.15	-	**2.13**	-	**2.13**	nr	**2.40**
pipes 55 - 110mm nominal size										
13mm thick	-	-	-	0.17	-	**2.42**	-	**2.42**	nr	**2.72**
19mm thick	-	-	-	0.23	-	**3.27**	-	**3.27**	nr	**3.68**
For pipes through existing plasterboard with skim finish; making good one side										
pipes not exceeding 55mm nominal size										
13mm thick	-	-	-	0.23	-	**3.27**	-	**3.27**	nr	**3.68**
pipes 55 - 110mm nominal size										
13mm thick	-	-	-	0.35	-	**4.97**	-	**4.97**	nr	**5.60**
Cutting chases for services										
In concrete; making good										
15mm nominal size pipes -1	-	-	-	0.35	-	**4.97**	-	**4.97**	m	**5.60**
22mm nominal size pipes -1	-	-	-	0.40	-	**5.68**	-	**5.68**	m	**6.39**
In brickwork										
15mm nominal size pipes -1	-	-	-	0.30	-	**4.26**	-	**4.26**	m	**4.80**
22mm nominal size pipes -1	-	-	-	0.35	-	**4.97**	-	**4.97**	m	**5.60**
In brickwork; making good fair face or facings										
15mm nominal size pipes -1	-	-	-	0.35	-	**4.97**	0.11	**5.08**	m	**5.71**
22mm nominal size pipes -1	-	-	-	0.40	-	**5.68**	0.16	**5.84**	m	**6.57**
In existing brickwork with plaster finish; making good										
15mm nominal size pipes -1	-	-	-	0.45	-	**6.39**	0.24	**6.64**	m	**7.47**
22mm nominal size pipes -1	-	-	-	0.55	-	**7.82**	0.39	**8.20**	m	**9.23**
In blockwork; making good										
15mm nominal size pipes -1	-	-	-	0.23	-	**3.27**	-	**3.27**	m	**3.68**
22mm nominal size pipes -1	-	-	-	0.29	-	**4.12**	-	**4.12**	m	**4.64**
Pipe and duct sleeves										
Steel pipes BS 1387 Table 2; casting into concrete; making good										
100mm long; to the following nominal size steel pipes										
15mm	0.58	5.00	**0.61**	0.12	0.06	**2.35**	-	**2.95**	nr	**3.32**
20mm	0.64	5.00	**0.67**	0.14	0.07	**2.74**	-	**3.41**	nr	**3.84**
25mm	0.79	5.00	**0.83**	0.15	0.08	**2.99**	-	**3.82**	nr	**4.29**
32mm	0.92	5.00	**0.97**	0.16	0.08	**3.13**	-	**4.10**	nr	**4.61**
40mm	1.06	5.00	**1.11**	0.18	0.09	**3.52**	-	**4.64**	nr	**5.22**
50mm	1.34	5.00	**1.41**	0.22	0.11	**4.30**	-	**5.72**	nr	**6.43**
150mm long; to the following nominal size steel pipes										
15mm	0.71	5.00	**0.75**	0.14	0.07	**2.74**	-	**3.49**	nr	**3.92**
20mm	0.80	5.00	**0.84**	0.16	0.08	**3.13**	-	**3.97**	nr	**4.46**
25mm	1.00	5.00	**1.05**	0.17	0.09	**3.38**	-	**4.43**	nr	**4.98**
32mm	1.18	5.00	**1.23**	0.18	0.09	**3.52**	-	**4.76**	nr	**5.35**
40mm	1.34	5.00	**1.41**	0.19	0.10	**3.77**	-	**5.18**	nr	**5.83**
50mm	1.74	5.00	**1.83**	0.23	0.12	**4.55**	-	**6.38**	nr	**7.18**
200mm long; to the following nominal size steel pipes										
15mm	0.84	5.00	**0.88**	0.15	0.08	**2.99**	-	**3.87**	nr	**4.35**
20mm	0.95	5.00	**0.99**	0.17	0.09	**3.38**	-	**4.37**	nr	**4.92**
25mm	1.20	5.00	**1.26**	0.18	0.09	**3.52**	-	**4.78**	nr	**5.38**
32mm	1.43	5.00	**1.50**	0.19	0.10	**3.77**	-	**5.27**	nr	**5.93**
40mm	1.64	5.00	**1.72**	0.20	0.10	**3.91**	-	**5.63**	nr	**6.34**
50mm	2.15	5.00	**2.26**	0.24	0.12	**4.70**	-	**6.96**	nr	**7.83**
250mm long; to the following nominal size steel pipes										
15mm	0.98	5.00	**1.03**	0.15	0.08	**2.99**	-	**4.01**	nr	**4.52**
20mm	1.09	5.00	**1.15**	0.19	0.10	**3.77**	-	**4.92**	nr	**5.53**
25mm	1.41	5.00	**1.48**	0.20	0.10	**3.91**	-	**5.39**	nr	**6.06**
32mm	1.69	5.00	**1.78**	0.21	0.11	**4.16**	-	**5.94**	nr	**6.68**
40mm	1.93	5.00	**2.03**	0.22	0.11	**4.30**	-	**6.33**	nr	**7.12**
50mm	2.56	5.00	**2.69**	0.26	0.13	**5.09**	-	**7.78**	nr	**8.75**
100mm long; to the following nominal size copper pipes										
15mm	0.64	5.00	**0.67**	0.12	0.06	**2.35**	-	**3.02**	nr	**3.40**
22mm	0.70	5.00	**0.74**	0.14	0.07	**2.74**	-	**3.48**	nr	**3.91**
28mm	0.87	5.00	**0.92**	0.15	0.08	**2.99**	-	**3.90**	nr	**4.39**
32mm	1.02	5.00	**1.07**	0.16	0.08	**3.13**	-	**4.20**	nr	**4.73**
45mm	1.17	5.00	**1.22**	0.18	0.09	**3.52**	-	**4.75**	nr	**5.34**
54mm	1.48	5.00	**1.56**	0.22	0.11	**4.30**	-	**5.86**	nr	**6.59**
150mm long; to the following nominal size copper pipes										
15mm	0.79	5.00	**0.83**	0.14	0.07	**2.74**	-	**3.57**	nr	**4.01**
22mm	0.88	5.00	**0.93**	0.16	0.08	**3.13**	-	**4.06**	nr	**4.56**
28mm	1.10	5.00	**1.16**	0.17	0.09	**3.38**	-	**4.54**	nr	**5.10**
32mm	1.29	5.00	**1.36**	0.18	0.09	**3.52**	-	**4.88**	nr	**5.49**
45mm	1.48	5.00	**1.56**	0.19	0.10	**3.77**	-	**5.33**	nr	**5.99**
54mm	1.92	5.00	**2.02**	0.23	0.12	**4.55**	-	**6.57**	nr	**7.39**

Labour hourly rates: (except Specialists) Craft Operatives £14.21 Labourer £10.71 Rates are national average prices. Refer to REGIONAL VARIATIONS for indicative levels of overall pricing in regions	MATERIALS			LABOUR				RATES		
	Del to Site £	Waste %	Material Cost £	Craft Optve Hrs	Lab Hrs	Labour Cost £	Sunds £	Nett Rate £	Unit	Gross Rate (+12.5%) £
1: HOLES/CHASES/COVERS/SUPPORTS FOR SERVICES; ILDERS WORK IN CONNECTION WITH MECHANICAL SERVICES nt.										
pe and duct sleeves Cont.										
teel pipes BS 1387 Table 2; casting into concrete; aking good Cont.										
200mm long; to the following nominal size copper pipes										
15mm	0.92	5.00	**0.97**	0.15	0.08	**2.99**	-	**3.96**	nr	**4.45**
22mm	1.04	5.00	**1.09**	0.17	0.09	**3.38**	-	**4.47**	nr	**5.03**
28mm	1.31	5.00	**1.38**	0.18	0.09	**3.52**	-	**4.90**	nr	**5.51**
32mm	1.58	5.00	**1.65**	0.19	0.10	**3.77**	-	**5.42**	nr	**6.10**
45mm	1.81	5.00	**1.90**	0.20	0.10	**3.91**	-	**5.81**	nr	**6.54**
54mm	2.37	5.00	**2.49**	0.24	0.12	**4.70**	-	**7.19**	nr	**8.09**
250mm long; to the following nominal size copper pipes										
15mm	1.08	5.00	**1.14**	0.17	0.09	**3.38**	-	**4.52**	nr	**5.08**
22mm	1.20	5.00	**1.26**	0.19	0.10	**3.77**	-	**5.03**	nr	**5.66**
28mm	1.54	5.00	**1.62**	0.20	0.10	**3.91**	-	**5.53**	nr	**6.23**
32mm	1.86	5.00	**1.95**	0.21	0.11	**4.16**	-	**6.11**	nr	**6.88**
45mm	2.11	5.00	**2.22**	0.22	0.11	**4.30**	-	**6.52**	nr	**7.34**
54mm	2.81	5.00	**2.95**	0.26	0.13	**5.09**	-	**8.04**	nr	**9.05**
teel pipes BS 1387 Table 2; building into blockwork; edding and pointing in cement mortar (1:3); making ood										
100mm long; to the following nominal size steel pipes										
15mm	0.58	5.00	**0.61**	0.17	0.09	**3.38**	-	**3.99**	nr	**4.48**
20mm	0.64	5.00	**0.67**	0.18	0.09	**3.52**	-	**4.19**	nr	**4.72**
25mm	0.79	5.00	**0.83**	0.19	0.10	**3.77**	-	**4.60**	nr	**5.17**
32mm	0.92	5.00	**0.97**	0.20	0.10	**3.91**	-	**4.88**	nr	**5.49**
40mm	1.06	5.00	**1.11**	0.21	0.11	**4.16**	-	**5.28**	nr	**5.94**
50mm	1.34	5.00	**1.41**	0.22	0.11	**4.30**	-	**5.72**	nr	**6.43**
140mm long; to the following nominal size steel pipes										
15mm	0.68	5.00	**0.72**	0.19	0.10	**3.77**	-	**4.49**	nr	**5.05**
20mm	0.77	5.00	**0.81**	0.20	0.10	**3.91**	-	**4.72**	nr	**5.31**
25mm	0.96	5.00	**1.00**	0.21	0.11	**4.16**	-	**5.17**	nr	**5.81**
32mm	1.12	5.00	**1.18**	0.22	0.11	**4.30**	-	**5.48**	nr	**6.17**
40mm	1.29	5.00	**1.36**	0.23	0.12	**4.55**	-	**5.91**	nr	**6.65**
50mm	1.67	5.00	**1.75**	0.24	0.12	**4.70**	-	**6.45**	nr	**7.26**
190mm long; to the following nominal size steel pipes										
15mm	0.82	5.00	**0.86**	0.21	0.11	**4.16**	-	**5.02**	nr	**5.65**
20mm	0.91	5.00	**0.96**	0.22	0.11	**4.30**	-	**5.26**	nr	**5.92**
25mm	1.17	5.00	**1.22**	0.23	0.12	**4.55**	-	**5.78**	nr	**6.50**
32mm	1.39	5.00	**1.46**	0.24	0.12	**4.70**	-	**6.15**	nr	**6.92**
40mm	1.58	5.00	**1.65**	0.25	0.13	**4.94**	-	**6.60**	nr	**7.42**
50mm	2.07	5.00	**2.17**	0.26	0.13	**5.09**	-	**7.26**	nr	**8.17**
215mm long; to the following nominal size steel pipes										
15mm	0.88	5.00	**0.93**	0.22	0.11	**4.30**	-	**5.23**	nr	**5.88**
20mm	0.99	5.00	**1.04**	0.23	0.12	**4.55**	-	**5.59**	nr	**6.29**
25mm	1.27	5.00	**1.33**	0.24	0.12	**4.70**	-	**6.03**	nr	**6.78**
32mm	1.51	5.00	**1.59**	0.25	0.13	**4.94**	-	**6.53**	nr	**7.35**
40mm	1.72	5.00	**1.81**	0.26	0.13	**5.09**	-	**6.90**	nr	**7.76**
50mm	2.28	5.00	**2.39**	0.27	0.14	**5.34**	-	**7.73**	nr	**8.70**
100mm long; to the following nominal size copper pipes										
15mm	0.64	5.00	**0.67**	0.17	0.09	**3.38**	-	**4.05**	nr	**4.56**
22mm	0.69	5.00	**0.73**	0.18	0.09	**3.52**	-	**4.25**	nr	**4.78**
28mm	0.87	5.00	**0.92**	0.19	0.10	**3.77**	-	**4.69**	nr	**5.27**
32mm	1.02	5.00	**1.07**	0.20	0.10	**3.91**	-	**4.98**	nr	**5.61**
45mm	1.06	5.00	**1.11**	0.21	0.11	**4.16**	-	**5.28**	nr	**5.94**
54mm	1.48	5.00	**1.56**	0.22	0.11	**4.30**	-	**5.86**	nr	**6.59**
140mm long; to the following nominal size copper pipes										
15mm	0.76	5.00	**0.79**	0.19	0.10	**3.77**	-	**4.56**	nr	**5.14**
22mm	0.84	5.00	**0.88**	0.20	0.10	**3.91**	-	**4.80**	nr	**5.39**
28mm	1.06	5.00	**1.11**	0.21	0.11	**4.16**	-	**5.28**	nr	**5.94**
32mm	1.24	5.00	**1.30**	0.22	0.11	**4.30**	-	**5.61**	nr	**6.31**
45mm	1.42	5.00	**1.49**	0.23	0.12	**4.55**	-	**6.04**	nr	**6.80**
54mm	1.84	5.00	**1.93**	0.24	0.12	**4.70**	-	**6.63**	nr	**7.45**
190mm long; to the following nominal size copper pipes										
15mm	0.90	5.00	**0.95**	0.21	0.11	**4.16**	-	**5.11**	nr	**5.75**
22mm	1.01	5.00	**1.06**	0.22	0.11	**4.30**	-	**5.36**	nr	**6.03**
28mm	1.28	5.00	**1.35**	0.23	0.12	**4.55**	-	**5.90**	nr	**6.64**
32mm	1.52	5.00	**1.60**	0.24	0.12	**4.70**	-	**6.29**	nr	**7.08**
45mm	1.73	5.00	**1.82**	0.25	0.13	**4.94**	-	**6.76**	nr	**7.61**
54mm	2.28	5.00	**2.39**	0.26	0.13	**5.09**	-	**7.48**	nr	**8.41**
215mm long; to the following nominal size copper pipes										
15mm	0.97	5.00	**1.01**	0.23	0.12	**4.55**	-	**5.57**	nr	**6.26**
22mm	1.08	5.00	**1.14**	0.24	0.12	**4.70**	-	**5.83**	nr	**6.56**
28mm	1.40	5.00	**1.47**	0.25	0.13	**4.94**	-	**6.41**	nr	**7.21**
32mm	1.66	5.00	**1.74**	0.26	0.13	**5.09**	-	**6.83**	nr	**7.68**
45mm	1.89	5.00	**1.98**	0.27	0.14	**5.34**	-	**7.32**	nr	**8.24**
54mm	2.51	5.00	**2.64**	0.28	0.14	**5.48**	-	**8.11**	nr	**9.13**

Labour hourly rates: (except Specialists) Craft Operatives £14.21 Labourer £10.71 Rates are national average prices. Refer to REGIONAL VARIATIONS for indicative levels of overall pricing in regions	MATERIALS			LABOUR				RATES		
	Del to Site £	Waste %	Material Cost £	Craft Optve Hrs	Lab Hrs	Labour Cost £	Sunds £	Nett Rate £	Unit	Gross Rate (+12.5%) £
P31: HOLES/CHASES/COVERS/SUPPORTS FOR SERVICES; BUILDERS WORK IN CONNECTION WITH MECHANICAL SERVICES Cont.										
Ends of supports for equipment, fittings, appliances and ancillaries										
Fix only; casting into concrete; making good										
holderbat or bracket	-	-	-	0.29	-	4.12	0.30	4.43	nr	4.98
Fix only; building into brickwork										
holderbat or bracket	-	-	-	0.12	-	1.71	-	1.71	nr	1.92
Fix only; cutting and pinning to brickwork; making good										
holderbat or bracket	-	-	-	0.14	-	1.99	0.09	2.08	nr	2.34
Fix only; cutting and pinning to brickwork; making good fair face or facings one side										
holderbat or bracket	-	-	-	0.17	-	2.42	0.15	2.56	nr	2.88
Cavity fixings for ends of supports for pipes and ducts										
Rawlnut Multi-purpose fixings, plated pan head screws; The Rawlplug Co. Ltd; fixing to soft building board, drilling										
M4 with 30mm long screw; product code 09 - 130										
at 300mm centres	1.40	5.00	1.47	0.19	-	2.70	-	4.17	m	4.69
at 450mm centres	0.93	5.00	0.98	0.12	-	1.71	-	2.69	m	3.02
isolated	0.42	5.00	0.44	0.06	-	0.85	-	1.29	nr	1.46
M5 with 30mm long screw; product code 09 - 235										
at 300mm centres	1.40	5.00	1.47	0.22	-	3.13	-	4.60	m	5.17
at 450mm centres	0.93	5.00	0.98	0.14	-	1.99	-	2.97	m	3.34
isolated	0.42	5.00	0.44	0.07	-	0.99	-	1.44	nr	1.62
M5 with 50mm long screw; product code 09 - 317										
at 300mm centres	1.88	5.00	1.97	0.22	-	3.13	-	5.10	m	5.74
at 450mm centres	1.25	5.00	1.32	0.14	-	1.99	-	3.31	m	3.72
isolated	0.56	5.00	0.59	0.07	-	0.99	-	1.59	nr	1.79
Rawlnut Multi-purpose fixings, plated pan head screws; The Rawlplug Co. Ltd; fixing to sheet metal, drilling										
M4 with 30mm long screw; product code 09 - 130										
at 300mm centres	1.40	5.00	1.47	0.36	-	5.12	-	6.59	m	7.41
at 450mm centres	0.93	5.00	0.98	0.24	-	3.41	-	4.39	m	4.94
isolated	0.42	5.00	0.44	0.11	-	1.56	-	2.00	nr	2.25
M5 with 30mm long screw; product code 09 - 235										
at 300mm centres	1.40	5.00	1.47	0.44	-	6.25	-	7.72	m	8.69
at 450mm centres	0.93	5.00	0.98	0.30	-	4.26	-	5.24	m	5.90
isolated	0.42	5.00	0.44	0.13	-	1.85	-	2.29	nr	2.57
M5 with 50mm long screw; product code 09 - 317										
at 300mm centres	1.88	5.00	1.97	0.44	-	6.25	-	8.23	m	9.25
at 450mm centres	1.25	5.00	1.32	0.30	-	4.26	-	5.58	m	6.28
isolated	0.56	5.00	0.59	0.13	-	1.85	-	2.44	nr	2.74
Interset high performance cavity fixings, plated pan head screws; The Rawlplug Co. Ltd; fixing to soft building board, drilling										
M4 with 40mm long screw; product code 41 - 620										
at 300mm centres	1.62	5.00	1.70	0.19	-	2.70	-	4.40	m	4.95
at 450mm centres	1.08	5.00	1.14	0.12	-	1.71	-	2.84	m	3.20
isolated	0.49	5.00	0.51	0.06	-	0.85	-	1.36	nr	1.53
M5 with 40mm long screw; product code 41 - 636										
at 300mm centres	1.94	5.00	2.04	0.22	-	3.13	-	5.16	m	5.81
at 450mm centres	1.29	5.00	1.36	0.14	-	1.99	-	3.35	m	3.77
isolated	0.58	5.00	0.61	0.07	-	0.99	-	1.61	nr	1.81
M5 with 55mm long screw; product code 41 - 652										
at 300mm centres	2.03	5.00	2.13	0.22	-	3.13	-	5.25	m	5.91
at 450mm centres	1.35	5.00	1.42	0.14	-	1.99	-	3.41	m	3.83
isolated	0.61	5.00	0.64	0.07	-	0.99	-	1.63	nr	1.84
Spring toggles plated pan head screws; The Rawlplug Co. Ltd; fixing to soft building board, drilling										
M5 with 80mm long screw; product code 94 - 439										
at 300mm centres	1.34	5.00	1.41	0.22	-	3.13	-	4.54	m	5.10
at 450mm centres	0.90	5.00	0.94	0.14	-	1.99	-	2.93	m	3.30
isolated	0.40	5.00	0.42	0.07	-	0.99	-	1.42	nr	1.60
M6 with 60mm long screw; product code 94 - 442										
at 300mm centres	1.57	5.00	1.64	0.22	-	3.13	-	4.77	m	5.37
at 450mm centres	1.04	5.00	1.10	0.14	-	1.99	-	3.09	m	3.47
isolated	0.47	5.00	0.49	0.07	-	0.99	-	1.49	nr	1.67
M6 with 80mm long screw; product code 94 - 464										
at 300mm centres	1.93	5.00	2.02	0.22	-	3.13	-	5.15	m	5.79
at 450mm centres	1.28	5.00	1.35	0.14	-	1.99	-	3.34	m	3.76
isolated	0.58	5.00	0.61	0.07	-	0.99	-	1.60	nr	1.80
Trench covers and frames										
Duct covers in cast iron, coated; continuous covers in 610mm lengths; steel bearers										
22mm deep, for pedestrian traffic; bedding frames to concrete in cement mortar (1:3); nominal width										
150mm	90.42	-	90.42	0.50	0.50	12.46	0.75	103.63	m	116.58
225mm	94.23	-	94.23	0.55	0.55	13.71	0.75	108.69	m	122.28
300mm	109.92	-	109.92	0.65	0.65	16.20	0.75	126.87	m	142.73
375mm	125.35	-	125.35	0.70	0.70	17.44	0.75	143.54	m	161.48

Labour hourly rates: (except Specialists) Craft Operatives £14.21 Labourer £10.71 Rates are national average prices. Refer to REGIONAL VARIATIONS for indicative levels of overall pricing in regions	MATERIALS			LABOUR				RATES		
	Del to Site £	Waste %	Material Cost £	Craft Optve Hrs	Lab Hrs	Labour Cost £	Sunds £	Nett Rate £	Unit	Gross Rate (+12.5%) £
P31: HOLES/CHASES/COVERS/SUPPORTS FOR SERVICES; BUILDERS WORK IN CONNECTION WITH MECHANICAL SERVICES Cont.										
Trench covers and frames Cont.										
Duct covers in cast iron, coated; continuous covers in 610mm lengths; steel bearers Cont.										
22mm deep, for pedestrian traffic; bedding frames to concrete in cement mortar (1:3); nominal width Cont.										
450mm	141.25	-	**141.25**	0.75	0.75	**18.69**	0.75	**160.69**	m	**180.77**
Pendock Profiles Ltd. floor ducting profiles; galvanized mild steel tray section with 12mm plywood cover board; fixing tray section to masonry with nails; fixing cover board with screws, countersinking										
For 50mm screeds										
reference FDT 100/50, 100mm wide	10.15	2.50	**10.41**	0.40	-	**5.68**	0.12	**16.21**	m	**18.24**
extra; stop end	2.89	2.50	**2.96**	0.12	-	**1.71**	-	**4.67**	nr	**5.25**
extra; corner	12.53	2.50	**12.84**	0.29	-	**4.12**	0.17	**17.14**	nr	**19.28**
extra; tee	12.53	2.50	**12.84**	0.35	-	**4.97**	0.17	**17.99**	nr	**20.24**
reference FDT 150/50, 150mm wide	13.42	2.50	**13.76**	0.46	-	**6.54**	0.12	**20.41**	m	**22.97**
extra; stop end	2.89	2.50	**2.96**	0.14	-	**1.99**	-	**4.95**	nr	**5.57**
extra; corner	13.95	2.50	**14.30**	0.35	-	**4.97**	0.17	**19.44**	nr	**21.87**
extra; tee	13.95	2.50	**14.30**	0.40	-	**5.68**	0.17	**20.15**	nr	**22.67**
reference FDT 200/50, 200mm wide	16.34	2.50	**16.75**	0.58	-	**8.24**	0.12	**25.11**	m	**28.25**
extra; stop end	2.89	2.50	**2.96**	0.17	-	**2.42**	-	**5.38**	nr	**6.05**
extra; corner	16.22	2.50	**16.63**	0.40	-	**5.68**	0.17	**22.48**	nr	**25.29**
extra; tee	16.22	2.50	**16.63**	0.46	-	**6.54**	0.17	**23.34**	nr	**26.25**
For 70mm screeds										
reference FDT 100/70, 100mm wide	10.97	2.50	**11.25**	0.40	-	**5.68**	0.12	**17.05**	m	**19.18**
extra; stop end	2.89	2.50	**2.96**	0.12	-	**1.71**	-	**4.67**	nr	**5.25**
extra; corner	12.53	2.50	**12.84**	0.29	-	**4.12**	0.17	**17.14**	nr	**19.28**
extra; tee	12.53	2.50	**12.84**	0.35	-	**4.97**	0.17	**17.99**	nr	**20.24**
reference FDT 150/70, 150mm wide	14.25	2.50	**14.60**	0.46	-	**6.54**	0.12	**21.26**	m	**23.92**
extra; stop end	2.89	2.50	**2.96**	0.14	-	**1.99**	-	**4.95**	nr	**5.57**
extra; corner	13.95	2.50	**14.30**	0.35	-	**4.97**	0.17	**19.44**	nr	**21.87**
extra; tee	13.95	2.50	**14.30**	0.40	-	**5.68**	0.17	**20.15**	nr	**22.67**
reference FDT 200/70, 200mm wide	17.05	2.50	**17.47**	0.58	-	**8.24**	0.12	**25.83**	m	**29.06**
extra; stop end	2.89	2.50	**2.96**	0.17	-	**2.42**	-	**5.38**	nr	**6.05**
extra; corner	16.22	2.50	**16.63**	0.40	-	**5.68**	0.17	**22.48**	nr	**25.29**
extra; tee	16.22	2.50	**16.63**	0.46	-	**6.54**	0.17	**23.34**	nr	**26.25**
Casings, bearers and supports for equipment										
Softwood, sawn										
25mm boarded platforms	5.71	10.00	**6.28**	1.25	0.16	**19.48**	0.17	**25.93**	m2	**29.17**
50 x 50mm bearers	0.52	10.00	**0.57**	0.18	0.02	**2.77**	0.02	**3.36**	m	**3.78**
50 x 75mm bearers	0.75	10.00	**0.82**	0.22	0.03	**3.45**	0.03	**4.30**	m	**4.83**
50 x 100mm bearers	1.03	10.00	**1.13**	0.26	0.03	**4.02**	0.03	**5.18**	m	**5.83**
38 x 50mm bearers; framed	0.42	10.00	**0.46**	0.29	0.04	**4.55**	0.04	**5.05**	m	**5.68**
Softwood, wrought										
25mm boarded platforms; tongued and grooved	11.12	10.00	**12.23**	1.40	0.17	**21.71**	0.20	**34.15**	m2	**38.42**
19mm boarded sides; tongued and grooved	8.43	10.00	**9.27**	1.25	0.16	**19.48**	0.17	**28.92**	m2	**32.53**
extra for holes for pipes not exceeding 55mm nominal size	-	-	-	0.30	-	**4.26**	-	**4.26**	nr	**4.80**
19mm boarded cover; tongued and grooved; ledged; sectional	15.98	10.00	**17.58**	1.40	0.17	**21.71**	0.20	**39.49**	m2	**44.43**
Fibreboard										
13mm sides	8.98	10.00	**9.87**	2.40	0.30	**37.32**	0.34	**47.53**	m2	**53.47**
extra for holes for pipes not exceeding 55mm nominal size	-	-	-	0.25	-	**3.55**	-	**3.55**	nr	**4.00**
Vermiculite insulation										
packing around tank	68.45	10.00	**75.30**	5.50	0.70	**85.65**	-	**160.95**	m3	**181.07**
Slag wool insulation										
packing around tank	34.06	10.00	**37.47**	7.00	0.85	**108.57**	-	**146.04**	m3	**164.30**
P31: HOLES/CHASES/COVERS/SUPPORTS FOR SERVICES; BUILDERS WORK IN CONNECTION WITH ELECTRICAL SERVICES										
Cutting or forming holes, mortices, sinkings and chases - New buildings										
Concealed steel conduits; making good										
luminaire points	-	-	-	0.80	0.40	**15.65**	0.40	**16.05**	nr	**18.06**
socket outlet points	-	-	-	0.70	0.35	**13.70**	0.35	**14.04**	nr	**15.80**
fitting outlet points	-	-	-	0.70	0.35	**13.70**	0.35	**14.04**	nr	**15.80**
equipment and control gear points	-	-	-	0.90	0.45	**17.61**	0.47	**18.08**	nr	**20.34**
Exposed p.v.c. conduits; making good										
luminaire points	-	-	-	0.28	0.14	**5.48**	0.14	**5.61**	nr	**6.32**
socket outlet points	-	-	-	0.23	0.12	**4.55**	0.12	**4.67**	nr	**5.25**
fitting outlet points	-	-	-	0.23	0.12	**4.55**	0.12	**4.67**	nr	**5.25**
equipment and control gear points	-	-	-	0.35	0.18	**6.90**	0.16	**7.06**	nr	**7.94**

Labour hourly rates: (except Specialists) Craft Operatives £14.21 Labourer £10.71 Rates are national average prices. Refer to REGIONAL VARIATIONS for indicative levels of overall pricing in regions	MATERIALS			LABOUR				RATES		
	Del to Site £	Waste %	Material Cost £	Craft Optve Hrs	Lab Hrs	Labour Cost £	Sunds £	Nett Rate £	Unit	Gross Rate (+12.5%) £
P31: HOLES/CHASES/COVERS/SUPPORTS FOR SERVICES; BUILDERS WORK IN CONNECTION WITH ELECTRICAL SERVICES Cont.										
Cutting mortices, sinking and the like for services - Existing buildings										
In existing concrete; making good										
75 x 75 x 35mm	-	-	-	0.50	0.18	**9.03**	0.16	**9.19**	nr	**10.34**
150 x 75 x 35mm	-	-	-	0.66	0.18	**11.31**	0.87	**12.18**	nr	**13.70**
In existing brickwork										
75 x 75 x 35mm	-	-	-	0.30	0.15	**5.87**	0.37	**6.24**	nr	**7.02**
150 x 75 x 35mm	-	-	-	0.44	0.15	**7.86**	0.53	**8.38**	nr	**9.43**
In existing brickwork with plaster finish; making good										
75 x 75 x 35mm	-	-	-	0.35	0.20	**7.12**	0.56	**7.67**	nr	**8.63**
150 x 75 x 35mm	-	-	-	0.50	0.23	**9.57**	0.81	**10.38**	nr	**11.67**
In existing blockwork; making good										
75 x 75 x 35mm	-	-	-	0.33	0.20	**6.83**	0.58	**7.41**	nr	**8.33**
150 x 75 x 35mm	-	-	-	0.39	0.12	**6.83**	0.27	**7.10**	nr	**7.99**
In existing blockwork with plaster finish; making good										
75 x 75 x 35mm	-	-	-	0.45	0.14	**7.89**	0.53	**8.42**	nr	**9.47**
150 x 75 x 35mm	-	-	-	0.50	0.18	**9.03**	0.46	**9.49**	nr	**10.68**
Cutting chases for services										
In existing concrete; making good										
20mm nominal size conduits -1	-	-	-	0.46	0.23	**9.00**	0.40	**9.40**	m	**10.57**
20mm nominal size conduits -3	-	-	-	0.58	0.29	**11.35**	0.47	**11.82**	m	**13.30**
20mm nominal size conduits -6	-	-	-	0.70	0.35	**13.70**	0.55	**14.24**	m	**16.02**
In existing brickwork										
20mm nominal size conduits -1	-	-	-	0.35	0.18	**6.90**	0.40	**7.30**	m	**8.21**
20mm nominal size conduits -3	-	-	-	0.46	0.23	**9.00**	0.47	**9.47**	m	**10.66**
20mm nominal size conduits -6	-	-	-	0.58	0.29	**11.35**	0.55	**11.89**	m	**13.38**
In existing brickwork with plaster finish; making good										
20mm nominal size conduits -1	-	-	-	0.52	0.26	**10.17**	0.53	**10.70**	m	**12.04**
20mm nominal size conduits -3	-	-	-	0.62	0.31	**12.13**	0.59	**12.72**	m	**14.31**
20mm nominal size conduits -6	-	-	-	0.75	0.38	**14.73**	0.65	**15.38**	m	**17.30**
In existing blockwork										
20mm nominal size conduits -1	-	-	-	0.26	0.13	**5.09**	0.40	**5.49**	m	**6.17**
20mm nominal size conduits -3	-	-	-	0.35	0.18	**6.90**	0.47	**7.37**	m	**8.30**
20mm nominal size conduits -6	-	-	-	0.44	0.22	**8.61**	0.55	**9.15**	m	**10.30**
In existing blockwork with plaster finish; making good										
20mm nominal size conduits -1	-	-	-	0.44	0.22	**8.61**	0.53	**9.13**	m	**10.28**
20mm nominal size conduits -3	-	-	-	0.52	0.26	**10.17**	0.59	**10.76**	m	**12.11**
20mm nominal size conduits -6	-	-	-	0.60	0.30	**11.74**	0.65	**12.39**	m	**13.94**
Lifting and replacing floorboards										
For cables or conduits										
in groups 1 - 3	-	-	-	0.18	0.09	**3.52**	0.12	**3.64**	m	**4.09**
in groups 3 - 6	-	-	-	0.22	0.11	**4.30**	0.13	**4.43**	m	**4.98**
in groups exceeding 6	-	-	-	0.28	0.14	**5.48**	0.14	**5.61**	m	**6.32**

Labour hourly rates: (except Specialists) Craft Operatives £14.21 Labourer £10.71 Rates are national average prices. Refer to REGIONAL VARIATIONS for indicative levels of overall pricing in regions	PLANT & TRANSPORT			LABOUR				RATES		
	Plant Cost £	Trans Cost £	P & T Cost £	Craft Optve Hrs	Lab Hrs	Labour Cost £	Sunds £	Nett Rate £	Unit	Gross Rate (+12.5%) £
Q10: KERBS/EDGINGS/CHANNELS/PAVING ACCESSORIES										
Excavating - by machine										
Trenches width not exceeding 0.30m										
maximum depth not exceeding 0.25m................	4.68	-	**4.68**	-	2.00	**21.42**	-	**26.10**	m3	**29.36**
maximum depth not exceeding 1.00m................	4.85	-	**4.85**	-	2.10	**22.49**	-	**27.34**	m3	**30.76**
Extra over any types of excavating irrespective of depth										
excavating below ground water level.............	4.15	-	**4.15**	-	-	**-**	-	**4.15**	m3	**4.66**
excavating in running silt, running sand or liquid mud	7.20	-	**7.20**	-	-	**-**	-	**7.20**	m3	**8.09**
Breaking out existing materials; extra over any types of excavating irrespective of depth										
hard rock	8.55	-	**8.55**	-	-	**-**	-	**8.55**	m3	**9.62**
concrete	7.10	-	**7.10**	-	-	**-**	-	**7.10**	m3	**7.98**
reinforced concrete	10.66	-	**10.66**	-	-	**-**	-	**10.66**	m3	**11.99**
brickwork, blockwork or stonework................	4.74	-	**4.74**	-	-	**-**	-	**4.74**	m3	**5.33**
100mm diameter drain and concrete bed under......	0.67	-	**0.67**	-	-	**-**	-	**0.67**	m	**0.75**
Breaking out existing hard pavings; extra over any types of excavating irrespective of depth										
concrete 150mm thick	1.12	-	**1.12**	-	-	**-**	-	**1.12**	m2	**1.26**
reinforced concrete 200mm thick..................	2.24	-	**2.24**	-	-	**-**	-	**2.24**	m2	**2.53**
brickwork, blockwork or stonework 100mm thick....	0.53	-	**0.53**	-	-	**-**	-	**0.53**	m2	**0.59**
coated macadam or asphalt 75mm thick............	0.19	-	**0.19**	-	-	**-**	-	**0.19**	m2	**0.21**
Disposal of excavated material										
depositing on site in temporary spoil heaps where directed										
25m..	-	2.04	**2.04**	-	0.12	**1.29**	-	**3.33**	m3	**3.74**
100m...	-	2.33	**2.33**	-	0.12	**1.29**	-	**3.61**	m3	**4.06**
800m...	-	3.58	**3.58**	-	0.12	**1.29**	-	**4.86**	m3	**5.47**
1600m..	-	4.05	**4.05**	-	0.12	**1.29**	-	**5.34**	m3	**6.00**
extra for each additional 1600m	-	0.63	**0.63**	-	-	**-**	-	**0.63**	m3	**0.71**
removing from site to tip a) Inert	-	17.00	**17.00**	-	-	**-**	-	**17.00**	m3	**19.13**
removing from site to tip b) Active	-	40.00	**40.00**	-	-	**-**	-	**40.00**	m3	**45.00**
removing from site to tip c) Contaminated (Guide price - always seek a quotation for specialist disposal costs.)	-	80.00	**80.00**	-	-	**-**	-	**80.00**	m3	**90.00**
Disposal of preserved top soil										
depositing on site in temporary spoil heaps where directed										
25m..	-	2.04	**2.04**	-	0.12	**1.29**	-	**3.33**	m3	**3.74**
50m..	-	2.13	**2.13**	-	0.12	**1.29**	-	**3.42**	m3	**3.84**
100m...	-	2.33	**2.33**	-	0.12	**1.29**	-	**3.61**	m3	**4.06**
Surface treatments										
Compacting										
bottoms of excavations	0.22	-	**0.22**	-	-	**-**	-	**0.22**	m2	**0.24**
Excavating - by hand										
Trenches width not exceeding 0.30m										
maximum depth not exceeding 0.25m................	-	-	**-**	-	4.70	**50.34**	-	**50.34**	m3	**56.63**
maximum depth not exceeding 1.00m................	-	-	**-**	-	4.95	**53.01**	-	**53.01**	m3	**59.64**
Extra over any types of excavating irrespective of depth										
excavating below ground water level.............	-	-	**-**	-	1.65	**17.67**	-	**17.67**	m3	**19.88**
excavating in running silt, running sand or liquid mud	-	-	**-**	-	6.60	**70.69**	-	**70.69**	m3	**79.52**
Breaking out existing materials; extra over any types of excavating irrespective of depth										
hard rock	-	-	**-**	-	12.00	**128.52**	-	**128.52**	m3	**144.59**
concrete	-	-	**-**	-	10.50	**112.46**	-	**112.46**	m3	**126.51**

Labour hourly rates: (except Specialists) Craft Operatives £14.21 Labourer £10.71 Rates are national average prices. Refer to REGIONAL VARIATIONS for indicative levels of overall pricing in regions	PLANT & TRANSPORT			LABOUR				RATES		
	Plant Cost £	Trans Cost £	P & T Cost £	Craft Optve Hrs	Lab Hrs	Labour Cost £	Sunds £	Nett Rate £	Unit	Gross Rate (+12.5%) £
Q10: KERBS/EDGINGS/CHANNELS/PAVING ACCESSORIES Cont.										
Excavating - by hand Cont.										
Breaking out existing materials; extra over any types of excavating irrespective of depth Cont.										
reinforced concrete	-	-	-	-	15.75	**168.68**	-	**168.68**	m3	**189.77**
brickwork, blockwork or stonework	-	-	-	-	6.30	**67.47**	-	**67.47**	m3	**75.91**
100mm diameter drain and concrete bed under	-	-	-	-	1.10	**11.78**	-	**11.78**	m	**13.25**
Breaking out existing hard pavings; extra over any types of excavating irrespective of depth										
concrete 150mm thick	-	-	-	-	1.65	**17.67**	-	**17.67**	m2	**19.88**
reinforced concrete 200mm thick	-	-	-	-	2.20	**23.56**	-	**23.56**	m2	**26.51**
brickwork, blockwork or stonework 100mm thick	-	-	-	-	1.00	**10.71**	-	**10.71**	m2	**12.05**
coated macadam or asphalt 75mm thick	-	-	-	-	0.30	**3.21**	-	**3.21**	m2	**3.61**
Disposal of excavated material										
depositing on site in temporary spoil heaps where directed										
25m	-	-	-	-	1.40	**14.99**	-	**14.99**	m3	**16.87**
50m	-	-	-	-	1.65	**17.67**	-	**17.67**	m3	**19.88**
100m	-	-	-	-	2.20	**23.56**	-	**23.56**	m3	**26.51**
extra for removing beyond 100m and not exceeding 400m	-	-	-	-	1.72	**18.42**	2.64	**21.06**	m3	**23.69**
extra for removing beyond 100m and not exceeding 800m	-	-	-	-	1.72	**18.42**	3.42	**21.84**	m3	**24.57**
depositing on site in permanent spoil heaps average 25m distant	-	-	-	-	1.40	**14.99**	-	**14.99**	m3	**16.87**
removing from site to tip										
a) Inert	-	-	-	-	1.50	**16.07**	17.00	**33.07**	m3	**37.20**
b) Active	-	-	-	-	1.50	**16.07**	44.00	**60.07**	m3	**67.57**
c) Contaminated (guide price - always seek a quote for specialist disposal cost	-	-	-	-	1.50	**16.07**	86.00	**102.07**	m3	**114.82**
Surface treatments										
Compacting										
bottoms of excavations	-	-	-	-	0.16	**1.71**	-	**1.71**	m2	**1.93**

This section continues on the next page

Labour hourly rates: (except Specialists) Craft Operatives £14.21 Labourer £10.71 Rates are national average prices. Refer to REGIONAL VARIATIONS for indicative levels of overall pricing in regions	MATERIALS			LABOUR				RATES		
	Del to Site £	Waste %	Material Cost £	Craft Optve Hrs	Lab Hrs	Labour Cost £	Sunds £	Nett Rate £	Unit	Gross Rate (+12.5%) £
Q10: KERBS/EDGINGS/CHANNELS/PAVING ACCESSORIES Cont.										
Earthwork support										
Earthwork support										
maximum depth not exceeding 1.00m; distance between opposing faces not exceeding 2.00m.......	1.10	-	**1.10**	-	0.21	**2.25**	0.03	**3.38**	m2	**3.81**
Earthwork support; curved										
Maximum depth not exceeding 1.00m; distance between opposing faces not exceeding 2.00m.......	1.76	-	**1.76**	-	0.33	**3.53**	0.07	**5.36**	m2	**6.03**
Earthwork support; unstable ground										
maximum depth not exceeding 1.00m; distance between opposing faces not exceeding 2.00m.......	1.87	-	**1.87**	-	0.35	**3.75**	0.03	**5.66**	m2	**6.36**
Earthwork support; next to roadways										
maximum depth not exceeding 1.00m; distance between opposing faces not exceeding 2.00m.......	2.18	-	**2.18**	-	0.44	**4.71**	0.08	**6.97**	m2	**7.84**
Precast concrete; standard or stock pattern units; BS 7263 Part 1; bedding, jointing and pointing in cement mortar (1:3); on plain in-situ concrete foundation; BS 5328 ordinary prescribed mix ST4, 20mm aggregate										
Kerbs; Figs. 1, 2, 6 and 7; concrete foundation and haunching; formwork										
125 x 255mm kerb; 400 x 200mm foundation.........	10.96	5.00	**11.65**	-	1.20	**12.85**	-	**24.51**	m	**27.57**
125 x 255mm kerb; 400 x 200mm foundation; curved 10.00m radius	14.44	5.00	**15.31**	-	1.28	**13.71**	-	**29.02**	m	**32.65**
150 x 305mm kerb; 400 x 200mm foundation.........	14.01	5.00	**14.85**	-	1.28	**13.71**	-	**28.56**	m	**32.13**
150 x 305mm kerb; 400 x 200mm foundation; curved 10.00m radius	21.18	5.00	**22.38**	-	1.36	**14.57**	-	**36.95**	m	**41.57**
Edgings; Figs. 10, 11, 12 and 13; concrete foundation and haunching; formwork										
50 x 150mm edging; 300 x 150mm foundation........	6.83	5.00	**7.26**	-	0.86	**9.21**	-	**16.47**	m	**18.53**
50 x 200mm edging; 300 x 150mm foundation........	7.53	5.00	**7.99**	-	0.99	**10.60**	-	**18.59**	m	**20.92**
50 x 250mm edging; 300 x 150mm foundation........	7.83	5.00	**8.30**	-	1.12	**12.00**	-	**20.30**	m	**22.83**
Channels; Fig. 8; concrete foundation and haunching; formwork										
255 x 125mm channel; 450 x 150mm foundation......	12.27	5.00	**13.00**	-	0.95	**10.17**	-	**23.17**	m	**26.07**
255 x 125mm channel; 450 x 150mm foundation; curved 10.00 radius	14.62	5.00	**15.47**	-	1.03	**11.03**	-	**26.50**	m	**29.81**
Quadrants; Fig. 14; concrete foundation and haunching; formwork										
305 x 305 x 150mm quadrant; 500 x 500 x 150mm foundation	11.79	5.00	**12.44**	-	0.67	**7.18**	-	**19.62**	nr	**22.07**
305 x 305 x 255mm quadrant; 500 x 500 x 150mm foundation	12.11	5.00	**12.78**	-	0.76	**8.14**	-	**20.92**	nr	**23.53**
455 x 455 x 150mm quadrant; 650 x 650 x 150mm foundation	15.35	5.00	**16.22**	-	0.94	**10.07**	-	**26.28**	nr	**29.57**
455 x 455 x 255mm quadrant; 650 x 650 x 150mm foundation	15.88	5.00	**16.77**	-	1.03	**11.03**	-	**27.80**	nr	**31.27**
Granite; standard units; BS EN 1343; bedding jointing and pointing in cement mortar (1:3); on plain in-situ concrete foundation; BS 5328 ordinary prescribed mix ST4, 20mm aggregate										
Edge kerb; concrete foundation and haunching; formwork										
150 x 300mm; 300 x 200mm foundation.............	19.76	2.50	**20.54**	0.55	0.72	**15.53**	-	**36.07**	m	**40.57**
150 x 300mm; 300 x 200mm foundation; curved external radius exceeding 1000mm................	29.98	2.50	**31.06**	0.88	0.72	**20.22**	-	**51.28**	m	**57.69**
200 x 300mm; 350 x 200mm foundation.............	35.55	2.50	**36.76**	0.66	0.80	**17.95**	-	**54.71**	m	**61.54**
200 x 300mm; 350 x 200mm foundation; curved external radius exceeding 1000mm................	38.65	2.50	**39.98**	0.99	0.80	**22.64**	-	**62.62**	m	**70.44**
Flat kerb; concrete foundation and haunching; formwork										
300 x 150mm; 450 x 200mm foundation.............	33.44	2.50	**34.65**	0.55	0.80	**16.38**	-	**51.04**	m	**57.41**
300 x 150mm; 450 x 200mm foundation; curved external radius exceeding 1000mm................	42.86	2.50	**44.36**	0.88	0.80	**21.07**	-	**65.43**	m	**73.61**
300 x 200mm; 450 x 200mm foundation.............	42.21	2.50	**43.65**	0.66	0.83	**18.27**	-	**61.92**	m	**69.66**
300 x 200mm; 450 x 200mm foundation; curved external radius exceeding 1000mm................	53.72	2.50	**55.48**	0.99	0.83	**22.96**	-	**78.44**	m	**88.25**
Q20: GRANULAR SUB-BASES TO ROADS/PAVINGS										
Hard, dry, broken brick or stone to be obtained off site; wheeling not exceeding 25m - by machine										
Filling to make up levels; depositing in layers 150mm maximum thickness										
average thickness not exceeding 0.25m...........	17.85	25.00	**22.31**	-	-	**-**	4.18	**26.49**	m3	**29.80**
average thickness exceeding 0.25m...............	17.85	25.00	**22.31**	-	-	**-**	3.47	**25.78**	m3	**29.00**
Surface packing to filling										
to vertical or battered faces....................	-	-	**-**	-	-	**-**	0.70	**0.70**	m2	**0.79**
Compacting with 680 kg vibratory roller										
filling; blinding with sand, ashes or similar fine material	0.18	40.00	**0.25**	-	0.05	**0.54**	0.26	**1.04**	m2	**1.18**

Labour hourly rates: (except Specialists) Craft Operatives £14.21 Labourer £10.71 Rates are national average prices. Refer to REGIONAL VARIATIONS for indicative levels of overall pricing in regions	MATERIALS			LABOUR				RATES		
	Del to Site £	Waste %	Material Cost £	Craft Optve Hrs	Lab Hrs	Labour Cost £	Sunds £	Nett Rate £	Unit	Gross Rate (+12.5%) £
Q20: GRANULAR SUB-BASES TO ROADS/PAVINGS Cont.										
Hard, dry, broken brick or stone to be obtained off site; wheeling not exceeding 25m - by machine Cont.										
Compacting with 6 - 8 tonnes smooth wheeled roller filling; blinding with sand, ashes or similar fine material	0.18	40.00	**0.25**	-	0.05	**0.54**	0.50	**1.29**	m2	**1.45**
Hard, dry, broken brick or stone to be obtained off site; wheeling not exceeding 50m - by machine										
Filling to make up levels; depositing in layers 150mm maximum thickness										
average thickness not exceeding 0.25m........	17.85	25.00	**22.31**	-	-	-	4.85	**27.17**	m3	**30.56**
average thickness exceeding 0.25m........	17.85	25.00	**22.31**	-	-	-	4.18	**26.49**	m3	**29.80**
Surface packing to filling to vertical or battered faces........	-	-	-	-	-	-	0.70	**0.70**	m2	**0.79**
Compacting with 680 kg vibratory roller filling; blinding with sand, ashes or similar fine material	0.18	40.00	**0.25**	-	0.05	**0.54**	0.26	**1.04**	m2	**1.18**
Compacting with 6 - 8 tonnes smooth wheeled roller filling; blinding with sand, ashes or similar fine material	0.18	40.00	**0.25**	-	0.05	**0.54**	0.50	**1.29**	m2	**1.45**
MOT Type 1 to be obtained off site; wheeling not exceeding 25m - by machine										
Filling to make up levels; depositing in layers 150mm maximum thickness										
average thickness not exceeding 0.25m........	20.95	25.00	**26.19**	-	-	-	4.18	**30.37**	m3	**34.16**
average thickness exceeding 0.25m........	20.95	25.00	**26.19**	-	-	-	3.47	**29.65**	m3	**33.36**
Surface packing to filling to vertical or battered faces........	-	-	-	-	-	-	0.70	**0.70**	m2	**0.79**
Compacting with 680 kg vibratory roller filling; blinding with sand, ashes or similar fine material	0.18	40.00	**0.25**	-	0.05	**0.54**	0.26	**1.04**	m2	**1.18**
Compacting with 6 - 8 tonnes smooth wheeled roller filling; blinding with sand, ashes or similar fine material	0.18	40.00	**0.25**	-	0.05	**0.54**	0.50	**1.29**	m2	**1.45**
MOT Type 1 to be obtained off site; wheeling not exceeding 50m - by machine										
Filling to make up levels; depositing in layers 150mm maximum thickness										
average thickness not exceeding 0.25m........	20.95	25.00	**26.19**	-	-	-	4.85	**31.04**	m3	**34.92**
average thickness exceeding 0.25m........	20.95	25.00	**26.19**	-	-	-	4.18	**30.37**	m3	**34.16**
Surface packing to filling to vertical or battered faces........	-	-	-	-	-	-	0.70	**0.70**	m2	**0.79**
Compacting with 680 kg vibratory roller filling; blinding with sand, ashes or similar fine material	0.18	40.00	**0.25**	-	0.05	**0.54**	0.26	**1.04**	m2	**1.18**
Compacting with 6 - 8 tonnes smooth wheeled roller filling; blinding with sand, ashes or similar fine material	0.18	40.00	**0.25**	-	0.05	**0.54**	0.50	**1.29**	m2	**1.45**
MOT Type 2 to be obtained off site; wheeling not exceeding 25m - by machine										
Filling to make up levels; depositing in layers 150mm maximum thickness										
average thickness not exceeding 0.25m........	19.40	25.00	**24.26**	-	-	-	4.18	**28.44**	m3	**31.99**
average thickness exceeding 0.25m........	19.40	25.00	**24.26**	-	-	-	3.47	**27.72**	m3	**31.19**
Surface packing to filling to vertical or battered faces........	-	-	-	-	-	-	0.70	**0.70**	m2	**0.79**
Compacting with 680 kg vibratory roller filling; blinding with sand, ashes or similar fine material	0.18	40.00	**0.25**	-	0.05	**0.54**	0.26	**1.04**	m2	**1.18**
Compacting with 6 - 8 tonnes smooth wheeled roller filling; blinding with sand, ashes or similar fine material	0.18	40.00	**0.25**	-	0.05	**0.54**	0.50	**1.29**	m2	**1.45**
MOT Type 2 to be obtained off site; wheeling not exceeding 50m - by machine										
Filling to make up levels; depositing in layers 150mm maximum thickness										
average thickness not exceeding 0.25m........	19.40	25.00	**24.26**	-	-	-	4.85	**29.11**	m3	**32.75**
average thickness exceeding 0.25m........	19.40	25.00	**24.26**	-	-	-	4.18	**28.44**	m3	**31.99**
Surface packing to filling to vertical or battered faces........	-	-	-	-	-	-	0.70	**0.70**	m2	**0.79**

Labour hourly rates: (except Specialists) Craft Operatives £14.21 Labourer £10.71 Rates are national average prices. Refer to REGIONAL VARIATIONS for indicative levels of overall pricing in regions	MATERIALS			LABOUR				RATES		
	Del to Site £	Waste %	Material Cost £	Craft Optve Hrs	Lab Hrs	Labour Cost £	Sunds £	Nett Rate £	Unit	Gross Rate (+12.5%) £
Q20: GRANULAR SUB-BASES TO ROADS/PAVINGS Cont.										
MOT Type 2 to be obtained off site; wheeling not exceeding 50m - by machine Cont.										
Compacting with 680 kg vibratory roller filling; blinding with sand, ashes or similar fine material	0.18	40.00	**0.25**	-	0.05	**0.54**	0.26	**1.04**	m2	**1.18**
Compacting with 6 - 8 tonnes smooth wheeled roller filling; blinding with sand, ashes or similar fine material	0.18	40.00	**0.25**	-	0.05	**0.54**	0.50	**1.29**	m2	**1.45**
Ashes to be obtained off site; wheeling not exceeding 25m - by machine										
Filling to make up levels; depositing in layers 150mm maximum thickness										
average thickness not exceeding 0.25m	12.00	40.00	**16.80**	-	-	-	4.43	**21.23**	m3	**23.88**
Compacting with 680 kg vibratory roller filling	-	-	-	-	-	-	0.26	**0.26**	m2	**0.29**
Compacting with 6 - 8 tonnes smooth wheeled roller filling	-	-	-	-	-	-	0.50	**0.50**	m2	**0.57**
Ashes to be obtained off site; wheeling not exceeding 50m - by machine										
Filling to make up levels; depositing in layers 150mm maximum thickness										
average thickness not exceeding 0.25m	12.00	40.00	**16.80**	-	-	-	5.14	**21.94**	m3	**24.68**
Compacting with 680 kg vibratory roller filling	-	-	-	-	-	-	0.26	**0.26**	m2	**0.29**
Compacting with 6 - 8 tonnes smooth wheeled roller filling	-	-	-	-	-	-	0.50	**0.50**	m2	**0.57**
Hoggin to be obtained off site; wheeling not exceeding 25m - by machine										
Filling to make up levels; depositing in layers 150mm maximum thickness										
average thickness not exceeding 0.25m	16.80	33.00	**22.34**	-	-	-	4.17	**26.51**	m3	**29.83**
average thickness exceeding 0.25m	16.80	33.00	**22.34**	-	-	-	3.47	**25.81**	m3	**29.04**
Compacting with 680 kg vibratory roller filling	-	-	-	-	-	-	0.26	**0.26**	m2	**0.29**
Compacting with 6 - 8 tonnes smooth wheeled roller filling	-	-	-	-	-	-	0.50	**0.50**	m2	**0.57**
Hoggin to be obtained off site; wheeling not exceeding 50m - by machine										
Filling to make up levels; depositing in layers 150mm maximum thickness										
average thickness not exceeding 0.25m	16.80	33.00	**22.34**	-	-	-	4.85	**27.20**	m3	**30.60**
average thickness exceeding 0.25m	16.80	33.00	**22.34**	-	-	-	4.17	**26.51**	m3	**29.83**
Compacting with 680 kg vibratory roller filling	-	-	-	-	-	-	0.26	**0.26**	m2	**0.29**
Compacting with 6 - 8 tonnes smooth wheeled roller filling	-	-	-	-	-	-	0.50	**0.50**	m2	**0.57**
Hard, dry, broken brick or stone to be obtained off site; wheeling not exceeding 25m - by hand										
Filling to make up levels; depositing in layers 150mm maximum thickness										
average thickness not exceeding 0.25m	17.85	25.00	**22.31**	-	1.58	**16.92**	-	**39.23**	m3	**44.14**
average thickness exceeding 0.25m	17.85	25.00	**22.31**	-	1.31	**14.03**	-	**36.34**	m3	**40.89**
Compacting with 680 kg vibratory roller filling; blinding with sand, ashes or similar fine material	0.18	40.00	**0.25**	-	0.05	**0.54**	0.26	**1.04**	m2	**1.18**
Compacting with 6 - 8 tonnes smooth wheeled roller filling; blinding with sand, ashes or similar fine material	0.18	40.00	**0.25**	-	0.05	**0.54**	0.50	**1.29**	m2	**1.45**
Hard, dry, broken brick or stone to be obtained off site; wheeling not exceeding 50m - by hand										
Filling to make up levels; depositing in layers 150mm maximum thickness										
average thickness not exceeding 0.25m	17.85	25.00	**22.31**	-	1.85	**19.81**	-	**42.13**	m3	**47.39**
average thickness exceeding 0.25m	17.85	25.00	**22.31**	-	1.58	**16.92**	-	**39.23**	m3	**44.14**
Compacting with 680 kg vibratory roller filling; blinding with sand, ashes or similar fine material	0.18	40.00	**0.25**	-	0.05	**0.54**	0.26	**1.04**	m2	**1.18**

Labour hourly rates: (except Specialists) Craft Operatives £14.21 Labourer £10.71 Rates are national average prices. Refer to REGIONAL VARIATIONS for indicative levels of overall pricing in regions	MATERIALS			LABOUR				RATES		
	Del to Site £	Waste %	Material Cost £	Craft Optve Hrs	Lab Hrs	Labour Cost £	Sunds £	Nett Rate £	Unit	Gross Rate (+12.5%) £
Q20: GRANULAR SUB-BASES TO ROADS/PAVINGS Cont.										
Hard, dry, broken brick or stone to be obtained off site; wheeling not exceeding 50m - by hand Cont.										
Compacting with 6 - 8 tonnes smooth wheeled roller filling; blinding with sand, ashes or similar fine material	0.18	40.00	**0.25**	-	0.05	**0.54**	0.50	**1.29**	m2	**1.45**
MOT Type 1 to be obtained off site; wheeling not exceeding 25m - by hand										
Filling to make up levels; depositing in layers 150mm maximum thickness										
average thickness not exceeding 0.25m	20.95	25.00	**26.19**	-	1.58	**16.92**	-	**43.11**	m3	**48.50**
average thickness exceeding 0.25m	20.95	25.00	**26.19**	-	1.31	**14.03**	-	**40.22**	m3	**45.24**
Compacting with 680 kg vibratory roller filling; blinding with sand, ashes or similar fine material	0.18	40.00	**0.25**	-	0.05	**0.54**	0.26	**1.04**	m2	**1.18**
Compacting with 6 - 8 tonnes smooth wheeled roller filling; blinding with sand, ashes or similar fine material	0.18	40.00	**0.25**	-	0.05	**0.54**	0.50	**1.29**	m2	**1.45**
MOT Type 1 to be obtained off site; wheeling not exceeding 50m - by hand										
Filling to make up levels; depositing in layers 150mm maximum thickness										
average thickness not exceeding 0.25m	20.95	25.00	**26.19**	-	1.85	**19.81**	-	**46.00**	m3	**51.75**
average thickness exceeding 0.25m	20.95	25.00	**26.19**	-	1.58	**16.92**	-	**43.11**	m3	**48.50**
Compacting with 680 kg vibratory roller filling; blinding with sand, ashes or similar fine material	0.18	40.00	**0.25**	-	0.05	**0.54**	0.26	**1.04**	m2	**1.18**
Compacting with 6 - 8 tonnes smooth wheeled roller filling; blinding with sand, ashes or similar fine material	0.18	40.00	**0.25**	-	0.05	**0.54**	0.50	**1.29**	m2	**1.45**
MOT Type 2 to be obtained off site; wheeling not exceeding 25m - by hand										
Filling to make up levels; depositing in layers 150mm maximum thickness										
average thickness not exceeding 0.25m	19.40	25.00	**24.26**	-	1.58	**16.92**	-	**41.18**	m3	**46.32**
average thickness exceeding 0.25m	19.40	25.00	**24.26**	-	1.31	**14.03**	-	**38.29**	m3	**43.07**
Compacting with 680 kg vibratory roller filling; blinding with sand, ashes or similar fine material	0.18	40.00	**0.25**	-	0.05	**0.54**	0.26	**1.04**	m2	**1.18**
Compacting with 6 - 8 tonnes smooth wheeled roller filling; blinding with sand, ashes or similar fine material	0.18	40.00	**0.25**	-	0.05	**0.54**	0.50	**1.29**	m2	**1.45**
MOT Type 2 to be obtained off site; wheeling not exceeding 50m - by hand										
Filling to make up levels; depositing in layers 150mm maximum thickness										
average thickness not exceeding 0.25m	19.40	25.00	**24.26**	-	1.85	**19.81**	-	**44.07**	m3	**49.58**
average thickness exceeding 0.25m	19.40	25.00	**24.26**	-	1.58	**16.92**	-	**41.18**	m3	**46.32**
Compacting with 680 kg vibratory roller filling; blinding with sand, ashes or similar fine material	0.18	40.00	**0.25**	-	0.05	**0.54**	0.26	**1.04**	m2	**1.18**
Compacting with 6 - 8 tonnes smooth wheeled roller filling; blinding with sand, ashes or similar fine material	0.18	40.00	**0.25**	-	0.05	**0.54**	0.50	**1.29**	m2	**1.45**
Ashes to be obtained off site; wheeling not exceeding 25m - by hand										
Filling to make up levels; depositing in layers 150mm maximum thickness										
average thickness not exceeding 0.25m	12.00	40.00	**16.80**	-	1.70	**18.21**	-	**35.01**	m3	**39.38**
Compacting with 680 kg vibratory roller filling	-	-	-	-	-	-	0.26	**0.26**	m2	**0.29**
Compacting with 6 - 8 tonnes smooth wheeled roller filling	-	-	-	-	-	-	0.50	**0.50**	m2	**0.57**
Ashes to be obtained off site; wheeling not exceeding 50m - by hand										
Filling to make up levels; depositing in layers 150mm maximum thickness										
average thickness not exceeding 0.25m	12.00	40.00	**16.80**	-	1.94	**20.78**	-	**37.58**	m3	**42.27**
Compacting with 680 kg vibratory roller filling	-	-	-	-	-	-	0.26	**0.26**	m2	**0.29**

Labour hourly rates: (except Specialists) Craft Operatives £14.21 Labourer £10.71 Rates are national average prices. Refer to REGIONAL VARIATIONS for indicative levels of overall pricing in regions	MATERIALS			LABOUR			RATES			
	Del to Site £	Waste %	Material Cost £	Craft Optve Hrs	Lab Hrs	Labour Cost £	Sunds £	Nett Rate £	Unit	Gross Rate (+12.5%) £
Q20: GRANULAR SUB-BASES TO ROADS/PAVINGS Cont.										
Ashes to be obtained off site; wheeling not exceeding 50m - by hand Cont.										
Compacting with 6 - 8 tonnes smooth wheeled roller										
filling	-	-	-	-	-	-	0.50	**0.50**	m2	**0.57**
Hoggin to be obtained off site; wheeling not exceeding 25m - by hand										
Filling to make up levels; depositing in layers 150mm maximum thickness										
average thickness not exceeding 0.25m	16.80	33.00	**22.34**	-	1.55	**16.60**	-	**38.94**	m3	**43.81**
average thickness exceeding 0.25m	16.80	33.00	**22.34**	-	1.30	**13.92**	-	**36.27**	m3	**40.80**
Compacting with 680 kg vibratory roller										
filling	-	-	-	-	-	-	0.26	**0.26**	m2	**0.29**
Compacting with 6 - 8 tonnes smooth wheeled roller										
filling	-	-	-	-	-	-	0.50	**0.50**	m2	**0.57**
Hoggin to be obtained off site; wheeling not exceeding 50m - by hand										
Filling to make up levels; depositing in layers 150mm maximum thickness										
average thickness not exceeding 0.25m	16.80	33.00	**22.34**	-	1.80	**19.28**	-	**41.62**	m3	**46.82**
average thickness exceeding 0.25m	16.80	33.00	**22.34**	-	1.55	**16.60**	-	**38.94**	m3	**43.81**
Compacting with 680 kg vibratory roller										
filling	-	-	-	-	-	-	0.26	**0.26**	m2	**0.29**
Compacting with 6 - 8 tonnes smooth wheeled roller										
filling	-	-	-	-	-	-	0.50	**0.50**	m2	**0.57**
Q21: IN-SITU CONCRETE ROADS/PAVINGS										
Plain in-situ concrete; mix 1:8, all in aggregate										
Beds; poured on or against earth or unblinded hardcore										
thickness not exceeding 150mm	70.66	7.50	**75.96**	-	4.68	**50.12**	1.80	**127.88**	m3	**143.87**
Plain in-situ concrete; mix 1:6; all in aggregate										
Beds; poured on or against earth or unblinded hardcore										
thickness 150 - 450mm	79.06	7.50	**84.99**	-	4.40	**47.12**	1.80	**133.92**	m3	**150.66**
thickness not exceeding 150mm	79.06	7.50	**84.99**	-	4.68	**50.12**	1.80	**136.92**	m3	**154.03**
Plain in-situ concrete; BS 5328, ordinary prescribed mix ST4, 20mm aggregate										
Beds										
thickness not exceeding 150mm	63.93	5.00	**67.13**	-	4.68	**50.12**	1.80	**119.05**	m3	**133.93**
Beds; poured on or against earth or unblinded hardcore										
thickness not exceeding 150mm	63.93	5.00	**67.13**	-	4.68	**50.12**	1.80	**119.05**	m3	**133.93**
Reinforced in-situ concrete; BS 5328, designed mix C20, 20mm aggregate, minimum cement content 240 kg/m3; vibrated										
Beds										
thickness 150 - 450mm	65.24	5.00	**68.51**	-	4.68	**50.12**	0.50	**119.13**	m3	**134.02**
thickness not exceeding 150mm	65.24	5.00	**68.51**	-	4.95	**53.01**	0.50	**122.02**	m3	**137.27**
Formwork and basic finish										
Edges of beds										
height not exceeding 250mm	1.08	12.50	**1.22**	0.58	0.12	**9.53**	0.24	**10.99**	m	**12.36**
height 250 - 500mm	1.90	12.50	**2.14**	1.06	0.21	**17.31**	0.45	**19.90**	m	**22.39**
Formwork steel forms										
Edges of beds										
height not exceeding 250mm	0.78	15.00	**0.90**	0.40	0.08	**6.54**	0.24	**7.68**	m	**8.64**
height 250 - 500mm	1.38	15.00	**1.58**	0.70	0.14	**11.45**	0.45	**13.48**	m	**15.17**
Reinforcement fabric; BS 4483, hard drawn plain round steel; welded; including laps, tying wire, all cutting and bending, and spacers and chairs which are at the discretion of the Contractor										
Reference C283, 2.61 kg/m2; 100mm side laps; 400mm end laps										
generally	1.27	17.50	**1.49**	0.06	0.02	**1.07**	0.07	**2.63**	m2	**2.95**
Reference C385, 3.41 kg/m2; 100mm side laps; 400mm end laps										
generally	1.65	17.50	**1.94**	0.06	0.02	**1.07**	0.09	**3.10**	m2	**3.49**
Reference C503, 4.34 kg/m2; 100mm side laps; 400mm end laps										
generally	2.04	17.50	**2.39**	0.07	0.02	**1.21**	0.11	**3.71**	m2	**4.18**

Labour hourly rates: (except Specialists) Craft Operatives £14.21 Labourer £10.71 Rates are national average prices. Refer to REGIONAL VARIATIONS for indicative levels of overall pricing in regions	MATERIALS			LABOUR				RATES		
	Del to Site £	Waste %	Material Cost £	Craft Optve Hrs	Lab Hrs	Labour Cost £	Sunds £	Nett Rate £	Unit	Gross Rate (+12.5%) £
Q21: IN-SITU CONCRETE ROADS/PAVINGS Cont.										
Reinforcement fabric; BS 4483, hard drawn plain round steel; welded; including laps, tying wire, all cutting and bending, and spacers and chairs which are at the discretion of the Contractor Cont.										
Reference C636, 5.55 kg/m2; 100mm side laps; 400mm end laps										
generally	2.72	17.50	**3.19**	0.07	0.02	**1.21**	0.14	**4.54**	m2	**5.11**
Reference C785, 6.72 kg/m2; 100mm side laps; 400mm end laps										
generally	3.28	17.50	**3.85**	0.08	0.02	**1.35**	0.17	**5.37**	m2	**6.05**
Formed joints										
Sealant to joint; Grace Construction Products. 'Servimastic 96' including preparation, cleaners, primers and sealers										
10 x 25mm	2.47	5.00	**2.60**	-	0.11	**1.18**	-	**3.77**	m	**4.25**
12.5 x 25mm	3.21	5.00	**3.37**	-	0.12	**1.29**	-	**4.66**	m	**5.24**
Worked finish on in-situ concrete										
Tamping by mechanical means										
level surfaces	-	-	-	-	0.16	**1.71**	0.29	**2.00**	m2	**2.25**
sloping surfaces	-	-	-	-	0.33	**3.53**	0.38	**3.91**	m2	**4.40**
surfaces to falls	-	-	-	-	0.44	**4.71**	0.38	**5.09**	m2	**5.73**
Trowelling										
level surfaces	-	-	-	-	0.36	**3.86**	-	**3.86**	m2	**4.34**
sloping surfaces	-	-	-	-	0.44	**4.71**	-	**4.71**	m2	**5.30**
surfaces to falls	-	-	-	-	0.55	**5.89**	-	**5.89**	m2	**6.63**
Rolling with an indenting roller										
level surfaces	-	-	-	-	0.11	**1.18**	0.28	**1.46**	m2	**1.64**
sloping surfaces	-	-	-	-	0.17	**1.82**	0.28	**2.10**	m2	**2.36**
surfaces to falls	-	-	-	-	0.33	**3.53**	0.36	**3.89**	m2	**4.38**
Accessories cast into in-situ concrete										
Mild steel dowels; half coated with bitumen										
16mm diameter x 400mm long	0.33	10.00	**0.37**	-	0.06	**0.64**	0.18	**1.19**	nr	**1.34**
20mm diameter x 500mm long; with plastic compression cap	0.64	10.00	**0.71**	-	0.08	**0.86**	0.18	**1.74**	nr	**1.96**
Q22: COATED MACADAM/ASPHALT ROADS/PAVINGS										
Coated macadam, BS 4987; base course 28mm nominal size aggregate, 50mm thick; wearing course 6mm nominal size aggregate, 20mm thick; rolled with 3 - 4 tonne roller; external										
70mm roads on concrete base; to falls and crossfalls, and slopes not exceeding 15 degrees from horizontal										
generally	-	-	**Spclist**	-	-	**Spclist**	-	**13.82**	m2	**15.55**
70mm pavings on concrete base; to falls and crossfalls, and slopes not exceeding 15 degrees from horizontal										
generally	-	-	**Spclist**	-	-	**Spclist**	-	**15.37**	m2	**17.29**
Coated macadam, BS 4987; base course 28mm nominal size aggregate, 50mm thick; wearing course 6mm nominal size aggregate, 20mm thick; rolled with 3 - 4 tonne roller; dressing surface with coated grit brushed on and lightly rolled; external										
70mm roads on concrete base; to falls and crossfalls, and slopes not exceeding 15 degrees from horizontal										
generally	-	-	**Spclist**	-	-	**Spclist**	-	**16.74**	m2	**18.83**
70mm pavings on concrete base; to falls and crossfalls, and slopes not exceeding 15 degrees from horizontal										
generally	-	-	**Spclist**	-	-	**Spclist**	-	**18.25**	m2	**20.53**
Coated macadam, BS 4987; base course 40mm nominal size aggregate, 65mm thick; wearing course 10mm nominal size aggregate, 25mm thick; rolled with 6 - 8 tonne roller; external										
90mm roads on concrete base; to falls and crossfalls, and slopes not exceeding 15 degrees from horizontal										
generally	-	-	**Spclist**	-	-	**Spclist**	-	**16.74**	m2	**18.83**
90mm pavings on concrete base; to falls and crossfalls, and slopes not exceeding 15 degrees from horizontal										
generally	-	-	**Spclist**	-	-	**Spclist**	-	**18.36**	m2	**20.66**

Labour hourly rates: (except Specialists) Craft Operatives £14.21 Labourer £10.71 Rates are national average prices. Refer to REGIONAL VARIATIONS for indicative levels of overall pricing in regions	MATERIALS			LABOUR				RATES		
	Del to Site £	Waste %	Material Cost £	Craft Optve Hrs	Lab Hrs	Labour Cost £	Sunds £	Nett Rate £	Unit	Gross Rate (+12.5%) £
Q22: COATED MACADAM/ASPHALT ROADS/PAVINGS Cont.										
Coated macadam, BS 4987; base course 40mm nominal size aggregate, 65mm thick; wearing course 10mm nominal size aggregate, 25mm thick; rolled with 6 - 8 tonne roller; dressing surface with coated grit brushed on and lightly rolled; external										
90mm roads on concrete base; to falls and crossfalls, and slopes not exceeding 15 degrees from horizontal										
generally	-	-	Spclist	-	-	Spclist	-	19.86	m2	22.34
90mm pavings on concrete base; to falls and crossfalls, and slopes not exceeding 15 degrees from horizontal										
generally	-	-	Spclist	-	-	Spclist	-	20.55	m2	23.12
Fine cold asphalt, BS 4987; single course 6mm nominal size aggregate; rolled with 3 - 4 tonne roller; external										
12mm roads on concrete base; to falls and crossfalls, and slopes not exceeding 15 degrees from horizontal										
generally	-	-	Spclist	-	-	Spclist	-	6.30	m2	7.09
19mm roads on concrete base; to falls and crossfalls, and slopes not exceeding 15 degrees from horizontal										
generally	-	-	Spclist	-	-	Spclist	-	7.30	m2	8.21
25mm roads on concrete base; to falls and crossfalls, and slopes not exceeding 15 degrees from horizontal										
generally	-	-	Spclist	-	-	Spclist	-	8.67	m2	9.75
12mm pavings on concrete base; to falls and crossfalls, and slopes not exceeding 15 degrees from horizontal										
generally	-	-	Spclist	-	-	Spclist	-	6.68	m2	7.52
19mm pavings on concrete base; to falls and crossfalls, and slopes not exceeding 15 degrees from horizontal										
generally	-	-	Spclist	-	-	Spclist	-	8.38	m2	9.43
25mm pavings on concrete base; to falls and crossfalls, and slopes not exceeding 15 degrees from horizontal										
generally	-	-	Spclist	-	-	Spclist	-	9.59	m2	10.79
Q22: GRAVEL/HOGGIN/WOODCHIP ROADS/PAVINGS										
Gravel paving; first layer clinker 75mm thick; intermediate layer coarse gravel 75mm thick; wearing layer blinding gravel 38mm thick; well water and roll										
188mm roads on blinded hardcore base; to falls and crossfalls and slopes not exceeding 15 degrees from horizontal										
generally	5.80	10.00	6.38	-	0.44	4.71	-	11.50	m2	12.94
Gravel paving; single layer blinding gravel; well water and roll										
50mm pavings on blinded hardcore base; to falls and crossfalls and slopes not exceeding 15 degrees from horizontal										
generally	2.11	10.00	2.32	-	0.15	1.61	-	4.03	m2	4.53
Gravel paving; first layer clinker 50mm thick; wearing layer blinding gravel 38mm thick; well water and roll										
88mm pavings on blinded hardcore base; to falls and crossfalls and slopes not exceeding 15 degrees from horizontal										
generally	2.68	10.00	2.95	-	0.26	2.78	-	5.84	m2	6.57
extra; 'Colas' and 10mm shingle dressing; one coat	1.19	5.00	1.25	-	0.17	1.82	-	3.07	m2	3.45
extra; 'Colas' and 10mm shingle dressing; two coats	2.03	5.00	2.13	-	0.27	2.89	-	5.02	m2	5.65
Q25: SLAB/BRICK/BLOCK/SETT/COBBLE PAVINGS										
York stone paving; 13mm thick bedding, jointing and pointing in cement mortar (1:3)										
Paving on blinded hardcore base										
50mm thick; to falls and crossfalls and to slopes not exceeding 15 degrees from horizontal	68.86	10.00	72.30	0.30	0.30	7.48	1.04	80.81	m2	90.91
75mm thick; to falls and crossfalls and to slopes not exceeding 15 degrees from horizontal	78.10	10.00	82.01	0.36	0.36	8.97	1.10	92.08	m2	103.59
extra; rubbed top surface	16.80	5.00	17.64	-	-	-	-	17.64	m2	19.85
Treads on concrete base										
50mm thick; 300mm wide	66.37	5.00	69.69	0.36	0.36	8.97	0.30	78.96	m	88.83

Labour hourly rates: (except Specialists) Craft Operatives £14.21 Labourer £10.71 Rates are national average prices. Refer to REGIONAL VARIATIONS for indicative levels of overall pricing in regions	MATERIALS			LABOUR			RATES			
	Del to Site £	Waste %	Material Cost £	Craft Optve Hrs	Lab Hrs	Labour Cost £	Sunds £	Nett Rate £	Unit	Gross Rate (+12.5%) £
Q25: SLAB/BRICK/BLOCK/SETT/COBBLE PAVINGS Cont.										
Precast concrete flags, BS 7263 Part 1, natural finish; 6mm joints, symmetrical layout; bedding in 13mm cement mortar (1:3); jointing and pointing with lime and sand (1:2)										
600 x 450 x 50mm units to pavings on sand, granular or blinded hardcore base										
50mm thick; to falls and crossfalls and to slopes not exceeding 15 degrees from horizontal.........	10.12	5.00	**10.57**	0.30	0.30	**7.48**	1.06	**19.10**	m2	**21.49**
600 x 600 x 50mm units to pavings on sand, granular or blinded hardcore base										
50mm thick; to falls and crossfalls and to slopes not exceeding 15 degrees from horizontal.........	8.46	5.00	**8.82**	0.28	0.28	**6.98**	1.00	**16.80**	m2	**18.90**
600 x 750 x 50mm units to pavings on sand, granular or blinded hardcore base										
50mm thick; to falls and crossfalls and to slopes not exceeding 15 degrees from horizontal.........	8.78	5.00	**9.15**	0.25	0.50	**8.91**	1.07	**19.12**	m2	**21.52**
600 x 900 x 50mm units to pavings on sand, granular or blinded hardcore base										
50mm thick; to falls and crossfalls and to slopes not exceeding 15 degrees from horizontal.........	7.89	5.00	**8.22**	0.23	0.46	**8.19**	1.08	**17.49**	m2	**19.68**
600 x 450 x 63mm units to pavings on sand, granular or blinded hardcore base										
63mm thick; to falls and crossfalls and to slopes not exceeding 15 degrees from horizontal.........	12.65	5.00	**13.21**	0.30	0.30	**7.48**	1.06	**21.75**	m2	**24.47**
600 x 600 x 63mm units to pavings on sand, granular or blinded hardcore base										
63mm thick; to falls and crossfalls and to slopes not exceeding 15 degrees from horizontal.........	11.40	5.00	**11.90**	0.28	0.28	**6.98**	1.00	**19.88**	m2	**22.36**
600 x 750 x 63mm units to pavings on sand , granular or blinded hardcore base										
63mm thick; to falls and crossfalls and to slopes not exceeding 15 degrees from horizontal.........	9.84	5.00	**10.27**	0.25	0.50	**8.91**	1.01	**20.18**	m2	**22.71**
600 x 900 x 63mm units to pavings on sand, granular or blinded hardcore base										
63mm thick; to falls and crossfalls and to slopes not exceeding 15 degrees from horizontal.........	10.89	5.00	**11.37**	0.23	0.46	**8.19**	1.02	**20.58**	m2	**23.15**
Precast concrete flags, BS 7263 Part 1, natural finish; 6mm joints, symmetrical layout; bedding in 13mm lime mortar (1:3); jointing and pointing with lime and sand (1:2)										
600 x 450 x 50mm units to pavings on sand, granular or blinded hardcore base										
50mm thick; to falls and crossfalls and to slopes not exceeding 15 degrees from horizontal.........	10.52	5.00	**10.97**	0.30	0.30	**7.48**	1.06	**19.50**	m2	**21.94**
600 x 600 x 50mm units to pavings on sand, granular or blinded hardcore base										
50mm thick; to falls and crossfalls and to slopes not exceeding 15 degrees from horizontal.........	8.86	5.00	**9.22**	0.28	0.28	**6.98**	1.00	**17.20**	m2	**19.35**
600 x 750 x 50mm units to pavings on sand, granular or blinded hardcore base										
50mm thick; to falls and crossfalls and to slopes not exceeding 15 degrees from horizontal.........	8.90	5.00	**9.27**	0.25	0.50	**8.91**	1.01	**19.19**	m2	**21.59**
600 x 900 x 50mm units to pavings on sand, granular or blinded hardcore base										
50mm thick; to falls and crossfalls and to slopes not exceeding 15 degrees from horizontal.........	8.02	5.00	**8.35**	0.23	0.46	**8.19**	1.06	**17.60**	m2	**19.80**
600 x 450 x 63mm units to pavings on sand, granular or blinded hardcore base										
63mm thick; to falls and crossfalls and to slopes not exceeding 15 degrees from horizontal.........	12.77	5.00	**13.34**	0.30	0.30	**7.48**	1.06	**21.87**	m2	**24.61**
600 x 600 x 63mm units to pavings on sand, granular or blinded hardcore base										
63mm thick; to falls and crossfalls and to slopes not exceeding 15 degrees from horizontal.........	11.53	5.00	**12.03**	0.28	0.28	**6.98**	1.00	**20.00**	m2	**22.51**
600 x 750 x 63mm units to pavings on sand, granular or blinded hardcore base										
63mm thick; to falls and crossfalls and to slopes not exceeding 15 degrees from horizontal.........	9.97	5.00	**10.39**	0.25	0.50	**8.91**	1.01	**20.31**	m2	**22.85**
600 x 900 x 63mm units to pavings on sand , granular or blinded hardcore base										
63mm thick; to falls and crossfalls and to slopes not exceeding 15 degrees from horizontal.........	11.02	5.00	**11.49**	0.23	0.46	**8.19**	1.02	**20.71**	m2	**23.30**

Labour hourly rates: (except Specialists) Craft Operatives £14.21 Labourer £10.71 Rates are national average prices. Refer to REGIONAL VARIATIONS for indicative levels of overall pricing in regions	MATERIALS			LABOUR				RATES		
	Del to Site £	Waste %	Material Cost £	Craft Optve Hrs	Lab Hrs	Labour Cost £	Sunds £	Nett Rate £	Unit	Gross Rate (+12.5%) £
Q25: SLAB/BRICK/BLOCK/SETT/COBBLE PAVINGS Cont.										
Precast concrete flags, BS 7263 Part 1, natural finish; 6mm joints, symmetrical layout; bedding in 25mm sand; jointing and pointing with lime and sand (1:2)										
600 x 450 x 50mm units to pavings on sand, granular or blinded hardcore base										
50mm thick; to falls and crossfalls and to slopes not exceeding 15 degrees from horizontal.........	9.27	5.00	**9.72**	0.30	0.30	**7.48**	1.06	**18.26**	m2	**20.54**
600 x 600 x 50mm units to pavings on sand, granular or blinded hardcore base										
50mm thick; to falls and crossfalls and to slopes not exceeding 15 degrees from horizontal.........	7.80	5.00	**8.18**	0.28	0.28	**6.98**	1.00	**16.15**	m2	**18.17**
600 x 750 x 50mm units to pavings on sand, granular or blinded hardcore base										
50mm thick; to falls and crossfalls and to slopes not exceeding 15 degrees from horizontal.........	7.66	5.00	**8.03**	0.25	0.50	**8.91**	1.01	**17.95**	m2	**20.19**
600 x 900 x 50mm units to pavings on sand, granular or blinded hardcore base										
50mm thick; to falls and crossfalls and to slopes not exceeding 15 degrees from horizontal.........	6.78	5.00	**7.10**	0.23	0.46	**8.19**	1.02	**16.32**	m2	**18.36**
Keyblok precast concrete paving blocks; Marshalls Mono Ltd Driveline 50; in standard units; bedding on sand; covering with sand, compacting with plate vibrator, sweeping off surplus										
200 x 100mm units to pavings on 50mm sand base; natural colour										
symmetrical half bond layout; level and to falls only										
50mm thick..................................	8.03	10.00	**8.83**	-	1.35	**14.46**	1.42	**24.71**	m2	**27.80**
laid in straight herringbone pattern; level and to falls only										
50mm thick..................................	8.03	10.00	**8.83**	-	1.50	**16.07**	1.42	**26.32**	m2	**29.61**
Brick paving on concrete base; 13mm thick bedding, jointing and pointing in cement mortar (1:3)										
Paving to falls and crossfalls and to slopes not exceeding 15 degrees from horizontal										
25mm thick Brick paviors P.C. £300.00 per 1000...	18.00	5.00	**18.90**	2.75	1.55	**55.68**	1.04	**75.62**	m2	**85.07**
65mm thick Facing bricks P.C. £210.00 per 1000....	12.60	5.00	**13.23**	2.75	1.55	**55.68**	1.04	**69.95**	m2	**78.69**
50mm thick Staffordshire blue chequered paviors..	18.00	5.00	**18.90**	2.75	1.55	**55.68**	1.04	**75.62**	m2	**85.07**
50mm thick Accrington non slip paviors...........	18.00	5.00	**18.90**	2.75	1.55	**55.68**	1.04	**75.62**	m2	**85.07**
Granite sett paving on concrete base; 13mm thick bedding and grouting in cement mortar (1:3)										
Paving to falls and crossfalls and to slopes not exceeding 15 degrees from horizontal										
125mm thick; new	35.53	2.50	**36.35**	1.80	2.20	**49.14**	2.31	**87.80**	m2	**98.78**
150mm thick; new	42.49	2.50	**43.48**	2.00	2.35	**53.59**	2.64	**99.71**	m2	**112.17**
100mm thick; reclaimed	39.79	2.50	**40.74**	1.65	2.00	**44.87**	1.98	**87.58**	m2	**98.53**
150mm thick; reclaimed	35.25	2.50	**36.07**	1.80	2.20	**49.14**	2.31	**87.52**	m2	**98.46**
200mm thick; reclaimed	46.54	2.50	**47.63**	2.00	2.35	**53.59**	2.64	**103.86**	m2	**116.84**
Cobble paving set in concrete bed (measured separately) tight butted, dry grouted with cement and sand (1:3) watered and brushed										
Paving level and to falls only										
plain	14.77	2.50	**15.14**	1.65	1.65	**41.12**	2.55	**58.81**	m2	**66.16**
set to pattern	14.77	2.50	**15.14**	2.00	2.00	**49.84**	2.55	**67.53**	m2	**75.97**
Crazy paving on blinded hardcore base, bedding in 38mm thick sand; pointing in cement mortar (1:4)										
Paving to falls and crossfalls and to slopes not exceeding 15 degrees from horizontal										
50mm thick precast concrete flag.................	5.17	2.50	**5.30**	0.40	0.40	**9.97**	2.15	**17.42**	m2	**19.59**
38-50mm thick York stone.........................	54.46	2.50	**55.82**	0.60	0.60	**14.95**	3.10	**73.87**	m2	**83.11**
38mm thick Westmorland green slate...............	112.75	2.50	**115.57**	0.60	0.60	**14.95**	2.15	**132.67**	m2	**149.25**
Grass concrete paving; voids filled with vegetable soil and sown with grass seed										
Precast concrete perforated slabs										
120mm thick units on 25mm thick sand bed; level and to falls only	18.22	2.50	**18.86**	0.32	0.32	**7.97**	-	**26.83**	m2	**30.19**
Q30: SEEDING/TURFING										
Cultivating										
Surfaces of natural ground										
digging over one spit deep; removing debris; weeding	-	-	**-**	-	0.28	**3.00**	-	**3.00**	m2	**3.37**
Surfaces of filling										
digging over one spit deep.......................	-	-	**-**	-	0.33	**3.53**	-	**3.53**	m2	**3.98**

Labour hourly rates: (except Specialists) Craft Operatives £14.21 Labourer £10.71 Rates are national average prices. Refer to REGIONAL VARIATIONS for indicative levels of overall pricing in regions	MATERIALS			LABOUR				RATES		
	Del to Site £	Waste %	Material Cost £	Craft Optve Hrs	Lab Hrs	Labour Cost £	Sunds £	Nett Rate £	Unit	Gross Rate (+12.5%) £
Q30: SEEDING/TURFING Cont.										
Surface applications										
Bone meal, 0.06 kg/m2; raking in general surfaces	0.03	5.00	**0.03**	-	0.06	**0.64**	-	**0.67**	m2	**0.76**
Selective weedkiller, 0.03 kg/m2; applying by spreader general surfaces	0.09	5.00	**0.09**	-	0.06	**0.64**	-	**0.74**	m2	**0.83**
Pre-seeding fertilizer at the rate of 0.05 kg/m2 general surfaces	0.08	5.00	**0.08**	-	0.06	**0.64**	-	**0.73**	m2	**0.82**
General grass fertilizer at the rate of 0.06 kg/m2 general surfaces	0.09	5.00	**0.09**	-	0.06	**0.64**	-	**0.74**	m2	**0.83**
Seeding										
Grass seed, 0.07 kg/m2, raking in; rolling; maintaining for 12 months after laying general surfaces	0.31	5.00	**0.33**	-	0.30	**3.21**	-	**3.54**	m2	**3.98**
Stone pick, roll and cut grass with a rotary cutter and remove arisings general surfaces	-	-	**-**	-	0.11	**1.18**	-	**1.18**	m2	**1.33**
Turfing										
Take turf from stack, wheel not exceeding 100m, lay, roll and water, maintaining for 12 months after laying general surfaces	-	-	**-**	-	0.70	**7.50**	-	**7.50**	m2	**8.43**
Rolawn® Medallion turf; cultivated; lay, roll and water; maintaining for 12 months after laying general surfaces	2.45	5.00	**2.57**	-	0.50	**5.36**	-	**7.93**	m2	**8.92**
Q31: PLANTING										
Planting trees, shrubs and hedge plants										
Excavate or form pit, hole or trench, dig over ground in bottom, spread and pack around roots with finely broken soil, refill with top soil with one third by volume of farmyard manure incorporated, water in, remove surplus excavated material and provide labelling										
small tree	1.02	-	**1.02**	-	0.85	**9.10**	0.39	**10.51**	nr	**11.83**
medium tree	1.29	-	**1.29**	-	1.65	**17.67**	0.56	**19.52**	nr	**21.96**
large tree	1.62	-	**1.62**	-	3.30	**35.34**	0.72	**37.68**	nr	**42.39**
shrub	0.77	-	**0.77**	-	0.55	**5.89**	0.16	**6.82**	nr	**7.67**
hedge plant	0.77	-	**0.77**	-	0.28	**3.00**	0.16	**3.92**	nr	**4.41**
60mm diameter treated softwood tree stake, pointed and driven into ground and with two PVC tree ties secured around tree and nailed to stake										
2100mm long	3.44	-	**3.44**	-	0.35	**3.75**	1.38	**8.57**	nr	**9.64**
2400mm long	3.66	-	**3.66**	-	0.35	**3.75**	1.38	**8.79**	nr	**9.89**
2700mm long	4.05	-	**4.05**	-	0.35	**3.75**	1.38	**9.18**	nr	**10.33**
Planting herbaceous plants, bulbs, corms and tubers										
Provide planting bed of screened and conditioned Rolawn® topsoil										
150mm thick	3.40	5.00	**3.57**	-	0.80	**8.57**	-	**12.14**	m2	**13.66**
225mm thick	5.10	5.00	**5.36**	-	1.25	**13.39**	-	**18.74**	m2	**21.09**
Form hole and plant										
herbaceous plants	0.45	5.00	**0.47**	-	0.22	**2.36**	-	**2.83**	nr	**3.18**
bulbs, corms and tubers	0.10	5.00	**0.11**	-	0.11	**1.18**	-	**1.28**	nr	**1.44**
Mulching after planting										
25mm peat	0.99	-	**0.99**	-	0.11	**1.18**	-	**2.17**	m2	**2.44**
75mm farmyard manure	0.48	-	**0.48**	-	0.11	**1.18**	-	**1.66**	m2	**1.87**
Q40: FENCING: POST AND WIRE FENCING										
Fencing; strained wire; BS 1722-2; concrete posts and struts; 4.00mm diameter galvanized mild steel line wires; galvanized steel fittings and accessories; backfilling around posts in concrete mix ST4										
900mm high fencing; type SC90; wires -3; posts with rounded tops										
posts at 2743mm centres; 600mm into ground	6.82	2.50	**6.95**	-	0.85	**9.10**	-	**16.05**	m	**18.06**
extra; end posts with struts -1	21.62	2.50	**22.08**	-	1.10	**11.78**	-	**33.86**	nr	**38.10**
extra; angle posts with struts -2	35.30	2.50	**36.06**	-	1.65	**17.67**	-	**53.73**	nr	**60.44**
1050mm high fencing; type SC105A; wires-5; posts rounded tops										
posts at 2743mm centres; 600mm into ground	7.27	2.50	**7.41**	-	0.90	**9.64**	-	**17.05**	m	**19.18**
extra; end posts with struts -1	23.35	2.50	**23.85**	-	1.20	**12.85**	-	**36.70**	nr	**41.29**
extra; angle posts with struts -2	38.11	2.50	**38.94**	-	1.80	**19.28**	-	**58.22**	nr	**65.49**

Labour hourly rates: (except Specialists) Craft Operatives £14.21 Labourer £10.71 Rates are national average prices. Refer to REGIONAL VARIATIONS for indicative levels of overall pricing in regions	MATERIALS			LABOUR				RATES		
	Del to Site £	Waste %	Material Cost £	Craft Optve Hrs	Lab Hrs	Labour Cost £	Sunds £	Nett Rate £	Unit	Gross Rate (+12.5%) £
Q40: FENCING: CHAIN LINK FENCING										
Fencing; chain link; BS 1722-1; concrete posts; galvanized mesh, line and tying wires; galvanized steel fittings and accessories; backfilling around posts in concrete mix ST4										
900mm high fencing; type GLC 90; posts with rounded tops										
posts at 2743mm centres; 600mm into ground.......	14.90	5.00	**15.56**	-	1.10	**11.78**	-	**27.34**	m	**30.76**
extra; end posts with struts -1.................	21.71	5.00	**22.63**	-	1.30	**13.92**	-	**36.56**	nr	**41.13**
extra; angle posts with struts -2...............	32.75	5.00	**34.14**	-	1.85	**19.81**	-	**53.95**	nr	**60.69**
1200mm high fencing; type GLC 120; posts with rounded tops										
posts at 2743mm centres; 600mm into ground.......	15.89	5.00	**16.60**	-	1.20	**12.85**	-	**29.45**	m	**33.13**
extra; end posts with struts -1.................	28.99	5.00	**30.27**	-	1.60	**17.14**	-	**47.41**	nr	**53.34**
extra; angle posts with struts -2...............	46.80	5.00	**48.89**	-	2.05	**21.96**	-	**70.85**	nr	**79.70**
1400mm high fencing; type GLC 140A; posts with rounded tops										
posts at 2743mm centres; 600mm into ground.......	16.39	5.00	**17.12**	-	1.30	**13.92**	-	**31.05**	m	**34.93**
extra; end posts with struts -1.................	28.82	5.00	**30.09**	-	1.80	**19.28**	-	**49.37**	nr	**55.54**
extra; angle posts with struts -2...............	47.84	5.00	**49.98**	-	2.30	**24.63**	-	**74.62**	nr	**83.94**
1800mm high fencing; type GLC 180; posts with rounded tops										
posts at 2743mm centres; 600mm into ground.......	18.61	5.00	**19.45**	-	1.60	**17.14**	-	**36.59**	m	**41.16**
extra; end posts with struts -1.................	35.68	5.00	**37.30**	-	1.90	**20.35**	-	**57.65**	nr	**64.85**
extra; angle posts with struts -2...............	56.84	5.00	**59.43**	-	2.45	**26.24**	-	**85.67**	nr	**96.38**
Fencing; chain link; BS 1722-1; concrete posts; plastics coated Grade A mesh, line and tying wire; galvanized steel fittings and accessories; backfilling around posts in concrete mix ST4										
900mm high fencing; type PLC 90A; posts with rounded tops										
posts at 2743mm centres; 600mm into ground.......	7.71	5.00	**8.02**	-	1.10	**11.78**	-	**19.80**	m	**22.27**
extra; end posts with struts -1.................	21.71	5.00	**22.63**	-	1.30	**13.92**	-	**36.56**	nr	**41.13**
extra; angle posts with struts -2...............	32.75	5.00	**34.14**	-	1.85	**19.81**	-	**53.95**	nr	**60.69**
1200mm high fencing; type PLC 120; posts with rounded tops										
posts at 2743mm centres; 600mm into ground.......	8.83	5.00	**9.19**	-	1.20	**12.85**	-	**22.04**	m	**24.80**
extra; end posts with struts -1.................	28.99	5.00	**30.27**	-	1.60	**17.14**	-	**47.41**	nr	**53.34**
extra; angle posts with struts -2...............	46.80	5.00	**48.89**	-	2.05	**21.96**	-	**70.85**	nr	**79.70**
1400mm high fencing; type PLC 140A; posts with rounded tops										
posts at 2743mm centres; 600mm into ground.......	10.06	5.00	**10.48**	-	1.30	**13.92**	-	**24.40**	m	**27.45**
extra; end posts with struts -1.................	28.82	5.00	**30.09**	-	1.80	**19.28**	-	**49.37**	nr	**55.54**
extra; angle posts with struts -2...............	47.84	5.00	**49.98**	-	2.30	**24.63**	-	**74.62**	nr	**83.94**
1800mm high fencing; type PLC 180; posts with rounded tops										
posts at 2743mm centres; 600mm into ground.......	12.77	5.00	**13.33**	-	1.60	**17.14**	-	**30.46**	m	**34.27**
extra; end posts with struts -1.................	35.68	5.00	**37.30**	-	1.90	**20.35**	-	**57.65**	nr	**64.85**
extra; angle posts with struts -2...............	56.84	5.00	**59.43**	-	2.45	**26.24**	-	**85.67**	nr	**96.38**
Fencing; chain link; BS 1722-1; rolled steel angle posts; galvanized mesh, line and tying wires; galvanized steel fittings and accessories; backfilling around posts in concrete mix ST4										
900mm high fencing; type GLS 90; posts with rounded tops										
posts at 2743mm centres; 600mm into ground.......	8.01	2.50	**8.16**	-	0.70	**7.50**	-	**15.66**	m	**17.62**
extra; end posts with struts -1.................	26.24	2.50	**26.80**	-	1.15	**12.32**	-	**39.11**	nr	**44.00**
extra; angle posts with struts -2...............	40.26	2.50	**41.12**	-	1.85	**19.81**	-	**60.93**	nr	**68.54**
1200mm high fencing; type GLS 120; posts with rounded tops										
posts at 2743mm centres; 600mm into ground.......	9.47	2.50	**9.66**	-	0.80	**8.57**	-	**18.23**	m	**20.51**
extra; end posts with struts -1.................	30.66	2.50	**31.33**	-	1.25	**13.39**	-	**44.72**	nr	**50.30**
extra; angle posts with struts -2...............	47.58	2.50	**48.62**	-	2.00	**21.42**	-	**70.04**	nr	**78.79**
1400mm high fencing; type GLS 140A; posts with rounded tops										
posts at 2743mm centres; 600mm into ground.......	10.16	2.50	**10.37**	-	0.90	**9.64**	-	**20.01**	m	**22.51**
extra; end posts with struts -1.................	32.97	2.50	**33.70**	-	1.70	**18.21**	-	**51.90**	nr	**58.39**
extra; angle posts with struts -2...............	51.18	2.50	**52.31**	-	2.20	**23.56**	-	**75.87**	nr	**85.35**
1800mm high fencing; type GLS 180; posts with rounded tops										
posts at 2743mm centres; 600mm into ground.......	12.08	2.50	**12.33**	-	1.10	**11.78**	-	**24.11**	m	**27.12**
extra; end posts with struts -1.................	42.66	2.50	**43.63**	-	1.85	**19.81**	-	**63.44**	nr	**71.37**
extra; angle posts with struts -2...............	63.87	2.50	**65.32**	-	2.40	**25.70**	-	**91.02**	nr	**102.40**

Labour hourly rates: (except Specialists) Craft Operatives £14.21 Labourer £10.71 Rates are national average prices. Refer to REGIONAL VARIATIONS for indicative levels of overall pricing in regions	MATERIALS			LABOUR			RATES			
	Del to Site £	Waste %	Material Cost £	Craft Optve Hrs	Lab Hrs	Labour Cost £	Sunds £	Nett Rate £	Unit	Gross Rate (+12.5%) £
Q40: FENCING: CHAIN LINK FENCING Cont.										
Fencing; chain link; BS 1722 Part 1; rolled steel angle posts; plastics coated Grade A mesh, line and tying wires; galvanized steel fittings and accessories; backfilling around posts in concrete mix ST4										
900mm high fencing; type PLS 90A; posts with rounded tops										
posts at 2743mm centres; 600mm into ground.......	12.57	5.00	**13.10**	-	0.70	**7.50**	-	**20.59**	m	**23.17**
extra; end posts with struts -1................	34.37	5.00	**35.89**	-	1.15	**12.32**	-	**48.20**	nr	**54.23**
extra; angle posts with struts -2................	51.32	5.00	**53.59**	-	1.85	**19.81**	-	**73.41**	nr	**82.58**
1200mm high fencing; type PLS 120; posts with rounded tops										
posts at 2743mm centres; 600mm into ground.......	13.66	5.00	**14.24**	-	0.80	**8.57**	-	**22.81**	m	**25.66**
extra; end posts with struts -1................	45.78	5.00	**47.87**	-	1.25	**13.39**	-	**61.26**	nr	**68.92**
extra; angle posts with struts -2................	61.47	5.00	**64.24**	-	2.00	**21.42**	-	**85.66**	nr	**96.37**
1400mm high fencing; type PLS 140A; posts with rounded tops										
posts at 2743mm centres; 600mm into ground.......	15.21	5.00	**15.87**	-	0.90	**9.64**	-	**25.51**	m	**28.70**
extra; end posts with struts -1................	49.39	5.00	**51.66**	-	1.70	**18.21**	-	**69.87**	nr	**78.61**
extra; angle posts with struts -2................	65.83	5.00	**68.82**	-	2.20	**23.56**	-	**92.38**	nr	**103.93**
1800mm high fencing; type PLS 180; posts with rounded tops										
posts at 2743mm centres; 600mm into ground.......	18.24	5.00	**19.05**	-	1.10	**11.78**	-	**30.84**	m	**34.69**
extra; end posts with struts -1................	66.29	5.00	**69.40**	-	1.85	**19.81**	-	**89.22**	nr	**100.37**
extra; angle posts with struts -2................	83.82	5.00	**87.72**	-	2.40	**25.70**	-	**113.42**	nr	**127.60**
Q40: FENCING: CHESTNUT FENCING										
Fencing; cleft chestnut pale; BS 1722 Part 4; sweet chestnut posts and struts; galvanized accessories; backfilling around posts in concrete mix ST4										
900mm high fencing; type CW90										
posts at 2000mm centres; 600mm into ground.......	10.39	5.00	**10.60**	-	0.28	**3.00**	-	**13.60**	m	**15.30**
extra; end posts with struts - 1................	18.65	5.00	**18.97**	-	0.35	**3.75**	-	**22.72**	nr	**25.56**
extra; angle posts with struts - 2...............	26.17	5.00	**26.57**	-	0.46	**4.93**	-	**31.50**	nr	**35.44**
1200mm high fencing; type CW 120										
posts at 2000mm centres; 600mm into ground.......	11.49	5.00	**11.76**	-	0.20	**2.14**	-	**13.90**	m	**15.64**
extra; end posts with struts - 1................	20.07	5.00	**20.46**	-	0.52	**5.57**	-	**26.03**	nr	**29.29**
extra; angle posts with struts - 2...............	28.01	5.00	**28.50**	-	0.70	**7.50**	-	**36.00**	nr	**40.50**
1500mm high fencing; type CW 150										
posts at 2000mm centres; 600mm into ground.......	14.45	5.00	**14.87**	-	0.25	**2.68**	-	**17.54**	m	**19.74**
extra; end posts with struts - 1................	21.38	5.00	**21.84**	-	0.58	**6.21**	-	**28.05**	nr	**31.56**
extra; angle posts with struts - 2...............	29.74	5.00	**30.32**	-	0.75	**8.03**	-	**38.35**	nr	**43.15**
Q40: FENCING: BOARDED FENCING										
Fencing; close boarded; BS 1722 Part 5; sawn softwood posts, rails, pales, gravel boards and centre stumps, pressure impregnated with preservative; backfilling around posts in concrete mix ST4										
1200mm high fencing; type BW 120										
posts at 3000mm centres; 600mm into ground.......	25.70	5.00	**26.58**	1.44		**20.46**	-	**47.05**	m	**52.93**
1500mm high fencing; type BW 150										
posts at 3000mm centres; 750mm into ground.......	30.79	5.00	**31.93**	1.73	-	**24.58**	-	**56.52**	m	**63.58**
1800mm high fencing; type BW 180A										
posts at 3000mm centres; 750mm into ground.......	33.43	5.00	**34.70**	1.92	-	**27.28**	-	**61.98**	m	**69.73**
Fencing; wooden palisade; BS 1722-5; softwood posts, rails, pales and stumps; pressure impregnated with preservative; backfilling around posts in concrete mix ST4										
1050mm high fencing; type WPW 105; 75 x 19mm rectangular pales with pointed tops										
posts at 3000mm centres; 600mm into ground.......	23.67	5.00	**24.46**	1.15	-	**16.34**	-	**40.80**	m	**45.90**
1200mm high fencing; type WPW 120; 75 x 19mm rectangular pales with pointed tops										
posts at 3000mm centres; 600mm into ground.......	25.85	5.00	**26.75**	1.44	-	**20.46**	-	**47.21**	m	**53.11**
Q40: FENCING: STEEL RAILINGS										
Ornamental steel railings; Ranalah Gates Ltd.; primed at works; fixing in brick openings										
Railing										
reference 200, 356mm high..........................	35.47	1.50	**36.01**	0.75	0.38	**14.73**	0.21	**50.94**	m	**57.31**
reference 220, 279mm high..........................	66.29	1.50	**67.29**	0.75	0.38	**14.73**	0.21	**82.23**	m	**92.50**
reference 230, 432mm high..........................	54.18	1.50	**54.99**	0.75	0.38	**14.73**	0.21	**69.93**	m	**78.67**
reference 240, 838mm high..........................	73.10	1.50	**74.20**	1.00	0.50	**19.57**	0.26	**94.02**	m	**105.77**

Labour hourly rates: (except Specialists) Craft Operatives £14.21 Labourer £10.71 Rates are national average prices. Refer to REGIONAL VARIATIONS for indicative levels of overall pricing in regions	MATERIALS			LABOUR				RATES		
	Del to Site £	Waste %	Material Cost £	Craft Optve Hrs	Lab Hrs	Labour Cost £	Sunds £	Nett Rate £	Unit	Gross Rate (+12.5%) £
Q40: FENCING: GUARD RAILS										
galvanized mild steel guard rail of tubing to BS 1387 medium grade with Kee Klamp fittings										
Rail or standard										
40mm diameter ref. 7-2-G	6.64	5.00	**6.98**	0.30	-	**4.26**	0.30	**11.54**	m	**12.98**
extra; flanged end	4.66	5.00	**4.90**	-	-	**-**	-	**4.90**	nr	**5.51**
extra; bend No.15-7	5.11	5.00	**5.36**	0.18	-	**2.56**	-	**7.92**	nr	**8.91**
extra; three-way intersection No.20-7	7.73	5.00	**8.12**	0.18	-	**2.56**	-	**10.68**	nr	**12.01**
extra; three-way intersection No.25-7	7.42	5.00	**7.79**	0.18	-	**2.56**	-	**10.35**	nr	**11.65**
extra; four-way intersection No.21-7	6.05	5.00	**6.36**	0.18	-	**2.56**	-	**8.91**	nr	**10.03**
extra; four-way intersection No.26-7	5.73	5.00	**6.02**	0.18	-	**2.56**	-	**8.58**	nr	**9.65**
extra; five-way intersection No.35-7	8.59	5.00	**9.02**	0.18	-	**2.56**	-	**11.58**	nr	**13.03**
extra; five-way intersection No.40-7	12.83	5.00	**13.47**	0.18	-	**2.56**	-	**16.03**	nr	**18.03**
extra; floor plate No.61-7	6.81	5.00	**7.15**	0.18	-	**2.56**	-	**9.71**	nr	**10.92**
Infill panel fixed with clips										
50 x 50mm welded mesh	21.97	5.00	**23.07**	0.90	-	**12.79**	0.21	**36.06**	m2	**40.57**
Q40: FENCING: PRECAST CONCRETE POSTS										
Precast concrete fence posts; BS 1722 excavating holes, backfilling around posts in concrete mix C20, disposing of surplus materials, earthwork support										
Fence posts for three wires, housing pattern										
100 x 100mm tapering intermediate post 1570mm long; 600mm into ground	7.74	2.50	**7.88**	0.29	0.29	**7.23**	-	**15.10**	nr	**16.99**
100 x 100mm square strainer post (end or intermediate) 1570mm long; 600mm into ground	8.95	2.50	**9.12**	0.29	0.29	**7.23**	-	**16.35**	nr	**18.39**
Clothes line post										
125 x 125mm tapering, 2670mm long	16.63	2.50	**16.98**	0.38	0.38	**9.47**	-	**26.45**	nr	**29.76**
Close boarded fence post										
94 x 100mm, 2745mm long	13.10	2.50	**13.37**	0.46	0.46	**11.46**	-	**24.83**	nr	**27.93**
Chain link fence post for 1800mm high fencing										
125 x 75mm tapering intermediate post, 2620mm long	12.11	2.50	**12.35**	0.38	0.38	**9.47**	-	**21.82**	nr	**24.55**
125 x 125mm end post, 1 strut, 2620mm long	18.18	2.50	**18.52**	0.72	0.72	**17.94**	-	**36.46**	nr	**41.02**
125 x 125mm angle post, 2 strut, 2620mm long	20.48	2.50	**20.82**	1.14	1.14	**28.41**	-	**49.23**	nr	**55.38**
125 x 125mm gate post, 2620mm long	18.72	2.50	**19.13**	0.38	0.38	**9.47**	-	**28.60**	nr	**32.18**
150 x 150mm gate post, 2620mm long	27.39	2.50	**28.01**	0.38	0.38	**9.47**	-	**37.48**	nr	**42.17**
Q40: FENCING: TIMBER GATES										
Gates; impregnated wrought softwood; featheredge pales; including ring latch and heavy hinges										
Gates (posts included elsewhere)										
900 x 1150mm	83.44	2.50	**85.53**	0.60	0.60	**14.95**	0.17	**100.65**	nr	**113.23**
900 x 1450mm	92.15	2.50	**94.45**	0.60	0.60	**14.95**	0.17	**109.57**	nr	**123.27**
Q40: FENCING: STEEL GATES										
Ornamental steel gates; Ranalah Gates Ltd.; primed at works; fixing in brick openings										
Popular, 1118mm high overall										
single gate reference 50, for 914mm wide opening	40.34	1.50	**40.95**	1.65	0.85	**32.55**	0.27	**73.77**	nr	**82.99**
Diana, 965mm high overall										
single gate reference 800, for 914mm wide opening	47.70	1.50	**48.41**	1.65	0.85	**32.55**	0.27	**81.23**	nr	**91.39**
single gate reference 801, for 1016mm wide opening	50.69	1.50	**51.45**	1.65	0.85	**32.55**	0.27	**84.27**	nr	**94.80**
double gates reference 810, for 2438mm wide opening	106.90	1.50	**108.50**	2.50	1.25	**48.91**	0.49	**157.90**	nr	**177.64**
Popular arch gate 1568mm high overall										
single gate reference 55, for 914mm wide opening	74.01	1.50	**75.12**	1.65	0.85	**32.55**	0.27	**107.94**	nr	**121.43**
Thrifty, 914mm high overall										
single gate reference 20, for 914mm wide opening	33.12	1.50	**33.61**	1.65	0.85	**32.55**	0.27	**66.43**	nr	**74.74**
Elite, 914mm high overall										
single gate reference 40, for 914mm wide opening	42.84	1.50	**43.48**	1.65	0.85	**32.55**	0.27	**76.30**	nr	**85.84**
double gates reference 41, for 2134mm wide opening	99.23	1.50	**100.72**	2.50	1.25	**48.91**	0.49	**150.12**	nr	**168.89**
double gates reference 42, for 2438mm wide opening	104.73	1.50	**106.31**	2.50	1.25	**48.91**	0.49	**155.71**	nr	**175.17**
Super, 965mm high overall										
single gate reference 60, for 914mm wide opening	49.91	1.50	**50.66**	1.65	0.85	**32.55**	0.27	**83.48**	nr	**93.91**
double gates reference 62, for 2286mm wide opening	126.84	1.50	**128.74**	2.50	1.25	**48.91**	0.49	**178.14**	nr	**200.41**
double gates reference 63, for 2438mm wide opening	131.47	1.50	**133.44**	2.50	1.25	**48.91**	0.49	**182.84**	nr	**205.69**
Concord, 914mm high overall										
single gate reference 160, for 914mm wide opening	56.48	1.50	**57.32**	1.65	0.85	**32.55**	0.27	**90.14**	nr	**101.41**
double gates reference 162, for 2286mm wide opening	129.77	1.50	**131.71**	2.50	1.25	**48.91**	0.49	**181.11**	nr	**203.75**
double gates reference 163, for 2438mm wide opening	137.66	1.50	**139.72**	2.50	1.25	**48.91**	0.49	**189.13**	nr	**212.77**
Regent, 914mm high overall										
single gate reference 150, for 914mm wide opening	63.38	1.50	**64.33**	1.65	0.85	**32.55**	0.27	**97.15**	nr	**109.29**
double gates reference 151, for 2134mm wide opening	137.89	1.50	**139.96**	2.50	1.25	**48.91**	0.49	**189.36**	nr	**213.03**

Labour hourly rates: (except Specialists) Craft Operatives £14.21 Labourer £10.71 Rates are national average prices. Refer to REGIONAL VARIATIONS for indicative levels of overall pricing in regions	MATERIALS			LABOUR				RATES		
	Del to Site £	Waste %	Material Cost £	Craft Optve Hrs	Lab Hrs	Labour Cost £	Sunds £	Nett Rate £	Unit	Gross Rate (+12.5%) £
Q40: FENCING: STEEL GATES Cont.										
Ornamental steel gates; Ranalah Gates Ltd.; primed at works; fixing in brick openings Cont.										
Regent, 914mm high overall Cont.										
double gates reference 152, for 2286mm wide opening	141.75	1.50	**143.88**	2.50	1.25	**48.91**	0.49	**193.28**	nr	**217.44**
double gates reference 153, for 2438mm wide opening	158.06	1.50	**160.43**	2.50	1.25	**48.91**	0.49	**209.84**	nr	**236.07**
Brighton, 991mm high overall										
single gate reference 600, for 914mm wide opening	71.72	1.50	**72.80**	1.65	0.85	**32.55**	0.27	**105.62**	nr	**118.82**
double gates reference 602, for 2286mm wide opening	189.83	1.50	**192.68**	2.50	1.25	**48.91**	0.49	**242.08**	nr	**272.34**
double gates reference 603, for 2438mm wide opening	194.51	1.50	**197.42**	2.50	1.25	**48.91**	0.49	**246.83**	nr	**277.68**
Preston, 1067mm high overall										
single gate reference 425, for 914mm wide opening	84.13	1.50	**85.39**	1.65	0.85	**32.55**	0.27	**118.21**	nr	**132.99**
single gate reference 426, for 1016mm wide opening	92.68	1.50	**94.07**	1.65	0.85	**32.55**	0.27	**126.89**	nr	**142.75**
double gates reference 430, for 2134mm wide opening	189.50	1.50	**192.34**	2.50	1.25	**48.91**	0.49	**241.74**	nr	**271.96**
double gates reference 431, for 2438mm wide opening	210.53	1.50	**213.69**	2.50	1.25	**48.91**	0.49	**263.09**	nr	**295.98**
Extra cost of fixing gates to and including pair of steel gate posts; backfilling around posts in concrete mix ST4										
for single or double gates; posts reference 901, size 40 x 40mm	39.71	1.50	**40.34**	0.35	0.20	**7.12**	-	**48.14**	nr	**54.16**
for single or double gates; posts reference 902, size 50 x 50mm	49.51	1.50	**50.29**	0.35	0.20	**7.12**	-	**57.75**	nr	**64.96**
for single or double gates; posts reference 903, size 70 x 70mm	66.11	1.50	**67.13**	0.75	0.35	**14.41**	-	**82.05**	nr	**92.31**
for single or double gates; posts reference 904, size 100 x 100mm	96.06	1.50	**97.55**	0.75	0.35	**14.41**	-	**112.47**	nr	**126.53**
Gates in chain link fencing										
Gate with 40 x 40 x 5mm painted angle iron framing, braces and rails, infilled with 50mm x 10 1/2 G mesh										
1200 x 1800mm; galvanized mesh	216.04	1.50	**219.28**	6.00	6.00	**149.52**	-	**368.80**	nr	**414.90**
1200 x 1800mm; plastic coated mesh	216.04	1.50	**219.28**	6.00	6.00	**149.52**	-	**368.80**	nr	**414.90**
Gate post of 80 x 80 x 6mm painted angle iron; backfilling around posts in concrete mix C30										
to suit 1800mm high gate	49.49	1.50	**50.18**	1.50	1.50	**37.38**	-	**87.56**	nr	**98.51**
Q50: SITE/STREET FURNITURE/EQUIPMENT										
Plain ordinary portland cement precast concrete bollards; setting in ground, excavating hole, removing surplus spoil, filling with concrete mix ST4; working around base										
230mm diameter										
455mm high above ground level; Woodhouse; Smooth Grey	68.00	-	**68.00**	0.65	0.65	**16.20**	3.10	**87.30**	nr	**98.21**
150mm diameter										
915mm high above ground level; Bridgford	76.00	-	**76.00**	1.05	1.05	**26.17**	3.10	**105.27**	nr	**118.42**
Precast concrete plant containers; setting in position										
Plant container (soil filling included elsewhere)										
710mm diameter x 500mm high; Strada	198.00	-	**198.00**	0.95	0.95	**23.67**	0.32	**221.99**	nr	**249.74**
1066mm diameter x 760mm high; Shirley	299.00	-	**299.00**	1.15	1.15	**28.66**	0.34	**328.00**	nr	**369.00**
500 x 500 x 500mm high; Strada	285.00	-	**285.00**	0.75	0.75	**18.69**	0.32	**304.01**	nr	**342.01**
Precast concrete litter bin with galvanized wire plastic coated inner basket Plastic Coated; setting in position										
Plain concrete										
65 litres capacity; Newstead	165.00	-	**165.00**	1.15	1.15	**28.66**	0.34	**194.00**	nr	**218.25**
190 litres capacity; Gransden	189.00	-	**189.00**	1.25	1.25	**31.15**	0.40	**220.55**	nr	**248.12**
Benches and seats with precast concrete supports and hardwood slats to seats and back rests; fixed or free-standing										
Bench; placing in position										
2000mm long; Kelvin; Smooth Grey	250.00	-	**250.00**	0.60	0.60	**14.95**	-	**264.95**	nr	**298.07**
2000mm long; Kelvin; Smooth White	250.00	-	**250.00**	0.60	0.60	**14.95**	-	**264.95**	nr	**298.07**
Steel cycle stands, Mawrob Co. (Engineers) Ltd; galvanized steel construction; free standing										
Single sided										
reference TR6 for 6 cycles	215.00	-	**215.00**	1.20	0.30	**20.27**	-	**235.27**	nr	**264.6[illegible]**
reference TR8 for 8 cycles	289.50	-	**289.50**	1.50	0.35	**25.06**	-	**314.56**	nr	**353.8[illegible]**
Double sided										
reference DTR10 for 10 cycles	260.00	-	**260.00**	1.80	0.45	**30.40**	-	**290.40**	nr	**326.7[illegible]**
reference DTR12 for 12 cycles	388.50	-	**388.50**	1.80	0.45	**30.40**	-	**418.90**	nr	**471.2[illegible]**

Labour hourly rates: (except Specialists) Craft Operatives £14.21 Labourer £10.71 Rates are national average prices. Refer to REGIONAL VARIATIONS for indicative levels of overall pricing in regions	MATERIALS			LABOUR				RATES		
	Del to Site £	Waste %	Material Cost £	Craft Optve Hrs	Lab Hrs	Labour Cost £	Sunds £	Nett Rate £	Unit	Gross Rate (+12.5%) £
Q50: SITE/STREET FURNITURE/EQUIPMENT Cont.										
Steel cycle stands, Mawrob Co. (Engineers) Ltd; galvanized steel construction; free standing Cont.										
Double sided Cont.										
reference DTR20 for 20 cycles	470.00	-	**470.00**	2.40	0.60	**40.53**	-	**510.53**	nr	**574.35**
Steel front wheel cycle supports; Mawrob Co. (Engineers) Ltd; galvanized steel construction, pillar mounted; bolting base plate to concrete										
Single sided										
reference MW/MC 6/S for 6 cycles	199.00	-	**199.00**	4.80	1.20	**81.06**	-	**280.06**	nr	**315.07**
reference MW/MC 9/S for 9 cycles	335.00	-	**335.00**	4.80	1.20	**81.06**	-	**416.06**	nr	**468.07**
reference MW/MC 12/S for 12 cycles	435.00	-	**435.00**	7.20	1.80	**121.59**	-	**556.59**	nr	**626.16**
Double sided										
reference MW/MC 6/D for 12 cycles	299.00	-	**299.00**	4.80	1.20	**81.06**	-	**380.06**	nr	**427.57**
reference MW/MC 9/D for 18 cycles	380.00	-	**380.00**	4.80	1.20	**81.06**	-	**461.06**	nr	**518.69**
reference MW/MC 12/D for 24 cycles	470.00	-	**470.00**	7.20	1.80	**121.59**	-	**591.59**	nr	**665.54**
Steel cycle shelters, Mawrob Co. (Engineers) Ltd; steel angle framing; with red oxide priming; corrugated galvanized steel roofing; assembling; bolting to concrete										
Horizontal loading, single sided										
reference MW/AV10 for 10 cycles	689.00	-	**689.00**	9.60	2.40	**162.12**	-	**851.12**	nr	**957.51**
reference MW/AK10 for 10 cycles	777.00	-	**777.00**	9.60	2.40	**162.12**	-	**939.12**	nr	**1056.51**
reference MW/AW8 for 8 cycles	970.00	-	**970.00**	7.20	1.80	**121.59**	-	**1091.59**	nr	**1228.04**
reference MW/AW10 for 10 cycles	1162.00	-	**1162.00**	9.60	2.40	**162.12**	-	**1324.12**	nr	**1489.64**

STREETWISE SUBBIE
Second Edition
Jack Russell

A guide to survival in the subcontracts jungle.

The author details a lifetime's experience of the construction industry, the last twenty-five in the services sector. He is a great believer in claims avoidance by the implementation and maintenance of basic routines on site, and believes that many of the disasters that befall sub-contractors could be avoided by a more systematic and "streetwise" approach.

Also included are the The Streetwise Subbie's site checklist - a list of basic reminders for display on the site cabin wall grouped under the main headings of Programme, Progress, Information, Claims and Records.

REVIEWS

"I am sure that the incalculable benefits which will accrue will save a vast amount of time, worry, trouble and money and enhance the WT Parker name further amongst our clients by promoting efficiency and expertise in our contracts to the benefit of both us as contractors and our clients. "

TG Morris, Director, WT Parker Group

£10.99 Paperback December 2000 160pp 210x148mm ISBN 075065029x

Five Ways to Order

ONLINE
http://books.elsevier.com/construction

E-MAIL
directorders@elsevier.com

FAX
+44 (0) 1865 474011

PHONE
+44 (0) 1865 474010

MAIL
Customer Services, Elsevier Limited
Linacre House, Jordan Hill, Oxford OX2 8DP

DISPOSAL SYSTEMS

Labour hourly rates: (except Specialists) Craft Operatives £17.38 Labourer £10.71 Rates are national average prices. Refer to REGIONAL VARIATIONS for indicative levels of overall pricing in regions	MATERIALS			LABOUR				RATES		
	Del to Site	Waste	Material Cost	Craft Optve	Lab	Labour Cost	Sunds	Nett Rate	Unit	Gross Rate (+12.5%)
	£	%	£	Hrs	Hrs	£	£	£		£
R10: RAINWATER PIPEWORK/GUTTERS: CAST IRON ROUND RAINWATER PIPES (BS 460)										
Cast iron pipes and fittings, BS 460 Type A sockets; dry joints										
Pipes; straight										
75mm; in or on supports (included elsewhere)	17.98	5.00	**18.88**	0.80	-	**13.90**	-	**32.78**	m	**36.88**
extra; shoes	14.86	2.00	**15.16**	1.00	-	**17.38**	-	**32.54**	nr	**36.60**
extra; bends	12.74	2.00	**12.99**	0.80	-	**13.90**	-	**26.90**	nr	**30.26**
extra; offset bends 75mm projection	16.07	2.00	**16.39**	0.80	-	**13.90**	-	**30.30**	nr	**34.08**
extra; offset bends 150mm projection	16.07	2.00	**16.39**	0.80	-	**13.90**	-	**30.30**	nr	**34.08**
extra; offset bends 225mm projection	18.70	2.00	**19.07**	0.80	-	**13.90**	-	**32.98**	nr	**37.10**
extra; offset bends 300mm projection	22.99	2.00	**23.45**	0.80	-	**13.90**	-	**37.35**	nr	**42.02**
extra; branches	22.31	2.00	**22.76**	0.80	-	**13.90**	-	**36.66**	nr	**41.24**
100mm; in or on supports (included elsewhere)	24.55	5.00	**25.78**	0.96	-	**16.68**	-	**42.46**	m	**47.77**
extra; shoes	20.03	2.00	**20.43**	1.20	-	**20.86**	-	**41.29**	nr	**46.45**
extra; bends	17.99	2.00	**18.35**	0.96	-	**16.68**	-	**35.03**	nr	**39.41**
extra; offset bends 75mm projection	30.30	2.00	**30.91**	0.96	-	**16.68**	-	**47.59**	nr	**53.54**
extra; offset bends 150mm projection	30.30	2.00	**30.91**	0.96	-	**16.68**	-	**47.59**	nr	**53.54**
extra; offset bends 225mm projection	36.70	2.00	**37.43**	0.96	-	**16.68**	-	**54.12**	nr	**60.88**
extra; offset bends 300mm projection	37.41	2.00	**38.16**	0.96	-	**16.68**	-	**54.84**	nr	**61.70**
extra; branches	26.51	2.00	**27.04**	0.96	-	**16.68**	-	**43.73**	nr	**49.19**
Cast iron pipes and fittings, BS 460 Type A sockets; ears cast on; dry joints										
Pipes; straight										
63mm; ears fixing to masonry with galvanized pipe nails and distance pieces	19.55	5.00	**20.53**	0.75	-	**13.04**	-	**33.57**	m	**37.76**
extra; shoes	17.14	2.00	**17.48**	0.85	-	**14.77**	-	**32.26**	nr	**36.29**
extra; bends	10.49	2.00	**10.70**	0.68	-	**11.82**	-	**22.52**	nr	**25.33**
extra; offset bends 75mm projection	16.07	2.00	**16.39**	0.68	-	**11.82**	-	**28.21**	nr	**31.74**
extra; offset bends 150mm projection	16.07	2.00	**16.39**	0.68	-	**11.82**	-	**28.21**	nr	**31.74**
extra; offset bends 225mm projection	18.70	2.00	**19.07**	0.68	-	**11.82**	-	**30.89**	nr	**34.75**
extra; offset bends 300mm projection	21.90	2.00	**22.34**	0.68	-	**11.82**	-	**34.16**	nr	**38.43**
extra; branches	20.63	2.00	**21.04**	0.68	-	**11.82**	-	**32.86**	nr	**36.97**
75mm; ears fixing to masonry with galvanized pipe nails and distance pieces	19.55	5.00	**20.53**	0.80	-	**13.90**	-	**34.44**	m	**38.74**
extra; shoes	17.14	2.00	**17.48**	1.00	-	**17.38**	-	**34.86**	nr	**39.22**
extra; bends	12.74	2.00	**12.99**	0.80	-	**13.90**	-	**26.90**	nr	**30.26**
extra; offset bends 75mm projection	16.07	2.00	**16.39**	0.80	-	**13.90**	-	**30.30**	nr	**34.08**
extra; offset bends 150mm projection	16.07	2.00	**16.39**	0.80	-	**13.90**	-	**30.30**	nr	**34.08**
extra; offset bends 225mm projection	18.70	2.00	**19.07**	0.80	-	**13.90**	-	**32.98**	nr	**37.10**
extra; offset bends 300mm projection	22.99	2.00	**23.45**	0.80	-	**13.90**	-	**37.35**	nr	**42.02**
extra; branches	22.31	2.00	**22.76**	0.80	-	**13.90**	-	**36.66**	nr	**41.24**
100mm; ears fixing to masonry with galvanized pipe nails and distance pieces	26.14	5.00	**27.45**	0.96	-	**16.68**	-	**44.14**	m	**49.65**
extra; shoes	22.76	2.00	**23.22**	1.20	-	**20.86**	-	**44.07**	nr	**49.58**
extra; bends	17.99	2.00	**18.35**	0.96	-	**16.68**	-	**35.03**	nr	**39.41**
extra; offset bends 75mm projection	30.30	2.00	**30.91**	0.96	-	**16.68**	-	**47.59**	nr	**53.54**
extra; offset bends 150mm projection	30.30	2.00	**30.91**	0.96	-	**16.68**	-	**47.59**	nr	**53.54**
extra; offset bends 225mm projection	36.70	2.00	**37.43**	0.96	-	**16.68**	-	**54.12**	nr	**60.88**
extra; offset bends 300mm projection	37.41	2.00	**38.16**	0.96	-	**16.68**	-	**54.84**	nr	**61.70**
extra; branches	26.51	2.00	**27.04**	0.96	-	**16.68**	-	**43.73**	nr	**49.19**
Pipe supports										
Steel holderbats										
for 75mm pipes; for building in (fixing included elsewhere)	7.55	2.00	**7.70**	-	-	**-**	-	**7.70**	nr	**8.66**
for 100mm pipes; for building in (fixing included elsewhere)	7.63	2.00	**7.78**	-	-	**-**	-	**7.78**	nr	**8.76**
for 75mm pipes; fixing to masonry with galvanized screws	6.08	2.00	**6.20**	0.20	-	**3.48**	-	**9.68**	nr	**10.89**
for 100mm pipes; fixing to masonry with galvanized screws	6.16	2.00	**6.28**	0.20	-	**3.48**	-	**9.76**	nr	**10.98**

Labour hourly rates: (except Specialists) Craft Operatives £17.38 Labourer £10.71 Rates are national average prices. Refer to REGIONAL VARIATIONS for indicative levels of overall pricing in regions	MATERIALS			LABOUR				RATES		
	Del to Site £	Waste %	Material Cost £	Craft Optve Hrs	Lab Hrs	Labour Cost £	Sunds £	Nett Rate £	Unit	Gross Rate (+12.5%) £
R10: RAINWATER PIPEWORK/GUTTERS: CAST IRON RECTANGULAR RAINWATER PIPES										
Cast iron rectangular pipes and fittings; ears cast on; dry joints										
Pipes; straight										
100 x 75mm; ears fixing to masonry with galvanized pipe nails and distance pieces	74.88	5.00	78.63	1.20	-	20.86	-	99.49	m	111.92
extra; shoes (front)	64.21	2.00	65.49	1.20	-	20.86	-	86.35	nr	97.14
extra; bends (front)	51.44	2.00	52.47	0.90	-	15.64	-	68.11	nr	76.62
extra; offset bends (front) 150 projection	66.85	2.00	68.19	0.90	-	15.64	-	83.83	nr	94.31
extra; offset bends (front) 300 projection	92.02	2.00	93.86	0.90	-	15.64	-	109.50	nr	123.19
R10: RAINWATER PIPEWORK/GUTTERS: ALUMINIUM RAINWATER PIPES (BS EN 612)										
Cast aluminium pipes and fittings BS EN 612 Section B, ears cast on; dry joints										
Pipes; straight										
63mm; ears fixing to masonry with galvanized screws	12.97	5.00	13.62	0.60	-	10.43	-	24.04	m	27.05
extra; shoes	8.24	2.00	8.40	0.75	-	13.04	-	21.44	nr	24.12
extra; bends	9.44	2.00	9.63	0.60	-	10.43	-	20.06	nr	22.56
extra; offset bends 75mm projection	18.13	2.00	18.49	0.60	-	10.43	-	28.92	nr	32.54
extra; offset bends 150mm projection	22.49	2.00	22.94	0.60	-	10.43	-	33.37	nr	37.54
extra; offset bends 225mm projection	25.48	2.00	25.99	0.60	-	10.43	-	36.42	nr	40.97
extra; offset bends 300mm projection	28.53	2.00	29.10	0.60	-	10.43	-	39.53	nr	44.47
75mm; ears fixing to masonry with galvanized pipe nails	15.08	5.00	15.83	0.80	-	13.90	-	29.74	m	33.45
extra; shoes	11.42	2.00	11.65	1.00	-	17.38	-	29.03	nr	32.66
extra; bends	12.36	2.00	12.61	0.80	-	13.90	-	26.51	nr	29.83
extra; offset bends 75mm projection	21.97	2.00	22.41	0.80	-	13.90	-	36.31	nr	40.85
extra; offset bends 150mm projection	25.25	2.00	25.76	0.80	-	13.90	-	39.66	nr	44.62
extra; offset bends 225mm projection	28.17	2.00	28.73	0.80	-	13.90	-	42.64	nr	47.97
extra; offset bends 300mm projection	31.27	2.00	31.90	0.80	-	13.90	-	45.80	nr	51.52
100mm; ears fixing to masonry with galvanized pipe nails	25.69	5.00	26.97	0.96	-	16.68	-	43.66	m	49.11
extra; shoes	13.65	2.00	13.92	1.20	-	20.86	-	34.78	nr	39.13
extra; bends	18.05	2.00	18.41	0.96	-	16.68	-	35.10	nr	39.48
extra; offset bends 75mm projection	24.84	2.00	25.34	0.96	-	16.68	-	42.02	nr	47.27
extra; offset bends 150mm projection	28.86	2.00	29.44	0.96	-	16.68	-	46.12	nr	51.89
extra; offset bends 225mm projection	32.65	2.00	33.30	0.96	-	16.68	-	49.99	nr	56.24
extra; offset bends 300mm projection	36.46	2.00	37.19	0.96	-	16.68	-	53.87	nr	60.61
R10: RAINWATER PIPEWORK/GUTTERS: UNPLASTICIZED PVC RAINWATER PIPES (BS EN 607)										
PVC-U pipes and fittings, BS EN 607; push fit joints; pipework and supports self coloured grey										
Pipes; straight										
68mm; in standard holderbats fixing to masonry with galvanized screws	4.98	5.00	5.20	0.48	-	8.34	-	13.54	m	15.23
extra; shoes	4.67	2.00	4.76	0.37	-	6.43	-	11.19	nr	12.59
extra; bends	5.41	2.00	5.52	0.30	-	5.21	-	10.73	nr	12.07
extra; offset bends (spigot)	2.99	2.00	3.05	0.30	-	5.21	-	8.26	nr	9.30
extra; offset bends (socket)	2.94	2.00	3.00	0.30	-	5.21	-	8.21	nr	9.24
extra; angled branches	10.76	2.00	10.98	0.30	-	5.21	-	16.19	nr	18.21
R10: RAINWATER PIPEWORK/GUTTERS: CAST IRON EAVES GUTTERS (BS 460)										
Cast iron half round gutters and fittings, BS 460; bolted and mastic joints										
Gutters; straight										
100mm; in standard fascia brackets fixing to timber with galvanized screws	10.94	5.00	11.46	0.60	-	10.43	-	21.88	m	24.62
extra; stopped ends	2.53	2.00	2.58	0.60	-	10.43	-	13.01	nr	14.63
extra; running outlets	7.33	2.00	7.48	0.60	-	10.43	-	17.90	nr	20.14
extra; stopped ends with outlet	8.60	2.00	8.77	0.60	-	10.43	-	19.20	nr	21.60
extra; angles	7.52	2.00	7.67	0.60	-	10.43	-	18.10	nr	20.36
114mm; in standard fascia brackets fixing to timber with galvanized screws	11.36	5.00	11.90	0.60	-	10.43	-	22.33	m	25.12
extra; stopped ends	3.28	2.00	3.35	0.60	-	10.43	-	13.77	nr	15.50
extra; running outlets	7.99	2.00	8.15	0.60	-	10.43	-	18.58	nr	20.90
extra; stopped ends with outlet	9.65	2.00	9.84	0.60	-	10.43	-	20.27	nr	22.80
extra; angles	7.75	2.00	7.91	0.60	-	10.43	-	18.33	nr	20.62
125mm; in standard fascia brackets fixing to timber with galvanized screws	13.11	5.00	13.74	0.68	-	11.82	-	25.55	m	28.75
extra; stopped ends	3.28	2.00	3.35	0.68	-	11.82	-	15.16	nr	17.06
extra; running outlets	9.14	2.00	9.32	0.68	-	11.82	-	21.14	nr	23.78
extra; stopped ends with outlet	10.79	2.00	11.01	0.68	-	11.82	-	22.82	nr	25.68
extra; angles	9.14	2.00	9.32	0.68	-	11.82	-	21.14	nr	23.78
150mm; in standard fascia brackets fixing to timber with galvanized screws	21.92	5.00	22.98	0.90	-	15.64	-	38.62	m	43.45
extra; stopped ends	4.55	2.00	4.64	0.90	-	15.64	-	20.28	nr	22.82
extra; running outlets	15.81	2.00	16.13	0.90	-	15.64	-	31.77	nr	35.74
extra; stopped ends with outlet	18.09	2.00	18.45	0.90	-	15.64	-	34.09	nr	38.36
extra; angles	16.69	2.00	17.02	0.90	-	15.64	-	32.67	nr	36.75

Labour hourly rates: (except Specialists) Craft Operatives £17.38 Labourer £10.71 Rates are national average prices. Refer to REGIONAL VARIATIONS for indicative levels of overall pricing in regions	MATERIALS			LABOUR				RATES		
	Del to Site £	Waste %	Material Cost £	Craft Optve Hrs	Lab Hrs	Labour Cost £	Sunds £	Nett Rate £	Unit	Gross Rate (+12.5%) £
R10: RAINWATER PIPEWORK/GUTTERS: CAST IRON EAVES GUTTERS (BS 460) Cont.										
Cast iron OG gutters and fittings, BS 460; bolted and mastic joints										
Gutters; straight										
100mm; fixing to timber with galvanized screws...	11.02	5.00	**11.57**	0.60	-	**10.43**	-	**22.00**	m	**24.75**
extra; stopped ends	2.59	2.00	**2.64**	0.60	-	**10.43**	-	**13.07**	nr	**14.70**
extra; running outlets	8.00	2.00	**8.16**	0.60	-	**10.43**	-	**18.59**	nr	**20.91**
extra; angles	7.86	2.00	**8.02**	0.60	-	**10.43**	-	**18.45**	nr	**20.75**
114mm; fixing to timber with galvanized screws...	12.12	5.00	**12.73**	0.60	-	**10.43**	-	**23.16**	m	**26.05**
extra; stopped ends	3.35	2.00	**3.42**	0.60	-	**10.43**	-	**13.85**	nr	**15.58**
extra; running outlets	8.51	2.00	**8.68**	0.60	-	**10.43**	-	**19.11**	nr	**21.50**
extra; angles	8.51	2.00	**8.68**	0.60	-	**10.43**	-	**19.11**	nr	**21.50**
125mm; fixing to timber with galvanized screws...	12.71	5.00	**13.35**	0.68	-	**11.82**	-	**25.17**	m	**28.31**
extra; stopped ends	3.35	2.00	**3.42**	0.68	-	**11.82**	-	**15.24**	nr	**17.14**
extra; running outlets	9.30	2.00	**9.49**	0.68	-	**11.82**	-	**21.30**	nr	**23.97**
extra; angles	9.30	2.00	**9.49**	0.68	-	**11.82**	-	**21.30**	nr	**23.97**
R10: RAINWATER PIPEWORK/GUTTERS: CAST IRON MOULDED SECTIONS (STOCK SECTIONS)										
Cast iron moulded (stock sections) gutters and fittings; bolted and mastic joints										
Gutters; straight										
100 x 75mm; in standard fascia brackets fixing to timber with galvanized screws.....................	20.85	5.00	**21.84**	0.90	-	**15.64**	-	**37.48**	m	**42.16**
extra; stopped ends	7.29	2.00	**7.44**	0.90	-	**15.64**	-	**23.08**	nr	**25.96**
extra; running outlets	19.49	2.00	**19.88**	0.90	-	**15.64**	-	**35.52**	nr	**39.96**
extra; angles	19.49	2.00	**19.88**	0.90	-	**15.64**	-	**35.52**	nr	**39.96**
extra; clips	8.15	2.00	**8.31**	0.90	-	**15.64**	-	**23.96**	nr	**26.95**
125 x 100mm; in standard fascia brackets fixing to timber with galvanized screws.....................	29.66	5.00	**31.09**	1.20	-	**20.86**	-	**51.94**	m	**58.44**
extra; stopped ends	9.45	2.00	**9.64**	1.20	-	**20.86**	-	**30.50**	nr	**34.31**
extra; running outlets	27.46	2.00	**28.01**	1.20	-	**20.86**	-	**48.87**	nr	**54.97**
extra; angles	27.46	2.00	**28.01**	1.20	-	**20.86**	-	**48.87**	nr	**54.97**
extra; clips	9.45	2.00	**9.64**	1.20	-	**20.86**	-	**30.50**	nr	**34.31**
R10: RAINWATER PIPEWORK/GUTTERS: CAST IRON BOX GUTTERS										
Cast iron box gutters and fittings; bolted and mastic joints										
Gutters; straight										
100 x 75mm; in or on brackets....................	33.28	5.00	**34.87**	0.90	-	**15.64**	-	**50.51**	m	**56.82**
extra; stopped ends	4.68	2.00	**4.77**	0.90	-	**15.64**	-	**20.42**	nr	**22.97**
extra; running outlets	22.50	2.00	**22.95**	0.90	-	**15.64**	-	**38.59**	nr	**43.42**
extra; angles	22.50	2.00	**22.95**	0.90	-	**15.64**	-	**38.59**	nr	**43.42**
extra; clips	5.92	2.00	**6.04**	0.90	-	**15.64**	-	**21.68**	nr	**24.39**
R10: RAINWATER PIPEWORK/GUTTERS: ALUMINIUM GUTTERS (BS EN 612)										
Cast aluminium half round gutters and fittings, BS EN 612 Section A; bolted and mastic joints										
Gutters; straight										
102mm; in standard fascia brackets fixing to timber with galvanized screws.....................	11.17	5.00	**11.70**	0.60	-	**10.43**	-	**22.12**	m	**24.89**
extra; stopped ends	3.03	2.00	**3.09**	0.60	-	**10.43**	-	**13.52**	nr	**15.21**
extra; running outlets	7.19	2.00	**7.33**	0.60	-	**10.43**	-	**17.76**	nr	**19.98**
extra; angles	6.25	2.00	**6.38**	0.60	-	**10.43**	-	**16.80**	nr	**18.90**
127mm; in standard fascia brackets fixing to timber with galvanized screws.....................	14.52	5.00	**15.19**	0.68	-	**11.82**	-	**27.01**	m	**30.38**
extra; stopped ends	4.02	2.00	**4.10**	0.68	-	**11.82**	-	**15.92**	nr	**17.91**
extra; running outlets	8.23	2.00	**8.39**	0.68	-	**11.82**	-	**20.21**	nr	**22.74**
extra; angles	7.99	2.00	**8.15**	0.40	-	**6.95**	-	**15.10**	nr	**16.99**
R10: RAINWATER PIPEWORK/GUTTERS: UNPLASTICISED PVC GUTTERS (BS EN 607)										
PVC-U half round gutters and fittings, BS EN 607; push fit connector joints; gutterwork and supports self coloured grey										
Gutters; straight										
114mm; in standard fascia brackets fixing to timber with galvanized screws.....................	5.29	2.00	**5.39**	0.40	-	**6.95**	-	**12.34**	m	**13.89**
extra; stopped ends	2.98	2.00	**3.04**	0.40	-	**6.95**	-	**9.99**	nr	**11.24**
extra; running outlets	5.54	2.00	**5.65**	0.40	-	**6.95**	-	**12.60**	nr	**14.18**
extra; angles	6.05	2.00	**6.17**	0.40	-	**6.95**	-	**13.12**	nr	**14.76**
R10: RAINWATER PIPEWORK/GUTTERS: ANCILLARIES										
Aluminium pipework ancillaries										
Rainwater heads; BS EN 612 square type										
250 x 180 x 180mm; 63mm outlet spigot; dry joint to pipe; fixing to masonry with galvanized screws	22.20	1.00	**22.42**	0.60	-	**10.43**	-	**32.85**	nr	**36.95**
250 x 180 x 180mm; 75mm outlet spigot; dry joint to pipe; fixing to masonry with galvanized screws	23.27	1.00	**23.50**	0.80	-	**13.90**	-	**37.41**	nr	**42.08**
250 x 180 x 180mm; 100mm outlet spigot; dry joint to pipe; fixing to masonry with galvanized screws	38.27	1.00	**38.65**	0.96	-	**16.68**	-	**55.34**	nr	**62.25**

Labour hourly rates: (except Specialists) Craft Operatives £17.38 Labourer £10.71 Rates are national average prices. Refer to REGIONAL VARIATIONS for indicative levels of overall pricing in regions	MATERIALS			LABOUR			RATES			
	Del to Site £	Waste %	Material Cost £	Craft Optve Hrs	Lab Hrs	Labour Cost £	Sunds £	Nett Rate £	Unit	Gross Rate (+12.5%) £
R10: RAINWATER PIPEWORK/GUTTERS: RAINWATER HEADS										
Cast iron pipework ancillaries										
Rainwater heads; BS 460 hopper type (flat pattern)										
210 x 160 x 185mm; 65mm outlet spigot; dry joint to pipe; fixing to masonry with galvanized screws	15.97	1.00	**16.13**	0.80	-	**13.90**	-	**30.03**	nr	**33.79**
210 x 160 x 185mm; 75mm outlet spigot; dry joint to pipe; fixing to masonry with galvanized screws	18.13	1.00	**18.31**	0.80	-	**13.90**	-	**32.21**	nr	**36.24**
250 x 180 x 175mm; 100mm outlet spigot; dry joint to pipe; fixing to masonry with galvanized screws	40.08	1.00	**40.48**	0.96	-	**16.68**	-	**57.16**	nr	**64.31**
Rainwater heads; BS 460 square type (flat pattern)										
250 x 180 x 175mm; 65mm outlet spigot; dry joint to pipe; fixing to masonry with galvanized screws	36.66	1.00	**37.03**	0.60	-	**10.43**	-	**47.45**	nr	**53.39**
250 x 180 x 175mm; 75mm outlet spigot; dry joint to pipe; fixing to masonry with galvanized screws	36.66	1.00	**37.03**	0.80	-	**13.90**	-	**50.93**	nr	**57.30**
250 x 180 x 175mm; 100mm outlet spigot; dry joint to pipe; fixing to masonry with galvanized screws	36.66	1.00	**37.03**	0.96	-	**16.68**	-	**53.71**	nr	**60.42**
Plastics pipework ancillaries self coloured black										
Rainwater heads; BS EN 607 square type										
280 x 155 x 230mm; 68mm outlet spigot; push fit joint to plastics pipe; fixing to masonry with galvanized screws	17.36	1.00	**17.53**	0.60	-	**10.43**	-	**27.96**	nr	**31.45**
252 x 195 x 210mm; 110mm outlet spigot; push fit joint to plastics pipe; fixing to masonry with galvanized screws	62.49	1.00	**63.11**	0.96	-	**16.68**	-	**79.80**	nr	**89.77**
R10: RAINWATER PIPEWORK/GUTTERS: ROOF OUTLETS										
Cast iron pipework ancillaries										
Roof outlets; luting flange for asphalt										
flat grating; 100mm outlet spigot; coupling joint to cast iron pipe	98.93	1.00	**99.92**	1.20	-	**20.86**	-	**120.78**	nr	**135.87**
domical grating; 100mm outlet spigot; coupling joint to cast iron pipe	100.15	1.00	**101.15**	1.20	-	**20.86**	-	**122.01**	nr	**137.26**
Plastics pipework ancillaries, self coloured grey										
Roof outlets; BS EN 607; luting flange for asphalt										
domical grating; 75mm outlet spigot; push fit joint to plastics pipe	29.94	1.00	**30.24**	0.45	-	**7.82**	-	**38.06**	nr	**42.82**
domical grating; 100mm outlet spigot; push fit joint to plastics pipe	29.94	1.00	**30.24**	0.45	-	**7.82**	-	**38.06**	nr	**42.82**
R11: FOUL DRAINAGE ABOVE GROUND: CAST IRON SOIL AND WASTE PIPES										
Cast iron pipes and fittings, BS 416; bolted and synthetic rubber gasket couplings										
Pipes; straight										
75mm; in or on supports (included elsewhere)	16.84	5.00	**17.61**	1.20	-	**20.86**	-	**38.46**	m	**43.27**
extra; pipes with inspection door bolted and sealed	42.98	2.00	**43.84**	0.80	-	**13.90**	-	**57.74**	nr	**64.96**
extra; short radius bends	26.47	2.00	**27.00**	0.80	-	**13.90**	-	**40.90**	nr	**46.02**
extra; short radius bends with inspection door bolted and sealed	43.66	2.00	**44.53**	0.80	-	**13.90**	-	**58.44**	nr	**65.74**
extra; offset bends 150mm projection	31.26	2.00	**31.89**	0.80	-	**13.90**	-	**45.79**	nr	**51.51**
extra; single angled branches	39.76	2.00	**40.56**	1.20	-	**20.86**	-	**61.41**	nr	**69.09**
extra; single angled branches with inspection door bolted and sealed	58.64	2.00	**59.81**	1.20	-	**20.86**	-	**80.67**	nr	**90.75**
extra; double angled branches	59.15	2.00	**60.33**	1.50	-	**26.07**	-	**86.40**	nr	**97.20**
boss pipes with one threaded boss	39.05	2.00	**39.83**	0.80	-	**13.90**	-	**53.74**	nr	**60.45**
100mm; in or on supports (included elsewhere)	20.59	2.00	**21.10**	0.96	-	**16.68**	-	**37.78**	m	**42.51**
extra; pipes with inspection door bolted and sealed	48.93	2.00	**50.49**	0.96	-	**16.68**	-	**67.17**	nr	**75.57**
extra; short radius bends	35.48	2.00	**36.77**	0.96	-	**16.68**	-	**53.45**	nr	**60.13**
extra; short radius bends with inspection door bolted and sealed	53.62	2.00	**55.27**	0.96	-	**16.68**	-	**71.95**	nr	**80.95**
extra; offset bends 150mm projection	41.61	2.00	**43.02**	0.96	-	**16.68**	-	**59.70**	nr	**67.17**
extra; offset bends 300mm projection	45.56	2.00	**47.05**	0.96	-	**16.68**	-	**63.73**	nr	**71.70**
extra; single angled branches	53.96	2.00	**55.90**	1.44	-	**25.03**	-	**80.93**	nr	**91.05**
extra; single angled branches with inspection door bolted and sealed	72.11	2.00	**74.42**	1.44	-	**25.03**	-	**99.44**	nr	**111.87**
extra; double angled branches	69.53	2.00	**72.07**	1.92	-	**33.37**	-	**105.44**	nr	**118.62**
extra; double angled branches with inspection door bolted and sealed	87.67	2.00	**90.58**	1.92	-	**33.37**	-	**123.95**	nr	**139.44**
extra; single anti-syphon branches, angled branch	55.17	2.00	**57.14**	1.44	-	**25.03**	-	**82.17**	nr	**92.44**
extra; straight wc connectors	26.04	2.00	**26.56**	0.96	-	**16.68**	-	**43.25**	nr	**48.65**
extra; roof connectors	28.08	2.00	**28.64**	0.96	-	**16.68**	-	**45.33**	nr	**50.99**
boss pipes with one boss	48.27	2.00	**49.81**	0.96	-	**16.68**	-	**66.50**	nr	**74.81**
boss pipes with two bosses	57.16	2.00	**58.88**	0.96	-	**16.68**	-	**75.56**	nr	**85.01**
50mm; in standard wall fixing bracket (fixing included elsewhere)	15.08	2.00	**15.38**	0.66	-	**11.47**	-	**26.85**	m	**30.21**
extra; short radius bends	25.09	2.00	**25.59**	0.66	-	**11.47**	-	**37.06**	nr	**41.70**
extra; short radius bends with inspection door bolted and sealed	42.28	2.00	**43.13**	0.66	-	**11.47**	-	**54.60**	nr	**61.42**
extra; single angled branches	37.69	2.00	**38.44**	1.00	-	**17.38**	-	**55.82**	nr	**62.80**
extra; single angled branches with inspection door bolted and sealed	56.57	2.00	**57.70**	1.00	-	**17.38**	-	**75.08**	nr	**84.47**

Labour hourly rates: (except Specialists) Craft Operatives £17.38 Labourer £10.71 Rates are national average prices. Refer to REGIONAL VARIATIONS for indicative levels of overall pricing in regions	MATERIALS			LABOUR				RATES		
	Del to Site £	Waste %	Material Cost £	Craft Optve Hrs	Lab Hrs	Labour Cost £	Sunds £	Nett Rate £	Unit	Gross Rate (+12.5%) £
R11: FOUL DRAINAGE ABOVE GROUND: CAST IRON SOIL AND WASTE PIPES Cont.										
Cast iron lightweight soil and waste pipes and fittings										
Pipes; straight										
50mm; in or on supports (included elsewhere).....	10.90	5.00	**11.40**	0.66	-	**11.47**	-	**22.87**	m	**25.73**
extra; short pipe with access door..............	28.21	2.00	**28.77**	0.66	-	**11.47**	-	**40.25**	nr	**45.28**
extra; bends	15.93	2.00	**16.25**	0.66	-	**11.47**	-	**27.72**	nr	**31.18**
extra; offset bend 130mm projection.............	21.05	2.00	**21.47**	0.66	-	**11.47**	-	**32.94**	nr	**37.06**
extra; single angled branch.....................	24.65	2.00	**25.14**	1.00	-	**17.38**	-	**42.52**	nr	**47.84**
extra; plug	7.24	2.00	**7.38**	0.33	-	**5.74**	-	**13.12**	nr	**14.76**
extra; universal plug with one inlet............	11.25	2.00	**11.48**	0.17	-	**2.95**	-	**14.43**	nr	**16.23**
extra; step coupling for connection to BS 416 pipe	6.02	2.00	**6.14**	0.50	-	**8.69**	-	**14.83**	nr	**16.68**
75mm; in or on supports (included elsewhere).....	12.53	5.00	**13.11**	0.80	-	**13.90**	-	**27.01**	m	**30.39**
extra; short pipe with access door..............	30.19	2.00	**30.79**	0.80	-	**13.90**	-	**44.70**	nr	**50.29**
extra; bends	17.69	2.00	**18.04**	0.80	-	**13.90**	-	**31.95**	nr	**35.94**
extra; offset bend 130mm projection.............	28.36	2.00	**28.93**	0.80	-	**13.90**	-	**42.83**	nr	**48.19**
extra; single angled branch.....................	26.57	2.00	**27.10**	1.20	-	**20.86**	-	**47.96**	nr	**53.95**
extra; diminishing piece 75/50..................	20.32	2.00	**20.73**	0.80	-	**13.90**	-	**34.63**	nr	**38.96**
extra; plug	7.82	2.00	**7.98**	0.40	-	**6.95**	-	**14.93**	nr	**16.79**
extra; universal plug with one inlet............	11.84	2.00	**12.08**	0.20	-	**3.48**	-	**15.55**	nr	**17.50**
extra; step coupling for connection to BS 416 pipe	7.39	2.00	**7.54**	0.70	-	**12.17**	-	**19.70**	nr	**22.17**
100mm; in or on supports (included elsewhere)....	15.08	5.00	**15.77**	1.00	-	**17.38**	-	**33.15**	m	**37.30**
extra; short pipe with access door..............	35.11	2.00	**35.81**	1.00	-	**17.38**	-	**53.19**	nr	**59.84**
extra; bends	22.08	2.00	**22.52**	1.00	-	**17.38**	-	**39.90**	nr	**44.89**
extra; long radius bend.........................	35.54	2.00	**36.25**	1.00	-	**17.38**	-	**53.63**	nr	**60.33**
extra; offset bend 75mm projection..............	27.37	2.00	**27.92**	1.00	-	**17.38**	-	**45.30**	nr	**50.96**
extra; offset bend 130mm projection.............	37.16	2.00	**37.90**	1.00	-	**17.38**	-	**55.28**	nr	**62.19**
extra; single angled branch.....................	34.29	2.00	**34.98**	1.50	-	**26.07**	-	**61.05**	nr	**68.68**
extra; double angled branch.....................	47.38	2.00	**48.33**	2.00	-	**34.76**	-	**83.09**	nr	**93.47**
extra; diminishing piece 100/50.................	23.74	2.00	**24.21**	1.00	-	**17.38**	-	**41.59**	nr	**46.79**
extra; diminishing piece 100/75.................	24.16	2.00	**24.64**	1.00	-	**17.38**	-	**42.02**	nr	**47.28**
extra; plug	9.77	2.00	**9.97**	0.50	-	**8.69**	-	**18.66**	nr	**20.99**
extra; universal plug with three inlets.........	12.80	2.00	**13.06**	0.50	-	**8.69**	-	**21.75**	nr	**24.46**
extra; step coupling for connection to BS 416 pipe	8.68	2.00	**8.85**	0.80	-	**13.90**	-	**22.76**	nr	**25.60**
extra; traditional joint connector..............	20.82	2.00	**21.24**	1.00	-	**17.38**	-	**38.62**	nr	**43.44**
extra; roof connector (asphalt).................	27.72	2.00	**28.27**	1.00	-	**17.38**	-	**45.65**	nr	**51.36**
Pipe supports										
galvanized steel vertical bracket										
for 50mm pipes; for fixing bolt(fixing included elsewhere)	4.46	2.00	**4.55**	-	-	-	-	**4.55**	nr	**5.12**
for 75mm pipes; for fixing bolt(fixing included elsewhere)	4.72	2.00	**4.81**	-	-	-	-	**4.81**	nr	**5.42**
for 100mm pipes; for fixing bolt(fixing included elsewhere)	5.01	2.00	**5.11**	-	-	-	-	**5.11**	nr	**5.75**
R11: FOUL DRAINAGE ABOVE GROUND: UNPLASTICIZED MPVC WASTE PIPES (SOLVENT WELDED)										
MuPVC pipes and fittings, BS EN 1329-1 Section Three; solvent welded joints										
Pipes; straight										
32mm; in standard plastics pipe brackets fixing to timber with screws	2.62	10.00	**2.82**	0.40	-	**6.95**	-	**9.77**	m	**11.00**
extra; straight expansion couplings.............	2.09	2.00	**2.13**	0.30	-	**5.21**	-	**7.35**	nr	**8.26**
extra; connections to plastics pipe socket; socket adaptor; solvent welded joint..................	2.23	2.00	**2.27**	0.30	-	**5.21**	-	**7.49**	nr	**8.42**
extra; connections to plastics pipe boss; boss adaptor; solvent welded joint....................	2.23	2.00	**2.27**	0.30	-	**5.21**	-	**7.49**	nr	**8.42**
extra; fittings; one end........................	2.29	2.00	**2.34**	0.15	-	**2.61**	-	**4.94**	nr	**5.56**
extra; fittings; two ends.......................	1.27	2.00	**1.30**	0.30	-	**5.21**	-	**6.51**	nr	**7.32**
extra; fittings; three ends.....................	1.62	2.00	**1.65**	0.30	-	**5.21**	-	**6.87**	nr	**7.72**
40mm; in standard plastics pipe brackets fixing to timber with screws	3.12	10.00	**3.36**	0.40	-	**6.95**	-	**10.32**	m	**11.60**
extra; straight expansion couplings.............	2.52	2.00	**2.57**	0.30	-	**5.21**	-	**7.78**	nr	**8.76**
extra; connections to plastics pipe socket; socket adaptor; solvent welded joint..................	2.23	2.00	**2.27**	0.30	-	**5.21**	-	**7.49**	nr	**8.42**
extra; connections to plastics pipe boss; boss adaptor; solvent welded joint....................	2.23	2.00	**2.27**	0.30	-	**5.21**	-	**7.49**	nr	**8.42**
extra; fittings; one end........................	2.28	2.00	**2.33**	0.15	-	**2.61**	-	**4.93**	nr	**5.55**
extra; fittings; two ends.......................	1.45	2.00	**1.48**	0.30	-	**5.21**	-	**6.69**	nr	**7.53**
extra; fittings; three ends.....................	2.06	2.00	**2.10**	0.30	-	**5.21**	-	**7.32**	nr	**8.23**
50mm; in standard plastics pipe brackets fixing to timber with screws	4.73	10.00	**5.09**	0.48	-	**8.34**	-	**13.44**	m	**15.12**
extra; straight expansion couplings.............	3.42	2.00	**3.49**	0.40	-	**6.95**	-	**10.44**	nr	**11.75**
extra; connections to plastics pipe socket; socket adaptor; solvent welded joint..................	3.18	2.00	**3.24**	0.40	-	**6.95**	-	**10.20**	nr	**11.47**
extra; connections to plastics pipe boss; boss adaptor; solvent welded joint....................	3.18	2.00	**3.24**	0.40	-	**6.95**	-	**10.20**	nr	**11.47**
extra; fittings; one end........................	3.75	2.00	**3.83**	0.20	-	**3.48**	-	**7.30**	nr	**8.21**
extra; fittings; two ends.......................	2.45	2.00	**2.50**	0.40	-	**6.95**	-	**9.45**	nr	**10.63**
extra; fittings; three ends.....................	4.01	2.00	**4.09**	0.40	-	**6.95**	-	**11.04**	nr	**12.42**
32mm; in standard plastics pipe brackets fixing to masonry with screws	2.65	10.00	**2.85**	0.48	-	**8.34**	-	**11.19**	m	**12.59**
40mm; in standard plastics pipe brackets fixing to masonry with screws	3.15	10.00	**3.39**	0.48	-	**8.34**	-	**11.74**	m	**13.20**
50mm; in standard plastics pipe brackets fixing to masonry with screws	4.76	10.00	**5.12**	0.56	-	**9.73**	-	**14.86**	m	**16.71**

Labour hourly rates: (except Specialists) Craft Operatives £17.38 Labourer £10.71 Rates are national average prices. Refer to REGIONAL VARIATIONS for indicative levels of overall pricing in regions	MATERIALS			LABOUR				RATES		
	Del to Site £	Waste %	Material Cost £	Craft Optve Hrs	Lab Hrs	Labour Cost £	Sunds £	Nett Rate £	Unit	Gross Rate (+12.5%) £
R11: FOUL DRAINAGE ABOVE GROUND: POLYPROPYLENE WASTE PIPES ('O' RING)										
Polypropylene pipes and fittings; butyl ring joints										
Pipes; straight										
32mm; in standard plastics pipe brackets fixing to timber with screws	1.55	10.00	**1.65**	0.40	-	**6.95**	-	**8.60**	m	**9.68**
extra; connections to plastics pipe socket; socket adaptor; solvent welded joint	2.41	2.00	**2.46**	0.30	-	**5.21**	-	**7.67**	nr	**8.63**
extra; connections to plastics pipe boss; boss adaptor; solvent welded joint	2.41	2.00	**2.46**	0.30	-	**5.21**	-	**7.67**	nr	**8.63**
extra; fittings; one end	1.79	2.00	**1.83**	0.15	-	**2.61**	-	**4.43**	nr	**4.99**
extra; fittings; two ends	1.03	2.00	**1.05**	0.30	-	**5.21**	-	**6.26**	nr	**7.05**
extra; fittings; three ends	1.59	2.00	**1.62**	0.30	-	**5.21**	-	**6.84**	nr	**7.69**
40mm; in standard plastics pipe brackets fixing to timber with screws	1.87	10.00	**1.99**	0.40	-	**6.95**	-	**8.95**	m	**10.06**
extra; connections to plastics pipe socket; socket adaptor; solvent welded joint	2.41	2.00	**2.46**	0.40	-	**6.95**	-	**9.41**	nr	**10.59**
extra; connections to plastics pipe boss; boss adaptor; solvent welded joint	2.41	2.00	**2.46**	0.40	-	**6.95**	-	**9.41**	nr	**10.59**
extra; fittings; one end	2.03	2.00	**2.07**	0.20	-	**3.48**	-	**5.55**	nr	**6.24**
extra; fittings; two ends	1.08	2.00	**1.10**	0.40	-	**6.95**	-	**8.05**	nr	**9.06**
extra; fittings; three ends	1.66	2.00	**1.69**	0.40	-	**6.95**	-	**8.65**	nr	**9.73**
32mm; in standard plastics pipe brackets fixing to masonry with screws	1.58	10.00	**1.68**	0.48	-	**8.34**	-	**10.02**	m	**11.28**
40mm; in standard plastics pipe brackets fixing to masonry with screws	1.90	10.00	**2.02**	0.48	-	**8.34**	-	**10.37**	m	**11.66**
R11: FOUL DRAINAGE ABOVE GROUND: UNPLASTICIZED SOIL AND VENTILATING PIPES										
PVC-U pipes and fittings, BS 4514; rubber ring joints; pipework self coloured grey										
Pipes; straight										
82mm; in or on supports (included elsewhere)	11.18	5.00	**11.68**	0.53	-	**9.21**	-	**20.90**	m	**23.51**
extra; pipes with inspection door	21.82	2.00	**22.26**	0.40	-	**6.95**	-	**29.21**	nr	**32.86**
extra; short radius bends	12.24	2.00	**12.48**	0.40	-	**6.95**	-	**19.44**	nr	**21.87**
extra; single angled branches	17.98	2.00	**18.34**	0.40	-	**6.95**	-	**25.29**	nr	**28.45**
pipes with one boss socket	9.73	2.00	**9.92**	0.40	-	**6.95**	-	**16.88**	nr	**18.99**
110mm; in or on supports (included elsewhere)	9.95	5.00	**10.39**	0.60	-	**10.43**	-	**20.82**	m	**23.42**
extra; pipes with inspection door	20.92	2.00	**21.34**	0.60	-	**10.43**	-	**31.77**	nr	**35.74**
extra; short radius bends	14.32	2.00	**14.61**	0.60	-	**10.43**	-	**25.03**	nr	**28.16**
extra; single angled branches	18.66	2.00	**19.03**	0.60	-	**10.43**	-	**29.46**	nr	**33.14**
extra; straight wc connectors	5.65	2.00	**5.76**	0.60	-	**10.43**	-	**16.19**	nr	**18.21**
pipes with one boss socket	10.48	2.00	**10.69**	0.60	-	**10.43**	-	**21.12**	nr	**23.76**
160mm; in or on supports (included elsewhere)	26.67	5.00	**27.87**	0.90	-	**15.64**	-	**43.51**	m	**48.95**
extra; pipes with inspection door	49.53	2.00	**50.52**	0.90	-	**15.64**	-	**66.16**	nr	**74.43**
extra; short radius bends	40.38	2.00	**41.19**	0.90	-	**15.64**	-	**56.83**	nr	**63.93**
extra; single angled branches	66.50	2.00	**67.83**	0.90	-	**15.64**	-	**83.47**	nr	**93.91**
pipes with one boss socket	20.34	2.00	**20.75**	0.90	-	**15.64**	-	**36.39**	nr	**40.94**
PVC-U pipes and fittings, BS 4514; solvent weld joints; pipework self coloured grey										
Pipes; straight										
110mm; in or on supports (included elsewhere)	9.95	5.00	**10.39**	0.63	-	**10.95**	-	**21.34**	m	**24.01**
extra; pipes with inspection door	20.92	2.00	**21.34**	0.63	-	**10.95**	-	**32.29**	nr	**36.32**
extra; short radius bends	14.32	2.00	**14.61**	0.63	-	**10.95**	-	**25.56**	nr	**28.75**
extra; single angled branches	18.66	2.00	**19.03**	0.63	-	**10.95**	-	**29.98**	nr	**33.73**
extra; straight wc connectors	5.65	2.00	**5.76**	0.63	-	**10.95**	-	**16.71**	nr	**18.80**
pipes with one boss socket	10.48	2.00	**10.69**	0.63	-	**10.95**	-	**21.64**	nr	**24.34**
160mm; in or on supports (included elsewhere)	26.67	5.00	**27.87**	0.95	-	**16.51**	-	**44.38**	m	**49.92**
extra; pipes with inspection door	49.53	2.00	**50.52**	0.95	-	**16.51**	-	**67.03**	nr	**75.41**
extra; short radius bends	40.38	2.00	**41.19**	0.95	-	**16.51**	-	**57.70**	nr	**64.91**
extra; single angled branches	66.50	2.00	**67.83**	0.95	-	**16.51**	-	**84.34**	nr	**94.88**
pipes with one boss socket	20.34	2.00	**20.75**	0.95	-	**16.51**	-	**37.26**	nr	**41.92**
Plastics pipework ancillaries, self coloured grey										
Weathering aprons										
fitted to 82mm pipes	4.47	2.00	**4.56**	0.40	-	**6.95**	-	**11.51**	nr	**12.95**
fitted to 110mm pipes	3.34	2.00	**3.41**	0.60	-	**10.43**	-	**13.83**	nr	**15.56**
Pipe supports for PVC-U soil and vent pipes, rubber ring joints; self coloured grey										
Plastics coated metal holderbats										
for 82mm pipes; for building in (fixing included elsewhere)	4.34	2.00	**4.43**	-	-	**-**	-	**4.43**	nr	**4.98**
for 110mm pipes; for building in (fixing included elsewhere)	6.29	2.00	**6.42**	-	-	**-**	-	**6.42**	nr	**7.22**
for 82mm pipes; fixing to timber with galvanized screws	3.54	2.00	**3.61**	0.20	-	**3.48**	-	**7.09**	nr	**7.98**
for 110mm pipes; fixing to timber with galvanized screws	4.72	2.00	**4.82**	0.20	-	**3.48**	-	**8.29**	nr	**9.33**
for 160mm pipes; fixing to timber with galvanized screws	8.84	2.00	**9.02**	0.20	-	**3.48**	-	**12.50**	nr	**14.06**
Pipe supports; solvent weld joints;										
Plastic brackets										
for 110mm pipes; for building in (fixing included elsewhere)	6.89	2.00	**7.03**	-	-	**-**	-	**7.03**	nr	**7.91**

Labour hourly rates: (except Specialists) Craft Operatives £17.38 Labourer £10.71 Rates are national average prices. Refer to REGIONAL VARIATIONS for indicative levels of overall pricing in regions	MATERIALS			LABOUR				RATES		
	Del to Site £	Waste %	Material Cost £	Craft Optve Hrs	Lab Hrs	Labour Cost £	Sunds £	Nett Rate £	Unit	Gross Rate (+12.5%) £
R11: FOUL DRAINAGE ABOVE GROUND: UNPLASTICIZED SOIL AND VENTILATING PIPES Cont.										
Pipe supports; solvent weld joints; Cont.										
Plastic brackets Cont.										
for 110mm pipes; fixing to timber with galvanized screws	3.36	2.00	**3.43**	0.30	-	**5.21**	-	**8.64**	nr	**9.72**
for 160mm pipes; fixing to timber with galvanized screws	8.02	2.00	**8.18**	0.30	-	**5.21**	-	**13.40**	nr	**15.07**
R11: FOUL DRAINAGE ABOVE GROUND: WIRE BALLOON GRATINGS										
Pipework ancillaries										
Wire balloon gratings										
fitted to 50mm pipes	2.26	3.00	**2.33**	0.12	-	**2.09**	-	**4.41**	nr	**4.97**
fitted to 82mm pipes	3.31	3.00	**3.41**	0.12	-	**2.09**	-	**5.49**	nr	**6.18**
fitted to 110mm pipes	3.44	3.00	**3.54**	0.12	-	**2.09**	-	**5.63**	nr	**6.33**
R11: FOUL DRAINAGE ABOVE GROUND: PLASTIC TRAPS										
Plastics pipework ancillaries self coloured white										
Traps; BS EN 274										
P, two piece, 75mm seal, inlet with coupling nut, outlet with seal ring socket										
32mm outlet	4.86	2.00	**4.96**	0.40	-	**6.95**	-	**11.91**	nr	**13.40**
40mm outlet	5.61	2.00	**5.72**	0.40	-	**6.95**	-	**12.67**	nr	**14.26**
S, two piece, 75mm seal, inlet with coupling nut, outlet with seal ring socket										
32mm outlet	6.16	2.00	**6.28**	0.40	-	**6.95**	-	**13.24**	nr	**14.89**
40mm outlet	7.20	2.00	**7.34**	0.40	-	**6.95**	-	**14.30**	nr	**16.08**
P, two piece, 75mm seal, with flexible polypropylene pipe for overflow connection, inlet with coupling nut, outlet with seal ring socket										
40mm outlet	10.40	2.00	**10.61**	0.60	-	**10.43**	-	**21.04**	nr	**23.67**
R11: FOUL DRAINAGE ABOVE GROUND: COPPER TRAPS										
Copper pipework ancillaries										
Traps; BS 1184 Section Two (Solid Drawn)										
P, two piece, 75mm seal, inlet with coupling nut, compression outlet										
35mm outlet	23.20	2.00	**23.66**	0.50	-	**8.69**	-	**32.35**	nr	**36.40**
42mm outlet	27.33	2.00	**27.88**	0.50	-	**8.69**	-	**36.57**	nr	**41.14**
54mm outlet	98.08	2.00	**100.04**	0.75	-	**13.04**	-	**113.08**	nr	**127.21**
S, two piece, 75mm seal, inlet with coupling nut, compression outlet										
35mm outlet	24.58	2.00	**25.07**	0.50	-	**8.69**	-	**33.76**	nr	**37.98**
42mm outlet	29.57	2.00	**30.16**	0.50	-	**8.69**	-	**38.85**	nr	**43.71**
54mm outlet	103.59	2.00	**105.66**	0.75	-	**13.04**	-	**118.70**	nr	**133.53**
bath, two piece, 75mm seal, overflow connection, cleaning eye and plug, inlet with coupling nut, compression outlet										
42mm outlet	38.80	2.00	**39.58**	0.80	-	**13.90**	-	**53.48**	nr	**60.17**

This section continues on the next page

Labour hourly rates: (except Specialists) Craft Operatives £14.21 Labourer £10.71 Rates are national average prices. Refer to REGIONAL VARIATIONS for indicative levels of overall pricing in regions	PLANT & TRANSPORT			LABOUR				RATES		
	Plant Cost £	Trans Cost £	P & T Cost £	Craft Optve Hrs	Lab Hrs	Labour Cost £	Sunds £	Nett Rate £	Unit	Gross Rate (+12.5%) £
R12: DRAINAGE BELOW GROUND: EXCAVATION FOR DRAINS - BY MACHINE										
Excavating trenches by machine to receive pipes not exceeding 200mm nominal size; disposing of surplus excavated material by removing from site										
Excavations commencing from natural ground level average depth										
500mm	1.56	3.06	**4.62**	-	0.24	**2.57**	0.93	**8.13**	m	**9.14**
750mm	2.49	3.06	**5.55**	-	0.35	**3.75**	1.41	**10.71**	m	**12.05**
1000mm	3.34	3.06	**6.40**	-	0.47	**5.03**	1.88	**13.31**	m	**14.97**
1250mm	4.28	3.06	**7.34**	-	0.64	**6.85**	2.34	**16.53**	m	**18.60**
1500mm	5.14	3.06	**8.20**	-	0.76	**8.14**	2.81	**19.15**	m	**21.54**
1750mm	6.08	3.06	**9.14**	-	0.90	**9.64**	3.29	**22.06**	m	**24.82**
2000mm	6.91	3.06	**9.97**	-	1.02	**10.92**	3.76	**24.65**	m	**27.74**
2250mm	8.84	3.06	**11.90**	-	1.77	**18.96**	4.22	**35.08**	m	**39.46**
2500mm	9.81	3.06	**12.87**	-	1.97	**21.10**	4.69	**38.66**	m	**43.50**
2750mm	10.84	3.06	**13.90**	-	2.16	**23.13**	5.17	**42.19**	m	**47.47**
3000mm	11.81	3.06	**14.87**	-	2.36	**25.28**	5.63	**45.77**	m	**51.49**
3250mm	12.86	3.06	**15.92**	-	2.56	**27.42**	6.10	**49.43**	m	**55.61**
3500mm	13.83	3.06	**16.89**	-	2.76	**29.56**	6.57	**53.02**	m	**59.64**
3750mm	14.85	3.06	**17.91**	-	2.96	**31.70**	7.03	**56.64**	m	**63.73**
4000mm	15.85	3.06	**18.91**	-	3.00	**32.13**	7.51	**58.54**	m	**65.86**
4250mm	16.87	3.06	**19.93**	-	4.46	**47.77**	7.98	**75.68**	m	**85.13**
4500mm	17.84	3.06	**20.90**	-	4.72	**50.55**	8.44	**79.89**	m	**89.88**
4750mm	18.86	3.06	**21.92**	-	4.98	**53.34**	8.91	**84.17**	m	**94.69**
5000mm	19.83	3.06	**22.89**	-	7.21	**77.22**	10.32	**110.43**	m	**124.24**
5250mm	20.88	3.06	**23.94**	-	6.88	**73.68**	9.85	**107.48**	m	**120.91**
5500mm	21.83	3.06	**24.89**	-	7.21	**77.22**	10.32	**112.43**	m	**126.48**
5750mm	22.88	3.06	**25.94**	-	7.53	**80.65**	10.79	**117.38**	m	**132.05**
6000mm	23.85	3.06	**26.91**	-	7.87	**84.29**	11.26	**122.46**	m	**137.77**
Excavating trenches by machine to receive pipes 225mm nominal size; disposing of surplus excavated material by removing from site										
Excavations commencing from natural ground level average depth										
500mm	1.68	3.23	**4.91**	-	0.25	**2.68**	0.93	**8.52**	m	**9.58**
750mm	2.62	3.23	**5.85**	-	0.37	**3.96**	1.41	**11.22**	m	**12.62**
1000mm	3.53	3.23	**6.76**	-	0.49	**5.25**	1.88	**13.89**	m	**15.63**
1250mm	4.46	3.23	**7.69**	-	0.67	**7.18**	2.34	**17.21**	m	**19.36**
1500mm	5.47	3.23	**8.70**	-	0.80	**8.57**	2.81	**20.08**	m	**22.59**
1750mm	6.39	3.23	**9.62**	-	0.94	**10.07**	3.29	**22.97**	m	**25.84**
2000mm	7.32	3.23	**10.55**	-	1.08	**11.57**	3.76	**25.87**	m	**29.11**
2250mm	9.30	3.23	**12.53**	-	1.85	**19.81**	4.22	**36.56**	m	**41.13**
2500mm	10.43	3.23	**13.66**	-	2.06	**22.06**	4.69	**40.42**	m	**45.47**
2750mm	11.46	3.23	**14.69**	-	2.26	**24.20**	5.17	**44.06**	m	**49.56**
3000mm	12.51	3.23	**15.74**	-	2.47	**26.45**	5.63	**47.82**	m	**53.79**
3250mm	13.56	3.23	**16.79**	-	2.68	**28.70**	6.10	**51.59**	m	**58.04**
3500mm	14.66	3.23	**17.89**	-	2.89	**30.95**	6.57	**55.42**	m	**62.34**
3750mm	15.71	3.23	**18.94**	-	3.10	**33.20**	7.03	**59.18**	m	**66.57**
4000mm	16.76	3.23	**19.99**	-	3.30	**35.34**	7.51	**62.84**	m	**70.70**
4250mm	17.81	3.23	**21.04**	-	4.68	**50.12**	7.98	**79.15**	m	**89.04**
4500mm	18.92	3.23	**22.15**	-	4.95	**53.01**	8.44	**83.60**	m	**94.05**
4750mm	19.97	3.23	**23.20**	-	5.22	**55.91**	8.91	**88.02**	m	**99.02**
5000mm	21.02	3.23	**24.25**	-	6.88	**73.68**	10.32	**108.25**	m	**121.79**
5250mm	22.04	3.23	**25.27**	-	7.23	**77.43**	9.85	**112.55**	m	**126.62**
5500mm	23.18	3.23	**26.41**	-	7.66	**82.04**	10.32	**118.76**	m	**133.61**
5750mm	24.23	3.23	**27.46**	-	7.91	**84.72**	10.79	**122.96**	m	**138.34**
6000mm	25.28	3.23	**28.51**	-	8.24	**88.25**	11.26	**128.02**	m	**144.02**
Excavating trenches by machine to receive pipes 300mm nominal size; disposing of surplus excavated material by removing from site										
Excavations commencing from natural ground level average depth										
500mm	1.85	3.57	**5.42**	-	0.26	**2.78**	0.93	**9.13**	m	**10.28**
750mm	2.90	3.57	**6.47**	-	0.44	**4.71**	1.41	**12.59**	m	**14.16**
1000mm	3.98	3.57	**7.55**	-	0.52	**5.57**	1.88	**15.00**	m	**16.87**
1250mm	5.06	3.57	**8.63**	-	0.70	**7.50**	2.34	**18.47**	m	**20.78**
1500mm	6.14	3.57	**9.71**	-	0.84	**9.00**	2.81	**21.52**	m	**24.21**
1750mm	7.19	3.57	**10.76**	-	0.99	**10.60**	3.29	**24.65**	m	**27.73**
2000mm	8.27	3.57	**11.84**	-	1.15	**12.32**	3.76	**27.92**	m	**31.41**
2250mm	10.54	3.57	**14.11**	-	1.95	**20.88**	4.22	**39.22**	m	**44.12**
2500mm	11.73	3.57	**15.30**	-	2.17	**23.24**	4.69	**43.23**	m	**48.63**
2750mm	12.94	3.57	**16.51**	-	2.36	**25.28**	5.17	**46.95**	m	**52.82**
3000mm	14.15	3.57	**17.72**	-	2.60	**27.85**	5.63	**51.19**	m	**57.59**
3250mm	15.34	3.57	**18.91**	-	2.81	**30.10**	6.10	**55.10**	m	**61.99**
3500mm	16.55	3.57	**20.12**	-	3.04	**32.56**	6.57	**59.25**	m	**66.66**
3750mm	17.76	3.57	**21.33**	-	3.25	**34.81**	7.03	**63.17**	m	**71.07**
4000mm	18.95	3.57	**22.52**	-	3.46	**37.06**	7.51	**67.08**	m	**75.47**
4250mm	20.16	3.57	**23.73**	-	4.91	**52.59**	7.98	**84.30**	m	**94.83**
4500mm	21.37	3.57	**24.94**	-	5.19	**55.58**	8.44	**88.97**	m	**100.09**
4750mm	22.56	3.57	**26.13**	-	5.49	**58.80**	8.91	**93.84**	m	**105.57**
5000mm	23.77	3.57	**27.34**	-	7.22	**77.33**	10.32	**114.99**	m	**129.36**
5250mm	24.98	3.57	**28.55**	-	7.58	**81.18**	9.85	**119.58**	m	**134.53**
5500mm	26.17	3.57	**29.74**	-	7.92	**84.82**	10.32	**124.88**	m	**140.49**
5750mm	27.38	3.57	**30.95**	-	8.29	**88.79**	10.79	**130.53**	m	**146.84**
6000mm	28.59	3.57	**32.16**	-	8.66	**92.75**	11.26	**136.18**	m	**153.20**

Labour hourly rates: (except Specialists) Craft Operatives £14.21 Labourer £10.71 Rates are national average prices. Refer to REGIONAL VARIATIONS for indicative levels of overall pricing in regions	PLANT & TRANSPORT			LABOUR				RATES		
	Plant Cost £	Trans Cost £	P & T Cost £	Craft Optve Hrs	Lab Hrs	Labour Cost £	Sunds £	Nett Rate £	Unit	Gross Rate (+12.5%) £
R12: DRAINAGE BELOW GROUND: EXCAVATION FOR DRAINS - BY MACHINE Cont.										
Excavating trenches by machine to receive pipes 400mm nominal size; disposing of surplus excavated material by removing from site										
Excavations commencing from natural ground level average depth										
2250mm	12.27	4.25	**16.52**	-	2.13	**22.81**	4.22	**43.55**	m	**49.00**
2500mm	13.64	4.25	**17.89**	-	2.37	**25.38**	4.69	**47.97**	m	**53.97**
2750mm	15.07	4.25	**19.32**	-	2.59	**27.74**	5.17	**52.23**	m	**58.76**
3000mm	16.45	4.25	**20.70**	-	2.83	**30.31**	5.63	**56.63**	m	**63.71**
3250mm	17.90	4.25	**22.15**	-	3.06	**32.77**	6.10	**61.02**	m	**68.65**
3500mm	19.25	4.25	**23.50**	-	3.31	**35.45**	6.57	**65.52**	m	**73.71**
3750mm	20.70	4.25	**24.95**	-	3.54	**37.91**	7.03	**69.90**	m	**78.64**
4000mm	22.08	4.25	**26.33**	-	3.78	**40.48**	7.51	**74.32**	m	**83.61**
4250mm	23.51	4.25	**27.76**	-	5.34	**57.19**	7.98	**92.93**	m	**104.54**
4500mm	24.88	4.25	**29.13**	-	5.67	**60.73**	8.44	**98.29**	m	**110.58**
4750mm	18.52	4.25	**22.77**	-	5.98	**64.05**	9.28	**96.10**	m	**108.12**
5000mm	27.68	4.25	**31.93**	-	7.87	**84.29**	10.32	**126.54**	m	**142.36**
5250mm	29.14	4.25	**33.39**	-	8.56	**91.68**	9.85	**134.91**	m	**151.77**
5500mm	30.48	4.25	**34.73**	-	8.66	**92.75**	10.32	**137.80**	m	**155.03**
5750mm	31.93	4.25	**36.18**	-	9.03	**96.71**	10.79	**143.68**	m	**161.64**
6000mm	23.31	4.25	**27.56**	-	9.45	**101.21**	11.74	**140.50**	m	**158.07**
Excavating trenches by machine to receive pipes 450mm nominal size; disposing of surplus excavated material by removing from site										
Excavations commencing from natural ground level average depth										
2250mm	13.13	4.42	**17.55**	-	2.21	**23.67**	4.22	**45.44**	m	**51.12**
2500mm	14.67	4.42	**19.09**	-	2.46	**26.35**	4.69	**50.13**	m	**56.39**
2750mm	16.10	4.42	**20.52**	-	2.70	**28.92**	5.17	**54.60**	m	**61.42**
3000mm	17.63	4.42	**22.05**	-	2.94	**31.49**	5.63	**59.17**	m	**66.56**
3250mm	18.10	4.42	**22.52**	-	3.20	**34.27**	6.10	**62.89**	m	**70.75**
3500mm	20.68	4.42	**25.10**	-	3.45	**36.95**	6.57	**68.62**	m	**77.20**
3750mm	22.13	4.42	**26.55**	-	3.69	**39.52**	7.03	**73.11**	m	**82.24**
4000mm	23.64	4.42	**28.06**	-	3.93	**42.09**	7.51	**77.66**	m	**87.36**
4250mm	25.18	4.42	**29.60**	-	5.56	**59.55**	7.98	**97.12**	m	**109.26**
4500mm	26.69	4.42	**31.11**	-	5.91	**63.30**	8.44	**102.84**	m	**115.70**
4750mm	28.14	4.42	**32.56**	-	6.24	**66.83**	8.91	**108.30**	m	**121.84**
5000mm	29.68	4.42	**34.10**	-	8.19	**87.71**	10.32	**132.13**	m	**148.65**
5250mm	31.19	4.42	**35.61**	-	8.61	**92.21**	9.85	**137.66**	m	**154.87**
5500mm	32.72	4.42	**37.14**	-	9.03	**96.71**	10.32	**144.17**	m	**162.19**
5750mm	34.15	4.42	**38.57**	-	9.39	**100.57**	10.79	**149.93**	m	**168.67**
6000mm	35.68	4.42	**40.10**	-	9.81	**105.07**	11.26	**156.43**	m	**175.99**
R12: DRAINAGE BELOW GROUND: EXCAVATION FOR DRAINS - BY HAND										
Excavating trenches by hand to receive pipes not exceeding 200mm nominal size; disposing of surplus excavated material on site										
Excavations commencing from natural ground level average depth										
500mm	-	-	**-**	-	1.20	**12.85**	0.93	**13.79**	m	**15.51**
750mm	-	-	**-**	-	2.15	**23.03**	1.41	**24.43**	m	**27.49**
1000mm	-	-	**-**	-	3.10	**33.20**	1.88	**35.08**	m	**39.47**
1250mm	-	-	**-**	-	3.90	**41.77**	2.34	**44.11**	m	**49.62**
1500mm	-	-	**-**	-	4.85	**51.94**	2.81	**54.76**	m	**61.60**
1750mm	-	-	**-**	-	5.80	**62.12**	3.29	**65.40**	m	**73.58**
2000mm	-	-	**-**	-	6.80	**72.83**	3.76	**76.59**	m	**86.16**
2250mm	-	-	**-**	-	8.00	**85.68**	4.22	**89.90**	m	**101.14**
2500mm	-	-	**-**	-	9.20	**98.53**	4.69	**103.22**	m	**116.13**
2750mm	-	-	**-**	-	10.40	**111.38**	5.17	**116.55**	m	**131.12**
3000mm	-	-	**-**	-	11.60	**124.24**	5.63	**129.86**	m	**146.10**
3250mm	-	-	**-**	-	12.80	**137.09**	6.10	**143.19**	m	**161.09**
3500mm	-	-	**-**	-	14.05	**150.48**	6.57	**157.05**	m	**176.68**
3750mm	-	-	**-**	-	15.25	**163.33**	7.03	**170.36**	m	**191.66**
4000mm	-	-	**-**	-	16.50	**176.72**	7.51	**184.22**	m	**207.25**
Excavating trenches by hand to receive pipes 225mm nominal size; disposing of surplus excavated material on site										
Excavations commencing from natural ground level average depth										
500mm	-	-	**-**	-	1.35	**14.46**	0.93	**15.39**	m	**17.32**
750mm	-	-	**-**	-	2.35	**25.17**	1.41	**26.58**	m	**29.90**
1000mm	-	-	**-**	-	3.35	**35.88**	1.88	**37.76**	m	**42.48**
1250mm	-	-	**-**	-	4.35	**46.59**	2.34	**48.93**	m	**55.05**
1500mm	-	-	**-**	-	5.34	**57.19**	2.81	**60.00**	m	**67.51**
1750mm	-	-	**-**	-	6.35	**68.01**	3.29	**71.29**	m	**80.21**
2000mm	-	-	**-**	-	7.35	**78.72**	3.76	**82.48**	m	**92.79**
2250mm	-	-	**-**	-	8.80	**94.25**	4.22	**98.47**	m	**110.78**
2500mm	-	-	**-**	-	10.10	**108.17**	4.69	**112.86**	m	**126.97**
2750mm	-	-	**-**	-	11.40	**122.09**	5.17	**127.26**	m	**143.17**
3000mm	-	-	**-**	-	12.70	**136.02**	5.63	**141.64**	m	**159.35**
3250mm	-	-	**-**	-	14.05	**150.48**	6.10	**156.57**	m	**176.15**
3500mm	-	-	**-**	-	15.40	**164.93**	6.57	**171.51**	m	**192.94**
3750mm	-	-	**-**	-	16.70	**178.86**	7.03	**185.89**	m	**209.13**
4000mm	-	-	**-**	-	18.10	**193.85**	7.51	**201.36**	m	**226.53**

Labour hourly rates: (except Specialists) Craft Operatives £14.21 Labourer £10.71 Rates are national average prices. Refer to REGIONAL VARIATIONS for indicative levels of overall pricing in regions	PLANT & TRANSPORT			LABOUR				RATES		
	Plant Cost £	Trans Cost £	P & T Cost £	Craft Optve Hrs	Lab Hrs	Labour Cost £	Sunds £	Nett Rate £	Unit	Gross Rate (+12.5%) £
R12: DRAINAGE BELOW GROUND: EXCAVATION FOR DRAINS - BY HAND Cont.										
Excavating trenches by hand to receive pipes 300mm nominal size; disposing of surplus material on site										
Excavations commencing from natural ground level average depth										
500mm	-	-	-	-	1.50	**16.07**	0.93	**17.00**	m	**19.12**
750mm	-	-	-	-	2.60	**27.85**	1.41	**29.25**	m	**32.91**
1000mm	-	-	-	-	3.65	**39.09**	1.88	**40.97**	m	**46.09**
1250mm	-	-	-	-	4.75	**50.87**	2.34	**53.21**	m	**59.86**
1500mm	-	-	-	-	5.80	**62.12**	2.81	**64.93**	m	**73.05**
1750mm	-	-	-	-	6.90	**73.90**	3.29	**77.19**	m	**86.83**
2000mm	-	-	-	-	7.95	**85.14**	3.76	**88.90**	m	**100.02**
2250mm	-	-	-	-	9.45	**101.21**	4.22	**105.43**	m	**118.61**
2500mm	-	-	-	-	10.85	**116.20**	4.69	**120.90**	m	**136.01**
2750mm	-	-	-	-	12.30	**131.73**	5.17	**136.90**	m	**154.01**
3000mm	-	-	-	-	13.70	**146.73**	5.63	**152.35**	m	**171.40**
3250mm	-	-	-	-	15.15	**162.26**	6.10	**168.36**	m	**189.40**
3500mm	-	-	-	-	16.60	**177.79**	6.57	**184.36**	m	**207.40**
3750mm	-	-	-	-	18.05	**193.32**	7.03	**200.35**	m	**225.39**
4000mm	-	-	-	-	19.50	**208.85**	7.51	**216.35**	m	**243.39**

This section continues on the next page

Labour hourly rates: (except Specialists) Craft Operatives £14.21 Labourer £10.71 Rates are national average prices. Refer to REGIONAL VARIATIONS for indicative levels of overall pricing in regions	MATERIALS			LABOUR			RATES			
	Del to Site £	Waste %	Material Cost £	Craft Optve Hrs	Lab Hrs	Labour Cost £	Sunds £	Nett Rate £	Unit	Gross Rate (+12.5%) £
R12: DRAINAGE BELOW GROUND: BEDS, BENCHINGS AND COVERS										
Granular material, 10mm nominal size pea shingle, to be obtained off site										
Beds										
400 x 150mm	1.59	10.00	**1.75**	-	0.18	**1.93**	-	**3.67**	m	**4.13**
450 x 150mm	1.77	10.00	**1.95**	-	0.19	**2.03**	-	**3.99**	m	**4.48**
525 x 150mm	2.06	10.00	**2.27**	-	0.20	**2.14**	-	**4.41**	m	**4.96**
600 x 150mm	2.38	10.00	**2.62**	-	0.26	**2.78**	-	**5.40**	m	**6.08**
700 x 150mm	2.78	10.00	**3.06**	-	0.29	**3.11**	-	**6.16**	m	**6.93**
750 x 150mm	2.99	10.00	**3.29**	-	0.32	**3.43**	-	**6.72**	m	**7.56**
Beds and surrounds										
400 x 150mm bed; 150mm thick surround to 100mm internal diameter pipes -1	4.23	10.00	**4.66**	-	0.55	**5.89**	-	**10.55**	m	**11.87**
450 x 150mm bed; 150mm thick surround to 150mm internal diameter pipes -1	5.34	10.00	**5.88**	-	0.62	**6.64**	-	**12.52**	m	**14.08**
525 x 150mm bed; 150mm thick surround to 225mm internal diameter pipes -1	7.28	10.00	**8.00**	-	0.82	**8.78**	-	**16.79**	m	**18.88**
600 x 150mm bed; 150mm thick surround to 300mm internal diameter pipes -1	9.53	10.00	**10.48**	-	1.01	**10.82**	-	**21.30**	m	**23.96**
700 x 150mm bed; 150mm thick surround to 400mm internal diameter pipes -1	12.97	10.00	**14.26**	-	1.20	**12.85**	-	**27.11**	m	**30.50**
750 x 150mm bed; 150mm thick surround to 450mm internal diameter pipes -1	14.82	10.00	**16.30**	-	1.39	**14.89**	-	**31.19**	m	**35.08**
Beds and filling to half pipe depth										
400 x 150mm overall; to 100mm internal diameter pipes - 1	3.70	10.00	**4.07**	-	0.14	**1.50**	-	**5.57**	m	**6.27**
450 x 225mm overall; to 150mm internal diameter pipes - 1	4.45	10.00	**4.89**	-	0.24	**2.57**	-	**7.46**	m	**8.39**
525 x 263mm overall; to 225mm internal diameter pipes - 1	6.77	10.00	**7.45**	-	0.34	**3.64**	-	**11.09**	m	**12.48**
600 x 300mm overall; to 300mm internal diameter pipes -1	9.53	10.00	**10.48**	-	0.43	**4.61**	-	**15.08**	m	**16.97**
700 x 350mm overall; to 400mm internal diameter pipes -1	13.89	10.00	**15.28**	-	0.59	**6.32**	-	**21.60**	m	**24.30**
750 x 375mm overall; to 450mm internal diameter pipes -1	16.35	10.00	**17.99**	-	0.67	**7.18**	-	**25.16**	m	**28.31**
Sand; to be obtained off site										
Beds										
400 x 50mm	0.31	5.00	**0.32**	-	0.05	**0.54**	-	**0.86**	m	**0.96**
400 x 100mm	0.61	5.00	**0.64**	-	0.10	**1.07**	-	**1.71**	m	**1.93**
450 x 50mm	0.34	5.00	**0.35**	-	0.06	**0.64**	-	**1.00**	m	**1.12**
450 x 100mm	0.69	5.00	**0.72**	-	0.11	**1.18**	-	**1.90**	m	**2.14**
525 x 50mm	0.40	5.00	**0.42**	-	0.07	**0.75**	-	**1.17**	m	**1.31**
525 x 100mm	0.80	5.00	**0.84**	-	0.13	**1.39**	-	**2.23**	m	**2.51**
600 x 50mm	0.46	5.00	**0.48**	-	0.07	**0.75**	-	**1.23**	m	**1.39**
600 x 100mm	0.92	5.00	**0.96**	-	0.14	**1.50**	-	**2.46**	m	**2.77**
700 x 50mm	0.54	5.00	**0.56**	-	0.08	**0.86**	-	**1.42**	m	**1.60**
700 x 100mm	1.07	5.00	**1.13**	-	0.17	**1.82**	-	**2.95**	m	**3.31**
750 x 50mm	0.57	5.00	**0.59**	-	0.10	**1.07**	-	**1.67**	m	**1.87**
750 x 100mm	1.15	5.00	**1.21**	-	0.18	**1.93**	-	**3.13**	m	**3.53**
Plain in-situ concrete; BS 5328, ordinary prescribed mix, ST3, 20mm aggregate										
Beds										
400 x 100mm	2.49	5.00	**2.61**	-	0.22	**2.36**	2.79	**7.76**	m	**8.73**
400 x 150mm	3.73	5.00	**3.92**	-	0.32	**3.43**	4.12	**11.47**	m	**12.90**
450 x 100mm	2.80	5.00	**2.94**	-	0.24	**2.57**	2.79	**8.30**	m	**9.34**
450 x 150mm	4.17	5.00	**4.38**	-	0.36	**3.86**	4.12	**12.35**	m	**13.90**
525 x 100mm	3.23	5.00	**3.40**	-	0.29	**3.11**	2.79	**9.30**	m	**10.46**
525 x 150mm	4.85	5.00	**5.09**	-	0.42	**4.50**	4.12	**13.71**	m	**15.43**
600 x 100mm	3.73	5.00	**3.92**	-	0.32	**3.43**	2.79	**10.14**	m	**11.41**
600 x 150mm	5.60	5.00	**5.88**	-	0.46	**4.93**	4.12	**14.93**	m	**16.79**
700 x 100mm	4.35	5.00	**4.57**	-	0.38	**4.07**	2.79	**11.44**	m	**12.87**
700 x 150mm	6.53	5.00	**6.86**	-	0.56	**6.00**	4.12	**16.98**	m	**19.10**
750 x 100mm	4.66	5.00	**4.90**	-	0.41	**4.39**	2.79	**12.08**	m	**13.59**
750 x 150mm	6.97	5.00	**7.31**	-	0.61	**6.53**	4.12	**17.97**	m	**20.21**
Beds and surrounds										
400 x 100mm bed; 150mm thick surround to 100mm internal diameter pipes -1	8.71	5.00	**9.14**	-	0.86	**9.21**	5.49	**23.84**	m	**26.82**
400 x 150mm bed; 150mm thick surround to 100mm internal diameter pipes -1	9.95	5.00	**10.45**	-	0.97	**10.39**	6.18	**27.02**	m	**30.40**
450 x 100mm bed; 150mm thick surround to 150mm internal diameter pipes -1	11.20	5.00	**11.76**	-	1.09	**11.67**	6.18	**29.61**	m	**33.31**
450 x 150mm bed; 150mm thick surround to 150mm internal diameter pipes -1	12.56	5.00	**13.19**	-	1.21	**12.96**	6.87	**33.03**	m	**37.15**
525 x 100mm bed; 150mm thick surround to 225mm internal diameter pipes -1	15.49	5.00	**16.26**	-	1.49	**15.96**	6.87	**39.09**	m	**43.98**
525 x 150mm bed; 150mm thick surround to 225mm internal diameter pipes -1	17.10	5.00	**17.96**	-	1.63	**17.46**	7.89	**43.31**	m	**48.73**
600 x 100mm bed; 150mm thick surround to 300mm internal diameter pipes -1	20.53	5.00	**21.55**	-	1.94	**20.78**	8.25	**50.58**	m	**56.90**
600 x 150mm bed; 150mm thick surround to 300mm internal diameter pipes -1	22.39	5.00	**23.51**	-	2.11	**22.60**	8.94	**55.04**	m	**61.93**
700 x 100mm bed; 150mm thick surround to 400mm internal diameter pipes -1	28.30	5.00	**29.72**	-	2.65	**28.38**	9.63	**67.73**	m	**76.19**
700 x 150mm bed; 150mm thick surround to 400mm internal diameter pipes -1	30.48	5.00	**32.00**	-	2.83	**30.31**	10.32	**72.63**	m	**81.71**

Labour hourly rates: (except Specialists) Craft Operatives £14.21 Labourer £10.71 Rates are national average prices. Refer to REGIONAL VARIATIONS for indicative levels of overall pricing in regions	MATERIALS			LABOUR				RATES		
	Del to Site £	Waste %	Material Cost £	Craft Optve Hrs	Lab Hrs	Labour Cost £	Sunds £	Nett Rate £	Unit	Gross Rate (+12.5%) £
R12: DRAINAGE BELOW GROUND: BEDS, BENCHINGS AND COVERS Cont.										
Plain in-situ concrete; BS 5328, ordinary prescribed mix, ST3, 20mm aggregate Cont.										
Beds and surrounds Cont.										
750 x 100mm bed; 150mm thick surround to 450mm internal diameter pipes -1	32.65	5.00	**34.29**	-	3.04	**32.56**	10.32	**77.17**	m	**86.81**
750 x 150mm bed; 150mm thick surround to 450mm internal diameter pipes -1	34.96	5.00	**36.70**	-	3.24	**34.70**	10.99	**82.39**	m	**92.69**
Beds and haunchings										
400 x 150mm; to 100mm internal diameter pipes - 1	6.22	5.00	**6.53**	-	0.48	**5.14**	4.12	**15.79**	m	**17.77**
450 x 150mm; to 150mm internal diameter pipes - 1	8.40	5.00	**8.82**	-	0.55	**5.89**	4.12	**18.83**	m	**21.18**
525 x 150mm; to 225mm internal diameter pipes - 1	12.19	5.00	**12.80**	-	0.74	**7.93**	4.12	**24.85**	m	**27.95**
600 x 150mm; to 300mm internal diameter pipes - 1	16.79	5.00	**17.63**	-	0.97	**10.39**	4.12	**32.14**	m	**36.16**
700 x 150mm; to 400mm internal diameter pipes - 1	23.95	5.00	**25.14**	-	1.32	**14.14**	4.12	**43.40**	m	**48.83**
750 x 150mm; to 450mm internal diameter pipes - 1	27.99	5.00	**29.39**	-	1.62	**17.35**	4.12	**50.86**	m	**57.22**
Vertical casings										
400 x 400mm to 100mm internal diameter pipes - 1	37.15	5.00	**37.65**	-	0.96	**10.28**	-	**47.93**	m	**53.92**
450 x 450mm to 150mm internal diameter pipes - 1	43.16	5.00	**43.79**	-	1.21	**12.96**	-	**56.75**	m	**63.85**
525 x 525mm to 225mm internal diameter pipes - 1	53.14	5.00	**54.00**	-	1.66	**17.78**	-	**71.78**	m	**80.75**
600 x 600mm to 300mm internal diameter pipes - 1	63.19	5.00	**64.31**	-	2.16	**23.13**	-	**87.44**	m	**98.38**
700 x 700mm to 400mm internal diameter pipes - 1	78.08	5.00	**79.60**	-	2.94	**31.49**	-	**111.09**	m	**124.97**
750 x 750mm to 450mm internal diameter pipes - 1	85.96	5.00	**87.70**	-	3.37	**36.09**	-	**123.80**	m	**139.27**
R12: DRAINAGE BELOW GROUND: VITRIFIED CLAY FLEXIBLE JOINT DRAINS										
Drains; vitrified clay pipes and fittings, BS EN 295, normal; flexible mechanical joints; Hepworth Hepseal										
Pipework in trenches										
400mm	99.52	5.00	**104.50**	0.66	0.22	**11.73**	-	**116.23**	m	**130.76**
extra; bends	373.97	5.00	**392.67**	0.47	-	**6.68**	-	**399.35**	nr	**449.26**
450mm	131.97	5.00	**138.57**	0.84	0.24	**14.51**	-	**153.07**	m	**172.21**
extra; bends	492.44	5.00	**517.07**	0.84	-	**11.94**	-	**529.00**	nr	**595.13**
Pipework in trenches; vertical										
400mm	99.52	5.00	**104.50**	0.72	0.22	**12.59**	-	**117.09**	m	**131.72**
450mm	131.97	5.00	**138.57**	0.92	0.24	**15.64**	-	**154.21**	m	**173.49**
R12: DRAINAGE BELOW GROUND: VITRIFIED CLAY SLEEVE JOINTS										
Drains; vitrified clay pipes and fittings, BS EN 295; sleeve joints, push-fit polypropylene standard ring flexible couplings; Hepworth Supersleeve										
Pipework in trenches										
100mm	6.60	5.00	**6.93**	0.19	0.07	**3.45**	-	**10.38**	m	**11.68**
extra; bends	8.98	5.00	**9.43**	0.10	-	**1.42**	-	**10.85**	nr	**12.21**
extra; branches	19.62	5.00	**20.60**	0.19	-	**2.70**	-	**23.30**	nr	**26.21**
150mm	12.59	5.00	**13.22**	0.22	0.10	**4.20**	-	**17.42**	m	**19.60**
extra; bends	18.61	5.00	**19.54**	0.11	-	**1.56**	-	**21.11**	nr	**23.74**
extra; branches	28.92	5.00	**30.36**	0.22	-	**3.13**	-	**33.49**	nr	**37.68**
Pipework in trenches; vertical										
100mm	6.60	5.00	**6.93**	0.22	0.07	**3.88**	-	**10.81**	m	**12.16**
150mm	12.59	5.00	**13.22**	0.24	0.10	**4.48**	-	**17.70**	m	**19.92**
Drains; vitrified clay pipes and fittings, BS EN 295; sleeve joints, push-fit polypropylene neoprene ring flexible couplings; Hepworth Supersleeve										
Pipework in trenches										
100mm	7.68	5.00	**8.07**	0.19	0.07	**3.45**	-	**11.52**	m	**12.96**
150mm	14.59	5.00	**15.32**	0.22	0.10	**4.20**	-	**19.52**	m	**21.96**
Pipework in trenches; vertical										
100mm	7.68	5.00	**8.07**	0.22	0.07	**3.88**	-	**11.95**	m	**13.44**
150mm	14.59	5.00	**15.32**	0.24	0.10	**4.48**	-	**19.80**	m	**22.28**
Drains; vitrified clay pipes and fittings, BS EN 295, normal; cement mortar (1:3)										
Pipework in trenches										
100mm	8.52	5.00	**8.95**	0.40	0.04	**6.11**	-	**15.06**	m	**16.95**
extra; double collars	10.25	5.00	**10.78**	0.12	-	**1.71**	-	**12.48**	nr	**14.04**
extra; bends	8.52	5.00	**8.95**	0.24	-	**3.41**	-	**12.36**	nr	**13.91**
extra; rest bends	13.94	5.00	**14.65**	0.24	-	**3.41**	-	**18.06**	nr	**20.32**
extra; branches	15.57	5.00	**16.36**	0.26	-	**3.69**	-	**20.05**	nr	**22.56**
150mm	8.52	5.00	**8.95**	0.52	0.05	**7.92**	-	**16.88**	m	**18.99**
extra; double collars	17.02	5.00	**17.88**	0.18	-	**2.56**	-	**20.44**	nr	**22.99**
extra; bends	13.05	5.00	**13.71**	0.32	-	**4.55**	-	**18.26**	nr	**20.54**
extra; rest bends	23.52	5.00	**24.70**	0.32	-	**4.55**	-	**29.25**	nr	**32.91**
extra; branches	25.73	5.00	**27.02**	0.34	-	**4.83**	-	**31.85**	nr	**35.83**
225mm	25.88	5.00	**27.19**	0.66	0.08	**10.24**	-	**37.43**	m	**42.11**
extra; double collars	39.79	5.00	**41.80**	0.28	-	**3.98**	-	**45.78**	nr	**51.50**
300mm	43.27	5.00	**45.45**	0.96	0.12	**14.93**	-	**60.38**	m	**67.92**
Pipework in trenches; vertical										
100mm	8.52	5.00	**8.95**	0.60	0.04	**8.95**	-	**17.91**	m	**20.14**
150mm	8.52	5.00	**8.95**	0.58	0.05	**8.78**	-	**17.73**	m	**19.94**

Labour hourly rates: (except Specialists) Craft Operatives £14.21 Labourer £10.71 Rates are national average prices. Refer to REGIONAL VARIATIONS for indicative levels of overall pricing in regions	MATERIALS			LABOUR				RATES		
	Del to Site £	Waste %	Material Cost £	Craft Optve Hrs	Lab Hrs	Labour Cost £	Sunds £	Nett Rate £	Unit	Gross Rate (+12.5%) £
R12: DRAINAGE BELOW GROUND: VITRIFIED CLAY SLEEVE JOINTS Cont.										
Drains; vitrified clay pipes and fittings, BS EN 295, normal; cement mortar (1:3) Cont.										
Pipework in trenches; vertical Cont.										
225mm	25.88	5.00	**27.19**	0.72	0.08	**11.09**	-	**38.28**	m	**43.07**
300mm	43.27	5.00	**45.45**	1.06	0.12	**16.35**	-	**61.80**	m	**69.52**
R12: DRAINAGE BELOW GROUND: PVC-U DRAINS										
Drains; Wavin; PVC-U pipes and fittings, BS EN 13598-1; ring seal joints										
Pipework in trenches										
110mm, ref. 4D.076	6.42	5.00	**6.74**	0.16	0.02	**2.49**	-	**9.23**	m	**10.38**
extra; connections to cast iron and clay sockets, ref. 4D.107	12.91	5.00	**13.55**	0.09	-	**1.28**	-	**14.83**	nr	**16.68**
extra; 45 degree bends, ref. 4D.163	11.00	5.00	**11.55**	0.15	0.02	**2.35**	-	**13.89**	nr	**15.63**
extra; branches, ref. 4D.210	16.20	5.00	**17.01**	0.15	0.02	**2.35**	-	**19.36**	nr	**21.78**
extra; connections cast iron and clay spigot, ref. 4D.128	12.91	5.00	**13.55**	0.09	-	**1.28**	-	**14.83**	nr	**16.68**
160mm, ref. 6D.076	14.13	5.00	**14.84**	0.22	0.03	**3.45**	-	**18.28**	m	**20.57**
extra; connections to cast iron and clay sockets, ref. 6D.107	25.92	5.00	**27.22**	0.11	-	**1.56**	-	**28.78**	nr	**32.38**
extra; 45 degree bends, ref. 6D.163	25.94	5.00	**27.24**	0.22	0.03	**3.45**	-	**30.69**	nr	**34.52**
extra; branches, ref. 6D.210	46.31	5.00	**48.63**	0.22	0.03	**3.45**	-	**52.08**	nr	**58.59**
extra; connections to cast iron and clay spigot, ref. 6D.128	25.92	5.00	**27.22**	0.11	-	**1.56**	-	**28.78**	nr	**32.38**
Pipework in trenches; vertical										
110mm, ref. 4D.076	6.42	5.00	**6.74**	0.31	0.02	**4.62**	-	**11.36**	m	**12.78**
160mm, ref. 6D.076	14.13	5.00	**14.84**	0.26	0.03	**4.02**	-	**18.85**	m	**21.21**
Drains; PVC-U solid wall concentric external rib-reinforced Wavin 'Ultra-Rib' pipes and fittings with sealing rings to joints										
Pipework in trenches										
150mm 6UR 046	8.21	5.00	**8.62**	0.25	0.02	**3.77**	-	**12.39**	m	**13.94**
extra; connectors 6UR 205	8.60	5.00	**9.03**	0.06	-	**0.85**	-	**9.89**	nr	**11.12**
extra; bends 6UR 563	11.58	5.00	**12.16**	0.25	-	**3.55**	-	**15.71**	nr	**17.68**
extra; branches 6UR 213	28.00	5.00	**29.40**	0.25	-	**3.55**	-	**32.95**	nr	**37.07**
extra; connections to clayware pipe ends 6UR 129	21.09	5.00	**22.14**	0.10	-	**1.42**	-	**23.57**	nr	**26.51**
225mm 9UR 046	18.16	5.00	**19.07**	0.35	0.03	**5.29**	-	**24.37**	m	**27.41**
extra; connectors 9UR 205	18.00	5.00	**18.90**	0.10	-	**1.42**	-	**20.32**	nr	**22.86**
extra; bends 9UR 563	46.60	5.00	**48.93**	0.35	-	**4.97**	-	**53.90**	nr	**60.64**
extra; branches 9UR 213	88.48	5.00	**92.90**	0.35	-	**4.97**	-	**97.87**	nr	**110.11**
extra; connections to clayware pipe ends 9UR 109	49.52	5.00	**51.99**	0.12	-	**1.71**	-	**53.70**	nr	**60.41**
300mm 12UR 043	30.47	5.00	**31.99**	0.45	0.05	**6.93**	-	**38.92**	m	**43.79**
extra; connectors 12UR 205	36.32	5.00	**38.14**	0.12	-	**1.71**	-	**39.85**	nr	**44.83**
extra; bends 12UR 563	80.57	5.00	**84.60**	0.45	-	**6.39**	-	**91.00**	nr	**102.37**
extra; branches 12UR 213	188.71	5.00	**198.15**	0.45	-	**6.39**	-	**204.54**	nr	**230.11**
extra; connections to clayware pipe ends 12UR 112	130.25	5.00	**136.77**	0.15	-	**2.13**	-	**138.90**	nr	**156.26**
Pipework in trenches; vertical										
150mm	8.21	5.00	**8.62**	0.50	0.02	**7.32**	-	**15.94**	m	**17.93**
225mm	18.16	5.00	**19.07**	0.45	0.03	**6.72**	-	**25.79**	m	**29.01**
300mm	30.47	5.00	**31.99**	0.55	0.05	**8.35**	-	**40.34**	m	**45.39**
R12: DRAINAGE BELOW GROUND: CONCRETE DRAINS										
Drains; unreinforced concrete pipes and fittings BS 5911, Class H, flexible joints										
Pipework in trenches										
300mm	12.65	5.00	**13.28**	0.71	0.29	**13.20**	-	**33.98**	m	**38.22**
extra; bends	126.50	5.00	**132.83**	0.36	0.14	**6.62**	-	**143.06**	nr	**160.94**
extra; 100mm branches	74.25	5.00	**77.96**	0.29	0.07	**4.87**	-	**84.64**	nr	**95.22**
extra; 150mm branches	74.25	5.00	**77.96**	0.48	0.13	**8.21**	-	**89.54**	nr	**100.73**
extra; 225mm branches	151.25	5.00	**158.81**	0.67	0.19	**11.56**	-	**175.28**	nr	**197.19**
extra; 300mm branches	173.25	5.00	**181.91**	0.86	0.26	**15.01**	-	**203.64**	nr	**229.10**
375mm	15.95	5.00	**16.75**	0.86	0.36	**16.08**	-	**42.13**	m	**47.40**
extra; bends	159.50	5.00	**167.48**	0.43	0.18	**8.04**	-	**180.17**	nr	**202.69**
extra; 100mm branches	82.50	5.00	**86.63**	0.29	0.07	**4.87**	-	**93.31**	nr	**104.97**
extra; 150mm branches	82.50	5.00	**86.63**	0.48	0.13	**8.21**	-	**98.20**	nr	**110.47**
extra; 225mm branches	159.50	5.00	**167.48**	0.67	0.19	**11.56**	-	**183.94**	nr	**206.94**
extra; 300mm branches	181.50	5.00	**190.58**	0.86	0.26	**15.01**	-	**212.30**	nr	**238.84**
extra; 375mm branches	203.50	5.00	**213.68**	1.06	0.32	**18.49**	-	**240.44**	nr	**270.49**
450mm	18.43	5.00	**19.35**	1.02	0.44	**19.21**	-	**49.93**	m	**56.17**
extra; bends	184.25	5.00	**193.46**	0.52	0.22	**9.75**	-	**208.90**	nr	**235.01**
extra; 100mm branches	93.50	5.00	**98.18**	0.29	0.07	**4.87**	-	**104.86**	nr	**117.96**
extra; 150mm branches	93.50	5.00	**98.18**	0.48	0.13	**8.21**	-	**109.75**	nr	**123.47**
extra; 225mm branches	170.50	5.00	**179.03**	0.67	0.19	**11.56**	-	**195.49**	nr	**219.93**
extra; 300mm branches	192.50	5.00	**202.13**	0.86	0.26	**15.01**	-	**223.85**	nr	**251.83**
extra; 375mm branches	214.50	5.00	**225.23**	1.06	0.32	**18.49**	-	**251.99**	nr	**283.49**
extra; 450mm branches	253.00	5.00	**265.65**	1.25	0.40	**22.05**	-	**298.04**	nr	**335.29**
525mm	22.28	5.00	**23.39**	1.18	0.52	**22.34**	-	**59.17**	m	**66.57**
extra; bends	222.75	5.00	**233.89**	0.59	0.26	**11.17**	-	**251.78**	nr	**283.25**
extra; 100mm branches	99.00	5.00	**103.95**	0.29	0.07	**4.87**	-	**110.63**	nr	**124.46**
extra; 150mm branches	99.00	5.00	**103.95**	0.48	0.13	**8.21**	-	**115.52**	nr	**129.97**
extra; 225mm branches	176.00	5.00	**184.80**	0.67	0.19	**11.56**	-	**201.27**	nr	**226.43**
extra; 300mm branches	198.00	5.00	**207.90**	0.86	0.26	**15.01**	-	**229.63**	nr	**258.33**
extra; 375mm branches	220.00	5.00	**231.00**	1.06	0.32	**18.49**	-	**257.76**	nr	**289.98**
extra; 450mm branches	258.50	5.00	**271.43**	1.25	0.40	**22.05**	-	**303.81**	nr	**341.79**

Labour hourly rates: (except Specialists) Craft Operatives £14.21 Labourer £10.71 Rates are national average prices. Refer to REGIONAL VARIATIONS for indicative levels of overall pricing in regions	MATERIALS			LABOUR				RATES		
	Del to Site £	Waste %	Material Cost £	Craft Optve Hrs	Lab Hrs	Labour Cost £	Sunds £	Nett Rate £	Unit	Gross Rate (+12.5%) £
R12: DRAINAGE BELOW GROUND: CONCRETE DRAINS Cont.										
Drains; unreinforced concrete pipes and fittings BS 5911, Class H, flexible joints Cont.										
Pipework in trenches Cont.										
600mm	27.23	5.00	**28.59**	1.32	0.60	**25.18**	-	**69.28**	m	**77.94**
extra; bends	272.25	5.00	**285.86**	0.66	0.30	**12.59**	-	**306.21**	nr	**344.49**
extra; 100mm branches	107.80	5.00	**113.19**	0.29	0.07	**4.87**	-	**119.87**	nr	**134.85**
extra; 150mm branches	107.80	5.00	**113.19**	0.48	0.13	**8.21**	-	**124.76**	nr	**140.36**
extra; 225mm branches	184.80	5.00	**194.04**	0.67	0.19	**11.56**	-	**210.51**	nr	**236.82**
extra; 300mm branches	206.80	5.00	**217.14**	0.86	0.26	**15.01**	-	**238.87**	nr	**268.73**
extra; 375mm branches	228.80	5.00	**240.24**	1.06	0.32	**18.49**	-	**267.00**	nr	**300.38**
extra; 450mm branches	267.30	5.00	**280.67**	1.25	0.40	**22.05**	-	**313.05**	nr	**352.19**
Drains; reinforced concrete pipes and fittings, BS 5911, Class H, flexible joints										
Pipework in trenches										
450mm	35.60	5.00	**37.38**	1.02	0.44	**19.21**	-	**67.96**	m	**76.46**
extra; bends	356.00	5.00	**373.80**	0.52	0.22	**9.75**	-	**389.23**	nr	**437.89**
extra; 100mm branches	99.32	5.00	**104.29**	0.29	0.07	**4.87**	-	**110.97**	nr	**124.84**
extra; 150mm branches	101.60	5.00	**106.68**	0.48	0.13	**8.21**	-	**118.25**	nr	**133.04**
extra; 225mm branches	105.01	5.00	**110.26**	0.67	0.19	**11.56**	-	**126.73**	nr	**142.57**
extra; 300mm branches	127.74	5.00	**134.13**	0.80	0.26	**14.15**	-	**155.00**	nr	**174.38**
extra; 375mm branches	144.79	5.00	**152.03**	1.06	0.32	**18.49**	-	**178.79**	nr	**201.14**
extra; 450mm branches	167.52	5.00	**175.90**	1.25	0.40	**22.05**	-	**208.29**	nr	**234.32**
525mm	43.51	5.00	**45.69**	1.18	0.52	**22.34**	-	**81.47**	m	**91.65**
extra; bends	435.12	5.00	**456.88**	0.59	0.26	**11.17**	-	**474.77**	nr	**534.11**
extra; 100mm branches	111.94	5.00	**117.54**	0.29	0.07	**4.87**	-	**124.22**	nr	**139.74**
extra; 150mm branches	114.21	5.00	**119.92**	0.48	0.13	**8.21**	-	**131.50**	nr	**147.93**
extra; 225mm branches	117.62	5.00	**123.50**	0.67	0.19	**11.56**	-	**139.97**	nr	**157.47**
extra; 300mm branches	134.67	5.00	**141.40**	0.86	0.26	**15.01**	-	**163.13**	nr	**183.52**
extra; 375mm branches	157.41	5.00	**165.28**	1.06	0.32	**18.49**	-	**192.04**	nr	**216.05**
extra; 450mm branches	180.14	5.00	**189.15**	1.25	0.40	**22.05**	-	**221.54**	nr	**249.23**
600mm	47.47	5.00	**49.84**	1.32	0.60	**25.18**	-	**90.54**	m	**101.86**
extra; bends	474.67	5.00	**498.40**	0.66	0.30	**12.59**	-	**518.75**	nr	**583.60**
extra; 100mm branches	101.48	5.00	**106.55**	0.29	0.07	**4.87**	-	**113.23**	nr	**127.39**
extra; 150mm branches	126.49	5.00	**132.81**	0.48	0.13	**8.21**	-	**144.39**	nr	**162.44**
extra; 225mm branches	129.90	5.00	**136.40**	0.67	0.19	**11.56**	-	**152.86**	nr	**171.97**
extra; 300mm branches	146.95	5.00	**154.30**	0.86	0.26	**15.01**	-	**176.03**	nr	**198.03**
extra; 375mm branches	169.68	5.00	**178.16**	1.06	0.32	**18.49**	-	**204.93**	nr	**230.54**
extra; 450mm branches	192.41	5.00	**202.03**	1.25	0.40	**22.05**	-	**234.42**	nr	**263.72**

This section continues on the next page

Labour hourly rates: (except Specialists) Craft Operatives £17.38 Labourer £10.71 Rates are national average prices. Refer to REGIONAL VARIATIONS for indicative levels of overall pricing in regions	MATERIALS			LABOUR				RATES		
	Del to Site £	Waste %	Material Cost £	Craft Optve Hrs	Lab Hrs	Labour Cost £	Sunds £	Nett Rate £	Unit	Gross Rate (+12.5%) £
R12: DRAINAGE BELOW GROUND: CAST IRON DRAINS										
Drains; cast iron pipes and cast iron fittings BS 437, coated; bolted cast iron coupling and synthetic rubber gasket joints										
Pipework in trenches										
100mm	28.01	5.00	**29.26**	1.20	-	**20.86**	-	**50.12**	m	**56.39**
extra; connectors, large socket for clayware	32.40	2.00	**33.05**	1.20	-	**20.86**	-	**53.90**	nr	**60.64**
extra; bends	57.36	2.00	**58.51**	1.20	-	**20.86**	-	**79.36**	nr	**89.28**
extra; branches	81.23	2.00	**82.85**	1.20	-	**20.86**	-	**103.71**	nr	**116.67**
extra; branches with access	129.28	2.00	**131.87**	1.20	-	**20.86**	-	**152.72**	nr	**171.81**
150mm	48.72	5.00	**50.98**	1.80	-	**31.28**	-	**82.27**	m	**92.55**
extra; connectors, large socket for clayware	48.37	2.00	**49.34**	1.80	-	**31.28**	-	**80.62**	nr	**90.70**
extra; diminishing pieces, reducing to 100mm	68.90	2.00	**70.28**	1.80	-	**31.28**	-	**101.56**	nr	**114.26**
extra; bends	99.67	2.00	**101.66**	1.80	-	**31.28**	-	**132.95**	nr	**149.57**
extra; branches	144.59	2.00	**147.48**	2.70	-	**46.93**	-	**194.41**	nr	**218.71**
extra; branches with access	210.43	2.00	**214.64**	2.70	-	**46.93**	-	**261.56**	nr	**294.26**
in runs not exceeding 3m										
100mm	37.94	5.00	**39.39**	1.80	-	**31.28**	-	**70.68**	m	**79.51**
150mm	60.75	5.00	**63.25**	2.70	-	**46.93**	-	**110.18**	m	**123.95**
Pipework in trenches; vertical										
100mm	37.94	5.00	**39.39**	1.44	-	**25.03**	-	**64.42**	m	**72.47**
150mm	60.75	5.00	**63.25**	2.16	-	**37.54**	-	**100.79**	m	**113.39**
R12: DRAINAGE BELOW GROUND: STAINLESS STEEL DRAINS										
Drains; stainless steel pipes and fittings, B.M. Stainless Steel Drains Ltd; ring seal push fit joints (AIS 1316)										
Pipework in trenches										
75mm	28.87	5.00	**30.31**	0.66	-	**11.47**	-	**41.78**	m	**47.01**
extra; bends	35.50	2.00	**36.21**	0.66	-	**11.47**	-	**47.68**	nr	**53.64**
extra; branches	41.25	2.00	**42.08**	0.99	-	**17.21**	-	**59.28**	nr	**66.69**
110mm	36.63	5.00	**38.46**	0.83	-	**14.43**	-	**52.89**	m	**59.50**
extra; diminishing pieces, reducing to 75mm	62.35	2.00	**63.60**	0.83	-	**14.43**	-	**78.02**	nr	**87.78**
extra; bends	45.32	2.00	**46.23**	0.83	-	**14.43**	-	**60.65**	nr	**68.23**
extra; branches	47.63	2.00	**48.58**	1.23	-	**21.38**	-	**69.96**	nr	**78.71**
in runs not exceeding 3m										
75mm	28.87	5.00	**30.31**	0.99	-	**17.21**	-	**47.52**	m	**53.46**
110mm	36.63	5.00	**38.46**	1.23	-	**21.38**	-	**59.84**	m	**67.32**
Pipework in trenches; vertical										
75mm	28.87	5.00	**30.31**	0.83	-	**14.43**	-	**44.74**	m	**50.33**
110mm	36.63	5.00	**38.46**	1.03	-	**17.90**	-	**56.36**	m	**63.41**
This section continues on the next page										

Labour hourly rates: (except Specialists) Craft Operatives £14.21 Labourer £10.71 Rates are national average prices. Refer to REGIONAL VARIATIONS for indicative levels of overall pricing in regions	MATERIALS			LABOUR				RATES		
	Del to Site £	Waste %	Material Cost £	Craft Optve Hrs	Lab Hrs	Labour Cost £	Sunds £	Nett Rate £	Unit	Gross Rate (+12.5%) £
R12: DRAINAGE BELOW GROUND: VITRIFIED CLAY GULLIES										
Vitrified clay accessories										
Gullies; Supersleve; joint to pipe; bedding and surrounding in concrete to BS 5328, ordinary prescribed mix ST3, 20mm aggregate; Hepworth Building Products										
100mm outlet, reference RG1/1, trapped, round; 255mm diameter top	44.21	5.00	**46.43**	0.66	0.09	**10.34**	-	**56.77**	nr	**63.86**
100mm outlet, reference SG2/1, trapped, square; 150 x 150 top	52.82	5.00	**55.46**	0.66	0.09	**10.34**	-	**65.80**	nr	**74.03**
100mm outlet, reference SG1/1, trapped, reversible, round; 100mm trap; hopper reference SH1, 100mm outlet, 150 x 150mm rebated top	80.42	5.00	**84.44**	1.10	0.11	**16.81**	-	**101.25**	nr	**113.91**
100mm outlet, reference SG1/1, trapped, reversible, round; 100mm trap; hopper reference SH2, 100mm outlet, 100mm horizontal inlet, 150 x 150mm rebated top	80.57	5.00	**84.59**	1.27	0.11	**19.22**	-	**103.82**	nr	**116.80**
100mm outlet, reference SG1/1, trapped, reversible, round; 100mm trap; raising pieces - 1, reference RRP2/2, 100mm diameter, 225mm high; hopper reference SH1, 100mm outlet, 150 x 150mm rebated top	129.58	5.00	**136.06**	1.54	0.17	**23.70**	-	**159.76**	nr	**179.73**
100mm outlet, reference SG1/1, trapped, reversible, round; 100mm trap; raising pieces - 1, reference RRP2/2, 100mm diameter, 225mm high; hopper reference SH2, 100mm outlet, 100mm horizontal inlet, 150 x 150mm rebated top	129.72	5.00	**136.21**	1.60	0.18	**24.66**	-	**160.87**	nr	**180.98**
100mm outlet, reference RGP5, trapped, round; 225mm diameter x 600mm deep; perforated galvanized steel bucket reference IBP3	154.08	5.00	**161.79**	2.75	0.28	**42.08**	-	**203.87**	nr	**229.35**
100mm outlet, reference RGR1, trapped, round; 300mm diameter x 600mm deep	110.97	5.00	**116.52**	2.75	0.36	**42.93**	-	**159.46**	nr	**179.39**
150mm outlet, reference RGR2, trapped, round; 300mm diameter x 600mm deep	112.89	5.00	**118.54**	2.75	0.36	**42.93**	-	**161.47**	nr	**181.65**
150mm outlet, reference RGR3, trapped, round; 400mm diameter x 750mm deep	125.96	5.00	**132.26**	4.40	0.50	**67.88**	-	**200.14**	nr	**225.15**
150mm outlet, reference RGR4, trapped, round; 450mm diameter x 900mm deep	159.45	5.00	**167.42**	6.05	0.55	**91.86**	-	**259.28**	nr	**291.69**
100mm outlet, reference RGU1, trapped; internal size 600 x 450 x 600mm deep; perforated tray and galvanized cover and frame	747.66	5.00	**785.04**	11.00	1.65	**173.98**	-	**959.03**	nr	**1078.90**
Rainwater shoes; Supersleve; joint to pipe; bedding and surrounding in concrete to BS 5328, ordinary prescribed mix ST3, 20mm aggregate										
100mm outlet, reference RRW/S3/1 trapless, round; 100mm vertical inlet, 250 x 150mm rectangular access opening	46.00	5.00	**48.30**	0.55	0.06	**8.46**	-	**56.75**	nr	**63.85**
R12: DRAINAGE BELOW GROUND: CONCRETE GULLIES										
Unreinforced concrete accessories										
Gullies; BS 5911; joint to pipe; bedding and surrounding in concrete to BS 5328 ordinary prescribed mix ST3, 20mm aggregate										
150mm outlet, trapped, round; 375mm diameter x 750mm deep internally, stopper	46.38	5.00	**48.69**	3.63	0.55	**57.47**	-	**106.17**	nr	**119.44**
150mm outlet, trapped, round; 375mm diameter x 900mm deep internally, stopper	57.50	5.00	**60.37**	3.85	0.55	**60.60**	-	**120.97**	nr	**136.10**
150mm outlet, trapped, round; 450mm diameter x 750mm deep internally, stopper	47.48	5.00	**49.85**	3.85	0.66	**61.78**	-	**111.63**	nr	**125.58**
150mm outlet, trapped, round; 450mm diameter x 900mm deep internally, stopper	58.60	5.00	**61.53**	4.07	0.66	**64.90**	-	**126.43**	nr	**142.24**
150mm outlet, trapped, round; 450mm diameter x 1050mm deep internally, stopper	65.92	5.00	**69.22**	4.40	0.66	**69.59**	-	**138.81**	nr	**156.16**
150mm outlet, trapped, round; 450mm diameter x 1200mm deep internally, stopper	80.97	5.00	**85.02**	4.95	0.66	**77.41**	-	**162.43**	nr	**182.73**
R12: DRAINAGE BELOW GROUND: PVC-U GULLIES										
PVC-U accessories										
Gullies; Osma; joint to pipe; bedding and surrounding in concrete to BS 5328, ordinary prescribed mix ST3, 20mm aggregate										
110mm outlet trap, base reference 4D.500, outlet bend with access reference 4D.569, inlet raising piece 300mm long, plain hopper reference 4D.503	62.99	5.00	**66.14**	1.20	0.20	**19.19**	-	**85.34**	nr	**96.00**
110mm outlet bottle gulley, reference 4D.900 trapped, round; 200mm diameter top with grating reference 4D.919	53.15	5.00	**55.81**	1.80	0.25	**28.26**	-	**84.06**	nr	**94.57**
Sealed rodding eyes; Osma; joint to pipe; surrounding in concrete to BS 5328; ordinary prescribed mix ST3, 20mm aggregate										
110mm outlet, reference 4D.316, with cover	50.14	5.00	**52.65**	2.25	0.45	**36.79**	-	**89.44**	nr	**100.62**

Labour hourly rates: (except Specialists) Craft Operatives £14.21 Labourer £10.71 Rates are national average prices. Refer to REGIONAL VARIATIONS for indicative levels of overall pricing in regions	MATERIALS			LABOUR				RATES		
	Del to Site £	Waste %	Material Cost £	Craft Optve Hrs	Lab Hrs	Labour Cost £	Sunds £	Nett Rate £	Unit	Gross Rate (+12.5%) £
R12: DRAINAGE BELOW GROUND: GULLY GRATING AND SEALING PLATES										
Cast iron accessories, painted black										
Gratings; Hepworth Building Products										
reference IG6C; 140mm diameter..................	3.27	-	**3.27**	-	0.03	**0.32**	-	**3.59**	nr	**4.04**
reference IG7C; 197mm diameter..................	4.99	-	**4.99**	-	0.03	**0.32**	-	**5.31**	nr	**5.97**
reference IG8C; 284mm diameter..................	10.88	-	**10.88**	-	0.03	**0.32**	-	**11.20**	nr	**12.60**
reference IG2C; 150 x 150mm.....................	3.27	-	**3.27**	-	0.03	**0.32**	-	**3.59**	nr	**4.04**
reference IG3C; 225 x 225mm.....................	9.72	-	**9.72**	-	0.03	**0.32**	-	**10.04**	nr	**11.30**
reference IG4C; 300 x 300mm.....................	21.90	-	**21.90**	-	0.03	**0.32**	-	**22.22**	nr	**25.00**
Gratings and frames; bedding frames in cement mortar (1:3); Hepworth Building Products										
reference IH5C; 100mm diameter..................	11.88	15.00	**13.66**	1.10	-	**15.63**	-	**29.29**	nr	**32.95**
reference IH6C; 150mm diameter..................	20.42	15.00	**23.48**	1.38	-	**19.61**	-	**43.09**	nr	**48.48**
reference IH7C; 225mm diameter..................	40.66	15.00	**46.76**	1.87	-	**26.57**	-	**73.33**	nr	**82.50**
reference IH2C; 150 x 150mm with lock and key....	13.18	15.00	**15.16**	1.38	-	**19.61**	-	**34.77**	nr	**39.12**
reference IH3C; 230 x 230mm with lock and key....	23.99	15.00	**27.59**	1.93	-	**27.43**	-	**55.02**	nr	**61.89**
reference IH4C; 316 x 316mm with lock and key....	63.39	15.00	**72.89**	2.09	-	**29.70**	-	**102.59**	nr	**115.42**
Sealing plates and frames; bedding frames in cement mortar (1:3); Hepworth Building Products										
reference IS5C; 140mm diameter..................	10.38	15.00	**11.93**	1.10	-	**15.63**	-	**27.56**	nr	**31.01**
reference IS6C; 197mm diameter..................	14.86	15.00	**17.09**	1.38	-	**19.61**	-	**36.70**	nr	**41.28**
reference IS7C; 273mm diameter..................	23.70	15.00	**27.25**	1.87	-	**26.57**	-	**53.82**	nr	**60.55**
reference IS2C; 150 x 150mm.....................	12.76	15.00	**14.68**	1.38	-	**19.61**	-	**34.28**	nr	**38.57**
reference IS3C; 225 x 225mm.....................	23.11	15.00	**26.58**	1.93	-	**27.43**	-	**54.00**	nr	**60.75**
Alloy accessories										
Gratings; Hepworth Building Products										
reference IG6; 140mm diameter...................	3.27	15.00	**3.76**	-	0.03	**0.32**	-	**4.08**	nr	**4.59**
reference IG7; 197mm diameter...................	4.99	15.00	**5.73**	-	0.03	**0.32**	-	**6.06**	nr	**6.81**
reference IG8; 284mm diameter...................	10.87	15.00	**12.50**	-	0.03	**0.32**	-	**12.82**	nr	**14.43**
reference IG2; 150 x 150mm......................	3.27	15.00	**3.76**	-	0.03	**0.32**	-	**4.08**	nr	**4.59**
reference IG3; 225 x 225mm......................	9.72	15.00	**11.18**	-	0.03	**0.32**	-	**11.50**	nr	**12.94**
reference IG4; 300 x 300mm......................	21.90	15.00	**25.18**	-	0.03	**0.32**	-	**25.50**	nr	**28.69**
Gratings and frames; bedding frames in cement mortar (1:3); Hepworth Building Products										
reference IH6; 193mm diameter...................	20.42	15.00	**23.48**	1.38	-	**19.61**	-	**43.09**	nr	**48.48**
reference IH2; 150 x 150mm with lock and key.....	13.18	15.00	**15.16**	1.38	-	**19.61**	-	**34.77**	nr	**39.12**
reference IH3; 225 x 225mm with lock and key.....	23.99	15.00	**27.59**	1.93	-	**27.43**	-	**55.02**	nr	**61.89**
Sealing plates and frames; bedding frames in cement mortar (1:3); Hepworth Building Products										
reference IS7; 273mm diameter...................	23.70	15.00	**27.25**	1.87	-	**26.57**	-	**53.82**	nr	**60.55**
reference IS2; 150 x 150mm......................	12.70	15.00	**14.60**	1.38	-	**19.61**	-	**34.21**	nr	**38.49**
Cast iron accessories, painted black										
Gratings and frames; Stanton PLC.; bedding frames in cement mortar (1:3)										
reference HY811 Watergate; 370 x 305mm...........	56.93	15.00	**65.47**	1.10	-	**15.63**	-	**81.10**	nr	**91.24**
reference HY812; 370 x 430mm.....................	56.93	15.00	**65.47**	2.75	-	**39.08**	-	**104.55**	nr	**117.62**
reference HY815; 510 x 360mm.....................	56.93	15.00	**65.47**	3.65	-	**51.87**	-	**117.34**	nr	**132.01**
reference HP894 Waterflow; 325 x 312mm...........	44.87	15.00	**51.60**	4.70	-	**66.79**	-	**118.39**	nr	**133.19**
reference HP895; 325 x 437mm.....................	51.63	15.00	**59.38**	3.05	-	**43.34**	-	**102.72**	nr	**115.56**
R12: DRAINAGE BELOW GROUND: SURFACE DRAINAGE SYSTEMS										
Precast polyester concrete ACO Drain surface drainage systems; Technologies Plc; bedding and haunching in concrete to BS 5328 ordinary prescribed mix ST5, 20mm aggregate										
N100K Channel system with resin composite 'Heelguard' non rusting lockable grating										
constant depth or complete with 0.6% fall........	25.74	5.00	**27.03**	1.65	2.15	**46.47**	-	**73.50**	m	**82.69**
extra; end cap	-	-	**-**	-	0.22	**2.36**	-	**2.36**	nr	**2.65**
ParkDrain Class one piece channel system with integral grating										
constant depth or complete with 0.6% fall........	25.74	5.00	**27.03**	1.87	2.31	**51.31**	-	**78.34**	m	**88.14**
extra; end cap	5.59	5.00	**5.87**	-	0.22	**2.36**	-	**8.23**	nr	**9.25**
Raindrain Class F900 one piece channel system with integral grating										
invert depth 115mm	39.87	5.00	**41.86**	1.54	1.85	**41.70**	-	**83.56**	m	**94.00**
extra; end cap	7.40	5.00	**7.77**	-	0.22	**2.36**	-	**10.13**	nr	**11.40**
extra; rodding access unit.......................	44.05	5.00	**46.25**	-	0.60	**6.43**	-	**52.68**	nr	**59.27**
extra; universal Gully assembly complete with bucket	192.06	5.00	**201.67**	1.10	1.10	**27.41**	-	**229.08**	nr	**257.71**
KerbDrain Class D400 combined kerb drainage system										
305mm x 150mm x 500mm	34.93	5.00	**36.67**	1.34	1.65	**36.71**	-	**73.38**	nr	**82.56**
extra; rodding access	68.60	5.00	**72.03**	-	0.60	**6.43**	-	**78.45**	nr	**88.26**
extra; Class D400 Gully..........................	184.79	5.00	**194.03**	1.10	1.10	**27.41**	-	**221.44**	nr	**249.12**

Labour hourly rates: (except Specialists) Craft Operatives £14.21 Labourer £10.71 Rates are national average prices. Refer to REGIONAL VARIATIONS for indicative levels of overall pricing in regions	PLANT & TRANSPORT			LABOUR				RATES		
	Plant Cost £	Trans Cost £	P & T Cost £	Craft Optve Hrs	Lab Hrs	Labour Cost £	Sunds £	Nett Rate £	Unit	Gross Rate (+12.5%) £
R12: DRAINAGE BELOW GROUND; MANHOLES AND SOAKAWAYS:										
Excavating - by machine										
Excavating pits commencing from natural ground level; maximum depth not exceeding										
0.25m	3.88	-	**3.88**	-	0.95	**10.17**	-	**14.05**	m3	**15.81**
1.00m	5.17	-	**5.17**	-	1.00	**10.71**	-	**15.88**	m3	**17.87**
2.00m	5.69	-	**5.69**	-	1.15	**12.32**	-	**18.01**	m3	**20.26**
4.00m	6.21	-	**6.21**	-	1.45	**15.53**	-	**21.74**	m3	**24.45**
Working space allowance to excavation pits	5.17	-	**5.17**	-	0.70	**7.50**	-	**12.67**	m2	**14.25**
Compacting bottoms of excavations	1.05	-	**1.05**	-	-	-	-	**1.05**	m2	**1.18**
Excavated material depositing on site in temporary spoil heaps where directed										
25m	2.07	-	**2.07**	-	-	-	-	**2.07**	m3	**2.33**
50m	2.59	-	**2.59**	-	-	-	-	**2.59**	m3	**2.91**
100m	3.10	-	**3.10**	-	-	-	-	**3.10**	m3	**3.49**
Excavating - by hand										
Excavating pits commencing from natural ground level; maximum depth not exceeding										
0.25m	-	-	-	-	3.70	**39.63**	-	**39.63**	m3	**44.58**
1.00m	-	-	-	-	3.85	**41.23**	-	**41.23**	m3	**46.39**
2.00m	-	-	-	-	4.40	**47.12**	-	**47.12**	m3	**53.01**
4.00m	-	-	-	-	5.60	**59.98**	-	**59.98**	m3	**67.47**
Working space allowance to excavation pits	-	-	-	-	3.45	**36.95**	-	**36.95**	m2	**41.57**
Compacting bottoms of excavations	-	-	-	-	0.13	**1.39**	-	**1.39**	m2	**1.57**
Excavated material depositing on site in temporary spoil heaps where directed										
25m	-	-	-	-	1.40	**14.99**	-	**14.99**	m3	**16.87**
50m	-	-	-	-	1.65	**17.67**	-	**17.67**	m3	**19.88**
100m	-	-	-	-	2.20	**23.56**	-	**23.56**	m3	**26.51**

This section continues on the next page

Labour hourly rates: (except Specialists) Craft Operatives £14.21 Labourer £10.71 Rates are national average prices. Refer to REGIONAL VARIATIONS for indicative levels of overall pricing in regions	MATERIALS			LABOUR				RATES		
	Del to Site £	Waste %	Material Cost £	Craft Optve Hrs	Lab Hrs	Labour Cost £	Sunds £	Nett Rate £	Unit	Gross Rate (+12.5%) £
R12: DRAINAGE BELOW GROUND; MANHOLES AND SOAKAWAYS: Cont.										
Earthwork support										
Earthwork support distance between opposing faces not exceeding 2.00m; maximum depth not exceeding										
1.00m	0.88	-	**0.88**	-	0.21	**2.25**	-	**3.13**	m2	**3.52**
2.00m	0.88	-	**0.88**	-	0.26	**2.78**	-	**3.66**	m2	**4.12**
4.00m	0.88	-	**0.88**	-	0.32	**3.43**	-	**4.31**	m2	**4.84**
distance between opposing faces 2.00 - 4.00m; maximum not exceeding										
1.00m	1.76	-	**1.76**	-	0.22	**2.36**	-	**4.11**	m2	**4.63**
2.00m	1.76	-	**1.76**	-	0.27	**2.89**	-	**4.65**	m2	**5.23**
4.00m	1.76	-	**1.76**	-	0.34	**3.64**	-	**5.40**	m2	**6.07**
Earthwork support; in unstable ground distance between opposing faces not exceeding 2.00m; maximum depth not exceeding										
1.00m	3.22	-	**3.22**	-	0.35	**3.75**	-	**6.97**	m2	**7.84**
2.00m	3.22	-	**3.22**	-	0.44	**4.71**	-	**7.93**	m2	**8.93**
4.00m	3.22	-	**3.22**	-	0.53	**5.68**	-	**8.90**	m2	**10.01**
distance between opposing faces 2.00 - 4.00m; maximum depth not exceeding										
1.00m	3.81	-	**3.81**	-	0.37	**3.96**	-	**7.77**	m2	**8.74**
2.00m	3.81	-	**3.81**	-	0.45	**4.82**	-	**8.63**	m2	**9.71**
4.00m	3.81	-	**3.81**	-	0.56	**6.00**	-	**9.81**	m2	**11.03**
R12: DRAINAGE BELOW GROUND; MANHOLES AND SOAKAWAYS: FILLING										
Excavated material arising from excavations										
Filling to excavation average thickness exceeding 0.25m	-	-	-	-	1.44	**15.42**	-	**15.42**	m3	**17.35**
Excavated material obtained from on site spoil heaps 25m distant										
Filling to excavation										
average thickness not exceeding 0.25m	-	-	-	-	1.85	**19.81**	-	**19.81**	m3	**22.29**
average thickness exceeding 0.25m	-	-	-	-	1.60	**17.14**	-	**17.14**	m3	**19.28**
Excavated material obtained from on site spoil heaps 50m distant										
Filling to excavation										
average thickness not exceeding 0.25m	-	-	-	-	2.10	**22.49**	-	**22.49**	m3	**25.30**
average thickness exceeding 0.25m	-	-	-	-	1.95	**20.88**	-	**20.88**	m3	**23.50**
Excavated material obtained from on site spoil heaps 100m distant										
Filling to excavation										
average thickness not exceeding 0.25m	-	-	-	-	4.40	**47.12**	-	**47.12**	m3	**53.01**
average thickness exceeding 0.25m	-	-	-	-	4.00	**42.84**	-	**42.84**	m3	**48.20**
Hard, dry broken brick or stone, 100 - 75mm gauge, to be obtained off site										
Filling to excavation average thickness exceeding 0.25m	17.85	25.00	**22.31**	-	1.50	**16.07**	-	**38.38**	m3	**43.17**
R12: DRAINAGE BELOW GROUND; MANHOLES AND SOAKAWAYS: CONCRETE WORK										
Plain in-situ concrete; BS 5328, ordinary prescribed mix ST3, 20mm aggregate										
Beds; poured on or against earth or unblinded hardcore										
thickness 150 - 450mm	62.20	5.00	**65.31**	-	4.45	**47.66**	-	**112.97**	m3	**127.09**
thickness not exceeding 150mm	62.20	5.00	**65.31**	-	4.75	**50.87**	-	**116.18**	m3	**130.70**
Benching in bottoms; rendering with 13mm cement and sand (1:2) before final set										
600 x 450 x average 225mm; trowelling	4.21	5.00	**4.46**	-	0.62	**6.64**	-	**11.10**	nr	**12.48**
675 x 675 x average 225mm; trowelling	7.09	5.00	**7.51**	-	1.06	**11.35**	-	**18.87**	nr	**21.23**
800 x 675 x average 225mm; trowelling	8.41	5.00	**8.91**	-	1.27	**13.60**	-	**22.52**	nr	**25.33**
800 x 800 x average 300mm; trowelling	13.81	5.00	**14.64**	-	2.00	**21.42**	-	**36.06**	nr	**40.57**
1025 x 800 x average 300mm; trowelling	17.26	5.00	**18.30**	-	2.54	**27.20**	-	**45.50**	nr	**51.19**
Reinforced in-situ concrete; BS 5328, designed mix ST4, 20mm aggregate, minimum cement content 240 kg/m3; vibrated										
Slabs										
thickness 150 - 450mm	63.93	5.00	**67.13**	-	3.95	**42.30**	-	**109.43**	m3	**123.11**
thickness not exceeding 150mm	63.93	5.00	**67.13**	-	4.40	**47.12**	-	**114.25**	m3	**128.53**
Formwork and basic finish										
Edges of suspended slabs; plain vertical height not exceeding 250mm	1.19	12.50	**1.34**	0.80	0.18	**13.30**	-	**14.63**	m	**16.46**

Labour hourly rates: (except Specialists) Craft Operatives £14.21 Labourer £10.71 Rates are national average prices. Refer to REGIONAL VARIATIONS for indicative levels of overall pricing in regions	MATERIALS			LABOUR			RATES			
	Del to Site £	Waste %	Material Cost £	Craft Optve Hrs	Lab Hrs	Labour Cost £	Sunds £	Nett Rate £	Unit	Gross Rate (+12.5%) £
R12: DRAINAGE BELOW GROUND; MANHOLES AND SOAKAWAYS: CONCRETE WORK Cont.										
Formwork and basic finish Cont.										
Soffits of slabs; horizontal										
slab thickness not exceeding 200mm; height to soffit not exceeding 1.50m	16.00	10.00	**17.60**	2.42	0.46	**39.31**	-	**56.91**	m2	**64.03**
slab thickness not exceeding 200mm; height to soffit 1.50 - 3.00m	16.00	10.00	**17.60**	2.20	0.46	**36.19**	-	**53.79**	m2	**60.51**
Holes; rectangular										
girth 1.00 - 2.00m; depth not exceeding 250mm	2.38	12.50	**2.67**	1.33	0.26	**21.68**	-	**24.36**	nr	**27.40**
girth 2.00 - 3.00m; depth not exceeding 250mm	3.56	12.50	**4.01**	1.73	0.34	**28.22**	-	**32.23**	nr	**36.26**
Reinforcement fabric; BS 4483; hard drawn plain round steel, welded; including laps, tying wire, all cutting and bending and spacers and chairs which are at the discretion of the Contractor										
Reference A142, 2.22kg/m2; 200mm side laps; 200mm end laps										
generally	1.18	15.00	**1.35**	0.30	0.02	**4.48**	-	**5.83**	m2	**6.56**
R12: DRAINAGE BELOW GROUND; MANHOLES AND SOAKAWAYS: BRICKWORK										
Common bricks, BS EN 772, Category M, 215 x 102.5 x 65mm, compressive strength 20.5 N/mm2; in cement mortar (1:3)										
Walls; vertical										
102mm thick; stretcher bond	14.02	5.00	**15.02**	1.50	1.65	**38.99**	-	**54.01**	m2	**60.76**
215mm thick; English bond	27.86	5.00	**29.86**	2.60	2.90	**68.01**	-	**97.86**	m2	**110.10**
Milton Hall Second hard stock bricks, BS EN 772, Category M, 215 x 102.5 x 65mm; in cement mortar (1:3)										
Walls; vertical										
102mm thick; stretcher bond	25.00	5.00	**26.55**	1.50	1.65	**38.99**	-	**65.54**	m2	**73.73**
215mm thick; English bond	49.64	5.00	**52.73**	2.60	2.90	**68.01**	-	**120.73**	m2	**135.82**
Engineering bricks, Category F, BS EN 772, 215 x 102.5 x 65mm, class B; in cement mortar (1:3)										
Walls; vertical										
102mm thick; stretcher bond	15.24	5.00	**16.30**	1.50	1.65	**38.99**	-	**55.29**	m2	**62.20**
215mm thick; English bond	30.28	5.00	**32.40**	2.60	2.90	**68.01**	-	**100.40**	m2	**112.95**
Common bricks, BS EN 772, Category M, 215 x 102.5 x 65mm, compressive strength 20.5 N/mm2; in cement mortar (1:3); flush smooth pointing as work proceeds										
Walls; vertical										
102mm thick; stretcher bond; facework one side	14.02	5.00	**15.02**	1.73	1.90	**44.93**	-	**59.96**	m2	**67.45**
215mm thick; English bond; facework one side	27.86	5.00	**29.86**	2.99	3.34	**78.26**	-	**108.12**	m2	**121.63**
Milton Hall Second hard stock bricks, BS EN 772, Category M, 215 x 102.5 x 65mm; in cement mortar (1:3); flush smooth pointing as work proceeds										
Walls; vertical										
102mm thick; stretcher bond; facework one side	25.00	5.00	**26.55**	1.73	1.90	**44.93**	-	**71.48**	m2	**80.42**
215mm thick; English bond; facework one side	49.64	5.00	**52.73**	2.99	3.34	**78.26**	-	**130.99**	m2	**147.36**
Engineering bricks, BS EN 772, Category F, 215 x 102.5 x 65mm, class B; in cement mortar (1:3); flush smooth pointing as work proceeds										
Walls; vertical										
102mm thick; stretcher bond; facework one side	15.24	5.00	**16.30**	1.73	1.90	**44.93**	-	**61.24**	m2	**68.89**
215mm thick; English bond; facework one side	30.28	5.00	**32.40**	2.50	3.34	**71.30**	-	**103.69**	m2	**116.66**
Accessories/sundry items for brick/block/stone walling										
Building in ends of pipes; 100mm diameter making good fair face one side										
102mm brickwork	0.08	15.00	**0.09**	0.10	0.10	**2.49**	-	**2.58**	nr	**2.90**
215mm brickwork	0.15	15.00	**0.17**	0.15	0.15	**3.74**	-	**3.91**	nr	**4.40**
Building in ends of pipes; 150mm diameter making good fair face one side										
102mm brickwork	0.08	15.00	**0.09**	0.15	0.15	**3.74**	-	**3.83**	nr	**4.30**
215mm brickwork	0.15	15.00	**0.17**	0.24	0.24	**5.98**	-	**6.16**	nr	**6.92**
Building in ends of pipes; 225mm diameter making good fair face one side										
102mm brickwork	0.15	15.00	**0.17**	0.24	0.24	**5.98**	-	**6.16**	nr	**6.92**
215mm brickwork	0.30	15.00	**0.35**	0.30	0.30	**7.48**	-	**7.83**	nr	**8.80**
Building in ends of pipes; 300mm diameter making good fair face one side										
102mm brickwork	0.23	15.00	**0.26**	0.30	0.30	**7.48**	-	**7.74**	nr	**8.71**
215mm brickwork	0.46	15.00	**0.52**	0.40	0.40	**9.97**	-	**10.49**	nr	**11.80**

Labour hourly rates: (except Specialists)
Craft Operatives £14.21 Labourer £10.71
Rates are national average prices.
Refer to REGIONAL VARIATIONS for indicative levels of overall pricing in regions

	MATERIALS			LABOUR				RATES		
	Del to Site £	Waste %	Material Cost £	Craft Optve Hrs	Lab Hrs	Labour Cost £	Sunds £	Nett Rate £	Unit	Gross Rate (+12.5%) £
R12: DRAINAGE BELOW GROUND; MANHOLES AND SOAKAWAYS: BRICKWORK Cont.										
Mortar, cement and sand (1:3)										
13mm work to walls on brickwork base; one coat; trowelled										
width exceeding 300mm	0.99	15.00	**1.14**	0.45	0.45	**11.21**	-	**12.35**	m2	**13.89**
13mm work to floors on concrete base; one coat; trowelled										
level and to falls only not exceeding 15 degrees from horizontal	0.99	15.00	**1.14**	0.70	0.70	**17.44**	-	**18.58**	m2	**20.90**
R12: DRAINAGE BELOW GROUND; MANHOLES AND SOAKAWAYS: CHANNELS, VITRIFIED CLAY										
Channels in bottoms; vitrified clay, BS 65, normal, glazed; cement mortar (1:3) joints; bedding in cement mortar (1:3)										
Half section										
straight; 600mm effective length										
100mm	3.76	5.00	**3.97**	0.35	0.05	**5.51**	-	**9.48**	nr	**10.66**
150mm	6.23	5.00	**6.58**	0.45	0.07	**7.14**	-	**13.72**	nr	**15.44**
straight; 1000mm effective length										
100mm	5.45	5.00	**5.77**	0.60	0.06	**9.17**	-	**14.94**	nr	**16.80**
150mm	8.93	5.00	**9.44**	0.75	0.08	**11.51**	-	**20.95**	nr	**23.57**
225mm	19.46	5.00	**20.51**	1.00	0.12	**15.50**	-	**36.01**	nr	**40.51**
300mm	39.30	5.00	**41.35**	1.50	0.18	**23.24**	-	**64.60**	nr	**72.67**
curved; 500mm effective length										
100mm	5.61	5.00	**5.91**	0.34	0.05	**5.37**	-	**11.28**	nr	**12.69**
150mm	9.19	5.00	**9.68**	0.45	0.07	**7.14**	-	**16.82**	nr	**18.92**
225mm	30.08	5.00	**31.63**	0.60	0.10	**9.60**	-	**41.23**	nr	**46.38**
300mm	61.02	5.00	**64.14**	0.90	0.10	**13.86**	-	**78.00**	nr	**87.75**
curved; 900mm effective length										
100mm	5.84	5.00	**6.18**	0.70	0.06	**10.59**	-	**16.76**	nr	**18.86**
150mm	9.49	5.00	**10.03**	0.90	0.08	**13.65**	-	**23.67**	nr	**26.63**
225mm	30.39	5.00	**31.98**	1.20	0.12	**18.34**	-	**50.32**	nr	**56.61**
300mm	61.33	5.00	**64.49**	1.80	0.20	**27.72**	-	**92.21**	nr	**103.73**
Three quarter section										
branch bends										
100mm	15.56	5.00	**16.79**	0.40	0.04	**6.11**	-	**22.90**	nr	**25.76**
150mm	24.11	5.00	**25.93**	0.60	0.07	**9.28**	-	**35.20**	nr	**39.60**
225mm	60.70	5.00	**63.81**	1.20	0.12	**18.34**	-	**82.15**	nr	**92.42**
Channels in bottoms; P.V.C.; 'O' ring joints; bedding in cement mortar (1:3)										
For 610mm clear opening										
straight										
110mm	30.60	5.00	**32.36**	0.45	0.04	**6.82**	-	**39.18**	nr	**44.08**
160mm	56.22	5.00	**59.34**	0.60	0.05	**9.06**	-	**68.40**	nr	**76.95**
short bend										
110mm	15.35	5.00	**16.34**	0.25	0.01	**3.66**	-	**20.00**	nr	**22.50**
160mm	33.83	5.00	**35.83**	0.35	0.02	**5.19**	-	**41.01**	nr	**46.14**
long radius bend										
110mm	50.99	5.00	**54.00**	0.40	0.02	**5.90**	-	**59.90**	nr	**67.38**
160mm	106.46	5.00	**112.39**	0.60	0.05	**9.06**	-	**121.46**	nr	**136.64**
R12: DRAINAGE BELOW GROUND; MANHOLES AND SOAKAWAYS: PRECAST CONCRETE INSPECTION CHAMBERS										
Precast concrete; standard or stock pattern units; BS 5911; jointing and pointing in cement mortar (1:3)										
Chamber or shaft sections										
900mm internal diameter										
250mm high	22.08	5.00	**23.19**	0.58	1.16	**20.67**	-	**43.85**	nr	**49.33**
500mm high	33.08	5.00	**34.74**	0.67	1.34	**23.87**	-	**58.61**	nr	**65.94**
750mm high	33.08	5.00	**34.74**	0.87	1.74	**31.00**	-	**65.74**	nr	**73.95**
1000mm high	44.08	5.00	**46.29**	0.93	1.86	**33.14**	-	**79.42**	nr	**89.35**
1050mm internal diameter										
305mm high	21.05	5.00	**22.12**	0.81	1.62	**28.86**	-	**50.98**	nr	**57.35**
458mm high	31.50	5.00	**33.09**	0.87	1.74	**31.00**	-	**64.09**	nr	**72.10**
610mm high	31.50	5.00	**33.09**	0.93	1.86	**33.14**	-	**66.23**	nr	**74.51**
762mm high	41.95	5.00	**44.06**	0.99	1.98	**35.27**	-	**79.34**	nr	**89.26**
1200mm internal diameter										
250mm high	26.55	5.00	**27.89**	0.99	1.98	**35.27**	-	**63.17**	nr	**71.06**
500mm high	39.75	5.00	**41.75**	1.04	2.08	**37.06**	-	**78.81**	nr	**88.66**
750mm high	39.75	5.00	**41.75**	1.22	2.44	**43.47**	-	**85.22**	nr	**95.88**
1000mm high	52.95	5.00	**55.61**	1.28	2.56	**45.61**	-	**101.22**	nr	**113.87**
1500mm internal diameter										
500mm high	39.75	5.00	**41.75**	1.04	2.08	**37.06**	-	**78.81**	nr	**88.66**
750mm high	39.75	5.00	**41.75**	1.22	2.44	**43.47**	-	**85.22**	nr	**95.88**
1000mm high	52.95	5.00	**55.61**	1.28	2.56	**45.61**	-	**101.22**	nr	**113.87**
1800mm internal diameter										
500mm high	39.75	5.00	**41.75**	1.04	2.08	**37.06**	-	**78.81**	nr	**88.66**
750mm high	39.75	5.00	**41.75**	1.22	2.44	**43.47**	-	**85.22**	nr	**95.88**
1000mm high	52.95	5.00	**55.61**	1.28	2.56	**45.61**	-	**101.22**	nr	**113.87**
2100mm internal diameter										
500mm high	39.75	5.00	**41.75**	1.04	2.08	**37.06**	-	**78.81**	nr	**88.66**
750mm high	39.75	5.00	**41.75**	1.22	2.44	**43.47**	-	**85.22**	nr	**95.88**
1000mm high	52.95	5.00	**55.61**	1.28	2.56	**45.61**	-	**101.22**	nr	**113.87**

Labour hourly rates: (except Specialists) Craft Operatives £14.21 Labourer £10.71 Rates are national average prices. Refer to REGIONAL VARIATIONS for indicative levels of overall pricing in regions	MATERIALS			LABOUR			RATES			
	Del to Site £	Waste %	Material Cost £	Craft Optve Hrs	Lab Hrs	Labour Cost £	Sunds £	Nett Rate £	Unit	Gross Rate (+12.5%) £
R12: DRAINAGE BELOW GROUND; MANHOLES AND SOAKAWAYS: PRECAST CONCRETE INSPECTION CHAMBERS Cont.										
Precast concrete; standard or stock pattern units; BS 5911; jointing and pointing in cement mortar (1:3) Cont.										
Chamber or shaft sections Cont.										
2400mm internal diameter										
750mm high	39.75	5.00	**41.75**	1.22	2.44	**43.47**	-	**85.22**	nr	**95.88**
1000mm high	52.95	5.00	**55.61**	1.28	2.56	**45.61**	-	**101.22**	nr	**113.87**
2700mm internal diameter										
750mm high	39.75	5.00	**41.75**	1.22	2.44	**43.47**	-	**85.22**	nr	**95.88**
1000mm high	52.95	5.00	**55.61**	1.28	2.56	**45.61**	-	**101.22**	nr	**113.87**
3000mm internal diameter										
750mm high	39.75	5.00	**41.75**	1.22	2.44	**43.47**	-	**85.22**	nr	**95.88**
1000mm high	52.95	5.00	**55.61**	1.28	2.56	**45.61**	-	**101.22**	nr	**113.87**
Cutting holes										
cutting holes for pipes; 100mm diameter; making good; pointing	30.15	-	**30.15**	-	0.09	**0.96**	-	**31.12**	nr	**35.01**
cutting holes for pipes; 150mm diameter; making good; pointing	30.23	-	**30.23**	-	0.11	**1.18**	-	**31.41**	nr	**35.33**
cutting holes for pipes; 225mm diameter; making good; pointing	45.38	-	**45.38**	-	0.13	**1.39**	-	**46.77**	nr	**52.62**
cutting holes for pipes; 300mm diameter; making good; pointing	60.53	-	**60.53**	-	0.18	**1.93**	-	**62.46**	nr	**70.27**
Step irons										
galvanized malleable cast iron step irons BS 1247, cast in	5.00	5.00	**5.25**	-	0.17	**1.82**	-	**7.07**	nr	**7.95**
Cover slabs, heavy duty										
for chamber or shaft sections 900mm internal diameter, 125mm thick; access opening 600mm x 600mm	48.84	5.00	**51.30**	1.70	2.55	**51.47**	-	**102.77**	nr	**115.61**
for chamber or shaft sections 1050mm internal diameter, 125mm thick; access opening 600mm x 600mm	56.04	5.00	**58.86**	2.25	3.35	**67.85**	-	**126.71**	nr	**142.55**
for chamber or shaft sections 1200mm internal diameter, 125mm thick; access opening 600mm x 600mm	74.07	5.00	**77.79**	2.80	4.20	**84.77**	-	**162.56**	nr	**182.88**
for chamber or shaft sections 1350mm internal diameter, 125mm thick; access opening 600mm x 600mm	108.59	5.00	**114.04**	3.35	5.05	**101.69**	-	**215.73**	nr	**242.69**
for chamber or shaft sections 1500mm internal diameter, 150mm thick; access opening 600mm x 600mm	134.14	5.00	**140.87**	3.95	5.90	**119.32**	-	**260.19**	nr	**292.71**
for chamber or shaft sections 1800mm internal diameter, 150mm thick access opening 600mm x 600mm	155.24	5.00	**163.02**	4.55	6.75	**136.95**	-	**299.97**	nr	**337.47**
for chamber or shaft sections 2100mm internal diameter, 150mm thick access opening 600mm x 600mm	320.88	5.00	**336.96**	5.15	7.60	**154.58**	-	**491.54**	nr	**552.98**
for chamber or shaft sections 2400mm internal diameter, 150mm thick access opening 600mm x 600mm	558.97	5.00	**586.96**	5.75	8.45	**172.21**	-	**759.16**	nr	**854.06**
for chamber or shaft sections 2700mm internal diameter, 150mm thick access opening 600mm x 600mm	723.60	5.00	**759.82**	6.35	9.30	**189.84**	-	**949.66**	nr	**1068.36**
for chamber or shaft sections 3000mm internal diameter, 150mm thick access opening 600mm x 600mm	852.17	5.00	**894.82**	6.95	10.15	**207.47**	-	**1102.29**	nr	**1240.07**
Precast concrete inspection chambers; BS 5911; jointing and pointing in cement mortar (1:3)										
Rectangular inspection chambers, internal size 457 x 610mm										
100mm high	6.57	5.00	**6.92**	0.50	0.35	**10.85**	-	**17.77**	nr	**19.99**
150mm high	7.54	5.00	**7.93**	0.65	0.45	**14.06**	-	**21.99**	nr	**24.74**
250mm high	10.11	5.00	**10.63**	0.75	0.53	**16.33**	-	**26.97**	nr	**30.34**
extra; galvanized malleable cast iron step irons BS 1247, built in	5.00	5.00	**5.25**	-	-	**-**	-	**5.25**	nr	**5.91**
Rectangular inspection chambers, internal size 760 x 610mm										
150mm high	10.65	5.00	**11.20**	1.05	0.75	**22.95**	-	**34.16**	nr	**38.43**
225mm high	13.57	5.00	**14.27**	1.21	0.88	**26.62**	-	**40.89**	nr	**46.00**
extra; galvanized malleable cast iron step irons BS 1247, built in	5.00	5.00	**5.25**	-	-	**-**	-	**5.25**	nr	**5.91**
Rectangular inspection chambers, internal size 990 x 610mm										
150mm high	12.88	5.00	**13.54**	1.05	0.75	**22.95**	-	**36.50**	nr	**41.06**
Rectangular inspection chambers, internal size 1200 x 750mm										
150mm high	17.78	5.00	**18.70**	1.20	0.85	**26.16**	-	**44.86**	nr	**50.47**
200mm high	21.85	5.00	**22.98**	1.38	1.00	**30.32**	-	**53.30**	nr	**59.96**
Base units										
450 x 610mm internal size; 360mm deep	30.28	5.00	**31.81**	1.05	0.75	**22.95**	-	**54.77**	nr	**61.61**
760 x 610mm internal size; 360mm deep	36.67	5.00	**38.54**	1.05	0.75	**22.95**	-	**61.49**	nr	**69.18**
990 x 610mm internal size; 360mm deep	48.83	5.00	**51.31**	1.05	0.75	**22.95**	-	**74.26**	nr	**83.55**
Cover slab with integral cover										
450 x 610mm internal size;	23.60	5.00	**24.80**	1.05	0.75	**22.95**	-	**47.75**	nr	**53.72**
760 x 610mm internal size;	43.78	5.00	**46.00**	1.05	0.75	**22.95**	-	**68.96**	nr	**77.58**
990 x 610mm internal size;	67.72	5.00	**71.14**	1.05	0.75	**22.95**	-	**94.10**	nr	**105.86**

Labour hourly rates: (except Specialists) Craft Operatives £14.21 Labourer £10.71 Rates are national average prices. Refer to REGIONAL VARIATIONS for indicative levels of overall pricing in regions	MATERIALS			LABOUR				RATES		
	Del to Site £	Waste %	Material Cost £	Craft Optve Hrs	Lab Hrs	Labour Cost £	Sunds £	Nett Rate £	Unit	Gross Rate (+12.5%) £
R12: DRAINAGE BELOW GROUND; MANHOLES AND SOAKAWAYS: GRP INSPECTION CHAMBERS										
Inspection chambers; Terrain; GRP with preformed benching; 4 nr pvc adaptors; A15 light duty single seal cast iron cover to BS EN 124 and plastic frame set in position										
Chamber or shaft sections and benching										
470mm internal diameter; depth to invert										
520mm	190.43	5.00	**199.95**	5.50	0.90	**87.79**	-	**287.74**	nr	**323.71**
755mm	227.02	5.00	**238.37**	5.50	0.90	**87.79**	-	**326.17**	nr	**366.94**
990mm	263.61	5.00	**276.79**	5.90	0.95	**94.01**	-	**370.81**	nr	**417.16**
1225mm	300.21	5.00	**315.22**	5.90	0.95	**94.01**	-	**409.23**	nr	**460.38**
R12: DRAINAGE BELOW GROUND; MANHOLES AND SOAKAWAYS: POLYPROPYLENE INSPECTION CHAMBERS										
Osma; Shallow inspection chamber ref. 4D.960 Polypropylene with preformed benching. Light duty single sealed cover and frame ref. 4D.961 set in position, 250mm diameter; for 100mm diameter pipes, depth to invert										
Chamber or shaft sections and benching										
600mm	62.13	5.00	**65.24**	2.50	1.44	**50.95**	-	**116.19**	nr	**130.71**
Osma; universal inspection chamber; Polypropylene; 450mm diameter										
preformed chamber base benching, ref. 4D.922; for 100mm pipes	89.81	5.00	**94.30**	1.50	0.10	**22.39**	-	**116.69**	nr	**131.27**
chamber shaft 230mm effective length, ref. 4D.925	22.90	5.00	**24.05**	-	0.60	**6.43**	-	**30.48**	nr	**34.28**
cast iron cover and frame; single seal light duty to suit, ref. 4D.942	72.15	5.00	**75.76**	0.60	-	**8.53**	-	**84.28**	nr	**94.82**
Polypropylene sealed cover and frame, ref. 4D.943	43.15	5.00	**45.31**	0.60	-	**8.53**	-	**53.83**	nr	**60.56**

This section continues on the next page

Labour hourly rates: (except Specialists) Craft Operatives £17.38 Labourer £10.71 Rates are national average prices. Refer to REGIONAL VARIATIONS for indicative levels of overall pricing in regions	MATERIALS			LABOUR				RATES		
	Del to Site £	Waste %	Material Cost £	Craft Optve Hrs	Lab Hrs	Labour Cost £	Sunds £	Nett Rate £	Unit	Gross Rate (+12.5%) £
R12: DRAINAGE BELOW GROUND; MANHOLES AND SOAKAWAYS: CAST IRON INSPECTION CHAMBERS										
Coated cast iron bolted inspection chambers										
100mm diameter straight chamber; coupling joints; bedding in cement mortar (1:3)										
with one branch one side	88.59	2.00	**90.37**	3.00	-	**52.14**	-	**142.51**	nr	**160.33**
with one branch each side	135.19	2.00	**137.90**	4.00	-	**69.52**	-	**207.42**	nr	**233.35**
with two branches each side	266.06	2.00	**271.39**	6.00	-	**104.28**	-	**375.67**	nr	**422.63**
150 x 100mm diameter straight chamber; coupling joints; bedding in cement mortar (1:3)										
with one branch one side	175.13	2.00	**178.64**	3.75	-	**65.18**	-	**243.82**	nr	**274.29**
150mm diameter straight chamber; coupling joints; bedding in cement mortar (1:3)										
with one branch each side	222.22	2.00	**226.68**	5.25	-	**91.25**	-	**317.92**	nr	**357.66**
with two branches each side	443.66	2.00	**452.55**	7.75	-	**134.70**	-	**587.24**	nr	**660.65**

This section continues on the next page

Labour hourly rates: (except Specialists) Craft Operatives £14.21 Labourer £10.71 Rates are national average prices. Refer to REGIONAL VARIATIONS for indicative levels of overall pricing in regions	MATERIALS			LABOUR				RATES		
	Del to Site £	Waste %	Material Cost £	Craft Optve Hrs	Lab Hrs	Labour Cost £	Sunds £	Nett Rate £	Unit	Gross Rate (+12.5%) £
R12: DRAINAGE BELOW GROUND; MANHOLES AND SOAKAWAYS: MANHOLE ACCESSORIES										
Manhole step irons										
Step irons; BS 1247 malleable cast iron; galvanized; building in to joints										
figure 1 general purpose pattern; 110mm tail	3.43	-	**3.43**	0.25	-	**3.55**	-	**6.98**	nr	**7.85**
figure 1 general purpose pattern; 225mm tail	4.29	-	**4.29**	0.27	-	**3.84**	-	**8.13**	nr	**9.14**
Manhole covers - A15 light duty										
Access covers; Durey Castings Ltd; BS EN 124; coated; bedding frame in cement mortar (1:3); bedding cover in grease and sand										
reference DU14B; clear opening 450mm dia	13.94	2.50	**14.30**	0.80	0.80	**19.94**	-	**34.23**	nr	**38.51**
reference DU10A; clear opening 450 x 450mm	18.56	2.50	**19.03**	0.80	0.80	**19.94**	-	**38.97**	nr	**43.84**
reference DU10B; clear opening 600 x 450mm	23.18	2.50	**23.77**	0.90	0.90	**22.43**	-	**46.20**	nr	**51.97**
reference DU10C; clear opening 600 x 600mm	32.42	2.50	**33.24**	1.40	1.40	**34.89**	-	**68.13**	nr	**76.64**
reference DU15A recessed; clear opening 600 x 450mm	35.19	2.50	**36.08**	1.30	1.30	**32.40**	-	**68.48**	nr	**77.04**
reference DU15A recessed; clear opening 600 x 600mm	43.97	2.50	**45.08**	1.60	1.60	**39.87**	-	**84.95**	nr	**95.57**
reference DU15A recessed; clear opening 750 x 600mm	225.53	2.50	**231.18**	1.75	1.75	**43.61**	-	**274.79**	nr	**309.14**
reference DU30A double seal; clear opening 450 x 450mm	240.32	2.50	**246.34**	1.45	1.45	**36.13**	-	**282.47**	nr	**317.78**
reference DU30A double seal; clear opening 600 x 450mm	253.90	2.50	**260.26**	1.60	1.60	**39.87**	-	**300.13**	nr	**337.65**
reference DU30A double seal; clear opening 600 x 600mm	282.13	2.50	**289.19**	1.75	1.75	**43.61**	-	**332.80**	nr	**374.40**
lifting keys	6.93	2.50	**7.10**	-	-	**-**	-	**7.10**	nr	**7.99**
Manhole covers - B125 medium duty										
Access covers; Stanton PLC; BS EN 124; coated; bedding frame in cement mortar (1:3); bedding cover in grease and sand										
reference H783 Trojan single seal; 600mm diameter	99.70	2.50	**102.38**	1.80	1.80	**44.86**	-	**147.24**	nr	**165.64**
reference H781; 600 x 450mm	93.21	2.50	**95.83**	1.58	1.58	**39.37**	-	**135.20**	nr	**152.10**
reference H782; 600 x 600mm	115.30	2.50	**118.47**	1.95	1.95	**48.59**	-	**167.06**	nr	**187.94**
reference HA08; 600 x 450mm	121.00	2.50	**124.21**	1.84	1.84	**45.85**	-	**170.06**	nr	**191.32**
reference HA09; 600 x 600mm	153.85	2.50	**158.08**	2.39	2.39	**59.56**	-	**217.63**	nr	**244.84**
lifting keys	4.94	2.50	**5.07**	-	-	**-**	-	**5.07**	nr	**5.70**
Manhole covers - D400 heavy duty										
Access covers; BS EN 124; coated; bedding frame in cement mortar (1:3); bedding cover in grease and sand										
reference H9330; 600mm diameter	125.24	2.50	**128.75**	2.04	2.04	**50.84**	-	**179.59**	nr	**202.04**
reference H9320; 600 x 600mm	108.09	2.50	**110.99**	2.04	2.04	**50.84**	-	**161.82**	nr	**182.05**
reference H9420; 750 x 750mm	261.43	2.50	**268.54**	2.40	2.40	**59.81**	-	**328.35**	nr	**369.39**
lifting keys	5.50	2.50	**5.64**	-	-	**-**	-	**5.64**	nr	**6.34**
R12: DRAINAGE BELOW GROUND; MANHOLES AND SOAKAWAYS: INTERCEPTING TRAPS										
Vitrified clay intercepting traps										
Intercepting traps; BS 65; joint to pipe; building into side of manhole; bedding and surrounding in concrete to BS 5328 ordinary prescribed mix C15P, 20mm aggregate										
cleaning arm and stopper										
100mm outlet, ref. R1 1/1	65.11	5.00	**68.36**	1.45	1.15	**32.92**	-	**101.28**	nr	**113.94**
150mm outlet, ref. R1 1/2	93.32	5.00	**97.99**	2.00	1.75	**47.16**	-	**145.15**	nr	**163.30**
225mm outlet, ref. R1 1/3	288.63	5.00	**303.06**	2.60	2.30	**61.58**	-	**364.64**	nr	**410.22**
R12: DRAINAGE BELOW GROUND; MANHOLES AND SOAKAWAYS: FRESH AIR INLETS										
Aluminium; fresh air inlets										
Air inlet valve, mice flap; set in cement mortar (1:3)										
100mm diameter	30.52	5.00	**32.06**	1.25	-	**17.76**	-	**49.82**	nr	**56.05**
R12: DRAINAGE BELOW GROUND; MANHOLES AND SOAKAWAYS: SETTLEMENT TANKS										
Fibreglass settlement tanks; including cover and frame										
Settlement tank; fibreglass; set in position										
1000mm deep to invert of inlet, standard grade										
2700 litres capacity	486.00	-	**486.00**	1.00	2.00	**35.63**	-	**521.63**	nr	**586.83**
3750 litres capacity	623.00	-	**623.00**	1.10	2.20	**39.19**	-	**662.19**	nr	**744.97**
4500 litres capacity	717.00	-	**717.00**	1.20	2.40	**42.76**	-	**759.76**	nr	**854.73**
7500 litres capacity	1080.00	-	**1080.00**	1.30	2.60	**46.32**	-	**1126.32**	nr	**1267.11**
9000 litres capacity	1155.00	-	**1155.00**	1.45	2.90	**51.66**	-	**1206.66**	nr	**1357.50**
1500mm deep to invert of inlet, heavy grade										
2700 litres capacity	548.00	-	**548.00**	1.00	2.00	**35.63**	-	**583.63**	nr	**656.58**
3750 litres capacity	686.00	-	**686.00**	1.10	2.20	**39.19**	-	**725.19**	nr	**815.84**
4500 litres capacity	780.00	-	**780.00**	1.20	2.40	**42.76**	-	**822.76**	nr	**925.60**
7500 litres capacity	1167.00	-	**1167.00**	1.30	2.60	**46.32**	-	**1213.32**	nr	**1364.98**
9000 litres capacity	1356.00	-	**1356.00**	1.45	2.90	**51.66**	-	**1407.66**	nr	**1583.62**

Labour hourly rates: (except Specialists) Craft Operatives £14.21 Labourer £10.71 Rates are national average prices. Refer to REGIONAL VARIATIONS for indicative levels of overall pricing in regions	MATERIALS			LABOUR				RATES		
	Del to Site £	Waste %	Material Cost £	Craft Optve Hrs	Lab Hrs	Labour Cost £	Sunds £	Nett Rate £	Unit	Gross Rate (+12.5%) £
R12: DRAINAGE BELOW GROUND; WORK TO EXISTING DRAINS										
Work to disused drains										
Fly ash filling to disused drain										
100mm diameter	0.46	-	**0.46**	0.40	-	**5.68**	-	**6.14**	m	**6.91**
150mm diameter	0.84	-	**0.84**	0.60	-	**8.53**	-	**9.37**	m	**10.54**
225mm diameter	1.80	-	**1.80**	0.90	-	**12.79**	-	**14.59**	m	**16.41**
Sealing end of disused drain with plain in-situ concrete										
100mm diameter	0.68	5.00	**0.72**	0.22	-	**3.13**	-	**3.84**	nr	**4.33**
150mm diameter	1.24	5.00	**1.31**	0.33	-	**4.69**	-	**6.00**	nr	**6.74**
225mm diameter	1.99	5.00	**2.09**	0.44	-	**6.25**	-	**8.34**	nr	**9.39**
R13: LAND DRAINAGE										
Drains; clayware pipes, BS 1196; butt joints										
Pipework in trenches										
75mm	3.21	5.00	**3.37**	-	0.26	**2.78**	-	**6.15**	m	**6.92**
100mm	5.51	5.00	**5.79**	-	0.29	**3.11**	-	**8.90**	m	**10.01**
150mm	11.30	5.00	**11.86**	-	0.36	**3.86**	-	**15.72**	m	**17.68**
Drains; porous concrete pipes, BS 5911; butt joints										
Pipework in trenches										
150mm	6.08	5.00	**6.38**	-	0.34	**3.64**	-	**10.03**	m	**11.28**
225mm	7.78	5.00	**8.17**	-	0.42	**4.50**	-	**12.67**	m	**14.25**
300mm	12.79	5.00	**13.43**	-	0.50	**5.36**	-	**18.78**	m	**21.13**
375mm	17.43	5.00	**18.30**	-	0.58	**6.21**	-	**24.51**	m	**27.58**
450mm	24.52	5.00	**25.75**	-	0.66	**7.07**	-	**32.81**	m	**36.92**
525mm	36.33	5.00	**38.15**	-	0.74	**7.93**	-	**46.07**	m	**51.83**
600mm	45.12	5.00	**47.38**	-	0.82	**8.78**	-	**56.16**	m	**63.18**
Drains; OGEE pipes, BS 5911 Part 110; butt joints										
Pipework in trenches										
150mm	4.96	5.00	**5.21**	-	0.34	**3.64**	-	**8.85**	m	**9.96**
225mm	8.03	5.00	**8.43**	-	0.42	**4.50**	-	**12.93**	m	**14.55**
300mm	13.18	5.00	**13.84**	-	0.50	**5.36**	-	**19.19**	m	**21.59**
375mm	17.90	5.00	**18.80**	-	0.58	**6.21**	-	**25.01**	m	**28.13**
450mm	22.74	5.00	**23.88**	-	0.66	**7.07**	-	**30.95**	m	**34.81**
525mm	30.72	5.00	**32.26**	-	0.74	**7.93**	-	**40.18**	m	**45.20**
600mm	41.31	5.00	**43.38**	-	0.82	**8.78**	-	**52.16**	m	**58.68**
750mm	61.67	5.00	**64.75**	-	0.90	**9.64**	-	**74.39**	m	**83.69**
Drains; clayware perforated pipes, butt joints; 150mm hardcore to sides and top										
Pipework in trenches										
100mm	11.81	5.00	**12.61**	-	0.44	**4.71**	-	**17.33**	m	**19.49**
150mm	19.06	5.00	**20.28**	-	0.65	**6.96**	-	**27.24**	m	**30.65**
225mm	32.59	5.00	**34.59**	-	0.70	**7.50**	-	**42.08**	m	**47.34**

PIPED SUPPLY SYSTEMS

Labour hourly rates: (except Specialists) Craft Operatives £17.38 Labourer £10.71 Rates are national average prices. Refer to REGIONAL VARIATIONS for indicative levels of overall pricing in regions	MATERIALS			LABOUR				RATES		
	Del to Site £	Waste %	Material Cost £	Craft Optve Hrs	Lab Hrs	Labour Cost £	Sunds £	Nett Rate £	Unit	Gross Rate (+12.5%) £
S115: HOT AND COLD WATER AND GAS: BLACK AND GALVANIZED STEEL TUBES AND FITTINGS										
Steel pipes, BS 1387 Table 2; steel fittings BS 1740 Part 1; screwed and PTFE tape joints; pipework black										
Pipes, straight in malleable iron pipe brackets fixing to timber with screws										
15mm	3.36	5.00	**3.53**	0.60	-	**10.43**	-	**13.96**	m	**15.70**
15mm: fittings										
extra; made bends	-	-	-	0.20	-	**3.48**	-	**3.48**	nr	**3.91**
extra; connections to copper pipe ends; compression joint	0.78	2.00	**0.80**	0.12	-	**2.09**	-	**2.88**	nr	**3.24**
extra; fittings; one end	0.59	2.00	**0.60**	0.15	-	**2.61**	-	**3.21**	nr	**3.61**
extra; fittings; two ends	0.60	2.00	**0.61**	0.30	-	**5.21**	-	**5.83**	nr	**6.55**
extra; fittings; three ends	0.82	2.00	**0.84**	0.30	-	**5.21**	-	**6.05**	nr	**6.81**
20mm	3.92	5.00	**4.12**	0.68	-	**11.82**	-	**15.94**	m	**17.93**
20mm; fittings										
extra; made bends	-	-	-	0.30	-	**5.21**	-	**5.21**	nr	**5.87**
extra; connections to copper pipe ends; compression joint	1.28	2.00	**1.31**	0.20	-	**3.48**	-	**4.78**	nr	**5.38**
extra; fittings; one end	0.67	2.00	**0.68**	0.20	-	**3.48**	-	**4.16**	nr	**4.68**
extra; fittings; two ends	0.82	2.00	**0.84**	0.40	-	**6.95**	-	**7.79**	nr	**8.76**
extra; fittings; three ends	1.20	2.00	**1.22**	0.40	-	**6.95**	-	**8.18**	nr	**9.20**
25mm	5.47	5.00	**5.74**	0.80	-	**13.90**	-	**19.65**	m	**22.10**
25mm; fittings										
extra; made bends	-	-	-	0.40	-	**6.95**	-	**6.95**	nr	**7.82**
extra; connections to copper pipe ends; compression joint	2.48	2.00	**2.53**	0.30	-	**5.21**	-	**7.74**	nr	**8.71**
extra; fittings; one end	0.84	2.00	**0.86**	0.30	-	**5.21**	-	**6.07**	nr	**6.83**
extra; fittings; two ends	1.27	2.00	**1.30**	0.60	-	**10.43**	-	**11.72**	nr	**13.19**
extra; fittings; three ends	1.72	2.00	**1.75**	0.60	-	**10.43**	-	**12.18**	nr	**13.71**
32mm	6.85	5.00	**7.19**	0.96	-	**16.68**	-	**23.88**	m	**26.86**
32mm; fittings										
extra; made bends	-	-	-	0.60	-	**10.43**	-	**10.43**	nr	**11.73**
extra; connections to copper pipe ends; compression joint	8.43	2.00	**8.60**	0.40	-	**6.95**	-	**15.55**	nr	**17.49**
extra; fittings; one end	1.36	2.00	**1.39**	0.40	-	**6.95**	-	**8.34**	nr	**9.38**
extra; fittings; two ends	2.15	2.00	**2.19**	0.80	-	**13.90**	-	**16.10**	nr	**18.11**
extra; fittings; three ends	2.92	2.00	**2.98**	0.80	-	**13.90**	-	**16.88**	nr	**18.99**
40mm	8.97	5.00	**9.39**	1.20	-	**20.86**	-	**30.25**	m	**34.03**
40mm; fittings										
extra; made bends	-	-	-	0.80	-	**13.90**	-	**13.90**	nr	**15.64**
extra; connections to copper pipe ends; compression joint	13.56	2.00	**13.83**	0.60	-	**10.43**	-	**24.26**	nr	**27.29**
extra; fittings; one end	1.61	2.00	**1.64**	0.50	-	**8.69**	-	**10.33**	nr	**11.62**
extra; fittings; two ends	3.61	2.00	**3.68**	1.00	-	**17.38**	-	**21.06**	nr	**23.69**
extra; fittings; three ends	4.48	2.00	**4.57**	1.00	-	**17.38**	-	**21.95**	nr	**24.69**
50mm	11.29	5.00	**11.86**	1.60	-	**27.81**	-	**39.67**	m	**44.63**
50mm; fittings										
extra; made bends	-	-	-	1.20	-	**20.86**	-	**20.86**	nr	**23.46**
extra; connections to copper pipe ends; compression joint	20.53	2.00	**20.94**	0.80	-	**13.90**	-	**34.84**	nr	**39.20**
extra; fittings; one end	3.36	2.00	**3.43**	0.60	-	**10.43**	-	**13.86**	nr	**15.59**
extra; fittings; two ends	4.23	2.00	**4.31**	1.20	-	**20.86**	-	**25.17**	nr	**28.32**
extra; fittings; three ends	6.46	2.00	**6.59**	1.20	-	**20.86**	-	**27.45**	nr	**30.88**
Pipes straight in malleable iron pipe brackets fixing to masonry with screws										
15mm	3.37	5.00	**3.54**	0.64	-	**11.12**	-	**14.67**	m	**16.50**
20mm	3.94	5.00	**4.14**	0.72	-	**12.51**	-	**16.65**	m	**18.73**
25mm	5.48	5.00	**5.76**	0.84	-	**14.60**	-	**20.36**	m	**22.90**
32mm	6.86	5.00	**7.21**	1.00	-	**17.38**	-	**24.59**	m	**27.66**
40mm	8.11	5.00	**8.52**	1.24	-	**21.55**	-	**30.07**	m	**33.83**
50mm	11.31	5.00	**11.88**	1.64	-	**28.50**	-	**40.38**	m	**45.43**
Pipes straight in malleable iron single rings, 165mm long screwed both ends steel tubes and malleable iron backplates fixing to timber with screws										
15mm	3.92	5.00	**4.12**	0.72	-	**12.51**	-	**16.63**	m	**18.71**
20mm	4.49	5.00	**4.72**	0.80	-	**13.90**	-	**18.62**	m	**20.95**

Labour hourly rates: (except Specialists) Craft Operatives £17.38 Labourer £10.71 Rates are national average prices. Refer to REGIONAL VARIATIONS for indicative levels of overall pricing in regions	MATERIALS			LABOUR				RATES		
	Del to Site £	Waste %	Material Cost £	Craft Optve Hrs	Lab Hrs	Labour Cost £	Sunds £	Nett Rate £	Unit	Gross Rate (+12.5%) £
S115: HOT AND COLD WATER AND GAS: BLACK AND GALVANIZED STEEL TUBES AND FITTINGS Cont.										
Steel pipes, BS 1387 Table 2; steel fittings BS 1740 Part 1; screwed and PTFE tape joints; pipework black Cont.										
Pipes straight in malleable iron single rings, 165mm long screwed both ends steel tubes and malleable iron backplates fixing to timber with screws Cont.										
25mm	6.01	5.00	**6.31**	0.92	-	**15.99**	-	**22.30**	m	**25.09**
32mm	7.19	5.00	**7.55**	1.08	-	**18.77**	-	**26.32**	m	**29.62**
40mm	8.36	5.00	**8.78**	1.32	-	**22.94**	-	**31.72**	m	**35.68**
50mm	11.40	5.00	**11.97**	1.72	-	**29.89**	-	**41.87**	m	**47.10**
Steel pipes, BS 1387 Table 3; steel fittings BS 1740 Part 1; screwed and PTFE tape joints; pipework black										
Pipes, straight in malleable iron pipe brackets fixing to timber with screws										
15mm	3.91	5.00	**4.10**	0.60	-	**10.43**	-	**14.53**	m	**16.35**
15mm; fittings										
extra; made bends	-	-	**-**	0.20	-	**3.48**	-	**3.48**	nr	**3.91**
extra; connections to copper pipe ends; compression joint	0.78	2.00	**0.80**	0.12	-	**2.09**	-	**2.88**	nr	**3.24**
extra; fittings; one end	0.59	2.00	**0.60**	0.15	-	**2.61**	-	**3.21**	nr	**3.61**
extra; fittings; two ends	0.60	2.00	**0.61**	0.30	-	**5.21**	-	**5.83**	nr	**6.55**
extra; fittings; three ends	0.82	2.00	**0.84**	0.30	-	**5.21**	-	**6.05**	nr	**6.81**
20mm	4.59	5.00	**4.82**	0.68	-	**11.82**	-	**16.64**	m	**18.72**
20mm; fittings										
extra; made bends	-	-	**-**	0.30	-	**5.21**	-	**5.21**	nr	**5.87**
extra; connections to copper pipe ends; compression joint	1.28	2.00	**1.31**	0.20	-	**3.48**	-	**4.78**	nr	**5.38**
extra; fittings; one end	0.67	2.00	**0.68**	0.20	-	**3.48**	-	**4.16**	nr	**4.68**
extra; fittings; two ends	0.82	2.00	**0.84**	0.40	-	**6.95**	-	**7.79**	nr	**8.76**
extra; fittings; three ends	1.20	2.00	**1.22**	0.40	-	**6.95**	-	**8.18**	nr	**9.20**
25mm	6.54	5.00	**6.87**	0.80	-	**13.90**	-	**20.77**	m	**23.37**
25mm; fittings										
extra; made bends	-	-	**-**	0.40	-	**6.95**	-	**6.95**	nr	**7.82**
extra; connections to copper pipe ends; compression joint	2.48	2.00	**2.53**	0.30	-	**5.21**	-	**7.74**	nr	**8.71**
extra; fittings; one end	0.84	2.00	**0.86**	0.30	-	**5.21**	-	**6.07**	nr	**6.83**
extra; fittings; two ends	1.27	2.00	**1.30**	0.60	-	**10.43**	-	**11.72**	nr	**13.19**
extra; fittings; three ends	1.72	2.00	**1.75**	0.60	-	**10.43**	-	**12.18**	nr	**13.71**
32mm	8.19	5.00	**8.60**	0.96	-	**16.68**	-	**25.28**	m	**28.44**
32mm; fittings										
extra; made bends	-	-	**-**	0.60	-	**10.43**	-	**10.43**	nr	**11.73**
extra; connections to copper pipe ends; compression joint	8.43	2.00	**8.60**	0.40	-	**6.95**	-	**15.55**	nr	**17.49**
extra; fittings; one end	1.36	2.00	**1.39**	0.40	-	**6.95**	-	**8.34**	nr	**9.38**
extra; fittings; two ends	2.15	2.00	**2.19**	0.80	-	**13.90**	-	**16.10**	nr	**18.11**
extra; fittings; three ends	2.92	2.00	**2.98**	0.80	-	**13.90**	-	**16.88**	nr	**18.99**
40mm	9.68	5.00	**10.16**	1.20	-	**20.86**	-	**31.02**	m	**34.90**
40mm; fittings										
extra; made bends	-	-	**-**	0.80	-	**13.90**	-	**13.90**	nr	**15.64**
extra; connections to copper pipe ends; compression joint	13.56	2.00	**13.83**	0.60	-	**10.43**	-	**24.26**	nr	**27.29**
extra; fittings; one end	1.61	2.00	**1.64**	0.50	-	**8.69**	-	**10.33**	nr	**11.62**
extra; fittings; two ends	3.61	2.00	**3.68**	1.00	-	**17.38**	-	**21.06**	nr	**23.69**
extra; fittings; three ends	4.48	2.00	**4.57**	1.00	-	**17.38**	-	**21.95**	nr	**24.69**
50mm	13.37	5.00	**14.04**	1.60	-	**27.81**	-	**41.85**	m	**47.08**
50mm; fittings										
extra; made bends	-	-	**-**	1.20	-	**20.86**	-	**20.86**	nr	**23.46**
extra; connections to copper pipe ends; compression joint	20.53	2.00	**20.94**	0.80	-	**13.90**	-	**34.84**	nr	**39.20**
extra; fittings; one end	3.36	2.00	**3.43**	0.60	-	**10.43**	-	**13.86**	nr	**15.59**
extra; fittings; two ends	4.23	2.00	**4.31**	1.20	-	**20.86**	-	**25.17**	nr	**28.32**
extra; fittings; three ends	6.46	2.00	**6.59**	1.20	-	**20.86**	-	**27.45**	nr	**30.88**
Pipes straight in malleable iron pipe brackets fixing to masonry with screws										
15mm	3.92	5.00	**4.12**	0.64	-	**11.12**	-	**15.24**	m	**17.15**
20mm	4.61	5.00	**4.84**	0.72	-	**12.51**	-	**17.35**	m	**19.52**
25mm	6.55	5.00	**6.88**	0.84	-	**14.60**	-	**21.48**	m	**24.17**
32mm	8.20	5.00	**8.61**	1.00	-	**17.38**	-	**25.99**	m	**29.24**
40mm	9.69	5.00	**10.18**	1.24	-	**21.55**	-	**31.73**	m	**35.70**
50mm	13.39	5.00	**14.06**	1.64	-	**28.50**	-	**42.56**	m	**47.88**
Pipes straight in malleable iron single rings, 165mm long screwed both ends steel tubes and malleable iron backplates fixing to timber with screws										
15mm	4.47	5.00	**4.70**	0.72	-	**12.51**	-	**17.21**	m	**19.36**
20mm	5.16	5.00	**5.42**	0.80	-	**13.90**	-	**19.33**	m	**21.74**
25mm	7.08	5.00	**7.43**	0.92	-	**15.99**	-	**23.42**	m	**26.35**
32mm	8.53	5.00	**8.96**	1.08	-	**18.77**	-	**27.73**	m	**31.20**
40mm	9.94	5.00	**10.44**	1.32	-	**22.94**	-	**33.38**	m	**37.55**
50mm	13.48	5.00	**14.16**	1.72	-	**29.89**	-	**44.05**	m	**49.56**
Steel pipes, BS 1387 Table 2; steel fittings BS 1740 Part 1; screwed and PTFE tape joints; pipework galvanized										
Pipes, straight in malleable iron pipe brackets fixing to timber with screws										
15mm	5.06	5.00	**5.31**	0.60	-	**10.43**	-	**15.74**	m	**17.71**

Labour hourly rates: (except Specialists) Craft Operatives £17.38 Labourer £10.71 Rates are national average prices. Refer to REGIONAL VARIATIONS for indicative levels of overall pricing in regions	MATERIALS			LABOUR				RATES		
	Del to Site £	Waste %	Material Cost £	Craft Optve Hrs	Lab Hrs	Labour Cost £	Sunds £	Nett Rate £	Unit	Gross Rate (+12.5%) £
S115: HOT AND COLD WATER AND GAS: BLACK AND GALVANIZED STEEL TUBES AND FITTINGS Cont.										
Steel pipes, BS 1387 Table 2; steel fittings BS 1740 Part 1; screwed and PTFE tape joints; pipework galvanized Cont.										
Pipes, straight in malleable iron pipe brackets fixing to timber with screws Cont.										
15mm; fittings										
extra; made bends	-	-	-	0.20	-	**3.48**	-	**3.48**	nr	**3.91**
extra; connections to copper pipe ends; compression joint	0.78	2.00	**0.80**	0.12	-	**2.09**	-	**2.88**	nr	**3.24**
extra; fittings; one end	0.82	2.00	**0.84**	0.15	-	**2.61**	-	**3.44**	nr	**3.87**
extra; fittings; two ends	0.83	2.00	**0.85**	0.30	-	**5.21**	-	**6.06**	nr	**6.82**
extra; fittings; three ends	1.14	2.00	**1.16**	0.30	-	**5.21**	-	**6.38**	nr	**7.17**
20mm	5.70	5.00	**5.99**	0.68	-	**11.82**	-	**17.81**	m	**20.03**
20mm; fittings										
extra; made bends	-	-	-	0.30	-	**5.21**	-	**5.21**	nr	**5.87**
extra; connections to copper pipe ends; compression joint	1.28	2.00	**1.31**	0.20	-	**3.48**	-	**4.78**	nr	**5.38**
extra; fittings; one end	0.94	2.00	**0.96**	0.20	-	**3.48**	-	**4.43**	nr	**4.99**
extra; fittings; two ends	1.14	2.00	**1.16**	0.40	-	**6.95**	-	**8.11**	nr	**9.13**
extra; fittings; three ends	1.66	2.00	**1.69**	0.40	-	**6.95**	-	**8.65**	nr	**9.73**
25mm	7.79	5.00	**8.18**	0.80	-	**13.90**	-	**22.08**	m	**24.84**
25mm; fittings										
extra; made bends	-	-	-	0.40	-	**6.95**	-	**6.95**	nr	**7.82**
extra; connections to copper pipe ends; compression joint	2.48	2.00	**2.53**	0.30	-	**5.21**	-	**7.74**	nr	**8.71**
extra; fittings; one end	1.26	2.00	**1.29**	0.30	-	**5.21**	-	**6.50**	nr	**7.31**
extra; fittings; two ends	1.76	2.00	**1.80**	0.60	-	**10.43**	-	**12.22**	nr	**13.75**
extra; fittings; three ends	2.39	2.00	**2.44**	0.60	-	**10.43**	-	**12.87**	nr	**14.47**
32mm	9.75	5.00	**10.24**	0.96	-	**16.68**	-	**26.92**	m	**30.29**
32mm; fittings										
extra; made bends	-	-	-	0.60	-	**10.43**	-	**10.43**	nr	**11.73**
extra; connections to copper pipe ends; compression joint	8.43	2.00	**8.60**	0.40	-	**6.95**	-	**15.55**	nr	**17.49**
extra; fittings; one end	1.94	2.00	**1.98**	0.40	-	**6.95**	-	**8.93**	nr	**10.05**
extra; fittings; two ends	3.30	2.00	**3.37**	0.80	-	**13.90**	-	**17.27**	nr	**19.43**
extra; fittings; three ends	4.53	2.00	**4.62**	0.80	-	**13.90**	-	**18.52**	nr	**20.84**
40mm	11.52	5.00	**12.10**	1.20	-	**20.86**	-	**32.95**	m	**37.07**
40mm; fittings										
extra; made bends	-	-	-	0.80	-	**13.90**	-	**13.90**	nr	**15.64**
extra; connections to copper pipe ends; compression joint	13.56	2.00	**13.83**	0.60	-	**10.43**	-	**24.26**	nr	**27.29**
extra; fittings; one end	2.47	2.00	**2.52**	0.50	-	**8.69**	-	**11.21**	nr	**12.61**
extra; fittings; two ends	4.95	2.00	**5.05**	1.00	-	**17.38**	-	**22.43**	nr	**25.23**
extra; fittings; three ends	6.20	2.00	**6.32**	1.00	-	**17.38**	-	**23.70**	nr	**26.67**
50mm	16.03	5.00	**16.83**	1.60	-	**27.81**	-	**44.64**	m	**50.22**
50mm; fittings										
extra; made bends	-	-	-	1.20	-	**20.86**	-	**20.86**	nr	**23.46**
extra; connections to copper pipe ends; compression joint	20.53	2.00	**20.94**	0.80	-	**13.90**	-	**34.84**	nr	**39.20**
extra; fittings; one end	4.61	2.00	**4.70**	0.60	-	**10.43**	-	**15.13**	nr	**17.02**
extra; fittings; two ends	5.79	2.00	**5.91**	1.20	-	**20.86**	-	**26.76**	nr	**30.11**
extra; fittings; three ends	8.93	2.00	**9.11**	1.20	-	**20.86**	-	**29.96**	nr	**33.71**
Pipes straight in malleable iron pipe brackets fixing to masonry with screws										
15mm	5.34	5.00	**5.60**	0.64	-	**11.12**	-	**16.72**	m	**18.81**
20mm	6.07	5.00	**6.36**	0.72	-	**12.51**	-	**18.87**	m	**21.23**
25mm	8.25	5.00	**8.65**	0.84	-	**14.60**	-	**23.25**	m	**26.15**
32mm	10.65	5.00	**11.15**	1.00	-	**17.38**	-	**28.53**	m	**32.10**
40mm	12.73	5.00	**13.33**	1.24	-	**21.55**	-	**34.88**	m	**39.24**
50mm	17.77	5.00	**18.61**	1.64	-	**28.50**	-	**47.11**	m	**53.00**
Pipes straight in malleable iron single rings, 165mm long screwed both ends steel tubes and malleable iron backplates fixing to timber with screws										
15mm	6.03	5.00	**6.33**	0.72	-	**12.51**	-	**18.84**	m	**21.20**
20mm	6.76	5.00	**7.09**	0.80	-	**13.90**	-	**20.99**	m	**23.62**
25mm	8.90	5.00	**9.33**	0.92	-	**15.99**	-	**25.32**	m	**28.49**
32mm	11.05	5.00	**11.57**	1.08	-	**18.77**	-	**30.34**	m	**34.13**
40mm	13.00	5.00	**13.62**	1.32	-	**22.94**	-	**36.56**	m	**41.13**
50mm	17.87	5.00	**18.71**	1.72	-	**29.89**	-	**48.60**	m	**54.68**
Steel pipes, BS 1387 Table 3; steel fittings BS 1740 Part 1; screwed and PTFE tape joints; pipework galvanized										
Pipes, straight in malleable iron pipe brackets fixing to timber with screws										
15mm	6.13	5.00	**6.43**	0.60	-	**10.43**	-	**16.86**	m	**18.97**
15mm; fittings										
extra; made bends	-	-	-	0.20	-	**3.48**	-	**3.48**	nr	**3.91**
extra; connections to copper pipe ends; compression joint	0.78	2.00	**0.80**	0.12	-	**2.09**	-	**2.88**	nr	**3.24**
extra; fittings; one end	0.82	2.00	**0.84**	0.15	-	**2.61**	-	**3.44**	nr	**3.87**
extra; fittings; two ends	0.83	2.00	**0.85**	0.30	-	**5.21**	-	**6.06**	nr	**6.82**
extra; fittings; three ends	1.14	2.00	**1.16**	0.30	-	**5.21**	-	**6.38**	nr	**7.17**
20mm	7.00	5.00	**7.34**	0.68	-	**11.82**	-	**19.16**	m	**21.55**
20mm; fittings										
extra; made bends	-	-	-	0.30	-	**5.21**	-	**5.21**	nr	**5.87**
extra; connections to copper pipe ends; compression joint	1.28	2.00	**1.31**	0.20	-	**3.48**	-	**4.78**	nr	**5.38**

Labour hourly rates: (except Specialists) Craft Operatives £17.38 Labourer £10.71 Rates are national average prices. Refer to REGIONAL VARIATIONS for indicative levels of overall pricing in regions	MATERIALS			LABOUR			RATES			
	Del to Site £	Waste %	Material Cost £	Craft Optve Hrs	Lab Hrs	Labour Cost £	Sunds £	Nett Rate £	Unit	Gross Rate (+12.5%) £
S115: HOT AND COLD WATER AND GAS: BLACK AND GALVANIZED STEEL TUBES AND FITTINGS Cont.										
Steel pipes, BS 1387 Table 3; steel fittings BS 1740 Part 1; screwed and PTFE tape joints; pipework galvanized Cont.										
Pipes, straight in malleable iron pipe brackets fixing to timber with screws Cont.										
20mm; fittings Cont.										
extra; fittings; one end	0.94	2.00	0.96	0.20	-	3.48	-	4.43	nr	4.99
extra; fittings; two ends	1.14	2.00	1.16	0.40	-	6.95	-	8.11	nr	9.13
extra; fittings; three ends	1.66	2.00	1.69	0.40	-	6.95	-	8.65	nr	9.73
25mm	9.72	5.00	10.19	0.80	-	13.90	-	24.10	m	27.11
25mm; fittings										
extra; made bends	-	-	-	0.40	-	6.95	-	6.95	nr	7.82
extra; connections to copper pipe ends; compression joint	2.48	2.00	2.53	0.30	-	5.21	-	7.74	nr	8.71
extra; fittings; one end	1.26	2.00	1.29	0.30	-	5.21	-	6.50	nr	7.31
extra; fittings; two ends	1.76	2.00	1.80	0.60	-	10.43	-	12.22	nr	13.75
extra; fittings; three ends	2.39	2.00	2.44	0.60	-	10.43	-	12.87	nr	14.47
32mm	12.49	5.00	13.09	0.96	-	16.68	-	29.77	m	33.49
32mm; fittings										
extra; made bends	-	-	-	0.60	-	10.43	-	10.43	nr	11.73
extra; connections to copper pipe ends; compression joint	8.43	2.00	8.60	0.40	-	6.95	-	15.55	nr	17.49
extra; fittings; one end	1.94	2.00	1.98	0.40	-	6.95	-	8.93	nr	10.05
extra; fittings; two ends	3.30	2.00	3.37	0.80	-	13.90	-	17.27	nr	19.43
extra; fittings; three ends	4.53	2.00	4.62	0.80	-	13.90	-	18.52	nr	20.84
40mm	14.93	5.00	15.64	1.20	-	20.86	-	36.50	m	41.06
40mm; fittings										
extra; made bends	-	-	-	0.80	-	13.90	-	13.90	nr	15.64
extra; connections to copper pipe ends; compression joint	13.56	2.00	13.83	0.60	-	10.43	-	24.26	nr	27.29
extra; fittings; one end	2.47	2.00	2.52	0.50	-	8.69	-	11.21	nr	12.61
extra; fittings; two ends	4.95	2.00	5.05	1.00	-	17.38	-	22.43	nr	25.23
extra; fittings; three ends	6.20	2.00	6.32	1.00	-	17.38	-	23.70	nr	26.67
50mm	20.66	5.00	21.64	1.60	-	27.81	-	49.44	m	55.62
50mm; fittings										
extra; made bends	-	-	-	1.20	-	20.86	-	20.86	nr	23.46
extra; connections to copper pipe ends; compression joint	20.53	2.00	20.94	0.80	-	13.90	-	34.84	nr	39.20
extra; fittings; one end	4.61	2.00	4.70	0.60	-	10.43	-	15.13	nr	17.02
extra; fittings; two ends	5.79	2.00	5.91	1.20	-	20.86	-	26.76	nr	30.11
extra; fittings; three ends	8.93	2.00	9.11	1.20	-	20.86	-	29.96	nr	33.71
Pipes straight in malleable iron pipe brackets fixing to masonry with screws										
15mm	6.15	5.00	6.45	0.64	-	11.12	-	17.57	m	19.77
20mm	7.02	5.00	7.36	0.72	-	12.51	-	19.87	m	22.35
25mm	9.74	5.00	10.21	0.84	-	14.60	-	24.81	m	27.91
32mm	12.51	5.00	13.10	1.00	-	17.38	-	30.48	m	34.30
40mm	14.95	5.00	15.66	1.24	-	21.55	-	37.21	m	41.86
50mm	20.67	5.00	21.65	1.64	-	28.50	-	50.16	m	56.42
Pipes straight in malleable iron single rings, 165mm long screwed both ends steel tubes and malleable iron backplates fixing to timber with screws										
15mm	6.84	5.00	7.18	0.72	-	12.51	-	19.69	m	22.15
20mm	7.71	5.00	8.09	0.80	-	13.90	-	21.99	m	24.74
25mm	10.39	5.00	10.90	0.92	-	15.99	-	26.89	m	30.25
32mm	12.91	5.00	13.52	1.08	-	18.77	-	32.29	m	36.33
40mm	15.22	5.00	15.95	1.32	-	22.94	-	38.89	m	43.75
50mm	20.77	5.00	21.75	1.72	-	29.89	-	51.65	m	58.10
S115: HOT AND COLD WATER AND GAS: LIGHT GAUGE STAINLESS STEEL TUBES AND FITTINGS										
Stainless steel pipes, BS EN 10312 Part 2; C.P. copper fittings, capillary, BS EN 1254 Part 2										
Pipes, straight in C.P. pipe brackets fixing to timber with screws										
15mm	5.07	5.00	5.32	0.40	-	6.95	-	12.28	m	13.81
15mm; fittings										
extra; fittings; one end	4.22	3.00	4.35	0.06	-	1.04	-	5.39	nr	6.06
extra; fittings; two ends	1.31	3.00	1.35	0.12	-	2.09	-	3.43	nr	3.86
extra; fittings; three ends	2.32	3.00	2.39	0.12	-	2.09	-	4.48	nr	5.03
22mm	7.95	5.00	8.33	0.43	-	7.47	-	15.81	m	17.78
22mm; fittings										
extra; fittings; one end	4.91	3.00	5.06	0.10	-	1.74	-	6.80	nr	7.64
extra; fittings; two ends	3.11	3.00	3.20	0.20	-	3.48	-	6.68	nr	7.51
extra; fittings; three ends	7.11	3.00	7.32	0.20	-	3.48	-	10.80	nr	12.15
28mm	10.74	5.00	11.25	0.48	-	8.34	-	19.59	m	22.04
28mm; fittings										
extra; fittings; one end	8.64	3.00	8.90	0.15	-	2.61	-	11.51	nr	12.94
extra; fittings; two ends	7.03	3.00	7.24	0.30	-	5.21	-	12.45	nr	14.01
extra; fittings; three ends	10.86	3.00	11.19	0.30	-	5.21	-	16.40	nr	18.45
35mm	13.76	5.00	14.40	0.60	-	10.43	-	24.83	m	27.93
35mm; fittings										
extra; fittings; one end	19.08	3.00	19.65	0.20	-	3.48	-	23.13	nr	26.02
extra; fittings; two ends	16.71	3.00	17.21	0.40	-	6.95	-	24.16	nr	27.18
extra; fittings; three ends	27.20	3.00	28.02	0.40	-	6.95	-	34.97	nr	39.34
42mm	19.20	5.00	20.07	0.80	-	13.90	-	33.98	m	38.22
42mm; fittings										
extra; fittings; one end	32.86	3.00	33.85	0.30	-	5.21	-	39.06	nr	43.94

Labour hourly rates: (except Specialists) Craft Operatives £17.38 Labourer £10.71 Rates are national average prices. Refer to REGIONAL VARIATIONS for indicative levels of overall pricing in regions	MATERIALS			LABOUR				RATES		
	Del to Site £	Waste %	Material Cost £	Craft Optve Hrs	Lab Hrs	Labour Cost £	Sunds £	Nett Rate £	Unit	Gross Rate (+12.5%) £
S115: HOT AND COLD WATER AND GAS: LIGHT GAUGE STAINLESS STEEL TUBES AND FITTINGS Cont.										
Stainless steel pipes, BS EN 10312 Part 2; C.P. copper fittings, capillary, BS EN 1254 Part 2 Cont.										
Pipes, straight in C.P. pipe brackets fixing to timber with screws Cont.										
42mm; fittings Cont.										
extra; fittings; two ends	27.62	3.00	**28.45**	0.60	-	**10.43**	-	**38.88**	nr	**43.74**
extra; fittings; three ends	43.66	3.00	**44.97**	0.60	-	**10.43**	-	**55.40**	nr	**62.32**
Pipes straight in C.P. pipe brackets fixing to masonry with screws										
15mm	5.09	5.00	**5.34**	0.45	-	**7.82**	-	**13.16**	m	**14.81**
22mm	7.96	5.00	**8.35**	0.48	-	**8.34**	-	**16.69**	m	**18.78**
28mm	10.75	5.00	**11.26**	0.53	-	**9.21**	-	**20.48**	m	**23.03**
35mm	13.78	5.00	**14.41**	0.65	-	**11.30**	-	**25.71**	m	**28.93**
42mm	19.21	5.00	**20.09**	0.85	-	**14.77**	-	**34.86**	m	**39.22**
Stainless steel pipes, BS EN 10312 Part 2; C.P. copper alloy fittings, compression, BS EN 1254 Part 2, Type A										
Pipes, straight in C.P. pipe brackets fixing to timber with screws										
15mm	5.39	5.00	**5.65**	0.40	-	**6.95**	-	**12.60**	m	**14.18**
15mm; fittings										
extra; fittings; one end	4.05	3.00	**4.17**	0.06	-	**1.04**	-	**5.21**	nr	**5.87**
extra; fittings; two ends	3.72	3.00	**3.83**	0.12	-	**2.09**	-	**5.92**	nr	**6.66**
extra; fittings; three ends	5.34	3.00	**5.50**	0.12	-	**2.09**	-	**7.59**	nr	**8.53**
22mm	8.21	5.00	**8.61**	0.43	-	**7.47**	-	**16.08**	m	**18.09**
22mm; fittings										
extra; fittings; one end	5.43	3.00	**5.59**	0.10	-	**1.74**	-	**7.33**	nr	**8.25**
extra; fittings; two ends	5.78	3.00	**5.95**	0.20	-	**3.48**	-	**9.43**	nr	**10.61**
extra; fittings; three ends	8.39	3.00	**8.64**	0.20	-	**3.48**	-	**12.12**	nr	**13.63**
28mm	11.11	5.00	**11.63**	0.48	-	**8.34**	-	**19.98**	m	**22.47**
28mm; fittings										
extra; fittings; one end	9.33	3.00	**9.61**	0.15	-	**2.61**	-	**12.22**	nr	**13.74**
extra; fittings; two ends	14.60	3.00	**15.04**	0.30	-	**5.21**	-	**20.25**	nr	**22.78**
extra; fittings; three ends	22.60	3.00	**23.28**	0.30	-	**5.21**	-	**28.49**	nr	**32.05**
35mm	14.85	5.00	**15.52**	0.60	-	**10.43**	-	**25.95**	m	**29.19**
35mm; fittings										
extra; fittings; one end	18.67	3.00	**19.23**	0.20	-	**3.48**	-	**22.71**	nr	**25.54**
extra; fittings; two ends	28.37	3.00	**29.22**	0.40	-	**6.95**	-	**36.17**	nr	**40.69**
extra; fittings; three ends	34.09	3.00	**35.11**	0.40	-	**6.95**	-	**42.06**	nr	**47.32**
42mm	20.51	5.00	**21.43**	0.80	-	**13.90**	-	**35.33**	m	**39.75**
42mm; fittings										
extra; fittings; one end	30.54	3.00	**31.46**	0.30	-	**5.21**	-	**36.67**	nr	**41.25**
extra; fittings; two ends	48.80	3.00	**50.26**	0.60	-	**10.43**	-	**60.69**	nr	**68.28**
extra; fittings; three ends	63.60	3.00	**65.51**	0.60	-	**10.43**	-	**75.94**	nr	**85.43**
Pipes straight in C.P. pipe brackets fixing to masonry with screws										
15mm	5.41	5.00	**5.66**	0.45	-	**7.82**	-	**13.49**	m	**15.17**
22mm	8.23	5.00	**8.62**	0.48	-	**8.34**	-	**16.97**	m	**19.09**
28mm	11.13	5.00	**11.65**	0.53	-	**9.21**	-	**20.86**	m	**23.47**
35mm	14.87	5.00	**15.54**	0.65	-	**11.30**	-	**26.84**	m	**30.19**
42mm	20.53	5.00	**21.44**	0.85	-	**14.77**	-	**36.22**	m	**40.74**
S115: HOT AND COLD WATER AND GAS: COPPER TUBING AND FITTINGS										
Copper pipes, BS EN 1057 Part 1 Table X; copper fittings, capillary, BS EN 1254 Part 2										
Pipes, straight in two piece copper spacing clips to timber with screws										
15mm	1.39	5.00	**1.46**	0.40	-	**6.95**	-	**8.41**	m	**9.46**
15mm; fitings										
extra; made bends	-	-	**-**	0.10	-	**1.74**	-	**1.74**	nr	**1.96**
extra; connections to iron pipe ends; screwed joint	2.15	3.00	**2.21**	0.12	-	**2.09**	-	**4.30**	nr	**4.84**
extra; fittings; one end	1.10	3.00	**1.13**	0.06	-	**1.04**	-	**2.18**	nr	**2.45**
extra; fittings; two ends	0.27	3.00	**0.28**	0.12	-	**2.09**	-	**2.36**	nr	**2.66**
extra; fittings; three ends	0.51	3.00	**0.53**	0.12	-	**2.09**	-	**2.61**	nr	**2.94**
22mm	2.69	5.00	**2.82**	0.43	-	**7.47**	-	**10.29**	m	**11.58**
22mm; fittings										
extra; made bends	-	-	**-**	0.16	-	**2.78**	-	**2.78**	nr	**3.13**
extra; connections to iron pipe ends; screwed joint	3.81	3.00	**3.92**	0.20	-	**3.48**	-	**7.40**	nr	**8.33**
extra; fittings; one end	2.06	3.00	**2.12**	0.10	-	**1.74**	-	**3.86**	nr	**4.34**
extra; fittings; two ends	0.71	3.00	**0.73**	0.20	-	**3.48**	-	**4.21**	nr	**4.73**
extra; fittings; three ends	1.64	3.00	**1.69**	0.20	-	**3.48**	-	**5.17**	nr	**5.81**
28mm	3.55	5.00	**3.73**	0.48	-	**8.34**	-	**12.07**	m	**13.58**
28mm; fittings										
extra; made bends	-	-	**-**	0.20	-	**3.48**	-	**3.48**	nr	**3.91**
extra; connections to iron pipe ends; screwed joint	6.15	3.00	**6.33**	0.30	-	**5.21**	-	**11.55**	nr	**12.99**
extra; fittings; one end	3.68	3.00	**3.79**	0.15	-	**2.61**	-	**6.40**	nr	**7.20**
extra; fittings; two ends	1.64	3.00	**1.69**	0.30	-	**5.21**	-	**6.90**	nr	**7.77**
extra; fittings; three ends	4.54	3.00	**4.68**	0.30	-	**5.21**	-	**9.89**	nr	**11.13**
35mm	7.79	5.00	**8.16**	0.60	-	**10.43**	-	**18.59**	m	**20.92**
35mm; fittings										
extra; made bends	-	-	**-**	0.30	-	**5.21**	-	**5.21**	nr	**5.87**
extra; connections to iron pipe ends; screwed joint	10.83	3.00	**11.15**	0.40	-	**6.95**	-	**18.11**	nr	**20.37**

Labour hourly rates: (except Specialists) Craft Operatives £17.38 Labourer £10.71 Rates are national average prices. Refer to REGIONAL VARIATIONS for indicative levels of overall pricing in regions	MATERIALS			LABOUR				RATES		
	Del to Site £	Waste %	Material Cost £	Craft Optve Hrs	Lab Hrs	Labour Cost £	Sunds £	Nett Rate £	Unit	Gross Rate (+12.5%) £
S115: HOT AND COLD WATER AND GAS: COPPER TUBING AND FITTINGS Cont.										
Copper pipes, BS EN 1057 Part 1 Table X; copper fittings, capillary, BS EN 1254 Part 2 Cont.										
Pipes, straight in two piece copper spacing clips to timber with screws Cont.										
35mm; fittings Cont.										
extra; fittings; one end	8.12	3.00	8.36	0.20	-	3.48	-	11.84	nr	13.32
extra; fittings; two ends	7.11	3.00	7.32	0.40	-	6.95	-	14.28	nr	16.06
extra; fittings; three ends	11.58	3.00	11.93	0.40	-	6.95	-	18.88	nr	21.24
42mm	9.89	5.00	10.37	0.80	-	13.90	-	24.27	m	27.30
42mm; fittings										
extra; made bends	-	-	-	0.40	-	6.95	-	6.95	nr	7.82
extra; connections to iron pipe ends; screwed joint	13.94	3.00	14.36	0.60	-	10.43	-	24.79	nr	27.88
extra; fittings; one end	13.98	3.00	14.40	0.30	-	5.21	-	19.61	nr	22.07
extra; fittings; two ends	11.75	3.00	12.10	0.60	-	10.43	-	22.53	nr	25.35
extra; fittings; three ends	18.58	3.00	19.14	0.60	-	10.43	-	29.57	nr	33.26
54mm	13.21	5.00	13.84	1.20	-	20.86	-	34.70	m	39.03
54mm; fittings										
extra; made bends	-	-	-	0.60	-	10.43	-	10.43	nr	11.73
extra; connections to iron pipe ends; screwed joint	21.18	3.00	21.82	0.80	-	13.90	-	35.72	nr	40.18
extra; fittings; one end	19.52	3.00	20.11	0.40	-	6.95	-	27.06	nr	30.44
extra; fittings; two ends	24.26	3.00	24.99	0.80	-	13.90	-	38.89	nr	43.75
extra; fittings; three ends	37.45	3.00	38.57	0.80	-	13.90	-	52.48	nr	59.04
Pipes straight in two piece copper spacing clips fixing to masonry with screws										
15mm	1.41	5.00	1.47	0.45	-	7.82	-	9.30	m	10.46
22mm	2.70	5.00	2.84	0.48	-	8.34	-	11.18	m	12.58
28mm	3.57	5.00	3.74	0.53	-	9.21	-	12.95	m	14.57
35mm	7.80	5.00	8.18	0.65	-	11.30	-	19.48	m	21.91
42mm	9.91	5.00	10.38	0.85	-	14.77	-	25.16	m	28.30
54mm	13.23	5.00	13.86	1.25	-	21.73	-	35.58	m	40.03
Pipes straight in pressed brass pipe brackets fixing to timber with screws										
15mm	1.80	5.00	1.89	0.40	-	6.95	-	8.85	m	9.95
22mm	3.17	5.00	3.32	0.43	-	7.47	-	10.80	m	12.15
28mm	4.12	5.00	4.32	0.48	-	8.34	-	12.66	m	14.25
35mm	8.37	5.00	8.77	0.60	-	10.43	-	19.20	m	21.60
42mm	10.68	5.00	11.20	0.80	-	13.90	-	25.10	m	28.24
54mm	14.26	5.00	14.94	1.20	-	20.86	-	35.80	m	40.27
Pipes straight in cast brass pipe brackets fixing to timber with screws										
15mm	1.91	5.00	2.00	0.40	-	6.95	-	8.96	m	10.08
22mm	3.38	5.00	3.54	0.43	-	7.47	-	11.02	m	12.40
28mm	4.42	5.00	4.63	0.48	-	8.34	-	12.98	m	14.60
35mm	9.10	5.00	9.55	0.60	-	10.43	-	19.97	m	22.47
42mm	11.55	5.00	12.11	0.80	-	13.90	-	26.01	m	29.27
54mm	15.70	5.00	16.45	1.20	-	20.86	-	37.31	m	41.97
Pipes straight In cast brass single rings and back plates fixing to timber with screws										
15mm	2.76	5.00	2.90	0.45	-	7.82	-	10.72	m	12.06
22mm	4.12	5.00	4.32	0.48	-	8.34	-	12.66	m	14.25
28mm	5.11	5.00	5.36	0.53	-	9.21	-	14.58	m	16.40
35mm	9.35	5.00	9.80	0.65	-	11.30	-	21.10	m	23.74
42mm	11.43	5.00	11.98	0.85	-	14.77	-	26.75	m	30.10
54mm	14.87	5.00	15.58	1.25	-	21.73	-	37.31	m	41.97
66.7mm	23.98	5.00	25.08	1.80	-	31.28	-	56.36	m	63.41
76.1mm	36.69	5.00	38.31	2.40	-	41.71	-	80.02	m	90.02
108mm	52.72	5.00	55.05	3.60	-	62.57	-	117.62	m	132.32

Copper pipes, BS EN 1057 Part 1 Table X; copper alloy fittings, compression, BS EN 1254 Part 2, Type A

	Del to Site £	Waste %	Material Cost £	Craft Optve Hrs	Lab Hrs	Labour Cost £	Sunds £	Nett Rate £	Unit	Gross Rate (+12.5%) £
Pipes, straight in C.P. pipe brackets fixing to timber with screws										
15mm	1.64	5.00	1.72	0.33	-	5.74	-	7.45	m	8.38
15mm; fittings										
extra; made bends	-	-	-	0.08	-	1.39	-	1.39	nr	1.56
extra; connections to iron pipe ends; screwed joint	0.78	3.00	0.80	0.12	-	2.09	-	2.89	nr	3.25
extra; fittings; one end	1.33	3.00	1.37	0.06	-	1.04	-	2.41	nr	2.71
extra; fittings; two ends	1.02	3.00	1.05	0.12	-	2.09	-	3.14	nr	3.53
extra; fittings; three ends	1.48	3.00	1.52	0.12	-	2.09	-	3.61	nr	4.06
22mm	3.10	5.00	3.24	0.43	-	7.47	-	10.72	m	12.06
22mm; fittings										
extra; made bends	-	-	-	0.16	-	2.78	-	2.78	nr	3.13
extra; connections to iron pipe ends; screwed joint	1.28	3.00	1.32	0.20	-	3.48	-	4.79	nr	5.39
extra; fittings; one end	1.59	3.00	1.64	0.10	-	1.74	-	3.38	nr	3.80
extra; fittings; two ends	1.73	3.00	1.78	0.20	-	3.48	-	5.26	nr	5.92
extra; fittings; three ends	2.42	3.00	2.49	0.20	-	3.48	-	5.97	nr	6.71
28mm	5.02	5.00	5.23	0.48	-	8.34	-	13.58	m	15.27
28mm; fittings										
extra; made bends	-	-	-	0.20	-	3.48	-	3.48	nr	3.91
extra; connections to iron pipe ends; screwed joint	2.48	3.00	2.55	0.30	-	5.21	-	7.77	nr	8.74
extra; fittings; one end	5.18	3.00	5.34	0.15	-	2.61	-	7.94	nr	8.94
extra; fittings; two ends	6.21	3.00	6.40	0.30	-	5.21	-	11.61	nr	13.06

Labour hourly rates: (except Specialists) Craft Operatives £17.38 Labourer £10.71 Rates are national average prices. Refer to REGIONAL VARIATIONS for indicative levels of overall pricing in regions	MATERIALS			LABOUR				RATES		
	Del to Site £	Waste %	Material Cost £	Craft Optve Hrs	Lab Hrs	Labour Cost £	Sunds £	Nett Rate £	Unit	Gross Rate (+12.5%) £
S115: HOT AND COLD WATER AND GAS: COPPER TUBING AND FITTINGS Cont.										
Copper pipes, BS EN 1057 Part 1 Table X; copper alloy fittings, compression, BS EN 1254 Part 2, Type A Cont.										
Pipes, straight in C.P. pipe brackets fixing to timber with screws Cont.										
28mm; fittings Cont.										
extra; fittings; three ends	11.30	3.00	**11.64**	0.30	-	**5.21**	-	**16.85**	nr	**18.96**
35mm	11.14	5.00	**11.62**	0.60	-	**10.43**	-	**22.05**	m	**24.80**
35mm; fittings										
extra; made bends	-	-	-	0.30	-	**5.21**	-	**5.21**	nr	**5.87**
extra; connections to iron pipe ends; screwed joint	8.43	3.00	**8.68**	0.40	-	**6.95**	-	**15.63**	nr	**17.59**
extra; fittings; one end	9.34	3.00	**9.62**	0.20	-	**3.48**	-	**13.10**	nr	**14.73**
extra; fittings; two ends	15.63	3.00	**16.10**	0.40	-	**6.95**	-	**23.05**	nr	**25.93**
extra; fittings; three ends	20.66	3.00	**21.28**	0.40	-	**6.95**	-	**28.23**	nr	**31.76**
42mm	14.39	5.00	**15.00**	0.80	-	**13.90**	-	**28.91**	m	**32.52**
42mm; fittings										
extra; made bends	-	-	-	0.40	-	**6.95**	-	**6.95**	nr	**7.82**
extra; connections to iron pipe ends; screwed joint	13.56	3.00	**13.97**	0.60	-	**10.43**	-	**24.39**	nr	**27.44**
extra; fittings; one end	15.27	3.00	**15.73**	0.30	-	**5.21**	-	**20.94**	nr	**23.56**
extra; fittings; two ends	21.87	3.00	**22.53**	0.60	-	**10.43**	-	**32.95**	nr	**37.07**
extra; fittings; three ends	31.79	3.00	**32.74**	0.60	-	**10.43**	-	**43.17**	nr	**48.57**
54mm	19.26	5.00	**20.07**	1.20	-	**20.86**	-	**40.93**	m	**46.04**
54mm; fittings										
extra; made bends	-	-	-	0.60	-	**10.43**	-	**10.43**	nr	**11.73**
extra; connections to iron pipe ends; screwed joint	20.53	3.00	**21.15**	0.80	-	**13.90**	-	**35.05**	nr	**39.43**
extra; fittings; one end	21.29	3.00	**21.93**	0.40	-	**6.95**	-	**28.88**	nr	**32.49**
extra; fittings; two ends	37.17	3.00	**38.29**	0.80	-	**13.90**	-	**52.19**	nr	**58.71**
extra; fittings; three ends	51.17	3.00	**52.71**	0.80	-	**13.90**	-	**66.61**	nr	**74.94**
Pipes; straight; in two piece copper spacing clips fixing to masonry with screws										
15mm	1.66	5.00	**1.73**	0.45	-	**7.82**	-	**9.56**	m	**10.75**
22mm	3.11	5.00	**3.26**	0.48	-	**8.34**	-	**11.60**	m	**13.05**
28mm	5.03	5.00	**5.25**	0.53	-	**9.21**	-	**14.46**	m	**16.27**
35mm	11.16	5.00	**11.64**	0.65	-	**11.30**	-	**22.93**	m	**25.80**
42mm	14.41	5.00	**15.02**	0.85	-	**14.77**	-	**29.79**	m	**33.52**
54mm	19.28	5.00	**20.09**	1.25	-	**21.73**	-	**41.81**	m	**47.04**
Pipes straight in pressed brass pipe brackets fixing to timber with screws										
15mm	2.06	5.00	**2.15**	0.40	-	**6.95**	-	**9.11**	m	**10.24**
22mm	3.58	5.00	**3.75**	0.43	-	**7.47**	-	**11.22**	m	**12.62**
28mm	5.58	5.00	**5.83**	0.48	-	**8.34**	-	**14.17**	m	**15.94**
35mm	11.72	5.00	**12.23**	0.60	-	**10.43**	-	**22.66**	m	**25.49**
42mm	15.18	5.00	**15.83**	0.80	-	**13.90**	-	**29.74**	m	**33.46**
54mm	20.31	5.00	**21.17**	1.20	-	**20.86**	-	**42.03**	m	**47.28**
Pipes straight in cast brass pipe brackets fixing to timber with screws										
15mm	2.16	5.00	**2.26**	0.40	-	**6.95**	-	**9.22**	m	**10.37**
22mm	3.79	5.00	**3.97**	0.43	-	**7.47**	-	**11.44**	m	**12.87**
28mm	5.88	5.00	**6.14**	0.48	-	**8.34**	-	**14.48**	m	**16.30**
35mm	12.46	5.00	**13.00**	0.60	-	**10.43**	-	**23.43**	m	**26.36**
42mm	16.05	5.00	**16.75**	0.80	-	**13.90**	-	**30.65**	m	**34.48**
54mm	21.75	5.00	**22.68**	1.20	-	**20.86**	-	**43.54**	m	**48.98**
Pipes straight in cast brass single rings and back plates fixing to timber with screws										
15mm	3.02	5.00	**3.16**	0.40	-	**6.95**	-	**10.11**	m	**11.38**
22mm	4.53	5.00	**4.74**	0.43	-	**7.47**	-	**12.22**	m	**13.75**
28mm	6.58	5.00	**6.87**	0.48	-	**8.34**	-	**15.21**	m	**17.12**
35mm	12.70	5.00	**13.26**	0.60	-	**10.43**	-	**23.69**	m	**26.65**
42mm	15.93	5.00	**16.62**	0.80	-	**13.90**	-	**30.52**	m	**34.34**
54mm	20.92	5.00	**21.81**	1.20	-	**20.86**	-	**42.67**	m	**48.00**
66.7mm	55.08	5.00	**57.11**	1.80	-	**31.28**	-	**88.39**	m	**99.44**
76.1mm	80.46	5.00	**83.39**	2.40	-	**41.71**	-	**125.10**	m	**140.73**
108mm	126.37	5.00	**130.91**	3.60	-	**62.57**	-	**193.48**	m	**217.67**
Copper pipes, BS EN 1057 Part 1 Table X; copper alloy fittings, compression, BS EN 1254 Part 2 Type B										
Pipes, straight in two piece copper spacing clips to timber with screws										
15mm	2.63	5.00	**2.74**	0.40	-	**6.95**	-	**9.69**	m	**10.90**
15mm; fittings										
extra; made bends	-	-	-	0.10	-	**1.74**	-	**1.74**	nr	**1.96**
extra; connections to iron pipe ends; screwed joint	3.51	3.00	**3.62**	0.16	-	**2.78**	-	**6.40**	nr	**7.20**
extra; fittings; one end	3.84	3.00	**3.96**	0.08	-	**1.39**	-	**5.35**	nr	**6.01**
extra; fittings; two ends	4.55	3.00	**4.69**	0.16	-	**2.78**	-	**7.47**	nr	**8.40**
extra; fittings; three ends	6.37	3.00	**6.56**	0.16	-	**2.78**	-	**9.34**	nr	**10.51**
22mm	4.68	5.00	**4.87**	0.43	-	**7.47**	-	**12.34**	m	**13.89**
22mm; fittings										
extra; made bends	-	-	-	0.16	-	**2.78**	-	**2.78**	nr	**3.13**
extra; connections to iron pipe ends; screwed joint	5.01	3.00	**5.16**	0.25	-	**4.35**	-	**9.51**	nr	**10.69**
extra; fittings; one end	4.68	3.00	**4.82**	0.13	-	**2.26**	-	**7.08**	nr	**7.96**
extra; fittings; two ends	7.41	3.00	**7.63**	0.25	-	**4.35**	-	**11.98**	nr	**13.47**
extra; fittings; three ends	10.73	3.00	**11.05**	0.25	-	**4.35**	-	**15.40**	nr	**17.32**
28mm	6.75	5.00	**7.02**	0.48	-	**8.34**	-	**15.36**	m	**17.28**

Labour hourly rates: (except Specialists) Craft Operatives £17.38 Labourer £10.71 Rates are national average prices. Refer to REGIONAL VARIATIONS for indicative levels of overall pricing in regions	MATERIALS			LABOUR				RATES		
	Del to Site £	Waste %	Material Cost £	Craft Optve Hrs	Lab Hrs	Labour Cost £	Sunds £	Nett Rate £	Unit	Gross Rate (+12.5%) £
S115: HOT AND COLD WATER AND GAS: COPPER TUBING AND FITTINGS Cont.										
Copper pipes, BS EN 1057 Part 1 Table X; copper alloy fittings, compression, BS EN 1254 Part 2 Type B Cont.										
Pipes, straight in two piece copper spacing clips to timber with screws Cont.										
28mm;fittings										
extra; made bends	-	-	-	0.20	-	3.48	-	3.48	nr	3.91
extra; connections to iron pipe ends; screwed joint	7.67	3.00	7.90	0.37	-	6.43	-	14.33	nr	16.12
extra; fittings; one end	8.78	3.00	9.04	0.19	-	3.30	-	12.35	nr	13.89
extra; fittings; two ends	12.55	3.00	12.93	0.37	-	6.43	-	19.36	nr	21.78
extra; fittings; three ends	18.07	3.00	18.61	0.37	-	6.43	-	25.04	nr	28.17
35mm	13.40	5.00	13.95	0.60	-	10.43	-	24.38	m	27.42
35mm; fittings										
extra; made bends	-	-	-	0.30	-	5.21	-	5.21	nr	5.87
extra; connections to iron pipe ends; screwed joint	15.15	3.00	15.60	0.48	-	8.34	-	23.95	nr	26.94
extra; fittings; one end	15.54	3.00	16.01	0.24	-	4.17	-	20.18	nr	22.70
extra; fittings; two ends	24.12	3.00	24.84	0.48	-	8.34	-	33.19	nr	37.33
extra; fittings; three ends	32.76	3.00	33.74	0.48	-	8.34	-	42.09	nr	47.35
42mm	17.36	5.00	18.06	0.80	-	13.90	-	31.96	m	35.96
42mm; fittings										
extra; made bends	-	-	-	0.40	-	6.95	-	6.95	nr	7.82
extra; connections to iron pipe ends; screwed joint	22.36	3.00	23.03	0.76	-	13.21	-	36.24	nr	40.77
extra; fittings; one end	24.44	3.00	25.17	0.38	-	6.60	-	31.78	nr	35.75
extra; fittings; two ends	36.14	3.00	37.22	0.76	-	13.21	-	50.43	nr	56.74
extra; fittings; three ends	53.30	3.00	54.90	0.76	-	13.21	-	68.11	nr	76.62
54mm	23.92	5.00	24.87	1.20	-	20.86	-	45.72	m	51.44
54mm; fittings										
extra; made bends	-	-	-	0.60	-	10.43	-	10.43	nr	11.73
extra; connections to iron pipe ends; screwed joint	32.37	3.00	33.34	1.00	-	17.38	-	50.72	nr	57.06
extra; fittings; one end	34.78	3.00	35.82	0.50	-	8.69	-	44.51	nr	50.08
extra; fittings; two ends	57.53	3.00	59.26	1.00	-	17.38	-	76.64	nr	86.22
extra; fittings; three ends	85.48	3.00	88.04	1.00	-	17.38	-	105.42	nr	118.60
Pipes straight in two piece copper spacing clips fixing to masonry with screws										
15mm	2.65	5.00	2.75	0.45	-	7.82	-	10.57	m	11.90
22mm	4.69	5.00	4.89	0.48	-	8.34	-	13.23	m	14.88
28mm	6.76	5.00	7.03	0.53	-	9.21	-	16.24	m	18.27
35mm	13.42	5.00	13.96	0.65	-	11.30	-	25.26	m	28.42
42mm	17.37	5.00	18.07	0.85	-	14.77	-	32.85	m	36.95
54mm	23.94	5.00	24.88	1.25	-	21.73	-	46.61	m	52.43
Pipes straight in pressed brass pipe brackets fixing to timber with screws										
15mm	3.05	5.00	3.17	0.45	-	7.82	-	10.99	m	12.37
22mm	5.16	5.00	5.38	0.48	-	8.34	-	13.72	m	15.43
28mm	7.31	5.00	7.61	0.53	-	9.21	-	16.82	m	18.92
35mm	13.98	5.00	14.56	0.65	-	11.30	-	25.85	m	29.09
42mm	18.15	5.00	18.89	0.85	-	14.77	-	33.66	m	37.87
54mm	24.97	5.00	25.97	1.25	-	21.73	-	47.69	m	53.66
Pipes straight in cast brass pipe brackets fixing to timber with screws										
15mm	3.15	5.00	3.28	0.45	-	7.82	-	11.10	m	12.49
22mm	5.37	5.00	5.60	0.48	-	8.34	-	13.94	m	15.68
28mm	7.61	5.00	7.92	0.53	-	9.21	-	17.13	m	19.28
35mm	14.72	5.00	15.33	0.65	-	11.30	-	26.63	m	29.95
42mm	19.02	5.00	19.80	0.85	-	14.77	-	34.57	m	38.89
54mm	26.41	5.00	27.48	1.25	-	21.73	-	49.20	m	55.35
Pipes straight in cast brass single rings and back plates fixing to timber with screws										
15mm	4.01	5.00	4.18	0.45	-	7.82	-	12.00	m	13.50
22mm	6.11	5.00	6.37	0.48	-	8.34	-	14.71	m	16.55
28mm	8.31	5.00	8.65	0.53	-	9.21	-	17.86	m	20.10
35mm	14.96	5.00	15.59	0.65	-	11.30	-	26.88	m	30.24
42mm	18.89	5.00	19.67	0.85	-	14.77	-	34.44	m	38.75
54mm	25.58	5.00	26.61	1.25	-	21.73	-	48.34	m	54.38
Copper pipes, BS EN 1057 Part 1 Table Y; copper fittings, capillary, BS EN 1254 Part 2										
Pipes; straight; in trenches										
15mm	2.66	5.00	2.79	0.20	-	3.48	-	6.27	m	7.05
15mm; fittings										
extra; made bends	-	-	-	0.10	-	1.74	-	1.74	nr	1.96
extra; connections to iron pipe ends; screwed joint	2.15	3.00	2.21	0.16	-	2.78	-	5.00	nr	5.62
extra; fittings; one end	1.10	3.00	1.13	0.08	-	1.39	-	2.52	nr	2.84
extra; fittings; two ends	0.27	3.00	0.28	0.16	-	2.78	-	3.06	nr	3.44
extra; fittings; three ends	0.51	3.00	0.53	0.16	-	2.78	-	3.31	nr	3.72
22mm	4.65	5.00	4.88	0.22	-	3.82	-	8.70	m	9.79
22mm; fittings										
extra; made bends	-	-	-	0.16	-	2.78	-	2.78	nr	3.13
extra; connections to iron pipe ends; screwed joint	3.81	3.00	3.92	0.25	-	4.35	-	8.27	nr	9.30
extra; fittings; one end	2.06	3.00	2.12	0.13	-	2.26	-	4.38	nr	4.93
extra; fittings; two ends	0.71	3.00	0.73	0.25	-	4.35	-	5.08	nr	5.71
extra; fittings; three ends	1.64	3.00	1.69	0.25	-	4.35	-	6.03	nr	6.7[illegible]

Labour hourly rates: (except Specialists) Craft Operatives £17.38 Labourer £10.71 Rates are national average prices. Refer to REGIONAL VARIATIONS for indicative levels of overall pricing in regions	MATERIALS			LABOUR				RATES		
	Del to Site £	Waste %	Material Cost £	Craft Optve Hrs	Lab Hrs	Labour Cost £	Sunds £	Nett Rate £	Unit	Gross Rate (+12.5%) £
S115: HOT AND COLD WATER AND GAS: COPPER TUBING AND FITTINGS Cont.										
Copper pipes, BS EN 1057 Part 1 Table Y; copper fittings, capillary, BS EN 1254 Part 2 Cont.										
Pipes; straight; in trenches Cont.										
22mm; fittings Cont.										
28mm	6.06	5.00	**6.36**	0.24	-	**4.17**	-	**10.53**	m	**11.85**
28mm; fittings										
extra; made bends	-	-	-	0.20	-	**3.48**	-	**3.48**	nr	**3.91**
extra; connections to iron pipe ends; screwed joint	6.15	3.00	**6.33**	0.37	-	**6.43**	-	**12.77**	nr	**14.36**
extra; fittings; one end	3.68	3.00	**3.79**	0.19	-	**3.30**	-	**7.09**	nr	**7.98**
extra; fittings; two ends	1.64	3.00	**1.69**	0.37	-	**6.43**	-	**8.12**	nr	**9.13**
extra; fittings; three ends	4.54	3.00	**4.68**	0.37	-	**6.43**	-	**11.11**	nr	**12.50**
35mm	9.26	5.00	**9.72**	0.27	-	**4.69**	-	**14.41**	m	**16.21**
35mm; fittings										
extra; made bends	-	-	-	0.30	-	**5.21**	-	**5.21**	nr	**5.87**
extra; connections to iron pipe ends; screwed joint	10.83	3.00	**11.15**	0.50	-	**8.69**	-	**19.84**	nr	**22.33**
extra; fittings; one end	8.12	3.00	**8.36**	0.25	-	**4.35**	-	**12.71**	nr	**14.30**
extra; fittings; two ends	7.11	3.00	**7.32**	0.50	-	**8.69**	-	**16.01**	nr	**18.01**
extra; fittings; three ends	11.58	3.00	**11.93**	0.50	-	**8.69**	-	**20.62**	nr	**23.19**
42mm	11.33	5.00	**11.89**	0.35	-	**6.08**	-	**17.97**	m	**20.22**
42mm; fittings										
extra; made bends	-	-	-	0.40	-	**6.95**	-	**6.95**	nr	**7.82**
extra; connections to iron pipe ends; screwed joint	13.94	3.00	**14.36**	0.76	-	**13.21**	-	**27.57**	nr	**31.01**
extra; fittings; one end	13.98	3.00	**14.40**	0.38	-	**6.60**	-	**21.00**	nr	**23.63**
extra; fittings; two ends	11.75	3.00	**12.10**	0.76	-	**13.21**	-	**25.31**	nr	**28.48**
extra; fittings; three ends	18.58	3.00	**19.14**	0.76	-	**13.21**	-	**32.35**	nr	**36.39**
54mm	19.40	5.00	**20.36**	0.39	-	**6.78**	-	**27.14**	m	**30.53**
54mm; fittings										
extra; made bends	-	-	-	0.60	-	**10.43**	-	**10.43**	nr	**11.73**
extra; connections to iron pipe ends; screwed joint	21.18	3.00	**21.82**	1.00	-	**17.38**	-	**39.20**	nr	**44.09**
extra; fittings; one end	19.52	3.00	**20.11**	0.50	-	**8.69**	-	**28.80**	nr	**32.40**
extra; fittings; two ends	24.26	3.00	**24.99**	1.00	-	**17.38**	-	**42.37**	nr	**47.66**
extra; fittings; three ends	37.45	3.00	**38.57**	1.00	-	**17.38**	-	**55.95**	nr	**62.95**
Copper pipes, BS EN 1057 Part 1 Table Y; non-dezincifiable fittings; compression, BS EN 1254 Part 2, Type B										
Pipes; straight; in trenches										
15mm	2.75	5.00	**2.89**	0.20	-	**3.48**	-	**6.36**	m	**7.16**
15mm; fittings										
extra; made bends	-	-	-	0.10	-	**1.74**	-	**1.74**	nr	**1.96**
extra; connections to iron pipe ends; screwed joint	1.32	3.00	**1.36**	0.16	-	**2.78**	-	**4.14**	nr	**4.66**
extra; fittings; one end	1.95	3.00	**2.01**	0.08	-	**1.39**	-	**3.40**	nr	**3.82**
extra; fittings; two ends	1.80	3.00	**1.85**	0.16	-	**2.78**	-	**4.63**	nr	**5.21**
extra; fittings; three ends	2.57	3.00	**2.65**	0.16	-	**2.78**	-	**5.43**	nr	**6.11**
22mm	4.78	5.00	**5.02**	0.22	-	**3.82**	-	**8.84**	m	**9.95**
22mm; fittings										
extra; made bends	-	-	-	0.16	-	**2.78**	-	**2.78**	nr	**3.13**
extra; connections to iron pipe ends; screwed joint	2.15	3.00	**2.21**	0.25	-	**4.35**	-	**6.56**	nr	**7.38**
extra; fittings; one end	2.63	3.00	**2.71**	0.13	-	**2.26**	-	**4.97**	nr	**5.59**
extra; fittings; two ends	2.79	3.00	**2.87**	0.25	-	**4.35**	-	**7.22**	nr	**8.12**
extra; fittings; three ends	4.05	3.00	**4.17**	0.25	-	**4.35**	-	**8.52**	nr	**9.58**
28mm	6.37	5.00	**6.68**	0.24	-	**4.17**	-	**10.85**	m	**12.21**
28mm; fittings										
extra; made bends	-	-	-	0.20	-	**3.48**	-	**3.48**	nr	**3.91**
extra; connections to iron pipe ends; screwed joint	3.47	3.00	**3.57**	0.37	-	**6.43**	-	**10.00**	nr	**11.26**
extra; fittings; one end	5.65	3.00	**5.82**	0.19	-	**3.30**	-	**9.12**	nr	**10.26**
extra; fittings; two ends	7.30	3.00	**7.52**	0.37	-	**6.43**	-	**13.95**	nr	**15.69**
extra; fittings; three ends	11.98	3.00	**12.34**	0.37	-	**6.43**	-	**18.77**	nr	**21.12**
35mm	9.91	5.00	**10.39**	0.27	-	**4.69**	-	**15.09**	m	**16.97**
35mm; fittings										
extra; made bends	-	-	-	0.30	-	**5.21**	-	**5.21**	nr	**5.87**
extra; connections to iron pipe ends; screwed joint	8.87	3.00	**9.14**	0.48	-	**8.34**	-	**17.48**	nr	**19.66**
extra; fittings; one end	9.34	3.00	**9.62**	0.24	-	**4.17**	-	**13.79**	nr	**15.52**
extra; fittings; two ends	17.19	3.00	**17.71**	0.48	-	**8.34**	-	**26.05**	nr	**29.30**
extra; fittings; three ends	20.66	3.00	**21.28**	0.48	-	**8.34**	-	**29.62**	nr	**33.32**
42mm	12.07	5.00	**12.65**	0.35	-	**6.08**	-	**18.73**	m	**21.07**
42mm; fittings										
extra; made bends	-	-	-	0.40	-	**6.95**	-	**6.95**	nr	**7.82**
extra; connections to iron pipe ends; screwed joint	14.84	3.00	**15.29**	0.76	-	**13.21**	-	**28.49**	nr	**32.06**
extra; fittings; one end	15.27	3.00	**15.73**	0.38	-	**6.60**	-	**22.33**	nr	**25.12**
extra; fittings; two ends	24.40	3.00	**25.13**	0.76	-	**13.21**	-	**38.34**	nr	**43.13**
extra; fittings; three ends	31.79	3.00	**32.74**	0.76	-	**13.21**	-	**45.95**	nr	**51.70**
54mm	20.30	5.00	**21.28**	0.39	-	**6.78**	-	**28.06**	m	**31.57**
54mm; fittings										
extra; made bends	-	-	-	0.60	-	**10.43**	-	**10.43**	nr	**11.73**
extra; connections to iron pipe ends; screwed joint	21.64	3.00	**22.29**	1.00	-	**17.38**	-	**39.67**	nr	**44.63**
extra; fittings; one end	21.29	3.00	**21.93**	0.50	-	**8.69**	-	**30.62**	nr	**34.45**
extra; fittings; two ends	40.84	3.00	**42.07**	1.00	-	**17.38**	-	**59.45**	nr	**66.88**
extra; fittings; three ends	51.17	3.00	**52.71**	1.00	-	**17.38**	-	**70.09**	nr	**78.85**

Labour hourly rates: (except Specialists) Craft Operatives £17.38 Labourer £10.71 Rates are national average prices. Refer to REGIONAL VARIATIONS for indicative levels of overall pricing in regions	MATERIALS			LABOUR			RATES			
	Del to Site £	Waste %	Material Cost £	Craft Optve Hrs	Lab Hrs	Labour Cost £	Sunds £	Nett Rate £	Unit	Gross Rate (+12.5%) £
S115: HOT AND COLD WATER AND GAS: POLYBUTYLENE TUBES AND FITTINGS										
Polybutylene pipes; Hepworth Hep20 flexible plumbing system BS 7291 Class H; polybutylene slimline fittings										
Pipes, flexible; in ducts in pipe clips to timber										
15mm	1.53	10.00	**1.64**	0.36	-	**6.26**	-	**7.90**	m	**8.89**
15mm; fittings										
extra; made bend	-	-	-	0.01	-	**0.17**	-	**0.17**	nr	**0.20**
extra; connection to copper pipe ends	1.29	2.00	**1.32**	0.12	-	**2.09**	-	**3.40**	nr	**3.83**
extra; fittings; one end	2.23	2.00	**2.27**	0.06	-	**1.04**	-	**3.32**	nr	**3.73**
extra; fittings; two ends	1.67	2.00	**1.70**	0.12	-	**2.09**	-	**3.79**	nr	**4.26**
extra; fittings; three ends	2.36	2.00	**2.41**	0.12	-	**2.09**	-	**4.49**	nr	**5.05**
22mm	2.17	10.00	**2.35**	0.40	-	**6.95**	-	**9.31**	m	**10.47**
22mm; fittings										
extra; made bend	-	-	-	0.01	-	**0.17**	-	**0.17**	nr	**0.20**
extra; connection to copper pipe ends	1.81	2.00	**1.85**	0.30	-	**5.21**	-	**7.06**	nr	**7.94**
extra; fittings; one end	2.99	2.00	**3.05**	0.15	-	**2.61**	-	**5.66**	nr	**6.36**
extra; fittings; two ends	2.36	2.00	**2.41**	0.30	-	**5.21**	-	**7.62**	nr	**8.57**
extra; fittings; three ends	2.82	2.00	**2.88**	0.30	-	**5.21**	-	**8.09**	nr	**9.10**
Straight pipe fittings in pipe clips fixing to masonry with screws										
15mm	1.62	10.00	**1.74**	0.36	-	**6.26**	-	**8.00**	m	**9.00**
22mm	2.21	10.00	**2.40**	0.40	-	**6.95**	-	**9.35**	m	**10.52**
Polybutylene pipes; Hepworth Hep20 flexible plumbing system BS 7291 Class H; polybutylene demountable fittings										
Pipes, flexible in pipe clips to timber with screws										
15mm	1.52	10.00	**1.64**	0.45	-	**7.82**	-	**9.46**	m	**10.64**
15mm; fittings										
extra; connection to copper pipe ends	1.26	2.00	**1.29**	0.12	-	**2.09**	-	**3.37**	nr	**3.79**
extra; fittings; one end	2.11	2.00	**2.15**	0.06	-	**1.04**	-	**3.20**	nr	**3.59**
extra; fittings; two ends	1.64	2.00	**1.67**	0.12	-	**2.09**	-	**3.76**	nr	**4.23**
extra; fittings; three ends	2.34	2.00	**2.39**	0.12	-	**2.09**	-	**4.47**	nr	**5.03**
22mm	2.15	10.00	**2.34**	0.48	-	**8.34**	-	**10.68**	m	**12.02**
22mm; fittings										
extra; connection to copper pipe ends	1.67	2.00	**1.70**	0.30	-	**5.21**	-	**6.92**	nr	**7.78**
extra; fittings; one end	2.83	2.00	**2.89**	0.15	-	**2.61**	-	**5.49**	nr	**6.18**
extra; fittings; two ends	2.32	2.00	**2.37**	0.30	-	**5.21**	-	**7.58**	nr	**8.53**
extra; fittings; three ends	2.97	2.00	**3.03**	0.30	-	**5.21**	-	**8.24**	nr	**9.27**
28mm	3.02	10.00	**3.27**	0.60	-	**10.43**	-	**13.70**	m	**15.41**
28mm; fittings										
extra; connection to copper pipe ends	3.97	2.00	**4.05**	0.40	-	**6.95**	-	**11.00**	nr	**12.38**
extra; fittings; one end	5.92	2.00	**6.04**	0.20	-	**3.48**	-	**9.51**	nr	**10.70**
extra; fittings; two ends	5.02	2.00	**5.12**	0.40	-	**6.95**	-	**12.07**	nr	**13.58**
extra; fittings; three ends	7.01	2.00	**7.15**	0.40	-	**6.95**	-	**14.10**	nr	**15.86**
Pipes flexible in pipe clips fixing to masonry with screws										
15mm	1.62	10.00	**1.74**	0.45	-	**7.82**	-	**9.56**	m	**10.75**
22mm	2.20	10.00	**2.39**	0.48	-	**8.34**	-	**10.73**	m	**12.07**
28mm	3.05	10.00	**3.31**	0.60	-	**10.43**	-	**13.73**	m	**15.45**
S115: HOT AND COLD WATER AND GAS: POLYETHYLENE TUBES AND FITTINGS										
Polythene pipes, BS 6572 Blue; copper alloy fittings, compression, BS EN 1254 Part 3, Type A										
Pipes; straight; in trenches										
20mm	0.64	10.00	**0.69**	0.17	-	**2.95**	-	**3.64**	m	**4.10**
20mm; fittings										
extra; connections to copper pipe ends; compression joint	5.38	3.00	**5.54**	0.16	-	**2.78**	-	**8.32**	nr	**9.36**
extra; fittings; one end	5.26	3.00	**5.42**	0.08	-	**1.39**	-	**6.81**	nr	**7.66**
extra; fittings; two ends	7.34	3.00	**7.56**	0.16	-	**2.78**	-	**10.34**	nr	**11.63**
extra; fittings; three ends	10.04	3.00	**10.34**	0.16	-	**2.78**	-	**13.12**	nr	**14.76**
25mm	0.88	10.00	**0.94**	0.20	-	**3.48**	-	**4.42**	m	**4.97**
25mm; fittings										
extra; connections to copper pipe ends; compression joint	8.20	3.00	**8.45**	0.25	-	**4.35**	-	**12.79**	nr	**14.39**
extra; fittings; one end	7.90	3.00	**8.14**	0.13	-	**2.26**	-	**10.40**	nr	**11.70**
extra; fittings; two ends	10.74	3.00	**11.06**	0.25	-	**4.35**	-	**15.41**	nr	**17.33**
extra; fittings; three ends	15.59	3.00	**16.06**	0.25	-	**4.35**	-	**20.40**	nr	**22.95**
32mm	1.42	10.00	**1.52**	0.24	-	**4.17**	-	**5.69**	m	**6.40**
32mm; fittings										
extra; connections to copper pipe ends; compression joint	14.05	3.00	**14.47**	0.37	-	**6.43**	-	**20.90**	nr	**23.51**
extra; fittings; one end	17.90	3.00	**18.44**	0.19	-	**3.30**	-	**21.74**	nr	**24.46**
extra; fittings; two ends	18.88	3.00	**19.45**	0.37	-	**6.43**	-	**25.88**	nr	**29.11**
extra; fittings; three ends	23.83	3.00	**24.54**	0.37	-	**6.43**	-	**30.98**	nr	**34.85**
50mm	3.52	10.00	**3.77**	0.40	-	**6.95**	-	**10.72**	m	**12.06**
50mm; fittings										
extra; fittings; two ends	44.07	3.00	**45.39**	0.60	-	**10.43**	-	**55.82**	nr	**62.80**
extra; fittings; three ends	58.58	3.00	**60.34**	0.60	-	**10.43**	-	**70.77**	nr	**79.61**
63mm	8.06	10.00	**8.53**	0.54	-	**9.39**	-	**17.91**	m	**20.15**
63mm; fittings										
extra; fittings; two ends	52.84	3.00	**54.43**	1.00	-	**17.38**	-	**71.81**	nr	**80.78**
extra; fittings; three ends	86.03	3.00	**88.61**	1.00	-	**17.38**	-	**105.99**	nr	**119.24**

Labour hourly rates: (except Specialists) Craft Operatives £17.38 Labourer £10.71 Rates are national average prices. Refer to REGIONAL VARIATIONS for indicative levels of overall pricing in regions	MATERIALS			LABOUR				RATES		
	Del to Site £	Waste %	Material Cost £	Craft Optve Hrs	Lab Hrs	Labour Cost £	Sunds £	Nett Rate £	Unit	Gross Rate (+12.5%) £
S115: HOT AND COLD WATER AND GAS: UNPLASTICISED PVC TUBES AND FITTINGS										
PVC-U pipes, BS EN 1452 Class E; fittings, solvent welded joints, BS 4346										
Pipes, straight in standard plastics pipe brackets fixing to timber with screws										
3/8''	4.07	5.00	**4.22**	0.40	-	**6.95**	-	**11.17**	m	**12.56**
3/8'; fittings										
extra; connections to iron pipe ends; screwed joint	1.23	2.00	**1.25**	0.12	-	**2.09**	-	**3.34**	nr	**3.76**
extra; fittings; one end	0.78	2.00	**0.80**	0.06	-	**1.04**	-	**1.84**	nr	**2.07**
extra; fittings; two ends	1.18	2.00	**1.20**	0.12	-	**2.09**	-	**3.29**	nr	**3.70**
extra; fittings; three ends	1.31	2.00	**1.34**	0.12	-	**2.09**	-	**3.42**	nr	**3.85**
1/2''	4.97	5.00	**5.16**	0.44	-	**7.65**	-	**12.81**	m	**14.41**
1/2'; fittings										
extra; connections to iron pipe ends; screwed joint	1.67	2.00	**1.70**	0.20	-	**3.48**	-	**5.18**	nr	**5.83**
extra; fittings; one end	0.95	2.00	**0.97**	0.10	-	**1.74**	-	**2.71**	nr	**3.05**
extra; fittings; two ends	1.34	2.00	**1.37**	0.20	-	**3.48**	-	**4.84**	nr	**5.45**
extra; fittings; three ends	1.54	2.00	**1.57**	0.20	-	**3.48**	-	**5.05**	nr	**5.68**
3/4''	5.28	5.00	**5.51**	0.52	-	**9.04**	-	**14.54**	m	**16.36**
3/4'; fittings										
extra; connections to iron pipe ends; screwed joint	1.81	2.00	**1.85**	0.30	-	**5.21**	-	**7.06**	nr	**7.94**
extra; fittings; one end	1.12	2.00	**1.14**	0.15	-	**2.61**	-	**3.75**	nr	**4.22**
extra; fittings; two ends	1.60	2.00	**1.63**	0.30	-	**5.21**	-	**6.85**	nr	**7.70**
extra; fittings; three ends	1.95	2.00	**1.99**	0.30	-	**5.21**	-	**7.20**	nr	**8.10**
1''	5.53	5.00	**5.78**	0.60	-	**10.43**	-	**16.20**	m	**18.23**
1'; fittings										
extra; connections to iron pipe ends; screwed joint	2.82	2.00	**2.88**	0.40	-	**6.95**	-	**9.83**	nr	**11.06**
extra; fittings; one end	1.25	2.00	**1.28**	0.20	-	**3.48**	-	**4.75**	nr	**5.34**
extra; fittings; two ends	2.22	2.00	**2.26**	0.40	-	**6.95**	-	**9.22**	nr	**10.37**
extra; fittings; three ends	2.95	2.00	**3.01**	0.40	-	**6.95**	-	**9.96**	nr	**11.21**
1 1/4''	7.93	5.00	**8.28**	0.72	-	**12.51**	-	**20.80**	m	**23.40**
11/4'; fittings										
extra; connections to iron pipe ends; screwed joint	3.96	2.00	**4.04**	0.60	-	**10.43**	-	**14.47**	nr	**16.28**
extra; fittings; one end	1.95	2.00	**1.99**	0.30	-	**5.21**	-	**7.20**	nr	**8.10**
extra; fittings; two ends	3.90	2.00	**3.98**	0.60	-	**10.43**	-	**14.41**	nr	**16.21**
extra; fittings; three ends	4.17	2.00	**4.25**	0.60	-	**10.43**	-	**14.68**	nr	**16.52**
1 1/2''	10.03	5.00	**10.49**	0.90	-	**15.64**	-	**26.13**	m	**29.40**
11/2'; fittings										
extra; connections to iron pipe ends; screwed joint	5.36	2.00	**5.47**	0.80	-	**13.90**	-	**19.37**	nr	**21.79**
extra; fittings; one end	3.29	2.00	**3.36**	0.40	-	**6.95**	-	**10.31**	nr	**11.60**
extra; fittings; two ends	5.03	2.00	**5.13**	0.80	-	**13.90**	-	**19.03**	nr	**21.41**
extra; fittings; three ends	6.03	2.00	**6.15**	0.80	-	**13.90**	-	**20.05**	nr	**22.56**
Pipes straight In standard plastics pipe brackets fixing to masonry with screws										
3/8''	4.19	5.00	**4.34**	0.46	-	**7.99**	-	**12.34**	m	**13.88**
1/2''	5.10	5.00	**5.29**	0.50	-	**8.69**	-	**13.98**	m	**15.72**
3/4''	5.35	5.00	**5.57**	0.58	-	**10.08**	-	**15.65**	m	**17.61**
1''	5.49	5.00	**5.74**	0.66	-	**11.47**	-	**17.21**	m	**19.36**
1 1/4''	7.97	5.00	**8.33**	0.78	-	**13.56**	-	**21.88**	m	**24.62**
1 1/2''	10.11	5.00	**10.57**	0.96	-	**16.68**	-	**27.25**	m	**30.66**
Pipes; straight; in trenches										
3/8''	2.20	5.00	**2.31**	0.20	-	**3.48**	-	**5.78**	m	**6.51**
1/2''	2.81	5.00	**2.95**	0.22	-	**3.82**	-	**6.77**	m	**7.62**
3/4''	3.97	5.00	**4.16**	0.24	-	**4.17**	-	**8.33**	m	**9.38**
1''	4.61	5.00	**4.84**	0.30	-	**5.21**	-	**10.05**	m	**11.31**
1 1/4''	6.83	5.00	**7.16**	0.35	-	**6.08**	-	**13.24**	m	**14.90**
1 1/2''	8.88	5.00	**9.31**	0.40	-	**6.95**	-	**16.26**	m	**18.29**
S115: HOT AND COLD WATER AND GAS: UNPLASTICISED OVERFLOW PIPES										
PVC-U pipes and fittings; solvent welded joints; pipework self coloured white										
Pipes, straight in standard plastics pipe brackets fixing to timber with screws										
19mm	1.68	5.00	**1.74**	0.30	-	**5.21**	-	**6.96**	m	**7.83**
19mm; fittings										
extra; fittings; two ends	1.35	2.00	**1.38**	0.25	-	**4.35**	-	**5.72**	nr	**6.44**
extra; fittings; three ends	1.46	2.00	**1.49**	0.25	-	**4.35**	-	**5.83**	nr	**6.56**
extra; fittings; tank connector	1.77	2.00	**1.81**	0.25	-	**4.35**	-	**6.15**	nr	**6.92**
S115: HOT AND COLD WATER AND GAS: DUCTILE IRON PIPES AND FITTINGS										
Ductile iron pipes and fittings, Tyton socketed flexible joints										
Pipes; straight; in trenches										
80mm	26.57	5.00	**27.90**	0.96	-	**16.68**	-	**44.58**	m	**50.16**
80mm; fittings										
extra; bends, 90 degree	65.00	2.00	**66.30**	0.96	-	**16.68**	-	**82.98**	nr	**93.36**
extra; duckfoot bends, 90 degree	148.17	2.00	**151.13**	0.96	-	**16.68**	-	**167.82**	nr	**188.80**
extra; tees	95.57	2.00	**97.48**	1.44	-	**25.03**	-	**122.51**	nr	**137.82**
extra; hydrant tees	130.51	2.00	**133.12**	1.44	-	**25.03**	-	**158.15**	nr	**177.92**
extra; branches, 45 degree	273.99	2.00	**279.47**	1.44	-	**25.03**	-	**304.50**	nr	**342.56**

Labour hourly rates: (except Specialists) Craft Operatives £17.38 Labourer £10.71 Rates are national average prices. Refer to REGIONAL VARIATIONS for indicative levels of overall pricing in regions	MATERIALS			LABOUR				RATES		
	Del to Site £	Waste %	Material Cost £	Craft Optve Hrs	Lab Hrs	Labour Cost £	Sunds £	Nett Rate £	Unit	Gross Rate (+12.5%) £
S115: HOT AND COLD WATER AND GAS: DUCTILE IRON PIPES AND FITTINGS Cont.										
Ductile iron pipes and fittings, Tyton socketed flexible joints Cont.										
Pipes; straight; in trenches Cont.										
80mm; fittings Cont.										
extra; flanged sockets	51.32	2.00	52.35	0.96	-	16.68	-	69.03	nr	77.66
extra; flanged spigots	48.00	2.00	48.96	0.96	-	16.68	-	65.64	nr	73.85
100mm	25.23	5.00	26.49	1.20	-	20.86	-	47.35	m	53.27
100mm; fittings										
extra; bends, 90 degree	68.43	2.00	69.80	1.20	-	20.86	-	90.65	nr	101.99
extra; duckfoot bends, 90 degree	160.56	2.00	163.77	1.20	-	20.86	-	184.63	nr	207.71
extra; tees	101.82	2.00	103.86	1.80	-	31.28	-	135.14	nr	152.03
extra; hydrant tees	96.28	2.00	98.21	1.80	-	31.28	-	129.49	nr	145.68
extra; branches, 45 degree	390.73	2.00	398.54	1.80	-	31.28	-	429.83	nr	483.56
extra; flanged sockets	55.52	2.00	56.63	1.20	-	20.86	-	77.49	nr	87.17
extra; flanged spigots	50.69	2.00	51.70	1.20	-	20.86	-	72.56	nr	81.63
150mm	32.15	5.00	33.76	1.80	-	31.28	-	65.04	m	73.17
150mm; fittings										
extra; bends, 90 degree	122.79	2.00	125.25	1.80	-	31.28	-	156.53	nr	176.10
extra; duckfoot bends, 90 degree	350.97	2.00	357.99	1.80	-	31.28	-	389.27	nr	437.93
extra; tees	189.67	2.00	193.46	2.70	-	46.93	-	240.39	nr	270.44
extra; hydrant tees	174.80	2.00	178.30	2.70	-	46.93	-	225.22	nr	253.37
extra; branches, 45 degree	494.31	2.00	504.20	2.70	-	46.93	-	551.12	nr	620.01
extra; flanged sockets	84.59	2.00	86.28	1.80	-	31.28	-	117.57	nr	132.26
extra; flanged spigots	58.79	2.00	59.97	1.80	-	31.28	-	91.25	nr	102.66
S115: HOT AND COLD WATER AND GAS: STOPCOCKS										
Brass stopcocks										
Stopcocks; BS 1010, crutch head; DZR; joints to polythene BS 6572										
each end										
20mm	24.31	2.00	24.80	0.40	-	6.95	-	31.75	nr	35.72
25mm	38.69	2.00	39.46	0.60	-	10.43	-	49.89	nr	56.13
32mm	51.92	2.00	52.96	0.80	-	13.90	-	66.86	nr	75.22
50mm	129.79	2.00	132.39	0.90	-	15.64	-	148.03	nr	166.53
Stopcocks BS 1010, crutch head; compression joints to copper (Type A)										
each end										
15mm	4.06	2.00	4.14	0.30	-	5.21	-	9.36	nr	10.52
22mm	7.05	2.00	7.19	0.40	-	6.95	-	14.14	nr	15.91
28mm	16.93	2.00	17.27	0.60	-	10.43	-	27.70	nr	31.16
35mm	50.87	2.00	51.89	0.80	-	13.90	-	65.79	nr	74.02
42mm	64.20	2.00	65.48	0.90	-	15.64	-	81.13	nr	91.27
54mm	96.90	2.00	98.84	1.20	-	20.86	-	119.69	nr	134.66
Stopcocks BS 1010, crutch head; DZR; compression joints to copper (Type A)										
each end										
15mm	14.86	2.00	15.16	0.30	-	5.21	-	20.37	nr	22.92
22mm	24.41	2.00	24.90	0.40	-	6.95	-	31.85	nr	35.83
28mm	40.38	2.00	41.19	0.60	-	10.43	-	51.62	nr	58.07
35mm	74.32	2.00	75.81	0.80	-	13.90	-	89.71	nr	100.92
42mm	106.38	2.00	108.51	0.90	-	15.64	-	124.15	nr	139.67
54mm	144.91	2.00	147.81	1.20	-	20.86	-	168.66	nr	189.75
Polybutylene stopcocks										
Stopcocks; fitted with Hep20 ends										
each end										
15mm	4.48	2.00	4.57	0.30	-	5.21	-	9.78	nr	11.01
22mm	5.37	2.00	5.48	0.40	-	6.95	-	12.43	nr	13.98
S115: HOT AND COLD WATER AND GAS: LEVER BALL VALVES										
Copper alloy ball valves										
Lever ball valves; screwed and PTFE joints										
each end threaded internally										
13mm	0.74	2.00	0.76	0.30	-	5.21	-	5.97	nr	6.72
19mm	0.78	2.00	0.79	0.40	-	6.95	-	7.74	nr	8.71
25mm	0.73	2.00	0.74	0.60	-	10.43	-	11.17	nr	12.56
32mm	0.63	2.00	0.64	0.80	-	13.90	-	14.54	nr	16.36
38mm	16.02	2.00	16.34	0.90	-	15.64	-	31.98	nr	35.98
51mm	21.79	2.00	22.23	1.20	-	20.86	-	43.08	nr	48.47
S115: HOT AND COLD WATER AND GAS: GATE VALVES										
Copper alloy gate valves										
Gate valves; BS EN 12288 series B; compression joints to copper (Type A)										
each end										
15mm	7.31	2.00	7.46	0.30	-	5.21	-	12.67	nr	14.25
22mm	8.88	2.00	9.06	0.40	-	6.95	-	16.01	nr	18.01
28mm	12.23	2.00	12.47	0.60	-	10.43	-	22.90	nr	25.77
35mm	23.02	2.00	23.48	0.80	-	13.90	-	37.38	nr	42.06
42mm	37.54	2.00	38.29	0.90	-	15.64	-	53.93	nr	60.67
54mm	58.14	2.00	59.30	1.20	-	20.86	-	80.16	nr	90.18

Labour hourly rates: (except Specialists) Craft Operatives £17.38 Labourer £10.71 Rates are national average prices. Refer to REGIONAL VARIATIONS for indicative levels of overall pricing in regions	MATERIALS			LABOUR				RATES		
	Del to Site £	Waste %	Material Cost £	Craft Optve Hrs	Lab Hrs	Labour Cost £	Sunds £	Nett Rate £	Unit	Gross Rate (+12.5%) £
S115: HOT AND COLD WATER AND GAS: GATE VALVES Cont.										
Copper alloy gate valves Cont.										
Gate valves; DZR; fitted with Hep20 ends each end										
15mm	4.19	2.00	**4.27**	0.30	-	**5.21**	-	**9.49**	nr	**10.67**
22mm	5.91	2.00	**6.03**	0.40	-	**6.95**	-	**12.98**	nr	**14.60**
S115: HOT AND COLD WATER AND GAS: BALL VALVES										
Copper alloy body float valves										
Float operated valves for low pressure; BS 1212; copper float inlet threaded externally; fixing to steel										
13mm; part 1	8.38	2.00	**8.55**	0.40	-	**6.95**	-	**15.50**	nr	**17.44**
13mm; part 2	9.90	2.00	**10.10**	0.40	-	**6.95**	-	**17.05**	nr	**19.18**
Float operated valves for low pressure; BS 1212; plastics float inlet threaded externally; fixing to steel										
13mm; part 1	5.60	2.00	**5.71**	0.40	-	**6.95**	-	**12.66**	nr	**14.25**
13mm; part 2	7.12	2.00	**7.26**	0.40	-	**6.95**	-	**14.21**	nr	**15.99**
Float operated valves for high pressure; BS 1212; copper float inlet threaded externally; fixing to steel										
13mm; part 1	6.72	2.00	**6.85**	0.40	-	**6.95**	-	**13.81**	nr	**15.53**
13mm; part 2	9.70	2.00	**9.89**	0.40	-	**6.95**	-	**16.85**	nr	**18.95**
19mm; part 1	19.67	2.00	**20.06**	0.60	-	**10.43**	-	**30.49**	nr	**34.30**
25mm; part 1	59.05	2.00	**60.23**	1.20	-	**20.86**	-	**81.09**	nr	**91.22**
Float operated valves for high pressure; BS 1212; plastics float inlet threaded externally; fixing to steel										
13mm; part 1	3.94	2.00	**4.02**	0.40	-	**6.95**	-	**10.97**	nr	**12.34**
13mm; part 2	6.92	2.00	**7.06**	0.40	-	**6.95**	-	**14.01**	nr	**15.76**
19mm; part 1	12.91	2.00	**13.17**	0.60	-	**10.43**	-	**23.60**	nr	**26.55**
25mm; part 1	49.68	2.00	**50.67**	1.20	-	**20.86**	-	**71.53**	nr	**80.47**
S115: HOT AND COLD WATER AND GAS: DRAINING TAPS										
Brass draining taps										
Drain cocks; BS 2879, square head Type2										
13mm	2.41	2.00	**2.46**	0.12	-	**2.09**	-	**4.54**	nr	**5.11**
19mm	18.88	2.00	**19.26**	0.16	-	**2.78**	-	**22.04**	nr	**24.79**
S115: HOT AND COLD WATER AND GAS: STORAGE TANKS										
GRP water storage tanks										
GRP Sarena mfg Ltd; pre insulated water storage tanks with cover										
reference SC10; 18 litres	76.00	-	**76.00**	1.80	-	**31.28**	-	**107.28**	nr	**120.69**
reference SC20; 55 litres	106.00	-	**106.00**	1.80	-	**31.28**	-	**137.28**	nr	**154.44**
reference SC30; 86 litres	116.00	-	**116.00**	2.10	-	**36.50**	-	**152.50**	nr	**171.56**
reference SC40; 114 litres	142.00	-	**142.00**	2.10	-	**36.50**	-	**178.50**	nr	**200.81**
reference SC50; 159 litres	161.00	-	**161.00**	2.40	-	**41.71**	-	**202.71**	nr	**228.05**
reference SC60; 191 litres	178.00	-	**178.00**	2.40	-	**41.71**	-	**219.71**	nr	**247.18**
reference SC70; 227 litres	198.00	-	**198.00**	3.00	-	**52.14**	-	**250.14**	nr	**281.41**
reference SC100/1; 327 litres	261.00	-	**261.00**	3.60	-	**62.57**	-	**323.57**	nr	**364.01**
reference SC100/2; 336 litres	254.00	-	**254.00**	3.60	-	**62.57**	-	**316.57**	nr	**356.14**
reference SC125; 423 litres	322.00	-	**322.00**	4.80	-	**83.42**	-	**405.42**	nr	**456.10**
reference SC150; 491 litres	383.00	-	**383.00**	4.80	-	**83.42**	-	**466.42**	nr	**524.73**
reference SC200; 709 litres	445.00	-	**445.00**	4.80	-	**83.42**	-	**528.42**	nr	**594.48**
reference SC250; 900 litres	510.00	-	**510.00**	7.20	-	**125.14**	-	**635.14**	nr	**714.53**
reference SC350; 1250 litres	650.00	-	**650.00**	8.40	-	**145.99**	-	**795.99**	nr	**895.49**
reference SC500; 1727 litres	850.00	-	**850.00**	9.60	-	**166.85**	-	**1016.85**	nr	**1143.95**
reference SC600; 2137 litres	980.00	-	**980.00**	10.80	-	**187.70**	-	**1167.70**	nr	**1313.67**
reference SC600H; 2350 litres	1100.00	-	**1100.00**	11.90	-	**206.82**	-	**1306.82**	nr	**1470.17**
reference SC1000; 3364 litres	1450.00	-	**1450.00**	14.40	-	**250.27**	-	**1700.27**	nr	**1912.81**
drilled holes for 13mm pipes	-	-	-	0.25	-	**4.35**	-	**4.35**	nr	**4.89**
drilled holes for 19mm pipes	-	-	-	0.25	-	**4.35**	-	**4.35**	nr	**4.89**
drilled holes for 25mm pipes	-	-	-	0.30	-	**5.21**	-	**5.21**	nr	**5.87**
drilled holed for 32mm pipes	-	-	-	0.30	-	**5.21**	-	**5.21**	nr	**5.87**
drilled holes for 38mm pipes	-	-	-	0.40	-	**6.95**	-	**6.95**	nr	**7.82**
drilled holes for 51mm pipes	-	-	-	0.40	-	**6.95**	-	**6.95**	nr	**7.82**
Plastics water storage cisterns										
Rectangular cold water storage cisterns; plastics; BS 4213 with sealed lid and byelaw 30 kit										
18 litres	15.15	2.00	**15.45**	1.50	-	**26.07**	-	**41.52**	nr	**46.71**
114 litres	50.55	2.00	**51.56**	2.10	-	**36.50**	-	**88.06**	nr	**99.07**
182 litres	79.90	2.00	**81.50**	2.40	-	**41.71**	-	**123.21**	nr	**138.61**
227 litres	86.95	2.00	**88.69**	3.00	-	**52.14**	-	**140.83**	nr	**158.43**
drilled holes for 13mm pipes	-	-	-	0.20	-	**3.48**	-	**3.48**	nr	**3.91**
drilled holes for 19mm pipes	-	-	-	0.20	-	**3.48**	-	**3.48**	nr	**3.91**
drilled holes for 25mm pipes	-	-	-	0.25	-	**4.35**	-	**4.35**	nr	**4.89**
drilled holed for 32mm pipes	-	-	-	0.25	-	**4.35**	-	**4.35**	nr	**4.89**
drilled holes for 38mm pipes	-	-	-	0.30	-	**5.21**	-	**5.21**	nr	**5.87**
drilled holes for 51mm pipes	-	-	-	0.30	-	**5.21**	-	**5.21**	nr	**5.87**

Labour hourly rates: (except Specialists) Craft Operatives £17.38 Labourer £10.71 Rates are national average prices. Refer to REGIONAL VARIATIONS for indicative levels of overall pricing in regions	MATERIALS			LABOUR				RATES		
	Del to Site £	Waste %	Material Cost £	Craft Optve Hrs	Lab Hrs	Labour Cost £	Sunds £	Nett Rate £	Unit	Gross Rate (+12.5%) £
S115: HOT AND COLD WATER AND GAS: STORAGE TANKS AND CYLINDERS										
Copper direct hot water cylinders; pre-insulated										
Direct hot water cylinders; copper cylinder, BS 699 Grade 3										
reference 3; 116 litres; four bosses	140.05	2.00	**142.85**	2.50	-	**43.45**	-	**186.30**	nr	**209.59**
reference 7; 120 litres; four bosses	125.57	2.00	**128.08**	2.50	-	**43.45**	-	**171.53**	nr	**192.97**
reference 8; 144 litres; four bosses	157.00	2.00	**160.14**	2.50	-	**43.45**	-	**203.59**	nr	**229.04**
reference 9; 166 litres; four bosses	202.50	2.00	**206.55**	2.50	-	**43.45**	-	**250.00**	nr	**281.25**
Copper indirect hot water cylinders; pre-insulated										
Double feed indirect hot water cylinders; copper cylinder, BS 1566 Part 1 Grade 3										
reference 3; 114 litres; four bosses	149.75	2.00	**152.75**	2.50	-	**43.45**	-	**196.20**	nr	**220.72**
reference 7; 117 litres; four bosses	128.25	2.00	**130.82**	2.50	-	**43.45**	-	**174.27**	nr	**196.05**
reference 8; 140 litres; four bosses	165.50	2.00	**168.81**	2.50	-	**43.45**	-	**212.26**	nr	**238.79**
reference 9; 162 litres; four bosses	207.50	2.00	**211.65**	2.50	-	**43.45**	-	**255.10**	nr	**286.99**
Single feed indirect hot water cylinders; copper cylinder, BS 1566 Part 2 Grade 3										
reference 7; 108 litres; four bosses	267.75	2.00	**273.11**	2.50	-	**43.45**	-	**316.56**	nr	**356.12**
reference 8; 130 litres; four bosses	315.50	2.00	**321.81**	2.50	-	**43.45**	-	**365.26**	nr	**410.92**
Copper combination hot water storage units; pre-insulated										
Combination direct hot water storage units; copper unit, BS 3198 with lid										
450mm diameter x 1050mm high; 115 litres hot water; 45 litres cold water	251.00	2.00	**256.02**	3.75	-	**65.18**	-	**321.20**	nr	**361.34**
450mm diameter x 1200mm high; 115 litres hot water; 115 litres cold water	278.00	2.00	**283.56**	3.75	-	**65.18**	-	**348.74**	nr	**392.33**
Combination double feed indirect hot water storage units; copper unit, BS 3198 with lid										
450mm diameter x 1050mm high; 115 litres hot water; 45 litres cold water	265.00	2.00	**270.30**	3.75	-	**65.18**	-	**335.48**	nr	**377.41**
450mm diameter x 1200mm high; 115 litres hot water; 115 litres cold water	281.75	2.00	**287.39**	3.75	-	**65.18**	-	**352.56**	nr	**396.63**
Combination single feed indirect hot water storage units; copper unit, BS 3198 with lid										
450mm diameter x 1200mm high; 115 litres hot water; 115 litres cold water	322.75	2.00	**329.21**	3.60	-	**62.57**	-	**391.77**	nr	**440.74**
S115: HOT AND COLD WATER AND GAS: PIPE INSULATION										
Thermal insulation; glass fibre preformed lagging, butt joints in the running length; secured with metal bands										
19mm thick insulation to copper pipework around one pipe										
15mm	2.94	3.00	**3.03**	0.20	-	**3.48**	-	**6.50**	m	**7.32**
22mm	3.10	3.00	**3.19**	0.26	-	**4.52**	-	**7.71**	m	**8.68**
28mm	3.29	3.00	**3.39**	0.30	-	**5.21**	-	**8.60**	m	**9.68**
35mm	3.52	3.00	**3.63**	0.36	-	**6.26**	-	**9.88**	m	**11.12**
42mm	3.89	3.00	**4.01**	0.40	-	**6.95**	-	**10.96**	m	**12.33**
54mm	4.42	3.00	**4.55**	0.50	-	**8.69**	-	**13.24**	m	**14.90**
25mm thick insulation to copper pipework around one pipe										
15mm	3.29	3.00	**3.39**	0.24	-	**4.17**	-	**7.56**	m	**8.50**
22mm	3.38	3.00	**3.48**	0.26	-	**4.52**	-	**8.00**	m	**9.00**
28mm	3.62	3.00	**3.73**	0.30	-	**5.21**	-	**8.94**	m	**10.06**
35mm	4.01	3.00	**4.13**	0.36	-	**6.26**	-	**10.39**	m	**11.69**
42mm	4.36	3.00	**4.49**	0.40	-	**6.95**	-	**11.44**	m	**12.87**
54mm	5.12	3.00	**5.27**	0.50	-	**8.69**	-	**13.96**	m	**15.71**
Thermal insulation; foamed polyurethane preformed lagging, butt joints in the running length; secured with adhesive bands										
13mm thick insulation to copper pipework around one pipe										
15mm	1.39	3.00	**1.43**	0.20	-	**3.48**	-	**4.91**	m	**5.52**
22mm	1.63	3.00	**1.68**	0.22	-	**3.82**	-	**5.50**	m	**6.19**
28mm	1.92	3.00	**1.98**	0.24	-	**4.17**	-	**6.15**	m	**6.92**
35mm	2.06	3.00	**2.12**	0.27	-	**4.69**	-	**6.81**	m	**7.67**
42mm	2.39	3.00	**2.46**	0.30	-	**5.21**	-	**7.68**	m	**8.64**
54mm	2.90	3.00	**2.99**	0.40	-	**6.95**	-	**9.94**	m	**11.18**
S115: HOT AND COLD WATER AND GAS: INSULATION JACKETS AND LAGGING UNITS FOR CYLINDERS AND TANKS										
Thermal insulation; glass fibre filled insulating jacket in strips, white pvc covering both sides; secured with metal straps and wire holder ring at top										
75mm thick insulation to equipment sides and tops of cylinders; overall size										
400mm diameter x 1050mm high	11.70	5.00	**12.29**	1.00	-	**17.38**	-	**29.67**	nr	**33.37**
450mm diameter x 900mm high	10.30	5.00	**10.82**	1.00	-	**17.38**	-	**28.20**	nr	**31.72**

Labour hourly rates: (except Specialists) Craft Operatives £17.38 Labourer £10.71 Rates are national average prices. Refer to REGIONAL VARIATIONS for indicative levels of overall pricing in regions	MATERIALS			LABOUR				RATES		
	Del to Site £	Waste %	Material Cost £	Craft Optve Hrs	Lab Hrs	Labour Cost £	Sunds £	Nett Rate £	Unit	Gross Rate (+12.5%) £
S115: HOT AND COLD WATER AND GAS: INSULATION JACKETS AND LAGGING UNITS FOR CYLINDERS AND TANKS Cont.										
Thermal insulation; glass fibre filled insulating jacket in strips, white pvc covering both sides; secured with metal straps and wire holder ring at top Cont.										
75mm thick insulation to equipment Cont.										
sides and tops of cylinders; overall size Cont.										
450mm diameter x 1050mm high	11.70	5.00	**12.29**	1.00	-	**17.38**	-	**29.67**	nr	**33.37**
450mm diameter x 1200mm high	31.00	5.00	**32.55**	1.00	-	**17.38**	-	**49.93**	nr	**56.17**
S115: HOT AND COLD WATER AND GAS: FUEL OIL STORAGE TANKS										
Steel fuel oil storage tanks										
Oil storage tanks; carbon steel tank, BS 799 Part 5 Type III, rectangular; 457mm diameter manhole cover, oil tight washer and screwed socket for fill, vent, sludge and draw off; painted one coat black bituminous paint										
14 gauge; 1520 x 610 x 1220mm; 1130 litre capacity	179.50	-	**179.50**	4.25	-	**73.87**	-	**253.37**	nr	**285.04**
14 gauge; 1830 x 610 x 1220mm; 1360 litre capacity	184.00	-	**184.00**	4.50	-	**78.21**	-	**262.21**	nr	**294.99**
14 gauge; 1830 x 1220 x 1220mm; 2730 litre capacity	452.50	-	**452.50**	5.00	-	**86.90**	-	**539.40**	nr	**606.83**

MECHANICAL HEATING/COOLING/REFRIGERATION SYSTEMS

Labour hourly rates: (except Specialists) Craft Operatives £17.38 Labourer £10.71 Rates are national average prices. Refer to REGIONAL VARIATIONS for indicative levels of overall pricing in regions	MATERIALS			LABOUR				RATES		
	Del to Site	Waste	Material Cost	Craft Optve	Lab	Labour Cost	Sunds	Nett Rate	Unit	Gross Rate (+12.5%)
	£	%	£	Hrs	Hrs	£	£	£		£
T10: GAS/OIL FIRED BOILERS										
Gas flue pipes comprising galvanised steel outer skin and aluminium inner skin with air space between, BS 715; socketed joints										
Pipes; straight										
100mm	21.36	5.00	**22.43**	0.80	-	**13.90**	-	**36.33**	m	**40.87**
extra; connections to appliance	4.39	2.00	**4.48**	0.60	-	**10.43**	-	**14.91**	nr	**16.77**
extra; adjustable pipes	12.50	2.00	**12.75**	0.60	-	**10.43**	-	**23.18**	nr	**26.08**
extra; terminals	16.30	2.00	**16.63**	0.60	-	**10.43**	-	**27.05**	nr	**30.44**
extra; elbows 0- 90 degrees	14.28	2.00	**14.57**	0.60	-	**10.43**	-	**24.99**	nr	**28.12**
extra; tees	31.68	2.00	**32.31**	0.60	-	**10.43**	-	**42.74**	nr	**48.08**
125mm	25.59	5.00	**26.87**	0.96	-	**16.68**	-	**43.55**	m	**49.00**
extra; connections to appliance	4.97	2.00	**5.07**	0.80	-	**13.90**	-	**18.97**	nr	**21.35**
extra; adjustable pipes	14.05	2.00	**14.33**	0.80	-	**13.90**	-	**28.24**	nr	**31.76**
extra; terminals	17.88	2.00	**18.24**	0.80	-	**13.90**	-	**32.14**	nr	**36.16**
extra; elbows 0- 90 degrees	16.86	2.00	**17.20**	0.80	-	**13.90**	-	**31.10**	nr	**34.99**
extra; tees	34.18	2.00	**34.86**	0.80	-	**13.90**	-	**48.77**	nr	**54.86**
150mm	28.57	5.00	**30.00**	1.20	-	**20.86**	-	**50.85**	m	**57.21**
extra; connections to appliance	5.39	2.00	**5.50**	1.20	-	**20.86**	-	**26.35**	nr	**29.65**
extra; adjustable pipes	17.70	2.00	**18.05**	1.20	-	**20.86**	-	**38.91**	nr	**43.77**
extra; terminals	22.95	2.00	**23.41**	1.20	-	**20.86**	-	**44.27**	nr	**49.80**
extra; elbows 0- 90 degrees	21.14	2.00	**21.56**	1.20	-	**20.86**	-	**42.42**	nr	**47.72**
extra; tees	35.57	2.00	**36.28**	1.20	-	**20.86**	-	**57.14**	nr	**64.28**
Flue supports; galvanised steel wall bands fixing to masonry with screws										
for 100mm pipes	6.99	2.00	**7.13**	0.20	-	**3.48**	-	**10.61**	nr	**11.93**
for 125mm pipes	7.44	2.00	**7.59**	0.20	-	**3.48**	-	**11.06**	nr	**12.45**
for 150mm pipes	9.43	2.00	**9.62**	0.20	-	**3.48**	-	**13.09**	nr	**14.73**
Fire stop spacers										
for 100mm pipes	3.04	2.00	**3.10**	0.30	-	**5.21**	-	**8.31**	nr	**9.35**
for 125mm pipes	3.04	2.00	**3.10**	0.40	-	**6.95**	-	**10.05**	nr	**11.31**
for 150mm pipes	3.46	2.00	**3.53**	0.60	-	**10.43**	-	**13.96**	nr	**15.70**
Gas fired boilers										
Gas fired boilers for central heating and hot water; automatically controlled by thermostat with electrical control, gas governor and flame failure device										
free standing boiler; conventional flue; approximate output rating B Th U per hour										
40000	466.75	1.00	**471.42**	6.00	-	**104.28**	-	**575.70**	nr	**647.66**
50000	488.25	1.00	**493.13**	6.00	-	**104.28**	-	**597.41**	nr	**672.09**
60000	519.50	1.00	**524.70**	6.00	-	**104.28**	-	**628.98**	nr	**707.60**
80000	674.00	1.00	**680.74**	7.00	-	**121.66**	-	**802.40**	nr	**902.70**
100000	864.00	1.00	**872.64**	9.00	-	**156.42**	-	**1029.06**	nr	**1157.69**
free standing boiler; balanced flue; approximate output rating B Th U per hour										
40000	609.00	1.00	**615.09**	7.00	-	**121.66**	-	**736.75**	nr	**828.84**
50000	631.00	1.00	**637.31**	7.00	-	**121.66**	-	**758.97**	nr	**853.84**
60000	675.00	1.00	**681.75**	7.00	-	**121.66**	-	**803.41**	nr	**903.84**
80000	913.00	1.00	**922.13**	7.00	-	**121.66**	-	**1043.79**	nr	**1174.26**
100000	1156.00	1.00	**1167.56**	9.00	-	**156.42**	-	**1323.98**	nr	**1489.48**
wall mounted boiler; balanced flue; approximate output rating B Th U per hour										
40000	520.00	1.00	**525.20**	7.50	-	**130.35**	-	**655.55**	nr	**737.49**
50000	563.00	1.00	**568.63**	7.50	-	**130.35**	-	**698.98**	nr	**786.35**
60000	606.00	1.00	**612.06**	7.50	-	**130.35**	-	**742.41**	nr	**835.21**
Oil fired boilers										
Oil fired boilers for central heating and hot water; automatically controlled by thermostat, with electrical control box										
free standing boiler; conventional flue; approximate output rating BThU per hour										
40,000 - 48,000	836.00	1.00	**844.36**	6.00	-	**104.28**	-	**948.64**	nr	**1067.22**
50,000 - 65,000	935.00	1.00	**944.35**	6.00	-	**104.28**	-	**1048.63**	nr	**1179.71**

Labour hourly rates: (except Specialists) Craft Operatives £17.38 Labourer £10.71 Rates are national average prices. Refer to REGIONAL VARIATIONS for indicative levels of overall pricing in regions	MATERIALS			LABOUR				RATES		
	Del to Site £	Waste %	Material Cost £	Craft Optve Hrs	Lab Hrs	Labour Cost £	Sunds £	Nett Rate £	Unit	Gross Rate (+12.5%) £
T10: GAS/OIL FIRED BOILERS Cont.										
Oil fired boilers Cont.										
Oil fired boilers for central heating and hot water; automatically controlled by thermostat, with electrical control box Cont.										
free standing boiler; conventional flue; approximate output rating BThU per hour Cont.										
70,000 - 85,000	1033.00	1.00	**1043.33**	6.00	-	**104.28**	-	**1147.61**	nr	**1291.06**
88,000 - 110,000	1158.00	1.00	**1169.58**	7.50	-	**130.35**	-	**1299.93**	nr	**1462.42**
T11: COAL FIRED BOILERS										
Equipment										
Solid fuel boilers for central heating and hot water; thermostatically controlled; with tools										
free standing boiler, gravity feed; approximate output rating BThU per hour										
45000	1672.00	1.00	**1688.72**	6.00	-	**104.28**	-	**1793.00**	nr	**2017.13**
60000	1948.00	1.00	**1967.48**	6.00	-	**104.28**	-	**2071.76**	nr	**2330.73**
80000	2167.00	1.00	**2188.67**	7.50	-	**130.35**	-	**2319.02**	nr	**2608.90**
T31: LOW TEMPERATURE HOT WATER HEATING										
Accelerator pumps										
Variable head accelerator pumps for forced central heating; small bore indirect systems										
BSP connections; with valves										
20mm	53.75	2.00	**54.83**	1.00	-	**17.38**	-	**72.21**	nr	**81.23**
25mm	54.50	2.00	**55.59**	1.00	-	**17.38**	-	**72.97**	nr	**82.09**
Radiators										
Pressed steel radiators single panel with convector; air cock, plain plug										
450mm high; fixing brackets to masonry with screws										
600mm long	23.69	1.00	**23.93**	3.75	-	**65.18**	-	**89.10**	nr	**100.24**
900mm long	35.51	1.00	**35.87**	3.75	-	**65.18**	-	**101.04**	nr	**113.67**
1200mm long	47.33	1.00	**47.80**	3.75	-	**65.18**	-	**112.98**	nr	**127.10**
1600mm long	72.25	1.00	**72.97**	3.75	-	**65.18**	-	**138.15**	nr	**155.42**
2400mm long	145.77	1.00	**147.23**	3.75	-	**65.18**	-	**212.40**	nr	**238.95**
600mm high; fixing brackets to masonry with screws										
600mm long	30.56	1.00	**30.87**	3.75	-	**65.18**	-	**96.04**	nr	**108.05**
900mm long	45.84	1.00	**46.30**	3.75	-	**65.18**	-	**111.47**	nr	**125.41**
1200mm long	61.13	1.00	**61.74**	3.75	-	**65.18**	-	**126.92**	nr	**142.78**
1600mm long	93.28	1.00	**94.21**	3.75	-	**65.18**	-	**159.39**	nr	**179.31**
2400mm long	188.20	1.00	**190.08**	3.75	-	**65.18**	-	**255.26**	nr	**287.16**
700mm high; fixing brackets to masonry with screws										
600mm long	35.05	1.00	**35.40**	3.75	-	**65.18**	-	**100.58**	nr	**113.15**
900mm long	60.11	1.00	**60.71**	3.75	-	**65.18**	-	**125.89**	nr	**141.62**
1200mm long	107.81	1.00	**108.89**	3.75	-	**65.18**	-	**174.06**	nr	**195.82**
1600mm long	143.77	1.00	**145.21**	3.75	-	**65.18**	-	**210.38**	nr	**236.68**
2400mm long	215.66	1.00	**217.82**	3.75	-	**65.18**	-	**282.99**	nr	**318.37**
Pressed steel radiators; double panel with single convector; air cock, plain plug										
450mm high; fixing brackets to masonry with screws										
600mm long	36.79	1.00	**37.16**	3.75	-	**65.18**	-	**102.33**	nr	**115.12**
900mm long	55.18	1.00	**55.73**	3.75	-	**65.18**	-	**120.91**	nr	**136.02**
1200mm long	84.20	1.00	**85.04**	3.75	-	**65.18**	-	**150.22**	nr	**168.99**
1600mm long	112.29	1.00	**113.41**	3.75	-	**65.18**	-	**178.59**	nr	**200.91**
2400mm long	226.31	1.00	**228.57**	3.75	-	**65.18**	-	**293.75**	nr	**330.47**
600mm high; fixing brackets to masonry with screws										
600mm long	46.36	1.00	**46.82**	3.75	-	**65.18**	-	**112.00**	nr	**126.00**
900mm long	69.53	1.00	**70.23**	3.75	-	**65.18**	-	**135.40**	nr	**152.33**
1200mm long	92.70	1.00	**93.63**	3.75	-	**65.18**	-	**158.80**	nr	**178.65**
1600mm long	141.47	1.00	**142.88**	3.75	-	**65.18**	-	**208.06**	nr	**234.07**
2400mm long	285.41	1.00	**288.26**	3.75	-	**65.18**	-	**353.44**	nr	**397.62**
700mm high; fixing brackets to masonry with screws										
600mm long	59.77	1.00	**60.37**	3.75	-	**65.18**	-	**125.54**	nr	**141.24**
900mm long	120.57	1.00	**121.78**	3.75	-	**65.18**	-	**186.95**	nr	**210.32**
1200mm long	160.72	1.00	**162.33**	3.75	-	**65.18**	-	**227.50**	nr	**255.94**
1600mm long	214.33	1.00	**216.47**	3.75	-	**65.18**	-	**281.65**	nr	**316.85**
2400mm long	321.48	1.00	**324.69**	3.75	-	**65.18**	-	**389.87**	nr	**438.60**
Brass valves for radiators; self colour										
Valves for radiators, BS 2767; wheel head										
inlet for copper; angle pattern										
15mm	6.81	2.00	**6.95**	0.30	-	**5.21**	-	**12.16**	nr	**13.68**
22mm	9.25	2.00	**9.44**	0.40	-	**6.95**	-	**16.39**	nr	**18.44**
Valves for radiators, BS 2767; lock shield										
inlet for copper; angle pattern										
15mm	6.81	2.00	**6.95**	0.30	-	**5.21**	-	**12.16**	nr	**13.68**
22mm	9.25	2.00	**9.44**	0.40	-	**6.95**	-	**16.39**	nr	**18.44**
Brass valves for radiators; chromium plated										
Valves for radiators, BS 2767; wheel head										
inlet for copper; angle pattern										
15mm	9.43	2.00	**9.62**	0.30	-	**5.21**	-	**14.83**	nr	**16.69**
22mm	12.33	2.00	**12.58**	0.40	-	**6.95**	-	**19.53**	nr	**21.97**

Labour hourly rates: (except Specialists) Craft Operatives £17.38 Labourer £10.71 Rates are national average prices. Refer to REGIONAL VARIATIONS for indicative levels of overall pricing in regions	MATERIALS			LABOUR				RATES		
	Del to Site £	Waste %	Material Cost £	Craft Optve Hrs	Lab Hrs	Labour Cost £	Sunds £	Nett Rate £	Unit	Gross Rate (+12.5%) £
T31: LOW TEMPERATURE HOT WATER HEATING Cont.										
Brass valves for radiators; chromium plated Cont.										
Valves for radiators, BS 2767; wheel head Cont.										
inlet for copper; straight pattern										
15mm	10.96	2.00	**11.18**	0.30	-	**5.21**	-	**16.39**	nr	**18.44**
22mm	14.26	2.00	**14.55**	0.40	-	**6.95**	-	**21.50**	nr	**24.18**
Valves for radiators, BS 2767; lock shield										
inlet for copper; angle pattern										
15mm	9.43	2.00	**9.62**	0.30	-	**5.21**	-	**14.83**	nr	**16.69**
22mm	12.33	2.00	**12.58**	0.40	-	**6.95**	-	**19.53**	nr	**21.97**
inlet for copper; straight pattern										
15mm	10.96	2.00	**11.18**	0.30	-	**5.21**	-	**16.39**	nr	**18.44**
22mm	14.26	2.00	**14.55**	0.40	-	**6.95**	-	**21.50**	nr	**24.18**
Brass thermostatic valves for radiators; chromium plated										
Thermostatic valves for radiators, one piece; ABS plastics head										
inlet for copper; angle pattern										
15mm	16.79	2.00	**17.13**	0.30	-	**5.21**	-	**22.34**	nr	**25.13**
inlet for copper; straight pattern										
15mm	16.79	2.00	**17.13**	0.30	-	**5.21**	-	**22.34**	nr	**25.13**
inlet for iron; angle pattern										
19mm	25.88	2.00	**26.40**	0.40	-	**6.95**	-	**33.35**	nr	**37.52**
Skirting heaters										
Pressed metal skirting heaters with fins on copper tube; straight										
900mm long	34.20	2.00	**34.88**	1.50	-	**26.07**	-	**60.95**	nr	**68.57**
1200mm long	45.00	2.00	**45.90**	2.00	-	**34.76**	-	**80.66**	nr	**90.74**
1500mm long	49.50	2.00	**50.49**	2.50	-	**43.45**	-	**93.94**	nr	**105.68**
1800mm long	54.00	2.00	**55.08**	3.00	-	**52.14**	-	**107.22**	nr	**120.62**
extra; internal corners	7.20	2.00	**7.34**	0.40	-	**6.95**	-	**14.30**	nr	**16.08**
extra; external corners	7.20	2.00	**7.34**	0.40	-	**6.95**	-	**14.30**	nr	**16.08**
extra; end stops	7.20	2.00	**7.34**	0.40	-	**6.95**	-	**14.30**	nr	**16.08**
extra; valve boxes	17.10	2.00	**17.44**	0.40	-	**6.95**	-	**24.39**	nr	**27.44**

This section continues on the next page

Labour hourly rates: (except Specialists) Craft Operatives £14.21 Labourer £10.71 Rates are national average prices. Refer to REGIONAL VARIATIONS for indicative levels of overall pricing in regions	MATERIALS			LABOUR				RATES		
	Del to Site £	Waste %	Material Cost £	Craft Optve Hrs	Lab Hrs	Labour Cost £	Sunds £	Nett Rate £	Unit	Gross Rate (+12.5%) £
T32: LOW TEMPERATURE HOT WATER HEATING (SMALL SCALE)										
Central heating installations - indicative prices										
Note the following are indicative prices for installation in two storey three bedroomed dwellings with a floor area of approximately 85m2										
Installation; boiler, copper piping, pressed steel radiators, heated towel rail; providing hot water to sink , bath, lavatory basin; complete with all necessary pumps, controls etc										
solid fuel fired	-	-	-	-	23.00	**246.33**	-	**3571.33**	it	**4017.75**
gas fired	-	-	-	-	22.00	**235.62**	-	**3335.62**	it	**3752.57**
oil fired, including oil storage tank	-	-	-	-	27.00	**289.17**	-	**4089.17**	it	**4600.32**

VENTILATION/AIR CONDITIONING SYSTEMS

Labour hourly rates: (except Specialists) Craft Operatives £20.54 Labourer £10.71 Rates are national average prices. Refer to REGIONAL VARIATIONS for indicative levels of overall pricing in regions	MATERIALS			LABOUR				RATES		
	Del to Site £	Waste %	Material Cost £	Craft Optve Hrs	Lab Hrs	Labour Cost £	Sunds £	Nett Rate £	Unit	Gross Rate (+12.5%) £
U10: GENERAL VENTILATION										
Ventilating fans										
Note: the following prices exclude cutting holes in glass										
Extract fans, window mounted; shutter and incorporated switch unit										
102mm diameter	78.40	2.50	**80.36**	1.50	-	**30.81**	-	**111.17**	nr	**125.07**
102mm diameter - low voltage	112.15	2.50	**114.95**	1.50	-	**30.81**	-	**145.76**	nr	**163.98**
257mm diameter	253.40	2.50	**259.74**	1.80	-	**36.97**	-	**296.71**	nr	**333.80**
324mm diameter	365.90	2.50	**375.05**	1.80	-	**36.97**	-	**412.02**	nr	**463.52**
Extract fans, wall mounted; shutter, wall duct and incorporated switch unit										
Note: the following prices exclude forming hole and making good										
102mm diameter	46.71	2.50	**47.88**	1.75	-	**35.95**	-	**83.82**	nr	**94.30**
102mm diameter - low voltage	80.46	2.50	**82.47**	1.75	-	**35.95**	-	**118.42**	nr	**133.22**
257mm diameter	221.71	2.50	**227.25**	2.30	-	**47.24**	-	**274.49**	nr	**308.81**
324mm diameter	334.21	2.50	**342.57**	2.30	-	**47.24**	-	**389.81**	nr	**438.53**
Extract fans, built in wall; shutter and incorporated switch unit										
290 x 321 mm	199.23	2.50	**204.21**	2.80	0.08	**58.37**	-	**262.58**	nr	**295.40**
409 x 433 mm	361.73	2.50	**370.77**	2.90	0.08	**60.42**	-	**431.20**	nr	**485.10**
473 x 476 mm	436.73	2.50	**447.65**	3.00	0.08	**62.48**	-	**510.13**	nr	**573.89**

ELECTRICAL SUPPLY/POWER/LIGHTING SYSTEMS

Labour hourly rates: (except Specialists) Craft Operatives £20.54 Labourer £10.71 Rates are national average prices. Refer to REGIONAL VARIATIONS for indicative levels of overall pricing in regions	MATERIALS			LABOUR				RATES		
	Del to Site £	Waste %	Material Cost £	Craft Optve Hrs	Lab Hrs	Labour Cost £	Sunds £	Nett Rate £	Unit	Gross Rate (+12.5%) £
V41: STREET/AREA/FLOOD LIGHTING										
Street lighting columns and set in ground; excavate hole, remove surplus spoil, filling with concrete mix ST4, working around base										
Aluminium; BS EN 40-1, nominal mounting height										
4.88m	271.61	-	**271.61**	5.50	10.95	**230.24**	22.97	**524.82**	nr	**590.43**
8.00m	669.05	-	**669.05**	6.50	12.65	**268.99**	35.73	**973.77**	nr	**1095.49**
10.00m	850.14	-	**850.14**	7.25	14.50	**304.21**	77.83	**1232.18**	nr	**1386.21**
12.00m	1191.79	-	**1191.79**	8.00	16.00	**335.68**	77.83	**1605.30**	nr	**1805.96**
Steel; BS EN 40-1, nominal mounting height										
4.00m	218.12	-	**218.12**	5.50	10.95	**230.24**	22.97	**471.33**	nr	**530.25**
5.00m	224.76	-	**224.76**	6.50	12.65	**268.99**	35.73	**529.48**	nr	**595.67**
6.00m	276.99	-	**276.99**	7.25	14.50	**304.21**	77.83	**659.03**	nr	**741.41**
8.00m	411.05	-	**411.05**	8.00	16.00	**335.68**	77.83	**824.56**	nr	**927.63**
V90: GENERAL LIGHTING AND POWER - COMMERCIAL										
Steel trunking										
Straight lighting trunking; PVC lid										
50 x 50mm	2.00	5.00	**2.10**	0.40	-	**8.22**	0.50	**10.81**	m	**12.17**
extra; end cap	2.32	-	**2.32**	0.08	-	**1.64**	0.25	**4.21**	nr	**4.74**
extra; internal angle	11.36	-	**11.36**	0.25	-	**5.14**	0.25	**16.74**	nr	**18.83**
extra; external angle	11.36	-	**11.36**	0.25	-	**5.14**	0.25	**16.74**	nr	**18.83**
extra; tee	12.17	-	**12.17**	0.25	-	**5.14**	0.25	**17.55**	nr	**19.75**
extra; four way intersection	12.57	-	**12.57**	0.40	-	**8.22**	0.25	**21.03**	nr	**23.66**
Straight dado trunking, two compartment; steel lid										
150 x 30mm	7.15	5.00	**7.51**	0.45	-	**9.24**	0.50	**17.25**	m	**19.40**
extra; end cap	5.17	-	**5.17**	0.08	-	**1.64**	0.25	**7.06**	nr	**7.94**
extra; internal angle	23.95	-	**23.95**	0.35	-	**7.19**	0.25	**31.39**	nr	**35.32**
extra; external angle	23.95	-	**23.95**	0.35	-	**7.19**	0.25	**31.39**	nr	**35.32**
extra; single socket plate	9.42	-	**9.42**	0.16	-	**3.29**	0.25	**12.96**	nr	**14.58**
extra; twin socket plate	10.01	-	**10.01**	0.16	-	**3.29**	0.25	**13.55**	nr	**15.24**
Straight dado trunking, three compartment; steel lid										
200 x 38mm	9.46	5.00	**9.93**	0.50	-	**10.27**	0.75	**20.95**	m	**23.57**
extra; end cap	5.65	-	**5.65**	0.08	-	**1.64**	0.25	**7.54**	nr	**8.48**
extra; internal angle	29.60	-	**29.60**	0.40	-	**8.22**	0.25	**38.07**	nr	**42.83**
extra; external angle	29.60	-	**29.60**	0.40	-	**8.22**	0.25	**38.07**	nr	**42.83**
extra; single socket plate	9.74	-	**9.74**	0.16	-	**3.29**	0.25	**13.27**	nr	**14.93**
extra; twin socket plate	10.44	-	**10.44**	0.16	-	**3.29**	0.25	**13.98**	nr	**15.73**
Straight skirting trunking; two compartments										
200 x 38mm	9.44	5.00	**9.92**	0.50	-	**10.27**	0.75	**20.94**	m	**23.55**
extra; internal bend	28.63	-	**28.63**	0.40	-	**8.22**	0.25	**37.09**	nr	**41.73**
extra; external bend	28.63	-	**28.63**	0.40	-	**8.22**	0.25	**37.09**	nr	**41.73**
Straight underfloor trunking; two compartments										
150 x 25mm	6.43	5.00	**6.75**	0.70	-	**14.38**	-	**21.13**	m	**23.77**
extra; floor junction boxes; adjustable frame and trap	77.12	-	**77.12**	1.85	-	**38.00**	-	**115.12**	nr	**129.51**
extra; horizontal bend	18.84	-	**18.84**	0.50	-	**10.27**	-	**29.11**	nr	**32.75**
PVC trunking; white										
Straight mini-trunking; clip on lid										
16 x 16mm	0.37	5.00	**0.39**	0.25	-	**5.14**	0.50	**6.03**	m	**6.78**
extra; end cap	0.33	-	**0.33**	0.06	-	**1.23**	0.13	**1.69**	nr	**1.90**
extra; internal bend	0.33	-	**0.33**	0.20	-	**4.11**	0.13	**4.57**	nr	**5.14**
extra; external bend	0.33	-	**0.33**	0.20	-	**4.11**	0.13	**4.57**	nr	**5.14**
extra; flat bend	0.33	-	**0.33**	0.20	-	**4.11**	0.13	**4.57**	nr	**5.14**
extra; equal tee	0.57	-	**0.57**	0.25	-	**5.14**	0.13	**5.84**	nr	**6.56**
25 x 16mm	0.42	5.00	**0.44**	0.25	-	**5.14**	0.13	**5.71**	m	**6.42**
extra; end cap	0.33	-	**0.33**	0.06	-	**1.23**	0.13	**1.69**	nr	**1.90**
extra; internal bend	0.33	-	**0.33**	0.20	-	**4.11**	0.13	**4.57**	nr	**5.14**
extra; external bend	0.33	-	**0.33**	0.20	-	**4.11**	0.13	**4.57**	nr	**5.14**
extra; flat bend	0.33	-	**0.33**	0.20	-	**4.11**	0.13	**4.57**	nr	**5.14**

Labour hourly rates: (except Specialists)
Craft Operatives £20.54 Labourer £10.71
Rates are national average prices.
Refer to REGIONAL VARIATIONS for indicative levels of overall pricing in regions

	MATERIALS			LABOUR				RATES		
	Del to Site £	Waste %	Material Cost £	Craft Optve Hrs	Lab Hrs	Labour Cost £	Sunds £	Nett Rate £	Unit	Gross Rate (+12.5%) £
V90: GENERAL LIGHTING AND POWER - COMMERCIAL Cont.										
PVC trunking; white Cont.										
Straight mini-trunking; clip on lid Cont.										
extra; equal tee	0.57	-	**0.57**	0.25	-	**5.14**	0.13	**5.84**	nr	**6.56**
40 x 16mm	0.63	5.00	**0.66**	0.30	-	**6.16**	0.13	**6.95**	m	**7.82**
extra; end cap	0.34	-	**0.34**	0.06	-	**1.23**	0.13	**1.70**	nr	**1.92**
extra; internal bend	0.34	-	**0.34**	0.25	-	**5.14**	0.13	**5.61**	nr	**6.31**
extra; external bend	0.34	-	**0.34**	0.25	-	**5.14**	0.13	**5.61**	nr	**6.31**
extra; flat bend	0.34	-	**0.34**	0.25	-	**5.14**	0.13	**5.61**	nr	**6.31**
extra; equal tee	0.60	-	**0.60**	0.30	-	**6.16**	0.13	**6.89**	nr	**7.75**
Straight dado trunking, three compartment; clip on lid										
145 x 40mm	21.81	5.00	**22.90**	0.40	-	**8.22**	0.50	**31.61**	m	**35.56**
extra; end caps	1.57	-	**1.57**	0.08	-	**1.64**	0.25	**3.46**	nr	**3.89**
extra; internal bend	5.36	-	**5.36**	0.30	-	**6.16**	0.25	**11.77**	nr	**13.24**
extra; external bend	5.36	-	**5.36**	0.30	-	**6.16**	0.25	**11.77**	nr	**13.24**
extra; tee	42.07	-	**42.07**	0.40	-	**8.22**	0.25	**50.54**	nr	**56.85**
extra; flat angle	36.06	-	**36.06**	0.30	-	**6.16**	0.25	**42.47**	nr	**47.78**
extra; mini trunking adaptor	1.52	-	**1.52**	0.20	-	**4.11**	0.25	**5.87**	nr	**6.61**
230 x 40mm	37.49	5.00	**39.36**	0.45	-	**9.24**	0.50	**49.10**	m	**55.24**
extra; end caps	4.97	-	**4.97**	0.08	-	**1.64**	0.25	**6.86**	nr	**7.72**
extra; internal bend	10.42	-	**10.42**	0.30	-	**6.16**	0.25	**16.83**	nr	**18.93**
extra; external bend	10.42	-	**10.42**	0.30	-	**6.16**	0.25	**16.83**	nr	**18.93**
extra; mini trunking adaptor	1.55	-	**1.55**	0.20	-	**4.11**	0.25	**5.91**	nr	**6.64**
Straight skirting trunking, three compartment; clip on lid										
100 x 40mm	3.84	5.00	**4.03**	0.35	-	**7.19**	0.50	**11.72**	m	**13.18**
extra; end cap	2.20	-	**2.20**	0.06	-	**1.23**	0.25	**3.68**	nr	**4.14**
extra; internal corner	2.82	-	**2.82**	0.25	-	**5.14**	0.25	**8.21**	nr	**9.24**
extra; external corner	2.82	-	**2.82**	0.25	-	**5.14**	0.25	**8.21**	nr	**9.24**
extra; mini trunking adaptor	1.99	-	**1.99**	0.16	-	**3.29**	0.25	**5.52**	nr	**6.21**
extra; socket adaptor, one gang	4.45	-	**4.45**	0.25	-	**5.14**	0.25	**9.83**	nr	**11.06**
extra; socket adaptor, two gang	6.38	-	**6.38**	0.25	-	**5.14**	0.25	**11.77**	nr	**13.24**
PVC heavy gauge conduit and fittings; push fit joints; spacer bar saddles at 600mm centres										
Conduits; straight										
20mm diameter; plugging to masonry; to surfaces	0.56	10.00	**0.62**	0.17	-	**3.49**	0.42	**4.53**	m	**5.10**
extra; small circular terminal boxes	1.00	2.50	**1.03**	0.10	-	**2.05**	0.27	**3.35**	nr	**3.77**
extra; small circular angle boxes	1.12	2.50	**1.15**	0.11	-	**2.26**	0.27	**3.68**	nr	**4.14**
extra; small circular three way boxes	1.21	2.50	**1.24**	0.13	-	**2.67**	0.27	**4.18**	nr	**4.70**
extra; small circular through way boxes	1.12	2.50	**1.15**	0.11	-	**2.26**	0.27	**3.68**	nr	**4.14**
25mm diameter; plugging to masonry; to surfaces	1.08	10.00	**1.18**	0.18	-	**3.70**	0.42	**5.30**	m	**5.96**
extra; small circular terminal boxes	1.54	2.50	**1.58**	0.10	-	**2.05**	0.27	**3.90**	nr	**4.39**
extra; small circular angle boxes	1.68	2.50	**1.72**	0.11	-	**2.26**	0.27	**4.25**	nr	**4.78**
extra; small circular three way boxes	1.86	2.50	**1.91**	0.13	-	**2.67**	0.27	**4.85**	nr	**5.45**
extra; small circular through way boxes	1.68	2.50	**1.72**	0.11	-	**2.26**	0.27	**4.25**	nr	**4.78**
20mm diameter; plugging to masonry; in chases	0.56	10.00	**0.62**	0.15	-	**3.08**	0.42	**4.12**	m	**4.64**
extra; small circular terminal boxes	1.00	2.50	**1.03**	0.10	-	**2.05**	0.27	**3.35**	nr	**3.77**
extra; small circular angle boxes	1.12	2.50	**1.15**	0.11	-	**2.26**	0.27	**3.68**	nr	**4.14**
extra; small circular three way boxes	1.21	2.50	**1.24**	0.13	-	**2.67**	0.27	**4.18**	nr	**4.70**
extra; small circular through way boxes	1.12	2.50	**1.15**	0.11	-	**2.26**	0.27	**3.68**	nr	**4.14**
25mm diameter; plugging to masonry; in chases	1.08	10.00	**1.18**	0.16	-	**3.29**	0.27	**4.74**	m	**5.33**
extra; small circular terminal boxes	1.54	2.50	**1.58**	0.10	-	**2.05**	0.27	**3.90**	nr	**4.39**
extra; small circular angle boxes	1.68	2.50	**1.72**	0.11	-	**2.26**	0.27	**4.25**	nr	**4.78**
extra; small circular three way boxes	1.86	2.50	**1.91**	0.13	-	**2.67**	0.27	**4.85**	nr	**5.45**
extra; small circular through way boxes	1.68	2.50	**1.72**	0.11	-	**2.26**	0.27	**4.25**	nr	**4.78**
Oval conduits; straight										
20mm (nominal size); plugging to masonry; in chases	0.32	10.00	**0.35**	0.15	-	**3.08**	0.42	**3.85**	m	**4.33**
25mm (nominal size); plugging to masonry in chases	0.34	10.00	**0.37**	0.16	-	**3.29**	0.42	**4.08**	m	**4.59**
20mm (nominal size); fixing to timber	0.32	10.00	**0.35**	0.13	-	**2.67**	0.21	**3.23**	m	**3.63**
25mm (nominal size); fixing to timber	0.34	10.00	**0.37**	0.15	-	**3.08**	0.21	**3.67**	m	**4.12**
Steel welded heavy gauge conduits and fittings, screwed joints; spacer bar saddles at 1000mm centres; enamelled black inside and outside by manufacturer										
Conduits; straight										
20mm diameter; plugging to masonry; to surfaces	0.80	10.00	**0.88**	0.33	-	**6.78**	0.25	**7.91**	m	**8.89**
extra; small circular terminal boxes	0.92	2.50	**0.94**	0.13	-	**2.67**	0.27	**3.88**	nr	**4.37**
extra; small circular angle boxes	1.04	2.50	**1.07**	0.16	-	**3.29**	0.27	**4.62**	nr	**5.20**
extra; small circular three way boxes	0.92	2.50	**0.94**	0.20	-	**4.11**	0.27	**5.32**	nr	**5.99**
extra; small circular through way boxes	0.98	2.50	**1.00**	0.16	-	**3.29**	0.27	**4.56**	nr	**5.13**
25mm diameter; plugging to masonry; to surfaces	1.04	10.00	**1.14**	0.42	-	**8.63**	0.27	**10.04**	m	**11.29**
extra; small circular terminal boxes	1.03	2.50	**1.06**	0.13	-	**2.67**	0.27	**4.00**	nr	**4.50**
extra; small circular angle boxes	1.16	2.50	**1.19**	0.16	-	**3.29**	0.27	**4.75**	nr	**5.34**
extra; small circular three way boxes	1.03	2.50	**1.06**	0.20	-	**4.11**	0.27	**5.43**	nr	**6.11**
extra; small circular through way boxes	1.10	2.50	**1.13**	0.16	-	**3.29**	0.27	**4.68**	nr	**5.27**
32mm diameter; plugging to masonry; to surfaces	2.43	10.00	**2.68**	0.50	-	**10.27**	0.27	**13.22**	m	**14.87**
Steel welded heavy gauge conduits and fittings, screwed joints; spacer bar saddles at 1000mm centres; galvanized inside and outside by manufacturer										
Conduits; straight										
20mm diameter; plugging to masonry; to surfaces	1.63	10.00	**1.79**	0.33	-	**6.78**	0.25	**8.82**	m	**9.93**
extra; small circular terminal boxes	1.09	2.50	**1.12**	0.13	-	**2.67**	0.27	**4.06**	nr	**4.56**
extra; small circular angle boxes	1.24	2.50	**1.27**	0.16	-	**3.29**	0.27	**4.83**	nr	**5.43**
extra; small circular three way boxes	1.09	2.50	**1.12**	0.20	-	**4.11**	0.27	**5.50**	nr	**6.18**
extra; small circular through way boxes	1.17	2.50	**1.20**	0.16	-	**3.29**	0.27	**4.76**	nr	**5.35**

Labour hourly rates: (except Specialists) Craft Operatives £20.54 Labourer £10.71 Rates are national average prices. Refer to REGIONAL VARIATIONS for indicative levels of overall pricing in regions	MATERIALS			LABOUR				RATES		
	Del to Site £	Waste %	Material Cost £	Craft Optve Hrs	Lab Hrs	Labour Cost £	Sunds £	Nett Rate £	Unit	Gross Rate (+12.5%) £
V90: GENERAL LIGHTING AND POWER - COMMERCIAL Cont.										
Steel welded heavy gauge conduits and fittings, screwed joints; spacer bar saddles at 1000mm centres; galvanized inside and outside by manufacturer Cont.										
Conduits; straight Cont.										
25mm diameter; plugging to masonry; to surfaces..	2.16	10.00	**2.37**	0.42	-	**8.63**	0.25	**11.25**	m	**12.65**
extra; small circular terminal boxes............	1.36	2.50	**1.39**	0.13	-	**2.67**	0.27	**4.33**	nr	**4.88**
extra; small circular angle boxes...............	1.53	2.50	**1.57**	0.16	-	**3.29**	0.27	**5.12**	nr	**5.77**
extra; small circular three way boxes...........	1.36	2.50	**1.39**	0.20	-	**4.11**	0.27	**5.77**	nr	**6.49**
extra; small circular through way boxes.........	1.45	2.50	**1.49**	0.16	-	**3.29**	0.27	**5.04**	nr	**5.67**
32mm diameter; plugging to masonry; to surfaces..	2.94	10.00	**3.23**	0.50	-	**10.27**	0.25	**13.75**	m	**15.47**
galvanized standard cable tray and fittings										
Light duty; straight										
50mm tray	1.22	5.00	**1.28**	0.25	-	**5.14**	0.09	**6.50**	m	**7.32**
extra; flat bend	7.23	-	**7.23**	0.20	-	**4.11**	0.25	**11.59**	nr	**13.03**
extra; tee	9.37	-	**9.37**	0.25	-	**5.14**	0.25	**14.76**	nr	**16.60**
extra; cross piece	15.82	-	**15.82**	0.40	-	**8.22**	0.50	**24.53**	nr	**27.60**
extra; riser bend	11.28	-	**11.28**	0.20	-	**4.11**	0.25	**15.63**	nr	**17.59**
100mm tray	1.76	5.00	**1.85**	0.33	-	**6.78**	0.09	**8.72**	m	**9.81**
extra; flat bend	7.80	-	**7.80**	0.23	-	**4.72**	0.25	**12.78**	nr	**14.37**
extra; tee	10.31	-	**10.31**	0.30	-	**6.16**	0.25	**16.73**	nr	**18.82**
extra; cross piece	16.86	-	**16.86**	0.41	-	**8.42**	0.50	**25.78**	nr	**29.01**
extra; riser bend	13.00	-	**13.00**	0.23	-	**4.72**	0.25	**17.98**	nr	**20.22**
extra; reducer 100 to 50mm......................	9.88	-	**9.88**	0.23	-	**4.72**	0.25	**14.86**	nr	**16.72**
150mm tray	2.58	5.00	**2.71**	0.42	-	**8.63**	0.09	**11.43**	m	**12.86**
extra; flat bend	9.11	-	**9.11**	0.25	-	**5.14**	0.25	**14.50**	nr	**16.31**
extra; tee	12.51	-	**12.51**	0.33	-	**6.78**	0.25	**19.54**	nr	**21.98**
extra; cross piece	19.79	-	**19.79**	0.42	-	**8.63**	0.50	**28.92**	nr	**32.53**
extra; riser bend	17.55	-	**17.55**	0.25	-	**5.14**	0.25	**22.94**	nr	**25.80**
extra; reducer 150 to 100mm.....................	13.02	-	**13.02**	0.25	-	**5.14**	0.25	**18.41**	nr	**20.71**
300mm tray	4.40	5.00	**4.62**	0.50	-	**10.27**	0.09	**14.98**	m	**16.85**
extra; flat bend	16.80	-	**16.80**	0.33	-	**6.78**	0.25	**23.83**	nr	**26.81**
extra; tee	23.02	-	**23.02**	0.50	-	**10.27**	0.25	**33.54**	nr	**37.74**
extra; cross piece	35.38	-	**35.38**	0.58	-	**11.91**	0.50	**47.79**	nr	**53.76**
extra; riser bend	30.89	-	**30.89**	0.33	-	**6.78**	0.25	**37.92**	nr	**42.66**
extra; reducer 300 to 150mm.....................	23.69	-	**23.69**	0.33	-	**6.78**	0.25	**30.72**	nr	**34.56**
450mm tray	8.88	5.00	**9.33**	0.58	-	**11.91**	0.09	**21.33**	m	**24.00**
extra; flat bend	26.89	-	**26.89**	0.50	-	**10.27**	0.25	**37.41**	nr	**42.09**
extra; tee	37.56	-	**37.56**	0.75	-	**15.41**	0.25	**53.22**	nr	**59.87**
extra; cross piece	58.39	-	**58.39**	1.00	-	**20.54**	0.50	**79.43**	nr	**89.36**
extra; riser bend	43.75	-	**43.75**	0.50	-	**10.27**	0.25	**54.27**	nr	**61.06**
extra; reducer 450 to 300mm.....................	39.33	-	**39.33**	0.50	-	**10.27**	0.25	**49.85**	nr	**56.08**
Medium duty; straight										
75mm tray	2.58	5.00	**2.71**	0.26	-	**5.34**	0.09	**8.14**	m	**9.16**
extra; flat bend	28.17	-	**28.17**	0.21	-	**4.31**	0.25	**32.73**	nr	**36.82**
extra; tee	37.16	-	**37.16**	0.26	-	**5.34**	0.25	**42.75**	nr	**48.10**
extra; cross piece	58.90	-	**58.90**	0.42	-	**8.63**	0.50	**68.03**	nr	**76.53**
extra; riser bend	15.31	-	**15.31**	0.21	-	**4.31**	0.25	**19.88**	nr	**22.36**
100mm tray	3.09	5.00	**3.24**	0.34	-	**6.98**	0.09	**10.32**	m	**11.61**
extra; flat bend	28.17	-	**28.17**	0.24	-	**4.93**	0.25	**33.35**	nr	**37.52**
extra; tee	37.16	-	**37.16**	0.31	-	**6.37**	0.25	**43.78**	nr	**49.25**
extra; cross piece	57.28	-	**57.28**	0.43	-	**8.83**	0.50	**66.61**	nr	**74.94**
extra; riser bend	14.01	-	**14.01**	0.24	-	**4.93**	0.25	**19.19**	nr	**21.58**
extra; reducer 100 to 75mm......................	13.08	-	**13.08**	0.24	-	**4.93**	0.25	**18.25**	nr	**20.54**
150mm tray	3.78	5.00	**3.97**	0.44	-	**9.04**	0.09	**13.10**	m	**14.74**
extra; flat bend	31.78	-	**31.78**	0.26	-	**5.34**	0.25	**37.37**	nr	**42.04**
extra; tee	42.29	-	**42.29**	0.34	-	**6.98**	0.25	**49.52**	nr	**55.71**
extra; cross piece	63.55	-	**63.55**	0.44	-	**9.04**	0.50	**73.09**	nr	**82.23**
extra; riser bend	18.40	-	**18.40**	0.26	-	**5.34**	0.25	**23.99**	nr	**26.99**
extra; reducer 150 to 100mm.....................	11.35	-	**11.35**	0.26	-	**5.34**	0.25	**16.94**	nr	**19.06**
300mm tray	4.96	5.00	**5.21**	0.52	-	**10.68**	0.09	**15.98**	m	**17.98**
extra; flat bend	43.60	-	**43.60**	0.34	-	**6.98**	0.25	**50.83**	nr	**57.18**
extra; tee	54.06	-	**54.06**	0.52	-	**10.68**	0.25	**64.99**	nr	**73.11**
extra; cross piece	83.22	-	**83.22**	0.61	-	**12.53**	0.50	**96.25**	nr	**108.28**
extra; riser bend	29.23	-	**29.23**	0.34	-	**6.98**	0.25	**36.46**	nr	**41.02**
extra; reducer 300 to 150mm.....................	23.90	-	**23.90**	0.34	-	**6.98**	0.25	**31.13**	nr	**35.03**
450mm tray	6.77	5.00	**7.11**	0.61	-	**12.53**	0.09	**19.73**	m	**22.20**
extra; flat bend	57.67	-	**57.67**	0.52	-	**10.68**	0.25	**68.60**	nr	**77.17**
extra; tee	75.55	-	**75.55**	0.78	-	**16.02**	0.25	**91.82**	nr	**103.30**
extra; cross piece	113.24	-	**113.24**	1.05	-	**21.57**	0.50	**135.31**	nr	**152.22**
extra; riser bend	44.08	-	**44.08**	0.52	-	**10.68**	0.25	**55.01**	nr	**61.89**
extra; reducer 450 to 300mm.....................	39.43	-	**39.43**	0.52	-	**10.68**	0.25	**50.37**	nr	**56.66**
Heavy duty; straight										
75mm tray	3.34	5.00	**3.51**	0.27	-	**5.55**	0.09	**9.14**	m	**10.29**
extra; flat bend	39.96	-	**39.96**	0.22	-	**4.52**	0.25	**44.73**	nr	**50.32**
extra; tee	49.11	-	**49.11**	0.27	-	**5.55**	0.25	**54.91**	nr	**61.77**
extra; cross piece	79.34	-	**79.34**	0.44	-	**9.04**	0.50	**88.88**	nr	**99.99**
extra; riser bend	27.46	-	**27.46**	0.22	-	**4.52**	0.25	**32.23**	nr	**36.25**
100mm tray	3.39	5.00	**3.56**	0.36	-	**7.39**	0.09	**11.05**	m	**12.43**
extra; flat bend	40.48	-	**40.48**	0.25	-	**5.14**	0.25	**45.87**	nr	**51.60**
extra; tee	49.53	-	**49.53**	0.33	-	**6.78**	0.25	**56.56**	nr	**63.63**
extra; cross piece	79.97	-	**79.97**	0.45	-	**9.24**	0.50	**89.71**	nr	**100.92**
extra; riser bend	28.14	-	**28.14**	0.25	-	**5.14**	0.25	**33.52**	nr	**37.71**
extra; reducer 100 to 75mm......................	18.31	-	**18.31**	0.25	-	**5.14**	0.25	**23.69**	nr	**26.65**
150mm tray	3.76	5.00	**3.95**	0.46	-	**9.45**	0.09	**13.48**	m	**15.17**
extra; flat bend	43.93	-	**43.93**	0.27	-	**5.55**	0.25	**49.73**	nr	**55.94**
extra; tee	54.44	-	**54.44**	0.36	-	**7.39**	0.25	**62.09**	nr	**69.85**
extra; cross piece	86.61	-	**86.61**	0.46	-	**9.45**	0.50	**96.56**	nr	**108.63**
extra; riser bend	30.54	-	**30.54**	0.27	-	**5.55**	0.25	**36.34**	nr	**40.88**
extra; reducer 150 to 100mm.....................	18.36	-	**18.36**	0.27	-	**5.55**	0.25	**24.15**	nr	**27.17**

Labour hourly rates: (except Specialists) Craft Operatives £20.54 Labourer £10.71 Rates are national average prices. Refer to REGIONAL VARIATIONS for indicative levels of overall pricing in regions	MATERIALS			LABOUR				RATES		
	Del to Site £	Waste %	Material Cost £	Craft Optve Hrs	Lab Hrs	Labour Cost £	Sunds £	Nett Rate £	Unit	Gross Rate (+12.5%) £
V90: GENERAL LIGHTING AND POWER - COMMERCIAL Cont.										
galvanized standard cable tray and fittings Cont.										
Heavy duty; straight Cont.										
300mm tray	5.47	5.00	**5.74**	0.55	-	**11.30**	0.09	**17.13**	m	**19.27**
extra; flat bend	50.42	-	**50.42**	0.36	-	**7.39**	0.25	**58.06**	nr	**65.32**
extra; tee	65.79	-	**65.79**	0.55	-	**11.30**	0.25	**77.34**	nr	**87.01**
extra; cross piece	104.13	-	**104.13**	0.64	-	**13.15**	0.50	**117.77**	nr	**132.50**
extra; riser bend	33.58	-	**33.58**	0.36	-	**7.39**	0.25	**41.22**	nr	**46.37**
extra; reducer 300 to 150mm	30.12	-	**30.12**	0.36	-	**7.39**	0.25	**37.77**	nr	**42.49**
450mm tray	9.46	5.00	**9.94**	0.64	-	**13.15**	0.09	**23.17**	m	**26.07**
extra; flat bend	85.51	-	**85.51**	0.55	-	**11.30**	0.25	**97.06**	nr	**109.19**
extra; tee	102.51	-	**102.51**	0.83	-	**17.05**	0.25	**119.81**	nr	**134.78**
extra; cross piece	160.87	-	**160.87**	1.10	-	**22.59**	0.25	**183.72**	nr	**206.68**
extra; riser bend	58.99	-	**58.99**	0.55	-	**11.30**	0.25	**70.54**	nr	**79.36**
extra; reducer 450 to 300mm	47.12	-	**47.12**	0.55	-	**11.30**	0.25	**58.67**	nr	**66.00**
Supports for cable tray										
Cantilever arms; mild steel; hot dip galvanized; fixing to masonry										
65mm wide	4.55	2.50	**4.66**	0.33	-	**6.78**	0.25	**11.69**	nr	**13.15**
90mm wide	4.55	2.50	**4.66**	0.33	-	**6.78**	0.25	**11.69**	nr	**13.15**
120mm wide	4.55	2.50	**4.66**	0.33	-	**6.78**	0.25	**11.69**	nr	**13.15**
Stand-off bracket; mild steel; hot dip galvanized; fixing to masonry										
75mm wide	4.55	2.50	**4.66**	0.33	-	**6.78**	0.25	**11.69**	nr	**13.15**
100mm wide	5.86	2.50	**6.00**	0.33	-	**6.78**	0.25	**13.03**	nr	**14.66**
150mm wide	5.86	2.50	**6.00**	0.33	-	**6.78**	0.25	**13.03**	nr	**14.66**
300mm wide	12.92	2.50	**13.24**	0.33	-	**6.78**	0.25	**20.27**	nr	**22.80**
450mm wide	17.68	2.50	**18.12**	0.33	-	**6.78**	0.25	**25.15**	nr	**28.29**
PVC insulated cables; single core; reference 6491X; stranded copper conductors; to BS 6004										
Drawn into conduits or ducts or laid or drawn into trunking										
1.5mm2	0.00	15.00	**0.00**	0.03	-	**0.62**	-	**0.62**	m	**0.69**
2.5mm2	0.00	15.00	**0.00**	0.03	-	**0.62**	-	**0.62**	m	**0.69**
4.0mm2	0.00	15.00	**0.00**	0.04	-	**0.82**	-	**0.82**	m	**0.93**
6.0mm2	0.00	15.00	**0.00**	0.05	-	**1.03**	-	**1.03**	m	**1.16**
10.0mm2	0.00	15.00	**0.00**	0.05	-	**1.03**	-	**1.03**	m	**1.16**
16.0mm2	0.01	15.00	**0.01**	0.08	-	**1.54**	-	**1.55**	m	**1.74**
PVC double insulated cables; single core; reference 6181Y; stranded copper conductors; to BS 6004										
Fixed to timber with clips										
6.0mm2	0.59	15.00	**0.68**	0.07	-	**1.44**	-	**2.12**	m	**2.38**
10.0mm2	0.65	15.00	**0.75**	0.08	-	**1.64**	-	**2.39**	m	**2.69**
16.0mm2	0.86	15.00	**0.99**	0.10	-	**2.05**	-	**3.04**	m	**3.42**
25.0mm2	1.09	15.00	**1.25**	0.15	-	**3.08**	-	**4.33**	m	**4.88**
35.0mm2	2.71	15.00	**3.12**	0.20	-	**4.11**	-	**7.22**	m	**8.13**
PVC insulated and PVC sheathed cables; multicore; copper conductors and bare earth continuity conductor 6242Y; to BS 6004										
Drawn into conduits or ducts or laid or drawn into trunking										
1.0mm2; twin with bare earth	0.00	15.00	**0.00**	0.05	-	**1.03**	-	**1.03**	m	**1.16**
1.5mm2; twin with bare earth	0.00	15.00	**0.00**	0.05	-	**1.03**	-	**1.03**	m	**1.16**
2.5mm2; twin with bare earth	0.00	15.00	**0.00**	0.05	-	**1.03**	-	**1.03**	m	**1.16**
4.0mm2; twin with bare earth	0.00	15.00	**0.01**	0.06	-	**1.23**	-	**1.24**	m	**1.39**
6.0mm2; twin with bare earth	0.01	15.00	**0.01**	0.07	-	**1.44**	-	**1.44**	m	**1.62**
Fixed to timber with clips										
1.0mm2; twin with bare earth	0.00	15.00	**0.00**	0.07	-	**1.44**	0.06	**1.50**	m	**1.69**
1.5mm2; twin with bare earth	0.00	15.00	**0.00**	0.07	-	**1.44**	0.06	**1.50**	m	**1.69**
2.5mm2; twin with bare earth	0.00	15.00	**0.00**	0.07	-	**1.44**	0.07	**1.51**	m	**1.70**
4.0mm2; twin with bare earth	0.00	15.00	**0.01**	0.08	-	**1.64**	0.08	**1.73**	m	**1.94**
6.0mm2; twin with bare earth	0.01	15.00	**0.01**	0.09	-	**1.85**	0.08	**1.93**	m	**2.18**
PVC insulated SWA armoured and PVC sheathed cables; multicore; copper conductors to BS 6346										
Laid and laced on cable tray										
16mm2; three core	3.54	5.00	**3.72**	0.17	-	**3.49**	0.11	**7.32**	m	**8.23**
Fixed to masonry with screwed clips at average 300mm centres; plugging										
16mm2; three core	3.54	5.00	**3.72**	0.25	-	**5.14**	0.94	**9.79**	m	**11.02**
Mineral insulated copper sheathed cables; PVC outer sheath; copper conductors; 600V light duty to BS EN 60702-1										
Fixed to masonry with screwed clips at average 300mm centres; plugging										
1.5mm2; two core	2.51	10.00	**2.76**	0.17	-	**3.49**	0.94	**7.19**	m	**8.09**
extra; termination including gland, seal and shroud	3.20	2.50	**3.28**	0.42	-	**8.63**	-	**11.91**	nr	**13.40**
2.5mm2; two core	3.68	10.00	**4.05**	0.17	-	**3.49**	0.94	**8.48**	m	**9.54**

Labour hourly rates: (except Specialists) Craft Operatives £20.54 Labourer £10.71 Rates are national average prices. Refer to REGIONAL VARIATIONS for indicative levels of overall pricing in regions	MATERIALS			LABOUR				RATES		
	Del to Site £	Waste %	Material Cost £	Craft Optve Hrs	Lab Hrs	Labour Cost £	Sunds £	Nett Rate £	Unit	Gross Rate (+12.5%) £
V90: GENERAL LIGHTING AND POWER - COMMERCIAL Cont.										
Mineral insulated copper sheathed cables; PVC outer sheath; copper conductors; 600V light duty to BS EN 60702-1 Cont.										
Fixed to masonry with screwed clips at average 300mm centres; plugging Cont.										
extra; termination including gland, seal and shroud	3.20	2.50	**3.28**	0.42	-	**8.63**	-	**11.91**	nr	**13.40**
4.0mm2; two core	4.66	10.00	**5.13**	0.20	-	**4.11**	0.94	**10.17**	m	**11.45**
extra; termination including gland, seal and shroud	3.20	2.50	**3.28**	0.50	-	**10.27**	-	**13.55**	nr	**15.24**
HV switchgear										
Switch fuse 500V, metal cased, fixed to masonry with screws; plugging										
32 amp; single pole and neutral; short circuit rating 6.4Ka	113.84	2.50	**116.68**	1.60	-	**32.86**	0.50	**150.05**	nr	**168.80**
Consumer control units, 250V, metal cased, fitted with 100 amp D.P. isolator and MCB's , fixed to masonry with screws; plugging										
eight way, single pole and neutral	94.41	2.50	**96.77**	1.60	-	**32.86**	0.50	**130.14**	nr	**146.40**
Distribution boards 500V, metal cased, MCB pattern, fitted with 100 amp D.P. isolator, fixed to masonry with screws; plugging										
nine way, single pole and neutral, 20 amp	112.71	2.50	**115.53**	1.80	-	**36.97**	0.50	**153.00**	nr	**172.12**
twelve way, triple pole and neutral, 32 amp	500.77	2.50	**513.29**	8.00	-	**164.32**	0.50	**678.11**	nr	**762.88**
Luminaires										
Note for lighting fittings and lamps see General Lighting and Power - Domestic Section										
Accessories; white plastics, with boxes, fixing to masonry with screws; plugging										
Flush plate switches										
5 amp; one gang; one way; single pole	1.55	2.50	**1.59**	0.42	-	**8.63**	0.25	**10.47**	nr	**11.77**
5 amp; two gang; one way; single pole	2.10	2.50	**2.15**	0.43	-	**8.83**	0.25	**11.23**	nr	**12.64**
Surface plate switches										
5 amp; one gang; one way; single pole	1.65	2.50	**1.69**	0.42	-	**8.63**	0.25	**10.57**	nr	**11.89**
5 amp; two gang; one way; single pole	2.20	2.50	**2.26**	0.43	-	**8.83**	0.25	**11.34**	nr	**12.75**
Ceiling switches										
5 amp; one gang; one way; single pole	2.10	2.50	**2.15**	0.43	-	**8.83**	0.25	**11.23**	nr	**12.64**
Flush switched socket outlets										
13 amp; one gang	3.00	2.50	**3.08**	0.43	-	**8.83**	0.25	**12.16**	nr	**13.68**
13 amp; two gang	3.50	2.50	**3.59**	0.43	-	**8.83**	0.25	**12.67**	nr	**14.25**
Surface switched socket outlets										
13 amp; one gang	3.15	2.50	**3.23**	0.43	-	**8.83**	0.25	**12.31**	nr	**13.85**
13 amp; two gang	3.80	2.50	**3.90**	0.43	-	**8.83**	0.25	**12.98**	nr	**14.60**
Flush switched socket outlets with RCCB (residual current device) protected at 30 mAmp										
13 amp; one gang	35.60	2.50	**36.49**	0.43	-	**8.83**	0.25	**45.57**	nr	**51.27**
Flush fused connection units										
13 amp; one gang; switched; flexible outlet	4.25	2.50	**4.36**	0.50	-	**10.27**	0.25	**14.88**	nr	**16.74**
13 amp; one gang; switched; flexible outlet; pilot lamp	4.65	2.50	**4.77**	0.50	-	**10.27**	0.25	**15.29**	nr	**17.20**
Accessories, metalclad, with boxes, fixing to masonry with screws; plugging										
Surface plate switches										
5 amp; one gang; two way; single pole	3.10	2.50	**3.18**	0.42	-	**8.63**	0.25	**12.05**	nr	**13.56**
5 amp; two gang; two way; single pole	4.00	2.50	**4.10**	0.43	-	**8.83**	0.25	**13.18**	nr	**14.83**
Surface switched socket outlets										
13 amp; one gang	4.50	2.50	**4.61**	0.43	-	**8.83**	0.25	**13.69**	nr	**15.41**
13 amp; two gang	6.50	2.50	**6.66**	0.43	-	**8.83**	0.25	**15.74**	nr	**17.71**
Weatherproof accessories, with boxes, fixing to masonry with screws; plugging										
Surface plate switches										
5 amp; one gang; two way; single pole	6.94	2.50	**7.11**	0.43	-	**8.83**	0.25	**16.19**	nr	**18.22**

Labour hourly rates: (except Specialists)
Craft Operatives £20.54 Labourer £10.71
Rates are national average prices.
Refer to REGIONAL VARIATIONS for indicative levels of overall pricing in regions

	MATERIALS			LABOUR				RATES		
	Del to Site £	Waste %	Material Cost £	Craft Optve Hrs	Lab Hrs	Labour Cost £	Sunds £	Nett Rate £	Unit	Gross Rate (+12.5%) £
V90: GENERAL LIGHTING AND POWER - DOMESTIC										
Electric wiring										
Note										
the following approximate prices of various types of installations are dependant on the number and disposition of points; lamps and fittings together with cutting and making good are excluded										
Electric wiring in new building										
Insulated Consumer unit, fixed to masonry withscrews; plugging										
2 usable ways 63a DP main Switch	17.58	2.50	18.02	1.10	-	22.59	0.25	40.86	nr	45.97
4 usable ways 100a DP main Switch	19.85	2.50	20.35	1.83	-	37.59	0.25	58.19	nr	65.46
6 usable ways 100a DP main Switch	23.13	2.50	23.71	2.20	-	45.19	0.25	69.15	nr	77.79
9 usable ways 100a DP main Switch	25.64	2.50	26.28	2.53	-	51.97	0.25	78.49	nr	88.30
13 usable ways 100a DP main Switch	32.53	2.50	33.35	3.03	-	62.24	0.25	95.83	nr	107.81
2 usable ways 63a 30ma RCD main Switch	57.10	2.50	58.53	1.28	-	26.29	0.25	85.07	nr	95.71
4 usable ways 80a 30ma RCD main Switch	63.62	2.50	65.21	1.38	-	28.35	0.25	93.81	nr	105.53
6 usable ways 80a 30ma RCD main Switch	70.22	2.50	71.98	1.65	-	33.89	0.25	106.12	nr	119.38
9 usable ways 80a 30ma RCD main Switch	76.34	2.50	78.25	1.83	-	37.59	0.25	116.09	nr	130.60
Miniature Circuit Breakers fitted in consumer unit; Bsen 60898										
6a Single Pole	6.24	2.50	6.40	0.14	-	2.88	-	9.27	nr	10.43
10a Single Pole	5.95	2.50	6.09	0.14	-	2.88	-	8.97	nr	10.09
16a Single Pole	5.95	2.50	6.09	0.14	-	2.88	-	8.97	nr	10.09
20a Single Pole	5.95	2.50	6.09	0.14	-	2.88	-	8.97	nr	10.09
32a Single Pole	5.95	2.50	6.09	0.14	-	2.88	-	8.97	nr	10.09
40a Single Pole	6.24	2.50	6.40	0.14	-	2.88	-	9.27	nr	10.43
Domestic Metal consumer units and accessories										
2 usable ways 63a DP main Switch	21.10	2.50	21.63	1.10	-	22.59	0.25	44.48	nr	50.04
4 usable ways 100a DP main Switch	22.62	2.50	23.19	1.83	-	37.59	0.25	61.03	nr	68.65
6 usable ways 100a DP main Switch	25.82	2.50	26.47	2.20	-	45.19	0.25	71.90	nr	80.89
9 usable ways 100a DP main Switch	31.11	2.50	31.89	2.53	-	51.97	0.25	84.10	nr	94.62
13 usable ways 100a DP main Switch	36.99	2.50	37.92	3.03	-	62.24	0.25	100.40	nr	112.95
Miniature Circuit Breakers fitted in consumer unit; Bsen 60898										
6a Single Pole	6.24	2.50	6.40	0.14	-	2.88	-	9.27	nr	10.43
10a Single Pole	5.95	2.50	6.09	0.14	-	2.88	-	8.97	nr	10.09
16a Single Pole	5.95	2.50	6.09	0.14	-	2.88	-	8.97	nr	10.09
20a Single Pole	5.95	2.50	6.09	0.14	-	2.88	-	8.97	nr	10.09
32a Single Pole	5.95	2.50	6.09	0.14	-	2.88	-	8.97	nr	10.09
40a Single Pole	6.24	2.50	6.40	0.14	-	2.88	-	9.27	nr	10.43
Installation with PVC insulated and sheathed cables										
lighting points	7.02	7.50	7.52	3.58	-	73.53	-	81.06	nr	91.19
socket outlets; 5A	7.66	7.50	8.21	4.84	-	99.41	-	107.62	nr	121.07
socket outlets; 13A ring main	4.20	7.50	4.49	4.07	-	83.60	-	88.09	nr	99.10
socket outlets; 13A radial circuit	4.33	7.50	4.63	4.24	-	87.09	-	91.72	nr	103.18
cooker points; 45A	14.25	7.50	15.27	6.93	-	142.34	-	157.61	nr	177.31
immersion heater points	6.54	7.50	7.00	5.06	-	103.93	-	110.94	nr	124.80
shaver sockets (transformer)	16.54	7.50	17.76	1.82	-	37.38	-	55.14	nr	62.03
Installation with mineral insulated copper sheathed cables										
lighting points	17.38	7.50	18.66	6.82	-	140.08	-	158.74	nr	178.58
socket outlets; 5A	32.09	7.50	34.44	5.34	-	109.68	-	144.12	nr	162.14
socket outlets; 13A ring main	22.24	7.50	23.88	4.57	-	93.87	-	117.75	nr	132.47
socket outlets; 13A radial circuit	28.29	7.50	30.38	6.11	-	125.50	-	155.88	nr	175.37
cooker points; 45A	79.12	7.50	85.01	10.12	-	207.86	-	292.87	nr	329.48
immersion heater points	36.74	7.50	39.47	7.26	-	149.12	-	188.59	nr	212.16
shaver sockets (transformer)	46.31	7.50	49.75	3.30	-	67.78	-	117.53	nr	132.23
Installation with black enamel heavy gauge conduit with PVC cables										
lighting points	33.01	7.50	35.23	5.12	-	105.16	-	140.39	nr	157.94
socket outlets; 5A	29.52	7.50	31.48	5.17	-	106.19	-	137.67	nr	154.88
socket outlets; 13A ring main	22.89	7.50	24.43	4.68	-	96.13	-	120.56	nr	135.63
socket outlets; 13A radial circuit	26.07	7.50	27.77	6.60	-	135.56	-	163.33	nr	183.75
cooker points; 45A	45.06	7.50	48.13	10.62	-	218.13	-	266.27	nr	299.55
immersion heater points	29.22	7.50	31.11	7.87	-	161.65	-	192.75	nr	216.85
shaver sockets (transformer)	46.35	7.50	49.15	2.53	-	51.97	-	101.12	nr	113.75
Installation with black enamel heavy gauge conduit with coaxial cable										
T.V. sockets	41.66	7.50	43.97	6.33	-	130.02	-	173.99	nr	195.74
Installation with black enamel heavy gauge conduit with draw wire										
telephone points	42.37	7.50	44.74	6.00	-	123.24	-	167.98	nr	188.97
Electric wiring in extension to existing building										
Installation with PVC insulated and sheathed cables										
lighting points	7.02	7.50	7.52	3.58	-	73.53	-	81.06	nr	91.19
socket outlets; 5A	7.66	7.50	8.21	4.84	-	99.41	-	107.62	nr	121.07
socket outlets; 13A ring main	4.20	7.50	4.49	4.07	-	83.60	-	88.09	nr	99.10
socket outlets; 13A radial circuit	4.33	7.50	4.63	4.24	-	87.09	-	91.72	nr	103.18
cooker points; 45A	14.25	7.50	15.27	6.93	-	142.34	-	157.61	nr	177.31
immersion heater points	6.54	7.50	7.00	5.06	-	103.93	-	110.94	nr	124.80
shaver sockets (transformer)	16.54	7.50	17.76	1.82	-	37.38	-	55.14	nr	62.03

Labour hourly rates: (except Specialists) Craft Operatives £20.54 Labourer £10.71 Rates are national average prices. Refer to REGIONAL VARIATIONS for indicative levels of overall pricing in regions	MATERIALS			LABOUR				RATES		
	Del to Site £	Waste %	Material Cost £	Craft Optve Hrs	Lab Hrs	Labour Cost £	Sunds £	Nett Rate £	Unit	Gross Rate (+12.5%) £
V90: GENERAL LIGHTING AND POWER - DOMESTIC Cont.										
Electric wiring in extension to existing building Cont.										
Installation with mineral insulated copper sheathed cables										
lighting points	17.38	7.50	**18.66**	6.82	-	**140.08**	-	**158.74**	nr	**178.58**
socket outlets; 5A	32.09	7.50	**34.44**	5.34	-	**109.68**	-	**144.12**	nr	**162.14**
socket outlets; 13A ring main	22.24	7.50	**23.88**	4.57	-	**93.87**	-	**117.75**	nr	**132.47**
socket outlets; 13A radial circuit	28.29	7.50	**30.38**	6.11	-	**125.50**	-	**155.88**	nr	**175.37**
cooker points; 45A	79.12	7.50	**85.01**	10.12	-	**207.86**	-	**292.87**	nr	**329.48**
immersion heater points	36.74	7.50	**39.47**	7.26	-	**149.12**	-	**188.59**	nr	**212.16**
shaver sockets (transformer)	46.31	7.50	**49.75**	3.30	-	**67.78**	-	**117.53**	nr	**132.23**
Installation with black enamel heavy gauge conduit with PVC cables										
lighting points	33.01	7.50	**35.23**	5.12	-	**105.16**	-	**140.39**	nr	**157.94**
socket outlets; 5A	29.52	7.50	**31.48**	5.17	-	**106.19**	-	**137.67**	nr	**154.88**
socket outlets; 13A ring main	22.89	7.50	**24.43**	4.68	-	**96.13**	-	**120.56**	nr	**135.63**
socket outlets; 13A radial circuit	26.07	7.50	**27.77**	6.60	-	**135.56**	-	**163.33**	nr	**183.75**
cooker points; 45A	45.06	7.50	**48.13**	10.62	-	**218.13**	-	**266.27**	nr	**299.55**
immersion heater points	29.22	7.50	**31.11**	7.87	-	**161.65**	-	**192.75**	nr	**216.85**
shaver sockets (transformer)	46.35	7.50	**49.15**	2.53	-	**51.97**	-	**101.12**	nr	**113.75**
Installation with black enamel heavy gauge conduit with coaxial cable										
T.V. sockets	41.66	7.50	**43.97**	6.33	-	**130.02**	-	**173.99**	nr	**195.74**
Installation with black enamel heavy gauge conduit with draw wire										
telephone points	42.37	7.50	**44.74**	6.00	-	**123.24**	-	**167.98**	nr	**188.97**
Electric wiring in extending existing installation in existing building										
Installation with PVC insulated and sheathed cables										
lighting points	7.02	7.50	**7.52**	4.13	-	**84.83**	-	**92.36**	nr	**103.90**
socket outlets; 5A	7.66	7.50	**8.21**	5.56	-	**114.20**	-	**122.41**	nr	**137.71**
socket outlets; 13A ring main	4.20	7.50	**4.49**	4.68	-	**96.13**	-	**100.62**	nr	**113.20**
socket outlets; 13A radial circuit	4.33	7.50	**4.63**	4.90	-	**100.65**	-	**105.27**	nr	**118.43**
cooker points; 45A	14.25	7.50	**15.27**	7.98	-	**163.91**	-	**179.18**	nr	**201.57**
immersion heater points	6.54	7.50	**7.00**	5.83	-	**119.75**	-	**126.75**	nr	**142.60**
shaver sockets (transformer)	16.54	7.50	**17.76**	2.09	-	**42.93**	-	**60.69**	nr	**68.27**
Installation with mineral insulated copper sheathed cables										
lighting points	17.38	7.50	**18.66**	7.87	-	**161.65**	-	**180.31**	nr	**202.85**
socket outlets; 5A	32.09	7.50	**34.44**	6.16	-	**126.53**	-	**160.97**	nr	**181.09**
socket outlets; 13A ring main	22.24	7.50	**23.88**	5.28	-	**108.45**	-	**132.33**	nr	**148.88**
socket outlets; 13A radial circuit	28.29	7.50	**30.38**	7.04	-	**144.60**	-	**174.99**	nr	**196.86**
cooker points; 45A	79.12	7.50	**85.01**	11.66	-	**239.50**	-	**324.50**	nr	**365.07**
immersion heater points	36.74	7.50	**39.47**	8.36	-	**171.71**	-	**211.18**	nr	**237.58**
shaver sockets (transformer)	46.31	7.50	**49.75**	3.80	-	**78.05**	-	**127.80**	nr	**143.78**
Installation with black enamel heavy gauge conduit with PVC cables										
lighting points	33.01	7.50	**35.23**	5.88	-	**120.78**	-	**156.00**	nr	**175.50**
socket outlets; 5A	29.52	7.50	**31.48**	5.94	-	**122.01**	-	**153.49**	nr	**172.67**
socket outlets; 13A ring main	22.89	7.50	**24.43**	5.39	-	**110.71**	-	**135.14**	nr	**152.03**
socket outlets; 13A radial circuit	26.07	7.50	**27.77**	7.60	-	**156.10**	-	**183.87**	nr	**206.86**
cooker points; 45A	45.06	7.50	**48.13**	12.21	-	**250.79**	-	**298.93**	nr	**336.29**
immersion heater points	29.22	7.50	**31.11**	9.08	-	**186.50**	-	**217.61**	nr	**244.81**
shaver sockets (transformer)	46.35	7.50	**49.15**	2.92	-	**59.98**	-	**109.13**	nr	**122.77**
Installation with black enamel heavy gauge conduit with coaxial cable										
T.V. sockets	41.66	7.50	**43.97**	7.26	-	**149.12**	-	**193.09**	nr	**217.23**
Installation with black enamel heavy gauge conduit with draw wire										
telephone points	42.37	7.50	**44.74**	6.88	-	**141.32**	-	**186.05**	nr	**209.31**
V90: LAMP FITTINGS, FANS, HEATERS etc., ALL INSTALLATIONS										
Fluorescent lamp fittings, mains voltage operations; switch start; including lamps										
Batten type; single tube										
1200mm; 36W	10.25	2.50	**10.51**	0.83	-	**17.05**	0.50	**28.05**	nr	**31.56**
1500mm; 58W	12.40	2.50	**12.71**	1.10	-	**22.59**	0.50	**35.80**	nr	**40.28**
1800mm; 70W	19.00	2.50	**19.48**	1.10	-	**22.59**	0.50	**42.57**	nr	**47.89**
2400mm; 100W	27.50	2.50	**28.19**	1.38	-	**28.35**	0.50	**57.03**	nr	**64.16**
Batten type; twin tube										
1200mm; 36W	16.88	2.50	**17.30**	0.83	-	**17.05**	0.50	**34.85**	nr	**39.21**
1500mm; 58W	20.00	2.50	**20.50**	1.10	-	**22.59**	0.50	**43.59**	nr	**49.04**
1800mm; 70W	30.00	2.50	**30.75**	1.10	-	**22.59**	0.50	**53.84**	nr	**60.57**
2400mm; 100W	40.00	2.50	**41.00**	1.38	-	**28.35**	0.50	**69.85**	nr	**78.58**
Metal trough reflector; single tube										
1200mm; 36W	34.41	2.50	**35.27**	1.10	-	**22.59**	0.50	**58.36**	nr	**65.66**
1500mm; 58W	37.69	2.50	**38.63**	1.32	-	**27.11**	0.50	**66.25**	nr	**74.53**
1800mm; 70W	44.41	2.50	**45.52**	1.32	-	**27.11**	0.50	**73.13**	nr	**82.27**
2400mm; 100W	63.21	2.50	**64.79**	1.60	-	**32.86**	0.50	**98.15**	nr	**110.42**

Labour hourly rates: (except Specialists) Craft Operatives £20.54 Labourer £10.71 Rates are national average prices. Refer to REGIONAL VARIATIONS for indicative levels of overall pricing in regions	MATERIALS			LABOUR				RATES		
	Del to Site £	Waste %	Material Cost £	Craft Optve Hrs	Lab Hrs	Labour Cost £	Sunds £	Nett Rate £	Unit	Gross Rate (+12.5%) £
V90: LAMP FITTINGS, FANS, HEATERS etc., ALL INSTALLATIONS Cont.										
Fluorescent lamp fittings, mains voltage operations; switch start; including lamps Cont.										
Metal trough reflector; twin tube										
1200mm; 36W	53.22	2.50	**54.55**	1.10	-	**22.59**	0.50	**77.64**	nr	**87.35**
1500mm; 58W	61.19	2.50	**62.72**	1.32	-	**27.11**	0.50	**90.33**	nr	**101.62**
1800mm; 70W	67.60	2.50	**69.29**	1.32	-	**27.11**	0.50	**96.90**	nr	**109.02**
2400mm; 100W	91.96	2.50	**94.26**	1.60	-	**32.86**	0.50	**127.62**	nr	**143.58**
Plastics diffused type; single tube										
1200mm; 36W	35.51	2.50	**36.40**	1.10	-	**22.59**	0.50	**59.49**	nr	**66.93**
1500mm; 58W	39.77	2.50	**40.76**	1.32	-	**27.11**	0.50	**68.38**	nr	**76.92**
1800mm; 70W	48.67	2.50	**49.89**	1.32	-	**27.11**	0.50	**77.50**	nr	**87.19**
2400mm; 100W	65.57	2.50	**67.21**	1.60	-	**32.86**	0.50	**100.57**	nr	**113.14**
Plastics diffused type; twin tube										
1200mm; 36W	66.28	2.50	**67.94**	1.00	-	**20.54**	0.50	**88.98**	nr	**100.10**
1500mm; 58W	80.52	2.50	**82.53**	1.32	-	**27.11**	0.50	**110.15**	nr	**123.91**
1800mm; 70W	90.95	2.50	**93.22**	1.32	-	**27.11**	0.50	**120.84**	nr	**135.94**
2400mm; 100W	116.15	2.50	**119.05**	1.60	-	**32.86**	0.50	**152.42**	nr	**171.47**
Lighting fittings complete with tungsten lamps										
Bulkhead; alloy										
100W G.L.S. lamp	10.50	2.50	**10.76**	0.72	-	**14.79**	0.25	**25.80**	nr	**29.03**
Wall glass; alloy; corner bracket										
100W G.L.S. lamp	27.87	2.50	**28.57**	0.83	-	**17.05**	0.25	**45.86**	nr	**51.60**
Ceiling sphere										
152mm; 60W G.L.S. lamp	14.85	2.50	**15.22**	0.72	-	**14.79**	0.25	**30.26**	nr	**34.04**
203mm; 100W G.L.S. lamp	21.64	2.50	**22.18**	0.72	-	**14.79**	0.25	**37.22**	nr	**41.87**
Down light; semi recessed										
175mm; 150W PAR38 lamp	18.54	2.50	**19.00**	0.72	-	**14.79**	0.25	**34.04**	nr	**38.30**
Lighting track; single circuit; surface mounted										
1250mm; starter pack	18.85	2.50	**19.32**	1.32	-	**27.11**	0.50	**46.93**	nr	**52.80**
plug-in fitting; 100W G.L.S. lamp	14.46	2.50	**14.82**	0.22	-	**4.52**	0.25	**19.59**	nr	**22.04**
Fan heaters										
3kw unit fan heaters for commercial application	159.66	2.50	**163.65**	1.65	-	**33.89**	0.50	**198.04**	nr	**222.80**
Thermostatic switch units comprising ON/OFF switch and selector switch between winter and summer (fan only) operation	54.45	2.50	**55.81**	1.10	-	**22.59**	0.25	**78.66**	nr	**88.49**
Tubular heaters										
60W per 305mm loading										
600mm long	11.34	2.50	**11.62**	1.10	-	**22.59**	0.75	**34.97**	nr	**39.34**
910mm long	12.48	2.50	**12.79**	1.10	-	**22.59**	0.75	**36.14**	nr	**40.65**
1220mm long	14.34	2.50	**14.70**	1.38	-	**28.35**	1.00	**44.04**	nr	**49.55**
1520mm long	18.74	2.50	**19.21**	1.38	-	**28.35**	1.00	**48.55**	nr	**54.62**
1830mm long	19.15	2.50	**19.63**	1.65	-	**33.89**	1.50	**55.02**	nr	**61.90**
Two way mounting brackets										
pair	5.45	2.50	**5.59**	0.44	-	**9.04**	0.50	**15.12**	nr	**17.01**
Night storage heaters; installed complete in new building including wiring										
Plastics insulated and sheathed cabled										
1.7KW	243.24	2.50	**249.32**	7.15	-	**146.86**	-	**396.18**	nr	**445.71**
2.55KW	320.27	2.50	**328.28**	7.70	-	**158.16**	-	**486.44**	nr	**547.25**
3.4KW	364.54	2.50	**373.66**	8.25	-	**169.46**	-	**543.12**	nr	**611.01**
Mineral insulated copper sheathed cables										
1.7KW	295.29	2.50	**302.67**	11.00	-	**225.94**	-	**528.61**	nr	**594.69**
2.55KW	372.29	2.50	**381.60**	11.55	-	**237.24**	-	**618.83**	nr	**696.19**
3.4KW	416.53	2.50	**426.94**	12.10	-	**248.53**	-	**675.48**	nr	**759.91**
Bell equipment										
Transformers										
3-5-8V	14.77	2.50	**15.14**	0.44	-	**9.04**	0.24	**24.42**	nr	**27.47**
12V	18.90	2.50	**19.37**	0.44	-	**9.04**	0.24	**28.65**	nr	**32.23**
Bells for transformer operation										
chime 2-note	10.61	2.50	**10.88**	0.55	-	**11.30**	0.24	**22.41**	nr	**25.21**
76mm domestic type	5.85	2.50	**6.00**	0.55	-	**11.30**	0.24	**17.53**	nr	**19.72**
150mm round bell type	21.27	2.50	**21.80**	0.44	-	**9.04**	0.24	**31.08**	nr	**34.96**
Bell pushes										
domestic	1.48	2.50	**1.52**	0.28	-	**5.75**	-	**7.27**	nr	**8.18**
industrial	5.84	2.50	**5.99**	0.33	-	**6.78**	-	**12.76**	nr	**14.36**

Labour hourly rates: (except Specialists) Craft Operatives £20.54 Labourer £10.71 Rates are national average prices. Refer to REGIONAL VARIATIONS for indicative levels of overall pricing in regions	MATERIALS			LABOUR				RATES		
	Del to Site £	Waste %	Material Cost £	Craft Optve Hrs	Lab Hrs	Labour Cost £	Sunds £	Nett Rate £	Unit	Gross Rate (+12.5%) £
V90: LAMP FITTINGS, FANS, HEATERS etc., ALL INSTALLATIONS Cont.										
Immersion heaters and thermostats										
Note: the following prices exclude Plumber's work										
2 or 3KW immersion heaters; without flanges										
305mm; non-withdrawable elements	23.37	2.50	**23.95**	0.72	-	**14.79**	4.65	**43.39**	nr	**48.82**
457mm; non-withdrawable elements	23.61	2.50	**24.20**	0.88	-	**18.08**	4.65	**46.93**	nr	**52.79**
686mm; non-withdrawable elements	23.72	2.50	**24.31**	1.05	-	**21.57**	4.65	**50.53**	nr	**56.85**
Thermostats										
immersion heater	10.24	2.50	**10.50**	0.55	-	**11.30**	0.25	**22.04**	nr	**24.80**
15 amp a.c. for air heating; without switch	15.47	2.50	**15.86**	0.44	-	**9.04**	0.25	**25.14**	nr	**28.29**
Water heaters										
Note: the following prices exclude Plumber's work										
Storage type units; free outlet										
7 litre	114.78	-	**114.78**	1.65	-	**33.89**	5.15	**153.82**	nr	**173.05**
23 litre	506.73	-	**506.73**	2.48	-	**50.94**	5.65	**563.32**	nr	**633.73**
55 litre	695.18	-	**695.18**	4.95	-	**101.67**	7.98	**804.83**	nr	**905.44**
Storage type units; multi-point										
50 litre	696.76	-	**696.76**	4.95	-	**101.67**	7.98	**806.41**	nr	**907.21**
75 litre	828.49	-	**828.49**	5.78	-	**118.72**	7.98	**955.19**	nr	**1074.59**
125 litre	948.19	-	**948.19**	6.60	-	**135.56**	10.30	**1094.05**	nr	**1230.81**
Shower units										
Note: the following prices exclude Plumber's work										
Instantaneous shower units										
complete with fittings	156.19	2.50	**160.09**	4.40	-	**90.38**	7.73	**258.20**	nr	**290.48**
V90: TESTING										
Testing; existing installations										
Point										
from	-	-	-	2.25	-	**46.22**	-	**46.22**	nr	**51.99**
to	-	-	-	3.85	-	**79.08**	-	**79.08**	nr	**88.96**
Complete installation; three bedroom house										
from	-	-	-	11.00	-	**225.94**	-	**225.94**	nr	**254.18**
to	-	-	-	17.50	-	**359.45**	-	**359.45**	nr	**404.38**

COMMUNICATIONS/SECURITY/CONTROL SYSTEMS

Labour hourly rates: (except Specialists) Craft Operatives £20.54 Labourer £10.71 Rates are national average prices. Refer to REGIONAL VARIATIONS for indicative levels of overall pricing in regions	MATERIALS			LABOUR				RATES		
	Del to Site £	Waste %	Material Cost £	Craft Optve Hrs	Lab Hrs	Labour Cost £	Sunds £	Nett Rate £	Unit	Gross Rate (+12.5%) £
W50: FIRE DETECTION AND ALARM										
Trunking, conduits, cables etc included in section V										
Fittings and equipment										
fire alarm 1 zone panel with batteries and charger	100.00	2.50	**102.50**	3.30	-	**67.78**	0.50	**170.78**	nr	**192.13**
fire alarm 4 zone panel with batteries and charger	150.00	2.50	**153.75**	4.40	-	**90.38**	0.50	**244.63**	nr	**275.20**
fire alarm 8 zone panel with batteries and charger	220.00	2.50	**225.50**	6.60	-	**135.56**	0.50	**361.56**	nr	**406.76**
Ionisation smoke detector........................	35.00	2.50	**35.88**	0.30	-	**6.16**	0.50	**42.54**	nr	**47.85**
Optical smoke detector	35.00	2.50	**35.88**	0.30	-	**6.16**	0.50	**42.54**	nr	**47.85**
Smoke detector mounting base with diode..........	8.00	2.50	**8.20**	0.85	-	**17.46**	0.50	**26.16**	nr	**29.43**
Manual surface mounted call point................	11.75	2.50	**12.04**	0.85	-	**17.46**	0.50	**30.00**	nr	**33.75**
Emergency lighting										
General purpose 8 watt non maintained............	25.00	2.50	**25.63**	1.10	-	**22.59**	0.50	**48.72**	nr	**54.81**
General purpose 8 watt maintained................	35.00	2.50	**35.88**	1.10	-	**22.59**	0.50	**58.97**	nr	**66.34**
General purpose 8 watt mains.....................	16.00	2.50	**16.40**	1.10	-	**22.59**	0.50	**39.49**	nr	**44.43**

This section continues on the next page

Labour hourly rates: (except Specialists) Craft Operatives £14.21 Labourer £10.71 Rates are national average prices. Refer to REGIONAL VARIATIONS for indicative levels of overall pricing in regions	MATERIALS			LABOUR				RATES		
	Del to Site £	Waste %	Material Cost £	Craft Optve Hrs	Lab Hrs	Labour Cost £	Sunds £	Nett Rate £	Unit	Gross Rate (+12.5%) £
W52: LIGHTNING PROTECTION										
Lightning conductors										
Note: the following are indicative average prices										
Air termination rods										
610mm long; in clips, fixing to masonry..........	-	-	-	-	-	-	-	**39.50**	nr	**44.44**
Air termination tape										
fixing to masonry..............................	-	-	-	-	-	-	-	**9.50**	m	**10.69**
Copper tape down conductors										
19 x 3mm; in clips, fixing to masonry...........	-	-	-	-	-	-	-	**14.35**	m	**16.14**
25 x 3mm; in clips, fixing to masonry...........	-	-	-	-	-	-	-	**15.35**	m	**17.27**
Test clamps										
with securing bolts	-	-	-	-	-	-	-	**22.00**	nr	**24.75**
Copper earth rods; tape connectors										
2438 x 16mm; driving into ground................	-	-	-	-	-	-	-	**140.00**	nr	**157.50**

TRANSPORT SYSTEMS

Labour hourly rates: (except Specialists) Craft Operatives £14.21 Labourer £10.71 Rates are national average prices. Refer to REGIONAL VARIATIONS for indicative levels of overall pricing in regions	MATERIALS			LABOUR				RATES		
	Del to Site £	Waste %	Material Cost £	Craft Optve Hrs	Lab Hrs	Labour Cost £	Sunds £	Nett Rate £	Unit	Gross Rate (+12.5%) £
X10: LIFTS										
Generally										
Note: the following are average indicative prices										
Light passenger lifts; standard range										
Electro Hydraulic drive; 630 kg, 8 person, 0.63m/s, 3 stop										
basic; primed car and entrances	-	-	-	-	-	-	-	15500.00	nr	17437.50
median; laminate walls, standard carpet floor, painted entrances	-	-	-	-	-	-	-	18200.00	nr	20475.00
ACVF drive; 1000 kg, 13 person, 0.63m/s, 4 stop										
basic; primed car and entrances	-	-	-	-	-	-	-	34600.00	nr	38925.00
median; laminate walls, standard carpet floor, painted entrances	-	-	-	-	-	-	-	35900.00	nr	40387.50
Extras on the above lifts										
800/900mm landing doors in cellulose paint finish	-	-	-	-	-	-	-	150.00	nr	168.75
800/900mm landing doors in brushed stainless steel	-	-	-	-	-	-	-	420.00	nr	472.50
800/900mm landing doors in brushed brass	-	-	-	-	-	-	-	1210.00	nr	1361.25
landing position indicators	-	-	-	-	-	-	-	237.00	nr	266.63
landing direction arrows and gongs	-	-	-	-	-	-	-	140.00	nr	157.50
car preference key switch	-	-	-	-	-	-	-	120.00	nr	135.00
car entrance safety (Progard L)	-	-	-	-	-	-	-	400.00	nr	450.00
fluorescent emergency light	-	-	-	-	-	-	-	372.00	nr	418.50
half height mirror to rear wall	-	-	-	-	-	-	-	420.00	nr	472.50
full height mirror to rear wall	-	-	-	-	-	-	-	610.00	nr	686.25
vandal resistant buttons; per car/entrance	-	-	-	-	-	-	-	136.00	nr	153.00
car bumper rail	-	-	-	-	-	-	-	820.00	nr	922.50
trailing cable for Warden Call System	-	-	-	-	-	-	-	349.00	nr	392.63
car telephone	-	-	-	-	-	-	-	274.00	nr	308.25
General purpose passenger lifts										
ACVF drive; 630 kg, 8 person, 1.0m/s, 4 stop										
median; laminate walls, standard carpet floor, painted entrances	-	-	-	-	-	-	-	35600.00	nr	40050.00
Variable speed a/c drive; 1000 kg, 13 person, 1.6m/s, 5 stop										
median; laminate walls, standard carpet floor, painted entrances	-	-	-	-	-	-	-	38200.00	nr	42975.00
Variable speed a/c drive; 1600 kg, 21 person, 1.6m/s, 6 stop										
median; laminate walls, standard carpet floor, painted entrances	-	-	-	-	-	-	-	92000.00	nr	103500.00
If a high quality architectural finish is required, e.g. veneered panelling walls, marble floors, special ceiling, etc, add to the above prices										
from	-	-	-	-	-	-	-	15000.00	nr	16875.00
to	-	-	-	-	-	-	-	42000.00	nr	47250.00
Bed/Passenger lifts										
Electro Hydraulic drive; 1800 kg, 24 person, 0.63m/s, 3 stop										
standard specification	-	-	-	-	-	-	-	48200.00	nr	54225.00
Variable speed a/c drive; 2000 kg, 26 person, 1.6m/s, 4 stop										
standard specification	-	-	-	-	-	-	-	87500.00	nr	98437.50

Labour hourly rates: (except Specialists) Craft Operatives £14.21 Labourer £10.71 Rates are national average prices. Refer to REGIONAL VARIATIONS for indicative levels of overall pricing in regions	MATERIALS			LABOUR				RATES		
	Del to Site £	Waste %	Material Cost £	Craft Optve Hrs	Lab Hrs	Labour Cost £	Sunds £	Nett Rate £	Unit	Gross Rate (+12.5%) £
X10: LIFTS Cont.										
Goods lifts (direct coupled)										
Electro Hydraulic drive; 1500 kg, 0.4m/s, 3 stop stainless steel car lining with chequer plate floor and galvanised shutters	-	-	-	-	-	-	-	**38200.00**	nr	**42975.00**
Electro Hydraulic drive; 2000 kg, 0.4m/s, 3 stop stainless steel car lining with chequer plate floor and galvanised shutters	-	-	-	-	-	-	-	**39400.00**	nr	**44325.00**
Service hoists										
Single speed a/c drive; 50 kg, 0.63m/s, 2 stop standard specification	-	-	-	-	-	-	-	**5400.00**	nr	**6075.00**
Single speed a/c drive; 250 kg, 0.4m/s, 2 stop standard specification	-	-	-	-	-	-	-	**6850.00**	nr	**7706.25**
Extra on the above hoists two hour fire resisting shutters; per landing	-	-	-	-	-	-	-	**227.00**	nr	**255.38**
X11: ESCALATORS										
Escalators										
30 degree inclination; 3.00m vertical rise department store specification	-	-	-	-	-	-	-	**58000.00**	nr	**65250.00**

BASIC PRICES OF MATERIALS

This section gives basic prices of materials delivered to site. Prices are exclusive of Value Added Tax

Excavation and Earthwork

Filling		
Broken brick or stone filling	17.85	m3
Clinker ashes filing	12.00	m3
MOT type 1 filling	20.95	m3
MOT type 2 filling	19.40	m3
Imported topsoil	15.00	m3
Sand	15.31	m3
Gravel		
Coarse gravel	17.34	m3
Blinding gravel	24.29	m3
Earthwork support timber	292.95	m3

Concrete Work

Materials for site mixed concrete		
40mm Aggregate	27.56	m3
20mm Aggregate	26.46	m3
10mm Aggregate	26.46	m3
White Portland Cement	350.00	tonne
Sharp sand	24.19	m3
White silver sand	28.35	m3
Ready mixed concrete, prescribed mix		
ST1 - 40mm aggregate	58.01	m3
ST3 - 20mm aggregate	62.20	m3
ST4 - 20mm aggregate	63.93	m3
ST5 - 20mm aggregate	66.12	m3
Ready mixed concrete, design mix		
C 15 - 15 N/mm2 - 20mm	64.17	m3
C 20 - 20 N/mm2 - 20mm	65.24	m3
C 25 - 25 N/mm2 - 20mm	67.58	m3
C 30 - 30 N/mm2 - 20mm	69.13	m3
Site mixed concretes, mortars, various mixes		
1:6, all in aggregate	79.06	m3
1:8, all in aggregate	70.66	m3
1:12, all in aggregate	62.26	m3
Lightweight concrete		
20.5 N/mm2; vibrated	88.00	m3
26.0 N/mm2; vibrated	92.23	m3
41.5 N/mm2; vibrated	100.70	m3
Mild steel rod reinforcement		
32mm diameter	399.00	tonne
25mm diameter	404.25	tonne
20mm diameter	404.25	tonne
16mm diameter	404.25	tonne
12mm diameter	409.50	tonne
10mm diameter	414.75	tonne
8mm diameter	425.25	tonne
6mm diameter	441.00	tonne
High yield rod reinforcement		
32mm diameter	409.50	tonne
25mm diameter	409.50	tonne
20mm diameter	409.50	tonne
16mm diameter	409.50	tonne
12mm diameter	420.00	tonne
10mm diameter	420.00	tonne
8mm diameter	430.50	tonne
6mm diameter	451.50	tonne
Stainless steel rod reinforcement		
25mm diameter	4250.00	tonne
20mm diameter	4250.00	tonne
16mm diameter	4250.00	tonne
12mm diameter	4250.00	tonne
10mm diameter	4250.00	tonne
8mm diameter	4250.00	tonne
Fabric reinforcement, BS Ref 4483		
A393	2.97	m2
A252	2.01	m2
A193	1.47	m2
A142	1.11	m2
A98	0.80	m2
B1131	5.35	m2
B785	3.96	m2
B503	2.88	m2
B385	2.22	m2
B283	1.82	m2
B196	1.50	m2
C785	3.28	m2
C503	2.04	m2
C385	1.65	m2
C283	1.27	m2
D98	0.80	m2
D49	0.77	m2
C636	2.72	m2
Formwork, basic finish		
Sides of foundations; height exceeding 1.00m	3.48	m2
Sides of foundations; height not exceeding 250mm	1.08	m
Sides of ground beams and edges of beds; height exceeding 1.00m	3.48	m2
Sides of ground beams and edges of beds; height not exceeding 250mm	1.08	m
Edges of suspended slabs; height not exceeding 250mm	1.30	m
Edges of suspended slabs; height 250 - 500mm	2.43	m
Sides of upstands; height not exceeding 250mm	1.54	m
Sides of upstands; height 250 - 500mm	1.81	m
Soffits of slabs; horizontal slab thickness not exceeding 200mm	10.53	m2
Soffits of slabs; horizontal slab thickness 200 - 300mm	11.59	m2
Soffits of slabs; sloping not exceeding 15 degrees slab thickness not exceeding 200mm	11.60	m2
Soffits of slabs; sloping exceeding 15 degrees slab thickness not exceeding 200mm	12.13	m2
Walls; plain	8.53	m2
Walls; plain; height exceeding 3.00m above floor level	7.87	m2
Beams; attached to slabs	7.36	m2
Beams; with 50mm wide chamfers	7.49	m2
Beams; isolated	7.72	m2
Beams; isolated with 50mm wide chamfers	7.87	m2
Columns; attached to walls	5.11	m2
Columns; attached to walls, height exceeding 3.00m above floor level	4.75	m2
Wall kickers straight	1.02	m
Wall kickers curved 2m radius	2.49	m
Wall kickers curved 10m radius	1.51	m
Stairflights; 1000mm wide; 155mm thick waist; 178mm risers	15.72	m
Stairflights; 1500mm wide; 180mm thick waist; 178mm risers	19.68	m
Stairflights; 1000mm wide; 155mm thick waist; 178mm undercut risers	15.72	m
Stairflights; 1500mm wide; 180mm thick waist; 178mm undercut risers	19.67	m
Claymaster expanded polystyrene permanent formwork		
2400 x 1200 x 75mm thick Reference 2611, 4.86 kg/m2	41.88	sheet
2400 x 1200 x 100mm thick Reference 2411, 6.34 kg/m2	62.63	sheet
2400 x 1200 x 150mm thick Reference 2811, 3.39 kg/m2	93.93	sheet
Liquid damp proof applications		
Synthaprufe bituminous emulsion	3.88	litre
Bituminous emulsion	3.88	litre
R.I.W. liquid asphaltic composition	4.57	litre
Flexible sheet damp proof membranes		
Polythene sheeting, 125mu	0.36	m2
Polythene sheeting, 250mu	0.42	m2
Polythene sheeting, 300mu	0.56	m2
Waterproof building paper to BS 1521		
British Sisalkraft Ltd 728 damp proof membrane	1.44	m2
Visqueen 1200 super damp proof membrane	0.56	m2
Precast concrete lintels		
75 x 150 x 1200mm lintel; reinf with 1.20 kg of 12mm lintel bars	7.05	nr

Concrete Work (Cont'd)

Precast concrete lintels (Cont'd)

102 x 150 x 1200mm lintel; reinf with 1.20 kg of 12mm lintel bars	9.59	nr
215 x 150 x 1200mm lintel; reinf with 2.40 kg of 12mm lintel bars	20.21	nr
327 x 150 x 1200mm lintel; reinf with 3.60 kg of 12mm lintel bars	30.74	nr
252 x 150 x 1200mm boot lintel; reinf with 3.60 kg of 12mm boot lintel bars and 0.56 kg of 6mm boot lintel links	29.13	nr
252 x 150 x 1800mm boot lintel; reinf with 5.19 kg of 12mm boot lintel bars and 0.83 kg of 6mm boot lintel links	43.70	nr

Prestressed precast lintels 100 x 700mm

900mm long	3.56	nr
1050mm long	4.15	nr
1200mm long	4.73	nr
1500mm long	5.92	nr
1800mm long	7.10	nr
2100mm long	8.28	nr
2400mm long	9.48	nr
2700mm long	10.65	nr
3000mm long	11.84	nr

Prestressed precast lintels 140 x 70mm

1200mm long	6.42	nr
3000mm long	16.09	nr

Prestressed precast lintels 225 x 70mm

1200mm long	9.91	nr
3000mm long	24.79	nr

Prestressed precast lintels 255 x 70mm

1200mm long	10.56	nr
3000mm long	26.38	nr

Precast concrete copings ; figure 1;

75 x 200mm; splayed; rebated joints	11.43	m
100 x 300mm; splayed; rebated joints	22.86	m
75 x 200mm; saddleback; rebated joints	12.19	m
100 x 300mm; saddleback; rebated joints	24.38	m

Brickwork and Blockwork

Bricks

Common bricks plain	204.00	1000
Common bricks keyed	211.00	1000
Selected regraded bricks	164.00	1000
Second hard stock bricks	402.00	1000
Engineering bricks class A	302.00	1000
Engineering bricks class B	214.00	1000
Staffordshire blue wirecut bricks	390.00	1000
Staffordshire blue pressed bricks	456.00	1000
Staffordshire blue splayed plinth or bullnose	1705.00	1000
Claydon Red Multi	249.70	1000
Tudors	281.60	1000
Milton Buff Ridgefaced	261.80	1000
Heathers	306.90	1000
Brecken Grey	240.90	1000
Leicester Red Stock	300.30	1000
West Hoathly Medium multi-stock	440.00	1000
Himley dark brown rustic	396.00	1000
Tonbridge hand made multi-coloured	638.00	1000
Waingroves smooth red	286.00	1000
Old English Russet	259.60	1000
First hard stocks	385.00	1000

Miscellaneous materials for brickwork and blockwork

Cement and sand mortar (1:3)	102.60	m3
Cement lime mortar (1:1:6)	110.87	m3
Hydrated Lime	210.19	tonne

Cavity wall insulation

Glass fibre slabs 50mm	3.85	m2
Glass fibre slabs 75mm	5.28	m2

Butterfly pattern stainless steel wall ties, 203 x 3mm

Butterfly pattern stainless steel	135.88	1000

Vertical twisted wall ties, 203 x 19 x 3mm

stainless steel	574.00	1000

Extension Profiles

Extension profiles 60-250mm ref C2KS	11.97	m
Extension profiles 100-150mm ref 1FUR100S	12.90	m
Extension profiles 190-250mm ref 1FUR215S	23.34	m

Thermabate insulated cavity closers for fixing to masonry with PVCU ties at 225mm centres

Thermabate 50; vertical	4.63	m3
Thermabate 75; vertical	4.84	m3

Thermalite Shield blocks

75mm thick blocks	7.22	m2
100mm thick blocks	9.49	m2
150mm thick blocks	14.07	m2
200mm thick blocks	18.62	m2

Thermalite Turbo blocks

100mm thick blocks	9.65	m2
115mm thick blocks	11.05	m2
130mm thick blocks	12.47	m2
190mm thick blocks	18.12	m2
215mm thick blocks	20.47	m2

Thermalite Party Wall blocks

215mm thick blocks	20.18	m2

Thermalite Hi-Strength blocks

100mm thick blocks	11.86	m2
140mm thick blocks	16.43	m2

Thermalite Trench blocks

255mm thick blocks	23.87	m2
275mm thick blocks	25.70	m2
305mm thick blocks	28.48	m2
355mm thick blocks	33.09	m2

Lignacite blocks solid Fair Face 7.0 N/mm2

100mm thick blocks	13.35	m2
140mm thick blocks	19.64	m2
190mm thick blocks	26.66	m2

Toplite solid Standard blocks 7.0 N/mm2

100mm thick blocks	9.32	m2
140mm thick blocks	13.05	m2
190mm thick blocks	17.70	m2
215mm thick blocks	20.04	m2

Hemelite solid Standard blocks 3.5 N/mm2

75mm thick blocks	5.91	m2
90mm thick blocks	6.58	m2
100mm thick blocks	7.16	m2
140mm thick blocks	10.52	m2
190mm thick blocks	14.27	m2
215mm thick blocks	16.14	m2

Hemelite solid Standard blocks 7.0 N/mm2

90mm thick blocks	6.91	m2
100mm thick blocks	7.59	m2
140mm thick blocks	11.16	m2
190mm thick blocks	15.15	m2
215mm thick blocks	17.13	m2

Damp proof course

BS 6398 Class A, bitumen with hessian base	7.60	m2
BS 6398 Class B, bitumen with fibre base	4.55	m2
BS 6398 Class D, bitumen with hessian base laminated with lead	15.07	m2
BS 6398 Class E, bitumen with fibre base laminated with lead	16.74	m2
Hyload Original, pitch polymer	6.08	m2
Synthaprufe bituminous latex emulsion	3.88	litre
Bituminous emulsion	3.88	litre
Slates 350 x 225mm	1474.00	1000

Type G Cavitray without lead

900mm long cavitray	8.96	nr
220 x 332mm external angle	7.94	nr
230 x 117mm internal angle	7.94	nr

Type X Cavitray with lead, 40 degree pitch

Intermediate tray	4.88	nr
Ridge tray	10.29	nr
Catchment tray	4.88	nr
Corner catchment angle tray	10.75	nr

Type W

Cavity weep ventilator	0.49	nr
Extension duct	0.82	nr

Stainless steel mesh reinforcement

61mm mesh	1.14	m
112mm mesh	1.74	m
178mm mesh	3.01	m
228mm mesh	3.93	m

Air bricks, terra cotta, red or buff, square hole

225 x 75mm	2.09	nr
225 x 150mm	3.99	nr
225 x 225mm	6.99	nr

Air bricks, iron, light

225 x 75mm	2.04	nr
225 x 150mm	3.29	nr
225 x 225mm	5.77	nr

Roofing

Profile 3 sheeting, standard grey colour

coverings	12.28	m2
eaves filler pieces	10.53	m
two piece plain angular adjustable ridge tiles	15.23	m
standard barge boards	8.63	m
apron flashings	10.58	m
standard ridge cap finials	9.53	nr

Profile 6 sheeting, standard coloured

coverings	11.73	m2
eaves filler pieces	8.73	m
two piece plain angular adjustable ridge tiles	16.06	m
standard barge boards	10.79	m
apron flashings	9.42	m
standard ridge cap finials	10.63	nr

Treated softwood roofing battens

25 x 19mm	0.24	m
32 x 19mm	0.25	m
32 x 25mm	0.36	m
38 x 19mm	0.28	m
50 x 19mm	0.42	m
50 x 25mm	0.52	m

Roofing underfelts, reinforced bitumen

type 1F 15kg/10m2	0.53	m2

Plain clay tiles, 268 x 165mm

Machine made	519.69	1000
hand made	806.44	1000

Redland plain tiles, 268 x 165mm

Granular faced	337.77	1000
Granular faced, tile and a half	67.50	100
Downland	376.88	1000
Downland, tile and a half	75.38	100

Roofing (Cont'd)

Fibre cement slates		
600 x 300mm	1107.00	1000
500 x 250mm	892.00	1000
400 x 200mm	644.00	1000
Ridges or hips	26.65	m
Welsh Blue/grey slates, 6.5mm thick		
600 x 300mm	5765.00	1000
500 x 250mm	2788.00	1000
400 x 200mm	1276.00	1000
Redland Cambrian interlocking riven textured slates		
300 x 336mm	1926.00	1000
slate and a half slates to match 300 x 336mm	3853.00	1000
150 x 6mm fibre cement undercloak	2.53	m
Stainless steel eaves clip to each slate	0.13	nr
Redland third round ridge or hip	3.38	nr
Redland half round ridge or hip	3.38	nr
Redland universal angle ridge or hip	3.67	m
Redland Dry Ridge system with half round ridge tiles	26.44	m
Redland Dry Ridge system with universal angle ridge tiles	27.12	m
Monoridge filler units	38.83	m
Redland Universal valley troughs	9.18	nr
Timloc uPVC ventilators		
Mark 3 eaves ventilators; reference 1126/40	4.14	m
Soffit ventilators; reference 1137	2.22	m
Polypropylene over-fascia ventilators; reference 3011	3.94	m
Redland single lap tiles		
Stonewold 430 x 380mm	1552.00	1000
Regent 418 x 332mm	815.00	1000
Grovebury 418 x 332mm	815.00	1000
Norfolk 381 x 227mm	575.00	1000
Redland 49, 381 x 227mm	562.00	1000
Renown 418 x 330mm	788.00	1000
Redland 50, 418 x 330mm (GRANULAR)	788.00	1000
Felt to BS 747		
Rubervent 3G underlay 2.6 kg/m2	1.74	m2
Ruberfort HP180 underlay 3.4 kg/m2	2.27	m2
Ruberfort HP180 mineral surface top sheet 3.6 kg/m2	3.15	m2
Type 3B -1.8 kg/m2	0.94	m2
Type 3E -2.8 kg/m2	1.88	m2
Hot bitumen bonding compound	0.53	kg
Insulating boards for felt roofing		
25mm Kingspan Thermaroof TR21 (PIR) Board	8.54	m2
30mm Kingspan Thermaroof TR21 (PIR) Board	9.79	m2
40mm Kingspan Thermaroof TR21 (PIR) Board	11.31	m2
50mm Kingspan Thermaroof TR21 (PIR) Board	12.92	m2
Lead sheet		
Code 3	16.84	m2
Code 4	22.91	m2
Code 5	28.51	m2
Aluminium roofing		
0.6 mm thick sheets	10.42	m2
150 mm girth flashings	2.75	m
180 mm girth flashings	3.63	m
240 mm girth flashings	4.13	m
300 mm girth flashings	4.13	m
450 mm girth flashings	6.38	m
180 x 180 mm soakers	0.65	nr
180 x 300 mm soakers	1.09	nr
360mm girth gutter linings	5.38	m
420mm girth gutter linings	5.38	nr
450mm girth gutter linings	5.38	nr
Copper roof sheeting		
0.55 mm thick sheets	25.80	m2
0.70 mm thick sheets	32.25	m2
Zinc roof sheeting		
0.80 mm thick	18.15	m2
Zinc flashings - 0.80 mm thick		
150 mm girth	2.72	m
180 mm girth	3.27	m
240 mm girth	4.36	m
300 mm girth	5.45	m
450 mm girth	8.17	m
Zinc soakers - 0.80 mm thick		
180 x 180 mm	0.59	nr
180 x 300 mm	0.98	nr
450 x 450 mm	3.68	nr

Woodwork

Sawn Softwood General Structural Grade SC3		
32 x 125mm	0.88	m
32 x 150mm	1.06	m
32 x 175mm	1.23	m
32 x 200mm	1.41	m
32 x 225mm	1.58	m
32 x 250mm	1.76	m
32 x 300mm	2.11	m
38 x 125mm	1.04	m
38 x 150mm	1.25	m
38 x 175mm	1.46	m
38 x 200mm	1.67	m
38 x 225mm	1.88	m
38 x 250mm	2.09	m
38 x 300mm	2.51	m
44 x 125mm	1.18	m
44 x 150mm	1.42	m
44 x 175mm	1.65	m
44 x 200mm	1.89	m
44 x 225mm	2.12	m
44 x 250mm	2.37	m
44 x 300mm	2.83	m
50 x 50mm	0.54	m
50 x 63mm	0.68	m
50 x 75mm	0.80	m
50 x 100mm	1.07	m
50 x 125mm	1.34	m
50 x 150mm	1.61	m
50 x 175mm	1.88	m
50 x 200mm	2.14	m
50 x 225mm	2.41	m
50 x 250mm	2.68	m
50 x 300mm	3.22	m
63 x 63mm	0.85	m
63 x 75mm	1.01	m
63 x 100mm	1.35	m
63 x 125mm	1.69	m
63 x 150mm	2.02	m
63 x 175mm	2.37	m
63 x 200mm	2.70	m
63 x 225mm	3.04	m
63 x 250mm	3.38	m
63 x 300mm	4.05	m
75 x 75mm	1.24	m
75 x 100mm	1.65	m
75 x 125mm	2.06	m
75 x 150mm	2.48	m
75 x 175mm	2.89	m
75 x 200mm	3.30	m
75 x 225mm	3.71	m
75 x 250mm	4.12	m
75 x 300mm	4.95	m
100 x 100mm	2.20	m
150 x 150mm	4.95	m
150 x 250mm	8.25	m
150 x 300mm	9.90	m
200 x 200mm	8.80	m
250 x 250mm	13.75	m
300 x 300mm	19.80	m
Sawn Softwood General Structural Grade SC3 Pressure impregnated		
32 x 125mm	0.99	m
32 x 150mm	1.19	m
32 x 175mm	1.39	m
32 x 200mm	1.58	m
32 x 225mm	1.78	m
32 x 250mm	1.98	m
32 x 300mm	2.38	m
38 x 125mm	1.18	m
38 x 150mm	1.41	m
38 x 175mm	1.65	m
38 x 200mm	1.88	m
38 x 225mm	2.11	m
38 x 250mm	2.35	m
38 x 300mm	2.82	m
44 x 125mm	1.33	m
44 x 150mm	1.60	m
44 x 175mm	1.86	m
44 x 200mm	2.13	m
44 x 225mm	2.40	m
44 x 250mm	2.66	m
44 x 300mm	3.19	m
50 x 50mm	0.60	m
50 x 63mm	0.76	m
50 x 75mm	0.91	m
50 x 100mm	1.21	m
50 x 125mm	1.51	m
50 x 150mm	1.82	m
50 x 175mm	2.12	m
50 x 200mm	2.42	m
50 x 225mm	2.72	m
50 x 250mm	3.02	m
50 x 300mm	3.63	m
63 x 63mm	0.96	m
63 x 75mm	1.14	m
63 x 100mm	1.52	m
63 x 125mm	1.91	m
63 x 150mm	2.29	m
63 x 175mm	2.67	m
63 x 200mm	3.05	m
63 x 225mm	3.43	m
63 x 250mm	3.81	m
63 x 300mm	4.57	m
75 x 75mm	1.39	m
75 x 100mm	1.86	m
75 x 125mm	2.32	m
75 x 150mm	2.78	m
75 x 175mm	3.25	m
75 x 200mm	3.71	m
75 x 225mm	4.18	m
75 x 250mm	4.64	m
75 x 300mm	5.57	m
100 x 100mm	2.48	m
150 x 150mm	5.57	m
150 x 225mm	8.35	m
150 x 250mm	9.28	m
150 x 300mm	11.14	m
200 x 200mm	9.90	m
250 x 250mm	15.47	m
300 x 300mm	22.28	m
Softwood boarding, wrot, impregnated		
19mm thick	7.74	m2
25mm thick	10.21	m2
Plywood BB/CC Far Eastern H/W		
18mm thick	11.21	m2
25mm thick	15.58	m2
Pressure Impregnated Softwood Boarding		
25 x 200mm	2.20	m
25 x 225mm	2.47	m
Softwood, wrought; - boarded flooring		
19 x 75mm, square edge	7.39	m2

Woodwork (Cont'd)

Softwood, wrought; - boarded flooring (Cont'd)

25 x 125mm, square edge.	9.25	m2
32 x 150mm, square edge.	11.76	m2
19 x 75mm, tongued and grooved joint	8.45	m2
25 x 125mm, tongued and grooved joint	10.13	m2

Western Red Cedar, wrought; - wall and ceiling boarding, tongued, grooved and veed joints

19mm thick, 100mm wide..	46.97	m2
25mm thick, 100mm wide..	58.96	m2

Knotty Pine, wrought, selected for transparent finish; - wall and ceiling boarding, tongued, grooved and veed joints

19mm thick, 100mm wide..	36.09	m2
25mm thick, 100mm wide..	45.40	m2

Sapele, wrought, selected for transparent finish; - wall and ceiling boarding, tongued, grooved and veed joints

19mm thick, 100mm wide..	58.29	m2
25mm thick, 100mm wide..	74.20	m2

Timber for First and Second Fixings

Iroko	1320.00	m3
Idigbo	1320.00	m3
African Mahogany	1540.00	m3
South American Mahogany	1540.00	m3
African Walnut	2200.00	m3
Beech	1320.00	m3
Afrormosia	1650.00	m3
American Oak	1650.00	m3
Sapele	1320.00	m3
British Columbain Pine	1010.00	m3
Teak	5225.00	m3

Blockboard Birch faced, Finnish manufacturer

16mm	10.91	m2
18mm	10.09	m2
25mm	16.58	m2

Blockboard Oak faced

18mm	17.73	m2

Blockboard Sapele faced

18mm	16.72	m2

Blockboard Beech faced

18mm	18.57	m2

Douglas Fir T&G plywood, unsanded select sheathing quality, WBP bonded

15mm thick	6.85	m2
18mm thick	8.68	m2

Birch faced T&G plywood, BB quality, WBP bonded

12mm thick	20.58	m2
15mm thick	24.71	m2
18mm thick	29.97	m2

Wood chipboard, BS EN 634, type II T&G flooring

18mm thick P4 T&G F/G ...	4.24	m2
22mm thick P4 T&G F/G ...	5.81	m2
18mm thick, moisture resisting, Furniture Grade.	3.21	m2

Standard hardboard, BS EN 622, type SHA; butt joints

3.2mm thick linings	1.37	m2
4.8mm thick linings	2.50	m2
6.4mm thick linings	3.12	m2

Low density medium board, BS EN 622, type LME; butt joints

6mm thick linings	8.59	m2
9mm thick linings	10.99	m2
12mm thick linings	14.02	m2

Medium density fibreboard, BS EN 622, type MDF; butt joints, linings to walls

9mm thick	3.61	m2
12mm thick	4.43	m2
18mm thick	5.82	m2
25mm thick	8.04	m2

Tacboard Class O fire resisting boards; butt joints

6mm thick linings	9.74	m2
9mm thick linings	17.94	m2
12mm thick linings	23.36	m2

Softwood standard skirtings, picture rails, architraves and the like; (finished sizes)

13 x 70mm	1.17	m
13 x 95mm	1.51	m
20 x 120mm	2.33	m
13 x 70mm	1.22	m
13 x 95mm	1.57	m
20 x 120mm	2.33	m

MDF skirtings, picture rails, architraves and the like; (finished sizes)

13 x 70mm	1.62	m
20 x 120mm	3.30	m

Structural Steelwork

Universal beams

914 x 419 x 343	675.00	tonne
914 x 419 x 388	675.00	tonne
914 x 305 x 201	670.00	tonne
914 x 305 x 224	670.00	tonne
914 x 305 x 253	670.00	tonne
914 x 305 x 289	670.00	tonne
838 x 292 x 176	665.00	tonne
838 x 292 x 194	665.00	tonne
838 x 292 x 226	665.00	tonne
762 x 267 x 134	665.00	tonne
762 x 267 x 147	665.00	tonne
762 x 267 x 173	665.00	tonne
762 x 267 x 197	665.00	tonne
686 x 254 x 125	665.00	tonne
686 x 254 x 140	665.00	tonne
686 x 254 x 152	665.00	tonne
686 x 254 x 170	665.00	tonne
610 x 305 x 238	655.00	tonne
610 x 305 x 179	655.00	tonne
610 x 305 x 149	655.00	tonne
610 x 229 x 140	655.00	tonne
610 x 229 x 125	655.00	tonne
610 x 229 x 113	655.00	tonne
610 x 229 x 101	655.00	tonne
533 x 210 x 122	640.00	tonne
533 x 210 x 109	640.00	tonne
533 x 210 x 101	640.00	tonne
533 x 210 x 92	640.00	tonne
533 x 210 x 82	640.00	tonne
457 x 191 x 98	630.00	tonne
457 x 191 x 89	630.00	tonne
457 x 191 x 82	630.00	tonne
457 x 191 x 74	630.00	tonne
457 x 191 x 67	630.00	tonne
457 x 152 x 82	630.00	tonne
457 x 152 x 74	630.00	tonne
457 x 152 x 67	630.00	tonne
457 x 152 x 60	630.00	tonne
457 x 152 x 52	630.00	tonne
406 x 178 x 74	635.00	tonne
406 x 178 x 67	635.00	tonne
406 x 178 x 60	635.00	tonne
406 x 178 x 54	635.00	tonne
406 x 140 x 46	635.00	tonne
406 x 140 x 39	635.00	tonne
356 x 171 x 67	635.00	tonne
356 x 171 x 57	635.00	tonne
356 x 171 x 51	635.00	tonne
356 x 171 x 45	635.00	tonne
356 x 127 x 39	635.00	tonne
356 x 127 x 33	635.00	tonne
305 x 165 x 54	630.00	tonne
305 x 165 x 46	630.00	tonne
305 x 165 x 40	630.00	tonne
305 x 127 x 48	630.00	tonne
305 x 127 x 42	630.00	tonne
305 x 127 x 37	630.00	tonne
305 x 102 x 33	630.00	tonne
305 x 102 x 28	630.00	tonne
305 x 102 x 25	630.00	tonne
254 x 102 x 28	625.00	tonne
254 x 102 x 25	625.00	tonne
254 x 102 x 22	625.00	tonne

Universal Columns

356 x 406 x 634	675.00	tonne
356 x 406 x 551	675.00	tonne
356 x 406 x 467	675.00	tonne
356 x 406 x 393	675.00	tonne
356 x 406 x 340	675.00	tonne
356 x 406 x 287	675.00	tonne
356 x 406 x 235	675.00	tonne
356 x 368 x 202	675.00	tonne
356 x 368 x 177	675.00	tonne
356 x 368 x 153	675.00	tonne
356 x 368 x 129	675.00	tonne
305 x 305 x 283	655.00	tonne
305 x 305 x 240	655.00	tonne
305 x 305 x 198	655.00	tonne
305 x 305 x 158	655.00	tonne
305 x 305 x 137	655.00	tonne
305 x 305 x 118	655.00	tonne
305 x 305 x 97	655.00	tonne
254 x 254 x 167	635.00	tonne
254 x 254 x 132	635.00	tonne
254 x 254 x 107	635.00	tonne
254 x 254 x 89	635.00	tonne
254 x 254 x 73	635.00	tonne
203 x 203 x 86	625.00	tonne
203 x 203 x 71	625.00	tonne
203 x 203 x 60	625.00	tonne
203 x 203 x 52	625.00	tonne
203 x 203 x 46	625.00	tonne
152 x 152 x 37	625.00	tonne
152 x 152 x 30	625.00	tonne
152 x 152 x 23	625.00	tonne

Quantity extras

10 tonnes and over	0.00	tonne
Under 10 tonnes to 5 tonnes	10.00	tonne
Under 5 tonnes to 2 tonnes	25.00	tonne
Under 2 tonnes	50.00	tonne

Parallel flange channels

430 x 100 x 64	685.00	tonne
330 x 100 x 54	685.00	tonne
300 x 100 x 46	655.00	tonne
300 x 90 x 41	655.00	tonne
260 x 90 x 35	655.00	tonne
260 x 75 x 28	655.00	tonne
230 x 90 x 32	655.00	tonne
230 x 75 x 26	655.00	tonne
200 x 90 x 30	655.00	tonne
180 x 90 x 26	655.00	tonne
150 x 90 x 24	655.00	tonne
Extra; Parallel flange channels, non-standard sizes	25.00	tonne

Equal and unequal angles

200 x 150 x 18	625.00	tonne
200 x 150 x 15	625.00	tonne
200 x 150 x 12	625.00	tonne
200 x 100 x 15	625.00	tonne
200 x 100 x 12	625.00	tonne
200 x 100 x 10	625.00	tonne
200 x 200 x 24	610.00	tonne
200 x 200 x 20	610.00	tonne
200 x 200 x 18	610.00	tonne
200 x 200 x 16	610.00	tonne
150 x 90 x 15	605.00	tonne
150 x 90 x 12	605.00	tonne
150 x 90 x 10	605.00	tonne
150 x 150 x 18	605.00	tonne
150 x 150 x 15	605.00	tonne
150 x 150 x 12	605.00	tonne
150 x 150 x 10	605.00	tonne
150 x 75 x 15	595.00	tonne
150 x 75 x 12	595.00	tonne
150 x 75 x 10	595.00	tonne
120 x 120 x 15	600.00	tonne
120 x 120 x 12	600.00	tonne
120 x 120 x 10	600.00	tonne
120 x 120 x 8	600.00	tonne
125 x 75 x 12	580.00	tonne

Structural Steelwork (Cont'd)

Equal and unequal angles (Cont'd)

125 x 75 x 10	580.00	tonne
125 x 75 x 8	580.00	tonne
100 x 75 x 12	575.00	tonne
100 x 75 x 10	575.00	tonne
100 x 75 x 8	575.00	tonne
100 x 65 x 10	575.00	tonne
100 x 65 x 8	575.00	tonne
100 x 65 x 7	575.00	tonne
100 x 100 x 15	565.00	tonne
100 x 100 x 12	565.00	tonne
100 x 100 x 10	565.00	tonne
100 x 100 x 8	565.00	tonne
90 x 90 x 12	565.00	tonne
90 x 90 x 10	565.00	tonne
90 x 90 x 8	565.00	tonne
90 x 90 x 7	565.00	tonne
90 x 90 x 6	565.00	tonne
Extra; angles non-standard sizes	25.00	tonne

Cold rolled zed purlins and cladding rails, Metsec purlin sleeved system

ref: 14214; 3.03 kg/m	14.79	m
ref: 14216; 3.47 kg/m	16.70	m
ref: 17214; 3.66 kg/m	17.50	m
ref: 17216; 4.11 kg/m	19.88	m
ref: 20216; 4.49 kg/m	21.71	m
ref: 20218; 5.03 kg/m	24.48	m
ref: 20220; 5.57 kg/m	26.89	m
ref: 23218; 5.73 kg/m	27.79	m
ref: 23220; 6.34 kg/m	30.82	m

Cold rolled zed purlins and cladding rails, Metsec purlin overlap system

ref: 142	9.94	m
ref: 172	10.56	m
ref: 202	11.51	m
ref: 232	12.45	m

Cold rolled zed purlins and cladding rails, Metsec purlin non continuous system

ref: 17215; 3.85 kg/m	18.61	m
ref: 17216; 4.11 kg/m	19.88	m
ref: 20216; 4.49 kg/m	21.71	m
ref: 20218; 5.03 kg/m	24.48	m
ref: 20220; 5.57 kg/m	26.89	m
ref: 23218; 5.73 kg/m	27.79	m
ref: 23223; 7.26 kg/m	35.61	m
ref: 26223; 7.92 kg/m	38.78	m
ref: 26229; 9.88 kg/m	48.09	m

Cold rolled zed purlins and cladding rails, Metsec galvanized purlin cleats

ref: 142 for 142mm deep purlins	5.17	nr
ref: 172 for 172mm deep purlins	5.89	nr
ref: 202 for 202mm deep purlins	6.43	nr
ref: 232 for 232mm deep purlins	7.08	nr
ref: 262 for 262mm deep purlins	7.73	nr

Cold rolled zed purlins and cladding rails, Metsec Round Lok anti-sag rods

1150mm centres	5.20	nr
1350mm centres	6.04	nr
1550mm centres	6.83	nr
1700mm centres	7.69	nr
1950mm centres	8.44	nr

Cold rolled zed purlins and cladding rails, Metsec side rail sleeved/single span system

ref: 14214; 3.03 kg/m	14.79	m
ref: 14216; 3.47 kg/m	16.70	m
ref: 17215; 3.85 kg/m	18.61	m
ref: 17216; 4.11 kg/m	19.88	m
ref: 20216; 4.49 kg/m	21.71	m
ref: 20218; 5.03 kg/m	24.48	m
ref: 20220; 5.57 kg/m	26.89	m
ref: 23218; 5.73 kg/m	27.79	m
ref: 23223; 7.26 kg/m	35.61	m

Cold rolled zed purlins and cladding rails, Metsec side rail cleats

ref: 142 for 142mm rails	5.18	nr
ref: 172 for 172mm rails	7.05	nr
ref: 202 for 202mm rails	8.68	nr
ref: 232 for 232mm rails	10.09	nr

Cold rolled zed purlins and cladding rails, Metsec galvanized side rail suports, 122-262 series

rail 1000mm long	15.20	nr
rail 1400mm long	16.59	nr
rail 1600mm long	18.40	nr
rail 1800mm long	22.14	nr

Cold rolled zed purlins and cladding rails, Metsec diagonal tie wire supports

1700mm long	29.66	nr
2200mm long	33.18	nr
2600mm long	36.70	nr
3600mm long	45.31	nr

Steel short span lattice joists primed at works

200mm deep, 8.2 kg/m	46.11	m
250mm deep, 8.5 kg/m	48.02	m
300mm deep, 10.7 kg/m	60.66	m
350mm deep, 11.6 kg/m	65.24	m
350mm deep, 12.8 kg/m	71.27	m

Steel intermediate span lattice joists primed at works

450mm deep, 12.2 kg/m	62.34	m
500mm deep, 15.8 kg/m	77.10	m
550mm deep, 19.4 kg/m	91.14	m
600mm deep, 22.5 kg/m	104.32	m
650mm deep, 29.7 kg/m	137.40	m

Steel long span lattice joists primed at works

700mm deep, 39.2 kg/m	177.69	m
800mm deep, 44.1 kg/m	190.25	m
900mm deep, 45.3 kg/m	191.32	m
1000mm deep, 46.1 kg/m	195.46	m
1500mm deep, 54.2 kg/m	231.15	m

Metalwork

Steel; galvanised - 30 x 2.5mm standard strapping

600mm long	0.88	nr
800mm long	1.29	nr
1000mm long	1.67	nr
1200mm long	2.01	nr
1600mm long	2.61	nr

Truss clips

for 38mm thick members	0.35	nr
for 50mm thick members	0.37	nr

Joist hangers; BAT Building Products Ltd

SPH type S

for 50 x 100mm joist	1.75	nr
for 50 x 150mm joist	1.78	nr
for 50 x 200mm joist	2.07	nr
for 50 x 250mm joist	2.83	nr
for 63 x 100mm joist	2.23	nr
for 63 x 150mm joist	2.59	nr
for 63 x 200mm joist	2.59	nr
for 63 x 250mm joist	2.96	nr

SPH type R

for 50 x 100mm joist	2.64	nr
for 50 x 150mm joist	2.78	nr
for 50 x 200mm joist	3.06	nr
for 50 x 250mm joist	3.98	nr
for 63 x 100mm joist	3.23	nr
for 63 x 150mm joist	3.64	nr
for 63 x 200mm joist	3.64	nr
for 63 x 250mm joist	4.11	nr

Steel; galvanised - truss plates and framing anchors

51 x 114mm	0.18	nr
76 x 254mm	0.68	nr
114 x 152mm	0.54	nr
114 x 254mm	0.93	nr
152 x 152mm	0.75	nr

Galvanized steel lintels; SUPERGALV (BIRTLEY) lintels reference CB 50

125 x 750mm	23.07	nr
125 x 1200mm	36.38	nr
125 x 1350mm	42.93	nr
125 x 1500mm	47.43	nr
125 x 1650mm	54.08	nr
165 x 1800mm	58.51	nr
165 x 1950mm	64.21	nr
165 x 2100mm	67.20	nr
165 x 2250mm	77.02	nr
165 x 2400mm	81.47	nr
215 x 2550mm	91.84	nr
215 x 2850mm	125.14	nr
215 x 3000mm	133.24	nr
215 x 3300mm	153.62	nr
215 x 3600mm	176.01	nr
215 x 3900mm	214.79	nr

Galvanized steel lintels; SUPERGALV (BIRTLEY) lintels reference CB 50 H.D

165 x 750mm	27.23	nr
165 x 1050mm	35.85	nr
165 x 1200mm	40.84	nr
165 x 1350mm	49.66	nr
165 x 1500mm	54.74	nr
215 x 1650mm	65.49	nr
215 x 1800mm	71.49	nr
215 x 2100mm	84.33	nr
215 x 2250mm	106.03	nr
215 x 2400mm	124.42	nr
215 x 2550mm	135.18	nr
215 x 3150mm	203.93	nr

Galvanized steel lintels; SUPERGALV (BIRTLEY) lintels reference CB 70

120 x 750mm	22.90	nr
120 x 1200mm	36.37	nr
120 x 1500mm	47.44	nr
160 x 1800mm	58.36	nr
215 x 2400mm	81.47	nr
215 x 3000mm	132.86	nr
215 x 3900mm	206.17	nr

Galvanized steel lintels; SUPERGALV (BIRTLEY) lintels reference CB 70 H.D

160 x 750mm	27.25	nr
160 x 1200mm	40.84	nr
160 x 1500mm	54.75	nr
215 x 1800mm	76.47	nr
215 x 2400mm	124.45	nr
215 x 3300mm	203.98	nr

Galvanized steel lintels; SUPERGALV (BIRTLEY) lintels reference CB 50/130

120 x 750mm	23.17	nr
120 x 1200mm	36.82	nr
120 x 1500mm	46.67	nr
165 x 1800mm	58.47	nr
165 x 2400mm	85.22	nr
215 x 3900mm	253.53	nr

Galvanized steel lintels; SUPERGALV (BIRTLEY) lintels reference CB 50/130 H.D

160 x 750mm	33.10	nr
160 x 1200mm	49.54	nr
160 x 1500mm	68.30	nr
215 x 1800mm	84.70	nr
215 x 2400mm	131.11	nr
215 x 3300mm	208.10	nr

Galvanized steel lintels; SUPERGALV (BIRTLEY) lintels reference CB 70/130

115 x 750mm	28.04	nr
115 x 1200mm	46.18	nr
115 x 1500mm	58.35	nr
155 x 1800mm	80.25	nr
155 x 2400mm	133.36	nr
210 x 3300mm	242.10	nr

Galvanized steel lintels; SUPERGALV (BIRTLEY) lintels reference HD 50

225 x 3900mm	252.15	nr
225 x 4200mm	298.21	nr
225 x 4500mm	329.63	nr
225 x 4800mm	351.57	nr

Metalwork (Cont'd)

Galvanized steel lintels; SUPERGALV (BIRTLEY) lintels reference HD 70

225 x 3000mm	204.75	nr
225 x 3300mm	225.21	nr
225 x 3600mm	246.83	nr

Galvanized steel lintels; SUPERGALV (BIRTLEY) lintels reference SB100

75 x 750mm	15.36	nr
75 x 1200mm	23.62	nr
140 x 1800mm	45.23	nr
140 x 2400mm	63.37	nr
215 x 3000mm	116.50	nr

Galvanized steel lintels; SUPERGALV (BIRTLEY) lintels reference SB140

140 x 750mm	19.41	nr
140 x 1200mm	31.57	nr
140 x 1800mm	53.10	nr
140 x 2400mm	75.77	nr
215 x 3000mm	128.29	nr

Galvanized steel lintels; SUPERGALV (BIRTLEY) lintels reference SBL200

142 x 750mm	26.83	nr
142 x 1200mm	42.18	nr
142 x 1800mm	66.67	nr
142 x 2400mm	93.99	nr
218 x 3000mm	158.53	nr

Galvanized steel lintels; SUPERGALV (BIRTLEY) internal door lintels reference INT100

100 x 900mm	12.18	nr
100 x 1050mm	14.24	nr
100 x 1200mm	16.29	nr

Mechanical Installations

Red Primer steel pipes, BS 1387 Table 2; steel fittings BS EN 10241

15mm pipes	2.85	m
15mm fittings; one end	0.59	nr
15mm fittings; two ends	0.60	nr
15mm fittings; three ends	0.82	nr
20mm pipes	3.36	m
20mm fittings; one end	0.67	nr
20mm fittings; two ends	0.82	nr
20mm fittings; three ends	1.20	nr
25mm pipes	4.82	m
25mm fittings; one end	0.84	nr
25mm fittings; two ends	1.27	nr
25mm fittings; three ends	1.72	nr
32mm pipes	5.97	m
32mm fittings; one end	1.36	nr
32mm fittings; two ends	2.15	nr
32mm fittings; three ends	2.92	nr
40mm pipes	6.94	m
40mm fittings; one end	1.61	nr
40mm fittings; two ends	3.61	nr
40mm fittings; three ends	4.48	nr
50mm pipes	9.76	m
50mm fittings; one end	3.36	nr
50mm fittings; two ends	4.23	nr
50mm fittings; three ends	6.46	nr

Red Primer steel pipes, BS 1387 Table 3; steel fittings BS EN 10241

15mm pipes	3.40	m
15mm fittings; one end	0.59	nr
15mm fittings; two ends	0.60	nr
15mm fittings; three ends	0.82	nr
20mm pipes	4.03	m
20mm fittings; one end	0.67	nr
20mm fittings; two ends	0.82	nr
20mm fittings; three ends	1.20	nr
25mm pipes	5.89	m
25mm fittings; one end	0.84	nr
25mm fittings; two ends	1.27	nr
25mm fittings; three ends	1.72	nr
32mm pipes	7.31	m
32mm fittings; one end	1.36	nr
32mm fittings; two ends	2.15	nr
32mm fittings; three ends	2.92	nr
40mm pipes	8.52	m
40mm fittings; one end	1.61	nr
40mm fittings; two ends	3.61	nr
40mm fittings; three ends	4.48	nr
50mm pipes	11.84	m
50mm fittings; one end	3.36	nr
50mm fittings; two ends	4.23	nr
50mm fittings; three ends	6.46	nr

Black steel pipe brackets

15mm malleable iron	0.97	m
20mm malleable iron	1.08	nr
25mm malleable iron	1.25	nr
32mm malleable iron	1.71	nr
40mm malleable iron	2.27	nr
50mm malleable iron	3.02	m

Galvanized steel pipes, BS 1387 Table 2; steel fittings BS EN 10241

15mm pipes	4.40	m
15mm pipe connectors	0.80	nr
15mm fittings; one end	0.82	nr
15mm fittings; two ends	0.83	nr
15mm fittings; three ends	1.14	nr
20mm pipes	4.97	m
20mm pipe connectors	1.05	nr
20mm fittings; one end	0.94	nr
20mm fittings; two ends	1.14	nr
20mm fittings; three ends	1.66	nr
25mm pipes	6.94	m
25mm pipe connectors	1.34	nr
25mm fittings; one end	1.26	nr
25mm fittings; two ends	1.76	nr
25mm fittings; three ends	2.39	nr
32mm pipes	8.60	m
32mm pipe connectors	2.67	nr
32mm fittings; one end	1.94	nr
32mm fittings; two ends	3.30	nr
32mm fittings; three ends	4.53	nr
40mm pipes	9.99	m
40mm pipe connectors	3.61	nr
40mm fittings; one end	2.47	nr
40mm fittings; two ends	4.95	nr
40mm fittings; three ends	6.20	nr
50mm pipes	14.02	m
50mm pipe connectors	5.23	nr
50mm fittings; one end	4.61	nr
50mm fittings; two ends	5.79	nr
50mm fittings; three ends	8.93	nr

Galvanized steel pipes, BS 1387 Table 3; steel fittings BS EN 10241

15mm pipes	5.21	m
15mm pipe connectors	0.80	nr
15mm fittings; one end	0.82	nr
15mm fittings; two ends	0.83	nr
15mm fittings; three ends	1.14	nr
20mm pipes	5.92	m
20mm pipe connectors	1.05	nr
20mm fittings; one end	0.94	nr
20mm fittings; two ends	1.14	nr
20mm fittings; three ends	1.66	nr
25mm pipes	8.43	m
25mm pipe connectors	1.34	nr
25mm fittings; one end	1.26	nr
25mm fittings; two ends	1.76	nr
25mm fittings; three ends	2.39	nr
32mm pipes	10.46	m
32mm pipe connectors	2.67	nr
32mm fittings; one end	1.94	nr
32mm fittings; two ends	3.30	nr
32mm fittings; three ends	4.53	nr
40mm pipes	12.21	m
40mm pipe connectors	3.61	nr
40mm fittings; one end	2.47	nr
40mm fittings; two ends	4.95	nr
40mm fittings; three ends	6.20	nr
50mm pipes	16.92	m
50mm pipe connectors	5.23	nr
50mm fittings; one end	4.61	nr
50mm fittings; two ends	5.79	nr
50mm fittings; three ends	8.93	nr

Galvanized steel pipe brackets

15mm malleable iron	1.27	m
20mm malleable iron	1.42	nr
25mm malleable iron	1.65	nr
32mm malleable iron	2.25	nr
40mm malleable iron	3.01	nr
50mm malleable iron	3.97	m

Stainless steel pipes, BS 4127 Part 2

15mm pipes	3.97	m
22mm pipes	6.37	m
28mm pipes	8.45	m
35mm pipes	10.01	m
42mm pipes	13.46	m

Copper pipes, BS 2871 Part 1 Table X

15mm pipes	1.26	m
22mm pipes	2.51	m
28mm pipes	3.24	m
35mm pipes	7.04	m
42mm pipes	8.62	m
54mm pipes	11.06	m
66.7mm pipes	14.94	m
76.1mm pipes	21.10	m
108mm pipes	30.58	m

Copper pipes, BS 2871 Part 1 Table Y

15mm pipes	2.65	m
22mm pipes	4.62	m
28mm pipes	5.99	m
35mm pipes	9.04	m
42mm pipes	10.96	m
54mm pipes	18.72	m

Capillary fittings for copper pipes

15mm fittings; one end	1.10	nr
15mm fittings; two ends	0.27	nr
15mm fittings; three ends	0.51	nr
22mm fittings; one end	2.06	nr
22mm fittings; two ends	0.71	nr
22mm fittings; three ends	1.64	nr
28mm fittings; one end	3.68	nr
28mm fittings; two ends	1.64	nr
28mm fittings; three ends	4.54	nr
35mm fittings; one end	8.12	nr
35mm fittings; two ends	7.11	nr
35mm fittings; three ends	11.58	nr
42mm fittings; one end	13.98	nr
42mm fittings; two ends	11.75	nr
42mm fittings; three ends	18.58	nr
54mm fittings; one end	19.52	nr
54mm fittings; two ends	24.26	nr
54mm fittings; three ends	37.45	nr

Compression fittings type A for copper pipes

15mm fittings; one end	1.33	nr
15mm fittings; two ends	1.02	nr
15mm fittings; three ends	1.48	nr
22mm fittings; one end	1.59	nr
22mm fittings; two ends	1.73	nr
22mm fittings; three ends	2.42	nr
28mm fittings; one end	5.18	nr
28mm fittings; two ends	6.21	nr
28mm fittings; three ends	11.30	nr
35mm fittings; one end	9.34	nr
35mm fittings; two ends	15.63	nr
35mm fittings; three ends	20.66	nr
42mm fittings; one end	15.27	nr
42mm fittings; two ends	21.87	nr
42mm fittings; three ends	31.79	nr
54mm fittings; one end	21.29	nr
54mm fittings; two ends	37.17	nr
54mm fittings; three ends	51.17	nr

Hepworth Hep20 polybutylene pipes

15mm pipes	0.85	m
22mm pipes	1.67	m
28mm pipes	2.30	m

Blue polythene pipes, BS EN 12201

20mm pipes	0.40	m
25mm pipes	0.54	m
32mm pipes	0.83	m
50mm pipes	2.09	m
63mm pipes	3.19	m

PVC-U pipes, BS 3505 Class E

3/8 pipes	2.05	m
1/2 pipes	2.64	m
3/4 pipes	3.78	m
1 pipes	4.39	m

Mechanical Installations (Cont'd)

PVC-U pipes, BS 3505 Class E (Cont'd)

1 1/4 pipes	6.43	m
1 1/2 pipes	8.41	m

PVC-U pipes self coloured white

19mm pipes	1.07	m

Ductile iron pipes

80mm pipes	26.57	m
100mm pipes	25.23	m
150mm pipes	32.15	m

Brass stopcocks

crutch head; DZR; joints to polythene

20mm	24.31	nr
25mm	38.69	nr
32mm	51.92	nr
50mm	129.79	nr

crutch head; compression joints to copper (Type A)

15mm	4.06	nr
22mm	7.05	nr
28mm	16.93	nr
35mm	50.87	nr
42mm	64.20	nr
54mm	96.90	nr

crutch head; DZR; compression joints to copper (Type A)

15mm	14.86	nr
22mm	24.41	nr
28mm	40.38	nr
35mm	74.32	nr
42mm	106.38	nr
54mm	144.91	nr

Polybutylene stopcocks fitted with Hep2O ends

15mm	4.48	nr
22mm	5.37	nr

Brass gate valves

BS 5154 series B; compression joints to copper (Type A)

15mm	7.31	nr
22mm	8.88	nr
28mm	12.23	nr
35mm	23.02	nr
42mm	37.54	nr
54mm	58.14	nr

DZR; fitted with Hep2O ends

15mm	4.19	nr
22mm	5.91	nr

Lever ball valves, screwed and PTFE joints

13mm	4.19	5.65
19mm	5.65	7.27
25mm	7.27	10.02
32mm	10.02	16.02
38mm	16.02	nr
51mm	21.79	nr

13mm Brass ball valves for low pressure; BS 1212;

part 1, copper float	8.38	nr
part 2, copper float	9.90	nr
part 1, plastics float	5.60	nr
part 2, plastics float	7.12	nr

Brass ball valves for high pressure; BS 1212

13mm; part 1, copper	6.72	nr
13mm; part 2, copper	9.70	nr
19mm; part 1, copper	19.67	nr
25mm; part 1, copper	59.05	nr
13mm; part 1, plastics	3.94	nr
13mm; part 2, plastics	6.92	nr
19mm; part 1, plastics	12.91	nr
25mm; part 1, plastics	49.68	nr

GRP Sarena mfg Ltd water storage tanks with cover

ref. SC10; 18 litres; pre insulated	76.00	nr
ref. SC20; 55 litres; pre insulated	106.00	nr
ref. SC30; 86 litres; pre insulated	116.00	nr
ref. SC40; 114 litres; pre insulated	142.00	nr
ref. SC50; 159 litres; pre insulated	161.00	nr
ref. SC60; 191 litres; pre insulated	178.00	nr
ref. SC70; 227 litres; pre insulated	198.00	nr
ref. SC100/1; 327 litres; pre insulated	261.00	nr
ref. SC100/2; 336 litres; pre insulated	254.00	nr
ref. SC125; 423 litres; pre insulated	322.00	nr
ref. SC150; 491 litres; pre insulated	383.00	nr
ref. SC200; 709 litres; pre insulated	445.00	nr
ref. SC250; 900 litres; pre insulated	510.00	nr
ref. SC350; 1250 litres; pre insulated	650.00	nr
ref. SC500; 1727 litres; pre insulated	850.00	nr
ref. SC600; 2137 litres; pre insulated	980.00	nr

Plastics water storage cisterns with sealed lid / Byelaw 30 thick

18 litres	15.15	nr
114 litres	50.55	nr
182 litres	79.90	nr
227 litres	86.95	nr

Copper direct hot water cylinders; pre-insulated

ref. 3; 116 litres; four bosses	140.05	nr
ref. 7; 120 litres; four bosses	125.57	nr
ref. 8; 144 litres; four bosses	157.00	nr
ref. 9; 166 litres; four bosses	202.50	nr

Copper indirect hot water cylinders; pre-insulated

ref. 7; 108 litres; four bosses	267.75	nr
ref. 8; 130 litres; four bosses	315.50	nr

Cast iron rainwater pipes and fittings, BS 460 Type A sockets; dry joints

75mm pipes	17.98	m
75mm shoes	14.86	nr
75mm bends	12.74	nr
75mm offset bends 150mm projection	16.07	nr
75mm offset bends 225mm projection	18.70	nr
75mm offset bends 300mm projection	22.99	nr
75mm angled branches	22.31	nr
100mm pipes	24.55	m
100mm shoes	20.03	nr
100mm bends	17.99	nr
100mm offset bends 75mm projection	30.30	nr
100mm offset bends 150mm projection	30.30	nr
100mm offset bends 225mm projection	36.70	nr
100mm offset bends 300mm projection	37.41	nr
100mm angled branches	26.51	nr

Cast iron rainwater pipes and fittings, BS 460 Type A sockets; ears cast on; dry joints

65mm pipes	19.21	m
65mm shoes	17.14	nr
65mm bends	10.49	nr
75mm pipes	19.21	m
75mm shoes	17.14	nr
75mm bends	12.74	nr
100mm pipes	25.80	m
100mm shoes	22.76	nr
100mm bends	17.99	nr

PVC-U rainwater pipes and fittings, BS EN 607; push fit joints; self coloured grey

68mm pipes	3.94	m
68mm pipe brackets	2.02	nr
68mm shoes	4.67	nr
68mm bends	5.41	nr

Cast iron half round gutters and fittings, BS 460

100mm gutters	9.87	m
100mm gutter brackets	2.12	nr
100mm stopped ends	2.53	nr
100mm running outlets	7.33	nr
100mm stopped ends with outlet	8.60	nr
100mm angles	7.52	nr
115mm gutters	10.29	m
115mm gutter brackets	2.12	nr
115mm stopped ends	3.28	nr
115mm running outlets	7.99	nr
115mm stopped ends with outlet	9.65	nr
115mm angles	7.75	nr
125mm gutters	12.04	m
125mm gutter brackets	2.12	nr
125mm stopped ends	3.28	nr
125mm running outlets	9.14	nr
125mm stopped ends with outlet	10.79	nr
125mm angles	9.14	nr
150mm gutters	20.57	m
150mm gutter brackets	2.68	nr
150mm stopped ends	4.55	nr
150mm running outlets	15.81	nr
150mm stopped ends with outlet	18.09	nr
150mm angles	16.69	nr

Cast iron box gutters

100 x 75mm gutters	30.79	m
100 x 75mm stopped ends	4.68	nr
100 x 75mm running outlets	22.50	nr
100 x 75mm gutter brackets	4.46	nr

Cast aluminium half round gutters and fittings, BS EN 612 Section A

100mm gutters	10.17	m
100mm gutter brackets	1.79	nr
100mm stopped ends	3.03	nr
100mm running outlets	7.19	nr
100mm angles	6.25	nr
125mm gutters	12.70	m
125mm gutter brackets	3.28	nr
125mm stopped ends	4.02	nr
125mm running outlets	8.23	nr
125mm angles	7.99	nr

PVC-U half round gutters and fittings, BS 4576, Part 1; push fit connector joints; self coloured grey

114mm gutters	3.82	nr
114mm Union	4.66	m
114mm gutter brackets	1.23	nr
114mm stopped ends	2.98	nr
114mm running outlets	5.54	nr
114mm angles	6.05	nr

Cast iron soil and waste pipes and fittings, BS 416

75mm pipes	14.41	m
75mm pipe couplings	7.36	nr
100mm pipes	17.42	m
100mm pipe couplings	9.61	nr
50mm pipes	12.88	m
50mm pipe couplings	6.67	nr

MuPVC waste pipes and fittings, BS EN 1329-1 Section Three; solvent welded joints

32mm pipes	1.81	m
32mm pipe couplings	1.19	nr
32mm pipe brackets	0.38	nr

Mechanical Installations (Cont'd)

MuPVC waste pipes and fittings, BS EN 1329-1 Section Three; solvent welded joints (Cont'd)

32mm fittings; three ends	1.62	nr
40mm pipes	2.24	m
40mm pipe couplings	1.18	nr
40mm pipe brackets	0.45	nr
40mm fittings; three ends	2.06	nr
50mm pipes	3.38	m
50mm pipe couplings	2.17	nr
50mm pipe brackets	0.59	nr
50mm fittings; three ends	4.01	nr

Polypropylene waste pipes and fittings; butyl ring joints

32mm pipes	0.87	m
32mm pipe couplings	1.03	nr
32mm fittings; three ends	1.59	nr
40mm pipes	1.10	m
40mm pipe couplings	1.14	nr
40mm pipe brackets	0.35	nr
40mm fittings; three ends	1.66	nr

PVC-U soil and vent pipes and fittings, BS 4514; rubber ring joints; self coloured grey

82mm pipes	9.25	m
82mm pipe couplings	5.86	nr
110mm pipes	7.92	m
110mm pipe couplings	6.16	nr
160mm pipes	22.01	m
160mm pipe couplings	14.13	nr

Pipe supports for PVC-U soil and vent pipes; for building in

for 82mm pipes	4.34	nr
for 110mm pipes	6.29	nr

Pipe supports for PVC-U soil and vent pipes; for fixing to timber

for 82mm pipes	3.50	nr
for 110mm pipes	4.68	nr
for 160mm pipes	8.80	nr

PVC balloon gratings

50mm pipes	2.26	nr
82mm pipes	3.31	nr
110mm pipes	3.44	nr

Plastic traps, two piece, 75mm seal, inlet with coupling nut, outlet with seal ring socket

32mm outlet, P	4.86	nr
40mm outlet, P	5.61	nr
32mm outlet, S	6.16	nr
40mm outlet, S	7.20	nr
40mm outlet, P, with flexible polypropylene pipe for overflow connection	10.40	nr

Copper traps, two piece, 75mm seal, inlet with coupling nut, compression outlet

35mm outlet, P	23.20	nr
42mm outlet, P	27.33	nr
54mm outlet, P	98.08	nr
35mm outlet, S	24.58	nr
42mm outlet, S	29.57	nr
54mm outlet, S	103.59	nr
42mm outlet, bath, overflow connection, cleaning eye and plug	38.80	nr

Pressed steel radiators single panel with convector; air cock, plain plug

450 x 600mm long	23.69	nr
450 x 900mm long	35.51	nr
450 x 1200mm long	47.33	nr
450 x 1600mm long	72.25	nr
450 x 2400mm long	145.77	nr
600 x 600mm long	30.56	nr
600 x 900mm long	45.84	nr
600 x 1200mm long	61.13	nr
600 x 1600mm long	93.28	nr
600 x 2400mm long	188.20	nr
700 x 600mm long	35.05	nr
700 x 900mm long	60.11	nr
700 x 1200mm long	107.81	nr
700 x 1600mm long	143.77	nr
700 x 2400mm long	215.66	nr

Electrical Installations

Lighting columns, aluminium; BS EN 40-1, nominal mounting height

4.88m	271.61	nr
8.00m	669.05	nr
10.00m	850.14	nr
12.00m	1191.79	nr

Lighting columns, steel; BS EN 40-1, nominal mounting height

4.00m	218.12	nr
5.00m	224.76	nr
6.00m	276.99	nr
8.00m	411.05	nr

Steel straight lighting trunking; PVC lid

50 x 50mm	6.00	3m
end cap	2.32	nr
internal angle	11.36	nr
external angle	11.36	nr
tee	12.17	nr
four way intersection	12.57	nr

Steel straight dado trunking, two compartment; steel lid

150 x 30mm	21.44	3m
end cap	5.17	nr
internal angle	23.95	nr
external angle	23.95	nr
single socket plate	9.42	nr
twin socket plate	10.01	nr

Steel straight dado trunking, three compartment; steel lid

200 x 38mm	28.38	3m
end cap	5.65	nr
internal angle	29.60	nr
external angle	29.60	nr
single socket plate	9.74	nr
twin socket plate	10.44	nr

Steel straight skirting trunking; two compartments

200 x 38mm	28.33	3m
internal bend	28.63	nr
external bend	28.63	nr

Steel straight underfloor trunking; two compartments

150 x 25mm	19.29	3m
floor junction boxes; adjustable frame and trap	77.12	nr
horizontal bend	18.84	nr

PVC trunking; white straight mini-trunking; clip on lid

16 x 16mm	1.12	3m
end cap	0.33	nr
internal bend	0.33	nr
external bend	0.33	nr
flat bend	0.33	nr
equal tee	0.57	nr
25 x 16mm	1.27	3m
end cap	0.33	nr
internal bend	0.33	nr
external bend	0.33	nr
flat bend	0.33	nr
equal tee	0.57	nr
40 x 16mm	1.88	3m
end cap	0.34	nr
internal bend	0.34	nr
external bend	0.34	nr
flat bend	0.34	nr
equal tee	0.60	nr

PVC trunking; white straight dado trunking, three compartment; clip on lid

145 x 40mm (3m)	65.42	3m
end caps	1.57	nr
internal bend	5.36	nr
external bend	5.36	nr
fabricated flat tee	42.07	nr
fabricated flat elbow	36.06	nr
mini trunking adaptor	1.52	nr
230 x 40mm (3m)	112.46	3m
end caps	4.97	nr
internal bend	10.42	nr
external bend	10.42	nr
mini trunking adaptor	1.55	nr

PVC trunking; white straight skirting trunking, three compartment; clip on lid

100 x 40mm	11.51	3m
end cap	2.20	nr
internal corner	2.82	nr
external corner	2.82	nr
mini trunking adaptor	1.99	nr
socket adaptor, one gang	4.45	nr
socket adaptor, two gang	6.38	nr

PVC heavy gauge conduits; straight with small circular boxes

20mm diameter conduit	1.69	3m
terminal boxes	1.00	nr
angle boxes	1.12	nr
three way boxes	1.21	nr
through way boxes	1.12	nr
25mm diameter conduit	3.23	3m
terminal boxes	1.54	nr
angle boxes	1.68	nr
three way boxes	1.86	nr
through way boxes	1.68	nr

PVC light gauge conduits; straight with small circular boxes

20mm diameter conduit	1.17	3m
terminal boxes	0.91	nr
angle boxes	1.01	nr
three way boxes	1.09	nr
through way boxes	1.01	nr
25mm diameter conduit	2.29	3m
terminal boxes	1.39	nr
angle boxes	1.51	nr
three way boxes	1.68	nr
through way boxes	1.51	nr

PVC heavy gauge oval conduits; straight

20mm (nominal size)	0.95	3m
25mm (nominal size)	1.02	3m

Steel welded black enamelled heavy gauge conduits; straight with small circular boxes

20mm diameter conduit	2.99	3.75m
terminal boxes	0.92	nr
angle boxes	1.04	nr
three way boxes	0.92	nr
through way boxes	0.98	nr
25mm diameter conduit	3.89	3.75m
terminal boxes	1.03	nr
angle boxes	1.16	nr
three way boxes	1.03	nr
through way boxes	1.10	nr
32mm diameter conduit	9.12	3.75m

Steel welded galvanised heavy gauge conduits; straight with small circular boxes

20mm diameter conduit	6.12	3.75m
terminal boxes	1.09	nr
angle boxes	1.24	nr
three way boxes	1.09	nr
through way boxes	1.17	nr
25mm diameter conduit	8.09	3.75m
terminal boxes	1.36	nr
angle boxes	1.53	nr
three way boxes	1.36	nr
through way boxes	1.45	nr
32mm diameter conduit	11.02	3.75m

Galvanised light duty standard cable tray and fittings; straight

50mm tray	3.65	3m
flat bend	7.23	nr
tee	9.37	nr
cross piece	15.82	nr
riser bend	11.28	nr
100mm tray	5.28	3m
flat bend	7.80	nr
tee	10.31	nr
cross piece	16.86	nr
riser bend	13.00	nr

Electrical Installations (Cont'd)

Galvanised light duty standard cable tray and fittings; straight (Cont'd)

reducer 100 to 50mm	9.88	nr
150mm tray	7.75	3m
flat bend	9.11	nr
tee	12.51	nr
cross piece	19.79	nr
riser bend	17.55	nr
reducer 150 to 100mm	13.02	nr
300mm tray	13.19	3m
flat bend	16.80	nr
tee	23.02	nr
cross piece	35.38	nr
riser bend	30.89	nr
reducer 300 to 150mm	23.69	nr
450mm tray	26.65	3m
flat bend	26.89	nr
tee	37.56	nr
cross piece	58.39	nr
riser bend	43.75	nr
reducer 450 to 300mm	39.33	nr

Galvanised medium duty standard cable tray and fittings; straight

75mm tray	7.75	3m
flat bend	28.17	nr
tee	37.16	nr
cross piece	58.90	nr
riser bend	15.31	nr
100mm tray	9.27	3m
flat bend	28.17	nr
tee	37.16	nr
cross piece	57.28	nr
riser bend	14.01	nr
reducer 100 to 75mm	13.08	nr
150mm tray	11.35	3m
flat bend	31.78	nr
tee	42.29	nr
cross piece	63.55	nr
riser bend	18.40	nr
reducer 150 to 100mm	11.35	nr
300mm tray	14.89	3m
flat bend	43.60	nr
tee	54.06	nr
cross piece	83.22	nr
riser bend	29.23	nr
reducer 300 to 150mm	23.90	nr
450mm tray	20.31	3m
flat bend	57.67	nr
tee	75.55	nr
cross piece	113.24	nr
riser bend	44.08	nr
reducer 450 to 300mm	39.43	nr

Galvanised heavy duty standard cable tray and fittings; straight

75mm tray	10.02	3m
flat bend	39.96	nr
tee	49.11	nr
cross piece	79.34	nr
riser bend	27.46	nr
100mm tray	10.18	3m
flat bend	40.48	nr
tee	49.53	nr
cross piece	79.97	nr
riser bend	28.14	nr
reducer 100 to 75mm	18.30	nr
150mm tray	11.28	3m
flat bend	43.93	nr
tee	54.44	nr
cross piece	86.61	nr
riser bend	30.54	nr
reducer 150 to 100mm	18.36	nr
300mm tray	16.41	3m
flat bend	50.42	nr
tee	65.79	nr
cross piece	104.13	nr
riser bend	33.58	nr
reducer 300 to 150mm	30.12	nr
450mm tray	28.39	3m
flat bend	85.51	nr
tee	102.51	nr
cross piece	160.87	nr
riser bend	58.99	nr
reducer 450 to 300mm	47.12	nr

Supports for cable tray; ms; galv

65mm cantilever arms	4.55	nr
90mm cantilever arms	4.55	nr
120mm cantilever arms	4.55	nr
75mm stand-off bracket	4.55	nr
100mm stand-off bracket	5.86	nr
150mm stand-off bracket	5.86	nr
300mm stand-off bracket	12.92	nr
450mm stand-off bracket	17.68	nr

PVC insulated cables; single core; reference 6491X; stranded copper conductors; to BS 6004

1.5mm2 (100m)	5.47	100m
2.5mm2 (100m)	8.72	100m
4.0mm2 (100m)	21.08	100m
6.0mm2 (100m)	25.52	100m
10.0mm2 (100m)	37.27	100m
16.0mm2 (100m)	61.77	100m

PVC double insulated cables; single core; reference 6181Y; stranded copper conductors; to BS 6004

16.0mm2 meter tails, 1m long	0.86	m
25.0mm2 meter tails, 1m long	1.09	m
35.0mm2 meter tails, 1m long	2.71	m

PVC insulated and PVC sheathed cables; multicore; copper conductors and bare earth continuity conductor 6242Y; to BS 6004

1.00mm2; twin and earth (100m)	11.96	100m
1.50mm2; twin and earth (100m)	15.89	100m
2.5mm2; twin and earth (100m)	17.10	100m
4.00mm2; twin and earth (100m)	48.46	100m
6.00mm2; twin and earth (100m)	52.79	100m

PVC insulated SWA armoured and LSF sheathed cables; multicore; copper conductors to BS 6724

16mm2; 3 core	3.54	m

PVC Sheathed, polyethylene insulated cables

Co-axial, braid screened, copper conductor 75 ohm (100m)	10.00	100m

Telephone cables

Mineral insulated copper sheathed cables; PVC outer sheath; copper conductors; 600V light duty to BS EN 60702-1

1.5mm2; two core cable	2.51	m
1.5mm2; termination including gland, seal and shroud	3.20	nr
2.5mm2; two core cable	3.68	m
2.5mm2; termination including gland, seal and shroud	3.20	nr
4.0mm2; two core cable	4.66	m
4.0mm2; termination including gland, seal and shroud	3.20	nr

HV switchgear

Switch fuse 500V, metal cased, 32 amp; single pole and neutral; short circuit rating 6.4Ka	113.84	nr
Consumer control units, 250V, metal cased, fitted with 100 amp D.P. isolator and MCB's, eight way, single pole and neutral	94.41	nr
Distribution boards 500V, metal cased, MCB pattern, fitted with 100 amp D.P. isolator, nine way, single pole and neutral, 20 amp	112.71	nr
Distribution boards 500V, metal cased, MCB pattern, fitted with 100 amp D.P. isolator, twelve way, triple pole and neutral, 32 amp	500.77	nr

Accessories; white plastics, with boxes

Flush plate switches 5 amp; one gang; one way; single pole	1.55	nr
Flush plate switches 5 amp; two gang; two way; single pole	2.10	nr
Surface plate switches 5 amp; one gang; one way; single pole	1.65	nr
Surface plate switches 5 amp; two gang; two way; single pole	2.20	nr
Ceiling switches 5 amp; one gang; one way; single pole	2.10	nr
Flush switched socket outlets 13 amp; one gang	3.00	nr
Flush switched socket outlets 13 amp; two gang	3.50	nr
Surface switched socket outlets 13 amp; one gang	3.15	nr
Surface switched socket outlets 13 amp; two gang	3.80	nr
Flush switched socket outlets with RCCB (residual current device) protected at 30 mAmp 13 amp; one gang	35.60	nr
Flush fused connection units 13 amp; one gang; switched; flexible outlet	4.25	nr
Flush fused connection units 13 amp; one gang; switched; flexible outlet; pilot lamp	4.65	nr

Accessories, metalclad, with boxes

Surface plate switches 5 amp; one gang; two way; single pole	3.10	nr
Surface plate switches 5 amp; two gang; two way; single pole	4.00	nr
Surface switched socket outlets 13 amp; one gang	4.50	nr
Surface switched socket outlets 13 amp; two gang	6.50	nr

Weatherproof accessories, with boxes

Surface plate switches 5 amp; one gang; two way; single pole	6.94	nr

Domestic electrical fittings, wiring and ancillaries

Ceiling rose	2.17	nr
Pendant	1.55	nr
Light switch	1.55	nr
Socket outlet 5 Amp	6.45	nr
Cooker point 4 SA	11.50	nr
Immersion heater point	4.65	nr
Shaver socket (transformer)	14.22	nr
Conduit PVC 16mm	0.72	3m
Grommets	0.02	nr
Clips	0.02	nr
MICS cable 1.00mm2	1.98	m
MICS cable 2.50mm2	2.96	m
MICS cable 6.00mm2	6.41	m
Gland	0.75	nr
20mm black enamel conduit	2.98	3.75m

Electrical Installations (Cont'd)

Domestic electrical fittings, wiring and ancillaries (Cont'd)

20mm black enamel conduit bends	2.10	nr
20mm black enamel conduit tees	2.93	nr
20mm black enamel conduit box terminals	3.75	nr
20mm black enamel conduit couplers	0.33	nr
20mm black enamel conduit saddles	0.27	nr
25mm black enamel conduit	3.89	3.75m
25mm black enamel conduit bends	3.02	nr
25mm black enamel conduit tees	4.48	nr
25mm black enamel conduit couplers	0.53	nr
25mm black enamel conduit saddles	0.38	nr
TV Socket and box	6.27	nr
Telephone point	4.00	nr
Draw wire	0.25	m

Fluorescent lamp fittings, mains voltage operations; switch start; including lamps

Batten type; single tube		
1200mm; 36W	10.25	nr
1500mm; 58W	12.40	nr
1800mm; 70W	19.00	nr
2400mm; 100W	27.50	nr
Batten type; twin tube		
1200mm; 36W	16.88	nr
1500mm; 58W	20.00	nr
1800mm; 70W	30.00	nr
2400mm; 100W	40.00	nr
Metal trough reflector; single tube		
1200mm; 36W	34.41	nr
1500mm; 58W	37.69	nr
1800mm; 70W	44.41	nr
2400mm; 100W	63.21	nr
Metal trough reflector; twin tube		
1200mm; 36W	53.22	nr
1500mm; 58W	61.19	nr
1800mm; 70W	67.60	nr
2400mm; 100W	91.96	nr
Plastics diffused type; single tube		
1200mm; 36W	35.51	nr
1500mm; 58W	39.77	nr
1800mm; 70W	48.67	nr
2400mm; 100W	65.57	nr
Plastics diffused type; twin tube		
1200mm; 36W	66.28	nr
1500mm; 58W	80.52	nr
1800mm; 70W	90.95	nr
2400mm; 100W	116.15	nr

Lighting fittings complete with tungsten lamps

Bulkhead; alloy 100W G.L.S. lamp	10.50	nr
Wall glass; alloy; corner bracket 100W G.L.S. lamp	27.87	nr
Ceiling sphere 152mm; 60W G.L.S. lamp	14.85	nr
Ceiling sphere 203mm; 100W G.L.S. lamp	21.64	nr
Down light; semi recessed 175mm; 150W PAR38 lamp	18.54	nr
Lighting track; single circuit; 1250mm; starter pack	18.85	nr
Lighting track; single circuit; plug-in fitting; 100W G.L.S. lamp	14.46	nr

Fan heaters

3kw unit fan heaters for commercial application	159.66	nr
Thermostatic switch units	54.45	nr

Turbular heaters (excluding wire guards), 60W per 305mm loading

600mm long	11.34	nr
910mm long	12.48	nr
1220mm long	14.34	nr
1520mm long	18.74	nr
1830mm long	19.15	nr
Pair two way mounting brackets	5.45	nr

Night storage heaters

1.7KW night storage heater	220.00	nr
2.55KW night storage heater	297.00	nr
3.4KW night storage heater	341.00	nr
MICS 1000v 2 core cable 4.0mm	5.35	m
20 Amp double pole switch	3.55	nr
8 way fuse board	92.11	nr

Bell equipment

Transformers 3-5-8V	14.77	nr
Transformers 12V	18.90	nr
Bells for transformer operation chime 2-note	10.61	nr
Bells for transformer operation 76mm domestic type	5.85	nr
Bells for transformer operation 150mm round bell type	21.27	nr
Bell pushes domestic	1.48	nr
Bell pushes industrial	5.84	nr

2 or 3KW immersion heaters and thermostats; non-withdrawable elements

without flanges 305mm	23.37	nr
without flanges 457mm	23.61	nr
without flanges 686mm	23.72	nr
Thermostats		
immersion heater	10.24	nr
15 amp a.c. for air heating; without switch	15.47	nr
Water heaters, Storage type units		
free outlet 7 litre	114.78	nr
free outlet 23 litre	506.73	nr
free outlet 55 litre	695.18	nr
multi-point 50 litre	696.76	nr
multi-point 75 litre	828.49	nr
multi-point 125 litre	948.19	nr
Instantaneous Shower units complete with fittings	156.19	nr

Extract fans, window mounted; shutter and incorporated switch unit

105mm dia; ref DX100	24.94	nr
257mm dia; ref GX9	199.94	nr
324mm dia; ref GX12	312.44	nr
Louvre shutter	6.24	nr
Window kit	37.93	nr

Extract fans, wall mounted; shutter and incorporated switch unit

105mm dia; ref DX100	24.94	nr
257mm dia; ref GX9	199.94	nr
324mm dia; ref GX12	312.44	nr
473 x 476mm; ref WX12	374.94	nr
Wall duct	6.24	nr
Louvre shutter	6.24	nr
Switch	9.29	nr
Fan controller	52.50	nr

Floor Wall and Ceiling Finishes

Screeds and toppings		
Cement and sand (1:3)	120.50	m3
Surface hardener for paving		
epoxy resin based	2.00	litre
polyurethane	6.70	litre
Nitoflor Lithurin	5.45	kg
Latex Flooring		
One coat levelling screeds	0.63	kg
Gypsum plasterboard to BS 1230 Part 2		
9.5mm handiboard	2.62	m2
9.5mm Duplex handiboard	3.10	m2
12.5mm handiboard	3.86	m2
9.5mm square edge wallboard	1.79	m2
12.5mm square edge wallboard	1.90	m2
15mm square edge wallboard	3.06	m2
Rendering and the like		
Cement and sand (1:3)	120.50	m3
Pea shingle for pebbledash	24.30	tonne
Cement for wet slurry	108.75	tonne
Plasters		
Board finish	139.01	tonne
Hardwall	204.60	tonne
Multi finish	152.13	tonne
Universal one coat	286.77	tonne
Thistle browning	188.76	tonne
Thistle finish	126.72	tonne
Thistle bonding	188.76	tonne
Thistle tough coat	193.71	tonne
Thistle renovating undercoat	227.04	tonne
Thistle renovating finish	199.65	tonne
Plaster beads and stops		
Angle ref: 550	0.56	m
Angle ref: 553	0.56	m
Angle ref: 554	0.73	m
Stop ref: 560	0.77	m
Stop ref: 561	0.77	m
Stop ref: 562	0.73	m
Stop ref: 563	0.73	m
Stop ref: 565	0.93	m
Stop ref: 566	0.93	m
Plasterboard edging ref: 567	1.18	m
Plasterboard edging ref: 568	1.18	m
Bonding fluid		
Bonding fluid	4.25	litre
Metal lathing		
Expanded metal lath, 9mm mesh x 0.500mm thick x 1.11 kg/m2	4.95	m2
Expanded metal lath 9mm mesh x 0.725mm thick x 1.61 kg/m2	5.80	m2
Expanded metal lath 9mm mesh x 0.950mm thick x 2.50 kg/m2	9.45	m2
Rib-Lath, 0.300mm thick x 1.16 kg/m2	6.20	m2
Rib-Lath, 0.500mm thick x 1.90 kg/m2	7.20	m2
Stainless steel Rib-Lath, 0.300mm thick x 1.16 kg/m2	16.55	m2
Expanded metal lath 9.2mmx 28.6mm mesh x 0.460mm thick x 1.15 kg/m2	8.75	m2
Red-rib lath, 0.500mm thick x 1.91 kg/m2	7.90	m2
Galvanized arch formers, corners		
372mm radius	15.35	nr
452mm radius	18.70	nr
602mm radius	23.30	nr
752mm radius	30.65	nr
Plasterboard coving		
Gyproc cove 100mm girth	1.13	m
Gyproc cove 127mm girth	1.00	m

Floor Wall and Ceiling Finishes (Cont'd)

Plasterboard coving (Cont'd)

Coving Adhesive	0.83	kg

Clay floor quarry tiles - terracotta - square edge

150 x 150 x 12.5mm	344.00	1000
150 x 150 x 20mm	516.00	1000
225 x 225 x 25mm	1170.00	1000

Ceramic floor tiles - red - square edge

100 x 100 x 9mm	15.55	m2
152 x 152 x 12mm	17.05	m2
200 x 200 x 12mm	18.35	m2

Ceramic floor tiles - cream - square edge

100 x 100 x 9mm	16.50	m2
152 x 152 x 12mm	17.85	m2
200 x 200 x 12mm	19.10	m2

Glazed ceramic wall tiles to BS 6431

108 x 108 x 6.5mm - white	16.75	m2
152 x 152 x 5.5mm - white	11.20	m2
108 x 108 x 6.5mm - light colours	16.75	m2
152 x 152 x 5.5mm - light colours	11.20	m2
108 x 108 x 6.5mm - dark colours	19.40	m2
152 x 152 x 5.5mm - dark colours	14.05	m2

Fixing materials for glazed ceramic wall tiles

Cement and sand (1:3)	120.50	m3
Tile adhesive	1.47	litre
White cement	1.05	kg

Terrazzo tiles

300 x 300 x 28mm units	31.00	m2
305 x 305 x 28mm units rounded nosing	15.00	m
305 x 165 x 28mm units	7.50	m

Terrazzo

Terrazzo mix 1:2	1000.00	m3

Wood block flooring, tongued and grooved, 70 x 230 x 20mm units

Maple wood blocks	28.75	m2
Merbau wood blocks	23.25	m2
Oak wood blocks	33.15	m2

PVC flooring, 300 x 300mm tiles

2mm tiles	7.85	m2
2.5mm tiles	9.50	m2
2mm heavy duty tiles	14.25	m2
2.5mm heavy duty tiles	15.75	m2

Linoleum flooring

500 x 500 x 2mm tiles	15.30	m2
2mm sheet flooring	14.25	m2
3.2mm sheet flooring	17.95	m2
Adhesive for linoleum flooring	2.90	litre

Burmatex carpeting

Adhesive for carpeting	12.25	litre

Edge fixed carpeting

Tredaire underlay	1.75	m2
Tufted wool/nylon	20.00	m2
Axminster wool/nylon	26.00	m2
Wilton wool/nylon	32.50	m2

Wallpapers and wall coverings

Lining paper	1.00	roll
Pulp paper	3.00	roll
Washable	4.00	roll
Vinyl coated	6.50	roll
Embossed	4.50	roll
Textured	5.00	roll
Woodchip	1.50	roll

Plastic faced plasterboard

Thermaline BASIC vapour check grade plasterboard

22mm Gyproc Thermaline	7.05	m2
30mm Gyproc Thermaline	7.24	m2
40mm Gyproc Thermaline	8.06	m2
50mm Gyproc Thermaline	10.21	m2

Cellular partitions

57mm panels	11.35	m2
37 x 19mm softwood battens	0.32	m
37 x 37mm softwood battens	0.41	m

Solidwall partitions plasterboard components

25 x 38mm softwood battens	0.29	m

Metal stud partitions

48mm studs	1.29	m
50mm floor and ceiling channels	1.44	m

Suspended ceiling tiles, bevelled, grooved and rebated

1200 x 600 x 9mm asbestos-free fire resisting	17.45	m2
1200 x 600 x 15.8mm plain mineral fibre	10.60	m2
600 x 600 x 15.8mm plain mineral fibre	10.60	m2

Glazing

Float glass

3mm GG quality	40.45	m2
4mm GG quality	40.45	m2
5mm GG quality	59.14	m2
6mm GG quality	59.14	m2
10mm GG quality	119.79	m2
3mm obscured ground	87.19	m2
4mm obscured ground	95.68	m2

Rough cast glass

6mm	53.28	m2

Pyroshield glass

6mm clear	95.71	m2
6mm clear, aligning panes	95.71	m2
6mm safety clear	112.70	m2
6mm safety clear, aligning panes	112.70	m2
7mm texture	44.28	m2
7mm texture, aligning panes	44.28	m2
7mm safety texture	57.61	m2
7mm safety texture, aligning panes	57.61	m2

Patterned glass

4mm white	37.52	m2
6mm white	56.47	m2
4mm tinted	55.40	m2
6mm tinted	62.69	m2

Antisun float glass

4mm grey	77.23	m2
6mm grey	111.45	m2
10mm grey	199.24	m2
12mm grey	275.01	m2
4mm bronze	77.23	m2
6mm bronze	111.45	m2
10mm bronze	199.24	m2
12mm bronze	275.01	m2
6mm green	142.84	m2

Thick float glass

10mm GG quality	119.79	m2
12mm GG quality	138.10	m2
15mm GG quality	229.77	m2
19mm GG quality	325.47	m2
25mm GG quality	516.65	m2

Toughened safety glass

4mm clear float glasses	56.94	m2
5mm clear float glasses	85.45	m2
6mm clear float glasses	85.45	m2
10mm clear float glasses	135.53	m2
12mm clear float glasses	248.19	m2
4mm white	114.79	m2
6mm white	119.86	m2
4mm tinted	114.77	m2
6mm tinted	119.86	m2

Toughened safety solar control glasses

4mm, antisun grey	104.41	m2
6mm, antisun grey	108.96	m2
10mm, antisun grey	215.55	m2
12mm, antisun grey	283.71	m2
4mm , antisun bronze	104.41	m2
6mm , antisun bronze	108.96	m2
10mm , antisun bronze	215.62	m2
12mm , antisun bronze	283.77	m2

Clear laminated glass - float quality

4.4mm safety glass	75.27	m2
6.4mm safety glass	63.11	m2
8.8mm safety glass	113.59	m2
10.8mm safety glass	132.84	m2
7.5mm anti-bandit glass	123.56	m2
9.5mm anti-bandit glass	154.54	m2
11.5mm anti-bandit glass	176.44	m2
11.5mm anti-bandit glass, large panes	211.84	m2

Georgian wired glass

7mm standard	44.28	m2
7mm safety	57.61	m2
6mm polished	95.71	m2
6mm polishedsafety	112.70	m2

Sundries

linseed oil putty	14.92	25kg
metal casement putty	15.69	25kg
washleather strip, 3mm glass	0.58	m
washleather strip, 6mm glass	0.70	m
rubber glazing strip, 3mm glass	0.54	m
rubber glazing strip, 6mm glass	0.65	m

Painting and Decorating

Timber treatments

Creosote	165.50	200L
Wood preservative	19.65	5L

Emulsion paint

matt brilliant white	7.40	5L
matt colours	8.15	5L
silk brilliant white	8.75	5L
silk colours	9.60	5L

Priming, paint

wood primer	17.35	5L
primer sealer	24.45	5L
aluminium sealer and wood primer	30.25	5L
alkali-resisting primer	26.60	5L
universal primer	27.45	5L
red oxide primer	21.20	5L
calcium plumbate primer	36.05	5L
zinc phosphate primer	28.00	5L
acrylic primer	16.15	5L

Undercoat paint

Brilliant white undercoat	12.25	5L
colours	16.60	5L

Gloss paint

Brilliant white gloss	12.25	5L
colours	16.60	5L

Eggshell paint

Brilliant white eggshell	15.80	5L
colours	17.45	5L

Plastic finish

coating compound	12.95	25kg

Waterproof cement paint

cream or white	46.10	5L
colours	46.10	5L
sealer	22.15	5L

Textured masonry paint

white	19.65	10L
colours	23.30	10L
Stabilising solution	52.35	20L

Mulitcolour painting

primer	14.15	5L

Painting and Decorating (Cont'd)

Mulitcolour painting (Cont'd)

basecoat	28.15	5L
finish	39.65	5L

Metallic paint

aluminium	39.00	5L
gold or bronze	57.90	5L

Bituminous paint

Bituminous paint	11.25	5L

Polyurethane lacquer

clear	28.25	5L

Linseed oil

raw	22.05	5L
boiled	26.10	5L

Fire retardant paints and varnishes

fire retardant paint	48.25	5L
fire retardant varnish	34.25	5L
overcoat varnish	53.65	5L

French polishing

white polish	8.00	litre
button polish	8.10	litre
garnet polish	7.60	litre
methylated spirit	1.90	litre

Road marking materials

solvent based paint	54.00	litre
50mm wide non-reflective self adhesive tape	1.60	m
100mm wide non-reflective self adhesive tape	2.95	m
50mm wide reflective self adhesive tape	2.24	m
100mm wide reflective self adhesive tape	4.13	m
road marking primer	3.60	litre

Surface treatments

silicone based water repellent	17.25	5L
fungicide	2.60	litre

Drainage

Drains; vitrified clay pipes and fittings, BS EN 295, normal; flexible mechanical joints; Hepworth Hepseal

400mm pipes	99.52	m
400mm bends	373.97	nr
450mm pipes	131.97	m
450mm bends	492.44	nr

Drains; vitrified clay pipes and fittings, BS EN 295; sleeve joints, push-fit polypropylene standard ring flexible couplings; Hepworth Supersleeve

100mm pipes	4.52	m
100mm pipe couplings	3.33	nr
100mm bends	8.98	nr
100mm branches	19.62	nr
150mm pipes	9.14	m
150mm pipe couplings	6.06	nr
150mm bends	18.61	nr
150mm branches	28.92	nr

Vitrified clay pipes and fittings, BS EN 295; sleeve joints, push-fit polypropylene neoprene ring flexible couplings; Hepworth Supersleeve

100mm pipes	4.52	m
100mm pipe couplings	5.07	nr
150mm pipes	9.14	m
150mm pipe couplings	9.56	nr

Drains; vitrified clay pipes and fittings, BS EN 295

100mm pipes	8.44	m
100mm double collars	10.18	nr
100mm bends	8.44	nr
100mm rest bends	13.87	nr
100mm branches	15.50	nr
150mm pipes	12.98	m
150mm double collars	16.95	nr
150mm bends	12.98	nr
150mm rest bends	23.44	nr
150mm branches	25.65	nr
225mm pipes	25.73	m
225mm double collars	39.64	nr
300mm pipes	43.12	m

Drains; Wavin; PVC-U pipes and fittings, BS EN 13598-1; ring seal joints

110mm pipes	4.92	m
110mm pipe couplings	4.54	nr
110mm connections to cast iron and clay sockets	12.91	nr
110mm 45 degree bends	11.00	nr
110mm branches	16.20	nr
110mm connections cast iron and clay spigot	12.91	nr
160mm pipes	11.11	m
160mm pipe couplings	9.14	nr
160mm connections to cast iron and clay sockets	25.92	nr
160mm 45 degree bends	25.94	nr
160mm branches	46.31	nr
160mm connections to cast iron and clay spigot	25.92	nr

PVC-U solid wall concentric external rib-reinforced Wavin Ultra-Rib pipes and fittings with sealing rings to joints

150mm 6UR 046 pipes	5.08	m
150mm pipe couplings	9.50	nr
150mm connectors 6UR 205	8.60	nr
150mm bends 6UR 563	11.58	nr
150mm branches 6UR 213	28.00	nr
150mm connections to clayware pipe ends 6UR 129	21.09	nr
225mm 9UR 046 pipes	11.64	m
225mm pipe couplings	19.76	nr
225mm connectors 9UR 205	18.00	nr
225mm bends 9UR 563	46.60	nr
225mm branches 9UR 213	88.48	nr
225mm connections to clayware pipe ends 9UR 109	49.52	nr
300mm 12UR 043 pipes	17.46	m
300mm pipe couplings	39.42	nr
300mm connectors 12UR 205	36.32	nr
300mm bends 12UR 563	80.57	nr
300mm branches 12UR 213	188.71	nr
300mm connections to clayware pipe ends 12UR 112	130.25	nr
Reducers 40/32mm	1.48	nr
32mm Pipe brackets - Metal	1.70	nr
40mm Pipe brackets - Metal	2.03	nr

Unreinforced concrete pipes and fittings BS 5911, Class H, flexible joints

300mm pipes	12.65	m
300mm bends	126.50	nr
300mm, 100mm branches	74.25	nr
300mm, 150mm branches	74.25	nr
300mm, 225mm branches	151.25	nr
300mm, 300mm branches	173.25	nr
375mm pipes	15.95	m
375mm bends	159.50	nr
375mm, 100mm branches	82.50	nr
375mm, 150mm branches	82.50	nr
375mm, 225mm branches	159.50	nr
375mm, 300mm branches	181.50	nr
375mm, 375mm branches	203.50	nr
450mm pipes	18.42	m
450mm bends	184.25	nr
450mm, 100mm branches	93.50	nr
450mm, 150mm branches	93.50	nr
450mm, 225mm branches	170.50	nr
450mm, 300mm branches	192.50	nr
450mm, 375mm branches	214.50	nr
450mm, 450mm branches	253.00	nr
525mm pipes	22.28	m
525mm bends	222.75	nr
525mm, 100mm branches	99.00	nr
525mm, 150mm branches	99.00	nr
525mm, 225mm branches	176.00	nr
525mm, 300mm branches	198.00	nr
525mm, 375mm branches	220.00	nr
525mm, 450mm branches	258.50	nr
600mm pipes	27.22	m
600mm bends	272.25	nr
600mm, 100mm branches	107.80	nr
600mm, 150mm branches	107.80	nr
600mm, 225mm branches	184.80	nr
600mm, 300mm branches	206.80	nr
600mm, 375mm branches	228.80	nr
600mm, 450mm branches	267.30	nr

Reinforced concrete pipes and fittings, BS 5911, Class H, flexible joints

450mm pipes	35.60	m
450mm bends	356.00	nr
450mm, 100mm branches	99.32	nr
450mm, 150mm branches	101.60	nr
450mm, 225mm branches	105.01	nr
450mm, 300mm branches	127.74	nr
450mm, 375mm branches	144.79	nr
450mm, 450mm branches	167.52	nr
525mm pipes	43.51	m
525mm bends	435.12	nr
525mm, 100mm branches	111.94	nr
525mm, 150mm branches	114.21	nr
525mm, 225mm branches	117.62	nr
525mm, 300mm branches	134.67	nr
525mm, 375mm branches	157.41	nr
525mm, 450mm branches	180.14	nr
600mm pipes	47.47	m
600mm bends	474.67	nr
600mm, 100mm branches	101.48	nr
600mm, 150mm branches	126.49	nr
600mm, 225mm branches	129.90	nr
600mm, 300mm branches	146.95	nr
600mm, 375mm branches	169.68	nr
600mm, 450mm branches	192.41	nr

Cast iron pipes and cast iron fittings BS 437, coated; bolted cast iron coupling and synthetic rubber gasket joints

100mm pipes	23.12	m
100mm pipe couplings	14.82	nr
100mm connectors, large socket for clayware	32.40	nr
100mm bends	27.72	nr
100mm branches	36.77	nr
150mm pipes	42.80	m
150mm pipe couplings	17.95	nr
150mm connectors, large socket for clayware	48.37	nr
150mm diminishing pieces, reducing to 100mm	36.13	nr
150mm bends	63.77	nr
150mm branches	90.74	nr

Stainless steel pipes and fittings, B.M. Stainless Steel Drains Ltd; ring seal push fit joints (AIS 1316)

75mm pipes	28.87	m
75mm pipe couplings	23.68	nr
75mm bends	35.50	nr
75mm branches	41.25	nr
110mm pipes	36.63	m
110mm pipe couplings	20.02	nr
110mm diminishing pieces, reducing to 75mm	62.35	nr
110mm bends	45.32	nr
110mm branches	47.63	nr

Precast circular concrete inspection chambers; standard or stock pattern units; BS 5911

250mm high, 900mm internal diameter	22.00	nr
500mm high, 900mm internal diameter	33.00	nr
750mm high, 900mm internal diameter	33.00	nr

Drainage (Cont'd)

Precast circular concrete inspection chambers; standard or stock pattern units; BS 5911 (Cont'd)

1000mm high, 900mm internal diameter	44.00	nr
250mm high, 1050mm internal diameter	20.90	nr
500mm high, 1050mm internal diameter	31.35	nr
750mm high, 1050mm internal diameter	31.35	nr
1000mm high, 1050mm internal diameter	41.80	nr
250mm high, 1200mm internal diameter	26.40	nr
500mm high, 1200mm internal diameter	39.60	nr
750mm high, 1200mm internal diameter	39.60	nr
1000mm high, 1200mm internal diameter	52.80	nr

Galvanized malleable cast iron steps

step irons BS 1247	5.00	nr

Cover slabs, heavy duty, 125mm thick, for chamber sections:

2400mm internal diameter	558.59	nr
2700mm internal diameter	723.14	nr
3000mm internal diameter	851.71	nr

Precast concrete inspection chambers; standard or stock pattern units; rectangular chamber sections

150mm high, internal size 1200 x 750mm	17.48	nr
200mm high, internal size 1200 x 750mm	21.55	nr

GRP inspection chambers with preformed benching; 475mm internal diameter

depth to invert 450mm	75.61	nr
depth to invert 585mm	36.59	nr
depth to invert 750mm	19.06	nr
depth to invert 930mm	78.22	nr

Polypropylene shallow inspection chamber with preformed benching. Light duty single sealed cover and frame ref. 4D.961

300mm diameter; for 100mm diameter pipes ref. 4D.960, depth to invert 600mm	62.13	nr

Polypropylene universal inspection chamber; 475mm diameter

preformed chamber base benching; for 100mm pipes, ref. 4D.922	89.81	nr
chamber shaft 230mm effective length, ref. 4D.925	22.90	nr
cast iron cover and frame; single seal light duty to suit, ref. 4D.942	72.15	nr
Polypropylene sealed cover and frame, ref. 4D.943	43.15	nr

Cast iron inspection chambers

100mm diameter straight with one branch	88.52	nr
100mm diameter straight with one branch	135.11	nr
100mm diameter straight with two branches	265.98	nr
150 x 100mm diameter straight with one branch	175.05	nr
150mm diameter straight with one branch	222.14	nr
150mm diameter straight with two branches	443.59	nr

Manhole covers, cast iron, A15 light duty, Single Seal

Solid top Circular 450mm dia	13.86	nr
Solid top 450mm x 450mm	18.48	nr
Solid top 600mm x 450mm	23.10	nr
Solid top 600mm x 600mm	32.34	nr
Recessed top 600mm x 450mm	35.11	nr
Recessed top 600mm x 600mm	43.89	nr
Recessed top 750mm x 600mm	225.46	nr

Manhole covers, cast iron, A15 light duty, Double Seal

Solid top 450mm x 450mm	240.24	nr
Solid top 600mm x 450mm	253.82	nr
Solid top 600mm x 600mm	282.05	nr
A15 light duty, lifting keys	6.93	nr

Manhole covers, cast iron, B125 medium duty

Ref. H783 Trojan single seal; 600mm diameter	98.18	nr
Ref. H781; 600 x 450mm	90.93	nr
Ref. H782; 600 x 600mm	113.02	nr
Ref. H743 Trojan S; 600 x 450mm	119.47	nr
Ref. H742; 600 x 600mm	150.81	nr
Ref. H741; 750 x 600mm	236.05	nr
B125 medium duty, lifting keys	4.94	nr

Manhole covers, cast iron, D400 heavy duty

Ref. H9330; 600mm diameter	122.20	nr
Ref. H9320; 600 x 600mm	106.57	nr
Ref. H9420; 750 x 750mm	256.87	nr
D400 heavy duty, lifting keys	5.50	nr

Clayware land drain pipes, BS 1196; butt joints

75mm	3.21	m
100mm	5.51	m
150mm	11.30	m

Porous concrete land drain pipes, BS 5911; butt joints

150mm	6.08	m
225mm	7.78	m
300mm	12.79	m
375mm	17.43	m
450mm	24.52	m
525mm	36.33	m
600mm	45.12	m

Fencing

Strained wire fencing - concrete posts

900mm intermediate post	5.15	nr
900mm end/angle post	6.30	nr
900mm strutt	5.02	nr
900mm stretcher bar	1.55	nr
1050mm intermediate post	5.58	nr
1050mm end/angle post	6.94	nr
1050mm strutt	5.58	nr
1050mm stretcher bar	2.08	nr
Ratchet winders	2.73	nr
2.5mm Stirrup wire	56.35	651m
4mm Galvanized line wire	37.34	254m

900mm Chain link fencing - concrete posts - galvanized mesh

intermediate post GLC	5.15	nr
end/angle post GLC	6.30	nr
strutt	5.02	nr
stretcher bar	1.64	nr
chain link	50.66	25m

1200mm Chain link fencing - concrete posts - galvanized mesh

intermediate post GLC	6.09	nr
end/angle post GLC	9.54	nr
strutt	5.78	nr
stretcher bar	2.20	nr
chain link	66.26	25m

1400mm Chain link fencing - concrete posts - galvanized mesh

intermediate post GLC	6.59	nr
end/angle post GLC	8.15	nr
strutt	6.59	nr
stretcher bar	2.60	nr
chain link	74.24	25m

1800mm Chain link fencing - concrete posts - galvanized mesh

intermediate post GLC	9.30	nr
end/angle post GLC	12.87	nr
strutt	8.01	nr
stretcher bar	3.32	nr
chain link	47.01	12.5m

Chain Link fencing sundries

Ratchet winders	2.73	nr
2.5mm Stirrup wire	56.35	651
PPC Line wire	5.72	26m
Ring fasteners	5.72	1000

900mm Chain link fencing - concrete posts - plastic coated mesh

intermediate post PLC	5.15	nr
end/angle post PLC	6.30	nr
strutt	5.02	nr
stretcher bar	1.64	nr
chain link 2.5 / 3.55mm	93.34	25m

1200mm Chain link fencing - concrete posts - plastic coated mesh

intermediate post PLC	6.09	nr
end/angle post PLC	9.54	nr
strutt	5.78	nr
stretcher bar	2.20	nr
chain link	112.19	25m

1400mm Chain link fencing - concrete posts - plastic coated mesh

intermediate post PLC	6.59	nr
end/angle post PLC	8.15	nr
strutt	6.59	nr
stretcher bar	2.60	nr
chain link	138.34	25m

1800mm Chain link fencing - concrete posts - plastic coated mesh

intermediate post PLC	9.30	nr
end/angle post PLC	12.87	nr
strutt	8.01	nr
stretcher bar	3.32	nr
chain link	175.72	25m

900mm Chain link fencing - steel posts - galvanized mesh

intermediate post GLS	9.57	nr
end post with strutts	22.29	nr
angle post with strutts	34.34	nr
chain link 3 x 50mm	50.66	25m

1200mm Chain link fencing - steel posts - galvanized mesh

intermediate post GLS	11.62	nr
end post with strutts	26.71	nr
angle post with strutts	41.65	nr
chain link	66.26	25m

1400mm Chain link fencing - steel posts - galvanized mesh

intermediate post GLS	12.64	nr
end post with strutts	29.02	nr
angle post with strutts	45.26	nr
chain link	74.24	25m

1800mm Chain link fencing - steel posts - galvanized mesh

intermediate post GLS	15.71	nr
end post with strutts	38.71	nr
angle post with strutts	57.95	nr
chain link	47.01	12.5m

900mm Chain link fencing - steel posts - plastic coated mesh

intermediate post PLS	14.85	nr
end post with strutts	30.42	nr
angle post with strutts	45.40	nr
chain link	93.34	25m

Fencing (Cont'd)

1200mm Chain link fencing - steel posts - plastic coated mesh		
intermediate post PLS	15.52	nr
end post with strutts	41.83	nr
angle post with strutts	55.54	nr
chain link	112.19	25m
1400mm Chain link fencing - steel posts - plastic coated mesh		
intermediate post PLS	16.92	nr
end post with strutts	45.44	nr
angle post with strutts	59.90	nr
chain link	138.34	25m
1800mm Chain link fencing - steel posts - plastic coated mesh		
intermediate post PLS	21.12	nr
end post with strutts	62.34	nr
angle post with strutts	77.90	nr
chain link	175.72	25m
Chestnut fencing		
900mm high fencing	33.02	9.1m
900mm intermediate post	1.36	nr
900mm end/angle post	5.04	nr
900mm strutts	1.47	nr
1200mm high fencing	41.79	9.1m
1200mm intermediate post	1.63	nr
1200mm end/angle post	6.04	nr
1200mm strutts	1.89	nr
1500mm high fencing	61.53	9.1m
1500mm intermediate post	3.20	nr
1500mm end/angle post	6.93	nr
1500mm strutts	2.31	nr
Close boarded fencing		
1200mm posts	10.82	nr
1200mm pales	0.63	nr
1500mm posts	14.38	nr
1500mm pales	0.80	nr
1800mm posts	16.17	nr
1800mm pales	0.97	nr
Rails	1.66	m
Gravel boards	1.79	m
Centre stumps	0.91	nr
50mm nails	46.50	25kg
75mm nails	46.50	25kg
Wooden palisade fencing		
1050mm posts	7.08	nr
1050mm pales	0.95	nr
1200mm posts	8.44	nr
1200mm pales	1.12	nr
Galvanized mild steel guard rail of tubing to BS 1387 medium grade with Kee Klamp fittings and intersections		
40mm diameter ref. 7-2-G	6.64	m
flanged end	4.66	nr
bend No.15-7	5.11	nr
three-way No.20-7	7.73	nr
three-way No.25-7	7.42	nr
four-way No.21-7	6.05	nr
four-way No.26-7	5.73	nr
five-way No.35-7	8.59	nr
five-way No.40-7	12.83	nr
floor plate No.62-7	6.81	nr

External works and landscaping

Precast concrete kerbs		
125 x 255 x 915mm	2.65	nr
125 x 255 x 915mm kerb; curved 10.00m radius	4.19	nr
150 x 305 x 915mm	5.44	nr
150 x 305 x 915mm kerb; curved 10.00m radius	10.36	nr
Precast concrete edgings		
50 x 150 x 915mm	1.26	m
50 x 200 x 915mm	1.90	m
50 x 250 x 915mm	2.17	m
Granite edge kerb		
150 x 300mm	13.00	m
200 x 300mm	28.14	m
York stone paving		
50mm thick	66.88	m2
75mm thick	75.54	m2
Precast concrete flags, BS 7263 Part 1, natural finish		
600 x 450 x 50mm	2.44	nr
600 x 600 x 50mm	2.65	nr
600 x 750 x 50mm	3.34	nr
600 x 900 x 50mm	3.53	nr
600 x 450 x 63mm	3.05	nr
Brick paving blocks		
25mm thick Brick paviors	300.00	1000
Granite sett paving		
125mm thick; new	33.14	m2
150mm thick; new	39.76	m2

COMPOSITE PRICES FOR APPROXIMATE ESTIMATING

The purpose of this section is to provide an easy reference to costs of "composite" items of work, which would usually be measured for Approximate Estimates. Rates for items, which are not included here, may be found by reference to the preceding pages of this volume. Allowance has been made in the rates for the cost of works incidental to and forming part of the described item although not described therein, i.e. the cost of formwork to edges of slabs has been allowed in the rates for slabs; the cost of finishings to reveals has been allowed in the rates for finishings to walls.

The rates are based on rates contained in the preceding Major Works section of the book and include the cost of materials, Labour fixing or laying and allow 7.5% for overheads and profit: Preliminaries are not included within the rates. The rates are related to a housing or simple commercial building contract in the range of £500,00 to £1,500,000. For contracts of a higher or lower value, rates should be adjusted by the percentage adjustments given in Essential Information at the beginning of this Volume. Landfill tax is included for inert waste.

The following cost plans have been included in both the Major Works and Small Works sections, and priced accordingly as examples on how to construct a Cost Plan, for information and to demonstrate how to use the Approximate Estimating Section.

COST PLAN FOR AN OFFICE DEVELOPMENT

The Cost Plan that follows gives an example of an Approximate Estimate for an office building as the attached sketch drawings. The estimate can be amended by using the examples of Composite Prices following or the prices included within the preceding pages. Details of the specification used are indicated within the elements. The cost plan is based on a fully measured bill of quantities priced using the detailed rates within this current edition of Laxton's. Preliminaries are shown separately.

Floor area **932** **m2**

Description	Item	Qty		Rate	Item Total	Cost per m2	Total
1.0 SUBSTRUCTURE							
Note: Rates allow for excavation by machine							
1.1 Substructure (271 m2)							
Clear site vegetation; excavate 150 topsoil and remove spoil.	Site clearance	288	m2	0.94	271.34	0.29	
300 x 600 reinforced concrete perimeter ground beams with attached 180 x 300 toes; damp proof membrane 50 concrete blinding; jablite; 102.5 thick common and facing bricks from toe to top of ground beam; 75 cavity filled with plain concrete; cavity weeps; damp proof courses (excavation included in ground slab excavation)	Ground beam	81	m	226.21	18,323.29	19.66	
600 x 450 reinforced concrete internal ground beams, mix C30; 50 thick concrete blinding; formwork; jablite protection (excavation included in ground slab excavation)	Ground beam	34	m	167.34	5,689.72	6.10	
Excavate to reduced levels; level and compact bottoms; 100 hardcore filling, blinded; damp proof membrane; 50 jablite insulation; 150 reinforced concrete ground slab 2 layers A252 fabric reinforcement	Ground floor slab	271	m2	80.99	21,948.26	23.55	
1900 x 1900 x 1150 lift pit; excavate; level and compact bottoms; working space; earthwork support; 50 concrete blinding; damp proof membrane; 200 reinforced concrete slab, 200 reinforced concrete walls, formwork; Bituthene tanking and protection board	Lift pit	1	nr	2,271.21	2,271.21	2.44	
1500 x 1500 x 1000 stanchion bases; excavate; level and compact bottoms; earthwork support; reinforced concrete foundations; 102.5 bricks from top of foundation to ground slab level	Stanchion base	13	nr	770.83	10,020.75	10.75	58,524.58
					58,524.58	62.79	
(Element unit rate = £ 215.96 m2)							

COST PLAN FOR AN OFFICE DEVELOPMENT (Cont'd)

Description	Item	Qty		Rate	Item Total	Cost per m2	Total
2.0 SUPERSTRUCTURE							
2.1 Frame (932 m2)							
300 x 300 reinforced concrete columns; formwork two coats RIW to exterior faces	Columns	111	m	103.76	11,517.40	12.36	
29.5m girth x 1.0m high steel frame to support roof over high level louvres	Frame	1	nr	6,879.26	6,879.26	7.38	
300 x 550 reinforced concrete beams; formwork	Beams	61	m	146.86	8,958.18	9.61	27,354.84
					27,354.84	29.35	
(Element unit rate = £ 29.35 m2)							
2.2 Upper floors (614 m2)							
275 reinforced concrete floor slab; formwork	Floor slab	614	m2	123.56	75,863.71	81.40	
300 x 400/675 x 175/300 x 150 reinforced concrete attached beams; formwork	Attached beams	273	m	113.15	30,889.35	33.14	106,753.06
					106,753.06	114.54	
(Element unit rate = £ 173.86 m2)							
2.3 Roof (403 m2 on plan)							
2.3.1 Roof Structure							
Pitched roof trusses, wall plates; 160 thick insulation and Netlon support	Roof structure	403	m2	47.59	19,180.04	20.58	
100 thick blockwork walls in roof space	Blockwork	27	m2	63.15	1,705.02	1.83	
Lift shaft capping internal size 1900 x 1900; 200 thick blockwork walls and 150 reinforced concrete slab	Lift shaft	1	nr	899.52	899.52	0.97	
2.3.2 Roof Coverings							
Concrete interlocking roof tiles, underlay and battens	Roof finish	465	m2	45.36	21,092.40	22.63	
Ridge	Ridge	22	m	18.10	398.20	0.43	
Hip	Hip	50	m	18.79	939.51	1.01	
Valley	Valley	6	m	40.59	243.54	0.26	
Abutments to cavity walls, lead flashings and cavity tray damp proof courses	Abutments	9	m	86.64	779.80	0.84	
Eaves framing, 25 x 225 softwood fascia 50 wide Resoplan soffit and ventilator; eaves tile ventilator and decoration; 164 x 127 pressed Aluminium gutters and fittings	Eaves	102	m	282.42	28,806.78	30.91	
2.3.3 Roof drainage							
100 x 100 Aluminium rainwater pipes and rainwater heads and hopper heads	Rain water pipes	58	m	59.07	3,425.84	3.68	77,470.64
					77,470.64	83.12	
(Element unit rate = £ 192.23 m2)							
2.4 Stairs							
Reinforced in-situ concrete staircase; blockwork support walls screed and vinyl floor finishes; plaster and paint to soffits; to wall							
3n2 1200 wide flights, 1nr half landing and 1nr quarter landing; 3100 rise; handrail one side	Stairs	1	nr	4,995.07	4,995.07	5.36	
4nr 1200 wide flights, 2nr half landings 620 rise; plastics coated steel balustrade one side and handrail one side	Stairs	1	nr	14,227.67	14,227.67	15.27	

COST PLAN FOR AN OFFICE DEVELOPMENT (Cont'd)

Description	Item	Qty		Rate	Item Total	Cost per m2	Total
2.0 SUPERSTRUCTURE (Cont'd)							
2.4 Stairs (Cont'd)							
4nr 1200 wide flights, 2nr half landings mix; 6200 rise; plastics coated steel and glass in fill panels balustrade one side and handrail one side	Stairs	1	nr	14,032.06	14,032.06	15.06	33,254.80
					33,254.80	35.68	
2.5 External Walls (614 m2)							
Cavity walls, 102.5 thick facing bricks (PC £250 per 1000), 75 wide insulated (PC £250 per cavity, 140 thick lightweight concrete blockwork	Facing brick wall	455	m2	135.38	61,597.90	66.09	
Cavity walls, 102.5 thick facing bricks (PC £250 per 1000), built against concrete beams and columns	Facing brick wall	159	m2	100.97	16,054.43	17.23	
880 x 725 x 150 cast stonework pier cappings	Stonework	9	nr	81.57	734.17	0.79	
780 x 1300 x 250 cast stonework spandrels	Stonework	9	nr	163.16	1,468.45	1.58	
6500 x 2500 attached main entrance canopy	Canopy	1	nr	7,201.69	7,201.69	7.73	
Milled lead sheet on impregnated softwood boarding and framing, and building paper	Lead sheet	4	m2	296.06	1,184.23	1.27	
Expansion joint in facing brickwork	Expansion joint	118	m	19.11	2,254.58	2.42	90,495.45
					90,495.45	97.10	
(Element unit rate = £ 147.39 m2)							
2.6 Windows and External Doors (214 m2)							
2.6.1 Windows (190 m2)							
Polyester powder coated Aluminium vertical pivot double glazed windows	Windows	190	m2	442.66	84,105.77	90.24	
Polyester powder coated Aluminium fixed louvre panels	Louvres	35	m2	185.75	6,501.18	6.98	
Heads 150 x 150 x 10 stainless steel angle bolted to concrete beam, cavity damp proof course, weep vents, plaster and decoration	Heads	98	m	101.43	9,940.39	10.67	
Reveals closing cavity with facing bricks, damp proof course, plaster and decoration	Reveals	99	m	36.33	3,597.03	3.86	
Works to reveals of openings comprising closing cavity with facing bricks, damp proof course, 13 thick cement and sand render with angle beads, and ceramic tile finish	Reveals	27	m	51.44	1,389.01	1.49	
Sills closing cavity with blockwork, damp proof course, and 19 x 150 Ash veneered MDF window boards	Sills	8	m	41.09	328.74	0.35	
Sills 230 x 150 cast stonework sills and damp proof course	Sills	36	m	53.89	1,940.18	2.08	
Sills 230 x 150 cast stonework sills, damp proof course, and 19 x 150 Ash veneered MDF window boards	Sills	36	m	301.94	10,869.98	11.66	
Sills closing cavity with blockwork, damp proof course, and ceramic tile sill finish	Sills	9	m	50.67	456.02	0.49	
Cavity wall lintels	Lintels	15	m	46.89	703.33	0.75	
Surrounds to 900 diameter circular openings; stainless steel radius lintels, 102.5 x 102.5 facing brick on edge surround, 13 thick plaster, and eggshell decoration	Lintels	8	m	161.97	1,295.77	1.39	
300 x 100 cast stonework surrounds to openings	Stonework	20	m	51.65	1,033.05	1.11	

COST PLAN FOR AN OFFICE DEVELOPMENT (Cont'd)

Description	Item	Qty		Rate	Item Total	Cost per m2	Total
2.0 SUPERSTRUCTURE (Cont'd)							
2.6 Windows and External Doors (214 m2) (Cont'd)							
2.6.2 External Doors (24 m2)							
Polyester powder coated Aluminium double glazed doors and any integral screens	External doors	24	m2	415.63	9,975.02	10.70	
Cavity wall lintels	Lintels	3	m	46.89	140.67	0.15	132,276.14
					132,276.14	141.93	
(Element unit rate = £ 618.11 m2)							
2.7 Internal Walls and Partitions (481 m2)							
100 thick blockwork walls	Block walls	231	m2	39.37	9,093.63	9.76	
200 thick blockwork walls	Block walls	113	m2	67.60	7,638.94	8.20	
215 thick common brickwork walls	Brick walls	84	m2	108.42	9,107.65	9.77	
Demountable partitions	Partitions	42	m2	27.30	1,146.63	1.23	
Extra over demountable partitions for single doors	Partitions	2	nr	26.26	52.52	0.06	
Extra over demountable partitions for double doors	Partitions	1	nr	31.52	31.52	0.03	
WC Cubicles	Cubicles	9	nr	126.48	1,138.33	1.22	
Screens in softwood glazed with 6 thick GWPP glass	Screens	11	m2	323.87	3,562.53	3.82	
Duct panels in Formica Beautyboard and 50 x 75 softwood framework	Ducts	27	m2	62.49	1,687.10	1.81	
Duct panels in veneered MDF board and 50 x 75 softwood framework	Ducts	26	m2	58.04	1,509.02	1.62	34,967.88
					34,967.88	37.52	
(Element unit rate = £ 72.70 m2)							
2.8 Internal Doors (84 m2)							
900 x 2100 half hour fire resistant single doorsets comprising doors, softwood frames, ironmongery and decoration	Doors	45	m2	354.37	15,946.48	17.11	
900 x 2100 one hour fire resistance single doorsets comprising doors, softwood frames, ironmongery and decoration	Doors	9	m2	535.83	4,822.45	5.17	
900 x 2100 half hour fire resistance single doorsets comprising doors with 2nr vision panels, softwood frames, ironmongery and decoration	Doors	11	m2	404.24	4,446.59	4.77	
1500 x 2100 half hour fire resistance double doorsets comprising Ash veneered doors with 2nr vision panels, Ash frames, ironmongery and decoration	Doors	19	m2	382.80	7,273.27	7.80	
100 wide lintels	Lintels	41	m	15.64	641.21	0.69	
200 wide lintels	Lintels	7	m	32.63	228.38	0.25	33,358.38
					33,358.38	35.79	
(Element unit rate = £ 397.12 m2)							
3.0 INTERNAL FINISHES							
3.1 Wall Finishes (1,086 m2)							
13 thick plaster to blockwork walls with eggshell paint finish	Plaster work	703	m2	20.79	14,615.26	15.68	
13 thick plaster to blockwork walls with lining paper and vinyl wall paper finish	Plaster work	124	m2	21.77	2,698.95	2.90	

COST PLAN FOR AN OFFICE DEVELOPMENT (Cont'd)

Description	Item	Qty		Rate	Item Total	Cost per m2	Total
3.0 INTERNAL FINISHES (Cont'd)							
3.1 Wall Finishes (1,086 m2) (Cont'd)							
10 thick plaster to concrete walls with eggshell paint finish	Plaster work	75	m2	31.44	2,358.13	2.53	
13 thick cement and sand render to blockwork walls with 150 x 150 coloured ceramic tiles finish	Tiling	184	m2	55.72	10,252.48	11.00	29,924.82
					29,924.82	32.11	
(Element unit rate = £ 27.56 m2)							
3.2 Floor Finishes (779 m2)							
50 thick sand and cement screed with 300 x 300 x 2.5 vinyl tile finish	Vinyl tiling	19	m2	46.24	878.51	0.94	
50 thick sand and cement screed with 2 thick sheet vinyl finish	Sheet vinyl	108	m2	41.92	4,527.22	4.86	
50 thick sand and cement screed with 250 x 250 x 6 carpet tile finish	Carpet tiling	587	m2	34.48	20,238.64	21.72	
50 thick sand and cement screed with carpet finish	Carpeting	17	m2	47.09	800.51	0.86	
50 thick sand and cement screed with 150 x 150 x 13 ceramic tile finish	Ceramic tiling	46	m2	65.12	2,995.74	3.21	
Entrance mats comprising brass Matt frame, 60 thick cement and sand screed and matting	Matting	2	m2	214.02	428.04	0.46	
150 high x 2.5 thick vinyl skirting	Skirting	45	m	7.41	333.45	0.36	
119 high x 2 thick sheet vinyl skirtings with 20 x 30 lacquered Hardwood cover fillet	Skirting	116	m	21.29	2,469.56	2.65	
25 x 50 Hardwood skirtings with lacquer finish	Skirting	214	m	16.22	3,471.51	3.72	
150 x 150 x 13 ceramic tile skirtings on 13 thick sand and cement render	Skirting	74	m	13.37	989.01	1.06	37,132.20
					37,132.20	39.84	
(Element unit rate = £ 47.67 m2)							
3.3 Ceiling Finishes (809 m2)							
Suspended ceiling systems to offices	Suspended	587	m2	39.06	22,928.22	24.60	
Suspended ceiling systems to toilets	Suspended	46	m2	31.29	1,439.33	1.54	
Suspended ceiling systems to reception area	Suspended	18	m2	31.29	563.21	0.60	
Suspended ceiling plain edge trims and 25 x 38 painted painted softwood shadow battens	Edge trim	300	m	15.25	4,574.55	4.91	
Gyproc MF suspended ceiling system with eggshell paint finish	Suspended	127	m2	27.39	3,478.85	3.73	
12.5 Gyproc Square edge plaster board and skim on softwood supports with eggshell paint finish	Plastering	29	m2	45.75	1,326.88	1.42	
10 thick plaster to concrete with eggshell paint finish	Plastering	2	m2	78.73	157.45	0.17	
Plaster coving	Cove	234	m	6.83	1,598.50	1.72	36,066.99
					36,066.99	38.70	
(Element unit rate = £ 44.58 m2)							
4.0 FITTINGS AND FURNISHINGS							
4.1 Fittings							
2100 x 500 x 900 reception counter in softwood	Counter	1	nr	1,020.08	1,020.08	1.09	
2700 x 500 x 900 vanity units in Formica faced chipboard	Units	6	nr	657.94	3,947.66	4.24	

COST PLAN FOR AN OFFICE DEVELOPMENT (Cont'd)

Description	Item	Qty		Rate	Item Total	Cost per m2	Total
4.0 FITTINGS AND FURNISHINGS (Cont'd)							
4.1 Fittings (Cont'd)							
600x 400 x 6 thick mirrors	Mirror	15	nr	81.93	1,228.95	1.32	
Vertical fabric blinds	Blinds	161	m2	49.55	7,977.63	8.56	
900 x 250 x 19 blockboard lipped shelves on spur support system	Shelves	12	nr	46.62	559.44	0.60	
1200 x 600 x 900 sink base units in melamine finished particle board	Sink base unit	3	nr	178.08	534.24	0.57	15,268.00
					15,268.00	16.38	
5.0 SERVICES							
5.1 Sanitary Appliances							
1200 x 600 stainless steel sink units with pillar taps	Sink	3	nr	165.37	496.11	0.53	
White glazed fireclay cleaners sink units with pillar taps	Cleaners sink	3	nr	310.93	932.79	1.00	
Vitreous china vanity basins with pillar taps	Vanity basins	12	nr	213.82	2,565.84	2.75	
Vitreous china wash hand basins and pedestal with pillar taps	Wash hand basin	1	nr	204.27	204.27	0.22	
Vitreous china wash hand basin with pillar taps	Wash hand basin	1	nr	132.19	132.19	0.14	
Vitreous china disabled wash hand basin with pillar taps	Wash hand basin	1	nr	179.00	179.00	0.19	
Close coupled vitreous china WC suites	WC suite	11	nr	239.57	2,635.27	2.83	
Showers comprising shower tray, mixer valve and fittings, and corner shower cubicle	Shower	1	nr	980.20	980.20	1.05	
Vitreous china urinals with automatic cistern	Urinals	3	nr	388.55	1,165.65	1.25	
Disabled grab rails set	Grab rails	1	nr	393.99	393.99	0.42	
Toilet roll holders	Accessories	11	nr	15.22	167.43	0.18	
Paper towel dispensers	Accessories	8	nr	25.02	200.15	0.21	
Paper towel waste bins	Accessories	8	nr	38.07	304.56	0.33	
Sanitary towel disposal units	Accessories	3	nr	717.91	2,153.73	2.31	
Hand dryers	Accessories	8	nr	250.23	2,001.81	2.15	14,512.98
					14,512.98	15.57	
5.2 Services Equipment		1	nr	2,092.59	2,092.59	2.15	2,092.59
5.3 Disposal Installations							
Upvc waste pipes and fittings to internal drainage	Internal Drainage	37	nr	149.22	5,521.30	5.92	5,521.30
5.4 Water Installations							
Hot and cold water supplies		37	nr	205.80	7,614.44	8.17	7,614.44
5.5 Heat Source							
Gas boiler for radiators and hot water from indirect storage tank		1	nr	80,488.80	80,488.80	86.36	80,488.80
5.6 Space Heating and Air treatment							
Low temperature hot water heating by radiators		1	nr	2,676.97	2,676.97	2.87	2,676.97
5.7 Ventilation Installations							
Extractor fan to toilet and kitchen areas		9	nr	658.81	5,929.31	6.36	5,929.31

COST PLAN FOR AN OFFICE DEVELOPMENT (Cont'd)

Description	Item	Qty		Rate	Item Total	Cost per m2	Total
5.0 SERVICES (Cont'd)							
5.8 Electrical Installations							
Lighting & Power		1	nr	92,204.16	92,204.16	98.93	92,204.16
5.9 Gas Installations							
To boiler room		1	nr	2,095.02	2,095.02	2.25	2,095.02
5.10 Lift and Conveyor Installations							
8 person 3 stop hydraulic lift		1	nr	24,864.76	24,864.76	26.68	24,864.76
5.11 Protective Installations							
Burglar Alarm System		1	nr	6,148.90	6,148.90	6.60	6,148.90
5.12 Communication Installations							
Cabling for telephones		1	nr	8,235.13	8,235.13	8.84	8,235.13
5.13 Special Installations							
Door entry to receptions on all floors		1	nr	1,360.18	1,360.18	1.46	1,360.18
5.14 Builders Work in Connection with Services							
Builders work in connection with plumbing, mechanical and electrical installations		1	nr	6,837.21	6,837.21	7.34	6,837.21
Builders Profit and Attendance on Services		1	nr	1,046.30	1,046.30	1.12	1,046.30
6.0 EXTERNAL WORKS							
6.1 Site Works							
6.1.1 Site Preparation							
Clearing site vegetation, filling root voids with excavated material	Site clearance	1716	m2	0.70	1,201.20	1.29	
Excavated topsoil 150 deep and preserve on site	Clear topsoil	214	m3	6.53	1,397.21	1.50	
Dispose of surplus topsoil off site	Disposal	70	m3	19.13	1,339.10	1.44	
6.1.2 Surface Treatments							
Excavated to reduce levels 400 deep; dispose of excavated material on site; level and compact bottoms; herbicide; 250 thick granular material filling, levelled, compacted and blinded; 70 thick two coat coated macadam pavings; white road lines and markings	Pavings	719	m2	37.81	27,188.96	29.17	
Excavate to reduce levels 400 deep; dispose of excavated material off site; level and compact bottoms; herbicide; 150 thick granular material filling, levelled compacted, and blinded; 200 x 100 x 65 thick clay paviors	Pavings	40	m2	46.58	1,863.34	2.00	
Excavate to reduce levels 400 deep; dispose of excavated material off site; level and compact bottoms; herbicide; 150 thick granular material filling, levelled, compacted and blinded; 900 x 600 x 50 thick concrete paving flags	Pavings	149	m2	34.47	5,136.34	5.51	
125 x 255 precast concrete road kerbs and foundations	Kerbs	265	m	27.99	7,418.41	7.96	
50 x 150 precast concrete edgings and foundations	Kerbs	48	m	18.53	889.44	0.95	
Excavate to reduce levels 400 deep; dispose of excavated material off site; prepare sub-soil; herbicide; 150 thick preserved topsoil filling; rotavate; turf	Landscaping	55	m2	26.66	1,466.53	1.57	

COST PLAN FOR AN OFFICE DEVELOPMENT (Cont'd)

Description	Item	Qty		Rate	Item Total	Cost per m2	Total
6.0 EXTERNAL WORKS (Cont'd)							
6.1 Site Works (Cont'd)							
6.1.2 Surface Treatments (Cont'd)							
Excavate to reduce levels 400 deep; dispose of excavated material off site; prepare sub-soil; herbicide; 450 thick preserved topsoil filling; cultivate; plant 450 - 600 size herbaceous plants - 3/m2, and 3m high trees - 1nr per 22m2; fertiliser; 25 thick peat mulch	Planting	397	m2	38.45	15,265.59	16.38	
6.1.3 Site Enclosure and Division							
Palisade fencing	Fencing	62	m	45.90	2,845.80	3.05	
1800 high fencing	Fencing	96	m	42.82	4,110.92	4.41	
Extra for 1000 x 2000 gate	Gate	1	nr	414.90	414.90	0.45	70,537.75
					70,537.75	75.68	
6.2 Drainage							
Excavate trench 250 - 500 deep; lay 100 diameter vitrified clay pipe and fittings bedded and surrounded in granular material	Pipes in trenches	44	m	43.39	1,909.05	2.05	
Excavate trench 500 - 750 deep; lay 100 diameter vitrified clay pipe and fittings bedded and surrounded in granular material	Pipes in trenches	59	m	46.55	2,746.40	2.95	
Excavate trench 750 - 1000 deep; lay 100 diameter vitrified clay pipe and fittings bedded and surrounded in granular material	Pipes in trenches	49	m	49.09	2,405.18	2.58	
Excavate trench 1000 - 1250 deep; lay 100 diameter vitrified clay pipe and fittings bedded and surrounded in granular material	Pipes in trenches	4	m	50.47	201.86	0.22	
Excavate trench 250 - 500 deep; lay 100 diameter vitrified clay pipe and fittings bedded and surrounded in in-situ concrete	Pipes in trenches	8	m	53.75	429.96	0.46	
Excavate trench 1000 - 1250 deep; lay 150 diameter vitrified clay pipe and fittings bedded and surrounded in in-situ concrete	Pipes in trenches	23	m	77.05	1,772.22	1.90	
Excavate trench 1500 - 1750 deep; lay 150 diameter vitrified clay pipe and fittings bedded and surrounded in in-situ concrete	Pipes in trenches	10	m	86.25	862.46	0.93	
100 diameter vertical vitrified clay pipe and fittings cased in in-situ concrete	Pipes in trenches	3	m	110.98	332.93	0.36	
100 outlet vitrified clay yard gully with cast iron grating, surrounded in in-situ concrete	Gully	1	nr	118.50	118.50	0.13	
100 outlet vitrified clay road gully with brick kerb and cast iron grating, surrounded in in-situ concrete	Gully	7	nr	204.47	1,431.29	1.54	
Inspection chambers 600 x 900 x 600 deep internally comprising excavation works, in-situ concrete bases, suspended slabs and benchings, engineering brick walls and kerbs, clay channels, galvanised iron step irons and grade A access covers and frames	Insp. chamber	4	nr	826.20	3,304.80	3.55	
Inspection chambers 600 x 900 x 600 deep internally comprising excavation works, in-situ concrete bases, suspended slabs and benchings, engineering brick walls and kerbs, clay channels, galvanised iron step irons and grade B access covers and frames	Insp. chamber	3	nr	827.16	2,481.47	2.66	17,996.12
					17,996.12	19.31	

COST PLAN FOR AN OFFICE DEVELOPMENT (Cont'd)

Description	Item	Qty		Rate	Item Total	Cost per m2	Total
6.0 EXTERNAL WORKS (Cont'd)							
6.3 External Services							
Excavate trench for 4nr ducts; lay 4nr 100 diameter UPVC ducts and fittings; lay warning tape; backfill trench		20	m	53.83	1,076.64	1.16	
Sewer connection charges		1	nr	5,231.48	5,231.48	5.61	
Gas main and meter charges		1	nr	4,185.18	4,185.18	4.49	
Electricity main and meter charges		1	nr	4,185.18	4,185.18	4.49	
Water main and meter charges		1	nr	3,138.89	3,138.89	3.37	
					17,817.37	19.12	17,817.37
Preliminaries							
Site staff, accommodation, lighting, water safety health and welfare, rubbish disposal, cleaning, drying, protection, security, plant, transport, temporary works and scaffolding with a contract period of 38 weeks		item			86,200.00	92.49	86,200.00
				Total			£ 1,167,027.05
				Cost per m2		**£ 1,252.17**	

COST PLAN FOR AN OFFICE DEVELOPMENT (Cont'd)

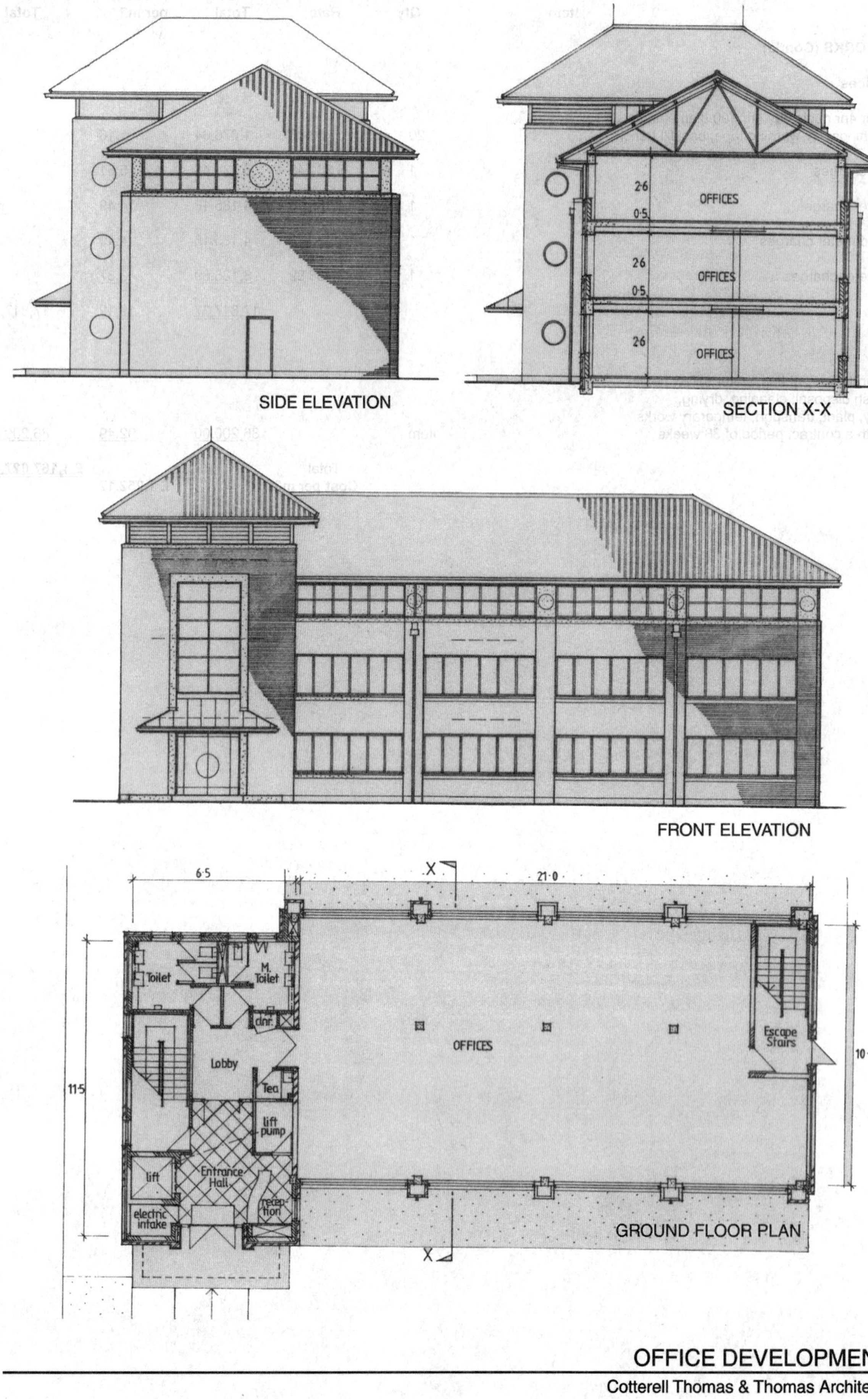

OFFICE DEVELOPMENT

Cotterell Thomas & Thomas Architects

COST PLAN FOR A DETACHED HOUSE

The Cost Plan that follows gives an example of an Approximate Estimate for a detached house with garage and porch as the attached sketch drawings. The estimate can be amended by using the examples of Composite Prices following or the prices included within the preceding pages. Details of the specification used are included within the elements. The cost plan is based on a fully measured bill of quantities priced using the detailed rates within this current edition of Laxton's. Preliminaries are shown separately

COST PLAN FOR A DETACHED HOUSE

Floor area

House (Including porch 5 m2)	120	m2
Garage	22	m2
Total	142	m2

Description	Item	Qty		Rate	Item Total	Cost per m2	Total
1.0 SUBSTRUCTURE							
Note: Rates allow for excavation by machine							
1.1 Substructure (86 m2)							
Clear site vegetation excavate 150 topsoil and remove spoil.	Site clearance	113	m2	3.94	445.16	3.13	
Excavate trench commencing at reduced level, average 800 deep, concrete to within 150 of ground level, construct cavity brick wall 300 high and lay pitch polymer damp proof course	Strip foundation	33	m	153.09	5,052.11	35.58	
Excavate trench commencing at reduced level, average 800mm deep, concrete to within 150 of ground level, construct 103 brick wall 300 high and lay pitch polymer damp proof course	Strip foundation	26	m	118.19	3,072.99	21.64	
Excavate trench commencing at reduced level, average 800mm deep, concrete to within 150 of ground level, construct 215 brick wall 300 high and lay pitch polymer damp proof course.	Strip foundation	2	m	268.66	537.31	3.78	
Excavate to reduce levels average 150 deep, lay 150mm bed of hardcore blinded, dpm and 150 concrete ground slab reinforced with Ref. A193 Fabric reinforcement.	Ground floor slab	85	m2	65.53	5,570.24	39.23	14,677.81
					14,677.81	103.36	
(Element unit rate = £ 170.67 m2)							
2.0 SUPERSTRUCTURE							
2.2 Upper Floors (55 m2)							
175 deep floor joists with 22 chipboard finish	Softwood floor	49	m2	43.08	2,110.96	14.87	
175 deep floor joists with 22 plywood finish	Softwood floor	6	m2	72.16	432.99	3.05	2,543.94
					2,543.94	17.92	
(Element unit rate = £ 46.25 m2)							
2.3 Roof (110 m2 on plan)							
2.3.1 Roof structure							
35 degree pitched roof with 50 x 100 rafters and ceiling joists, 100 glass fibre insulation.	Pitched roof	110	m2	62.49	6,873.71	48.41	
Gable end 275 cavity brick and blockwork facings PC £250/1000	Gable wall	29	m2	119.40	3,462.49	24.38	
Gable end 102 brick wall facings £250/1000	Gable wall	8	m2	87.45	699.61	4.93	

COST PLAN FOR A DETACHED HOUSE (Cont'd)

Description	Item	Qty	Rate	Item Total	Cost per m2	Total
2.0 SUPERSTRUCTURE (Cont'd)						
2.3 Roof (110 m2 on plan) (Cont'd)						
2.3.2 Roof coverings						
Interlocking concrete tiles, felt underlay, battens.	Roof tiling	143 m2	22.68	3,243.24	22.84	
Verge, 150 under cloak.	Verge	38 m	13.55	514.90	3.63	
Ridge, half round	Ridge	14 m	18.10	253.40	1.78	
Valley trough tiles.	Valley	2 m	40.59	81.18	0.57	
Abutments, lead flashings cavity trays.	Abutment	11 m	117.73	1,295.05	9.12	
Sheet lead cladding.	Lead roof	2 m2	129.85	259.70	1.83	
Eaves soffit, fascia and 100 PVC gutter	Eaves	29 m	63.76	1,849.13	13.02	
2.3.3 Roof Drainage						
68 PVCU rainwater pipe.	Rainwater pipe	24 m	28.48	683.55	4.81	19,215.96
				19,215.96	135.32	
(Element unit rate = £ 174.69 m2)						
2.4 Stairs (1nr)						
Softwood staircase 2670 rise, balustrade one side, 910 wide two flights with half landing	Stairs	1 nr	1,537.72	1,537.72	10.83	1,537.72
				1,537.72	10.83	
2.5 External Walls (76 m2)						
Hollow wall, one skin light weight blocks, one skin facings PC £250/1000, ties.	Hollow wall	127 m2	124.66	15,832.27	111.49	
Facing brick wall PC £250/1000 102 thick.	1/2B Facing wall	25 m2	85.56	2,138.90	15.06	
Facing brick wall PC £250/1000 215 thick.	1B Facing wall	3 m2	149.02	447.06	3.15	18,418.23
				18,418.23	129.71	
(Element unit rate = £ 121.17 m2)						
2.6 Windows and External Doors (35 m2)						
2.6.1 Windows (16 m2)						
Hardwood double glazed casement windows.	HW windows	16 m2	471.36	7,541.69	53.11	
Galvanised steel lintel, dpc, brick soldier course.	Lintels	16 m	53.51	856.11	6.03	
Softwood window board, dpc close cavity.	Window board	15 m	11.38	170.70	1.20	
Close cavity at jamb, dpc facings to reveal.	Jamb	32 m	16.95	542.30	3.82	
2.6.2 External Doors (19 m2)						
Hardwood glazed doors, frames, cills, ironmongery	Hardwood doors	6 m2	627.64	3,765.84	26.52	
Hardwood patio doors.	Patio doors	9 m2	430.45	3,874.04	27.28	
Galvanised steel garage door.	Garage door	4 m2	182.38	729.53	5.14	
Galvanised steel lintel, dpc, brick soldier course.	Lintels	12 m	54.94	659.34	4.64	
Close cavity at jamb, dpc, facings to reveal.	Jamb	4 m	16.93	67.72	0.48	18,207.27
				18,207.27	128.22	
(Element unit rate = £ 520.21 m2)						

COST PLAN FOR A DETACHED HOUSE (Cont'd)

Description	Item	Qty		Rate	Item Total	Cost per m2	Total
2.0 SUPERSTRUCTURE (Cont'd)							
2.7 Internal Walls and Partitions (153 m2)							
Lightweight concrete block 100 thick.	Block walls	104	m2	41.64	4,330.11	30.49	
75 timber stud partition plasterboard and skim both sides.	Stud partitions	53	m2	71.73	3,801.69	26.77	8,131.80
					8,131.80	57.27	
(Element unit rate = £ 53.15 m2)							
2.8 Internal doors (40 m2)							
12 hardboard embossed panel door, linings ironmongery PC £20.00	Flush doors	18	m2	291.20	5,241.61	36.91	
3 softwood "GG glazed door, linings ironmongery PC £20.00	Glazed doors	6	m2	306.15	1,836.92	12.94	
6 Plywood faced door, linings, ironmongery PC £20.00	Flush doors	12	m2	267.81	3,213.68	22.63	
1 plywood faced half hour fire check flush door, frame, ironmongery PC £60.00	Fire doors	2	m2	218.22	436.43	3.07	
Plywood ceiling hatch	Hatch	1	nr	88.76	88.76	0.63	
Precast concrete lintels.	Lintels	10	m	21.63	216.25	1.52	11,033.65
					11,033.65	77.70	
(Element unit rate = £ 275.84 m2)							
3.0 INTERNAL FINISHES							
3.1 Wall finishes (264 M2)							
13 light weight plaster.	Plaster work	264	m2	20.01	5,282.27	37.20	
152 x 152 wall tiling coloured PC £20.00 supply	Tiling	13	m2	43.74	568.62	4.00	5,850.89
					5,850.89	41.20	
(Element unit rate = £ 22.16 m2)							
3.2 Floor Finishes (63 M2)							
22 Chipboard, 50 insulation.	Board flooring	63	m2	31.62	1,992.06	14.03	
3 PVC sheet flooring.	Sheet vinyl	18	m2	22.71	408.78	2.88	
25 x 125 softwood skirting.	Skirting	131	m	12.86	1,684.66	11.86	4,085.50
					4,085.50	28.77	
(Element unit rate = £ 64.85 m2)							
3.3 Ceiling Finishes (115 M2)							
9.5 Plasterboard and set.	Plasterboard	109	m2	18.02	1,964.18	13.83	
Plywood finish	Plywood	6	m2	51.12	306.72	2.16	2,270.90
					2,270.90	15.99	
(Element unit rate = £ 19.75 m2)							
4.0 FITTINGS AND FURNITURE							
4.1 Fittings							
Kitchen units :-							
600 base unit		2	nr	231.56	463.12	3.26	
1000 base unit		6	nr	371.28	2,227.68	15.69	
500 broom cupboard		1	nr	318.91	318.91	2.25	
600 wall unit		3	nr	136.07	408.21	2.87	
1000 wall unit		2	nr	202.66	405.32	2.85	
Work tops		10	m	63.97	639.70	4.50	
Gas fire surround		1	nr	618.75	618.75	4.36	5,081.69
					5,081.69	35.79	

COST PLAN FOR A DETACHED HOUSE (Cont'd)

Description	Item	Qty		Rate	Item Total	Cost per m2	Total
5.0 SERVICES							
5.1 Sanitary Appliances							
Lavatory basin with pedestal.		3	nr	171.69	515.07	3.63	
Bath with side panel.		1	nr	297.74	297.74	2.10	
WC suite.		3	nr	286.44	859.32	6.05	
Stainless steel sink and drainer. (500)		1	nr	165.37	165.37	1.16	
Stainless steel sink and drainer. (600)		1	nr	182.65	182.65	1.29	
shower.		1	nr	775.98	775.98	5.46	
5.2 Services Equipment							
Electric oven built in.		1	nr	794.03	794.03	5.59	
Gas hob built in.		1	nr	387.48	387.48	2.73	
Kitchen extract unit		1	nr	317.03	317.03	2.23	
Gas fire with balanced flue.		1	nr	397.01	397.01	2.80	
5.3 Disposal Installations							
Upvc soil and vent pipe.		15	m	102.19	1,532.87	10.79	
Automatic air admittance valve.		3	nr	83.73	251.19	1.77	
5.4 Water Installations							
Cold water, tanks, 9 draw off points		item		1,064.90	1,064.90	7.50	
Hot water cylinder, 7 draw off points		item		1,474.23	1,474.23	10.38	
5.5 Heat Source							
Boiler, pump, and controls.		item		1,012.50	1,012.50	7.13	
5.6 Space heating							
Low temperature hot water radiator system.		item		4,387.50	4,387.50	30.90	
5.7 Ventilating Systems							
Extract fan units.		3	nr	138.04	414.12	2.92	
5.8 Electrical Installations							
Electrical Installation comprising: 16 Nr lighting points, 30 Nr outlet points, cooker points, immersion heater point and 9 Nr fittings points to include smoke alarms, bells, telephone and aerial installation.		item		4,905.56	4,905.56	34.55	
5.9 Gas installation							
Installation for boiler, hob and fire		item		587.09	587.09	4.13	
5.10 Builders work							
Building work for service installation.		item		1,631.50	1,631.50	11.49	21,953.14
					21,953.14	154.60	
(Element unit rate = £ 154.60 m2)							
6.0 EXTERNAL WORKS							
6.1 Site work							
Clear site vegetation excavate 150 topsoil and remove spoil.	Site clearance	479	m2	3.94	1,887.02	13.29	
Excavate and lay 50 precast concrete flag paving on hardcore.	Paving	120	m2	61.98	7,437.80	52.38	
Brick wall, excavation 600 deep, concrete foundation, 215 brickwork facing. PC £200/1000 damp proof course brick on edge coping 900 high	Garden wall	13	m	144.93	1,884.10	13.27	
Plastic coated chain link fencing 900 high on angle iron posts	Fencing	92	m	48.39	4,451.98	31.35	
Turfed Areas	Landscaping	103	m2	38.78	3,993.88	28.13	

COST PLAN FOR A DETACHED HOUSE (Cont'd)

Description	Item	Qty		Rate	Item Total	Cost per m2	Total
6.0 EXTERNAL WORKS (Cont'd)							
6.2 Drainage							
100 diameter clay pipes, excavation, 150 bed and surround, pea shingle, backfill 750 deep.		85	m	59.45	5,053.01	35.58	
Polypropylene inspection chambers; universal inspection chamber; polypropylene; 475 diameter; preformed chamber base benching; for 100mm pipe; chamber shaft, 175mm effective length; cast iron cover and frame; single seal A15 light duty to suit.		6	nr	260.37	1,562.22	11.00	
6.3 External Services							
100 diameter clay duct, excavation backfill, polythene warning tape, 500 deep.		38	m	21.16	804.24	5.66	
Provisional sums for External Services Gas, electric, water, telephone.		item		4,000.00	4,000.00	28.17	31,074.25
					31,074.25	218.83	
Preliminaries							
Site staff, accommodation, lighting, water, safety and health and welfare, rubbish disposal cleaning, drying, protection, security, plant, transport, temporary works and scaffolding with a contract period of 28 weeks					13,000.00	91.55	13,000.00
				Total			**£ 177,082.75**
				Cost per m2		**£ 1,247.06**	

COST PLAN FOR A DETACHED HOUSE (cont'd)

REAR ELEVATION

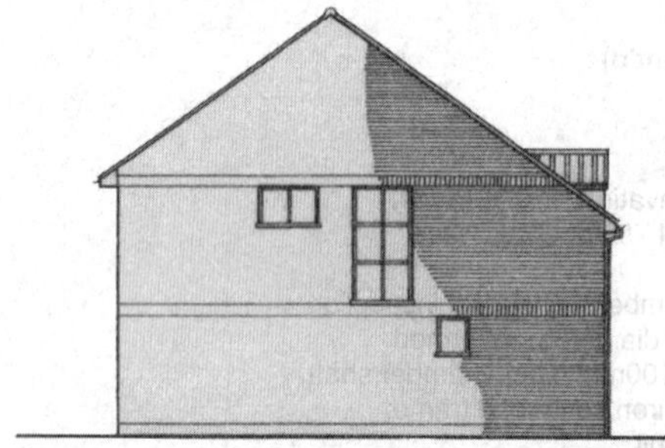

SIDE ELEVATION

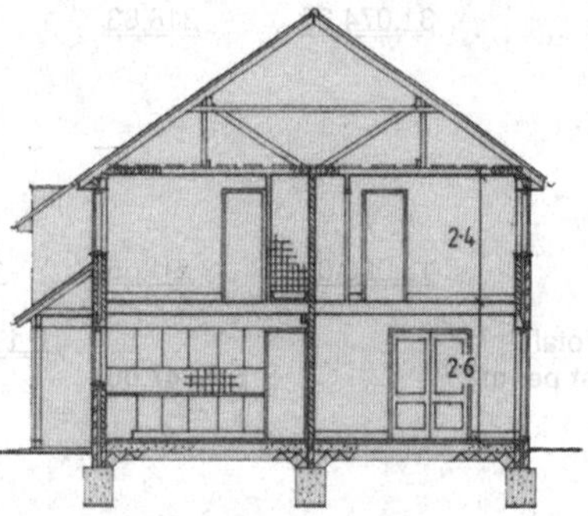

SECTION X-X

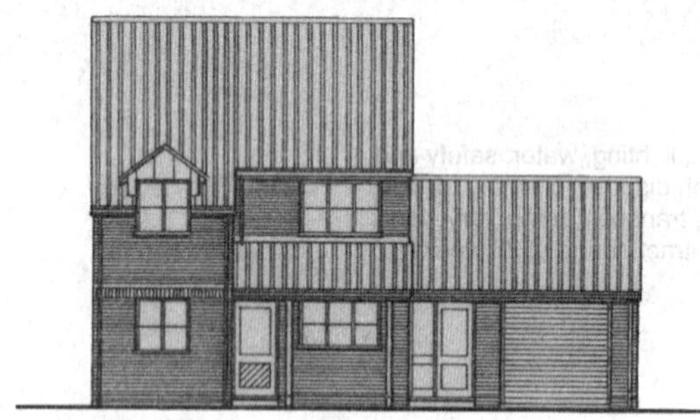

FRONT ELEVATION

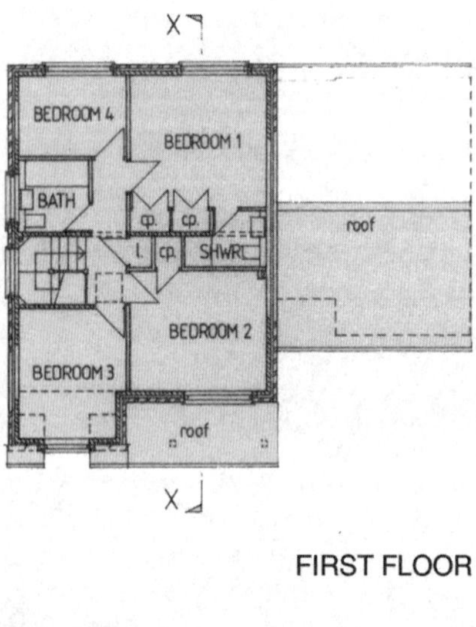

FIRST FLOOR PLAN

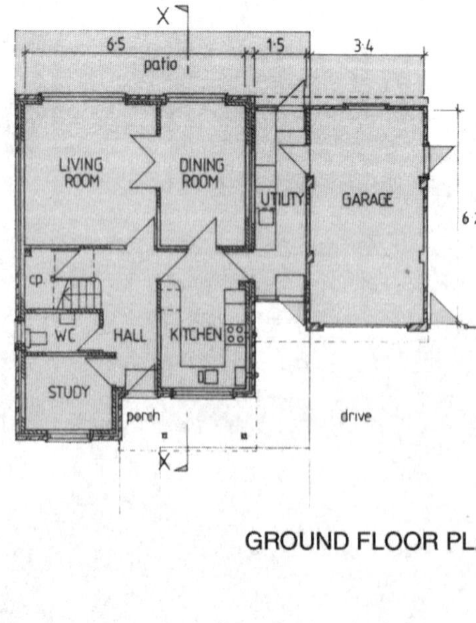

GROUND FLOOR PLAN

4 - BEDROOM DETACHED HOUSE

Cotterell Thomas & Thomas Architects

COMPOSITE PRICES

The following composite prices are based on the detailed rates herein and may be used to amend the previous example cost plans or to create new approximate budget estimates.
The composite prices can be further extended and amended by reference to the various sections
Allowances should be made for location factors, tender values and price changes.

PRELIMINARIES

The following preliminary items need to be added to the composite prices that follow.
A calculation can be made by reference to the preliminaries section.

		£
Typical allowances (small works)		
Management and staff - average each	week	520.00
Site accommodation	m²	26.00
Power and lighting	week	13.00
Heating	week	18.00
Water	item	140.00
Telephone	week	20.00
Safety, health and welfare	week	12.00
Rubbish disposal and cleaning site	week	30.00
Final clean	item	200.00
Drying out building	item	100.00
Security	week	30.00
Small plant and tools	%	1.50
Insurance of works	%	0.30
Travelling time and fares	week	40.00
Miscellaneous		
Rates on temporary buildings	week	10.00
Defects	item	730.00
Mechanical plant		
Hoist	item	340.00
Pumping	item	140.00
Transport	item	500.00
Temporary access	item	175.00
Scaffolding		
Mobile tower	item	720.00

DEMOLITIONS AND ALTERATIONS

Demolition		
Demolish timber outbuilding size 2.50 x 2.00 x 3.00 high	nr	211.86
Demolish brick lean-to and make good size 1.50 x 1.00 x 2.00 high	nr	267.93
Demolishing parts of structures		
Substructures		
incl. Excavation and removal by hand		
break out brickwork	m³	169.74
break out concrete	m³	209.50
break out reinforced concrete	m³	275.76
break up concrete bed		
100 thick	m²	20.95
150 thick	m²	31.42
200 thick	m²	41.90
break up reinforced concrete bed		
100 thick	m²	27.58
150 thick	m²	41.36
200 thick	m²	51.30
Walls		
brick internal		
102 thick	m²	16.07
215 thick	m²	33.12
327 thick	m²	50.73
brick external		
102 thick	m²	12.22
250 thick cavity wall	m²	25.41
Floors		
concrete reinforced; screed		
150 thick; 50 screed	m²	54.85
200 thick; 50 screed	m²	68.63
timber boarded on joists		
225 x 50 joists; floor boards	m²	12.71
Roofs		
timber flat roofs		
150 x 50 joists, 25 boards; felt	m²	13.37
timber pitched roof		
100 x 50 rafters; joists; plain tile	m²	16.54
100 x 50 rafters; joists; interlocking tiles	m²	14.38
100 x 50 rafters; joists; slates	m²	16.45
Alterations		
Includes make good finishes		
Cut opening through reinforced concrete floor 150 thick x 900 x 2134 for staircase	nr	908.87
Cut opening through block wall 100 thick x 838 x 2032 for doorway	nr	568.72
Cut opening through brick wall		
215 thick x 914 x 2082 for doorway	nr	831.82
250 thick cavity wall x 900 x 900 for window	nr	538.39
Cut opening through stud partition 100 thick x 838 x 2032 for doorway	nr	399.99
Fill opening in block wall 100 thick x 838 x 2032; door removed	nr	401.04
Fill opening in brick wall		
102 thick x 838 x 2032; door removed	nr	476.69
215 thick x 914 x 2082; external door removed	nr	766.68
250 thick cavity wall x 914 x 2082; external door removed	nr	715.74
Fill opening in stud partition 100 thick x 838 x 2032; door removed	nr	360.34
Removal		
Remove window		
1486 x 923 metal	nr	22.48
1225 x 1070 wood casement; frame	nr	25.03
915 x 1525 cased frame; sashes	nr	37.34
Remove door and frame		
single internal	nr	18.31
single external	nr	20.39
Remove roof coverings		
3 layer felt	m²	4.31
asphalt	m²	5.39
slates	m²	12.12
tiles	m²	12.21

COMPOSITE PRICES (Cont'd)

SUBSTRUCTURE

Note: Rates allow for excavation by machine.

Strip Foundations

Excavating, 225x450 concrete, brickwork, dpc, common bricks			
800	deep for 103 wall	m	71.43
1000	deep for 103 wall	m	86.92
1200	deep for 103 wall	m	109.55
800	deep for 215 wall	m	107.54
1000	deep for 215 wall	m	132.65
1200	deep for 215 wall	m	157.61
800	deep for 275 wall	m	127.40
1000	deep for 275 wall	m	159.61
1200	deep for 275 wall	m	190.03
Extra 100mm depth concrete foundation; excavation; concrete		m	7.91
Extra 100mm depth 103 brick wall in foundation; excavation; brickwork		m	7.84
Extra 100mm depth 215 brick wall in foundation; excavation; brickwork		m	12.57
Extra 100mm depth 275 brick wall in foundation; excavation; brickwork		m	14.11
Extra for four courses facing bricks P.C. £250.00 per 1000		m	3.49
Excavating, 225x600 concrete, brickwork, dpc, common bricks			
800	deep for 275 wall	m	133.39
1000	deep for 275 wall	m	166.16
1200	deep for 275 wall	m	193.86
Extra 100mm depth 275 brick wall in foundation; excavation; brickwork		m	14.29
Extra for four courses facing bricks P.C. £250.00 per 1000		m	3.49
Excavating, 225x450 concrete, brickwork, dpc, Engineering bricks Class B			
800	deep for 215 wall	m	113.68
1000	deep for 215 wall	m	140.47
1200	deep for 215 wall	m	167.12
800	deep for 275 wall	m	134.51
1000	deep for 275 wall	m	168.68
1200	deep for 275 wall	m	201.05
Extra 100mm depth concrete foundation; excavation; concrete		m	7.91
Extra 100mm depth 103 brick wall in foundation; excavation; brickwork		m	8.33
Extra 100mm depth 215 brick wall in foundation; excavation; brickwork		m	13.41
Extra 100mm depth 275 brick wall in foundation; excavation; brickwork		m	15.09
Extra for four courses facing bricks P.C. £250.00 per 1000		m	2.02
Excavating, 225x450 concrete, blockwork, dpc, Concrete blocks			
800	deep for 100 wall	m	57.77
1000	deep for 100 wall	m	64.73
1200	deep for 100 wall	m	75.45
Extra 100mm depth concrete foundation; excavation; concrete		m	7.91
Extra 100mm depth 100 block wall in foundation; excavation; blockwork		m	5.45
800	deep for 140 wall	m	61.01
1000	deep for 140 wall	m	73.55
1200	deep for 140 wall	m	86.10
Extra 100mm depth concrete foundation; excavation; concrete		m	7.91
Extra 100mm depth 140 block wall in foundation; excavation; blockwork		m	6.35
800	deep for 190 wall	m	67.59
1000	deep for 190 wall	m	81.76
1200	deep for 190 wall	m	95.94
Extra 100mm depth concrete foundation; excavation; concrete		m	7.91
Extra 100mm depth 190 block wall in foundation; excavation; blockwork		m	7.16
Excavating, 225 x 450 concrete, blockwork, dpc, Thermalite Trenchblocks			
800	deep for 275 wall	m	86.51
1000	deep for 275 wall	m	105.78
1200	deep for 275 wall	m	125.04
Extra 100mm depth concrete foundation; excavation; concrete		m	7.91
Extra 100mm depth 190 block wall in foundation; excavation; blockwork		m	9.77
Excavating, 225x600 concrete, blockwork, dpc, Thermalite Trenchblocks			
800	deep for 275 wall	m	93.20
1000	deep for 275 wall	m	113.02
1200	deep for 275 wall	m	132.69
Extra 100mm depth concrete foundation; excavation; concrete;		m	7.91
Extra 100mm depth 275 brick wall in foundation; excavation; blockwork		m	9.77

Fabric reinforcement ref. A193 (no allowance for working space or formwork)		
350 wide	m	1.25
600 wide	m	2.15
750 wide	m	3.17

Trench Fill Foundations

Excavating 450 wide concrete fill to 150 of ground level, brickwork, dpc			
Common bricks			
800	deep for 102 wall	m	70.56
1000	deep for 102 wall	m	84.95
1200	deep for 102 wall	m	99.34
800	deep for 215 wall	m	84.31
1000	deep for 215 wall	m	98.70
1200	deep for 215 wall	m	113.09
800	deep for 275 wall	m	91.71
1000	deep for 275 wall	m	106.10
1200	deep for 275 wall	m	120.49
Extra 100mm depth concrete foundation; excavation; concrete;		m	7.91
Extra for four courses facing bricks PC £250 per 1000		m	3.49

Column bases

Excavating, 450 concrete		
1000 x 1000 x 1000 deep	nr	89.99
1000 x 1000 x 1200 deep	nr	102.65
1000 x 1000 x 1500 deep	nr	114.25
1200 x 1200 x 1000 deep	nr	125.93
1200 x 1200 x 1200 deep	nr	301.87
1200 x 1200 x 1500 deep	nr	157.70
1500 x 1500 x 1000 deep	nr	191.06
1500 x 1500 x 1200 deep	nr	213.90
1500 x 1500 x 1500 deep	nr	235.74
2000 x 2000 x 1000 deep	nr	329.49
2000 x 2000 x 1200 deep	nr	365.09
2000 x 2000 x 1500 deep	nr	400.13
Excavating, 750 concrete		
1000 x 1000 x 1000 deep	nr	127.32
1000 x 1000 x 1200 deep	nr	139.98
1000 x 1000 x 1500 deep	nr	151.58
1000 x 1000 x 2000 deep	nr	170.92
1200 x 1200 x 1000 deep	nr	179.69
1200 x 1200 x 1200 deep	nr	196.11
1200 x 1200 x 1500 deep	nr	211.45
1200 x 1200 x 2000 deep	nr	237.03
1500 x 1500 x 1000 deep	nr	275.05
1500 x 1500 x 1200 deep	nr	297.89
1500 x 1500 x 1500 deep	nr	319.73
1500 x 1500 x 2000 deep	nr	356.14
2000 x 2000 x 1000 deep	nr	478.81
2000 x 2000 x 1200 deep	nr	514.41
2000 x 2000 x 1500 deep	nr	549.45
2000 x 2000 x 2000 deep	nr	607.85
3000 x 3000 x 1000 deep	nr	1,054.46
3000 x 3000 x 1200 deep	nr	1,123.29
3000 x 3000 x 1500 deep	nr	1,193.60

COMPOSITE PRICES (Cont'd)

SUBSTRUCTURE (Cont'd)

Column Bases (Cont'd)

Excavating, 1000 concrete		
1000 x 1000 x 1000 deep	nr	158.43
1000 x 1000 x 1200 deep	nr	171.09
1000 x 1000 x 1500 deep	nr	182.69
1000 x 1000 x 2000 deep	nr	202.03
1200 x 1200 x 1000 deep	nr	224.48
1200 x 1200 x 1200 deep	nr	240.90
1200 x 1200 x 1500 deep	nr	256.25
1200 x 1200 x 2000 deep	nr	281.82
1500 x 1500 x 1000 deep	nr	345.04
1500 x 1500 x 1200 deep	nr	367.88
1500 x 1500 x 1500 deep	nr	389.72
1500 x 1500 x 2000 deep	nr	426.13
2000 x 2000 x 1000 deep	nr	603.24
2000 x 2000 x 1200 deep	nr	638.84
2000 x 2000 x 1500 deep	nr	673.88
2000 x 2000 x 2000 deep	nr	732.28
3000 x 3000 x 1000 deep	nr	1,334.43
3000 x 3000 x 1200 deep	nr	1,403.26
3000 x 3000 x 1500 deep	nr	1,473.57
3000 x 3000 x 2000 deep	nr	1,590.75
Excavating, 1500 concrete		
1000 x 1000 x 1500 deep	nr	244.91
1000 x 1000 x 2000 deep	nr	264.25
1500 x 1500 x 1500 deep	nr	529.71
1500 x 1500 x 2000 deep	nr	566.11
2000 x 2000 x 1500 deep	nr	922.74
2000 x 2000 x 2000 deep	nr	981.14
3000 x 3000 x 1500 deep	nr	2,033.51
3000 x 3000 x 2000 deep	nr	2,150.69
Excavating, 450 reinforced concrete		
1000 x 1000 x 1000 deep	nr	266.63
1000 x 1000 x 1200 deep	nr	290.46
1000 x 1000 x 1500 deep	nr	317.22
1000 x 1000 x 2000 deep	nr	361.82
1200 x 1200 x 1000 deep	nr	360.00
1200 x 1200 x 1200 deep	nr	388.55
1200 x 1200 x 1500 deep	nr	422.08
1200 x 1200 x 2000 deep	nr	477.97
Excavating, 750 reinforced concrete		
1000 x 1000 x 1000 deep	nr	447.94
1000 x 1000 x 1200 deep	nr	470.70
1000 x 1000 x 1500 deep	nr	497.46
1000 x 1000 x 2000 deep	nr	542.06
1200 x 1200 x 1000 deep	nr	599.11
1200 x 1200 x 1200 deep	nr	627.66
1200 x 1200 x 1500 deep	nr	661.19
1200 x 1200 x 2000 deep	nr	717.08

Ground beams

Excavation, blinding, reinforced concrete		
300 x 450	m	86.61
300 x 600	m	135.56
300 x 450	m	117.89
300 x 600	m	175.76

Ground slabs

Excavation, 150 hardcore, blinding, concrete		
150 ground slab	m²	38.04
200 ground slab	m²	42.96
225 ground slab	m²	46.41
Extra: 50 blinding	m²	7.53
Extra: Fabric reinforcement A142	m²	3.08
Extra: Fabric reinforcement A193	m²	3.58
Slab thickening 450 wide x 150 thick	m²	9.32
Damp proof membrane		
250mu Polythene sheeting	m²	1.65
2 coats Bitumen emulsion	m²	15.62
Extra: 50mm Expanded polystyrene insulation	m²	7.05
Extra: 50mm Concrete thickness	m²	5.87
Extra: 100mm Reduced level excavation	m²	2.06
Extra: 100mm hardcore filling	m²	2.98

Beam and block floors

(excludes excavation, blinding etc.)		
155mm beam floor span 3.9m	m²	29.12
Extra: 150mm Reduced level excavation; 50 concrete blinding	m²	9.21

Piling

On/off site charge	nr	4,616.05
450 bored cast piles n/e 6m deep	m	27.04
610 bored cast piles n/e 6m deep	m	40.88
Cutting off tops of piles	nr	9.25

INSIDE EXISTING BUILDING

Note: rates allow for excavation by hand.

Strip foundations

Excavating, 225 concrete, brickwork, dpc		
Common bricks		
800 deep for 103 wall	m	92.50
1000 deep for 103 wall	m	112.47
1200 deep for 103 wall	m	156.92
800 deep for 215 wall	m	130.34
1000 deep for 215 wall	m	160.85
1200 deep for 215 wall	m	198.70
Extra 100mm depth concrete foundation; excavation; concrete	m	11.57

Trench Fill Foundations

Excavating 450 wide concrete fill to 150 of ground level, brickwork, dpc		
Common bricks		
800 deep for 102 wall	m	96.45
1000 deep for 102 wall	m	117.44
800 deep for 215 wall	m	110.20
1000 deep for 215 wall	m	112.34
Extra 100mm depth concrete foundation; excavation; concrete	m	11.57

Ground slabs

Excavation, 150 hardcore, blinding, concrete		
150 ground slab	m²	61.03
200 ground slab	m²	71.84
Slab thickening 450 wide x 150 thick	m	14.59
Damp proof membrane		
250 mu polythene sheeting	m²	1.65
2 coats bitumen emulsion	m²	15.62

FRAME £

Reinforced concrete

200 x 200 column	m	59.14
225 x 225 column	m	66.53
300 x 300 column	m	97.04
200 x 200 attached beam	m	60.43
225 x 250 attached beam	m	74.80
300 x 450 attached beam	m	134.14
400 x 650 attached beam	m	212.45
Casing to 152 x 152 steel column	m	49.17
Casing to 203 x 203 steel column	m	63.47
Casing to 146 x 254 steel beam	m	63.07
Casing to 165 x 305 steel beam	m	74.08
Extra: wrought formwork	m²	4.70

Steel

Columns and beams	tonne	1,558.41
Roof trusses	tonne	2,205.29
Purlins and cladding rails	tonne	1,575.29
Small sections	tonne	1,158.75
Two coats primer at works	tonne	211.68
Prime, undercoat and gloss paint	m²	8.15

APPROXIMATE ESTIMATING

COMPOSITE PRICES (Cont'd)

FRAME (Cont'd)

Glued laminated softwood beams

65x150x4000 long	nr	86.88
65x250x4000 long	nr	148.97
65x300x6000 long	nr	253.52
65x325x8000 long	nr	359.75
90x200x4000 long	nr	150.93
90x325x8000 long	nr	453.00
90x400x10000 long	nr	686.98
90x450x12000 long	nr	916.61
115x250x4000 long	nr	236.26
115x300x4000 long	nr	281.38
115x350x8000 long	nr	621.80
115x425x12000 long	nr	1,074.89
115x450x8000 long	nr	778.62
115x500x14000 long	nr	1,478.58
140x350x8000 long	nr	752.27
140x450x12000 long	nr	1,401.58
165x450x12000 long	nr	1,633.90
190x475x14000 long	nr	2,311.27

UPPER FLOORS

Concrete

In-situ reinforced concrete		
150 thick	m^2	76.54
200 thick	m^2	82.63
225 thick	m^2	90.26
In-situ reinforced concrete with permanent Expamet Hy-rib formwork ref 2411 and A142 mesh reinforcement		
100 thick	m^2	77.00
125 thick	m^2	80.99
150 thick	m^2	84.94
Hollow prestressed concrete		
125 thick	m^2	44.67
150 thick	m^2	48.90
200 thick	m^2	53.35
Beam and block		
200 thick	m^2	29.12

Timber

Softwood joist		
100 thick	m^2	18.86
150 thick	m^2	22.46
175 thick	m^2	24.13
200 thick	m^2	26.25
225 thick	m^2	28.72
250 thick	m^2	31.59
Floor boarding		
19 square edge softwood	m^2	23.78
25 square edge softwood	m^2	25.22
19 tongued and grooved softwood	m^2	26.90
25 tongued and grooved softwood	m^2	28.01
18 tongued and grooved chipboard	m^2	9.38
22 tongued and grooved chipboard	m^2	10.15
15 tongued and grooved plywood Douglas Fir unsanded	m^2	14.41
18 tongued and grooved plywood Douglas Fir unsanded	m^2	16.78
15 tongued and grooved plywood Birch Faced BB quality	m^2	36.00
18 tongued and grooved plywood Birch Faced BB quality	m^2	42.52

ROOF

Roof structure

Reinforced in-situ concrete		
100 thick	m^2	65.30
150 thick	m^2	74.48
200 thick	m^2	75.24
50 lightweight screed to falls	m^2	26.53
50 cement sand screed to falls	m^2	16.59
50 vermiculite screed to falls	m^2	24.51
50 flat glass fibre insulation board	m^2	10.92
Extra for vapour barrier	m^2	4.07
Softwood flat roof joists		
50 x 100 joists @ 400 centres	m^2	23.95
50 x 175 joists @ 400 centres	m^2	26.94
50 x 200 joists @ 400 centres	m^2	29.39
50 x 225 joists @ 400 centres	m^2	31.97
50 x 100 joists @ 600 centres	m^2	17.31
50 x 175 joists @ 600 centres	m^2	19.60
50 x 200 joists @ 600 centres	m^2	20.88
50 x 225 joists @ 600 centres	m^2	22.62
Boarding		
18 chipboard	m^2	10.07
25 softwood tongued grooved	m^2	16.78
18 plywood decking	m^2	20.89
24 plywood decking	m^2	27.22
Pitched roof with 50 x 100 rafters and ceiling joists at 400 centres	m^2	43.00
Pitched roof with trussed rafters at 600 centres	m2	26.10

Roof coverings

Three layer fibre based felt	m^2	21.63
Three layer high performance felt	m^2	55.07
20 two coat mastic asphalt	m^2	27.88
50 lightweight screed to falls	m^2	26.53
50 cement sand screed to falls	m^2	16.59
60 Vermiculite screed to falls	m^2	29.33
50 flat glass fibre insulation board	m^2	10.92
Extra for vapour barrier	m^2	4.07
Profile 3 fibre cement sheeting (on slope)	m^2	24.28
Profile 6 fibre cement sheeting (on slope)	m^2	21.07
Colour coated galvanised steel sheeting (on slope)	m^2	30.75
381 x 227 concrete interlocking tiles (on slope)	m^2	28.36
268 x 165 plain concrete tiles (on slope)	m^2	58.59
268 x 165 machine made clay tiles (on slope)	m^2	59.88
265 x 165 hand made tiles (on slope)	m^2	77.42
400 x 200 fibre cement slates (on slope)	m^2	46.52
500 x 250 fibre cement slates (on slope)	m^2	40.22
600 x 300 fibre cement slates (on slope)	m^2	34.08
400 x 200 natural slates (on slope)	m^2	68.65
500 x 250 natural slates (on slope)	m^2	80.40
600 x 300 natural slates (on slope)	m^2	101.61
100 Fibreglass insulation quilt	m^2	7.69
100 Rockwool insulation quilt	m^2	6.01
150 Rockwool insulation quilt	m^2	9.27

Eaves, soffits, fascias and gutters

Eaves soffit, fascia, gutter and decoration		
225 x 150, 114 pvc-u	m	49.03
225 x 150, 100 half round Cast Iron gutter	m	65.88
225 x 150, 114 half round Cast Iron gutter	m	67.25
225 x 150, 125 half round Cast Iron gutter	m	72.44
225 x 150, 150 half round Cast Iron gutter	m	92.90
225 x 150, 100 og Cast Iron gutter	m	66.08
225 x 150, 114 og Cast Iron gutter	m	68.26
225 x 150, 125 og Cast Iron gutter	m	72.08
400 x 200, 114 pvc-u	m	57.07
400 x 200, 100 half round Cast Iron gutter	m	73.92
400 x 200, 114 half round Cast Iron gutter	m	75.29
400 x 200, 125 half round Cast Iron gutter	m	80.48
400 x 200, 150 half round Cast Iron gutter	m	100.94
400 x 200, 100 og Cast Iron gutter	m	74.12
400 x 200, 114 og Cast Iron gutter	m	76.30
400 x 200, 125 og Cast Iron gutter	m	80.12
400 x 200, 102 half round cast aluminium gutter	m	72.04
400 x 200, 127 half round cast aluminium gutter	m	80.23

COMPOSITE PRICES (Cont'd)

ROOF (Cont'd)

Verges, hips, ridges and valleys

Verge undercloak, barge board and decoration		
225 x 25	m	30.67
Hip rafter and capping		
225 x 50, concrete capping tiles	m	28.13
225 x 50, clay machine made capping tiles	m	40.97
225 x 50, clay hand made capping tiles	m	38.77
225 x 50, concrete bonnet hip tiles	m	40.97
225 x 50, clay machine made bonnet hip tiles	m	40.97
225 x 50, clay hand made bonnet hip tiles	m	40.97
Ridge board and capping		
175 x 32, concrete capping tile	m	23.44
175 x 32, machine made clay tile	m	36.28
175 x 32, hand made clay tile	m	34.08
Valley rafter and tile		
225 x 32, concrete angular valley tile	m	27.18
225 x 32, concrete valley trough	m	30.59
225 x 32, machine made clay valley tile	m	40.79
225 x 32, hand made clay valley tile	m	39.57
Valley rafter, sole boards and leadwork		
225 x 32, code 4 lead	m	101.53
225 x 32, code 5 lead	m	108.54

Rain water pipes

Rainwater pipe			
68	pvc-u pipe	m	27.30
63	cast iron pipe and decoration	m	65.82
75	cast iron pipe and decoration	m	71.73
100	cast iron pipe and decoration	m	91.79
63	cast aluminium pipe	m	49.61
75	cast aluminium pipe	m	63.28
100	cast aluminium pipe	m	88.59

STAIRS

Concrete

Reinforced concrete 2670 rise, mild steel balustrade one side, straight flight		
1000 wide granolithic finish	nr	2,849.09
1500 wide granolithic finish	nr	3,714.87
1000 wide pvc sheet finish non slip nosings	nr	3,272.47
1500 wide pvc sheet finish non slip nosings	nr	3,643.75
1000 wide terrazzo finish	nr	3,544.36
1500 wide terrazzo finish	nr	3,946.71
Reinforced concrete 2670 rise, mild steel balustrade one side in two flights with half landing		
1000 wide granolithic finish	nr	3,375.37
1500 wide granolithic finish	nr	3,831.27
1000 wide pvc sheet finish non slip nosings	nr	3,816.58
1500 wide pvc sheet finish non slip nosings	nr	4,157.01
1000 wide terrazzo finish	nr	4,127.87
1500 wide terrazzo finish	nr	5,068.54

Timber

Softwood staircase 2670 rise, balustrade one side.		
910 wide straight flight	nr	1,072.67
910 wide 1 flight with 3 winders	nr	1,322.25
910 wide 2 flight with half landing	nr	1,322.25

Steel

Mild steel staircase 2670 rise, balustrade, both sides.		
910 wide straight flight	nr	3,117.94
910 wide 2 flights with quarter landing	nr	5,802.47
910 wide 2 flights with half landing	nr	7,025.06
Spiral steel staircase, balustrade, decorated		
1568 diameter 2640 rise, domestic	nr	2,565.67
1568 diameter 3000 rise, shop, office, factory and domestic	nr	3,108.84
2072 diameter 3000 rise, shop, office, factory and domestic	nr	3,460.08
1568 diameter, 3000 rise, shop, office, factory and domestic; galvanised	nr	3,355.07
2072 diameter, 3000 rise, shop, office, factory and domestic; galvanised	nr	3,689.58

EXTERNAL WALLS

Brickwork

Common bricks		
102 wall	m²	59.94
215 wall	m²	105.78
327 wall	m²	140.42
Hollow wall with stainless steel ties		
275 common bricks both skins	m²	123.36
275 facings one side P.C. £150/1000	m²	128.01
275 facings one side P.C. £200/1000	m²	131.49
275 facings one side P.C. £250/1000	m²	134.98
275 facings one side P.C. £300/1000	m²	138.46
Engineering bricks Class B		
102 wall	m²	64.84
215 wall	m²	114.19
327 wall	m²	150.94
Engineering bricks Class A		
102 wall	m²	70.98
215 wall	m²	126.56
327 wall	m²	169.55
Facing bricks		
102 wall P.C. £150 per 1000	m²	64.59
102 wall P.C. £200 per 1000	m²	68.07
102 wall P.C. £250 per 1000	m²	71.56
102 wall P.C. £300 per 1000	m²	75.04
215 wall P.C. £150 per 1000	m²	106.60
215 wall P.C. £200 per 1000	m²	113.63
215 wall P.C. £250 per 1000	m²	120.66
215 wall P.C. £300 per 1000	m²	127.68

Blockwork

Aerated concrete block 4.0n/mm²		
100 wall	m²	36.09
140 wall	m²	44.46
190 wall	m²	51.85
Aerated concrete block 7.0n/mm²		
100 wall	m²	38.89
140 wall	m²	48.32
190 wall	m²	57.11
215 wall	m²	63.11
Dense concrete block 7.0n/mm²		
100 wall	m²	48.31
140 wall	m²	56.15
190 wall	m²	65.06
215 wall	m²	71.25
Fair face concrete block		
100 wall	m²	51.67
140 wall	m²	63.93
190 wall	m²	74.24

Composite

Hollow wall in light weight blocks with galvanised ties.		
275 facings one side P.C. £150/1000	m²	104.16
275 facings one side P.C. £200/1000	m²	107.64
275 facings one side P.C. £250/1000	m²	111.13
275 facings one side P.C. £300/1000	m²	114.61
Extra for:		
50 mm glass fibre slab insulation	m²	9.10
13 mm two coat plain render painted	m²	11.98
15 mm pebbled dash render	m²	21.11

COMPOSITE PRICES (Cont'd)

WINDOWS AND EXTERNAL DOORS

Windows

Softwood single glazed casement window		
type N09V size 488 x 900	nr	176.91
type 110V size 631 x 1050	nr	227.19
type 2N10W size 915 x 1050	nr	279.97
type 212CV size 1220 x 1220	nr	396.01
type 312CVC size 1769 x 1050	nr	590.71
average cost	m²	332.76
purpose made in panes 0.10-0.50m²	m²	283.54
Softwood double glazed casement window		
type N09V size 488 x 900	nr	249.55
type 110V size 631 x 1050	nr	319.06
type 2N10W size 915 x 1050	nr	377.49
type 212CV size 1220 x 1220	nr	533.59
type 310CVC size 1769 x 1050	nr	715.53
type 312CVC size 1769 x 1050	nr	784.29
average cost	m²	447.06
purpose made in panes 0.10-0.50m²	m²	407.28
Hardwood double glazed standard casement windows		
type W107VH size 626 x 750	nr	381.16
type W112VH size 626 x 1200	nr	457.55
type W2N10CVH size 913 x 1050	nr	530.19
type WH212CH size 1220 x 1220	nr	616.74
average cost	m²	610.98
Afrormosia double glazed casement windows		
purpose made in panes 0.l0-0.50m²	m²	594.88
European Oak double glazed casement windows		
purpose made in panes 0.10-0.50m²	m²	708.74
Softwood glazed double hung sash window		
type GS3 size 825 x 1094	nr	710.50
type GS4 size 825 x 1394	nr	778.59
type GSW4 size 1051 x 1394	nr	946.79
average cost	m²	782.49
Softwood glazed double hung sash window 38 thick; cased frames; oak cills; sash chords		
size 900 x 1200	nr	748.83
size 1050 x 1200	nr	903.69
size 1200 x 1500	nr	1,063.33
size 1200 x 1800	nr	1,293.83
average cost	m²	719.20
Softwood glazed double hung sash window 50 thick; cased frames; oak cills; sash chords		
size 1050 x 1200	nr	957.98
size 1200 x 1500	nr	1,135.46
size 1200 x 1800	nr	1,377.86
size 1500 x 1800	nr	1,610.18
average cost	m²	725.01
Galvanised steel windows;4mm glass		
type NC6F size 279 x 923	nr	131.93
type NC05F size 508 x 1067	nr	174.71
type NE2 size 997 x 628	nr	199.20
type NC02F size 997 x 1067	nr	367.02
type ND2F size 997 x 1218	nr	359.82
type NC4F size 1486 x 923	nr	459.21
type NC04F size 1486 x 1067	nr	496.95
type ND4 size 1486 x 1218	nr	483.52
type NC11F size 1994 x 923	nr	533.16
type ND11F size 1994 x 1218	nr	614.91
average cost	m²	338.86
Aluminium double glazed; vertical sliders; cill		
size 600 x 800	nr	494.96
size 600 x 1200	nr	656.82
size 800 x 800	nr	806.13
size 800 x 1200	nr	1,029.43
size 1200 x 800	nr	1,581.47
size 1200 x 1200	nr	1,819.05
Aluminium double glazed; horizontal sliders; cill		
size 1200 x 800	nr	506.21
size 1200 x 1400	nr	592.70
size 1400 x 800	nr	541.14
size 1400 x 1400	nr	631.12
size 1600 x 800	nr	580.02
size 1600 x 1600	nr	708.32
Upvc windows; opening/fixed lights; factory double glazed		
size 600 x 1350 window	nr	438.04
size 750 x 1350 window	nr	473.13
size 900 x 1350 window	nr	510.61
size 1200 x 1350 window	nr	574.26
Upvc windows tilt / turn factory double glazed		
size 600 x 600 window	nr	603.27
size 600 x 1050 window	nr	730.21
size 750 x 1050 window	nr	758.59
size 900 x 1200 window	nr	850.24
size 900 x 1500 window	nr	947.70
size 1200 x 1050 window	nr	901.66
Upvc windows tilt / turn; fixed light; factory double glazed		
size 600 x 1500 window	nr	840.28
size 750 x 1500 window	nr	890.85
size 900 x 1500 window	nr	947.70
size 1200 x 1500 window	nr	1,070.65
Upvc windows tilt / turn; fixed side light; factory double glazed		
size 1800 x 1050 window	nr	1,080.43
size 1800 x 1350 window	nr	1,285.26
Upvc windows tilt / turn; fixed centre light; factory double glazed		
size 2400 x 1050 window	nr	1,657.06
size 2400 x 1350 window	nr	1,857.58

Doors

44 External ledged and braced door, 50 x 100 softwood frame, ironmongery P.C. £50.00 per leaf and decorate		
762 x 1981 door	nr	342.16
838 x 1981 door	nr	365.47
1372 x 1981 pair doors	nr	576.16
44 External framed, ledged and braced door, 50 x 100 softwood frame, ironmongery P.C. £50.00 per leaf and decorate		
762 x 1981 door	nr	342.52
838 x 1981 door	nr	382.86
1372 x 1981 pair doors	nr	636.98
44 External plywood flush door, 50 x 100 SW frame, architrave one side, ironmongery P.C. £50.00 per leaf and decorate		
762 x 1981 door	nr	371.60
838 x 1981 door	nr	379.89
1372 x 1981 pair doors	nr	606.62
44 Softwood external panel door glazed panels, 50 x 100 softwood frame, architrave one side, ironmongery P.C. £50.00 decorated.		
2XG size 762 x 1981	nr	430.57
2XG size 838 x 1981	nr	374.18
2XGG size 762 x 1981	nr	366.45
2XGG size 838 x 1981	nr	377.48
KXT size 762 x 1981	nr	467.40
KXT size 838 x 1981	nr	480.60
44 Hardwood Magnet Carolina door, clear glass 50 x 100 hardwood frame architrave one side, ironmongery P.C. £50.00 decorated.		
size 838 x 1981	nr	512.54
size 813 x 2032	nr	517.42
44 Hardwood Magnet Alicante door, bevelled glass 50 x 100 hardwood frame architrave one side, ironmongery P.C. £50.00 decorated.		
size 838 x 1981	nr	601.69
size 813 x 2032	nr	606.57

Garage doors

Galvanised steel garage door		
size 2135 x 1980	nr	866.70
size 4265 x 1980	nr	1,911.59
Western Red Cedar garage door		
size 2135 x 1980 side hung pair	nr	959.53
size 2135 x 1980 up and over	nr	1,189.82

Opening in cavity wall

Galvanised steel lintel, DPC, brick soldier course	m	108.93
Concrete boot lintel		
size 252x150x1200	nr	56.34
size 252x225x1200	nr	75.36
size 252x150x1800	nr	84.48
size 252x225x1800	nr	113.23
Softwood window board, DPC, close cavity	m	28.93
Close cavity at jamb, DPC, facings to reveal	m	17.88

COMPOSITE PRICES (Cont'd)

INTERNAL WALLS

Brickwork

Common bricks		
102 wall	m²	59.94
215 wall	m²	105.78
327 wall	m²	140.42
Engineering bricks		
102 wall	m²	64.84
215 wall	m²	114.19
327 wall	m²	150.94

Blockwork

Light weight concrete block 4.0 n/m²		
75 wall	m²	30.35
90 wall	m²	35.01
100 wall	m²	36.09
140 wall	m²	44.46
190 wall	m²	51.85
Dense concrete block 7.0 n/m²		
90 wall	m²	45.79
100 wall	m²	48.31
140 wall	m²	56.15
190 wall	m²	65.06
215 wall	m²	71.25
Hollow glass blocks		
80 wall - 190 x 190 blocks	m²	368.03
80 wall - 240 x 240 blocks	m²	301.55

Partitions

Timber stud partition with plasterboard both sides		
75 partition, direct decoration	m²	63.75
100 partition direct decoration	m²	69.18
75 partition with skim coat plaster and decorate	m²	79.69
100 partition with skim coat plaster and decorate	m²	85.12
75 partition with plastic faced plasterboard	m²	63.58
100 partition with plastic faced plasterboard	m²	69.01
Extra for 75 insulation	m²	10.35
Cellular core partition		
57 partition direct decoration	m²	55.78
57 partition with skim coat plaster and decorate	m²	71.72
Gyproc laminated partition		
50 partition direct decoration	m²	58.33
65 partition direct decoration	m²	66.07
50 partition with skim coat plaster and decorate	m²	74.27
65 partition with skim coat plaster and decorate	m²	82.01
Metal stud partition with one layer plasterboard each side		
75 partition direct decoration	m²	50.13
75 partition with skim coat plaster and decorate	m²	66.07
Metal stud partition with two layers plasterboard each side		
100 partition direct decoration	m²	58.74
100 partition with skim coat plaster and decorate	m²	74.68

INTERNAL DOORS

Panel doors

34 softwood panelled type SA door; bevelled glass; 38 x 150 lining, stop, architrave both sides, ironmongery PC £20.00 per leaf; decorated		
size 762 x 1981	nr	512.13
34 Magnet Winsoe hardwood six panel door 38 x 150 lining, stop, architrave both sides, ironmongery PC £20.00 per leaf; decorated		
size 686 x 1981	nr	427.27
size 762 x 1981	nr	438.64
34 Magnet Victorian hardwood four panel door; 38 x 150 lining, stop, architrave both sides, ironmongery PC £20.00 per leaf; decorated		
size 686 x 1981	nr	423.36
size 762 x 1981	nr	436.04
34 softwood panelled pattern 2G door; 4mm toughened glass top panel; 38 x 150 lining, stop, architrave both sides, ironmongery PC £20.00 per leaf; decorated		
size 762 x 1981	nr	413.68
34 softwood panelled pattern 2GG door; 4mm toughened glass both panels; 38 x 150 lining, stop, architrave both sides, ironmongery PC £20.00 per leaf; decorated		
size 762 x 1981	nr	434.15
34 softwood panelled pattern 10 door; 4mm toughened glass to one large panel; 38 x 150 lining, stop, architrave both sides, ironmongery PC £20.00 per leaf; decorated		
size 762 x 1981	nr	408.63
Softwood panelled door		
38 x 150 lining, stop, architrave both sides, ironmongery PC £20.00 per leaf; decorated		
38 thick x 762 x 1981 two panel	nr	391.02
38 thick x 762 x 1981 four panel	nr	405.16
38 thick x 762 x 1981 six panel	nr	419.82
50 thick x 838 x 1981 six panel	nr	438.97
Afromosia panelled door		
38 x 150 swd lining, stop, architrave both sides, ironmongery PC £20.00 per leaf; decorated		
38 thick x 762 x 1981 two panel	nr	641.45
38 thick x 762 x 1981 four panel	nr	665.61
38 thick x 762 x 1981 six panel	nr	689.75
50 thick x 838 x 1981 six panel	nr	733.11

Flush doors

34 flush door 38 x 150 lining, stop, architrave both sides, ironmongery PC £20.00 per leaf, decorated		
762 x 1981 hardboard faced	nr	304.94
838 x 1981 hardboard faced	nr	316.66
1372 x 1981 pair hardboard faced	nr	440.22
762 x 1981 plywood faced	nr	304.94
838 x 1981 plywood faced	nr	326.43
1372 x 1981 pair plywood faced	nr	440.22
762 x 1981 Sapele faced	nr	279.57
838 x 1981 Sapele faced	nr	296.56
1372 x 1981 pair Sapele faced	nr	401.09
44 half hour fire check flush door 50 x 100 frame, architrave both sides, ironmongery PC £60.00 per leaf, decorated.		
762 x 1981 plywood faced	nr	327.34
838 x 1981 plywood faced	nr	359.93
1372 x 1981 pair plywood faced	nr	567.45
762 x 1981 Sapele faced	nr	354.80
838 x 1981 Sapele faced	nr	369.57
Extra 6mm Georgian polished wired glass		
508 x 508 panel	nr	45.34
584 x 584 panel	nr	52.25

Lintels

Precast prestressed concrete lintels		
100 x 70 x 900	nr	10.58
100 x 70 x 1200	nr	13.79
100 x 70 x 1800	nr	20.02
100 x 70 x 2400	nr	25.58
100 x 70 x 3000	nr	31.24
150 x 70 x 1200	nr	16.02
150 x 70 x 3000	nr	36.60
225 x 70 x 1200	nr	20.20
225 x 70 x 3000	nr	46.92
255 x 70 x 1200	nr	20.98
255 x 70 x 3000	nr	49.19
100 x 145 x 1800	nr	28.90
100 x 145 x 3000	nr	45.95
Precast concrete lintels		
100x150x1200	nr	18.53
100x150x1650	nr	25.24
215x150x1200	nr	38.07
215x150x1650	nr	53.29

COMPOSITE PRICES (Cont'd)

INTERNAL DOORS (Cont'd)

Lintels (Cont'd)

Galvanized steel lintels - Birtley		
ref INT100; 100 wide x 900 long	nr	24.70
ref INT100; 100 wide x 1050 long	nr	27.91
ref INT100; 100 wide x 1200 long	nr	31.12
ref SB100; 75 deep x 1200 long	nr	44.62
ref SB100; 140 deep x 1800 long	nr	73.18
ref SB100; 215 deep x 3000 long	nr	164.90
ref SB140; 140 deep x 1200 long	nr	53.79
ref SB140; 140 deep x 1800 long	nr	82.26
ref SB140; 215 deep x 3000 long	nr	179.33
ref SBL200; 142 deep x 1200 long	nr	66.02
ref SBL200; 142 deep x 1800 long	nr	100.15
ref SBL200; 218 deep x 3000 long	nr	213.37
ref HD50; 225 deep x 3900 long	nr	323.56
ref HD50; 225 deep x 4800 long	nr	441.57
ref HD70; 225 deep x 3600 long	nr	314.34

WALL FINISHES

Plaster

Hardwall plaster in two coats		
13 thick	m²	15.80
13 thick, emulsion	m²	19.21
13 thick, oil paint	m²	21.67
13 thick textured plastic coating, stipple	m²	19.77
Lightweight plaster in two coats		
13 thick	m²	12.91
13 thick, emulsion	m²	16.32
13 thick, oil paint	m²	18.78
13 thick textured plastic coating, stipple	m²	16.88

Plasterboard

Plasterboard for direct decoration		
9.5 thick	m²	11.40
9.5 thick, emulsion	m²	14.81
9.5 thick, oil paint	m²	17.27
9.5 thick textured plastic coating, stipple	m²	15.37
12.5 thick	m²	13.07
12.5 thick, emulsion	m²	16.48
12.5 thick, oil paint	m²	18.94
12.5 thick textured plastic coating, stipple	m²	17.04
15 thick	m²	16.02
15 thick, emulsion	m²	19.43
15 thick, oil paint	m²	21.89
15 thick textured plastic coating, stipple	m²	19.99
two layer 9.5 thick total 19mm thick	m²	20.35
two layer 9.5 + 12.5; total 22mm thick	m²	22.68
two layer 12.5; total 25mm thick	m²	23.91
Plasterboard with 3 plaster skim coat finish		
9.5 thick plasterboard; skim	m²	16.99
9.5 thick plasterboard; skim, emulsion	m²	20.40
9.5 thick plasterboard; skim, oil paint	m²	22.86
9.5 thick plasterboard; skim, textured plastic coating, stipple	m²	12.99
12.5 thick plasterboard and skim	m²	18.44
12.5 thick plasterboard and skim, emulsion	m²	21.85
12.5 thick plasterboard and skim, oil paint	m²	24.31
12.5 thick plasterboard and skim, textured plastic coating; stipple	m²	22.41
15 thick plasterboard and skim	m²	22.37
15 thick plasterboard and skim, emulsion	m²	25.78
15 thick plasterboard and skim, oil paint	m²	28.24
15 thick plasterboard and skim, textured plastic coating; stipple	m²	26.34
two layer 9.5 thick total 21mm thick including skim	m²	26.08
two layer 9.5 + 12.5; total 25mm thick including skim	m²	27.37
two layer 12.5; total 28mm thick including skim	m²	28.54
Gyproc plaster cove		
100 girth	m	6.51
127 girth	m	6.34

Extra

lining paper PC £1.00/roll	m²	2.43
wall paper PC £4.00/roll	m²	5.11
vinyl coated wall paper PC £7.00/roll	m²	5.96
embossed paper PC £4.50/roll	m²	8.90
wall tiling 108 x 108 x 6.5 white glazed	m²	58.15
wall tiling 152 x 152 x 5.5 white glazed	m²	43.74
wall tiling 152 x 152 x 5.5 light colour	m²	43.74

Redecorations

Stripping existing paper and lining with paper PC £1.00 per roll	m²	5.87
Two coats emulsion paint on existing	m²	4.81
one undercoat; one gloss paint on existing	m²	5.92
one undercoat; one eggshell paint on existing	m²	5.97
Textured plastic coating on existing painted plaster	m²	5.92

FLOOR FINISHES

Cement; sand bed

25 thick	m²	10.55
32 thick	m²	13.35
38 thick	m²	13.56
50 thick	m²	16.59
25 thick, quarry tile paving	m²	57.70
32 thick, quarry tile paving	m²	60.50
38 thick, quarry tile paving	m²	60.71
50 thick, quarry tile paving	m²	63.74
Cement; sand bed and flooring		
25 thick, 3 pvc sheet	m²	39.61
32 thick, 3 pvc sheet	m²	42.41
38 thick, 3 pvc sheet	m²	42.62
50 thick, 3 pvc sheet	m²	45.65
25 thick, 2.5 pvc tile	m²	39.61
32 thick, 2.5 pvc tile	m²	42.41
38 thick, 2.5 pvc tile	m²	42.62
50 thick, 2.5 pvc tile	m²	45.65
25 thick, 20 oak wood block	m²	100.13
32 thick, 20 oak wood block	m²	102.93
38 thick, 20 oak wood block	m²	103.14
50 thick, 20 oak wood block	m²	106.17
25 thick, 20 maple wood block	m²	94.94
32 thick, 20 maple wood block	m²	97.74
38 thick, 20 maple wood block	m²	97.95
50 thick, 20 maple wood block	m²	100.98
25 thick, 20 Merbau wood block	m²	88.44
32 thick, 20 Merbau wood block	m²	91.24
38 thick, 20 Merbau wood block	m²	91.45
50 thick, 20 Merbau wood block	m²	94.48
Cement; sand bed; Carpet and underlay		
25 thick, carpet PC £28/m²	m²	57.93
32 thick, carpet PC £28/m²	m²	60.73
38 thick, carpet PC £28/m²	m²	60.94
50 thick, carpet PC £28/m²	m²	63.97

Granolithic pavings

25 thick	m²	15.13
32 thick	m²	17.00
38 thick	m²	18.98
50 thick	m²	22.48
Granolithic skirtings		
19 thick x 150 high	m	15.75
25 thick x 150 high	m	16.93
32 thick x 150 high	m	17.92

Levelling screed

Latex levelling screed; one coat		
3 thick average to concrete	m²	12.09

Terrazzo

300x300x28 tiles; 40 bed; polished	m²	71.43
16 thick laid insitu in bays	m²	78.33
6 thick skirtings laid insitu 75 high	m	27.09
16 thick skirtings laid insitu 150 high	m	38.95

COMPOSITE PRICES (Cont'd)

FLOOR FINISHES (Cont'd)

PVC sheet flooring		
2 thick	m²	22.71
2.5 thick	m²	24.66
PVC coved skirting		
100 high set in	m	7.00
100 high sit on	m	7.41
Softwood flooring		
Floor boarding and insulation to concrete		
22 tongued and grooved chipboard and 50 insulation board	m²	31.62
Softwood skirting decorated		
13 x 95 to BS 1186 ref. 13RS95	m	10.34
20 x 70 to BS 1186 ref. 20CA70	m	10.32
20 x 120 to BS 1186 ref. 20CS120	m	12.10
19 x 100 square	m	11.78
19 x 150 square	m	13.37
25 x 100 square	m	12.86
25 x 150 square	m	15.23
15 x 125 moulded	m	11.45
19 x 125 moulded	m	12.43

CEILING FINISHES

Plaster		
Hardwall plaster in two coats		
13 thick	m²	17.72
13 thick, emulsion	m²	20.35
13 thick, oil paint	m²	24.08
13 thick textured plastic coatings, stipple	m²	23.20
Lightweight plaster in two coats		
10 thick	m²	14.51
10 thick, emulsion	m²	17.14
10 thick, oil paint	m²	20.87
10 thick textured plastic coatings, stipple	m²	19.99
Plasterboard		
Plasterboard for direct decoration		
9.5 thick	m²	12.70
9.5 thick, emulsion	m²	15.33
9.5 thick, oil paint	m²	19.06
9.5 thick textured plastic coatings, stipple	m²	18.18
12.5 thick	m²	15.37
12.5 thick, emulsion	m²	18.00
12.5 thick, oil paint	m²	21.73
12.5 thick textured plastic coating, stipple	m²	20.85
two layer 9.5 thick total 19mm thick	m²	22.60
two layer 9.5 + 12.5; total 22mm thick	m²	25.25
two layer 12.5; total 25mm thick	m²	26.70
Plasterboard with 3 plaster skim coat finish		
9.5 thick plasterboard and skim	m²	19.06
9.5 thick plasterboard and skim, emulsion	m²	21.69
9.5 thick plasterboard and skim, oil paint	m²	25.42
9.5 thick plasterboard; skim, textured plastic coating, stipple	m²	24.54
12.5 thick plasterboard and skim	m²	20.67
12.5 thick plasterboard and skim, emulsion	m²	23.30
12.5 thick plasterboard and skim, oil paint	m²	27.03
12.5 thick plasterboard; skim, textured plastic coating, stipple	m²	26.15
two layer 9.5 thick total 21mm thick including skim	m²	29.12
two layer 9.5 + 12.5; total 25mm thick including skim	m²	30.57
two layer 12.5; total 28mm thick including skim	m²	31.90
Suspended ceilings		
300 x 300 x 15.8 plain mineral fibre tile	m²	31.18
300 x 300 x 15.8 textured mineral fibre tile	m²	34.31
300 x 300 patterned tile P.C. £15.00	m²	37.27
600 x 600 x 15.8 plain mineral fibre tile	m²	31.69
600 x 600 x 15.8 drilled mineral fibre tile	m²	36.11
600 x 600 patterned tile P.C. £15.00	m²	39.06

FITTINGS AND FURNISHINGS

Mirrors		
6mm bevelled edges; silvered; copper backing		
250 x 400	nr	50.12
460 x 600	nr	81.93
600 x 900	nr	123.43
Venetian blinds		
25mm slat; plain colours		
1000 wide x 1200 drop	nr	56.83
2000 wide x 1200 drop	nr	86.52
3000 wide x 1200 drop	nr	116.18
50mm slat; plain colours		
1000 wide x 1200 drop	nr	56.88
2000 wide x 1200 drop	nr	86.71
3000 wide x 1200 drop	nr	116.37
Roller blinds		
self acting roller		
1000 wide x 1200 drop	nr	84.81
2000 wide x 1200 drop	nr	109.21
3000 wide x 1200 drop	nr	133.02
Vertical louvre blinds		
89mm wide louvres		
1000 wide x 1200 drop	nr	92.61
2000 wide x 1200 drop	nr	158.02
3000 wide x 1200 drop	nr	223.44
127mm wide louvres		
1000 wide x 1200 drop	nr	85.92
2000 wide x 1200 drop	nr	144.66
3000 wide x 1200 drop	nr	203.91
Kitchen units		
Pre-assembled domestic kitchen units		
400 floor unit	nr	195.37
500 floor unit	nr	216.02
600 floor unit	nr	231.56
1000 floor unit	nr	371.28
1200 floor unit	nr	403.14
1200 sink unit	nr	285.87
1500 sink unit	nr	413.47
500 floor one drawer unit	nr	258.53
600 floor one drawer unit	nr	274.07
1000 floor one drawer unit	nr	432.96
1200 sink one drawer unit	nr	352.58
500 floor four drawer unit	nr	342.99
600 floor four drawer unit	nr	363.45
500 x 600 high wall unit	nr	127.08
600 x 600 high wall unit	nr	136.07
1000 x 600 high wall unit	nr	202.66
1200 x 600 high wall unit	nr	219.47
500 x 1950 tall store unit	nr	336.26
600 x 1950 tall store unit	nr	370.93
28 work top	m	63.97
40 work top	m	63.97
plinth	m	9.82
cornice	m	13.62
pelmet	m	10.37

SANITARY APPLIANCES

Sanitary appliance with allowance for waste fittings, traps, overflows, taps and builder's work		
Domestic building, with plastic wastes		
lavatory basin	nr	163.86
lavatory basin with pedestal	nr	185.09
bath with panels to front and one end	nr	306.49
low level w.c. suite	nr	206.39
stainless steel sink and drainer	nr	179.63
shower	nr	664.63

COMPOSITE PRICES (Cont'd)

SANITARY APPLIANCES (Cont'd)

Commercial building, with copper wastes		
lavatory basin	nr	206.98
low level w.c. suite	nr	206.39
bowl type urinal	nr	429.69
stainless steel sink	nr	206.51
drinking fountain	nr	443.23
shower	nr	650.13

DISPOSAL INSTALLATIONS

Soil and vent pipe for ground and first floor		
plastic wastes and SVP	nr	340.17
copper wastes and cast iron SVP	nr	777.95

WATER INSTALLATIONS

Hot and cold water services		
domestic building	nr	1,838.58
gas instantaneous sink water heater	nr	722.56
electric instantaneous sink water heater	nr	655.35
undersink water heater	nr	710.34
electric shower unit	nr	503.88

HEATING INSTALLATIONS

Boiler, radiators, pipework, controls and pumps		
100m² domestic building, solid fuel	m²	39.21
100m² domestic building, gas	m²	35.89
100m² domestic building, oil	m²	44.74
150m² domestic building, solid fuel	m²	35.00
150m² domestic building, gas	m²	31.48
150m² domestic building, oil	m²	39.12
500m² commercial building, gas	m²	52.30
500m² commercial building, oil	m²	60.86
1000m² commercial building, gas	m²	62.04
1000m² commercial building, oil	m²	73.57
1500m² commercial building, gas	m²	72.12
1500m² commercial building, oil	m²	84.51
2000m² commercial building, gas	m²	78.35
2000m² commercial building, oil	m²	92.47
Underfloor heating with polymer pipework and insulation - excludes screed.	m²	27.12

ELECTRICAL INSTALLATIONS

Concealed installation with lighting, power, and ancillary cooker, immersion heater etc. circuit.		
100m² domestic building		
PVC insulated		
PVC insulated and sheathed cables	m²	43.82
MICS cables	m²	60.25
steel conduit	m²	70.20
150m² domestic building,		
PVC insulated and sheathed cables	m²	39.32
MICS cables	m²	53.21
steel conduit	m²	61.42
500m² commercial building, light fittings, steel conduit	m²	87.74
1000m² commercial building, light fittings, steel conduit	m²	93.23
1500m² commercial building, light fittings, steel conduit	m²	98.71
2000m² commercial building, light fittings, steel conduit	m²	104.19

LIFT AND ESCALATOR INSTALLATIONS

Light passenger lift, laminate walls, stainless steel doors.		
3 stop, 8 person	nr	23,811.77
4 stop, 13 person	nr	44,463.40
General purpose passenger lift, laminate walls stainless steel doors.		
4 stop, 8 person	nr	44,125.90
5 stop, 13 person	nr	47,790.03
6 stop, 21 person	nr	109,054.16
Service hoist		
2 stop, 50 kg	nr	6,559.65
Escalator 30 degree inclination department store specification		
3.00 m rise	nr	65,250.00

SITE WORK

Kerbs, edgings channels

Precast concrete kerb and foundation		
125 x 255 kerb	m	34.02
150 x 305 kerb	m	39.63
Precast concrete edging and foundation		
50 x 150 edging	m	24.30
50 x 200 edging	m	27.96
50 x 250 edging	m	31.04
Precast concrete channel and foundation		
255 x 125 channel	m	32.58
Granite kerb on edge and foundation		
150 x 300 kerb	m	47.75
200 x 300 kerb	m	69.41
Granite kerb laid flat and foundation		
300 x 150 kerb	m	63.28
300 x 200 kerb	m	76.63

Pavings

Concrete; 250 excavation, 150 hardcore		
150 concrete road tamped	m²	41.78
200 concrete road tamped	m²	47.16
150 concrete and 70 bitumen macadam	m²	54.28
200 concrete and 70 bitumen macadam	m²	59.66
Extra for Ref. A252 fabric reinforcement	m²	4.50
Extra for Ref. C636 fabric reinforcement	m²	5.66
Gravel; 250 excavation; 100 hardcore; 75 clinker; 75 course gravel; 38 blinding gravel	m²	21.66
Gravel; 150 excavation; 100 hardcore; 50 blinding gravel	m²	11.05
Gravel; 150 excavation; 100 hardcore; 50 clinker;		
38 blinding gravel	m²	13.09
38 blinding gravel with 'Colas'; Shingle dressing	m²	16.54
York Stone; 200 excavation, 150 hardcore		
50 paving	m²	100.02
75 paving	m²	112.70
Precast concrete; 200 excavation, 150 hardcore		
600 x 450 x 50 slabs	m²	30.60
600 x 600 x 50 slabs	m²	28.01
600 x 750 x 50 slabs	m²	30.63
600 x 900 x 50 slabs	m²	28.79
600 x 450 x 63 slabs	m²	33.58
600 x 600 x 63 slabs	m²	31.47
600 x 750 x 63 slabs	m²	31.82
600 x 900 x 63 slabs	m²	32.26
50 thick Crazy paving	m²	28.70
Precast concrete; 200 excavation, 150 hardcore		
200 x 100 x 50 Blocks on 50 sand	m²	36.91
Granite Sets; 300 excavation, 100 hardcore		
125 thick on 100 concrete	m²	119.34
150 thick on 100 concrete	m²	132.73
Cobbles; 200 excavation, 150 hardcore		
Cobbles set in 100 concrete	m²	84.52
Westmorland Green slate; 250 excavation, 150 hardcore		
38 slate on 100 concrete	m²	168.71

Seeding/Turfing

Seeding; Cultivating;digging over one spit deep; removing debris; weeding; weedkiller		
Grass seed; fertilizer; maintain	m²	9.00
Seeding; 150 excavation; 150 topsoil; weedkiller		
Grass seed; fertilizer; maintain	m²	13.59
Turfing; Cultivating;digging over one spit deep; removing debris; weeding; weedkiller		
Turf; maintain	m²	12.63
Turfing; 150 excavation; 150 topsoil; weedkiller		
Turf; maintain	m²	17.22

Walls

Brick wall, excavation 600 deep concrete foundation, 215 wall with piers at 3m centres in facings PC £200 / 1000, damp proof course.		
900 high brick on edge coping	m	186.99
1800 high brick on edge coping	m	302.12
900 high 300 x 100 PCC coping	m	212.60
1800 high 300 x 100 PCC coping	m	327.73

COMPOSITE PRICES (Cont'd)

SITE WORK (Cont'd)

Fences

Plastic coated chain link fencing, posts at 2743 centres.		
900 high 38 x 38 angle iron posts, two line wires	m	23.17
1200 high 38 x 38 angle iron posts, two line wires	m	25.66
1400 high 38 x 38 angle iron posts, three line wires	m	28.70
1800 high 38 x 38 angle iron posts, three line wires	m	34.69
900 high concrete posts two line wires	m	22.27
1200 high PCC posts two line wires	m	24.80
1400 high PCC posts three line wires	m	27.45
1800 high PCC posts three line wires	m	34.27
Gates; chain link; posts		
1200 x 1800	nr	611.92
Close boarded fencing, 100 x 100 posts at 2743 centres 25 x 100 pales, 25 x 150 gravel boards, creosote, post holes filled with concrete		
1200 high two arris rails	m	52.93
1500 high three arris rails	m	63.58
1800 high three arris rails	m	69.73
Gates; timber		
900 x 1150	nr	177.59
900 x 1450	nr	187.63

DRAINAGE

Drains

Vitrified clay sleeve joint drain, excavation and backfill.		
100 diameter 500 deep	m	22.15
100 diameter 750 deep	m	25.68
100 diameter 1000 deep	m	29.23
100 diameter 1500 deep	m	37.04
150 diameter 500 deep	m	30.35
150 diameter 750 deep	m	33.96
150 diameter 1000 deep	m	37.58
150 diameter 1500 deep	m	45.55
Vitrified clay sleeve joint drain, excavation 150 bed and surround pea shingle and backfill.		
100 diameter 500 deep	m	36.00
100 diameter 750 deep	m	39.53
100 diameter 1000 deep	m	43.07
100 diameter 1500 deep	m	50.89
150 diameter 500 deep	m	46.83
150 diameter 750 deep	m	50.44
150 diameter 1000 deep	m	54.06
150 diameter 1500 deep	m	62.03
Vitrified clay sleeve joint drain, excavation 150 bed and haunch concrete and backfill.		
100 diameter 500 deep	m	40.77
100 diameter 750 deep	m	44.30
100 diameter 1000 deep	m	47.85
100 diameter 1500 deep	m	56.13
150 diameter 500 deep	m	49.07
150 diameter 750 deep	m	52.68
150 diameter 1000 deep	m	56.30
150 diameter 1500 deep	m	64.27
Vitrified clay sleeve joint drain, excavation 150 bed and surround concrete and back fill.		
100 diameter 500 deep	m	50.95
100 diameter 750 deep	m	54.48
100 diameter 1000 deep	m	58.02
100 diameter 1500 deep	m	65.84
150 diameter 500 deep	m	59.57
150 diameter 750 deep	m	63.18
150 diameter 1000 deep	m	66.80
150 diameter 1500 deep	m	74.77
Upvc drain, excavation 50 mm bed sand and backfill.		
110 diameter 500 deep	m	22.10
110 diameter 750 deep	m	25.63
110 diameter 1000 deep	m	29.18
110 diameter 1500 deep	m	36.99
160 diameter 500 deep	m	32.80
160 diameter 750 deep	m	36.41
160 diameter 1000 deep	m	40.03
160 diameter 1500 deep	m	47.99
Upvc drain, excavation bed, and surround pea shingle and back fill.		
110 diameter 500 deep	m	26.96
110 diameter 750 deep	m	30.49
110 diameter 1000 deep	m	34.03
110 diameter 1500 deep	m	41.85
160 diameter 500 deep	m	38.22
160 diameter 750 deep	m	41.83
160 diameter 1000 deep	m	45.45
160 diameter 1500 deep	m	53.42
Upvc drain, excavation bed and haunch concrete and back fill.		
110 diameter 500 deep	m	48.66
110 diameter 750 deep	m	52.19
110 diameter 1000 deep	m	55.73
110 diameter 1500 deep	m	63.54
160 diameter 500 deep	m	50.40
160 diameter 750 deep	m	54.01
160 diameter 1000 deep	m	57.63
160 diameter 1500 deep	m	65.60
Upvc drain, excavation bed and surround concrete and back fill.		
110 diameter 500 deep	m	49.65
110 diameter 750 deep	m	53.18
110 diameter 1000 deep	m	56.72
110 diameter 1500 deep	m	64.54
160 diameter 500 deep	m	67.05
160 diameter 750 deep	m	70.66
160 diameter 1000 deep	m	74.28
160 diameter 1500 deep	m	82.25
Cast iron drain, excavation and back fill.		
100 diameter 500 deep	m	66.86
100 diameter 750 deep	m	70.39
100 diameter 1000 deep	m	73.94
100 diameter 1500 deep	m	81.75
150 diameter 500 deep	m	103.30
150 diameter 750 deep	m	106.91
150 diameter 1000 deep	m	110.53
150 diameter 1500 deep	m	118.50
Cast iron drain, excavation, bed and haunch concrete and back fill.		
100 diameter 500 deep	m	94.63
100 diameter 750 deep	m	98.16
100 diameter 1000 deep	m	101.70
100 diameter 1500 deep	m	109.51
150 diameter 500 deep	m	122.33
150 diameter 750 deep	m	125.94
150 diameter 1000 deep	m	129.56
150 diameter 1500 deep	m	137.53
Cast iron drain, excavation, bed and surround concrete and back fill.		
100 diameter 500 deep	m	95.66
100 diameter 750 deep	m	99.19
100 diameter 1000 deep	m	102.73
100 diameter 1500 deep	m	110.55
150 diameter 500 deep	m	139.03
150 diameter 750 deep	m	142.64
150 diameter 1000 deep	m	146.26
150 diameter 1500 deep	m	154.23

Brick manhole

Excavation; 150 concrete bed; 215 class B engineering brick walls; 100 clayware main channel; concrete benching; step irons and cast iron manhole cover		
600 x 450, two branches 750 deep	nr	642.41
600 x 450, two branches 1500 deep	nr	1,039.50
750 x 450, three branches 750 deep	nr	961.97
750 x 450, three branches 1000 deep	nr	1,102.86
750 x 450, three branches 1500 deep	nr	1,367.75
900 x 600, five branches 750 deep	nr	991.32
900 x 600, five branches 1000 deep	nr	1,167.93
900 x 600, five branches 1500 deep	nr	1,483.80

COMPOSITE PRICES (Cont'd)

DRAINAGE (Cont'd)

Concrete circular manhole
Excavation; 150 concrete bed; manhole rings; cover slab; 100 clayware main channel; concrete benching; step irons; cast iron manhole cover

900 dia., two branches 750 deep	nr	719.53
900 dia., two branches 1000 deep	nr	861.35
900 dia., two branches 1500 deep	nr	1,035.06
1050 dia., three branches 750 deep	nr	826.31
1050 dia., three branches 1000 deep	nr	956.51
1050 dia., three branches 1500 deep	nr	1,194.28
1200 dia., five branches 750 deep	nr	944.21
1200 dia., five branches 1000 deep	nr	1,085.50
1200 dia., five branches 1500 deep	nr	1,306.66

Concrete rectangular manhole
Excavation; concrete bed; manhole sections; base unit; cover slab with integral cover

600 x 450 x 500 deep	nr	246.33
600 x 450 x 750 deep	nr	320.79
600 x 450 x 1000 deep	nr	383.44

GRP Inspection chamber
Excavation; GRP Inspection chamber cast iron manhole cover

470 diameter, 520 deep	nr	378.06
470 diameter, 990 deep	nr	510.38

Polypropolene Inspection chamber
Excavation; Polypropolene Inspection chamber cast iron manhole cover

250 diameter, 600 deep	nr	165.03
450 diameter, 600 deep	nr	348.66
450 diameter, 1000 deep	nr	446.53

Concrete gullies
Excavation; concrete bed; gully; bricks;cast iron road grating

375 diameter x 750 deep	nr	323.18
375 diameter x 900 deep	nr	349.82
450 diameter x 750 deep	nr	348.85
450 diameter x 900 deep	nr	375.59

Vitrified clay gullies
Excavation; concrete bed; gully; grating

trapped round; 100 outlet	nr	97.40

Pvc-u gullies
Excavation; concrete bed; bottle gully; grating

200 diameter top; trapped round	nr	118.40

ALTERATION WORKS

The measured rates in this section are intended to apply to Works of a minor nature up to the value of £15,000 based on National Average prices.

REGIONAL VARIATIONS

Laxton's Building Price Book is based upon national average prices (=1.00). Indicative levels of pricing in regions at the second quarter 2005 are given on this map and the factors shown may be applied to adjust overall pricing.

MAJOR WORKS - TENDER VALUES

The measured rates are based on contracts valued in the range of £500,000 to £1,500,000.

As a guide to pricing works of larger value and for cost planning or budgetary purposes, the following adjustments may be made to overall project values:

Project value:
£1,500,000 to £2,500,000 ... deduct 2.5%
£2,500,000 to £5,000,000 ... deduct 5%

SMALL WORKS - TENDER VALUES

The measured rates are based on contracts valued in the range of £50,000 to £100,000.

As a guide to pricing works of smaller or larger value and for cost planning or budgetary purposes, the following adjustments may be made to overall project values:

Project value:
£15,000 to £25,000.......... add 10 %
£25,000 to £50,000.......... add 5%
£50,000 to £100,000........ rates as shown
£100,000 to £200,000....... deduct 5%
£200,000 to £500,000....... deduct 7.5%

ALTERATIONS/REPAIRS/MAINTENANCE

The measured rates are based on contracts valued in the range of £5,000 to £15,000.
The rates may be used for orders of lower value however due allowances must be added for disproportionate waste factors, working in occupied premises and preliminary items.

EXISTING SITE/BUILDINGS/SERVICES

Labour hourly rates: (except Specialists) Craft Operatives £14.21 Labourer £10.71 Rates are national average prices. Refer to REGIONAL VARIATIONS for indicative levels of overall pricing in regions	MATERIALS			LABOUR				RATES		
	Del to Site	Waste	Material Cost	Craft Optve	Lab	Labour Cost	Sunds	Nett Rate	Unit	Gross Rate (+20%)
	£	%	£	Hrs	Hrs	£	£	£		£
C20: DEMOLITION										
Pulling down outbuildings										
Demolishing individual structures										
timber outbuilding 2.50 x 2.00 x 3.00m maximum high	-	-	-	-	12.10	129.59	82.98	212.57	nr	255.09
outbuilding with half brick walls, one brick piers and felt covered timber roof, 2.50 x 2.00 x 3.00m maximum high	-	-	-	-	24.20	259.18	165.97	425.16	nr	510.19
greenhouse with half brick dwarf walls and timber framing to upper walls and roof, 3.00 x 2.00 x 3.00m maximum high	-	-	-	-	19.80	212.06	142.26	354.32	nr	425.19
Demolishing individual structures; setting aside materials for re-use										
metal framed greenhouse, 3.00 x 2.00 x 3.00m maximum high	-	-	-	4.40	7.70	144.99	-	144.99	nr	173.99
prefabricated concrete garage, 5.40 x 2.60 x 2.40m maximum high	-	-	-	6.60	12.10	223.38	-	223.38	nr	268.05
Demolishing individual structures; making good structures										
timber framed lean-to outbuilding and remove flashings and make good facing brick wall at ridge and vertical abutments, 3.00 x 1.90 x 2.40m maximum high	-	-	-	1.65	13.20	164.82	87.87	252.69	nr	303.22
Unit rates for pricing the above and similar work										
pull down building 3.00 x 1.90 x 2.40m maximum high	-	-	-	-	11.00	117.81	82.98	200.79	nr	240.95
remove flashing	-	-	-	-	0.13	1.41	0.47	1.88	m	2.26
make good facing brick wall at ridge	-	-	-	0.22	0.22	5.48	0.43	5.91	m	7.10
make good facing brick wall at vertical abutment	-	-	-	0.19	0.19	4.66	0.43	5.09	m	6.11
make good rendered wall at ridge	-	-	-	0.22	0.22	5.48	1.00	6.49	m	7.78
make good rendered wall at vertical abutment	-	-	-	0.22	0.22	5.48	1.00	6.49	m	7.78
Demolishing individual structures; making good structures										
lean-to outbuilding with half brick walls and slate covered timber roof and remove flashings, hack off plaster to house wall and make good with rendering to match existing, 1.50 x 1.00 x 2.00m maximum high	-	-	-	4.40	13.75	209.79	55.84	265.63	nr	318.75
Unit rates for pricing the above and similar work										
pull down building with half brick walls, 1.50 x 1.00 x 2.00m maximum high	-	-	-	-	5.83	62.44	37.93	100.37	nr	120.45
pull down building with one brick walls, 1.50 x 1.00 x 2.00m maximum high	-	-	-	-	11.00	117.81	71.13	188.94	nr	226.73
hack off plaster or rendering and make good to match existing brick facings	-	-	-	1.10	2.20	39.19	1.42	40.61	m2	48.74
hack off plaster or rendering and make good to match existing rendering	-	-	-	1.43	2.53	47.42	5.74	53.15	m2	63.78
Removal of old work										
Demolishing parts of structures										
in-situ plain concrete bed										
75mm thick	-	-	-	-	0.99	10.60	3.56	14.16	m2	16.99
100mm thick	-	-	-	-	1.32	14.14	4.74	18.88	m2	22.66
150mm thick	-	-	-	-	1.98	21.21	7.12	28.32	m2	33.99
200mm thick	-	-	-	-	2.64	28.27	9.49	37.76	m2	45.32
in-situ reinforced concrete flat roof										
100mm thick	-	-	-	-	2.09	22.38	4.74	27.13	m2	32.55
150mm thick	-	-	-	-	3.14	33.58	7.12	40.69	m2	48.83
200mm thick	-	-	-	-	4.18	44.77	9.49	54.26	m2	65.11
225mm thick	-	-	-	-	4.71	50.42	10.68	61.10	m2	73.32
in-situ reinforced concrete upper floor										
100mm thick	-	-	-	-	2.09	22.38	4.74	27.13	m2	32.55
150mm thick	-	-	-	-	3.14	33.58	7.12	40.69	m2	48.83

Labour hourly rates: (except Specialists) Craft Operatives £14.21 Labourer £10.71 Rates are national average prices. Refer to REGIONAL VARIATIONS for indicative levels of overall pricing in regions	MATERIALS			LABOUR				RATES		
	Del to Site £	Waste %	Material Cost £	Craft Optve Hrs	Lab Hrs	Labour Cost £	Sunds £	Nett Rate £	Unit	Gross Rate (+20%) £
C20: DEMOLITION Cont.										
Removal of old work Cont.										
Demolishing parts of structures Cont.										
in-situ reinforced concrete upper floor Cont.										
200mm thick	-	-	-	-	4.18	**44.77**	9.49	**54.26**	m2	**65.11**
225mm thick	-	-	-	-	4.71	**50.42**	10.68	**61.10**	m2	**73.32**
in-situ reinforced concrete beam	-	-	-	-	20.90	**223.84**	47.42	**271.26**	m3	**325.51**
in-situ reinforced concrete column	-	-	-	-	19.80	**212.06**	47.42	**259.48**	m3	**311.38**
in-situ reinforced concrete wall										
100mm thick	-	-	-	-	1.98	**21.21**	4.74	**25.95**	m2	**31.14**
150mm thick	-	-	-	-	2.97	**31.81**	7.12	**38.93**	m2	**46.71**
200mm thick	-	-	-	-	3.96	**42.41**	9.49	**51.90**	m2	**62.28**
225mm thick	-	-	-	-	4.46	**47.71**	10.68	**58.39**	m2	**70.07**
in-situ reinforced concrete casing to beam	-	-	-	-	17.60	**188.50**	47.42	**235.92**	m3	**283.10**
in-situ reinforced concrete casing to column	-	-	-	-	16.78	**179.66**	47.42	**227.08**	m3	**272.50**
brick internal walls in lime mortar										
102mm thick	-	-	-	-	0.69	**7.42**	4.84	**12.26**	m2	**14.71**
215mm thick	-	-	-	-	1.43	**15.32**	10.19	**25.51**	m2	**30.61**
327mm thick	-	-	-	-	2.20	**23.56**	15.51	**39.07**	m2	**46.89**
brick internal walls in cement mortar										
102mm thick	-	-	-	-	1.05	**11.19**	4.84	**16.03**	m2	**19.23**
215mm thick	-	-	-	-	2.13	**22.86**	10.19	**33.05**	m2	**39.66**
327mm thick	-	-	-	-	3.28	**35.11**	15.51	**50.62**	m2	**60.74**
reinforced brick internal walls in cement lime mortar										
102mm thick	-	-	-	-	1.05	**11.19**	4.84	**16.03**	m2	**19.23**
215mm thick	-	-	-	-	2.13	**22.86**	10.19	**33.05**	m2	**39.66**
hollow clay block internal walls in cement-lime mortar										
50mm thick	-	-	-	-	0.33	**3.53**	2.37	**5.91**	m2	**7.09**
75mm thick	-	-	-	-	0.44	**4.71**	3.56	**8.27**	m2	**9.93**
100mm thick	-	-	-	-	0.57	**6.13**	4.74	**10.87**	m2	**13.05**
hollow clay block internal walls in cement mortar										
50mm thick	-	-	-	-	0.35	**3.77**	2.37	**6.14**	m2	**7.37**
75mm thick	-	-	-	-	0.47	**5.07**	3.56	**8.62**	m2	**10.35**
100mm thick	-	-	-	-	0.63	**6.72**	4.74	**11.46**	m2	**13.75**
concrete block internal walls in cement lime mortar										
75mm thick	-	-	-	-	0.39	**4.12**	3.56	**7.68**	m2	**9.22**
100mm thick	-	-	-	-	0.58	**6.24**	4.74	**10.99**	m2	**13.19**
190mm thick	-	-	-	-	0.77	**8.25**	9.01	**17.25**	m2	**20.70**
215mm thick	-	-	-	-	1.55	**16.61**	10.19	**26.80**	m2	**32.17**
concrete block internal walls in cement mortar										
75mm thick	-	-	-	-	0.62	**6.60**	3.56	**10.16**	m2	**12.19**
100mm thick	-	-	-	-	0.80	**8.60**	4.74	**13.34**	m2	**16.01**
190mm thick	-	-	-	-	1.00	**10.72**	9.01	**19.73**	m2	**23.67**
215mm thick	-	-	-	-	1.84	**19.67**	10.19	**29.87**	m2	**35.84**
if internal walls plastered, add per side	-	-	-	-	0.13	**1.41**	0.95	**2.37**	m2	**2.84**
if internal walls rendered, add per side	-	-	-	-	0.20	**2.12**	0.95	**3.07**	m2	**3.69**
brick external walls in lime mortar										
102mm thick	-	-	-	-	0.53	**5.65**	4.84	**10.49**	m2	**12.59**
215mm thick	-	-	-	-	1.08	**11.55**	10.19	**21.74**	m2	**26.09**
327mm thick	-	-	-	-	1.66	**17.79**	15.51	**33.30**	m2	**39.96**
brick external walls in cement mortar										
102mm thick	-	-	-	-	0.69	**7.42**	4.84	**12.26**	m2	**14.71**
215mm thick	-	-	-	-	1.43	**15.32**	10.19	**25.51**	m2	**30.61**
327mm thick	-	-	-	-	2.20	**23.56**	15.51	**39.07**	m2	**46.89**
reinforced brick external walls in cement lime mortar										
102mm thick	-	-	-	-	0.69	**7.42**	4.84	**12.26**	m2	**14.71**
215mm thick	-	-	-	-	1.43	**15.32**	10.19	**25.51**	m2	**30.61**
if external walls plastered, add per side	-	-	-	-	0.10	**1.06**	0.95	**2.01**	m2	**2.41**
if external walls rendered or rough cast, add per side	-	-	-	-	0.13	**1.41**	0.95	**2.37**	m2	**2.84**
clean old bricks in lime mortar and stack for re-use, per thousand	-	-	-	-	16.50	**176.72**	-	**176.72**	nr	**212.06**
clean old bricks in cement mortar and stack for re-use, per thousand	-	-	-	-	27.50	**294.53**	-	**294.53**	nr	**353.43**
rough rubble walling 600mm thick, in lime mortar	-	-	-	-	4.13	**44.18**	28.46	**72.63**	m2	**87.16**
random rubble walling 350mm thick, in lime mortar	-	-	-	-	2.42	**25.92**	16.59	**42.51**	m2	**51.01**
random rubble walling 500mm thick, in lime mortar	-	-	-	-	3.41	**36.52**	23.71	**60.23**	m2	**72.28**
dressed stone walling 100mm thick, in gauged mortar	-	-	-	2.20	2.53	**58.36**	4.74	**63.10**	m2	**75.72**
dressed stone walling 200mm thick, in gauged mortar	-	-	-	4.29	4.95	**113.98**	9.49	**123.46**	m2	**148.16**
stone copings 300 x 50mm	-	-	-	0.22	0.29	**6.19**	0.72	**6.91**	m	**8.29**
stone copings 375 x 100mm	-	-	-	0.44	0.57	**12.38**	1.79	**14.16**	m	**17.00**
stone staircases	-	-	-	-	16.50	**176.72**	47.42	**224.14**	m3	**268.96**
stone steps	-	-	-	-	16.23	**173.77**	47.42	**221.19**	m3	**265.43**
structural timbers										
50 x 75mm	-	-	-	-	0.14	**1.53**	0.18	**1.71**	m	**2.06**
50 x 100mm	-	-	-	-	0.17	**1.77**	0.23	**2.00**	m	**2.40**
50 x 150mm	-	-	-	-	0.20	**2.12**	0.36	**2.49**	m	**2.98**
50 x 225mm	-	-	-	-	0.26	**2.83**	0.53	**3.36**	m	**4.03**
75 x 100mm	-	-	-	-	0.20	**2.12**	0.36	**2.49**	m	**2.98**
75 x 150mm	-	-	-	-	0.22	**2.36**	0.53	**2.89**	m	**3.47**
75 x 225mm	-	-	-	-	0.34	**3.65**	0.80	**4.45**	m	**5.34**
steelwork										
steel beams, joists and lintels not exceeding 10 kg/m	-	-	-	55.00	55.00	**1370.60**	47.42	**1418.02**	t	**1701.63**
steel beams, joists and lintels 10 - 20 kg/m	-	-	-	50.60	50.60	**1260.95**	47.42	**1308.37**	t	**1570.05**

Labour hourly rates: (except Specialists) Craft Operatives £14.21 Labourer £10.71 Rates are national average prices. Refer to REGIONAL VARIATIONS for indicative levels of overall pricing in regions	MATERIALS			LABOUR				RATES		
	Del to Site £	Waste %	Material Cost £	Craft Optve Hrs	Lab Hrs	Labour Cost £	Sunds £	Nett Rate £	Unit	Gross Rate (+20%) £
C20: DEMOLITION Cont.										
Removal of old work Cont.										
Demolishing parts of structures Cont.										
steelwork Cont.										
steel beams, joists and lintels 20 - 50 kg/m .	-	-	-	44.00	44.00	**1096.48**	47.42	**1143.90**	t	**1372.68**
steel columns and stanchions not exceeding 10 kg/m	-	-	-	66.00	66.00	**1644.72**	47.42	**1692.14**	t	**2030.57**
steel columns and stanchions 10 - 20 kg/m	-	-	-	61.60	61.60	**1535.07**	47.42	**1582.49**	t	**1898.99**
steel columns and stanchions 20 - 50 kg/m	-	-	-	55.00	55.00	**1370.60**	47.42	**1418.02**	t	**1701.63**
steel purlins and rails not exceeding 10 kg/m	-	-	-	55.00	55.00	**1370.60**	47.42	**1418.02**	t	**1701.63**
steel purlins and rails 10 - 20 kg/m	-	-	-	55.00	55.00	**1370.60**	47.42	**1418.02**	t	**1701.63**
C30: SHORING/FACADE RETENTION										
Support of structures not to be demolished										
Provide and erect timber raking shore complete with sole piece, cleats, needles and 25mm brace boarding with										
two 150 x 150mm rakers (total length 8.00m) and 50 x 175mm wall piece	134.20	10.00	**147.62**	4.40	4.40	**109.65**	26.05	**283.32**	nr	**339.98**
weekly cost of maintaining last................	-	-	-	1.10	1.10	**27.41**	1.31	**28.72**	nr	**34.46**
three 225 x 225mm rakers (total length 21.00m) and 50 x 250mm wall piece	507.10	10.00	**557.81**	25.52	25.52	**635.96**	113.23	**1307.00**	nr	**1568.40**
weekly cost of maintaining last................	-	-	-	1.65	1.65	**41.12**	5.59	**46.70**	nr	**56.04**
Provide and erect timber flying shore of 100 x 150mm main member, 50 x 175mm wall pieces and 50 x 100mm straining pieces and 75 x 100mm struts, distance between wall faces										
4.00m	136.40	10.00	**150.04**	16.50	16.50	**411.18**	5.59	**566.81**	nr	**680.17**
weekly cost of maintaining last................	-	-	-	2.75	2.75	**68.53**	1.31	**69.84**	nr	**83.80**
5.00m	158.40	10.00	**174.24**	19.58	19.58	**487.93**	6.51	**668.69**	nr	**802.42**
weekly cost of maintaining last................	-	-	-	3.30	3.30	**82.24**	1.50	**83.73**	nr	**100.48**
6.00m	172.70	10.00	**189.97**	5.61	22.66	**322.41**	7.22	**519.60**	nr	**623.52**
weekly cost of maintaining last................	-	-	-	3.85	3.85	**95.94**	1.71	**97.66**	nr	**117.19**
C40: CLEANING MASONRY/CONCRETE										
Cleaning surfaces										
Thoroughly clean existing concrete surfaces prior to applying damp proof membrane										
floors	-	-	-	-	0.21	**2.24**	0.03	**2.26**	m2	**2.72**
walls	-	-	-	-	0.23	**2.47**	0.03	**2.50**	m2	**3.00**
Thoroughly clean existing concrete surfaces, fill in nail holes and small surface imperfections and leave smooth										
walls	-	-	-	0.39	0.22	**7.83**	0.05	**7.88**	m2	**9.46**
soffits	-	-	-	0.55	0.32	**11.23**	0.05	**11.28**	m2	**13.54**
Thoroughly clean existing brick or block surfaces prior to applying damp proof membrane										
walls	-	-	-	-	0.28	**2.95**	-	**2.95**	m2	**3.53**
Clean out air brick										
225 x 150mm	-	-	-	-	0.22	**2.36**	-	**2.36**	nr	**2.83**
C41: REPAIRING/RENOVATING/CONSERVING MASONRY										
Cut out decayed bricks and replace with new bricks in gauged mortar (1:2:9) and make good surrounding work										
Fair faced common bricks										
singly	0.24	10.00	**0.27**	0.19	0.19	**4.66**	0.17	**5.09**	nr	**6.11**
small patches	14.52	10.00	**15.97**	6.38	6.38	**158.99**	7.07	**182.03**	m2	**218.44**
Picked stock facing bricks										
singly	0.51	10.00	**0.56**	0.19	0.19	**4.66**	0.17	**5.38**	nr	**6.46**
small patches	30.36	10.00	**33.40**	6.38	6.38	**158.99**	7.07	**199.46**	m2	**239.35**
Cut out defective wall and re-build in gauged mortar (1:2:9) and tooth and bond to surrounding work (25% new bricks allowed)										
Half brick wall; stretcher bond										
common bricks	3.63	10.00	**3.99**	4.40	4.40	**109.65**	4.06	**117.70**	m2	**141.24**
One brick wall; English bond										
common bricks	7.26	10.00	**7.99**	8.69	8.69	**216.55**	8.72	**233.26**	m2	**279.91**
common bricks faced one side with picked stock facings	13.20	10.00	**14.52**	9.13	9.13	**227.52**	8.72	**250.75**	m2	**300.91**
picked stock facing bricks faced both sides......	15.18	10.00	**16.70**	9.68	9.68	**241.23**	8.72	**266.64**	m2	**319.97**
One brick wall; Flemish bond										
common bricks	7.26	10.00	**7.99**	8.69	8.69	**216.55**	4.06	**228.60**	m2	**274.32**
common bricks faced one side with picked stock facings	13.20	10.00	**14.52**	9.13	9.13	**227.52**	8.72	**250.75**	m2	**300.91**
picked stock facing bricks faced both sides......	15.18	10.00	**16.70**	9.68	9.68	**241.23**	8.72	**266.64**	m2	**319.97**
275mm hollow wall; skins in stretcher bond										
two half brick skins in common bricks...........	7.26	10.00	**7.99**	8.98	8.98	**223.68**	8.88	**240.55**	m2	**288.66**

Labour hourly rates: (except Specialists) Craft Operatives £14.21 Labourer £10.71 Rates are national average prices. Refer to REGIONAL VARIATIONS for indicative levels of overall pricing in regions	MATERIALS			LABOUR				RATES		
	Del to Site £	Waste %	Material Cost £	Craft Optve Hrs	Lab Hrs	Labour Cost £	Sunds £	Nett Rate £	Unit	Gross Rate (+20%) £
C41: REPAIRING/RENOVATING/CONSERVING MASONRY Cont.										
Cut out defective wall and re-build in gauged mortar (1:2:9) and tooth and bond to surrounding work (25% new bricks allowed) Cont.										
275mm hollow wall; skins in stretcher bond Cont.										
one half brick skin in common bricks and one half brick skin in picked stock facings	11.22	10.00	**12.34**	9.46	9.46	**235.74**	8.88	**256.97**	m2	**308.36**
one 100mm skin in concrete blocks and one half brick skin in picked stock facings	10.37	10.00	**11.40**	9.09	9.09	**226.42**	8.30	**246.13**	m2	**295.35**
Cut out crack in brickwork and stitch across with new bricks in gauged mortar(1:2:9); average 450mm wide										
Half brick wall										
common bricks	6.29	10.00	**6.92**	2.09	2.09	**52.08**	3.94	**62.94**	m	**75.53**
common bricks fair faced one side	6.29	10.00	**6.92**	2.53	2.53	**63.05**	3.94	**73.91**	m	**88.69**
picked stock facing bricks faced both sides	13.16	10.00	**14.47**	2.64	2.64	**65.79**	3.94	**84.20**	m	**101.04**
One brick wall										
common bricks	12.58	10.00	**13.84**	4.29	4.29	**106.91**	8.48	**129.23**	m	**155.07**
common bricks fair faced one side	12.58	10.00	**13.84**	4.84	4.84	**120.61**	8.48	**142.93**	m	**171.52**
common bricks faced one side with picked stock facings	22.88	10.00	**25.17**	5.06	5.06	**126.10**	8.48	**159.74**	m	**191.69**
picked stock facing bricks faced both sides	26.31	10.00	**28.94**	5.17	5.17	**128.84**	8.48	**166.26**	m	**199.51**
Cut out defective arch; re-build in picked stock facing bricks in gauged mortar (1:2:9) including centering										
Brick-on-edge flat arch										
102mm on soffit	3.54	10.00	**3.90**	1.10	1.10	**27.41**	0.99	**32.30**	m	**38.76**
215mm on soffit	6.74	10.00	**7.42**	1.05	1.05	**26.04**	1.74	**35.20**	m	**42.24**
Brick-on-end flat arch										
102mm on soffit	6.74	10.00	**7.42**	1.05	1.05	**26.04**	1.74	**35.20**	m	**42.24**
215mm on soffit	13.49	10.00	**14.84**	2.09	2.09	**52.08**	3.48	**70.40**	m	**84.48**
Segmental arch in two half brick rings										
102mm on soffit	6.74	10.00	**7.42**	3.19	3.19	**79.49**	1.74	**88.65**	m	**106.38**
215mm on soffit	13.49	10.00	**14.84**	3.19	3.19	**79.49**	3.48	**97.81**	m	**117.37**
Semi-circular arch in two half brick rings										
102mm on soffit	6.74	10.00	**7.42**	3.19	3.19	**79.49**	1.74	**88.65**	m	**106.38**
215mm on soffit	13.49	10.00	**14.84**	3.19	3.19	**79.49**	3.48	**97.81**	m	**117.37**
Cut out defective arch; replace with precast concrete, BS 5328, designed mix C20, 20mm aggregate lintel; wedge and pin up to brickwork over with slates in cement mortar (1:3) and extend external rendering over										
Remove defective brick-on-end flat arch replace with 102 x 215mm precast concrete lintel reinforced with one 12mm mild steel bar	30.00	10.00	**33.00**	2.09	2.09	**52.08**	2.73	**87.81**	m	**105.37**
Cut out defective external sill and re-build										
Shaped brick-on-edge sill in picked stock facing bricks in gauged mortar (1:2:9)										
225mm wide	6.74	10.00	**7.42**	1.10	1.10	**27.41**	2.42	**37.26**	m	**44.71**
Roofing tile sill set weathering and projecting in cement mortar (1:3)										
two courses	6.76	10.00	**7.43**	1.32	1.32	**32.89**	1.25	**41.58**	m	**49.90**
Take down defective brick-on-edge coping and re-build in picked stock facing bricks in gauged mortar (1:2:9)										
Coping										
to one brick wall	6.74	10.00	**7.42**	0.66	0.66	**16.45**	2.42	**26.29**	m	**31.55**
Coping with cement fillets both sides										
to one brick wall; with oversailing course	10.12	10.00	**11.13**	0.74	0.74	**18.37**	3.64	**33.14**	m	**39.77**
to one brick wall; with single course tile creasing	10.12	10.00	**11.14**	1.20	1.20	**29.88**	3.05	**44.07**	m	**52.88**
to one brick wall; with double course tile creasing	13.50	10.00	**14.85**	1.52	1.52	**37.83**	4.30	**56.99**	m	**68.38**
Cut out defective air brick and replace with new; bed and point in gauged mortar (1:2:9)										
Cast iron, light, square hole										
225 x 75mm	4.35	-	**4.35**	0.50	0.50	**12.34**	0.19	**16.87**	nr	**20.24**
225 x 150mm	7.98	-	**7.98**	0.50	0.50	**12.34**	0.31	**20.62**	nr	**24.74**
Terra cotta										
215 x 65mm	1.46	-	**1.46**	0.50	0.50	**12.34**	0.19	**13.99**	nr	**16.79**
215 x 140mm	2.04	-	**2.04**	0.50	0.50	**12.34**	0.31	**14.68**	nr	**17.61**
Rake out joints of old brickwork 20mm deep and re-point in cement mortar (1:3)										
For turned in edge of lead flashing										
flush pointing; horizontal	-	-	**-**	0.55	0.55	**13.71**	0.28	**13.99**	m	**16.79**
flush pointing; stepped	-	-	**-**	0.94	0.94	**23.30**	0.40	**23.70**	m	**28.44**

Labour hourly rates: (except Specialists) Craft Operatives £14.21 Labourer £10.71 Rates are national average prices. Refer to REGIONAL VARIATIONS for indicative levels of overall pricing in regions	MATERIALS			LABOUR				RATES		
	Del to Site £	Waste %	Material Cost £	Craft Optve Hrs	Lab Hrs	Labour Cost £	Sunds £	Nett Rate £	Unit	Gross Rate (+20%) £
C41: REPAIRING/RENOVATING/CONSERVING MASONRY Cont.										
Rake out joints of old brickwork 20mm deep and re-point in cement mortar (1:3) Cont.										
For turned in edge of lead flashing Cont.										
weathered pointing; horizontal	-	-	-	0.61	0.61	**15.08**	0.28	**15.36**	m	**18.43**
weathered pointing; stepped	-	-	-	0.99	0.99	**24.67**	0.40	**25.07**	m	**30.09**
ironed in pointing; horizontal	-	-	-	0.61	0.61	**15.08**	0.28	**15.36**	m	**18.43**
ironed in pointing; stepped	-	-	-	0.99	0.99	**24.67**	0.40	**25.07**	m	**30.09**
Rake out joints of old brickwork 20mm deep and re-point in cement-lime mortar(1:1:6)										
Generally; English bond										
flush pointing	-	-	-	1.10	0.73	**23.41**	0.82	**24.22**	m2	**29.07**
weathered pointing	-	-	-	1.10	0.73	**23.41**	0.82	**24.22**	m2	**29.07**
ironed in pointing	-	-	-	1.32	0.88	**28.18**	0.82	**29.00**	m2	**34.80**
Generally; Flemish bond										
flush pointing	-	-	-	1.10	0.73	**23.41**	0.82	**24.22**	m2	**29.07**
weathered pointing	-	-	-	1.10	0.73	**23.41**	0.82	**24.22**	m2	**29.07**
ironed in pointing	-	-	-	1.32	0.88	**28.18**	0.82	**29.00**	m2	**34.80**
Isolated areas not exceeding 1.00m2; English bond										
flush pointing	-	-	-	1.65	1.10	**35.23**	0.82	**36.04**	m2	**43.25**
weathered pointing	-	-	-	1.65	1.10	**35.23**	0.82	**36.04**	m2	**43.25**
ironed in pointing	-	-	-	1.98	1.32	**42.27**	0.82	**43.09**	m2	**51.71**
Isolated areas not exceeding 1.00m2; Flemish bond										
flush pointing	-	-	-	1.65	1.10	**35.23**	0.82	**36.04**	m2	**43.25**
weathered pointing	-	-	-	1.65	1.10	**35.23**	0.82	**36.04**	m2	**43.25**
ironed in pointing	-	-	-	1.98	1.32	**42.27**	0.82	**43.09**	m2	**51.71**
For using coloured mortar										
add	-	-	-	-	-	-	0.59	**0.59**	m2	**0.71**
Rake out joints of old brickwork 20mm deep and re-point in cement mortar (1:3)										
Generally; English bond										
flush pointing	-	-	-	1.32	0.88	**28.18**	0.82	**29.00**	m2	**34.80**
weathered pointing	-	-	-	1.32	0.88	**28.18**	0.82	**29.00**	m2	**34.80**
Generally; Flemish bond										
flush pointing	-	-	-	1.32	0.88	**28.18**	0.82	**29.00**	m2	**34.80**
weathered pointing	-	-	-	1.32	0.88	**28.18**	0.82	**29.00**	m2	**34.80**
Isolated areas not exceeding 1.00m2; English bond										
flush pointing	-	-	-	1.98	1.32	**42.27**	0.82	**43.09**	m2	**51.71**
weathered pointing	-	-	-	1.98	1.32	**42.27**	0.82	**43.09**	m2	**51.71**
Isolated areas not exceeding 1.00m2; Flemish bond										
flush pointing	-	-	-	1.98	1.32	**42.27**	0.82	**43.09**	m2	**51.71**
weathered pointing	-	-	-	1.98	1.32	**42.27**	0.82	**43.09**	m2	**51.71**
Repairing/Renovating stone										
Note restoration and repair of stonework is work for a specialist; prices being obtained for specific projects. Some firms that specialise in this class of work are included within the list at the end of section F-Masonry										
Repairs in Portland stone work and set in gauged mortar (1:2:9) and point to match existing										
Cut out decayed stones in facings of walls, piers or the like, prepare for and set new 75mm thick stone facings										
single stone	392.70	-	**392.70**	16.50	16.50	**411.18**	7.62	**811.50**	m2	**973.80**
areas of two or more adjacent stones	392.70	-	**392.70**	13.20	13.20	**328.94**	7.62	**729.26**	m2	**875.11**
Take out decayed stone sill, prepare for and set new sunk weathered and throated sill										
175 x 100mm	210.10	-	**210.10**	1.98	1.98	**49.34**	1.03	**260.47**	m	**312.56**
250 x 125mm	286.00	-	**286.00**	2.75	2.75	**68.53**	1.74	**356.27**	m	**427.52**
Take out sections of decayed stone coping, prepare for and set new weathered and twice throated coping										
300 x 75mm	255.20	-	**255.20**	1.54	1.54	**38.38**	1.29	**294.86**	m	**353.84**
300 x 100mm	291.50	-	**291.50**	1.87	1.87	**46.60**	1.70	**339.80**	m	**407.77**
Repairs in York stone work and set in gauged mortar (1:2:9) and point to match existing										
Cut out decayed stones in facings of walls, piers or the like, prepare for and set new 75mm thick stone facings										
single stone	294.80	-	**294.80**	18.70	18.70	**466.00**	7.62	**768.42**	m2	**922.11**
areas of two or more adjacent stones	294.80	-	**294.80**	15.13	15.13	**376.92**	7.62	**679.33**	m2	**815.20**
Take out decayed stone sill, prepare for and set new sunk weathered and throated sill										
175 x 100mm	130.90	-	**130.90**	1.98	1.98	**49.34**	1.29	**181.53**	m	**217.83**
250 x 125mm	203.50	-	**203.50**	2.75	2.75	**68.53**	1.70	**273.73**	m	**328.48**

Labour hourly rates: (except Specialists) Craft Operatives £14.21 Labourer £10.71 Rates are national average prices. Refer to REGIONAL VARIATIONS for indicative levels of overall pricing in regions	MATERIALS			LABOUR				RATES		
	Del to Site £	Waste %	Material Cost £	Craft Optve Hrs	Lab Hrs	Labour Cost £	Sunds £	Nett Rate £	Unit	Gross Rate (+20%) £
C41: REPAIRING/RENOVATING/CONSERVING MASONRY Cont.										
Repairs in York stone work and set in gauged mortar (1:2:9) and point to match existing Cont.										
Take out sections of decayed stone coping, prepare for and set new weathered and twice throated coping										
300 x 75mm	143.00	-	**143.00**	1.54	1.54	**38.38**	1.29	**182.66**	m	**219.20**
300 x 100mm	169.40	-	**169.40**	1.87	1.87	**46.60**	1.70	**217.70**	m	**261.25**
Rake out decayed mortar joints and re-point in gauged mortar (1:2:9)										
Re-point rubble walling										
coursed	-	-	-	1.38	1.38	**34.27**	1.26	**35.53**	m2	**42.63**
uncoursed	-	-	-	1.45	1.45	**36.18**	1.26	**37.45**	m2	**44.94**
squared	-	-	-	1.38	1.38	**34.27**	1.26	**35.53**	m2	**42.63**
Re-point stonework										
ashlar	-	-	-	1.32	1.32	**32.89**	1.26	**34.16**	m2	**40.99**
coursed block	-	-	-	1.32	1.32	**32.89**	1.26	**34.16**	m2	**40.99**
C42: REPAIRING/ RENOVATING/CONSERVING CONCRETE										
Re-bed loose copings										
Take off loose concrete coping and clean and remove old bedding mortar										
re-bed, joint and point in cement mortar (1:3)	-	-	-	1.10	1.10	**27.41**	1.34	**28.76**	m	**34.51**
Repairing cracks in concrete										
Repair cracks; clean out all dust and debris and fill with mortar mixed with a bonding agent										
up to 5mm wide	-	-	-	0.55	0.55	**13.71**	0.65	**14.35**	m	**17.22**
Repair cracks; cutting out to form groove, treat with bonding agent and fill with fine concrete mixed with a bonding agent										
25 x 25mm deep	-	-	-	0.55	0.55	**13.71**	1.17	**14.88**	m	**17.85**
40 x 40mm deep	-	-	-	0.66	0.66	**16.45**	2.77	**19.22**	m	**23.06**
C45: DAMP PROOF COURSE RENEWAL/INSERTION										
Insert damp proof course in old wall by cutting out one course of brickwork in alternate lengths not exceeding 1.00m and replace old bricks with 50mm bricks in cement mortar (1:3)										
B.S.6398 Class D, bitumen with hessian base laminated with lead in										
one brick wall	4.24	10.00	**4.66**	3.30	3.30	**82.24**	6.54	**93.44**	m	**112.13**
one and a half brick wall	6.37	10.00	**7.01**	4.40	4.40	**109.65**	9.83	**126.48**	m	**151.78**
254mm hollow wall	4.24	10.00	**4.66**	3.85	3.85	**95.94**	6.54	**107.14**	m	**128.57**
Hyload, pitch polymer in										
one brick wall	2.03	10.00	**2.24**	3.30	3.30	**82.24**	6.54	**91.01**	m	**109.21**
one and a half brick wall	3.05	10.00	**3.36**	4.40	4.40	**109.65**	9.83	**122.83**	m	**147.40**
254mm hollow wall	2.03	10.00	**2.24**	3.85	3.85	**95.94**	6.54	**104.72**	m	**125.66**
Two courses slates in cement mortar (1:3) in										
one brick wall	6.64	10.00	**7.31**	3.30	3.30	**82.24**	6.54	**96.08**	m	**115.30**
one and a half brick wall	9.97	10.00	**10.96**	4.40	4.40	**109.65**	9.83	**130.44**	m	**156.52**
254mm hollow wall	6.64	10.00	**7.31**	3.85	3.85	**95.94**	6.54	**109.79**	m	**131.75**
0.6mm bitumen coated copper in										
one brick wall	12.58	10.00	**13.84**	3.30	3.30	**82.24**	6.54	**102.61**	m	**123.14**
one and a half brick wall	18.90	10.00	**20.79**	4.40	4.40	**109.65**	9.83	**140.26**	m	**168.32**
254mm hollow wall	12.58	10.00	**13.84**	3.85	3.85	**95.94**	6.54	**116.32**	m	**139.58**
Insert damp proof course in old wall by hand sawing in 600mm lengths										
B.S.6398 Class D, bitumen with hessian base laminated with lead in										
one brick wall	4.24	10.00	**4.66**	2.75	2.75	**68.53**	9.78	**82.97**	m	**99.57**
one and a half brick wall	6.37	10.00	**7.01**	3.85	3.85	**95.94**	14.67	**117.62**	m	**141.15**
254mm hollow wall	4.24	10.00	**4.66**	3.30	3.30	**82.24**	9.78	**96.68**	m	**116.01**
Hyload, pitch polymer in										
one brick wall	2.03	10.00	**2.24**	2.75	2.75	**68.53**	9.78	**80.54**	m	**96.65**
one and a half brick wall	3.05	10.00	**3.36**	3.85	3.85	**95.94**	14.67	**113.97**	m	**136.77**
254mm hollow wall	2.03	10.00	**2.24**	3.30	3.30	**82.24**	9.78	**94.25**	m	**113.10**
Two courses slates in cement mortar (1:3) in										
one brick wall	6.64	10.00	**7.31**	2.75	2.75	**68.53**	9.78	**85.62**	m	**102.74**
one and a half brick wall	9.97	10.00	**10.96**	3.85	3.85	**95.94**	14.67	**121.58**	m	**145.90**
254mm hollow wall	3.32	10.00	**3.65**	3.30	3.30	**82.24**	9.78	**95.67**	m	**114.80**
0.6mm bitumen coated copper in										
one brick wall	12.58	10.00	**13.84**	2.75	2.75	**68.53**	9.78	**92.15**	m	**110.58**
one and a half brick wall	18.90	10.00	**20.79**	3.85	3.85	**95.94**	14.67	**131.41**	m	**157.69**
254mm hollow wall	12.58	10.00	**13.84**	3.30	3.30	**82.24**	9.78	**105.85**	m	**127.02**

Labour hourly rates: (except Specialists) Craft Operatives £14.21 Labourer £10.71 Rates are national average prices. Refer to REGIONAL VARIATIONS for indicative levels of overall pricing in regions	MATERIALS			LABOUR				RATES		
	Del to Site £	Waste %	Material Cost £	Craft Optve Hrs	Lab Hrs	Labour Cost £	Sunds £	Nett Rate £	Unit	Gross Rate (+20%) £
C45: DAMP PROOF COURSE RENEWAL/INSERTION Cont.										
Insert damp proof course in old wall by machine sawing in 600mm lengths										
B.S.6398 Class D, bitumen with hessian base laminated with lead in										
one brick wall	4.24	10.00	**4.66**	2.20	2.20	**54.82**	12.98	**72.47**	m	**86.97**
one and a half brick wall	6.37	10.00	**7.01**	3.30	3.30	**82.24**	19.46	**108.71**	m	**130.45**
254mm hollow wall	4.24	10.00	**4.66**	2.75	2.75	**68.53**	12.98	**86.18**	m	**103.41**
Hyload, pitch polymer in										
one brick wall	2.03	10.00	**2.24**	2.20	2.20	**54.82**	12.98	**70.04**	m	**84.05**
one and a half brick wall	3.05	10.00	**3.36**	3.30	3.30	**82.24**	19.46	**105.06**	m	**126.07**
254mm hollow wall	2.03	10.00	**2.24**	2.75	2.75	**68.53**	12.98	**83.75**	m	**100.50**
Two courses slates in cement mortar (1:3) in										
one brick wall	6.64	10.00	**7.31**	2.20	2.20	**54.82**	12.98	**75.12**	m	**90.14**
one and a half brick wall	9.97	10.00	**10.96**	3.30	3.30	**82.24**	19.46	**112.66**	m	**135.20**
254mm hollow wall	3.32	10.00	**3.65**	2.75	2.75	**68.53**	12.98	**85.17**	m	**102.20**
0.6mm bitumen coated copper in										
one brick wall	12.58	10.00	**13.84**	2.20	2.20	**54.82**	12.98	**81.65**	m	**97.98**
one and a half brick wall	18.90	10.00	**20.79**	3.30	3.30	**82.24**	19.46	**122.49**	m	**146.99**
254mm hollow wall	12.58	10.00	**13.84**	2.75	2.75	**68.53**	12.98	**95.35**	m	**114.42**
Insert cavity tray in old wall by cutting out by hand in short lengths; insert individual trays and make good with picked stock facing bricks in gauged mortar (1:2:9) including pinning up with slates										
Type E Cavitray by Cavity Trays Ltd in										
half brick skin of cavity wall	9.96	10.00	**10.96**	1.93	1.93	**47.97**	8.30	**67.23**	m	**80.68**
external angle	6.88	10.00	**7.56**	1.10	1.10	**27.41**	7.63	**42.60**	nr	**51.12**
internal angle	6.88	10.00	**7.56**	1.38	1.38	**34.27**	7.63	**49.45**	nr	**59.35**
Type X Cavity Trays Ltd to suit 40 degree pitched roof complete with attached code 4 lead flashing and dress over tiles in										
half brick skin of cavity wall	41.42	10.00	**45.56**	4.68	4.68	**116.50**	16.61	**178.68**	m	**214.41**
ridge tray	10.29	10.00	**11.31**	1.10	1.10	**27.41**	3.87	**42.59**	nr	**51.11**
catchment tray	4.87	10.00	**5.36**	0.55	0.55	**13.71**	2.03	**21.10**	nr	**25.32**
corner catchment tray	10.75	10.00	**11.82**	1.10	1.10	**27.41**	3.87	**43.10**	nr	**51.72**
Damp proofing old brick wall by silicone injection method (excluding removal and reinstatement of plaster, etc.)										
Damp proofing										
one brick wall	8.78	-	**8.78**	1.10	1.10	**27.41**	0.89	**37.08**	m	**44.49**
one and a half brick wall	13.11	-	**13.11**	1.65	1.65	**41.12**	1.32	**55.55**	m	**66.67**
254mm hollow wall	8.78	-	**8.78**	1.49	1.49	**37.01**	0.89	**46.67**	m	**56.01**
C50: REPAIRING/RENOVATING/CONSERVING METAL										
Replace defective arch bar										
Take out defective arch bar including cutting away brickwork as necessary, replace with new mild steel arch bar primed all round and make good all work disturbed										
32 x 6mm flat arch bar	1.71	-	**1.71**	1.13	1.13	**28.23**	0.82	**30.76**	m	**36.91**
50 x 6mm flat arch bar	2.64	-	**2.64**	1.23	1.23	**30.70**	1.20	**34.54**	m	**41.45**
50 x 50 x 6mm angle arch bar	4.90	-	**4.90**	1.40	1.40	**34.81**	2.14	**41.84**	m	**50.21**
75 x 50 x 6mm angle arch bar	6.11	-	**6.11**	1.58	1.58	**39.47**	2.69	**48.27**	m	**57.92**
Replace defective baluster										
Cut out defective baluster, clean core and bottom rails, provide new mild steel baluster and weld on										
13mm diameter x 686mm long	0.99	-	**0.99**	1.01	1.01	**25.22**	1.71	**27.92**	nr	**33.50**
13mm diameter x 762mm long	0.99	-	**0.99**	1.03	1.03	**25.77**	1.71	**28.47**	nr	**34.16**
13 x 13 x 686mm long	1.32	-	**1.32**	1.03	1.03	**25.77**	2.02	**29.11**	nr	**34.93**
13 x 13 x 762mm long	1.38	-	**1.38**	1.05	1.05	**26.04**	2.02	**29.44**	nr	**35.33**
Repair damaged weld										
Clean off damaged weld including removing remains of weld and cleaning off surrounding paint back to bright metal and re-weld connection between 10mm rail and										
13mm diameter rod	-	-	**-**	0.55	0.55	**13.71**	2.52	**16.22**	nr	**19.47**
13 x 13mm bar	-	-	**-**	0.58	0.58	**14.53**	2.99	**17.52**	nr	**21.02**
25 x 25mm bar	-	-	**-**	0.66	0.66	**16.45**	3.75	**20.20**	nr	**24.24**
C51: REPAIRING/RENOVATING/CONSERVING TIMBER										
Cut out rotten or infected structural timber members including cutting away wall, etc. and shoring up adjacent work and replace with new pressure impregnated timber										
Plates and bed in cement mortar (1:3)										
50 x 75mm	1.39	10.00	**1.52**	0.33	0.33	**8.22**	0.19	**9.94**	m	**11.92**
75 x 100mm	2.97	10.00	**3.27**	0.55	0.55	**13.71**	0.36	**17.34**	m	**20.80**

Labour hourly rates: (except Specialists) Craft Operatives £14.21 Labourer £10.71 Rates are national average prices. Refer to REGIONAL VARIATIONS for indicative levels of overall pricing in regions	MATERIALS			LABOUR				RATES		
	Del to Site £	Waste %	Material Cost £	Craft Optve Hrs	Lab Hrs	Labour Cost £	Sunds £	Nett Rate £	Unit	Gross Rate (+20%) £
C51: REPAIRING/RENOVATING/CONSERVING TIMBER Cont.										
Cut out rotten or infected structural timber members including cutting away wall, etc. and shoring up adjacent work and replace with new pressure impregnated timber Cont.										
Floor or roof joists										
50 x 175mm	3.22	10.00	**3.55**	0.63	0.63	**15.62**	0.42	**19.59**	m	**23.51**
50 x 225mm	4.15	10.00	**4.56**	0.74	0.74	**18.37**	0.54	**23.47**	m	**28.16**
75 x 175mm	5.21	10.00	**5.74**	0.87	0.87	**21.66**	0.64	**28.03**	m	**33.63**
Rafters										
38 x 100mm	1.80	10.00	**1.98**	0.39	0.39	**9.59**	0.19	**11.77**	m	**14.12**
50 x 100mm	1.84	10.00	**2.02**	0.42	0.42	**10.42**	0.24	**12.68**	m	**15.22**
Ceiling joists and collars										
50 x 100mm	1.84	10.00	**2.02**	0.42	0.42	**10.42**	0.24	**12.68**	m	**15.22**
50 x 150mm	2.76	10.00	**3.04**	0.57	0.57	**14.25**	0.36	**17.65**	m	**21.19**
Purlins, ceiling beams and struts										
75 x 100mm	2.97	10.00	**3.27**	0.61	0.61	**15.08**	0.36	**18.71**	m	**22.45**
100 x 225mm	9.37	10.00	**10.31**	1.35	1.35	**33.72**	1.09	**45.11**	m	**54.14**
Roof trusses										
75 x 100mm	2.97	10.00	**3.27**	1.01	1.01	**25.22**	0.36	**28.85**	m	**34.62**
75 x 150mm	4.46	10.00	**4.90**	1.51	1.51	**37.55**	0.54	**42.99**	m	**51.59**
Hangers in roof										
25 x 75mm	0.86	10.00	**0.94**	0.33	0.33	**8.22**	0.10	**9.27**	m	**11.12**
38 x 100mm	1.80	10.00	**1.98**	0.50	0.50	**12.34**	0.19	**14.51**	m	**17.41**
Cut out rotten or infected timber and prepare timber under to receive new										
Roof boarding; provide new 25mm pressure impregnated softwood boarding										
flat boarding; areas not exceeding 0.50m2	12.10	10.00	**13.31**	2.16	2.16	**53.73**	1.20	**68.24**	m2	**81.88**
flat boarding; areas 0.50 - 5.00m2	12.10	10.00	**13.31**	1.79	1.79	**44.68**	1.20	**59.19**	m2	**71.03**
sloping boarding; areas not exceeding 0.50m2	12.10	10.00	**13.31**	2.57	2.57	**64.14**	1.20	**78.65**	m2	**94.38**
sloping boarding; areas 0.50 - 5.00m2	12.10	10.00	**13.31**	2.08	2.08	**51.81**	1.20	**66.32**	m2	**79.58**
Gutter boarding and bearers; provide new 25mm pressure impregnated softwood boarding and 50 x 50mm bearers										
over 300mm wide	15.80	10.00	**17.38**	2.27	2.27	**56.47**	1.44	**75.29**	m2	**90.34**
200mm	4.27	10.00	**4.69**	0.61	0.61	**15.08**	0.48	**20.26**	m	**24.31**
225mm	4.57	10.00	**5.03**	0.68	0.68	**17.00**	0.53	**22.55**	m	**27.06**
Gutter sides; provide new 19mm pressure impregnated softwood boarding										
150mm	1.40	15.00	**1.61**	0.51	0.51	**12.61**	0.13	**14.35**	m	**17.22**
200mm	1.94	15.00	**2.23**	0.55	0.55	**13.71**	0.19	**16.12**	m	**19.34**
225mm	2.37	15.00	**2.72**	0.63	0.63	**15.62**	0.21	**18.55**	m	**22.26**
Eaves or verge soffit and bearers and provide new 25mm pressure impregnated softwood boarding and 25 x 50mm bearers										
over 300mm wide	17.39	15.00	**19.95**	2.31	2.31	**57.57**	1.44	**78.95**	m2	**94.74**
225mm	4.81	15.00	**5.48**	0.74	0.74	**18.37**	0.53	**24.37**	m	**29.24**
Fascia; provide new pressure impregnated softwood										
25 x 150mm	2.51	15.00	**2.88**	0.51	0.51	**12.61**	0.19	**15.68**	m	**18.82**
25 x 200mm	3.30	15.00	**3.80**	0.56	0.56	**13.98**	0.24	**18.02**	m	**21.62**
Barge board; provide new pressure impregnated softwood										
25 x 225mm	3.75	15.00	**4.31**	0.61	0.61	**15.08**	0.28	**19.67**	m	**23.60**
25 x 250mm	4.13	15.00	**4.74**	0.63	0.63	**15.62**	0.31	**20.68**	m	**24.81**
Take up floor boarding, remove all nails and clean joists under, re-lay boarding, clear away and replace damaged boards and sand surface (20% new softwood boards allowed)										
Square edged boarding										
25mm	2.97	15.00	**3.42**	1.90	1.90	**47.42**	0.48	**51.32**	m2	**61.59**
32mm	3.94	15.00	**4.53**	1.94	1.94	**48.25**	0.61	**53.38**	m2	**64.05**
Tongued and grooved boarding										
25mm	3.15	20.00	**3.78**	2.06	2.06	**51.26**	0.48	**55.52**	m2	**66.63**
32mm	4.17	20.00	**5.01**	2.08	2.08	**51.81**	0.61	**57.42**	m2	**68.91**
Take up floor boarding, remove all nails and clean joists under, re-lay boarding, destroy and replace infected boards with new pressure impregnated boards and sand surface (20% new softwood boards allowed)										
Square edged boarding										
25mm	3.27	15.00	**3.76**	1.99	1.99	**49.62**	0.48	**53.86**	m2	**64.63**
32mm	4.33	15.00	**4.98**	2.01	2.01	**50.16**	0.61	**55.75**	m2	**66.90**
Tongued and grooved boarding										
25mm	3.47	20.00	**4.16**	2.13	2.13	**53.18**	0.48	**57.82**	m2	**69.39**
32mm	4.59	20.00	**5.50**	2.15	2.15	**53.45**	0.61	**59.56**	m2	**71.48**

Labour hourly rates: (except Specialists) Craft Operatives £14.21 Labourer £10.71 Rates are national average prices. Refer to REGIONAL VARIATIONS for indicative levels of overall pricing in regions	MATERIALS			LABOUR				RATES		
	Del to Site £	Waste %	Material Cost £	Craft Optve Hrs	Lab Hrs	Labour Cost £	Sunds £	Nett Rate £	Unit	Gross Rate (+20%) £
C51: REPAIRING/RENOVATING/CONSERVING TIMBER Cont.										
Take up worn or damaged floor boarding, remove all nails and clean joists under and replace with new softwood boarding										
Square edged boarding										
25mm in areas not exceeding 0.50m2	14.85	15.00	**17.08**	2.09	2.09	**52.08**	1.20	**70.36**	m2	**84.43**
32mm in areas not exceeding 0.50m2	19.69	15.00	**22.64**	2.13	2.13	**53.18**	1.53	**77.35**	m2	**92.82**
25mm in areas 0.50 - 3.00m2	14.85	15.00	**17.08**	1.73	1.73	**43.04**	1.20	**61.31**	m2	**73.58**
32mm in areas 0.50 - 3.00m2	19.69	15.00	**22.64**	1.77	1.77	**44.13**	1.53	**68.31**	m2	**81.97**
Tongued and grooved boarding										
25mm in areas not exceeding 0.50m2	15.74	20.00	**18.89**	2.20	2.20	**54.82**	1.20	**74.91**	m2	**89.89**
32mm in areas not exceeding 0.50m2	20.87	20.00	**25.04**	2.27	2.27	**56.47**	1.53	**83.04**	m2	**99.65**
25mm in areas 0.50 - 3.00m2	15.74	20.00	**18.89**	2.00	2.00	**49.89**	1.20	**69.98**	m2	**83.97**
32mm in areas 0.50 - 3.00m2	20.87	20.00	**25.04**	2.06	2.06	**51.26**	1.53	**77.83**	m2	**93.40**
Punch down all nails, remove all tacks, etc. and fill nail holes										
Surface of existing flooring										
generally	-	-	-	0.55	0.11	**8.99**	0.03	**9.03**	m2	**10.83**
generally and machine sand in addition	-	-	-	1.10	0.33	**19.17**	1.30	**20.46**	m2	**24.56**
Note										
removal of old floor coverings included elsewhere										
Easing and overhauling doors										
Ease and adjust door and oil ironmongery										
762 x 1981mm	-	-	-	0.36	0.07	**5.87**	0.02	**5.89**	nr	**7.06**
914 x 2134mm	-	-	-	0.43	0.07	**6.80**	0.02	**6.82**	nr	**8.19**
Take off door, shave 12mm off bottom edge and re-hang										
762 x 1981mm	-	-	-	0.73	0.12	**11.61**	-	**11.61**	nr	**13.93**
914 x 2134mm	-	-	-	0.87	0.14	**13.88**	-	**13.88**	nr	**16.66**
Adapting doors										
Take off door and re-hang on opposite hand, piece out as necessary and oil ironmongery										
762 x 1981mm	-	-	-	2.75	0.28	**42.02**	0.11	**42.13**	nr	**50.56**
914 x 2134mm	-	-	-	2.92	0.48	**46.61**	0.11	**46.72**	nr	**56.06**
Refixing removed doors and frames										
Doors										
762 x 1981mm	-	-	-	2.20	0.37	**35.27**	0.15	**35.42**	nr	**42.51**
914 x 2134mm	-	-	-	2.48	0.42	**39.65**	0.15	**39.80**	nr	**47.76**
Doors and frames; oil existing ironmongery										
762 x 1981mm	-	-	-	4.40	0.73	**70.30**	0.39	**70.68**	nr	**84.82**
914 x 2134mm	-	-	-	4.95	0.83	**79.18**	0.39	**79.56**	nr	**95.47**
Repairing doors and frames										
Piece in door where ironmongery removed										
rim lock	-	-	-	1.10	0.17	**17.40**	0.18	**17.57**	nr	**21.09**
rim lock or latch and furniture	-	-	-	1.65	0.28	**26.39**	0.43	**26.82**	nr	**32.18**
mortice lock	-	-	-	1.65	0.28	**26.39**	0.43	**26.82**	nr	**32.18**
mortice lock or latch and furniture	-	-	-	2.20	0.35	**35.03**	0.51	**35.54**	nr	**42.65**
butt hinge	-	-	-	0.55	0.09	**8.76**	0.24	**9.00**	nr	**10.80**
Piece in frame where ironmongery removed										
lock or latch keep	-	-	-	0.55	0.09	**8.76**	0.24	**9.00**	nr	**10.80**
butt hinge	-	-	-	0.55	0.09	**8.76**	0.24	**9.00**	nr	**10.80**
Take off door, dismantle as necessary, cramp and re-wedge with new glued wedges, pin and re-hang										
762 x 1981mm, panelled	-	-	-	5.50	0.92	**88.05**	0.18	**88.23**	nr	**105.87**
914 x 2134mm, panelled	-	-	-	6.60	1.10	**105.57**	0.18	**105.74**	nr	**126.89**
762 x 1981mm, glazed panelled	-	-	-	6.05	0.99	**96.57**	0.18	**96.75**	nr	**116.10**
914 x 2134mm, glazed panelled	-	-	-	7.15	1.19	**114.32**	0.18	**114.50**	nr	**137.40**
Take off softwood panelled door 762 x 1981mm, dismantle as necessary and replace the following damaged members and re-hang										
38 x 125mm top rail	5.93	10.00	**6.52**	2.37	0.40	**37.85**	0.50	**44.86**	nr	**53.83**
38 x 125mm hanging stile	15.41	10.00	**16.95**	2.60	0.43	**41.48**	1.25	**59.68**	nr	**71.62**
38 x 125mm locking stile	15.41	10.00	**16.95**	2.60	0.43	**41.48**	1.25	**59.68**	nr	**71.62**
38 x 200mm middle rail	9.26	10.00	**10.19**	4.65	0.78	**74.48**	0.81	**85.49**	nr	**102.58**
38 x 200mm bottom rail	9.26	10.00	**10.19**	2.42	0.41	**38.75**	0.81	**49.75**	nr	**59.70**
Take off afromosia panelled door 762 x 1981mm, dismantle as necessary and replace the following damaged members and re-hang										
38 x 125mm top rail	30.53	10.00	**33.58**	5.72	0.96	**91.53**	0.50	**125.61**	nr	**150.73**
38 x 125mm hanging stile	79.36	10.00	**87.30**	5.90	0.98	**94.27**	1.25	**182.82**	nr	**219.38**
38 x 125mm locking stile	79.36	10.00	**87.30**	5.90	0.98	**94.27**	1.25	**182.82**	nr	**219.38**
38 x 200mm middle rail	48.32	10.00	**53.15**	11.44	1.90	**182.94**	0.81	**236.91**	nr	**284.29**
38 x 200mm bottom rail	48.32	10.00	**53.15**	5.83	0.97	**93.21**	0.81	**147.18**	nr	**176.62**
Take off damaged softwood stop and replace with new										
12 x 38mm	0.51	10.00	**0.56**	0.39	0.07	**6.18**	0.11	**6.84**	m	**8.21**
25 (fin) x 50mm glue and screw on	0.76	10.00	**0.83**	0.50	0.09	**7.98**	0.13	**8.94**	m	**10.73**

Labour hourly rates: (except Specialists) Craft Operatives £14.21 Labourer £10.71 Rates are national average prices. Refer to REGIONAL VARIATIONS for indicative levels of overall pricing in regions	MATERIALS			LABOUR				RATES		
	Del to Site £	Waste %	Material Cost £	Craft Optve Hrs	Lab Hrs	Labour Cost £	Sunds £	Nett Rate £	Unit	Gross Rate (+20%) £
C51: REPAIRING/RENOVATING/CONSERVING TIMBER Cont.										
Repairing doors and frames Cont.										
Take off damaged afromosia stop and replace with new										
12 x 38mm	2.04	10.00	**2.24**	0.50	0.09	**7.98**	0.11	**10.32**	m	**12.39**
25 (fin) x 50mm glue and screw on	4.07	10.00	**4.48**	0.76	0.13	**12.20**	0.13	**16.81**	m	**20.17**
Flush facing old doors										
Take off panelled door, remove ironmongery, cover both sides with 3.2mm hardboard; refix ironmongery, adjust stops and re-hang										
762 x 1981mm	7.92	10.00	**8.71**	3.41	0.55	**54.35**	0.28	**63.33**	nr	**76.00**
838 x 1981mm	7.92	10.00	**8.71**	3.47	0.56	**55.25**	0.28	**64.23**	nr	**77.08**
Take off panelled door, remove ironmongery, cover one side with 3.2mm hardboard and other side with 6mm fire resistant sheeting screwed on; refix ironmongery, adjust stops and re-hang										
762 x 1981mm	49.83	10.00	**54.81**	4.40	0.74	**70.42**	0.90	**126.13**	nr	**151.36**
838 x 1981mm	49.83	10.00	**54.81**	4.46	0.75	**71.32**	0.90	**127.03**	nr	**152.44**
Take off panelled door, remove ironmongery, cover one side with 3.2mm hardboard, fill panels on other side with 6mm fire resistant sheeting and cover with 6mm fire resistant sheeting screwed on; refix ironmongery, adjust stops and re-hang										
762 x 1981mm	72.77	10.00	**80.04**	5.50	0.91	**87.93**	1.16	**169.13**	nr	**202.96**
838 x 1981mm	72.77	10.00	**80.04**	5.56	0.92	**88.83**	1.16	**170.03**	nr	**204.03**
Take off flush door, remove ironmongery, cover one side with 6mm fire resistant sheeting screwed on; refix ironmongery and re-hang										
762 x 1981mm	45.87	10.00	**50.46**	2.75	0.46	**44.03**	0.58	**95.07**	nr	**114.08**
838 x 1981mm	45.87	10.00	**50.46**	2.75	0.47	**44.14**	0.58	**95.18**	nr	**114.22**
Easing and overhauling windows										
Ease and adjust and oil ironmongery										
casement opening light	-	-	-	0.45	0.08	**7.23**	0.02	**7.26**	nr	**8.71**
double hung sash and renew sash lines with best quality flax cord	1.30	10.00	**1.43**	1.80	0.30	**28.82**	0.06	**30.30**	nr	**36.36**
Refixing removed windows										
Windows										
casement opening light and oil ironmongery	-	-	-	1.13	0.19	**18.10**	0.04	**18.15**	nr	**21.78**
double hung sash and provide new best quality flax sash cord	1.30	10.00	**1.43**	1.87	0.31	**29.87**	0.06	**31.36**	nr	**37.63**
Single or multi-light timber casements and frames										
440 x 920mm	-	-	-	4.40	0.73	**70.30**	0.04	**70.34**	nr	**84.41**
1225 x 1070mm	-	-	-	6.60	1.10	**105.57**	0.06	**105.62**	nr	**126.75**
2395 x 1225mm	-	-	-	8.80	1.46	**140.72**	0.07	**140.78**	nr	**168.94**
Repairing windows										
Remove damaged members of cased frame and replace with new softwood										
parting bead	0.64	10.00	**0.70**	1.28	0.21	**20.37**	0.06	**21.13**	m	**25.35**
pulley stile	2.48	10.00	**2.72**	1.85	0.31	**29.56**	0.15	**32.44**	m	**38.92**
stop bead	0.83	10.00	**0.91**	0.68	0.11	**10.87**	0.06	**11.83**	m	**14.20**
Take out sash not exceeding 50mm thick, dismantle as necessary, replace the following damaged members in softwood and re-hang with new best quality flax sash cord										
glazing bar	3.08	10.00	**3.39**	3.45	0.57	**55.21**	0.06	**58.66**	m	**70.39**
stile	3.78	10.00	**4.16**	3.33	0.55	**53.25**	0.10	**57.51**	m	**69.01**
rail	5.75	10.00	**6.32**	3.08	0.50	**49.07**	0.28	**55.66**	m	**66.80**
Take out casement opening light not exceeding 50mm thick, dismantle as necessary, replace the following damaged members in softwood and re-hang										
glazing bar	1.78	10.00	**1.96**	2.17	0.35	**34.56**	0.06	**36.58**	m	**43.89**
hanging stile	2.48	10.00	**2.72**	2.04	0.33	**32.45**	0.11	**35.28**	m	**42.34**
shutting stile	2.48	10.00	**2.72**	2.04	0.33	**32.45**	0.11	**35.28**	m	**42.34**
rail	4.44	10.00	**4.89**	1.77	0.30	**28.35**	0.29	**33.52**	m	**40.23**
Take out sash, cramp and re-wedge with new glued wedges, pin and re-hang										
sliding sash	-	-	-	6.22	1.05	**99.51**	0.96	**100.46**	nr	**120.56**
opening casement light	-	-	-	5.50	0.94	**88.17**	0.96	**89.13**	nr	**106.95**
Repair corner of sash with 100 x 100mm angle repair plate										
let in flush	0.46	-	**0.46**	1.10	0.19	**17.63**	0.19	**18.28**	nr	**21.94**
Carefully rake out and dry crack in sill and fill with two part wood repair system	-	-	-	0.55	0.09	**8.76**	1.65	**10.41**	m	**12.49**

Labour hourly rates: (except Specialists) Craft Operatives £14.21 Labourer £10.71 Rates are national average prices. Refer to REGIONAL VARIATIONS for indicative levels of overall pricing in regions	MATERIALS			LABOUR				RATES		
	Del to Site £	Waste %	Material Cost £	Craft Optve Hrs	Lab Hrs	Labour Cost £	Sunds £	Nett Rate £	Unit	Gross Rate (+20%) £
C51: REPAIRING/RENOVATING/CONSERVING TIMBER Cont.										
Repairs to softwood skirtings, picture rails, architraves, etc.										
Take off and refix										
skirting	-	-	-	0.44	0.08	**7.08**	0.04	**7.12**	m	**8.55**
dado or picture rail	-	-	-	0.39	0.07	**6.18**	0.03	**6.21**	m	**7.45**
architrave	-	-	-	0.29	0.06	**4.65**	0.03	**4.69**	m	**5.62**
Cut out damaged section and piece in new										
19 x 100mm square skirting	1.14	10.00	**1.26**	0.76	0.12	**12.08**	0.22	**13.56**	m	**16.27**
25 x 150mm square skirting	2.23	10.00	**2.46**	0.78	0.13	**12.51**	0.26	**15.23**	m	**18.28**
19 x 125mm moulded skirting	5.06	10.00	**5.57**	0.77	0.12	**12.24**	0.25	**18.06**	m	**21.67**
19 x 50mm moulded dado or picture rail	2.72	10.00	**2.99**	0.81	0.13	**12.98**	0.11	**16.08**	m	**19.30**
25 x 50mm moulded or splayed architrave	3.66	10.00	**4.03**	0.56	0.10	**9.03**	0.13	**13.19**	m	**15.83**
38 x 75mm moulded or splayed architrave	5.27	10.00	**5.80**	0.57	0.10	**9.19**	0.25	**15.24**	m	**18.28**
Cut out rotten or infected skirting and grounds and replace with new treated softwood										
19 x 100mm square skirting	1.63	10.00	**1.79**	1.10	0.19	**17.63**	0.30	**19.72**	m	**23.67**
25 x 150mm square skirting	2.72	10.00	**2.99**	1.18	0.20	**18.85**	0.41	**22.24**	m	**26.69**
19 x 125mm moulded skirting	5.54	10.00	**6.10**	1.13	0.19	**18.10**	0.34	**24.54**	m	**29.45**
Repairs to staircase members in softwood										
Cut out damaged members of staircase and piece in new										
32 x 275mm tread	8.68	10.00	**9.55**	4.76	0.79	**76.16**	0.51	**86.22**	m	**103.46**
38 x 275mm tread	10.32	10.00	**11.35**	5.34	0.89	**85.35**	0.58	**97.29**	m	**116.74**
19 x 200mm riser	4.22	10.00	**4.65**	2.48	0.42	**39.65**	0.20	**44.49**	m	**53.39**
25 x 200mm riser	5.45	10.00	**5.99**	3.33	0.56	**53.37**	0.30	**59.66**	m	**71.59**
Cut back old tread for a width of 125mm and provide and fix new nosing with glued and dowelled joints to tread										
32mm with rounded nosing	4.66	10.00	**5.13**	2.38	0.40	**38.00**	0.84	**43.97**	m	**52.76**
38mm with rounded nosing	5.39	10.00	**5.93**	2.81	0.47	**44.92**	0.94	**51.79**	m	**62.15**
32mm with moulded nosing	5.54	10.00	**6.10**	2.79	0.46	**44.65**	0.84	**51.59**	m	**61.90**
38mm with moulded nosing	6.29	10.00	**6.92**	3.37	0.56	**53.84**	0.99	**61.75**	m	**74.10**
Replace missing square bar baluster 914mm high; clean out old mortices										
25 x 25mm	0.83	-	**0.83**	0.62	0.10	**9.81**	-	**10.64**	nr	**12.77**
38 x 38mm	1.50	-	**1.50**	0.83	0.14	**13.25**	-	**14.75**	nr	**17.70**
Cut out damaged square bar baluster 914mm high and piece in new										
25 x 25mm	0.83	-	**0.83**	0.83	0.03	**12.08**	0.03	**12.93**	nr	**15.52**
38 x 38mm	1.50	-	**1.50**	1.00	0.17	**15.99**	0.06	**17.54**	nr	**21.05**
Cut out damaged handrail and piece in new										
50mm mopstick	27.73	10.00	**30.50**	2.15	0.36	**34.37**	3.52	**68.39**	m	**82.07**
75 x 100mm moulded	46.24	10.00	**50.87**	4.37	0.73	**69.83**	3.82	**124.52**	m	**149.42**
Refixing removed handrail										
to new balustrade	-	-	-	0.76	0.13	**12.20**	-	**12.20**	m	**14.64**
Repairs to staircase members in oak										
Cut out damaged members of staircase and piece in new										
32 x 275mm tread	49.94	10.00	**54.93**	6.70	1.12	**107.21**	0.51	**162.65**	m	**195.18**
38 x 275mm tread	60.18	10.00	**66.20**	7.90	1.32	**126.37**	0.58	**193.15**	m	**231.78**
19 x 200mm riser	26.90	10.00	**29.58**	3.67	0.62	**58.80**	0.20	**88.59**	m	**106.30**
25 x 200mm riser	33.30	10.00	**36.63**	5.02	0.84	**80.23**	0.32	**117.18**	m	**140.61**
Cut back old tread for a width of 125mm and provide and fix new nosing with glued and dowelled joints to tread										
32mm with rounded nosing	29.44	10.00	**32.38**	2.45	0.41	**39.22**	0.84	**72.43**	m	**86.92**
38mm with rounded nosing	33.30	10.00	**36.63**	3.21	0.54	**51.42**	0.99	**89.03**	m	**106.84**
32mm with moulded nosing	30.73	10.00	**33.81**	3.52	0.58	**56.26**	0.84	**90.91**	m	**109.09**
38mm with moulded nosing	34.57	10.00	**38.03**	4.35	0.73	**69.52**	0.99	**108.54**	m	**130.25**
Replace missing square bar baluster 914mm high; clean out old mortices										
25 x 25mm	4.35	-	**4.35**	1.03	0.18	**16.58**	-	**20.92**	nr	**25.11**
38 x 38mm	8.83	-	**8.83**	1.09	0.19	**17.48**	-	**26.31**	nr	**31.57**
Cut out damaged square bar baluster 914mm high and piece in new										
25 x 25mm	4.35	-	**4.35**	1.33	0.22	**21.27**	0.03	**25.65**	nr	**30.78**
38 x 38mm	8.83	-	**8.83**	1.45	0.24	**23.22**	0.06	**32.11**	nr	**38.54**
Cut out damaged handrail and piece in new										
50mm mopstick	61.04	10.00	**67.14**	4.00	0.67	**64.08**	3.52	**134.75**	m	**161.70**
75 x 100mm moulded	101.72	10.00	**111.89**	6.05	1.01	**96.81**	3.82	**212.51**	m	**255.02**
Refixing removed handrail										
to new balustrade	-	-	-	1.49	0.25	**23.81**	-	**23.81**	m	**28.57**
Remove broken ironmongery and fix only new to softwood										
Door ironmongery										
75mm butt hinges	-	-	-	0.25	0.04	**4.07**	0.01	**4.08**	nr	**4.89**
100mm butt hinges	-	-	-	0.25	0.04	**4.07**	0.01	**4.08**	nr	**4.89**
tee hinges up to 300mm	-	-	-	0.68	0.11	**10.87**	0.01	**10.88**	nr	**13.06**

Labour hourly rates: (except Specialists) Craft Operatives £14.21 Labourer £10.71 Rates are national average prices. Refer to REGIONAL VARIATIONS for indicative levels of overall pricing in regions	MATERIALS			LABOUR				RATES		
	Del to Site £	Waste %	Material Cost £	Craft Optve Hrs	Lab Hrs	Labour Cost £	Sunds £	Nett Rate £	Unit	Gross Rate (+20%) £
C51: REPAIRING/RENOVATING/CONSERVING TIMBER Cont.										
Remove broken ironmongery and fix only new to softwood Cont.										
Door ironmongery Cont.										
single action floor spring hinges	-	-	-	3.30	0.55	**52.78**	1.43	**54.21**	nr	**65.06**
barrel bolt up to 300mm	-	-	-	0.45	0.08	**7.23**	0.01	**7.24**	nr	**8.69**
lock or latch furniture	-	-	-	0.48	0.08	**7.70**	0.01	**7.71**	nr	**9.26**
rim lock	-	-	-	0.88	0.14	**14.04**	0.01	**14.05**	nr	**16.86**
rim latch	-	-	-	0.88	0.14	**14.04**	0.01	**14.05**	nr	**16.86**
mortice lock	-	-	-	1.34	0.22	**21.43**	0.01	**21.44**	nr	**25.72**
mortice latch	-	-	-	1.19	0.20	**19.00**	0.01	**19.01**	nr	**22.82**
pull handle	-	-	-	0.44	0.08	**7.08**	0.03	**7.11**	nr	**8.53**
Window ironmongery										
sash fastener	-	-	-	1.20	0.20	**19.16**	0.01	**19.17**	nr	**23.00**
sash lift	-	-	-	0.25	0.04	**4.07**	0.01	**4.08**	nr	**4.89**
casement stay	-	-	-	0.44	0.08	**7.08**	0.01	**7.09**	nr	**8.51**
casement fastener	-	-	-	0.44	0.08	**7.08**	0.01	**7.09**	nr	**8.51**
Sundry ironmongery										
hat and coat hook	-	-	-	0.32	0.06	**5.12**	0.01	**5.13**	nr	**6.16**
toilet roll holder	-	-	-	0.32	0.06	**5.12**	0.01	**5.13**	nr	**6.16**
Remove broken ironmongery and fix only new to hardwood										
Door ironmongery										
75mm butt hinges	-	-	-	0.33	0.06	**5.28**	0.01	**5.29**	nr	**6.35**
100mm butt hinges	-	-	-	0.33	0.06	**5.28**	0.01	**5.29**	nr	**6.35**
tee hinges up to 300mm	-	-	-	0.95	0.15	**15.09**	0.01	**15.10**	nr	**18.12**
single action floor spring hinges	-	-	-	4.62	0.77	**73.90**	1.43	**75.33**	nr	**90.39**
barrel bolt up to 300mm	-	-	-	0.66	0.11	**10.56**	0.01	**10.57**	nr	**12.68**
lock or latch furniture	-	-	-	0.67	0.11	**10.71**	0.01	**10.72**	nr	**12.87**
rim lock	-	-	-	1.27	0.21	**20.21**	0.01	**20.23**	nr	**24.27**
rim latch	-	-	-	1.27	0.21	**20.21**	0.01	**20.23**	nr	**24.27**
mortice lock	-	-	-	1.95	0.33	**31.20**	0.01	**31.21**	nr	**37.45**
mortice latch	-	-	-	1.74	0.29	**27.76**	0.01	**27.77**	nr	**33.33**
pull handle	-	-	-	0.63	0.11	**10.09**	0.03	**10.12**	nr	**12.14**
Window ironmongery										
sash fastener	-	-	-	1.73	0.29	**27.60**	0.01	**27.61**	nr	**33.14**
sash lift	-	-	-	0.34	0.06	**5.43**	0.01	**5.45**	nr	**6.53**
casement stay	-	-	-	0.62	0.10	**9.81**	0.01	**9.82**	nr	**11.79**
casement fastener	-	-	-	0.62	0.10	**9.81**	0.01	**9.82**	nr	**11.79**
Sundry ironmongery										
hat and coat hook	-	-	-	0.46	0.08	**7.39**	0.01	**7.40**	nr	**8.88**
toilet roll holder	-	-	-	0.46	0.08	**7.39**	0.01	**7.40**	nr	**8.88**
C52: FUNGUS/BEETLE ERADICATION										
Protective treatment of existing timbers										
Treat with two coats of spray applied preservative										
boarding	0.66	10.00	**0.72**	-	0.17	**1.77**	0.04	**2.53**	m2	**3.04**
structural timbers	0.98	10.00	**1.08**	-	0.22	**2.36**	0.06	**3.49**	m2	**4.19**
Treat with two coats of brush applied preservative										
boarding	0.98	10.00	**1.08**	-	0.29	**3.06**	0.03	**4.18**	m2	**5.01**
structural timbers	1.51	10.00	**1.66**	-	0.88	**9.42**	0.06	**11.14**	m2	**13.36**
Insecticide treatment of existing timbers										
Treat worm infected timbers with spray applied proprietary insecticide										
boarding	0.98	10.00	**1.08**	-	0.26	**2.83**	0.06	**3.96**	m2	**4.76**
structural timbers	1.31	10.00	**1.44**	-	0.31	**3.30**	0.07	**4.81**	m2	**5.77**
joinery timbers	1.31	10.00	**1.44**	-	0.35	**3.77**	0.07	**5.28**	m2	**6.33**
Treat worm infected timbers with brush applied proprietary insecticide										
boarding	1.31	10.00	**1.44**	-	0.40	**4.24**	0.03	**5.72**	m2	**6.86**
structural timbers	1.57	10.00	**1.73**	-	0.46	**4.95**	0.06	**6.73**	m2	**8.08**
joinery timbers	1.57	10.00	**1.73**	-	0.53	**5.65**	0.06	**7.44**	m2	**8.93**
Treatment of wall surfaces										
Treat surfaces of concrete and brickwork adjoining areas where infected timbers removed										
with a blow lamp	-	-	-	-	0.28	**2.95**	0.03	**2.98**	m2	**3.57**
C90: ALTERATIONS - SPOT ITEMS										
Removal of fittings and fixtures										
Removing fittings and fixtures										
cupboard with doors, frames, architraves, etc	-	-	-	-	0.65	**6.95**	1.36	**8.31**	m2	**9.97**
worktop with legs and bearers	-	-	-	-	0.62	**6.60**	1.81	**8.41**	m2	**10.09**
shelving exceeding 300mm wide with bearers and brackets	-	-	-	-	0.50	**5.30**	1.15	**6.45**	m2	**7.74**
shelving not exceeding 300mm wide with bearers and brackets	-	-	-	-	0.22	**2.36**	0.36	**2.72**	m	**3.27**
draining board and bearers										
500 x 600mm	-	-	-	-	0.31	**3.30**	0.69	**3.99**	nr	**4.79**
600 x 900mm	-	-	-	-	0.41	**4.36**	1.23	**5.58**	nr	**6.70**

Labour hourly rates: (except Specialists) Craft Operatives £14.21 Labourer £10.71 Rates are national average prices. Refer to REGIONAL VARIATIONS for indicative levels of overall pricing in regions	MATERIALS			LABOUR				RATES		
	Del to Site £	Waste %	Material Cost £	Craft Optve Hrs	Lab Hrs	Labour Cost £	Sunds £	Nett Rate £	Unit	Gross Rate (+20%) £
C90: ALTERATIONS - SPOT ITEMS Cont.										
Removal of fittings and fixtures Cont.										
Removing fittings and fixtures Cont.										
wall cupboard unit										
510 x 305 x 305mm	-	-	-	-	0.52	**5.54**	1.09	**6.63**	nr	**7.96**
510 x 305 x 610mm	-	-	-	-	0.72	**7.66**	2.12	**9.78**	nr	**11.74**
1020 x 305 x 610mm	-	-	-	-	1.03	**11.07**	4.31	**15.39**	nr	**18.47**
1220 x 305 x 610mm	-	-	-	-	1.27	**13.55**	5.14	**18.68**	nr	**22.42**
floor cupboard unit										
510 x 510 x 915mm	-	-	-	-	1.01	**10.84**	5.40	**16.24**	nr	**19.48**
610 x 510 x 915mm	-	-	-	-	1.33	**14.26**	6.44	**20.69**	nr	**24.83**
1020 x 510 x 915mm	-	-	-	-	1.82	**19.44**	10.79	**30.23**	nr	**36.28**
1220 x 510 x 915mm	-	-	-	-	2.12	**22.74**	12.93	**35.67**	nr	**42.80**
sink base unit										
585 x 510 x 895mm	-	-	-	-	1.01	**10.84**	6.09	**16.93**	nr	**20.31**
1070 x 510 x 895mm	-	-	-	-	1.82	**19.44**	11.08	**30.52**	nr	**36.62**
tall cupboard unit										
510 x 510 x 1505mm	-	-	-	-	1.72	**18.38**	8.89	**27.27**	nr	**32.72**
610 x 510 x 1505mm	-	-	-	-	1.95	**20.85**	10.62	**31.48**	nr	**37.77**
bath panel and bearers	-	-	-	-	0.52	**5.54**	1.15	**6.68**	nr	**8.02**
steel or concrete clothes line post with concrete base	-	-	-	-	0.55	**5.89**	2.27	**8.16**	nr	**9.79**

This section continues on the next page

Labour hourly rates: (except Specialists) Craft Operatives £17.38 Labourer £10.71 Rates are national average prices. Refer to REGIONAL VARIATIONS for indicative levels of overall pricing in regions	MATERIALS			LABOUR				RATES		
	Del to Site	Waste	Material Cost	Craft Optve	Lab	Labour Cost	Sunds	Nett Rate	Unit	Gross Rate (+20%)
	£	%	£	Hrs	Hrs	£	£	£		£
C90: ALTERATIONS - SPOT ITEMS										
Removal of plumbing and electrical installations										
Removing plumbing and electrical installations										
100mm cast iron rainwater gutters	-	-	-	0.22	-	3.82	0.22	4.05	m	4.85
75mm cast iron rainwater pipes	-	-	-	0.28	-	4.78	0.25	5.03	m	6.03
100mm cast iron rainwater pipes	-	-	-	0.28	-	4.78	0.42	5.20	m	6.24
cast iron rainwater head	-	-	-	0.33	-	5.74	0.69	6.43	nr	7.71
cast iron soil and vent pipe with caulked joints										
50mm	-	-	-	0.39	-	6.69	0.13	6.82	m	8.19
75mm	-	-	-	0.39	-	6.69	0.25	6.94	m	8.33
100mm	-	-	-	0.39	-	6.69	0.42	7.11	m	8.53
150mm	-	-	-	0.55	-	9.56	0.99	10.55	m	12.66
100mm asbestos cement soil and vent pipe	-	-	-	0.28	-	4.78	0.42	5.20	m	6.24
150mm asbestos cement soil and vent pipe	-	-	-	0.28	-	4.78	0.99	5.77	m	6.92
100mm p.v.c. soil and vent pipe	-	-	-	0.28	-	4.78	0.42	5.20	m	6.24
150mm p.v.c. soil and vent pipe	-	-	-	0.28	-	4.78	0.99	5.77	m	6.92
100mm lead soil and vent pipe	-	-	-	0.39	-	6.69	0.42	7.11	m	8.53
150mm lead soil and vent pipe	-	-	-	0.39	-	6.69	0.99	7.68	m	9.22
w.c. suites	-	-	-	1.71	-	29.63	4.25	33.88	nr	40.66
lavatory basins	-	-	-	1.49	-	25.81	2.55	28.36	nr	34.04
baths	-	-	-	1.71	-	29.63	8.51	38.14	nr	45.77
glazed ware sinks	-	-	-	1.49	-	25.81	4.25	30.06	nr	36.07
stainless steel sinks	-	-	-	1.49	-	25.81	2.55	28.36	nr	34.04
stainless steel sinks with single drainer	-	-	-	1.49	-	25.81	3.42	29.22	nr	35.07
stainless steel sinks with double drainer	-	-	-	1.49	-	25.81	4.25	30.06	nr	36.07
hot and cold water, waste pipes, etc. 8 - 25mm diameter										
copper	-	-	-	0.17	-	2.87	0.05	2.92	m	3.50
lead	-	-	-	0.17	-	2.87	0.05	2.92	m	3.50
polythene or p.v.c.	-	-	-	0.17	-	2.87	0.05	2.92	m	3.50
stainless steel	-	-	-	0.28	-	4.78	0.05	4.83	m	5.80
steel	-	-	-	0.28	-	4.78	0.05	4.83	m	5.80
hot and cold water, waste pipes, etc. 32 - 50mm diameter										
copper	-	-	-	0.22	-	3.82	0.13	3.95	m	4.74
lead	-	-	-	0.22	-	3.82	0.13	3.95	m	4.74
polythene or p.v.c.	-	-	-	0.22	-	3.82	0.13	3.95	m	4.74
stainless steel	-	-	-	0.33	-	5.74	0.13	5.87	m	7.04
steel	-	-	-	0.33	-	5.74	0.13	5.87	m	7.04
cold water cisterns up to 454 litres capacity	-	-	-	2.20	-	38.24	19.32	57.55	nr	69.06
hot water tanks or cylinders up to 227 litre capacity	-	-	-	2.20	-	38.24	9.65	47.88	nr	57.46
solid fuel boilers up to 66000 BTU output	-	-	-	4.13	-	71.69	-	71.69	nr	86.03
gas or oil fired boilers up to 66000 BTU output	-	-	-	4.13	-	71.69	-	71.69	nr	86.03
wall mounted water heaters	-	-	-	2.20	-	38.24	-	38.24	nr	45.88
wall type radiators up to 914mm long, up to 610mm high	-	-	-	1.10	-	19.12	0.86	19.98	nr	23.97
wall type radiators up to 914mm long, 610 - 914mm high	-	-	-	1.49	-	25.81	1.71	27.52	nr	33.02
wall type radiators 914 - 1829mm long, up to 610mm high	-	-	-	1.71	-	29.63	2.61	32.24	nr	38.69
wall type radiators 914 - 1829mm long, 610 - 914mm high	-	-	-	2.15	-	37.28	3.42	40.70	nr	48.83
column type radiators up to 914mm long, up to 610mm high	-	-	-	1.10	-	19.12	4.47	23.59	nr	28.31
column type radiators up to 914mm long, 610 - 914mm high	-	-	-	1.49	-	25.81	8.93	34.74	nr	41.69
column type radiators 914 - 1829mm long, up to 610mm high	-	-	-	1.71	-	29.63	13.40	43.03	nr	51.64
column type radiators 914 - 1829mm long, 610 - 914mm high	-	-	-	2.15	-	37.28	17.86	55.14	nr	66.17
pipe insulation from pipes 8 - 25mm diameter	-	-	-	0.11	-	1.91	0.13	2.04	m	2.45
pipe insulation from pipes 32 - 50mm diameter	-	-	-	0.14	-	2.49	0.25	2.73	m	3.28
water tank or calorifier insulation from tanks up to 227 litres	-	-	-	0.55	-	9.56	9.65	19.20	nr	23.05
water tank or calorifier insulation from tanks 227 - 454 litres	-	-	-	0.83	-	14.34	19.32	33.66	nr	40.39
Removing plumbing and electrical installations; setting aside for re-use										
cast iron soil and vent pipe with caulked joints										
50mm	-	-	-	0.61	-	10.51	-	10.51	m	12.62
75mm	-	-	-	0.66	-	11.47	-	11.47	m	13.76
100mm	-	-	-	0.88	-	15.29	-	15.29	m	18.35
150mm	-	-	-	1.10	-	19.12	-	19.12	m	22.94

This section continues on the next page

Labour hourly rates: (except Specialists) Craft Operatives £20.54 Labourer £10.71 Rates are national average prices. Refer to REGIONAL VARIATIONS for indicative levels of overall pricing in regions	MATERIALS			LABOUR				RATES		
	Del to Site £	Waste %	Material Cost £	Craft Optve Hrs	Lab Hrs	Labour Cost £	Sunds £	Nett Rate £	Unit	Gross Rate (+20%) £
C90: ALTERATIONS - SPOT ITEMS Cont.										
Removal of plumbing and electrical installations Cont.										
Removing plumbing and electrical installations										
wall mounted electric fire	-	-	-	2.20	-	**45.19**	2.27	**47.46**	nr	**56.95**
1.5 kW night storage heater	-	-	-	2.75	-	**56.49**	6.80	**63.29**	nr	**75.95**
2 KW night storage heater	-	-	-	2.75	-	**56.49**	9.07	**65.56**	nr	**78.67**
3 KW night storage heater	-	-	-	2.75	-	**56.49**	13.61	**70.09**	nr	**84.11**
Removing plumbing and electrical installations; extending and making good finishings										
lighting points	-	-	-	1.38	-	**28.24**	1.85	**30.09**	nr	**36.11**
flush type switches	-	-	-	1.38	-	**28.24**	1.85	**30.09**	nr	**36.11**
surface mounted type switches	-	-	-	1.10	-	**22.59**	1.49	**24.08**	nr	**28.90**
flush type socket outlets	-	-	-	1.38	-	**28.24**	1.85	**30.09**	nr	**36.11**
surface mounted type socket outlets	-	-	-	1.10	-	**22.59**	1.49	**24.08**	nr	**28.90**
flush type fitting points	-	-	-	1.38	-	**28.24**	1.85	**30.09**	nr	**36.11**
surface mounted type fitting points	-	-	-	1.10	-	**22.59**	1.49	**24.08**	nr	**28.90**
surface mounted p.v.c. insulated and sheathed cables	-	-	-	0.04	-	**0.90**	0.08	**0.98**	m	**1.18**
surface mounted mineral insulated copper sheathed cables	-	-	-	0.11	-	**2.26**	0.16	**2.42**	m	**2.90**
conduits up to 25mm diameter with junction boxes	-	-	-	0.17	-	**3.39**	0.22	**3.61**	m	**4.33**
surface mounted cable trunking up to 100 x 100mm	-	-	-	0.61	-	**12.43**	0.81	**13.23**	m	**15.88**

This section continues on the next page

EXISTING SITE/BUILDINGS/SERVICES

Labour hourly rates: (except Specialists) Craft Operatives £14.21 Labourer £10.71 Rates are national average prices. Refer to REGIONAL VARIATIONS for indicative levels of overall pricing in regions	MATERIALS			LABOUR				RATES		
	Del to Site £	Waste %	Material Cost £	Craft Optve Hrs	Lab Hrs	Labour Cost £	Sunds £	Nett Rate £	Unit	Gross Rate (+20%) £
C90: ALTERATIONS - SPOT ITEMS Cont.										
Removal of floor, wall and ceiling finishings										
Removing finishings										
cement and sand to floors	-	-	-	-	0.95	**10.13**	2.27	**12.40**	m2	**14.88**
granolithic to floors	-	-	-	-	1.01	**10.84**	2.27	**13.11**	m2	**15.73**
granolithic to treads and risers	-	-	-	-	1.57	**16.85**	2.27	**19.11**	m2	**22.94**
plastic or similar tiles to floors	-	-	-	-	0.22	**2.36**	0.23	**2.59**	m2	**3.11**
plastic or similar tiles to floors; cleaning off for new	-	-	-	-	0.52	**5.54**	0.23	**5.77**	m2	**6.93**
plastic or similar tiles to floors and screed under	-	-	-	-	1.16	**12.37**	2.50	**14.87**	m2	**17.85**
linoleum and underlay to floors	-	-	-	-	0.08	**0.82**	2.27	**3.09**	m2	**3.71**
ceramic or quarry tiles to floors	-	-	-	-	1.33	**14.26**	1.36	**15.61**	m2	**18.73**
ceramic or quarry tiles to floors and screed under	-	-	-	-	2.63	**28.16**	3.62	**31.78**	m2	**38.14**
extra; cleaning and keying surface of concrete under	-	-	-	-	0.58	**6.24**	0.05	**6.30**	m2	**7.56**
asphalt to floors on loose underlay	-	-	-	-	0.39	**4.12**	0.91	**5.04**	m2	**6.04**
asphalt to floors keyed to concrete or screed	-	-	-	-	0.84	**8.95**	0.91	**9.87**	m2	**11.84**
granolithic skirtings	-	-	-	-	0.36	**3.89**	0.23	**4.12**	m	**4.95**
plastic or similar skirtings; cleaning off for new	-	-	-	-	0.14	**1.53**	0.05	**1.58**	m	**1.90**
timber skirtings	-	-	-	-	0.13	**1.41**	0.23	**1.65**	m	**1.98**
ceramic or quarry tile skirtings	-	-	-	-	0.47	**5.07**	0.26	**5.33**	m	**6.39**
plaster to walls	-	-	-	-	0.66	**7.07**	0.91	**7.98**	m2	**9.58**
rendering to walls	-	-	-	-	0.96	**10.25**	0.91	**11.16**	m2	**13.39**
rough cast to walls	-	-	-	-	0.99	**10.60**	1.15	**11.75**	m2	**14.10**
match boarding linings to walls with battens	-	-	-	-	0.42	**4.48**	1.36	**5.83**	m2	**7.00**
plywood or similar sheet linings to walls with battens	-	-	-	-	0.34	**3.65**	1.36	**5.01**	m2	**6.01**
insulating board linings to walls with battens	-	-	-	-	0.34	**3.65**	1.36	**5.01**	m2	**6.01**
plasterboard to walls	-	-	-	-	0.34	**3.65**	0.69	**4.34**	m2	**5.21**
plasterboard dry linings to walls	-	-	-	-	0.55	**5.89**	0.69	**6.58**	m2	**7.90**
plasterboard and skim to walls	-	-	-	-	0.66	**7.07**	0.91	**7.98**	m2	**9.58**
lath and plaster to walls	-	-	-	-	0.48	**5.18**	1.15	**6.33**	m2	**7.60**
metal lath and plaster to walls	-	-	-	-	0.36	**3.89**	1.15	**5.03**	m2	**6.04**
ceramic tiles to walls	-	-	-	-	0.83	**8.84**	0.69	**9.53**	m2	**11.43**
ceramic tiles to walls and backing under	-	-	-	-	1.22	**13.08**	1.15	**14.22**	m2	**17.07**
asphalt coverings to walls keyed to concrete or brickwork	-	-	-	-	0.73	**7.78**	0.91	**8.69**	m2	**10.43**
plaster to ceilings	-	-	-	-	0.81	**8.72**	0.91	**9.63**	m2	**11.56**
match boarding linings to ceilings with battens	-	-	-	-	0.59	**6.36**	1.36	**7.72**	m2	**9.26**
plywood or similar sheet linings to ceilings with battens	-	-	-	-	0.42	**4.48**	1.36	**5.83**	m2	**7.00**
insulating board linings to ceilings with battens	-	-	-	-	0.29	**3.06**	1.36	**4.42**	m2	**5.30**
plasterboard to ceilings	-	-	-	-	0.36	**3.89**	0.69	**4.58**	m2	**5.49**
plasterboard and skim to ceilings	-	-	-	-	0.44	**4.71**	0.91	**5.62**	m2	**6.75**
lath and plaster to ceilings	-	-	-	-	0.65	**6.95**	1.15	**8.10**	m2	**9.72**
metal lath and plaster to ceilings	-	-	-	-	0.52	**5.54**	1.15	**6.68**	m2	**8.02**
Hacking surfaces of concrete as key for new finishings										
ceiling	-	-	-	-	0.74	**7.89**	0.05	**7.95**	m2	**9.53**
floor	-	-	-	-	0.58	**6.24**	0.05	**6.30**	m2	**7.56**
wall	-	-	-	-	0.58	**6.24**	0.05	**6.30**	m2	**7.56**
Hacking surfaces of brick wall and raking out joints as key for new finishings										
wall	-	-	-	-	0.66	**7.07**	0.05	**7.12**	m2	**8.54**
Removing finishings; extending and making good finishings										
plaster cornices up to 100mm girth on face	0.38	10.00	**0.42**	0.86	0.86	**21.38**	0.09	**21.89**	m	**26.27**
plaster cornices 100 - 200mm girth on face	0.62	10.00	**0.69**	0.97	0.97	**24.12**	0.23	**25.04**	m	**30.05**
plaster cornices 200 - 300mm girth on face	1.10	10.00	**1.21**	1.07	1.07	**26.59**	0.69	**28.49**	m	**34.19**
plaster ceiling roses up to 300mm diameter	0.19	10.00	**0.21**	0.45	0.45	**11.24**	0.18	**11.63**	nr	**13.96**
plaster ceiling roses 300 - 450mm diameter	0.38	10.00	**0.42**	0.70	0.70	**17.54**	0.36	**18.33**	nr	**21.99**
plaster ceiling roses 450 - 600mm diameter	0.75	10.00	**0.82**	0.99	0.99	**24.67**	0.72	**26.21**	nr	**31.45**
Removing finishings; carefully handling and disposing toxic or other special waste by approved method										
asbestos cement sheet linings to walls	-	-	-	2.20	2.20	**54.82**	3.62	**58.45**	m2	**70.14**
asbestos cement sheet linings to ceilings	-	-	-	3.30	3.30	**82.24**	3.62	**85.86**	m2	**103.03**
Removal of roof coverings										
Removing coverings										
felt to roofs	-	-	-	-	0.31	**3.30**	0.91	**4.21**	m2	**5.05**
felt skirtings to roofs	-	-	-	-	0.36	**3.89**	0.09	**3.98**	m	**4.77**
asphalt to roofs	-	-	-	-	0.39	**4.12**	1.15	**5.27**	m2	**6.32**
asphalt skirtings to roofs; on expanded metal reinforcement	-	-	-	-	0.22	**2.36**	0.18	**2.54**	m	**3.05**
asphalt skirtings to roofs; keyed to concrete, brickwork, etc.	-	-	-	-	0.33	**3.53**	0.18	**3.72**	m	**4.46**
asphalt coverings to roofs; on expanded metal reinforcement; per 100mm of width	-	-	-	-	0.14	**1.53**	0.14	**1.67**	m	**2.01**
asphalt coverings to roofs; keyed to concrete, brickwork, etc.; per 100mm of width	-	-	-	-	0.22	**2.36**	0.14	**2.50**	m	**3.00**
slate to roofs	-	-	-	-	0.37	**4.01**	0.91	**4.92**	m2	**5.90**
tiles to roofs	-	-	-	-	0.34	**3.65**	1.36	**5.01**	m2	**6.01**
extra; removing battens	-	-	-	-	0.31	**3.30**	0.91	**4.21**	m2	**5.05**
extra; removing counter battens	-	-	-	-	0.11	**1.18**	0.18	**1.36**	m2	**1.63**
extra; removing underfelt	-	-	-	-	0.22	**2.36**	0.36	**2.72**	m2	**3.27**
lead to roofs	-	-	-	-	0.53	**5.65**	0.91	**6.57**	m2	**7.88**
lead flashings to roofs; per 25mm of girth	-	-	-	-	0.06	**0.59**	0.05	**0.64**	m	**0.77**
zinc to roofs	-	-	-	-	0.37	**4.01**	0.91	**4.92**	m2	**5.90**
zinc flashings to roofs; per 25mm of girth	-	-	-	-	0.03	**0.35**	0.05	**0.41**	m	**0.49**

Labour hourly rates: (except Specialists) Craft Operatives £14.21 Labourer £10.71 Rates are national average prices. Refer to REGIONAL VARIATIONS for indicative levels of overall pricing in regions	MATERIALS			LABOUR			RATES			
	Del to Site £	Waste %	Material Cost £	Craft Optve Hrs	Lab Hrs	Labour Cost £	Sunds £	Nett Rate £	Unit	Gross Rate (+20%) £
C90: ALTERATIONS - SPOT ITEMS Cont.										
Removal of roof coverings Cont.										
Removing coverings Cont.										
copper to roofs	-	-	-	-	0.37	4.01	0.91	4.92	m2	5.90
copper flashings to roofs; per 25mm of girth	-	-	-	-	0.03	0.35	0.05	0.41	m	0.49
corrugated metal sheeting to roofs	-	-	-	-	0.34	3.65	1.36	5.01	m2	6.01
corrugated translucent sheeting to roofs	-	-	-	-	0.34	3.65	1.36	5.01	m2	6.01
board roof decking	-	-	-	-	0.34	3.65	1.36	5.01	m2	6.01
woodwool roof decking	-	-	-	-	0.43	4.59	1.36	5.95	m2	7.14
Removing coverings; setting aside for re-use										
slate to roofs	-	-	-	-	0.66	7.07	-	7.07	m2	8.48
tiles to roofs	-	-	-	-	0.66	7.07	-	7.07	m2	8.48
clean and stack 405 x 205mm slates; per 100	-	-	-	-	1.65	17.67	-	17.67	nr	21.21
clean and stack 510 x 255mm slates; per 100	-	-	-	-	2.08	22.27	-	22.27	nr	26.72
clean and stack 610 x 305mm slates; per 100	-	-	-	-	2.46	26.39	-	26.39	nr	31.67
clean and stack concrete tiles; per 100	-	-	-	-	2.46	26.39	-	26.39	nr	31.67
clean and stack plain tiles; per 100	-	-	-	-	1.65	17.67	-	17.67	nr	21.21
Removing coverings; carefully handling and disposing toxic or other special waste by approved method										
asbestos cement sheeting to roofs	-	-	-	2.20	2.20	54.82	3.62	58.45	m2	70.14
asbestos cement roof decking	-	-	-	2.20	2.20	54.82	3.62	58.45	m2	70.14
Removal of woodwork										
Removing										
stud partitions plastered both sides	-	-	-	-	0.73	7.78	3.42	11.19	m2	13.43
roof boarding	-	-	-	-	0.21	2.24	1.15	3.39	m2	4.06
roof boarding; prepare joists for new	-	-	-	-	0.50	5.30	1.15	6.45	m2	7.74
gutter boarding and the like	-	-	-	-	0.21	2.24	1.15	3.39	m2	4.06
weather boarding and battens	-	-	-	-	0.26	2.83	1.36	4.18	m2	5.02
tilting fillets, angle fillets and the like	-	-	-	-	0.10	1.06	0.09	1.15	m	1.38
fascia boards 150 mm wide	-	-	-	-	0.11	1.18	0.18	1.36	m	1.63
barge boards 200mm wide	-	-	-	-	0.11	1.18	0.23	1.41	m	1.70
soffit boards 300mm wide	-	-	-	-	0.11	1.18	0.36	1.54	m	1.85
floor boarding	-	-	-	-	0.36	3.89	1.15	5.03	m2	6.04
handrails and brackets	-	-	-	-	0.11	1.18	0.72	1.90	m	2.27
balustrades complete down to and with cappings to aprons or strings	-	-	-	0.64	0.11	10.24	1.81	12.06	m	14.47
newel posts; cut off flush with landing or string	-	-	-	0.53	0.09	8.45	0.46	8.90	nr	10.68
ends of treads projecting beyond face of cut outer string including scotia under; make good treads where balusters removed	-	-	-	0.99	0.17	15.84	1.36	17.19	m	20.63
dado or picture rails with grounds	-	-	-	-	0.11	1.18	0.18	1.36	m	1.63
architrave	-	-	-	-	0.07	0.71	0.18	0.89	m	1.07
window boards and bearers	-	-	-	-	0.13	1.41	0.36	1.78	m	2.13
Removing; setting aside for reuse										
handrails and brackets	-	-	-	0.28	0.04	4.38	-	4.38	m	5.25
Removal of windows and doors										
Removing										
metal windows with internal and external sills; in conjunction with demolition										
997 x 923mm	-	-	-	-	0.44	4.71	2.18	6.89	nr	8.27
1486 x 923mm	-	-	-	-	0.55	5.89	3.13	9.02	nr	10.82
1486 x 1513mm	-	-	-	-	0.66	7.07	5.14	12.20	nr	14.65
1994 x 1513mm	-	-	-	-	0.77	8.25	6.86	15.10	nr	18.12
metal windows with internal and external sills; preparatory to filling openings; cut out lugs										
997 x 923mm	-	-	-	-	1.43	15.32	2.18	17.49	nr	20.99
1486 x 923mm	-	-	-	-	1.76	18.85	3.13	21.98	nr	26.37
1486 x 1513mm	-	-	-	-	2.20	23.56	5.14	28.70	nr	34.44
1994 x 1513mm	-	-	-	-	2.53	27.10	6.86	33.95	nr	40.74
wood single or multi-light casements and frames with internal and external sills; in conjunction with demolition										
440 x 920mm	-	-	-	-	0.31	3.30	1.41	4.71	nr	5.65
1225 x 1070mm	-	-	-	-	0.32	3.42	4.44	7.86	nr	9.43
2395 x 1225mm	-	-	-	-	0.33	3.53	9.98	13.52	nr	16.22
wood single or multi-light casements and frames with internal and external sills; preparatory to filling openings; cut out fixing cramps										
440 x 920mm	-	-	-	-	1.38	14.73	1.41	16.13	nr	19.36
1225 x 1070mm	-	-	-	-	1.87	20.03	4.44	24.47	nr	29.37
2395 x 1225mm	-	-	-	-	2.77	29.69	9.98	39.67	nr	47.61
wood cased frames and sashes with internal and external sills complete with accessories and weights; in conjunction with demolition										
610 x 1225mm	-	-	-	-	0.67	7.19	3.42	10.60	nr	12.72
915 x 1525mm	-	-	-	-	0.73	7.78	6.35	14.12	nr	16.95
1370 x 1525mm	-	-	-	-	0.83	8.84	9.48	18.31	nr	21.97
wood cased frames and sashes with internal and external sills complete with accessories and weights; preparatory to filling openings; cut out fixing cramps										
610 x 1225mm	-	-	-	-	1.97	21.09	3.42	24.50	nr	29.40
915 x 1525mm	-	-	-	-	2.82	30.16	6.35	36.51	nr	43.81
1370 x 1525mm	-	-	-	-	4.52	48.42	9.48	57.90	nr	69.48
wood doors										
single internal doors	-	-	-	-	0.25	2.71	2.72	5.43	nr	6.52

Labour hourly rates: (except Specialists) Craft Operatives £14.21 Labourer £10.71 Rates are national average prices. Refer to REGIONAL VARIATIONS for indicative levels of overall pricing in regions	MATERIALS			LABOUR				RATES		
	Del to Site £	Waste %	Material Cost £	Craft Optve Hrs	Lab Hrs	Labour Cost £	Sunds £	Nett Rate £	Unit	Gross Rate (+20%) £
C90: ALTERATIONS - SPOT ITEMS Cont.										
Removal of windows and doors Cont.										
Removing Cont.										
wood doors Cont.										
single internal doors and frames; in conjunction with demolition	-	-	-	-	0.20	**2.12**	6.13	**8.25**	nr	**9.90**
single internal doors and frames; preparatory to filling openings; cut out fixing cramps	-	-	-	-	1.10	**11.78**	6.13	**17.91**	nr	**21.49**
double internal doors	-	-	-	-	0.52	**5.54**	5.45	**10.99**	nr	**13.18**
double internal doors and frames; in conjunction with demolition	-	-	-	-	0.21	**2.24**	9.31	**11.55**	nr	**13.85**
double internal doors and frames; preparatory to filling openings; cut out fixing cramps	-	-	-	-	1.32	**14.14**	9.31	**23.44**	nr	**28.13**
single external doors	-	-	-	-	0.28	**2.95**	4.54	**7.48**	nr	**8.98**
single external doors and frames; in conjunction with demolition	-	-	-	-	0.20	**2.12**	8.16	**10.28**	nr	**12.34**
single external doors and frames; preparatory to filling openings; cut out fixing cramps	-	-	-	-	1.10	**11.78**	8.16	**19.94**	nr	**23.93**
double external doors	-	-	-	-	0.52	**5.54**	5.45	**10.99**	nr	**13.18**
double external doors and frames; in conjunction with demolition	-	-	-	-	0.21	**2.24**	9.31	**11.55**	nr	**13.85**
double external doors and frames; preparatory to filling openings; cut out fixing cramps	-	-	-	-	1.32	**14.14**	9.31	**23.44**	nr	**28.13**
single door frames; in conjunction with demolition	-	-	-	-	0.19	**2.00**	3.42	**5.42**	nr	**6.50**
single door frames; preparatory to filling openings; cut out fixing cramps	-	-	-	-	0.85	**9.07**	3.42	**12.49**	nr	**14.98**
double door frames; in conjunction with demolition	-	-	-	-	0.20	**2.12**	3.86	**5.98**	nr	**7.17**
double door frames; preparatory to filling openings; cut out fixing cramps	-	-	-	-	0.90	**9.66**	3.86	**13.52**	nr	**16.22**
Removing; setting aside for reuse										
metal windows with internal and external sills; in conjunction with demolition										
997 x 923mm	-	-	-	0.77	0.55	**16.83**	-	**16.83**	nr	**20.20**
1486 x 923mm	-	-	-	0.94	0.72	**20.94**	-	**20.94**	nr	**25.13**
1486 x 1513mm	-	-	-	0.99	0.83	**22.90**	-	**22.90**	nr	**27.48**
1994 x 1513mm	-	-	-	1.10	0.94	**25.64**	-	**25.64**	nr	**30.77**
metal windows with internal and external sills; preparatory to filling openings; cut out lugs										
997 x 923mm	-	-	-	0.77	1.54	**27.44**	-	**27.44**	nr	**32.92**
1486 x 923mm	-	-	-	0.94	1.93	**33.90**	-	**33.90**	nr	**40.68**
1486 x 1513mm	-	-	-	0.99	2.37	**39.40**	-	**39.40**	nr	**47.28**
1994 x 1513mm	-	-	-	1.10	2.70	**44.49**	-	**44.49**	nr	**53.39**
wood single or multi-light casements and frames with internal and external sills; in conjunction with demolition										
440 x 920mm	-	-	-	0.55	0.40	**12.06**	-	**12.06**	nr	**14.47**
1225 x 1070mm	-	-	-	0.91	0.47	**18.04**	-	**18.04**	nr	**21.65**
2395 x 1225mm	-	-	-	1.10	0.51	**21.05**	-	**21.05**	nr	**25.26**
wood single or multi-light casements and frames with internal and external sills; preparatory to filling openings; cut out fixing cramps										
440 x 920mm	-	-	-	0.55	1.46	**23.48**	-	**23.48**	nr	**28.18**
1225 x 1070mm	-	-	-	0.91	2.02	**34.65**	-	**34.65**	nr	**41.58**
2395 x 1225mm	-	-	-	1.10	2.95	**47.20**	-	**47.20**	nr	**56.64**
wood cased frames and sashes with internal and external sills complete with accessories and weights; in conjunction with demolition										
610 x 1225mm	-	-	-	1.07	0.86	**24.35**	-	**24.35**	nr	**29.22**
915 x 1525mm	-	-	-	1.79	1.02	**36.43**	-	**36.43**	nr	**43.72**
1370 x 1525mm	-	-	-	2.13	1.18	**42.93**	-	**42.93**	nr	**51.52**
wood cased frames and sashes with internal and external sills complete with accessories and weights; preparatory to filling openings; cut out fixing cramps										
610 x 1225mm	-	-	-	1.07	2.16	**38.25**	-	**38.25**	nr	**45.90**
915 x 1525mm	-	-	-	1.79	3.11	**58.82**	-	**58.82**	nr	**70.58**
1370 x 1525mm	-	-	-	2.13	4.87	**82.51**	-	**82.51**	nr	**99.02**
note: the above rates assume that windows and sashes will be re-glazed when re-used										
wood doors										
single internal doors	-	-	-	0.55	0.34	**11.47**	-	**11.47**	nr	**13.76**
single internal doors and frames; in conjunction with demolition	-	-	-	1.10	0.37	**19.64**	-	**19.64**	nr	**23.56**
single internal doors and frames; preparatory to filling openings; cut out fixing cramps	-	-	-	1.10	1.28	**29.30**	-	**29.30**	nr	**35.16**
double internal doors	-	-	-	1.10	0.69	**23.05**	-	**23.05**	nr	**27.66**
double internal doors and frames; in conjunction with demolition	-	-	-	1.62	0.47	**28.04**	-	**28.04**	nr	**33.65**
double internal doors and frames; preparatory to filling openings; cut out fixing cramps	-	-	-	1.62	1.60	**40.06**	-	**40.06**	nr	**48.07**
single external doors	-	-	-	0.55	0.36	**11.70**	-	**11.70**	nr	**14.04**
single external doors and frames; in conjunction with demolition	-	-	-	1.10	0.37	**19.64**	-	**19.64**	nr	**23.56**
single external doors and frames; preparatory to filling openings; cut out fixing cramps	-	-	-	1.10	1.28	**29.30**	-	**29.30**	nr	**35.16**
double external doors	-	-	-	1.10	0.69	**23.05**	-	**23.05**	nr	**27.66**
double external doors and frames; in conjunction with demolition	-	-	-	1.62	0.48	**28.16**	-	**28.16**	nr	**33.79**
double external doors and frames; preparatory to filling openings; cut out fixing cramps	-	-	-	1.62	1.60	**40.06**	-	**40.06**	nr	**48.07**

Labour hourly rates: (except Specialists) Craft Operatives £14.21 Labourer £10.71 Rates are national average prices. Refer to REGIONAL VARIATIONS for indicative levels of overall pricing in regions	MATERIALS			LABOUR				RATES		
	Del to Site £	Waste %	Material Cost £	Craft Optve Hrs	Lab Hrs	Labour Cost £	Sunds £	Nett Rate £	Unit	Gross Rate (+20%) £
C90: ALTERATIONS - SPOT ITEMS Cont.										
Removal of windows and doors Cont.										
Removing; setting aside for reuse Cont.										
wood doors Cont.										
single door frames; in conjunction with demolition	-	-	-	0.55	0.28	10.76	-	10.76	nr	12.91
single door frames; preparatory to filling openings; cut out fixing cramps	-	-	-	0.55	0.94	17.83	-	17.83	nr	21.40
double door frames; in conjunction with demolition	-	-	-	0.55	0.29	10.88	-	10.88	nr	13.05
double door frames; preparatory to filling openings; cut out fixing cramps	-	-	-	0.55	0.99	18.42	-	18.42	nr	22.10
Removal of ironmongery										
Removing door ironmongery (piecing in doors and frames included elsewhere)										
butt hinges	-	-	-	-	0.07	0.71	0.01	0.72	nr	0.86
tee hinges up to 300mm	-	-	-	-	0.22	2.36	0.01	2.37	nr	2.84
floor spring hinges and top centres	-	-	-	-	1.01	10.84	0.26	11.10	nr	13.32
barrel bolts up to 200mm	-	-	-	-	0.14	1.53	0.01	1.54	nr	1.85
flush bolts up to 200mm	-	-	-	-	0.19	2.00	0.01	2.02	nr	2.42
indicating bolts	-	-	-	-	0.25	2.71	0.01	2.72	nr	3.27
double panic bolts	-	-	-	-	0.76	8.13	0.14	8.27	nr	9.93
single panic bolts	-	-	-	-	0.64	6.83	0.09	6.92	nr	8.31
Norfolk or Suffolk latches	-	-	-	-	0.25	2.71	0.01	2.72	nr	3.27
cylinder rim night latches	-	-	-	-	0.25	2.71	0.01	2.72	nr	3.27
rim locks or latches and furniture	-	-	-	-	0.25	2.71	0.01	2.72	nr	3.27
mortice dead locks	-	-	-	-	0.19	2.00	0.01	2.02	nr	2.42
mortice locks or latches and furniture	-	-	-	-	0.32	3.42	0.01	3.43	nr	4.12
Bales catches	-	-	-	-	0.13	1.41	0.01	1.43	nr	1.71
overhead door closers, surface fixed	-	-	-	-	0.32	3.42	0.14	3.56	nr	4.27
pull handles	-	-	-	-	0.13	1.41	0.01	1.43	nr	1.71
push plates	-	-	-	-	0.13	1.41	0.01	1.43	nr	1.71
kicking plates	-	-	-	-	0.19	2.00	0.05	2.05	nr	2.47
letter plates	-	-	-	-	0.13	1.41	0.01	1.43	nr	1.71
Removing window ironmongery										
sash centres	-	-	-	-	0.13	1.41	0.01	1.43	nr	1.71
sash fasteners	-	-	-	-	0.19	2.00	0.01	2.02	nr	2.42
sash lifts	-	-	-	-	0.08	0.82	0.01	0.84	nr	1.01
sash screws	-	-	-	-	0.19	2.00	0.01	2.02	nr	2.42
casement fasteners	-	-	-	-	0.13	1.41	0.01	1.43	nr	1.71
casement stays	-	-	-	-	0.13	1.41	0.01	1.43	nr	1.71
quadrant stays	-	-	-	-	0.13	1.41	0.01	1.43	nr	1.71
fanlight catches	-	-	-	-	0.14	1.53	0.01	1.54	nr	1.85
curtain tracks	-	-	-	-	0.17	1.77	0.09	1.86	m	2.23
Removing sundry ironmongery										
hat and coat hooks	-	-	-	-	0.08	0.82	0.01	0.84	nr	1.01
cabin hooks and eyes	-	-	-	-	0.13	1.41	0.01	1.43	nr	1.71
shelf brackets	-	-	-	-	0.08	0.82	0.05	0.88	nr	1.05
toilet roll holders	-	-	-	-	0.10	1.06	0.05	1.11	nr	1.33
towel rollers	-	-	-	-	0.13	1.41	0.09	1.50	nr	1.81
Removal of metalwork										
Removing										
balustrades 1067mm high	-	-	-	-	0.89	9.54	1.17	10.72	m	12.86
wire mesh screens with timber or metal beads screwed on										
generally	-	-	-	-	0.68	7.30	1.15	8.45	m2	10.14
225 x 225mm	-	-	-	-	0.22	2.36	0.09	2.45	nr	2.94
305 x 305mm	-	-	-	-	0.24	2.59	0.14	2.74	nr	3.28
guard bars of vertical bars at 100mm centres welded to horizontal fixing bars at 450mm centres, fixed with bolts	-	-	-	-	0.68	7.30	1.15	8.45	m2	10.14
guard bars of vertical bars at 100mm centres welded to horizontal fixing bars at 450mm centres, fixed with screws	-	-	-	-	0.43	4.59	1.15	5.74	m2	6.89
extra; piecing in softwood after removal of bolts	-	-	-	0.28	0.04	4.38	0.12	4.50	nr	5.40
small bracket	-	-	-		0.20	2.12	0.09	2.21	nr	2.65
Removing; making good finishings										
balustrades 1067mm high; making good mortices in treads	-	-	-	1.07	1.07	26.59	1.49	28.08	m	33.69
guard bars of vertical bars at 100mm centres welded to horizontal fixing bars at 450mm centres fixed in mortices; making good concrete or brickwork and plaster	-	-	-	0.66	1.56	26.11	1.55	27.66	m2	33.19
small bracket built or cast in; making good concrete or brickwork and plaster	-	-	-	0.28	0.28	6.85	0.61	7.47	nr	8.96
Removal of manholes										
Removing										
manholes overall size; remove cover and frame; break up brick sides and concrete bottom; fill in void with hardcore										
overall size 914 x 1067mm and 600mm deep to invert	2.20	30.00	2.86	-	11.00	117.81	4.54	125.21	nr	150.25
overall size 914 x 1219mm and 900mm deep to invert	4.40	30.00	5.72	-	12.65	135.48	4.99	146.19	nr	175.43

Labour hourly rates: (except Specialists) Craft Operatives £14.21 Labourer £10.71 Rates are national average prices. Refer to REGIONAL VARIATIONS for indicative levels of overall pricing in regions	MATERIALS			LABOUR				RATES		
	Del to Site £	Waste %	Material Cost £	Craft Optve Hrs	Lab Hrs	Labour Cost £	Sunds £	Nett Rate £	Unit	Gross Rate (+20%) £
C90: ALTERATIONS - SPOT ITEMS Cont.										
Removal of manholes Cont.										
Removing Cont.										
manholes overall size; remove cover and frame; break up brick sides and concrete bottom; fill in void with hardcore Cont.										
manhole covers and frames; clean off brickwork	-	-	-	-	0.55	**5.89**	0.81	**6.70**	nr	**8.04**
fresh air inlets; clean out pipe socket	-	-	-	-	0.55	**5.89**	0.26	**6.15**	nr	**7.38**
Removal of fencing										
Removing										
chestnut pale fencing with posts										
610mm high	-	-	-	-	0.33	**3.53**	0.72	**4.25**	m	**5.10**
914mm high	-	-	-	-	0.37	**4.01**	1.09	**5.10**	m	**6.12**
1219mm high	-	-	-	-	0.42	**4.48**	1.46	**5.94**	m	**7.12**
close boarded fencing with posts										
1219mm high	-	-	-	-	0.75	**8.01**	1.67	**9.68**	m	**11.62**
1524mm high	-	-	-	-	0.89	**9.54**	2.09	**11.63**	m	**13.95**
1829mm high	-	-	-	-	1.00	**10.72**	2.50	**13.22**	m	**15.87**
chain link fencing with posts										
914mm high	-	-	-	-	0.32	**3.42**	0.81	**4.22**	m	**5.07**
1219mm high	-	-	-	-	0.37	**4.01**	3.36	**7.37**	m	**8.84**
1524mm high	-	-	-	-	0.42	**4.48**	1.36	**5.83**	m	**7.00**
1829mm high	-	-	-	-	0.46	**4.95**	1.67	**6.62**	m	**7.94**
timber gate and posts	-	-	-	-	1.08	**11.55**	3.42	**14.96**	nr	**17.95**
Strutting for forming openings										
Strutting generally										
the cost of strutting in connection with forming openings is to be added to the cost of forming openings; the following are examples of costs for strutting for various sizes and locations of openings. Costs are based on three uses of timber										
Strutting for forming small openings in internal load bearing walls										
opening size 900 x 2000mm	9.80	5.00	**10.29**	2.20	0.36	**35.15**	0.86	**46.30**	nr	**55.56**
opening size 2000 x 2300mm	21.78	5.00	**22.87**	5.50	0.91	**87.93**	1.86	**112.67**	nr	**135.20**
Strutting for forming openings in external one brick walls										
opening size 900 x 2000mm	14.71	5.00	**15.44**	3.30	0.55	**52.78**	1.49	**69.71**	nr	**83.65**
opening size 1500 x 1200mm	14.71	5.00	**15.44**	3.30	0.55	**52.78**	1.49	**69.71**	nr	**83.65**
Strutting for forming openings in external 280mm brick cavity walls										
opening size 900 x 2000mm	20.15	5.00	**21.16**	4.40	0.73	**70.30**	1.67	**93.13**	nr	**111.75**
opening size 1500 x 1200mm	20.15	5.00	**21.16**	4.40	0.73	**70.30**	1.67	**93.13**	nr	**111.75**
Strutting for forming large openings in internal load bearing walls on the ground floor of an average two storey building with timber floors and pitched roof; with load bearing surface 450mm below ground floor; including cutting holes and making good										
opening size 3700 x 2300mm	136.13	5.00	**142.93**	22.00	22.00	**548.24**	19.76	**710.93**	nr	**853.12**
Unit rates for pricing the above and similar work										
plates, struts, braces and wedges in supports to floor and roof	1.90	5.00	**2.00**	0.31	0.31	**7.68**	0.29	**9.96**	m	**11.95**
dead shore and needle, sole plates, braces and wedges including cutting holes and making good ...	68.61	5.00	**72.04**	11.00	11.00	**274.12**	9.84	**356.00**	nr	**427.20**
Strutting for forming large openings in external one brick walls on the ground floor of an average two storey building as described above										
opening size 6000 x 2500mm	275.52	5.00	**289.29**	44.00	44.00	**1096.48**	39.20	**1424.97**	nr	**1709.96**
Unit rates for pricing the above and similar work										
strutting to window opening over new opening	13.72	5.00	**14.40**	2.20	2.20	**54.82**	1.94	**71.17**	nr	**85.40**
plates, struts, braces and wedges in supports to floor and roof	1.85	5.00	**1.94**	0.31	0.31	**7.68**	0.29	**9.90**	m	**11.88**
dead shore and needle, sole plates, braces and wedges including cutting holes and making good ...	107.92	5.00	**113.32**	17.27	17.27	**430.37**	15.45	**559.13**	nr	**670.96**
set of two raking shores with 50mm wall piece, wedges and dogs	27.56	5.00	**28.93**	4.40	4.40	**109.65**	3.90	**142.48**	nr	**170.97**
Strutting for forming large openings in external 280mm brick cavity walls on the ground floor of an average two storey building as above described										
opening size 6000 x 2500mm	301.98	5.00	**317.08**	48.29	48.29	**1203.39**	43.09	**1563.56**	nr	**1876.28**
Unit rates for pricing the above and similar work										
strutting to window opening over new opening	27.45	5.00	**28.82**	4.40	4.40	**109.65**	3.90	**142.36**	nr	**170.84**
plates, struts, braces and wedges in supports to floor and roof	1.85	5.00	**1.94**	0.31	0.31	**7.68**	0.29	**9.90**	m	**11.88**
dead shore and needle, sole plates, braces and wedges including cutting holes and making good ...	114.35	5.00	**120.06**	18.37	18.37	**457.78**	16.46	**594.31**	nr	**713.17**
set of two raking shores with 50mm wall piece, wedges and dogs	27.87	5.00	**29.27**	4.40	4.40	**109.65**	3.90	**142.81**	nr	**171.38**

Labour hourly rates: (except Specialists) Craft Operatives £14.21 Labourer £10.71 Rates are national average prices. Refer to REGIONAL VARIATIONS for indicative levels of overall pricing in regions	MATERIALS			LABOUR				RATES		
	Del to Site £	Waste %	Material Cost £	Craft Optve Hrs	Lab Hrs	Labour Cost £	Sunds £	Nett Rate £	Unit	Gross Rate (+20%) £
C90: ALTERATIONS - SPOT ITEMS Cont.										
Spot items										
Spot items generally the following items are usually specified as spot items where all work in all trades is included in one item; to assist in adapting these items for varying circumstances, unit rates have been given for the component operation where possible while these items are not in accordance with SMM7 either by description or measurement, it is felt that the pricing information given will be of value to the reader										
Breaking up and reinstatement of concrete floors for excavations										
Break up concrete bed and 150mm hardcore bed under for a width of 760mm for excavation of trench and reinstate with new hardcore and in-situ concrete, BS 5328, designed mix C20, 20mm aggregate; make good up to new wall and existing concrete bed both side										
with 100mm plain concrete bed....................	11.91	10.00	**13.42**	-	2.97	**31.81**	8.42	**53.65**	m	**64.37**
with 150mm plain concrete bed....................	17.08	10.00	**19.11**	-	5.50	**58.91**	10.19	**88.20**	m	**105.84**
with 100mm concrete bed reinforced with steel fabric to BS 4483, Reference A193................	13.46	10.00	**15.12**	-	4.46	**47.71**	8.42	**71.25**	m	**85.50**
with 150mm concrete bed reinforced with steel fabric to BS 4483, Reference A193................	18.63	10.00	**20.81**	-	8.06	**86.35**	10.19	**117.35**	m	**140.83**
Openings through concrete										
Form opening for door frame through 150mm reinforced concrete wall plastered both sides and with skirting both sides; take off skirtings, cut opening through wall, square up reveals, make good existing plaster up to new frame both sides and form fitted ends on existing skirtings up to new frame; extend cement and sand floor screed and vinyl floor covering through opening and make good up to existing										
838 x 2032mm	20.68	-	**20.68**	24.75	30.80	**681.57**	39.37	**741.61**	nr	**889.93**
Unit rates for pricing the above and similar work										
cut opening through wall										
100mm thick....................................	-	-	-	-	5.83	**62.44**	9.27	**71.71**	m2	**86.05**
150mm thick....................................	-	-	-	-	8.69	**93.07**	13.86	**106.93**	m2	**128.31**
200mm thick....................................	-	-	-	-	11.55	**123.70**	18.44	**142.15**	m2	**170.57**
make good fair face around opening	-	-	-	-	0.88	**9.42**	0.65	**10.08**	m	**12.09**
square up reveals to opening										
100mm wide.....................................	-	-	-	0.83	0.83	**20.56**	1.62	**22.18**	m	**26.61**
150mm wide.....................................	-	-	-	1.18	1.18	**29.33**	2.28	**31.61**	m	**37.93**
200mm wide.....................................	-	-	-	1.52	1.52	**37.83**	3.62	**41.45**	m	**49.74**
make good existing plaster up to new frame ...	0.35	10.00	**0.39**	0.61	0.61	**15.08**	-	**15.46**	m	**18.56**
13mm two coat hardwall plaster to reveal not exceeding 300mm wide	0.70	10.00	**0.77**	0.70	0.70	**17.54**	-	**18.32**	m	**21.98**
take off old skirting	-	-	-	0.24		**3.44**	0.13	**3.57**	m	**4.28**
form fitted end on existing skirting up to new frame..	-	-	-	0.42		**5.94**	-	**5.94**	nr	**7.13**
short length of old skirting up to 100mm long with mitre with existing one end	-	-	-	0.64		**9.07**	0.05	**9.12**	nr	**10.94**
short length of old skirting up to 200mm long with mitres with existing both ends	-	-	-	1.27	-	**17.98**	0.08	**18.05**	nr	**21.66**
38mm cement and sand (1:3) screeded bed in opening...	5.43	10.00	**5.97**	1.69	1.69	**42.21**	-	**48.18**	m2	**57.82**
vinyl or similar floor covering to match existing, fixed with adhesive, not exceeding 300mm wide......................................	7.09	10.00	**7.79**	0.17	0.17	**4.11**	0.25	**12.15**	m	**14.58**
iroko threshold; twice oiled; plugged and screwed										
15 x 100mm.....................................	5.51	10.00	**6.06**	0.99	-	**14.07**	0.07	**20.20**	m	**24.23**
15 x 150mm.....................................	8.06	10.00	**8.87**	1.32	-	**18.76**	0.08	**27.70**	m	**33.25**
15 x 200mm.....................................	9.75	10.00	**10.72**	1.71	-	**24.23**	0.12	**35.07**	m	**42.08**
Form opening for staircase through 150mm reinforced concrete suspended floor plastered on soffit and with screed and vinyl floor covering on top; take up floor covering and screed, cut opening through floor, square up edges of slab, make good ceiling plaster up to lining and make good screed and floor covering to lining										
900 x 2134mm	81.39	-	**81.39**	20.90	42.90	**756.45**	50.84	**888.67**	nr	**1066.41**
Unit rates for pricing the above and similar work										
take up vinyl or similar floor covering..........	-	-	-	-	0.61	**6.48**	0.18	**6.66**	m2	**7.99**
hack up screed	-	-	-	-	1.65	**17.67**	1.68	**19.35**	m2	**23.22**
cut opening through 150mm thick slab............	-	-	-	-	8.69	**93.07**	13.86	**106.93**	m2	**128.31**
cut opening through 200mm thick slab............	-	-	-	-	11.55	**123.70**	18.44	**142.15**	m2	**170.57**
make good fair face around opening...............	-	-	-	-	0.88	**9.42**	0.61	**10.04**	m	**12.04**
square up edges of 150mm thick slab............	-	-	-	1.18	1.18	**29.33**	2.41	**31.74**	m	**38.09**
square up edges of 200mm thick slab............	-	-	-	1.52	1.52	**37.83**	3.22	**41.05**	m	**49.26**
make good existing plaster up to new lining......	0.35	10.00	**0.39**	0.61	0.61	**15.08**	-	**15.46**	m	**18.56**
make good existing 38mm floor screed up to new lining	0.86	10.00	**0.94**	0.50	0.50	**12.34**	-	**13.28**	m	**15.93**
make good existing vinyl or similar floor covering up to new lining	9.45	10.00	**10.39**	0.74	0.74	**18.37**	0.39	**29.15**	m	**34.98**

Labour hourly rates: (except Specialists) Craft Operatives £14.21 Labourer £10.71 Rates are national average prices. Refer to REGIONAL VARIATIONS for indicative levels of overall pricing in regions	MATERIALS			LABOUR				RATES		
	Del to Site	Waste	Material Cost	Craft Optve	Lab	Labour Cost	Sunds	Nett Rate	Unit	Gross Rate (+20%)
	£	%	£	Hrs	Hrs	£	£	£		£
C90: ALTERATIONS - SPOT ITEMS Cont.										
Filling holes in concrete										
Fill holes in concrete structure where pipes removed with concrete to BS 5328, designed mix C20, 20mm aggregate; formwork										
wall; thickness 100mm; pipe diameter										
50mm	-	-	-	0.66	0.33	**12.91**	1.99	**14.91**	nr	**17.89**
100mm	-	-	-	0.86	0.43	**16.79**	2.63	**19.42**	nr	**23.30**
150mm	-	-	-	1.08	0.54	**21.09**	3.17	**24.26**	nr	**29.11**
wall; thickness 200mm; pipe diameter										
50mm	-	-	-	0.86	0.43	**16.79**	2.63	**19.42**	nr	**23.30**
100mm	-	-	-	1.08	0.54	**21.09**	3.17	**24.26**	nr	**29.11**
150mm	-	-	-	1.10	0.85	**24.70**	4.07	**28.77**	nr	**34.52**
floor; thickness 200mm; pipe diameter										
50mm	-	-	-	0.81	0.41	**15.93**	2.39	**18.31**	nr	**21.97**
100mm	-	-	-	1.02	0.51	**19.96**	3.05	**23.01**	nr	**27.61**
150mm	-	-	-	1.05	0.53	**20.50**	3.86	**24.36**	nr	**29.24**
Make good fair face to filling and surrounding work										
to one side of wall; pipe diameter										
50mm	-	-	-	0.11	0.11	**2.74**	0.27	**3.01**	nr	**3.62**
100mm	-	-	-	0.22	0.22	**5.48**	0.42	**5.90**	nr	**7.08**
150mm	-	-	-	0.28	0.28	**6.85**	0.51	**7.36**	nr	**8.83**
to soffit of floor; pipe diameter										
50mm	-	-	-	0.11	0.11	**2.74**	0.27	**3.01**	nr	**3.62**
100mm	-	-	-	0.22	0.22	**5.48**	0.42	**5.90**	nr	**7.08**
150mm	-	-	-	0.28	0.28	**6.85**	0.51	**7.36**	nr	**8.83**
Make good plaster to filling and surrounding work										
to one side of wall; pipe diameter										
50mm	-	-	-	0.21	0.21	**5.21**	0.27	**5.48**	nr	**6.58**
100mm	-	-	-	0.40	0.40	**9.87**	0.42	**10.29**	nr	**12.34**
150mm	-	-	-	0.50	0.50	**12.34**	0.51	**12.84**	nr	**15.41**
to soffit of floor; pipe diameter										
50mm	-	-	-	0.22	0.22	**5.48**	0.27	**5.76**	nr	**6.91**
100mm	-	-	-	0.44	0.44	**10.96**	0.42	**11.38**	nr	**13.66**
150mm	-	-	-	0.55	0.55	**13.71**	0.51	**14.21**	nr	**17.06**
Make good floor screed and surrounding work										
pipe diameter 50mm	-	-	-	0.21	0.21	**5.21**	0.12	**5.33**	nr	**6.39**
pipe diameter 100mm	-	-	-	0.40	0.40	**9.87**	0.25	**10.12**	nr	**12.14**
pipe diameter 150mm	-	-	-	0.50	0.50	**12.34**	0.34	**12.67**	nr	**15.21**
Make good vinyl or similar floor covering and surrounding work										
pipe diameter 50mm	-	-	-	0.55	0.55	**13.71**	2.03	**15.74**	nr	**18.89**
pipe diameter 100mm	-	-	-	1.10	1.10	**27.41**	2.72	**30.14**	nr	**36.16**
pipe diameter 150mm	-	-	-	1.32	1.32	**32.89**	3.82	**36.71**	nr	**44.06**
Fill holes in concrete structure where metal sections removed with concrete to BS 5328, designed mix C20, 20mm aggregate; formwork										
wall; thickness 100mm; section not exceeding 250mm deep	-	-	-	0.66	0.66	**16.45**	2.42	**18.87**	nr	**22.65**
wall; thickness 200mm; section not exceeding 250mm deep	-	-	-	0.83	0.83	**20.56**	3.53	**24.09**	nr	**28.91**
wall; thickness 100mm; section 250 - 500mm deep	-	-	-	0.83	0.83	**20.56**	3.53	**24.09**	nr	**28.91**
wall; thickness 200mm; section 250 - 500mm deep	-	-	-	0.99	0.99	**24.67**	4.26	**28.93**	nr	**34.72**
Make good fair face to filling and surrounding work										
to one side of wall; section not exceeding 250mm deep	-	-	-	0.24	0.24	**6.03**	0.23	**6.27**	nr	**7.52**
to one side of wall; section 250 - 500mm deep	-	-	-	0.29	0.29	**7.13**	0.36	**7.49**	nr	**8.99**
Make good plaster to filling and surrounding work										
to one side of wall; section not exceeding 250mm deep	-	-	-	0.39	0.39	**9.59**	0.64	**10.23**	nr	**12.28**
to one side of wall; section 250 - 500mm deep	-	-	-	0.47	0.47	**11.79**	0.96	**12.75**	nr	**15.30**
Filling openings in concrete										
Fill opening where door removed in 150mm reinforced concrete wall with concrete reinforced with 12mm mild steel bars at 300mm centres both ways tied to existing structure; hack off plaster to reveals, hack up floor covering and screed in opening, prepare edges of opening to form joint with filling, drill edges of opening and tie in new reinforcement, plaster filling both sides and extended skirting both sides; making good junction of new and existing plaster and skirtings and make good floor screed and vinyl covering up to filling both sides										
838 x 2032mm	136.49	-	**136.49**	24.75	37.95	**758.14**	12.55	**907.18**	nr	**1088.62**
Unit rates for pricing the above and similar work										
hack off plaster to reveal not exceeding 100mm wide	-	-	-	-	0.26	**2.83**	0.09	**2.92**	m	**3.50**
hack off plaster to reveal 100 - 200mm wide	-	-	-	-	0.36	**3.89**	0.18	**4.07**	m	**4.88**
take up vinyl or similar floor covering and screed	-	-	-	-	2.20	**23.56**	1.86	**25.43**	m2	**30.51**
prepare edge of opening 100mm wide to form joint with filling	-	-	-	0.28	0.28	**6.85**	-	**6.85**	m	**8.22**
prepare edge of opening 200mm wide to form joint with filling	-	-	-	0.39	0.39	**9.59**	-	**9.59**	m	**11.51**

Labour hourly rates: (except Specialists) Craft Operatives £14.21 Labourer £10.71 Rates are national average prices. Refer to REGIONAL VARIATIONS for indicative levels of overall pricing in regions	MATERIALS			LABOUR			RATES			
	Del to Site £	Waste %	Material Cost £	Craft Optve Hrs	Lab Hrs	Labour Cost £	Sunds £	Nett Rate £	Unit	Gross Rate (+20%) £
C90: ALTERATIONS - SPOT ITEMS Cont.										
Filling openings in concrete Cont.										
Unit rates for pricing the above and similar work Cont.										
concrete to BS 5328, designed mix C20, 20mm aggregate in walls 150 - 450m thick	121.90	5.00	128.00	-	13.20	141.37	0.53	269.90	m3	323.88
concrete to BS 5328, designed mix C20, 20mm aggregate in walls not exceeding 150mm thick	121.90	5.00	128.00	-	13.20	141.37	0.53	269.90	m3	323.88
12mm mild steel bar reinforcement; straight	1064.80	10.00	1171.28	93.50	-	1328.64	5.88	2505.79	t	3006.95
drill edge of opening and tie in new reinforcement	-	-	-	-	1.10	11.78	0.53	12.32	m	14.78
formwork and basic finish to wall, per side	9.41	10.00	10.35	2.75	2.75	68.53	1.28	80.16	m2	96.19
formwork and fine formed finish to wall, per side	9.41	10.00	10.35	3.03	3.03	75.38	1.43	87.17	m2	104.60
junction between new and existing fair face	-	-	-	-	0.33	3.53	-	3.53	m	4.24
13mm two coat hardwall plaster over 300mm wide	2.36	10.00	2.59	1.43	1.43	35.64	-	38.23	m2	45.88
25 x 100mm softwood chamfered skirting, primed all round	2.20	10.00	2.42	0.62	0.10	9.81	0.14	12.38	m	14.85
junction with existing	-	-	-	0.55	0.09	8.76	-	8.76	nr	10.51
make good floor screed and vinyl or similar floor covering up to filling not exceeding 300mm wide	8.66	10.00	9.52	1.23	1.23	30.70	0.42	40.64	m	48.77
Fill opening where staircase removed in 150mm reinforced concrete floor with concrete reinforced with 12mm mild steel bars at 225mm centres both ways tied to existing structure; remove timber lining, prepare edges of opening to form joint with filling, drill edges of opening tie in new reinforcement, plaster filling on soffit and extend cement and sand screed and vinyl floor covering to match existing; make good junction of new and existing plaster and vinyl floor covering										
900 x 2134mm	154.54	-	154.54	16.23	24.53	493.27	10.43	658.24	nr	789.89
Unit rates for pricing the above and similar work										
take off timber lining	-	-	-	-	0.26	2.83	0.13	2.96	m	3.55
prepare edge of opening 150mm wide to form joint with filling	-	-	-	0.33	0.33	8.22	-	8.22	m	9.87
prepare edge of opening 200mm wide to form joint with filling	-	-	-	0.39	0.39	9.59	-	9.59	m	11.51
concrete to BS 5328, designed mix C20, 20mm aggregate in slabs 150 - 450mm thick	121.90	5.00	128.00	-	13.20	141.37	0.53	269.90	m3	323.88
concrete to BS 5328, designed mix C20, 20mm aggregate in slabs not exceeding 150mm thick	121.90	5.00	128.00	-	13.20	141.37	0.53	269.90	m3	323.88
12mm mild steel bar reinforcement; straight	1064.80	10.00	1171.28	93.50	-	1328.64	5.88	2505.79	t	3006.95
drill edge of opening and tie in new reinforcement	-	-	-	-	1.10	11.78	0.53	12.32	m	14.78
formwork and basic finish to soffit, slab thickness not exceeding 200mm; height to soffit 1.50 - 3.00m	10.45	10.00	11.50	2.48	2.48	61.68	1.26	74.44	m2	89.32
formwork and basic finish to soffit, slab thickness 200 - 300mm; height to soffit 1.50 - 3.00m	11.55	10.00	12.71	2.75	2.75	68.53	1.43	82.67	m2	99.20
formwork and fine formed finish to soffit, slab thickness not exceeding 200mm; height to soffit 1.50 - 3.00m	10.45	10.00	11.50	2.75	2.75	68.53	1.46	81.48	m2	97.78
formwork and fine formed finish to soffit, slab thickness 200 - 300mm; height to soffit 1.50 - 3.00m	11.55	10.00	12.71	3.03	3.03	75.38	1.51	89.60	m2	107.52
junction between new and existing fair face	-	-	-	-	0.33	3.53	-	3.53	m	4.24
10mm two coat lightweight plaster over 300mm wide	2.36	10.00	2.59	1.43	1.43	35.64	-	38.23	m2	45.88
38mm cement and sand (1:3) trowelled bed	5.43	10.00	5.97	0.69	0.69	17.27	-	23.24	m2	27.89
vinyl or similar floor covering to match existing fixed with adhesive exceeding 300mm wide	23.62	10.00	25.98	1.38	0.55	25.43	0.73	52.14	m2	62.57
Openings through brickwork or blockwork										
Form openings for door frames through 100mm block partition plastered both sides and with skirting both sides; take off skirtings, cut opening through partition, insert 100 x 150mm precast concrete lintel and wedge and pin up over, quoin up jambs, extend plaster to faces of lintel and make good junction with existing plaster; make good existing plaster up to new frame both sides and form fitted ends on existing skirtings up to new frame; extend softwood board flooring through opening on and with bearers and make good up to existing										
838 x 2032mm	24.28	-	24.28	22.00	18.92	515.25	16.55	556.08	nr	667.30
Unit rates for pricing the above and similar work										
cut opening through 75mm block partition plastered both sides	-	-	-	1.21	1.43	32.51	5.10	37.61	m2	45.13
cut opening through 100mm block partition plastered both sides	-	-	-	1.43	1.72	38.70	5.98	44.68	m2	53.62
make good fair face around opening	-	-	-	0.28	0.28	6.85	0.13	6.98	m	8.38
precast concrete, BS 5328, designed mix C20, 20mm aggregate lintel 75 x 150 x 1200mm, reinforced with 1.01 kg of 12mm mild steel bars	9.57	10.00	10.52	0.88	0.88	21.93	0.33	32.78	nr	39.33
precast concrete, BS 5328, designed mix C20, 20mm aggregate lintel 100 x 150 x 1200mm, reinforced with 1.01 kg of 12mm mild steel bars	18.89	10.00	20.78	1.10	1.10	27.41	0.43	48.62	nr	58.34
precast concrete, BS 5328 designed mix C20, 20mm aggregate lintel 75 x 150 x 1200mm, reinforced with 1.01 kg of 12mm mild steel bars, fair finish all faces	9.57	10.00	10.53	0.88	0.88	21.93	0.33	32.78	nr	39.34

Labour hourly rates: (except Specialists) Craft Operatives £14.21 Labourer £10.71 Rates are national average prices. Refer to REGIONAL VARIATIONS for indicative levels of overall pricing in regions	MATERIALS			LABOUR				RATES		
	Del to Site	Waste	Material Cost	Craft Optve	Lab	Labour Cost	Sunds	Nett Rate	Unit	Gross Rate (+20%)
	£	%	£	Hrs	Hrs	£	£	£		£
C90: ALTERATIONS - SPOT ITEMS Cont.										
Openings through brickwork or blockwork Cont.										
Unit rates for pricing the above and similar work Cont.										
precast concrete, BS 5328 designed mix C20, 20mm aggregate lintel 100 x 150 x 1200mm, reinforced with 1.01 kg of 12mm mild steel bars, fair finish all faces	18.89	10.00	**20.78**	1.10	1.10	**27.41**	0.43	**48.62**	nr	**58.34**
wedge and pin up over lintel 75mm wide	-	-	-	0.55	0.55	**13.71**	0.81	**14.51**	m	**17.42**
wedge and pin up over lintel 100mm wide	-	-	-	0.73	0.73	**18.09**	1.07	**19.16**	m	**22.99**
quoin up 75mm wide jambs	-	-	-	0.57	0.57	**14.25**	0.81	**15.06**	m	**18.07**
quoin up 100mm wide jambs	-	-	-	0.75	0.75	**18.64**	1.07	**19.71**	m	**23.65**
extend 13mm hardwall plaster to face of lintel	2.36	10.00	**2.59**	2.34	2.34	**58.39**	0.08	**61.06**	m2	**73.27**
make good existing plaster up to new frame	0.35	10.00	**0.39**	0.61	0.61	**15.08**	-	**15.46**	m	**18.56**
take off old skirting	-	-	-	0.24		**3.44**	0.13	**3.57**	m	**4.28**
form fitted end on existing skirting up to new frame	-	-	-	0.42		**5.94**	-	**5.94**	nr	**7.13**
38mm cement and sand (1:3) screeded bed in opening	5.43	10.00	**5.97**	1.69	1.69	**42.21**	-	**48.18**	m2	**57.82**
extend 25mm softwood board flooring through opening on and with bearers	34.22	20.00	**39.22**	17.82	2.97	**285.03**	0.70	**324.95**	m2	**389.94**
iroko threshold; twice oiled; plugged and screwed										
19 x 75mm	4.87	10.00	**5.36**	0.88	-	**12.50**	0.09	**17.96**	m	**21.55**
19 x 100mm	5.51	10.00	**6.06**	0.99	-	**14.07**	0.10	**20.23**	m	**24.28**
Form openings for door frames through one brick wall plastered one side and with external rendering other side; take off skirting, cut opening through wall, insert 215 x 150mm precast concrete lintel and wedge and pin up over; quoin up jambs; extend plaster to face of lintel and make good junction with existing plaster; make good existing plaster up to new frame one side; extend external rendering to face of lintel and to reveals and make good up to existing rendering and new frame the other side; form fitted ends on existing skirting up to new frame										
914 x 2082mm	39.80	-	**39.80**	30.80	27.94	**736.91**	36.63	**813.33**	nr	**976.00**
Unit rates for pricing the above and similar work										
cut opening through half brick wall	-	-	-	2.15	2.64	**58.75**	6.20	**64.96**	m2	**77.95**
cut opening through one brick wall	-	-	-	4.18	5.17	**114.77**	11.74	**126.51**	m2	**151.82**
make good fair face around opening	-	-	-	0.55	0.55	**13.71**	0.81	**14.51**	m	**17.42**
make good facings to match existing around opening	1.77	10.00	**1.95**	0.55	0.55	**13.71**	0.81	**16.46**	m	**19.75**
precast concrete, BS 5328, designed mix C20, 20mm aggregate lintel 102 x 150 x 1200mm, reinforced with 1.07 kg of 12mm mild steel bars	18.89	10.00	**20.78**	1.10	1.10	**27.41**	0.47	**48.66**	nr	**58.39**
precast concrete, BS 5328, designed mix C20, 20mm aggregate lintel 215 x 150 x 1200mm, reinforced with 2.13 kg of 12mm mild steel bars	38.70	10.00	**42.57**	1.65	1.65	**41.12**	0.96	**84.65**	nr	**101.58**
precast concrete, BS 5328, designed mix C20, 20mm aggregate lintel 102 x 150 x 1200mm, reinforced with 1.07 kg of 12mm mild steel bars, fair finish all faces	18.89	10.00	**20.78**	1.10	1.10	**27.41**	0.47	**48.66**	nr	**58.39**
precast concrete, BS 5328, designed mix C20, 20mm aggregate lintel 215 x 150 x 1200mm, reinforced with 2.13 kg of 12mm mild steel bars, fair finish all faces	40.85	10.00	**44.94**	1.65	1.65	**41.12**	0.94	**87.00**	nr	**104.40**
wedge and pin up over lintel 102mm wide	-	-	-	0.73	0.73	**18.09**	1.07	**19.16**	m	**22.99**
wedge and pin up over lintel 215mm wide	-	-	-	1.10	1.10	**27.41**	2.28	**29.69**	m	**35.63**
quoin up half brick jambs	-	-	-	0.77	0.77	**19.19**	1.07	**20.26**	m	**24.31**
quoin up one brick jambs	-	-	-	1.40	1.40	**34.81**	2.28	**37.09**	m	**44.51**
facings to match existing to margin	1.77	10.00	**1.95**	0.74	0.74	**18.37**	1.07	**21.38**	m	**25.66**
make good existing plaster up to new frame	0.35	10.00	**0.39**	0.61	0.61	**15.08**	-	**15.46**	m	**18.56**
13mm two coat hardwall plaster to reveal not exceeding 300mm wide	0.70	10.00	**0.77**	0.86	0.86	**21.38**	0.03	**22.18**	m	**26.62**
make good existing external rendering up to new frame	0.27	10.00	**0.29**	0.52	0.52	**12.88**	-	**13.18**	m	**15.81**
13mm two coat cement and sand (1:3) external rendering to reveal not exceeding 300mm wide	0.54	10.00	**0.59**	0.81	0.81	**20.28**	0.03	**20.90**	m	**25.08**
take off old skirting	-	-	-	0.25	-	**3.60**	0.12	**3.71**	m	**4.45**
short length of old skirting up to 100mm long with mitre with existing one end	-	-	-	0.64		**9.07**	0.05	**9.12**	nr	**10.94**
short length of old skirting up to 300mm long with mitres with existing both ends	-	-	-	1.27		**17.98**	0.07	**18.04**	nr	**21.65**
38mm cement and sand (1:3) screeded bed in opening	5.43	10.00	**5.97**	1.69	1.69	**42.21**	-	**48.18**	m2	**57.82**
extend 25mm softwood board flooring through opening on and with bearers	6.67	15.00	**7.48**	4.02	0.67	**64.24**	0.18	**71.90**	nr	**86.28**
iroko threshold, twice oiled, plugged and screwed										
15 x 113mm	5.65	10.00	**6.22**	0.99	-	**14.07**	0.07	**20.35**	m	**24.42**
15 x 225mm	10.45	10.00	**11.50**	1.82	-	**25.79**	0.12	**37.40**	m	**44.88**
Form opening for window through 275mm hollow wall with two half brick skins, plastered one side and faced with picked stock facings the other; cut opening through wall, insert 275 x 225mm precast concrete boot lintel and wedge and pin up over; insert damp proof course and cavity gutter; quoin up jambs, close cavity at jambs with brickwork bonded to existing and with vertical damp proof course and close cavity at sill with one course of slates; face margin externally to match existing and extend plaster to face of lintel and to reveals and make good up to existing plaster and new frame										
900 x 900mm	60.94	-	**60.94**	17.49	18.48	**446.45**	19.03	**526.42**	nr	**631.7**

Labour hourly rates: (except Specialists) Craft Operatives £14.21 Labourer £10.71 Rates are national average prices. Refer to REGIONAL VARIATIONS for indicative levels of overall pricing in regions	MATERIALS			LABOUR				RATES		
	Del to Site £	Waste %	Material Cost £	Craft Optve Hrs	Lab Hrs	Labour Cost £	Sunds £	Nett Rate £	Unit	Gross Rate (+20%) £
C90: ALTERATIONS - SPOT ITEMS Cont.										
Openings through brickwork or blockwork Cont.										
Unit rates for pricing the above and similar work										
cut opening through wall with two half brick skins	-	-	-	4.29	5.28	**117.51**	10.64	**128.15**	m2	**153.78**
cut opening through wall with one half brick and one 100mm block skins	-	-	-	3.58	4.36	**97.45**	10.64	**108.09**	m2	**129.71**
precast concrete, BS 5328, designed mix C20, 20mm aggregate boot lintel 275 x 225 x 1200mm long reinforced with 3.60 kg of 12mm mild steel bars and 0.67 kg of 6mm mild steel links.............	62.91	10.00	**69.20**	2.53	2.53	**63.05**	1.07	**133.32**	nr	**159.98**
close cavity at jambs with blockwork bonded to existing and with 112.5mm wide bituminous hessian based damp proof course, BS 6398, Class A........	2.67	10.00	**2.93**	0.55	0.55	**13.71**	0.27	**16.91**	m	**20.29**
close cavity at jambs with brickwork bonded to existing and with 112.5mm wide bituminous hessian based damp proof course, BS 6398, Class A........	2.69	10.00	**2.96**	0.77	0.77	**19.19**	1.07	**23.22**	m	**27.86**
close cavity at jambs with slates in cement mortar (1:3) set vertically	3.32	10.00	**3.65**	0.48	0.48	**12.06**	0.27	**15.99**	m	**19.19**
close cavity at sill with one course of slates in cement mortar (1:3)	3.32	10.00	**3.65**	0.48	0.48	**12.06**	0.27	**15.99**	m	**19.19**
lead cored bituminous hessian based damp proof course and cavity tray, BS 6398, Class F, width exceeding 225mm	18.84	10.00	**20.73**	0.88	0.88	**21.93**	0.27	**42.93**	m2	**51.52**
precast concrete, BS 5328, designed mix C20, 20mm aggregate lintel 160 x 225 x 1200mm long reinforced with 2.13 kg of 12mm mild steel bars..	43.20	10.00	**47.52**	1.65	1.65	**41.12**	0.81	**89.44**	nr	**107.33**
76 x 76 x 6mm mild steel angle arch bar, primed..	7.37	5.00	**7.74**	0.55	0.55	**13.71**	0.81	**22.25**	m	**26.70**
take out three courses of facing bricks and build brick-on-end flat arch in picked stock facing bricks	6.74	10.00	**7.42**	1.10	1.10	**27.41**	2.36	**37.19**	m	**44.63**
For unit rates for finishes, quoining up jambs, etc see previous items										
Form opening for window where door removed, with window head at same level as old door head, through one brick wall plastered one side and faced the other with picked stock facings; remove old lintel and flat arch, cut away for and insert 102 x 225mm precast concrete lintel and wedge and pin up over; build brick-on-end flat arch in facing bricks on 76 x 76 x 6mm mild steel angle arch bar and point to match existing; cut away jambs of old door opening to extend width; quoin up jambs and face margins externally to match existing; fill in lower part of old opening with brickwork in gauged mortar (1:2:9), bond to existing and face externally and point to match existing 1200 x 900mm; old door opening 914 x 2082mm; extend plaster to face of lintel, reveals and filling and make good up to existing plaster and new frame; extend skirting one side and make good junction with existing	131.21	-	**131.21**	24.20	23.65	**597.17**	32.59	**760.97**	nr	**913.16**
Unit rates for pricing the above and similar work										
take out old lintel	-	-	-	1.10	1.43	**30.95**	4.43	**35.38**	m	**42.45**
cut away one brick wall to increase width of opening..	-	-	-	4.18	5.17	**114.77**	11.08	**125.85**	m2	**151.02**
cut away 275mm hollow wall of two half brick skins to increase width of opening.....................	-	-	-	4.29	5.28	**117.51**	11.08	**128.59**	m2	**154.31**
For other unit rates for pricing the above and similar work see previous items and items in 'Filling openings in Brickwork and Blockwork' section										
Form opening for door frame where old window removed, with door head at same level as old window head, through one brick wall plastered one side and faced the other with picked stock facings; remove old lintel and flat arch, cut away for and insert 102 x 225mm precast concrete lintel and wedge and pin up over; build brick-on-end flat arch in facing bricks on 76 x 76 x 6mm mild steel angle arch bar and point, fill old opening at sides of new opening to reduce width with brickwork in gauged mortar (1:2:9), bond to existing, wedge and pin up at soffit, face externally and to margins and point to match existing; take off skirtings, cut away wall below sill of old opening, quion up jambs and face margins externally 914 x 2082mm, old window opening 1200 x 900mm; extend plaster to face of lintel and filling and make good up to existing plaster and new frame; form fitted ends on existing skirting up to new frame	79.33	-	**79.33**	26.40	26.18	**655.53**	37.15	**772.01**	nr	**926.42**
For unit rates for pricing the above and similar work see previous items and items in 'Filling openings in Brickwork and Blockwork' section										
Air bricks in old walls										
Cut opening through old one brick wall, render all round in cement mortar (1:3); build in clay air brick externally and fibrous plaster ventilator internally and make good facings and plaster										
225 x 150mm	10.09	-	**10.09**	1.93	1.93	**47.97**	0.72	**58.77**	nr	**70.53**
225 x 225mm	8.23	-	**8.23**	2.56	2.56	**63.87**	0.90	**73.00**	nr	**87.60**

Labour hourly rates: (except Specialists) Craft Operatives £14.21 Labourer £10.71 Rates are national average prices. Refer to REGIONAL VARIATIONS for indicative levels of overall pricing in regions	MATERIALS			LABOUR			RATES			
	Del to Site £	Waste %	Material Cost £	Craft Optve Hrs	Lab Hrs	Labour Cost £	Sunds £	Nett Rate £	Unit	Gross Rate (+20%) £
C90: ALTERATIONS - SPOT ITEMS Cont.										
Air bricks in old walls Cont.										
Cut opening through old 275mm hollow brick wall and seal cavity with slates; build in clay air brick externally and fibrous plaster ventilator internally and make good facings and plaster										
225 x 150mm	13.41	10.00	**13.74**	2.35	2.35	**58.66**	0.72	**73.12**	nr	**87.74**
225 x 225mm	11.55	10.00	**11.88**	3.09	3.09	**77.03**	0.90	**89.81**	nr	**107.77**
Openings through rubble walling										
Form opening for door frame through 600mm rough rubble wall faced both sides; cut opening through wall, insert 600 x 150mm precast concrete lintel finished fair on all exposed faces and wedge and pin up over; quoin and face up jambs and extend softwood board flooring through opening on and with bearers and make good up to existing										
838 x 2032mm	90.88	-	**90.88**	45.43	48.07	**1160.39**	79.64	**1330.92**	nr	**1597.10**
Unit rates for pricing the above and similar work										
cut opening through 350mm random rubble wall	-	-	-	6.60	7.70	**176.25**	15.51	**191.76**	m2	**230.12**
cut opening through 600mm rough rubble wall	-	-	-	11.00	13.20	**297.68**	26.59	**324.27**	m2	**389.13**
precast concrete, BS 5328, designed mix C20, 20mm aggregate lintel 350 x 150 x 1140mm reinforced with 3.04 kg of 12mm mild steel bars; fair finish all faces	63.17	10.00	**69.49**	2.48	2.48	**61.68**	0.94	**132.11**	nr	**158.53**
precast concrete, BS 5328, designed mix C20, 20mm aggregate lintel 600 x 150 x 1140mm reinforced with 5.00 kg of 12mm mild steel bars; fair finish all faces	108.30	10.00	**119.12**	4.02	4.02	**100.05**	1.47	**220.65**	nr	**264.78**
wedge and pin up over lintel 350mm wide	-	-	-	2.20	2.20	**54.82**	2.69	**57.51**	m	**69.01**
wedge and pin up over lintel 600mm wide	-	-	-	3.30	3.30	**82.24**	4.71	**86.94**	m	**104.33**
quoin up 350mm wide jambs	-	-	-	2.20	2.20	**54.82**	2.69	**57.51**	m	**69.01**
quoin up 600mm wide jambs	-	-	-	3.30	3.30	**82.24**	4.71	**86.94**	m	**104.33**
extend 25mm softwood board flooring through opening on and with bearers	16.67	15.00	**18.70**	10.67	1.76	**170.47**	0.51	**189.68**	nr	**227.62**
Openings through stone faced walls										
Form opening for window through wall of half brick backing, plastered, and 100mm stone facing; cut opening through wall, insert 102 x 150mm precast concrete lintel and wedge and pin up over and build in 76 x 76 x 6mm mild steel angle arch bar to support stone facing, clean soffit of stone and point; quoin up jambs and face up and point externally; extend plaster to face of lintel and to reveals and make good up to existing plaster and new frame										
900 x 900mm	29.92	-	**29.92**	21.12	21.78	**533.38**	24.64	**587.94**	nr	**705.52**
Unit rates for pricing the above and similar work										
cut opening through wall	-	-	-	5.67	6.44	**149.42**	10.64	**160.06**	m2	**192.07**
precast concrete, BS 5328, designed mix C20, 20mm aggregate lintel 102 x 150 x 1200mm reinforced with 1.01 kg of 12mm mild steel bars	18.89	10.00	**20.78**	1.16	1.16	**28.78**	0.40	**49.96**	nr	**59.95**
76 x 76 x 6mm mild steel angle arch bar, primed	7.37	5.00	**7.74**	0.55	0.55	**13.71**	0.81	**22.25**	m	**26.70**
quoin up jambs	-	-	-	0.77	0.77	**19.19**	1.21	**20.40**	m	**24.48**
face stone jambs and point	-	-	-	2.31	2.31	**57.57**	3.36	**60.93**	m	**73.11**
clean stone head and point	-	-	-	2.31	2.31	**57.57**	0.40	**57.97**	m	**69.56**
For unit rates for finishes, etc. see previous items										
Filling openings in brickwork or blockwork										
Fill opening where door removed in 100mm block wall with concrete blocks in gauged mortar (1:2:9); provide 50mm preservative treated softwood sole plate on timber floor in opening, bond to blockwork at jambs and wedge and pin up at head; plaster filling both sides and extend skirting both sides; make good junction of new and existing plaster and skirtings										
838 x 2032mm	64.66	-	**64.66**	15.51	8.97	**316.41**	11.05	**392.12**	nr	**470.55**
Unit rates for pricing the above and similar work										
75mm blockwork in filling	9.26	10.00	**10.19**	1.10	1.38	**30.36**	1.47	**42.02**	m2	**50.42**
100mm blockwork in filling	11.11	10.00	**12.22**	1.32	1.65	**36.43**	2.02	**50.67**	m2	**60.80**
75mm blockwork in filling and fair face and flush smooth pointing one side	9.26	10.00	**10.19**	1.38	1.65	**37.21**	1.75	**49.15**	m2	**58.97**
100mm blockwork in filling and fair face and flush smooth pointing one side	11.11	10.00	**12.22**	1.60	1.93	**43.28**	2.15	**57.65**	m2	**69.18**
preservative treated softwood sole plate										
50 x 75mm	1.39	10.00	**1.52**	0.33	0.06	**5.28**	0.27	**7.08**	m	**8.49**
50 x 100mm	1.84	10.00	**2.02**	0.36	0.07	**5.87**	0.36	**8.25**	m	**9.90**
lead cored bituminous hessian based damp proof course, BS 6398, Class F, width not exceeding 225mm	18.84	10.00	**20.73**	0.88	0.88	**21.93**	0.53	**43.19**	m2	**51.83**
wedge and pin up at head 75mm wide	-	-	-	0.11	0.11	**2.74**	0.53	**3.28**	m	**3.93**
wedge and pin up at head 100mm wide	-	-	-	0.13	0.13	**3.29**	0.81	**4.10**	m	**4.92**
cut pockets and bond 75mm filling to existing blockwork at jambs	-	-	-	0.44	0.44	**10.96**	0.94	**11.90**	m	**14.28**
cut pockets and bond 100mm filling to existing blockwork at jambs	-	-	-	0.55	0.55	**13.71**	1.33	**15.04**	m	**18.04**

Labour hourly rates: (except Specialists) Craft Operatives £14.21 Labourer £10.71 Rates are national average prices. Refer to REGIONAL VARIATIONS for indicative levels of overall pricing in regions	MATERIALS			LABOUR			RATES			
	Del to Site £	Waste %	Material Cost £	Craft Optve Hrs	Lab Hrs	Labour Cost £	Sunds £	Nett Rate £	Unit	Gross Rate (+20%) £
C90: ALTERATIONS - SPOT ITEMS Cont.										
Filling openings in brickwork or blockwork Cont.										
Unit rates for pricing the above and similar work Cont.										
preservative treated softwood sole plate Cont.										
12mm (average) cement and sand dubbing over 300mm wide on filling	1.71	10.00	**1.89**	0.59	0.59	**14.80**	-	**16.69**	m2	**20.03**
13mm two coat lightweight plaster over 300mm wide on filling	2.36	10.00	**2.59**	1.40	1.40	**34.81**	0.03	**37.43**	m2	**44.92**
19 x 100mm softwood chamfered skirting, primed all round	1.79	10.00	**1.97**	0.55	0.09	**8.76**	0.14	**10.87**	m	**13.05**
junction with existing	-	-	-	0.55	0.09	**8.76**	-	**8.76**	nr	**10.51**
make good vinyl or similar floor covering up to filling	7.86	10.00	**8.65**	0.55	0.55	**13.71**	0.18	**22.54**	m	**27.05**
Fill opening where door removed in half brick wall with brickwork in gauged mortar (1:2:9); provide lead cored damp proof course in opening, lapped with existing, bond to existing brickwork at jambs and wedge and pin up at head; plaster filling both sides and extend skirting both sides; make good junction of new and existing plaster and skirtings										
838 x 2032mm	56.93	-	**56.93**	17.27	14.03	**395.61**	13.56	**466.10**	nr	**559.32**
Unit rates for pricing the above and similar work										
half brick filling in common bricks	14.52	10.00	**15.97**	2.20	2.20	**54.82**	3.36	**74.16**	m2	**88.99**
half brick filling in common bricks fair faced and flush pointed one side	14.52	10.00	**15.97**	2.56	2.56	**63.87**	3.62	**83.47**	m2	**100.16**
50 x 100mm softwood sole plate	1.84	10.00	**2.02**	0.36	0.07	**5.87**	0.33	**8.21**	m	**9.85**
lead cored bituminous hessian based damp proof course, BS 6398, Class F, width not exceeding 225mm	18.84	10.00	**20.73**	0.88	0.88	**21.93**	0.53	**43.19**	m2	**51.83**
wedge and pin up at head 102mm wide	-	-	-	0.13	0.13	**3.29**	0.81	**4.10**	m	**4.92**
cut pockets and bond half brick filling to existing brickwork at jambs	-	-	-	0.55	0.55	**13.71**	1.34	**15.05**	m	**18.06**
For unit rates for finishes, etc see previous items										
Fill opening where door removed in one brick external wall with brickwork in gauged mortar (1:2:9) faced externally with picked stock facings pointed to match existing; remove old lintel and arch, provide lead cored damp proof course in opening lapped with existing, bond to existing brickwork at jambs and wedge and pin up at head; plaster filling one side and extend skirting; make good junction of new and existing plaster and skirting										
914 x 2082mm	186.92	-	**186.92**	21.62	20.74	**529.22**	33.50	**749.64**	nr	**899.57**
Unit rates for pricing the above and similar work										
remove old lintel and arch	-	-	-	1.61	1.80	**42.14**	2.66	**44.80**	m	**53.76**
one brick filling in common bricks	29.04	10.00	**31.94**	4.40	4.40	**109.65**	8.06	**149.65**	m2	**179.58**
one brick filling in common bricks faced one side with picked stock facings and point to match existing	52.80	10.00	**58.08**	5.50	5.50	**137.06**	8.06	**203.20**	m2	**243.83**
one brick filling in common bricks fair faced and flush smooth pointing one side	29.04	10.00	**31.94**	4.76	4.76	**118.69**	8.06	**158.69**	m2	**190.43**
two course slate damp proof course, width not exceeding 225mm	26.58	10.00	**29.23**	1.12	1.12	**27.96**	0.94	**58.13**	m2	**69.76**
lead cored bituminous hessian based damp proof course, BS 6398, Class F, width not exceeding 225mm	18.84	10.00	**20.73**	0.88	0.88	**21.93**	0.40	**43.06**	m2	**51.67**
wedge and pin up at head 215mm wide	-	-	-	0.22	0.22	**5.48**	1.88	**7.36**	m	**8.83**
cut pockets and bond one brick filling to existing brickwork at jambs	-	-	-	1.10	1.10	**27.41**	3.09	**30.50**	m	**36.60**
12mm cement and sand (1:3) two coat external rendering over 300mm wide on filling	1.71	10.00	**1.89**	1.38	1.38	**34.27**	0.05	**36.20**	m2	**43.44**
For unit rates for internal finishes, etc. see previous items										
Fill opening where door removed in 275mm external hollow wall with inner skin in common bricks and outer skin in picked stock facings pointed to match existing, in gauged mortar (1:2:9); remove old lintel and arch, provide lead cored combined damp proof course and cavity gutter lapped with existing; form cavity with ties; bond to existing brickwork at jambs and wedge and pin up at head; plaster filling one side and extend skirting; make good junction of new and existing plaster and skirting										
914 x 2082mm	129.55	-	**129.55**	21.89	21.01	**536.07**	34.22	**699.84**	nr	**839.81**
Unit rates for pricing the above and similar work										
half brick filling in common bricks	14.52	10.00	**15.97**	2.20	2.20	**54.82**	3.36	**74.16**	m2	**88.99**
half brick filling in common bricks and fair face and flush smooth pointing one side	14.52	10.00	**15.97**	2.56	2.56	**63.87**	3.62	**83.47**	m2	**100.16**
half brick filling in picked stock facings and point to match existing	30.36	10.00	**33.40**	3.25	3.25	**80.87**	3.62	**117.89**	m2	**141.46**
form 50mm wide cavity with BS 1243 Fig. 3 galvanised steel twisted wall ties built in	1.40	10.00	**1.54**	0.22	0.22	**5.48**	-	**7.02**	m2	**8.43**
lead cored bituminous hessian based damp proof course and cavity tray, BS 6398, Class F, width exceeding 225mm	18.84	10.00	**20.73**	0.88	0.88	**21.93**	0.40	**43.06**	m2	**51.67**
wedge and pin up at head 102mm wide	-	-	-	0.13	0.13	**3.29**	0.81	**4.10**	m	**4.92**

Labour hourly rates: (except Specialists) Craft Operatives £14.21 Labourer £10.71 Rates are national average prices. Refer to REGIONAL VARIATIONS for indicative levels of overall pricing in regions	MATERIALS			LABOUR				RATES		
	Del to Site £	Waste %	Material Cost £	Craft Optve Hrs	Lab Hrs	Labour Cost £	Sunds £	Nett Rate £	Unit	Gross Rate (+20%) £
C90: ALTERATIONS - SPOT ITEMS Cont.										
Filling openings in brickwork or blockwork Cont.										
Unit rates for pricing the above and similar work Cont.										
cut pockets and bond half brick wall in common bricks to existing brickwork at jambs	-	-	-	0.55	0.55	**13.71**	1.34	**15.05**	m	**18.06**
cut pockets and bond half brick wall in facing bricks to existing brickwork at jambs and make good facings	-	-	-	0.66	0.66	**16.45**	2.82	**19.26**	m	**23.12**
For unit rates for internal finishes, etc. see previous items										
Fillings openings in rubble walling										
Fill opening where window removed in 600mm rough rubble wall with rubble walling in lime mortar (1:3) pointed to match existing; bond to existing walling at jambs and wedge and pin up at head; plaster filling one side and make good junction of new and existing plaster										
900 x 900mm	128.99	-	**128.99**	13.20	13.20	**328.94**	20.73	**478.66**	nr	**574.39**
Unit rates for pricing the above and similar work										
rough rubble walling in filling										
600mm thick	133.10	10.00	**146.41**	6.60	6.60	**164.47**	16.11	**326.99**	m2	**392.39**
random rubble walling in filling										
300mm thick	96.80	10.00	**106.48**	6.38	6.38	**158.99**	8.06	**273.53**	m2	**328.23**
500mm thick	143.00	10.00	**157.30**	11.99	11.99	**298.79**	13.43	**469.52**	m2	**563.42**
wedge and pin up at head										
300mm wide	-	-	-	0.31	0.31	**7.68**	2.69	**10.36**	m	**12.43**
500mm wide	-	-	-	0.52	0.52	**12.88**	4.56	**17.45**	m	**20.94**
600mm wide	-	-	-	0.62	0.62	**15.35**	5.37	**20.72**	m	**24.87**
cut pockets and bond walling to existing at jambs										
300mm	7.26	10.00	**7.99**	0.91	0.91	**22.75**	0.81	**31.55**	m	**37.86**
500mm	10.73	10.00	**11.80**	1.52	1.52	**37.83**	1.21	**50.84**	m	**61.01**
600mm	9.98	10.00	**10.98**	1.93	1.93	**47.97**	1.47	**60.42**	m	**72.51**
For unit rates for internal finishes, etc. see previous items										
Work to chimney stacks										
SEAL AND VENTILATE TOP OF FLUE - remove chimney pot and flaunching and provide 50mm precast concrete, BS 5328, designed mix C20, 20mm aggregate, weathered and throated capping and bed in cement mortar (1:3) to seal top of stack; cut opening through brick side of stack and build in 225 x 225mm clay air brick to ventilate flue and make good facings										
stack size 600 x 600mm with one flue	23.39	10.00	**25.16**	5.39	5.39	**134.32**	5.63	**165.11**	nr	**198.14**
stack size 825 x 600mm with two flues	35.66	10.00	**38.11**	9.90	9.90	**246.71**	10.38	**295.19**	nr	**354.23**
stack size 1050 x 600mm with three flues	47.94	10.00	**51.05**	14.41	14.41	**359.10**	14.22	**424.37**	nr	**509.24**
RENEW DEFECTIVE CHIMNEY POT - remove chimney pot and provide new clay chimney pot set and flaunched in cement mortar (1:3)										
150mm diameter x 600mm high	28.49	-	**28.49**	2.75	2.75	**68.53**	5.10	**102.12**	nr	**122.54**
REBUILD DEFECTIVE CHIMNEY STACK - pull down defective stack to below roof level for a height of 2.00m; prepare for raising and rebuild in common brickwork faced with picked stock facings in gauged mortar (1:2:9) pointed to match existing; parge and core flues; provide No. 4 lead flashings and soakers; provide 150mm diameter clay chimney pots 600mm high set and flaunched in cement mortar (1:3); make good roof tiling and all other work disturbed										
600 x 600mm with one flue	308.85	10.00	**336.88**	38.83	41.31	**994.15**	76.91	**1407.94**	nr	**1689.53**
825 x 600mm with two flues	404.69	10.00	**439.46**	51.87	52.69	**1301.31**	106.89	**1847.66**	nr	**2217.19**
REMOVE DEFECTIVE CHIMNEY STACK - pull down stack to below roof level; remove all flashings and soakers; piece in 50 x 100mm rafters; extend roof tiling with machine made plain tiles to 100mm gauge nailed every fourth course with galvanised nails to and with 25 x 19mm battens; bond new tiling to existing										
850 x 600mm for a height of 2.00m	39.62	10.00	**43.58**	28.60	24.53	**669.12**	46.93	**759.63**	nr	**911.56**
Work to chimney breasts and fireplaces										
TAKE OUT FIREPLACE AND FILL IN OPENING - remove fire surround, fire back and hearth; hack up screed to hearth and extend 25mm tongued and grooved softwood floor boarding over hearth on and with bearers; fill in opening where fireplace removed with 75mm concrete blocks in gauged mortar (1:2:9) bonded to existing at jambs and wedged and pinned up to soffits; form 225 x 225mm opening and provide and build in 225 x 225mm fibrous plaster louvered air vent; plaster filling with 12mm two coat lightweight plaster on 12mm (average) dubbing plaster; extend 19 x 100mm softwood chamfered skirting over filling and join with existing; make good all new finishings to existing										
fireplace opening 570 x 685mm, hearth 1030 x 405mm	43.77	-	**43.77**	8.80	8.20	**212.82**	13.43	**270.01**	nr	**324.01**
fireplace opening 800 x 760mm, hearth 1260 x 405mm	53.50	-	**53.50**	10.62	9.85	**256.28**	15.64	**325.43**	nr	**390.51**

Labour hourly rates: (except Specialists) Craft Operatives £14.21 Labourer £10.71 Rates are national average prices. Refer to REGIONAL VARIATIONS for indicative levels of overall pricing in regions	MATERIALS			LABOUR				RATES		
	Del to Site £	Waste %	Material Cost £	Craft Optve Hrs	Lab Hrs	Labour Cost £	Sunds £	Nett Rate £	Unit	Gross Rate (+20%) £
C90: ALTERATIONS - SPOT ITEMS Cont.										
Work to chimney breasts and fireplaces Cont.										
REMOVE CHIMNEY BREAST - pull down chimney breast for full height from ground to roof level (two storeys) including removing two fire surrounds and hearths complete; make out brickwork where flues removed and make out brickwork where breasts removed; extend 50 x 175mm floor joists and 25mm tongued and grooved softwood floor boarding (ground and first floors); extend 50 x 100mm ceiling joists (first floor); plaster walls with 12mm two coat lightweight plaster on 12mm (average) plaster dubbing; extend ceiling plaster with expanded metal lathing and 19mm three coat lightweight plaster; run 19 x 100mm softwood chamfered skirting to walls to join with existing; make good all new finishings up to existing										
chimney breast 2.00m wide and 7.50m high including gathering in roof space	180.29	-	**180.29**	68.20	92.40	**1958.73**	187.70	**2326.72**	nr	**2792.06**
Form trap door in ceiling										
Form new trap door in ceiling with 100mm deep joists and lath and plaster finish; cut away and trim ceiling joists around opening and insert new trimming and trimmer joists; provide 32x 127mm softwood lining with 15 x 25mm stop and 20 x 70mm architrave; provide and place in position 18mm blockboard trap door, lipped all round; make good existing ceiling plaster around openings										
600 x 900mm	47.96	-	**47.96**	24.04	6.16	**407.51**	11.86	**467.33**	nr	**560.80**
Unit rates for pricing the above and similar work										
cut away plastered plasterboard ceiling	-	-	-	1.65	0.36	**27.33**	1.11	**28.44**	m2	**34.13**
cut away lath and plaster ceiling	-	-	-	2.20	0.45	**36.09**	1.33	**37.42**	m2	**44.91**
cut away and trim existing 100mm deep joists around opening										
664 x 964mm	-	-	-	1.05	0.18	**16.73**	0.93	**17.66**	nr	**21.19**
964 x 1444mm	-	-	-	2.09	0.36	**33.59**	1.86	**35.45**	nr	**42.54**
cut away and trim existing 150mm deep joists around opening										
664 x 964mm	-	-	-	1.57	0.26	**25.18**	1.42	**26.60**	nr	**31.92**
964 x 1444mm	-	-	-	3.15	0.53	**50.36**	2.79	**53.15**	nr	**63.78**
softwood trimming or trimmer joists										
75 x 100mm	2.97	10.00	**3.27**	1.05	0.18	**16.73**	0.47	**20.47**	m	**24.56**
75 x 150mm	4.46	10.00	**4.90**	1.10	0.19	**17.63**	0.51	**23.04**	m	**27.65**
softwood lining										
32 x 125mm	2.31	10.00	**2.54**	1.65	0.28	**26.39**	0.14	**29.08**	m	**34.89**
32 x 150mm	2.40	10.00	**2.64**	1.76	0.30	**28.19**	0.10	**30.93**	m	**37.12**
15 x 25mm softwood stop	0.50	10.00	**0.54**	0.36	0.06	**5.75**	0.07	**6.36**	m	**7.63**
19 x 50mm softwood twice rounded architrave to Patt. 18	0.90	10.00	**0.99**	0.36	0.06	**5.75**	0.07	**6.80**	m	**8.17**
18mm blockboard trap door lipped all round										
600 x 900mm	14.23	10.00	**15.66**	3.30	0.55	**52.78**	0.22	**68.66**	nr	**82.39**
900 x 1280mm	32.56	10.00	**35.82**	6.60	1.10	**105.57**	0.39	**141.77**	nr	**170.13**
75mm steel butt hinge to softwood	0.39	-	**0.39**	0.73	0.12	**11.61**	0.22	**12.22**	nr	**14.66**
make good plastered plasterboard ceiling around opening	-	-	-	0.83	0.83	**20.56**	1.60	**22.16**	m	**26.59**
make good lath and plaster ceiling around opening	-	-	-	0.91	0.91	**22.75**	1.97	**24.72**	m	**29.66**
Openings through stud partitions										
Form opening for door frame through lath and plaster finished 50 x 100mm stud partition; take off skirtings, cut away plaster, cut away and trim studding around opening and insert new studding at head and jambs of opening; make good existing lath and plaster up to new frame both sides and form fitted ends on existing skirtings up to new frame; extend softwood board flooring through opening on and with bearers and make good up to existing										
838 x 2032mm	15.33	-	**15.33**	19.91	6.71	**354.79**	20.99	**391.11**	nr	**469.33**
Unit rates for pricing the above and similar work										
cut away plastered plasterboard	-	-	-	1.10	0.28	**18.58**	1.11	**19.68**	m2	**23.62**
cut away lath and plaster	-	-	-	1.21	0.29	**20.26**	1.33	**21.59**	m2	**25.90**
cut away and trim existing 75mm studding around opening										
838 x 2032mm	-	-	-	2.26	0.40	**36.28**	2.84	**39.13**	nr	**46.95**
914 x 2032mm	-	-	-	2.26	0.40	**36.28**	2.84	**39.13**	nr	**46.95**
cut away and trim existing 100mm studding around opening										
838 x 2032mm	-	-	-	2.81	0.50	**45.16**	3.71	**48.88**	nr	**58.65**
914 x 2032mm	-	-	-	2.81	0.50	**45.16**	3.71	**48.88**	nr	**58.65**
50 x 75mm softwood studding	1.39	10.00	**1.52**	0.55	0.09	**8.76**	0.14	**10.43**	m	**12.51**
50 x 100mm softwood studding	1.84	10.00	**2.02**	0.66	0.11	**10.56**	0.16	**12.73**	m	**15.28**
make good plastered plasterboard around opening	-	-	-	0.83	0.83	**20.56**	1.60	**22.16**	m	**26.59**
make good lath and plaster around opening	-	-	-	0.91	0.91	**22.75**	1.97	**24.72**	m	**29.66**
take off skirting	-	-	-	0.23	-	**3.28**	0.23	**3.52**	m	**4.22**
form fitted end on existing skirting	-	-	-	0.63	-	**8.91**	-	**8.91**	nr	**10.69**
extend 25mm softwood tongued and grooved board flooring through opening on and with bearers	3.42	20.00	**3.92**	1.65	0.28	**26.39**	0.08	**30.39**	nr	**36.47**

Labour hourly rates: (except Specialists) Craft Operatives £14.21 Labourer £10.71 Rates are national average prices. Refer to REGIONAL VARIATIONS for indicative levels of overall pricing in regions	MATERIALS			LABOUR				RATES		
	Del to Site £	Waste %	Material Cost £	Craft Optve Hrs	Lab Hrs	Labour Cost £	Sunds £	Nett Rate £	Unit	Gross Rate (+20%) £
C90: ALTERATIONS - SPOT ITEMS Cont.										
Filling openings in stud partitions										
Fill opening where door removed in stud partition with 50 x 100mm studding and sole plate, covered on both sides with 9.5mm plasterboard baseboard with 5mm one coat board finish plaster; extend skirting on both sides; make good junction of new and existing plaster and skirtings										
838 x 2032mm	35.18	-	**35.18**	17.60	5.90	**313.24**	3.91	**352.33**	nr	**422.80**
Unit rates for pricing the above and similar work										
50 x 75mm softwood sole plate	1.39	10.00	**1.52**	0.55	0.09	**8.76**	0.14	**10.43**	m	**12.51**
50 x 100mm softwood sole plate	1.84	10.00	**2.02**	0.66	0.11	**10.56**	0.18	**12.76**	m	**15.31**
50 x 75mm softwood studding	1.39	10.00	**1.52**	0.55	0.09	**8.76**	0.14	**10.43**	m	**12.51**
50 x 100mm softwood studding	1.84	10.00	**2.02**	0.66	0.11	**10.56**	0.18	**12.76**	m	**15.31**
9.5mm plasterboard baseboard to filling	2.39	10.00	**2.63**	0.55	0.11	**8.99**	0.36	**11.98**	m2	**14.38**
9.5mm plasterboard baseboard to filling including packing out not exceeding 10mm	2.39	10.00	**2.63**	1.10	0.19	**17.63**	0.48	**20.74**	m2	**24.89**
5mm one coat board finish plaster on plasterboard to filling	1.25	10.00	**1.38**	1.00	1.00	**24.94**	0.04	**26.36**	m2	**31.64**
10mm two coat hardwall plaster on plasterboard to filling	1.93	10.00	**2.12**	1.40	1.40	**34.81**	0.04	**36.98**	m2	**44.37**
19 x 100mm softwood chamfered skirting, primed all round	1.79	10.00	**1.97**	0.59	0.10	**9.50**	0.23	**11.71**	m	**14.05**
junction with existing	-	-	-	0.63	-	**8.91**	-	**8.91**	nr	**10.69**
25 x 150mm softwood chamfered skirting, primed all round	3.21	10.00	**3.53**	0.61	0.10	**9.66**	0.31	**13.50**	m	**16.20**
junction with existing	-	-	-	0.63	-	**8.91**	-	**8.91**	nr	**10.69**
Work to existing drains										
BREAK UP CONCRETE PAVING FOR EXCAVATING DRAIN TRENCH AND REINSTATE - break up 100mm concrete paving and 150mm hardcore bed under for excavation of drain trench and reinstate with new hardcore and concrete, BS 5328, designed mix C20, 20mm aggregate and make good up to existing paving both sides										
400mm wide	6.27	30.00	**7.06**	-	3.60	**38.52**	4.43	**50.02**	m	**60.02**
TAKE UP FLAG PAVING FOR EXCAVATING DRAIN TRENCH AND REINSTATE - carefully take up 50mm precast concrete flag paving and set aside for re-use and break up 150mm hardcore bed under for excavation of drain trench and reinstate with new hardcore and re-lay salvaged flag paving on and with 25mm sand bed and grout in lime sand (1:3) and make good up to existing paving on both sides										
400mm wide	2.17	30.00	**2.80**	1.62	1.62	**40.30**	4.08	**47.18**	m	**56.62**
INSERT NEW JUNCTION IN EXISTING DRAIN -excavate for and trace and expose existing drain, break into glazed vitrified clay drain and insert junction with 100mm branch, short length of new pipe and double collar and joint to existing drain; support earthwork, make good concrete bed and haunching and backfill										
existing 100mm diameter drain, invert depth										
600mm	32.73	10.00	**36.00**	3.30	13.20	**188.27**	11.28	**235.54**	nr	**282.65**
750mm	32.73	10.00	**36.00**	3.30	15.68	**214.77**	12.19	**262.96**	nr	**315.55**
900mm	32.73	10.00	**36.00**	3.30	18.15	**241.28**	13.36	**290.64**	nr	**348.77**
existing 150mm diameter drain invert depth										
750mm	59.54	10.00	**65.50**	4.95	17.33	**255.89**	14.08	**335.47**	nr	**402.56**
900mm	59.54	10.00	**65.50**	4.95	19.80	**282.40**	15.97	**363.86**	nr	**436.64**
1050mm	59.54	10.00	**65.50**	4.95	22.28	**308.90**	17.47	**391.87**	nr	**470.24**
1200mm	59.54	10.00	**65.50**	4.95	24.75	**335.41**	19.10	**420.01**	nr	**504.01**
REPAIR DEFECTIVE DRAIN - excavate for and trace and expose existing drain, break out fractured glazed vitrified clay drain pipe, replace with new pipe and double collars and joint to existing drain; support earthwork, make good concrete bed and haunching and backfill										
existing 100mm diameter drain; single pipe length; invert depth										
450mm	27.62	10.00	**30.38**	3.30	12.65	**182.37**	8.54	**221.30**	nr	**265.55**
600mm	27.62	10.00	**30.38**	3.30	14.85	**205.94**	10.30	**246.62**	nr	**295.94**
750mm	27.62	10.00	**30.38**	3.30	17.60	**235.39**	12.58	**278.35**	nr	**334.02**
add for each additional pipe length; invert depth										
450mm	17.85	10.00	**19.64**	2.20	7.43	**110.78**	8.15	**138.57**	nr	**166.28**
600mm	17.85	10.00	**19.64**	2.20	8.36	**120.80**	9.84	**150.28**	nr	**180.33**
750mm	17.85	10.00	**19.64**	2.20	9.68	**134.93**	11.80	**166.37**	nr	**199.64**
Work to existing manholes										
RAISE TOP OF EXISTING MANHOLE - take off cover and frame and set aside for re-use, prepare level bed on existing one brick sides of manhole internal size 610 x 457mm and raise with Class B engineering brickwork in cement mortar (1:3) finished with a fair face and flush pointed; refix salvaged cover and frame, bed frame in cement mortar and cover in grease and sand										
raising 150mm high	18.41	10.00	**20.26**	5.83	5.83	**145.28**	4.13	**169.67**	nr	**203.61**
raising 225mm high	27.62	10.00	**30.38**	7.43	7.43	**185.03**	6.13	**221.54**	nr	**265.85**
raising 300mm high	36.83	10.00	**40.51**	9.35	9.35	**233.00**	7.85	**281.36**	nr	**337.63**

Labour hourly rates: (except Specialists) Craft Operatives £14.21 Labourer £10.71 Rates are national average prices. Refer to REGIONAL VARIATIONS for indicative levels of overall pricing in regions	MATERIALS			LABOUR				RATES		
	Del to Site £	Waste %	Material Cost £	Craft Optve Hrs	Lab Hrs	Labour Cost £	Sunds £	Nett Rate £	Unit	Gross Rate (+20%) £
C90: ALTERATIONS - SPOT ITEMS Cont.										
Work to existing manholes Cont.										
INSERT NEW BRANCH BEND IN BOTTOM OF EXISTING MANHOLE - break into bottom and one brick side of manhole, insert new glazed vitrified clay three quarter section branch bend to discharge over existing main channel, build in end of new drain and make good benching and side of manhole to match existing										
100mm diameter branch bend	11.64	10.00	**12.80**	6.33	6.33	**157.62**	2.96	**173.38**	nr	**208.06**
EXTEND EXISTING MANHOLE - take off cover and frame and set aside; break up one brick end wall of 457mm wide (internal) manhole and excavate for and extend manhole 225mm; support earthwork, level and compact bottom, part backfill and remove surplus spoil from site; extend bottom with 150mm concrete, BS 5328, ordinary prescribed mix C15P, 20mmaggregate bed; extend one brick sides and end with Class B engineering brickwork in cement mortar (1:3) fair faced and flush pointed and bond to existing; extend 100mm main channel and insert 100mm three quarter section branch channel bend; extend benching and build in end of new 100mm drain to one brick side, corbel over end of manhole manholes of the following invert depths; refix salvaged cover and frame, bed frame in cement mortar and cover in grease and sand										
450mm	56.17	10.00	**61.78**	19.36	19.69	**485.99**	11.42	**559.19**	nr	**671.02**
600mm	61.83	10.00	**68.01**	21.78	22.22	**547.47**	14.68	**630.16**	nr	**756.19**
750mm	67.68	10.00	**74.45**	24.20	24.75	**608.95**	17.94	**701.34**	nr	**841.61**
900mm	73.43	10.00	**80.77**	26.62	27.28	**670.44**	21.19	**772.40**	nr	**926.88**
Temporary weatherproof coverings										
Providing and erecting and clearing away on completion; for scaffold tube framework see preliminaries										
corrugated iron sheeting	6.95	10.00	**7.65**	0.17	0.55	**8.24**	1.03	**16.91**	m2	**20.29**
flexible reinforced plastic sheeting	3.00	10.00	**3.30**	0.11	0.28	**4.51**	0.83	**8.65**	m2	**10.38**
tarpaulins	6.38	10.00	**7.02**	0.11	0.28	**4.51**	0.83	**12.36**	m2	**14.83**
Temporary internal screens										
Providing and erecting and clearing away on completion; temporary screen of 50 x 50mm softwood framing and cover with										
reinforced building paper	2.86	10.00	**3.15**	0.47	0.33	**10.26**	0.29	**13.69**	m2	**16.43**
heavy duty polythene sheeting	1.96	10.00	**2.15**	0.47	0.33	**10.26**	0.26	**12.67**	m2	**15.20**
3mm hardboard	2.95	10.00	**3.24**	0.58	0.35	**12.05**	0.43	**15.73**	m2	**18.87**
Providing and erecting and clearing away on completion; temporary dustproof screen of 50 x 75mm softwood framing securely fixed to walls, floor and ceiling, the joints and edges of lining sealed with masking tape and cover with										
3mm hardboard lining one side	3.80	10.00	**4.17**	1.21	0.20	**19.31**	0.57	**24.06**	m2	**28.88**
3mm hardboard lining both sides	5.24	10.00	**5.76**	1.43	0.24	**22.91**	0.72	**29.39**	m2	**35.27**
6mm plywood lining one side	7.41	10.00	**8.16**	1.27	0.21	**20.21**	0.64	**29.01**	m2	**34.81**
6mm plywood lining both sides	12.41	10.00	**13.65**	1.54	0.25	**24.59**	1.00	**39.25**	m2	**47.09**
providing 35mm hardboard faced flush door size 838 x 1981mm in dustproof screen with 25 x 87mm softwood frame, 18 x 25mm softwood stop, pair of 100mm butts, pull handle and ball catch	151.10	10.00	**166.21**	1.65	0.28	**26.39**	3.62	**196.22**	nr	**235.47**
emporary dustproof corridors										
ote										
for prices of walls and doors, see dustproof screens above										
roviding and erecting and clearing away on ompletion; ceiling of 50 x 75mm softwood joists at 50mm centres, the joints and edges sealed with asking tape and cover with										
3mm hardboard lining to soffit	4.09	10.00	**4.50**	1.21	0.20	**19.31**	0.57	**24.39**	m2	**29.27**
6mm plywood lining to soffit	7.68	10.00	**8.45**	1.27	0.21	**20.21**	0.72	**29.38**	m2	**35.25**
emporary timber balustrades										
roviding and erecting and clearing away on ompletion; temporary softwood balustrade consisting f 50 x 75mm plate, 50 x 50mm standards at 900mm entres, four 25 x 150mm intermediate rails and 50 x 0mm handrail										
1150mm high	7.51	10.00	**8.26**	1.10	0.29	**18.69**	1.64	**28.60**	m	**34.32**
mporary steel balustrades										
oviding and erecting and clearing away on mpletion; temporary steel balustrade constructed of mm diameter galvanised scaffold tubing and fittings th standards at 900mm centres intermediate rail, ndrail, plastic mesh infill and with plates on ends standards fixed to floor										
1150mm high	7.48	-	**7.48**	0.20	0.99	**13.42**	0.39	**21.29**	m	**25.55**
weekly cost of hire and maintenance	0.74	-	**0.74**	0.02	0.11	**1.49**	0.05	**2.28**	m	**2.74**

Labour hourly rates: (except Specialists) Craft Operatives £14.21 Labourer £10.71 Rates are national average prices. Refer to REGIONAL VARIATIONS for indicative levels of overall pricing in regions	MATERIALS			LABOUR			RATES			
	Del to Site	Waste	Material Cost	Craft Optve	Lab	Labour Cost	Sunds	Nett Rate	Unit	Gross Rate (+20%)
	£	%	£	Hrs	Hrs	£	£	£		£
C90: ALTERATIONS - SPOT ITEMS Cont.										
Temporary fillings to openings in external walls										
Providing and erecting and clearing away on completion; temporary filling to window or door opening of 50 x 100mm softwood framing and cover with										
corrugated iron sheeting	15.31	10.00	**16.84**	0.50	0.50	**12.34**	1.71	**30.89**	m2	**37.06**
25mm softwood boarding	13.66	10.00	**15.03**	0.66	0.66	**16.45**	1.99	**33.47**	m2	**40.16**
12mm external quality plywood	12.45	10.00	**13.70**	0.55	0.55	**13.71**	1.29	**28.69**	m2	**34.43**
C90: REPAIRING/RENOVATING ASPHALT WORK										
Cut out crack in old covering and make good with new material to match existing										
Floor or tanking										
20mm thick in two coats, horizontal	-	-	-	0.73	0.73	**18.09**	0.65	**18.74**	m	**22.49**
30mm thick in three coats, horizontal	-	-	-	0.99	0.99	**24.67**	0.98	**25.65**	m	**30.78**
30mm thick in three coats, vertical	-	-	-	1.31	1.31	**32.62**	0.98	**33.60**	m	**40.32**
Cut out detached blister in old covering and make good with new material to match existing										
Floor or tanking										
20mm thick in two coats, horizontal	-	-	-	0.33	0.33	**8.22**	0.33	**8.55**	nr	**10.26**
30mm thick in three coats, horizontal	-	-	-	0.39	0.39	**9.59**	0.46	**10.05**	nr	**12.06**
30mm thick in three coats, vertical	-	-	-	0.53	0.53	**13.16**	0.46	**13.61**	nr	**16.34**
Cut out crack in old covering and make good with new material to match existing										
Roof covering										
20mm thick in two coats	-	-	-	0.73	0.73	**18.09**	0.65	**18.74**	m	**22.49**
Cut out detached blister in old covering and make good with new material to match existing										
20mm thick in two coats	-	-	-	0.33	0.33	**8.22**	0.33	**8.55**	nr	**10.26**
C90: REPAIRING/RENOVATING ROOF COVERINGS										
Slate roofing repairs to roofs covered with 610 x 300mm slates										
Remove damaged slates and replace with new; sloping or vertical										
one slate	6.30	5.00	**6.62**	0.35	0.35	**8.77**	0.12	**15.51**	nr	**18.61**
patch of 10 slates	63.03	5.00	**66.18**	1.21	1.21	**30.15**	0.91	**97.25**	nr	**116.70**
patch of slates 1.00 - 3.00m2	77.21	5.00	**81.07**	1.38	1.38	**34.27**	0.99	**116.33**	m2	**139.59**
Examine battens, remove defective and provide 20% new										
19 x 38mm	0.23	10.00	**0.25**	0.10	0.10	**2.47**	0.07	**2.79**	m2	**3.34**
25 x 50mm	0.40	10.00	**0.44**	0.10	0.10	**2.47**	0.09	**2.99**	m2	**3.59**
Re-cover roof with slates previously removed and stacked and fix with slate nails										
with 75mm lap	-	-	-	0.85	0.85	**21.11**	0.12	**21.22**	m2	**25.47**
extra for providing 20% new slates	15.44	5.00	**16.21**	-	-	-	0.23	**16.45**	m2	**19.74**
Remove double course at eaves and										
refix	-	-	-	0.26	0.26	**6.58**	0.40	**6.98**	m	**8.38**
extra for providing 20% new slates	5.36	5.00	**5.63**	-	-	-	0.12	**5.74**	m	**6.89**
Remove double course at verge and replace with new slates and bed and point in										
mortar	23.57	5.00	**24.75**	0.32	0.32	**7.95**	0.76	**33.46**	m	**40.15**
Slate roofing repairs for roofs covered with 510 x 250mm slates										
Remove damaged slates and replace with new; sloping or vertical										
one slate	3.04	5.00	**3.19**	0.35	0.35	**8.77**	0.12	**12.08**	nr	**14.49**
patch of 10 slates	30.36	5.00	**31.88**	1.10	1.10	**27.41**	0.77	**60.06**	nr	**72.07**
patch of slates 1.00 - 3.00m2	54.65	5.00	**57.38**	1.64	1.64	**40.84**	0.98	**99.20**	m2	**119.04**
Examine battens, remove defective and provide 20% new										
19 x 38mm	0.28	10.00	**0.31**	0.12	0.12	**3.02**	0.09	**3.42**	m2	**4.10**
25 x 50mm	0.49	10.00	**0.53**	0.12	0.12	**3.02**	0.12	**3.67**	m2	**4.40**
Re-cover roof with slates previously removed and stacked and fix with slate nails										
with 75mm lap	-	-	-	1.10	1.10	**27.41**	0.17	**27.58**	m2	**33.1**
extra for providing 20% new slates	10.93	5.00	**11.48**	-	-	-	0.23	**11.71**	m2	**14.05**
Remove double course at eaves and										
refix	-	-	-	0.28	0.28	**6.85**	0.69	**7.54**	m	**9.05**
extra for providing 20% new slates	3.40	5.00	**3.57**	-	-	-	0.10	**3.67**	m	**4.4**
Remove double course at verge and replace with new slates and bed and point in										
mortar	13.97	5.00	**14.66**	0.32	0.32	**7.95**	0.74	**23.36**	m	**28.0**

Labour hourly rates: (except Specialists) Craft Operatives £14.21 Labourer £10.71 Rates are national average prices. Refer to REGIONAL VARIATIONS for indicative levels of overall pricing in regions	MATERIALS			LABOUR			RATES			
	Del to Site £	Waste %	Material Cost £	Craft Optve Hrs	Lab Hrs	Labour Cost £	Sunds £	Nett Rate £	Unit	Gross Rate (+20%) £
C90: REPAIRING/RENOVATING ROOF COVERINGS Cont.										
Slate roofing repairs for roofs covered with 405 x 200mm slates										
Remove damaged slates and replace with new; sloping or vertical										
one slate	1.50	5.00	1.57	0.35	0.35	8.77	0.10	10.45	nr	12.54
patch of 10 slates	14.96	5.00	15.71	0.99	0.99	24.67	0.66	41.04	nr	49.25
patch of slates 1.00 - 3.00m2	44.88	5.00	47.12	2.28	2.28	56.74	1.08	104.95	m2	125.94
Examine battens, remove defective and provide 20% new										
19 x 38mm	0.38	10.00	0.42	0.14	0.14	3.56	0.10	4.08	m2	4.90
25 x 50mm	0.65	10.00	0.71	0.14	0.14	3.56	0.12	4.40	m2	5.27
Re-cover roof with slates previously removed and stacked and fix with slate nails										
with 75mm lap	-	-	-	1.25	1.25	31.25	0.25	31.50	m2	37.80
extra for providing 20% new slates	8.98	5.00	9.42	-	-	-	0.23	9.66	m2	11.59
Remove double course at eaves and refix	-	-	-	0.29	0.29	7.13	0.68	7.80	m	9.37
extra for providing 20% new slates	2.03	5.00	2.14	-	-	-	0.10	2.24	m	2.69
Remove double course at verge and replace with new slates and bed and point in mortar	8.72	5.00	9.16	0.33	0.33	8.22	0.82	18.20	m	21.84
Tile roofing repairs for roofs covered with machine made clay plain tiles laid to 100mm gauge										
Remove damaged tiles and replace with new, sloping or vertical										
one tile	0.56	10.00	0.62	0.35	0.35	8.77	0.04	9.43	nr	11.32
patch of 10 tiles	5.63	10.00	6.20	0.99	0.99	24.67	0.33	31.19	nr	37.43
patch of tiles 1.00 - 3.00m2	34.36	10.00	37.79	1.89	1.89	47.15	2.09	87.02	m2	104.43
Examine battens, remove defective and provide 20% new										
19 x 25mm	0.51	10.00	0.56	0.14	0.14	3.56	0.10	4.22	m2	5.07
19 x 38mm	0.62	10.00	0.68	0.15	0.15	3.84	0.12	4.63	m2	5.56
19 x 50mm	0.92	10.00	1.02	0.15	0.15	3.84	0.14	5.00	m2	6.00
Re-cover roof with tiles previously removed and stacked										
nail every fourth course with galvanised nails	-	-	-	1.10	1.10	27.41	0.22	27.63	m2	33.16
extra for nailing every course	-	-	-	0.22	0.22	5.48	0.59	6.07	m2	7.28
extra for providing 20% new tiles	6.87	10.00	7.56	-	-	-	0.23	7.79	m2	9.35
Tile roofing repairs to roofs covered with machine made clay plain tiles laid to 90mm gauge										
Remove damaged tiles and replace with new, sloping or vertical										
one tile	0.56	10.00	0.62	0.35	0.35	8.77	0.04	9.43	nr	11.32
patch of 10 tiles	5.63	10.00	6.20	0.99	0.99	24.67	0.33	31.19	nr	37.43
patch of tiles 1.00 - 3.00m2	38.30	10.00	42.13	2.26	2.26	56.19	2.32	100.64	m2	120.77
Examine battens, remove defective and provide 20% new										
19 x 25mm	0.56	10.00	0.62	0.15	0.15	3.84	0.12	4.57	m2	5.49
19 x 38mm	0.68	10.00	0.75	0.17	0.17	4.11	0.14	5.00	m2	6.01
19 x 50mm	1.03	10.00	1.13	0.17	0.17	4.11	0.16	5.40	m2	6.48
Re-cover roof with tiles previously removed and stacked										
nail every fourth course with galvanised nails	-	-	-	1.27	1.27	31.52	0.25	31.77	m2	38.13
extra for nailing every course	-	-	-	0.25	0.25	6.30	0.61	6.92	m2	8.30
extra for providing 20% new tiles	7.66	10.00	8.43	-	-	-	0.23	8.66	m2	10.39
Tile roofing repairs for roofs covered with 381 x 227mm concrete interlocking tiles laid to 75mm laps										
Remove damaged tiles and replace with new, sloping or vertical										
one tile	0.69	10.00	0.76	0.35	0.35	8.77	0.09	9.62	nr	11.54
patch of 10 tiles	6.86	10.00	7.55	0.99	0.99	24.67	0.95	33.17	nr	39.81
patch of tiles 1.00 - 3.00m2	11.67	10.00	12.84	0.94	0.94	23.30	1.51	37.65	m2	45.18
Examine battens, remove defective and provide 20% new										
19 x 25mm	0.18	10.00	0.20	0.06	0.06	1.37	0.09	1.66	m2	1.99
19 x 38mm	0.22	10.00	0.24	0.07	0.07	1.64	0.09	1.98	m2	2.38
19 x 50mm	0.33	10.00	0.37	0.08	0.08	1.92	0.10	2.39	m2	2.87
Re-cover roof with tiles previously removed and stacked										
nail every fourth course with galvanised nails	-	-	-	0.55	0.55	13.71	0.03	13.73	m2	16.48
extra for nailing every course	-	-	-	0.06	0.06	1.37	0.08	1.45	m2	1.74
extra for providing 20% new tiles	2.75	10.00	3.02	-	-	-	0.29	3.31	m2	3.97
Tile roofing repairs generally										
Remove damaged clay tiles and provide new										
half round ridge or hip and bed in mortar	21.71	15.00	24.96	0.66	0.66	16.45	1.42	42.83	m	51.40
bonnet hip	30.36	5.00	31.88	0.83	0.83	20.56	0.39	52.83	m	63.39
trough valley	30.36	5.00	31.88	0.83	0.83	20.56	0.39	52.83	m	63.39
vertical angle	30.36	10.00	33.40	0.83	0.83	20.56	0.39	54.35	m	65.22

Labour hourly rates: (except Specialists) Craft Operatives £14.21 Labourer £10.71 Rates are national average prices. Refer to REGIONAL VARIATIONS for indicative levels of overall pricing in regions	MATERIALS			LABOUR			RATES			
	Del to Site £	Waste %	Material Cost £	Craft Optve Hrs	Lab Hrs	Labour Cost £	Sunds £	Nett Rate £	Unit	Gross Rate (+20%) £
C90: REPAIRING/RENOVATING ROOF COVERINGS Cont.										
Tile roofing repairs generally Cont.										
Remove double course and replace with new clay tiles and bed and point in mortar										
at eaves	6.76	10.00	**7.43**	0.25	0.25	**6.30**	0.59	**14.33**	m	**17.19**
at verge	14.08	10.00	**15.49**	0.42	0.42	**10.42**	1.81	**27.72**	m	**33.26**
Rake out defective pointing and re-point in cement mortar (1:3)										
ridge or hip tiles	-	-	-	0.37	0.37	**9.32**	0.22	**9.54**	m	**11.45**
Hack off defective cement mortar fillet and renew in cement mortar (1:3)	-	-	-	0.37	0.37	**9.32**	0.44	**9.76**	m	**11.72**
Remove defective hip hook and replace with new	2.28	5.00	**2.39**	0.37	0.37	**9.32**	0.09	**11.80**	nr	**14.16**
Corrugated fibre cement roofing repairs										
Remove damaged sheets and provide and fix new with screws and washers to timber purlins and ease and adjust edges of adjoining sheets as necessary										
one sheet; 1825mm long Profile 3 natural grey	18.16	10.00	**19.98**	1.35	1.35	**33.72**	2.59	**56.29**	nr	**67.55**
one sheet; 1825mm long Profile 3 standard colour	21.79	10.00	**23.97**	1.35	1.35	**33.72**	2.59	**60.28**	nr	**72.34**
one sheet; 1825mm long Profile 6 natural grey	22.79	10.00	**25.07**	1.96	1.96	**48.79**	3.83	**77.70**	nr	**93.24**
one sheet; 1825mm long Profile 6 standard colour	27.36	10.00	**30.09**	1.96	1.96	**48.79**	3.83	**82.72**	nr	**99.26**
over one sheet; Profile 3 natural grey	16.51	10.00	**18.16**	0.98	0.98	**24.40**	2.45	**45.01**	m2	**54.01**
over one sheet; Profile 3 standard colour	19.81	10.00	**21.79**	0.98	0.98	**24.40**	2.45	**48.64**	m2	**58.37**
over one sheet; Profile 6 natural grey	13.41	10.00	**14.75**	0.98	0.98	**24.40**	3.61	**42.76**	m2	**51.31**
over one sheet; Profile 6 standard colour	16.09	10.00	**17.70**	0.98	0.98	**24.40**	3.61	**45.71**	m2	**54.85**
Remove damaged sheets and provide and fix new with hook bolts and washers to steel purlins and ease and adjust edges of adjoining sheets as necessary										
one sheet; 1825mm long Profile 3 natural grey	18.16	10.00	**19.98**	1.51	1.51	**37.55**	2.59	**60.13**	nr	**72.15**
one sheet; 1825mm long Profile 3 standard colour	21.79	10.00	**23.97**	1.51	1.51	**37.55**	2.59	**64.12**	nr	**76.94**
one sheet; 1825mm long Profile 6 natural grey	22.79	10.00	**25.07**	2.18	2.18	**54.28**	3.83	**83.18**	nr	**99.82**
one sheet; 1825mm long Profile 6 standard colour	27.36	10.00	**30.09**	2.18	2.18	**54.28**	3.83	**88.20**	nr	**105.84**
over one sheet; Profile 3 natural grey	16.51	10.00	**18.16**	1.13	1.13	**28.23**	2.45	**48.85**	m2	**58.62**
over one sheet; Profile 3 standard colour	19.81	10.00	**21.79**	1.13	1.13	**28.23**	2.45	**52.48**	m2	**62.97**
over one sheet; Profile 6 natural grey	13.41	10.00	**14.75**	1.13	1.13	**28.23**	3.61	**46.59**	m2	**55.91**
over one sheet; Profile 6 standard colour	16.09	10.00	**17.70**	1.13	1.13	**28.23**	3.61	**49.55**	m2	**59.46**
Felt roofing repairs for roofs covered with bituminous felt roofing, BS 747										
Sweep clean, cut out defective layer, re-bond adjacent felt and cover with one layer of felt bonded in hot bitumen; patches not exceeding 1.00m2										
fibre based roofing felt type 1B	1.56	15.00	**1.80**	0.97	0.97	**24.12**	3.10	**29.02**	m2	**34.83**
for each defective underlayer cut out and replaced, add	1.56	15.00	**1.80**	0.97	0.97	**24.12**	3.10	**29.02**	m2	**34.83**
fibre based mineral surfaced roofing felt type 1E	2.97	15.00	**3.41**	1.16	1.16	**28.78**	3.10	**35.30**	m2	**42.36**
high performance polyester based roofing felt type 5B	6.05	15.00	**6.96**	1.12	1.12	**27.96**	3.10	**38.02**	m2	**45.62**
for each defective underlayer cut out and replaced, add	6.05	15.00	**6.96**	1.12	1.12	**27.96**	3.10	**38.02**	m2	**45.62**
high performance polyester based mineral surfaced roofing felt type 5E	4.29	15.00	**4.93**	1.24	1.24	**30.98**	3.10	**39.01**	m2	**46.81**
Sweep clean, cut out defective layer, re-bond adjacent felt and cover with one layer of felt bonded in hot bitumen; patches 1.00 - 3.00m2										
fibre based roofing felt type 1B	1.56	15.00	**1.80**	0.47	0.47	**11.79**	3.10	**16.69**	m2	**20.02**
for each defective underlayer cut out and replaced, add	1.56	15.00	**1.80**	0.47	0.47	**11.79**	3.10	**16.69**	m2	**20.02**
fibre based mineral surfaced roofing felt type 1E	2.97	15.00	**3.41**	0.62	0.62	**15.35**	3.10	**21.86**	m2	**26.24**
high performance polyester based roofing felt type 5B	6.05	15.00	**6.96**	0.58	0.58	**14.53**	3.10	**24.59**	m2	**29.51**
for each defective underlayer cut out and replaced, add	6.05	15.00	**6.96**	0.58	0.58	**14.53**	3.10	**24.59**	m2	**29.51**
high performance polyester based mineral surfaced roofing felt type 5E	4.29	15.00	**4.93**	0.72	0.72	**17.82**	3.10	**25.85**	m2	**31.02**
Sweep clean surface of existing roof, prime with hot bitumen and dress with										
13mm layer of limestone or granite chippings	1.98	10.00	**2.18**	-	0.25	**2.71**	1.73	**6.62**	m2	**7.95**
13mm layer of pea shingle	1.38	10.00	**1.51**	-	0.40	**4.24**	1.73	**7.49**	m2	**8.98**
extra; removing existing chippings or shingle	-	-	-	-	0.26	**2.83**	0.52	**3.35**	m2	**4.02**
Liquid bitumen proofing (black) brush applied on old roof covering including cleaning old covering										
Corrugated asbestos roofing										
one coat	2.61	10.00	**2.87**	-	0.23	**2.47**	-	**5.34**	m2	**6.41**
two coats	4.86	10.00	**5.35**	-	0.28	**2.95**	-	**8.29**	m2	**9.95**
Felt roofing										
one coat	2.26	10.00	**2.48**	-	0.19	**2.00**	-	**4.48**	m2	**5.38**
two coats	3.62	10.00	**3.98**	-	0.23	**2.47**	-	**6.45**	m2	**7.75**
For top coat in green in lieu black										
add	0.69	10.00	**0.76**	-	-	-	-	**0.76**	m2	**0.91**

Labour hourly rates: (except Specialists) Craft Operatives £17.38 Labourer £10.71 Rates are national average prices. Refer to REGIONAL VARIATIONS for indicative levels of overall pricing in regions	MATERIALS			LABOUR			RATES			
	Del to Site	Waste	Material Cost	Craft Optve	Lab	Labour Cost	Sunds	Nett Rate	Unit	Gross Rate (+20%)
	£	%	£	Hrs	Hrs	£	£	£		£
C90: REPAIRING/RENOVATING ROOF COVERINGS Cont.										
Repairs to lead roofing										
Repair crack										
clean out and fill with copper bit solder	-	-	-	0.88	-	15.29	0.39	15.69	m	18.82
Turn back flashing										
and re-dress	-	-	-	0.53	-	9.18	-	9.18	m	11.01
Repairs to lead flashings, etc.; replace with new lead										
Remove flashing and provide new code Nr4 lead; wedge into groove										
150mm girth	3.78	5.00	3.97	1.17	-	20.27	0.34	24.58	m	29.49
240mm girth	6.05	5.00	6.35	1.64	-	28.49	0.34	35.18	m	42.21
300mm girth	7.57	5.00	7.95	2.00	-	34.79	0.34	43.08	m	51.70
Remove stepped flashing and provide new code Nr 4 lead; wedge into groove										
180mm girth	4.54	5.00	4.77	1.69	-	29.44	0.34	34.55	m	41.46
240mm girth	6.05	5.00	6.35	2.13	-	37.09	0.34	43.78	m	52.54
300mm girth	7.57	5.00	7.95	2.57	-	44.74	0.34	53.02	m	63.63
Remove apron flashing and provide new code Nr 5 lead; wedge into groove										
150mm girth	4.71	5.00	4.94	1.17	-	20.27	0.34	25.55	m	30.66
300mm girth	9.42	5.00	9.89	2.00	-	34.79	0.34	45.02	m	54.02
450mm girth	14.11	5.00	14.82	2.84	-	49.32	0.52	64.66	m	77.60
Remove lining to valley gutter and provide new code Nr 5 lead										
240mm girth	7.52	5.00	7.90	1.64	-	28.49	-	36.39	m	43.66
300mm girth	9.42	5.00	9.89	2.00	-	34.79	-	44.68	m	53.62
450mm girth	14.11	5.00	14.82	2.84	-	49.32	-	64.14	m	76.97
Remove gutter lining and provide new code Nr 6 lead										
360mm girth	13.39	5.00	14.06	2.33	-	40.53	-	54.59	m	65.50
420mm girth	15.61	5.00	16.39	2.67	-	46.46	-	62.85	m	75.42
450mm girth	16.72	5.00	17.56	2.84	-	49.32	-	66.88	m	80.26
Repairs to zinc flashings, etc.; replace with 0.8mm zinc										
Remove flashing and provide new; wedge into groove										
150mm girth	2.99	5.00	3.14	1.23	-	21.41	0.23	24.79	m	29.75
240mm girth	4.80	5.00	5.04	1.76	-	30.59	0.23	35.86	m	43.03
300mm girth	6.00	5.00	6.29	2.13	-	37.09	0.34	43.72	m	52.47
Remove stepped flashing and provide new; wedge into groove										
180mm girth	3.60	5.00	3.78	1.82	-	31.54	0.23	35.56	m	42.67
240mm girth	4.80	5.00	5.04	2.29	-	39.77	0.23	45.04	m	54.04
300mm girth	6.00	5.00	6.29	2.77	-	48.18	0.34	54.81	m	65.77
Remove apron flashing and provide new; wedge into groove										
150mm girth	2.99	5.00	3.14	1.23	-	21.41	0.23	24.79	m	29.75
300mm girth	6.00	5.00	6.29	2.13	-	37.09	0.34	43.72	m	52.47
450mm girth	8.99	5.00	9.44	3.04	-	52.77	0.42	62.62	m	75.14
Remove lining to valley gutter and provide new										
240mm girth	4.80	5.00	5.04	1.76	-	30.59	-	35.62	m	42.75
300mm girth	6.00	5.00	6.29	2.13	-	37.09	-	43.38	m	52.06
450mm girth	8.99	5.00	9.44	3.04	-	52.77	-	62.20	m	74.64
Remove gutter lining and provide new										
360mm girth	7.18	5.00	7.54	2.50	-	43.40	-	50.94	m	61.13
420mm girth	8.38	5.00	8.80	2.87	-	49.90	-	58.70	m	70.44
450mm girth	8.99	5.00	9.44	3.04	-	52.77	-	62.20	m	74.64
Repairs to copper flashings, etc.; replace with 0.55mm copper										
Remove flashing and provide new; wedge into groove										
150mm girth	4.26	5.00	4.47	1.17	-	20.27	0.47	25.20	m	30.25
240mm girth	6.81	5.00	7.15	1.64	-	28.49	0.47	36.10	m	43.33
300mm girth	8.51	5.00	8.94	2.00	-	34.79	0.47	44.20	m	53.04
Remove stepped flashing and provide new; wedge into groove										
180mm girth	5.10	5.00	5.36	1.69	-	29.44	0.47	35.27	m	42.32
240mm girth	6.81	5.00	7.15	2.13	-	37.09	0.47	44.71	m	53.65
300mm girth	8.51	5.00	8.94	2.57	-	44.74	0.47	54.15	m	64.97
Remove apron flashing and provide new; wedge into groove										
150mm girth	4.26	5.00	4.47	1.17	-	20.27	0.47	25.20	m	30.25
300mm girth	8.51	5.00	8.94	2.00	-	34.79	0.47	44.20	m	53.04
450mm girth	12.77	5.00	13.41	2.84	-	49.32	-	62.73	m	75.28
Remove lining to valley gutter and provide new										
240mm girth	6.81	5.00	7.15	1.64	-	28.49	-	35.64	m	42.76
300mm girth	8.51	5.00	8.94	2.00	-	34.79	-	43.73	m	52.48
450mm girth	12.77	5.00	13.41	2.84	-	49.32	-	62.73	m	75.28

Labour hourly rates: (except Specialists) Craft Operatives £17.38 Labourer £10.71 Rates are national average prices. Refer to REGIONAL VARIATIONS for indicative levels of overall pricing in regions	MATERIALS			LABOUR				RATES		
	Del to Site	Waste	Material Cost	Craft Optve	Lab	Labour Cost	Sunds	Nett Rate	Unit	Gross Rate (+20%)
	£	%	£	Hrs	Hrs	£	£	£		£
C90: REPAIRING/RENOVATING ROOF COVERINGS Cont.										
Repairs to copper flashings, etc.; replace with 0.55mm copper Cont.										
Remove gutter lining and provide new										
360mm girth	10.22	5.00	10.73	2.33	-	40.53	-	51.26	m	61.51
420mm girth	11.92	5.00	12.52	2.67	-	46.46	-	58.98	m	70.77
450mm girth	12.77	5.00	13.41	2.84	-	49.32	-	62.73	m	75.28
Repairs to aluminium flashings etc; replace with 0.60mm aluminium										
Remove flashings and provide new; wedge into groove										
150mm girth	3.03	5.00	3.18	1.17	-	20.27	0.21	23.65	m	28.38
240mm girth	4.54	5.00	4.77	1.64	-	28.49	0.21	33.46	m	40.16
300mm girth	4.54	5.00	4.77	2.00	-	34.79	0.21	39.77	m	47.73
Remove stepped flashing and provide new; wedge into groove										
180mm girth	3.99	5.00	4.19	1.69	-	29.44	0.21	33.84	m	40.61
240mm girth	4.54	5.00	4.77	2.13	-	37.09	0.21	42.07	m	50.48
300mm girth	4.54	5.00	4.77	2.57	-	44.74	0.21	49.71	m	59.66
Remove apron flashing and provide new; wedge into groove										
150mm girth	3.03	5.00	3.18	1.17	-	20.27	0.21	23.65	m	28.38
300mm girth	4.54	5.00	4.77	2.00	-	34.79	0.21	39.77	m	47.73
450mm girth	7.02	5.00	7.37	2.82	-	48.94	3.91	60.22	m	72.27
Remove lining to valley gutter and provide new										
240mm girth	4.54	5.00	4.77	1.64	-	28.49	-	33.26	m	39.91
300mm girth	4.54	5.00	4.77	2.00	-	34.79	-	39.56	m	47.48
450mm girth	7.02	5.00	7.37	2.84	-	49.32	-	56.69	m	68.03
Remove gutter lining and provide new										
360mm girth	5.92	5.00	6.21	2.33	-	40.53	-	46.74	m	56.09
420mm girth	5.92	5.00	6.21	2.67	-	46.46	-	52.67	m	63.20
450mm girth	5.92	5.00	6.21	2.84	-	49.32	-	55.54	m	66.65
C90: REPAIRING/RENOVATING PLUMBING										
Repairs to rainwater goods										
Clean out and make good loose fixings										
eaves gutter	-	-	-	0.44	-	7.65	-	7.65	m	9.18
rainwater pipe	-	-	-	0.44	-	7.65	-	7.65	m	9.18
rainwater head	-	-	-	0.83	-	14.34	-	14.34	nr	17.21
Clean out defective joint to existing eaves gutter and re-make joint										
with jointing compound and new bolt	1.10	-	1.10	0.83	-	14.34	-	15.44	nr	18.53
Remove remains of old balloon grating and provide new plastic grating in outlet or end of pipe										
50mm diameter	3.15	2.00	3.21	0.55	-	9.56	-	12.77	nr	15.32
82mm diameter	3.19	2.00	3.25	0.55	-	9.56	-	12.81	nr	15.38
110mm diameter	3.21	2.00	3.28	0.55	-	9.56	-	12.84	nr	15.40
Repair or replacement of stopcocks, valves, etc.										
Turn off water supply drain down as necessary and re-new washer to the following up to 19mm										
main stopcock	0.55	-	0.55	2.20	-	38.24	-	38.79	nr	46.54
service stopcock	0.55	-	0.55	2.20	-	38.24	-	38.79	nr	46.54
bib or pillar cock	0.55	-	0.55	2.20	-	38.24	-	38.79	nr	46.54
supatap	0.55	-	0.55	2.20	-	38.24	-	38.79	nr	46.54
draining tap	0.55	-	0.55	2.20	-	38.24	-	38.79	nr	46.54
Turn off water supply, drain down as necessary, take out old valves and re-new with joints to copper										
13mm brass stopcock (BS 1010)	4.47	2.00	4.56	2.20	-	38.24	-	42.79	nr	51.35
13mm chromium plated sink pillar cock (BS 1010)	16.12	2.00	16.44	2.20	-	38.24	-	54.67	nr	65.61
13mm brass bib cock (BS 1010)	9.68	2.00	9.87	2.20	-	38.24	-	48.11	nr	57.73
13mm drain tap (BS 2879) type 2	2.65	2.00	2.70	2.20	-	38.24	-	40.94	nr	49.13
19mm drain tap (BS 2879) type 2	20.77	2.00	21.18	2.20	-	38.24	-	59.42	nr	71.30
13mm high pressure ball valve (BS 1212) piston type with copper ball	7.39	2.00	7.54	2.20	-	38.24	-	45.78	nr	54.93
13mm high pressure ball valve (BS 1212) piston type with plastic ball	4.33	2.00	4.42	2.20	-	38.24	-	42.66	nr	51.19
13mm low pressure ball valve (BS 1212) piston type with copper ball	9.22	2.00	9.40	2.20	-	38.24	-	47.64	nr	57.17
13mm low pressure ball valve (BS 1212) piston type with plastic ball	6.16	2.00	6.28	2.20	-	38.24	-	44.52	nr	53.42
19mm high pressure ball valve (BS 1212) piston type with copper ball	21.64	2.00	22.07	2.75	-	47.80	-	69.86	nr	83.84
19mm high pressure ball valve (BS 1212) piston type with plastic ball	20.96	2.00	21.37	2.75	-	47.80	-	69.17	nr	83.00
Replacement of traps										
Take out old plastic trap and replace with new 76mm seal trap (BS EN 274) with 'O'' ring joint outlet										
36mm diameter P trap	5.35	2.00	5.45	1.38	-	23.90	-	29.35	nr	35.22
42mm diameter P trap	6.17	2.00	6.29	1.65	-	28.68	-	34.97	nr	41.97
36mm diameter S trap	6.78	2.00	6.91	1.38	-	23.90	-	30.81	nr	36.97

Labour hourly rates: (except Specialists) Craft Operatives £17.38 Labourer £10.71 Rates are national average prices. Refer to REGIONAL VARIATIONS for indicative levels of overall pricing in regions	MATERIALS			LABOUR				RATES		
	Del to Site	Waste	Material Cost	Craft Optve	Lab	Labour Cost	Sunds	Nett Rate	Unit	Gross Rate (+20%)
	£	%	£	Hrs	Hrs	£	£	£		£
C90: REPAIRING/RENOVATING PLUMBING Cont.										
Replacement of traps Cont.										
Take out old plastic trap and replace with new 76mm seal trap (BS EN 274) with 'O'' ring joint outlet Cont.										
42mm diameter S trap	7.92	2.00	8.08	1.65		28.68		36.76	nr	44.11
42mm diameter bath trap with overflow connection	11.44	2.00	11.67	1.93	-	33.46	-	45.13	nr	54.15
Take out old copper trap and replace with new 76mm seal two piece trap (BS 1184) with compression joint										
35mm diameter P trap	25.52	2.00	26.03	1.38	-	23.90	-	49.93	nr	59.91
42mm diameter P trap	30.06	2.00	30.66	1.65	-	28.68	-	59.34	nr	71.21
35mm diameter S trap	27.04	2.00	27.58	1.38	-	23.90	-	51.48	nr	61.77
42mm diameter S trap	32.53	2.00	33.18	1.65	-	28.68	-	61.85	nr	74.23
42mm diameter bath trap with male iron overflow connection	42.68	2.00	43.53	1.93	-	33.46	-	76.99	nr	92.39
Preparation of pipe for insertion of new pipe or fitting										
Cut into cast iron pipe and take out a length up to 1.00m for insertion of new pipe or fitting										
50mm diameter	-	-	-	4.07	-	70.74	-	70.74	nr	84.88
64 and 76mm diameter	-	-	-	4.73	-	82.21	-	82.21	nr	98.65
89 and 102mm diameter	-	-	-	5.39	-	93.68	-	93.68	nr	112.41
Cut into cast iron pipe with caulked joint and take out a length up to 1.00m for insertion of new pipe or fitting										
50mm diameter	-	-	-	4.07	-	70.74	-	70.74	nr	84.88
64 and 76mm diameter	-	-	-	4.73	-	82.21	-	82.21	nr	98.65
89 and 102mm diameter	-	-	-	5.39	-	93.68	-	93.68	nr	112.41
Cut into asbestos cement pipe and take out a length up to 1.00m for insertion of new pipe or fitting										
50mm diameter	-	-	-	3.47	-	60.22	-	60.22	nr	72.27
64 and 76mm diameter	-	-	-	4.13	-	71.69	-	71.69	nr	86.03
89 and 102mm diameter	-	-	-	4.79	-	83.16	-	83.16	nr	99.80
Cut into polythene or p.v.c. pipe and take out a length up to 1.00m for insertion of new pipe or fitting										
up to 25mm diameter	-	-	-	2.75	-	47.80	-	47.80	nr	57.35
32 - 50mm diameter	-	-	-	3.47	-	60.22	-	60.22	nr	72.27
64 and 76mm diameter	-	-	-	4.13	-	71.69	-	71.69	nr	86.03
89 and 102mm diameter	-	-	-	4.84	-	84.12	-	84.12	nr	100.94
Cut into steel pipe and take out a length up to 1.00m for insertion of new pipe or fitting										
up to 25mm diameter	-	-	-	3.36	-	58.31	-	58.31	nr	69.97
32 - 50mm diameter	-	-	-	4.07	-	70.74	-	70.74	nr	84.88
64 and 76mm diameter	-	-	-	4.73	-	82.21	-	82.21	nr	98.65
89 and 102mm diameter	-	-	-	5.39	-	93.68	-	93.68	nr	112.41
Cut into stainless steel pipe and take out a length up to 1.00m for insertion of new pipe or fitting										
up to 25mm diameter	-	-	-	3.19	-	55.44	-	55.44	nr	66.53
32 - 50mm diameter	-	-	-	3.85	-	66.91	-	66.91	nr	80.30
64 and 76mm diameter	-	-	-	4.51	-	78.38	-	78.38	nr	94.06
89 and 102mm diameter	-	-	-	5.23	-	90.81	-	90.81	nr	108.97
Cut into copper pipe and take out a length up to 1.00m for insertion of new pipe or fitting										
up to 25mm diameter	-	-	-	3.19	-	55.44	-	55.44	nr	66.53
32 - 50mm diameter	-	-	-	3.85	-	66.91	-	66.91	nr	80.30
64 and 76mm diameter	-	-	-	4.51	-	78.38	-	78.38	nr	94.06
89 and 102mm diameter	-	-	-	5.23	-	90.81	-	90.81	nr	108.97
Cut into lead pipe and take out a length up to 1.00m for insertion of new pipe or fitting										
up to 25mm diameter	-	-	-	3.36	-	58.31	-	58.31	nr	69.97
32 - 50mm diameter	-	-	-	4.07	-	70.74	-	70.74	nr	84.88
64 and 76mm diameter	-	-	-	4.73	-	82.21	-	82.21	nr	98.65
89 and 102mm diameter	-	-	-	5.39	-	93.68	-	93.68	nr	112.41
Removal of traps, valves, etc., for re-use										
Take out the following, including unscrewing or uncoupling, for re-use and including draining down as necessary										
trap up to 25mm diameter	-	-	-	0.77	-	13.38	-	13.38	nr	16.06
trap 32 - 50mm diameter	-	-	-	0.94	-	16.25	-	16.25	nr	19.50
stop valve up to 25mm diameter	-	-	-	2.20	-	38.24	-	38.24	nr	45.88
stop valve 32 - 50mm diameter	-	-	-	2.48	-	43.02	-	43.02	nr	51.62
radiator valve up to 25mm diameter	-	-	-	2.20	-	38.24	-	38.24	nr	45.88
radiator valve 32 - 50mm diameter	-	-	-	2.48	-	43.02	-	43.02	nr	51.62
tap up to 25mm diameter	-	-	-	2.20	-	38.24	-	38.24	nr	45.88
tap 32 - 50mm diameter	-	-	-	2.48	-	43.02	-	43.02	nr	51.62
extra; burning out one soldered joint up to 25mm diameter	-	-	-	0.55	-	9.56	-	9.56	nr	11.47
extra; burning out one soldered joint 32 - 50mm diameter	-	-	-	0.88	-	15.29	-	15.29	nr	18.35

Labour hourly rates: (except Specialists) Craft Operatives £17.38 Labourer £10.71 Rates are national average prices. Refer to REGIONAL VARIATIONS for indicative levels of overall pricing in regions	MATERIALS			LABOUR				RATES		
	Del to Site £	Waste %	Material Cost £	Craft Optve Hrs	Lab Hrs	Labour Cost £	Sunds £	Nett Rate £	Unit	Gross Rate (+20%) £
C90: REPAIRING/RENOVATING PLUMBING Cont.										
Replacement of old sanitary fittings										
Disconnect trap, valves and services pipes, remove old fitting and provide and fix new including re-connecting trap, valves and services pipes										
560 x 405mm vitreous china lavatory basin with 32mm chromium plated waste outlet, plug, chain and stay and pair of cast iron brackets	59.40	5.00	**62.37**	6.60	-	**114.71**	-	**177.08**	nr	**212.49**
600 x 1000mm stainless steel sink with single drainer with 38mm chromium plated waste outlet, plug, chain and stay	100.93	5.00	**105.97**	6.60	-	**114.71**	-	**220.68**	nr	**264.82**
600 x 1500mm stainless steel sink with double drainer with 38mm chromium plated waste outlet, overflow plug, chain and stay	123.48	5.00	**129.65**	7.15	-	**124.27**	-	**253.92**	nr	**304.70**
1500mm pressed steel vitreous enamelled rectangular top bath with cradles and with 38mm chromium plated waste outlet, plug, chain and stay and 32mm chromium plated overflow and including removing old cast iron bath	106.98	5.00	**112.32**	8.80	-	**152.94**	-	**265.27**	nr	**318.32**
1700mm pressed steel vitreous enamelled rectangular top bath with cradles and with 38mm chromium plated waste outlet, plug, chain and stay and 32mm chromium plated overflow and including removing old cast iron bath	106.98	5.00	**112.32**	8.80	-	**152.94**	-	**265.27**	nr	**318.32**
Take off old W.C. seat and provide and fix new ring pattern black plastic seat and cover	13.75	5.00	**14.44**	1.65	-	**28.68**	-	**43.11**	nr	**51.74**
Disconnect supply pipe, overflow and flush pipe and take out old W.C. cistern and provide and fix new 9 litre black plastic W.C. cistern with cover and connect pipe and ball valve										
low level	52.53	5.00	**55.15**	5.50	-	**95.59**	-	**150.74**	nr	**180.89**
high level	64.08	5.00	**67.28**	5.50	-	**95.59**	-	**162.87**	nr	**195.44**
Disconnect flush pipe, take off seat and remove old W.C. pan, provide and fix new white china pan, connect flush pipe, form cement joint with drain and refix old seat										
S or P trap pan	30.80	5.00	**32.34**	5.50	-	**95.59**	-	**127.93**	nr	**153.52**
Disconnect and remove old high level W.C. suite complete and provide and fix new suite with P or S trap pan, plastic ring seat and cover, plastic flush pipe or bend, 9 litre plastic W.C. cistern with cover, chain and pull or lever handle and 13mm low pressure ball valve; adapt supply pipe and overflow as necessary and connect to new suite and cement joint to drain										
new high level suite	112.20	5.00	**117.81**	9.35	-	**162.50**	-	**280.31**	nr	**336.38**
new low level suite (vitreous china cistern)	105.05	5.00	**110.30**	9.90	-	**172.06**	-	**282.36**	nr	**338.84**
Disconnect and remove old low level W.C. suite complete and provide and fix new suite with P or S trap pan, plastic ring seat and cover, plastic flush bend, 9 litre vitreous china W.C. cistern with cover, lever handle and 13mm low pressure ball valve with plastic ball; re-connect supply pipe and overflow and cement joint to drain										
new low level suite	105.05	5.00	**110.30**	9.35	-	**162.50**	-	**272.81**	nr	**327.37**
Replacement of old water tanks										
Disconnect all pipework, set aside ball valve and take out cold water storage cistern in roof space; replace with new GRP water storage tank (BS EN 13280); reconnect all pipework and ball valve; remove metal filings and clean inside of tank										
type SC70 size 915 x 610 x 580mm (227 litre)	217.80	2.00	**222.16**	9.90	-	**172.06**	0.56	**394.78**	nr	**473.73**

This section continues on the next page

Labour hourly rates: (except Specialists) Craft Operatives £20.54 Labourer £10.71 Rates are national average prices. Refer to REGIONAL VARIATIONS for indicative levels of overall pricing in regions	MATERIALS			LABOUR				RATES		
	Del to Site £	Waste %	Material Cost £	Craft Optve Hrs	Lab Hrs	Labour Cost £	Sunds £	Nett Rate £	Unit	Gross Rate (+20%) £
C90: REPAIRING/RENOVATING ELECTRICAL INSTALLATIONS										
Generally										
Note										
the following approximate estimates of costs are dependent on the number and disposition of the points. Lamps and fittings together with cutting and making good are excluded										
Strip out and re-wire complete										
Provide new lamp holders, flush switches, flush socket outlets, fitting outlets and consumer unit with circuit breakers; a three bedroom house with ten lighting points and associated switches, sixteen 13 amp socket outlets, three 13 amp fitting outlets, one 30 amp cooker control panel, three earth bonding points and one eight way consumer unit										
PVC cables in existing conduit	1101.79	7.50	**1184.43**	58.14	-	**1194.09**	33.00	**2411.52**	nr	**2893.82**
PVC cables and conduit	1221.68	7.50	**1313.31**	108.90	-	**2236.81**	33.00	**3583.11**	nr	**4299.74**
Strip out and re-wire with PVC insulated and sheathed cable in existing conduit										
Point to junction box, re-using existing lamp holder, socket outlet etc										
5 amp socket outlet	9.90	7.50	**10.64**	0.94	-	**19.20**	0.50	**30.34**	nr	**36.41**
13 amp socket or fitting outlet	13.20	7.50	**14.19**	0.94	-	**19.20**	0.50	**33.89**	nr	**40.67**
lighting point or switch	5.81	7.50	**6.24**	1.08	-	**22.14**	0.50	**28.88**	nr	**34.66**
Strip out and re-wire with PVC insulated and sheathed cable and new conduit										
Point to junction box, re-using existing lamp holder, socket outlet etc										
5 amp socket outlet	10.91	7.50	**11.73**	1.32	-	**27.11**	0.55	**39.39**	nr	**47.27**
13 amp socket outlet	14.21	7.50	**15.28**	1.32	-	**27.11**	0.55	**42.94**	nr	**51.53**
lighting point or switch	8.12	7.50	**8.73**	1.43	-	**29.37**	0.33	**38.43**	nr	**46.11**
Luminaires and accessories										
Take out damaged and renew										
lamp holder, batten type	2.92	7.50	**3.13**	0.62	-	**12.65**	0.28	**16.06**	nr	**19.27**
lamp holder, pendant type with rose	1.43	7.50	**1.54**	1.17	-	**23.95**	0.28	**25.76**	nr	**30.91**
switch, flush type	1.27	7.50	**1.36**	0.83	-	**16.95**	0.50	**18.80**	nr	**22.56**
switch, surface mounted type	2.20	7.50	**2.37**	0.83	-	**16.95**	0.50	**19.81**	nr	**23.77**
13 amp socket outlet, flush type	2.48	7.50	**2.66**	0.83	-	**16.95**	0.50	**20.10**	nr	**24.12**
13 amp socket outlet, surface mounted type	4.40	7.50	**4.73**	0.97	-	**19.88**	0.50	**25.11**	nr	**30.13**
5 amp socket outlet, flush type	5.94	7.50	**6.39**	0.77	-	**15.82**	0.50	**22.70**	nr	**27.24**
5 amp socket outlet, surface mounted type	6.71	7.50	**7.21**	0.97	-	**19.88**	0.50	**27.59**	nr	**33.11**

This section continues on the next page

Labour hourly rates: (except Specialists) Craft Operatives £14.21 Labourer £10.71 Rates are national average prices. Refer to REGIONAL VARIATIONS for indicative levels of overall pricing in regions	MATERIALS			LABOUR			RATES			
	Del to Site £	Waste %	Material Cost £	Craft Optve Hrs	Lab Hrs	Labour Cost £	Sunds £	Nett Rate £	Unit	Gross Rate (+20%) £
C90: REPAIRING/RENOVATING FINISHINGS										
Repairs to old wall and ceiling finishings										
Cut out damaged two coat plaster to wall in patches and make out in new plaster to match existing										
under 1.00m2	2.98	10.00	**3.27**	2.34	2.34	**58.39**	0.32	**61.98**	m2	**74.38**
1.00 - 2.00m2	2.98	10.00	**3.27**	1.94	1.94	**48.25**	0.32	**51.84**	m2	**62.21**
2.00 - 4.00m2	2.98	10.00	**3.27**	1.53	1.53	**38.10**	0.32	**41.70**	m2	**50.03**
Cut out damaged three coat plaster to wall in patches and make out in new plaster to match existing										
under 1.00m2	3.95	10.00	**4.35**	2.75	2.75	**68.53**	0.49	**73.37**	m2	**88.05**
1.00 - 2.00m2	3.95	10.00	**4.35**	2.27	2.27	**56.47**	0.49	**61.31**	m2	**73.57**
2.00 - 4.00m2	3.95	10.00	**4.35**	1.78	1.78	**44.41**	0.49	**49.25**	m2	**59.10**
Cut out damaged two coat rendering to wall in patches and make out in new rendering to match existing										
under 1.00m2	1.74	10.00	**1.92**	2.12	2.12	**52.91**	0.32	**55.14**	m2	**66.17**
1.00 - 2.00m2	1.74	10.00	**1.92**	1.76	1.76	**43.86**	0.32	**46.10**	m2	**55.32**
2.00 - 4.00m2	1.74	10.00	**1.92**	1.39	1.39	**34.54**	0.32	**36.78**	m2	**44.13**
Cut out damaged three coat rendering to wall in patches and make out in new rendering to match existing										
under 1.00m2	2.68	10.00	**2.95**	2.51	2.51	**62.50**	0.49	**65.94**	m2	**79.13**
1.00 - 2.00m2	2.68	10.00	**2.95**	2.06	2.06	**51.26**	0.49	**54.70**	m2	**65.64**
2.00 - 4.00m2	2.68	10.00	**2.95**	1.62	1.62	**40.30**	0.49	**43.74**	m2	**52.49**
Cut out damaged two coat plaster to ceiling in patches and make out in new plaster to match existing										
under 1.00m2	2.98	10.00	**3.27**	2.62	2.62	**65.24**	0.32	**68.83**	m2	**82.60**
1.00 - 2.00m2	2.98	10.00	**3.27**	2.16	2.16	**53.73**	0.32	**57.32**	m2	**68.78**
2.00 - 4.00m2	2.98	10.00	**3.27**	1.69	1.69	**42.21**	0.32	**45.81**	m2	**54.97**
Cut out damaged three coat plaster to ceiling in patches and make out in new plaster to match existing										
under 1.00m2	3.95	10.00	**4.35**	3.07	3.07	**76.48**	0.49	**81.32**	m2	**97.59**
1.00 - 2.00m2	3.95	10.00	**4.35**	2.53	2.53	**63.05**	0.49	**67.89**	m2	**81.47**
2.00 - 4.00m2	3.95	10.00	**4.35**	1.99	1.99	**49.62**	0.49	**54.46**	m2	**65.35**
Cut out damaged 9.5mm plasterboard and two coat plaster to wall in patches and make out in new plasterboard and plaster to match existing										
under 1.00m2	4.96	10.00	**5.46**	2.45	2.45	**61.13**	0.55	**67.13**	m2	**80.56**
1.00 - 2.00m2	4.96	10.00	**5.46**	1.96	1.96	**48.79**	0.55	**54.80**	m2	**65.76**
2.00 - 4.00m2	4.96	10.00	**5.46**	1.50	1.50	**37.28**	0.55	**43.29**	m2	**51.94**
Cut out damaged 12.5mm plasterboard and two coat plaster to wall in patches and make out in new plasterboard and plaster to match existing										
under 1.00m2	5.17	10.00	**5.69**	2.51	2.51	**62.50**	0.61	**68.80**	m2	**82.56**
1.00 - 2.00m2	5.17	10.00	**5.69**	2.01	2.01	**50.16**	0.61	**56.46**	m2	**67.75**
2.00 - 4.00m2	5.17	10.00	**5.69**	1.52	1.52	**37.83**	0.61	**44.13**	m2	**52.95**
Cut out damaged 9.5mm plasterboard and two coat plaster to ceiling in patches and make out in new plasterboard and plaster to match existing										
under 1.00m2	4.96	10.00	**5.46**	2.79	2.79	**69.63**	0.55	**75.63**	m2	**90.76**
1.00 - 2.00m2	4.96	10.00	**5.46**	2.23	2.23	**55.65**	0.55	**61.65**	m2	**73.98**
2.00 - 4.00m2	4.96	10.00	**5.46**	1.71	1.71	**42.49**	0.55	**48.49**	m2	**58.19**
Cut out damaged 12.5mm plasterboard and two coat plaster to ceiling in patches and make out in new plasterboard and plaster to match existing										
under 1.00m2	5.17	10.00	**5.69**	2.82	2.82	**70.17**	0.61	**76.47**	m2	**91.77**
1.00 - 2.00m2	5.17	10.00	**5.69**	2.27	2.27	**56.47**	0.61	**62.77**	m2	**75.32**
2.00 - 4.00m2	5.17	10.00	**5.69**	1.73	1.73	**43.04**	0.61	**49.33**	m2	**59.20**
Cut out damaged lath and plaster to wall in patches and make out with new metal lathing and three coat plaster to match existing										
under 1.00m2	7.69	10.00	**8.46**	2.78	2.78	**69.35**	0.61	**78.43**	m2	**94.1[illegible]**
1.00 - 2.00m2	7.69	10.00	**8.46**	2.35	2.35	**58.66**	0.61	**67.73**	m2	**81.2[illegible]**
2.00 - 4.00m2	7.69	10.00	**8.46**	1.90	1.90	**47.42**	0.61	**56.50**	m2	**67.7[illegible]**
Cut out damaged lath and plaster to ceiling in patches and make out with new metal lathing and three coat plaster to match existing										
under 1.00m2	7.69	10.00	**8.46**	3.16	3.16	**78.67**	0.61	**87.75**	m2	**105.2[illegible]**
1.00 - 2.00m2	7.69	10.00	**8.46**	2.67	2.67	**66.61**	0.61	**75.68**	m2	**90.8[illegible]**
2.00 - 4.00m2	7.69	10.00	**8.46**	2.15	2.15	**53.45**	0.61	**62.53**	m2	**75.0[illegible]**
Cut out crack in plaster, form dovetailed key and make good with plaster										
not exceeding 50mm wide	0.14	10.00	**0.15**	0.22	0.22	**5.48**	0.02	**5.66**	m	**6.7[illegible]**
Make good plaster where the following removed										
small pipe	-	-	-	0.18	0.18	**4.39**	0.02	**4.40**	nr	**5.2[illegible]**
small steel section	-	-	-	0.23	0.23	**5.76**	0.02	**5.78**	nr	**6.9[illegible]**
Cut out damaged plaster moulding or cornice and make out with new to match existing										
up to 100mm girth on face	0.92	10.00	**1.02**	1.38	0.69	**26.96**	0.13	**28.11**	m	**33.7[illegible]**
100 - 150mm girth on face	1.54	10.00	**1.69**	2.05	1.02	**40.03**	0.25	**41.98**	m	**50.3[illegible]**
150 - 200mm girth on face	3.08	10.00	**3.39**	2.48	1.24	**48.48**	0.46	**52.33**	m	**62.7[illegible]**

Labour hourly rates: (except Specialists) Craft Operatives £14.21 Labourer £10.71 Rates are national average prices. Refer to REGIONAL VARIATIONS for indicative levels of overall pricing in regions	MATERIALS			LABOUR				RATES		
	Del to Site £	Waste %	Material Cost £	Craft Optve Hrs	Lab Hrs	Labour Cost £	Sunds £	Nett Rate £	Unit	Gross Rate (+20%) £
C90: REPAIRING/RENOVATING FINISHINGS Cont.										
Repairs to old wall and ceiling finishings Cont.										
Remove damaged fibrous plaster louvered ventilator, replace with new and make good										
229 x 75mm	1.76	5.00	**1.85**	0.55	0.55	**13.71**	0.06	**15.62**	nr	**18.74**
229 x 150mm	2.11	-	**2.11**	0.66	0.66	**16.45**	0.10	**18.66**	nr	**22.40**
Take out damaged 152 x 152 x 5.5mm white glazed wall tiles and renew to match existing										
isolated tile	0.49	5.00	**0.52**	0.35	0.35	**8.77**	0.04	**9.33**	nr	**11.19**
patch of 5 tiles	2.46	5.00	**2.59**	0.66	0.66	**16.45**	0.18	**19.22**	nr	**23.06**
patch 0.50 - 1.00m2	21.68	5.00	**22.77**	2.55	2.55	**63.60**	1.45	**87.81**	m2	**105.37**
patch 1.00 - 2.00m2	21.68	5.00	**22.77**	1.83	1.83	**45.50**	1.45	**69.72**	m2	**83.66**
Repairs to old floor finishings										
Take out damaged 150 x 150 x 12.5mm clay floor tiles and renew to match existing										
isolated tile	0.35	5.00	**0.37**	0.35	0.23	**7.48**	0.04	**7.88**	nr	**9.46**
patch of 5 tiles	1.76	5.00	**1.85**	0.72	0.50	**15.46**	0.18	**17.49**	nr	**20.99**
patch 0.50 - 1.00m2	15.84	5.00	**16.63**	2.64	1.77	**56.48**	1.45	**74.56**	m2	**89.47**
patch 1.00 - 2.00m2	15.84	5.00	**16.63**	2.09	1.40	**44.66**	1.45	**62.74**	m2	**75.29**
Cut out damaged 25mm granolithic paving in patches and make out in new granolithic to match existing										
under 1.00m2	7.89	5.00	**8.29**	1.65	1.65	**41.12**	0.68	**50.08**	m2	**60.10**
1.00 - 2.00m2	7.89	5.00	**8.29**	1.38	1.38	**34.27**	0.68	**43.23**	m2	**51.88**
2.00 - 4.00m2	7.89	5.00	**8.29**	1.18	1.18	**29.33**	0.68	**38.30**	m2	**45.95**
Cut out damaged 32mm granolithic paving in patches and make out in new granolithic to match existing										
under 1.00m2	10.10	5.00	**10.61**	1.85	1.85	**46.05**	0.79	**57.45**	m2	**68.95**
1.00 - 2.00m2	10.10	5.00	**10.61**	1.52	1.52	**37.83**	0.79	**49.23**	m2	**59.08**
2.00 - 4.00m2	10.10	5.00	**10.61**	1.27	1.27	**31.52**	0.79	**42.93**	m2	**51.51**
Cut out damaged 38mm granolithic paving in patches and make out in new granolithic to match existing										
under 1.00m2	12.00	5.00	**12.60**	1.98	1.98	**49.34**	0.94	**62.87**	m2	**75.45**
1.00 - 2.00m2	12.00	5.00	**12.60**	1.65	1.65	**41.12**	0.94	**54.65**	m2	**65.58**
2.00 - 4.00m2	12.00	5.00	**12.60**	1.38	1.38	**34.27**	0.94	**47.80**	m2	**57.36**
Cut out damaged 50mm granolithic paving in patches and make out in new granolithic to match existing										
under 1.00m2	15.79	5.00	**16.57**	2.33	2.33	**58.11**	1.14	**75.83**	m2	**90.99**
1.00 - 2.00m2	15.79	5.00	**16.57**	1.90	1.90	**47.42**	1.14	**65.14**	m2	**78.16**
2.00 - 4.00m2	15.79	5.00	**16.57**	1.60	1.60	**39.75**	1.14	**57.46**	m2	**68.95**
Cut out damaged granolithic paving to tread 275mm wide and make out in new granolithic to match existing										
25mm thick	2.21	5.00	**2.32**	0.48	0.48	**12.06**	0.19	**14.57**	m	**17.49**
32mm thick	2.84	5.00	**2.98**	0.53	0.53	**13.16**	0.22	**16.36**	m	**19.63**
38mm thick	3.16	5.00	**3.31**	0.55	0.55	**13.71**	0.25	**17.27**	m	**20.73**
50mm thick	4.42	5.00	**4.64**	0.65	0.65	**16.17**	0.31	**21.13**	m	**25.35**
Cut out damaged granolithic covering to plain riser 175mm wide and make out in new granolithic to match existing										
13mm thick	0.63	5.00	**0.66**	0.88	0.88	**21.93**	1.74	**24.33**	m	**29.20**
19mm thick	0.95	5.00	**0.99**	0.92	0.92	**23.03**	1.75	**25.77**	m	**30.92**
25mm thick	1.26	5.00	**1.33**	0.97	0.97	**24.12**	1.77	**27.22**	m	**32.66**
32mm thick	1.89	5.00	**1.99**	1.02	1.02	**25.49**	1.78	**29.26**	m	**35.11**
Cut out damaged granolithic covering to undercut riser 175mm wide and make out in new granolithic to match existing										
13mm thick	0.63	5.00	**0.66**	0.96	0.96	**23.85**	1.74	**26.25**	m	**31.50**
19mm thick	0.95	5.00	**0.99**	1.02	1.02	**25.49**	1.75	**28.24**	m	**33.88**
25mm thick	1.26	5.00	**1.33**	1.08	1.08	**26.86**	1.77	**29.96**	m	**35.95**
32mm thick	1.89	5.00	**1.99**	1.14	1.14	**28.51**	1.78	**32.28**	m	**38.73**
Cut out damaged granolithic skirting 150mm wide and make out in new granolithic to match existing										
13mm thick	0.63	5.00	**0.66**	0.79	0.79	**19.74**	1.48	**21.88**	m	**26.26**
19mm thick	0.95	5.00	**0.99**	0.84	0.84	**20.83**	1.50	**23.33**	m	**28.00**
25mm thick	1.26	5.00	**1.33**	0.88	0.88	**21.93**	1.52	**24.77**	m	**29.73**
32mm thick	1.58	5.00	**1.66**	0.90	0.90	**22.48**	1.53	**25.67**	m	**30.80**
Cut out crack in granolithic paving not exceeding 50mm wide, form dovetailed key and make good to match existing										
25mm thick	0.32	5.00	**0.33**	0.66	0.66	**16.45**	0.03	**16.81**	m	**20.17**
32mm thick	0.63	5.00	**0.66**	0.73	0.73	**18.09**	0.04	**18.79**	m	**22.55**
38mm thick	0.63	5.00	**0.66**	0.76	0.76	**18.91**	0.05	**19.63**	m	**23.55**
50mm thick	0.95	5.00	**0.99**	0.87	0.87	**21.66**	0.06	**22.71**	m	**27.25**
C90: REPAIRING GLAZING										
Resecure glass										
Remove decayed putties, paint one coat on edge of rebate and re-putty										
wood window	0.15	10.00	**0.16**	0.76	-	**10.79**	0.02	**10.97**	m	**13.17**
metal window	0.15	10.00	**0.16**	0.84	-	**11.88**	0.02	**12.07**	m	**14.48**

Labour hourly rates: (except Specialists) Craft Operatives £14.21 Labourer £10.71 Rates are national average prices. Refer to REGIONAL VARIATIONS for indicative levels of overall pricing in regions	MATERIALS			LABOUR			RATES			
	Del to Site £	Waste %	Material Cost £	Craft Optve Hrs	Lab Hrs	Labour Cost £	Sunds £	Nett Rate £	Unit	Gross Rate (+20%) £
C90: REPAIRING GLAZING Cont.										
Resecure glass Cont.										
Remove beads, remove decayed bedding materials, paint one coat on edge of rebate, re-bed glass and re-fix beads										
wood window	0.15	10.00	**0.16**	1.34	-	**19.07**	0.02	**19.26**	m	**23.11**
metal window	0.15	10.00	**0.16**	1.45	-	**20.63**	0.02	**20.82**	m	**24.98**
C90: REPAIRING/RENOVATING FENCING										
Repairs to softwood fencing										
Remove defective timber fence post and replace with new 127 x 102mm post; letting post 450mm into ground; backfilling around in concrete mix C20; securing ends of existing arris rails and gravel boards to new post										
to suit fencing 914mm high	14.85	10.00	**15.46**	2.24	2.24	**55.92**	0.45	**71.83**	nr	**86.20**
to suit fencing 1219mm high	15.79	10.00	**16.41**	2.27	2.27	**56.47**	0.52	**73.39**	nr	**88.07**
to suit fencing 1524mm high	16.78	10.00	**17.40**	2.29	2.29	**57.02**	0.61	**75.02**	nr	**90.02**
to suit fencing 1829mm high	18.28	10.00	**18.89**	2.31	2.31	**57.57**	0.70	**77.16**	nr	**92.59**
Remove defective pale from palisade fence and replace with new 19 x 75mm pointed pale; fixing with galvanised nails										
764mm long	0.79	-	**0.79**	0.61	0.10	**9.66**	0.20	**10.65**	nr	**12.78**
1069mm long	1.11	-	**1.11**	0.62	0.10	**9.81**	0.24	**11.17**	nr	**13.40**
1374mm long	1.43	-	**1.43**	0.64	0.11	**10.24**	0.28	**11.95**	nr	**14.34**
Remove defective pales from close boarded fence and replace with new 100mm wide feather edged pales; fixing with galvanised nails										
singly	1.79	-	**1.79**	0.28	0.04	**4.38**	0.23	**6.40**	m	**7.68**
areas up to 1.00m2	10.93	20.00	**13.12**	1.45	0.24	**23.22**	1.93	**38.27**	m2	**45.92**
areas over 1.00m2	10.93	20.00	**13.12**	1.08	0.18	**17.20**	1.93	**32.25**	m2	**38.70**
Remove defective gravel board and centre stump and replace with new 25 x 150mm gravel board and 50 x 50mm centre stump; fixing with galvanised nails										
securing to fence posts	2.68	5.00	**2.82**	0.58	0.10	**9.34**	0.23	**12.39**	m	**14.87**
Repairs to oak fencing										
Remove defective timber fence post and replace with new 127 x 102mm post; letting post 450mm into ground; backfilling around in concrete mix C20; securing ends of existing arris rails and gravel boards to new post										
to suit fencing 914mm high	25.61	10.00	**26.22**	2.39	2.39	**59.48**	0.45	**86.15**	nr	**103.38**
to suit fencing 1219mm high	29.80	10.00	**30.41**	2.43	2.43	**60.58**	0.52	**91.51**	nr	**109.81**
to suit fencing 1524mm high	33.98	10.00	**34.59**	2.48	2.48	**61.68**	0.61	**96.87**	nr	**116.25**
to suit fencing 1829mm high	38.17	10.00	**38.78**	2.52	2.52	**62.77**	0.70	**102.26**	nr	**122.71**
Remove defective pale from palisade fence and replace with new 19 x 75mm pointed pale; fixing with galvanised nails										
764mm long	2.09	-	**2.09**	0.77	0.13	**12.36**	0.20	**14.64**	nr	**17.57**
1069mm long	2.90	-	**2.90**	0.84	0.14	**13.41**	0.24	**16.56**	nr	**19.87**
1374mm long	3.75	-	**3.75**	0.91	0.15	**14.62**	0.28	**18.65**	nr	**22.38**
Remove defective pales from close boarded fence and replace with new 100mm wide feather edged pales; fixing with galvanised nails										
singly	4.70	-	**4.70**	0.30	0.06	**4.81**	0.23	**9.74**	m	**11.68**
areas up to 1.00m2	28.66	20.00	**34.39**	2.82	0.47	**45.08**	1.93	**81.39**	m2	**97.67**
areas over 1.00m2	28.66	20.00	**34.39**	2.28	0.39	**36.48**	1.93	**72.79**	m2	**87.35**
Remove defective gravel board and centre stump and replace with new 25 x 150mm gravel board and 50 x 50mm centre stump; fixing with galvanised nails										
securing to fence posts	7.00	5.00	**7.35**	1.21	0.20	**19.31**	0.23	**26.89**	m	**32.27**
Repairs to posts										
100 x 100mm precast concrete spur 1219mm long; setting into ground; backfilling around in concrete mix C20										
bolting to existing timber post	15.42	10.00	**16.03**	0.84	0.84	**20.83**	1.83	**38.69**	nr	**46.43**
C90: REPAIRING/RENOVATING EXTERNAL PAVINGS										
Repairs to precast concrete flag paving										
Take up uneven paving and set aside for re-use; level up and consolidate existing hardcore and relay salvaged paving										
bedding on 25mm sand bed; grouting in lime mortar (1:3)	1.03	33.00	**1.34**	0.74	1.02	**21.43**	0.13	**22.91**	m2	**27.49**
Repairs to York stone paving										
Take up old paving; re-square; relay random in random sizes										
bedding, jointing and pointing in cement mortar (1:3)	4.83	10.00	**5.31**	1.21	1.21	**30.15**	0.99	**36.45**	m2	**43.74**

Labour hourly rates: (except Specialists) Craft Operatives £14.21 Labourer £10.71 Rates are national average prices. Refer to REGIONAL VARIATIONS for indicative levels of overall pricing in regions	MATERIALS			LABOUR				RATES		
	Del to Site £	Waste %	Material Cost £	Craft Optve Hrs	Lab Hrs	Labour Cost £	Sunds £	Nett Rate £	Unit	Gross Rate (+20%) £
C90: REPAIRING/RENOVATING EXTERNAL PAVINGS Cont.										
Repairs to granite sett paving										
Take up, clean and stack old paving	-	-	-	-	0.99	10.60	-	10.60	m2	12.72
Take up old paving, clean and re-lay bedding and grouting in cement mortar (1:3)	-	-	-	1.82	2.81	55.83	1.32	57.15	m2	68.58
Clean out joints of old paving and re-grout in cement mortar (1:3)	-	-	-	-	0.61	6.48	0.57	7.05	m2	8.46
Repairs to bitumen paving										
Cut out damaged 50mm paving to footpath and fill with bitumen macadam under 1.00m2	39.60	10.00	43.56	1.82	2.81	55.83	1.32	100.71	m2	120.86
Cut out damaged 75mm paving to road fill with macadam and seal all around with hot bitumen under 1.00m2	59.40	10.00	65.34	1.82	2.81	55.83	1.32	122.49	m2	146.99

GROUNDWORK

Labour hourly rates: (except Specialists) Craft Operatives £14.21 Labourer £10.71 Rates are national average prices. Refer to REGIONAL VARIATIONS for indicative levels of overall pricing in regions	PLANT & TRANSPORT			LABOUR				RATES		
	Plant Cost £	Trans Cost £	P & T Cost £	Craft Optve Hrs	Lab Hrs	Labour Cost £	Sunds £	Nett Rate £	Unit	Gross Rate (+20%) £
D20: EXCAVATING AND FILLING										
Site preparation										
Removing trees; filling voids left by removal of roots with selected material arising from excavation										
girth 1.50 - 3.00m	154.44	-	**154.44**	-	25.30	**270.96**	-	**425.40**	nr	**510.48**
Excavating - by machine										
Top soil for preservation										
average 150mm deep	0.36	-	**0.36**	-	-	-	-	**0.36**	m2	**0.43**
To reduce levels										
maximum depth not exceeding										
0.25m	2.79	-	**2.79**	-	-	-	-	**2.79**	m3	**3.35**
1.00m	1.47	-	**1.47**	-	-	-	-	**1.47**	m3	**1.76**
2.00m	1.69	-	**1.69**	-	-	-	-	**1.69**	m3	**2.02**
Basements and the like										
maximum depth not exceeding										
0.25m	2.79	-	**2.79**	-	0.50	**5.30**	-	**8.09**	m3	**9.71**
1.00m	1.76	-	**1.76**	-	0.55	**5.89**	-	**7.65**	m3	**9.18**
2.00m	2.06	-	**2.06**	-	0.61	**6.48**	-	**8.54**	m3	**10.25**
Trenches width not exceeding 0.30m										
maximum depth not exceeding										
0.25m	7.21	-	**7.21**	-	2.20	**23.56**	-	**30.77**	m3	**36.92**
1.00m	7.46	-	**7.46**	-	2.31	**24.74**	-	**32.20**	m3	**38.65**
Trenches width exceeding 0.30m										
maximum depth not exceeding										
0.25m	7.19	-	**7.19**	-	0.72	**7.66**	-	**14.85**	m3	**17.82**
1.00m	6.16	-	**6.16**	-	0.77	**8.25**	-	**14.41**	m3	**17.29**
2.00m	6.16	-	**6.16**	-	0.88	**9.42**	-	**15.59**	m3	**18.71**
Pits										
maximum depth not exceeding										
0.25m	6.31	-	**6.31**	-	1.05	**11.19**	-	**17.50**	m3	**21.00**
1.00m	7.77	-	**7.77**	-	1.10	**11.78**	-	**19.55**	m3	**23.47**
2.00m	7.77	-	**7.77**	-	1.27	**13.55**	-	**21.32**	m3	**25.59**
Extra over any types of excavating irrespective of depth										
excavating below ground water level	6.38	-	**6.38**	-	-	-	-	**6.38**	m3	**7.66**
excavating in heavy soil or clay	1.42	-	**1.42**	-	-	-	-	**1.42**	m3	**1.70**
excavating in gravel	2.21	-	**2.21**	-	-	-	-	**2.21**	m3	**2.66**
Breaking out existing materials; extra over any types of excavating irrespective of depth										
concrete	9.01	-	**9.01**	-	-	-	-	**9.01**	m3	**10.81**
reinforced concrete	13.53	-	**13.53**	-	-	-	-	**13.53**	m3	**16.24**
brickwork, blockwork or stonework	6.04	-	**6.04**	-	-	-	-	**6.04**	m3	**7.25**
drain and concrete bed under										
100mm diameter	0.84	-	**0.84**	-	-	-	-	**0.84**	m	**1.01**
150mm diameter	1.02	-	**1.02**	-	-	-	-	**1.02**	m	**1.22**
Breaking out existing hard pavings; extra over any types of excavating irrespective of depth										
concrete 150mm thick	1.42	-	**1.42**	-	-	-	-	**1.42**	m2	**1.71**
reinforced concrete 200mm thick	2.85	-	**2.85**	-	-	-	-	**2.85**	m2	**3.42**
coated macadam or asphalt 75mm thick	0.24	-	**0.24**	-	-	-	-	**0.24**	m2	**0.29**
Working space allowance to excavation; including additional earthwork support, disposal (excluding landfill tax) and backfilling										
reduce levels, basements and the like	6.46	-	**6.46**	-	0.33	**3.53**	-	**10.00**	m2	**11.99**
pits	7.34	-	**7.34**	-	0.77	**8.25**	-	**15.58**	m2	**18.70**
trenches	6.46	-	**6.46**	-	0.55	**5.89**	-	**12.35**	m2	**14.82**

Labour hourly rates: (except Specialists) Craft Operatives £14.21 Labourer £10.71 Rates are national average prices. Refer to REGIONAL VARIATIONS for indicative levels of overall pricing in regions	PLANT & TRANSPORT			LABOUR				RATES		
	Plant Cost £	Trans Cost £	P & T Cost £	Craft Optve Hrs	Lab Hrs	Labour Cost £	Sunds £	Nett Rate £	Unit	Gross Rate (+20%) £
D20: EXCAVATING AND FILLING Cont.										
Excavating - by machine Cont.										
Compacting										
bottoms of excavations	0.24	-	**0.24**	-	-	-	-	**0.24**	m2	**0.28**
Compacting with 680 kg vibratory roller										
ground	0.25	-	**0.25**	-	-	-	-	**0.25**	m2	**0.29**
Trimming										
sides of cuttings; vertical or battered	0.62	-	**0.62**	-	-	-	-	**0.62**	m2	**0.74**
sides of embankments; vertical or battered	0.62	-	**0.62**	-	-	-	-	**0.62**	m2	**0.74**
Disposal of excavated material										
depositing on site in temporary spoil heaps where directed										
25m	-	2.25	**2.25**	-	0.13	**1.41**	-	**3.66**	m3	**4.39**
removing from site to tip a) Inert	-	18.70	**18.70**	-	-	-	-	**18.70**	m3	**22.44**
removing from site to tip b) Active	-	44.00	**44.00**	-	-	-	-	**44.00**	m3	**52.80**
removing from site to tip c) Contaminated (Guide price - always seek a quotation for specialist disposal costs.)	-	88.00	**88.00**	-	-	-	-	**88.00**	m3	**105.60**
Disposal of preserved top soil										
depositing on site in temporary spoil heaps where directed										
25m	-	2.25	**2.25**	-	0.13	**1.41**	-	**3.66**	m3	**4.39**
Excavating - by hand										
Top soil for preservation										
average 150mm deep	-	-	-	-	0.36	**3.89**	-	**3.89**	m2	**4.67**
To reduce levels										
maximum depth not exceeding										
0.25m	-	-	-	-	3.03	**32.40**	-	**32.40**	m3	**38.88**
1.00m	-	-	-	-	3.19	**34.16**	-	**34.16**	m3	**41.00**
2.00m	-	-	-	-	3.63	**38.88**	-	**38.88**	m3	**46.65**
Basements and the like										
maximum depth not exceeding										
0.25m	-	-	-	-	3.14	**33.58**	-	**33.58**	m3	**40.29**
1.00m	-	-	-	-	3.36	**35.93**	-	**35.93**	m3	**43.12**
2.00m	-	-	-	-	3.74	**40.06**	-	**40.06**	m3	**48.07**
Trenches width not exceeding 0.30m										
maximum depth not exceeding										
0.25m	-	-	-	-	5.17	**55.37**	-	**55.37**	m3	**66.44**
1.00m	-	-	-	-	5.45	**58.32**	-	**58.32**	m3	**69.98**
Trenches width exceeding 0.30m										
maximum depth not exceeding										
0.25m	-	-	-	-	3.45	**36.99**	-	**36.99**	m3	**44.39**
1.00m	-	-	-	-	3.63	**38.88**	-	**38.88**	m3	**46.65**
2.00m	-	-	-	-	4.24	**45.36**	-	**45.36**	m3	**54.43**
Pits										
maximum depth not exceeding										
0.25m	-	-	-	-	4.07	**43.59**	-	**43.59**	m3	**52.31**
1.00m	-	-	-	-	4.24	**45.36**	-	**45.36**	m3	**54.43**
2.00m	-	-	-	-	4.84	**51.84**	-	**51.84**	m3	**62.20**
Extra over any types of excavating irrespective of depth										
excavating below ground water level	-	-	-	-	1.82	**19.44**	-	**19.44**	m3	**23.33**
excavating in heavy soil or clay	-	-	-	-	0.77	**8.25**	-	**8.25**	m3	**9.90**
excavating in gravel	-	-	-	-	1.98	**21.21**	-	**21.21**	m3	**25.45**
Breaking out existing materials; extra over any types of excavating irrespective of depth										
concrete	-	-	-	-	12.10	**129.59**	-	**129.59**	m3	**155.51**
reinforced concrete	-	-	-	-	18.15	**194.39**	-	**194.39**	m3	**233.26**
brickwork, blockwork or stonework	-	-	-	-	8.47	**90.71**	-	**90.71**	m3	**108.86**
drain and concrete bed under										
100mm diameter	-	-	-	-	0.99	**10.60**	-	**10.60**	m	**12.72**
150mm diameter	-	-	-	-	1.21	**12.96**	-	**12.96**	m	**15.55**
Breaking out existing hard pavings; extra over any types of excavating irrespective of depth										
concrete 150mm thick	-	-	-	-	1.82	**19.44**	-	**19.44**	m2	**23.33**
reinforced concrete 200mm thick	-	-	-	-	3.63	**38.88**	-	**38.88**	m2	**46.65**
coated macadam or asphalt 75mm thick	-	-	-	-	0.33	**3.53**	-	**3.53**	m2	**4.24**
Working space allowance to excavation; including additional earthwork support, disposal (excluding landfill tax) and backfilling										
reduce levels, basements and the like	-	-	-	-	3.14	**33.58**	-	**33.58**	m2	**40.29**
pits	-	-	-	-	3.80	**40.64**	-	**40.64**	m2	**48.77**
trenches	-	-	-	-	3.41	**36.52**	-	**36.52**	m2	**43.83**
Compacting										
bottoms of excavations	-	-	-	-	0.17	**1.77**	-	**1.77**	m2	**2.12**
Trimming										
sides of cuttings; vertical or battered	-	-	-	-	0.24	**2.59**	-	**2.59**	m2	**3.11**

Labour hourly rates: (except Specialists) Craft Operatives £14.21 Labourer £10.71 Rates are national average prices. Refer to REGIONAL VARIATIONS for indicative levels of overall pricing in regions	PLANT & TRANSPORT			LABOUR				RATES		
	Plant Cost £	Trans Cost £	P & T Cost £	Craft Optve Hrs	Lab Hrs	Labour Cost £	Sunds £	Nett Rate £	Unit	Gross Rate (+20%) £
D20: EXCAVATING AND FILLING Cont.										
Excavating - by hand Cont.										
Trimming Cont.										
sides of embankments; vertical or battered.......	-	-	-	-	0.24	2.59		2.59	m2	3.11
Disposal of excavated material										
depositing on site in temporary spoil heaps where directed										
25m..	-	-	-	-	1.54	16.49		16.49	m3	19.79
removing from site to tip a) Inert	-	18.70	18.70	-	1.65	17.67		36.37	m3	43.65
removing from site to tip b) Active	-	44.00	44.00	-	1.65	17.67		61.67	m3	74.01
removing from site to tip c) Contaminated (Guide price - always seek a quotation for specialist disposal costs.)	-	88.00	88.00	-	1.65	17.67		105.67	m3	126.81
Disposal of preserved top soil										
depositing on site in temporary spoil heaps where directed										
25m..	-	-	-	-	1.54	16.49		16.49	m3	19.79
Excavating inside existing building - by hand										
To reduce levels										
maximum depth not exceeding										
0.25m..	-	-	-	-	4.24	45.36	-	45.36	m3	54.43
1.00m..	-	-	-	-	4.51	48.30	-	48.30	m3	57.96
2.00m..	-	-	-	-	5.12	54.78	-	54.78	m3	65.74
Trenches width not exceeding 0.30m										
maximum depth not exceeding										
0.25m..	-	-	-	-	7.32	78.34	-	78.34	m3	94.01
1.00m..	-	-	-	-	7.70	82.47	-	82.47	m3	98.96
Trenches width exceeding 0.30m										
maximum depth not exceeding										
0.25m..	-	-	-	-	4.90	52.43	-	52.43	m3	62.91
1.00m..	-	-	-	-	5.17	55.37	-	55.37	m3	66.44
2.00m..	-	-	-	-	6.05	64.80	-	64.80	m3	77.75
Pits										
maximum depth not exceeding										
0.25m..	-	-	-	-	5.72	61.26	-	61.26	m3	73.51
1.00m..	-	-	-	-	5.94	63.62	-	63.62	m3	76.34
2.00m..	-	-	-	-	6.82	73.04	-	73.04	m3	87.65
Extra over any types of excavating irrespective of depth										
excavating below ground water level.............	-	-	-	-	2.59	27.69	-	27.69	m3	33.22
Breaking out existing materials; extra over any types of excavating irrespective of depth										
concrete..	-	-	-	-	13.20	141.37	-	141.37	m3	169.65
reinforced concrete	-	-	-	-	19.03	203.81	-	203.81	m3	244.57
brickwork, blockwork or stonework..............	-	-	-	-	8.58	91.89	-	91.89	m3	110.27
drain and concrete bed under										
100mm diameter...............................	-	-	-	-	1.38	14.73	-	14.73	m	17.67
150mm diameter...............................	-	-	-	-	1.65	17.67	-	17.67	m	21.21
Breaking out existing hard pavings; extra over any types of excavating irrespective of depth										
concrete 150mm thick	-	-	-	-	1.82	19.44		19.44	m2	23.33
reinforced concrete 200mm thick................	-	-	-	-	3.63	38.88		38.88	m2	46.65
Working space allowance to excavation; including additional earthwork support, disposal (excluding landfill tax) and backfilling										
pits	-	-	-	-	4.84	51.84		51.84	m2	62.20
trenches	-	-	-	-	4.51	48.30	-	48.30	m2	57.96
Compacting										
bottoms of excavations	-	-	-	-	0.17	1.77	-	1.77	m2	2.12
Disposal of excavated material										
depositing on site in temporary spoil heaps where directed										
25m..	-	-	-	-	2.15	22.97		22.97	m3	27.57
removing from site to tip a) Inert	-	18.70	18.70	-	2.75	29.45	-	48.15	m3	57.78
removing from site to tip b) Active	-	44.00	44.00	-	2.75	29.45	-	73.45	m3	88.14
removing from site to tip c) Contaminated (Guide price - always seek a quotation for specialist disposal costs.)	-	88.00	88.00	-	2.75	29.45	-	117.45	m3	140.94

This section continues on the next page

Labour hourly rates: (except Specialists) Craft Operatives £14.21 Labourer £10.71 Rates are national average prices. Refer to REGIONAL VARIATIONS for indicative levels of overall pricing in regions	MATERIALS Del to Site £	 Waste %	 Material Cost £	LABOUR Craft Optve Hrs	 Lab Hrs	 Labour Cost £	 Sunds £	RATES Nett Rate £	 Unit	 Gross Rate (+20%) £
D20: EXCAVATING AND FILLING Cont.										
Earthwork support										
Earthwork support										
distance between opposing faces not exceeding 2.00m; maximum depth not exceeding										
1.00m	1.21	-	**1.21**	-	0.23	**2.47**	0.04	**3.72**	m2	**4.47**
2.00m	1.52	-	**1.52**	-	0.29	**3.06**	0.05	**4.63**	m2	**5.56**
distance between opposing faces 2.00 - 4.00m; maximum depth not exceeding										
1.00m	1.31	-	**1.31**	-	0.24	**2.59**	0.05	**3.95**	m2	**4.75**
2.00m	1.61	-	**1.61**	-	0.30	**3.18**	0.05	**4.84**	m2	**5.81**
distance between opposing faces exceeding 4.00m; maximum depth not exceeding										
1.00m	1.50	-	**1.50**	-	0.25	**2.71**	0.05	**4.25**	m2	**5.10**
2.00m	1.74	-	**1.74**	-	0.32	**3.42**	0.06	**5.22**	m2	**6.26**
Earthwork support; next to roadways										
distance between opposing faces not exceeding 2.00m; maximum depth not exceeding										
1.00m	2.40	-	**2.40**	-	0.46	**4.95**	0.07	**7.42**	m2	**8.90**
2.00m	2.91	-	**2.91**	-	0.58	**6.24**	0.10	**9.25**	m2	**11.10**
distance between opposing faces 2.00 - 4.00m; maximum depth not exceeding										
1.00m	2.71	-	**2.71**	-	0.48	**5.18**	0.08	**7.97**	m2	**9.57**
2.00m	3.24	-	**3.24**	-	0.61	**6.48**	0.11	**9.82**	m2	**11.79**
distance between opposing faces exceeding 4.00m; maximum depth not exceeding										
1.00m	2.91	-	**2.91**	-	0.51	**5.42**	0.10	**8.43**	m2	**10.11**
2.00m	3.53	-	**3.53**	-	0.65	**6.95**	0.11	**10.59**	m2	**12.71**
Earthwork support; left in										
distance between opposing faces not exceeding 2.00m; maximum depth not exceeding										
1.00m	4.36	-	**4.36**	-	0.19	**2.00**	0.07	**6.43**	m2	**7.72**
2.00m	4.36	-	**4.36**	-	0.23	**2.47**	0.08	**6.92**	m2	**8.30**
distance between opposing faces 2.00 - 4.00m; maximum depth not exceeding										
1.00m	4.36	-	**4.36**	-	0.20	**2.12**	0.07	**6.55**	m2	**7.86**
2.00m	4.36	-	**4.36**	-	0.24	**2.59**	0.10	**7.04**	m2	**8.45**
distance between opposing faces exceeding 4.00m; maximum depth not exceeding										
1.00m	5.80	-	**5.80**	-	0.21	**2.24**	0.08	**8.12**	m2	**9.75**
2.00m	5.80	-	**5.80**	-	0.25	**2.71**	0.10	**8.61**	m2	**10.33**
Earthwork support; next to roadways; left in										
distance between opposing faces not exceeding 2.00m; maximum depth not exceeding										
1.00m	14.50	-	**14.50**	-	0.37	**4.01**	0.13	**18.64**	m2	**22.37**
2.00m	14.50	-	**14.50**	-	0.46	**4.95**	0.17	**19.62**	m2	**23.54**
distance between opposing faces 2.00 - 4.00m; maximum depth not exceeding										
1.00m	17.40	-	**17.40**	-	0.40	**4.24**	0.16	**21.80**	m2	**26.16**
2.00m	17.40	-	**17.40**	-	0.48	**5.18**	0.19	**22.78**	m2	**27.33**
Earthwork support; inside existing building										
distance between opposing faces not exceeding 2.00m; maximum depth not exceeding										
1.00m	1.26	-	**1.26**	-	0.25	**2.71**	0.04	**4.01**	m2	**4.81**
2.00m	1.57	-	**1.57**	-	0.32	**3.42**	0.05	**5.04**	m2	**6.04**
distance between opposing faces 2.00 - 4.00m; maximum depth not exceeding										
1.00m	1.39	-	**1.39**	-	0.26	**2.83**	0.05	**4.27**	m2	**5.12**
2.00m	1.73	-	**1.73**	-	0.33	**3.53**	0.06	**5.32**	m2	**6.39**
distance between opposing faces exceeding 4.00m; maximum depth not exceeding										
1.00m	1.56	-	**1.56**	-	0.28	**2.95**	0.05	**4.55**	m2	**5.46**
2.00m	1.87	-	**1.87**	-	0.36	**3.89**	0.06	**5.82**	m2	**6.98**
Excavated material arising from excavation - by machine										
Filling to excavation										
average thickness exceeding 0.25m	-	-	-	-	-	-	3.70	**3.70**	m3	**4.43**
Compacting										
filling	-	-	-	-	-	-	0.39	**0.39**	m2	**0.47**
Selected excavated material obtained from on site spoil heaps 25m distant - by machine										
Filling to make up levels; depositing in layers 150mm maximum thickness										
average thickness not exceeding 0.25m	-	-	-	-	-	-	5.34	**5.34**	m3	**6.41**
average thickness exceeding 0.25m	-	-	-	-	-	-	4.60	**4.60**	m3	**5.52**
Compacting with 680 kg vibratory roller										
filling	-	-	-	-	-	-	0.28	**0.28**	m2	**0.34**
Trimming										
sides of cuttings; vertical or battered	-	-	-	-	-	-	0.77	**0.77**	m2	**0.92**
sides of embankments; vertical or battered	-	-	-	-	-	-	0.77	**0.77**	m2	**0.92**

Labour hourly rates: (except Specialists) Craft Operatives £14.21 Labourer £10.71 Rates are national average prices. Refer to REGIONAL VARIATIONS for indicative levels of overall pricing in regions	MATERIALS			LABOUR				RATES		
	Del to Site £	Waste %	Material Cost £	Craft Optve Hrs	Lab Hrs	Labour Cost £	Sunds £	Nett Rate £	Unit	Gross Rate (+20%) £
D20: EXCAVATING AND FILLING Cont.										
Preserved topsoil obtained from on site spoil heaps not exceeding 100m distant - by machine										
Filling to make up levels										
average thickness not exceeding 0.25m............	-	-	-	-	-	-	8.85	**8.85**	m3	**10.62**
Imported topsoil; wheeling not exceeding 50m - by machine										
Filling to make up levels										
average thickness not exceeding 0.25m............	16.50	10.00	**18.15**	-	-	-	10.68	**28.83**	m3	**34.60**
Hard, dry, broken brick or stone to be obtained off site; wheeling not exceeding 25m - by machine										
Filling to make up levels; depositing in layers 150mm maximum thickness										
average thickness not exceeding 0.25m............	19.64	25.00	**24.54**	-	-	-	4.60	**29.14**	m3	**34.97**
average thickness exceeding 0.25m................	19.64	25.00	**24.54**	-	-	-	3.82	**28.36**	m3	**34.03**
Surface packing to filling										
to vertical or battered faces....................	-	-	-	-	-	-	0.77	**0.77**	m2	**0.92**
Compacting with 680 kg vibratory roller										
filling; blinding with sand, ashes or similar fine material	0.20	40.00	**0.28**	-	0.06	**0.59**	0.28	**1.15**	m2	**1.38**
MOT Type 1 to be obtained off site; wheeling not exceeding 25m - by machine										
Filling to make up levels; depositing in layers 150mm maximum thickness										
average thickness not exceeding 0.25m............	23.04	25.00	**28.80**	-	-	-	4.60	**33.40**	m3	**40.08**
average thickness exceeding 0.25m................	23.04	25.00	**28.80**	-	-	-	3.82	**32.62**	m3	**39.14**
Surface packing to filling										
to vertical or battered faces....................	-	-	-	-	-	-	0.77	**0.77**	m2	**0.92**
Compacting with 680 kg vibratory roller										
filling; blinding with sand, ashes or similar fine material	0.20	40.00	**0.28**	-	0.06	**0.59**	0.28	**1.15**	m2	**1.38**
MOT Type 2 to be obtained off site; wheeling not exceeding 25m - by machine										
Filling to make up levels; depositing in layers 150mm maximum thickness										
average thickness not exceeding 0.25m............	21.34	25.00	**26.68**	-	-	-	4.60	**31.28**	m3	**37.54**
average thickness exceeding 0.25m................	21.34	25.00	**26.68**	-	-	-	3.82	**30.50**	m3	**36.60**
Surface packing to filling										
to vertical or battered faces....................	-	-	-	-	-	-	0.77	**0.77**	m2	**0.92**
Compacting with 680 kg vibratory roller										
filling; blinding with sand, ashes or similar fine material	0.20	40.00	**0.28**	-	0.06	**0.59**	0.28	**1.15**	m2	**1.38**
Ashes to be obtained off site; wheeling not exceeding 25m - by machine										
Filling to make up levels; depositing in layers 150mm maximum thickness										
average thickness not exceeding 0.25m............	13.20	40.00	**18.48**	-	-	-	4.87	**23.35**	m3	**28.02**
Compacting with 680 kg vibratory roller										
filling	-	-	-	-	-	-	0.28	**0.28**	m2	**0.34**
Sand to be obtained off site; wheeling not exceeding 25m - by machine										
Filling to make up levels; depositing in layers 150mm maximum thickness										
average thickness not exceeding 0.25m............	16.84	33.00	**22.40**	-	-	-	5.34	**27.74**	m3	**33.29**
average thickness exceeding 0.25m................	16.84	33.00	**22.40**	-	-	-	4.59	**26.98**	m3	**32.38**
Compacting with 680 kg vibratory roller										
filling	-	-	-	-	-	-	0.28	**0.28**	m2	**0.34**
Hoggin to be obtained off site; wheeling not exceeding 25m - by machine										
Filling to make up levels; depositing in layers 150mm maximum thickness										
average thickness not exceeding 0.25m............	18.48	33.00	**24.58**	-	-	-	4.59	**29.16**	m3	**35.00**
average thickness exceeding 0.25m................	18.48	33.00	**24.58**	-	-	-	3.82	**28.39**	m3	**34.07**
Compacting with 680 kg vibratory roller										
filling	-	-	-	-	-	-	0.28	**0.28**	m2	**0.34**
Excavated material arising from excavation - by hand										
Filling to excavation										
average thickness exceeding 0.25m................	-	-	-	-	1.58	**16.96**	-	**16.96**	m3	**20.36**

Labour hourly rates: (except Specialists) Craft Operatives £14.21 Labourer £10.71 Rates are national average prices. Refer to REGIONAL VARIATIONS for indicative levels of overall pricing in regions	MATERIALS			LABOUR				RATES		
	Del to Site £	Waste %	Material Cost £	Craft Optve Hrs	Lab Hrs	Labour Cost £	Sunds £	Nett Rate £	Unit	Gross Rate (+20%) £
D20: EXCAVATING AND FILLING Cont.										
Excavated material arising from excavation - by hand Cont.										
Compacting										
filling	-	-	-	-	-		0.55	**0.55**	m2	**0.66**
Selected excavated material obtained from on site spoil heaps 25m distant - by hand										
Filling to make up levels; depositing in layers 150mm maximum thickness										
average thickness not exceeding 0.25m	-	-	-	-	2.04	**21.79**	-	**21.79**	m3	**26.15**
average thickness exceeding 0.25m	-	-	-	-	1.76	**18.85**	-	**18.85**	m3	**22.62**
Compacting with 680 kg vibratory roller										
filling	-	-	-	-			0.28	**0.28**	m2	**0.34**
Trimming										
sides of cuttings; vertical or battered	-	-	-	-	0.24	**2.59**	-	**2.59**	m2	**3.11**
sides of embankments; vertical or battered	-	-	-	-	0.24	**2.59**	-	**2.59**	m2	**3.11**
Preserved topsoil obtained from on site spoil heaps not exceeding 100m distant - by hand										
Filling to make up levels										
average thickness not exceeding 0.25m	-	-	-	-	4.84	**51.84**	-	**51.84**	m3	**62.20**
Filling to external planters										
average thickness not exceeding 0.25m	-	-	-	-	6.05	**64.80**	-	**64.80**	m3	**77.75**
Imported topsoil; wheeling not exceeding 50m - by hand										
Filling to make up levels										
average thickness not exceeding 0.25m	16.50	10.00	**18.15**	-	4.24	**45.36**	-	**63.51**	m3	**76.21**
Filling to external planters										
average thickness exceeding 0.25m	16.50	10.00	**18.15**	-	5.45	**58.32**	-	**76.47**	m3	**91.76**
Hard, dry, broken brick or stone to be obtained off site; wheeling not exceeding 25m - by hand										
Filling to make up levels; depositing in layers 150mm maximum thickness										
average thickness not exceeding 0.25m	19.64	25.00	**24.54**	-	1.74	**18.61**	-	**43.16**	m3	**51.79**
average thickness exceeding 0.25m	19.64	25.00	**24.54**	-	1.44	**15.43**	-	**39.98**	m3	**47.97**
Surface packing to filling										
to vertical or battered faces	-	-	-	-	0.24	**2.59**	-	**2.59**	m2	**3.11**
Compacting with 680 kg vibratory roller										
filling; blinding with sand, ashes or similar fine material	0.20	40.00	**0.28**	-	0.06	**0.59**	0.28	**1.15**	m2	**1.38**
MOT Type 1 to be obtained off site; wheeling not exceeding 25m - by hand										
Filling to make up levels; depositing in layers 150mm maximum thickness										
average thickness not exceeding 0.25m	23.04	25.00	**28.80**	-	1.74	**18.61**	-	**47.42**	m3	**56.90**
average thickness exceeding 0.25m	23.04	25.00	**28.80**	-	1.44	**15.43**	-	**44.24**	m3	**53.08**
Surface packing to filling										
to vertical or battered faces	-	-	-	-	0.24	**2.59**	-	**2.59**	m2	**3.11**
Compacting with 680 kg vibratory roller										
filling; blinding with sand, ashes or similar fine material	0.20	40.00	**0.28**	-	0.06	**0.59**	0.28	**1.15**	m2	**1.38**
MOT Type 2 to be obtained off site; wheeling not exceeding 25m - by hand										
Filling to make up levels; depositing in layers 150mm maximum thickness										
average thickness not exceeding 0.25m	21.34	25.00	**26.68**	-	1.74	**18.61**	-	**45.29**	m3	**54.35**
average thickness exceeding 0.25m	21.34	25.00	**26.68**	-	1.44	**15.43**	-	**42.11**	m3	**50.54**
Surface packing to filling										
to vertical or battered faces	-	-	-	-	0.24	**2.59**	-	**2.59**	m2	**3.11**
Compacting with 680 kg vibratory roller										
filling; blinding with sand, ashes or similar fine material	0.20	40.00	**0.28**	-	0.06	**0.59**	0.28	**1.15**	m2	**1.38**
Ashes to be obtained off site; wheeling not exceeding 25m - by hand										
Filling to make up levels; depositing in layers 150mm maximum thickness										
average thickness not exceeding 0.25m	13.20	40.00	**18.48**	-	1.87	**20.03**	-	**38.51**	m3	**46.21**
Compacting with 680 kg vibratory roller										
filling	-	-	-	-	-	-	0.28	**0.28**	m2	**0.34**

Labour hourly rates: (except Specialists) Craft Operatives £14.21 Labourer £10.71 Rates are national average prices. Refer to REGIONAL VARIATIONS for indicative levels of overall pricing in regions	MATERIALS			LABOUR				RATES		
	Del to Site £	Waste %	Material Cost £	Craft Optve Hrs	Lab Hrs	Labour Cost £	Sunds £	Nett Rate £	Unit	Gross Rate (+20%) £
D20: EXCAVATING AND FILLING Cont.										
Sand to be obtained off site; wheeling not exceeding 25m - by hand										
Filling to make up levels; depositing in layers 150mm maximum thickness										
average thickness not exceeding 0.25m...........	16.84	33.00	**22.40**	-	2.02	**21.68**	-	**44.07**	m3	**52.89**
average thickness exceeding 0.25m...............	16.84	33.00	**22.40**	-	1.74	**18.61**	-	**41.01**	m3	**49.21**
Compacting with 680 kg vibratory roller										
filling	-	-	**-**	-	-	**-**	0.28	**0.28**	m2	**0.34**
Hoggin to be obtained off site; wheeling not exceeding 25m - by hand										
Filling to make up levels; depositing in layers 150mm maximum thickness										
average thickness not exceeding 0.25m...........	18.48	33.00	**24.58**	-	1.71	**18.26**	-	**42.84**	m3	**51.41**
average thickness exceeding 0.25m...............	18.48	33.00	**24.58**	-	1.43	**15.32**	-	**39.89**	m3	**47.87**
Compacting with 680 kg vibratory roller										
filling	-	-	**-**	-	-	**-**	0.28	**0.28**	m2	**0.34**
Excavated material arising from excavation; work inside existing building - by hand										
Filling to excavation										
average thickness exceeding 0.25m...............	-	-	**-**	-	3.08	**32.99**	-	**32.99**	m3	**39.58**
Compacting										
filling	-	-	**-**	-	-	**-**	0.28	**0.28**	m2	**0.34**
Selected excavated material obtained from on site spoil heaps 25m distant; work inside existing building - by hand										
Filling to make up levels; depositing in layers 150mm maximum thickness										
average thickness not exceeding 0.25m...........	-	-	**-**	-	3.59	**38.41**	-	**38.41**	m3	**46.09**
average thickness exceeding 0.25m...............	-	-	**-**	-	3.39	**36.29**	-	**36.29**	m3	**43.54**
Compacting										
filling	-	-	**-**	-	-	**-**	0.28	**0.28**	m2	**0.34**
Hard, dry, broken brick or stone to be obtained off site; wheeling not exceeding 25m; work inside existing building - by hand										
Filling to make up levels; depositing in layers 150mm maximum thickness										
average thickness not exceeding 0.25m...........	19.64	25.00	**24.54**	-	3.29	**35.23**	-	**59.77**	m3	**71.72**
average thickness exceeding 0.25m...............	19.64	25.00	**24.54**	-	3.09	**33.10**	-	**57.65**	m3	**69.18**
Surface packing to filling										
to vertical or battered faces...................	-	-	**-**	-	0.24	**2.59**	-	**2.59**	m2	**3.11**
Compacting										
filling; blinding with sand, ashes or similar fine material	0.20	40.00	**0.28**	-	0.06	**0.59**	0.28	**1.15**	m2	**1.38**
Ashes to be obtained off site; wheeling not exceeding 25m; work inside existing building - by hand										
Filling to make up levels; depositing in layers 150mm maximum thickness										
average thickness not exceeding 0.25m...........	13.20	40.00	**18.48**	-	3.41	**36.52**	-	**55.00**	m3	**66.00**
Compacting										
filling	-	-	**-**	-	-	**-**	0.28	**0.28**	m2	**0.34**
Sand to be obtained off site; wheeling not exceeding 25m; work inside existing building - by hand										
Filling to make up levels; depositing in layers 150mm maximum thickness										
average thickness not exceeding 0.25m...........	16.84	33.00	**22.40**	-	3.59	**38.41**	-	**60.80**	m3	**72.96**
average thickness exceeding 0.25m...............	16.84	33.00	**22.40**	-	3.29	**35.23**	-	**57.62**	m3	**69.15**
Compacting										
filling	-	-	**-**	-	-	**-**	0.28	**0.28**	m2	**0.34**

IN-SITU CONCRETE/LARGE PRECAST CONCRETE

Labour hourly rates: (except Specialists) Craft Operatives £14.21 Labourer £10.71 Rates are national average prices. Refer to REGIONAL VARIATIONS for indicative levels of overall pricing in regions	MATERIALS			LABOUR			RATES			
	Del to Site	Waste	Material Cost	Craft Optve	Lab	Labour Cost	Sunds	Nett Rate	Unit	Gross Rate (+20%)
	£	%	£	Hrs	Hrs	£	£	£		£
E10: MIXING/CASTING/CURING IN-SITU CONCRETE - (READY MIXED)										
Plain in-situ concrete; BS 5328, ordinary prescribed mix ST3, 20mm aggregate										
Foundations poured on or against earth or unblinded hardcore										
generally	68.42	7.50	**73.55**	-	2.53	**27.10**	-	**100.65**	m3	**120.78**
Isolated foundations; poured on or against earth or unblinded hardcore										
generally	68.42	7.50	**73.55**	-	3.03	**32.40**	-	**105.95**	m3	**127.14**
Beds; poured on or against earth or unblinded hardcore										
thickness 150 - 450mm	68.42	7.50	**73.55**	-	2.75	**29.45**	-	**103.00**	m3	**123.60**
thickness not exceeding 150mm	68.42	7.50	**73.55**	-	3.41	**36.52**	-	**110.07**	m3	**132.09**
Filling hollow walls										
thickness not exceeding 150mm	68.42	5.00	**71.84**	-	6.05	**64.80**	-	**136.64**	m3	**163.96**
Plain in-situ concrete; BS 5328, ordinary prescribed mix ST4, 20mm aggregate										
Foundations poured on or against earth or unblinded hardcore										
generally	70.33	7.50	**75.60**	-	2.53	**27.10**	-	**102.70**	m3	**123.24**
Isolated foundations; poured on or against earth or unblinded hardcore										
generally	70.33	7.50	**75.60**	-	3.03	**32.40**	-	**108.00**	m3	**129.60**
Beds; poured on or against earth or unblinded hardcore										
thickness 150 - 450mm	70.33	7.50	**75.60**	-	2.75	**29.45**	-	**105.05**	m3	**126.06**
thickness not exceeding 150mm	70.33	7.50	**75.60**	-	3.41	**36.52**	-	**112.12**	m3	**134.55**
Filling hollow walls										
thickness not exceeding 150mm	70.33	5.00	**73.84**	-	6.05	**64.80**	-	**138.64**	m3	**166.37**
Reinforced in-situ concrete; BS 5328, designed mix C20, 20mm aggregate, minimum cement content 240 kg/m3; vibrated										
Foundations										
generally	71.77	5.00	**75.36**	-	2.86	**30.63**	0.67	**106.66**	m3	**127.99**
Ground beams										
generally	71.77	5.00	**75.36**	-	3.96	**42.41**	0.67	**118.44**	m3	**142.13**
Isolated foundations										
generally	71.77	5.00	**75.36**	-	3.30	**35.34**	0.67	**111.37**	m3	**133.64**
Beds										
thickness 150 - 450mm	71.77	5.00	**75.36**	-	3.03	**32.40**	0.67	**108.43**	m3	**130.11**
thickness not exceeding 150mm	71.77	5.00	**75.36**	-	3.85	**41.23**	0.67	**117.26**	m3	**140.71**
Slabs										
thickness 150 - 450mm	71.77	2.50	**73.56**	-	3.80	**40.64**	0.67	**114.88**	m3	**137.85**
thickness not exceeding 150mm	71.77	2.50	**73.56**	-	5.01	**53.60**	0.67	**127.84**	m3	**153.40**
Walls										
thickness 150 - 450mm	71.77	2.50	**73.56**	-	4.29	**45.95**	0.67	**120.18**	m3	**144.22**
thickness not exceeding 150mm	71.77	2.50	**73.56**	-	5.50	**58.91**	0.67	**133.14**	m3	**159.77**
Beams										
isolated	71.77	2.50	**73.56**	-	5.17	**55.37**	0.67	**129.60**	m3	**155.53**
isolated deep	71.77	2.50	**73.56**	-	5.28	**56.55**	0.67	**130.78**	m3	**156.94**
attached deep	71.77	2.50	**73.56**	-	5.17	**55.37**	0.67	**129.60**	m3	**155.53**
Columns										
generally	71.77	2.50	**73.56**	-	6.93	**74.22**	0.67	**148.45**	m3	**178.14**

Labour hourly rates: (except Specialists) Craft Operatives £14.21 Labourer £10.71 Rates are national average prices. Refer to REGIONAL VARIATIONS for indicative levels of overall pricing in regions	MATERIALS			LABOUR				RATES		
	Del to Site £	Waste %	Material Cost £	Craft Optve Hrs	Lab Hrs	Labour Cost £	Sunds £	Nett Rate £	Unit	Gross Rate (+20%) £
E10: MIXING/CASTING/CURING IN-SITU CONCRETE - (READY MIXED) Cont.										
Reinforced in-situ concrete; BS 5328, designed mix C20, 20mm aggregate, minimum cement content 240 kg/m3; vibrated Cont.										
Staircases										
generally	71.77	2.50	**73.56**	-	6.05	**64.80**	0.67	**139.03**	m3	**166.83**
Upstands										
generally	71.77	2.50	**73.56**	-	8.47	**90.71**	0.67	**164.95**	m3	**197.94**
Reinforced in-situ concrete; BS 5328, designed mix C30, 20mm aggregate, minimum cement content 290 kg/m3; vibrated										
Foundations										
generally	76.04	5.00	**79.84**	-	2.86	**30.63**	0.67	**111.14**	m3	**133.37**
Ground beams										
generally	76.04	5.00	**79.84**	-	3.96	**42.41**	0.67	**122.92**	m3	**147.51**
Isolated foundations										
generally	76.04	5.00	**79.84**	-	3.30	**35.34**	0.67	**115.86**	m3	**139.03**
Beds										
thickness 150 - 450mm	76.04	5.00	**79.84**	-	3.03	**32.40**	0.67	**112.91**	m3	**135.49**
thickness not exceeding 150mm	76.04	5.00	**79.84**	-	3.85	**41.23**	0.67	**121.75**	m3	**146.10**
Slabs										
thickness 150 - 450mm	76.04	2.50	**77.94**	-	3.80	**40.64**	0.67	**119.26**	m3	**143.11**
thickness not exceeding 150mm	76.04	2.50	**77.94**	-	5.01	**53.60**	0.67	**132.22**	m3	**158.66**
Walls										
thickness 150 - 450mm	76.04	2.50	**77.94**	-	4.29	**45.95**	0.67	**124.56**	m3	**149.47**
thickness not exceeding 150mm	76.04	2.50	**77.94**	-	5.50	**58.91**	0.67	**137.52**	m3	**165.02**
Beams										
isolated	76.04	2.50	**77.94**	-	5.17	**55.37**	0.67	**133.98**	m3	**160.78**
isolated deep	76.04	2.50	**77.94**	-	5.28	**56.55**	0.67	**135.16**	m3	**162.19**
attached deep	76.04	2.50	**77.94**	-	5.17	**55.37**	0.67	**133.98**	m3	**160.78**
Columns										
generally	76.04	2.50	**77.94**	-	6.93	**74.22**	0.67	**152.83**	m3	**183.40**
Upstands										
generally	76.04	2.50	**77.94**	-	8.47	**90.71**	0.67	**169.33**	m3	**203.19**
E10: MIXING/CASTING/CURING IN-SITU CONCRETE - (SITE MIXED)										
Plain in-situ concrete; mix 1:6, all in aggregate										
Foundations poured on or against earth or unblinded hardcore										
generally	86.97	7.50	**93.49**	-	3.74	**40.06**	2.42	**135.97**	m3	**163.16**
Isolated foundations; poured on or against earth or unblinded hardcore										
generally	86.97	7.50	**93.49**	-	4.29	**45.95**	2.42	**141.86**	m3	**170.23**
Beds; poured on or against earth or unblinded hardcore										
thickness 150 - 450mm	86.97	7.50	**93.49**	-	3.96	**42.41**	2.42	**138.32**	m3	**165.99**
thickness not exceeding 150mm	86.97	7.50	**93.49**	-	4.68	**50.07**	2.42	**145.98**	m3	**175.18**
Beds; sloping not exceeding 15 degrees; poured on or against earth or unblinded hardcore										
thickness 150 - 450mm	86.97	7.50	**93.49**	-	4.18	**44.77**	2.42	**140.68**	m3	**168.82**
thickness not exceeding 150mm	86.97	7.50	**93.49**	-	4.90	**52.43**	2.42	**148.34**	m3	**178.00**
Plain in-situ concrete; BS 5328, ordinary prescribed mix C20P, 20mm aggregate										
Foundations poured on or against earth or unblinded hardcore										
generally	90.99	7.50	**97.81**	-	3.74	**40.06**	2.42	**140.29**	m3	**168.34**
Isolated foundations; poured on or against earth or unblinded hardcore										
generally	90.99	7.50	**97.81**	-	4.29	**45.95**	2.42	**146.18**	m3	**175.41**
Beds; poured on or against earth or unblinded hardcore										
thickness 150 - 450mm	90.99	7.50	**97.81**	-	3.96	**42.41**	2.42	**142.64**	m3	**171.17**
thickness not exceeding 150mm	90.99	7.50	**97.81**	-	4.68	**50.07**	2.42	**150.30**	m3	**180.36**
Filling hollow walls										
thickness not exceeding 150mm	90.99	7.50	**97.81**	-	7.26	**77.75**	2.42	**177.99**	m3	**213.58**
Reinforced in-situ concrete; BS 5328, designed mix C25, 20mm aggregate, minimum cement content 290 kg/m3; vibrated										
Foundations										
generally	88.94	5.00	**93.39**	-	4.07	**43.59**	3.09	**140.07**	m3	**168.09**

Labour hourly rates: (except Specialists) Craft Operatives £14.21 Labourer £10.71 Rates are national average prices. Refer to REGIONAL VARIATIONS for indicative levels of overall pricing in regions	MATERIALS			LABOUR				RATES		
	Del to Site £	Waste %	Material Cost £	Craft Optve Hrs	Lab Hrs	Labour Cost £	Sunds £	Nett Rate £	Unit	Gross Rate (+20%) £
E10: MIXING/CASTING/CURING IN-SITU CONCRETE - (SITE MIXED) Cont.										
Reinforced in-situ concrete; BS 5328, designed mix C25, 20mm aggregate, minimum cement content 290 kg/m3; vibrated Cont.										
Ground beams										
generally	88.94	5.00	**93.39**	-	5.17	**55.37**	3.09	**151.85**	m3	**182.23**
Isolated foundations										
generally	88.94	5.00	**93.39**	-	4.51	**48.30**	3.09	**144.79**	m3	**173.74**
Beds										
thickness 150 - 450mm	88.94	5.00	**93.39**	-	4.29	**45.95**	3.09	**142.43**	m3	**170.92**
thickness not exceeding 150mm	88.94	5.00	**93.39**	-	4.95	**53.01**	3.09	**149.50**	m3	**179.40**
Cement and sand (1:3); for grouting and the like										
Grouting.										
stanchion bases.	2.74	10.00	**3.01**	-	0.50	**5.30**	1.41	**9.72**	nr	**11.67**
E10: MIXING/CASTING/CURING IN-SITU CONCRETE - (SUNDRIES)										
The foregoing concrete is based on the use of Portland cement. For other cements and waterproofers ADD as follows										
Concrete 1:6										
rapid hardening cement	4.36	-	**4.36**	-	-	-	-	**4.36**	m3	**5.23**
sulphate resisting cement	5.46	-	**5.46**	-	-	-	-	**5.46**	m3	**6.56**
waterproofing powder	12.60	-	**12.60**	-	-	-	-	**12.60**	m3	**15.12**
waterproofing liquid	6.42	-	**6.42**	-	-	-	-	**6.42**	m3	**7.70**
Concrete ST3 and C15										
rapid hardening cement	2.91	-	**2.91**	-	-	-	-	**2.91**	m3	**3.50**
sulphate resisting cement	3.63	-	**3.63**	-	-	-	-	**3.63**	m3	**4.36**
waterproofing powder	8.40	-	**8.40**	-	-	-	-	**8.40**	m3	**10.08**
waterproofing liquid	6.42	-	**6.42**	-	-	-	-	**6.42**	m3	**7.70**
Concrete ST4 and C20										
rapid hardening cement	4.31	-	**4.31**	-	-	-	-	**4.31**	m3	**5.17**
sulphate resisting cement	5.40	-	**5.40**	-	-	-	-	**5.40**	m3	**6.48**
waterproofing powder	12.42	-	**12.42**	-	-	-	-	**12.42**	m3	**14.91**
waterproofing liquid	6.42	-	**6.42**	-	-	-	-	**6.42**	m3	**7.70**
Concrete ST5 and C25										
rapid hardening cement	5.82	-	**5.82**	-	-	-	-	**5.82**	m3	**6.98**
sulphate resisting cement	7.27	-	**7.27**	-	-	-	-	**7.27**	m3	**8.73**
waterproofing powder	16.78	-	**16.78**	-	-	-	-	**16.78**	m3	**20.14**
waterproofing liquid	6.42	-	**6.42**	-	-	-	-	**6.42**	m3	**7.70**
Polythene sheeting as temporary protection to surface of concrete (use and waste)										
Polythene sheeting and laying										
125mu	0.38	10.00	**0.42**	-	0.03	**0.35**	-	**0.77**	m2	**0.93**
Waterproof building paper, BS 1521; laying on hardcore to receive concrete										
Waterproof building paper and laying										
Grade B1F	0.38	25.00	**0.47**	-	0.07	**0.71**	-	**1.18**	m2	**1.41**
Grade B2	1.47	25.00	**1.83**	-	0.07	**0.71**	-	**2.54**	m2	**3.05**
Polythene sheeting, laying on hardcore to receive concrete										
Polythene sheeting and laying										
300mu	0.62	20.00	**0.74**	-	0.07	**0.71**	-	**1.45**	m2	**1.74**
E20: FORMWORK FOR IN-SITU CONCRETE										
Formwork and basic finish										
Sides of foundations; plain vertical										
height exceeding 1.00m	3.83	12.50	**4.31**	2.12	0.43	**34.76**	1.00	**40.08**	m2	**48.09**
height not exceeding 250mm	1.19	12.50	**1.34**	0.64	0.13	**10.48**	0.31	**12.13**	m	**14.56**
height 250 - 500mm	2.09	12.50	**2.35**	1.17	0.23	**19.04**	0.58	**21.98**	m	**26.37**
height 0.50m - 1.00m	3.92	12.50	**4.41**	2.17	0.44	**35.51**	1.00	**40.92**	m	**49.10**
Sides of ground beams and edges of beds; plain vertical										
height exceeding 1.00m	3.83	12.50	**4.31**	2.12	0.43	**34.76**	1.00	**40.08**	m2	**48.09**
height not exceeding 250mm	1.19	12.50	**1.34**	0.64	0.13	**10.48**	0.31	**12.13**	m	**14.56**
height 250 - 500mm	2.09	12.50	**2.35**	1.17	0.23	**19.04**	0.58	**21.98**	m	**26.37**
height 0.50m - 1.00m	3.92	12.50	**4.41**	2.17	0.44	**35.51**	1.00	**40.92**	m	**49.10**
Edges of suspended slabs; plain vertical										
height not exceeding 250mm	1.43	12.50	**1.61**	0.80	0.15	**13.06**	0.35	**15.02**	m	**18.03**
height 250 - 500mm	2.68	12.50	**3.01**	1.45	0.29	**23.70**	0.69	**27.40**	m	**32.88**
extra; recesses; plain rectangular										
25 x 50mm	0.05	-	**0.05**	0.08	-	**1.09**	-	**1.15**	m	**1.38**
50 x 100mm	0.30	-	**0.30**	0.13	-	**1.88**	-	**2.18**	m	**2.62**

Labour hourly rates: (except Specialists) Craft Operatives £14.21 Labourer £10.71 Rates are national average prices. Refer to REGIONAL VARIATIONS for indicative levels of overall pricing in regions	MATERIALS			LABOUR				RATES		
	Del to Site £	Waste %	Material Cost £	Craft Optve Hrs	Lab Hrs	Labour Cost £	Sunds £	Nett Rate £	Unit	Gross Rate (+20%) £
E20: FORMWORK FOR IN-SITU CONCRETE Cont.										
Formwork and basic finish Cont.										
Edges of suspended slabs; plain vertical Cont.										
extra; rebates; plain rectangular										
25 x 50mm	0.05	-	**0.05**	0.09	-	**1.25**	-	**1.30**	m	**1.56**
50 x 100mm	0.32	-	**0.32**	0.15	-	**2.19**	-	**2.50**	m	**3.00**
extra; chamfers										
60mm wide	0.09	-	**0.09**	0.06	-	**0.78**	-	**0.87**	m	**1.04**
Sides of upstands; plain vertical										
height not exceeding 250mm	1.70	12.50	**1.91**	0.87	0.19	**14.35**	0.39	**16.65**	m	**19.98**
height 250 - 500mm	1.99	12.50	**2.24**	1.01	0.22	**16.74**	0.51	**19.49**	m	**23.39**
extra; recesses; plain rectangular										
25 x 50mm	0.04	-	**0.04**	0.09	-	**1.25**	-	**1.29**	m	**1.55**
Machine bases and plinths; plain vertical										
height exceeding 1.00m	4.22	12.50	**4.75**	2.33	0.46	**38.09**	1.12	**43.96**	m2	**52.75**
height 250 - 500mm	2.31	12.50	**2.60**	1.29	0.24	**20.88**	0.60	**24.08**	m	**28.90**
Soffits of slabs; horizontal										
slab thickness not exceeding 200mm; height to soffit										
1.50 - 3.00m	11.59	10.00	**12.74**	1.82	0.36	**29.68**	1.40	**43.83**	m2	**52.59**
3.00 - 4.50m	11.59	10.00	**12.74**	2.08	0.43	**34.14**	1.92	**48.80**	m2	**58.56**
slab thickness 200 - 300mm; height to soffit										
1.50 - 3.00m	12.75	10.00	**14.03**	2.00	0.40	**32.69**	1.58	**48.30**	m2	**57.96**
3.00 - 4.50m	12.75	10.00	**14.03**	2.27	0.46	**37.15**	2.07	**53.25**	m2	**63.90**
Soffits of slabs; sloping not exceeding 15 degrees										
slab thickness not exceeding 200mm; height to soffit										
1.50 - 3.00m	12.76	10.00	**14.04**	2.00	0.40	**32.69**	1.57	**48.30**	m2	**57.96**
3.00 - 4.50m	12.76	10.00	**14.04**	2.27	0.46	**37.15**	2.05	**53.23**	m2	**63.88**
Soffits of landings; horizontal										
slab thickness not exceeding 200mm; height to soffit										
not exceeding 1.50m	11.59	10.00	**12.74**	2.18	0.44	**35.66**	1.40	**49.81**	m2	**59.77**
1.50 - 3.00m	11.59	10.00	**12.74**	2.44	0.50	**40.00**	1.92	**54.66**	m2	**65.60**
3.00 - 4.50m	11.59	10.00	**12.74**	2.71	0.56	**44.46**	2.45	**59.65**	m2	**71.58**
Walls; vertical										
plain	9.38	15.00	**10.79**	1.94	0.39	**31.63**	1.30	**43.72**	m2	**52.47**
extra; recesses; plain rectangular										
12 x 25mm	0.03	-	**0.03**	0.07	-	**0.94**	-	**0.97**	m	**1.16**
50 x 75mm	0.22	-	**0.22**	0.11	-	**1.56**	-	**1.78**	m	**2.14**
Walls; vertical; curved 1m radius										
plain	19.19	15.00	**22.07**	4.84	0.97	**79.14**	3.24	**104.46**	m2	**125.35**
Walls; vertical; curved 10m radius										
plain	11.55	15.00	**13.28**	2.90	0.58	**47.51**	1.98	**62.77**	m2	**75.33**
Beams; attached to slabs										
regular shaped; rectangular; height to soffit										
1.50 - 3.00m	8.10	10.00	**8.91**	2.73	0.50	**44.07**	1.25	**54.22**	m2	**65.06**
3.00 - 4.50m	8.10	10.00	**8.91**	2.73	0.50	**44.07**	1.53	**54.50**	m2	**65.40**
Beams; isolated										
regular shaped; rectangular; height to soffit										
1.50 - 3.00m	8.49	10.00	**9.34**	2.85	0.52	**46.02**	1.25	**56.61**	m2	**67.93**
3.00 - 4.50m	8.49	10.00	**9.34**	2.85	0.52	**46.02**	1.53	**56.89**	m2	**68.27**
Columns; attached to walls										
regular shaped; rectangular	5.62	10.00	**6.18**	2.78	0.51	**44.97**	1.17	**52.32**	m2	**62.79**
Columns; isolated										
regular shaped; rectangular	5.59	10.00	**6.15**	2.73	0.50	**44.07**	1.17	**51.39**	m2	**61.66**
Extra over a basic finish for a fine formed finish										
slabs	-	-	-	-	0.43	**4.59**	-	**4.59**	m2	**5.51**
walls	-	-	-	-	0.43	**4.59**	-	**4.59**	m2	**5.51**
beams	-	-	-	-	0.43	**4.59**	-	**4.59**	m2	**5.51**
columns	-	-	-	-	0.43	**4.59**	-	**4.59**	m2	**5.51**
Wall kickers										
straight	1.12	15.00	**1.29**	0.32	0.09	**5.48**	0.23	**6.99**	m	**8.39**
Wall ends, soffits and steps in walls; plain										
width not exceeding 250mm	2.27	15.00	**2.61**	0.64	0.13	**10.48**	0.41	**13.50**	m	**16.20**
Openings in walls; plain										
width not exceeding 250mm	2.27	15.00	**2.61**	0.64	0.13	**10.48**	0.41	**13.50**	m	**16.20**
Stairflights; 1000mm wide; 155mm thick waist; 178mm undercut risers; includes formwork to soffits, risers and strings										
strings 300mm wide	17.29	10.00	**19.02**	7.33	1.46	**119.77**	2.96	**141.75**	m	**170.10**
Mortices; circular										
girth not exceeding 500mm; depth not exceeding 250mm	2.07	15.00	**2.38**	1.17	0.24	**19.16**	0.55	**22.09**	nr	**26.51**

Labour hourly rates: (except Specialists) Craft Operatives £14.21 Labourer £10.71 Rates are national average prices. Refer to REGIONAL VARIATIONS for indicative levels of overall pricing in regions	MATERIALS			LABOUR				RATES		
	Del to Site £	Waste %	Material Cost £	Craft Optve Hrs	Lab Hrs	Labour Cost £	Sunds £	Nett Rate £	Unit	Gross Rate (+20%) £
E20: FORMWORK FOR IN-SITU CONCRETE Cont.										
Formwork and basic finish Cont.										
Holes; rectangular										
girth not exceeding 500mm; depth not exceeding 250mm	1.03	15.00	**1.18**	0.58	0.12	**9.58**	0.28	**11.04**	nr	**13.25**
Holes; circular										
girth not exceeding 500mm; depth not exceeding 250mm	2.07	15.00	**2.38**	1.17	0.24	**19.16**	0.55	**22.09**	nr	**26.51**
Formwork and basic finish; coating with retarding agent										
Sides of foundations; plain vertical										
height exceeding 1.00m	4.63	12.50	**5.21**	2.12	0.72	**37.83**	1.00	**44.04**	m2	**52.85**
height not exceeding 250mm	1.27	12.50	**1.43**	0.64	0.17	**10.83**	0.31	**12.57**	m	**15.08**
height 250 - 500mm	2.40	12.50	**2.70**	1.17	0.34	**20.22**	0.57	**23.49**	m	**28.18**
height 0.50m - 1.00m	4.54	12.50	**5.11**	2.17	0.65	**37.74**	1.00	**43.86**	m	**52.63**
Sides of ground beams and edges of beds; plain vertical										
height exceeding 1.00m	4.66	12.50	**5.24**	2.12	0.72	**37.83**	1.00	**44.07**	m2	**52.88**
height not exceeding 250mm	1.27	12.50	**1.43**	0.64	0.17	**10.83**	0.31	**12.57**	m	**15.08**
height 250 - 500mm	2.40	12.50	**2.70**	1.17	0.34	**20.22**	0.57	**23.49**	m	**28.18**
height 0.50m - 1.00m	4.54	12.50	**5.11**	2.17	0.65	**37.74**	1.00	**43.86**	m	**52.63**
Edges of suspended slabs; plain vertical										
height not exceeding 250mm	1.51	12.50	**1.70**	0.80	0.20	**13.53**	0.35	**15.58**	m	**18.69**
height 250 - 500mm	2.93	12.50	**3.30**	1.45	0.43	**25.23**	0.69	**29.22**	m	**35.06**
Soffits of slabs; horizontal										
slab thickness not exceeding 200mm; height to soffit										
1.50 - 3.00m	12.39	10.00	**13.63**	1.82	0.65	**32.74**	1.61	**47.98**	m2	**57.58**
3.00 - 4.50m	12.39	10.00	**13.63**	2.08	0.72	**37.20**	1.92	**52.75**	m2	**63.30**
Walls; vertical										
plain	8.70	15.00	**10.00**	1.94	0.68	**34.81**	1.30	**46.12**	m2	**55.34**
Beams; attached to slabs										
regular shaped; rectangular; height to soffit 1.50 - 3.00m	9.00	10.00	**9.90**	2.73	0.84	**47.72**	1.25	**58.87**	m2	**70.64**
Beams; isolated										
regular shaped; rectangular; height to soffit 1.50 - 3.00m	9.35	10.00	**10.29**	2.85	0.86	**49.67**	1.25	**61.21**	m2	**73.45**
Columns; attached to walls										
regular shaped; rectangular	6.57	10.00	**7.23**	2.78	0.85	**48.62**	1.17	**57.02**	m2	**68.42**
Columns; isolated										
regular shaped; rectangular	6.57	10.00	**7.23**	2.73	0.84	**47.72**	1.25	**56.20**	m2	**67.44**
Claymaster low density expanded polystyrene permanent formwork; fixing with Clayfix hooks - 3/m2										
Sides of foundations; plain vertical 75mm thick										
height exceeding 1.00m	16.00	5.00	**16.80**	-	0.08	**0.82**	0.81	**18.43**	m2	**22.12**
height not exceeding 250mm	4.00	5.00	**4.20**	-	0.02	**0.24**	0.27	**4.70**	m	**5.65**
height 250-500mm	8.00	5.00	**8.40**	-	0.09	**0.94**	0.27	**9.61**	m	**11.53**
height 0.50m-1.00m	12.00	5.00	**12.60**	-	0.09	**0.94**	0.54	**14.08**	m	**16.90**
Sides of foundations; plain vertical 150mm thick										
height exceeding 1.00m	35.87	5.00	**37.67**	-	0.12	**1.30**	0.81	**39.77**	m2	**47.73**
height not exceeding 250mm	8.97	5.00	**9.42**	-	0.03	**0.35**	0.27	**10.04**	m	**12.05**
height 250-500mm	17.94	5.00	**18.83**	-	0.08	**0.82**	0.27	**19.93**	m	**23.91**
height 0.50m-1.00m	26.91	5.00	**28.25**	-	0.13	**1.41**	0.54	**30.21**	m	**36.25**
To underside of foundations; laid on earth or hardcore; 75mm thick										
horizontal	16.00	5.00	**16.80**	-	0.06	**0.59**	-	**17.38**	m2	**20.86**
To underside of foundations; laid on earth or hardcore; 150mm thick										
horizontal	35.87	5.00	**37.67**	-	0.09	**0.94**	-	**38.61**	m2	**46.33**
E30: REINFORCEMENT FOR IN-SITU CONCRETE										
Reinforcement bars; BS 4449, hot rolled plain round mild steel including hooks and tying wire, and spacers and chairs which are at the discretion of the Contractor										
Straight										
6mm	485.10	7.50	**521.48**	56.10	5.50	**856.09**	31.25	**1408.81**	t	**1690.58**
8mm	467.78	7.50	**502.86**	48.40	5.50	**746.67**	27.67	**1277.19**	t	**1532.63**
10mm	456.23	7.50	**490.44**	39.60	5.50	**621.62**	22.46	**1134.52**	t	**1361.43**
12mm	450.45	7.50	**484.23**	34.10	5.50	**543.47**	19.20	**1046.90**	t	**1256.28**
16mm	444.68	7.50	**478.03**	26.40	5.50	**434.05**	15.79	**927.87**	t	**1113.44**
20mm	444.68	7.50	**478.03**	22.00	5.50	**371.53**	8.63	**858.18**	t	**1029.82**
25mm	444.68	7.50	**478.03**	22.00	5.50	**371.53**	8.63	**858.18**	t	**1029.82**
Bent										
6mm	485.10	7.50	**521.48**	56.10	5.50	**856.09**	31.25	**1408.81**	t	**1690.58**

Labour hourly rates: (except Specialists) Craft Operatives £14.21 Labourer £10.71 Rates are national average prices. Refer to REGIONAL VARIATIONS for indicative levels of overall pricing in regions	MATERIALS			LABOUR				RATES		
	Del to Site £	Waste %	Material Cost £	Craft Optve Hrs	Lab Hrs	Labour Cost £	Sunds £	Nett Rate £	Unit	Gross Rate (+20%) £
E30: REINFORCEMENT FOR IN-SITU CONCRETE Cont.										
Reinforcement bars; BS 4449, hot rolled plain round mild steel including hooks and tying wire, and spacers and chairs which are at the discretion of the Contractor Cont.										
Bent Cont.										
8mm	467.78	7.50	**502.86**	48.40	5.50	**746.67**	27.67	**1277.19**	t	**1532.63**
10mm	456.23	7.50	**490.44**	39.60	5.50	**621.62**	22.46	**1134.52**	t	**1361.43**
12mm	450.45	7.50	**484.23**	34.10	5.50	**543.47**	19.20	**1046.90**	t	**1256.28**
16mm	444.68	7.50	**478.03**	26.40	5.50	**434.05**	15.79	**927.87**	t	**1113.44**
20mm	444.68	7.50	**478.03**	22.00	5.50	**371.53**	8.63	**858.18**	t	**1029.82**
25mm	444.68	7.50	**478.03**	22.00	5.50	**371.53**	8.63	**858.18**	t	**1029.82**
Links										
6mm	485.10	7.50	**521.48**	72.60	5.50	**1090.55**	31.25	**1643.28**	t	**1971.94**
8mm	467.78	7.50	**502.86**	72.60	5.50	**1090.55**	27.67	**1621.07**	t	**1945.29**
Reinforcement bars; BS 4449, hot rolled deformed high yield steel including hooks and tying wire, and spacers and chairs which are at the discretion of the Contractor										
Straight										
6mm	496.65	7.50	**533.90**	56.10	5.50	**856.09**	31.25	**1421.23**	t	**1705.48**
8mm	473.55	7.50	**509.07**	48.40	5.50	**746.67**	27.67	**1283.40**	t	**1540.08**
10mm	462.00	7.50	**496.65**	39.60	5.50	**621.62**	22.46	**1140.73**	t	**1368.87**
12mm	462.00	7.50	**496.65**	34.10	5.50	**543.47**	19.20	**1059.32**	t	**1271.18**
16mm	450.45	7.50	**484.23**	26.40	5.50	**434.05**	15.79	**934.08**	t	**1120.89**
20mm	450.45	7.50	**484.23**	22.00	5.50	**371.53**	8.63	**864.39**	t	**1037.27**
25mm	450.45	7.50	**484.23**	22.00	5.50	**371.53**	8.63	**864.39**	t	**1037.27**
Bent										
6mm	496.65	7.50	**533.90**	56.10	5.50	**856.09**	31.25	**1421.23**	t	**1705.48**
8mm	473.55	7.50	**509.07**	48.40	5.50	**746.67**	27.67	**1283.40**	t	**1540.08**
10mm	462.00	7.50	**496.65**	39.60	5.50	**621.62**	22.46	**1140.73**	t	**1368.87**
12mm	462.00	7.50	**496.65**	34.10	5.50	**543.47**	19.20	**1059.32**	t	**1271.18**
16mm	450.45	7.50	**484.23**	26.40	5.50	**434.05**	15.79	**934.08**	t	**1120.89**
20mm	450.45	7.50	**484.23**	22.00	5.50	**371.53**	8.63	**864.39**	t	**1037.27**
25mm	450.45	7.50	**484.23**	22.00	5.50	**371.53**	8.63	**864.39**	t	**1037.27**
Links										
6mm	496.65	7.50	**533.90**	72.60	5.50	**1090.55**	31.25	**1655.70**	t	**1986.83**
8mm	473.55	7.50	**509.07**	72.60	5.50	**1090.55**	27.67	**1627.28**	t	**1952.74**
Reinforcement fabric; BS 4483, hard drawn plain round steel; welded; including laps, tying wire, all cutting and bending, and spacers and chairs which are at the discretion of the Contractor										
Reference A142, 2.22kg/m2; 200mm side laps; 200mm end laps										
generally	1.22	17.50	**1.44**	0.09	0.02	**1.49**	0.09	**3.02**	m2	**3.62**
strips in one width										
750mm wide	1.22	17.50	**1.44**	0.13	0.02	**2.11**	0.09	**3.64**	m2	**4.37**
Reference A193, 3.02kg/m2; 200mm side laps; 200mm end laps										
generally	1.61	17.50	**1.89**	0.09	0.02	**1.49**	0.12	**3.50**	m2	**4.21**
strips in one width										
750mm wide	1.61	17.50	**1.89**	0.13	0.02	**2.11**	0.12	**4.13**	m2	**4.96**
Reference A393, 6.16 kg/m2; 200mm side laps; 200mm end laps										
generally	3.26	17.50	**3.84**	0.11	0.02	**1.80**	0.23	**5.87**	m2	**7.04**
strips in one width										
750mm wide	3.26	17.50	**3.84**	0.17	0.02	**2.58**	0.23	**6.65**	m2	**7.98**
Reference B1131, 10.90 kg/m2; 100mm side laps; 200mm end laps										
generally	5.89	17.50	**6.92**	0.15	0.03	**2.54**	0.42	**9.88**	m2	**11.86**
strips in one width										
750mm wide	5.89	17.50	**6.92**	0.23	0.03	**3.64**	0.42	**10.97**	m2	**13.17**
Reference B385, 4.53 kg/m2; 100mm side laps; 200mm end laps										
generally	2.45	17.50	**2.87**	0.10	0.02	**1.64**	0.17	**4.69**	m2	**5.62**
strips in one width										
750mm wide	2.45	17.50	**2.87**	0.15	0.02	**2.42**	0.17	**5.47**	m2	**6.56**
Reference B503, 5.93 kg/m2; 100mm side laps; 200mm end laps										
generally	3.17	17.50	**3.72**	0.11	0.02	**1.80**	0.23	**5.76**	m2	**6.91**
strips in one width										
750mm wide	3.17	17.50	**3.72**	0.17	0.02	**2.58**	0.23	**6.54**	m2	**7.85**
Reference B785, 8.14 kg/m2; 100mm side laps; 200mm end laps										
generally	4.36	17.50	**5.12**	0.12	0.03	**2.07**	0.31	**7.50**	m2	**9.00**
strips in one width										
750mm wide	4.36	17.50	**5.12**	0.18	0.03	**2.85**	0.31	**8.29**	m2	**9.94**
Reference C385, 3.41 kg/m2; 100mm side laps; 400mm end laps										
generally	1.82	17.50	**2.14**	0.09	0.02	**1.49**	0.14	**3.76**	m2	**4.51**

Labour hourly rates: (except Specialists) Craft Operatives £14.21 Labourer £10.71 Rates are national average prices. Refer to REGIONAL VARIATIONS for indicative levels of overall pricing in regions	MATERIALS			LABOUR				RATES		
	Del to Site £	Waste %	Material Cost £	Craft Optve Hrs	Lab Hrs	Labour Cost £	Sunds £	Nett Rate £	Unit	Gross Rate (+20%) £
E30: REINFORCEMENT FOR IN-SITU CONCRETE Cont.										
Reinforcement fabric; BS 4483, hard drawn plain round steel; welded; including laps, tying wire, all cutting and bending, and spacers and chairs which are at the discretion of the Contractor Cont.										
Reference C385, 3.41 kg/m2; 100mm side laps; 400mm end laps Cont.										
strips in one width										
750mm wide	1.82	17.50	**2.14**	0.13	0.02	**2.11**	0.14	**4.39**	m2	**5.26**
Reference C503, 4.34 kg/m2; 100mm side laps; 400mm end laps										
generally	2.24	17.50	**2.63**	0.10	0.02	**1.64**	0.17	**4.44**	m2	**5.33**
strips in one width										
750mm wide	2.24	17.50	**2.63**	0.15	0.02	**2.42**	0.17	**5.23**	m2	**6.27**
Reference D49, 0.77 kg/m2; 100mm side laps; 100mm end laps										
bent	0.85	17.50	**0.99**	0.04	0.01	**0.74**	0.09	**1.83**	m2	**2.20**
E40: DESIGNED JOINTS IN IN-SITU CONCRETE										
Formed joints										
Incorporating 10mm thick Korkpak, Grace Construction Products Limited - Servicised; formwork; reinforcement laid continuously across joint										
in concrete, depth not exceeding 150mm; horizontal	4.16	5.00	**4.36**	-	0.09	**0.94**	6.30	**11.61**	m	**13.93**
in concrete, depth 150 - 300mm; horizontal	7.27	5.00	**7.64**	-	0.10	**1.06**	11.04	**19.73**	m	**23.68**
Incorporating 19mm thick Korkpak, Grace Construction Products Limited - Servicised; formwork; reinforcement laid continuously across joint										
in concrete, depth not exceeding 150mm; horizontal	5.98	5.00	**6.28**	-	0.09	**0.94**	6.30	**13.52**	m	**16.23**
in concrete, depth 150 - 300mm; horizontal	10.46	5.00	**10.99**	-	0.10	**1.06**	11.04	**23.08**	m	**27.70**
E41: WORKED FINISHES/ CUTTING ON IN-SITU CONCRETE										
Worked finishes on in-situ concrete										
Tamping unset concrete										
surfaces	-	-	-	-	0.24	**2.59**	0.39	**2.98**	m2	**3.58**
Trowelling										
surfaces	-	-	-	-	0.40	**4.24**	-	**4.24**	m2	**5.09**
Hacking; by hand										
surfaces	-	-	-	-	0.61	**6.48**	-	**6.48**	m2	**7.78**
to soffits	-	-	-	-	0.79	**8.48**	-	**8.48**	m2	**10.18**
Hacking; by machine										
surfaces	-	-	-	-	0.24	**2.59**	0.31	**2.90**	m2	**3.48**
to soffits	-	-	-	-	0.31	**3.30**	0.31	**3.61**	m2	**4.33**
Bush hammering										
surfaces	-	-	-	1.57	-	**22.35**	0.87	**23.23**	m2	**27.87**
to soffits	-	-	-	2.42	-	**34.39**	1.16	**35.54**	m2	**42.65**
Cutting on in-situ concrete										
Cutting chases										
depth not exceeding 50mm	-	-	-	-	0.21	**2.24**	-	**2.24**	m	**2.69**
Cutting chases; making good										
depth not exceeding 50mm	-	-	-	-	0.24	**2.59**	0.34	**2.93**	m	**3.51**
Cutting mortices										
50 x 50mm; depth not exceeding 100mm	-	-	-	-	0.48	**5.18**	-	**5.18**	nr	**6.22**
Cutting holes										
225 x 150mm; depth not exceeding 100mm	-	-	-	-	1.38	**14.73**	0.46	**15.18**	nr	**18.22**
225 x 150mm; depth 200 - 300mm	-	-	-	-	2.42	**25.92**	0.75	**26.67**	nr	**32.01**
50mm diameter; depth not exceeding 100mm	-	-	-	-	0.94	**10.01**	0.27	**10.28**	nr	**12.34**
50mm diameter; depth 100 - 200mm	-	-	-	-	1.51	**16.14**	0.44	**16.58**	nr	**19.90**
E42: ACCESSORIES CAST INTO IN-SITU CONCRETE										
Cast in accessories										
Galvanized steel dowels										
12mm diameter x 150mm long	0.77	10.00	**0.85**	-	0.10	**1.06**	-	**1.91**	nr	**2.29**
Galvanized expanded metal tie										
300 x 50mm; bending and temporarily fixing to formwork	0.92	10.00	**1.01**	-	0.14	**1.53**	-	**2.54**	nr	**3.05**
Steel rag bolts										
M 10 x 100mm long	1.54	5.00	**1.62**	0.11	-	**1.56**	-	**3.18**	nr	**3.82**
M 12 x 160mm long	1.98	5.00	**2.08**	0.13	-	**1.88**	-	**3.96**	nr	**4.75**
M 16 x 200mm long	3.39	5.00	**3.56**	0.15	-	**2.19**	-	**5.75**	nr	**6.90**

Labour hourly rates: (except Specialists) Craft Operatives £14.21 Labourer £10.71 Rates are national average prices. Refer to REGIONAL VARIATIONS for indicative levels of overall pricing in regions	MATERIALS			LABOUR				RATES		
	Del to Site £	Waste %	Material Cost £	Craft Optve Hrs	Lab Hrs	Labour Cost £	Sunds £	Nett Rate £	Unit	Gross Rate (+20%) £
E60: PRECAST/COMPOSITE CONCRETE DECKING										
Prestressed concrete beams and 100mm building blocks; hoist bed and grout										
155mm thick floors (Bison 'House Beam')										
beams at 510mm centres	13.97	2.50	**14.32**	-	1.27	**13.55**	0.60	**28.47**	m2	**34.17**
double beams at 624mm centres....................	19.10	2.50	**19.58**	-	1.10	**11.78**	0.60	**31.96**	m2	**38.35**
beams at 285mm centres	20.33	2.50	**20.84**	-	1.27	**13.55**	0.60	**34.99**	m2	**41.99**
beams at 114mm centres with concrete between.....	39.88	2.50	**40.88**	-	1.10	**11.78**	0.60	**53.27**	m2	**63.92**

MASONRY

Labour hourly rates: (except Specialists) Craft Operatives £14.21 Labourer £10.71 Rates are national average prices. Refer to REGIONAL VARIATIONS for indicative levels of overall pricing in regions	MATERIALS			LABOUR				RATES		
	Del to Site £	Waste %	Material Cost £	Craft Optve Hrs	Lab Hrs	Labour Cost £	Sunds £	Nett Rate £	Unit	Gross Rate (+20%) £
F10: BRICK/BLOCK WALLING										
Common bricks, BS EN 772, Category M, 215 x 102.5 x 65mm, compressive strength 20.5 N/mm2; in cement-lime mortar (1:2:9)										
Walls; vertical										
102mm thick; stretcher bond	13.24	5.00	**13.90**	1.60	1.76	**41.51**	3.19	**58.61**	m2	**70.33**
215mm thick; English bond	26.70	5.00	**28.04**	2.59	2.92	**67.95**	7.44	**103.43**	m2	**124.12**
Walls; building against old brickwork; tie new to old with 3mm diameter galvanized twisted wire ties - 6/m2; vertical										
102mm thick; stretcher bond	14.14	10.00	**14.89**	1.76	1.93	**45.63**	3.19	**63.71**	m2	**76.45**
215mm thick; English bond	27.60	10.00	**29.03**	2.92	3.25	**76.18**	7.44	**112.64**	m2	**135.17**
Isolated piers; vertical										
215mm thick; English bond	26.70	5.00	**28.04**	5.17	5.50	**132.37**	7.44	**167.85**	m2	**201.42**
327mm thick; English bond	40.17	5.00	**42.18**	6.27	6.77	**161.55**	11.69	**215.42**	m2	**258.50**
Projections; vertical										
215mm wide x 112mm projection	2.99	5.00	**3.14**	0.33	0.39	**8.81**	0.85	**12.80**	m	**15.37**
215mm wide x 215mm projection	5.98	5.00	**6.28**	0.55	0.66	**14.88**	1.70	**22.87**	m	**27.44**
327mm wide x 112mm projection	4.49	5.00	**4.71**	0.50	0.61	**13.51**	1.28	**19.50**	m	**23.40**
327mm wide x 215mm projection	8.98	5.00	**9.42**	0.83	1.05	**22.92**	2.55	**34.89**	m	**41.87**
Projections; bonding to old brickwork; cutting pockets; including extra material; vertical										
215mm wide x 112mm projection	4.98	5.00	**5.23**	1.02	1.08	**26.08**	1.42	**32.73**	m	**39.28**
215mm wide x 215mm projection	7.98	5.00	**8.37**	1.24	1.35	**32.15**	2.27	**42.80**	m	**51.36**
327mm wide x 112mm projection	7.48	5.00	**7.85**	1.54	1.65	**39.55**	2.13	**49.54**	m	**59.44**
327mm wide x 215mm projection	10.97	5.00	**11.52**	1.87	2.09	**48.96**	3.12	**63.60**	m	**76.31**
Arches including centering; flat										
112mm high on face; 215mm thick; width of exposed soffit 215mm; bricks-on-edge	2.99	5.00	**3.14**	1.65	1.38	**38.17**	0.85	**42.16**	m	**50.60**
215mm high on face; 112mm thick; width of exposed soffit 112mm; bricks-on-end	2.99	5.00	**3.14**	1.60	1.38	**37.39**	0.85	**41.38**	m	**49.66**
Arches including centering; semi-circular										
112mm high on face; 215mm thick; width of exposed soffit 215mm; bricks-on-edge	4.70	5.00	**4.93**	2.37	1.76	**52.46**	1.34	**58.72**	m	**70.47**
215mm high on face; 215mm thick; width of exposed soffit 215mm; bricks-on-edge	9.40	5.00	**9.87**	3.25	2.70	**74.97**	2.67	**87.51**	m	**105.02**
Closing cavities; vertical										
50mm wide with brickwork 102mm thick	1.46	5.00	**1.53**	0.44	0.44	**10.96**	0.41	**12.91**	m	**15.49**
75mm wide with brickwork 102mm thick	1.46	5.00	**1.53**	0.44	0.44	**10.96**	0.41	**12.91**	m	**15.49**
Closing cavities; horizontal										
50mm wide with brickwork 102mm thick	1.46	5.00	**1.53**	0.44	0.44	**10.96**	0.41	**12.91**	m	**15.49**
75mm wide with brickwork 102mm thick	1.46	5.00	**1.53**	0.44	0.44	**10.96**	0.41	**12.91**	m	**15.49**
Bonding ends to existing common brickwork; cutting pockets; extra material										
walls; bonding every third course										
102mm thick	1.00	5.00	**1.05**	0.41	0.41	**10.14**	0.28	**11.47**	m	**13.77**
215mm thick	1.99	5.00	**2.09**	0.73	0.73	**18.09**	0.57	**20.75**	m	**24.90**
327mm thick	2.99	5.00	**3.14**	1.05	1.05	**26.04**	0.85	**30.03**	m	**36.04**
Second hard stock bricks, BS EN 772, Category M, 215 x 102.5 x 65mm; in cement-lime mortar (1:2:9)										
Walls; vertical										
102mm thick; stretcher bond	26.09	5.00	**27.39**	1.60	1.76	**41.51**	3.19	**72.10**	m2	**86.52**
215mm thick; English bond	52.62	5.00	**55.25**	2.59	2.92	**67.95**	7.44	**130.64**	m2	**156.77**
Isolated piers; vertical										
215mm thick; English bond	52.62	5.00	**55.25**	5.17	5.50	**132.37**	7.44	**195.06**	m2	**234.07**
327mm thick; English bond	79.15	5.00	**83.11**	6.27	6.77	**161.55**	11.69	**256.36**	m2	**307.63**

Labour hourly rates: (except Specialists) Craft Operatives £14.21 Labourer £10.71 Rates are national average prices. Refer to REGIONAL VARIATIONS for indicative levels of overall pricing in regions	MATERIALS			LABOUR				RATES		
	Del to Site £	Waste %	Material Cost £	Craft Optve Hrs	Lab Hrs	Labour Cost £	Sunds £	Nett Rate £	Unit	Gross Rate (+20%) £
F10: BRICK/BLOCK WALLING Cont.										
Second hard stock bricks, BS EN 772, Category M, 215 x 102.5 x 65mm; in cement-lime mortar (1:2:9) Cont.										
Chimney stacks; vertical										
440mm thick; English bond	105.24	5.00	**110.51**	7.26	7.92	**187.99**	15.94	**314.43**	m2	**377.32**
890mm thick; English bond	210.49	5.00	**221.01**	11.44	12.76	**299.22**	31.88	**552.11**	m2	**662.54**
Projections; vertical										
215mm wide x 112mm projection	5.89	5.00	**6.19**	0.33	0.39	**8.81**	0.85	**15.85**	m	**19.02**
215mm wide x 215mm projection	11.79	5.00	**12.38**	0.55	0.66	**14.88**	1.70	**28.97**	m	**34.76**
327mm wide x 112mm projection	8.84	5.00	**9.29**	0.50	0.61	**13.51**	1.28	**24.08**	m	**28.89**
327mm wide x 215mm projection	17.69	5.00	**18.57**	0.83	1.05	**22.92**	2.55	**44.04**	m	**52.85**
Projections; bonding to old brickwork; cutting pockets; including extra material; vertical										
215mm wide x 112mm projection	9.82	5.00	**10.31**	1.02	1.08	**26.08**	1.42	**37.81**	m	**45.38**
215mm wide x 215mm projection	15.72	5.00	**16.50**	1.24	1.35	**32.15**	2.27	**50.93**	m	**61.11**
327mm wide x 112mm projection	14.74	5.00	**15.48**	1.54	1.65	**39.55**	2.13	**57.16**	m	**68.59**
327mm wide x 215mm projection	21.61	5.00	**22.70**	1.87	2.09	**48.96**	3.12	**74.77**	m	**89.73**
Arches including centering; flat										
112mm high on face; 215mm thick; width of exposed soffit 215mm; bricks-on-edge	5.89	5.00	**6.19**	1.65	1.38	**38.17**	0.85	**45.21**	m	**54.26**
215mm high on face; 112mm thick; width of exposed soffit 112mm; bricks-on-end	5.89	5.00	**6.19**	1.60	1.38	**37.39**	0.85	**44.43**	m	**53.32**
Arches including centering; semi-circular										
112mm high on face; 215mm thick; width of exposed soffit 215mm; bricks-on-edge	9.26	5.00	**9.72**	2.37	1.76	**52.46**	1.34	**63.51**	m	**76.21**
215mm high on face; 215mm thick; width of exposed soffit 215mm; bricks-on-edge	18.51	5.00	**19.44**	3.25	2.70	**74.97**	2.67	**97.09**	m	**116.51**
Closing cavities; vertical										
50mm wide with brickwork 102mm thick	2.87	5.00	**3.02**	0.44	0.44	**10.96**	0.41	**14.40**	m	**17.28**
75mm wide with brickwork 102mm thick	2.87	5.00	**3.02**	0.44	0.44	**10.96**	0.41	**14.40**	m	**17.28**
Closing cavities; horizontal										
50mm wide with brickwork 102mm thick	2.87	5.00	**3.02**	0.44	0.44	**10.96**	0.41	**14.40**	m	**17.28**
75mm wide with brickwork 102mm thick	2.87	5.00	**3.02**	0.44	0.44	**10.96**	0.41	**14.40**	m	**17.28**
Bonding ends to existing common brickwork; cutting pockets; extra material										
walls; bonding every third course										
102mm thick	1.96	5.00	**2.06**	0.41	0.41	**10.14**	0.28	**12.49**	m	**14.99**
215mm thick	3.93	5.00	**4.12**	0.73	0.73	**18.09**	0.57	**22.78**	m	**27.34**
Engineering bricks, BS EN 772, Category F, 215 x 102.5 x 65, Class B; in cement-lime mortar (1:2:9)										
Isolated piers; vertical										
215mm thick; English bond	28.01	5.00	**29.41**	5.67	6.00	**144.71**	7.44	**181.56**	m2	**217.87**
327mm thick; English bond	42.14	5.00	**44.24**	6.88	7.37	**176.63**	11.69	**232.56**	m2	**279.08**
Chimney stacks; vertical										
440mm thick; English bond	56.03	5.00	**58.83**	7.98	8.64	**205.81**	15.94	**280.57**	m2	**336.69**
890mm thick; English bond	112.05	5.00	**117.65**	12.60	13.92	**328.00**	31.88	**477.54**	m2	**573.05**
Projections; vertical										
215mm wide x 112mm projection	3.14	5.00	**3.29**	0.36	0.42	**9.64**	0.85	**13.78**	m	**16.54**
215mm wide x 215mm projection	6.28	5.00	**6.59**	0.61	0.72	**16.25**	1.70	**24.55**	m	**29.46**
327mm wide x 112mm projection	4.71	5.00	**4.94**	0.55	0.66	**14.88**	1.28	**21.10**	m	**25.33**
327mm wide x 215mm projection	9.42	5.00	**9.89**	0.91	1.13	**25.11**	2.55	**37.55**	m	**45.06**
Projections; bonding to old brickwork; cutting pockets; including extra material; vertical										
215mm wide x 112mm projection	5.23	5.00	**5.49**	1.12	1.18	**28.55**	1.42	**35.46**	m	**42.55**
215mm wide x 215mm projection	8.37	5.00	**8.78**	1.36	1.47	**35.17**	2.27	**46.22**	m	**55.47**
327mm wide x 112mm projection	7.85	5.00	**8.24**	1.69	1.80	**43.39**	2.13	**53.76**	m	**64.51**
327mm wide x 215mm projection	11.51	5.00	**12.08**	2.06	2.28	**53.62**	3.12	**68.82**	m	**82.58**
Arches including centering; flat										
112mm high on face; 215mm thick; width of exposed soffit 215mm; bricks-on-edge	3.14	5.00	**3.29**	1.76	1.43	**40.32**	0.85	**44.47**	m	**53.37**
215mm high on face; 112mm thick; width of exposed soffit 112mm; bricks-on-end	3.14	5.00	**3.29**	1.71	1.49	**40.13**	0.85	**44.28**	m	**53.13**
Arches including centering; semi-circular										
112mm high on face; 215mm thick; width of exposed soffit 215mm; bricks-on-edge	4.93	5.00	**5.17**	2.48	1.82	**54.61**	1.34	**61.12**	m	**73.34**
215mm high on face; 215mm thick; width of exposed soffit 215mm; bricks-on-edge	9.86	5.00	**10.35**	3.41	2.81	**78.50**	2.67	**91.52**	m	**109.82**
Closing cavities; vertical										
50mm wide with brickwork 102mm thick	1.53	5.00	**1.61**	0.48	0.48	**12.06**	0.41	**14.08**	m	**16.90**
75mm wide with brickwork 102mm thick	1.53	5.00	**1.61**	0.48	0.48	**12.06**	0.41	**14.08**	m	**16.90**
Closing cavities; horizontal										
50mm wide with brickwork 102mm thick	1.53	5.00	**1.61**	0.48	0.48	**12.06**	0.41	**14.08**	m	**16.90**
75mm wide with brickwork 102mm thick	1.53	5.00	**1.61**	0.48	0.48	**12.06**	0.41	**14.08**	m	**16.90**
Bonding ends to existing common brickwork; cutting pockets; extra material										
walls; bonding every third course										
102mm thick	1.05	5.00	**1.10**	0.45	0.45	**11.24**	0.28	**12.62**	m	**15.14**

Labour hourly rates: (except Specialists) Craft Operatives £14.21 Labourer £10.71 Rates are national average prices. Refer to REGIONAL VARIATIONS for indicative levels of overall pricing in regions	MATERIALS			LABOUR			RATES			
	Del to Site £	Waste %	Material Cost £	Craft Optve Hrs	Lab Hrs	Labour Cost £	Sunds £	Nett Rate £	Unit	Gross Rate (+20%) £
F10: BRICK/BLOCK WALLING Cont.										
Engineering bricks, BS EN 772, Category F, 215 x 102.5 x 65, Class B; in cement-lime mortar (1:2:9) Cont.										
Bonding ends to existing common brickwork; cutting pockets; extra material Cont.										
walls; bonding every third course Cont.										
215mm thick	2.09	5.00	**2.19**	0.80	0.80	**20.01**	0.57	**22.77**	m	**27.33**
For using Sulphate Resisting cement in lieu of Portland cement, ADD as follows										
For each half brick thickness										
1:2:9 cement-lime mortar	-	-	-	-	-	-	0.13	**0.13**	m2	**0.16**
1:1:6 cement-lime mortar	-	-	-	-	-	-	0.19	**0.19**	m2	**0.23**
1:4 cement mortar	-	-	-	-	-	-	0.30	**0.30**	m2	**0.35**
1:3 cement mortar	-	-	-	-	-	-	0.38	**0.38**	m2	**0.45**
Facing bricks p.c. £200.00 per 1000, 215 x 102.5 x 65mm; in cement-lime mortar (1:2:9); flush smooth pointing as work proceeds										
Walls; vertical										
stretcher bond; facework one side										
102mm thick	12.98	5.00	**13.63**	1.93	2.09	**49.74**	3.19	**66.56**	m2	**79.87**
stretcher bond; facework both sides										
102mm thick	12.98	5.00	**13.63**	2.26	2.42	**57.96**	3.19	**74.78**	m2	**89.74**
English bond; facework both sides										
215mm thick	26.18	5.00	**27.49**	2.92	3.25	**76.18**	7.44	**111.10**	m2	**133.32**
Flemish bond; facework both sides										
215mm thick	26.18	5.00	**27.49**	3.25	3.58	**84.40**	7.44	**119.33**	m2	**143.19**
Walls; building overhand; vertical										
stretcher bond; facework one side										
102mm thick	12.98	5.00	**13.63**	2.37	2.53	**60.70**	3.19	**77.52**	m2	**93.03**
Arches including centering; flat										
112mm high on face; 215mm thick; width of exposed soffit 112mm; bricks-on-edge	2.93	5.00	**3.08**	1.16	1.16	**28.78**	0.85	**32.71**	m	**39.26**
112mm high on face; 215mm thick; width of exposed soffit 215mm; bricks-on-edge	2.93	5.00	**3.08**	1.65	1.38	**38.17**	0.85	**42.10**	m	**50.52**
215mm high on face; 112mm thick; width of exposed soffit 112mm; bricks-on-end	2.93	5.00	**3.08**	1.60	1.38	**37.39**	0.85	**41.32**	m	**49.59**
Arches including centering; segmental										
215mm high on face; 215mm thick; width of exposed soffit 112mm; bricks-on-edge	5.87	5.00	**6.16**	2.64	2.37	**62.84**	1.70	**70.70**	m	**84.85**
215mm high on face; 215mm thick; width of exposed soffit 215mm; bricks-on-edge	5.87	5.00	**6.16**	3.25	2.70	**74.97**	1.70	**82.84**	m	**99.40**
Arches including centering; semi-circular										
215mm high on face; 215mm thick; width of exposed soffit 112mm; bricks-on-edge	5.87	5.00	**6.16**	2.64	2.37	**62.84**	1.70	**70.70**	m	**84.85**
215mm high on face; 215mm thick; width of exposed soffit 215mm; bricks-on-edge	5.87	5.00	**6.16**	3.25	2.70	**74.97**	1.70	**82.84**	m	**99.40**
Facework copings; bricks-on-edge; pointing top and each side										
215 x 102.5mm; horizontal	2.93	5.00	**3.08**	0.77	0.83	**19.78**	0.85	**23.71**	m	**28.45**
215 x 102.5mm; with oversailing course and cement fillets both sides; horizontal	2.93	5.00	**3.08**	1.21	1.27	**30.74**	0.85	**34.67**	m	**41.61**
327 x 102.5mm; with oversailing course and cement fillets both sides; horizontal	4.40	5.00	**4.62**	1.65	1.71	**41.71**	1.28	**47.60**	m	**57.12**
extra; galvanized iron coping cramp at angle or end	2.57	5.00	**2.70**	0.04	0.04	**1.10**	-	**3.80**	nr	**4.56**
Facing bricks p.c. £300.00 per 1000, 215 x 102.5 x 65mm; in cement-lime mortar (1:2:9); flush smooth pointing as work proceeds										
Walls; vertical										
stretcher bond; facework one side										
102mm thick	19.47	5.00	**20.44**	1.93	2.09	**49.74**	3.19	**73.37**	m2	**88.05**
stretcher bond; facework both sides										
102mm thick	19.47	5.00	**20.44**	2.26	2.42	**57.96**	3.19	**81.60**	m2	**97.92**
English bond; facework both sides										
215mm thick	39.27	5.00	**41.23**	2.92	3.25	**76.18**	7.44	**124.85**	m2	**149.82**
Flemish bond; facework both sides										
215mm thick	39.27	5.00	**41.23**	3.25	3.58	**84.40**	7.44	**133.07**	m2	**159.69**
Walls; building overhand; vertical										
stretcher bond; facework one side										
102mm thick	19.47	5.00	**20.44**	2.37	2.53	**60.70**	3.19	**84.34**	m2	**101.21**
Arches including centering; flat										
112mm high on face; 215mm thick; width of exposed soffit 112mm; bricks-on-edge	4.40	5.00	**4.62**	1.16	1.16	**28.78**	0.85	**34.25**	m	**41.10**
112mm high on face; 215mm thick; width of exposed soffit 215mm; bricks-on-edge	4.40	5.00	**4.62**	1.65	1.38	**38.17**	0.85	**43.64**	m	**52.37**
215mm high on face; 112mm thick; width of exposed soffit 112mm; bricks-on-end	4.40	5.00	**4.62**	1.60	1.38	**37.39**	0.85	**42.86**	m	**51.43**
Arches including centering; segmental										
215mm high on face; 215mm thick; width of exposed soffit 112mm; bricks-on-edge	8.80	5.00	**9.24**	2.64	2.37	**62.84**	1.70	**73.78**	m	**88.54**

Labour hourly rates: (except Specialists) Craft Operatives £14.21 Labourer £10.71 Rates are national average prices. Refer to REGIONAL VARIATIONS for indicative levels of overall pricing in regions	MATERIALS			LABOUR				RATES		
	Del to Site £	Waste %	Material Cost £	Craft Optve Hrs	Lab Hrs	Labour Cost £	Sunds £	Nett Rate £	Unit	Gross Rate (+20%) £
F10: BRICK/BLOCK WALLING Cont.										
Facing bricks p.c. £300.00 per 1000, 215 x 102.5 x 65mm; in cement-lime mortar (1:2:9); flush smooth pointing as work proceeds Cont.										
Arches including centering; segmental Cont.										
215mm high on face; 215mm thick; width of exposed soffit 215mm; bricks-on-edge	8.80	5.00	**9.24**	3.25	2.70	**74.97**	1.70	**85.92**	m	**103.10**
Arches including centering; semi-circular										
215mm high on face; 215mm thick; width of exposed soffit 112mm; bricks-on-edge	8.80	5.00	**9.24**	2.64	2.37	**62.84**	1.70	**73.78**	m	**88.54**
215mm high on face; 215mm thick; width of exposed soffit 215mm; bricks-on-edge	8.80	5.00	**9.24**	3.25	2.70	**74.97**	1.70	**85.92**	m	**103.10**
Facework copings; bricks-on-edge; pointing top and each side										
215 x 102.5mm; horizontal	4.40	5.00	**4.62**	0.77	0.83	**19.78**	0.85	**25.25**	m	**30.30**
215 x 102.5mm; with oversailing course and cement fillets both sides; horizontal	4.40	5.00	**4.62**	1.21	1.27	**30.74**	0.85	**36.21**	m	**43.45**
327 x 102.5mm; with oversailing course and cement fillets both sides; horizontal	6.60	5.00	**6.93**	1.65	1.71	**41.71**	1.28	**49.91**	m	**59.90**
extra; galvanized iron coping cramp at angle or end	2.57	5.00	**2.70**	0.04	0.04	**1.10**	-	**3.80**	nr	**4.56**
For each £10.00 difference in cost of 1000 facing bricks, add or deduct the following										
Wall in facing bricks										
102mm thick; English bond; facework one side	0.98	5.00	**1.03**	-	-	**-**	-	**1.03**	m2	**1.23**
215mm thick; facework both sides	1.31	5.00	**1.37**	-	-	**-**	-	**1.37**	m2	**1.65**
Extra over flush smooth pointing for the following types of pointing										
Pointing as work proceeds										
ironing in joints	-	-	**-**	0.03	0.03	**0.82**	-	**0.82**	m2	**0.99**
weathered pointing	-	-	**-**	0.03	0.03	**0.82**	-	**0.82**	m2	**0.99**
recessed pointing	-	-	**-**	0.04	0.04	**1.10**	-	**1.10**	m2	**1.32**
Raking out joints and pointing on completion										
flush smooth pointing with cement-lime mortar (1:2:9)	-	-	**-**	0.83	0.83	**20.56**	0.21	**20.77**	m2	**24.93**
weathered pointing with cement-lime mortar (1:2:9)	-	-	**-**	0.91	0.91	**22.75**	0.21	**22.96**	m2	**27.56**
Aerated concrete blocks, Thermalite, 440 x 215mm, Shield blocks (4.0 N/mm2);in cement-lime mortar (1:1:6)										
Walls; vertical										
75mm thick; stretcher bond	7.94	5.00	**8.34**	0.77	0.88	**20.37**	0.97	**29.68**	m2	**35.61**
100mm thick; stretcher bond	10.44	5.00	**10.96**	0.88	0.99	**23.11**	1.22	**35.29**	m2	**42.34**
140mm thick; stretcher bond	14.48	5.00	**15.20**	0.99	1.16	**26.44**	1.83	**43.47**	m2	**52.16**
190mm thick; stretcher bond	19.48	5.00	**20.46**	1.05	1.21	**27.81**	2.44	**50.70**	m2	**60.84**
Bonding ends to common brickwork; forming pockets; extra material										
walls; bonding every third course										
75mm	1.99	5.00	**2.08**	0.22	0.22	**5.48**	-	**7.57**	m	**9.08**
100mm	2.61	5.00	**2.74**	0.28	0.28	**6.85**	-	**9.59**	m	**11.51**
140mm	3.62	5.00	**3.80**	0.39	0.39	**9.59**	-	**13.39**	m	**16.07**
Bonding ends to existing common brickwork; cutting pockets; extra material										
walls; bonding every third course										
75mm	1.99	5.00	**2.08**	0.44	0.44	**10.96**	-	**13.05**	m	**15.66**
100mm	2.61	5.00	**2.74**	0.55	0.55	**13.71**	-	**16.45**	m	**19.74**
140mm	3.62	5.00	**3.80**	0.77	0.77	**19.19**	-	**22.99**	m	**27.59**
190mm	4.87	5.00	**5.11**	0.88	0.88	**21.93**	-	**27.04**	m	**32.45**
Bonding ends to existing concrete blockwork; cutting pockets; extra material										
walls; bonding every third course										
75mm	1.99	5.00	**2.08**	0.33	0.33	**8.22**	-	**10.31**	m	**12.37**
100mm	2.61	5.00	**2.74**	0.44	0.44	**10.96**	-	**13.71**	m	**16.45**
140mm	3.62	5.00	**3.80**	0.61	0.61	**15.08**	-	**18.88**	m	**22.65**
190mm	4.87	5.00	**5.11**	0.72	0.72	**17.82**	-	**22.93**	m	**27.52**
Aerated concrete blocks, Thermalite, 440 x 215mm, Turbo blocks (2.8 N/mm2); in cement-lime mortar (1:1:6)										
Walls; vertical										
100mm thick; stretcher bond	10.62	5.00	**11.15**	0.88	0.88	**21.93**	1.22	**34.29**	m2	**41.15**
150mm thick; stretcher bond	15.80	5.00	**16.59**	0.99	1.16	**26.44**	1.83	**44.86**	m2	**53.83**
200mm thick; stretcher bond	20.96	5.00	**22.00**	1.10	1.32	**29.77**	2.44	**54.21**	m2	**65.05**
Closing cavities; vertical										
50mm wide with blockwork 100mm thick	2.65	5.00	**2.79**	0.28	0.28	**6.85**	0.26	**9.90**	m	**11.88**
75mm wide with blockwork 100mm thick	2.65	5.00	**2.79**	0.28	0.28	**6.85**	0.26	**9.90**	m	**11.88**

Labour hourly rates: (except Specialists) Craft Operatives £14.21 Labourer £10.71 Rates are national average prices. Refer to REGIONAL VARIATIONS for indicative levels of overall pricing in regions	MATERIALS			LABOUR				RATES		
	Del to Site £	Waste %	Material Cost £	Craft Optve Hrs	Lab Hrs	Labour Cost £	Sunds £	Nett Rate £	Unit	Gross Rate (+20%) £
F10: BRICK/BLOCK WALLING Cont.										
Aerated concrete blocks, Thermalite, 440 x 215mm, Hi-Strength 7 blocks (7.0 N/mm2); in cement-lime mortar (1:1:6)										
Walls; vertical										
100mm thick; stretcher bond	13.05	5.00	**13.70**	0.88	0.99	**23.11**	1.22	**38.02**	m2	**45.63**
140mm thick; stretcher bond	18.07	5.00	**18.98**	0.99	1.16	**26.44**	1.83	**47.25**	m2	**56.70**
200mm thick; stretcher bond	25.63	5.00	**26.91**	1.10	1.32	**29.77**	2.44	**59.12**	m2	**70.94**
Closing cavities; vertical										
50mm wide with blockwork 100mm thick	3.26	5.00	**3.42**	0.28	0.28	**6.85**	0.26	**10.54**	m	**12.65**
75mm wide with blockwork 100mm thick	3.26	5.00	**3.42**	0.28	0.28	**6.85**	0.26	**10.54**	m	**12.65**
Aerated concrete blocks, Thermalite, 440 x 215mm, Trenchblocks (4.0 N/mm2); in cement mortar (1:4)										
Walls; vertical										
255mm thick; stretcher bond	26.26	5.00	**27.57**	1.43	1.65	**37.99**	2.53	**68.09**	m2	**81.71**
275mm thick; stretcher bond	28.27	5.00	**29.68**	1.60	1.82	**42.10**	3.04	**74.83**	m2	**89.79**
305mm thick; stretcher bond	31.33	5.00	**32.89**	1.82	2.04	**47.59**	3.04	**83.52**	m2	**100.22**
355mm thick; stretcher bond	36.40	5.00	**38.22**	2.15	2.37	**55.81**	3.55	**97.58**	m2	**117.09**
Tarmac Topblock, medium density concrete block, Hemelite, 440 x 215mm, solid Standard blocks (3.5 N/mm2); in cement-lime mortar (1:1:6)										
Walls; vertical										
75mm thick; stretcher bond	6.50	5.00	**6.83**	0.83	0.99	**22.33**	0.97	**30.12**	m2	**36.15**
100mm thick; stretcher bond	7.88	5.00	**8.27**	0.99	1.16	**26.44**	1.22	**35.93**	m2	**43.11**
140mm thick; stretcher bond	11.57	5.00	**12.15**	1.10	1.32	**29.77**	1.83	**43.75**	m2	**52.50**
190mm thick; stretcher bond	15.70	5.00	**16.48**	1.16	1.38	**31.14**	2.44	**50.06**	m2	**60.07**
Closing cavities; vertical										
50mm wide with blockwork 100mm thick	1.97	5.00	**2.07**	0.28	0.28	**6.85**	0.26	**9.18**	m	**11.02**
75mm wide with blockwork 100mm thick	1.97	5.00	**2.07**	0.28	0.28	**6.85**	0.26	**9.18**	m	**11.02**
Bonding ends to common brickwork; forming pockets; extra material										
walls; bonding every third course										
75mm	1.63	5.00	**1.71**	0.28	0.28	**6.85**	-	**8.56**	m	**10.27**
100mm	1.97	5.00	**2.07**	0.33	0.33	**8.22**	-	**10.29**	m	**12.35**
140mm	2.89	5.00	**3.04**	0.44	0.44	**10.96**	-	**14.00**	m	**16.80**
190mm	3.92	5.00	**4.12**	0.50	0.50	**12.34**	-	**16.46**	m	**19.75**
Bonding ends to existing common brickwork; cutting pockets; extra material										
walls; bonding every third course										
75mm	1.63	5.00	**1.71**	0.55	0.55	**13.71**	-	**15.41**	m	**18.50**
100mm	1.97	5.00	**2.07**	0.66	0.66	**16.45**	-	**18.51**	m	**22.22**
140mm	2.89	5.00	**3.04**	0.88	0.88	**21.93**	-	**24.97**	m	**29.96**
190mm	3.92	5.00	**4.12**	0.99	0.99	**24.67**	-	**28.79**	m	**34.55**
Tarmac Topblock, medium density concrete block, Hemelite, 440 x 215mm, solid Standard blocks (7.0 N/mm2); in cement-lime mortar (1:1:6)										
Walls; vertical										
100mm thick; stretcher bond	8.35	5.00	**8.77**	0.99	1.16	**26.44**	1.22	**36.42**	m2	**43.71**
140mm thick; stretcher bond	12.28	5.00	**12.89**	1.10	1.32	**29.77**	1.83	**44.49**	m2	**53.39**
190mm thick; stretcher bond	16.67	5.00	**17.50**	1.16	1.38	**31.14**	2.44	**51.07**	m2	**61.29**
Closing cavities; vertical										
50mm wide with blockwork 100mm thick	2.09	5.00	**2.19**	0.28	0.28	**6.85**	0.26	**9.30**	m	**11.17**
75mm wide with blockwork 100mm thick	2.09	5.00	**2.19**	0.28	0.28	**6.85**	0.26	**9.30**	m	**11.17**
Tarmac Topblock, fair face concrete blocks, Lignacite, 440 x 215mm, solid Standard blocks (7.0 N/mm2); in cement-lime mortar (1:1:6)										
Walls; vertical										
100mm thick; stretcher bond	14.69	5.00	**15.42**	1.05	1.21	**27.81**	1.22	**44.45**	m2	**53.33**
140mm thick; stretcher bond	21.60	5.00	**22.68**	1.21	1.38	**31.92**	1.83	**56.44**	m2	**67.72**
190mm thick; stretcher bond	29.33	5.00	**30.79**	1.27	1.49	**33.88**	2.44	**67.11**	m2	**80.53**
Closing cavities; vertical										
50mm wide with blockwork 100mm thick	3.67	5.00	**3.85**	0.28	0.28	**6.85**	0.26	**10.97**	m	**13.16**
75mm wide with blockwork 100mm thick	3.67	5.00	**3.85**	0.28	0.28	**6.85**	0.26	**10.97**	m	**13.16**
Aerated concrete blocks, Thermalite, 440 x 215mm, Smooth Face blocks (4.0 N/mm2); in cement-lime mortar (1:1:6); flush smooth pointing as work proceeds										
Walls; vertical										
stretcher bond; facework one side										
100mm thick	17.47	5.00	**18.34**	1.65	1.72	**41.82**	1.22	**61.38**	m2	**73.66**
140mm thick	24.29	5.00	**25.50**	1.21	1.38	**31.92**	1.83	**59.26**	m2	**71.11**
190mm thick	32.81	5.00	**34.45**	1.27	1.49	**33.88**	2.44	**70.77**	m2	**84.92**
stretcher bond; facework both sides										
100mm thick	17.47	5.00	**18.34**	1.32	1.49	**34.66**	1.22	**54.22**	m2	**65.07**
140mm thick	24.29	5.00	**25.50**	1.43	1.60	**37.40**	1.83	**64.74**	m2	**77.69**
190mm thick	32.81	5.00	**34.45**	1.49	1.71	**39.36**	2.44	**76.25**	m2	**91.50**

Labour hourly rates: (except Specialists) Craft Operatives £14.21 Labourer £10.71 Rates are national average prices. Refer to REGIONAL VARIATIONS for indicative levels of overall pricing in regions	MATERIALS			LABOUR				RATES		
	Del to Site £	Waste %	Material Cost £	Craft Optve Hrs	Lab Hrs	Labour Cost £	Sunds £	Nett Rate £	Unit	Gross Rate (+20%) £
F30: ACCESSORIES/SUNDRY ITEMS FOR BRICK/BLOCK/STONE WALLING										
Forming cavities in hollow walls										
Cavity with BS 1243 Fig 1 stainless steel wire butterfly wall ties, -4/m2 built in										
width of cavity 50mm	0.60	10.00	0.66	0.11	0.11	2.74	-	3.40	m2	4.08
Cavity with 200mm stainless steel vertical twisted wall ties -4/m2 built in; 50 fibreglass resin bonded slab cavity insulation										
width of cavity 50mm	6.76	10.00	7.44	0.39	0.39	9.59	1.28	18.31	m2	21.97
Cavity with stainless steel vertical twisted wall ties -4/m2 built in; 75 fibreglass resin bonded slab cavity insulation										
width of cavity 75mm	8.33	10.00	9.17	0.44	0.44	10.96	1.28	21.41	m2	25.69
Damp proof courses										
BS 6398 Class A, bitumen with hessian base; 100mm laps; bedding in cement mortar (1:3); no allowance made for laps										
width not exceeding 225mm; vertical	8.36	5.00	8.78	0.55	-	7.82	-	16.59	m2	19.91
width exceeding 225mm; vertical	8.36	5.00	8.78	0.50	-	7.03	-	15.81	m2	18.97
width not exceeding 225mm; horizontal	8.36	5.00	8.78	0.39	-	5.47	-	14.25	m2	17.10
width exceeding 225mm; horizontal	8.36	5.00	8.78	0.33	-	4.69	-	13.47	m2	16.16
cavity trays; width not exceeding 225mm; horizontal	8.36	5.00	8.78	0.61	-	8.60	-	17.38	m2	20.85
cavity trays; width exceeding 225mm; horizontal	8.36	5.00	8.78	0.55	-	7.82	-	16.59	m2	19.91
Hyload, pitch polymer; 100mm laps; bedding in cement lime mortar (1:1:6); no allowance made for laps										
width not exceeding 225mm; vertical	6.69	5.00	7.02	0.55	-	7.82	-	14.84	m2	17.81
width exceeding 225mm; vertical	6.69	5.00	7.02	0.50	-	7.03	-	14.06	m2	16.87
width not exceeding 225mm; horizontal	6.69	5.00	7.02	0.39	-	5.47	-	12.49	m2	14.99
width exceeding 225mm; horizontal	6.69	5.00	7.02	0.33	-	4.69	-	11.71	m2	14.05
Two courses slates in cement mortar (1:3)										
width not exceeding 225mm; horizontal	38.91	15.00	44.75	1.43	1.43	35.64	-	80.39	m2	96.46
width exceeding 225mm; horizontal	38.91	15.00	44.75	1.32	1.32	32.89	-	77.65	m2	93.17
Cavity trays										
Type G Cavitray by Cavity Trays Ltd in										
half brick skin of cavity wall	10.94	5.00	11.49	0.22	-	3.13	-	14.61	m	17.54
external angle	8.73	5.00	9.17	0.11	-	1.56	-	10.73	nr	12.88
internal angle	8.73	5.00	9.17	0.11	-	1.56	-	10.73	nr	12.88
Type X Cavitray by Cavity Trays Ltd to suit 40 degree pitched roof complete with attached code 4 lead flashing and dress over tiles in										
half brick skin of cavity wall	45.63	5.00	47.91	0.22	-	3.13	-	51.04	m	61.24
ridge tray	11.32	5.00	11.88	0.11	-	1.56	-	13.45	nr	16.14
catchment tray	5.37	5.00	5.64	0.11	-	1.56	-	7.20	nr	8.64
corner catchment angle tray	11.83	5.00	12.42	0.11	-	1.56	-	13.98	nr	16.78
Type W Cavity weep/ventilator by Cavity Trays Ltd in										
half brick skin of cavity wall	0.60	5.00	0.63	0.11	-	1.56	-	2.19	nr	2.63
extension duct	0.90	5.00	0.95	0.11	-	1.56	-	2.51	nr	3.01
Joint reinforcement										
Expamet grade 304/S15 Exmet reinforcement, stainless steel; 150mm laps;										
65mm wide	1.25	5.00	1.32	0.04	0.04	1.10	0.04	2.45	m	2.94
115mm wide	1.91	5.00	2.01	0.06	0.06	1.37	0.04	3.42	m	4.10
175mm wide	3.31	5.00	3.48	0.07	0.07	1.64	0.06	5.18	m	6.22
225mm wide	4.32	5.00	4.54	0.08	0.08	1.92	0.07	6.53	m	7.83
Pointing in flashings										
Cement mortar (1:3)										
horizontal	-	-	-	0.50	-	7.03	0.28	7.32	m	8.78
horizontal; in old wall	-	-	-	0.55	-	7.82	0.28	8.10	m	9.72
stepped	-	-	-	0.88	-	12.50	0.45	12.95	m	15.54
stepped; in old wall	-	-	-	0.94	-	13.29	0.45	13.74	m	16.48
Wedging and pinning										
Two courses slates in cement mortar (1:3)										
width of wall 215mm	9.73	15.00	11.19	0.55	0.55	13.71	0.17	25.06	m	30.07
Joints										
Expansion joints in facing brickwork										
13mm wide; vertical; filling with Servicised Ltd, Fibrepack filler, Vertiseal compound pointing one side; including preparation, cleaners, primers and sealers										
102mm thick wall	3.78	7.50	4.07	0.11	0.22	3.92	-	7.99	m	9.58
215mm thick wall	6.01	7.50	6.46	0.12	0.24	4.31	-	10.77	m	12.92

Labour hourly rates: (except Specialists) Craft Operatives £14.21 Labourer £10.71 Rates are national average prices. Refer to REGIONAL VARIATIONS for indicative levels of overall pricing in regions	MATERIALS			LABOUR				RATES		
	Del to Site	Waste	Material Cost	Craft Optve	Lab	Labour Cost	Sunds	Nett Rate	Unit	Gross Rate (+20%)
	£	%	£	Hrs	Hrs	£	£	£		£
F30: ACCESSORIES/SUNDRY ITEMS FOR BRICK/BLOCK/STONE WALLING Cont.										
Slates and tiles for creasing										
Nibless creasing tiles, red, machine made, 265 x 165 x 10mm; in cement-lime mortar (1:1:6)										
two courses 253mm wide	5.86	5.00	6.15	0.94	0.94	23.30	0.61	30.07	m	36.08
two courses 365mm wide	8.79	5.00	9.23	1.27	1.27	31.52	0.85	41.61	m	49.93
Slate and tile sills										
Clay plain roofing tiles, red, BS EN 1304, machine made, 265 x 165 x 13mm; in cement-lime mortar										
one course 150mm wide; set weathering	1.95	5.00	2.05	0.50	0.50	12.34	0.28	14.67	m	17.60
two courses 150mm wide; set weathering	3.91	5.00	4.10	0.83	0.83	20.56	0.57	25.23	m	30.27
Chimney pots										
Clay chimney pots; set and flaunched in cement mortar (1:3)										
tapered roll top; 600mm high	42.83	5.00	44.98	0.88	1.05	23.70	0.50	69.17	nr	83.00
tapered roll top; 900mm high	76.92	5.00	80.77	1.21	1.54	33.69	0.74	115.20	nr	138.24
Air bricks										
Clay air bricks, BS 493, square hole pattern; opening with slate lintel over										
common brick wall 102mm thick; opening size										
225 x 75mm	2.30	-	2.30	0.22	0.22	5.48	0.74	8.53	nr	10.23
225 x 150mm	4.39	-	4.39	0.28	0.28	6.85	0.79	12.03	nr	14.44
225 x 225mm	7.69	-	7.69	0.33	0.33	8.22	0.85	16.76	nr	20.12
cavity wall 252mm thick with 102mm facing brick outer skin, 100mm block inner skin and 50mm cavity; sealing cavity with slates in cement mortar (1:3); rendering all round with cement mortar (1:3); opening size										
225 x 75mm	2.30	-	2.30	0.66	0.66	16.45	2.72	21.47	nr	25.76
225 x 150mm	4.39	-	4.39	0.77	0.77	19.19	3.35	26.92	nr	32.31
225 x 225mm	7.69	-	7.69	0.88	0.88	21.93	3.91	33.53	nr	40.24
Building in										
Building in metal windows; building in lugs; bedding in cement mortar (1:3), pointing with Secomastic standard mastic one side										
900mm high; lugs to brick jambs; plugging and screwing to brick sill and concrete head										
500mm wide	-	-	-	0.98	0.52	19.45	2.59	22.04	nr	26.45
900mm wide	-	-	-	1.24	0.66	24.73	3.36	28.09	nr	33.71
1500mm wide	-	-	-	1.64	0.87	32.60	4.49	37.09	nr	44.51
1800mm wide	-	-	-	1.83	0.96	36.20	5.03	41.22	nr	49.47
1300mm high; lugs to brick jambs; plugging and screwing to brick sill and concrete head										
500mm wide	-	-	-	1.23	0.63	24.22	3.32	27.54	nr	33.05
900mm wide	-	-	-	1.51	0.76	29.54	4.08	33.62	nr	40.35
1500mm wide	-	-	-	1.90	0.99	37.64	5.24	42.88	nr	51.46
1800mm wide	-	-	-	2.11	1.09	41.67	5.79	47.47	nr	56.96
Building in factory glazed metal windows; screwing with galvanized screws; bedding in cement mortar (1:3), pointing with Secomastic standard mastic one side										
900mm high; plugging and screwing lugs to brick jambs and sill and concrete head										
600mm wide	-	-	-	1.25	0.78	26.18	2.80	28.99	nr	34.78
1200mm wide	-	-	-	1.82	1.14	38.04	3.91	41.96	nr	50.35
1800mm wide	-	-	-	2.38	1.51	49.90	5.03	54.93	nr	65.91
1300mm high; plugging and screwing lugs to brick jambs and sill and concrete head										
600mm wide	-	-	-	1.55	0.94	32.05	3.72	35.78	nr	42.93
1200mm wide	-	-	-	2.27	1.47	47.99	4.67	52.66	nr	63.19
1800mm wide	-	-	-	2.99	1.99	63.84	5.79	69.63	nr	83.56
Building in wood windows; building in lugs; bedding in cement mortar (1:3), pointing with Secomastic standard mastic one side										
920mm high; lugs to brick jambs										
438mm wide	-	-	-	1.89	0.31	30.18	2.52	32.70	nr	39.24
1225mm wide	-	-	-	2.87	0.39	44.92	4.08	49.00	nr	58.80
1809mm wide	-	-	-	3.60	0.45	55.94	5.24	61.18	nr	73.42
1225mm high; lugs to brick jambs										
438mm wide	-	-	-	2.31	0.35	36.60	3.12	39.72	nr	47.66
1225mm wide	-	-	-	3.37	0.45	52.66	4.65	57.31	nr	68.77
1809mm wide	-	-	-	4.13	0.55	64.51	5.75	70.25	nr	84.30
Proprietary items										
Precast prestressed concrete lintels; Stressline Ltd; Roughcast; bedding in cement-lime mortar (1:1:6)										
100 x 70mm										
900mm long	3.92	2.50	4.01	0.31	0.15	6.03	0.31	10.35	nr	12.42
1200mm long	5.20	2.50	5.33	0.40	0.20	7.75	0.40	13.48	nr	16.18
1500mm long	6.51	2.50	6.67	0.48	0.24	9.47	0.48	16.63	nr	19.95
1800mm long	7.81	2.50	8.01	0.56	0.29	11.03	0.53	19.57	nr	23.49
2400mm long	10.43	2.50	10.69	0.69	0.35	13.62	0.71	25.02	nr	30.02

Labour hourly rates: (except Specialists) Craft Operatives £14.21 Labourer £10.71 Rates are national average prices. Refer to REGIONAL VARIATIONS for indicative levels of overall pricing in regions	MATERIALS			LABOUR				RATES		
	Del to Site £	Waste %	Material Cost £	Craft Optve Hrs	Lab Hrs	Labour Cost £	Sunds £	Nett Rate £	Unit	Gross Rate (+20%) £
F30: ACCESSORIES/SUNDRY ITEMS FOR BRICK/BLOCK/STONE WALLING Cont.										
Proprietary items Cont.										
Precast prestressed concrete lintels; Stressline Ltd; Roughcast; bedding in cement-lime mortar (1:1:6) Cont.										
150 x 70mm										
900mm long	5.31	2.50	**5.45**	0.32	0.17	**6.30**	0.31	**12.05**	nr	**14.46**
1200mm long	7.06	2.50	**7.24**	0.41	0.21	**8.02**	0.40	**15.66**	nr	**18.79**
1500mm long	8.84	2.50	**9.07**	0.50	0.25	**9.74**	0.50	**19.31**	nr	**23.17**
1800mm long	10.63	2.50	**10.89**	0.57	0.29	**11.19**	0.54	**22.63**	nr	**27.15**
2400mm long	14.16	2.50	**14.51**	0.73	0.36	**14.20**	0.71	**29.42**	nr	**35.31**
225 x 70mm										
900mm long	8.17	2.50	**8.38**	0.33	0.17	**6.46**	0.31	**15.14**	nr	**18.17**
1200mm long	10.90	2.50	**11.17**	0.42	0.21	**8.18**	0.40	**19.75**	nr	**23.70**
1500mm long	13.64	2.50	**13.98**	0.51	0.25	**9.90**	0.50	**24.38**	nr	**29.25**
1800mm long	16.37	2.50	**16.78**	0.58	0.30	**11.47**	0.54	**28.79**	nr	**34.54**
2400mm long	21.81	2.50	**22.36**	0.74	0.37	**14.48**	0.74	**37.58**	nr	**45.10**
100 x 145mm										
900mm long	7.90	2.50	**8.10**	0.33	0.17	**6.46**	0.31	**14.86**	nr	**17.83**
1200mm long	10.55	2.50	**10.81**	0.42	0.21	**8.18**	0.43	**19.42**	nr	**23.30**
1500mm long	13.18	2.50	**13.51**	0.51	0.25	**9.90**	0.50	**23.90**	nr	**28.68**
1800mm long	15.81	2.50	**16.20**	0.58	0.30	**11.47**	0.59	**28.26**	nr	**33.91**
2400mm long	21.09	2.50	**21.61**	0.74	0.37	**14.48**	0.74	**36.84**	nr	**44.20**
Galvanized steel lintels; SUPERGALV (BIRTLEY) lintels reference CB 50; bedding in cement-lime mortar (1:1:6)										
125mm deep x 750mm long	25.38	2.50	**26.01**	0.35	0.35	**8.77**	-	**34.78**	nr	**41.74**
125mm deep x 1200mm long	40.02	2.50	**41.02**	0.48	0.48	**12.06**	-	**53.08**	nr	**63.70**
125mm deep x 1500mm long	52.17	2.50	**53.48**	0.56	0.56	**13.98**	-	**67.46**	nr	**80.95**
165mm deep x 2100mm long	73.92	2.50	**75.77**	0.72	0.72	**17.82**	-	**93.59**	nr	**112.30**
Galvanized steel lintels; SUPERGALV (BIRTLEY) lintels reference CB 50 H.D.; bedding in cement-lime mortar (1:1:6)										
165mm deep x 750mm long	29.95	2.50	**30.70**	0.36	0.36	**9.05**	-	**39.75**	nr	**47.70**
165mm deep x 1200mm long	44.92	2.50	**46.05**	0.48	0.48	**12.06**	-	**58.11**	nr	**69.73**
165mm deep x 1350mm long	54.63	2.50	**55.99**	0.53	0.53	**13.16**	-	**69.15**	nr	**82.98**
215mm deep x 1800mm long	78.64	2.50	**80.60**	0.64	0.64	**15.90**	-	**96.50**	nr	**115.80**
215mm deep x 2100mm long	92.76	2.50	**95.08**	0.72	0.72	**17.82**	-	**112.90**	nr	**135.48**
Galvanized steel lintels; SUPERGALV (BIRTLEY) lintels reference SBL200; bedding in cement-lime mortar (1:1:6)										
142mm deep x 750mm long	29.51	2.50	**30.25**	0.57	0.57	**14.25**	-	**44.50**	nr	**53.40**
142mm deep x 1050mm long	41.02	2.50	**42.04**	0.65	0.65	**16.17**	-	**58.22**	nr	**69.86**
142mm deep x 1350mm long	52.53	2.50	**53.85**	0.73	0.73	**18.09**	-	**71.94**	nr	**86.33**
142mm deep x 1650mm long	67.19	2.50	**68.87**	0.85	0.85	**21.11**	-	**89.98**	nr	**107.97**
Galvanized steel lintels; SUPERGALV (BIRTLEY) internal door lintels reference INT100; bedding in cement-lime mortar (1:1:6)										
100mm wide, 900mm long	13.40	2.50	**13.74**	0.42	0.42	**10.42**	-	**24.16**	nr	**28.99**
100mm wide, 1050mm long	15.66	2.50	**16.06**	0.45	0.45	**11.24**	-	**27.29**	nr	**32.75**
100mm wide, 1200mm long	17.92	2.50	**18.37**	0.48	0.48	**12.06**	-	**30.43**	nr	**36.52**
Stainless steel Furfix wall extension profiles; single flange; plugging and screwing to brickwork and building ties in to joints of new walls										
100mm	13.17	5.00	**13.83**	0.33	-	**4.69**	-	**18.51**	m	**22.22**
150mm	14.19	5.00	**14.90**	0.39	-	**5.47**	-	**20.37**	m	**24.44**
215mm	25.67	5.00	**26.96**	0.44	-	**6.25**	-	**33.21**	m	**39.85**
Thermabate insulated cavity closers; fixing to masonry with PVC-U ties at 225mm centres										
Thermabate 50; vertical	5.09	5.00	**5.35**	0.44	0.44	**10.96**	-	**16.31**	m	**19.57**
Thermabate 75; vertical	5.32	5.00	**5.59**	0.44	0.44	**10.96**	-	**16.56**	m	**19.87**
Thermabate 50; horizontal	5.09	5.00	**5.35**	0.44	0.44	**10.96**	-	**16.31**	m	**19.57**
Thermabate 75; horizontal	5.32	5.00	**5.59**	0.44	0.44	**10.96**	-	**16.56**	m	**19.87**
F31: PRECAST CONCRETE SILLS/LINTELS/COPINGS/FEATURES										
Precast concrete sills; BS 5642 Part 1; bedding in cement lime mortar (1:1:6)										
Sills; figure 2 or figure 4										
75 x 150 x 300mm splayed and grooved	3.77	2.50	**3.87**	0.21	0.21	**5.21**	0.14	**9.22**	nr	**11.06**
75 x 150 x 700mm splayed and grooved	8.80	2.50	**9.02**	0.42	0.42	**10.42**	0.25	**19.69**	nr	**23.62**
75 x 150 x 1300mm splayed and grooved	16.34	2.50	**16.75**	0.66	0.66	**16.45**	0.38	**33.58**	nr	**40.29**
extra for stooled end	3.56	-	**3.56**	-	-	**-**	-	**3.56**	nr	**4.27**
Precast concrete copings BS 5642 Part 2; bedding in cement-lime mortar (1:1:6)										
Copings; figure 1; horizontal										
100 x 300mm; splayed; rebated joints	25.15	2.50	**25.77**	0.61	0.31	**11.90**	0.34	**38.01**	m	**45.62**
extra; stopped ends	2.14	-	**2.14**	-	-	**-**	-	**2.14**	nr	**2.56**
extra; internal angles	3.56	-	**3.56**	-	-	**-**	-	**3.56**	nr	**4.27**
100 x 300mm; saddleback; rebated joints	26.82	2.50	**27.49**	0.61	0.31	**11.90**	0.34	**39.73**	m	**47.68**
extra; hipped ends	2.14	-	**2.14**	-	-	**-**	-	**2.14**	nr	**2.56**
extra; internal angles	3.56	-	**3.56**	-	-	**-**	-	**3.56**	nr	**4.27**

Labour hourly rates: (except Specialists) Craft Operatives £14.21 Labourer £10.71 Rates are national average prices. Refer to REGIONAL VARIATIONS for indicative levels of overall pricing in regions	MATERIALS			LABOUR				RATES		
	Del to Site £	Waste %	Material Cost £	Craft Optve Hrs	Lab Hrs	Labour Cost £	Sunds £	Nett Rate £	Unit	Gross Rate (+20%) £
F31: PRECAST CONCRETE SILLS/LINTELS/COPINGS/FEATURES Cont.										
Precast concrete lintels; BS 5328, designed mix C25, 20mm aggregate minimum cement content 360 kg/m3; vibrated; reinforcement bars BS 4449 hot rolled plain round mild steel; bedding in cement-lime mortar (1:1:6)										
Lintels										
75 x 150 x 1200mm; reinforced with 1.20 kg of 12mm bars	7.76	2.50	**7.95**	0.22	0.11	**4.30**	0.14	**12.40**	nr	**14.87**
102 x 150 x 1200mm; reinforced with 1.20 kg of 12mm bars	10.55	2.50	**10.81**	0.36	0.19	**7.16**	0.14	**18.11**	nr	**21.74**
215 x 150 x 1200mm; reinforced with 2.40 kg of 12mm bars	22.23	2.50	**22.79**	0.73	0.36	**14.20**	0.24	**37.23**	nr	**44.67**
215 x 150 x 1650mm; reinforced with 3.20 kg of 12mm bars	30.57	2.50	**31.33**	1.00	0.58	**20.47**	0.31	**52.11**	nr	**62.53**
215 x 225 x 1200mm; reinforced with 2.40 kg of 12mm bars	33.35	2.50	**34.18**	0.95	0.47	**18.51**	0.25	**52.94**	nr	**63.52**
327 x 150 x 1200mm; reinforced with 3.60 kg of 12mm bars	33.81	2.50	**34.66**	0.95	0.47	**18.51**	0.27	**53.44**	nr	**64.13**
75 x 150 x 1200mm; reinforced with 1.20 kg of 12mm bars; fair finish two faces	9.54	2.50	**9.78**	0.22	0.11	**4.30**	0.14	**14.22**	nr	**17.07**
Boot lintels										
215 x 150 x 1200mm; reinforced with 3.60 kg of 12mm bars and 0.51 kg of 6mm links	27.34	2.50	**28.03**	0.73	0.36	**14.20**	0.22	**42.45**	nr	**50.94**
252 x 150 x 1200mm; reinforced with 3.60 kg of 12mm bars and 0.56 kg of 6mm links	32.05	2.50	**32.85**	0.87	0.44	**17.06**	0.26	**50.17**	nr	**60.20**
Precast concrete padstones; BS 5328, designed mix C25, 20mm aggregate, minimum cement content 360 kg/m3; vibrated; reinforced at contractor's discretion; bedding in cement-lime mortar (1:1:6)										
Padstones										
215 x 215 x 150mm	5.03	2.50	**5.16**	0.10	0.04	**1.88**	0.13	**7.17**	nr	**8.60**
327 x 327 x 150mm	11.65	2.50	**11.94**	0.21	0.10	**4.03**	0.15	**16.12**	nr	**19.34**

NATURAL STONEWORK

Natural stonework including restoration and repair is work for the specialist and the cost depends on many factors, prices should be obtained for specific projects.

There are a number of reputable firms who specialize in this class of work and who will submit quotations. Among these are the following:-

STONEWEST LTD.
Lamberts Place, St James Road, Croydon, Surrey, CR9 2HX. Tel: 020 8684 6646
Fax: 020 8684 9323

A. F. JONES (STONEMASONS).
33 Bedford Road, Reading, Berkshire, RG1 7EX. Tel: 0118 957 3537
Fax: 0118 957 4334

DIESPEKER MARBLE & TERRAZZO LTD.
132-136 Ormside Street, Peckham, London, SE15 1TF Tel: 020 7358 0160
Fax: 020 7369 1695

C.W.O. LIMITED.
Terminus Road, Chichester, West Sussex, PO19 2TX. Tel: 01243 784225
Fax: 01243 813700

LAING STONEMASONRY
Dalston Road, Carlisle, Cumbria, CA2 5NR. Tel: 01228 521404
Fax: 01228 511306

MACC STONE CO.
196-198 New Street, Biddulph Moor, Staffordshire, Moorlands, ST8 7NW. Tel: 01782 514353
Fax: 01782 516783

STONEGUARD (LONDON) LTD.
St Martins House, The Runway, South Ruislip, Middlesex, H04 6SG. Tel: 020 8839 9977
Fax: 020 8839 9988

APS MASONRY LTD.
Osney Mead, Oxford, OX2 0EQ. Tel: 01865 246677
Fax: 01865 254617

DUNHOUSE QUARRY CO. LTD.
Staine Drop, Darlington, Co. Durham DL2 3QU Tel: 01833 660208
Fax: 01833 660748

STRUCTURAL/CARCASSING METAL/TIMBER

Labour hourly rates: (except Specialists) Craft Operatives £14.21 Labourer £10.71 Rates are national average prices. Refer to REGIONAL VARIATIONS for indicative levels of overall pricing in regions	MATERIALS			LABOUR				RATES		
	Del to Site £	Waste %	Material Cost £	Craft Optve Hrs	Lab Hrs	Labour Cost £	Sunds £	Nett Rate £	Unit	Gross Rate (+20%) £
G12: ISOLATED STRUCTURAL METAL MEMBERS										
Isolated unfabricated structural members										
Weldable steel, BS EN 10025 : 1993 Grade S275 (formerly BS 7668 Grade 43A), hot rolled sections										
plain member; beam										
weight less than 40 Kg/m	1045.00	-	**1045.00**	55.00	-	**781.55**	-	**1826.55**	t	**2191.86**
weight 40-100 Kg/m	1045.00	-	**1045.00**	51.76	-	**735.44**	-	**1780.44**	t	**2136.53**
weight exceeding 100 Kg/m	1045.00	-	**1045.00**	45.32	-	**644.00**	-	**1689.00**	t	**2026.80**
Fixing bolts; steel										
Black bolts, BS 4190 grade 4.6										
M 12; with nuts and washers										
50mm long	0.37	5.00	**0.38**	0.09	-	**1.25**	-	**1.64**	nr	**1.96**
100mm long	0.56	5.00	**0.59**	0.10	-	**1.41**	-	**2.00**	nr	**2.40**
150mm long	1.05	5.00	**1.10**	0.13	-	**1.88**	-	**2.98**	nr	**3.57**
Black bolts, BS 4190 grade 4.6; galvanized										
M 12; with nuts and washers										
50mm long	0.60	5.00	**0.63**	0.09	-	**1.25**	-	**1.88**	nr	**2.25**
100mm long	0.88	5.00	**0.92**	0.10	-	**1.41**	-	**2.33**	nr	**2.80**
150mm long	1.61	5.00	**1.69**	0.13	-	**1.88**	-	**3.57**	nr	**4.28**
G20: CARPENTRY/TIMBER FRAMING/FIRST FIXING										
Softwood, sawn, BS 4978, GS grade - carcassing										
Floor members										
38 x 75mm	0.69	15.00	**0.79**	0.15	0.02	**2.42**	0.01	**3.22**	m	**3.87**
50 x 100mm	1.18	15.00	**1.36**	0.21	0.02	**3.21**	0.01	**4.57**	m	**5.49**
50 x 200mm	2.36	15.00	**2.71**	0.35	0.04	**5.47**	0.02	**8.20**	m	**9.85**
50 x 250mm	2.95	15.00	**3.39**	0.42	0.06	**6.53**	0.02	**9.94**	m	**11.93**
75 x 150mm	2.72	15.00	**3.13**	0.39	0.04	**5.94**	0.02	**9.09**	m	**10.91**
75 x 200mm	3.63	15.00	**4.17**	0.46	0.06	**7.15**	0.03	**11.35**	m	**13.62**
75 x 250mm	4.54	15.00	**5.22**	0.59	0.08	**9.27**	0.03	**14.52**	m	**17.42**
Wall or partition members										
38 x 75mm	0.69	15.00	**0.79**	0.24	0.03	**3.79**	0.01	**4.60**	m	**5.52**
50 x 100mm	1.18	15.00	**1.36**	0.32	0.04	**5.00**	0.02	**6.38**	m	**7.65**
75 x 100mm	1.82	15.00	**2.09**	0.40	0.04	**6.10**	0.02	**8.21**	m	**9.85**
38 x 75mm; fixing to masonry	0.69	15.00	**0.79**	0.47	0.06	**7.31**	0.03	**8.13**	m	**9.75**
50 x 100mm; fixing to masonry	1.18	15.00	**1.36**	0.57	0.08	**8.95**	0.03	**10.34**	m	**12.41**
75 x 100mm; fixing to masonry	1.82	15.00	**2.09**	0.55	0.08	**8.64**	0.03	**10.76**	m	**12.91**
Plates										
38 x 75mm	0.69	15.00	**0.79**	0.24	0.03	**3.79**	0.01	**4.60**	m	**5.52**
50 x 100mm	1.18	15.00	**1.36**	0.32	0.04	**5.00**	0.02	**6.38**	m	**7.65**
50 x 100mm; fixing by bolting	1.18	15.00	**1.36**	0.39	0.04	**5.94**	-	**7.30**	m	**8.76**
Roof members; flat										
50 x 175mm	2.06	15.00	**2.37**	0.32	0.04	**5.00**	0.02	**7.40**	m	**8.88**
50 x 200mm	2.36	15.00	**2.71**	0.35	0.04	**5.47**	0.02	**8.20**	m	**9.85**
50 x 250mm	2.95	15.00	**3.39**	0.42	0.06	**6.53**	0.02	**9.94**	m	**11.93**
75 x 175mm	3.18	15.00	**3.65**	0.44	0.06	**6.84**	0.02	**10.52**	m	**12.62**
75 x 200mm	3.63	15.00	**4.17**	0.46	0.06	**7.15**	0.03	**11.35**	m	**13.62**
75 x 250mm	4.54	15.00	**5.22**	0.59	0.08	**9.27**	0.03	**14.52**	m	**17.42**
Roof members; pitched										
25 x 125mm	0.76	15.00	**0.87**	0.21	0.02	**3.21**	0.01	**4.09**	m	**4.90**
32 x 175mm	1.36	15.00	**1.56**	0.29	0.03	**4.42**	0.01	**5.99**	m	**7.19**
38 x 75mm	0.69	15.00	**0.79**	0.20	0.02	**3.05**	0.01	**3.85**	m	**4.62**
38 x 100mm	0.92	15.00	**1.06**	0.24	0.03	**3.79**	0.01	**4.86**	m	**5.84**
50 x 75mm	0.88	15.00	**1.02**	0.24	0.03	**3.79**	0.01	**4.82**	m	**5.79**
50 x 100mm	1.18	15.00	**1.36**	0.26	0.03	**4.10**	0.01	**5.48**	m	**6.57**
50 x 150mm	1.77	15.00	**2.04**	0.34	0.04	**5.32**	0.02	**7.37**	m	**8.84**
50 x 225mm	2.65	15.00	**3.05**	0.47	0.06	**7.31**	0.03	**10.39**	m	**12.47**
75 x 100mm	1.82	15.00	**2.09**	0.34	0.04	**5.32**	0.02	**7.42**	m	**8.91**

Labour hourly rates: (except Specialists) Craft Operatives £14.21 Labourer £10.71 Rates are national average prices. Refer to REGIONAL VARIATIONS for indicative levels of overall pricing in regions	MATERIALS			LABOUR				RATES		
	Del to Site £	Waste %	Material Cost £	Craft Optve Hrs	Lab Hrs	Labour Cost £	Sunds £	Nett Rate £	Unit	Gross Rate (+20%) £
G20: CARPENTRY/TIMBER FRAMING/FIRST FIXING Cont.										
Softwood, sawn, BS 4978, GS grade - carcassing Cont.										
Joist strutting; herringbone; depth of joist 175mm										
50 x 50mm	1.06	15.00	**1.22**	0.86	0.11	**13.37**	0.05	**14.64**	m	**17.57**
Joist strutting; herringbone; depth of joist 225mm										
50 x 50mm	1.42	15.00	**1.63**	0.86	0.11	**13.37**	0.05	**15.05**	m	**18.06**
Joist strutting; block; depth of joist 200mm										
50 x 200mm	2.36	15.00	**2.71**	0.92	0.12	**14.43**	0.05	**17.19**	m	**20.63**
Noggings to joists										
50 x 50mm	0.59	15.00	**0.68**	0.37	0.04	**5.79**	0.02	**6.48**	m	**7.78**
50 x 75mm	0.88	15.00	**1.02**	0.44	0.06	**6.84**	0.02	**7.88**	m	**9.46**
Softwood, sawn, BS 4978, GS grade, impregnated - carcassing										
Floor members										
38 x 75mm	0.78	15.00	**0.89**	0.15	0.02	**2.42**	0.01	**3.32**	m	**3.99**
50 x 100mm	1.33	15.00	**1.53**	0.21	-	**2.97**	0.01	**4.51**	m	**5.41**
50 x 200mm	2.66	15.00	**3.06**	0.35	0.04	**5.47**	0.02	**8.55**	m	**10.26**
50 x 250mm	3.33	15.00	**3.83**	0.42	0.06	**6.53**	0.02	**10.38**	m	**12.45**
75 x 150mm	3.06	15.00	**3.52**	0.39	0.04	**5.94**	0.02	**9.49**	m	**11.38**
75 x 200mm	4.08	15.00	**4.70**	0.46	0.06	**7.15**	0.03	**11.88**	m	**14.25**
75 x 250mm	5.10	15.00	**5.87**	0.59	0.08	**9.27**	0.03	**15.17**	m	**18.20**
Wall or partition members										
38 x 75mm	0.78	15.00	**0.89**	0.24	0.03	**3.79**	0.01	**4.70**	m	**5.64**
50 x 100mm	1.33	15.00	**1.53**	0.32	0.04	**5.00**	0.02	**6.55**	m	**7.86**
75 x 100mm	2.04	15.00	**2.35**	0.40	0.04	**6.10**	0.02	**8.47**	m	**10.16**
38 x 75mm; fixing to masonry	0.78	15.00	**0.89**	0.47	0.06	**7.31**	0.03	**8.23**	m	**9.87**
50 x 100mm; fixing to masonry	1.33	15.00	**1.53**	0.57	0.08	**8.95**	0.03	**10.52**	m	**12.62**
75 x 100mm; fixing to masonry	2.04	15.00	**2.35**	0.55	0.08	**8.64**	0.03	**11.02**	m	**13.22**
Plates										
38 x 75mm	0.78	15.00	**0.89**	0.24	0.03	**3.79**	0.01	**4.70**	m	**5.64**
50 x 100mm	1.33	15.00	**1.53**	0.32	0.04	**5.00**	0.02	**6.55**	m	**7.86**
50 x 100mm; fixing by bolting	1.33	15.00	**1.53**	0.39	0.04	**5.94**	-	**7.47**	m	**8.97**
Roof members; flat										
50 x 175mm	2.33	15.00	**2.68**	0.32	0.04	**5.00**	0.02	**7.70**	m	**9.24**
50 x 200mm	2.66	15.00	**3.06**	0.35	0.04	**5.47**	0.02	**8.55**	m	**10.26**
50 x 250mm	3.33	15.00	**3.83**	0.42	0.06	**6.53**	0.02	**10.38**	m	**12.45**
75 x 175mm	3.57	15.00	**4.11**	0.44	0.06	**6.84**	0.02	**10.97**	m	**13.17**
75 x 200mm	4.08	15.00	**4.70**	0.46	0.06	**7.15**	0.03	**11.88**	m	**14.25**
75 x 250mm	5.10	15.00	**5.87**	0.59	0.08	**9.27**	0.03	**15.17**	m	**18.20**
Roof members; pitched										
25 x 125mm	0.85	15.00	**0.98**	0.21	0.02	**3.21**	0.01	**4.19**	m	**5.03**
32 x 175mm	1.52	15.00	**1.75**	0.29	0.03	**4.42**	0.01	**6.19**	m	**7.42**
38 x 75mm	0.78	15.00	**0.89**	0.20	0.02	**3.05**	0.01	**3.95**	m	**4.74**
38 x 100mm	1.03	15.00	**1.19**	0.24	0.03	**3.79**	0.01	**5.00**	m	**6.00**
50 x 75mm	1.00	15.00	**1.15**	0.24	0.03	**3.79**	0.01	**4.95**	m	**5.95**
50 x 100mm	1.33	15.00	**1.53**	0.26	0.03	**4.10**	0.01	**5.65**	m	**6.78**
50 x 150mm	2.00	15.00	**2.30**	0.34	0.04	**5.32**	0.02	**7.63**	m	**9.16**
50 x 225mm	2.99	15.00	**3.44**	0.47	0.06	**7.31**	0.03	**10.78**	m	**12.94**
75 x 100mm	2.04	15.00	**2.35**	0.34	0.04	**5.32**	0.02	**7.68**	m	**9.22**
75 x 150mm	3.06	15.00	**3.52**	0.46	0.06	**7.15**	0.03	**10.70**	m	**12.84**
Joist strutting; herringbone; depth of joist 175mm										
50 x 50mm	1.20	15.00	**1.38**	0.86	0.11	**13.37**	0.05	**14.80**	m	**17.75**
Joist strutting; herringbone; depth of joist 225mm										
50 x 50mm	1.60	15.00	**1.84**	0.86	0.11	**13.37**	0.05	**15.25**	m	**18.31**
Joist strutting; block; depth of joist 175mm										
50 x 175mm	2.33	15.00	**2.68**	0.92	0.12	**14.43**	0.05	**17.16**	m	**20.59**
Noggings to joists										
50 x 50mm	0.67	15.00	**0.77**	0.37	0.04	**5.79**	0.02	**6.57**	m	**7.88**
50 x 75mm	1.00	15.00	**1.15**	0.44	0.06	**6.84**	0.02	**8.01**	m	**9.61**
Stress grading softwood										
General Structural (GS) grade included in rates										
Special Structural (SS) grade add 10% to materials prices										
Flame proofing treatment to softwood										
For timbers treated with proofing process to Class 1 add 100% to materials prices										
Softwood, sawn - supports										
Butt jointed supports										
width exceeding 300mm; 25 x 50mm at 300mm centres; fixing to masonry	1.01	10.00	**1.11**	1.43	0.18	**22.21**	0.08	**23.39**	m2	**28.0**
Framed supports										
width exceeding 300mm; 50 x 50mm members at 400mm centres one way; 50 x 50mm subsidiary members at 400mm centres one way	3.03	10.00	**3.33**	1.43	0.18	**22.21**	0.08	**25.61**	m2	**30.7**

Labour hourly rates: (except Specialists) Craft Operatives £14.21 Labourer £10.71 Rates are national average prices. Refer to REGIONAL VARIATIONS for indicative levels of overall pricing in regions	MATERIALS			LABOUR				RATES		
	Del to Site £	Waste %	Material Cost £	Craft Optve Hrs	Lab Hrs	Labour Cost £	Sunds £	Nett Rate £	Unit	Gross Rate (+20%) £
G20: CARPENTRY/TIMBER FRAMING/FIRST FIXING Cont.										
Softwood, sawn - supports Cont.										
Framed supports Cont.										
width not exceeding 300mm; 50 x 50mm members longitudinally; 50 x 50mm subsidiary members at 400mm centres laterally	3.03	10.00	**3.33**	0.55	0.07	**8.52**	0.03	**11.88**	m	**14.25**
Individual supports										
12 x 38mm	0.15	10.00	**0.16**	0.11	0.01	**1.68**	0.01	**1.85**	m	**2.22**
25 x 50mm	0.30	10.00	**0.33**	0.13	0.02	**2.11**	0.01	**2.45**	m	**2.94**
50 x 50mm	0.61	10.00	**0.67**	0.15	0.02	**2.42**	0.01	**3.10**	m	**3.72**
50 x 150mm	1.82	10.00	**2.00**	0.22	0.03	**3.48**	0.01	**5.49**	m	**6.58**
12 x 38mm; fixing to masonry	0.15	10.00	**0.16**	0.31	0.04	**4.85**	0.02	**5.03**	m	**6.03**
25 x 50mm; fixing to masonry	0.30	10.00	**0.33**	0.33	0.04	**5.16**	0.02	**5.51**	m	**6.61**
50 x 50mm; fixing to masonry	0.61	10.00	**0.67**	0.37	0.04	**5.79**	0.02	**6.47**	m	**7.76**
Softwood, sawn, impregnated - supports										
Butt jointed supports										
width exceeding 300mm; 50mm wide x 50mm average depth at 450mm centres	3.02	10.00	**3.32**	0.47	0.06	**7.31**	0.03	**10.66**	m2	**12.79**
width exceeding 300mm; 50mm wide x 50mm average depth at 600mm centres	2.26	10.00	**2.49**	0.35	0.04	**5.47**	0.02	**7.98**	m2	**9.57**
Individual supports										
25 x 25mm; 1 labours	0.21	10.00	**0.23**	0.14	0.02	**2.27**	0.01	**2.51**	m	**3.01**
25 x 38mm; 1 labours	0.32	10.00	**0.36**	0.14	0.02	**2.27**	0.01	**2.63**	m	**3.16**
triangular; extreme dimensions										
25 x 25mm	0.21	10.00	**0.23**	0.14	0.02	**2.27**	0.01	**2.51**	m	**3.01**
50 x 75mm	1.28	10.00	**1.40**	0.20	0.02	**3.05**	0.01	**4.46**	m	**5.36**
75 x 100mm	2.55	10.00	**2.81**	0.25	0.03	**3.95**	0.01	**6.77**	m	**8.12**
rounded roll for lead										
50 x 50mm	2.04	10.00	**2.25**	0.29	0.03	**4.42**	0.01	**6.68**	m	**8.01**
Softwood, wrot, impregnated - gutter boarding										
Gutter boards including sides, tongued and grooved joints										
width exceeding 300mm; 25mm thick	11.23	10.00	**12.35**	2.97	0.37	**46.21**	0.16	**58.72**	m2	**70.46**
width not exceeding 300mm; 25mm thick										
150mm wide	1.68	10.00	**1.85**	0.44	0.06	**6.84**	0.02	**8.72**	m	**10.46**
225mm wide	2.53	10.00	**2.78**	0.58	0.08	**9.11**	0.03	**11.92**	m	**14.30**
cesspool 225 x 225 x 150mm deep	2.08	10.00	**2.28**	1.98	0.25	**30.85**	0.11	**33.24**	nr	**39.88**
Lier boards; tongued and grooved joints										
width exceeding 300mm; 25mm thick	11.23	10.00	**12.35**	1.32	0.17	**20.52**	0.07	**32.95**	m2	**39.54**
Valley sole boards; butt joints										
width not exceeding 300mm; 25mm thick										
150mm wide	1.68	10.00	**1.85**	0.66	0.09	**10.32**	0.04	**12.21**	m	**14.65**
Plywood BS EN 636, II/III grade, WBP bonded, butt joints - gutter boarding										
Gutter boards including sides, butt joints										
width exceeding 300mm; 18mm thick	12.33	10.00	**13.56**	1.98	0.25	**30.85**	0.11	**44.51**	m2	**53.42**
width exceeding 300mm; 25mm thick	17.14	10.00	**18.85**	2.20	0.28	**34.21**	0.12	**53.18**	m2	**63.81**
width not exceeding 300mm; 18mm thick										
150mm wide	1.85	10.00	**2.03**	0.26	0.03	**4.10**	0.01	**6.15**	m	**7.38**
width not exceeding 300mm; 25mm thick										
150mm wide	2.57	10.00	**2.83**	0.31	0.04	**4.85**	0.02	**7.69**	m	**9.23**
225mm wide	3.86	10.00	**4.24**	0.33	0.04	**5.16**	0.02	**9.42**	m	**11.30**
cesspool 225 x 225 x 150mm deep	3.17	10.00	**3.49**	1.98	0.25	**30.85**	0.11	**34.44**	nr	**41.33**
Softwood, wrot, impregnated - eaves and verge boarding										
Fascia and barge boards										
width not exceeding 300mm; 25mm thick										
150mm wide	1.68	10.00	**1.85**	0.29	0.03	**4.42**	0.01	**6.28**	m	**7.54**
200mm wide	2.25	10.00	**2.47**	0.30	0.03	**4.57**	0.01	**7.06**	m	**8.47**
Fascia and barge boards, tongued, grooved and veed joints										
width not exceeding 300mm; 25mm thick										
225mm wide	2.53	10.00	**2.78**	0.40	0.06	**6.22**	0.02	**9.02**	m	**10.82**
250mm wide	2.81	10.00	**3.09**	0.42	0.06	**6.53**	0.02	**9.64**	m	**11.57**
Eaves or verge soffit boards										
width exceeding 300mm; 25mm thick	11.23	10.00	**12.35**	1.32	0.17	**20.52**	0.07	**32.95**	m2	**39.54**
width not exceeding 300mm; 25mm thick										
225mm wide	2.53	10.00	**2.78**	0.40	0.06	**6.22**	0.02	**9.02**	m	**10.82**
Plywood BS EN 636, II/III grade, WBP bonded, butt joints - eaves and verge boarding										
Fascia and barge boards										
width not exceeding 300mm; 18mm thick										
150mm wide	1.85	10.00	**2.03**	0.29	0.03	**4.42**	0.01	**6.47**	m	**7.76**
250mm wide	3.08	10.00	**3.39**	0.31	0.04	**4.85**	0.02	**8.26**	m	**9.91**
Eaves or verge soffit boards										
width exceeding 300mm; 18mm thick	12.33	10.00	**13.56**	0.99	0.12	**15.36**	0.05	**28.98**	m2	**34.78**
width not exceeding 300mm; 18mm thick										
225mm wide	2.77	10.00	**3.05**	0.33	0.04	**5.16**	0.02	**8.23**	m	**9.88**

Labour hourly rates: (except Specialists) Craft Operatives £14.21 Labourer £10.71 Rates are national average prices. Refer to REGIONAL VARIATIONS for indicative levels of overall pricing in regions	MATERIALS			LABOUR			RATES			
	Del to Site £	Waste %	Material Cost £	Craft Optve Hrs	Lab Hrs	Labour Cost £	Sunds £	Nett Rate £	Unit	Gross Rate (+20%) £
G20: CARPENTRY/TIMBER FRAMING/FIRST FIXING Cont.										
Fascia and bargeboards; Caradon Celuform Ltd										
Fascia reference 4247; butt joint with joint trims reference 4560; fixing to timber with Polytop screws										
150mm wide	13.47	5.00	**14.14**	0.26	0.03	**4.10**	0.19	**18.44**	m	**22.12**
250mm wide	21.29	5.00	**22.36**	0.26	0.03	**4.10**	0.38	**26.85**	m	**32.22**
Elite solid fascia board reference 4245; butt joint with joint trims reference 4560; fixing to timber with Polytop screws										
150mm wide	13.54	5.00	**14.22**	0.26	0.03	**4.10**	0.19	**18.51**	m	**22.22**
250mm wide	22.48	5.00	**23.60**	0.26	0.03	**4.10**	0.38	**28.09**	m	**33.71**
Supaliner fascia board reference 4270; butt joint with joint trims reference 4586; fixing to timber with Polytop screws										
150mm wide	10.91	5.00	**11.45**	0.39	0.04	**5.94**	0.19	**17.59**	m	**21.10**
250mm wide	17.61	5.00	**18.49**	0.44	0.07	**6.96**	0.38	**25.83**	m	**31.00**
Celuvent multi purpose/soffite board reference 4258; butt joint with panel joint trim reference 4570; fixing to timber with stainless steel nails; fixing soffite board channel reference 4572 to masonry with Polytop screws										
100mm wide	13.23	5.00	**13.89**	0.50	0.08	**7.86**	0.19	**21.94**	m	**26.33**
325mm wide	29.24	5.00	**30.70**	0.77	0.10	**12.00**	0.57	**43.28**	m	**51.93**
Multi purpose/soffite board reference 4558; butt joint with panel joint trim reference 4570; fixing to timber with stainless steel nails; fixing soffite board channel reference 4572 to masonry with Polytop screws										
125mm wide	14.83	5.00	**15.57**	0.50	0.08	**7.86**	0.19	**23.62**	m	**28.34**
200mm wide	18.81	5.00	**19.75**	0.50	0.08	**7.86**	0.38	**27.99**	m	**33.59**
300mm wide	27.38	5.00	**28.75**	0.77	0.10	**12.00**	0.57	**41.32**	m	**49.59**
Steel - tie rods and straps										
Tie rods										
19mm diameter; threaded	6.92	2.50	**7.10**	0.33	0.04	**5.16**	-	**12.26**	m	**14.71**
extra; with nut and washer	0.34	2.50	**0.35**	0.28	0.03	**4.26**	-	**4.61**	nr	**5.53**
Straps										
3 x 38; holes -4; fixing with bolts (bolts included elsewhere)										
500mm long	1.44	2.50	**1.48**	0.18	0.02	**2.74**	-	**4.22**	nr	**5.06**
1000mm long	2.89	2.50	**2.96**	0.33	0.04	**5.16**	-	**8.12**	nr	**9.74**
6 x 50; holes -4; fixing with bolts (bolts included elsewhere)										
500mm long	3.16	2.50	**3.24**	0.21	0.02	**3.21**	-	**6.44**	nr	**7.73**
1000mm long	6.27	2.50	**6.43**	0.40	0.04	**6.10**	-	**12.52**	nr	**15.03**
Steel; galvanized - straps										
Straps and clips; BAT Building Products Ltd.										
30 x 2.5mm standard strapping; fixing with nails										
600mm long	0.97	5.00	**1.02**	0.20	0.02	**3.05**	0.01	**4.08**	nr	**4.89**
1000mm long	1.84	5.00	**1.93**	0.24	0.03	**3.79**	0.01	**5.74**	nr	**6.88**
30 x 2.5mm standard strapping; twists -1; fixing with nails										
600mm long	1.07	5.00	**1.12**	0.20	0.02	**3.05**	0.01	**4.18**	nr	**5.02**
1000mm long	1.93	5.00	**2.02**	0.24	0.03	**3.79**	0.01	**5.83**	nr	**7.00**
1600mm long	2.98	5.00	**3.13**	0.26	0.03	**4.10**	0.01	**7.24**	nr	**8.69**
30 x 5mm M 305 strapping; bends -1; nailing one end (building in included elsewhere)										
600mm long	2.06	5.00	**2.17**	0.11	0.01	**1.68**	0.01	**3.85**	nr	**4.63**
1000mm long	3.36	5.00	**3.53**	0.12	0.01	**1.84**	0.01	**5.37**	nr	**6.45**
30 x 5mm M305 strapping; bends -1; twists -1; nailing one end (building in included elsewhere)										
600mm long	2.14	5.00	**2.25**	0.11	0.01	**1.68**	0.01	**3.94**	nr	**4.73**
1000mm long	3.44	5.00	**3.61**	0.12	0.01	**1.84**	0.01	**5.46**	nr	**6.55**
1600mm long	5.41	5.00	**5.68**	0.13	0.02	**2.11**	0.01	**7.80**	nr	**9.36**
Steel; galvanized - joist hangers										
Joist hangers; BAT Building Products Ltd; (building in where required included elsewhere)										
SPH type S, for 50 x 100mm joist	1.93	2.50	**1.98**	-	-	-	-	**1.98**	nr	**2.37**
SPH type S, for 50 x 125mm joist	1.96	2.50	**2.01**	-	-	-	-	**2.01**	nr	**2.41**
SPH type S, for 50 x 150mm joist	1.96	2.50	**2.01**	-	-	-	-	**2.01**	nr	**2.41**
SPH type S, for 50 x 200mm joist	2.28	2.50	**2.33**	-	-	-	-	**2.33**	nr	**2.80**
SPH type S, for 50 x 225mm joist	2.44	2.50	**2.50**	-	-	-	-	**2.50**	nr	**3.00**
SPH type S, for 63 x 250mm joist	3.26	2.50	**3.34**	-	-	-	-	**3.34**	nr	**4.01**
SPH type R, for 50 x 100mm joist	2.91	2.50	**2.98**	-	-	-	-	**2.98**	nr	**3.58**
SPH type R, for 50 x 125mm joist	2.95	2.50	**3.03**	-	-	-	-	**3.03**	nr	**3.63**
SPH type R, for 63 x 200mm joist	4.00	2.50	**4.10**	-	-	-	-	**4.10**	nr	**4.92**
SPH type R, for 63 x 250mm joist	4.52	2.50	**4.63**	-	-	-	-	**4.63**	nr	**5.56**
SPH type ST for 50 x 100mm joist	5.10	2.50	**5.22**	-	-	-	-	**5.22**	nr	**6.27**
SPH type ST for 50 x 125mm joist	5.24	2.50	**5.37**	-	-	-	-	**5.37**	nr	**6.45**
SPH type ST for 50 x 150mm joist	4.98	2.50	**5.11**	-	-	-	-	**5.11**	nr	**6.13**
SPH type ST for 50 x 200mm joist	5.96	2.50	**6.11**	-	-	-	-	**6.11**	nr	**7.34**
SPH type ST for 50 x 225mm joist	6.53	2.50	**6.69**	-	-	-	-	**6.69**	nr	**8.03**

Labour hourly rates: (except Specialists) Craft Operatives £14.21 Labourer £10.71 Rates are national average prices. Refer to REGIONAL VARIATIONS for indicative levels of overall pricing in regions	MATERIALS			LABOUR				RATES		
	Del to Site £	Waste %	Material Cost £	Craft Optve Hrs	Lab Hrs	Labour Cost £	Sunds £	Nett Rate £	Unit	Gross Rate (+20%) £
G20: CARPENTRY/TIMBER FRAMING/FIRST FIXING Cont.										
Steel; galvanized - joist hangers Cont.										
Speedy Minor type for the following size joists; fixing with nails										
38 x 100mm	0.54	2.50	**0.55**	0.20	0.02	**3.05**	0.01	**3.61**	nr	**4.34**
50 x 100mm	0.54	2.50	**0.55**	0.20	0.02	**3.05**	0.01	**3.61**	nr	**4.34**
Speedy short leg type for the following size joists fixing with nails										
38 x 100mm	0.99	2.50	**1.02**	0.20	0.02	**3.05**	0.01	**4.08**	nr	**4.89**
50 x 175mm	0.99	2.50	**1.02**	0.20	0.02	**3.05**	0.01	**4.08**	nr	**4.89**
Speedy Standard Leg type for the following size joists; fixing with nails										
50 x 175mm	1.08	2.50	**1.11**	0.20	0.02	**3.05**	0.01	**4.17**	nr	**5.00**
75 x 225mm	1.13	2.50	**1.16**	0.22	0.03	**3.48**	0.01	**4.65**	nr	**5.58**
Steel; galvanized - joist struts										
Herringbone joist struts; BAT Building Products Ltd. to suit joists at the following centres; fixing with nails										
400mm	0.38	5.00	**0.40**	0.13	0.02	**2.11**	0.01	**2.52**	m	**3.03**
Steel; galvanized - truss plates and framing anchors										
Framing anchors; BAT Building Products Ltd.										
type A; fixing with nails	0.51	5.00	**0.53**	0.28	0.03	**4.26**	0.01	**4.81**	nr	**5.77**
type B; fixing with nails	0.51	5.00	**0.53**	0.28	0.03	**4.26**	0.01	**4.81**	nr	**5.77**
type C; fixing with nails	0.51	5.00	**0.53**	0.28	0.03	**4.26**	0.01	**4.81**	nr	**5.77**
Steel - timber connectors										
Connectors BS EN 912 double sided round toothed-plate connector, table 4										
51mm diameter	0.35	5.00	**0.37**	0.11	0.01	**1.68**	-	**2.05**	nr	**2.46**
76mm diameter	0.63	5.00	**0.66**	0.11	0.01	**1.68**	-	**2.34**	nr	**2.81**
Steel - bolts and nuts										
Black bolts, BS 4190 grade 4.6, hexagon head										
M 12; with nuts and washers										
100mm long	0.60	5.00	**0.63**	0.12	0.01	**1.84**	-	**2.47**	nr	**2.96**
160mm long	1.58	5.00	**1.65**	0.14	0.02	**2.27**	-	**3.92**	nr	**4.71**
M 16; with nuts and washers										
100mm long	0.83	5.00	**0.87**	0.12	0.01	**1.84**	-	**2.71**	nr	**3.25**
160mm long	2.10	5.00	**2.20**	0.15	0.02	**2.42**	-	**4.63**	nr	**5.55**
M 20; with nuts and washers										
100mm long	1.82	5.00	**1.91**	0.13	0.02	**2.11**	-	**4.02**	nr	**4.83**
160mm long	3.13	5.00	**3.28**	0.19	0.02	**2.89**	-	**6.18**	nr	**7.41**
M 12; with nuts and 38 x 38 x 3mm square plate washers										
100mm long	0.61	5.00	**0.64**	0.12	0.01	**1.84**	-	**2.48**	nr	**2.97**
160mm long	1.56	5.00	**1.64**	0.14	0.02	**2.27**	-	**3.91**	nr	**4.69**
Steel; galvanized - bolts and nuts										
Black bolts, BS 4190 grade 4.6, hexagon head										
M 12; with nuts and washers										
100mm long	0.92	5.00	**0.97**	0.12	0.01	**1.84**	-	**2.81**	nr	**3.37**
160mm long	2.38	5.00	**2.50**	0.14	0.02	**2.27**	-	**4.77**	nr	**5.73**
M 16; with nuts and washers										
100mm long	1.54	5.00	**1.62**	0.12	0.01	**1.84**	-	**3.45**	nr	**4.14**
160mm long	3.21	5.00	**3.37**	0.15	0.02	**2.42**	-	**5.79**	nr	**6.95**
M 20; with nuts and washers										
100mm long	2.82	5.00	**2.96**	0.13	0.02	**2.11**	-	**5.07**	nr	**6.09**
160mm long	4.79	5.00	**5.03**	0.19	0.02	**2.89**	-	**7.93**	nr	**9.51**
M 12; with nuts and 50 x 50 x 3mm square plate washers										
100mm long	1.08	5.00	**1.13**	0.12	0.01	**1.84**	-	**2.97**	nr	**3.56**
160mm long	2.54	5.00	**2.67**	0.14	0.02	**2.27**	-	**4.93**	nr	**5.92**
Steel - expanding bolts										
Expanding bolts; bolt projecting Rawlbolts; drilling masonry with nuts and washers; reference										
44 - 505 (M6 10P)	1.03	5.00	**1.08**	0.26	0.03	**4.10**	-	**5.19**	nr	**6.23**
44 - 555 (M8 10P)	1.30	5.00	**1.36**	0.29	0.03	**4.42**	-	**5.78**	nr	**6.94**
44 - 660 (M12 30P)	3.05	5.00	**3.20**	0.35	0.04	**5.47**	-	**8.67**	nr	**10.41**
Expanding bolts; loose bolt Rawlbolts; drilling masonry with washers; reference										
44 - 015 (M6 10L)	1.03	5.00	**1.09**	0.26	0.03	**4.10**	-	**5.19**	nr	**6.23**
44 - 060 (M8 25L)	1.34	5.00	**1.41**	0.29	0.03	**4.42**	-	**5.83**	nr	**6.99**
44 - 160 (M12 25L)	2.84	5.00	**2.98**	0.35	0.04	**5.47**	-	**8.45**	nr	**10.14**

Labour hourly rates: (except Specialists) Craft Operatives £14.21 Labourer £10.71 Rates are national average prices. Refer to REGIONAL VARIATIONS for indicative levels of overall pricing in regions	MATERIALS			LABOUR				RATES		
	Del to Site £	Waste %	Material Cost £	Craft Optve Hrs	Lab Hrs	Labour Cost £	Sunds £	Nett Rate £	Unit	Gross Rate (+20%) £
G20: CARPENTRY/TIMBER FRAMING/FIRST FIXING Cont.										
Chemical anchors										
Chemical anchors; Kemfix capsules and standard studs; drilling masonry										
capsule reference 60-428; stud reference 60-649; with nuts and washers (M8 x 110mm)..............	2.10	5.00	**2.21**	0.32	0.04	**5.00**	-	**7.21**	nr	**8.65**
capsule reference 60-432; stud reference 60-661; with nuts and washers (M12 x 160mm).............	2.89	5.00	**3.03**	0.40	0.06	**6.22**	-	**9.25**	nr	**11.10**

Labour hourly rates: (except Specialists) Craft Operatives £14.21 Labourer £10.71 Rates are national average prices. Refer to REGIONAL VARIATIONS for indicative levels of overall pricing in regions	MATERIALS			LABOUR				RATES		
	Del to Site £	Waste %	Material Cost £	Craft Optve Hrs	Lab Hrs	Labour Cost £	Sunds £	Nett Rate £	Unit	Gross Rate (+20%) £
H21: TIMBER WEATHERBOARDING										
Softwood, wrought, impregnated										
Boarding to walls, shiplapped joints; 25mm thick, 150mm wide boards										
width exceeding 300mm; fixing to timber..........	11.62	10.00	**12.78**	0.99	0.12	**15.36**	0.17	**28.31**	m2	**33.97**
Abutments										
25mm thick softwood boarding										
raking cutting	1.16	10.00	**1.28**	0.09	-	**1.25**	-	**2.53**	m	**3.03**
curved cutting	1.74	10.00	**1.92**	0.13	-	**1.88**	-	**3.79**	m	**4.55**
H30: FIBRE CEMENT PROFILED SHEET CLADDING/COVERING/SIDING										
Roof coverings; corrugated reinforced cement Profile 3 sheeting, standard grey colour; lapped one and a half corrugations at sides and 150mm at ends										
Coverings; fixing to timber joists at 900mm general spacing with galvanized mild steel drive screws and washers; drilling holes										
pitch 30 degrees from horizontal................	13.51	15.00	**15.53**	0.31	0.31	**7.68**	0.53	**23.74**	m2	**28.49**
Coverings; fixing to steel purlins at 900mm general spacing with galvanized hook bolts and washers; drilling holes										
pitch 30 degrees from horizontal................	13.51	15.00	**15.53**	0.42	0.42	**10.42**	1.08	**27.03**	m2	**32.43**
Eaves										
eaves filler pieces	11.58	10.00	**12.74**	0.13	0.13	**3.29**	-	**16.03**	m	**19.24**
Ridges										
two piece plain angular adjustable ridge tiles...	16.75	10.00	**18.43**	0.25	0.25	**6.30**	-	**24.73**	m	**29.68**
Barge boards										
standard barge boards	9.49	10.00	**10.44**	0.13	0.13	**3.29**	-	**13.73**	m	**16.48**
Aprons/sills										
apron flashings	11.64	10.00	**12.80**	0.17	0.17	**4.11**	-	**16.91**	m	**20.30**
Finials										
standard ridge cap finials.......................	10.48	10.00	**11.53**	0.19	0.19	**4.66**	-	**16.19**	nr	**19.43**
Cutting										
raking	2.70	15.00	**3.11**	0.31	0.31	**7.68**	-	**10.78**	m	**12.94**
Holes										
for pipes, standards or the like................	-	-	**-**	0.64	-	**9.07**	-	**9.07**	nr	**10.88**
Roof coverings; corrugated reinforced cement Profile 3 sheeting, standard coloured; lapped one and a half corrugations at sides and 150mm at ends										
Coverings; fixing to timber joists at 900mm general spacing with galvanized mild steel drive screws and washers; drilling holes										
pitch 30 degrees from horizontal................	16.21	15.00	**18.65**	0.31	0.31	**7.68**	0.70	**27.03**	m2	**32.43**
Coverings; fixing to steel purlins at 900mm general spacing with galvanized hook bolts and washers; drilling holes										
pitch 30 degrees from horizontal................	16.21	15.00	**18.65**	0.42	0.42	**10.42**	1.39	**30.45**	m2	**36.54**
Eaves										
eaves filler pieces	14.49	10.00	**15.94**	0.13	0.13	**3.29**	-	**19.23**	m	**23.07**
Ridges										
two piece plain angular adjustable ridge tiles...	20.21	10.00	**22.23**	0.25	0.25	**6.30**	-	**28.53**	m	**34.24**

Labour hourly rates: (except Specialists) Craft Operatives £14.21 Labourer £10.71 Rates are national average prices. Refer to REGIONAL VARIATIONS for indicative levels of overall pricing in regions	MATERIALS			LABOUR				RATES		
	Del to Site £	Waste %	Material Cost £	Craft Optve Hrs	Lab Hrs	Labour Cost £	Sunds £	Nett Rate £	Unit	Gross Rate (+20%) £
H30: FIBRE CEMENT PROFILED SHEET CLADDING/COVERING/SIDING Cont.										
Roof coverings; corrugated reinforced cement Profile 3 sheeting, standard coloured; lapped one and a half corrugations at sides and 150mm at ends Cont.										
Barge boards										
standard barge boards	11.87	10.00	**13.06**	0.13	0.13	**3.29**	-	**16.35**	m	**19.61**
Aprons/sills										
apron flashings	14.53	10.00	**15.98**	0.17	0.17	**4.11**	-	**20.10**	m	**24.12**
Finials										
standard ridge cap finials	11.91	10.00	**13.10**	0.19	0.19	**4.66**	-	**17.76**	nr	**21.32**
Cutting										
raking	3.24	15.00	**3.73**	0.31	0.31	**7.68**	-	**11.40**	m	**13.69**
Holes										
for pipes, standards or the like	-	-	**-**	0.64	-	**9.07**	-	**9.07**	nr	**10.88**
Roof coverings; corrugated reinforced cement Profile 6 sheeting, standard coloured; lapped half a corrugation at sides and 150mm at ends										
Coverings; fixing to timber joists at 900mm general spacing with galvanized mild steel drive screws and washers; drilling holes										
pitch 30 degrees from horizontal	12.90	15.00	**14.84**	0.31	0.31	**7.68**	0.69	**23.21**	m2	**27.85**
Coverings; fixing to steel purlins at 900mm general spacing with galvanized hook bolts and washers; drilling holes										
pitch 30 degrees from horizontal	12.90	15.00	**14.84**	0.42	0.42	**10.42**	1.39	**26.64**	m2	**31.97**
Eaves										
eaves filler pieces	9.60	10.00	**10.56**	0.13	0.13	**3.29**	-	**13.85**	m	**16.62**
Ridges										
two piece plain angular adjustable ridge tiles	17.67	10.00	**19.43**	0.25	0.25	**6.30**	-	**25.74**	m	**30.88**
Barge boards										
standard barge boards	11.87	10.00	**13.06**	0.13	0.13	**3.29**	-	**16.35**	m	**19.61**
Aprons/sills										
apron flashings	10.36	10.00	**11.40**	0.17	0.17	**4.11**	-	**15.51**	m	**18.61**
Finials										
standard ridge cap finials	11.69	10.00	**12.86**	0.19	0.19	**4.66**	-	**17.52**	nr	**21.03**
Cutting										
raking	2.58	15.00	**2.97**	0.31	0.31	**7.68**	-	**10.64**	m	**12.77**
Holes										
for pipes, standards or the like	-	-	**-**	0.64	-	**9.07**	-	**9.07**	nr	**10.88**
H31: METAL PROFILED/FLAT SHEET CLADDING/COVERING/SIDING										
Wall cladding; PVC colour coated both sides galvanized steel profiled sheeting 0.70mm thick and with sheets secured at seams and laps										
Coverings; fixing to steel rails at 900mm general spacing with galvanized hook bolts and washers; drilling holes										
pitch 90 degrees from horizontal	18.22	5.00	**19.13**	0.44	0.44	**10.96**	1.30	**31.39**	m2	**37.67**
Vertical angles										
vertical corner flashings	6.81	5.00	**7.15**	0.19	0.19	**4.66**	-	**11.81**	m	**14.17**
Filler blocks										
polyethylene	0.77	5.00	**0.81**	0.13	0.13	**3.29**	-	**4.10**	m	**4.92**
PVC	2.17	5.00	**2.28**	0.13	0.13	**3.29**	-	**5.56**	m	**6.68**
black synthetic rubber	2.41	5.00	**2.53**	0.13	0.13	**3.29**	-	**5.82**	m	**6.98**
Cutting										
raking	3.64	5.00	**3.83**	0.31	0.31	**7.68**	-	**11.50**	m	**13.80**
Roof coverings; PVC colour coated both sides galvanized steel profiled sheeting 0.70mm thick and with sheets secured at seams and laps										
Coverings; fixing to steel purlins at 900mm general spacing with galvanized hook bolts and washers; drilling holes										
pitch 30 degrees from horizontal	18.22	5.00	**19.13**	0.39	0.39	**9.59**	1.34	**30.06**	m2	**36.08**
Ridges										
ridge cappings	6.81	5.00	**7.15**	0.19	0.19	**4.66**	-	**11.81**	m	**14.17**
Flashings										
gable flashings	6.81	5.00	**7.15**	0.19	0.19	**4.66**	-	**11.81**	m	**14.17**
eaves flashings	4.53	5.00	**4.76**	0.19	0.19	**4.66**	-	**9.42**	m	**11.30**

Labour hourly rates: (except Specialists) Craft Operatives £14.21 Labourer £10.71 Rates are national average prices. Refer to REGIONAL VARIATIONS for indicative levels of overall pricing in regions	MATERIALS			LABOUR				RATES		
	Del to Site £	Waste %	Material Cost £	Craft Optve Hrs	Lab Hrs	Labour Cost £	Sunds £	Nett Rate £	Unit	Gross Rate (+20%) £
H31: METAL PROFILED/FLAT SHEET CLADDING/COVERING/SIDING Cont.										
Roof coverings; PVC colour coated both sides galvanized steel profiled sheeting 0.70mm thick and with sheets secured at seams and laps Cont.										
Filler blocks										
polyethylene	0.77	5.00	**0.81**	0.13	0.13	**3.29**	-	**4.10**	m	**4.92**
PVC	2.17	5.00	**2.28**	0.13	0.13	**3.29**	-	**5.56**	m	**6.68**
black synthetic rubber	2.41	5.00	**2.53**	0.13	0.13	**3.29**	-	**5.82**	m	**6.98**
Cutting										
raking	3.64	5.00	**3.83**	0.31	0.31	**7.68**	-	**11.50**	m	**13.80**
H60: PLAIN ROOF TILING										
Quantities required for 1m2 of tiling										
Plain tiles to 65mm lap										
tiles, 60nr										
battens, 10m										
Plain tiles to 85mm lap										
tiles, 70nr										
battens, 11m										
Roof coverings; clayware Machine made plain tiles BS EN 1304, red, 265 x 165mm; fixing every fourth course with two galvanized nails per tile to 65mm lap; 38 x 19mm pressure impregnated softwood battens fixed with galvanized nails; underlay BS 747 type 1F reinforced bitumen felt; 150mm laps; fixing with galvanized steel clout nails										
Pitched 50 degrees from horizontal										
generally	37.96	5.00	**40.07**	0.79	0.40	**15.50**	1.72	**57.28**	m2	**68.74**
holes	-	-	-	0.64	0.32	**12.48**	-	**12.48**	nr	**14.98**
abutments; square	3.43	5.00	**3.60**	0.31	0.15	**6.03**	-	**9.63**	m	**11.55**
abutments; raking	4.86	5.00	**5.10**	0.46	0.23	**9.04**	-	**14.14**	m	**16.97**
double course at eaves; Machine made eaves tile	3.50	5.00	**3.67**	0.39	0.20	**7.59**	0.61	**11.87**	m	**14.24**
verges; bed and point in coloured cement-lime mortar (1:1:6)	3.43	5.00	**3.60**	0.44	0.22	**8.61**	0.46	**12.67**	m	**15.21**
verges; single extra undercloak course plain tiles; bed and point in coloured cement-lime mortar (1:1:6)	4.57	5.00	**4.80**	0.64	0.32	**12.48**	0.46	**17.75**	m	**21.30**
Machine made ridge tiles, half round; in 300mm effective lengths; butt jointed; bedding and pointing in coloured cement-lime mortar (1:1:6)	20.22	5.00	**21.23**	0.39	0.20	**7.59**	0.46	**29.28**	m	**35.14**
Machine made hip tiles, half round; in 300mm effective lengths; butt jointed; bedding and pointing in coloured cement-lime mortar (1:1:6)	20.22	5.00	**21.23**	0.39	0.20	**7.59**	0.46	**29.28**	m	**35.14**
Machine made hip tiles; angular	20.22	5.00	**21.23**	0.39	0.20	**7.59**	0.46	**29.28**	m	**35.14**
Machine made hip tiles; bonnet pattern; bedding and pointing in coloured cement-lime mortar (1:1:6)	20.22	5.00	**21.23**	0.39	0.20	**7.59**	0.46	**29.28**	m	**35.14**
Machine made valley tiles; angular	20.22	5.00	**21.23**	0.55	0.28	**10.76**	0.46	**32.45**	m	**38.94**
Wall coverings; clayware Machine made plain tiles BS EN 1304, red, 265 x 165mm;fixing every course with two galvanized nails per tile to 38mm lap; 38 x 19mm pressure impregnated softwood battens fixed with galvanized nails; underlay BS 747 type 1F reinforced bitumen felt; 150mm laps; fixing with galvanized steel clout nails										
Vertical										
generally	33.79	5.00	**35.69**	0.94	0.47	**18.35**	2.51	**56.55**	m2	**67.86**
Fittings										
Hip irons										
galvanized mild steel; 32 x 3 x 380mm girth; scrolled; fixing with galvanized steel screws to timber	2.28	5.00	**2.39**	0.13	0.07	**2.58**	-	**4.97**	nr	**5.97**
Lead soakers										
fixing only	-	-	-	0.39	0.20	**7.59**	-	**7.59**	nr	**9.11**
Roof coverings; Redland Plain granular faced tiles, 268 x 165mm; fixing every fifth course with two aluminium nails to each tile to 65mm lap; 38 x 19mm pressure impregnated softwood battens fixed with galvanized nails; underlay BS 747 type 1F bitumen reinforced felt; 150mm laps; fixing with galvanized steel clout nails										
Pitched 40 degrees from horizontal										
generally	25.63	5.00	**27.10**	0.77	0.39	**15.07**	1.27	**43.43**	m2	**52.12**
double course at eaves and nailing each tile with two aluminium nails	2.23	5.00	**2.34**	0.39	0.20	**7.59**	0.56	**10.49**	m	**12.59**
verges; Redland Dry Verge system with plain tiles and tile-and-a-half tiles in alternate courses, clips and aluminium nails	12.93	5.00	**13.58**	0.31	0.15	**6.03**	1.19	**20.79**	m	**24.95**

Labour hourly rates: (except Specialists) Craft Operatives £14.21 Labourer £10.71 Rates are national average prices. Refer to REGIONAL VARIATIONS for indicative levels of overall pricing in regions	MATERIALS			LABOUR				RATES		
	Del to Site £	Waste %	Material Cost £	Craft Optve Hrs	Lab Hrs	Labour Cost £	Sunds £	Nett Rate £	Unit	Gross Rate (+20%) £
H60: PLAIN ROOF TILING Cont.										
Roof coverings; Redland Plain granular faced tiles, 268 x 165mm; fixing every fifth course with two aluminium nails to each tile to 65mm lap; 38 x 19mm pressure impregnated softwood battens fixed with galvanized nails; underlay BS 747 type 1F bitumen reinforced felt; 150mm laps; fixing with galvanized steel clout nails Cont.										
Pitched 40 degrees from horizontal Cont.										
verges; plain tile undercloak; tiles and undercloak bedded and pointed in tinted cement mortar (1:3) and with plain tiles and tile-and-a-half tiles in alternate courses and aluminium nails	1.11	5.00	**1.17**	0.31	0.15	**6.03**	0.50	**7.69**	m	**9.23**
double course at top edges with clips and nailing each tile with two aluminium nails	2.25	5.00	**2.36**	0.37	0.19	**7.32**	4.19	**13.87**	m	**16.65**
ridge or hip tiles, Redland Plain angle; butt jointed; bedding and pointing in tinted cement mortar (1:3)	19.85	5.00	**20.85**	0.39	0.20	**7.59**	0.46	**28.90**	m	**34.68**
ridge or hip tiles, Redland half round; butt jointed; bedding and pointing in tinted cement mortar (1:3)	8.26	5.00	**8.67**	0.39	0.20	**7.59**	0.46	**16.73**	m	**20.07**
valley tiles , Redland Plain valley tiles	10.56	5.00	**11.09**	0.39	0.20	**7.59**	0.46	**19.14**	m	**22.97**
Fittings										
Ventilator tiles										
Red Vent ridge ventilation terminal 450mm long for half round ridge	55.94	5.00	**58.73**	1.60	0.80	**31.27**	0.73	**90.72**	nr	**108.87**
Redvent eaves ventilators fixing with aluminium nails	10.10	5.00	**10.60**	0.62	0.31	**12.05**	0.03	**22.69**	m	**27.23**
Red Vent Thruvent tiles for Redland plain tiles	49.62	5.00	**52.10**	0.62	0.31	**12.05**	0.28	**64.43**	nr	**77.31**
Hip irons										
galvanized mild steel; 32 x 3 x 380mm girth; scrolled; fixing with galvanized steel screws to timber	2.28	5.00	**2.39**	0.12	0.07	**2.43**	-	**4.82**	nr	**5.78**
Lead soakers										
fixing only	-	-	-	0.37	0.19	**7.32**	-	**7.32**	nr	**8.78**
H61: FIBRE CEMENT SLATING										
Quantities required for 1m2 of slating										
Slates to 70mm lap										
400 x 240mm, 25.3nr										
Slates to 102mm lap										
600 x 300mm, 13nr										
500 x 250mm, 19.5nr										
400 x 200mm, 32nr										
Roof coverings; asbestos-free cement slates, 600 x 300mm; centre fixing with copper nails and copper disc rivets to 102mm lap; 38 x 19mm pressure impregnated softwood battens fixed with galvanized nails; underlay; BS 747 type 1F bitumen reinforced felt; 150mm laps; fixing with galvanized steel clout nails										
Pitched 30 degrees from horizontal										
generally	17.65	5.00	**18.65**	0.66	0.33	**12.91**	1.76	**33.32**	m2	**39.98**
holes	-	-	-	0.64	0.32	**12.48**	-	**12.48**	nr	**14.98**
abutments; square	1.58	5.00	**1.66**	0.25	0.13	**5.01**	-	**6.67**	m	**8.01**
abutments; raking	2.37	5.00	**2.49**	0.31	0.15	**6.03**	-	**8.52**	m	**10.22**
abutments; curved to 3000mm radius	3.17	5.00	**3.32**	0.39	0.20	**7.59**	-	**10.92**	m	**13.10**
double course at eaves	4.26	5.00	**4.48**	0.19	0.10	**3.72**	0.80	**9.00**	m	**10.79**
verges; slate undercloak and point in cement mortar (1:3)	2.44	5.00	**2.56**	0.10	0.06	**2.00**	0.66	**5.21**	m	**6.26**
ridges or hips; asbestos free cement; fixing with nails	29.32	5.00	**30.78**	0.25	0.13	**5.01**	0.46	**36.25**	m	**43.50**
Roof coverings; asbestos-free cement slates, 500 x 250mm; centre fixing with copper nails and copper disc rivets to 102mm lap; 50 x 25mm pressure impregnated softwood battens fixed with galvanized nails; underlay; BS 747 type 1F bitumen reinforced felt; 150mm laps; fixing with galvanized steel clout nails										
Pitched 30 degrees from horizontal										
generally	22.58	5.00	**23.91**	0.79	0.40	**15.50**	2.40	**41.80**	m2	**50.1**
Fittings										
Hip irons										
galvanized mild steel; 32 x 3 x 380mm girth; scrolled; fixing with galvanized steel screws to timber	2.28	5.00	**2.39**	0.12	0.07	**2.43**	-	**4.82**	nr	**5.7**
Lead soakers										
fixing only	-	-	-	0.37	0.19	**7.32**	-	**7.32**	nr	**8.7**

Labour hourly rates: (except Specialists) Craft Operatives £14.21 Labourer £10.71 Rates are national average prices. Refer to REGIONAL VARIATIONS for indicative levels of overall pricing in regions	MATERIALS			LABOUR				RATES		
	Del to Site £	Waste %	Material Cost £	Craft Optve Hrs	Lab Hrs	Labour Cost £	Sunds £	Nett Rate £	Unit	Gross Rate (+20%) £
H62: NATURAL SLATING										
Quantities required for 1m2 of slating										
Slates to 75mm lap										
600 x 300mm, 12.65nr										
500 x 250mm, 18.9nr										
400 x 200mm, 31nr										
Roof coverings; blue/grey slates, 600 x 300mm, 6.5mm thick; fixing with slate nails to 75mm lap; 38 x 19mm pressure impregnated softwood battens fixed with galvanized nails; underlay; BS 747 type 1F bitumen reinforced felt; 150mm laps; fixing with galvanized steel clout nails										
Pitched 30 degrees from horizontal										
generally	82.03	5.00	**86.26**	0.64	0.32	**12.48**	0.62	**99.36**	m2	**119.23**
holes	-	-	-	0.64	0.32	**12.48**		**12.48**	nr	**14.98**
abutments; square	7.80	5.00	**8.19**	0.19	0.10	**3.72**	-	**11.91**	m	**14.29**
abutments; raking	11.73	5.00	**12.32**	0.29	0.14	**5.60**	-	**17.91**	m	**21.50**
double course at eaves	22.20	5.00	**23.31**	0.18	0.09	**3.44**	0.55	**27.30**	m	**32.76**
verges; slate undercloak and point in cement mortar (1:3)	12.68	5.00	**13.32**	0.10	0.06	**2.00**	0.51	**15.82**	m	**18.98**
Fittings										
Hip irons										
galvanized mild steel; 32 x 3 x 380mm girth; scrolled; fixing with galvanized steel screws to timber	2.28	5.00	**2.39**	0.12	0.07	**2.43**	-	**4.82**	nr	**5.78**
Lead soakers										
fixing only	-	-	-	0.37	0.19	**7.32**	-	**7.32**	nr	**8.78**
H65: SINGLE LAP ROOF TILING										
Roof coverings; Redland Stonewold through coloured slates, 430 x 380mm; fixing every fourth course with galvanized nails to 75mm lap; 38 x 19mm pressure impregnated softwood battens fixed with galvanized nails; underlay; BS 747 type 1F reinforced bitumen felt; 150mm laps; fixing with galvanized steel clout nails										
Pitched 40 degrees from horizontal										
generally	15.51	5.00	**16.39**	0.53	0.26	**10.33**	0.89	**27.61**	m2	**33.13**
abutments; square	1.40	5.00	**1.47**	0.13	0.07	**2.58**	-	**4.05**	m	**4.86**
abutments; raking	2.10	5.00	**2.20**	0.19	0.10	**3.72**	-	**5.92**	m	**7.11**
abutments; curved to 3000mm radius	2.80	5.00	**2.94**	0.25	0.13	**5.01**	-	**7.95**	m	**9.54**
supplementary fixing eaves course with one clip per slate	0.40	5.00	**0.42**	0.03	0.02	**0.70**	-	**1.12**	m	**1.34**
verges; Redland Dry Verge system with half slates, full slates and clips	16.94	5.00	**17.78**	0.31	0.15	**6.03**	2.11	**25.92**	m	**31.10**
verges; Redland Dry Verge system with half slates, verge slates and clips	17.45	5.00	**18.32**	0.31	0.15	**6.03**	2.11	**26.46**	m	**31.75**
ridge tiles, Redland universal angle; butt jointed; bedding and pointing in tinted cement mortar (1:3)	9.29	5.00	**9.75**	0.25	0.13	**5.01**	0.46	**15.22**	m	**18.26**
ridge tiles, Redland Dry Ridge system with half round ridge tiles; fixing with stainless steel batten straps and ring shanked fixing nails with neoprene washers and sleeves; polypropylene ridge seals and uPVC profile filler units	30.68	5.00	**32.21**	0.50	0.25	**9.74**	-	**41.96**	m	**50.35**
ridge tiles, Redland Dry Vent Ridge system with half round ridge tiles; fixing with stainless steel batten straps and ring shanked nails with neoprene washers and sleeves; polypropylene ridge seals, PVC air flow control units and uPVC ventilated profile filler units	31.87	5.00	**33.46**	0.50	0.25	**9.74**	-	**43.20**	m	**51.84**
valley tiles, Redland Universal valley troughs; laid with 100mm laps	30.39	5.00	**31.91**	0.62	0.31	**12.05**	0.46	**44.43**	m	**53.31**
Roof coverings; Redland Grovebury granular faced double pantiles, 418 x 332mm; fixing every fourth course with galvanized nails to 75mm lap; 38 x 22mm pressure impregnated softwood battens fixed with galvanized nails; underlay; BS 747 type 1F bitumen reinforced felt; 150mm laps; fixing with galvanized steel clout nails										
Pitched 40 degrees from horizontal										
generally	10.37	5.00	**11.00**	0.52	0.26	**10.17**	1.00	**22.17**	m2	**26.61**
reform eaves filler unit and eaves clip to each eaves tile	0.40	5.00	**0.42**	0.08	0.04	**1.57**	0.78	**2.76**	m	**3.31**
verges; cloaked verge tiles and aluminium nails	10.96	5.00	**11.50**	0.19	0.10	**3.72**	0.08	**15.30**	m	**18.36**
verges; half tiles, cloaked verge tiles and aluminium nails	12.75	5.00	**13.39**	0.25	0.13	**5.01**	0.12	**18.52**	m	**22.22**
verges; 150 x 6mm fibre cement undercloak; tiles and undercloak bedded and pointed in tinted cement mortar (1:3) and with standard tiles and clips	5.47	5.00	**5.75**	0.25	0.13	**5.01**	2.50	**13.25**	m	**15.90**
verges; 150 x 6mm fibre cement undercloak; tiles and undercloak bedded and pointed in tinted cement mortar (1:3) and with treble roll verge tiles and clips	5.47	5.00	**5.75**	0.25	0.13	**5.01**	2.50	**13.25**	m	**15.90**

Labour hourly rates: (except Specialists) Craft Operatives £14.21 Labourer £10.71 Rates are national average prices. Refer to REGIONAL VARIATIONS for indicative levels of overall pricing in regions	MATERIALS			LABOUR				RATES		
	Del to Site £	Waste %	Material Cost £	Craft Optve Hrs	Lab Hrs	Labour Cost £	Sunds £	Nett Rate £	Unit	Gross Rate (+20%) £
H65: SINGLE LAP ROOF TILING Cont.										
Roof coverings; Redland Grovebury granular faced double pantiles, 418 x 332mm; fixing every fourth course with galvanized nails to 75mm lap; 38 x 22mm pressure impregnated softwood battens fixed with galvanized nails; underlay; BS 747 type 1F bitumen reinforced felt; 150mm laps; fixing with galvanized steel clout nails Cont.										
Pitched 40 degrees from horizontal Cont.										
ridge or hip tiles, Redland third round; butt jointed; bedding and pointing in tinted cement mortar (1:3)	8.18	5.00	**8.59**	0.37	0.19	**7.32**	2.50	**18.40**	m	**22.08**
ridge or hip tiles, Redland half round; butt jointed; bedding and pointing in tinted cement mortar (1:3)	8.18	5.00	**8.59**	0.37	0.19	**7.32**	2.50	**18.40**	m	**22.08**
valley tiles, Redland Universal valley troughs; laid with 100mm laps	30.39	5.00	**31.91**	0.37	0.19	**7.32**	0.46	**39.69**	m	**47.63**
Roof coverings; Redland Norfolk through coloured pantiles, 381 x 227mm; fixing every fourth course with galvanized nails to 75mm lap; 38 x 25mm pressure impregnated softwood battens fixed with galvanized nails; underlay; BS 747 type 1F bitumen reinforced felt; 150mm laps; fixing with galvanized steel clout nails										
Pitched 40 degrees from horizontal										
generally	12.34	5.00	**13.09**	0.67	0.34	**13.19**	1.45	**27.73**	m2	**33.28**
abutments; square	1.03	5.00	**1.08**	0.12	0.07	**2.43**	-	**3.51**	m	**4.21**
abutments; raking	1.54	5.00	**1.62**	0.19	0.10	**3.72**	-	**5.34**	m	**6.41**
abutments; curved to 3000mm radius	2.06	5.00	**2.17**	0.25	0.13	**5.01**	-	**7.17**	m	**8.61**
reform eaves filler unit and aluminium nails to each eaves tile	0.40	5.00	**0.42**	0.08	0.04	**1.57**	0.78	**2.76**	m	**3.31**
verges; plain tile undercloak; tiles and undercloak bedded and pointed in tinted cement mortar (1:3) and with standard tiles and aluminium nails	1.90	5.00	**1.99**	0.12	0.07	**2.43**	0.50	**4.91**	m	**5.90**
verges; plain tile undercloak; tiles and undercloak bedded and pointed in tinted cement mortar (1:3) and with standard tiles and clips	1.90	5.00	**1.99**	0.12	0.07	**2.43**	3.80	**8.21**	m	**9.86**
verges; 150 x 6mm fibre cement undercloak; tiles and undercloak bedded and pointed in tinted cement mortar (1:3) and with standard tiles and aluminium nails	4.68	5.00	**4.91**	0.19	0.10	**3.72**	0.53	**9.16**	m	**10.99**
verges; 150 x 6mm fibre cement undercloak; tiles and undercloak bedded and pointed in tinted cement mortar (1:3) and with standard tiles and clips	4.68	5.00	**4.91**	0.19	0.10	**3.72**	3.80	**12.43**	m	**14.91**
ridge or hip tiles, Redland third round; butt jointed; bedding and pointing in tinted cement mortar (1:3)	8.55	5.00	**8.98**	0.37	0.19	**7.32**	0.46	**16.76**	m	**20.11**
ridge or hip tiles, Redland half round; butt jointed; bedding and pointing in tinted cement mortar (1:3)	8.55	5.00	**8.98**	0.37	0.19	**7.32**	0.46	**16.76**	m	**20.11**
valley tiles, Redland Universal valley troughs; laid with 100mm laps	30.39	5.00	**31.91**	0.37	0.19	**7.32**	0.46	**39.69**	m	**47.63**
Fittings										
Ventilator tiles										
Red Vent ridge ventilation terminal 450mm long for half round ridge	55.94	5.00	**58.73**	1.60	0.80	**31.27**	0.73	**90.72**	nr	**108.87**
Red Vent Thruvent tiles for										
Stonewold slates	42.69	5.00	**44.83**	0.62	0.31	**12.05**	0.28	**57.15**	nr	**68.58**
Grovebury double pantiles	42.69	5.00	**44.83**	0.62	0.31	**12.05**	0.28	**57.15**	nr	**68.58**
Norfolk pantiles	42.69	5.00	**44.83**	0.62	0.31	**12.05**	0.28	**57.15**	nr	**68.58**
Red line ventilation tile complete with underlay seal and fixing clips for										
Stonewold slates	52.90	5.00	**55.54**	1.23	0.62	**24.10**	0.28	**79.92**	nr	**95.91**
Grovebury double pantiles	52.90	5.00	**55.54**	1.23	0.62	**24.10**	0.28	**79.92**	nr	**95.91**
Gas terminals										
Gas Flue ridge terminal Mark III, 450mm long with sealing gasket and fixing brackets										
half round ridge	54.82	5.00	**57.57**	1.86	0.92	**36.31**	0.73	**94.60**	nr	**113.52**
extra for 150mm extension adaptor	20.67	5.00	**21.70**	0.37	0.19	**7.32**	0.28	**29.29**	nr	**35.15**
Hip irons										
galvanized mild steel; 32 x 3 x 380mm girth; scrolled; fixing with galvanized steel screws to timber	2.28	5.00	**2.39**	0.12	0.07	**2.43**	-	**4.82**	nr	**5.78**
Lead soakers										
fixing only	-	-	-	0.37	0.19	**7.32**	-	**7.32**	nr	**8.78**

This section continues on the next page

Labour hourly rates: (except Specialists) Craft Operatives £17.38 Labourer £10.71 Rates are national average prices. Refer to REGIONAL VARIATIONS for indicative levels of overall pricing in regions	MATERIALS			LABOUR				RATES		
	Del to Site £	Waste %	Material Cost £	Craft Optve Hrs	Lab Hrs	Labour Cost £	Sunds £	Nett Rate £	Unit	Gross Rate (+20%) £
H71: LEAD SHEET COVERINGS/FLASHINGS										
Sheet lead										
Technical data										
Code 3, 1.32mm thick, 14.97 kg/m2, colour code green										
Code 4, 1.80mm thick, 20.41 kg/m2, colour code blue										
Code 5, 2.24mm thick, 25.40 kg/m2, colour code red										
Code 6, 2.65mm thick, 30.05 kg/m2, colour code black										
Code 7, 3.15mm thick, 35.72 kg/m2, colour code white										
Code 8, 3.55mm thick, 40.26 kg/m2, colour code orange										
Milled lead sheet, BS EN 12588										
Nr 5 roof coverings; fixing to timber with milled lead cleats and galvanized screws										
pitch 7.5 degrees from horizontal	31.36	5.00	32.93	6.27	-	108.97	0.40	142.31	m2	170.77
extra; oil patination to surfaces	2.04	10.00	2.24	0.25	-	4.40	-	6.64	m2	7.96
Flashings; horizontal										
Nr 4; 150mm lapped joints; fixing to masonry with milled lead clips and lead wedges										
150mm girth	3.78	5.00	3.97	0.55	-	9.56	0.34	13.87	m	16.64
240mm girth	6.05	5.00	6.35	0.90	-	15.68	0.34	22.37	m	26.84
300mm girth	7.57	5.00	7.95	1.12	-	19.50	0.34	27.78	m	33.34
Nr 5; 150mm lapped joints; fixing to masonry with milled lead clips and lead wedges										
150mm girth	4.71	5.00	4.94	0.64	-	11.09	0.40	16.44	m	19.72
240mm girth	7.52	5.00	7.90	1.01	-	17.59	0.40	25.89	m	31.07
300mm girth	9.42	5.00	9.89	1.27	-	21.99	0.40	32.28	m	38.73
Flashings; stepped										
Nr 4; 150mm lapped joints; fixing to masonry with milled lead clips and lead wedges										
180mm girth	4.54	5.00	4.77	0.90	-	15.68	0.52	20.96	m	25.16
240mm girth	6.05	5.00	6.35	1.21	-	21.03	0.52	27.90	m	33.48
300mm girth	7.57	5.00	7.95	1.50	-	26.00	0.52	34.46	m	41.36
Nr 5; 150mm lapped joints; fixing to masonry with milled lead clips and lead wedges										
180mm girth	5.64	5.00	5.93	1.01	-	17.59	0.62	24.13	m	28.96
240mm girth	7.52	5.00	7.90	1.34	-	23.32	0.62	31.84	m	38.21
300mm girth	9.42	5.00	9.89	1.67	-	29.06	0.62	39.56	m	47.48
Hips; sloping; dressing over slating and tiling										
Nr 4; 150mm lapped joints; fixing to timber with milled lead clips										
300mm girth	7.57	5.00	7.95	1.12	-	19.50	0.34	27.78	m	33.34
Nr 5; 150mm lapped joints; fixing to timber with milled lead clips										
300mm girth	9.42	5.00	9.89	1.27	-	21.99	0.40	32.28	m	38.73
Kerbs; horizontal										
Nr 4; 150mm lapped joints; fixing to timber with copper nails										
300mm girth	7.57	5.00	7.95	1.12	-	19.50	0.34	27.78	m	33.34
Ridges; horizontal; dressing over slating and tiling										
Nr 4; 150mm lapped joints; fixing to timber with milled lead clips										
300mm girth	7.57	5.00	7.95	1.12	-	19.50	0.34	27.78	m	33.34
Nr 5; 150mm lapped joints; fixing to timber with milled lead clips										
300mm girth	9.42	5.00	9.89	1.27	-	21.99	0.40	32.28	m	38.73
Valleys; sloping										
Nr 4; dressing over tilting fillets -1; 150mm lapped joints; fixing to timber with copper nails										
300mm girth	7.57	5.00	7.95	1.12	-	19.50	0.34	27.78	m	33.34
450mm girth	11.33	5.00	11.90	1.67	-	29.06	0.52	41.47	m	49.77
Cavity gutters										
Nr 4 dressings										
corrugated roofing; fibre cement; down corrugations	-	-	-	0.32	-	5.54	-	5.54	m	6.65
corrugated roofing; fibre cement; across corrugations	-	-	-	0.32	-	5.54	-	5.54	m	6.65
glass and glazing bars; timber	-	-	-	0.32	-	5.54	-	5.54	m	6.65
Nr 5 dressings										
corrugated roofing; fibre cement; down corrugations	-	-	-	0.32	-	5.54	-	5.54	m	6.65

Labour hourly rates: (except Specialists) Craft Operatives £17.38 Labourer £10.71 Rates are national average prices. Refer to REGIONAL VARIATIONS for indicative levels of overall pricing in regions	MATERIALS			LABOUR				RATES		
	Del to Site £	Waste %	Material Cost £	Craft Optve Hrs	Lab Hrs	Labour Cost £	Sunds £	Nett Rate £	Unit	Gross Rate (+20%) £
H71: LEAD SHEET COVERINGS/FLASHINGS Cont.										
Cavity gutters Cont.										
Nr 5 dressings Cont.										
corrugated roofing; fibre cement; across corrugations	-	-	-	0.32	-	5.54	-	5.54	m	6.65
glass and glazing bars; timber	-	-	-	0.32	-	5.54	-	5.54	m	6.65
Nr 4 soakers and slates handed to others for fixing										
180 x 300mm	1.36	5.00	1.43	0.25	-	4.40	-	5.83	nr	7.00
Nr 5 soakers and slates handed to others for fixing										
180 x 300mm	1.69	5.00	1.78	0.25	-	4.40	-	6.18	nr	7.41

WATERPROOFING

Labour hourly rates: (except Specialists) Craft Operatives £14.21 Labourer £10.71 Rates are national average prices. Refer to REGIONAL VARIATIONS for indicative levels of overall pricing in regions	MATERIALS			LABOUR				RATES		
	Del to Site £	Waste %	Material Cost £	Craft Optve Hrs	Lab Hrs	Labour Cost £	Sunds £	Nett Rate £	Unit	Gross Rate (+20%) £
J20: MASTIC ASPHALT TANKING/ DAMP PROOF MEMBRANES										
Tanking and damp proofing BS 6925 (limestone aggregate)										
Tanking and damp proofing 20mm thick; coats of asphalt -2; to concrete base; horizontal work subsequently covered										
width not exceeding 150mm	-	-	-	-	-	-	-	7.76	lm	9.31
width 150 - 225mm	-	-	-	-	-	-	-	9.74	lm	11.68
width 225 - 300mm	-	-	-	-	-	-	-	12.02	lm	14.43
width exceeding 300mm	-	-	-	-	-	-	-	25.31	m2	30.37
Tanking and damp proofing 20mm thick; coats of asphalt -2; to concrete base; vertical work subsequently covered										
width not exceeding 150mm	-	-	-	-	-	-	-	15.27	lm	18.32
width 150 - 225mm	-	-	-	-	-	-	-	17.85	lm,	21.42
width 225 - 300mm	-	-	-	-	-	-	-	20.36	lm	24.43
width exceeding 300mm	-	-	-	-	-	-	-	52.60	m2	63.12
J21: MASTIC ASPHALT ROOFING/INSULATION/FINISHES										
Roofing; BS 6925 (limestone aggregate); sand rubbing; BS 747 type 4A sheathing felt isolating membrane, butt joints										
Roofing 20mm thick; coats of asphalt -2; to concrete base; pitch not exceeding 6 degrees from horizontal										
width exceeding 300mm	-	-	-	-	-	-	-	27.26	m2	32.71
extra; solid water check roll	-	-	-	-	-	-	-	13.67	m	16.41
Roofing 20mm thick; coats of asphalt -2; to concrete base; pitch 30 degrees from horizontal										
width exceeding 300mm	-	-	-	-	-	-	-	41.69	m2	50.03
Paint two coats of Solar reflective roof paint on surfaces of asphalt										
width exceeding 300mm	-	-	-	-	-	-	-	7.43	m2	8.91
Skirtings 13mm thick; coats of asphalt -2; to brickwork base										
girth not exceeding 150mm	-	-	-	-	-	-	-	16.19	m	19.43
Coverings to kerbs 20mm thick; coats of asphalt -2; to concrete base										
girth 600mm	-	-	-	-	-	-	-	34.33	m	41.20
Jointing new roofing to existing										
20mm thick	-	-	-	-	-	-	-	20.52	m	24.62
Edge trim; aluminium; silver anodized priming with bituminous primer; butt joints with internal jointing sleeves; fixing with aluminium alloy screws to timber; working two coat asphalt into grooves -1										
63.5mm wide x 44.4mm face depth	8.64	5.00	9.07	0.44	-	6.25	1.52	16.84	m	20.20
Roof Ventilators										
aluminium; setting in position	10.16	2.50	10.42	0.94	-	13.29	0.59	24.30	nr	29.16
Roofing; BS 6925 (limestone aggregate); covering with 13mm white spar chippings in hot bitumen; BS 747 type 4A sheathing felt isolating membrane; butt joints										
Roofing 20mm thick; coats of asphalt -2; to concrete base; pitch not exceeding 6 degrees from horizontal										
width exceeding 300mm	-	-	-	-	-	-	-	32.77	m2	39.32

Labour hourly rates: (except Specialists) Craft Operatives £14.21 Labourer £10.71 Rates are national average prices. Refer to REGIONAL VARIATIONS for indicative levels of overall pricing in regions	MATERIALS			LABOUR				RATES		
	Del to Site £	Waste %	Material Cost £	Craft Optve Hrs	Lab Hrs	Labour Cost £	Sunds £	Nett Rate £	Unit	Gross Rate (+20%) £
J21: MASTIC ASPHALT ROOFING/INSULATION/FINISHES Cont.										
Roofing; BS 6925 (limestone aggregate); covering with 13mm white spar chippings in hot bitumen; BS 747 type 4A sheathing felt isolating membrane; butt joints Cont.										
Roofing 20mm thick; coats of asphalt -2; to concrete base; pitch 30 degrees from horizontal										
width exceeding 300mm	-	-	-	-	-	-	-	47.18	m2	56.61
J30: LIQUID APPLIED TANKING/DAMP PROOF MEMBRANES										
Synthaprufe bituminous emulsion; blinding with sand										
Coverings, coats -2; to concrete base; horizontal										
width not exceeding 150mm	5.72	10.00	6.29	-	0.51	5.42	0.15	11.86	m2	14.24
width 150 - 225mm	5.72	10.00	6.29	-	0.42	4.48	0.15	10.92	m2	13.11
width 225 - 300mm	5.72	10.00	6.29	-	0.33	3.53	0.15	9.98	m2	11.98
width exceeding 300mm	5.72	10.00	6.29	-	0.25	2.71	0.15	9.15	m2	10.99
Coverings, coats -3; to concrete base; horizontal										
width exceeding 300mm	8.54	10.00	9.39	-	0.35	3.77	0.15	13.31	m2	15.98
Coverings; coats -2; to concrete base; vertical										
width not exceeding 150mm	5.72	10.00	6.29	-	0.76	8.13	0.15	14.57	m2	17.49
width 150 - 225mm	5.72	10.00	6.29	-	0.63	6.72	0.15	13.16	m2	15.79
width 225 - 300mm	5.72	10.00	6.29	-	0.50	5.30	0.15	11.75	m2	14.10
width exceeding 300mm	5.72	10.00	6.29	-	0.39	4.12	0.15	10.57	m2	12.68
Coverings; coats -3; to concrete base; vertical										
width exceeding 300mm	8.54	10.00	9.39	-	0.53	5.65	0.15	15.20	m2	18.24
R.I.W. liquid asphaltic composition										
Coverings, coats -2; to concrete base; horizontal										
width not exceeding 150mm	8.85	10.00	9.73	-	0.51	5.42	0.15	15.31	m2	18.37
width 150 - 225mm	8.85	10.00	9.73	-	0.42	4.48	0.15	14.36	m2	17.24
width 225 - 300mm	8.85	10.00	9.73	-	0.33	3.53	0.15	13.42	m2	16.10
width exceeding 300mm	8.85	10.00	9.73	-	0.25	2.71	0.15	12.60	m2	15.12
Coverings, coats -3; to concrete base; horizontal										
width exceeding 300mm	13.32	10.00	14.65	-	0.35	3.77	0.15	18.58	m2	22.29
Coverings; coats -2; to concrete base; vertical										
width not exceeding 150mm	8.85	10.00	9.73	-	0.76	8.13	0.15	18.02	m2	21.62
width 150 - 225mm	8.85	10.00	9.73	-	0.63	6.72	0.15	16.60	m2	19.92
width 225 - 300mm	8.85	10.00	9.73	-	0.50	5.30	0.15	15.19	m2	18.23
width exceeding 300mm	8.85	10.00	9.73	-	0.39	4.12	0.15	14.01	m2	16.81
Coverings; coats -3; to concrete base; vertical										
width exceeding 300mm	13.32	10.00	14.65	-	0.53	5.65	0.15	20.46	m2	24.56
J40: FLEXIBLE SHEET TANKING/DAMP PROOFING MEMBRANES										
Polythene sheeting; 100mm welted laps										
Tanking and damp proofing; 125mu										
horizontal; on concrete base	0.40	20.00	0.48	-	0.07	0.71	-	1.18	m2	1.42
vertical; on concrete base	0.40	20.00	0.48	-	0.10	1.06	-	1.54	m2	1.84
Bituthene 500X self adhesive damp proof membrane; 75mm lapped joints										
Tanking and damp proofing										
horizontal; on concrete base	5.58	20.00	6.69	-	0.25	2.71	-	9.40	m2	11.28
vertical; on concrete base; priming with Servicised B primer	5.58	20.00	6.69	-	0.39	4.12	0.29	11.10	m2	13.32
extra; Bituthene internal angle fillet; 40 x 40mm	6.27	10.00	6.90	-	0.13	1.41	-	8.31	m	9.97
Visqueen 1200 super damp proof membrane; 150mm laps sealed with tape and mastic										
Tanking and damp proofing										
horizontal; on concrete base	0.62	20.00	0.74	-	0.09	0.94	0.12	1.80	m2	2.16
J41: BUILT UP FELT ROOF COVERINGS										
Roofing; felt BS 747, comprising 3G Rubervent underlay 2.6 kg/m2, 1 layer Ruberfort HP180 underlay 3.4 kg/m2, 1 layer Ruberfort HP180 mineral surface top sheet 3.6 kg/m2										
Roof coverings; layers of felt -3; bonding with hot bitumen compound to timber base										
pitch 7.5 degrees from horizontal	11.37	15.00	13.26	0.51	0.07	7.90	-	21.15	m2	25.38
pitch 40 degrees from horizontal	11.37	15.00	13.26	0.84	0.10	12.94	-	26.19	m2	31.43
Roof coverings; layers of felt -3; bonding with hot bitumen compound to cement and sand or concrete base										
pitch 7.5 degrees from horizontal	11.37	15.00	13.26	0.51	0.07	7.90	-	21.15	m2	25.38
pitch 40 degrees from horizontal	11.37	15.00	13.26	0.84	0.10	12.94	-	26.19	m2	31.43
Skirtings; layers of felt - 3; bonding with hot bitumen compound to brickwork base										
girth 0 - 200mm	2.27	15.00	2.65	0.31	0.03	4.73	-	7.38	m	8.8

Labour hourly rates: (except Specialists) Craft Operatives £14.21 Labourer £10.71 Rates are national average prices. Refer to REGIONAL VARIATIONS for indicative levels of overall pricing in regions	MATERIALS			LABOUR				RATES		
	Del to Site £	Waste %	Material Cost £	Craft Optve Hrs	Lab Hrs	Labour Cost £	Sunds £	Nett Rate £	Unit	Gross Rate (+20%) £
J41: BUILT UP FELT ROOF COVERINGS Cont.										
Roofing; felt BS 747, comprising 3G Rubervent underlay 2.6 kg/m2, 1 layer Ruberfort HP180 underlay 3.4 kg/m2, 1 layer Ruberfort HP180 mineral surface top sheet 3.6 kg/m2 Cont.										
Skirtings; layers of felt - 3; bonding with hot bitumen compound to brickwork base Cont.										
girth 200 - 400mm	4.55	15.00	**5.30**	0.51	0.07	**7.90**	-	**13.20**	m	**15.84**
Collars around pipes; standard and the like; layers of felt -1; bonding with hot bitumen compound to metal base										
50mm diameter x 150mm long; hole in 3 layer covering	0.95	15.00	**1.10**	0.19	0.02	**2.89**	-	**3.99**	nr	**4.79**
100mm diameter x 150mm long; hole in 3 layer covering	1.21	15.00	**1.41**	0.24	0.03	**3.79**	-	**5.20**	nr	**6.24**
Eaves trim; aluminium; silver anodized butt joints over matching sleeve pieces 200mm long; fixing to timber base with aluminium alloy screws; bedding in mastic										
63.5mm wide x 44.4mm face depth	8.64	5.00	**9.07**	0.44	-	**6.25**	1.65	**16.97**	m	**20.36**
Roofing; felt BS 747, bottom and intermediate layers type 3B -1.8 kg/m2, top layer type 3E -2.8 kg/m2, 75mm laps; fully bonding layers with hot bitumen bonding compound										
Roof coverings; layers of felt -2; bonding with hot bitumen compound to timber base										
pitch 7.5 degrees from horizontal	5.43	15.00	**6.37**	0.42	0.06	**6.53**	-	**12.89**	m2	**15.47**
pitch 40 degrees from horizontal	5.43	15.00	**6.37**	0.69	0.09	**10.79**	-	**17.16**	m2	**20.59**
Roof coverings; layers of felt -2; bonding with hot bitumen compound to cement and sand or concrete base										
pitch 7.5 degrees from horizontal	5.43	15.00	**6.37**	0.42	0.06	**6.53**	-	**12.89**	m2	**15.47**
pitch 40 degrees from horizontal	5.43	15.00	**6.37**	0.69	0.09	**10.79**	-	**17.16**	m2	**20.59**
Roof coverings; layers of felt -3; bonding with hot bitumen compound to timber base										
pitch 7.5 degrees from horizontal	7.63	15.00	**8.95**	0.61	0.08	**9.42**	-	**18.38**	m2	**22.05**
pitch 40 degrees from horizontal	7.63	15.00	**8.95**	1.01	0.13	**15.79**	-	**24.75**	m2	**29.70**
Roof coverings; layers of felt -3; bonding with hot bitumen compound to cement and sand or concrete base										
pitch 7.5 degrees from horizontal	7.63	15.00	**8.95**	0.61	0.08	**9.42**	-	**18.38**	m2	**22.05**
pitch 40 degrees from horizontal	7.63	15.00	**8.95**	1.01	0.13	**15.79**	-	**24.75**	m2	**29.70**
Skirtings; layers of felt - 3; bonding with hot bitumen compound to brickwork base										
girth 0 - 200mm	1.53	15.00	**1.79**	0.31	0.03	**4.73**	-	**6.52**	m	**7.83**
girth 200 - 400mm	3.05	15.00	**3.58**	0.51	0.07	**7.90**	-	**11.48**	m	**13.77**
Collars around pipes; standard and the like; layers of felt -1; bonding with hot bitumen compound to metal base										
50mm diameter x 150mm long; hole in 3 layer covering	0.60	15.00	**0.70**	0.19	0.02	**2.89**	-	**3.59**	nr	**4.31**
100mm diameter x 150mm long; hole in 3 layer covering	0.80	15.00	**0.92**	0.24	0.03	**3.79**	-	**4.72**	nr	**5.66**
Roofing; sheet, Ruberoid Building Products Superflex Ultrabond system, bottom layer Rubervent BS 747 type 3G; intermediate layer Superbase Ultrabond; top layer Superflex Ultrabond slate surfaced cap sheet, 50mm side laps, 75mm end laps; partially bonding bottom layer, fully bonding other layers in Ruberoid 95/25 hot bonding bitumen										
Roof coverings; layers of sheet -3; bonding with hot bitumen compound to timber base										
pitch not exceeding 5.0 degrees from horizontal	17.58	15.00	**20.39**	0.61	0.31	**11.90**	-	**32.29**	m2	**38.74**
Roof coverings; layers of sheet -3; bonding with hot bitumen compound to cement and sand or concrete base										
pitch not exceeding 5.0 degrees from horizontal	17.58	15.00	**20.39**	0.61	0.31	**11.90**	-	**32.29**	m2	**38.74**
Roof coverings; layers of sheet -3; bonding with hot bitumen compound to metal base										
pitch not exceeding 5.0 degrees from horizontal	17.58	15.00	**20.39**	0.61	0.31	**11.90**	-	**32.29**	m2	**38.74**
Skirtings; layers of sheet -2; bonding with hot bitumen compound to brickwork base										
girth 0 - 200mm	2.15	15.00	**2.50**	0.31	0.06	**4.97**	-	**7.46**	m	**8.96**
girth 200 - 400mm	4.30	15.00	**5.00**	0.51	0.07	**7.90**	-	**12.89**	m	**15.47**

WATERPROOFING

Labour hourly rates: (except Specialists) Craft Operatives £14.21 Labourer £10.71 Rates are national average prices. Refer to REGIONAL VARIATIONS for indicative levels of overall pricing in regions	MATERIALS			LABOUR				RATES		
	Del to Site	Waste	Material Cost	Craft Optve	Lab	Labour Cost	Sunds	Nett Rate	Unit	Gross Rate (+20%)
	£	%	£	Hrs	Hrs	£	£	£		£
J41: BUILT UP FELT ROOF COVERINGS Cont.										
Roofing; sheet, Ruberoid Building Products Superflex Ultrabond system, bottom layer Rubervent BS 747 type 3G; intermediate layer Superbase Ultrabond; top layer Superflex Ultrabond slate surfaced cap sheet, 50mm side laps, 75mm end laps; partially bonding bottom layer, fully bonding other layers in Ruberoid 95/25 hot bonding bitumen; 25mm Kingspan Thermaroof TR21 (PIR) Board										
Roof coverings; layers of sheet -3; bonding with hot bitumen compound to timber base										
pitch not exceeding 5.0 degrees from horizontal ..	26.97	15.00	**31.19**	1.21	0.40	**21.44**	1.22	**53.85**	m2	**64.62**
Roof coverings; layers of sheet -3; bonding with hot bitumen compound to cement and sand or concrete base										
pitch not exceeding 5.0 degrees from horizontal ..	26.97	15.00	**31.19**	1.28	0.40	**22.37**	1.22	**54.79**	m2	**65.74**
Roof coverings; layers of sheet -3; bonding with hot bitumen compound to metal base										
pitch not exceeding 5.0 degrees from horizontal ..	26.97	15.00	**31.19**	1.25	0.40	**22.06**	1.22	**54.47**	m2	**65.37**
Roofing; sheet, Ruberoid Building Products Superflex Ultratorch system, bottom layer Superbase Ultratorch; top layer Superflex Ultratorch slate surfaced cap sheet, 75mm side and end laps; fully bonding both layers (by torching); SuperRock Torch, mineral wool insulation with bitumen primed surface; Superbar vapour control layer										
Roof coverings; layers of sheet -3; bonding to timber base										
pitch not exceeding 5.0 degrees from horizontal ..	44.39	15.00	**51.04**	0.61	0.31	**11.90**	3.16	**66.10**	m2	**79.31**
Roof coverings; layers of sheet -3; bonding to cement and sand or concrete base										
pitch not exceeding 5.0 degrees from horizontal ..	44.39	15.00	**51.04**	0.61	0.31	**11.90**	3.16	**66.10**	m2	**79.31**
Roof coverings; layers of sheet -3; bonding to metal base										
pitch not exceeding 5.0 degrees from horizontal ..	44.39	15.00	**51.04**	0.65	0.31	**12.52**	3.16	**66.72**	m2	**80.06**
Skirtings; layers of sheet -2; bonding with hot bitumen compound to brickwork base										
girth 0 - 200mm	3.34	15.00	**3.84**	0.31	0.06	**4.97**	0.64	**9.44**	m	**11.33**
girth 200 - 400mm	6.67	15.00	**7.67**	0.51	0.07	**7.90**	1.28	**16.84**	m	**20.21**

LININGS/SHEATHING/DRY PARTITIONING

Labour hourly rates: (except Specialists) Craft Operatives £14.21 Labourer £10.71 Rates are national average prices. Refer to REGIONAL VARIATIONS for indicative levels of overall pricing in regions	MATERIALS			LABOUR				RATES		
	Del to Site £	Waste %	Material Cost £	Craft Optve Hrs	Lab Hrs	Labour Cost £	Sunds £	Nett Rate £	Unit	Gross Rate (+20%) £
K10: PLASTERBOARD DRY LINING/PARTITIONS/CEILINGS										
Linings; tapered edge sheeting of one layer 9.5mm thick Gyproc wall board for direct decoration, BS 1230, butt joints; fixing with galvanized nails to timber base; butt joints filled with joint filler tape and joint finish, spot filling										
Walls										
height 2100 - 2400mm	5.81	5.00	**6.10**	1.06	0.53	**20.66**	-	**26.76**	m	**32.11**
height 2400 - 2700mm	6.54	5.00	**6.86**	1.19	0.59	**23.24**	-	**30.11**	m	**36.13**
height 2700 - 3000mm	7.17	5.00	**7.53**	1.32	0.66	**25.83**	-	**33.36**	m	**40.03**
Reveals and soffits of openings and recesses										
width not exceeding 300mm	0.79	10.00	**0.86**	0.26	0.07	**4.46**	-	**5.32**	m	**6.38**
Ceilings										
generally	2.44	5.00	**2.56**	0.53	0.22	**9.86**	-	**12.42**	m2	**14.90**
Linings; tapered edge sheeting of one layer 12.5mm thick Gyproc wallboard for direct decoration, BS 1230 butt joints; fixing with galvanized nails to timber base; butt joints filled with joint filler tape and joint finish, spot filling										
Walls										
height 2100 - 2400mm	6.10	5.00	**6.41**	1.24	0.62	**24.26**	-	**30.67**	m	**36.80**
height 2400 - 2700mm	6.86	5.00	**7.21**	1.40	0.69	**27.27**	-	**34.48**	m	**41.38**
height 2700 - 3000mm	7.62	5.00	**8.00**	1.55	0.78	**30.40**	-	**38.41**	m	**46.09**
Reveals and soffits of openings and recesses										
width not exceeding 300mm	0.83	10.00	**0.90**	0.31	0.08	**5.20**	-	**6.10**	m	**7.32**
Ceilings										
generally	2.56	5.00	**2.68**	0.62	0.26	**11.58**	-	**14.27**	m2	**17.12**
Linings; tapered edge sheeting of one layer 15mm thick Gyproc wall board for direct decoration, BS 1230, butt joints; fixing with galvanized nails to timber base; butt joints filled with joint filler tape and joint finish, spot filling										
Walls										
height 2100 - 2400mm	9.16	5.00	**9.62**	1.43	0.72	**27.98**	-	**37.60**	m	**45.12**
height 2400 - 2700mm	10.31	5.00	**10.82**	1.61	0.80	**31.42**	-	**42.25**	m	**50.69**
height 2700 - 3000mm	11.36	5.00	**11.93**	1.78	0.89	**34.86**	-	**46.79**	m	**56.15**
Reveals and soffits of openings and recesses										
width not exceeding 300mm	1.21	10.00	**1.32**	0.35	0.09	**5.94**	-	**7.26**	m	**8.72**
Ceilings										
generally	3.83	5.00	**4.02**	0.70	0.31	**13.30**	-	**17.33**	m2	**20.79**
Linings; tapered edge sheeting of one layer 9.5mm thick Gyproc Duplex wallboard for direct decoration, BS 1230 butt joints; fixing with galvanized nails to timber base; butt joints filled with joint filler tape and joint finish, spot filling										
Walls										
height 2100 - 2400mm	7.42	5.00	**7.79**	1.06	0.53	**20.66**	-	**28.45**	m	**34.14**
height 2400 - 2700mm	8.35	5.00	**8.77**	1.19	0.59	**23.24**	-	**32.01**	m	**38.41**
height 2700 - 3000mm	9.18	5.00	**9.64**	1.32	0.66	**25.83**	-	**35.47**	m	**42.56**
Reveals and soffits of openings and recesses										
width not exceeding 300mm	0.99	10.00	**1.08**	0.26	0.07	**4.46**	-	**5.54**	m	**6.65**
Ceilings										
generally	3.11	5.00	**3.26**	0.53	0.22	**9.86**	-	**13.12**	m2	**15.75**

Labour hourly rates: (except Specialists) Craft Operatives £14.21 Labourer £10.71 Rates are national average prices. Refer to REGIONAL VARIATIONS for indicative levels of overall pricing in regions	MATERIALS			LABOUR			RATES			
	Del to Site £	Waste %	Material Cost £	Craft Optve Hrs	Lab Hrs	Labour Cost £	Sunds £	Nett Rate £	Unit	Gross Rate (+20%) £
K10: PLASTERBOARD DRY LINING/PARTITIONS/CEILINGS Cont.										
Linings; two layers of Gypsum wallboard, first layer square edge sheeting 9.5mm thick, second layer tapered edge sheeting 12.5mm thick for direct decoration, BS 1230, butt joints; fixing with galvanized nails to timber base; butt joints of second layer filled with joint filler tape and joint finish, spot filling										
Walls										
height 2100 - 2400mm	11.33	5.00	**11.90**	2.11	1.06	**41.32**	-	**53.22**	m	**63.87**
height 2400 - 2700mm	12.75	5.00	**13.39**	2.38	1.19	**46.49**	-	**59.88**	m	**71.85**
height 2700 - 3000mm	14.08	5.00	**14.78**	2.64	1.32	**51.65**	-	**66.43**	m	**79.72**
Reveals and soffits of openings and recesses										
width not exceeding 300mm	1.48	10.00	**1.59**	0.53	0.13	**8.92**	-	**10.50**	m	**12.60**
Ceilings										
generally	4.74	5.00	**4.97**	1.06	0.44	**19.72**	-	**24.69**	m2	**29.63**
K10: PLASTERBOARD DRY LINING/PARTITIONS/CEILINGS: METAL STUD PARTITIONS										
Proprietary partitions; Gyproc Gypwall metal stud; tapered edge sheeting of one layer 12.5mm thick Gyproc wallboard for direct decoration, BS 1230, butt joints; applying continuous beads of Gyproc acoustical sealant to perimeters of construction both sides, fixing with Pozidriv head screws to steel frame comprising 50mm head and floor channels, 48mm vertical studs at 600mm centres; butt joints filled with joint filler tape and joint finish, spot filling; fixing to masonry with cartridge fired nails										
Height 2100 - 2400mm, 75mm thick										
boarded both sides	22.12	5.00	**23.23**	3.14	1.57	**61.40**	-	**84.62**	m	**101.55**
Proprietary partitions; Gyproc Gypwall metal stud; tapered edge sheeting of two layers 12.5mm thick Gyproc wallboard for direct decoration, BS 1230, butt joints; applying continuous beads of Gyproc acoustical sealant to perimeters of construction both sides, fixing with Pozidriv head screws to steel frame comprising 50mm head and floor channels, 48mm vertical studs at 600mm centres; butt joints filled with joint filler tape and joint finish, spot filling; fixing to masonry with cartridge fired nails										
Height 2100 - 2400mm, 100mm thick										
boarded both sides	33.16	5.00	**34.82**	3.58	1.79	**70.00**	-	**104.83**	m	**125.79**
Height 2400 - 2700mm, 100mm thick										
boarded both sides	36.86	5.00	**38.70**	4.02	2.01	**78.61**	-	**117.31**	m	**140.77**
Height 2700 - 3000mm, 100mm thick										
boarded both sides	40.55	5.00	**42.58**	4.46	2.23	**87.22**	-	**129.80**	m	**155.76**
Angles to partitions; fixing end studs together through partition; extending sheeting to one side across end of partition; joints filled with joint filler tape and joint finish										
Plain										
75mm thick partitions	1.58	10.00	**1.66**	0.14	-	**2.03**	-	**3.69**	m	**4.43**
100mm thick partitions	1.63	10.00	**1.72**	0.53	-	**7.50**	-	**9.22**	m	**11.07**
K11: RIGID SHEET FLOORING/SHEATHING/LININGS/CASINGS										
Douglas Fir plywood, unsanded select sheathing quality, WBP bonded; tongued and grooved joints long edges										
18mm thick sheeting to floors										
width exceeding 300mm	9.55	7.50	**10.27**	0.40	0.04	**6.10**	0.04	**16.41**	m2	**19.69**
Birch faced plywood, BB quality, WBP bonded; tongued and grooved joints all edges										
12mm thick sheeting to floors										
width exceeding 300mm	22.64	7.50	**24.34**	0.35	0.04	**5.47**	0.04	**29.85**	m2	**35.82**
Wood chipboard, BS EN 312, type II flooring; tongued and grooved joints										
22mm thick sheeting to floors										
width exceeding 300mm	6.39	7.50	**6.87**	0.40	0.04	**6.10**	0.04	**13.01**	m2	**15.61**
K11: RIGID SHEET FLOORING/SHEATHING/LININGS/CASINGS: FIBREBOARDS AND HARDBOARDS										
Standard hardboard, BS EN 622; butt joints										
3.2mm thick sheeting to floors										
width exceeding 300mm	1.51	5.00	**1.59**	0.21	0.02	**3.21**	0.07	**4.86**	m2	**5.83**

Labour hourly rates: (except Specialists) Craft Operatives £14.21 Labourer £10.71 Rates are national average prices. Refer to REGIONAL VARIATIONS for indicative levels of overall pricing in regions	MATERIALS			LABOUR				RATES		
	Del to Site £	Waste %	Material Cost £	Craft Optve Hrs	Lab Hrs	Labour Cost £	Sunds £	Nett Rate £	Unit	Gross Rate (+20%) £
K11: RIGID SHEET FLOORING/SHEATHING/LININGS/CASINGS: FIBREBOARDS AND HARDBOARDS Cont.										
Tempered hardboard, BS EN 622; butt joints										
3.2mm thick linings to walls										
width exceeding 300mm	2.17	5.00	**2.28**	0.42	0.06	**6.53**	0.14	**8.95**	m2	**10.74**
6.4mm thick linings to walls										
width exceeding 300mm	4.00	5.00	**4.20**	0.45	0.06	**7.00**	0.15	**11.34**	m2	**13.61**
Medium density fibreboard, BS EN 622, type MDF; butt joints										
9mm thick linings to walls										
width exceeding 300mm	3.97	5.00	**4.17**	0.45	0.06	**7.00**	0.15	**11.32**	m2	**13.58**
K11: RIGID SHEET FLOORING/SHEATHING/LININGS/CASINGS: ASBESTOS FREE BOARD										
Masterboard Class 0 fire resisting boards; butt joints										
6mm thick linings to walls										
width exceeding 300mm	12.44	5.00	**13.06**	0.79	0.10	**12.31**	0.26	**25.64**	m2	**30.77**
Supalux fire resisting boards, Sanded; butt joints; fixing to timber with screws, countersinking										
6mm thick linings to walls										
width exceeding 300mm	17.23	5.00	**18.09**	0.79	0.10	**12.31**	0.26	**30.66**	m2	**36.80**
Vermiculux fire resisting boards; butt joints; fixing to steel with self tapping screws, countersinking										
20mm thick casings to isolated beams or the like										
total girth not exceeding 600mm.................	26.57	10.00	**29.23**	2.07	0.25	**32.10**	0.68	**62.00**	m2	**74.41**
total girth 600 - 1200mm........................	26.57	5.00	**27.90**	1.24	0.15	**19.31**	0.41	**47.62**	m2	**57.14**
40mm thick casings to isolated beams or the like										
total girth not exceeding 600mm.................	65.88	10.00	**72.47**	2.46	0.31	**38.31**	0.81	**111.60**	m2	**133.92**
total girth 600 - 1200mm........................	65.88	5.00	**69.18**	1.51	0.19	**23.42**	0.50	**93.09**	m2	**111.71**
K11: RIGID SHEET FLOORING/SHEATHING/LININGS/CASINGS: DUCT FRONTS										
Plywood BS EN 636, II/III grade, MR bonded; butt joints; fixing to timber with screws, countersinking										
12mm thick linings to walls										
width exceeding 300mm	15.22	20.00	**18.27**	1.27	0.15	**19.62**	0.07	**37.96**	m2	**45.56**
K11: RIGID SHEET FLOORING/SHEATHING/LININGS/CASINGS: PIPE CASINGS										
Plywood BS EN 636, II/III grade, MR bonded; butt joints; fixing to timber with screws, countersinking										
12mm thick linings to pipe ducts and casings										
width exceeding 300mm	15.22	20.00	**18.27**	1.87	0.23	**29.05**	0.10	**47.42**	m2	**56.90**
width not exceeding 300mm.......................	4.57	20.00	**5.48**	0.66	0.09	**10.32**	0.04	**15.84**	m	**19.01**
K11: RIGID SHEET FLOORING/SHEATHING/LININGS/CASINGS: ROOF BOARDING										
Plywood BS EN 636, II/III grade, WBP bonded; butt joints										
18mm thick sheeting to roofs; external										
width exceeding 300mm	12.33	10.00	**13.56**	0.44	0.06	**6.84**	0.02	**20.43**	m2	**24.51**
extra; cesspool 300 x 300 x 150mm...............	3.39	10.00	**3.73**	1.98	0.24	**30.73**	0.11	**34.56**	nr	**41.48**
K20: TIMBER BOARD FLOORING/SHEATHING/LININGS/CASINGS										
Softwood, wrought; fixing to timber - boarded flooring										
Boarding to floors, tongued and grooved joints; 25mm thick, 125mm wide boards										
width exceeding 300mm	11.14	7.50	**11.97**	0.99	0.12	**15.36**	0.05	**27.39**	m2	**32.87**
K20: TIMBER BOARD FLOORING/SHEATHING/LININGS/CASINGS: SURFACE TREATMENT OF EXISTING WOOD FLOORING										
Sanding and sealing existing wood flooring										
Machine sanding										
width exceeding 300mm	-	-	**-**	0.61	-	**8.60**	0.80	**9.40**	m2	**11.27**
Prepare, one priming coat and one finish coat of seal										
width exceeding 300mm	-	-	**-**	0.20	-	**2.81**	0.40	**3.21**	m2	**3.86**

Labour hourly rates: (except Specialists) Craft Operatives £14.21 Labourer £10.71 Rates are national average prices. Refer to REGIONAL VARIATIONS for indicative levels of overall pricing in regions	MATERIALS			LABOUR				RATES		
	Del to Site £	Waste %	Material Cost £	Craft Optve Hrs	Lab Hrs	Labour Cost £	Sunds £	Nett Rate £	Unit	Gross Rate (+20%) £
K20: TIMBER BOARD FLOORING/SHEATHING/LININGS/CASINGS: WALL AND CEILING BOARDING										
Softwood, wrought; fixing to timber										
Boarding to walls, tongued, grooved and veed joints; 19mm thick, 100mm wide boards										
width exceeding 300mm	13.36	10.00	**14.69**	1.32	0.17	**20.52**	0.07	**35.29**	m2	**42.35**
Boarding to ceilings, tongued, grooved and veed joints; 19mm thick, 150mm wide boards										
width exceeding 300mm	12.20	10.00	**13.42**	1.32	0.17	**20.52**	0.07	**34.01**	m2	**40.82**
Boarding to ceilings, tongued, grooved and veed joints; 25mm thick, 100mm wide boards										
width exceeding 300mm	17.42	10.00	**19.17**	2.04	0.25	**31.63**	0.11	**50.90**	m2	**61.08**
Finished angles										
External; 19mm thick softwood boarding										
tongued and mitred	-	-	-	0.88	-	**12.50**	-	**12.50**	m	**15.01**
K40: DEMOUNTABLE SUSPENDED CEILINGS										
Suspended ceilings; 600 x 600mm bevelled, grooved and rebated patterned tiles p.c. £15.00/m2; laying in position in metal suspension system of main channel members on wire or rod hangers										
Depth of suspension 150 - 500mm										
15.8mm thick linings; fixing hangers to masonry..	17.33	5.00	**18.19**	0.79	0.40	**15.50**	4.51	**38.20**	m2	**45.84**
Edge trims										
Plain										
angle section; white stove enamelled aluminium; fixing to masonry with screws at 450mm centres...	1.38	5.00	**1.44**	0.55	0.28	**10.76**	-	**12.20**	m	**14.65**

WINDOWS/DOORS/STAIRS

Labour hourly rates: (except Specialists) Craft Operatives £14.21 Labourer £10.71 Rates are national average prices. Refer to REGIONAL VARIATIONS for indicative levels of overall pricing in regions	MATERIALS			LABOUR				RATES		
	Del to Site £	Waste %	Material Cost £	Craft Optve Hrs	Lab Hrs	Labour Cost £	Sunds £	Nett Rate £	Unit	Gross Rate (+20%) £
L10: WINDOWS/ROOFLIGHTS/SCREENS/LOUVRES: TIMBER WINDOWS										
Casements in softwood, base coat stain by manufacturer										
Magnet Trade; softwood Statesman Plus casement windows without bars; softwood sub-sills; hinges; fasteners										
631 x 750mm overall; 107V	105.16	2.50	**107.79**	1.71	0.85	**33.30**	1.78	**142.87**	nr	**171.45**
631 x 1050mm overall; 110V	113.83	2.50	**116.67**	1.98	0.99	**38.74**	1.78	**157.19**	nr	**188.63**
915 x 1050mm overall; 2N10W	134.15	2.50	**137.51**	2.26	1.12	**44.06**	1.78	**183.35**	nr	**220.02**
1200 x 900mm overall; 209W	160.03	2.50	**164.03**	2.37	1.18	**46.21**	1.78	**212.02**	nr	**254.42**
1200 x 1050mm overall; 210W	163.03	2.50	**167.10**	2.53	1.27	**49.50**	1.78	**218.39**	nr	**262.06**
1524 x 1050mm overall; 3NN10WW	209.34	2.50	**214.58**	2.75	1.38	**53.80**	1.78	**270.16**	nr	**324.20**
Fully reversible windows in softwood, wrought; base coat stain by manufacturer										
Magnet Trade Hi-Profile; softwood horizontal tilt windows; locking catch; high security espagnolette bolt; fully weatherstripped and Factory Double Glazed; clear low E glass										
600 x 900mm overall; HP0906	323.47	2.50	**331.56**	1.76	0.88	**34.43**	1.78	**367.77**	nr	**441.33**
900 x 1050mm overall; HP0910	372.39	2.50	**381.70**	2.20	1.10	**43.04**	1.78	**426.53**	nr	**511.84**
1200 x 1050mm overall; HP1210	425.31	2.50	**435.94**	2.48	1.23	**48.36**	1.78	**486.08**	nr	**583.30**
Windows in softwood, base coat stained by manufacturer										
Jeld-Wen Ltd. Softwood Sovereign windows plain casement non bar type side hung; glazing beads; weatherstripping, hinges; fasteners; base coat stain by manufacturer										
630 x 900mm overall; W109C	122.38	2.50	**125.43**	2.42	1.21	**47.35**	1.78	**174.56**	nr	**209.48**
1200 x 900mm overall; W209C	154.06	2.50	**157.91**	3.19	1.60	**62.41**	1.78	**222.10**	nr	**266.52**
Windows in hardwood, base coat stained by manufacturer										
Jeld-Wen Ltd. Hardwood Sovereign windows casement with vents; glazing beads; weatherstripping, hinges; fasteners; base coat stain by manufacturer										
630 x 900mm overall; W109VH	214.62	2.50	**219.99**	2.92	1.19	**54.15**	1.78	**275.91**	nr	**331.10**
915 x 1050mm overall; W2N10CVH	263.71	2.50	**270.30**	3.14	1.56	**61.28**	1.78	**333.36**	nr	**400.03**
1200 x 1500mm overall; W215TH	518.09	2.50	**531.04**	3.96	1.98	**77.48**	1.78	**610.30**	nr	**732.36**
Purpose made windows in softwood, wrought										
38mm moulded casements or fanlights										
in one pane	116.76	2.50	**119.68**	1.93	0.24	**29.95**	0.31	**149.93**	m2	**179.92**
divided into panes 0.10 - 0.50m2	134.27	2.50	**137.63**	1.93	0.24	**29.95**	0.31	**167.88**	m2	**201.46**
divided into panes not exceeding 0.10m2	151.79	2.50	**155.58**	1.93	0.24	**29.95**	0.31	**185.84**	m2	**223.00**
50mm moulded casements or fanlights										
in one pane	134.27	2.50	**137.63**	2.04	0.25	**31.63**	0.33	**169.59**	m2	**203.50**
divided into panes 0.10 - 0.50m2	151.79	2.50	**155.58**	2.04	0.25	**31.63**	0.33	**187.54**	m2	**225.05**
divided into panes not exceeding 0.10m2	169.30	2.50	**173.53**	2.04	0.25	**31.63**	0.33	**205.49**	m2	**246.59**
Labours										
fitting and hanging casement or fanlight on butts (included elsewhere)										
38mm	-	-	**-**	0.83	0.10	**12.78**	-	**12.78**	nr	**15.34**
50mm	-	-	**-**	0.88	0.11	**13.68**	-	**13.68**	nr	**16.42**
fitting and hanging casement or fanlight on sash centres (included elsewhere)										
38mm	-	-	**-**	1.82	0.23	**28.27**	-	**28.27**	nr	**33.92**
50mm	-	-	**-**	1.98	0.24	**30.73**	-	**30.73**	nr	**36.87**
urpose made windows in Afromosia, wrought										
0mm moulded casements or fanlights										
in one pane	258.95	2.50	**265.42**	3.03	0.37	**46.99**	0.50	**312.90**	m2	**375.49**
divided into panes 0.10 - 0.50m2	292.72	2.50	**300.04**	3.03	0.37	**46.99**	0.50	**347.52**	m2	**417.03**
divided into panes not exceeding 0.10m2	326.50	2.50	**334.66**	3.03	0.37	**46.99**	0.50	**382.14**	m2	**458.57**

Labour hourly rates: (except Specialists) Craft Operatives £14.21 Labourer £10.71 Rates are national average prices. Refer to REGIONAL VARIATIONS for indicative levels of overall pricing in regions	MATERIALS			LABOUR				RATES		
	Del to Site £	Waste %	Material Cost £	Craft Optve Hrs	Lab Hrs	Labour Cost £	Sunds £	Nett Rate £	Unit	Gross Rate (+20%) £
L10: WINDOWS/ROOFLIGHTS/SCREENS/LOUVRES: TIMBER WINDOWS Cont.										
Sash windows in softwood, wrought, preservative treated; priming or staining by manufacturer										
Magnet Trade; Georgian, open rebated for glass; weatherstripping; spiral balances; brass plated fittings; fixing to masonry with galvanized steel cramps -4 nr, 25 x 3 x 150mm girth, flat section, holes -2										
825 x 1394mm overall; GS4	420.58	2.50	431.10	3.30	1.65	64.56	1.71	497.37	nr	596.84
L10: WINDOWS/ROOFLIGHTS/SCREENS/LOUVRES: TIMBER ROOFLIGHTS										
Roof windows in Nordic red pine, wrought, treated										
Velux roof windows GGL/GHL 3000 Range, factory glazed clear float double glazed sealed unit; type EDL 0000; fixing to timber with screws; dressing flashings										
778 x 550mm overall, C02 GGL Range	166.11	2.50	170.26	6.05	3.03	118.37	-	288.63	nr	346.36
1178 x 1140mm overall, S06 GGL Range	278.65	2.50	285.61	8.80	4.40	172.17	-	457.78	nr	549.34
L10: WINDOWS/ROOFLIGHTS/SCREENS/LOUVRES: METAL SCREENS, BORROWED LIGHTS, FRAMES AND GRILLES										
Grilles in mild steel										
Grilles; 13 x 51mm flat bar framing to perimeter; 13mm diameter vertical infill bars at 150mm centres; welded connections										
1000 x 1000mm overall; fixing to masonry with screws	425.84	5.00	447.13	4.95	2.48	96.85	-	543.98	nr	652.77
Bedding and pointing frames										
Pointing with Secomastic standard metal frames one side	0.80	10.00	0.88	0.15	-	2.19	-	3.07	m	3.69
Bedding in cement mortar (1:3); pointing with Secomastic standard mastic one side metal frames	0.99	5.00	1.07	0.22	-	3.13	-	4.20	m	5.04
L10 - WINDOWS/ROOFLIGHTS/SCREENS/LOUVRES: PLASTICS WINDOWS										
Windows in uPVC; white										
Windows; factory glazed 20mm double glazed units; hinges; fastenings										
fixed light; fixing to masonry with cleats and screws; overall size										
600 x 1050mm	200.14	2.50	205.14	3.91	1.95	76.34	1.71	283.19	nr	339.83
1200 x 900mm	259.66	2.50	266.15	4.73	2.37	92.54	1.71	360.40	nr	432.48
L10 - WINDOWS/ROOFLIGHTS/SCREENS/LOUVRES: PLASTIC ROOFLIGHTS										
Roof domelights in glass reinforced plastics										
Domelights, translucent										
plain; fixing to masonry with screws										
914 x 914mm	131.63	2.50	134.92	2.53	2.53	63.05	0.69	198.66	nr	238.40
controlled ventilator; fixing to masonry with screws										
914 x 914mm	277.47	2.50	284.41	2.53	2.53	63.05	0.69	348.15	nr	417.78
L20: DOORS/SHUTTERS/HATCHES: TIMBER DOORS										
Doors in softwood, wrought										
Doors in softwood, wrought, preservative treated										
External ledged and braced doors; L & B Magnet Trade										
762 x 1981 x 44mm (f sizes)	82.76	2.50	84.83	1.65	0.18	25.33	-	110.16	nr	132.20
External panelled doors; 2XG, glazed with 4mm clear toughend glass; Magnet Trade										
762 x 1981 x 44mm (f sizes)	99.07	2.50	101.55	2.15	0.26	33.31	-	134.86	nr	161.83
Doors in Hardwood (solid, laminated or veneered), wrought										
External panelled doors; Elizabethan; Magnet Trade										
838 x 1981 x 44mm (f sizes)	173.93	2.50	178.28	2.92	0.36	45.31	-	223.59	nr	268.31
Flush doors										
Internal; hardboard facings; Magnet Trade										
762 x 1981 x 34mm	30.98	2.50	31.75	2.15	0.26	33.31	-	65.06	nr	78.07
Internal; Magnaseal; Magnet Trade										
762 x 1981 x 34mm	24.10	2.50	24.70	2.42	0.31	37.69	-	62.39	nr	74.87

Labour hourly rates: (except Specialists) Craft Operatives £14.21 Labourer £10.71 Rates are national average prices. Refer to REGIONAL VARIATIONS for indicative levels of overall pricing in regions	MATERIALS			LABOUR				RATES		
	Del to Site £	Waste %	Material Cost £	Craft Optve Hrs	Lab Hrs	Labour Cost £	Sunds £	Nett Rate £	Unit	Gross Rate (+20%) £
L20: DOORS/SHUTTERS/HATCHES: TIMBER DOORS Cont.										
Flush doors Cont.										
Internal; plywood facings; Magnet Trade										
762 x 1981 x 34mm	30.98	2.50	**31.75**	2.15	0.26	**33.31**	-	**65.06**	nr	**78.07**
Internal; Sapele Showpiece; Magnet Trade										
762 x 1981 x 34mm	35.12	2.50	**35.99**	2.64	0.33	**41.05**	-	**77.04**	nr	**92.45**
Internal; fire resisting; Plywood facings; FD30; Magnet Trade										
762 x 1981 x 44mm	73.03	2.50	**74.86**	2.42	0.31	**37.69**	-	**112.54**	nr	**135.05**
Panelled doors										
Panelled doors in softwood, wrought										
50mm square framed (or chamfered or moulded one or both sides)										
two panel	88.64	2.50	**90.85**	1.27	0.15	**19.62**	-	**110.48**	m2	**132.57**
four panel	97.90	2.50	**100.34**	1.27	0.15	**19.62**	-	**119.97**	m2	**143.96**
six panel	107.18	2.50	**109.86**	1.27	0.15	**19.62**	-	**129.49**	m2	**155.38**
add if upper panels open moulded in small squares for glass	19.24	2.50	**19.72**	-	-	**-**	-	**19.72**	m2	**23.67**
Panelled doors in Sapele, wrought										
50mm square framed (or chamfered or moulded one or both sides)										
two panel	215.65	2.50	**221.04**	1.93	0.24	**29.95**	-	**250.99**	m2	**301.19**
four panel	231.29	2.50	**237.07**	1.93	0.24	**29.95**	-	**267.02**	m2	**320.42**
six panel	246.91	2.50	**253.09**	1.93	0.24	**29.95**	-	**283.03**	m2	**339.64**
add if upper panels open moulded in small squares for glass	37.51	2.50	**38.44**	-	-	**-**	-	**38.44**	m2	**46.13**
Patio doors in hardwood, wrought										
Double glazed clear toughened glass, Magnet Trade, one opening leaf, doors and frames treated with base coat stains, fixing frame to masonry with screws										
2073 X 2387mm overall, HP 8	1539.93	2.50	**1578.43**	13.20	1.65	**205.24**	-	**1783.67**	nr	**2140.41**
extra; ventilator head, for HP 8	24.20	2.50	**24.81**	-	-	**-**	-	**24.81**	nr	**29.77**
Doorsets mainly in softwood, wrought										
Doorsets; 28mm thick jambs, head and transome with 12mm thick stop to suit 100mm thick wall; 14mm thick hardwood threshold; honeycomb core plywood faced flush door lipped two long edges; 6mm plywood transome panel fixed with pinned beads; 65mm snap in hinges; 57mm backset mortice latch; fixing frame to masonry with screws										
726 x 2040 x 40mm flush door -1 nr; basic dimensions 800 x 2100mm	146.05	2.50	**149.70**	1.38	0.18	**21.42**	-	**171.12**	nr	**205.35**
Note actual frame size is 14mm narrower and 14mm shorter than basic size dimensions										
L20: DOORS/SHUTTERS/HATCHES: ROLLER SHUTTERS										
Roller shutters in wood										
Note roller shutters are always purpose made to order and the following prices are indicative only. Firm quotations should always be obtained										
Pole and hook operation										
2134 x 2134mm	1650.00	-	**1650.00**	13.20	13.20	**328.94**	27.50	**2006.44**	nr	**2407.73**
L20: DOORS/SHUTTERS/HATCHES: LININGS										
Door frames and door lining sets in softwood, wrought										
Sets										
linings; fixing to masonry with screws										
32 x 113mm	6.03	2.50	**6.18**	0.50	0.07	**7.74**	0.10	**14.02**	m	**16.83**
32 x 225mm	10.83	2.50	**11.10**	0.77	0.09	**11.88**	0.14	**23.13**	m	**27.75**
Internal door lining sets; supplied unassembled; fixing to masonry with screws										
32 x 115mm rebated linings; assembling for 762 x 1981mm doors	26.88	2.50	**27.55**	2.53	0.32	**39.37**	0.41	**67.33**	nr	**80.80**
32 x 140mm rebated linings; assembling for 762 x 1981mm doors	29.33	2.50	**30.06**	2.53	0.32	**39.37**	0.41	**69.83**	nr	**83.80**
Door frames and door lining sets in Sapele, wrought, selected for transparent finish										
Sets										
linings; fixing to masonry with screws										
32 x 113mm	12.68	2.50	**12.99**	0.72	0.09	**11.10**	0.13	**24.23**	m	**29.07**
38 x 225mm	26.07	2.50	**26.72**	1.01	0.12	**15.68**	0.19	**42.59**	m	**51.10**

Labour hourly rates: (except Specialists) Craft Operatives £14.21 Labourer £10.71 Rates are national average prices. Refer to REGIONAL VARIATIONS for indicative levels of overall pricing in regions	MATERIALS			LABOUR				RATES		
	Del to Site £	Waste %	Material Cost £	Craft Optve Hrs	Lab Hrs	Labour Cost £	Sunds £	Nett Rate £	Unit	Gross Rate (+20%) £
L20: DOORS/SHUTTERS/HATCHES: FRAMES										
Door frames and door lining sets in softwood, wrought - frames										
Sets										
50 x 75mm jambs	6.53	2.50	**6.69**	0.25	0.03	**3.95**	0.04	**10.69**	m	12.82
38 x 75mm jambs; labours -1	5.78	2.50	**5.93**	0.23	0.03	**3.64**	0.04	**9.61**	m	11.53
38 x 75mm heads	5.65	2.50	**5.79**	0.23	0.03	**3.64**	0.04	**9.47**	m	11.36
50 x 75mm heads	6.53	2.50	**6.69**	0.25	0.03	**3.95**	0.04	**10.69**	m	12.82
38 x 75mm heads; labours -1	5.78	2.50	**5.93**	0.23	0.03	**3.64**	0.03	**9.60**	m	11.52
63 x 125mm sills	12.86	2.50	**13.18**	0.29	0.03	**4.42**	0.04	**17.64**	m	21.17
75 x 125mm sills	14.66	2.50	**15.02**	0.31	0.04	**4.85**	0.06	**19.93**	m	23.91
63 x 125mm sills; labours -2	13.09	2.50	**13.41**	0.29	0.03	**4.42**	0.04	**17.87**	m	21.45
Internal door frame sets; supplied unassembled										
38 x 75mm jambs and head; stops (supplied loose); assembling										
for 762 x 1981mm doors	26.06	2.50	**26.71**	1.27	0.15	**19.62**	0.21	**46.55**	nr	55.85
50 x 100mm jambs and head; stops (supplied loose); assembling										
for 762 x 1981mm doors	40.72	2.50	**41.73**	1.27	0.15	**19.62**	0.21	**61.57**	nr	73.88
External door frame sets; one coat external primer before delivery to site										
63 x 75mm jambs and head, rebates -1 nr; hardwood sill										
for 838 x 1981 x 50mm doors	99.36	2.50	**101.84**	1.60	0.20	**24.79**	0.26	**126.89**	nr	152.27
Door frames and door lining sets in Sapele, wrought, selected for transparent finish - frames										
Sets										
38 x 75mm jambs	11.88	2.50	**12.17**	0.34	0.04	**5.32**	0.06	**17.55**	m	21.05
50 x 100mm jambs	15.93	2.50	**16.33**	0.39	0.04	**5.94**	0.07	**22.34**	m	26.81
38 x 75mm heads	11.88	2.50	**12.17**	0.34	0.04	**5.32**	0.06	**17.55**	m	21.05
50 x 100mm heads	15.93	2.50	**16.33**	0.39	0.04	**5.94**	0.07	**22.34**	m	26.81
38 x 75mm heads; labours -1	12.10	2.50	**12.40**	0.34	0.04	**5.32**	0.06	**17.77**	m	21.33
75 x 150mm sills	34.76	2.50	**35.63**	0.47	0.06	**7.31**	0.08	**43.02**	m	51.62
Intumescent strips and smoke seals										
Albi-Flex, white PVC sleeved self adhesive; Rentokil Ltd.										
12 x 4mm intumescent strip, half hour application; setting into groove in timber frame or door	4.45	10.00	**4.89**	0.19	-	**2.66**	-	**7.55**	m	9.06
18 x 4mm intumescent strip, one hour application; setting into groove in timber frame or door	6.24	10.00	**6.87**	0.19	-	**2.66**	-	**9.52**	m	11.43
L30 STAIRS/WALKWAYS/BALUSTRADES: TIMBER STAIRS										
Stairs in softwood, wrought										
Straight flight staircase and balustrade; 25mm treads; 19mm risers, glued; wedged and blocked; 25mm wall string; 38mm outer string; 75 x 75mm newels; 25 x 25mm balusters; 38 x 75mm hardwood handrail										
864mm wide x 2600mm rise overall; balustrade to one side; fixing to masonry with screws	770.10	2.50	**789.35**	16.50	2.09	**256.85**	2.64	**1048.84**	nr	1258.61
864mm wide x 2600mm rise overall with 3 nr winders at bottom; balustrade to one side; fixing to masonry with screws	924.12	2.50	**947.22**	22.00	2.75	**342.07**	3.58	**1292.87**	nr	1551.44
L40: GENERAL GLAZING										
Glass; BS 952, clear float										
4mm thick to wood rebates with BS 544 putty										
not exceeding 0.15m2	44.50	5.00	**46.72**	1.06	0.11	**16.18**	0.39	**63.29**	m2	75.95
0.15 - 4.00m2	44.50	5.00	**46.72**	0.53	0.06	**8.09**	0.19	**55.01**	m2	66.01
6mm thick to wood rebates with BS 544 putty										
not exceeding 0.15m2	65.05	5.00	**68.31**	1.19	0.11	**18.06**	0.39	**86.75**	m2	104.10
0.15 - 4.00m2	65.05	5.00	**68.31**	0.59	0.06	**9.03**	0.19	**77.53**	m2	93.04
4mm thick to wood rebates with bradded wood beads (included elsewhere) and BS 544 putty										
not exceeding 0.15m2	44.50	5.00	**46.72**	1.32	0.11	**19.94**	0.39	**67.04**	m2	80.45
0.15 - 4.00m2	44.50	5.00	**46.72**	0.66	0.06	**9.97**	0.19	**56.88**	m2	68.26
6mm thick to wood rebates with bradded wood beads (included elsewhere) and BS 544 putty										
not exceeding 0.15m2	65.05	5.00	**68.31**	1.45	0.11	**21.81**	0.39	**90.50**	m2	108.61
0.15 - 4.00m2	65.05	5.00	**68.31**	0.73	0.06	**10.91**	0.19	**79.41**	m2	95.29
4mm thick to wood rebates with screwed wood beads and glazing strip (included elsewhere)										
not exceeding 0.15m2	44.50	5.00	**46.72**	1.85	0.11	**27.44**	0.39	**74.55**	m2	89.45
0.15 - 4.00m2	44.50	5.00	**46.72**	0.92	0.06	**13.72**	0.19	**60.63**	m2	72.76
3mm thick to metal rebates with metal casement glazing compound										
not exceeding 0.15m2	44.50	5.00	**46.72**	1.19	0.11	**18.06**	0.45	**65.23**	m2	78.27
0.15 - 4.00m2	44.50	5.00	**46.72**	0.59	0.06	**9.03**	0.22	**55.97**	m2	67.16

Labour hourly rates: (except Specialists) Craft Operatives £14.21 Labourer £10.71 Rates are national average prices. Refer to REGIONAL VARIATIONS for indicative levels of overall pricing in regions	MATERIALS			LABOUR				RATES		
	Del to Site £	Waste %	Material Cost £	Craft Optve Hrs	Lab Hrs	Labour Cost £	Sunds £	Nett Rate £	Unit	Gross Rate (+20%) £
L40: GENERAL GLAZING Cont.										
Glass; BS 952, clear float Cont.										
4mm thick to metal rebates with metal casement glazing compound										
not exceeding 0.15m2	44.50	5.00	**46.72**	1.19	0.11	**18.06**	0.45	**65.23**	m2	**78.27**
0.15 - 4.00m2	44.50	5.00	**46.72**	0.59	0.06	**9.03**	0.22	**55.97**	m2	**67.16**
4mm thick to metal rebates with clipped metal beads and gaskets (included elsewhere)										
not exceeding 0.15m2	44.50	5.00	**46.72**	1.58	0.11	**23.69**	0.39	**70.79**	m2	**84.95**
0.15 - 4.00m2	44.50	5.00	**46.72**	0.79	0.06	**11.84**	0.19	**58.76**	m2	**70.51**
Glass; BS 952, rough cast										
6mm thick to wood rebates with screwed wood beads and glazing strip (included elsewhere)										
not exceeding 0.15m2	58.61	5.00	**61.54**	1.98	0.11	**29.31**	0.39	**91.24**	m2	**109.49**
0.15 - 4.00m2	58.61	5.00	**61.54**	0.99	0.06	**14.66**	0.19	**76.39**	m2	**91.67**
6mm thick to metal rebates with clipped metal beads and gaskets (included elsewhere)										
not exceeding 0.15m2	58.61	5.00	**61.54**	1.72	0.11	**25.56**	0.45	**87.55**	m2	**105.06**
0.15 - 4.00m2	58.61	5.00	**61.54**	0.86	0.06	**12.78**	0.22	**74.54**	m2	**89.44**
Glass; BS 952, Pyroshield clear										
6mm thick to wood rebates with screwed beads and intumescent glazing strip (included elsewhere)										
not exceeding 0.15m2	105.28	5.00	**110.55**	2.38	0.11	**34.94**	-	**145.49**	m2	**174.58**
not exceeding 0.15m2; aligning panes with adjacent panes	105.28	8.50	**114.23**	2.64	0.11	**38.69**	-	**152.92**	m2	**183.51**
0.15 - 4.00m2	105.28	5.00	**110.55**	1.19	0.06	**17.47**	-	**128.02**	m2	**153.62**
Glass; BS 952, Pyroshield safety clear										
6mm thick to wood rebates with screwed beads and intumescent glazing strip (included elsewhere)										
not exceeding 0.15m2	123.97	5.00	**130.17**	2.38	0.11	**34.94**	-	**165.11**	m2	**198.13**
0.15 - 4.00m2	123.97	5.00	**130.17**	1.19	0.06	**17.47**	-	**147.64**	m2	**177.17**
Glass; BS 952, Pyroshield safety texture										
7mm thick to wood rebates with screwed beads and intumescent glazing strip (included elsewhere)										
not exceeding 0.15m2	63.37	5.00	**66.54**	2.38	0.11	**34.94**	-	**101.48**	m2	**121.78**
0.15 - 4.00m2	63.37	5.00	**66.54**	1.19	0.06	**17.47**	-	**84.01**	m2	**100.81**
Glass; BS 952, white patterned										
4mm thick to wood rebates with bradded wood beads (included elsewhere) and BS 544 putty										
not exceeding 0.15m2	41.27	5.00	**43.34**	1.32	0.11	**19.94**	0.39	**63.66**	m2	**76.39**
0.15 - 4.00m2	41.27	5.00	**43.34**	0.66	0.06	**9.97**	0.19	**53.50**	m2	**64.20**
Patterns available										
4mm white										
Arctic										
4 and 6mm white										
Autumn; Cotswold; Driftwood; Everglade; Flemish; Linkon; Mayflower; Reeded; Stippolyte										
Glass; BS 952, antisun float; grey										
4mm thick to wood rebates with bradded wood beads (included elsewhere) and BS 544 putty										
not exceeding 2400 x 1200mm	84.95	5.00	**89.20**	0.92	0.13	**14.54**	0.28	**104.02**	m2	**124.83**
6mm thick to wood rebates with bradded wood beads (included elsewhere) and BS 544 putty										
not exceeding 5950 x 3150mm	122.60	5.00	**128.72**	0.99	0.13	**15.48**	0.28	**144.48**	m2	**173.38**
Glass; BS 952, antisun float; bronze										
4mm thick to wood rebates with bradded wood beads (included elsewhere) and BS 544 putty										
not exceeding 2400 x 1200mm	84.95	5.00	**89.20**	0.92	0.13	**14.54**	0.28	**104.02**	m2	**124.83**
6mm thick to wood rebates with bradded wood beads (included elsewhere) and BS 544 putty										
not exceeding 5950 x 3150mm	122.60	5.00	**128.72**	0.99	0.13	**15.48**	0.28	**144.48**	m2	**173.38**
Glass; BS 952, toughened clear float										
4mm thick to metal rebates with screwed metal beads and gaskets (included elsewhere)										
not exceeding 2400 x 1300mm	62.63	5.00	**65.77**	0.79	0.13	**12.67**	-	**78.43**	m2	**94.12**
5mm thick to metal rebates with screwed metal beads and gaskets (included elsewhere)										
not exceeding 2500 x 1520mm	94.00	5.00	**98.69**	0.86	0.13	**13.61**	-	**112.30**	m2	**134.76**

Labour hourly rates: (except Specialists) Craft Operatives £14.21 Labourer £10.71 Rates are national average prices. Refer to REGIONAL VARIATIONS for indicative levels of overall pricing in regions	MATERIALS			LABOUR				RATES		
	Del to Site £	Waste %	Material Cost £	Craft Optve Hrs	Lab Hrs	Labour Cost £	Sunds £	Nett Rate £	Unit	Gross Rate (+20%) £
L40: GENERAL GLAZING Cont.										
Glass; BS 952, laminated safety, clear float										
4.4mm thick to metal rebates with screwed metal beads and gaskets (included elsewhere)										
not exceeding 2100 x 1200mm	82.80	5.00	86.94	2.11	0.13	31.43	-	118.36	m2	142.03
6.4mm thick to metal rebates with screwed metal beads and gaskets (included elsewhere)										
not exceeding 3210 x 2000mm	69.42	5.00	72.89	2.64	0.13	38.93	-	111.82	m2	134.18
Materials resembling glass; UVA stabilised polycarbonate sheet latex paper masked both sides										
4mm thick to metal rebates with screwed metal beads and gaskets (included elsewhere)										
standard grade	60.04	5.00	63.04	1.32	0.13	20.17	-	83.21	m2	99.85
Materials resembling glass; Meshlite vandal resistant glazing; Georgian, clear, smooth both faces or crinkle one face										
4mm thick to wood rebates with screwed wood beads (included elsewhere) and BS 544 putty										
general purpose grade	43.57	5.00	45.75	1.27	0.13	19.39	0.18	65.32	m2	78.38
fire retardant grade, class 2	52.61	5.00	55.24	1.27	0.13	19.39	0.18	74.81	m2	89.78
fire retardant grade, class 0	61.94	5.00	65.04	1.27	0.13	19.39	0.18	84.61	m2	101.53
Factory made double glazed hermetically sealed units										
Two panes BS 952, clear float 3 or 4mm thick; to metal rebates with screwed metal beads and gaskets (included elsewhere)										
521mm wide x 421mm high	20.99	5.00	22.04	1.27	-	17.98	-	40.01	nr	48.02
740mm wide x 740mm high	52.38	5.00	55.00	2.20	-	31.26	-	86.26	nr	103.52
Two panes BS 952, clear float 5 or 6mm thick; to metal rebates with screwed metal beads and gaskets (included elsewhere)										
521mm wide x 421mm high	30.90	5.00	32.44	1.38	-	19.54	-	51.98	nr	62.38
740mm wide x 740mm high	77.19	5.00	81.05	2.42	-	34.39	-	115.43	nr	138.52
Inner pane BS 952, clear float 4mm thick; outer pane BS 952, white patterned 4mm thick; to metal rebates with screwed metal beads and gaskets (included elsewhere)										
521mm wide x 421mm high	25.77	5.00	27.06	1.27	-	17.98	-	45.04	nr	54.04
740mm wide x 740mm high	64.04	5.00	67.24	2.20	-	31.26	-	98.51	nr	118.21
Polyester window films; Durable Berkeley Company Ltd.,										
3M Scotchtint Solar Control films; applying to glass										
type P18, silver	-	-	-	-	-	-	-	34.99	m2	41.99
type RE15S1X, external	-	-	-	-	-	-	-	44.53	m2	53.43
Drilling										
Hole through sheet or float glass; not exceeding 6mm thick										
6 - 15mm diameter	4.38	-	4.38	-	-	-	-	4.38	nr	5.25
16 - 38mm diameter	6.18	-	6.18	-	-	-	-	6.18	nr	7.42
exceeding 38mm diameter	12.48	-	12.48	-	-	-	-	12.48	nr	14.97
Bedding edges of panes										
Rubber glazing strips										
to edges of 3mm thick glass or the like	0.59	5.00	0.62	0.07	-	0.94	-	1.56	m	1.87
to edges of 6mm thick glass or the like	0.72	5.00	0.75	0.08	-	1.09	-	1.84	m	2.21
Hacking out existing glass; preparing for re-glazing										
Float glass										
wood rebates	-	-	-	0.48	0.48	12.06	0.21	12.27	m	14.72
wood rebates and screwed wood beads; storing beads for re-use	-	-	-	0.57	0.57	14.25	0.21	14.46	m	17.35
Float glass behind guard bars in position										
wood rebates and screwed wood beads; storing beads for re-use	-	-	-	0.94	0.94	23.30	0.21	23.51	m	28.21

SURFACE FINISHES

Labour hourly rates: (except Specialists) Craft Operatives £14.21 Labourer £10.71 Rates are national average prices. Refer to REGIONAL VARIATIONS for indicative levels of overall pricing in regions	MATERIALS			LABOUR			RATES			
	Del to Site £	Waste %	Material Cost £	Craft Optve Hrs	Lab Hrs	Labour Cost £	Sunds £	Nett Rate £	Unit	Gross Rate (+20%) £
M10: CEMENT: SAND/CONCRETE SCREEDS/TOPPINGS										
Mortar, cement and sand (1:3) - screeds										
13mm work to walls on brickwork or blockwork base; one coat; screeded										
width exceeding 300mm	1.72	5.00	**1.81**	0.51	0.25	**9.90**	-	**11.71**	m2	**14.05**
25mm work to floors on concrete base; one coat; screeded										
level and to falls only not exceeding 15 degrees from horizontal	3.50	5.00	**3.66**	0.32	0.20	**6.65**	-	**10.32**	m2	**12.38**
50mm work to floors on concrete base; one coat; screeded										
level and to falls only not exceeding 15 degrees from horizontal	6.81	5.00	**7.14**	0.44	0.26	**9.08**	-	**16.22**	m2	**19.47**
Granolithic screed; steel troweled										
25mm work to floors on concrete base; one coat level and to falls only not exceeding 15 degrees from horizontal	4.61	5.00	**4.82**	0.53	0.23	**9.98**	-	**14.79**	m2	**17.75**
50mm work to floors on concrete base; one coat level and to falls only not exceeding 15 degrees from horizontal	8.84	5.00	**9.26**	0.67	0.30	**12.72**	-	**21.98**	m2	**26.38**
to slopes exceeding 15 degrees from horizontal	8.84	5.00	**9.26**	0.87	0.30	**15.53**	-	**24.79**	m2	**29.75**
If carborundum troweled into surface										
add	4.24	10.00	**4.66**	0.11	0.06	**2.15**	-	**6.81**	m2	**8.17**
25mm work to skirtings on concrete base										
height 150mm	1.02	10.00	**1.12**	1.05	0.06	**15.44**	-	**16.56**	m	**19.87**
Rounded angles and intersections										
10 - 100mm radius	-	-	**-**	0.22	-	**3.13**	-	**3.13**	m	**3.75**
Surface hardeners on paving										
Proprietary surface hardener (Epoxy Resin Based)										
two coats	1.98	10.00	**2.18**	-	0.25	**2.71**	-	**4.89**	m2	**5.87**
Polyurethane floor sealer										
two coats	2.28	10.00	**2.51**	-	0.25	**2.71**	-	**5.22**	m2	**6.27**
Nitoflor Lithurin concrete surface dressing										
one coat	0.45	10.00	**0.49**	-	0.23	**2.47**	-	**2.97**	m2	**3.56**
two coats	0.90	10.00	**0.99**	-	0.33	**3.53**	-	**4.52**	m2	**5.43**
If on old floors										
add for cleaning and degreasing	0.39	5.00	**0.41**	-	0.44	**4.71**	-	**5.12**	m2	**6.15**
Vermiculite screed consisting of cement and Vermiculite aggregate finished with 20mm cement and sand (1:4) screeded bed										
70mm work to roofs on concrete base; two coats to falls and crossfalls and to slopes not exceeding 15 degrees from horizontal	13.22	5.00	**13.87**	0.72	0.77	**18.41**	-	**32.28**	m2	**38.73**
Screed reinforcement										
Reinforcement; galvanized wire netting, BS EN 10223-2, 25mm mesh, 20 gauge wire; 150mm laps; placing in position										
floors	1.88	5.00	**1.98**	0.06	-	**0.78**	-	**2.76**	m2	**3.31**

SURFACE FINISHES

Labour hourly rates: (except Specialists) Craft Operatives £14.21 Labourer £10.71 Rates are national average prices. Refer to REGIONAL VARIATIONS for indicative levels of overall pricing in regions	MATERIALS			LABOUR				RATES		
	Del to Site £	Waste %	Material Cost £	Craft Optve Hrs	Lab Hrs	Labour Cost £	Sunds £	Nett Rate £	Unit	Gross Rate (+20%) £
M10: CEMENT: SAND/CONCRETE SCREEDS/TOPPINGS Cont.										
Division strips										
Aluminium dividing strips; setting in bed										
5 x 25mm; flat section	5.34	5.00	**5.60**	0.23	-	**3.28**	-	**8.88**	m	**10.66**
Brass dividing strips; setting in bed										
5 x 25mm; flat section	6.71	5.00	**7.05**	0.23	-	**3.28**	-	**10.33**	m	**12.39**
M10: CEMENT: SAND/CONCRETE SCREEDS/TOPPINGS; STAIRCASE AREAS										
Granolithic screed; steel troweled										
25mm work to floors on concrete base; one coat level and to falls only not exceeding 15 degrees from horizontal	4.61	5.00	**4.82**	0.79	0.23	**13.73**	-	**18.55**	m2	**22.26**
25mm work to treads on concrete base width 275mm	1.79	10.00	**1.96**	0.33	0.07	**5.40**	-	**7.35**	m	**8.82**
13mm work to undercut risers on concrete base height 175mm	0.51	10.00	**0.56**	0.66	0.06	**9.97**	-	**10.53**	m	**12.63**
13mm work to strings on concrete base height 300mm	1.02	10.00	**1.12**	0.83	0.06	**12.31**	-	**13.43**	m	**16.12**
M12: TROWELED BITUMEN/RESIN/RUBBER LATEX FLOORING										
One coat levelling screeds										
3mm work to floors on existing timber boarded base; one coat level and to falls only not exceeding 15 degrees from horizontal	3.85	5.00	**4.04**	0.64	0.07	**9.77**	-	**13.81**	m2	**16.57**
M20: PLASTERED/RENDERED/ROUGHCAST COATINGS - PLASTERBOARD										
Baseboarding; Gyproc Handiboard; 5mm joints, filled with plaster and scrimmed; fixing with Galvanized nails, 30mm long at 400mm ccs; internal										
9.5mm work to walls on timber base										
width exceeding 300mm	3.30	5.00	**3.47**	0.32	0.17	**6.30**	-	**9.77**	m2	**11.73**
width not exceeding 300mm	1.08	10.00	**1.17**	0.19	0.06	**3.25**	-	**4.42**	m	**5.31**
9.5mm work to ceilings on timber base										
width exceeding 300mm	3.30	5.00	**3.47**	0.39	0.17	**7.24**	-	**10.71**	m2	**12.85**
9.5mm work to isolated beams on timber base										
width exceeding 300mm	3.30	5.00	**3.47**	0.48	0.17	**8.64**	-	**12.12**	m2	**14.54**
Baseboarding; Gyproc Handiboard; 5mm joints, filled with plaster and scrimmed; fixing with adhesive dabs; internal										
9.5mm work to walls on masonry base										
width exceeding 300mm	3.27	5.00	**3.44**	0.32	0.17	**6.30**	-	**9.74**	m2	**11.69**
width not exceeding 300mm	1.06	10.00	**1.16**	0.19	0.06	**3.25**	-	**4.40**	m	**5.28**
Baseboarding; two layers 9.5mm Gyproc square edge wallboard; second layer with 3mm joints filled with plaster and scrimmed; fixing with Galvanized nails, 30mm long at 400mm ccs; internal										
19mm work to walls on timber base										
width exceeding 300mm	4.57	5.00	**4.79**	0.66	0.33	**12.91**	-	**17.71**	m2	**21.25**
width not exceeding 300mm	1.52	10.00	**1.65**	0.40	0.10	**6.69**	-	**8.34**	m	**10.01**
19mm work to ceilings on timber base										
width exceeding 300mm	4.57	5.00	**4.79**	0.79	0.33	**14.79**	-	**19.58**	m2	**23.50**
19mm work to isolated beams on timber base										
width exceeding 300mm	4.57	5.00	**4.79**	0.88	0.33	**16.04**	-	**20.83**	m2	**25.00**
width not exceeding 300mm	1.52	10.00	**1.65**	0.53	0.10	**8.56**	-	**10.21**	m	**12.26**
Baseboarding; two layers; first layer 9.5mm Gyproc Duplex wallboard, second layer 9.5mm Gyproc wallboard both layers with square edge boards; second layer with 3mm joints filled with plaster and scrimmed; fixing with Galvanized nails, 30mm long at 400mm ccs; internal										
19mm work to walls on timber base										
width exceeding 300mm	5.24	5.00	**5.50**	0.66	0.33	**12.91**	-	**18.41**	m2	**22.09**
width not exceeding 300mm	1.72	10.00	**1.87**	0.40	0.10	**6.69**	-	**8.56**	m	**10.27**
19mm work to ceilings on timber base										
width exceeding 300mm	5.24	5.00	**5.50**	0.79	0.33	**14.79**	-	**20.29**	m2	**24.35**
width not exceeding 300mm	1.72	10.00	**1.87**	0.47	0.10	**7.78**	-	**9.65**	m	**11.58**
19mm work to isolated beams on timber base										
width exceeding 300mm	5.24	5.00	**5.50**	0.88	0.33	**16.04**	-	**21.54**	m2	**25.85**
width not exceeding 300mm	1.72	10.00	**1.87**	0.53	0.10	**8.56**	-	**10.43**	m	**12.52**

Labour hourly rates: (except Specialists) Craft Operatives £14.21 Labourer £10.71 Rates are national average prices. Refer to REGIONAL VARIATIONS for indicative levels of overall pricing in regions	MATERIALS			LABOUR				RATES		
	Del to Site £	Waste %	Material Cost £	Craft Optve Hrs	Lab Hrs	Labour Cost £	Sunds £	Nett Rate £	Unit	Gross Rate (+20%) £
M20: PLASTERED/RENDERED/ROUGHCAST COATINGS: DAMP WALL TREATMENT										
Newlath; fixing with masonry nails; internal										
Work to walls on brickwork or blockwork base										
width exceeding 300mm	13.59	5.00	**14.13**	0.44	0.22	**8.61**	-	**22.74**	m2	**27.28**
width not exceeding 300mm	4.08	5.00	**4.24**	0.26	0.07	**4.46**	-	**8.70**	m	**10.44**
M20: PLASTERED/RENDERED/ROUGHCAST COATINGS: HARDWALL PLASTERING										
Plaster, BS 1191, Part 1, Class B; finishing coat of board finish, 3mm thick; steel troweled; internal										
3mm work to walls on concrete or plasterboard base										
width exceeding 300mm	0.59	5.00	**0.62**	0.36	0.19	**7.16**	0.02	**7.79**	m2	**9.35**
width not exceeding 300mm	0.31	10.00	**0.34**	0.22	0.06	**3.72**	0.00	**4.06**	m	**4.87**
3mm work to ceilings on concrete or plasterboard base										
width exceeding 300mm	0.59	5.00	**0.62**	0.44	0.19	**8.26**	0.02	**8.89**	m2	**10.67**
width exceeding 300mm; 3.50 - 5.00m above floor	0.59	5.00	**0.62**	0.47	0.19	**8.72**	0.02	**9.36**	m2	**11.23**
width not exceeding 300mm	0.31	10.00	**0.34**	0.26	0.06	**4.34**	0.00	**4.68**	m	**5.62**
Plaster; first coat of hardwall, 11mm thick; finishing coat of multi-finish 2mm thick; steel troweled; internal										
13mm work to walls on brickwork or blockwork base										
width exceeding 300mm	2.36	5.00	**2.48**	0.66	0.33	**12.91**	0.06	**15.45**	m2	**18.54**
width not exceeding 300mm	1.07	10.00	**1.17**	0.40	0.10	**6.69**	0.02	**7.88**	m	**9.46**
13mm work to ceilings on concrete base										
width exceeding 300mm	2.36	5.00	**2.48**	0.79	0.33	**14.79**	0.06	**17.33**	m2	**20.79**
width exceeding 300mm; 3.50 - 5.00m above floor	2.36	5.00	**2.48**	0.86	0.33	**15.73**	0.06	**18.26**	m2	**21.92**
Rounded angles										
radius 10 - 100mm	-	-	-	0.33	-	**4.69**	-	**4.69**	m	**5.63**
M20: PLASTERED/RENDERED/ROUGHCAST COATINGS: LIGHTWEIGHT PLASTERING										
Plaster, Thistle; pre-mixed; floating coat of browning, 11mm thick; finishing coat of finish, 2mm thick; steel troweled; internal										
13mm work to walls on brickwork or blockwork base										
width exceeding 300mm	1.73	5.00	**1.82**	0.55	0.28	**10.76**	0.04	**12.62**	m2	**15.15**
width exceeding 300mm; dubbing average 12mm thick	3.19	5.00	**3.34**	0.86	0.53	**17.85**	0.08	**21.27**	m2	**25.53**
width not exceeding 300mm	0.76	10.00	**0.84**	0.33	0.09	**5.63**	0.01	**6.48**	m	**7.78**
width not exceeding 300mm; dubbing average 12mm thick	1.39	10.00	**1.52**	0.52	0.17	**9.11**	0.02	**10.66**	m	**12.79**
M20: PLASTERED/RENDERED/ROUGHCAST COATINGS: RENOVATING PLASTER										
Renovating plaster, Thistle; undercoat 11mm thick; finishing coat 2mm thick; steel troweled; internal										
13mm work to walls on existing brickwork or blockwork base										
width exceeding 300mm	2.69	5.00	**2.82**	0.72	0.36	**14.05**	0.07	**16.94**	m2	**20.32**
width exceeding 300mm; dubbing average 12mm thick	4.93	5.00	**5.18**	1.02	0.76	**22.67**	0.13	**27.98**	m2	**33.57**
width not exceeding 300mm	1.22	10.00	**1.34**	0.43	0.11	**7.27**	0.02	**8.64**	m	**10.36**
width not exceeding 300mm; dubbing average 12mm thick	2.22	10.00	**2.44**	0.62	0.23	**11.23**	0.04	**13.70**	m	**16.45**
M20: PLASTERED/RENDERED/ROUGHCAST COATINGS: PLASTER BEADS AND STOPS										
Accessories										
galvanized steel angle beads										
Expamet fixing to brickwork or blockwork with masonry nails										
Reference 550	0.65	10.00	**0.71**	0.11	-	**1.56**	-	**2.27**	m	**2.73**
galvanized steel stop beads										
Expamet fixing to brickwork or blockwork with masonry nails										
Reference 560 for 3mm	0.86	10.00	**0.95**	0.11	-	**1.56**	-	**2.51**	m	**3.01**
Accessories; external										
Stainless steel angle beads										
Expamet fixing to brickwork or blockwork with masonry nails										
Reference 545	2.17	10.00	**2.38**	0.11	-	**1.56**	-	**3.94**	m	**4.73**
Stainless steel stop beads										
Expamet fixing to brickwork or blockwork with masonry nails										
Reference 547	2.10	10.00	**2.30**	0.11	-	**1.56**	-	**3.87**	m	**4.64**

Labour hourly rates: (except Specialists) Craft Operatives £14.21 Labourer £10.71 Rates are national average prices. Refer to REGIONAL VARIATIONS for indicative levels of overall pricing in regions	MATERIALS			LABOUR				RATES		
	Del to Site £	Waste %	Material Cost £	Craft Optve Hrs	Lab Hrs	Labour Cost £	Sunds £	Nett Rate £	Unit	Gross Rate (+20%) £
M20: PLASTERED/RENDERED/ROUGHCAST COATINGS: ANTI-CRACK STRIPS										
Accessories										
Lathing; BB galvanized expanded metal lath, 9mm mesh X 0.500mm thick x 1.11 kg/m2; butt joints fixing with nails										
100mm wide to walls; one edge to timber; one edge to brickwork or blockwork	0.54	5.00	**0.57**	0.11	-	**1.56**	0.04	**2.17**	m	**2.61**
M20: PLASTERED/RENDERED/ROUGHCAST COATINGS: BONDING FLUID										
Prepare and apply bonding fluid to receive plaster or cement rendering										
Work to walls on existing cement and sand base										
width exceeding 300mm	0.62	10.00	**0.68**	0.28	-	**3.91**	-	**4.59**	m2	**5.51**
M30: METAL MESH LATHING/ANCHORED REINFORCEMENT FOR PLASTERED COATINGS										
Lathing; BB galvanized expanded metal lath, 9mm mesh x 0.500mm thick x 1.11 kg/m2; butt joints; fixing with galvanized staples										
Work to walls										
width exceeding 300mm	5.45	5.00	**5.72**	0.20	0.10	**3.87**	0.07	**9.66**	m2	**11.59**
Work to ceilings										
width exceeding 300mm	5.45	5.00	**5.72**	0.25	0.10	**4.66**	0.08	**10.46**	m2	**12.55**
Work to isolated beams										
width exceeding 300mm	5.45	5.00	**5.72**	0.30	0.10	**5.28**	0.10	**11.10**	m2	**13.32**
width not exceeding 300mm	1.91	10.00	**2.10**	0.18	0.03	**2.85**	0.06	**5.01**	m	**6.01**
Lathing; Expamet stainless steel Rib-Lath, 0.300mm thick x 1.16 kg/m2; butt joints; fixing with stainless steel staples										
Work to walls										
width exceeding 300mm	18.21	5.00	**19.12**	0.20	0.10	**3.87**	0.07	**23.05**	m2	**27.67**
Work to ceilings										
width exceeding 300mm	18.21	5.00	**19.12**	0.25	0.10	**4.66**	0.08	**23.85**	m2	**28.62**
Work to isolated beams										
width exceeding 300mm	18.21	5.00	**19.12**	0.30	0.10	**5.28**	0.10	**24.49**	m2	**29.39**
width not exceeding 300mm	6.37	10.00	**7.01**	0.18	0.03	**2.85**	0.06	**9.92**	m	**11.91**
Extra cost of fixing lathing to brickwork, blockwork or concrete with cartridge fired nails in lieu of to timber with staples										
Work to walls										
width exceeding 300mm	-	-	-	0.13	-	**1.88**	0.18	**2.05**	m2	**2.46**
width not exceeding 300mm	-	-	-	0.08	-	**1.09**	0.10	**1.20**	m	**1.44**
M31: FIBROUS PLASTER: PLASTERBOARD COVE										
Coves										
Gyproc cove; fixing with adhesive										
100mm girth	1.41	10.00	**1.55**	0.20	0.10	**3.87**	-	**5.42**	m	**6.51**
extra; ends	-	-	-	0.07	-	**0.94**	-	**0.94**	nr	**1.13**
extra; internal angles	-	-	-	0.33	-	**4.69**	-	**4.69**	nr	**5.63**
extra; external angles	-	-	-	0.33	-	**4.69**	-	**4.69**	nr	**5.63**
127mm girth	1.26	10.00	**1.39**	0.20	0.10	**3.87**	-	**5.26**	m	**6.32**
extra; ends	-	-	-	0.07	-	**0.94**	-	**0.94**	nr	**1.13**
extra; internal angles	-	-	-	0.33	-	**4.69**	-	**4.69**	nr	**5.63**
extra; external angles	-	-	-	0.33	-	**4.69**	-	**4.69**	nr	**5.63**
M40: STONE/CONCRETE/QUARRY/CERAMIC TILING/MOSAIC: CLAY TILE PAVING										
Clay floor quarries, BS 6431, terracotta; 3mm joints, symmetrical layout; bedding in 10mm cement mortar (1:3); pointing in cement mortar (1:3); on cement and sand base										
150 x 150 x 12.5mm units to floors level or to falls only not exceeding 15 degrees from horizontal										
plain	18.26	5.00	**19.17**	1.05	0.66	**21.92**	-	**41.09**	m2	**49.31**
150 x 150 x 12.5mm units to skirtings on brickwork or blockwork base										
height 150mm; rounded top edge and cove at bottom	5.19	10.00	**5.69**	0.40	0.10	**6.69**	-	**12.37**	m	**14.85**

Labour hourly rates: (except Specialists) Craft Operatives £14.21 Labourer £10.71 Rates are national average prices. Refer to REGIONAL VARIATIONS for indicative levels of overall pricing in regions	MATERIALS			LABOUR				RATES		
	Del to Site	Waste	Material Cost	Craft Optve	Lab	Labour Cost	Sunds	Nett Rate	Unit	Gross Rate (+20%)
	£	%	£	Hrs	Hrs	£	£	£		£
M40: STONE/CONCRETE/QUARRY/CERAMIC TILING/MOSAIC: CLAY TILE PAVING Cont.										
Ceramic floor tiles, BS 6431, fully vitrified, cream; 3mm joints, symmetrical layout; bedding in 10mm cement mortar (1:3); pointing in cement mortar (1:3); on cement and sand base										
152 x 152 x 12mm units to floors level or to falls only not exceeding 15 degrees from horizontal										
plain	21.62	5.00	**22.70**	1.21	0.75	**25.21**	-	**47.91**	m2	**57.49**
Extra for pointing with tinted mortar										
152 x 152 x 12mm tiles	1.03	10.00	**1.13**	0.13	-	**1.88**	-	**3.00**	m2	**3.60**
152 x 152 x 12mm units to skirtings on brickwork or blockwork base										
height 152mm; rounded top edge and cove at bottom	5.81	10.00	**6.37**	0.40	0.11	**6.81**	-	**13.17**	m	**15.81**
M40: STONE/CONCRETE/QUARRY/CERAMIC TILING/MOSAIC: TILE WALL LININGS										
Ceramic tiles, BS 6431, glazed white; 2mm joints; symmetrical layout; fixing with thin bed adhesive; pointing with neat white cement; on plaster base										
108 x 108 x 6.5mm units to walls										
plain, width exceeding 300mm	21.33	5.00	**22.54**	1.76	0.99	**35.61**	-	**58.15**	m2	**69.78**
plain, width not exceeding 300mm	7.39	10.00	**8.13**	1.05	0.63	**21.56**	-	**29.69**	m	**35.63**
152 x 152 x 5.5mm units to walls										
plain, width exceeding 300mm	15.04	5.00	**15.93**	1.38	0.80	**28.14**	-	**44.07**	m2	**52.88**
plain, width not exceeding 300mm	5.02	10.00	**5.52**	0.83	0.50	**17.02**	-	**22.55**	m	**27.06**
Ceramic tiles, BS 6431, glazed light colour; 2mm joints, symmetrical layout; fixing with thin bed adhesive; pointing with neat white cement; on plaster base										
108 x 108 x 6.5mm units to walls										
plain, width exceeding 300mm	21.33	5.00	**22.54**	1.76	0.99	**35.61**	-	**58.15**	m2	**69.78**
plain, width not exceeding 300mm	7.39	10.00	**8.13**	1.05	0.63	**21.56**	-	**29.69**	m	**35.63**
152 x 152 x 5.5mm units to walls										
plain, width exceeding 300mm	15.04	5.00	**15.93**	1.38	0.80	**28.14**	-	**44.07**	m2	**52.88**
plain, width not exceeding 300mm	5.02	10.00	**5.52**	0.83	0.50	**17.02**	-	**22.55**	m	**27.06**
Ceramic tiles, BS 6431, glazed dark colour; 2mm joints, symmetrical layout; bedding in 10mm cement mortar (1:3); pointing with neat white cement; on brickwork or blockwork base										
108 x 108 x 6.5mm units to walls										
plain, width exceeding 300mm	23.13	5.00	**24.31**	1.76	0.99	**35.61**	-	**59.92**	m2	**71.90**
plain, width not exceeding 300mm	7.65	10.00	**8.39**	1.05	0.63	**21.56**	-	**29.96**	m	**35.95**
152 x 152 x 5.5mm units to walls										
plain, width exceeding 300mm	17.06	5.00	**17.92**	1.38	0.80	**28.14**	-	**46.06**	m2	**55.28**
plain, width not exceeding 300mm	5.28	10.00	**5.79**	0.83	0.50	**17.02**	-	**22.82**	m	**27.38**
M40: STONE/CONCRETE/QUARRY/CERAMIC TILING/MOSAIC: TILE CILLS										
Ceramic tiles, BS 6431, glazed white; 2mm joints, symmetrical layout; bedding in 10mm cement mortar (1:3); pointing with neat white cement; on brickwork or blockwork base - cills										
152 x 152 x 5.5mm units to cills on brickwork or blockwork base										
width 150mm; rounded angle	3.13	5.00	**3.29**	0.40	0.12	**6.92**	-	**10.21**	m	**12.25**
width 225mm; rounded angle	4.23	5.00	**4.44**	0.55	0.19	**9.82**	-	**14.26**	m	**17.11**
M42: WOOD BLOCK/COMPOSITION BLOCK/ PARQUET FLOORING										
Maple wood blocks, tongued and grooved joints; symmetrical herringbone pattern layout, two block plain borders; fixing with adhesive; sanding, one coat sealer										
70 x 230 x 20mm units to floors on cement and sand base; level or to falls only not exceeding 15 degrees from horizontal										
plain	35.52	5.00	**37.29**	1.98	1.60	**45.22**	-	**82.51**	m2	**99.01**
M50: RUBBER/PLASTICS/CORK/LINO/CARPET TILING/SHEETING: PVC FLOORING										
P.V.C. floor tiles, BS EN 654; butt joints, symmetrical layout; fixing with adhesive; two coats sealer; on cement and sand base										
300 x 300 x 2mm units to floors level or to falls only not exceeding 15 degrees from horizontal										
width exceeding 300mm	12.76	5.00	**13.60**	0.66	0.33	**12.91**	-	**26.51**	m2	**31.82**
width not exceeding 300mm	4.45	10.00	**4.89**	0.40	0.10	**6.69**	-	**11.58**	m	**13.89**

SURFACE FINISHES

Labour hourly rates: (except Specialists) Craft Operatives £14.21 Labourer £10.71 Rates are national average prices. Refer to REGIONAL VARIATIONS for indicative levels of overall pricing in regions	MATERIALS			LABOUR				RATES		
	Del to Site £	Waste %	Material Cost £	Craft Optve Hrs	Lab Hrs	Labour Cost £	Sunds £	Nett Rate £	Unit	Gross Rate (+20%) £
M50: RUBBER/PLASTICS/CORK/LINO/CARPET TILING/SHEETING: PVC FLOORING Cont.										
Fully flexible PVC heavy duty floor tiles, BS EN 649 type A; butt joints, symmetrical layout; fixing with adhesive; two coats sealer; on cement and sand base										
610 x 610 x 2.5mm units to floors level or to falls only not exceeding 15 degrees from horizontal										
width exceeding 300mm	21.45	5.00	**22.72**	0.66	0.33	**12.91**	-	**35.64**	m2	**42.76**
width not exceeding 300mm	7.05	10.00	**7.76**	0.40	0.10	**6.69**	-	**14.45**	m	**17.34**
Fully flexible P.V.C. heavy duty sheet,BS EN 649 Type A; butt joints; welded; fixing with adhesive; two coats sealer; on cement and sand base										
P.V.C. coved skirting; fixing with adhesive										
Skirtings on plaster base										
height 100mm - Sit on	2.93	5.00	**3.11**	0.28	0.02	**4.14**	-	**7.25**	m	**8.70**
M50: RUBBER/PLASTICS/CORK/LINO/CARPET TILING/SHEETING: LINOLEUM FLOORING										
Linoleum tiles, BS EN 548 & 670 butt joints, symmetrical layout; fixing with adhesive; on cement and sand base										
500 x 500 x 2mm units to floors level or to falls only not exceeding 15 degrees from horizontal										
width exceeding 300mm	17.88	5.00	**18.83**	0.50	0.31	**10.33**	-	**29.16**	m2	**34.99**
width not exceeding 300mm	5.53	10.00	**6.08**	0.30	0.09	**5.16**	-	**11.24**	m	**13.49**
Linoleum sheet, BS EN 548 & 670 butt joints; laying loose; on cement and sand base										
3.2mm work to floors level or to falls only not exceeding 15 degrees from horizontal										
width exceeding 300mm	19.75	5.00	**20.73**	0.22	0.17	**4.89**	-	**25.63**	m2	**30.75**
width not exceeding 300mm	5.92	10.00	**6.52**	0.13	0.06	**2.46**	-	**8.98**	m	**10.78**
M50: RUBBER/PLASTICS/CORK/LINO/CARPET TILING/SHEETING; STAIRCASE AREAS - STAIR NOSINGS AND TREADS										
Accessories										
Heavy duty aluminium alloy stair nosings with anti-slip inserts; 80mm wide with two line inserts; fixing to masonry with screws										
22mm drop	15.74	5.00	**16.53**	0.99	-	**14.07**	0.13	**30.73**	m	**36.87**
51mm drop	22.03	5.00	**23.14**	0.99	-	**14.07**	0.13	**37.33**	m	**44.80**
M51: EDGE FIXED CARPETING										
Underlay to carpeting; Tredaire; fixing with tacks; on timber base										
Work to floors; level or to falls only not exceeding 15 degrees from horizontal										
width exceeding 300mm	1.93	5.00	**2.02**	0.17	0.09	**3.29**	0.18	**5.49**	m2	**6.59**
Fitted carpeting; Tufted wool/nylon, p.c. £20.00 m2; fixing with tackless grippers; on timber base										
Work to floors; level or to falls only not exceeding 15 degrees from horizontal										
width exceeding 300mm	22.00	5.00	**23.10**	0.44	0.22	**8.61**	0.18	**31.89**	m2	**38.27**
M52: DECORATIVE PAPERS/FABRICS										
Pulp paper (24' drop pattern match) p.c. £4.00 roll; sizing; applying adhesive; hanging; butt joints										
Plaster walls and columns										
exceeding 0.50m2	0.94	40.00	**1.29**	0.34	-	**4.85**	0.09	**6.23**	m2	**7.47**
Plaster ceilings and beams										
exceeding 0.50m2	0.94	35.00	**1.25**	0.36	-	**5.16**	0.10	**6.50**	m2	**7.80**
Vinyl coated paper p.c. £7.00 roll; sizing; applying adhesive; hanging; butt joints										
Plaster walls and columns										
exceeding 0.50m2	1.62	25.00	**2.00**	0.26	-	**3.75**	0.07	**5.82**	m2	**6.99**
Plaster ceilings and beams										
exceeding 0.50m2	1.62	20.00	**1.92**	0.29	-	**4.06**	0.08	**6.07**	m2	**7.28**
Woodchip paper p.c. £1.50 roll; sizing; applying adhesive; hanging; butt joints										
Plaster walls and columns										
exceeding 0.50m2	0.39	25.00	**0.47**	0.26	-	**3.75**	0.07	**4.29**	m2	**5.14**
Plaster ceilings and beams										
exceeding 0.50m2	0.39	20.00	**0.45**	0.29	-	**4.06**	0.07	**4.59**	m2	**5.51**

Labour hourly rates: (except Specialists) Craft Operatives £14.21 Labourer £10.71 Rates are national average prices. Refer to REGIONAL VARIATIONS for indicative levels of overall pricing in regions	MATERIALS			LABOUR				RATES		
	Del to Site £	Waste %	Material Cost £	Craft Optve Hrs	Lab Hrs	Labour Cost £	Sunds £	Nett Rate £	Unit	Gross Rate (+20%) £
M52: DECORATIVE PAPERS/FABRICS Cont.										
Paper border strips p.c. £0.50m; applying adhesive; hanging; butt joints										
Border strips										
75mm wide to papered walls and columns	0.57	10.00	0.63	0.11	-	1.56	0.03	2.22	m	2.67
Hessian wall covering p.c. £4.00m2; sizing; applying adhesive; hanging; butt joints										
Plaster walls and columns exceeding 0.50m2	4.63	30.00	5.95	0.26	-	3.75	0.07	9.77	m2	11.73
M52: DECORATIVE PAPERS/FABRICS - REDECORATIONS										
Stripping existing paper; pulp paper p.c. £3.00 roll; sizing; applying adhesive; hanging; butt joints										
Plaster walls and columns exceeding 0.50m2	0.72	30.00	0.92	0.43	-	6.10	0.29	7.30	m2	8.76
Plaster ceilings and beams exceeding 0.50m2	0.72	25.00	0.88	0.46	-	6.57	0.31	7.76	m2	9.31
Stripping existing washable or vinyl coated paper; vinyl coated paper P.C. £8.00 roll; sizing; applying adhesive; hanging; butt joints										
Plaster walls and columns exceeding 0.50m2	1.84	25.00	2.28	0.44	-	6.25	0.29	8.82	m2	10.59
Plaster ceilings and beams exceeding 0.50m2	1.84	20.00	2.19	0.47	-	6.72	0.32	9.23	m2	11.07
Washing down old distempered or painted surfaces; washable paper p.c. £4.00 roll; sizing; applying adhesive; hanging; butt joints										
Plaster walls and columns exceeding 0.50m2	0.96	25.00	1.18	0.36	-	5.16	0.10	6.43	m2	7.72
Stripping existing paper; textile hessian paper backed wall covering P.C. £7.50m2; sizing; applying adhesive; hanging; butt joints										
Plaster walls and columns exceeding 0.50m2	4.63	30.00	5.95	0.56	-	7.97	0.38	14.30	m2	17.16
M60: PAINTING/CLEAR FINISHING - EMULSION PAINTING										
Mist coat, one full coat emulsion paint										
Concrete general surfaces girth exceeding 300mm	0.20	10.00	0.22	0.17	-	2.34	0.04	2.61	m2	3.14
Concrete general surfaces 3.50 - 5.00m above floor girth exceeding 300mm	0.20	10.00	0.22	0.19	-	2.66	0.05	2.93	m2	3.52
Plaster general surfaces girth exceeding 300mm	0.16	10.00	0.18	0.14	-	2.03	0.04	2.25	m2	2.70
Brickwork general surfaces girth exceeding 300mm	0.20	10.00	0.22	0.19	-	2.66	0.05	2.93	m2	3.52
Mist coat, two full coats emulsion paint										
Concrete general surfaces girth exceeding 300mm	0.34	10.00	0.38	0.23	-	3.28	0.06	3.72	m2	4.47
Concrete general surfaces 3.50 - 5.00m above floor girth exceeding 300mm	0.34	10.00	0.38	0.26	-	3.75	0.07	4.20	m2	5.04
Plaster general surfaces girth exceeding 300mm	0.28	10.00	0.30	0.21	-	2.97	0.06	3.33	m2	4.00
Brickwork general surfaces girth exceeding 300mm	0.34	10.00	0.38	0.26	-	3.75	0.07	4.20	m2	5.04
M60: PAINTING/CLEAR FINISHING - CEMENT PAINTING										
Two coats sealer, one coat waterproof cement paint; external work										
Cement rendered general surfaces girth exceeding 300mm	3.40	10.00	3.74	0.37	-	5.31	0.10	9.16	m2	10.99
Two coats textured masonry paint; external work										
Cement rendered general surfaces girth exceeding 300mm	0.61	10.00	0.67	0.26	-	3.75	0.07	4.49	m2	5.39
isolated surfaces, girth not exceeding 300mm	0.17	10.00	0.19	0.10	-	1.41	0.03	1.62	m	1.95

SURFACE FINISHES

Labour hourly rates: (except Specialists) Craft Operatives £14.21 Labourer £10.71 Rates are national average prices. Refer to REGIONAL VARIATIONS for indicative levels of overall pricing in regions	MATERIALS			LABOUR				RATES		
	Del to Site £	Waste %	Material Cost £	Craft Optve Hrs	Lab Hrs	Labour Cost £	Sunds £	Nett Rate £	Unit	Gross Rate (+20%) £
M60: PAINTING/CLEAR FINISHING - CEMENT PAINTING Cont.										
One coat stabilizing solution, one coat textured masonry paint; external work										
Cement rendered general surfaces										
girth exceeding 300mm	0.59	10.00	**0.65**	0.22	-	**3.13**	0.06	**3.84**	m2	**4.60**
isolated surfaces, girth not exceeding 300mm	0.17	10.00	**0.19**	0.08	-	**1.09**	0.02	**1.31**	m	**1.57**
M60: PAINTING/CLEAR FINISHING - PRESERVATIVE TREATMENT										
One coat wood preservative, internal work (sawn timber)										
Wood general surfaces										
girth exceeding 300mm	0.35	10.00	**0.38**	0.13	-	**1.88**	0.04	**2.29**	m2	**2.75**
isolated surfaces, girth not exceeding 300mm	0.11	10.00	**0.12**	0.04	-	**0.63**	0.01	**0.76**	m	**0.91**
One coat wood preservative, internal work (wrought timber)										
Wood general surfaces										
girth exceeding 300mm	0.29	10.00	**0.32**	0.11	-	**1.56**	0.03	**1.91**	m2	**2.29**
isolated surfaces, girth not exceeding 300mm	0.09	10.00	**0.10**	0.03	-	**0.47**	0.01	**0.57**	m	**0.69**
Two coats wood preservative, external work (sawn timber)										
Wood general surfaces										
girth exceeding 300mm	0.69	10.00	**0.76**	0.24	-	**3.44**	0.07	**4.27**	m2	**5.12**
isolated surfaces, girth not exceeding 300mm	0.22	10.00	**0.24**	0.09	-	**1.25**	0.02	**1.51**	m	**1.81**
Two coats Sadolins High Performance Varnish										
Wood general surfaces										
girth exceeding 300mm	1.32	10.00	**1.45**	0.33	-	**4.69**	0.09	**6.23**	m2	**7.48**
isolated surfaces, girth not exceeding 300mm	0.40	10.00	**0.44**	0.12	-	**1.72**	0.03	**2.19**	m	**2.63**
Wood partially glazed doors										
girth exceeding 300mm; panes, area 0.50 - 1.00m2	0.72	10.00	**0.80**	0.66	-	**9.38**	0.18	**10.35**	m2	**12.42**
girth exceeding 300mm; panes, area exceeding 1.00m2	0.72	10.00	**0.80**	0.64	-	**9.07**	0.17	**10.03**	m2	**12.04**
Wood windows and screens										
girth exceeding 300mm; panes, area not exceeding 0.10m2	0.76	10.00	**0.83**	1.20	-	**17.04**	0.32	**18.19**	m2	**21.83**
girth exceeding 300mm; panes, area 0.10 - 0.50m2	0.53	10.00	**0.59**	0.86	-	**12.19**	0.23	**13.01**	m2	**15.61**
girth exceeding 300mm; panes, area 0.50 - 1.00m2	0.47	10.00	**0.52**	0.73	-	**10.32**	0.20	**11.04**	m2	**13.25**
girth exceeding 300mm; panes, area exceeding 1.00m2	0.47	10.00	**0.52**	0.64	-	**9.07**	0.17	**9.76**	m2	**11.71**
One coat Sadolins Classic; two coats Sadolins Extra; external work										
Wood general surfaces										
girth exceeding 300mm	2.15	10.00	**2.36**	0.61	-	**8.60**	0.16	**11.12**	m2	**13.35**
isolated surfaces, girth not exceeding 300mm	0.63	10.00	**0.70**	0.21	-	**2.97**	0.06	**3.72**	m	**4.47**
Wood partially glazed doors										
girth exceeding 300mm; panes, area 0.50 - 1.00m2	1.18	10.00	**1.29**	0.57	-	**8.13**	0.15	**9.58**	m2	**11.49**
girth exceeding 300mm; panes, area exceeding 1.00m2	1.18	10.00	**1.29**	0.55	-	**7.82**	0.15	**9.26**	m2	**11.11**
Wood windows and screens										
girth exceeding 300mm; panes, area not exceeding 0.10m2	1.18	10.00	**1.29**	1.08	-	**15.32**	0.29	**16.90**	m2	**20.28**
girth exceeding 300mm; panes, area 0.10 - 0.50m2	0.84	10.00	**0.93**	0.79	-	**11.25**	0.21	**12.40**	m2	**14.88**
girth exceeding 300mm; panes, area 0.50 - 1.00m2	0.76	10.00	**0.84**	0.67	-	**9.53**	0.18	**10.55**	m2	**12.66**
girth exceeding 300mm; panes, area exceeding 1.00m2	0.76	10.00	**0.84**	0.59	-	**8.44**	0.16	**9.44**	m2	**11.33**
M60: PAINTING/CLEAR FINISHING - OIL PAINTING WALLS AND CEILINGS										
One coat primer, one undercoat, one coat full gloss finish										
Concrete general surfaces										
girth exceeding 300mm	0.96	10.00	**1.05**	0.36	-	**5.16**	0.10	**6.31**	m2	**7.57**
Concrete general surfaces 3.50 - 5.00m above floor										
girth exceeding 300mm	0.96	10.00	**1.05**	0.40	-	**5.63**	0.11	**6.79**	m2	**8.14**
Plaster general surfaces										
girth exceeding 300mm	0.87	10.00	**0.96**	0.33	-	**4.69**	0.09	**5.74**	m2	**6.89**
isolated surfaces, girth not exceeding 300mm	1.72	10.00	**1.89**	0.14	-	**2.03**	0.04	**3.96**	m	**4.76**
Plaster general surfaces 3.50 - 5.00m above floor										
girth exceeding 300mm	0.87	10.00	**0.96**	0.36	-	**5.16**	0.10	**6.22**	m2	**7.46**
One coat primer, one undercoat, one coat eggshell finish										
Plaster general surfaces										
girth exceeding 300mm	0.92	10.00	**1.01**	0.33	-	**4.69**	0.09	**5.79**	m2	**6.95**

Labour hourly rates: (except Specialists) Craft Operatives £14.21 Labourer £10.71 Rates are national average prices. Refer to REGIONAL VARIATIONS for indicative levels of overall pricing in regions	MATERIALS			LABOUR				RATES		
	Del to Site £	Waste %	Material Cost £	Craft Optve Hrs	Lab Hrs	Labour Cost £	Sunds £	Nett Rate £	Unit	Gross Rate (+20%) £
M60: PAINTING/CLEAR FINISHING - OIL PAINTING WALLS AND CEILINGS Cont.										
One coat primer, two undercoats, one coat eggshell finish										
Brickwork general surfaces										
girth exceeding 300mm	1.44	10.00	**1.58**	0.55	-	**7.82**	0.15	**9.54**	m2	**11.45**
Plaster general surfaces										
girth exceeding 300mm	1.10	10.00	**1.21**	0.44	-	**6.25**	0.12	**7.58**	m2	**9.09**
isolated surfaces, girth not exceeding 300mm	0.34	10.00	**0.37**	0.20	-	**2.81**	0.05	**3.24**	m	**3.89**
Plaster general surfaces 3.50 - 5.00m above floor										
girth exceeding 300mm	1.10	10.00	**1.21**	0.48	-	**6.88**	0.02	**8.10**	m2	**9.73**
M60: PAINTING/CLEAR FINISHING - PLASTIC FINISH										
Textured plastic coating - Combed Finish										
Plaster general surfaces										
girth exceeding 300mm	0.41	10.00	**0.45**	0.31	-	**4.38**	0.08	**4.91**	m2	**5.89**
Plaster general surfaces 3.50 - 5.00m above floor										
girth exceeding 300mm	0.41	10.00	**0.45**	0.33	-	**4.69**	0.09	**5.23**	m2	**6.28**
M60: PAINTING/CLEAR FINISHING - OIL PAINTING METALWORK										
One undercoat, one coat full gloss finish on ready primed metal surfaces										
Iron or steel structural work										
girth exceeding 300mm	0.35	10.00	**0.39**	0.34	-	**4.85**	0.09	**5.32**	m2	**6.39**
isolated surfaces, girth not exceeding 300mm	0.11	10.00	**0.12**	0.14	-	**2.03**	0.04	**2.19**	m	**2.63**
Iron or steel structural members of roof trusses, lattice girders, purlins and the like										
girth exceeding 300mm	0.35	10.00	**0.39**	0.44	-	**6.25**	0.12	**6.76**	m2	**8.11**
isolated surfaces, girth not exceeding 300mm	0.11	10.00	**0.12**	0.14	-	**2.03**	0.04	**2.19**	m	**2.63**
One coat primer, one undercoat, one coat full gloss finish on metal surfaces										
Iron or steel general surfaces										
girth exceeding 300mm	1.01	10.00	**1.12**	0.47	-	**6.72**	0.13	**7.97**	m2	**9.56**
isolated surfaces, girth not exceeding 300mm	0.32	10.00	**0.35**	0.15	-	**2.19**	0.04	**2.58**	m	**3.10**
Galvanized glazed doors, windows or screens										
girth exceeding 300mm; panes, area not exceeding 0.10m2	0.55	10.00	**0.61**	0.87	-	**12.35**	0.23	**13.19**	m2	**15.83**
girth exceeding 300mm; panes, area 0.10 - 0.50m2	0.41	10.00	**0.45**	0.62	-	**8.75**	0.17	**9.37**	m2	**11.24**
girth exceeding 300mm; panes, area 0.50 - 1.00m2	0.35	10.00	**0.38**	0.52	-	**7.35**	0.14	**7.87**	m2	**9.44**
girth exceeding 300mm; panes, area exceeding 1.00m2	0.35	10.00	**0.38**	0.47	-	**6.72**	0.13	**7.23**	m2	**8.68**
Iron or steel structural work										
girth exceeding 300mm	1.01	10.00	**1.12**	0.51	-	**7.19**	0.14	**8.44**	m2	**10.13**
isolated surfaces, girth not exceeding 300mm	0.32	10.00	**0.35**	0.22	-	**3.13**	0.06	**3.54**	m	**4.24**
Iron or steel structural members of roof trusses, lattice girders, purlins and the like										
girth exceeding 300mm	1.01	10.00	**1.12**	0.65	-	**9.22**	0.18	**10.51**	m2	**12.62**
isolated surfaces, girth not exceeding 300mm	0.32	10.00	**0.35**	0.22	-	**3.13**	0.06	**3.54**	m	**4.24**
Iron or steel services										
girth exceeding 300mm	1.01	10.00	**1.12**	0.65	-	**9.22**	0.18	**10.51**	m2	**12.62**
isolated surfaces, girth not exceeding 300mm	0.32	10.00	**0.35**	0.22	-	**3.13**	0.06	**3.54**	m	**4.24**
isolated areas not exceeding 0.50m2 irrespective of girth	0.52	10.00	**0.57**	0.65	-	**9.22**	0.18	**9.97**	nr	**11.96**
Copper services										
girth exceeding 300mm	1.01	10.00	**1.12**	0.65	-	**9.22**	0.18	**10.51**	m2	**12.62**
isolated surfaces, girth not exceeding 300mm	0.32	10.00	**0.35**	0.22	-	**3.13**	0.06	**3.54**	m	**4.24**
Galvanized services										
girth exceeding 300mm	1.01	10.00	**1.12**	0.65	-	**9.22**	0.18	**10.51**	m2	**12.62**
isolated surfaces, girth not exceeding 300mm	0.32	10.00	**0.35**	0.22	-	**3.13**	0.06	**3.54**	m	**4.24**
One coat primer, one undercoat, one coat full gloss finish on metal surfaces ; external work										
Iron or steel general surfaces										
girth exceeding 300mm	1.01	10.00	**1.12**	0.51	-	**7.19**	0.14	**8.44**	m2	**10.13**
isolated surfaces, girth not exceeding 300mm	0.32	10.00	**0.35**	0.20	-	**2.81**	0.05	**3.22**	m	**3.86**
Iron or steel railings, fences and gates; plain open type										
girth exceeding 300mm	0.78	10.00	**0.86**	0.47	-	**6.72**	0.13	**7.71**	m2	**9.25**
isolated surfaces, girth not exceeding 300mm	0.32	10.00	**0.35**	0.15	-	**2.19**	0.04	**2.58**	m	**3.10**
Iron or steel railings, fences and gates; close type										
girth exceeding 300mm	1.01	10.00	**1.12**	0.40	-	**5.63**	0.11	**6.85**	m2	**8.22**
Iron or steel railings, fences and gates; ornamental type										
girth exceeding 300mm	0.78	10.00	**0.86**	0.80	-	**11.41**	0.22	**12.49**	m2	**14.99**

Labour hourly rates: (except Specialists) Craft Operatives £14.21 Labourer £10.71 Rates are national average prices. Refer to REGIONAL VARIATIONS for indicative levels of overall pricing in regions	MATERIALS			LABOUR				RATES		
	Del to Site	Waste	Material Cost	Craft Optve	Lab	Labour Cost	Sunds	Nett Rate	Unit	Gross Rate (+20%)
	£	%	£	Hrs	Hrs	£	£	£		£
M60: PAINTING/CLEAR FINISHING - OIL PAINTING METALWORK Cont.										
One coat primer, one undercoat, one coat full gloss finish on metal surfaces ; external work Cont.										
Iron or steel eaves gutters										
girth exceeding 300mm	1.01	10.00	**1.12**	0.55	-	**7.82**	0.15	**9.08**	m2	**10.90**
isolated surfaces, girth not exceeding 300mm	0.32	10.00	**0.35**	0.19	-	**2.66**	0.05	**3.06**	m	**3.67**
One coat primer, two undercoats, one coat full gloss finish on metal surfaces										
Iron or steel general surfaces										
girth exceeding 300mm	1.19	10.00	**1.31**	0.63	-	**8.91**	0.17	**10.39**	m2	**12.47**
One coat primer, two undercoats, one coat full gloss finish on metal surfaces; external work										
Iron or steel general surfaces										
girth exceeding 300mm	1.19	10.00	**1.31**	0.68	-	**9.69**	0.18	**11.18**	m2	**13.42**
isolated surfaces, girth not exceeding 300mm	0.37	10.00	**0.41**	0.23	-	**3.28**	0.06	**3.76**	m	**4.51**
One coat primer, one undercoat, two coats full gloss finish on metal surfaces										
Iron or steel general surfaces										
girth exceeding 300mm	1.19	10.00	**1.31**	0.63	-	**8.91**	0.17	**10.39**	m2	**12.47**
isolated surfaces, girth not exceeding 300mm	0.37	10.00	**0.41**	0.21	-	**2.97**	0.06	**3.44**	m	**4.12**
M60: PAINTING/CLEAR FINISHING - BITUMINOUS PAINT										
Two coats black bitumen paint; external work										
Iron or steel general surfaces										
girth exceeding 300mm	0.54	10.00	**0.60**	0.32		**4.53**	0.09	**5.22**	m2	**6.26**
M60: PAINTING/CLEAR FINISHING - OIL PAINTING WOODWORK										
One coat primer; carried out on site before fixing members										
Wood general surfaces										
girth exceeding 300mm	0.27	10.00	**0.29**	0.20	-	**2.81**	0.05	**3.16**	m2	**3.79**
isolated surfaces, girth not exceeding 300mm	0.08	10.00	**0.08**	0.08	-	**1.09**	0.02	**1.20**	m	**1.44**
One undercoat, one coat full gloss finish on ready primed wood surfaces										
Wood general surfaces										
girth exceeding 300mm	0.35	10.00	**0.39**	0.39	-	**5.47**	0.10	**5.96**	m2	**7.15**
isolated surfaces, girth not exceeding 300mm	0.11	10.00	**0.12**	0.13	-	**1.88**	0.04	**2.03**	m	**2.44**
Wood partially glazed doors										
girth exceeding 300mm; panes, area not exceeding 0.10m2	0.27	10.00	**0.30**	0.52	-	**7.35**	0.14	**7.78**	m2	**9.34**
girth exceeding 300mm; panes, area 0.10 - 0.50m2	0.22	10.00	**0.24**	0.43	-	**6.10**	0.12	**6.45**	m2	**7.74**
girth exceeding 300mm; panes, area 0.50 - 1.00m2	0.19	10.00	**0.21**	0.39	-	**5.47**	0.10	**5.78**	m2	**6.94**
girth exceeding 300mm; panes, area exceeding 1.00m2	0.19	10.00	**0.21**	0.37	-	**5.31**	0.10	**5.62**	m2	**6.75**
Wood windows and screens										
girth exceeding 300mm; panes, area not exceeding 0.10m2	0.27	10.00	**0.30**	0.72	-	**10.16**	0.19	**10.65**	m2	**12.78**
girth exceeding 300mm; panes, area 0.10 - 0.50m2	0.22	10.00	**0.24**	0.51	-	**7.19**	0.14	**7.56**	m2	**9.08**
girth exceeding 300mm; panes, area 0.50 - 1.00m2	0.19	10.00	**0.21**	0.43	-	**6.10**	0.12	**6.42**	m2	**7.70**
girth exceeding 300mm; panes, area exceeding 1.00m2	0.19	10.00	**0.21**	0.39	-	**5.47**	0.10	**5.78**	m2	**6.94**
One undercoat, one coat full gloss finish on ready primed wood surfaces; external work										
Wood general surfaces										
girth exceeding 300mm	0.35	10.00	**0.39**	0.41	-	**5.78**	0.11	**6.28**	m2	**7.53**
isolated surfaces, girth not exceeding 300mm	0.11	10.00	**0.12**	0.14	-	**2.03**	0.04	**2.19**	m	**2.63**
Wood railings fences and gates; open type										
girth exceeding 300mm	0.27	10.00	**0.30**	0.31	-	**4.38**	0.08	**4.76**	m2	**5.71**
isolated surfaces, girth not exceeding 300mm	0.11	10.00	**0.12**	0.10	-	**1.41**	0.03	**1.55**	m	**1.86**
Wood railings fences and gates; close type										
girth exceeding 300mm	0.35	10.00	**0.39**	0.26	-	**3.75**	0.07	**4.21**	m2	**5.05**
Two undercoats, one coat full gloss finish on ready primed wood surfaces										
Wood general surfaces										
girth exceeding 300mm	0.53	10.00	**0.58**	0.56	-	**7.97**	0.15	**8.70**	m2	**10.44**
isolated surfaces, girth not exceeding 300mm	0.16	10.00	**0.18**	0.19	-	**2.66**	0.05	**2.89**	m	**3.46**
Two undercoats, one coat full gloss finish on ready primed wood surfaces; external work										
Wood general surfaces										
girth exceeding 300mm	0.53	10.00	**0.58**	0.59	-	**8.44**	0.16	**9.18**	m2	**11.02**
isolated surfaces, girth not exceeding 300mm	0.16	10.00	**0.18**	0.21	-	**2.97**	0.06	**3.20**	m	**3.85**

Labour hourly rates: (except Specialists) Craft Operatives £14.21 Labourer £10.71 Rates are national average prices. Refer to REGIONAL VARIATIONS for indicative levels of overall pricing in regions	MATERIALS			LABOUR				RATES		
	Del to Site	Waste	Material Cost	Craft Optve	Lab	Labour Cost	Sunds	Nett Rate	Unit	Gross Rate (+20%)
	£	%	£	Hrs	Hrs	£	£	£		£
M60: PAINTING/CLEAR FINISHING - OIL PAINTING WOODWORK Cont.										
One coat primer, one undercoat, one coat full gloss finish on wood surfaces										
Wood general surfaces										
girth exceeding 300mm	0.62	10.00	**0.68**	0.56	-	**7.97**	0.15	**8.80**	m2	**10.56**
isolated surfaces, girth not exceeding 300mm	0.18	10.00	**0.20**	0.19	-	**2.66**	0.05	**2.91**	m	**3.49**
Wood partially glazed doors										
girth exceeding 300mm; panes, area not exceeding 0.10m2	0.48	10.00	**0.53**	0.74	-	**10.47**	0.20	**11.20**	m2	**13.44**
girth exceeding 300mm; panes, area 0.10 - 0.50m2	0.37	10.00	**0.41**	0.62	-	**8.75**	0.17	**9.32**	m2	**11.19**
girth exceeding 300mm; panes, area 0.50 - 1.00m2	0.34	10.00	**0.38**	0.56	-	**7.97**	0.15	**8.50**	m2	**10.20**
girth exceeding 300mm; panes, area exceeding 1.00m2	0.34	10.00	**0.38**	0.53	-	**7.50**	0.14	**8.02**	m2	**9.63**
Wood windows and screens										
girth exceeding 300mm; panes, area not exceeding 0.10m2	0.48	10.00	**0.53**	1.01	-	**14.38**	0.27	**15.18**	m2	**18.22**
girth exceeding 300mm; panes, area 0.10 - 0.50m2	0.37	10.00	**0.41**	0.74	-	**10.47**	0.20	**11.08**	m2	**13.29**
girth exceeding 300mm; panes, area 0.50 - 1.00m2	0.34	10.00	**0.38**	0.62	-	**8.75**	0.17	**9.30**	m2	**11.15**
girth exceeding 300mm; panes, area exceeding 1.00m2	0.34	10.00	**0.38**	0.55	-	**7.82**	0.15	**8.34**	m2	**10.01**
One coat primer, one undercoat, one coat full gloss finish on wood surfaces; external work										
Wood general surfaces										
girth exceeding 300mm	0.62	10.00	**0.68**	0.59	-	**8.44**	0.16	**9.28**	m2	**11.14**
isolated surfaces, girth not exceeding 300mm	0.18	10.00	**0.20**	0.21	-	**2.97**	0.06	**3.23**	m	**3.87**
One coat primer, two undercoats, one coat full gloss finish on wood surfaces										
Wood general surfaces										
girth exceeding 300mm	0.79	10.00	**0.87**	0.73	-	**10.32**	0.20	**11.38**	m2	**13.66**
isolated surfaces, girth not exceeding 300mm	0.24	10.00	**0.26**	0.24	-	**3.44**	0.07	**3.77**	m	**4.52**
One coat primer, two undercoats, one coat full gloss finish on wood surfaces; external work										
Wood general surfaces										
girth exceeding 300mm	0.79	10.00	**0.87**	0.77	-	**10.94**	0.21	**12.02**	m2	**14.43**
isolated surfaces, girth not exceeding 300mm	0.24	10.00	**0.26**	0.26	-	**3.75**	0.07	**4.08**	m2	**4.90**
One coat primer, one undercoat, two coats full gloss finish on wood surfaces										
Wood general surfaces										
girth exceeding 300mm	0.79	10.00	**0.87**	0.73	-	**10.32**	0.20	**11.38**	m2	**13.66**
isolated surfaces, girth not exceeding 300mm	0.24	10.00	**0.26**	0.24	-	**3.44**	0.07	**3.77**	m	**4.52**
M60: PAINTING/CLEAR FINISHING - POLYURETHANE LACQUER										
Two coats polyurethane lacquer										
Wood general surfaces										
girth exceeding 300mm	0.93	10.00	**1.03**	0.48	-	**6.88**	0.13	**8.03**	m2	**9.64**
isolated surfaces, girth not exceeding 300mm	0.28	10.00	**0.31**	0.15	-	**2.19**	0.04	**2.54**	m	**3.05**
Wood partially glazed doors										
girth exceeding 300mm; panes, area not exceeding 0.10m2	0.75	10.00	**0.82**	0.65	-	**9.22**	0.18	**10.22**	m2	**12.26**
girth exceeding 300mm; panes, area 0.10 - 0.50m2	0.56	10.00	**0.62**	0.53	-	**7.50**	0.14	**8.26**	m2	**9.91**
girth exceeding 300mm; panes, area 0.50 - 1.00m2	0.53	10.00	**0.58**	0.48	-	**6.88**	0.13	**7.59**	m2	**9.11**
girth exceeding 300mm; panes, area exceeding 1.00m2	0.53	10.00	**0.58**	0.46	-	**6.57**	0.12	**7.27**	m2	**8.73**
Wood windows and screens										
girth exceeding 300mm; panes, area not exceeding 0.10m2	0.75	10.00	**0.82**	0.89	-	**12.66**	0.24	**13.72**	m2	**16.47**
girth exceeding 300mm; panes, area 0.10 - 0.50m2	0.56	10.00	**0.62**	0.64	-	**9.07**	0.17	**9.85**	m2	**11.82**
girth exceeding 300mm; panes, area 0.50 - 1.00m2	0.53	10.00	**0.58**	0.53	-	**7.50**	0.14	**8.23**	m2	**9.87**
girth exceeding 300mm; panes, area exceeding 1.00m2	0.53	10.00	**0.58**	0.48	-	**6.88**	0.13	**7.59**	m2	**9.11**
Two coats polyurethane lacquer; external work										
Wood general surfaces										
girth exceeding 300mm	0.93	10.00	**1.03**	0.51	-	**7.19**	0.14	**8.35**	m2	**10.02**
isolated surfaces, girth not exceeding 300mm	0.28	10.00	**0.31**	-	-	**-**	0.04	**0.35**	m	**0.42**
M60: PAINTING/CLEAR FINISHING - FIRE RETARDANT PAINTS AND VARNISHES										
Two coats fire retardant paint										
Wood general surfaces										
girth exceeding 300mm	3.50	10.00	**3.85**	0.48	-	**6.88**	0.13	**10.86**	m2	**13.03**
isolated surfaces, girth not exceeding 300mm	1.06	10.00	**1.17**	0.15	-	**2.19**	0.04	**3.40**	m	**4.08**

Labour hourly rates: (except Specialists) Craft Operatives £14.21 Labourer £10.71 Rates are national average prices. Refer to REGIONAL VARIATIONS for indicative levels of overall pricing in regions	MATERIALS			LABOUR				RATES		
	Del to Site £	Waste %	Material Cost £	Craft Optve Hrs	Lab Hrs	Labour Cost £	Sunds £	Nett Rate £	Unit	Gross Rate (+20%) £
M60: PAINTING/CLEAR FINISHING - FIRE RETARDANT PAINTS AND VARNISHES Cont.										
Two coats fire retardant varnish, one overcoat varnish										
Wood general surfaces										
girth exceeding 300mm	3.78	10.00	**4.16**	0.65	-	**9.22**	0.18	**13.56**	m2	**16.27**
isolated surfaces, girth not exceeding 300mm.....	1.17	10.00	**1.28**	0.22	-	**3.13**	0.06	**4.47**	m	**5.36**
M60: PAINTING/CLEAR FINISHING - FRENCH AND WAX POLISHING										
Stain; two coats white polish										
Wood general surfaces										
girth exceeding 300mm	3.04	10.00	**3.35**	0.62	-	**8.75**	0.17	**12.27**	m2	**14.72**
isolated surfaces, girth not exceeding 300mm.....	0.96	10.00	**1.05**	0.21	-	**2.97**	0.06	**4.08**	m	**4.90**
Stain; body in; fully French polish										
Wood general surfaces										
girth exceeding 300mm	5.16	10.00	**5.68**	4.36	-	**61.90**	1.18	**68.75**	m2	**82.50**
isolated surfaces, girth not exceeding 300mm.....	1.63	10.00	**1.79**	1.45	-	**20.63**	0.39	**22.81**	m	**27.38**
M60: PAINTING/CLEAR FINISHING - ROAD MARKINGS										
Two coats solvent based road marking paint; external work										
Concrete general surfaces										
isolated surfaces, 50mm wide.....................	2.97	5.00	**3.12**	0.20	-	**2.81**	0.05	**5.99**	m	**7.18**
isolated surfaces, 100mm wide....................	5.94	5.00	**6.24**	0.23	-	**3.28**	0.06	**9.58**	m	**11.50**
M60: PAINTING/CLEAR FINISHING (REDECORATIONS) - EMULSION PAINTING										
Generally										
Note: the following rates include for the cost of all preparatory work, e.g. washing down, etc.										
Two coats emulsion paint, existing emulsion painted surfaces										
Plaster general surfaces										
girth exceeding 300mm	0.22	10.00	**0.24**	0.31	-	**4.38**	0.08	**4.70**	m2	**5.64**
Brickwork general surfaces										
girth exceeding 300mm	0.28	10.00	**0.30**	0.35	-	**5.00**	0.10	**5.40**	m2	**6.48**
Two full coats emulsion paint, existing textured plastic coating surfaces										
Plaster general surfaces										
girth exceeding 300mm	0.22	10.00	**0.24**	0.35	-	**5.00**	0.10	**5.34**	m2	**6.41**
Two full coats emulsion paint, existing gloss painted surfaces										
Plaster general surfaces										
girth exceeding 300mm	0.22	10.00	**0.24**	0.32	-	**4.53**	0.09	**4.86**	m2	**5.83**
M60: PAINTING/CLEAR FINISHING (REDECORATIONS) - CEMENT PAINTING										
Generally										
Note: the following rates include for the cost of all preparatory work, e.g. washing down, etc.										
One coat sealer, one coat waterproof cement paint; external work										
Cement rendered general surfaces										
girth exceeding 300mm	2.70	10.00	**2.97**	0.40	-	**5.63**	0.11	**8.70**	m2	**10.44**
Brickwork general surfaces										
girth exceeding 300mm	2.49	10.00	**2.74**	0.47	-	**6.72**	0.13	**9.59**	m2	**11.51**
Rough cast general surfaces										
girth exceeding 300mm	2.70	10.00	**2.97**	0.57	-	**8.13**	0.15	**11.25**	m2	**13.50**
Two coats textured masonry paint, existing cement painted surfaces; external work										
Cement rendered general surfaces										
girth exceeding 300mm	0.61	10.00	**0.67**	0.43	-	**6.10**	0.12	**6.88**	m2	**8.25**
Brickwork general surfaces										
girth exceeding 300mm	1.08	10.00	**1.19**	0.53	-	**7.50**	0.14	**8.83**	m2	**10.60**
Rough cast general surfaces										
girth exceeding 300mm	1.43	10.00	**1.57**	0.68	-	**9.69**	0.18	**11.44**	m2	**13.73**

Labour hourly rates: (except Specialists) Craft Operatives £14.21 Labourer £10.71 Rates are national average prices. Refer to REGIONAL VARIATIONS for indicative levels of overall pricing in regions	MATERIALS			LABOUR				RATES		
	Del to Site	Waste	Material Cost	Craft Optve	Lab	Labour Cost	Sunds	Nett Rate	Unit	Gross Rate (+20%)
	£	%	£	Hrs	Hrs	£	£	£		£
M60: PAINTING/CLEAR FINISHING (REDECORATIONS) - OIL PAINTING WALLS AND CEILINGS										
Generally										
Note: the following rates include for the cost of all preparatory work, e.g. washing down, etc.										
One undercoat, one coat full gloss finish, existing gloss painted surfaces										
Brickwork general surfaces										
girth exceeding 300mm	0.39	10.00	0.43	0.43	-	6.10	0.12	6.64	m2	7.97
Plaster general surfaces										
girth exceeding 300mm	0.34	10.00	0.37	0.37	-	5.31	0.10	5.79	m2	6.94
Two undercoats, one coat full gloss finish, existing gloss painted surfaces										
Plaster general surfaces										
girth exceeding 300mm	0.51	10.00	0.56	0.48	-	6.88	0.13	7.57	m2	9.09
Two undercoats, one coat eggshell finish, existing gloss painted surfaces										
Brickwork general surfaces										
girth exceeding 300mm	0.71	10.00	0.78	0.53	-	7.50	0.14	8.43	m2	10.11
Plaster general surfaces										
girth exceeding 300mm	0.56	10.00	0.61	0.48	-	6.88	0.13	7.62	m2	9.15
One coat primer, one undercoat, one coat full gloss finish, existing washable distempered or emulsion painted surfaces										
Plaster general surfaces										
girth exceeding 300mm	0.94	10.00	1.03	0.48	-	6.88	0.13	8.04	m2	9.65
M60: PAINTING/CLEAR FINISHING (REDECORATIONS) - PLASTIC FINISH										
Generally										
Note: the following rates include for the cost of all preparatory work, e.g. washing down, etc.										
Textured plastic coating - combed finish, existing washable distempered or emulsion painted surfaces										
Brickwork general surfaces										
girth exceeding 300mm	0.55	10.00	0.60	0.55	-	7.82	0.15	8.57	m2	10.28
Plaster general surfaces										
girth exceeding 300mm	0.41	10.00	0.45	0.44	-	6.25	0.12	6.82	m2	8.19
M60: PAINTING/CLEAR FINISHING (REDECORATIONS) - CLEAN OUT GUTTERS										
Generally										
Note: the following rates include for the cost of all preparatory work, e.g. washing down, etc.										
Clean out gutters prior to repainting, staunch joints with red lead, bituminous compound or mastic										
Iron or steel eaves gutters										
generally	-	-	-	0.07	-	0.94	0.07	1.00	m	1.20
M60: PAINTING/CLEAR FINISHING (REDECORATIONS) - OIL PAINTING METALWORK										
Generally										
Note: the following rates include for the cost of all preparatory work, e.g. washing down, etc.										
One undercoat, one coat full gloss finish, existing gloss painted metal surfaces										
Iron or steel general surfaces										
girth exceeding 300mm	0.35	10.00	0.39	0.52	-	7.35	0.14	7.87	m2	9.45
isolated surfaces, girth not exceeding 300mm	0.11	10.00	0.12	0.18	-	2.50	0.05	2.67	m	3.20
Galvanized glazed doors, windows or screens										
girth exceeding 300mm; panes, area not exceeding 0.10m2	0.19	10.00	0.21	0.97	-	13.76	0.26	14.22	m2	17.07
girth exceeding 300mm; panes, area 0.10 - 0.50m2	0.13	10.00	0.15	0.70	-	10.00	0.19	10.34	m2	12.41
girth exceeding 300mm; panes, area 0.50 - 1.00m2	0.11	10.00	0.12	0.58	-	8.28	0.16	8.56	m2	10.27
girth exceeding 300mm; panes, area exceeding 1.00m2	0.11	10.00	0.12	0.51	-	7.19	0.14	7.45	m2	8.93
Iron or steel structural work										
girth exceeding 300mm	0.35	10.00	0.39	0.57	-	8.13	0.15	8.67	m2	10.40
isolated surfaces, girth not exceeding 300mm	0.11	10.00	0.12	0.23	-	3.28	0.06	3.46	m	4.16
Iron or steel structural members of roof trusses, lattice girders, purlins and the like										
girth exceeding 300mm	0.35	10.00	0.39	0.72	-	10.16	0.19	10.74	m2	12.89

Labour hourly rates: (except Specialists) Craft Operatives £14.21 Labourer £10.71 Rates are national average prices. Refer to REGIONAL VARIATIONS for indicative levels of overall pricing in regions	MATERIALS			LABOUR				RATES		
	Del to Site £	Waste %	Material Cost £	Craft Optve Hrs	Lab Hrs	Labour Cost £	Sunds £	Nett Rate £	Unit	Gross Rate (+20%) £
M60: PAINTING/CLEAR FINISHING (REDECORATIONS) - OIL PAINTING METALWORK Cont.										
One undercoat, one coat full gloss finish, existing gloss painted metal surfaces Cont.										
Iron or steel structural members of roof trusses, lattice girders, purlins and the like Cont.										
isolated surfaces, girth not exceeding 300mm.....	0.11	10.00	**0.12**	0.24	-	**3.44**	0.07	**3.62**	m	**4.35**
Iron or steel services										
girth exceeding 300mm	0.35	10.00	**0.39**	0.72	-	**10.16**	0.19	**10.74**	m2	**12.89**
isolated surfaces, girth not exceeding 300mm.....	0.11	10.00	**0.12**	0.24	-	**3.44**	0.07	**3.62**	m	**4.35**
Copper services										
isolated surfaces, girth not exceeding 300mm.....	0.11	10.00	**0.12**	0.24	-	**3.44**	0.07	**3.62**	m	**4.35**
Galvanized services										
isolated surfaces, girth not exceeding 300mm.....	0.11	10.00	**0.12**	0.24	-	**3.44**	0.07	**3.62**	m	**4.35**
One undercoat, one coat full gloss finish, existing gloss painted metal surfaces; external work										
Iron or steel general surfaces										
girth exceeding 300mm	0.35	10.00	**0.39**	0.55	-	**7.82**	0.15	**8.35**	m2	**10.02**
isolated surfaces, girth not exceeding 300mm.....	0.11	10.00	**0.12**	0.19	-	**2.66**	0.05	**2.83**	m	**3.39**
Iron or steel railings, fences and gates; plain open type										
girth exceeding 300mm	0.27	10.00	**0.30**	0.54	-	**7.66**	0.15	**8.10**	m2	**9.72**
isolated surfaces, girth not exceeding 300mm.....	0.11	10.00	**0.12**	0.18	-	**2.50**	0.05	**2.67**	m	**3.20**
Iron or steel railings, fences and gates; close type										
girth exceeding 300mm	0.35	10.00	**0.39**	0.50	-	**7.03**	0.13	**7.55**	m2	**9.06**
Iron or steel railings, fences and gates; ornamental type										
girth exceeding 300mm	0.27	10.00	**0.30**	0.99	-	**14.07**	0.27	**14.63**	m2	**17.56**
Iron or steel eaves gutters										
girth exceeding 300mm	0.35	10.00	**0.39**	0.58	-	**8.28**	0.16	**8.83**	m2	**10.59**
isolated surfaces, girth not exceeding 300mm.....	0.11	10.00	**0.12**	0.19	-	**2.66**	0.05	**2.83**	m	**3.39**
Two undercoats, one coat full gloss finish, existing gloss painted metal surfaces										
Iron or steel general surfaces										
girth exceeding 300mm	0.53	10.00	**0.58**	0.68	-	**9.69**	0.18	**10.45**	m2	**12.54**
isolated surfaces, girth not exceeding 300mm.....	0.16	10.00	**0.18**	0.23	-	**3.28**	0.06	**3.52**	m	**4.23**
Two undercoats, one coat full gloss finish, existing gloss painted metal surfaces; external work										
Iron or steel general surfaces										
girth exceeding 300mm	0.53	10.00	**0.58**	0.72	-	**10.16**	0.19	**10.93**	m2	**13.12**
isolated surfaces, girth not exceeding 300mm.....	0.16	10.00	**0.18**	0.24	-	**3.44**	0.07	**3.68**	m	**4.42**
M60: PAINTING/CLEAR FINISHING (REDECORATIONS) - OIL PAINTING WOODWORK										
Generally										
Note: the following rates include for the cost of all preparatory work, e.g. washing down, etc.										
One undercoat, one coat full gloss finish, existing gloss painted wood surfaces										
Wood general surfaces										
girth exceeding 300mm	0.35	10.00	**0.39**	0.50	-	**7.03**	0.13	**7.55**	m2	**9.06**
isolated surfaces, girth not exceeding 300mm.....	0.11	10.00	**0.12**	0.17	-	**2.34**	0.04	**2.51**	m	**3.01**
Wood partially glazed doors										
girth exceeding 300mm; panes, area not exceeding 0.10m2	0.27	10.00	**0.30**	0.74	-	**10.47**	0.20	**10.97**	m2	**13.16**
girth exceeding 300mm; panes, area 0.10 - 0.50m2.	0.22	10.00	**0.24**	0.59	-	**8.44**	0.16	**8.84**	m2	**10.61**
girth exceeding 300mm; panes, area 0.50 - 1.00m2.	0.19	10.00	**0.21**	0.53	-	**7.50**	0.14	**7.85**	m2	**9.42**
girth exceeding 300mm; panes, area exceeding 1.00m2	0.19	10.00	**0.21**	0.47	-	**6.72**	0.13	**7.06**	m2	**8.47**
Wood windows and screens										
girth exceeding 300mm; panes, area not exceeding 0.10m2	0.27	10.00	**0.30**	0.94	-	**13.29**	0.25	**13.84**	m2	**16.60**
girth exceeding 300mm; panes, area 0.10 - 0.50m2.	0.22	10.00	**0.24**	0.67	-	**9.53**	0.18	**9.95**	m2	**11.94**
girth exceeding 300mm; panes, area 0.50 - 1.00m2.	0.19	10.00	**0.21**	0.57	-	**8.13**	0.15	**8.49**	m2	**10.19**
girth exceeding 300mm; panes, area exceeding 1.00m2	0.19	10.00	**0.21**	0.48	-	**6.88**	0.13	**7.22**	m2	**8.66**
One undercoat, one coat full gloss finish, existing gloss painted wood surfaces; external work										
Wood general surfaces										
girth exceeding 300mm	0.35	10.00	**0.39**	0.53	-	**7.50**	0.14	**8.03**	m2	**9.64**
isolated surfaces, girth not exceeding 300mm.....	0.11	10.00	**0.12**	0.18	-	**2.50**	0.05	**2.67**	m	**3.20**
Wood railings fences and gates; open type										
girth exceeding 300mm	0.27	10.00	**0.30**	0.45	-	**6.41**	0.12	**6.83**	m2	**8.19**

Labour hourly rates: (except Specialists) Craft Operatives £14.21 Labourer £10.71 Rates are national average prices. Refer to REGIONAL VARIATIONS for indicative levels of overall pricing in regions	MATERIALS			LABOUR				RATES		
	Del to Site £	Waste %	Material Cost £	Craft Optve Hrs	Lab Hrs	Labour Cost £	Sunds £	Nett Rate £	Unit	Gross Rate (+20%) £
60: PAINTING/CLEAR FINISHING (REDECORATIONS) - OIL AINTING WOODWORK Cont.										
ne undercoat, one coat full gloss finish, existing loss painted wood surfaces; external work Cont.										
ood railings fences and gates; open type Cont.										
isolated surfaces, girth not exceeding 300mm.....	0.11	10.00	**0.12**	0.14	-	**2.03**	0.04	**2.19**	m	**2.63**
ood railings fences and gates; close type										
girth exceeding 300mm	0.35	10.00	**0.39**	0.41	-	**5.78**	0.11	**6.28**	m2	**7.53**
urn off, one coat primer, one undercoat, one coat ull gloss finish, existing painted wood surfaces										
ood general surfaces										
girth exceeding 300mm	0.62	10.00	**0.68**	0.99	-	**14.07**	0.66	**15.41**	m2	**18.49**
isolated surfaces, girth not exceeding 300mm.....	0.18	10.00	**0.20**	0.29	-	**4.06**	0.19	**4.46**	m	**5.35**
ood partially glazed doors										
girth exceeding 300mm; panes, area not exceeding 0.10m2	0.48	10.00	**0.53**	1.32	-	**18.76**	0.88	**20.17**	m2	**24.20**
girth exceeding 300mm; panes, area 0.10 - 0.50m2.	0.37	10.00	**0.41**	1.09	-	**15.47**	0.73	**16.61**	m2	**19.93**
girth exceeding 300mm; panes, area 0.50 - 1.00m2.	0.34	10.00	**0.38**	0.99	-	**14.07**	0.66	**15.11**	m2	**18.13**
girth exceeding 300mm; panes, area exceeding 1.00m2	0.34	10.00	**0.38**	0.95	-	**13.44**	0.63	**14.45**	m2	**17.34**
ood windows and screens										
girth exceeding 300mm; panes, area not exceeding 0.10m2	0.48	10.00	**0.53**	1.82	-	**25.79**	1.22	**27.53**	m2	**33.04**
girth exceeding 300mm; panes, area 0.10 - 0.50m2.	0.37	10.00	**0.41**	1.30	-	**18.44**	0.87	**19.72**	m2	**23.66**
girth exceeding 300mm; panes, area 0.50 - 1.00m2.	0.34	10.00	**0.38**	1.09	-	**15.47**	0.73	**16.58**	m2	**19.90**
girth exceeding 300mm; panes, area exceeding 1.00m2	0.34	10.00	**0.38**	0.97	-	**13.76**	0.65	**14.78**	m2	**17.74**
urn off, one coat primer, one undercoat, one coat ull gloss finish, existing painted wood surfaces; xternal work										
ood general surfaces										
girth exceeding 300mm	0.62	10.00	**0.68**	1.05	-	**14.85**	0.70	**16.23**	m2	**19.47**
isolated surfaces, girth not exceeding 300mm.....	0.18	10.00	**0.20**	0.32	-	**4.53**	0.21	**4.95**	m	**5.94**
ood railings fences and gates; open type										
girth exceeding 300mm	0.48	10.00	**0.53**	0.89	-	**12.66**	0.60	**13.79**	m2	**16.54**
isolated surfaces, girth not exceeding 300mm.....	0.18	10.00	**0.20**	0.26	-	**3.75**	0.18	**4.13**	m	**4.96**
ood railings fences and gates; close type										
girth exceeding 300mm	0.62	10.00	**0.68**	0.84	-	**11.88**	0.56	**13.12**	m2	**15.74**
urn off, one coat primer, two undercoats, one coat ull gloss finish, existing painted wood surfaces										
ood general surfaces										
girth exceeding 300mm	0.79	10.00	**0.87**	1.17	-	**16.57**	0.78	**18.22**	m2	**21.87**
isolated surfaces, girth not exceeding 300mm.....	0.24	10.00	**0.26**	0.35	-	**5.00**	0.24	**5.50**	m	**6.60**
urn off, one coat primer, two undercoats, one coat ull gloss finish, existing painted wood surfaces; xternal work										
ood general surfaces										
girth exceeding 300mm	0.79	10.00	**0.87**	1.22	-	**17.35**	0.82	**19.04**	m2	**22.85**
isolated surfaces, girth not exceeding 300mm.....	0.24	10.00	**0.26**	0.39	-	**5.47**	0.26	**5.99**	m	**7.19**
60: PAINTING/CLEAR FINISHING (REDECORATIONS) - OLYURETHANE LACQUER										
enerally										
ote: the following rates include for the cost of all reparatory work, e.g. washing down, etc.										
wo coats polyurethane lacquer, existing lacquered urfaces										
ood general surfaces										
girth exceeding 300mm	0.93	10.00	**1.03**	0.59	-	**8.44**	0.16	**9.63**	m2	**11.55**
isolated surfaces, girth not exceeding 300mm.....	0.28	10.00	**0.31**	0.20	-	**2.81**	0.05	**3.17**	m	**3.81**
ood partially glazed doors										
girth exceeding 300mm; panes, area not exceeding 0.10m2	0.75	10.00	**0.82**	0.85	-	**12.04**	0.23	**13.08**	m2	**15.70**
girth exceeding 300mm; panes, area 0.10 - 0.50m2.	0.56	10.00	**0.62**	0.68	-	**9.69**	0.18	**10.49**	m2	**12.59**
girth exceeding 300mm; panes, area 0.50 - 1.00m2.	0.53	10.00	**0.58**	0.61	-	**8.60**	0.16	**9.34**	m2	**11.21**
girth exceeding 300mm; panes, area exceeding 1.00m2	0.53	10.00	**0.58**	0.55	-	**7.82**	0.15	**8.55**	m2	**10.25**
ood windows and screens										
girth exceeding 300mm; panes, area not exceeding 0.10m2	0.75	10.00	**0.82**	1.09	-	**15.47**	0.29	**16.59**	m2	**19.91**
girth exceeding 300mm; panes, area 0.10 - 0.50m2.	0.56	10.00	**0.62**	0.72	-	**10.16**	0.19	**10.97**	m2	**13.16**
girth exceeding 300mm; panes, area 0.50 - 1.00m2.	0.53	10.00	**0.58**	0.65	-	**9.22**	0.18	**9.98**	m2	**11.97**
girth exceeding 300mm; panes, area exceeding 1.00m2	0.53	10.00	**0.58**	0.57	-	**8.13**	0.15	**8.86**	m2	**10.64**

Labour hourly rates: (except Specialists) Craft Operatives £14.21 Labourer £10.71 Rates are national average prices. Refer to REGIONAL VARIATIONS for indicative levels of overall pricing in regions	MATERIALS			LABOUR				RATES		
	Del to Site £	Waste %	Material Cost £	Craft Optve Hrs	Lab Hrs	Labour Cost £	Sunds £	Nett Rate £	Unit	Gross Rate (+20%) £
M60: PAINTING/CLEAR FINISHING (REDECORATIONS) - POLYURETHANE LACQUER Cont.										
Two coats polyurethane lacquer, existing lacquered surfaces; external work										
Wood general surfaces										
girth exceeding 300mm	0.93	10.00	**1.03**	0.63	-	**8.91**	0.17	**10.10**	m2	**12.13**
isolated surfaces, girth not exceeding 300mm	0.28	10.00	**0.31**	0.21	-	**2.97**	0.06	**3.33**	m	**4.00**
M60: PAINTING/CLEAR FINISHING (REDECORATIONS) - FRENCH AND WAX POLISHING										
Generally										
Note: the following rates include for the cost of all preparatory work, e.g. washing down, etc.										
Stain; two coats white polish										
Wood general surfaces										
girth exceeding 300mm	3.03	10.00	**3.34**	0.44	-	**6.25**	0.12	**9.71**	m2	**11.65**
isolated surfaces, girth not exceeding 300mm	0.95	10.00	**1.05**	0.15	-	**2.19**	0.04	**3.28**	m	**3.93**
Fully French polish existing polished surfaces										
Wood general surfaces										
girth exceeding 300mm	5.16	10.00	**5.68**	1.69	-	**24.07**	0.46	**30.21**	m2	**36.25**
isolated surfaces, girth not exceeding 300mm	1.63	10.00	**1.79**	0.57	-	**8.13**	0.15	**10.07**	m	**12.09**
M60: PAINTING/CLEAR FINISHING (REDECORATIONS) - WATER REPELLENT										
One coat solvent based water repellent, existing surfaces										
Brickwork general surfaces										
girth exceeding 300mm	0.21	5.00	**0.22**	0.14	-	**2.03**	0.04	**2.29**	m2	**2.75**
M60: PAINTING/CLEAR FINISHING (REDECORATIONS) - REMOVAL OF MOULD GROWTH										
Apply fungicide to existing decorated and infected surfaces										
Cement rendered general surfaces										
girth exceeding 300mm	0.13	5.00	**0.14**	0.24	-	**3.44**	0.07	**3.64**	m2	**4.37**
Plaster general surfaces										
girth exceeding 300mm	0.11	5.00	**0.12**	0.24	-	**3.44**	0.07	**3.62**	m2	**4.35**

FURNITURE/EQUIPMENT

Labour hourly rates: (except Specialists) Craft Operatives £14.21 Labourer £10.71 Rates are national average prices. Refer to REGIONAL VARIATIONS for indicative levels of overall pricing in regions	MATERIALS Del to Site £	Waste %	Material Cost £	LABOUR Craft Optve Hrs	Lab Hrs	Labour Cost £	Sunds £	RATES Nett Rate £	Unit	Gross Rate (+20%) £
N10: GENERAL FIXTURES/FURNISHING/EQUIPMENT: MIRRORS										
BS 952, clear float, SG, silvered and protected with copper backing										
6mm thick; fixing to masonry with brass screws, chromium plated dome covers, rubber sleeves and washers										
holes 6mm diameter -4; edges polished										
460 x 600mm	52.99	5.00	**55.64**	0.72	-	**10.16**	0.50	**66.29**	nr	**79.55**
holes 6mm diameter -4; edges bevelled										
460 x 600mm	66.14	5.00	**69.45**	0.72	-	**10.16**	0.50	**80.11**	nr	**96.13**
N10: GENERAL FIXTURES/FURNISHING/EQUIPMENT: BLINDS										
Internal blinds										
Venetian blinds; stove enamelled aluminium alloy slats 25mm wide; plain colours with single cord control										
1200mm drop; fixing to timber with screws										
1000mm wide	50.87	2.50	**52.14**	1.21	0.15	**18.84**	0.07	**71.05**	nr	**85.26**
2000mm wide	79.49	2.50	**81.48**	1.82	0.23	**28.27**	0.10	**109.84**	nr	**131.81**
N10: GENERAL FIXTURES/FURNISHING/EQUIPMENT: SHELVING SYSTEMS - SHELVING SUPPORT SYSTEMS										
Spur patent shelf supports in steel with white enamelled finish; Spur Shelving Ltd; Spur Steel-Lok components										
Wall uprights										
type 9013, 1000mm long; fixing to masonry with screws	3.58	2.50	**3.66**	0.50	-	**7.03**	-	**10.70**	nr	**12.84**
Straight brackets										
type 9003, 270mm long	1.54	2.50	**1.58**	0.06	-	**0.78**	-	**2.36**	nr	**2.83**
N10: GENERAL FIXTURES/FURNISHING/EQUIPMENT: MAT RIMS										
Matwells in polished brass										
Mat frames; brazed fabrication										
38 x 38 x 6mm angle section; angles mitred; plain lugs -4, welded on; mat space										
914 x 610mm	201.96	2.50	**207.01**	2.09	2.09	**52.08**	-	**259.09**	nr	**310.91**
N10: GENERAL FIXTURES/FURNISHING/EQUIPMENT: HAT AND COAT RAILS										
Hat and coat rails										
Rails in softwood, wrought										
25 x 75mm; chamfers -2	1.56	10.00	**1.71**	0.32	0.04	**5.00**	0.02	**6.74**	m	**8.08**
Hat and coat hooks										
B.M.A. finish	2.42	2.50	**2.48**	0.44	-	**6.25**	-	**8.73**	nr	**10.48**
SAA finish	1.21	2.50	**1.24**	0.44	-	**6.25**	-	**7.49**	nr	**8.99**

This section continues on the next page

Labour hourly rates: (except Specialists) Craft Operatives £17.38 Labourer £10.71 Rates are national average prices. Refer to REGIONAL VARIATIONS for indicative levels of overall pricing in regions	MATERIALS			LABOUR				RATES		
	Del to Site £	Waste %	Material Cost £	Craft Optve Hrs	Lab Hrs	Labour Cost £	Sunds £	Nett Rate £	Unit	Gross Rate (+20%) £
N13: SANITARY APPLIANCES/FITTINGS										
Supply and fix appliances - sinks										
Note: the following prices include joints to copper services and wastes but do not include traps										
Sink units; single bowl with single drainer and back ledge; stainless steel; BS EN 13310 Type A; 38mm waste, plug and chain to BS EN 274 Part 1 with combined overflow; pair 13mm pillar taps to BS 1010 Part 2										
fixing on base unit with metal clips										
1000 x 600mm	100.93	2.50	103.45	3.30	-	57.35	0.90	161.70	nr	194.04
Supply and fix appliances - lavatory basins										
Wash basins; coloured vitreous china; BS 5506 Part 3; 32mm slotted waste, chain, stay and plug to BS EN 274 Part 1; painted build-in brackets; pair 13mm pillar taps to BS 1010 Part 2										
building in brackets to masonry; sealing at back with mastic sealant										
560 x 406mm	86.90	2.50	89.07	4.62	-	80.30	1.80	171.16	nr	205.40
Supply and fix appliances - baths										
Baths; coloured acrylic; BS 4305; 32mm overflow with front grid and 38mm waste, chain and plug to BS EN 274 Part 1 with combined overflow; metal cradles with adjustable feet and wall fixing brackets; pair 19mm pillar taps to BS 1010 Part 2										
700 x 1700 x 570mm overall; fixing brackets to masonry with screws; sealing at walls with mastic sealant	136.40	2.50	139.81	6.60	-	114.71	3.93	258.45	nr	310.13
Supply and fix appliances - W.C. suites										
Low level W.C. suites; coloured vitreous china pan to BS 5503; 9 litre coloured vitreous china cistern with valveless fittings and plastics flush bend to BS 1125; 13mm piston type high pressure ball valve to BS 1212 Part 1 with 127mm plastics float to BS 2456; coloured plastics seat and cover to BS 1254 type 1										
pan with S or P trap conversion bend to BS 5627; fixing pan and cistern brackets to masonry with screws; bedding pan in mastic; jointing pan to soil pipe with Multikwik connector	170.23	2.50	174.48	5.94	-	103.24	2.36	280.07	nr	336.09
Supply and fix appliances - urinals										
Wall urinals; range of 3; white vitreous china bowls to BS 5520; white vitreous china divisions; 14 litre white plastics automatic cistern to BS 1876; stainless steel flush pipes and spreaders; 32mm plastics waste outlets										
fixing bowls, divisions, cistern and pipe brackets to masonry with screws	774.95	2.50	794.32	17.82	-	309.71	3.65	1107.68	nr	1329.22
Supply and fix appliances - showers										
Showers; coloured glazed acrylic tray, 760 x 760 x 180mm; 38mm waste outlet; recessed thermostatic valve and swivel spray head										
bedding tray in cement mortar (1:4); fixing valve and head to masonry with screws; sealing at walls with mastic sealant	522.23	2.50	535.28	6.27	-	108.97	5.61	649.86	nr	779.84
Shower curtains; heavy duty plastic; anodised aluminium rail, glider hooks, end and suspension fitting; fixing supports to masonry with screws										
straight, 914mm long x 1829mm drop	46.20	2.50	47.36	1.32	-	22.94	0.56	70.86	nr	85.03
angled, 1676mm girth x 1829mm drop	62.43	2.50	63.99	1.98	-	34.41	0.73	99.13	nr	118.95

BUILDING FABRIC SUNDRIES

Labour hourly rates: (except Specialists) Craft Operatives £14.21 Labourer £10.71 Rates are national average prices. Refer to REGIONAL VARIATIONS for indicative levels of overall pricing in regions	MATERIALS			LABOUR				RATES		
	Del to Site £	Waste %	Material Cost £	Craft Optve Hrs	Lab Hrs	Labour Cost £	Sunds £	Nett Rate £	Unit	Gross Rate (+20%) £
P10: SUNDRY INSULATION/PROOFING WORK/ FIRE STOPS										
Waterproof reinforced building paper BS 1521; grade A1F; 150mm lapped joints										
Across members at 450mm centres vertical	0.91	5.00	**0.96**	0.22	0.03	**3.48**	0.03	**4.47**	m2	**5.36**
British Sisalkraft Ltd SK860 vapour control layer; 100mm laps sealed with tape										
Plain areas horizontal; laid loose	1.56	10.00	**1.72**	0.09	0.01	**1.37**	0.04	**3.13**	m2	**3.76**
Paper faced mineral wool insulation quilt; butt joints										
60mm thick; between members at 450mm centres vertical	5.07	15.00	**5.83**	0.28	0.03	**4.26**	0.03	**10.12**	m2	**12.15**
Glass fibre insulation quilt; butt joints										
100mm thick; across members at 450mm centres horizontal; laid loose	4.71	15.00	**5.41**	0.13	0.02	**2.11**	-	**7.52**	m2	**9.03**
100mm thick; between members at 450mm centres horizontal; laid loose	4.71	10.00	**5.18**	0.20	0.02	**3.05**	-	**8.23**	m2	**9.87**
Rockwool Rollbatts; insulation quilt; butt joints										
100mm thick; between members at 450mm centres horizontal; laid loose	3.57	10.00	**3.92**	0.12	0.02	**1.96**	-	**5.88**	m2	**7.05**
150mm thick; between members at 450mm centres horizontal; laid loose	5.51	10.00	**6.06**	0.19	0.03	**3.01**	-	**9.07**	m2	**10.88**
Dow Construction Products; Styrofoam Floormate 200; butt joints										
50mm thick; plain areas horizontal; laid loose	12.73	10.00	**14.01**	0.24	0.04	**3.91**	-	**17.92**	m2	**21.50**
Expanded polystyrene sheeting; butt joints										
25mm thick; plain areas										
horizontal; laid loose	1.87	5.00	**1.96**	0.13	0.07	**2.58**	-	**4.54**	m2	**5.45**
vertical; fixing with adhesive	1.87	5.00	**1.96**	0.33	0.17	**6.46**	1.21	**9.63**	m2	**11.55**
P20: UNFRAMED ISOLATED TRIMS/SKIRTINGS/SUNDRY ITEMS										
Softwood, wrought - skirtings, architraves, picture rails and cover fillets										
Skirtings, picture rails, architraves and the like; (finished sizes)										
19 x 150mm	3.43	10.00	**3.78**	0.35	0.04	**5.47**	0.06	**9.31**	m	**11.17**
25 x 75mm; splays -1	2.64	10.00	**2.91**	0.32	0.04	**5.00**	0.06	**7.97**	m	**9.56**
25 x 100mm; chamfers -2	3.34	10.00	**3.67**	0.33	0.04	**5.16**	0.06	**8.89**	m	**10.66**
Cover fillets, stops, trims, beads, nosings and the like; (finished sizes)										
19 x 50mm; mouldings -1	1.47	10.00	**1.61**	0.28	0.03	**4.26**	0.04	**5.92**	m	**7.10**
25 x 38mm; chamfers -1	1.39	10.00	**1.53**	0.28	0.03	**4.26**	0.04	**5.83**	m	**7.00**
Afrormosia, wrought - skirtings, architraves, picture rails and cover fillets										
Skirtings, picture rails, architraves and the like; (finished sizes)										
19 x 100mm	7.30	10.00	**8.03**	0.53	0.07	**8.21**	0.09	**16.32**	m	**19.59**
25 x 150mm	12.17	10.00	**13.39**	0.61	0.08	**9.42**	0.10	**22.91**	m	**27.49**
extra; ends	1.86	10.00	**2.05**	0.07	0.01	**1.06**	0.01	**3.12**	nr	**3.74**

Labour hourly rates: (except Specialists) Craft Operatives £14.21 Labourer £10.71 Rates are national average prices. Refer to REGIONAL VARIATIONS for indicative levels of overall pricing in regions	MATERIALS			LABOUR			RATES			
	Del to Site £	Waste %	Material Cost £	Craft Optve Hrs	Lab Hrs	Labour Cost £	Sunds £	Nett Rate £	Unit	Gross Rate (+20%) £
P20: UNFRAMED ISOLATED TRIMS/SKIRTINGS/SUNDRY ITEMS Cont.										
Afrormosia, wrought - skirtings, architraves, picture rails and cover fillets Cont.										
Skirtings, picture rails, architraves and the like; (finished sizes) Cont.										
extra; angles	3.03	-	**3.03**	0.12	0.02	**1.96**	0.02	**5.00**	nr	**6.00**
extra; mitres	3.03	-	**3.03**	0.12	0.02	**1.96**	0.02	**5.00**	nr	**6.00**
19 x 125mm; mouldings -1	8.22	10.00	**9.04**	0.58	0.08	**9.11**	0.10	**18.25**	m	**21.89**
Cover fillets, stops, trims, beads, nosings and the like; (finished sizes)										
19 x 50mm; mouldings -1	4.27	10.00	**4.70**	0.47	0.06	**7.31**	0.08	**12.09**	m	**14.50**
25 x 38mm; chamfers -1	3.94	10.00	**4.34**	0.47	0.06	**7.31**	0.08	**11.73**	m	**14.07**
Medium density fibreboard, BS EN 622, skirtings, architraves, picture rails and cover fillets to BS 1186 Part 3										
Skirtings, picture rails, architraves and the like; (finished sizes)										
20 x 70mm; reference 20CA70	2.21	10.00	**2.43**	0.37	0.06	**5.90**	0.08	**8.41**	m	**10.10**
13 x 45mm; reference 13RA45	1.41	10.00	**1.55**	0.36	0.06	**5.75**	0.08	**7.38**	m	**8.85**
Cover fillets, stops, trims, beads, nosings and the like; (finished sizes)										
20 x 45mm; reference HR45	1.68	10.00	**1.84**	0.32	0.04	**5.00**	0.07	**6.91**	m	**8.30**
12 x 95mm; reference 12HRCS95	2.62	10.00	**2.88**	0.39	0.06	**6.06**	0.07	**9.00**	m	**10.81**
Softwood, wrought - window boards										
Window boards										
25 x 150mm; rounded edges -1	4.51	10.00	**4.97**	0.37	0.04	**5.79**	0.06	**10.81**	m	**12.97**
Sapele; wrought - window boards										
Window boards										
25 x 150mm; rounded edges -1	10.21	10.00	**11.23**	0.57	0.08	**8.95**	0.10	**20.28**	m	**24.34**
Moisture resistant medium density fibreboard, BS EN 622										
Window boards										
25 x 225mm; nosed and tongued	6.90	10.00	**7.59**	0.37	0.06	**5.90**	0.07	**13.55**	m	**16.27**
25 x 250mm; nosed and tongued	7.66	10.00	**8.43**	0.40	0.06	**6.22**	0.07	**14.71**	m	**17.66**
Softwood, wrought - isolated shelves and worktops										
Isolated shelves and worktops										
25 x 150mm	2.69	10.00	**2.96**	0.25	0.03	**3.95**	0.04	**6.95**	m	**8.34**
25 x 300mm	5.66	10.00	**6.22**	0.37	0.04	**5.79**	0.06	**12.06**	m	**14.47**
25 x 450mm overall; 25 x 50mm slats spaced 25mm apart	6.26	10.00	**6.88**	1.27	0.15	**19.62**	0.21	**26.72**	m	**32.06**
Afrormosia, wrought - isolated shelves and worktops										
Isolated shelves and worktops										
25 x 150mm	9.27	10.00	**10.20**	0.44	0.06	**6.84**	0.07	**17.10**	m	**20.52**
25 x 300mm	18.55	10.00	**20.40**	0.67	0.09	**10.48**	0.11	**30.99**	m	**37.19**
25 x 450mm; cross tongued	38.51	2.50	**39.48**	0.77	0.10	**12.00**	0.12	**51.60**	m	**61.92**
Plywood BS EN 636, II/III grade, INT bonded; butt joints - isolated shelves and worktops										
Isolated shelves and worktops										
12 x 150mm	1.83	20.00	**2.20**	0.24	0.03	**3.79**	0.07	**6.06**	m	**7.27**
12 x 300mm	3.68	20.00	**4.41**	0.36	0.04	**5.63**	0.07	**10.11**	m	**12.13**
15 x 600mm	8.83	20.00	**10.60**	0.51	0.07	**7.90**	0.07	**18.56**	m	**22.27**
Blockboard, BS 3444, 2/2 grade, INT bonded, faced with 1.5mm laminated plastic sheet, BS EN 438, classified HGS, with 1.2mm laminated plastic sheet balance veneer; butt joints										
Isolated shelves and worktops										
18 x 150mm	34.76	20.00	**41.72**	0.51	0.07	**7.90**	0.13	**49.75**	m	**59.69**
18 x 300mm	57.99	20.00	**69.59**	0.76	0.10	**11.85**	0.13	**81.57**	m	**97.88**
18 x 600mm	91.96	20.00	**110.36**	1.01	0.13	**15.79**	0.13	**126.28**	m	**151.54**
P21: IRONMONGERY										
Fix only ironmongery (prime cost sum for supply included elsewhere)										
To hardwood or the like										
butt hinges; 50mm	-	-	-	0.28	-	**3.91**	-	**3.91**	nr	**4.69**
butt hinges; 100mm	-	-	-	0.28	-	**3.91**	-	**3.91**	nr	**4.69**
rising hinges; 100mm	-	-	-	0.44	-	**6.25**	-	**6.25**	nr	**7.50**
tee hinges; 150mm	-	-	-	0.66	-	**9.38**	-	**9.38**	nr	**11.25**
single action floor springs and top centres	-	-	-	4.51	-	**64.09**	-	**64.09**	nr	**76.90**
double action floor springs and top centres	-	-	-	5.39	-	**76.59**	-	**76.59**	nr	**91.91**
overhead door closers	-	-	-	2.70	-	**38.30**	-	**38.30**	nr	**45.96**
concealed overhead door closers	-	-	-	4.51	-	**64.09**	-	**64.09**	nr	**76.90**
Perko door closers	-	-	-	2.70	-	**38.30**	-	**38.30**	nr	**45.96**

Labour hourly rates: (except Specialists) Craft Operatives £14.21 Labourer £10.71 Rates are national average prices. Refer to REGIONAL VARIATIONS for indicative levels of overall pricing in regions	MATERIALS			LABOUR				RATES		
	Del to Site £	Waste %	Material Cost £	Craft Optve Hrs	Lab Hrs	Labour Cost £	Sunds £	Nett Rate £	Unit	Gross Rate (+20%) £
P21: IRONMONGERY Cont.										
Fix only ironmongery (prime cost sum for supply included elsewhere) Cont.										
To hardwood or the like Cont.										
cabin hooks	-	-	-	0.39	-	**5.47**	-	**5.47**	nr	**6.57**
casement fasteners	-	-	-	0.61	-	**8.60**	-	**8.60**	nr	**10.32**
sash screws	-	-	-	0.88	-	**12.50**	-	**12.50**	nr	**15.01**
mortice latches	-	-	-	1.21	-	**17.19**	-	**17.19**	nr	**20.63**
night latches	-	-	-	1.82	-	**25.79**	-	**25.79**	nr	**30.95**
rim latches	-	-	-	0.88	-	**12.50**	-	**12.50**	nr	**15.01**
Suffolk latches	-	-	-	1.43	-	**20.32**	-	**20.32**	nr	**24.38**
cylinder locks	-	-	-	1.82	-	**25.79**	-	**25.79**	nr	**30.95**
mortice locks	-	-	-	1.82	-	**25.79**	-	**25.79**	nr	**30.95**
casement stays	-	-	-	0.44	-	**6.25**	-	**6.25**	nr	**7.50**
barrel bolts; 150mm	-	-	-	0.50	-	**7.03**	-	**7.03**	nr	**8.44**
monkey tail bolts										
flush bolts; 200mm	-	-	-	1.21	-	**17.19**	-	**17.19**	nr	**20.63**
indicating bolts	-	-	-	1.82	-	**25.79**	-	**25.79**	nr	**30.95**
panic bolts; to single door	-	-	-	2.70	-	**38.30**	-	**38.30**	nr	**45.96**
panic bolts; to double doors	-	-	-	3.63	-	**51.58**	-	**51.58**	nr	**61.90**
knobs	-	-	-	0.39	-	**5.47**	-	**5.47**	nr	**6.57**
lever handles	-	-	-	0.39	-	**5.47**	-	**5.47**	nr	**6.57**
pull handles; 225mm	-	-	-	0.39	-	**5.47**	-	**5.47**	nr	**6.57**
back plates	-	-	-	0.44	-	**6.25**	-	**6.25**	nr	**7.50**
kicking plates	-	-	-	0.88	-	**12.50**	-	**12.50**	nr	**15.01**
letter plates	-	-	-	2.70	-	**38.30**	-	**38.30**	nr	**45.96**
push plates; 225mm	-	-	-	0.39	-	**5.47**	-	**5.47**	nr	**6.57**
shelf brackets	-	-	-	0.50	-	**7.03**	-	**7.03**	nr	**8.44**
To hardwood and brickwork										
cabin hooks	-	-	-	0.44	-	**6.25**	-	**6.25**	nr	**7.50**
Water bars; steel; galvanized										
Water bars; to concrete										
flat section; setting in groove in mastic										
25 x 3 x 900mm long	2.55	5.00	**2.67**	0.83	0.10	**12.78**	0.50	**15.95**	nr	**19.14**
Drawer pulls; standard quality										
To hardwood or the like										
drawer pulls; B.M.A.; 100mm	2.24	2.50	**2.30**	0.33	-	**4.69**	-	**6.99**	nr	**8.39**
drawer pulls; S.A.A.; 100mm	1.32	2.50	**1.35**	0.33	-	**4.69**	-	**6.04**	nr	**7.25**
Hooks; standard quality										
To hardwood or the like										
cup hooks; polished brass; 25mm	0.20	2.50	**0.20**	0.11	-	**1.56**	-	**1.77**	nr	**2.12**
hat and coat hooks; B.M.A.	2.10	2.50	**2.15**	0.50	-	**7.03**	-	**9.19**	nr	**11.02**
hat and coat hooks; S.A.A.	0.92	2.50	**0.95**	0.50	-	**7.03**	-	**7.98**	nr	**9.58**

This section continues on the next page

Labour hourly rates: (except Specialists) Craft Operatives £14.21 Labourer £10.71 Rates are national average prices. Refer to REGIONAL VARIATIONS for indicative levels of overall pricing in regions	PLANT & TRANSPORT			LABOUR			RATES			
	Plant Cost £	Trans Cost £	P & T Cost £	Craft Optve Hrs	Lab Hrs	Labour Cost £	Sunds £	Nett Rate £	Unit	Gross Rate (+20%) £
P30: TRENCHES/PIPEWAYS/PITS FOR BURIED ENGINEERING SERVICES										
Site preparation										
Lifting turf for preservation										
stacking on site average 50 metres distance for immediate use; watering	0.00	-	**0.00**	-	0.66	**7.07**	-	**7.07**	m2	**8.49**
Re-laying turf										
taking from stack average 50 metres distance, laying on prepared bed, watering, maintaining	-	-	-	-	0.66	**7.07**	-	**7.07**	m2	**8.48**
Excavating trenches to receive services not exceeding 200mm nominal size - by machine										
Excavations commencing from existing ground level; levelling and grading backfilling to receive turf										
not exceeding 1m deep; average 750mm deep	2.09	0.56	**2.65**	-	0.28	**2.95**	4.10	**9.69**	m	**11.63**
Extra over excavating trenches irrespective of depth; breaking out existing materials										
concrete	15.56	-	**15.56**	-	-	-	-	**15.56**	m3	**18.67**
brickwork, blockwork or stonework	10.37	-	**10.37**	-	-	-	-	**10.37**	m3	**12.45**
Extra over excavating trenches irrespective of depth; breaking out existing hard pavings										
concrete 150mm thick	2.42	-	**2.42**	-	-	-	-	**2.42**	m2	**2.90**
reinforced concrete 200mm thick	5.01	-	**5.01**	-	-	-	-	**5.01**	m2	**6.02**
concrete 150mm thick; reinstating	2.94	-	**2.94**	-	-	-	15.08	**18.02**	m2	**21.63**
macadam paving 75mm thick	0.43	-	**0.43**	-	-	-	-	**0.43**	m2	**0.52**
macadam paving 75mm thick; reinstating	0.43	-	**0.43**	-	-	-	19.61	**20.04**	m2	**24.05**
concrete flag paving 50mm thick	-	-	-	-	0.33	**3.53**	-	**3.53**	m2	**4.24**
concrete flag paving 50mm thick; re-instating	-	-	-	0.66	0.99	**19.98**	12.67	**32.65**	m2	**39.18**
Excavating trenches to receive services not exceeding 200mm nominal size - by hand										
Excavations commencing from natural ground level; compacting; backfilling with excavated material										
not exceeding 1m deep; average 750mm deep	-	-	-	-	1.32	**14.14**	8.86	**23.00**	m	**27.60**

This section continues on the next page

PAVING/PLANTING/FENCING/SITE FURNITURE

Labour hourly rates: (except Specialists) Craft Operatives £14.21 Labourer £10.71 Rates are national average prices. Refer to REGIONAL VARIATIONS for indicative levels of overall pricing in regions	PLANT & TRANSPORT			LABOUR				RATES		
	Plant Cost £	Trans Cost £	P & T Cost £	Craft Optve Hrs	Lab Hrs	Labour Cost £	Sunds £	Nett Rate £	Unit	Gross Rate (+20%) £
Q10: KERBS/EDGINGS/CHANNELS/PAVING ACCESSORIES										
Excavating - by machine										
Trenches width not exceeding 0.30m										
maximum depth not exceeding 0.25m................	5.15	-	**5.15**	-	2.20	**23.56**	-	**28.71**	m3	**34.45**
maximum depth not exceeding 1.00m................	5.33	-	**5.33**	-	2.31	**24.74**	-	**30.07**	m3	**36.09**
Breaking out existing materials; extra over any types of excavating irrespective of depth										
concrete	7.81	-	**7.81**	-	-	-	-	**7.81**	m3	**9.37**
reinforced concrete	11.73	-	**11.73**	-	-	-	-	**11.73**	m3	**14.07**
brickwork, blockwork or stonework................	5.21	-	**5.21**	-	-	-	-	**5.21**	m3	**6.25**
100mm diameter drain and concrete bed under......	0.73	-	**0.73**	-	-	-	-	**0.73**	m	**0.88**
Breaking out existing hard pavings; extra over any types of excavating irrespective of depth										
concrete 150mm thick	1.23	-	**1.23**	-	-	-	-	**1.23**	m2	**1.48**
reinforced concrete 200mm thick..................	2.47	-	**2.47**	-	-	-	-	**2.47**	m2	**2.96**
brickwork, blockwork or stonework 100mm thick....	0.58	-	**0.58**	-	-	-	-	**0.58**	m2	**0.70**
coated macadam or asphalt 75mm thick.............	0.21	-	**0.21**	-	-	-	-	**0.21**	m2	**0.25**
Disposal of excavated material										
depositing on site in temporary spoil heaps where directed										
25m..	-	2.25	**2.25**	-	0.13	**1.41**	-	**3.66**	m3	**4.39**
removing from site to tip a) Inert	-	18.70	**18.70**	-	-	-	-	**18.70**	m3	**22.44**
removing from site to tip b) Active	-	44.00	**44.00**	-	-	-	-	**44.00**	m3	**52.80**
removing from site to tip c) Contaminated (Guide price - always seek a quotation for specialist disposal costs.)	-	88.00	**88.00**	-	-	-	-	**88.00**	m3	**105.60**
Surface treatments										
Compacting										
bottoms of excavations	0.24	-	**0.24**	-	-	-	-	**0.24**	m2	**0.29**
Excavating - by hand										
Trenches width not exceeding 0.30m										
maximum depth not exceeding 0.25m................	-	-	**-**	-	5.17	**55.37**	-	**55.37**	m3	**66.44**
maximum depth not exceeding 1.00m................	-	-	**-**	-	5.45	**58.32**	-	**58.32**	m3	**69.98**
Breaking out existing materials; extra over any types of excavating irrespective of depth										
hard rock	-	-	**-**	-	13.20	**141.37**	-	**141.37**	m3	**169.65**
concrete	-	-	**-**	-	11.55	**123.70**	-	**123.70**	m3	**148.44**
reinforced concrete	-	-	**-**	-	17.33	**185.55**	-	**185.55**	m3	**222.66**
brickwork, blockwork or stonework................	-	-	**-**	-	6.93	**74.22**	-	**74.22**	m3	**89.06**
100mm diameter drain and concrete bed under......	-	-	**-**	-	1.21	**12.96**	-	**12.96**	m	**15.55**
Breaking out existing hard pavings; extra over any types of excavating irrespective of depth										
concrete 150mm thick	-	-	**-**	-	1.82	**19.44**	-	**19.44**	m2	**23.33**
reinforced concrete 200mm thick..................	-	-	**-**	-	2.42	**25.92**	-	**25.92**	m2	**31.10**
brickwork, blockwork or stonework 100mm thick....	-	-	**-**	-	1.10	**11.78**	-	**11.78**	m2	**14.14**
coated macadam or asphalt 75mm thick.............	-	-	**-**	-	0.33	**3.53**	-	**3.53**	m2	**4.24**
Disposal of excavated material										
depositing on site in temporary spoil heaps where directed										
25m..	-	-	**-**	-	1.54	**16.49**	-	**16.49**	m3	**19.79**
removing from site to tip										
a) Inert.....................................	-	-	**-**	-	1.65	**17.67**	18.70	**36.37**	m3	**43.65**
b) Active....................................	-	-	**-**	-	1.65	**17.67**	48.40	**66.07**	m3	**79.29**
c) Contaminated (guide price - always seek a quote for specialist disposal cost	-	-	**-**	-	1.65	**17.67**	94.60	**112.27**	m3	**134.73**

Labour hourly rates: (except Specialists) Craft Operatives £14.21 Labourer £10.71 Rates are national average prices. Refer to REGIONAL VARIATIONS for indicative levels of overall pricing in regions	PLANT & TRANSPORT			LABOUR				RATES		
	Plant Cost £	Trans Cost £	P & T Cost £	Craft Optve Hrs	Lab Hrs	Labour Cost £	Sunds £	Nett Rate £	Unit	Gross Rate (+20%) £
Q10: KERBS/EDGINGS/CHANNELS/PAVING ACCESSORIES Cont.										
Surface treatments										
Compacting										
bottoms of excavations	-	-	-	-	0.18	**1.88**	-	**1.88**	m2	**2.26**

This section continues on the next page

Labour hourly rates: (except Specialists) Craft Operatives £14.21 Labourer £10.71 Rates are national average prices. Refer to REGIONAL VARIATIONS for indicative levels of overall pricing in regions	MATERIALS			LABOUR			RATES			
	Del to Site £	Waste %	Material Cost £	Craft Optve Hrs	Lab Hrs	Labour Cost £	Sunds £	Nett Rate £	Unit	Gross Rate (+20%) £
Q10: KERBS/EDGINGS/CHANNELS/PAVING ACCESSORIES Cont.										
Earthwork support										
Earthwork support maximum depth not exceeding 1.00m; distance between opposing faces not exceeding 2.00m.......	1.21	-	1.21	-	0.23	2.47	0.04	3.72	m2	4.47
Earthwork support; curved Maximum depth not exceeding 1.00m; distance between opposing faces not exceeding 2.00m.......	1.93	-	1.93	-	0.36	3.89	0.07	5.89	m2	7.07
Earthwork support; next to roadways maximum depth not exceeding 1.00m; distance between opposing faces not exceeding 2.00m.......	2.40	-	2.40	-	0.48	5.18	0.08	7.67	m2	9.20
Precast concrete; standard or stock pattern units; BS 7263 Part 1; bedding, jointing and pointing in cement mortar (1:3); on plain in-situ concrete foundation; BS 5328 ordinary prescribed mix ST4, 20mm aggregate										
Kerbs; Figs. 1, 2, 6 and 7; concrete foundation and haunching; formwork										
125 x 255mm kerb; 400 x 200mm foundation.........	12.06	5.00	12.82	-	1.32	14.14	-	26.96	m	32.35
125 x 255mm kerb; 400 x 200mm foundation; curved 10.00m radius	15.89	5.00	16.84	-	1.41	15.08	-	31.92	m	38.31
Edgings; Figs. 10, 11, 12 and 13; concrete foundation and haunching; formwork										
50 x 150mm edging; 300 x 150mm foundation........	7.52	5.00	7.98	-	0.95	10.13	-	18.12	m	21.74
Channels; Fig. 8; concrete foundation and haunching; formwork										
255 x 125mm channel; 450 x 150mm foundation......	13.50	5.00	14.30	-	1.05	11.19	-	25.49	m	30.59
255 x 125mm channel; 450 x 150mm foundation; curved 10.00 radius	16.09	5.00	17.02	-	1.13	12.13	-	29.15	m	34.98
Quadrants; Fig. 14; concrete foundation and haunching; formwork										
305 x 305 x 150mm quadrant; 500 x 500 x 150mm foundation	12.97	5.00	13.69	-	0.74	7.89	-	21.58	nr	25.90
Q20: GRANULAR SUB-BASES TO ROADS/PAVINGS										
Hard, dry, broken brick or stone to be obtained off site; wheeling not exceeding 25m - by machine										
Filling to make up levels; depositing in layers 150mm maximum thickness										
average thickness not exceeding 0.25m...........	19.64	25.00	24.54	-	-	-	4.60	29.14	m3	34.97
average thickness exceeding 0.25m...............	19.64	25.00	24.54	-	-	-	3.82	28.36	m3	34.03
Surface packing to filling to vertical or battered faces....................	-	-	-	-	-	-	0.77	0.77	m2	0.92
Compacting with 680 kg vibratory roller filling; blinding with sand, ashes or similar fine material	0.20	40.00	0.28	-	0.06	0.59	0.28	1.15	m2	1.38
Hard, dry, broken brick or stone to be obtained off site; wheeling not exceeding 50m - by machine										
Filling to make up levels; depositing in layers 150mm maximum thickness										
average thickness not exceeding 0.25m...........	19.64	25.00	24.54	-	-	-	5.34	29.88	m3	35.86
average thickness exceeding 0.25m...............	19.64	25.00	24.54	-	-	-	4.60	29.14	m3	34.97
Surface packing to filling to vertical or battered faces....................	-	-	-	-	-	-	0.77	0.77	m2	0.92
Compacting with 680 kg vibratory roller filling; blinding with sand, ashes or similar fine material	0.20	40.00	0.28	-	0.06	0.59	0.28	1.15	m2	1.38
MOT Type 2 to be obtained off site; wheeling not exceeding 25m - by machine										
Filling to make up levels; depositing in layers 150mm maximum thickness										
average thickness not exceeding 0.25m...........	21.34	25.00	26.68	-	-	-	4.60	31.28	m3	37.54
average thickness exceeding 0.25m...............	21.34	25.00	26.68	-	-	-	3.82	30.50	m3	36.60
Surface packing to filling to vertical or battered faces....................	-	-	-	-	-	-	0.77	0.77	m2	0.92
Compacting with 680 kg vibratory roller filling; blinding with sand, ashes or similar fine material	0.20	40.00	0.28	-	0.06	0.59	0.28	1.15	m2	1.38
Ashes to be obtained off site; wheeling not exceeding 25m - by machine										
Filling to make up levels; depositing in layers 150mm maximum thickness										
average thickness not exceeding 0.25m...........	13.20	40.00	18.48	-	-	-	4.87	23.35	m3	28.02

Labour hourly rates: (except Specialists) Craft Operatives £14.21 Labourer £10.71 Rates are national average prices. Refer to REGIONAL VARIATIONS for indicative levels of overall pricing in regions	MATERIALS			LABOUR				RATES		
	Del to Site	Waste	Material Cost	Craft Optve	Lab	Labour Cost	Sunds	Nett Rate	Unit	Gross Rate (+20%)
	£	%	£	Hrs	Hrs	£	£	£		£
Q20: GRANULAR SUB-BASES TO ROADS/PAVINGS Cont.										
Ashes to be obtained off site; wheeling not exceeding 25m - by machine Cont.										
Compacting with 680 kg vibratory roller										
filling	-	-	-	-		-	0.28	**0.28**	m2	**0.34**
Hoggin to be obtained off site; wheeling not exceeding 25m - by machine										
Filling to make up levels; depositing in layers 150mm maximum thickness										
average thickness not exceeding 0.25m...........	18.48	33.00	**24.58**	-		-	4.59	**29.16**	m3	**35.00**
average thickness exceeding 0.25m..................	18.48	33.00	**24.58**	-		-	3.82	**28.39**	m3	**34.07**
Compacting with 680 kg vibratory roller										
filling	-	-	-	-		-	0.28	**0.28**	m2	**0.34**
Hard, dry, broken brick or stone to be obtained off site; wheeling not exceeding 25m - by hand										
Filling to make up levels; depositing in layers 150mm maximum thickness										
average thickness not exceeding 0.25m...........	19.64	25.00	**24.54**	-	1.74	**18.61**	-	**43.16**	m3	**51.79**
average thickness exceeding 0.25m..................	19.64	25.00	**24.54**	-	1.44	**15.43**	-	**39.98**	m3	**47.97**
Compacting with 680 kg vibratory roller										
filling; blinding with sand, ashes or similar fine material	0.20	40.00	**0.28**	-	0.06	**0.59**	0.28	**1.15**	m2	**1.38**
MOT Type 1 to be obtained off site; wheeling not exceeding 25m - by hand										
Filling to make up levels; depositing in layers 150mm maximum thickness										
average thickness not exceeding 0.25m...........	23.04	25.00	**28.80**	-	1.74	**18.61**	-	**47.42**	m3	**56.90**
average thickness exceeding 0.25m..................	23.04	25.00	**28.80**	-	1.44	**15.43**	-	**44.24**	m3	**53.08**
Compacting with 680 kg vibratory roller										
filling; blinding with sand, ashes or similar fine material	0.20	40.00	**0.28**	-	0.06	**0.59**	0.28	**1.15**	m2	**1.38**
MOT Type 2 to be obtained off site; wheeling not exceeding 25m - by hand										
Filling to make up levels; depositing in layers 150mm maximum thickness										
average thickness not exceeding 0.25m...........	21.34	25.00	**26.68**	-	1.74	**18.61**	-	**45.29**	m3	**54.35**
average thickness exceeding 0.25m..................	21.34	25.00	**26.68**	-	1.44	**15.43**	-	**42.11**	m3	**50.54**
Compacting with 680 kg vibratory roller										
filling; blinding with sand, ashes or similar fine material	0.20	40.00	**0.28**	-	0.06	**0.59**	0.28	**1.15**	m2	**1.38**
Ashes to be obtained off site; wheeling not exceeding 25m - by hand										
Filling to make up levels; depositing in layers 150mm maximum thickness										
average thickness not exceeding 0.25m...........	13.20	40.00	**18.48**	-	1.87	**20.03**	-	**38.51**	m3	**46.21**
Hoggin to be obtained off site; wheeling not exceeding 25m - by hand										
Filling to make up levels; depositing in layers 150mm maximum thickness										
average thickness not exceeding 0.25m...........	18.48	33.00	**24.58**	-	1.71	**18.26**	-	**42.84**	m3	**51.41**
average thickness exceeding 0.25m..................	18.48	33.00	**24.58**	-	1.43	**15.32**	-	**39.89**	m3	**47.87**
Q21: IN-SITU CONCRETE ROADS/PAVINGS										
Plain in-situ concrete; mix 1:6; all in aggregate										
Beds; poured on or against earth or unblinded hardcore										
thickness 150 - 450mm	86.97	7.50	**93.49**	-	4.84	**51.84**	1.98	**147.31**	m3	**176.77**
thickness not exceeding 150mm....................	86.97	7.50	**93.49**	-	5.15	**55.14**	1.98	**150.61**	m3	**180.73**
Plain in-situ concrete; BS 5328, ordinary prescribed mix ST4, 20mm aggregate										
Beds										
thickness not exceeding 150mm....................	70.33	5.00	**73.84**	-	5.15	**55.14**	1.98	**130.96**	m3	**157.15**
Beds; poured on or against earth or unblinded hardcore										
thickness not exceeding 150mm....................	70.33	5.00	**73.84**	-	5.15	**55.14**	1.98	**130.96**	m3	**157.1**
Reinforced in-situ concrete; BS 5328, designed mix C20, 20mm aggregate, minimum cement content 240 kg/m3; vibrated										
Beds										
thickness 150 - 450mm	71.77	5.00	**75.36**	-	5.15	**55.14**	0.55	**131.04**	m3	**157.2**
thickness not exceeding 150mm....................	71.77	5.00	**75.36**	-	5.45	**58.32**	0.55	**134.22**	m3	**161.0**

Labour hourly rates: (except Specialists) Craft Operatives £14.21 Labourer £10.71 Rates are national average prices. Refer to REGIONAL VARIATIONS for indicative levels of overall pricing in regions	MATERIALS			LABOUR			RATES			
	Del to Site £	Waste %	Material Cost £	Craft Optve Hrs	Lab Hrs	Labour Cost £	Sunds £	Nett Rate £	Unit	Gross Rate (+20%) £
Q21: IN-SITU CONCRETE ROADS/PAVINGS Cont.										
Formwork and basic finish										
Edges of beds										
height not exceeding 250mm.......................	1.19	12.50	**1.34**	0.64	0.13	**10.48**	0.26	**12.09**	m	**14.50**
height 250 - 500mm	2.09	12.50	**2.35**	1.17	0.23	**19.04**	0.50	**21.89**	m	**26.27**
Formwork steel forms										
Edges of beds										
height not exceeding 250mm.......................	0.86	15.00	**0.99**	0.44	0.09	**7.19**	0.26	**8.45**	m	**10.14**
height 250 - 500mm	1.52	15.00	**1.74**	0.77	0.15	**12.59**	0.50	**14.83**	m	**17.80**
Reinforcement fabric; BS 4483, hard drawn plain round steel; welded; including laps, tying wire, all cutting and bending, and spacers and chairs which are at the discretion of the Contractor										
Reference C283, 2.61 kg/m2; 100mm side laps; 400mm end laps										
generally	1.39	17.50	**1.64**	0.07	0.02	**1.17**	0.08	**2.89**	m2	**3.47**
Reference C636, 5.55 kg/m2; 100mm side laps; 400mm end laps										
generally	2.99	17.50	**3.51**	0.08	0.02	**1.33**	0.15	**5.00**	m2	**6.00**
Formed joints										
Sealant to joint; Grace Construction Products. ' Servimastic 96' including preparation, cleaners, primers and sealers										
10 x 25mm	2.72	5.00	**2.86**	-	0.12	**1.30**	-	**4.15**	m	**4.98**
Worked finish on in-situ concrete										
Tamping by mechanical means										
level surfaces	-	-	**-**	-	0.18	**1.88**	0.32	**2.20**	m2	**2.64**
surfaces to falls	-	-	**-**	-	0.48	**5.18**	0.42	**5.60**	m2	**6.72**
Trowelling										
level surfaces	-	-	**-**	-	0.40	**4.24**	-	**4.24**	m2	**5.09**
surfaces to falls	-	-	**-**	-	0.61	**6.48**	-	**6.48**	m2	**7.78**
Accessories cast into in-situ concrete										
Mild steel dowels; half coated with bitumen										
20mm diameter x 500mm long; with plastic compression cap	0.71	10.00	**0.78**	-	0.09	**0.94**	0.20	**1.92**	nr	**2.30**
Q22: COATED MACADAM/ASPHALT ROADS/PAVINGS										
Coated macadam, BS 4987; base course 28mm nominal size aggregate, 50mm thick; wearing course 6mm nominal size aggregate, 20mm thick; rolled with 3 - 4 tonne roller; external										
70mm roads on concrete base; to falls and crossfalls, and slopes not exceeding 15 degrees from horizontal										
generally	-	-	**-**	-	-	**-**	-	**15.20**	m2	**18.24**
70mm pavings on concrete base; to falls and crossfalls, and slopes not exceeding 15 degrees from horizontal										
generally	-	-	**-**	-	-	**-**	-	**16.91**	m2	**20.29**
Q22: GRAVEL/HOGGIN/WOODCHIP ROADS/PAVINGS										
Gravel paving; single layer blinding gravel; well water and roll										
50mm pavings on blinded hardcore base; to falls and crossfalls and slopes not exceeding 15 degrees from horizontal										
generally	2.32	10.00	**2.55**	-	0.17	**1.77**	-	**4.43**	m2	**5.31**
Q25: SLAB/BRICK/BLOCK/SETT/COBBLE PAVINGS										
Precast concrete flags, BS 7263 Part 1, natural finish; 6mm joints, symmetrical layout; bedding in 13mm cement mortar (1:3); jointing and pointing with lime and sand (1:2)										
600 x 450 x 50mm units to pavings on sand, granular or blinded hardcore base										
50mm thick; to falls and crossfalls and to slopes not exceeding 15 degrees from horizontal.........	11.13	5.00	**11.63**	0.33	0.33	**8.22**	1.17	**21.02**	m2	**25.22**

Labour hourly rates: (except Specialists) Craft Operatives £14.21 Labourer £10.71 Rates are national average prices. Refer to REGIONAL VARIATIONS for indicative levels of overall pricing in regions	MATERIALS			LABOUR			RATES			
	Del to Site £	Waste %	Material Cost £	Craft Optve Hrs	Lab Hrs	Labour Cost £	Sunds £	Nett Rate £	Unit	Gross Rate (+20%) £
Q25: SLAB/BRICK/BLOCK/SETT/COBBLE PAVINGS Cont.										
Precast concrete flags, BS 7263 Part 1, natural finish; 6mm joints, symmetrical layout; bedding in 13mm lime mortar (1:3); jointing and pointing with lime and sand (1:2)										
600 x 450 x 50mm units to pavings on sand, granular or blinded hardcore base										
50mm thick; to falls and crossfalls and to slopes not exceeding 15 degrees from horizontal.........	11.57	5.00	**12.07**	0.33	0.33	**8.22**	1.17	**21.46**	m2	**25.75**
Keyblok precast concrete paving blocks; Marshalls Mono Ltd Driveline 50; in standard units; bedding on sand; covering with sand, compacting with plate vibrator, sweeping off surplus										
200 x 100mm units to pavings on 50mm sand base; natural colour										
symmetrical half bond layout; level and to falls only										
50mm thick..................................	8.83	10.00	**9.72**	-	1.49	**15.90**	1.56	**27.18**	m2	**32.62**
laid in straight herringbone pattern; level and to falls only										
50mm thick..................................	8.83	10.00	**9.72**	-	1.65	**17.67**	1.56	**28.95**	m2	**34.74**
Granite sett paving on concrete base; 13mm thick bedding and grouting in cement mortar (1:3)										
Paving to falls and crossfalls and to slopes not exceeding 15 degrees from horizontal										
150mm thick; new	46.74	2.50	**47.83**	2.20	2.59	**58.95**	2.90	**109.68**	m2	**131.62**
150mm thick; reclaimed	38.78	2.50	**39.68**	1.98	2.42	**54.05**	2.54	**96.27**	m2	**115.53**
Cobble paving set in concrete bed (measured separately) tight butted, dry grouted with cement and sand (1:3) watered and brushed										
Paving level and to falls only										
plain	16.25	2.50	**16.66**	1.82	1.82	**45.23**	2.81	**64.69**	m2	**77.63**
Grass concrete paving; voids filled with vegetable soil and sown with grass seed										
Precast concrete perforated slabs										
120mm thick units on 25mm thick sand bed; level and to falls only	20.04	2.50	**20.74**	0.35	0.35	**8.77**	-	**29.52**	m2	**35.42**
Q30: SEEDING/TURFING										
Cultivating										
Surfaces of natural ground										
digging over one spit deep; removing debris; weeding	-	-	-	-	0.31	**3.30**	-	**3.30**	m2	**3.96**
Surfaces of filling										
digging over one spit deep....................	-	-	-	-	0.36	**3.89**	-	**3.89**	m2	**4.67**
Surface applications										
Bone meal, 0.06 kg/m2; raking in										
general surfaces	0.03	5.00	**0.03**	-	0.07	**0.71**	-	**0.74**	m2	**0.89**
Selective weedkiller, 0.03 kg/m2; applying by spreader										
general surfaces	0.10	5.00	**0.10**	-	0.07	**0.71**	-	**0.81**	m2	**0.97**
Pre-seeding fertilizer at the rate of 0.05 kg/m2										
general surfaces	0.09	5.00	**0.09**	-	0.07	**0.71**	-	**0.80**	m2	**0.96**
Seeding										
Grass seed, 0.07 kg/m2, raking in; rolling; maintaining for 12 months after laying										
general surfaces	0.34	5.00	**0.36**	-	0.33	**3.53**	-	**3.89**	m2	**4.67**
Stone pick, roll and cut grass with a rotary cutter and remove arisings										
general surfaces	-	-	-	-	0.12	**1.30**	-	**1.30**	m2	**1.56**
Turfing										
Rolawn® Medallion turf; cultivated; lay, roll and water; maintaining for 12 months after laying										
general surfaces	2.69	5.00	**2.83**	-	0.55	**5.89**	-	**8.72**	m2	**10.46**

Labour hourly rates: (except Specialists) Craft Operatives £14.21 Labourer £10.71 Rates are national average prices. Refer to REGIONAL VARIATIONS for indicative levels of overall pricing in regions	MATERIALS			LABOUR				RATES		
	Del to Site £	Waste %	Material Cost £	Craft Optve Hrs	Lab Hrs	Labour Cost £	Sunds £	Nett Rate £	Unit	Gross Rate (+20%) £
Q31: PLANTING										
Planting trees, shrubs and hedge plants										
Excavate or form pit, hole or trench, dig over ground in bottom, spread and pack around roots with finely broken soil, refill with top soil with one third by volume of farmyard manure incorporated, water in, remove surplus excavated material and provide labelling										
small tree	1.12	-	**1.12**	-	0.94	**10.01**	0.43	**11.56**	nr	**13.88**
shrub	0.84	-	**0.84**	-	0.61	**6.48**	0.18	**7.50**	nr	**9.00**
hedge plant	0.84	-	**0.84**	-	0.31	**3.30**	0.18	**4.32**	nr	**5.18**
60mm diameter treated softwood tree stake, pointed and driven into ground and with two PVC tree ties secured around tree and nailed to stake										
2100mm long	3.78	-	**3.78**	-	0.39	**4.12**	1.52	**9.43**	nr	**11.31**
Q40: FENCING: CHAIN LINK FENCING										
Fencing; chain link; BS 1722-1; rolled steel angle posts; galvanized mesh, line and tying wires; galvanized steel fittings and accessories; backfilling around posts in concrete mix ST4										
1200mm high fencing; type GLS 120; posts with rounded tops										
posts at 2743mm centres; 600mm into ground	10.42	2.50	**10.63**	-	0.88	**9.42**	-	**20.05**	m	**24.06**
extra; end posts with struts -1	33.73	2.50	**34.46**	-	1.38	**14.73**	-	**49.19**	nr	**59.02**
extra; angle posts with struts -2	52.33	2.50	**53.48**	-	2.20	**23.56**	-	**77.04**	nr	**92.45**
1800mm high fencing; type GLS 180; posts with rounded tops										
posts at 2743mm centres; 600mm into ground	13.28	2.50	**13.56**	-	1.21	**12.96**	-	**26.52**	m	**31.82**
extra; end posts with struts -1	46.93	2.50	**47.99**	-	2.04	**21.79**	-	**69.79**	nr	**83.75**
extra; angle posts with struts -2	70.26	2.50	**71.85**	-	2.64	**28.27**	-	**100.13**	nr	**120.15**
Fencing; chain link; BS 1722 Part 1; rolled steel angle posts; plastics coated Grade A mesh, line and tying wires; galvanized steel fittings and accessories; backfilling around posts in concrete mix ST4										
1200mm high fencing; type PLS 120; posts with rounded tops										
posts at 2743mm centres; 600mm into ground	15.02	5.00	**15.67**	-	0.88	**9.42**	-	**25.09**	m	**30.11**
extra; end posts with struts -1	50.36	5.00	**52.66**	-	1.38	**14.73**	-	**67.39**	nr	**80.86**
extra; angle posts with struts -2	67.61	5.00	**70.67**	-	2.20	**23.56**	-	**94.23**	nr	**113.08**
1800mm high fencing; type PLS 180; posts with rounded tops										
posts at 2743mm centres; 600mm into ground	20.07	5.00	**20.96**	-	1.21	**12.96**	-	**33.92**	m	**40.70**
extra; end posts with struts -1	72.92	5.00	**76.34**	-	2.04	**21.79**	-	**98.14**	nr	**117.77**
extra; angle posts with struts -2	92.20	5.00	**96.49**	-	2.64	**28.27**	-	**124.76**	nr	**149.72**
Q40: FENCING: BOARDED FENCING										
Fencing; close boarded; BS 1722 Part 5; sawn softwood posts, rails, pales, gravel boards and centre stumps, pressure impregnated with preservative; backfilling around posts in concrete mix ST4										
1800mm high fencing; type BW 180A										
posts at 3000mm centres; 750mm into ground	36.77	5.00	**38.17**	2.11	-	**30.01**	-	**68.18**	m	**81.82**
Q40: FENCING: GUARD RAILS										
galvanized mild steel guard rail of tubing to BS 1387 medium grade with Kee Klamp fittings										
Rail or standard										
40mm diameter ref. 7-2-G	7.31	5.00	**7.67**	0.33	-	**4.69**	0.33	**12.69**	m	**15.23**
extra; flanged end	5.13	5.00	**5.39**	-	-	-	-	**5.39**	nr	**6.46**
extra; bend No.15-7	5.62	5.00	**5.90**	0.20	-	**2.81**	-	**8.71**	nr	**10.46**
extra; three-way intersection No.20-7	8.51	5.00	**8.93**	0.20	-	**2.81**	-	**11.75**	nr	**14.10**
extra; three-way intersection No.25-7	8.17	5.00	**8.57**	0.20	-	**2.81**	-	**11.39**	nr	**13.66**
extra; four-way intersection No.21-7	6.66	5.00	**6.99**	0.20	-	**2.81**	-	**9.81**	nr	**11.77**
extra; floor plate No.61-7	7.49	5.00	**7.87**	0.20	-	**2.81**	-	**10.68**	nr	**12.82**
Infill panel fixed with clips										
50 x 50mm welded mesh	24.16	5.00	**25.37**	0.99	-	**14.07**	0.23	**39.67**	m2	**47.60**
Q40: FENCING: STEEL GATES										
Gates in chain link fencing										
Gate with 40 x 40 x 5mm painted angle iron framing, braces and rails, infilled with 50mm x 10 1/2 G mesh										
1200 x 1800mm; plastic coated mesh	237.64	1.50	**241.21**	6.60	6.60	**164.47**	-	**405.68**	nr	**486.81**
Gate post of 80 x 80 x 6mm painted angle iron; backfilling around posts in concrete mix C30										
to suit 1800mm high gate	54.44	1.50	**55.20**	1.65	1.65	**41.12**	-	**96.32**	nr	**115.58**

DISPOSAL SYSTEMS

Labour hourly rates: (except Specialists) Craft Operatives £17.38 Labourer £10.71 Rates are national average prices. Refer to REGIONAL VARIATIONS for indicative levels of overall pricing in regions	MATERIALS			LABOUR				RATES		
	Del to Site £	Waste %	Material Cost £	Craft Optve Hrs	Lab Hrs	Labour Cost £	Sunds £	Nett Rate £	Unit	Gross Rate (+20%) £
R10: RAINWATER PIPEWORK/GUTTERS: CAST IRON ROUND RAINWATER PIPES (BS 460)										
Cast iron pipes and fittings, BS 460 Type A sockets; ears cast on; dry joints										
Pipes; straight										
75mm; ears fixing to masonry with galvanized pipe nails and distance pieces	21.51	5.00	**22.59**	0.88	-	**15.29**	-	37.88	m	**45.46**
extra; shoes	18.85	2.00	**19.23**	1.10	-	**19.12**	-	38.35	nr	**46.02**
extra; bends	14.01	2.00	**14.29**	0.88	-	**15.29**	-	29.59	nr	**35.51**
extra; offset bends 150mm projection	17.68	2.00	**18.03**	0.88	-	**15.29**	-	33.32	nr	**39.99**
extra; offset bends 300mm projection	25.29	2.00	**25.79**	0.88	-	**15.29**	-	41.09	nr	**49.31**
extra; branches	24.54	2.00	**25.03**	0.88	-	**15.29**	-	40.33	nr	**48.39**
100mm; ears fixing to masonry with galvanized pipe nails and distance pieces	28.75	5.00	**30.20**	1.06	-	**18.35**	-	48.55	m	**58.26**
extra; shoes	25.04	2.00	**25.54**	1.32	-	**22.94**	-	48.48	nr	**58.17**
extra; bends	19.79	2.00	**20.18**	1.06	-	**18.35**	-	38.54	nr	**46.25**
extra; offset bends 150mm projection	33.33	2.00	**34.00**	1.06	-	**18.35**	-	52.35	nr	**62.82**
extra; offset bends 300mm projection	41.15	2.00	**41.97**	1.06	-	**18.35**	-	60.33	nr	**72.39**
R10: RAINWATER PIPEWORK/GUTTERS: UNPLASTICIZED PVC RAINWATER PIPES (BS EN 607)										
PVC-U pipes and fittings, BS EN 607; push fit joints; pipework and supports self coloured grey										
Pipes; straight										
68mm; in standard holderbats fixing to masonry with galvanized screws	5.48	5.00	**5.72**	0.53	-	**9.18**	-	14.90	m	**17.88**
extra; shoes	5.14	2.00	**5.24**	0.41	-	**7.07**	-	12.31	nr	**14.78**
extra; bends	5.95	2.00	**6.07**	0.33	-	**5.74**	-	11.81	nr	**14.17**
extra; offset bends (spigot)	3.29	2.00	**3.35**	0.33	-	**5.74**	-	9.09	nr	**10.91**
extra; offset bends (socket)	3.23	2.00	**3.30**	0.33	-	**5.74**	-	9.03	nr	**10.84**
extra; angled branches	11.84	2.00	**12.07**	0.33	-	**5.74**	-	17.81	nr	**21.37**
R10: RAINWATER PIPEWORK/GUTTERS: CAST IRON EAVES GUTTERS (BS 460)										
Cast iron half round gutters and fittings, BS 460; bolted and mastic joints										
Gutters; straight										
100mm; in standard fascia brackets fixing to timber with galvanized screws	12.04	5.00	**12.60**	0.66	-	**11.47**	-	24.07	m	**28.89**
extra; stopped ends	2.78	2.00	**2.84**	0.66	-	**11.47**	-	14.31	nr	**17.17**
extra; running outlets	8.06	2.00	**8.22**	0.66	-	**11.47**	-	19.70	nr	**23.63**
extra; stopped ends with outlet	9.46	2.00	**9.65**	0.66	-	**11.47**	-	21.12	nr	**25.34**
extra; angles	8.27	2.00	**8.44**	0.66	-	**11.47**	-	19.91	nr	**23.89**
R10: RAINWATER PIPEWORK/GUTTERS: UNPLASTICISED PVC GUTTERS (BS EN 607)										
PVC-U half round gutters and fittings, BS EN 607; push fit connector joints; gutterwork and supports self coloured grey										
Gutters; straight										
114mm; in standard fascia brackets fixing to timber with galvanized screws	5.81	2.00	**5.93**	0.44	-	**7.65**	-	13.58	m	**16.29**
extra; stopped ends	3.28	2.00	**3.34**	0.44	-	**7.65**	-	10.99	nr	**13.19**
extra; running outlets	6.09	2.00	**6.22**	0.44	-	**7.65**	-	13.86	nr	**16.64**
extra; angles	6.66	2.00	**6.79**	0.44	-	**7.65**	-	14.44	nr	**17.32**
R10: RAINWATER PIPEWORK/GUTTERS: RAINWATER HEADS										
Cast iron pipework ancillaries										
Rainwater heads; BS 460 square type (flat pattern)										
250 x 180 x 175mm; 65mm outlet spigot; dry joint to pipe; fixing to masonry with galvanized screws	40.32	1.00	**40.73**	0.66	-	**11.47**	-	52.20	nr	**62.64**

Labour hourly rates: (except Specialists) Craft Operatives £17.38 Labourer £10.71 Rates are national average prices. Refer to REGIONAL VARIATIONS for indicative levels of overall pricing in regions	MATERIALS			LABOUR				RATES		
	Del to Site £	Waste %	Material Cost £	Craft Optve Hrs	Lab Hrs	Labour Cost £	Sunds £	Nett Rate £	Unit	Gross Rate (+20%) £
R10: RAINWATER PIPEWORK/GUTTERS: RAINWATER HEADS Cont.										
Plastics pipework ancillaries self coloured black										
Rainwater heads; BS EN 607 square type										
280 x 155 x 230mm; 68mm outlet spigot; push fit joint to plastics pipe; fixing to masonry with galvanized screws	19.09	1.00	**19.28**	0.66	-	**11.47**	-	**30.76**	nr	**36.91**
R11: FOUL DRAINAGE ABOVE GROUND: CAST IRON SOIL AND WASTE PIPES										
Cast iron pipes and fittings, BS 416; bolted and synthetic rubber gasket couplings										
Pipes; straight										
100mm; in or on supports (included elsewhere)	22.65	2.00	**23.21**	1.06	-	**18.35**	-	**41.56**	m	**49.87**
extra; pipes with inspection door bolted and sealed	53.82	2.00	**55.53**	1.06	-	**18.35**	-	**73.89**	nr	**88.66**
extra; short radius bends	39.03	2.00	**40.44**	1.06	-	**18.35**	-	**58.80**	nr	**70.56**
extra; short radius bends with inspection door bolted and sealed	58.98	2.00	**60.80**	1.06	-	**18.35**	-	**79.15**	nr	**94.98**
extra; offset bends 150mm projection	45.77	2.00	**47.32**	1.06	-	**18.35**	-	**65.67**	nr	**78.81**
extra; offset bends 300mm projection	50.12	2.00	**51.75**	1.06	-	**18.35**	-	**70.11**	nr	**84.13**
extra; single angled branches	59.36	2.00	**61.49**	1.58	-	**27.53**	-	**89.02**	nr	**106.83**
extra; single angled branches with inspection door bolted and sealed	79.32	2.00	**81.86**	1.58	-	**27.53**	-	**109.39**	nr	**131.27**
extra; double angled branches	76.48	2.00	**79.28**	2.11	-	**36.71**	-	**115.99**	nr	**139.19**
extra; double angled branches with inspection door bolted and sealed	96.44	2.00	**99.63**	2.11	-	**36.71**	-	**136.34**	nr	**163.61**
extra; straight wc connectors	28.64	2.00	**29.22**	1.06	-	**18.35**	-	**47.57**	nr	**57.08**
extra; roof connectors	30.89	2.00	**31.51**	1.06	-	**18.35**	-	**49.86**	nr	**59.83**
boss pipes with one boss	53.10	2.00	**54.79**	1.06	-	**18.35**	-	**73.15**	nr	**87.78**
boss pipes with two bosses	62.88	2.00	**64.77**	1.06	-	**18.35**	-	**83.12**	nr	**99.75**
Pipe supports										
galvanized steel vertical bracket										
for 100mm pipes; for fixing bolt(fixing included elsewhere)	5.51	2.00	**5.62**	-	-	-	-	**5.62**	nr	**6.75**
R11: FOUL DRAINAGE ABOVE GROUND: UNPLASTICIZED MPVC WASTE PIPES (SOLVENT WELDED)										
MuPVC pipes and fittings, BS EN 1329-1 Section Three; solvent welded joints										
Pipes; straight										
32mm; in standard plastics pipe brackets fixing to timber with screws	2.89	10.00	**3.10**	0.44	-	**7.65**	-	**10.75**	m	**12.90**
extra; straight expansion couplings	2.30	2.00	**2.34**	0.33	-	**5.74**	-	**8.08**	nr	**9.70**
extra; connections to plastics pipe socket; socket adaptor; solvent welded joint	2.45	2.00	**2.50**	0.33	-	**5.74**	-	**8.24**	nr	**9.88**
40mm; in standard plastics pipe brackets fixing to timber with screws	3.43	10.00	**3.70**	0.44	-	**7.65**	-	**11.35**	m	**13.62**
extra; straight expansion couplings	2.77	2.00	**2.83**	0.33	-	**5.74**	-	**8.56**	nr	**10.28**
extra; connections to plastics pipe socket; socket adaptor; solvent welded joint	2.45	2.00	**2.50**	0.33	-	**5.74**	-	**8.24**	nr	**9.88**
extra; fittings; one end	2.51	2.00	**2.56**	0.17	-	**2.87**	-	**5.43**	nr	**6.51**
extra; fittings; two ends	1.60	2.00	**1.63**	0.33	-	**5.74**	-	**7.36**	nr	**8.83**
extra; fittings; three ends	2.27	2.00	**2.31**	0.33	-	**5.74**	-	**8.05**	nr	**9.66**
50mm; in standard plastics pipe brackets fixing to timber with screws	5.20	10.00	**5.60**	0.53	-	**9.18**	-	**14.78**	m	**17.74**
extra; straight expansion couplings	3.76	2.00	**3.84**	0.44	-	**7.65**	-	**11.48**	nr	**13.78**
extra; connections to plastics pipe socket; socket adaptor; solvent welded joint	3.50	2.00	**3.57**	0.44	-	**7.65**	-	**11.22**	nr	**13.46**
extra; fittings; one end	4.13	2.00	**4.21**	0.22	-	**3.82**	-	**8.03**	nr	**9.64**
extra; fittings; two ends	2.70	2.00	**2.75**	0.44	-	**7.65**	-	**10.40**	nr	**12.48**
extra; fittings; three ends	4.41	2.00	**4.50**	0.44	-	**7.65**	-	**12.15**	nr	**14.58**
32mm; in standard plastics pipe brackets fixing to masonry with screws	2.92	10.00	**3.14**	0.53	-	**9.18**	-	**12.31**	m	**14.78**
40mm; in standard plastics pipe brackets fixing to masonry with screws	3.46	10.00	**3.73**	0.53	-	**9.18**	-	**12.91**	m	**15.49**
50mm; in standard plastics pipe brackets fixing to masonry with screws	5.23	10.00	**5.63**	0.62	-	**10.71**	-	**16.34**	m	**19.61**
R11: FOUL DRAINAGE ABOVE GROUND: UNPLASTICIZED SOIL AND VENTILATING PIPES										
PVC-U pipes and fittings, BS 4514; solvent weld joints; pipework self coloured grey										
Pipes; straight										
110mm; in or on supports (included elsewhere)	10.95	5.00	**11.43**	0.69	-	**12.04**	-	**23.47**	m	**28.17**
extra; pipes with inspection door	23.01	2.00	**23.47**	0.69	-	**12.04**	-	**35.52**	nr	**42.62**
extra; short radius bends	15.75	2.00	**16.07**	0.69	-	**12.04**	-	**28.11**	nr	**33.73**
extra; single angled branches	20.53	2.00	**20.94**	0.69	-	**12.04**	-	**32.98**	nr	**39.58**
extra; straight wc connectors	6.22	2.00	**6.34**	0.69	-	**12.04**	-	**18.38**	nr	**22.06**
pipes with one boss socket	11.53	2.00	**11.76**	0.69	-	**12.04**	-	**23.80**	nr	**28.56**
160mm; in or on supports (included elsewhere)	29.34	5.00	**30.65**	1.05	-	**18.16**	-	**48.82**	m	**58.58**
extra; pipes with inspection door	54.48	2.00	**55.57**	1.05	-	**18.16**	-	**73.73**	nr	**88.48**
extra; short radius bends	44.42	2.00	**45.31**	1.05	-	**18.16**	-	**63.47**	nr	**76.16**
extra; single angled branches	73.15	2.00	**74.61**	1.05	-	**18.16**	-	**92.78**	nr	**111.33**
pipes with one boss socket	22.37	2.00	**22.82**	1.05	-	**18.16**	-	**40.98**	nr	**49.18**

Labour hourly rates: (except Specialists) Craft Operatives £17.38 Labourer £10.71 Rates are national average prices. Refer to REGIONAL VARIATIONS for indicative levels of overall pricing in regions	MATERIALS			LABOUR			RATES			
	Del to Site £	Waste %	Material Cost £	Craft Optve Hrs	Lab Hrs	Labour Cost £	Sunds £	Nett Rate £	Unit	Gross Rate (+20%) £
R11: FOUL DRAINAGE ABOVE GROUND: UNPLASTICIZED SOIL AND VENTILATING PIPES Cont.										
Plastics pipework ancillaries, self coloured grey										
Weathering aprons										
fitted to 110mm pipes	3.67	2.00	**3.75**	0.66	-	**11.47**	-	**15.22**	nr	**18.26**
Pipe supports; solvent weld joints;										
Plastic brackets										
for 110mm pipes; for building in (fixing included elsewhere)	7.58	2.00	**7.73**	-	-	-	-	**7.73**	nr	**9.28**
for 110mm pipes; fixing to timber with galvanized screws	3.70	2.00	**3.77**	0.33	-	**5.74**	-	**9.51**	nr	**11.41**
for 160mm pipes; fixing to timber with galvanized screws	8.82	2.00	**9.00**	0.33	-	**5.74**	-	**14.74**	nr	**17.68**
R11: FOUL DRAINAGE ABOVE GROUND: WIRE BALLOON GRATINGS										
Pipework ancillaries										
Wire balloon gratings										
fitted to 82mm pipes	3.64	3.00	**3.75**	0.13	-	**2.29**	-	**6.04**	nr	**7.25**
fitted to 110mm pipes	3.78	3.00	**3.90**	0.13	-	**2.29**	-	**6.19**	nr	**7.43**
R11: FOUL DRAINAGE ABOVE GROUND: PLASTIC TRAPS										
Plastics pipework ancillaries self coloured white										
Traps; BS EN 274										
P, two piece, 75mm seal, inlet with coupling nut, outlet with seal ring socket										
32mm outlet	5.35	2.00	**5.45**	0.44	-	**7.65**	-	**13.10**	nr	**15.72**
40mm outlet	6.17	2.00	**6.29**	0.44	-	**7.65**	-	**13.94**	nr	**16.73**
S, two piece, 75mm seal, inlet with coupling nut, outlet with seal ring socket										
32mm outlet	6.78	2.00	**6.91**	0.44	-	**7.65**	-	**14.56**	nr	**17.47**
40mm outlet	7.92	2.00	**8.08**	0.44	-	**7.65**	-	**15.73**	nr	**18.87**
P, two piece, 75mm seal, with flexible polypropylene pipe for overflow connection, inlet with coupling nut, outlet with seal ring socket										
40mm outlet	11.44	2.00	**11.67**	0.66	-	**11.47**	-	**23.14**	nr	**27.77**
R11: FOUL DRAINAGE ABOVE GROUND: COPPER TRAPS										
Copper pipework ancillaries										
Traps; BS 1184 Section Two (Solid Drawn)										
P, two piece, 75mm seal, inlet with coupling nut, compression outlet										
35mm outlet	25.52	2.00	**26.03**	0.55	-	**9.56**	-	**35.59**	nr	**42.71**
42mm outlet	30.06	2.00	**30.66**	0.55	-	**9.56**	-	**40.22**	nr	**48.27**

This section continues on the next page

DISPOSAL SYSTEMS

Labour hourly rates: (except Specialists) Craft Operatives £14.21 Labourer £10.71 Rates are national average prices. Refer to REGIONAL VARIATIONS for indicative levels of overall pricing in regions	PLANT & TRANSPORT			LABOUR			RATES			
	Plant Cost £	Trans Cost £	P & T Cost £	Craft Optve Hrs	Lab Hrs	Labour Cost £	Sunds £	Nett Rate £	Unit	Gross Rate (+20%) £
R12: DRAINAGE BELOW GROUND: EXCAVATION FOR DRAINS - BY MACHINE										
Excavating trenches by machine to receive pipes not exceeding 200mm nominal size; disposing of surplus excavated material by removing from site										
Excavations commencing from natural ground level average depth										
500mm	1.72	3.37	**5.08**	-	0.26	**2.83**	1.03	**8.94**	m	**10.73**
750mm	2.74	3.37	**6.11**	-	0.39	**4.12**	1.55	**11.78**	m	**14.13**
1000mm	3.67	3.37	**7.04**	-	0.52	**5.54**	2.07	**14.64**	m	**17.57**
1500mm	5.65	3.37	**9.02**	-	0.84	**8.95**	3.09	**21.07**	m	**25.28**
2000mm	7.60	3.37	**10.97**	-	1.12	**12.02**	4.13	**27.12**	m	**32.54**
2500mm	10.79	3.37	**14.16**	-	2.17	**23.21**	5.16	**42.53**	m	**51.03**
3000mm	12.99	3.37	**16.35**	-	2.60	**27.80**	6.19	**50.34**	m	**60.41**
3500mm	15.21	3.37	**18.57**	-	3.04	**32.52**	7.23	**58.32**	m	**69.98**
Excavating trenches by machine to receive pipes 225mm nominal size; disposing of surplus excavated material by removing from site										
Excavations commencing from natural ground level average depth										
500mm	1.84	3.55	**5.40**	-	0.28	**2.95**	1.03	**9.37**	m	**11.24**
1000mm	3.88	3.55	**7.44**	-	0.54	**5.77**	2.07	**15.28**	m	**18.33**
1500mm	6.02	3.55	**9.57**	-	0.88	**9.42**	3.09	**22.09**	m	**26.51**
2000mm	8.05	3.55	**11.60**	-	1.19	**12.72**	4.13	**28.46**	m	**34.15**
2500mm	11.47	3.55	**15.03**	-	2.27	**24.27**	5.16	**44.46**	m	**53.35**
3000mm	13.76	3.55	**17.31**	-	2.72	**29.10**	6.19	**52.60**	m	**63.12**
3500mm	16.13	3.55	**19.68**	-	3.18	**34.05**	7.23	**60.96**	m	**73.15**
R12: DRAINAGE BELOW GROUND: EXCAVATION FOR DRAINS - BY HAND										
Excavating trenches by hand to receive pipes not exceeding 200mm nominal size; disposing of surplus excavated material on site										
Excavations commencing from natural ground level average depth										
500mm	-	-	-	-	1.32	**14.14**	1.03	**15.16**	m	**18.20**
750mm	-	-	-	-	2.37	**25.33**	1.55	**26.88**	m	**32.25**
1000mm	-	-	-	-	3.41	**36.52**	2.07	**38.59**	m	**46.31**
1500mm	-	-	-	-	5.34	**57.14**	3.09	**60.23**	m	**72.28**
2000mm	-	-	-	-	7.48	**80.11**	4.13	**84.25**	m	**101.09**
2500mm	-	-	-	-	10.12	**108.39**	5.16	**113.55**	m	**136.26**
3000mm	-	-	-	-	12.76	**136.66**	6.19	**142.85**	m	**171.42**
3500mm	-	-	-	-	15.46	**165.52**	7.23	**172.75**	m	**207.30**
Excavating trenches by hand to receive pipes 225mm nominal size; disposing of surplus excavated material on site										
Excavations commencing from natural ground level average depth										
500mm	-	-	-	-	1.49	**15.90**	1.03	**16.93**	m	**20.32**
750mm	-	-	-	-	2.59	**27.69**	1.55	**29.23**	m	**35.08**
1000mm	-	-	-	-	3.69	**39.47**	2.07	**41.53**	m	**49.84**
1500mm	-	-	-	-	5.87	**62.91**	3.09	**66.01**	m	**79.21**
2000mm	-	-	-	-	8.09	**86.59**	4.13	**90.73**	m	**108.87**
2500mm	-	-	-	-	11.11	**118.99**	5.16	**124.15**	m	**148.98**
3000mm	-	-	-	-	13.97	**149.62**	6.19	**155.81**	m	**186.97**

This section continues on the next page

Labour hourly rates: (except Specialists) Craft Operatives £14.21 Labourer £10.71 Rates are national average prices. Refer to REGIONAL VARIATIONS for indicative levels of overall pricing in regions	MATERIALS			LABOUR				RATES		
	Del to Site £	Waste %	Material Cost £	Craft Optve Hrs	Lab Hrs	Labour Cost £	Sunds £	Nett Rate £	Unit	Gross Rate (+20%) £
R12: DRAINAGE BELOW GROUND: BEDS, BENCHINGS AND COVERS										
Granular material, 10mm nominal size pea shingle, to be obtained off site										
Beds										
400 x 150mm	1.75	10.00	**1.92**	-	0.20	**2.12**	-	**4.04**	m	**4.85**
450 x 150mm	1.95	10.00	**2.15**	-	0.21	**2.24**	-	**4.38**	m	**5.26**
525 x 150mm	2.27	10.00	**2.50**	-	0.22	**2.36**	-	**4.85**	m	**5.82**
Beds and surrounds										
400 x 150mm bed; 150mm thick surround to 100mm internal diameter pipes -1	4.66	10.00	**5.12**	-	0.61	**6.48**	-	**11.60**	m	**13.92**
450 x 150mm bed; 150mm thick surround to 150mm internal diameter pipes -1	5.88	10.00	**6.47**	-	0.68	**7.30**	-	**13.77**	m	**16.53**
525 x 150mm bed; 150mm thick surround to 225mm internal diameter pipes -1	8.00	10.00	**8.80**	-	0.90	**9.66**	-	**18.46**	m	**22.16**
Beds and filling to half pipe depth										
400 x 150mm overall; to 100mm internal diameter pipes - 1	4.07	10.00	**4.48**	-	0.15	**1.65**	-	**6.13**	m	**7.36**
450 x 225mm overall; to 150mm internal diameter pipes - 1	4.89	10.00	**5.38**	-	0.26	**2.83**	-	**8.21**	m	**9.85**
525 x 263mm overall; to 225mm internal diameter pipes - 1	7.45	10.00	**8.20**	-	0.37	**4.01**	-	**12.20**	m	**14.64**
Plain in-situ concrete; BS 5328, ordinary prescribed mix, ST3, 20mm aggregate										
Beds										
400 x 100mm	2.74	5.00	**2.87**	-	0.24	**2.59**	3.07	**8.54**	m	**10.25**
450 x 100mm	3.08	5.00	**3.23**	-	0.26	**2.83**	3.07	**9.13**	m	**10.96**
525 x 100mm	3.56	5.00	**3.74**	-	0.32	**3.42**	3.07	**10.23**	m	**12.27**
Beds and surrounds										
400 x 100mm bed; 150mm thick surround to 100mm internal diameter pipes -1	9.58	5.00	**10.06**	-	0.95	**10.13**	6.04	**26.23**	m	**31.47**
450 x 100mm bed; 150mm thick surround to 150mm internal diameter pipes -1	12.32	5.00	**12.93**	-	1.20	**12.84**	6.80	**32.57**	m	**39.09**
525 x 150mm bed; 150mm thick surround to 225mm internal diameter pipes -1	18.82	5.00	**19.76**	-	1.79	**19.20**	8.68	**47.64**	m	**57.17**
Beds and haunchings										
400 x 150mm; to 100mm internal diameter pipes - 1	6.84	5.00	**7.18**	-	0.53	**5.65**	4.53	**17.37**	m	**20.85**
450 x 150mm; to 150mm internal diameter pipes - 1	9.24	5.00	**9.70**	-	0.61	**6.48**	4.53	**20.71**	m	**24.85**
525 x 150mm; to 225mm internal diameter pipes - 1	13.41	5.00	**14.08**	-	0.81	**8.72**	4.53	**27.33**	m	**32.80**
Vertical casings										
400 x 400mm to 100mm internal diameter pipes - 1.	40.87	5.00	**41.41**	-	1.06	**11.31**	-	**52.72**	m	**63.27**
450 x 450mm to 150mm internal diameter pipes - 1.	47.48	5.00	**48.17**	-	1.33	**14.26**	-	**62.43**	m	**74.91**
525 x 525mm to 225mm internal diameter pipes - 1.	58.46	5.00	**59.40**	-	1.83	**19.56**	-	**78.96**	m	**94.75**
R12: DRAINAGE BELOW GROUND: VITRIFIED CLAY FLEXIBLE JOINT DRAINS										
Drains; vitrified clay pipes and fittings, BS EN 295, normal; flexible mechanical joints; Hepworth Hepseal										
Pipework in trenches										
Pipework in trenches; vertical										
400mm	109.47	5.00	**114.95**	0.79	0.24	**13.85**	-	**128.79**	m	**154.55**
R12: DRAINAGE BELOW GROUND: VITRIFIED CLAY SLEEVE JOINTS										
Drains; vitrified clay pipes and fittings, BS EN 295; sleeve joints, push-fit polypropylene standard ring flexible couplings; Hepworth Supersleeve										
Pipework in trenches										
100mm	7.26	5.00	**7.62**	0.21	0.08	**3.79**	-	**11.42**	m	**13.70**
extra; bends	9.88	5.00	**10.37**	0.11	-	**1.56**	-	**11.94**	nr	**14.32**
extra; branches	21.58	5.00	**22.66**	0.21	-	**2.97**	-	**25.63**	nr	**30.76**
150mm	13.85	5.00	**14.55**	0.24	0.11	**4.62**	-	**19.16**	m	**22.99**
extra; bends	20.47	5.00	**21.50**	0.12	-	**1.72**	-	**23.22**	nr	**27.86**
extra; branches	31.81	5.00	**33.40**	0.24	-	**3.44**	-	**36.84**	nr	**44.21**
Pipework in trenches; vertical										
100mm	7.26	5.00	**7.62**	0.24	0.08	**4.26**	-	**11.89**	m	**14.26**
150mm	13.85	5.00	**14.55**	0.26	0.11	**4.93**	-	**19.48**	m	**23.37**
Drains; vitrified clay pipes and fittings, BS EN 295, normal; cement mortar (1:3)										
Pipework in trenches										
100mm	9.37	5.00	**9.85**	0.44	0.04	**6.72**	-	**16.57**	m	**19.88**
extra; double collars	11.28	5.00	**11.85**	0.13	-	**1.88**	-	**13.73**	nr	**16.47**
extra; bends	9.37	5.00	**9.85**	0.26	-	**3.75**	-	**13.60**	nr	**16.32**
extra; rest bends	15.34	5.00	**16.11**	0.26	-	**3.75**	-	**19.87**	nr	**23.84**
extra; branches	17.13	5.00	**18.00**	0.29	-	**4.06**	-	**22.06**	nr	**26.47**
150mm	9.37	5.00	**9.85**	0.57	0.06	**8.72**	-	**18.56**	m	**22.28**
extra; double collars	18.73	5.00	**19.67**	0.20	-	**2.81**	-	**22.48**	nr	**26.98**
extra; bends	14.36	5.00	**15.09**	0.35	-	**5.00**	-	**20.09**	nr	**24.11**
extra; rest bends	25.87	5.00	**27.18**	0.35	-	**5.00**	-	**32.18**	nr	**38.61**

Labour hourly rates: (except Specialists) Craft Operatives £14.21 Labourer £10.71 Rates are national average prices. Refer to REGIONAL VARIATIONS for indicative levels of overall pricing in regions	MATERIALS			LABOUR				RATES		
	Del to Site	Waste	Material Cost	Craft Optve	Lab	Labour Cost	Sunds	Nett Rate	Unit	Gross Rate (+20%)
	£	%	£	Hrs	Hrs	£	£	£		£
R12: DRAINAGE BELOW GROUND: VITRIFIED CLAY SLEEVE JOINTS Cont.										
Drains; vitrified clay pipes and fittings, BS EN 295, normal; cement mortar (1:3) Cont.										
Pipework in trenches; vertical										
100mm	9.37	5.00	**9.85**	0.66	0.04	**9.85**	-	**19.70**	m	**23.64**
150mm	9.37	5.00	**9.85**	0.64	0.06	**9.66**	-	**19.50**	m	**23.40**
R12: DRAINAGE BELOW GROUND: VITRIFIED CLAY GULLIES										
Vitrified clay accessories										
Gullies; Supersleve; joint to pipe; bedding and surrounding in concrete to BS 5328, ordinary prescribed mix ST3, 20mm aggregate; Hepworth Building Products										
100mm outlet, reference RG1/1, trapped, round; 255mm diameter top	48.64	5.00	**51.07**	0.73	0.10	**11.38**	-	**62.44**	nr	**74.93**
100mm outlet, reference SG1/1, trapped, reversible, round; 100mm trap; hopper reference SH1, 100mm outlet, 150 x 150mm rebated top	88.46	5.00	**92.89**	1.21	0.12	**18.49**	-	**111.38**	nr	**133.65**
Rainwater shoes; Supersleve; joint to pipe; bedding and surrounding in concrete to BS 5328, ordinary prescribed mix ST3, 20mm aggregate										
100mm outlet, reference RRW/S3/1 trapless, round; 100mm vertical inlet, 250 x 150mm rectangular access opening	50.60	5.00	**53.13**	0.61	0.07	**9.30**	-	**62.43**	nr	**74.92**
R12: DRAINAGE BELOW GROUND: CONCRETE GULLIES										
Unreinforced concrete accessories										
Gullies; BS 5911; joint to pipe; bedding and surrounding in concrete to BS 5328 ordinary prescribed mix ST3, 20mm aggregate										
150mm outlet, trapped, round; 375mm diameter x 900mm deep internally, stopper	63.25	5.00	**66.41**	4.24	0.61	**66.66**	-	**133.07**	nr	**159.68**
R12: DRAINAGE BELOW GROUND: PVC-U GULLIES										
PVC-U accessories										
Gullies; Osma; joint to pipe; bedding and surrounding in concrete to BS 5328, ordinary prescribed mix ST3, 20mm aggregate										
110mm outlet trap, base reference 4D.500, outlet bend with access reference 4D.569, inlet raising piece 300mm long, plain hopper reference 4D.503	69.29	5.00	**72.76**	1.32	0.22	**21.11**	-	**93.87**	nr	**112.64**
110mm outlet bottle gulley, reference 4D.900 trapped, round; 200mm diameter top with grating reference 4D.919	58.47	5.00	**61.39**	1.98	0.28	**31.08**	-	**92.47**	nr	**110.97**
Sealed rodding eyes; Osma; joint to pipe; surrounding in concrete to BS 5328; ordinary prescribed mix ST3, 20mm aggregate										
110mm outlet, reference 4D.316, with cover	55.15	5.00	**57.91**	2.48	0.50	**40.47**	-	**98.38**	nr	**118.06**
R12: DRAINAGE BELOW GROUND: GULLY GRATING AND SEALING PLATES										
Cast iron accessories, painted black										
Gratings; Hepworth Building Products										
reference IG7C; 197mm diameter	5.48	-	**5.48**	-	0.03	**0.35**	-	**5.84**	nr	**7.01**
Sealing plates and frames; bedding frames in cement mortar (1:3); Hepworth Building Products										
reference IS6C; 197mm diameter	16.34	15.00	**18.80**	1.52	-	**21.57**	-	**40.37**	nr	**48.44**
Cast iron accessories, painted black										
Gratings and frames; Stanton PLC.; bedding frames in cement mortar (1:3)										
reference HY815; 510 x 360mm	62.63	15.00	**72.02**	4.02	-	**57.05**	-	**129.07**	nr	**154.89**

This section continues on the next page

Labour hourly rates: (except Specialists) Craft Operatives £14.21 Labourer £10.71 Rates are national average prices. Refer to REGIONAL VARIATIONS for indicative levels of overall pricing in regions	PLANT & TRANSPORT			LABOUR			RATES			
	Plant Cost £	Trans Cost £	P & T Cost £	Craft Optve Hrs	Lab Hrs	Labour Cost £	Sunds £	Nett Rate £	Unit	Gross Rate (+20%) £
R12: DRAINAGE BELOW GROUND; MANHOLES AND SOAKAWAYS:										
Excavating - by machine										
Excavating pits										
commencing from natural ground level; maximum depth not exceeding										
1.00m	5.69	-	**5.69**	-	1.10	**11.78**	-	**17.47**	m3	**20.96**
2.00m	6.26	-	**6.26**	-	1.27	**13.55**	-	**19.81**	m3	**23.77**
Working space allowance to excavation pits	5.69	-	**5.69**	-	0.77	**8.25**	-	**13.94**	m2	**16.72**
Compacting bottoms of excavations	1.16	-	**1.16**	-	-	-	-	**1.16**	m2	**1.39**
Excavated material depositing on site in temporary spoil heaps where directed 25m	2.28	-	**2.28**	-	-	-	-	**2.28**	m3	**2.73**
Excavating - by hand										
Excavating pits										
commencing from natural ground level; maximum depth not exceeding										
0.25m	-	-	-	-	4.07	**43.59**	-	**43.59**	m3	**52.31**
1.00m	-	-	-	-	4.24	**45.36**	-	**45.36**	m3	**54.43**
Working space allowance to excavation pits	-	-	-	-	3.80	**40.64**	-	**40.64**	m2	**48.77**
Compacting bottoms of excavations	-	-	-	-	0.14	**1.53**	-	**1.53**	m2	**1.84**
Excavated material depositing on site in temporary spoil heaps where directed 25m	-	-	-	-	1.54	**16.49**	-	**16.49**	m3	**19.79**

This section continues on the next page

Labour hourly rates: (except Specialists) Craft Operatives £14.21 Labourer £10.71 Rates are national average prices. Refer to REGIONAL VARIATIONS for indicative levels of overall pricing in regions	MATERIALS			LABOUR				RATES		
	Del to Site	Waste	Material Cost	Craft Optve	Lab	Labour Cost	Sunds	Nett Rate	Unit	Gross Rate (+20%)
	£	%	£	Hrs	Hrs	£	£	£		£
R12: DRAINAGE BELOW GROUND; MANHOLES AND SOAKAWAYS: Cont.										
Earthwork support										
Earthwork support distance between opposing faces not exceeding 2.00m; maximum depth not exceeding										
1.00m	0.97	-	**0.97**	-	0.23	**2.47**	-	**3.44**	m2	**4.13**
2.00m	0.97	-	**0.97**	-	0.29	**3.06**	-	**4.03**	m2	**4.84**
R12: DRAINAGE BELOW GROUND; MANHOLES AND SOAKAWAYS: FILLING										
Excavated material arising from excavations										
Filling to excavation average thickness exceeding 0.25m	-	-	-		1.58	**16.96**	-	**16.96**	m3	**20.36**
Excavated material obtained from on site spoil heaps 25m distant										
Filling to excavation average thickness exceeding 0.25m	-	-	-	-	1.76	**18.85**	-	**18.85**	m3	**22.62**
R12: DRAINAGE BELOW GROUND; MANHOLES AND SOAKAWAYS: CONCRETE WORK										
Plain in-situ concrete; BS 5328, ordinary prescribed mix ST3, 20mm aggregate										
Beds; poured on or against earth or unblinded hardcore thickness not exceeding 150mm	68.42	5.00	**71.84**	-	5.23	**55.96**	-	**127.80**	m3	**153.36**
Benching in bottoms; rendering with 13mm cement and sand (1:2) before final set										
600 x 450 x average 225mm; trowelling	4.63	5.00	**4.90**	-	0.68	**7.30**	-	**12.21**	nr	**14.65**
800 x 675 x average 225mm; trowelling	9.25	5.00	**9.80**	-	1.40	**14.96**	-	**24.77**	nr	**29.72**
Reinforced in-situ concrete; BS 5328, designed mix ST4, 20mm aggregate, minimum cement content 240 kg/m3; vibrated										
Slabs thickness not exceeding 150mm	70.32	5.00	**73.84**	-	4.84	**51.84**	-	**125.68**	m3	**150.81**
Formwork and basic finish										
Edges of suspended slabs; plain vertical height not exceeding 250mm	1.31	12.50	**1.47**	0.88	0.20	**14.63**	-	**16.09**	m	**19.31**
Soffits of slabs; horizontal slab thickness not exceeding 200mm; height to soffit not exceeding 1.50m	17.60	10.00	**19.36**	2.66	0.51	**43.25**	-	**62.61**	m2	**75.13**
Holes; rectangular girth 2.00 - 3.00m; depth not exceeding 250mm	3.92	12.50	**4.41**	1.90	0.37	**31.05**	-	**35.46**	nr	**42.55**
Reinforcement fabric; BS 4483; hard drawn plain round steel, welded; including laps, tying wire, all cutting and bending and spacers and chairs which are at the discretion of the Contractor										
Reference A142, 2.22kg/m2; 200mm side laps; 200mm end laps generally	1.29	15.00	**1.49**	0.33	0.02	**4.92**	-	**6.41**	m2	**7.70**
R12: DRAINAGE BELOW GROUND; MANHOLES AND SOAKAWAYS: BRICKWORK										
Engineering bricks, Category F, BS EN 772, 215 x 102.5 x 65mm, class B; in cement mortar (1:3)										
Common bricks, BS EN 772, Category M, 215 x 102.5 x 65mm, compressive strength 20.5 N/mm2; in cement mortar (1:3); flush smooth pointing as work proceeds										
Walls; vertical 215mm thick; English bond; facework one side	30.64	5.00	**32.84**	3.29	3.67	**86.09**	-	**118.93**	m2	**142.71**
Accessories/sundry items for brick/block/stone walling										
Building in ends of pipes; 100mm diameter making good fair face one side 215mm brickwork	0.17	15.00	**0.19**	0.17	0.17	**4.11**	-	**4.30**	nr	**5.16**
Building in ends of pipes; 150mm diameter making good fair face one side 215mm brickwork	0.17	15.00	**0.19**	0.26	0.26	**6.58**	-	**6.77**	nr	**8.13**
Mortar, cement and sand (1:3)										
13mm work to walls on brickwork base; one coat; trowelled width exceeding 300mm	1.09	15.00	**1.25**	0.50	0.50	**12.34**	-	**13.58**	m2	**16.30**

Labour hourly rates: (except Specialists) Craft Operatives £14.21 Labourer £10.71 Rates are national average prices. Refer to REGIONAL VARIATIONS for indicative levels of overall pricing in regions	MATERIALS			LABOUR			RATES			
	Del to Site	Waste	Material Cost	Craft Optve	Lab	Labour Cost	Sunds	Nett Rate	Unit	Gross Rate (+20%)
	£	%	£	Hrs	Hrs	£	£	£		£
R12: DRAINAGE BELOW GROUND; MANHOLES AND SOAKAWAYS: CHANNELS, VITRIFIED CLAY										
Channels in bottoms; vitrified clay, BS 65, normal, glazed; cement mortar (1:3) joints; bedding in cement mortar (1:3)										
Half section										
straight; 600mm effective length										
100mm	4.13	5.00	**4.36**	0.39	0.06	**6.06**	-	**10.42**	nr	**12.51**
150mm	6.86	5.00	**7.23**	0.50	0.08	**7.86**	-	**15.09**	nr	**18.11**
straight; 1000mm effective length										
100mm	6.00	5.00	**6.35**	0.66	0.07	**10.09**	-	**16.43**	nr	**19.72**
150mm	9.83	5.00	**10.38**	0.83	0.09	**12.67**	-	**23.05**	nr	**27.66**
curved; 500mm effective length										
100mm	6.17	5.00	**6.50**	0.37	0.06	**5.90**	-	**12.41**	nr	**14.89**
150mm	10.11	5.00	**10.64**	0.50	0.08	**7.86**	-	**18.50**	nr	**22.20**
Three quarter section										
branch bends										
100mm	17.11	5.00	**18.47**	0.44	0.04	**6.72**	-	**25.19**	nr	**30.23**
150mm	26.52	5.00	**28.52**	0.66	0.08	**10.20**	-	**38.72**	nr	**46.47**
R12: DRAINAGE BELOW GROUND; MANHOLES AND SOAKAWAYS: PRECAST CONCRETE INSPECTION CHAMBERS										
Precast concrete; standard or stock pattern units; BS 5911; jointing and pointing in cement mortar (1:3)										
Chamber or shaft sections										
900mm internal diameter										
250mm high	24.28	5.00	**25.51**	0.64	1.28	**22.73**	-	**48.24**	nr	**57.89**
500mm high	36.38	5.00	**38.21**	0.74	1.47	**26.26**	-	**64.47**	nr	**77.36**
1050mm internal diameter										
305mm high	23.16	5.00	**24.33**	0.89	1.78	**31.75**	-	**56.08**	nr	**67.29**
Cutting holes										
cutting holes for pipes; 100mm diameter; making good; pointing	33.17	-	**33.17**	-	0.10	**1.06**	-	**34.23**	nr	**41.07**
cutting holes for pipes; 150mm diameter; making good; pointing	33.25	-	**33.25**	-	0.12	**1.30**	-	**34.55**	nr	**41.46**
Step irons										
galvanized malleable cast iron step irons BS 1247, cast in	5.50	5.00	**5.78**	-	0.19	**2.00**	-	**7.78**	nr	**9.33**
Cover slabs, heavy duty										
for chamber or shaft sections 900mm internal diameter, 125mm thick; access opening 600mm x 600mm	53.73	5.00	**56.43**	1.87	2.81	**56.61**	-	**113.04**	nr	**135.65**
for chamber or shaft sections 1050mm internal diameter, 125mm thick; access opening 600mm x 600mm	61.65	5.00	**64.75**	2.48	3.69	**74.64**	-	**139.38**	nr	**167.26**
Precast concrete inspection chambers; BS 5911; jointing and pointing in cement mortar (1:3)										
Rectangular inspection chambers, internal size 457 x 610mm										
150mm high	8.30	5.00	**8.73**	0.72	0.50	**15.46**	-	**24.19**	nr	**29.03**
250mm high	11.12	5.00	**11.70**	0.83	0.58	**17.97**	-	**29.66**	nr	**35.60**
extra; galvanized malleable cast iron step irons BS 1247, built in	5.50	5.00	**5.78**	-	-	-	-	**5.78**	nr	**6.93**
Rectangular inspection chambers, internal size 760 x 610mm										
150mm high	11.71	5.00	**12.32**	1.16	0.83	**25.25**	-	**37.57**	nr	**45.09**
225mm high	14.92	5.00	**15.70**	1.33	0.97	**29.28**	-	**44.98**	nr	**53.97**
extra; galvanized malleable cast iron step irons BS 1247, built in	5.50	5.00	**5.78**	-	-	-	-	**5.78**	nr	**6.93**
Base units										
450 x 610mm internal size; 360mm deep	33.31	5.00	**35.00**	1.16	0.83	**25.25**	-	**60.24**	nr	**72.29**
760 x 610mm internal size; 360mm deep	40.34	5.00	**42.39**	1.16	0.83	**25.25**	-	**67.64**	nr	**81.17**
R12: DRAINAGE BELOW GROUND; MANHOLES AND SOAKAWAYS: POLYPROPYLENE INSPECTION CHAMBERS										
Osma; Shallow inspection chamber ref. 4D.960 polypropylene with preformed benching. Light duty single sealed cover and frame ref. 4D.961 set in position, 250mm diameter; for 100mm diameter pipes, depth to invert										
Chamber or shaft sections and benching										
600mm	68.35	5.00	**71.76**	2.75	1.58	**56.04**	-	**127.80**	nr	**153.37**
Osma; universal inspection chamber; Polypropylene; 450mm diameter										
preformed chamber base benching, ref. 4D.922; for 100mm pipes	98.79	5.00	**103.73**	1.65	0.11	**24.62**	-	**128.36**	nr	**154.03**
chamber shaft 230mm effective length, ref. 4D.925	25.19	5.00	**26.45**	-	0.66	**7.07**	-	**33.52**	nr	**40.23**
cast iron cover and frame; single seal light duty to suit, ref. 4D.942	79.36	5.00	**83.33**	0.66	-	**9.38**	-	**92.71**	nr	**111.25**

Labour hourly rates: (except Specialists) Craft Operatives £14.21 Labourer £10.71 Rates are national average prices. Refer to REGIONAL VARIATIONS for indicative levels of overall pricing in regions	MATERIALS			LABOUR				RATES		
	Del to Site	Waste	Material Cost	Craft Optve	Lab	Labour Cost	Sunds	Nett Rate	Unit	Gross Rate (+20%)
	£	%	£	Hrs	Hrs	£	£	£		£
R12: DRAINAGE BELOW GROUND; MANHOLES AND SOAKAWAYS: MANHOLE ACCESSORIES										
Manhole step irons										
Step irons; BS 1247 malleable cast iron; galvanized; building in to joints										
figure 1 general purpose pattern; 225mm tail.....	4.72	-	**4.72**	0.30	-	**4.22**	-	**8.94**	nr	**10.73**
Manhole covers - A15 light duty										
Access covers; Durey Castings Ltd; BS EN 124; coated; bedding frame in cement mortar (1:3); bedding cover in grease and sand										
reference DU10B; clear opening 600 x 450mm.......	25.50	2.50	**26.14**	0.99	0.99	**24.67**	-	**50.82**	nr	**60.98**
reference DU15A recessed; clear opening 600 x 450mm	38.71	2.50	**39.69**	1.43	1.43	**35.64**	-	**75.32**	nr	**90.39**
reference DU15A recessed; clear opening 750 x 600mm	248.09	2.50	**254.30**	1.93	1.93	**47.97**	-	**302.27**	nr	**362.73**
reference DU30A double seal; clear opening 600 x 450mm	279.29	2.50	**286.28**	1.76	1.76	**43.86**	-	**330.14**	nr	**396.17**
lifting keys	7.62	2.50	**7.81**	-	-	**-**	-	**7.81**	nr	**9.38**
Manhole covers - B125 medium duty										
Access covers; Stanton PLC; BS EN 124; coated; bedding frame in cement mortar (1:3); bedding cover in grease and sand										
reference H783 Trojan single seal; 600mm diameter	109.67	2.50	**112.62**	1.98	1.98	**49.34**	-	**161.96**	nr	**194.35**
reference H781; 600 x 450mm......................	102.54	2.50	**105.41**	1.74	1.74	**43.31**	-	**148.72**	nr	**178.47**
reference H782; 600 x 600mm......................	126.83	2.50	**130.31**	2.15	2.15	**53.45**	-	**183.77**	nr	**220.52**
lifting keys	5.44	2.50	**5.57**	-	-	**-**	-	**5.57**	nr	**6.69**
Manhole covers - D400 heavy duty										
Access covers; BS EN 124; coated; bedding frame in cement mortar (1:3); bedding cover in grease and sand										
reference H9330; 600mm diameter.................	137.76	2.50	**141.63**	2.24	2.24	**55.92**	-	**197.55**	nr	**237.06**
reference H9320; 600 x 600mm......................	118.90	2.50	**122.08**	2.24	2.24	**55.92**	-	**178.01**	nr	**213.61**
lifting keys	6.05	2.50	**6.20**	-	-	**-**	-	**6.20**	nr	**7.44**
R12: DRAINAGE BELOW GROUND; WORK TO EXISTING DRAINS										
Work to disused drains										
Fly ash filling to disused drain										
100mm diameter	0.50	-	**0.50**	0.44	-	**6.25**	-	**6.75**	m	**8.10**
150mm diameter	0.92	-	**0.92**	0.66	-	**9.38**	-	**10.30**	m	**12.36**
225mm diameter	1.98	-	**1.98**	0.99	-	**14.07**	-	**16.05**	m	**19.26**
Sealing end of disused drain with plain in-situ concrete										
100mm diameter	0.75	5.00	**0.79**	0.24	-	**3.44**	-	**4.23**	nr	**5.07**
150mm diameter	1.37	5.00	**1.44**	0.36	-	**5.16**	-	**6.60**	nr	**7.91**
225mm diameter	2.19	5.00	**2.30**	0.48	-	**6.88**	-	**9.18**	nr	**11.01**
R13: LAND DRAINAGE										
Drains; clayware perforated pipes, butt joints; 150mm hardcore to sides and top										
Pipework in trenches										
100mm	12.99	5.00	**13.88**	-	0.48	**5.18**	-	**19.06**	m	**22.87**
150mm	20.97	5.00	**22.31**	-	0.72	**7.66**	-	**29.97**	m	**35.96**
225mm	35.85	5.00	**38.05**	-	0.77	**8.25**	-	**46.29**	m	**55.55**

PIPED SUPPLY SYSTEMS

Labour hourly rates: (except Specialists) Craft Operatives £17.38 Labourer £10.71 Rates are national average prices. Refer to REGIONAL VARIATIONS for indicative levels of overall pricing in regions	MATERIALS			LABOUR				RATES		
	Del to Site	Waste	Material Cost	Craft Optve	Lab	Labour Cost	Sunds	Nett Rate	Unit	Gross Rate (+20%)
	£	%	£	Hrs	Hrs	£	£	£		£
S115: HOT AND COLD WATER AND GAS: BLACK AND GALVANIZED STEEL TUBES AND FITTINGS										
Steel pipes, BS 1387 Table 3; steel fittings BS 1740 Part 1; screwed and PTFE tape joints; pipework black										
Pipes, straight in malleable iron pipe brackets fixing to timber with screws										
15mm	4.30	5.00	**4.52**	0.66	-	**11.47**	-	**15.99**	m	**19.18**
15mm; fittings										
extra; made bends	-	-	-	0.22	-	**3.82**	-	**3.82**	nr	**4.59**
extra; connections to copper pipe ends; compression joint	0.86	2.00	**0.88**	0.13	-	**2.29**	-	**3.17**	nr	**3.80**
extra; fittings; one end	0.65	2.00	**0.66**	0.17	-	**2.87**	-	**3.53**	nr	**4.24**
extra; fittings; two ends	0.66	2.00	**0.67**	0.33	-	**5.74**	-	**6.41**	nr	**7.69**
extra; fittings; three ends	0.90	2.00	**0.92**	0.33	-	**5.74**	-	**6.66**	nr	**7.99**
20mm	5.05	5.00	**5.31**	0.75	-	**13.00**	-	**18.31**	m	**21.97**
Pipes straight in malleable iron pipe brackets fixing to masonry with screws										
15mm	4.32	5.00	**4.53**	0.70	-	**12.24**	-	**16.77**	m	**20.12**
20mm	5.07	5.00	**5.32**	0.79	-	**13.76**	-	**19.09**	m	**22.91**
S115: HOT AND COLD WATER AND GAS: COPPER TUBING AND FITTINGS										
Copper pipes, BS EN 1057 Part 1 Table X; copper fittings, capillary, BS EN 1254 Part 2										
Pipes, straight in two piece copper spacing clips to timber with screws										
15mm	1.53	5.00	**1.60**	0.44	-	**7.65**	-	**9.25**	m	**11.10**
15mm; fitings										
extra; made bends	-	-	-	0.11	-	**1.91**	-	**1.91**	nr	**2.29**
extra; connections to iron pipe ends; screwed joint	2.37	3.00	**2.44**	0.13	-	**2.29**	-	**4.73**	nr	**5.68**
extra; fittings; one end	1.21	3.00	**1.25**	0.07	-	**1.15**	-	**2.39**	nr	**2.87**
extra; fittings; two ends	0.30	3.00	**0.31**	0.13	-	**2.29**	-	**2.60**	nr	**3.12**
extra; fittings; three ends	0.56	3.00	**0.58**	0.13	-	**2.29**	-	**2.87**	nr	**3.45**
22mm	2.96	5.00	**3.10**	0.47	-	**8.22**	-	**11.32**	m	**13.59**
Pipes straight in two piece copper spacing clips fixing to masonry with screws										
15mm	1.55	5.00	**1.62**	0.50	-	**8.60**	-	**10.23**	m	**12.27**
22mm	2.97	5.00	**3.12**	0.53	-	**9.18**	-	**12.30**	m	**14.76**
28mm	3.92	5.00	**4.12**	0.58	-	**10.13**	-	**14.25**	m	**17.10**
S115: HOT AND COLD WATER AND GAS: POLYBUTYLENE TUBES AND FITTINGS										
Polybutylene pipes; Hepworth Hep20 flexible plumbing system BS 7291 Class H; polybutylene demountable fittings										
Pipes, flexible in pipe clips to timber with screws										
15mm	1.68	10.00	**1.80**	0.50	-	**8.60**	-	**10.41**	m	**12.49**
15mm; fittings										
extra; connection to copper pipe ends	1.39	2.00	**1.41**	0.13	-	**2.29**	-	**3.71**	nr	**4.45**
extra; fittings; one end	2.32	2.00	**2.37**	0.07	-	**1.15**	-	**3.51**	nr	**4.22**
extra; fittings; two ends	1.80	2.00	**1.84**	0.13	-	**2.29**	-	**4.13**	nr	**4.96**
extra; fittings; three ends	2.57	2.00	**2.63**	0.13	-	**2.29**	-	**4.92**	nr	**5.90**
22mm	2.37	10.00	**2.57**	0.53	-	**9.18**	-	**11.75**	m	**14.10**
Pipes flexible in pipe clips fixing to masonry with screws										
15mm	1.78	10.00	**1.91**	0.50	-	**8.60**	-	**10.51**	m	**12.62**
22mm	2.42	10.00	**2.63**	0.53	-	**9.18**	-	**11.80**	m	**14.16**

Labour hourly rates: (except Specialists) Craft Operatives £17.38 Labourer £10.71 Rates are national average prices. Refer to REGIONAL VARIATIONS for indicative levels of overall pricing in regions	MATERIALS			LABOUR				RATES		
	Del to Site	Waste	Material Cost	Craft Optve	Lab	Labour Cost	Sunds	Nett Rate	Unit	Gross Rate (+20%)
	£	%	£	Hrs	Hrs	£	£	£		£
S115: HOT AND COLD WATER AND GAS: POLYETHYLENE TUBES AND FITTINGS										
Polythene pipes, BS 6572 Blue; copper alloy fittings, compression, BS EN 1254 Part 3, Type A										
Pipes; straight; in trenches										
20mm	0.70	10.00	**0.75**	0.19	-	**3.25**	-	**4.00**	m	**4.81**
20mm; fittings										
extra; connections to copper pipe ends; compression joint	5.92	3.00	**6.10**	0.18	-	**3.06**	-	**9.15**	nr	**10.99**
extra; fittings; one end	5.79	3.00	**5.96**	0.09	-	**1.53**	-	**7.49**	nr	**8.99**
extra; fittings; two ends	8.07	3.00	**8.32**	0.18	-	**3.06**	-	**11.38**	nr	**13.65**
extra; fittings; three ends	11.04	3.00	**11.38**	0.18	-	**3.06**	-	**14.43**	nr	**17.32**
25mm	0.97	10.00	**1.04**	0.22	-	**3.82**	-	**4.86**	m	**5.83**
S115: HOT AND COLD WATER AND GAS: STOPCOCKS										
Brass stopcocks										
Stopcocks BS 1010, crutch head; DZR; compression joints to copper (Type A)										
each end										
15mm	16.35	2.00	**16.67**	0.33	-	**5.74**	-	**22.41**	nr	**26.89**
22mm	26.85	2.00	**27.39**	0.44	-	**7.65**	-	**35.04**	nr	**42.04**
28mm	44.42	2.00	**45.31**	0.66	-	**11.47**	-	**56.78**	nr	**68.13**
S115: HOT AND COLD WATER AND GAS: GATE VALVES										
Copper alloy gate valves										
Gate valves; BS EN 12288 series B; compression joints to copper (Type A)										
each end										
15mm	8.04	2.00	**8.20**	0.33	-	**5.74**	-	**13.94**	nr	**16.72**
22mm	9.77	2.00	**9.96**	0.44	-	**7.65**	-	**17.61**	nr	**21.13**
28mm	13.45	2.00	**13.72**	0.66	-	**11.47**	-	**25.19**	nr	**30.23**
S115: HOT AND COLD WATER AND GAS: BALL VALVES										
Copper alloy body float valves										
Float operated valves for high pressure; BS 1212; plastics float										
inlet threaded externally; fixing to steel										
13mm; part 1	4.33	2.00	**4.42**	0.44	-	**7.65**	-	**12.07**	nr	**14.48**
19mm; part 1	14.20	2.00	**14.49**	0.66	-	**11.47**	-	**25.96**	nr	**31.15**
S115: HOT AND COLD WATER AND GAS: DRAINING TAPS										
Brass draining taps										
Drain cocks; BS 2879, square head Type2										
13mm	2.65	2.00	**2.70**	0.13	-	**2.29**	-	**5.00**	nr	**6.00**
S115: HOT AND COLD WATER AND GAS: STORAGE TANKS										
Plastics water storage cisterns										
Rectangular cold water storage cisterns; plastics; BS 4213 with sealed lid and byelaw 30 kit										
18 litres	16.67	2.00	**17.00**	1.65	-	**28.68**	-	**45.68**	nr	**54.81**
227 litres	95.65	2.00	**97.56**	3.30	-	**57.35**	-	**154.91**	nr	**185.89**
drilled holes for 13mm pipes	-	-	-	0.22	-	**3.82**	-	**3.82**	nr	**4.59**
S115: HOT AND COLD WATER AND GAS: PIPE INSULATION										
Thermal insulation; glass fibre preformed lagging, butt joints in the running length; secured with metal bands										
19mm thick insulation to copper pipework										
around one pipe										
15mm	3.23	3.00	**3.33**	0.22	-	**3.82**	-	**7.15**	m	**8.59**
28mm	3.62	3.00	**3.73**	0.33	-	**5.74**	-	**9.46**	m	**11.36**
Thermal insulation; foamed polyurethane preformed lagging, butt joints in the running length; secured with adhesive bands										
13mm thick insulation to copper pipework										
around one pipe										
15mm	1.53	3.00	**1.57**	0.22	-	**3.82**	-	**5.40**	m	**6.48**
28mm	2.11	3.00	**2.18**	0.26	-	**4.59**	-	**6.76**	m	**8.12**

MECHANICAL HEATING/COOLING/REFRIGERATION SYSTEMS

Labour hourly rates: (except Specialists) Craft Operatives £17.38 Labourer £10.71 Rates are national average prices. Refer to REGIONAL VARIATIONS for indicative levels of overall pricing in regions	MATERIALS			LABOUR				RATES		
	Del to Site £	Waste %	Material Cost £	Craft Optve Hrs	Lab Hrs	Labour Cost £	Sunds £	Nett Rate £	Unit	Gross Rate (+20%) £
T10: GAS/OIL FIRED BOILERS										
Gas flue pipes comprising galvanised steel outer skin and aluminium inner skin with air space between, BS 715; socketed joints										
Pipes; straight										
100mm	23.50	5.00	**24.67**	0.88	-	**15.29**	-	**39.97**	m	**47.96**
extra; connections to appliance	4.83	2.00	**4.93**	0.66	-	**11.47**	-	**16.40**	nr	**19.68**
extra; adjustable pipes	13.75	2.00	**14.03**	0.66	-	**11.47**	-	**25.50**	nr	**30.59**
extra; terminals	17.93	2.00	**18.29**	0.66	-	**11.47**	-	**29.76**	nr	**35.71**
extra; elbows 0- 90 degrees	15.71	2.00	**16.02**	0.66	-	**11.47**	-	**27.49**	nr	**32.99**
extra; tees	34.85	2.00	**35.54**	0.66	-	**11.47**	-	**47.02**	nr	**56.42**
Flue supports; galvanised steel wall bands fixing to masonry with screws										
for 100mm pipes	7.69	2.00	**7.84**	0.22	-	**3.82**	-	**11.67**	nr	**14.00**
Gas fired boilers										
Gas fired boilers for central heating and hot water; automatically controlled by thermostat with electrical control, gas governor and flame failure device										
free standing boiler; conventional flue; approximate output rating B Th U per hour										
40000	513.43	1.00	**518.56**	6.60	-	**114.71**	-	**633.27**	nr	**759.92**
80000	741.40	1.00	**748.81**	7.70	-	**133.83**	-	**882.64**	nr	**1059.17**
wall mounted boiler; balanced flue; approximate output rating B Th U per hour										
40000	572.00	1.00	**577.72**	8.25	-	**143.39**	-	**721.11**	nr	**865.33**
60000	666.60	1.00	**673.27**	8.25	-	**143.39**	-	**816.65**	nr	**979.98**
T31: LOW TEMPERATURE HOT WATER HEATING										
Accelerator pumps										
Variable head accelerator pumps for forced central heating; small bore indirect systems										
BSP connections; with valves										
20mm	59.13	2.00	**60.31**	1.10	-	**19.12**	-	**79.43**	nr	**95.31**
25mm	59.95	2.00	**61.15**	1.10	-	**19.12**	-	**80.27**	nr	**96.32**
Radiators										
Pressed steel radiators single panel with convector; air cock, plain plug										
450mm high; fixing brackets to masonry with screws										
600mm long	26.06	1.00	**26.32**	4.13	-	**71.69**	-	**98.01**	nr	**117.61**
900mm long	39.06	1.00	**39.45**	4.13	-	**71.69**	-	**111.14**	nr	**133.37**
1200mm long	52.06	1.00	**52.58**	4.13	-	**71.69**	-	**124.28**	nr	**149.13**
1600mm long	79.48	1.00	**80.27**	4.13	-	**71.69**	-	**151.96**	nr	**182.35**
2400mm long	160.35	1.00	**161.95**	4.13	-	**71.69**	-	**233.64**	nr	**280.37**
600mm high; fixing brackets to masonry with screws										
600mm long	33.62	1.00	**33.95**	4.13	-	**71.69**	-	**105.64**	nr	**126.77**
900mm long	50.42	1.00	**50.93**	4.13	-	**71.69**	-	**122.62**	nr	**147.14**
1200mm long	67.24	1.00	**67.92**	4.13	-	**71.69**	-	**139.61**	nr	**167.53**
1600mm long	102.61	1.00	**103.63**	4.13	-	**71.69**	-	**175.33**	nr	**210.39**
2400mm long	207.02	1.00	**209.09**	4.13	-	**71.69**	-	**280.78**	nr	**336.94**
700mm high; fixing brackets to masonry with screws										
600mm long	38.56	1.00	**38.94**	4.13	-	**71.69**	-	**110.63**	nr	**132.76**
900mm long	66.12	1.00	**66.78**	4.13	-	**71.69**	-	**138.47**	nr	**166.17**
1200mm long	118.59	1.00	**119.78**	4.13	-	**71.69**	-	**191.47**	nr	**229.76**
1600mm long	158.15	1.00	**159.73**	4.13	-	**71.69**	-	**231.42**	nr	**277.71**
2400mm long	237.23	1.00	**239.60**	4.13	-	**71.69**	-	**311.29**	nr	**373.55**
Pressed steel radiators; double panel with single convector; air cock, plain plug										
450mm high; fixing brackets to masonry with screws										
600mm long	40.47	1.00	**40.87**	4.13	-	**71.69**	-	**112.57**	nr	**135.08**
900mm long	60.70	1.00	**61.30**	4.13	-	**71.69**	-	**133.00**	nr	**159.60**
1200mm long	92.62	1.00	**93.55**	4.13	-	**71.69**	-	**165.24**	nr	**198.29**

Labour hourly rates: (except Specialists) Craft Operatives £17.38 Labourer £10.71 Rates are national average prices. Refer to REGIONAL VARIATIONS for indicative levels of overall pricing in regions	MATERIALS			LABOUR				RATES		
	Del to Site £	Waste %	Material Cost £	Craft Optve Hrs	Lab Hrs	Labour Cost £	Sunds £	Nett Rate £	Unit	Gross Rate (+20%) £
T31: LOW TEMPERATURE HOT WATER HEATING Cont.										
Radiators Cont.										
Pressed steel radiators; double panel with single convector; air cock, plain plug Cont.										
450mm high; fixing brackets to masonry with screws Cont.										
1600mm long	123.52	1.00	**124.75**	4.13	-	**71.69**	-	**196.45**	nr	**235.74**
2400mm long	248.94	1.00	**251.43**	4.13	-	**71.69**	-	**323.12**	nr	**387.75**
600mm high; fixing brackets to masonry with screws										
600mm long	51.00	1.00	**51.51**	4.13	-	**71.69**	-	**123.20**	nr	**147.84**
900mm long	76.48	1.00	**77.25**	4.13	-	**71.69**	-	**148.94**	nr	**178.73**
1200mm long	101.97	1.00	**102.99**	4.13	-	**71.69**	-	**174.68**	nr	**209.62**
1600mm long	155.62	1.00	**157.17**	4.13	-	**71.69**	-	**228.87**	nr	**274.64**
2400mm long	313.95	1.00	**317.09**	4.13	-	**71.69**	-	**388.78**	nr	**466.54**
700mm high; fixing brackets to masonry with screws										
600mm long	65.75	1.00	**66.40**	4.13	-	**71.69**	-	**138.10**	nr	**165.72**
900mm long	132.63	1.00	**133.95**	4.13	-	**71.69**	-	**205.65**	nr	**246.77**
1200mm long	176.79	1.00	**178.56**	4.13	-	**71.69**	-	**250.25**	nr	**300.30**
1600mm long	235.76	1.00	**238.12**	4.13	-	**71.69**	-	**309.81**	nr	**371.78**
2400mm long	353.63	1.00	**357.16**	4.13	-	**71.69**	-	**428.86**	nr	**514.63**
Brass valves for radiators; self colour										
Valves for radiators, BS 2767; wheel head										
inlet for copper; angle pattern										
15mm	7.49	2.00	**7.64**	0.33	-	**5.74**	-	**13.38**	nr	**16.05**
22mm	10.18	2.00	**10.38**	0.44	-	**7.65**	-	**18.03**	nr	**21.63**
Valves for radiators, BS 2767; lock shield										
inlet for copper; angle pattern										
15mm	7.49	2.00	**7.64**	0.33	-	**5.74**	-	**13.38**	nr	**16.05**
22mm	10.18	2.00	**10.38**	0.44	-	**7.65**	-	**18.03**	nr	**21.63**
Brass valves for radiators; chromium plated										
Valves for radiators, BS 2767; wheel head										
inlet for copper; angle pattern										
15mm	10.37	2.00	**10.58**	0.33	-	**5.74**	-	**16.32**	nr	**19.58**
22mm	13.56	2.00	**13.83**	0.44	-	**7.65**	-	**21.48**	nr	**25.78**
Brass thermostatic valves for radiators; chromium plated										
Thermostatic valves for radiators, one piece; ABS plastics head										
inlet for copper; angle pattern										
15mm	18.47	2.00	**18.84**	0.33	-	**5.74**	-	**24.57**	nr	**29.49**
inlet for copper; straight pattern										
15mm	18.47	2.00	**18.84**	0.33	-	**5.74**	-	**24.57**	nr	**29.49**
inlet for iron; angle pattern										
19mm	28.47	2.00	**29.04**	0.44	-	**7.65**	-	**36.68**	nr	**44.02**

VENTILATION/AIR CONDITIONING SYSTEMS

Labour hourly rates: (except Specialists) Craft Operatives £20.54 Labourer £10.71 Rates are national average prices. Refer to REGIONAL VARIATIONS for indicative levels of overall pricing in regions	MATERIALS			LABOUR				RATES		
	Del to Site £	Waste %	Material Cost £	Craft Optve Hrs	Lab Hrs	Labour Cost £	Sunds £	Nett Rate £	Unit	Gross Rate (+20%) £
U10: GENERAL VENTILATION										
Ventilating fans										
Note: the following prices exclude cutting holes in glass										
Extract fans, window mounted; shutter and incorporated switch unit										
102mm diameter	86.24	2.50	**88.40**	1.65	-	**33.89**	-	**122.29**	nr	**146.74**
102mm diameter - low voltage....................	123.37	2.50	**126.45**	1.65	-	**33.89**	-	**160.34**	nr	**192.41**
257mm diameter	278.74	2.50	**285.71**	1.98	-	**40.67**	-	**326.38**	nr	**391.65**
324mm diameter	402.49	2.50	**412.55**	1.98	-	**40.67**	-	**453.22**	nr	**543.87**
Extract fans, wall mounted; shutter, wall duct and incorporated switch unit										
Note: the following prices exclude forming hole and making good										
102mm diameter	51.38	2.50	**52.67**	1.93	-	**39.54**	-	**92.21**	nr	**110.65**
102mm diameter - low voltage....................	88.51	2.50	**90.72**	1.93	-	**39.54**	-	**130.26**	nr	**156.31**
257mm diameter	243.88	2.50	**249.98**	2.53	-	**51.97**	-	**301.94**	nr	**362.33**
324mm diameter	367.63	2.50	**376.82**	2.53	-	**51.97**	-	**428.79**	nr	**514.55**
Extract fans, built in wall; shutter and incorporated switch unit										
290 x 321 mm	219.15	2.50	**224.63**	3.08	0.09	**64.21**	-	**288.84**	nr	**346.61**
409 x 433 mm	397.90	2.50	**407.85**	3.19	0.09	**66.47**	-	**474.32**	nr	**569.18**
473 x 476 mm	480.40	2.50	**492.41**	3.30	0.09	**68.72**	-	**561.14**	nr	**673.37**

ELECTRICAL SUPPLY/POWER/LIGHTING SYSTEMS

Labour hourly rates: (except Specialists) Craft Operatives £20.54 Labourer £10.71 Rates are national average prices. Refer to REGIONAL VARIATIONS for indicative levels of overall pricing in regions	MATERIALS			LABOUR				RATES		
	Del to Site £	Waste %	Material Cost £	Craft Optve Hrs	Lab Hrs	Labour Cost £	Sunds £	Nett Rate £	Unit	Gross Rate (+20%) £
V90: GENERAL LIGHTING AND POWER - COMMERCIAL										
Steel trunking										
Straight lighting trunking; PVC lid										
50 x 50mm	2.20	5.00	2.31	0.44	-	9.04	0.55	11.90	m	14.28
extra; end cap	2.55	-	2.55	0.09	-	1.81	0.28	4.63	nr	5.56
extra; internal angle	12.49	-	12.49	0.28	-	5.65	0.28	18.42	nr	22.10
extra; external angle	12.49	-	12.49	0.28	-	5.65	0.28	18.42	nr	22.10
extra; tee	13.38	-	13.38	0.28	-	5.65	0.28	19.31	nr	23.17
Straight dado trunking, three compartment; steel lid										
200 x 38mm	10.41	5.00	10.93	0.55	-	11.30	0.83	23.05	m	27.66
extra; end cap	6.21	-	6.21	0.09	-	1.81	0.28	8.30	nr	9.95
extra; internal angle	32.56	-	32.56	0.44	-	9.04	0.28	41.87	nr	50.25
extra; external angle	32.56	-	32.56	0.44	-	9.04	0.28	41.87	nr	50.25
extra; single socket plate	10.71	-	10.71	0.18	-	3.62	0.28	14.60	nr	17.52
extra; twin socket plate	11.49	-	11.49	0.18	-	3.62	0.28	15.38	nr	18.46
Straight underfloor trunking; two compartments										
150 x 25mm	7.07	5.00	7.43	0.77	-	15.82	-	23.24	m	27.89
extra; floor junction boxes; adjustable frame and trap	84.83	-	84.83	2.04	-	41.80	-	126.63	nr	151.96
extra; horizontal bend	20.72	-	20.72	0.55	-	11.30	-	32.02	nr	38.42
PVC trunking; white										
Straight mini-trunking; clip on lid										
25 x 16mm	0.47	5.00	0.49	0.28	-	5.65	0.14	6.28	m	7.54
extra; end cap	0.36	-	0.36	0.07	-	1.36	0.14	1.86	nr	2.23
extra; internal bend	0.36	-	0.36	0.22	-	4.52	0.14	5.02	nr	6.03
extra; external bend	0.36	-	0.36	0.22	-	4.52	0.14	5.02	nr	6.03
extra; equal tee	0.63	-	0.63	0.28	-	5.65	0.14	6.42	nr	7.70
Straight dado trunking, three compartment; clip on lid										
145 x 40mm	23.99	5.00	25.19	0.44	-	9.04	0.55	34.77	m	41.73
extra; end caps	1.73	-	1.73	0.09	-	1.81	0.28	3.81	nr	4.57
extra; internal bend	5.90	-	5.90	0.33	-	6.78	0.28	12.95	nr	15.54
extra; external bend	5.90	-	5.90	0.33	-	6.78	0.28	12.95	nr	15.54
extra; tee	46.28	-	46.28	0.44	-	9.04	0.28	55.59	nr	66.71
extra; flat angle	39.67	-	39.67	0.33	-	6.78	0.28	46.72	nr	56.06
extra; mini trunking adaptor	1.67	-	1.67	0.22	-	4.52	0.28	6.46	nr	7.75
230 x 40mm	41.24	5.00	43.30	0.50	-	10.17	0.55	54.01	m	64.82
extra; end caps	5.47	-	5.47	0.09	-	1.81	0.28	7.55	nr	9.06
extra; internal bend	11.46	-	11.46	0.33	-	6.78	0.28	18.51	nr	22.22
extra; external bend	11.46	-	11.46	0.33	-	6.78	0.28	18.51	nr	22.22
extra; mini trunking adaptor	1.70	-	1.70	0.22	-	4.52	0.28	6.50	nr	7.80
Straight skirting trunking, three compartment; clip on lid										
100 x 40mm	4.22	5.00	4.43	0.39	-	7.91	0.55	12.89	m	15.47
extra; end cap	2.42	-	2.42	0.07	-	1.36	0.28	4.05	nr	4.86
extra; internal corner	3.11	-	3.11	0.28	-	5.65	0.28	9.03	nr	10.84
extra; external corner	3.11	-	3.11	0.28	-	5.65	0.28	9.03	nr	10.84
extra; mini trunking adaptor	2.19	-	2.19	0.18	-	3.62	0.28	6.08	nr	7.29
extra; socket adaptor, one gang	4.89	-	4.89	0.28	-	5.65	0.28	10.81	nr	12.98
extra; socket adaptor, two gang	7.02	-	7.02	0.28	-	5.65	0.28	12.94	nr	15.53
PVC heavy gauge conduit and fittings; push fit joints; spacer bar saddles at 600mm centres										
Conduits; straight										
20mm diameter; plugging to masonry; to surfaces	0.62	10.00	0.68	0.19	-	3.84	0.46	4.98	m	5.98
extra; small circular terminal boxes	1.10	2.50	1.13	0.11	-	2.26	0.30	3.68	nr	4.42
extra; small circular angle boxes	1.23	2.50	1.26	0.12	-	2.49	0.30	4.05	nr	4.85
extra; small circular three way boxes	1.33	2.50	1.36	0.14	-	2.94	0.30	4.60	nr	5.52
25mm diameter; plugging to masonry; to surfaces	1.18	10.00	1.30	0.20	-	4.07	0.46	5.83	m	7.00
extra; small circular terminal boxes	1.69	2.50	1.74	0.11	-	2.26	0.30	4.29	nr	5.15
extra; small circular angle boxes	1.85	2.50	1.89	0.12	-	2.49	0.30	4.68	nr	5.61
20mm diameter; plugging to masonry; in chases	0.62	10.00	0.68	0.17	-	3.39	0.46	4.53	m	5.44
25mm diameter; plugging to masonry; in chases	1.18	10.00	1.30	0.18	-	3.62	0.30	5.21	m	6.26

Labour hourly rates: (except Specialists) Craft Operatives £20.54 Labourer £10.71 Rates are national average prices. Refer to REGIONAL VARIATIONS for indicative levels of overall pricing in regions	MATERIALS			LABOUR				RATES		
	Del to Site £	Waste %	Material Cost £	Craft Optve Hrs	Lab Hrs	Labour Cost £	Sunds £	Nett Rate £	Unit	Gross Rate (+20%) £
V90: GENERAL LIGHTING AND POWER - COMMERCIAL Cont.										
Steel welded heavy gauge conduits and fittings, screwed joints; spacer bar saddles at 1000mm centres; enamelled black inside and outside by manufacturer										
Conduits; straight										
20mm diameter; plugging to masonry; to surfaces..	0.88	10.00	**0.96**	0.36	-	**7.46**	0.28	**8.70**	m	**10.43**
extra; small circular terminal boxes............	1.01	2.50	**1.04**	0.14	-	**2.94**	0.30	**4.27**	nr	**5.13**
extra; small circular angle boxes...............	1.14	2.50	**1.17**	0.18	-	**3.62**	0.30	**5.08**	nr	**6.10**
extra; small circular three way boxes...........	1.01	2.50	**1.04**	0.22	-	**4.52**	0.30	**5.85**	nr	**7.02**
extra; small circular through way boxes.........	1.08	2.50	**1.10**	0.18	-	**3.62**	0.30	**5.02**	nr	**6.02**
25mm diameter; plugging to masonry; to surfaces..	1.14	10.00	**1.26**	0.46	-	**9.49**	0.30	**11.04**	m	**13.25**
extra; small circular terminal boxes............	1.13	2.50	**1.16**	0.14	-	**2.94**	0.30	**4.40**	nr	**5.27**
extra; small circular angle boxes...............	1.28	2.50	**1.31**	0.18	-	**3.62**	0.30	**5.22**	nr	**6.26**
Steel welded heavy gauge conduits and fittings, screwed joints; spacer bar saddles at 1000mm centres; galvanized inside and outside by manufacturer										
Conduits; straight										
20mm diameter; plugging to masonry; to surfaces..	1.79	10.00	**1.97**	0.36	-	**7.46**	0.28	**9.71**	m	**11.65**
extra; small circular terminal boxes............	1.20	2.50	**1.23**	0.14	-	**2.94**	0.30	**4.46**	nr	**5.36**
extra; small circular three way boxes...........	1.20	2.50	**1.23**	0.22	-	**4.52**	0.30	**6.04**	nr	**7.25**
25mm diameter; plugging to masonry; to surfaces..	2.37	10.00	**2.61**	0.46	-	**9.49**	0.28	**12.37**	m	**14.85**
extra; small circular terminal boxes............	1.50	2.50	**1.53**	0.14	-	**2.94**	0.30	**4.77**	nr	**5.72**
extra; small circular angle boxes...............	1.68	2.50	**1.73**	0.18	-	**3.62**	0.30	**5.64**	nr	**6.76**
extra; small circular three way boxes...........	1.50	2.50	**1.53**	0.22	-	**4.52**	0.30	**6.35**	nr	**7.62**
galvanized standard cable tray and fittings										
Medium duty; straight										
100mm tray	3.40	5.00	**3.57**	0.37	-	**7.68**	0.10	**11.35**	m	**13.62**
extra; flat bend	30.99	-	**30.99**	0.26	-	**5.42**	0.28	**36.68**	nr	**44.02**
extra; tee	40.88	-	**40.88**	0.34	-	**7.00**	0.28	**48.16**	nr	**57.79**
extra; cross piece	63.01	-	**63.01**	0.47	-	**9.72**	0.55	**73.27**	nr	**87.93**
extra; riser bend	15.41	-	**15.41**	0.26	-	**5.42**	0.28	**21.10**	nr	**25.32**
150mm tray	4.16	5.00	**4.37**	0.48	-	**9.94**	0.10	**14.41**	m	**17.29**
extra; flat bend	34.96	-	**34.96**	0.29	-	**5.87**	0.28	**41.10**	nr	**49.33**
extra; tee	46.52	-	**46.52**	0.37	-	**7.68**	0.28	**54.48**	nr	**65.37**
Supports for cable tray										
Cantilever arms; mild steel; hot dip galvanized; fixing to masonry										
120mm wide	5.01	2.50	**5.13**	0.36	-	**7.46**	0.28	**12.86**	nr	**15.43**
Stand-off bracket; mild steel; hot dip galvanized; fixing to masonry										
100mm wide	6.44	2.50	**6.60**	0.36	-	**7.46**	0.28	**14.34**	nr	**17.20**
150mm wide	6.44	2.50	**6.60**	0.36	-	**7.46**	0.28	**14.34**	nr	**17.20**
PVC insulated cables; single core; reference 6491X; stranded copper conductors; to BS 6004										
Drawn into conduits or ducts or laid or drawn into trunking										
1.5mm2	0.00	15.00	**0.00**	0.03	-	**0.68**	-	**0.68**	m	**0.81**
2.5mm2	0.00	15.00	**0.00**	0.03	-	**0.68**	-	**0.68**	m	**0.81**
PVC insulated and PVC sheathed cables; multicore; copper conductors and bare earth continuity conductor 6242Y; to BS 6004										
Drawn into conduits or ducts or laid or drawn into trunking										
1.0mm2; twin with bare earth.....................	0.00	15.00	**0.00**	0.06	-	**1.13**	-	**1.13**	m	**1.36**
1.5mm2; twin with bare earth.....................	0.00	15.00	**0.00**	0.06	-	**1.13**	-	**1.13**	m	**1.36**
2.5mm2; twin with bare earth.....................	0.00	15.00	**0.00**	0.06	-	**1.13**	-	**1.13**	m	**1.36**
PVC insulated SWA armoured and PVC sheathed cables; multicore; copper conductors to BS 6346										
Fixed to masonry with screwed clips at average 300mm centres; plugging										
16mm2; three core	3.89	5.00	**4.09**	0.28	-	**5.65**	1.03	**10.77**	m	**12.93**
Mineral insulated copper sheathed cables; PVC outer sheath; copper conductors; 600V light duty to BS EN 60702-1										
Fixed to masonry with screwed clips at average 300mm centres; plugging										
2.5mm2; two core	4.05	10.00	**4.45**	0.19	-	**3.84**	1.03	**9.33**	m	**11.19**
extra; termination including gland, seal and shroud	3.52	2.50	**3.61**	0.46	-	**9.49**	-	**13.10**	nr	**15.72**
HV switchgear										
Switch fuse 500V, metal cased, fixed to masonry with screws; plugging										
32 amp; single pole and neutral; short circuit rating 6.4Ka	125.22	2.50	**128.35**	1.76	-	**36.15**	0.55	**165.05**	nr	**198.06**

Labour hourly rates: (except Specialists) Craft Operatives £20.54 Labourer £10.71 Rates are national average prices. Refer to REGIONAL VARIATIONS for indicative levels of overall pricing in regions	MATERIALS			LABOUR				RATES		
	Del to Site	Waste	Material Cost	Craft Optve	Lab	Labour Cost	Sunds	Nett Rate	Unit	Gross Rate (+20%)
	£	%	£	Hrs	Hrs	£	£	£		£
V90: GENERAL LIGHTING AND POWER - COMMERCIAL Cont.										
HV switchgear Cont.										
Consumer control units, 250V, metal cased, fitted with 100 amp D.P. isolator and MCB's , fixed to masonry with screws; plugging										
eight way, single pole and neutral	103.85	2.50	**106.45**	1.76	-	**36.15**	0.55	**143.15**	nr	**171.78**
Luminaires										
Note for lighting fittings and lamps see General Lighting and Power - Domestic Section										
Accessories; white plastics, with boxes, fixing to masonry with screws; plugging										
Flush plate switches										
5 amp; one gang; one way; single pole	1.71	2.50	**1.75**	0.46	-	**9.49**	0.28	**11.51**	nr	**13.81**
5 amp; two gang; one way; single pole	2.31	2.50	**2.37**	0.47	-	**9.72**	0.28	**12.36**	nr	**14.83**
Surface plate switches										
5 amp; one gang; one way; single pole	1.82	2.50	**1.86**	0.46	-	**9.49**	0.28	**11.62**	nr	**13.95**
5 amp; two gang; one way; single pole	2.42	2.50	**2.48**	0.47	-	**9.72**	0.28	**12.47**	nr	**14.97**
Ceiling switches										
5 amp; one gang; one way; single pole	2.31	2.50	**2.37**	0.47	-	**9.72**	0.28	**12.36**	nr	**14.83**
Flush switched socket outlets										
13 amp; one gang	3.30	2.50	**3.38**	0.47	-	**9.72**	0.28	**13.37**	nr	**16.05**
13 amp; two gang	3.85	2.50	**3.95**	0.47	-	**9.72**	0.28	**13.94**	nr	**16.72**
Surface switched socket outlets										
13 amp; one gang	3.47	2.50	**3.55**	0.47	-	**9.72**	0.28	**13.54**	nr	**16.25**
13 amp; two gang	4.18	2.50	**4.28**	0.47	-	**9.72**	0.28	**14.27**	nr	**17.13**
Flush switched socket outlets with RCCB (residual current device) protected at 30 mAmp										
13 amp; one gang	39.16	2.50	**40.14**	0.47	-	**9.72**	0.28	**50.13**	nr	**60.16**
Flush fused connection units										
13 amp; one gang; switched; flexible outlet	4.68	2.50	**4.79**	0.55	-	**11.30**	0.28	**16.36**	nr	**19.64**
13 amp; one gang; switched; flexible outlet; pilot lamp	5.12	2.50	**5.24**	0.55	-	**11.30**	0.28	**16.81**	nr	**20.18**
Accessories, metalclad, with boxes, fixing to masonry with screws; plugging										
Surface switched socket outlets										
13 amp; one gang	4.95	2.50	**5.07**	0.47	-	**9.72**	0.28	**15.06**	nr	**18.08**
13 amp; two gang	7.15	2.50	**7.33**	0.47	-	**9.72**	0.28	**17.32**	nr	**20.78**
V90: GENERAL LIGHTING AND POWER - DOMESTIC										
Electric wiring										
Note the following approximate prices of various types of installations are dependant on the number and disposition of points; lamps and fittings together with cutting and making good are excluded										
Electric wiring in new building										
Insulated Consumer unit, fixed to masonry withscrews; plugging										
9 usable ways 100a DP main Switch	28.20	2.50	**28.90**	2.78	-	**57.16**	0.28	**86.34**	nr	**103.61**
Miniature Circuit Breakers fitted in consumer unit; Bsen 60898										
6a Single Pole	6.87	2.50	**7.04**	0.15	-	**3.16**	-	**10.20**	nr	**12.24**
10a Single Pole	6.54	2.50	**6.70**	0.15	-	**3.16**	-	**9.87**	nr	**11.84**
Installation with PVC insulated and sheathed cables										
lighting points	7.73	7.50	**8.28**	3.94	-	**80.89**	-	**89.16**	nr	**107.00**
socket outlets; 13A ring main	4.62	7.50	**4.94**	4.48	-	**91.96**	-	**96.90**	nr	**116.28**
socket outlets; 13A radial circuit	4.76	7.50	**5.09**	4.66	-	**95.80**	-	**100.89**	nr	**121.07**
cooker points; 45A	15.67	7.50	**16.79**	7.62	-	**156.58**	-	**173.37**	nr	**208.04**
immersion heater points	7.19	7.50	**7.70**	5.57	-	**114.33**	-	**122.03**	nr	**146.44**
shaver sockets (transformer)	18.20	7.50	**19.53**	2.00	-	**41.12**	-	**60.66**	nr	**72.79**
Installation with black enamel heavy gauge conduit with PVC cables										
lighting points	36.31	7.50	**38.75**	5.63	-	**115.68**	-	**154.43**	nr	**185.32**
socket outlets; 13A ring main	25.18	7.50	**26.87**	5.15	-	**105.74**	-	**132.61**	nr	**159.13**
socket outlets; 13A radial circuit	28.68	7.50	**30.55**	7.26	-	**149.12**	-	**179.67**	nr	**215.60**
cooker points; 45A	49.57	7.50	**52.95**	11.68	-	**239.95**	-	**292.90**	nr	**351.47**
immersion heater points	32.15	7.50	**34.22**	8.66	-	**177.81**	-	**212.03**	nr	**254.44**
shaver sockets (transformer)	50.99	7.50	**54.06**	2.78	-	**57.16**	-	**111.23**	nr	**133.47**
Electric wiring in extension to existing building										
Installation with PVC insulated and sheathed cables										
lighting points	7.73	7.50	**8.28**	3.94	-	**80.89**	-	**89.16**	nr	**107.00**
socket outlets; 13A ring main	4.62	7.50	**4.94**	4.48	-	**91.96**	-	**96.90**	nr	**116.28**
socket outlets; 13A radial circuit	4.76	7.50	**5.09**	4.66	-	**95.80**	-	**100.89**	nr	**121.07**
cooker points; 45A	15.67	7.50	**16.79**	7.62	-	**156.58**	-	**173.37**	nr	**208.04**

Labour hourly rates: (except Specialists) Craft Operatives £20.54 Labourer £10.71 Rates are national average prices. Refer to REGIONAL VARIATIONS for indicative levels of overall pricing in regions	MATERIALS			LABOUR				RATES		
	Del to Site £	Waste %	Material Cost £	Craft Optve Hrs	Lab Hrs	Labour Cost £	Sunds £	Nett Rate £	Unit	Gross Rate (+20%) £
V90: GENERAL LIGHTING AND POWER - DOMESTIC Cont.										
Electric wiring in extension to existing building Cont.										
Installation with PVC insulated and sheathed cables Cont.										
immersion heater points	7.19	7.50	**7.70**	5.57	-	**114.33**	-	**122.03**	nr	**146.44**
shaver sockets (transformer)	18.20	7.50	**19.53**	2.00	-	**41.12**	-	**60.66**	nr	**72.79**
Installation with black enamel heavy gauge conduit with PVC cables										
lighting points	36.31	7.50	**38.75**	5.63	-	**115.68**	-	**154.43**	nr	**185.32**
socket outlets; 13A ring main	25.18	7.50	**26.87**	5.15	-	**105.74**	-	**132.61**	nr	**159.13**
socket outlets; 13A radial circuit	28.68	7.50	**30.55**	7.26	-	**149.12**	-	**179.67**	nr	**215.60**
cooker points; 45A	49.57	7.50	**52.95**	11.68	-	**239.95**	-	**292.90**	nr	**351.47**
immersion heater points	32.15	7.50	**34.22**	8.66	-	**177.81**	-	**212.03**	nr	**254.44**
shaver sockets (transformer)	50.99	7.50	**54.06**	2.78	-	**57.16**	-	**111.23**	nr	**133.47**
V90: LAMP FITTINGS, FANS, HEATERS etc., ALL INSTALLATIONS										
Fluorescent lamp fittings, mains voltage operations; switch start; including lamps										
Batten type; single tube										
1200mm; 36W	11.28	2.50	**11.56**	0.91	-	**18.75**	0.55	**30.86**	nr	**37.03**
Batten type; twin tube										
1200mm; 36W	18.57	2.50	**19.03**	0.91	-	**18.75**	0.55	**38.34**	nr	**46.00**
Plastics diffused type; single tube										
1200mm; 36W	39.06	2.50	**40.04**	1.21	-	**24.85**	0.55	**65.44**	nr	**78.53**
Plastics diffused type; twin tube										
1200mm; 36W	72.91	2.50	**74.73**	1.10	-	**22.59**	0.55	**97.87**	nr	**117.45**
Lighting fittings complete with tungsten lamps										
Bulkhead; alloy										
100W G.L.S. lamp	11.55	2.50	**11.84**	0.79	-	**16.27**	0.28	**28.38**	nr	**34.06**
Down light; semi recessed										
175mm; 150W PAR38 lamp	20.39	2.50	**20.90**	0.79	-	**16.27**	0.28	**37.45**	nr	**44.94**
Immersion heaters and thermostats										
Note: the following prices exclude Plumber's work										
2 or 3KW immersion heaters; without flanges										
457mm; non-withdrawable elements	25.97	2.50	**26.62**	0.97	-	**19.88**	5.12	**51.62**	nr	**61.94**
Thermostats										
immersion heater	11.26	2.50	**11.55**	0.61	-	**12.43**	0.28	**24.25**	nr	**29.10**
Water heaters										
Note: the following prices exclude Plumber's work										
Storage type units; free outlet										
23 litre	557.40	-	**557.40**	2.73	-	**56.03**	6.22	**619.65**	nr	**743.58**
Shower units										
Note: the following prices exclude Plumber's work										
Instantaneous shower units										
complete with fittings	171.81	2.50	**176.10**	4.84	-	**99.41**	8.50	**284.02**	nr	**340.82**
V90: TESTING										
Testing; existing installations										
Point										
from	-	-	-	2.48	-	**50.84**	-	**50.84**	nr	**61.00**
to	-	-	-	4.24	-	**86.99**	-	**86.99**	nr	**104.38**
Complete installation; three bedroom house										
from	-	-	-	12.10	-	**248.53**	-	**248.53**	nr	**298.24**
to	-	-	-	19.25	-	**395.40**	-	**395.40**	nr	**474.47**

ALTERATIONS INDEX

S

STANDARD RATES OF WAGES

WAGES

Note: Wage agreements and amendments to the following made after the end of June 2005 and notified to us prior to publication will be noted at the end of the Preliminaries section or as noted herein.

ENGLAND, SCOTLAND AND WALES

Following negotiations between the parties to the Construction Industry Joint Council in June 2003 the Council adopted recommendations for a three year agreement on pay and conditions

With effect on and from Monday, 27th June, 2005 the following basic rates of pay, allowances and additional payments will be applicable:

Note: The rates shown below are weekly rates. The hourly equivalents for a 39 hour weel are shown in brackets.

1. ENTITLEMENT TO BASIC AND ADDITIONAL RATES - WR.1

		Rate per week £	per hour £
Craft Operative		351.00	(9.00p)
Skill Rate	1	334.62	(8.58p)
	2	322.14	(8.26p)
	3	301.47	(7.73p)
	4	284.31	(7.29p)
General Building Operative		264.03	(6.77p)

Additional Payment for Skilled Work – WR.1.2.2

Skilled Operative Additional Rate:

	1	54.60	(1.40p)
	2	33.15	(0.85p)
	3	9.36	(0.24p)

Apprentice Rates – WR. 1.5.2

	Rate per week £	per hour £
Year 1	146.25	(3.75)
Year 2	188.76	(4.84)
Year 3 without NVQ2	220.74	(5.66)
With NVQ2	280.80	(7.20)
Year 3 with NVQ3	351.00	(9.00)
On completion of apprenticeship with NVQ2	351.00	(9.00)

2. BONUS - WR.2

It shall be open to employers and employees on any job to agree a bonus scheme based on measured output and productivity for any operation or operations on that particular job.

3. STORAGE OF TOOLS - WR.12

Employers maximum liability shall be £400.00

4. LOSS OF CLOTHING - WR.13

Employers maximum liability shall be £30.00

5. SUBSISTENCE ALLOWANCE - WR.15

£ 26.47 per night

6. SICK PAY - WR.20

Effective on and from 27th June 2005 employers whose terms and conditions of employment incorporate the Working Rule Agreement of the CIJC should make a payment to operatives, who are absent from work due to sickness or injury, of £93.60 per week.

7. BENEFIT SCHEMES WR 21

Death benefit – WR 21 1(a) £20,000.00

Accident cover – up to £16,000.00

EasyBuild pension contributions – WR 21.3
With affect from 4th August 2003, the minimum employer contribution shall be £2.50 per week. Where the operative contributes between £2.51 and £10.00 per week the employer shall increase the minimum contribution to match that of the operative up to a maximum of £10.00 per week

SCHEDULE 2

Entitlement to additional rates of pay for intermittent skill, responsibility or working in adverse conditions as Schedule 2.

Extra Rate	P/hr	Weekly Rate based on 39 hours
A	(15p)	05.85
B	(24p)	09.36
C	(32p)	12.48
D	(37p)	14.43
E	(56p)	21.84

ENGLAND, SCOTLAND AND WALES (Cont'd)

RATES OF WAGES SINCE 1987

The following tables give the rates of wages since 1987.

	Grade A and Scotland District Craft Operatives per week £	Labourers per week £	London & Liverpool District Craft Operatives per week £	Labourers per week £
June 29th, 1987	103.35	88.14	103.545	88.335
June 27th, 1988	110.565	94.185	110.76	94.38
June 26th, 1989	120.12	102.375	120.315	102.57
June 25th, 1990	131.82	112.32	132.015	112.515
June 24th, 1991	139.035	118.365	139.23	118.56
June 29th, 1992	159.51	135.915	159.705	136.11
June 27th, 1994	163.41	139.23	163.605	139.425
	All areas			
June 26th, 1995	173.55	143.52		
June 24th, 1996	178.62	147.81		
August 18th, 1997	188.37	156.00		

	Craft Operative	Skilled Operative Rate: 1	2	3	4	General Building Operative
June 29th, 1998	214.50	209.04	201.24	188.76	177.84	164.97
June 28th, 1999	235.95	224.64	216.45	202.80	191.10	177.45
June 26th, 2000	247.65	235.95	227.37	212.94	200.85	186.42
June 25th, 2001	261.30	248.82	239.85	224.64	211.77	196.56.
June 24th, 2002	284.70	271.05	261.30	244.53	230.49	214.11
June 30th, 2003	299.13	284.70	274.56	257.01	242.19	225.03
June 28th 2004	320.38	305.37	294.06	273.34	259.35	241.02
June 27th, 2005	351.00	334.62	322.14	301.47	284.31	264.03

GUARANTEED MINIMUM BONUS

	Grade A and Scotland District Craft Operatives per week £	Labourers per week £	London and Liverpool District Craft Operatives per week £	Labourers per week £
June 27th, 1988	15.015	12.675	15.015	12.675
June 26th, 1989	15.015	12.675	15.015	12.675
June 25th, 1990	16.38	14.04	16.38	14.04
June 24th, 1991	16.575	14.235	16.575	14.235
June 29th, 1992	-	-	-	-

HOLIDAYS WITH PAY

Employers Contributions for Annual Holiday Stamps (per week)

Date effective from	Adults £	Under 18 £	* Contribution included to Retirement and Death Benefit Scheme £
August 1st, 1987	13.60*	9.30	1.20
August 1st, 1988	14.45*	9.85	1.30
July 31st, 1989	16.10*	11.05	1.40
July 30th, 1990	17.70*	12.15	1.55
July 29th, 1991	18.70*	12.75	1.75
August 3rd, 1992	19.20*	13.05	1.85
August 1st, 1994	19.30*	13.05	1.95
July 29th, 1996	19.60*	13.05	2.05
August 4th, 1997	20.00*	13.45	2.25
August 3rd, 1998	20.80*	14.25	2.45

"Template" introduced to meet the industry's Working Rule Agreements and the Working Time Regulations.
Holiday pay annual contribution calculated by multiplying number of hours holiday by hourly rate.
Employers Contributions (per week)

	Holiday Hours	Craft Operative £	General Operative £	Retirement Benefit £
August 2nd 1999	233	27.11	20.38	1.90
June 26th 2000	226	27.60	20.78	1.90

	Holiday Hours	Craft Operative £	General Operative £	Retirement and Death Benefit £
June 25th 2001	226	29.12	20.78	3.20 (July 2nd 2001)
June 24th 2002	226	31.73	23.86	5.40 (July 1st 2002)
June 30th 2003	226	33.34	25.08	8.40 (August 4th 2003)
June 28th 2004	226	35.73	26.86	10.90
June 27th 2005	226	39.12	29.42	10.90

NORTHERN IRELAND

The Joint Council for the Building and Civil Engineering Industry in Northern Ireland has agreed a three year wage award for the Building and Civil Engineering Industry (Northern Ireland):

Effective from	2nd August 2004 £	1st August 2005 £	31st July 2006 £
Advanced Craftsmen	322.14	340.08	360.75
Craftsmen	299.52	315.90	335.40
Skilled Operatives	278.07	290.55	303.42
General Building Operative/Basic Operatives	257.40	269.10	280.80

RATES OF WAGES SINCE 1990

The following tables give the rates of wages since 1990

	Craftsmen Per Week	Labourers Per Week	GUARANTEED MINIMUM BONUS Craftsmen Per Week	GUARANTEED MINIMUM BONUS Labourers Per Week
July 30th 1990	£146.25	£127.14	7.00	7.00
July 29th 1991	£159.51	£134.84	7.00	7.00
July 27th 1992	£165.75	£144.30	7.00	7.00
July 26th 1993	£177.06	£155.22	-	-
July 25th 1994	£180.96	£158.73	-	-
July 31st 1995	£188.37	£164.97	-	-

	Craftsmen Per Week	General Building Operative
August 6th 1996	£194.22	£170.04
February 3rd 1997	£200.07	£175.11
September 1st 1997	£205.92	£180.18
February 2nd 1998	£219.96	£192.27
August 3rd 1998	£229.71	£200.85
January 4th 1999	£236.73	£207.09
August 2nd 1999	£244.92	£214.50
July 31st 2000	£255.84	£220.74

	Advanced Craftsmen Per Week	Craftsmen Per Week	Skilled Operative Per Week	General Building Operative
July 30th 2001	-	£264.81	£246.87	£228.54
August 5th 2002	£296.40	£275.34	£256.62	£237.51
August 4th 2003	£306.93	£285.09	£265.59	£245.70

HOLIDAYS WITH PAY

Employers Contributions for Operatives:
consolidated holiday credit
Date effective (including benefit contribution)

May 7th, 1990	£20.80
May 6th, 1991	£22.90
May 4th, 1992	£24.20
May 3rd, 1993	£25.00
May 2nd, 1994	£25.20
May 1st, 1995	£26.00
April 29th, 1996	£27.15
April 28th, 1997	£29.25
April 27th, 1998	£32.25
April 26th, 1999	£34.35

From 31st July 2000 Contributions provide holiday pay equivalent to basic wages to comply with the Working Time Regulations.

PENSION AND DEATH BENIFITS CONTRIBUTION

From 13th September 2004 pension and death benifits contribution will increase from £7.00 per week to £8.00 per week.

From April 2006 it is proposed to bring in a new scheme with employers contributions of £11.00 per week

HEATING, VENTILATING AND DOMESTIC ENGINEERING CONTRACTS

Rates agreed between the Heating and Ventilating Contractors' Association, and AMICUS.

RATES OF WAGES

All districts of the United Kingdom
Note : Ductwork Erection Operatives, are entitled to the same rates and allowances as the parallel fitter grades shown.
Three year wage agreement effective from 3rd November 2003

RATES AND ALLOWANCES	From 7 October 2002		From 3 November 2003		From 4 October 2004		From 3 October 2005	
Hourly Rate								
	£		£		£		£	
Foreman	11.61		12.02		12.56		13.20	
Senior Craftsman(RAS+RAW)	10.80		11.16		11.67		12.25	
Senior Craftsman(RAS)	10.40		10.75		11.24		11.80	
Senior Craftsman(RAW)	10.00		10.34		10.81		11.35	
Senior Craftsman	9.60		9.93		10.38		10.90	
Craftsman(+2RA)	9.60		9.93		10.38		10.90	
Craftsman(+1RA)	9.20		9.52		9.95		10.45	
Craftsman	8.80		9.11		9.52		10.00	
Installer	7.98		8.26		8.63		9.07	
Adult Trainee	6.72		6.96		7.27		7.64	
Mate (over 18)	6.72		6.96		7.27		7.64	
Mate (17-18)	4.32		4.47		4.67		4.91	
Mate (under 17)	3.12		3.23		3.38		3.55	
Modern Apprentices								
Junior	4.38		4.53		4.73		4.97	
Intermediate	6.18		6.40		6.69		7.03	
Senior	7.98		8.26		8.63		9.07	
Responsibility Allowance (Craftsmen)	P		p		p		P	
1 Unit	40		41		43		45	
2 Units	80		82		86		90	
Responsibility Allowance (Senior Craftsmen)								
Second welding skill	40		41		43		45	
Supervisory responsibility	80		82		86		90	
Second welding skill and supervisory responsibility	1.20		1.23		1.29		1.35	
DAILY ABNORMAL CONDITIONS MONEY £ per day	2.99		No change					
LODGING ALLOWANCE £ per night	25.75		26.50		27.10		To be promulgated separately	
DAILY TRAVELLING ALLOWANCE Scale 2 – (Outside M25) C = Craftsmen including installers M + A = Mates, Apprentices, Adult Trainees.								
Direct distance from centre to job in miles	C	M & A	C	M & A	C	M & A	C	M & A
Over-Not exceeding	£	£	£	£	£	£	£	£
15-20	1.92	1.65	1.99	1.71	2.08	1.79	2.19	1.88
20-30	4.97	4.29	5.14	4.44	5.37	4.64	5.64	4.88
30-40	7.15	6.17	7.40	6.39	7.73	6.68	8.12	7.02
40-50	9.40	8.05	9.73	8.33	10.17	8.71	10.69	9.15
* N.B. from 7/10/2002 distance to be over 15 not exceeding 20 miles								

WEEKLY HOLIDAY CREDIT/WELFARE CONTRIBUTIONS

From 6th October 2003			From 4th October 2004		From 3rd October 2005	
Credit Value Category (Note 1)	Weekly Holiday Credit £	Combined Weekly Credit and Welfare Contribution £	Weekly Holiday Credit £	Combined Weekly Credit and Welfare Contribution £	Weekly Holiday Credit	Combined Weekly Credit and Welfare Contribution £
a	57.50	62.46	60.08	65.30	63.13	68.61
b	53.48	58.44	55.89	61.11	58.73	64.21
c	51.51	56.47	53.82	59.04	56.55	62.03
d	49.52	54.48	51.75	56.97	54.38	59.86
e	47.54	52.50	49.68	54.90	52.20	57.68
f	45.57	50.53	47.61	52.38	50.03	55.51
g	43.58	48.54	45.54	50.76	47.85	53.33
h	39.52	44.48	41.30	46.52	43.40	48.88
i	33.28	38.24	34.78	40.00	36.55	42.03
j	30.61	35.57	31.99	37.21	33.62	39.10
k	21.70	26.66	22.68	27.90	23.83	29.31
l	21.40	26.36	22.36	27.58	23.50	28.98
m	15.46	20.42	16.15	21.37	16.97	22.45

HEATING, VENTILATING AND DOMESTIC ENGINEERING CONTRACTS (Cont'd)
NOTES TO THE APPENDIX ON WAGE RATES, ALLOWANCES AND OTHER PROVISIONS

1. From 6th October 2003, the grades of H&V Employees covered by the range of credit values and entitled to the different rates of Sickness and Accident Benefit, and weekly Holiday Credit and Welfare Contributions are as follows:

a	Foreman
b	Senior Craftsman (RAS & RAW)
c	Senior Craftsman(RAS)
d	Senior Craftsman(RAW)
e	Senior Craftsman
	Craftsman (+2RA)
f	Craftsmen(+1 RA)
g	Craftsman
h	Installer
	Senior Modern Apprentice
i	Adult Trainee
	Mate(over 18)
j	Intermediate Modern Apprentice
k	Junior Modern Apprentice
l	Mate (aged 17 - 18)
m	Mate (under 17)

For premium rates explanations, fuller details and general conditions the reader is advised to obtain a copy of the National Working Rule Agreement available from the Heating and Ventilation Contractors' Association Publications Department, Old Mansions House, Eamont Bridge, Penrith, Cumbria, CA10 2BX www.hvca.org.uk www.amicustheunion.org

JOINT INDUSTRY BOARD FOR THE ELECTRICAL CONTRACTING INDUSTRY

JIB INDUSTRIAL DETERMINATION 2005/2006/2007

Promulgated by the JIB National Board to be effective from the dates shown.

1. WAGE RATES

From **7th January 2002** two different wage rates have applied to JIB Graded operatives working on site, depending on whether the employer transports them to site or whether they provide their own transport. The two categories are:

Job Employed (Transport Provided)
Payable to an operatives who are transported to and from the job by their employer. Operatives shall also be entitled to payment for Travel Time, travelling in their own time, as detailed in the appropriate scale.

Job Employed (Own Transport)
Payable to operatives who travel by their own means to and from the job. Operatives under this category shall be entitled to payment for Travel Allowance and also when travelling in their own times to Travel Time, as detailed in the appropriate scale.

There are separate rate to accommodate those few operatives who are employed permanently at the shop and who do not work on site. These rates are shown under section 6 of this Determination.

2. WAGES 2005

From and including **10th January 2005,** the JIB hourly rates for Job Employed operatives shall be:

(i) National Standard Rates

Grade	Transport Provided	Own Transport
Technician (or equivalent specialist grade)	£12.29	£12.91
Approved Electrician (or equivalent specialist grade)	£10.84	£11.46
Electrician (or equivalent specialist grade	£9.94	£10.57
Electrical Improver	£8.95	£9.51
Labourer	£7.89	£8.48
Senior Graded Electrical Trainee	£8.95	£9.51
Adult Trainee	£7.89	£8.48
Adult Trainee (under 21)	£5.92	£6.36

(ii) **London Rate**

for Operatives engaged upon work falling within the terms of JIB National Working Rule 6.2 will be:

Grade	Transport Provided	Own Transport
Technician (or equivalent specialist grade)	£13.76	£14.46
Approved Electrician (or equivalent specialist grade)	£12.14	£12.84
Electrician (or equivalent specialist grade	£11.13	£11.84
Electrical Improver	£10.02	£10.65
Labourer	£8.84	£9.50
Senior Graded Electrical Trainee	£10.02	£10.65
Adult Trainee	£8.84	£9.50
Adult Trainee (under 21)	£6.63	£7.12

3 WAGES 2006

From and including 9th **January 2006,** the JIB hourly rates for Job Employed operatives shall be:

(i) National Standard Rates

Grade	Transport Provided	Own Transport
Technician (or equivalent specialist grade)	£12.84	£13.49
Approved Electrician (or equivalent specialist grade)	£11.33	£11.98
Electrician (or equivalent specialist grade	£10.39	£11.05
Electrical Improver	£9.35	£9.94
Labourer	£8.25	£8.86
Senior Graded Electrical Trainee	£9.35	£9.94
Adult Trainee	£8.25	£8.86
Adult Trainee (under 21)	£6.19	£6.65

JOINT INDUSTRY BOARD FOR THE ELECTRICAL CONTRACTING INDUSTRY (Cont'd)
WAGES 2006(Cont'd)

(ii) **London Rate**

for Operatives engaged upon work falling within the terms of JIB National Working Rule 6.2 will be:

Grade	Transport Provided	Own Transport
Technician (or equivalent specialist grade)	£14.38	£15.11
Approved Electrician (or equivalent specialist grade)	£12.69	£13.42
Electrician (or equivalent specialist grade	£11.64	£12/38
Electrical Improver	£10.47	£11.13
Labourer	£9.24	£9.92
Senior Graded Electrical Trainee	£10.47	£11.13
Adult Trainee	£9.24	£9.92
Adult Trainee (under 21)	£6.93	£7.45

4 WAGES 2007

From and including 8th **January 2007,** the JIB hourly rates for Job Employed operatives shall be:

(i) **National Standard Rates**

Grade	Transport Provided	Own Transport
Technician (or equivalent specialist grade)	£13.48	£14.16
Approved Electrician (or equivalent specialist grade)	£11.90	£12.58
Electrician (or equivalent specialist grade	£10.91	£11.60
Electrical Improver	£9.82	£10.44
Labourer	£8.66	£9.30
Senior Graded Electrical Trainee	£9.82	£10.44
Adult Trainee	£8.66	£9.30
Adult Trainee (under 21)	£6.50	£6.98

(ii) **London Rate**

for Operatives engaged upon work falling within the terms of JIB National Working Rule 6.2 will be:

Grade	Transport Provided	Own Transport
Technician (or equivalent specialist grade)	£15.10	£15.86
Approved Electrician (or equivalent specialist grade)	£13.33	£14.09
Electrician (or equivalent specialist grade	£12.22	£12.99
Electrical Improver	£11.00	£11.69
Labourer	£9.70	£10.42
Senior Graded Electrical Trainee	£11.00	£11.69
Adult Trainee	£9.70	£10.42
Adult Trainee (under 21)	£7.28	£7.82

6. OPERATIVES EMPLOYED PERMANENTLY AT THE SHOP

From and including 10th January 2005, 9th January 2006 and 8th January 2007 respectively, the JIB hourly rates for operatives employed permanently at the shop and nowhere else shall be:.

(i) **National Standard Rates**

Grade	2005	2006	2007
Technician (or equivalent specialist grade)	£11.49	£12.01	£12.61
Approved Electrician (or equivalent specialist grade)	£10.06	£10.51	£11.04
Electrician (or equivalent specialist grade	£9.17	£9.58	£10.06
Labourer	£7.09	£7.41	£7.78

(ii) **London Rate**

for Operatives engaged upon work falling within the terms of JIB National Working Rule 6(b) will be:

Grade	2005	2006	2007
Technician (or equivalent specialist grade)	£12.87	£13.45	£14.12
Approved Electrician (or equivalent specialist grade)	£11.27	£11.77	£12.36
Electrician (or equivalent specialist grade	£10.27	£10.73	£11.27
Labourer	£7.94	£8.30	£8.71

Further details of the wage agreement, allowances and working Rules may be obtained on written application to the Joint Industry Board for the Electrical Contracting Industry, Kingswood House, 47/51 Sidcup Hill, Sidcup, Kent DA14 6HP.

THE NATIONAL JOINT COUNCIL FOR THE LAYING SIDE OF THE MASTIC ASPHALT INDUSTRY

Rates of wages as approved by the National Joint Council for the Laying Side of the Mastic Asphalt Industry.

Rates from Monday 21st June, 2004	Basic hourly rate (incl. GMPP) £	Guaranteed Basic Weekly Pay £
Chargehands Responsibility Allowance..............	10.50 (per day)	
Spreaders Basic (39 hours)	7.72 / hour	301.08 / week
Bonus	6.70 / hour	
Mixermen/Potmen Basic (39 hours)....................	7.14 / hour	278.46 / week
Bonus	6.21 / hour	
Classified Labourers Basic (39 hours)................	7.14 / hour	278.46 / week
Bonus	5.69 / hour	
Unclassified Labourers.......................................	Building Trade Labourer's Rate	

Apprentice/Trainee rates of pay

0-6 months	182.33 per week
Adult Trainees..	194.38 per week
6-12 months	197.53 per week minimum – 25% production
12-18 months	212.72 per week minimum – 40% production
18-24 months	227.91 per week minimum – 55% production
24-30 months	251.61 per week minimum – 70% production
30-36 months	275.92 per week minimum – 85% production
36 months +	300.98 – 100% production

Further information and details can be obtained from:-
Employers' Secretary, Mastic Asphalt Council Ltd, P O Box 77, Hastings. TN35 4WL
Tel: 01424 814400 www.masticasphaltcouncil.co.uk

THE JOINT INDUSTRY BOARD FOR PLUMBING MECHANICAL ENGINEERING SERVICES IN ENGLAND AND WALES

Basic hourly rates of pay inclusive of tool allowance shall be:

		from 2nd Dec 2002 £	from 5th January 2004 £	from 3rd January 2005
(a)	**Operatives**			
	Technical Plumber and Gas Service Technician	10.70	11.34	12.02
	Advanced Plumber and Gas Service Engineer	9.63	10.21	10.82
	Trained Plumber and Gas Service Fitter	8.25	8.75	9.28
(b)	**Apprentices**			
	4th year of training with NVQ level 3*	8.00		8.99
	4th year of training with NVQ level 2*..	7.24		8.13
	4th year of training............................	6.37	6.75	7.16
	3rd year of training with NVQ level 2*..	6.30		7.08
	3rd year of training............................	5.18	5.49	5.82
	2nd year of training...........................	4.59	4.87	5.16
	1st year of training............................	4.01	4.25	4.51
(c)	**Adult Trainees**			
	3rd 6 months of employment	7.21	7.64	8.10
	2nd 6 months of employment	6.91	7.32	7.76
	1st 6 months of employment	6.45	6.84	7.25

* Note Where apprentices have achieved NVQ's, the appropriate rate is payable from the date of attainment except that it shall not be any earlier than the commencement of the promulgated year of Training in which it applies.

WORKING HOURS AND OVERTIME

The normal working weeks shall be 37.5 hours
Overtime
45 hours at normal rates are to be worked (Monday to Friday) before overtime rates shall apply
Note overtime hours worked over time Monday to Friday shall be paid at time and a half up to 8.00pm.
After 8.00pm overtime shall be paid at double time.
Saturday at time and a half up to 1.00pm when double time then applies up until normal starting time on Monday
From Monday 2nd December 2002 the Guaranteed Minimum Payment shall no longer apply.

Further information can be obtained from The Joint Industry Board for Plumbing Mechanical Engineering Services in England and Wales, Brook House, Brook Street, St. Neots, Huntingdon, Cambs. PE19 2HW. Tel: 01480 476925 Fax: 01480 403081

Responsibility/Incentive Pay allowance

Since Monday 3rd September 2003, Employers may, In consultation with the employees concerned, enhance the basis graded rates of the payment of an additional amount, as per the brands shown below, where it is agreed that their work involves extra responsibility, productivity or flexibility.

Band 1	an additional rate of	24p per hour
Band 2	an additional rate of	44p per hour
Band 3	an additional rate of	64p per hour
Band 4	an additional rate of	84p per hour

This allowance forms part of an operative's basic rate of pay and shall be used to calculate premium payments. It is payable **EITHER** on a contract-by-contract basis **OR** on an annual review basis.

NATIONAL INSURANCES

Employers Contribution:

April 1992	4.6% of employee's earnings where £54.00 per week and not exceeding £89.99 per week. 6.6% of employee's earnings where £90.00 per week and not exceeding £134.99 per week. 8.6% of employee's earnings where £135.00 per week and not exceeding £189.99 per week. 10.4% of employee's earnings where £190.00 per week and over.
April 1993	4.6% of employee's earnings where £56.00 per week and not exceeding £94.99 per week. 6.6% of employee's earnings where £95.00 per week and not exceeding £139.99 per week. 8.6% of employee's earnings where £140.00 per week and not exceeding £194.99 per week. 10.4% of employee's earnings where £195.00 per week and over.
April 1994	3.6% of employee's earnings where £57.00 per week and not exceeding £99.99 per week. 5.6% of employee's earnings where £100.00 per week and not exceeding £144.99 per week. 7.6% of employee's earnings where £145.00 per week and not exceeding £199.99 per week. 10.2% of employee's earnings where £200.00 per week and over.
April 1995	3% of employee's earnings where £58.00 per week and not exceeding £104.99 per week. 5% of employee's earnings where £105.00 per week and not exceeding £149.99 per week. 7% of employee's earnings where £150.00 per week and not exceeding £204.99 per week. 10.2% of employee's earnings where £205.00 per week and over.
April 1996	3% of employee's earnings where £61.00 per week and not exceeding £109.99 per week. 5% of employee's earnings where £110.00 per week and not exceeding £154.99 per week. 7% of employee's earnings where £155.00 per week and not exceeding £209.99 per week. 10.2% of employee's earnings where £210.00 per week and not exceeding £445.00 per week. 10.2% of employee's earnings where £445.00 per week and over.
April 1997	3% of employee's earnings where £62.00 per week and not exceeding £109.99 per week. 5% of employee's earnings where £110.00 per week and not exceeding £154.99 per week. 7% of employee's earnings where £155.00 per week and not exceeding £209.99 per week. 10% of employee's earnings where £210.00 per week and not exceeding £465.00 per week. 10% of employee's earnings where £465.00 per week and over.
April 1998	3% of employee's earnings where £64.00 per week and not exceeding £109.99 per week. 5% of employee's earnings where £110.00 per week and not exceeding £154.99 per week. 7% of employee's earnings where £155.00 per week and not exceeding £309.99 per week. 10% of employee's earnings where £210.00 per week and not exceeding £485.00 per week. 10% of employee's earnings where £485.00 per week and over.
April 1999	12.2% of employee's earnings above the earnings threshold where £83.00 per Week or more
April 2000	12.2% of employee's earnings above the earnings threshold where £84.00 per Week or more
April 2001	11.9% of employee's earnings above the earnings threshold where £87.01 per Week or more
April 2002	11.8% of employee's earnings above the earnings threshold where £89.01 per Week or more
April 2003	12.8% of employee's earnings above the earnings threshold where £89.01 per Week or more
April 2004	12.8% of employee's earnings above the earnings threshold where £91.01 per Week or more
April 2005	12.8% of employee's earnings above the earnings threshold where £94.01 per week or more

CITB LEVY

The CITB levies applicable following The Industrial Training Levy (Construction Board) orders are as follows:

Date came into force	Payroll	Labour only
1st February 1994	0.25%	2%
13th March 1998	0.29%	2.28%
2nd February 1999	0.38%	2.28%
23rd February 2000	0.5%	2.28%
19th March 2001	0.5%	1.5%

Subsequent Industrial Training Levy (Construction Board) Orders keep the PAYE rate at 0.5% and the labour only subcontractor rate at 1.5%

BUILDERS AND CONTRACTORS PLANT

MECHANICAL PLANT

The following rates are inclusive of operator as noted, fuel, consumables,repairs and maintenance, but excludes transport.

	£	
Excavators including Driver and Banksman		
Hydraulic, full circle slew, crawler mounted with single equipment (shovel)		
0.70 m3	23.71	hour
1.00 m3	24.86	hour
1.50 m3	26.57	hour
Hydraulic, offset or centre post, half circle slew, wheeled, dual purpose (back loader)		
0.60 m3	21.43	hour
0.80 m3	21.43	hour
1.00 m3	22.29	hour
1.20 m3	22.57	hour
Mini excavators, full circle slew, crawler mounted with single equipment (back actor)		
1.4 tonne capacity	22.29	hour
2.4 tonne capacity	24.14	hour
3.6 tonne capacity	25.14	hour
Compactor including operator		
Vibrating plate		
90kg	9.71	hour
120kg	9.71	hour
176kg	9.86	hour
310kg	9.99	hour
Rollers including driver		
Road roller dead-weight		
2.54 tonnes	11.34	hour
6.10 tonnes	12.51	hour
8.13 tonnes	13.29	hour
10.20 tonnes	13.86	hour
Vibratory pedestrian operated		
single roller 550kg	10.00	hour
single roller 680kg	10.14	hour
twin roller 650kg	10.43	hour
twin roller 950kg	10.57	hour
twin roller 1300kg	10.71	hour
Self propelled vibratory tandem roller		
1.07 tonne	11.86	hour
2.03 tonne	12.29	hour
3.05 tonne	12.71	hour
6.10 tonne	13.43	hour
Pumps including attendant quarter time		
Single diphragm (excluding hoses)		
51mm	4.68	hour
76mm	4.88	hour
102mm	6.10	hour
Double diphragm (excluding hoses)		
51mm	4.88	hour
76mm	5.18	hour
102mm	6.40	hour
Suction or delivery hose, flexible, including coupling, valve and strainer, per 7.62m length		
51mm	0.22	hour
76mm	0.35	hour
102mm	0.53	hour
Compressors including driver quarter time		
Portable, normal delivery of free air at 7kg/cm3		
2.0m3/min	3.14	hour
4.0m3/min	3.14	hour
4.0m3/min	3.57	hour
Compressor tools with 15.3m hose (excluding operator)		
breaker including steels	0.71	hour
light pick including steels	0.71	hour
clay spade including blade	0.71	hour
Compressor tool sundries		
Muffler	0.10	hour
Concrete Mixers Including Driver		
Closed drum		
0.20/0.15m3	11.35	hour
0.30/0.20m3	12.15	hour
Closed drum with swing batch weighing gear		
0.40/0.30m3	14.35	hour
Mixer with batch weighing gear and power loading shovel		
0.30/0.20m3	14.35	hour
0.40/0.30m3	16.55	hour
Mortar Mixers Including Driver		
Mortar Mixer		
0.15/0.10m3	11.05	hour
Concrete Equipment including Driver/Operator		
Concrete pump, skid mounted, exclusive of piping		
20/26m3	34.40	hour
30/45m3	36.40	hour
46/54m3	38.40	hour
Piping per 3.05m length		
102mm diameter	0.35	hour
127mm diameter	0.40	hour
152mm diameter	0.45	hour
Vibrator poker		
petrol	12.40	hour
diesel	12.70	hour
electric	11.80	hour
air	11.50	hour
Vibrator, external type, clamp on		
small	11.50	hour
medium	12.40	hour
large	13.00	hour

MECHANICAL PLANT (Cont'd)

Lorries including Driver

	£	
Ordinary of the following plated gross vehicle weights		
3.56 tonnes	14.52	hour
5.69 tonnes	17.02	hour
8.74 tonnes	18.02	hour
10.16 tonnes	21.57	hour
Tipper of the following plated gross vehicle weights		
6.60 tonnes	16.02	hour
8.74 tonnes	19.27	hour
9.65 tonnes	21.57	hour
Vans, including driver, carrying capacity		
0.4 tonnes	9.52	hour
1.0 tonnes	10.52	hour
1.25 tonnes	11.02	hour

Dumpers including Driver

Manual gravity tipping 2 wheel drive		
750kg	11.43	hour
1000kg	11.51	hour
2 tonnes	11.63	hour
3 tonnes	11.71	hour
Hydraulic tipping 4 wheel drive		
2 tonnes	11.86	hour
3 tonnes	12.14	hour
4 tonnes	13.29	hour
5 tonnes	14.14	hour

Hoists

Scaffold type hoist excluding operator		
250 kg	95.00	week
Ladder type hoist excluding operator		
150 kg electric	120.00	week
Goods type hoist up to 10m height including operator, excluding erecting and dismantling		
mobile type 500 kg	355.00	week
Tied-in type 750 kg	375.00	week

CHARGES FOR HIRE

The following Trade List Prices have been provided by HSS Hire Shops, 25 Willow Lane, Mitcham, Surrey, CR4 4TS, Tel: 020 8260 3100 to whom acknowledgement is hereby given. All hire charges are per week, day rates available on application. Guide prices only and subject to alteration without notice. All carriage to and from site charged extra.

Builders' Ladders

		£
Extension Ladders (Aluminium rope separated)		
2 part, up to 6.2m	each	25.20
2 part, up to 9.0m	each	36.00
2 part, up to 11.0m	each	63.00
3 part, up to 13.7m	each	72.00
3 part, up to 16.0m	each	82.80
Roof Ladders		
5.9 single	each	37.80
4.6 single	each	36.00
7.6m double	each	41.40

Adjustable Steel Trestles, 4 boards wide (fixed leg)

	closed	extended		
no 1	0.5m	0.8m	each	2.70
no 2	0.8m	1.2m	each	2.70
no 3	1.1m	1.6m	each	2.70

Aluminium Work Towers

Span Towers (Multilevel Work Platforms) narrow width 0.85m x 1.8m (or 2.5m) and full width 1.45m x 1.8m (or 2.5m)		
2.2m height	each	70.20
4.2m height	each	102.60
6.2m height	each	135.00
8.2m height	each	165.60
10.2m height	each	196.20
Additional 1m	each	15.30

Stagings

Lightweight staging 600mm wide deck		
2.4m	each	27.00
3.6m	each	30.60
4.8m	each	41.40
6.0m	each	43.20
6.6m	each	46.80
Guard rail...2.4m long	each	18.00
Scissor Lifts (2.44 x 0.81 x 1.73m)		
Narrow Aisle up to 10m		319.50

PURCHASE PRICES

LIGHT ALLOY INDUSTRIAL LADDERS TO DIN 4568/EN 131

The following Trade Prices have been provided by Light Alloy Ltd., Dales Road, Ipswich, Suffolk, IP1 4JR, Tel 01473 740445, to whom acknowledgement is hereby given.

Single part ladders

Alloy ladder with rungs		
Ref. 41511 overall length 1.9m	each	54.65
Ref. 41512 overall length 2.5m	each	68.69
Ref. 41513 overall length 3.0m	each	81.97
Ref. 41514 overall length 3.6m	each	97.06
Ref. 41515 overall length 4.2m	each	114.95
Ref. 41516 overall length 4.7m	each	141.33
Ref. 41517 overall length 5.8m	each	186.55
Ref. 41518 overall length 7.0m	each	228.96
GRP ladder		
Ref. 41251 overall length 1.9m	each	146.99
Ref. 41253 overall length 3.0m	each	213.88
Ref. 41255 overall length 4.2m	each	292.08

LIGHT ALLOY INDUSTRIAL LADDERS TO DIN 4568/EN 131 (Cont'd)
PURCHASE PRICES (Cont'd)

Single part ladders (Cont'd)

	£
GRP ladder with aluminium alloy rungs	
Ref. 41151 overall length 1.9m..... each	116.84
Ref. 41153 overall length 3.0m..... each	179.03
Ref. 41155 overall length 4.2m..... each	243.08
Alloy ladder with double depth rungs	
Ref. 40451 overall length 1.8m..... each	84.80
Ref. 40452 overall length 2.4m..... each	104.59
Ref. 40453 overall length 2.9m..... each	127.20
Ref. 40454 overall length 3.5m..... each	149.80
Ref. 40455 overall length 4.0m..... each	173.36
Ref. 40456 overall length 4.6m..... each	206.34

Double sided step ladders

Alloy trestle steps	
Ref. 40311 vertical height 1.75m... each	132.85
Ref. 40312 vertical height 2.29m... each	163.95
Ref. 40313 vertical height 2.83m... each	201.63
Ref. 40314 vertical height 3.37m... each	240.26
Ref. 40315 vertical height 3.91m... each	280.78
Ref. 40316 vertical height 4.45m... each	339.19
Ref. 40317 vertical height 5.57m... each	479.59
GRP trestle steps	
Ref. 41256 overall height 1.82m.... each	309.05
Ref. 41257 overall height 2.35m.... each	386.31
Ref. 41258 overall height 2.89m.... each	474.87
Ref. 41260 overall height 3.95m.... each	662.37
Alloy trestle steps with double depth rungs	
Ref. 40461 vertical height 1.71m....each	184.67
Ref. 40462 vertical height 2.23m....each	235.55
Ref. 40463 vertical height 2.76m....each	288.31
Ref. 40464 vertical height 3.28m....each	337.32
Ref. 40465 vertical height 3.81m....each	411.74
Ref. 40466 vertical height 4.34m....each	489.01

Push-up and rope operated ladders

Alloy two part push-up ladder	
Ref. 40245 extended length 3.0m...each	104.59
Ref. 40246 extended length 4.1m...each	131.91
Ref. 40247 extended length 4.9m...each	170.53
Ref. 40248 extended length 6.1m...each	215.76
Ref. 40249 extended length 7.2m...each	257.22
Ref. 40214 extended length 8.3m...each	304.34
Ref. 40215 extended length 9.2m...each	375.95
Ref. 40216 extended length 10.3m.each	459.80
Ref. 40217 extended length 12.2m.each	574.74
Two part push-up ladder in GRP with aluminium alloy rungs	
Ref. 41163 extended length 4.9m...each	408.91
Ref. 41164 extended length 6.0m...each	471.10
Alloy two part rope operated ladder	
Ref. 40206 extended length 7.2m...each	340.14
Ref. 40207 extended length 8.3m...each	389.14
Ref. 40208 extended length 9.2m...each	465.44
Ref. 40209 extended length 10.3m.each	505.02
Ref. 40210 extended length 12.2m.each	613.38

Prices and full catalogue available on application.

RAMSAY INDUSTRIAL CLASS 1 LADDERS: BS 2037 1984

The following Trade Prices have been provided by Ramsay & Sons (Forfar) Ltd., 61, West High Street, Forfar, Angus, DD8 1BH, Tel: 01307 462255, to whom acknowledgement is hereby given.

Aluminium Ladders

Single (parallel sides)	length (m)	rungs	approx. weight (kg)		£
SA3. OMK	3.05	12	6	each	54.80
SA3. 5MK	3.55	14	7	each	63.00
SA4. OMK	4.05	16	8	each	71.60
SA4. 5MK	4.55	18	9	each	81.20
SA5. OML	5.05	20	12	each	98.40
SA5. 5MF	5.55	22	16	each	130.00
SA6. OMF	6.05	24	18	each	146.00

Double Extending (push up type)	closed height (m)	extended height (m)	rungs	approx. weight (kg)		
DE3. OML	3.05	5.55	24	13	each	140.00
DE3. 5ML	3.55	6.30	28	17	each	156.80
DE4. OMF	4.05	7.55	32	27	each	237.60
DE4. 5MF	4.55	8.55	36	30	each	264.00
DE5. OMF	5.05	9.30	40	33	each	288.00
DE5. 5MF	5.55	10.05	44	36	each	313.60
DE6. OMJ	6.05	11.05	47	52	each	480.00
DE6. 5MJ	6.55	11.75	51	56	each	514.40
DE7. OMJ	7.05	12.75	55	63	each	549.60

Non-skid rubber suction grips	per pair	14.40
Ropes and pulleys	Per Double Ladder	40.00

PURCHASE PRICES (Cont'd)
RAMSAY INDUSTRIAL CLASS 1 LADDERS: BS 2037 1984 (Cont'd)

Ramsay Standard Range Ladders

	closed height	extended height		approx. weight		
Triple Extending (push up type)	(m)	(m)	rungs	(kg)		£
TE4. OMF	4.05	10.30	47	42	each	396.00
TE4. 5MF	4.50	11.80	53	47	each	439.00
TE5. OMF	5.05	13.30	59	52	each	485.80
Triple Extending (rope operated)						£
TE6. OMJR	6.05	15.80	69	87	each	860.00
TE7. OMJR	7.05	18.80	81	101	each	1000.00

Non-skid rubber suction grips	per pair	14.40
Ropes and pulleys	Per Triple Ladder	61.60

Loft Ladders

AL1	2.29m to 2.54m	each	174.80
AL2	2.57m to 2.82m	each	183.20
AL3	2.84m to 3.10m	each	188.00
AL4	3.12m to 3.38m	each	204.40
AL5	3.40m to 3.66m	each	222.40

N.B. The above prices are inclusive of one safety handrail per ladder

Additional safety handrail	per ladder	16.80

Other sizes quoted for on request.

Aluminium Industrial Platforms

	height	approx. weight		
Single Sided	(m)	(kg)		
DPS 01	0.240	6	each	71.60
DPS 02	0.475	7	each	89.20
DPS 03	0.710	9	each	107.60
DPS 04	0.945	10	each	124.80
DPS 05	1.180	11	each	142.40
DPS 06	1.415	12	each	160.80
Double Sided				
DPD 02	0.475	8	each	133.20
DPD 03	0.710	10	each	147.20
DPD 04	0.945	12	each	160.80
DPD 05	1.180	15	each	174.00
DPD 06	1.415	17	each	188.00
DPD 07	1.645	19	each	200.80

Standard trade discounts available on all the above items.

GUIDE PRICES TO BUILDING TYPES PER SQUARE METRE

The prices throughout this section are ranged over the most commonly occurring average building prices and are based upon the total floor area of all storeys measured inside external walls and over internal walls and partitions. They exclude contingencies, external works, furniture, fittings and equipment, professional fees and VAT.

Building prices are influenced significantly by location, market, local conditions, size and specification. The prices given are intended only as an indicative guide to representative building types in a U.K. location and must be used with caution, professional skill and judgement. Prices outside the ranges stated can, of course, be encountered.

New Build

Industrial Buildings	per m2 £
Agricultural Shed	350-700
Factory, light industrial for letting	
incoming mains only	360-700
including lighting and power, toilets and basic office accommodation ...	400-800
for owner occupation	
including all services and basic accommodation ...	400-900
high tech: with controlled environment	500-2200
Garage/Showroom	450-1100
Workshop	400-1000
Warehouse/Store	350-780
Administrative Buildings	
Council Offices	880-1300
Magistrates Court	1000-1760
Offices, low rise (for letting)	
non air conditioned	770-1200
air conditioned	1000-1600
Offices, High rise (for owner occupation)	
air conditioned	1000-2000
Bank (branch)	1200-1700
Shops	450-920
Retail Warehouse	350-1600
Shopping Mall air-conditioned	1000-2000
Police Station	1150-1700
Ambulance Station.	710-1040
Fire Station	1140-1630

Health and Welfare Buildings	
General Hospital	1100-1750
Private Hospital	1200-1900
Day Hospital	1200-1800
Health Centre	860-1250
Hospice	1000-1460
Day Centre	970-1380
Home for the Mentally Ill	820-1220
Children's Home	880-1380
Home for the Aged	800-1200
Refreshment, Entertainment and Recreation Buildings	
Restaurant	1200-1700
Public House	920-1440
Kitchen with equipment	1300-2250
Theatres	1450-1850
Community Centre	780-1220
Swimming Pool (covered) ...	1450-3000
Sports Hall	700-1230
Squash Courts	740-940
Sports Pavilion	720-1250
Religious Buildings	
Church	1000-1600
Church Hall	900-1550
Educational, Cultural and Scientific Buildings	
Nursery School	950-1550
Primary School	830-1300
Middle School	680-1110
Secondary School	770-1230
Special School	850-1260
Universities	900-1500
College	830-1250
Training Centre	800-1200

Laboratory building	1000-1750
Computer Building	1200-2720
Libraries	950-1400
Residential Buildings	
Local Authority and Housing Association Schemes	
low rise flats	650-870
4/5 person two storey houses	470-820
Private Developments	
low rise flats	650-900
4/5 person two storey houses	500-900
Private Houses, detached...	760-1890
Sheltered Housing	
Bungalows	670-960
low rise	670-910
flats with lifts and Warden's accommodation, etc.....	770-980
Student's Residence	860-1400
Hotel	920-2000
Refurbishment	
Factories	190-880
Offices (basic refurbishment)	380-1170
Offices (high quality including air conditioning)	1000-1800
Banks	610-1200
Shops	400-1400
Hospitals	640-1270
Public Houses	600-1140
Theatres	720-1600
Churches	430-1120
Schools	330-930
Housing & Flats	220-1200
Hotels	530-1100

COST ALLOWANCES

HOUSING ASSOCIATION SCHEMES

The following is based on the circular for Total Cost Indicators, grant rates and administration allowances for 2004/2005 and the Guidance notes for 2004/2005 and 2005/2006 effective from 1st April 2004 reproduced here in part by kind permission of the Housing Corporation.

Tables in full are no longer included in Laxton's. Those specifically requiring this detail may view the tables on-line at www. www.housingcorplibrary.org.uk

CIRCULAR 09/03

Advises registered social landlords of the Total Cost Indicators, discount amounts, and administration allowances that will apply from 1 April 2004. This circular will, from 1 April 2004, replace circular F2-22/02 24/10/2003
Note: 'housing association' is used as a generic term for registered social landlords.
'Corporation' means Housing Corporation.
This circular is for the attention of:
Registered Social Landlords; more than 250 properties
Registered Social Landlords; less than 250 properties
Local Authorities
Registered Social Landlords' Auditors

1 Legislation, Regulation and Policy Requirements

n/a

2 Definitions

TCIs are the Housing Corporation's estimate of the norm total cost of providing different types of housing in different parts of the country
The TCI system is the Housing Corporation's benchmark for assessing the value for money of estimated scheme costs at the point at which grant is formally confirmed.

3 Corporation Expectations

Introduction

The Total Cost Indicators, value limits, discount amounts and administration allowances which will apply to registered social landlord (RSL) schemes sponsored by the Housing Corporation or by local authorities and receiving grant confirmation on or after 1 April 2004 have now been determined. The grant rate calculator calculates grant rates for individual schemes. Details of and guidance on Total Cost Indicators, grant rate calculator and administration allowances for 2004/2005 and 2005/2006, effective from 1 April 2004, are available on the Corporation's website at:
http://www.housingcorplibrary.org.uk/housingcorp.nsf/Programmes/A4.

Total Cost Indicators (TCIs)

TCIs are the Housing Corporation's estimate of the norm total cost of providing different types of housing in different parts of the country.
The TCI system is the Housing Corporation's benchmark for assessing the value for money of estimated scheme costs at the point at which grant is formally confirmed.
The Corporation's discretion to approve schemes continues to be limited to 110% TCI.
The changes from the 2003/2004 TCIs are set out below:

The TCI framework outlined in the TCI guidance notes and accompanying tables form the basis of assessment for both 2004/2005 and 2005/2006. No intermediate adjustments are planned.

The 2003/2004 TCI table levels have been retained as the basis for future adjustments. Cost groupings will remain unchanged and TCI base tables, key and supplementary multipliers and key and supplementary on-costs areat unchanged levels. However, specific uplifts of between 5% and 15% apply to the TCI base tables for each individual local authority.
The categories of Homebuy have been expanded. Value limits no longer apply to Homebuy categories.
Administrative allowance and expenses remain unchanged and are now fixed at levels set for 2003/2004.

Grant rate calculator

The grant rates as determined by the new grant rate calculator represent for each type of scheme in each area the maximum percentage of total qualifying scheme costs that may be funded with public capital subsidy. The grant rate calculator is available on the Corporation's website
The grant rate calculator has been adapted this year to enable both social rent and intermediate rent grant rate calculations using one and the same model. The differences between the calculations is explained in the Guide to the Allocation Process available on the website.

LCHO, Homebuy, VPG, RTA

For LCHO grant rates see the grant rate calculator on the website.

The composition of the VPG/RTA cost groupings remain as 2003/4. The number of cost groups remains at six.

Discount amounts for the Voluntary Purchase Grant (VPG) and the Right to Acquire (RTA) schemes remain unchanged.

The categories of Homebuy have been expanded. Value limits no longer apply to Homebuy categories.

THE HOUSING CORPORATION (Cont'd)

Administration allowances

These allowances remain the same as for 2003/2004 and are available on the Corporation's website

Limit on the management element of the service charge in leasehold schemes for the elderly (LSE)
The limit for 2004/2005 will be announced in due course

Further Information

The Housing Corporation publishes details of the methodology and data sources for TCI reviews and grant rate modelling. This is available on the Corporation's website
The details will be amended shortly to take account of the change in approach adopted for 2004/2005 and 2005/2006.

4 Assessing Compliance

The application of the guidance associated with this circular and the information provided by RSLs will be subject to audit by the Housing Corporation.

5 Enquiries about this circular should be directed to:

The appropriate office of the Housing Corporation.

THE HOUSING CORPORATION

Total Cost Indicators, grant rates and administration allowances 2004/2005 and 2005/2006.

Guidance notes effective from 1st April 2004

FUNDAMENTAL CHANGES FROM 2003/2004 TCI

Two year applicability

The Corporation has invited new bids for the two-year period 2004/2005 and 2005/2006. Therefore the TCI framework outlined in these TCI guidance notes and the accompanying TCI tables will form the basis of assessment for both 2004/2005 and 2005/2006. No intermediate adjustments are planned.

TCI enhancements for individual local authorities

Following detailed discussions with Government Departments, including H M Treasury, concerning the need for additional outputs and the desire to maintain a more stable cost framework, it has been decided to retain the 2003/2004 TCI tables as the basis for future adjustments. Cost groupings will remain unchanged and TCI base tables, key and supplementary multipliers and key and supplementary on-costs are published at unchanged levels. The opportunity has been taken to correct a number of minor errors contained in the 2003/2004 TCI tables, but otherwise the tables remain as last year.

However, to take account of forecast cost movements in 2004/2005, that may differ not only by cost group but also by local authority, the TCI base tables 1.4 and 1.5 are subject to specific uplifts of between +5% and +15%. The relevant enhancement percentages applicable to individual local authorities are contained within table 1.12.

Homebuy

The principle changes are as follows:
1. Value Limits no longer apply to Homebuy purchases
2. Two new categories of Homebuy have been introduced to complement the traditional Homebuy option, namely:
   - Key Worker Homebuy (Market Purchase).
   - London Challenge Key Teacher Homebuy (Market Purchase)
3. Following piloting within the 2003/2004 Challenge Fund programme, a related option of Newbuild Homebuy has been mainstreamed to operate within the TCI framework.
4. The eligibility criteria, operating procedures and allowances vary according to the category.

VPG and RTA discount levels

The discount levels remain unchanged. The relevant cost group headings have been amended to reflect the fact that the unchanged six cost groups now apply only to VPG/RTA discount levels and no longer apply to Homebuy.

LA DIYSO

Withdrawn.

Administrative allowances and expenses

These remain unchanged and fixed. The local authority enhancements to the TCI base tables do not apply to administrative allowances and expenses.

EXPLANATION OF TCI

A key objective of the funding system is to achieve value for money in return for grant, and to ensure the correct level of grant is paid. TCIs form the basis of this system, and are divided into unit type and cost group area categories.

TCIs apply equally to units funded with Social Housing Grant (SHG) by the Housing Corporation, or those sponsored by a local authority.

TCIs represent the basis for a cost evaluation of SHG funded units. TCIs are also used to calculate the maximum level of grant or other public subsidy payable.

Key and supplementary multipliers are applied to the base TCIs to allow for scheme variations as outlined in the multiplier tables (tables 1.6 to 1.9 and table 1.12).

TYPES OF ACCOMMODATION

Different types of accommodation other than self contained housing for general needs are classified as follows:-

Accommodation for older people

i) Category 1 - self-contained accommodation for the more active older person, which may include an element of support and/or additional communal facilities;

ii) Category 2 - self-contained accommodation for the less active older person, which includes an element of support and the full range of communal facilities;
The term 'sheltered' is used generally to describe Category 1 and Category 2 schemes;

iii) Frail older people – extra care accommodation, which may be either shared or self-contained, for the frail older person. Includes the full range of communal facilities, plus additional special features, including wheelchair user environments and supportive management.

THE HOUSING CORPORATION (Cont'd)
Total Cost Indicators, grant rates and administration allowances 2004/2005 and 2005/2006. (Cont'd)

Shared Accommodation

Accommodation predominantly for single persons, which includes a degree of sharing between tenants of some facilities (e.g. kitchens, bathrooms, living room) and may (optional) include an element of support and/or additional communal facilities.

Supported housing

Accommodation, which may be either shared or self-contained, designed to meet the needs of particular user groups for intensive housing management (see the Housing Corporation's publication The Guide to Supported Housing) Such accommodation may also include additional communal facilities.

Accommodation for wheelchair users

Accommodation, which may be either shared or self-contained, designed for independent living by people with physical disabilities and wheelchair users. Where such accommodation is incorporated within schemes containing communal facilities, these facilities should be wheelchair accessible.

Communal Facilities

Ancillary communal accommodation, the range of which comprises:

i) Common room - consisting of common room/s of adequate size to accommodate tenants and occasional visitors, chair store and kitchenette for tea-making.

ii) Associated communal facilities - consisting of warden's office, laundry room and guest room.

TREATMENT OF COMBINED SUPPORTED HOUSING AND GENERAL NEEDS SCHEMES

Arrangements exist which allow the combination of supported housing and general needs units within a single scheme. Further guidance with regards to supported housing schemes is set out under the heading "Selection of Supplementary Multipliers".

TEMPORARY HOUSING

Temporary Social Housing (TSH) Grant is a term used to describe SHG paid to registered social landlords (RSLs) to cover the cost of bringing properties into temporary use.

Properties are eligible for TSH Grant if they are available for use by the RSL for a period of time covered by a lease or licence for not less than 2 years and not more than 29 years

A capital grant contribution will be available towards initial acquisition costs or periodic lease charges up to a grant maximum. The maxima are based upon the Housing Corporation's own assessment of what constitutes a reasonable contribution to the capitalised lease value. This is calculated using capitalised lease premium factors.

The *TSH multipliers and capitalised lease premium factor tables for 2004/2005* (tables 1.9 and 1.10) are included in this guidance.

THE COMPOSITION OF TCI

TCI comprise the following elements:

Acquisition
i) Purchase price of land/property
Works
i) Main works contract costs (including where applicable adjustments for additional claims and fluctuations, but excluding any costs defined as on-costs) (see note below);

ii) Major site development works (where applicable). These include piling, soil stabilisation, road/sewer construction, major demolition;

iii) Major pre-works (rehabilitation) where applicable;

iv) Statutory agreements, associated bonds and party wall agreements (including all fees and charges directly attributable to such works) where applicable;

v) Additional costs associated with complying with archaeological works and party wall agreement awards (including all fees, charges and claims attributable to such works) where applicable;

vi) Home loss and associated costs. This applies to new build only;

vii) VAT on the above, where applicable.

On-costs
i) Legal fees, disbursements and expenses;

ii) Stamp duty;

iii) Net gains/losses via interest charges on development period loans;

iv) Building society or other valuation and administration fees;

v) Fees for building control and planning permission;

vi) Fees and charges associated with compliance with European Community directives, and the Housing Corporation's requirements relating to energy rating of dwellings, EcoHomes certification and Housing Quality Indicators;

vii) In house or external consultants' fees, disbursements and expenses (where the development contract is a design and build contract) (see note below);

viii) Insurance premiums including building warranty and defects/liability insurance (except contract insurance included in works costs);

ix) Contract performance bond premiums;

x) Borrowing administration charges (including associated legal and valuation fees);

xi) An appropriate proportion of the RSL's development and administration costs (formerly Acquisition and Development allowances), excluding Co-operative Promotional Allowance (CPA) and Special Projects Promotion Allowances (SPPA) and including an appropriate proportion of any abortive scheme costs;

xii) Furniture, loose fittings and furnishings;

xiii) Home loss and disturbance payments for rehabilitation;

xiv) Preliminary minor site development works (new build), pre-works (rehabilitation), and minor works (off the shelf) and minor works and repairs in connection with existing satisfactory purchases;

xv) Marketing costs - for sale schemes only;

xvi) Post completion interest - for sale schemes only;

xvii) Legal, administrative and related fees and costs associated with negotiating and arranging leases – for TSH only;

xviii) VAT on the above, where applicable.

THE HOUSING CORPORATION (Cont'd)
Total Cost Indicators, grant rates and administration allowances 2004/2005 and 2005/2006. (Cont'd)

Note:
Where the development contract is design and build, the on-costs are deemed to include the builder's design fee element of the contract sum. Therefore the amount included by the builder for design fees should be deducted from the works cost element submitted by the RSL to the Housing Corporation.

Similarly, other non-works costs that may be included by the builder such as fees for building and planning permission, building warranty and defects/liability insurance, contract performance bond and energy rating of dwellings should also be deducted from the works cost element submitted by the RSL to the Housing Corporation.

The Housing Corporation will subsequently check compliance through its compliance audit framework.

EXPLANATION OF ON-COSTS

TCIs are inclusive of on-costs contained in the relevant on-cost table. The on-costs vary according to TCI cost group and the general purpose of the scheme. TCI levels are set with the assumption that the RSL's development and administrative costs will be contained within the percentages in the relevant on-cost table.

In order to allow a proper comparison between the total eligible costs and the relevant TCI it is necessary to add the percentage on-cost (from the relevant table) to the estimated eligible final costs of acquisition and works.

Major repairs schemes and adaptation schemes are not measured against TCI. However, on-costs (from the relevant table) should be added to such schemes. Supplementary on-costs may not be used in any circumstances in connection with major repairs or adaptation schemes.

SELECTION OF ON-COSTS

One key on-cost will apply per scheme. To this should be added any appropriate supplementary on-cost.

Supplementary on-costs may be used when the accommodation is designed to meet the relevant standards set out in the Housing Corporation publication 'Scheme Development Standards' (April 2003) including the enhancements, applicable from April 2004, notified in the Guide to the Allocation process (GTAP) 2004/2005.

The appropriate key or supplementary on-cost is determined by the predominant dwelling type in a scheme. Predominance is established, where necessary, by the largest number of persons in total. Where two key on-costs or two supplementary on-costs are equally applicable (e.g. supported housing and shared), the higher should be used.

TRANCHES

For units developed under the Housing Act 1996, a set percentage of approved grant can be paid once a unit reaches certain key development stages. These grant payments are known as tranches.

The key stages are:

i) exchange of purchase contracts (ACQ);

ii) start on site of main contract works (SOS); this is deemed to be the date when the contractor took possession of the site/property in accordance with the signed main building contract;

iii) practical completion of the scheme (PC).

Where a *public subsidy* is given by way of discounted land, (acquisition public subsidy), the whole subsidy is deducted from the first tranche; any excess balance should be deducted from the second tranche. This is to ensure that grant is not paid in advance of need. In other circumstances any other public subsidy will be deducted from each tranche on a pro-rata basis.

In the case of *land inclusive packages* the first and second tranches are paid together at start on site stage.

Tranche details for Special Projects Promotion Allowances (SPPA) and Co-op Promotion Allowances (CPA) together with further guidance relating to tranches for management contracting arrangements can be found in the Capital Funding Guide.

The grant on outstanding mortgages for reimprovement schemes will be paid in accordance with the tranche percentage for the scheme.

All tranche payments may be paid directly to RSL's rather than via solicitors.

Table 1.1 KEY ON-COSTS 2004/05 and 2005/06 BY COST GROUP

	Key On-costs	A1 & A2 %	A3 %	A4 %	B1 %	B2 %	B3 %	C1 %	C2 %	C3 %	D1 %	D2 %	E1 %
		Cost Group											
a)	New Build												
	i) acquisition and works	11	12	12	12	13	13	14	15	15	16	16	16
	ii) off the Shelf	8	8	8	8	8	8	8	8	8	8	8	8
	iii) works only	17	17	17	17	17	17	17	17	17	17	17	17
b)	Rehabilitation												
	i) acquisition and works (vacant)	11	11	12	12	13	13	14	15	15	16	17	17
	ii) acquisition and works (tenanted)	13	14	14	15	16	16	17	18	19	20	21	21
	iii) existing satisfactory	9	9	9	9	9	9	9	9	9	9	9	9
	iv) purchase and repair	10	10	10	10	10	10	10	10	10	10	10	10
	v) RH works only (vacant)	22	22	22	22	22	22	22	22	22	22	22	22
	vi) RH works only (tenanted)	26	26	26	26	26	26	26	26	26	26	26	26
	vii) reimprovements	22	22	22	22	22	22	22	22	22	22	22	22
	viii) TSH (unimproved vacant)	21	21	21	21	21	21	21	21	21	21	21	21
	ix) TSH (improved vacant)	10	10	10	10	10	10	10	10	10	10	10	10
c)	Major repairs and miscellaneous works	25	25	25	25	25	25	25	25	25	25	25	25
d)	Adaptation works	13	13	13	13	13	13	13	13	13	13	13	13

THE HOUSING CORPORATION (Cont'd)

Total Cost Indicators, grant rates and administration allowances 2004/2005 and 2005/2006. (Cont'd)

Table 1.2 SUPPLEMENTARY ON-COSTS 2004/05 and 2005/06

	Supplementary on-costs (All Cost Groups)	Purchase & repair and Acquisition & Works %	Works Only and Reimprovements %	Off-the-Shelf and Existing Satisfactory %
a)	Sheltered with common room or communal facilities	2	3	1
b)	Frail older persons	4	6	3
c)	(i) Supported housing	4	6	3
	(ii) with common room or common facilities	5	7	4
d)	Shared	5	7	5
e)	Construction Clients Charter	1	1	0
f)	Housing for Sale	4	6	5
g)	Comunity self build	4	0	0
h)	i) TSH shared unimproved	7		
	ii) TSH shared improved	5		

Table 1.3 GRANT PAYMENTS FOR TRANCHES 2004/05 and 2005/06

TRANCHE PERCENTAGES

		Acquisition and Works			Works Only & Reimprovement		Off-the-shelf/ existing satisfac.	Purchase & repair	
Scheme type	**Cost groups**	**(i)**	**(ii)**	**(iii)**	**(i)**	**(ii)**	**(i)**	**(i)+(ii)**	**(iii)**
Housing for rent									
Mixed funded: New Build	ALL	40	40	20	65	35	100		
Mixed funded Rehabilitation	ALL	50	30	20	60	40	100	80	20
Supported housing: New Build	As,Bs and Cs	30	35	35	50	50	100		
Supported housing: New Build	Ds and Es	20	35	45	45	55	100		
Supported housing: Rehabilitation	As,Bs, and Cs	50	20	30	40	60	100	80	20
Supported housing: Rehabilitation	Ds and Es	45	20	35	35	65	100	80	20
TSH unimproved					65	35			
TSH improved							100		
Housing for sale									
Mixed funded: New build and rehabilitation	ALL	50	45	5	95	5	100	95	5

THE USE OF THE TCI BASE TABLE

The unit size in square metres shown in the *TCI base table 2003/2004; self-contained accommodation* (table 1.4) relates to the total floor area of the unit. The probable occupancy figure in the tables is only a guideline figure. The number of occupants is derived from the total number of bedspaces provided.

The TCI for a unit where the total floor area exceeds 120m2 will be the cost of a unit of 115-120m2 plus for each additional 5m2 or part thereof, the difference between the cost of a 110-115m2 unit and the cost of a 115-120m2 unit. For practical purposes the unit size calculation limit has been set at a 345-350m2 unit.

In selecting the appropriate TCI floor area band the actual floor area should first be rounded to the nearest whole number. This rounded number must be used to select the TCI floor area band.

For example

actual floor area = 40.2m²
actual floor area rounded = 40m²
selected floor area band = >35 – 40m²

You must not make the wrong assumption that 40.2m² falls into the range ">40 – 45m² because 40.2m²" must be first rounded to 40m² which falls into the range ">35 – 40m².

For Self-Contained Accommodation

Self contained units provide each household, defined as a tenancy, with all their basic facilities behind their own lockable front door.

For self-contained units the base TCI is determined by its total floor area and the cost group in which it is located. The dwelling floor area is determined by the area of each unit for the private use of a single household. Communal areas or any facilities shared by two or more households should be excluded. The total floor area of self-contained accommodation is measured to the finished internal faces of the main containing walls on each floor of the accommodation and includes the space, on plan, taken up by private staircases, partitions, internal walls (but not `party' or similar walls), chimney breasts, flues and heating appliances. It includes the area of tenant's internal and/or external essential storage space.
It excludes:

i) any space where the height to the ceiling is less than 1.5m (e.g. areas in rooms with sloping ceilings, external dustbin enclosures);

ii) any porch, covered way, etc, open to air;

iii) all balconies (private, escape and access) and decks

iv) non-habitable basements, attics, thermal buffer zones, conservatories, or sheds;

v) external storage space in excess of 2.5m2;

vi) all space for purposes other than housing (e.g. garages, commercial premises, etc).

The TCI for Frail older persons dwellings should always be calculated as *self-contained* units even if they have some characteristics which are more typical of shared accommodation.

THE HOUSING CORPORATION (Cont'd)

Total Cost Indicators, grant rates and administration allowances 2004/2005 and 2005/2006. (Cont'd

For Shared Accommodation

Shared accommodation is defined as one household (i.e. one tenancy or licence) which shares facilities (i.e. bathroom, kitchen) with other households. Each household sharing such accommodation may comprise more than one person. The base TCI for shared accommodation should be calculated separately to any self-contained accommodation in the scheme. The base TCI for shared accommodation is calculated on a per bed space basis and may include any staff with a residential tenancy for shared accommodation. Staff sleepover accommodation which is not subject to a tenancy is not regarded as a bedspace for TCI purposes. The relevant base TCI cost for the cost group from the TCI *base table* 2004/2005; *shared accommodation* (table 1.5) is used.

For TSH shared accommodation

The relevant floor area, for TCI and capitalised lease premium calculation purposes only, should be taken as the overall building gross floor area measured to the finished internal faces of the main containing walls all as otherwise described for self-contained units above, divided by the number of people sharing. This calculation will result in the correct band size per person sharing which will then be used to select the base figure from the TCI base table: self-contained accommodation (table 1.4). This figure should then be multiplied by the number of people sharing prior to applying the relevant TSH key and supplementary multipliers or capitalised lease premium factor.

Table 1.4 BASE TABLE 2004/05 and 2005/06: SELF-CONTAINED ACCOMMODATION

Total unit costs		ALL SELF-CONTAINED ACCOMMODATION (including all frail older persons and TSH) £ per unit Cost Group											
Unit floor area m2	Probable occupancy (Persons)	A1 & A2	A3	A4	B1	B2	B3	C1	C2	C3	D1	D2	E1
Up to 25m^2	1	78,400	70,100	65,200	60,600	57,300	53,100	49,900	47,300	43,900	41,800	39,700	36,900
Exceeding/not exceeding													
25/30m^2	1	89,200	79,300	73,400	67,800	63,900	59,100	55,200	52,100	48,300	45,600	43,200	40,100
30/35m^2	1 and 2	100,000	88,500	81,600	75,100	70.600	65,100	60,600	56,900	52,600	49,400	46,600	43,200
35/40m^2	1 and 2	110,800	97,600	89,800	82,400	77,300	71,100	65,900	61,700	56,900	53,300	50,100	46,400
40/45m^2	2	121,600	106,800	98,000	89,700	84,100	77,000	71,200	66,400	61,200	57,100	53,600	49,600
45/50m^2	2	132,400	116,000	106,200	97,000	90,600	83,000	76,600	71,200	65,500	60,900	57,100	52,800
50/55m^2	2 and 3	143,200	125,100	114,500	104,300	97,300	89,000	81,900	76,000	69,800	64,700	60,600	55,900
55/60m^2	2 and 3	154,000	134,300	122,700	111,600	104,000	95,000	87,200	80,800	74,100	68,500	64,100	59,100
60/65m^2	3 and 4	164,800	143,500	130,900	118,900	110,600	101,000	92,500	85,500	78,400	72,300	67,600	62,300
65/70m^2	3 and 4	175,600	152,600	139,100	126,200	117,300	106,900	97,900	90,300	82,700	76,100	71,100	65,500
70/75m^2	3, 4 and 5	186,400	161,800	147,300	133,500	124,000	112,900	103,200	95,100	87,000	79,900	74,500	68,600
75/80m^2	3, 4 and 5	197,200	171,000	155,500	140,800	130,700	118,900	108,500	99,900	91,300	83,700	78,000	71,800
80/85m^2	4, 5 and 6	208,000	180,100	163,700	148,100	137,300	124,900	113,900	104,600	95,600	87,600	81,500	75,000
85/90m^2	4, 5 and 6	218,800	189,300	171,900	155,400	144,000	130,900	119,200	109,400	99,900	91,400	85,000	78,100
90/95m^2	5 and 6	229,600	198,500	180,200	162,700	150,700	136,800	124.500	114,200	104,200	95,200	88,500	81,300
95/100m^2	5 and 6	240,400	207,600	188,400	169,900	157,400	142,800	129,800	119,000	108,500	99,000	92,000	84,500
100/105m^2	6 and 7	251,200	216,800	196,600	177,200	164,000	148,800	135,200	123,700	112,800	102,800	95,500	87,700
105/110m^2	6 and 7	262,000	226,000	204,800	184,500	170,700	154,800	140,500	128,500	117,100	106,600	99,000	90,800
110/115m^2	6, 7 and 8	272,700	235,100	213,000	191,800	177,400	160,800	145,800	133,300	121,400	110,400	102,400	94,000
115/120m^2	6, 7 and 8	283,500	244,300	221,200	199,100	184,100	166,700	151,200	138,100	125,700	114,200	105,900	97,200

Note:
In selecting the appropriate TCI floor area band the actual floor area should first be rounded to the nearest whole number.
This rounded figure must then be used to select the TCI floor area band.
For example:
Actual floor area = 40.2M^2
Actual floor area rounded = 40.2M^2
Selected floor area band = >35-40M^2
You must not make the wrong assumption that 40.2m^2 falls into the range because 40.2m^2 must be first rounded to 40m^2 which falls in the band 35-40m^2.

Table 1.5 TCI BASE TABLE 2004/05 and 2005/06: SHARED ACCOMMODATION

	All shared accommodation (except frail older person and TSH) £ per person sharing Cost group											
TCI per person	A1 & A2	A3	A4	B1	B2	B3	C1	C2	C3	D1	D2	E1
Each person bedspace	115,400	100,000	90,900	82,400	76,400	69,600	63,500	58,500	53,400	49,000	45,700	42,000

SELECTION OF KEY MULTIPLIERS

Only one key multiplier can be used per unit. (See table 1.6)
New build acquisition and works is the basic key multiplier, hence its neutral value.
The *off-the-shelf* multiplier is used where new dwellings to a standard suitable for social housing letting are purchased, following inspection, from contractors/developers or their agents. The cost of any minor works required should be set against the on-cost allowance.

The *existing satisfactory* multiplier is used where existing dwellings of a standard and in a condition suitable for social housing letting are purchased, following inspection, from the second hand property market. The cost of any minor works required should be set against the on-cost allowance.

THE HOUSING CORPORATION (Cont'd)
Total Cost Indicators, grant rates and administration allowances 2004/2005 and 2005/2006. (Cont'd)
SELECTION OF KEY MULTIPLIERS (Cont'd)

The *purchase and repair* multiplier is used where existing dwellings are purchased, following inspection, from the property market which necessitate a degree of repair to bring them to a standard and condition suitable for social housing but not full rehabilitation. Purchase and repair classification will apply where the estimated repair/improvement costs of each dwelling exceed £1,500 but are less than £10,000.

The *works only* multiplier is used for accommodation which involves the development of land or property already in the RSL's ownership and for which no acquisition costs (other than legal charges) apply.

The *reimprovement* multiplier is used for dwellings which have already received some form of public/grant subsidy and are now being rehabilitated. Reimprovement schemes are generally expected to be submitted no less than 15 years after a rehabilitation or 30 years after construction. It should be noted that the outstanding mortgage of the original works can be considered an eligible cost and will attract the grant rate applicable to the new scheme as a whole. The reimprovement key multiplier is linked to that of the Rehabilitation works only (vacant) multiplier rather than the rehabilitation works only (tenanted) multiplier in order to reflect the average cost of provision.

The TSH *improved and unimproved vacant* multiplier is used for accommodation with a lease of between 2 and 29 years.

Table 1.6 KEY MULTIPLIER 2004/2005 and 2005/2006 BY COST GROUP

Note: Only one of the following to be used.

	TCI calc. form	Cost group											
Scheme type	**Line No.**	**A1 &A2**	**A3**	**A4**	**B1**	**B2**	**B3**	**C1**	**C2**	**C3**	**D1**	**D2**	**E1**
a) New Build													
i) acquisition and works	0010	1.00	1.00	1.00	1.00	1.00	1.00	1.00	1.00	1.00	1.00	1.00	1.00
ii) off-the-shelf	0020	1.00	1.00	1.00	1.00	1.00	1.00	1.00	1.00	0.97	0.90	0.90	0.90
iii) works only	0090	0.46	0.50	0.53	0.58	0.60	0.64	0.68	0.73	0.76	0.83	0.86	0.89
b) Rehabilitation													
i) acquisition and works (vacant)	0100	1.18	1.16	1.17	1.17	1.16	1.15	1.15	1.15	1.15	1.14	1.15	1.15
acquisition and works (tenanted)	0105	1.08	1.08	1.08	1.08	1.07	1.07	1.07	1.06	1.07	1.09	1.09	1.09
ii) existing satisfactory	0120	1.08	1.24	1.17	1.20	1.15	1.07	1.01	0.97	0.95	0.94	0.94	0.94
iii) purchase and repair	0110	1.09	1.25	1.18	1.21	1.16	1.08	1.02	0.98	0.96	0.95	0.95	0.95
iv) works only (vacant)	0125	0.43	0.47	0.50	0.54	0.57	0.60	0.64	0.69	0.72	0.78	0.81	0.83
works only (tenanted)	0130	0.44	0.48	0.51	0.56	0.58	0.62	0.66	0.71	0.74	0.80	0.84	0.86
v) reimprovements	0135	0.43	0.47	0.50	0.54	0.57	0.60	0.64	0.69	0.72	0.78	0.81	0.83
c) TSH													
i) vacant improved	0030	1.18	1.16	1.17	1.17	1.16	1.15	1.15	1.15	1.15	1.14	1.15	1.15
ii) vacant unimproved	0035	1.18	1.16	1.17	1.17	1.16	1.15	1.15	1.15	1.15	1.14	1.15	1.15

SELECTION OF SUPPLEMENTARY MULTIPLIERS

Table 1.7 is used for acquisition and works, off-the-shelf, existing satisfactory and purchase and repair schemes. Works only and reimprovement schemes types use the table 1.8. None of these supplementary multipliers are applicable to TSH schemes. Supplementary multipliers for TSH schemes are in table 1.9.

Supplementary multipliers can be applied to new build and rehabilitation units when the accommodation is designed to meet the relevant standards set out in the Housing Corporation publication 'Scheme Development Standards'. (April 2003) including the enhancements, applicable from April 2004, notified in the Guide to the Allocation Process (GTAP) 2004/2005.

More than one supplementary multiplier can be used per unit. However certain combinations of multipliers are invalid. Multipliers for sheltered, frail older people, supported housing and extended families cannot be combined. The matrix (table 1.11) gives a comprehensive list of valid combinations of multipliers.

The main reason for combinations of multipliers being invalid is that the combination of those multipliers would lead to a duplication of the financial provision for the facilities accounted for in the multipliers. e.g. category 2 includes allowance for new lifts or single storey. This means that in these circumstances the other relevant multiplier does not apply i.e. a new lift or single storey multiplier cannot be used with a category 2 multiplier.

The *supported housing* multiplier and on-costs will only apply to schemes approved within the funding framework introduced in 1995. This multiplier should only be used where no Supported People Grant is required or where written confirmation has been received from the Administering Authority Supporting People Team that Supported People Grant is available for the scheme.

Supported housing schemes with shared facilities cannot be combined with either of the three common room multipliers in of the supplementary multiplier table.

The supported housing multiplier cannot be used for staff units.

The appropriate *shared* supplementary multiplier is determined by the total number of bedspaces provided within all of the households sharing facilities in the scheme. Where a cluster (i.e. more than one) of shared accommodation is provided the appropriate shared supplementary multiplier is determined by the total number of bedspaces contained within the households comprising each independent and self sufficient shared accommodation arrangement. The *TCI base table: shared accommodation* (table 1.5) is deemed to include all communal and ancillary facilities.

The shared multiplier does not apply to either sheltered or frail old people schemes.

The *extended families* multiplier is used when a self contained dwelling is to cater for eight or more persons and additional or duplicate facilities i.e. kitchen and/or sanitary fittings/equipment are provided. The additional space requirement is accounted for in the size band selection. Where significant additional facilities are not provided the RSL should contact the relevant regional office of the corporation to receive confirmation that the extended multiplier may be applied.

The *served by new lifts* multiplier is used when new vertical passenger lift provision is incorporated for access to dwelling entrances and communal accommodation. The multiplier does not apply to dwellings with entrances at ground floor level unless, exceptionally, the scheme includes essential communal accommodation provided at a level other than at ground floor level.

THE HOUSING CORPORATION (Cont'd)

Total Cost Indicators, grant rates and administration allowances 2004/2005 and 2005/2006. (Cont'd)

SELECTION OF SUPPLEMENTARY MULTIPLIERS (Cont'd)

The *wheelchair with individual carport* multiplier should be used only with individual self-contained dwellings designed in accordance with the relevant standards set out in the Housing Corporation publication Scheme Development Standards. Where non-individual self-contained wheelchair user dwellings are provided, e.g. on some category 2 schemes, or where a waiver for non-compliance with the carport provisions has been obtained from the regional office, the wheelchair without individual carport multiplier should be used. The *wheelchair* multipliers are intended to cover the costs of providing housing to suit a range of wheelchair users. The multipliers does not allow for the costs of any fixed additional equipment for people with disabilities; these needs should be met via the adaptations funding framework.

The *parent with children refuges* multiplier applies to shared accommodation specifically designed to meet the needs of parents with children, including vulnerable women with babies and women and children at risk of domestic violence. The TCI is calculated on a per bed space basis with children counted as full bedspaces. The provision of additional bunk-bedspaces should be disregarded for TCI calculation purposes.

The *housing for sale* multiplier is used for all sale schemes. It cannot be used with the supported housing or shared multipliers.

The *rehabilitation to pre-1919 properties* multiplier is used where the scope of the refurbishment or conversion work is carried out on a property or properties originally constructed prior to 1919. It cannot be used in connection with existing satisfactory or purchase and repair multipliers.

The *No VAT rehabilitation* multiplier is used when the scope of the refurbishment or conversion work is such that the relevant local Customs & Excise office determines that VAT is not chargeable on the Works.

The *rural housing* multiplier is used for schemes identified by rural investment codes R, G and F. Rural areas mainly comprise those with 1,000 or less inhabitants. Exceptionally, the limit may be extended to 3,000 inhabitants on a case by case basis by the relevant local office of the Housing Corporation. (See the Housing Corporation's Guide to the Allocations Process for details of population settlements, the Housing Corporation's Rural Settlements Gazetteer, last published in 1998)

The *national park* multiplier may be applied to any scheme within a formally designated national park or formally designated area of outstanding beauty. It may be used in conjunction with the rural housing multiplier. A listing of the recognised areas is set out in table 1.13.

The *sustainability* multiplier is intended to encourage RSLs towards greater sustainability. In order to qualify for the multiplier, two separate aspects must be addressed:

i) greening – in pursuit of the government's stated policy of increasing energy efficiency and reducing levels of CO_2 in the atmosphere, an 'EcoHome' rating of 'Good' (second level) must be certified. From April 2005 a rating of 'Very Good' will be required to qualify;

ii) Security – Secured by Design certification must be obtained for the scheme.

The *adaptability* multiplier is intended to encourage RSLs to design homes with lofts that may be readily adapted to provide additional habitable space. In order to qualify for the multiplier the following features must be present:

- a clear loft area of a size and height that will satisfy planning requirements to accommodate a single bedroom;
- loft floor joist sized for floor loading;
- loft floor trimmed for new staircase;
- landing layout and size that will accommodate a new access to the converted loft;
- a bedroom sized gable window, dormer window or opening rooflight.

The *construction clients' charter* multiplier is intended to recognise the additional expense involved in setting up and maintaining 'Chartered' client status as supported by Government and encouraged for all agencies and associated organisations in receipt of public funds. In order to qualify for the supplementary on-cost and associated multiplier, the RSL must be able to satisfy and comply with either the RSL group 'Clients Charter' or the new 'mini-charter' as appropriate. The mini-charter is intended for RSLs with small or occasional development programmes. Details of qualifying criteria and useful links are set out in the Guide to the Allocation Process(GTAP) 2004/2005. (see also www.clientsuccess.org).

The *P&R/ESP* multiplier is intended to address the extremes of any imbalance between base TCI and dwelling prices. The level of the ESP and P&R key multipliers are linked to the average TCI/dwelling price difference in each cost group. The supplementary multipliers adjusts TCIs at an individual local authority level. The P&R/ESP multiplier for each local authority area is et out in Table 1.12.

The *TSH* supplementary multiplier is used for all schemes with a lease length of between two and 29 years requiring works, (see table 1.9) and it can only be used with the TSH shared supplementary multiplier. It does not apply to improved TSH schemes.

Note: Where a RSL is in any doubt concerning the appropriateness of the application of any supplementary multiplier, it is advised to seek advice from the relevant local office of the Corporation. RSL's should also ensure that relevant information relating to the appropriateness of the application of supplementary multipliers is maintained on file for compliance audit purposes.

CALCULATION OF MAXIMUM GRANT CONTRIBUTION TO LEASE COSTS

The factors in table 1.10 are used to calculate maximum grant contributions for lease premiums outside the normal TCI framework. Whilst the factors are applied to the appropriate figures in the TCI base table and the relevant TSH key multiplier to determine the maximum contribution, any grant paid is additional to that paid in respect of works costs, which is subject to separate value for money assessment using the TCI multipliers The grant payment will be calculated according to the actual cost of the lease premium.

The above factors should be used only in conjunction with the *TCI Base table for 2004/2005: self contained accommodation* (table 1.4) and the relevant TSH key multiplier. The resultant value reflects the maximum grant contribution towards any capitalised lease premium payable to acquire the lease. No other factors or supplementary multipliers apply. The on-costs associated with setting up the lease are included within the TCI element of eligible costs. The different principles underlying the calculation of TCI for self-contained accommodation and for shared accommodation as outlined apply equally to the calculation of capitalised lease premiums.

Tables 1.7 to 1.11 are no longer included in Laxton's. Those specifically requiring this detail may view the tables on-line at www. www.housingcorplibrary.org.uk

Table 1.12 follows, however the LA Enhancement column is not shown and requires a percentage to be added to the figures indicated.

THE HOUSING CORPORATION (Cont'd)

Total Cost Indicators, grant rates and administration allowances 2004/2005 and 2005/2006. (Cont'd)

Table 1.12; TCI grouping, ESP and P&R supplementary multipliers and Homebuy and RTA groupings 2003/2004

In alphabetical order – commencing with the London Boroughs

LA code	LA name	TCI grouping	P&R multiplier	Homebuy grouping	RTA grouping
202	City of London	A1	1.00	H1	H1
203	Barking and Dagenham	B3	0.95	H3	H3
204	Barnet	A2	1.15	H1	H1
205	Bexley	A4	0.90	H1	H1
206	Brent	A3	1.05	H1	H1
207	Bromley	A2	1.00	H1	H1
208	Camden	A2	1.15	H1	H1
209	Croydon	A3	0.85	H1	H1
210	Ealing	A3	0.95	H1	H1
211	Enfield	A2	1.15	H1	H1
212	Greenwich	B1	1.00	H1	H1
213	Hackney	A3	1.00	H1	H1
214	Hammersmith	A2	1.15	H1	H1
215	Haringey	A3	1.00	H1	H1
216	Harrow	A3	1.00	H1	H1
217	Havering	B1	0.90	H2	H2
218	Hillingdon	A3	0.85	H1	H1
219	Hounslow	A2	1.00	H1	H1
220	Islington	A2	1.15	H1	H1
221	Kensington and Chelsea	A1	1.15	H1	H1
222	Kingston upon Thames	A4	1.15	H1	H1
223	Lambeth	A2	1.10	H1	H1
224	Lewisham	A4	1.00	H1	H1
225	Merton	A3	1.15	H1	H1
226	Newham	A4	0.90	H2	H2
227	Redbridge	B1	0.95	H1	H2
228	Richmond upon Thames	A3	1.10	H1	H1
229	Southwark	A2	1.00	H1	H1
230	Sutton	A3	0.85	H1	H1
231	Tower Hamlets	A4	1.15	H1	H1
232	Waltham Forest	B1	1.00	H2	H2
233	Wandsworth	A3	1.15	H1	H1
234	Westminster	A1	1.15	H1	H1
272	Adur	C2	1.15	H2	H2
638	Allerdale	D2	0.90	H6	H6
594	Alnwick	D2	1.00	H6	H6
482	Amber Valley	D1	0.85	H6	H6
273	Arun	B2	0.90	H3	H3
474	Ashfield	D2	0.90	H6	H6
251	Ashford	C1	1.00	H3	H4
316	Aylesbury Vale	B3	1.00	H2	H2
348	Babergh	C3	1.00	H4	H4
551	Barnsley	E1	0.90	H6	H6
639	Barrow-in-Furness	D1	0.85	H6	H6
355	Basildon	C1	1.00	H2	H2
279	Basingstoke and Deane	B1	0.95	H2	H2
475	Basset Law	D1	0.80	H6	H6
701	Bath and North East Somerset	B1	0.85	H3	H3
331	Bedford	C2	1.10	H3	H3
595	Berwick-upon-Tweed	D2	0.90	H6	H6
515	Birmingham	C2	1.00	H3	H3
451	Blaby	C3	0.95	H5	H5
624	Blackburn	D2	0.90	H6	H6
625	Blackpool	D1	0.85	H6	H6
596	Blyth Valley	E1	0.90	H6	H6
483	Bolsover	D2	0.90	H6	H6
606	Bolton	D1	0.90	H5	H6
460	Boston	D2	0.90	H6	H6
401	Bournemouth	B2	0.85	H3	H3
294	Bracknell Forest	B2	1.00	H2	H2
555	Bradford	E1	1.00	H5	H6
357	Braintree	C1	1.00	H3	H4
341	Breckland	D1	0.90	H5	H6
356	Brentwood	B2	1.05	H1	H1
531	Bridgnorth	C3	1.10	H3	H4
706	Brighton & Hove	B1	1.00	H1	H1
436	Bristol	B2	0.95	H2	H2
342	Broadland	C3	0.95	H5	H5
501	Bromsgrove	C2	0.85	H4	H4
321	Broxbourne	B1	1.00	H1	H1
476	Broxtowe	D1	0.95	H5	H6
626	Burnley	D2	0.90	H6	H6
607	Bury	D2	0.95	H5	H6
556	Calderdale	E1	0.90	H6	H6
335	Cambridge	A4	1.15	H1	H1
522	Cannock Chase	C3	0.85	H6	H6
252	Canterbury	B2	0.85	H3	H4
429	Caradon	D1	1.00	H4	H4
640	Carlisle	D2	1.00	H5	H5
430	Carrick	C3	1.15	H3	H3
597	Castle Morpeth	D2	1.00	H5	H5
358	Castle Point	C1	1.00	H3	H3
452	Charnwood	C2	0.90	H5	H5
359	Chelmsford	B3	1.00	H2	H2
441	Cheltenham	B3	1.05	H2	H2
311	Cherwell	B3	1.00	H3	H3
616	Chester	C3	1.00	H4	H4
484	Chesterfield	D1	0.90	H5	H5
581	Chester-le-Street	D2	0.90	H6	H6
484	Chichester	B1	0.95	H1	H2
318	Chiltern	B1	1.00	H1	H1
627	Chorley	C3	0.85	H5	H6
402	Christchurch	B2	0.90	H2	H2
360	Colchester	C2	1.00	H3	H3
617	Congleton	D2	1.00	H4	H4
641	Copeland	D2	0.90	H6	H6
467	Corby	D2	0.90	H6	H6
442	Cotswold	C1	1.15	H2	H2
516	Coventry	D1	1.00	H4	H4
560	Craven	D2	1.10	H3	H4
275	Crawley	B3	1.05	H2	H2
618	Crewe & Nantwich	D1	1.00	H5	H5
322	Dacorum	A3	1.00	H1	H1
582	Darlington	D1	0.85	H5	H5
253	Dartford	C1	1.00	H3	H3
468	Daventry	C2	0.95	H4	H4
485	Derby	D1	0.85	H6	H6
490	Derbyshire Dales	C3	1.00	H4	H4
583	Derwentside	D1	0.85	H6	H6
552	Doncaster	D1	0.85	H6	H6
254	Dover	C2	1.05	H3	H4
517	Dudley	C3	1.00	H5	H5
584	Durham	D1	1.00	H4	H5
585	Easington	D2	0.90	H6	H6
336	East Cambridgeshire	C1	1.05	H2	H3
419	East Devon	C1	1.00	H3	H4
408	East Dorset	B3	1.00	H2	H2
280	East Hampshire	B2	1.00	H2	H2
323	East Hertfordshire	B2	1.00	H1	H2
461	East Lindsey	D2	0.90	H6	H6
469	East Northamptonshire	C1	0.85	H5	H5
708	East Riding	D2	0.90	H6	H6
523	East Staffordshire	D2	1.00	H5	H6
266	Eastbourne	C1	1.15	H1	H1
281	Eastleigh	B3	1.00	H2	H3
642	Eden	D1	1.00	H4	H5
619	Ellesmere Port & Neston	D2	0.90	H6	H6
300	Elmbridge	B1	1.15	H1	H1
361	Epping Forest	B1	1.15	H1	H1
301	Epsom and Ewell	B2	1.15	H1	H1
486	Erewash	D1	0.90	H5	H6
420	Exeter	C1	1.00	H3	H4
282	Fareham	B3	1.00	H2	H3
337	Fenland	D1	1.00	H4	H5
349	Forest Heath	C2	1.00	H4	H4
443	Forest of Dean	C3	1.05	H3	H4
628	Flyde	C2	0.85	H5	H5
589	Gateshead	D2	0.95	H5	H5
477	Gedling	D1	0.95	H5	H5
444	Gloucester	C2	1.00	H3	H4
283	Gosport	C1	1.00	H3	H3
256	Gravesham	C2	1.00	H3	H3
343	Great Yarmouth	D2	0.95	H5	H6
302	Guildford	B1	1.10	H1	H1

THE HOUSING CORPORATION (Cont'd)

Total Cost Indicators, grant rates and administration allowances 2004/2005 and 2005/2006. (Cont'd)

Table 1.12 (Cont'd)

LA code	LA name	TCI grouping	ESP multiplier	Homebuy grouping	
620	Halton	C3	0.85	H6	H6
561	Hambleton	C3	1.00	H4	H4
453	Harborough	C1	0.85	H4	H4
362	Harlow	C1	1.15	H2	H2
562	Harrogate	C2	1.10	H3	H3
284	Hart	A4	1.00	H1	H1
577	Hartlepool	D2	0.90	H6	H6
267	Hastings	C3	1.15	H3	H3
285	Havant	B3	0.95	H3	H3
712	Herefordshire	C1	0.85	H4	H4
324	Hertsmere	A3	1.00	H1	H1
487	High Peak	C3	1.00	H4	H4
454	Hinckley and Boxsworth	C3	0.95	H5	H5
276	Horsham	B2	1.00	H1	H2
338	Huntingdonshire	C1	1.00	H3	H3
629	Hyndburn	E1	0.90	H6	H6
350	Ipswich	C2	1.00	H4	H4
707	Isle of Wight	C3	1.00	H4	H4
447	Isles of Scilly	A2	1.00	H1	H1
409	Kennet	B3	1.00	H3	H3
431	Kerrier	D2	1.10	H4	H4
470	Kettering	C2	0.85	H5	H5
347	King's Lynn and West Norfolk	D1	0.90	H5	H6
574	Kingston upon Hull	D2	0.90	H6	H6
557	Kirklees	D2	0.90	H5	H5
601	Knowsley	D2	0.90	H6	H6
630	Lancaster	C3	0.95	H5	H5
558	Leeds	D1	1.00	H4	H4
455	Leicester	C3	0.85	H5	H6
269	Lewes	B2	0.85	H3	H3
524	Lichfield	C2	1.00	H4	H4
462	Lincoln	D2	0.90	H5	H5
602	Liverpool	C3	0.85	H5	H5
332	Luton	C2	1.00	H3	H4
621	Macclesfield	C2	1.10	H3	H3
257	Maidstone	B3	1.00	H3	H3
363	Maldon	C2	1.10	H3	H3
713	Malvern Hills District	C1	1.00	H3	H3
608	Manchester	C2	1.00	H4	H4
478	Mansfield	D2	0.90	H6	H6
456	Melton	C2	0.85	H5	H6
414	Mendip	D1	1.00	H4	H4
333	Mid Bedfordshire	C1	1.10	H2	H3
425	Mid Devon	C2	1.00	H4	H4
351	Mid Suffolk	C3	1.00	H4	H4
277	Mid Sussex	B2	1.00	H1	H2
579	Middlesborough	D2	0.90	H6	H6
319	Milton Keynes	C1	0.95	H3	H4
303	Mole Valley	B2	1.15	H1	H1
286	New Forest	B2	1.00	H2	H3
479	Newark & Sherwood	D1	0.90	H5	H6
590	Newcastle upon Tyne	C3	1.00	H4	H5
525	Newcastle-under-Lyme	D2	0.90	H6	H6
432	North Cornwall	D1	1.15	H3	H4
421	North Devon	C2	1.00	H4	H4
403	North Dorset	C1	1.00	H3	H3
488	North East Derbyshire	D1	1.00	H5	H5
709	N.E. Lincolnshire	E1	0.90	H6	H6
325	North Hertfordshire	B3	1.15	H1	H2
463	North Kesteven	D1	0.85	H6	H6
710	North Lincolnshire	E1	0.90	H6	H6
344	North Norfolk	D1	0.95	H5	H5
532	North Shropshire	D2	1.00	H5	H5
440	North Somerset	C1	1.00	H2	H2
591	North Tyneside	D1	1.00	H5	H5
510	North Warwickshire	C3	1.00	H5	H5
457	North West Leicestershire	C3	0.85	H5	H6
410	North Wiltshire	B3	0.95	H3	H3
471	Northampton	C1	0.90	H4	H4
345	Norwich	C1	0.85	H4	H5
480	Nottingham City	C3	0.85	H5	H6
511	Nuneaton and Bedworth	C3	0.85	H5	H5
458	Oadby and Wigston	C3	1.00	H4	H5
609	Oldham	D1	0.85	H5	H6
533	Oswestry	D1	0.85	H6	H6
312	Oxford	A2	0.95	H1	H1
631	Pendle	D2	0.90	H6	H6
433	Penwith	D1	1.05	H3	H4
339	Peterborough	C3	1.00	H4	H5
422	Plymouth	C3	1.00	H4	H5
404	Poole	B2	0.85	H3	H3
287	Portsmouth	C1	1.00	H3	H3
632	Preston	C3	0.85	H5	H5
405	Purbeck	C1	1.00	H3	H3
296	Reading	A4	0.90	H1	H1
578	Redcar and Cleveland	E1	0.90	H6	H6
505	Redditch	C3	1.00	H4	H4
304	Reigate and Banstead	B2	1.10	H1	H1
434	Restormel	D1	1.05	H4	H4
633	Ribble Valley	C3	1.00	H4	H4
563	Richmondshire	D2	1.05	H4	H4
610	Rochdale	D1	0.90	H5	H5
364	Rochford	C1	1.00	H2	H3
634	Rossendale	D2	1.00	H5	H6
270	Rother	C1	1.00	H3	H3
553	Rotherham	E1	0.90	H6	H6
512	Rugby	C3	0.95	H5	H5
305	Runnymede	B2	1.05	H1	H1
481	Rushcliffe	C3	1.00	H4	H4
288	Rushmoor	B2	1.00	H1	H1
459	Rutland	C1	0.95	H4	H4
564	Ryedale	D1	1.00	H4	H4
611	Salford	D1	0.85	H6	H6
411	Salisbury	B3	1.00	H2	H2
518	Sandwell	D1	0.90	H5	H5
566	Scarborough	D2	0.95	H5	H5
586	Sedgefield	D2	0.90	H6	H6
415	Sedgemoor	D1	1.05	H4	H4
604	Sefton	D1	0.95	H5	H5
565	Selby	D2	1.10	H4	H4
259	Sevenoaks	B2	1.05	H1	H1
554	Sheffield	D2	1.00	H4	H4
260	Shepway	C3	1.05	H4	H4
534	Shrewsbury and Atcham	D1	1.00	H4	H4
297	Slough	B2	1.00	H1	H1
519	Solihull	B2	0.95	H2	H2
334	South Bedfordshire	C2	1.05	H3	H3
317	South Buckinghamshire	B1	1.10	H1	H1
340	South Cambridgeshire	B3	1.15	H1	H2
489	South Derbyshire	D2	0.90	H6	H6
702	South Gloucestershire	B2	0.85	H3	H3
423	South Hams	C3	1.15	H2	H3
464	South Holland	D2	0.90	H6	H6
465	South Kesteven	D1	0.95	H5	H5
643	South Lakeland	C3	1.00	H4	H4
346	South Norfolk	C3	0.95	H5	H5
472	South Northamptonshire	B3	0.85	H4	H4
313	South Oxfordshire	B2	1.00	H1	H2
635	South Ribble	C3	0.85	H5	H5
535	South Shropshire	D1	1.00	H4	H4
418	South Somerset	C2	1.00	H4	H4
526	South Staffordshire	C2	0.85	H5	H5
592	South Tyneside	D2	1.00	H5	H5
289	Southampton	B3	1.00	H3	H3
365	Southend-on-Sea	C2	1.00	H3	H3
306	Spelthorne	B2	1.15	H1	H1
326	St. Albans	A3	1.15	H1	H1
352	St Edmundsbury	C2	1.00	H3	H3
603	St. Helens	C3	0.85	H6	H6
527	Stafford	D1	0.90	H5	H5
528	Staffordshire Moorlands	D2	0.90	H6	H6
327	Stevenage	C1	1.15	H1	H2
612	Stockport	C1	1.00	H3	H3
580	Stockton-on-Tees	D2	0.90	H6	H6
529	Stoke-on-Trent	D2	0.90	H6	H6
513	Stratford-upon-Avon	B1	1.00	H1	H1
445	Stroud	C2	1.00	H4	H4
353	Suffolk Coastal	C3	1.00	H4	H4

THE HOUSING CORPORATION (Cont'd)

Total Cost Indicators, grant rates and administration allowances 2004/2005 and 2005/2006. (Cont'd)

Table 1.12 (Cont'd)

LA code	LA name	TCI grouping	ESP multiplier	Homebuy grouping	
593	Sunderland	D1	1.00	H5	H5
307	Surrey Heath	B2	1.05	H1	H1
261	Swale	C2	1.10	H4	H4
412	Swindon	C1	1.00	H3	H3
613	Tameside	D1	0.85	H6	H6
530	Tamworth	D1	1.00	H5	H5
308	Tandridge	B2	1.00	H1	H1
416	Taunton Deane	C1	1.00	H3	H3
587	Teesdale	D2	0.90	H5	H5
424	Teignbridge	C1	0.95	H4	H4
366	Tendring	D1	1.05	H3	H4
290	Test Valley	C1	1.00	H3	H3
446	Tewkesbury	C2	1.00	H4	H4
262	Thanet	C2	1.00	H3	H4
711	The Meadway Towns	C2	1.00	H3	H4
536	The Wrekin	D1	0.95	H5	H6
328	Three Rivers	A3	1.05	H1	H1
367	Thurrock	C2	1.10	H3	H3
263	Tonbridge and Malling	B2	1.00	H2	H2
426	Torbay	C2	1.00	H4	H4
427	Torridge	C3	1.00	H4	H4
614	Trafford	B3	0.85	H3	H4
264	Tunbridge Wells	B3	1.00	H2	H2
598	Tynedale	D1	1.00	H5	H5
368	Uttlesford	B2	1.00	H1	H2
314	Vale of White Horse	B3	1.00	H2	H2
622	Vale Royal	C3	1.00	H4	H4
559	Wakefield	D1	0.85	H5	H5
520	Walsall	C3	0.85	H5	H6
599	Wansbeck	D2	0.90	H6	H6
623	Warrington	C3	0.85	H6	H6
514	Warwick	B3	1.05	H1	H2
329	Watford	A3	0.90	H1	H1
354	Waveney	D2	0.95	H5	H5
309	Waverley	B1	1.10	H1	H1
271	Wealden	C1	1.15	H2	H2
588	Wear Valley	D2	0.90	H6	H6
473	Wellingborough	C3	0.90	H5	H5
330	Welwyn Hatfield	B1	1.00	H1	H1
295	West Berkshire	B1	0.95	H2	H2
428	West Devon	C3	1.15	H2	H3
406	West Dorset	B3	1.00	H3	H3
636	West Lancashire	C2	0.90	H5	H5
466	West Lindsey	E1	0.90	H6	H6
315	West Oxfordshire	B1	0.90	H2	H2
417	West Somerset	C3	1.00	H4	H4
413	West Wiltshire	C1	0.95	H3	H4
407	Weymouth & Portland	C1	0.85	H3	H3
615	Wigan	D1	0.90	H5	H6
291	Winchester	A4	1.00	H1	H1
298	Windsor and Maidenhead	B1	1.15	H1	H1
605	Wirral	D2	0.90	H6	H6
310	Woking	B1	1.00	H1	H1
299	Wokingham	B2	1.05	H1	H1
521	Wolverhampton	D2	0.90	H5	H6
507	Worcester City	C2	1.00	H3	H3
278	Worthing	C1	1.10	H2	H3
508	Wychavon	C2	1.00	H3	H3
320	Wycombe	B1	0.90	H2	H2
637	Wyre	C3	0.85	H5	H5
509	Wyre Forest	C2	0.90	H4	H5
567	York	C3	1.05	H3	H4

Table 1.13: National parks and areas of outstanding beauty in England

National parks

The Broads
Dartmoor
Exmoor
Lake District
The New Forest
North Yorkshire Moor
Northumberland
Peak District
Yorkshire Dales

Areas of outstanding beauty

Arnside & Silverdale
Blackdown Hills
Cannock Chase
Chichester Harbour
Chilterns
Cornwall
Cotswolds
Cranbourne Chase and West Wiltshire Downs
Dedham Vale
Dorset
East Devon
East Hampshire
Forest of Bowland
High Weald
Howardian Hills
Isle of Wight
Isles of Scilly
Kent Downs
Lincolnshire Wolds
Malvern Hills
Mendip Hills
Nidderdale
Norfolk Coast
North Devon
North Pennines
Northumberland Coast
North Wessex Downs
Quantock Hills
Shropshire Hills
Solway Coast
South Devon
South Hampshire Coast
Suffolk Coast and Heaths
Surrey Hills
Sussex Downs
Tamar Valley
Wye Valley

THE HOUSING CORPORATION (Cont'd)

The Housing Corporation's Website:-
www.housingcorp.gov.uk

The Housing Corporation's Offices

Corporate Services
Maple House
149 Tottenham Court Road
London
W1T 7BN
Tel. (020) 7393 2000
Fax. (020) 7393 2111
e-mail. enquiries@housingcorp.gsx.gov.uk

Regions:

London
Waverley House
7-12 Noel Street
London
W1F 8BA
Tel. (020) 7292 4400
Fax. (020) 7292 4401

North

North Eastern
Leeds
St Paul's House
23 Park Square South
Leeds
LS1 2ND
Tel. (0113) 233 7100
Fax. (0113) 233 7101

North West
Manchester
Elisabeth House
16 St Peter's Square
Manchester
M2 3DF
Tel. (0161) 242 2000
Fax. (0161) 242 2001

Merseyside
Liverpool
Colonial Chambers
3-11 Temple Street
Liverpool
L2 5RH
Tel. (0151) 242 1200
Fax. (0151) 242 1201

Central

Leicester
East
Attenborough House
109/119 Charles Street
Leicester
LE1 1FQ
Tel. (0116) 242 4800
Fax. (0116) 242 4801

West Midlands
Wolverhampton
31 Waterloo Road
Wolverhampton
WV1 4DJ
Tel. (01902) 795000
Fax. (01902) 795001

South

South East
Croydon
Leon House
High Street
Croydon
Surrey
CR9 1UH
Tel. (020) 8253 1400
Fax. (020) 8253 1444

South West
Exeter
Beaufort House
51 New North Road
Exeter
EX4 4EP
Tel. (01392) 428200
Fax. (01392) 428201

NATIONAL WORKING RULE AGREEMENT

The following are extracts from the Working Rule Agreement for the Construction Industry as approved by the Construction Industry Joint Council (Revised 30th June 2003)

WR.1 ENTITLEMENT TO BASIC AND ADDITIONAL RATES OF PAY

Operatives employed to carry out work in the Building and Civil Engineering Industry are entitled to basic pay in accordance with this Working Rule (W. R. 1) Rates of pay are set out in a separate schedule, published periodically by the Council.

Classification of basic and additional pay rates of operatives:

General operative
Skilled Operative Rate:- 4
3
2
1
Craft Operative.

1.1 **General Operatives**

1.1.1 General Operatives employed to carry out general building and/or civil engineering work are entitled to receive the "General Operatives Basic Rate of Pay".

Payment for Occasional Skilled Work

1.1.2 General Operatives, employed as such, who are required to carry out building and/or civil engineering work defined in Schedule 1, on an occasional basis, are entitled to receive the "General Operative Basic Rate of Pay" increased to the rate of pay specified in Schedule 1 for the hours they are engaged to carry out the defined work.

1.2 **Skilled Operatives**

1.2.1 Skilled Operatives engaged and employed whole time as such, who are required to carry out skilled building and/or civil engineering work defined in Schedule 1. on a continuous basis, are entitled to the "Basic Rate of Pay" specified in Schedule 1.

Additional payment for Skilled Work

1.2.2 In addition to the "Basic Rate of Pay" Skilled Operatives who are required to carry out skilled work defined in Schedule 1 are entitled to an additional payment in accordance with the following scale:

Skilled Operative Additional Rate iii
ii
I

1.3 **Craft operatives**

Craft operatives employed to carry out craft building and/or civil engineering work are entitled to receive the "Craft Operative Basic Rate of Pay".

1.4 **Additional payments for Intermittent Skill, Responsibility or Working in Adverse Conditions.**

1.4.1. While carrying out duties specified in Schedule 2 of the Working Rule Agreement, an operative shall be entitled, during the hour he is so engaged, to an additional payment as specified in Schedule 2.

1.5 **Conditions of employment of apprentices.**

1.5.1 **Conditions.**

1.5.2 **Wages**

. Rates of pay are set out in a separate schedule, published periodically by the Council. Payment under the schedule is due from the date of entry into employment as an apprentice, whether the apprentice is working on site or undergoing full-time training on an approved course, subject to the provisions of WR1.5.3. Payment under the scale is due from the beginning of the pay week during which the specified period starts.

1.5.3 **PAYMENT DURING OFF THE JOB TRAINING**

. Apprentices are entitled to be paid during normal working hours to attend approved courses off-the-job training in accordance with the requirement of their apprenticeship. payment during such attendance shall be at their normal rate of pay, but the employer may withhold payment for hours during which an apprentice, without authorisation fails to attend the course

1.5.4 **Overtime**

. The working of overtime by apprentices under 18 years of age shall not be permitted. Where an apprentice age 18 or over is required to work overtime payment shall be in accordance with WR.4.

1.5.5 **Daily Fare and Travel Allowances**

. The apprentice shall be entitled to fare and travel allowances in accordance with WR. 5.

1.5.6 **Other terms and conditions of Engagement.**

. The apprentice shall be subject to all other provisions and entitlements contained within the Working Rule Agreement.

WR.1 ENTITLEMENT TO BASIC AND ADDITIONAL RATES OF PAY (Cont'd)

Note: Normal hourly rate
The expression 'normal hourly rate' in this agreement means the craft, skilled or general operatives' weekly basic rate of pay as above, divided by the hours defined in Working Rule. 3 "Working Hours". Additional payments for intermittent skill or responsibility, occasional skilled work, work at heights, work in adverse conditions or bonus payments are not taken into account for calculating the "normal hourly rate".

WR.2 BONUS

It shall be open to employers and employees on any job to agree a bonus scheme based on measured output and productivity for any operation or operations on that particular job.

WR.3 WORKING HOURS

Working Hours
The normal working hours shall be:

Monday to Thursday	8 hours per day
Friday	7 hours per day
Total	39 hours per week

except for operatives working shifts whose working hours shall continue to be 8 hours per weekday. 40 hours per week.

The expression "normal working hours" means the number of hours prescribed above for any day (or night) when work is actually undertaken reckoned from the starting time fixed by the employer.

3.1 **Rest/Meal Breaks**

3.1.1. **Meal/Refreshment Breaks**

At each site or job there shall be a break or breaks for rest and/or refreshment at times to be fixed by the employer. The breaks, which shall not exceed one hour per day on aggregate, shall include a meal break of not less than a half an hour.

3.1.2. **Daily/Weekly Rest Breaks**

Where there are objective or technical reasons concerning the organisation of work the application of the Working Time Regulations 1998 – Regulations 10(1) Daily Rest Period – 11(1) and 11(2) Weekly Rest Period is excluded.

3.2. **Average Weekly Working Hours**

The average working time for each 7 day period will be calculated by reference to the formula set out in the Working Time Regulations 1998–Regulation 4–(6) and (7)

WR.4 OVERTIME RATES

The employer may require overtime to be worked and the operative may not unreasonably refuse to work overtime.

Overtime will be calculated on a daily basis, but overtime premium rates will not be payable until the normal hours (39 hours-WR3) have been worked in the pay week unless the short time is authorised by the employer on compassionate or other grounds or is a certified absence due to sickness or injury.
Note: The number of hours worked in excess of normal hours will be reduced by the number of hours at unauthorised absence before the overtime premium is calculated.

Overtime shall be calculated as follows:

Monday to Friday
For the first four hours after completion of the normal working hours of the day time and a half; thereafter at the rate of double time until starting time the following day.

Saturday
Time and a half, until completion of the first four hours, and thereafter at double time.

Sunday
At the rate of double time, until starting time on Monday morning.

When an operative is called out after completing the normal working hours of the day or night, he shall be paid at overtime rates for the additional time worked as if he had worked continuously. Any intervening period shall count to determine the rate, but shall not be paid.

Overtime shall be calculated on the normal hourly rate. Additional payments for intermittent skill or responsibility or adverse conditions and bonus shall not be included when calculating overtime payments.

In no case shall payment exceed double time.

WR.5 DAILY FARE AND TRAVEL ALLOWANCES

5.1 **Extent of Payment**

Operatives are entitled to a daily fare and travel allowance, measured one way from their home to the job/site. The allowances will be paid in accordance with the table published periodically by the Council. There is no entitlement to allowances under this Rule for operatives who report, at normal starting time, to a fixed establishment, such as the employers yard, shop or permanent depot. The distance travelled will be calculated by reference to Working Rule 5.2. There is no entitlement under this Rule to allowances for distances less than 15 kilometres or in excess of 75 kilometres. The Employer will, on production of a receipt, reimburse the cost of ferries and tolls incurred by the operative.

Notes: Part of a kilometre counts as a full kilometre. Time spent in daily travelling (except in transfer during working hours) is not to be reckoned as part of the normal working hours.

5.2 **Measurement of Distance**

All distances shall be measured on an Ordnance Survey map in the 1: 50,000 series in a straight line from the centre of the kilometre square in which his home lies to the centre of the kilometre square where his place of work is located (job centre).
Where special physical conditions such as river, mountain range or motorway make straight-line measurement inequitable, the distance shall be measured from the job centre to the nearest crossing place or exit/entry point, then in a straight line across the crossing place where applicable and then in a straight line to the home centre.

An operative's home is the address at which he is living while employed on the work to which he travels daily.

5.3 **Transport provided free by the employer**

Where the employer provides free transport, the operative shall not he entitled to fare allowance. However, operatives who travel to the pick up point for transport provided free by the employer are entitled to fare allowance for that part of the journey, in accordance with the table.

WR.5 DAILY FARE AND TRAVEL ALLOWANCES (Cont'd)

5.4 **Transfer during Working Day**

An operative transferred to another place of work during work hours shall, on the day of the transfer only, be paid any fares actually incurred:

(i) in the transfer, and

(ii)in travelling home from the place where he finishes work if this differs from the place where he reported for work at starting time, subject to deduction of half the daily fare allowance.

5.5 **Emergency Work**

An operative called from the home (or temporary place of residence) outside normal working hours to carry out emergency work shall be paid travelling expenses and his normal hourly rate for the time spent travelling to and from the job.

WR.6 ROTARY SHIFT WORKING

Shift working means a situation in which more than one shift of not less than eight hours is worked on a job in a 24 hour period and such shifts do not overlap.

On all work which is carried out on two or more shifts in a 24 hour period the following provisions shall apply:

The first shift in the week shall be the first shift that ends after midnight on Sunday

The normal hours of a shift shall be eight hours, excluding meal breaks, notwithstanding which, the hours to be worked on any particular shift shall be established by the employer.

The rate payable for the normal hours of the shift shall be the operative's normal hourly rate plus, in the case of an operative completing a shift, a shift allowance of 14% of the normal rate.

An operative required to work continuously for over eight hours on a shift or shifts shall be paid at the rate of time and a half, plus a shift allowance of 14% of his normal hourly rate, for the first four hours beyond eight and thereafter at double time but such double time payment shall not be enhanced by the 14% shift allowance (i.e. the maximum rate in any circumstance shall be double the normal hourly rate).

After having worked five complete scheduled shifts in a week an operative shall on the first shift thereafter be paid at the rate of time and a half of normal rate plus 12.5% shift allowance for the first eight hours of the shift and thereafter and on any subsequent shift in that week at the rate of double time but with no shift allowance.

Where the work so requires, an operative shall be responsible for taking over from and handing over to his counterpart at commencement and on completion of duty unless otherwise instructed by his employer.

Where the nature of the work is such as to require an operative to remain at his workplace and remain available for work during mealtimes a shift allowance of 20% shall apply instead of the 14% or 12.5% otherwise referred to in this rule.

The shift allowance shall be regarded as a conditions payment and shall not be included when calculating overtime payments.

Under this rule the first five complete shifts worked by an operative shall count as meeting the requirements of the Guaranteed Minimum Rule.

This Rule does not apply to operatives employed on night work, continuous working or tunnel work.

Employers and operatives may agree alternative shift working arrangements and rates of pay where, at any job or site, flexibility is essential to effect completion of the work.

WR.7 NIGHT WORK

7.1 **Night Work Allowance and Overtime**

Where work is carried out at night by a separate gang of operatives from those working during daytimbe, operatives so working shall be paid at their normal hourly rate plus an allowance of 25% of the normal hourly rate.

Overtime shall be calculated on the normal hourly rate provided that in no case shall the total rate exceed double the normal hourly rate. Overtime shall therefore be paid as follows:

(a) Monday to Friday:
after completion of the normal working hours at the rate of time and a half plus the night work allowance (i.e. time and a half plus 25% of normal hourly rate) for the next four working hours and thereafter at double time.

(b) Weekends:
All hours worked on Saturday or Sunday night at double time until the start of working hours on Monday.

This rule does not apply to operatives employed on shift work, tunnel work or continuous working.

7.2 **Average Weekly Working Hours**

Where there are objectives or technical reasons concerning the organisation of work, average weekly working hours will be calculated by reference to a twelve month period subject to the employer complying with the general principles relating to the protection of health and safety of workers and providing equivalent compensatory rest periods or, in exceptional cases where it is not possible for objective reasons to grant such periods, ensuring appropriate protection for the operatives concerned.

7.3 **Health and Risk Assessments**

Under the terms of this agreement night workers will:

a) Be provided with a free health assessment before taking up night work and at intervals of not less than three months during which night work is undertaken.

b) Be provided on request with a written risk assessment identifying any special hazards or heavy physical or mental strain and the control measures, which apply to eliminate or adequately reduce such risks.

WR.8 CONTINUOUS WORKING

An operative whose normal duties are such as to require him to be available for work during mealtimes and consequently has no regular mealtime, shall be deemed a "continuous worker" and shall be responsible for taking over from and handing over to his counterpart at commencement and completion of duty unless otherwise instructed by his employer. He will be paid for the number of hours on duty on the job at normal hourly rate plus 20% as a continuous working allowance. If in the normal cycle of operations for the particular job he is required to be on duty between 10 p.m. Saturday and 10 p.m. Sunday, he will during these hours be paid the rate of time and a half, plus 20% of his normal hourly rate as a continuous working allowance, provided that the continuous working allowance of 20% of normal hourly rate shall be deemed to be a conditions payment and shall not be reckoned for the purpose of calculating overtime payments. If work between 10 p.m. Saturday to 10 p.m. Sunday is not within the normal cycle of operations for the particular job, then no continuous working allowance shall be paid but the rate of payment shall be double the normal hourly rate.

This rule does not apply to operatives on night-work, shift work or tunnel work.

WR.9 TIDE WORK

9.1 Where work governed by tidal conditions is carried out during part only of the normal working hours, and an operative is employed on other work for the remainder of the normal working hours, the normal hourly rate (with the addition of any additional rate payable in respect of the conditions under which work is done), shall be paid during the normal working hours a thereafter shall be in accordance with the rule on Overtime Rates.

9.2 Where work governed by tidal conditions necessitates operatives turning out for each tide and they are not employed on other work, they shall be paid a minimum for each tide of six hours pay at ordinary rates, provided they do not work more than eight hours in the two tides. Payment for hours worked beyond a total of eight on two tides shall be calculated proportionately i.e. those hours worked in excess of eight multiplied by the total hours worked and the result divided by eight, to give the number of hours to be paid in addition to the twelve paid hours (six for each tide) provided for above. Work done on Saturday after 4 p.m. and all Sunday shall be paid at the rate of double time. Operatives shall be guaranteed eight hours at ordinary rate for time work between 4 p.m. and midnight on Saturday and 16 hours at ordinary rates for two tides work on Sunday. Payment under the rule on Overtime Rates does not apply to this sub-paragraph.

WR.10 TUNNEL WORK

The long-standing custom of the industry that tunnel work is normally carried out by day and by night is reaffirmed. Where shifts are being worked within and in connection with the driving tunnels the first period of a shift equivalent to the normal working hours specified in the Working Hours rule for that day shall be deemed to be the ordinary working day. Thereafter the next four working hours shall be paid at time and a half and thereafter at double time provided that

(a) In the case of shifts worked on Saturday the first four hours shall be paid at time and a half and thereafter at double time.

(b) In the case of shifts worked wholly on Sunday payment shall be made double time.

(c) In the case of shifts commencing on Saturday but continuing into Sunday, payment shall be made for all hours worked at double time.

(d) In case of shifts commencing on Sunday but continuing into Monday, hours worked before midnight shall be paid for at double time and thereafter four working hours calculated from midnight shall be at time and a half and thereafter at double time.

This rule does not apply to an operative employed under the Continuous Working rule.

WR.11 REFUELLING, SERVICING, MAINTENANCE AND REPAIRS

Operators of mechanical plant, such as excavators, cranes, derricks, rollers, locomotives, compressors or concrete mixers and boiler attendants shall, if required, work and be paid at their normal hourly rate for half an hour before and half an hour after the working hours prescribed by the employer for preparatory or finishing work such as refuelling firing up, oiling, greasing, getting out, starting up, banking down, checking over, cleaning out and parking and securing the machine or equipment.

Refuelling, Servicing, Maintenance and Repair work carried out on Saturday and Sundays shall be paid in accordance with the rule on Overtime Rates.

WR.12 STORAGE OF TOOLS

When practicable and reasonable on a site, job or in a shop the employer shall provide an adequate lock-up or lock-up boxes, where tools can be left at the owner's risk, provided always that the employer shall accept liability up to a maximum amount specified by the Council for any loss caused by fire or theft of tools properly secured by an operative in such lock-up or lock-up boxes, on the understanding that the employee shall comply with all the employer's requirements as regards the storage of tools. At all times an operative shall take good care of his tools and personal property and act in a responsible manner to ensure their reasonable safety.

WR.13 LOSS OF CLOTHING

Where an operative leaves clothing in accommodation provided by the employer the Employer shall be liable up to a maximum amount specified by the Council for loss of such clothing through fire.

WR.14 TRANSFER ARRANGEMENTS

14.1 At any time during the period of employment, the operative may, at the discretion of the employer be transferred from one job to another.

14.2 The employer shall have the right to transfer an operative to any site within daily travelling distance of where the operative is living. A site is within daily travelling distance if:

(a) when transport is provided free by the employer the operative can normally get from where he is living to the pickup point designated by the employer within one hour, using public transport if necessary or

(b) in any other case the operative, by using the available public transport on the most direct surface route, can normally get to the site within two hours.

WR.14 TRANSFER ARRANGEMENTS (Cont'd)

14.3 Transfer to a job which requires the operative to live away from home shall be by mutual consent. The consent shall not be necessary where the operative has been in receipt of subsistence allowance in accordance with the Subsistence Allowance Rule from the same employer at any time within the preceding 12 months.

WR.15 SUBSISTENCE ALLOWANCE

When an operative is recruited on the job or site and employment commences on arrival at job or site he shall not be entitled to payment of subsistence allowance. An operative necessarily living away from the place in which he normally resides shall be entitled to a subsistence allowance of an amount specified by the Council.

Subsistence allowance shall not be paid in respect of any day on which an operative is absent from work except when that absence is due to sickness or industrial injury and he continues to live in the temporary accommodation and meets the industry sick pay requirements.

Alternatively, the employer may make suitable arrangements for a sick or injured operative to return home, the cost of which shall be met in full by the employer

An operative in receipt of subsistence allowance shall only be entitled to daily fare allowance under the Daily Fare Allowance rule between his accommodation and the job if he satisfies the employer that he is living as near to the job as there is a accommodation available.

WR.16 PERIODIC LEAVE

16.1 When an operative is recruited on the job or site and employment commences on arrival at the job or site he shall not be entitled to the periodic leave allowances in this rule. In other cases when an operative is recruited or a sent to a job which necessitates his living away from the place in which he normally resides, he shall he entitled to:

Payment of his fares or conveyance in transport provided by the employer as follows:-

(a) from a convenient centre to the job at commencement

(b) to the convenient centre and back to the job at the following periodic leave intervals

(i) for jobs up to 128 kilometres from the convenient centre (measured in a straight line), every four weeks

(ii) for jobs over 128 kilometres from the convenient centre (measured in a straight line) at an interval fixed by mutual arrangement between the employer and operative before he goes to the job.

(c) from the job to the convenient centre at completion.

16.2 **Payment of fares**

Where an employer does not exercise the option to provide free transport, the obligation to pay fares may at the employer's option be discharged by the provision of a free railway or bus ticket or travel voucher or the rail fare.

Payment for the time spent travelling between the convenient centre and the job as follows:

(a) On commencement of his employment at the job, the time travelling from the convenient centre to the job, provided that an operative shall not be entitled to such payment if within one month from the date of commencement of his employment on the job he discharges himself voluntarily or is discharged for misconduct.

(b) When returning to the job (i.e. one way only) after periodic leave, provided that he returns to the job at the time specified by the employer and provided also that an operative shall not be entitled to such payment if, within one month from the date of his return to the jobs, he discharges himself voluntarily or is discharged for misconduct.

(c) On termination of his employment on the job by his employer, the time spent travelling from the job to the convenient centre, provided that he is not discharged for misconduct.

(d) Time spent in periodic travelling is not to be reckoned as part of the normal working hours; periodic travelling time payments shall in all cases be at the operative's normal hourly rate to the nearest quarter of an hour and shall not exceed payment for eight hours per journey.

16.3 **Convenient centre**

The convenient centre shall be a railway station, bus station or other similar suitable place in the area in which the operative normally resides.

WR.17 GUARANTEED MINIMUM WEEKLY EARNINGS

Where an operative, who is employed under the Working Rules of the Council, has been available for work throughout the normal working hours for the week whether or not work has been provided by the employer. He shall be entitled to guaranteed minimum weekly earnings as defined in Rule 1.

17.1 **Loss of Guarantee**

There shall be no entitlement to guaranteed minimum weekly earnings where the employer is unable to provide continuity of work due to industrial action or for Tide Work Rule 9 or work paid by shift Rule,6.

17.2 **Proportional Reduction**

Where an operative is absent for part of normal working hours due to certified sickness or injury or for one or more days of annual or recognised public holiday the requirement for the operative to be available for work will be deemed to be met and the payment of Guaranteed Minimum Weekly Earnings will be proportionately reduced. The proportionate reduction will not apply where the employer authorises the absence on compassionate or other grounds.

17.3 **Availability for Work**

An operative has satisfied the requirements to keep himself available for work during normal working hours if he complies with the following conditions:

That, unless otherwise instructed by the employer, he has presented himself for work at the starting time and location prescribed by the employer and has remained available for work during normal working hours.

That he carries out satisfactorily the work for which he was engaged or suitable alternative work if instructed by the employer and

He complies with the instructions of the employer as to when, during normal working hours, work is to be carried out, interrupted or resumed.

WR.17 GUARANTEED MINIMUM WEEKLY EARNINGS (Cont'd)

17.4 **Temporary Lay-off**

17.4.1 Where work is temporarily stopped or is not provided by the employer the operative may be temporarily laid off. The operative shall, subject to the provisions of Rule 17.4.2, be paid one fifth of his "Guaranteed Minimum Weekly Earnings" as defined in Rule 17 for the day on which he is notified of the lay-off and for each of the first five days of temporary lay-off. While the stoppage of work continues and the operative is prevented from actually working, he will be required by the employer to register as available for work at the operatives local job centre.

17.4.2 The payment described in Rule. 17.4.1 will be made, provided that in the three months prior to any lay-off there has not been a previous period or periods of lay-off in respect of which a guaranteed payment was made for 5 consecutive days or 5 days cumulative, excluding the day or days of notification of lay-off. In any such case the operative will not be entitled to a further guaranteed payment until a total of three months has elapsed from the last day of the period covered by the previous payment. Thereafter and for so long as the stoppage lasts, he shall be entitled to a further guaranteed payment of up to five days.

17.5 **Disputes**

A dispute arising under this agreement concerning guaranteed minimum payment due may, at the option of the claimant, be referred to ACAS and/or an industrial tribunal in the event of no decision by the Council.

WR.18 ANNUAL HOLIDAYS

The Holiday Year will run from the second Monday in January each Year. Operatives are entitled to twenty-one days annual holiday with pay as follows:

18.1 **The Winter Holiday** shall be seven working days taken in conjunction with Christmas Day, Boxing Day and New Year's Day, to give a Winter Holiday of two calendar weeks. The Council shall publish the dates of each Winter Holiday.

18.2 **The Easter (Spring) Holiday** shall be the four working days immediately following Easter Monday, to give an Easter Holiday of one calendar week.

It shall be open to employers and operatives to agree that all or some of the 'Winter Holiday' and/or the 'Easter (Spring) Holiday' will be taken on alternative dates.

18.3 **The Summer Holiday** shall be two calendar weeks, by mutual agreement not necessarily consecutive, to be granted in the "summer period" - i.e. between 1 May and 31 October.

18.4 **Payment for Annual Holiday**
Payment for annual holiday which shall be made on the last pay-day preceding the commencement of each holiday period shall be made at either:

18.4.1 **Where the operatives pay does not vary with the amount of work done**
A weeks' pay is simply the normal weekly wage for the contractual weekly hours as defined by the contract of employment.

18.4.2 **Where the operatives pay varies with the amount of work done.**
Where earnings vary because of piece work or productivity bonus arrangements, then a weeks' pay is arrived at calculating the earnings during the normal working week as defined by the contract of employment, averaged over the 12 complete weeks worked immediately prior to the holiday week.

18.4.3 One day's pay is calculated by dividing a week's pay as defined by either WR 18.4.1 or 18.4.2 by the number of hours in the normal working week and multiplying by the normal hours in the particular day, both as defined by the contract of employment.

18.5 **Leavers**

18.5.1 Operatives who leave the employment of the employer during a leave year are entitled to a compensatory payment calculated as follows:

(A÷52) x 21 - B

Where;
A is the number of complete weeks of service in the leave year.

B is the number of days leave taken by the operative in the leave year excluding public/bank holidays.

18.5.2 Where the number of days leave taken exceeds the operative's entitlement the employer has the right to make a deduction from payment made to the operative leaving the employment of the employer in respect of any overpayment of holiday pay. Such deduction will be calculated in accordance with WR 18.6

18.6 **Amount of the compensatory payment**
The operative is entitled to a compensatory payment for each day, or part of a day, of entitlement calculated by reference to the hourly rate of pay under **WR.18.4** multiplied by the normal contractual working hours.

18.7 **General Provisions related to Annual Holiday.**
Where employment commences after the start of the leave year the operatives will be entitled to the proportion of the twenty-one days annual holiday equivalent to the proportion of the leave year calculated from the first week of employment to the last week of the leave year.

An operative has no entitlement to payment for holidays not taken during the holiday leave year or to carry forward entitlement to holiday from one holiday year to the subsequent holiday year.

WR.19 PUBLIC HOLIDAYS

19.1 The following are recognised as public holiday for the purpose of this agreement:

(a) **England and Wales**
Christmas Day, Boxing Day, New Year's Day, Good Friday, Easter Monday, the May Day, Bank Holiday, the Spring Bank Holiday, the Summer Bank Holiday shall be recognised as public holidays in England and Wales, provided that such days are generally recognised as holidays in the locality in which the work is being done.

(b) **Scotland**
Christmas Day, Boxing Day, New Year's Day, Easter Monday, the first Monday in May, the Friday immediately preceding the Annual Summer Local Trades Holiday and the Friday and Monday at the Autumn Holiday, as fixed by the competent Local Authority.

WR.19 PUBLIC HOLIDAYS (Cont'd)

19.1 (Cont'd)

(c) **Local Variations**

Where, in any locality, any of the above public holidays is generally worked and another day is recognised instead as a general holiday, such other day shall be recognised as the alternative holiday.

(d) **Alternative Days**

When Christmas Day, Boxing Day or New Year's Day falls on a Saturday or Sunday an alternative day or days of public holiday will be promulgated. Any reference in this Rule to Christmas Day, Boxing Day or New Year's Day shall be taken to apply to the alternative day so fixed.

19.2 **Payment in respect of Public Holidays**

Payment for days of public holiday recognised under this rule shall be made by the employer to an operative in his employment at the time of each such holiday on the pay day in respect of the payweek in which such holiday occurs, except that payment for Christmas Day, Boxing Day and New Year's Day shall be made on the last pay day before the Winter Holiday. The amount of payment for each day of public holiday shall be in accordance with WR.18.4.3.

19.3 **General provisions related to payment for Public Holidays**

19.3.1 The entitlement to take leave on Public Holidays and to receive payment for those days is subject to the operative being available for work, in accordance with Rule 17.3, on the last normal working day preceding and the first normal working day following the day or days of public holiday.

19.3.2 An operative who has completed 4 weeks in employment, but who is no longer in the employment of employer on such a public holiday shall, nevertheless, be entitled to payment in respect of such holiday if he was in the employer's employment on the last working day preceding the day or days of public holiday but the employment was terminated by the employer (otherwise than for misconduct) before the holiday occurred. Where the holiday is Christmas Day, payment shall also be made for Boxing Day and Year's Day. Where the holiday is Good Friday payment shall also be made for Easter Monday. Payment shall be made on termination of employment.

19.3 **General provisions related to payment for Public Holidays**

19.3.3 An operative who is required to work on Christmas Day, Boxing Day, Good Friday or Easter Monday has the option by arrangement with the employer of an alternative day of holiday as soon thereafter as is mutually convenient, in which case the payment prescribed by this rule shall be made in respect of such alternative day instead of the public holiday. When the employment is terminated before such alternative day occurs, the operative shall receive such payment on the termination of employment.

19.3.4 There is no entitlement to payment for public holidays for an operative who has been unavailable for work, due to sickness or injury for a continuous period of ten weeks or more and who remains unavailable for work on the last working day preceding the or days of public holiday.

19.4 **Payment for Work on a Public Holiday**

All hours worked on a day designated as a public holiday shall be paid for at double time.

WR.20 AMOUNT AND DURATION OF PAYMENT OF INDUSTRY SICK PAY

Absence from Work due to Sickness or Injury

20.1 **Relationship of Industry Sick Pay with Statutory Sick Pay**

Under existing legislation there is an entitlement to statutory sick pay. Any payment due under this rule shall be increased by an amount equivalent to any statutory sick pay that may be payable in respect of the same day of incapacity for work under the Regulations made under that Act. These are referred to elsewhere in this Rule as "SSP Regulations".

20.2 **Scope**

This Rule applies to all adult operatives i.e. all operatives aged 18 years and over.

20.3 **Qualifying Days**

For the purpose of both this Rule and the SSP Regulations, the "qualifying days" that shall generally apply in the industry are Monday to Friday in each week.

While the qualifying days referred to above shall generally be the same five days as those which form the normal week of guaranteed employment under this agreement, it is accepted that there might be certain exceptions, e.g. where the particular circumstances of the workplace require continuous six or seven day working. In these situations it is in order, where there is mutual agreement, for other days to be regarded as 'qualifying days' for the purpose of this Rule and SSP.

20.4 **Amount and duration of Payment**

(i) An operative who during employment with an employer is absent from work on account of sickness or injury shall, subject to satisfying all the conditions set out in this Rule be paid the appropriate proportion of a weekly amount specified by the Council for each qualifying day of incapacity for work. For this purpose, the appropriate proportion due for a day shall be the weekly rate divided by the number of qualifying days specified under Rule 20.3 above.

(ii) During the first four continuous weeks of employment with a new employer the Operative shall be entitled to Statutory Sick Pay for absence which the employer is satisfied is due to genuine sickness or injury.

(iii) After four continuous weeks of employment the Operative shall be entitled to a total of four weeks Industry sick pay in addition to SSP in respect of absence that starts after these four weeks.

(iv) After fifteen continuous weeks of employment the Operative shall be entitled to a total of seven weeks Industry sick pay in addition to SSP (less any sick pay received) in respect of absence that starts after these fifteen weeks.

(v) After twenty six continuous weeks of employment the Operative shall be entitled to a total of ten weeks Industry sick pay in addition to SSP (less any sick pay received) in respect of absence that starts after these twenty six weeks. This entitlement is based on a rolling 12 month period or single period of absence, whichever is the longer.

WR.20 AMOUNT AND DURATION OF PAYMENT OF INDUSTRY SICK PAY (Cont'd)

20.5 **Notification of Incapacity for Work**

An operative shall not be entitled to payment under this Rule unless, during the first qualifying day in the period of incapacity, his employer is notified that he is unable to work due to sickness or injury and when the incapacity for work started. Thereafter, the operative shall, at intervals not exceeding one week throughout the whole period of absence, keep the employer informed of his continuing incapacity for work. Where the employer is notified later than this rule requires, he may nevertheless make payment under the rule if satisfied that there was good cause for the delay.

20.6 **Certification of Incapacity for Work**

The whole period of absence from work shall be covered by a certificate or certificates of incapacity for work to the satisfaction of the employer. For the first seven consecutive days of sickness absence, including weekends and public holidays, a self certificate will normally suffice for this purpose. Any additional days of the same period of absence must be covered by a certificate or certificates given by a registered medical practitioner.

NOTE: For the purpose of this paragraph a self certificate means a signed statement made by the operative in a form that is approved by the employer, that he has been unable to work due to sickness/injury for the whole period specified in the statement.

20.7 **Qualifying Conditions for Payment**

An operative shall not be entitled to the payment prescribed in this rule unless the following conditions are satisfied:

20.7.1 That incapacity has been notified to the employer in accordance with Rule 20.5 above

20.7.2 That the requirements of Rule 20.6 above to supply certificate(s) of incapacity for work have been complied with.

20.7.3 That the first three qualifying days (for which no payment shall be due) have elapsed each period of absence.

20.7.4 That none of the qualifying days concerned is a day of annual or public holiday granted in accordance with the provisions of this Working Rule Agreement.

20.7.5 That the incapacity does not arise directly or indirectly from insurrection or war, attempted suicide or self-inflicted injury, the operative's own misconduct, any gainful occupation outside working hours or participation as a professional in sports or games.

20.7.6 That the limit of payment has not been reached.

20.8 The employer shall be responsible for keeping records of absence and payments made to the operatives.

WR.21 BENEFIT SCHEMES

21.1 An operative is entitled to be provided by his employer with cover for:

(a) death benefit, of an amount promulgated from time to time by the Trustees of the Benefits Scheme, with an additional discretionary amount payable if the cause of death is an accident at the place of work or while travelling to or from work, all such amounts payable at the discretion of the Trustees.

There is no entitlement to cover in the case of an operative over the age of 65 years, except for death benefit if the if the cause of death is an accident at the place of work or an accident while travelling to or from work when the operative has obtained his 65th birthday; or over and

(b) accidental injury benefit for a specified injury (or injuries) sustained as a result of an accident at the place of work or an accident while travelling to or from work, the amount being dependent on the nature of the qualifying injury or injuries, all such payments being payable at the discretion of the Trustees of the Building and Civil Engineering Accident Benefit Scheme.

In each case in accordance with the Trust Deed and Rules of the relevant Scheme, which are published separately.

21.2 Cover for the above benefits is to be provided by payment made by the employer to the Building and Civil Engineering Benefits Scheme Trustee Limited.

21.3 An employer is required to make payment on behalf of the operative aged between 16 years and 64 years of such minimum amount or amounts as is promulgated from time to time by the relevant parties to this agreement for the purpose of providing a retirement benefit. The normal retirement age of an operative shall be 65. Such benefit is an entitlement under this agreement and the employer should normally make payment of such amount to the B & CE EasyBuild Stakeholder Pension Scheme. However, if the operative and the employer shall agree in writing , payment may be made to an alternative approved pension arrangement provider. In any event the parties shall not agree to contribute less than the minimum amounts promulgated from time to time,

21.4 Where an operative wishes to make voluntary contributions on his own account for the purposes of providing further retirement benifit, and requests his employer to do so, the employer shall deduct from the operative's weekly pay an amount duly authorised by the operative, and shall arrange for that amount to be transmitted to either B&CE EasyBuild Stakeholder Pension Scheme, or to the employer's approved pension arrangement, by no later than the nineteenth day of the following month that such amount is deducted from the operative's pay.

WR.22 GRIEVANCE PROCEDURE

Procedure for dealing with grievances

Any issue which may give rise to or has given rise to a grievance (including issues relating to discipline) affecting the employer's workplace and operatives employed by that employer at that workplace shall be dealt with in accordance with the following procedure.

There shall be no stoppage of work, either partial or general, including a 'go-slow', strike, lock out or any other kind of disruption or restriction in output or departure from normal working, in relation to any grievance unless the grievance procedure has been fully used and exhausted at all levels.

Every effort should be made by all concerned to resolve any issue at the earliest stage.

A written record shall be kept of meetings held and conclusions reached or decisions taken. The appropriate management or union representative should indicate at each stage of the procedure when an answer to questions arising is likely to be given, which should be as quickly as practicable.

WR.22 GRIEVANCE PROCEDURE (Cont'd)
Procedure for dealing with grievances (Cont'd)

Stage 1
Any operative who has a grievance concerning his employment shall raise the matter orally or in writing, with his immediate supervisor.

Stage 2
If the issue remains unresolved, the operative or his immediate supervisor shall bring it to the attention of the next level of supervision. The operative may involve an appointed steward of a recognised trade union from this stage. A steward may also be involved if a grievance affects a number of operatives.

Stage 3
Failing resolution of the issue at stage 2, the appropriate supervisor or the operative(s) concerned or the steward (if appointed) shall, if there are good grounds for so doing, refer the matter to the agent or his nominee. If the matter then remains unresolved, the steward shall report the matter to the appropriate full time union official who shall, if he considers it appropriate, pursue any outstanding issue with the agent or his nominee after advising him in writing of the issue(s) he wishes to pursue.

Stage 4
Failing resolution of the issue at stage 3, the full time local union official shall report the matter up to the appropriate full time union official and the agent (or nominee) shall report the matter to an appropriate representative of the employer. Such union official, if there are good grounds for so doing shall pursue the issue with the appropriate representative of the employer.

Stage 5
Failing resolution of the issue at stage 4, the union official concerned shall, if it is decided to pursue the matter further, put the issue in writing to the employer and it is the duty of such official and/or the employer to submit the matter, as quickly as practicable, to the Construction Industry Joint Council for settlement.

The decisions of the Construction Industry Joint Council shall be accepted and implemented by all concerned.

WR.23 DISCIPLINARY PROCEDURE

It is recognised that, in order to maintain good morale, the employer has the right to discipline those:

- who fail to fulfil competently and to instructions of the company the duties and responsibilities called for by the position they hold; and/or
- whose behaviour is unsatisfactory; and/or
- who fail to make appropriate use of the disputes procedure for the resolution of questions arising without recourse to strike or other industrial action.

It is equally recognised that the employer must exercise this right consistently and with justice and care.

Where there is cause to take disciplinary action, the employer should give the operative adequate notice of the date on which the disciplinary hearing will be held. The notice must contain details of the complaint against the operative and advise of the right to be accompanied at the hearing.

Discipline shall be applied in accordance with the following procedure:

23.1 An oral warning shall be given to the operative of the employer's dissatisfaction and the improvement called for within a stated period. The steward, if appointed, may be present and the warning shall be entered in the operative's record.

23.2 If the required improvement does not occur, or a new complaint arises, a written warning shall be issued stating that it is a final warning failure to improve within a stated period will result in dismissal.

The steward, if appointed, shall be given a copy of the written warning.

23.3 On continued failure, the operative shall be dismissed with appropriate notice in writing, stating, if the operative so requests, the reason for dismissal. A copy of the written notice shall be given to the steward, if appointed.

23.4 If, at any stage, alleged gross misconduct arises, the case shall be investigated promptly and a decision taken by the employer, after a fair hearing of the operative and his steward, if appointed. Such decision may be summary dismissal; return to normal working; or transfer to another workplace. Where the decision is dismissal, it shall be in writing stating, if the operative so requests, the reason for dismissal. A copy of the written decision will be given to the steward, if appointed.

At the discretion of the employer, the operative(s) may be warned of the investigation and/or suspended from the place of work on pay. Such pay will be forfeited if the investigation confirms that dismissal is appropriate.

23.5 An operative has the right of appeal from a disciplinary decision made by the employer. The employer must advise the operative of the right of appeal, the appeal procedure, and any time limit applying to the right to appeal.

WR.24 TERMINATION OF EMPLOYMENT

24.1 The minimum period of notice of termination of employment that an employer shall give to an employee is:-

(i) During the first four weeks	One day's notice
(ii) After four weeks' continuous employment but less than two years	One week's notice
(iii) After two years' continuous employment but less than twelve years	One weeks notice for each full year of employment.
(iv) Twelve years' continuous employment or more	Twelve weeks' notice

24.2 **The minimum period of notice of termination of employment that an employee shall give an employer is:**

(i) During the first four weeks -	one day's notice
(ii) After four weeks' continuous employment -	one week's notice.

24.3 The employment may be terminated at any time by mutual consent which should preferably be expressed in writing.

24.4 All outstanding wages are to be paid at the expiration of the period of notice and the employee advised of his entitlement to holiday credits and PAYE certificates or, in lieu thereof, a written statement that they will be forwarded as soon as possible.

24.5 In the event of gross misconduct, an operative may be summarily discharged at any time subject to a sufficient investigation into the circumstances being carried out in accordance with the recommended procedures.

WR.24 TERMINATION OF EMPLOYMENT (Cont'd)

24.5 (Cont'd)

Where gross misconduct is alleged which may result in dismissal without previous warning, the case shall be investigated, a hearing shall be arranged and a decision taken by the employer as quickly as practicable.

At the discretion of the employer; the operative(s) may be suspended on full pay whilst the case is investigated. Such pay will be forfeited if the investigation confirms that dismissal is appropriate.

WR.25 TRADE UNIONS

25.1 The Construction Confederation recognises the signatory Trade Unions within the Construction Industry Joint Council for the purposes of collective bargaining. Both parties are fully committed to the Working Rule Agreement and strongly urge employers to:

25.1.1 recognise the trade unions who are signatories to the Agreement;

25.1.2 ensure that all operatives are in the direct employment of the company or its sub contractors and are engaged under the terms and conditions of the Working Rule Agreement.

25.2 **Deduction of Union Subscriptions**

When requested by a signatory union, employers should not unreasonably refuse facilities for the deduction of union subscriptions (check-off) from the pay of union members.

25.3 **Full Time Trade Union Officials**

A full time official of a trade union which is party to the Agreement shall be entitled, by prior arrangement with the employer's agent or other senior representative in charge and on presenting his credentials, to visit a workplace to carry out trade union duties and to see that the Working Rule Agreement is being properly observed.

25.4 **Trade Union Stewards**

An operative is eligible for appointment as a steward on completion of not less than four weeks continuous work in the employment of the employer. Where an operative has been properly appointed as a steward in accordance with the rules of his trade union (being a trade union signatory to the Agreement) and issued with written credentials by the trade union concerned, the trade union shall notify the employer's agent or representative of the appointment, for formal recognition by the employer of the steward. On completion of this procedure the employer will recognise the steward, unless the employer has any objection to granting recognition, in which case he shall immediately notify the trade union with a view to resolving the question.

An employer shall not be required to recognise for the purposes of representation more than one officially accredited steward for each trade or union at any one site or workplace.

25.5 **Convenor Stewards**

Where it is jointly agreed by the employer and the unions, having regard to the number of operatives employed by the employer at the workplace and/or the size of the workplace the recognised trade unions may appoint a convenor steward, who should normally be in the employment of the main contractor, from among the stewards and such appointment shall be confirmed in writing by the Operatives' side. On completion of this procedure the employer will recognise the Convenor Steward, unless the employer has any objection to granting recognition in which case he shall immediately notify the trade union with a view to resolving the question.

25.6 **Duties and Responsibilities of Stewards and Convenor Stewards.**

25.6.1 The duties of the Stewards shall be:

i) To represent the members of their union engaged on the site/factory/depot,
ii) to ensure that the Working rule is observed,
iii) to recruit members on the site/factory/depot into a signatory union,
iv) to participate in the Grievance Procedure at the appropriate stage under WR. 22,
v) to assist in the resolution of disputes in accordance with the Working Rule Agreement.

25.6.2 The duties of the Convenor Steward, in addition to those set out in WR 25.6.1 shall be:

i) to represent the operatives on matters concerning members of more than one union,
ii) to co-operate with Management and to assist individual Shop Stewards
iii) to ensure that disputes are resolved by negotiation in accordance with the Working Rule Agreement.

25.6.3 No Steward or Convenor Steward shall leave their place of work to conduct union business without prior permission of their immediate supervisor. Such permission should not be unreasonably withheld but should only be given where such business is urgent and relevant to the site, job or shop.

25.7 **Steward Training**

To assist them in carrying out their functions, Shop Stewards will be allowed reasonable release to attend training courses approved by their union.

25.8 **Stewards Facilities**

Management shall give recognised union officials and/or convenor stewards reasonable facilities for exercising their proper duties and functions. These facilities, which must not be unreasonably withheld, must not be abused. The facilities should include (a) use of a meeting room, (b) access to a telephone and (c) the use of a notice board on site. If the convenor steward so requests, the employer shall provide him regularly with the names of operatives newly engaged by that contractor for work on that site or job or in that shop.

25.9 **Meetings**

Meetings of operatives, stewards or convenor stewards may be held during working hours only with the prior consent of Management.

WR.26 SCAFFOLDERS

The following provisions are to be read in conjunction with the provisions of the Construction Industry Scaffolders' Record Scheme published separately by the Council and the Construction Industry Training Board. Special attention is drawn to the provision in the Scheme that if a scaffolder or trainee scaffolder is not in possession of a CITB training record card at the time of engagement, application should be made immediately by the new employer on the prescribed form. Any difficulties should be referred to the Joint Secretaries for action in accordance with the Scheme.

WR.26 SCAFFOLDERS (Cont'd)

26.1 Scaffolders employed whole time as such are to be in one of three categories, as defined below:

Trainee Scaffolder: an operative who can produce one of the following types of CITB training record card:

(i) a valid Trainee Scaffolder's card, or
(ii) a Basic Scaffolder's card with post-dated validity.

Basic Scaffolder: an operative who has at least one year's whole-time experience of scaffolding and who can produce one of the following types of CITB training record cards:

(i) a valid Basic Scaffolder's card, or

(ii) an Advanced Scaffolder's card with post-dated validity.

Advanced Scaffolder: an operative who has at least two years' whole-time experience of scaffolding and at least one year's experience as a Basic Scaffolder and who can produce an Advanced Scaffolder's CITB training record card.

26.2 No operative other than a Basic or Advanced Scaffolder, as defined above, may be employed on basic scaffolding operations and no operative other than an Advanced Scaffolder may be employed on advanced scaffolding operations (see schedule Rule 26.5) unless

(i) under adequate supervision, or
(ii) working together with a Basic or Advanced Scaffolder, or
(iii) erecting, altering or dismantling simple access scaffolding with a working platform no higher than 5 metres, or
(iv) a craft operative who is normally required to erect, alter or dismantle scaffolding in the course of his work and who has received adequate training in the scaffolding operation(s) concerned.

These exceptions do not affect the application to such scaffolding work of the provisions of safety legislation and, in particular, the Construction (Health, Safety and Welfare) Regulations 1996.

26.3 The onus of proof of training and experience required under this Working Rule is on the operative concerned and the onus of checking the proof submitted is on the employer.

26.4 Scaffolders covered by Rule 26.1 are entitled to the "Skilled Operative Basic Rate of Pay" 4,3,2,1 or Craft Rate as follows:

	Rate
Trainee Scaffolder	4.
Basic Scaffolder with less than $2^1/_2$ years whole time experience in scaffolding and less than $1^1/_2$ years whole time experience as a Basic Scaffolder	2.
Basic Scaffolder with at least $2^1/_2$ years whole time experience in scaffolding and at least $1^1/_2$ years whole time experience as a Basic Scaffolder and Advanced Scaffolder	Craft Rate

26.5 Approved list of scaffolding operations,

Basic:
Erecting, adapting and dismantling:
Independent, putlog and birdcage scaffolds and static and mobile towers.
Beams to form gantries and openings, correctly braced.
Hoist frameworks and guides
Protective fans.
Stack scaffolds.
Roof scaffolds
Proprietary systems.
Fixing wire netting to hoist framework or scaffold framework.
Fixing sheeting to scaffold framework.
Interpreting simple design layout drawing for scaffolding detailed above.
Applying knowledge of Construction Regulations to operations listed above.

Advanced:
All work in list of Basic Operations:
Erecting, adapting and dismantling:
Scaffolding to form truss-out scaffolds.
Slung scaffolds including use of lifting equipment, wire ropes, chains and shackling.
Raking and flying shores.
Other forms of designed structures, e.g. larger truss-outs, cantilevers, lifting structures, ramps, footbridges and temporary roofs.
Scaffolding or standard props (including all bracing) to form a dead shore, including adjustable bases and forkheads.
Scaffolding and proprietary systems (including levelling to within reasonable tolerances) to support formwork as laid out in engineering scaffold drawings.
Interpreting scaffold design drawings.
Applying knowledge of Construction Regulations to operations listed above.

WR.27 HEALTH SAFETY AND WELFARE

27.1 The nature of work in the building and civil engineering industry presents particular hazards to those engaged in it. The employers and operatives Organisation who are signatories to the Working Rule Agreement shall take all reasonable steps to ensure that employers and employees, whether operatives or management, comply with the requirement of legislation dealing with health, safety and welfare.

Safety Representatives

27.2 Legislation provides that recognised trade unions may appoint safety representatives to represent operatives. Provision is also made for the establishment of safety committees where a formal request, in writing, is made to an employer by at least two safety representatives who have been appointed in accordance with legislation.

WR. 28 REFERENCE PERIODS AND DEFINITIONS

28.1 For the purpose of compliance with the Working Rule Agreement, statutory definitions, entitlements and calculations the reference period shall, subject to the requirements of WR 3.2 and WR.7.2 - "AVERAGE WEEKLY WORKING HOURS", be twelve months, in accordance with the Working time regulations 1998

28.2 A night worker is defined as a worker who works at least 3 hours of his daily working time between the hours of 11.00pm and 6.00am on the majority of days worked.

WR.29 SUPPLEMENTARY AGREEMENTS

When it is agreed between the employer(s) and operative(s) that at any particular workplace it would be appropriate to enter into an agreement specifically for that workplace, any such agreement shall be supplementary to and not in conflict with this Working Rule Agreement. Where any dispute arises in this respect the National Working Rule Agreement takes precedence.

WR.30 DURATION OF AGREEMENT

This Agreement shall continue in force and the parties to it agree to honour its terms until the expiration of three calendar month's notice to withdraw from it, given by either the Employers' side or the Operatives' side.

SCHEDULE 1

Specified Work Establishing Entitlement to the Skilled Operative Pay Rate 4,3,2 1. or Craft Rate

	Rates of Pay Basic Additional
QUALIFIED BAR BENDERS AND REINFORCEMENT FIXERS	
Bender and fixer of Concrete Reinforcement capable of reading and understanding drawings and bending schedules and able to set out work	Craft Rate
CONCRETE	
Concrete Leveller or Vibrator Operator Screeder and Concrete Surface Finisher man working off forms or other datum (e.g. road-form or concrete haunch, or edge beam or wire) man required by his employer to use trowel or hand or powered float to produce high quality finished concrete	4
DRILLING AND BLASTING	
Drills, rotary or percussive: mobile rigs, operator of	3.
Operative attending drill rig	4.
Shotfirer, operative in control of and responsible for explosives, including placing, connecting and detonating charges	3.
Operatives attending on shotfirer including stemming	4.
FORMWORK CARPENTERS	
1st year trainee	4.
2nd year trainee	3.
Formwork Carpenters	Craft Rate
GANGERS AND TRADE CHARGEHANDS	
(Higher Grade payments may be made at the employer's discretion)	4.
GAS DISTRIBUTION	
Operatives who have successfully completed approved training to the standard of: Gas Distribution Operative	
Category 1 and 2	4.
Category 3	3.
Team Leader - Service layer	2.
Team Leader-Main Layer	1.
Team Leader - Main and Service Layer	1.
LINESMEN - ERECTORS	
1st Grade	2.
(Skilled in all works associated with the erection of O.H. Transmission Lines including Assembly and Erection of Steel Towers; Concrete and Wood Poles of all types; Insulators and O.H.Line Switchgear; Stringing, tensioning, sagging, jointing, clamping and making off all types of O.H.conductors; making off stay wires; also similar structures and other works appertaining thereto.)	
2nd Grade	3.
(As above but lesser degree of skill - or competent and fully skilled either in assembly and erection of all types of supports or assembly and erection of the insulators, conductors, stay wires - all as specified above.)	
LINESMEN - ERECTORS' MATE.	
(Semi-skilled in works specified above and a general helper)	4.
SKILLED MECHANICS	
Maintenance Mechanic on site, i.e., man capable of carrying out field service, maintenance, and minor repairs ancillary to civil engineering works	2.
Contractors' Plant Mechanic in shop, depot or on site, i.e., man capable of carrying out major repairs and overhauls including welding work, operating turning lathe or similar machine and/or carrying out highly skilled work ancillary to civil engineering works	1.
Contractors' Plant Mechanics' Mate on site or in depot	4.
Greaser on site, servicing machines	4.
Tyre Fitter, heavy earth mover tyres	2.
MECHANICAL PLANT	
Compressors and Generators	
Air compressors diesel or petrol or generators over 10 KW operator or Driver of	4.
Concrete Placing Equipment	
Portable or static concrete pumps; pneumatic concrete placers; concrete placing booms; operator of	4.
Self-propelled Mobile Concrete Pump, with or without boom, mounted on lorry or lorry chassis; driver/operator of	3.
Mobile Cranes, Hoists or Fork-Lift Trucks	
Self-propelled Mobile Crane on road wheels or caterpillar tracks including lorry mounted:	
Capacity up to and including 5 tons (or metric tonnes); driver of	4.
Capacity up to and including 10 tons (or metric tonnes); driver of	3.
Capacity over 10 tons (or metric tonnes) and up to and including 120 tons (or metric tonnes); driver of	Craft Rate
Capacity over 120 tons (or metric tonnes); driver of	Craft Rate iii
Where grabs are attached to Cranes the next higher skilled basic rate of pay applies except over 120 tons (or metric tonnes) where the rate is at the employer's discretion.	
Tower Cranes (including travelling or climbing)	
Up to and including 2 tons (or metric tonnes) maximum lifting capacity at minimum radius; driver of	4.
Over 2 tons (or metric tonnes) up to and including 10 tons (or metric tonnes) lifting capacity at minimum radius; driver of	3.
Over 10 tons (or metric tonnes) up to and including 20 tons (or metric tonnes) lifting capacity at minimum radius; driver of	2.
Over 20 tons (or metric tonnes) lifting capacity at minimum radius; driver of	1.

SCHEDULE 1 (Cont'd) — Rates of Pay: Basic / Additional

Miscellaneous Cranes, etc.

Overhead traveller or gantry up to and including 10 tons (or metric tonnes) capacity; driver of — 3.

Loco traveller up to and including 10 tons (or metric tonnes) capacity; driver of — 3.

Power driven Hoist/crane; driver of — 4.

Where grabs are attached to Cranes — 3.

Derricks (Stationary or Travelling)
Power-driven derrick: capacity up to and including 20 tons (or metric tonnes); driver of — 3.

Capacity over 20 tons (or metric tonnes); driver of — 2.

Where grabs are attached to Derricks the next higher skilled Operative Basic Rate of pay applies.

Banksmen appointed to attend Crane, hoist or Derrick and be responsible for fastening or slinging loads and generally direct Crane hoist or Derrick Driver — 4.

Smooth or Rough Terrain fork lift trucks or side loader up to and including 3 tons (or metric tonnes) capacity driver of. — 4.

Over 3 tons (or metric tonnes) capacity driver of — 3.

Dumpers and Dump Trucks
Up to and including 7 tons (or metric tonnes) carrying capacity; driver of — 4.

Over 7 tons (or metric tonnes) and up to and including 16 tons (or metric tonnes) carrying capacity; driver of — 3.

Over 16 tons (or metric tonnes) and up to and including 60 tons (or metric tonnes) carrying capacity; driver of — 2.

Over 60 tons (or metric tonnes) and up to and including 125 tons (or metric tonnes) carrying capacity; driver of — 1.

Over 125 tons (or metric tonnes) carrying capacity; driver of — Craft Rate

Excavators (360 degree slewing, rope or hydraulic)
Excavators with rated bucket capacity up to and including cu. yd. (0.6 cu. metres); driver of — 3.

Excavator with rated bucket capacity over - cu. yd. (0.6 cu. metres) and up to and including 5 cu. yds (3.85 cu. metres); driver of — 2.

Excavator with rated bucket capacity over 5 cu. yds. (3.85 cu. metres) and up to and including 10 cu. yds. (7.65 cu. metres);driver of — 1.

Excavator with rated bucket capacity over 10 cu. yds. (765 cu. metres); and up to 20 cu. yds (15.28 cu metres) driver of — Craft Rate (iii)

Excavator with rated bucket capacity over 20 cu. Yds (15.28 cu. metres) and up to and including 40 cu. Yds (30.57 cu. metres) driver of — Craft Rate (ii)

Excavator with rated bucket capacity over 40 cu. Yds (30.57cu. metres) driver of — Craft Rate (i)

Operative attending excavator or responsible for positioning vehicles during loading of tipping — 4.

Shovel Loaders, Wheeled or Tracked
See Tractors and Equipments.

Locos all; driver of — 4.

Mixers
Mortar Pan or Concrete Mixer up to but not including 21/4 (or 400 litres) wet capacity, to applyto one man only per machine; man employed on, and actually responsible for, operating. — 4.

Mixer Concrete 21/14 and up to and including 2 cu. yds. (or 400 litres and up to but not including 1,500 litres) wet capacity to apply to one man only per machine; man employed on, and actually responsible for, operating — 3.

Mixer Concrete of Over 2 cu, yds. (or 1,500 litres) wet capacity, to apply to one man only per machine; man employed on, and actually responsible for, operating — 2

Mixer Concrete, mobile self-loading and batching up to 2,500 litres wet capacity, to apply to one man only per machine; man employed on, and actually responsible for, operating — 2.

Drag Shovel: operative of — 4.

Motor Graders
Motor Grader; driver of — 2.

Motor Vehicles (Road Licensed Vehicles within the Construction and Use Regulations)
Vehicle up to and including 7.5 tons (or metric tonnes) Gross Vehicle Weight: driver of — 4.

Operations requiring a LGV Licence of either or both Classes C & E; driver of — 1.

Motorised Scrapers
Motorised Scraper; driver of — 2.

Power Driven Tools
Operatives using power-driven tools such as breakers, percussive drills, picks and spades, rammers and tamping machines. — 4.

Pumps
Power-driven pump(s); attendant of — 4.

Power Rollers
Roller up to but not including 4 tons (or metric tonnes);driver of — 4.

Roller, 4 tons (or metric tonnes) and upwards; driver of — 3.

Tractors (Agricultural)

Tractor, rubber-tyred agricultural type, when used to tow trailer and/or with mounted S compressor; driver of. — 4.

Tractors (Wheeled or Tracked - with or without equipment such as buckets, blades, loaders, backhoes)

Tractor, up to and including 100 h.p.; driver of — 3.

Tractor, over 100 h.p. up to and including 400 h.p.; driver of — 2.

Tractor, over 400 h.p. up to and including 650 h.p. driver of — 1.

Tractor, over 650 h.p.; driver of — Craft Rate

SCHEDULE 1 (Cont'd)

	Rates of Pay Basic	Additional
Trenching Machine (Multi Bucket)		
Trenching Machine, up to and including 30 h.p, driver of		4.
Trenching Machine, over 30 h.p. and up to and including 70 h.p.; driver of		3.
Trenching Machine, over 70 h.p., driver of		2.
Winches		
Power driven winch, driver of		4.
PAVIORS, MASONS, ETC.		
Paviors Rammerman		4.
Kerb & Paving Jointer		4.
Operative engaged in face pitching or dry walling		3.
PILING		
General Skilled Piling Operative		4.
Piling Ganger/Chargehand		3.
Pile Frame Winch Driver		3.
Rotary or Specialist Mobile Piling Rig Driver		2.
PIPE JOINTERS		
Jointers, Stoneware or Concrete Pipes		4.
Jointers working on flexible joints or setting up or caulking lead joints under 12 in. (or 300 mm.) dia		4.
Jointers working on flexible joints or setting up or caulking lead joints of 12in. (or 300 mm.) dia and over		4.
Jointers, Cast Iron or Steel Pipes using lead EXCEPT		4.
(a) in mains of 12 in. dia. up to and including 21in. dia. (or 300 mm. dia, up to and including 535 mm. dia.) when experienced in all or any of the following operations: i) Jointing with run lead; ii) Jointing with cold strip lead, wool or similar type		3.
(b) on mains of over 21 in. (or 535 mm.) dia. when so experienced		2.
PIPELAYERS		
Men preparing the bed and laying pipes up to 12 in (or 300 mm.) dia		4.
Men preparing the bed and laying pipes over 12in (or 300mm.) dia.		4
PLATE LAYER not labourer in gang		3.
POST TENSIONING AND PRE-STRESSING CONCRETE		
Operative in control of and responsible for hydraulic jack and other tensioning devices engaged in post-tensioning and/or pre-stressing whilst so employed (not labourer in gang)		3.
SCAFFOLDERS See WR. 26		
STEELWORK CONSTRUCTION		
Men fully skilled in and engaged in the assembly erection and fixing of steel-framed construction of a permanent nature ancillary to civil engineering works		1.
Operative capable of and engaged in fixing simple steelwork and structures ancillary to civil engineering works		3.
TAR SPRAYING, MANUAL OR MECHANICAL		
Chargehand		4.
Spraybar Operators		4.
Gritter Operators		4.
TIMBERMAN		
Timberman		3.
Except in cases where the timbering required calls for a special degree of skill, as, for example, in running sand and/or cofferdams, or where walings, struts and/or rakers are of 10 in. by 5 in, (or 250 mm, by 125 mm.) nominal or large timbers, in which cases the rates shall be		2.
Operative attending		4.
TUNNELS		
Operative working below ground on the construction or reconstruction of Tunnels or sinking shafts for Tunnels in hard rock or soft ground:		
Face Tunnelling machine; operator of		2.
Tunnel Miner (the skilled operative working at the face, or machineman working a drifter type of machine).		3.
Tunnel Miner's mates (the operative who assists the miner at the face or break-up, including such men who work pneumatic tools breakers, and/or in a soft tunnel, prepare the timber).		4.
(The above tunnel rates are exclusive of plus rate payments prescribed for work in active foul or surface water sewers but inclusive of any other plus rate payment for conditions which might otherwise apply)		
Other operatives engaged in driving headings over 2 metres in length from the entrance, in connection with drain cable and main laying		4.
WELDERS		
Gas or electric arc welder capable of welding mild steel and building up to normal welds, ancillary to civil engineering work, when so required		2.
Electric arc welder able and required to weld to highest standards for structural fabrication and simple pressure vessels (air receivers), including C02 processes, ancillary to civil engineering works		1.
Electric arc welder able and required to undertake all welding processes on all weldable materials, ancillary to civil engineering works, including working on his own initiative from drawings		Craft Rate
Opencast Coal Washeries and Screening Plants (where work is ancillary to civil engineering works).		
Screening Plant Machine Operators, Operatives employed and responsible for controlling:		
Mechanical equipment in a dry coal plant, or a washing plant machine		4
ROLLED ASPHALT ROAD SURFACING WORK, TAR AND/OR BITUMEN MACADAM SURFACING WORK		
Operatives employed on this class of work to be paid as follows:		
Raker		3.
Tamperman		4.
Chipper		4.

SCHEDULE 1 (Cont'd)

	Rates of Pay Basic	Additional
Mixing Platform Chargehand	4.	
Power Roller, 4 tons (or metric tonnes) and upwards driver of	3.	
Operator of Mechanical Spreader of the Barber-Greene or similar type	3.	
Leveller on Mechanical Spreader of the Barber Greene or similar type	3.	

The hours of work as set in this Working Rule Agreement shall apply to this class of work, subject to the proviso that due to the exceptional circumstances of the work and the desirability of keeping work going continuously, it is necessary to limit intervals for meals to a period of half-hour each.

* In the event of a breakdown of plant (including haulage plant) or of conditions arising which make it necessary or desirable in the interests of the job to cease work on the job for any period exceeding one hour, the following arrangements shall apply:

(a) If the operatives who have to cease work are given alternative work, hours worked either on that work, or on Rolled Asphalt Road Surfacing Work after the end of the normal working day they shall be paid at the overtime rates in accordance with WR.4 Overtime Rates.

(b) If the operatives who have to cease work are not given alternative work but are notwithstanding paid for the period of such stoppage, work may be continued beyond what would otherwise be the normal working day for a period equal to the total of the periods of such stoppages, but not exceeding three hours at ordinary hourly rates, and after a period of three hours, at the overtime rate prescribed by WR.4 Overtime Rates as if the day had ended at the time at which ordinary hourly rates cease to be payable.

* NOTE Para 4 does not apply to Tar Macadam Surfacing Work,

Operatives below 18 years of age.

Operatives below 18 years of age will receive payment of a percentage of the General Operative Basic Rate:
At 16 years of age 50% of the relevant rate
At 17 years of age 75% of the relevant rate
At 18 years of age or over 100% of the relevant rate.

Dry-liners

Operatives undergoing approved training in dry-lining operations	4.	
Operatives who can produce a certificate of training achievement indicating satisfactory completion of approved dry-lining training module (iii) (iv) or (v)	3.	

SCHEDULE 2

Entitlement to additional rates of pay for intermittent skill, responsibility or working in adverse conditions.

Extra Rate

1.4.1 **STONE CLEANING**

Operatives other than craft operatives employed on dry-cleaning stone-work by mechanical process for the removal of protective material and/or discoloration — E

1.4.2 **TUNNELS**

Operatives (other than Tunnel Machine Operators, Tunnel Miners and Tunnel Miners' Mates) wholly or mainly engaged in work of actual construction including the removal anddumping of mined materials but excluding operatives whose employment in the tunnel isoccasional and temporary:
In unlined tunnels — A

(The above tunnels are exclusive of extra payments prescribed for work in active foul orsurface water sewers, but inclusive of any other extra payment for conditions which might otherwise apply).

1.4.3 **SEWER WORK**

Operatives working within a totally enclosed active foul sewer of any nature or condition — D

Operatives working within a totally enclosed active surface water sewer of any nature or condition . — B

Operatives working outside existing sewers excavating or removing foul materials emanating from existing sewers. — C

1.4.4 **WORKING AT HEIGHT**

Operatives (including drivers of tower cranes, but excluding the drivers of power driven derricks on high stages and linesmen-erectors and their mates and scaffolders) employed on "detached work" shall receive in respect of conditions extra payments on the following scale calculated from the "point of departure".

Above 45m and up to 90m — C

Thereafter a further extra payment at the same rate for each additional 45 metres

For the purpose of this rule "detached work" shall be deemed to comprise work on detached "tower like structures" whether completely detached or rising above and being connected to a main structure. In the latter case the "point of departure" shall be the level at which the detached structure leaves the top of the main structure.

CONSTRUCTION INDUSTRY JOINT COUNCIL

Joint Secretaries Guidance Notes on the Working Rule Agreement of the Construction Industry Joint Council.

Introduction

These Notes of Guidance, whilst not forming part of the Working Rules, are intended to assist employers and operatives to understand and implement the Working Rule Agreement.

It is the intent of all parties to this Agreement that operatives employed in the building and civil engineering industry are engaged under the terms and conditions of the CIJC Working Rule Agreement.

Requests for definitions, clarification or resolution of disputes in relation to this Agreement should be addressed to the Construction Industry Joint Council at the address set out on the front cover of the Working Rule Agreement.

WR. 1 Entitlement to Basic Pay and Additional Rates of Pay.

WR. 1 sets out the entitlement to the basic rate of pay, additional payments for skilled work and intermittent skill, responsibility or for working in adverse conditions.

There are six basic rates of pay under this Agreement; General Operative rate, four rates for Skilled Operatives and a rate for a Craft Operative.

CONSTRUCTION INDUSTRY JOINT COUNCIL (Cont'd)

Joint Secretaries Guidance Notes on the Working Rule Agreement of the Construction Industry Joint Council. (Cont'd)

Payment for Occasional Skilled Work

WR. 1.1.2 deals with the payment for occasional skilled work and provides that general operatives who are required to carry out work defined in Schedule 1 on an occasional basis should receive an increased rate of pay commensurate with the work they are carrying out for the period such work is undertaken. This sets out the flexibility to enable enhanced payment to be made to general operatives undertaking skilled work for a limited amount of time but should not be used where the operative is engaged whole time on skilled work.

Skilled Operatives

WR.1.2.1 sets out a permanent rate of pay for skilled operatives who are engaged whole time on the skilled activity and does not permit the operative engaged on whole time skilled work to have his pay reduced to the General Operative basic rate when occasional alternative work is undertaken.

Additional Payment for Skilled Work.
WR.1.2.2 provides additional payment for skilled work for those operatives engaged in the work defined in Schedule 1. These payments are permanent payments for skilled operatives carrying out the specified work.

WR. 1.4. Additional Payments for Intermittent Skill, Responsibility or Working in Adverse Conditions.

WR 1.4.1 Provides additional payments for intermittent skill, responsibility or working in adverse conditions. These payments are made for the duration of the requirement for the skilled work or responsibility or to the extent the work is undertaken in adverse conditions and are not permanent payments.

N.B. For definition of normal hourly rate see the notes contained in the Working Rule Agreement.

WR.3.2 Average Weekly Working Hours.

Working Rules 3.2.1 to 3.2.4 provide, where there are objective or technical reasons, for the calculation of average weekly working hours by reference to a period of twelve months.

Whilst it is open to employers and employees to agree to work additional hours over the "normal working hours," Rule 3.2 does not give the employer the unilateral right to introduce excessive hours on a job or site. The 12 month averaging period referred to may only be applied where, for objective reasons, it is necessary to ensure completion of the work efficiently and effectively or where there are technical reasons that require additional hours to be worked.

Examples of objective and/or technical reasons which may require average weekly working hours to be calculated using a twelve month reference period are set out below. The list is not exhaustive other objective or technical reasons may apply.

Objective Reasons:

Work on infrastructure, roads, bridges, tunnels and tide work etc.

Client requirements for work to be completed within a tightly defined period, work undertaken for exhibitions, schools, retail outlets, shopfitting and banks etc.

Emergency work, glazing and structural safety etc.

Technical Reasons;

Work requiring a continuous concrete pour surfacing and coating work, tunnelling etc.

N.B. Any disputes regarding the validity of objective and/or technical reasons may be referred to the National Conciliation Panel of the Construction Industry Joint Council.

W.R. 4 Overtime Rates.

Where an operative who has worked overtime fails without authorisation to be available for work during normal weekly working hours he may suffer a reduction in or may not be entitled to premium payments in respect of overtime worked.

To calculate the number of hours paid at premium rate (overtime) you subtract the number of hours of unauthorised absence from the total number of hours overtime worked. This is in effect using a part of the hours of overtime worked to make the hours paid at the normal hourly rate up to 39 hours per week, the balance of overtime is paid at the appropriate premium rate.

CONSTRUCTION INDUSTRY JOINT COUNCIL (Cont'd)
Joint Secretaries Guidance Notes on the Working Rule Agreement of the Construction Industry Joint Council. (Cont'd)

EXAMPLES

Example 1.

An operative works three hours overtime on Monday. Tuesday and Wednesday works normal hours on, Thursday and is unavailable for work on Friday due to unauthorised absence. In these circumstances overtime premia will be calculated as follows:

	Normal Hours	Overtime Hours
Monday	8	3
Tuesday	8	3
Wednesday	8	3
Thursday	8	-
Friday	-	-
Total hours worked	(A) 32	(B) 9

Normal weekly working hours	(C) 39
Less total normal hours worked	(B) 32
Hours required to make up to 39	(D) 7
Total overtime hours	(B) 9
Less	(D) 7
Hours to be paid at premium rate	(E) 2

The operative is therefore, entitled to be paid:

(A+D)	39	Hours at "Normal Hourly Rate"
(E)	2	Hours at premium rate (time and a half)
Total	41 Hours	

Example 2.

An operative works four hours overtime on a Monday, three hours on Tuesday five hours (one hour double time) on the Wednesday, no overtime on either Thursday or Friday and absents himself from work on Friday at 12 noon without authorisation and then works six hours on Saturday. This calculation would be as follows:

	Normal hours	**Overtime Hours**
Monday	8	4
Tuesday	8	3
Wednesday	8	5 (1hour double time)
Thursday	8	-
Friday	3	-
Saturday	-	6 (2hours double time)
Total Hours worked	(A) $35\frac{1}{2}$	(B) 18

Normal weekly working hours	(C)	39
Less total normal hours worked	(B)	$35\frac{1}{2}$
Hours required to make up to 39	(D)	$3\frac{1}{2}$
Total overtime hours	(B)	18
Less	(D)	$3\frac{1}{2}$
Hours to be paid at premium rate	(E)	$14\frac{1}{2}$

The operative is, therefore, entitled to be paid:

(A+D)	39	Hours at "Normal Hourly Rate"
(E)	$14\frac{1}{2}$	Hours at premium rate (time and a half)
Total	$53\frac{1}{2}$	Hours

The entitlement to 3 hours pay at double time is extinguished in this example by the hours of unauthorised absence. If the number of hours worked at double time exceeds the number of hours of unauthorised absence the balance must be paid at the rate of double time.

NB. There shall be no reduction in overtime premium payments for operatives who are absent from work with the permission of the employer or who are absent due to sickness or injury

W.R. 17.4 Temporary Lay-off.

The temporary lay-off provisions may only be used when the employer has a reasonable expectation of being able to provide work within a reasonable time.

In this context an example of an employer who has a reasonable expectation to be able to provide work may be where a tender has been accepted but commencement delayed, where work is temporarily stopped due to weather conditions or for some other reason outside the employer's control. Reasonable time is not legally defined, however, an employee who has been temporarily laid off for four or more consecutive weeks or six weeks cumulative in any thirteen week period may claim a redundancy payment.

In no circumstances may the temporary lay-off rule be used where a genuine redundancy situation exists or to evade statutory obligations.

An employee who is temporarily laid off is entitled to payment of one fifth of his guaranteed minimum earnings for the day of notification of lay-off and for each of the first five days of the layoff subject to the limitations in WR 17.4.2.

WR 22. Grievance Procedure.

The Grievance Procedure allows companies and operatives to resolve problems at a local level and in the event of a failure to agree to refer the dispute to a CIJC National Conciliation Panel. The Working Rule Agreement sets out a five stage procedure for dealing with grievances; Stage 4 permits employer representatives and full time trade union officials to hold a local meeting to discuss resolution of the grievance or dispute. In the rare event that such a meeting does not resolve the problem a reference may be made to the National Conciliation Panel. Both operatives and employers have the right of appeal to the Construction Industry Joint Council from the decisions made by the National Conciliation Panel.

N.B. For full details of the conciliation dispute machinery please see the Constitution and Rules of the CIJC.

Any question relating to Working Rules not covered by the notes of guidance should be referred to the:-

Construction Industry Joint Council
Construction House
56-64 Leonard Street
London
EC2A 4JX

Building Construction Handbook

Fifth Edition

R. Chudley & R. Greeno

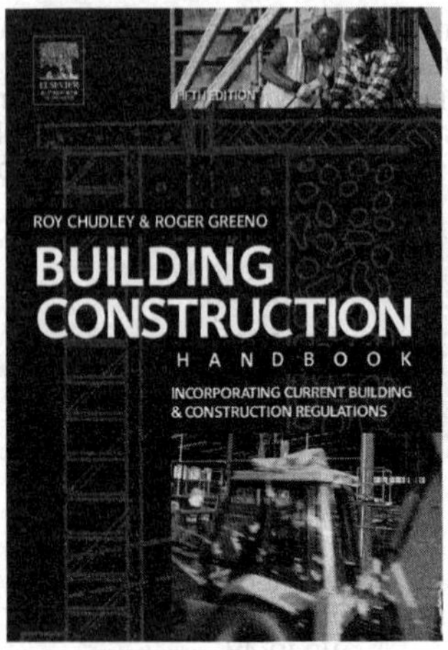

Building Construction Handbook is an essential text for all building construction students up to, and including, undergraduate level. It summarises all elements of building construction practice, techniques, components and procedures.

The comprehensive coverage and numerous references to the latest Building Regulations, Water Supply Regulations, British and European Standards also makes this book an invaluable reference tool for construction professionals.

The fifth edition consolidates recent amendments to the Building Regulations, including: Part A – Structure; Part C - Site preparation; Part E - Resistance to the passage of sound; Part L - Conservation of fuel and power; and Part M - Access to and use of buildings.

REVIEWS OF THE PREVIOUS EDITION:

'Book of the Year...the ideal handy reference for the site office.'
Professional Builder

'Pages of clear diagrams and brief text packed full of top tips and typical details.'
Building Design

£19.99 **Paperback March 2004** **736pp** **234 x 165mm** **ISBN 0750661968**

Five Ways to Order

ONLINE
http://books.elsevier.com/construction

E-MAIL
directorders@elsevier.com

FAX
+44 (0) 1865 474011

PHONE
+44 (0) 1865 474010

MAIL
Customer Services, Elsevier Limited
Linacre House, Jordan Hill, Oxford OX2 8DP

DAYWORK CHARGES

CONTRACT WORK

DEFINITION OF PRIME COST OF DAYWORK CARRIED OUT UNDER A BUILDING CONTRACT

Reproduced by permission of the Royal Institution of Chartered Surveyors.

This Definition of Prime Cost is published by The Royal Institution of Chartered Surveyors and the National Federation of Building Trades Employers, for convenience and for use by people who choose to use it. Members of the National Federation of Building Trades Employers are not in any way debarred from defining Prime Cost and rendering their accounts for work carried out on that basis in any way they choose. Building owners are advised to reach agreement with contractors on the Definition of Prime Cost to be used prior to issuing instructions.

SECTION 1

Application

1.1 This definition provides a basis for the valuation of daywork executed under such building contracts as provide for its use (e.g. contracts embodying the Standard Forms issued by the Joint Contracts Tribunal).

1.2 It is not applicable in any other circumstances, such as jobbing or other work carried out as a separate or main contract nor in the case of daywork executed during the Defects Liability Period of contracts embodying the above mentioned Standard Forms.

SECTION 2

Composition of total charges

2.1 The prime cost of daywork comprises the sum of the following costs:

(a) Labour as defined in Section 3.

(b) Materials and goods as defined in Section 4.

(c) Plant as defined in Section 5.

2.2 Incidental costs, overheads and profit as defined in Section 6, as provided in the building contract and expressed therein as percentage adjustments, are applicable to each of 2.1 (a)-(c).

SECTION 3

Labour

3.1 The standard wage rates, emoluments and expenses referred to below and the standard working hours referred to in 3.2 are those laid down for the time being in the rules or decisions of the National Joint Council for the Building Industry and the terms of the Building and Civil Engineering Annual and Public Holiday Agreements applicable to the works, or the rules or decisions or agreements of such body, other than the National Joint Council for the Building Industry, as may be applicable relating to the class of labour concerned at the time when and in the area where the daywork is executed.

3.2 Hourly base rates for labour are computed by dividing the annual prime cost of labour, based upon standard working hours and as defined in 3.4 (a)-(i), by the number of standard working hours per annum (see examples)

3.3 The hourly rates computed in accordance with 3.2 shall be applied in respect of the time spent by operatives directly engaged on daywork, including those operating mechanical plant and transport and erecting and dismantling other plant (unless otherwise expressly provided in the building contract).

3.4 The annual prime cost of labour comprises the following:

(a) Guaranteed minimum weekly earnings (e.g. Standard Basic Rate of Wages, Joint Board Supplement and Guaranteed Minimum Bonus Payment in the case of NJCBI rules).

(b) All other guaranteed minimum payments (unless included in Section 6).

(c) Differentials or extra payments in respect of skill, responsibility, discomfort, inconvenience or risk (excluding those in respect of supervisory responsibility - see 3.5).

(d) Payments in respect of public holidays.

(e) Any amounts which may become payable by the Contractor to or in respect of operatives arising from the operation of the rules referred to in 3.1 which are not provided for in 3.4 (a)-(d) or in Section 6.

(f) Employer's National Insurance contributions applicable to 3.4 (a)-(e).

(g) Employer's contributions to annual holiday credits.

(h) Employer's contributions to death benefit scheme.

DEFINITION OF PRIME COST OF DAYWORK CARRIED OUT UNDER A BUILDING CONTRACT (Cont'd)

CONTRACT WORK (Cont'd)

SECTION 3 (Cont'd)

3.4 (Cont'd)

(i) Any contribution, levy or tax imposed by statute, payable by the Contractor in his capacity as an employer.

3.5 Note

Differentials or extra payments in respect of supervisory responsibility are excluded from the annual prime cost (see Section 6). The time of principals, foremen, gangers, leading hands and similar categories, when working manually, is admissible under this Section at the appropriate rates for the trades concerned.

SECTION 4

Materials and Goods

4.1 The prime cost of materials and goods obtained from stockists or manufacturers is the invoice cost after deduction of all trade discounts but including cash discounts not exceeding 5 per cent and includes the cost of delivery to site

4.2 The prime cost of materials and goods supplied from the Contractor's stock is based upon the current market prices plus any appropriate handling charges.

4.3 Any Value Added Tax which is treated, or is capable of being treated, as input tax (as defined in the Finance Act, 1972) by the Contractor is excluded.

SECTION 5

Plant

5.1 The rates for plant shall be as provided in the building contract.

5.2 The costs included in this Section comprise the following:

(a) Use of mechanical plant and transport for the time employed on daywork.

(b) Use of non-mechanical plant (excluding non-mechanical hand tools) for the time employed on daywork.

5.3 Note: The use of non-mechanical hand tools and of erected scaffolding, staging, trestles or the like is excluded (see Section 6).

SECTION 6

Incidental Costs, Overheads and Profit

6.1 The percentage adjustments provided in the building contract, which are applicable to each of the totals of Sections 3, 4 and 5, comprise the following:

(a) Head Office charges

(b) Site staff including site supervision.

(c) The additional cost of overtime (other than that referred to in 6.2).

(d) Time lost due to inclement weather.

(e) The additional cost of bonuses and all other incentive payments in excess of any guaranteed minimum included in 3.4 (a).

(f) Apprentices study time.

(g) Subsistence and periodic allowances.

(h) Fares and travelling allowances.

(i) Sick pay or insurance in respect thereof.

(j) Third party and employer's liability insurance.

(k) Liability in respect of redundancy payments to employees.

(l) Employer's National Insurance contributions not included in Section 3.4

(m) Tool allowances.

(n) Use, repair and sharpening of non-mechanical hand tools.

(o) Use of erected scaffolding, staging, trestles or the like.

(p) Use of tarpaulins, protective clothing, artificial lighting, safety and welfare facilities, storage and the like that may be available on the site.

(q) Any variation to basic rates required by the Contractor in cases where the building contract provides for the use of a specified schedule of basic plant charges (to the extent that no other provision is made for such variation).

(r) All other liabilities and obligations whatsoever not specifically referred to in this Section nor chargeable under any other section.

(s) Profit.

6.2 Note: The additional cost of overtime, where specifically ordered by the Architect/Supervising Officer, shall only be chargeable in the terms of prior written agreement between the parties to the building contract

DEFINITION OF PRIME COST OF DAYWORK CARRIED OUT UNDER A BUILDING CONTRACT (Cont'd)
CONTRACT WORK (Cont'd)

Example of calculation of typical standard hourly base rate (as defined in Section 3) for NJCBI building craft operative and labour in Grade A areas **based upon rates ruling at 1st July 1975.**

		Rate	Craft Operative	Rate	Labourer
Guaranteed minimum weekly earnings		£	£	£	£
Standard Basic Rate	49 wks	37.00	1813.00	31.40	1538.60
Joint Board Supplement	49 wks	5.00	245.00	4.20	205.80
Guaranteed Minimum Bonus	49 wks	4.00	196.00	3.60	176.40
			2254.00		1920.80
Employer's National Insurance Contribution at 8.5%			191.59		63.27
			2445.59		2084.07
Employer's Contributions to:					
CITB annual levy			15.00		3.00
Annual holiday credits	49 wks	2.80	137.20	2.80	137.20
Public holidays (included in guaranteed minimum weekly earnings above)			-		-
Death benefit scheme	49 wks	0.10	4.90	0.10	4.90
Annual labour cost as defined in Section 3		£	2602.69	£	2229.17

	Craft operative	Labourer
Hourly base rates as defined in Section 3, clause 3.2	£ 2602.69 / 1904 = £ 1.37	£2229.17 / 1904 = £ 1.17

Note: (1) Standard working hours per annum calculated as follows:

	52 weeks at 40 hours	=		2080
Less	3 weeks holiday at 40hours	=	120	
	7 days public holidays at 8 hours	=	56	176
				1904

(2) It should be noted that all labour costs incurred by the Contractor in his capacity as an employer, other than those contained in the hourly base rate, are to be taken into account under Section 6.

(3) The above example is for the convenience of users only and does not form part of the Definition; all the basic costs are subject to re-examination according to the time when and in the area where the daywork is executed.

BUILD UP OF STANDARD HOURLY BASE RATES - RATES APPLICABLE AT 30th JUNE 2004

Under the JCT Standard Form, dayworks are calculated in accordance with the Definition of the Prime Cost of Dayworks carried out under a Building Contract published by the RICS and the NFBTE. The following build-up has been calculated from information provided by the Building Cost Information Service (BCIS) in liaison with the NFBTE. The example is for the calculation of the standard hourly base rates and is for convenience only and does not form part of the Definition; all basic rates are subject to re-examination according to when and where the dayworks are executed.

Standard working hours per annum			
52 weeks at 39 hours		2028	hours
Less - 21 days annual holiday			
16 days at 8 hours	128		
5 days at 7 hours	35		
8 days public holiday			
7 days at 8 hours	56		
1 day at 7 hours	7	226	hours
		1802	hours

	Craft Rate	General Operative
Guaranteed Minimum weekly earnings	£	£
Standard basic rate	299.13	225.03

GENERAL INFORMATION

DEFINITION OF PRIME COST OF DAYWORK CARRIED OUT UNDER A BUILDING CONTRACT (Cont'd)
CONTRACT WORK (Cont'd)

BUILD UP OF STANDARD HOURLY BASE RATES - RATES APPLICABLE AT 30th JUNE 2004 (Cont'd)

Hourly Base Rate	Rate £	Craftsman £	Rate £	General Operative £
Guaranteed minimum weekly earnings 46.205. weeks at	320.58	14,812.44(£8.22)	241.02	11,136.36(£6.18)
Extra payments for skill, responsibility, discomfort, inconvenience or risk* - 1802 hours		-		-
Employer's National Insurance Contribution** - 12.8% of	10,607.78	1357.80	6931.70	887.26
Holidays with Pay – 226 Hours	8.22	1857.72	6.18	1396.68
*** CITB Annual Levy - 0.5% of	16,670.16	83.35	12533.04	62.67
Welfare benefit – 52 stamps at £10.00		520.00		520.00
Public holidays (included with Holidays with Pay above).		-		-
		£18,631.31		£14,002.96

Hourly Base Rate	Craftsman	General Operative
	$\frac{£18631.31}{1802}$ = £10.34	$\frac{£14002.96}{1802}$ = £7.77

Notes

* Only included in hourly base rate for operatives receiving such payments.

** National Insurance Contribution based on 12.8% above the Earnings Threshold for 46.2 weeks at £91.00 (=£4204.20).

*** From 1st April 2002, the CITB levy is calculated at 0.5% of the PAYE payroll of each employee plus 1.5% of payments made under a labour only agreement. Building contractors having a combined payroll of £61,000 or less are exempt. The levy is included in the daywork rate of each working operative, the remainder being a constituent part of the overheads percentage.

**** From 29th June 1992, the Guaranteed Minimum Bonus ceased and is consolidated in the Guaranteed Minimum Earnings.

Holidays – The winter holiday is 2 calendar weeks taken in conjunction with Christmas, Boxing and New Years Day.
The Easter Holiday shall be four working days immediately following Easter Monday
The Summer Holiday shall be two calendar weeks.

DAYWORK, OVERHEADS AND PROFIT

Fixed percentage additions to cover overheads and profit no longer form part of the published Daywork schedules issued by the Royal Institution of Chartered Surveyors and Contractors are usually asked to state the percentages they require as part of their tender. These will vary from firm to firm and before adding a percentage, the list of items included in Section 6 of the above Schedule should be studied and the cost of each item assessed and the percentage overall additions worked out.

As a guide, the following percentages are extracted from recent tenders for Contract Work:
On Labour costs 80-200%
On Material costs 10-20%
On Plant costs 5-30%
Much higher percentages may occur on small projects or where the amount of daywork envisaged is low.

JOBBING WORK

DEFINITION OF PRIME COST OF BUILDING WORKS OF A JOBBING OR MAINTENANCE CHARACTER

Reproduced by permission of the Royal Institution of Chartered Surveyors.

This Definition of Prime Cost is published by the Royal Institution of Chartered Surveyors and the National Federation of Building Trades Employers for convenience and for use by people who choose to use it. Members of the National Federation of Building Trades Employers are not in any way debarred from defining prime cost and rendering their accounts for work carried out on that basis in any way they choose.

Building owners are advised to reach agreement with contractors on the Definition of Prime Cost to be used prior to issuing instructions.

SECTION 1

Application

1.1 This definition provides a basis for the valuation of work of a jobbing or maintenance character executed under such building contracts as provide for its use.

1.2 It is not applicable in any other circumstances such as daywork executed under or incidental to a building contract.

SECTION 2

Composition of Total Charges

2.1.1 The prime cost of jobbing work comprises the sum of the following costs:

(a) Labour as defined in Section 3.

(b) Materials and goods as defined in Section 4.

(c) Plant, consumable stores and services as defined in Section 5.

(d) Sub-contracts as defined in Section 6.

2.2 Incidental costs, overhead and profit as defined in Section 7 and expressed as percentage adjustments are applicable to each of 2.1 (a)-(d).

JOBBING WORK (Cont'd)
DEFINITION OF PRIME COST OF BUILDING WORKS OF A JOBBING OR MAINTENANCE CHARACTER (Cont'd)

SECTION 3

Labour

3.1 Labour costs comprise all payments made to or in respect of all persons directly engaged upon the work, whether on or off the site, except those included in Section 7.

3.2 Such payments are based upon the standard wage rates, emoluments and expenses as laid down for the time being in the rules or decisions of the National Joint Council for the Building Industry and the terms of the Building and Civil Engineering Annual and Public Holiday Agreements applying to the works, or the rules or decisions or agreements of such other body as may relate to the class of labour concerned, at the time when and in the area where the work is executed, together with the Contractor's statutory obligations, including:

(a) Guaranteed minimum weekly earnings (e.g. Standard Basic Rate of Wages and Guaranteed Minimum Bonus Payment in the case of NJCBI rules).

(b) All other guaranteed minimum payments (unless included in Section 7).

(c) Payments in respect of incentive schemes or productivity agreements applicable to the works.

(d) Payments in respect of overtime normally worked; or necessitated by the particular circumstances of the work; or as otherwise agreed between parties.

(e) Differential or extra payments in respect of skill responsibility, discomfort, inconvenience or risk.

(f) Tool allowance.

(g) Subsistence and periodic allowances.

(h) Fares, travelling and lodging allowances.

(j) Employer's contributions to annual holiday credits.

(k) Employer's contributions to death benefit schemes.

(l) Any amounts which may become payable by the Contractor to or in respect of operatives arising from the operation of the rules referred to in 3.2 which are not provided for in 3.2 (a)-(k) or in Section 7.

(m) Employer's National Insurance contributions and any contributions, levy or tax imposed by statute, payable by the Contractor in his capacity as employer.

Note: Any payments normally made by the Contractor which are of a similar character to those described in 3.2 (a)-(c) but which are not within the terms of the rules and decisions referred to above are applicable subject to the prior agreement of the parties, as an alternative to 3.2(a)-(c).

3.3 The wages or salaries of supervisory staff, timekeepers, storekeepers, and the like, employed on or regularly visiting site, where the standard wage rates etc. are not applicable, are those normally paid by the Contractor together with any incidental payments of a similar character to 3.2 (c)-(k).

3.4 Where principals are working manually their time is chargeable, in respect of the trades practised, in accordance with 3.2.

SECTION 4

Materials and Goods

4.1 The prime cost of materials and goods obtained by the Contractor from stockists or manufacturers is the invoice cost after deduction of all trade discounts but including cash discounts not exceeding 5 per cent, and includes the cost of delivery to site.

4.2 The prime cost of materials and goods supplied from the Contractor's stock is based upon the current market prices plus any appropriate handling charges.

4.3 The prime cost under 4.1 and 4.2 also includes any costs of:

(a) non-returnable crates or other packaging.

(b) returning crates and other packaging less any credit obtainable.

4.4 Any Value Added Tax which is treated, or is capable of being treated, as input tax (as defined in the Finance Act, 1972 or any re-enactment thereof) by the Contractor is excluded.

SECTION 5

Plant, Consumable Stores and Services

5.1 The prime cost of plant and consumable stores as listed below is the cost at hire rates agreed between the parties or in the absence of prior agreement at rates not exceeding those normally applied in the locality at the time when the works are carried out, or on a use and waste basis where applicable:

(a) Machinery in workshops.

(b) Mechanical plant and power-operated tools.

5.1 (c) Scaffolding and scaffold boards.

(d) Non-Mechanical plant excluding hand tools.

(e) Transport including collection and disposal of rubbish.

(f) Tarpaulins and dust sheets.

(g) Temporary roadways, shoring, planking and strutting, hoarding, centering, formwork, temporary fans, partitions or the like.

(h) Fuel and consumable stores for plant and power-operated tools unless included in 5.1 (a), (b), (d) or (e) above.

(j) Fuel and equipment for drying out the works and fuel for testing mechanical services.

5.2 The prime cost also includes the net cost incurred by the Contractor of the following services, excluding any such cost included under Sections 3, 4 or 7:

(a) Charges for temporary water supply including the use of temporary plumbing and storage

(b) Charges for temporary electricity or other power and lighting including the use of temporary installations.

(c) Charges arising from work carried out by local authorities or public undertakings.

(d) Fee, royalties and similar charges.

(e) Testing of materials

JOBBING WORK (Cont'd)

DEFINITION OF PRIME COST OF BUILDING WORKS OF A JOBBING OR MAINTENANCE CHARACTER (Cont'd)

SECTION 5 (Cont'd)

5.2 (Cont'd)

(f) The use of temporary buildings including rates and telephone and including heating and lighting not charged under (b) above.

(g) The use of canteens, sanitary accommodation, protective clothing and other provision for the welfare of persons engaged in the work in accordance with the current Working Rule Agreement and any Act of Parliament, statutory instrument, rule, order, regulation or bye-law.

(h) The provision of safety measures necessary to comply with any Act of Parliament.

(j) Premiums or charges for any performance bonds or insurances which are required by the Building Owner and which are not referred to elsewhere in this Definition.

SECTION 6

Sub-Contracts

6.1 The prime cost of work executed by sub-contractors, whether nominated by the Building Owner or appointed by the Contractor is the amount which is due from the Contractor to the sub-contractors in accordance with the terms of the sub-contracts after deduction of all discounts except any cash discount offered by any sub-contractor to the Contractor not exceeding 2 1/2 per cent.

SECTION 7

Incidental Costs, Overheads and Profit

7.1 The percentages adjustments provided in the building contract, which are applicable to each of the totals of Sections 3-6, provide for the following:-

(a) Head Office Charges.

(b) Off-site staff including supervisory and other administrative staff in the Contractor's workshops and yard.

(c) Payments in respect of public holidays.

(d) Payments in respect of apprentices' study time.

(e) Sick pay or insurance in respect thereof.

(f) Third Party and employer's liability insurance.

(g) Liability in respect of redundancy payments made to employees.

(h) Use, repair and sharpening of non-mechanical hand tools.

(j) Any variation to basic rates required by the Contractor in cases where the building contract provides for the use of a specified schedule of basic plant charges (to the extent that no other provision is made for such variation).

(k) All other liabilities and obligations whatsoever not specifically referred to in this Section nor chargeable under any other section.

(l) Profit.

SPECIMEN ACCOUNT FORMAT

If this Definition of Prime Cost is followed the Contractor's account could be in the following format:- £

Labour (as defined in Section 3)
Add % (see Section 7)
Materials and goods (as defined in Section 4)
Add % (see Section 7)
Plant, consumable stores and services (as defined in Section 5)
Add % (see Section 7)
Sub-contracts (as defined in Section 6)
Add % (see Section 7)

£

VAT to be added if applicable

SCHEDULE OF BASIC PLANT CHARGES

For use in connection with Dayworks under a Building Contract (Fifth revision - 1st May 2001)
This schedule is reproduced by permission of the Royal Institution of Chartered Surveyors.

Explanatory Notes

1. The rates in the Schedule are intended to apply solely to daywork carried out under and incidental to a Building Contract. They are NOT intended to apply to:-
(i) Jobbing or any other work carried out as a main or separate contract; or
(ii) Work carried out after the date of commencement of the Defects Liability Period.

2. The rates apply only to plant and machinery already on site, whether hired or owned by the Contractor.

3. The rates, unless otherwise stated, include the cost of fuel and power of every description, lubricating oils, grease, maintenance, sharpening of tools, replacement of spare parts, all consumable stores and for licences and insurances applicable to items of plant.

4. The rates, unless otherwise stated, do not include the cost of drivers and attendants.

5. The rates are base costs and may be subject to the overall adjustment for price movement, overheads and profit, quoted by the Contractor prior to the placing of the Contract.

6. The rates should be applied to the time during which the plant is actually engaged in daywork.

7. Whether or not plant is chargeable on daywork depends on the daywork agreement in use and the inclusion of an item of plant in this schedule does not necessarily indicate that that item is chargeable.

8. Rates for plant not included in the Schedule or which is not on site and is specifically provided or hired for daywork shall be settled at prices which are reasonably related to the rates in the Schedule having regard to any overall adjustment quoted by the Contractor in the Conditions of Contract.

Item of Plant	Size/Rating	Unit	Rate/Hr
PUMPS			
Mobile Pumps			
Including pump hoses, valves and strainers etc.			
Diaphragm	50mm dia.	Each	0.87
Diaphragm	76mm dia	Each	1.29
Submersible	50mm dia	Each	1.18
Induced flow	50mm dia	Each	1.54
Induced flow	76mm dia	Each	2.05
Centrifugal, self priming	50mm dia	Each	1.96
Centrifugal, self priming	102mm dia	Each	2.52
Centrifugal, self priming	152mm dia	Each	3.87
SCAFFOLDING, SHORING, FENCING			
Complete Scaffolding			
Mobile working towers, single width	1.8m x 0.8m base x 7m high	Each	2.00
Mobile working towers, single width	1.8m x 0.8m base x 9m high	Each	2.80
Mobile working towers, double width	1.8m x 1.4m base x 7m high	Each	2.15
Mobile working towers, double width	1.8m x 1.4m base x 15m high	Each	5.10
Chimney scaffold, single unit		Each	1.79
Chimney scaffold, twin unit		Each	2.05
Chimney scaffold, four unit		Each	3.59
Trestle			
Trestle, adjustable	Any height	Pair	0.10
Trestle, Painters	1.8m high	Pair	0.21
Trestle, Painters	2.4m high	Pair	0.26
Shoring, Planking and Strutting			
'Acrow adjustable prop	Sizes up to 4.9m (open)	Each	0.10
'Strong boy' support attachment		Each	0.15
Adjustable trench struts	Sizes up to 1.67m (open)	Each	0.10
Trench sheet		Metre	0.01
Backbone trench box		Each	1.00
Temporary Fencing			
Including block and coupler			
Site fencing steel grid panel	3.5m x 2.0m	Each	0.08
Anti-climb site steel grid fence panel	3.5m x 2.0m	Each	0.08

SCHEDULE OF BASIC PLANT CHARGES (Cont'd)

Item of Plant	Size/Rating		Unit	Rate/Hr
LIFTING APPLIANCES AND CONVEYORS				
Cranes				
Mobile Cranes				
Rates are inclusive of drivers				
Lorry mounted, telescopic jib				
Two wheel drive	6 tonnes		Each	24.40
Two wheel drive	7 tonnes		Each	25.00
Two wheel drive	8 tonnes		Each	25.62
Two wheel drive	10 tonnes		Each	26.90
Two wheel drive	12 tonnes		Each	28.25
Two wheel drive	15 tonnes		Each	29.66
Two wheel drive	18 tonnes		Each	31.14
Two wheel drive	20 tonnes		Each	32.70
Two wheel drive	25 tonnes		Each	34.33
Four wheel drive	10 tonnes		Each	27.44
Four wheel drive	12 tonnes		Each	28.81
Four wheel drive	15 tonnes		Each	30.25
Four wheel drive	20 tonnes		Each	33.35
Four wheel drive	25 tonnes		Each	35.19
Four wheel drive	30 tonnes		Each	37.12
Four wheel drive	45 tonnes		Each	39.16
Four wheel drive	50 tonnes		Each	41.32
Track-mounted tower crane				
Rates inclusive of driver				
Note: Capacity equals maximum lift tonnes times maximum radius at which it can be lifted				
	Capacity (metre/tonnes) up to	Height under hook above ground (m) up to		
Tower crane	10	17	Each	7.99
Tower crane	15	17	Each	8.59
Tower crane	20	18	Each	9.18
Tower crane	25	20	Each	11.56
Tower crane	30	22	Each	13.78
Tower crane	40	22	Each	18.09
Tower crane	50	22	Each	22.20
Tower crane	60	22	Each	24.32
Tower crane	70	22	Each	23.00
Tower crane	80	22	Each	25.91
Tower crane	110	22	Each	26.45
Tower crane	125	30	Each	29.38
Tower crane	150	30	Each	32.35
Static tower cranes				
Rates inclusive of driver				
To be charged at 90% of the above rates for track mounted tower cranes				
Crane Equipment				
Muck tipping skip	up to 0.25m3		Each	0.56
Muck tipping skip	0.5m		Each	0.67
Muck tipping skip	0.75m3		Each	0.82
Muck tipping skip	1.0m3		Each	1.03
Muck tipping skip	1.5m3		Each	1.18
Muck tipping skip	2.0m3		Each	1.38
Mortar Skips	up to 0.38m3		Each	0.41
Boat skips	1.0m3		Each	1.08
Boat skips	1.5m3		Each	1.33
Boat skips	2.0m3		Each	1.59
Concrete skips. Hand levered	0.5m3		Each	1.00
Concrete skips. Hand levered	0.75m3		Each	1.10
Concrete skips. Hand levered	1.0m3		Each	1.25
Concrete skips. Hand levered	1.5m3		Each	1.50
Concrete skips. Hand levered	2.0m3		Each	1.65
Concrete skips, geared	0.5m3		Each	1.30
Concrete skips, geared	0.75m3		Each	1.40
Concrete skips, geared	1.0m3		Each	1.55
Concrete skips, geared	1.5m3		Each	1.80
Concrete skips, geared	2.0m3		Each	2.05

SCHEDULE OF BASIC PLANT CHARGES (Cont'd)

Item of Plant	Size/Rating		Unit	Rate/Hr
LIFTING APPLIANCES AND CONVEYORS (Cont'd)				
Hoists				
Scaffold hoists	200kg		Each	1.92
Rack and pinion (goods only)	500kg		Each	3.31
Rack and pinion (goods only)	1100kg		Each	4.28
Rack and pinion goods and passenger	15 person, 1200kg		Each	5.62
Wheelbarrow chain sling			Each	0.31
Conveyors				
Belt conveyors				
Conveyor	7.5m long x 400mm wide		Each	6.41
Miniveyor, control box and loading hopper	3m unit		Each	3.59
Other Conveying Equipment				
Wheelbarrow			Each	0.21
Hydraulic superlift			Each	2.95
Pavac slab lifter			Each	1.03
Hand pad and hose attachment			Each	0.26
Lifting Trucks				
Fork lift, two wheel drive	Payload	Maximum Lift		
Fork lift, two wheel drive	1100kg	up to 3.01m	Each	4.87
Fork lift, two wheel drive	2540kg	up to 3.7m	Each	5.12
Fork lift, four wheel drive	1524kg	up to 6.0m	Each	6.04
Fork lift, four wheel drive	2600kg	up to 5.4m	Each	7.69
Lifting Platforms				
Hydraulic platforms (Cherry picker)	7.5m		Each	4.23
Hydraulic platforms (Cherry picker)	13m		Each	9.23
Scissors lift	7.8m		Each	7.56
Telescopic handlers	7m, 2 tonne		Each	7.18
Telescopic handlers	13m, 3 tonne		Each	8.72
Lifting and Jacking gear				
Pipe winch including gantry	1 tonne		Sets	1.92
Pipe winch including gantry	3 tonne		Sets	3.21
Chain block	1 tonne		Each	0.45
Chain block	2 tonne		Each	0.71
Chain block	5 tonnes		Each	1.22
Pull lift (Tirfor winch)	1 tonne		Each	0.64
Pull lift (Tirfor winch)	1.6 tonnes		Each	0.90
Pull lift (Tirfor winch)	3.2 tonnes		Each	1.15
Brother or chain slings, two legs	not exceeding 4.2 tonnes		Set	0.35
Brother or chain slings, two legs	not exceeding 7.5 tonnes		Set	0.45
Brother or chain slings, four legs	not exceeding 3.1 tonnes		Set	0.41
Brother or chain slings, four legs	not exceeding 11.2 tonnes		Set	1.28
CONSTRUCTION VEHICLES				
Lorries				
Plated lorries				
Rates are inclusive of driver				
Platform lorries	7.5 tonnes		Each	19.00
Platform lorries	17 tonnes		Each	21.00
Platform lorries	24 tonnes		Each	26.00
Platform lorries with winch and skid	7.5 tonnes		Each	21.40
Platform lorries with crane	17 tonnes		Each	27.50
Platform lorries with crane	24 tonnes		Each	32.10
Tipper Lorries				
Rates are inclusive of driver				
Tipper lorries	15/17 tonnes		Each	19.50
Tipper lorries	24 tonnes		Each	21.40
Tipper lorries	30 tonnes		Each	27.10
Dumpers				
Site use only (excluding tax, insurance and extra Cost of DERV etc, when operating on highway)				
	Makers capacity			
Two wheel drive	0.8 tonnes		Each	1.20
Two wheel drive	1 tonnes		Each	1.30
Two wheel drive	1.2 tonnes		Each	1.60
Four wheel drive	2 tonnes		Each	2.50
Four wheel drive	3 tonnes		Each	3.00

SCHEDULE OF BASIC PLANT CHARGES (Cont'd)

Item of Plant	Size/Rating	Unit	Rate/Hr
CONSTRUCTION VEHICLES (Cont'd)			
Dumpers (Cont'd)			
Four wheel drive	4 tonnes	Each	3.50
Four wheel drive	5 tonnes	Each	4.00
Four wheel drive	6 tonnes	Each	4.50
Dumper Trucks			
Rates are inclusive of drivers			
Dumper trucks	10/13 tonnes	Each	20.00
Dumper trucks	18/20 tonnes	Each	20.40
Dumper trucks	22/25 tonnes	Each	26.30
Dumper trucks	35/40 tonnes	Each	36.60
Tractors			
Agricultural type			
Wheeled, rubber-clad tyred			
Light	48 h.p.	Each	4.65
Heavy	65 h.p.	Each	5.15
Crawler Tractors			
With bull or angle dozer	80/90 h.p.	Each	21.40
With bull or angle dozer	115/130 h.p.	Each	25.10
With bull or angle dozer	130/150 h.p.	Each	26.00
With bull or angle dozer	155/175 h.p.	Each	27.74
With bull or angle dozer	210/230 h.p.	Each	28.00
With bull or angle dozer	300/340 h.p.	Each	31.10
With bull or angle dozer	400/440 h.p.	Each	46.90
With loading shovel	0.8m3	Each	25.00
With loading shovel	1.0m3	Each	28.00
With loading shovel	1.2m3	Each	32.00
With loading shovel	1.4m3	Each	36.00
With loading shovel	1.8m3	Each	45.00
Light Vans			
Ford Escort or the like		Each	4.74
Ford Transit or the like	1.0 tonnes	Each	6.79
Luton Box Van or the like	1.8 tonnes	Each	8.33
Water/Fuel Storage			
Mobile water container	110 litres	Each	0.28
Water bowser	1100 litres	Each	0.55
Water bowser	3000 litres	Each	0.74
Mobile fuel container	110 litres	Each	0.28
Fuel bowser	1100 litres	Each	0.65
Fuel bowser	3000 litres	Each	1.02
EXCAVATORS AND LOADERS			
Excavators			
Wheeled hydraulic	7/10 tonnes	Each	12.00
Wheeled hydraulic	11/13 tonnes	Each	12.70
Wheeled hydraulic	15/16 tonnes	Each	14.80
Wheeled hydraulic	17/18 tonnes	Each	16.70
Wheeled hydraulic	20/23 tonnes	Each	16.70
Crawler, hydraulic	12/14 tonnes	Each	12.00
Crawler, hydraulic	15/17 tonnes	Each	14.00
Crawler, hydraulic	20/23 tonnes	Each	16.00
Crawler, hydraulic	25/30 tonnes	Each	21.00
Crawler, hydraulic	30/35 tonnes	Each	30.00
Mini excavators	1000/1500kg	Each	4.50
Mini excavators	2150/2400kg	Each	5.50
Mini excavators	2700/3500kg	Each	6.50
Mini excavators	3500/4500kg	Each	8.50
Mini excavators	4500/6000kg	Each	9.50
Loaders			
Wheeled skip loader		Each	4.50
Shovel loaders, four wheel drive	1.6m3	Each	12.00
Shovel loaders, four wheel drive	2.4m3	Each	19.00
Shovel loaders, four wheel drive	3.6m3	Each	22.00
Shovel loaders, four wheel drive	4.4m3	Each	23.00
Shovel loaders, crawlers	0.8m3	Each	11.00

SCHEDULE OF BASIC PLANT CHARGES (Cont'd)

Item of Plant	Size/Rating	Unit	Rate/Hr
EXCAVATORS AND LOADERS (Cont'd)			
Loaders (Cont'd)			
Shovel loaders, crawlers	1.2m3	Each	14.00
Shovel loaders, crawlers	1.6m3	Each	16.00
Shovel loaders, crawlers	2m3	Each	17.00
Skid steer loaders wheeled	300/400kg payload	Each	6.00
Excavators loaders			
Wheeled tractor type with back-hoe excavator			
Four wheel drive	2.5/3.5 tonnes	Each	7.00
Four wheel drive, 2 wheel steer	7/8 tonnes	Each	9.00
Four wheel drive, 4 wheel steer	7/8 tonnes	Each	10.00
Crawler hydraulic	12 tonnes	Each	20.00
Crawler hydraulic	20 tonnes	Each	16.00
Crawler hydraulic	30 tonnes	Each	35.00
Crawler hydraulic	40 tonnes	Each	38.00
Attachments			
Breakers for excavation		Each	7.50
Breakers for mini excavation		Each	3.60
Breakers for back-hoe excavators/loaders		Each	6.00
COMPACTION EQUIPMENT			
Rollers			
Vibrating roller	368kg-420kg	Each	1.68
Single roller	533kg	Each	1.92
Single roller	750kg	Each	2.41
Twin roller	698kg	Each	1.93
Twin roller	851kg	Each	2.41
Twin roller with seat end steering wheel	1067kg	Each	3.03
Twin roller with seat end steering wheel	1397kg	Each	3.17
Pavement rollers	3-4 tonnes dead weight	Each	3.18
Pavement rollers	4-6 tonnes	Each	4.13
Pavement rollers	6-10 tonnes	Each	4.84
Rammers			
Tamper rammer 2 stroke-petrol	225mm-275mm	Each	1.59
Soil Compactors			
Plate compactor	375mm-400mm	Each	1.20
Plate compactor rubber pad	375mm-1400mm	Each	0.33
Plate compactor reversible plate-petrol	400mm	Each	2.20
CONCRETE EQUIPMENT			
Concrete/Mortar Mixes			
Open drum without hopper	0.09/0.06m3	Each	0.62
Open drum without hopper	0.12/0.09m3	Each	0.68
Open drum without hopper	0.15/0.10m3	Each	0.72
Open drum with hopper	0.20/0.15m3	Each	0.80
Concrete/Mortar Transport Equipment			
Concrete pump including hose, valve and couplers			
Lorry mounted concrete pump	23m max distance	Each	36.00
Lorry mounted concrete pump	50m max distance	Each	46.00
Concrete Equipment			
Vibrator. Poker, petrol beam	up to 75mm dia	Each	1.62
Air vibrator (excluding compressor and hose)	up to 75mm dia	Each	0.79
Extra poker heads	25/36/60mm dia	Each	0.77
Vibrating screed unit with beam	5m	Each	1.77
Vibrating screed unit with adjustable beam	3m-5m	Each	2.18
Power float	725mm-900mm	Each	1.72
Power grouter		Each	0.92
TESTING EQUIPMENT			
Pipe Testing Equipment			
Pressure testing pump, electric		Sets	1.87
Pipe pressure-testing equipment, hydraulic		Sets	2.46
Pressure test pump		Sets	0.64

SCHEDULE OF BASIC PLANT CHARGES (Cont'd)

Item of Plant	Size/Rating	Unit	Rate/Hr
SITE ACCOMMODATION AND TEMPORARY SERVICES			
Heating Equipment			
Space heaters-propane	80,000Btu/hr	Each	0.77
Space heaters-propane/electric	125,000Btu/hr	Each	1.56
Space heaters-propane/electric	250,000Btu/hr	Each	1.79
Space heaters-propane	125,000Btu/hr	Each	1.33
Space heaters-propane	260,000Btu/hr	Each	1.64
Cabinet heaters		Each	0.41
Cabinet heaters catalytic		Each	0.46
Electric halogen heaters		Each	1.28
Ceramic heaters	3kW	Each	0.79
Fan heaters	3kW	Each	0.41
Cooling fan		Each	1.15
Mobile cooling unit-small		Each	1.38
Mobile cooling unit-large		Each	1.54
Air-conditioning unit		Each	2.62
Site Lighting and Equipment			
Tripod floodlight	500W	Each	0.36
Tripod floodlight	1000W	Each	0.34
Towable floodlight	4x1000W	Each	2.00
Hand held floodlight	500W	Each	0.22
Rechargeable light		Each	0.62
Inspection light		Each	0.15
Plasterers light		Each	0.56
Lighting mast		Each	0.92
Festoon light string	33m	Each	0.31
Site Electrical Equipment			
Extension leads	240V/14m	Each	0.20
Extension leads	110V/14m	Each	0.20
Cable reel	25m110V/24V	Each	0.28
Cable reel	50m110V/24V	Each	0.33
4 way junction box	110V	Each	0.17
Power Generating Units			
Generator petrol	2kVA	Each	1.08
Generator silenced petrol	2kVA	Each	1.54
Generator petrol	3kVA	Each	1.38
Generator diesel	5kVA	Each	1.92
Generator silenced diesel	8kVA	Each	3.59
Generator silenced diesel	15kVA	Each	7.69
Tail adaptor	240V	Each	0.20
Transformers			
Transformer	3kVA	Each	0.36
Transformer	5kVA	Each	0.51
Transformer	7.5kVA	Each	0.82
Transformer	10kVA	Each	0.87
Rubbish Collection and Disposal Equipment			
Rubbish Chutes			
Standard plastic module	1m section	Each	0.18
Steel liner insert		Each	0.26
Steel top hopper		Each	0.20
Plastic side entry hopper		Each	0.20
Plastic side entry hopper liner		Each	0.20
Dust Extraction Plant			
Dust extraction unit, light duty		Each	1.03
Dust extraction unit, heavy duty		Each	1.64
SITE EQUIPMENT			
Welding Equipment			
Arc-(Electric) complete with Leads.			
Welder generator-petrol	200amp	Each	2.26
Welder generator-diesel	300/350amp	Each	3.33
Welder generator-diesel	4000amp	Each	4.74
Extra welding lead sets		Each	0.29

SCHEDULE OF BASIC PLANT CHARGES (Cont'd)

Item of Plant	Size/Rating	Unit	Rate/Hr
SITE EQUIPMENT (Cont'd)			
Welding Equipment (Cont'd)			
Gas-Oxy Welder			
Welding and cutting set (including oxygen and acetylene, excluding underwater equipment and thermic boring)			
Small		Each	1.41
Large		Each	2.00
Mig welder		Each	1.00
Fume extractor		Each	0.92
Road works Equipment			
Traffic lights, mains/generator	2-way	Set	4.01
Traffic lights, mains/generator	3-way	Set	7.92
Traffic lights, mains/generator	4-way	Set	9.81
Traffic lights, mains/generator-trailer mounted	2-way	Set	3.98
Flashing light		Each	0.20
Road safety cone	450mm	10	0.26
Safety cone	750mm	10	0.38
Safety barrier plank	1.25m`	Each	0.03
Safety barrier plank	2m`	Each	0.04
Road sign		Each	0.26
DPC Equipment			
Damp proofing injection machine		Each	1.49
Cleaning Equipment			
Vacuum cleaner (industrial wet) single motor		Each	0.62
Vacuum cleaner (industrial wet) twin motor		Each	1.23
Vacuum cleaner (industrial wet) triple motor		Each	1.44
Vacuum cleaner (industrial wet) back pack		Each	0.97
Pressure washer, light duty, electric	1450PSI	Each	0.97
Pressure washer. Heavy duty, Diesel	2500PSI	Each	2.69
Cold pressure washer. Electric		Each	1.79
Hot pressure washer, petrol		Each	2.92
Cold pressure washer. petrol		Each	2.00
Sandblast attachment to last washer		Each	0.54
Drain cleaning attachment to last washer		Each	0.31
Surface Preparation Equipment			
Rotavators	5 h.p.	Each	1.67
Scabbler, up to three heads		Each	1.15
Scabbler, pole		Each	1.50
Scabbler, multi-headed floor		Each	4.00
Floor preparation machine		Each	2.82
Compressors and Equipment			
Portable Compressors			
Compressors – electric	0.23m3/min	Each	1.59
Compressors – petrol	0.28m3/min	Each	1.74
Compressors – petrol	0.71m3/min	Each	2.00
Compressors – diesel	up to 2.83m3/min	Each	1.24
Compressors – diesel	up to 3.68m3/min	Each	1.49
Compressors – diesel	up to 4.25m3/min	Each	1.60
Compressors – diesel	up to 4.81m3/min	Each	1.92
Compressors – diesel	up to 7.64m3/min	Each	3.08
Compressors – diesel	up to 11.32m3/min	Each	4.23
Compressors – diesel	up to 18.40m3/min	Each	5.73
Mobile Compressors			
Lorry mounted compressors *(machine plus lorry only)*	2.86-4.24m3/min	Each	12.50
Tractor mounted compressors *(Machine plus rubber tyred tractor)*	2.86-3.40m3/min	Each	13.50
Accessories (Pneumatic Tools)			
(with and including up to 15m of air hose)			
Demolition pick		Each	1.03
Breakers (with six steels)light	up to 150kg	Each	0.79
Breakers (with six steels)medium	295kg	Each	1.08
Breakers (with six steels)heavy	386kg	Each	1.44
Rock drill (for use with compressor) hand held		Each	0.90
Additional hoses	15m	Each	0.16
Muffler, tool silencer		Each	0.14

SCHEDULE OF BASIC PLANT CHARGES (Cont'd)

Item of Plant	Size/Rating	Unit	Rate/Hr
SITE EQUIPMENT (Cont'd)			
Compressors and Equipment (Cont'd)			
Breakers			
Demolition hammer drill, heavy duty, electric		Each	1.00
Road breaker, electric		Each	1.65
Road breaker. 2 stroke, petrol		Each	2.05
Hydraulic breaker unit, light duty, petrol		Each	2.05
Hydraulic breaker unit, heavy duty, petrol		Each	2.60
Hydraulic breaker unit, heavy duty, diesel		Each	2.95
Quarrying and Tooling Equipment			
Block and stone splitter, hydraulic	600mm x 600mm	Each	1.35
Block and slab splitter, manual		Each	1.10
Steel Reinforcement Equipment			
Bar bending machine – manual	up to 13mm dia.rods	Each	0.90
Bar bending machine – manual	up to 20mm dia.rods	Each	1.28
Bar shearing machine – electric	up to 38mm dia.rods	Each	2.82
Bar shearing machine – electric	up to 40mm dia.rods	Each	3.85
Bar cropper machine – electric	up to 13mm dia.rods	Each	1.54
Bar cropper machine – electric	up to 20mm dia.rods	Each	2.05
Bar cropper machine – electrical	up to 40mm dia.rods	Each	2.82
Bar cropper machine – 3 phase	up to 40mm dia.rods	Each	3.85
Dehumidifiers			
110/240v Water	68 litres extraction per 24 hrs	Each	1.28
110/240v Water	90 litres extraction per 24 hrs	Each	1.85
SMALL TOOLS			
Saws			
Masonry bench saw	350mm – 500mm dia.	Each	2.80
Floor saw	350mm dia., 125mm max.cut	Each	1.90
Floor saw	450mm dia., 150mm max.cut	Each	2.60
Floor saw, reversible	Max. cut 300mm	Each	13.00
Chop/cut off saw, electric	350mm dia.	Each	1.33
Circular saw, electric	230mm dia.	Each	0.60
Tyrannosaw		Each	1.20
Reciprocating saw		Each	0.60
Door trimmer		Each	0.90
Chainsaw, petrol	500mm	Each	2.13
Full chainsaw safety kit		Each	0.50
Worktop jig		Each	0.60
Pipework Equipment			
Pipe bender	15mm-22mm	Each	0.33
Pipe bender, hydraulic	50mm	Each	0.60
Pipe bender, electric	50mm-150mm dia	Each	1.35
Pipe cutter, hydraulic		Each	1.84
Tripod pipe vice		Set	0.40
Ratchet threader	12mm-32mm	Each	0.55
Pipe threading machine, electric	12mm-75mm	Each	2.40
Pipe threading machine, electric	12mm-100mm	Each	3.00
Impact wrench, electric		Each	0.54
Impact wrench, two stroke, petrol		Each	4.49
Impact wrench, heavy duty, electric		Each	1.13
Plumber's furnace, calor gas or similar		Each	2.16
Hand-held Drills and Equipment			
Impact or hammer drill	up to 25mm dia.	Each	0.50
Impact or hammer drill	35mm dia.	Each	0.90
Angle head drills		Each	0.70
Stirrer. Mixer drills		Each	0.70
Paint, Insulation Application Equipment			
Airless spray unit		Each	4.20
Portaspray unit		Each	1.65
HVLP turbine spray unit		Each	1.65
Compressor and spray gun		Each	2.20

SCHEDULE OF BASIC PLANT CHARGES (Cont'd)

Item of Plant	Size/Rating	Unit	Rate/Hr
SMALL TOOLS (Cont'd)			
Other Handtools			
Screwing machine	13mm – 50mm dia.	Each	0.77
Screwing machine	25mm – 100mm dia.	Each	1.57
Staple gun		Each	0.33
Air nail gun	110V	Each	3.33
Cartridge hammer		Each	1.00
Tongue and groove nailer complete with mallet		Each	0.93
Chasing machine	152mm	Each	1.72
Chasing machine	76mm – 203mm	Each	5.99
Floor grinder		Each	3.00
Floor plane		Each	3.67
Diamond concrete planer		Each	2.05
Autofeed screwdriver, electric		Each	1.13
Laminate trimmer		Each	0.64
Biscuit jointer		Each	0.87
Random orbital sander		Each	0.72
Floor sander		Each	1.33
Palm, delta, flap or belt sander		Each	0.38
Saw cutter, 2 stroke, petrol	300mm	Each	1.26
Grinder, angle or cutter	up to 225mm	Each	0.60
Grinder, angle or cutter	300 mm	Each	1.10
Mortar raking tool attachment		Each	0.15
Floor/ polisher scrubber	325mm	Each	1.03
Floor tile stripper		Each	1.74
Wallpaper stripper, electric		Each	0.56
Electric scrapper		Each	0.51
Hot air paint stripper		Each	0.38
Electric diamond tile cutter	All sizes	Each	1.38
Hand tile cutter		Each	0.36
Electric needle gun		Each	1.08
Needle chipping gun		Each	0.72
Pedestrian floor sweeper	1.2m wide	Each	0.87

ELECTRICAL CONTRACTORS

A "Definition of Prime Cost of Daywork carried out under an electrical contract" agreed between the Royal Institution of Chartered Surveyors and the Electrical Contractors Associations can be obtained from the RICS, 12 Great George Street, London SW1P 3AD, Tel. 020 7222 7000

TERRAZZO MARBLE AND MOSAIC SPECIALISTS

Daywork rates approved by the National Federation of Terrazzo Marble & Mosaic Specialists, P.O. Box 2843, London W1A 5PG

Minimum Rates for Dayworks

		£
Terrazzo Marble or Mosaic Craftsman..........	per hour	32.00
Labourers..	per hour	27.80

Certain marble works may involve higher rates than the above
All travelling and waiting time to be paid at the above rates.

The above rates cover Wages, National Insurance, Graduated Pension Contribution, Workmen's Compensation and Third Party Insurance, Annual and Public Holidays Contributions, Head Office supervision, Overheads and Profit. Value added Tax is not included. All known wage increases up to 20th June, 2005 are covered

Non-productive overtime or overtime premiums should be charged at £19.00 per hour for all classes of labour.

Increased labour costs should be plus 20% on basis of clause 31 (e) of R.I.B.A. Standard Form of Building Contract.

Floor polishing machines (single head) £50.00 per day or part day. Dry sanding machines £50.00 per day or part day. Multi-head floor polishing machines, £200 per day or part day. Exclusive of cost of fuel and abrasive stones and discs.

Fares to and from the site, lodging allowances and materials will be charged at cost plus 25%.
Delivery and collection of materials/plant not included in any of the above rates.

HEATING, VENTILATING, AIR CONDITIONING, REFRIGERATION, PIPEWORK AND/OR DOMESTIC ENGINEERING CONTRACTS

DEFINITION OF PRIME COST OF DAYWORK CARRIED OUT UNDER A HEATING, VENTILATING, AIR CONDITIONING, REFRIGERATION, PIPEWORK AND/OR DOMESTIC ENGINEERING CONTRACT

Reproduced by permission of the Royal Institution of Chartered Surveyors and the Heating and Ventilating Contractors' Association.

This definition of Prime Cost is published by the Royal Institution of Chartered Surveyors and the Heating and Ventilating Contractors' Association for convenience and for use by people who choose to use it. Members of the Heating and Ventilating Contractors' Association are not in any way debarred from defining Prime Cost and rendering their accounts for work carried out on that basis in any way they choose. Building owners are advised to reach agreement with contractors on the Definition of Prime Cost to be used prior to entering into a contract or sub-contract.

SECTION 1

Application

1.1 This definition provides a basis for the valuation of daywork executed under such heating, ventilating, air conditioning, refrigeration, pipework, and/or domestic engineering contracts as provide for its use.

1.2 It is not applicable in any other circumstances, such as jobbing or other work carried out as a separate or main contract nor in the case of daywork executed after a date of practical completion.

1.3 The terms "contract" and "contractor" herein shall be read as "sub-contract" and "sub-contractor" as applicable.

SECTION 2

Composition of Total Charges

2.1 The Prime Cost of daywork comprises the sum of the following costs:

(a) Labour as defined in Section 3.

(b) Materials and goods as defined in Section 4.

(c) Plant as defined in Section 5.

2.2 Incidental costs, overheads and profit as defined in Section 6, as provided in the contract and expressed therein as percentage adjustments, are applicable to each of 2.1 (a)-(c).

SECTION 3

Labour

3.1 The standard wage rates, emoluments and expenses referred to below and the standard working hours referred to in 3.2 are those laid down for the time being in the rules or decisions or agreements of the Joint Conciliation Committee of the Heating, Ventilating and Domestic Engineering Industry applicable to the Works (or those of such other body as may be appropriate) and to the grade of operative concerned at the time when and the area where the daywork is executed.

3.2 Hourly base rates for labour are computed by dividing the annual prime cost of labour, based upon the standard working hours and as defined in 3.4, by the number of standard working hours per annum. See example at end of Section 6.

3.3 The hourly rates computed in accordance with 3.2 shall be applied in respect of the time spent by operatives directly engaged on daywork, including those operating mechanical plant and transport and erecting and dismantling other plant (unless otherwise expressly provided in the contract) and handling and distributing the materials and goods used in the daywork.

3.4 The annual prime cost of labour comprises the following:

(a) Standard weekly earnings (i.e. the standard working week as determined at the appropriate rate for the operative concerned).

(b) Any supplemental payments.

(c) Any guaranteed minimum payments (unless included in Section 6.1 (a)-(p)).

(d) Merit money.

(e) Differentials or extra payments in respect of skill, responsibility, discomfort, inconvenience or risk (excluding those in respect of supervisory responsibility - see 3.5).

(f) Payments in respect of public holidays.

(g) Any amounts which may become payable by the contractor to or in respect of operatives arising from the rules etc. referred to in 3.1 which are not provided for in 3.4 (a)-(f) nor in Section 6.1 (a)-(p).

HEATING, VENTILATING, ETC (Cont'd)
Section 3 (Cont'd)

3.4 The annual prime cost of labour comprises the following:-(Cont'd)

(h) Employer's contributions to the Annual Holiday with Pay and Welfare Benefits Scheme or payments in lieu thereof.

(i) Employer's National Insurance contributions as applicable to 3.4 (a)-(h).

(j) Any contribution, levy or tax imposed by Statute, payable by the contractor in his capacity as an employer.

3.5 Differentials or extra payments in respect of supervisory responsibility are excluded from the annual prime cost (see Section 6). The time of principals, staff, foremen, chargehands and the like when working manually is admissible under this Section at the rates for the appropriate grades.

SECTION 4

Materials and Goods

4.1 The prime cost of materials and goods obtained specifically for the daywork is the invoice cost after deducting all trade discounts and any portion of cash discounts in excess of 5 per cent.

4.2 The prime cost of all other materials and goods used in the daywork is based upon the current market prices plus any appropriate handling charges.

4.3 The prime cost referred to in 4.1 and 4.2 includes the cost of delivery to site.

4.4 Any Value Added Tax which is treated, or is capable of being treated, as input tax (as defined by the Finance Act 1972, or any re-enactment or amendment thereof or substitution therefor) by the contractor is excluded.

SECTION 5

Plant

5.1 Unless otherwise stated in the contract, the prime cost of plant comprises the cost of the following:

(a) use or hire of mechanically-operated plant and transport for the time employed on and/or provided or retained for the daywork;

(b) use of non-mechanical plant (excluding non-mechanical hand tools) for the time employed on and/or provided or retained for the daywork;

(c) transport to and from site and erection and dismantling where applicable.

5.2 The use of non-mechanical hand tools and of erected scaffolding, staging, trestles or the like is excluded (see Section 6), unless specifically retained for the daywork.

SECTION 6

Incidental Costs, Overheads and Profit

6.1 The percentage adjustments provided in the contract which are applicable to each of the totals of Sections 3, 4 and 5, comprise the following:

(a) Head office charges.

(b) Site staff including site supervision.

(c) The additional cost of overtime (other than that referred to in 6.2).

(d) Time lost due to inclement weather.

(e) The additional cost of bonuses and all other incentive payments in excess of any included in 3.4.

(f) Apprentices' study time.

(g) Fares and travelling allowances.

(h) Country, lodging and periodic allowances.

(i) Sick pay or insurances in respect thereof, other than as included in 3.4.

(j) Third party and employers' liability insurance.

(k) Liability in respect of redundancy payments to employees.

(l) Employer's National Insurance contributions not included in 3.4.

(m) Use and maintenance of non-mechanical hand tools.

(n) Use of erected scaffolding, staging, trestles or the like (but see 5.2).

(o) Use of tarpaulins, protective clothing, artificial lighting, safety and welfare facilities, storage and the like that may be available on site.

(p) Any variation to basic rates required by the contractor in cases where the contract provides for the use of a specified schedule of basic plant charges (to the extent that no other provision is made for such variation - see 5.1).

(q) In the case of a sub-contract which provides that the sub-contractor shall allow a cash discount, such provision as is necessary for the allowance of the prescribed rate of discount.

(r) All other liabilities and obligations whatsoever not specifically referred to in this Section nor chargeable under any other Section.

(s) Profit.

6.2 The additional cost of overtime where specifically ordered by the Architect/Supervising Officer shall only be chargeable in the terms of a prior written agreement between the parties.

HEATING, VENTILATING, ETC (Cont'd)

Example of calculation of typical standard hourly base rate

Example of calculation of typical standard hourly base rate (as defined in section 3) for Advanced Fitter (qualified gas and arc) and mate employed in the Heating, Ventilating, Air Conditioning, Piping and Domestic Engineering Industry under NJIC Rules based upon rates ruling at 4th February 1980.

	Rate £		Advanced Fitter (qual. gas & arc) £	Rate £	Mate £
Standard Weekly Earnings: 48 weeks x 38 hours	2.26		4122.24	1.64	2991.36
Welding Supplement (gas and arc): 48 weeks x 38 hours	.20		364.80	-	-
Merit Money and other variables as applicable			*		*
			4487.04		2991.36
Employer's National Insurance Contribution at 13.5%			605.75		403.83
			5092.79		3395.19
CITB Annual Levy			60.00		6.00
Weekly Holiday Credit and Welfare Stamp: 48 weeks	9.83		471.84	8.08	387.84
Annual Labour Cost as Defined in Section 3			£ 5624.63		£ 3789.03
Hourly base rates as defined in Section 3 clause 3.2	5624.63 / 1763.2 =	£3.19	£3789.03 / 1763.2 =		£2.15

Note: (1) Standard working hours per annum calculated as follows:

52 weeks at 38 hours	=		1976
Less			
4 weeks holiday at 38 hours	=	152	
8 days Public Holidays at (average) 7.6 hours per day	=	60.8	212.8
			1763.2

(2) Where applicable, Merit Money and other variables (e.g. Daily Abnormal Conditions Money), which attract Employer's National Insurance Contribution, should be included at*.

It should be noted that all labour costs incurred by the Contractor in his capacity as an Employer, other than those contained in the hourly base rate, as defined under Section 3, are to be taken into account under Section 6.

(3) The above example is for the convenience of users only and does not form part of the Definition; all the basic costs are subject to re-examination according to the time when and the area where the daywork is executed.

THE ASSOCIATION OF CIVIL ENGINEERING CONTRACTORS

I.C.E. SCHEDULES of DAYWORKS CARRIED OUT INCIDENTAL TO CONTRACT WORK
August 2003

1. **LABOUR**
2. **MATERIALS**
3. **SUPPLEMENTARY CHARGES**
4. **PLANT**

These Schedules are the Schedules referred to in the I.C.E. Conditions of Contract and have been prepared for use in connection with Dayworks carried out incidental to contract work where no other rates have been agreed. They are not intended to be applicable for Dayworks ordered to be carried out after the contract works have been substantially completed or to a contract to be carried out wholly on a daywork basis. The circumstances of such works vary so widely that the rates applicable call for special consideration and agreement between contractor and employing authority.

1st July 1998

1. LABOUR

Add to the amount of wages paid to operatives................148%

1. "Amount of wages" means:-
   Actual wages and bonus paid, daily travelling allowances (fare and/or time) tool allowance and all prescribed payments including those in respect of time lost due to inclement weather paid to operatives at plain time rates and/or at overtime rates.
2. The percentage addition provides for all statutory charges at the date of publication and other charges including:-

   National Insurance payments
   Normal Contract Works, Third Party, & Employer's Liability Insurances.
   Annual and Public Holidays with Pay.
   Statutory and industry sick pay.
   Welfare benifits
   Industrial Training Levy.
   Redundancy Payments.
   The Employment Rights Act 1996
   Employment Relations Act 1999
   Site supervision and staff including foremen and walking gangers, but the time of the gangers or charge handsworking their gangs is to be paid for as for operatives. Small Tools - such as Picks, Shovels, Barrows, Trowels, Hand Saws, Buckets, Trestles, Hammers, Chisels and all items of a like nature.
   Protective Clothing.
   Head Office charges and profit.
3. The time spent in training, mobilisation, demobilisation etc. for the Dayworks operation is chargeable.
4. All hired plant drivers and labour sub-contractors' accounts to be charged in full (without deduction of any cash discounts not exceeding 2.5%) plus 88%.
5. Subsistence or lodging allowances and periodic travel allowances (fare and/or time) paid to or incurred on behalf of operatives are chargeable at cost plus 12.5%.

2. MATERIALS

Add to cost of materials delivered to site..........................12.5%

1. The percentage addition provides for Head Offices charges and Profit.
2. The cost of materials means the invoiced price of materials including delivery to site without deduction of any cash discounts not exceeding 2.5%.
3. Unloading of materials:-

   The percentage added to the cost of materials excludes the cost of handling which shall be charged in addition. An Allowance for unloading into site stock or storage including wastage should be added where materials are taken from existing stock.

3. SUPPLEMENTARY CHARGES

1. Transport provided by contractors for operatives to, from in and around the site to be charged at the appropriate Schedule Rates.
2. Any other charges incurred in respect of any Dayworks operation including, tipping charges, professional fees, sub-contractor's accounts and the like shall be paid for in full plus 12.5% (without deduction of cash discounts not exceeding 2.5%). Labour subcontractor's being dealt with in Section 1.
3. The cost of operating welfare facilities to be charged by the contractor at cost plus 12.5%.
4. The cost of additional insurance premiums for abnormal contract work or special site conditions to be charged at cost plus 12.5%.
5. The cost of watching and lighting specially necessitated by Dayworks is to be paid for separately at Schedule Rates.

4. PLANT

1. These rates apply <u>only</u> to plant already on site, exclusive of drivers and attendants, but inclusive of fuel and consumable stores unless stated to be charged in addition, repairs and maintenance, insurance of plant but excluding time spent in general servicing.
2. Where plant is hired in specifically for dayworks: plant hire (exclusive of drivers and attendants), fuel, oil and grease, insurance, transport etc., to be charged at full amount of invoice (without deduction of any cash discount not exceeding 2.5%) to which should be added consumables where supplied by the Contractor, all plus 12.5%.
3. Fuel distribution, mobilisation and demobilisation are not included in the rates quoted which shall be an additional charge.
4. Metric capacities are adopted and these are not necessarily exact conversions from their imperial equivalents, but cater for the variations arising from comparison of plant manufacturing firms' ratings.
5. Minimum hire charge will be for the period quoted.
6. Hire rates for plant not included below, shall be settled at prices reasonably related to the rates quoted.
7. The rates provide for Head Office charges and Profit.

CIVIL ENGINEERING - PLANT (Cont'd)

Section & Item No.	Description	Nominal Size or Capacity		Hire Rate	Remarks
	1. ACCESS PLATFORMS	Capacity Kg	Platform Height m		
(1)	Scissor lift self-propelled - yard travel..................	Up to 400	Up to 5.0	4.88	
(2)	Ditto...	Up to 400	Up to 10.0	5.66	
(3)	Ditto...	Up to 800	Up to 9.0	5.59	
(4)	Ditto...	Up to 800	Up to 15.0	8.54	
(5)	Ditto...	Over 800	any height	8.88	
(6)	Scissor lift self-propelled - rough terrain travel.......	Up to 800	Up to 9.0	10.35	
(7)	Ditto...	Over 800	any height	16.11	
(8)	Scissor lift - hand-propelled or towed..................	Any Capacity	Up to 7.0	4.70	
(9)	Ditto...	-	Up to 10.0	4.79	
(10)	Telescopic self-propelled - yard travel..................	Up to 400	Up to 10.0	8.46	
(11)	Ditto...	Up to 400	Up to 15.0	18.81	
(12)	Telescopic self-propelled - rough terrain travel........			+ 95% on above	
(13)	Vehicle mounted platform - telescopic/ articulated (Excluding Vehicle)..	Any Capacity	Up to 10.0	7.33	
(14)	Ditto...	"	Up to 20.0	16.30	
(15)	Ditto...	"	Up to 30.0	34.92	
(16)	Ditto...	"	Up to 40.0	55.23	
	2. ASPHALT EQUIPMENT				
(1)	Asphalt and/or coated macadam spreader crawler or wheeled..	Up to 37kW		24.46	
(2)	Ditto...	Up to 56kW		32.19	
(3)	Ditto...	Up to 82kW		43.17	
(4)	Ditto...	Up to 150kW		60.73	
(5)	Extra for above m/c's used with Dry Lean Concrete..			+ 14%	
(6)	Joint matcher. Separately or in conjunction with longitudinal beams (extra over machine)..................			4.91	
(7)	Joint matcher with addition of transverse control devices or fully automatic equipment capable of control from reference guides (extra over machine)..	Up to 150kW		7.02	
(8)	Self-propelled, metered coated chipping applicator for asphalt work (including trailer)........................	Up to 3.5m width		21.66	
(9)	Ditto...	Up to 4.0m width		21.91	
(10)	Ditto...	Up to 4.5m width		22.69	
(11)	Cold planer..	Up to 0.5m width incl. picks		35.33	
(12)	Ditto...	Up to 1.00m incl. picks		55.13	
(13)	Ditto...	Up to 2.25m incl. picks		157.96	
(14)	Asphalt road burner (portable)........................	-		11.55	
(15)	Ditto................(self propelled)	-		118.39	
	3. BAR BENDING AND BAR SHEARING MACHINES	For mild steel rods Up to mm. diameter			
(1)	Bar bending machine,......................hand operated	Up to 25		4.36	Per day
(2)	Ditto...power driven	Up to 40		2.96	
(3)	Ditto...Elect	Up to 50		3.69	
(4)	Ditto...Diesel	Up to 50		6.20	
(5)	Bar shearing machine,..................hand operated	Up to 25		3.24	Per day
(6)	Ditto...power driven	Up to 40		2.71	
(7)	Ditto.. Elect	Up to 50		3.47	
(8)	Ditto... Diesel	Up to 50		4.57	
		For high tensile steel rods			
(9)	Bar bending machine,......................power driven	Up to 50		5.92	
(10)	Bar shearing machine,.................................Ditto	Up to 50		5.73	
	Note: Jaws to be charged as consumables				
	4. COMPRESSORS	Normal delivery of free air per minute at 7kg per cm2 (for compressors with higher working pressure, rates to be negotiated)			
(1)	Compressor (silenced) (machine only)	Up to 2.5m³/min		6.98	
(2)	Ditto...	Up to 3.0m³/min		7.91	
(3)	Ditto...	Up to 4.0m³/min		8.35	
(4)	Ditto...	Up to 5.0m³/min		9.33	
(5)	Ditto...	Up to 6.0m³/min		9.89	
(6)	Ditto...	Up to 8.0m³/min		11.85	
(7)	Ditto...	Up to 11.0m³/min		15.91	
(8)	Ditto...	Up to 14.0m³/min		21.58	
(9)	Ditto...	Up to 17.0m³/min		28.23	

The rates shown above are per unit machine per hour unless otherwise stated.

CIVIL ENGINEERING - PLANT (Cont'd)

Section & Item No.	Description	Nominal Size or Capacity	Hire Rate	Remarks
	4. COMPRESSORS (Cont'd)			
(10)	Ditto	Up to 20.0m³/min	37.15	
(11)	Ditto	Up to 26.0m³/min	52.65	
(12)	Lorry mounted silenced compressor (machine + lorry only) (site and public highway use - including Tax, Insurance and extra cost of D.E.R.V.)	Up to 3.0m³/min	35.32	
(13)	Ditto	Up to 5.0m³/min	41.96	
(14)	Ditto	Up to 7.5m³/min	50.37	
(15)	Tractor mounted compressor (machine plus rubber tyred tractor with or without front loading bucket) (site use - excluding Tax, Insurance and extra cost D.E.R.V.)	Up to 4.0m³/min Normal delivery of free air per minute at 7 bar (for compressors with higher working pressure, rates to be negotiated.	19.26	
(16)	Air receiver	Up to 2.0m³	8.97	Per day
(17)	Ditto	Up to 4.0m³	10.73	Per day
(18)	Ditto	Up to 8.0m³	17.39	Per day

Note (i) "Super Silenced" compressors (less than 81 dBA at 7m in any direction) would carry a premium of 50% at the above rates.

(ii) No allowance has been made in the above rates for "*written schemes for inspectors*" under the "Pressure Systems Safety Regulations 2000."

Section & Item No.	Description	Nominal Size or Capacity	Hire Rate	Remarks
	5. CONCRETE MIXERS	Wet Capacity		
(1)	Tilting Drum mixer	Up to 90 litres	1.75	
(2)	Ditto	Up to 100 litres	1.85	
(3)	Ditto	Up to 150 litres	2.07	
(4)	Ditto	Up to 185 litres	2.67	
(5)	Ditto	Up to 200 litres	2.72	
(6)	Non-Tilting mixer with integral batch weighing	Up to 300 litres	3.30	
(7)	Ditto	Up to 400 litres	3.82	
(8)	Ditto	Up to 500 litres	4.34	
(9)	Extra for power driven loading shovel (items (1)-(8) inclusive)	-	1.98	
(10)	Extra for aggregate feed apron	2 compartment	0.58	
(11)	Ditto	3 compartment	0.61	
(12)	Tilting or non-tilting drum mixers or pan mixers incorporated into a structure with cement, etc., silos at high or low level and built in weighing and batching equipment	-	Rates to be negotiated	
	Capacity			
(13)	Cement Silo - low level	Up to 13 tonnes	2.43	
(14)	Ditto	Up to 21 tonnes	3.93	
(15)	Ditto	Up to 31 tonnes	4.68	
(16)	Ditto	Up to 51 tonnes	5.43	
(17)	Ditto	Up to 100 tonnes	9.18	
(18)	Extra for autofeed, hydraulic cement weighgear and screw discharge	-	2.98	
(19)	Ditto, with electronic weighgear	-	3.14	
		Maximum Output litre/min		
(20)	Grout mixer(Single drum)	Up to 60	2.94	
(21)	Ditto(Double drum)	Up to 110	3.75	
(22)	Ditto(ditto)	Up to 170	5.23	
(23)	Grout mixer, Roller type	Up to 10 litre/min	2.93	
(24)	Ditto	Up to 22 litre/min	4.16	
(25)	Ditto	Up to 43 litre/min	6.37	
(26)	Grout mixer/pump c/w loading hopper and weighgear and double drum	Up to 22	4.43	
(27)	Grouting machine including pump and hopper	Up to 40	3.34	
(28)	Ditto	Up to 70	4.16	
(29)	Ditto	Up to 110	4.66	
(30)	Grout pump (electric)	Up to 310	5.70	
(31)	Ditto	Up to 480	5.77	
(32)	Agitating tank for use with grout pumps	--	3.32	
(33)	Grout pump hand operated - 38mm	Up to 25	1.55	
(34)	Concrete truck/mixer/agitator (including separate engine with mechanical or hydraulic drive to drum)	Capacity Up to 5.0m³	35.63	
(35)	Ditto	Up to 6.0m³	37.60	
(36)	Ditto	Up to 7.0m³	48.28	

The rates shown above are per unit machine per hour unless otherwise stated.

CIVIL ENGINEERING - PLANT (Cont'd)

Section & Item No.	Description	Nominal Size or Capacity	Hire Rate	Remarks
	6. CONCRETE EQUIPMENT	Maximum Output per hour		
(1)	Concrete pump, skid or trailer mounted including pipe cleaning equipment (excluding piping)........	Up to 26m³	14.98	
(2)	Ditto........	Up to 40m³	19.35	
(3)	Ditto........	Up to 55m³	23.98	
(4)	Ditto........	Up to 85m³	41.24	
(5)	Ditto........	Up to 110m³	65.04	
(6)	Concrete pump (lorry mounted) (exclusive of piping) (including boom and pipe cleaning equipment)........	Up to 50m³	66.82	
(7)	Ditto........	Up to 80m³	156.05	
(8)	Ditto........	Up to 120m³	271.74	
		Diameter		
(9)	Piping per 3m. straight length........	Up to 102mm/per 3m	1.01	Per day
(10)	Ditto........	Up to 127mm/per 3m	1.42	Per day
(11)	Ditto........	Up to 152mm/per 3m	2.43	Per day
(12)	Bend........	Up to 102mm/per 3m	1.49	Per day
(13)	Ditto........	Up to 127mm/per 3m	1.97	Per day
(14)	Ditto........	Up to 152mm/per 3m	2.75	Per day
(15)	Flexible distributor hose per 4m length........	Up to 102mm/per 3m	6.01	Per day
(16)	Ditto........	Up to 127mm/per 3m	10.08	Per day
(17)	Ditto........	Up to 152mm/per 3m	27.62	Per day
(18)	Shut off pipe section........	Up to 102mm/per 3m	8.79	Per day
(19)	Ditto........	Up to 127mm/per 3m	9.50	Per day
(20)	Ditto........	Up to 152mm/per 3m	22.50	Per day
(21)	Vibrator poker(petrol driven)........	-	2.63	
(22)	Ditto(diesel driven)........	-	2.58	
(23)	Ditto(electric)........	-	1.57	
(24)	Ditto (air) (excluding compressor)........	-	1.02	
(25)	Alternator/frequency converter........	Up to 2.5 kVA	1.14	
(26)	Ditto........	Up to 4.5 kVA	1.27	
(27)	Vibrator poker (H/F motor in head type)........	-	1.18	
(28)	Vibrator external type clamp-on electric........	small	1.31	
(29)	Ditto........	medium	1.42	
(30)	Ditto........	large	1.52	
(31)	Ditto air (excluding compressor)........	-	1.11	
(32)	Vibrator tamper type including single timber or metal screed board........	screed length Up to 3.25m	1.77	
(33)	Extra for additional single screed length........	additional per 300mm	0.13	
(34)	Vibrator tamper type including double screed board........	screed length Up to 3.25m	2.16	
(35)	Extra for additional double screed length........	additional per 300mm	0.13	
(36)	Power float(petrol driven)........	-	2.28	
(37)	Ditto(electric driven)........	-	1.76	
(38)	Concrete saw (exclusive of blades and water supply to be charged in addition) manually propelled........	blade diameter Up to 350mm	2.75	
(39)	Ditto........	Up to 450mm	3.41	
(40)	Self-propelled........	Up to 500mm	3.60	
(41)	Ditto(with hydraulic system for blade movement)	Up to 600mm	6.96	
(42)	Ditto........	Up to 900mm	7.89	
(43)	Joint former hand-propelled with vibratory blade........	width Up to 2.4m	2.29	
(44)	Ditto........	Up to 8.2m	2.92	
(45)	Grinder pedestrian operated........	-	2.04	
		Struck Capacity		
(46)	Concrete pram dobbin rubber tyred tipping.	Up to 120 litre	8.29	per day
(47)	Concrete pram dobbin tyred tipping........	Up to 200 litre	8.43	per day
(48)	Ditto........	Up to 280 litre	8.54	per day
(49)	Power driven barrow........	Up to 170 litre	1.59	
	7. CRANES			
(1)	Mobile rubber tyred all terrain type (site use only *excluding* Tax, Insurance and extra cost of DERV)	Maximum S.W.L Up to 13 tonnes	36.68	
(2)	Ditto........	Up to 18 tonnes	41.24	
(3)	Ditto........	Up to 25 tonnes	46.39	
(4)	Ditto........	Up to 40 tonnes	62.56	
(5)	Ditto........	Up to 55 tonnes	81.16	

The rates shown above are per unit machine per hour unless otherwise stated

CIVIL ENGINEERING - PLANT (Cont'd)

Section & Item No.	Description	Nominal Size or Capacity	Hire Rate	Remarks
	7. CRANES			
(6)	Ditto...	Up to 70 tonnes	107.63	
(7)	Crawler mounted cranes (site use only)...	Up to 28 tonnes	45.30	
(8)	Ditto..	Up to 36 tonnes	54.58	
(9)	Ditto..	Up to 56 tonnes	76.88	
(10)	Ditto..	Up to 75 tonnes	96.82	
(11)	Ditto..	Up to 85 tonnes	107.93	
(12)	Ditto..	Up to 115 tonnes	136.41	
(13)	Ditto..	Up to 150 tonnes	166.37	
(14)	Lorry mounted or all terrain Crane (including) Tax, Insurance and extra cost of D.E.R.V.).	Up to 12 tonnes	70.91	
(15)	Ditto..	Up to 20 tonnes	82.41	
(16)	Ditto..	Up to 26 tonnes	92.68	
(17)	Ditto..	Up to 30 tonnes	101.15	
(18)	Ditto..	Up to 45 tonnes	123.52	
(19)	Ditto..	Up to 60 tonnes	146.60	
(20)	Lorry mounted or all terrain Crane (including) Tax, Insurance and extra cost of D.E.R.V.).	Up to 90 tonnes	188.21	
(21)	Ditto..	Up to 110 tonnes	217.07	
(22)	Ditto..	Up to 135 tonnes	255.78	
(23)	Ditto..	Up to 200 tonnes	339.91	
(24)	Ditto..	Up to 300 tonnes	474.59	
(25)	Tower crane – electric – (erected complete with ballast). Standard trolley or luffing jib, rail mounted (static or travelling) or fixed on sill plates or fixing angles. Capacity (in metre tonnes) is the maximum lift in tonnes x maximum radius at which that load can be lifted when the crane is fitted with its maximum jib length. Up to 30m.t	Height under hook above ground for trolley jib cranes (jib pivot height for luffing jib cranes) Up to 25m	20.45	
(26)	Ditto up to 40m.t	Up to 30m	24.55	
(27)	Ditto up to 50m.t	Up to 30m	26.91	
(28)	Ditto up to 65m.t	Up to 30m	30.62	
(29)	As above at (25) but excluding luffing jib cranes 65 - 80m.t..	Up to 36m	36.20	
(30)	Ditto up to 110m.t..	Up to 36m	44.16	
(31)	Ditto up to 140m.t..	Up to 36m	51.76	
(32)	Ditto up to 170m.t..	Up to 36m	59.57	
(33)	Ditto up to 200m.t..	Up to 36m	66.85	
(34)	Ditto up to 250m.t..	Up to 36m	78.75	
(35)	Luffing jib cranes exeeding 65m.t rating to be charged at the following percentage of the above rates		126%	
(36)	Extra for extend height of tower (including accessories) for trolley jib cranes up to 65m.t	per 1m of height	1.10	
(36)	Ditto up to 140m.t..	per 1m of height	1.15	
(38)	Ditto up to 170m.t..	per 1m of height	1.17	
(39)	Ditto up to 250m.t..	per 1m of height	1.22	
(40)	Extra for extended height of tower (including accessories) for luffing jib cranes up to 60m.t	per 1m of height	1.10	
(41)	Ditto up to 100m.t..	per 1m of height	1.15	
(42)	Ditto up to 170m.t..	per 1m of height	1.17	
(43)	Tower crane track complete with Fish plates, bolts, concr. sleeper pads, and gauge holders (excl. installation)	Up to 61m.t./per m	4.29	per day
(44)	Ditto..	Up to 300m.t./per m	4.52	per day
	8. CRANE EQUIPMENT	struck capacity		
(1)	Crane Grab, all types, excavating (normal weight)..	Up to 0.50m³	4.52	
(2)	Ditto..	Up to 0.75m³	4.99	
(3)	Ditto..	Up to 1.25m³	5.58	
(4)	Ditto (Ditto)..	Up to 1.50m³	5.80	
(5)	Ditto (Ditto)..	Up to 2.00m³	6.13	
(6)	Ditto (heavy duty)............................	Up to 0.50m³	5.91	
(7)	Ditto...	Up to 0.75m³	6.40	
(8)	Ditto..	Up to 1.25m³	7.01	
(9)	Ditto..	Up to 1.50m³	7.23	
(10)	Ditto (Ditto)............................	Up to 2.00m³	7.57	
(11)	Crane Grab, all types, rehandling (normal weight)..	Up to 0.50m³	3.59	

The rates shown above are per unit machine per hour unless otherwise stated.

GENERAL INFORMATION

CIVIL ENGINEERING - PLANT (Cont'd)

Section & Item No.	Description	Nominal Size or Capacity		Hire Rate	Remarks
	8. CRANE EQUIPMENT (Cont'd)	struck capacity			
(12)	Ditto (Ditto)......................................	Up to 0.75m³		3.71	
(13)	Ditto (Ditto)......................................	Up to 1.25m³		3.88	
(14)	Ditto (Ditto)......................................	Up to 1.50m³		3.94	
(15)	Ditto (Ditto)......................................	Up to 2.00m³		4.04	
(16)	Ditto (heavy duty)...............................	Up to 0.50m³		3.75	
(17)	Ditto (Ditto)......................................	Up to 0.75m³		4.11	
(18)	Ditto (Ditto)......................................	Up to 1.25m³		4.55	
(19)	Ditto (Ditto)......................................	Up to 1.50m³		4.71	
(20)	Ditto (Ditto)......................................	Up to 2.00m³		4.96	
(21)	Skip, muck tipping circular.....................	Up to 0.10m³		0.61	
(22)	Ditto..	Up to 0.15m³		0.64	
(23)	Ditto..	Up to 0.20m³		0.66	
(24)	Ditto..	Up to 0.50m³		0.78	
(25)	Ditto..	Up to 0.75m³		0.88	
(26)	Ditto..	Up to 1.00m³		0.98	
(27)	Ditto..	Up to 1.50m³		1.19	
(28)	Ditto..	Up to 2.00m³		1.40	
(29)	Skip, concrete lay down or roll over standard front or bottom discharge..................	Up to 0.50m³		1.23	
(30)	Ditto..	Up to 0.75m³		1.38	
(31)	Ditto..	Up to 1.00m³		1.48	
(32)	Ditto..	Up to 1.50m³		1.62	
(33)	Ditto..	Up to 2.00m³		1.73	
(34)	Skip, concrete roll over geared or hydraulic hand operated clamshell.....................	Up to 0.50m³		1.20	
(35)	Ditto..	Up to 0.75m³		1.54	
(36)	Ditto..	Up to 1.00m³		1.79	
(37)	Ditto..	Up to 1.50m³		2.13	
(38)	Ditto..	Up to 2.00m³		2.38	
		Chain diameter mm.	Safe working load tonnes		
(39)	Chain slings or brothers - Single...............	6	1.12	1.06	per day
(40)	Double..(at 90 degrees)......................	6	1.60	1.35	per day
(41)	Ditto - Single...	10	3.15	1.22	per day
(42)	Ditto - Double(Ditto)...........................	10	4.25	1.55	per day
(43)	Ditto - Single...	13	5.30	1.41	per day
(44)	Ditto - Double(Ditto)............................	13	7.50	1.91	per day
(45)	Ditto - Single...	16	8.00	1.66	per day
(46)	Ditto - Double(Ditto)............................	16	11.20	2.27	per day
(47)	Ditto - Single...	19	11.20	2.05	per day
(48)	Ditto - Double(Ditto)..........................	19	16.00	2.77	per day
(49)	Ditto - Single...	22	15.00	2.75	per day
(50)	Ditto - Double(Ditto)............................	22	21.20	4.20	per day
(51)	Grab tag lines..	Up to 1.20m³ grab		6.29	per day
(52)	Ditto..	Over 2.50m³ grab		9.49	per day
		Dead Weight			
(53)	Demolition Ball..	Up to 260Kg		1.45	per day
(54)	Ditto..	Up to 510Kg		1.94	per day
(55)	Ditto..	Up to 770Kg		2.36	per day
(56)	Ditto..	Up to 1020Kg		2.84	per day
(57)	Ditto..	Up to 1300Kg		3.30	per day
(58)	Ditto..	Up to 1550Kg		3.98	per day
(59)	Ditto..	Up to 1800Kg		4.12	per day
(60)	Ditto..	Up to 2050Kg		4.78	per day
	9. DUMPERS				
(1)	Small Dumper (site use only - excluding Tax, Insurance and extra cost of D.E.R.V.) 2 wheel drive – hydraulic tipping	Maker's Rated payload Up to 1500Kg		5.29	
(2)	Ditto..	Up to 2150Kg		5.98	
(3)	Ditto - 4 wheel drive, hydraulic tipping......	Up to 2100Kg		6.08	
(4)	Ditto..	Up to 2550Kg		6.71	
(5)	Ditto..	Up to 3100Kg		7.45	
(6)	Ditto..	Up to 4100Kg		9.01	
(7)	Ditto..	Up to 5100Kg		10.66	
(8)	Ditto..	Up to 6100Kg		11.67	
(9)	Ditto..	Up to 10000Kg		13.82	
(10)	Extra for high discharge on equivalent standard dumper rate..............................			+ 15%	

The rates shown above are per unit machine per hour unless otherwise stated

CIVIL ENGINEERING - PLANT (Cont'd)

Section & Item No.	Description	Nominal Size or Capacity	Hire Rate	Remarks
	9. DUMPERS (Cont'd)			
(11)	Extra for Turntable Skip on equivalent standard dumper rate.............................		+ 22%	
(12)	Rear Dump Truck..................................	Up to 15500Kg	23.60	
(13)	Ditto..	Up to 17000Kg	27.16	
(14)	Ditto..	Up to 22000Kg	35.15	
(15)	Ditto..	Up to 25000Kg	42.38	
(16)	Ditto..	Up to 28000Kg	51.36	
(17)	Ditto..	Up to 32000Kg	60.72	
(18)	Ditto..	Up to 38000Kg	76.23	
(19)	Ditto..	Up to 45000Kg	95.07	
(20)	Ditto..	Up to 50000Kg	114.97	
(21)	Articulated Dump Truck..........................	Up to 12200Kg	30.90	
(22)	Ditto..	Up to 18500Kg	40.12	
(23)	Ditto..	Up to 23000Kg	44.49	
(24)	Ditto..	Up to 30000Kg	52.70	
(25)	Ditto..	Up to 35000Kg	64.62	
(26)	Ditto..	Up to 40000Kg	76.99	
	10. EXCAVATORS			
(1)	Rope operated, full circle slew, crawler mounted, with single equipment(dragline)	Maker's Rated Dragline Capacity Up to .70m³	24.42	
(2)	Ditto..	Up to .80m³	31.00	
(3)	Ditto..	Up to 1.00m³	42.00	
(4)	Ditto..	Up to 1.20m³	50.76	
(5)	Ditto..	Up to 1.40m³	59.35	
(6)	Ditto..	Up to 1.50m³	64.35	
(7)	Ditto..	Up to 2.00m³	79.47	
(8)	Ditto..	Up to 2.30m³	85.75	
(9)	Ditto..	Up to 2.70m³	97.35	
(10)	Roper operated, full circle slew, crawler Mounted, with single equipment (face shovel)................................	Maker's Rated Face Shovel Capacity Up to .70m³	23.81	
(11)	Ditto..	Up to 1.00m³	41.94	
(12)	Ditto..	Up to 1.40m³	61.39	
(13)	Ditto..	Up to 1.50m³	65.01	
(14)	Ditto..	Up to 2.50m³	90.36	
(15)	Hydraulic, full circle slew, crawler or wheel mounted, with single equipment	Maker's Rated Nominal weight of machine Up to 2.0 tonnes	10.35	
(16)	Ditto..	Up to 3.0 tonnes	12.95	
(17)	Ditto..	Up to 4.0 tonnes	14.81	
(18)	Ditto..	Up to 6.0 tonnes	17.53	
(19)	Ditto..	Up to 11.0 tonnes	21.89	
(20)	Ditto..	Up to 14.0 tonnes	26.22	
(21)	Ditto..	Up to 17.0 tonnes	29.47	
(22)	Ditto..	Up to 21.0 tonnes	33.43	
(23)	Ditto..	Up to 25.0 tonnes	37.65	
(24)	Ditto..	Up to 30.0 tonnes	47.94	
(25)	Ditto..	Up to 38.0 tonnes	62.70	
(26)	Ditto..	Up to 55.0 tonnes	84.53	
(27)	Ditto..	Up to 75.0 tonnes	111.82	
(28)	Extra for m/c mounted percussion breaker	Unit weight less cradle Up to 100Kg	5.34	
(29)	Ditto..	Up to 500Kg	8.55	
(30)	Ditto..	Up to 1000Kg	12.57	
(31)	Ditto..	Up to 1500Kg	16.59	
(32)	Extra for m/c mounted percussion breaker	Up to 2000Kg	20.61	
(33)	Ditto..	Up to 3000Kg	28.65	
(34)	Hydraulic, offset or centre post, half circle slew, wheeled, dual purpose (back hoe/loader)......................................	Maker's Rated Loader Bucket Capacity Up to .60m³	14.94	
(35)	Ditto..	Up to .80m³	19.87	
(36)	Ditto..	Up to 1.00m³	21.02	
(37)	Extra on above rated for low ground pressure		+15%	
(37)	Excavator mats................................		To be negotiated	

The rates shown above are per unit machine per hour unless otherwise stated.

CIVIL ENGINEERING - PLANT (Cont'd)

Section & Item No.	Description	Nominal Size or Capacity	Hire Rate	Remarks
	11. GENERATING SETS AND TRANSFORMERS	Nominal rating		
(1)	Generating set......................................	Up to 2.0kVa	3.54	
(2)	Ditto...	Up to 4.0kVa	4.29	
(3)	Ditto...	Up to 10.0kVa	4.16	
(4)	Ditto...	Up to 16.0kVa	4.79	
(5)	Ditto...	Up to 30.0kVa	6.15	
(6)	Ditto...	Up to 50.0kVa	7.88	
(7)	Ditto...	Up to 70.0kVa	9.77	
(8)	Ditto...	Up to 90.0kVa	10.82	
(9)	Ditto...	Up to 100.0kVa	12.16	
(10)	Ditto...	Up to 120.0kVa	13.36	
(11)	Ditto...	Up to 160.0kVa	16.40	
(12)	Ditto...	Up to 210.0kVa	19.95	
(13)	Ditto...	Up to 280.0kVa	24.00	
(14)	Ditto...	Up to 330.0kVa	27.95	
(15)	Transformer, stationary...(air cooled)	Up to 1.0kVva	0.48	
(16)	Ditto (Ditto)....................................	Up to 2.5kVa	0.49	
(17)	Ditto (Ditto)....................................	Up to 5.0kVa	0.55	
(18)	Ditto (Ditto)....................................	Up to 7.5kVa	0.61	
(19)	Ditto (Ditto)....................................	Up to 10.0kVa	0.67	
(20)	Ditto (Ditto)....................................	Up to 12.5kVa	0.71	
(21)	Ditto (Ditto)....................................	Up to 15.0kVa	0.76	
(22)	Ditto (oil immersed)..........................	Up to 35.0kVa	1.16	
(23)	Ditto (Ditto)....................................	Up to 50.0kVa	1.19	
(24)	Ditto (Ditto)....................................	Up to 60.0kVa	1.22	
(25)	Ditto (Ditto)....................................	Up to 100.0kVa	1.31	
(26)	Ditto (Ditto)....................................	Up to 200.0kVa	1.54	
(27)	Ditto (Ditto)....................................	Up to 300.0kVa	1.77	
(28)	Ditto (Ditto)....................................	Up to 500.0kVa	2.20	
(29)	Mobile lighting unit, 2 light tungsten halogen including 1 1/2kVa generator.......	Tower/mast height Up to 8.0m	4.46	
(30)	Ditto, 4 light mercury vapour including 6 1/4kVa generator................................	Up to 16.0m	6.51	
(31)	Ditto, 4 light tungsten halogen including 7 1/2kVa generator................................	Up to 20.0m	6.88	
	Note: All the above rates apply to medium voltage only. For high voltage rates are to be negotiated.			
	12. HOISTS			
(1)	Goods hoist, cantilever or rope suspended c/w safety devices and limits, bottom gate and one landing gate, interlocked. Erected to 9m top landing height..............................	Up to 500Kg	8.15	
(2)	Extra for mast extension including ties.......	per 1m	0.08	
(3)	Extra for additional landing gate................		0.10	
(4)	Goods hoists, rack and pinion complete with safety devices and limits, bottom gate and one landing gate interlocked. Erected to 50m landing height...................................	Up to 400Kg	8.41	
(5)	Extra for mast extension for item (4) including cable, ties etc............................	per 1m	0.06	
(6)	Extra for additional interlocking landing gate for item (4)..		0.07	
(7)	As item (4) - Single cage...........................	Up to 800Kg	10.68	
(8)	Ditto..	Up to 1500Kg	12.11	
(9)	Item (7) or (8) when fitted with twin cage....		+ 66% on above rates	
(10)	Extra for mast extension including cable, ties, etc. for items (7) or (8)	per 1m	0.07	
(11)	Ditto (for item (9))......................................	per 1m	0.12	
(12)	Extra for additional interlocking landing gate for item (7) or (8).......................................		0.07	
(13)	Ditto (for item (9))......................................		0.14	
(14)	Passenger/Goods hoist, rack and pinion complete with base unit, bottom gate and one set interlocking landing gate, cable and mast to 60m top landing height(Single Cage)	Up to 1000Kg or 12 men	17.81	
(15)	Ditto..	Up to 1600Kg or 20 men	19.44	
(16)	Ditto(Twin Cage)	Up to 1000Kg or 12 men	32.25	
(17)	Ditto..	Up to 1600Kg or 20 men	32.78	

The rates shown above are per unit machine per hour unless otherwise stated.

CIVIL ENGINEERING - PLANT (Cont'd)

Section & Item No.	Description	Nominal Size or Capacity		Hire Rate	Remarks
	12. HOISTS (Cont'd)				
(18)	Extra for mast extension for single cage Pass/Goods hoist (including cable, ties etc.)..	per 1m		0.13	
(19)	Extra for additional interlocking landing gates for single cage pass/goods hoist......			0.39	
(20)	Extra for mast extension for Twin cage Pass/Goods hoist (including cable, ties etc.)..	per 1m		0.19	
(21)	Extra for additional interlocking landing gates for Twin cage Pass/Goods hoist......			0.81	
(22)	Scaffold Hoist..	Up to 75Kg		3.13	
(23)	Ditto..	Up to 200Kg		3.23	
	13. LIFTING AND JACKING GEAR				
(1)	Shear legs, steel, 5 metres.......................	Up to 1 tonne		6.49	per day
(2)	Pipe gantry..	Up to 3 tonnes		10.97	per day
(3)	Chain blocks..	Up to 1 tonne		4.89	per day
(4)	Ditto..	Up to 2 tonnes		5.21	per day
(5)	Ditto..	Up to 3.5 tonnes		5.75	per day
(6)	Ditto..	Up to 4.5 tonnes		6.15	per day
(7)	Ditto..	Up to 5.5 tonnes		6.58	per day
(8)	Ditto..	Up to 6.5 tonnes		7.04	per day
(9)	Ditto..	Up to 8 tonnes		7.80	per day
(10)	Ditto..	Up to 10 tonnes		8.92	per day
(11)	Ditto..	Up to 15 tonnes		12.28	per day
(12)	Ditto..	Up to 20 tonnes		16.45	per day
(13)	Hydraulic Jack (Hand operated)................	Up to 10.00 tonnes		5.33	per day
(14)	Ditto..	Up to 15.00 tonnes		5.44	per day
(15)	Ditto..	Up to 20.00 tonnes		5.56	per day
(16)	Ditto..	Up to 30.00 tonnes		5.84	per day
(17)	Ditto..	Up to 60.00 tonnes		7.00	per day
(18)	Ditto..	Up to 100.00 tonnes		9.25	per day
(19)	Ratchet Jack..	Up to 5.50 tonnes		5.23	per day
(20)	Ditto..	Up to 10.50 tonnes		5.88	per day
(21)	Ditto..	Up to 16.00 tonnes		6.36	per day
(22)	Ditto..	Up to 21.00 tonnes		6.71	per day
		Safe working load			
		Lifting	Pulling		
(23)	Lifting and pulling machine.......................	800Kg	1200Kg	5.32	per day
(24)	Ditto..	1600Kg	2500Kg	5.70	per day
(25)	Ditto..	3000Kg	5000Kg	6.53	per day
	14. LORRIES, VANS, ETC.				
(1)	Lorry, ordinary (site use and public highway use - including Tax, Insurance and extra cost of petrol or D.E.R.V.).	Plated Gross Vehicle Weight Up to 6.5 tonnes		13.03	
(2)	Ditto..	Up to 7.5 tonnes		15.09	
(3)	Ditto..	Up to 12.0 tonnes		18.16	
(4)	Ditto..	Up to 14.0 tonnes		20.07	
(5)	Ditto..	Up to 17.0 tonnes		21.85	
(6)	Ditto..	Up to 24.5 tonnes		29.16	
(7)	Ditto..	Up to 30.0 tonnes		35.56	
(8)	Ditto (Articulated).....................................	Up to 33.0 tonnes		28.61	
(9)	Ditto (Articulated).....................................	Up to 39.0 tonnes		38.86	
(10)	Extra for lorry fitted with crane attachment (whether used or not).............	Maximum load Up to 1.5 tonnes		1.97	
(11)	Ditto..	Up to 2.5 tonnes		2.75	
(12)	Ditto..	Up to 3.5 tonnes		3.64	
(13)	Ditto..	Up to 5.0 tonnes		5.13	
(14)	Ditto..	Over 5.1 tonnes		7.95	
(15)	Lorry, tipper (site use and public highway use - including Tax, Insurance and extra cost of D.E.R.V.)...................................	Plated Gross Vehicle Weight Up to 11.0 tonnes		14.51	
(16)	Ditto..	Up to 17.0 tonnes		22.57	
(17)	Ditto..	Up to 25.0 tonnes		29.39	
(18)	Ditto..	Up to 31.0 tonnes		35.64	
(19)	Extra for Side Tipping.............................	-		+ 20%	
		Carrying Capacity			
(20)	Van, pick up or similar utility vehicle.........	Up to .60 tonnes		6.95	
(21)	Ditto..	Up to 1.10 tonnes		12.39	

The rates shown above are per unit machine per hour unless otherwise stated.

CIVIL ENGINEERING - PLANT (Cont'd)

Section & Item No.	Description	Nominal Size or Capacity	Hire Rate	Remarks
	14. LORRIES, VANS, ETC. (cont'd)			
		Carrying Capacity		
(22)	Ditto..	Up to 1.30 tonnes	14.44	
(23)	Ditto..	Up to 1.60 tonnes	16.62	
(24)	Ditto..	Up to 2.10 tonnes	19.46	
(25)	Ditto..	Up to 2.60 tonnes	21.51	
(26)	Passenger/goods, cross country.............	Wheelbase up to 2.4m	14.04	
(27)	Ditto..	Wheelbase 2.40m and over	16.33	
(28)	Ditto, Station wagon................................	Up to 7 seater	16.53	
(29)	Ditto..	Up to 12 seater	18.51	
(30)	Personnel carrier/coach/bus....................	Up to 9/13 seater	18.93	
(31)	Ditto..	Up to 14/17 seater	20.62	
		Hopper Capacity		
(32)	Road sweeper/cleaner self-propelled........	Up to 1.0m³	26.60	
(33)	Ditto..	Up to 2.0m³	32.22	
(34)	Road sweeper complete with gulley emptier (site and public highway use including Tax, Insurance and use of D.E.R.V.)..	Up to 3.0m³	35.77	
(35)	Ditto..	Up to 6.0m³	41.19	
(36)	Tractor mounted brush (front or rear) non-collector (excluding tractor)...............	-	3.01	
	15. MATERIALS HANDLING			
(1)	Trailer flat (towed with knock on brakes) (site use only excluding Tax and Insurance)..	Carrying Capacity Up to 2.0 tonnes	12.30	per day
(2)	Ditto..	Up to 3.0 tonnes	13.45	per day
(3)	Ditto..	Up to 4.0 tonnes	14.31	per day
(4)	Extra over items (1) to (3) for powered tipping	Up to 4 tonnes	+15%	
(5)	4 wheeled flat trailer with Hydraulic or air brakes...	Up to 8 tonnes	18.27	per day
(6)	Ditto..	Up to 10 tonnes	21.13	per day
(7)	Ditto..	Up to 12 tonnes	25.20	per day
		Maximum Lifting Capacity (forward loading)		
(8)	Fork lift truck (yard type)..........................	Up to 1000Kg	5.14	
(9)	Ditto..	Up to 1250Kg	5.57	
(10)	Ditto..	Up to 1500Kg	6.00	
(11)	Ditto..	Up to 2000Kg	6.64	
(12)	Ditto..	Up to 2500Kg	7.18	
(13)	Fork lift truck (rough terrain type) (2 wheel drive)..	Up to 1000Kg	4.61	
(14)	Ditto..	Up to 1250Kg	5.60	
(15)	Ditto..	Up to 1500Kg	6.62	
(16)	Ditto..	Up to 2000Kg	8.91	
(17)	Ditto..	Up to 2500Kg	10.64	
(18)	Ditto..	Up to 3000Kg	11.35	
(19)	Ditto..	Up to 4000Kg	14.01	
(20)	Ditto..	Up to 5000Kg	15.86	
(21)	Fork lift truck (rough terrain) (4 wheel drive)..	Up to 1500Kg	8.76	
(22)	Ditto..	Up to 2000Kg	9.28	
(23)	Ditto..	Up to 2500Kg	10.55	
(24)	Ditto..	Up to 3000Kg	12.02	
(25)	Ditto..	Up to 3500Kg	13.88	
(26)	Ditto..	Up to 4000Kg	14.59	
(27)	Ditto..	Up to 5000Kg	17.27	
(28)	Ditto..	Up to 6000Kg	18.72	
(29)	Ditto..	Up to 7000Kg	21.71	
(30)	Ditto..	Up to 8000Kg	24.38	
(31)	Telescopic Site Handling Machines (2 wheel drive)..	Up to 2500Kg	11.63	
(32)	Ditto..	Up to 4000Kg	16.10	
(33)	Telescopic Site Handing Machines (4 wheel drive)..	Up to 2500Kg	14.70	
(34)	Ditto..	Up to 4000kg	18.79	

Note: If man-riding basket is required, rates to be negotiated.

The rates shown above are per unit machine per hour unless otherwise stated.

CIVIL ENGINEERING - PLANT (Cont'd)

Section & Item No.	Description	Nominal Size or Capacity	Hire Rate	Remarks
	16. OFFICES, STORES, ETC.			
(1)	Offices on site with usual fittings ie, heating and lighting, desks, tables, chairs, plan chests, etc......................................	Per 10m^2 floor area	45.71	per week
(2)	Mess room on site with usual fittings, ie, heating and lighting, tables, forms etc. but excluding kitchen equipment.....................	"	41.04	per week
(3)	Stores on site with usual fittings ie, counter, desk, chairs, racks, shelves or bins, heating and lighting..	"	37.41	per week
(4)	Toilet unit, single chemical type, (excluding servicing and consumables)........................		20.35	per week
(5)	Toilet unit, single mains flushing type, with wash basin, excluding consumables............		25.43	per week
	Note:- Rate for multiple toilet units: Add 70% of single unit rate for each additional unit.			
(6)	Mobile office including heating and lighting (mains) and usual fittings,		38.97	per week
(7)	Ditto..		62.76	per week
(8)	I.S.O. Containers used as Tool Stores, Fitting shop etc.	Up to 5m long	86.16	per week
		Up to 7m long	112.32	per week
		Up to 13m long	5.45	per day
	17. PILING PLANT			
	(Excluding Compressor)	Hammer weight		
(1)	Piling Hammer double-acting (air)................	Up to 155Kg	5.70	
(2)	Ditto...	Up to 305Kg	6.45	
(3)	Ditto...	Up to 1143Kg	7.44	
(4)	Piling Hammer double-acting (air)................	Up to 2132Kg	8.88	
(5)	Ditto...	Up to 3006Kg	9.55	
(6)	Ditto...	Up to 4334Kg	10.56	
(7)	Ditto...	Up to 6350Kg	11.63	
(8)	Hydraulically operated drop hammer c/w hydraulic power pack...................................	Up to 3000Kg	52.26	
(9)	Ditto...	Up to 5000kg	54.27	
(10)	Ditto...	Up to 7000Kg	56.29	
(11)	Air driven impulse hammer...........................	Total weight Up to 5 tonnes	13.99	
(12)	Drop Hammer (bare) (suitable for use with channel or tubular leaders fitted with leather guides and rubber inserts)............................	Hammer weight Up to 1016Kg	1.07	
(13)	Ditto..	Up to 2032Kg	1.22	
(14)	Ditto..	Up to 3048Kg	1.37	
(15)	Internal drop hammer (for use with cased piles):			
(16)	Internal diameter of cased pile. Up to 305mm	Up to 1270Kg	0.92	
(16)	Ditto.......................................Up to 406mm	Up to 3556Kg	1.04	
(17)	Ditto.......................................Up to 508mm	Up to 5588Kg	1.19	
(18)	Air operated Extractor (excluding compressor)..	Unit weight Up to 764Kg	5.65	
(19)	Ditto...	Up to 1705Kg	6.60	
(20)	Ditto...	Up to 3000Kg	7.31	
(21)	Ditto...	Up to 4590Kg	8.71	
(22)	Flexible Reinforced Rubber Hose (for compressed air), per metre........................	Diameter Up to 25mm/per m	0.94	per day
(23)	Ditto..	Up to 32mm/per m	0.95	per day
(24)	Ditto..	Up to 38mm/per m	0.96	per day
(25)	Ditto..	Up to 51mm/per m	0.98	per day
(26)	Ditto..	Up to 64mm/per m	1.01	per day
(27)	Diesel Hammer (including complete set of guide equipment) (single acting).................	Piston weight Up to 500Kg	8.74	
(28)	Ditto..	Up to 800Kg	11.49	
(29)	Ditto..	Up to 1300Kg	15.94	
(30)	Ditto..	Up to 1500Kg	17.69	
(31)	Ditto..	Up to 1900Kg	21.12	
(32)	Ditto..	Up to 2500Kg	26.09	
(33)	Ditto..	Up to 3000Kg	30.09	
(34)	Ditto..	Up to 3500Kg	33.95	
(35)	Ditto..	Up to 4600Kg	36.79	
(36)	Ditto..	Up to 6200Kg	66.96	
(37)	Ditto..	Up to 8000Kg	77.09	
(38)	Ditto..	Up to10000Kg	85.75	

The rates shown above are per unit machine per hour unless otherwise stated.

CIVIL ENGINEERING - PLANT (Cont'd)

Section & Item No.	Description	Nominal Size or Capacity		Hire Rate	Remarks
	17. PILING PLANT (Cont'd)				
		Centrifugal force Tonnes			
(39)	Hydraulic vibrating hammer/extractor including power pack................................	Up to 21		29.24	
(40)	Ditto..	Up to 38		32.99	
(41)	Ditto..	Up to 62		43.29	
(42)	Ditto..	Up to 143		65.33	
(43)	Ditto..	Up to 207		113.97	
(44)	Pile Helmet for Pile....................................	203mm x 203mm		1.18	
(45)	Ditto..	254mm x 254mm		1.47	
(46)	Ditto..	305mm x 305mm		1.73	
(47)	Ditto..	356mm x 356mm		1.93	
(48)	Ditto..	406mm x 406mm		2.09	
(49)	Ditto..	457mm x 457mm		2.20	
(50)	Pile helmet for pile.............................. (Note:Plastic Dollies or equivalent to be paid for in addition)	508mm x 508mm		2.27	
		Length of jib			
(51)	Hanging Leaders Channel type (for use with drop hammer.....................................	12.2m		3.24	
(52)	Ditto..	15.3m		3.24	
(53)	Hanging Leaders, rectangular section .61m x .61m (for use with drop or diesel hammers)...	Crane Boom 9.2m	Nominal Length 15.3m	4.87	
(54)	Ditto..	15.3m	24.4m	5.44	
(55)	Hanging Leaders, rectangular section .84 x .84m..	12.2m	21.4m	6.51	
(56)	Ditto..	18.3m	29.9m	7.67	
(57)	Ditto..	24.4m	36.6m	8.45	
(58)	Auger Piling and diaphragm Walling machines and other specialist equipment			To be Negotiated	
	18. PILING Temporary steel piling and steel trench sheeting to be charged as a material. The residual value of recovered steel piling and steel trench sheeting to be the subject of special agreement.				
	19. PIPE BENDING EQUIPMENT				
(1)	Special purpose pipe winches (for standard winches - See Section 36)........................	All sizes		To be negotiated	
		Nominal pipe diameter			
(2)	Pipe bending machine (Hand operated).....	Up to 50mm		0.95	
(3)	Ditto..	Up to 75mm		1.17	
(4)	Ditto..	Up to 150mm		2.04	
(5)	Pipe bending machine (motorised)............	Up to 50mm		1.11	
(6)	Ditto..	Up to 75mm		1.82	
(7)	Ditto..	Up to 100mm		2.79	
(8)	Ditto..	Up to 150mm		3.49	

Section & Item No.	Remarks Description	Nominal Size or Capacity	Basic Rate per Day	Extra Working Hour	per
	20. PUMPS, PORTABLE (exclusive of all hoses)				
(1)	Semi rotary (hand)......................................	Up to 19mm	3.29	0.31	
(2)	Ditto..	Up to 25mm	4.61	0.32	
(3)	Single diaphragm......................................	Up to 51mm	8.30	1.20	
(4)	Ditto..	Up to 76mm	13.63	1.21	
(5)	Ditto..	Up to 102mm	25.83	1.47	
(6)	Double diaphragm......................................	Up to 51mm	6.51	1.00	
(7)	Ditto..	Up to 76mm	11.81	1.23	
(8)	Ditto..	Up to 102mm	29.98	1.67	
(9)	Self priming centrifugal..............................	Up to 38mm	9.28	1.29	
(10)	Ditto..	Up to 51mm	13.80	1.44	
(11)	Ditto..	Up to 76mm	19.94	1.76	
(12)	Ditto..	Up to 102mm	24.47	2.17	
(13)	Self priming centrifugal..........................	Up to 152mm	30.61	3.19	

The rates shown above are per unit machine per hour unless otherwise stated.

CIVIL ENGINEERING - PLANT (Cont'd)

Section & Item No.	Description	Nominal Size or Capacity	Basic Rate per Day	Extra per Working Hour	Remarks
	20. PUMPS, PORTABLE (Cont'd) (exclusive of all hoses)				
(14)	Sludge and Sewage.................................	Up to 76mm	22.23	1.63	
(15)	Ditto..	Up to 102mm	26.22	1.69	
(16)	Ditto..	Up to 152mm	37.72	3.04	
(17)	Ditto..	Up to 204mm	55.03	3.62	
(18)	Sump pneumatic (excluding compressor)..	Up to 51mm	6.88	0.98	
(19)	Ditto..	Up to 64mm	7.44	1.00	
(20)	Ditto..	Up to 76mm	8.03	1.01	
(21)	Electric submersible................................	Up to 38mm	4.99	0.84	
(22)	Ditto..	Up to 51mm	5.27	0.94	
(23)	Ditto..	Up to 64mm	5.64	0.93	
(24)	Ditto..	Up to 76mm	6.07	0.10	
(25)	Ditto..	Up to 102mm	7.30	1.46	

Section & Item No.	Description	Nominal Size or Capacity	Hire Rate	Remarks
	21. PUMPING EQUIPMENT			
	Pump hoses, flexible, suction or delivery, including coupling, valve and strainer	Diameter		per day
(1)	Suction per metre..................................	Up to 38mm	0.05	per day
(2)	Ditto..	Up to 51mm	0.07	per day
(3)	Ditto..	Up to 76mm	0.10	per day
(4)	Ditto..	Up to 102mm	0.13	per day
(5)	Ditto..	Up to 152mm	0.20	per day
(6)	Ditto..	Up to 204mm	0.26	per day
(7)	Delivery per metre..................................	Up to 38mm	0.08	per day
(8)	Ditto..	Up to 51mm	0.10	per day
(9)	Ditto..	Up to 76mm	0.15	per day
(10)	Ditto..	Up to 102mm	0.20	per day
(11)	Ditto..	Up to 152mm	0.30	per day
(12)	Ditto..	Up to 204mm	0.41	per day
(13)	Steel suction or delivery, including flanges, bolts and joint rings (excluding valve and strainer)per metre..................................	Up to 76mm	0.05	per day
(14)	Ditto..	Up to 152mm	0.10	per day
(15)	Ditto..	Up to 204mm	0.14	per day
(16)	Bend to be charged as 9.0m length...........	-		
(17)	Valve to be charged as 9.0m length..........	-		
	Note: Attendance to be charged additionally			
22. RAMMERS AND COMPACTORS				
(1)	Vibro or vibration rammer........................	Up to 60Kg	1.89	
(2)	Ditto..	Up to 70Kg	1.99	
(3)	Ditto..	Up to 90Kg	2.01	
(4)	Ditto..	Up to 115Kg	2.06	
(5)	Ditto..	Up to 200Kg	2.19	
(6)	Vibrating plate compactor.......................	Up to 80Kg	2.02	
(7)	Ditto..	Up to 110Kg	2.15	
(8)	Ditto..	Up to 150Kg	1.94	
(9)	Ditto..	Up to 200Kg	2.08	
(10)	Ditto..	Up to 300Kg	2.45	
(11)	Ditto..	Up to 400Kg	2.93	
(12)	Ditto..	Up to 500Kg	3.60	
(13)	Ditto..	Up to 600Kg	4.26	

The rates shown above are per unit machine per hour unless otherwise stated.

CIVIL ENGINEERING - PLANT (Cont'd)

Section & Item No.	Description	Nominal Size or Capacity	Hire Rate	Remarks
	23. ROLLERS	Unballasted weight		
(1)	Road Deadweight (steel 3 wheel/3 roll)	Up to 6.10 tonnes	16.32	
(2)	Ditto..	Up to 8.50 tonnes	18.88	
(3)	Ditto..	Up to 10.50 tonnes	20.26	
(4)	Ditto..	Up to 13.00 tonnes	22.60	
(5)	Ditto(tandem) Diesel........................	Up to 8.50 tonnes	19.60	
		Average Weight		
(6)	Rubber tyred self propelled....................	Up to 1.0 tonne/wheel	21.80	
(7)	Ditto(Ditto).................................	Up to 2.0 tonnes/wheel	24.17	
(8)	Ditto(Ditto)..................................	Up to 3.0 tonnes/wheel	25.61	
(9)	Vibratory pedestrian operated................ ....(single roller)................................	Makers weight Up to 550Kg	3.99	
(10)	Ditto..	Up to 700Kg	4.44	
(11)	Vibratory pedestrian operated................ ....(twin roller)	Up to 650Kg	5.23	
(12)	Ditto..	Up to 950Kg	5.76	
(13)	Ditto..	Up to 1300Kg	6.77	
(14)	Ditto..	Up to 1750Kg	8.61	
(15)	Towed, Vibratory trailer..........................	Up to 1500Kg	11.98	
(16)	Ditto..	Up to 6500Kg	14.06	
(17)	Ditto..	Up to 8800Kg	15.13	
(18)	Ditto..	Up to 11700Kg	16.63	
(19)	Ditto..	Up to 13300Kg	17.51	
(20)	Self propelled vibratory tandem (seated control)...	Up to 1.4 tonnes	6.66	
(21)	Ditto..	Up to 2.5 tonnes	9.25	
(22)	Ditto..	Up to 5.0 tonnes	14.18	
(23)	Ditto..	Up to 9.0 tonnes	21.96	
(24)	Ditto..	Up to 12.0 tonnes	27.60	
(25)	Ditto, single roll, rubber tyred driving wheels..	Up to 8.2 tonnes	23.16	
(26)	Ditto..	Up to 10.5 tonnes	26.44	
(27)	Scarifier (working time) extra over Roller including sharpening tines.....................	1 Tine	2.90	
(28)	Ditto..	2 Tine Hydraulic	3.25	
(29)	Ditto..	3 Tine Hydraulic	3.63	
(30)	Compactor sheeps foot self-propelled Tamping foot heavy duty wheeled...........	Maker's rated flywheel Up to 130Kw	52.99	
(31)	Ditto..	Up to 200Kw	81.47	
(32)	Ditto..	Up to 300Kw	116.45	
	24. SAWS MECHANICAL			
(1)	Chain Saw including Heavy Duty protective clothing..................................	Guide bar length Up to 310mm	1.77	
(2)	Ditto..	Up to 410mm	1.94	
(3)	Ditto..	Up to 510mm	2.09	
(4)	Ditto..	Up to 650mm	2.24	
(5)	Ditto..	Up to 800mm	2.37	
(6)	Ditto..	Up to 950mm	2.47	
(7)	Ditto..	Up to 1300mm	2.67	
		Saw diameter		
(8)	Portable Saw Bench..............................	Up to 260mm	1.23	
(9)	Ditto..	Up to 310mm	1.53	
(10)	Ditto..	Up to 410mm	1.99	
(11)	Ditto..	Up to 460mm	2.19	
(12)	Ditto..	Up to 510mm	2.36	
(13)	Ditto..	Up to 610mm	2.66	
(14)	Ditto..	Up to 660mm	2.79	
(15)	Ditto..	Up to 760mm	3.03	
(16)	Band Saw..	-	3.69	

The rates shown above are per unit machine per hour unless otherwise stated.

CIVIL ENGINEERING - PLANT (Cont'd)

Section & Item No.	Description	Nominal Size or Capacity	Hire Rate	Remarks
	25. SCAFFOLDING			
(1)	Tubular steel..	51mm dia. Nominal per m	0.03	per week
(2)	Ditto, alloy..	51mm dia. Nominal per m	0.05	per week
(3)	Putlog steel..	1.52m long Nominal	0.07	per week
(4)	Ditto, alloy..	1.52m long Nominal	0.13	per week
(5)	Ditto, steel..	1.83m long Nominal	0.08	per week
(6)	Putlog alloy..	1.83m long Nominal	0.15	per week
(7)	Fitting steel, single double or swivel coupler, joint pin, fixed base plate...........	-	0.02	per week
(8)	Ditto, Adjustable Base Plate....................	-	0.06	per week
(9)	Ditto, Hop up Bracket.............................	-	0.16	per week
(10)	Ditto, Split Head trestle folding type/adjustable......................................	0.48m - 0.84m	1.07	per week
(11)	Ditto...	0.76m - 1.37m	1.07	per week
(12)	Ditto...	1.07m - 1.83m	1.08	per week
(13)	Ditto...	1.37m - 2.44m	1.09	per week
(14)	Castor..	-	1.13	per week
(15)	Ditto, rubber tyred................................		1.16	per week
(16)	Ditto, nylon tyred..................................	-	1.38	per week
(17)	Jenny Wheel including 30m rope...........	Up to 254mm diameter	1.57	per week
(18)	Ditto...	Up to 305mm diameter	1.64	per week
(19)	Board..	Up to 4.0m long	0.07	per week
	Rates for special items of prefabricated scaffold units owned by the main contractor to be similar to those charge by Proprietary Firms, plus 12.5%			
	26. SHORING, PLANKING AND STRUTTING			
(1)	Baulk timber, use and waste (excluding nails, dogs, wedges, etc., to be charged in addition as consumables)................. (Minimum charge 18 days hire)	Per m³	3.78	per day
(2)	Timber for planking and strutting use and waste (excluding nails, dogs, wedges, etc., to be charged in addition as consumables).................................. (Minimum charge 12 days hire)	Per m³	6.86	per day
(3)	Adjustable Steel Strut, extending...........	457mm - 711mm	1.21	per week
(4)	Ditto..	686mm - 1.09m	1.25	per week
(5)	Ditto..	1.04m - 1.70m	1.32	per week
		Length / Height		
(6)	Dragbox...	Up to 3.1m / Up to 1.9m	15.39	
(7)	Ditto...	Up to 5.0m / Up to 2.5m	22.42	
(8)	Ditto...	Up to 7.1m / Up to 2.0m	31.16	
(9)	Top extension to item 7..................	Up to 5.0m / Up to 1.5m	19.08	
	27. SHUTTERING			
(1)	Steel Shutters, all types. Rates for items of steel shuttering to be based on those chargeable by Proprietory Firms. (Wedges, keys and consumables to be added) plus 12.5%................................			
(2)	Steel road forms..................................	102mm per 10 metres	0.61	per day
(3)	Ditto..	152mm per 10 metres	0.67	per day
(4)	Ditto..	203mm per 10 metres	0.78	per day
(5)	Ditto..	203mm (heavy section) per 10 metres	0.78	per day
(6)	Ditto..	203mm (heavy section with rail) per 10 metres	1.07	per day
(7)	Ditto..	254mm (heavy section with rail) per 10 metres	1.32	per day
(8)	Ditto..	305mm (heavy section with rail) per 10 metres	1.45	per day
(9)	Telescopic steel floor centre, extending lattice girder type.................................	Span up to 1.8m	2.57	per week
(10)	Ditto..	Span up to 2.1m	2.69	per week
(11)	Telescopic steel floor centre, extending lattice girder type	Span up to 2.7m	2.93	per week
(12)	Ditto..	Span up to 3.6m	3.29	per week
(13)	Telescopic steel prop extending...........	1.04m - 1.83m	2.21	per week
(14)	Ditto..	1.75m - 3.12m	2.29	per week

The rates shown above are per unit machine per hour unless otherwise stated.

CIVIL ENGINEERING - PLANT (Cont'd)

Section & Item No.	Description	Nominal Size or Capacity	Hire Rate	Remarks
	27. SHUTTERING (Cont'd)			
(15)	Ditto	1.98m - 3.35m	2.33	per week
(16)	Ditto	2.44m - 3.96m	2.35	per week
(17)	Column Shutter Clamps, extending	254mm - 508mm per set	2.41	per week
(18)	Ditto	406mm - 813mm per set	2.55	per week
(19)	Ditto	0.61m - 1.22m per set	2.81	per week
(20)	Beam Shutter Clamps, clamping width 114mm to 850mm	305mm arm per set	2.63	per week
(21)	Ditto	457mm arm per set	2.67	per week
(22)	Ditto	610mm arm per set	2.71	per week
(23)	Wall Shutter Clamps, concrete thickness	102mm to 305mm per set	2.04	per week
(24)	Ditto	102mm to 610mm per set	2.13	per week
(25)	Ditto	102mm to 915mm per set	2.23	per week
(26)	Timber used for shutter. Rough (minimum charge 10 days hire) (excluding wedges, nails, screws, bolts, etc., to be charged in addition as consumables + 12.5%)	per m³	12.47	per day
(27)	Ditto, Wrot (minimum charge 10 days hire)	per m³	12.47	per day
(28)	Douglas Fir Plywood Sheeting (Excluding timber and steel or timber supports to be charged in addition to + 12.5%) (Thickness 13mm (minimum charge 12 days Hire)	Good 1 side per 3m²	1.16	per day
(29)	Ditto	Good 2 sides per 3m²	1.34	per day
(30)	Thickness 19mm (minimum charge 12 days hire)	Good 1 sides per 3m²	1.59	per day
(31)	Ditto	Good 2 sides per 3m²	1.83	per day
(32)	Thickness 25mm (minimum charge 12 days hire)	Good 1 sides per 3m²	2.24	per day
(33)	Ditto	Good 2 sides per 3m²	2.49	per day
	Plastic Faced			
(34)	Thickness 13mm (minimum charge 12 days hire)	per 3m²	1.53	per day
(35)	Thickness 19mm (minimum charge 12 days hire)	per 3m²	2.14	per day

Item (28) to (35) exclude wedges, nails screws, bolts etc., to be charged in addition as consumables at cost plus 12 1/2%

Section & Item No.	Description	Nominal Size or Capacity	Hire Rate	Remarks
	28. SURVEYING INSTRUMENTS	Standard Deviation		
(1)	Engineer's level c/w Tripod and Staff	2.5mm/km	3.22	per day
(2)	Engineer's Auto level c/w Tripod and Staff	2.0mm/km	3.64	per day
(3)	Engineers Precise Auto level c/w Tripod and Staff	1.5mm/km	4.78	per day
(4)	Engineers Precise Auto level with parallel plate c/w Tripod and Staff	0.3mm/km	6.80	per day
(5)	Digital Theodolite	1.5mm/km	13.60	per day
(6)	Electronic Theodolite	20 seconds	11.10	per day
(7)	Electronic Total Station	10 seconds	15.51	per day
(8)	Ditto	5 seconds	21.80	per day
(9)	Ditto	3 seconds	30.54	per day
(10)	Ditto	0.5 seconds	70.02	per day
(11)	Precise Optical Plummet	1mm/30m	10.55	per day
(12)	Ranging rod or pole, timber	-	0.05	per day
(13)	Ranging rod or pole, alloy	-	0.05	per day
(14)	Rates for GPS or LAS to be charged at proprietary prices +12%	-		
	29. TAR SPRAYING AND COLD EMULSION PLANT			
(1)	Tar Boiler and Sprayer (including firing) hand operated	Up to 1000 litres	5.70	
(2)	Ditto, power operated	Up to 2000 litres	6.81	
(3)	Gritter (attached to lorry) extra over lorry	-	4.87	
(4)	Gritter (towed hopper type) ditto	-	5.35	
(5)	Cold Emulsion Sprayer hand operated	Up to 250 litres	1.02	
(6)	Ditto, power operated	Over 700 litres	1.91	

The rates shown above per unit machine per hour unless otherwise stated.

CIVIL ENGINEERING - PLANT (Cont'd)

Section & Item No.	Description	Nominal Size or Capacity	Hire Rate	Remarks
	30 TOOLS PNEUMATIC			
	(Excluding compressors but including 15m of hose)			
(1)	Breaker including steels.........................	-	1.75	
(2)	Light pneumatic pick including steels.....	-	1.55	
(3)	Pneumatic clay Spade including blade...	-	1.64	
(4)	Chipping/Scaling/Caulking hammer.........	-	1.38	
	Note: The rates for the above items includes the sharpening of the steel or blade.			
(5)	Hand-held Rock Drill without Drill Rods or detachable Bits................light weight	16-20Kg	0.32	
(6)	Ditto..................................middle weight	Up to 24Kg	1.52	
(7)	Ditto...................................heavy weight	Up to 32Kg	1.92	
	Note: Drill Rods and Bits for Hand-held Rock Drill to be paid for in addition + 12 1/2%			
(9)	Additional hoses.....................................	Per 15m	0.06	
(10)	Drill (excluding consumables).............	Up to 13mm chuck	0.50	
(11)	Reversible drill (Ditto).......................	Up to 13mm chuck	0.51	
(12)	Grinder (lightweight)(Ditto)..............	76mm diameter	0.51	
(13)	Ditto (heavy weight)(Ditto).............	117mm diameter	0.61	
(14)	Sander (excluding pad and disc)(Ditto)...	-	0.49	
(15)	Riveting hammer (Ditto)......................	-	0.97	
(16)	Pneumatic paint scraper tool...................	-	1.18	
(17)	Polisher..	-	0.52	
	Consumables to be charged in addition + 12.5% except where otherwise stated.			
	31. TOOLS, PORTABLE ELECTRIC			
	(Excluding generator or power) (consumables, to be charged in addition + 12 1/2%)			
(1)	Drill...	Up to 10mm diameter	0.32	
(2)	Ditto..	Up to 13mm diameter	0.36	
(3)	Ditto..	Up to 19mm diameter	0.42	
(4)	Ditto..	Up to 32mm diameter	0.57	
(5)	Extra for Stand...	-	0.43	
(6)	Ditto, magnetic Stand................................	-	1.25	
(7)	Bench Grinder and pedestal......................	Up to 152mm diameter	0.26	
(8)	Ditto..	Up to 204mm diameter	0.29	
(9)	Ditto..	Up to 250mm diameter	0.31	
(10)	Angle Grinder...	Up to 230mm diameter	0.37	
(11)	Straight Grinder..	Up to 152mm diameter	0.47	
(12)	Sander..	Up to 180mm diameter	0.36	
(13)	Ditto..	Up to 230mm diameter	0.39	
(14)	Polisher/Sander...	-	0.39	
(15)	Electric Demolition Hammer/Drill Light weight...	Up to 25mm diameter	0.76	
(16)	Ditto (heavy weight)...............................	Up to 51mm diameter	1.01	
(17)	Portable electric Saw (circular)...............	Up to 250mm diameter	0.50	
(18)	Extra for power to be added to above items where applicable............................		+ 50%	
	32. TRACTORS, SCRAPERS, ETC.			
(1)	Tractor (crawler) with bull or angle dozer (hydraulically or winch operated)............	Maker's rated fly wheel Kw Up to 70.0Kw	26.54	
(2)	Ditto..	Up to 85.0Kw	34.44	
(3)	Ditto..	Up to 100.0Kw	47.10	
(4)	Ditto..	Up to 115.0Kw	49.95	
(5)	Ditto..	Up to 135.0Kw	53.83	
(6)	Ditto..	Up to 185.0Kw	68.11	
(7)	Ditto..	Up to 215.0Kw	85.86	
(8)	Ditto..	Up to 250.0Kw	105.10	
(9)	Ditto..	Up to 350.0Kw	143.27	
(10)	Ditto..	Up to 450.0Kw	197.30	
	Ripper attachments to items (1) to (10) subject to negotiation	S.A.E. rated capacity		
(11)	Tractor Loading Shovel (Crawler)............	Up to .80m³	24.24	
(12)	Ditto..	Up to 1.00m³	29.22	
(13)	Ditto..	Up to 1.20m³	33.66	
(14)	Ditto..	Up to 1.40m³	38.36	
(15)	Ditto..	Up to 1.80m³	46.71	
(16)	Ditto..	Up to 2.00m³	51.10	
(17)	Ditto..	Up to 2.10m³	53.57	
(18)	Ditto..	Up to 3.50m³	78.29	

The rates shown above per unit machine per hour unless otherwise stated.

CIVIL ENGINEERING - PLANT (Cont'd)

Section & Item No.	Description	Nominal Size or Capacity	Hire Rate	Remarks
	32. TRACTORS, SCRAPERS, ETC. (Cont'd)			
(19)	Tractor Loading Shovel (Crawler) with Back Hoe Equipment	Add to above rates	+ 17%	
(20)	Tractor Loading Shovel (Crawler) with 4 in 1 attachment	Add to above rates	+ 10%	
(21)	Tractor Loading Shovel (Crawler) with additional hydraulic mounted ripper	S.A.E. rated capacity Up to .60m³	18.70	
(22)	Ditto	Up to .80m³	32.34	
(23)	Ditto	Up to 1.00m³	43.02	
(24)	Ditto	Up to 1.20m³	51.78	
(25)	Ditto	Up to 1.40m³	59.20	
(26)	Ditto	Up to 1.60m³	65.59	
(27)	Ditto	Up to 1.80m³	71.12	
(28)	Tractor Loading Shovel (wheeled) with 4 wheel drive - Articulated	S.A.E. rated capacity Up to 1.00m³	16.36	
(29)	Ditto	Up to 1.20m³	18.44	
(30)	Ditto	Up to 1.40m³	20.75	
(31)	Ditto	Up to 1.50m³	21.91	
(32)	Ditto	Up to 1.60m³	23.00	
(33)	Ditto	Up to 1.80m³	26.00	
(34)	Ditto	Up to 2.00m³	28.13	
(35)	Ditto	Up to 2.30m³	32.66	
(36)	Ditto	Up to 2.70m³	37.87	
(37)	Ditto	Up to 3.10m³	43.15	
(38)	Ditto	Up to 3.50m³	49.01	
(39)	Ditto	Up to 3.85m³	54.48	
(40)	Ditto	Up to 5.00m³	73.55	
(41)	Ditto	Up to 6.00m³	146.49	
(42)	Tractor Loading Shovel (wheeled) with 4 wheel drive - Articulated with 4 in 1 attachment	S.A.E. rated capacity Up to 1.00m³	19.79	
(43)	Ditto	Up to 1.40m³	22.11	
(44)	Ditto	Up to 1.60m³	24.24	
(45)	Ditto	Up to 2.10m³	36.19	
(46)	Ditto	Up to 2.70m³	48.97	
(47)	Ditto	Up to 3.50m³	59.94	
(48)	Tractor Loading Shovel (wheeled, skid steer loader)	S.A.E. rated capacity Up to 0.15m³	6.95	
(49)	Ditto	Up to 0.30m³	8.68	
(50)	Ditto	Up to 0.40m³	10.25	
(51)	Tractor Loading Shovel (wheeled) with 2 wheel drive	S.A.E. rated capacity Up to 1.00m³	13.75	
(52)	Ditto	Up to 1.10m³	14.71	
(53)	Ditto	Up to 1.20m³	15.81	
		Variable Blade/Flywheel Kw		
(54)	Motor Grader	Up to 80Kw	29.62	
(55)	Ditto	Up to 110Kw	42.58	
(56)	Ditto	Up to 120Kw	48.16	
(57)	Ditto	Up to 160Kw	77.12	
(58)	Ditto	Up to 200Kw	125.78	
(59)	Wheeled Tractor (rubber tyred)	Up to 40Kw	6.98	
(60)	Ditto	Up to 50Kw	10.19	
(61)	Ditto	Up to 60Kw	13.95	
(62)	Ditto	Up to 70Kw	16.64	
(63)	Ditto	Up to 75Kw	18.22	
(64)	Ditto	Up to 90Kw	21.53	
(65)	Ditto	Up to 95Kw	22.95	
(66)	Ditto	Up to 105Kw	24.90	
(67)	Ditto	Up to 120Kw	27.75	
(68)	Ditto	Up to 140Kw	31.07	
(69)	Motorised Scraper (rubber tyred) (single engine)	Heaped Capacity (S.A.E.) Up to 16.0m³	84.93	
(70)	Ditto	Up to 23.0m³	117.21	
(71)	Motorised Scraper (rubber tyred) (twin engine)	Up to 16.0m³	106.79	
(72)	Ditto	Up to 25.0m³	210.39	
(73)	Ditto	Up to 35.0m³	251.42	
(73)	Motorised elevating Scraper (rubber tyred)	Up to 9.0m³	68.25	
(75)	Ditto	Up to 15.0m³	102.35	
(76)	Ditto	Up to 20.0m³	135.79	

The rates shown above are per unit machine per hour unless otherwise stated.

CIVIL ENGINEERING - PLANT (Cont'd)

Section & Item No.	Description	Nominal Size or Capacity	Hire Rate	Remarks
	33. TRENCHERS	Flywheel Kw		
(1)	Chain bucket or wheel type....................	Up to 10 Kw	7.32	
(2)	Ditto...	Up to 30 Kw	14.62	
(3)	Ditto...	Up to 50 Kw	21.49	
(4)	Ditto...	Up to 70 Kw	25.69	
(5)	Ditto...	Up to 100 Kw	35.90	
(6)	Ditto...	Up to 200 Kw	68.80	
(7)	Ditto...	Up to 300 Kw	96.27	
	34. WATER AND FUEL SUPPLY	(Exclusive of supporting structure)		
(1)	Static water storage tank........................	Up to 1500 litres	2.02	per day
(2)	Ditto...	Up to 3000 litres	3.67	per day
(3)	Ditto...	Up to 6000 litres	6.64	per day
(4)	Ditto...	Up to 12000 litres	9.43	per day
(5)	Static fuel storage tank............................	Up to 1500 litres	7.06	per day
(6)	Ditto...	Up to 3000 litres	9.34	per day
(7)	Ditto...	Up to 6000 litres	13.42	per day
(8)	Ditto...	Up to 12000 litres	17.24	per day
(9)	Add 50% to above rates for bunding.........			per day
(10)	Mobile water storage tank - towed...........	Up to 1500 litres	4.15	
(11)	Ditto...	Up to 3000 litres	5.15	
(12)	Ditto...	Up to 6000 litres	7.14	per day
(13)	Ditto...	Up to 12000 litres	11.10	per day
(14)	Mobile fuel storage tank - towed..............	Up to 1500 litres	7.22	per day
(15)	Ditto...	Up to 3000 litres	8.45	per day
(16)	Ditto...	Up to 6000 litres	10.76	per day
(17)	Ditto...	Up to 12000 litres	11.93	per day
(18)	Bunded mobile fuel storage tank – towed..	Up to 1500 litres	9.23	per day
(19)	Ditto...	Up to 3000 litres	10.28	per day
(20)	Mobile water storage tank – self propelled	Up to 4500 litres	13.43	
(21)	Ditto...	Up to 7000 litres	14.27	
(22)	Mobile water storage tank – self propelled	Up to 4500 litres	14.39	
(23)	Ditto...	Up to 7000 litres	18.94	
	35. WELDING AND CUTTING SETS			
(1)	Oxy-acetylene cutting and welding set (inclusive of oxygen and acetylene) (excluding under-water equipment)........	-	2.39	
(2)	Welding set, diesel.................................. (exclusive of electrodes to be charged in addition +12.5%)	150 amp single operator	2.26	
(3)	Ditto........................Ditto.........................	300 amp single operator	3.46	
(4)	Ditto........................Ditto.........................	480 amp double operator	5.06	
(5)	Ditto..........................transformer electric	150 amp single operator	1.83	
(6)	Ditto........................Ditto.........................	300 amp single operator	2.29	
(7)	Head screen...	-	1.85	per day
(8)	Helmet..	-	2.77	per day
(9)	Standard kit 300 amp. set.........................	-	5.54	per day
	36. WINCHES	Line pull tonnes		
(1)	Double drum diesel friction winch...........	Up to 1.00 tonnes	5.19	
(2)	Ditto...	Up to 3.00 tonnes	6.62	
(3)	Ditto...	Up to 4.00 tonnes	7.32	
(4)	Single drum winch, diesel hydraulic........	Up to 10.00 tonnes	10.75	
(5)	Ditto...	Up to 15.00 tonnes	14.07	
(6)	Single drum winch, electric......................	Up to 1.00 tonnes	1.11	
(7)	Ditto...	Up to 2.00 tonnes	1.64	
(8)	Ditto...	Up to 3.00 tonnes	2.29	
(9)	Ditto...	Up to 5.00 tonnes	3.35	
(10)	Ditto...	Up to 10.00 tonnes	5.57	
(11)	Ditto...	Up to 15.00 tonnes	7.40	
(12)	Hand operated winch...............................	Up to 1.00 tonnes	0.61	
(13)	Ditto...	Up to 2.00 tonnes	1.45	
(14)	Ditto...	Up to 3.00 tonnes	2.09	
(15)	Ditto...	Up to 5.00 tonnes	2.13	
(16)	Ditto, mounted on lorry (power ex lorry battery)..	Up to 2.00 tonnes	3.47	
(17)	Air powered winches (excluding supply of compressed air).................................	Up to 2.00 tonnes	7.05	
(18)	Ditto...	Up to 4.00 tonnes	11.37	
(19)	Ditto...	Up to 5.00 tonnes	12.78	

Note: Special rope requirements to be charged extra.

The rates shown above are per unit machine per hour unless otherwise stated.

CIVIL ENGINEERING - PLANT (Cont'd)

Section & Item No.	Description	Nominal Size or Capacity	Hire Rate	Remarks
	37. MISCELLANEOUS PLANT AND CONSUMABLE STORES			
(1)	Air testing machine for drains..................	-	2.29	per day
(2)	Fencing Chestnut....................................	per metre	1.87	per day
(3)	Fencing, post and rail or similar...............	per metre	1.88	per day
(4)	L.P.G. for tar boiler..................................	-	3.35	
(5)	Watchmans heater including LPG............	-	0.46	
(6)	Road barrier...	per metre	0.12	per day
(7)	Tarpaulin..	per m^2	0.05	per day
(8)	Tilley Lamp (flood) including LPG............	-	2.62	per night
(9)	Traffic signals, two way (excluding generator and including mains power)....	per set	93.12	per week
(10)	Watchman's lamp including paraffin and attendance..	-	4.60	per night
		BTU's hour		
(11)	Space Heater (paraffin/electric)..............	Up to 30000	1.42	
(12)	Ditto..........................Ditto........................	Up to 63000	1.75	
(13)	Ditto..........................Ditto........................	Up to 100000	2.16	
(14)	Ditto..........................Ditto........................	Up to 150000	2.77	
(15)	Ditto..........................Ditto........................	Up to 240000	4.10	
(16)	Ditto..........................Ditto........................	Up to 325000	5.69	
(17)	Extra for space heater using LPG in lieu of paraffin..		add 24%	
(18)	Brickwork and masonry saw (excluding abrasive discs to be charged at cost +12.5%)	Blade diameter Up to 356mm	1.13	
(19)	Ditto...	Up to 457mm	1.31	
(20)	Crossing plates (steel sheets)................	Thickness 20mm per m^2	0.25	per day
(21)	Flashing traffic warning lamp unit including batteries..................................	-	1.50	per day
(22)	Flashing traffic warning lamp unit including batteries, bollard or tripod	-	3.54	per day
(23)	Pendent barrier marker including fencing pins...	per 26m	2.10	per day
(24)	Drainage rods including accessories......	102 to 152mm diameter per 10m	0.08	per day
(25)	Ditto, extra per rod.................................	.91m length each	0.01	per day
(26)	Drain stopper (expanding)......................	Up to 102mm diameter each	0.05	per day
(27)	Ditto...	Up to 152mm diameter each	0.08	per day
(28)	Ditto...	Up to 305mm diameter each	0.24	per day
(29)	Ditto...	Up to 457mm diameter each	0.49	per day
(30)	Ditto...	Up to 610mm diameter each	0.54	per day
(31)	Ditto...	Up to 762mm diameter each	0.72	per day
(32)	Ditto...	Up to 914mm diameter each	1.04	per day
(33)	Ditto...	Up to 1220mm diameter each	1.37	per day
		Water extraction rate per 24 hours		
(34)	Dehumidifier...	Up to 68 litres	7.39	per day
(35)	Ditto...	Up to 91 litres	11.62	per day
(36)	Ditto...	Up to 160 litres	17.51	per day
(37)	Ditto, with temperature control...............	Up to 160 litres	20.09	per day
(38)	Ladders - all types and lengths.............	-	to be negotiated	
		Height		
(39)	Traffic warning cone (plastic)..............	Up to 500mm	0.05	per day
(40)	Ditto...	Up to 600mm	0.07	per day
(41)	Ditto...	Up to 800mm	0.11	per day
(42)	Ditto...	Up to 1000mm	0.14	per day
(43)	Traffic warning signs (circular)............	Up to 600mm diameter	0.21	per day
(44)	Ditto...	Up to 750mm diameter	0.33	per day
(45)	Ditto...	Up to 900mm diameter	0.49	per day
(46)	Ditto...	Up to 1200mm diameter	0.80	per day
(47)	Traffic warning sign (triangular)...........	Up to 600mm height	0.18	per day
(48)	Ditto...	Up to 700mm height	0.25	per day
(49)	Ditto...	Up to 900mm height	0.37	per day
(50)	Ditto...	Up to 1200mm height	0.69	per day
(51)	Traffic warning board (rectangular).....	1050 x 750mm	0.65	per day
(52)	Paint spraying machine, 1 gun type with 7.5m lengths of air and fluid hose and 10.0 litre pressure tank..................	(Excluding compressor)	0.53	
(53)	Ditto (power driven)..............................	(Electric or petrol)	2.16	
(54)	Paint spraying machine, 2 gun type with 15.0m lengths of air and fluid hose and 25 litre pressure tank.....................	(Excluding compressor)	0.85	
(55)	Ditto (power driven)..............................	(Electric, petrol or diesel)	3.04	

The rates shown above are per unit machine per hour unless otherwise stated .

FEES

ROYAL INSTITUTE OF BRITISH ARCHITECTS

The following extracts from 'A guide to RIBA Standard Forms of Appointment 1999 and other Architect's Appointments', 'Standard Form of Agreement for the Appointment of an Architect (SFA/99) (Updated April 2000)', 'A Client's Guide to Engaging an Architect including guidance on fees' and 'Conditions of Engagement for the appointment of an Architect (CE/99)' are printed by permission of the Royal Institute of British Architects, 66 Portland Place, London, W1N 4AD from whom full copies can be obtained.

Note: 'Standard Form of Agreement for the Appointment of an Architect (SFA/99) (Updated April 2004) is now available and should be referred to for full details of amendments

A guide to RIBA Standard Forms of Appointment 1999 and other Architect's Appointments

The guide provides Architects with advice on the application and completion of the RIBA Forms of Appointment1999 and their various supplements, with worked examples.

Clients may also find the guide useful to understand the Architect's role and the responsibilities of the parties.

RIBA Forms of Appointment 1999

SFA/99 - Standard Form of Agreement for the appointment of an Architect

The core document from which all other forms are derived. It is suitable for use where an Architect is to provide services for a fully designed building project in a wide range of sizes or complexity and/or to provide other professional services. It is used with Articles of Agreement and includes notes on completion and an optional Services Supplement

CE/99 - Conditions of Engagement for the Appointment of an Architect

Suitable for use where an architect is to provide services for a fully designed building project and/or to provide other professional services where a letter of Appointment is preferred to the Articles of Agreement in SFA/99.It includes notes on completion, a draft Model Letter and an optional (modified) Services Supplement.

SW/99 - Small Works

Suitable for use where an architect is to provide services of a relatively straightforward nature where the cost of construction works is not expected to exceed £150,000 and use of the JCT Agreement for Minor Works is appropriate. It is used with a Letter of Appointment and includes notes on completion a draft Model Letter and an optional Schedule of Services for Small Works

SC/99 - Form of appointment as Sub-consultant

Suitable for use where a consultant wishes another consultant (sub-consultant) to perform a part of his or her responsibility but not for use where the intention is for the client to appoint consultants directly. It is used With Articles of Agreement and includes notes on completion and a draft form of Warranty to the Client

DB1/99 - Employers Requirements

An amendment for SFA/99 and CE/99 where an architect is appointed by the employer client to prepare Employer's requirements for a design and build contract. It includes a replacement Services Supplement and notes on completion for initial appointment where a change to design and build occurs later or where 'consultant switch' or 'Novation' is contemplated

DB2/99 - Contractors proposals

An amendment for SFA/99 where an architect is appointed by the contractor client to prepare Contractor's Proposals under a design and build contract. (Use with CE/99 is not recommended.) It includes replacement Articles, Appendix and Services Supplement and notes on completion for initial appointment and for 'consultant switch' and 'Novation'.

PS/99 - Form of Appointment as Planning Supervisor

Used for the appointment as planning supervisor under the CDM Regulations I994 of suitably qualified construction professionals. An appointment as planning supervisor is distinct from the provisions of architectural services under other RIBA Forms of Appointment It is used with Articles of Agreement and includes notes on completion. Advice on use with a Letter of Appointment is given in this Guide

PM/99 - Form of Appointment as Project Manager

Suitable for a Wide range of projects where the client wishes to appoint a Project Manager to provide a management service and/or other professional services. Does not duplicate or conflict with an architect's services under other RIBA Forms. It is used with Articles of Agreement and includes notes on completion. Not yet published, at the time of going to press.

STANDARD FORM OF AGREEMENT
For the appointment of an Architect (SFA/99)

The Standard Form of Agreement for the appointment of an Architect (SFA/99) has been issued to replace the previous Architects Appointment (1992).

The Standard Form of Agreement for the Appointment of an Architect (SFA/99) consists of:-

Articles of Agreement

Schedule One: - Project Description

Schedule Two: - Services

Schedule Three: - Fees and Expenses

Schedule Four: - Other Appointments - made under separate Agreements by the Client).

Services Supplement - Optional

Notes for Architects on the use and completion of SFA/99

Conditions of Engagement

Conditions of Engagement - Extracts

CONDITIONS

1 General

Interpretation 1.1 The headings and notes to these conditions are for convenience only and do not affect the interpretation of the conditions.

1.2 Words denoting the masculine gender include the feminine gender and words denoting natural persons include corporations and firms and shall be construed interchangeably in that manner.

Applicable law 1.3 The law applicable to this Agreement shall be that stated in the letter of Appointment and, if not so stated, shall be the law of England and Wales.

Communications 1.4 Communications, between the Client and the Architect, including any notice or other document required under the Agreement, shall be in writing and given or served by any effective means. Communications that are not in writing shall be of no effect unless and until confirmed in writing by the sender or the other party. Communications shall take effect when received at an agreed address of the recipient. Communications sent by recorded or registered fist-class post shall be presumed to have arrived at the address to which they are posted on the second day after posting.

Public Holidays 1.5 Where under this Agreement an action is required within a specific period of days, that period shall exclude any day which is a Public Holiday.

Services Variation 1.6 In relation to the Services, either party shall advise the other upon becoming aware of:

.1 a need to vary the Services, the Project Timetable and/or the fees and/or any other part of the Agreement;

.2 any incompatibility in or between any of the Client's requirements in the Brief, any Client's instruction, the Construction Cost, the Timetable and/or the approved design; or any need to vary any part of them;

.3 any issues affecting or likely to affect the progress, quality or cost of the Project;

.4 any information or decisions required from the Client or others in connection with performance of the Services;

and the parties shall agree how to deal with the matter.

2 Obligations & authority of the Architect

Duty of care 2.1 The Architect shall in performing the Services and discharging all the obligations under this part 2 of these Conditions, exercise reasonable skill and care in conformity with the normal standards of the Architect's profession.

Architect's authority 2.2 The Architect shall act on behalf of the Client in the matters set out or necessarily implied in the Agreement.

2.3 In relation to the Services, the Architect shall obtain the authority of the Client before proceeding with the Services or initiating any Work Stage. The Architect shall confirm such authority in writing.

Appointment of Consultants or other persons 2.4 The Architect shall advise the Client on the appointment of Consultants or other persons, other than those named in Schedule 4, to design and/or carry out certain parts of the Works or to provide specialists advice if required in connection with the Project.

Appointment of Site Inspectors 2.5 The Architect shall advise the Client on the appointment of full- or part-time Site Inspectors, other than those named in Schedule 4, under separate agreements where the Architect considers that the execution of the Works warrants such appointment.

Co-operation etc 2.6 The Architect in performing the Services shall when reasonably required by any of the persons identified in Schedule 4:

.1 Co-operate with them as reasonably necessary for carrying out their services;

.2 provide them with information concerning the Services for carrying out their services;

.3 consider and where appropriate comment on their work so that they may consider making any necessary changes to their work;

.4 integrate into his work relevant information provided by them.

No alteration to Services or design 2.7 The Architect shall make no material alteration or addition to or omission from the Services or the approved design without the knowledge and consent of the Client, which consent shall be confirmed in writing by the Architect. In an emergency the Architect may make such alteration, addition or omission without the knowledge and consent of the Client but shall inform the Client without delay and subsequently confirm such action.

STANDARD FORM OF AGREEMENT
For the appointment of an Architect (SFA/99) (Cont'd)
Conditions of Engagement (Cont'd)
Obligations & authority of the Client (Cont'd)

Visits to the Works — 2.8 The Architect shall in providing the Services make such visits to the Works as the Architect at the date of the appointment reasonably expected to be necessary.

3 Obligations and authority of the Client

Client's representative — 3.1 The Client shall name the person who shall exercise the powers of the Client under the Agreement and through whom all instructions shall be given.

Information, decisions, approvals and instructions — 3.2 The Client shall supply, free of charge, accurate information as necessary for the proper and timely performance of the Services and to comply with CDM Regulation 11.

3.3 The Client, when requested by the Architect, shall give decisions and approvals as necessary for the proper and timely performance of the Services.

3.4 The Client shall advise the Architect of the relative priorities of the Brief, the Construction Cost and the Timetable.

3.5 The Client shall have authority to issue instructions to the Architect, subject to the Architect's right of reasonable objection. Such instructions and all instructions to any Consultants or Contractors or other persons providing services in connection with the project shall be issued through the Lead Consultant.

Statutory and other consents required — 3.6 The Client shall instruct the making of applications for consents under planning legislation, building acts, regulations or other statutory requirements and by freeholders and others having an interest in the Project. The Client shall pay any statutory charges and any fees, expenses and disbursements in respect of such applications.

CDM Regulations — 3.7 The Client Shall, *where required by CDM Regulations:*
.1 comply with his obligations under the CDM Regulations and in any conflict between the obligations under the Regulations and this Agreement, the former shall take precedence;
.2 appoint a competent planning Supervisor;
.3 appoint a competent Principal Contractor.

Appointments and payment others — 3.8 Where it is agreed Consultants, or other persons, are to be appointed, the Client shall appoint and pay them under separate agreements and shall confirm in writing to the Architect the services to be performed by such persons so appointed.

Nominations — 3.9 Either the Client or the Architect may propose the appointment of such Consultants or other persons, at any time, subject to acceptance by each party.

Site Inspectors — 3.10 Where it is agreed Site inspectors shall be appointed they shall be under the direction of the Lead Consultant and the Client shall appoint and pay them under separate agreements and shall confirm in writing to the Architect the services to be performed, their disciplines and the expected duration of their employment.

Responsibilities of others — 3.11 The Client, in respect of any work or services in connection with the Project performed or to be performed by any person other than the Architect, shall:
.1 hold such person responsible for the competence and performance of his services and for visits to the site in connection with the work undertaken by him;
.2 ensure that such person shall co-operate with the Architect and provide to the Architect drawings and information reasonably needed for the proper and timely performance of the Services;
.3 ensure that such person shall, when requested by the Architect, consider and comment on work of the Architect in relation to their own work so that the Architect may consider making any necessary change to his work.

3.12 The client shall hold the Principle Contractor and/or other contractors appointed to undertake construction works and not the Architect responsible for their management and operational methods, for the proper carrying out and completion of the Works in compliance with the building contract and for health and safety provisions on the Site.

Legal advice — 3.13 The Client shall procure such legal advice and provide such information and evidence as required for the resolution of any dispute between the Client and any parties providing services in connection with the Project.

4 Assignment and sub-letting

Assignment — 4.1 Neither the Architect nor the Client shall assign the whole or any part of the Agreement without the consent of the other in writing.

Sub-letting — 4.2 The Architect shall not appoint and Sub-Consultants to perform any part of the Services without the consent of the Client, which consent shall not be unreasonably withheld. The Architect shall confirm such consent in writing.

5 Payment

Fees for performance of the Services — 5.1 The Fees for performance of the Services, including the anticipated Visits to the Works, shall be calculated and charged as set out in Schedule 3.
Works Stage Fees shall be:
.1 as a percentage of the Construction Costs calculated in accordance with clause 5.2.2; or
.2 as lumps sums in accordance with clause 5.3; or
.3 as time charges in accordance with clause 5.4; or
.4 other agreed method.

STANDARD FORM OF AGREEMENT
For the appointment of an Architect (SFA/99) (Cont'd)
Conditions of Engagement (Cont'd)
Payment (Cont'd)

Percentage fees 5.2 .1 Percentage fees shall be in accordance with clause 5.2.2 or 5.3.1 as stated in Schedule 3.

.2 Where this clause 5.2.2 is stated to apply in Schedule 3 the percentage or percentages stated therein shall be applied to the Construction Costs. Until the final Construction Cost has been ascertained interim fee calculations shall be based on:

(a) before tenders have been obtained - the lowest acceptable estimate of the Construction Cost;

(b) after tenders have been obtained - the lowest acceptable tender;

(c) after the contract is let either the certified value or the anticipated final account.

The final fee shall be calculated on the ascertained gross final cost of all Works included in the Construction Cost, excluding any adjustment for loss and/or expense payable to or liquidated damages recoverable from a contractor by the Client. Where the Client is the Contractor, the final cost shall include an allowance for the Contractor's profit and overheads.

Lump sums 5.3 A lump sum fee or fees shall be in accordance with clause 5.3.1 or 5.3.2 as stated in Schedule 3 and subject to clause 5.5. Such fee or fees shall be:

.1 calculated by applying the percentage stated in Schedule 3, to be in accordance with this clause 5.3.1, to create:

(a) a lump sum or sums based on the Construction Cost approved by the Client at the end of Work Stage D; or

(b) separate lump sums for each Work Stages based on the Construction Cost approved by the Client at the end of the previous stage; or

.2 a fixed lump sum or sums stated in Schedule 3 to be in accordance with this clause 5.3.2, which shall be adjusted in accordance with clause 5.6 if substantial amendments are made to the Brief and/or the Construction Cost and/or the Timetable.

Time charges 5.4 A time based fee shall be ascertained by multiplying the time reasonably spent in the performance of the Services by the relevant hourly rate set out in Schedule 3. Time 'reasonably spent' shall include the time spent in connection with performance of the Services in travelling from and returning to the Architect's office.

Revision of lump sums and time charge and other rates 5.5 Lump sums complying with clause 5.3.1a or 5.3.2, rates for time charges mileage and printing carried out in the Architect 's office shall be revised every 12 months in accordance with changes in the Retail Price Index*. Each 12-month period shall commence on the anniversary of the Effective Date of the Agreement, or the date of calculation of lump sums complying with clause 5.3.1a, whichever is the later.

**Retail Price Index is set out in Table 6.1 (All items) to 'Labour Market trends' published by the Office for National Statistics.*

Additional Fees 5.6 If the Architect, for reasons beyond his control is involved in extra work or incurs extra expense, for which he will not otherwise be remunerated, the Architect shall be entitled to additional fees calculated on a time basis unless otherwise agreed. Reasons for such entitlement include, but shall not be limited to:

.1 the scope of the Services or the Timetable or the period specified for any work stage is varied by the Client;

.2 the nature of the Project requires that substantial parts of the design can not be completed or must be specified provisionally or approximately before construction commences;

.3 the Architect is required to vary any item of work commenced or completed pursuant to the Agreement or to provide a new design after the Client has authorised the Architect to develop an approved design;

.4 delay or disruption by others;

.5 prolongation of any building contract(s) relating to the Project;

.6 the Architect consents to enter into any third party agreement the form or beneficiary of which had not been agreed by the Architect at the date of the Agreement;

.7 the cost of any work designed by the Architect or the cost of special equipment is excluded from the Construction Cost.

This clause 5.6 shall not apply where the extra work and/or expense to which it refers is due to a breach of the Agreement by the Architect.

5.7 Where for any reason the Architect completed provides only part of the Services specified in Schedule 2, the Architect shall be entitled to fees calculated as follows:

.1 for completed Services, as described for those Services in Schedule 3;

.2 for completed Work Stages, as apportioned for those Work Stages in Schedule 3;

.3 for services or Work Stages not completed, a fee proportionate to that described or apportioned in Schedule 3 based on the Architect's estimate of the percentage of completion.

Expenses and disbursements 5.8 The Client shall reimburse at net cost plus the handling charge stated in Schedule 3:

.1 expenses specified in Schedule 3;

.2 expenses other than those specified and incurred with the prior authorisation of the Client;

.3 any disbursements made on the Client's behalf.

Maintain records 5.9 If the Architect is entitled to reimbursement of time spent on Services performed on a time basis, and of expenses and disbursements, the Architect shall maintain records and shall make these available to the Client on reasonable request.

STANDARD FORM OF AGREEMENT
For the appointment of an Architect (SFA/99) (Cont'd)
Conditions of Engagement (Cont'd)
Payment (Cont'd)

Payments

5.10 Payments under the Agreement shall become due to the Architect on issue of the Architect's accounts. The final date for such payments by the Client shall be 30 days from the date of issue of an account.

The Architect's account shall be issued at intervals of not less than one month and shall include any additional fees, expenses or disbursements and state the basis of calculation of the amounts due.

Instalments of the fees shall be calculated on the basis of the Architect's estimate of the percentage of completion of the Work Stage or other Services or such other method specified in Schedule 3.

5.11 The Client may not withhold payment of any part of an account for a sum or sums due to the Architect under the Agreement by reason of claims or alleged claims against the Architect unless the amount to be withheld has been agreed by the Architect as due to the Client, or has been awarded in adjudication, arbitration or litigation in favour of the Client and arises out of or under the Agreement. Save as aforesaid, all rights of set-off at common law or in equity which the Client would otherwise be entitled to exercise are hereby expressly excluded.

Payment notices

5.12 A written notice from the Client to the Architect:

.1 may be given within 5 days of the date of issue of an account specifying the amount the Client proposes to pay and the basis of calculation of that amount; and/or

.2 shall be given not later than 5 days before the final date for payment of any amount due to the Architect if the Client intends to withhold payment of any part of that amount stating the amount proposed to be withheld and the ground for doing so or, if there is more than one ground, each ground and the amount attributable to it.

If no such notices are given the amount due shall be the amount stated as due in the account. The Client shall not delay payment of any undisputed part of an account.

Late payment

5.13 Any sums due and remaining unpaid at the expiry of 30 days after the date of issue of an account from the Architect shall bear interest. Interest shall be payable at 8% over Bank of England base rate current at the date of issue of the account.

For the avoidance of doubt the Architect's entitlement to interest at the expiry of 30 days after the date of issue of an account shall also apply to any amount which an Adjudicator decides should be paid to the Architect.

Payment on suspension or determination

5.15 If the Client or the Architect suspends or determines performance of the Services, the Architect shall be entitled to payment of any part of the fee or other amounts due at the date of suspension or determination on issue of the Architect's account in accordance with clause 5.10.

5.16 Where the performance of the Services is suspended or determined by the Client, or suspended or determined by the Architect because of breach of the Agreement by the Client, the Architect shall be entitled to payment of all expenses and other costs necessarily incurred as a result of any suspension and any resumption or determination. On issue of the Architect's account in accordance with clause 5.10.

VAT

5.17 Fees, expenses and disbursements arising under the Agreement do not include Value Added Tax. The Client shall pay any Value Added Tax chargeable on the net value of the Architect's fees and expenses.

6 Copyright

Copyright

6.1 The Architect owns the copyright on the work produced by him in performing the Services and generally asserts the right to be identified as the author of the artistic work/work of architecture comprising the Project.

Licence

6.2 The Client shall have a licence to copy and use and allow other Consultants and contractors providing services to the Project to use and copy drawings, documents and bespoke software produced by the Architect, in performing the Services hereinafter called 'the Material', but only for purposes related to the Project on the Site or part of the Site to which the design relates.

Such purposes shall include its operation, maintenance, repairs, reinstatement, alteration, extending, promotion, leasing and/or sale, but shall exclude the reproduction of the Architect's design for any part of any extension of the Project and/or for any other project unless a licence fee in respect of any identified part of the Architect's design stated in Schedule 3.

Provided that:

.1 the Architect shall not be liable if the Material is modified other than by or with the consent of the Architect, or used for any purpose other than that for which it was prepared; or used for any unauthorised purpose;

.2 in the event of any permitted use occurring after the date of the last Service performed under the Agreement and prior to practical completion of the completion of the construction of the project, the Client shall:

(a) where the Architect has not completed Detailed Proposals (Work Stage D), obtain the Architect's consent, which consent shall not be unreasonably withheld; and/or

(b) pay to the Architect a reasonable licence fee where no licence fee is specified in Schedule 3;

STANDARD FORM OF AGREEMENT
For the appointment of an Architect (SFA/99) (Cont'd)
Conditions of Engagement (Cont'd)
Copyright (Cont'd)

.3 in the event of the Client being in default of payment of any fees or other amounts due, the Architect may suspend further use of the licence on giving 7 day's notice of the intention of doing so. Use of the licence maybe resumed on receipt of outstanding amounts.

7 Liabilities and Insurance

Limitation of warranty by Architect

7.1 Subject always to the provisions of clause 2.1, the Architect does not warrant:
.1 that the Services will be completed in accordance with the Timetable;
.2 that planning permission and other statutory approvals will be granted;
.3 the performance, work or the products of others;
.4 the solvency of any other appointed body whether or not such appointment was made on the advice of the Architect.

Time limit for action or proceeding

7.2 No action or proceedings whatsoever for any breach of this Agreement or arising out of or in connection with this Agreement whether in contract, negligence, tort or howsoever shall be commenced against the Architect after the expiry of the period stated in the Letter of Appointment from the date of the Last Services performed under the Agreement or (if earlier) practical completion of the construction of the project.

Architect's liability

7.3 In any action or proceedings brought against the Architect under or in connection with the Agreement whether in contract, negligence, tort or howsoever the Architect's liability for loss or damage in respect of any one occurrence or series of occurrences arising out of one event shall be limited to whichever is the lesser of the sum:
.1 stated in the Appendix; or
.2 such sum as it is just and equitable for the Architect to pay having regard to the extent of his responsibility for the loss and/or damage in question when compared with the responsibilities of contractors, sub-contractors, Consultants and other persons for that loss and/or damage. Such sums to be assessed on the basis that such persons are deemed to have provided contractual undertakings to the Client no less onerous than those of the Architect under the Agreement and had paid to the Client such sums as it would be just and equitable for them to pay having regard to the extent of their responsibility for that loss and/or damage.

Professional Indemnity Insurance

7.4 The Architect shall maintain Professional Indemnity Insurance cover in the amount stated in the Appendix for any one occurrence or series of occurrences arising out of any one event until at least the expiry of the period stated in the Appendix from the date of the Last Services, performed under the Agreement or (if earlier) practical completion of the construction of the Project provided such insurance is available at commercially reasonable rates and generally available in the insurance market to the Architect.

The Architect when requested by the Client, shall produce for inspection documentary evidence that the Professional Indemnity Insurance required under the Agreement is being maintained.

The Architect shall inform the Client if such insurance ceases to be available at commercially reasonable rates in order that the Architect and Client can discuss the best means of protecting their respective positions in respect of the project in the absence if such insurance.

Third Party Agreements

7.5 Where the Client has notified, prior to the signing of this Agreement, that he will require the Architect to enter into an agreement with a third party or third parties, the terms of which and the names or categories of other parties who will sign similar agreements are set out in an annex to this Agreement, then the Architect shall enter such agreement or agreements within a reasonable period of being requested to do so by the Client.

Rights of Third Parties

7.6 For the avoidance of doubt nothing in this Agreement shall confer or purport to confer on any third party any benefit or the right to enforce any term of this Agreement.

8 Suspension and determination

Suspension

8.1 The Client may suspend the performance of any or all of the Services by giving at least 7-days' notice to the Architect. The notice shall specify the Services affected.

8.2 The Architect may suspend performance of the Services and his obligations under the Agreement on giving at least 7-days' notice to the Client of his intention and the grounds for doing so in the event that the Client:
Is in default of any fees or other amounts due; or
Fails to comply with the requirements of the CDM Regulations.
The Architect shall resume performance of his obligations on receipt of the outstanding amounts.

8.3 If any period of suspension arising from a valid notice given under clause 8.1 or clause 8.2 exceeds 6 months the Architect shall request the Client to issue instructions. If written instructions have not been received within 30 days of the date of such request the Architect shall have the right to treat performance of any Service or his obligations affected as determined.

8.4 Any period of suspension arising from a valid notice given under clause 8.1 or clause 8.2 shall be disregarded in computing any contractual date for completion of the Services.

Determination

8.5 The Client or the Architect may by giving 14-days' notice in writing to the other determine performance of any or all of the Services and the Architect's obligations under Part 2 of these conditions stating the grounds for doing so and the Services and obligations affected.

STANDARD FORM OF AGREEMENT
For the appointment of an Architect (SFA/99) (Cont'd)
Conditions of Engagement (Cont'd)
Suspension and determination (Cont'd)

8.6 Performance of the Services and the Architect's obligations under Part 2 of these Conditions may be determined immediately by notice from either party in the event of:
.1 insolvency of the Client or the Architect; or
.2 the Architect becoming unable to provide the Services through death or incapacity.

8.7 On determination on performance of the Services or the Architect's obligations under Part 2 of these conditions, a copy of the Material referred to in clause 6.2 shall be delivered on demand to the Client by the Architect, subject to the terms of the licence under clause 6.2 and payment of the Architect's reasonable copying charges.

8.8 Determination of the performance of the Services or the Architect's obligations shall be without prejudice to the accrued rights and remedies of either party.

9 Dispute resolution*

Negotiation or conciliation

9.1 In the event of any dispute or difference arising under the Agreement, the Client and the Architect may attempt to settle such dispute or difference by negotiation or in accordance with the RIBA Conciliation Procedure.

Adjudication: England and Wales

9.2 Where the law of England and Wales is England the applicable law, any dispute or difference arising out of this Agreement may be referred to adjudication by the Client or the Architect at any time. The adjudication procedures and the Agreement for the Appointment of an Adjudicator shall be as set out in the "Model Adjudication Procedures" published by the *Construction Industry Council* current at the date of the reference. Clause 28 of the 'Model Adjudication Procedures' shall be deleted and replaced as follows: 'The adjudicator may in his discretion direct the payment of legal costs and expenses of one party by another as part of his decision. The adjudicator may determine the amount of costs to be paid or may delegate the task to an independent costs draftsman.

Adjudication: Scotland

9.3 .1 Where the law of Scotland is the applicable law, any dispute or difference touching or concerning any matter or thing arising out of this Agreement (other than with regard to the meaning or construction of this Agreement) may be referred to some independent and fit persons within 7 days of the application of the Client or the Architect and any fees which may become payable to the persons so appointed shall be within the award of that person.

.2 Any such Adjudicator appointed in terms of clause 9.3.1 hereof shall have 28 days from the date of referral within which to reach a decision on the dispute, or such longer period as is agreed between the parties after the dispute has been referred but without prejudice, to the foregoing the Adjudicator shall be permitted to extend the said period of 28 days by up to 14 days, with the consent of the party by whom the dispute was referred.

The Adjudicator shall act impartially and shall be entitled to take the initiative in ascertaining the facts and the law relating to the dispute. The decision of the Adjudicator shall be binding on both parties until the dispute is finally determined by arbitration pursuant to clause 9.5 hereof.

The Adjudicator shall not be liable for anything done or omitted in the discharge or purported discharge of his functions as Adjudicator unless the act or omission is in bad faith, and all employees or agents of the Adjudicator are similarly protected from liability subject to the same proviso.

Naming or nomination of an Adjudicator

9.2 Where no Adjudicator is named in the Agreement and the parties are unable to agree on a person to act as Adjudicator, the Adjudicator shall be a person to be nominated at the request of either party by the nominator identified in the Letter of Appointment.

Arbitration

9.5 When in accordance with the Letter of Appointment either the Client or the Architect require any dispute or difference to be referred to Arbitration the requiring party shall give notice to the other to such effect and the dispute or difference shall or referred to the arbitration and final decision of a person to be agreed between the parties or, failing agreement within 14 days of the date of the notice, the Appointor shall be the person identified in the Appendix.

Provided that where the law of England and Wales is applicable to the Agreement:
.1 the Client or the Architect may litigate any claim for a pecuniary remedy which does not exceed £5,000 or such other sum as is provided by statute pursuant to section 91 of the Arbitration Act 1996;
.2 the Client or the Architect may litigate the enforcement of any decision of an Adjudicator;
.3 where and to the extent that the claimant in any dispute which is referred to arbitration is the Architect, the arbitrator shall not have the power referred to in Section 38(3) of the Arbitration Act 1996.

Costs

9.6 The Client shall indemnify the Architect in respect of his legal and other costs in any action or proceedings, together with a reasonable sum in respect of his time spent in connection with such action or proceedings or any part thereof, if:
.1 the Architect obtains a judgement of the court or an Arbitrator's award on his favour for the recovery of fees and/or expenses under the Agreement; or
.2 the Client fails to obtain a judgement of the court or an Arbitrator's award in the Client's favour for any claim or any part of any claim against the Architect.

- *Architects are subject to the disciplinary sanction of the Architects Registration Board in relation to complaints of unacceptable professional conduct or serious professional incompetence.*

STANDARD FORM OF AGREEMENT
For the appointment of an Architect (SFA/99) (Cont'd)

Schedule Two **Services**

The Architect shall:

1. Perform the Services as designer, design leader, lead consultant during pre-construction and construction Work Stages.
2. Perform the Services for the Work Stages indicated below and in accordance with the Services Supplement.
3. Make visits to the Works in accordance with clause 2.8.
4. Perform any other Services identified below.

Feasibility	**A** Appraisal	**B** Strategic Brief
Pre-Construction	**C** Outline Proposals **D** Detailed Proposals **E** Final Proposals	**F** Production Information **G** Tender Documentation **H** Tender Action
Construction	**J** Mobilisation **K** To Practical Completion	**L** After Practical Completion

Other Services *If identifying any other services the description should be sufficient to identify its scope.*

Other Activities

These activities do not form part of the services unless identified under 'Other Services'

Performance of any of these activities will attract additional fees in accordance with clause 5.6.

Activities - *These may be required in connection with:*

Sites and Buildings

- o Selection of Consultants
- o Options approval
- o Compiling, revising and editing:
  - (a) strategic brief
  - (b) detailed (written) brief
  - (c) room data sheets
- o Selection of site and/or buildings
- o Outline planning submission
- o Environmental studies
- o Surveys, inspection or specialist investigations
- o Party wall matters
- o Two-stage tendering
- o Negotiating a price with a contractor (in-lieu tendering)
- o Use of energy in new or existing buildings
- o Value management services
- o Compiling maintenance and operational manuals
- o Specially prepared drawings of a building as built
- o Submission of plans for proposed works for approval of landlords, founders, freeholders, tenants etc
- o Applications or negotiations for statutory and other grants

Design Skills

- o Interior design services
- o Selection of furniture and fittings
- o Design of furniture and fittings
- o Landscape design services
- o Special drawings, photographs, models or technical information produced at the Client's request

Historic Buildings and Conservation

- o Detailed inspection and report
- o Historical research and archaeological records
- o Listed building consent
- o Conservation area consents
- o Grant aided works

Special Activities - These may be required in connection with:

- o Exceptional negotiations with planning or other statutory authorities
- o Revision of documents to
  - (a) comply with requirements of planning or statutory authorities, or landlords, etc
  - (b) comply with changes in interpretation or enactment or revisions to laws or statutory regulations
  - (c) make corrections not arising from any failure of the Architect
- o Ascertainment of contractor's claims
- o Investigations and instructions relating to work not in accordance with the building contract
- o Assessment of alternative designs, materials or products proposed by the contractor
- o Dispute resolution services on behalf of the Client
- o Damage to or destruction of a building in construction or to existing buildings
- o Determination of any contract or agreement with any other party providing services to the project
- o Insolvency of any other party providing services to the project
- o Valuations for mortgage or other purposes
- o Easements or other legal agreements
- o Investigation of building failures
- o Feed Back – post-completion evaluation

STANDARD FORM OF AGREEMENT
For the appointment of an Architect (SFA/99) (Cont'd)

Services Supplement: Design and Management.

Architect's design services

All Commissions

1.1 Receive Client's instructions

1.2 Advise Client on the need to obtain statutory approvals and of the duties of the Client under the CDM Regulations

1.3 Receive information about the site from the Client (CDM Reg 11)

1.4 Where applicable, co-operate with and pass information to the Planning Supervisor

1.5 Visit the site and carry out an initial appraisal

A Appraisal

1 Carry out studies to determine the feasibility of the Client's Requirements

2A Review with Client alternative design and construction approaches and the cost implications, *or*

2B Provide information for report on cost implications

B Strategic Brief

1 (Strategic Brief prepared by or for the Client)

C Outline Proposals

1 Receive Strategic Brief and Commence development into Project Brief

2 Prepare Outline Proposals

3A Prepare an approximation of construction cost, *or*

3B Provide information for cost planning

4 Submit to Client Outline Proposals and approximate construction cost

D Detailed Proposals

1 Complete development of Project Brief

2 Develop the Detailed Proposals from approved Outline Proposals

3A Prepare a cost estimate, *or*

3B Provide information for preparation of cost estimate

4 Consult statutory authorities

5 Submit to Client the Detailed Proposals, showing spatial arrangements, materials and appearance, and a cost estimate

6 Prepare and submit application for full planning permission

E Final Proposals

1 Develop Final Proposals from approved Detailed Proposals

2A Revise cost estimate, *or*

2B Provide information for revision of cost estimate

3 Consult statutory authorities on developed design proposals

4 Submit to Client type of construction, quality of materials, standard of workmanship and revised cost estimate

5 Advise on consequences of any subsequent changes on cost and programme

F Production Information

1 Prepare production information for tender purposes

2A Prepare schedules of rates and/or quantities and/or schedules of works for tendering purposes and revise cost estimate, *or*

2B Provide information for preparation of tender pricing documents and revision of cost estimate

3A Prepare and make submissions under building acts and/or regulations or other statutory requirements, *or*

3B Prepare and give building notice under building acts and/or regulations*
**Not applicable in Scotland*

3 Prepare further production information for construction purposes

G Tender Documentation

1 Prepare and collate tender documents in sufficient detail to enable a tender or tenders to be obtained

2 Where applicable pass final information to Planning Supervisor for pre-tender Health and Safety Plan

3A Prepare pre-tender cost estimate, *or*

3B Provide information for preparation of pre-tender cost estimate

H Tender Action

1 Contribute to appraisal and report on tenders/ negotiations

2 *If instructed, revise production information to meet adjustments in the tender sum*

J Mobilisation

1 Provide production information as required for the building contract and for construction

K Construction to Practical Completion

1 Make Visits to the Works in connection with the Architect's design

2 Provide further information reasonably required for construction

3 Review design information from contractors or specialists

4 Provide drawings showing the building and main lines of drainage and other information for the Health and Safety File

4 Give general advice on operation and maintenance of the building

L After Practical Completion

1 Identify defects and make final inspections

2A Settle Final Account, *or*

2B Provide information required by others for settling final account

> If the Architect is to provide cost advice Alternative **A** applies
>
> If a Quantity Surveyor is appointed Alternative **B** applies

STANDARD FORM OF AGREEMENT
For the appointment of an Architect (SFA/99) (Cont'd)
Services Supplement (Cont'd)

Architect's Management Services

As Design Leader

The Architect has authority and responsibility for:

1 directing the design process;

2 co-ordinating design of all constructional elements, including work by any Consultants, specialists or suppliers;

3 establishing the form and content of design outputs, their interfaces and a verification procedure;

5 communicating with the Client on significant design issues.

As Lead Consultant (pre-construction)

The Architect has authority and responsibility in the pre-construction Work Stages for:

1 co-ordinating and monitoring design work;

2 communications between the Client and the Consultants; except that communications on significant design matters are dealt with as design leader;

3 advising on the need for and the scope of services by Consultants, specialists, sub-contractors or suppliers;

4 advising on methods of procuring construction;

5 receiving regular status reports from each Consultant, including the design leader;

6 developing and managing change control procedures, making or obtaining decisions necessary for time and cost control;

7 reporting to the Client as appropriate.

As Lead Consultant and Contract Administrator

The Architect has authority and responsibility in the tender and construction Work Stages A-G for:

1 inviting and appraising a tender or tenders, including:
- considering with the Client a tenderer or a list of tenderers for the Works;
- considering with the Client appointment of a contractor, and the responsibilities of the parties and the Architect/Contract Administrator under the building contract;
- preparing the building contract and arranging for signatures;

2 administering the building contract, including:
- monitoring the progress of the Works against the Contractor's programme;
- issuing information, instructions, etc;
- preparing and certifying valuations of work carried out or completed and preparing financial reports for the Client; or
- certifying valuations of work prepared by others, and presenting financial reports prepared by others to the Client;
- collating record information including the Health and Safety File;

3 co-ordinating and monitoring the work of Consultants and Site Inspectors, if any, to the extent required for the administration of the building contract, including:
- receiving reports from such Consultants and Site Inspectors to enable decisions to be made in respect of the administration of the building contract;
- consulting any Consultant or other person whose design or specification may be affected by a Client instruction relating to the construction contract, obtaining any information required and issuing any necessary instructions to the Contractor;
- managing change control procedures and making or obtaining decisions as necessary for cost control during the construction period;
- providing information obtained during the administration of the building contract to the Consultants and Site Inspectors;

4 communications between the Client and the Consultants;

5 reporting to the Client as appropriate.

ROYAL INSTITUTE OF BRITISH ARCHITECTS

A Client's guide to Engaging an Architect including guidance on fees

The Royal Institute of British Architects Guide provides guidance for Clients on engaging an Architect, and on professional services, conditions of engagement and fees, together with information on :-

Architect's Services
Other services
Appointment
RIBA Standard Forms
Design and Build
Historic buildings
CDM Regulations

The Client – Architect Agreement
The architect
The client
Statutory requirements
Other appointments
Copyright
Liability and insurance
Suspension and determination
Dispute Resolution

Architect's Fees and Expenses
Percentage basis
Lump sums
Time basis
Expenses
Payment

Architect's Fees and Expenses

RIBA forms of appointment provide a number options for the calculation of the architect's fee, i.e.:

- a quoted percentage of the final cost of the building work;
- a fixed or calculated lump sum or sums;
- time charge; or
- an other agreed basis.

There is no 'standard' or 'recommended' basis for calculation of the fee, but the relative cost usually incurred in providing architects' normal services is indicated in the tables below. The tables recognise that the additional complexity of the work in alterations or extensions or to historic buildings may justify a higher fee than for new work. Figures 1 and 2 show indicative percentage rates in relation to construction cost for new works and works to existing buildings respectively for five classes of building (Fig 5). Figure 6 and 7 provide similar information for historic buildings.

Before fees can be agreed, client and architect should establish the requirements, the services to be provided, the approximate construction cost and the timetable. It is usually helpful to identify the proportion of the fee for each work Stage (Fig 3) and any other fees payable for additional services such as a survey.

In proposing a fee the architect will take account of the requirement that a client must be *'reasonably satisfied'* that designers will allocate adequate resources for the purposes of CDM Regulations.

Percentage basis

The architect's fees for normal services are expressed as a percentage of the final construction cost executed under the architect's direction. This method is best used for clearly defined building projects. Figures 1 and 2 may be a helpful point of reference in selecting the percentage(s).

Lump sums

Fixed lump sums – may be suitable where the parameters of the services, i.e. the requirements, time and cost, can be clearly defined from the outset. Then, if these are varied by more than a stated amount the lump sum itself may be varied.

Calculated lump sums – with this option the lump sums for the Work Stages are calculated by applying quoted percentages to the latest approved cost either, when the design and estimated cost have been settled at the end of Stage D or at the commencement of each stage. This method will be particularly appropriate where the project parameters cannot be pre-determined with the necessary accuracy or there are inflationary or deflationary pressures in the market place.

Time basis

This basis is best used where the scope of the work cannot be reasonably foreseen or where services do not relate to the cost of construction, for example, for:

- varied or additional services to an otherwise basic service;
- open-ended exploration work, particularly for repair and conservation work;
- feasibility studies;
- protracted planning negotiations;
- party wall services, etc.

The time spent on such services by principals and technical staff is charged at agreed rates, usually expressed as hourly rates, which may be revised at 12-monthly intervals. A range of indicative values is given in Figure 4. Time spent in travelling in conjunction with the time-charged services is chargeable, but the time of secretarial and administrative staff is usually excluded.

Expenses

In addition to the fees, expenses identified in the Agreement or Letter of Appointment for reimbursement may include:

- the cost of copies of drawings and other documents;
- the purchase of maps and printed contract documentation;
- the cost of photography;
- postage and fax charges;
- travel costs and accommodation where relevant.

Additionally, the relevant fee must accompany applications for building regulations or planning approval. If the architect agrees to make such payments on behalf of the client, these disbursements will also be chargeable, perhaps with a handling charge.

Payment

Under RIBA forms of appointment accounts for instalments of the fees and expenses will normally be issued monthly for payment within 30 days. If required, regular payments can be budgeted over a period with review at completion of say the pre-construction Work Stages. Alternatively, fees may be paid at the completion of each Work Stage.

The agreed arrangements will be set out in the Agreement or Letter of Appointment, showing the basis on which the fee is calculated and the rates that are to apply. Statutory provision apply to the payment terms.

Architects will keep and make available for inspection records of expenditure on expenses and disbursements and time spent on services carried out on a time-charged basis. Where applicable, VAT is due on architects' fees and expenses at the appropriate rate.

A Clients guide to Engaging an Architect (cont'd.)
Architect's Fees and Expenses (Cont'd)

Indicative Percentage Fee Scales – Normal Services

Figure 1

New Works

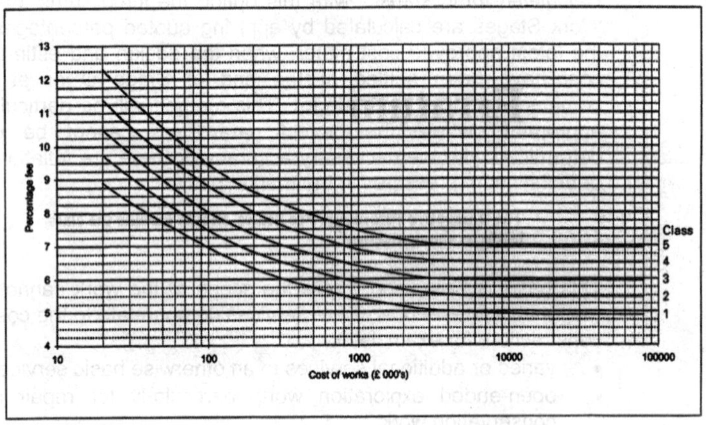

Figure 3

Proportion of fee by Work stage

Proportion of fee			%
Work stage	C	Outline proposals	20
	D	Detailed Proposals	20
	E	Final Proposals	15
	F	Production Information	20
	G	Tender Documentation	2
	H	Tender Action	1
	J	Mobilisation	1
	K	Construction to Practical Completion	20
	L	After Practical Completion	1
			100%

Stages A and B. Appraisal and strategic brief, are normally carried out on a time-charged basis and are therefore not included above.

Figure 2

Works to existing buildings

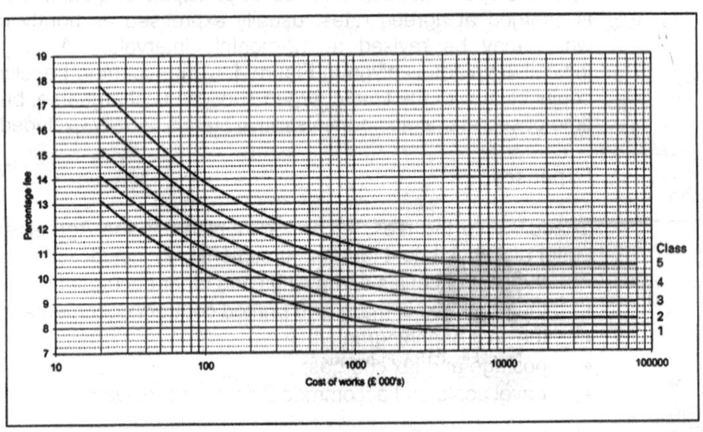

Figure 4

Indicative hourly rates (June 1999)

Type of work	*General*	*Complex*	*Specialist*
Partner/director Or equivalent	£95	£140	£180
Senior Architect	£75	£105	£140
Architect	£55	£75	£95

Rates are purely indicative and may be negotiated above or below the figures suggested to reflect the nature of the work.

Figure 6

Historic Buildings

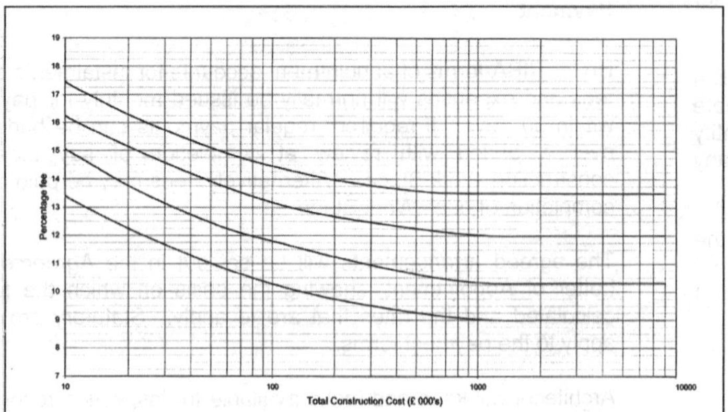

Figure 7

Classification of work to Historic Buildings

Class HB1	Work of a straight forward nature where one repair element predominates (eg large areas of re-roofing).
Class HB2	Work of general repair not including a great deal; of internal alteration.
Class HB3	Work of general repair and alteration, including substantial alteration work. Repair work where the services installations are to be renewed or replaced.
Class HB4	Work including specialist conservation work, requiring an exceptional amount of detailed specification and inspection.

A Clients guide to Engaging an Architect (Cont'd.)

Figure 5 **Classification by building type**

Type	Class 1	Class 2	Class3	Class 4	Class 5
Industrial	• Storage sheds	• Speculative factories and warehouse • Assembly and machine workshops • Transport garages	• Purpose-built factories and warehouses		
Agricultural	• Barns and sheds	• Stables	• Animal breeding units		
Commercial	• Speculative shops • Surface car parks	• Multi-storey and underground car parks	• Supermarkets • Banks • Purpose-built shops • Office developments • Retail warehouses • Garages/ showrooms	• Department stores • Shopping centres • Food processing units • Breweries • Telecommunications and computer buildings • Restaurants • Public houses	• High risk research and production buildings • Research and development laboratories • Radio, TV and recording studios
Community		• Community halls	• Community centres • Branch libraries • Ambulance and fire stations • Bus stations • Railway stations • Airports • Police stations • Prisons • Postal Buildings • Broadcasting	• Civic centres • Churches and crematoria • Specialist libraries • Museums and art galleries • Magistrates/ County Courts	• Theatres • Opera houses • Concert halls • Cinemas • Crown Courts
Residential		• Dormitory hostels	• Estates housing and flats • Barracks • Sheltered housing • Housing for single people • Student housing	• Parsonages/ manses • Apartment blocks • Hotels • Housing for the handicapped • Housing for the frail elderly	• Houses and flats for individual clients
Education			• Primary/ nursery/ first schools	• Other schools including middle and secondary • University complexes	• University laboratories
Recreation			• Sports halls • Squash courts	• Swimming pools • Leisure complexes	• Leisure pools • Specialised complexes
Medical/Social services			• Clinics	• Health Centres • General hospitals • Nursing homes • Surgeries	• Teaching hospitals • Hospital laboratories • Dental surgeries

CONDITIONS OF ENGAGEMENT FOR THE APPOINTMENT OF AN ARCHITECT (CE/99)

The RIBA *Conditions of Engagement for the Appointment of an Architect (CE/99)* is suitable for use where an architect is to provide services for fully designed building project in a wide range of sizes or complexity and/or to provide other professional services. It is fully compliant with the requirements of the Housing Grants, Construction and Regeneration Act 1996.

The Agreement is completed by the exchange of a Letter of Appointment to which a copy of the form is attached, after completion of the Schedules.

'Notes for Architects' and a draft Model letter is included in CE/99 and may assist in identifying those matters which will affect the scope and terms of the Agreement.

It should be noted that the appointment of an architect as Planning Supervisor, where required by the CDM Regulations 1994, is separate from the provision of architectural services under CE/99 and should be made using the RIBA *Form of Appointment as Planning Supervisor (PS/99)*

A series of guides addressed directly to Clients is also published on topics associated with the appointment of an Architect, which Architects may wish to send to their Clients as appropriate and also refer to themselves. The guides can be obtained from RIBA Publications by mail order (020) 7251 0791 or from RIBA bookshops in London and the regions.

+THE ROYAL INSTITUTION OF CHARTERED SURVEYORS

Appointing a Quantity Surveyor

The Scales of Professional Charges for Quantity Surveying Services have been withdrawn and replaced by the RICS information document "Appointing a Quantity Surveyor: A guide for Clients and Surveyors". The document gives no indication of the level of fees to be anticipated although a Form of Enquiry, Schedule of Services, Fee Offer, Form of Agreement and Terms of Appointment are included. The view being that consultants are now appointed less often by reputation and recommendation and more often after a process of competition. To assist employers, quantity surveyors and those procuring professional services and where consultants are being appointed by reputation or recommendation the following will assist in assessing the reasonableness of any fee proposal.

Extracts from Appointing a Quantity Surveyor: A guide for Clients and Surveyors
Scales of Professional Charges
Example of calculation of fees for lump sum and % basis as clause 5.6 of the guide
Anticipated Percentage fees by project and value based on calculations as example.

Appointing a Quantity Surveyor

The following extracts are for information and the full document should be referred to for further advice.

Contents

SECTION 1

Selection and Appointment Advice

This Section outlines for Clients and Surveyors when and how to select and appoint a Quantity Surveyor

1. Introduction

1.1 The Royal Institution of Chartered Surveyors ('the RICS') has published this Guide to assist Clients when they wish to appoint a chartered Quantity Surveyor. The contents of the Guide should also assist Quantity Surveyors when concluding an agreement with their Client.

1.2 Over the last few years changes have taken place in the market place for professional services. Consultants are now appointed less often by reputation and recommendation and more often after a process of competition. Competition takes place in the private and public sectors and is supported by European Union requirements and UK Compulsory Competitive Tendering legislation.

1.3 This Guide is applicable to a range of situations - whether details of the services required are proposed to or by the Quantity Surveyor; or whether appointment is of a single Quantity Surveyor or following selection from a limited list. The Guide sets out principles to be considered and applied rather than a detailed step-by-step procedure. Details often vary from project to project depending on the type of Client and the nature of the project. A detailed procedure that attempts to cover all projects is not therefore considered appropriate.

1.4 Quantity Surveyors may be appointed directly by a Client or as part of a multidisciplinary team. The RICS recommends that the quantity surveying appointment is made directly by a Client so as to ensure that independent financial advice is made directly to that Client by the Quantity Surveyor

1.5 The Guide provides a basis for the appointment of a Quantity Surveyor so that those concerned are clear about:

- the services being requested and/or offered
- the terms and conditions of the contract
- the fee payable and the method of payment.

THE ROYAL INSTITUTION OF CHARTERED SURVEYORS
Appointing a Quantity Surveyor (Cont'd)
Selection and Appointment Advice (Cont'd)

A Client and his Quantity Surveyor should set out clearly their requirements in their agreement so that the services to be provided and the conditions of engagement are certain. It is necessary for a Client to set out his requirements with the same clarity as when seeking tenders from contractors on construction projects. It is now compulsory for Members of the RICS to provide written notification to their Client of the terms and conditions of the appointment. This documentation is intended to form a good basis upon which this certainty can be achieved and is recommended for use between the Client and his Quantity Surveyor.

2. **Notes on use and completion of documents**

2.1 **Form of Enquiry**
The Form of Enquiry sets out the details of the services a Client wishes a Quantity Surveyor to provide and should be attached to the Form of Agreement and identified as belonging to it.

2.2 **Form of Agreement and Terms of Appointment**
The Form of Agreement should be completed and signed only when the services, fees and expenses have been agreed between a Client and his Quantity Surveyor. The Form of Agreement states the names of the parties, their intentions and their arrangement.

The names and addresses of the parties should be inserted, a brief description of the project given and any site identified. If the Form of Agreement is to made as a simple contract it should be signed by both parties and the date entered. If it is to be executed as a Deed the alternative version should be used. See notes *1 and *2.

3. **When to appoint a Quantity Surveyor**

In order that maximum benefit can be gained from his skills a Quantity Surveyor should be appointed by a Client as soon as possible in the life of a project, preferably at the inception of a scheme, so that advice can be provided on:

- the costs of the project so that a realistic budget can be set at inception and cost management can be applied throughout
- the procurement method best suited to the requirements of the Client
- the implications of the appointment of other consultants and contractors.

It is recommended that a Client and his prospective Quantity Surveyor should meet and discuss the appointment before an agreement is reached, unless the services of the Quantity Surveyor are to be restricted only to some of those from the range available and shown in the Form of Enquiry.

4. **How to select and appoint a Quantity Surveyor**

Detailed guidance on the selection and appointment of Chartered Surveyors is given in the RICS publication A Guide to Securing the Services of a Chartered Surveyor. Chapters 2, 3 and 4 also provide useful information on preparing a brief for a Client's requirements, producing tender and contract documentation, organising competitions and evaluating offers.

Methods of selecting a Quantity Surveyor include:-

Selection based on existing knowledge
A Client may select and then appoint a Quantity Surveyor using existing knowledge of that Quantity Surveyor's performance and reputation. This knowledge may arise from a previous working relationship or be based on the recommendation of others

Selection from a panel maintained by a Client
A Client may maintain a panel of Quantity Surveyors. He will have records of their experience which will enable him to make his selection and appointment

Selection from an ad hoc list produced by a Client
If a Client is unable to make an appointment based on knowledge or reputation or by selection from a standing panel it may be more appropriate for an ad hoc list to be prepared.

4.3 Whichever of the above methods of selection is used it is important for the selection criteria to include the following:

- the financial standing of the Quantity Surveyor under consideration
- the experience, competence and reputation of each Quantity Surveyor in the area of the project/skill being considered so that selection is made using comparable standards between firms.
- the ability of each Quantity Surveyor to provide the required service at the relevant time.

There is no one correct way to use competitive tendering as a basis for appointment and no single way of complying with legislation. It is important, however, that the procedure is always fair and open. Where legislation means that competition is compulsory, reference should be made to Clause 4.8 below and any other guidance available.

4.4 If competition is to be used in the selection process the following should be borne in mind:

- a Client should choose between the various methods of competition from either open, selective or restrictive lists, or single- or two-stage tenders, according to the complexity of the project which is to be the subject of competition
- the basis of selection can be on quality, competence or price, or a combination of all three. Whichever basis is chosen it must be clearly stated in the documentation inviting interest in the commission
- it is recommended that price is never regarded in isolation, as will be clear from the detailed guidance set out below.

4.5 To ensure that any competition is well organised the key ingredients required are that:

- specification of the service required is clear
- definition of the skills and competencies required to deliver the services is clear
- criteria on which offers will be evaluated will be made available with the tender documents
- any weighting to be applied to skills, competencies and price is made available with the tender documents
- each stage of the tender process is documented and the procedures for selection and invitation to tender match the requirements of the commission.

4.6 Every effort should be made to ensure that contract documentation is complete. If circumstances change to the extent that criteria are altered all tenderers should immediately be advised and given the opportunity to respond and confirm a willingness to continue.

THE ROYAL INSTITUTION OF CHARTERED SURVEYORS
Appointing a Quantity Surveyor (Cont'd)
Selection and Appointment Advice (Cont'd)

4.7 **Private contracts**
In the case of private contracts there is no upper or lower limit to the number of private sector firms that may be invited to tender, nor formal rules as to their selection. Nevertheless, the RICS recommends that not less than 3 and not more than 6 firms are invited to submit proposals.

4.8 **Public contracts**
In the case of public sector contracts Compulsory Competitive Tendering rules and European Union rules, together with relevant Treasury regulations and DETR guidance, need to be followed. More information is available in the RICS publication A Guide to Securing the Services of a Chartered Surveyor. Chapter 2, pages 3 to 6, provides details on regulations but it is emphasised that current requirements should always be verified before seeking public sector offers.

5. **Specifying the quantity surveying service and determining the fee options and expense costs**

5.1 It is emphasised that the fee for a project can either be negotiated, sought as a sole offer or sought in competition. The following considerations may apply to any of the methods of selecting a Quantity Surveyor and determining his fee in relation to the services to be provided.

5.2 When fee quotations are to be sought from more than one Quantity Surveyor it is of paramount importance that the enquiry and/or the submissions relate precisely to the same details of service(s) required. Proper use of the Form of Enquiry should ensure that this is achieved. In particular, the services required should be specified in detail, by using the Schedule of Services provided and/or by adding or deleting any special services not listed or required in that Schedule.

5.3 Information provided by a Client in an offer document should include, as a minimum:

- the complete scope of the project
- a full and precise description of the quantity surveying services to be provided
- the terms and conditions which will apply, preferably the Form of Agreement and Terms of Appointment contained in this Guide, together with any further requirements that will apply
- the anticipated time scales that will apply, both for the quantity surveying service and for the project
- information on the inclusion or exclusion of expense costs
- the basis of the fee upon which the offer is being invited or offered including:
  - where a fee submission is to deal separately with different components of the service, the way in which any components will be evaluated in order to give comparable totals for selection purposes; and
  - wherever a percentage is to be quoted by a Quantity Surveyor; the definition of the total construction cost to which that percentage is to be applied
- provisions for stage or instalment payments
- where the nature, size, scope, time or value of the project or of the quantity surveying service(s) are likely to vary, the method(s) of appropriately adjusting the fee.

5.4 In determining the basis of the services and the fees required the following should be considered:

- where quantity surveying services are incapable of precise definition at the time of appointment, or where they could change substantially, enough information should be contained in an agreement to enable possible variations in those services to be negotiated
- where services are likely to be executed over a particularly long period, or where the service and/or the project might be delayed or postponed, the method and timing of any increase(s) in the fees should be stated
- the basis of the fees to be offered might be one of, or a combination of:
  - a single percentage encompassing all the described services
  - separate percentages for individually defined components of services
  - a single lump sum encompassing all the described services
  - separate lump sums for individually defined components of services
  - a time charge with either hourly rates quoted or a multiplier or adjustment factor to be applied to a specified definition of hourly cost.

5.5 Where the nature or scale of the project warrants it Quantity Surveyors may be invited to:

- explain in outline their proposed method of operation for the project, and/or
- demonstrate their ability to provide the services required from their own resources, and/or
- state the extent of partner/director involvement, and/or
- name the key personnel to be allocated to the project with details of relevant experience.

5.6 **Fee options**
Before offering any fee proposal a Quantity Surveyor will need to evaluate the costs of carrying out the service being offered. Options for charging fees are given below and may be used alone or in combination.

5.6.1 Percentage fees
This method of charging fees is appropriate when a construction project is reasonably straightforward and the quantity surveying service can be clearly defined but the exact amount of the total construction cost cannot be determined with much certainty

In determining a fee, judgements need to be made by the Quantity Surveyor on the size of the project, its complexity, degree of repetition, method of procurement and contract arrangements. This method may be seen as a 'broad brush' way of assessment which sometimes may be vulnerable to market forces and their influence on contractors' tenders.

Fees are expressed as a percentage of a sum, which is invariably the total construction cost, for the provision of a defined service. This sum will generally be calculated from the value of all construction contracts and other items of work carried out directly by a Client on a project.

THE ROYAL INSTITUTION OF CHARTERED SURVEYORS
Appointing a Quantity Surveyor (Cont'd)
Selection and Appointment Advice (Cont'd)

It is advisable to have a definition of total construction cost agreed between a Client and his Quantity Surveyor Exclusions from the total construction cost should also be defined.

5.6.2 Lump sum fee
This method of charging fees is appropriate when a construction project is reasonably certain in programme time, project size and construction cost and the quantity surveying service can be clearly defined. A total fee is agreed for providing a defined service or amount of work. If appropriate, percentage-based or time-based fees may be converted to a lump sum fee once a project has become sufficiently defined.

If time, size, cost or circumstances of the project are significantly varied by more than specified amounts the lump sum fee may be varied, perhaps according to a formula contained in the Agreement. Otherwise the fee remains fixed. It should be borne in mind that a lump sum fee, whilst giving certainty to both parties at the start of a commission, may not always be appropriate unless the circumstances of the project remain significantly as stated at the time that the Agreement was signed.

5.6.3 Time charge fees
This method of charging fees is appropriate where the scope and/or the extent of the services to be provided cannot reasonably be foreseen and/or cannot reasonably be related to the cost of construction. It is often appropriate where open-ended services are required on feasibility studies, negotiation, claims consultancy etc. It is also a method that allows for additional or varied services to be provided, in addition to the provision of basic services quoted for in a fee agreement.

Fees are calculated from all the time expended on a project by partners/directors and staff and charged at rates previously agreed. The rates may be increased periodically to allow for the effects of inflation.

It is advisable to have previously agreed rates for grades of staff and partners/directors and methods of revising those rates periodically to reflect subsequent changes in salaries and costs. The inclusion of overhead allowances, generally including secretarial and administrative staff, within the rates needs careful calculation by the Quantity Surveyor Principles for the reimbursement of the time involved also need to have been agreed in advance. Sometimes this method may appear to be open ended, perhaps with little incentive for working efficiently. Periodic review of the charges is important so that progress is monitored against budgets.

5.6.4 Target cost fees
This method is appropriate where time charge fees are the basis for the Agreement but a Client wishes to have a guaranteed maximum target for fees.

Fees are recovered on a time charge basis but a 'capped' or guaranteed target cost is agreed between the parties before the work is carried out by the Quantity Surveyor If the target cost is exceeded then the Quantity Surveyor may not receive all of his costs, for instance part of his overhead allowances. If alternatively the total amount of the time charges falls below the budget cost then the Quantity Surveyor may receive an additional payment, a 'bonus' to reflect or share in the 'saving' to his Client. A formula for this should be agreed at the outset.

5.7 **Expense costs**

Expense costs are the costs to a firm such as the provision of cars, mileage-based payments to staff for travel, other travel costs, hotels, subsistence, meals, reproduction of documents, photocopying, postage and phone/fax charges, purchase costs of items bought specifically for a project, expenses in connection with appointing and engaging site-based staff, exceptional time spent by staff in travelling outside of usual hours and/or beyond usual distances etc.

In any agreement on fees it is advisable to be clear about the inclusion or exclusion of any category of expense costs within the fee charges, to identify which expenses are chargeable and to have a machinery for adjustment where necessary.

5.7.1 Recovering expense costs in fee arrangements.
Expense costs in fee bids and fee arrangements may be dealt with in a number of ways. Tender enquiries may prescribe the way required or there may be flexibility in how expense costs are included within the fee figures or shown separately from them. Three options exist for the recovery of expense costs in fee arrangements:

- **Expense costs may be included within a lump sum fee**
  Expense costs of any variety can be assessed and included within a fee and not shown separately. The fee offer thus includes a lump sum for all expense costs. This is often referred to as a 'rolled-up' offer in that the costs have not been shown separately in the bid. Although expense costs have been so included it is often necessary to consider how any adjustment to those costs may be made if changed circumstances require. For instance, if travel costs have been included in a 'rolled-up' fee and much more travel becomes necessary, it may then be required to demonstrate the basis of travel expenses that had been included in the fee offer in order to agree an adjustment

- **Expense costs may be converted to a percentage of the total construction cost of the project**
  An assessment is made of the likely expense costs that may be incurred and the amount is then converted into a percentage of the total construction cost. This percentage may be added to the percentage shown as a fee offer for the quantity surveying service or it may be shown separately. The relationship between construction costs, to which a percentage fee would be applied, and the amount of expense costs that will be incurred is often tenuous and this method should only be entered into after careful consideration by a Client and his Quantity Surveyor

- **Expense costs may be paid as a separate lump sum**
  A lump sum can be calculated for expense costs and shown separately from any other lump sum, percentage-based or time charge fee for the project. Whilst this method will give a Client certainty at the outset, provision should be made for adjustment by the parties if circumstances change.

THE ROYAL INSTITUTION OF CHARTERED SURVEYORS
Appointing a Quantity Surveyor (Cont'd)
Selection and Appointment Advice (Cont'd)

5.7.2 Constituents of expense costs and payment considerations
If expense costs are not to be included in a lump sum, percentage-based or time charge fee they may be recovered by the Quantity Surveyor on the submission of the cost records. If this method is used it is important to establish how expense costs will be calculated and reimbursed to a Quantity Surveyor by his Client. There are three methods for this:

- **Recovery of actual expense costs of disbursements incurred by a Quantity Surveyor with the authority of his Client**
This category of cost would normally include items such as public transport costs, hotels and subsistence, and sundry expenses incurred, for instance, in obtaining financial reports on companies when considering prospective tender lists, or in establishing a site office
- **Recovery of expense costs for in-house resources provided by an organisation**
This would include, for example, photocopying and the reproduction of tender and contract documentation. The amounts due under this category of costs are not always as clearly demonstrable as the costs referred to in the previous category. It may be helpful to set up an agreement between the parties that allows the Quantity Surveyor to recover an amount in addition to the actual material costs incurred so that some staff and overhead costs can be recovered in addition to the basic costs
- **Recovery of expense costs at market rates**
This category is for resources provided by an organisation from its in-house facilities, for example photocopying or drawing reproduction, travel carried out by personnel using company cars etc. These costs would be recovered at what is agreed to be the market rate for the service provided.

5.8 **Implication of termination or suspension**
If a project is terminated or suspended it may become necessary to consider the implications on fees and on expenses, for instance for staff, offices, cars, rented accommodation or leased equipment taken on especially for a project. It may be that some of these costs cannot readily be avoided if termination or suspension occurs and a fee agreement should contain provisions to allow adjustments to be made, particularly to lump sum, 'rolled-up' or percentage fees.

6. Submission and comparison of fee offers and selection and notification of results

6.1 **Submission**

6.1.1 Clients are advised to invite fee offers only from firms of comparable capability and to make a selection taking into account value as well as the fee bid.

6.1.2 Where competitive offers have been sought, instructions on the submission of offers should be clear; giving the date, time and place for their delivery. Offers should all be opened at the same time and treated as confidential until notification of results is possible.

6.2 **Comparison and selection**

6.2.1 Competitive offers should be analysed and compared (including where there are different components of a fee evaluation, as referred to above). Comparison of offers should be on a basis which incorporates all the component parts of an offer in order to indicate the lowest offer and its relationship with other offers.

6.2.2 If two or more submissions give identical or very close results then it may be appropriate to apply a sensitivity test, namely to check the impact of possible or probable changes in the scope, size or value of the project on the fee bids.

6.2.3 In comparing offers it is advisable to weigh quality criteria against the offer price. Further detailed guidance is given in Appendix 10 of the RICS publication A Guide to Securing the Services of a Chartered Surveyor. It is usual to weigh quality of service criteria at a minimum of 60 per cent relative to price criteria at a maximum of 40 per cent.

6.3 **Notification**

6.3.1 Once a decision has been taken to appoint a Quantity Surveyor; the successful firm should be notified. Unsuccessful firms should also be notified of the decision and given information on the bids received (where appropriate this can be by the use of 'indices' which do not link the names of each firm to its offer). All notifications should be made in writing and be sent as soon as possible after the decision to appoint has been made.

7. Confirmation of the Agreement between the Client and the Quantity Surveyor

7.1 The Agreement between a Client and his Quantity Surveyor should be effected by either:

- using the Form of Agreement and Terms of Appointment contained in this Guide, with cross-reference to the Form of Enquiry, Schedule of Services and Fee Offer
- using a separate form of agreement, terms and conditions, with cross reference to the services to be provided and to the fee offer
- a simple exchange of letters incorporating the information given above, making reference to the fee offer.

The 1996 Housing Grants, Construction and Regeneration Act ('the HGCR Act') introduced a right to refer any dispute to adjudication and certain provisions relating to payment. This documentation complies with the HGCR Act and includes the necessary adjudication and payment provisions. Other forms of agreement will need to comply with the HGCR Act or the adjudication and payment provisions of the Statutory Scheme for Construction Contracts will apply. An exchange of letters is unlikely to comply and the Scheme will therefore also apply to these.

THE ROYAL INSTITUTION OF CHARTERED SURVEYORS
Appointing a Quantity Surveyor (Cont'd)
Selection and Appointment Advice (Cont'd)

8. Complaints handling and disputes resolution

8.1 Quantity Surveyors who are partners or directors in firms providing surveying services must operate an internal complaints handling procedure, which applies to disputes less than £50,000, under the RICS Bye-Laws. The RICS also sets a minimum standard of complaints handling, as laid out in its Professional Conduct – Rules of Conduct and Disciplinary Procedures. If the complaint cannot be resolved internally by the firm then the matter must go to final resolution by a third party. A reference to adjudication under the HGCR Act would be sufficient for the purposes of satisfying RICS regulations, pending any final resolution of the dispute at the end of the contract period. See also Clause 11 of the Terms of Appointment.

8.2 The HGCR Act introduced a compulsory scheme of third party neutral dispute resolution – which can occur at any time during the construction process – called adjudication. Any party may refer any dispute to an adjudicator at any time.

SECTION 2
Form of Enquiry
Schedule of Services

1. Category One: General services

The following services may be provided on any project, whatever its nature and whatever the method of procurement adopted.

☐ The following services shall, where relevant, also apply to environmental engineering service (mechanical and electrical engineering) if indicated by placing a cross in this box.

1.1 Inception and feasibility

1.1.1 ☐ Liaise with Client and other consultants to determine Client's initial requirements and subsequent development of the full brief

1.1.2 ☐ Advise on selection of other consultants if not already appointed

1.1.3 ☐ Advise on implications of proposed project and liase with other experts in developing such advice

1.1.4 ☐ Advise on feasibility of procurement options

1.1.5 ☐ Establish Client's order of priorities for quality, time and cost

1.1.6 ☐ Prepare initial budget estimate from feasibility proposals

1.1.7 ☐ Prepare overall project cost calculation and cash flow projections

1.2 Pre-contract cost control

1.2.1. ☐ Prepare and develop preliminary cost plan

1.2.2 ☐ Advise on cost of design team's proposals, including effects of site usage, shape of buildings, alternative forms of design and construction as design develops

1.2.3 ☐ Monitor cost implications during detailed design stage

1.2.4 ☐ Maintain and develop cost plan, and prepare periodic reports and updated cash flow forecasts

1.3 Tender and contractual documentation

1.3.1 ☐ Advise on tendering and contractual arrangements taking into account the Client's priorities and information available from designers

1.3.2 ☐ Advise on insurance responsibilities and liase with Client's insurance adviser

1.3.3 ☐ Advise on warranties

1.3.4 ☐ Advise on bonds for performance and other purposes

1.3.5 ☐ Prepare tender and contract documentation in conjunction with the Client and members of the design team

1.3.6 ☐ Provide copies of documentation as agreed

1.3.7 ☐ Advise on use and/or amendment of standard forms of contract or contribute to drafting of particular requirements in association with Client's legal advisers

1.3.8 ☐ Draw up forms of contract, obtain contract drawings from members of design team and prepare and deliver to both parties contract copies of all documents

1.4 Tender selection and appraisal

1.4.1 ☐ Advise on short listing prospective tenderers

1.4.2 ☐ Investigate prospective tenderers and advise Client on financial status and experience

1.4.3 ☐ Attend interviews with tenderers

1.4.4 ☐ Arrange delivery of documents to selected tenderers

1.4.5 ☐ Check tender submissions for accuracy, level of pricing, pricing policy etc.

1.4.6 ☐ Advise on errors and qualifications and, if necessary, negotiate thereon

1.4.7 ☐ Advise on submission of programme of work and method statement

THE ROYAL INSTITUTION OF CHARTERED SURVEYORS
Appointing a Quantity Surveyor (Cont'd)
Schedule of Services (Cont'd)
Category One: General services (Cont'd)

1.4.8 ☐ Prepare appropriate documentation, if required, to adjust the tender received to an acceptable contract sum

1.4.9 ☐ Review financial budget in view of tenders received and prepare revised cash flow

1.4.10 ☐ Prepare report on tenders with appropriate recommendations

1.4.11 ☐ Advise on letters of intent and issue in conjunction with Client's advisers

1.5 Interim valuations

1.5.1 ☐ Prepare recommendations for interim payments to contractors, subcontractors and suppliers in accordance with contract requirements

1.6 Post-contract cost control

1.6.1 ☐ Value designers' draft instructions for varying the project before issue

1.6.2 ☐ Prepare periodic cost reports in agreed format at specified intervals including any allocations of cost and/or copies as requested by third parties

1.7 Final account

1.7.1 ☐ Prepare the final account

1.8 Attendance at meetings

1.8.1 ☐ Attend meetings as provided for under this Agreement

1.9 Provision of printing/reproduction/copying of documents and the like

1.9.1 ☐ Provide copies of documentation as provided for under this Agreement

2. Category Two: Services particular to non-traditional methods of procurement

These services relate to particular methods of procurement and contract arrangement and should be incorporated into the Agreement as required in conjunction with services from Categories One and Three of the Schedule.

2.1 Services particular to prime cost and management or construction management contracts

2.1.1 ☐ Obtain agreement of a contractor to the amount of the approximate estimate and confirm the amount of the fee for the contract

2.1.2 ☐ Prepare recommendations for interim payments to contractor based on contractor's prime costs

2.1.3 ☐ Adjust the approximate estimate to take account of variations and price fluctuations

2.1.4 ☐ Check the final amounts due to contractors, subcontractors and suppliers

2.2 Services particular to management and construction management contracts

The terms 'management contracting' and 'construction management' mean contractual arrangements where a firm is employed for a fee to manage, organise, supervise and secure the carrying out of the work by other contractors.

2.2.1 ☐ If required, assist in drafting special forms of contract

2.2.2 ☐ Prepare tender documents for the appointment of a management contractor or construction manager

2.2.3 ☐ Attend interviews of prospective contractors or managers

2.2.4 ☐ Obtain manager's agreement to contract cost plan and confirm amount of manager's fee

2.2.5 ☐ Assist in allocation of cost plan into work packages

2.2.6 ☐ Assist in preparation of tender and contract documents

2.2.7 ☐ Price tender documents to provide an estimate comparable with tenders

2.2.8 ☐ Review cost plan as tenders are obtained and prepare revised forecasts of cash flow

2.2.9 ☐ Prepare periodic cost reports to show effect of variations, tenders let and prime costs

2.2.10 ☐ Check the final amounts due to managers, contractors, subcontractors or works contractors and suppliers

2.3 Services particular to design and build contracts - services available to a Client

2.3.1 ☐ Draft the Client's brief, in association with the Client and his designers

2.3.2 ☐ Prepare tender documents incorporating the Client's requirements

2.3.3 ☐ Prepare contract documentation, taking into account any changes arising from the contractor's proposals

2.3.4 ☐ Prepare recommendations for interim and final payments to the contractor; including compliance with statutory requirements of the 1996 Housing Grants, Construction and Regeneration Act

THE ROYAL INSTITUTION OF CHARTERED SURVEYORS
Appointing a Quantity Surveyor (Cont'd)
Schedule of Services (Cont'd)

2.3.5 ☐ Assist in agreement of settlement of the contractor's final account

2.4 Services particular to design and build contracts - services available to a contractor

2.4.1 ☐ Prepare bills of quantities to assist in the preparation of a contractor's tender

2.4.2 ☐ Prepare alternative cost studies to assist in determining the optimum scheme for a contractor's submission

2.4.3 ☐ Draft specifications forming the contractor's proposals

2.4.4 ☐ Assist with specialist enquiries in compiling the contractor's tender

2.4.5 ☐ Measure and price variations for submission to the Client's representative

2.4.6 ☐ Prepare applications for interim payments

2.4.7 ☐ Agree final account with Client's representative

2.5 Services particular to measured term contracts

2.5.1 ☐ Take measurements, price from agreed schedule of rates and agree totals with contractor

2.5.2 ☐ Check final amounts due to contractor(s)

3. Category Three: Services not always required in Categories One and Two

These services should be incorporated into the Agreement in conjunction with services from Categories One and Two of the Schedule as required. Where points are left blank, the Client and/or the Quantity Surveyor should specify their own service as required.

3.1 Bill of quantities

3.1.1 ☐ Provide bills of quantities for mechanical and engineering services

3.1.2 ☐ Price bills of quantities to provide an estimate comparable with tenders

3.2 Cost analysis

3.2.1 ☐ Prepare cost analysis based on agreed format or special requirement

3.3 Advise on financial implications as follows:

3.3.1 ☐ Cost options of developing different sites

3.3.2 ☐ Preparation of development appraisals

3.3.3 ☐ Cost implications of alternative development programmes

3.3.4 ☐ Effect of capital and revenue expenditure

3.3.5 ☐ Life cycle cost studies and estimate of annual running costs

3.3.6 ☐ Availability of grants

3.3.7 ☐ Assist in applications for grants and documentation for these

3.3.8 ☐ Evaluation of items for capital allowances, grant payments or other such matters

3.4 Advice on use of areas and provide:

3.4.1 ☐ Measurement of gross floor areas

3.4.2 ☐ Measurement of net lettable floor areas

3.5 Provide advice on contractual matters affecting the following:

3.5.1 ☐ Entitlement to liquidated and ascertained damages

3.5.2 ☐ Final assessment of VAT

3.5.3 ☐ Opinion on delays and/or disruptions and requests for extensions of time

3.5.4 ☐ Consequences of acceleration

3.5.5 ☐ Assessment of the amount of loss and expense or other such matters and if instructed carrying out negotiations with contractors to reach a settlement

3.6 Provide value management and value engineering services as follows:

3.6.1 ☐ (state)____________________

3.6.2 ☐ (state)____________________

3.7 Provide risk assessment and management services as follows:

3.7.1 ☐ (state)____________________

3.7.2 ☐ (state)____________________

3.8 Adjudication services

3.8.1 ☐ Provide services acting as an adjudicator in construction disputes

3.8.2 ☐ Provide services in connection with advising the Client in relation to active or threatened adjudication proceedings

3.9 Provide services in connection with arbitration and/or litigationas follows:

3.9.1 ☐ (state)____________________

3.9.2 ☐ (state)____________________

THE ROYAL INSTITUTION OF CHARTERED SURVEYORS
Appointing a Quantity Surveyor (Cont'd)
Schedule of Services (Cont'd)
Category Three: Services not always required (Cont'd)

3.10 Provide services arising from fire or other damage to buildings including preparing and negotiating claims with loss adjusters as follows:

3.10.1 ☐ (state)__________

3.10.2 ☐ (state)__________

3.11 Provide services to a contractor in connection with negotiation of claims as follows:

3.11.1 ☐ (state)__________

3.11.2 ☐ (state)__________

3.12 Project Management

Project management services are available from Quantity Surveyors and are set out in detail (together with a Form of Agreement and Guidance Notes) in the RICS publication Project Management Agreement and Conditions of Engagement.

Clients are referred to the Project Management Agreement if the service is mainly for project management. This section is intended to be used if ancillary project management is required to a mainly quantity surveying service.

Provide project management services as follows:

3.12.1 ☐ (state)__________

3.12.2 ☐ (state)__________

3.13 Provide programme co-ordination and monitoring services as follows:

3.13.1 ☐ (state)__________

3.13.2 ☐ (state)__________

3.14 Provide planning supervisor services as follows:

3.14.1 ☐ (state)__________

3.14.2 ☐ (state)__________

3.15 Provide information for use in future management and / or maintenance of the building as follows:

3.15.1 ☐ (state)__________

3.15.2 ☐ (state)__________

3.16 Any other services not listed elsewhere in the Form of Enquiry:

3.16.1 ☐ (state)__________

3.16.2 ☐ (state)__________

3.16.3 ☐ (state)__________

Other Quantity Surveying Services

Other services required or provided by quantity surveyors include:-

Planning Supervisor duties
Contractors Services – Measurement
Pricing
Procurement advice -
Subcontractors
Suppliers
Post Contract services

Adjudication
Mediation
Arbitration
Expert Witness

Example - Quantity Surveying Fee Calculation (see clause 5.6 of Guide)

Example - Complex Project Resource Schedule

Project Value £1,500,000 (Including alteration works value £150,000)

	Hours	Rate	Cost
1. COST PLANNING			
Client Liaison	20	25.00	500.00
Attendance at meetings	20	25.00	500.00
Pre tender cost planning	72	20.00	1440.00
			2440.00
2. BILLS OF QUANTITIES			
Client Liaison	12	25.00	300.00
Attendance at meetings	20	25.00	500.00
Demolitions & Alterations	32	16.00	512.00
Substructures	24	16.00	384.00
Frame	16	16.00	256.00
Upper floors	12	16.00	192.00
Roof	16	16.00	256.00
Stairs	8	16.00	128.00
External walls	16	16.00	256.00
Windows & Doors	32	16.00	512.00
Internal Doors	16	16.00	256.00
Wall Finishes	16	16.00	256.00
Floor Finishes	12	16.00	192.00
Ceiling Finishes	8	16.00	128.00
Fittings & Furnishings	8	16.00	128.00
Sanitary appliances	8	16.00	128.00
Services equipment	8	16.00	128.00
Disposal installations	12	16.00	192.00
Water installations	8	16.00	128.00
Heat source	8	16.00	128.00
Space heating & air treatment	12	16.00	192.00
Ventilation installations	8	16.00	128.00
Electrical installations	24	16.00	384.00
Gas Installations	8	16.00	128.00
Lift installations	8	16.00	128.00
Protective installations	8	16.00	128.00
Communication installations	8	16.00	128.00
Special installations	8	16.00	128.00
Builders work in connection	24	16.00	384.00
External works	42	16.00	672.00
Drainage	16	16.00	256.00
External services	8	16.00	128.00
Preliminaries & Preambles	24	16.00	384.00
Pre-tender estimates	24	16.00	384.00
Tender report & Cost Analysis	16	16.00	256.00
Contract documentation	8	20.00	160.00
			9120.00

Example - Quantity Surveying Fee Calculation (Cont'd)
Example - Complex Project Resource Schedule (Cont'd)

	Hours	Rate	Cost
3. POST CONTRACT			
Client liaison	24	25.00	600.00
Attendance at meetings	100	17.00	1700.00
Valuations	80	17.00	1360.00
Cost Reports	80	17.00	1360.00
Pre Estimate Instructions	40	17.00	680.00
Final Account	80	17.00	1360.00
			7060.00
Total estimated cost			8620.00
Add overheads		50 %	9310.00
			27930.00
Add profit		25 %	6982.00
			£34,913.00
Cost Planning only			
Estimated Cost			2440.00
Add Overheads		50 %	1220.00
			3660.00
Add Profit		25 %	915.00
			4575.00
Bills of Quantities only			
Estimated Cost			9120.00
Add Overheads		50%	4560.00
			13,680.00
Add Profit		25%	3420.00
			17,100.00
Post Contract only			
Estimated Cost			7060.00
Add Overheads		50%	3530.00
			10,590.00
Add Profit		25%	2648.00
			13,238.00
Total	£34,913.00		

The above example excludes all disbursements – travel, reproduction of documents, forms of contracts and the like.

The rates indicated equate to:

Partner	cost rate £25.00	Charge Rate £46.87
Senior Surveyor	cost rate £20.00	Charge Rate £37.50
Taker off	cost rate £16.00	Charge Rate £30.00
Post Contract Surveyor	cost rate £17.00	Charge Rate £31.87

The hours indicated and rates used should be adjusted as appropriate to suit each individual project and the costs, overheads and profit requirements of the practice bidding for work.

Other factors that should be considered when calculating fees include timescale, workload, relationship with client and other consultants, prestige, likely ongoing programme of work, availability of labour

Typical Quantity Surveying fees

Typical fees calculated as above for provision of basic services

	Cost Planning	Bills of Quantities	Post Contract services
Complex Works with little repetition			
Work Value			
50,000	250	1100	700
150,000	650	2850	1900
300,000	1250	4900	3400
600,000	2500	8100	6300
1,500,000	4600	16,200	13,300
3,000,000	8200	7000	24,000
6,000,000	14,500	6800	43,800
10,000,000	21,500	70,800	67,800
Less Complex work with some repetition			
Work Value			
50,000	250	1000	700
150,000	650	2700	1900
300,000	1250	4450	3400
600,000	2500	7100	6100
1,500,000	4600	13,000	12,000
3,000,000	8200	22,000	21,000
6,000,000	14,500	38,200	37,200
10,000,000	21,500	57,400	56,400
Simple works or works with a substantial amount of repetition			
Work Value			
50,000	200	900	600
150,000	500	2400	1600
300,000	1000	4000	2900
600,000	1800	6100	5400
1,500,000	3400	10,900	11,300
3,000,000	6100	18,100	19,400
6,000,000	10,600	30,700	33,800
10,000,000	15,400	45,000	50,600

Alteration Works – Additional Fee on value of alteration works

Work Value	
50,000	300
150,000	900

Decoration Works – Additional Fee on value of alteration works

Work Value	
50,000	450
150,000	1350

Where Provisional Sums, Prime Cost Sums, Subcontracts, Dayworks contingencies form a substantial part of the project the above figures may require adjustment accordingly.

Mechanical and Electrical Services – Additional Fee for provision of Bills of Quantities

Value of M & E work	Bills of Quantities	Post Contract services
120,000	1800	1500
240,000	3420	2600
600,000	7500	5300
1,000,000	11,000	8250
4,000,000	31,700	26,200
6,000,000	45,500	44,500

Negotiating Tenders – Additional Fee

Work Value	
150,000	450
600,000	1250
1,200,000	1950
3,000,000	3000

THE ROYAL INSTITUTION OF CHARTERED SURVEYORS

The following Scales of Professional Charges printed by permission of The Royal Institution of Chartered Surveyors, 12 Great George Street, Westminster, London SW1P 3AD, are now obsolete and are reproduced for information only.

SCALE NO. 36

INCLUSIVE SCALE OF PROFESSIONAL CHARGES FOR QUANTITY SURVEYING SERVICES FOR BUILDING WORKS

(Effective from 29th July, 1988)

This Scale has been determined by the Fees Committee appointed by the Quantity Surveyors Divisional Council of The Royal Institution of Chartered Surveyors. The Scale is for guidance and is not mandatory.

1.0 GENERALLY

1.1 This scale is for use when an inclusive scale of professional charges is considered to be appropriate by mutual agreement between the employer and the quantity surveyor.

1.2 This scale does not apply to civil engineering works, housing schemes financed by local authorities and the Housing Corporation and housing improvement work for which separate scales of fees have been published.

1.3 The fees cover quantity surveying services as may be required in connection with a building project irrespective of the type of contract from initial appointment to final certification of the contractor's account such as:

(a) Budget estimating; cost planning and advice on tendering procedures and contract arrangements.

(b) Preparing tendering documents for main contract and specialist sub-contracts; examining tenders received and reporting thereon or negotiating tenders and pricing with a selected contractor and/or sub-contractors.

(c) Preparing recommendations for interim payments on account to the contractor; preparing periodic assessments of anticipated final cost and reporting thereon; measuring work and adjusting variations in accordance with the terms of the contract and preparing final account, pricing same and agreeing totals with the contractor.

(d) Providing a reasonable number of copies of bills of quantities and other documents; normal travelling and other expenses. Additional copies of documents, abnormal travelling and other expenses (e.g. in remote areas or overseas) and the provision of checkers on site shall be charged in addition by prior arrangement with the employer.

1.4 If any of the materials used in the works are supplied by the employer or charged at a preferential rate, then the actual or estimated market value thereof shall be included in the amounts upon which fees are to be calculated.

1.5 If the quantity surveyor incurs additional costs due to exceptional delays in building operations or any other cause beyond the control of the quantity surveyor then the fees shall be adjusted by agreement between the employer and the quantity surveyor to cover the reimbursement of these additional costs.

1.6 The fees and charges are in all cases exclusive of value added tax which will be applied in accordance with legislation.

1.7 Copyright in bills of quantities and other documents prepared by the quantity surveyor is reserved to the quantity surveyor.

2.0 INCLUSIVE SCALE

2.1 The fees for the services outlined in paragraph 1.3, subject to the provision of paragraph 2.2, shall be as follows:

(a) CATEGORY A. Relatively complex works and/or works with little or no repetition.

Examples:
Ambulance and fire stations; banks; cinemas; clubs; computer buildings; council offices; crematoria; fitting out of existing buildings; homes for the elderly; hospitals and nursing homes; laboratories; law courts; libraries; 'one off' houses; petrol stations; places of religious worship; police stations; public houses, licensed premises; restaurants; sheltered housing; sports pavilions; theatres; town halls; universities, polytechnics and colleges of further education (other than halls of residence and hostels); and the like.

Value of Work			Category A Fee	
£			£	£
Up to		150000	380 + 6.0% (Minimum Fee £3380)	
150,000	-	300,000	9380 + 5.0% on balance over	150,000
300,000	-	600,000	16,880 + 4.3% on balance over	300,000
600,000	-	1,500,000	29,780 + 3.4% on balance over	600,000
1,500,000	-	3,000,000	60,380 + 3.0% on balance over	1,500,000
3,000,000	-	6,000,000	105,380 + 2.8% on balance over	3,000,000
Over		6,000,000	189,380 + 2.4% on balance over	6,000,000

THE ROYAL INSTITUTION OF CHARTERED SURVEYORS (Cont'd)
SCALE NO. 36 (Cont'd)
INCLUSIVE SCALE (Cont'd)

(b) CATEGORY B. Less complex works and/or works with some element of repetition.

Examples:
Adult education facilities; canteens; church halls; community centres; departmental stores; enclosed sports stadia and swimming baths; halls of residence; hostels; motels; offices other than those included in Categories A and C; railway stations; recreation and leisure centres; residential hotels; schools; self-contained flats and maisonettes; shops and shopping centres; supermarkets and hypermarkets; telephone exchanges; and the like.

Value of Work		Category B Fee	
	£	£	£
Up to	150,000	360 + 5.8% (Minimum Fee £3260)	
150000	- 300,000	9060 + 4.7% on balance over	150,000
300000	- 600,000	16,110 + 3.9% on balance over	300,000
600000	- 1,500,000	27,810 + 2.8% on balance over	600,000
1500000	- 3,000,000	53,010 + 2.6% on balance over	1,500,000
3000000	- 6,000,000	92,010 + 2.4% on balance over	3,000,000
Over	6,000,000	64,010 + 2.0% on balance over	6,000,000

(c) CATEGORY C. Simple works and/or works with a substantial element of repetition.

Examples:
Factories; garages; multi-storey car parks; open-air sports stadia; structural shell offices not fitted out; warehouses; workshops; and the like.

Value of Work		Category C Fee	
	£	£	£
Up to	150,000	300 + 4.9% (Minimum Fee £2750)	
150,000	- 300,000	7650 + 4.1% on balance over	150,000
300,000	- 600,000	13,800 + 3.3% on balance over	300,,000
600,000	- 1,500,000	23,700 + 2.5% on balance over	600000
1,500,000	- 3,000,000	46,200 + 2.2% on balance over	1,500,000
3,000,000	- 6,000,000	79,200 + 2.0% on balance over	3,000,000
Over	6,000,000	139,200 + 1.6% on balance over	6,000,000

2.1 (d) Fees shall be calculated upon the total of the final account for the whole of the work including all nominated sub-contractors' and nominated suppliers' accounts. When work normally included in a building contract is the subject of a separate contract for which the quantity surveyor has not been paid fees under any other clause hereof, the value of such work shall be included in the amount upon which fees are charged.

(e) When a contract comprises buildings which fall into more than one category, the fee shall be calculated as follows:

(i) The amount upon which fees are chargeable shall be allocated to the categories of work applicable and the amounts so allocated expressed as percentages of the total amount upon which fees are chargeable.

(ii) Fees shall then be calculated for each category on the total amount upon which fees are chargeable.

(iii) The fee chargeable shall then be calculated by applying the percentages of work in each category to the appropriate total fee and adding the resultant amounts.

(iv) A consolidated percentage fee applicable to the total value of the work may be charged by prior agreement between the employer and the quantity surveyor. Such a percentage shall be based on this scale and on the estimated cost of the various categories of work and calculated in accordance with the principles stated above.

(f) When a project is the subject of a number of contracts then, for the purpose of calculating fees, the values of such contracts shall not be aggregated but each contract shall be taken separately and the scale of charges (paragraphs 2.1 (a) to (e)) applied as appropriate.

THE ROYAL INSTITUTION OF CHARTERED SURVEYORS (Cont'd)
SCALE NO. 36 (Cont'd)
INCLUSIVE SCALE (Cont'd)

2.2 Air Conditioning, Heating, Ventilating and Electrical Services

(a) When the services outlined in paragraph 1.3 are provided by the quantity surveyor for the air conditioning, heating, ventilating and electrical services there shall be a fee for these services in addition to the fee calculated in accordance with paragraph 2.1 as follows:

Value of Work			Additional Fee	
		£	£	£
Up to		120,000	5.0%	
120,000	-	240,000	6000 + 4.7% on balance over	120,000
240,000	-	480,000	11,640 + 4.0% on balance over	240,000
480,000	-	750,000	21,240 + 3.6% on balance over	480,000
750,000	-	1,000,000	30,960 + 3.0% on balance over	750,000
1,000,000	-	4,000,000	38,460 + 2.7% on balance over	1,000,000
Over		4,000,000	119,460 + 2.4% on balance over	4,000,000

(b) The value of such services, whether the subject of separate tenders or not, shall be aggregated and the total value of work so obtained used for the purpose of calculating the additional fee chargeable in accordance with paragraph (a). (Except that when more than one firm of consulting engineers is engaged on the design of these services, the separate values for which each such firm is responsible shall be aggregated and the additional fees charged shall be calculated independently on each such total value so obtained.)

(c) Fees shall be calculated upon the basis of the account for the whole of the air conditioning, heating, ventilating and electrical services for which bills of quantities and final accounts have been prepared by the quantity surveyor.

2.3 Works of Alteration

On works of alteration or repair, or on those sections of the work which are mainly works of alteration or repair, there shall be a fee of 1.0% in addition to the fee calculated in accordance with paragraphs 2.1 and 2.2.

2.4 Works of Redecoration and Associated Minor Repairs

On works of redecoration and associated minor repairs, there shall be a fee of 1.5% in addition to the fee calculated in accordance with paragraphs 2.1 and 2.2.

2.5 Generally

If the works are substantially varied at any stage or if the quantity surveyor is involved in an excessive amount of abortive work, then the fees shall be adjusted by agreement between the employer and the quantity surveyor.

3.0 ADDITIONAL SERVICES

3.1 For additional services not normally necessary, such as those arising as a result of the termination of a contract before completion, liquidation, fire damage to the buildings, services in connection with arbitration, litigation and investigation of the validity of contractors' claims, services in connection with taxation matters and all similar services where the employer specifically instructs the quantity surveyor, the charges shall be in accordance with paragraph 4.0 below.

4.0 TIME CHARGES

4.1 (a) For consultancy and other services performed by a principal, a fee by arrangement according to the circumstances including the professional status and qualifications of the quantity surveyor.

(b) When a principal does work which would normally be done by a member of staff, the charge shall be calculated as paragraph 4.2 below.

4.2 (a) For services by a member of staff, the charges for which are to be based on the time involved, such charges shall be calculated on the hourly cost of the individual involved plus 145%.

4.2 (b) A member of staff shall include a principal doing work normally done by an employee (as paragraph 4.1 (b) above), technical and supporting staff, but shall exclude secretarial staff or staff engaged upon general administration.

(c) For the purpose of paragraph 4.2 (b) above, a principal's time shall be taken at the rate applicable to a senior assistant in the firm.

(d) The supervisory duties of a principal shall be deemed to be included in the addition of 145% as paragraph 4.2 (a) above and shall not be charged separately.

(e) The hourly cost to the employer shall be calculated by taking the sum of the annual cost of the member of staff of:

(i) Salary and bonus but excluding expenses;

THE ROYAL INSTITUTION OF CHARTERED SURVEYORS (Cont'd)
SCALE NO. 36 (Cont'd)
INCLUSIVE SCALE (Cont'd)
TIME CHARGES (Cont'd)

(ii) Employer's contributions payable under any Pension and Life Assurance Schemes;
(iii) Employer's contributions made under the National Insurance Acts, the Redundancy Payments Act and any other payments made in respect of the employee by virtue of any statutory requirements; and
(iv) Any other payments or benefits made or granted by the employer in pursuance of the terms of employment of the member of staff; and dividing by 1650.

5.0 INSTALMENT PAYMENTS

5.1 In the absence of agreement to the contrary, fees shall be paid by instalments as follows:

(a) Upon acceptance by the employer of a tender for the works, one half of the fee calculated on the amount of the accepted tender.

(b) The balance by instalments at intervals to be agreed between the date of the first certificate and one month after final certification of the contractor's account.

5.2 (a) In the event of no tender being accepted, one half of the fee shall be paid within three months of completion of the tender documents. The fee shall be calculated upon the basis of the lowest original bona fide tender received. In the event of no tender being received, the fee shall be calculated upon a reasonable valuation of the works based upon the tender documents.
Note: In the foregoing context "bona fide tender" shall be deemed to mean a tender submitted in good faith without major errors of computation and not subsequently withdrawn by the tenderer.

(b) In the event of the project being abandoned at any stage other than those covered by the foregoing, the proportion of the fee payable shall be by agreement between the employer and the quantity surveyor.

SCALE NO. 37

ITEMISED SCALE OF PROFESSIONAL CHARGES FOR QUANTITY SURVEYING SERVICES FOR BUILDING WORKS
(29th July, 1988)

This Scale has been determined by the Fees Committee appointed by the Quantity Surveyors Divisional Council of The Royal Institution of Chartered Surveyors. The Scale is for guidance and is not mandatory.

1.0 GENERALLY

1.1 The fees are in all cases exclusive of travelling and other expenses (for which the actual disbursement is recoverable unless there is some prior arrangement for such charges) and of the cost of reproduction of bills of quantities and other documents, which are chargeable in addition at net cost.

1.2 The fees are in all cases exclusive of services in connection with the allocation of the cost of the works for purposes of calculating value added tax for which there shall be an additional fee based on the time involved (see paragraphs 19.1 and 19.2)

1.3 If any of the materials used in the works are supplied by the employer or charged at a preferential rate, then the actual or estimated market value thereof shall be included in the amounts upon which fees are to be calculated.

1.4 The fees are in all cases exclusive of preparing a specification of the materials to be used and the works to be done, but the fees for preparing bills of quantities and similar documents do include for incorporating preamble clauses describing the materials and workmanship (from instructions given by the architect and/or consulting engineer).

1.5 If the quantity surveyor incurs additional costs due to exceptional delays in building operations or any other cause beyond the control of the quantity surveyor then the fees may be adjusted by agreement between the employer and the quantity surveyor to cover the reimbursement of these additional costs.

1.6 The fees and charges are in all cases exclusive of value added tax which will be applied in accordance with legislation.

1.7 Copyright in bills of quantities and other documents prepared by the quantity surveyor is reserved to the quantity surveyor.

THE ROYAL INSTITUTION OF CHARTERED SURVEYORS (Cont'd)
SCALE NO. 37 (Cont'd)
CONTRACTS BASED ON BILLS OF QUANTITIES:
PRE-CONTRACT SERVICES

2.0 BILLS OF QUANTITIES

2.1 Basic Scale: For preparing bills of quantities and examining tenders received and reporting thereon.

2.1 (a) CATEGORY A: Relatively complex works and/or works with little or no repetition.

Examples:

Ambulance and fire stations; banks; cinemas; clubs; computer buildings; council offices; crematoria; fitting out of existing buildings; homes for the elderly; hospitals and nursing homes; laboratories; law courts; libraries; 'one off' houses; petrol stations; places of religious worship; police stations; public houses, licensed premises; restaurants; sheltered housing; sports pavilions; theatres; town halls; universities, polytechnics and colleges of further education (other than halls of residence and hostels); and the like.

Value of Work			Category A Fee	
£			£	£
Up to 150,000			230 + 3.0% (minimum fee £1730)	
150,000	-	300,000	4730 + 2.3% on balance over	150,000
300,000	-	600,000	8180 + 1.8% on balance over	300,000
600,000	-	1,500,000	13,580 + 1.5% on balance over	600,000
1,500,000	-	3,000,000	27,080 + 1.2% on balance over	1,500,000
3,000,000	-	6,000,000	45,080 + 1.1% on balance over	3,000,000
Over	-	6,000,000	78,080 + 1.0% on balance over	6,000,000

(b) CATEGORY B: Less complex works and/or works with some element of repetition.

Examples:

Adult education facilities; canteens; church halls; community centres; departmental stores; enclosed sports stadia and swimming baths; halls of residence; hostels; motels; offices other than those included in Categories A and C; railway stations; recreation and leisure centres; residential hotels; schools; self-contained flats and maisonettes; shops and shopping centres; supermarkets and hypermarkets; telephone exchanges; and the like.

Value of Work			Category B Fee	
£			£	£
Up to150,000			210 + 2.8% (minimum fee £1610)	
150,000	-	300,000	4410 + 2.0% on balance over	150,000
300,000	-	600,000	7410 + 1.5% on balance over	300,000
600,000	-	1,500,000	11,910 + 1.1% on balance over	600,000
1,500,000	-	3,000,000	21,810 + 1.0% on balance over	1,500,000
3,000,000	-	6,000,000	36,810 + 0.9% on balance over	3,000,000
Over		6,000,000	63,810 + 0.8% on balance over	6,000,000

(c) CATEGORY C: Simple works and/or works with a substantial element of repetition.

Examples:

Factories; garages; multi-storey car parks; open-air sports stadia; structural shell offices not fitted out; warehouses; workshops; and the like.

Value of Work			Category C Fee	
£			£	£
Up to 150,000			180 + 2.5% (minimum fee £1430)	
150,000	-	300,000	3930 + 1.8% on balance over	150,000
300,000	-	600,000	6630 + 1.2% on balance over	300,000
600,000	-	1,500,000	10,230 + 0.9% on balance over	600,000
1,500,000	-	3,000,000	18,330 + 0.8% on balance over	1,500,000
3,000,000	-	6,000,000	30,330 + 0.7% on balance over	3,000,000
Over		6,000,000	51,330 + 0.6% on balance over	6,000,000

THE ROYAL INSTITUTION OF CHARTERED SURVEYORS (Cont'd)
SCALE NO. 37 (Cont'd)
BILLS OF QUANTITIES (Cont'd)

2.1 (d) The scales of fees for preparing bills of quantities (paragraphs 2.1 (a) to (c)) are overall scales based upon the inclusion of all provisional and prime cost items, subject to the provision of paragraph 2.1 (g). When work normally included in a building contract is the subject of a separate contract for which the quantity surveyor has not been paid fees under any other clause hereof, the value of such work shall be included in the amount upon which fees are charged.

(e) Fees shall be calculated upon the accepted tender for the whole of the work subject to the provisions of paragraph 2.6. In the event of no tender being accepted, fees shall be calculated upon the basis of the lowest original bona fide tender received. In the even of no such tender being received, the fees shall be calculated upon a reasonable valuation of the works based upon the original bills of quantities.

Note: In the foregoing context "bona fide tender" shall be deemed to mean a tender submitted in good faith without major errors of computation and not subsequently withdrawn by the tenderer.

(f) In calculating the amount upon which fees are charged the total of any credits and the totals of any alternative bills shall be aggregated and added to the amount described above. The value of any omission or addition forming part of an alternative bill shall not be added unless measurement or abstraction from the original dimension sheets was necessary.

(g) Where the value of the air conditioning, heating, ventilating and electrical services included in the tender documents together exceeds 25% of the amount calculated as described in paragraphs 2.1 (d) and (e), then, subject to the provisions of paragraph 2.2, no fee is chargeable on the amount by which the value of these services exceeds the said 25% in this context the term "value" excludes general contractor's profit, attendance, builder's work in connection with the services, preliminaries and any similar additions.

(h) When a contract comprises buildings which fall into more than one category, the fee shall be calculated as follows:

(i) The amount upon which fees are chargeable shall be allocated to the categories of work applicable and the amounts so allocated expressed as percentages of the total amount upon which fees are chargeable.

(ii) Fees shall then be calculated for each category on the total amount upon which fees are chargeable.

(iii) The fee chargeable shall then be calculated by applying the percentages of work in each category to the appropriate total fee and adding the resultant amounts.

(j) When a project is the subject of a number of contracts then, for the purpose of calculating fees, the values of such contracts shall not be aggregated but each contract shall be taken separately and the scale of charges (paragraphs 2.1 (a) to (h)) applied as appropriate.

(k) Where the quantity surveyor is specifically instructed to provide cost planning services the fee calculated in accordance with paragraphs 2.1 (a) to (j) shall be increased by a sum calculated in accordance with the following table and based upon the same value of work as that upon which the aforementioned fee has been calculated:

2.1 (k) CATEGORIES A & B: (as defined in paragraphs 2.1 (a) and (b)).

Value of Work		Fee	
	£	£	£
Up to	600,000	0.7%	
600,000	- 3,000,000	4200 + 0.4% on balance over	600,000
3000,000	- 6,000,000	13,800 + 0.35% on balance over	3,000,000
Over	6,000,000	24,300 + 0.3% on balance over	6,000,000

CATEGORY C: (as defined in paragraph 2.1 (c)).

Value of Work		Fee	
	£	£	£
Up to	600,000	0.5%	
600,000	- 3,000,000	3000 + 0.3% on balance over	600,000
3,000,000	- 6,000,000	10200 + 0.25% on balance over	3,000,000
Over	6,000,000	17700 + 0.2% on balance over	6,000,000

THE ROYAL INSTITUTION OF CHARTERED SURVEYORS (Cont'd)
SCALE NO. 37 (Cont'd)
BILLS OF QUANTITIES (Cont'd)

2.2 Air Conditioning, Heating, Ventilating and Electrical Services

(a) Where bills of quantities are prepared by the quantity surveyor for the air conditioning, heating, ventilating and electrical services there shall be a fee for these services (which shall include examining tenders received and reporting thereon), in addition to the fee calculated in accordance with paragraph 2.1 as follows:

Value of Work			Additional Fee	
£	£	£		£
Up to	120,000		2.5%	
120,000	- 240,000	3000 + 2.25%	on balance over	120,000
240,000	- 480,000	5700 + 2.0%	on balance over	240,000
480,000	- 750,000	10,500 + 1.75%	on balance over	480,000
750,000	- 1,000,000	15,225 + 1.25%	on balance over	750,000
Over	1,000,000	18,350 + 1.15%	on balance over	1,000,000

(b) The values of such services, whether the subject of separate tenders or not, shall be aggregated and the total value of work so obtained used for the purpose of calculating the additional fee chargeable in accordance with paragraph (a).

(Except that when more than one firm of consulting engineers is engaged on the design of these services, the separate values for which each such firm is responsible shall be aggregated and the additional fees charged shall be calculated independently on each such total value so obtained).

(c) Fees shall be calculated upon the accepted tender for the whole of the air conditioning, heating, ventilating and electrical services for which bills of quantities have been prepared by the quantity surveyor. In the event of no tender being accepted, fees shall be calculated upon the basis of the lowest original bona fide tender received. In the event of no such tender being received, the fees shall be calculated upon a reasonable valuation of the services based upon the original bills of quantities.

Note: In the foregoing context "bona fide tender" shall be deemed to mean a tender submitted in good faith without major errors of computation and not subsequently withdrawn by the tenderer.

(d) When cost planning services are provided by the quantity surveyor for air conditioning, heating, ventilating and electrical services (or for any part of such services) there shall be an additional fee based on the time involved (see paragraphs 19.1 and 19.2). Alternatively the fee may be on a lump sum or percentage basis agreed between the employer and the quantity surveyor.

Note: The incorporation of figures for air conditioning, heating, ventilating and electrical services provided by the consulting engineer is deemed to be included in the quantity surveyor's services under paragraph 2.1.

2.3 Works of Alteration

On works of alteration or repair, or on those sections of the works which are mainly works of alteration or repair, there shall be a fee of 1.0% in addition to the fee calculated in accordance with paragraphs 2.1 and 2.2.

2.4 Works of Redecoration and Associated Minor Repairs
On works of redecoration and associated minor repairs, there shall be a fee of 1.5% in addition to the fee calculated in accordance with paragraphs 2.1 and 2.2.

2.5 Bills of Quantities Prepared in Special Forms

Fees calculated in accordance with paragraphs 2.1, 2.2, 2.3 and 2.4 include for the preparation of bills of quantities on a normal trade basis. If the employer requires additional information to be provided in the bills of quantities or the bills to be prepared in an elemental, operational or similar form, then the fee may be adjusted by agreement between the employer and the quantity surveyor.

2.6 Action of Tenders

(a) When cost planning services have been provided by the quantity surveyor and a tender, when received, is reduced before acceptance and if the reductions are not necessitated by amended instructions of the employer or by the inclusion in the bills of quantities of items which the quantity surveyor has indicated could not be contained within the approved estimate, then in such a case no charge shall be made by the quantity surveyor for the preparation of bills of reductions and the fee for the preparation of the bills of quantities shall be based on the amount of the reduced tender.

(b) When cost planning services have not been provided by the quantity surveyor and if a tender, when received, is reduced before acceptance, fees are to be calculated upon the amount of the unreduced tender. When the preparation of bills of reductions is required, a fee is chargeable for preparing such bills of reductions as follows:

(i) 2.0% upon the gross amount of all omissions requiring measurement or abstraction from original dimension sheets.

(ii) 3.0% upon the gross amount of all additions requiring measurement.

(iii) 0.5% upon the gross amount of all remaining additions.

THE ROYAL INSTITUTION OF CHARTERED SURVEYORS (Cont'd)
SCALE NO. 37 (Cont'd)
BILLS OF QUANTITIES (Cont'd)

Note: The above scale for the preparation of bills of reductions applies to work in all categories.

2.7 Generally

If the works are substantially varied at any stage or if the quantity surveyor is involved in an excessive amount of abortive work, then the fees shall be adjusted by agreement between the employer and the quantity surveyor.

3.0 NEGOTIATING TENDERS

3.1 (a) For negotiating and agreeing prices with a contractor:

Value of Work			Fee	
		£	£	£
Up to		150,000	0.5%	
150,000	-	600,000	750 + 0.3% on balance over	150,000
600,000	-	1,200,000	2100 + 0.2% on balance over	600,000
Over		1,200,000	3300 + 0.1% on balance over	1,200,000

3.1 (b) The fee shall be calculated on the total value of the works as defined in paragraphs 2.1 (d), (e), (f), (g) and (j).

(c) For negotiating and agreeing prices with a contractor for air conditioning, heating, ventilating and electrical services there shall be an additional fee as paragraph 3.1 (a) calculated on the total value of such services as defined in paragraph 2.2 (b).

4.0 CONSULTATIVE SERVICES AND PRICING BILLS OF QUANTITIES

4.1 Consultative Services

Where the quantity surveyor is appointed to prepare approximate estimates, feasibility studies or submissions for the approval of financial grants or similar services, then the fee shall be based on the time involved (see paragraphs 19.1 and 19.2) or, alternatively, on a lump sum or percentage basis agreed between the employer and the quantity surveyor.

4.2 Pricing Bills of Quantities

(a) For pricing bills of quantities, if instructed, to provide an estimate comparable with tenders, the fee shall be one-third (33 1/3%) of the fee for negotiating and agreeing prices with a contractor, calculated in accordance with paragraphs 3.1 (a) and (b).

(b) For pricing bills of quantities, if instructed, to provide an estimate comparable with tenders for air conditioning, heating, ventilating and electrical services the fee shall be one-third (33 1/3%) of the fee calculated in accordance with paragraph 3.1 (c).

CONTRACTS BASED ON BILLS OF QUANTITIES: POST-CONTRACT SERVICES

Alternative Scales (I and II) for post-contract services are set out below to be used at the quantity surveyor's discretion by prior agreement with the employer.

ALTERNATIVE 1

5.0 OVERALL SCALE OF CHARGES FOR POST-CONTRACT SERVICES

5.1 If the quantity surveyor appointed to carry out the post-contract services did not prepare the bills of quantities then the fees in paragraphs 5.2 and 5.3 shall be increased to cover the additional services undertaken by the quantity surveyor.

5.2 Basic Scale: For taking particulars and reporting valuations for interim certificates for payments on account to the contractor, preparing periodic assessments of anticipated final cost and reporting thereon, measuring and making up bills of variations including pricing and agreeing totals with the contractor, and adjusting fluctuations in the cost of labour and materials if required by the contract.

THE ROYAL INSTITUTION OF CHARTERED SURVEYORS (Cont'd)
SCALE NO. 37 (Cont'd)
OVERALL SCALE OF CHARGES FOR POST-CONTRACT SERVICES (Cont'd)

(a) CATEGORY A: Relatively complex works and/or works with little or no repetition.

Examples:
Ambulance and fire stations; banks; cinemas; clubs; computer buildings; council offices; crematoria; fitting out of existing buildings; homes for the elderly; hospitals and nursing homes; laboratories; law courts; libraries; 'one-off' houses; petrol stations; places of religious worship; police stations; public houses, licensed premises; restaurants; sheltered housing; sports pavilions; theatres; town halls; universities, polytechnics and colleges of further education (other than halls of residence and hostels); and the like.

Value of Work			Category A Fee	
	£		£	£
Up to 150,000			150 + 2.0% (minimum fee £1150)	
150,000	-	300,000	3150 + 1.7% on balance over	150,000
300,000	-	600,000	5700 + 1.6% on balance over	300,000
600,000	-	1,500,000	10,500 + 1.3% on balance over	600,000
1,500,000	-	3,000,000	22,200 + 1.2% on balance over	1,500,000
3,000,000	-	6,000,000	40,200 + 1.1% on balance over	3,000,000
Over		6,000,000	73,200 + 1.0% on balance over	6,000,000

(b) CATEGORY B: Less complex works and/or works with some element of repetition.

Examples:
Adult education facilities; canteens; church halls; community centres; departmental stores; enclosed sports stadia and swimming baths; halls of residence; hostels; motels; offices other than those included in Categories A and C; railway stations; recreation and leisure centres; residential hotels; schools; self-contained flats and maisonettes; shops and shopping centres; supermarkets and hypermarkets; telephone exchanges; and the like.

Value of Work			Category B Fee	
	£		£	£
Up to150,000			150 + 2.0% (minimum fee £1150)	
150,000	-	300,000	3150 + 1.7% on balance over	150,000
300,000	-	600,000	5700 + 1.5% on balance over	300,000
600,000	-	1,500,000	10,200 + 1.1% on balance over	600,000
1,500,000	-	3,000,000	20,100 + 1.0% on balance over	1,500,000
3,000,000	-	6,000,000	35,100 + 0.9% on balance over	3,000,000
Over		6,000,000	62,100 + 0.8% on balance over	6,000,000

(c) CATEGORY C: Simple works and/or works with a substantial element of repetition.

Examples:
Factories; garages; multi-storey car parks; open-air sports stadia; structural shell offices not fitted out; warehouses; workshops; and the like.

Value of Work			Category C Fee	
	£		£	£
Up to 150,000			120 + 1.6% (minimum fee £920)	
150,000	-	300000	2520 + 1.5% on balance over	150,000
300,000	-	600,000	4770 + 1.4% on balance over	300,000
600,000	-	1,500,000	8970 + 1.1% on balance over	600,000
1,500,000	-	3,000,000	18,870 + 0.9% on balance over	1,500,000
3,000,000	-	6,000,000	32,370 + 0.8% on balance over	3,000,000
Over		6,000,000	56,370 + 0.7% on balance over	6,000,000

THE ROYAL INSTITUTION OF CHARTERED SURVEYORS (Cont'd)
SCALE NO. 37 (Cont'd)
OVERALL SCALE OF CHARGES FOR POST-CONTRACT SERVICES (Cont'd)

(d) The scales of fees for post-contract services (paragraphs 5.2 (a) and (c) are overall scales, based upon the inclusion of all nominated sub-contractors' and nominated suppliers' accounts, subject to the provision of paragraph 5.2 (g). When work normally included in a building contract is the subject of a separate contract for which the quantity surveyor has not been paid fees under any other clause hereof, the value of such work shall be included in the amount on which fees are charged.

(e) Fees shall be calculated upon the basis of the account for the whole of the work, subject to the provisions of paragraph 5.3.

(f) In calculating the amount on which fees are charged the total of any credits is to be added to the amount described above.

(g) Where the value of air conditioning, heating, ventilating and electrical services included in the tender documents together exceeds 25% of the amount calculated as described in paragraphs 5.2 (d) and (e) above, then, subject to the provisions of paragraph 5.3, no fee is chargeable on the amount by which the value of these services exceeds the said 25%. In this context the term "value" excludes general contractor's profit, attendance, builders work in connection with the services, preliminaries and any other similar additions.

(h) When a contract comprises buildings which fall into more than one category, the fee shall be calculated as follows:

(i) The amount upon which fees are chargeable shall be allocated to the categories of work applicable and the amounts so allocated expressed as percentages of the total amount upon which fees are chargeable.

(ii) Fees shall then be calculated for each category on the total amount upon which fees are chargeable.

(iii) The fee chargeable shall then be calculated by applying the percentages of work in each category to the appropriate total fee and adding the resultant amounts.

(j) When a project is the subject of a number of contracts then, for the purposes of calculating fees, the values of such contracts shall not be aggregated but each contract shall be taken separately and the scale of charges (paragraphs 5.2 (a) to (h)) applied as appropriate.

(k) When the quantity surveyor is required to prepare valuations of materials or goods off site, an additional fee shall be charged based on the time involved (see paragraphs 19.1 and 19.2).

(l) The basic scale for post-contract services includes for a simple routine of periodically estimating final costs. When the employer specifically requests a cost monitoring service which involves the quantity surveyor in additional or abortive measurement an additional fee shall be charged based on the time involved (see paragraphs 19.1 and 19.2), or alternatively on a lump sum or percentage basis agreed between the employer and the quantity surveyor.

(m) The above overall scales of charges for post contract services assume normal conditions when the bills of quantities are based on drawings accurately depicting the building work the employer requires. If the works are materially varied to the extent that substantial remeasurement is necessary then the fee for post-contract services shall be adjusted by agreement between the employer and the quantity surveyor.

5.3 Air Conditioning, Heating, Ventilating and Electrical Services

(a) Where final accounts are prepared by the quantity surveyor for the air conditioning, heating, ventilating and electrical services there shall be a fee for these services, in addition to the fee calculated in accordance with paragraph 5.2, as follows:

Value of Work		Additional Fee		
	£	£		£
Up to	120,000		2.0%	
120,000	- 240,000	2400 + 1.6%	on balance over	120,000
240,000	- 1,000,000	4320 + 1.25%	on balance over	240,000
1,000,000	- 4,000,000	13,820 + 1.0%	on balance over	1,000,000
Over	4,000,000	43,820 + 0.9%	on balance over	4,000,000

GENERAL INFORMATION

THE ROYAL INSTITUTION OF CHARTERED SURVEYORS (Cont'd)
SCALE NO. 37 (Cont'd)
OVERALL SCALE OF CHARGES FOR POST-CONTRACT SERVICES (Cont'd)
Air Conditioning, Heating, Ventilating and Electrical Services (Cont'd)

(b) The values of such services, whether the subject of separate tenders or not, shall be aggregated and the total value of work so obtained used for the purpose of calculating the additional fee chargeable in accordance with paragraph (a).

(Except that when more than one firm of consulting engineers is engaged on the design of these services the separate values for which each such firm is responsible shall be aggregated and the additional fee charged shall be calculated independently on each such total value so obtained).

(c) The scope of the services to be provided by the quantity surveyor under paragraph (a) above shall be deemed to be equivalent to those described for the basic scale for post-contract services.

(d) When the quantity surveyor is required to prepare periodic valuations of materials or goods off site, an additional fee shall be charged based on the time involved (see paragraphs 19.1 and 19.2).

(e) The basic scale for post-contract services includes for a simple routine of periodically estimating final costs. When the employer specifically requests a cost monitoring service which involves the quantity surveyor in additional or abortive measurement an additional fee shall be charged based on the time involved (see paragraph 19.1 and 19.2), or alternatively on a lump sum or percentage basis agreed between the employer and the quantity surveyor.

(f) Fees shall be calculated upon the basis of the account for the whole of the air conditioning, heating, ventilating and electrical services for which final accounts have been prepared by the quantity surveyor.

Scale 37 Alternative II; - Separate Stages of Post Contract Services and Scales 38 - 47 are not reproduced here.

THE BUILDING (LOCAL AUTHORITY CHARGES) REGULATIONS 1998

Notes

Before you build , extend or convert, you or your agent must advise your local authority either by submitting Full Plans or a Building Notice. The fee payable depends on the type of work. The number of dwellings in a building and the total floor area. The following tables may be used in conjunction with the Fees Regulations to calculate the fees. If you have any difficulties calculating the fees please consult your local authority building control office.

Should you submit Full Plans you will pay a plan fee at the time of submission to cover their passing or rejection.

For further advice and to check current fees consult your local authority.

With Full Plans submissions, for most types of work, an inspection fee covering all necessary site visits will be payable following the first inspection.

Should you submit a Building Notice, the appropriate Building Notice fee is payable at the time of submission and covers all necessary checks and site visits. The Building Notice fee is equivalent to the sum of the relevant plan fee and inspection fee.

Should you apply for a regularisation certificate, in respect of unauthorised building work, commenced on or after 11 November 1985, you will pay a regularisation fee to cover the cost of assessing your application and all inspections. The fee equivalent to the Building Notice fee plus 20%.

Total estimated cost means an estimate accepted by the local authority of a reasonable cost that would be charged by a person in business to carry out the work shown or described in the application excluding VAT and any fees paid to an architect, engineer or surveyor etc. and also excluding land acquisition costs.

THE BUILDING (LOCAL AUTHORITY CHARGES) REGULATIONS 1998 (Cont'd)
FULL PLAN AND INSPECTION CHARGES Charges Scheme No.1 1999 Annex A

Description of work		**Plan charge**	**VAT @ 17.5%**	**Total plan charge**	**Inspection charge**	**VAT @ 17.5%**	**Total inspection charge payable**
		£	**£**	**£**	**£**	**£**	**£**
Domestic extensions							
(not exceeding 3	10m2 & under	200	35	235	-	-	-
storeys incl. basement)	40m2 & under	75	13.13	88.13	225	39.38	264.38
Total areas to be aggregated	60m2 & under	100	17.5	117.5	300	52.5	352.50
	Over 60m2	Estimate required see table below for other work					
Detached garages/ Carports	40m & under	25	4.38	29.38	75	13.13	88.13
(not otherwise exempt)	Over 40m	Estimate required see table below for other work					
New house and flats	**No. of dwellings**						
(small domestic	1.	140	24.50	164.50	160	28.00	188.00
buildings not	2.	205	35.88	240.88	300	52.50	352.50
exceeding 300m2 floor	3.	270	47.25	317.25	435	76.13	511.13
area and no more than	4.	335	58.63	393.63	575	100.63	675.63
3 storeys incl. Basement)	5.	405	70.88	475.88	710	124.25	834.25
	6.	475	83.13	558.13	785	137.38	922.38
	7.	495	86.63	581.63	945	165.38	1,110.38
	8.	515	90.13	605.13	1,105	193.38	1,298.38
	9.	535	93.63	628.63	1,265	221.38	1,486.38
	10.	540	94.50	634.50	1,440	252.00	1,692.00
	11.	545	95.38	640.38	1,580	276.50	1,856.50
	12.	550	96.25	646.25	1,720	301.00	2,021.00
	13.	555	97.13	652.13	1,860	325.50	2,185.50
	14.	560	98.00	658.00	2,000	350.00	2,350.00
	15.	565	98.88	663.88	2,140	374.5	2,514.50
	16.	570	99.75	669.75	2,280	399.00	2,679.00
	17.	575	100.63	675.63	2,420	423.50	2,843.50
	18.	580	101.50	681.50	2,560	448.00	3,008.00
	19.	585	102.38	687.38	2,700	472.50	3,172.50
	20.	590	103.25	693.25	2,840	497.00	3,337.00
	21 to 30	600	105	705.00	2,940	514.50	3,454.50
	Plus £10 + £1.75 VAT for each additional dwelling above 21				**Plus £100 + £17.50 VAT for each additional dwelling above 21**		
	31 & over	700	122.50	822.50	3,940	689.50	4,629.50
	Plus £5 + £0.88 VAT for each additional dwelling above 31				**Plus £75 + £13.13 VAT for each additional dwelling above 31**		

OTHER WORK

Estimated cost of building work (excluding VAT) (£.)	**Plan charge**	**VAT @ 17.5%**	**Total plan charge payable**	**Inspection charge**	**VAT @ 17.5%**	**Total inspection charge payable**
	£	£	£	£	£	£
2,000 or less	100	17.50	117.50	-	-	-
Exceeding 2,000 but not exceeding 5,000	165	28.88	193.88	-	-	-
Exceeding 5,000 but not exceeding 20,000	41.25	7.22	48.47	123.75	21.66	145.41
	Plus for every 1,000 (or part thereof) by which the cost exceeds £5,000					
	2.25	0.39	2.64	6.75	1.18	7.93
Exceeding 20,000 but not exceeding 100,000	75	13.13	88.13	225	39.38	264.38
	Plus for every £1,000 (or part thereof) by which the cost exceeds £20,000					
	2	0.35	235	6	1.05	7.05
Exceeding 100,000 but not exceeding 1,000,000	235	41.13.	276.13	7.05	123.38	828.38
	Plus for every £1,000 (or part thereof) by which the cost exceeds £100,000:					
	0.88	0.15	1.03	2.62	0.46	3.08
Exceeding 1,000,000 but not exceeding 10,000,000	1,022.50	178.94	1,201.44	3,067.50	536.81	3,604.31
	Plus for every £1,000 (or part thereof) by which the cost exceeds £1,000,000:					
	0.69	0.12	0.81	2.06	0.36	2.42
Exceeding 10,000,000	7,210	1,261.75	8,471.75	21,630	3,785.25	25,415.25
	Plus for every £1,000 (or part thereof) by which the cost exceeds £10,000,000:					
	0.50	0.09	0.59	1.50	0.26	1.76

THE BUILDING (PRESCRIBED FEES) REGULATIONS 1994 (Cont'd)
BUILDING NOTICE & REVERSION CHARGES **REGULARISATION CHARGES**

Description of Work		Charge £.	VAT @ 17.5% £	Total payable £	Charge £	VAT @ 0% £	Total payable £
Domestic extensions							
(not exceeding 3	10m2 & under	200	35.00	235.00	240	-	240
storeys incl.	40m2 & under	300	52.50	352.50	360	-	360
Basement). Total floor	60m2 & under	400	70.00	470.00	480	-	480
areas to be aggregated	Over 60m2	Estimate required see table below					
Detached garages/carports (not otherwise exempt)	40m2 & under	100	17.50		120	-	120
	Over 40m						
		Estimate required see table below			Estimate required see table below		
New houses and flats	**No. of dwellings**						
(small domestic	1.	300	52.50	352.50	360	-	360
buildings) not	2.	505	88.38	593.38	606	-	606
exceeding 300m2 floor	3.	705	123.38	828.38	846	-	846
area and no more than	4.	910	159.25	1,069.25	1,092	-	1,092
3 storeys incl.	5.	1,115	195.13	1,310.13	1,338	-	1,338
Basement.	6.	1,260	220.50	1,480.50	1,512	-	1,512
	7.	1,440	252.00	1,692.00	1,728	-	1,728
	8.	1,620	283.50	1,903.50	1,944	-	1,944
	9.	1,800	315.00	2,115.00	2,160	-	2,160
	10.	1,980	346.50	2,326.50	2,376,	-	2,376
	11.	2,125	371.88	2,496.88	2,550	-	2,550
	12.	2,270	397.25	2,667.25	2,724	-	2,724
	13.	2,415	422.63	2,837.63	2,898	-	2,898
	14.	2,560	448.00	3,008.00	3,072	-	3,072
	15.	2,705	473.38	3,178.38	3,246	-	3,246
	16.	2,850	498.75	3,348.75	3,420	-	3,420
	17.	2,995	524.13	3,519.13	3,594	-	3,594
	18.	3,140	549.50	3,689.50	3,768	-	3,768
	19.	3,285	574.88	3,859.88	3,942	-	3,942
	20.	3,430	600.25	4,030.25	4,116	-	4,116
	21 to 30.	3,540	619.50	4,159.50	4,248	-	4,248
	Plus £110 + £19.25 VAT or each additional dwelling above 21				**Plus £132 for each additional dwelling above 21**		
	31 & over	4,640	812.00	5,452.00	5,568	-	5,568
	Plus £80 + £14.00 VAT for each additional dwelling over 31				**Plus £96 for each additional dwelling over 31**		

OTHER WORK

Estimated cost of building work (excluding VAT) (£)	Charge £	VAT@ 17.5%. £	Total Payable £	Charge £	VAT@ 0% £	Total Payable £
2,000 or less	100	17.50	117.50	120	-	120
Exceeding 2,000 but not exceeding 5,000	165	28.88	193.88	198	-	198
Exceeding 5,000 but not exceeding 20,000	165	28.88	193.88	198	-	198
	Plus for every £1,000 (or part thereof) by which the cost exceeds £5,000					
	9	1.58	10.58	10.80	-	10.80
Exceeding 20,000 but not exceeding 100,000	300	52.50	352.50	360	-	360
	Plus for every £1,000 (or part thereof) by which the cost exceeds £20,000					
	8	1.40	9.40	9.60	-	9.60
Exceeding 100,000 but not exceeding 1,000,000	940	164.50	1,104.50	1,128	-	1,128
	Plus for every £1,000 (or part thereof) by which the cost exceeds £100,000					
	3.50	0.61	4.11	4.20	-	4.20
Exceeding 1,000,000 but not exceeding 10,000,000	4,090	715.75	4,805.75	4,908	-	4,908
	Plus for every £1,000 (or part thereof) by which the cost exceeds £1,000,000					
	2.75	0.48	3.23	3.30	-	4.20
Exceeding 10,000,000	28,840	5,047	33,887	34,608	-	34,608
	Plus for every £1,000 (or part thereof) by which the cost exceeds £10,000,000					
	2.00	0.35	2.35	2.4	-	2.4

CONSTRUCTION (DESIGN AND MANAGEMENT) REGULATIONS 1994

CONSTRUCTION (DESIGN AND MANAGEMENT) REGULATIONS 1994

The Construction (Design and Management) Regulations 1994 came into force on 31st March 1995. The CDM Regulations and Amendments provide a framework in health and safety terms for the design, construction, maintenance and demolition of a structure.

Due regard must be made of The Management of Health and Safety at Work Regulations 1999 and The Construction (Design and Management) (Amendment) Regulations 2000 which provides clarification of the definition of the "Designer", both of which are not reproduced here. Managing health and safety in construction: Approved Code of Practice: Construction (Design and Management) Regulations 1994 (HSG224) became enforceable on 1st February 2002 and replaces the previous Code of Practice, further details from HSE (www.hse.gov.uk or www.hsebooks.co.uk)

If you have not followed the relevant provisions of the code you may be prosecuted for breach of health and safety law.

The regulations place duties upon clients, client's agents, designers and contractors so that health and safety is taken into account and then co-ordinated and managed effectively throughout all stages of a construction project; from conception to execution of works on site and subsequent maintenance and repair and demolitions.

The following is an outline of the requirements of the code including excerpts together with extracts from the HSE leaflet "CDM Regulations - How the regulations affect you". No attempt has been made to cover all the detail and readers are advised to study the code in full.

The Client

Clients in particular should note the obligations put on them, Key duties are:-

a) To appoint a health and safety planning supervisor and principal contractor.

b) Be satisfied that the planning supervisor and principal contractor are competent and will allocate adequate resources for health and safety.

c) Be satisfied that designers and contractors are also competent and will allocate adequate resources.

d) Provide the planning supervisor with information relevant to health and safety on the project.

e) Ensure construction work does not start until the principal contractor has prepared a satisfactory health and safety plan.

f) Ensure the health and safety file is available for inspection after the project is completed.

Duties on clients do not apply to domestic households having construction projects carried out or very minor works carried out in occupied premises.

The CDM Regulations will generally apply to construction work which is notifiable i.e. lasts for more than 30 days or will involve more than 500 person days of work. CDM will apply to any design which involves five people or more on site at any one time or if the work includes demolition.

The Planning Supervisor

The planning supervisor has to co-ordinate the health and safety aspects project design and the initial planning to ensure as much as they can that:

a) designers comply with their duties - in particular, the avoidance and reduction of risk;

b) designers co-operate with each other for the purposes of health and safety;

c) a health and safety plan is prepared before arrangements are made for a principal contractor to be appointed;

d) they are able to give advice, if requested, to the client on the competence and allocation of resources by designers and all contractors; advise contractors appointing designers; and also advise the client on the health and safety plan before the construction phase starts;

e) the project is notified to the Health and Safety Executive;

f) the health and safety file is prepared and deliver it to the client at the end of a project.

The Designer

Designers must design in a way which avoids, reduces, or controls risks to health and safety as far as is reasonably practicable so that projects they design can be constructed and maintained safely. Where risks remain, they have to be stated to the extent necessary to enable reliable performance by a competent contractor.

The designer's key duties are, as far as reasonably practicable, to:

a) alert clients to their duties;

b) consider during the development of designs the hazards and risks which may arise to those constructing and maintaining the structure;

The Designer (Cont'd)

c) design to avoid risks to health and safety so far as is reasonable practicable;

d) reduce risks at source if avoidance is not possible;

e) consider measures which will protect all workers if neither avoidance nor reduction to a safe level is possible;

f)

(f) ensure that the design includes adequate information on health and safety.

(g) pass this information on to the planning supervisor so that it can be included in the health and safety plan; and ensure that it is given on drawings or in specifications etc.;

h) co-operate with the planning supervisor and, where necessary, other designers involved in the project.

Note:

Designer means any person who prepares a design.

Design includes drawings, design details, specifications and bills of quantities (Quantity Surveyors Note!)

The Principal Contractor

The principal contractor should take account of the specific requirements of a project when preparing and presenting tenders or similar documents, take over and develop the health and safety plan and co-ordinate the activities of all contractors so that they comply with health and safety law.

The principal contractor's key duties are to:

a) develop and implement the health and safety plan;

b) arrange for competent and adequately resourced contractors to carry out the work where it is subcontracted;

c) ensure the co-ordination and co-operation of contractors;

d) obtain from contractors the main findings of their risk assessments and details of how they intend to carry out high risk operations;

e) ensure that contractors have information about risks on site;

f) ensure that workers on site have been given adequate training;

g) ensure that contractors and workers comply with any site rules which may have been set out in the health and safety plan;

h) monitor health and safety performance;

I) ensure that all workers are properly informed and consulted;

j) make sure only authorised people are allowed onto the site;

k) display the notification of the project to HSE;

l) pass information to the planning supervisor for the Health and Safety file;

Contractors and the Self-employed

Contractors in general have duties to play their part in the successful management of health and safety during construction work. The key duties are to;

a) provide information for the health and safety plan about risks to health and safety arising from their work and the steps they will take to control and manage the risks;

b) manage their work so that they comply with rules in the health and safety plan and directions from the principal contractor;

c) provide information for the health and safety file, and about injuries, dangerous occurrences and ill health;

d) provide information to their employees.

The self-employed also have these duties when they act as contractors.

Employees

Under CDM employees will benefit by being better informed and more able to play an active part in health and safety.

Employees are:

a) entitled to information about health and safety during the construction phase;

b) able to express their views about health and safety to the principal contractor.

The principal contractor has to check that employees have been provided with adequate information and training.

The pre-tender health and safety plan

The pre-tender plan which the planning supervisor has to ensure is prepared, should include:

a) a general description of the work;

b) details of timings within the project;

c) details of risks to workers as far as possible at that stage;

d) information required by potential principal contractors to demonstrate competence or adequacy of resources;

e) information for preparing a health and safety plan for the construction phase and information for welfare provision.

The health and safety plan for the construction phase

The plan developed by the principal contractor is the foundation on which health and safety management of construction work is based. It should include:

a) arrangements for ensuring the health and safety of all who may be affected by the construction work;

b) arrangements for the management of health and safety of construction work and monitoring of compliance with health and safety law;

c) information about welfare arrangements.

Health and safety file

The health and safety file amounts to a normal maintenance manual enlarged to alert those who will be responsible for a structure after handover to risks that must be managed when the structure and associated plant is maintained, repaired, renovated or demolished. It is a record of information to inform future decisions on the management of health and safety.

Competence and provision for health and safety

Checking on competence and on the allocation of adequate resources for health and safety may be done in a way that fits in with established procedures, with the degree of detail suited for the project.

The Approved Code of Practice

The Construction (Design and Management) Regulations 1994 are reproduced in italics in the Code with each part of the Code printed after the relevant regulation. The Code covers the following topics.

The following is an outline of the Appendices for information

Outline of Appendix 1 — Summary of the Management of Health and Safety at Work Regulations 1992

1 The Regulations aim to improve health and safety management and make more explicit what is required of employers under the Health and Safety at Work etc. Act 1974 with a more systematic and better organised approach to dealing with health and safety.

2 The Regulations place duties on employers (and in some cases, the self-employed) whether they are clients, designers, planning supervisors, principal contractors or other contractors.

3 Employers have to:

(a) assess the risks to the health and safety of their employees and others who may be affected by the work activity. This is to identify the preventive and protective measures necessary. Employers with five or more employees must record the significant findings of the assessment (regulation 3);

(b) ensure that arrangements are made for implementing the measures following the risk assessment. This will cover planning, organisation, control, monitoring and review, i.e. the management of health and safety (regulation 4);

(c) provide health surveillance for employees whenever the risk assessment determines it to be needed (regulation 5);

(d) appoint competent people to help devise and apply the measures (regulation 6);

(e) set up emergency procedures (regulation 7);

(f) provide employees with relevant information on health and safety (regulation 8);

(g) co-operate with other employers sharing the same work site and co-ordinate the preventive and protective measures needed (regulations 9 and 10);

(h) make sure employees have adequate health and safety training (regulation 11);

(I) provide temporary workers with health and safety information to meet special needs (regulation 13);

4 Employees also have duties (regulation 12) to:

(a) use equipment in accordance with training and instruction;

(b) report dangerous situations;

(c) report any shortcomings in health and safety arrangements.

Outline of Appendix 2 — The principles of prevention and protection

The principles of Prevention and Protection are:

(a) **If possible avoid the risk completely.**

(b) **Combat risks at source.**

(c) **Wherever possible, adapt work to the individual,** particularly in the choice of work equipment and methods of work.

(d) **Take advantage of technological progress.**

(e) **Incorporate the prevention measures into a coherent plan** to reduce progressively those risks which cannot altogether be avoided and which takes into account working conditions, organisational factors, the working environment and social factors. On individual projects, the health and safety plan (regulation 15) will act as the focus for bringing together and co-ordinating the individual policies of everyone involved. Where an employer is required under section 2(3) of the Health and Safety at Work etc. Act 1974 to have a health and safety policy, this should be prepared and applied by reference to these principles.

(f) **Give priority to those measures which protect the whole workforce or activity**, i.e. give collective protective measures, such as suitable working platforms with edge protection, priority over individual measures, such as safety harnesses.

(g) **Employees and the self-employed need to understand what they need to do,** e.g. by training, instruction, and communication of plans and risk assessments.

(h) **The existence of an active safety culture affecting the organisations responsible for developing and executing the project needs to be assured.**

The Approved Code of Practice (Cont,d)

Appendix 3 Where the local authority is the enforcing authority under the Health and Safety (Enforcing Authority) Regulations 1989
(Not reproduced here)

Note:

For construction work, HSE is the enforcing authority in all premises for which it is responsible and in all premises which would be inspected by the local authority except in respect of minor construction works as noted in the regulations.

Appendix 4 The Health and Safety plan prepared under Regulation 15(1)-(3)

1 Nature of the project

(a) Name of client.

(b) Location.

(c) Nature of construction work to be carried out.

(d) Timescale for completion of the construction work.

2 The existing environment

(a) Surrounding land uses and related restrictions, e.g. premises (schools, shops or factories) adjacent to proposed construction site, planning restrictions which might affect health and safety.

(b) Existing services, e.g. underground and overheads lines.

(c) Existing traffic systems and restrictions, e.g. access for fire appliances, times of delivery, ease of delivery and parking.

(d) Existing structures, e.g. special health problems from materials in existing structures which are being demolished or refurbished, any fragile materials which require special safety precautions or instability problems.

(e) Ground conditions, e.g. contamination, gross instability, possible subsidence, old mine workings or underground obstructions.

3 Existing drawings

(a) Available drawings of structure(s) to be demolished or incorporated in the proposed structure(s) (this may include a health and safety file prepared for the structure(s) and held by the client).

4 The design

(a) Significant hazards or work sequences identified by designers which cannot be avoided or designed out and, where appropriate, a broad indication of the precautions assumed for dealing with them.

(b) The principles of the structural design and any precautions that might be needed or sequences of assembly that need to be followed during construction.

(c) Detailed reference to specific problems where contractors will be required to explain their proposals for managing these problems.

5 Construction materials

(a) Health hazards arising from construction materials where particular precautions are required either because of their nature or the manner of their intended use. These will have been identified by designers as hazards which cannot be avoided or designed out. They should be specified as far as is necessary to ensure reliable performance by a competent contractor who may be assumed to know the precautionary information that suppliers are, by law, required to provide.

6 Site-wide elements

(a) Positioning of site access and egress points (e.g. for deliveries and emergencies).

(b) Location of temporary site accommodation.

(c) Location of unloading, layout and storage areas.

(d) Traffic/pedestrian routes.

7 Overlap with client's undertaking

(a) Consideration of the health and safety issues which arise when the project is to be located in premises occupied or partially occupied by the client.

8 Site rules

(a) Specific site rules which the client or the planning supervisor may wish to lay down as a result of points 2 to 7 above or for other reasons, e.g. specific permit-to-work rules, emergency procedures.

9 Continuing liaison

(a) Procedures for considering the health and safety implications of design elements of the principal contractor's and other contractor's packages.

(b) Procedures for dealing with unforeseen eventualities during project execution resulting in substantial design change and which might affect resources.

Appendix 5 The health and safety file prepared under regulation 14(d)-(f)

Information contained in the file needs to include that which will assist persons carrying out construction work on the structure at any time after completion of the current project and may include:

(a) record or 'as built' drawings and plans used and produced throughout the construction process along with the design criteria;

(b) general details of the construction methods and materials used;

(c) details of the structure's equipment and maintenance facilities;

(d) maintenance procedures and requirements for the structure;

(e) manuals produced by specialist contractors and suppliers which outline operating and maintenance procedures and schedules for plant and equipment installed as part of the structure; and

(f) details on the location and nature of utilities and services, including emergency and fire-fighting systems.

ARRANGEMENT OF REGULATIONS

1. Citation and commencement
2. Interpretation
3. Application of regulations
4. Clients and agents of clients
5. Requirements on developer
6. Appointments of planning supervisor and principal contractor

The Approved Code of Practice (Cont,d)
ARRANGEMENT OF REGULATIONS (Cont'd)

7. Notification of project
8. Competence of planning supervisor, designer and contractors
9. Provision for health and safety
10. Start of construction phase
11. Client to ensure information is available
12. Client to ensure health and safety file is available for inspection
13. Requirements on designer
14. Requirements on planning supervisor
15. Requirements relating to the health and safety plan
16. Requirements on and powers of principal contractor
17. Information and training
18. Advice from, and views of, persons at work
19. Requirements and prohibitions on contractors
20. Extension outside Great Britain
21. Exclusion of civil liability
22. Enforcement
23. Transitional provisions
24. Repeals, revocations and modifications

Schedule 1 Particulars to be notified to the Executive

Schedule 2 Transitional provisions

Schedule 1 Particulars to be notified to the Executive Regulation 7

1 Date of forwarding

2 Exact address of the construction site

3 Name and address of the client or clients (see note)

3 Type of project

5 Name and address of the planning supervisor

6 A declaration signed by or on behalf of the planning supervisor that he has been appointed as such.

7 Name and address of the principal contractor

7 A declaration signed by or on behalf of the principal that he has been appointed as such

9 Date planned for start of the construction phase

10 Planned duration of the construction phase

11 Estimated maximum number of people at work on the construction site

12 Planned number of contractors on the construction site

13 Name and address of any contractor or contractors already chosen

Note: Where a declaration has been made in accordance with regulation 4(4), item 3 above refers to the client or clients on the basis that that declaration has not yet taken effect.

Acknowledgement

Excerpts reproduced from HSE publications -

(a) CDM Regulations - How the regulations affect you!

(b) Managing Construction for Health and Safety, Construction (Design and Management) Regulation 1994, Approved Code of Practice.

(c) Statutory Instruments 1994 No.3140 HEALTH AND SAFETY the Construction (Design and Management) Regulations 1994.

New regulations will come into force in October 2006. The consultation document is available at:- HTTP://consultations.HSE.gov.uk/CONSULT.TI/CONREGS/CONSULTATIONHOME

ARBITRATION

ADJUDICATION

A more 'immediate' and statutory method of resolving disputes was introduced by The Housing Grants, Construction and Regeneration Act 1996 (commonly known as 'The Construction Act'), which became law on 1st May 1998. It provides for the Act's "Scheme for Construction Contracts" to be read into all defined construction contracts, entered into after 1st May 1998, which do not already contain similar provisions.

Included in the provisions is a statutory right for the parties to a construction contract to refer all disputes to adjudication by a third party - the adjudicator.

The aim of the provisions is to resolve disputes speedily and effectively as they occur. The adjudicator is to be appointed either by agreement or by appointment by an 'adjudicator nominating body'.

The object is to ensure the appointment of the adjudicator and the referral of the dispute to him within 7 days of a notice being given by either party that he intends to refer a dispute to adjudication.

The adjudicator has then to reach a decision within 28 days of referral or within such longer period as is agreed by the parties. If the parties fail to agree on an extension of time, requested by the adjudicator, an extension of 14 days may be granted with the sole consent of the party by whom the dispute was referred.

The adjudicator has to act impartially but is not bound to follow any set procedure. He is expected to take the initiative in ascertaining the facts and the law.

The parties may agree to accept the decision of the adjudicator as final. However, the contract may provide that his decision is binding until the dispute is eventually determined by arbitration, by legal proceedings or by agreement.

Previously, the standard forms of sub-contract provided adjudication only as a means of establishing the allocation of money in set-off disputes. The current provisions, which are statutory and apply to all defined construction contracts (including consultancy agreements), can be invoked for all types of dispute at any time.

An adjudicator can be named in the contract. If no name appears in the contract documents, application can be made by a party to an 'adjudicator nominating body' for an appointment to be made. A variety of organisations offer the service of selecting a suitable adjudicator and run training programmes for 'would be' adjudicators. The 'adjudicator nominating bodies' include the Chartered Institute of Building, Confederation of Construction Specialists, Construction Confederation, Construction Industry Council, ICE, RIBA, RICS and the Chartered Institute of Arbitrators.

ARBITRATION

Most standard forms of building contract and subcontract conditions contain an arbitration clause. This means that the parties agree to refer any dispute, arising under the contract, to a person to be agreed between the parties, to act as arbitrator.

If the parties are unable to agree on a person then the president or vice-president of one of the professional institutes will appoint a person from a panel of experienced arbitrators. Depending upon the wording in the conditions of contract, the request for an appointment by either party may be made to the president of the Royal Institute of British Architects, The Royal Institution of Chartered Surveyors, the Institution of Civil Engineers or the Chartered Institute of Arbitrators.

Compared with High Court procedures, arbitration has the benefit of being private.

It is recommended that a negotiated settlement of a dispute should be pursued as far as possible before the attempt at an agreement is abandoned and arbitration commenced. Once arbitration is started it may be necessary to search files for all relevant documents and witnesses may have to appear to give evidence. The arbitrator will need to see or hear all the necessary evidence upon which to base his award. The process can be very demanding in management and staff time. Sometimes even the "winner" gets little financial reward for the amount of time and trouble spent on preparing for the arbitration, particularly where small amounts of money are involved.

The latest Arbitration Act gives the parties, in an arbitration, the opportunity to avoid unnecessary delay and expense.

The substantive provisions of the Act - The Arbitration Act 1996 - came into force on 31 January 1997. The Act is printed on the following pages (except Schedules 1-4).

The underlying philosophy of the Act is that the parties should be free to agree between themselves how their dispute should be resolved. The Act is founded on three principles- a) the object of arbitration is to obtain the fair resolution of disputes by an impartial tribunal without unnecessary delay and expense; b) the parties should be free to agree how their disputes are resolved, subject only to safeguards as are necessary in the public interest; c) the Court should not intervene except in the very limited circumstances provided by the Act.

The arbitrator now has the power to decide all procedural and evidential matters, subject to the right of the parties to agree any matter if they so wish. The strict rules of evidence, adhered to by the Courts, may be ignored. The arbitrator may take the initiative in ascertaining the facts and law instead of relying upon the parties or their representatives to ascertain them by examination and cross-examination. The absolute right, of either party, to an oral hearing has been swept away.

ARBITRATION (Cont'd)

The arbitrator will seek to establish the wishes of the parties at a Preliminary Meeting. He will be made aware of the nature of the dispute and assist the parties in agreeing the most suitable procedure for the resolution of the dispute. Subsequently he will issue an order giving the timetable for the arbitration process.

The 1996 Act gives arbitrators a completely new power. Unless otherwise agreed between the parties, the arbitrator may direct that the recoverable costs of the arbitration or any part of the arbitral proceedings shall be limited to a specified amount. The direction has to be given sufficiently in advance of the costs being incurred for the limit to be taken into account.

The Act also gives the arbitrator immunity from liability for anything he may do or omit to do in the discharge of his duties except acting in bad faith.

The right to challenge an award for the technical misconduct of the arbitrator is limited and the Court will only intervene in those cases of serious irregularity where substantial injustice has occurred. The kind of irregularity is narrowly defined by the Act.

The Act also contains guidelines concerning those circumstances in which the Court will give leave to appeal on a point of law. It generally needs to be on a point of general public importance for the Court to give leave. Thus it will be very difficult to overturn the award of the arbitrator.

The arbitration process, prior to the Act coming into force, had become almost a copy of Court procedures. Arbitration had become expensive and time consuming. It is hoped that the Act will enable parties to make agreements on procedure and costs and arbitrators to use their discretion for the sake of the fair and swift resolution of damaging disputes.

Following the enactment of the Arbitration Act, the Society of Construction Arbitrators initiated the production of Model Arbitration Rules for adoption by all construction institutions and other bodies having interests in construction arbitration.

The outcome was The Construction Industry Model Arbitration Rules (CIMAR). These rules together with supplementary and advisory procedures were published by the Joint Contracts Tribunal in 1998 and are incorporated in all current editions of the JCT standard forms of building contract.

The rules provide procedures and time limits for document only, sort hearing or full procedure arbitrations. Sections of the Act, which need to be read with the Rules, are printed in the published text and other sections are reproduced in the Appendix.

Where arbitration agreements are incorporated in non-standard contract conditions, without any rules being stated, it might be helpful to the parties, in the interest of speed and cost effectiveness, to adopt the Model Rules, by agreement, when the arbitration commences.

ARBITRATION ACT 1996

CHAPTER 23

ARRANGEMENT OF SECTIONS

PART I

ARBITRATION PURSUANT TO AN ARBITRATION AGREEMENT

ARBITRATION ACT 1996 (Cont'd)

ARRANGEMENT OF SECTIONS (Cont'd)

Arbitration Act 1996

ELIZABETH II

Arbitration Act 1996

1996 CHAPTER 23

An Act to restate and improve the law relating to arbitration pursuant to an arbitration agreement; to make other provision relating to arbitration and arbitration awards; and for connected purposes. [17th June 1996]

BE IT ENACTED by the Queen's most Excellent Majesty, by and with the advice and consent of the Lords Spiritual and Temporal, and Commons, in this present Parliament assembled, and by the authority of the same, as follows:-

PART I

ARBITRATION PURSUANT TO AN ARBITRATION AGREEMENT

Introductory

General principles

1. The provisions of this Part are founded on the following principles, and shall be construed accordingly-

(a) the object of arbitration is to obtain the fair resolution of disputes by an impartial tribunal without unnecessary delay or expense;

(b) the parties should be free to agree how their disputes are resolved, subject only to such safeguards as are necessary in the public interest;

(c) in matters governed by this Part the court should not intervene except as provided by this Part.

Scope of application of provisions

2. (l) The provisions of this Part apply where the seat of the arbitration is in England and Wales or Northern Ireland.

(2) The following sections apply even if the seat of the arbitration is outside England and Wales or Northern Ireland or no seat has been designated or determined-

(a) sections 9 to 11 (stay of legal proceedings, &c.), and

(b) section 66 (enforcement of arbitral awards).

(3) The powers conferred by the following sections apply even if the seat of the arbitration is outside England and Wales or Northern Ireland or no seat has been designated or determined-

(a) section 43 (securing the attendance of witness), and

(b) section 44 (court powers exercisable in support of arbitral proceedings);

but the court may refuse to exercise any such power, if in the opinion of the court, the fact that the seat of the arbitration is outside England and Wales or Northern Ireland, or that when designated or determined the seat is likely to be outside England and Wales or Northern Ireland, makes it inappropriate to do so.

(4) The court may exercise a power conferred by any provision of this Part not mentioned in subsections (2) or (3) for the purpose of supporting the arbitral process where-

(a) no seat of the arbitration has been designated or determined, and

(b) by reason of a connection with England and Wales or Northern Ireland the court is satisfied that it is appropriate to do so.

(5) Section 7 (separability of arbitration agreement) and section 8 (death of a party) apply where the law of England and Wales or Northern Ireland even if the seat of the arbitration is outside England and Wales or Northern Ireland or has not been designated or determined.

The seat of the arbitration

3. In this Part "the seat of the arbitration" means the juridical seat of the arbitration designated-

(a) by the parties to the arbitration agreement, or

(b) by any arbitral or other institution or person vested by the parties with power in that regard, or

(c) by the arbitral tribunal if so authorised by the parties,

or determined, in the absence of any such designation, having regard to the parties' agreement and all the relevant circumstances.

Mandatory and non-mandatory provisions

4.(1) The mandatory provisions of this Part are listed in Schedule 1 and have effect notwithstanding any agreement to the contrary,

(2) The other provisions of this Part (the "non-mandatory provisions") allow the parties to make their own arrangements by agreement but provide rules which apply in the absence of such agreement

(3) The parties may make such arrangements by agreeing to the application of institutional rules or providing any other means by which a matter may be decided.

(4) It is immaterial whether or not the law applicable to the parties' agreement is the law of England and Wales or, as the case may be, Northern Ireland.

(5) The choice of a law other than the law of England and Wales or Northern Ireland as the applicable law in respect of a matter provided for by a non-mandatory provision of this Part is equivalent to an agreement making provision about that matter.

For this purpose an applicable law determined in accordance with the parties' agreement, or which is objectively determined in the absence of any express or implied choice, shall be treated as chosen by the parties.

Agreements to be in writing

5.(1) The provisions of this Part apply only where the arbitration agreement is in writing, and any other agreement between the parties as to any matter is effective for the purposes of this Part only if in writing.

Arbitration Act 1996

Part I

The expressions "agreement", "agree" and "agreed" shall be construed accordingly.

(2) There is an agreement in writing-

(a) if the agreement is made in writing (whether or not it is signed by the parties),

(b) if the agreement is made by exchange of communications in writing, or

(c) if the agreement is evidenced in writing.

(3) Where parties agree otherwise than in writing by reference to terms which are in writing, they make an agreement in writing.

(4) An agreement is evidenced in writing if an agreement made otherwise than in writing is recorded by one of the parties, or by a third party, with the authority of the parties to the agreement.

(5) An exchange of written submissions in arbitral or legal proceedings in which the existence of an agreement otherwise than in writing is alleged by one party against another party and not denied by the other party in his response constitutes as between those parties an agreement in writing to the effect alleged.

(6) References in this Part to anything being written or in writing include its being recorded by any means.

The arbitration agreement

Definition of arbitration agreement

6.(1) In this Part an "arbitration agreement" means an agreement to submit to arbitration present or future disputes (whether they are contractual or not).

(2) The reference in an agreement to a written form of arbitration clause or to a document containing an arbitration clause constitutes an arbitration agreement if the reference is such as to make the clause part of the agreement.

Separability or arbitration agreement

7. Unless otherwise agreed by the parties, an arbitration agreement which forms or was intended to form part of another agreement (whether or not in writing) shall not be regarded as invalid, non-existent or ineffective because that other agreement is invalid, or did not come into existence or has become ineffective, and it shall for that purpose be treated as a distinct agreement.

Whether an agreement discharged by death of a party

8.(1) Unless otherwise agreed by the parties, arbitration agreement is not discharged by the death of a party and may be enforced by or against the personal representatives of that party.

(2) Subsection (1) does not affect the operation of any enactment or rule of law by virtue of which a substantive right or obligation is extinguished by death.

Stay of legal proceedings

Stay of Legal proceedings

9.(1) A party to an arbitration agreement against whom legal proceedings are brought (whether by way of claim or counterclaim in respect of a matter which under the agreement is to be referred to arbitration may (upon notice to the other parties to the proceedings) apply to the court in which the proceedings have been brought to stay the proceedings so far as they concern that matter.

Arbitration Act 1996

(2) An application may be made notwithstanding that the matter is to be referred to arbitration only after the exhaustion of other dispute resolution procedures.

(3) An application may not be made by a person before taking the appropriate procedural step (if any) to acknowledge the legal proceedings against him or after he has taken any step in those proceedings to answer the substantive claim.

(4) On an application under this section the court shall grant a stay unless satisfied that the arbitration agreement is null and void, inoperative, or incapable of being performed.

(5) If the court refuses to stay the legal proceedings, any provision that an award is a condition precedent to the bringing of legal proceedings in respect of any matter is of no effect in relation to those proceedings.

Reference of interpleader issue to arbitration

10.(1)Where in legal proceedings relief by way of interpleader is granted and any issue between the claimants is one in respect of which there is an arbitration agreement between them, the court granting the relief shall direct that the issue be determined in accordance with the agreement unless the circumstances are such that proceedings brought by a claimant in respect of the matter would not be stayed.

(2) Where subsection (1) applies but the court does not direct that the issue be determined in accordance with the arbitration agreement, any provision that an award is a condition precedent to the bringing of legal proceedings in respect of any matter shall not affect the determination of that issue by the court.

Retention of security where Admiralty proceedings stayed

11.(1)Where Admiralty proceedings are stayed on the ground that the dispute in question should be submitted to arbitration, the court granting the stay may, if in those proceedings property has been arrested or bail or other security has been given to prevent or obtain release from arrest.

(a) order that the property arrested be retained as security for the satisfaction of any award given in the arbitration in respect of that dispute, or

(b) order that the stay of those proceedings be conditional on the provision of equivalent security for the satisfaction of any such award.

(2) Subject to any provision made by rules of court and to any necessary modifications, the same law and practice shall apply in relation to property retained in pursuance of an order as would apply if it were held for the purposes of proceedings in the court making the order.

Arbitration Act 1996

Commencement of arbitral proceedings

Part I

Power of court to extend time for beginning arbitral proceedings &c

12.(1)Where an arbitration agreement to refer future disputes to arbitration provides that a claim shall be barred, or the claimant's right extinguished, unless the claimant takes within a time fixed by the agreement some step.

(a) to begin arbitral proceedings, or

(b) to begin other dispute resolution procedures which must be exhausted before arbitral proceedings can be begun.

the court may by order extend the time for taking that step.

(2) Any party to the arbitration agreement may apply for such an order (upon notice to the other parties), but only after a claim has arisen and after exhausting any available arbitral process for obtaining an extension of time.

(3) The court shall make an order only if satisfied.

(a) that the circumstances are such as were outside the reasonable contemplation of the parties when they agreed the provision in question, and that it would be just to extend the time, or

(b) that the conduct of one party makes it unjust to hold the other party to the strict terms of the provision in question.

(4) The court may extend the time for such period and on such terms as it thinks fit, and may do so whether or not the time previously fixed (by agreement or by a previous order) has expired.

(5) An order under this section does not affect the operation of the Limitation Acts (see section 13).

(6) The leave of the court is required for any appeal from a decision of the court under this section.

Application of Limitation Acts

13.(1)The Limitation Acts apply to arbitral proceedings as they apply to legal proceedings.

(2) The court may order that in computing the time prescribed by the Limitation Acts for the commencement of proceedings (including arbitral proceedings) in respect of a dispute which was the subject matter.

(a) of an award which the court orders to be set aside or declares to be of no effect, or

(b) of the affected part of an award which the court orders to be set aside in part, or declares to be in part of no effect,

the period between the commencement of the arbitration and the date of the order referred to in paragraph (a) or (b) shall be excluded.

(3) In determining for the purposes of the Limitation Acts when a cause of action accrued, any provision that an award is a condition precedent to the bringing of legal proceedings in respect of a matter to which an arbitration agreement applies shall be disregarded.

Arbitration Act 1996

(4) In this Part "the Limitation Acts" means-

1980 c.58
1984 c.16

(a) in England and Wales, the Limitation Act 1980, the Foreign Limitation Periods Act 1984 and any other enactment (whenever passed) relating to the limitation of actions;

S.I. 1989/1339 (N.I.11)
S.I.1985/754 (N.I.5)

(b) in Northern Ireland, the Limitation (Northern Ireland) Order 1989, the Foreign Limitation Periods (Northern Ireland) Order 1985 and any other enactment (whenever passed) relating to the limitation of actions.

Commencement of arbitral

14.(1)The parties are free to agree when arbitral proceedings are to be regarded as proceedings commenced for the purposes of this Part and for the purposes of the Limitations Acts.

(2) If there is no such agreement the following provisions apply.

(3) Where the arbitrator is named or designated in the arbitration agreement, arbitral proceedings are commenced in respect of a matter when one party serves on the other party or parties a notice in writing requiring him or them to submit that matter to the person so named or designated.

(4) Where the arbitrator or arbitrators are to be appointed by the parties, arbitral proceedings are commenced in respect of a matter when one party serves on the other party or parties notice in writing requiring him or them to appoint an arbitrator or to agree to the appointment of an arbitrator in respect of that matter.

(5) Where the arbitrator or arbitrators are to be appointed by a person other than a party to the proceedings, arbitral proceedings are commenced in respect of a matter when one party gives notice in writing to that person requesting him to make the appointment in respect of that matter.

The arbitral tribunal

The arbitral tribunal

15.(1)The parties are free to agree on the number of arbitrators to form the tribunal and whether there is to be a chairman or umpire.

(2) Unless otherwise agreed by the parties, an agreement that the number of arbitrators shall be two or any other even number shall be understood as requiring the appointment of an additional arbitrator as chairman of the tribunal.

(3) If there is no agreement as to the number of arbitrators, the tribunal shall consist of a sole arbitrator.

Procedure for appointment of arbitrators

16.(1)The parties are free to agree on the procedure for appointing the arbitrator or arbitrators including the procedure for appointing any chairman or umpire.

(2) If or to the extent that there is no such agreement, the following provisions apply.

Arbitration Act 1996

Part I

(3) If the tribunal is to consist of a sole arbitrator, the parties shall jointly appoint the arbitrator not later than 28 days after service of a request in writing by either party to do so.

(4) If the tribunal is to consist of two arbitrators, each party shall appoint one arbitrator not later than 14 days after service of a request in writing by either party to do so.

(5) If the tribunal is to consist of three arbitrators-

(a) each party shall appoint one arbitrator not later than 14 days after service of a request in writing by either party to do so, and

(b) the two so appointed shall forthwith appoint a third arbitrator as the chairman of the tribunal.

(6) If the tribunal is to consist of two arbitrators and an umpire-

(a) each party shall appoint one arbitrator not later than 14 days after service of a request in writing by either party to do so, and

(b) the two so appointed may appoint an umpire at any time after they themselves are appointed and shall do so before any substantive hearing or forthwith if they cannot agree on a matter relating to the arbitration.

(7) In any other case (in particular, if there are more than two parties) section 18 applies as in the case of a failure of the agreed appointment procedure.

Power in case of default to appoint sole arbitrator

17.(1)Unless the parties otherwise agree, where each of two parties to an arbitration agreement is to appoint an arbitrator and one party ("the party in default") refuses to do so, or fails to do so within the time specified, the other party, having duly appointed his arbitrator, may give notice in writing to the party in default that he proposes to appoint his arbitrator to act as sole arbitrator.

(2) If the party in default does not within 7 clear days of that notice being given-

(a) make the required appointment, and

(b) notify the other party that he has done so,

the other party may appoint his arbitrator as sole arbitrator whose award shall be binding on both parties as if he had been so appointed by agreement.

(3) Where a sole arbitrator has been appointed under subsection (2), the party in default may (upon notice to the appointing party) apply to the court which may set aside the appointment.

(4) The leave of the court is required for any appeal from a decision of the court under this section.

Arbitration Act 1996

Failure of appointment procedure

18.(1)The parties are free to agree what is to happen in the event of a failure of the procedure for the appointment of the arbitral tribunal.
There is no failure if an appointment is duly made under section 17 (power in case of default to appoint sole arbitrator), unless that appointment is set aside.

(2) If or to the extent that there is no such agreement any party to the arbitration agreement may (upon notice to the other parties) apply to the court to exercise its powers under this section.

(3) Those powers are-

(a) to give directions as to the making of any necessary appointments;

(b) to direct that the tribunal shall be constituted by such appointments (or any one or more of them) as have been made;

(c) to revoke any appointments already made;

(d) to make any necessary appointments itself.

(4) An appointment made by the court under this section has effect as if made with the agreement of the parties.

(5) The leave of the court is required for any appeal from a decision of the court under this section.

Court to have regard to agreed qualifications

19. In deciding whether to exercise, and in considering how to exercise, any of its powers under section 16 (procedure for appointment of arbitrators) or section 18 (failure of appointment procedure), the court shall have due regard to any agreement of the parties as to the qualifications required of the arbitrators.

Chairman

20.(1)Where the parties have agreed that there is to be a chairman, they are free to agree what the functions of the chairman are to be in relation to the making of decisions, orders and awards.

(2) If or to the extent that there is no such agreement, the following provisions apply.

(3) Decisions, orders and awards shall be made by all or a majority of the arbitrators (including the chairman).

(4) The view of the chairman shall prevail in relation to a decision, order or award in respect of which there is neither unanimity nor a majority under subsection (3).

Umpire

21.(1)Where the parties have agreed that there is to be an umpire, they are free to agree what the functions of the umpire are to be, and in particular-

(a) whether he is to attend the proceedings, and

(b) when he is to replace the other arbitrators as the tribunal with power to make decisions, orders and awards.

(2) If or to the extent that there is no such agreement, the following provisions apply.

Arbitration Act 1996

Part I

(3) The umpire shall attend the proceedings and be supplied with the same documents and other materials as are supplied to the other arbitrators.

(4) Decisions, orders and awards shall be made by the other arbitrators unless and until they cannot agree on a matter relating to the arbitration.

In that event they shall forthwith give notice in writing to the parties and the umpire, whereupon the umpire shall replace them as the tribunal with power to make decisions, orders and awards as if he were sole arbitrator.

(5) If the arbitrators cannot agree but fail to give notice of that fact, or if any of them fails to join in the giving of notice, any party to the arbitral proceedings may (upon notice to the other parties and to the tribunal) apply to the court which may order that the umpire shall replace the other arbitrators as the tribunal with power to make decisions, orders and awards as if he were sole arbitrator.

(6) The leave of the court is required for any appeal from a decision of the court under this section.

Decision-making where no chairman or umpire awards.

22.(1)Where the parties agreed that there shall be two or more arbitrators with no chairman or umpire, the parties are free to agree how the tribunal is to make decisions, orders and

(2) If there is no such agreement, decisions, orders and awards shall be made by all or a majority of the arbitrators.

Revocation arbitrator's authority

23.(1)The parties are free to agree in what circumstances the authority of an arbitrator may be revoked.

(2) If or to the extent that there is no such agreement the following provisions apply.

(3) The authority of an arbitrator may not be revoked except-

(a) by the parties acting jointly, or

(b) by an arbitral or other institution or person vested by the parties with powers in that regard.

(4) Revocation of the authority of an arbitrator by the parties acting jointly must be agreed in writing unless the parties also agree (whether or not in writing) to terminate the arbitration agreement.

(5) Nothing in this section affects the power of the court-

(a) to revoke an appointment under section 18 (powers exercisable in case of failure of appointment procedure), or

(b) to remove an arbitrator on the grounds specified in section 24.

Power of court to remove arbitrator

24.(1)A party to arbitral proceedings may (upon notice to the other parties, to the arbitrator concerned and to any other arbitrator) apply to the court to remove an arbitrator on any of the following grounds-

Arbitration Act 1996

(a) that circumstances exist that give rise to justifiable doubts as to his impartiality;

(b) that he does not possess the qualifications required by the arbitration agreement;

(c) that he physically or mentally incapable of conducting the proceedings or there are justifiable doubts as to his capacity to do so;

(d) that he has refused or failed-

(i) properly to conduct the proceedings, or

(ii) to use all reasonable despatch in conducting the proceedings or making an award, and that substantial injustice has been or will caused to the applicant.

24(2) If there is arbitral or other institution or person vested by the parties with power to remove an arbitrator, the court shall not exercise its power of removal unless satisfied that the applicant has first exhausted any available recourse to that institution or person.

(3) The arbitral tribunal may continue the arbitral proceedings and make an award while an application to the court under this section is pending.

(4) Where the court removes an arbitrator, it may make such order as it thinks fit with respect to his entitlement (if any) to fees or expenses, or the repayment of any fees or expenses already paid.

(5) The arbitrator concerned is entitled to appear and be heard by the court before it makes any order under this section.

(6) The leave of the court is required for any appeal from a decision of the court under this section.

Resignation of arbitrator

25.(1)The parties are free to agree with an arbitrator as to the consequences of his resignation as regards-

(a) his entitlement (if any) to fees or expenses, and
(b) any liability thereby incurred by him.

(2) If or to the extent that there is no such agreement the following provisions apply.

(3) An arbitrator who resigns his appointment may (upon notice to the parties) apply to the court-

(a) to grant him relief from any liability thereby incurred by him, and

(b) to make such order as it thinks fit with respect to his entitlement (if any) to fees or expenses or the repayment of any fees or expenses already paid.

(4) If the court is satisfied that in all the circumstances it was reasonable for the arbitrator to resign, it may grant such relief as it mentioned in subsection (3)(a) on such terms as it thinks fit.

Arbitration Act 1996

Part I

(5) The leave of the court is required for any appeal from a decision of the court under this section.

Death of arbitrator or person appointing him

26.(1)The authority of an arbitrator is personal and ceases on his death.

(2) Unless otherwise agreed by the parties, the death of the person by whom an arbitrator was appointed does not revoke the arbitrator's authority.

Filling of Vacancy &c

27.(1)Where an arbitrator ceases to hold office, the parties are free to agree-

(a) whether and if so how the vacancy is to be filled,

(b) whether and if so to what extent the previous proceedings should stand, and

(c) what effect (if any) his ceasing to hold office has on any appointment made by him (alone or jointly).

(2) If or to the extent that there is no such agreement, the following provisions apply.

(3) The provisions of sections 16 (procedure for appointment of arbitrators) and 18 (failure of appointment procedure) apply in relation to the filling of the vacancy as in relation to an original appointment.

(4) The tribunal (when reconstituted) shall determine whether and if so to what extent the previous proceedings should stand.

This does not affect any right of a party to challenge those proceedings on any ground which had arisen before the arbitrator ceased to hold office.

(5) His ceasing to hold office does not affect any appointment by him (alone or jointly) of another arbitrator, in particular any appointment of a chairman or umpire.

Joint and several liability of parties to arbitrators for fees and expenses

28.(1)The parties are jointly and severally liable to pay to the arbitrators such reasonable fees and expenses (if any) as are appropriate in the circumstances.

(2) Any party may apply to the court (upon notice to the other parties and to the arbitrators) which may order that the amount of the arbitrators' fees and expenses shall be considered and adjusted by such means and upon such terms as it may direct.

(3) If the application is made after any amount has been paid to the arbitrators by way of fees or expenses, the court may order the repayment of such amount (if any) as is shown to be excessive, but shall not do so unless it is shown that it is reasonable in the circumstances to order repayment.

(4) The above provisions have effect subject to any order of the court under section 24(4) or 25(3)(b) (order as to entitlement to fees or expenses in case of removal or resignation of arbitrator).

(5) Nothing in this section affects any liability of a party to any other party to pay all or any of the costs of arbitration (see sections 59 to 65) or

any contractual right of an arbitrator to payment of his fees and expenses.

(6) In this section references to arbitrators include an arbitrator who has ceased to act and an umpire who has not replaced the other arbitrators.

Immunity of arbitrator

29.(1)An arbitrator is not liable for anything done or omitted in the discharge or purported discharge of his functions as arbitrator unless the act or omission is shown to have been in bad faith.

(2) Subsection (1) applies to an employee or agent of an arbitrator as it applies to the arbitrator himself.

(3) This section does not affect any liability incurred by an arbitrator by reason of his resigning (but see section 25).

Jurisdiction of the arbitral tribunal

Competence of tribunal to rule on its own jurisdiction

30. (1)Unless otherwise agreed by the parties, the arbitral tribunal may rule on its own substantive jurisdiction, that is, as to-

(a) whether there is a valid arbitration agreement

(b) whether the tribunal is properly constituted, and

(c) what matters have been submitted to arbitration in accordance with the arbitration agreement.

(2) Any such ruling may be challenged by any available arbitral process of appeal or review or in accordance with the provisions of this Part.

Objection to substantive jurisdiction of tribunal

31.(1)An objection that the arbitral tribunal lacks substantive jurisdiction at the outset of the proceedings must be raised by a party not later than the time he takes the first step in the proceedings to contest the merits of any matter in relation to which he challenges the tribunal's jurisdiction.

A party is not precluded from raising such an objection by the fact that he has appointed or participated in the appointment of an arbitrator.

(2) Any objection during the course of the arbitral proceedings that the arbitral tribunal is exceeding its substantive jurisdiction must be made as soon as possible after the matter alleged to be beyond its jurisdiction is raised.

(3) The arbitral tribunal may admit an objection later than the time specified in subsection (1) or (2) if it considers the delay justified.

(4) Where an objection is duly taken to the tribunal's substantive jurisdiction and the tribunal has power to rule on its own jurisdiction, it may-

(a) rule on the matter in an award as to jurisdiction, or
(b) deal with the objection in its award on the merits.

If the parties agree which of these courses the tribunal should take, the tribunal shall proceed accordingly.

31.(5) The tribunal may in any case, and shall if the parties so agree, stay proceedings whilst an application is made to the court under section 32 (determination of preliminary point of jurisdiction).

Determination of preliminary point of jurisdiction

32.(1)The court may, on the application of a party to arbitral proceedings (upon notice to the other parties), determine any question as to the substantive jurisdiction of the tribunal.

A party may lose the right to object (see section 73).

(2) An application under this section shall not be considered unless-

(a) it is made with the agreement in writing of all the other parties to the proceedings, or

(b) it is made with the permission of the tribunal and the court is satisfied-

(i) that the determination of the question is likely to produce substantial savings in costs,

(ii) that the application was made without delay, and

(iii) that there is good reason why the matter should be decided by the court.

(3) An application under this section, unless made with the agreement of all the other parties to the proceedings, shall state the grounds on which it is said that the matter should be decided by the court.

(4) Unless otherwise agreed by the parties, the arbitral tribunal may continue the arbitral proceedings and make an award while an application to the court under this section is pending.

(5) Unless the court gives leave, no appeal lies from a decision of the court whether the conditions specified in subsection (2) are met.

(6) The decision of the court on the question of jurisdiction shall be treated as a judgement of the court for the purposes of an appeal.

But no appeal lies without the leave of the court which shall not be given unless the court considers that the question involves a point of law which is one of general importance or is one which for some other special reason should be considered by the Court of Appeal.

The arbitral proceedings

General duty of the tribunal

33.(1)The tribunal shall-

(a) act fairly and impartially as between the parties, giving each party a reasonable opportunity of putting his case and dealing with that of his opponent, and

(b) adopt procedures suitable to the circumstances of the particular case, avoiding unnecessary delay or expense, so as to provide a fair means for the resolution of the matters falling to be determined.

Arbitration Act 1996

(2) The tribunal shall comply with that general duty in conducting the arbitral proceedings, in its decisions on matters of procedure and evidence and in the exercise of all other powers conferred on it.

Procedural and evidential matters

34.(1)It shall be for the tribunal to decide all procedural and evidential matters, subject to the right of the parties to agree any matter.

(2) Procedural and evidential matters include-

(a) when and where any part of the proceedings is to be held;

(b) the language or languages to be used in the proceedings and whether translations of any relevant documents are to be supplied;

(c) whether any and if so what form of written statements of claim and defence are to be used, when these should be supplied and the extent to which such statements can be later amended;

(d) whether any and if so which documents or classes of documents should be disclosed between and produced by the parties and at what stage;

(e) whether any and if so what questions should be put to and answered by the respective parties and when and in what form this should be done;

(f) whether to apply strict rules of evidence (or any other rules) as to the admissibility, relevance or weight of any material (oral, written or other) sought to be tendered on any matters of fact or opinion, and the time, manner and form in which such material should be exchanged and presented;

(g) whether and to what extent the tribunal should itself take the initiative in ascertaining the facts and the law;

(h) whether and to what extent there should be oral or written evidence or submissions.

(3) The tribunal may fix the time within which any directions given by it are to be complied with, and may if it thinks fit extend the time so fixed (whether or not it has expired).

Consolidation of proceedings and concurrent hearings

35.(1)The parties are free to agree-

(a) that the arbitral proceedings shall be consolidated with other arbitral proceedings, or

(b) that concurrent hearings shall be held, on such terms as may be agreed.

(2) Unless the parties agree to confer such power on the tribunal, the tribunal has no power to order consolidation of proceedings or concurrent hearings.

Legal or other representation

36. Unless otherwise agreed by the parties, a party to arbitral proceedings may be represented in the proceedings by a lawyer or other person chosen by him.

Arbitration Act 1996

Power to appoint experts, legal advisers or assessors

37.(1)Unless otherwise agreed by the parties-

(a) the tribunal may-

(i) appoint experts or legal advisers to report to it and the parties, or

(ii) appoint assessors to assist it on technical matters,

and may allow any such expert, legal adviser or assessor to attend the proceedings; and

(b) the parties shall be given a reasonable opportunity to comment on any information, opinion or advice offered by any such person.

(2) The fees and expenses of an expert, legal adviser or assessor appointed by the tribunal for which the arbitrators are liable are expenses of the arbitrators for the purposes of this Part.

General powers exercisable by the tribunal

38.(1)The parties are free to agree on the powers exercisable by the arbitral tribunal for the purposes of and in relation to the proceedings.

(2) Unless otherwise agreed by the parties the tribunal has the following powers.

(3) The tribunal may order a claimant to provide security for the costs of the arbitration.

This power shall not be exercised on the ground that the claimant is-

(a) an individual ordinarily resident outside the United Kingdom, or

(b) a corporation or association incorporated or formed under the law of a country outside the United Kingdom, or whose central management and control is exercised outside the United Kingdom.

(4) The tribunal may give directions in relation to any property which is the subject of the proceedings or as to which any question arises in the proceedings, and which is owned by or is in the possession of a party to the proceedings-

(a) for the inspection, photographing, preservation, custody or detention of the property by the tribunal, an expert or a party, or

(b) ordering that samples be taken from, or any observation be made of or experiment conducted upon, the property.

38.(5) The tribunal made direct that a party or witness shall be examined on oath or affirmation, and may for that purpose administer any necessary oath or take any necessary affirmation.

(6) The tribunal may give directions to a party for the preservation for the purposes of the proceedings of any evidence in his custody or control.

Power to make provisional awards

39.(1) The parties are free to agree that the tribunal shall have power to order on a provisional basis any relief which it would have power to grant in a final award.

Arbitration Act 1996

(2) This includes, for instance, making-

(a) a provisional order for the payment of money or the disposition of property as between the parties, or

(b) an order to make an interim payment on account of the costs of the arbitration.

(3) Any such order shall be subject to the tribunal's final adjudication; and the tribunal's final award, on the merits or as to costs, shall take account of any such order.

(4) Unless the parties agree to confer such power on the tribunal, the tribunal has no such power. This does not affect its powers under section 47 (awards on different issues, &c).

General duty of parties

40.(1)The parties shall do all things necessary for the proper and expeditious conduct of the arbitral proceedings.

(2) This includes-

(a) complying without delay with any determination of the tribunal as to procedural or evidential matters, or with any order or directions of the tribunal, and

(b) where appropriate, taking without delay any necessary steps to obtain a decision of the court on a preliminary question of jurisdiction or law (see sections 32 and 45).

Powers of tribunal in case of party's default

41.(1)The parties are free to agree on the powers of the tribunal in case of a party's failure to do something necessary for the proper and expeditious conduct of the arbitration.

(2) Unless otherwise agreed by the parties, the following provisions apply.

(3) If the tribunal is satisfied that there has been inordinate and inexcusable delay on the part of the claimant in pursuing his claim and that the delay-

(a) gives rise, or is likely to give rise, to a substantial risk that it is not possible to have a fair resolution of the issues in that claim, or

(b) has caused, or is likely to cause, serious prejudice to the respondent,

the tribunal may make an award dismissing the claim.

(4) If without showing sufficient cause a party-

(a) fails to attend or be represented at an oral hearing of which due notice was given, or

(b) where matters are to be dealt with in writing, fails after due notice to submit written evidence or make written submissions,

the tribunal may continue the proceedings in the absence of that party of, as the case may be, without any written evidence or submissions on his behalf, and may make an award on the basis of the evidence before it.

Arbitration Act 1996

Arbitration Act 1996

Part I

(5) If without showing sufficient cause a party fails to comply with any order or directions of the tribunal, the tribunal may make a peremptory order to the same effect, prescribing such time for compliance with it as the tribunal considers appropriate.

(6) If a claimant fails to comply with a peremptory offer of the tribunal to provide security for costs, the tribunal may make an award dismissing his claim.

(7) If a party fails to comply with any other kind of peremptory order, then, without prejudice to section 42 (enforcement by court of tribunal's peremptory orders), the tribunal may do any of the following-

(a) direct that the party in default shall not be entitled to rely upon any allegation or material which was the subject matter of the order;

(b) draw such adverse inferences from the act of non-compliance as the circumstances justify;

(c) proceed to an award on the basis of such materials as have been properly provided to it;

(d) make such order as it thinks fit as to the payment of costs of the arbitration incurred in consequence of the non-compliance.

Powers of court in relation to arbitral proceedings

Enforcement of peremptory orders of tribunal

42.(1)Unless otherwise agreed by the parties, the court may make an order requiring a party to comply with a peremptory order made by the tribunal.

(2) An application for an order under this section may be made-

(a) by the tribunal (upon notice to the parties),

(b) by a party to the arbitral proceedings with the permission of the tribunal (and upon notice to the other parties), or

42.(2).(c) where the parties have agreed that the powers of the court under this section shall be available.

(3) The court shall not act unless it is satisfied that the applicant has exhausted any available arbitral process in respect of failure to comply with the tribunal's order.

(4) No order shall be made under this section unless the court is satisfied that the person to whom the tribunal's order was directed has failed to comply with it within the time prescribed in the order, or, if no time was prescribed, within a reasonable time.

(5) The leave of the court is required for any appeal from a decision of the court under this section.

Securing the attendance of witnesses

43.(1)A party to arbitral proceedings may use the same court procedures as are available in relation to legal proceedings to secure the attendance before the tribunal of a witness in order to give oral testimony or to produce documents or other material evidence.

(2) This may only be done with the permission of the tribunal or the agreement of the other parties.

(3) The court procedures may only be used if-

(a) the witness is in the United Kingdom, and

(b) the arbitral proceedings are being conducted in England and Wales or, as the case may be, Northern Ireland.

(4) A person shall not be compelled by virtue of this section to produce any document or other material evidence which he could not be compelled to produce in legal proceedings.

Court powers exercisable in support of arbitral proceedings

44.(1)Unless otherwise agreed by the parties, the court has for the purposes of and in relation to arbitral proceedings the same power of making orders about the matters listed below as it has for the purposes of and in relation to legal proceedings.

(2) Those matters are-

(a) the taking of the evidence of witnesses;

(b) the preservation of evidence;

(c) making orders relating to property which is the subject of the proceedings or as to which any question arises in the proceedings-
(i) for the inspection, photographing, preservation, custody or detention of the property, or
(ii) ordering that samples to be taken from, or any observation be made of or experiment conducted upon, the property;

and for that purpose authorising any person to enter any premises in the possession or control of a party to the arbitration;

(d) the sale of any goods the subject of the proceedings;

(e) the granting of an interim injunction or the appointment of a receiver.

(3) If the case is one of urgency, the court may, on the application of a party or proposed party to the arbitral proceedings, make such orders as it thinks necessary for the purpose of preserving evidence or assets.

(4) If the case is not one of urgency, the court shall act only on the application of a party to the arbitral proceedings (upon notice to the other parties and to the tribunal) made with the permission of the tribunal or the agreement in writing of the other parties.

(5) In any case the court shall act only if or to the extent that the arbitral tribunal, and any arbitral or other institution or person vested by the parties with power in that regard, has no power or is unable for the time being to act effectively.

Arbitration Act 1996

Part I

(6) If the court so orders, an order made by it under this section shall cease to have effect in whole or in part on the order of the tribunal or of any such arbitral or other institution or person having power to act in relation to the subject-matter of the order.

(7) The leave of the court is required for any appeal from a decision of the court under this section.

Determination of preliminary point of law

45.(1)Unless otherwise agreed by the parties, the court may on the application of a party to arbitral proceedings (upon notice to the other parties) determine any question of law arising in the course of the proceedings which the court is satisfied substantially affects the rights of one or more of the parties.

An agreement to dispense with reasons for the tribunal's award shall be considered an agreement to exclude the court's jurisdiction under this section.

(2) An application under this section shall not be considered unless-

(a) it is made with the agreement of all the other parties to the proceedings, or

(b) it is made with the permission of the tribunal and the court is satisfied-

(i) that the determination of the question is likely to produce substantial savings in costs, and

(ii) that the application was made without delay.

(3) The application shall identify the question of law to be determined and, unless made with the agreement of all the other parties to the proceedings, shall state the grounds on which it is said that the question should be decided by the court.

(4) Unless otherwise agreed by the parties, the arbitral tribunal may continue the arbitral proceedings and made an award while an application to the court under this section is pending.

(5) Unless the court gives leave, no appeal lies from a decision of the court whether the conditions specified in subsection (2) are met.

(6) The decision of the court on the question of law shall be treated as a judgement of the court for the purposes of an appeal.

But no appeal lies without the leave of the court which shall not be given unless the court considers that the question is one of general importance, or is one which for some other special reason should be considered by the Court of Appeal.

The award

Rules applicable to substance of dispute

46.(1)The arbitral tribunal shall decide the dispute-

(a) in accordance with the law chosen by the parties as applicable to the substance of the dispute, or

Arbitration Act 1996

(b) if the parties so agree, in accordance with such other considerations as are agreed by them or determined by the tribunal.

(2) For this purpose the choice of the laws of a country shall be understood to refer to the substantive laws of that country and not its conflict of laws rules.

(3) If or to the extent that there is no such choice or agreement, the tribunal shall apply the law determined by the conflict of laws rules which it considers applicable.

Awards on different issues, &c.

47.(1)Unless otherwise agreed by the parties, the tribunal may make more than one award at different times on different aspects of the matters to be determined.

(2) The tribunal may, in particular, make an award relating-

(a) to an issue affecting the whole claim, or

(b) to a part only of the claims or cross-claims submitted to it for decision.

(3) If the tribunal does so, it shall specify in its award the issue, or the claim or part of a claim, which is the subject matter of the award.

Remedies

48.(1)The parties are free to agree on the powers exercisable by the arbitral tribunal as regards remedies.

(2) Unless otherwise agreed by the parties, the tribunal has the following powers.

(3) The tribunal may make a declaration as to any matter to be determined in the proceedings.

48.(4) The tribunal may order the payment of a sum of money, in any currency.

(5) The tribunal has the same powers as the court-

(a) to order a party to do or refrain from doing anything;

(b) to order specific performance of a contract (other than a contract relating to land);

(c) to order the rectification, setting aside or cancellation of a deed or other document.

Interest

49.(1)The parties are free to agree on the powers of the tribunal as regards the award of interest.

(2) Unless otherwise agreed by the parties the following provisions apply.

(3) The tribunal may award simple or compound interest from such dates at such rates and with such rests as it considers meets the justice of the case-

(a) on the whole or part of any amount awarded by the tribunal, in respect of any period up to the date of the award;

Arbitration Act 1996

Part I

(b) on the whole or part of any amount claimed in the arbitration and outstanding at the commencement of the arbitral proceedings but paid before the award was made, in respect of any period up to the date of payment.

(4) The tribunal may award simple or compound interest from the date of the award (or any later date) until payment, at such rates and with such rests as it considers meets the justice of the case, on the outstanding amount of any award (including any award of interest under subsection (3) and any award as to costs).

(5) References in this section to an amount awarded by the tribunal include an amount payable in consequence of a declaratory award by the tribunal.

(6) The above provisions do not affect any other power of the tribunal to award interest.

Extension of time for making award

50.(1)Where the time for making an award is limited by or in pursuance of the arbitration agreement, then, unless otherwise agreed by the parties, the court may in accordance with the following provisions by order extend that time.

(2) An application for an order under this section may be made-

(a) by the tribunal (upon notice to the parties), or

(b) by any party to the proceedings upon notice to the tribunal and the other parties),

but only after exhausting any available arbitral process for obtaining an extension of time.

50.(3) The court shall only make an order if satisfied that a substantial injustice would otherwise be done.

(4) The court may extend the time for such period and on such terms as it thinks fit, and may do so whether or not the time previously fixed (by or under the agreement or by a previous order) has expired.

(5) The leave of the court is required for any appeal from a decision of the court under this section.

Settlement

51.(1)If during arbitral proceedings the parties settle the dispute, the following provisions apply unless otherwise agreed by the parties.

(2) The tribunal shall terminate the substantive proceedings and, if so requested by the parties and not objected to by the tribunal, shall record the settlement in the form of an agreed award.

(3) An agreed award shall state that it is an award of the tribunal and shall have the same status and effect as any other award on the merits of the case.

(4) The following provisions of this Part relating to awards (sections 52 to 58) apply to an agreed award.

(5) Unless the parties have also settled the matter of the payment of the costs of the arbitration, the provisions of this Part relating to costs (sections 59 to 65) continue to apply.

Arbitration Act 1996

Form of award

52.(1)The parties are free to agree on the form of an award.

(2) If or to the extent that there is no such agreement, the following provisions apply.

(3) The award shall be in writing signed by all the arbitrators or all those assenting to the award.

(3) The award shall contain the reasons for the award unless it is an agreed award or the parties have agreed to dispense with reasons.

(5) The award shall state the seat of the arbitration and the date when the award is made.

Place where award treated as made

53. Unless otherwise agreed by the parties, where the seat of the arbitration is in England and Wales or Northern Ireland, any award in the proceedings shall be treated as made there, regardless of where it was signed, despatched or delivered to any of the parties.

Date of award

54.(1)Unless otherwise agreed by the parties, the tribunal may decide what is to be taken to be the date on which the award was made.

(2) In the absence of any such decision, the date of the award shall be taken to be the date on which it is signed by the arbitrator or, where more than one arbitrator signs the award, by the last of them.

Notification of award

55.(1)The parties are free to agree on the requirements as to notification of the award to the parties.

(2) If there is no such agreement, the award shall be notified to the parties by service on them of copies of the award, which shall be done without delay after the award is made.

(3) Nothing in this section affects section 56 (power to withhold award in case of non-payment).

Power to withhold award in case of Non-payment

56.(1)The tribunal may refuse to deliver an award to the parties except upon full payment of fees and expenses of the arbitrators.

(2) If the tribunal refuses on that ground to deliver an award, a party to the arbitral proceedings may (upon notice to the other parties and the tribunal) apply to the court, which may order that-

(a) the tribunal shall deliver the award on the payment into court by the applicant of the fees and expenses demanded, or such lesser amount as the court may specify,

(b) the amount of the fees and expenses properly payable shall be determined by such means and upon such terms as the court may direct, and

(c) out of the money paid into court there shall be paid out such fees and expenses as may be found to be properly payable and the balance of the money (if any) shall be paid out to the applicant.

(3) For this purpose the amount of fees and expenses properly payable is the amount the applicant is liable to pay under section 28 or any agreement relating to the payment of the arbitrators.

(4) No application to the court may be made where there is any available arbitral process for appeal or review of the amount of the fees or expenses demanded.

(5) References in this section to arbitrators include an arbitrator who has ceased to act and an umpire who has not replaced the other arbitrators.

(6) The above provisions of this section also apply in relation to any arbitral or other institution or person vested by the parties with powers in relation to the delivery of the tribunal's award.

As they so apply, the references to the fees and expenses of the arbitrators shall be construed as including the fees and expenses of that institution or person.

(7) The leave of the court is required for any appeal from a decision of the court under this section.

(8) Nothing in this section shall be construed as excluding an application under section 28 where payment has been made to the arbitrators in order to obtain the award.

Correction of award or additional award

57.(1) The parties are free to agree on the powers of the tribunal to correct an award or make an additional award.

(2) If or to the extent there is no such agreement, the following provisions apply.

(3) The tribunal may on its own initiative or on the application of a party-

(a) correct an award so as to remove any clerical mistake or error arising from an accidental slip or omission or clarify or remove any ambiguity in the award, or

(b) make an additional award in respect of any claim (including a claim for interest or costs) which was presented to the tribunal but was not dealt with in the award.

These powers shall not be exercised without first affording the other parties a reasonable opportunity to make representations to the tribunal.

(4) Any application for the exercise of those powers must be made within 28 days of the date of the award or such longer period as the parties may agree.

(5) Any correction of an award shall be made within 28 days of the date the application was received by the tribunal or, where the correction is made by the tribunal on its own initiative, within 28 days of the date of the award or, in either case, such longer period as the parties may agree.

(6) Any additional award shall be made within 56 days of the date of the original award or such longer period as the parties may agree.

(7) Any correction of an award shall form part of the award.

Effect of award

58.(1) Unless otherwise agreed by the parties, an award made by the tribunal pursuant to an arbitration agreement is final and binding both on the parties and on any persons claiming through or under them.

(2) This does not affect the right of a person the challenge the award by any available arbitral process of appeal or review or in accordance with the provisions of this Part.

Costs of the arbitration

Costs of the arbitration

59.(1) References in this Part to the costs of the arbitration are to-

(a) the arbitrators' fees and expenses,

(b) the fees and expenses of any arbitral institution concerned, and

(c) the legal or other costs of the parties.

(2) Any such reference includes the costs of or incidental to any proceedings to determine the amount of the recoverable costs of the arbitration (see section 63).

Agreement to pay costs in any event

60. An agreement which has the effect that a party is to pay the whole or part of the costs of the arbitration in any event is only valid if made after the dispute in question has arisen.

Award of costs

61.(1) The tribunal may make an award allocating the costs of the arbitration as between the parties, subject to any agreement of the parties.

(2) Unless the parties otherwise agree, the tribunal shall award costs on the general principle that costs should follow the event except where it appears to the tribunal that in the circumstances this is not appropriate in relation to the whole or part of the costs.

Effect of agreement or award about costs

62. Unless the parties otherwise agree, any obligation under an agreement between them as to how the costs of the arbitration are to be borne, or under an award allocating the costs of the arbitration, extends only to such costs as are recoverable.

The recoverable costs of the Arbitration

63.(1) The parties are free to agree what costs of arbitration are recoverable.

(2) If or to the extent there is no such agreement, the following provisions apply.

(3) The tribunal may determine by award the recoverable costs of the arbitration on such basis as it thinks fit.

If it does so, it shall specify-

(a) the basis on which it has acted, and

Arbitration Act 1996

Part I

(b) the items of recoverable costs and the amount referable to each.

(4) If the tribunal does not determine the recoverable costs of the arbitration, any party to the arbitral proceedings may apply to the court (upon notice to the other parties) which may-

(a) determine the recoverable costs of the arbitration on such basis as it thinks fit, or

(b) order that they shall be determined by such means and upon such terms as it may specify.

(5) Unless the tribunal or the court determines otherwise-

(a) the recoverable costs of the arbitration shall be determined on the basis that there shall be allowed a reasonable amount in respect of all costs reasonably incurred, and

(b) any doubt as to whether costs were reasonably incurred or were reasonable in amount shall be resolved in favour of the paying party.

(6) The above provisions have effect subjection to section 64 (recoverable fees and expenses of arbitrators).

63.(7) Nothing in this section affects any right of the arbitrators, any expert, legal adviser or assessor appointed by the tribunal, or any arbitral institution, to payment of their fees and expenses.

Recoverable fees and expenses of arbitrators

64.(1)Unless otherwise agreed by the parties, the recoverable costs of the arbitration shall include in respect of the fees and expenses of the arbitrators only such reasonable fees and expenses as are appropriate in the circumstances.

(2) If there is any question as to what reasonable fees and expenses are appropriate in the circumstances, and the matter is not already before the court on an application under section 63(4), the court may on the application of any party (upon notice to other parties)-

(a) determine the matter, or

(b) order that it be determined by such means and upon such terms as the court may specify.

(3) Subsection (1) has effect subject to any order of the court under section 24(4) or 25(3)(b) (order as to entitlement to fees or expenses in case of removal or resignation of arbitrator).

(4) Nothing in this section affects any right of the arbitrator to payment of his fees and expenses.

Power to limit recoverable costs

65.(1)Unless otherwise agreed by the parties, the tribunal may direct that the recoverable costs of the arbitration, or of any part of the arbitral proceedings, shall be limited to a specified amount.

(2) Any direction may be made or varied at any stage, but this must be done sufficiently in advance of the incurring of costs to which is relates, or the taking of any steps in the proceedings which may be affected by it, for the limit to be taken into account.

Arbitration Act 1996

Powers of the court in relation to award

Enforcement of the award

66.(1)An award made by the tribunal pursuant to an arbitration agreement may, by leave of the court, be enforced in the same manner as a judgement or order of the court to the same effect.

(2) Where leave is so given, judgement may be entered in terms of the award.

(3) Leave to enforce an award shall not be given where, or to the extent that, the person against whom it is sought to be enforced shows that the tribunal lacked substantive jurisdiction to make the award.

The right to raise such an objection may have been lost (see section 73).

1950 c.27

66.(4) Nothing in this section affects the recognition or enforcement of an award under any other enactment or rule of law, in particular under Part II of the Arbitration Act 1950 (enforcement of awards under Geneva Convention) or the provisions of Part III of this Act relating to the recognition and enforcement of awards under the New York Convention or by an action on the award.

Challenging the award: substantive jurisdiction

67.(1)A party to arbitral proceedings may (upon notice to the other parties and to the tribunal) apply to the court-

(a) challenging any award of the arbitral tribunal as to its substantive jurisdiction; or

(b) for an order declaring an award made by the tribunal on the merits to be of no effect, in whole or in part, because the tribunal did not have substantive jurisdiction.

A party may lose the right to object (see section 73) and the right to apply is subject to the restrictions in section 70(2) and (3).

(2) The arbitral tribunal may continue the arbitral proceedings and make a further award while an application to the court under this section is pending in relation to an award as to jurisdiction.

(3) On an application under this section challenging an award of the arbitral tribunal as to its substantive jurisdiction, the court may by order-

(a) confirm the award,

(b) vary the award, or

(c) set aside the award in whole or in part.

(4) The leave of the court is required for any appeal from a decision of the court under this section.

Challenging the award: serious irregularity

68.(1) A party to arbitral proceedings may (upon notice to the other parties and to the tribunal) apply to the court challenging an award in the proceedings on the ground of serious irregularity affecting the tribunal, the proceedings or the award.

Arbitration Act 1996

Part I

A party may lose the right to object (see section 73) and the right to apply is subject to the restrictions in section 70(2) and (3).

(2) Serious irregularity means an irregularity of one or more of the following kinds which the court considers has caused or will cause substantial injustice to the applicant-

(a) failure by the tribunal to comply with section 33 (general duty of tribunal);

(b) the tribunal exceeding its powers (otherwise then by exceeding its substantive jurisdiction: see section 67);

(c) failure by the tribunal to conduct the proceedings in accordance with the procedure agreed by the parties;

68.(2).(d)failure by the tribunal to deal with all the issues that were put to it;

(e) any arbitral or other institution or person vested by the parties with powers in relation to the proceedings or the award exceeding its powers;

(f) uncertainty or ambiguity as to the effect of the award;

(g) the award being obtained by fraud or the award or the way in which it was procured being contrary to public policy;

(h) failure to comply with the requirements as to the form of the award; or

(i) any irregularity in the conduct of the proceedings or in the award which is admitted by the tribunal or by any arbitral or other institution or person vested by the parties with powers in relation to the proceedings or the award.

(3) If there is shown to be serious irregularity affecting the tribunal, the proceedings or the award, the court may-

(a) remit the award to the tribunal, in whole or in part, for reconsideration,

(b) set the award aside in whole or in part, or

(c) declare the award to be of no effect, in whole or in part.

The court shall not exercise its power to set

aside or to declare an award to be of no effect, in whole or in part, unless it is satisfied that it would be inappropriate to remit the matters in question to the tribunal for reconsideration.

(4) The leave of the court is required for any appeal from a decision of the court under this section.

Appeal on point of law

69.(1) Unless otherwise agreed by the parties, a party to arbitral proceedings may (upon notice to the other parties and to the tribunal) appeal to the court on a question of law arising out of an award made in the proceedings.

An agreement to dispense with reasons for the tribunal's award shall be considered an agreement to exclude the court's jurisdiction under this section.

Arbitration Act 1996

(2) An appeal shall not be brought under this section except-

(a) with the agreement of all the other parties to the proceedings, or

(b) with the leave of the court.

The right to appeal is also subject to the restrictions in section 70(2) and (3).

(3) Leave to appeal shall be given only if the court is satisfied-

69.(3).(a)that the determination of the question will substantially affect the rights of one or more of the parties,

(b) that the question is one which the tribunal was asked to determine,

(c) that, on the basis of the findings of fact in the award-

(i) the decision of the tribunal on the question is obviously wrong, or

(ii) the question is one of general public importance and the decision of the tribunal is at least open to serious doubt, and

(d) that, despite the agreement of the parties to resolve the matter by arbitration, it is just and proper in all the circumstances for the court to determine the question.

(4) An application for leave to appeal under this section shall identify the question of law to be determined and state the grounds on which it is alleged that leave to appeal should be granted.

(5) The court shall determine an application for leave to appeal under this section without a hearing unless it appears to the court that a hearing is required.

(6) The leave of the court is required for any appeal from a decision of the court under this section to grant or refuse leave to appeal.

(7) On an appeal under this section the court may by order-

(a) confirm the award,

(b) vary the award,

(c) remit the award to the tribunal, in whole or in part, for reconsideration in the light of the court's determination, or

(d) set aside the award in whole or in party.

The court shall not exercise its power to set aside an award, in whole or in part, unless it is satisfied that it would be inappropriate to remit the matters in question to the tribunal for reconsideration.

(8) The decision of the court on an appeal under this section shall be treated as a judgement of the court for the purposes of a further appeal.

Arbitration Act 1996

Part I

But no such appeal lies without the leave of the court which shall not be given unless the court considers that the question is one of general importance or is one which for some other special reason should be considered by the Court of Appeal.

Challenge or appeal: supplementary provisions

70.(1) The following provisions apply to an application or appeal under section 67, 68 or 69.

(2) An application or appeal may not be brought if the applicant or appellant has not first exhausted-

70.(2).(a)any available arbitral process of appeal or review, and

(b) any available recourse under section 57 (correction of award or additional award).

(3) Any application or appeal must be brought within 28 days of the date of the award or, if there has been any arbitral process of appeal or review, of the date when the applicant or appellant was notified of the result of that process.

(4) If on an application or appeal it appears to the court that the award-

(a) does not contain the tribunal's reasons, or

(b) does not set out the tribunal's reasons in sufficient detail to enable the court properly to consider the application or appeal,

the court may order the tribunal to state the reasons for its award in sufficient detail for that purpose.

(5) Where the court makes an order under subsection (4), it may make such further order as it thinks fit with respect to any additional costs of the arbitration resulting from its order.

(6) The court may order the applicant or appellant to provide security for the costs of the application or appeal, and may direct that the application or appeal be dismissed if the order is not complied with.

The power to order security for costs shall not be exercised on the ground that the applicant or appellant is-

(a) an individual ordinarily resident outside the United Kingdom, or

(b) a corporation or association incorporated or formed under the law of a country outside the United Kingdom, or whose central management and control is exercised outside the United Kingdom.

(7) The court may order that any money payable under the award shall be brought into court or otherwise secured pending the determination of the application or appeal, and may direct that the application or appeal be dismissed if the order is not complied with.

(8) The court may grant leave to appeal subject to conditions to the same or similar effect as an order under subsection (6) or (7).

Arbitration Act 1996

This does not affect the general discretion of the court to grant leave subject to conditions.

Challenge or appeal: effect of order of court

71.(1) The following provisions have effect where the court makes an order under section 67, 68 or 69 with respect to an award.

(2) Where the award is varied, the variation has effect as part of the tribunal's award.

73(3)Where the award is remitted to the tribunal, in whole or in part, for reconsideration, the tribunal shall make a fresh award in respect of the matters remitted within three months of the date of the order for remission or such longer or shorter period as the court may direct.

(4) Where the award is set aside or declared to be of no effect, in whole or in part, the court may also order that any provision that an award is a condition precedent to the bringing of legal proceedings in respect of a matter to which the arbitration agreement applies, is of no effect as regards the subject matter of the award or, as the case may be, the relevant part of the award.

Miscellaneous

Saving for rights of person who takes no part in proceedings

72.(1) A person alleged to be a party to arbitral proceedings but who takes no part in the proceedings may question-

(a) whether there is a valid arbitration agreement,

(b) whether the tribunal is properly constituted, or

I

(c) what matters have been submitted to arbitration in accordance with the arbitration agreement,

by proceedings in the court for a declaration or injunction or other appropriate relief.

(2) He also has the same right as a party to the arbitral proceedings to challenge an award-

(a) by an application under section 67 on the ground of lack of substantive jurisdiction in relation to him, or

(b) by an application under section 68 on the ground of serious irregularity (within the meaning of that section) affecting him;

and section 70(2) (duty to exhaust arbitral procedures) does not apply in his case.

Loss of right to object

73.(1)If a party to arbitral proceedings takes part, or continues to take part, in the proceedings without making, either forthwith or within such time as is allowed by the arbitration agreement or the tribunal or by any provision of this Part, any objection-

(a) that the tribunal lacks substantive jurisdiction,

(b) that the proceedings have been improperly conducted,

(c) that there has been a failure to comply with the arbitration agreement or with any provision of this Part, or

Arbitration Act 1996

Part I

(d) that there has been any other irregularity affecting the tribunal or the proceedings.

73.(1) (Cont'd) He may not raise that objection later, before the tribunal or the court, unless he shows that, at the time he took part or continued to take part in the proceedings, he did not know and could not with reasonable diligence have discovered the grounds for the objection.

(2) Where the arbitral tribunal rules that it has substantive jurisdiction and a party to arbitral proceedings who could have questions that ruling-

(a) by any available arbitral process of appeal or review, or

(b) by challenging the award,

does not do so, or does not do so within the time allowed by the arbitration agreement or any provision of this Part, he may not object later to the tribunal's substantive jurisdiction on any ground which was the subject of that ruling.

Immunity of arbitral institutions, &c

74.(1) An arbitral or other institution or person designated or requested by the parties to appoint or nominate an arbitrator is not liable for anything done or omitted in the discharge or purported discharge of that function unless the act or omission is shown to have been in bad faith.

(2) An arbitral or other institution or person by whom an arbitrator is appointed or nominated is not liable, by reason of having appointed or nominated him, for anything done or omitted by the arbitrator (or his employees or agents) in the discharge or purported discharge of his functions as arbitrator).

(3) The above provisions apply to an employee or agent of an arbitral or other institution or person as they apply to the institution or person himself.

Charge to secure payment of solicitors' costs
1974 c.47
S.I. 1976/582 (N.I.12)

75. The powers of the court to make declarations and orders under section 73 of the Solicitors Act 1974 or Article 71H of the Solicitors (Northern Ireland) Order 1976 (power to charge property recovered in the proceedings with the payment of solicitors' costs) may be exercised in relation to arbitral proceedings as if those proceedings were proceedings in the court.

Supplementary

Service of notices, &c

76.(1) The parties are free to agree on the manner of service of any notice or other document required or authorised to be given or served in pursuance of the arbitration agreement or for the purposes of the arbitral proceedings.

(2) If or to the extent that there is no such agreement the following provisions apply.

(3) A notice or other document may be served on a person by any effective means.

(4) If a notice or other document is addressed, pre-paid and delivered by post-

Arbitration Act 1996

(a) to the addressee's last known principal residence or, if he is or has been carrying on a trade, profession or business, his last known principal business address, or

(b) where the addressee is a body corporate, to the body's registered or principal office,

it shall be treated as effectively served.

(5) This section does not apply to the service of documents for the purposes of legal proceedings, for which provision is made by rules of court.

(6) References in this Part to a notice or other document include any form of communication in writing and references to giving or serving a notice or other document shall be construed accordingly.

Powers of court in relation to service of documents

77.(1) this section applies where service of a document on a person in the manner agreed by the parties, or in accordance with provisions of section 76 having effect in default of agreement, is not reasonably practicable.

(2) Unless otherwise agreed by the parties, the court may make such order as it thinks fit-

(a) for service in such manner as the court may direct, or

(b) dispensing with service of the document.

(3) Any party to the arbitration agreement may apply for an order, but only after exhausting any available arbitral process for resolving the matter.

(4) The leave of the court is required for any appeal from a decision of the court under this section.

Reckoning periods of time

78.(1) The parties are free to agree on the method of reckoning periods of time for the purposes of any provision agreed by them or any provision of this Part having effect in default of such agreement.

(2) If or to the extent there is no such agreement, periods of time shall be reckoned in accordance with the following provisions.

(3) Where the act is required to be done within a specified period after or from a specified date, the period begins immediately after that date.

(4) Where the act is required to be done a specified number of clear days after a specified date, at least that number of days must intervene between the day on which the act is done and that date.

(5) Where the period is a period of seven days or less which would include a Saturday, Sunday or a public holiday in the place where anything which has to be done within the period falls to be done, that day shall be excluded.

1971 c.80

In relation to England and Wales or Northern Ireland, a "public holiday" means Christmas Day, Good Friday or a day which under the Banking and Financial Dealings Act 1971 is a bank holiday.

Arbitration Act 1996

Part I

Power of court to extend time limits relating to arbitral proceedings

79.(1) Unless the parties otherwise agree, the court may by order extend any time limit agreed by them in relation to any matter relating to the arbitral proceedings or specified in any provision of this Part having effect in default of such agreement.

This section does not apply to a time limit to which section 12 applies (power of court to extend time for beginning arbitral proceedings, &c).

(2) An application for an order may be made-

(a) by any party to the arbitral proceedings (upon notice to the other parties and to the tribunal), or

(b) by the arbitral tribunal (upon notice to the parties).

(3) The court shall not exercise its power to extend a time limit unless it is satisfied-

(a) that any available recourse to the tribunal, or to any arbitral or other institution or person vested by the parties with power in that regard, has first been exhausted, and

(b) that a substantial injustice would otherwise be done.

(4) The court's power under this section may be exercised whether or not the time has already expired.

(5) An order under this section may be made on such terms as the court thinks fit.

(6) The leave of the court is required for any appeal from a decision of the court under this section.

Notice and other requirements in connection with legal proceedings

80.(1) References in this Part to an application, appeal or other step in relation to legal proceedings being taken "upon notice" to the other parties to the arbitral proceedings, or to the tribunal, are to such notice of the originating process as is required by rules of court and do not impose any separate requirement.

(2) Rules of court shall be made-

(a) requiring such notice to be given as indicated by any provision of this Part, and

(b) as to the manner, form and content of any such notice.

(3) Subject to any provision made by rules of court, a requirement to give notice to the tribunal of legal proceedings shall be construed-

(a) if there is more than one arbitrator, as a requirement to give notice to each of them; and

(b) if the tribunal is not fully constituted, as a requirement to give notice to any arbitrator who has been appointed.

(4) References in this Part to making an application or appeal to the court within a specified period are to the issue within that period of the appropriate originating process in accordance with rules of court.

Arbitration Act 1996

(5) Where any provision of this Part requires an application or appeal to be made to the court within a specified time, the rules of the court relating to the reckoning of periods, the extending or abridging of periods, and the consequences of not taking a step within the period prescribed by the rules, apply in relation to that requirement.

(6) Provision may be made by rules of court amending the provisions of this Part-

(a) with respect to the time within which any application or appeal to the court must be made,

(b) so as to keep any provision made by this Part in relation to arbitral proceedings in step with the corresponding provision of rules of court applying in relation to proceedings in the court, or

(c) so as to keep any provision made by this Part in relation to legal proceedings in step with the corresponding provision of rules of court applying generally in relation to proceedings in the court.

(7) Nothing in this section affects the generality of the power to make rules of court.

Saving for certain matters governed by common law

81.(1) Nothing in this Part shall be construed as excluding the operation of any rule of law consistent with the provisions of this Part, in particular, any rule of law as to-

(a) matters which are not capable of settlement by arbitration;

(b) the effect of an oral arbitration agreement; or

(c) the refusal of recognition or enforcement of an arbitral award on grounds of public policy.

(2) Nothing in this Act shall be construed as reviving any jurisdiction of the court to set aside or remit an award on the ground of errors of fact or law on the face of the award.

Minor definitions

82.(1) In this Part-

"arbitrator", unless the context otherwise requires, includes an umpire;

"available arbitral process", in relation to any matter, includes any process of appeal to or review by an arbitral or other institution or person vested by the parties with powers in relation to that matter;

"claimant", unless the context otherwise requires, includes a counter claimant, and related expressions shall be construed accordingly;

"dispute" includes any difference;

"enactment" includes an enactment contained in Northern Ireland legislation;

"legal proceedings" means civil proceedings in the High Court or a county court;

Arbitration Act 1996

Part I

"peremptory order" means an order made under section 41(5) or made in exercise of any corresponding power conferred by the parties;

"premises" includes land, buildings, moveable structures, vehicles, vessels, aircraft and hovercraft;

"question of law" means-

(a) for a court in England and Wales, a question of the law of England and Wales, and

(b) for a court in Northern Ireland, a question of the law of Northern Ireland;

"substantive jurisdiction", in relation to an arbitral tribunal, refers to the matters specified in section 30(1)(a) to (c), and references to the tribunal exceeding its substantive jurisdiction shall be construed accordingly.

(2) References in this Part to a party to an arbitration agreement include any person claiming under or through a party to the agreement.

Index of defined expressions: Part I

83. In this Part the expressions listed below are defined or otherwise explained by the provisions indicated-

agreement, agree and agreed	section 5(1)
agreement in writing	section 5(2) to (5)
arbitration agreement	sections 6 and 5(1)
arbitrator	section 82(1)
available arbitral process	section 82(1)
claimant	section 82(1)
commencement (in relation to arbitral proceedings)	section 14
costs of the arbitration	section 59
the court	section 105
dispute	section 82(1)
enactment	section 82(1)
legal proceedings	section 82(1)
Limitation Acts	section 13(4)
notice (or other document)	section 76(6)
party-	
- in relation to an arbitration agreement	section 82(2)
- where section 106(2) or (3) applies	section 106(4)
peremptory order	section 82(1) (and see section 41(5)
premises	section 82(1)
question of law	section 82(1)
recoverable costs	sections 63 and 64
seat of the arbitration	section 3
serve and service (of notice or other document)	section 76(6)
substantive jurisdiction (in relation to an arbitral tribunal)	section 82(1) (and see section 30(1)(a) to (c))
upon notice (to the parties or the tribunal)	section 80
written and in writing	section 5(6)

Arbitration Act 1996

Transitional provisions

84.(1)The provisions of this Part do not apply to arbitral proceedings commenced before the date on which this Part comes into force.

(2) They apply to arbitral proceedings commenced on or after that date under an arbitration agreement whenever made.

(3) The above provisions have effect subject to any transitional provision made by an order under section 109(2) (power to include transitional provisions in commencement order).

PART II

OTHER PROVISIONS RELATING TO ARBITRATION

Part II

Domestic arbitration agreements
(Note Sections 85 to 87 are not yet in force)

Modification of Part I in relation to domestic arbitration agreement

85.(1)In the case of a domestic arbitration agreement the provisions of Part I are modified in accordance with the following sections.

(2) For this purpose a "domestic arbitration agreement" means an arbitration agreement to which none of the parties is-

(a) an individual who is a national of, or habitually resident in, a state other than the United Kingdom, or

(b) a body corporate which is incorporated in, or whose central control and management is exercised in, a state other than the United Kingdom,

and under which the seat of the arbitration (if the seat has been designated or determined) is in the United Kingdom.

(3) In subsection (2) "arbitration agreement" and "seat of the arbitration" have the same meaning as in Part I (see sections 3, 5(1) and 6).

Staying of legal proceedings

86.(1)In section 9 (stay of legal proceedings), subsection (4) (stay unless the arbitration agreement is null and void, inoperative, or incapable of being performed) does not apply to a domestic arbitration agreement.

(2) On an application under that section in relation to a domestic arbitration agreement the court shall grant a stay unless satisfied-

(a) that the arbitration agreement is null and void, inoperative, or incapable of being performed, or

(b) that there are other sufficient grounds for not requiring the parties to abide by the arbitration agreement.

(3) The court may treat as a sufficient ground under subsection (2)(b) the fact that the applicant is or was at any material time not ready and willing to do all things necessary for the proper conduct of the arbitration or of any other dispute resolution procedures required to be exhausted before resorting to arbitration.

(4) For the purposes of this section the question whether an arbitration agreement is a domestic arbitration agreement shall be determined by reference to the facts at the time the legal proceedings are commenced.

Part II

Effectiveness of agreement to exclude court's jurisdiction

87.(1)In the case of a domestic arbitration agreement any agreement to exclude the jurisdiction of the court under-

(a) section 45 (determination of preliminary point of law), or

(b) section 69 (challenging the award: appeal on point of law),

is not effective unless entered into after the commencement of the arbitral proceedings in which the question arises or the award is made.

(2) For this purpose the commencement of the arbitral proceedings has the same meaning as in Part I (see section 14).

(3) For the purposes of this section the question whether an arbitration agreement is a domestic arbitration agreement shall be determined by reference to the facts at the time the agreement is entered into.

Power to repeal or amend sections 85 to 87

88.(1)The Secretary of State may by order repeal or amend the provisions of sections 85 to 87.

(2) An order under this section may contain such supplementary, incidental and transitional provisions as appear to the Secretary of State to be appropriate.

(3) An order under this section shall be made by statutory instrument and no such order shall be made unless a draft of it has been laid before and approved by a resolution of each House of Parliament.

Consumer arbitration agreements

Application of unfair terms regulations to consumer arbitration agreements S.I.1994/3159

89.(1)The following sections extend the application of the Unfair Terms in Consumer Contracts Regulations 1994 in relation to a term which constitutes an arbitration agreement

For this purpose "arbitration agreement" means an agreement to submit to arbitration present or future disputes or differences (whether or not contractual).

(2) In those sections "the Regulations" means those regulations and includes any regulations amending or replacing those regulations.

(3) Those sections apply whatever the law applicable to the arbitration agreement.

Regulations apply where consumer is a legal person

90. The Regulations apply where the consumer is a legal person as they apply where the consumer is a natural person.

Arbitration agreement unfair where modest amount sought

91.(1)A term which constitutes an arbitration agreement is unfair for the purposes of the Regulations so far as it relates to a claim for a pecuniary remedy which does not exceed the amount specified by order for the purposes of this section.

(2) Orders under this section may make different provision for different cases and for different purposes.

(3) The power to make orders under this section is exercisable-

Arbitration Act 1996

(a) for England and Wales, by the Secretary of State with the concurrence of the Lord Chancellor

(b) for Scotland, by the Secretary of State with the concurrence of the Lord Advocate, and

(c) for Northern Ireland, by the Department of Economic Development for Northern Ireland with the concurrence of the Lord Chancellor.

(4) Any such order for England and Wales or Scotland shall be made by statutory instrument which shall be subject to annulment in pursuance of a resolution of either House of Parliament.

S.I.1979/1573 (N.I.12) 1954 c.33 (N.I.)

(5) Any such order for Northern Ireland shall be a statutory rule for the purposes of the Statutory Rules (Northern Ireland) Order 1979 and shall be subject to negative resolution, within the meaning of section 41(6) of the Interpretation Act (Northern Ireland) 1954.

Small claims arbitration in the county court

Exclusion of Part I in relation to small claims arbitration in the county court 1984 c.28

92. Nothing in Part I of this Act applies to arbitration under section 64 of the County Courts Act 1984.

Appointment of judges as arbitrators

Part II

Appointment of judges as arbitrators

93.(1)A judge of the Commercial Court or an official referee may, if in all the circumstances he thinks fit, accept appointment as a sole arbitrator or as umpire by or by virtue of an arbitration agreement.

(2) A judge of the Commercial Court shall not do so unless the Lord Chief Justice has informed him that, having regard to the state of business in the High Court and the Crown Court, he can be made available.

(3) An official referee shall not do so unless the Lord Chief Justice has informed him that, having regard to the state of official referees' business, he can be made available.

(4) The fees payable for the services of a judge of the Commercial Court or official referee as arbitrator or umpire shall be taken in the High Court.

(5) In this section-

"arbitration agreement" has the same meaning as in Part I; and

1981 c.54

"official referee" means a person nominated under section 68(1)(a) of the Supreme Court Act 1981 to deal with official referees' business.

(6) The provisions of Part I of this Act apply to arbitration before a person appointed under this section with the modifications specified in Schedule 2.

Statutory arbitrations

Application of

94.(1)The provisions of Part I apply to every

Arbitration Act 1996

Part II
Part I to statutory arbitrations

arbitration under an enactment (a "statutory arbitration"), whether the enactment was passed or made before or after the commencement of this Act, subject to the adaptations and exclusions specified in sections 95 to 98.

(2) The provisions of Part I do not apply to a statutory arbitration if or to the extent that their application-

(a) is inconsistent with the provisions of the enactment concerned, with any rules or procedure authorised or recognised by it, or

(b) is excluded by any other enactment.

(3) In this section and the following provisions of this Part "enactment"-

1978 c.30

(a) in England and Wales, includes an enactment contained in subordinate legislation within the meaning of the Interpretation Act 1978;

1954 c.33 (N.I.)

(b) in Northern Ireland, means a statutory provision within the meaning of section 1(f) of the Interpretation Act (Northern Ireland) 1954.

General adaptations of provisions in relation to statutory arbitrations

95.(1)The provisions of Part I apply to a statutory arbitration-

(a) as if the arbitration were pursuant to an arbitration agreement and as if the enactment were that agreement, and

(b) as if the persons by and against whom a claim subject to arbitration in pursuance of the enactment may be or has been made were parties to that agreement.

(2) Every statutory arbitration shall be taken to have its seat in England and Wales or, as the case may be, Northern Ireland.

Specific adaptations of provisions in relation to statutory arbitrations

96.(1)The following provisions of Part I apply to a statutory arbitration with the following adaptations.

(2) In section 30(1) (competence of tribunal to rule on its own jurisdiction), the reference in paragraph (a) to whether there is a valid arbitration agreement shall be construed as a reference to whether the enactment applies to the dispute or difference in question.

(3) Section 35 (consolidation of proceedings and concurrent hearings) applies only so as to authorise the consolidation of proceedings, or concurrent hearings in proceedings, under the same enactment.

(4) Section 46 (rules applicable to substance of dispute) applies with the omission of subsection (1)(b) (determination in accordance with considerations agreed by parties).

Provisions excluded from applying to statutory arbitrations

97. The following provisions of Part I do not apply in relation to a statutory arbitration-

(a) section 8 (whether agreement discharged by death of a party);

Arbitration Act 1996

(b) section 12 (power of court to extend agreed time limits;

(c) sections 9(5), 10(2) and 71(4) (restrictions on effect of provision that award condition precedent to right to bring legal proceedings).

Power to make further provision by regulations

98.(1)The Secretary of State may make provision by regulations for adapting or excluding any provision of Part I in relation to statutory arbitrations in general or statutory arbitrations of any particular description.

(2) The power is exercisable whether the enactment concerned is passed or made before or after the commencement of this Act.

(3) Regulations under this section shall be made by statutory instrument which shall be subject to annulment in pursuance of a resolution of either House of Parliament.

PART III

RECOGNITION AND ENFORCEMENT OF CERTAIN FOREIGN AWARDS

Enforcement of Geneva Convention awards

Continuation of Part II of the Arbitration Act 1950

1950 c.27

99. Part II of the Arbitration Act 1950 (enforcement of certain foreign awards) continues to apply in relation to foreign awards within the meaning of that Part which are not also New York Convention awards

Recognition and enforcement of New York Convention awards

New York Convention awards

100.(1)In this Part a "New York Convention award" means an award made, in pursuance of an arbitration agreement, in the territory of a state (other than the United Kingdom) which is a party to the New York Convention.

(2) For the purposes of subsection (1) and of the provisions of this Part relating to such awards-

(a) "arbitration agreement" means an arbitration agreement in writing, and

(b) an award shall be treated as made at the seat of arbitration, regardless of where it was signed, despatched or delivered to any of the parties.

In this subsection "agreement in writing" and "seat of the arbitration" have the same meaning as in Part I.

(3) If Her Majesty by Order in Council declares that a state specified in the Order is a party to the New York Convention, or is a party in respect of any territory so specified, the Order shall, while in force, be conclusive evidence of that fact.

(4) In this section, "the New York Convention" means the Convention on the Recognition and Enforcement of Foreign Arbitral Awards adopted by the United Nations Conference on International Commercial Arbitration on 10th June 1958.

Arbitration Act 1996

Part III

Recognition and enforcement of awards

101.(1)A New York Convention award shall be recognised as binding on the persons as between whom it was made, and may accordingly be relied on by those persons by way of defence, set-off or otherwise in any legal proceedings in England and Wales or Northern Ireland.

(2) A New York Convention award may, by leave of the court, be enforced in the same manner as a judgement or order of the court to the same effect.

As to the meaning of "the court" see section 105.

(3) Where leave is so given, judgement may be entered in terms of the award.

Evidence to be produced by party seeking recognition or enforcement

102.(1)A party seeking the recognition or enforcement of a New York Convention award must produce-

(a) the duly authenticated original award or a duly certified copy of it, and

(b) the original arbitration agreement or a duly certified copy of it.

(2) If the award or agreement is in a foreign language, the party must also produce a translation of it certified by an official or sworn translator or by a diplomatic or consular agent.

Refusal of recognition or enforcement

103.(1)Recognition or enforcement of a New York Convention award shall not be refused except in the following cases.

(2) Recognition or enforcement of the award may be refused if the person against whom it invoked proves-

(a) that a party to the arbitration agreement was (under the law applicable to him) under some incapacity;

(b) that the arbitration agreement was not valid under the law to which the parties subjected it or, failing any indication thereon, under the law of the country where the award was made;

(c) that he was not given proper notice of the appointment of the arbitrator or of the arbitration proceedings or was otherwise unable to present his case;

(d) that the award deals with a difference not contemplated by or not falling within the terms of the submission to arbitration or contains decisions on matters beyond the scope of the submission to arbitration (but see subsection (4));

(e) that the composition of the arbitral tribunal or the arbitral procedure was not in accordance with the agreement of the parties or, failing such agreement, with the law of the country in which the arbitration took place;

(f) that the award has not yet become binding on the parties, or has been set aside or suspended by a competent authority of the country in which, or under the law of which, it was made.

(3) Recognition or enforcement of the award may also be refused if the award is in respect of a matter which is not capable of settlement by arbitration, or if it would be contrary to public policy to recognise or enforce the award.

(4) An award which contains decisions on matters not submitted to arbitration may be recognised or enforced to the extent that it contains decisions on matters submitted to arbitration which can be separated from those on matters not so submitted.

(5) Where an application for the setting aside or suspension of the award has been made to such a competent authority as is mentioned in subsection (2)(f), the court before which the award is sought to be relied upon, may if it considers it proper, adjourn the decision on the recognition or enforcement of the award.

It may also on the application of the party claiming recognition or enforcement of the award order the other party to give suitable security.

Saving for other bases of recognition or enforcement

104 Nothing in the preceding provisions of this Part affects any right to rely upon or enforce a New York Convention award at common law or under section 66.

PART IV

GENERAL PROVISIONS

Meaning of "the court": jurisdiction of High Court and county court

105.(1)In this Act "the court" means the High Court or a county court, subject to the following provisions.

(2) The Lord Chancellor may be order make provision-

(a) allocating proceedings under this Act to the High Court or to county courts; or

(b) specifying proceedings under this Act which may be commenced or taken only in the High Court or in a county court.

(3) The Lord Chancellor may by order make provision requiring proceedings of any specified description under this Act in relation to which a county court has jurisdiction to be commenced or taken in one or more specified county courts.

Any jurisdiction so exercisable by a specified county court is exercisable throughout England and Wales or, as the case may be, Northern Ireland.

(4) An order under this section-

(a) may differentiate between categories of proceedings by reference to such criteria as the Lord Chancellor sees fit to specify, and

(b) may make such incidental or transitional provision as the Lord Chancellor considers necessary or expedient.

(5) An order under this section for England and Wales shall be made by statutory instrument which shall be subject to annulment in pursuance of a resolution of either House of Parliament.

Arbitration Act 1996

Part IV

.S.I. 1979/1573 (N.I. 12) 1946 c.36

(6) An order under this section for Northern Ireland shall be a statutory rule for the purposes of the Statutory Rules (Northern Ireland) Order 1979 which shall be subject to annulment in pursuance of a resolution of either House of Parliament in like manner as a statutory instrument and section 5 of the Statutory Instruments Act 1946 shall apply accordingly.

Crown application

106.(1)Part I of this Act applies to any arbitration agreement to which Her Majesty, either in right of the Crown or of the Duchy of Lancaster or otherwise, or the Duke of Cornwall is a party.

(2) Where Her Majesty is party to an arbitration agreement otherwise than in right of the Crown, Her Majesty shall be represented for the purposes of any arbitral proceedings-

(a) where the agreement was entered into by Her Majesty in right of the Duchy of Lancaster, by the Chancellor of the Duchy or such person as he may appoint, and

(b) in any other case, by such person as Her Majesty may appoint in writing under the Royal Sign Manual.

(3) Where the Duke of Cornwall is party to an arbitration agreement, he shall be represented for the purposes of any arbitral proceedings by such person as he may appoint.

(4) References in Part I to a party or the parties to the arbitration agreement or to arbitral proceedings shall be construed, where subsection (2) or (3) applies, as references to the person representing Her Majesty or the Duke of Cornwall.

Consequential amendments and repeals

107.(1)The enactments specified in Schedule 3 are amended in accordance with that Schedule, the amendments being consequential on the provisions of this Act.

(2) The enactments specified in Schedule 4 are repealed to the extent specified.

Extent

108.(1)The provisions of this Act extend to England and Wales and, except as mentioned below, to Northern Ireland.

(2) The following provisions of Part II do not extend to Northern Ireland-

section 92 (exclusion of Part I in relation to small claims arbitration in the county court), and

section 93 and Schedule 2 (appointment of judges as arbitrators).

(3) Sections 89, 90 and 91 (consumer arbitration agreements) extend to Scotland and the provisions of Schedules 3 and 4 (consequential amendments and repeals) extend to Scotland so far as they relate to enactments which so extend, subject as follows.

Arbitration Act 1996

1975 c.3.

(4) The repeal of the Arbitration Act 1975 extends only to England and Wales and Northern Ireland.

Commencement

109.(1)The provisions of this Act come into force on such day as the Secretary of State may appoint by order made by statutory instrument, and different days may be appointed for different purposes.

(2) An order under subsection (1) may contain such transitional provisions as appear to the Secretary of State to be appropriate.

Short title

110. This Act may be cited as the Arbitration Act 1996.

Please note: **THE SCHEDULES TO THE ARBITRATION ACT ARE NOT REPRODUCED HERE.**

The Housing Grants, Construction and Regeneration Act 1996

This act, provides a new right to fast impartial ADJUDICATION usually within 28 days.

The Act requires relevant contracts to provide for-

Adjudication for disputes which may be binding until the matter is resolved by litigation, arbitration or agreement.

Periodic payments for contracts greater than 45 days.

Notice to be given of intention to withhold payments.

A right to suspend performance for non payment.

The "Scheme for Construction Contracts" will provide for the matters raised by the act including supplementing contracts that do not provide for the Act.

The following extracts from the Act should be noted-

PART II

CONSTRUCTION CONTRACTS

Introductory provisions

104. Construction contracts.
105. Meaning of "construction operations".
106. Provisions not applicable to contract with residential occupier.
107. Provisions applicable only to agreements in writing.

Adjudication

108. Right to refer disputes to adjudication.

Payment

109. Entitlement to stage payments.
110. Dates for payment.
111. Notice of intention to withhold payment.
112. Right to suspend performance for non-payment.
113. Prohibition of conditional payment provisions.

Supplementary provisions

114. The Scheme for Construction Contracts.
115. Service of notices, etc.
116. Reckoning periods of time.
117. Crown application.

Adjudication

Right to refer disputes to adjudication

108.(1) A party to a construction contract has the right to refer a dispute arising under the contract for adjudication under a procedure complying with this section.

1996 c.23.

For this purpose "dispute" includes any difference.

(2) The contract shall-

(a) enable a party to give notice at any time of his intention to refer a dispute to adjudication;

(b) provide a timetable with the object of securing the appointment of the adjudicator and referral of the dispute to him within 7 days of such notice;

(c) require the adjudicator to reach a decision within 28 days of referral or such longer period as is agreed by the parties after the dispute has been referred;

(d) allow the adjudicator to extend the period of 28 days by up to 14 days, with the consent of the party by whom the dispute was referred;

(e) impose a duty on the adjudicator to act impartially; and

(f) enable the adjudicator to take the initiative in ascertaining the facts and the law.

(3) The contract shall provide that the decision of the adjudicator is binding until the dispute is finally determined by legal proceedings, by arbitration (if the contract provides for arbitration or the parties otherwise agree to arbitration) or by agreement.

The parties may agree to accept the decision of the adjudicator as finally determining the dispute.

(4) The contract shall also provide that the adjudicator is not liable for anything done or omitted in the discharge or purported discharge of his functions as adjudicator unless the act or omission is in bad faith, and that any employee or agent of the adjudicator is similarly protected from liability.

(5) If the contract does not comply with the requirements of subsections (1) to (4), the adjudication provisions of the Scheme for Construction Contracts apply.

(6) For England and Wales, the Scheme may apply the provisions of the Arbitration Act 1996 with such adaptations and modifications as appear to the Minister making the scheme to be appropriate.

For Scotland, the Scheme may include provision conferring powers on courts in relation to adjudication and provision relating to the enforcement of the adjudicator's decision.

TABLES AND MEMORANDA

AVERAGE COVERAGE OF PAINTS

The following information has been provided by the British Decorators Association, whose permission to publish is hereby acknowledged.

In this revision a range of spreading capacities is given. Figures are in square metres per litre, except for oil-bound water paint and cement-based paint which are in square metres per kilogram.

For comparative purposes figures are given for a single coat, but users are recommended to follow manufactures' recommendations as to when to use single or multicoat systems.

It is emphasised that the figures quoted in the schedule are practical figures for brush application, achieved in big scale painting work and take into account losses and wastage. They are not optima figures based upon ideal conditions of surface, nor minima figures reflecting the reverses of these conditions.

There will be instances when the figures indicated by paint manufacturers in their literature will be higher than those shown in the schedule. The Committee realise that under ideal conditions of application, and depending on such factors as the skill of the applicator, type and quality of the product, substantially better covering figures can be achieved.

The figures given below are for application by brush and to appropriate systems on each surface. They are given for guidance and are qualified to allow for variation depending on certain factors.

SCHEDULE OF AVERAGE COVERAGE OF PAINTS IN SQUARE METRES PER LITRE

Coating per litre	Finishing plaster	Wood floated rendering	Smooth concrete/ cement	Fair faced brick-work	Block-work	Roughcast pebble-dash	Hardboard	Soft fibre insulating board
Water thinned primer/undercoat								
as primer........................	13-15	-	-	-	-	-	10-12	7-10
as undercoat....................	-	-	-	-	-	-	-	10-12
Plaster primer (including building board)................	9-11	8-12	9-11	7-9	5-7	2-4	8-10	7-9
Alkali resistant primer.......	7-11	6-8	7-11	6-8	4-6	2-4	-	-
External wall primer sealer.	6-8	6-7	6-8	5-7	4-6	2-4	-	-
Undercoat.......................	11-14	7-9	7-9	6-8	6-8	3-4	11-14	10-12
Oil based thixotropic finish	Figures should be obtained from individual manufacturers							
Eggshell/semi-gloss finish (oil based)........................	11-14	9-11	11-14	8-10	7-9	-	10-13	10-12
Acrylic eggshell	11-14	10-12	11-14	8-11	7-10	-	11-14	10-12
Emulsion paint								
standard..........................	12-15	8-12	11-14	8-12	6-10	2-4	12-15	8-10
contract...........................	10-12	7-11	10-12	7-10	5-9	2-4	10-12	7-9
Glossy emulsion.	Figures should be obtained from individual manufacturers							
Heavy textured coating....	2-4	2-4	2-4	2-4	2-4	-	2-4	2-4
Masonry paint...............	5-7	4-6	5-7	4-6	3-5	2-4	-	-
per kilogram								
Oil bound water paint......	7-9	6-8	7-9	6-8	5-7	-	-	-
Cement based paint..........	-	4-6	6-7	3-6	3-6	2-3	-	-

AVERAGE COVERAGE OF PAINTS (cont'd)

SCHEDULE OF AVERAGE COVERAGE OF PAINTS IN SQUARE METRES PER LITRE (cont'd)

Coating per litre	Fire retardent fibre insulating board	Smooth paper faced board	Hard asbestos sheet **	Structural steel work	Metal sheeting	Joinery	Smooth primed surfaces	Smooth under-coated surfaces
Woodprimer (oil based)......	-	-	-	-	-	8-11	-	-
Water thinned primer/undercoat								
as primer..........................	-	8-11	* *	-	-	10-14	-	-
as undercoat....................	-	10-12	-	-	-	12-15	12-15	-
Aluminium sealer +								
spirit based......................	-	-	-	-	-	7-9	-	-
oil based..........................	-	-	-	-	9-13	9-13	-	-
Metal primer								
conventional....................	-	-	-	7-10	10-13	-	-	-
Specialised......................	Figures should be obtained from individual manufacturers							
Plaster primer (including building board)...............	8-10	10-12	10-12	-	-	-	-	-
Alkali resistant primer....	-	-	8-10	-	-	-	-	-
External wall primer sealer.	-	-	6-8	-	-	-	-	-
Undercoat.....................	10-12	11-14	10-12	10-12	10-12	10-12	11-14	-
Gloss finish..................	10-12	11-14	10-12	10-12	10-12	10-12	11-14	11-14
Eggshell/semi-gloss finish (oil based)..........................	10-12	11-14	10-12	10-12	10-12	10-12	11-14	11-14
Acrylic eggshell	10-12	11-14	-	-	11-14	11-13	11-14	-
Emulsion paint								
standard...........................	8-10	12-15	10-12	-	-	10-12	12-15	12-15
contract............................	-	10-12	8-10	-	-	10-12	10-12	10-12
Heavy textured coating.....	2-4	2-4	2-4	2-4	2-4	2-4	2-4	2-4
Masonry paint..................	-	-	5-7	-	-	-	6-8	6-8
Coating per kilogram								
Oil bound water paint.......	-	-	7-9	-	-	-	-	-
Cement based paint.........	-	-	4-6	-	-	-	-	-
Glossy emulsion.	Figures should be obtained from individual manufacturers							

+ Aluminium primer/sealer is normally used over "bitumen" painted surfaces.

* On some roughcast/pebbledash surfaces appreciably lower coverage may be obtained.

In many instances the coverages achieved will be affected by the suction and texture to the backing; for example the suction and texture of brickwork can vary to such an extent that coverages outside those quoted may on occasions be obtained.

It is necessary to take these factors into account when using this table.

* * Owing to new legislation (COSHH) further advice regarding the encapsulation of asbestos should be sought.

APPROXIMATE WEIGHT OF SUNDRY MATERIALS

IMPERIAL

	lb per ft2	lb per ft3
Asbestos-cement sheeting		
1/4 in corrugated...............	3 1/4	-
1/4 in flat..........................	2 1/4	-
Asbestos-cement slating		
Diamond...........................	3	-
Rectangular......................	4	-
Asphalt 1 in thick........................	12	-
Ballast, river..................................	-	120
Bituminous felt roofing..................	1	-
Brickwork, Commons.....................	-	113
Cement, Portland..........................	-	90
Cement screeding (1:3) 1/2 in.......	6	-
Concrete, cement		
Ballast............................	-	140
Brick...............................	-	115
Clinker............................	-	95
Pumice...........................	-	70
Reinforced (about 2% steel).		150
Cork slabs 1 in thick.....................	1	-
Fibre boards		
1/2 in thick.....................	3/4	-
Compressed 1/4 in thick.	2/3	-
Fibrous plaster 5/8 in thick...........	3	-
Flooring		
1 in Magnesium Oxychloride (sawdust filler)................	7 1/2	-
1 in Magnesium Oxychloride (mineral filler).................	11 1/2	-
1/4 in Rubber..................	2 3/4	-
1 in (Nominal) softwood.. .	2 1/4	-
1 in (Nominal) pitchpine...	3	-
1 in (Nominal) hardwood..	3 1/4	-
Glass, 1/4 in plate........................	3 1/2	-
Gravel..	-	115
Lime, chalk................................	-	44

APPROXIMATE WEIGHT OF SUNDRY MATERIALS IMPERIAL (Cont'd)

Partition slabs (solid)			
	2 in Coke breeze or pumice	11 1/2	-
	2 in Clinker.......................	15	-
	2 in Terrazzo.....................	25	-
Partition slabs (hollow)			
	2 in Coke breeze or pumice	9 1/4	-
	2 in Clinker......................	11 1/2	-
	2 in Moler.........................	2 1/2	-
Plastering			
	3/4 in Lime or gypsum......	7 1/2	-
	on lathing..........................	8 3/4	-
Roof boarding, 1 in........................		2 1/2	-
Reconstructed stone......................		-	145
Sand,	pit..................................	-	90
	river................................	-	120
Shingle.......................................		-	90
Slag wool..................................		-	17
Slate	1 in slab........................	15	-
Slating, 3 in lap			
	Cornish (medium)...........	7 1/2	-
	Welsh (medium).............	8 1/2	-
	Westmorland (medium)...	11 1/2	-
	Westmorland (thin).........	9	-
Terrazzo pavings 5/8 in................		7	-
Tiling, roof			
	Machine made 4 in gauge.	13	-
	Hand made 4 in gauge......	14	-
Tiling	floor (1/2 in).....................	5 3/4	-
	wall (3/8 in).......................	4	-
Timber, seasoned			
	Elm.....................	-	39
	Baltic Fir..........................	-	35-38
	Red Pine...........................	-	40
	Yellow Pine........................	-	28
	Douglas Fir........................	-	33
	Canadian Spruce............	-	29
	White Pine.........................	-	27
	Yellow Birch.......................	-	44
	Canadian Maple.............	-	47
	Honduras Mahogany.........	-	34
	Spanish Mahogany..........	-	44
	English Oak.......................	-	45
	American Oak..................	-	47
	Baltic Oak.........................	-	46
	Indian Teak........................	-	41
	African Teak.......................	-	60
	Blackbean..........................	-	40-47
Water	....	-	62 1/2
	rain..................................	-	62 1/2
	sea..................................	-	64

WEIGHTS OF VARIOUS METALS - IMPERIAL

Weight in lb per Square Foot

	Thickness								
Inches	1/16	1/8	1/4	3/8	1/2	5/8	3/4	7/8	1
Steel	2.55	5.1	10.2	15.3	20.4	25.5	30.6	35.7	40.8
Wrought Iron	2.50	5.0	10.0	15.0	20.0	25.0	30.0	35.0	40.0
Cast iron	2.35	4.69	9.37	14.06	18.75	23.44	28.12	32.81	37.5
Brass	2.75	5.48	10.94	16.42	21.88	27.34	32.81	38.29	43.76
Copper	2.89	5.78	11.56	17.34	23.12	28.99	36.68	40.46	46.24
Cast lead	3.70	7.39	14.78	22.17	29.56	36.95	44.34	51.73	59.12

Weights of Steel Flat Bar in lb per Foot Lineal

Width	Thickness									
Inches	1/8	1/4	5/16	3/8	7/16	1/2	5/8	3/4	7/8	1
1	0.43	0.85	1.06	1.28	1.49	1.70	2.13	2.55	2.98	3.40
1 1/8	0.48	0.96	1.20	1.43	1.67	1.91	2.39	2.87	3.35	3.83
1 1/4	0.53	1.06	1.33	1.59	1.86	2.13	2.66	3.19	3.72	4.25
1 3/8	0.58	1.17	1.46	1.75	2.05	2.34	2.92	3.51	4.09	4.68
1 1/2	0.64	1.28	1.59	1.91	2.23	2.55	3.19	3.83	4.46	5.10
1 5/8	0.69	1.38	1.73	2.07	2.42	2.76	3.45	4.14	4.83	5.53
1 3/4	0.74	1.49	1.86	2.23	2.60	2.98	3.72	4.46	5.21	5.95
1 7/8	0.80	1.59	1.99	2.39	2.79	3.19	3.98	4.78	5.58	6.38
2	0.85	1.70	2.13	2.55	2.98	3.40	4.25	5.10	5.95	6.80
2 1/4	0.96	1.91	2.39	2.87	3.35	3.83	4.78	5.74	6.69	7.65
2 1/2	1.06	2.13	2.66	3.19	3.72	4.25	5.31	6.38	7.44	8.50
2 3/4	1.17	2.34	2.92	3.51	4.09	4.68	5.84	7.01	8.18	9.35
3	1.28	2.55	3.19	3.83	4.46	5.10	6.38	7.65	8.93	10.20
3 1/4	1.38	2.76	3.45	4.14	4.83	5.53	6.91	8.29	9.67	11.05
3 1/2	1.49	2.98	3.72	4.46	5.21	5.95	7.44	8.93	10.41	11.90
3 3/4	1.59	3.19	3.98	4.78	5.58	6.38	7.97	9.56	11.16	12.75
4	1.70	3.40	4.25	5.10	5.95	6.80	8.50	10.20	11.90	13.60
4 1/4	1.81	3.61	4.52	5.42	6.32	7.23	9.03	10.84	12.64	14.45
4 1/2	1.91	3.83	4.78	5.74	6.69	7.65	9.56	11.48	13.39	15.30
4 3/4	2.02	4.04	5.05	6.06	7.07	8.08	10.09	12.11	14.13	16.15
5	2.13	4.25	5.31	6.38	7.44	8.50	10.63	12.75	14.88	17.00
5 1/2	2.34	4.68	5.84	7.01	8.18	9.35	11.69	14.03	16.36	18.70
6	2.55	5.10	6.38	7.56	8.93	10.20	12.75	15.30	17.85	20.40
6 1/2	2.76	5.53	6.91	8.29	9.67	11.05	13.81	16.58	19.34	22.10
7	2.98	5.95	7.44	8.93	10.41	11.90	14.88	17.85	20.83	23.80
7 1/2	3.19	6.38	7.97	9.56	11.16	12.75	15.94	19.13	22.31	25.50
8	3.40	6.80	8.50	10.20	11.90	13.60	17.00	20.40	23.80	27.20
9	3.83	7.65	9.56	11.48	13.39	15.30	19.13	22.95	26.78	30.60
10	4.25	8.50	10.63	12.75	14.88	17.00	21.25	25.50	29.75	34.00
11	4.68	9.35	11.69	14.03	16.36	18.70	23.38	28.05	32.73	37.40
12	5.10	10.20	12.75	15.30	17.85	20.40	25.50	30.60	35.70	40.80

Weights of Steel Round and Square Bars in lb per Foot Lineal

Inches	Round	Square	Inches	Round	Square	Inches	Round	Square	Inches	Round	Square
1/8	0.042	0.053	7/8	2.044	2.603	2 3/8	15.06	19.18	5	66.76	85.00
3/16	0.094	0.120	15/16	2.347	2.988	2 1/2	16.69	21.25	5 1/4	73.60	93.71
1/4	0.167	0.213	1	2.670	3.400	2 5/8	18.40	23.43	5 1/2	80.78	102.85
5/16	0.261	0.332	1 1/8	3.380	4.303	2 3/4	20.19	25.71	5 3/4	88.29	112.41
3/8	0.376	0.478	1 1/4	4.172	5.312	2 7/8	22.07	28.10	6	96.13	122.40
7/16	0.511	0.651	1 3/8	5.049	6.428	3	24.03	30.60	6 1/4	104.31	138.81
1/2	0.668	0.849	1 1/2	6.008	7.650	3 1/4	28.21	35.91	6 1/2	112.82	143.65
9/16	0.845	1.076	1 5/8	7.051	8.978	3 1/2	32.71	41.65	6 3/4	121.67	154.88
5/8	1.043	1.328	1 3/4	8.178	10.412	3 3/4	37.55	47.81	7	130.85	-
11/16	1.262	1.607	1 7/8	9.388	11.953	4	42.73	54.40	7 1/2	150.21	-
3/4	1.502	1.912	2	10.681	13.600	4 1/4	48.23	61.41	8	170.90	-
13/16	1.763	2.245	2 1/8	12.06	15.35	4 1/2	54.07	68.85	9	216.00	-
			2 1/4	13.52	17.21	4 3/4	60.25	76.71	10	267.00	-

WEIGHTS OF STEEL JOISTS TO BS 4: PART 1: 1980

Size		Weight		Size		Weight	
inches	millimetres	lb/ft	kg/m	inches	millimetres	lb/ft	kg/m
10 x 8	254 x 203	55.0	81.85	5 x 4.5	127 x 114	18.0	26.79
10 x 4.5	254 x 114	25.0	37.20	4.5 x 4.5	114 x 114	18.0	26.79
8 x 6	203 x 152	35.0	52.09	4 x 4	102 x 102	15.5	23.07
6 x 5	152 x 127	25.0	37.20	3.5 x 3.5	89 x 89	13.0	19.35
5 x 4.5	127 x 114	20.0	29.76	3 x 3	76 x 76	8.5	12.65

WEIGHTS OF STEEL CHANNELS TO BS 4: PART 1: 1980

Size of channel		Weight		Size of channel		Weight	
in	mm	lb/ft	kg/m	in	mm	lb/ft	kg/m
17 x 4	432 x 102	44	65.54	8 x 3.5	203 x 89	20	29.78
15 x 4	381 x 102	37	55.10	8 x 3	203 x 76	16	23.82
12 x 4	305 x 102	31	46.18	7 x 3.5	178 x 89	18	26.81
12 x 3.5	305 x 89	28	41.69	7 x 3	178 x 76	14	20.84
10 x 3.5	254 x 89	24	35.74	6 x 3.5	152 x 89	16	23.84
10 x 3	254 x 76	19	28.29	6 x 3	152 x 76	12	17.88
9 x 3.5	229 x 89	22	32.76	5 x 2.5	127 x 64	10	14.90
9 x 3	229 x 76	17.5	26.06				

WEIGHTS OF STRUCTURAL STEEL TEE BARS SPLIT FROM UNIVERSAL BEAMS

Size of Tee		Weight		Size of Tee		Weight	
in	mm	lb/ft	kg/m	in	mm	lb/ft	kg/m
12 x 12	305 x 305	60.0	90	7 x 8	178 x 203	18.0	27
10 x 13.5	254 x 343	42.0	63	5.5 x 8	140 x 203	15.5	23
9 x 12	229 x 305	47.0	70	6.75 x 7	171 x 178	22.5	34
9 x 12	229 x 305	34.0	51	6.75 x 7	171 x 178	15.0	23
8.25 x 10.5	210 x 267	41.0	61	5 x 7	127 x 178	13.0	20
8.25 x 10.5	210 x 267	27.5	41	6.5 x 6	165 x 152	18.0	27
7 x 9	191 x 229	33.0	49	5 x 6	127 x 152	16.0	24
7 x 9	191 x 229	22.5	34	4 x 6	102 x 152	11.0	17
6 x 9	152 x 229	27.5	41	5.75 x 5	146 x 127	14.5	22
6 x 9	152 x 229	17.5	26	4 x 5	102 x 127	9.5	14
7 x 8	178 x 203	25.0	37	5.25 x 4	133 x 102	10.0	15

WEIGHTS OF STRUCTURAL STEEL TEE BARS SPLIT FROM UNIVERSAL COLUMNS

Size of Tee		Weight		Size of Tee		Weight	
in	mm	lb/ft	kg/m	in	mm	lb/ft	kg/m
12 x 6	305 x 152	53.0	79	8 x 4	203 x 102	29.0	43
12 x 6	305 x 152	32.5	49	8 x 4	203 x 102	15.5	23
10 x 5	254 x 127	44.5	66	6 x 3	152 x 76	12.5	19
10 x 5	254 x 127	24.5	37	6 x 3	152 x 76	7.85	12

WEIGHTS OF STEEL EQUAL ANGLES TO B.S. 4848 PART 4: 1972

Size	Thickness	Weight	Size	Thickness	Weight	Size	Thickness	Weight
mm	mm	kg/m	mm	mm	kg/m	mm	mm	kg/m
250 x 250	35	128.0	150 x 150	10	23.0	110 x 110	12	19.8
	32	118.0					10	16.7
	28	104.0	130 x 130	16	30.8		8	13.5
	25	93.6		12	23.5			
				10	19.8	100 x 100	15	21.9
200 x 200	24	71.1		8	16.0		12	17.8
	20	59.9					10	15.0
	18	54.2	120 x 120	15	26.6		8	12.2
	16	48.5		12	21.6			
				10	18.2	90 x 90	12	15.9
150 x 150	18	40.1		8	14.7		10	13.4
	15	33.8					8	10.9
	12	27.3	110 x 110	16	25.8		6	8.3

METRIC SYSTEM

SYMBOLS USED

Length	m =	linear metre	Area	m2 =	square metre
	mm =	linear millimetre		mm2 =	square millimetre
Volume	m3 =	cubic metre	Mass	kg =	kilogramme
	mm3 =	cubic millimetre		g =	gramme
Pressure	N/m2 =	Newton per square metre	Density	kg/m =	kilogrammes per linear metre
				kg/m3 =	kilogrammes per cubic metre

CONVERSION FACTORS

To convert Imperial weights and measures into Metric or vice versa, multiply by the following conversion factors.

Length	Factor	Length	Factor
Inches to millimetres (mm)............................	25.4000	Millimetres to inches..............................	0.0394
Feet to metres (m)..	0.3048	Metres to feet..	3.281
Yards to metres (m)....................................	0.9144	Metres to yards....................................	1.094
Area			
Square feet to square metres (m2)..............	0.0929	Square metres to square feet.................	10.764
Square yards to ditto...................................	0.8361	Ditto to square yards.............................	1.196
Volume			
Cubic feet to cubic metres (m3)...................	0.0283	Cubic metres to cubic feet.....................	35.315
Cubic yards to ditto....................................	0.7645	Ditto to cubic yards................................	1.308
Mass			
Tons (2240 lbs) to tonnes...........................	1.0161	Tonnes (1000 kg) to tons.......................	0.9842
Cwts to kilogrammes..................................	50.802	Kilogrammes to cwts..............................	0.0196
Pounds to kilogrammes (kg).......................	0.454	Kilogrammes to pounds..........................	2.205
Ounces to grammes (g)..............................	28.350	Grammes to ounces...............................	0.0353
Pressure			
Pounds per square inch to Newtons per square metre (N/mm2)...............................	6894.800	Newtons per square metre to pounds per square inch.......................................	0.000145
Density			
Pounds per linear foot to kilogrammes per linear metre (kg/m)...............................	1.488	Kilogrammes per linear metre to Pounds per linear foot..............................	0.6720
Pounds per square foot to kilogrammes per metre (k/m2)..	4.882	Kilogrammes per square metre to Pounds per square foot...........................	0.2048
Pounds per square yard to ditto.................	0.542	Ditto to pounds per square yard..............	1.845
Pounds per cubic foot to kilogrammes per cubic metre (kg/m3)..................................	16.019	Kilogrammes per cubic metre to Pounds per cubic foot..............................	0.0624
Capacity			
Gallons to litres..	4.546	Litres to gallons......................................	0.220

CONVERSION FACTORS (Cont'd)

To convert items priced in Imperial units into Metric or vice versa, mulitply by the following conversion factors.

	£ Factor		£ Factor
Linear feet to linear metres	3.281	Linear metres to linear feet	0.3048
Linear yards to linear metres	1.094	Linear metres to linear yards	0.9144
Square feet to square metres	10.764	Square metres to square feet	0.0929
Square yards to square metres	1.196	Square metres to square yards	0.8361
Cubic feet to cubic metres	35.315	Cubic metres to cubic feet	0.028
Cubic yards to cubic metres	1.308	Cubic metres to cubic yards	0.7645
Pounds to kilogrammes	2.2046	Kilogrammes to pounds	0.454
Cwts to kilogrammes	0.0196	Kilogrammes to cwts	50.802
Cwts to tonnes	19.684	Tonnes to cwts	0.051
Tons to tonnes	0.9842	Tonnes to tons	1.0161
Gallons to litres	0.220	Litres to gallons	4.54

METRIC EQUIVALENTS

Linear inches to linear millimetres 12 in = 304.80 millimetres

Inches	0	1	2	3	4	5	6	7	8	9	10	11
0		25.40	50.80	76.20	101.60	127.00	152.40	177.80	203.20	228.60	154.00	279.40
1/32	0.79	26.19	51.59	76.99	102.39	127.79	153.19	178.59	203.99	229.39	254.79	280.19
1/16	1.59	26.99	52.39	77.79	103.19	128.59	153.99	179.39	204.79	230.19	255.59	280.99
3/32	2.38	27.78	53.18	78.58	103.98	129.38	154.78	180.18	205.58	230.98	256.38	281.78
1/8	3.18	28.58	53.98	79.38	104.78	130.18	155.58	180.98	206.38	231.78	257.18	282.58
5/32	3.97	29.37	54.77	80.17	105.57	130.97	156.37	181.77	207.17	232.57	257.97	283.37
6/16	4.76	30.16	55.56	80.96	106.36	131.76	157.16	182.56	207.96	233.36	258.76	284.16
7/32	5.56	30.96	56.36	81.76	107.16	132.56	157.96	183.36	208.76	234.16	259.56	284.96
1/4	6.35	31.75	57.15	82.55	107.95	133.35	158.75	184.15	209.55	234.95	260.35	285.75
3/8	9.53	34.93	60.33	85.73	111.13	136.53	161.93	187.33	212.73	238.13	263.53	288.93
1/2	12.70	38.10	63.50	88.90	114.30	139.70	165.10	190.50	215.90	241.30	266.70	292.10
5/8	15.88	41.28	66.68	92.08	117.48	142.88	168.28	193.68	219.08	244.48	269.88	295.28
3/4	19.05	44.45	69.85	95.25	120.65	146.05	171.45	196.85	222.25	247.65	273.05	298.45
7/8	22.23	47.63	98.43	98.43	123.83	149.23	174.63	200.03	225.43	150.83	276.23	301.63

Linear feet and inches to linear millimetres

Feet	1	2	3	4	5	6	7	8	9	10	11	12
0 in	304.8	609.6	914.4	1219.2	1524.0	1828.8	2133.6	2438.4	2743.2	3048.0	3352.8	3657.6
1 in	330.2	635.0	939.8	1244.6	1549.4	1854.2	2159.0	2463.8	2768.6	3073.4	3378.2	2683.0
2 in	355.6	660.4	965.2	1270.0	1574.8	1879.6	2184.4	2489.2	2794.0	3098.8	3403.6	3708.4
3 in	381.0	385.8	990.6	1295.4	1600.2	1905.0	2209.8	2514.6	2819.4	3124.2	3429.0	3733.8
4 in	406.4	711.2	1016.0	1320.8	1625.6	1930.4	2235.2	2540.0	2844.8	3149.6	3454.4	3759.2
5 in	431.8	736.6	1041.4	1346.4	1651.0	1955.8	2260.6	2565.4	2870.2	3175.0	3479.8	3784.6
6 in	457.2	762.0	1066.8	1371.6	1676.4	1981.2	2286.0	2590.8	2895.6	3200.4	3505.2	3810.0
7 in	482.6	787.4	1092.2	1397.0	1701.8	2006.6	2311.4	2616.2	2921.0	3225.8	3530.6	3835.4
8 in	508.0	812.8	1117.6	1422.4	1727.2	2032.0	2336.8	2641.6	2946.4	3251.2	3556.0	3860.8
9 in	533.4	838.2	1143.0	1447.8	1752.6	2057.4	2362.2	2667.0	2971.8	3276.6	3581.4	3886.2
10 in	558.8	863.6	1168.4	1473.2	1778.0	2082.8	2387.6	2692.4	2997.2	3302.0	3606.8	3911.6
11 in	584.2	889.0	1193.8	1498.6	1803.4	2108.2	2413.0	2717.8	3022.6	3327.4	3632.2	3937.0

METRIC VALUES

POWER

1000	microwatts	=	1 milliwatt
1000	milliwatts	=	1 watt
1000	watts	=	1 kilowatt
1000	kilowatts	=	1 megawatt
1000	gigawatts	=	1 terawatt

VOLUME AND CAPACITY

1000	cu millimetres	=	1 cu centimetre
1000	cu centimetres	=	1 cu decimetre
1000	cu decimetres	=	1 cu metre
1000	cu metres	=	1 cu dekametre
1000	cu dekametres	=	1 cu hectometre
1000	cu hectometre	=	1 cu kilometre
1000	microlitres	=	1 millilitres (1cc)
10	millilitres	=	1 centilitre
10	centilitres	=	1 decilitre
1000	millilitres	=	1 litre
100	centilitres	=	1 litre
1000	litres	=	1 kilolitre
1000	litres	=	1 cu metre

PRESSURE AND STRESS

1000	micropascals	=	1 millipascal
100	millipascals	=	1 microbar
1000	millipascals	=	1 pascal
100	pascals	=	1 millibar
1000	pascals	=	1 kilopascal
1000	millibars	=	1 bar
1000	kilopascals	=	1 megapascal
1000	bars	=	1 kilobar
1000	megapascals	=	1 gigapascal

VELOCITY

36	km/hour	=	1 metre/second
3600	km/hour	=	1 km/second

ELECTRICITY AND MAGNETISM

1000	picoamperes	=	1 nanoampere
1000	nanoamperes	=	1 microampere
1000	microamperes	=	1 milliampere
1000	milliamperes	=	1 ampere
1000	amperes	=	1 kiloampere
1000	millicoulombs	=	1 coulomb
1000	coulombs	=	1 kilocoulomb
1000	microvolts	=	1 millivolt
1000	millivolts	=	1 volt
1000	volts	=	1 kilovolt
1000	kilovolts	=	1 megavolt
1000	microhms	=	1 milliohm
1000	milliohms	=	1 ohm
1000	ohms	=	1 kilohms
1000	kilohms	=	1 megaohm
1000	millisiemens	=	1 siemen
1000	millihenrys	=	1 henry
1000	milliteslas	=	1 tesla

LENGTH

1000	picometres	=	1 nanometre
1000	nanometres	=	1 micrometre
1000	micrometres	=	1 millimetre
10	millimetres	=	1 centimetre
10	centimetres	=	1 decimetre
1000	millimetres	=	1 metre
100	centimetres	=	1 metre
100	metres	=	1 hectometre
1000	metres	=	1 kilometre
1000	kilometres	=	1 megametre
1852	nautical metres	=	1 international nautical mile

METRIC VALUES(Cont'd)

AREA

100	sq millimetres	=	1 sq centimetre
100	sq centimetres	=	1 sq decimetre
10000	sq centimetres	=	1 sq metre
100	sq metres	=	1 are
10	ares	=	1 dekare
10000	sq metres	=	1 hectare
100	hectares	=	1 sq kilometre

MASS

1000	nanograms	=	1 microgram
1000	micrograms	=	1 milligram
200	milligrams	=	1 metric carat
1000	milligrams	=	1 gram
25	grams	=	1 metric ounce
100	grams	=	1 hectogram
1000	grams	=	1 kilogram
100	kilograms	=	1 quintal
1000	kilograms	=	1 megagram

ENERGY (WORK AND HEAT)

10000	ergs	=	1 millijoule
1000	millijoules	=	1 joule
1000	joules	=	1 kilojoule
1000	kilojoules	=	1 megajoule
3.6	megajoules	=	1 kilowatt hour
1000	megajoules	=	1 gigajoule
1000	gigajoules	=	1 terajoule

FORCE

10	micronewtons	=	1 dyne
1000	micronewtons	=	1 millinewton
10	millinewtons	=	1 centinewton
1000	millinewtons	=	1 newton
1000	newtons	=	1 kilonewton
1000	kilonewtons	=	1 meganewton

FREQUENCY

1000	hertz	=	1 kilohertz
1000	kilohertz	=	1 megahertz
1000	megahertz	=	1 gigahertz
1000	gigahertz	=	1 terahertz
1000	terahertz	=	1 petahertz
1000	petahertz	=	1 exahertz

METRIC WEIGHTS OF MATERIALS

Material	kg/m3
Aerated concrete	800-960
Aluminium	
cast	2550
rolled	2700
Asphalt	
natural	1000
paving	2080
Ballast, loose, graded	600
Beech	700
Birch	
American	640
yellow	700
Brass	
cast	8330
rolled	8570
Bricks (common burnt clay)	
stacked	1600-1920
sand cement	1840
sand lime	2080
ballast	1200
brickwork	1920
Cement	
bags	1280
bulk	1281-1442
casks	960
Concrete (cement, plain)	
brick aggregate	1840
clinker	1440
stone ballast	2240
Concrete (cement, reinforced)	
1% steel	2370
2% steel	2420
5% steel	2580
Copper	
cast	8760
drawn or sheet	8940
Cork	130-240
Deal, yellow	430
Ebony	1180-1330
Elm	
American	670
Dutch	580
English	580
Wych	690
Fir	
Douglas	530
Silver	480
Flint	2560
Foam slag	700
Freestone	2243-4280
masonry, dressed	2400
rubble	2240
Glass	
bottle	2720
flint, best	3080
flint, heavy	5000-6000
plate	2800

Material	kg/m3
Lime	
grey chalk, lump	700
grey stone, lump	800
hydrate, bags	510
Macadam	2100
Mahogany	
African	560
Honduras	540
Spanish	690
Marble	2590-2830
Mortar	
cement, set	1920-2080
lime, set	1600-1760
Oak	
African	960
American red	720
English	800-880
Padouk	780
Paint	
aluminium	1200
bituminous emulsion	1120
red lead	3120
white lead	2800
zinc	2400
Pine	
American red	530
British Columbian	530
Christina	690
Oregon	530
Pitch	650
Plywood	480-640
Polyvinyl chloride acetate	1200-1350
Poplar	450
Portland cement	
loose	1200-1360
bags	1120-1280
drums	1200
Redwood	
American	530
Baltic	500
non-graded	430
Rhodesian	910
Slate	
Welsh	2800
Westmorland	3000
Steel	7830
Stone	
Ancaster	2500
Bath	2080
Darley Dale	2370
Forest of Dean	2430
Granite	2640
Hopton Wood	2530
Kentish rag	2670
Mansfield	2260
Portland	2240
Purbeck	2700

METRIC WEIGHTS OF MATERIALS (Cont'd)

Granolithic	2240
Hardcore	1920
Hoggin	1760
Iroko	600
Iron	
cast	7200
malleable cast	7370-7500
wrought	7690
Lead	
cast or rolled	11300
Lime	
acetate or, bags	1280
blue lias, ground	850
carbonate of, barrels	1280
chloride of, lead-lined cases	450
York	2240
Tarmacadam	2080
Teak	
Burma, African	650
Walnut	660
Water	
fresh	1001
salt	1009-1201
Whitewood	460
Zinc	
cast	6840
rolled	7190
sheets, packed	900

WEIGHTS OF VARIOUS METALS

in kilogrammes per square metre

mm	in	Steel kg	Wt. iron kg	Cast iron kg	Brass kg	Copper kg	Cast Lead kg
1.59	1/16	12.45	12.21	11.47	13.43	14.11	18.07
3.18	1/8	24.90	24.41	22.90	26.76	28.22	36.08
6.35	1/4	49.80	48.82	45.75	53.41	56.44	72.16
9.53	3/8	74.70	73.24	68.65	80.17	84.66	108.24
12.70	1/2	99.60	97.65	91.55	106.83	112.88	144.32
15.88	5/8	124.50	122.06	144.44	133.49	141.54	180.41
19.05	3/4	149.40	146.47	137.29	160.19	169.32	216.49
22.23	7/8	174.30	170.88	160.19	186.95	197.54	252.57
25.40	1	199.20	195.30	183.09	213.65	225.76	288.65

WEIGHTS OF ROUND AND SQUARE STEEL BARS

mm	Round kg/m	lb/lin ft	Square kg/m	lb/lin ft	mm	Round kg/m	lb/lin ft	Square kg/m	lb/lin ft
6	0.222	0.149	0.283	0.190	20	2.466	1.657	3.139	2.110
8	0.395	0.265	0.503	0.338	25	3.854	2.590	4.905	3.296
10	0.616	0.414	0.784	0.527	32	6.313	4.243	8.035	5.400
12	0.888	0.597	1.130	0.759	40	9.864	6.629	12.554	8.437
16	1.579	1.061	2.010	1.351	50	15.413	10.358	19.617	13.183

WEIGHTS OF FLAT STEEL BARS

in kilogrammes per linear metre

Width mm	in	Thickness in millimetres and inches 3.18 (1/8 in)	6.35 (1/4 in)	7.94 (5/16 in)	9.53 (3/8 in)	11.11 (7/16 in)	12.70 (1/2 in)	15.88 (5/8 in)	19.05 (3/4 in)	22.23 (7/8 in)	25.40 (1 in)
25.40	1	0.64	1.27	1.58	1.91	2.22	2.53	3.17	3.80	4.44	5.06
28.58	1 1/8	0.71	1.43	1.79	2.13	2.49	2.84	3.56	4.27	4.99	5.70
31.75	1 1/4	0.79	1.58	1.98	2.37	2.77	3.17	3.96	4.75	5.54	6.33
34.93	1 3/8	0.86	1.74	2.17	2.60	3.05	3.48	4.35	5.22	6.09	6.97
38.10	1 1/2	0.95	1.91	2.37	2.84	3.32	3.80	4.75	5.70	6.64	7.59
41.28	1 5/8	1.03	2.05	2.58	3.08	3.60	4.11	5.13	6.16	7.19	8.23
44.45	1 3/4	1.10	2.22	2.77	3.32	3.87	4.44	5.54	6.64	7.75	8.86
50.80	2	1.27	2.53	3.17	3.80	4.44	5.06	6.33	7.59	8.86	10.12
57.15	2 1/4	1.44	2.84	3.56	4.27	4.99	5.70	7.11	8.54	9.96	11.39
63.50	2 1/2	1.58	3.17	3.96	4.75	5.54	6.33	7.90	9.50	11.07	12.68
69.85	2 3/4	1.74	3.48	4.35	5.22	6.09	6.97	8.69	10.43	12.17	13.92
76.20	3	1.91	3.80	4.75	5.70	6.64	7.59	9.50	11.39	13.29	15.18
82.55	3 1/4	2.05	4.11	5.13	6.16	7.19	8.23	10.28	12.34	14.39	16.44
88.90	3 1/2	2.22	4.44	5.54	6.64	7.75	8.86	11.07	13.29	15.49	17.71
95.25	3 3/8	2.37	4.75	5.92	7.11	8.30	9.50	11.86	14.23	16.61	18.97
101.60	4	2.53	5.06	6.33	7.59	8.86	10.12	12.65	15.18	17.71	20.24
107.95	4 1/4	2.69	5.37	6.73	8.07	9.41	10.76	13.44	16.13	18.81	21.50
114.30	4 1/2	2.84	5.70	7.11	8.54	9.96	11.39	14.23	17.08	19.93	22.77
120.65	4 3/4	3.01	6.01	7.52	9.02	10.52	12.02	15.02	18.02	21.03	24.03
127.00	5	3.17	6.33	7.90	9.50	11.07	12.65	15.82	18.97	22.14	25.30
139.70	5 1/2	3.48	6.97	8.69	10.43	12.17	13.92	17.40	20.88	24.35	27.83
152.40	6	3.80	7.59	9.50	11.39	13.29	15.18	18.97	22.77	26.56	30.36
165.10	6 1.2	4.11	8.23	10.28	12.34	14.39	16.44	20.55	24.67	28.78	32.89
177.80	7	4.44	8.86	11.07	13.29	15.49	17.71	22.14	26.56	31.00	35.42
190.50	7 1/2	4.75	9.50	11.86	14.23	16.61	18.97	23.72	28.47	33.20	37.95
203.20	8	5.06	10.12	12.65	15.18	17.71	20.24	25.30	30.36	35.42	40.48
228.60	9	5.70	11.39	14.23	17.08	19.93	22.77	28.47	34.15	39.85	45.54
254.00	10	6.73	12.65	15.82	18.97	22.14	25.30	31.62	37.95	44.27	50.50
279.40	11	6.97	13.92	17.40	20.88	24.35	27.83	34.79	41.74	48.71	55.66
304.80	12	7.59	15.18	18.97	22.77	26.56	30.36	37.95	45.54	53.13	60.72

BRANDS AND TRADE NAMES

202 - Sapa Building Systems (monarch and glostal)
238 - Chalmit Lighting
242 - Crawford Amber Ltd
325 - Sapa Building Systems (monarch and glostal)
542 - Crawford Amber Ltd
1200 - GE Bayer Silicones
2000 - Healthmatic Ltd
(IPAF Training) - Turner Access Ltd
(MC) - Barber Wilsons & Co Ltd
(PASMA Training) - Turner Access Ltd
+8 Scaffolding System - Turner Access Ltd
>B< - IBP Ltd
>B< 3000 - IBP Ltd
>B< 8000 - IBP Ltd
>B< Plan - IBP Ltd
>B< Press - IBP Ltd
£1 coin locks for coats - Locking Systems Ltd
100 High - Hudevad Britain
1890's (GA) - Barber Wilsons & Co Ltd
1890's Regent Mastersraft - Barber Wilsons & Co Ltd
1900 Series Cutting & Grinding Discs - Acton & Borman
20/20 Interlocking clay plain (tile) - Sandtoft Roof Tiles Ltd
25/40/56 Series - Albany Standard Pumps
2Wire Systems - BPT Security Systems (UK) Ltd
3 Step Mat - Dycem Ltd
316 Range - James Gibbons Format Ltd
3230 Classic Fibre Bonded Carpet - Burmatex Ltd
3m Commercial Graphics - Northgate Solar Controls
3M Matting - Floor Protection Services Ltd
4000AC - GE Bayer Silicones
425 Range - Horstmann Controls Ltd
4250 Alarm Confirmation - ADT Fire and Security Plc
4th Connection - Horne Engineering Ltd
500 Series - Opella Ltd
506 NCE - Otis Ltd
513 NPE - Otis Ltd
606 NCT - Otis Ltd
800 Series Call System - C-TEC (Computionics) Ltd
900 Series - Girdlestone Pumps Ltd
9235 Waterproof membrane - Laticrete UK
A + K - Anders + Kern UK Ltd
A Division of Castle House Joinery. - Orchard Street Furniture
A100 - Auld Valves Ltd
AAF LtFd - McQuay UK Ltd
Aalco - Aalco Hull Amari Metal Ltd
Aaztec - Aaztec Cubicles
Abacus - Smart F & G (Shopfittings)Ltd
Abacus Lighting Ltd - Abacus Lighting Ltd
Abanaki - Furmanite International Ltd
ABB I.C. Moller Pipe System - ABB Industrial Systems Ltd
Abbcol - The Abbseal Group incorporating Everseal (Thermovitrine Ltd)
Abbey - Abbey Building Supplies Co
Abbotsford - Brintons Ltd
Abbotsford - Brintons Ltd
Abco Metal Refiner - ABCO Products Ltd.
Abco Metal Refiner - ABCO Products Ltd.
Abet - Viking Laminates Ltd
ABG - Ingersoll-Rand European Sales Ltd
Abigail - Trent Bathrooms
Abingdon - Carpets International (UK) Ltd
ABK - Arnull Bernard J & Co Ltd
Able and Martin roberts - Martin Roberts
Abloy Disklock Pro - Abloy Security Ltd
Abloy Exec - Abloy Security Ltd
Abloy Exec - Abloy Security Ltd
Abloy High Profile - Abloy Security Ltd
Abloy Protec - Abloy Security Ltd
Abloy® - Abloy Security Ltd
Abmags - Abloy Security Ltd
Abrafract - Carborundum Abrasives G.B. Ltd
Abrobility - Gabriel & Co Ltd
Abroclamps - Gabriel & Co Ltd
Abrotube - Gabriel & Co Ltd
ABS Floors - Optiroc Ltd
ABS Pumps - ABS Pumps Ltd
AbSence/ PreSense - Philips Lighting Solutions
Abtex - ABG Ltd
AC - Howden Buffalo
ACAS - Advisory Council and Arbitration Service
Access - Gradus Ltd
Access 1000 - CEM Systems Ltd
Access 1000 Enterprise - CEM Systems Ltd
Access 2000 - CEM Systems Ltd
Access Ramp - Sportmark™ Ltd
Access Steps - Sportmark™ Ltd
Accessories and Cabinets - Jendico
Acclaim - Chalmit Lighting
Acclaim - Dermide Ltd
Acclaim - Dudley Thomas Ltd
Accoflex - DLW Flooring
Accolade - Gledhill Water Storage Ltd
Accolade - Heatrae Sadia Heating Ltd
Accolade - Jaga Heating Poducts (UK) Ltd
Accord - Stocksigns Ltd
Accorroni - May William (Ashton) Ltd
Acc-q-line - Metalrax Ltd
Accrodal - Smithbrook Building Products Ltd
Accuro - Draeger Safety UK Ltd
Accu-Tune - IAC
ACE - Association of Cunsulting Engineers
Ace & Ace High - Halstead Boilers Ltd
ACH - Acoustic Ceiling Hangers - Christie & Grey Ltd
Ackermann Spider - Ackermann Ltd
Acklaim - Ackermann Ltd
Acknowledge - Ackermann Ltd
Ackwire - Ackermann Ltd
Ackwire - Ackermann Ltd
Acme - Eternit Building Materials
ACO - ACO Technologies Plc
ACO - ACO Technologies Plc
ACO Drain - ACO Technologies Plc
ACO Fulbora - ACO Technologies Plc
ACO vinyl seal - ACO Technologies Plc
Acorn - Oakdale (Contracts) Ltd
Acousti Shock - Acousticabs Industrial Noise Control Ltd
Acoustic Vents - Simon R W Ltd
Acousticel M20AD - Sound Service (Oxford) Ltd
Acousticel R10 - Sound Service (Oxford) Ltd
Acousticurtain - Acousticabs Industrial Noise Control Ltd
Acoustifoam - Acousticabs Industrial Noise Control Ltd
Acoustilouvre - Acousticabs Industrial Noise Control Ltd
Acoustilouvre - Acousticabs Industrial Noise Control Ltd
Acoustislab - Acousticabs Industrial Noise Control Ltd
Acrylacote - Dacrylate Paints Ltd
ACT - Artur Fischer (UK) Ltd
Actfast - Lycetts Industrial Doors
Actimatic - Wade International Ltd
Actionair - Ruskin Air Management
Activ - Pilkington Birmingham
Activa - Full Spectrum Fluorescent Tubes - Sylvania Lighting International INC.
Active 8 - Gas Measurement Instruments Ltd
Activ-Ox - Feedwater Ltd
Acumina - Orsogril Spa UK Distributor
Aczent - Tarkett Sommer Ltd
Adams Rite - Permclose Door Services Ltd
Adanced counters & Serveries - Viscount Catering Ltd
Adapta Tap - Airstream Products Ltd
Adaptalok - Adaptaflex Ltd
Adaptaring - Adaptaflex Ltd
Adaptaseal - Adaptaflex Ltd
Adaptasteel - Adaptaflex Ltd
Adaptasteel - Adaptaflex Ltd
Adcor - Grace Construction Products Ltd
Addablock - Addagrip Surface Treatments Ltd
Addacrete - Addagrip Surface Treatments Ltd
Addaflor - Addagrip Surface Treatments Ltd
Addagrip1000 System - Addagrip Surface Treatments Ltd
Addalevel - Addagrip Surface Treatments Ltd
Addapave - Addagrip Surface Treatments Ltd
Addaprime - Addagrip Surface Treatments Ltd
Addaprime - Addagrip Surface Treatments Ltd
Addaprint - Addagrip Surface Treatments Ltd
Addastone - Addagrip Surface Treatments Ltd
Addastone Matrix Paver - Addagrip Surface Treatments Ltd
Addastone TP - Addagrip Surface Treatments Ltd
Adesilex Range - Mapei (UK) Ltd
Adhesin - Henkel Industrial Adhesives
Adjustabars - Baddeley
Adjustabars - Baddeley
Adjustagate - Baddeley
Admatic - Potter & Soar Ltd
Admix - Sheardown Engineering Ltd
Admont - McLoughlin Wood Ltd
Adorn - Valor
adpro axiam - Vision Sytems (Europe) Ltd
adpro fast scan series III - Vision Sytems (Europe) Ltd
adpro fast trace - Vision Sytems (Europe) Ltd
adpro fast vu - Vision Sytems (Europe) Ltd
adpro pir detectors - Vision Sytems (Europe) Ltd
Adria - Armstrong World Industries Ltd
ADT Modern - ADT Modern Security Systems Ltd
ADT Modern - ADT Modern Security Systems Ltd
Adva - Grace Construction Products Ltd
Advance - Nilfisk Advance Ltd
Advance Matt Gaffa - Advance Tapes International Ltd
Advance Premier Closure Plate Tape - Advance Tapes International Ltd
Advance Sectional Buildings - UK Cabin Co
Advanced Interior Solutions Ltd - Advanced Interior Solutions Ltd
Advantage - Dauphin PLC
Advantage - Dixon Turner Wallcoverings
Advantage - Saint-Gobain Ecophon Ltd
Advantage Respirators - MSA (Britain) Ltd
Advent range - Caradon Bathrooms Ltd
Advisor & Infoboard - Arbory Group Limited
AEG - Electrolux Domestic Appliances
AEG - Parkinson Cowan and Tricity Bendix Co, The
AEL Video - ADT Modern Security Systems Ltd
Aercon - Aercon Wiring Systems Ltd
Aercon - Aercon Wiring Systems Ltd
Aero - Blakey John H Ltd
Aero - Twyford Bushboard
Aerodyne cowl - Brewer Metalcraft
Aerodyne range - Saint Gobain pipelines PLC
Aeroguard - Celuform
Aeroguard - Manton Industrial Seals Ltd
Aerovent - ABB Control Ltd
Aerovent - Celuform
AET - AET.gb Ltd
AEX - The Salex Group Ltd.
Affinity Disposable Respirtiors - MSA (Britain) Ltd
Afghan - Carpets of Worth Ltd
AFP Analogue Addressable Fire Panels - C-TEC (Computionics) Ltd
Aga - Aga-Rayburn
Aga Companion - Aga-Rayburn
Aga Cook Shop Collection - Aga-Rayburn
Aga Masterchef - Aga-Rayburn
Agabekov - Crescent Lighting
Agastat - Thomas & Betts Manufacturing Services Ltd
Agc 200 - Stewart Wales, Somerville Ltd
Agglio Conglomerate Marble - Reed Harris - Reed Harris
Agglosimplex Conglomerate Marble - Marble Flooring Spec. Ltd
AGP - Creteco Sales Ltd
Agrement certificates - British Board of Agrement
Agri drain - Naylor Drainage Ltd
Agrikote AF - Pearl Paints Ltd
Agru - Capper Plastics
Aidalarm - Hoyles Electronic Developments Ltd
Ainsworth OSB - Wood International Agency Ltd
Ainsworth Pourform 107 - Wood International Agency Ltd
Ainsworth Pourform H00 - Wood International Agency Ltd
Air Marshall - Rapaway Energy Ltd
Air Miser - Rapaway Energy Ltd
Aircare 2000 series - Aircare Europe Ltd
Aircare 2000 series - Aircare Europe Ltd
Aircore - Ryton's Building Products Ltd
Airdale - Airedale International Air Conditioning Ltd
Airdale Service - Airedale International Air Conditioning Ltd
Airdor - S & P Coil Products Ltd
Airdry - Hills Industries Ltd
AirDuct - Airflow (Nicoll Ventilators) Ltd
Airetronix - Airedale International Air Conditioning Ltd
Airfix - Flamco UK Ltd
Airguard - Aquastat Ltd
Airking - IMI Air Conditioning Ltd
Airline - Ves Andover Ltd
Airliner - Ryton's Building Products Ltd
Airmat - Camas Aggregates Ltd - Aggregate Industries
Airseal - Fillcrete Ltd
Airspeed - Crosbie Coatings
Airspeed - Warner Howard Group Ltd
Airsprung Beds Ltd - Airsprung Beds Ltd
Airstar - Rapaway Energy Ltd
Airstrip - Simon R W Ltd
Airswitch - Setsquare
Airworld - Warner Howard Group Ltd
Aitersall - Deva Tap Co Ltd
Akademy Classroom System - Portakabin Ltd
Akor - ABB Control Ltd
Akord - DW Windsor Lighting
Aktivated Showers - Hansgrohe
AL 50 - SG System Products Ltd
Alabastine - Polycell Products Ltd
Alag Aggregate - Lafarge Aluminates Ltd
Alarm-a-fence® - Jackson H S & Son (Fencing) Ltd
Alarming UK - UK Fire International Ltd
Alarmline - Linear Heat Detection - Kidde Fire Protection Ltd
AlarmSense - Apollo Fire Detectors Ltd
Alba De-Fence Systems - Albion Fencing Ltd
Albany - Gloster Furniture Ltd
Albany Paints - Brewer C & Sons Ltd
Albany Paints - Brewer C & Sons Ltd
Albany Plus - MK (MK Electric Ltd)
Albany Range - Caradon Bathrooms Ltd
Albany Wallcoverings - Brewer C & Sons Ltd
Albi-Clear - Rentokil Initial U.K. Ltd.
Albi-Grips - Albion Manufacturing
Albi-guard - Rentokil Initial U.K. Ltd.
Albi-Max - Rentokil Initial U.K. Ltd.
Albion - Swedecor Ltd
Albi-Pruf - Rentokil Initial U.K. Ltd.
Albi-Slings - Albion Manufacturing
Albi-Steel 90 - Rentokil Initial U.K. Ltd.
Albi-Traps - Albion Manufacturing
Alcione - Orsogril Spa UK Distributor
Alclad - Solair Ltd
Alcoa Extrusions - Alcoa Europe-Extrusions & Endproducts (UK)
Alcoclad - Alcover UK Ltd
Alcoglass - Alcover UK Ltd
Alcoglass - Alcover UK Ltd
Aldwark - York Handmade Brick Co Ltd
Alenti - Arjo Ltd
Aleonard - Smithbrook Building Products Ltd
Alertcall - C-TEC (Computionics) Ltd
Alexandra - BRB Industrial Services Ltd
Alfab - The Safety Tread Ltd
Alfacryl - Alfas Industries Ltd
Alfacryl FR - Alfas Industries Ltd
Alfapower - MJ Electronics Services (International) Ltd
Alfas Bond - Alfas Industries Ltd
Alfas Bond FR - Alfas Industries Ltd
Alfas Bond FR - Compriband Ltd
Alfas 'C' seal - Alfas Industries Ltd

Alfas GBT - Alfas Industries Ltd
Alfas SGT & Alfas SGT super - Alfas Industries Ltd
Alfas TMT - Alfas Industries Ltd
Alfasil - Alfas Industries Ltd
Alfatherm - AEL
Alfed - Aluminium Federation (Alfed) Ltd
Alfed - Aluminium Federation (Alfed) Ltd
Algopilot - Control Terminal - Siemens Building Technologies Ltd,
Algorex - Fire Detection System - Siemens Building Technologies Ltd,
Ali - Satchwell Control Systems Ltd
Aliflex - Flexible Ducting Ltd
Aligator ® - Marley Plumbing and Drainage
Alimaster Shelving - Beford & Soar Ltd
A-Line - Desking Systems Ltd
Alitherm - Smart Systems Ltd
Alkorflex - Alkor Draka Ltd
Alkorplan - Alkor Draka Ltd
Alkorplan - Alkor Draka Ltd
Alkortop - Alkor Draka Ltd
Alkysil - Dacrylate Paints Ltd
All Abrasive Strip - The Safety Tread Ltd
All in one sealer - Sealocrete Ltd
Allart Deco - Allart, Frank, & Co Ltd
Allart Deco - Allart, Frank, & Co Ltd
Allegra - Hansgrohe
Allegro - Erlau AG
Allergi Guard - Nationwide Filter Co Ltd
Allgood Hardware - Allgood plc
Allgood Secure - Allgood plc
Allibert - Allibert (UK) Ltd
alligator entrance matting - Jaymart Rubber & Plastics Ltd
Allite - The Safety Tread Ltd
Allmat - Allmat (East Surrey) Ltd
Allmat - Allmat (East Surrey) Ltd
Alno - Alno (United Kingdom) Ltd
Alno 2000 - Alno (United Kingdom) Ltd
Alnova - Alno (United Kingdom) Ltd
Alnova - Alno (United Kingdom) Ltd
AL-OR - Harbro Supplies Ltd
Aloxite - Carborundum Abrasives G.B. Ltd
Alpha - Brett Paving
Alpha - Titan Pollution Control Ltd
Alpha 24 - Premdor
Alpha 24 Secure - Premdor
Alpha Cool - Airedale International Air Conditioning Ltd
Alpha scan - Milliken Carpet
Alphaline - ABG Ltd
Alpine - EBC UK Ltd
Alpine Automation - Alpine Autimation Ltd.
Alpine seat - Furnitubes International Ltd
Althon - Allpipe Ltd
Althon - Allpipe Ltd
Althon-Lite - Allpipe Ltd
Altimus - Envoplan
Alto - Alto Cleaning Systems (UK) Ltd
Alton Green Houses - Compton Buildings Ltd
Alton Greenhouses - Phillip Ader Ltd
Altro PU - Altro Floors
Altroscreed Quartz - Altro Floors
Altrosmooth - Altro Floors
Altrotimber - Altro Floors
ALU - Osram Ltd
Alucobond - Booth Muirie
Alucobond - Booth Muirie
Alucobond ACM - Technology Telford Ltd
Aluglas - Clow Group Ltd
Aluglaze - Panel Systems Ltd
Alu-lift - Metreel Ltd
Alulux - Brown F plc
Alulux - Brown F plc
Alumasc - Brooks Belfast
Alumasc - Drainage Systems
Alumatilt - Caldwell Hardware (UK) Ltd
Aluminex - Heywood Williams Architectural
Aluminium A Vent - Areco Roofing Supplies Company
Aluminium Systems - Barlow Group
Alu-Steps - Quantum Profile Systems Ltd
Alutec - Marley Plumbing and Drainage
Aluzink - SSAB Dobel Coated Steel Ltd
Alvas - Trent Bathrooms
Alwitra - ICB (International Construction Bureau) Ltd
Ama Drainer - KSB Ltd
Ama-Drainer Box - KSB Ltd
Amal - Bailey Birkett Sales Division
Ama-porter - KSB Ltd
Ama-Porter ICS - KSB Ltd
Amarex - KSB Ltd
Amazon - Potter & Soar Ltd
Ambassador - Metlex
Amber Booth - Designed for Sound Ltd
Amberclems - Ambersil Ltd
Amberklene - Ambersil Ltd
Ambiance Lumiere - Crescent Lighting
Ambi-deck - Redman Fisher Engineering
Ambi-dry verge - Lafarge Roofing Ltd
Ambience - Helvar Electrosonic
Ambi-Rad - Ambi-Rad Ltd
Amcor - Amcor (Appliances) Ltd
Amdega - Amdega Ltd
Amercoat - Kenyon Group Ltd
American Lincoln - Alto Cleaning Systems (UK) Ltd
American Sanders - Alto Cleaning Systems (UK) Ltd
Amerlock - Kenyon Group Ltd
Ameron - Andrews Coatings Ltd
Amie - Tunstal Telecom Ltd
Amiran Anti-Reflective Glass - SCHOTT UK Ltd
Amor metals international - Aalco Hull Amari Metal Ltd
Amore - Brockway Carpets Ltd
Amphistoma - Adams- Hydraulics Ltd
Amphistoma - Adams- Hydraulics Ltd
Ancaster - Realstone Ltd
Ancaster Limestone - Rare Stone Group, The
Ancestry - Ibstock Brick Ltd
Anchorlite slate - Forticrete Ltd
Anchorlite slate - Forticrete Roofing Products
Ancon - Ancon Building Products
Ancon - Ancon Building Products
Ancon Optima - Ancon Building Products
and many more - Eternit Building Materials
Anda-Crib - PHI Group Ltd
Anders & Kern - Anders + Kern UK Ltd
Anders & Kern - Anders + Kern UK Ltd
Andersen ® - Andersen/ Black Millwork
Anderson Art Glass - Andersen/ Black Millwork
Andersons - Andersons Ltd
Anderton interloc - Anderton Concrete Products Ltd
Andrews - Andrews Sykes Hire Ltd
Andrews - Andrews Sykes Hire Ltd
Andrews Air Conditioning - Andrews Sykes Hire Ltd
Andrews Heat for Hire - Andrews Sykes Hire Ltd
Andura - Andura Textured Masonry Coatings Ltd
Andura - Andura Textured Masonry Coatings Ltd
Anglepoise - Anglepoise Lighting Ltd
Angletie - Redifix Ltd
Anglo - Caradon Bathrooms Ltd
Anglo Abrasives - Carborundum Abrasives G.B. Ltd
Angolar - Provex Products Ltd
Animo - Provex Products Ltd
Ankalok Revetment - Ruthin Precast Concrete Ltd
Anki - Sparkes K & L
Annabelle Classique - Brintons Ltd
Annable - Trent Bathrooms
Ansell - BRB Industrial Services Ltd
Antares - Optelma Lighting Ltd
Antel - Andrews Coatings Ltd
Anti Bird Line - Birdscarer Products & Net Supplies Ltd
Anti Bird Line - Birdscarer Products & Net Supplies Ltd
Anti Bird Matting - Birdscarer Products & Net Supplies Ltd
Anti Bird Mesh - Birdscarer Products & Net Supplies Ltd
Anti Bird Net - Birdscarer Products & Net Supplies Ltd
Anti- Bird-Curtain - Birdscarer Products & Net Supplies Ltd
Anti Climb - Andura Textured Masonry Coatings Ltd
Anti fatigue mat - Dycem Ltd
Anti Nesting Cage - Birdscarer Products & Net Supplies Ltd
Anti-Bird- Spikes - Birdscarer Products & Net Supplies Ltd
Anti-block Mfish - Birdscarer Products & Net Supplies Ltd
Anticarb - Andura Textured Masonry Coatings Ltd
Antifreeze - Flexelec(UK) Ltd
Antium - Decra Ltd
Apa plywood - Wood International Agency Ltd
Apache - Ellis Patents
apachestat soft foot - Jaymart Rubber & Plastics Ltd
Aperture - Briggs Roofing & Cladding Ltd, Roofing Contract
Apex - Gaskell Carpet Tiles
Apex - Vulcanite Limited
Apex Doors - Regency Garage Door Services Ltd
Apex heritage - Alumasc Exterior Building Products Ltd
Apex Piling - Apex Piling Ltd
Apex range - Dixon Turner Wallcoverings
APF International forest machine exhibition - Forestry and Timber Association
APHC - Association of Plumbing and Heating Contractors
Apollo - Brierkrete Ltd
Apollo - Guardall Ltd
Apollo - Iles Waste Systems
Apollo bathroom Furniture - Adam Furniture Group PLC
Apollo incinerators - Combustion Linings Ltd
Apollo Windows - Heywood Williams Architectural
APP Liner - Acousticabs Industrial Noise Control Ltd
Appro - Robot (UK) Ltd
Aprimatic automatic gates - Automatic Doors & Gates
Apropos - Clear Span Ltd
APT Barriers - APT Controls Ltd
APT Barriers - APT Controls Ltd
APT Bollards (Hydraulic) - APT Controls Ltd
Apton - Dexion Ltd
Apton Partitions - Adex Interiors for industry Ltd
APW Ltd - Associated Perforators & Weavers Ltd
Aqua - Clear - Aldous & Stamp Services Ltd
Aqua Blue Designs - Aqua-Blue Ltd
Aqua Blue Designs - Aqua-Blue Ltd
Aqua Drolics - Aqua Design Ltd
Aqua Tackseal - Charcon Tunnels (Division of Tarmac plc)
Aqua TP - Pearl Paints Ltd
Aquacatch - Design & Display (Sales) Ltd
Aquachill - J & E Hall Ltd
Aquadish - Jones of Oswestry Ltd
Aqua-epoxeel - IFT Supplies Ltd
Aquaflex - Richard Hose Ltd
Aquaflow - Armorex Ltd
Aquaflow - Formpave Ltd
Aquaflow ML - Formpave Ltd
Aquagard - Marley Waterproofing
Aquaglide - Design & Display (Sales) Ltd
Aquagrip - Viking Johnson
Aquaguard - Aquastat Ltd
Aquaguard - Aquastat Ltd
Aquahib - Aquastat Ltd
Aqua-ifthane - IFT Supplies Ltd
Aqualine - Alumasc Exterior Building Products Ltd
Aqualine - Alumasc Exterior Building Products Ltd
Aqualine - Rangemaster
Aqua-Lite - Naylor Drainage Ltd
Aqualock Surface DPM - Isocrete Floor Screeds Ltd
Aquamaster - ABB Instrumentation Ltd.
Aquameter - Channel Electronics
Aquaminder - Setsquare
Aquamix - Martin Orgee Ltd
Aquamixa - Aqualisa Products
Aquaprobe - ABB Instrumentation Ltd.
Aquarelle - Tarkett Sommer Ltd
Aquarian - Aqualisa Products
Aquasave - HRS Hevac Ltd
Aquasave BIO - HRS Hevac Ltd
Aquaseal - Degussa Construction Chemicals (UK)
Aquasentry - Aquasentry
Aquasil - Aldous & Stamp Services Ltd
Aquaslab - Formpave Ltd
Aquaslique - Midland Stom Ltd
Aquaslot - Jones of Oswestry Ltd
Aquasol - Conren Ltd
Aquaspeed - Crosbie Coatings
Aquastore - CST Vulcan
Aquastream - Aqualisa Products
Aquastyle - Aqualisa Products
Aquasystems - George Fischer Sales Ltd
Aquatec - Loblite Ltd
Aquatec Beluga bathlift - Bison Bede Ltd
Aquatec Elan bathlift - Bison Bede Ltd
Aquatec Fortuna bathlift - Bison Bede Ltd
Aquatech - ICI Dulux Trade Paints
Aquatique - Aqualisa Products
Aquatread - Tor Coatings Ltd
Aquavalve - Aqualisa Products
Aquavalve - Aqualisa Products
Arbo - Adshead Ratcliffe & Co Ltd
Arbo - Adshead Ratcliffe & Co Ltd
Arbocrylic - Adshead Ratcliffe & Co Ltd
Arbofoam - Adshead Ratcliffe & Co Ltd
Arbokol - Adshead Ratcliffe & Co Ltd
Arbomast - Adshead Ratcliffe & Co Ltd
Arborslot - Jones of Oswestry Ltd
Arboseal - Adshead Ratcliffe & Co Ltd
Arbosil - Adshead Ratcliffe & Co Ltd
Arbothane - Adshead Ratcliffe & Co Ltd
Arc - Illuma Lighting Ltd
ARCA - Asbestos Removal Contractors' Association
Arcade - Shackerley (Holdings) Group Ltd incorporating Designer ceramics
Arcadia Pantile - Sandtoft Roof Tiles Ltd
Arcadian - Haddonstone Ltd
Arcadian garden features - Haddonstone Ltd
Arcal - Hanson Aggregates
Archco-Rigidon - Winn & Coales (Denso) Ltd
Archform - Helmsman
Architectural - Protim Solignum Ltd
Architectural Concrete Ltd - Marble Mosaic Co Ltd, The
Architectural coping - Alifabs (Woking) Ltd
Architectural Facing Masonry - Lignacite Ltd
Architectural Vinyls - Architectural Textiles Ltd
Architectural Window Films - Northgate Solar Controls
Archmaster® - Simpson Strong-Tie®
Archtec - Cintec International Ltd
Arcitectural Textiles - Architectural Textiles Ltd
Arclime - Hanson Aggregates
Arclion - Warner H & Son Ltd
Arco - Provex Products Ltd
Arcus - Erlau AG
Arcus - Laidlaw Architectural Hardware
Arcus - The HB Sign Company Ltd
Arden lighting - Arden Manufacturing Birmingham Ltd
Arden lighting - Arden Manufacturing Birmingham Ltd
Ardenbrite - Tor Coatings Ltd
Ardesia - EBC UK Ltd
Ardex - Ardex UK Ltd
Ardex DMP - Ardex UK Ltd
Ardex Feather Finish - Ardex UK Ltd
Ardex UK Ltd - Ardex UK Ltd
Ardex-flex - Ardex UK Ltd
Arditex - Ardex UK Ltd
Areana Job Costing - Arena Software Ltd
Areana Job Costing - Arena Software Ltd
Areco Polycarbonate - Areco Roofing Supplies Company
Arecotrim - Areco Roofing Supplies Company
Arena - Hille Ltd
Arena Accounts - Arena Software Ltd
ARES - Commercial Lighting Systems
Argenta - Imperial Machine Co Ltd
Argo - TSE Brownson
Argo-Braze - Johnson Matthey PLC - Metal Joining
Aria - Ellis J T & Co Ltd
Ariane Spiral - Loft Centre Products
Aries - Fitzgerald Lighting Ltd
Aristocast - Troika Architectural Mouldings Ltd
Aristocrat - Jaga Heating Poducts (UK) Ltd
Aristom - Diamond Merchants
Arito - Top Office Equipment Ltd
Ark - Vulcanite Limited
Arlon - DLW Flooring
Armacolour - Yonder Hill Ltd
Armadillo Products - Yonder Hill Ltd
Armadillo Range - Wybone Ltd
Armaflex - Davant Products Ltd
Armaglaze - Yonder Hill Ltd
Armamesh - Yonder Hill Ltd
Armapod - CSI Security LTD
Armarr Sectional + Roller Shutters - Regency Garage Door Services Ltd
Armaseam - Alumasc Exterior Building Products Ltd
Armatherm - Crosbie Coatings
Armbond - Armstrong (Concrete Blocks), Thomas, Ltd
Armbond - Armstrong (Concrete Blocks), Thomas, Ltd
Armco - Asset International Ltd
Armitage Shanks - Barflow-LCA
Armorcote - Armorex Ltd
Armordon - Attwater & Sons Ltd
Armordon - Attwater & Sons Ltd
Armorsol - Armorex Ltd
Armourcast - Gillespie (UK) Ltd
armourclad - HMG Paints
Armourdoor - Piper Double Glazing Ltd
Armspan - Armfibre Ltd
Armspan - Armfibre Ltd
Armstrong - Kitsons Insulation Products Ltd
Armstrong - Nevill Long Limited
Armstrong - New Forest Ceilings Ltd
Arnold Montrose - Denmans Montrose Ltd
Aro Pumps - Ingersoll-Rand European Sales Ltd
Arosa - Gloster Furniture Ltd
Arpa - Deralam Laminates Ltd
Arpax - Leigh's Paints
Arpeggio - Rawson Carpets Ltd
Arresta Rail - Unistrut Limited
Arroclip - Barlow Group
Arrogard - Barlow Group
Arrolok - Barlow Group
Arrone - HOPPE (UK) Ltd
Arrow - Walls & Ceilings International Ltd
Arstyl - nmc (UK) Ltd
Art Deco Suite - Brass Art Ltd
Arteco - British Gypsum Ltd
Artek - Astron Buildings Parker Hannifin PLC
Artel - Air Improvement Centre Ltd
Artemide - Quip Lighting, Div of Bruce & Macintyre Ltd
Artemis - Desking Systems Ltd
Artemis - Stocksigns Ltd
Artesit - Erlau AG
Artificial Grass - Sportmark™ Ltd
Artikula - Anglepoise Lighting Ltd
Artista Fusing Glass - SCHOTT UK Ltd
Artizan - Sovereign Brush Co Ltd, The
Artoleum - Forbo - Nairn Ltd
Artworks - Gaskell Carpet Tiles
Arunhithe - Falco
AS Aqua Seal - Consult Lighting
Asbestolok® - Liquid Plastics Ltd
Asbestoscheck Interactive - National Britannia Ltd
Asco - Asco Extinguishers Co Ltd T/A The Asco Fire Group
Asco - Asco Extinguishers Co Ltd T/A The Asco Fire Group
Ascott Designs - Desney Products Ltd
Ashbury - Paragon by Heckmondwike
Ashdowne - Eternit Building Materials
Ashfab - Ash & Lacy Building Products Ltd
Ashfix - Ash & Lacy Building Products Ltd
Ashflow - Ash & Lacy Building Products Ltd
Ashgrid - Ash & Lacy Building Products Ltd
Ashgrid - Ash & Lacy Building Products Ltd
Ashjack - Ash & Lacy Building Products Ltd
Ashley - Hager Ltd
Ashtherm - Ash & Lacy Building Products Ltd
Ashworth Speedy tester - GE Protimeter
Ashworth Weyrite - Phoenix Scales Ltd
ASI Glass Building Integrated Photovoltaics - SCHOTT UK Ltd
ASK - RPS Data Products (UK) Ltd
Aslux - Yorian Garden Products
Asmentic Collection - Masterframe Windows Ltd
Aspiromatic - Docherty New Vent Chimney Group
Aspirotor - Brewer Metalcraft
Aspirotor - Marflex Chimney Systems
ASR 683 - Monowa Manufacturing (UK) Ltd
ASSA - ASSA Ltd
ASSA - ASSA Ltd
Associated Asphalt - Stenoak Associated Services PLC
Associated Asphalt - Stenoak Associated Services PLC
Associated International - Stenoak Associated Services PLC
Associated Metal - Pland Stainless Ltd
Associated Solutions - Stenoak Associated Services PLC
Association of Safety Fencing Contractors (ASFA) - Fencing Contractors Association
Assured Data Sheets - British Board of Agrement
Asthetics - ARC Partitioning
Asthetics - ARC Partitioning
Astoria - Muraspec

Astra - Guardall Ltd
Astra - Hometex Trading Ltd
Astra Micra - Guardall Ltd
Astracast - Astracast PLC
Astral Frame - James Smellie Fabrications Ltd
Astralite (minor surgical lamps) - Brandon Medical
Astraroof - Stoakes Systems Ltd
Astrawall - Stoakes Systems Ltd
Astrawall SSG - Stoakes Systems Ltd
Astro - Zellweger Analytic Ltd, Sieger Division
Astron - Astron Buildings Parker Hannifin PLC
Astron - Astron Buildings Parker Hannifin PLC
Astron MSB - Astron Buildings Parker Hannifin PLC
Astronet - Astron Buildings Parker Hannifin PLC
Astrotec - Astron Buildings Parker Hannifin PLC
Astura - Trent Bathrooms
ASX - Hochiki Europe (UK) Ltd
AT ™ - Maestro International Ltd
ATA - PCT Group Sales Ltd
Atelier Sedap - Optelma Lighting Ltd
Athens - Metlex
Athlon - Trespa UK Ltd
Athmer - Strand Hardware Ltd
Atlantic - Signs & Labels Ltd
Atlas - Ellis Patents
Atlas - T & D Packaging Ltd
Atlas (Smart & Brown) - Bernlite Ltd
Atlas Buildings - Atlas Ward Structures Ltd
Atlas Buildings - Atlas Ward Structures Ltd
Atlas Concorde - Swedecor Ltd
Atlas Solar - Sundwel Solar Ltd
Atlaz - Woolliscroft Tiles Ltd
Atox - Aldous & Stamp Services Ltd
Audio System 200 - BPT Security Systems (UK) Ltd
Audiopath' Voice Alarm Systems - PEL Services Limited
Aura - Eaton Electric Limited
Aurora - Novellini Bathroom Products Ltd
Ausmark - Ramset Fasteners Ltd
Auto Flow - Hattersley Newman Hender Ltd
Auto M - Airmaster Engineering Ltd
Auto Pile - Cementation Foundations SKANSKA Ltd
Auto Seal - Exitex Ltd
Autodor - Garador Ltd
Autodos - Golden Coast Ltd
Autoflow - Pillinger G C & Co (Engineers) Ltd
Auto-Freway - Tuke & Bell Ltd
Autoglide - Cardale Doors Ltd
Autoglide - Silent Gliss Ltd
AutoKote - Grace Construction Products Ltd
Automat - Elequip Projects Ltd
Automax - Flowserve Flow Control
Automet - Metalrax Ltd
Auton - Harrison Drape, Divsion of Newell Ltd
Autopa - Autopa Ltd
AutoStamp™ Franking & Weighing System - Neopost Ltd
Autowalk - Kone Escalators Ltd
Autowaste - Fordwater Pumping Supplies
Autowaste - Fordwater Pumping Supplies Ltd
Avalon - Formed Steel Radiators - Clyde Combustions Ltd
Avalon Glass Tiles - Kirkstone
Avance - Douglas Industrial Ltd
Avanti Partitions - Davies John Interiors Ltd
Avanti Partitions - Logic Office Contracts
Avanza Benching System - Envopak Group Ltd
Avdel® - Textron Fastening Systems Ltd
Avdelok® - Textron Fastening Systems Ltd
Avenue - Neptune Outdoor Furniture Ltd
Aveseal® - Textron Fastening Systems Ltd
Avex® - Textron Fastening Systems Ltd
Avon - Top Office Equipment Ltd
avon gutter system - Dales Fabrications Ltd
Avon Industrial Doors Ltd - Avon Industrial Doors Ltd
Avonova - Avon Manufacturing Ltd
Axcess - Rackline Systems Storage Ltd
Axedstone - Hampton Stone Ltd
Axim - Permclose Door Services Ltd
Axiom - Armstrong World Industries Ltd
Axis - Aqualisa Products
Axis - Dixon Turner Wallcoverings
Axjet - Howden Buffalo
Axminster - Tankard
Axona Hydrotherapy pool - Arjo Ltd
Axor - Hansgrohe
Axpet - Amari Plastics Plc
Axtehal GX Fire Resistant Plastics - Plasticable Ltd
Azimex - Azimex Fabrications Ltd
Aztec Boilers - Trianco Redfyre Ltd
Aztec Europe Ltd - Flexel International Ltd
B & G Bertha - Kair ventilation Ltd
B&R - Power Breaker
B.P. - Anderson Gibb & Wilson
Babcock Transformers - Babcock Transformers
Babcock Transformers - Babcock Transformers
Baby Ben - General Time Europe
Baco Alcan, now A Line Systems - Alframes Ltd
BACS - British Association for Chemical Specialities
Bacticell - Nationwide Filter Co Ltd
Bacti-clean - Liquid Technology Ltd
Bacticoat - Nationwide Filter Co Ltd
Bacti-coat floor coating - Liquid Technology Ltd
Bacti-coat wall / ceiling coating - Liquid Technology Ltd
Bactiguard - Liquid Technology Ltd
Baffles - Rockfon Ltd
Baggeridge - Baggeridge Brick PLC
Baggeridge - Baggeridge Brick PLC
Bailey - Bailey Birkett Sales Division
Bailey - Cash - Bailey Birkett Sales Division
Baka - Industrial Devices Ltd
Bakaware - Brettell & Shaw Ltd
Bal - Stokes R J & Co Ltd
Bal - Admix AD1/GT1 - Building Adhesives Ltd
Bal - Easypoxy - Building Adhesives Ltd
Bal - Floor Epoxy - Building Adhesives Ltd
Bal - Grout - Building Adhesives Ltd
Bal - Pourable Thick Bed - Building Adhesives Ltd
Bal - Rapidset - Building Adhesives Ltd
Bal - Superflex Wall & Wide Joint Grouts - Building Adhesives Ltd
Bal - Wall Green/Blue/White Star - Building Adhesives Ltd
Bal Gold Star - Building Adhesives Ltd
Bal Microseal - Building Adhesives Ltd
Bal Supergrout - Building Adhesives Ltd
Bal- Wide Joint Grout - Building Adhesives Ltd
Ball - Millar Dennis (1998) Ltd
Balley Birkett - The BSS Group PLC
Ball-O-Star Ball Valves - Klinger Fluid Instrumentation Ltd
Ball-O-Top - Klinger Fluid Instrumentation Ltd
Balmoral - Davant Products Ltd
Balmoral - Sandtoft Roof Tiles Ltd
Balmoral bunded tanks - Balmoral Tanks
Balmoral Porcelain - Balmoral Porcelain Limited/Longmead Ceramics
Balmoral Porcelain - Balmoral Porcelain Limited/Longmead Ceramics
Balmoral spaceformer modular buildings - Balmoral Tanks
Balmoral superfill - Balmoral Tanks
Balmoral Tanks - Balmoral Tanks
Baltic - Neptune Outdoor Furniture Ltd
Baltimore - Townscape Products Ltd
Bal-Trol - Metreel Ltd
Balzard - Girsberger London
Bandit - Brissco Equipment Ltd
Bandit - Brissco Equipment Ltd
Bandit - Surmak Products Ltd
Bandseal - Ashmead Buidling Supplies Ltd
Band-seal - Naylor Drainage Ltd
Baraflor - MGL Carpet Finishers Ltd
Barbican - Carpets of Worth Ltd
Barbican ® - Jackson H S & Son (Fencing) Ltd
Barbican Steel Doors - Bradbury Group Ltd
Barclay Kellett - Albany Standard Pumps
Barco Pantile - Blyth William
Barco Pantile - Blyth William
Barcooler - Imperial Machine Co Ltd
Bardeau - Smithbrook Building Products Ltd
Bardoline - Onduline Building Products Ltd
Barduct BL - Barduct Ltd
Barduct CL - Barduct Ltd
Barduct ML - Barduct Ltd
Barduct UL - Barduct Ltd
Barduct XL - Barduct Ltd
Barduct XL - Barduct Ltd
Barent - Neptune Outdoor Furniture Ltd
Baripex - IBP Ltd
Barkeller - Imperial Machine Co Ltd
Barkers engineering Ltd - Chestnut Products Limited
Barkwith engineering - Ermine Engineering Co. Ltd
Barlo - Diamond Merchants
Barlo Radiators - Barlo Radiators Limited
Barlow metal fabrications - Barlow Group
Barlow metal fabrications - Barlow Group
Barnwood - Barnwood Shopfitting Ltd
Barnwood - Barnwood Shopfitting Ltd
Baron - Bartlett Catering Equipment Ltd
Barrier Floor Seal SB/30 - Tank Storage & Services Ltd
Barrier Foil - Thatching Advisory Services Ltd
Barspeed - BRC Ltd
Bartangle - Barton Storage Systems Ltd
Bartender - Imperial Machine Co Ltd
Bartisan - Clenaware Systems Ltd
Bartoline - Bartoline Ltd
Bartoline - Bartoline Ltd
Barton & Merton - Anthony de Grey Trellises
Bartrak - Barton Storage Systems Ltd
Bartspan - Barton Storage Systems Ltd
Barweld - BRC Ltd
Barwil - Barber Wilsons & Co Ltd
Barwil - Barber Wilsons & Co Ltd
Bas 2800+ - Satchwell Control Systems Ltd
Basalton Revetment - Ruthin Precast Concrete Ltd
Baseline - Twyford Bushboard
Baseline - Twyford Bushboard
Basepak - Flamco UK Ltd
Basis - Flexiform Business Furniture Ltd
Bat - Allmat (East Surrey) Ltd
Bat - Expamet Building Products
Batchmatic - BSA Machine Tools Ltd
Baths/Brassware/Panels/Seats/Urinals/Low Level - Topravit (UK) Ltd
Bathscreens by Coram - Coram Showers Ltd
Batiguard - Nationwide Filter Co Ltd
Bauclad - CEP Claddings Ltd
Baufix - Pump Technical Services Ltd
Bax- net - bespoke network - Baxall Ltd
Bayferrox - Bayer Chemicals Inoegate Pigment Group
Bayferrox - Bayer Chemicals Inoegate Pigment Group
Baysilone - GE Bayer Silicones
BBA - British Board of Agrement
BBA - British Board of Agrement
BCA - British Cement Association
BCC - British Ceramic Confederation
BCF - British Coatings Federation
BCIS - Building Cost Information Service
BCIS Online - Building Cost Information Service
BCMA - The Carpet Foundation
Beach Glass Mosaics - Kirkstone
BEAMA - British Electrotechnical and Allied Manufacturers Association
Beamform - Rom Ltd
Beamrider - Metreel Ltd
Beany Block - Marshalls Mono Ltd
Beaufort - Bassett & Findley Ltd
Beaufort - Neptune Outdoor Furniture Ltd
Beaumont Chimneys - Beaumont Ltd F E
Beaumont Chimneys - Beaumont Ltd F E
Beauport - Bennett Windows & Doors
Beauty of marble collection - Emsworth Fireplaces Ltd
Beaver - Hipkiss, H, & Co Ltd
Becker - F. W. Harris & Co Ltd
Becker Acroma - Anderson Gibb & Wilson
Bedale - Oakdale (Contracts) Ltd
Bee Bumps - B&C Roadmarkings Ltd
Beeline - Halls Beeline Ltd
Beemul - Emusol Products (LEICS) Ltd
Bekaert - Betafence Limited
Bekaert - Hi-Tec Window Films
Bekaert fencing - Chestnut Products Limited
Bekaert Fencing 10 yr guarantee system - Warefence Limited
Bekassure - Termstall Limited
BEKassure 10 Year Guarantee - Albion Fencing Ltd
Belair - Draeger Safety UK Ltd
Belcor - May William (Ashton) Ltd
Belgravia - S & P Coil Products Ltd
Belgravia Mixture - Cranleigh Brick & Tile Co Ltd
BELL - British Electric Lamps Ltd
Bell Bollard - Furnitubes International Ltd
Bell Fireplaces - Bell & Co Ltd
Bell Fireplaces - Bell & Co Ltd
Bell fires - Bell & Co Ltd
Bell Twist - Brintons Ltd
Bella - Kermi (UK) Ltd
Belle - Chippindale Plant Ltd
Bendix - Electrolux Domestic Appliances
Bendix - Parkinson Cowan and Tricity Bendix Co, The
Benga W - Jotun Henry Clark Ltd Decorative Division
Beniedale - Chiltern Invadex
Bennet Doors - Bennett Windows & Doors
Bennet Windows - Bennett Windows & Doors
Berber Twist Elite - Brockway Carpets Ltd
Berber Twist Forte - Brockway Carpets Ltd
Beresford - Baresford Pumps Ltd
Beresford - Baresford Pumps Ltd
Bergen - Loft Centre Products
Bergen - Scandinavian Window Systems Ltd
Berkeley - Beeston Heating Group Ltd
Berkeley - Carpets of Worth Ltd
Berkeley - Crittall Steel Windows Ltd
Berkeley Variations - Carpets of Worth Ltd
Berkshire Multi - Charnwood Brick Group Ltd
Berry - Glen Dimplex UK Ltd
Berry - Seaque Technologies Ltd
Berry Beam - Berry Systems
BESA - British Electrical Systems Association
BESA - British Electrical Systems Association
Best In the World - Stockten Heath Forge (Caldwells) Ltd
Besta - LSA Projects Ltd
Beta - Beta Lighting Ltd
Beta - Beta Lighting Ltd
Beta - Brett Paving
Beta Naco - Air Diffusion Ltd
Beta Naco - Air Diffusion Ltd
Betec - Beton Construction Ltd
Betec - Forma Lighting Ltd
Betoatlas - Grass Concrete Ltd
Betoconcept - Grass Concrete Ltd
Betokem - Optiroc Ltd
Betonap - Grass Concrete Ltd
Betonap - Grass Concrete Ltd
Betotitan - Grass Concrete Ltd
Beverley - Beeston Heating Group Ltd
Bexor - BRB Industrial Services Ltd
BFA - Cast Metals Federation
BFCMA - British Flue and Chimney Manufacturers Association
BFRC - British Fenestration Rating Council
BFT - Crimeguard Systems Ltd
BFT automatic gates - Automatic Doors & Gates
BHHMA - British Hardware and Housewares Manufacturers Association
Bianca - Arjo Ltd
Bianco - Kermi (UK) Ltd
Bicester Court Floors - Bicester Products Ltd
Bicester Play Wall System - Bicester Products Ltd
Bicester Play Wall System - Bicester Products Ltd
BICTA - Cast Metals Federation
BiDi Safe - Safetell Ltd
Bidston Door - AS Newbould Ltd
Bieri - Biral Pumps Ltd
Bifjet - Howden Buffalo
Biflo - Grohe Ltd
Big Ben - General Time Europe
Big Green Chart - Sportmark™ Ltd
Bighead ® - BigHead Bonding Fasteners Ltd
Bighead ® - BigHead Bonding Fasteners Ltd
Bigwood Stokers - Hodgkinson Bennis Ltd
Bilco - Bilco UK Ltd
Bilco - Bilco UK Ltd
Bimagrip - Clare R S & Co Ltd
Binnlock Mesh Fencing - Binns Fencing Ltd
Binnlock Mesh Fencing - Binns Fencing Ltd
Binwall - Grass Concrete Ltd
Bio rolls - Rawtex Ltd
BioDisc ® - Klargester Environmental Ltd
Biogard - Leigh's Paints
Biogaurd - Armstrong World Industries Ltd
Biogene - Corroless Corrosion Control
Biomass Grease Traps - Progressive Product Developments Ltd
Biometric - FC Frost Ltd
Biostat - Tuke & Bell Ltd
Biotank - Biotank Ltd
Biotec - Titan Pollution Control Ltd
Biral - Beeston Heating Group Ltd
Biral - Biral Pumps Ltd
Biral - Biral Pumps Ltd
Birco - Marshalls Mono Ltd
Birkdale Bathroom Suite - Ideal-Standard Ltd
Birkett - Bailey Birkett Sales Division
Birthday - Hoskins Medical Equipment Ltd
Bisley - Beeston Heating Group Ltd
Bisley - Beeston Heating Group Ltd
Bisley - Bisley Office Equipment
Bisley - Bisley Office Equipment
Bison Beams - ACP (Concrete) Ltd
Bitite - Callenders Ltd
Bitite - Callenders Ltd
Bitron - Interphone Security Group PLC
Bitubond Cork - SCP Concrete Sealing Technology Ltd
Bitufor - Bekaert Building Products
Bituline - Onduline Building Products Ltd
Bitumastic - Pearl Paints Ltd
Bitusheet - SCP Concrete Sealing Technology Ltd
Bituthene - Grace Construction Products Ltd
Biwater - Biwater Industries Ltd
Biwater - Biwater Industries Ltd
Biz - Biz Contracts
Biz - Biz Contracts
BK - Imperial Machine Co Ltd
BL Bollard - Consult Lighting
Black Beauty - Valor
Black Ebony - Manders Paint Ltd
Black Prince - Pearl Paints Ltd
Blackbox Domestic Safe - Chubb Safe Equipment Company
Blackfriar - Blackfriar Paints Ltd
Blackfriar - Blackfriar Paints Ltd
Blackfriar - Tor Coatings Ltd
Blackheat ® - Roberts-Gordon
Blackjack Square Spiral - Loft Centre Products
Blake - Tuke & Bell Ltd
Blakely Electrics - Blakley Electrics Ltd
Blakley - Blakley Electrics Ltd
Blanc de Bierges - Blanc de Bierges
Blanc de Bierges - Blanc de Bierges
Blanc de Bierges - Milner Delvaux Ltd
Blast Coolers - Turnbull & Scott (Engineers) Ltd
Blastrac - USF Surface Preparation
Blaw-Knox - Ingersoll-Rand European Sales Ltd
Blenheim - Adam Carpets Ltd
Blenheim - Bassett & Findley Ltd
Blika - LSA Projects Ltd
Blind Bolt - Henry Venables Interiors Ltd
B-Line - Bartlett Catering Equipment Ltd
Blink - Shackerley (Holdings) Group Ltd incorporating Designer ceramics
Blockmaster - Aggregate Industries UK Ltd
Blockmaster - Aggregate Industries UK Ltd
Blockmaster - CAEC Howard (Holdings) Ltd
Blocnail - Glasgow Steel Nail Co Ltd
Blok-Kade 'D' - Manton Industrial Seals Ltd
Bloomsbury - Ellis J T & Co Ltd
BLP - BLP (Hamble) Ltd
BLP - BLP (Hamble) Ltd
BLP - BLP Group
BLP - BLP Group
BluBat - Junckers Limited
Blue 92 - Laticrete UK
Blue ridge - Campbell Marson & Co
Blue ridge - Campbell Marson & Co
BLW - BLW Associates Ltd
BLW - BLW Associates Ltd
BM 310 - Monarflex Anderson Ltd
BMA - Bathroom Manufacturers Association, The
BMCA - Cast Metals Federation
BMF - Builders Merchants Federation
BMI - Building Cost Information Service
BMI Building Maintenance Price Book - Building Maintenance Information
BMI Bulletin Service - Building Maintenance Information
Bn - Wilo Samson Pumps Ltd
Boatman Plastics - Boatman Plastics Ltd
BOB Stevenson - BOB Stevenson Ltd
BOB Stevensons - BOB Stevenson Ltd
Bobi Mail Boxes - JB Architectural Ironmongery Ltd
Boda - Industrial Devices Ltd
Boge - F. W. Harris & Co Ltd
Boiler Crown Valves - Millar Dennis (1998) Ltd
Boilermate - Gledhill Water Storage Ltd
Bold roll - Forticrete Roofing Products
Bolderaja - Boardcraft Ltd
Boldroll - Britmet Tileform Limited
Bolero Frame - James Smellie Fabrications Ltd
Bolivar - IBP Ltd
Bolom Fire Protection - The Bollom Group

Bolton brady - Konematic Bolton Brady
Bolton Wood Yorkstone - Pickard Group
Bomag - Chippindale Plant Ltd
Bonakemi - Campbell Marson & Co
Bond It - Avocet Hardware Ltd
Bonda Contact Adhesive - Bondaglass Voss Ltd
Bonda PU Adhesive - Bondaglass Voss Ltd
Bonda PVA Adhesive - Bondaglass Voss Ltd
Bonda Woodfill - Bondaglass Voss Ltd
Bonda Woodstop - Bondaglass Voss Ltd
Bonda Woodstop - Bondaglass Voss Ltd
Bondax - Carpets of Worth Ltd
Bondel - Boardcraft Ltd
Bondite - Trade Sealants
Bondmaster - National Starch & Chemical Ltd
Bondwave - Flexible Reinforcements Ltd
Bonsai - Satchwell Control Systems Ltd
Bonus by Coram - Coram Showers Ltd
Bonus Plus by Coram - Coram Showers Ltd
Booths - Booth Samuel & Co Ltd
Booths - Booth Samuel & Co Ltd
Bordalok - Rawson Fillings Ltd
Border Range - Esk Building Products
Border Stone - Aggregate Industries UK Ltd
Borems - Luxo UK Ltd
Bosch - Arrow Supply Company Ltd
Bosch - Crimeguard Systems Ltd
Bosch - Robert Bosch Ltd
Boss - SGB Youngman
Boss White - The BSS Group PLC
Boulevard - Fitzgerald Lighting Ltd
Boulevard - Marshalls Mono Ltd
Boulton & Paul - Jeld-Wen UK Ltd
Bourne Parking - Bourne Steel Ltd
Bourne Special Projects - Bourne Steel Ltd
Bowland Stone - Concrete Products (Lincoln) 1980 Ltd
Bowls Bumper - Sportmark™ Ltd
BowTie - Helifix Ltd
Box culverts - Johnston Pipes Ltd
BPD - Zled (DPC) Co Ltd
BPF - British Plastics Federation, The
BPMA - British Pump Manufacturers Association
BPT - Crimeguard Systems Ltd
Brad - NBS
Bradco - Sissons W & G Ltd
Bradley - Relcross Ltd
Bradstone - Aggregate Industries UK Ltd
Bradstone - Ashmead Buidling Supplies Ltd
Bradstone Paving - B & M Fencing Limited
Bradstone slabs - B & M Fencing Limited
Braemar suite - Armitage Shanks Ltd
Brahams - Bitmen Products Ltd
Brailliant Touch - Signs & Labels Ltd
Bramah - Bramah Security Equipment Ltd
Bramah - Bramah Security Equipment Ltd
Brampton - Anthony de Grey Trellises
Bran + Luebbe - SPX Process Equipment
Brandy Crag - Burlington Slate Ltd
Brathay Blue/Black Slate - Kirkstone
Brazilian Pinc plywood - Wood International Agency Ltd
BRC - BRC Ltd
BRC - BRC Ltd
BRE - Building Reserch Establishment
Breathalyzer - Draeger Safety UK Ltd
Brecon - Sashless Window Co Ltd
Breeam - BRE
Breeam - BRE
Breeza - London Fan Company Ltd
Breezair - Radiant Services Ltd
Breezax - London Fan Company Ltd
Breeze - Rycroft Ltd
Bretshaw - Brettell & Shaw Ltd
Bretshaw - Brettell & Shaw Ltd
Brett Landscaping and Building Products - B & M Fencing Limited
Brett martin - FTW Windows
Brewer - Brewer Metalcraft
Brewer - Brewer Metalcraft
BRFC - British Flat Roofing Council (BFRC)
Brick & Stone - Optiroc Ltd
Brickfill - Fillcrete Ltd
Brickforce® - BRC Special Products
Brickforce® - BRC Special Products
Brickhouse - Drainage Systems
Bricklifter - Benton Co Ltd Edward
Bricklifter - Edward Benton & Co Ltd
Brickseal - Fillcrete Ltd
Brickslotdrain - ACO Technologies Plc
Bricktie® - BRC Special Products
Bricktor® - BRC Special Products
Bridge - Checkmate Industries Ltd
Bridge - Townscape Products Ltd
Bridge Quarry - Bridge Quarry
Bridge-Coat - Andura Textured Masonry Coatings Ltd
Bridgestone - Braithaite Engineers Ltd
Bridgestone - Braithaite Engineers Ltd
Bridgman Doors - Industrial Building Componants Ltd
Bridgwater - Sandtoft Roof Tiles Ltd
Briercom - Brierkrete Ltd
Brierkrete - Brierkrete Ltd
Brierkrete - Brierkrete Ltd
Briermarine - Brierkrete Ltd
Brifen - Stenoak Associated Services PLC
Briflo - Brickhouse
Briflo - Brickhouse
Briflor - Brickhouse
Brig - Henry Venables Interiors Ltd
Brig - Henry Venables Ltd
Brigadier - Richard Hose Ltd
Brillite - Witham Oil & Paint Ltd
Brimar Tanks - Brimar Plastics Ltd
Brimar Tanks - Brimar Plastics Ltd
Brimax - Brickhouse
Bripave - Brickhouse
Briseal - Brickhouse
Brisol - British Solvent Oils, A Division of Careless
Brisol - British Solvent Oils, A Division of Careless
Bristlex - Exitex Ltd
Bristol - Gloster Furniture Ltd
Bristol Armour - Meggitt Armour Systems
Bristol Maid - Hospital Metalcraft Ltd
Bristowes - Bitmen Products Ltd
Bristowes - Bitmen Products Ltd
Brit Clips - Kem Edwards Ltd
Britannia - Diamond Merchants
Britannia - Marflow Eng Ltd
Britannia Extinguishers - UK Fire International Ltd
Britclips - Emlux Holdings Ltd
Britelite - Cooper Lighting and Security Ltd
Britelux - Metal Halide Lamps - Sylvania Lighting International INC.
British Decorators Association - Painting & Decorating association
British Electric Lamps Ltd - British Electric Lamps Ltd
British Electric Lamps Ltd - British Electric Lamps Ltd
British Glass - British Glass Manufacturers Confederation
British Gypsum - Brikenden (Builders Merchants) Ltd
British Gypsum - Hatmet Ltd
British Gypsum - Kitsons Insulation Products Ltd
British Gypsum - Nevill Long Limited
British Monorail - Morris Material Handling Ltd
British Standards Online - British Standards Institution
Britlock - Sandtoft Roof Tiles Ltd
Briton - IR Security and Safety
Briton - Laidlaw Architectural Hardware
Briton - Laidlaw Architectural Hardware Ltd
Briton - Stanton plc
Britool - Arrow Supply Company Ltd
Britosterope - Orsogril Spa UK Distributor
Briv® - Textron Fastening Systems Ltd
Briwax - The Bollom Group
BRIX - Federation of Master Builders
Broadmead - Broadmead Cast Stone
Broadmead - Broadmead Cast Stone
Broadstel - Brickhouse
Broadway - Atlas Stone Products
Broadway - Atlas Stone Products
Broag-Remeha - Broag Ltd
Broag-Remeha - Broag Ltd
Brodclad - Broderick Structures Limited
Brodclad - Broderick Structures Limited
Brodeck - Broderick Structures Limited
Brodform - Broderick Structures Limited
Brodscrean - Broderick Structures Limited
Brodseam - Broderick Structures Limited
Broflame - The Bollom Group
Broflor - The Bollom Group
Brolac - Crown Trade Paints
Bromel - The Bollom Group
Bromel - The Bollom Group
Bromgard - Feedwater Ltd
Bron-Lock - Booth Muirie
Bronz-Steps - Quantum Profile Systems Ltd
Brook Air Changer - Brook Airchanger
Brook vent - Brook Airchanger
Brook Vent - Brook Airchanger
Brookes - Elequip Projects Ltd
Brooking - Clement Steel Windows
Brooking Collection - Brass Art Ltd
Brooks Deck - Brooks Belfast
Brooks Roof Units - Matthews & Yates Ltd
Brosteel - The Bollom Group
Broughton Controls - Konematic Bolton Brady
Broughton Moor - Burlington Slate Ltd
Broughton Moor - Burlington Slate Ltd
Broxap ® - Broxap Ltd
Broxley - Beeston Heating Group Ltd
Brummer - Clam-Brummer Ltd
Brunel (External bin) - Glasdon UK Ltd
Brunner Mono - Anderson Gibb & Wilson
Bruynzeel - Dexion Ltd
BSA - BSA Machine Tools Ltd
BSA - BSA Machine Tools Ltd
BSI British Standards - British Standards Institution
BSI Business information - British Standards Institution
BSI Business Soloution - British Standards Institution
BSI Information - British Standards Institution
BSI Inspectorate - British Standards Institution
BSI Management System - British Standards Institution
BSI Product Services - British Standards Institution
BSIA - British Security Industry Association
BSSA - British Stainless Steel Association
BSWA - British Structural Waterproofing Association, The
BTA PrintMachine Eazimark™ Laserprint Management - Neopost Ltd
Buccaneer - Brickhouse
Buchanan - Thomas & Betts Manufacturing Services Ltd
Buckden - Anthony de Grey Trellises
Buckingham - Metlex
Budget - Spur Shelving Ltd
Buell - Campbell Marson & Co
Buggie - Manitou (Site Lift) Ltd
Buggiescopic - Manitou (Site Lift) Ltd
Buildaid - Armorex Ltd
Builders Guild - Guild of Builders & Contractors, The
Buildex - ITW Spit
Building Bookshop - The Building Centre Group Ltd
Building Centre - The Building Centre Group Ltd
Building Watch - Advanced Interior Solutions Ltd
Bulletin Board - Forbo - Nairn Ltd
bullnose fascia soffit - Dales Fabrications Ltd
Bunnie Dryers - Wandsworth Elecrtrical Ltd
Burkle - Lyncrete Burkle Services Ltd
Burley - Beeston Heating Group Ltd
Buroflex - Metalliform Products Plc
Burr Bloc anti-ram raid bollard - Furnitubes International Ltd
Bursting Stone - Burlington Slate Ltd
Bush Circulators - Turney Turbines (pumps) Ltd
Bushboard - Anderson C F & Son Ltd
Butler MR24 Standing seam roofing systems - CA Group Ltd - MR-24 Division
Butterley - Hanson Brick Ltd, Northern Regonal Sales Office
Button diamond - Lionweld Kennedy Ltd
Butyl-Nek - Creteco Sales Ltd
Butzbach - Envirodoor Markus Ltd
BVC - Bivac - Barloworld Vacuum Technology Plc
Bvent Universal Gas Vent - Rite-Vent Ltd
BWA - Bitumen Waterproofing Association
BWF - British Woodworking Federation
BX2000 - Horstmann Controls Ltd
Bygone Collection - Masterframe Windows Ltd
Byland - York Handmade Brick Co Ltd
C & R - Accomodex
C raftwork - Bennett Windows & Doors
C.A.T. - Radiodetection Ltd
C.D.S. Car Deck Systems - Waterproofing Multi-Syorey Car Parks - Ferguson G A & Co Ltd
C.F.P - Westpile Ltd
C.R. window systems - BLW Associates Ltd
C+/C plywood - Wood International Agency Ltd
C4000 Combination - Elster Metering Ltd
C6L - McMullen Architectural Systems
CA Profiles - CA Group Ltd - MR-24 Division
CA Profiles - CA Group Ltd - MR-24 Division
Cabaret - Qualitas Bathrooms
Cabeline Ambassador - Mita (UK) Ltd
Cabeline Classic - Mita (UK) Ltd
Cabeline Powertrack - Mita (UK) Ltd
Cable - Cable Lift Installations Ltd
Cable - Cable Lift Installations Ltd
Cable Tidy - Tower Manufacturing
Cableline Ambassador Duo - Mita (UK) Ltd
Cablelink-Rapide - Ackermann Ltd
Cablemann - Ackermann Ltd
Cablofil - Mita (UK) Ltd
CAB-RAD - Property Mechanical Products Ltd
Cabsys - Mita (UK) Ltd
Cada - Kermi (UK) Ltd
Cadeby Limestone - Rare Stone Group, The
Cadenza - Laird Security Hardware
Cadet - Selbix Joinery Ltd
Cadet Kerb - Lafaroe Redland Precast
Cadweld ® - Erico Europa (GB) Ltd
Cafco - Firebarrier Services Ltd
Caimi - Lesco Products Ltd
Calder - Calder Industrial Materials Ltd
Calderdale - Sandtoft Roof Tiles Ltd
Caldwells - Stockten Heath Forge (Caldwells) Ltd
Caldy Door - AS Newbould Ltd
Caleffi - Altecnic Ltd
Caleffi - Altecnic Ltd
Calibre - BRE
Call Centre - Paragon Business Furniture
Callendrite - Callenders Ltd
Calor Autogas - Calor Gas Ltd
Calor Gas - Calor Gas Ltd
Calypso - Eltron Chromalox
Calypso Range - Evergreens Uk
Camargue - Carpets of Worth Ltd
Camargue - Qualitas Bathrooms
Cambrian - Brickhouse
Cambridge Gazebo - Anthony de Grey Trellises
Came - Crimeguard Systems Ltd
Came automatic gates - Automatic Doors & Gates
Camer - Anthony de Grey Trellises
Camer - Anthony de Grey Trellises
Camer Arbour seat - Anthony de Grey Trellises
Camer Rose Arch - Anthony de Grey Trellises
Camer Rose Walk - Anthony de Grey Trellises
Camfine Thermsaver - Kenilworth Daylight Centre Ltd
Camflex - Claydon Architectural Metalwork Ltd
Cam-Jack Technology - Apex Piling Ltd
Camjoint - Claydon Architectural Metalwork Ltd
Camlok - Camlok Lifting Clamps Ltd
Camlok - Camlok Lifting Clamps Ltd
Campden Pitched Stone - Atlas Stone Products
Campden Stone - Atlas Stone Products
Camray - Boulter Boilers Ltd
Camray - Boulter Boilers Ltd
Camrian slates - Lafarge Roofing Ltd
Camtwin - Claydon Architectural Metalwork Ltd
Canal - Optelma Lighting Ltd
Can-Corp - Bush Nelson PLC
Can-Corp - Bush Nelson PLC
Candy - British Ceramic Tile Ltd
Candy - British Ceramic Tile Ltd
Canigo' - Harbro Supplies Ltd
Cannon Bollard - Furnitubes International Ltd
Cannon Consumables - Cannon Hygiene Ltd
Cannon Horticulture - Cannon Hygiene Ltd
Cannon Pest Control - Cannon Hygiene Ltd
Cannon Textile CARE - Cannon Hygiene Ltd
Canon - RPS Data Products (UK) Ltd
Canonbury - Canonbury Asphalte Co. Ltd.
Canonbury - Canonbury Asphalte Co. Ltd.
Canply Cofi plywood - Wood International Agency Ltd
Canterbury - Atlas Stone Products
Canti-Bolt - Hi-Store Ltd
Canti-Clad - Hi-Store Ltd
Canti-Frame - Hi-Store Ltd
Canti-Guide - Hi-Store Ltd
Canti-Lec - Hi-Store Ltd
Canti-Lock - Hi-Store Ltd
Canti-Track - Hi-Store Ltd
Canti-Weld - Hi-Store Ltd
Capco - Capco Test Equipment Ltd
Capco - Capco Test Equipment Ltd
Cape - Firebarrier Services Ltd
Cape Boards - Kitsons Insulation Products Ltd
Capella - Ward Building Components Ltd
Capex - Exitex Ltd
Caplock - Ibstock Building Products Ltd
Capoplastic GP - Thermica Ltd
Cappair - Swish Building Products
Cappit - Swish Building Products
Capricorn Range - Caradon Bathrooms Ltd
Capstat - Heat Trace Ltd
Capyt SS100 composition - Thermica Ltd
Carbital SB - Imerys Minerals Ltd
Carboline - Nullifire Ltd
Carborundum Abrasives - Carborundum Abrasives G.B. Ltd
Carborundum Abrasives - Carborundum Abrasives G.B. Ltd
Cardale - Cardale Doors Ltd
Cardale - Cardale Doors Ltd
Cardale - Chiltern Invadex
Cardale - Moffett Thallon & Co Ltd
Cardale Doors - Regency Garage Door Services Ltd
Cardax - Vidionics Security Systems Ltd
Cardoc - Kent & Co (Twines) Ltd
Care LST - Hudevad Britain
Carefree Buffable - Johnson Wax Professional
Carefree Emulsion - Johnson Wax Professional
Carefree Eternum - Johnson Wax Professional
Carefree Gloss Restorer - Johnson Wax Professional
Carefree Maintainer - Johnson Wax Professional
Carefree Satin - Johnson Wax Professional
Carefree Speed Stripper - Johnson Wax Professional
Carefree Stride 1000 - Johnson Wax Professional
Carefree Stride 2000 - Johnson Wax Professional
Carefree Stride 3000 - Johnson Wax Professional
Carefree Stripper - Johnson Wax Professional
Carefree Undercoat - Johnson Wax Professional
Caremix - Altecnic Ltd
Carendo - Arjo Ltd
Carerscreen - Contour Showers Ltd
Caretaker III - Rapaway Energy Ltd
Care™ - Jack Moody Ltd
Cargo 2000 - Otis Ltd
Cargotrak - Rackline Systems Storage Ltd
Caribe - Holophane Europe Ltd
Caribe ® - Roberts-Gordon
Carlton - Metlex
Carlton - Mita (UK) Ltd
Carlton - Tuke & Bell Ltd
Carlton Brick - Carlton Main Brickworks Ltd
Carnival - Gaskell Carpet Tiles
Carousel - Heatrae Sadia Heating Ltd
Carpetguard - Gaskell Carpet Tiles
Carr - Carr Gymnasium Equipment Ltd
Carr of Nottingham - Carr Gymnasium Equipment Ltd
Carrier air - Toshiba Carrier UK
Carron Phoenix - Carron Phoenix Ltd
Carter Coldstores - Carter Retail Equipment Ltd
Carter Parratt - Railex Systems Ltd
Carter Refridgeration Display - Carter Retail Equipment Ltd
Carter Refridgeration Services - Carter Retail Equipment Ltd
Casana - Air Improvement Centre Ltd
Cascade - Calomax Ltd
Cascade - Calomax Ltd
Cascade - Potter & Soar Ltd
Cascade Wallwash Luminaire - Profile Lighting Services Ltd
Cascamite - Humbrol Ltd
Cascamite - Wessex Resins & Adhesives Ltd
Cascofil - Humbrol Ltd
Cascophen - Humbrol Ltd
Cascophen - Wessex Resins & Adhesives Ltd
Cascorez - Humbrol Ltd
Casco-Tape - Wessex Resins & Adhesives Ltd
Cashfast - CSI Security LTD
Casscom II - Telelarm care Ltd
Cast in situ linings - Cico Chimney Linings Ltd
Castalia cistern - Opella Ltd
Castalia Filter Clear - Opella Ltd
Castalia Tap Range - Opella Ltd
Castell - Castell Safety International
Castell 150 - Lok - Castell Safety International
Castelli - Haworth UK Ltd
Castle Bricks - Castle Brick (Wales) Ltd
Castle Multicem - Castle Cement Ltd
Castlemead Twist - Adam Carpets Ltd
Cataphas CR Paint - B&C Roadmarkings Ltd
Caterpillar - Briggs Industrial Footwear Ltd
Caterpillar - FG Wilson

Cathedral - Lafaroe Redland Precast
Cathedral Axminster - Mackay Carpets, Hugh
Cathedral Contract - Mackay Carpets, Hugh
Catnic - Catnic
Catnic - Wade Building Services Ltd
Catnic Classic - Catnic
Catnic Unique - Garador Ltd
Catseye - Reflecting Roadstuds Ltd
CATV & Telephony Cabinets - H&H Electronics Ltd
Caviarch - Cavity Trays Ltd
Cavicloser - Cavity Trays Ltd
Caviflash - Cavity Trays Ltd
Cavilintel - Cavity Trays Ltd
Caviroll - Cavity Trays Ltd
Cavitray - Cavity Trays Ltd
Cavity Trays of Yeovil - Cavity Trays Ltd
Cavity Wall Batts - Rockwool Ltd
Cavity Wallboard - Callenders Ltd
Cavivent - Cavity Trays Ltd
Caviweep - Cavity Trays Ltd
CBA - Concrete Block Association
Cbi - Worcester Heat Systems Ltd
CCL - CCL. Veloduct Ltd
CD 9000 Cameras - Baxall Ltd
CD 9000 Cameras - Baxall Ltd
CD Series - IMI Air Conditioning Ltd
CDC - Clark Door Ltd
CDI - USF Permutit/Ionpure
CDI - Worcester Heat Systems Ltd
CDR - CSW Coldform Ltd
CDR Cameras - Baxall Ltd
CDSP 9000 cameras - Baxall Ltd
CDW Products - Chindwell Co Ltd
CDX - Hochiki Europe (UK) Ltd
CE SI Wall Tiles - Stokes R J & Co Ltd
CECA - Civil Engineering Contractors Association
Cedar Green Houses - Compton Buildings Ltd
Cedec footpath gravel - CED Ltd
CeeJay - Simpson Strong-Tie®
Cefndy Healthcare - Cefndy Healthcare
Cego - Laird Security Hardware
Celafelt - Davant Products Ltd
Celazole® PBI - DSM Engineering Plastic Products (UK) Ltd
Celbor - Rentokil Ltd
Celbrite - Rentokil Ltd
Celbronze - Rentokil Ltd
Celco - Helvar Electrosonic
Celcure - Rentokil Ltd
Celeno - Orsogril Spa UK Distributor
Celeste - Ubbink (UK) Ltd
Celina- klinder Germany - Padipa Ltd
Cellarcote - Sealocrete Ltd
Cellarguard - Trade Sealants
Cellcore - Cordek Ltd
Cellcote - SGL Acotec Ltd
Cellform - Cordek Ltd
Cellio - Armstrong World Industries Ltd
Celmac - Polypipe Bathroom & Kitchen Products Ltd
Celpruf - Rentokil Ltd
Celtic Pantile - Blyth William
Celtite - Exchem Mining and Construction
Celuform - Caradon Duraflex Systems Ltd
Celuform - Celuform
Celuguard - Celuform
Cementitious Coating 851 - Flexcrete Ltd
Cemglaze - Cement Glaze Decorators Ltd
Cemglaze Specialist Wall Finish - Cement Glaze Decorators Ltd
Cemlevel - Armorex Ltd
Cempatch - Don Construction Products Ltd
Cemset - Cementation Foundations SKANSKA Ltd
CemTie - Helifix Ltd
Cemwhite - Cembrit Blunn Ltd
Centaur Plus - Horstmann Controls Ltd
Centaurstat - Horstmann Controls Ltd
Centemmial - Midland Stom Ltd
Centerline - Crane Fluid Systems
Centra - Tully
CentrAlert - Geoquip Ltd
CentralScotland Plumbers - Ogilvie Construction Ltd
Centrel - Emergi-Lite Safety Systems
Centurian - Laidlaw Architectural Hardware Ltd
Centurion - Boulter Boilers Ltd
Centurion - Heat Trace Ltd
Centurion - Laidlaw Architectural Hardware
Centurion - Shackerley (Holdings) Group Ltd incorporating Designer ceramics
Centurion 12.5 - Forticrete Roofing Products
Century - Yale Security Products Ltd
Cepac - Emusol Products (LEICS) Ltd
Cepeldor® - Vicaima Ltd
Ceramic Wall & Floor Tiles - Topravit (UK) Ltd
Ceramica - Golden Coast Ltd
Ceramitz Sprayed Stone - Textured European Finishes Ltd
Cermamica Impex - Stokes R J & Co Ltd
Certificates of Conformity - British Board of Agrement
Certikin - Certikin International Ltd
Cesi - Arnull Bernard J & Co Ltd
Cestilene® - DSM Engineering Plastic Products (UK) Ltd
Cetemodule DHW generator - HRS Hevac Ltd
CFP 2,4 and 8 Zone Fire Alarm Control Panels - C-TEC (Computionics) Ltd
CGL - CGL Cometec Ltd
CGL Totalwall - CGL Cometec Ltd
Chainex - BRB Industrial Services Ltd
Challenger - Abacus Lighting Ltd
Challenger - Dunhams of Norwich
Challenger - Flamco UK Ltd
Chameleon - Ulster Carpet Mills Ltd
Chamelneon - Oldham Lighting Ltd
Champion - Rawson Carpets Ltd
Chamry Surface Range - Lumitron Ltd
Chancery - Muraspec
Channel Plus - Horstmann Controls Ltd
Chantal Range - Caradon Bathrooms Ltd
Charcon - Aggregate Industries UK Ltd
Charisma - Laird Security Hardware
Charles Yorke - Symphony Group PLC, The
Charlotte - Qualitas Bathrooms
Charlotte Space Saver Spiral - Loft Centre Products
Charlton Gates - B & M Fencing Limited
Charter panel - Dunhams of Norwich
Chartered Institution of Waste Management - Chartered Institute of Wastes Management
Chartres - Formpave Ltd
Chase - Chase Equipment Ltd
Chateau - Muraspec
Check - Millar Dennis (1998) Ltd
Checkmate Servicing Package - Smoke Control Services Ltd
Checkstat - Checkmate Industries Ltd
Chelsea - Gloster Furniture Ltd
Cheltenham - Chase James & Son (Furnishing) Ltd
Cheltenham - Ellis J T & Co Ltd
Chelwood - Chelwood Brick Ltd
Chemco - Brash John & Co Ltd
Chem-crete - Cross-Guard International Ltd
Chemflo - CPV Ltd
Chemi-epoxeel - IFT Supplies Ltd
Chemset - Ramset Fasteners Ltd
Cheriton - Hamilton R & Co Ltd
Cherwell PV - Portacel
Cheshire Garden Tools - Stockten Heath Forge (Caldwells) Ltd
Cheshire Heritage - Optiroc Ltd
Chess - Flexiform Business Furniture Ltd
Chesterflex - Chesterfelt Ltd
Chestermeric - Chesterfelt Ltd
Chesterstar - Chesterfelt Ltd
Chevin - Cooper Lighting and Security Ltd
Chieftain - Falcon Food Services Equipment
Childers - Kingspan Industrial Ins. Ltd
Childrey seat - Anthony de Grey Trellises
Chilled ceilings - Clestra Hauserman Ltd
Chilstone Architectural Stonework - Chilstone
Chilstone Garden Ornaments - Chilstone
Chiltern - Maceman + Amstad Ltd
Chimaster MF - Docherty New Vent Chimney Group
Chimflex LW/SB - Rite-Vent Ltd
Chimliner - Marflex Chimney Systems
Chimney Capper - Brewer Metalcraft
Chimney Cowls - Birdscarer Products & Net Supplies Ltd
Chindwell Doors - Chindwell Co Ltd
Chippy - Owlett - Jaton
Chislestone - Hampton Stone Ltd
C-Hitlite - BLV Licht-und Vakuumtecnik
Chocflex - Bonar Floors Ltd
Chrome bond - Howse Thomas Ltd
Chrome Plus - MK (MK Electric Ltd)
Churchill - BSA Machine Tools Ltd
Churchill - Northcot Brick Ltd
CIArb - Chartered Institute of Arbitrators
Cico - Cico Chimney Linings Ltd
Cico flex - Cico Chimney Linings Ltd
Cig-Arrete - Airstream Products Ltd
CIJC - Construction Industry Joint Council
Cilplan - CIL International Ltd
Cimberio - SAV United Kingdom Ltd
Ciment Fondu Lafarge - Lafarge Aluminates Ltd
Cimm - SAV United Kingdom Ltd
CINCA Mosaic & Tiles - Reed Harris
Cintec - Cintec International Ltd
CIPEC JEP Asphaltic plug joints - Pitchmastic Pmb Ltd
Circulux - Poselco Lighting
CIRIA - Construction Industry Research & Information Association
Cirrus - Dermide Ltd
Cirrus - S & P Coil Products Ltd
Cirrus Gold - Dermide Ltd
CIS - Crack Inducing System - Compriband Ltd
Cissell Dryers - Warner Howard Group Ltd
Cistermiser - Cistermiser Ltd
Citadel - CSI Security LTD
Citadel - Fitzpatrick Doors Ltd
Citadel - SG System Products Ltd
Citax - Henkel Industrial Adhesives
CITB - Construction Industry Training Board
Citipave - Camas Aggregates Ltd - Aggregate Industries
Citizen - Townscape Products Ltd
Citycabin - Wernick S & Sons Ltd
Cityspace - Kone PLC
Citywall - CSI Security LTD
Civa - Optelma Lighting Ltd
CK Station - KSB Ltd
Clarendon - Legrand Electric Ltd
Clarke - Alto Cleaning Systems (UK) Ltd
Clarke "C" spray - Clarke UK Ltd
Class I Chimney Systems - Dunbrik (Yorks) Ltd
Classic - Alumasc Construction Products Ltd
Classic - Dunhams of Norwich
Classic - Hill Leigh Ltd
Classic - Kermi (UK) Ltd
Classic - Kompan Ltd
Classic - Paddock Fabrications Ltd
Classic - Pellfold Parthos Ltd.
Classic - Rangemaster
Classic - Vicaima Ltd
Classic - Vision Neon & Display Ltd
Classic 25 - Altro Floors
Classic 62 - Spectus Systems
Classic Coarse and Fine Texture - Andura Textured Masonry Coatings Ltd
Classic Collection - Masterframe Windows Ltd
Classic Collection - Masterframe Windows Ltd
Classic Line - Schneider (UK)
Classic Range - Eternit Building Materials
Classic Trolleys - Envopak Group Ltd
Classic® Kerb System - Brett Paving
Classica - Matki plc
Classicair louvres - Grille Diffuser & Louvre Co Ltd The
Classical - Saint-Gobain Pipelines PLC
Classical express - Saint-Gobain Pipelines PLC
Classical plus - Saint-Gobain Pipelines PLC
Classics - Muraspec
Classidur - Blackfriar Paints Ltd
Classique - Balmoral Porcelain Limited/Longmead Ceramics
Classmate - Rawson Carpets Ltd
Claudgen - Consort Equipment Products Ltd
Claverton - Gloster Furniture Ltd
Claymaster - Cordek Ltd
Claypenny - Oakdale (Contracts) Ltd
Clean steam - Fulton Boiler Works (Great Britain) Ltd
Cleanflow - Johnson & Starley
Cleanline - Neslo Partitioning & Interiors
Cleanline - Welconstruct Co Ltd
Cleanroom - Neslo Partitioning & Interiors
Cleanshield ™ - Accent Hansen
Cleanshield ™ - Accent Hansen
Clearfire - McKenzie-Martin Ltd
Clearline - Dyer Environmental Controls
Clearline - Neslo Partitioning & Interiors
Clearline - Teleflex - Morse
Clearstor Shelving - Moresecure Ltd
Clearview Glazed Overhead Doors - Avon Industrial Doors Ltd
Clenaglass - Clenaware Systems Ltd
Clenewall - Brown F plc
Clereflo - Conder Products Ltd
Cleveland - Sashless Window Co Ltd
Clic - Signs & Labels Ltd
Clic - Signs & Labels Ltd
Clickfix coping - Alifabs (Woking) Ltd
Clickfix coping - Alifabs (Woking) Ltd
Clients own brand if required - FG Wilson
Clifton - Formpave Ltd
Clifton - Maceman + Amstad Ltd
Clifton Landscape and design - Clifton Nurseries
Clifton little Venice - Clifton Nurseries
Climadur-Dabo - EJOT Ecofast Ltd
Climaflex - Davant Products Ltd
Climaflex - nmc (UK) Ltd
Climatronic - Satchwell Control Systems Ltd
Climet 100 Series - H&H Electronics Ltd
Clinicall - Tunstal Telecom Ltd
Clino21 - Ackermann Ltd
ClinoOpt 99 - Ackermann Ltd
Clipclad Panel Management - Grant Westfield Ltd
Clipfix 750 - European Profiles
Clipfold - Lycetts Industrial Doors
Clip-In - PFP Electrical Products Ltd
CliplaN10 - CIL International Ltd
Clipper - Calomax Ltd
Clipper - Norton Diamond Products
Clipstrip - Ryton's Building Products Ltd
Clip-Top - Gradus Ltd
Clivet porducts - Clivet UK Ltd
Clorocote - Witham Oil & Paint Ltd
Cloudburst - T & D Packaging Ltd
Club - Lappset UK Ltd
Clyde - IBP Ltd
Clyde - Portacel
Clyde Rail Spikes - Glasgow Steel Nail Co Ltd
cmf - Cast Metals Federation
CO2 Monitor - Gas Measurement Instruments Ltd
Coalbrookdale Stove - Aga-Rayburn
Coalmaster - Docherty New Vent Chimney Group
Coarse buff - Realstone Ltd
Cobra - Allmat (East Surrey) Ltd
Cobra - MSA (Britain) Ltd
Coburn - Hillaldam Coburn Ltd
Coex - Alumasc Construction Products Ltd
Cofast adhesive - Combustion Linings Ltd
Cofax castables - Combustion Linings Ltd
Cofax cements - Combustion Linings Ltd
Cofferlite - Indirect Lighting Ltd
Cog - Indirect Lighting Ltd
Coir - Land Wood & Water Co. Ltd
Coir - Rawtex Ltd
Cokeglow - Interoven Ltd
Colaquex - Emusol Products (LEICS) Ltd
Cold Cathode - Oldham Lighting Ltd
Coldblocker - Space - Ray UK
Coldfusion Bonded built up roofing system - Liquid Plastics Ltd
College Cycle Stand - Furnitubes International Ltd
Colman - Senior Air Systems
Colne - Portacel
Colonade - Matki plc
Colonial - Intersolar Group Ltd
Colorail - Rothley Burn Ltd
Colordril - EJOT Ecofast Ltd
Colorduct 2 - Hotchkiss Air Supply
Colorette - DLW Flooring Ltd
Colorex - Forbo - Nairn Ltd
Colorfast - EJOT Ecofast Ltd
Colorfects - Crown Trade Paints
Colorlite - BLV Licht-und Vakuumtecnik
Colormate - EJOT Ecofast Ltd
Colorpan - CEP Claddings Ltd
Colorslat - Barker W H Ltd
Colorslat - Barker W H Ltd
Colortone - Armstrong World Industries Ltd
Colortread - EMS Entrance Matting Systems Ltd
Colosseum - Metalliform Products Plc
Colour dimensions - ICI Dulux Trade Paints
Colour Express - Milliken Carpet
Colour Index - Muraspec
Colour Palette - ICI Dulux Trade Paints
Colour Palette - Tektura Plc
Colour Range - Ibstock Brick Ltd
Colour Select Glazed Bricks - Ibstock Brick Ltd
Colourcast - Procter Bros Limited
Colourfast - The Bollom Group
Colourmark - Armorex Ltd
Colouroute - Clare R S & Co Ltd
Colourpave - Clare R S & Co Ltd
Colourplex™ - Johnson Matthey Pigments & Dispersions
Colourtex - Bardon (England) Ltd
Colson - Avocet Hardware Ltd
Colt Houses - Colt W H & Co Ltd
Coltec - Cadisch MDA Ltd Ltd
Coltlite - Colt International
Columbia taping tools - Lafarge Plasterboard Ltd
Columbus timers - Elkay Electrical Manufacturing Co Ltd
Columns - Roman Mosaic Contracts Ltd
Comax - RFA Group Ltd
Combat - Range Cylinders
Combat ® - Roberts-Gordon
Combi - IR Security and Safety
Combi LNHI - Babcock Wanson Uk Ltd
Combi LNHI - Babcock Wanson Uk Ltd
Combiflex - Sika Ltd
Combimate - Cistermiser Ltd
Come Back to Carpit - The Carpet Foundation
Comelit - Crimeguard Systems Ltd
Comet - Laird Security Hardware
Cometec - CGL Cometec Ltd
Comfort Plus - Milliken Carpet
Comforto - Haworth UK Ltd
Commander - ABB Instrumentation Ltd.
Commando - Richard Hose Ltd
Communicall - Tunstal Telecom Ltd
Community - Checkmate Industries Ltd
Compac - Rycroft Ltd
Compact - ASSA Ltd
Compact - Calomax Ltd
Compact - Dewhurst plc
Compact & Origin Switches - Ensto Briticent Ltd
Compact Activ - Alno (United Kingdom) Ltd
Compact Sack - Assi Doman Sacks (UK)
Compact Sack - Assi Doman Sacks (UK)
Compactdoor - Henry Venables Ltd
Compak - Fordwater Pumping Supplies Ltd
Companion 90 - Telelarm care Ltd
Compas Sort Units - Envopak Group Ltd
Complete Bathrooms - Topravit (UK) Ltd
Compriband - Compriband Ltd
Comprisil - Compriband Ltd
Compton - Dawson McDonald & Dawson Ltd
Compton Buildings Ltd - Phillip Ader Ltd
Compton Garages - Compton Buildings Ltd
Computaire - Denco Ltd
Conbextra - Fosroc Ltd
Concept - Kermi (UK) Ltd
Concept 2000 bath - Chiltern Invadex
Concept Suite - Armitage Shanks Ltd
Concept Suite - Armitage Shanks Ltd
Concierge - Vision Neon & Display Ltd
Concord - Concord Lighting
Concord - Dunhams of Norwich
Concrete Bookshop - Concrete Information Ltd
Concrete Contracts - Concrete Information Ltd
Concrete current awareness - Concrete Information Ltd
Concrete info - Concrete Information Ltd
Concrex - Watco UK Ltd
Concryl - IFT Supplies Ltd
Condense - Condensing Boilers - Clyde Combustions Ltd
Condensor - Keddy-Poujoulat (UK) Ltd
Conderbrake - Conder Products Ltd
Conductite - Laybond Products Ltd
Cone Crushers - Parker Plant Ltd
Conex - Hipkin F W Ltd
Conex - IBP Ltd
Congress - Project Office Furniture PLC
Congrip - Conren Ltd
Conidec Waterproofing System - Dundec Contracts Ltd
Conipox 772 - Dundec Contracts Ltd
Conipur 255 - Dundec Contracts Ltd
Conipur 258 - Dundec Contracts Ltd
Conipur 275 - Dundec Contracts Ltd
Conipur 58 - Dundec Contracts Ltd
Conipur 79 - Dundec Contracts Ltd
Coniroof Systems - Dundec Contracts Ltd
Conision - Potter & Soar Ltd
Connaught - Trent Bathrooms
Connect grid - Saint-Gobain Ecophon Ltd
Connexions Steel Partitioning - Bradbury Group Ltd
Connoisseur Handles - Fab & Fix Ltd
Conquest - Celuform
Consolite - Cheshire Concrete Products
Consort - Clenaware Systems Ltd
Consort - Consort Equipment Products Ltd

Consort - Marflow Eng Ltd
Consort - Mita (UK) Ltd
Consort - Vantrunk Engineering Ltd
Constable suite - Brass Art Ltd
Constant Force™ - Latchways plc
Constructa - Taylor & Portway Ltd
Construction and Building Abstracts - NBS
Construction Information Service - NBS
Constructionline - Chestnut Products Limited
Constructionline - Constructionline
Constructor - Dexion Ltd
Conta-clip terminals - Elkay Electrical Manufacturing Co Ltd
Contact - Girsberger London
Containafire - Lydney Containers Ltd
Contain-It - George Fischer Sales Ltd
Contect Duck oil - Deb Ltd
Contetch - IFT Supplies Ltd
Contigym - Continental Sports Ltd
Contimat - Continental Sports Ltd
Continental Range - Pedley Furniture International Ltd
Continuum - Twyford Bushboard
Contitramp - Continental Sports Ltd
Contour - Calomax Ltd
Contour - Hambleside Danelaw Ltd
Contour - Valor
Contour 2 - Armitage Shanks Ltd
Contour Cisterns - Dudley Thomas Ltd
Contour Counters - Sealine Business Products
Contour Factor 20 - Hambleside Danelaw Ltd
Contour OTT - Hambleside Danelaw Ltd
Contour Stepsafe - Hambleside Danelaw Ltd
Contour Vinyl Wallcoverings - Forbo (Lancaster) Ltd
Contourail - Jendico
Contract - Owlett - Jaton
Contract 'E' - Eltron Chromalox
Contract Journal - Contract Journal
Contracters - GE Bayer Silicones
Contravent - Contravent Regal Ltd
Control Packs - Horstmann Controls Ltd
Controlmaster - ABB Control Ltd
Conturan Display Enhancement Glass - SCHOTT UK Ltd
Convectastream - Falcon Food Services Equipment
Cool-fit - George Fischer Sales Ltd
Coolfit - Low voltage Halogen Lamps - Sylvania Lighting International INC.
Coolflow - Biddle Air Systems Ltd
Coolguard - Feedwater Ltd
Coolkote - Bonwyke Ltd
Coolplex - Feedwater Ltd
CoolView (examination lamps) - Brandon Medical
coopers-uk.com - Cooper Group Ltd
Coopertiva D'imola - Arnull Bernard J & Co Ltd
Coo-Var ® - Coo-Var Ltd
Coo-Var ® - Teal & Mackrill Ltd
Copal - Fernox Alpha Fry Technologies Europe
Copar 700 - Monowa Manufacturing (UK) Ltd
Copeland - Copeland Corporation Ltd
Copenhagen - Loft Centre Products
Copley Decor - Copley Decor Ltd
Copon - Andrews Coatings Ltd
Copon - Wood E Ltd
Copper Bot 2000 Aantifouling - Wessex Resins & Adhesives Ltd
Copperad - The BSS Group PLC
Copperfield - Swintex Ltd
Copper-Flo - Johnson Matthey PLC - Metal Joining
Copydex - Henkel Consumer Adhesives
Coradoor Preliminary Zone - Tufton Ltd
Coragrip Preliminary Zone - Tufton Ltd
Coral Brush - Bonar Floors Ltd
Coral Classic - Bonar Floors Ltd
Coral Clean Off Zone - Tufton Ltd
Coral Door - Bonar Floors Ltd
Coral Door - Bonar Floors Ltd
Coral Grip - Bonar Floors Ltd
Coral Luxe - Bonar Floors Ltd
Coral Plus Clean Off Zone - Tufton Ltd
CoRayVac ® - Roberts-Gordon
Corbreak - Cordek Ltd
Corflam - British Cork Mills Ltd
Corflam - British Cork Mills Ltd
CORGI - Council for Registered Gas Installers, The
Corian - Du - pont - Decra Ltd
Corinhian - Trent Bathrooms
Corium - Baggeridge Brick PLC
Corkfast - Western Cork Ltd
Corklok - Gradient Insulations (UK) Ltd
Corkmaster - Amorim (UK) Ltd
Corkmaster - Amorim (UK) Ltd
Cornice - Amilake Southern Ltd
Cornish - Formpave Ltd
Corocord - Lawrence Surfaces Plc, Charles
Corofil - Firebarrier Services Ltd
Corofix - Corofil Woodall
Coroflow - Corofil Woodall
Coroline - Ariel Plastics Ltd
Corolla - AEL
Corolux - Ariel Plastics Ltd
Corolux 2000 - Ariel Plastics Ltd
Corona - SCHUCO International KG
Coroseal - Corofil Woodall
Corotherm - Ariel Plastics Ltd
Corpoate W20 - Crittall Steel Windows Ltd
Corporate 2000 - Crittall Steel Windows Ltd
Corporate 2000 - Crittall Steel Windows Ltd
Corporate W20 - Crittall Steel Windows Ltd
Corralux - Rada Lighting
Correx - Cordek Ltd
Corridor - Saint-Gobain Ecophon Ltd
Corroban - Feedwater Ltd
Corrocure - Corroless Corrosion Control
Corrocure - Kenyon Group Ltd
Corrogiene - Corroless Corrosion Control
Corroguard - Corroless Corrosion Control
Corroless - Corroless Corrosion Control
Corroshield - Corroless Corrosion Control
Corrosperse - Feedwater Ltd
Corruspan - Compriband Ltd
Cortal - Haworth UK Ltd
Co-Seal Liquid Plastic Dressing - Stewart Wales, Somerville Ltd
Cosmofin - Beton Construction Ltd
Cosmopolitan - Firth Interface Europe Ltd
Cosybug - Harton Heating Appliances
Cottage Roof Tiles - Tudor Roof Tile Co Ltd
Cottarstone - Camas Building Materials
Cottarstone - Fyfe John Ltd
Cougar - Catnic
Countdown - Checkmate Industries Ltd
Counterline - Celuform
Counterpoint - Mackay Carpets, Hugh
Countershield - Safetell Ltd
Country- CLAD - Hadley Industries Plc
Country collection - Emsworth Fireplaces Ltd
Countryman - Bennett Windows & Doors
County Collection - Taylor & Portway Ltd
County Kerb - Lafaroe Redland Precast
County pantile - Sandtoft Roof Tiles Ltd
County Walling - Atlas Stone Products
Courtaulds performance films - Banafix Limited
Courtaulds performance films - Banafix Limited
Courtrai - Smithbrook Building Products Ltd
Cove red - Realstone Ltd
Coverad (Polymer) - Coverad Ltd
Coverad Estate (Steel) - Coverad Ltd
Coverlife - EBC UK Ltd
Coverpipe - Coverad Ltd
Coverwalk Paving System - Coverite Ltd
Coveseal - Armorex Ltd
Coxdome T.P.X - Cox Building Products Ltd
Coxdome Trade - Cox Building Products Ltd
Coxdone 2000 - Cox Building Products Ltd
Coxdone Mark 5 - Cox Building Products Ltd
Coxspan Baselock - Cox Building Products Ltd
Coxspan GlassPlank - Cox Building Products Ltd
Coxspan Modular roof glazing - Cox Building Products Ltd
CP Canopy - Consult Lighting
CPA - Construction Plant-hire Association
CPL Fineliner - Composite Panels Ltd
CPL Finerib - Composite Panels Ltd
CPL Fireclad - Composite Panels Ltd
CPL Hipertec - Composite Panels Ltd
CPL Micro - V - Composite Panels Ltd
CPL Super 4 - Composite Panels Ltd
Cpv - df - CPV Ltd
Crack Bond - Helifix Ltd
Cractie - Redifix Ltd
Craftsman Elite - Brockway Carpets Ltd
Craftwall - Neslo Partitioning & Interiors
Craftwork Collection - Spur Shelving Ltd
Crane in a Box - Metreel Ltd
Cranleigh Light Multi - Cranleigh Brick & Tile Co Ltd
Cranleigh Red Multi - Cranleigh Brick & Tile Co Ltd
Craven - Morris Material Handling Ltd
Creativline - Sissons W & G Ltd
Credifon® Remote Postage Meter Recrediting System - Neopost Ltd
Credo - Kermi (UK) Ltd
Creighton Thomson Associates - Apex Piling Ltd
Crescent - Crescent Lighting
Cresco - Creteco Sales Ltd
Cresent Cl - Crescent Lighting
Cresfinex - Exitex Ltd
Cresset - Adams- Hydraulics Ltd
Cresta - Brockway Carpets Ltd
Cresta - MSA (Britain) Ltd
Cresta Supreme - Brockway Carpets Ltd
Cresta Supreme Heathers - Brockway Carpets Ltd
Creteangle - Benton Co Ltd Edward
Creteangle - Benton Co Ltd Edward
Creteangle - Edward Benton & Co Ltd
Cricket Gas Detectors - MSA (Britain) Ltd
Crimegard Systems - Crimeguard Systems Ltd
Cristal - H & R Johnson Tiles Ltd
Critic - Surge Protection Products - Erico Europa (GB) Ltd
Crittall Composite - Crittall Steel Windows Ltd
Crocodile - Simpson Strong-Tie®
Croda - Andrews Coatings Ltd
Cromleigh - Allan Harris & Sons Ltd
Cromleigh - Allan Harris & Sons Ltd
Cromleigh S.V.K - Allan Harris & Sons Ltd
Crompack - Cooper Lighting and Security Ltd
Cromp-arc stud welding - CSW Coldform Ltd
Crompton - ERA Products
Crompton Lighting - Cooper Lighting and Security Ltd
Cromwell Canopy - James Smellie Fabrications Ltd
Crosinox - Strand Hardware Ltd
Cross-cote - Cross-Guard International Ltd
Crossgrip - Plastic Extruders Ltd
Crossguard - Bradbury Group Ltd
Crossguard - Bradbury Group Ltd
Crossley - Carpets International (UK) Ltd
Crossvent - Howden Buffalo
Crown - Albany Standard Pumps
Crown Trade - Crown Trade Paints
Crown Wool - Knauf Insulation Ltd
CRT - Range Cylinders
Cryclad - Cryotherm Insulation Ltd
Cryosil - Cryotherm Insulation Ltd
Cryostop - Cryotherm Insulation Ltd
Cryotherme - Firebarrier Services Ltd
Cubase1 - Foster W H & Sons Ltd
Cubase2 - Foster W H & Sons Ltd
Cubase3 - Foster W H & Sons Ltd
Cube wall - Grant Westfield Ltd
Cubes Range - Pedley Furniture International Ltd
Cubic - Alumasc Construction Products Ltd
CubicleTrack - Integra Products
Cubit - Packaged Modular Steam Generators - Eaton-Williams Group Ltd
Cubo enclosures - Ensto Briticent Ltd
Cullamix Tyrolean - Weber Building Solutions
Cummins - Cummins Power Generation Ltd
Cummins Power Generation - Cummins Power Generation Ltd
Cumpasso - USG (UK) Ltd
Cupa slates - Brooks Belfast
Cupa slates - Brooks Belfast
Cuplok - SGB Youngman
Cupola - Hughes Safety Showers Limited
Cuprinol - ICI Paints
Cuprofit - IBP Ltd
Cuproright 70 - Righton Ltd
Cuproright 90 - Righton Ltd
Curvatura - USG (UK) Ltd
Curved Tempered Glass CTG Ltd - Darby Glass Ltd
Curvetech - Stocksigns Ltd
Cushion Coil - Kleen-Tex Industries Ltd
Cutting Edge Garden Tools - Stockten Heath Forge (Caldwells) Ltd
Cwt y Bugail - McAlpine Slate Ltd, Alfred
CX 2 - President Office Furniture Ltd
Cyclong - Colt International
Cypherlok - Abloy Security Ltd
Cyra - Whitecroft Lighting Limited
d Line - Allgood plc
D & H - Dyer Environmental Controls
D S 90 - Salex Interiors Ltd
D.F.D - Dudley Factory Doors Ltd - Dudley Factory Doors Limited
D.F.D. - (Dudley Factory Doors Ltd.) - Shutter Door Repair & Maintenance Ltd
D.F.D.-(Dudley Factory Doors Ltd.) - Priory Shutter & Door Co Ltd
Dabo - EJOT Ecofast Ltd
Dac Chlor - Dacrylate Paints Ltd
Dac Crete - Dacrylate Paints Ltd
Dac flex - Dacrylate Paints Ltd
Dac Roc - Dacrylate Paints Ltd
Dac Shield - Dacrylate Paints Ltd
Dacatie - Quantum Profile Systems Ltd
Dacpol - Dacrylate Paints Ltd
Daikin - HV Air Conditioning Ltd
Daikin - Purified Air Ltd
Dalamatic - Torit DCE
Dal-equipe - Dalsouple Direct Ltd
Dales Decor - Copley Decor Ltd
Dalesauna - Dalesauna Ltd
Dalkit - Dalsouple Direct Ltd
Dal-lastic - Dalsouple Direct Ltd
Dalsouple Magnetic - Dalsouple Direct Ltd
Dalsport - Dalsouple Direct Ltd
Damcor - TDI (UK) Ltd
Damixa - Berglen Group Ltd
Damixa - Berglen Group Ltd
Dampa 10 - Dampa (UK) Ltd
Dampa 100 - Dampa (UK) Ltd
Dampa 200 - Dampa (UK) Ltd
Dampa 300 - Dampa (UK) Ltd
Dampa Chess - Dampa (UK) Ltd
Dampa Interval - Dampa (UK) Ltd
Dampa Lamel - Dampa (UK) Ltd
Dampa Sport - Dampa (UK) Ltd
Dampa Tile - Dampa (UK) Ltd
Dampa X3 - Dampa (UK) Ltd
Dampcoursing 132 - Kiltox Damp Free Solutions
Dampseal - Marley Waterproofing
Danelaw - Drainage Systems
Danelaw - Hambleside Danelaw Ltd
Danesmoor - Worcester Heat Systems Ltd
Danfoss - Danfoss Randall Ltd
Daniel Platt - Stokes R J & Co Ltd
Danish Oil - Anglo Building Products Ltd
Danpalon - EBC UK Ltd
Dantherm - Hayward Air
Darbo Particleboard - Willamette Europe Ltd
Darbyseal' Insulating Glass Unit - Darby Glass Ltd
Darbytuf' Tempered Safety Glass - Darby Glass Ltd
Dark Victorian Red - Charnwood Brick Group Ltd
Darkroom safelights - Encapsulite International Ltd
Darksky - Cooper Lighting and Security Ltd
Daryl - Daryl Industries Ltd
Dasphalte - IFT Supplies Ltd
Datacall Aquarius - Blick Communication Systems Ltd
Datacall Aquarius - Blick Communication Systems Ltd
Datacall Gemini - Blick Communication Systems Ltd
Datacall Minder - Blick Communication Systems Ltd
Dataworks - MK (MK Electric Ltd)
Datlux - Setsquare
Dauphin Trendline - Dauphin PLC
Davenset - Elequip Projects Ltd
David Ball Test Sands - David Ball Group Plc
David Mclean Contractors Ltd - David Mclean Group
Davis - Legrant Electric Ltd
Daylight Simulation - Full Spectrum Lighting Ltd
Day-Lite - Emergi-Lite Safety Systems
dBan - Interfloor Limited
D-Bus - Eaton Electric Limited
D-Bus - Home Automation Ltd
DCE - Torit DCE
DCE10000 - Torit DCE
Dea Madore Range - Brass Art Ltd
Deanlite - Deans Blinds & Awnings (UK) Ltd
Deanmaster - Deans Blinds & Awnings (UK) Ltd
Deanox - Elementis Pigments Ltd
Deans - Deans Blinds & Awnings (UK) Ltd
Deans signs - Deans Blinds & Awnings (UK) Ltd
Deb Lime - Deb Ltd
Deb Natural - Deb Ltd
Deb Protect - Deb Ltd
Deb Restore - Deb Ltd
Deben Systems - Phoenix Scales Ltd
Debotec - Chesterfelt Ltd
Debut - Ness Furniture Ltd
Debut Range - Caradon Bathrooms Ltd
Dec - Nology - Owlett - Jaton
Decadex - Liquid Plastics Ltd
Decamel - Formica Ltd
Decapo - Concord: Marlin
Decathlon - Hodkin Jones (Sheffield) Ltd
Deck drain - ABG Ltd
Deckshield - Flowcrete Plc
Deco - Jaga Heating Poducts (UK) Ltd
Decoflair - nmc (UK) Ltd
Decoform - Celuform
Decoline - Deceuninck Ltd
Decolit - STO Ltd
Deconyl - Plascoat Systems Ltd
Deconyl - Plastic Coatings Ltd
Decor - Deceuninck Ltd
Décor Profile - Sarnafil Ltd
Decorative - Lafaroe Redland Precast
Decorfix - Decorfix Ltd
Decoroc - Deceuninck Ltd
Decorscreed - Conren Ltd
Decostar - Osram Ltd
Decothane - Liquid Plastics Ltd
Decra Stratos - Decra Roofing Systems
Decra tiles - Decra Roofing Systems
Decra Vent - Decra Roofing Systems
Deepflow - Marley Plumbing and Drainage
Deeplane Yorkstone - Pickard Group
Deeplas - Deceuninck Ltd
Deepstor - Dexion Ltd
Defender - Marshall J A & Co (Brushes) Ltd
Defender - Saint Gobain pipelines PLC
Defender Egress - Securistyle Ltd
Defender Standard - Securistyle Ltd
Defendor - Kaba Door Systems Ltd
Defensor - Geoquip Ltd
Deflecta - Brierkrete Ltd
Degreasol - IFT Supplies Ltd
Degussa - Anderson Gibb & Wilson
Deha - Halfen Ltd
Dekguard - Fosroc Ltd
Dekor - Rockfon Ltd
Dekordor ® - Vicaima Ltd
Dektite - ITW Construction Products
Delabole - The Delabole Slate Co. Ltd
Delabole Slate - The Delabole Slate Co. Ltd
Delbraze - IBP Ltd
Delchem - Opella Ltd
Delflo - Hattersley Newman Hender Ltd
Delight - Glamox Electric (UK) Ltd
Deliplan - DLW Flooring Ltd
Delphis Glass Mosaics - Kirkstone
Delstar - Bolivar Stamping Ltd
Delstar - Bolivar Stamping Ltd
Delta - Brett Paving
Delta - Harbro Supplies Ltd
Delta - Kaba Door Systems Ltd
Delta downlights - LB Lighting Ltd
delta gutter system - Dales Fabrications Ltd
Deltascreen - Quartet-GBC UK Ltd
DelTherm - Hattersley Newman Hender Ltd
Deltos - Lightfoot Charles Ltd
Delvo - Feb MBT
DEM Tip-Up Seating System - Race Furniture Ltd
Demidekk - Jotun Henry Clark Ltd Decorative Division
Deminpac - Feedwater Ltd
Denby - Rawson Carpets Ltd
Denchem - Naylor Drainage Ltd
Denduct - Naylor Drainage Ltd
Denka Lifts - Denka International
Denline - Naylor Drainage Ltd
Denlok - Naylor Drainage Ltd
Dennis Ruabon - Stokes R J & Co Ltd
Denrod - Naylor Drainage Ltd
Denseal - Ashmead Buidling Supplies Ltd
Denseal - Naylor Drainage Ltd
Densleeve - Ashmead Buidling Supplies Ltd
Densleeve - Ashmead Buidling Supplies Ltd
Densleeve - Naylor Drainage Ltd
Denso - Winn & Coales (Denso) Ltd
Derbigum - Alumasc Exterior Building Products Ltd
Dernier & Hamilyn - Denmans Montrose Ltd
Design 2001 - Sportmark™ Ltd
Design Line - Railex Systems Ltd
Designer - Clark Door Ltd

Designer - Pilkington's Tiles Ltd
Designer 2000 - Bassett & Findley Ltd
Designer Collection - Brintons Ltd
Designer Range - Leaderflush + Shapland Ltd
Desimpel - Hanson Brick Ltd, Northern Regonal Sales Office
DeskTop - Forbo - Nairn Ltd
Desmo - ASSA Ltd
Desney Products Ltd - Desney Products Ltd
DespatchLink™ Despatch Management Software - Neopost Ltd
Destiny - CCTV over ATM - Baxall Ltd
Destiny-IP-CCTV Ethernet System - Baxall Ltd
Detan tension rod system - Halfen Ltd
Detectamesh - Boddingtons Ltd
Detectatape - Boddingtons Ltd
Developer - Checkmate Industries Ltd
Devweld structural - Kenyon Group Ltd
Dexlux - Chalmit Lighting
DF.D- (Dudley Factory Doors Ltd.) - Neway Doors Ltd
Dialock - Haffele UK Ltd
Diamard - Emusol Products (LEICS) Ltd
Diamond - Iles Waste Systems
Diamond 600 & 750 Raised Access Flooring - Durabella System Floors Ltd
Diamond Access Floor - Stone Raised Access Floor - Chelsea Artisans Ltd
Diamond Décor - Impact Resistant Decorative Glass - Chelsea Artisans Ltd
Diamond Glaze - ICI Dulux Trade Paints
Diamond Mirror - Impact Resistant Safety Mirror - Chelsea Artisans Ltd
Diamond Optic® - DW Windsor Lighting
Diamond Sign System - Drakard & Humble Ltd
Diamond Ultralite Internal/External - Chelsea Artisans Ltd
Dibond ® - Righton Ltd
Digital System - BPT Security Systems (UK) Ltd
Dignity (Series 750) - Parker Bath Limited
Diklon Sunshade fabric - Dallas & Forrest Ltd
Dimensions - Anderson C F & Son Ltd
Dimple Marker - Marking Machines International
Dimplex - Glen Dimplex UK Ltd
Dinella - Provex Products Ltd
Diplomat - Dudley Thomas Ltd
Diplomat Range - Lumitron Ltd
Direct Worktops - Viking Laminates Ltd
Directions - Desking Systems Ltd
Direx - Universal Components Ltd
Disbocrete - PermaRock Products Ltd
Disc Fan - Domus Ventilation Ltd.
Discovery - Apollo Fire Detectors Ltd
Discreet - Integra Products
Discreet Lighting - Vision Neon & Display Ltd
Disposapad - Sissons W & G Ltd
DK - Harbro Supplies Ltd
DMP Bonder - Isocrete Floor Screeds Ltd
DNA - Bernlite Ltd
Dobel 200 XT - SSAB Dobel Coated Steel Ltd
Dobelshield stucco Aluzink - SSAB Dobel Coated Steel Ltd
DOCM Packs - Armitage Shanks Ltd
Dolphin - Dolphin Showers Ltd
Dominator - Crane Fluid Systems
Domino - OWA UK Ltd
Domo - Dalesauna Ltd
Domostyl - nmc (UK) Ltd
Don - Gazco Ltd
DON - Record & Parkes Ltd
Don - The Safety Tread Ltd
Don Burners - Don Heating Products Ltd
Donaldson filter components - Encon Air Systems Ltd
Doncrete Dense Concrete Blocks - Broome Bros (Doncaster) Ltd
Donio - Dalesauna Ltd
Donlite Building Blocks - Broome Bros (Doncaster) Ltd
Donlite Building Blocks - Broome Bros (Doncaster) Ltd
Donlite Foundation Blocks - Broome Bros (Doncaster) Ltd
Donlite light weight concrete blocks - Broome Bros (Doncaster) Ltd
Donn - USG (UK) Ltd
Donn - Brown - Brown F plc
Donn DX Grid - USG (UK) Ltd
Doorfit - Doorfit Products Ltd
Dooria - Scandinavian Window Systems Ltd
Dorchester - Hamworthy Heating Ltd
Doric Bollard - Furnitubes International Ltd
Dorma - Permclose Door Services Ltd
Dorma ED800 - Dorma Door Services Ltd
Dorma ED800 - Dorma Door Services Ltd
Dormers Direct - Arbory Group Limited
Dorothea ® - Broxap Ltd
Dorset - Pilkington's Tiles Ltd
Dorus - Henkel Industrial Adhesives
Double Diamond - Chelwood Brick Ltd
Double Duty - Integra Products
Double-R - Celotex Ltd
Doublestore - Railex Systems Ltd
Doubleswing - Blair Joinery
Doubleswing - Blair Joinery
Dove - Contour Showers Ltd
Dove gutter system - Dales Fabrications Ltd
Dovedale - Chiltern Invadex
DowCorning - Dow Corning Ltd
Down To Earth Garden Tools - Stockten Heath Forge (Caldwells) Ltd
Downflo - Torit DCE
Dowty Spats - Corofil Woodall
DP5 - Kenilworth Daylight Centre Ltd
DP91 - Kenilworth Daylight Centre Ltd
DPn - Wilo Samson Pumps Ltd
Dr Marten - Briggs Industrial Footwear Ltd
Draeger-Tubes - Draeger Safety UK Ltd
Dragged Stone - Hampton Stone Ltd
Dragon - Fulton Boiler Works (Great Britain) Ltd
Dragon HLO - Elan-Dragonair
Dragon HNG - Elan-Dragonair
Dragonair Hlo - Dragonair Ltd
Dragonair Hng - Dragonair Ltd
Drainmaster - Pump Technical Services Ltd
Dramix- Steelwire Fibres - Bekaert Building Products
Dramix- Steelwire Fibres - Bekaert Building Products
Drape - Harrison Drape, Divsion of Newell Ltd
Draper - Arrow Supply Company Ltd
Draper - Kem Edwards Ltd
Draught Buster - Airflow (Nicoll Ventilators) Ltd
Draughtbuster - Timloc Expamet Building Products
Dravo - Dravo Environmental Services
Dravo AE unit heater - Dravo Environmental Services
Dravo Air Handling Units - Dravo Environmental Services
Dravo DF - Dravo Environmental Services
Dravo Directflow - Dravo Environmental Services
Dravo Euroflow - Dravo Environmental Services
Dravo FS Directflow - Dravo Environmental Services
Dravo Heat Exchange Modules - Dravo Environmental Services
Dravo Heaters - Dravo Environmental Services
Dravo Heatmiser - Dravo Environmental Services
Dravo NDS - Dravo Environmental Services
Dravo Ventilation - Dravo Environmental Services
Drawmet - Drawn Metal Ltd
Dreadnought - Baydale Architectural Systems Ltd
Dreadnought Tiles - Hinton Perry & Davenhill Ltd
Dream - Orsogril Spa UK Distributor
Dream Fireslide - Valor
Dremel - Robert Bosch Ltd
Dricon - Arch Timber Protection
Dri-Marker - Magiboards Ltd
Dritherm - Knauf Insulation Ltd
Driva - ITW Construction Products
Driveway - Lafaroe Redland Precast
Drum Form - Scapa Tapes UK Ltd.
Dry Fix - Helifix Ltd
Dry Shake Armorshield - Armorex Ltd
Dryangles - Atlas Stone Products
Dryfast - Hayward Air
Dryflo - Torit DCE
Dryflow - Johnson & Starley
Dryseal - Hambleside Danelaw Ltd
Drystone - Atlas Stone Products
Drywall Academy - British Gypsum Ltd
DS - 3 - Fernox Alpha Fry Technologies Europe
DS - 40 - Fernox Alpha Fry Technologies Europe
DS-3 - Cookson Electronics
DS-40 - Cookson Electronics
DS90 - The Salex Group Ltd.
DSA - Drilling & Sawing Association, The
DSMA - Door and Shutter Manufacturers Association
DSS - The HB Sign Company Ltd
D-stain - Milliken Carpet
DTFA - Whitecroft Lighting Limited
DualAir ® - Roberts-Gordon
Dualcase - Universal Components Ltd
Dualframe - Sapa Building Systems (monarch and glostal)
Dualtec - C & K Intellisense
Dublo - Range Cylinders
Duck Tapes - Henkel Consumer Adhesives
Ducksback - Cuprinol Ltd
Ductex - Rega Ductex Ltd
Ductex - Ubbink (UK) Ltd
Duets - Forticrete Ltd
Duke - Iles Waste Systems
Dukes - Realstone Ltd
Dulastone - Polypipe Bathroom & Kitchen Products Ltd
Dulux Trade - ICI Dulux Trade Paints
Dulux Trade Emulsions - ICI Dulux Trade Paints
Dulux Trade Gloss - ICI Dulux Trade Paints
Dulux Trade Natural Wood finishes - ICI Dulux Trade Paints
DUNA - Reiner Fixing Devices (Ex: Hardo Fixing Devices)
Dundec Car Park Rejuvenation - Dundec Contracts Ltd
Dundec Seamless Waterproofing - Dundec Contracts Ltd
Dune Premier - Armstrong World Industries Ltd
Dunelm - Mackay Carpets, Hugh
Dunham Strip - Dunham Bush Ltd
Dunlop - BRB Industrial Services Ltd
Dunsfold Selected Dark - Cranleigh Brick & Tile Co Ltd
Dunvent - Dunbrik (Yorks) Ltd
Duo - Metlex
Duo-Fast - Duo-Fast (UK) Ltd
Duoform - Celuform
Duotherm - Smart Systems Ltd
Dupar - Dewhurst plc
Duplex - Airsprung Beds Ltd
Duplex - Airsprung Beds Ltd
Duplex - Fillcrete Ltd
Duplex Building System - Portakabin Ltd
Duplo - Becker (SLIDING PARTITIONS) Ltd
Dupox - Andrews Coatings Ltd
Duquesa Slate - Cembrit Blunn Ltd
Durable - Durable
Duraflex - Caradon Duraflex Systems Ltd
Duraflex - Caradon Duraflex Systems Ltd
Duraflon - Plastic Coatings Ltd
Durafort - Dixon Turner Wallcoverings
Durafort - Tektura Plc
Duraglas - Amari Plastics Plc
Duraglas - Flexitallic Ltd
Duraglass - Leigh's Paints
Duraguard - Plastic Coatings Ltd
Duralay - Interfloor Limited
Duralife Polyester Powder Coat - Crittall Steel Windows Ltd
Duralite - Pryorsign
Duralock range post and rails - Duralock (UK) Ltd
Durama X - Quartet-GBC UK Ltd
Durapipe - Capper Plastics
Durapipe - Capper Plastics
Durashade Blockout blinds - Dallas & Forrest Ltd
Duraspring - Airsprung Beds Ltd
Durastat - Heat Trace Ltd
Durasteel - Promat UK Ltd
Durham carpets - Mackay Carpets, Hugh
Durham county twist - Mackay Carpets, Hugh
Durham edition - Mackay Carpets, Hugh
Duropal - Anderson C F & Son Ltd
Duropal - Deralam Laminates Ltd
Durox - Tarmac Topblock Ltd
Dursax - Mackay Carpets, Hugh
Dustguard - Conren Ltd
DV - Whipp & Bourne
DVS & DVSR Switchers - Baxall Ltd
DWM Copeland - Copeland Corporation Ltd
Dycel Revetment - Ruthin Precast Concrete Ltd
Dycem Clean-Zone - Dycem Ltd
Dycem Protectamat - Dycem Ltd
Dycem Work - Zone - Dycem Ltd
Dyke- Roof Solar - Dyke Chemicals Ltd
Dyke-Aluminium - Dyke Chemicals Ltd
Dyke-Flashing - Dyke Chemicals Ltd
Dyke-Glass - Dyke Chemicals Ltd
Dyke-Mastic - Dyke Chemicals Ltd
Dyke-Roof - Dyke Chemicals Ltd
Dyke-Roof colour finish - Dyke Chemicals Ltd
Dyke-Roof universal primer - Dyke Chemicals Ltd
Dyke-Seal - Dyke Chemicals Ltd
Dyke-Sil - Dyke Chemicals Ltd
Dyke-Silver - Dyke Chemicals Ltd
Dynabolt - Ramset Fasteners Ltd
Dynaset - Ramset Fasteners Ltd
Dyno-locks - Dyno Rod PLC
Dyno-rod - Dyno Rod PLC
Dytap Revetment - Ruthin Precast Concrete Ltd
E Range - James Gibbons Format Ltd
E. F.S - Fulton Boiler Works (Great Britain) Ltd
E3000/Birkdale - Eternit Building Materials
E7BX - Horstmann Controls Ltd
EAC - Harbro Supplies Ltd
EAC - Unicorn Abrasives (UK) Ltd
Earth square - Milliken Carpet
Easby - Oakdale (Contracts) Ltd
Easi joist - Wyckham Blackwell Ltd
Easi-away - Easi-Fall International Ltd
Easicheck - Cooper Lighting and Security Ltd
Easiclamp - Viking Johnson
Easicollar - Viking Johnson
EasiCool - Airedale International Air Conditioning Ltd
Easi-Fall - Easi-Fall International Ltd
Easifiles - Railex Systems Ltd
Easi-fill - Easi-Fall International Ltd
Easifix - Decra Ltd
Easi-flex - Easi-Fall International Ltd
Easifoam - Vita Liquid Polymers Ltd
Easigrip - Uponor Ltd
Easi-mend - Easi-Fall International Ltd
Easipipe 100, 125, 150 - Domus Ventilation Ltd.
Easi-rect - Harkness Hall Ltd
Easitap - Viking Johnson
Easitee - Viking Johnson
Easiwork - Metreel Ltd
Eastwood Cookers - Don Heating Products Ltd
Easy Change - Tapworks Water Softeners
Easy Change - Vision Neon & Display Ltd
Easyarch - Simpson Strong-Tie®
Easybolt - Tinsley & Co Ltd, Eliza
Easyclad - Advanced Hygienic Contracting Ltd
Easyclad - Advanced Hygienic Contracting Ltd
Easyclean - Vicaima Ltd
EasyFinish - Readymix Limited
Easyfix - Tinsley & Co Ltd, Eliza
Easy-Flo - Johnson Matthey PLC - Metal Joining
Easyhang - Spur Shelving Ltd
Easymark - Marking Machines International
Easy-slide - Hill Leigh Ltd
Easyway - Sapa Building Systems (monarch and glostal)
Eauzone - Matki plc
Eavesguard - Ryton's Building Products Ltd
Eavesguard - Vulcanite Limited
Eazicoat - Howse Thomas Ltd
Eazimark™ OMR Coding Software - Neopost Ltd
EB - Clement Steel Windows
EBC UK Ltd - EBC UK Ltd
Ebco - Barber Edward & Co Ltd
Ebco - Barber Edward & Co Ltd
EBL - Emporium Brazil Ltd (EBL)
EC 2000 - ABB Control Ltd
ECB - Torit DCE
ECCO 5000 - Reznor UK Ltd
Eclipse - Amilake Southern Ltd
Eclipse - Chalmit Lighting
Eclipse - CSI Security LTD
Eclipse - Eclipse Blind Systems Ltd
Eclipse - Fitzgerald Lighting Ltd
Eclipse - Northern Incinerators (GB) Ltd
Eclipse - Safetell Ltd
Eclipse - Spear & Jackson
Eclipse PUR - Tarkett-Marley Floors Ltd
Eclipse Spiralux - Spear & Jackson
Eclispe - Grace Construction Products Ltd
ECO - Kair ventilation Ltd
Eco - Townscape Products Ltd
Eco Countryman - Bennett Windows & Doors
Ecoarc - Improved Retrofit for Mercury Vapour Lamp - Sylvania Lighting International INC.
Ecodisc - Kone PLC
Ecofibre - Gilmour Ecometal
Ecoflex - Uponor Housing Solutions
Ecohome - Gilmour Ecometal
Ecojoist - Gang-Nail Systems Ltd
Ecoliner - BPB Paperboard
Ecomax Reduc - Trim Acoustics
Ecomax Soundslab - Trim Acoustics
Ecometal - George Gilmour (Metals) Ltd
Ecometal - Gilmour Ecometal
Econocoat - Stewart Wales, Somerville Ltd
Econoflame - Stokvis R S & Sons Ltd
Econoloc - Arvin Motion Control Ltd
Economaire - Johnson & Starley
Economix - Sheardown Engineering Ltd
Economy 7 QTZ - Horstmann Controls Ltd
Econoplas Voge - Stewart Wales, Somerville Ltd
Econoroll - Crawford Amber Ltd
Econospan - Gilmour Ecometal
Econotex - Stewart Wales, Somerville Ltd
Econovent - AEL
Ecopack - Airedale International Air Conditioning Ltd
Ecopanel - Gilmour Ecometal
Ecophon - Nevill Long Limited
Ecophon - New Forest Ceilings Ltd
Eco-rax - Barton Storage Systems Ltd
Ecospan - Gilmour Ecometal
Ecotel - Airedale International Air Conditioning Ltd
Ecotherm - Prestoplan
Ecotray - Gilmour Ecometal
Ecovent - Ves Andover Ltd
Ecowall - Gilmour Ecometal
Ecozip - Gilmour Ecometal
ECS Reset - Philips Lighting Solutions
Ecurse TS - Interlux Ltd
Edale - Chiltern Invadex
Edelman Leather (Europe) - Mach & Garbarino
Eden - Rawson Carpets Ltd
Eden CBX 24 - Halstead Boilers Ltd
Eden SBX 30 - Halstead Boilers Ltd
Eden VBX 18 - Halstead Boilers Ltd
Eden VBX 30 - Halstead Boilers Ltd
Edenaire - Air Conditioning Units - Eaton-Williams Group Ltd
Edge - MK (MK Electric Ltd)
Edge Slot - Spur Shelving Ltd
Edilkamin - Emsworth Fireplaces Ltd
Edwin clarke stairways - Ermine Engineering Co. Ltd
EEKO - Electric Elements Co, The
Eeto - Davant Products Ltd
Eetofoam - Davant Products Ltd
Eff Eff - Viking Security Systems Ltd
Effebi Scale - Higginson Staircases Ltd
E-fix - Artur Fischer (UK) Ltd
EFP1 Single Zone Fire Alarm Control Panel - C-TEC (Computionics) Ltd
EFP1 Single Zone Fire Alarm Control Panel - C-TEC (Computionics) Ltd
Egger - Anderson C F & Son Ltd
Egger - Montague L Meyer
Egg-shpae liners - Johnston Pipes Ltd
Egoluce - Forma Lighting Ltd
EJ - Zon International
Elastaseal - Tor Coatings Ltd
Elasto - Fillcrete Ltd
Elcock - Purdie Elcock Ltd
Electrace - Flexelec(UK) Ltd
Electrak - Electrak International Ltd
Electramate - Gledhill Water Storage Ltd
Electric Security Fencing Federartion (ESFF) - Fencing Contractors Association
Electrisaver - Horstmann Controls Ltd
Electro-Fence - Advanced Perimeter Systems Ltd
Electroline - Jebron Ltd
Electrolux - Electrolux Domestic Appliances
Electrolux - Electrolux Laundry Systems
Electrolux - Parkinson Cowan and Tricity Bendix Co, The
Electrolux proffessional service - Electrolux Foodservice
Electromatic - Tuke & Bell Ltd
Electronic 7 - Horstmann Controls Ltd
Electronic Range - Biral Pumps
Electropack - Fulton Boiler Works (Great Britain) Ltd
Electrosonic - Helvar Electrosonic
Electrospot - Illuma Lighting Ltd
Electroway - Viscount Catering Ltd
ElectroWire - Geoquip Ltd
Elegance - Provex Products Ltd
Elegance Vetro - Unilock Ltd
Eleganza - Matki plc
Element - RB UK Ltd
Elementis Pigments - Elementis Pigments
Elements - H & R Johnson Tiles Ltd
Eleptec - Hubbard Architectural Metalwork Ltd

Elequant Cost Modelling Software - Johnson V B LLP
Elequip - Elequip Projects Ltd
Elite - Cardale Doors Ltd
Elite - Celuform
Elite - Dudley Thomas Ltd
Elite - Ellis Patents
Elite - Fitzgerald Lighting Ltd
Elite - Gradus Ltd
Elite - Nevill Long Limited
Elite - Shackerley (Holdings) Group Ltd incorporating Designer ceramics
Elite PUR - Tarkett-Marley Floors Ltd
Elixir - Matki plc
Eliza Tinley gate fittings - B & M Fencing Limited
Eliza Tinsley - Challenge Fencing Ltd
Elizabethan The Word of Cushions - Elizabethan Textiles
ElJan - Johnson & Starley
Elkington Gatic - Alumasc Construction Products Ltd
Elkosta - Elkosta; (UK) Ltd
Elkosta - Fernden Construction (South Western) Limited
Ellatec Lighting - Woodhouse UK Plc
Elliott Group Ltd - FineLine
Elliott Linked Modules - Elliott Group Ltd
Elliott Lucas - Spear & Jackson
Elliptipar - Crescent Lighting
Ellisse - Hille Ltd
Elolegic - Kaba (UK) Ltd
Elson - Elsy & Gibbons Ltd
Elterwater - Burlington Slate Ltd
Eltex - Brettell & Shaw Ltd
Elvaco - Elvaco UK Ltd
Emaco - Feb MBT
EmaW - Degussa Construction Chemicals (UK)
Embedded Retaining Walls - Westpile Ltd
Emerald - En-tout-cas plc
Emerald - En-tout-cas plc (Scotland)
Emerald - Polypipe Bathroom & Kitchen Products Ltd
Emergi - Man - Emergi-Lite Safety Systems
Emergi-Lite - Thomas & Betts Manufacturing Services Ltd
Emil Ceramica - Stokes R J & Co Ltd
Emir - Emmerich (Berlon) Ltd
Emit - Allgood plc
Emperor - Ellis Patents
Ems - Envirodoor Markus Ltd
Emtec - RPS Data Products (UK) Ltd
Enamelcoat - Conren Ltd
Encore - Arjo Ltd
Energy Beta - Beta Lighting Ltd
Energymaster - Lumitron Ltd
Engerseal - Callenders Ltd
Engineered Wall - Gang-Nail Systems Ltd
Enreo - IMI Air Conditioning Ltd
Ensign - Drainage Systems
Ensign - Saint-Gobain Pipelines PLC
Ensto - Ensto Briticent Ltd
Ensudisc - Ensor Building Products
Enterprise - Accomodex
Enterprise - Dunsley Heat Ltd
Entice Range - Caradon Bathrooms Ltd
Entro - Turnquest UK Ltd
Envireau - Titan Pollution Control Ltd
Envirofast - Envirodoor Markus Ltd
Envirogard - Leigh's Paints
Envirograf - Firebarrier Services Ltd
Envirolite - Envirodoor Markus Ltd
Enviromantal Noise Barrier Association (ASFC) - Fencing Contractors Association
Enviromat - Linatex Ltd
Enviropave - Linatex Ltd
Enviroplast - Envirodoor Markus Ltd
Envirotector - Conder Products Ltd
Enviva - Durable
Enyiroseal - Mann McGowan Group
EP - E. H. S. Roofing Ltd
E-Pack - Entrelec MTE Ltd
Epcon - ITW Construction Products
Epicol - Pearl Paints Ltd
Epidek - Leigh's Paints
Epidox - Witham Oil & Paint Ltd
Epigrip - Leigh's Paints
Epocem - Sika Ltd
Epoch Soventless Plastic - Stewart Wales, Somerville Ltd
Epo-dur - ICI Dulux Trade Paints
Epoxeel - IFT Supplies Ltd
Epoxy Verti-Patch - Anglo Building Products Ltd
Epoxy Wall Shield - Anglo Building Products Ltd
Epson - RPS Data Products (UK) Ltd
Equator - Marley Plumbing and Drainage
ERA - ERA Products
Erasmus - Hille Ltd
Erdu - Hughes Safety Showers Limited
Ergo - Top Office Equipment Ltd
Ergoflex - BRB Industrial Services Ltd
Ergolite (bedhead lamps) - Brandon Medical
Erico - Kem Edwards Ltd
Erico-caddy ® - Erico Europa (GB) Ltd
Ericson - Sunrite Blinds Ltd
Erik Joergensen - Zon International
Eritech - Ground Rods - Erico Europa (GB) Ltd
Ermine engineering - Ermine Engineering Co. Ltd
Erosamat - ABG Ltd
Erskines - Newton John & Co Ltd
Ertacetal® C PC 1000 - DSM Engineering Plastic Products (UK) Ltd
Ertalon® 6SA - DSM Engineering Plastic Products (UK) Ltd
Ertalyte® - DSM Engineering Plastic Products (UK) Ltd
Esam - Barloworld Vacuum Technology Plc
Esavian - 1200 - Jewers Doors Ltd
Esavian - 126 - Jewers Doors Ltd
Esavian - 127 - Jewers Doors Ltd
Esavian - 128 - Jewers Doors Ltd
Escalator SafetyStrip - Kleeneze Sealtech Ltd
Escana - Denco Ltd
Escofet - Woodhouse UK Plc
Escol - Escol Products Ltd
ESP - Hochiki Europe (UK) Ltd
Espero - Dividers Modernfold Ltd
Esprit - Celuform
Esprit - Logic Office Group
Essar Aquacoat SP - Smith & Rodger Ltd
Essar Aquacoat Xtra - Smith & Rodger Ltd
Essar Bar Top Lacquer - Smith & Rodger Ltd
Essar French Polish - Smith & Rodger Ltd
Essar Precatalysed Lacquers - Smith & Rodger Ltd
Essar Woodshield - Smith & Rodger Ltd
Essential - Vicaima Ltd
Essex - Essex
Essex - Railex Systems Ltd
Esteem - Tarkett-Marley Floors Ltd
Estoryl Sunshade fabric - Dallas & Forrest Ltd
Etalbond - CEP Claddings Ltd
Eternit-Promat - Firebarrier Services Ltd
EtherProx - CEM Systems Ltd
Ethos - Gaskell Carpet Tiles
EU 2000 - ABB Control Ltd
Euracom - Ackermann Ltd
Eureka - Eureka Products Ltd
Eurisol - UK Mineral Wool Association,The
Euro Shelving - Moresecure Ltd
Eurobar - Imperial Machine Co Ltd
Eurobench - Q.M.P.
Eurocheck - Saint Gobain pipelines PLC
Eurocord 2000 - Rawson Carpets Ltd
Eurodekor - Egger (UK) Ltd
Eurodoor 2000 - James & Bloom Ltd
Eurofix - Lastite Building Products Ltd
Euroflood - Whitecroft Lighting Limited
Eurofoil - Roof Units Group
Eurofold Folding Shutter - Bolton Gate Co. Ltd
Euroglas - Clow Group Ltd
Euro-Guard ® - Jackson H S & Son (Fencing) Ltd
Eurogullies - ACO Technologies Plc
Eurohose - Flexible Ducting Ltd
Eurolag - The Salex Group Ltd.
Euroline - Rangemaster
Eurolon Noise Absorber - The Salex Group Ltd.
Euro-meridian - Castle Care-Tech
Euromesh - Boddingtons Ltd
Euronox - Wellman Robey Ltd
Europa - Sandtoft Roof Tiles Ltd
Europa - Tarkett-Marley Floors Ltd
Europak - Reznor UK Ltd
Europak - Roof Units Group
Europave - Bardon (England) Ltd
Europave 2000 - Camas Aggregates Ltd - Aggregate Industries
Europclip - Bolivar Stamping Ltd
European - Wellman Robey Ltd
European Technical Approvals - British Board of Agrement
Europipe Drainage Pipework System - Blucher UK Ltd
Europipe Drainage Pipework System - Blucher UK Ltd
Europitch - Roof Units Group
Eurosafe - Chubb Safe Equipment Company
Euroseries - Roof Units Group
Euroslide - Q.M.P.
Eurospan - Egger (UK) Ltd
Eurostar - BLV Licht-und Vakuumtecnik
Eurostar Boilers - Trianco Redfyre Ltd
Eurotape - Boddingtons Ltd
Eurotec - Astron Buildings Parker Hannifin PLC
Eurotherm - Martin Orgee Ltd
Evac+Chair - Evac+chair international Ltd
Evacast facing masonary - Hanson Concrete Products
Evalast Blocks - Hanson Concrete Products
Evalastic - ICB (International Construction Bureau) Ltd
Evalon - ICB (International Construction Bureau) Ltd
Evamatic - KSB Ltd
Evamatic Box - KSB Ltd
Evans - Evans Vanodine International PLC
Evans Pourform - Wood International Agency Ltd
Evaporative Coolers - Radiant Services Ltd
Evelite - Amari Plastics Plc
Evenlode - Gloster Furniture Ltd
Everclad - The Abbseal Group incorporating Everseal (Thermovitrine Ltd)
Everdry - Timloc Expamet Building Products
Everest - Caradon Everest Ltd
Everest - Caradon Everest Ltd
Everflex - Everlac (GB) Ltd
Everlast Paint grade Masonary - Hanson Concrete Products
Everlite - Amari Plastics Plc
Everlite Facing Masonary - Hanson Concrete Products
Everlite paint Grade masonary - Hanson Concrete Products
Everseal - Everlac (GB) Ltd
Evertaut - Evertaut Ltd
Evertex - Everlac (GB) Ltd
Everywear - Firth Interface Europe Ltd
Evolution - Carter Retail Equipment Ltd
Evolution - Chalmit Lighting
Evolution radiant strip - Dunham Bush Ltd
Evolve - Flexiform Business Furniture Ltd
Evo-Stik - Evode Ltd
EWT - Glen Dimplex UK Ltd
Excalibar - Kenrick Archibald & Sons Ltd
Excalibur - Bartlett Catering Equipment Ltd
Excel - Axter Ltd
Excel - Axter Ltd
Excel - James Gibbons Format Ltd
Excel - Kleeneze Sealtech Ltd
Excel 5000 - Honeywell Control Systems Ltd
Excel Life Safety - Honeywell Control Systems Ltd
Excel Pro - Marlow Ropes Ltd
Excel Security Manager - Honeywell Control Systems Ltd
Excelastic - Geosynthetic Technology Ltd
Excelflex - Axter Ltd
Exceliner - Geosynthetic Technology Ltd
Excell - OWA UK Ltd
Excelsior - ADT Fire and Security Plc
Excem - Exchem Mining and Construction
Excludoor - Sunray Engineering Ltd
Exclusive - Vicaima Ltd
Executive - Midland Alloy Ltd
Executive - Rollalong Ltd
Exedra - BPT Security Systems (UK) Ltd
E-xel - Logic Office Group
Exhausterl blowers - BVC
Exhausterl blowers - BVC
Exide - MJ Electronics Services (International) Ltd
Exit Box - Fitzgerald Lighting Ltd
Exit:Way guidance - Hoyles Electronic Developments Ltd
Exitguard - Hoyles Electronic Developments Ltd
Exmet - Expamet Building Products
Exodus - Hille Ltd
Exodus - Orsogril Spa UK Distributor
Exos 8000 - Kaba (UK) Ltd
Expament - Allmat (East Surrey) Ltd
Expamet - Advanced Fencing Systems Ltd
Expamet - Expamet Building Products
Expamet - M & M Fencing Limited
Expamet fencing - Chestnut Products Limited
Expanda -PAC - C & R Building Systems Ltd
Expandafoam - Fosroc Ltd
Expelex - Exitex Ltd
Explorer and Harlequin Range - Lumitron Ltd
Exposit - Erlau AG
Expowall - Deceuninck Ltd
Express - Heatrae Sadia Heating Ltd
Express Evans - Express Evans Lift
Express Evans Lifts - Express Evans Lifts
Extended Performance - Leaderflush + Shapland Ltd
Extra - Everlac (GB) Ltd
Extra Aqua Vent - Areco Roofing Supplies Company
Extral - Rada Lighting
Extraproof - Briggs Roofing & Cladding Ltd, Roofing Contract
Extrusions Direct - Extrusions Direct
Ezelift - Maswell Engineering Ltd
Ezi - lift - Didsbury Engineering Co Ltd
F.S.B. - Haffele UK Ltd
F`Light - Glamox Electric (UK) Ltd
F2P - Artur Fischer (UK) Ltd
F2P-G - Artur Fischer (UK) Ltd
FAAC - Crimeguard Systems Ltd
FAAC automatic gates - Automatic Doors & Gates
Faber - Faber Blinds UK Ltd
Faber 1800 - Faber Blinds UK Ltd
Faber 2000 - Faber Blinds UK Ltd
Faber Autostop - Faber Blinds UK Ltd
Faber Maximatic - Faber Blinds UK Ltd
Faber Metalet - Faber Blinds UK Ltd
Faber Metamatic - Faber Blinds UK Ltd
Faber Midimatic - Faber Blinds UK Ltd
Faber Minimatic - Faber Blinds UK Ltd
Faber Multistop - Faber Blinds UK Ltd
Faber Softline - Faber Blinds UK Ltd
Fabresa - Stokes R J & Co Ltd
Fabrex - Tuke & Bell Ltd
Façade - Hillaldam Coburn Ltd
Facet - Zon International
Facom - Arrow Supply Company Ltd
Factor 4 - Interface Europe Ltd
Fade shield - Madico Film Products Division Inc
Fadini - Automatic Doors & Gates
Fagley Yorkstone - Pickard Group
Fairmile - Advanced Fencing Systems Ltd
Fairmile - Fairmile Fencing Ltd
Fairmile A - Pale - Fairmile Fencing Ltd
Fairmile Crownguard - Fairmile Fencing Ltd
Fairmile Maxigaurd '358' - Fairmile Fencing Ltd
Fairmile Palisade - Fairmile Fencing Ltd
Fairmile permagaurd - Fairmile Fencing Ltd
Fairmile Roadguard - Fairmile Fencing Ltd
Fairmile Safeguard - Fairmile Fencing Ltd
Fairmile Siteguard - Fairmile Fencing Ltd
Fairmile Sovereign Grille - Fairmile Fencing Ltd
Fairmile Stylegaurd - Fairmile Fencing Ltd
Fairmile Vertical Bar - Fairmile Fencing Ltd
Fairmitre - Fairmitre Ltd
Fairoaks Red Multi - Cranleigh Brick & Tile Co Ltd
Fairway - Metalrax Ltd
Faiveley - Automatic Doors & Gates
Faiveley - Automatic Doors & Gates
Falco - Falco
Falcon - James Gibbons Format Ltd
Falcon - Swintex Ltd
Falcon 3.6.9 - Falcon Food Services Equipment
Falcon 350 - Falcon Food Services Equipment
Falcon warewash - Falcon Food Services Equipment
Fall Detector - Telelarm care Ltd
Faltec - Crawford Amber Ltd
Fanaire - Fantasia Distribution Ltd
Fantasia - Lappset UK Ltd
Fantasia ceiling fans - Fantasia Distribution Ltd
Faral Etal - Townhive Ltd
Faral Green - Townhive Ltd
Faral Lampo - Townhive Ltd
Faral Lineal - Townhive Ltd
Faral Longo - Townhive Ltd
Faral Mera - Townhive Ltd
Faral Tondo 2 - Townhive Ltd
Faral Tropical - Townhive Ltd
Faral Tropical 140 - Townhive Ltd
Faral Vega - Townhive Ltd
Farmington - Farmington Natural Stone Ltd
FarmPave - Readymix Limited
Farr 3D/3D - Farr Filtration Ltd
FAS - Fabricated Aluminium Services Ltd
FAS Ltd - Fabricated Aluminium Services Ltd
FAS Windows & Doors - Fabricated Aluminium Services Ltd
Fascinating finishes - Kingstonian Point Ltd
Fascinations - Carpets of Worth Ltd
Fascut Thin Wheels - Unicorn Abrasives (UK) Ltd
Fasset - CEP Claddings Ltd
Fast build barriers - Bell & Webster Concrete Ltd
fast build platforms - Bell & Webster Concrete Ltd
fast build rooms - Bell & Webster Concrete Ltd
Fast build rooms - Bell & Webster Concrete Ltd
Fast build tendering - Bell & Webster Concrete Ltd
fast build walls - Bell & Webster Concrete Ltd
Fast fit - Panel & Louvre Co Ltd
Fast rack - Legrant Electric Ltd
Fastcall - Tunstal Telecom Ltd
Fastel - P4 Limited
Fastel - P4 Limited
Fastfill - Flexcrete Ltd
FastFinish - Readymix Limited
Fastfloor - Tarmac Top Floor
Fastfloor - Tarmac Top Floor
Fastie - Redifix Ltd
Fastline - IFT Supplies Ltd
Fast-Track - Tektura Plc
Fast-track - Tully
FastTrack™ Folder-Inserter System - Neopost Ltd
Fastwall - Fastwall Ltd
FaTec - Artur Fischer (UK) Ltd
Faulhaber - Electro Mechanical Systems Ltd
Favourite - Valor
FCA member - Chestnut Products Limited
FEB - Degussa Construction Chemicals (UK)
FEB - Strand Hardware Ltd
Fecon Baffle Filtors - Vianen Kitchen Ventilation (UK) Ltd.
Federal - Schneider Electric Ltd
Federation of Scottish Fencing Contractors (FSFC) - Fencing Contractors Association
Feedmaster - CAEC Howard (Holdings) Ltd
Felix Dance Floor - Felix Design Ltd
Felix Riser Stage - Felix Design Ltd
Felix Staging - Felix Design Ltd
Feltex - Rawson Fillings Ltd
Feltguard - Campbell Marson & Co
Fence Decor - Jotun Henry Clark Ltd Decorative Division
Fenceguard - Guardall Ltd
Fencemaker - Fleetwood Fengate Co Ltd
Fencetone - Protim Solignum Ltd
Fencing Division - May Gurney (Technical Services) Ltd - Fencing Division
Fenlite Blocks - Hanson Concrete Products
Fentox - Aldous & Stamp Services Ltd
Fep 'O' rings - Ashton Seals Ltd
Fep 'O' rings - Ashton Seals Ltd
Ferham - Titan plastech
Fermacell dry flooring elements - Fels UK
Fermacell wallboards - Fels UK
Fermec International Ltd - Fermec Holdings Ltd
Fermec Manufacturing Ltd - Fermec Holdings Ltd
Ferndown - Hamworthy Heating Ltd
Feron - Clayton-Munroe Ltd
Fersina - Cestrum Building Products Ltd
Ffestiniog - McAlpine Slate Ltd, Alfred
FG Contracts - FG Contracts Limited
FG Wilson - FG Wilson
Fibaform - Fibaform Products Ltd.
Fibarack - Mita (UK) Ltd
Fibastrut - Mita (UK) Ltd
Fibatray - Mita (UK) Ltd
Fibercill - Fibercill
Fibermat - EBC UK Ltd
Fibermat - Vandex (UK) Ltd
Fibermesh MD - S.I. Concrete Systems
Fiberskirt - Fibercill
Fiberstars - Crescent Lighting
Fibertrave - Fibercill
Fibo - Lignacite (North London) Ltd
Fibo Lightweight Insulation Blocks - Stowell Concrete Ltd
Fibral - Rockfon Ltd
Fibreroll E240 Rolling Shutter - Bolton Gate Co. Ltd
Fibreroll E30 Rolling Shutter - Bolton Gate Co. Ltd

Fibreshield Rolling Fire Curtain - Bolton Gate Co. Ltd
Fibrespan - Armfibre Ltd
Fibrotrace ® - Maestro International Ltd
Fido bin (Dog waste bin) - Glasdon UK Ltd
Fiesta - Gaskell Carpet Tiles
Fife Alarms - Fife Fire Engineers & Consultants Ltd
Fife Fire - Fife Fire Engineers & Consultants Ltd
Fife Stone - Aggregate Industries UK Ltd
Fiji 2 - Rangemaster
Filatrap - Conder Products Ltd
Fildek - Don Construction Products Ltd
Filetrak - Rackline Systems Storage Ltd
Fillaboard - Fillcrete Ltd
Fillcrete - Fillcrete Ltd
Filmatic - Martin Orgee Ltd
Filon - Filion Products Ltd
Filon Citadel - Filion Products Ltd
Filon DR - Filion Products Ltd
Filon Monarch Barrellights - Filion Products Ltd
Filon Multiclad - Filion Products Ltd
Filon Overproof with Profix - Filion Products Ltd
Filon Plate - Filion Products Ltd
Filon SupaSafe - Filion Products Ltd
Filto-Bench - Horizon International Ltd
Filtrair - Auchard Development Co Ltd
Filtrair - Auchard Development Co Ltd
Filtrasol ® - Hunter Douglas Ltd (Contracts Division) Luxaflex Blinds ®
Fimap - Barloworld Vacuum Technology Plc
Fin Cycle Stand - Furnitubes International Ltd
Fine surface treatment - Fels UK
Fine Textured Renovation Blend - Charnwood Brick Group Ltd
Fine Uplands Twist - Adam Carpets Ltd
Fine Worcester Twist - Adam Carpets Ltd
Finebar Glazing system - Blair Joinery
Finelight - Andersen/ Black Millwork
FineLine - FineLine
Fineline - Jebron Ltd
Fineline - Matki plc
Fineline - Neslo Partitioning & Interiors
Finesse - ABG Ltd
Finesse - ABG Ltd
Finesse - Nevill Long Limited
Finest Platinum - Halstead Boilers Ltd
FingerSafe - Dorma Door Services Ltd
Finned Tube - Turnbull & Scott (Engineers) Ltd
Finned Tube Radiators - Turnbull & Scott (Engineers) Ltd
Finno - Lappset UK Ltd
Finsa - Boardcraft Ltd
Finvector - Dunham Bush Ltd
Firaqua - Firwood Paints Ltd
Fire Curtain - Colt International
Fire Flexmaster - Docherty New Vent Chimney Group
Fire master - Paddock Fabrications Ltd
Fire Path - Colt International
Fire Protection Prducts - Smith & Rodger Ltd
Fire Safe - Kingspan Ltd
Fire secur - Ide T & W Ltd
Fireact - Rea Metal Windows Ltd
fireates - Macron Safety Systems (UK) Ltd
Firebeam - Photain Controls Plc
Firebrand - Hart Door Systems Ltd
Firebreak - Rea Metal Windows Ltd
Firecel - Fitzpatrick Doors Ltd
Firecheck 8000 Series - Tann Synchronome Ltd
Firecheck concept - Tann Synchronome Ltd
Firecheck Interactive - National Britannia Ltd
Firecheck MX4000series - Tann Synchronome Ltd
Firecheck PK3000 series - Tann Synchronome Ltd
Firecheck synchro - Tann Synchronome Ltd
Firechef - Profab Access Ltd
Fireclamps® - Abesco Ltd
Firecom' Fire Telephone Systems - PEL Services Limited
Fired Earth - Fired Earth
Fired Earth H2O (The Miracle of Water) - Fired Earth
Firedex - Cooper Lighting and Security Ltd
Firedex - JSB Electrical PLC
Firedex 3302 - JSB Electrical PLC
Firedex Biwire - JSB Electrical PLC
FireFoam - Sealmaster
Fireframe - Sapa Building Systems (monarch and glostal)
FireGlaze - Sealmaster
Fireguard - TCW Services (Control) Ltd
Firelite - Valor
Firelok Cladding Panel - Gradient Insulations (UK) Ltd
Firelok Deck - Gradient Insulations (UK) Ltd
Firemac - McKenzie-Martin Ltd
Fireman - Clark Door Ltd
FireMaster - Cooper Group Ltd
Firemaster - Docherty New Vent Chimney Group
Firemaster - Firemaster Extinguisher Ltd
Firemaster - Premdor
Fireplan - Internal Partitions Systems
Firepoint - Hoyles Electronic Developments Ltd
Firesecur - Twide-Paragon Ltd
Fireshield - Premdor
Fireshield - The Bollom Group
Fireshield ™ - Accent Hansen
Firestile - Mann McGowan Group
Firestop - Dow Corning Ltd
Firestop - George Fischer Sales Ltd
Firestop - Yonder Hill Ltd
Firestore - Franklin Hodge Industries Ltd
Fireswiss - C. G. I. International Ltd
Fireswiss Foam - C. G. I. International Ltd
Firetainer - Franklin Hodge Industries Ltd
Firetec - AEI Cables Limited
Firetec - AEI Cables Limited
Firetech - Knauf Insulation Ltd
Firetex - Kaba Door Systems Ltd
Firetex - Leigh's Paints
Firewall - Neslo Partitioning & Interiors
Firewarn' Fire Alarm Systems - PEL Services Limited
Firewatch - TCW Services (Control) Ltd
Firex - James Smellie Fabrications Ltd
Firglo - Firwood Paints Ltd
Firlene - Firwood Paints Ltd
Firlex - Firwood Paints Ltd
Firpavar - Firwood Paints Ltd
First Alert - BRK Brands Europe Ltd
First Alert - BRK Brands Europe Ltd
First class - Lappset UK Ltd
First-Rol - Reflex-Rol (UK)
Firsyn - Firwood Paints Ltd
Firth Carpets - Interface Europe Ltd
Firwood - Firwood Paints Ltd
Fischer - Arrow Supply Company Ltd
Fischer - Artur Fischer (UK) Ltd
Fischer - Artur Fischer (UK) Ltd
Fischer - Kem Edwards Ltd
Fitex - Airsprung Beds Ltd
Fitt box - PFP Electrical Products Ltd
Fitter & Poulton - PFP Electrical Products Ltd
Five 5 Star signs - Arbory Group Limited
Five Brazilian Slates - Kirkstone
Five Star - HIS Ltd
Fix - Grorud Industries Ltd
Fix-A-Tap - Essex
FKR - Cast Iron Column Radiators - Clyde Combustions Ltd
FL Façade - Consult Lighting
Flagmaster - Zephyr The Visual Communicators
Flagship Range - Chelwood Brick Ltd
Flair - Taylor & Portway Ltd
Flairline - Hadley Industries Plc
Flame Bar - Fire Protection Ltd
Flamecheck - Arch Timber Protection
Flameguard - Bonar Floors Ltd
Flameguard Hinged Fire Door - Bolton Gate Co. Ltd
Flamenco - MMP International
Flamestat - Flexelec(UK) Ltd
Flametex - Airsprung Beds Ltd
Flammtex - Vita Liquid Polymers Ltd
Flamrad - Infraglo (Sheffield) Ltd
Flanders - Forticrete Roofing Products
Flapmaster - Paddock Fabrications Ltd
Flare - Indirect Lighting Ltd
Flashpoint infra red - Emergi-Lite Safety Systems
Flavel - Flavel - Leisure Glynwed Consumer & Construction Products Ltd
Flaviker - Arnull Bernard J & Co Ltd
Flaviker Tiles - Reed Harris
Fleck Coat - Stewart Wales, Somerville Ltd
Flectoline - Rockfon Ltd
Flex - Harbro Supplies Ltd
Flex - seal - Flex-Seal Couplings Ltd
Flex line - Schneider (UK)
Flexbeam - Asset International Ltd
Flexbond - Rawson Fillings Ltd
Flexcase - Universal Components Ltd
Flexcon - Flamco UK Ltd
Flexconpak - Flamco UK Ltd
Flex-face signs - Vision Neon & Display Ltd
Flexfelt - Rawson Fillings Ltd
Flexfiller - Flamco UK Ltd
Flexflyte - Flexible Ducting Ltd
Flexguard - C & K Intellisense
Flexibar 2000 - Manton Industrial Seals Ltd
Flexiboard - Salex Interiors Ltd
Flexi-bollard - Berry Systems
Flexibuild Sort Units - Envopak Group Ltd
Flexiburo - Flexiform Business Furniture Ltd
Flexicon - Polypipe Bathroom & Kitchen Products Ltd
Flexicool - ABB Control Ltd
Flexideck 2000 - Flexcrete Ltd
Flexideck Interdeck system - Flexcrete Ltd
Flexideck Sealant - Flexcrete Ltd
Flexidoor - Envirodoor Markus Ltd
Flexidoor - Telford Doors and Barriers Ltd
Flexiframe ® - Andersen/ Black Millwork
Flexiglide - Flexiform Business Furniture Ltd
Flexiguard - Advanced Perimeter Systems Ltd
Flexiguard - Advanced Perimeter Systems Ltd
Flexiline - B&C Roadmarkings Ltd
Fleximetric - Flexiform Business Furniture Ltd
Flexipark 2 - Tully
Flexi-post - Berry Systems
Flexisound - Salex Interiors Ltd
Flexistor - Flexiform Business Furniture Ltd
Flexithane - Tor Coatings Ltd
Flexitrack - Beta Lighting Ltd
Flexiwall acrylic wall coating - Altro Floors
Flexlite - Ventlane Ltd
Flexlock - Viking Johnson
Flex-net - Metway Electrical Industries Ltd
Flex-net II - Metway Electrical Industries Ltd
Flexocel - Baxenden Chemicals Ltd
Flextight seccoral systems - P.C.I Construction Systems Ltd
Flextract 2000 - Flexible Ducting Ltd
Flexvent - Flamco UK Ltd
Flexyl - STO Ltd
Flipflap - Envirodoor Markus Ltd
FLo2 - Ves Andover Ltd
Floating Solar Fountain - Intersolar Group Ltd
Flocoat - Hub Tubes
FloGuard - FloPlast Ltd
Flo-jo ® - Maestro International Ltd
Flolift - Conder Products Ltd
Floodline - Illuma Lighting Ltd
Floor Muffle - TDI (UK) Ltd
Floor to Floor - Engineering & Lift Ltd
Floorcote - Coo-Var Ltd
Floorfast - P.C.I Construction Systems Ltd
Floor-fast - Lindapter International
Floorline - Egger (UK) Ltd
Floormaker - Western Cork Ltd
Floormaster - ABB Control Ltd
Floormaster - Interfloor Limited
Floorshield - Springvale E P S
Floorworks - Interfloor Limited
Flopast - Allmat (East Surrey) Ltd
Florad - FloRad Heating Systems
Floralsilk - Pouliot Designs by Floralsilk Ltd
FlorAwall - PHI Group Ltd
Florentine - Forticrete Ltd
Florette - Carpets of Worth Ltd
Florex - Tor Coatings Ltd
Floristoon - Blakley Electrics Ltd
Flos/Arteluce - Quip Lighting, Div of Bruce & Macintyre Ltd
FloSaver - FloPlast Ltd
Flotex - Bonar Floors Ltd
Floway - CPV Ltd
Flowcrete - Flowcrete Plc
Flowdeck - Redman Fisher Engineering
Flowforge® - Redman Fisher Engineering
Flowgrid - Redman Fisher Engineering
FlowGRiP TM - Redman Fisher Engineering
Flowguard - Isopad Ltd
Flowlok - Redman Fisher Engineering
Flowmaster - Titan plastech
Flowmax - Range Cylinders
Flowpac - Barr & Wray Ltd
Flowtop - Watco UK Ltd
Flowtread - Redman Fisher Engineering
Flueflex - Ventlane Ltd
Fluemaster - Docherty New Vent Chimney Group
Fluidair - F. W. Harris & Co Ltd
Fluidair Economy drive system - Fluidair International Ltd
Fluidair rotary - Fluidair International Ltd
Fluorosint® - DSM Engineering Plastic Products (UK) Ltd
Flush glazing - Clestra Hauserman Ltd
Flushplan - Internal Partitions Systems
Flushwall - Neslo Partitioning & Interiors
FLX Floodlight - Consult Lighting
Flying end fabric - BRC Ltd
FMB - Federation of Master Builders
FM-Zoo® - Hygood Ltd
Foamalux ® - Righton Ltd
Foamglas - Pittsburg Corning (UK) Ltd
Focal Point - Brintons Ltd
Focal point - nmc (UK) Ltd
Focus - James Gibbons Format Ltd
Focus - Saint-Gobain Ecophon Ltd
Fogo Montanha - Emsworth Fireplaces Ltd
Foilboard - Callenders Ltd
Foldaway - Tilley International Ltd
Fondaline - Onduline Building Products Ltd
Fondu - Lafarge Aluminates Ltd
Foodcheck Interactive - National Britannia Ltd
Footlights - Paragon by Heckmondwike
Footnotes - Woolliscroft Tiles Ltd
Footprint - Footprint Tools Ltd
Force - Axter Ltd
Forceflow - Biddle Air Systems Ltd
Forceflow - Biddle Air Systems Ltd
Fordham - Fordham
Forest - Lappset UK Ltd
Forest-Saver - Earth Anchors Ltd
Forgebrand - Stockten Heath Forge (Caldwells) Ltd
Forgecrete - Cheshire Concrete Products
Forgecrete plus - Cheshire Concrete Products
Forgemaster Hammers - Stockten Heath Forge (Caldwells) Ltd
Form3Fast - Square D, Electrical Distribution & Building applicators Division
Forma D400 - Allpipe Ltd
Form-a-duct - Airflow (Nicoll Ventilators) Ltd
Format - James Gibbons Format Ltd
Format 50 - James Gibbons Format Ltd
Formawall - European Profiles
Formica - Decra Ltd
Formica - Formica Ltd
Formica - Viking Laminates Ltd
Formica Spawall - Formica Ltd
Formica Cubewall - Formica Ltd
Formica Firewall - Formica Ltd
Formica Lifeseal - Formica Ltd
Formica Prima - Formica Ltd
Formica Unipanel - Formica Ltd
Formost - Flexiform Business Furniture Ltd
Formwork Tape - Cordek Ltd
Forpave - Bardon (England) Ltd
Forpave - Camas Aggregates Ltd - Aggregate Industries
Fortic - Range Cylinders
Foscarini - Quip Lighting, Div of Bruce & Macintyre Ltd
Fospro - Hellermann Tyton
Foulmaster - Pump Technical Services Ltd
Foundation 2000 - Computer Foundations Ltd
Foundation Flow - Readymix Limited
Four-D - LSA Projects Ltd
Four-hour fire door - Rolflex Doors Ltd
Foxxx swimming pool kits - Fox Pool (UK) Ltd
FP 2-14 Zone Panel - C-TEC (Computionics) Ltd
fpdc - Federation of Plastering and Drywall Contractors
FPS - Federation of Piling Specialists
FPS Matting + Washroom Services - Floor Protection Services Ltd
Framefast - HW Systems Ltd
Framelights - Illuma Lighting Ltd
Frametherm - Davant Products Ltd
Franch - Zon International
Frank - Creteco Sales Ltd
Franke Eco System - Franke UK Ltd
Franke Kitchen Systems - Franke UK Ltd
Franke Sinks - Franke UK Ltd
Franke Taps - Franke UK Ltd
Franke Triflow - Franke UK Ltd
Franki - Cementation Foundations SKANSKA Ltd
Franklin Hodge Industries - Denco Ltd
Freedom 5 Vertical Lift - Brooks Stairlifts Ltd
Freeflex - William Freeman Ltd
FreeFlush - Cistermiser Ltd
Freespace - Project Office Furniture PLC
Freestanding Systems - Spur Shelving Ltd
Freestyle - Swedecor Ltd
Freeway - Rawson Carpets Ltd
Freeway - Vantrunk Engineering Ltd
Freeway 11 - Cistermiser Ltd
FreeWeigh™ In-Motion Electronic Weighing System - Neopost Ltd
Freezstop - Heat Trace Ltd
Frenchwood ® - Andersen/ Black Millwork
Frico - shearflow LTD
Fridgewatch - J & E Hall Ltd
Friedland - Friedland Ltd
Friedland - Friedland Ltd
Frimeda Rapid Lift - Halfen Ltd
Fritztile Resilient Terrazzo - Marble Flooring Spec. Ltd
Froggo (Frog bin) - Glasdon UK Ltd
Frontier - Boddingtons Ltd
Frontrunner - Plastic Extruders Ltd
Frost - FC Frost Ltd
Frostar - Imperial Machine Co Ltd
Frostguard - Isopad Ltd
FRT 80 - Thatching Advisory Services Ltd
FRT Exterior® - Brash John & Co Ltd
FRX (Fluidair rotary Screw) - Fluidair International Ltd
FS Screed - Conren Ltd
FSB - Allgood plc
FSB Door Furniture - JB Architectural Inronmongery Ltd
FSC - Montague L Meyer
FSS - The HB Sign Company Ltd
FST (Flexible Security Topping) - Binns Fencing Ltd
Fuel bank bunded storage units - Cookson and Zinn (PTL) Ltd
Fufil - Springvale E P S
Fujitsu - HV Air Conditioning Ltd
Fujitsu - Pro Display
Fujitsu - Purified Air Ltd
Fulcrum - Paddock Fabrications Ltd
Full Spectrum - Full Spectrum Lighting Ltd
Fullflow - Fullflow Group Ltd
Fulton-Series E - Fulton Boiler Works (Great Britain) Ltd
Fungishield - Glixtone Ltd
Funtheme - Design & Display (Sales) Ltd
Funxion - ASSA Ltd
Furfix Wall Profiles - Simpson Strong-Tie®
Furmanite Flange Covers - Furmanite International Ltd
Furmanite Sheet/Jointing Compound - Furmanite International Ltd
Furmanite Tube Plug - Furmanite International Ltd
Fuseal - George Fischer Sales Ltd
Futura - Paragon Business Furniture
FW 50 + - SCHUCO International KG
FW 60 + - SCHUCO International KG
Fyfestone - Camas Building Materials
Fyfestone - Camas Building Materials
Fyfestone - Fyfe John Ltd
Fyfestone - Fyfe John Ltd
Fyrebag 240 - Mann McGowan Group
Fyrex - Iles Waste Systems
G.B. Wendland - Caradon Duraflex Systems Ltd
G146 Gutter - Alifabs (Woking) Ltd
G3 Gates - Albion Fencing Ltd
G3 Gates - Albion Fencing Ltd
G3000 - Girsberger London
G3300 - Girsberger London
G55 Gutter - Alifabs (Woking) Ltd
Gabions - PHI Group Ltd
GAI - Guild of Architectural Ironmongers
Galaxy - Kompan Ltd
Galaxy - Metreel Ltd
Galaxy Ultra (operating theatre lamps) - Brandon Medical
Galaxy Ultra (operating theatre lamps) - Brandon Medical
Galena Spencer Ltd - Kidde Fire Protection Services Ltd
Galerie design range - Caradon Bathrooms Ltd
Galleria - Brintons Ltd
Galparket - Deralam Laminates Ltd
Galtres - York Handmade Brick Co Ltd
Galvashield - Fosroc Ltd
Gang-Nail Connectors - Gang-Nail Systems Ltd
Garador - Garador Ltd
Garbina - Unicorn Containers Ltd
Garden Classic - Titan plastech

Gardex - Palgrave Brown Timber & Fireproofing Co
Gardwall - Brown F plc
Garfa - Zon International
Garmex - Metalrax Ltd
Garnier sign - Burnham Signs
Gas 210 ECO - Broag Ltd
Gas zips - Power Plastics Ltd
Gascogne Limestone - Kirkstone
Gascool - Birdsall Services Ltd
Gascoseeker - Gas Measurement Instruments Ltd
Gasgard - Marley Waterproofing
Gasgard - MSA (Britain) Ltd
Gasgard gas detector - MSA (Britain) Ltd
Gass - SGB Youngman
GasSeal - Marley Waterproofing
Gastech - Birdsall Services Ltd
Gastech - Birdsall Services Ltd
Gastyle II - Dunbrik (Yorks) Ltd
Gasurveyor - Gas Measurement Instruments Ltd
Gate - Millar Dennis (1998) Ltd
Gate Automation and access Barrier Association (EA - Fencing Contractors Association
Gatic 2000 - Alumasc Construction Products Ltd
Gatic Easylift - Alumasc Construction Products Ltd
Gatic Halliday - Alumasc Construction Products Ltd
Gatic Slotdrain - Alumasc Construction Products Ltd
Gatorband - Simplex Signs Limited
Gayton Door - AS Newbould Ltd
GEA - Garage Equipment Association Ltd
GEC Anderson - Anderson GEC Ltd
GEC Anderson - Anderson GEC Ltd
Gedina - Saint-Gobain Ecophon Ltd
Geepee - Swish Building Products
Gem - Crane Fluid Systems
Gem - Tunstal Telecom Ltd
Gemini - Forticrete Roofing Products
Gemini - Hoyles Electronic Developments Ltd
Gemini - Rangemaster
Gemini - Roof Units Group
Gemini Shelving - Sealine Business Products
Gemini tile - Forticrete Ltd
GEN2 - Otis Ltd
Genesis - Dixon Turner Wallcoverings
Genesis - Eltron Chromalox
Genie - Genie Europe
Genny - Radiodetection Ltd
Genovent - Gebhardt Kiloheat
Gent - Gent Limited
Gent Services - Gent Limited
Geo - Jaga Heating Poducts (UK) Ltd
GEO - Woodhouse UK Plc
Geodome - CST Vulcan
Geofix - Degussa Construction Chemicals (UK)
Geojute - Land Wood & Water Co. Ltd
Geomat - Land Wood & Water Co. Ltd
George Fischer - Capper Plastics
George Fischer - The BSS Group PLC
Georgian & Victorian Brass - Eaton Electric Limited
Geotextiles - Land Wood & Water Co. Ltd
Gerard - Decra Roofing Systems
Gerbergraphix - Green Brothers Ltd
Gerni - Nilfisk Advance Ltd
Geryon - Kaba Door Systems Ltd
Geschwender - LSA Projects Ltd
Gestra - Flowserve Flow Control
Gethal FSC Certiply - Wood International Agency Ltd
Geze - Haffele UK Ltd
G-Flex - Johnston Pipes Ltd
GH 310 ECO - Broag Ltd
GIAS - Profile Lighting Services Ltd
Gifcrete - Don Construction Products Ltd
Gildeloc - HCL Verta
Gillair smoke vent products - Matthews & Yates Ltd
Gilt - Baggeridge Brick PLC
Girdlestone - Girdlestone Pumps Ltd
GKD - Cadisch MDA Ltd Ltd
GL Globe - Consult Lighting
GL1000 - Gleno Industries Ltd
GL24 - Gleno Industries Ltd
GL24W - Gleno Industries Ltd
GL252 - Gleno Industries Ltd
GL3 - Gleno Industries Ltd
GL32 - Gleno Industries Ltd
GL32W - Gleno Industries Ltd
GL38 - Gleno Industries Ltd
GL38W - Gleno Industries Ltd
GL6 - Gleno Industries Ltd
GL914 - Gleno Industries Ltd
Glada - Novellini Bathroom Products Ltd
Glamox - Glamox Electric (UK) Ltd
Glamur - Jotun Henry Clark Ltd Decorative Division
Glasal - Eternit Building Materials
Glasbord - Kemlite Ltd
Glasflex - Flexible Ducting Ltd
Glasgrid - Stenoak Associated Services PLC
Glasket - Alfas Industries Ltd
Glasroc - British Gypsum Ltd
Glasroc - Nevill Long Limited
Glass 97 - CST Vulcan
Glass Blocks - Swedecor Ltd
Glass chalkboards - Lightfoot Charles Ltd
Glass fibre Manufacturors - Hynes Aden K Sculpture Studios
Glass Houses - Haddonstone Ltd
Glass louvre - Colt International
Glassgard - Bonwyke Ltd
Glasstrim GRP - Areco Roofing Supplies Company
Glasstrim GRP - Areco Roofing Supplies Company
Glasswhite - Cardale Doors Ltd
Glasswood - Cardale Doors Ltd
Glasurit - HMG Paints
Glazed Bricks - Ibstock Hathernware Ltd
Glazing Accessories - Simon R W Ltd
Glen - Glen Dimplex UK Ltd
Glendale - Lycetts Industrial Doors
Glendyne Slate - Cembrit Blunn Ltd
Glendyne Slate - Cembrit Blunn Ltd
Glenfield - Bennett Windows & Doors
Glenium - Feb MBT
Gleno - Therm - Gleno Industries Ltd
Glidden - ICI Dulux Trade Paints
Glidebolt - Paddock Fabrications Ltd
Glidemaster - Hillaldam Coburn Ltd
Glidetrak - Rackline Systems Storage Ltd
Glidevale - Zled (DPC) Co Ltd
Glipp-loc - Johnston Pipes Ltd
Glo Lights - Brett Paving
Global - EBC UK Ltd
Global Glass - CST Vulcan
Global Sign Sales - Signs International
Globe - Millar Dennis (1998) Ltd
Globelite - LB Lighting Ltd
Glocote - Coo-Var Ltd
Glo-dac - Dacrylate Paints Ltd
Gloria - Kermi (UK) Ltd
Glosspile - Marshall J A & Co (Brushes) Ltd
Glostal - Glostal - Monarch
Glostal - Sapa Building Systems (monarch and glostal)
Gloster - Gloster Furniture Ltd
Gloucester - Atlas Stone Products
Glover Series 2000 Extinguishers - Fife Fire Engineers & Consultants Ltd
Glow-Worm - Hepworth Heating Ltd
Gloy - Henkel Consumer Adhesives
Gluegunsdirect - Kenyon Performance Adhesives
Glulam - Structural Timbers Ltd
Glynn-Johnson - Relcross Ltd
GM & RD - Townhive Ltd
Go - Atmospheric Boilers - Clyde Combustions Ltd
Go Green - ABG Ltd
Godwin Pumps - Goodwin HJ Ltd
Gold - Wilo Samson Pumps Ltd
Goldbach - Mauser Interiors (UK) Ltd
Goldseal - Marley Waterproofing
Goldstar - Wilo Samson Pumps Ltd
Golf Range - Helmsman
Goods-rol - Lycetts Industrial Doors
Gordon Russell - Harvey Furniture
Gordon Russell - Steelcase Strafor Plc
Goslo - Fyfe John Ltd
Gothic - Oswestry Reinforced Plastics Ltd
Goxhill Plain Tile - Sandtoft Roof Tiles Ltd
GR 1 Graffiti Remover - Stewart Wales, Somerville Ltd
Grabber screws - Lafarge Plasterboard Ltd
Grada - Panel & Louvre Co Ltd
Gradelux - Laidlaw Architectural Hardware Ltd
Graduat - AEL
Gradus - Gradus Ltd
Graffiti-Gard - Andura Textured Masonry Coatings Ltd
Gragreen - En-tout-cas plc
Grampian - Sashless Window Co Ltd
Granclene - Harbro Supplies Ltd
Grando UK - Buckingham Swimming Pools Ltd
Grando UK - Buckingham Swimming Pools Ltd
Granette - DLW Flooring Ltd
Granguard Aqua - Granwood Flooring Ltd
Graniplast Trowelled Marble - Textured European Finishes Ltd
Granirapid - Mapei (UK) Ltd
Granit - Tarkett Sommer Ltd
Granline - Granwood Flooring Ltd
Gransprung - Granwood Flooring Ltd
Gransprung Hi-load - Granwood Flooring Ltd
Granwax - Atkinson & Kirby Ltd
Granwood - Granwood Flooring Ltd
Granyte Woodfinishes - Atkinson & Kirby Ltd
Graphic Inlay - Kleen-Tex Industries Ltd
Grassblock - Grass Concrete Ltd
Grasscrete - Grass Concrete Ltd
Grasshopper - Lawrence Surfaces Plc, Charles
Grasskerb - Grass Concrete Ltd
Grasspave - ABG Ltd
Grassroad - Grass Concrete Ltd
Grate-fast - Lindapter International
Grateglow - Robinson Willey Ltd
Graviner - Explosion Protection Systems - Kidde Fire Protection Ltd
Grease Bugs - Progressive Product Developments Ltd
Greaves Portmadoc Slate - Greaves Welsh Slate Co Ltd
Greenmoor sandstone - Lindley George & Sons Ltd
Greenstar HE - Worcester Heat Systems Ltd
Greenwood - Sandtoft Roof Tiles Ltd
Gres de Valles - Stokes R J & Co Ltd
Grid Plus - MK (MK Electric Ltd)
Gridframe - Sapa Building Systems (monarch and glostal)
Gridliner - Sportmark™ Ltd
Grid-lok - Vantrunk Engineering Ltd
Gridplex - Cameron & Moore (Gridplex) Ltd
Gridplex - Cameron & Moore (Gridplex) Ltd
Griffin - Tinsley & Co Ltd, Eliza
Griffin Fire - Griffin and General Fire Services Ltd
grime grabber entrance matting - Jaymart Rubber & Plastics Ltd
Gripfalt SMA - Johnston Roadstone Ltd
Gripfast - Vallance
Gripfill - Laybond Products Ltd
Grippa - Talbot
Gripper handlamps - Ensto Briticent Ltd
Gripperrods - Interfloor Limited
Gripple - Kem Edwards Ltd
Griptight Plus Woodscrews - UK Fasteners Ltd
Griptop - Don Construction Products Ltd
Groheart - Grohe Ltd
Grohedal - Grohe Ltd
Grohetec - Grohe Ltd
Grohetec special fittings - Grohe Ltd
Gro-Lux - Plant House /Greenhouse & Aquarium Lamps - Sylvania Lighting International INC.
Ground Anchors - PHI Group Ltd
Groundform - Cordek Ltd
Groundsman's Bible - Sportmark™ Ltd
Groutfast - P.C.I Construction Systems Ltd
Grovesnor - ciptek - Grosvenor Pumps Ltd
GRP 4 - Group Four Glassfibre Ltd
Grundfos - The BSS Group PLC
Grundy Kuppersbusch - Grundy Catering
Grundy Olner Toms - Grundy Catering
Grundy Shopfitting - Grundy Catering
Grundy Survice - Grundy Catering
Grundy Systems - Grundy Catering
GT Trolleys - Envopak Group Ltd
Guard - Dixon Turner Wallcoverings
Guard Dog - Poselco Lighting
Guardian - Alumasc Exterior Building Products Ltd
Guardian - Intersolar Group Ltd
Guardian - Raised Floor Systems Ltd
Guardian - Sissons W & G Ltd
Guardmaster - Docherty New Vent Chimney Group
Guardstation - Guardall Ltd
Guardwire - Geoquip Ltd
Guideline - The Building Centre Group Ltd
Guidograph - Marking Machines International
Guiraud Freres - Smithbrook Building Products Ltd
Guiting (Cotswold Stone) - Hanson Bath & Portland Stone
Gulfstream - Gledhill Water Storage Ltd
Gullwing See-Saw - Record Playground Equipment Ltd
Gunk - Unicorn Abrasives (UK) Ltd
Gun-point - Gun-Point Ltd
Gustafs - LSA Projects Ltd
GVR - Recloser - Whipp & Bourne
GW Axial - Matthews & Yates Ltd
Gwent Weighing Systems - Phoenix Scales Ltd
Gwerthiannau Arwyddion Bydeang - Signs International
Gymnasium Side Mounting Range - Lumitron Ltd
Gypframe - British Gypsum Ltd
Gypframe - Metsec Building Products Ltd
Gypglas - Nevill Long Limited
Gyproc - British Gypsum Ltd
Gyproc - British Gypsum Ltd
Gyproc - Nevill Long Limited
Gyvlon - Flowcrete Plc
Gyvlon - Isocrete Floor Screeds Ltd
H & R Johnson - Stokes R J & Co Ltd
H & R Johnson Tiles Ltd - H & R Johnson Tiles Ltd
H.S.M - Tony Team Ltd
H20 - E grp pipes - Johnston Pipes Ltd
H2Ocheck Interactive - National Britannia Ltd
H30000 - Reznor UK Ltd
H4000 Helix - Elster Metering Ltd
Haddon - Tecstone - Haddonstone Ltd
Haddoncraft - Haddonstone Ltd
Haddonstone - Haddonstone Ltd
Hadrian - Relcross Ltd
Hadron slates - Forticrete Roofing Products
Hafa - Crawford Amber Ltd
Haffele UK Ltd - Haffele UK Ltd
Hager - Hager Ltd
Hagley - Welconstruct Co Ltd
Hairexcel Granite - Haironville TAC Ltd
Hairfelt - Davant Products Ltd
Hairplan - Haironville TAC Ltd
Hairy snake - Manton Industrial Seals Ltd
Halcyon - Rixonway Kitchens Ltd
Halcyon - Trent Bathrooms
Hale Hamilton Ltd - Hale Hamilton Ltd
Halex - AEI Cables Limited
Halfen Channels - Halfen Ltd
Hallamshire - Phoenix Scales Ltd
Hallmark - Forticrete Ltd
Hallmark - J & E Hall Ltd
Hallmark - Marcrist International Ltd
Halls - Halls Beeline Ltd
Hallscrew - J & E Hall Ltd
Halo - Bowater Containers Ltd
Halo - Bowater Containers Ltd
Halolux - JSB Electrical PLC
Halopar - Osram Ltd
Halstead - Diamond Merchants
Hambleton - Sashless Window Co Ltd
Hambleton - York Handmade Brick Co Ltd
Hamilton Litestat - Hamilton R & Co Ltd
Handi-Access - Profab Access Ltd
Handiclamp - Viking Johnson
Handitap - Viking Johnson
Handiune - Hills Industries Ltd
Handmade Hadley Red - Cranleigh Brick & Tile Co Ltd
Handmade Haven Red - Cranleigh Brick & Tile Co Ltd
Handmade Light Multi - Cranleigh Brick & Tile Co Ltd
Handmade Loxwood Multi - Cranleigh Brick & Tile Co Ltd
Handmade Red Multi - Cranleigh Brick & Tile Co Ltd
Handmade Victorian Buff - Cranleigh Brick & Tile Co Ltd
Handy Angle - Link 51 (Storage Products)
Handy Tube - Link 51 (Storage Products)
Hangmann - Ackermann Ltd
Hansgrohe - Hansgrohe
Hanson Bricks - Brikenden (Builders Merchants) Ltd
Hanson Bricks - Brikenden (Builders Merchants) Ltd
Harbex - Harris & Bailey Ltd
Harbourite - S.I. Concrete Systems
HARD DECK Chemical Concrete Dustproofing Treatmet - Ferguson G A & Co Ltd
Hard Hat - Andrews Coatings Ltd
Hardrive - Westpile Ltd
Hardrow - Forticrete Ltd
Hardside - Salex Interiors Ltd
Hardyhemp - Marlow Ropes Ltd
Harland Fire Extinguishers - Nu-Swift International Ltd
Harmen - Alumasc Exterior Building Products Ltd
Harmer - Drainage Systems
Harmonica-in-vinyl - Becker (SLIDING PARTITIONS) Ltd
Harmonika-on-wood - Becker (SLIDING PARTITIONS) Ltd
Harmonise - Paragon by Heckmondwike
Harmony - Bartlett Catering Equipment Ltd
Harpoon - BRB Industrial Services Ltd
Harrier - Whitecroft Lighting Limited
Hart - Hart Door Systems Ltd
Hart - High Sensitivity Smoke Detection - Kidde Fire Protection Ltd
Hartam Park Ground Stone - bath stone - Hanson Bath & Portland Stone
Hartland - Hamilton R & Co Ltd
Hartleybotanic - Clear Span Ltd
Harton and Harco Pak. - Harton Heating Appliances
Harton Boost - Harton Heating Appliances
Harvey - Harvey Furniture
Harwood Hardware - RAP Group Ltd
HAS seal - Hotchkiss Air Supply
HAScheck Interactive - National Britannia Ltd
Hasflex - Hotchkiss Air Supply
Hatcrete - Hatfield Ltd, Roy
Hathernware - Ibstock Brick Ltd
Hathernware architectural faience - Ibstock Hathernware Ltd
Hathernware architectural terracotta - Ibstock Hathernware Ltd
Hathernware faience - Ibstock Hathernware Ltd
Hathernware terracotta - Ibstock Hathernware Ltd
Hattersley - Pegler Hattersley
Hattersley - The BSS Group PLC
Hatz - Chippindale Plant Ltd
Haucphalt - Bardon (England) Ltd
Haucphalt - Camas Aggregates Ltd - Aggregate Industries
Haven Hinge - Fab & Fix Ltd
Hawa - Haffele UK Ltd
Hawk White - Cookson Electronics
Hawkins - Eternit Building Materials
Haworth - Haworth UK Ltd
Hazguard - Thurston Building Systems
HBS flagpoles - Harrision Flagpoles
HC Slingsby PLC - HC Slingsby PLC
HCI - Osram Ltd
HCS (HOPPE Compact System) - HOPPE (UK) Ltd
HDS High Density Stacking Chair - Race Furniture Ltd
Healthcheck Interactive - National Britannia Ltd
Healthguard® - Earth Anchors Ltd
Heartbeat - Valor
Heartland - Neptune Outdoor Furniture Ltd
Heat Tracer - Heat Trace Ltd
Heatbusters - SGB Youngman
Heatec - Loblite Ltd
Heathland - Checkmate Industries Ltd
Heatkeeper Homes - Walker Timber Ltd
Heatovent - Seaque Technologies Ltd
Heat-Rad - Horizon International Ltd
Heatsave - Tarmac Top Floor
Heat-Saver - Horizon International Ltd
Heatslave - Worcester Heat Systems Ltd
Heavyside - Oakdale (Contracts) Ltd
Hebefix - Pump Technical Services Ltd
Helagrip - Hellermann Tyton
Helatemp - Hellermann Tyton
Heldite - Heldite Ltd
Heli Bar - Helifix Ltd
HeliBond - Helifix Ltd
Helit - Lesco Products Ltd
Hellerman - Hellermann Tyton
Hellermann - Kem Edwards Ltd
Hellermark - Hellermann Tyton
Helmsman - Helmsman
Helo - Finnish Sauna Baths (Sales) Ltd
Helsyn - Hellermann Tyton
Helvar - Bernlite Ltd
Helvar - Helvar Electrosonic

Helvin - Hellermann Tyton
Hemelite - Tarmac Topblock Ltd
Hemelite Blocks - CAEC Howard (Holdings) Ltd
Hemsec - Hemsec Manufacturing Ltd
Henlow - Hill Leigh Ltd
Henry Nuttall - Viscount Catering Ltd
Henslowe & Fox - Commodore Kitchens Ltd
Hentage Collection - Masterframe Windows Ltd
Hep v0 - Hepworth Building Products Ltd
Hep20 - Hepworth Building Products Ltd
Hep30 - Hepworth Building Products Ltd
HepBio - Hepworth Building Products Ltd
Hepflow - Hepworth Building Products Ltd
Hepseal - Hepworth Building Products Ltd
Hepsleve - Hepworth Building Products Ltd
Hepsure - Hepworth Building Products Ltd
Heraklith - Skanda
Herakustic star - Skanda
Herakustik - Skanda
Heraperm - Isocrete Floor Screeds Ltd
Heras - Advanced Fencing Systems Ltd
Heras - Advanced Fencing Systems Ltd
Heratekta - Skanda
Herbert - BSA Machine Tools Ltd
Herbol - HMG Paints
Hercal - Range Cylinders
Hercules - Berry Systems
Heritage - Alumasc Exterior Building Products Ltd
Heritage - Bennett Windows & Doors
Heritage - Marshalls Mono Ltd
Heritage Brass - Marcus M Ltd
Heritage Range - Eternit Building Materials
Heritage Range - Pedley Furniture International Ltd
Heritage System - Hyperion Wall Furniture Ltd
Hermastic - Pearl Paints Ltd
Herz - Ellis Miller Ltd
Hessian 51 - Muraspec
Heswell Door - AS Newbould Ltd
Heuga - Interface Europe Ltd
HEVAC - Heating Ventilating and Air Conditioning Manufacturers Association
Hevilifts - Hydratight Sweeney Limited
HEWI - Turnquest UK Ltd
Hewlett Packard - RPS Data Products (UK) Ltd
Hexadeck - Turner Access Ltd
Hexcell - Interlux Ltd
Hexsert® - Textron Fastening Systems Ltd
Hexslot - Alumasc Construction Products Ltd
Heydal - Heydal Ltd
Heydal Clean Systems - Heydal Ltd
Heydal Concept - Heydal Ltd
Heydal Ventialted Ceiling System - Heydal Ltd
Heyes - PFP Electrical Products Ltd
Hi Grip Excel Decking - CTS Ltd
Hi Grip Plus Decking - CTS Ltd
Hi Grip Standard Decking - CTS Ltd
HI Hibay - Consult Lighting
Hi Load Masonry - Simpson Strong-Tie®
Hi Spot - Mains Voltage Halogen Lamps - Sylvania Lighting International INC.
Hi Watt Batteries - J.F. Poynter Ltd
HiBuild - Everlac (GB) Ltd
Hic units - Johnston Pipes Ltd
Hi-Cap Rainwater System - FloPlast Ltd
Hid - Viking Security Systems Ltd
Hideaway - Hilladam Coburn Ltd
Hideaway - Polypipe Bathroom & Kitchen Products Ltd
Hi-Deck - Hi-Store Ltd
Hi-Frame - Hi-Store Ltd
High Impact ABS - Regency Garage Door Services Ltd
High Performance Sports 2-part Aqueous Lacquer - Junckers Limited
high street entrance matting - Jaymart Rubber & Plastics Ltd
Highbeam Refractor - DW Windsor Lighting
Higherpower - Blakley Electrics Ltd
Highlander - Richard Hose Ltd
High-Performance - Andersen/ Black Millwork
Highplow - Worcester Heat Systems Ltd
Highways Agency Products - International Protective Coatings,
HiJan - Johnson & Starley
Hiline - CPV Ltd
Hill & Knowles - CrowsonFabrics Ltd
Hills Supadry - Hills Industries Ltd
Hilltop - McAlpine Slate Ltd, Alfred
Hilmor - Hilmor
Hilo - MSA (Britain) Ltd
Hiltons Italian Plaster Coving - Hilton Banks Ltd
Hilume AAA - Stocksigns Ltd
Himalayan - Briggs Industrial Footwear Ltd
Hinton Pavilion - Anthony de Grey Trellises
Hip line - Schneider (UK)
Hiper Bar - Exitex Ltd
Hiperlan - Alcatel Cablenet
Hiperlan-E - Alcatel Cablenet
Hiperlink - Alcatel Cablenet
Hipernet - Alcatel Cablenet
Hiperway - Alcatel Cablenet
Hiperway - Alcatel Cablenet
Hippo - Horizon International Ltd
Hirecabin - UK Cabin Co
Hirsch - Viking Security Systems Ltd
Hisec - Vidionics Security Systems Ltd
Hi-Slim - Oldham Lighting Ltd
Hispec - Johnson & Starley
Hit - BLV Licht-und Vakuumtecnik
Hitachi - Crimeguard Systems Ltd
Hitachi - HV Air Conditioning Ltd
Hitachi - RPS Data Products (UK) Ltd
Hitex - Camas Aggregates Ltd - Aggregate Industries
Hitherm - Kingspan Industrial Ins. Ltd
Hit-Lite - BLV Licht-und Vakuumtecnik
Hi-Trolley - Hi-Store Ltd
Hi-way - Vantrunk Engineering Ltd
Hi-Wide - Welconstruct Co Ltd
HOCOPLAST Profile - Juno Roplasto Ltd
Hodge Close - Burlington Slate Ltd
Hodgkinson Bennis Stokers - Hodgkinson Bennis Ltd
Hodgkisson - TSE Brownson
Hollo-bolt - Lindapter International
Home Automation - Home Automation Ltd
Home 'n' dry - Johnson & Starley
Home of scaffold used in official "World Record" - Turner Plus Eight Ltd
Homefit - Doorfit Products Ltd
Homegard - Marley Waterproofing
Homelift - Wessex Medical
Homelight - Crittall Steel Windows Ltd
Homelight - Crittall Steel Windows Ltd
Home-Loc 6000 - Timloc Expamet Building Products
Homelux - Homelux Nenplas
Homeqest - RB UK Ltd
Homer - Interoven Ltd
Homerette - Interoven Ltd
Hometrim - LB Plastics Ltd
Honeywell - The BSS Group PLC
Hoppe (UK) LTD - Moffett Thallon & Co Ltd
Hoppings decking products - B & M Fencing Limited
Hopton Wood Limcstone - Kirkstone
Horizon - Hudevad Britain
Horizon - Telelarm care Ltd
Hormann - Crimeguard Systems Ltd
Hormann - Hormann (UK) Ltd
Horne - Horne Engineering Ltd
Horne 15 - Horne Engineering Ltd
Horne 20 - Horne Engineering Ltd
Hoskins Medical Equipment - Hoskins Medical Equipment Ltd
Hot Haus - Intercraft Designs Ltd
Hotflo - Heatrae Sadia Heating Ltd
Hotfoil - Isopad Ltd
Hotrace - Flexelec(UK) Ltd
Hotwat - Heat Trace Ltd
Hoylake Door - AS Newbould Ltd
HQI - Osram Ltd
HR nails, tacks & pins - John Reynolds & Sons (Birmingham) Ltd
HRV - Kair ventilation Ltd
HRV - Kiltox Damp Free Solutions
HS/FR Armouring Compound - Thermica Ltd
HSS Hire - Weld - HSS Hire Service Group
HSS Lift & Shift - HSS Hire Service Group
HSS One call - HSS Hire Service Group
HSS Safe & Sure - HSS Hire Service Group
HSS Workshops - HSS Hire Service Group
HT Elastomeric - Anderson D & Son
HT Elastomeric - Anderson D & Son
HT Polyflex - Anderson D & Son
HT Range - Rycroft Ltd
HTH Door Systems - Carter Retail Equipment Ltd
HTV - Babcock Wanson Uk Ltd
Hufcor - Kaba Door Systems Ltd
Hughes & Lancaster - Satec Ltd
Humber Plain Tile - Sandtoft Roof Tiles Ltd
Humidity Vents - Simon R W Ltd
Humidvent - Airflow Developments Ltd
Humidvent - Airflow Developments Ltd
Hunter - Capper Plastics
Huntonit - Norske Interiors (UK) Ltd
Hurricane - Norske Interiors (UK) Ltd
Huwil - RB UK Ltd
HW70 - HW Systems Ltd
HX Range - Biral Pumps
HXD Range - Biral Pumps
Hyarain - KSB Ltd
Hyclad - AvestaPolarit Ltd
Hyclean - Potter & Soar Ltd
Hyclene - Welconstruct Co Ltd
Hycor System 3000/F - TDI (UK) Ltd
Hyde - Hyde Brian Ltd
Hydome - Axter Ltd
Hydrasenta - Daryl Industries Ltd
Hydroban - TRC Service Group Ltd
Hydroban ® - TRC (Midlands) Ltd
Hydrocol - Bayer Chemicals Inoegate Pigment Group
Hydrocol - W. Hawley & Son Ltd
Hydrodeck - Golden Coast Ltd
Hydroduct - Grace Construction Products Ltd
Hydroferrox - W. Hawley & Son Ltd
Hydroflow - Hydropath (UK) Ltd
Hydron Cyklon - Aldous & Stamp Services Ltd
Hydrophilic Joint Seals - Charcon Tunnels (Division of Tarmac plc)
Hydrophilic Sealant - Charcon Tunnels (Division of Tarmac plc)
Hydrotech - Alumasc Exterior Building Products Ltd
Hydrothane - Conren Ltd
Hydrotite - Charcon Tunnels (Division of Tarmac plc)
Hydrotite - Laybond Products Ltd
Hyflek - Flowcrete Plc
Hyflex - Briggs Roofing & Cladding Ltd, Roofing Contract
Hyflex 15 - Hyflex Roofing
Hyflex 5 - Hyflex Roofing
Hyflex BW - Hyflex Roofing
Hyflex Corricover - Hyflex Roofing
Hyflex CP - Hyflex Roofing
Hyflex Flexstep - Hyflex Roofing
Hyflex10 - Hyflex Roofing
Hyflex10 C.A - Hyflex Roofing
Hygiene - Saint-Gobain Ecophon Ltd
Hygiene Safe - Kingspan Ltd
Hygienic - Rockfon Ltd
Hygood - Macron Safety Systems (UK) Ltd
HyGuard - Hygood Ltd
Hyload - Ashmead Buidling Supplies Ltd
Hypercab - AEI Cables Limited
Hyperscan - Robot (UK) Ltd
Hyranger - Axter Ltd
Hy-rib - Expamet Building Products
HySense - Hygood Ltd
Hyslide - Welconstruct Co Ltd
Hytech - Welconstruct Co Ltd
Hytex - Potter & Soar Ltd
Hytherm - Axter Ltd
Hyvodex - Hyflex Roofing
Hyweld - Potter & Soar Ltd
I Tre - Forma Lighting Ltd
I.C.I. - Anderson Gibb & Wilson
I.C.P - Rockfon Ltd
IAT - Institute of Ashphalt Technology, the
IBC - Industrial Building Componants Ltd
iBet- Transeat - Evac+chair international Ltd
Ibiza - Kermi (UK) Ltd
IBM - INMAC (a division of microwarehouse Ltd)
Iboflor - Checkmate Industries Ltd
Ibotuft - Checkmate Industries Ltd
Ibstock Hathernware - Ibstock Building Products Ltd
Ibstock Scottish brick - Ibstock Building Products Ltd
Ibstock Scottish Stone - Ibstock Scotish Brick
Icelert - Findlay Irvine Ltd
Icemate - CSSP (Construction Software Services Partnership)
Iceni - Selbix Joinery Ltd
Iceni Extinguishers - UK Fire International Ltd
Icepac - CSSP (Construction Software Services Partnership)
ICI - Anderson Gibb & Wilson
ICI - Anderson Gibb & Wilson
Icon - Valor
ICS chimney system - Rite-Vent Ltd
ICS5000 chimney system - Rite-Vent Ltd
ICT - Institute of Concrete Technology The
Ideal - Ideal Furniture Enterprises Ltd
Ideal - The BSS Group PLC
Ideal - Tuke & Bell Ltd
Ideal Caradon - Brikenden (Builders Merchants) Ltd
IDM - Kone PLC
Ifthane - IFT Supplies Ltd
IG - IG Limited
Ikos Encaustic Plaster - Textured European Finishes Ltd
I-Line - Square D, Electrical Distribution & Building applicators Division
Illsonic - Salex Interiors Ltd
Illusion - Matki plc
IMA Pro spec and insolvency guarantees - Anderson D & Son
Image - Interface Europe Ltd
Imagine - Helvar Electrosonic
Imera body tinted coloured glass - SCHOTT UK Ltd
Imit - Altecnic Ltd
Immersion Paint Stripper - Howse Thomas Ltd
Impact Crusher - Parker Plant Ltd
Impactor - Geoquip Ltd
Impactor - Imperial Machine Co Ltd
Impera - Sissons W & G Ltd
Imperial - Miele Co Ltd
Imperial - Smart Systems Ltd
Imperial Collection - Masterframe Windows Ltd
Imperial locks - Guardian Lock and Engineering Co Ltd
Impermia - Callenders Ltd
Impex - Dexion Ltd
Impression - Komfort Office Environments Plc
Impressionist II - Altro Floors
Impressions Kitchen Collection - Adam Furniture Group PLC
Imprint - Prismo Ltd
Impro Proximity Access Control - BPT Security Systems (UK) Ltd
Impulse - ITW Construction Products
Impulse - Mode Lighting (UK) Ltd
Index cable glands - Elkay Electrical Manufacturing Co Ltd
Index of Quality Names - The Carpet Foundation
Indicator Mesh - Stenoak Associated Services PLC
Industrial premier strip - Manton Industrial Seals Ltd
Inergen - ADT Fire and Security Plc
Infil - Infil Ltd
Infinim - Golden Coast Ltd
infiniti - H & R Johnson Tiles Ltd
Infinity - Shackerley (Holdings) Group Ltd incorporating Designer ceramics
infopanel - Stocksigns Ltd
InfoProx - CEM Systems Ltd
Inforail ® - Broxap Ltd
Infrapod - Setsquare
ingenius entrance matting - Jaymart Rubber & Plastics Ltd
Ingersoll - Yale Security Products Ltd
Ingersoll Rand - F. W. Harris & Co Ltd
Ingersol-Rand - Ingersoll-Rand European Sales Ltd
Initial Sherrock Monitoring - Initial Electronic Security Systems Ltd
Initial Sherrork Surveilance - Initial Electronic Security Systems Ltd
Initial Shorrock Fire - Initial Electronic Security Systems Ltd
Initial Shorrock Monitoring - Ingersoll-Rand European Sales Ltd
Initial Shorrock Security - Initial Electronic Security Systems Ltd
Initial Shorrock Surveillance - Ingersoll-Rand European Sales Ltd
Initial Shorrock Systems - Initial Electronic Security Systems Ltd
Injection Plastics - Hines & Sons, P E, Ltd
Inlay Range - Pedley Furniture International Ltd
INMP/ NDF - Babcock Wanson Uk Ltd
Innerspace Storage Wall - ARC Partitioning
Innova - Premdor
Innovair - Biddle Air Systems Ltd
Innovation - Rawson Carpets Ltd
Innovators in Clay - Ibstock Brick Ltd
Inperim - TDI (UK) Ltd
Insitex - Creteco Sales Ltd
InSkew - Helifix Ltd
Inspirations - Trespa UK Ltd
Instacem - Underwater mortar - David Ball Group Plc
Instacom - Rollalong Ltd
Instacoustic - InstaGroup Ltd
Instafibre - InstaGroup Ltd
Instaflex - George Fischer Sales Ltd
Instamac - instant repair macadam - David Ball Group Plc
Instant Adhesive - Loctite UK Ltd
Instapruf - Polymerflooring topping - David Ball Group Plc
Insudoor - Stokvis R S & Sons Ltd
Insugard - Kaba Door Systems Ltd
Insuglaze - Panel Systems Ltd
Insulath® - BRC Special Products
Insulator - Hambleside Danelaw Ltd
Insulderm - Crosbie Coatings
Insulite - European Profiles
Insulite FR - European Profiles
Insuloid - Hellermann Tyton
Insul-Sheet - nmc (UK) Ltd
Insultube - Davant Products Ltd
Insul-Tube - nmc (UK) Ltd
Insulux - Madico Film Products Division Inc
Insupex - IPPEC Sytsems
Insuroll - Stokvis R S & Sons Ltd
Intec - Creteco Sales Ltd
INTEC Special Finishes - Quiligotti Access Flooring Limited
Integra - Bayer Chemicals Inoegate Pigment Group
Integra - Dauphin PLC
Integra - Durable
Integra - Fitzgerald Lighting Ltd
Integra - Integra Products
integra - W. Hawley & Son Ltd
Integra Monitor - BPT Security Systems (UK) Ltd
Intejan - Johnson & Starley
Intelekt - Eclipse Blind Systems Ltd
Intellect - Cooper Lighting and Security Ltd
Intellem - Cooper Lighting and Security Ltd
Intelligent Brick Work - Ibstock Brick Ltd
Intelligent home control - Square D, Electrical Distribution & Building applicators Division
Intellisense - C & K Intellisense
Intelock - Vantrunk Engineering Ltd
Interact - Ackermann Ltd
Interbond FP thin film intumescent coating - International Protective Coatings,
Intercell - Interface Europe Ltd
Intercure rapid curing epoxies - International Protective Coatings,
Interface - Interface Europe Ltd
Interface Fabrics - Interface Europe Ltd
Interfine 629 iso cyonate free finishes - International Protective Coatings,
Interfine Polysiloxane - International Protective Coatings,
Interfloor Ltd. - Interfloor Limited
Interline Tank linings - International Protective Coatings,
Interlock - Howarth Timber (Elland) Ltd
Interlock - Interface Europe Ltd
International Waterproofing Association - Bitumen Waterproofing Association
Interplan - Internal Partitions Systems
Interplan maintanance Painting - International Protective Coatings,
Interplus Surface Tolerant - International Protective Coatings,
Interpretations - Paragon by Heckmondwike
Interscope - Allen P W & Co Ltd
Interscreen - Safetell Ltd
Interseal Pir - Hemsec Manufacturing Ltd
Interspan - Compriband Ltd
Interspan All Metal - Compriband Ltd
Interstrip - Shackerley (Holdings) Group Ltd incorporating Designer ceramics
Interzinc 72 epoxy zinc rich - International Protective Coatings,
Interzone high solid expoxies - International Protective Coatings,
Intex - Pearl Paints Ltd
Intrad - Fixatrad Ltd
Intralux - Indirect Lighting Ltd
Intravent - Gebhardt Kiloheat
Intuclear - The Bollom Group
Intufoam - Quelfire

Intulac - The Bollom Group
intusteel WB - The Bollom Group
Intuwrap - Quelfire
Invadex - Chiltern Invadex
Invaflo Shower Drain - Chiltern Invadex
Invalux - Chiltern Invadex
Invicta - Invicita Plastics Ltd
Invis - JKO Ltd
Invisidor - Biddle Air Systems Ltd
Invisirung - Metreel Ltd
Ionpure - USF Permutit/Ionpure
IP64 Lampholders - Jerrards Bros. PLC Division
IP67 Ballast - Jerrards Bros. PLC Division
Ipe - Wilo Samson Pumps Ltd
Ipn - Wilo Samson Pumps Ltd
IPS washroom systems - Armitage Shanks Ltd
IQ - Trend Control Systems Ltd
Iquana - Jaga Heating Poducts (UK) Ltd
Irby Door - AS Newbould Ltd
IRC - Osram Ltd
Iresident - Pellfold Parthos Ltd.
Iron-Horse - Kleen-Tex Industries Ltd
IRT - Isopad Ltd
IRTS - Institute of Remedial Treatment Surveyors, The
Isis - Stuart Turner Ltd
Isis suite - Brass Art Ltd
Island - Neptune Outdoor Furniture Ltd
ISO - Metlex
Isocrete - Flowcrete Plc
Isocrete Self Level - Isocrete Floor Screeds Ltd
Isodrain - Haironville TAC Ltd
Isofast - SFS Intec LTD
Isoflex - Isolated Systems Ltd
Isofoam - Baxenden Chemicals Ltd
Isofoam - Baxenden Chemicals Ltd
Isogran - Isocrete Floor Screeds Ltd
Isokern - ISOKERN (UK) Ltd
Isokoat - ISOKERN (UK) Ltd
Isola Marble & Granite - Reed Harris
Isolgomma - Sound Service (Oxford) Ltd
IsoSystems Original - IsoSystems
IsoSystems Traditional - IsoSystems
Isotape - Isopad Ltd
Isotherm - Biddle Air Systems Ltd
Isothin - Isocrete Floor Screeds Ltd
Isoval - Clivet UK Ltd
Isowool - British Gypsum Ltd
Isowool - Nevill Long Limited
Ispotherm - Alumasc Exterior Building Products Ltd
Italian Travertines - Kirkstone
Italiana Camini - Emsworth Fireplaces Ltd
Itconi - Lewes Design Contracts Ltd
ITS (Interior Transformation Services) - ITS Ceilings & Partitions
IVE - Institute of Vitreous Enamellers, The
J & S Pumps - ABS Pumps Ltd
J & S Pumps - ABS Pumps Ltd
J R Pearson - Drawn Metal Ltd
J R Pearson Revolvers - Pearsons J R (Birmingham) Ltd
J/S Warm Flow - Johnson & Starley
Jabcore - Vencel Resil Ltd
Jabdec - Vencel Resil Ltd
Jabfill - Vencel Resil Ltd
Jabfloor - Vencel Resil Ltd
Jablite - Vencel Resil Ltd
Jablok - Vencel Resil Ltd
Jabroof Board - Vencel Resil Ltd
Jabroof Panel - Vencel Resil Ltd
Jabroof Slimfix - Vencel Resil Ltd
Jabsqueeze - Vencel Resil Ltd
Jabtherm - Vencel Resil Ltd
Jabvent - Vencel Resil Ltd
Jackson - Jebron Ltd
Jackson - Viscount Catering Ltd
Jacksons fencing ® - Jackson H S & Son (Fencing) Ltd
Jacksons finefencing ® - Jackson H S & Son (Fencing) Ltd
Jacuzzi - Jacuzzi UK Ltd, Division of Jacuzzi Inc
Jakcure ® - Jackson H S & Son (Fencing) Ltd
Jaktop ® - Jackson H S & Son (Fencing) Ltd
JALI - Jali Ltd
Janitor - Deb Ltd
Janssen-Dings - Smithbrook Building Products Ltd
Janus - Johnson & Starley
Jasper - Project Office Furniture PLC
Javelin - Whitecroft Lighting Limited
Javelin 2 - CIL International Ltd
Jazz Tones - Gaskell Carpet Tiles
JCB - J C Bamford (Excavators) Ltd
JCB - J C Bamford (Excavators) Ltd
JCP - Ankerit - Owlett - Jaton
Jebron - Jebron Ltd
Jeka - Smithbrook Building Products Ltd
Jet Print - Kleen-Tex Industries Ltd
Jetdye Digital Printing - Zephyr The Visual Communicators
Jetfloor - Hanson Concrete Products Ltd
Jet-rite - Flamco UK Ltd
Jetslab - Hanson Concrete Products Ltd
Jetstream Ventilation Fan - Colt International
JIB - Joint Industry Board for the Electrical Contracting Industry
Jimi-Heat - Flexelec(UK) Ltd
Jizer - Deb Ltd
Jobomulch - ABG Ltd
John Carr - Jeld-Wen UK Ltd
John Tann - Rosengerns Tann Ltd
Johnson - H & R Johnson Tiles Ltd
Johnstone's Paints - Kalon Decorative Products
Johnstones Performance Coatings - Kalon Decorative Products
Johnstone's Woodworks - Kalon Decorative Products
Joltec - Ronacrete Ltd
Jorgian Wire - Oswestry Reinforced Plastics Ltd
Joseph Bramah - Bramah Security Equipment Ltd
Jotalakk - Jotun Henry Clark Ltd Decorative Division
Jotaplast - Jotun Henry Clark Ltd Decorative Division
Jotapro - Jotun Henry Clark Ltd Decorative Division
Jotawall - Jotun Henry Clark Ltd Decorative Division
Joy - Zon International
JP - Halls Beeline Ltd
JPA - Pulsford Associates Ltd, John
JPM - Abloy Security Ltd
JSB - Cooper Lighting and Security Ltd
Jubilee - JSB Electrical PLC
Jubilee - Wade International Ltd
Jubilee Gold - Howse Thomas Ltd
Juggenaut Poles - Harrision Flagpoles
Juliette Spiral - Loft Centre Products
Jumbo - Swish Building Products
Jumbotec - Swish Building Products
Jumbovent - Swish Building Products
junckers flooring - Brooks Belfast
Jung - Pump Technical Services Ltd
Jung Pumpen - Pump Technical Services Ltd
Jupiter - Fabrikat (Nottingham) Ltd
Jupiter - Guardall Ltd
Jupiter - Metreel Ltd
Jupiter - Ubbink (UK) Ltd
Jupiter Range - Caradon Bathrooms Ltd
Just Shelving - Moresecure Ltd
Jute - Rawtex Ltd
Jutland - Cembrit Blunn Ltd
JVC - Vidionics Security Systems Ltd
JW Green Swimming Pools Ltd - Regency Swimming Pools Ltd
K -Rend - Kilwaughter Chemical Co Ltd
K Screed - Isocrete Floor Screeds Ltd
K4 - Komfort Office Environments Plc
K40 - Kaba Door Systems Ltd
K500 / K700 Delivery hose - UK Fire International Ltd
K7 - Komfort Office Environments Plc
Kaba 20 - Kaba (UK) Ltd
Kaba Benzing - Kaba (UK) Ltd
Kaba garog - Kaba Garog
Kaba GeGe - Kaba (UK) Ltd
Kaba Modular - Kaba (UK) Ltd
Kadett - Dermide Ltd
Kair - Kair ventilation Ltd
Kair - Kiltox Damp Free Solutions
Kalahari - Kiltox Damp Free Solutions
Kalea - Wessex Medical
Kaloric - Kaloric Heater Co Ltd
Kalwall - Stoakes Systems Ltd
Kalzip - E. H. S. Roofing Ltd
Kameleon - Colt International
Kameo - Komfort Office Environments Plc
Kami - Link Build Products Ltd
Kami metal roofing - Lafarge Roofing Ltd
Kanaline - Flexible Ducting Ltd
Kappa - Rockwell Sheet Sales Ltd
Kappa Curvo - Rockwell Sheet Sales Ltd
Kara - Neptune Outdoor Furniture Ltd
Karbo - Stanton-Thompson (Agencies) Ltd
Karibia - Kompan Ltd
Kasbah Twist - Adam Carpets Ltd
Kassel Kerb - Lafaroe Redland Precast
Kawneer - EBS Services Ltd
K-Bond - Kenyon Group Ltd
K-Bond - Kenyon Performance Adhesives
Kee Anchor™ - Kee Klamp Ltd
Kee Guard ® - Kee Klamp Ltd
Kee Klamp ® - Kee Klamp Ltd
Kee Koat® - Kee Klamp Ltd
Kee Line ® - Kee Klamp Ltd
Kee Lite® - Kee Klamp Ltd
Kee nect ™ - Kee Klamp Ltd
Keep Kool - Airedale International Air Conditioning Ltd
Kemira - Anderson Gibb & Wilson
Kemira - Andrews Coatings Ltd
Kemnay Ashlar - Camas Building Materials
Kemnay Ashlar - Fyfe John Ltd
Kemperol - Granflex Roofing Ltd
Kempston - Hanson Brick Ltd, Northern Regonal Sales Office
Kenco - The Kenco Coffee Company
Kenco In-Cup - The Kenco Coffee Company
Kenco Singles - The Kenco Coffee Company
Kenngott international - Ermine Engineering Co. Ltd
Kenrick - Kenrick Archibald & Sons Ltd
KensalFlame - Gazco Ltd
Kenstack - McKenzie-Martin Ltd
Kentish Range - Esk Building Products
Kenyon floor paint - Kenyon Group Ltd
Keps - Springvale E P S
Kerabond - Mapei (UK) Ltd
Keracolor - Mapei (UK) Ltd
Kerafloor - Mapei (UK) Ltd
Keralastic - Mapei (UK) Ltd
Keraquick - Mapei (UK) Ltd
Kerastar - H & R Johnson Tiles Ltd
Kerbdrain - ACO Technologies Plc
Kerbmaster - Benton Co Ltd Edward
Kerbmaster - Edward Benton & Co Ltd
Kernal - Range Cylinders
Kerridge - Macc Stone Ltd
Kerridge Stone - Macclesfield Stone Quarries Ltd
Kestrel - Owlett - Jaton
Kestrel series - Falcon Food Services Equipment
Ketron® PEEK - DSM Engineering Plastic Products (UK) Ltd
Kew - Alto Cleaning Systems (UK) Ltd
Kew - Alto Cleaning Systems (UK) Ltd
Key 4 Architects - NBS
Keyblok - Marshalls Mono Ltd
Keyguard - Hoyles Electronic Developments Ltd
Keylex - Relcross Ltd
Keymer - Keymer Hand Made Clay Tiles
Keymer Clayton - Keymer Hand Made Clay Tiles
Keymer Inline Ventilation System - Keymer Hand Made Clay Tiles
Keymer Shire - Keymer Hand Made Clay Tiles
Keymer Traditional - Keymer Hand Made Clay Tiles
Keypac - Wernick S & Sons Ltd
K-Flex - Isopad Ltd
Khrvf100 - Kair ventilation Ltd
Khrvisu - Kair ventilation Ltd
KiddEx - Explosion Suppression agent - Kidde Fire Protection Ltd
Kiddi-Ride - Record Playground Equipment Ltd
Killaspray - Hozelock Ltd
Kiltox - Kiltox Damp Free Solutions
Kinetic Floors - The Salex Group Ltd.
Kinetics - Envoplan
kinetics rubber floorings - Jaymart Rubber & Plastics Ltd
King Hoists - PCT Group Sales Ltd
King Sliding Door Gear - Lakeside Group
Kingfisher - MacBee Manufacturing Ltd
Kingfisher - Zled (DPC) Co Ltd
Kingley - Flowflex Componants Ltd
Kingshield - Power Breaker
Kingspan - E. H. S. Roofing Ltd
Kingston - Balmoral Porcelain Limited/Longmead Ceramics
Kingston - Gloster Furniture Ltd
Kingstonian - Kingstonian Point Ltd
Kingswood - Allied Manufacturing Co (London) Ltd
Kinmar ® - Kinnings Marlow Ltd
Kirkby - Burlington Slate Ltd
Kirkstone Volcanic Stone - Kirkstone
Kitchen Aspects - Adam Furniture Group PLC
Kitchen Collection - Ellis J T & Co Ltd
Kitchen Confidence - Symphony Group PLC, The
Kito Bitumen Emultions - South Western Tar Distilleries
Kito Chain Hoists - PCT Group Sales Ltd
Kiz - Interlux Ltd
Kizfireguard - Interlux Ltd
Klassic - Komfort Office Environments Plc
Kleen Thru Plus - Kleen-Tex Industries Ltd
Kleenscrape - Kleen-Tex Industries Ltd
Kleen-Stat - Kleen-Tex Industries Ltd
Kleerlok - Nettlefolds Ltd
Kleertred - Exitex Ltd
Klemm - Ingersoll-Rand European Sales Ltd
Klik - Envoplan
Klik - Hager Ltd
KM 3 - Komfort Office Environments Plc
Knauf - Firebarrier Services Ltd
Knauf - Kitsons Insulation Products Ltd
Knauf Insulation Polyfoam - Knauf Insulation Ltd
Knight - Iles Waste Systems
Knight - Marshall J A & Co (Brushes) Ltd
Knobs Range - Pedley Furniture International Ltd
Knorr - The Kenco Coffee Company
Knotless Netting - Sportmark™ Ltd
Kodak - RPS Data Products (UK) Ltd
Kohlangaz - Flavel - Leisure Glynwed Consumer & Construction Products Ltd
Koi-sen - Lotus Water Garden Products Ltd
Kolorband - Kronospan Ltd
Kolorcourt - Emusol Products (LEICS) Ltd
Komfire - Komfort Office Environments Plc
Komfort - HEM Interiors Group Ltd
Komfort - New Forest Ceilings Ltd
Komfort Partitions - Davies John Interiors Ltd
Kompan - Kompan Ltd
Kone Eco3000 - Kone PLC
Konematic auto door division - Konematic Bolton Brady
Konexion - Kone PLC
Konfigure - Komfort Office Environments Plc
Kontrakt - Pedley Furniture International Ltd
Konvertex - Pearl Paints Ltd
KoolDuct - Kingspan Industrial Ins. Ltd
Koolshade - Cooper Group Ltd
Kooltherm - Kingspan Industrial Ins. Ltd
Kooltherm - Kingspan Insulation Ltd
Kooltherm - Kitsons Insulation Products Ltd
Kopex - Kem Edwards Ltd
Koral - Rockfon Ltd
Korean plywood - Wood International Agency Ltd
Korifit - Adaptaflex Ltd
Koroseal - Tektura Plc
Kosset - Carpets International (UK) Ltd
Koverflor - Witham Oil & Paint Ltd
Kraftex - Package Products Ltd
Krantz - Designed for Sound Ltd
Kromax - Booth Samuel & Co Ltd
Kromital Trowelled Stone - Textured European Finishes Ltd
Kromofin Multicolour Spray - Textured European Finishes Ltd
Krono Country - Kronospan Ltd
Krono LAQ - Kronospan Ltd
Kronofix - Kronospan Ltd
Kronofloor - Kronospan Ltd
Kronoplus - Kronospan Ltd
Kronosilent - Kronospan Ltd
Kronospan - Anderson C F & Son Ltd
Kronospan - Boardcraft Ltd
Kronospan - Boardcraft Ltd
KS1000 - Kingspan Ltd
K-Seal patio range - Kingfisher Building Products Ltd.
K-stone - Blocks - Esk Building Products
KT3 - Marlow Ropes Ltd
Kudos - Calomax Ltd
Kudos - Calor Gas Ltd
Kue System 918 - Cleanstone Co (Bradford) Ltd
Kuterlex - Yorkshire Copper Tube Ltd
Kuterlex plus - Yorkshire Copper Tube Ltd
Kuterlon - Yorkshire Copper Tube Ltd
Kvent - Rite-Vent Ltd
Kwikastrip - Halfen Ltd
kwikfix fascia soffit system - Dales Fabrications Ltd
Kwikpoint - Kingfisher Building Products Ltd.
Kwikroll - Kaba Door Systems Ltd
Kwiktie - Fix Fast (Fasteners) Wales Ltd
KX10 wood stains - Kingfisher Building Products Ltd.
Kyomi Bathroom Suite - Ideal-Standard Ltd
L&H Polymer - William Freeman Ltd
L.A Range - Balmoral Porcelain Limited/Longmead Ceramics
L2 Highflow - Armorex Ltd
Label controls - Kaba Garog
Labline 2 - Norwood Partition Systems Ltd
Lachat - Zellweger Analytic Ltd, Sieger Division
Lacuna - Opto International Ltd
Ladder Up Safety Post - Bilco UK Ltd
LadderLatch™ - Latchways plc
Laddermatic - Clow Group Ltd
Lafage Homespan - Lafarge Plasterboard Ltd
Lafarge db check - Lafarge Plasterboard Ltd
Lafarge Deco Systems - Lafarge Plasterboard Ltd
Lafarge firecheck - Lafarge Plasterboard Ltd
Lafarge fresco - Lafarge Plasterboard Ltd
Lafarge Moisturecheck - Lafarge Plasterboard Ltd
Lafarge thermalcheck - Lafarge Plasterboard Ltd
Lafarge Toughcheck - Lafarge Plasterboard Ltd
Laguna - Kompan Ltd
Lakeshore - Halls Beeline Ltd
Lakeside Group - Lakeside Group
Lakeside Lifestyle.com - Lakeside Group
Lakeside Sercurity - Lakeside Group
Lakshield - Degussa Ltd
Lamax - Booth Muirie
Lamelad - JKO Ltd
Lamella - McLoughlin Wood Ltd
Lami - Lami Doors UK Ltd
Lamina Filtertrap - Progressive Product Developments Ltd
Laminated Supplies Ltd - Halmark Panels Ltd
Lampro products by Kemlite Ltd - Kemlite Ltd
LamproSystem hygenic grp cladding - Kemlite Ltd
Lamson - Barloworld Vacuum Technology Plc
Lamwood - Laminated Wood Ltd
Lancaster - Bassett & Findley Ltd
Lancaster - Bassett & Findley Ltd
Lancaster - Carpets International (UK) Ltd
Lancer - Forson Design & Engineering Ltd
Lanchester - Mackay Carpets, Hugh
Landmark (slate & double pantile) - Lafarge Roofing Ltd
Landsurveyor - Gas Measurement Instruments Ltd
Langford - Langford Joinery Ltd
Langley By Shavrin - Shavrin Levatap Co Ltd
Lanka - Swedecor Ltd
LANmark - Alcatel Cablenet
Lantana - Brockway Carpets Ltd
Lapidolith - Conren Ltd
Larfarge Acoustic Homespan - Lafarge Plasterboard Ltd
Larfarge Cormet systems - Lafarge Plasterboard Ltd
Laser - Flexiform Business Furniture Ltd
Lastogum - P.C.I Construction Systems Ltd
Latapoxy - Laticrete UK
Lateral Restraint Tie (LRT.) - Redifix Ltd
Latham Clad - James Latham plc
Laticrete - Laticrete UK
Laumans - Smithbrook Building Products Ltd
Laura Ashley -British Ceramic Tile Ltd - British Ceramic Tile Ltd
Lavanda - Saint-Gobain Ecophon Ltd

Laybond adhesives - Laybond Products Ltd
Laytrad - FG Contracts Limited
Laytrad contract furnishers - Laytrad Ltd
Lbofall - Accomodex
LC-35 Barrier - APT Controls Ltd
LCN - Relcross Ltd
LCS - Kenilworth Daylight Centre Ltd
LD Barrier - APT Controls Ltd
Le Farge - Brikenden (Builders Merchants) Ltd
Le Grand - Schneider Electric Ltd
Leakguard - Isopad Ltd

Leakmaster - Charcon Tunnels (Division of Tarmac plc)
Lean-to - CIL International Ltd
Lecol - Campbell Marson & Co
LED Digital - Vision Neon & Display Ltd
Ledcor - Callenders Ltd
Ledder Flush Shapland - Moffett Thallon & Co Ltd
Ledu - Luxo UK Ltd
Ledumite - Callenders Ltd
Lee Bishop - Viscount Catering Ltd
Legge - IR Security and Safety
Legge - Laidlaw Architectural Hardware
Legrand - Legrand Electric Ltd
LEIA - Lift and Escalator Industry Association,The
Leisure - Flavel - Leisure Glynwed Consumer & Construction Products Ltd
Leisure Sinks - Rangemaster
Leisuretex - Bardon (England) Ltd
Leisuretex - Camas Aggregates Ltd - Aggregate Industries
Lentex - Dixon Turner Wallcoverings
Lenton ® - Erico Europa (GB) Ltd
Leromur - Grass Concrete Ltd
Les Actuels de Lucien Gau - Myddleton Hall Lighting, Division of Peerless Design Ltd
Lesco - Lesco Products Ltd
Lessman - Hyde Brian Ltd
Letterboxes @ Signs of the times - Signs Of The Times Ltd
Levelay HO - Conren Ltd
Leveldale - Chiltern Invadex
Levolux - Levolux A.T. Limited
Lexin - Rangemaster
Lexis - Concord: Marlin
Leyland Paints - Kalon Decorative Products
LFH - Hellermann Tyton
Liberator - Allart, Frank, & Co Ltd
Liberty - Amilake Southern Ltd
Liberty - Satchwell Control Systems Ltd
Liberty (Series 120) - Parker Bath Limited
Liberty range of platform lifts - Wessex Medical
Libra - Sealine Business Products
Libra - Sealine Business Products
Liebig - Arrow Supply Company Ltd
Lieutenant - Neptune Outdoor Furniture Ltd
Lifestyle Showers - Hansgrohe
Li-Flat - Aremco Products
Li-Flat - Aremco Products
Lift 'n' Lock - Airsprung Beds Ltd
Liftingclamps.com - Camlok Lifting Clamps Ltd
Lift-Slab - Hydratight Sweeney Limited
Light & shade - Beta Lighting Ltd
Light Coffers - Saint-Gobain Ecophon Ltd
Light Fantastic - Intersolar Group Ltd
Light Russet - Charnwood Brick Group Ltd
Light Victorian Red - Charnwood Brick Group Ltd
Lighting Distribution Centres - Home Automation Ltd
Lightmanager - Philips Lighting Solutions
Lightmaster - Philips Lighting Solutions
Lightmatic - Optex (Europe) Ltd
Lightnin - SPX Process Equipment
Lightpack - Fitzgerald Lighting Ltd
Lightrak - Electrak International Ltd
Lightseal - Illuma Lighting Ltd
Lightspan - Duplus Architectural Systems Ltd
Lignacite - Lignacite (North London) Ltd
Lignacite - Lignacite Ltd
Lignacrete - Lignacite (North London) Ltd
Lignacrete - Lignacite Ltd
Lignastone Architectural Dressings - Lignacite Ltd
Lignoform - Salex Interiors Ltd
Lignum - Opto International Ltd
Lilliput - Hamworthy Heating Ltd
Lilliput Nursery - Portakabin Ltd
Lime quick - Legge Thompson F & Co Ltd
Limebacker cable ramps - Lewden Electrical Industries
Limoges - Qualitas Bathrooms
Limpet BD6 self setting composition - Thermica Ltd
Limpetite Rubber Coatings - Bristol Metal Spraying & Protective Coatings Ltd
Linbide - Marshall J A & Co (Brushes) Ltd
Lincoln Model Stairlift - Brooks Stairlifts Ltd
Lincoln Pantile - Blyth William
Lincoln Stairlifts - Brooks Stairlifts Ltd
Lincoln Stairlifts - Brooks Stairlifts Ltd
Lincolnshire Limestone - Rare Stone Group, The
Lindapter - Kem Edwards Ltd
Lindapters - Arrow Supply Company Ltd
Lindapters - Arrow Supply Company Ltd
Lindiclip - Lindapter International
Lindum - York Handmade Brick Co Ltd
Line Lazer - Marking Machines International
Linea - Hamilton R & Co Ltd
Linea - Jaga Heating Poducts (UK) Ltd
Linear - Rangemaster
Liner Hydratow Fast Tow Concrete Mixer - Multi Marque Production Engineering Ltd
Liner Major Concrete Mixers - Multi Marque Production Engineering Ltd
Liner Rolpanit Pan Mixers - Multi Marque Production Engineering Ltd
Liner Roughrider Dumpers - Multi Marque Production Engineering Ltd
Linergrip - Viking Johnson
Linesman - Marking Machines International
Linesman - MSA (Britain) Ltd
Link 51 - Libraco
Linkex - AEI Cables Limited
Linkguard - Bonar Floors Ltd
Linklight - Beta Lighting Ltd
Linolite - Concord Lighting
Linosom - Tarkett Sommer Ltd
Lintels - Tarmac Top Floor
Lintie - Redifix Ltd
Linton - Anthony de Grey Trellises
Linurette - DLW Flooring Ltd
Linx Railing system - Furnitubes International Ltd
Lion - Boardcraft Ltd
Lion Abrasives - Acton & Borman
Lion Abrasives - Acton & Borman
Lion Brand - Cottam Bros Ltd
Lion Brand - Davis Burrow & Sons
Liquid 99 - British Nova Works Ltd
Liquid Nails - Vallance
Liquid Trent - Portacel
Liquistore - Franklin Hodge Industries Ltd
Liquitainer - Franklin Hodge Industries Ltd
Liquitite - Readymix Limited
Lisbon - Swedecor Ltd
Lisson ® - Cooper H W & Co Ltd
Lister-Petter - Chippindale Plant Ltd
Lister-Petter - Martin (Marine) Ltd Alec
Liteglaze - Ariel Plastics Ltd
Liteglaze - Ariel Plastics Ltd
Litelink LSC - Eaton Electric Limited
Liteminder - Setsquare
Litetile - Catnic
Lithos Venetian Stucco - Textured European Finishes Ltd
Litta Pikka (Litter collection apparatus) - Glasdon UK Ltd
Little crown - Kompan Ltd
Livonett Wood - Tarkett Sommer Ltd
LM Series - Baxall Ltd
LMS Modular - Intergrated System Terminal - Siemens Building Technologies Ltd,
LO Lobay - Consult Lighting
Load Pro - Nico Manufacturing Ltd
Loadbank - Dorman Smith Switchgear Ltd
LoadCentre KQ - Square D, Electrical Distribution & Building applicators Division
Loadframe - Dorman Smith Switchgear Ltd
Loadlimiter - Dorman Smith Switchgear Ltd
Loadline - Dorman Smith Switchgear Ltd
Loadmaster - Cooke Brothers Ltd
Loadswitch - Dorman Smith Switchgear Ltd
Loba - Campbell Marson & Co
Loblite - Loblite Ltd
LOC Strip - Marble Flooring Spec. Ltd
Locate - Universal Components Ltd
Lochrin ® - Wm. Bain Fencing Ltd
Lochrin ® - Wm. Bain Fencing Ltd
Lochrin Classic Rivetless Palisade - Wm. Bain Fencing Ltd
Lochrin® Bain - Wm. Bain Fencing Ltd
Lockblock - Camas Building Materials
Lockblock - Fyfe John Ltd
Lockhart Project services - Lockhart Catering Equipment
Lockmaster - Paddock Fabrications Ltd
Lockstone - PHI Group Ltd
Loctite - Henkel Consumer Adhesives
Loctite - Loctite UK Ltd
Lodlifta - Engineering & Lift Ltd
Log walls - PHI Group Ltd
Logic Office Contracts - Logic Office Group
Logic Plus - MK (MK Electric Ltd)
Logicwall Cupboard Storage - Logic Office Contracts
Lokfix - Fosroc Ltd
Lokmesh - Potter & Soar Ltd
Lomond - Combat Polystyrene Group Ltd
London - Hanson Brick
London - Hanson Brick Ltd, Northern Regonal Sales Office
Longden - Leaderflush + Shapland Ltd
Longglow Photoluminescent PVC - Signs International
Longley - Longley James & Co Ltd
Longlife - Nicholson Plastics Ltd
Longline - Nevill Long Limited
Longmead Ceramics - Balmoral Porcelain Limited/Longmead Ceramics
Longmore Conduit - Tyco European Tubes
Longspan Shelving - Moresecure Ltd
Lonox Pulse - Fulton Boiler Works (Great Britain) Ltd
Lonsdale - EBS Services Ltd
Loovent - Airflow Developments Ltd
Lotamax - Lotus Water Garden Products Ltd
Lotrak geotextiles - Don & Low Ltd
Lotrak pavelay - Don & Low Ltd
Lotus - Lotus Water Garden Products Ltd
Lotus - Qualitas Bathrooms
Louver-Lite Ltd - Louver-Lite Ltd
Louvolite - Dallas & Forrest Ltd
Louvolite - Louver-Lite Ltd
Louvrelite - Durable Berkeley Co Ltd
Louvrestyle - Hallmark Blinds Ltd
Low Energy Lighting Systems - Connect Lighting Systems (UK) Ltd
Low H2O - Jaga Heating Poducts (UK) Ltd
Low Voltage Neon - Vision Neon & Display Ltd
LP 1000 - Wessex Medical
LP 550 - Wessex Medical
Lucas Furniture Systems - Carleton Furniture Group Ltd
Lucas Furniture Systems - Carleton Furniture Group Ltd
Luce Plan - Quip Lighting, Div of Bruce & Macintyre Ltd
Lulworth - Hamworthy Heating Ltd
Lumaseal - Illuma Lighting Ltd
Lumaspot - Illuma Lighting Ltd
Lumatrac - Illuma Lighting Ltd
Lumax - light shelf blind - Hallmark Blinds Ltd
Lumiere Polychromatic Coating - Textured European Finishes Ltd
Lumilux - Osram Ltd
Luminex - Interlux Ltd
Lumisty - Northgate Solar Controls
Luna - Rada Lighting
Lutz - Halfen Ltd
Lux Classic - Kronospan Ltd
Lux Lift - Park Products
Lux Lok - Kronospan Ltd
Luxaflex ® - Hunter Douglas Ltd (Contracts Division) Luxaflex Blinds ®
Luxblocks - Luxcrete Ltd
Luxcrete - Luxcrete Ltd
Luxe - Rangemaster
Luxfibre - Luxcrete Ltd
Luxfix - Luxcrete Ltd
Luxflex - Indirect Lighting Ltd
Luxfloor - Kronospan Ltd
Luxline Plus - Triphosphor Fluorescent Tubes - Sylvania Lighting International INC.
Luxmix - Luxcrete Ltd
Luxo - Luxo UK Ltd
LX - L Range - Biral Pumps
LXD / LD Range - Biral Pumps
Lydney Lock and Leave® - Lydney Containers Ltd
Lydney Modular Space - Lydney Containers Ltd
Lyncrete - Lyncrete Burkle Services Ltd
Lyndon - Trent Bathrooms
Lynx - Compact Fluorescent Lamps - Sylvania Lighting International INC.
Lyric - President Office Furniture Ltd
Lysander - Brockway Carpets Ltd
Lyssand - Broxwood (scotland Ltd)
Lytag - CEMEX
Lytag - Lytag Ltd
M Range - Biral Pumps
M11 - Maby Hire Ltd
M2 - Guardall Ltd
M7 - Maby Hire Ltd
Mab Door Closers - JB Architectural Inronmongery Ltd
MAC - Maestro International Ltd
Macalloy - MacAlloy Limited
Macaw - Paragon by Heckmondwike
Maccast - Macc Stone Ltd
Macclex - Exitex Ltd
Maccstone - Macclesfield Stone Quarries Ltd
Macdee - Polypipe Bathroom & Kitchen Products Ltd
Machin - Amdega Ltd
Mackay Engineering - Mackay Engineering
Mackridge - McKenzie-Martin Ltd
MacLarenline - Johnson Controls Systems Ltd
Macmount - MacLellan Rubber Ltd
Macpherson - Akzo Coatings Plc
Macpherson - Akzo Coatings Plc
Macphersons - Crown Trade Paints
Macromelt - Henkel Industrial Adhesives
Mactac - Northgate Solar Controls
Mactie - Redifix Ltd
Macvent - MacLellan Rubber Ltd
Madico - Invicta Window Films Ltd
Madico - Northgate Solar Controls
Madico Inc - Madico Film Products Division Inc
Madico/ M.S.C. Specialty Films - Hi-Tec Window Films
Madison - Gloster Furniture Ltd
Maestro ® - Maestro International Ltd
Maestroflo ® - Maestro International Ltd
Maestropac ® - Maestro International Ltd
Mag Master - ABB Instrumentation Ltd.
Mag Master - ABB Instrumentation Ltd.
Magiboards - Magiboards Ltd
MagiRail - Magiboards Ltd
Magma - Thatching Advisory Services Ltd
Magnapleat - Farr Filtration Ltd
Magnaseal - Magnet Ltd
Magnastar - Magnet Ltd
Magnastorm - Magnet Ltd
Magnetek - Bernlite Ltd
Magnetic accessories - Clestra Hauserman Ltd
Magnetic Auto Control - Electro Mechanical Systems Ltd
Magsi - Elliott John H (Monostamp) Ltd
Maharam - Tektura Plc
Mahtal - James Latham plc
Mail master letterboxes - Fab & Fix Ltd
MailManager™ Mail Management Software - Neopost Ltd
Main - Myson Radiators Ltd
Maine Reception System - Hyperion Wall Furniture Ltd
Maintain - MMP International
Majestic - Brintons Ltd
Majestic - Majestic Shower Co Ltd
Majestic - Metlex
Makita - Arrow Supply Company Ltd
Makrolon - Amari Plastics Plc
Makroswing - Envirodoor Markus Ltd
Maldives - Novellini Bathroom Products Ltd
Malibu - Arjo Ltd
Mamoreete - DLW Flooring
Manade - Lesco Products Ltd
Managed Garment range - Helmsman
Managers Decorating Sundries - Kalon Decorative Products
Manby - Mandor Engineering Ltd
Manchester Bollard - Furnitubes International Ltd
Mancuna - Mandor Engineering Ltd
Mandolite - Cafco International
Mandolite - Cafco International
Mandor - Mandor Engineering Ltd
Mangers - Davant Products Ltd
Mangers - Vallance
Manhattan - DW Windsor Lighting
Manhattan - JSB Electrical PLC
Manhattan Furniture - Dennis & Robinson Ltd
Maniaccess - Manitou (Site Lift) Ltd
Manifix - Manitou (Site Lift) Ltd
Manireach - Manitou (Site Lift) Ltd
Maniscopic - Manitou (Site Lift) Ltd
Manitransit - Manitou (Site Lift) Ltd
Manntech - Cento Engineering Company Ltd
Manor - Metlex
Mansafe - Unistrut Limited
ManSafe® - Latchways plc
Mansion - Metlex
Manton - Manton Industrial Seals Ltd
Mapecem - Mapei (UK) Ltd
Mapelastic - Mapei (UK) Ltd
Marblex - Swish Building Products
Marcaddy - WF Electrical Distributors
Marchrist - Marcrist International Ltd
Margard - Dacrylate Paints Ltd
Marghestone Assimilated Granite - Marble Flooring Spec. Ltd
Marine - Richard Hose Ltd
Marineflex - Trade Sealants
Marineseal - Trade Sealants
Mark 1 - Channel Electronics
Markar - Relcross Ltd
Markus - Envirodoor Markus Ltd
Marlborough - Adam Carpets Ltd
Marlborough - Ellis J T & Co Ltd
Marlborough - Gloster Furniture Ltd
Marlborough - Metlex
Marley - Capper Plastics
Marley Roofing - Marley Building Materials Ltd
Marleyflex - Tarkett-Marley Floors Ltd
Marleyflor Plus PU - Tarkett-Marley Floors Ltd
Marleyseal - Marley Waterproofing
Marleytorch - Marley Waterproofing
Marlinea - Concord: Marlin
Marlit - STO Ltd
Marlow - Bartlett Catering Equipment Ltd
Marlux ® - Marshall Contracts Ltd
Marmarette - DLW Flooring Ltd
Marmoleum - Forbo - Nairn Ltd
Marmolit - Harbro Supplies Ltd
Marples - Record Tools Ltd
Marquis - Brintons Ltd
Marrakesh - Brintons Ltd
Marshalite - Marshalls Mono Ltd
Marstair - IMI Air Conditioning Ltd
Marston - Bailey Birkett Sales Division
Marstron - Marlow Ropes Ltd
Martin Lunel - WF Electrical Distributors
Martin Sectional Doors - Regency Garage Door Services Ltd
Marutex Magnum stainless drilling screws - UK Fasteners Ltd
Marvac - Bailey Birkett Sales Division
Marxco - Marcrist International Ltd
Masonite - Boardcraft Ltd
Masons Timber Products - Mason FW & Sons Ltd
Masonstone - Hampton Stone Ltd
Mass - Asset International Ltd
Mastascrew - Nettlefolds Ltd
Master - Norwood Partition Systems Ltd
Master - Securikey Ltd
Master Craftsman Forte - Brockway Carpets Ltd
Masterboard - Promat UK Ltd
Masterclass - Elliott Group Ltd
Mastercraft - Barber Wilsons & Co Ltd
Masterpiece - Checkmate Industries Ltd
Masterseal - Degussa Construction Chemicals (UK)
Masterseal - MK (MK Electric Ltd)
Masterspray - Hills Industries Ltd
Mastertop - Degussa Construction Chemicals (UK)
Mastertop - Feb MBT
Matador - Crosbie Coatings
Matchplay - En-tout-cas plc (Scotland)
Mathys - Andrews Coatings Ltd
Matki - Matki plc
Matrex - Terrapin Ltd
Matrix - Dixon Turner Wallcoverings
Matrix - Tarkett-Marley Floors Ltd
Matrix Naturals - Tarkett-Marley Floors Ltd
Matterson Cranes - PCT Group Sales Ltd
Matthews & Yates Centrifugal Fans - Matthews & Yates Ltd
Mattseal - Armorex Ltd
Maurice Hill - Briggs Roofing & Cladding Ltd, Roofing Contract
Mauser - Mauser Interiors (UK) Ltd
Mawrob ® - Broxap Ltd
Max - Jotun Henry Clark Ltd Decorative Division
Max - Ves Andover Ltd
Max ST - Max Appliances Ltd
Maxam - Race Furniture Ltd
Max-Econ - Horizon International Ltd
Maxi - Dexion Ltd
Maxi - Isolated Systems Ltd
Maxi - Jaga Heating Poducts (UK) Ltd
Maxicryl - STO Ltd
Maxifit - Viking Johnson
Maxigard - Lycetts Industrial Doors
Maxilift - Stannah Lifts Ltd
Maxim - Entrelec MTE Ltd

Maxim - J.F. Poynter Ltd
Maxim - Laird Security Hardware
Maxima - Norwood Partition Systems Ltd
Maximair - McKenzie-Martin Ltd
Maximin - Max Appliances Ltd
Maximiser - Rycroft Ltd
Maximixam - O'Brian Manufacturing Ltd
Maxi-Mixer - Cardale Doors Ltd
Maxinex - Chalmit Lighting
Maxivent - Airflow Developments Ltd
Maxmatic 1000 - Max Appliances Ltd
Maxmatic 1500 - Max Appliances Ltd
Maxmatic 2000 - Max Appliances Ltd
Maxmatic 3000 - Max Appliances Ltd
Maxmatic 4000 - Max Appliances Ltd
Maxmatic 5000 - Max Appliances Ltd
Maxol 12D - Burco Maxol
Maxol M10D - Burco Maxol
Maxol M15 - Burco Maxol
Maxol Microturbo - Burco Maxol
Maxol Mirage - Burco Maxol
Maxol Montana - Burco Maxol
Maxol Morocco - Burco Maxol
Maxol Morocco - Burco Maxol
Maxpax - Craft Jacobs Suchard Ltd
Maxum R ™ - Dezurik International Ltd
Maxum V ™ - Dezurik International Ltd
Maxweld - Twin County Fan Company Ltd
Maxwell House - The Kenco Coffee Company
Mayfair - Gower Furniture Ltd
Mayfair - Kaloric Heater Co Ltd
Mayfair Chrome Accessories - Allibert (UK) Ltd
Mayfair Chrome Accessories - Allibert (UK) Ltd
Mayfield Brol - Kingstonian Point Ltd
MB - 1 - Fernox Alpha Fry Technologies Europe
Mb1 - Cookson Electronics
M-Bond - Isocrete Floor Screeds Ltd
M-Bond Xtra - Isocrete Floor Screeds Ltd
Mbrace - Degussa Construction Chemicals (UK)
Mbrace - Feb MBT
MBT - Degussa Construction Chemicals (UK)
MBT-Feb - Ashmead Buidling Supplies Ltd
MC - Howden Buffalo
McAlpine - Hipkin F W Ltd
Mcnnesmann - Cento Engineering Company Ltd
McQuay UK Ltd - McQuay UK Ltd
MCR Barrier - APT Controls Ltd
MCS Barrier - APT Controls Ltd
Mc-Wall - Smart Systems Ltd
MDA Scientific - Zellweger Analytic Ltd, Sieger Division
Meadrain - Mea UK Ltd
Meagard - Mea UK Ltd
Meagard S - Mea UK Ltd
Meakerb - Mea UK Ltd
Mealine - Mea UK Ltd
Meaplas - Mea UK Ltd
Mearin - Mea UK Ltd
Mechbond - Vulcanite Limited
Meckel - HCL Verta
Mecury - Fabrikat (Nottingham) Ltd
Medici - Forticrete Ltd
Medicool (bedhead lamps) - Brandon Medical
Mediland MDF - Willamette Europe Ltd
Medite - Boardcraft Ltd
Medite - Boardcraft Ltd
Medite - Montague L Meyer
Medite 313 MR MDF - Willamette Europe Ltd
Medite Exterior - Willamette Europe Ltd
Medite FQ - Willamette Europe Ltd
Medite FR Class 0 - Willamette Europe Ltd
Medite FR Class 1 - Willamette Europe Ltd
Medite HD - Willamette Europe Ltd
Medite MDF - Willamette Europe Ltd
Medite ZF - Willamette Europe Ltd
Medway - Portacel
Medway - Potter & Soar Ltd
Meg - Abet Ltd
Meg - Abet Ltd
Mega Vent - Areco Roofing Supplies Company
Megadoor - Crawford Amber Ltd
MegaDuct 220 - Domus Ventilation Ltd.
Megafit - Viking Johnson
Megaflo - Diamond Merchants
Megaflo - Heatrae Sadia Heating Ltd
MegaLife - Heatrae Sadia Heating Ltd
Megasys - Advanced Perimeter Systems Ltd
MegaTEE - Uponor Ltd
Megatower - Turner Access Ltd
Meleto - Panel Systems Ltd
meltaway - Uponor Housing Solutions
Meltone - Lafaroe Redland Precast
Memera 2000 - Eaton Electric Limited
Memera 2000 AD - Eaton Electric Limited
Memlite BC3 - Eaton Electric Limited
Memstyle - Eaton Electric Limited
Menvier - Cooper Lighting and Security Ltd
Menvier Security - Cooper Security Ltd
Merchant Hardware - Avocet Hardware Ltd
Mercury - Eltron Chromalox
Mercury and Eclipse, Versailles - Metlex
Mercury Litestat - Hamilton R & Co Ltd
Meridian - CSI Security LTD
Meridian Garage Doors - Regency Garage Door Services Ltd
Merino - Gaskell Carpet Tiles
Merlin - Swintex Ltd
Merlin - T & D Packaging Ltd
Merlin - Whitecroft Lighting Limited
Merlin Gerin - Schneider Electric Ltd
Mermaid - Norske Interiors (UK) Ltd
Merriott Radiators - Barlo Radiators Limited
Mervene - Firwood Paints Ltd
Meshlite - Oswestry Reinforced Plastics Ltd
Meshtec - Cadisch MDA Ltd Ltd
META Cubicles - Detlef Muller Ltd
Metabin - Metalrax Ltd
Metabolt - Metalrax Ltd
Metaclad Systems - Haironville TAC Ltd
Metaclip - Metalrax Ltd
Metagard - Leigh's Paints
Metal Halide Lamps - Sylvania Lighting International INC.
Metalarc - Metal Halide Lamps - Sylvania Lighting International INC.
Metalclad Plus - MK (MK Electric Ltd)
Metalliform - Metalliform Products Plc
Metalphoto - Pryorsign
Metalset - Loctite UK Ltd
Metasys - Johnson Controls Systems Ltd
Metavent - Cembrit Blunn Ltd
Metbar - Metreel Ltd
Metbar - Mid Essex Trading Co Ltd
Meteon - Trespa UK Ltd
Meterheat - Flexelec(UK) Ltd
Metframe - Metsec Building Products Ltd
Metiflash - Metra Non Ferrous Metals Ltd
Metis - Paragon Business Furniture
Metizinc - Metra Non Ferrous Metals Ltd
Metlex - Triton PLC
Metpost - Challenge Fencing Ltd
Metpost - Expamet Building Products
Metpost products - B & M Fencing Limited
Metreboard - Callenders Ltd
Metreisk - Callenders Ltd
Metrex - Hattersley Newman Hender Ltd
Metro - Harton Heating Appliances
Metro - Townscape Products Ltd
Metro Therm - Harton Heating Appliances
Metro Walling - CSI Security LTD
Metrobond - Britmet Tileform Limited
Metropole - Silent Gliss Ltd
Metropolitan - Silent Gliss Ltd
Metrowel - Mid Essex Trading Co Ltd
Metsec - Metsec Building Products Ltd
Metsec floor beams - Metsec Building Products Ltd
Metsec lattice joists - Metsec Building Products Ltd
Metsec Zed purlins - Metsec Building Products Ltd
Metspan - Metsec Building Products Ltd
Metstrut - Metsec Building Products Ltd
Met-track - Metreel Ltd
Metway - Metway Electrical Industries Ltd
Meva Screens - Adams- Hydraulics Ltd
Meyden - Montague L Meyer
Meyer - Commercial Lighting Systems
Meynell - Caradon Plumbing Solutions
Meynell - Caradon Plumbing Solutions
Meynell - Kohler Mira Ltd
MFP 4-28 Zone Microprocessor Fire Alarm Control - C-TEC (Computionics) Ltd
MGB - Titan plastech
Micafil - Dupre Vermiculite
MICE Kaymar Ltd - MICE Kaymar Limited
Michelmersh - Michelmersh Brick & Tile Co Ltd
Mico - ASSA Ltd
Micro-Air - Feb MBT
Microbore - Wednesbury Tube
Microcard Readers - Time and Data Systems International Ltd (TDSI)
Microflon - Plastic Coatings Ltd
Microfloor 600 - Quiligotti Access Flooring Limited
MicroFlo™ - Johnson Matthey Pigments & Dispersions
Microlock Systems - Time and Data Systems International Ltd (TDSI)
MicroNet - Satchwell Control Systems Ltd
Micronex - Chalmit Lighting
MicroPac - Draeger Safety UK Ltd
MicroPoint - Paragon by Heckmondwike
Microtracer - Heat Trace Ltd
Midland Portable Buildings Ltd - Phillip Ader Ltd
Midlift - Stannah Lifts Ltd
Miele - Miele Co Ltd
Miko - Optelma Lighting Ltd
Miko - The HB Sign Company Ltd
Milano - Symphony Group PLC, The
Milestone - Forticrete Ltd
Millars - Bitmen Products Ltd
Millenium - Stocksigns Ltd
Millenium - Troika Architectural Mouldings Ltd
Millenium Bed Range - Sidhil Care
Millenium Freedom Bed Range - Sidhil Care
Millennium Aluplast Aluminium - Signs International
Milliken - Hattersley Newman Hender Ltd
Milliken Colours - Milliken Carpet
Milliken Obex Matting - Floor Protection Services Ltd
Millwood - Project Office Furniture PLC
Milner - Chubb Safe Equipment Company
Milton - Ashmead Buidling Supplies Ltd
Milton Precast - Milton Pipes Ltd
Mimo - Helvar Electrosonic
Minerac - Treetex Acoustics Ltd
Minerit - Cembrit Blunn Ltd
Minerva MX - ADT Fire and Security Plc
Mingardi - Dyer Environmental Controls
Mini Clearflow - Dunbrik (Yorks) Ltd
Mini Compack - Fulton Boiler Works (Great Britain) Ltd
Mini Locker - Helmsman
Mini Loovent - Airflow Developments Ltd
Mini Seal - Sheardown Engineering Ltd
Mini slate - Forticrete Ltd
Minibore - Yorkshire Copper Tube Ltd
Mini-compacta - KSB Ltd
Miniguard - Asset International Ltd
Minilift - Didsbury Engineering Co Ltd
Mini-Lynx - Compact Fluorescent Lamps - Sylvania Lighting International INC.
Minima - Fastwall Ltd
Minimaster - ABB Control Ltd
Minipack - Ramset Fasteners Ltd
Minislate - Forticrete Roofing Products
Minisoccer Goals - Sportmark™ Ltd
Minispace - Kone PLC
Minitracer - Heat Trace Ltd
Miniwarn - Draeger Safety UK Ltd
Minor - Norwood Partition Systems Ltd
Minster - Minsterstone Ltd
Minton Hollins - H & R Johnson Tiles Ltd
Miodox - Witham Oil & Paint Ltd
Mipolam - Brighton (Handrails), W
Mira - Kohler Mira Ltd
Mira Design - Kohler Mira Ltd
Mira showers - Caradon Plumbing Solutions
Miracle wood - Cuprinol Ltd
Miraflex - Knauf Insulation Ltd
Mirage - Dudley Thomas Ltd
Mirage - Komfort Office Environments Plc
Mirage - Matki plc
Mirage - Mode Lighting (UK) Ltd
Mirage - Muraspec
Mirage - Shackerley (Holdings) Group Ltd incorporating Designer ceramics
Mirage - Whitecroft Lighting Limited
Mirage flooring - Brooks Belfast
Miralyte - Harkness Hall Ltd
Miranti - Arjo Ltd
Mirazzo - Camas Building Materials
Mirazzo - Fyfe John Ltd
Mirka abrasives - Campbell Marson & Co
MIRO - Alanod Ltd
Mirogard Anti-Reflective Glass - SCHOTT UK Ltd
Mirror-Flex - Opals (Mirror-Flex) Co Ltd
Mistral - Airedale International Air Conditioning Ltd
Mistral - Elizabethan Textiles
Mistral - HEM Interiors Group Ltd
Mistral - Imperial Machine Co Ltd
Mistral - Unilock Ltd
Mistral (Series 550) - Parker Bath Limited
Mitix - Howse Thomas Ltd
Mitlite - Oswestry Reinforced Plastics Ltd
Mitraclad - Oswestry Reinforced Plastics Ltd
Mitsubishi - Deralam Laminates Ltd
Mitsubishi - HV Air Conditioning Ltd
Mitsubishi - Purified Air Ltd
Mitx - Oswestry Reinforced Plastics Ltd
Mixcal - Altecnic Ltd
Mixed Golden Russet - Charnwood Brick Group Ltd
Mixed Hampshire Red - Charnwood Brick Group Ltd
Mixed Regency Red - Charnwood Brick Group Ltd
Mixmaster - CAEC Howard (Holdings) Ltd
MJC - Airmaster Engineering Ltd
MJX - Airmaster Engineering Ltd
MJX - Airmaster Engineering Ltd
MK - MK (MK Electric Ltd)
MM2 - Helifix Ltd
MM74 - Whipp & Bourne
MM81 - Whipp & Bourne
MMS - Kone PLC
Mobile 2000 - Lappset UK Ltd
Mobilier International - Haworth UK Ltd
Mobilspace - C & R Building Systems Ltd
Mobilspace - C & R Building Systems Ltd
Mock Birds of Prey - Birdscarer Products & Net Supplies Ltd
Mod 4 - Accomodex
Mode - Allgood plc
Mode - Bernlite Ltd
Modelmark - Elliott John H (Monostamp) Ltd
Modernfold - Dividers Modernfold Ltd
Modicon - Schneider Electric Ltd
Modric - Allgood plc
Modric - Allgood plc
Moducel - Senior Air Systems
Modulair - Biddle Air Systems Ltd
Modular - ASSA Ltd
Modular - Futimis Ltd
Modular - Helmsman
Modular 2000 - ACO Technologies Plc
Modular Glazing - Heywood Williams Architectural
Module 4 - Harrison Drape, Divsion of Newell Ltd
Modulight - Mermet U.K
Moduline - IAC
Modulite - Cheshire Concrete Products
Modulite plus - Cheshire Concrete Products
Moelven - Moelven Laminated Timber Structures Ltd
Mogaspaan - Metra Non Ferrous Metals Ltd
Mogat - Metra Non Ferrous Metals Ltd
Molar Major - Manton Industrial Seals Ltd
Molar Minor - Manton Industrial Seals Ltd
Molyneux - LB Plastics Ltd
Molyneux concealed weepholes - Molyneux G (Products) Ltd
Moment - Tektura Plc
Momento - Hille Ltd
Monafloor - Monarflex Anderson Ltd
Monaframe - Glostal - Monarch
Monaframe - Sapa Building Systems (monarch and glostal)
Monarch - Glostal - Monarch
Monarch - Sapa Building Systems (monarch and glostal)
Monarch aluminium, now GrangiuosSystems - Alframes Ltd
Monarfol 250 - Monarflex Anderson Ltd
Monarperm 450 - Monarflex Anderson Ltd
Mondial - Uponor Ltd
Mondo Sportflex - Altro Floors
Mondo Sportflex Athletic Surface - Bernhards Landscapes Ltd
Mondo Sportflex Athletic Surface - Bernhards Landscapes Ltd
Mondopave - Altro Floors
Monitohm - Heat Trace Ltd
Monitor - Concord: Marlin
Monks Park - Bath Stone - Hanson Bath & Portland Stone
Monmouth Canopy - James Smellie Fabrications Ltd
Mono - Pellfold Parthos Ltd.
Monobolt® - Textron Fastening Systems Ltd
Monoceram - Stokes R J & Co Ltd
monocoque support system - Dales Fabrications Ltd
Monodraught suncatchers - Sunpipe @ Monodraught Ltd
Monodraught sunpipes - Sunpipe @ Monodraught Ltd
Monodraught vertical balanced flus - Sunpipe @ Monodraught Ltd
Monodraught windcatchers - Sunpipe @ Monodraught Ltd
Monofilament 250 - Monarflex Anderson Ltd
Monolevel FC - Flexcrete Ltd
Monolite - Flexcrete Ltd
Monolux - Flexcrete Ltd
Monolux - Promat UK Ltd
Monomix - Flexcrete Ltd
Monoplan - Becker (SLIDING PARTITIONS) Ltd
Monoplan - Becker (SLIDING PARTITIONS) Ltd
Monoprofe - Ronacrete Ltd
Monoscape - Marshalls Mono Ltd
Monoset - Ronacrete Ltd
Monoshake - Don Construction Products Ltd
Monosil - Flexcrete Ltd
Monospace - Kone PLC
Monotrak - Rackline Systems Storage Ltd
Monowall5oo - Monowa Manufacturing (UK) Ltd
Monsoon - Stuart Turner Ltd
Monsoon Extra - Stuart Turner Ltd
Montage - Satchwell Control Systems Ltd
Montana - Armitage Shanks Ltd
Montanna - Libraco
Monument - Monument Tools Ltd
Moonlight Marker - Intersolar Group Ltd
Moorwood Vulcan - Viscount Catering Ltd
Morris - Morris Material Handling Ltd
Morris and Co. - Sanderson
Morrland - Trent Bathrooms
Mosaiq - Kompan Ltd
Mosstex - Rawtex Ltd
Motif - Paragon Business Furniture
Mould Making - Hynes Aden K Sculpture Studios
Mouldguard - Trade Sealants
Mouldshield - ICI Dulux Trade Paints
Moventi - Zon International
Movitex - Vision Neon & Display Ltd
Moxon - Stenoak Associated Services PLC
MRS 7 - MR Limited
MSR Columns - Consult Lighting
MSS 0.5 - The HB Sign Company Ltd
MT70 IP67 waterproof fitting - Encapsulite International Ltd
MTM - Railex Systems Ltd
Muftilag - The Salex Group Ltd.
Mufti-Lag - Salex Interiors Ltd
Mulchmat 2000 - Land Wood & Water Co. Ltd
Multi - Glamox Electric (UK) Ltd
Multi Block - Nico Manufacturing Ltd
Multi Brindle - Charnwood Brick Group Ltd
Multi Red Brown - Charnwood Brick Group Ltd
Multi Versatile - Woolliscroft Tiles Ltd
Multi Wall Paneling - Combat Polystyrene Group Ltd
Multibeam - Ward Building Components Ltd
Multibond - Loctite UK Ltd
Multibond Gold - Laybond Products Ltd
Multichannel - Ward Building Components Ltd
Multicolour - Andura Textured Masonry Coatings Ltd
Multicryl - Laybond Products Ltd
Multideck - Ward Building Components Ltd
Multiduct - Ward Building Components Ltd
Multifire - McKenzie-Martin Ltd
Multifix Airbrick - Ryton's Building Products Ltd
Multiflex - Marflex Chimney Systems
Multiflex - Multikwik Ltd
MULTIFLEX - Slab formwork - PERI Ltd
Multiflush - Multikwik Ltd
Multi-form - Square D, Electrical Distribution & Building applicators Division
Multiguard - Hoyles Electronic Developments Ltd
Multiguard - Manton Industrial Seals Ltd
Multikwik - Hipkin F W Ltd
Multikwik - Multikwik Ltd
Multilab laboratory furniture - Grant Westfield Ltd
Multiline - Opto International Ltd
Multiplate - Asset International Ltd
Multiplate - Asset International Ltd
Multipoint - Heatrae Sadia Heating Ltd
Multipot - TSE Brownson
Multirend - Thermica Ltd
Multishield ™ - Accent Hansen

Multistore - Readymix Limited
Multi-system Less-abled Products - Armitage Shanks Ltd
Multitack - Laybond Products Ltd
Multitex - Firth Interface Europe Ltd
Multitex - Multitex GRP & GRG Products
Multitrak - Rackline Systems Storage Ltd
Multi-User - Healthmatic Ltd
Multivap - Ubbink (UK) Ltd
Multivent - Domus Ventilation Ltd.
Multivent - Ubbink (UK) Ltd
Multiwarn - Draeger Safety UK Ltd
Municiple - Fitzgerald Lighting Ltd
MUPRO Pipe Supports - Industrial Hangers Ltd
Muralon - Muraspec
Murek - Muraspec
Murfor-Masonry Reinforcement - Bekaert Building Products
Mvi - Wilo Samson Pumps Ltd
MW Motorway - Consult Lighting
MWS35 - Manton Industrial Seals Ltd
MX - Carborundum Abrasives G.B. Ltd
Myflo - Ves Andover Ltd
Myotex hp - Eclipse Blind Systems Ltd
Myotex sp - Eclipse Blind Systems Ltd
Myrtha - Aqua Design Ltd
Myrtha - Aqua Design Ltd
Myson - The BSS Group PLC
N T Access - Permclose Door Services Ltd
N.T.Master & Minor - Nevill Long Limited
Nailor - Advanced Air (UK) Ltd
Nailor - Advanced Air (UK) Ltd
Nanu - Girsberger London
Nappigon - Imperial Machine Co Ltd
Nappigon - Sissons W & G Ltd
Narcis Mirror Radiator - Jaga Heating Poducts (UK) Ltd
Narrowline ® - Andersen/ Black Millwork
Narrowline ® - Andersen/ Black Millwork
Narvik - Smithbrook Building Products Ltd
NAS - National Association of Shopfitters
Nathan Range - Claydon Architectural Metalwork Ltd
National - Kenyon Performance Adhesives
National bondmaster - Kenyon Group Ltd
National Range - Lumitron Ltd
National Register of Warranted Builders - Federation of Master Builders
Natur - Vicaima Ltd
Natura - Eternit Building Materials
Naturefloor - Pilkington's Tiles Ltd
Nautilus - Aldous & Stamp Services Ltd
Navigator - Project Office Furniture PLC
Naylor - Ashmead Buidling Supplies Ltd
Naylor Drainage - Brikenden (Builders Merchants) Ltd
NBS - NBS
NBS Access - NBS
NBS Landscape - NBS
NBS Scheduler - NBS
NBS Specification Manager - NBS
NBS Specification Writer - NBS
Neaco - Norton Engineering Alloys Co Ltd
Neaco Support Systems - Norton Engineering Alloys Co Ltd
Nearly neon led - Vision Neon & Display Ltd
Neata Lever - Sheardown Engineering Ltd
Neata Seal - Sheardown Engineering Ltd
Neata Tap - Sheardown Engineering Ltd
Neata Toggle - Sheardown Engineering Ltd
Neata Turn - Sheardown Engineering Ltd
Neatdek 2 - Norton Engineering Alloys Co Ltd
Neatform/match/rout - Montague L Meyer
Neatgrille - Norton Engineering Alloys Co Ltd
Neatmat - Norton Engineering Alloys Co Ltd
NEC - Pro Display
Nedair - Ellis Miller Ltd
Neff - Diamond Merchants
Nelson - Marlow Ropes Ltd
Nelton - Tensar International Group LTD
Neo - Desking Systems Ltd
Neopost™ Mailroom Equipment and Supplies - Neopost Ltd
Neotex - Eclipse Blind Systems Ltd
Neotran - Mode Lighting (UK) Ltd
Neotronics - Zellweger Analytic Ltd, Sieger Division
Neptune - Neptune Outdoor Furniture Ltd
Neptune - Neptune Showers Ltd
Nestor (Grit storage bin) - Glasdon UK Ltd
Neta - Harrison Drape, Divsion of Newell Ltd
NetComms - ADT Fire and Security Plc
Net-works - Metway Electrical Industries Ltd
NeutraZone - Airstream Products Ltd
Nevada Western - Thomas and Betts
New c3m/p5. Moisture Resistant Carcase - Rixonway Kitchens Ltd
New Century - Hoskins Medical Equipment Ltd
New concept doors - Kind, J B, Ltd
New Era - Bennett Windows & Doors
New Era Leveling Systems - Atkinson & Kirby Ltd
New Kent Kerb - Brett Paving
New Ultomat - Project Office Furniture PLC
Neway - Dudley Factory Doors Limited
Neway - Neway Doors Ltd
Neway - (Newway Doors Ltd.) - Shutter Door Repair & Maintenance Ltd
Neway-(Neway Doors Ltd) - Priory Shutter & Door Co Ltd
Newdawn - Newdawn & Sun Ltd
Newdome (Rooflights) - Gardner & Newton
Newland - Cembrit Blunn Ltd
Newlath - Newton John & Co Ltd
Newline - Tunstal Telecom Ltd

Newman Monmore - Tyco European Tubes
Newspack - Accomodex
Newspack - Accomodex
Newteam - NewTeam Ltd
Newton Door - AS Newbould Ltd
Newwood - Montague L Meyer
Next Generation - Spur Shelving Ltd
Nextspeed cat 6 - Hubbell Premise Wiring
Nextstep - Otis Ltd
Nexus - Astracast PLC
Nexxus - Chalmit Lighting
NFDC - National federation of Demolition Contractors
NFRC - National Federation of Roofing Contractors, The
NHBC - National House Building Council
Niagara rainwater system - FloPlast Ltd
Nic-cool - Nicholson Plastics Ltd
NICEIC - National Inspection Council for Electrical Installation Contracting
Nickleby - Swintex Ltd
Nico - Nico Manufacturing Ltd
Nicobond - Nicholls & Clarke Ltd
Nico-O-Grout - Nicholls & Clarke Ltd
Night 'n' Day - Eclipse Blind Systems Ltd
Nika3 - Industrial Devices Ltd
Nilflam - Kingspan Industrial Ins. Ltd
Nine thousand 9000 Series - Jebron Ltd
Nine thousand 900SX series - Falcon Food Services Equipment
Niroflex - BRB Industrial Services Ltd
Nite Site - Leigh's Paints
Nitobond - Fosroc Ltd
Nitofil - Fosroc Ltd
Nitoflor - Fosroc Ltd
Nitoseal - Fosroc Ltd
Nitromatric - Pillinger G C & Co (Engineers) Ltd
Nitromors - Henkel Consumer Adhesives
Noble - Marshall J A & Co (Brushes) Ltd
Noeplast Formliners - Grayhill Trading Ltd
Noise Lock - IAC
Nomafoam - nmc (UK) Ltd
Nomapack - nmc (UK) Ltd
Nomastyl - nmc (UK) Ltd
Non Com X - Arch Timber Protection
Non-slip Bench Mat - Dycem Ltd
Norbo - Booth Muirie
Norboflow - Booth Muirie
Norbord - Montague L Meyer
Norbro - Flowserve Flow Control
Nordal rainwater system - Dales Fabrications Ltd
Nordic Shelving - Sealine Business Products
Nordica - Deceuninck Ltd
Nordyl - Dixon Turner Wallcoverings
Norges Vinduet - Norlandia Ltd
Nori - Accrington Brick Tile Co Ltd
Nori - Marshalls Mono Ltd
Norit - Mauser Interiors (UK) Ltd
Norlyn - Smart F & G (Shopfittings)Ltd
Normban - Laidlaw Architectural Hardware Ltd
Normbau - IR Security and Safety
Normet - Burlington Engineers Ltd
Norslo - Power Breaker
North - BRB Industrial Services Ltd
North Star - Tapworks Water Softeners
Norusto - Protim Services Ltd
Norwegian 'H' - Scandinavian Window Systems Ltd
NOS - Strand Hardware Ltd
Nostalgia - Scholes Windows Ltd
Notifier Fire Detection - Fife Fire Engineers & Consultants Ltd
Nouvelle - Qualitas Bathrooms
Nova - BPT Security Systems (UK) Ltd
Nova - Rawson Fillings Ltd
Nova - Zon International
Nova / Novaplex - British Nova Works Ltd
Nova / Novaplex - British Nova Works Ltd
Novacare - British Nova Works Ltd
Novacryl - British Nova Works Ltd
Novaglaze - Swedecor Ltd
Novalift - British Nova Works Ltd
Novalin - Dixon Turner Wallcoverings
Novalite - British Nova Works Ltd
Novaract - British Nova Works Ltd
Novashield - British Nova Works Ltd
Novastet - British Nova Works Ltd
Novastone - Forticrete Ltd
Novastone - Swedecor Ltd
Novatone - USG (UK) Ltd
Novatreet - British Nova Works Ltd
Novatron - British Nova Works Ltd
Novaways - British Nova Works Ltd
Novellini - Novellini Bathroom Products Ltd
Novera - Regency Garage Door Services Ltd
Noviasol - Dalesauna Ltd
Novoferm - Envirodoor Markus Ltd
Novojet - Reznor UK Ltd
Novoment - P.C.I Construction Systems Ltd
Novus - Harris T A Ltd
NP100 Slurry - Imerys Minerals Ltd
NRWB - Federation of Master Builders
Nuance - Twyford Bushboard
Nuastyle taps - Armitage Shanks Ltd
Nucana - Ward's Flexible Rod Company Ltd
Nufins - Hines & Sons, P E, Ltd
Nuflex - Ward's Flexible Rod Company Ltd
Nuflex 95 - Lastite Building Products Ltd
Nu-klad - Kenyon Group Ltd
Nullifire - Firebarrier Services Ltd
Nullifire - Nullifire Ltd
Nullifire - Nullifire Ltd
Nuralite - Nuralite (UK) Ltd
Nuraply - Nuralite (UK) Ltd

Nursecall 800 Panel - C-TEC (Computionics) Ltd
Nutrim - Areco Roofing Supplies Company
Nutsert® - Textron Fastening Systems Ltd
Nuway - Bonar Floors Ltd
Nylatron® MC901 - DSM Engineering Plastic Products (UK) Ltd
Oakey - Unicorn Abrasives (UK) Ltd
Oasis - Kompan Ltd
Oasis - Rovacabin
Oasis - Whitecroft Lighting Limited
Oberflex - Abet Ltd
Oberon - Decra Roofing Systems
Ocean - Signs & Labels Ltd
Ocean - Whitecroft Lighting Limited
Octadoor - Righton Ltd
Octet - Mode Lighting (UK) Ltd
OCTO 250 - Turner Access Ltd
Odeon Fireplace - James Smellie Fabrications Ltd
Odyssey worksurfaces - Twyford Bushboard
OEP contract - OEP Furniture Group PLC
OEP Engineering - OEP Furniture Group PLC
OEP matrix - OEP Furniture Group PLC
Office Watch - Advanced Interior Solutions Ltd
Offsigns - Vision Neon & Display Ltd
Ogilive Sealants - Ogilvie Construction Ltd
Ogilvie Construction Ltd - Ogilvie Construction Ltd
Oil slurps - IFT Supplies Ltd
Oilmaster - Docherty New Vent Chimney Group
Old Clamp - York Handmade Brick Co Ltd
Old colonial brick - Fyfe John Ltd
Old English pantile - Sandtoft Roof Tiles Ltd
Oligo - Futimis Ltd
Olsen - Scandinavian Window Systems Ltd
Olympia - Erlau AG
Olympian - FG Wilson
Olympic - Swedecor Ltd
Olympus - Carpets of Worth Ltd
OMA - Harbro Supplies Ltd
Omega - Anderson C F & Son Ltd
Omega - Viking Laminates Ltd
Omega 4 - JSB Electrical PLC
Omega Duo - Brett Paving
Omega worksurfaces - Twyford Bushboard
Omega® - Brett Paving
Omegadeck - Turner Access Ltd
Omicron - Auchard Development Co Ltd
Omni-Pro - En-tout-cas plc
Omnirack - Drainage Systems
Onyx - Pland Stainless Ltd
Onyx - Project Office Furniture PLC
Opalika White Flashed Opal Glass - SCHOTT UK Ltd
Opaline - Concord: Marlin
Opalux - Northgate Solar Controls
Open Lok - Unifix Ltd
Open Options - Project Office Furniture PLC
Opera - AEL
Opera - Arjo Ltd
Opiocolour Mosaics - Reed Harris
Optelma - Optelma Lighting Ltd
Optiflame - Glen Dimplex UK Ltd
Optiflex - Anglepoise Lighting Ltd
Optiflex - Hughes Safety Showers Limited
Optifloat - Pilkington Birmingham
Optiguard window screens - Bradbury Group Ltd
Optilam - Pilkington Birmingham
Optima - Cardale Doors Ltd
Optima - DW Windsor Lighting
Optima - FDB Electrical Ltd
Optima - Tarkett Sommer Ltd
Optimirror - Pilkington Birmingham
Optimum - Gaskell Carpet Tiles
Options - Bartlett Catering Equipment Ltd
Options - Hands of Wycombe
Options - Sanderson
Options Toilet Cubicles - Grant Westfield Ltd
Options worksurfaces - Twyford Bushboard
Optisneen - Aalco Hull Amari Metal Ltd
Optisneen - Aalco Hull Amari Metal Ltd
Optitherm - Pilkington Birmingham
Optivent - ABB Control Ltd
Optix - Indirect Lighting Ltd
Opto Contour Showcases - Opto International Ltd
Optorex - **Wideband Optical Smoke Detector** - Siemens Building Technologies Ltd,
OPV 2000 - Colt International
Orator - Gent Limited
Orbik - Bernlite Ltd
Orbis - Apollo Fire Detectors Ltd
Orbis - Laidlaw Architectural Hardware
Orbis - Laidlaw Architectural Hardware Ltd
Orcal - Armstrong World Industries Ltd
Orcon - Fillcrete Ltd
Originals LTD - Troika Architectural Mouldings Ltd
Origins - Baggeridge Brick PLC
Orion - Rada Lighting
Orion - Ubbink (UK) Ltd
Orion Range of Recorders - Computertel Ltd
Orsogril - Orsogril Spa UK Distributor
Orsogril - Squires Metal Fabrications Ltd
Orsopanel - Orsogril Spa UK Distributor
Ortho Mat - Interface Europe Ltd
OSLO - Scandinavian Window Systems Ltd
Osma - Wavin Plastics Ltd
Osma Amazon - Wavin Plastics Ltd
Osma DeepLine - Wavin Plastics Ltd
Osma RoofLine - Wavin Plastics Ltd
Osma Roundline - Wavin Plastics Ltd
Osma Squareline - Wavin Plastics Ltd
Osma SuperLine - Wavin Plastics Ltd
Osma UltraRib - Wavin Plastics Ltd
OsmaDrain - Wavin Plastics Ltd

OsmaGold - Wavin Plastics Ltd
OsmaSoil - Wavin Plastics Ltd
OsmaWeld - Wavin Plastics Ltd
OSMO - OSMO UK Ltd
OSMO Color - OSMO UK Ltd
OSMO floor - OSMO UK Ltd
OSMOgard - OSMO UK Ltd
Osprey - Swintex Ltd
Osram - Bernlite Ltd
Osram Dulux - Osram Ltd
OT Medipower (isolated power systems) - Brandon Medical
Otis 2000 E - Otis Ltd
Otis 2000 H - Otis Ltd
Otto Klein Singen Germany - Padipa Ltd
Ovalgrip - Hellermann Tyton
Overlord - Boddingtons Ltd
OWAcontruct - OWA UK Ltd
OWAcoustic - OWA UK Ltd
OWAlux - OWA UK Ltd
OWAspectra - OWA UK Ltd
OWAtecta - OWA UK Ltd
Owens Corning - Kitsons Insulation Products Ltd
Oxan WP - Jotun Henry Clark Ltd Decorative Division
Oxylene - Palgrave Brown Timber & Fireproofing Co
P. S. Fan units - Elan-Dragonair
P.B.C (Prestressed Beam Construction) - Hanson Building Products
P.P.P - Wade International Ltd
P5 - Hudevad Britain
P5K - Hudevad Britain
P6R and P3 Sheeting - Eternit Building Materials
PAC - Vidionics Security Systems Ltd
Pace - Caldwell Hardware (UK) Ltd
Pacemaker - Portakabin Ltd
Pacesottca - Fyfe John Ltd
Pacific - Deceuninck Ltd
Pacifyre Fire Stops - Industrial Hangers Ltd
Padipa - Padipa Ltd
Pagette - Polypipe Bathroom & Kitchen Products Ltd
Pahlens - Golden Coast Ltd
Pakyderm - Crosbie Coatings
Palace - Pellfold Parthos Ltd.
Palace Switches - Wandsworth Elecrtrical Ltd
Palletstor - Moresecure Ltd
Palmyco - Palmyra Ltd
Panasonic - Connought Communication Systems Ltd
Panasonic - Pro Display
Panasonic - Vidionics Security Systems Ltd
Panatrim - Universal Components Ltd
Panda - Horizon International Ltd
Pandiut - Brissco Equipment Ltd
Panelmaster - Timeguard Ltd
Panic Hardware - Dorma Door Services Ltd
Pannell Signs - Pannell Signs Ltd
Panorama - Muraspec
Panther - Eclipse Blind Systems Ltd
Panther - Hilladam Coburn Ltd
Panther - Simpson Strong-Tie®
Pantile 2000 - Britmet Tileform Limited
Paptrim Aluminium - Pitchmastic Pmb Ltd
Paptrim Roof Edge Trime - Pitchmastic Pmb Ltd
Parabolt - Allman Fasteners Ltd
Parabolt - Arrow Supply Company Ltd
Parabolt - Tucker Fasteners Ltd
Parade - Iles Waste Systems
Parade - Welconstruct Co Ltd
Paraflam - C. G. I. International Ltd
Parafon - Armstrong World Industries Ltd
Paragon - Maceman + Amstad Ltd
Paragon - Twide-Paragon Ltd
Paragon AV - Maceman + Amstad Ltd
Paraline - Marlow Ropes Ltd
Paraline - Twyford Bushboard
Parallel - Securistyle Ltd
Parallel Flooring Accessories - McLoughlin Wood Ltd
Paralock - Locking Systems Ltd
Paralon membranes - Intergrated Polymer Systems (UK) Ltd
Paramount - Maceman + Amstad Ltd
Parasquare - Interlux Ltd
Parat C - Draeger Safety UK Ltd
Paray - Arnull Bernard J & Co Ltd
Parcon ® - Composite Structures
Parctile - Decra Roofing Systems
Parflu - May William (Ashton) Ltd
parge - Pellfold Parthos Ltd.
Pargon - Fitzgerald Lighting Ltd
Parigi - Altecnic Ltd
Parinox - Altecnic Ltd
Park Home Coatings - Everlac (GB) Ltd
Park Street Furniture - Lancashire Steel Fabrication Co Ltd
Parker Bath - Arjo Ltd
Parker Hygiene System (Series 610) - Parker Bath Limited
Parkiflex - Western Cork Ltd
Parkray - Hepworth Heating Ltd
Parmet - May William (Ashton) Ltd
Paroc - Hemsec Manufacturing Ltd
Paroc - Knauf Insulation Ltd
Parqcolor - Abet Ltd
Parquet Countryhouse - DLW Flooring Ltd
Parquet Professional - DLW Flooring Ltd
Parquet Shipsdeck - DLW Flooring Ltd
Parquet Virginia - DLW Flooring Ltd
Parthos - Pellfold Parthos Ltd.
Partitions - Roman Mosaic Contracts Ltd
Pasivent - Passivent Ltd

Paslite Aggregates - CAEC Howard (Holdings) Ltd
Paslode - ITW Spit
Passivent - Zled (DPC) Co Ltd
Passivent Tricklevents - Passivent Ltd
Pastoral - Alno (United Kingdom) Ltd
Patay Pumps - Pump International Ltd
Patchfast - Conren Ltd
Patented energy control - Fluidair International Ltd
Patex - Henkel Industrial Adhesives
Pathfinder - Boulter Boilers Ltd
Pathfinder - Intersolar Group Ltd
Patina - Shadbolt F R & Sons Ltd
Patio - Rawson Carpets Ltd
PAVAC - Park Products
Pavia - Pilkington's Tiles Ltd
Pavia Toscana - Kirkstone
Pavigres - Stokes R J & Co Ltd
Pay on foot - Tully
PC99 - Pearl Paints Ltd
PDA Induction Loop Amplifiers - C-TEC (Computionics) Ltd
PE100 - Uponor Ltd
PE80 - Uponor Ltd
Peabody Water Services - Satec Ltd
Peakmoor - Realstone Ltd
Peakstone - CEMEX
Pearce Homes - Pearce Construction(Barnstaple)Ltd
Pearce Property Services - Pearce Construction(Barnstaple)Ltd
Pecafil formwork - BRC Ltd
Pedalo - Erlau AG
Pedestal Tip-Up Seating System - Race Furniture Ltd
Pedmon - CEM Systems Ltd
Pegler - Pegler Hattersley
Pelican - Tuke & Bell Ltd
Pelicolor - Eternit Building Materials
Pel-Job - Chippindale Plant Ltd
Pella - Pellfold Parthos Ltd.
Pemberley - CPV Ltd
Pemko - Relcross Ltd
Pendock Casings and Enclosures - Alumasc Interior Building Products Limited
Pendock Casings and Enclosures - Alumasc Interior Building Products Limited
Pendock profiles - Alumasc Interior Building Products Limited
Penn - Johnson Controls Systems Ltd
Pennine - Laidlaw Architectural Hardware
Pennine - Laidlaw Architectural Hardware Ltd
Pennine Stone - Haddonstone Ltd
Penrhyn - McAlpine Slate Ltd, Alfred
Pensby Door - AS Newbould Ltd
Pepex Pipes - Uponor Housing Solutions
Perfecta Circulators - Turney Turbines (pumps) Ltd
Perfector - Crane Fluid Systems
Performing Art - Brintons Ltd
Perftec - Cadisch MDA Ltd Ltd
Pergola - Panel & Louvre Co Ltd
PERI UP - System scaffolding - PERI Ltd
Period suite - Brass Art Ltd
Periplan - P.C.I Construction Systems Ltd
Periscope - Ryton's Building Products Ltd
Periwarm - Unilock Ltd
Perko powermatic® - Samual Heath
Perko® - Samual Heath
Perkomatic® - Samual Heath
Perlux - Harkness Hall Ltd
Permabond - Kenyon Group Ltd
Perma-clean - Andersen/ Black Millwork
Permaclick - Haze Cladding Systems
Permaclick - Haze Cladding Systems
Permacrib - PHI Group Ltd
Permafoam - Isocrete Floor Screeds Ltd
Permagard - Andura Textured Masonry Coatings Ltd
Permagold - Deva Tap Co Ltd
Permaguard - PermaRock Products Ltd
Permalath - PermaRock Products Ltd
Permalux Sealed Range - Lumitron Ltd
Permarend - PermaRock Products Ltd
PermaRock - PermaRock Products Ltd
Permascreed - DPC Screeding Ltd
Permashield ® - Andersen/ Black Millwork
Permashine - Polypipe Bathroom & Kitchen Products Ltd
Permasint - Plastic Coatings Ltd
Permatite - Readymix Limited
Permatred - Permali Dehoplast Ltd
Permoglaze - Akzo Coatings Plc
Permoglaze - Crown Trade Paints
Permutit - USF Permutit/Ionpure
Personal Surveyor - Gas Measurement Instruments Ltd
Perspectives - Gradus Ltd
Perstorp - Viking Laminates Ltd
Petrarch - CEP Claddings Ltd
Petrel - PFP Electrical Products Ltd
Petrelux - PFP Electrical Products Ltd
Petro fast modular forecourt systems - Cookson and Zinn (PTL) Ltd
Pexapipe - IPPEC Sytsems
Pexatherm - IPPEC Sytsems
Pfleiderer - Anderson C F & Son Ltd
PH Products - PH Products Ltd
Phantom - Dudley Thomas Ltd
Pharo - Hansgrohe
Phenblox Pipe Support Inserts - Industrial Hangers Ltd
Phenomenon - Paragon by Heckmondwike
Philips - RPS Data Products (UK) Ltd
Phillips - Bernlite Ltd
Phlexicare - Nicholls & Clarke Ltd
Phoenix - Bennett Windows & Doors
Phoenix - Cooke Brothers Ltd
Phoenix - Intersolar Group Ltd
Phoenix - Swintex Ltd
Phoenix - Kingfisher - Jewers Doors Ltd
Phoenix - Osprey - Jewers Doors Ltd
Phoenix - Swift - Jewers Doors Ltd
Phonic - Pellfold Parthos Ltd.
Phosco - CU Phosco Lighting
Piazza - Erlau AG
Piccola - Draeger Safety UK Ltd
Piccolo lift - Stannah Lifts Ltd
Pickering Europe Ltd - Pickerings Ltd
Pickwick - Swintex Ltd
Pieri - Grace Construction Products Ltd
Pikes - Pike Signals Ltd
Pilkington - FTW Windows
Pilkington Acoustic Laminate - Pilkington UK Ltd
Pilkington Insulight - Pilkington UK Ltd
Pilkington K Glass - Pilkington UK Ltd
Pilkington 'K' Glass - Pilkington Birmingham
Pilkington Optifloat - Pilkington UK Ltd
Pilkington Optimirror Plus - Pilkington UK Ltd
Pilkington Planar Structural Glazing - Pilkington UK Ltd
Pilkington Planarclad - Pilkington UK Ltd
Pilkington Pyrodur - Pilkington UK Ltd
Pilkington Pyroshield Safety - Pilkington UK Ltd
Pilkington Pyrostop - Pilkington UK Ltd
Pilkington Satin - Pilkington UK Ltd
Pilkington Suncool - Pilkington UK Ltd
Pilkington Tiles - Stokes R J & Co Ltd
Pimlico - Ellis J T & Co Ltd
Pinnacle Envioromental Plastic - Signs International
Pinpoint - Stocksigns Ltd
Pipegard - Leigh's Paints
Piper Lifeline - Tunstal Telecom Ltd
Pirelli - Kem Edwards Ltd
Pirouette - Rackline Systems Storage Ltd
Pivette - Leaderflush + Shapland Ltd
PivotSafe - Dorma Door Services Ltd
Placon - Plastic Construction Fabrications Ltd
Plain Tile - Decra Roofing Systems
Plan - Hudevad Britain
Planal - Axter Ltd
Planar - Ide T & W Ltd
Planar - Pilkington Birmingham
Planet - Pland Stainless Ltd
Planet Peripherals - Planet Communications Group Ltd
Planet Projects - Planet Communications Group Ltd
Planet Videoconferenceing - Planet Communications Group Ltd
Plano - Rockfon Ltd
Planrad - AEL
PLAN-RAD - Property Mechanical Products Ltd
Planta - Erlau AG
Plantec - Cadisch MDA Ltd Ltd
Plascoat - Plastic Coatings Ltd
Plasflow - Fullflow Group Ltd
Plasinter - Plastic Coatings Ltd
Plasmeta - Broxwood (scotland Ltd)
Plassim - Brett Martin Ltd
Plastalux - Clare R S & Co Ltd
Plaster Arris Protector - Areco Roofing Supplies Company
Plastex - Plastic Extruders Ltd
Plasti Drain - Hepworth Building Products Ltd
Plastic Padding - Henkel Consumer Adhesives
Plastic padding - Loctite UK Ltd
Plastilac - Trimite Ltd
Plastimetal - EBC UK Ltd
Plastite - Nettlefolds Ltd
Platinum 600 & 750 Raised Access Flooring - Durabella System Floors Ltd
platinum Floorshield - Springvale E P S
Platinum Fulfil - Springvale E P S
Platinum Wallshield - Springvale E P S
Platinum Warm shark - Springvale E P S
Platinum Warm Squeez - Springvale E P S
Playdek - En-tout-cas plc
Playtime - Record Playground Equipment Ltd
Playtime® - Jackson H S & Son (Fencing) Ltd
Playtop - Lawrence Surfaces Plc, Charles
Plaza Range - Pedley Furniture International Ltd
Plazgard - Lonsdale Metal Company
Pleatex - Sampson T F Ltd
Pleione - Orsogril Spa UK Distributor
Plenty - SPX Process Equipment
Plenty mirrlees - SPX Process Equipment
Plexiglas - Amari Plastics Plc
Plugless - Millar Dennis (1998) Ltd
Plumbers bits - Traps & wc connectors - Hunter Plastics Ltd
Plumbfit - Brett Martin Ltd
Plumbfit - Davant Products Ltd
Plumbing-flexi hoses - Guest (Speedfit) Ltd, John,
Plumbuild - Airflow (Nicoll Ventilators) Ltd
Plus Clips - Tower Manufacturing
Plus Eight - Turner Plus Eight Ltd
Plus Eight Universal System - Turner Plus Eight Ltd
Plus Gas - Unicorn Abrasives (UK) Ltd
Plycorapid - Laybond Products Ltd
Plylok Deck - Gradient Insulations (UK) Ltd
Plymovent Limited - Encon Air Systems Ltd
PM2000 - Gas Measurement Instruments Ltd
PmB : Waterproofing - Pitchmastic Pmb Ltd
Pmb waterproofing Systmes - Pitchmastic Pmb Ltd
PMF - E. H. S. Roofing Ltd
PMP-RAD - Property Mechanical Products Ltd
PNC 3 Vision - Tunstal Telecom Ltd
Pneu - Compact - Polypipe Bathroom & Kitchen Products Ltd
Podium - Barlow Group
Podium Steps - Turner Access Ltd
Polagard - Manders Paint Ltd
Polar - Komfort Office Environments Plc
Polar - Rockfon Ltd
Polcarb 40 S - Imerys Minerals Ltd
Polcarb 45 Slurry - Imerys Minerals Ltd
Polcarb 60 (S) - Imerys Minerals Ltd
Polcarb 60 Slurry - Imerys Minerals Ltd
Polcarb 90 - Imerys Minerals Ltd
Polcarb SB - Imerys Minerals Ltd
Policor C.C. - TDI (UK) Ltd
Polo - Swish Building Products
Polow - Troika Architectural Mouldings Ltd
Poly Buytl - Alfas Industries Ltd
Polycell - ICI Paints
Polycell - Polycell Products Ltd
Polycote - Cross-Guard International Ltd
Polycup - Rockwell Sheet Sales Ltd
Polyduct - Polypipe Civils Ltd
Polyfence - Danbury Fencing PLC
Polyfilla - Polycell Products Ltd
Polyflex - Polyflor Ltd
Polyflor - Polyflor Ltd
Polylath® - BRC Special Products
PolyLay - Junckers Limited
Polymate - Nettlefolds Ltd
Polymatic - CPV Ltd
Polymer Modified Bitumen Emultions and cutbacks. - South Western Tar Distilleries
Polymetron - Zellweger Analytic Ltd, Sieger Division
Polypile - Marshall J A & Co (Brushes) Ltd
Polypipe - Ashmead Buidling Supplies Ltd
Polypipe - Brikenden (Builders Merchants) Ltd
Polypipe - Drainage Systems
Polypipe - Hipkin F W Ltd
PolyPlus - Helifix Ltd
Polyrex - Neural Smoke Detector - Siemens Building Technologies Ltd,
Polyrey - Anderson C F & Son Ltd
Polyrey - Decra Ltd
Polyrey - Viking Laminates Ltd
Polysafe - Polyflor Ltd
Polyscreed - Cross-Guard International Ltd
Polyshield - Anglo Building Products Ltd
Polyshield - Feedwater Ltd
Polystar - Rockwell Sheet Sales Ltd
Polytan - Feedwater Ltd
Polytek - Notcutt Ltd
Polytherm - Smart Systems Ltd
Polytred - Polyflor Ltd
Polytron - Draeger Safety UK Ltd
Polyu - Rockwell Sheet Sales Ltd
Polyvent 225, 300 - Domus Ventilation Ltd.
Pond liner - Firestone Building Products Europe
Pool Lifts - Arjo Ltd
Pop - Tucker Fasteners Ltd
POP Rivets - Allman Fasteners Ltd
POP Rivets - Allman Fasteners Ltd
Popstar - BLV Licht-und Vakuumtecnik
Pop-up - Ackermann Ltd
Portadec - Hughes Safety Showers Limited
Portaflex - Hughes Safety Showers Limited
Portaheater - Hughes Safety Showers Limited
Portaro® - Vicaima Ltd
Portaspray - Dawson McDonald & Dawson Ltd
Portland - Hamworthy Heating Ltd
Portway - Taylor & Portway Ltd
Poselco - Poselco Lighting
Posicote - Cross-Guard International Ltd
Posilok - Exitex Ltd
Posi-Strut - MiTek Industries Ltd
Posi-Web - MiTek Industries Ltd
Posners - Posners the floorstore
Post master - Paddock Fabrications Ltd
Potissimum - Orsogril Spa UK Distributor
Pottelberg - Smithbrook Building Products Ltd
Pouliot - Pouliot Designs by Floralsilk Ltd
Poulot Designs - Desney Products Ltd
Powa glide - Moresecure Ltd
Powdagalv - Howse Thomas Ltd
Powder Prime - Howse Thomas Ltd
Powdermark - Marcrist International Ltd
Power - Kenyon Performance Adhesives
Power breaker - Power Breaker
Power hotmelts - Kenyon Group Ltd
Power Nails - Atkinson & Kirby Ltd
Powerball - Osram Ltd
Powerbond - Gilmour Ecometal
Powerbrace - Maby Hire Ltd
Powerbrake - Entrelec MTE Ltd
Powercyl - Gledhill Water Storage Ltd
Powerdrain - ACO Technologies Plc
Powerflow Flux - Cookson Electronics
Powerflow Flux - Fernox Alpha Fry Technologies Europe
Powerflow Solder - Cookson Electronics
Powerflow Solder - Fernox Alpha Fry Technologies Europe
Powerform 25 - Dorman Smith Switchgear Ltd
Powerform 63 - Dorman Smith Switchgear Ltd
Powerframe - Sapa Building Systems (monarch and glostal)
Powerheat - Heat Trace Ltd
Powerline - Roof Units Group
Powerlink Plus - MK (MK Electric Ltd)
Powerlok - Nettlefolds Ltd
PowerPlas 520 - Power Plastics Ltd
PowerPlas 541 - Power Plastics Ltd
Powerpole - Mita (UK) Ltd
Powerrac - Dezurik International Ltd
Powersafe - Hawker Ltd
Powershade - Eclipse Blind Systems Ltd
Powershare - Maby Hire Ltd
Powersoft - Entrelec MTE Ltd
Powerstar - Osram Ltd
Powerstart - Entrelec MTE Ltd
Powerstation Contol Panels - Ves Andover Ltd
Powerstock - Hamworthy Heating Ltd
Power-Style - Square D, Electrical Distribution & Building applicators Division
Powerswitch - Entrelec MTE Ltd
Powertrack - Arena-Walsall Ltd
Powertrak - Metreel Ltd
Powertrak - Rackline Systems Storage Ltd
Powertrim - Heat Trace Ltd
Powervent - Kiltox Damp Free Solutions
Powervent - Martin Orgee Ltd
Powrmatic - The BSS Group PLC
Pozi GP - Nettlefolds Ltd
Pozidrain - ABG Ltd
Pozzolith - Feb MBT
PPA 571 - Plascoat Systems Ltd
PPC - Brighton (Handrails), W
PRA - P R A Coating Technology
Precedence - Paragon by Heckmondwike
Precious Gems - Carpets of Worth Ltd
Precision - Komfort Office Environments Plc
Prelude - Ellis J T & Co Ltd
Premier - Dixon Turner Wallcoverings
Premier - Fitzgerald Lighting Ltd
Premier - Integra Products
Premier - Manton Industrial Seals Ltd
Premier - Owlett - Jaton
Premier - Securikey Ltd
Premier - Xpelair Ltd
Premier by Coram - Coram Showers Ltd
Premier Collection - Tarkett-Marley Floors Ltd
Premier Electric/Manual (Series 410/420) - Parker Bath Limited
Premier Fire Extinguishers - Nu-Swift International Ltd
Premier Modular Buildings - Premier Group
Premier Service - Flexiform Business Furniture Ltd
Premier Transline Hire - Premier Group
Premium - Everlac (GB) Ltd
Premix - Hanson Aggregates
Premix Concrete - Hanson Building Materials Europe
Prepakt - Westpile Ltd
Preprufe - Grace Construction Products Ltd
Prescor - Flamco UK Ltd
PreSense/ AbSence - Philips Lighting Solutions
President - Deans Blinds & Awnings (UK) Ltd
President - Securistyle Ltd
Presspak - Flamco UK Ltd
Pressure Jet Marker - Marking Machines International
Pressure Pak - Fordwater Pumping Supplies
Pressurepak - Fordwater Pumping Supplies Ltd
Prestcold - Copeland Corporation Ltd
Prestige - Balmoral Porcelain Limited/Longmead Ceramics
Prestige - Gent Limited
Prestige - Novellini Bathroom Products Ltd
Prestige - Stocksigns Ltd
Prestige Plus - MK (MK Electric Ltd)
Prestige Sports Surfaces Ltd - Pro-Sport - Hepworth Minerals & Chemicals Ltd
Prestige Sports Surfaces Ltd - TopSport - Hepworth Minerals & Chemicals Ltd
Presto - Girsberger London
Presweb - Prestoplan
prialpas rubber floorings - Jaymart Rubber & Plastics Ltd
Prima - Armstrong World Industries Ltd
Prima - Armstrong World Industries Ltd
Prima - Cardale Doors Ltd
Prima - Dermide Ltd
Prima - Laird Security Hardware
Prima - Prima Security & Fencing Products
Prima - Sampson & Partners Fencing
Prima - The Salex Group Ltd.
Prima 2 Gas Vent - Rite-Vent Ltd
Prima Plus Single Wall System - Rite-Vent Ltd
Prima SW Single Wall System - Rite-Vent Ltd
Primacalc - Fullflow Group Ltd
Primaflow - Fullflow Group Ltd
Primary Options - Project Office Furniture PLC
Primatic - Range Cylinders
Primera - Gaskell Carpet Tiles
Primo - Industrial Devices Ltd
Primofit - George Fischer Sales Ltd
Primus Washers/ Ironers - Warner Howard Group Ltd
Prince - Pellfold Parthos Ltd.
Princess suite - Brass Art Ltd
Princess suite - Brass Art Ltd
Prinmuls Lite LMP90K Polymer Emultion - South Western Tar Distilleries
Prinmuls Mac H.A.U.C. Binder - South Western Tar Distilleries
Prinmuls MP90X Polymer Emultion - South Western Tar Distilleries
Print HPL - Abet Ltd
PrintMachine™ Laserprinting Management Software - Neopost Ltd
Priory - CEMEX
Priory - Dudley Factory Doors Limited

Priory - Priory Shutter & Door Co Ltd
Priory - (The Priory Shutter & Door Co. Ltd.) - Shutter Door Repair & Maintenance Ltd
Priory-(The Priory Shutter & Door Co. Ltd.) - Neway Doors Ltd
Prisma Lighting - Futimis Ltd
Prismafit - H & R Johnson Tiles Ltd
Prismatics - H & R Johnson Tiles Ltd
Pritt - Henkel Consumer Adhesives
Prixmat - Camas Aggregates Ltd - Aggregate Industries
Pro - Net Contractor - Termstall Limited
Pro - Select - Owlett - Jaton
Pro Bounce - En-tout-cas plc (Scotland)
Pro line - Schneider (UK)
Proarc - PCT Group Sales Ltd
Pro-Balance - Crane Fluid Systems
Pro-Bounce - En-tout-cas plc
Probuild - Vallance
Procast - Procter Bros Limited
PROcheck - National Britannia Ltd
Procor - Grace Construction Products Ltd
procter cast stone - Procter Cast stone & Concrete Products
procter concrete products - Procter Cast stone & Concrete Products
procter fencing products - Procter Cast stone & Concrete Products
Procut - PCT Group Sales Ltd
Prodec coatings - Intergrated Polymer Systems (UK) Ltd
Prodigy - Oldham Lighting Ltd
Prodisplay™ - Pro Display
Prodrive - Camas Aggregates Ltd - Aggregate Industries
Profil - Fillcrete Ltd
Profile - Rackline Systems Storage Ltd
Profile - Railex Systems Ltd
Profile 22 - Plastal (SBP Ltd)
Profile 49 - Britmet Tileform Limited
Profile Paving - Readymix Limited
Profiles - Anderson D & Son
Profiles - Twyford Bushboard
Profiles solo - Anderson D & Son
Profilia - Henry Venables Interiors Ltd
Profilia - Henry Venables Ltd
Profilptin - DLW Flooring Ltd
Pro-flite Suspended Aerofoil - Profile Lighting Services Ltd
Proform - Rackline Systems Storage Ltd
Proframe - Sapa Building Systems (monarch and glostal)
ProFuse - Uponor Ltd
Progeny - Viking Security Systems Ltd
Proglas - Marley Waterproofing
Programastat - Timeguard Ltd
Progress - Pellfold Parthos Ltd.
Progressive - Paragon Business Furniture
Prohire - PCT Group Sales Ltd
Projects - Concord: Marlin
Prolift - PCT Group Sales Ltd
Proline - Rangemaster
Promacase - Promat Fire Protection Ltd
Promalit - Promat Fire Protection Ltd
Promalit - Promat UK Ltd
Promalit60 - Promat UK Ltd
Promaseal - Promat Fire Protection Ltd
Promat - Kitsons Insulation Products Ltd
Promat TL Board - Promat UK Ltd
Promatecet FW - Promat UK Ltd
Promatect - Promat Fire Protection Ltd
Promatect 250 - Promat UK Ltd
Promatherm Lining Panel - Promat Fire Protection Ltd
Promenade - Lappset UK Ltd
Promesh - Procter Bros Limited
Promfire - Haironville TAC Ltd
Promisol Bi-Modular Flat Panel - Haironville TAC Ltd
Promlouvre - Haironville TAC Ltd
Promplan - Haironville TAC Ltd
Pro-Net Contractor - Albion Fencing Ltd
Pronto - Girsberger London
Proofex - Fosroc Ltd
Propadek - Raised Floor Systems Ltd
Proplene - CPV Ltd
Propower - Blakley Electrics Ltd
Propower - PCT Group Sales Ltd
Proseal - Armorex Ltd
Proseal - Illuma Lighting Ltd
Proseal - Sika Ltd
Pro-Set - Wessex Resins & Adhesives Ltd
Prospex - Terrapin Ltd
Protal - Winn & Coales (Denso) Ltd
Protan - E. H. S. Roofing Ltd
Protecta - Chalmit Lighting
Protectacap - Interlux Ltd
Protector - Amber Doors Ltd
Protector - Crawford Amber Ltd
Protector - Securikey Ltd
Protector Frames - H&H Electronics Ltd
Protectosie 300 E - Degussa Ltd
Protectosie 800 E - Degussa Ltd
Protectosie Antigraffiti - Degussa Ltd
Protectosie WS 405 - Degussa Ltd
Pro-Tee - Crane Fluid Systems
Protekt - Madico Film Products Division Inc
Proteus - Golden Coast Ltd
Protim - Protim Solignum Ltd
Protim- Prevac - Protim Solignum Ltd
Protimeter Humilog - GE Protimeter
Protimeter Hygromaster - GE Protimeter
Protimeter mini - GE Protimeter
Protimeter MMS - GE Protimeter
Protimeter Surveymaster SM - GE Protimeter
Protimeter Timbermaster - GE Protimeter
Proton - Paddock Fabrications Ltd
Providence - Trent Bathrooms
PS2000 - Gas Measurement Instruments Ltd
Psicon - Geoquip Ltd
PSM - Elster Metering Ltd
Public - Maceman + Amstad Ltd
Pudlo - David Ball Group Plc
Puf - Maceman + Amstad Ltd
Pullman - Portakabin Ltd
Pulsa - ITW Construction Products
Pulsacoil - Gledhill Water Storage Ltd
Pulsar - Rada Lighting
PULSAR12 - CSW Coldform Ltd
pulse secure electric fencing - Chestnut Products Limited
Puma - Mode Lighting (UK) Ltd
Puma - Resdev Ltd
Pumadur - Resdev Ltd
Pumaflor - Resdev Ltd
Pumalite - Fyfe John Ltd
Pumpen - Pump Technical Services Ltd
Puntoluce - Forma Lighting Ltd
Pure - Isolated Systems Ltd
Pure Air - Consort Equipment Products Ltd
Purewell - Hamworthy Heating Ltd
Purilan - Ubbink (UK) Ltd
Puriton - Uponor Ltd
Purmelt - Henkel Industrial Adhesives
Purmo Radiators - Rettig Heating Ltd
Purmo Underfloor Heating - Rettig Heating Ltd
Purogene - Hertel Services
Purogene - Verna Ltd
Push and Lock - Ryton's Building Products Ltd
Push Flush - Sheardown Engineering Ltd
Pushfit - Talbot
PushLock™ - Latchways plc
Pushtite - Nettlefolds Ltd
PV 80 - Lafarge Roofing Ltd
PV 800 - Lafarge Roofing Ltd
PVCU fencing systems - Danbury Fencing PLC
PVCU-Steps - Quantum Profile Systems Ltd
PX Control Panel - Guardall Ltd
Pygme - Grosvenor Pumps Ltd
Pyramid - Baxall Ltd
Pyramid - Vulcanite Limited
Pyran S Fire-Resistant Glass - SCHOTT UK Ltd
Pyranova Insulated Fire-resistant Glass - SCHOTT UK Ltd
Pyratox - Aldous & Stamp Services Ltd
Pyratox - Aldous & Stamp Services Ltd
Pyrocoil - Mann McGowan Group
Pyrocryl - Vallance
Pyrodur - Ide T & W Ltd
Pyrodur - Pilkington Birmingham
Pyrodur - Twide-Paragon Ltd
Pyrodur plus - Ide T & W Ltd
Pyrodur plus - Twide-Paragon Ltd
Pyroglaze - Mann McGowan Group
Pyrogrille - Mann McGowan Group
Pyroguard - C. G. I. International Ltd
Pyroguard - C. G. I. International Ltd
Pyro-Guard - Gamma Resources Ltd
Pyroguard extra - C. G. I. International Ltd
Pyromas - Mann McGowan Group
Pyropol - Vallance
Pyroputty - Mann McGowan Group
Pyroshield - Pilkington Birmingham
Pyrosleeve - Mann McGowan Group
Pyrospan - Mann McGowan Group
Pyrostem - C. G. I. International Ltd
Pyrostop - Ide T & W Ltd
Pyrostop - Pilkington Birmingham
Pyrostop - Twide-Paragon Ltd
Pyrostrip - Mann McGowan Group
Pyrotenax - Tyeo Thermal Control
Pyrotenax - Tyeo Thermal Control
Pyrovista - Mann McGowan Group
Pyruma - Purimachos Ltd
Pyxel - Grosvenor Pumps Ltd
Q Ply - Intergrated Polymer Systems (UK) Ltd
QB - Welconstruct Co Ltd
Qbs - Metalrax Ltd
QD 90 - Blackfriar Paints Ltd
QED - Raised Floor Systems Ltd
Q-Rail 2000 - Quartet-GBC UK Ltd
Quadbreak - Square D, Electrical Distribution & Building applicators Division
Quadra line - Schneider (UK)
Quadrascott - Fyfe John Ltd
Quadrasett - Camas Building Materials
Quadro 4 - Rockfon Ltd
Qualceram - Shires Bathrooms
Qualitair - Air Conditioning Units - Eaton-Williams Group Ltd
Quality Mark - The Carpet Foundation
Qualube - Witham Oil & Paint Ltd
Quantak Stoves - Don Heating Products Ltd
Quantec Call System - C-TEC (Computionics) Ltd
Quantum - Marley Plumbing and Drainage
Quantum - Valor
Quarry Pak - Aggregate Industries UK Ltd
Quarrycast - Troika Architectural Mouldings Ltd
Quartermaster Shelving - Beford & Soar Ltd
Quartermaster Shelving - Beford & Soar Ltd
Quartet - Mode Lighting (UK) Ltd
Quartsan - Sissons W & G Ltd
Quartz - Aqualisa Products
Quartzstone Quartz Tiles - Marble Flooring Spec. Ltd
Quasar - Fitzgerald Lighting Ltd
Quasilan - Baxenden Chemicals Ltd
Queensfil 240 - Imerys Minerals Ltd
Queensfil 25 - Imerys Minerals Ltd
Queensfil 300 - Imerys Minerals Ltd
Quelcote - Quelfire
Quelfire - Firebarrier Services Ltd
Quelfire - Quelfire
Quelfire QF1 Mortar - Quelfire
Quelfire QF2 Mortar - Quelfire
Quelfire QF4 Mortar - Quelfire
Quicbuilt - Brierkrete Ltd
Quick Clip - Junckers Limited
Quickfit - Vantrunk Engineering Ltd
Quickfit - Viking Johnson
Quickflex - Sealocrete Ltd
Quickframe - IPPEC Sytsems
Quickheat - IPPEC Sytsems
Quickline - Dyer Environmental Controls
Quickmast - Don Construction Products Ltd
Quickpipe - IPPEC Sytsems
Quickpost - Catnic
Quickscreed - Laticrete UK
Quickshelf - Barton Storage Systems Ltd
Quickshelf-Plus - Barton Storage Systems Ltd
Quickstrip - Laybond Products Ltd
Quicktronic - Osram Ltd
Quiclean - Hansgrohe
Quiclok - Brierkrete Ltd
Quiet Mat - Salex Interiors Ltd
Quietpack - Roof Units Group
Quikaboard - Quinton & Kaines (Holdings) Ltd
Quilfloor 600 - Quiligotti Access Flooring Limited
Quinta - Broag Ltd
Quintesse - Matki plc
Quip - Quip Lighting, Div of Bruce & Macintyre Ltd
Quitetite - Auld Valves Ltd
Quorndon - Lafaroe Redland Precast
Qwikline II - Square D, Electrical Distribution & Building applicators Division
Qx Control Panel - Guardall Ltd
Racklight - IMP Lighting Ltd
Rada - Kohler Mira Ltd
Rada - Rada Lighting
Rada Controls - Caradon Plumbing Solutions
Radbalc - PVC-u Balcony System - Radway, Door and Windows Ltd.
Radcliffe King of Diamonds - Radcliffe & Sons (London)
Radcliffe London Warrented - Radcliffe & Sons (London)
Radiance - Matki plc
Radipex - Uponor Housing Solutions
Radius - Hudevad Britain
Radius - Montague L Meyer
Radson Radiators - Rettig Heating Ltd
Radway - PVC-u widows doors and fenestraction Products - Radway, Door and Windows Ltd.
Rafid - Geoquip Ltd
Raftex - Exitex Ltd
Ragdale - Lafaroe Redland Precast
Rahau - FTW Windows
Rail Track Specialists - Southern Zone - GJ Durafencing Limited
Rail, Roll & Ride - Record Playground Equipment Ltd
Railcraft - Hills Industries Ltd
Railex - Railex Systems Ltd
Railtrack Propducts - International Protective Coatings,
Raincoat - Tor Coatings Ltd
Raindrain - ACO Technologies Plc
Rainstopper - Andura Textured Masonry Coatings Ltd
Rallibondite - Honeywell Speciality Wax and Additives Ltd
Rallikol - Honeywell Speciality Wax and Additives Ltd
Rallisil - Honeywell Speciality Wax and Additives Ltd
Rallisil - Trade Sealants
Rallithane - Honeywell Speciality Wax and Additives Ltd
Rall-on - Wessex Medical
Rambolt - Ramset Fasteners Ltd
Ramesis - Oldham Lighting Ltd
Ramplug - Ramset Fasteners Ltd
Ramsay Access - Ramsay & Sons (Forfar) Ltd
Ramsay Ladders - Ramsay & Sons (Forfar) Ltd
Ramset - Arrow Supply Company Ltd
Ramset - ITW Spit
Ramset - Ramset Fasteners Ltd
Randi - IR Security and Safety
Randi - Laidlaw Architectural Hardware Ltd
Randi-line - Laidlaw Architectural Hardware
Range Cooker Co - Diamond Merchants
Rangemaster - Rangemaster
Rapid - Clark Door Ltd
Rapid - Fillcrete Ltd
Rapidac - Dacrylate Paints Ltd
Rapide - Project Office Furniture PLC
Rapide Shelving - Sealine Business Products
Rapide solo - Anderson D & Son
Rapidplan - Wernick S & Sons Ltd
Rapid-Roll - Stokvis R S & Sons Ltd
Rapier - Kenrick Archibald & Sons Ltd
Rapier 2 - CIL International Ltd
Rapier Marine - CIL International Ltd
Rapier rising screen - Pearce Security Systems
Rapitech - Stokvis R S & Sons Ltd
RAS Attenuator - The Salex Group Ltd.
Ravenna - Paragon by Heckmondwike
Ravenna Bathroom Suite - Ideal-Standard Ltd
Ravier - Taskworthy Ltd
Rawl - British Gypsum Ltd
Rawlplug - Arrow Supply Company Ltd
Rawlplug - Kem Edwards Ltd
Rayburn - Aga-Rayburn
Rayburn - Aga-Rayburn
Rayburn Cookmaster - Aga-Rayburn
Rayburn Heatranger - Aga-Rayburn
Raypak - AEL
Raypak - AEL
RB - R-B International Ltd
RB Horizontal - Fulton Boiler Works (Great Britain) Ltd
RBS - Race Furniture Ltd
RCC - RCC, Division of Tarmac Precast Concrete Ltd
RCC Safety Barriers - Tarmac Precast Concrete Limited
RCC-retaining Walls - Tarmac Precast Concrete Limited
RD Road - Consult Lighting
RD. - Hille Ltd
RD400 - Radiodetection Ltd
RDBA - Rural design and Building Association
Rea frame - Rea Metal Windows Ltd
Rea W20 - Rea Metal Windows Ltd
Rea W40 - Rea Metal Windows Ltd
Ready block - CEMEX
Ready floor - CEMEX
Ready pave - CEMEX
Ready Plumbed Modules - Twyford Bushboard
Ready span - CEMEX
Readymix - CEMEX
Readyscreed - Readymix Limited
ReadySpread - Readymix Limited
Realstone Ltd - Realstone Ltd
Realwood - Montague L Meyer
Reclaim - Reclaim.UK.Com
Rectaleen - Flexible Reinforcements Ltd
Recupovent - Thermal Technology (sales) Ltd
Red Box Wipes - Deb Ltd
Reddicord - Reddiglaze Ltd
Reddifoam - Reddiglaze Ltd
Reddihinges - Reddiglaze Ltd
Reddilock - Reddiglaze Ltd
Reddipile - Reddiglaze Ltd
Reddiprene - Reddiglaze Ltd
Redfyre - Gazco Ltd
Redfyre Cookers - Trianco Redfyre Ltd
Redgra - Hanson Aggregates
Rediguard - Hoyles Electronic Developments Ltd
Redipress - Hoyles Electronic Developments Ltd
Redi-Resin - Redifix Ltd
Redline - Lafarge Roofing Ltd
Redvent - Lafarge Roofing Ltd
Refatex - Astron Buildings Parker Hannifin PLC
Reflections - Tarkett-Marley Floors Ltd
Reflections Bathroom Suite - Ideal-Standard Ltd
Reflecto shield - Madico Film Products Division Inc
RefleKto - BLV Licht-und Vakuumtecnik
Reflex 275 - Monarflex Anderson Ltd
Reflex Flameproof - Monarflex Anderson Ltd
ReflexSuper - Monarflex Anderson Ltd
Reflux - Lumitron Ltd
Regaflex S - Rega Ductex Ltd
Regaflex W - Rega Ductex Ltd
Regal - Contravent Regal Ltd
Regal Doors - Kind, J B, Ltd
Regalsafe - Cooper Security Ltd
Regency - Forticrete Ltd
Regency Buff - Cranleigh Brick & Tile Co Ltd
Regency Divestone - Fyfe John Ltd
Regency Drestone - Camas Building Materials
Regency Swimming Pools - Regency Swimming Pools Ltd
Regent - Clenaware Systems Ltd
Regent - JSB Electrical PLC
Regent (R) - Barber Wilsons & Co Ltd
Registered Specialists - The Carpet Foundation
Regular - Integra Products
Regulator - Checkmate Industries Ltd
Rehall - Brighton (Handrails), W
Rehau - Architectural Plastics (Handrail) Ltd
Rehau - Architectural Plastics (Handrail) Ltd
Reid Span - Reid John & Sons (Strucsteel) Ltd
Reinforced Earth™ TerraBlock - Freyssinet Ltd
Relax - Provex Products Ltd
Relite - Fantasia Distribution Ltd
Remclamp - LPA Industries PLC
Remelt Facility - Alcoa Europe-Extrusions & Endproducts (UK)
Remstrap - LPA Industries PLC
Rendabrick - MR Limited
Rendalath® - BRC Special Products
Renderoc - Fosroc Ltd
Renfor - Super Cement Ltd
Reno - Johnson & Starley
Renovaction - Aqua Design Ltd
Renovent - Ubbink (UK) Ltd
Renovisions - Interface Europe Ltd
Rentokil LFB30/60 - Rentokil Initial U.K. Ltd.
rentokil tas fire wall - Thatching Advisory Services Ltd
Renz Mail Boxes - JB Architectural Inronmongery Ltd
Repairite - MMP International
Repcoat - Don Construction Products Ltd
Reportline - National Britannia Ltd
Resdev - Resdev Ltd
Resiblcok F.R.I - Resiblock Ltd
Resiblock 22 - Resiblock Ltd
Resiblock ER - Resiblock Ltd
Resiblock OR - Resiblock Ltd
Resiblock Superior - Resiblock Ltd
Resiblock Superior Matt - Resiblock Ltd
Resiblock Ultra - Resiblock Ltd

Resiblock Ultra Matt - Resiblock Ltd
Resiblock W.J.C. - Resiblock Ltd
Resibond - Exchem Mining and Construction
Resiclean - Resiblock Ltd
Resicote - Exchem Mining and Construction
Resicrete - Exchem Mining and Construction
Resiecco - Resiblock Ltd
Resifix 3P Anchor Grout - Exchem Mining and Construction
Resilay - Exchem Mining and Construction
Resipatch - Exchem Mining and Construction
Resistex - Leigh's Paints
Resistoid - Harrison Drape, Divsion of Newell Ltd
ResiTie - Helifix Ltd
Reso Pal - Montague L Meyer
Respond - Elizabethan Textiles
Responseline - National Britannia Ltd
Restol - Arch Timber Protection
Restover restoration glass - SCHOTT UK Ltd
Restwall - Architectural Textiles Ltd
Retriever 55 bing (Dog waste bin) - Glasdon UK Ltd
Retroflame - Gamma Resources Ltd
retroflect - HMG Paints
Retrofloor - Redman Fisher Engineering
RetroTie - Helifix Ltd
Retrotread - Redman Fisher Engineering
Revato - Amilake Southern Ltd
Revelation - Spectus Systems
Revo Actuators - Alfa Laval Saunders Ltd
Revo Actuators - Alfa Laval Saunders Ltd
Revue Bathroom Suite - Ideal-Standard Ltd
Rex - Halls Beeline Ltd
Rex - Ves Andover Ltd
Rexovent - ABB Control Ltd
Reynobond - Booth Muirie
Reynobond 33 - Amari Plastics Plc
Reynobond ACM - Technology Telford Ltd
reynobond cassette panels - Dales Fabrications Ltd
RH 'U' Tube Heater - Radiant Services Ltd
Rhapsody - Arjo Ltd
Rhapsody - Arjo Ltd
Rhapsody Range - Caradon Bathrooms Ltd
RHD Double Linear Heater - Radiant Services Ltd
Rheobuild - Feb MBT
Rheocell - Feb MBT
Rheofinish - Feb MBT
Rheomix - Feb MBT
Rhino Doors - European Profiles
Rhino Flex - DLW Flooring
Rhino Quartz - DLW Flooring
Rhino Tex - DLW Flooring
Rhino Tough - Wybone Ltd
RHL Linear Heater - Radiant Services Ltd
Rhodius - Hyde Brian Ltd
Rialto - Kermi (UK) Ltd
RIAS Bookshops - Royal Incorporation of Architects in Scotland
RIAS Competitions - Royal Incorporation of Architects in Scotland
RIAS Consultancy - Royal Incorporation of Architects in Scotland
RIAS CPD - Royal Incorporation of Architects in Scotland
RIAS Gallery - Royal Incorporation of Architects in Scotland
RIAS Insurance Services - Royal Incorporation of Architects in Scotland
RIAS Practice Services - Royal Incorporation of Architects in Scotland
Ribblelite - Richard Hose Ltd
Ribloc - Forticrete Ltd
Richardson - Amdega Ltd
Richardson - Amdega Ltd
Ridgidrain Advanced Drainage System - Polypipe Civils Ltd
Ridgiduct - Polypipe Civils Ltd
Rigaflo - Farr Filtration Ltd
Righton PQ - Righton Ltd
Rigibeam - Amilake Southern Ltd
Rigidal Lokrol - Rigidal Industries Ltd.
Rigidal Lokroll - Rigidal Systems LTD
Rigidal Profiled Cladding - Rigidal Systems LTD
Rigidal Thermocore - Rigidal Industries Ltd.
Rigidal Thermowall - Rigidal Industries Ltd.
Rigidal Trocaldek - Rigidal Industries Ltd.
Rigidal Ziplok - Rigidal Industries Ltd.
Rigidal Ziplok - Rigidal Systems LTD
Rigid-Cor/F - TDI (UK) Ltd
Rigidline - Phoenix Metal Products Ltd
Rigidlock 15 - Phoenix Metal Products Ltd
Rigidlock 24 - Phoenix Metal Products Ltd
Rigifix - Huntley & Sparks
Rilass V - Farr Filtration Ltd
Rinaldi - Atkinson & Kirby Ltd
Rior - Ward's Flexible Rod Company Ltd
Ripac - CSSP (Construction Software Services Partnership)
Ripon - Ellis J T & Co Ltd
Riposeal - Shackerley (Holdings) Group Ltd incorporating Designer ceramics
Ritz - New Haden Pumps Ltd
River Gate - Claydon Architectural Metalwork Ltd
Riverdale - Eternit Building Materials
Rivscrew® - Textron Fastening Systems Ltd
RMB 300 - Monarflex Anderson Ltd
RMC Concrete Products - CEMEX
Road Drain - ACO Technologies Plc
Roadline - Leigh's Paints
Roadside - Maceman + Amstad Ltd
Robec - Tuke & Bell Ltd
Robert Lynam collection - Emsworth Fireplaces Ltd
Robette - Tuke & Bell Ltd
Robey - Wellman Robey Ltd
Robinhood - Beeston Heating Group Ltd
Robinlight - Torvale Building Products, A Division of Stadium Group PLC
Robinson Willey - Robinson Willey Ltd
Robinsons Greenhouses - Compton Buildings Ltd
Robocal - Altecnic Ltd
Robofil - Altecnic Ltd
Robokit - Altecnic Ltd
Robolink - Altecnic Ltd
Robot - Robot (UK) Ltd
Rockcem - Cryotherm Insulation Ltd
Rocket - Illuma Lighting Ltd
Rockfon - Nevill Long Limited
Rockfon - New Forest Ceilings Ltd
Rockliner - Cryotherm Insulation Ltd
Rockranger Primary Crushing Outfits - Parker Plant Ltd
Rocksil - Knauf Insulation Ltd
Rockwell - Rockwell Sheet Sales Ltd
Rockwood - Rockwell Sheet Sales Ltd
Rockwool - Firebarrier Services Ltd
Rockwool - Rockwool Ltd
Rofatop Membrane - Cembrit Blunn Ltd
Rola - Bramah Security Equipment Ltd
Rollable - Cuprinol Ltd
Rollajack - Rollalong Ltd
Rollalong Linx - Rollalong Ltd
Rollaskid - Rollalong Ltd
Rollatape - Advance Tapes International Ltd
Rollawheel - Rollalong Ltd
Rollercash - Safetell Ltd
Rollerseal - Kleeneze Sealtech Ltd
Roll-fix ridge - Klober Ltd
Rolls Rotary - Tuke & Bell Ltd
Rollstick - Axter Ltd
Rolux - Ubbink (UK) Ltd
Rolyat - Elsy & Gibbons Ltd
Roma Transport - Phoenix Scales Ltd
Romil - Anderson Gibb & Wilson
Ronabond - Ronacrete Ltd
RonAdeck - Ronacrete Ltd
Ronafix - Ronacrete Ltd
Ronascreed - Ronacrete Ltd
Ronascreed DB Fast Floor - Ronacrete Ltd
Ronaset - Ronacrete Ltd
Rondelux - Poselco Lighting
Ronseal - Campbell Marson & Co
Roofite - Watco UK Ltd
Roofmate - Dow Construction Products
Rooftex - Conren Ltd
Room Service - Brintons Ltd
Roomstat - Horstmann Controls Ltd
Roomvent - Airflow Developments Ltd
Rootfast® - Earth Anchors Ltd
Rope and Bow - Balmoral Porcelain Limited/Longmead Ceramics
Ropetwist Bathroom Suite - Ideal-Standard Ltd
Rosamon - Heating Equipment Ltd
Roscal - Higginson Staircases Ltd
Rosconi Benches - Detlef Muller Ltd
Rosconi Cloakroom Systems - Detlef Muller Ltd
Rosebury - Qualitas Bathrooms
Rosemary clay - Lafarge Roofing Ltd
Rosengerns - Rosengerns Tann Ltd
Rossella - Dermide Ltd
Rossetti - Symphony Group PLC, The
Rossflex - Dermide Ltd
Rota Guard - Geoquip Ltd
Rota spike ® - Jackson H S & Son (Fencing) Ltd
Rotaflow - Gebhardt Kiloheat
Rotapak - Fluidair International Ltd
Rotasoc - Electrak International Ltd
Rotastar - Fluidair International Ltd
Rotavent - Gebhardt Kiloheat
Rotaworm - Ward's Flexible Rod Company Ltd
Rotherbar - Civil Engineered Products Ldt
Rothwell - Deva Tap Co Ltd
Roto - Roto Frank Ltd
RotoLatch™ - Latchways plc
Rougeite - CED Ltd
Rough at the edges - Clayton-Munroe Ltd
Roulcabin - UK Cabin Co
Roulston - UK Cabin Co
Rovacabin - Rovacabin
Rovalink - Rovacabin
Rovaspan - Rovacabin
Roxy Fireplace - James Smellie Fabrications Ltd
Royair - Royair Ltd
Royal - Pellfold Parthos Ltd.
Royal Forest - Formpave Ltd
Royal Granite - Pilkington's Tiles Ltd
Royal S - SCHUCO International KG
Royale - Hill Leigh Ltd
Royale - Midland Stom Ltd
Royale Doors - Kind, J B, Ltd
Royalux - JSB Electrical PLC
Royce - Morris Material Handling Ltd
RPM Salvage Ltd - Reclaim.UK.Com
RPS - RPS Data Products (UK) Ltd
RPV 2000 - Reznor UK Ltd
RS Rotary screw compressor - Fluidair International Ltd
RSS - Rouse Security Service
Rubberfuse - E. H. S. Roofing Ltd
Rubberfuse - Intergrated Polymer Systems (UK) Ltd
Rubbergard - Firestone Building Products Europe
Rubberguard EPDM - Firestone Building Products Europe
Rubberline - Dudley Thomas Ltd
Rubberwell - Dudley Thomas Ltd
Rubboseal - Trade Sealants
Ruberoid - Ashmead Buidling Supplies Ltd
Rubit - Hansgrohe
Rudgwick - Baggeridge Brick PLC
Rugby cement - Brikenden (Builders Merchants) Ltd
Rugby Hydrated Lime - Rugby Cement
Rugby Masonry - Rugby Cement
Rugby Oilwell - Rugby Cement
Rugby Portland - Rugby Cement
Rugby Rapid - Rugby Cement
Rugby Special - Rugby Cement
Rugby Sulfate - Rugby Cement
Rugby White - Rugby Cement
Ruko - ASSA Ltd
RUNDFLEX - Circular wall formwork - PERI Ltd
Runtal CX - Zehnder Ltd
Runtal Jet - Zehnder Ltd
Runtal LST - Zehnder Ltd
Runtal RX - Zehnder Ltd
Russell Hobbs - Pifco Ltd
Rustic Kerb - Brett Paving
Rustins - Rustins Ltd
Rustoleum - Andrews Coatings Ltd
Rustoleum - Andrews Coatings Ltd
Rustoleum - Kenyon Group Ltd
Rutland Range - Evergreens Uk
RV Active Inra Red Beams - Geoquip Ltd
RW Linquartz - Robinson Willey Ltd
Ryt Comb - Ryton's Building Products Ltd
Ryt Hose - Ryton's Building Products Ltd
Rytweep - Ryton's Building Products Ltd
S.D.R.M. - Shutter Door Repair & Maintenance Ltd - Dudley Factory Doors Limited
S.D.R.M.-(Shutter & Door Repare & Maintenance Ltd) - Shutter Door Repair & Maintenance Ltd
S.D.R.M.-(Shutter Door Repair & Maintenance Ltd.) - Priory Shutter & Door Co Ltd
S.E.F.T. Lighting - Vision Neon & Display Ltd
S.M.I.T.M. - Telelarm care Ltd
S.R.- EPO RESIN FLOOR SCREED - For upgrading - Ferguson G A & Co Ltd
S.V.K. Cromleigh - Allan Harris & Sons Ltd
S2000 CSJ - Elster Metering Ltd
Sabre - CIL International Ltd
Sabre - Viking Security Systems Ltd
Sadia Refrigeration - Viscount Catering Ltd
Sadolin Classic - Sadolin UK Ltd
Sadolin Extra - Sadolin UK Ltd
Sadolin Fencing Wood Stain - Sadolin UK Ltd
Sadolin High Performance Varnish - Sadolin UK Ltd
Sadolin Prestige - Sadolin UK Ltd
Sadolin PV67 Heavy Duty Floor Varnish - Sadolin UK Ltd
Sadolin Quick Drying Woodstain - Sadolin UK Ltd
Sadolin Shed and Fence Protection - Sadolin UK Ltd
Sadolin Supercoat - Sadolin UK Ltd
Sadolin Superdec - Sadolin UK Ltd
Saf 2 - The Safety Tread Ltd
Safe Call - Raidio-Based Nurse Call System - Wandsworth Elecrtrical Ltd
Safe guard - DLW Flooring
Safe 'n' Sound - Premdor
Safe T Epoxy - Anglo Building Products Ltd
Safe T Epoxy - Anglo Building Products Ltd
safe-deko vinyl floorings - Jaymart Rubber & Plastics Ltd
Safedress - Stenoak Associated Services PLC
Safegard - James & Bloom Ltd
Safegrate - Lionweld Kennedy Ltd
Safegrid - Lionweld Kennedy Ltd
Safeguard - Securikey Ltd
Safeguard - Unicorn Abrasives (UK) Ltd
SafeGuard - Resident Monitoring/Patient Tagging - Wandsworth Elecrtrical Ltd
Safelock - Lionweld Kennedy Ltd
Safepave - Stenoak Associated Services PLC
Safepower - Blakley Electrics Ltd
Saferail - Lionweld Kennedy Ltd
Safestep - Forbo - Nairn Ltd
Safestore - Barton Storage Systems Ltd
Safe-T-Bar - Fab & Fix Ltd
Safetread - Lionweld Kennedy Ltd
Safetred Range - Tarkett-Marley Floors Ltd
Safe-T-Store - Elliott Group Ltd
Safety Speed cut - JKO Ltd
Saffire - Baydale Architectural Systems Ltd
Saffire - Baydale Architectural Systems Ltd
Salamandre - Legrant Electric Ltd
Salex Acoustic Products - The Salex Group Ltd.
Salient - Paragon Business Furniture
Salmen's - Fine H & Son Ltd
Salter - Phoenix Scales Ltd
Salter Abbey - Phoenix Scales Ltd
Sam - Ves Andover Ltd
SaMontec - Artur Fischer (UK) Ltd
sampsonsironmongery.co.uk - Sampson T F Ltd
Samson - Rockfon Ltd
Samsumo TM Safety Access Systems - Redman Fisher Engineering
Samsung - Connought Communication Systems Ltd
Samtor - Axter Ltd
Samuel Booth collection - Booth Samuel & Co Ltd
Sanador - EDS Shutter Door Systems Ltd
Sand Drain - ABG Ltd
Sanddresser Turf - Lawrence Surfaces Plc, Charles
Sandene - Intergrated Polymer Systems (UK) Ltd
Sanderson - Sanderson
Sandfilled Turf - Lawrence Surfaces Plc, Charles
Sandringham - Fab & Fix Ltd
Sandringham - Loft Centre Products
Sandringham suite - Armitage Shanks Ltd
Sandtex - Akzo Coatings Plc
Sandtex - Crown Trade Paints
Sandwell - Chiltern Invadex
Sani Panel - Jaga Heating Poducts (UK) Ltd
Sani Ronda Towel Rail - Jaga Heating Poducts (UK) Ltd
Sani-fem - Unicorn Containers Ltd
Sanistrel - Imperial Machine Co Ltd
Sanistrel - Sissons W & G Ltd
Sanitas - OWA UK Ltd
Santric - Pland Stainless Ltd
Saphir - EJOT Ecofast Ltd
Sapphire 600 & 750 Raised Access Flooring - Durabella System Floors Ltd
Saracen - Laird Security Hardware
Sarel - Schneider Electric Ltd
Sarena Housings - Sarena MFG Ltd
Sarena Ladders - Sarena MFG Ltd
Sarena Mouldings - Sarena MFG Ltd
Sarena Tanks - Sarena MFG Ltd
Sarina rubber floorings - Jaymart Rubber & Plastics Ltd
Sarnafast - Sarnafil Ltd
Sarnafil - Sarnafil Ltd
Sarnalite - Sarnafil Ltd
Sarnatherm - Sarnafil Ltd
Sarnatred - Sarnafil Ltd
Sartec - Anderson D & Son
Sartec vaperm breather membrane - Anderson D & Son
Sash Jammers - Fab & Fix Ltd
Sasmox - McLoughlin Wood Ltd
Satalico - Satec Ltd
Satchwell Sigma - Satchwell Control Systems Ltd
Satellite - Paragon Business Furniture
Saturn - Elizabethan Textiles
Saturn - Fabrikat (Nottingham) Ltd
Saunier Duval - Hepworth Heating Ltd
Savanna - En-tout-cas plc
Savanna - En-tout-cas plc (Scotland)
Saver CF - Draeger Safety UK Ltd
Saver PP - Draeger Safety UK Ltd
Sawco - The Saw Centre Group
Sawcut - The Saw Centre Group
Sawflex - nmc (UK) Ltd
SB - Wilo Samson Pumps Ltd
SBS - Hawker Ltd
SC - Hudevad Britain
SC500 Beam Mounted Seating System - Race Furniture Ltd
Scala - DLW Flooring
Scala - DLW Flooring Ltd
Scalpture Products - Hynes Aden K Sculpture Studios
Scalpture Services - Hynes Aden K Sculpture Studios
Scan Pumps - ABS Pumps Ltd
Scanco - McMullen Architectural Systems
Scan-Q - Zon International
Scantronic - Cooper Security Ltd
Scenario - Mode Lighting (UK) Ltd
Sceneset - Helvar Electrosonic
Sceneview - Helvar Electrosonic
Sceptre - Paragon Business Furniture
Schafer Lockers - Detlef Muller Ltd
Schlage - Relcross Ltd
Scholar - Ellis J T & Co Ltd
SCHOTT Termofrost Cold Room Doors - SCHOTT UK Ltd
Schuco - EBS Services Ltd
Schwan - Hyde Brian Ltd
Schwarzwald - Erlau AG
Schwing - Burlington Engineers Ltd
Schwing - Burlington Engineers Ltd
Scion - USF Permutit/Ionpure
Scotchshield - Durable
Scotchtint - Durable
Scoth Buff - Realstone Ltd
Scottish Brick - Ibstock Brick Ltd
Screedmaster - Laybond Products Ltd
Screening Outfits - Parker Plant Ltd
Screenmaster - ABB Instrumentation Ltd.
Screwlok - Camlok Lifting Clamps Ltd
SD.RM- (Shutter Door Repair & Maintenance Ltd.) - Neway Doors Ltd
SDR - Stoves PLC
Seagull - Gledhill Water Storage Ltd
Sealdeck - Laybond Products Ltd
Sealex - Klinger Fluid Instrumentation Ltd
Sealmaster - Sealmaster
Sealobond - Sealocrete Ltd
Sealoflash - Sealocrete Ltd
Sealofoam - Sealocrete Ltd
Sealomix - Sealocrete Ltd
Sealopol - Sealocrete Ltd
Sealoproof - Sealocrete Ltd
Sealoshield Asbestos - Sealocrete Ltd
Sealotone - Sealocrete Ltd
Seamaker - Barr & Wray Ltd
Seamaker - Barr & Wray Ltd
Seating - Paragon Business Furniture
SEC - Race Furniture Ltd
Secar Cements - Lafarge Aluminates Ltd
Seculate - Pearl Paints Ltd
Securaglide insulated roller shutters - Regency Garage Door Services Ltd
Secure-Mix - FC Frost Ltd
Securesheild ™ - Accent Hansen
Securicel - Fitzpatrick Doors Ltd

Securigard - Solaglas Laminated
Securiguard window shutter - Bolton Gate Co. Ltd
Securitech - C & K Intellisense
Security Plus® - Nico Manufacturing Ltd
Securus - Hill Leigh Ltd
Sedia - Erlau AG
Seedmat 2000 - Land Wood & Water Co. Ltd
Seefire - Colt International
Seekure - Cordek Ltd
Sefac - Alumasc Construction Products Ltd
Seip Remote Control - Regency Garage Door Services Ltd
Sela - SFS Intec LTD
Seldex - Haworth UK Ltd
Seleceta - Broag Ltd
Select Range - Pedley Furniture International Ltd
Selecta Shower Screens - Shires Bathrooms
Selectarail - Rothley Burn Ltd
Selectascreen - Pearce Security Systems
Selectos - Nu-Way Ltd
Self Serve - Unicorn Containers Ltd
Selfix Capsules - Exchem Mining and Construction
Selfix Carbofibe - Exchem Mining and Construction
Selfix Style 3 Bolts - Exchem Mining and Construction
Sellite - Sellite Blocks Ltd
Sellite Blocks - CAEC Howard (Holdings) Ltd
Sellite Blocks - CAEC Howard (Holdings) Ltd
Selux Lighting - Woodhouse UK Plc
Sembla - Allgood plc
Semitron® ESD - DSM Engineering Plastic Products (UK) Ltd
Senator - Chubb Safe Equipment Company
Senator - Deans Blinds & Awnings (UK) Ltd
Senator - Dudley Thomas Ltd
Senator - Forticrete Roofing Products
Senior Aluminium systems - EBS Services Ltd
Sensalux - Setsquare
sensations - H & R Johnson Tiles Ltd
Sensatronic - Ecowater Systems Ltd
Sensor - Legrand Electric Ltd
Sensor Coil 600 - Geoquip Ltd
Sensor flow & electronic taps - Armitage Shanks Ltd
Sentinal - Advanced Fencing Systems Ltd
sentinal ATM enclosure - Pearce Security Systems
Sentinel - CSI Security LTD
Sentinel - Fitzgerald Lighting Ltd
Sentinel - Nettlefolds Ltd
Sentinel - Race Furniture Ltd
Sentinel - Shackerley (Holdings) Group Ltd incorporating Designer ceramics
Sentree - Boddingtons Ltd
Sentrilock - Kenrick Archibald & Sons Ltd
Sentry - MK (MK Electric Ltd)
Sentry - Montague L Meyer
Sentry ® - Jackson H S & Son (Fencing) Ltd
SEP - Satchwell Control Systems Ltd
SerckAudio - Flowserve Flow Control
Series 2100 - James Gibbons Format Ltd
Series 55 Hose Reel - Macron Safety Systems (UK) Ltd
Series 60 - Apollo Fire Detectors Ltd
Series 60 - Apollo Fire Detectors Ltd
Series 60 Low Voltage - Apollo Fire Detectors Ltd
Series 65 - Apollo Fire Detectors Ltd
Series 90 Heavy Duty Benching - Envopak Group Ltd
Series AM - Dunham Bush Ltd
Series BM - Dunham Bush Ltd
Series CM - Dunham Bush Ltd
Series F - Dunham Bush Ltd
Series UH - Dunham Bush Ltd
Serina - DW Windsor Lighting
Serpo RG - Optiroc Ltd
Serpo Rock - Optiroc Ltd
Serristrap - Grace Construction Products Ltd
Servery - Cooper Group Ltd
Servicare - Evac+chair international Ltd
Servicised - Grace Construction Products Ltd
Serviflex - Uponor Ltd
Serviseal - Grace Construction Products Ltd
Servisol - Ambersil Ltd
Servisol - Ambersil Ltd
Servitherm - Servitherm
Servowarm - Servowarm
SES - Shepherd Engineering Services Ltd
Sesam - Dividers Modernfold Ltd
Sesame - Profab Access Ltd
Seth Thomas - General Time Europe
Setlite - Setsquare
Seventh Heaven - Interfloor Limited
severn gutter system - Dales Fabrications Ltd
Sewer Drain - Hepworth Building Products Ltd
Sewtec - Saint Gobain pipelines PLC
SGB Events - SGB Hire Plus
SGB Formwork - SGB Hire Plus
SGB Taskforce - SGB Hire Plus
SGG Antelio - Solaglas
SGG Contour - Solaglas
SGG Cool-Lite - Solaglas
SGG Diamant - Solaglas
SGG Eko Plus - Solaglas
SGG Planilux - Solaglas
SGG Priva-Lite - Solaglas
SGG Selvrit - Solaglas
SGG Seralit - Solaglas
SGG Spider Glass - Solaglas
SGG Stadip Silence - Solaglas
SGG Stadis - Solaglas
Shadacrete - W. Hawley & Son Ltd
Shadeacrete - Bayer Chemicals Inoegate Pigment Group
Shades of light - Eclipse Blind Systems Ltd
shadex shading system - Dales Fabrications Ltd
Shadflam - Shadbolt F R & Sons Ltd
Shadmaster - Shadbolt F R & Sons Ltd
Shaftbrace - Maby Hire Ltd
Shaftesbury - Hamworthy Heating Ltd
Shallovent - Therminal Technology (sales) Ltd
Shape shield - Madico Film Products Division Inc
Shapphire - Heatrae Sadia Heating Ltd
Sharp - Pro Display
Sharps - Inter Public Urban Systems Uk Ltd
Shavrin - Shavrin Levatap Co Ltd
Shavrin Bijoux - Shavrin Levatap Co Ltd
Shavrin Levamixa - Shavrin Levatap Co Ltd
Shavrin Levatap - Shavrin Levatap Co Ltd
Shavrin Modesta - Shavrin Levatap Co Ltd
Shaws glazed bricks - Ibstock Building Products Ltd
Shear flow - shearflow LTD
Sheer - Hamilton R & Co Ltd
Sheerframe - LB Plastics Ltd
Sheerline - LB Plastics Ltd
Sheffield - Kaba Door Systems Ltd
Shell - Balmoral Porcelain Limited/Longmead Ceramics
Shelter - Checkmate Industries Ltd
Shepherd - Kenrick Archibald & Sons Ltd
Sherborne - Hamworthy Heating Ltd
Sheriden - Ulster Carpet Mills Ltd
Sherwood Shelving - Sealine Business Products
Shev - Dunsley Heat Ltd
Shield - Stocksigns Ltd
Shield LST - Hudevad Britain
Shieldoor - Norwood Partition Systems Ltd
Shinsei seiki - Kaba Garog
Shopline - Smart Systems Ltd
Shoptrack - CIL International Ltd
Shoreseal - Shackerley (Holdings) Group Ltd incorporating Designer ceramics
Shower Panels - Hansgrohe
Shower Temples - Hansgrohe
Shower Trays - Roman Mosaic Contracts Ltd
Showercare - Contour Showers Ltd
Showercyl - Gledhill Water Storage Ltd
Showerforce - NewTeam Ltd
Showerhouse Shower Canopies - Jendico
Showerhouse Shower Curtains - Jendico
Showermate - Stuart Turner Ltd
Showermax - Range Cylinders
Showerminder - Setsquare
ShowerSport - ShowerSport Ltd
Showertime - Neptune Showers Ltd
Showertrays by Coram - Coram Showers Ltd
Shuttering - Cordek Ltd
Si - Worcester Heat Systems Ltd
Sibe Scale - Higginson Staircases Ltd
Sidetrak - Rackline Systems Storage Ltd
Sidewinder - Clarke UK Ltd
Siedle - Viking Security Systems Ltd
Sieger - Zellweger Analytic Ltd, Sieger Division
Sienna - President Office Furniture Ltd
Siesta - Erlau AG
Siesta - Kermi (UK) Ltd
Sightline 70 - Spectus Systems
Sigma - Andrews Coatings Ltd
Sigma - Brett Paving
Sigmaclad - Sigma Coatings Ltd
Sigmacover - Sigma Coatings Ltd
Sigmadeck - Turner Access Ltd
Sigmadur - Sigma Coatings Ltd
Sigmafast - Sigma Coatings Ltd
Sigmaferro - Sigma Coatings Ltd
Sigmaguard - Sigma Coatings Ltd
Sigmarite - Sigma Coatings Ltd
Sigmatherm - Sigma Coatings Ltd
Sigmaweld - Sigma Coatings Ltd
Sigmetal - Sigma Coatings Ltd
Signal - Shadbolt F R & Sons Ltd
Signature - Unilock Ltd
Signe - Silent Gliss Ltd
Signet - George Fischer Sales Ltd
Signfix modular systems - Green Brothers Ltd
Signicolour - Amari Plastics Plc
Signsystems modular systems - Green Brothers Ltd
Sika-1 - Sika Ltd
SikaBond - Sika Ltd
SikaDur - Sika Ltd
Sikaflex - Sika Ltd
SikaFloor - Sika Ltd
Sikaplan - Sika Ltd
Sikatop - Sika Ltd
Sikkens - Akzo Coatings Plc
Silbond - Vallance
Silbralloy - Johnson Matthey PLC - Metal Joining
Silco Color - STO Ltd
Silent floor - Brooks Belfast
Silent Gliss - Silent Gliss Ltd
Silfos - Johnson Matthey PLC - Metal Joining
Silglaze - GE Bayer Silicones
Silica Paint Potassium Silicate - Textured European Finishes Ltd
Silk Machines - Furmanite International Ltd
Silovil - Campbell Marson & Co
Silpruf - GE Bayer Silicones
Silver-Flo - Johnson Matthey PLC - Metal Joining
Silverline - Aluminium RW Supplies Ltd
Silvia - Townhive Ltd
Simon Fiesta - Simon R W Ltd
Simpak compactors - Randalls Fabrications Ltd
Simplan-3 - Booth Muirie
Simply Elegant - Hodkin Jones (Sheffield) Ltd
Simpson - Ashmead Buidling Supplies Ltd
Simpson Strong-Tie® - Simpson Strong-Tie®
Single Point Gas Alarm - Gas Measurement Instruments Ltd
Single Touch Communication Systems - BPT Security Systems (UK) Ltd
Sintamatic - Torit DCE
Sintesi - Tubular Steel Radiators - Clyde Combustions Ltd
Sirius - Poselco Lighting
Sirius - Valor
Sirocco - Paragon by Heckmondwike
Sirplus - GE Bayer Silicones
Sirrus - Gummers A & J Ltd
Sisal - Rawtex Ltd
Sisalite - British Sisalkraft Ltd
Sisalite - British Sisalkraft Ltd
Sisalkraft - British Sisalkraft Ltd
SissonsCF - Sissons W & G Ltd
Sitecop - Rediweld Rubber & Plastics Ltd
Siteguard - GE Capital Modular Space
Sitesealer Membrane - Cavity Trays Ltd
Six hundred 600 Series - Komfort Office Environments Plc
Skanform - Norske Interiors (UK) Ltd
Skarsten - Skarsten Tools
SKF - Magnetic - Electro Mechanical Systems Ltd
Skil - Robert Bosch Ltd
SKIL - Skil Controls Ltd
Skinner for Fencing - since 1920 - WA Skinner & Co Limited
Skirtings - Roman Mosaic Contracts Ltd
SKYDECK - Aluminum panel slab formwork - PERI Ltd
Skygard - Lonsdale Metal Company
Skyline - Alumasc Exterior Building Products Ltd
Skyline - Concord: Marlin
Skyline - Ves Andover Ltd
Skyway - Scholes Windows Ltd
Slab Stress - Freyssinet Ltd
Slate2000 - Britmet Tileform Limited
Slatevent II - Cembrit Blunn Ltd
Sleepsafe - PH Products Ltd
Slik Bar - Furmanite International Ltd
Slim Vent - Ryton's Building Products Ltd
Slimglide - Integra Products
Slimline - Cooke Brothers Ltd
Slimline - Dudley Thomas Ltd
Slimlinedrain - ACO Technologies Plc
Slimpack - Roof Units Group
Slimstyle - Hallmark Blinds Ltd
Slimstyle Groundfloor Treatments - Heywood Williams Architectural
Slingsby - HC Slingsby PLC
Slite - Indirect Lighting Ltd
Sloan & Davidson - Longbottom J & J W Ltd
Slotline - Royair Ltd
Slotz - Winther Browne & co Ltd
SM Range - Becker (SLIDING PARTITIONS) Ltd
SM4 - C & R Building Systems Ltd
Smallstore - Lydney Containers Ltd
Smaragd - Forbo - Nairn Ltd
Smart systems - Alframes Ltd
Smartank Duplux Pump Sets - Metcraft Ltd
Smartank Overfill Alarm Unit - Metcraft Ltd
Smartank Pump Control Pannels - Metcraft Ltd
Smartroof - Wyckham Blackwell Ltd
Smartscreen - Smart Systems Ltd
smartsigns - Stocksigns Ltd
Smartstar - Roof Units Group
Smatex - Bardon (England) Ltd
Smatex - Bardon (England) Ltd
Smatex - Camas Aggregates Ltd - Aggregate Industries
Smeed Dean - Chelwood Brick Ltd
Smiths - Timeguard Ltd
Smog-Eater - Horizon International Ltd
Smog-Mobile - Horizon International Ltd
Smog-Rambler - Horizon International Ltd
Smoke Ventilation Louvres - Matthews & Yates Ltd
Smokemaster curtain - Colt International
SmokeStop - Cooper Group Ltd
Smokex - Keddy-Poujoulat (UK) Ltd
Smometa - Photain Controls Plc
Smooth - On - Notcutt Ltd
Smoothline - Rega Ductex Ltd
Smoothline - WP Metals Ltd
SMU - Stanton plc
Snake Way - Marshall-Tufflex Ltd
Snaptyte - Newdawn & Sun Ltd
snipef - Scottish & Northern Ireland Plumbing Employers' Federation
Sofco ® - Broxap Ltd
Softbac - Gaskell Carpet Tiles
Softcell - Whitecroft Lighting Limited
Softgrip Tools - Tower Manufacturing
Soil nails - PHI Group Ltd
Soil Panel - PHI Group Ltd
Soilcrete - Keller Ground Engineering
Soilfrac - Keller Ground Engineering
soilmaster entrance matting - Jaymart Rubber & Plastics Ltd
Solair - Solair Ltd
Solar C Brise soleil - Colt International
Solar Fin - Colt International
Solar Knight - Intersolar Group Ltd
Solar Moler - Intersolar Group Ltd
Solar Protect - Eclipse Blind Systems Ltd
Solar Pumpkit - Intersolar Group Ltd
Solargard - Hi-Tec Window Films
Solaris - Shackerley (Holdings) Group Ltd incorporating Designer ceramics
Solarvent - Intersolar Group Ltd
Solarvent Turbo - Intersolar Group Ltd
Solatherm - IPPEC Sytsems
Solenium - Interface Europe Ltd
Solga - Harbro Supplies Ltd
Solid - ASSA Ltd
Solidbase - Foster W H & Sons Ltd
Solifer - Industrial Devices Ltd
Solignum - Protim Solignum Ltd
Solignum - Tor Coatings Ltd
Solitaire - Mode Lighting (UK) Ltd
Solitex Plus - Fillcrete Ltd
Solitex Wa - Fillcrete Ltd
Solitude - Cep Ceilings Ltd
Soll - HCL Verta
Solo - Ellis J T & Co Ltd
Solo - Helvar Electrosonic
Solo - Paragon Business Furniture
Solo (Series 210) - Parker Bath Limited
solo choice - Paragon Business Furniture
Solomat - Zellweger Analytic Ltd, Sieger Division
Solos - Forticrete Roofing Products
Solosun - Ubbink (UK) Ltd
Solray - Comyn Ching & Co Ltd (SOLRAY)
Solutions - Hands of Wycombe
Solutions 2000 SHR - Beeston Heating Group Ltd
Solvite - Henkel Consumer Adhesives
Sombra - Saint-Gobain Ecophon Ltd
Somdrain - Geosynthetic Technology Ltd
Somerset Doorset Range - Hill Leigh Ltd
Somerton - Top Office Equipment Ltd
Sommer - Tarkett Sommer Ltd
Somtube - Geosynthetic Technology Ltd
Sonae - Anderson C F & Son Ltd
Sonae - Boardcraft Ltd
Sonae - Deralam Laminates Ltd
Sonafold Acoustic Folding Door - Bolton Gate Co. Ltd
Sonar - Rockfon Ltd
Sonata - Laird Security Hardware
Sonata - Quartet-GBC UK Ltd
Sonata BS - Polypipe Bathroom & Kitchen Products Ltd
Sonatone - USG (UK) Ltd
Sonex - Salex Interiors Ltd
Sonneborn - Feb MBT
Sony - Crimeguard Systems Ltd
Sony - RPS Data Products (UK) Ltd
Sony - Vidionics Security Systems Ltd
Sorba - LSA Projects Ltd
Sorrus - Lumitron Ltd
SOS - Maestro International Ltd
Sottini bathrooms - Ideal-Standard Ltd
Sound Control Ear Muffs - MSA (Britain) Ltd
Sound Masking - The Salex Group Ltd.
Sound Reduction Systems Ltd Acoustilay - Trim Acoustics
Sound secure - Premdor
Soundcel - Fitzpatrick Doors Ltd
Soundcote - Cafco International
Soundcote - Calder Industrial Materials Ltd
Soundmaster - Monowa Manufacturing (UK) Ltd
Soundpac - Hotchkiss Air Supply
Soundsafe - Emergi-Lite Safety Systems
Soundshield ™ - Accent Hansen
Southampton - Neptune Outdoor Furniture Ltd
Sovereign - Baggeridge Brick PLC
Sovereign - Clenaware Systems Ltd
Sovereign - Commercial Lighting Systems
Sovereign - Sovereign Brush Co Ltd, The
Sovereign Doors - Jeld-Wen Ltd
Sovereign Paving - Thakeham Tiles Ltd
Sovereign Standard, Fixed and Hi-Lo - Parker Bath Limited
Sovereign Windows - Jeld-Wen Ltd
Sovereign Windows - Jeld-Wen Ltd
SP100 - Laticrete UK
SPAC - Brighton (Handrails), W
SPAC - Brighton (Handrails), W
Space Bathroom Suite - Ideal-Standard Ltd
Space -Ray - Space - Ray UK
Spacebuild - Wernick S & Sons Ltd
Spacedeck - Turner Access Ltd
Spacedek - Raised Floor Systems Ltd
Spacemaster Bollard - Dee-Organ Ltd
Spacepac - Wernick S & Sons Ltd
Spacer - Pellfold Parthos Ltd.
Spacesaver Bollard - Dee-Organ Ltd
Spacestopper Major - Dee-Organ Ltd
Spacestor - Envoplan
Spacewhirl - Record Playground Equipment Ltd
Spalex - Vita Liquid Polymers Ltd
Spandex - Dexion Ltd
Spandex Modular System - Green Brothers Ltd
Spangard - Lonsdale Metal Company
Spania - Marshalls Mono Ltd
Spanseal - Compriband Ltd
Spantile Callette Ventilated Ceilling System - Vianen Kitchen Ventilation (UK) Ltd.
Spanwall - McMullen Architectural Systems
Spanwell - Welconstruct Co Ltd
Spartacus - Metalliform Products Plc
Spartan - Spartan Tiles Ltd
Spawall wall lining - Grant Westfield Ltd
SPC - Furmanite International Ltd
Spear & Jackson - Spear & Jackson
Specialist facades - Haironville TAC Ltd
Specialist Formwork 3D - Cordek Ltd
Specify it (online service) - Technical Indexes Ltd
Specktralock non stain grout - Laticrete UK
Spectral - Commercial Lighting Systems
Spectral 240 3D - Harkness Hall Ltd
Spectrum - Norton Engineering Alloys Co Ltd

Spectrum Paint - Sanderson
Specula - Shadbolt F R & Sons Ltd
Spedec - SFS Intec LTD
Speed - Speed Saws Ltd.
Speed Patch - Anglo Building Products Ltd
Speed Screed - Anglo Building Products Ltd
Speed Set - Premdor
Speedal ® - Righton Ltd
Speedbolt® - Nico Manufacturing Ltd
Speedchannel cat 5E - Hubbell Premise Wiring
Speeddeck - E. H. S. Roofing Ltd
Speedfit - Guest (Speedfit) Ltd, John,
Speedframe - Dexion Ltd
Speedlock - Dexion Ltd
Speedor - Hart Door Systems Ltd
Speedplug - Ramset Fasteners Ltd
Speedrax - Metalrax Ltd
Speedturn - BSA Machine Tools Ltd
Speedway - Vantrunk Engineering Ltd
Spelsburg enclosures - Elkay Electrical Manufacturing Co Ltd
Spencers - Anderson Gibb & Wilson
Sphere - Silent Gliss Ltd
Sphinx - Royal Dutch Sphinx Ltd
Spike - SFS Intec LTD
Spikemaster - Rawson Carpets Ltd
Spiketop - Lawrence Surfaces Plc, Charles
Spiral construction - Consculpt Spiral Stairs Ltd
Spiralift - Caldwell Hardware (UK) Ltd
Spiralite - Cryotherm Insulation Ltd
Spiratube - Flexible Ducting Ltd
Spiratube - **HRS Heat Exchangers Ltd** - HRS Hevac Ltd
Spirex - Caldwell Hardware (UK) Ltd
Spirit - Bartlett Catering Equipment Ltd
Spirotred - Pedley Furniture International Ltd
Spirtech underlay - Lafarge Roofing Ltd
Spit - ITW Spit
Splash (dolphin bin) - Glasdon UK Ltd
Splitline - Lewden Electrical Industries
Sports panel - Horne Engineering Ltd
Sports-Cote Excel - Everlac (GB) Ltd
Sports-Cote Premium - Everlac (GB) Ltd
Sportswall - Record Playground Equipment Ltd
Sporturf - En-tout-cas plc (Scotland)
SPRA - Single Ply Roofing Association
Spraydon - Cafco International
Sprayed Limpet Mineral Wool-GP - Thermica Ltd
Sprayed Limpet Mineral Wool-HT - Thermica Ltd
Sprayed Limpet Mineral Wool-TI - Thermica Ltd
Sprayed Limpet Vermiculite-external - Thermica Ltd
Sprayed Limpet Vermiculite-Internal - Thermica Ltd
Spraymar - Marking Machines International
Spring steel buffer - Berry Systems
Springflex - Rawson Fillings Ltd
Springflow - Astracast PLC
Springflow - Fordham
Springlok - Rawson Fillings Ltd
Springvin - Flexible Ducting Ltd
Sprint - Paddock Fabrications Ltd
Sprint - SGB Youngman
Sprint Truss Equipment - Gang-Nail Systems Ltd
Sproughton Arbour Seat - Anthony de Grey Trellises
Spyhawk - Hunting Engineering Ltd
Spyscan - Hunting Engineering Ltd
Square D - Schneider Electric Ltd
Squire - Squire Henry & Sons
Squire Dlike - Iles Waste Systems
SRS acoustic floating floor - Christie & Grey Ltd
SS System - Quelfire
SSI Schaefer Noell - SSI Schaefer Ltd
SSI Schaefer Peem - SSI Schaefer Ltd
St James - Marflow Eng Ltd
STA - Solar Trade Association, The
Sta - Tie - Fix Fast (Fasteners) Wales Ltd
Stabil Binders - Huntley & Sparks
Stabila - Hyde Brian Ltd
Stabox - Creteco Sales Ltd
Staffline - CSI Security LTD
Staifix - Ancon Building Products
Stainless Steel - Swift & Sure - Arvin Motion Control Ltd
Staircases - Tarmac Top Floor
Stairiser - Stannah Lifts Ltd
Sta-Lok Rigging - Scotia Rigging Services (Industrial Stainless Steel Rigging Specialists)
Stalwart - CSI Security LTD
Stamford - Trent Bathrooms
Standard Patent Glazing - Standard Patent Glazing Co Ltd
Standard Plus - Square D, Electrical Distribution & Building applicators Division
Standfast - Auld Valves Ltd
Standfast - Auld Valves Ltd
Stang Monitors - UK Fire International Ltd
Stanhope - Albany Standard Pumps
Stanley - Relcross Ltd
Stanley Cookers - Diamond Merchants
Stanton - Drainage Systems
Stanweld - Lionweld Kennedy Ltd
Stanwin - Albany Standard Pumps
Stanza - Laird Security Hardware
Stanza - Mackay Carpets, Hugh
Stapoco - Standby Power Company
Star - Metreel Ltd
Star - Novellini Bathroom Products Ltd
Star - Rada Lighting
Star Premium PVC - Signs International
Starbloc - Holden Brooke Pullen
Starcel - E+R Polymers Ltd
Starfix Twin - Holden Brooke Pullen
Starflex - Holden Brooke Pullen
Stargard - SG System Products Ltd
Starkan - Holden Brooke Pullen
Starlight - IMP Lighting Ltd
Starline - Deceuninck Ltd
Starline - Holden Brooke Pullen
Starnorm - Holden Brooke Pullen
Starquest - Rawson Carpets Ltd
Starsystem - Watco UK Ltd
Startwin - Holden Brooke Pullen
Statements of Conformity - British Board of Agrement
Statesman - Magnet Ltd
Statesman - Project Office Furniture PLC
Stayflex - Adaptaflex Ltd
Stealth - S.I. Concrete Systems
Steden - IR Security and Safety
Steden - Laidlaw Architectural Hardware
Steel Clad - Elliott Group Ltd
Steel -Lok - Spur Shelving Ltd
Steel Mobile Units - Lydney Containers Ltd
Steel monoblock - Clestra Hauserman Ltd
Steelaccom - Lydney Containers Ltd
Steelcase Strafor - Steelcase Strafor Plc
Steelcoat - Winn & Coales (Denso) Ltd
Steelcote - Quelfire
Steelex - AvestaPolarit Ltd
Steelex/Hyclad - AvestaPolarit Ltd
Steelite - Marlow Ropes Ltd
Steelkane - Ward's Flexible Rod Company Ltd
Steel-line - Cardale Doors Ltd
Steelplan - Elliott Group Ltd
Steelstore - Lydney Containers Ltd
Steeltone - Henry Venables Ltd
Steelwall - Brown F plc
Steelwall - Neslo Partitioning & Interiors
Stelduct - Brickhouse
Steni colour - CEP Claddings Ltd
Steni nature - CEP Claddings Ltd
Stenoak Fencing - Stenoak Associated Services PLC
Stent - Stent Foundations Ltd
Sterisept - Liquid Plastics Ltd
Sterisheen - Liquid Plastics Ltd
Steristeel - Associated Metal (Stainless) Ltd
Steristeel - Associated Metal (Stainless) Ltd
Sterling - Chalmit Lighting
Sterling - Marlow Ropes Ltd
Sterling - Securistyle Ltd
Sterling range of cable management systems - Marshall-Tufflex Ltd
Sterope - Orsogril Spa UK Distributor
Sterreberg - Smithbrook Building Products Ltd
Stetter - Burlington Engineers Ltd
Stewart Film Screen - Anders + Kern UK Ltd
Stick-Lite - Encapsulite International Ltd
Stik - Iles Waste Systems
Stile - Campbell Marson & Co
Stilpro - Allpipe Ltd
Sto Marmorino - STO Ltd
Stockfloor - Readymix Limited
Stockmaster Shelving - Beford & Soar Ltd
Stocks Blocks - Brikenden (Builders Merchants) Ltd
Stolastic Color - STO Ltd
Stolevell Uni - STO Ltd
Stolit - STO Ltd
Stom - Polypipe Bathroom & Kitchen Products Ltd
Stomflow - Midland Stom Ltd
Stomstar - Midland Stom Ltd
Stone slate - Forticrete Roofing Products
Stonecor - TDI (UK) Ltd
Stoneraise red - Realstone Ltd
Stontex - Atlas Stone Products
Stopdust - Don Construction Products Ltd
Stopgap - Ball F & Co Ltd
Stopgap - Ball F & Co Ltd
Stoplite - Cryotherm Insulation Ltd
Stoprep Contact - STO Ltd
Stoprim Micro - STO Ltd
Storage Wall - Davies John Interiors Ltd
Storbox - Moresecure Ltd
Store Watch - Advanced Interior Solutions Ltd
Storemaster - Airedale International Air Conditioning Ltd
Storemaster - T & D Packaging Ltd
Storesafe - Lydney Containers Ltd
Storeshield - Lycetts Industrial Doors
StoRFP - STO Ltd
Stork - Race Furniture Ltd
Storm - Blakey John H Ltd
Storm - Blakey John H Ltd
Storm - Concord: Marlin
Storm - Securistyle Ltd
Stormflo Rainwater Sytems - Hunter Plastics Ltd
Stormfox - Johnston Pipes Ltd
Stormframe - Sapa Building Systems (monarch and glostal)
Stormgard - Corroless Corrosion Control
Stormor Clearstor - Link 51 (Storage Products)
Stormor Drawerstor - Link 51 (Storage Products)
Stormor Euro - Link 51 (Storage Products)
Stormor Longspan - Link 51 (Storage Products)
Stormor Roller Edge - Link 51 (Storage Products)
Stormor XL Pallet - Link 51 (Storage Products)
Stormproof - Premdor
Stortech - Rothley Burn Ltd
Stothert & Pitt - Bitmen Products Ltd
Stoves - Stoves PLC
Stoves Newhome - Stoves PLC
Stowell - Ashmead Buidling Supplies Ltd
Stowlite Medium Blocks - Stowell Concrete Ltd
Stra - Optiroc Ltd
Strading - SG System Products Ltd
Strand - DW Windsor Lighting
Strand FEB - Strand Hardware Ltd
Strata - Stokvis R S & Sons Ltd
Stratabord - European Profiles
Stratalite - LB Lighting Ltd
Strate-Grip - Andura Textured Masonry Coatings Ltd
Stratogrid - HT Martingale Ltd
Stratos - Rada Lighting
Stratotrack - HT Martingale Ltd
Stratton Canopy - James Smellie Fabrications Ltd
Streamline - Clenaware Systems Ltd
Streamline - Heatrae Sadia Heating Ltd
Streamline - Wade International Ltd
street king entrance matting - Jaymart Rubber & Plastics Ltd
Streetcrete ™ - Broxap Ltd
Streetiron ™ - Broxap Ltd
Streetscene ® - Broxap Ltd
Streetscene ® - Broxap Ltd
Stretch - Andura Textured Masonry Coatings Ltd
Strip Form - Scapa Tapes UK Ltd.
Stroke - Bernhards Landscapes Ltd
Strong Water Based Lacquer - Junckers Limited
Strongbeam - Spur Shelving Ltd
Strongcast concrete - Challenge Fencing Ltd
Stronghold - Catnic
Stronghold - SG System Products Ltd
Strongoat - Don Construction Products Ltd
Structural Soils (Northern) Ltd - Structural Soils Ltd
Strypit - Rustins Ltd
Stuart - Stuart Turner Ltd
Studflex vibration isolation mat - Christie & Grey Ltd
Studio 3 - Eaton Electric Limited
Studio 3 - Home Automation Ltd
Studio 5 - Eaton Electric Limited
Studio 6 - Home Automation Ltd
Studio Bathroom Suite - Ideal-Standard Ltd
Studio Sanderson - Sanderson
Studio System - Hyperion Wall Furniture Ltd
Studrail system acoustic floating floor - Christie & Grey Ltd
Sturdee - Marlow Ropes Ltd
Styccobond - Ball F & Co Ltd
Stylemaster - Checkmate Industries Ltd
Styrene - Ariel Plastics Ltd
Styroclad - Panel Systems Ltd
Styrodar - Axter Ltd
Styrodur - Kingspan Insulation Ltd
Styrofloor - Panel Systems Ltd
Styroglaze - Panel Systems Ltd
Styroliner - Panel Systems Ltd
Suba Care - William Freeman Ltd
Suba Seal - William Freeman Ltd
Suelux - Luxo UK Ltd
Sufer Rusmin - Mackay Carpets, Hugh
Suffolk - Dixon Turner Wallcoverings
Suhner - Harbro Supplies Ltd
Sumlock - Sumlock J.H.P.
Summit - Rada Lighting
Sumo - Pump Technical Services Ltd
Sumo can crushers - Randalls Fabrications Ltd
Sumpmaster - Pump Technical Services Ltd
Suncatcher - Sunpipe @ Monodraught Ltd
Suncell - CPV Ltd
Suncool - Pilkington Birmingham
Sundeala A Grade - Sundeala Division of Celotex Ltd
Sundeala K Grade - Sundeala Division of Celotex Ltd
Sundeala superspec FR Class 0 - Sundeala Division of Celotex Ltd
Sundeala Surespec FR Class 1 - Sundeala Division of Celotex Ltd
Sundial Fireplace - James Smellie Fabrications Ltd
Sunfilm Anti fade blinds - Dallas & Forrest Ltd
Sunflex ® - Hunter Douglas Ltd (Contracts Division) Luxaflex Blinds ®
Sunfold - Becker (SLIDING PARTITIONS) Ltd
Sungard - Bonwyke Ltd
Sungard - Bonwyke Ltd
Sungard - Invicta Window Films Ltd
Sunrite - Sunrite Blinds Ltd
Sunscreen - Mermet U.K
Sunstar - Space - Ray UK
Suntech - Deans Blinds & Awnings (UK) Ltd
Suntech systems - Deans Blinds & Awnings (UK) Ltd
Suntime - Faber Blinds UK Ltd
sunvector brie soleil - Hallmark Blinds Ltd
Sunway ® - Hunter Douglas Ltd (Contracts Division) Luxaflex Blinds ®
Supa Glide - Moresecure Ltd
Supaboard - Celuform
Supabore - Avesta Sheffield Ltd
Supabore - Avesta Sheffield Ltd
Supachute - Watkins
Supacom - Lydney Containers Ltd
Supadriv - Nettlefolds Ltd
Supafil - Knauf Insulation Ltd
Supaflo - Readymix Limited
Supaliner - Celuform
Supalite - Danbury Fencing PLC
Supalux - Promat UK Ltd
Supamate - Watkins
Supapac - Rycroft Ltd
Supapile - Marshall J A & Co (Brushes) Ltd
Supascrew - Nettlefolds Ltd
Supastor - Flamco UK Ltd
Supastrike box striking plate - Guardian Lock and Engineering Co Ltd
Supatube - Avesta Sheffield Ltd
Supaxpress - Nettlefolds Ltd
Super Clean - Dunsley Heat Ltd
Super Seal - Hepworth Building Products Ltd
Super Shaftbrace - Maby Hire Ltd
Super Slot - Versatile Fittings Ltd
Super Top - Armorex Ltd
Super Top - Armorex Ltd
Super Twosome - Integra Products
Superbolt - Furmanite International Ltd
Supercal - Range Cylinders
Superchill - Heatrae Sadia Heating Ltd
Supercoil - Airsprung Beds Ltd
Supercoil Rolling Smoke Curtain - Bolton Gate Co. Ltd
Supercomfort - Dunham Bush Ltd
Superconcentrate Boiler Noise Silencer - Cookson Electronics
Superconcentrate Boiler Noise Silencer - Fernox Alpha Fry Technologies Europe
Superconcentrate Leaksealer - Cookson Electronics
Superconcentrate Leaksealer - Fernox Alpha Fry Technologies Europe
Superconcentrate Limescale preventer - Cookson Electronics
Superconcentrate Limescale Preventer - Fernox Alpha Fry Technologies Europe
Superconcentrate Protector - Cookson Electronics
Superconcentrate Protector - Cookson Electronics
Superconcentrate Protector - Fernox Alpha Fry Technologies Europe
Superconcentrate Restorer - Cookson Electronics
Superconcentrate Restorer - Fernox Alpha Fry Technologies Europe
Supercote - Witham Oil & Paint Ltd
Superdeluxe - Premdor
Superfine - Integra Products
Superfloc - Fernox Alpha Fry Technologies Europe
Superfold 4000 high-security folding door - Bolton Gate Co. Ltd
Superfoot - Bernhards Landscapes Ltd
Superfoot Acrylic Paint - Sportmark™ Ltd
Superfoot plus - Bernhards Landscapes Ltd
SuperG - Saint-Gobain Ecophon Ltd
Supergalv Lintels - Birtley Building Products Ltd
Supergully - Johnston Pipes Ltd
Superior - Beacon Machine Tools Ltd
Superior - Beacon Machine Tools Ltd
Superior Folding Shutter - Bolton Gate Co. Ltd
SuperJan - Johnson & Starley
Superline - BLV Licht-und Vakuumtecnik
Superline - Marlow Ropes Ltd
Superlintel - Jones of Oswestry Ltd
Superlite Blocks - Hanson Concrete Products
Superlite masonary - Hanson Concrete Products
Superloo - Elliott Group Ltd
Supermat - Kleen-Tex Industries Ltd
Supermix - Marlow Ropes Ltd
Supermix products - B & M Fencing Limited
Supersafe - Hawker Ltd
Super-Saphir - EJOT Ecofast Ltd
Supersave - Hemsec Manufacturing Ltd
Superseal - Kleeneze Sealtech Ltd
Superseal-pipe insert - Guest (Speedfit) Ltd, John,
Supersleve House Drain - Hepworth Building Products Ltd
Superspan - Compriband Ltd
Superstore - Elliott Group Ltd
Superstrong - Boddingtons Ltd
Superstrong - Boddingtons Ltd
Superswing - Record Playground Equipment Ltd
Superswitch - Friedland Ltd
Supertube 125 - Domus Ventilation Ltd.
Supervent - Airflow Developments Ltd
SuperVigilant - Auld Valves Ltd
Supervisor - Gent Limited
Supex hoist - Hills Industries Ltd
Supplyline - National Britannia Ltd
Supplymaster - Timeguard Ltd
Supra-Arch - Jones of Oswestry Ltd
Suprabloc - Jones of Oswestry Ltd
Supragate - Jones of Oswestry Ltd
Supragrid - Jones of Oswestry Ltd
Suprasteel - Jones of Oswestry Ltd
Supratrim - Jones of Oswestry Ltd
Supreme Boiling Water Products - Heatrae Sadia Heating Ltd
Suprosol - Pearl Paints Ltd
Surefire - Thurston Building Systems
Surefire Adhesive - Advance Tapes International Ltd
Surefit Rainwater Sytems - Hunter Plastics Ltd
Suregrip - Wincro Metal Industires Ltd
Suregrip ® - Coo-Var Ltd
Suregrip ® - Teal & Mackrill Ltd
Sureguard - Thurston Building Systems
Surell - Decra Ltd
Surell - Formica Ltd
Surelock - Laird Security Hardware
Sureloo - Thurston Building Systems
Sureplay - En-tout-cas plc (Scotland)
Surespace - Thurston Building Systems
Surespan - Thurston Building Systems
Suresport - Thurston Building Systems
Surestep - Forbo - Nairn Ltd
Suretex - Halls Beeline Ltd
Surface Mounted Road Blockers - APT Controls Ltd

Surfking - Barr & Wray Ltd
Surmabond - Surmak Products Ltd
Surmaglaze - Surmak Products Ltd
Surmaseal - Surmak Products Ltd
Surrey-Sac - Surrey Sackholders Ltd.
Suwide - Tektura Plc
Swaledale - Oakdale (Contracts) Ltd
Swaledale - Realstone Ltd
Swan - Boardcraft Ltd
Swanglide - Integra Products
Swarfega - Deb Ltd
Swarfega Orange - Deb Ltd
Swarfega Power - Deb Ltd
Swash - Clenaware Systems Ltd
Swedeglaze - Swedecor Ltd
Swedhouse - Swedhouse Ltd
Swedish Plain roller blind fabric - Dallas & Forrest Ltd
Swedsign - Signs & Labels Ltd
Swift - Kermi (UK) Ltd
Swift-and-Sure - Arvin Motion Control Ltd
Swift-and-Sure - Arvin Motion Control Ltd
Swiftlay - Swifts Of Scarborough Ltd
Swiftplan - Wernick S & Sons Ltd
Swiftrack - Legrant Electric Ltd
Swiftrack - Swifts Of Scarborough Ltd
Swifts - Legrant Electric Ltd
Swimmer - Golden Coast Ltd
Swimtile - Pilkington's Tiles Ltd
Swing-Boom - Horizon International Ltd
Swisslab - MR Limited
Swisslab/lax/pan/tech/therm - Alumasc Exterior Building Products Ltd
Swissplan - MR Limited
Switchplan - Legrand Electric Ltd
Sws Envirocoat - Stewart Wales, Somerville Ltd
Sygnette - Wellman Robey Ltd
Sykes Pumps - Andrews Sykes Hire Ltd
SylvaFoam - Junckers Limited
SylvaKet - Junckers Limited
Sylvania Lamps - Concord Lighting
SylvaRed - Junckers Limited
SylvaSport Club - Junckers Limited
SylvaSport Premium - Junckers Limited
SylvaSquash - Junckers Limited
SylvaThene - Junckers Limited
Sylvatone - Henry Venables Interiors Ltd
Sylvatone - Henry Venables Ltd
Symbol - Concord: Marlin
Syncrolux - Rada Lighting
Syncros - Floor Protection Services Ltd
Synergy - H. B. Fuller Coatings Ltd
Synkoflex - Creteco Sales Ltd
Syntha Pulvin - H. B. Fuller Coatings Ltd
Syntha Pulvin - HB Fuller Coatings Ltd
Syntha Pulvin Plus - H. B. Fuller Coatings Ltd
Synthascreed - DPC Screeding Ltd
Synthatel - HB Fuller Coatings Ltd
Synthtec Plus - H. B. Fuller Coatings Ltd
Syntropal - ICB (International Construction Bureau) Ltd
System 100 - BPT Security Systems (UK) Ltd
System 100 - Domus Ventilation Ltd.
System 160 - Maby Hire Ltd
System 2000 - nmc (UK) Ltd
System 25 - Metalrax Ltd
System 300 - BPT Security Systems (UK) Ltd
System 4 - Simplex Signs Limited
System 45 - Hamilton R & Co Ltd
System 500 - Newton John & Co Ltd
System 9000 - Timloc Expamet Building Products
System Access - Saint-Gobain Ecophon Ltd
System J - Nullifire Ltd
System one - Twyford Bushboard
System S - Nullifire Ltd
System W - Nullifire Ltd
System2000 range - Helmsman
Systemate - Gledhill Water Storage Ltd
Systemform - Leaderflush + Shapland Ltd
Systemglas - Promat Fire Protection Ltd
SystemLogic - Range Cylinders
Systems 600 - Marley Waterproofing
Syston doors - Syston Rolling Shutters Ltd
Syvac - Torit DCE
T Grade Seal - IFT Supplies Ltd
T line counters - Platonoff & Harris Plc
T Vent - Areco Roofing Supplies Company
T.D.I. Firestop - TDI (UK) Ltd
T.I.P.S. - Hellermann Tyton
T14 storage wall - Gifford Grant Ltd
T2 - Flexiform Business Furniture Ltd
T2000 - Reznor UK Ltd
T5 Lighting - Encapsulite International Ltd
Table Portfolio - Envoplan
Tablo - Dermide Ltd
Tacdeck 508 - Haironville TAC Ltd
tactile - H & R Johnson Tiles Ltd
Tactray 90 - Haironville TAC Ltd
Tactual - Stocksigns Ltd
TAF - Pearl Paints Ltd
Tahiti - Qualitas Bathrooms
Takpave - Rediweld Rubber & Plastics Ltd
Talia - Orsogril Spa UK Distributor
Talisman - Tilley International Ltd
Taller Uno - Optelma Lighting Ltd
Tanalith - Arch Timber Protection
Tanalith Extra - Arch Timber Protection
Tanatone - Arch Timber Protection
Tanda - Drawn Metal Ltd
Tangye Pumps - Satec Ltd
Tanking 134 - Kiltox Damp Free Solutions
Tantofex Bathroom taps and mixers - Ideal-Standard Ltd
Tantofex Kitchen taps and mixers - Ideal-Standard Ltd
Tapes4 Builders - Advance Tapes International Ltd
Tapetex - Architectural Textiles Ltd
Tapetex - Architectural Textiles Ltd
Tappex - Uponor Housing Solutions
Taptite - Nettlefolds Ltd
Tapworks - Ecowater Systems Ltd
Tapworks - Tapworks Water Softeners
Taranis - Girsberger London
Targha - BPT Security Systems (UK) Ltd
Tarkett - Quiligotti Access Flooring Limited
Tarkett - Tarkett Sommer Ltd
Tarkett Sommer - Tarkett Sommer Ltd
TAS Attenuator - The Salex Group Ltd.
Taskmaster - Doors & Hardware Ltd
Taskmaster S series - Doors & Hardware Ltd
Taskmaster Select - Lumitron Ltd
Taskmaster T series - Doors & Hardware Ltd
Taskworthy - Taskworthy Ltd
Tasso - Property Mechanical Products Ltd
Tate Automation - Tate Fencing
Tate Floors - Hatmet Ltd
Tayler tools - Hyde Brian Ltd
T-Box-H-Matic - Healthmatic Ltd
TBS Baffle - Acousticabs Industrial Noise Control Ltd
TBS Panels - Acousticabs Industrial Noise Control Ltd
TC 16 - CSW Coldform Ltd
TC 20 - CSW Coldform Ltd
TC 25 - CSW Coldform Ltd
TD - Torit DCE
TD Board - Cafco International
TD Board - Caird Environmental Ltd, Industrial Services Division
TDI 0.35 Loft Match - TDI (UK) Ltd
TDI Cavity stop - TDI (UK) Ltd
Teacast - Haddonstone Ltd
Tealite - Haddonstone Ltd
Teamac - Teal & Mackrill Ltd
TEBRAX - Tebrax Ltd
Tecbond - Kenyon Performance Adhesives
Tecfix - CCL. Veloduct Ltd
Tecflex / Tectherm - CCL. Veloduct Ltd
Techdek - Norton Engineering Alloys Co Ltd
Technal - Cantifix of London Limited
Technal - Cantifix of London Limited
Technics - Interfloor Limited
Techniflam - Eclipse Blind Systems Ltd
Techno - Concord: Marlin
Technomelt - Henkel Industrial Adhesives
Technostone Ltd - Technostone Ltd
TechSpan® - Freyssinet Ltd
Techtonic rooflight blinds - Hallmark Blinds Ltd
Techtron® HP V PPS PPSU PEI PSU PVDF - DSM Engineering Plastic Products (UK) Ltd
Tecroc - Drainage Systems
Tecseal - CCL. Veloduct Ltd
Tecsom - Tarkett Sommer Ltd
Tecstone - Haddonstone Ltd
Tecstore - CST Vulcan
Tecstrut - CCL. Veloduct Ltd
Tectape - CCL. Veloduct Ltd
Tectite - CCL. Veloduct Ltd
Tectrix Fins - Hallmark Blinds Ltd
Tedlar - Tektura Plc
Tee Gee - Swish Building Products
Teen Scene - Record Playground Equipment Ltd
Teetector Golf - Evergreens Uk
Tego Security Products - McNeill-McManus Ltd
Tegola - Matthew Hebden
Tegometall - RB UK Ltd
Tegometall - Smart F & G (Shopfittings)Ltd
Tegula - Marshalls Mono Ltd
Tehalit - Hager Ltd
Tekfix - Tektura Plc
Tektalan - Skanda
Tektaroot - Watkins
Telabon ACM - Technology Telford Ltd
Telebloc - Jones of Oswestry Ltd
Tele-Call - bedside TV/Telephone system - Wandsworth Elecrtrical Ltd
Telecom Security Ltd - ADT Modern Security Systems Ltd
Teleflex - Dyer Environmental Controls
Telemecanique - Schneider Electric Ltd
Telford Flip-Chart - Magiboards Ltd
Tema - Optelma Lighting Ltd
Tem-Energy - May William (Ashton) Ltd
Tempcheck deck - Celotex Ltd
Tempest - Rada Lighting
Tempo - Jaga Heating Poducts (UK) Ltd
Tempower - Blakley Electrics Ltd
Tempus - Zon International
Ten Brazilian Granites - Kirkstone
Ten Plus - Kompan Ltd
Tenby - Legrand Electric Ltd
Tenon - HEM Interiors Group Ltd
Tensabarrier - Tensator Ltd
Tensaguide - Tensator Ltd
Tensar - Tensar International Group LTD
Terminex - Peter Cox Ltd
Terodem - Henkel Industrial Adhesives
Terokal - Henkel Industrial Adhesives
Terophon - Henkel Industrial Adhesives
Teroson - Beton Construction Ltd
Teroson - Henkel Industrial Adhesives
Terostat - Henkel Industrial Adhesives
TerraClass™ - Freyssinet Ltd
Terraforce - Keller Ground Engineering
Terraglass - Diespeker Marble & Terrazzo Ltd
Terrain - Capper Plastics
Terrain - Drainage Systems
Terraneo - Legrand Electric Ltd
Terranova - Pilkington's Tiles Ltd
Terrapin - Terrapin Ltd
Terraplegel - Link Build Products Ltd
Terratone - Andersen/ Black Millwork
TerraTrel™ - Freyssinet Ltd
Terrex - ABG Ltd
Terrier - Pegler Hattersley
Terrings - Compriband Ltd
Terry's - The Kenco Coffee Company
Tesa - Beiersdorf (UK) Ltd
Tesa - Beiersdorf (UK) Ltd
Tesi - Tubular Steel Column Radiators - Clyde Combustions Ltd
Teviot - Gaskell Carpet Tiles
Tex - Ves Andover Ltd
Texitone - Lafaroe Redland Precast
Texsol - Trade Sealants
Texspra - En-tout-cas plc (Scotland)
Textomur - Keller Ground Engineering
Textura - Gradus Ltd
Texture - Pilkington Birmingham
Texwood - McLoughlin Wood Ltd
Thames - Portacel
Thatchbatt - Thatching Advisory Services Ltd
The Carpet Foundation - The Carpet Foundation
The Carpet Foundation Quality Mark - The Carpet Foundation
The CAST Rooflight - Clement Steel Windows
The contry range - Cheshire Concrete Products
The Cube Collection - Cube Collection Ltd
The Design Archives - CrowsonFrabrics Ltd
The Diamond System - Chelsea Artisans Ltd
The Gallery - Symphony Group PLC, The
The Invisible Lightswitch - Forbes & Lomax Ltd
The rapid range - Lafarge Roofing Ltd
The Rutland Press - Royal Incorporation of Architects in Scotland
The White Group - ADT Modern Security Systems Ltd
The window centre - BLW Associates Ltd
Theben - Timeguard Ltd
Themerend - Thermica Ltd
Therma - Kingspan Industrial Ins. Ltd
Thermac - McKenzie-Martin Ltd
Thermaclip - Flexible Ducting Ltd
Thermacool - Reznor UK Ltd
Thermadoor - Stokvis R S & Sons Ltd
Thermafibre - Davant Products Ltd
Thermafit - Haze Cladding Systems
Thermaflex - Flexible Ducting Ltd
Thermafloor - Kingspan Insulation Ltd
Thermafold Insulated Folding Door - Bolton Gate Co. Ltd
Thermaglide - Cardale Doors Ltd
Thermal Mitx - Oswestry Reinforced Plastics Ltd
Thermal Safe - Kingspan Ltd
Thermalath® - BRC Special Products
Thermaline - Harton Heating Appliances
Thermaliner - Kingspan Insulation Ltd
Thermalite - Marley Building Materials Ltd
Thermalsave Heat Re-circulators - Turnbull & Scott (Engineers) Ltd
Thermalux - Hill Leigh Ltd
Thermapipe 100, 125, 150 - Domus Ventilation Ltd.
Thermapitch - Kingspan Insulation Ltd
Thermaroof - Kingspan Insulation Ltd
Thermashield - Glixtone Ltd
Thermataper - Kingspan Insulation Ltd
Thermatic - Martin Orgee Ltd
Thermawall - Kingspan Insulation Ltd
Thermaweld - Anderson D & Son
Thermax - Birdsall Services Ltd
Thermgard - Lonsdale Metal Company
Therminox - Keddy-Poujoulat (UK) Ltd
Thermoclear - Kenilworth Daylight Centre Ltd
Thermocool - Thermoscreens Ltd
Thermocor C.C. - TDI (UK) Ltd
Thermoflue - Marflex Chimney Systems
Thermoframe - HIS Ltd
Thermoheat - Tyeo Thermal Control
Thermolier unit heaters - Turnbull & Scott (Engineers) Ltd
Thermonda - Rockwell Sheet Sales Ltd
Thermorex - Heat Detector - Siemens Building Technologies Ltd,
Thermoscreens - Thermoscreens Ltd
Thermoson - Axter Ltd
Thermotapes - Flexelec(UK) Ltd
Thermotone air curtains - shearflow LTD
Theta Duo - Brett Paving
Theta® - Brett Paving
Thin Joint Brickforce® - BRC Special Products
Thinpave - Camas Aggregates Ltd - Aggregate Industries
Thioflex - Fosroc Ltd
Thirkleby - York Handmade Brick Co Ltd
Thistle - British Gypsum Ltd
Tholoflow - Fullflow Group Ltd
Thomas Dudley - Hipkin F W Ltd
Thomas Lowe Joinery - Thomas Lowe Joinery
Thor - Thor Hammer Co Ltd
Thorace - Thor Hammer Co Ltd
Thorex - Thor Hammer Co Ltd
Thorlite - Thor Hammer Co Ltd
Thoroseal - Thoro Systems Products Ltd
Thortex - Wood E Ltd
Three 3G Vulcavent - Vulcanite Limited
Three 3K - Boulter Boilers Ltd
Three 3M Scotchtint - Durable Berkeley Co Ltd
Three M 3M Scotchshield - Durable Berkeley Co Ltd
ThreeC 3C Tricontrol - Martin Orgee Ltd
ThreeM - Clean Step Ltd
Threshex - Exitex Ltd
thunderbolt - Unifix Ltd
Thwaites - Chippindale Plant Ltd
Tiara - Hometex Trading Ltd
Tibmix - Don Construction Products Ltd
TICA - Thermal Insulation Contractors Association
Tico - Tiflex Ltd
Tidy Bin - Unicorn Containers Ltd
Tie-Sleeve - Redifix Ltd
Tiger Diamond Blades - Unicorn Abrasives (UK) Ltd
Tikkurila - Andrews Coatings Ltd
Tile Association - The Tile Association
Tile-A-Door - Howe-Green Ltd
Tilefast - P.C.I Construction Systems Ltd
Tileflex - Compriband Ltd
Tilene - Visqueen Building Products
Tilevent PT - Cembrit Blunn Ltd
Tilley Lamp-Kerosene - Tilley International Ltd
Tillwite - Till & Whitehead Ltd, t/a L. Birkinshaw
Tilt Wash - Andersen/ Black Millwork
Timbacrib - Keller Ground Engineering
Timbasol™ - Johnson Matthey Pigments & Dispersions
Timb-a-Tilt - Caldwell Hardware (UK) Ltd
Timber care - Cuprinol Ltd
Timbercare - Manders Paint Ltd
Timberflex - P.C.I Construction Systems Ltd
Timberland - Briggs Industrial Footwear Ltd
timberline - DLW Flooring
TimberLine - Premdor
Timbersound noise barrier - Chris Wheeler Construction Limited
Timbertie - Redifix Ltd
Timbertone - Protim Solignum Ltd
Timberwrap - Monarflex Anderson Ltd
Timelox - ASSA Ltd
Time-Piece - Arena Software Ltd
Timesaver - Drainage Systems
Timesaver - Saint-Gobain Pipelines PLC
Timloc - Allmat (East Surrey) Ltd
Timonox - Crown Trade Paints
Tipform - Cordek Ltd
Tisca - Hometex Trading Ltd
Titan - Fabrikat (Nottingham) Ltd
Titan - Ingersoll-Rand European Sales Ltd
Titan - Osram Ltd
Titan - PHI Group Ltd
Titan - Titan Ladders Ltd
Titan Environmental - Titan plastech
Titan Garden Buildings - Phillip Ader Ltd
Titazel - The Safety Tread Ltd
TJM - Brooks Belfast
TLC Control - Telelarm care Ltd
TMC Fine Texture - Andura Textured Masonry Coatings Ltd
Toby Dustbins - Iles Waste Systems
Toesavers - Briggs Industrial Footwear Ltd
Tokstrip - Winn & Coales (Denso) Ltd
Toledo - Townscape Products Ltd
Tonewood - Henry Venables Interiors Ltd
Tonewood - Henry Venables Ltd
Tonon Fan Coil Units - Therminal Technology (sales) Ltd
Tony Team - Tony Team Ltd
Top office - OEP Furniture Group PLC
Top Plan - Top Office Equipment Ltd
Top Star Rooflights - Klober Ltd
Top Tech - Top Office Equipment Ltd
Top Track - Lab Furnishings (UK) Ltd
Top Vent - Klober Ltd
Topcrete - Tarmac Topblock Ltd
Topdek - Ward Building Components Ltd
Topflood - BLV Licht-und Vakuumtecnik
Topfoam - Tarmac Northern Ltd
TopForm - SFS Intec LTD
Toplab Plus - Trespa UK Ltd
Toplite - Tarmac Topblock Ltd
Topmesh - Barton Storage Systems Ltd
Topmix - Tarmac Northern Ltd
Topper Range - Lumitron Ltd
Toprax - Barton Storage Systems Ltd
Toproc - Tarmac Northern Ltd
Topshelf - Barton Storage Systems Ltd
Topspot - BLV Licht-und Vakuumtecnik
Topstore - Barton Storage Systems Ltd
Topstore - Barton Storage Systems Ltd
Topsy - Glasdon UK Ltd
Topsy Jubilee (bins) - Glasdon UK Ltd
Toptek - Barlow Group
Torbeck - Opella Ltd
Torbeck Variflush - Opella Ltd
Torclad / Tordeck / Torlife / Torrex / Torshield / Tortread - Tor Coatings Ltd
Torcure MC - Tor Coatings Ltd
Toreboda - Moelven Laminated Timber Structures Ltd
Torit - Torit DCE
Torlife WB - Tor Coatings Ltd
Torlite - Torvale Building Products, A Division of Stadium Group PLC
Torlon® - DSM Engineering Plastic Products (UK) Ltd
Tornado - Arjo Ltd
Toro - Hille Ltd
Torprufe CRC / Torprufe EMF - Tor Coatings Ltd
Torrent - Gledhill Water Storage Ltd
Torso - Caldwell Hardware (UK) Ltd
Toscana - Denco Ltd
Toscana - Marshalls Mono Ltd
Toscana - Paragon by Heckmondwike
Toshiba - HV Air Conditioning Ltd

Toshiba - Purified Air Ltd
Toshiba air conditioning - Toshiba Carrier UK
Tough-Cote™ High Build Texture - Glixtone Ltd
Toughliner - Lotus Water Garden Products Ltd
Toughness - BRB Industrial Services Ltd
Toughness - BRB Industrial Services Ltd
Tower - Pifco Ltd
Towerfill - Tower Manufacturing
Towerfire - Tower Manufacturing
Towerfoam - Tower Manufacturing
Towergrip - Tower Manufacturing
Towerpak - Tower Manufacturing
Towerseal - Tower Manufacturing
TPC LN/B - Babcock Wanson Uk Ltd
Tplus - Monarflex Anderson Ltd
T-Pren - Matthew Hebden
TQ - Owlett - Jaton
Trackelast - Tiflex Ltd
Trackmaster - Rawson Carpets Ltd
Trackranger Crawler Primary Crushing Outfits - Parker Plant Ltd
Trademark - Marcrist International Ltd
Tradesman - Ibstock Brick Ltd
Tradesman - Turner Access Ltd
Traditional - Northcot Brick Ltd
Traditional Marble Granite & Limestone - Chelsea Artisans Ltd
Traffic - Kompan Ltd
Traffic 2 - Woolliscroft Tiles Ltd
Traficop - Rediweld Rubber & Plastics Ltd
Trakway - Eve Trakway
Trammel fin drain - Don & Low Ltd
Trane - HV Air Conditioning Ltd
Transaction - Rackline Systems Storage Ltd
Transcab - AEI Cables Limited
Transfastener™ - Latchways plc
Translite - Harkness Hall Ltd
Transmit - Tunstal Telecom Ltd
Trans-Oxide® - Johnson Matthey Pigments & Dispersions
Transplex - EDS Shutter Door Systems Ltd
Transpower - Blakley Electrics Ltd
Transtor - Desking Systems Ltd
Transvario - Kone Escalators Ltd
Trashmaster - Pump Technical Services Ltd
Travel master hoists - Wessex Medical
Travertin - Skanda
Trav-O-lators ® - Otis Ltd
Treacle Taps - Millar Dennis (1998) Ltd
Tread Safe - Anglo Building Products Ltd
Treadmaster - Tiflex Ltd
Treads & Risers - Roman Mosaic Contracts Ltd
Treadspire - Consculpt Spiral Stairs Ltd
Trebitt - Jotun Henry Clark Ltd Decorative Division
Trebitt Opaque - Jotun Henry Clark Ltd Decorative Division
Tredaire - Interfloor Limited
Trellidor - Telford Doors and Barriers Ltd
Trench Heating - Turnbull & Scott (Engineers) Ltd
Trent - Woolliscroft Tiles Ltd
Trent Bathrooms - Shires Bathrooms
trent gutter system - Dales Fabrications Ltd
Trentobond - Tremco Ltd
Trentocrete - Tremco Ltd
Trentodel - Tremco Ltd
Trentoflex - Tremco Ltd
Trentoshield - Tremco Ltd
Trentothane - Tremco Ltd
Trespa - Amari Plastics Plc
Trespa - Decra Ltd
Trespa - Trespa UK Ltd
Tretol - Viking Laminates Ltd
Trevi Showers - Armitage Shanks Ltd
Trevi showers - Ideal-Standard Ltd
Trewor - Broxwood (scotland Ltd)
Triaqua - Multikwik Ltd
Tribol Limited - MMP International
Tribond - Vulcanite Limited
Tribond"E" - Vulcanite Limited
Tricity - Electrolux Domestic Appliances
Tricity - Parkinson Cowan and Tricity Bendix Co, The
Triclamp - Opto International Ltd
Tricosal - Beton Construction Ltd
Tricosal - Beton Construction Ltd
Tricosal - Beton Construction Ltd
Trident Water Gardens - Baresford Pumps Ltd
Tridonic - Bernlite Ltd
Tridonic - Bernlite Ltd
Trief Kerb - Lafaroe Redland Precast
Triflow - IBP Ltd
Triglide - Brickhouse
Tri-Guard ® - Jackson H S & Son (Fencing) Ltd
Trilax - Girsberger London
Triline - Smart Systems Ltd
Tri-lock - Grass Concrete Ltd
Trim Acoustic Cradle System - Trim Acoustics
Trim Defender Door Bars - Trim Acoustics
Trim Defender Wall - Trim Acoustics
Trimalac - Trimite Ltd
Trimapanel - European Profiles
Trimawall - European Profiles
Trimguard - Manton Industrial Seals Ltd
Trimite - Andrews Coatings Ltd
Trimite - Trimite Ltd
Trimloc Interchange directories - Drakard & Humble Ltd
Trimpak - Whitecroft Lighting Limited
Trimtop - Swish Building Products
Trimwall - Neslo Partitioning & Interiors
Trio - Panel & Louvre Co Ltd
TRIO - Panel wall formwork - PERI Ltd
Triple Seven - Dunbrik (Yorks) Ltd
Tri-Plugs - Tower Manufacturing
Tripower - Blakley Electrics Ltd
TripTrap Oils - McLoughlin Wood Ltd
Tri-Shell - Dudley Thomas Ltd
Trisomet - European Profiles
Tritainer - Franklin Hodge Industries Ltd
Triton - Albany Standard Pumps
Triton - Albany Standard Pumps
Triton - Triton PLC
Tritorch - Vulcanite Limited
Trixene - Baxenden Chemicals Ltd
Trizone - Airsprung Beds Ltd
Troikafix - Troika Architectural Mouldings Ltd
Trojan - Bitmen Products Ltd
Trojan - Crane Fluid Systems
Trojan - Hallmark Blinds Ltd
Trojan - Neptune Showers Ltd
Trojan - Piper Double Glazing Ltd
Trojan - Shackerley (Holdings) Group Ltd incorporating Designer ceramics
Tropical Fish - Hayward Air
Tropicano - Xpelair Ltd
Trubolt - Ramset Fasteners Ltd
Trueflue - Marflex Chimney Systems
True-lite - Full Spectrum Lighting Ltd
Trufrule - Rabone Chesterman Ltd
Truline - Simpson Strong-Tie®
Trulok - Armstrong World Industries Ltd
Trumeter - Marking Machines International
Truss Tray - European Profiles
Truth - Caldwell Hardware (UK) Ltd
TS60P - Ramset Fasteners Ltd
TS750P - Ramset Fasteners Ltd
TSV1 surface mounted shower pannel - Horne Engineering Ltd
TT - Optelma Lighting Ltd
T-T controls - T-T Pumps & T-T controls
TTA - The Tile Association
TTF - Simon R W Ltd
TTF - Timber Trade Federation
TUBA-RAD - Property Mechanical Products Ltd
Tub-box - Creteco Sales Ltd
Tubeclamps - Tubeclamps Ltd
Tubela - Tubela Engineering Co Ltd
Tubelite - LB Lighting Ltd
Tube-U-Fit Hanging Rail & Supports - Daro Factors Ltd
Tubinox - Keddy-Poujoulat (UK) Ltd
Tucker - Arrow Supply Company Ltd
Tudor Roof Tiles - Tudor Roof Tile Co Ltd
Tuf Light (Roof Sheets) - Wilsons Fibreglass Ltd
Tuf Light roof sheets - Alpha Glass Fibres Ltd
Tufanega - Deb Ltd
Tufcon 80 - Conren Ltd
Tuff Stuff - Twyford Bushboard
Tuffblocks - RFA Group Ltd
Tuffgrip - Hanson Aggregates
Tuffgrip - Hanson Building Materials Europe
Tuffmate - Titan plastech
Tuff-R - Celotex Ltd
Tuffwall - Brown F plc
Tuflok - Rabone Chesterman Ltd
Tuftex - Vita Liquid Polymers Ltd
Tuftiguard - Bonar Floors Ltd
Tuhsu - Elizabethan Textiles
Tully range - Tully
Tunnel Mortar - Vandex (UK) Ltd
TUO - Optelma Lighting Ltd
Tuplex - McLoughlin Wood Ltd
Turbo - Roof Units Group
Turbo cast 300 (Grit Spreader) - Glasdon UK Ltd
TurboFast - Helifix Ltd
Turbo-Flo - MSA (Britain) Ltd
Turbolite - Marcrist International Ltd
TurboSyphons - Dudley Thomas Ltd
Turkish room - Dalesauna Ltd
Turmix - Air Improvement Centre Ltd
Turnerised - TRC Service Group Ltd
Turnquest - Turnquest UK Ltd
Tusc - Unistrut Limited
Tuscany - Muraspec
TVSS - Eaton Electric Limited
Tweeny - Engineering & Lift Ltd
Tweeny - The Haigh Tweeny Co Ltd
Twenty-one Limestones - Kirkstone
Twiclad - Ide T & W Ltd
Twilfix - Betafence Limited
Twill - Advanced Fencing Systems Ltd
Twill Wire products - Challenge Fencing Ltd
Twilweld - United Wire Ltd
Twin & Big Twin - Helmsman
Twin Traps - Midland Stom Ltd
Twinaplate - Alpine Automation Ltd.
Twinaplate - Alpine Automation Ltd.
Twinarc - Two Arc Tube High Pressure Sodium Lamps - Sylvania Lighting International INC.
Twinbreak - Square D, Electrical Distribution & Building applicators Division
Twinfix - Amari Plastics Plc
Twinfold Fine Twist - Brockway Carpets Ltd
Twinimum - Ves Andover Ltd
Twinlok Self Sealing Test Plugs - Industrial Hangers Ltd
Twinlux - Illuma Lighting Ltd
Twinqwik - Owlett - Jaton
Twisco - Manitou (Site Lift) Ltd
Twist - Kermi (UK) Ltd
Twizzler - Record Playground Equipment Ltd
Twosome - Integra Products
TW-Tec Metal cubicles - Detlef Muller Ltd
Twyfords - Caradon Bathrooms Ltd
Twyfords - Caradon Bathrooms Ltd
Tycho - Kronospan Ltd
Tyfords - Barflow-LCA
Tyfords - Barflow-LCA
Tyglas - Fothergill Engineered Fabrics Ltd
Tylo - Golden Coast Ltd
Type E - Cavity Trays Ltd
Type X - Cavity Trays Ltd
Typhoo - The Kenco Coffee Company
Tyrespan - Metalrax Ltd
Tyton - Hellermann Tyton
Tyveks - Klober Ltd
Tyzack - Spear & Jackson
U.S.G Donn - New Forest Ceilings Ltd
U+ - Uponor Ltd
U45 Insulated Sectional Doors - Avon Industrial Doors Ltd
UbiSoil - Ubbink (UK) Ltd
Ubivent - Ubbink (UK) Ltd
Ucrete - Degussa Construction Chemicals (UK)
Ucrete - Thoro Systems Products Ltd
UL-M - Johnston Roadstone Ltd
Ulster Velvet - Ulster Carpet Mills Ltd
Ultima - Airedale International Air Conditioning Ltd
Ultima - Armstrong World Industries Ltd
Ultima - Laird Security Hardware
Ultima - Marley Waterproofing
Ultima - Midland Stom Ltd
Ultima - Polypipe Bathroom & Kitchen Products Ltd
Ultima - For the less abled. - Rixonway Kitchens Ltd
Ultimate - Owlett - Jaton
Ultimate 99 - Sportmark™ Ltd
Ultimate Collection - Ball William Ltd
Ultimate Collection - Ball William Ltd
Ultimate flue outlet - Brewer Metalcraft
Ultimate Range - Valor
Ultra - Eaton Electric Limited
Ultra - Monarflex Anderson Ltd
Ultra 80 - Norwood Partition Systems Ltd
Ultra Bathscreens by Coram - Coram Showers Ltd
Ultrabeam - Hadley Industries Plc
Ultracal - Range Cylinders
Ultracolor - Mapei (UK) Ltd
Ultra-Cool - Beta Lighting Ltd
Ultraflex - Dyer Environmental Controls
Ultraflex - P.C.I Construction Systems Ltd
Ultraframe - FTW Windows
Ultraframe - Ultraframe PLC
Ultrafresh - Elizabethan Textiles
Ultragard Software - Time and Data Systems International Ltd (TDSI)
Ultragard Vision-ID - Time and Data Systems International Ltd (TDSI)
Ultraglaze - GE Bayer Silicones
Ultragrain ® - Righton Ltd
Ultralife - BLV Licht-und Vakuumtecnik
Ultralift - Caldwell Hardware (UK) Ltd
Ultralift - Caldwell Hardware (UK) Ltd
Ultralite 500 - Ultraframe PLC
Ultralite Blocks - Hanson Concrete Products
Ultramastic - Mapei (UK) Ltd
Ultraplan - Mapei (UK) Ltd
Ultra-Rib - Brett Martin Ltd
Ultra-Rib - Uponor Ltd
Ultraspec reversible windows - Dale Joinery Ltd
Ultrasteel - Hadley Industries Plc
Ultrastor - Norwood Partition Systems Ltd
Ultrastrong - Boddingtons Ltd
Ultratile - Britmet Tileform Limited
Ultraturn - Righton Ltd
Ultrazed - Hadley Industries Plc
Uni- Ecoloc - CEMEX
Uni -Walten - DLW Flooring Ltd
Unibond - Henkel Consumer Adhesives
Unibond No More Nails - Henkel Consumer Adhesives
Unibond No More Sealent Guns - Henkel Consumer Adhesives
Unibond Unifilla - Henkel Consumer Adhesives
Unica - Hansgrohe
Unica - Novellini Bathroom Products Ltd
Unican - Abloy Security Ltd
Unicell - Ronacrete Ltd
Unicell - Torit DCE
Unidare - Glen Dimplex UK Ltd
Unidare - Seaque Technologies Ltd
Unidox - Witham Oil & Paint Ltd
Unieck - DLW Flooring
Unifact - Satchwell Control Systems Ltd
Unifit - NewTeam Ltd
Uni-Fix Hi-Load - Unifix Ltd
Uni-Fix Klip-Lok - Unifix Ltd
Unifloor Flooring System - Hanson Concrete Products
Uniframe - Hanson Concrete Products
Unigas - GP Burners (CIB) Ltd
Uni-Junction - Johnston Pipes Ltd
Unikrame - Hanson Concrete Products
Unilight - Universal Components Ltd
Uniline - Rangemaster
Uniliner - BPB Paperboard
Unimaster - Torit DCE
Unimate - Quartet-GBC UK Ltd
Unimotor - Dyer Environmental Controls
Union - Parkes Josiah & Sons Ltd
Union Jack Label Fittings - Daro Factors Ltd
Uniperf - United Wire Ltd
Unipipe - Uponor Housing Solutions
Uniplant - Witham Oil & Paint Ltd
Unipleat - Eclipse Blind Systems Ltd
Uniply - Marley Waterproofing
Unistrut - Kem Edwards Ltd
Unistrut - Unistrut Limited
Unit Swimming Pools - Regency Swimming Pools Ltd
Unitas - Witham Oil & Paint Ltd
Uni-Trex - Terrapin Ltd
Univent smoke relief chimney - Smoke Control Services Ltd
Universal - EBS Services Ltd
Universal - Matki plc
Universal - Unicorn Abrasives (UK) Ltd
Universal Gulley Adaptor - Johnston Pipes Ltd
Universal louvre - Colt International
Universal Metricove - Pilkington's Tiles Ltd
Universal range of rainwater products - Marshall-Tufflex Ltd
Universal roofboard - Callenders Ltd
Uniwall - Hanson Concrete Products
Uniweave® - Ulster Carpet Mills Ltd
Uno - Becker (SLIDING PARTITIONS) Ltd
Upat - Artur Fischer (UK) Ltd
UPE - Talbot
Upline - Illuma Lighting Ltd
Upton Door - AS Newbould Ltd
Uptraply TPO - Firestone Building Products Europe
USP Intruder Guard - United Safety Products UK Ltd
Utrain - Uponor Ltd
UV Dermagard - Bonwyke Ltd
UV Free Lighting - Encapsulite International Ltd
V Trim GRP - Areco Roofing Supplies Company
V/ Slider - Hill Leigh Ltd
V100 PSM - Elster Metering Ltd
V200 - Elster Metering Ltd
V210 - Elster Metering Ltd
V3 - President Office Furniture Ltd
V3 - Reznor UK Ltd
V300 Master - Elster Metering Ltd
Vac Vac - Arch Timber Protection
Vacsol - Arch Timber Protection
Valiant - Laird Security Hardware
Vallance - Vallance
Valley - Iles Waste Systems
Valley Range - Dale Joinery Ltd
Valspar - Akzo Coatings Plc
Valtek - Flowserve Flow Control
Value Seating - Sandler Seating Ltd
Valves - Horstmann Controls Ltd
Van Besouw - E. H. S. Roofing Ltd
Van Erven - McLoughlin Wood Ltd
Vandal Resistant Entry Panels - BPT Security Systems (UK) Ltd
Vandalene ® - Coo-Var Ltd
Vandex BB75 - Vandex (UK) Ltd
Vandex BB75E - Vandex (UK) Ltd
Vandex Injection Mortar - Vandex (UK) Ltd
Vandex Plug - Vandex (UK) Ltd
Vandex Premix - Vandex (UK) Ltd
Vandex Super - Vandex (UK) Ltd
Vandex Unimortar - Vandex (UK) Ltd
Vandgard - Anti-Climb Guards Ltd
Vandgard - Anti-Climb Guards Ltd
Vanguard - Forticrete Roofing Products
Vanguard - Top Office Equipment Ltd
Vangvord - Bennett Windows & Doors
Vanity flair - Ellis J T & Co Ltd
Vanity Tops - Roman Mosaic Contracts Ltd
Vanodine - Evans Vanodine International PLC
VAPAC - Humidifiers - Eaton-Williams Group Ltd
Vapourfoil - Vulcanite Limited
Vapourguard - Trade Sealants
Varedplan - Edmonds A & Co Ltd
Vari-Ceat - Ellis Patents
Variflow - Grosvenor Pumps Ltd
Vari-Level - Wade International Ltd
Varilift - Arvin Motion Control Ltd
Vario - Kompan Ltd
VARIO - Wall formwork - PERI Ltd
Varioguard - Asset International Ltd
Variplan - Arena-Walsall Ltd
Vari-Purpose - FC Frost Ltd
Varispace - Harkness Hall Ltd
Varitone - IAC
Varley TopJet - Marking Machines International
Vasco - The Salex Group Ltd.
Vasura - Erlau AG
VC - Variable Counterbalance Excavators - R-B International Ltd
VCD - Air Diffusion Ltd
Vector - Securistyle Ltd
VECTOR - Vantrunk Engineering Ltd
Vector Excluder (ancillary security device) - Securistyle Ltd
Vector plus - Securistyle Ltd
Vedette - Helmsman
Vedette Specialized range - Helmsman
Vee - Project Office Furniture PLC
Vega - Guardall Ltd
Veha Radiators - Barlo Radiators Limited
Veha Radiators - Brikenden (Builders Merchants) Ltd
VEKA - ABB Control Ltd
Veka - FTW Windows
Velfac - Velfac Ltd
Velopa - Autopa Ltd
Veltitech underlay - Lafarge Roofing Ltd
Velux - VELUX Co Ltd, The
Vemo - RFA Group Ltd
Venetian - Symphony Group PLC, The
Venetian Silk - Dermide Ltd
Vent A Matic - Simon R W Ltd
Ventform - Cordek Ltd
VentLight - Domus Ventilation Ltd.
Ventlite - Ventlane Ltd
Ventmaster AG - Docherty New Vent Chimney Group
Ventraflow - Jones of Oswestry Ltd

Ventrolla - Durable
Ventrust - Callenders Ltd
Ventura - McKenzie-Martin Ltd
Venture Parry - Bernlite Ltd
Venturer - Record Playground Equipment Ltd
Venus - Novellini Bathroom Products Ltd
Verco - Verco Office Furniture Ltd
Verena - Deceuninck Ltd
Verges - Zon International
Vermeer - B-Trac Equipment Ltd
Vermeer - B-Trac Equipment Ltd
Vermiculite - Dupre Vermiculite
Vermiculux - Promat UK Ltd
Vermont - Deceuninck Ltd
Vernacare - Verna Ltd
Vernagene - Verna Ltd
Verona Range - Caradon Bathrooms Ltd
Veronastone Assimilated Limestone - Marble Flooring Spec. Ltd
Versailles - Firth Interface Europe Ltd
Versailles - Qualitas Bathrooms
Versaline - Lewden Electrical Industries
Versalite - Righton Ltd
Versalux - JSB Electrical PLC
Versa-nut insert® - Textron Fastening Systems Ltd
Versatemp - Clivet UK Ltd
Versiplan - Becker (SLIDING PARTITIONS) Ltd
Versitex - Camas Aggregates Ltd - Aggregate Industries
Vertalux - Rada Lighting
Verti-Frame - Hi-Store Ltd
vesda laser compact + vesda aspirating pipe & fiti - Vision Sytems (Europe) Ltd
vesda laser plus - Vision Sytems (Europe) Ltd
Vespel® - DSM Engineering Plastic Products (UK) Ltd
Vesselpak - Flamco UK Ltd
Vestos - Arch Timber Protection
VET Agent Assessment System - Computertel Ltd
Vetcare™ - Steel Line Ltd
Vetisol - Axter Ltd
Vex - Guardall Ltd
V-Gard - MSA (Britain) Ltd
Viajoint 'D' Expansion Ioint - Pitchmastic Pmb Ltd
Vibro - Keller Ground Engineering
Vicrtex - Tektura Plc
Victaulic - Midland Tube and Fabrications
Victaulic - Victaulic Systems
Victorglos - The Bollom Group
Victorian - Property Mechanical Products Ltd
Victoriana - AEL
Victory - Novellini Bathroom Products Ltd
Victory - PCT Group Sales Ltd
Vicucase - Promat Fire Protection Ltd
Vicuclad - Promat Fire Protection Ltd
Vicuclad - Promat UK Ltd
Vicutube - Promat Fire Protection Ltd
Vicutube - Promat UK Ltd
Video System 200 - BPT Security Systems (UK) Ltd
Videx - Crimeguard Systems Ltd
Viero Decorative Coatings - Textured European Finishes Ltd
Vieroquartz Masonry Paint - Textured European Finishes Ltd
Viewguard - Hadley Industries Plc
Viewpoint - Pryorsign
Vigilant - Auld Valves Ltd
Vigilante - Blick Communication Systems Ltd
Vigilon - Gent Limited
Viking - Macron Safety Systems (UK) Ltd
Viking - Roof Units Group
Viking - Viking Laundry Equipment Ltd
Viking - Viking Laundry Equipment Ltd
Villa - Kompan Ltd
Village doors - Kind, J B, Ltd
Village Stone - CEMEX
Vinadac - Dacrylate Paints Ltd
Vintage Collection - Masterframe Windows Ltd
Vinyllon - Dixon Turner Wallcoverings
Vinylock - HW Systems Ltd
Viper spike ® - Jackson H S & Son (Fencing) Ltd
Vipps - CEM Systems Ltd
Virage (handle) - Securistyle Ltd
Virgil - Interoven Ltd
Viridian - Tully
Virtuon - Trespa UK Ltd
Visage - Valor
Viscacid - Kenyon Group Ltd
Viscount - Forticrete Roofing Products
Viscount (handle) - Securistyle Ltd
Visedge - Howe-Green Ltd
Visibin - Barton Storage Systems Ltd
Vision (handle) - Securistyle Ltd
Visions Bathrooms - Shires Bathrooms
Visofold - Smart Systems Ltd
Visoglide - Smart Systems Ltd
Visoline - Smart Systems Ltd
Visolplast textured coatings - Textured European Finishes Ltd
Visorguard - CSI Security LTD
Visqueen - Anaplast (Greenock)
Visqueen - Anaplast (Greenock)
Visqueen - Visqueen Building Products
Vista - Dallas & Forrest Ltd
Vista - Fabrikat (Nottingham) Ltd
Vista - Stocksigns Ltd
Vista - Welconstruct Co Ltd
Vista-Fix - Vista Engineering Ltd
Vistalux - Ariel Plastics Ltd
Vista-Plas - Vista Engineering Ltd
Visual Builder - Opto International Ltd
Vitachem - Vita Liquid Polymers Ltd
Vitaplas - Vita Liquid Polymers Ltd
Viternus - Crosbie Coatings
Vitogres - Arnull Bernard J & Co Ltd
Vitopan Steel Clean - Kleen-Tex Industries Ltd
Vitra - Relcross Ltd
Vitra - Swedecor Ltd
Vitra - Vitra (UK) Ltd
Vitraclad - Panel Systems Ltd
Vitraglaze - Panel Systems Ltd
Vitral - EBS Services Ltd
Vitram - Evans Concrete Products Ltd
Vitrium - CST Vulcan
Vitros - Crosbie Coatings
Viva - Arnull Bernard J & Co Ltd
Vivak - Amari Plastics Plc
Vivant - Whitecroft Lighting Limited
Vivante 2 - Envoplan
Vixalit Lime-based paint - Textured European Finishes Ltd
V-Kool - Northgate Solar Controls
VLT ® - Danfoss Ltd
Vodafone - Vodafone Paging Ltd
Vogue - Dermide Ltd
Vogue - Novellini Bathroom Products Ltd
Voidak - Dunham Bush Ltd
Vokera - Diamond Merchants
Vokes - Vokes
Volacel - E+R Polymers Ltd
Voltan - Townscape Products Ltd
Volumeter - Gebhardt Kiloheat
Volvo Compact Equipment - Chippindale Plant Ltd
Von Duprin - Relcross Ltd
Vortax - Axial - Howden Buffalo
Vossloh Schwabe - Bernlite Ltd
VPX LN/RR - Babcock Wanson Uk Ltd
Vtech - Vantrunk Engineering Ltd
Vulcadome - Vulcan Plastics Ltd
Vulcaflex - Vulcanite Limited
Vulcalap - Vulcan Plastics Ltd
Vulcalon Spg - Vulcan Plastics Ltd
Vulcan - Ellis Patents
Vulcan - Tuke & Bell Ltd
Vulcaplast Spg - Vulcan Plastics Ltd
Vulcathene - Capper Plastics
Vulcatorch - Vulcanite Limited
Vulcatuf - Vulcan Plastics Ltd
Vylon Plus - Tarkett-Marley Floors Ltd
Vynarac - Hispack Limited
W Range - Biral Pumps
W Schneider - Schneider (UK)
W/A formwork plywood - Wood International Agency Ltd
W/A pine doors - Wood International Agency Ltd
W20 - Clement Steel Windows
W40 - Clement Steel Windows
Wacker Silicones - Notcutt Ltd
Wade - Drainage Systems
Wade - Wade International Ltd
Waldor - Howe-Green Ltd
wales and wales - Wales & Wales
Walk On Ceiling - Norwood Partition Systems Ltd
Walker Timber Frame - Walker Timber Ltd
Walksafe - McKenzie-Martin Ltd
Wall panels - Saint-Gobain Ecophon Ltd
Wall Shield - Springvale E P S
Wallflex - Lastite Building Products Ltd
Wallforce® - BRC Special Products
Wallis - Bitmen Products Ltd
Wallmaster - Hilladam Coburn Ltd
Wallmate - Dow Construction Products
Wallstor - Desking Systems Ltd
Walltalker - Tektura Plc
Walraven - Kem Edwards Ltd
Walsall - Arena-Walsall Ltd
Walsall - Arena-Walsall Ltd
Walther - Barloworld Vacuum Technology Plc
Walton - Forbo - Nairn Ltd
Walton Sheds - Challenge Fencing Ltd
Ward - E. H. S. Roofing Ltd
Warm Air Cabinet Heaters - Radiant Services Ltd
Warm Air Unit Heaters - Radiant Services Ltd
Warm Squeez - Springvale E P S
Warmsafe LST - Dunham Bush Ltd
Warmskark - Springvale E P S
Warmwell - Hamworthy Heating Ltd
Wascator - Electrolux Laundry Systems
Wash-Horse - Kleen-Tex Industries Ltd
Waste King - Berglen Group Ltd
Wastematic Bench Mounted - Max Appliances Ltd
Wastematic Free Standing - Max Appliances Ltd
Wastematic Sink Mounted - Max Appliances Ltd
Watchdog - Heat Trace Ltd
Watchkeepers - Rapaway Energy Ltd
Watchman - Hygood Ltd
Water Hawk - Cookson Electronics
Waterguard Showertrays by Coram - Coram Showers Ltd
Water-Horse - Kleen-Tex Industries Ltd
Waterless - Relcross Ltd
Watermatic - Harcostar Drums
WaterPoint - Airstream Products Ltd
Waterproof Surface Range - Lumitron Ltd
Watersaver - Sheardown Engineering Ltd
Waterseal - Sealocrete Ltd
Watertight - Laticrete UK
Watertimer - Sheardown Engineering Ltd
Watrous - FC Frost Ltd
Watts cliff - Realstone Ltd
Wauderer-Castle care-tech - Castle Care-Tech
Wave litter Bin - Furnitubes International Ltd
Wave Seat - Furnitubes International Ltd
Wavelay - Boddingtons Ltd
Waveley - Trent Bathrooms
Wavespan - Flexible Reinforcements Ltd
Wayland - Advanced Fencing Systems Ltd
Waylite 800 - JSB Electrical PLC
Ways with Doorways - Kern (Design Products) J T Ltd
Weather Master - Ves Andover Ltd
Weather Safe - Kronospan Ltd
Weather Ten - Exitex Ltd
Weatherbeaters - Owlett - Jaton
Weathercor/F - TDI (UK) Ltd
Weathershield - ICI Dulux Trade Paints
Weatherwhite - Cardale Doors Ltd
Weatherwood - Cardale Doors Ltd
Weavespread - Flexible Reinforcements Ltd
Weavetop - Flexible Reinforcements Ltd
Weavetop Sonshade fabric - Dallas & Forrest Ltd
Web wall - ABG Ltd
Webbseal - Alfas Industries Ltd
Weber Building solutions - Weber Building Solutions
Weber cem - Weber Building Solutions
Weber certite - Weber Building Solutions
Weber cote - Weber Building Solutions
Weber fix - Weber Building Solutions
Weber Joint - Weber Building Solutions
Weber mulsifix - Weber Building Solutions
Weber Pral - Weber Building Solutions
Weber rent - Weber Building Solutions
Weber Set - Weber Building Solutions
Weber tec - Weber Building Solutions
Weep - Airflow (Nicoll Ventilators) Ltd
Weger Air Handling Units - Therminal Technology (sales) Ltd
WEHA - Harbro Supplies Ltd
Weholite - Asset International Ltd
Weigh-tronix - Phoenix Scales Ltd
Weldmesh - Betafence Limited
Welex - Welconstruct Co Ltd
Wellington - Bassett & Findley Ltd
Welltec - Cadisch MDA Ltd Ltd
Welmade - Q.M.P.
Wenmore Rooflights - Wenmore Rooflights, A Division of W.F.Joy & Co. Ltd
Wensleydale - Carpets of Worth Ltd
Wernick - Wernick S & Sons Ltd
Wescol Glosford - Wescol Glosford
Wessex - Hamworthy Heating Ltd
Wessex - Healthmatic Ltd
Wessex - Multitex GRP & GRG Products
Wessex Doors - Regency Garage Door Services Ltd
West System - Wessex Resins & Adhesives Ltd
West System Epoxy - Wessex Resins & Adhesives Ltd
Westbond - Hometex Trading Ltd
Westbourne Acoustic - Durabella System Floors Ltd
Westbrick - Ibstock Brick Ltd
Westbury Canopy - James Smellie Fabrications Ltd
Westclox - General Time Europe
Westdale - Chiltern Invadex
Westerland - Cembrit Blunn Ltd
Western Cork Ltd - Westco Group Ltd
Westman Systems - MJ Electronics Services (International) Ltd
Westminster - Atlas Stone Products
Westminster - Branson Leisure
Westminster - Branson Leisure
Westminster - Sheardown Engineering Ltd
Weston - Gloster Furniture Ltd
Westwood Ground - Bath Stone - Hanson Bath & Portland Stone
Weyroc - Egger (UK) Ltd
Weyroc HDX - Egger (UK) Ltd
Weyroc Project - Egger (UK) Ltd
Weyroc V313 - Egger (UK) Ltd
WF - Westbury Filters Limited
WF1 - Westbury Filters Limited
WFB - Westbury Filters Limited
WFC - Westbury Filters Limited
WFF - Westbury Filters Limited
WFG - Westbury Filters Limited
WFH HEPA - Westbury Filters Limited
WFK - Westbury Filters Limited
WFV - Westbury Filters Limited
WH Collier Ltd - Chelwood Brick Ltd
WhatRisk.com - National Britannia Ltd
Whirl-A-Waste - Sissons W & G Ltd
Whirline - Hipkin F W Ltd
Whirlpool & Spa Bathsystems - Jendico
Whirlwind - Envirodoor Markus Ltd
Whispair - Xpelair Ltd
White Book - British Gypsum Ltd
Whitehill - Whitehill Spindle Tools Ltd
Whiterock Doors - Altro Floors
Whiterock PVC Ceiling system - Altro Floors
Whiterock PVC Sport Cladding - Altro Floors
Whitestar - BLV Licht-und Vakuumtecnik
Whiting & Davis safety - BRB Industrial Services Ltd
Whizard - BRB Industrial Services Ltd
WHS - Spear & Jackson
Wicanders Cork-o-Floor - Amorim (UK) Ltd
Wides Lab - Tarmac Top Floor
Widespan - Metalrax Ltd
Wilclo - Clow Group Ltd
Wildfire - Marshall J A & Co (Brushes) Ltd
Wildgoose - Derby Timber Supplies
Wilka Locks & Cylinders - JB Architectural Inronmongery Ltd
Wilkinson's Furniture - Carleton Furniture Group Ltd
Wilo SE - Wilo Samson Pumps Ltd
Wilo Top E - Wilo Samson Pumps Ltd
Wilo Top S - Wilo Samson Pumps Ltd
Wilo Top SD - Wilo Samson Pumps Ltd
Wilson Art - Montague L Meyer
Wilsonart - Anderson C F & Son Ltd
Wilsons Fibreglass ltd - Alpha Glass Fibres Ltd
Wilton - Tankard
Wilton Royal - Carpets International (UK) Ltd
Wimbledon - Gloster Furniture Ltd
Wimborne - Hamworthy Heating Ltd
Winchester - BSA Machine Tools Ltd
Windcatcher - Sunpipe @ Monodraught Ltd
Window Care Systems - Window Care Systems Limited
Window Files - Railex Systems Ltd
Windowgard - Kaba Door Systems Ltd
Windowmaster Screws - UK Fasteners Ltd
Windsor - DW Windsor Lighting
Windsor - Guardall Ltd
Windsor - Cast Iron Column Radiators - Clyde Combustions Ltd
Windsor handles - Fab & Fix Ltd
Windsor Water Fittings - Deva Tap Co Ltd
Wings - Bisley Office Equipment
Winther Browne & Co Ltd - Winther Browne & co Ltd
Wintun - Caradon Duraflex Systems Ltd
Wiremaster - Philips Lighting Solutions
WirralRange - AS Newbould Ltd
WirralRange Doors - AS Newbould Ltd
Wisa-Deck - Schauman (UK) Ltd
Wispa Hoist - Chiltern Invadex
Wizard - T & D Packaging Ltd
WMA - Wallcovering Manufacturers Association
WMS - Avocet Hardware Ltd
WMS - Avocet Hardware Ltd
Woco - Witham Oil & Paint Ltd
Wolfin - Beton Construction Ltd
Wonderex - Optex (Europe) Ltd
Wondertrack - Optex (Europe) Ltd
Woodacoustic - Henry Venables Ltd
Woodcelip - Torvale Building Products, A Division of Stadium Group PLC
Woodcemair - Torvale Building Products, A Division of Stadium Group PLC
Woodcemax - Torvale Building Products, A Division of Stadium Group PLC
Woodcraft - Dallas & Forrest Ltd
Woodcraft - Design Line
Wooden Lockers - Helmsman
Woodflex - Polycell Products Ltd
Woodgrip - Pedley Furniture International Ltd
Woodland - Neptune Outdoor Furniture Ltd
Woodland System - Hyperion Wall Furniture Ltd
Woodman Twist - Brockway Carpets Ltd
Wood-n-Play - Record Playground Equipment Ltd
Wood-o-Cork - Amorim (UK) Ltd
Wood-o-Floor - Amorim (UK) Ltd
Woodro - Hille Ltd
Woodscape - Woodscape Ltd
Woodstock Accent - Adam Carpets Ltd
Woodstock Classic - Adam Carpets Ltd
Woodstock Classic - Adam Carpets Ltd
Woodstock Studio - Adam Carpets Ltd
Woodstyle - Hallmark Blinds Ltd
Woodworks - Interfloor Limited
Woodwright - Andersen/ Black Millwork
Worcester - The BSS Group PLC
Working Walls - Muraspec
Workspace - Paragon by Heckmondwike
World hand dryers - Warner Howard Group Ltd
Worldspan - Clow Group Ltd
Worldspan - Clow Group Ltd
WP Wallpack - Consult Lighting
Write Easy Chalkboards - Signs International
WSG Donn Floors - Hatmet Ltd
WSM - Welconstruct Co Ltd
WTIF - Wall Tie Installers Federation
Wulch - Rawtex Ltd
www.electrolux.com/laundrysystems - Electrolux Laundry Systems
www.frostlandscapes.co.uk - Frosts Landscape Construction Ltd
www.sales@firebarrier.co.uk - Firebarrier Services Ltd
Wyrem - Flexible Ducting Ltd
Wyvern Fireplaces - Wyvern Marlborough Ltd
X1000 - Reznor UK Ltd
Xen - Optelma Lighting Ltd
Xenex - CEMEX
Xenex - Gent Limited
Xerra - Girsberger London
Xetex - Electrix International Ltd
Xodus - Xpelair Ltd
XP95 - Apollo Fire Detectors Ltd
XP95 Intrinsically safe - Apollo Fire Detectors Ltd
X-Plas PVC - Signs International
Xplorer - Apollo Fire Detectors Ltd
Xtra load DPC - Anderson D & Son
Xtra load elite - Anderson D & Son
Xtra load preformed cavity trays - Anderson D & Son
Xtraflex - Adaptaflex Ltd
Xtraflex Sanding Pads - Acton & Borman
Xypex - Fullstop Technology Ltd
Yakcam Wilton - Mackay Carpets, Hugh
Yale - Yale Security Products Ltd
Yanmar - Martin (Marine) Ltd Alec
Yeodrain Sheild Total Surface Protection. - Moffett Thallon & Co Ltd
Yeoman - Bartlett Catering Equipment Ltd

COMPANY INFORMATION

Aalco Hull Amari Metal Ltd The Hersham Centre The Green Hersham Surrey KT12 4HP tel:(020) 8481 3300 fax:(020) 8481 3301 marketing@amari-metals.com www.aalco

Aardee Security Shutters Ltd 16 Boswell Square Hillington Ind. Estate Glasgow G52 4BQ tel:(0141) 810 3444 fax:(0141) 810 3777 sales@aardee.co.uk www.aardeesecurity.co.uk

Aaztec Cubicles Becklands Close Bar Lane, Roecliffe Boroughbridge North Yorkshire YO51 9NR tel:(01423) 326400 fax:(01423) 325115 sales@aaztec.com www.aaztec.com

Abacus Integrated Systems Ltd Unit 3, Redland Centre Redlands Coulsdon Surrey CR5 2UT tel:(020) 8763 8800 fax:(020) 8763 9996

Abacus Lighting Ltd Sutton-in-Ashfield Nottinghamshire NG17 5FT tel:(01623) 511111 fax:(01623) 552133 sales@abacuslighting.com www.abacuslighting.com

Abacus Signs Unit L4A, Colchester Industrial Estate Colchester Avenue Cardiff Glamorgan CF23 9AP tel:(029) 20 465030 fax:(029) 20 487376 sales@abacussigns.co.uk www.abacussigns.co.uk

ABB Control Ltd Grovelands House Longford Road Gravelands Ind Est Coventry CV7 9ND tel:(024) 76 368500 fax:(024) 76 364499 www.abb.com/uk

ABB Industrial Systems Ltd District Heating Division Stafford Park 12 Telford Shropshire TF3 3BJ tel:(01952) 205450 fax:(01852) 205451

ABB Instrumentation Ltd. Howard Road Eaton Socon St Neots Huntingdon PE19 3EU tel:(01480) 475321 fax:(01480) 218361 info@gb.abb.com www.abb.com

Abbex Fire Protection Ltd 17 Redchurch London E2 7DJ tel:(020) 7739 6251 fax:(020) 7336 0667

Abbey Building Supplies Co 213 Stourbridge Road Halesowen W. Midlands B63 3QY tel:(0121) 550 7674 fax:(0121) 585 5031

Abbey Craftsmen Ltd 56 High Street Haslemere Surrey GU27 2LA tel:(01428) 652666 fax:(01428) 642125

ABC Studios (Plaster Mouldings) Ltd. Oxford Lane City Road Routh Cardiff CF2 3DU tel:(029) 20 482886

ABCO Products Ltd. Bldg. 45, Membury Airfield Lambourne Newbury Berks. RG16 7TJ tel:(01488) 72414

Abel Alarm Co Ltd Detection House 4 Vaughan Way Leicester LE1 4ST tel:(0116) 265 4200 fax:(0116) 251 5341 sales@abelalarm.co.uk www.abelalarm.co.uk

Abesco Ltd Abesco House Laurencekirk Business Park Laurencekirk AB30 1EY tel:(01561) 377766 fax:(01561) 378887 technical@abesco.co.uk www.abesco.co.uk

Abet Ltd 70 Roding Road London Industrial Park London E6 4LS tel:(020) 7473 6910 fax:(020) 7476 6935 sales@abet.ltd.uk www.abet-ltd.co.uk

ABG Ltd Unit E7, Meltham Mills Meltham Mills Road Meltham W. Yorks. HD9 4DS tel:(01484) 852096 fax:(01484) 851562 geo@abgltd.com www.abg-geosynthetics.com

Ableson L & D Ltd Thwaite Gate Low Road Leeds LS10 2SL tel:(01532) 713527 fax:(01532) 711352 sales@ablesons.co.uk www.ablesons.co.uk

Abloy Security Ltd 2-3 Hatters Lane Croxley Business Park Watford Herts. WD18 8QY tel:(01923) 255066 fax:(01923) 655001 sales@abloysecurity.co.uk www.abloysecurity.co.uk

ABS Pumps Ltd Elm Park Court, Tilgate Forest Business Centre Brighton Road W. Sussex RH11 8BP tel:(01293) 821975 fax:(01293) 821976

Abstracta Construction Ltd 187 Brent Cresent London NW10 7RX tel:(020) 8965 8845 fax:(020) 8965 5120

Accent Hansen Greengate Industrial Park Greengate Middleton Manchester M24 1SW tel:(0161) 284 4100 fax:(0161) 655 3119 sales@accenthansen.co.uk www.hansengroup.co.uk

Access Industries Group Ltd Unit 2K Cherrycourt Way Leighton Buzzard Beds. LU7 8UH tel:(01525) 383101 fax:(01525) 384381

Access Piling Ltd Lyngarth Sutton Field Road Sutton Doncaster DN6 9JX tel:(01302) 707506 fax:(01302) 707506

Accomodex Leofric House Ryton-on-Dunsmore Coventry CV8 3ED tel:(024) 76 301301 fax:(024) 76 301148 100577.2713@compuserve.com

Accrington Brick Tile Co Ltd Acrrington Nori Factory Whinney Hill Accrington Lancs. BB5 6NR tel:(0870) 2383631 fax:(01254) 351769 nori_acid@marshalls.co.uk www.nori-acidbrick.com

Ackermann Ltd The Arnold Centre Paycocke Road Basildon Essex SS14 3EA 01268 563252 01268 563437 general.enquiries@ackermann.co.uk www.ackermann.co.uk

Acmeflooring Ltd, Marketing Department St Peters Road Huntington Cambs. PE18 7DN tel:(01480) 52101

Acmexd Doors Ltd Unit 2 Stone Close West Drayton Middx. UB7 8JU tel:(01895) 444044

ACO Technologies Plc ACO Business Park Hitchin Road Shefford Beds. SG17 5TE tel:(01462) 816666 fax:(01462) 815895 drainsales@aco.co.uk www.aco.co.uk

Acousticabs Industrial Noise Control Ltd Unit 52 Pocklington Ind Estate Pocklington Yorks. YO42 1NR tel:(01759) 305266 fax:(01759) 305268 info@acousticabs.com www.acousticabs.com

ACP (Concrete) Ltd Risehow Industrial Estate Firmby, Maryprot Cumbria CA15 8PD tel:(01900) 814659 fax:(01900) 816200 acp@thosarmstrong.co.uk www.acp-concrete.co.uk

ACP Concrete (Contracts) Limited Unit 11 Westside Industrial Estate Jackson Street St Helens Merseyside WA9 3AT tel:(01744) 24600 fax:(01744) 451815

ACR Heat Transfer Manufacturing Ltd Rollesby Road, Hardwick Industrial Estate Kings Lynn Norfolk PE30 4CN tel:(01553) 763371 fax:(01553) 771322

Active Carbon Filters Ltd Unit 4, Vickers Industrial Estate Mellishaw Lane Morecambe Lancs. LA3 3EN tel:(01524) 383200 fax:(01524) 389438 info@activecarbonfilters.com www.fengroup.com

Acton & Borman Cavendish Road Stevenage Herts. SG1 2EG tel:(01438) 312243 fax:(01438) 741335

Adam Carpets Ltd Greenhill Works Birmingham Road Kidderminster Worcs DY10 2SH tel:(01562) 822247 fax:(01562) 751471 ahardwic@adamcarpets.com www.adamcarpets.com

Adam Furniture Group PLC Fairfield Road Droylsden Manchester M43 6AR tel:(0161) 370 8317 fax:(0161) 370 4307 sales@adamfurniture.co.uk

Adamant Engineering Co Ltd Headley Road East Woodley Reading RG5 4SN tel:(01734) 690980 fax:(01734) 442093

Adams- Hydraulics Ltd PO Box 15 York YO30 4TA tel:(01904) 695695 fax:(01904) 695600 sales@adamshydraulics.co.uk www.adamshydraulics.co.uk

Adaptaflex Ltd Station Road Coleshill Birmingham B46 1HT tel:(01675) 468200 fax:(01675) 464930 sales@adaptaflex.co.uk www.adaptaflex.com

Addacabin Ltd Southend Thornton Kircaldy Fife KY1 4ED tel:(01592) 774401 fax:(01592) 774088

Addagrip Surface Treatments Ltd Addagrip House Bell Lane Industrial Estate Uckfield E. Sussex TN22 1QL tel:(01825) 761333 fax:(01825) 768566 sales@addagrip.co.uk www.addagrip.co.uk

Adex Interiors for industry Ltd Unit 5, Avebury Court Mark Road Hemel Hempstead Herts. HP2 7TA tel:(01442) 232327 fax:(01442) 262713 adex@msn.com www.Adex.co.uk

Adshead Ratcliffe & Co Ltd Derby Road Belper Derbys. DE56 1WJ tel:(01773) 826661 fax:(01773) 821215 arbo@arbo.co.uk www.arbo.co.uk

ADT Fire and Security Plc Security House The Sunmmitt, Hanworth Road Sunbury-on-Thames Middx TW16 5DB tel:(01932) 743333 fax:(01932) 743155 www.adt.co.uk

ADT Modern Security Systems Ltd The Clock House The Campus Hemel Hempstead Herts. HP2 7TL tel:(01442) 234123 fax:(01442) 252815

Advance Tapes International Ltd P.O Box 122 Abbey Meadows Leics. LE4 5RA tel:(0116) 251 0191 fax:(0116) 2652046 sales@advancetapes.com www.advancetapes.com

Advance Technical Panels Ltd Longland Industrial Estate Milner Way Ossett W. Yorks. WF5 9JE tel:(01924) 263655 fax:(01924) 280193

Advanced Air (UK) Ltd Burell Way Thetford Norfolk Suffolk IP24 3WB tel:(01284) 701356 fax:(01284) 701357 sales@advancedair.co.uk www.advancedair.co.uk

Advanced Fencing Systems Ltd 104-120 Blackburn Road Wincobank Sheffield S. Yorks. S61 2DN tel:(0114) 2891891 fax:(0114) 2891892

Advanced Hygienic Contracting Ltd Hammerain House Hookstone Avenue Harrogate N. Yorks. HG2 8ER tel:(01423) 870049 fax:(01423) 870051 advanced@cladding.co.uk www.cladding.co.uk

Advanced Interior Solutions Ltd West Richardson Street High Wycombe Bucks. HP11 2SB tel:(01494) 450722 fax:(01494) 448998 info@advancedinteriors.net www.advancedinteriors.net

Advanced Panels & Products Grosvenor Road, Gillingham Business Park Gillingham Kent -ME8 0SA tel:(01634) 378880 fax:(01634) 378381

Advanced Perimeter Systems Ltd 16 Cunningham Road, Springkerse Ind Est Stirling FK7 7TP tel:(01786) 479862 fax:(01786) 470331 admin@apsltd.net www.apsltd.net

Advanta Seed UK Sleaford Lincolnshire England NG 34 7HA tel:(01480) 459500 fax:(01480) 459990

Advisory Council and Arbitration Service Brandon House 180 Borough High Street London SE1 1LW 020 7210 3613 www.acas.org.uk

AEI Cables Limited Birtley Chester-Le-Street Durham DH3 2RA tel:(0191) 4103111 fax:(0191) 4108312 homesales@aeicables.co.uk www.aeicables.co.uk

AEL 7 Berkeley Court Manor Park Runcorn Cheshire WA7 1TQ tel:(01928) 579068 fax:(01928) 579523 ael@totalise.co.uk www.aelheating.com

AEL Furniture Group A27 Hastingwood Trading Estate Harbet Road Edmonton London N18 3LP tel:(020) 8807 7476 fax:(020) 8807 7579 www.yell.co.uk

Aercon Ltd Woodside Road Eastleigh Hampshire SO5 4DZ tel:(023) 80614322 fax:(023) 80620649

Aercon Wiring Systems Ltd Aercon Works Alfred Road Gravesend Kent DA11 7QF tel:(01268) 418822 fax:(01268) 418822 sales@aerconwiringsystems.co.uk www.aerconwiringsystems.co.uk

AET.gb Ltd 23 Cranford Drive Alton Hants GU34 4HJ tel:(01420) 80937 fax:(01420) 85478 sales@aet.gb.com www.aet.gb.com

Aga-Rayburn Station Road Ketley Telford Shropshire TF1 5AQ tel:(01952) 642000 fax:(01952) 222048 info@aga-rayburn.co.uk www.aga.rayburn.co.uk

Aggregate Industries UK Ltd Hulland Ward Ashbourne Derbys. DE6 3ET tel:(01335) 372222 fax:(01335) 370074 sales.charcon@aggregate.com www.charcon.com

AGN Fencing Limited Plumtree Farm Industrial Estate Bawtry Road, Harworth Doncaster S. Yorks. DN11 8EW tel:(01302) 710247 fax:(01302) 711751

Ainsworth B R (Southern) Ltd Scarisbrick House Brunnel Way Fareham Hampshire PO15 5TX tel:(01489) 885565 fax:(01489) 885258

Air Diffusion Ltd Stourbridge Road Bridgnorth Shropshire WV15 5BB tel:(01746) 761921 fax:(01746) 766450

Air Heating Ltd Murray Street Paisley Renfrewshire tel:(0141) 889 4802

Air Improvement Centre Ltd 23 Denbigh Street London SW1V 2HF tel:(020) 7834 2834 fax:(020) 7630 8485 tlc@air-improvement.co.uk www.air-improvement.co.uk

Air Safety Ltd 47 Weir Road Wimbledon London SW19 8UG tel:(020) 8944 8383 fax:(020) 8944 1716 info@airsafe.demon.co.uk

Aircare Europe Ltd Unit 27, Tatton Court Kingsland Grange, Woolston Warrington Cheshire WA1 4RR tel:(08707) 445588 fax:(01925) 850325 info@aircareurope.com www.aircareurope.com

Airdri Ltd Technology House, Oakfield Industrial Est Eynsham Oxford OX8 1TH tel:(01865) 882330 fax:(01865) 881647 sales@airdri.com www.airdri.com

Airedale International Air Conditioning Ltd Leeds Road Rawdon Leeds LS19 6JY tel:(0113) 239 1000 fax:(0113) 250 7219 enquiries@airedale.co.uk www.Airedale.co.uk

Airflow (Nicoll Ventilators) Ltd Unit 3 Hamilton Business Park Gore Road Industrial Estate New Milton Hampshire BH25 6AX tel:(01425) 611547 fax:(01425) 638912 sales@airflow-vent.co.uk www.airflow-vent.co.uk

Airflow Air Bricks Ltd Oxhey Lane Watford WD1 4RQ tel:(01923) 232736

Airflow Developments Ltd Lancaster Road High Wycombe Bucks. HP12 3QP tel:(01494) 525252 fax:(01494) 461073 info@airflow.co.uk www.airflow.co.uk

Airmaster Engineering Ltd Limewood Approach Seacroft Leeds LS14 1NG tel:(0113) 273 3333 fax:(0113) 265 0735

Airsprung Beds Ltd Canal Road Trowbridge Wilts. BA14 8RQ tel:(01225) 779101 fax:(01225) 779123 contracts@airsprungbeds.co.uk www.airsprungbeds.co.uk

Airstream Products Ltd PHS Group PLC Western Ind Est Caerphilly CF83 1XH tel:(0161) 866 9292 fax:(0161) 866 6207 info@airstream-products.co.uk www.airstream-products.co.uk

Akzo Coatings Plc 135 Milton Park Abingdon Oxon. OX14 4SB tel:(01235) 862226 fax:(01235) 862236

Alanod Ltd Chippenham Drive Kingston Milton Keynes MK10 0AN tel:(01908) 282044 fax:(01908) 282032 alanod@alanod.co.uk www.alanod.co.uk

Albany Engineering Co Ltd Church Road Lydney Glos. GL15 5EQ tel:(01594) 842275 fax:(01594) 842574 sales@albany.pumps.co.uk www.albany.pumps.co.uk

Albany Standard Pumps Richter Works Garnett Street Bradford W. Yorks. BD3 9HB tel:(01274) 725351 fax:(01274) 742467 sales@albany-pumps.co.uk www.albany-pumps.co.uk

Albion Fencing Ltd Albion House 2239 London Road Glasgow G32 8XL tel:(0141) 778 1672 fax:(0141) 778 6688 info@albionfencing.co.uk www.albionfencing.co.uk

Albion Manufacturing The Granary Silfield Road Wymondham Norfolk NR18 9AU tel:(01953) 605983 fax:(01953) 606764 sales@albionmanufacturing.com www.albionmanufacturing.com

Alcatel Cablenet Tandy House Felixstowe Road Abbey Wood London SE2 9AA tel:(020) 8557 3131 fax:(020) 8557 3535

Alcoa Europe-Extrusions & Endproducts (UK) Southam Road Banbury Oxon. OX16 7SN tel:(01295) 454545 fax:(01295) 454683 extrusions.uk@alcoa.com www.alcoa.com

Alcover UK Ltd 110 Gloucester Avenue Primrose Hill London NW1 8JA tel:(020) 7722 0857 fax:(020) 7209 5095 cmi@cmiltd.demon.co.uk www.cmiltd.co.uk

Aldous & Stamp Services Ltd 86-90 Avenue Road Beckenham Kent BR3 4SA tel:(020) 8659 1833 fax:(020) 8676 9676 sales@aldous-stamp.co.uk www.aldous-stamp.co.uk

Alfa Laval Saunders Ltd Grange Road Cwmbran Gwent NP44 3XX tel:(01633) 486666 fax:(01633) 486777

Alfas Industries Ltd Bentall Business Park Glover, District 11 Washington Tyne & Wear NE37 3JD tel:(0191) 419 0505 fax:(0191) 419 2200 sales@alfas.com www.alfas.com

Alframes Ltd 1A Arnold Road Tooting London SW17 9HU tel:(020) 8648 9394 fax:(020) 8648 4985 denisepurdy@btconnect.com

Alifabs (Woking) Ltd Kernel Court Walnut Tree Close Guildford Surrey GU1 4UD tel:(01483) 546547 fax:(01483) 546548 info@alifabs.com www.alifabs.com

Alkor Draka Ltd Axis 5 Rhodes Way Watford Herts. WD24 4TN tel:(01923) 249511 fax:(01923) 227427 david.davis@solvay.com www.alkorproof.com

Allan Harris & Sons Ltd Station Road St Georges Weston-super-Mare N. Somerset BS22 7XN tel:(01934) 511166 fax:(01934) 513066 allanharrisslate@aol.com www.allanharrisslate.co.uk

Allart, Frank, & Co Ltd 15-35 Great Tindal Street Ladywood Birmingham B16 8DR tel:(0121) 454 2977 fax:(0121) 456 2234 sales@allart.co.uk www.allart.co.uk

Allaway Acoustics Ltd 1 Queens Road Hertford SG14 1EN tel:(01992) 550825 fax:(01992) 554982 enquiries@allawayacoustics.demon.co.uk www.allawayacoustics.co.uk

Allen (Concrete) Ltd 38 Willow Lane Micham Surrey CR4 4NA tel:(020) 86872222 fax:(020) 86875400 www.allenconcrete.co.uk

Allen (Fencing) Ltd Birch Walk West Byfleet Surrey KT14 6EJ tel:(01932) 349607 fax:(01932) 354868 sales@allenfencing.co.uk www.allenfencing.co.uk

Allen P W & Co Ltd Allen House Ashchurch Business Center Tewkesbury Glos GL20 8TD tel:(01684) 851100 fax:(01684) 851101 sales@pwallen.co.uk www.inspection.pwallen.co.uk

Allermuir Contract Furniture Ltd Branch Road Lower Darwen Lancs. BB3 0PR tel:(01254) 682421 fax:(01254) 673793 sales@allermuir.co.uk.

Allgood plc 297 Euston Road London NW1 3AQ tel:(020) 7387 9951 fax:(020) 7380 1232 info@allgood.co.uk www.allgood.co.uk

Allibert (UK) Ltd PO Box 66 St Andrews Square Droitwich Spa Worcs. WR9 8XE tel:(01905) 770469 fax:(01905) 771959

Allied Guilds Unit 19, Reddicap Trading Estate Coleshill Road Sutton Coalfield W. Midlands B75 7 BU tel:(0121) 329 2874 fax:(0121) 311 1883

Allied Manufacturing Co (London) Ltd Sarena House Grove Park Colindale London NW9 0EB tel:(020) 8205 8844 fax:(020) 8200 9510

Allman Fasteners Ltd Unit 42, Heaton Mersey Industrial Estate Station Road Stockport Cheshire SK4 3QT tel:(0161) 442 6366 fax:(0161) 443 1755

Allmat (East Surrey) Ltd Kenley Treatment Works Godstone Road Kenley Surrey CR8 5AE tel:(020) 8668 6666 fax:(020) 8763 2110 info@allmat.co.uk www.allmat.co.uk

Allpipe Ltd Vulcan Road South Norwich NR6 6AF tel:(01603) 488700 fax:(01603) 488598 info@althon.net www.althon.net

Alltons (Structural Steel) Ltd PO Box 4 Ure Bank Rippon Yorks. HG4 1JE tel:(01765) 604351

Alltype Fencing Specialists Ye Wentes Wayes, High Road Langdon Hills Basildon Essex SS16 6HY tel:(01268) 545192 fax:(01268) 545192

Alno (United Kingdom) Ltd Unit 10, Hampton Farm Industrial Estate Hampton Road West Hanworth Middx. TW13 6DB tel:(020) 8898 4781 fax:(020) 8898 0268

Alpha Glass Fibres Ltd Fitzherbert Road Farlington Industrial Estate Portsmouth Hampshire PO6 1RU tel:(02392) 379990 fax:(02392) 210716

Alpha M & T Ltd Unit 2, Munro Drive Cline Road London N11 2LZ tel:(020) 8368 2230 fax:(020) 8368 2301

Alpha Therm Ltd United House Goldsel Road Swanley Kent BR8 8EX tel:(01322) 613924 fax:(01322) 662313 info@alphatherm.co.uk www.alpha-boilers.com

Alpine Autimation Ltd. Threemilestone Industrial Estate Threemilestone Truro Cornwall TR4 9LD tel:(01872) 245450 fax:(01872) 245451 www.alpineuk.com

Altecnic Ltd Airfield Industrial Estate Hixon Staffs. ST18 0PF tel:(01889) 207200 fax:(01889) 270577 sales@altecnic.co.uk www.altecnic.co.uk

Alto Cleaning Systems (UK) Ltd Bowerbank Way Gilwilly Industrial Estate Penrith Cumbria CA11 9BN tel:(01768) 868995 fax:(01768) 864713 sales@alto.uk.com

Altro Floors Works Road Letchworth, Garden City Herts. SG6 1NW tel:(01462) 480480 fax:(01462) 480010 info@altro.co.uk www.altro.co.uk

Alumasc Construction Products Ltd Elkington Gatic, Hammond House Holmestone Road, Poulton Close Dover Kent CT17 0UF tel:(01304) 203545 fax:(01304) 215001 acp@gaticdover.co.uk www.alumasc-construction.co.uk

Alumasc Exterior Building Products Ltd White House Works Bold Road Sutton, St Helens Merseyside WA9 4JG tel:(01744) 648400 fax:(01744) 648401 info@alumasc-exteriors.co.uk www.alumasc-exteriors.co.uk

Alumasc Interior Building Products Limited Halesfield 19 Telford Shropshire TF7 4QT tel:(01952) 580590 fax:(01952) 587805 sales@alumascinteriors.com www.pendock.co.uk

Aluminium Federation (Alfed) Ltd Broadway House Calthorpe Road Five Ways Birmingham B15 1TN tel:(0121) 456 1103 fax:(0870) 138 9714 alfed@alfed.org.uk www.alfed.org.uk

Aluminium RW Supplies Ltd Ryan House, Unit 6 Dumballs Road Cardiff CF10 5DF tel:(029) 20 390576 fax:(029) 20 238410 arws@compuserve.com www.arwsltd.com

Amari Plastics Plc Holmes House 24-30 Baker Street Weybridge Surrey KT13 8AU tel:(01932) 835000 fax:(01932) 835002 ho.marketing@amariplastics.com www.amariplastics.com

Ambec Fencing The Farm Depot Bentley Lane Heskin Chorley PR7 5PY tel:(01257) 451412 fax:(01257) 450094

Amber Doors Ltd Mason Way, Platts Common Industrial Estate Hoyland, Barnsley S. Yorks. S74 9TG tel:(01226) 351135 fax:(01226) 350176 sales@amber-doors.co.uk www.amber-doors.co.uk

Amberol Ltd The Plantation King Street Alfreton Derbys. DE55 7TT tel:(01773) 830930 fax:(01773) 834191 www.amberol.co.uk

Ambersil Ltd Wylds Road, Castlefield Industrial Estate Bridgwater Somerset TA6 4DD tel:(01278) 424200 fax:(01278) 425644 ambersil@btinternet.com www.ukindustry.co.uk/ambersil

Ambi-Rad Ltd Fens Pool Avenue Wallows Industrial Estate Brierley Hill W. Midlands DY5 1QA tel:(01384) 489700 fax:(01384) 489707 sales@ambirad.co.uk www.ambirad.co.uk.

Amcor (Appliances) Ltd Unit 2, Canal Court 152-156 High Street Brentford Middx. TW8 8JA tel:(020) 8560 4141 fax:(020) 8232 8814

Amdega Ltd Faverdale Darlington Co. Durham DL3 0PW tel:(01325) 468522 fax:(01325) 489209 info@amdega.com www.amegda.co.uk

Amey Grilles & Doors Ltd Westminster Road Vauxhall Industrial Estate Canterbury Kent CT1 1TX tel:(01227) 456081 fax:(01227) 450977

Amilake Southern Ltd PO BOX 4790 Estate Ferndown Dorset RH22 9WB tel:(01202) 890 279 fax:(01202) 891 716 amilake@freeuk.com www.amilake.co.uk

Amitico 18 Hanover Square London W1R tel:(020) 7629 6258

AMK Fence-In Limited Wallace Road Parkwood Springs Sheffield Yorks. S3 9SR tel:(0114) 2739372 fax:(0114) 2739372 amkfence/in@aol.com www.amkfence-in.co.uk

Amorim (UK) Ltd Bishops Weald House Albion Way Horsham W. Sussex RH13 8RA tel:(01403) 710001 fax:(01403) 710003 www.amorim.com

Anaplast (Greenock) 96 Port Glasgow Road Greenock Renfrewshire PA15 2RP tel:(01475) 501100 fax:(01475) 501080

Ancon Building Products President Way President Park Sheffield S. Yorks. S4 7UR tel:(0114) 275 5224 fax:(0114) 276 8543 info@ancon.co.uk www.ancon.co.uk

Anders + Kern UK Ltd Norderstedt House James Carter Road Mildenhall Suffolk IP28 7RQ tel:(01638) 510900 fax:(01638) 510901 sales@anders-kern.co.uk www.anders-kern.co.uk

Andersen/ Black Millwork Andersen House Dallow Street Burton-on-Trent Staffs. DE14 2PQ tel:(01283) 511122 fax:(01283) 510863 info@blackmillwork.co.uk www.blackmillwork.co.uk

Anderson C F & Son Ltd 228 London Road, Marks Tey Colchester Essex CO6 1HD tel:(01206) 211666 fax:(01206) 212450 info@cfanderson.co.uk www.cfanderson.co.uk

Anderson D & Son Barton Dock Road Stretford Manchester M32 0YL tel:(0161) 865 4444 fax:(0161) 864 1178 marketing@anderson-roofing.co.uk www.anderson-roofing.co.uk

Anderson GEC Ltd Oakengrove Shire Lane Hastoe Herts HP23 6LY tel:(01442) 826999 fax:(01442) 825999 email@gecanderson.co.uk www.gecanderson.co.uk

Anderson Gibb & Wilson A Division of Charles Tennant & Co Ltd 543 Gorgie Road Edinburgh E11 3AR tel:(0131) 443 4556 fax:(0131) 455 7608

Andersons Ltd Hillview Enerprise Park Belfast BT14 7BZ tel:(028) 90 741222 fax:(028) 90 351440 anderson.doors.co.ltd@itnet

Anderton Concrete Products Ltd Anderton Wharf Soot Hill Anderton, Northwich Cheshire CW9 6AA tel:(01606) 79436 fax:(01606) 871590 sales@andertonconcrete.co.uk www.andertonconcrete.co.uk

Andrews Coatings Ltd Carver Building Littles Lane Wolverhampton W. Midlands WV1 1JY tel:(01902) 429190 fax:(01902) 426574 info@andrewscoatings.co.uk www.antigraffiti.co.uk

Andrews Sykes Hire Ltd Premier House Darlington Street Wolverhampton W. Midlands WV1 4JJ tel:(01902) 328700 fax:(01902) 422466 info@andrews-sykes.com www.andrews-sykes.com

Andura Textured Masonry Coatings Ltd 20 Murdock Road Bicester Oxon. OX26 4PP tel:(01869) 240374 fax:(01869) 240375 ideacon.andura@business.ntl.com www.andura.com

Andy Thornton Ltd Ainleys Industrial Estate Elland W. Yorks. HX5 9JP tel:(01422) 375595 fax:(01422) 377455 email@ataa.co.uk www.andythornton.com

Angle Ring Co Ltd Bloomfield Road Tipton W. Midlands DY4 9EH tel:(0121) 557 7241 fax:(0121) 522 4555 sales@anglering.co.uk www.anglering.co.uk

Anglepoise Lighting Ltd Unit 51 Enfield Industrial Area Redditch Worcs. B97 6DR tel:(01527) 63771 fax:(01527) 61232 jt@anglepoise.co.uk www.anglepoise.co.uk

Anglia Lead Ltd 49 Barker Street Norwich Norfolk NR2 4TN tel:(01603) 630979 fax:(01603) 619171

Anglian Home Improvments Unit 6 Kingsford Business Park Onslow Basingstoke RG24 8QL tel:(01256) 472247 fax:(01256) 346697

Anglo Building Products Ltd Branksome House Filmer Grove Godalming Surrey GU7 3AB tel:(01483) 427777 fax:(01483) 428888 sales@anglobuild.co.uk www.anglobuild.co.uk

Angus Fire Thame Park Road Thame Oxon. OX9 3RT tel:(01844) 214545 fax:(01844) 213511

Anixter (UK) Ltd 1 York Road Uxbridge Middx. UB8 1RN tel:(01895) 818181 fax:(01895) 818182

Anthony de Grey Trellises Broadhinton Yard 77a North Street London SW4 0HQ tel:(020) 7738 8866 fax:(020) 7498 9075 sales@anthonydegrey.com www.anthonydegrey.com

Anti Corrosion Services Ltd Carrington Business Park Carrington Urmston Manchester M31 4QW tel:(0161) 776 4019 fax:(0161) 775 8995

Anti-Climb Guards Ltd PO Box 51 Edenbridge Kent TN8 6WY tel:(01732) 865901 fax:(01732) 867567 acg@vandgard.co.uk www.vandgard.co.uk

Antocks Lairn Ltd Lancaster Road, Cressex Industrial Estate High Wycombe Bucks. HP12 3HZ tel:(01494) 465454 fax:(01494) 465901

APA Youlditch Barns Peter Tavy Tavistock Devon PL19 9LY tel:(01822) 810187/8 fax:(01822) 810189 sales@alanpow.co.uk www.alanpow.co.uk

APE Crossley Ltd PO Box 1 Manchester M11 2DP tel:(01616) 223 1353

Apex Piling Ltd Orchard House 22 Nethergate Street Clare Suffolk CO10 8NP tel:(01787) 277888 fax:(01787) 277876 svt@apexpiling.co.uk www.apexpiling.co.uk

Apex Roofing Contractors Victor Lane Heavitree Exeter Devon EX1 3BJ tel:(01392) 466297 fax:(01392) 432202

Apollo Fire Detectors Ltd 36 Brookside Road Havant Hampshire PO9 1JR tel:(023) 92492412 fax:(023) 9249 2754 enquiries@apollo-fire.co.uk www.apollo-fire.co.uk

Apollo Space Systems Ltd Apollo House Wharf Road Pinxton Notts. NG16 6LF tel:(01773) 812800 fax:(01773) 861607

Applied Chemicals Ltd Applied House Wilsons Lane Coventry CV6 6JA tel:(024) 76 363575 fax:(024) 76 366639

Applied Felts Ltd Castle Bank Mills Portobello Road Wakefied West Yorkshire WF1 5PS Tel:(01924) 200535 Fax:(01924) 366951 sales@appliedfelts.co.uk www.appliedfelts.co.uk

APT Controls Ltd The Power House, Chantry Place Headstone Lane Harrow Middx. HA3 6NY tel:(020) 8421 2411 fax:(020) 8421 3951 info@aptcontrols.co.uk www.aptcontrols.co.uk

APV UK Ltd Gatwick Road Crawley W. Sussex RH10 2JP tel:(01293) 527777 fax:(01293) 552640

Aqua Design Ltd Gratton Way Roundswell Industrial Estate Barnstaple Devon EX31 3NL tel:(01271) 325825 fax:(01271) 371699 aquadesign@goldenc.com www.goldenc.com

Aqua-Blue Ltd Wessex House, Suite 3B12 Upper Market Street Eastleigh Hampshire SO50 9FD tel:(02380) 643888 fax:(02380) 643888

Aqualisa Products The Flyer's Way Westerham Kent TN16 1DE tel:(01959) 560000 fax:(01959) 560030 marketing@aqualisa.co.uk www.aqualisa.co.uk

Aquasentry 3 London Road Alderley Edge Cheshire SK9 7JT tel:(01625) 585685 fax:(01625) 585943 sales@aquasentry.co.uk www.aquasentry.co.uk

Aquastat Ltd Aquastat House, Tudor Works Beaconsfield Road Hayes Middx. UB4 0SL tel:(020) 8848 8811 fax:(020) 8756 0841 david_a.garland@msn.com

Aquatanks Ltd Unit 3, Westcombe Trading Estate Station Road Illminster Somerset TA19 9DW tel:(01460) 55664 fax:(01460) 53338 aquatanks@aol.com

Arbory Group Limited Holker Business Centre Burnley Road Colne Lancs. BB8 8EG tel:(01254) 394417 fax:(01254) 382278 sales@trojanproducts.co.uk www.trojanproducts.co.uk

ARC Partitioning 212-214 Great Portland Street London W1N 5HG tel:(020) 7299 0300 fax:(020) 7631 3721 arc.partitions@pentos-plc.co.uk www.arc-partitioning.co.uk

ARC Northern Clifford House York Road Wetherby W. Yorks. LS22 4NS tel:(01937) 581977 fax:(01937) 581610

Arch Timber Protection Wheldon Road Castleford W. Yorks. WF10 2JT tel:(01977)71400 fax:(01977) 714001 advice@archchemicals.com www.archtp.com

Architectural & Building Products Ltd Unit 4, Ponders End Indstrial Estate Ducks Lees Lane Enfield Middx. EN3 7SP tel:(020) 8805 4444 fax:(020) 8805 0022

Architectural Association 6 Bedford Square London WCIB 3ES tel:(0171)636 0974

Architectural Components Ltd 4-8 Exibition Road South Kensington London SW7 2HF tel:(020) 7581 2401 fax:(020) 7589 4928 sales@knobs.co.uk www.doorhandles.co.uk

Architectural Panel Products (2004) Ltd Threxton Road Industrial estate Watton Norfolk IP25 6NG 01953 884949 01953 884950 www.app-uk.net

Architectural Plastics (Handrail) Ltd Unit One 2 Robert Street Harrogate N. Yorks. HG1 1HP tel:(01423) 561852 fax:(01423) 520728

Architectural Textiles Ltd Brook Farm Barnston Essex CM6 3NK tel:(01763) 268474 fax:(01371)874255 sales@architecturaltextiles.co.uk

Arden Manufacturing Birmingham Ltd 47-50 Tenby Street North Birmingam B1 3EG tel:(0121) 693 1818 fax:(0121) 693 1819 info@arden-lighting.co.uk www.arden-lighting.co.uk

Ardex UK Ltd Homefield Road Haverhill Suffolk CB9 8QP tel:(01440) 714939 fax:(01440) 716660 info@ardex.co.uk www.ardex.co.uk

Areco Roofing Supplies Company Unit 3, Weston Works Weston Lane Tyseley Birmingham B11 3RP tel:(0121) 706 4909 fax:(0121) 707 3031 sales@areco.co.uk www.areco.co.uk

Aremco Products Foxoak Street Cradley Heath W. Midlands B64 5DQ tel:(01384) 568566 fax:(01384) 638919

Arena Software Ltd 3 Rayleigh Close Cambridge CB2 2AZ tel:(01223) 302220 fax:(01223) 460920 arena@arenasoftware.co.uk. www.arenasoftware.co.uk.

Arena-Walsall Ltd Dial Lane Hill Top Bromwich W. Midlands B70 0EB tel:(0121) 5222222 fax:(0121) 5222251 sales@arena-walsall.co.uk

Ariel Plastics Ltd Speedwell Ind. Est. Staveley Derbys. S34 3JP tel:(01246) 561122 fax:(01246) 561111

Arjo Ltd St Catherine Street Gloucester GL1 2SL tel:(08702) 430430 fax:(01452) 428344 uksales@arjo.co.uk www.arjo.com

Arkinstall Galvanizing Ltd 38 Coventry Street Birmingham W. Midlands B5 5NQ tel:(0121) 643 6455 fax:(0121) 643 0192 info@galvanizing.co.uk www.galvanizing.co.uk

Armes, Williams, Ltd Armes Trading Estate Cronard Road Sudbury Suffolk CO10 6XB tel:(01787) 372988 fax:(01787) 379383

Armfibre Ltd Drove Road Everton Sandy Beds. SG19 2HX tel:(01767) 651811 fax:(01767) 651901 sales@armfibre.u-net.com www.armfibre.u-net.com

Armitage Shanks Ltd Armitage Rugeley Staffs. WS15 4BT tel:(01543) 490253 fax:(01543) 491677 www.armitage.shanks.co.uk

Armorex Ltd Riverside House Bury Road Lavenham Suffolk CO10 9QD tel:(01787) 248482 fax:(01787) 248277 sales@armorex.co.uk www.armorex.com

Armstrong (Concrete Blocks), Thomas, Ltd Whinfield Industrial Estate Rowlands Gill Tyne & Wear NA39 1EH tel:(01207) 544214 fax:(01207) 542761 blocks@thomasarmstrong.co.uk www.xlomasarmstrong.co.uk

Armstrong (Timber), Thomas, Ltd Workington Road Flimby, Maryport Cumbria CA15 8RY tel:(01900) 68226 fax:(01900) 870800

Armstrong Pumps Ltd Peartree Road Stanway, Colchester Essex CO3 OLP tel:(01206) 579491 fax:(01206) 760532 apl@armlink.com www.armstrongpumps.com
Armstrong World Industries Ltd Armstrong House 38 Market Square Uxbridge Middx. UB8 1NG tel:(0800) 371849 fax:(01895) 274287 sales_support@armstrong.com www.armstrong-ceilings.co.uk
Arnold Wragg Ltd. Unit 2, parkway Parkway Drive Sheffield S. Yorkshire S9 4WU tel:(0115) 939 4646 fax:(0115) 939 8321
Arnull Bernard J & Co Ltd 17-21 Sunbeam Road Park Royal London NW10 6JP tel:(020) 8965 6094 fax:(020) 8961 1585 bernard.arnull@easynet.co.uk. www.bernardarnull.co.uk
Arrow Supply Company Ltd Sunbeam Road Woburn Road Industrial Estate Kempston Beds. MK42 7BZ tel:(01234) 840404 fax:(01234) 840374 information@arrow-supply.co.uk www.arrow-supply.co.uk
Artex Ltd Artex Avenue Newhaven E. Sussex BN9 9DD tel:(01273) 513100 fax:(01273) 513100
Artistic Plastercraft Ltd Lyndhurst Studios 16-18 Lyndhurst Road, Oldfield Park Bath Avon BA2 3JH tel:(01225) 315404 fax:(01225) 315404 www.artisticplastercraft.com
Arts Council of England 14 Great Peter Street London SW1P 3NQ tel:(020) 7333 0100 fax:(020) 7973 6590 www.artscouncil.org.uk
Artur Fischer (UK) Ltd Whiteley Road Hithercroft Trading Est Willingford Oxfordshire OX10 9AT tel:(01491) 827900 fax:(01491) 827953 info@fischer.co.uk www.fischer.co.uk
Arvin Motion Control Ltd 15 New Star Road Leicester LE4 9JD tel:(0116) 274 3600 fax:(0116) 274 3620 info.amcl@arvinmeritor.com www.camloc.com
AS Newbould Ltd 19 Tarran Way West Tarran Way Industrial Estate Moreton Wirral CH46 4TT tel:(0151) 677 6906 fax:(0151) 678 0680 www.asnewbouldltd.co.uk
Asbestos Removal Contractors' Association Arca House 237 Branston Road Burton upon Trent Staffs DE14 3BT tel:(01283) 531126 fax:(01283) 568228 info@arca.org.uk www.arca.org.uk
Asco Extinguishers Co Ltd T/A The Asco Fire Group Morgan Works 47A West Bowling Green Street Edinburgh EH6 5NX tel:(0131) 553 3838 fax:(0131) 555 2969 sales@asco.uk.com www.asco.uk.com
Ascot Industrial Doors Ltd Britannia Way Industrial Park Union Road Bolton Lancs. BL2 2HE tel:(0990) 556644 fax:(01204) 545800 100613.3133@compuserv.com
Ash & Lacy Building Products Ltd Bromford Lane West Bromwich W.Mids B70 3JJ tel:(0121) 525 1444 fax:(0121) 524 3444 firstname.lastname@ashandlacybp.co.uk www.ashandlacybp.co.uk
Ash Joseph Galvanizing PO Box 16 Charles Henry Street Birmingham B12 0SP tel:(0121) 622 4661 fax:(0121) 622 4317
Ashford Marine & Industrial Ltd Ashford Works 9 Ashford Road Fordingbridge Hampshire SP6 1DA tel:(019425) 52422
Ashmead Buidling Supplies Ltd Portview Road Avonmouth Bristol BS11 9LD tel:(0117) 982 8281 fax:(0117) 982 0135 avon@ashmead.co.uk www.ashmead.co.uk
Ashton Seals Ltd PO Box 269 Cortonwood Drive, Cortonwood Business Park Brampton, Barnsley S. Yorks. S73 0YP tel:(01226) 273700 fax:(01226) 756774
ASSA Ltd 75 Sumner Road Croydon Surrey CR0 3LN tel:(020) 8688 5191 fax:(020) 8688 0285 sales@assa.co.uk www.assa.co.uk
Asset International Ltd Stephenson Street Newport South Wales NP19 4XH tel:(01633) 271906 fax:(01633) 290519 postbox@assetint.co.uk www.assetint.co.uk
Assi Doman Sacks (UK) Medway House New Hythe Lane Aylesford Kent ME20 6SH tel:(01622) 717855 fax:(01622) 716360 sales.sacksuk@asdo.com www.asdo.com
Associated Metal (Stainless) Ltd 101 Brook Street Glasgow G40 3AP tel:(0141) 551 0707 fax:(0141) 551 0690 info@assoc-metal.co.uk www.assoc-metal.co.uk
Associated Perforators & Weavers Ltd 75 Church Street Warrington Cheshire WA1 2SR tel:(01925) 632402 fax:(01925) 413810 sales@apw.co.uk www.apw.co.uk
Association of Cost Engineers Administrative Office Lea House, 5 Middlewich Road Sandbach Cheshire CW11 1XL tel:(01270) 764798 fax:(01270) 766180 enquiries@acoste.org.uk www.acoste.org.uk
Association of Cunsulting Engineers Alliance House 12 Caxton Street London SW1H 0QL 020 7222 6557 020 7222 0750 consult@acenet.co.uk www.acenet.co.uk
Association of Interior Specialists Olton Bridge 245 Warwick Road Solihull West Midlands B92 7AH 0121 707 0077 0121 706 1949 info@ais-interiors.org.uk www.ais-interiors.org.uk
Association of Plumbing and Heating Contractors 14 Ensign House Ensign Business Centre, Westwood Way Coventry CV4 8JA 024 7647 0626 024 7647 0942 www.aphc.co.uk

Astracast PLC PO Box 20 Spring Ram Business Park Birstall W. Yorks. WF17 9XD tel:(01924) 477466 fax:(01924) 475801 marketing@astracast.co.uk www.astracast.com
Astrofade Ltd Kyle Road Gateshead Tyne & Wear NE8 2YE tel:(0191) 420 0515 fax:(0191) 460 4185 astrofade@aol.com www.astrofade.co.uk
Astrolighting Ltd 21 River Way Mead Ind Park Harlow Essex CM20 2SE tel:(01279) 422540 fax:(01279) 421570 uksales@maychristie-uk.com
Astron Buildings Parker Hannifin PLC Tachbrook Park Drive Tachbrook Park Warwick CV34 6TU tel:(01926) 888080 fax:(01926) 885088 astron.uk@btinternet.com www.astron-buildings.com
Atcost Ltd Spa House Faircrouch Lane Wadhurst Business Park East Sussex TN5 6PT tel:(01892) 526288 fax:(01892) 515348
Atkinson & Kirby Ltd 81 Wigan Road Ormskirk Lancs. L39 2AR tel:(01695) 573234 fax:(01695) 586902 sales@akirby.co.uk www.akirby.co.uk
Atlas Fire Engineering Ltd Serendib House 67A Boston Manor Rd. Brentford Middx. TW8 9JQ tel:(020) 8232 8866 fax:(020) 8568 0097 gen@atlasfire.co.uk www.atlas-fire.com
Atlas Stone Co Ltd Harbour Road Rye E. Sussex TN31 7TE tel:(01797) 223955
Atlas Stone Products Westington Quarry Chipping Campden Glos. GL55 6EG tel:(01386) 841104 fax:(01386) 841356 sales@atlasstone.co.uk www.atlasstone.co.uk
Atlas Ward Structures Ltd Sherburn, Malton N. Yorks. YO17 8PZ tel:(01944) 710421 fax:(01944) 710759 enquiries@atlasward.com www.atlasward.com
Atofina UK Ltd Globe House Bayley Street Stalybridge Cheshire SK15 1PY tel:(0161) 338 4411 fax:(0161) 303 1908 www.atofina.com
Attwater & Sons Ltd Hopwood Street Mills Preston Lancs. PR1 1TH tel:(01772) 258245 fax:(01772) 203361 into@attwater.co.uk
Auchard Development Co Ltd Old Road Southam, Leamington Spa Warwks. CV33 0HP tel:(01926) 812419 fax:(01926) 817425 sales@auchard.co.uk www.auchard.co.uk
Auckland Construction Ltd Mill Lane Arlesey Beds SG15 6RF
Auld Valves Ltd Cowlairs Industrial Estate Finlas Street Glasgow Scotland G22 5DQ tel:(0141) 557 0515 fax:(0141) 558 1059 sales@auldvalves.com www.auldvalves.com
Auto-Klean Filtration Ltd Lascar Works Hounslow Middx. TW3 3JL tel:(020) 8570 7722 fax:(020) 8570 4438
Automatic Doors & Gates 18a St Johns Road Isleworth Middx. TW7 6NW tel:(020) 8568 6781 fax:(020) 8847 2682
Autopa Ltd Cottage Leap Rugby Warwicks CV21 3XP tel:(01788) 550 556 fax:(01788) 550 265 info@autopa.co.uk www.autopa.co.uk
Aveling-Barford (Machines) plc Houghton Road Grantham Lincs. NG31 6JE tel:(01476) 515500 fax:(01476) 79947
Avesta Sheffield Ltd Stelco Hardy Blaenrhondda Treorchy M. Glam. CF42 5BY tel:(01443) 778621 fax:(01443) 778626 (Sales) lyn.curtis@avestasheffield.com
AvestaPolarit Ltd Precision Strip, PO Box 54 Meadowhall Sheffield S. Yorks. S9 1HU tel:(0114) 243 7272 fax:(0114) 243 1277 malvina.telfer@avestapolarit.com www.avestapolarit.com
Avocet Hardware Ltd Brookfoot Mills Elland Road Brighouse W. Yorks. HD6 2RW tel:(01484) 711700 fax:(01484) 720124
Avon Industrial Doors Ltd Armstrong Way Yate Bristol BS37 5NG tel:(01454) 273110 fax:(01454) 323224 sales@avondoors.co.uk www.avondoors.co.uk
Avon Manufacturing Ltd Montague Road Warwick Warwickshire CV34 5LS tel:(01926) 496331 fax:(01926) 400291 stewartl@avonova.co.uk www.avonova.co.uk
Axter Ltd Cliff Road Ipswich Suffolk IP3 0AY tel:(01473) 217154 fax:(01473) 232118 sales@axterltd.co.uk www.axter.fr
Azimex Fabrications Ltd Cartwright House 44 Cartwright Road Northampton NN2 6HF tel:(01604) 717712 fax:(01604) 791087
B & D Clays & Chemicals Ltd 10 Wandle Way Willow Lane Trading Estate Mitcham Surrey CR4 4NB tel:(020) 8640 9221 fax:(020) 8648 5033
B & M Fencing Limited Reading Road Hook Basingstoke Hampshire RG27 9DB tel:(01256) 762739 fax:(01256) 766891 bmfencing@totalise.co.uk www.bmfencing.co.uk
B & M Laminaites PO Box 9 Alfreton Derbys. DE55 5ZZ tel:(01773) 812608 fax:(01773) 812608
B&C Roadmarkings Ltd Unit 4 Lansbury Estate Lower Guilford Road Woking Surrey GU21 2EP 01483 488310 01483 487 919 sales@sportsmark.net www.sportsmark.net
Babcock Joinery Rosyth Royal Dockyard Bell Road Rosyth Fife KY11 2YD tel:(01383) 423049 fax:(01383) 423727
Babcock Transformers Oxford Street Bilston W. Midlands WV14 7DL tel:(01902) 492681 fax:(01902) 491116

Babcock Wanson Uk Ltd 7 Elstree Way Borehamwood Herts. WD6 1SA tel:(020) 8953 7111 fax:(020) 8207 5177 wanson@babcock-wanson.co.uk www.wanson.com
BAC Ltd Fanningdon Avenue Romford Essex RM3 8SP tel:(01708) 724824 fax:(01708) 382285 www.bacwindows.co.uk
Bacon WM & RW Ltd Walter House, Wickford Business Park Hodgson Way Wickford Essex SS11 8YG tel:(01268) 561035 fax:(01268) 561036 sales@wmbacon.co.uk www.wmbacon.co.uk
Baddeley Marshmoor Works Great North Road North Mymms, Hatfield Herts. AL9 5SD tel:(01707) 257689 fax:(01707) 257690 stella@baddeley.freeserve.co.uk
Baggeridge Brick PLC Fir Street Sedgley Dudley W. Midlands DY3 4AA tel:(01902) 880555 fax:(01902) 880432 marketing@baggeridge.co.uk www.baggeridge.co.uk
Bailey Birkett Sales Division Sharp Street Worsley Manchester M28 3NA tel:(0161) 790 7741 fax:(0161) 703 8451 support@baileybirkett.com www.baileybirkett.com
Bainbridge Engineering Ltd Woodhill Road Bury Lancs. BL8 1BW tel:(0161) 764 5034 fax:(0161) 764 5020
Baker Fencing Limited Unit 14 Wonastow Road, Ind. Estate West Monmouth Monmouthshire NP25 4PQ tel:(01600) 710820 fax:(01600) 710827 sales@bakerfencing.co.uk www.bakerfencing.co.uk
Ball F & Co Ltd Churnetside Business Park Station Road Cheddleton, Leek Staffs. ST13 7RS tel:(01538) 361633 fax:(01538) 361622 webmaster@f-ball.co.uk. www.f-ball.co.uk
Ball William Ltd Ultimate House London Road Grays Essex RM20 4WB tel:(01375) 375151 fax:(01375) 393355 marketing@wball.co.uk www.wball.co.uk
Ballantine, Bo Ness Iron Co Ltd Links Road Bo'ness Scotland EH51 9PW tel:(01506) 822721 fax:(01506) 827326 sales@ballantineboness.co.uk www.creativeironworks.co.uk
Ballofix Valves Ltd Bishops Gate Works 68 Lower City Road Tividale, Warley W. Midlands B69 2HF tel:(0121) 552 5281 fax:(0121) 552 6895
Balmoral Porcelain Limited/Longmead Ceramics Millwey ind. Est. Axminster Devon EX13 5HU tel:(01297) 32578 fax:(01297) 32710
Balmoral Tanks Balmoral Park Loirston Aberdeen AB12 3GY tel:(01224) 859100 fax:(01224) 859123 tanks@balmoral.co.uk www.balmoraltanks.com
Balsam (UK) Ltd 515a Mansfield Road Nottingham Nottinghamshire Leicestershire NG5 2JL tel:(01664) 410022 fax:(01664) 410745 sales@balsam.co.uk www.balsam.co.uk
Banafix Limited Banafix House Hillbottom Road High Wycombe Bucks. HP12 4HJ tel:(01494) 539898 fax:(01494) 539191
Banbury Windows Ltd Alton Works Long Bank Bewdley Worcs. DY12 2UJ tel:(01299) 266332 fax:(01299) 266676
Barber Edward & Co Ltd Paxton Road Tottenham London N17 0BS tel:(020) 8808 5161 fax:(020) 8801 5718
Barber Wilsons & Co Ltd Crawley Road Wood Green London N22 6AH tel:(020) 8888 3461 fax:(020) 88882041 sales@barwil.co.uk
Barbour Enquiry Service New Lodge Drift Road Windsor SL4 4RQ tel:(01344) 884999 fax:(01344) 899377 ann.mansfield@barbour-index.co.uk www.barbour-index.co.uk
Bardon (England) Ltd Bardon Hill Coalville Leics. LE67 1TL tel:(01530) 510066 fax:(01530) 510123
Barduct Ltd Gatehouse Close Aylesbury Bucks HP19 3DJ tel:(01296) 339388 fax:(01296) 339969 elainec@barduct.co.uk. www.barduct.co.uk
Baresford Pumps Ltd Carlton Road Foleshill Coventry CV6 7FL tel:(024) 76 669016 fax:(024) 76 637891 www.beresfordpumps.co.uk
Barflow-LCA 116 London Road Hailsham E. Sussex BN27 3AL tel:(01765) 690690
Barker & Co (Leeds) Ltd George Highfield Works Highfield Road Bradford Yorks. BD10 8RF tel:(01274) 611141
Barker & Geary Limited The Yard, Romsey Road, Kings Somborne Nr Stockbridge Hampshire SO20 6PW tel:(01794) 388205 fax:(01794) 388205
Barker Bros Aggregates Ltd The Green Downham Market Norfolk PE38 9DY tel:(01366) 382525 fax:(01366) 383002
Barker W H Ltd Etna Works Duke Street Fenton Stoke-on-Trent ST4 3NS tel:(01782) 319264 fax:(01782) 599724
Barlo Radiators Limited Spinning Jenny Way Leigh Lancs. WN7 4PE tel:(01942) 671105 fax:(01942) 261801 marketing@barlo.co.uk www.barlo.co.uk
Barlow Group 136 London Road Sheffield S. Yorks. S2 4NX tel:(0114) 280 3000 fax:(0114) 280 3001 barlow@barlowgroup.co.uk www.barlowgroup.co.uk
Barlow Tyrie Ltd Braintree Essex CM7 2RN tel:(01376) 557600 fax:(01376) 557610 sales@teak.com www.teak.com

Barloworld Vacuum Technology Plc Harbour Road Gosport Hampshire PO12 1BG tel:(0870) 0106929 fax:(0870) 0106916 marketing@barloworldvt.com www.barloworldvt.com
Barnwood Shopfitting Ltd 203 Barnwood Road Gloucester GL4 3HT tel:(01452) 614124 fax:(01452) 372933 mail@barnwoodshopfitting.com www.barnwoodshopfitting.com
Barr & Wray Ltd 324 Drumoyne Road Glasgow G51 4DY tel:(0141) 882 9991 fax:(0141) 882 3690 carey@barlandwray.com
Barrow Hepburn Sala Ltd 4 Old Mill Road Portishead Avon BS20 9BX tel:(01275) 846119 fax:(01275) 8409114
Bartlett Catering Equipment Ltd Maylands Avenue Hemel Hempsted Herts. HP2 7EN tel:(01442) 284284 fax:(01442) 231265 sales@bartlettcatering.com www.bartlettcatering.com
Bartoline Ltd Barmston Close Beverley E. Yorks. HU17 0LG tel:(01482) 678710 fax:(01482) 872606 info@bartoline.co.uk www.bartoline.co.uk
Barton B C & Son Ltd PO Box 67 No 1 Hainge Road Tividale, Oldbury W. Midlands B69 2NJ tel:(0121) 557 2272 fax:(0121) 557 2276 sales@b-c-b.co.uk www.b-c-b.co.uk
Barton Engineering Birchills Walsall WS2 8QE tel:(01922) 626581 fax:(01922) 646675 sales@barton-engineering.co.uk www.barton-engineering.co.uk
Barton Storage Systems Ltd Barton Industrial Park Mount Pleasant, Bilston W. Midlands WV14 7NG tel:(01902) 499500 fax:(01902) 353098 enquiries@barton-storage-systems.co.uk www.barton-storage-systems.co.uk
Bassaire Ltd Duncan Road Park Gate Southampton SO31 1ZS tel:(01489) 885111 fax:(01489) 885211 bassaire@compuserve.com www.bassaire.co.uk
Bassett & Findley Ltd Talbot Road North Wellingborough Northamptonshire NN8 1QS tel:(01933) 224898 fax:(01933) 227731 info@basseltandfinley.ltd.uk www.bassettandfindley.co.uk
Basta Parsons Alma Street Wolverhampton W. Midlands WV10 OEY tel:(01902) 877770 fax:(01902) 877771
Bathroom Manufacturers Association, The Federation House Station Road Stoke on Trent ST4 2RT 01782 747123 01782 747161 info@bathroom-association.org.uk www.bathroom-association.org
Bauer Foundations Limited Bauer House Woodrow Way, Fairhill Industrial Estate Irlam Manchester M44 6ZQ tel:(0161) 777 4400 fax:(0161) 776 2446 sales@bauerfoundations.co.uk www.bauerfoundations.co.uk
Baxall Ltd Unit 1, Castlehill Horsfield Way, Bredbury Park Industrial Estate Stockport Cheshire SK6 2SU tel:(0161) 406 6611 fax:(0161) 406 8988 info@baxall.com www.baxall.com
Baxenden Chemicals Ltd Paragon Works Baxenden Nr Accrington Lancs. BB5 2SL tel:(01254) 872278 fax:(01284) 871 247 mail@baxchem.co.uk www.baxchem.co.uk.
Baydale Architectural Systems Ltd 14 Northfield Way, Aycliffe Industrial Estate Newton Aycliffe Co. Durham DL5 6EJ tel:(01325) 307030 fax:(01325) 308030 sales@baydale.co.uk www.baydale.co.uk
Bayer Chemicals Inoegate Pigment Group W.Hawley & Son Ltd Lichfield Road Branston Staffs DE14 3WH tel:(01283) 714200 fax:(01283) 714210 info@hawley.edi.co.uk www.bayferrox.de
BB & EA Ltd 63-65 London Road Sandy Bedfordshire SG19 1DJ tel:(01767) 680291 fax:(01767) 691288
Beacon Machine Tools Ltd Mission Works Purdy Road Bilston W. Midlands WV4 8UB tel:(01902) 493331 fax:(01902) 493241 dheng@btclick.com www.home.btclick.com/dheng
Beaker Acroma Ltd Rookwood Way HeavenHill Suffolk CB9 8PQ tel:(01922) 410101 fax:(01992) 419563
Beama Ltd Westminster Tower 3 Albert Embankment London SE1 7SL tel:(020) 7793 3000 fax:(020) 7793 3003
Beasley Joiners Ltd Farnley Low Mills Whitehall Road Leeds LS12 5PS tel:(0113) 263 0524 fax:(0113) 279 2389 info@beasleyjoiners.co.uk www.beasleyjoiners.co.uk
Beaumont Ltd F E Woodlands Road Mere Wilts. BA12 6BT tel:(01747) 860481 fax:(01747) 861076 sales@beaumont-chimneys.co.uk www.beaumont-chimneys.co.uk
Becker (SLIDING PARTITIONS) Ltd Wemco House 477 Whippendell Road Watford Herts WD1 7PS tel:(01923) 236906 fax:(01923) 230149 sales@becker.uk.com www.becker.uk.com
Bedford Fencing Co. Limited Unit 8 Sargeant Turner Ind. Estate, Bromley Street , Lye Stourbridge W. Midlands DY9 8HZ tel:(01384) 422668 fax:(01384) 422688
Beeston Heating Group Ltd Derwentside Industrial Park Derby Road Belper Derbys. DE5 1UX tel:(01773) 828383 fax:(01773) 829091 info@beestonheating.co.uk

Beford & Soar Ltd 6 Kingfisher Court Farnham Road Slough Berks. SL2 1JF tel:(01753) 680666 fax:(01753) 680520 bedfords@btinternet.com www.bedfordshelf.co.uk

Beiersdorf (UK) Ltd tesa Division Yeomans Drive Blakelands, Milton Keynes Bucks. MK14 5LS tel:(01908) 211333 fax:(01908) 211555

Bekaert Building Products PO Box 119 Shepcote Lane Sheffield S9 1TY tel:(0114) 2244487 fax:(0114) 2244564 buildingproducts@bekaert.com www.bekaert.com/building

Beldam Lascar Seals Ltd Lascar Works Hounslow Middx. TW3 3JL tel:(020) 8570 7722 fax:(020) 8570 4438

Bell & Co Ltd Kingsthorpe Road Northampton NN2 6LT tel:(01604) 777500 fax:(01604) 777501 sales@abell.co.uk www.abell.co.uk

Bell & Webster Concrete Ltd Alma Park Road Grantham Lincs. NG31 9SE tel:(01476) 562277 fax:(01476) 562944 sales@bellandwebster.co.uk www.bellandwebster.co.uk

Belzona Ltd Claro Road Harrogate N. Yorks. HG1 4AY tel:(01423) 567641 fax:(01423) 505967

Be-Modern Ltd Western Approach South Shields Tyne & Wear NE33 5QZ tel:(0191) 455 3571 fax:(0191) 456 5556

Benbow Interiors Bradley Mill Newton Abbott Devon TQ12 1NF tel:(01626) 367861 fax:(01626) 355591 mail@benbow-interiors.co.uk

Benfield Fencing Limited Unit 22 Philadelphia Workshops Houghton-Le-Spring Tyne & Wear DH4 4TG tel:(0191) 584 7272 fax:(0191) 584 9222

Bennett Windows & Doors Park Road Ratby Leics. LE6 0JL tel:(0116) 239 5353 fax:(0116) 238 7295 sales@bennettwindows.co.uk www.bennettwindows.co.uk

Bennett Jr Ltd Ben Lisle Road Rotherham Yorks. S60 2RL tel:(01709) 382251 fax:(01706) 369206 limestone@benbennettjr.co.uk www.benbennettjr.co.uk

Benson Industries Ltd Unit 5, Norcroft Industrial Estate Norcroft Street Bradford W. Yorks. BD7 1JA tel:(01274) 722204 fax:(01274) 306319

Benton Co Ltd Edward Creteangle Works Brook Lane Ferring, Worthing W. Sussex BN12 5LP tel:(01903) 241349 fax:(01903) 700213 sales@creteangle.com www.creteangle.com

Be-Plas Marketing Ltd Old Hall Industrial Estate Bromborough Merseyside L62 3QA tel:(0151) 334 5133 fax:(0151) 334 9399 be-plas@demon.co.uk www.be-plas.co.uk

Berglen Group Ltd Watford Bus Park Caxton Way Herts London WD18 78F tel:(020) 8205 1133 fax:(020) 8200 0074

Bernhards Landscapes Ltd Bilton Road Rugby Warwks CV22 7DT tel:(01788) 811500 fax:(01788) 816803 bernhards@btconnect.com www.bernhards.co.uk

Bernlite Ltd 3 Brookside Colne Way Watford Herts. WD24 7QJ tel:(01923) 200160 fax:(01923) 246057 sales@bernlite.co.uk www.bernlite.co.uk

Bernstein Group PLC Silburn House Great Bank Road Westhoughton Bolton BL5 3XU tel:(01942) 840840 fax:(01942) 840084

Berry Systems A Memberof Hill & Smith Infrastructure Group Springvale Business and Ind. Park Bilston, Wolverhampton West Midlands WV14 0QL tel:(01902) 491100 fax:(01902) 494080 sales@berrysystems.co.uk www.berrysystems.co.uk

Besam Ltd Washington House Brooklands House Sunbury on Thames Middx. TW16 7EQ tel:(01932) 765888 fax:(01932) 812235

Best & Lloyd Ltd Cambray Works William Street West Smethwick, Warley W. Midlands B66 2NX tel:(0121) 558 1191 fax:(0121) 565 3547

Beta Lighting Ltd 383 /387 Leeds Road Bradford W. Yorks. BD3 9LZ tel:(01274) 721129 fax:(01274) 305007 beta@betalighting.com www.betalighting.com

Betafence Limited PO Box 119 Shepcote Lane Sheffield South Yorks. S9 1TY tel:0870 120 3252 fax:0870 120 3242 info.uk@betafence.com www.betafence.com

Beton Construction Ltd PO Box 11 Basingstoke Hampshire RG21 8EL tel:(01256) 353146 fax:(01256) 840621 info@betonconmat.co.uk www.betonconmat.co.uk

BetzDearborn Ltd Foundry Lane Widnes Cheshire WA8 8UD tel:(0151) 424 5351 fax:(0151) 423 2722

Bevan Funnell Ltd Beach Road Newhaven E. Sussex BN9 0BZ tel:(01273) 513762 fax:(01273) 516735

BFRC Services Ltd 177 Bagnall Road Basford Nottingham NG6 8SJ tel:(0115) 942 4200 fax:(0115) 942 4488 jim@bfrc.demon.co.uk www.roofinguk.com

Bicester Products Ltd 55 West End Witney Oxon. OX8 6NJ tel:(01993) 774426 fax:(01993) 779569 graham@bicpro.co.uk www.squashcourt.co.uk

Biddle Air Systems Ltd St Mary's Road Nuneaton Warwks. CV11 5AU tel:(024) 76 384233 fax:(024) 76 373621 sales@biddle-air.co.uk www.biddle.air.co.uk

BigHead Bonding Fasteners Ltd Unit 15-16 Elliott Road West Howe Ind Estate Bournemouth Dorset BH11 8LZ tel:(01202) 574601 fax:(01202) 578300 info@bighead.co.uk www.bighead.co.uk

Bilco UK Ltd Park Farm Business Centre Fornham St Genevieve Bury St Edmunds Suffolk IP28 6TS tel:(01284) 701696 fax:(01284) 702531 bilcouk@bilco.com www.bilco.com

Bill Switchgear Reddings Lane Birmingham B11 3EZ tel:(0121) 685 2080 fax:(0121) 685 2184 sales@bill-switchgear.com www.bill-switchgear.com

Billericay Fencing Ltd Morbec Farm Arterial Road Wickford Essex SS12 9JF tel:(01268) 727712 fax:(01268) 590225

Billington Structures Ltd Barnsley Road Wombwell, Barnsley S Yorks. S73 8DS tel:(01226) 340666 fax:(01266) 755947 postroom@billington-modern.co.uk www.billington-structures.co.uk

Binder Old Ipswich Road Claydon, Ipswich IP6 0AG tel:(01473) 830582 fax:(01473) 832175

Binns Fencing Ltd Harvest House Cranborne Road Potters Bar Hertfordshire EN6 3JF tel:(01707) 855555 fax:(01707) 857565 contracts@binns-fencing.com www.binns-fencing.com

Biotank Ltd Radar House Willingale Road Fyfield, Ongar Essex CM5 0SD 01277 898 000 01277 899 001 chris.white@biotank.co.uk www.biotank.co.uk

Biral Pumps Sole distributor: Hamworthy Heating Fleets corner Poole Dorset BH17 0HH tel:(01202) 662 500 fax:(01202) 665 111 sales@hamworthy-heating.com

Biral Pumps Ltd Derwentside Industrial Park Derby Road Belper Derbys. DE56 1UX tel:(01773) 821020 fax:(01773) 829091 info@beeston-heating.co.uk

Birdsall Services Ltd 6 Frogmore Road Hemel Hempstead Herts. HP3 9RW tel:(01442) 212501 fax:(01442) 248989

Birdscarer Products & Net Supplies Ltd Oak House 24 Lovedean Lane Horndean Hampshire P08 8HJ tel:(02392) 363208 fax:(02392) 796300 info@birdscarer.freeserve.co.uk www.netsuppliesyou.com

Birmingham Guild Ltd Guild House Cradley Road Netherton, Dudley W. Midlands DY2 9TH tel:(01384) 411511 fax:(01384) 411234

Birtley Building Products Ltd Mary Avenue Birtley Co. Durham DH3 1JF tel:(0191) 410 6631 fax:(0191) 410 0650 info@birtley-building.co.uk www.birtley-building.co.uk

Bisley Office Equipment Queens Road Bisley, Woking Surrey GU24 9BJ tel:(01483) 474577 fax:(01483) 489962

Bison Bede Ltd Unit 9 No.1 Ind. Estate Consett Co. Durham DH8 6ST tel:(01207) 585000 fax:(01207) 585085 sales@bisonbede.co.uk www.bisonbede.com

Bison Concrete Products Ltd Millennium Court First Avenue, Centrum 100 Burton-Upon-Trent Staffs DE14 2WR tel:(01283) 495000 fax:(01283) 544900 concrete@bison.co.uk www.bison.co.uk

Bison Structures Ltd London Road Tetbury Glos. GL8 8HH tel:(01666) 502792 fax:(01666) 504246 enquiry@bison.co.uk www.bison.co.uk

Bitmen Products Ltd PO Box 339 Over Cambridge Cambs. CB4 5TU tel:(01954) 231315 fax:(01954) 231512 sales@bitmen.force9.co.uk www.bitmen.co.uk

Bitumen Waterproofing Association 19 Regina Crescent Ravenshead Nottingham NG15 9AE 01623 430574 01623 798098 info@bwa-europe.com www.bwa-europe.com

Biwater Industries Ltd Clay Cross Chesterfield Derbys. S45 9NG tel:(01246) 250740 fax:(01246) 250741

Biz Contracts 50 St. Andrews Street Hertford Herts SG14 1JA tel:(01992) 538032 fax:(01992) 538035

Blackfriar Paints Ltd Blackfriars Road Nailsea Nr. Bristol BS48 4DJ tel:(01275) 854911 fax:(01275) 858108 sales@blackfriar.co.uk www.blackfriar.co.uk

Blagg & Johnson Ltd Massey Street Newark Notts. NG24 1PF tel:(01636) 703137 fax:(01636) 701914

Blair Joinery 9 Baker Street Greenock Strathclyde PA15 4TU tel:(01475) 721256 fax:(01475) 787364 sales@blairsofscotland.co.uk www.blairsofscotland.co.uk

Blake John Clayton le Moors Accrington BB5 5LP tel:(01254) 235441 fax:(01254) 382899

Blakell Europlacer Ltd Blanford Heights Blandford Dorset DT11 7TE tel:(01258) 451353 fax:(01258) 480183

Blakey John H Ltd New Bridge Mill Lord Street Industrial Estate Radcliffe Manchester M26 3BZ tel:(0161) 723 2414 fax:(0161) 725 9148

Blakley Electrics Ltd Conington Road Lewisham London SE13 7LJ tel:(020) 8852 4383 fax:(020) 8318 5284 sales@blakley.co.uk www.blakley.co.uk

Blanc de Bierges Eastrea Road Whittlesey, Peterborough Cambs. PE7 2AG tel:(01733) 202566 fax:(01733) 205405 info@blancdebierges.com www.blancdebierges.com

Blaze Neon Ltd Patricia Way Pysons Road Broadstairs Kent CT10 2XZ tel:(01843)601075 fax:(01843) 867924 sales.dev@blazeneon.com www.blazeneon.co.uk

Blick Communication Systems Ltd Blick House Bramble Road Swindon Wilts. SN2 8ER tel:(01793) 692401 fax:(01793) 615848 mtweedie@blick.co.uk www.blick.co.uk

Blindcraft Industries 12 Edgefauld Avenue Glasgow G21 4BB tel:(0141) 287 0800 fax:(0141) 287 0802

Blockleys Brick Ltd Sommerfield Road Trench Lock, Telford Shropshire TF1 4RY tel:(01952) 251933 fax:(01952) 641900 info@blockleys.com

Blount Shutters Ltd Unit B 734 London Road West Thurrock Essex RM16 1NL tel:(01708) 860000 fax:(01708) 861272

BLP (Hamble) Ltd Race Shop Hamble Point Quay, School Lane Hamble Southampton SO31 5NB tel:(023) 80455537 fax:(023) 80456466

BLP Group Sandallstones Road Kirk Sandall, Doncaster S. Yorks. DN3 1QR tel:(01302) 890555 fax:(01302) 886724 marketing@blpuk.com www.blpuk.com

Blucher UK Ltd Station Road Industrial Estate Tadcaster N. Yorks. LS24 9SG tel:(01937) 838000 fax:(01937) 832454 mail@blucher.co.uk www.blucher.co.uk

BLV Licht-und Vakuumtecnik Units 25 & 26 Rabans Close, Rabans Lane Industrial Estate Aylesbury Bucks. HP19 3RS tel:(01296) 399334 fax:(01296) 393422 info@blv.co.uk www.blv.co.uk

BLW Associates Ltd Caledonian Works Alexandra Drive Lockerbie Dumfries & Galloway DG11 2PD tel:(01576) 203595 fax:(01576) 202276

Blyth William Hoe Hill Barton-on-Humber North Lincolnshire DN18 5RB tel:(01652) 632175 fax:(01652) 660966

BMK Ltd Burnside Street Kilmarnock Ayrshire KA1 1SX tel:(01563) 578000 fax:(01563) 578011

Boardcraft Ltd Howard Road Eaton Socon St Neots, Huntingdon Cambs. PE19 3ET tel:(01480) 213266 fax:(01480) 219095 info@boardcraft.co.uk www.boardcraft.co.uk

Boatman Plastics Ltd GT Hales Street Market Drayton Shropshire TF9 2AA tel:(01630) 657286 fax:(01630) 655545

BOB Stevenson Ltd Coleman Street Derby DE24 8NN tel:(01332) 574112 fax:(01332) 757286

Boddingtons Ltd Unit 10 Chelmsford Road Industrial Estate Gt Dunmow Essex CM6 1HF tel:(01371) 875101 fax:(01371) 874906 sales@boddingtons-ltd.com www.boddingtons-ltd.com

Boilden MKM Ltd Middlemore Lane Aldridge Walsall WS9 8ND tel:(01922) 743321 fax:(01922) 459317 enqs@bolidenmkm.co.uk www.bolidenbrass.com

Bolivar Stamping Ltd Crown Works Luton Street Keighley W. Yorks. BD21 2LB tel:(01535) 605766 fax:(01535) 609915

Bollom Fire Protection Ltd Croydon Road Elmers End, Beckenham Kent BR3 4BL tel:(020) 8658 2299 fax:(020) 8658 8672 fire@bollom.com www.bollom.com

Bolton Brady Ltd Unit J3, Colchester Avenue Factory Estate Colchester Avenue Cardiff CF3 7XG tel:(029) 20 494771 fax:(029) 20 494191

Bolton Gate Co. Ltd Waterloo Street Bolton BL1 2SP tel:(01204) 871000 fax:(01204) 871049 sales@boltongate.co.uk www.boltongate.co.uk

Bolton, Thomas, Ltd PO Box 1 Froghall nr Cheadle Stock-on-Trent ST10 2HF tel:(01538) 752241 fax:(01538) 756715

Bonar Floors Ltd High Holborn Road Ripley Derbys. DE5 3NT tel:(01773) 744121 fax:(01773) 744142 bonarfloors.uk@lowandbonar.com www.nuwaymats.com

Bondaglass Voss Ltd 158-160 Ravenscroft Road Beckenham Kent BR3 4TW tel:(020) 8778 0071 fax:(020) 8659 5297 bondaglass@btconnect.com

Bonwyke Ltd Bonwyke House 41-43 Redlands Lane Fareham Hampshire PO14 1HL tel:(01329) 289621 fax:(01329) 822768 sungard@bonwyke.demon.co.uk. www.bonwyke.demon.co.u.k.

Booth Industries Ltd PO Box 50 Nelson Street Bolton Lancashire BL3 ZAP tel:(01204) 366333 fax:(01204) 380888 marketing@blast-systems.co.uk www.booth-industries.co.uk

Booth Muirie 870 South Street Glasgow G14 0SY tel:(0141) 959 1183 fax:(0141) 958 1173 info@booth-muirie.co.uk www.booth-muirie.co.uk

Booth Samuel & Co Ltd Cheapside Works Cheapside, Birmingham B12 0PS tel:(0121) 772 2717 fax:(0121) 766 6962 www.samuelbooth.co.uk

Border Concrete Products Ltd Jedburgh Road Kelso Roxburghshire TD5 8 JG tel:(01573) 224393 fax:(01573) 226360 sales@borderconcrete.co.uk www.borderconcrete.co.uk

Bostik Ltd Ulverscroft Road Leicester LE4 6BW tel:(0116) 2510051 fax:(0116) 2513943

Boston Chemical Co Ltd Thorpe Arch Trading Estate Avenue 4 Wetherby W. Yorks. LS23 7BZ tel:(01937) 843413 fax:(01937) 841458

Bostwick Doors (UK) Ltd Grove Works, Mersey Industrial Estate Heaton Mersey Stcokport Gtr. Manchester SK4 3ED tel:(0161) 442 7227 fax:(0161) 431 6406

Boulter Boilers Ltd Magnet House White House Road Ipswich Suffolk IP1 5JA tel:(01473) 241555 fax:(01473) 241321

Bourne Steel Ltd St Clements House St Clements Road Poole Dorset BH12 4GP tel:(01202) 746666 fax:(01202) 732002 sales@bourne-steel.co.uk www.bournesteel.co.uk

Bovingdon Brickworks Ltd Leyhill Road Hemel Hempstead Herts. HP3 0NW tel:(01442) 833176 fax:(01442) 834539 les@bovingdonbrickworks.co.uk www.bovingdonbrickworks.co.uk

Bowater Containers Ltd Monmouthshire Broad Mills Alexandra Docks Newport Gwent NPT 2WE tel:(01633) 246666

Bowden Fencing Ltd Leicester Lane Great Bowden Market Harborough Leics. LE16 7HA tel:(01858) 410660 fax:(01858) 433957 www.bowdenfencing.co.uk

Bowthorpe EMP Ltd Stevenson Road Brighton Sussex BN2 2DF tel:(01273) 692591 fax:(01273) 676637

BPB Paperboard Davidson Mill Mugiemoss Bucksburn Aberdeen AB21 9AA tel:(01224) 712821 fax:(01224) 712453

BPT Automation Ltd Unit 16, Sovereign Park Cleveland Way Hemel Hempstead Herts. HP2 7DA tel:(01442) 235355 fax:(01442) 244729 sales@bptautomation.co.uk www.bptautomation.co.uk

BPT Security Systems (UK) Ltd Unit 16, Sovereign Park Cleveland Way Hemel Hempstead Herts. HP2 7DA tel:(01442) 230800 fax:(01442) 244729 sales@bpt.co.uk www.bpt.co.uk

Bradbury Group Ltd Dunlop Way, Queensway Enterprise Estate Scunthorpe N. Lincolnshire DN16 3RN tel:(01724) 271999 fax:(01724) 271888 sales@bradburyuk.com www.bradburyuk.com

Brads Fencing Co. Limited 22 Hare Lane Farncombe Godalming Surrey GU7 3EE tel:(01483) 414745 fax:(01483) 419394 www.bradsfencing.co.uk

Braithaite Engineers Ltd Neptune Works Usk Way Newport S. Wales NP9 2UY tel:(01633) 262141 fax:(01633) 250631 tanks@braithwaite.co.uk

Bramah Security Equipment Ltd 31 Oldbury Place London W1U 5PT 08700 272 624 fax:(020) 7935 2779 lock.sales@bramah.co.uk www.bramah.co.uk

Brandon Medical Holme Well Road Leeds West Yorkshire LS10 4TQ tel:(0113) 2777393 fax:(0113) 2728844 enquiries@brandon-medical.com www.brandon-medical.com

Brannan S & Sons Ltd Leconfield Ind Est Cleator Moor Cumbria CA25 5QE tel:(01946) 816600 fax:(01946) 816625 sales@brannan.co.uk www.brannan.co.uk

Brannan Thermometers & Gauges Leconfield Industrial Estate Cleator Moor Cumbria CA25 5QE tel:(01946) 816624 fax:(01946) 816625 sales@brannan.co.uk www.brannan.co.uk

Branson Leisure Roman House, Temple Bank River Way Harlow Essex CM20 2DY tel:(01279) 432151 fax:(01279) 450542 sales@bransonleisure.co.uk www.bransonleisure.co.uk

Brash John & Co Ltd The Old ShipYard Gainsborough Lincs. DN21 1NG tel:(01427) 613858 fax:(01427) 810218 info@johnbrash.co.uk www.johnbrash.co.uk

Brass Art Ltd Regent Works Attwood Street Lye, Stourbridge W. Midlands DY9 8Ry tel:(01384) 894814 fax:(01384) 423824 sales@brassart.co.uk www.brassart.co.uk

Brass Tracks Hardware Ltd Unit 8 Kidare Close East Court Middx HA4 9LG tel:(020) 8566 9669 fax:(020) 8566 9339

BRB Industrial Services Ltd Douro Place Off Edge Lane Liverpool Merseyside L13 1AG tel:(0151) 259 6161 fax:(0151) 220 4410 (0151) 259 7097 - Zetafax sales @BRB-industrial.co.uk www.BRB-industrial.co.uk

BRC Ltd 79 - 81 Station Road Mansfield Nottinghamshire NG17 5FR tel:(01623) 555111 fax:(01623) 440932 sales@brc_eastmidlands.co.uk www.brc-uk.co.uk

BRC Special Products Carver Road, Astonfields Industrial Estate Stafford ST16 3BP tel:(01785)222288 fax:(01785)240029 stafford.sales@brc-special-products.co.uk www.brc-special-products.co.uk

BRE Bucknalls Lane Garston Watford Herts WD25 9XX tel:(01923) 664000 fax:(01923) 664010 enquiries@bre.co.uk www.bre.co.uk

Brett Martin Ltd Speedwell Industrial Estate Staveley Chesterfield Derbys. S43 3JP tel:(01246) 280000 fax:(01246) 280001 gsmith@brettmartin.co.uk www.brettmartin.co.uk

Brett Paving Salt Lane Cliffe, Rochester Kent ME3 7SZ tel:(01634) 221801 fax:(01634) 222186 sales@brettpaving.co.uk www.brettpaving.com

Brettell & Shaw Ltd Vulcan House Heyes Lane LYE W. Midlands DY9 8QT 01384 898911 01384 899100 sales@bretshaweltex.com www.bretshaweltex.com

Brewer C & Sons Ltd Albany House Ashford Road Eastbourne E. Sussex BN21 3TR tel:(01323) 411080 fax:(01323) 721435 www.brewers.co.uk

Brewer Metalcraft Thicket Lane Halnaker near Chichester W. Sussex PO18 0QS tel:(01243) 539639 fax:(01243) 533184 sales@brewer.co.uk www.brewercowl.co.uk

Brewer T. & Co Ltd Old Station Yard Springbank Road Hither Green London SE13 6SS tel:(020) 8461 2471 fax:(020) 8461 4822 sales@tbrewer.com

Brick Development Association Woodside House Winkfield Windsor Berks. SL4 2DX tel:(01344) 885651 fax:(01344) 890129 brick@brick.org.uk. www.brick.org.uk

Brickhouse Pontymister Industrial Estate Risca Gwent NP1 6YL tel:(01633) 612833 fax:(01633) 601593 brickhse@celtic.co.uk

Bridge Quarry Windmill Lane Kerridge Macclesfield Cheshire SK10 5AZ tel:(01625) 572700 fax:(01625) 572700 davidtooth@tiscali.co.uk www.bridgequarry.co.uk

Bridon International Ltd Carr Hill Doncaster S. Yorks. DN4 8DG tel:(01302) 344010 fax:(01302) 382593

Brierkrete Ltd Catterall Gates Lane Garstang, nr Preston Lancs. PR3 1XP tel:(01995) 603431 fax:(01995) 604234 general@brierkrete.co.uk www.brierkrete.co.uk

Briggs Industrial Footwear Ltd Edwin House Cornwall Road South Wigston Leicester LE18 4YU tel:(01162) 444700 fax:(01162) 444744 sales@briggsfootwear.co.uk

Briggs Roofing & Cladding Ltd, Roofing Contract Halfords Lane Smethwick West Midlands B66 1BJ tel:(0121) 565 2336 fax:(0121) 565 2721 jtyler@briggsroofing.co.uk www.briggsroofing.co.uk

Brighton (Handrails), W 55 Quarry Hill Tamworth Staffs. B77 5BW tel:(01827) 284488 fax:(01827) 250907 wbrightonhandrails@ntlworld.com www.wbrightonhandrails.com

Bright's of London Ltd Westgate Business Park Pickering North Yorkshire YO18 8LX tel:(01751) 474333

Brikenden (Builders Merchants) Ltd Grange Lane North Scunthorpe North Lincolnshire DN16 1DP tel:(01724) 282829 fax:(01274) 281375

Brimar Plastics Ltd North Road Yate Bristol BS37 7PR tel:(01454) 322111 fax:(01454) 316955 brimar@brimarplastics.co.uk www.brimar.co.uk

Brintons Ltd PO Box 16 Exchange Street Kidderminster Worcs. DY10 1AG tel:(01562) 820000 fax:(01562) 515597 solutions@brintons.co.uk www.brintons.net

Brissco Equipment Ltd 13/15 Carter Road Bishopsworth Bristol BS13 7TX tel:(0117) 964 3700 fax:(0117) 964 7906 sales@brissco.co.uk

Bristol Metal Spraying & Protective Coatings Ltd Paynes Shipyard Coronation Road Bristol BS3 1RP tel:(01272) 662206 fax:(01272) 661158 sales@bmspc.co.uk www.bmspc.co.uk

British Adhesives and Sealants Association 5 Alderson Road Worksop Notts J80 1UZ 01909 480 888 01909 473 834 secretary@basaonline.org www.basaonline.org

British Association for Chemical Specialities The Gatehouse White Cross Lancaster Lancashire LA1 4XQ 01524 849606 01524 894194 enquiries@bacsnet.org www.bacsnet.org

British Board of Agrement PO Box 195 Bucknalls Lane Garston , Watford Herts WD25 9BA tel:(01923) 665300 tel: (01923) 665301 mail@bba.star.co.uk www.bbacerts.co.uk

British Building & Engineering Appliances plc 63-65 London Road Sandy Beds SG19 1DJ tel:(01767) 680291 fax:(01767) 691288

British Cement Association Riverside House 4 Meadows Business Park, Station Approach Blackwater, Camberly Surrey GU17 9AB 01276 608700 01276 608701 info@bca.org.uk www.bca.org.uk

British Ceramic Confederation Federation House Station Road Stoke on Trent ST4 2SA 01782 744631 01782 744102 bcc@ceramfed.co.uk www.ceramfed.co.uk

British Ceramic Research Ltd. Queens Road Penkhull, Stoke-on-Trent Staffordshire ST4 7LQ 01782 454331 01782 412331

British Ceramic Tile Ltd Millenn Tile Works Heathfield Newton Abbot Devon TQ12 6RF tel:(01626) 834774 fax:(01626) 834775 info@bctltd.co.uk www.bctltd.co.uk

British Coatings Federation James House Bridge Street Leatherhead Surrey KT22 7EP 01372 360660 01372 376069 enquiry@bcf.co.uk www.coatings.org.uk

British Constructional Steelwork Association Ltd 4 Whitehall Court Westminster London SW1A 2ES tel:(020) 7839 8566 fax:(020) 7976 1634 postroom@steelconstruction.org www.steelconstruction.org

British Cork Mills Ltd Gerona House 17 Merton Road Bootle Merseyside L20 2PG tel:(0151) 922 1917 fax:(0151) 922 5843

British Electric Lamps Ltd Spencer Hill Road London SW19 4EN tel:(020) 8946 5035 fax:(020) 88790602

British Electrical Systems Association Granville Chambers 2 Radford Street Stone Staffs. ST15 8DA tel:(01785) 812426 fax:(01785) 818157 besacabman@webfactory.co.uk

British Electrotechnical and Allied Manufacturers Association Westminster Tower 3 Albert Embankment London SE1 7SL 020 7793 3000 020 7793 3003 info@beama.org.uk www.beama.org.uk

British Fenestration Rating Council PO Box 24 Hitchen Herts SG5 2FP 08700 278 494 08700 278 493 info@bfrc.org www.bfrc.org

British Fire Protection Systems Asssociation Neville House 55 Eden Street Kingston upon Thames Surrey KT1 1BW tel:(020) 8549 5855 fax:(020) 8547 1564 bfpsa@abft.orq.uk www.bfpsa.org.uk

British Flat Roofing Council (BFRC) 186 Beardall Street Hucknall Notts NG15 7JU 0115 956 6666 0115 963 3444

British Flue and Chimney Manufacturers Association Henry Road Medmenham Marlow Bucks SL7 2ER 01491 578674 01491 575024 info@feta.co.uk www.feta.co.uk/bfcma/index.htm

British Glass Manufacturers Confederation 9 Churchill Road Chapeltown Sheffield S35 2PY 0114 290 1850 0114 290 1851 info@britglass.co.uk www.britglass.co.uk

British Gypsum Ltd Drywall Acadamy Advice Centre East Leake Loughborough Leicestershire LE12 6JT tel:(08705) 456123 fax:(08705) 456356 bgtechnical.enquiries@bpb.com www.british-gypsum.com

British Hardware and Housewares Manufacturers Association Brooke House 4 The Lakes, Bedford Road Northampton NN4 7YD 01604 622023 01604 631252 bhhma@brookehouse.co.uk www.bhhma.co.uk

British Lead Mills Peartree Lane Welwyn Garden City Herts AL7 3UB tel:(01707) 324595 fax:(01707) 328941 sales@britishlead.co.uk www.britishlead.co.uk

British Library Business Information Service The British Library 96 Euston Road London NW1 2DB tel:(020) 7412 7454 fax:(020) 7412 7453 business-information@bl.uk www.bl.uk/bis

British Nova Works Ltd Neville House Beautmont Road Banbury Oxfordshire OX16 1RB tel:(020) 8574 6531 fax:(020) 8571 7572 britishnova@yahoo.com

British Plastics Federation, The 6 Bath Place Rivington Street London EC2A 3JE 020 7457 5000 020 7457 5045 bpf@bpf.co.uk www.bpf.co.uk

British Precast Concrete Federation 60 Charles Street Leicester Leicestershire LE1 1FB 0116 253 6161 0116 251 4568 info@britishprecast.org www.britishprecast.org

British Pump Manufacturers Association The McLaren Building 35 Dale End Birmingham B4 7LN 0121 200 1299 0121 200 1306 enquiry@bpma.org.uk www.bpma.org.uk

British Security Industry Association Security House Barbourne Road Worcester WR1 1RS 01905 21464 01905 613625 info@bsia.co.uk www.bsia.co.uk

British Sisalkraft Ltd Commissioners Road Strood Kent ME2 4ED tel:(01634) 290505 fax:(01634) 291029

British Slate Association Construction House 56-64 Leonard Street London EC2A 4JX 020 7608 5094 020 7608 5081 enquiries@stone-federationgb.org.uk www.stone-federationgb.org.uk

British Solvent Oils, A Division of Careless PO Box 6 Metcalf Drive, Altham Industrial Estate Altham, nr Accrington Lancs. BB5 5TU tel:(01282) 79112 fax:(01282) 78660

British Stainless Steel Association Broomgrove 59 Clarkehouse Road Sheffield S10 2LE 0114 267 1260 0114 266 1252 enquiry@bssa.org.uk www.bssa.org.uk

British Standards Institution 389 Chiswick High Road London W4 4AL tel:(020) 8996 9000 fax:(020) 8996 7001 orders@bsi-global.com www.bsi-global.com

British Steel Special Sections PO Box 9, Carr House Whessoe Road Darlington DL3 0RG tel:(01325) 382382 fax:(01325) 380038

British Structural Waterproofing Association, The Westcott House Catlins Lane Pinner Middlesex HA5 2EZ 020 8866 8339 020 8868 9971 enquiries@bswa.org.uk www.bswa.org.uk

British Wall Tie Association PO Box 22 Goring Reading RG8 9YX tel:(01189) 842674 fax:(01189) 845396

British Water 1 Queen Anne's Gate London SW1H 9BT tel:(020) 7957 4554 fax:(020) 7957 4565 info@britishwater.co.uk www.britishwater.co.uk

British Wood Preserving & Damp Proofing Association 1 Glaneagles House Vernon Gate South Street Derbyshire DE1 1UP 01332 225100 01332 255101 info@bwpda.co.uk www.bwpda.co.uk

British Woodworking Federation 55 Tufton Street London SW1P 3QL 0870 458 6939 0870 458 6949 bwf@bwf.org.uk www.bwf.org.uk

Britmet Tileform Limited Spital Farm Grimsbury Banbury Oxon. OX16 4RZ tel:(01295) 250998 fax:(01295) 271068 BritmetTileform@hotmail.com www.britmet.co.uk

BRK Brands Europe Ltd Fountain House Canal View Road Newbury Berks. RG14 5XF tel:(01635) 528080 fax:(01635) 521635 sales@brk-firstalert.co.uk

Broadbent Machine Tools Ltd Quarry Works Burnly Road Sowerby Bridge Yorks. HX6 2TF tel:(01422) 882816

Broadcrown Ltd Alliance Works Airfield Industrial Estate Hixon Stafford ST18 0PF 08794 122200 fax:(01889) 272220 info@broadcrown.co.uk www.broadcrown.com

Broadmead Cast Stone Broadmead Works Hart Street Maidstone Kent ME16 8RE tel:(01622) 690960 fax:(01622) 765484 sales@broadmead.datanet.co.uk www.kier.co.uk

Broag Ltd Remeha House Molly Millars Lane Wokingham Berks. RG41 2QP tel:(0118) 978 3434 fax:(0118) 978 6977 boilers@broag-remeha.com www.uk.remeha.com

Brockway Carpets Ltd Hoobrook Kidderminster Worcs. DY10 1XW tel:(01562) 824737 fax:(01562) 752010 sales@brockway.co.uk

Broderick Structures Limited Forsyth Road Sheerwater Woking Surrey GU21 5RR tel:(01483) 750207 fax:(01483) 750209 sales_wok@broderickstructures.com

Broen Valves Ltd 7 Clecton Street Business Park Clecton Street Tipton W. Midlands DY4 7TR tel:(0121) 522 4515 fax:(0121) 552 4535

Bromag Structures Ltd Monarch Court House 2 Mill Lane Benson Wallingford, Oxon. OX10 6SA tel:(01491) 838808 fax:(01491) 834183 jo@bromag.co.uk www.bromag.co.uk

Bromford Iron & Steel Co Ltd Bromford Lane West Bromwich W. Midlands B70 7JJ tel:(0121) 525 3110 fax:(0121) 525 4673 enquiries@bromfordsteels.co.uk www.bromfordsteels.co.uk

Brook Airchanger Brook House Dummarry Industrial Estate Dummarry Belfast BT17 9HU tel:(028) 90 616505 fax:(028) 90 616518

Brooks & Walker Ltd Century House 82 Tanner Street London SE1 3JP tel:(020) 7231 4030 fax:(020) 7231 5369

Brooks Belfast 27 Duncrue Street Belfast Antrim BT3 9AR tel:(028) 9074 4201 fax:(028) 9074 8952 brooks.belfast@upm-kymmere.com www.brooksgroup.uk.com

Brooks Stairlifts Ltd Hawksworth Hall Hawksworth Lane Guiseley West Yorks LS20 8NU tel:(01943) 877000 fax:(01943) 871491 help@stairlifts.co.uk www.stairlifts.co.uk

Broome Bros (Doncaster) Ltd Line Side York Road Doncaster South Yorkshire DN5 8AR tel:(01302) 361733 fax:(01302) 328536 broomebros@hotmail.com

Broughton Controls Ltd Shaw Road Oldham OL1 4AW tel:(0161) 627 0060 fax:(0161) 627 1362 info@broughton_controls.co.uk www.broughton_controls.co.uk

Brown & Tawse Ltd Crown Works Mucklow Hill Halesowen W. Midlands B62 8DB tel:(0121) 550 9921 fax:(0121) 501 1013

Brown F plc Unit 17 Block 1 Vale of Leven Industrial Estate Dumbarton G82 3PD tel:(01389) 759911 fax:(01389) 758986

Brown N C (Storage Equipment) Ltd Hutchinson House, Firwood Industrial Estate Thicketford Road Bolton BL2 3TR tel:(01204) 596777 fax:(01204) 596555

Broxap Ltd Rowhurst Industrial Estate Chesteron Newcastle-under-Lyme Staffs. ST5 6BD tel:(01782) 564411 fax:(01782) 565357 e-mail@broxap.co.uk www.broxap.co.uk

Broxwood (scotland Ltd) Invekalmond Way Invekalmond Industrial Estate Perth PH1 3VQ tel:(01738) 444456 fax:(01738) 444452 sales@broxwood.com www.broxwood.com

Bruntons (Musselburgh) Ltd Musselburgh Scotland EH21 7UG tel:(0131) 665 3888 fax:(0131) 653 2236

BSA Machine Tools Ltd Mackadown Lane Kitts Green Birmingham B33 0LE tel:(0121) 783 4071 fax:(0121) 784 5921 ch&ceo@bsatools.co.uk www.bsamachinetools.co.uk

BSS Zenith 981 Scotswood Road Newcastle-upon-Tyne NE99 1CS tel:(0191) 273 5938 fax:(0191) 272 3487

B-Trac Equipment Ltd 45-51 Rixon Road Wellingborough Northamptonshire NN8 4BA tel:(01933) 274400 fax:(01933) 274403 sales@b-trac.co.uk www.b-trac.co.uk

Buckingham Swimming Pools Ltd Dalehouse Lane Kenilworth Warwks. CV32 5UX tel:(01926) 852351 fax:(01926) 512387 info@buckswimpools.co.uk www.buckinghampools.com

BUFCA Ltd PO Box 23 Kenilworth Warwks. CV8 2YZ tel:(01926) 513187 fax:(01926) 513187 www.bufca.co.uk

Builder's Equipment (Norwich) Limited 24 City Road Norwich Norfolk NR1 3AN tel:(01603) 612211 fax:(01603) 630408

Builders Merchants Federation 15 Soho Square London W1D 3HL 0870 901 3380 020 7734 2766 info@bmf.org.uk www.bmf.org.uk

Building Additions Ltd Unit 1 Vallis Road Frome Somerset BA11 3HT tel:(01373) 454577 fax:(01373) 454578 partitions@buildingadditions.co.uk www.buildingadditions.co.uk

Building Adhesives Ltd Longton Road Trentham Stoke-on-Trent ST4 8JB tel:(01782) 591100 fax:(01782) 591101 info@building_adhesives.com www.building_adhesives.com

Building Cost Information Service BCIS 3 Cadogan Gate London SW1X 0AS tel:(020) 7695 1500 fax:(020) 7695 1501 bcis@bcis.co.uk www.bcis.co.uk

Building Maintenance Information BCIS 3 Cadogan Gate London SW1X 0AS tel:(020) 7695 1500 fax:(020) 7695 1501 bmi@bcis.co.uk www.bcis.co.uk

Building Reserch Establishment BRE Garston Watford WD25 9XX 01923 664000 enquiries@bre.co.uk www.bre.co.uk

Bull Group of Companies Ltd Masonry Works Priors Haw Road, North Weldon Industrial Estate Corby Northants NN17 1JG tel:(01536) 69579 fax:(01536) 240265

Bullock & Driffill Ltd Staunton Works Newark Road Staunton in the Vale, Nottingham Nottinghamshire NG13 9PF tel:(01400) 280000 fax:(01400) 280010 bullock.driffill@btopenworld.com

Bulthaup 37 Wigmore Street London W1U 1PP tel:(020) 7495 3663 fax:(020) 7495 0139 info.wi@bulthaup.co.uk www.bulthaup.com

Burco Maxol Rosegrove Burnley Lancs. BB12 6AL tel:(01282) 427241 fax:(01282) 831206

Burgess Architectural Products Ltd PO Box 2 Brookfield Road Hinkley Leics. LE10 2LL tel:(01455) 618787 fax:(01455) 251061 info@burgessceilings.co.uk www.burgessceilings.co.uk

Burlington Engineers Ltd Unit 11, Perivale Industrial Park Horsenden Lane South Greenford Middx. UB6 7RL tel:(020) 8997 1515 fax:(020) 8998 3517 info@burlington-engineers.co.uk www.burlington-engineers.co.uk

Burlington Slate Ltd Cavendish House Kirkby in Furness Cumbria LA17 7UN tel:(01229) 889661 fax:(01229) 889466 sales@burlingtonstone.co.uk www.burlingtonstone.co.uk

Burmatex Ltd Victoria Mills The Green Ossett W. Yorks. WF5 0AN tel:(01924) 262525 fax:(01924) 280033 info@burmatex.co.uk www.burmatex.co.uk

Burn Fencing Limited West End Farm, West End Lane Balne Goole E. Riding DN14 0EH tel:(01302) 708706 fax:(01302) 707377

Burndept Electronics Ltd Tom Cribb Road Thamesmead London SE28 0BM tel:(020) 8316 4477

Burnett N R Ltd Union Building Clarence Street Hull HU9 1DH tel:(01482) 20648 fax:(01482) 219600

Burnham Signs Burnham Way Kangley Bridge Road London SE26 5AE tel:(020) 8659 1525 fax:(020) 8659 4707 john.waites@burnhamsigns.com www.burnhamsigns.com

Bush Nelson PLC Stephenson Way Three Bridges Crawley W. Sussex RH10 1TN tel:(01293) 547361 fax:(01293) 531432 sales@bush-nelson.com www.bush-nelson.com

Butler Davies (Fencing) Limited 24 Mount Pleasant Road Northam Southampton Hampshire SO14 0QA tel:(023) 80222686 fax:(023) 80231201

Butler Europe Limited Mitchelston Industrial Estate Kirkcaldy Fife KY1 3LL tel:(01592) 652300 fax:(01592) 653135 butler-europe@t-online.de www.butler-europe.de

Butterworth-Heinemann Linacre House Jordan Hill Oxford OX2 8DP tel:(01865) 314091 www.bh.com

BVC Harbour Road Gosport Hampshire PO12 1BG tel:(023) 92584281 fax:(023) 92504648 marketing@bvc.co.uk www.bvc.co.uk

C & K Intellisense 2 Redwood Crescent Peel Park Campus East Killbride G74 5PA tel:(01527) 68111 fax:(01527) 68222 www.cksys.com

C & R Building Systems Ltd Great Scausby Manor Scausby Halifax W. Yorks. HX2 9UR tel:(01422) 243500 fax:(01422) 243501 schippenda@aol.com

C & W Fencing Ltd The Steelworks Bradfield Road, Wix Manningtree Essex CO11 2SG tel:(01255) 871300 fax:(01255) 871309 Sales@cw-fencing.co.uk www.cw-fencing.co.uk

C. G. I. International Ltd International House Millford Lane Haydock Merseyside WA11 9GA tel:(01924) 710720 fax:(01924) 710730 info@cgii.co.uk www.cgii.co.uk

C.R. Longley & Co. Ltd Ravensthorpe Road Thornhill Lees Dewsbury W. Yorks. WF12 9EF tel:(01924) 464283 fax:(01924) 459183 sales@longley.uk.com www.longley.uk.com

CA Group Ltd - MR-24 Division No 1 The Parade Lodge Drive Culcheth, nr Warrington Cheshire WA3 4ES tel:(01925) 764335 fax:(01925) 763445 camr24@cagroup.ltd.uk www.cagroup.ltd.uk

Cable Lift Installations Ltd Unit 1, Gemini Business Centre South Crescent Cody Road London E16 4TL tel:(020) 7540 2671 fax:(020) 7540 2671 cableifcs@kone.com

Cadisch MDA Ltd Ltd Unit 1, Finchley Industrial Centre 879 High Road Finchley London N12 8QA tel:(020) 8492 0444 fax:(020) 8492 0333 info@cadisch.com www.cadisch.com
CAEC Howard (Holdings) Ltd Grove Crescent House 18 Grove Place Bedford Beds. MK40 3JJ tel:(01234) 363171 fax:(01234) 357160
Cafco International Lawn Road Carlton In Lindrick Worsop Nottingham S81 9LB tel:(0115) 9464454 fax:(0115) 9464405
Caird Environmental Ltd, Industrial Services Division Unit 21, Freshwharf Highbridge Road Barking Essex IG11 7BW tel:(020) 8594 5099 fax:(020) 8594 8499
Calder Industrial Materials Ltd 1 Derwent Court Earlsway, Team Valley Trading Estate Gateshead Tyne & Wear NE11 0TF tel:(0191) 482 7350 fax:(0191) 482 7351 info@caldergroup.co.uk www.caldergroup.co.uk
Caldwell Hardware (UK) Ltd Herald Way Binley Industrial Estate Coventry Warwks. CV3 2RQ tel:(024) 7643 7900 fax:(024) 7643 7969 specifier@caldwell.co.uk www.caldwell.co.uk
Callenders Ltd Harvey Road Burnt Mills Industrial Estate Basildon Essex SS13 1QJ tel:(01268) 591155 fax:(01268) 591165 technical@callendars.co.uk www.callenders.co.uk
Calomax Ltd Lupton Avenue Leeds W. Yorks. LS9 7DD tel:(0113) 249 6681 fax:(0113) 235 0358 sales@calomax.co.uk www.calomax.co.uk
Calor Gas Ltd Athena Drive Tachbrook Park Warwick CV34 6RL tel:(0800) 216659 fax:(01926) 318989 enquirestele@calorgas.co.uk www.calorgas.co.uk
Calpeda Limited Wedgwood Road Industrial Estate Bicester Oxon. OX26 4UL tel:(01869) 241441 fax:(01869) 240681 pumps@calpeda.co.uk www.calpeda.co.uk
Camas Aggregates Ltd - Aggregate Industries Croft House Huncote Road Croft Leics. LE9 3GS tel:(0845) 600 0860 fax:(01455) 288601 roger.warrington@aggregate.com www.aggregate-uk.com
Camas Building Materials Kemnay Quarry Aquithie Road Kemnay Aberdeenshire AB51 9PD tel:(01467) 643861 fax:(01467) 642342
Cambridge Asphalte Co Ltd Ely Road Waterbeach Cambridge CB5 9PG tel:(01223) 863000 fax:(01223) 440006
Cambridge Structures Ltd 1 Avenue Business Park Elsworth Cambridgeshire CB3 8HY tel:(01954) 267676 fax:(01954) 267677
Cameron & Moore (Gridplex) Ltd Rannock Industrial Estate Geddington Road Corby Northamptonshire NN18 8AA tel:(01536) 260261 fax:(01536) 400208
Camlok Lifting Clamps Ltd Knutsford Way Sealand Industrial Estate Chester Cheshire CH14NZ tel:(01244) 375375 fax:(01244) 377403 camlok@deeweld.u-net.com www.camlok.co.uk
Campbell Marson & Co Unit 34, Wimbledon Business Centre Riverside Road London SW17 0BA tel:(020) 88791909 fax:(020) 8946 9395 sales@campbell-marson.co.uk www.campbell-marson.co.uk
Camtwix Engineering (Division of Terrapan Ltd) Unit 1, Bond Trading Estate Bond Avenue Bletchley Milton Keynes MK1 1RE tel:(01908) 372238 fax:(01908) 373277 camtwix@terrapin-ltd.co.uk
Cannon Hygiene Ltd Northgate White Lund Industrial Estate Morecambe Lancs. LA3 3BJ 0870 444 1988 0807 444 3938 hygiene@cannonhygiene.com www.cannonhygiene.co.uk
Canonbury Asphalte Co. Ltd. 451 Wick Lane Bow London E3 2TB tel:(020) 8980 6501 fax:(020) 8981 2956 canonburyasphalt@cs.com
Cantifix of London Limited Unit 9, Garrick Industrial Centre Irving Way London NW9 6AQ tel:(020) 8203 6203 fax:(020) 8203 6454 e-mail@cantifix.co.uk www.cantifix.co.uk
Capco Test Equipment Ltd Riverside View Wickham Market Woodbridge Suffolk IP13 0TA tel:(01728) 747407 fax:(01728) 747599 sales@capco.co.uk www.capco.co.uk
Cape Ceilings Ltd Veralum Road Stafford ST16 3EA tel:(01785) 223435 fax:(01785) 51309
Capel Fencing Contractors Ltd. 22 Sychem Lane Five Oak Green Tonbridge Kent TNl2 6TR tel:(01892) 836036 fax:(01892) 834844 info@capelfencing.co.uk www.capelfencing.co.uk
Capper Plastics Plastiques house Roman Way Coleshill Birmingham B46 1HQ tel:(01675) 467557 fax:(01675) 465972 apitt@capperplastics.com www.capperplastics.com
Caradon Bathrooms Ltd Lawton Road Alsager Stoke ST7 2DF tel:(01270) 879777 fax:(01270) 873864 www.caradon.com
Caradon Duraflex Systems Ltd Customer Centre, Central way Arle Road Cheltenham Glos. GL51 8LX tel:(0990) 351351 fax:(01242) 239604
Caradon Everest Ltd Everest House Sopers Road Cuffley, Potters Bar Herts. EN6 4SG tel:(01707) 875700 fax:(01707) 875621
Caradon Jones Ltd Whittington Road Oswestry Shropshire SY11 1HZ tel:(01691) 653251 fax:(01691) 658222
Caradon Plumbing Solutions Cromwell Road Cheltenham Glos GL52 5EP tel:(01242) 221221 fax:(0870) 8400052
Caradon Terrain Ltd Aylesford,Maidstone Kent ME20 7PJ tel:(01622) 717811 fax:(01622) 716920
Carborundum Abrasives G.B. Ltd Trafford Park Road Trafford Park Manchester M17 1HP tel:(0161) 872 2381 fax:(0161) 953 2982 marketing@carbogb.co.uk www.carbogb.co.uk
Cardale Doors Ltd Buckingham Road Industrial Estate Brackley Northants NN13 7EA tel:(01280) 703022 fax:(01280) 701138 marketing@cardale.co.uk www.cardale.com
Cardale Engineering Ltd Selbourne Road Luton Beds. LU4 8NT tel:(01582) 572546 fax:(01582) 505503
Cardinal Fencing (Northern) Limited Unit 3A, Royce Industrial Estate Ashburton Road West Trafford Park Manchester M17 1RY tel:(0161) 872 0725 fax:(0161) 872 0107 fencing@cardinal.u-nett.com
Carleton Furniture Group Ltd Mill Dam Lane Monkill Pontefract W. Yorks. WF8 2NS tel:(01977) 700770 fax:(01977) 708740 cfgit@compuserve.com
Carlton Main Brickworks Ltd Grimethorpe Barnsley S. Yorks. S72 7BG tel:(01226) 711521 fax:(01226) 780417 sales@carltonbrick.co.uk www.carltonbrick.co.uk
Carnock Building Services 24 Norwood Drive Giffnock Glasgow G46 8JE tel:(0141) 638 1830 fax:(0141) 620 3006
Carpets International (UK) Ltd PO Box 255 Toftshaw Lane Bradford W. Yorks. BD4 6QW tel:(01274) 651155 fax:(01274) 652491
Carpets of Worth Ltd Townshed Works Puxton Lane Kidederminster Worcs. DY11 5DF tel:(01562) 745000 fax:(01562) 732827 sales@carpetsofworth.co.uk www.carpetsofworth.co.uk
Carr Gymnasium Equipment Ltd Ronald Street Radford Nottingham NG7 3GY tel:(0115) 942 2252 fax:(0115) 942 2276 carrofnottm@btconnect.com www.carrofnottm.co.uk
Carrier Holdings Ltd Cameron Court Cameron Street Hillington Industrial Estate Glasgow G52 4JH tel:(0141) 882 9692 fax:(0141) 882 9291
Carrington Tom & Co Ltd EGA-KUT Works Willenhall Lane Bloxwich Walsall WS3 2XN tel:(01922) 406611 fax:(01922) 493493
Carron Phoenix Ltd Stenhouse Road Carron Falkirk FK2 8DW tel:(01324) 638321 fax:(01324) 620978 www.carron.com
Carrs Paper Ltd Shirley Solihull B90 4LJ tel:(0121) 733 3030 fax:(0121) 733 3811
Carson Office Furniture Systems 29-31 Great Portland Street London W1N 5DD tel:(020) 74361771 fax:(020) 3230156
Carter Concrete Ltd Briton's lane Beeston Regis Sheringham Norfolk NR26 8TP tel:(01263) 823434 fax:(01263) 825678 mail@carter-concrete.co.uk www.carter-concrete.co.uk
Carter Industrial Products Ltd Bedford Road Birmingham B11 1AY tel:(0121) 772 4300 fax:(0121) 772 3672
Carter Retail Equipment Ltd Redhill Road Birmingham B25 8EY tel:(0121) 250 1000 fax:(0121) 250 1005 info@cre-ltd.co.uk www.cre-ltd.co.uk
Casaire Ltd Raebarn House Northolt Road Harrow Middlesex HA2 0DY tel:(020) 8423 2323 fax:(020) 8864 2952
Casella CEL Regent House Wolseley Road Kempston Bedford MK42 7JY tel:(01462) 422411 fax:(01462) 422511
Cast Metals Federation The National Metal Forming Centre 47 Birmingham Road West Bromwich West Midlands B70 6PY 0121 601 6390 0121 601 6391 admin@cmfed.co.uk www.cmfed.com
Castell Safety International Kingsbury Works Kingsbury Road Kingsbury London NW9 8UR tel:(020) 8200 1200 fax:(020) 8205 0055 sales@castell.co.uk www.castell.co.uk
Castle Brick (Wales) Ltd Waterloo Road Penygroes Camarthenshire SA147NR tel:(01269) 831003 fax:(01269) 831003
Castle Care-Tech North Street Winkfield, Windsor Berks. SL4 4SY tel:(01344) 887788 fax:(01344) 890024 sales@castle-caretech.com www.castle-caretech.com
Castle Cement Ltd Park Square, 3160 Solihull Parkway Brimingham Business Park Birmingham West Midlands B37 7YN tel:(0121) 779 7771 fax:(0121) 779 7609 www.castlecement.co.uk
Castle Wood Floor Showroom & Warehouse 36 Lombard Road Battersea London SW11 3RP tel: (020) 7564 2315 fax:(020) 7564 2314
Castrol (UK) Ltd Burmah House Pipers Way Swindon SB3 1RE tel:(01793) 512712 fax:(01793) 513506
Caswell & Co Ltd 6 Princewood Road Earlstrees Ind. Estate Corby Northants NN17 4AP tel:(01536) 464800 fax:(01536) 464801 sales@caswell-adhesives.co.uk www.caswell-adhesives.co.uk
Catering Equipment Suppliers Association 235/237 Vauxhall Bridge Road London SW1V 1EJ 020 7233 7724 020 7828 0667 enquiries@cesa.org.uk www.cesa.org.uk
Catnic Pontypandy Industrial Estate Caerphilly M. Glam. CF83 3GL tel:(0290) 337900 fax:(0290) 863178 www.catnic.com
Catomance Ltd 4 Claxton Place Stevenage Herts. AL7 1JW tel:(01707) 324373 fax:(01707) 372191
Caunton Engineering Ltd Moorgreen Industrial Park Moorgreen Nottingham NG16 3QU tel:(01773) 531111 fax:(01773) 532020 sales@caunton.co.uk www.caunton.co.uk
Cave Tab Plc 1 Sovereign Court South Portway Close, Round Spinney Northampton Northamptonshire NN3 8RH tel:(01604) 643677 fax:(01604) 648542 cavetab@compuserve.com www.cavetab.plc.uk
Cavity Trays Ltd Administration Centre Boundary Road Yeovil Somerset BA22 8HU tel:(01935) 474769 fax:(01935) 428223 sales@cavitytrays.co.uk www.cavitytrays.com
Cavtie Ltd Conway Tolhurst Works Termorfa Cardiff
CCL. Veloduct Ltd 10 Woodall Road, Redburn Industrial Estate Enfield Middx. EN3 4LE tel:(020) 8805 3656 fax:(020) 8805 0558 sales@cclveloduct.co.uk www.cclveloduct.co.uk
CCTCO Europe Ltd Scots Quays East Street Birkenhead Merseyside L41 1FB tel:(0151) 606 5900
CED Ltd 728 London Road West Thurrock, Grays Essex RM20 3LU tel:(01708) 867237 fax:(01708) 867230 sales@ced.ltd.uk www.ced.ltd.uk
Cefndy Healthcare Cefndy Road Rhyl Denbighshire LL18 2HG tel:(01745) 343877 fax:(01745) 355806 cefndy.sales@denbighshire.gov.uk www.cefndy.com
Cegelec Industrial Control Ltd Kidsgrove Stoke-on-Trent ST7 1TW tel:(01784) 163551
Celmac Ltd Saxon Street Denton Manchester M34 3AB tel:(0161) 336 4401 fax:(0161) 335 0026
Celotex Ltd Lady Lane Industrial Estate Hadleigh Ipswich Suffolk IP7 6BA tel:(01473) 822093 fax:(01473) 820880 info@celotex.co.uk www.celotex.co.uk
Celuform New Hythe Business Park New Hythe Lane Aylesford Kent ME20 7SX tel:(01622) 719199 fax:(01622) 882258 celuform@cdw.co.uk www.celuform.co.uk
CEM Systems Ltd Unit 4, Ravenhill Business Park Ravenhill Road Belfast BT6 8AW tel:(028) 90 456767 fax:(028) 90 454535 sales@cemsys.com www.cemsys.com
Cembrit Blunn Ltd 57 Kellner Road London SE28 0AX tel:(020) 8301 8900 fax:(020) 8301 8901 sales@cembritblunn.co.uk www.cembritbluunn.co.uk
Cement Admixtures Association 38A Tilehouse Green Lane Knowle West Midlands B93 9EY 01564 776362 01564 776362 jmd@btclick.com
Cement Glaze Decorators Ltd 5 Barry Parade Barry Road London SE22 OJA tel:(020) 8299 2553 fax:(020) 8299 2346
Cementation Foundations SKANSKA Ltd Denham Way Maple Cross Rickmansworth Herts. WD3 9AS tel:(01923) 423100 fax:(01923) 423681 cementation.foundations@skanska.co.uk www.cementationfoundations.skanska.co.uk
CEMEX CEMEX House Coldharbour Lane Thorpe, Egham Surrey TW20 8TD tel:(01932 583600 fax:(01932 583611
Cento Engineering Company Ltd unit 6, Baddow Park Great Baddow Chelmsford Essex CM2 7SY tel:(01245) 477708 fax:(01245) 477748 info@cento.co.uk www.cento.co.uk
Cep Ceilings Ltd Veralum Road Common Road Industrial Estate Stafford ST16 3EA tel:(01785) 223435 fax:(01785) 251309 cep@cep-ms.demon.co.uk www.cep-ms.demon.co.uk
CEP Claddings Ltd Wainwright Close Churchfields Hastings E. Sussex TN38 9PP tel:(01424) 852641 fax:(01424) 852797 claddings@cepgroup.co.uk www.copcladdings.com
Ceram Building Technology Queens Road Penkhull Stoke-on-Trent ST4 7LQ tel:(01782) 746476 fax:(01782) 764458 info@ceram.co.uk www.ceram.com
Cerro (Manganese Bronze) Ltd PO Box 22, Hanford Works Hadleigh Road Ipswich Suffolk IP2 0EG tel:(01473) 252127 fax:(01473) 218229 salescmt@aol.com www.cerromb.com
Certifield Laboratories (Div or CPS Industries Ltd) Landchard House Victoria Street West Bromwich W. Midlands B70 8ER tel:(0121) 525 6678 fax:(0121) 500 5386
Certikin International Ltd Unit 9, Witan Park, Avenue 2 Station Lane Industrial Estate Witney Oxon. OX8 6FH tel:(01993) 778855 fax:(01993) 778620 info@certikin.co.uk www.certikin.co.uk
Cestrum Building Products Ltd Carlton Industrial Estate Carlton Barnsley S. Yorks. S70 3NH tel:(01226) 728310 fax:(01226) 709200 indo@cestrum.com www.cestrum.com
CFS Carpets Unit 3, Nexus Park Lysons Avenue Ash Vale Surrey GU12 5QE tel:(0870) 6074321 fax:(01252) 529400 sales@cfscarpets.com www.cfscarpets.com
CGL Cometec Ltd 2 Young Place Kelvin Hill East Kilbridge Glasgow G75 0TD tel:(01355) 235561 fax:(01355) 247189 sales@systems.co.uk www.cglsystems.co.uk
Chaffoteaux et Maury Ltd 6a Derriford Park Plymouth Devon PL6 5QZ tel:(01852) 222727 fax:(01952) 243493
Challenge Fencing Ltd Downside Road, Cobham Surrey KT11 3LY tel:(01932) 866555 fax:(01932) 866445 sales@challengefencing.com www.challengefencing.com
Chalmit Lighting 388 Hillington Road Glasgow Lanarkshire G52 4BL tel:(0141) 882 5555 fax:(0141) 883 3704 info@chalmit.com www.chalmit.com
Channel Electronics 23a Cradle Hill Industrial Estate Seaford E. Sussex BN25 3JE tel:(01323) 894961 fax:(01323) 893149
Channel Safety Systems Ltd 9 Petersfield Business Park Bedford Road Petersfield Hants. GU32 3QA tel:(01730) 268231 fax:(01730) 265552 sales@channelsafety.co.uk www.channelsafety.co.uk
Channel Woodcraft Ltd Bowles Well Gardens Folkestone Kent CT19 6NP tel:(01303) 850231 fax:(01303) 850734 enquiry@cwcltd.demon.co.uk www.channelwoodcraft.com
Chapel Studio Bridge Road Hunton Bridge Kings Langley Herts. WD4 8RE tel:(01923) 226386 fax:(01923) 269707 chapelstudio@btinternet.com www.btinternet.com/~chapelstudio
Charcon Tunnels (Division of Tarmac plc) PO Box 1 Southwell Lane Kirby-in-Ashfield Notts. NG17 8GQ tel:(01623) 754493 fax:(01623) 759825 charcon@tarmac.co.uk www.tarmacprecastconcrete.co.uk
Charles Collinge (Architectural Ironmongery) Ltd 44 Loman Street London SE1 0EH tel:(020) 7928 3541 fax:(020) 7620 1326
Charles Homes (Halesowen) Ltd Burrell Way Thetford Norfolk IP24 3QT
Charnwood Brick Group Ltd Old Station Close Shepshed nr Loughborough Leics. LE12 9RJ tel:(01509) 503203 fax:(01509) 507566
Charnwood Fencing Limited Beveridge Lane Bardon Hill Leicester Leics. LE67 1TB tel:(01530) 835835 fax:(01530) 814545
Chartered Building Company Scheme Englemere Kings Ride Ascot Berks. SL5 7TB tel:(01344) 630 743 fax:(01344) 630771 cbcinfo@ciob.org.uk www.cbcscheme.org.uk
Chartered Building Consultancy Scheme Englemere Kings Ride Ascot Berks SL5 7TB tel:(01344) 630700 fax:(01344) 630777 cbcinfo@ciob.org.uk www.cbcscheme.org.uk
Chartered Institute of Arbitrators International Arbitration Centre 12 Bloomsbury Square London WC1A 2LP 020 7412 7444 020 7404 4023 info@arbitrators.org www.arbitrators.org
Chartered Institute of Building Englemere Kings Ride Ascot Berks. SL5 7TB tel:(01344) 630700 fax:(01344) 630777 reception@ciob.org.uk www.ciob.org.uk
Chartered Institute of Wastes Management 9 Saxon Court St Peters Gardens Northampton Northamptonshire NN1 1SX tel:(01604) 620426 fax:(01604) 621339 ciwm@ciwm.co.uk www.ciwm.co.uk
Chartered Institution of Building Services Engineers (CIBSE) Delta House 222 Balham High Road London SW12 9BS tel:(020) 8675 5211 fax:(020) 8675 5449 info@cibse.org www.cibse.org
Chase Equipment Ltd Wellington House Sangwin Road Coseley, Bilston West Midlands WV14 9EE tel:(01902) 675835 fax:(01902) 674998 sales@chaseequipment.com www.chaseequipment.com
Chase James & Son (Furnishing) Ltd Thornton Rd. Bradford W. Yorks BD1 2JT tel:(01274) 738 282 fax:(01274) 736641 sales@james-chase.co.uk www.james-chase.co.uk
Checkmate Industries Ltd Bridge House Bridge Street Halstead Essex CO9 1HT tel:(01787) 477272 fax:(01787) 476334 checkmatecarpets@btopenworld.com
Chelsea Artisans Ltd Unit C2, Sandown Industrial Park Mill Road Esher Surrey KT10 8BL tel:(01372) 469301 fax:(01372) 470590 info@chelsea-artisans.co.uk www.chelsea-artisans.co.uk
Chelwood Brick Ltd Adswood Road Cheadle Hulme Cheadle Cheshire SK8 5QY tel:(0161) 485 8211 fax:(0161) 488 4351 marketing@chelwood.co.uk www.chelwood.co.uk
Chemsearch (Div of NCH UK Ltd) Lanchard House Victoria Street West Bromwich W. Midlands B70 8ER tel:(0121) 525 1666 fax:(0121) 500 5386
Cheshire Concrete Products Brooks Lane Middlewich Cheshire CW10 0JQ tel:(01606) 837364 fax:(01606) 837365 jim.woodyer@pochinsplc.co.uk www.cheshireconcrete.co.uk
Chesterfelt Ltd Foxwood Way Sheepbridge Chesterfield Derbys. S41 9RX tel:(01246) 268000 fax:(01246) 268001 chesterfelt@tauk21.com www.chesterfelt.co.uk
Chestnut Products Limited Unit 15, Gaza Trading Estate Hildenborough Tonbridge Kent TN11 8PL tel:(01732) 463 777 fax:(01732) 454 636 sales@chestnutproducts.fsnet.co.uk www.chestnutproducts.co.uk

Chilstone Victoria Park Fordcombe Road, Langton Green Tunbridge Wells Kent TN3 0RD tel:(01892) 740866 fax:(01892) 740249 office@chilstone.com www.chilstone.com
Chiltern Concrete and Stone Ltd Faldo Road Barton - Le - Clay Bedfordshire MK45 4RH tel:(01582) 881414 fax:(01582) 881855
Chiltern Invadex 66-68 Manchester Road Chapel-en-Le-Firth High Peak SK23 9TP tel:(01298) 816366 fax:(01298) 816106 marketing@chilterninvadex.co.uk www.chilterninvadex.co.uk
Chindwell Co Ltd Hyde House Edgware Road The Hyde London NW9 6JT tel:(020) 8205 6171 fax:(020) 8205 8800 chindwell_co_ltd@compuserve.com www.chindwell.com
Chiorino UK Ltd Highlands Road Shirley, Solihull W. Midlands B90 4HN tel:(0121) 705 8271 fax:(0121) 711 3741
Chippindale Plant Ltd Butterbowl Works Lower Wortley Ring Road Leeds W. Yorkshire LS12 5AJ tel:(0113) 263 2344 fax:(0113) 279 1710 nigel.chippindale@chippindale-plant.co.uk www.chippindale-plant.co.uk
Chris Wheeler Construction Limited Church Farm Burbage Nr Marlborough Wilts. SN8 3AT tel:(01672) 810315 fax:(01672) 810309
Christie & Grey Ltd Morley Road Tonbridge Kent TN9 1RA tel:(01732) 371100 fax:(01732) 359666 sales@christiegrey.com www.christiegrey.com
Chromalox (UK) L td Eltron House 20-28 Whitehorse Road Croydon CR0 2JA tel: (020) 86658900 fax: (020) 86890571 www.chromolox.co.uk
Chubb Fire Ltd Chubb House Sunbury-on-Thames Surrey TW16 7AR tel:(0800) 321666 fax:(01932) 765630 info@chubb.co.uk www.chubb.co.uk
Chubb Safe Equipment Company PO Box 61 Wednesfield Road Wolverhampton WV10 0EW tel:(01902) 455111 fax:(01902) 351961 info@chubb-safes.com www.chubb-safes.com
Cico Chimney Linings Ltd The Street Westleton Saxmundham Suffolk IP17 3AG tel:(01728) 648608 fax:(01728) 648428 cico@chimney-problems.co.uk www.chimney-problems.co.uk
CIDDA Wyatt House 72 Francis Road Edgbaston Birmingham B16 8SP tel:(0)121 693 9909 fax:(0)121 455 9785 cidda@wyatt-inter.co.uk www.cidda.com
CIL International Ltd CIL Trading Estate Fonthill Road London N4 3HN tel:(020) 7272 0222 fax:(020) 7272 6402 info@cil_international.com www.cil_international.com
Cila Landscapes 15 Cave Street Cwmdu, Swansea Swansea SA5 8JY tel:(017920 585545 fax:(01792) 585545
Cintec International Ltd Factory Road Newport Gwent S. Wales NP20 5FA tel:(01633) 246614 fax:(01633) 246110 cintec@aol.com www.cintec.com
Cistermiser Ltd Unit 1, Woodley Park Estate 59-69 Reading Raod Woodley Reading RG5 3AN tel:(0118) 969 1611 fax:(0118) 944 1426 cistermiser@btinternet.com www.cistermiser.co.uk
Civic Trust 17 Carlton House Terrace London SW1Y 5AW tel:(020) 7930 0914 pride@civictrust.org.uk www.civictrust.org.uk
Civil & Marine Slag Cement Ltd London Road Grays Essex RM20 3NL tel:(01708) 864813 fax:(01708) 865907 enquiries@civilmarine.co.uk www.civilmarine.co.uk
Civil Engineered Products Ldt Meadowbank Works Meadowbank Road Rotherham S61 2ND tel:(01709) 555 258 fax:(01709) 555 316 cep_ltd@lineone.net www.cep_ltd.co.uk
Civil Engineering Contractors Association Construction House 55 Tufton Street London SW1P 3QL 020 7227 4620 020 7227 4621 enquiries@ceca.co.uk www.ceca.co.uk
Cladcolour Profiling Whitehead Estate Docks Way Newport Gwent tel:(01633) 252191
Cladding Consultants (UK) Ltd Burton End Stanstead Essex CM24 8UE tel:(01279) 812224 fax:(01279) 812693
Clam-Brummer Ltd London Road Spellbook, Nr. Bishops Stortford Herts. CM23 4BA tel:(020) 7476 3171 fax:(020) 7474 0098 sales@brummer.ltd.com www.brummer.ltd.com
Clare R S & Co Ltd Stanhope Street Liverpool Merseyside L8 5RQ tel:(0151) 709 2902 fax:(0151) 709 0518 kbannon@rsclare.co.uk www.rsclare.com
Clark & Fenn Ltd Unit 19, Mitcham Industrial Estate Streatham Road Mitcham Surrey CR4 2AP tel:(020) 8685 5000 fax:(020) 8401986
Clark Door Ltd Willowholme Carlisle Cumbria CA2 5RR tel:(01228) 522321 fax:(01228) 401854 mail@clarkdoor.com www.clarkdoor.com
Clarke Instruments Ltd Distloc House, Old Sarum Airfield The Portway Salisbury Wilts. SP4 6DZ tel:(01722) 323451 fax:(01722) 335154 sales@clarke-inst.com www.clarke-inst.com
Clarke UK Ltd Grange Works Lomond Road Coatbridge Lanarkshire ML5 2NN tel:(01236) 422075 fax:(01236) 427274
Clarksteel Ltd Station Works Station Road Yaxley Peterborough PE7 3EG tel:(01733) 240811 fax:(01733) 240201 sales@clarksteel.com www.clarksteel.com
Clay Pipe Development Association Ltd Copsham House 53 Broad Street Chesham Bucks. HP5 3EA tel:(01494) 791456 fax:(01494) 792378 cpda@aol.com www.cpda.co.uk
Claydon Architectural Metalwork Ltd Units 11/12 Claydon Industrial Park Great Blakenham Suffolk IP6 0NL tel:(01473) 831000 fax:(01473) 832154 sales@cam-ltd.co.uk www.cam-ltd.co.uk
Clayton-Munroe Ltd West Drive Kingston Staverton, Totnes Devon TQ9 6AR tel:(01803) 762626 fax:(01803) 762584 mail@claytonmunroe.co.uk www.claytonmunroe.co.uk
Clean Step Ltd Unit 2, Ravenseft Park Cheney Manor Industrial Estate Swindon SN2 2QJ tel:(01793) 420088 fax:(01793) 431113 sales@cleanstepmats.co.uk www.cleanstepmats.co.uk
Cleaner Surfaces (Repair & Restoration) Ltd Basford Works Egypt Road Basford Nottingham NG7 7GL tel:(01602)783177 fax:(01602) 422798
Cleanstone Co (Bradford) Ltd Dick Lane Bradford BD4 8JW tel:(01274) 669516 fax:(01274) 665356
Clear Span Ltd Wellington Road Greenfield Oldham Lancs. OL3 7AG tel:(01457) 873244 fax:(01457) 870151 www.clearspan.co.uk
Clearwater Environmental Engineering Ltd Little London Spalding Lincs. PE11 2UE tel:(01775) 768694 fax:(01775) 710294
Clement Steel Windows Clement House Haslemere Surrey GU27 1HR tel:(01428) 647700 fax:(01428) 661369 info@clementwg.co.uk www.clementwg.co.uk
Clenaware Systems Ltd The Trading Estate Farnham Surrey GU9 9PQ tel:(01252) 712789 fax:(01252) 723719 info@cleanaware.com www.clenaware.com
Clestra Hauserman Ltd Hamilton House 3 North Street Carshalton Surrey SM5 2HZ tel:(020) 8773 2121 fax:(020) 8773 4793
Clifford Partitioning Co Ltd Champion House Burlington Road New Malden KT3 4NB tel:(020) 8942 6646 fax:(020) 8336 2369
Clifton Nursories 5a Clifton Villas Little Venice London W9 2PH tel:(020) 7289 6851 fax:(020) 7286 4215 e-mail@clifton.co.uk www.clifton.co.uk
Clinicon Fire 5 Ren House Bristol BS16 1WB tel:(07000) 240240 fax:(07000) 241242 sales@aca-tv.demon.co.uk
Clivet UK Ltd 4 Kingdom Close Segensworth East Hampshire PO15 5TJ tel:(01489) 572238 fax:(01489) 573033 l.joy@clivet-uk.co.uk www.clivet.com
Clough (Croydon) Ltd Suprete House 85 Manor Road Wallington Surrey SM6 0DH tel:(020) 8395 8787 fax:(020) 8647 9235
Clow Group Ltd Diamond Ladder Factory 562-584 Lea Bridge Road Leyton London E10 7DW tel:(020) 8558 0300 fax:(020) 8558 0301 clow@ladders-direct.co.uk www.ladders-direct.co.uk
Clyde Canvas Fabric Structures Ltd 50 Lindsay Road Leith Edinburgh EH6 6NW tel:(0131) 554 1331 fax:(0131) 553 7655
Clyde Combustions Ltd Cox Lane Chessington Surrey KT9 1SL tel:(020) 8391 2020 fax:(020) 8397 4598 info@clyde4heat.co.uk www.clyde4heat.co.uk
Coates Fencing Limited Unit 3, Barhams Close Wylds Road Bridgwater Somerset TA6 4DS tel:(01278) 423577 fax:(01278) 427760 info@coatesfencing.co.uk www.coatesfencing.co.uk
Cobham Composites Ltd Davey House Gelders Hall Road Shepshed, Loughborough Leics. LE12 9NH tel:(01509) 504541 fax:(01509) 507563
Coblands Landscapes Ltd South Farm Barn South Farm Lane, Langton Green Tunbridge Wells Kent TN3 9JN tel:(01892) 863535 fax:(01892) 863778 jamieharper@coblandslandscapes.com www.coblandslandscapes.com
Colas Roofing Harvey Road Basildon Essex SS13 1ES tel:(01268) 591155 fax:(01268) 591165
Cold Rolled Sections Association Robson Rhodes, Centre City Tower 7 Hill Street Birmingham B5 4UU tel:(0121) 697 6000 fax:(0121) 697 6113
Colebrand Ltd Cloebrand House 18-20 Warwick Street London W1R 6BE tel:(020) 7439 1000 fax:(020) 7734 3358 contracting@colebrand.com www.colebrand.com
Coleford Brick & Tile Co Ltd Hawkwell Green Cinderford Glos. GL14 3JJ tel:(01594) 822160 fax:(01594) 826655
Collier W H Ltd Church Lane Marks Tey Colchester Essex CO6 1LN tel:(01206) 230301 fax:(01206) 212540 chelwood_marketing@compuserv.com www.chelwood.co.uk
Collins Walker Ltd, Elec Steam & Hot Water Boilers Unit 7a, Nottingham S & Wilford Ind. Est. Ruddington Lane Nottingham NG11 7EP tel:(01159) 818044 fax:(01159) 455376 malcolm@collins-walker.co.uk www.collins/walker.co.uk
Colman Greeves Unit 3 Rutland Street Ashton under Lyne OL6 6TX tel:(0161) 330 9316 fax:(0161) 339 5016
Colortype Ltd Poughill Bude Cornwall EX23 9EZ tel:(01288) 352879
Colt International New Lane Havant Hants. PO9 2LY tel:(023) 92451111 fax:(023) 92454220 info@coltgroup.com www.coltgroup.com
Colt W H & Co Ltd Prestige House, Landews Meadow Green Lane, Challock Ashford Kent TN25 4BL tel:(01233) 740074 fax:(01233) 740123 mail@colthouses.co.uk www.colthouses.co.uk
Coltman Precast Concrete Ltd London Road Canwell, Sutton Coldfield W. Midlands B75 5SX tel:(01543) 480482 fax:(01543) 481587 general@coltman.co.uk www.coltman.co.uk
Colton Electrical Equipment Co 329 Front Lane Cranham, Upminster Essex RM14 1LN tel:(014022) 24454 fax:(014022) 21191
Comar Architectural Aluminuim Systems Unit 5, The Willow Centre 17 Willow Lane Mitcham Surrey CR4 4NX tel:(020) 8685 9685 fax: (020) 8646 5096 technical@parksidegrp.co.uk www.comar-alu.co.uk
Combat Polystyrene Group Ltd Unit 5c, Grange Road Houston Industrial Estate Livingston EH54 5DE tel:(01506) 441188 fax:(01506) 431133
Comber Models 17 London Lane London E8 3PR tel:(020) 8533 6592 fax:(020) 8533 5333 models@combermodels.demon.co.uk
Combustion Linings Ltd Jacaidem Works Walley Street Burslem Stoke-on-Trent ST6 2AH tel:(01782) 266766 fax:(01782) 266890 jeff.hurst@combustionlinings.com www.combustionlinings.com
Commercial Floorings Exhibition 19 Woodland Rise Ascot Berks. SL5 9HP tel:(01344) 876123 fax:(01344) 28838
Commercial Lighting Systems Units 16 &17 Chandlers Way Swanick Southampton SO31 1FQ tel: (01489) 581002 fax:(01489) 576262 sales@commercial-lighting.co.uk www.commercial-lighting.co.uk
Commission For Architecture and the Built Environment 7 St James's Square London SW1Y 4JU tel:(020) 7839 6537 fax:(020) 8398475 enquiries@cabe.org.uk www.cabe.org.uk
Commodore Kitchens Ltd Acorn House Gumley Road Grays Essex RM20 4XP tel:(01375) 382323 fax:(01375) 394955 info@commodorekitchens.co.uk www.commodorekitchens.co.uk
Complete Laboratory Installations Unit 5, The techology Centre Epinal Way Loughborough Leicestershire LE11 0QE tel:(01509) 611322 fax:(01509) 611416 completelabs@lboro.ac.uk www.completelabs.co.uk
Component Developments Halesfield 10 Telford Shropshire TF7 4QP tel:(01952) 588488 fax:(01952) 684395 info@componentdevelopments.com www.componentdevelopments.com
Composite Panels Ltd CPS House Clay Street Hull North Humberside HU8 8HA tel:(0141) 812 6866 fax:(0141) 812 7721 info@compositepanels.co.uk www.compositepanels.co.uk
Composite Structures Eastleigh House Upper Market Street Eastleigh Hants. SO50 9FD tel:(023) 80616712 fax:(023) 80643665 paull@mthojgaard.co.uk www.mthojgaard.co.uk
Compriband Ltd Bentall Business Park Glover, District II Washington Tyne & Wear NE37 3JD tel:(0191) 4196860 fax:(0191) 4196861 sales@compribrand.co.uk www.compribrand.co.uk
Compton Buildings Ltd Station Works Fenny Crompton Southam Warwks. CV47 2XB tel:(01295) 770291 fax:(01295) 770748 sales@compton-buildings.co.uk www.comptongarages.co.uk
Computer Foundations Ltd Foundation House 75- 77 London Road Lexden, Colchester Essex CO3 5AL tel:(01206) 541455 fax:(01206) 761930 info@foundation.ltd.uk
Computertel Ltd CTL House, The Maltings, Unit 53 Bath Street Gravesend Kent DA11 0DF tel:(01474) 561111 fax:(01474) 561122 sales@computertel.co.uk www.computertel.co.uk
Comyn Ching & Co Ltd (SOLRAY) 7 Skylines Village Lime Harbour London E14 9TS tel:(020) 7987 8787 fax:(020) 7987 9898
Concord Lighting Avis Way Newhaven E. Sussex BN9 0ED tel:(01273) 515811 fax:(01273) 611101
Concord Lighting Ltd 174 High Holborn London WC1V 7AA tel:(020) 7497 1400 fax:(020) 7497 1404
Concord: Marlin Hampton Road West Hanworth Trading Estate Feltham Middx. TW13 6DR tel:(020) 8894 5522 fax:(020) 8898 8480 www.marlinlighting.com
Concordia Electric Wire & Cable Co. Ltd Derwent Street Long Eaton Nottingham NG10 3LP tel:(0115) 946 7400 fax:(0115) 946 1026 sales@concordia.ltd.uk www.concordia.ltd.uk
Concrete Block Association 60 Charles Street Leicester LE1 1FB 0116 253 6161 0116 251 4568 enquiries@cba-blocks.org.uk www.cba-blocks.org.uk
Concrete Information Ltd Century House Telford Avenue Crowthorne Berks RG45 6YS tel:(01344) 762676 tel: (01344) 727202 enquiries@concreteinfo.org www.concreteinfo.org
Concrete Pipe Association 60 Charles Street Leicester Leicestershire LE1 1FB 0116 253 6161 0116 251 4568 briprecast@aol.com
Concrete Products (Lincoln) 1980 Ltd Riverside Industrial Estate Skellingthorpe Road Saxilby Lincoln LN1 2LR tel:(01522) 704158 fax:(01522) 704233
Concrete Repairs Ltd Cathite House 23a Willow Lane Mitcham Surrey CR4 4TU tel:(020) 8288 4848 fax:(020) 8288 4847 jdrewett@concrete-repairs.co.uk www.concrete.repairs.co.uk
Concrete Society Century House Telford Avenue Crowthorne Berkshire RG45 6YS 01344 466007 01344 466008 concsoc@concrete.org.uk www.concrete.org.uk
Conder Products Ltd Pullman House Barton Park, Chickenhall Lane Eastleigh Hampshire SO50 6RR tel:(023) 8068 7100 fax:(023) 8068 7101 sales@conderproducts.com.co.uk
Connect Lighting Systems (UK) Ltd 1B Wessex Gate Portsmouth Road Horndean Hants. PO8 9LP tel:(023) 9257 0098 fax:(023) 9257 0097 sales@connectlighting.co.uk www.connectlighting.co.uk
Connoleys (Blackley) Ltd Delaunays Road Manchester M9 3FP tel:(01928) 722727 fax:(01928) 762486
Connought Communication Systems Ltd Systems House Reddicap Estate Sutton Coalfield Birmingham B75 7BU tel:(0121) 311 1010 fax:(0121) 311 1890
Conren Ltd Redwither Road Wrexham County Borough of Wrexham LL13 9RD tel:(01978) 611991 fax:(01978) 661120 info@conren.com www.conren.com
Consculpt Spiral Stairs Ltd Queensway Goldsithney Penzance Cornwall TR20 9LE tel:(01736) 719444 fax:(01736) 719440 info@spiralstairs.uk.com www.spiralstairs.uk.com
Conservatory Roof Systems Ltd Bedwas Road Caerphilly M. Glam. CF83 1WG tel:(029) 20 862900 fax:(029) 20 884552 conservatoryroofsystems@ukbusiness.com www.ukbusiness.com/conservatoryroofsystems
Consort Equipment Products Ltd Thornton Industrial Estate Milford Haven Pembrokeshire SA73 2RT tel:(01646) 692172 fax:(01646) 695195 enquiries@consortepl.com www.consortepl.com
Construction Confederation Construction House 56 - 64 Leonard Street London EC2A 4JX tel:(020) 7608 5000 fax:(020) 7608 5001 enquiries@thecc.org.uk www.thecc.org.uk
Construction Employers Federation 143 Malone Road Belfast BT9 6SU tel:(028) 90 877143 fax:(028) 90 877155 mail@cefni.co.uk www.cefni.co.uk
Construction Industry Joint Council 55 Tufton Street London SW1P 3QL tel:(020) 7227 4522 fax:(020) 7227 4521 www.thecc.org.uk
Construction Industry Research & Information Association Classic House 174 - 180 Old Street London EC1V 9BP 020 7549 3300 020 7253 0523 enquiries@ciria.org www.ciria.org
Construction Industry Training Board CITB Bircham Newton Kings Lynn Norfolk PE31 6RH 01485 577577 01485 577793 information.centre@citb.co.uk www.citb.org.uk
Construction Plant-hire Association 27/28 Newbury Street Barbican London EC1A 7HU 020 7796 3366 020 7796 3399 enquiries@cpa.uk.net www.cpa.uk.net
Construction Specialist (UK) Ltd Conspec House Bicester Road Aylesbury Bucks. HP19 3AF tel:(01296) 399 700
Constructionline PO Box 6441 Basingstoke Hampshire RG21 7FN 0870 240 0152 01256 383899 constructionline@capita.co.uk www.constructionline.co.uk
Consult Lighting Consult House 92 Grange Lane Barnsley S. Yorkshire S71 5QQ tel:(01226) 246445 fax:(01226) 246101 john@consultlighting.co.uk www.consultlighting.co.uk
Contactum Ltd Edgware Road London NW2 6LF tel:(020) 8452 6366 fax:(020) 8208 3340 general@contactum.co.uk www.contactum.co.uk
Continental Shutters Ltd Unit 1 Heatway Industrial Estate Manchester Way, Wantz Road Dagenham Essex RH10 8PN tel:(020) 8517 8877 fax:(020) 8593 5721
Continental Sports Ltd Hill Top Road Paddock Huddersfield HD1 4SD tel:(01484) 542051 fax:(01484) 539148 sales@contisports.co.uk www.contisports.co.uk
Contour Showers Ltd Siddorn Street Winsford Cheshire CW7 2BA tel:(01606) 592586 fax:(01606) 861260 sales@contour-shower.co.uk www.contour-showers.co.uk
Contract Flooring Association 4c St Marys Place The Lace Market Nottingham NG1 1PH tel:(0115) 941 1126 fax:(0115) 941 2238 info@cfa.org.uk www.cfa.org.uk

Contract Journal Reed Business Information, Quadrant House The Quadrant Sutton Surrey SM2 5AS tel:(020) 8652 4805 fax:(020) 8652 4804 contract.journale@rbi.co.uk www.contractjournal.com

Contract Maintenance Ltd Newgrow Avenue Dublin Ireland

Contravent Regal Ltd 5 Cowley Mill Trading Estate Longbridge Way Uxbridge Middx. UB8 2YG tel:(01895) 257766 fax:(01895) 257860 info@contravent.co.uk www.contravent.co.uk

Cooke Brothers Ltd Northgate Aldridge Walsall W. Midlands WS9 8TL tel:(01922) 740001 fax:(01922) 456227 sales@cookebrothers.co.uk www.cookebrothers.co.uk

Cookery & Food Association 1 Victoria Parade by 331 Sandycombe Road Richmond Surrey TW9 3NB 020 8948 3870

Cookson and Zinn (PTL) Ltd Station Road Works Hadleigh, Ipswich Suffolk IP7 5PN tel:(01473) 825200 fax:(01473) 824164 info@czltd.com www.czltd.com

Cookson Electronics Forsyth Road Sheerwater Woking Surrey GU21 5RZ tel:(01483) 793200 fax:(01483) 793201 sales@fernox.com www.fernox.com

Cooper & Turner Sheffield Road Sheffield S9 1RS tel:(0114) 256 0057 fax:(0114) 244 5529

Cooper Group Ltd Unit 18, The Tanneries Brockhampton Lane Havant Hants. PO9 1JB tel:(02392) 454405 fax:(02392) 492732 coopergroup@msn.com www.coopersblinds.co.uk

Cooper H W & Co Ltd Unit 5 27 - 29 Vauxhall Grove London SW8 1SY tel:(020) 7582 1874 fax:(020) 7840 0272

Cooper Lighting and Security Ltd Wheatley Hall Road Doncaster S. Yorks. DN2 4NB tel:(01302) 321541 fax:(01302) 303220 sales@cooper-ls.com www.cooper-ls.com

Cooper Security Ltd Security House Vantage Point Business Village Mitcheldean Glos. GL17 0SZ tel:(01594) 545400 fax:(01594) 545401 enquiries@coopersecurity.co.uk www.coopersecurity.co.uk

Coo-Var Ltd Elenshaw Works Lockwood Street Hull HU2 0HN tel:(01482) 328053 fax:(01482) 219266 info@coo-var.co.uk www.coo-var.co.uk

Cope & Trimmins Ltd Angel Road Works Edmonton London N18 3AY tel:(020) 8803 6481 fax:(020) 8884 232

Copeland Corporation Ltd Colthrop Way Colthrop, Thatcham Berks. RG19 4NQ tel:(01635) 876161 fax:(01635) 877111

Copley Decor Ltd Unit 1 Leyburn Business Park Leyburn N. Yorks. DL8 5QA tel:(01969) 623410 fax:(01969) 624398 mouldings@copleydecor.co.uk www.copleydecor.com

Coppa Cutta Ltd 11 - 15 Tailsman Business Centre Duncan Road Park Gate Hants. SO31 7GA tel:(01489) 554769 fax:(01489) 577858 coppagutta@good-directions.co.uk www.coppagutta.com

Copper Development Association 5 Grovelands Business Centre Boundary Way Hemel Hempstead Herts. HP2 7TE tel:(01442) 275700 fax:(01442) 275716 helpline@copperdev.co.uk www.cda.org.uk

Coram Showers Ltd Stanmore Industrial Estate Bridgnorth Shropshire WV15 5HP tel:(01746) 766466 fax:(01746) 764140 sales@coram.co.uk www.coram.co.uk

Corbett & Co (Galvanising) W Ltd Alexandra Works Hollies Road Wellington,Telford Shropshire TF1 1QJ tel:(01952) 412777 fax:(01952) 412888

Cordek Ltd Spring Copse Business Park Slinfold W. Sussex RH13 0SZ tel:(01403) 799600 fax:(01403) 791718 sales@cordek.com www.cordek.com

Corofil Woodall Unit 2 Navigation Drive Hurst Business park Brierley Hill, West Midlands DY5 1UT tel:(01384) 489400 fax:(01384) 489401 sales@corofilwoodall.co.uk www.corofilwoodall.co.uk

Corroless Corrosion Control Kelvin Way West Bromwich West Midlands B70 7JZ tel:(0121) 524 2235 fax:(0121) 553 2787 contact@corroless.com www.corroless.com

Corrugated Sheets & Profiles Ltd Ridgecare Road Black Lake West Bromwich W. Midlands B71 1BB tel:(0121) 553 6671 fax:(0121) 500 6133

Corus PO Box 1 Frodingham House Scunthorpe S. Humberside DN16 1BP tel:(01724) 280280 fax:(01724) 282040

Cottage Craft Spirals The Barn, Gorsty Low Farm The Wash Chapel-en-le-Firth High Peak SK23 0QL tel:(01663) 750716 fax:(01663) 751093 sales@cottspirals.u-net.com www.castspiralstairs.com

Cottam & Preedy Ltd 68 Lower City Road Tividale W. Midlands B69 2HF tel:(0121) 552 5281 fax:(0121) 454 9613

Cottam Bros Ltd Wilson Street North Sheepfolds Industrial Estate Sunderland SR5 1BB tel:(0191) 567 1091 fax:(0191) 510 8187 sales@cottambros.com www.cottambros.com

Council for Aluminium in Building River View House Bond's Mill Stonehouse Glouchestershire GL10 3RF tel:(01453) 828851 fax:(01453) 828861 enquiries@c-a-b.org.uk www.c-a-b.org.uk

Council for Registered Gas Installers, The 1 Elmwood, Chineham Park Crockford Lane Basingstoke Hants RG24 8WG 0870 401 2200 0870 401 2600 enquiries@corgi-gas.com www.corgi-gas.com

Country Manor Bricks The Brickyard East Wall Road Dublin 1 Ireland tel:(00 353) 1836 6901 fax:(00 353) 1855 4743 sales@cmb.ie www.cmb.ie

Courtaulds/Celanese Foleshill Road Coventry CV6 5NL tel:(024) 76 688771

Coverad Ltd 6-7 Leapale Road Guildford GU1 4JX tel:(01483) 43205 fax:(01483) 506712 coverad.uk@virgin.net www.coverad.co.uk

Coverite Ltd Bridge Road Wood Green London N22 7SP tel:(020) 8888 7821 fax:(020) 8889 0731 gavin.hamblett@coverite.co.uk www.coverite.co.uk

Covers Timber Structures Ltd Sussex House Quarry Lane Industrial Estate Chichester W. Sussex PO19 2PE tel:(01243) 531818 fax:(01243) 537373 chichester@covers-timber-structures.ltd.uk www.covers-timber-structures.ltd.uk

Coverscreen Unit 174 78 Marylebone High Street London W1U 5AP tel:(07905) 292101 fax:(01923) 451600

Cowley Structural Timberwork Ltd The Quarry Grantham Road Waddington Lincoln LN5 9NT tel:(01522) 803800 fax:(01522) 803801 admin@cowleytimberwork.co.uk

Cox Building Products Ltd Unit 1 Shaw Road Bushbury Wolverhampton WV10 9LA tel:(01902) 371800 fax:(01902 371810 sales@coxdome.co.uk www.coxbp.com

Cox Long Ltd Airfield Industrial Estate Hixon Stafford Staffordshire ST18 0PA tel:(01889) 270166 fax:(01889) 271041 info@coxlong.com

CPV Ltd Woodington Mill East Wellow Romsey Hants SO51 6DQ tel:(01794) 322884 fax:(01794) 322885 sales@cpv.co.uk www.cpv.co.uk

Cradley Castings Ltd Mill Street Cradley Halesowen W. Midlands B63 2UB tel:(01384) 560601 fax:(01384) 639181 sales@cradleycastings.co.uk www.cradleycastings.co.uk

Craft Jacobs Suchard Ltd Banbury Oxon. OX16 7QU tel:(01295) 264433 fax:(01295) 259018

Crane Fluid Systems Nacton Road Ipswich Suffolk IP3 9QH tel:(01473) 277400 fax:(01473) 277401 enquiries@cranefs.com www.cranefs.com

Cranleigh Brick & Tile Co Ltd Baynards Rudgwick Nr Horsham W. Sussex RH12 3AG tel:(01403) 823251 fax:(01403) 823351

Crawford Amber Ltd Whittle Road Meir,Stoke on Trent Staffs. ST3 7QA tel:(01782) 599899 fax:(01782) 599989 sales@crawfordamber.co.uk www.crawfordamber.co.uk

Crescent Lighting 8 Rivermead, Pipers Lane Thatcham Berks. RG19 4EP tel:(01635) 878888 fax:(01635) 873888 sales@crescent.co.uk www.crescent.co.uk

Crescent of Cambridge Ltd Edison Road St. Ives Cambs. PE27 3LG tel:(01480) 301522 fax:(01480) 494001 info@crescentstairs.co.uk www.crescentstairs.co.uk

Creteco Sales Ltd 17 St Martin's Street Wallingford Oxon. OX10 0EA tel:(01491) 839488 fax:(01491) 833879 www.creteco.co.uk

Crickhowell Furniture Co. Ltd The Granary Standard Street Crickhowell, Powys Wales NP8 1BP tel:(01873) 810973 fax:(01873) 812270 sales@crickhowellfurniture.co.uk www.crickhowellfurniture.co.uk

Crimeguard Systems Ltd Unit 44, Carlyon Road School Road Wembley Middlesex HA0 1JE tel:(020) 810 9933 fax:(020) 810 8899 info@crimeguardsystems.com www.crimeguardsystems.com

Crittall Steel Windows Ltd 39 Durham Street Glasgow G41 1BS tel:(0141) 427 4931 fax:(0141) 427 1463 hq@crittall-windows.co.uk www.crittall-windows.co.uk

Crittall Steel Windows Ltd Springwood Drive Braintree Essex CM7 2YN tel:(01376) 324106 fax:(01376) 349662 hq@crittall-windows.co.uk www.crittall-windows.co.uk

Crosbie Coatings Walsall Street Wolverhampton West Midlands WV1 3LP tel:(01902) 352020 fax:(01902) 456392 enquiries@crosbie-casco.co.uk www.crosbie-casco.co.uk

Cross-Guard International Ltd Bridge House Severn Bridge Riverside North Bewdley Worcestershire DY12 1AB tel:(01299) 406022 fax:(01299) 406023 info@cross-guard.co.uk www.cross-guard.co.uk

Crown Nail Co Ltd Commercial Road Wolverhampton WV1 3QS tel:(01902) 351806 fax:(01902) 871212

Crown Trade Paints Akzo Nobel Decorative Coatings Ltd Crown House, P O Box 37, Hollins Road Darwen Lancs BB3 0BG tel:(01254) 704951 fax:(01254) 870155

CrowsonFabrics Ltd Crowson House Bolton Close Bellbrook Ind Est East Sussex TN22 1QZ tel:(01202) 753277 fax:(01202) 762582 contracts@monkwell.co.uk www.monkwell.com

Crowthorne Fencing Englemere Sawmill London Road Ascot Berks. SL5 8DG tel:(01344) 885451 fax:(01344) 893101

Croxton + Garry Ltd Curtis Road Dorking Surrey RH4 1XA tel:(01306) 886688

Cryotherm Insulation Ltd Hirst Wood Works Hirst Wood Road Shipley W. Yorks. BD18 4BU tel:(01274) 589175 fax:(01274) 593315

Cryselco Ltd Cryselco House 274 Ampthill Road Bedford MK42 9QL tel:(01234) 273355 fax:(01234) 210867

CSI Security LTD Ronald Close Kempston Bedford MK42 7SH tel:(01234) 840840 fax:(01234) 840333 sales@csisec.com www.csisec.com

CSL (Services) Limited Malt Cottages Basford Nottingham NG7 7DX tel:(0115) 978 3177 fax:(0115) 942 2798

CSSP (Construction Software Services Partnership) 29 London Road Bromley Kent BR1 1DG tel:(020) 8460 0022 fax:(020) 8460 1196 enq@cssp.co.uk www.cssp.co.uk

CST Vulcan Cates Park Lane Coates Park Industrial Estate Alfreton Derbys. DE55 4NJ tel:(01773) 835321 fax:(01773) 836578 sales@vulcantanks.co.uk www.vulcantanks.co.uk

CSW Coldform Ltd Valley Way Market Harborough Leicester LE16 7PS tel:(01858) 410600 fax:(01858) 466536 sales@coldform.co.uk

C-TEC (Computionics) Ltd Stephens Way Goose Green, Wigan Lancs. WN3 6PH tel:(01942) 322744 fax:(01942) 829867 sales@c-tec.co.uk www.c-tec.co.uk

CTS Ltd Abbey Road Shepley, Huddersfield W. Yorks. HD8 8BX tel:(01484) 606416 fax:(01484) 608763 ctsltd@clara.net www.ctsbridges.co.uk

CU Lighting Ltd Great Amwell Ware Herts. SG12 9AT tel:(01920) 462272 fax:(01920) 461370

CU Phosco Lighting Charles House, Lower Road Great Amwell Ware Herts. SG12 9TA tel:(01920) 860600 fax:(01920) 485915 sales@cuphosco.co.uk www.cuphosco.com

Cube Collection Ltd 14 Holland Street London W8 4LT tel:(020) 7938 2244 fax:(020) 7938 1920 info@thecube.co.uk www.thecube.co.uk

Cumberland Construction Units 6c-6f Albright Ind Estate Ferry Lane Rainham Essex RM13 9YH tel:(01708) 555226 fax:(01708) 520231 contracts@cumberland-construction.co.uk www.cumberlandconstruction.co.uk

Cummins Power Generation Ltd Manston Park Columbus Avenue, Manston Ramsgate Kent CT12 5BF tel:(01843) 255000 fax:(01843) 255913 cpg.uk@cummins.com www.cumminspower.com

Cuprinol Ltd Wexham Road Slough Berkshire SL2 5DS tel:(01753) 550555 www.cuprinol.co.uk

Curtis Steel Ltd Brookside Works Mill Road Radstock Bath BA3 5TX tel:(01761) 432841 fax:(01761) 433919

Cutting R C & Co Arcadia Avenue London N3 2JU tel:(020) 8371 0001 fax:(020) 8371 0003 info@rccutting.co.uk www.rccutting.co.uk

Cyclone Stripper Ltd Phoenix Works 8 Avery Hill Road New Eltham London SE9 2BD tel:(020) 8850 1458

Dacrylate Paints Ltd Lime Street Kirkby in Ashfield Nottingham NG17 8AL tel:(01623) 753845 fax:(01623) 757151 sales@dacrylate.co.uk www.dacrylate.co.uk

Dalair Ltd Southern Way Wednesbury W. Midlands WS10 7BU tel:(0121) 556 9944 fax:(0121) 502 3124

Dale Joinery Ltd Unit 4A -D Chase Park Ind Est Ring Road Burntwood, Staffordshire WS7 3JQ tel:(01706) 350350 fax:(01706) 350351

Dale Power Systems Electricity Buildings Filey N. Yorks. YO14 9PJ tel:(01723) 514141 fax:(01723) 515723

Dales Fabrications Ltd Crompton Road Industrial Estate Ilkeston Derbys. DE7 4BG tel:(0115) 930 1521 fax:(0115) 930 7625 sales@dales-eaves.co.uk www.dales-eaves.co.uk

Dalesauna Ltd Chatsworth Road Harrogate N. Yorks. NG1 5HS tel:(01423) 522241 fax:(01423) 509450

Dallas & Forrest Ltd 21 Old Castle Road Cathcart Glasgow G44 4BM tel:(0141) 637 3073 fax:(0141) 637 1582

Dalsouple Direct Ltd P O Box 140 Bridgwater Somerset TA5 6ZR tel:(01278) 727777 fax:(01278) 727788 info@dalsouple.com www.dalsouple.com

Dampa (UK) Ltd Berinsfield Wallingford Oxford OX10 7LZ tel:(01865) 342200 fax:(01865) 341482

Dampcoursing Limited 10-12 Dorset Road Tottenham London N15 5AJ tel:(020) 8802 2233 fax:(020) 8809 1839 dampcoursingltd@london.com

Danbury Fencing PLC Olivers Farm Maldon Road Witham Essex CM8 3HY tel:(01376) 502020 fax:(01376) 520500 sales@danburyfencing.com www.danburyfencing.com

Danfoss Ltd Capswood Oxforc Street Denham Bucks UB9 4LH tel:(0870) 241 7030 fax:(0870) 241 7035 uk.controls.sales@danfloss.com www.danfoss.co.uk

Danfoss Randall Ltd Ampthill Road Bedford Beds. MK42 9ER tel:(01234) 364621 fax:(01234) 219705 danfossrandall@danfoss.com www.danfoss-randall.co.uk

Daniel Platt Ltd Canal Lane Tunstall Stoke-on-Trent Staffs ST6 4NY tel:(01782) 577187 fax:(01782) 577877 sales@danielplatt.co.uk www.danielplatt.co.uk

Dantherm Ltd Hither Green Clevedon Avon BS21 6XT tel:(01275) 876851 fax:(01275) 343086 ikf.ltd.uk@dantherm.com

Darby Glass Ltd Darby House Sunningdale Road Scunthorpe North Lincolnshire DN17 2SS tel:(01724) 280044 fax:(01724) 868295 sales@darbyglass.co.uk www.darbyglass.co.uk

Darfen Durafencing Unit 12 Smallford Works Smallford Lane St. Albans Herts AL4 0SA tel:(01727) 828290 fax:(01727) 828299 darfen@stalbans.co.uk www.darfen.co.uk

Darfen Durafencing, Scottish Region Units 7-8 Rochsolloch Road Industrial Estate Airdrie Scotland ML6 9BA tel:(01236) 755001 fax:(01236) 747012

Daro Factors Ltd 80-84 Wallis Road London E9 5LW tel:(020) 8510 4000 fax:(020) 8510 4001 sales@daro.com www.daro.com

Dartford Portable Buildings Ltd 389-397 Princes Road Dartford Kent DA1 1JU tel:(01322) 229521 fax:(01322) 221948 mark@dpbl.co.uk www.dpbl.co.uk

Daryl Industries Ltd Alfred Road Wallasey Wirral L44 7HY tel:(0151) 606 5000 fax:(0151) 638 0303 daryl@daryl-showers.co.uk

Dauphin PLC Peter Street Blackburn Lancs. BB1 5HL tel:(01254) 52220 fax:(01254) 680401 sb@dauphin.plc.uk www.dauphin.plc.uk

Davant Products Ltd Davant House, Jugs Green Business Park Jugs Green Staplow, nr Ledbury Herefordshire HR8 1NR tel:(01531) 640880 fax:(01531) 640827 info@davant.co.uk www.davant.co.uk

David Ball Group Plc David Ball Buildings Huntington Road Bar Hill Cambridge CB3 8HN tel:(01954) 780687 fax:(01954) 782912 sales@davidballgroup.com www.davidballgroup.com

David Mclean Group Enterprise House 28 Parkway Industrial Park Deeside Flintshire CH5 2NS tel:(01244) 285 500 fax:(01244) 285 501 bill.addy@davidmclean.co.uk www.davidmclean.co.uk

Davies John Interiors Ltd Century House Century Road Oldbury W. Midlands B69 4AD tel:(0121) 511 1000 fax:(0121) 511 1116 jdi@jdinteriors.co.uk www.johndavisinteriors.co.uk

Davis Burrow & Sons Wilson Street North Sunderland Tyne & Wear SR5 1BB tel:(0191) 567 5889 fax:(0191) 510 8187 sales@cottambros.com www.cottambros.com

Dawson McDonald & Dawson Ltd Compton Works Ashbourne Derbys. DE6 1DB tel:(01335) 343184 fax:(01335) 342540 sales@dmd.org.uk www.dmd.org.uk

Dawson-Keith Ltd Dekay House Brookhampton Lane Havant Hants. PO9 1HQ tel:(023) 92474122 fax:(023) 92470921

Day Marketing Ltd 64A Guildford Street Chertsey Surrey KT16 9BD tel:(01932) 567341 fax:(01932) 568858

De Longhi Radiators srl 19 Intake Lane Dunnington York YO1 5NX tel:(01904) 488212

Deanes Furniture Ltd Wardsgrove Lane Chalgrove Oxfordshire OX44 7TM tel:(01865) 891444 fax:(01865) 891427

Deans Blinds & Awnings (UK) Ltd Unit 4, Haslemere Industrial Estate Ravensbury Terrace London SW18 4SE tel:(020) 8947 8931 fax:(020) 8947 8336 info@deansblinds.com www.deansblinds.com

Deb Ltd Spencer Road Belper Derbys. DE56 1JX tel:(01773) 596700 fax:(01773) 822548 enquiry@deb.co.uk www.deb.co.uk

Deborah Services Limited 10 South Parade Wakefield W. Yorks. WF1 1LS tel:(01924) 378222 fax:(01924) 366250 enquiries@deborahservices.co.uk www.deborahservices.co.uk

Deceuninck Ltd Unit 2 Stanier Road, Porte Marsh Industrial Estate Calne Wilts. SN11 9PX tel:(01249) 816969 fax:(01249) 815234 deceuninck.ltd@deceuninck.com www.deceuninck.com

Décor Ireland 58 Trench Road Mallusk Newtown Abbey BT36 4TY tel:(028) 90 834910

Decorfix Ltd Chapel Works Chapel Lane Lower Halstow, Sittingbourne Kent ME9 7AB tel:(01795) 843124 fax:(01795) 842465

Decra Ltd 34 Forest Business Park Argoll Avenue Leyton London E10 7FB tel:(020) 8520 4371 fax:(020) 8521 0605 sales@decraltd.co.uk www.decraltd.co.uk

Decra Roofing Systems Unit 3, Faraday Center Faraday Road Crawley W. Sussex RH10 2PX tel:(01293) 545058 fax:(01293) 562709 technical@decra.co.uk www.decra.co.uk/

Dee-Organ Ltd 5 Sandyford Road Paisley Renfrewshire PA3 4HP tel:(0141) 889 7000 fax:(0141) 889 7764 signs@dee-organ.co.uk

Deepdale Engineering Co Ltd Pedmore Road Dudley W. Midlands DY2 0RD tel:(01384) 480022 fax:(01384) 480489 sales@deepdale-eng.co.uk www.deepdale-eng.co.uk

Degussa Construction Chemicals (UK) Albany House Swinton Hall Road Swinton Manchester M27 4DT tel:(0161) 794 7411 fax:(0161) 727 8547 mbtfeb@degussa.com www.degussa-cc.co.uk

Degussa Ltd Winterton House Winterton Way Macclesfield Cheshire SK11 0LP tel:(01625) 503050 fax:(01625) 502096
DEL Piling Contractors Park Road Holmewood Industrial Park Chesterfield Derbys. S42 5UY tel:(01246) 855855 fax:(01246) 854909
Delbridge Lifts Ltd Steetley Industrial Estate Sangwin Road Corseley, Bilston W. Midlands WV14 9EE tel:(01902) 662998 fax:(01902) 674998
Deloro Satellite Ltd Unit 103 Cheney Monor Swindon Wilts. SN2 2PW tel:(01793) 822451 fax:(01793) 823814
Delta Crompton Cables Ltd Milmarsh Lane Brimsdown Enfield Middx. EN3 7QD tel:(020) 8804 2468 fax:(020) 8443 2281
Denco Ltd Holmer Road Hereford HR4 9SJ tel:(01432) 277277 fax:(01432) 268005 www.denco.co.uk
Denka International Red Roofs Chinnor Road Thame Oxon. OX9 3RF tel:(01844) 216754 fax:(01844) 216141 101336.1667@compuserve.com
Denmans Montrose Ltd 28-30 South Bank Business Centre Ponton Road Vauxhall London SW8 5BL tel:(020) 7622 1221 fax:(020) 7622 5152
Dennis & Robinson Ltd Blenhiem Road Churchill Industrial Estate Lancing W. Sussex BN15 8HU tel:(01903) 524300 fax:(01903) 750679 manhattan@manhattan.co.uk www.manhattan.co.uk
Department of Environment Romney House 43 Marsham Street London SW1B 3EB tel:(020) 7276 3000
Deralam Laminates Ltd West Coast Park Bradley Lane Standish Wigan Lancs WN6 OYR tel:(01257) 478540 fax:(01257) 478550 sales@deralam.co.uk
Derby Timber Supplies 3 John Street Derby DE1 2LU tel:(01332) 348340 fax:(01332) 385573
Dermide Ltd Westfield Mill Carr Road Barnoldswick Lancs. BB18 5UU tel:(01282) 812581 fax:(01282) 812366 sales@dermide.freeserve.co.uk www.dermide.sagenet.co.uk
Design & Display (Sales) Ltd Dunstable Studios Dunstable Road Richmond Surrey TW9 1UH tel:(020)8940 2256 fax:(020) 8948 7997 grp@design-and-display.co.uk www.design-and-display.co.uk
Design Line 45 Lawley Middleway Aston Birmingham B4 7XH tel:(0121) 359 6941 fax:(0121) 359 5758
Design Windows (South West) Ltd PO Box 165 Fishponds Bristol BS16 7YP tel:(0117) 9702585 fax:(0117) 9567608 www.designrooflights.co.uk
Designed for Sound Ltd 61-67 Rectory Road Wivenhoe Essex CO7 9ES tel:(01206) 827171 fax:(01206) 826936 rs@d4s.co.uk www.d4s.co.uk
Designplan Lighting Ltd Wealdstone Road Kimpton Industrial Estate Sutton Surrey SM3 9RW tel:(020) 8254 2020 fax:(020) 8644 4253 sales@designplan.co.uk www.designplan.co.uk
Desking Systems Ltd Warpsgrove Lane Chalgrove Oxfordshire OX44 7TH tel:(01865) 891444 fax:(01865) 891427 sales@desking.co.uk www.desking.co.uk
Desney Products Ltd Units 5-6 Marley Industrial Estate Southam Road Banbury Oxon. OX16 7RL tel:(01295) 270848 fax:(01295) 270965 sales@desney.demon.co.uk
Detlef Muller Ltd 82 Chobham Road Frimley Camberley Surrey GU16 5PP tel:(01276) 61967 fax:(01276) 64711 detlef.muller@btclick.com
Deva Tap Co Ltd Brooklands Mill English Street Leigh Lancs. WN7 3EH tel:(01942) 680177 fax:(01942) 680190 sales@devatap.co.uk www.devatap.co.uk
Dew Construction Limited 1st Floor, 10 Lake End Court Taplow Road Taplow, Maidenhead Berks. SL9 0JQ tel:(01628) 669888 fax:(01628) 669800
Dewey Waters Ltd Cox Green Wrington Bristol BS40 5QS tel:(01934) 862601 fax:(01934) 862604 sales@deweywaters.co.uk www.deweywaters.co.uk
Dewhurst plc Inverness Road Hounslow Middx. TW3 3LT tel:(020) 8607 7300 fax:(020) 8572 5986 info@dewhurst.co.uk www.dewhurst.co.uk
Dewplan Ltd Maple Court 17-21 Queens Road High Wycombe Bucks. HP13 6AQ tel:(01494) 557300 fax:(01494) 465489 post@dewplan.co.uk
Dexine Rubber Co Ltd Spotland Bridge Works Rochdale Lancs. OL12 6AU tel:(01706) 40011 fax:(01706) 527714
Dexion Ltd Maylands Avenue Hemel Hempstead Herts. HP2 7DE tel:(01442) 242261 fax:(01442) 217145 enquiries@dexion.co.uk www.dexion.com
Dexter Paints Ltd Dexter House Trafalgar Street Burnley Lancs. B11 1RE tel:(01282) 23361 fax:(01282) 414573 sales@dexter-paints.co.uk www.dexter-paints.co.uk
Dezurik International Ltd Nelson Way Nelson Industrial Estate Cramlington Northumberland NE23 9BJ tel:(01670) 714111 fax:(01670) 737813
Diamond Merchants 43 Acre Lane Brixton London SW2 5TN tel:(020) 7274 6624 fax:(020) 7978 8370 www.diamondmerchants.co.uk
Dicon Safety Products (UK) Ltd P.O. Box 402 Gloucester GL2 9YB tel:(01452) 714999 fax:(01452 713103 sales@dicon.w-net.com www.diconsafety.co.uk
Didsbury Engineering Co Ltd Clifton works Manor Road Levenshulme Manchester M19 3EJ tel:(0161) 224 6224 fax:(0161) 224 2098 sales@didsbury.com
Diespeker Marble & Terrazzo Ltd 132-136 Ormside Street London SE15 1TF tel:(020) 7358 0160 fax:(020) 7639 1695 sales@diespeker.co.uk www.diespeker.co.uk
Dinnington Fencing Co. Limited North Hill Dinnington Newcastle-upon-Tyne NE13 7LG tel:(01661) 824046 fax:(01661) 872234 info@dinningtonfencing.co.uk www.dinningtonfencing.co.uk
Direction Group Ltd 4 Riverside Business Centre Walnut Tree Close Guildford Surrey GU1 4UG tel:(01483) 455555 fax:(01483) 451590
Distribution tube Oxford Street Bilston West Midlands WV14 7DS tel:(01902) 499700 fax:(01902 405838 wtube@globalnet www.globalnet.co.uk/~wtube/
Dividers Modernfold Ltd Great Gutter Lane Willerby Hull East Yorkshire HU10 6BS tel:(01482) 651331 fax:(01482) 651497 sales@esperowalls.com www.esperowalls.com
Dixon Turner Wallcoverings Brecon Works Henfaes Lane Welshpool Powys SY21 7BE tel:(0870) 606 1237 fax:(0870) 606 1239 enquiries@dixon-turner.co.uk www.dixon-turner.co.uk
DJ Profiles Stafford Park 15 Telford Shropshire TF3 3BB tel:(01952) 290303 fax:(01952) 290226 sales@djprofiles.co.uk www.djprofiles.co.uk
DLW Flooring St Centurion Court Milton Park Addington Oxon OX14 4RY tel:(01642) 763224 fax:(01642) 750213 sales@dlw.co.uk www.dlw.co.uk
DLW Flooring Ltd Centurian Court Milton Park Abingdon Oxon. OX14 4RY tel:(01235) 831296 fax:(01235) 861016 sales@dlw.co.uk www.dlw.co.uk
Docherty New Vent Chimney Group Unit 3 Sawmill Road, Redshute Hill Indusrial Estate Hermitage, Newbury Berks. RG18 9QL tel:(01635) 200145 fax:(01635) 201737 sales@docherty.co.uk
Doherty Medical Ltd Doherty House 278 Alma Road Enfield Middx. EN3 7BH tel:(020) 8804 1244 fax:(020) 8804 9314
Dolphin Showers Ltd Bromwich Road Worcester WR2 4DB tel:(01905) 748500 fax:(01905) 429034
Dom-Nemef-Corbin Unit 1, Harolds Close Harolds Road Harlow Essex CM19 5TH tel:(01279) 454709 fax:(01279) 454711
Domus Tiles Ltd Cantebury Court Brixton Road Battersea London SW9 6TA tel:(020) 7223 5555 fax:(020) 7924 2556 info@domustiles.co.uk www.domustiles.co.uk
Domus Ventilation Ltd. Bearwalden House, Bearwalden Business Park Royston Road Wendens Ambo, Saffron Waldon Essex CB11 4JX tel:(01799) 540602 fax:(01799) 541143 info@domusventilation.co.uk www.domusventilation.com
Don & Low Ltd Newfordpark House Glamis Road Forfar Angus DD8 IFR tel:(01307) 452200 fax:(01307) 452300 lotrak@donlow.co.uk www.donlow.co.uk
Don Construction Products Ltd Churnetside Business Park Station Road Cheddleton, Leek Staffordshire ST13 7RS tel:(01538) 361799 fax:(01538) 361899 info@donconstruction.co.uk www.donconstruction.co.uk
Don Heating Products Ltd Sowton Ind Est Osprey Road Exeter Somerset TA21 8SS tel:(01823) 663181 fax:(01823) 665889 eastwood@don-products.co.uk www.don-products.co.uk
Door and Shutter Manufacturers Association 42 Heath Street Tamworth Staffs B79 7JH 01827 52337 01827 310827 info@dsma.org.uk www.dsma.org.uk
Door Panels PLC Rectory Road Upton-upon-Severn Worcs. WR8 0LX tel:(01684) 594561 fax:(01684) 593431
Doorfit Products Ltd Icknield House Heaton Street, Hockley Birmingham W. Midlands B18 5BA tel:(0121) 523 4171 fax:(0121) 554 3859 enquiries@doorfit.co.uk www.doorfit.co.uk
Doors & Hardware Ltd Taskmaster Works Maybrook Road Minworth, Sutton Coldfield W. Midlands B76 1AL tel:(0121) 351 5276 fax:(0121) 313 1228 sales@doors-and-hardware.com www.doors-and-hardware.com
Dorma Door Services Ltd Woodside Industrial Park Works Road Letchworth Herts SG6 1LA tel:(01462) 472502 fax:(01462) 472503 service@dorma-uk.co.uk www.dorma-uk.co.uk
Dorman Smith Switchgear Ltd Blackpool Road Preston Lancs. PR2 2DQ tel:(01772) 325400 fax:(01772) 726276 brian.anderson@dormansmith.co.uk www.dormansmith.co.uk
Douglas Industrial Ltd Wards Farm Industial Estate Greenmoore, Woodcote Reading RG8 0RB tel:(01491) 682100 fax:(01491) 682037 sales@douglasindustrial.co.uk
Dover Engineering Works Ltd Hammond House Poulton Road Holnestone Dover CT17 0UF tel:(01304) 203545
Dover Trussed Roof Co Ltd Shelvin Manor Shelvin, Canterbury Kent CT4 6RL tel:(01303) 844303 fax:(01303) 844342 sales@dovertruss.co.uk www.dovertruss.co.uk
Dow Construction Products Lakeside House Stockley Park Uxbridge Middx. UB11 1BE tel:(020) 8848 5050 fax:(020) 8848 5413
Dow Corning Ltd Meriden Business Park Copse Drive, Allesley Coventry CV5 9RG tel:(01676) 528000 fax:(01676) 528001 www.dowcorning.com
Down & Francis Ltd, Division of Metalrax PLC Ardath Road Kings Norton Birmingham B38 9PN tel:(0121) 433 3300 fax:(0121) 459 9222 enquiries@downandfrances.co.uk www.downandfrances.co.uk
DP Sercurity Ryecroft Buildings Cireon Street Green Road Kent DA2 8DX tel:(01322) 278178 fax:(01322) 284315 dpsecurity@btinternet.com www.dpsecurity.co.uk
DPC Screeding Ltd Brunwick Industrial Estate Brunswick Village Newcastle-upon-Tyne NE13 7BA tel:(0191) 236 4226 fax:(0191) 236 2242 dpcscreeding@btconnect.com
Draeger Safety UK Ltd Ullswater Close Kitty Brewster Industrial Estate Blyth Northumberland NE24 4RG tel:(01670) 352891 fax:(01670) 356266 marketing@draeger.ltd.uk www.draeger-safety.co.uk
Dragonair Ltd Fitzherbert Road Units 7-8, Dragon Industrial Centre Fallington, Portsmouth Hants. PO6 1SQ tel:(023) 92376451 fax:(023) 92370411
Drainage Systems Brickhouse Lane West Bromwich West Midlands B70 0DX tel:(0121) 521 2900 fax:(0121) 522 2197 drainage.system@dial.pipex.com
Drakard & Humble Ltd Brunleys Kiln Farm Milton Keynes MK11 3EP tel:(01908) 567909 fax:(01908) 563527 sales@drakardsigns.co.uk www.drakardsigns.co.uk
Draught Proofing Advisory Association PO Box 12 Haslemere Surrey GU27 3AH tel:(01428) 654011 fax:(01428) 651401 dpaaassociation@aol.com www.insulationassociation.org.uk
Dravo Environmental Services Sir Henry Parkes Road Canley Coventry CV5 6BN tel:(024) 76717170 fax:(024) 76714978 dravo@covrad.co.uk
Drawn Metal Ltd Swinnow Lane Bramley Leeds LS13 4NE tel:(0113) 256 5661 fax:(0113) 239 3194 drawmet@btinternet.com www.drawmet.com
Drayton Controls (Engineering) Ltd Chantry Close West Drayton Middx. UB7 7SP tel:(01895) 444012 fax:(01895) 441288
Drilling & Sawing Association, The PO Box 16 Belper Derbyshire DE56 4WF 01773 820000 01773 821284 dsa@drillandsaw.org.uk www.drillandsaw.org.uk
Drugasar Ltd Deans Road Swinton Manchester M27 3JH tel:(0161) 793 8700 fax:(0161) 727 8057
Dry Stone Walling Association of Great Britain Lane Farm Crooklands Milnthorpe Cumbria LA7 7NH tel:(015395) 67953 information@dswa.org.uk www.dswa.org.uk
DSM Engineering Plastic Products (UK) Ltd 83 Bridge Road East Welwyn Garden City Hertfordshire. AL7 1LA tel:(01707) 361800 fax:(01707) 361801 info@dsmepp.co.uk www.dsmepp.com
DSM UK Ltd DSM House Papermill Drive Redditch Worcs. B98 8QJ tel:(01527) 590590 fax:(01527) 590555
Du Pont (UK) Ltd Maylands Avenue Hemel Hempstead Herts. HP2 7DP tel:(01442) 61251 fax:(01442) 249463
Ducal Ltd Andover Hants. SP10 5AZ tel:(01264) 332720 fax:(01264) 334046
Ducana Surrey Ltd 5 Sandrock Hill Road Farnham Surrey GU10 4NR tel:(01252) 715768 fax:(01252) 737152
Duct Access Covers Ltd Newtown Industrial Estate Crosskeys Newport Gwent NP1 7PZ tel:(01495) 272727 fax:(01495) 270978
Ductform Engineering Ltd 141 Barkby Road Leicester LE4 7LW tel:(0116) 276 6636 fax:(0116) 240 0426
Ductile Sections Planetary Road Willenhall W. Midlands WV13 3SW tel:(01902) 739739 fax:(01902) 862841
Dudley Factory Doors Limited Unit G6 Grice Street West Bromwich West Midlands B70 7EZ tel:(0121) 558 8989 fax:(0121) 558 4616
Dudley Thomas Ltd PO Box 28 Birmingham New Road Dudley DY1 4SN tel:(0121) 557 5411 fax:(0121) 557 5345 info@thomasdudley.co.uk
Dufaylite Developments Ltd Cromwell Road St Neots Huntingdon Cambs. PE19 1QW tel:(01480) 215000 fax:(01480) 405526
Dunbrik (Yorks) Ltd Ferry Lane Stanley Ferry Wakefield West Yorkshire WF3 4LT tel:(01924) 373694 fax:(01924) 383459 Flues@dunbrik.co.uk www.dunbrik.co.uk
Dundec Contracts Ltd 10 Boundary Business Court Church Road Mitcham Surrey CR4 3TD tel:(020) 86467272 fax:(020) 86465100 marketing@dundec.com www.dundec.com
Dunham Bush Ltd European Headquaters Downley Road Havant Hants. PO9 2JD tel:(02392) 477700 fax:(012392) 450396 info@dunham-bush.co.uk
Dunhams of Norwich Hellesdon Park Road Norwich NR6 5DR tel:(01603) 424855
Dunlop Adhesives Building Adhesives Longton Road Trentham Stroke-on-Trent ST4 8JB tel:(0121) 373 8101 fax:(0121) 384 2826
Dunphy Combustion Ltd Queensway Rochdale Lancs. OL11 2SL tel:(01706) 649217 fax:(01706) 655512 www.dunphy.co.uk
Dunsley Heat Ltd Fearnough Huddersfield Road Holmfirth W. Yorks. HD7 2TU tel:(01484) 682635 fax:(01484) 688428 www.i-inetdunsleyht
Duo-Fast (UK) Ltd Gazely House Rockingham Drive Milton Keynes Bucks MK14 6PD tel:(01908) 667788 fax:(01908) 672689 info@duo-fast.co.uk www.paslode.co.uk
Duplus Architectural Systems Ltd 370 Melton Road Leicester LE4 7SL 0116 2610710 0116 2610539 sales@duplus.co.uk www.duplus.co.uk
Duplus Domes Ltd 370 Melton Road Leicester Leicestershire LE4 7SL tel:(0116) 261 0710 fax:(0116) 261 0539 sales@duplus.co.uk www.duplus.co.uk
Dupre Vermiculite Tamworth Road Hertford Herts SG13 7DL tel:(01992) 582541 fax:(01992) 553436 dupre@microfine.co.uk www.dupre.microfine.co.uk
Durabella System Floors Ltd Talisman Square Kenilworth Warwickshire CV8 1JB tel:(0870) 7894000 fax:(0870) 7894100 info@durabella-system-floors.co.uk www.durabella-system-floors.co.uk
Durable 10-11 Southview Park Marsack Street Caversham Reading RG4 5AF tel:(0118) 9483500 fax:(01734) 462114 durable_co@msn.com www.durable.co.uk
Durable Berkeley Co Ltd 10-13 Southview Park Marsack Street Caversham Reading RG4 5AF tel:(0118) 948 3500 fax:(0118) 946 2114 durable_co@msn.com www.durable.co.uk
Durable Contracts Ltd Durable House Crabtree Manorway Belvedere Kent DA17 6AB tel:(020) 8311 1211 fax:(020) 8310 7893 sales@durable-online.com www.durable-online.com
Duralock (UK) Ltd Unit 12 Tamar View Ind Est Saltash Cornwall PL12 6LD tel:(01752) 484085 fax:(01752) 484086 duralock.ltd@virgin.net www.duralock.com
Duratan Ltd West Street Buckingham MK18 1HE tel:(01280) 814048 fax:(01280) 817842
Durey Casting Ltd Hawley Road Dartford Kent DA1 1PU tel:(01322) 272424 fax:(01322) 288073 sales@dureycastings.co.uk www.dureycastings.co.uk
Durox Building Products Ltd Linford Stanford le Hope Essex SS17 0PY tel:(01375) 673344 fax:(01735) 642460
Duroy Fibreglass Mouldings Ltd Mercury Yacht Harbour Satchell Lane Hamble Southampton SO31 4HQ tel:(023) 80453781 fax:(023) 80455538 duroygrp@aol.com
DW Windsor Lighting Pindar Road Hoddesdon Herts. EN11 0DX tel:(01992) 474600 fax:(01992) 474601 info@dwwindsor.co.uk www.dwwindsor.co.uk
Dycem Ltd Ashley Trading Estate Bristol BS2 9BB tel:(0117) 955 9921 fax:(0117) 954 1194 uk@dycem.com www.dycem.com
Dyer Environmental Controls 1 Brooklyn Road Cheadle Cheshire SK8 1BS tel:(0161) 4914840 fax:(0161) 4914841
Dyke Chemicals Ltd PO Box 158 Cardiff CF1 9RU tel:(01932) 866096 fax:(01932) 866097
Dynamic Plastics Ltd Egerton Works Egerton Street Farnworth, Bolton Lancashire BL4 7ER tel:(01204 575151 fax:(01204) 861219
Dynaspan (UK) Ltd Parkhouse High Street Colnbrook, Slough Berks. SL3 0LX tel:(01753) 689977 fax:(01753) 689970
Dyno Rod PLC Zockoll House 143 Maple Road Surbiton Surrey KT6 4BJ tel:(020) 8481 2200 fax:(020) 8481 2288 postmaster@dyno.com www.dyno.com
E. H. S. Roofing Ltd 42 Flint Green Road Acocks Green Birmingham B27 6QA tel:(0121) 764 4500 fax:(0121) 706 2650 sales@ehsroofing.com www.ehsroofing.com
E+R Polymers Ltd V.C. Works High Street East Glossop Derbys. SK13 8QB tel:(01457) 853254 fax:(01457) 869470 info@eandrpolmers.com www.eandrpolymers.com
Eagles, William, Ltd Liverpool Street Salford Manchester M5 4LP tel:(0161) 736 1661 fax:(0161) 745 7765
Earth Anchors Ltd 15 Campbell Road Croydon Surrey CR0 2SQ tel:(020) 8684 9601 fax:(020) 8684 2230 www.earth-anchors.com
Ease-E-Load Trolleys Ltd Saunders House Moor Lane Wiltan Birmingham B6 7HH tel:(0121) 356 7411 fax:(0121) 344 3358
Easi-Fall International Ltd Brindley House Brindley Avenue Sale Cheshire M33 7BE tel:(0161) 973 0304 fax:(0161) 969 5009 e.fallint@zen.co.uk www.easifall.co.uk
Easy Arches Ltd Lodgefield Road Halesowen W. Midlands B62 8RT tel:(0121) 559 5613 fax:(0121) 561 2895
Eaton Electric Limited Grimshaw Lane Middleton Manchester M23 1GQ tel:(0161) 655 8900 08700 507 525 ukresiorders@eaton.com www.mem250.com
Eaton Elizabeth 85 Bourne Street London SW1W 8HF tel:(020) 7730 2262 fax:(020) 7730 7294

Eaton-Williams Group Ltd Station Road Edenbridge Kent TN8 6EG tel:(01732) 866055 fax:(01732) 863461 www.eaton-williams.com

EBC UK Ltd Unit 16, Old brewery Yard Kilton Road Worksop Notts. S80 2DE tel:(01909) 479276 fax:(01909) 479278 rob@ebc.force9.co.uk www.e-b-c-uk.com

Eberle GmbH 25 Gosforth Close Sandy Business Park Sandy Beds. SG19 1RB tel:(01767) 692323 fax:(01767) 692333

Ebm-papst (UK) Limited Chelmsford Business Park Chelmsford Essex CM2 5EZ tel:(01245) 468555 fax:(01245) 466336 sales@ebmpapst.com www.ebm-papst.co.uk

Ebor Concretes Ltd PO Box 4 Ure Bank Ripon N. Yorks. HG4 1JE tel:(01765) 604351 fax:(01765) 690065 sales@eborconcrete.co.uk www.eborconcrete.co.uk

EBS Services Ltd Thames House Longreach Road Barking Essex IG11 0JR tel:(020) 8594 5255 fax:(020) 8594 5404

Eclipse Blind Systems Ltd Inchinnan Business Park !0 Founain Cress Renfrew Renfkewshire PA4 9RE tel:(0141) 812 3322 fax:(0141) 812 5253 email@compuserv.com www.eclipse-blinds.com

Eclipse Sprayers Ltd 120 Beakes Road Smethwick W. Midlands B67 5AB tel:(0121) 420 2494 fax:(0121) 429 1668

Ecolec Sharrocks Street Wolverhampton W. Midlands WV1 3RP tel:(01902) 457575 fax:(01902) 457797 ecolec@thama.co.uk www.ecolec.co.uk

Ecowater Systems Ltd Mill Road Stockenchurch Bucks. HP14 3TP tel:(01494) 484000 fax:(01494) 484396 info@ecowater.co.uk www.ecowater.co.uk

Edenaire Ltd Station Road Edenbridge Kent TN8 6EG tel:(01732) 866066 fax:(01732) 866653

Edmonds A & Co Ltd 91 Constitution Hill Birmingham B19 3JY tel:(0121) 236 8351 fax:(0121) 236 4793 edmondsmark@tesco.net ww.edmondsa.co.uk

EDS Shutter Door Systems Ltd Sava Works Barlow Way, Fairview Industrial Park Rainham Essex RM13 8UE tel:(01708) 551964 fax:(01708) 551964

Edward Benton & Co Ltd Creteangle Works Brook Lane, Ferring Worthing West Sussex BN12 5LP tel:(01903) 241349 fax:(01903) 700213 sales@creteangle.com www.creteangle.com

EFG Business Furniture (UK) Ltd 3 Finway Road Hemel Hempstead Herts. HP2 7PT tel:(01442) 216666 fax:(01442) 217023

Ega Ltd St Asaph Clwyd N. Wales LI17 0ER tel:(01745) 584927

Egetaepper (UK) Ltd Ege House Drumhead Road, Chorley North Business Park Chorley Lancs. PR6 7BZ tel:(01257) 239000 fax:(01257) 239001

Egger (UK) Ltd Anick Range Road Hexham Northumberland NE46 4JS tel:(01434) 602191 fax:(01434) 605103 building.uk@egger.com www.egger.co.uk

EITE & Associates 58 Carsington Cresent Allestree Derby DE22 2QZ tel:(01332) 559929 fax:(01332) 559929 petereite@associates83.freeserve.co.uk

EJOT Ecofast Ltd EJOT House Fox Way Hunslet Leeds LS10 IPS tel:(0113) 2470880 fax:(0113) 2470882 sales@ejot.co.uk www.ejot.co.uk

Elan-Dragonair Units 7-8, Dragon Industrial Centre Fitzherbert Road Farlington Portsmouth PO6 1SQ tel:(023) 92376451 fax:(023) 92370411

Electrak International Ltd Unit 12, No.1 Industrial Estate Medomsley Road Consett Co. Durham DH8 6SR tel:(01207) 503400 fax:(01207) 501799 sales@electrak.co.uk www.electrak.co.uk

Electric Elements Co, The Tokenhouse Yard Bridlesmith Gate Nottingham NG1 2HH tel:(0115) 9505253 fax:(0115) 9588283

Electrical Contractors Association, The Esca House 34 Palace Court London W2 4HY tel:(020) 7313 4800 fax:(020) 7221 7344 electricalcontractors@eca.co.uk www.eca.co.uk

Electrical Review Quadrant House The Quadrant Sutton SM2 tel:(020) 8652 3492

Electrical Times Quadrant House The Quadrant Sutton SM2 tel:(020) 8652 8735

Electrix International Ltd Dovecot Hill South Church Enterprise Park Bishop Auckland Co. Durham DL14 6XP tel:(01388) 774455 fax:(01388) 777359 enquires@electrix.co.uk www.electrix.co.uk

Electro Mechanical Systems Ltd Eros House Calleva Industrial Park Aldermaston Reading RG7 8LN tel:(0118) 981 7391 fax:(0118) 981 7613 barriers@ems-ltd.com www.ems-limited.co.uk

Electrolux Domestic Appliances PO Box 545 55-57 High Street Slough Berks. SL1 9BG tel:(0990) 146146 fax:(01753) 872381

Electrolux Foodservice Crystal Court, Aston Cross Business Park Rocky Lane, Aston Birmingham West Midlands B6 5RQ tel:(0121)220 2800 fax:(0121) 220 2801 fse.info@electrolux.co.uk www.foodservice.electrolux.com

Electrolux Laundry Systems Unit 3A Humphrys Road, Woodside Estate Dunstable Bedfordshire LU5 4TP tel:(0870) 0604118 fax:(0870) 0604113 elsinfo@electrolux.co.uk www.electrolux-wascator.co.uk

Electrosonic Ltd Hawley Mill Hawley Road Dartford Kent DA2 7SY tel:(01322) 222211 fax:(01322) 282282

Elementis Pigments Liliput Road Bracksmills Industrial Estate Northampton NN4 7DT tel:(01604) 827403 fax:(01604) 827400 pigmentsinfo.eu@elementis.com www.elementis.com

Elequip Projects Ltd Power Supplies Division Gloucester Crescent Wigston, Leicester Leicestershire LE18 4YN tel:(01162) 222 000 fax:(01162) 779369 sales@elequip.co.uk www.elequip.com

Elga Ltd High Street Lane End High Wycombe Bucks. HP14 3JH tel:(01494) 887700 fax:(01494) 881007 elga@elga.demon.co.uk www.elga.co.uk

Elite Trade Kitchens Ltd 90 Willesden Lane Kiburn London NW6 7TA tel:(020) 7328 1243 fax:(020) 7328 1243 sales@elitekitchens.co.uk www.elitekitchens.co.uk

Elizabethan Textiles Castle Bank Mills Portobello Road Wakefield West Yorkshire WF1 5PS Tel:(01924) 290711 Fax:(01924) 290334

Elkay Electrical Manufacturing Co Ltd Unit 18, Mochdre Industrial Estate Newtown Powys SY16 4LF tel:(01686) 627000 fax:(01686) 628276 elkay@co.uk

Elkosta; (UK) Ltd Bishops Hull, Taunton Somerset TA1 5EA tel:(01823) 271911 fax:(01823) 335763 info@elkosta.co.uk www.elkosta.com

Elliott Construction Division Braemar House Snelsins Road Cleckheaton W. Yorks. BD19 3UE tel:(01274) 863221 fax:(01274) 861582 construction@elliott-group.co.uk www.elliottconstruction.co.uk

Elliott Group Ltd Delta Way Cannock Staffordshire WS11 3BE tel:(01543) 404040 fax:(01543) 572710 sales@elliott-group.co.uk www.elliott-group.co.uk

Elliott John H (Monostamp) Ltd 4 Evans Street Off Milton Street Sheffield S3 7WE tel:(0114) 272 6141 fax:(0114) 272 3218

Ellis & McDougall Lifts Ltd Argill Works 86 Broad Street Bridgestone Glasgow G40 2PX tel:(0141) 554 7604 fax:(0141) 554 6762

Ellis J T & Co Ltd Kilner Bank Industrial Esatae Silver Street Huddersfield W. Yorks. HD5 9BA tel:(01484) 514212 fax:(01484) 456433/533454 sales@ellisfurniture.co.uk www.ellisfurniture.co.uk

Ellis Miller Ltd Market House 62 St Judes Road Englefield Green Surrey TW20 0BU tel:(01784) 435302 fax:(01784) 471846 sales@ellismiller.com

Ellis Patents High Street Rillington, Malton N. Yorks. YO17 8LA tel:(01944) 758395 fax:(01944) 758808 sales@ellispatents.co.uk www.ellispatents.co.uk

Elster Metering Ltd Pondwicks Road Luton Beds. LU1 3LJ tel:(01582) 402020 fax:(01582) 438052 water.metering@gb.elster.com www.elstermetering.com

Elsy & Gibbons Ltd Simonside South Shields Tyne & Wear N34 9PE tel:(0191) 427 0777 fax:(0191) 427 0888

Elta Fans Ltd 15 Barnes Wallis Road Segensworth East Industrial Estate Fareham Hants. PO15 5ST tel:(01489) 583044 fax:(01489) 584699

Eltron Chromalox Eltron House 28 Whitehorse Road Croydon Surrey CR0 2JA tel:(020) 8665 8900 fax:(020) 8689 0571 uksales@chromalox.com www.chromalox.co.uk

Elvaco UK Ltd Cambridge House Gogmore Lane Chertsey Surrey KT16 9AP tel:(01932) 564444 fax:(01932) 570707 pab@elvaco.jsbusiness.co.uk

Elvet Structures Low Willington Industrial Estate Willington Crook Co. Durham DL5 0UH tel:(01388) 747120

Elwell Buildings Ltd 204 Oldbury Road West Bromwich W. Midlands B70 9DE tel:(0121) 553 5723 fax:(0121) 553 5723

Elwell Sections Ltd Phoenix Street West Bromwich W. Midlands B70 0AQ tel:(0121) 553 4274 fax:(0121) 553 4272

Emcol International Ltd Westfordmill Wellington Somerset TA21 0DT tel:(01823) 660769 fax:(01823) 665719

Emergi-Lite Safety Systems (Thomas & Betts Ltd) Bruntcliffe Lane Morley, Leeds W. Yorks. LS27 9LL tel:(0113) 2810600 fax:(0113) 2810601 emergi-lite_sales@tnb.com www.emergi-lite.co.uk

Emlux Holdings Ltd Industrial Estate Black Bourton Road Carterton Oxford OX18 3LY tel:(01993) 841574 fax:(01993) 843186 info@britclips.co.uk www.britclips.co.uk

Emmerich (Berlon) Ltd Kingsnorth Industrial Estate Wotton Road Ashford Kent TN23 6JY tel:(01233) 622684 fax:(01233) 645801 lax@emir.co.uk www.emir.co.uk

Emmott George (Pawsons) Ltd Wadsworth Mill Oxlenhope Keighley W. Yorks. BD22 9NE tel:(01535) 643733 fax:(01535) 642108

Emporium Brazil Ltd (EBL) Unit 4 - Arianne Business Centre Townsend Industrial Estate - Blackburn Road Houghton Regis Bedfordshire LU5 5DZ 01582 601 000 01582 661 616 info@emporium-brazil.co.uk www.emporium-brazil.co.uk

EMS Entrance Matting Systems Ltd Northbridge Roadf Berkhampstead Herts HP4 1EH tel:(01442) 875273 fax:(01442) 874242 info@entrance-matting.com www.entrance-matting.com

Emsworth Fireplaces Ltd Unit 3 Station Approach North Street Emsworth Hampshire PO10 7PW 01243 373 431 fax:(01243) 371023 info@emsworth.co.uk www.emsworth.co.uk

Emusol Products (LEICS) Ltd 10 Mandervell Road Oadby Leicester LE2 5LQ tel:(0116) 271 3763 fax:(0116) 271 2133 emusol@aol.com

Enamelled Signs (UK) Ltd 76 Windmill Hill Colley Gate Halesowen W. Midlands B63 2BZ tel:(0121) 558 6060 fax:(0121) 555 5532

Encapsulite International Ltd 17 Chartwell Business Park Chartmoor Road Leighton Buzzard Beds. LU7 4WG tel:(01525) 376974 fax:(01525) 850306 info@encapsulite.co.uk www.encapsulite.co.uk

Encon Air Systems Ltd 31 Quarry Park Close Charter Gate, Moulton Park Industrial Estate Northampton NN3 6QB tel:(01604) 494187 fax:(01604) 645848 sales@encon-air.co.uk www.encon-air.co.uk

Enfield Speciality Doors Alexandra Road The Ride Enfield Middx. EN3 7EH tel:(020) 8805 6662 fax:(020) 8443 1290 sales@enfielddoors.co.uk www.enfielddoors.co.uk

Engineering & Lift Ltd 5 Feelix Av Luton Bedfordshire LU2 7LE tel:(0114) 243 4241 fax:(0114) 256 0991

English Abrasives & Chemicals Co Ltd Doxey Road Stafford ST16 1EA tel:(01785) 251288 fax:(01785) 259428 info@english-abrasives.co.uk www.unicorn-abrasives.co.uk

English Architectural Glazing Ltd Chiskwick Avenue Mildenhall Suffolk IP28 7AY tel:(01638) 510000 fax:(01638) 510400 www.eag.uk.com

English Heritage Fortress House 23 Savile Road London W1S 2ET tel:(020) 7973 3000

Ensor Building Products Blackmoor Road, Guide Lancashire BB1 2LQ tel:(0161) 427 2746 fax:(0161) 427 3074 sales@ensormetal.co.uk www.ensormetal.co.uk

Ensto Briticent Ltd Unit 6, Priory Industrial park Airspeed Road Christchurch Dorset BH23 4HD tel:(01425) 283300 fax:(01425) 280480 sales@ensto.co.uk www.ensto.com

En-tout-cas plc Unit 11 Bartlett Road Lynx Trading Estate Yeovil Summerset BA20 2PJ tel:(01935) 425372 fax:(01935) 432413 sales@en-tout-cas-cas.co.uk www.en-tout-cas-cas.co.uk

En-tout-cas plc (Scotland) Axwel House East Mains Broxburn W. Lothian EH52 5AU tel:(01506) 857044 fax:(01506) 857250

Entrelec MTE Ltd Stephenson Road Eastwood Leith-on-Sea Essex SS9 5LS tel:(01702) 421124 fax:(01702) 420365 info@entrelec-mte.co.uk www.entrelec-mte.co.uk

Envair Ltd York Avenue Haslingden Rossendale Lancs. BB4 4HX tel:(01706) 228416 fax:(01706) 229577 info@envair.co.uk www.envair.co.uk

Envirodoor Markus Ltd Viking Close Willerby Hull East Yorkshire HU10 6BS tel:(01482) 659375 fax:(01482) 655131 sales@envirodoor.com www.envirodoor.com

Environmental Acoustics Ltd Goose Lane Barwell Leics. LE9 8DH tel:(01455) 843678 fax:(01455) 841740 riklewis@clara.net

Environmental Lining Systems Ltd Westland Square Leeds LS11 5SS tel:(0113 277 5635 fax:(0113) 277 5454

Envirtec 11 Sherwood Close Whitstable Kent CT5 4PE tel:(01227) 275015 fax:(01227) 275015

Envopak Group Ltd Edgington Way Sidcup Kent DA14 5EF tel:(020) 8308 8000 fax:(020) 8300 3832 sales@envopak.co.uk www.envopak.com

Envoplan 5 Aintree Road Perivale Greenford Middlesex UB6 7LA tel:(020) 8997 9656 fax:(020) 8991 1442

ERA Products Straight Road Short Heath, Willenhall W. Midlands WV12 5RA tel:(01922) 490000 fax:(01922) 490044 info@era-security.com www.era-security.com

Erico Europa (GB) Ltd 52 Milford Road Reading RG1 8LJ tel:(01189) 588386 fax:(01189) 594856 info@erico.com www.erico.com

Erisco Bauder Limited Broughton House Broughton Road Ipswich Suffolk IP1 3QR tel:(01473) 257671 fax:(01473) 230761 systems@erisco-bauder.co.uk www.erisco-bauder.com

Erlau AG UK Sales Office 42 Macclesfield Road Hazel Grove Stockport SK7 6BE tel:(01625) 877277 fax:(01625) 850242 erlau@zuppinger.u-net.com www.erlau.de

Ermine Engineering Co. Ltd Freancis House, Silver Birch Park Great Northern Terrace Lincoln Lincolnshre LN5 8LG tel:(01522) 510977 fax:(01522) 510929 info@ermineengineering.co.uk www.ermineengineering.co.uk

Escol Products Ltd Paisley Works Windover Road Huntingdon Cambridgeshire PE29 7EB tel:(01933) 276136 fax:(01933) 440121 info@escolpanels.co.uk www.escolpanels.co.uk

Esk Building Products Calsil Brickworks Brisco Carlisle Cumbria CA4 0QY tel:(01228) 527621 fax:(01228) 514019 alan@bell984.tsbusiness.co.uk www.hyeoak.co.uk

Essex 73 Park Lane Liverpool L1 5EX tel:(0151) 709 6636 fax:(0151) 709 2109

Estec Environmental Ltd Old Pump House, Elmer Works Hawks Hill Leatherhead Surrey KT22 9DA tel:(01372) 361451 fax:(01372) 361453

Estel Limited UK 31-33 Priory Park Road London NW6 7UP 020 7372 2549 020 7372 7324 information@estel.co.uk www.estel.co.uk

Eternit Building Materials Whaddon Road Meldreth Nr Royston Herts. SG8 5RL tel:(01763) 264600 fax:(01763) 262338 marketing@eternit.co.uk www.eternit.co.uk

Eureka Products Ltd Norfolk Street Nelson Lancs. BB9 7SY tel:(01282) 615661 fax:(01282) 699542

Euro Materials Handling Ltd Emec Buildings Forge Lane, Mucklow Hill Halesowen W. Midlands B62 8EA tel:(0121) 585 5818 fax:(0121) 585 6127

Eurogrid Ltd Halesfield 19 Telford Shropshire TF7 4QT tel:(01952) 581988 fax:(01952) 586285

European Profiles Llandybie Ammanford Carmarthenshire SA18 3JG tel:(01269) 850691 fax:(01269) 851081 info@europeanprofiles.co.uk www.europeanprofiles.co.uk

Evac+chair international Ltd Paraid House Weston Lane Birmingham B11 3RS tel:(0121) 706 6744 fax:(0121) 706 6746 info@paraid.co.uk www.paraid.co.uk

Evans Concrete Products Ltd Pease Hill Road Ripley Derbys. DE5 3HZ tel:(01773) 748026 fax:(01773) 570354 evans@concrete.fsnet.co.uk www.vitram.co.uk

Evans Howard Roofing Ltd Tyburn Road Erdington Birmingham B24 8NB tel:(0121) 327 1336 fax:(0121) 327 3423 roofherl@ukbusiness.com

Evans Lifts Unit 46 Oak Hill Trading Estate Devonshire Road Worsley Manchester M28 5PT tel:(01204) 861234 fax:(01204) 861242

Evans Vanodine International PLC Brierley Road Walton Summit Centre Bamber Bridge Preston PR5 8AH tel:(01772) 322200 fax:(01772) 626000 sales@evansvanodine.co.uk www.evansvanodine.co.uk

Eve Trakway Coxmoor Road Sutton-in-ashfield Nottinghamshire NG17 5LA tel:(08700) 767676 fax:(08700) 737373 mail@evetrakway.co.uk www.evegroup.co.uk

Evergreens Uk No. 2 Extons Units MKT overton Ind. Est. Market Overton Nr. Oakham Leics LE15 7PP tel:(01572) 768208 fax:(01572) 768261 sales@evergreensuk.com www.evergreensuk.com

Everlac (GB) Ltd The Maltings Fordham Road Newmarket Suffolk CB8 7AA tel:(01638) 664241 fax:(01638) 560015 l.fraser@everlac.co.uk www.everlac.co.uk

Evertaut Ltd Lions Drive Shadsworth Business Park Blackburn Lancs. BB1 2QS 01254 297 880 01254 274 859 sales@evertaut.co.uk www.evertaut.co.uk

Evode Ltd Common Road Stafford ST16 3EH tel:(01785) 257755 fax:(01785) 252337 evo-stik@evode.co.uk www.evode.co.uk

Exchem Mining and Construction PO Box 7 Venture Cresent Alfreton Derbys. DE55 7RE tel:(01773) 540440 fax:(01773) 607638 emc@exchem.com www.exchem.com

Exitex Ltd Dundalk Ireland tel:(00 353) 4293 71244 fax:(00 353) 4293 71221 info@exitex.co.uk

Expamet Building Products PO Box 52 Longhill Industrial Estate (North) Hartlepool TS25 1PR tel:(01429) 866611 fax:(01429) 866633 admin@expamet.net www.expamet.co.uk

Expanded Piling Co Ltd, The Cheapside Works Waltham Grimsby N. E. Lincolnshire DN37 0JD tel:(01472) 822552 fax:(01472) 220675 marketing@expandedpiling.com www.expandedpiling.com

Express Evans Lift 123 Abbey Lane Abbey Lane Leicester LE4 5QX tel:(0116) 201 1200 fax:(0116) 266 8592

Express Evans Lifts Unit 1E Wavetree Boulevard South Wavetree Technology Park Liverpool L7 9PF tel:(0151) 472 1500 fax:(0151) 472 1520

Extrusions Direct Crittall Road Witham Essex CM8 3AW tel:(01376) 503504 fax:(0800) 526005 extrusions.direct@binternet.com

Eyre & Baxter (Stampcraft) Ltd 229 Derbyshire Lane Sheffield S8 8SD tel:(0114) 2500153 fax:(0114) 2580856 sales@eryeandbaxter.co.uk www.eryeandbaxter.co.uk

F. W. Harris & Co Ltd Moorland Road Burslem Stoke-on-Trent ST6 1DR tel:(01782) 575181 fax:(01782) 839439

Fab & Fix Ltd Shilton Industial Estate Shilton Coventry CV7 9JY tel:(024) 76618182 fax:(024) 76619913 www.fabnif.co.uk

Faber Blinds UK Ltd Kilvey Road Brackmills Northampton NN4 7BQ tel:(01604) 766251 fax:(01604) 705209 contracts-uk@faber.dk www.faberblinds.co.uk
Fabricated Aluminium Services Ltd 97-99 Beddington Lane Croydon Surrey CR0 4TD tel:(020) 8664 3333 fax:(020) 8664 3300 sales@faslimited.co.uk www.faslimited.co.uk
Fabrikat (Nottingham) Ltd Hamilton Road Sutton-in-Ashfield Nottingham NG17 5LN tel:(01623) 442200 fax:(01623) 442233
Fairfield Displays & Lighting Ltd 127 Albert Street Fleet Hampshire GU51 3SN tel:(01252) 812211 fax:(01252) 812123 info@fairfielddisplays.co.uk www.fairfielddisplays.co.uk
Fairmile Fencing Ltd Gower Fencing St Georges Teleford Shropshire TF2 9DB tel:(01952) 620062 fax:(01952) 610011 enquiries@fairmilefencing.co.uk www.fairmilefencing.co.uk
Fairmitre Ltd Ashburton Road East Trafford Park Manchester M17 1BB tel:(0161) 877 7200 fax:(0161) 877 7300
Falco Arunhithe Ltd, Barnfield Close Barnfield Industrial Estate Leek Staffs ST13 5EG tel:(01538) 380080 fax:(01538) 386421 sales@falco.co.uk www.falco.co.uk
Falcon Food Services Equipment Wallace Veiw Hill Foots Road Blairlogie Stirlingshire FK9 5PY tel:(01324) 554221 fax:(01324) 552211
Fantasia Distribution Ltd Unit B The Flyers Way Westerham Kent TN16 1DE tel:(01959) 564440 fax:(01959) 564829
Farefence Limited Pinfold House Pinfold Road Worsley Manchester M28 5DZ tel:(0161) 799 4925 fax:(0161) 703 8542 info@farefence.co.uk www.farefence.co.uk
Farmington Natural Stone Ltd Northleach Cheltenham Glos. GL54 3NZ tel:(01451) 860280 fax:(01451) 860115 cotswold.stone@farmington.co.uk
Farr Filtration Ltd 272 Kings Road Tyseley Birmingham B11 2AB tel:(0121) 707 8211 fax:(0121) 706 9986
Fastclean Scotland Ltd Burngrange Lodge Dalziel Estate Motherwell Strathclyde ML1 2BL tel:(01689) 263963 fax:(01698) 263963
Fastwall Ltd PO Box 500 Epsom Surrey KT18 6YB tel:(01372) 278111 fax:(01372) 279222 sales@fastwall.com www.fastwall.com
Faversham Fencing Ltd 62-66 Leatherhead Road Chessington Surrey KT9 2HZ tel:(020) 8397 6164 fax:(020) 8397 5115
FC Frost Ltd Bankside Works Benfield Way Braintree Essex CM7 3YS tel:(01376) 329 111 fax:(01376) 347 002 info@fcfrost.com www.fcfrost.com
FCY Plc Bankwood Lane Industrial Estate Rossington Doncaster S. Yorks. DN11 0PS tel:(01302) 868567 fax:(01302) 864249 fencingconstruct@compuserve.com
FDB Electrical Ltd Reynard Business Park Windmill Road Brentford Middx. TW8 9NZ tel:(020) 8568 4621 fax:(020) 8569 7899 fdbelectricalltd@btinternet.com
FDI- Samborn Ltd 820 Yeoville Road Slough Berks. tel:(01753) 6923557 fax:(01753) 692358
Feb MBT Albany House Swinton Hall Road Swinton Manchester M27 4DT tel:(0161) 794 7411 fax:(0161) 727 8547
Federation of Master Builders Gordon Fisher House 14-15 Great James Street London WC1N 3DP tel:(020) 7242 7583 fax:(020) 7404 0296 central@fmb.org.uk www.findabuilder.co.uk
Federation of Piling Specialists Forum Court 83 Copers Cope Road Beckenham Kent BR3 1NR 020 8663 0947 020 8663 0949 fps@fps.org.uk www.fps.org.uk
Federation of Plastering and Drywall Contractors The Building Centre 26 Store Street London WC1E 7BT 020 7580 3545 020 7580 3288 enquiries@fpdc.org www.fpdc.org
Feedwater Ltd Tarran Way Moreton Wirral L46 4TP tel:(0151) 606 0808 fax:(0151) 678 5459 enquiries@feedwater.co.uk www.feedwater.co.uk
Felix Design Ltd Unit 15, Tiverton Way Tiverton Business Park Tiverton Devon EX16 6SR tel:(01884) 255420 fax:(01884) 242613 sales@felixstaging.co.uk www.felixstaging.co.uk
Fels UK Trinity Place Midland Drive Sutton Coalfield W. Midlands B72 1TX tel:(0121) 321 1155 fax:(0121) 321 1018 uksales@fels.de www.fels.de
Fence Hire Limited Stubs Industrial Estate Holly Bush Lane Aldershot Hants. GU11 2PX
Fencelines Ltd Unit 16 Westbrook Road Trafford Park Manchester M17 1AY tel:(0161) 848 8311 fax:(0161) 872 9643
Fencing Contractors Association Hillside Grange Warren Road Trellech Monmouthshire NP25 4PQ tel:(07000) 560722 fax:(01600) 860888 info@fencingcontractors.org www.fencingcontractors.org
Fencing Design & Development Co Limited Netherhall Farm Netherhall Road Roydon Harlow Essex CM19 5JP tel:(01279) 793529 fax:(01279) 793667
Fenlock-Hansen Ltd Heworth House William Street Felling, Gateshead Tyne & Wear NE10 0JP tel:(0191) 438 3222 fax:(0191) 438 1686 sales@fendorhansen.co.uk www.hansongroup.biz
Ferfa Association House 99 West Street Farnham Surrey GU9 7EN 01252 739149 01252 739140 ferfa@associationhouse.org.uk www.ferfa.org.uk
Ferguson G A & Co Ltd Western Road Oldbury Warley W. Midlands B69 4LY tel:(0121) 552 3674 Fax: (0121) 552 3676
Fermec Holdings Ltd Barton Dock Road Stretford Manchester M32 0YH tel:(0161) 865 4400 fax:(0161) 875 5210 www.fermec.com
Fernden Construction (South Western) Limited Bishops Hull Taunton Somerset TA1 5EA tel:(01823) 271911 fax:(01823) 335763 info@fernden.com www.fernden.com
Fernden Construction Limited Barfield Close Winchester Hampshire SO23 9SQ tel:(01962) 866400 fax:(01962) 864139 sales@ferndenwin.co.uk www.ferndenwin.co.uk
Fernox Alpha Fry Technologies Europe Tandem House, Marlowe Way Beddington Farm Road Croydon Surrey CRO 4XS tel:(020) 8665 6666 fax:(020) 8665 4695 sales@frytechnology.cookson.com www.fernox.com
Ferrograph Limited New York Way New York Industrial Park Newcastle-upon-Tyne NE27 OQF tel:(0191) 280 8800 fax:(0191) 280 8810 ferrograph@btinternet.com www.ferrograph.com
FG Contracts Limited PO Box 3 Macclesfield Cheshire SK11 8XY tel:(01625) 820181 fax:(08700) 522836 info@fgcontracts.co.uk www.laytrad.co.uk
FG Wilson Old Glenarm Road Larne Co. Antrim BT40 1EJ tel:(028) 28261000 fax:(028) 28261111 sales@fgwilson.com www.fgwilson.com
FGF (Ashton) Ltd Shadwell House Shadwell Street Birmingham B4 6LJ tel:(0121) 233 1144 fax:(0121) 212 2539
Fibaflo Ltd Industrial Estate Holton Heath Poole Dorset BH16 6LP tel:(01202) 624141 fax:(01202) 631589 sales@fibaflo.com www.fibaflo.com
Fibaform Products Ltd. Unit 22a Lansil Industrial Estate Caton Road Lancaster Lancs LA1 3PQ tel:(01524) 60182 fax:(01524) 389289 info@fibaform.co.uk www.fibaform.co.uk
Fibercill The Moorings Hurst Business Park Brierley Hill W. Midlands DY5 1UX tel:(01384) 482221 fax:(01384) 482212 mail@fibercill.com www.fibercill.com
Fibre Reinforced Products Ltd 2 Whitehouse Way South West Industrial Estate Paterlee Co. Durham SR8 2HZ tel:(0191) 586 5311 fax:(0191) 586 1274
Fibremesh Synthetic Industries Europe Ltd Fibremesh House Smeckley Wood Close Sheepsbridge, Chesterfield Derbys. S41 9PZ tel:(01246) 453102 fax:(01246) 455841 fibremesh@aol.com
Fiche (UK) Ltd Security House Acrewood Way St Albans Herts. AL4 0JL tel:(01727) 863863 fax:(01727) 41079
Fieldmount (Terrazzo) Ltd 152 West End Lane West Hampstead London NW6 1SD tel:(020) 7624 8866 fax:(020) 7328 1836
Fife Fire Engineers & Consultants Ltd Waverley Road, Mitchelston Industrial Estate Kirkcaldy Fife KY1 3NH tel:(01592) 653661 fax:(01592) 653990
Fighting Films 1 Triangle House 2 Broomhill Road Wandsworth London SW18 4HX tel: +44 (0)20 8877 1441 fax: +44 (0)20 8874 8590 info@fightingfilms.com
Filion Products Ltd Aldridge Road Streetly Sutton Coalfield W. Midlands B74 2DZ tel:(0121) 353 0814 fax:(0121) 352 0886
Fillcrete Ltd Maple House 5 Over Minnis New Ash Green Kent DA3 8JA tel:(01474) 872444 fax:(01474) 872426 sales@fillcrete.com www.fillcrete.com
Findlay Irvine Ltd Bog Road Penicuik Midlothian EH26 9BU tel:(01968) 671200 fax:(01968) 671237 sales@findlayirvine.com www.findlayirvine.com
Fine H & Son Ltd Victoria House 93 Manor Farm Road Wembley Middx. HA0 1XB tel:(020) 8997 5055 fax:(020) 8997 8410 sales@hfine.co.uk www.hfine.co.uk
Finecraft Joinery Ltd Arundel House Arundel Road Uxbridge Middx. UB8 2RX tel:(01895) 233101 fax:(01895) 231933
FineLine Whitewall Road, Midway Estate Rochester Kent ME2 4EW tel:(01634) 719701 fax:(01634) 716394 info@finelinewindows.co.uk www.finelinewindows.co.uk
Finlay Breton Concrete Products Tor Hill Concrete Works Tor Hill Wells Somerset BA5 3NT tel:(01749) 675757 fax:(01749) 677750 tim.butler@finlayconcrete.com www.finlaybreton.com
Finlock Gutters Unit 3 Ladycross Business Park Hollow Lane Dormansland, Lingfield Surrey RH7 6 PB tel:(01342) 871992 fax:(01342) 870939 ahbltd@aol.com www.finlock-concretegutters.co.uk
Finn Forest Old Golf Course Fishtoft Road Boston Lincs. PE21 OBJ tel:(01205) 362461 fax:(01205) 354488
Finnish Fibreboard UK Ltd Viking House Swallowdale Lane Hemel Hampstead Hertfordshire HP2 7EA tel:(01438) 798746 fax:(01438) 798305
Finnish Sauna Baths (Sales) Ltd 4 Shelbourne Close Pinner Middx. HA5 3AF tel:(020) 8868 7170
Finwood Leisure Ltd 41 Feus Auchterarder Tayside PH3 1EP tel:(01764) 664920 fax:(01764) 664923 sales@finwood.co.uk www.finwood.co.uk
Firco Construction Ltd Unit1, Denton Slipways Wharf Road, Norfolk Road Industrial Estate Gravesend Kent DA12 2RW tel:(01474) 351644 fax:(01474) 358936 firco@gravesend.demon.co.uk
Fire Escapes & Fabrications (UK) Ltd Foldhead Mills Newgate Mirfield W. Yorks. WF14 8DD tel:(01924) 498787 fax:(01924) 497424 info@fireescapes.co.uk
Fire Protection Ltd Millars 3 Southmill Road Bishops Stortford Herts. CM23 3DH tel:(01279) 467077 fax:(01279) 466994 fire.protection@btinternet.com
Fire Security (Sprinkler Installations) Ltd Homefield Road Haverhill Suffolk CB9 8QP tel:(01440) 705815 fax:(01440) 704352 info@firesecurity.co.uk www.firesecurity.co.uk
Firebarrier Services Ltd The Old Bank Chambers Brunswick Square Torquay Devon TQ1 4UT tel:(01803) 291185 fax:(01803) 290026 sales@firebarrier.co.uk www.firebarrier.co.uk
Fired Earth Twyford Mill Oxford Road Adderbury Oxon. OX17 3HP tel:(01295) 812088 fax:(01295) 810832 enquiries@firedearth.com www.firedearth.com
Firemaster Extinguisher Ltd Firex House 174-176 Hither Green Lane London SE13 6QB tel:(020) 8852 8585 fax:(020) 8297 8020 sales@firemaster.co.uk www.firemaster.co.uk
Firestone Building Products Europe Meridian House Road One Winsford Cheshire CW7 3QG tel:(01606) 552026 fax:(01606) 592666 info@fbpl.co.uk www.firestonebpe.com
Firetecnics Systems Ltd Southbank House Black Prince Road London SE1 7SJ tel:(020) 7587 1493 fax:(020) 7582 3496 sales@firetecnics.bdx.co.uk www.firetecnics.com
First Stop Builders Merchants Ltd Queens Drive Kilmarnock Ayrshire KA1 3XA tel:(01563) 534818 fax:(01563) 537848 sales@firststopbm.co.uk www.firststopbm.co.uk
Firth Interface Europe Ltd PO Box 17 Clifton Mills Brighouse W. Yorks. HD6 4EJ tel:(0870) 600 2008 fax:(01484) 728040 firthcontract@eu.interfaceinc.com www.firthcarpets.com
Firth-Vickers Special Steels Ltd Staybrite Works Weedon Street Sheffield S9 2FU tel:(01742) 449955 fax:(01742) 445390
Firwood Paints Ltd Victoria Works Oakenbottom Road Bolton Lancs. BL2 6DP tel:(01204) 525231 fax:(01204) 362522 sales@firwood.co.uk www.firwood.co.uk
Fisher Scientific (UK) Ltd Bishop Meadow Road Loughborough Leics. LE11 5RG tel:(01509) 231166 fax:(01509) 231893 info@fisher.co.uk www.fisher.co.uk
Fitchett & Woolacott Ltd Willow Road Lenton Lane Nottingham NG7 2PR tel:(01602) 700691 fax:(01602) 420912
Fitzgerald Lighting Ltd Normandy Way Bodmin Cornwall PL31 1HH tel:(01208) 262200 fax:(01208) 262334 sales@flg.co.uk www.flg.co.uk
Fitzpatrick Doors Ltd Rushey Lane Tyseley Birmingham B11 2BL tel:(0121) 706 6363 fax:(0121) 708 2250 fitzuk1@aol.com www.fitzpatrickmetaldoors.com
Fix Fast (Fasteners) Wales Ltd Penallta Industrial Estate Ystrad Mynach Hengoed M. Glam. CF82 7QZ tel:(01443) 815441 fax:(01443) 815021 fixfast@whsmithnet.co.uk
Fixatrad Ltd Unit 29, Robert Cort Industrial Estate Britten Road Reading Berkshire RG2 0AW tel:(0118) 921 2100 fax:(0118) 921 0634
Fixing Delivery 6 Three Arch Road Redhill RH1 5SS tel:(01737) 767656 fax:(01737) 772867
Flamco UK Ltd Unit 4, St Micheals Road Lea Green Industrial Estate St. Helens Merseyside WA9 4WZ tel:(01744) 818100 fax:(01744) 830400 info@flamco.co.uk www.flamco.co.uk
Flavel - Leisure Glynwed Consumer & Construction Products Ltd Clarence Street Leamington Spa Warwks. CV31 2AD tel:(01926) 427027 fax:(01926) 450526 flavel-leisure.market@dial.pipex.com
Fleetwood Fengate Co Ltd 62-66 Leatherhead Road Chessington Surrey KT9 2HZ tel:(020) 8397 6161 fax:(020) 8397 5115
Fleetwood Paints (Sales) Ltd 117 Wood Lane Isleworth Middx. TW7 5EG tel:(020) 8560 7634
Fleming Buildings Ltd 23 Auchinloch Road Lenzie Glasgow G66 5ET tel:(0141) 776 1181 fax:(0141) 775 1394
Fleming Homes Ltd Coldstream Road Duns Berwickshire TD11 3HS tel:(01361) 883785 fax:(01361) 883898 timberframe@fleminghomes.fsnet.co.uk www.fleminghomes.co.uk
Flexcrete Ltd Iotech House, PO Box 7 Miller Street Preston Lancs. PR1 1EA tel:(01772) 255074 fax:(01772) 255670 info@flexcrete.co.uk www.flexcrete.co.uk
Flexel International Ltd Queensway Industrial Estate Glenrothes Fife KY7 5QF tel:(01592) 757313 fax:(01592) 754535 sales@flexel.co.uk www.flexel.co.uk
Flexelec(UK) Ltd 200 Rickmansworth Road Watford Herts. WD18 7JS tel:(01923) 234477 fax:(01923) 240264 sales@flexelec.co.uk. www.flexelec.com
Flexello Limited 136 Edinburgh Avenue Slough Berks. SL1 4SS tel:(01753) 775200 fax:(01753) 775300 sales@flexello.co.uk
Flexible Ducting Ltd Cloberfield Milngavie Glasgow G62 7LW tel:(0141) 956 4551 fax:(0141) 956 4847 info@flexibleducting.co.uk www.flexibleducting.co.uk
Flexible Reinforcements Ltd Queensway House Queensway Clitheroe Lancs. BB7 1AU tel:(01200) 442266 fax:(01200) 452010
Flexicote Ltd Grinstead House Partridge Green Horsham W. Sussex RH13 8EJ tel:(01306) 741599 fax:(01306) 885901
Flexiform Business Furniture Ltd The Business Furniture Centre 1392 Leeds Road Bradford W. Yorks. BD3 7AE tel:(01274) 706206 fax:(01274) 660867 flexiform@aol.com www.flexiform.co.uk
Flexi-Plan Partitions Ltd Unit J1 Halesfield 19 Telford Shropshire TF7 4QT tel:(01952) 586126 fax:(01952) 581174 flexi.planpartitions@barclays.net
Flexitallic Ltd Marsh Works Dewsbury Rd. Cleckheaton W. Yorks. BD19 5BT tel:(01274) 851273 fax:(01274) 851386 sales@flexitallic.com www.flexitallic.com
Flex-Seal Couplings Ltd Endeavour Works Newlands Way, Valley Park Wombwell, Barnsley South Yorkshire S73 OUW tel:(01226) 340888 fax:(01226) 340999 sales@flexseal.co.uk www.flexseal.co.uk
Flexwood Veneers 9 Egsign Drive Palmers Green London N13 5AF tel:(020) 8447 8479 fax:(020) 8886 2483
Float Glass Industries Float Glass House Floats Road Roundthorn Manchester M23 9QA tel:(0161) 946 8000 fax:(0161) 946 8092
Floor Protection Services Ltd Triumph Way Kempston Beds. MK42 7QP tel:(01234) 303090 fax:(01234) 303080 sales@fpsmatting.co.uk www.fpsmatting.co.uk
FloPlast Ltd Eurolink Business Park Castle Road Sittingbourne Kent ME10 3FP tel:(01795) 431731 fax:(01795) 431188 sales@floplast.co.uk www.floplast.co.uk
FloRad Heating Systems Unit 1, Horseshoe Business Park Lye Lane Bricket Wood, St. Albans Herts. AL2 3TA tel:(01923) 893025 fax:(01923) 670723 floradhs@hotmail.com www.florad.co.uk
Flowcrete Plc Flowcrete Business Park Booth Lane Moston, Sandbach Cheshire CW11 3QF tel:(01270) 753000 fax:(01270) 753333 marketing@flowcrete.co.uk www.flowcrete.co.uk
Flowflex Componants Ltd Samuel Blaser Works Tounge Lane Industrial Estate Buxton Derbys. SK17 7LR tel:(01298) 77211 fax:(01298) 72362
Flowline Grilles Block 3 Narrow Boat Way Dudley W. Midlands DY2 0XQ tel:(01384) 459260 fax:(01384) 458381
Flowserve Flow Control Burrell Road Haywards Heath W Sussex RH16 1TL tel:(01444) 314400 fax:(01444) 314401 wvukinfo@flowserve.com www.flowserve.com
Fluidair International Ltd School Hill Works Kent Street Bolton Manchester BL1 2LN tel:(0204) 559955 fax:(01204) 559966 sales@fluidair.co.uk www.fluidair.co.uk
Fluorel Ltd 312 Broadmead Road Woodford Green Essex IG8 8PG tel:(020) 8504 9691 fax:(020) 8506 1792 info@fluorel.co.uk www.fluorel.co.uk
Foamseal Ltd New Street House New Street Petworth West Sussex GU28 0AS tel:(01798) 345400 fax:(01798) 344093 mail@foamseal.co.uk www.foamseal.co.uk
Focal Signs Ltd 12 Wandle Way Mitcham Surrey CR4 4NB tel:(020) 8687 5300 fax:(020) 8687 5301 sales@focalsigns.co.uk www.focalsigns.co.uk
Focus SB Napier Road Castleham Industrial Estate St Leonards-on-Sea E. Sussex TN38 9NY tel:(01424) 440734 fax:(01424) 853862 sales@focus-sb.co.uk www.focus-sb.co.uk
Footprint Tools Ltd PO Box 19 Hollis Croft Sheffield S. Yorks. S1 3HY tel:(01142) 753200 fax:(01142) 759613 sales@footprint-tools.co.uk www.footprint-tools.co.uk
Forbes & Lomax Ltd 205a St John's Hill London SW11 1TH tel:(020) 7738 0202 fax:(020) 7738 9224 sales@forbesandlomax.co.uk www.forbesandlomax.co.uk
Forbo - Nairn Ltd PO Box 1 Den Road Kirkcaldy Fife KY1 2SB tel:(01592) 643777 fax:(01592) 643999 headoffice@forbo-nairn.co.uk www.forbo-nairn.co.uk
Forbo (Lancaster) Ltd Lune Mills Lancaster LA1 5QN tel:(01524) 65222 fax:(01524) 61638
Fordham PO Box 20 Spring Ram Business Park Birstal W. Yorks. WF17 9XD tel:(01924) 351351 fax:(01924) 351333 marketing@astracast.co.uk www.fordham-sinks.co.uk
Fordwater Pumping Supplies 11 Lea Road Abingdon Northampton NN1 4PE tel:(01604) 39805 fax:(01604) 20143

Fordwater Pumping Supplies Ltd Unit 32 Forestvale Ind Estate Cinderford Glos. GL14 2PH tel:(01594) 826780 fax:(01594) 826780 info@pumpsuk.com www.pumpsuk.com

Forest of Dean Stone Firms Ltd Bixslade Stone Works Parkend Nr Lydney Glos. GL15 4JS tel:(01594) 562304 fax:(01594) 564184 paul@fodstonefirms.co.uk

Forestry and Timber Association 5 Dublin Street Lane South Edinburgh EH1 3PX tel:(0131) 5387111 fax:(0131) 5387222 info@forestryandtimber.org www.forestryandtimber.org

Forma Lighting Ltd Unit 3, Mitcham Industrial Estate 85 Streatham Road Mitcham Surrey CR4 2AP tel:(020) 8640 6811 fax:(020) 8640 6910

Formica Ltd Coast Road North Shields Tyne & Wear NE29 8RE tel:(0191) 259 3000 fax:(0191) 258 2719 www.Formica-europe.com

Formpave Ltd Tufthorne Avenue Coleford Glos. GL16 8PR tel:(01594) 836999 fax:(01594) 810577 sales@formpave.co.uk www.formpave.co.uk

Forson Design & Engineering Ltd Commerce Way Lancing W. Sussex BN15 8TQ tel:(01903) 752835 fax:(01903) 756677 forsons@mistral.co.uk

Fortafix Ltd First Drove Fengate Peterborough Cambs. PE1 5BJ tel:(01733) 566136 fax:(01733) 315393 sales@fortafix.com www.ourworld.compuserve.com/homepages/fortafix

Forticrete Ltd Bridle Way Bootle Merseyside L30 4UA tel:(0151) 521 3545 fax:(0151) 521 5696 mormesher@forticrete.com www.forticrete.com

Forticrete Roofing Products Heath Road Leighton Buzzard Beds. LU7 3ER tel:(01525) 244900 fax:(01525) 850432 roofing@forticrete.com www.forticrete.co.uk

Fosroc Ltd Coleshill Road Tamworth Staffs. B78 3TL tel:(01827) 262222 fax:(01827) 262444 sales@fosrocuk.com www.fosrocuk.com

Foster C G & Sons, (Contract Soft Furnishers) Sherbourne House 247 Humber Avenue Coventry CV1 2AQ tel:(024) 76 224229 fax:(024) 76 222786

Foster W H & Sons Ltd 3 Cardale Street Rowley Regis W. Midlands B65 0LX tel:(0121) 561 1103 fax:(0121) 559 2620 sales@whfoster.co.uk www.whfoster.co.uk

Fosters Upholstery Services More Hall Ibstock Road Ellistown Leics. LE67 1ED tel:(01530) 260211

Fothergill Engineered Fabrics Ltd PO Box 1 Summit Littleborough Lancs. OL15 OLU tel:(01706) 372414 fax:(01706) 376422 sales@fothergill.co.uk www.fothergill.co.uk

Fox Pool (UK) Ltd Mere House Stow Lincoln Lincs. LN1 2BZ tel:(01427) 788662 fax:(01427) 788526 foxpooluk@aol.com www.fox-pool.co.uk

Francis Frith Collection Frith's Barn Teffont Salisbury Wilts SP3 5QP tel:(01722) 716376 fax:(01722) 716 881 sales@francisfrith.co.uk www.francisfrith.co.uk

Franke UK Ltd West Park, Manchester Int'l Office Centre Styal Road Manchester Greater Manchester M22 5WB tel:(0161) 436 6280 fax:(0161) 436 2180 info.uk@franke.com www.franke.co.uk

Franklin Hodge Industries Ltd P. O. Box 11, Holmer Road Hereford Herefordshire HR4 9SJ tel:(01432) 277277 fax:(01432) 277454 fhisales@denco.co.uk

Fray Design Ltd The Bradley Centre Bradley, Keighley W. Yorks. BD20 9LE tel:(01535) 636433 fax:(01535) 634642

Frazer Safety & Construction Station Road Hebburn Tyne & Wear NE31 1BD tel:(0191) 428 0242 fax:(0191) 483 3628

Freeman T R Beadle Trading Estate Ditton Walk Cambridge CB5 8PD tel:(01223) 410600 fax:(01223) 411061 t.r.freeman@dial.pipex.com

Freshfield Lane Brickworks Ltd Freshfield Lane Dane Hill, nr Haywards Heath W. Sussex RH17 7HH tel:(01825) 790350 fax:(01825) 790779

Freyssinet Ltd 7 Hollinswood Court StaffordPark 1 Telford Shropshire TF3 3DE tel:(01952) 201901 fax:(01925) 201753 info@freyssinet.co.uk www.freyssinet.co.uk

Friedland Ltd Houldsworth Street Reddish, Stockport Gtr. Manchester SK5 6BP tel:(0161) 432 0277 fax:(0161) 975 1300 www.friedland.co.uk

Friends of the Earth 26 Underwood Street London N1 7JQ tel:(020) 7490 1555 fax: (020) 7490 0881 info@foe.co.uk www.foe.co.uk

Frosts Landscape Construction Ltd Newport Road Woburn Sands Milton Keynes Bucks MK17 8UZ tel:(01908) 583611 fax:(01908) 585238 landscapes@frostgroup.com www.frostsgroup.com

FTW Windows Blue Diamonds Lawn Lane Sutton Cambridgeshire CB6 ZRE tel:(0800) 294 94 54 fax:(01353) 778021 sales@ftwwindows.com www.ftwwindows.com

Full Spectrum Lighting Ltd Unit 48 Marlow Road Stokenchurch High Wycombe Bucks. HP14 3QJ tel:(01494) 484851 fax:(01494) 484855

Fullflow Group Ltd Fullflow House Holbrook Avenue Holbrook, Sheffield S. Yorks. S20 3FF tel:(0114) 247 3655 fax:(0114) 247 7805 info@uk.fullflow.com www.fullflow.com

Fullstop Technology Ltd Caidan House Canal Road Timperley Cheshire WA14 1TE tel:(0161) 962 8093 fax:(0161) 905 1436 xypexfullstop@aol.com

Fulton Boiler Works (Great Britain) Ltd Broomhill Road Bristol BS4 4TU tel:(0117) 972 3322 fax:(0117) 972 3358 uk-info@fulton.com www.fulton.com

Furmanite International Ltd Furman House Shap Road Kendal Cumbria LA9 6RU tel:(01539) 729009 fax:(01539) 729359 uk.enquiry@furmanite.com www.furmanite.com

Furnitubes International Ltd Meridian House Royal Hill Greenwich London SE10 8RT tel:(020) 8378 3200 fax:(020) 8378 3250 sales@furnitubes.com www.furnitubes.com

Furse Wilford Road Nottingham Nottinghamshire NG2 1EB tel:(0115) 964 3700 fax:(0115) 986 0538 enquiry@furse.com www.furse.com

Futimis Ltd Futimis House 11 Mead Park, River Way Harlow Essex CM20 2SE tel:(01279) 411131 fax:(01279) 453107 info@futimus.co.uk www.futimus.com

Fyfe John Ltd Westhill Industrial Estate Westhill Skene Aberdeen AB32 6TQ tel:(01224) 744144 fax:(01224) 744500

G. Miccoli & Son Ltd 2A Selhurst Road South Norwood London SE25 5QF tel:(020) 8684 3816 fax:(020) 8689 6514

Gabriel & Co Ltd Abro Works 10 Hay Hall Road Tyseley Birmingham B11 2AU tel:(0121) 248 3333 fax:(0121) 248 3330 john@gabrielco.com www.gabrielco.com

Galvanizers Association Wrens Court 56 Victoria Road Sutton Coldfield W. Midlands B72 1SY tel:(0121) 355 8838 fax:(0121) 355 8727 ga@hdg.org.uk www.galvanizing.org.uk

Gamma Resources Ltd 8 Western Business Centre Winterstroke Road Weston-Super-Mare N. Somerset BS24 9AB tel:(01934) 632651 fax:(01934) 623849 pyroguard@aol.com www.timberwise

Gang-Nail Systems Ltd Christy Estate Ivy Road Aldershot Hants. GU12 4XG tel:(01252) 334691 fax:(01252) 334562 info@gang-nail.co.uk www.gangnail.co.uk

Garador Ltd Bunford Lane Yeovil Somerset BA20 2YA tel:(01935) 443700 fax:(01935) 443744 enquiries@garador.co.uk www.garador.co.uk

Garage Equipment Association Ltd 2-3 Church Walk Daventry NN11 4BL 01327 312616 01327 312606 info@gea.co.uk www.gea.co.uk

Gardner & Newton PO Box 45 Hall Street St Helens Merseyside WA10 1ER tel:(01744) 22189/28661 fax:(01744) 26934 info@glassbending.co.uk www.glassbending.co.uk

Garlick E & Sons Ltd 78 Hoyle Street Sheffield Yorks. S3 7EX tel:(01742) 725412 fax:(01742) 725412

Garran Lockers Ltd Garran House Nantgarw Road Caerphilly CF83 1AQ tel:(029) 20 869924 fax:(029) 20 882841 sales@garran-lockers.com www.garran-lockers.com

Gartec Ltd Wellington House Lower Icknield Way Longwick Princess Risborough Bucks HP27 9RZ tel:(01844) 345100 fax:(01844) 345 104 eduncomb@gartec.com

Gas Measurement Instruments Ltd Inchinnan Business Park Renfrewshire PA4 9RG tel:(0141) 812 3211 fax:(0141) 812 7820 sales@gmiuk.com www.gmiuk.com

Gaskell Carpet Tiles Clayton Park Clayton - Le - Moors Accrington Lancs. BB5 5GT tel:(01254) 724300 fax:(01254) 724301 tiles@gaskellcarpettiles.co.uk www.gaskellcarpettiles.co.uk

Gazco Ltd Osprey Road Sowton Industrial Estate Exeter EX2 7JG tel:(01392) 216999 fax:(01392) 444148 info@gazco www.gazco.com

GB Metals Ltd Grange Works, Cromwell Centre Sellina's Lane Dagenham Essex RM8 1QH tel:(020) 8984 7470 fax:(020) 8984 7970

GBG Fences Limited 25 Barns Lane Rushall Walsall Staffs. WS4 1HQ tel:(01922) 623207 fax:(01922)722110 113105.1311@compuserve.com

GE Bayer Silicones Old Hall Road Sale Gtr. Manchester M33 2HG tel:(01257) 473867 fax:(01257) 473868 alan.lacey@gepex.ge.com

GE Capital Modular Space G E House Ten Pound Walk Doncaster S. Yorks. DN4 5HW tel:(01427) 871141 fax:(01427) 871156 nicholas.bell@gecapitol.com www.modularspace.co.uk

GE Lighting Ltd Lincoln Road Enfield Surrey EN1 1SB tel:(020) 8626 8500 fax:(020) 8626 8501

GE Plastics Ltd Old Hall Road Sale Gtr. Manchester M33 2HG tel:(0161) 905 5000 fax:(0161) 905 5160

GE Protimeter Meter House Fieldhouse Lane Marlow Bucks. SL7 1LW tel:(01628) 472722 fax:(01628) 474312 protimeter@ge.com www.protimeter.com

Gearing F T Landscape Services Ltd Crompton Road Depot Stevenage Herts. SG1 2EE tel:(01438) 369321 fax:(01483) 353039 Sales@ft-gearing.co.uk www.ft-gearing.co.uk

Geberit UK Metcalf House 25 Kirkgate Rippon N. Yorks. HG4 1PB tel:(01765) 602082 fax:(01765) 640765

Gebhardt Kiloheat Kiloheat House Enterprise Way Edenbridge Kent TN8 6HF tel:(01732) 866000 fax:(01732) 866370

GEC Anderson Ltd Oakengrove Hastoe Tring Hertfordshire HP23 6LY tel:(020) 8950 1826 fax:(020) 8950 3626

Gee Joinery Ltd Heanor Road Smalley Derbyshire DE7 6DW tel:(01332) 780601 fax:(01332) 780605

Geemat 1 Triangle houseq 2 Broomhill Road Wandsworth London SW18 4HX tel: (020) 8877 1441 fax:(020) 8874 8590 info@fightingfilms.com www.fightingfilms.com

General Combustion Ltd Genlor - Beverley Brookers Road Billinghurst W. Sussex RH14 9RZ tel:(01403) 782091 fax:(01403) 782087 info@genlor-beverley.co.uk www.gencor-beverley.co.uk

General Time Europe 8 Heathcote Way Warwick CV34 6TE tel:(01926) 885400 fax:(01926) 885723 Westcloxeurope@generaltime.com www.westclox.co.uk

Genie Europe The Maltings Wharf Road Grantham Lincs NG31 6BH tel:(01476) 584333 fax:(01476) 584334 info@genieind.com www.genielift.com

Gent Limited 140 Waterside Road Hamilton Industrial Park Leicester LE5 1TN tel:(0116) 246 2000 fax:(0116) 246 2300 gent_enquiry@gent.co.uk www.gent.co.uk

Geoquip Ltd Unit 6, Kingsfield Industrial Estate Derby Road Wirksworth, Matlock Derbys. DE4 4BG tel:(01629) 824891 fax:(01429) 824896 info@geoquip.com www.geoquip.com

George Fischer Sales Ltd Coventry Walsgrave Triangle, Paradise Way Hinckley Road Coventry West Midlands CV2 2ST tel:(024) 765 35535 fax:(024) 765 30450 info@georgefischer.co.uk www.georgefischer.co.uk

George Gilmour (Metals) Ltd 245 Govan Road Glasgow G51 2SQ tel:(0141) 427 1264 fax:(0141) 427 2205

George James Partnership Tincklers House Tincklers Lane Eccleston, Chorley Lancs. PR7 5QX tel:(01257) 452151 fax:(01257) 450094

Geosynthetic Technology Ltd Little Bulmer Farm Wiston Road Nayland, Colchester Essex CO6 4LT tel:(01206) 262676 fax:(01206) 262998 sales@geosynthetic.co.uk www.geosynthetic.co.uk

Gerflor Ltd Wedgnock House Wedgnock Lane Warwick Warwickshire CV34 5AP tel:(020) 7723 6601 fax:(020) 7723 9557

Geze UK Chelmsford Business Park Colchester Road Chelmsford Essex CM2 5LA tel:(01245) 451093 fax:(01245) 451108

Gifford Grant Ltd 144 Clarence Road Fleet Hants. GU13 9RS tel:(01252) 816188 fax:(01252) 625980 giffgrant@aol.com

Gillespie (UK) Ltd Alma House 38 Crimea Road Aldershot Hampshire GU11 1UD tel:(01252) 323311 fax:(01252) 336836 davidbland@gillespieuk.co.uk www.gillespie.co.uk

Gilmer Limited Southdown Works Laughton Road Ringmer, Nr Lewes E. Sussex BN8 5PN tel:(01273) 813370 fax:(01273) 813963

Gilmour Ecometal 245 Govan Road Glasgow G51 2SQ tel:(0141) 427 7000 fax:(0141) 427 5345 www.gilmour.ecometal.co.uk

Giraffe Products Ltd Units 1-3, Fowler Industrial Park Chorley New Road Horwich Bolton BL6 5LU tel:(01204) 690052 fax:(01204) 695636

Girdlestone Pumps Ltd Station Road Melton, Woodbridge Suffolk IP12 1ER tel:(01394) 383777 fax:(01394) 386733 girdlestone.enquiries@girdlestone.co.uk www.girdlestone.co.uk

Girsberger London 140 Old Street London EC1V 9BJ tel:(020) 7490 3223 fax:(020) 7490 5665 mail@girsberger-uk.com www.girsberger.com

Giscol Bricks 20 Clifton Street Glasgow G3 7XS tel:(0141) 332 3786 fax:(0141) 332 7588

GJ Durafencing Limited Silverlands Park Nursery Holloway Hill Chertsey Surrey KT16 0AE tel:(01932) 568727 fax:(01932) 567799 johnb@gavinjones.co.uk www.gavinjones.co.uk

Glamox Electric (UK) Ltd Unit 3E Howden Green Industrial Estate Wallsend Tyne & Wear NE28 6SX tel:(0191) 262 7126 fax:(0191) 262 4118 info@glamox.com www.glamox.com

Glasdon UK Ltd Preston New Road Blackpool FY4 4UL tel:(01253) 600410 fax:(01253) 792558 sales@glasdon.uk.co.uk www.glasdon.com

Glasgow Steel Nail Co Ltd Lowmoss Bishopbriggs Glasgow G64 2HX tel:(0141) 762 3355 fax:(0141) 762 0914 glasgowsteelnail@compuserve.com ourworld.compuserve.com/homepages/glasgowsteelnail

Glass and Glazing Federation The 44-48 Borough High Street London SE1 1XB tel:(020) 7403 7177 fax:(020) 7357 7458 info@ggf.org.uk www.ggf.org.uk

Glaverbel (UK) Ltd Chesnut Field Regents Place Rugby Warwks. CV21 2XH tel:(01788) 535353 fax:(01788) 560853

Glazzard (Dudley) Ltd The Washington Centre Netherton, Dudley W. Midlands DY2 9RE tel:(01384) 233151 fax:(01384) 250224 gdl@glazzard.co.uk www.glazzard.co.uk

Gledhill Water Storage Ltd Sycamore Trading Estate Squires Gate Lane Blackpool Lancs. FY4 3RL tel:(01253) 474444 fax:(01253) 474445 sales@gledhill.net www.gledhill.net

Glen Dimplex UK Ltd Millbrook House Grange Drive Hedge End Southampton SO30 2DF tel:(0870) 077 7117 fax:(0870) 727 0109 customer.services@glendimplexheating.co.uk www.dimplex.com

Glenigan Ltd 41-47 Seabourne Road Bournemouth Dorset BH5 2HU tel:(0800) 373771 fax:(01202) 423441 info@glenigan.emap.co.uk www.glenigan.com

Gleno Industries Ltd Grange Road Houstoun Industrial Estate Livingston W. Lothian EH54 5DH tel:(01506) 32551 fax:(01506) 34386 glenosales@aol.com

Glixtone Ltd Westminster Works Alvechurch Road West Heath Birmingham B31 3PG tel:(0121) 243 1122 fax:(0121) 243 1123 excellence@carrspaints.com www.carrspaints.com

Gloddfa Ganol Slate Mine Blaenau Ffestiniog Gwynedd LL41 3NB tel:(01766) 830664 fax:(01766) 831527

Glostal - Monarch Ashchurch Tewkesbury Glos. GL20 8NB tel:(01684) 853500 fax:(01684) 851850 enquire@granges.co.uk www.granges.co.uk

Gloster Furniture Ltd Concorde Road Patchway Bristol BS34 5TB tel:(0117) 931 5335 fax:(0117) 931 5334 enquiries@gloster.co.uk www.gloster.co.uk

Gold & Wassall (Hinges) Ltd Castle Works Tamworth Staffs. B79 7TH tel:(01827) 63391 fax:(01827) 310819

Golden Coast Ltd Gratton Way Roundswell Industrial Estate Barnstaple Devon EX31 3NL tel:(01271) 378100 fax:(01271) 371699 swimmer@goldenc.com www.goldenc.com

Goodacres Fencing Shoby Lodge Farm Loughborough Road Melton Mowbray Leics. LE14 3PF tel:(01664) 813989 fax:(01664) 813989 kgoodacre1@aol.com

Goodman Croggon Ltd 6-8 Clothier Road Brislington Trading Estate Bristol BS4 5PE tel:(01272) 770721 fax:(01272) 723676

Goodwin HJ Ltd Quenington, nr Cirencester Glos. GL7 5BX tel:(01285) 750271 fax:(01285) 750352

Goodwin Tanks Ltd Pontefract Street Derby DE24 8JD tel:(01332) 363112 fax:(01332) 294 683 info@goodwintanks.co.uk www.goodwintanks.co.uk

Gower Furniture Ltd Holmfield Industrial Estate Holmefield Halifax W. Yorks. tel:(01422) 246201 fax:(01422) 249932

GP Burners (CIB) Ltd 2d Hargreaves Road Groundwell Industrial Estate Swindon Wilts. SN25 5AZ tel:(01793) 709050 fax:(01793) 709060 info@gpburners.co.uk www.gpburners.co.uk

Grace Construction Products Ltd Ajax Avenue Slough Berkshire SL1 4BH tel:(01753) 692929 fax:(01753) 691623 uksales@grace.com www.graceconstruction.com

Gradient Insulations (UK) Ltd Station Road Four Ashes Wolverhampton WV10 7DB tel:(01902) 791888 fax:(01902) 791886 sales@gradientuk.com

Gradus Ltd Park Green Macclesfield Cheshire SK11 7LZ tel:(01625) 428922 fax:(01625) 433949 sales@gradusworld.com www.gradusworld.com

Gradwood Ltd Landsdowne House 85 BuxtonRoad Stockport Cheshire SK2 6LR tel:(0161) 480 9629 fax:(0161) 474 7433

Grando (UK) Ltd Cotton Drive Kenilworth Warwks. CV8 2EB tel:(01926) 854977 fax:(01926) 856772 grando-uk@hotmail.com www.grando.co.uk

Granflex Roofing Ltd Brick Kiln Lane Basford Stoke-on-Trent ST4 7BT tel:(01782) 202208 fax:(01782) 273601 sales@granflexroofing.co.uk www.granflexroofing.co.uk

Grange Fencing Erectors Limited Halesfield 21 Telford Shropshire TF7 4PA tel:(01952) 587892 fax:(01952) 684461

Grant & Livingston Ltd Kings Road Charfleets Industrial Estate Canvey Island Essex SS8 0RA tel:(01268) 696855 fax:(01268) 697018 gandi.canvey@btconnect.com www.grantandlivingstonltd.com

Grant Westfield Ltd Westfield Avenue Edinburgh Lothian EH11 2QH tel:(0131) 337 6262 fax:(0131) 337 2859 sales@grantwestfield.co.uk www.grantwestfield.co.uk

Granwood Flooring Ltd PO Box 60 Alfreton DE55 4ZX tel:(01773) 606060 fax:(01773) 606030 sales@granwood.co.uk www.granwood.co.uk

Grass Concrete Ltd Walker House 22 Bond Street Wakefield W. Yorkshire WF1 2QP tel:(01924) 379443 fax:(01924) 290289 info@grasscrete.com www.grasscrete.com

Gravesend Fencing Limited The Fencing Works Codrington Crescent Gravesend Kent DA12 5DD tel:(01474) 326016 fax:(01474) 324562

Gray J. W. & Son Ltd 163 Wantz Road Maldon Essex CM9 5DD tel:(01621) 859950 fax:(01621) 852713
Grayhill Trading Ltd PO Box 521 Great Dunmow Essex CM6 1UY tel:(01376) 348664 fax:(01376) 348664
Great Metropolitan Flooring Co Ltd 83 Kinnerton Street Knightsbridge London SW1X 8ER tel:(020) 7235 1161 fax:(020) 7235 9027
Greaves Welsh Slate Co Ltd Llechwedd Slate Mines Blaenau Ffestiniog Gwynedd LL41 3NB tel:(01766) 830522 fax:(01766) 830711 llechwedd@aol.com www.llechwedd.co.uk
Green A P Refractories Ltd Dock Road South Bromborough Wirral Merseyside L62 4SP tel:(0151) 645 0701 fax:(0151) 645 8261
Green Alexander Lough Road Lurgan Co. Armagh BT66 6LY tel:(028) 38322086 fax:(028) 38325502
Green Bros (Geebro) Ltd Hailsham E. Sussex BN27 3DT tel:(01323) 840771 fax:(01323) 440109
Green Brothers Ltd Shentonfield Road Sharston Industrial Estate Manchester M22 4TJ tel:(0161) 741 7270 fax:(0161) 741 7272 sales@greensigns.co.uk www.greensigns.co.uk
Green Gerald 211 Hinckley Road Nuneaton Warwks. CV11 6LL tel:(02476) 325059 fax:(02476) 325059 gerald@ggarts.co.uk www.ggarts.demon.co.uk
Greenwood Air Management Brookside Industrial Estate Rustington W. Sussex BN16 3LH tel:(01903) 771021 fax:(01903) 782398 info@greenwood.co.uk www.greenwood.co.uk
Greif UK Ltd Merseyside Works Ellesmere Port Cheshire CH65 4EZ tel:(0151) 373 2000 fax:(0151) 373 2072 info.uk@greif.com www.greif.com
Griff Chains Ltd Quarry Road Dudley Wood Dudley West Midlands DY2 0ED tel:(01384) 569415 fax:(01384) 410580 sales@griffchairs.co.uk www.griffchairs.co.uk
Griffin and General Fire Services Ltd Unit F 7 Willow Street London EC2A 4BH tel:(020) 7251 9379 fax:(020) 7729 5652 headoffice@griffinfire.co.uk www.griffinfire.co.uk
Grille Diffuser & Louvre Co Ltd The Air Diffusion Works Woolley Bridge Road Hollingworth, Hyde Cheshire SK14 7BW tel:(01457) 861538 fax:(01457) 866010
Gripperrods Ltd Wyrley Brook Park Walkmill Lane Bridgtown, Cannock Staffs. WS11 3RX tel:(01922) 417777 fax:(01922) 419411
Grohe Ltd 1 River Road Barking Essex IG11 0HD tel:(020) 8594 7292 fax:(020) 8594 8898
Grorud Industries Ltd Castleside Industrial Estate Consett Co. Durham DH8 8HG tel:(01207) 581485 fax:(01207) 580036
Grosvenor Pumps Ltd Trevoole Praze Cambourne Cornwall TR14 0PJ tel:(01242) 227400 fax:(01242) 227404 gravesnor.pumps@virgin.net
Grothkarst UK Ltd Unit 28, Balfour Business Centre Balfour Road Southall Middx. UB2 5BD tel:(020) 8574 6464 fax:(020) 8573 1189
Group Four Glassfibre Ltd Church Road Business Centre Murston Sittingbourne Kent ME10 3RS tel:(01795) 429424 fax:(01795) 476248 info@groupfourglassfibre.co.uk www.groupfourglassfibre.co.uk
Groupco Ltd 18 Thresham Road Orton Southgate Peterborough PE2 6SG tel:(01733) 234750 fax:(01733) 235246
Grundfos Pumps Ltd Groveberry Road Leighton Buzzard Beds. LU7 8TL tel:(01525) 850000 fax:(01525) 850011
Grundy Catering Grundy House Ascot Road Bedfont, Feltham Middx. TW14 8QH tel:(01784) 251244 fax:(01784) 246137
Guardall Ltd Queen Ann Drive Lochend Industrial Estate Newbridge Edinburgh EH28 8PL tel:(0131) 333 2900 fax:(0131) 333 4919 sales@guardall.co.uk www.guardall.co.uk
Guardian Lock and Engineering Co Ltd Imperial Works Wednesfield Road Willenhall W. Midlands WV13 1AL tel:(01902) 635964 fax:(01902) 630675
Guardian Wire Ltd Guardian Works Stock Lane Chadderton Lancs. OL9 9EY tel:(0161) 624 6020 fax:(0161) 620 2880
Guest (Speedfit) Ltd, John, Horton Road West Drayton Middx. UB7 8JL tel:(01895) 449233 fax:(01895) 420321
Guild of Architectural Ironmongers 8 Stepney Green London E1 3JU 020 7790 3431 www.martex.co.uk/ace/ace0291.htm
Guild of Builders & Contractors, The Crest House 102-104 Church Road Teddington Middlesex TW11 8PY 020 8977 1105 020 8943 3151 info@buildersguild.co.uk www.buildersguild.co.uk
Gummers A & J Ltd Unit H, Redfern Park Way Tysley Birmingham West Midlands B11 2DN tel:(0121) 706 2241 fax:(0121) 706 2960 Sales@Gummers.co.uk www.gummers.co.uk
Gunnebo Ltd Bell Brook Business Park Uckfield E. Sussex TN22 1QQ tel:(01252) 373700 fax:(01252) 373707
Gun-Point Ltd Thavies Inn House 3-4 Holborn Circus London EC1N 2PL tel:(020) 7353 1759 fax:(020) 7583 7259
Gunton Leisure Ltd David Longley House East Park Crawley W. Sussex RH10 6AP tel:(01293) 564777 fax:(01293) 564789
Guthrie Allesbrook & Co Ltd The Old Hall Works Arborfield Road Shinfield Reading RG2 9DP tel:(01734) 882022 fax:(01734) 884678
Guthrie Douglas Ltd 12 Heathcote Way Heathcote Industrial Estate Warwick CV34 6TE tel:(01926) 452452 fax:(01926) 336417
Guttercrest Victoria Road Oswestry Shropshire SY11 2HX tel:(01691) 663300 fax:(01691) 663311 info@guttercrest.co.uk www.guttercrest.co.uk
H & R Johnson Tiles Ltd Harewood Street Tunstall Stoke-on-Trent ST6 5JZ tel:(01782) 575575 fax:(01782) 524 138 sales@johnson-tiles.com www.johnson-ceramics.com
H Pickup Mechanical & Electrical Services Ltd Durham House Lower Clark Street Scarborough N. Yorks. YO12 7PW tel:(01723) 369191 fax:(01723) 362044 pickup@hpickup.demon.co.uk www.hpickup.demon.co.uk
H&H Electronics Ltd Port Marsh Industrial Estate Port Marsh Road Calne Wilts. SN11 9BW tel:(01249) 813412 fax:(01249) 815184 sales@handh.uk.com www.handh.uk.com
H. B. Fuller Coatings Ltd 95 Aston Church Road Birmingham B7 5RQ tel:(0121) 322 6900 fax:(0121) 322 6901 www.hbfuller.com
Haddonstone Ltd The Forge House Church Lane East Haddon Northants NN6 8DB tel:(01604) 770711 fax:(01604) 770027 info@haddonstone.co.uk www.haddonstone.co.uk
Hadley Industries Plc PO Box 92 Downing Street Smethwick W. Midlands B66 2PA tel:(0121) 555 1300 fax:(0121) 555 1301 sales@hadleygroup.co.uk www.hadleygroup.co.uk
Haffele UK Ltd Swift Valley Industrial Estate Rugby Warwks. CV21 1RD tel:(01788) 542020 fax:(01788) 544440 email@hafele.co.uk www.hafele.co.uk
Hager Ltd Hortonwood 50 Telford Shropshire TF1 7FT tel: (0870) 2402400 fax: (0870) 2400400 info@hager.co.uk www.hager.co.uk
Hago Products Ltd Durban Road, South Berstead Industrial Estate Bogner Regis W. Sussex PO22 9RD tel:(01243) 863131 fax:(01243) 827687
Haironville TAC Ltd Abbotsfield Road Abbotsfield Industrial Park St Helens Merseyside WA9 4HU tel:(01744) 818181 fax:(01744) 851555 sales@haironvilletac.co.uk www.haironvilletac.co.uk
Hale Hamilton Ltd Frays Mills Works Cowley Road Uxbridge Middx. UB8 2AF tel:(01895) 236525 fax:(01895) 231407 gpearson@halehamilton.com
Halfen Ltd Humphreys Road, Woodside Estate Dunstable Beds. LU5 4TP tel:(08705) 316300 fax:(08705) 316304 info@halfen.co.uk www.halfen.co.uk
Hall & Watts Ltd 266 Hatfield Road St Albans Herts. AL1 4UN tel:(01727) 859288 fax:(01727) 835683
Hallidays UK Ltd High Street Dorchester on Thames Oxon OX10 7HL tel:(01865) 340028 fax:(01865) 341149 www.hallidays.com
Hallmark Blinds Ltd Hallmark House 173 Caladonian Road Islington London N1 0SL tel:(020) 7837 0964 fax:(020) 7833 1693 info@hallmarkblindsltd.co.uk www.hallmarkblindsltd.co.uk
Halls Beeline Ltd Northgate White Lund Estate Morecambe Lancs. LA3 3PA tel:(01524) 63233 fax:(01524) 65792 info@beeline-rex.co.uk www.beeline-rex.co.uk
Halls Group Ltd Riverside Road Gorleston Great Yarmouth Norfolk NR31 6PX tel:(01493) 656672 fax:(01493) 440225 reception@hallsgroup.uk.com www.hallsgroup.uk.com
Halmark Panels Ltd Valletta House, Valletta Street Hedon Road Hull HU9 5NP tel:(01482) 703222 fax:(01482) 714714 johnrolland@hallmark-panels.demon.co.uk
Halmatic Ltd Saxon Wharf Lower York Street Northam Southampton SO14 5QF tel:(023) 80337477 fax:(023) 80337478
Halo, W H S Division Of Bowater Industries 1 Cranfield Road, Lostock Industrial Estate Lostock Bolton BL6 4SB tel:(01204) 699211 fax:(01204) 699559
Halstead Boilers Ltd 20/22 First Avenue, Bluebridge Industrial Estate Halstead Essex CO9 2EX tel:(01787) 475557 fax:(01787) 474588 sales@halsteadboilers.co.uk www.halsteadboilers.co.uk
Hambleside Danelaw Ltd 2-8 Bentley Way Royal Oak Industrial Estate Daventry Northants NN11 5QH tel:(01327) 701900 fax:(01327) 701909 marketing@hambleside-danelaw.co.uk www.hambleside-danelaw.co.uk
Hamilton Acorn Ltd Halford Road Attleborough Norfolk N17 2HZ tel:(01953) 453201 fax:(01953) 454943 info@hamilton-acorn.co.uk www.hamilton-acorn.co.uk
Hamilton R & Co Ltd Unit G Quarry Industrial Estate Mere Wilts. BA12 6LA tel:(01747) 860088 fax:(01747) 861032 info@hamilton-litestat.com www.hamilton-litestat.com
Hammond & Champness Ltd 4-5 St Johns Square London EC1M 4HS tel:(020) 7490 2316 fax:(020) 7250 3012
Hammonds - New Home Division Fleming Road Harrowbrook Industrial Estate Hinckley Leicestershire LE10 3DT tel:(01455) 251451 fax:(01455) 633981 www.hammonds-uk.com
Hampton Stone Ltd The Hampton Centre Cirencester Road Chalford Glos. GL6 8PE tel:(01453) 882180 fax:(01453) 886975 info@hamptonstone.co.uk www.hamptonstone.co.uk
Hamworthy Heating Ltd Fleets Corner Poole Dorset BH17 0HH tel:(01202) 662500 fax:(01202) 665111 sales@hamworthy-heating.com www.hamworthy-heating.com
Hand Tools Ltd Wreakes Lane Dronfeild Sheffield S18 6PN tel:(01246) 413139 fax:(01246) 415208
Hands of Wycombe 36 Dashwood Avenue High Wycombe Bucks. HP12 3DX tel:(01494) 524222 fax:(01494) 526508 info@hands.co.uk www.hands.co.uk
Hanovia Ltd 145 Farnham Road Slough Berks. SL1 4XB tel:(01753) 515300 fax:(01753) 534277 sales@hanovia.co.uk www.hanovia.co.uk
Hansen Windows Ltd 6 and 7 The Rickyard Clifton Reynes Olney Bucks MK46 5LQ tel:(01234) 713151 fax:(01234) 713094 jrmayhew@btinternet.com
Hansgrohe Units D1 & D2 Sandown Park Trading Estate Royal Mills, Esher Surrey KT10 8BL tel:(0870) 7701972 fax:(0870) 7701973 sales@hansgrohe.co.uk www.hansgrohe.co.uk
Hanson Aggregates Venture Way Priorswood Ind Est Taunton Somerset TA2 8DG tel:(01373) 463211 fax:(01373) 465843 www.hanson-aggregates.com
Hanson Bath & Portland Stone Bumpers Lane Wakeham Portland Dorset DT5 1HY tel:(01305) 820207 fax:(01305) 860275 colin.goble@hansonpk.com www.hanson-europe.com
Hanson Brick Stewartby Bedford MK43 9LZ tel:(08705) 258258 fax:(01234) 762040 info@hansonbrick.com www.hansonbrick.com
Hanson Brick Ltd, Northern Regonal Sales Office Unicorn House Wellington Street Ripley Derbys. DE5 3DZ tel:(0990) 258258 fax:(01733) 206040
Hanson Building Materials Europe The Ridge Chipping Sodbury Bristol BS37 6AY tel:(01454) 316000 fax:(01454) 332791 www.quarryproducts.com
Hanson Building Products Quarry Lane Woodkirk Dewsbury W Yorkshire WF12 7JJ tel:(01773) 602432 fax:(01733) 603134 robert.drayton@hansonbp.com www.hansonprecast.com
Hanson Concrete Products PO Box 14 Appleford Road Sutton Courtenay, Abbingdon Oxon. OX14 4UB tel:(01235) 848877 fax:(01235) 848767 sales@hansonconcreteproducts.com www.hansonplc.com
Hanson Concrete Products Mells Road Mells Frome Somerset BA11 3PD tel:(01179) 812791 fax:(01179) 814516 sales@hanson-pipes.com www.hanson-pipes.com
Hanson Concrete Products Alfreton Road Derby DE21 4BN tel:(01332) 364314 fax:(01332) 372208 reception@hansonplc.com www.hanson.biz
Hanson Concrete Products Ltd Hoveringham Nottingham NG14 7JX tel:(01636) 832000 fax:(01636) 832020
Harbro Supplies Ltd Morland Street Bishop Auckland Co. Durham DL14 6JQ tel:(01388) 605363 fax:(01388) 603263 harbrosupplies@hotmail.com www.habrosupplies.com
Harcostar Drums Harcostar garden products Windover Road Huntingdon Cambs. PE29 7EE tel:(01480) 445113 fax:(01480) 413203
Hardall International Ltd 34 Clarke Road Mount Farm Milton Keynes MK1 1LG tel:(01908) 274441 fax:(01908) 367265
Hardman John Studios Lightwood House, Lightwood Park Hagley Road West Bearwood, Warley W. Midlands B67 5DP tel:(0121) 429 7609
Hargreaves Quarries Ltd Moota Qyarry Cockermouth Cumbria CA13 0QE tel:(016973) 20234 fax:(016973) 20179
Harkness Hall Ltd The Gate Studios Station Road Borehamwood Herts. WD6 1DQ tel:(020) 8953 3611 fax:(020) 8207 3657 Sales@harknesshall.com www.harknesshall.com
Harris & Bailey Ltd 50 Hastings Road Croydon Surrey CR9 6BR tel:(020) 8654 3181 fax:(020) 8656 9369 mail@harris-bailey.co.uk www.harris-bailey.co.uk
Harris Road Roofing & Flooring Ltd Unit 56 Youngs Industrial Estate Paices Hill, Aldermaston Reading Berks. RG7 4PW tel:(01189) 811466 fax:(0118) 981 9021
Harris T A Ltd Hargreen House 134 New Kent Road London SE1 6TY tel:(020) 7703 3842 fax:(020) 7708 2596
Harrision Flagpoles Borough Road Darlington Co. Durham DL1 1SW tel:(01325) 355433 fax:(01325) 461726 sales@flagpoles.co.uk www.flagpoles.co.uk
Harrison Drape, Divsion of Newell Ltd PO Box 233 Bradford Street Birmingham B12 0PE tel:(0121) 766 6111 fax:(0121) 772 0696 mlaydon@harrisondrape.com
Harrison Thompson & Co Ltd Yeoman House, Whitehall Estate Whitehall Road Leeds LS12 5JB tel:(0113) 279 5854 fax:(0113) 231 0406 info@yeomanshield.com www.yeomanshield.com
Hart Door Systems Ltd Redburn Road Westerhope Industrial Estate Newcastle-upon-Tyne Tyne & Wear NE5 1PJ tel:(0191) 214 0404 fax:(0191) 271 1611 response@hart-speedor.co.uk www.hart-speedor.co.uk
Hartley & Sugden Atlas Works Gibbet Street Halifax W. Yorks. HX1 4DB tel:(01422) 355651 fax:(01422) 359636 hartleyandsugden@dial.pipex.com
Hartley Clear Span Ltd Wellington Road Greenfield, Oldham Lancs. OL3 7AG tel:(01457) 873244 fax:(01457) 870151
Hartley John W Ltd Lunds Fields Carnforth Lancs. LA5 9AB tel:(01455) 272457 fax:(01455) 274564
Harton Heating Appliances Unit 6, Thustlebrook Industrial Est. Eynsham Drive Abbey Wood SE2 9RB tel:(020) 8310 0421 fax:(020) 8310 6785
Harty Holdings Ltd Crossbeg Industrial Estate Upper Ballymount Road Tallaght Dublin 24 Ireland tel:(00 353) 1 4566482 fax:(00 353) 1 4501214
Harvey Fabrication Ltd Hancock Road London E3 3DA tel:(020) 8981 7811 fax:(020) 8981 7815
Harvey Furniture Newlands Drive Poyle Berks. SL3 0DX tel:(01753) 680200 fax:(01753) 683673
Hatema (UK) Ltd St Pancras Commercial Centre 63 Prat Street London NW1 0BY tel:(020) 7482 4466 fax:(020) 7267 3239
Hatfield Ltd, Roy Fullerton Road Rotherham S. Yorks. S60 1DL tel:(01709) 820855 fax:(01709) 374062
Hatmet Ltd Interiors House Lynton Road Crouch End London N8 8SL tel:(020) 8348 9262 fax:(020) 8341 9878 trevorh@hatmet.co.uk
Hattersley & Davidson Ltd 71 Greystock Street Shefield S4 7WA tel:(01742) 728957 fax:(01742) 729330
Hattersley Newman Hender Ltd Burscough Road Ormskirk Lancs. L39 2XG tel:(01695)577199 fax:(01695)578775 uksales@hattersley.com www.hattersley.com
Havering Fencing Co. 237 Chase Cross Road Collier Row Romford Essex RM5 3XS tel:(01708) 747855 fax:(01708) 721010 haveringfencingco@hfco.fsbusiness.co.uk www.haveringfencing.co.uk
Hawker Ltd Rake Lake Clifton Junction Swinton Manchester M27 8LR tel:(0161) 794 4611 fax:(0161) 793 6606 hawkeruk.sales@hawker.invensys.com www.hawker.invensys.com
Hawkins Insulation Ltd Central House, 101Central Park Petherton Road Hengrave Bristol BS14 9BZ tel:(01275) 839500 fax:(01275)835555 asbestos.equiries@fsmail.net www.hawkinsinsulation.co.uk
Haworth UK Ltd Cannon CT 5 Brew House Yard St Johns London WC1A 1EE tel:(020) 7404 1617 fax:(020) 7404 1607 www.haworth.com
Hay Joinery Chapel Road Lingwood Norwich NR13 4NY tel:(01603) 712392 fax:(01603) 714248 build@js-hay.co.uk
Haydon & Hagan Construction Ltd Palfrey House 71A Palfrey Place London SW8 1AR tel:(020) 7587 1150 fax:(020) 7820 9080 h.haydon@haydonandhagan.demon.co.uk
Haymills (Contractors) Ltd Wesley House 1 -7 Wesley Av London NW10 7BZ tel:(020) 8991 4200 fax:(020) 8991 4201
Hayward Air 120 Avon Road Bournemouth Dorset BH8 8SF tel:(01202) 390333 fax:(01202) 390666 sales@hayward-air.co.uk www.hayward-air.co.uk
Hazard Safety Products Ltd 55-57 Bristol Road Edgbaston Birmingham B5 7TU tel:(0121) 446 4433 fax:(0121) 446 4230
Haze Cladding Systems Briding Road Hatfield Road Cardiff CF11 8TL tel:(01443) 813814 fax:(01443) 815813 sales@brohome.com www.brohome.com
HB Fuller Coatings Ltd 95 Aston Church Road Nechells Birmingham B7 5RQ tel:(0121) 322 6900 fax:(0121) 322 6902
HC Slingsby PLC Preston Street Bradford W. Yorks. BD7 1JF tel:(01274) 721591 fax:(01274) 723044 sales@slingsby.com www.slingsby.com
HCL Contracts Ltd Bridge House Commerce Road Brentford Middx. TW8 8LQ tel:(0800) 212867 fax:(020) 8847 9003 info@hclgroup.co.uk www.hclgroup.co.uk
HCL Verta Unit 6 Wharfeilde Business Centre Trafford Wharf Road Manchester M17 1EX tel:(0161) 877 6619 fax:(0161) 877 7197 hclverta@compuserve.com
HE Services (Plant Hire) Ltd Ashford Depot Ellingham Industrial Estate Ashford Kent TN22 2LZ tel:(01233) 639831 fax:(01233) 639831 mail@heservices.co.uk www.heservices.co.uk
Healthmatic Ltd 3 Wellington Square Ayr Ayrshire KA7 1EN Tel:(01292) 265 456 Fax:(01292) 287 869 sales@healthmatic.com www.healthmatic.com
Heap Joshua & Co Ltd Boodle Street Ashton-under-Lyne Lancs. OL6 8NG tel:(0161) 330 1868 fax:(0161) 339 2040
Heat Trace Ltd Tracer House Cromwell Road Bredbury, Stockport Gtr. Manchester SK6 2RF tel:(0161) 430 8333 fax:(0161) 430 8654 sales@heat-trace.ltd.uk www.heat-trace.ltd.uk

Heath Samuel & Sons PLC Cobden Works Leopold Street Birmingham B12 0UJ tel:(0121) 772 2303 fax:(0121) 772 3334
Heating & Ventilating Contractors Association Esca House 34 Palace Court London W2 4JG tel:(020) 7313 4900 fax:(020) 7727 9268 contact@hvca.org.uk www.hvca.org.uk
Heating Equipment Ltd Hughenden Manor Petersfinger Salisbury Wilts. SP5 3DF tel:(01722) 335139 fax:(01722) 323592
Heating Ventilating and Air Conditioning Manufacturers Association Henley Road Medenham Marlow Bucks SL7 2ER 01491 578674 01491 575024 info@feta.co.uk www.feta.co.uk/hevac/index.htm
Heatrae Sadia Heating Ltd Hurricane Way Norwich NR6 6EA tel:(01603) 420110 fax:(01603) 420149 sales@heatraesadia.com www.heatraesadia.com
Heldite Ltd Heldite Centre Bristow Road Hounslow Middx. TW3 1UP tel:(020) 8577 9157/9257 fax:(020) 8577 9057
Helifix Ltd Second Floor Suite 21 Warple Way London W3 0RX tel:(020) 8735 5200 fax:(020) 8735 5201 sales@helifix.co.uk www.helifix.co.uk
Hellermann Tyton Sharston Green Buisness Park , 1 Robeson Way Altrincham Road Wythenshawe Manchester M22 4TY tel:(0161) 945 41811 fax:(0161) 954 3708 www.hellermanntyton.co.uk
Helmsman Northern Way Bury St Edmunds Suffolk IP32 6NH tel:(01284) 727600 fax:(01284) 727601 sales@helmsman.co.uk www.helmsman.co.uk
Helvar Electrosonic Hawley Mill Hawley Road Dartford Kent DA2 7SY tel:(01322) 222211 fax:(01322) 282282 www.helvar.com
HEM Interiors Group Ltd HEM House Kirkstall Road Leeds LS4 2BT tel:(0113) 263 2222 fax:(0113) 231 0237 heminteriors@btinternet.com
Hemsec Manufacturing Ltd Stoney Lane Rainhill Prescot Merseyside L35 9LL tel:(0151) 426 7171 fax:(0151) 493 1331 sales@hemsec.com www.hemsec.com
Henkel Consumer Adhesives Road 5 Winsford Ind Est Winsford Cheshire CW1 3QY tel:(020) 8804 3343 fax:(020) 8443 4321 www.unibond.co.uk
Henkel Industrial Adhesives Marketing and Sales Department Watchmead Welwyn Garden City Hertfordshire AL7 1JB tel:(01707) 355303 fax:(01707) 355302 aisalesoff@henkel.co.uk www.henkel.com
Henkel Industrial Adhesives (AIV) Cromwell Road St. Neots Cambs. PE19 2JW tel:(01480) 404505 fax:(01480) 406111
Henry Venables Interiors Ltd Tollgate Drive Tollgate Industrial Estate Beaconside, Stafford Staffs ST16 3HS tel:(01785) 270610 fax:(01785) 270616 enquiries@henryvenables.co.uk www.henryvenables.co.uk
Henry Venables Ltd Castletown Stafford Staffordshire ST16 2EN tel:(01785) 259131 fax:(01785) 215087 enquiries@henryvenables.co.uk www.henryvenables.co.uk
Hepworth Building Products Ltd Hazlehead Crow Edge Sheffield S. Yorks. S36 4HG tel:(01226) 763561 fax:(01226) 764827 info@hepworthbp.co.uk www.hepworthbp.co.uk
Hepworth Heating Ltd Nottingham Road Belper Derbys. DE56 1JT tel:(01773) 824141 fax:(01773) 820569 www.hep-heat.co.uk
Hepworth Minerals & Chemicals Ltd Brookside Hall Sandbach Cheshire CW11 4TF tel:(01270) 752 752 fax:(01270) 752753 sales@h-m-c.co.uk www.h-m-c.co.uk
Herberts, Division of HPG Freshwater Road Dagenham Essex RM8 1RU tel:(020) 8590 6030 fax:(020) 8599 3527
Herga Electric Ltd Northern Way Bury St Edmunds Suffolk IP32 6NN tel:(01284) 701422 fax:(01284) 753112
Hertel Services Hertel (UK) Ltd Poolspringe Llanwarne Hereford HR2 8JJ tel:(01981) 540020 fax:(01981) 540853
Hettich UK Ltd Unit 200, Metroplex Business Park Broadway Salford Manchester M5 2UW tel:(0161) 872 9552 fax:(0161) 848 7605
Heydal Ltd Unit 25 Shepley South Industrial Estate Audenshaw Manchester M34 5DW tel:(0161) 336 0900 fax:(0161) 337 0710 info@heydal.com www.heydal.com
Heywood Williams Architectural Birds Royd Lane Brighouse W. Yorks. HD6 1NG tel:(01484) 717677 fax:(01484) 400148
HF (GB) Ltd York House Vicarage Lane Bowdon, Altrincham Cheshire WA14 3BA tel:(0161) 928 4572 fax:(0161) 929 8513
Hickton & Co Ltd, John PO Box 10 Stourbridge Road Holesowen W. Midlands B63 3QY tel:(0121) 550 1671 fax:(0121) 585 5031
Higginson Staircases Ltd Unit 1 Carlisle Road London NW9 0HD tel:(020) 8200 4848 fax:(020) 8200 8249 sales@higginson.co.uk www.higginson.co.uk
Hill Leigh Ltd Brue Way Walrow Industrial Estate Highbridge Somerset TA9 4AW tel:(01278) 789156 fax:(01278) 781768 enquiries@hill-leigh.co.uk www.hill-leigh.co.uk
Hill Top Sections Ltd Ridgacre Road West Bromwich W. Midlands B71 1BB tel:(0121) 5551470 fax:(0121)5551471 hts@hadleygroup.co.uk www.hadleygroup.co.uk
Hillaldam Coburn Ltd Unit 6 Wyvern Estate Beverley Way New Malden Surrey KT3 4PH tel:(020) 8336 1515 fax:(020) 8336 1414 sales@hillaldam.co.uk www.coburn.co.uk
Hillday 1 Haverscroft Industrial Estate Attleborough Norfolk NR17 1YE tel:(01953) 454014 fax:(01953) 454014
Hille Ltd Two Gates Mill Angus Street Darwen BB3 3QA tel:(01942) 262283 fax:(01942) 603232 hille.sales@hille.co.uk www.hille.co.uk
Hills Industries Ltd Pontgwindy Industrial Estate Caerphilly M. Glam. CF8 3HU tel:(029) 20 883951 fax:(029) 20 886102
Hilmor Caxton Way Stevenage Herts. SG1 2DQ tel:(01438) 312466 fax:(01438) 728327 sales@hilmar.co.uk www.hilmar.co.uk
Hilti (Great Britain) Ltd 1 Trafford Wharf Road Manchester M17 1BY tel:(0800) 886100 fax:(0800) 886200 www.hilti.com
Hilton Banks Ltd 74 Oldfield Road Hampton Middx. TW12 2HR tel:(08452) 300 404 fax:(020) 8779 8294 hilton banks@btinternet.com
Hines & Sons, P E, Ltd Whitbridge Lane Stone Staffs. ST15 8LU tel:(01785) 814921 fax:(01785) 818808
Hinton Perry & Davenhill Ltd Dreadnought Works Pensnett Brierley Hill Staffs. DY5 4TH tel:(01384) 77405 fax:(01384) 74553 office@dreadnought.freeuk.com www.dreadnought-tiles.co.uk
Hipkin F W Ltd Victory Works Maitland Road Stratford London E15 4HF tel:(020) 8555 8201 fax:(020) 8519 6606
Hipkiss, H, & Co Ltd 40-45 George Street Birmingham B3 1QA tel:(0121) 236 5342 fax:(0121) 236 4174 info@hipkiss.co.uk www.hipkiss.co.uk
Hird Hastie Paints Ltd 73 Milnpark Street Glasgow G41 1EH tel:(0141) 429 2775 fax:(0141) 429 1407
HIS Ltd Pearl House Church Street Bridgwater Somerset TA6 5AT tel:(01278) 450600 fax:(01278) 452160
Hispack Limited 8 School Road Downham, Billericay Essex CM11 1QU tel:(01268) 711499 fax:(01268) 711068 hispack@btinternet.com www.hispack.co.uk
Hi-Store Ltd Station Approach Four Marks Alton Hants. GU34 5HN tel:(01420) 562522 fax:(01420) 564420 hi-store@msn.com www.hi-store.com
Hi-Tec Window Films The Stone House 79 Mount Road Tettenhall Wood Wolverhampton WV6 8HQ tel:(01902) 761303 fax:(01902) 766225 hitec@aol.com www.hitecwindowfilms.com
Hi-Vee Ltd Southdown Western Road Crowborough E. Sussex TN6 3EW tel:(01892) 662177 fax:(01892) 667225
HMG Paints Riverside Works Collyhurst Road Collyhurst Manchester M40 7RU tel:(0161) 205 7631 fax:(0161) 205 4829 sales@hmgpaint.com www.hmgpaint.com
HMSO Publications Office Parliamentary Press Mandela Way London SE1 tel:(020) 7394 4200
Hochiki Europe (UK) Ltd Grosvenor Road, Gillingham Business Park Gillingham Kent ME8 0SA tel:(01634) 260133 fax:(01634) 260132 www.hochikieurope.com
Hodgkinson Bennis Ltd Unit 7A Highfield Road Little Hulton, Worsley Manchester M38 9SS tel:(0161) 790 4411 fax:(0161) 703 8505
Hodkin Jones (Sheffield) Ltd Callywhite Lane Dronfield Sheffield S18 2XP tel:(01246) 290890 fax:(01246) 290292 info@hodkin-jones.co.uk www.hodkin-jones.co.uk
Holden Brooke Pullen Wenlock Way Manchester M12 5JL tel:(0161) 223 2223 fax:(0161) 220 9660 sales@holdenbrookepullen.com www.holdenbrookepullen.com
Hollaender Rainer PO Box 52 Walsall W. Midlands WS2 7PL tel:(01922) 495777 fax:(01922) 497943
Holophane Europe Ltd Bond Avenue Bletchley Milton Keynes MK1 1JG tel:(01908) 649292 fax:(01908) 270006
Homa Castors (Cosby) Ltd Cambridge Road Cosby Leicester LE9 5SH tel:(01533) 848266 fax:(01533) 867692
Home Automation Ltd Bumpers Way Chippenham Wilts. SN14 6LF tel:(01249) 443422 fax:(01249) 443315 sales@homeautomation.co.uk www.homeautomation.co.uk
Homelux Nenplas Airfield Industrial Airfield Industrial Estate Ashbourne Derbys. DE6 1HA tel:(01335) 347300 fax:(01335) 340333 www.tiletransfers.com
Hometex Trading Ltd 13 Wheatley Road Ilkey W. Yorks. LS29 8TS tel:(01943) 608197 fax:(01943) 816370
Honeywell Control Systems Ltd Honeywell House Arlington Business Park Bracknell Berks. RG12 1EB tel:(01344) 656000 fax:(01344) 656240 uk.infocentre@honeywell.com www.honeywell.com/uk
Honeywell Speciality Wax and Additives Ltd Tavistock Road West Drayton Middx. UB7 7RA tel:(01895) 427000 fax:(01895) 449199 www.honeywell.com
Hoover European Appliances Group Pentrbach, Merthyr Tydfil M. Glam. CF48 4TU tel:(01685) 721222 fax:(01685) 382946
Hopkinsons Ltd PO Box B27 Britania Works Huddersfield HD2 2UR tel:(01484) 422171 fax:(01484) 518092
HOPPE (UK) Ltd Gailey Park Gravelly Way Standeford Wolverhampton WV10 7GW tel:(01902) 484400 fax:(01902) 484406 info.uk@hoppe.com www.hoppe.co.uk
Horizon International Ltd Willment Way Avonmouth Bristol BS11 8DJ tel:(0117) 982 1415 fax:(0117) 982 0630 sales@horizon-int.com www.horizon-int.com
Hormann (UK) Ltd Gee Road Coalville Leicestershire LE67 4JW tel:(01530) 513000 fax:(01530) 513001
Horne Engineering Ltd PO Box 7 Rankine Street Johnstone Renfrewshire PA5 8BD tel:(01505) 321455 fax:(01505) 336287 sales@horne.co.uk www.horne.co.uk
Horstmann Controls Ltd South Bristol Business Park Roman Farm Road Bristol BS4 1UP tel:(0117) 978 8700 fax:(0117) 978 8701 sales@horstmann.co.uk www.horstmann.co.uk
Hoskins Medical Equipment Ltd Admail 1001 Birmingham B1 1HJ tel:(0121) 707 6600 fax:(0121) 707 6688
Hospital Metalcraft Ltd Blandford Heights Blandford Forum Dorset DT11 7TG tel:(01258) 451338 fax:(01258) 455056 sales@bristolmaid.com www.bristolmaid.com
Hotchkiss Air Supply Heath Mill Road Wombourne Wolverhampton WV5 8AP tel:(01903) 895161 fax:(01903) 892045 sales@hotchkissairsupply.co.uk www.hotchkissairsupply.co.uk
Housing Corporation 149 Tottenham Court Road London W1T 7BN tel:(020) 7393 2000 fax:(020) 7393 2111 enquiries@housingcare.gov.uk www.housingcorp.gov.uk
Hovair Systems Ltd Unit 3, Kingstons Ind Est Eastern Road Aldershot Hampshire GU12 4TD tel:(01252) 319922 fax:(01252) 341872 sales@hovair.co.uk www.hovair.co.uk
Hoval Ltd Northgate Newark-on-Trent Notts. NG24 1JN tel:(01636) 72711 fax:(01636) 73532
Howard Bros Joinery Ltd Station Approach Battle E. Sussex TN33 0DE tel:(01424) 773272 fax:(01424) 773836 sales@howard-bros-joinery.com www.howard-bros-Joinery.com
Howard Evans Roofing Limited 75 Tyburn Road Erdington Birmingham B24 8NB tel:(0121) 327 1336 fax:(0121) 327 3423 roofherl@ukbusiness.com
Howard Home Improvements Ltd 45 Mawney Road Romford Essex RM7 7HL tel:(01708) 722245 / 720048 fax:(01708) 736090
Howarth Timber (Elland) Ltd Elland Lane Elland Halifax Yorks. HX5 9DZ tel:(01422) 376012 fax:(01422) 310107
Howden Buffalo Old Govan Road Renfrew Renfrewshire PA4 8XJ tel:(0141) 885 7500 fax:(0141) 885 7555 buffalo.marketing@howden.com www.howden.com
Howe Fencing & Sectional Buildings Horse Cross, Standon Road Standon Nr Ware Herts. SG11 2PU tel:(01920) 822055 fax:(01920) 822871 sales@howefencing.co.uk www.howefencing.co.uk
Howe-Green Ltd Merchant Drive Mead Lane Industrial Estate Hertford Herts. SG13 7BH tel:(01992) 554388 fax:(01992) 584612 info@howegreen.co.uk www.howegreen.co.uk
Howse Thomas Ltd Cakemore Road Rowley Regis West Midlands B65 0RD tel:(0121) 559 1451 fax:(0121) 559 2722 sales@howsepaints.co.uk www.howsepaints.co.uk
Hoyles Electronic Developments Ltd Unit 3 Milbrook Business Park Rainford Industrial Estate Rainford, St. Helens Merseyside WA11 8LZ tel:(01744) 886600 fax:(01744) 886607 info@hoyles.com www.hoyles.com
Hozelock Ltd Haddenham Aylesbury Bucks. HP17 8JD tel:(01844) 291881 fax:(01844) 290344
HRS Hevac Ltd PO Box 230 Watford Herts. WD1 8TX tel:(01923) 232335 fax:(01923) 230266 mail@hrs.co.uk www.hrshevac.co.uk
HSS Hire Service Group 25 Willow Lane Mitcham Surrey CR4 4TS tel:(020) 8260 3100 fax:(020) 8687 5005 hire@hss.co.uk www.hss.co.uk
HT Martingale Ltd Ridgeway Industrial Estate Iver Bucks. SL0 9HU tel:(01753) 654411 fax:(01753) 630002
Hub Tubes 14-16 McDowall Street Paisley Strathclyde PA3 2NB tel:(0141) 848 6767 fax:(0141) 848 6991
Hubbard Architectural Metalwork Ltd 3 Hurricane Way Norwich Norfolk NR6 6HS tel:(01603) 424817 fax:(01603) 487158 tony.hubbard@hubbardsmetalwork.co.uk www.hubbardsmetalwork.co.uk
Hubbell Premise Wiring Chantry Avenue Kempston Bedford MK42 7RR tel:(01234) 848559 fax:(01234) 856190 smcilroy@hubbell.premise.com www.hubbell-premise.com
Hudevad Britain Hudevad House, Walton Lodge Bridge Street Walton-on-Thames Surrey KT12 1BT tel:(01932) 247835 fax:(01932) 247694 sales@hudevad.co.uk www.hudevad.co.uk
Hufcor (Partitions) Ltd Trent Lane Castle Donington Derby DE74 2NP tel:(01332) 810576 fax:(01332) 811059 lnonbury@kst.kaba.com
Hughes Safety Showers Limited Whitefield Road Bredbury, Stockport Cheshire SK6 2SS tel:(0161) 430 6618 fax:(0161) 4307928 sales@hughes-safety-showers.co.uk www.hughes-safety-showers.co.uk
Humbrol Ltd Marfleet Hull HU9 5NE tel:(01482) 701191 fax:(01482) 712908
Humphrey & Stretton PLC Pindar Road Hoddesdon Herts. EN11 0EU tel:(01992) 462965 fax:(01992) 463996 firedors@aol.com
Hunt Europe Ltd Chester Hall Lane Basildon Essex SS14 3BG tel:(01268) 530331 fax:(01268) 527211
Hunter Douglas Ltd (Contracts Division) Luxaflex Blinds ® Unit 8A Business Centre 17 London Road Swanscombe Kent DA10 0LH tel:(01322) 624580 fax:(01322) 624558 info.contract@luxaflex-sunway.co.uk www.luxaflexprojects.com
Hunter Flemming Associates Ltd 3 Tythe Barn Brumstead Road Stalham Norwich NR12 9DH tel:(01692) 581545 fax:(01692) 582893
Hunter Plastics Ltd Nathan Way London SE28 0AE tel:(020) 8855 9851 fax:(020) 8317 7764 hunplas@hunterplastics.co.uk www.hunterplastics.co.uk.
Hunter Timber (Scotland) Earls Road Grangemouth Saw Mills Grangemouth Stirlingshire FK3 8XF tel:(01324) 483294
Hunting Engineering Ltd Reddings Wood Ampthill Bedford MK45 2HD tel:(01525) 840162 fax:(01525) 843730
Hunting Industrial Coatings Ltd Derby Road Widnes Cheshire WA8 9ND tel:(0151) 495 3505 fax:(0151) 495 3522
Huntley & Sparks Sterling House Blacknell Lane Crewkerne Somerset TA18 8LL tel:(01460) 72222 fax:(01460) 76402 mail@sterling-hydraulics.co.uk www.sterling-hydraulics.co.uk
Huntree Fencing Ltd 104 Green End Road Great Barford Beds. MK44 3HD tel:(01234) 870864 fax:(01480) 471082
HV Air Conditioning Ltd Coach Fold Works Coach Fold Road Haley Hill, Halifax W. Yorks. HX3 6ED tel:(01422) 366631 fax:(01422) 366061
HW Systems Ltd Olympus Park Quedgeley Gloucester GL2 4NF tel:(01452) 722227 fax:(01452) 727600
Hyde Brian Ltd Stirling Road Shirley Solihull W. Midlands B90 4LZ tel:(0121) 705 7987 fax:(0121) 711 2465 sales@brianhyde.co.uk www.brianhyde.co.uk
Hydraseeders Ltd Coxbench Derby Derbys. DE21 5BH tel:(01332) 880364 fax:(01332) 883241 enquiries@hydraseeders.demon.co.uk www.hydraseeders.uk.com
Hydratight Sweeney Limited Bently Road South Darlaston W. Midlands WS10 8LQ tel:(0121) 5050 600 fax:(0121) 5050 800 enquiry@hydratight.com www.hevilifts.com
Hydro Polymers Ltd Newton Acliffe Co. Durham DL5 6EA tel:(01325) 300555 fax:(01325) 300215
Hydropath (UK) Ltd Unit F Acorn Park Ind Est Nottinghamshire NG7 2TR tel:(01923) 210028 fax:(01923) 800243
Hydropure Group Ltd Hydropure House Alington Road St Neots, Huntingdon Cambs. PE19 2RB tel:(01480) 407900 fax:(01480) 407911
Hydrovane Compressor Co Ltd Claybrook Drive Washford Industrial Estate Redditch B98 0DS tel:(01527) 525522 fax:(01527) 510862
Hyflex Roofing Halfords Lane Smethwick W. Midlands B66 1BJ tel:(0121) 555 6464 fax:(0121) 555 5862 enquiries@hyflex.co.uk www.hyflex.co.uk
Hygood Ltd Woodlands Place Woodlands Road Guilford Surrey GU1 1RN tel:(01483) 572222 fax:(01483) 302180 sales@hygood.co.uk www.hygood.co.uk
Hymo Ltd 35 Cornwell Business Park Salthouse Road Brackmills Northampton NN4 7EX tel:(01604) 661601 fax:(01604) 660166 info@hymo.ltd.uk www.hymo.com
Hynes Aden K Sculpture Studios 3 Hornsby Square Southfield Industrial Park Laindon, Basildon Essex SS15 6SD tel:(01268) 418837 fax:(01268) 414118 www.sculpturestudios.co.uk
Hyperion Wall Furniture Ltd Business Park 7 Brook Way Leatherhead Surrey KT22 7NA tel:(01372) 378279 fax:(01372) 362004 enquiries@hyperionfurniture.co.uk www.hyperionfurniture.co.uk
Hy-Ten Reinforcement Co Ltd Whitewall Road Medway City Industrial Estate Strood, Rochester Kent ME2 4DZ tel:(01634) 290102 fax:(01634) 290574 sales@hy-ten.co.uk
I R Sercurity Bascot Crescent Walsall West Midlands WS1 4DL tel:(01902) 366332 fax:(01902) 603935

IAC IAC House Moorside Road Winchester Hamp. SO23 7US tel:(01962) 873 000 fax:(01962) 873111 info@iacl.co.uk www.iacl.co.uk

IBL Lighting Unit C60, Barwell Business Park Leatherhead Road Chessington Surrey KT9 2NY tel:(020) 8391 7500 fax:(020) 8974 1629 info@ibl.co.uk www.ibl.co.uk

IBP Ltd Whitehall Road Tipton W. Midlands DY4 7JU tel:(0121) 557 2831 fax:(0121) 520 8778 salesuk@ibpgroup.com www.ibp.com

Ibstock Brick Ltd Leicester Road Ibstock Leics. LE67 6HS tel:(01530) 261999 fax:(01530) 261888 marketing@ibstock.co.uk www.ibstock.co.uk

Ibstock Building Products Ltd Anstone Office Kiveton Park Station, Kiveton Park Sheffield S. Yorks. S26 6NP tel:(01909) 771122 fax:(01909) 515281

Ibstock Hathernware Ltd Station Works, Rempstone Road Normanton on Soar Loughborough Leics. LE12 5EW tel:(01509) 842273 fax:(01509) 843629 hathernware@lbstock.co.uk www.hathernware.co.uk

Ibstock Scotish Brick Tannochside Works Old Edingburgh Road Tanockside, Uddingston Strathclyde G71 6HL tel:(01698) 810686 fax:(01698) 812364 www.ibstock.co.uk

ICB (International Construction Bureau) Ltd Unit 9, Elliott Road West Howe Industrial Estate Bournemouth Dorset BH11 8JX tel:(01202) 579208 fax:(01202) 581748 www.alwitra.co.uk

ICI Chemicals & Polymers Ltd S & T Department PO Box 90 Wiltan Middlesborough TS6 8JE tel:(01928) 514444 fax:(01928) 572296

ICI Dulux Trade Paints Wexham Road Slough Berks. SL2 5DS tel:(01753) 691690 fax:(01753) 530336 www.duluxtrade.co.uk

ICI Paints Wexham Road Slough Berks SL1 5HU tel:(01753) 550000 fax:(01753) 530336 www.dulux.com

ICI Polyurethanes Hitchen Lane Shepton Mallet BA4 5TZ tel:(01749) 343061 fax:(01749) 346283

Ide T & W Ltd Glasshouse Fields London E1W 3JA tel:(020) 7790 2333 fax:(020) 7790 0201 sales@twigroup.co.uk www.twigroup.co.uk

Ideal Building Systems Ltd Lancaster Road Carnaby Industrial Estate Bridlington E. Yorks. YO15 3QY tel:(01262) 606750 fax:(01262) 671960 www.idealbuildingsystems.co.uk

Ideal Furniture Enterprises Ltd Broadmeadow Industrial Estate Dumbarton G82 2RG tel:(01389) 42588 fax:(01389) 61819

Ideal-Standard Ltd The Bathroom Works National Avenue Kingston Upon Hull HU5 4HS tel:(01482) 346461 fax:(01482) 445886 www.ideal-standard.co.uk

IFT Supplies Ltd Unit 39, Cromwell Estate Staffa Road Leyton London E10 7QZ tel:(020) 8556 0678 fax:(020) 8556 0678

IG Limited Avondale Road Cwmbran Gwent NP44 1XY tel:(01633) 486486 fax:(01633) 486465 info@igetd.co.uk www.igltd.co.uk

Iles Waste Systems Valley Mills Valley Road Bradford W. Yorks. BD1 4RU tel:(01274) 728837 fax:(01274) 734351 wastesystems@trevoriles.co.uk www.trevoriles.co.uk

Illuma Lighting Ltd Illuma House Gelders Hall Road Shepshed Leicestershire LE12 9NH tel:(01509) 601050 fax:(01509) 601036 info@illuma.co.uk www.illuma.co.uk

Images Storage & Partitioning 68 Iron Mill Lane Crayford Kent DA1 4RR tel:(01322) 525975 fax:(01322 558032

Imagetrim (UK) Ltd Unit 3, Teakcroft Centre, Fairview Industrial Park Marsh Way Rainham Essex RM13 8UH tel:(01708) 630472 fax:(01708) 630470

Imerys Minerals Ltd Westwood Beverley N. Humberside HU17 8RQ tel:(01482) 881234 fax:(01482) 872301

IMI Air Conditioning Ltd Armytage Road Brighouse W. Yorks. HD6 1QF tel:(01484) 714361 fax:(01484) 400543 info@imi-airconditioning.co.uk www.imi-airconditioning.co.uk

IMI Stanton Ltd Somerton Works Newport Gwent NP9 0XU tel:(01633) 277711 fax:(01633) 276004

IMP Lighting Ltd Spring Lane Malven Link Worcester WR14 1AT tel:(01684) 891211 fax:(01684) 891046 enquiries@implighting.co.uk www.implighting.co.uk

Imperial Electrical Wholesalers 199 Watling Street Radlett Herts WD7 7NW tel:(01923) 850333

Imperial Machine Co Ltd Harvey Road Croxley Green Herts. WD3 3AX tel:(01923) 718000 fax:(01923) 777273 mail@imc.co.uk www.imco.co.uk

IMS UK Arley Road Saltley Birmingham B8 1BB tel:(0121) 326 5000 fax:(0121) 326 5005 imsuk-south@ims-group.com www.ims-uk.com

Incinerator The Co Ltd Howard Road Eaton Socon St Neots Huntington PE19 3ER tel:(01480) 213171

Indalex Ltd Kingsditch Road Cheltenham Glos. GL51 9PD tel:(01242) 521641 fax:(01242) 513304

Indirect Lighting Ltd Unit7A, The Seedbed Business Centre Vangaurd Way Shoeburyness, Southend-on-Sea Essex SS3 9QX tel:(01702) 382321 fax:(01702) 382323 info@indirectlighting.co.uk

Industrial Brushware Ltd 77 Malt Mill Lane Halesowen West Midlands B62 8JJ tel:(0121) 559 3862 fax:(0121) 559 9404 brushware@compuserve.com

Industrial Building Componants Ltd PO Box 36 Longhill Industrial Estate (North) Hartlepool Cleveland TS25 1QR tel:(01429) 221111 fax:(01429) 274035 ibc_ltd@compuserve.com

Industrial Devices Ltd 309 West End Lane London NW6 1RG tel:(020) 7431 1118 fax:(020) 7794 3913

Industrial Hangers Ltd IHL House Thorpe Close, Thorpe Way Industrial Estate Banbury Oxon. OX16 8UU tel:(01295) 753400 fax:(01295) 753428 sales@ihl.co.uk www.ihl.co.uk

Industrial Lighting Ltd Pinnacle House Stanley Road Bootle Merseyside L20 7JF tel:(0151) 922 0052 fax:(0151) 922 3959

Infil Ltd Infil House Hadnall Shrewsbury Shropshire SY4 4AG tel:(01939) 210320 fax:(01939) 210690

Infraglo (Sheffield) Ltd Dannemora Drive Greenland Road Industrial Park Sheffield S. Yorks. S9 5DF tel:(0114) 2495445 fax:(0114) 2495066 info@infraglo.co.uk www.infraglo.co.uk

Ingersol Locks Ltd Wood Street Willenhall W. Midlands WV13 1LA tel:(01902) 366911 fax:(01902) 363535

Ingersoll-Rand European Sales Ltd PO Box 2 Chorley New Road Horwich Bolton BL6 6JN tel:(01204) 690690 fax:(01204) 690388

Initial Electronic Security Systems Ltd Shadsworth Road Blackburn BB1 2PR tel:(01254) 688688 fax:(01254) 662571

Initial Plant Services Ltd, Redispace and Johnson Valletta Street Hedon Road Hull HU9 5NP tel:(01482) 781202 fax:(01482) 712157

INMAC (a division of microwarehouse Ltd) Harizon 1 Studio Way Borehamwood Herts. WD6 5WH tel:(01923) 471288 fax:(01923) 227215 www.inmac.co.uk

Insituform Permaline Ltd Roundwood Industrial Estate Ossett W. Yorks. WF5 9SQ tel:(01924) 277076 fax:(01924) 265107

InstaGroup Ltd Insta House Ivanhoe Road, Hogwood Business Park Finchampstead, Wokingham Berks. RG40 4PZ tel:(0118) 932 8811 fax:(0118) 932 8314 101470.3501@compuserve.com www.instagroup.co.uk

Instant Zip-Up Ltd Audley Avenue Newport Shropshire TF10 7DP tel:(01952) 825815 fax:(01952) 825635

Institute of Ashphalt Technology, the Paper Mews Place 290 High street Dorking Surrey RH4 1QT 01306 742792 01306 888902 secretary@instofasphalt.demon.co.uk www.instofasphalt.org

Institute of Concrete Technology The PO Box 7827 Crowthorne Berks RG45 6FR tel:(01344) 752096 fax:(01344) 752096 ict@ictech.org www.ictech.org

Institute of Plumbing The 64 Station Lane Hornchurch Essex RM12 6NB tel:(01708) 472791 fax:(01708) 448987 info@plumbers.org.uk www.plumbers.org.uk

Institute of Quarrying 7 Regent Street Nottingham Nottinghamshire HG1 5BS 0115 941 1315 0115 948 4035

Institute of Remedial Treatment Surveyors, The Essex House High Street Chipping Ongar Essex CM5 9EB 0800 915 6363 0870 755 5432 info@irts.co.uk www.irts.co.uk

Institute of Vitreous Enamellers, The 39 Sweetbriar Way Heath Hayes Cannock Staffs WS12 2US 01543 450596 08700 941237 info@ive.org.uk www.ive.org.uk

Institution of Civil Engineers One Great George Street London SW1P 3AA tel:(020) 7222 7722 fax:(020) 7222 7500 shana.cooper@ice.org.uk www.ice.org.uk

Institution of Electrical Engineers Savoy Place London WC2R OBL tel:(020) 7240 1871 fax:(020) 7240 7735 postmaster@iee.org.uk www.iee.org.uk

Institution of Incorporated Engineers (IIE) Savoy Hill House Savoy Hill London WC2R 0BS tel:(020) 7836 3357 fax:(020) 7497 9006 info@iie.org.uk www.iie.org.uk

Institution of Mechanical Engineers 1 Birdcage Walk Westminster London SW1H 9JJ tel:(020) 7222 7899 fax:(020) 7222 4557 enquiries@imeche.org.uk www.imeche.org.uk

Institution of Structural Engineers 11 Upper Belgrave Street London SW1X 8BH tel:(020) 7235 4535 fax:(020) 7235 4294 www.istructe.org.uk

Insulated Render & Cladding Association PO Box 12 Haslemere Surrey GU27 3AH tel:(01428) 654011 fax:(01428) 651401 incaassociation@aol.com www.inca-ltd.org.uk

Integra Products Eastern Avenue Lichfield Staffs. WS13 7SB tel:(01543) 267100 fax:(01543) 267104 cuscare@integra-products.co.uk www.integra-proucts.co.uk

Inter Public Urban Systems Uk Ltd Suite 16, Horsehay House Horsehay Telford Shropshire TF4 3PY tel:(01952) 502012 fax:(01952) 502022 sales@interpublicurbansystems.co.uk www.interpublicurbansystems.co.uk

Intercraft Designs Ltd The Arenson Centre Arenson Way Dunstable Beds LU5 5UL tel:(01582) 678300 fax:(01582) 678333 info@intercraft.co.uk www.intercraft.co.uk

Interface Europe Ltd Shelf Mills Shelf Halifax West Yorkshire HX3 7PA tel:(01274) 690690 fax:(01274) 694095 enquiries@eu.interfaceinc.com www.interfaceeurope.com

Interfloor Limited Edinburgh Road Heathhall Dumfries DG1 1QA tel:(01387) 253111 fax:(01387) 255726 sales@interfloor.com www.interfloor.com

Interframe Ltd Aspen Way Yalberton Industrial Estate Paignton Devon TQ4 7QR tel:(01803) 666633 fax:(01803) 663030

Interglass Ltd Greenfields Industrial Estate Tindale Crescent Bishop Auckland Co. Durham DL1 9TF tel:(01388) 603667 fax:(01388) 609744

Intergrated Polymer Systems (UK) Ltd Allen House Harmby Road Ley Burn North Yorks DL8 5WS 01969 625000 01969 623000 ips@zoom.co.uk www.ips-uk-ltd.com

Interlink Group Interlink House Commerse Way, Lancing Industrial Estate Lancing W. Sussex BN15 8TA tel:(01903) 763663 fax:(01903) 762621

Interlux Ltd 204 Oldbury Road West Bromwich W. Midlands B70 9DE tel:(0121) 5537551 fax:(0121) 580 0238 sales@interlux.co.uk www.interlux.co.uk

Internal Partitions Systems Interplan House, Dunmow Industrial Estate Chelmsford Road Great Dunmow Essex CM6 1HD tel:(01371) 874241 fax:(01371) 873848 contact@ipsinteriors.co.uk www.ips-interiors.co.uk

International Protective Coatings, Stoneygate Lane Felling on Tyne Tyne & Wear NE10 0JY tel:(0191) 469 6111 fax:(0191) 495 0676 www.international-pc.com

Interoven Ltd 9 Shaftesbury Parade South Harrow Middx. HA2 0AJ tel:(020) 8422 2541 fax:(020) 8864 7514

Interpave 60 Charles Street Leicester Leicestershire LE1 1FB 0116 253 6161 0116 251 4568 info@paving.org.uk www.paving.org.uk

Interphone Security Group PLC Interphone House 12-22 Herga Road Wealdstone Harrow, Middx HA3 5AS tel:(020) 8621 6000 fax:(020) 8621 6100 www.interphone.co.uk

Intersolar Group Ltd Magdalen Center The Oxford Science Park Oxford Oxon. OX4 4GA tel:(01865) 784670 fax:(01865) 784681 intersolar@intersolar.com www.intersolar.com

Invicita Plastics Ltd Harborough Road Oadby Leics. LE2 4LB tel:(0116) 272 0555 fax:(0116) 272 0626 invicta.sales@dial.pipex.com www.invictagroup.co.uk

Invicta Window Films Ltd Invicta House 7 Western Parade Woodhatch, Reigate Surrey RH2 8AU tel:(0800) 393380 fax:(01737) 240845 invictafilms@btconnect.com www.invicawindowfilms.com

IPPEC Sytsems 66 Rea Street South Digbeth Birmingham B5 6LB tel:(0121) 622 4333 fax:(0121) 622 5768 www.co.ippec.co.uk

IR Security and Safety Bescot Crescent Walsall W. Midlands WS1 4DL tel:(01922) 707400 fax:(01922) 646303 info@irsecurityandsafety.co.uk www.irsecurityandsafety.co.uk

Island Cottages Ltd 2nd floor 2 St Thomas Square Newport Isle of Wight P030 1SN tel:(01983) 532531 fax:(01983) 532531

Isocrete Floor Screeds Ltd Flowcrete Business Park Booth Lane Moston Sandbach, Cheshire CW11 3QF tel:(01270)753753 fax:(01270) 753333 help@flowcrete.co.uk www.flowcrete.co.uk

ISOKERN (UK) Ltd 14 Haviland Road Ferndown Industrial Estate Wimborne Dorset BH21 7RF tel:(01202) 861650 fax:(01202) 861632 sales@isokern.co.uk www.isokern.co.uk

Isolated Systems Ltd Adams Close, Heanor Gate Industrial Estate Heanor Derbys. DE75 7SW tel:(01773) 761226 fax:(01773) 760408 sales@isolatedsystems.com www.isolatedsystems.com

Isopad Ltd Oldmixon Industrial Estate Oldmixon Crescent Weston super Mare N. Somerset BS24 9AY tel:(01934) 629273 fax:(01934) 626125 sales@isopad.co.uk www.isopad.co.uk

IsoSystems Unit 145 Hartlebury Trading Estate Hartlebury, Kidderminster Worcs. DY10 4JB tel:(01299) 254300 fax:(01299) 254335 sales@isosystems.co.uk www.isosystems.co.uk

Isowall UK) Ltd Winchester Hill, Romsey Hants. SO51 7UT tel:(01794) 522910 fax:(01794) 523243

ITM Primary (uk) Ltd Unit 4 Focus 303 Bus Centre Focus Way Andover Hampshire SP10 5NY tel:(01252) 28512 fax:(01252) 343217

ITS Ceilings & Partitions 44 Portman Road Reading Berks. RG30 1EA tel:(0118) 9500225 fax:(0118) 9503267

ITT Flygt Ltd Colwick Nottingham NG4 2AN tel:(0115) 940 0111 fax:(0115) 940 0444 www.flygt.com

ITW Automotive Finishing UK Anchorbrook Industrial Estate Lockside Aldridge, Walsall WS9 8EG tel:(01922) 423700 fax:(01922) 423705 info@itwautomotivefinishing.co.uk www.itwautomotivefinishing.co.uk

ITW Construction Products Crompton Way Crawley W.Sussex RH10 2QR tel:(01293) 441353 fax:(01293) 515186

ITW Spit Fleming Way Crawley W. Sussex RH10 2QR tel:(01293) 523372 fax:(01293) 515186 www.itwcp.co.uk

J & E Hall Ltd Invicta House Sir Thomas Longley Rd Rochester Kent ME2 4DP tel:(01634) 731400 fax:(01634) 731401 helpline@jehall.co.uk www.jehall.co.uk

J & M Fencing Services Unit 3, Wrexham Road, Laindon Basildon Essex SS15 6PX tel:(01268) 415233 fax:(01268) 417357

J C Bamford (Excavators) Ltd Rocester Staffs. ST14 5JP tel:(01889) 590312 fax:(01889) 590588 enquiries@jcb.com www.jcb.com

J Pugh-Lewis Limited Bushypark Farm Pilsley Chesterfield Derbys. S45 8HW 01773 872362 01773 874763 info@pugh-lewis.co.uk www.pugh-lewis.co.uk

J. B. Corrie & Co Limited Frenchmans Road Petersfield Hampshire GU32 3AP tel:(01730) 237100 fax:(01730) 264915 sales@jbcorrie.co.uk www.jbcorrie.co.uk

J. Scott (Thrapston) Ltd Bridge Street Thrapston Northants NN14 4LR tel:(01832) 732366 fax:(01832) 733703

J.F. Poynter Ltd Maxim Lamp Works Cuckfield Hurstpierpoint W. Sussex BN6 9RW tel:(01273) 834890 fax:(01273) 834979 maximlamp@aol.com

Jack Moody Ltd Hollybush Farm Warstone Road Shareshill Wolverhampton WV10 7LX tel:(01922) 417648 fax:(01922) 413420 jackmoody@mail.com

Jackson & Sons, G, ITD Unit 19, Mitcham Ind Estate Streatham Road Surrey CR4 2AP tel:(020) 8685 5000 fax:(020)8 640 1986

Jackson H S & Son (Fencing) Ltd Stowting Common Ashford Kent TN25 6BN tel:(01233) 750393 fax:(01233) 750403 sales€@jacksons-fencing.co.uk www.jacksons-fencing.co.uk

Jackson James & Co (London) Ltd 76-89 Alscot Road Bermondsey London SE1 3AW tel:(020) 7237 2862 fax:(020) 7232 2362

Jacuzzi UK Ltd, Division of Jacuzzi Inc Silverdale Road Newcastle under Lyme Staffs ST5 6EL tel:(01782) 717175 fax:(01782) 717166 www.jacuzziuk.com

Jaga Heating Poducts (UK) Ltd Jaga House, Orchard Business Park Bromyard Road Ledbury Herefordshire HR8 1LG tel:(01531) 631533 fax:(01531) 631534 admin@jaga.co.uk www.theradiatorfactory.com

Jakem Timbers Ltd The Old Malt House 125 High Street Uckfield E. Sussex TN22 1EG tel:(01825) 768555 fax:(01825) 768483 richard@jakem.demon.co.uk

Jali Ltd Albion Works Church Lane Barham, Canterbury Kent CT4 6QS tel:(01227 833333 fax:(01227 831950 sales@jali.co.uk www.jali.co.uk

Jamak Fabrications (Europe) Ltd Unit H1 & H2, Europa Trading Estate Stoneclough Road Radcliffe Manchester M26 1GG tel:(01204) 794554 fax:(01204) 574521 sales@jamak.co.uk www.jamak.co.uk

James & Bloom Ltd Crossley Park Crossley Road Stockport Cheshire SK4 5DF tel:(0161) 432 5555 fax:(0161) 432 5312 jamesandbloom.co@btinternet.com

James D Cocks (Enterprises) St Michaels, Brightlingsea Road Thorrington Colchester Essex CO7 8JJ tel:(01206) 302285 fax:(01206) 302285

James Gibbons Format Ltd Vulcon Road Bilston WV14 7JG tel:(01902) 458585 fax:(01902) 351336 deptname@jgf.co.uk www.jgf.co.uk

James Latham plc Unit 2 Swallow Park Finley Road Hemel Hampstead HP2 7QU tel:(020) 8806 3333 fax:(020) 8806 7249 marketing@lathams.co.uk www.lathamtimber.co.uk

James Smellie Fabrications Ltd Unit F Leona Industrial Estate Nimmings Road Halesowen W. Midlands B62 9JQ tel:(0121) 561 1167 fax:(0121) 559 0336

James Southerton Ltd 24a Reddicap Trading Est Sutton Coldfield W. Midlands B75 7BY tel:(0121) 378 0194 fax:(0121) 378 3438

Jandor Architectural Ltd Unit 26, Perryvale Industrial Park Horsenden Lane South Perryvale Middx. UB6 7RJ tel:(020) 8998 4321 fax:(020) 8991 1374

Jarvis Construction (UK) Ltd Suite 7/6, Clydeway Skypark 8 Elliott Place Glasgow G3 8EP tel:(0141) 248 7733 fax:(0141) 248 8779

Jay Interiors Corbine House Owen Road Diss Norfolk IP22 4ER tel:(01379) 652 212 fax:(01379) 652 961 info@jay-interiors.co.uk www.jay-innteriors.co.uk

Jaymart Rubber & Plastics Ltd Woodland Trading Estate Edenvale Road Westbury Wilts. BA13 3QS tel:(01373) 864926 01373 858454 maltings@jaymert.net www.jaymart.net

JB Architectural Inronmongery Ltd Avis Way Newhaven E. Sussex BN9 0DU tel:(01273) 514961 fax:(01273) 516764 info@jbai.co.uk

Jebron Ltd Bright Street Wednesbury W. Midlands WS10 9HY tel:(0121) 526 2212 fax:(0121) 526 5322

Jeld-Wen Ltd Watch House Watch House Lane Stadium Est Doncaster DN5 9LR tel:(01603) 706000 fax:(01603) 706050

Jeld-Wen UK Ltd Watch House Lane Doncaster S. Yorks. DN5 9LR tel:(01302) 394000 fax:(01302) 787383 www.jeld-wen.co.uk

Jendico Jendico House 1 Cork Lane Glen Parva Leicester LE2 9JR tel:(0116) 277 0474 fax:(0116) 277 2941 sales@jendico.co.uk www.jendico.co.uk

Jenkins Newell Dunford Ltd Thrumpton Lane Retford Notts. DN22 7AN tel:(01777) 706777 fax:(01777) 708141

Jerrards Bros. PLC Division Arcadia House Cairo New Road Croydon London CR0 1XP tel:(020) 8251 5555 fax:(020) 8251 5500 enquiries@jerrards.com www.jerrards.com

Jewers Doors Ltd Stratton Business Park Normandy Lane Biggleswade Beds. SG18 8QB tel:(01767) 317090 fax:(01767) 312305 postroom@jewersdoors.co.uk www.jewersdoors.co.uk

Jewson Ltd High Street Brasted Westerham Kent TN16 1NG tel:(01959) 563856 fax:(01959) 564666

JKO Ltd Hughenden Avenue High Wycombe Bucks. HP13 5SQ tel:(01494) 521051 fax:(01494) 461176 mike@jkd.co.uk www.Jkd.co.uk

JLC Pumps & Engineering Co Ltd PO Box 225 Barton-le-Clay Beds. MK45 4PN tel:(01582) 881946 fax:(01582) 881951

John Davidson (Pipes) Ltd Townfoot Industrial Estate Longtown Carlisle Cumbria CA6 5LY tel:(01228) 791503 fax:(01228) 792051 headoffice@jdpipes.co.uk www.jdpipes.co.uk

John Reynolds & Sons (Birmingham) Ltd Church Lane West Bromwich W. Midlands B71 1DJ tel:(0121) 553 1287 / 2754 fax:(0121) 500 5460

Johnson & Johnson PLC Units 12-19, Guiness Road Trading Estate Trafford Park Manchester M17 1SB tel:(0161) 872 7041 fax:(0161) 872 7351

Johnson & Starley Rhosili Road Brackmills Northampton NN4 7LZ tel:(01604) 762881 fax:(01604) 767408 marketing@johnsonandstarley.co.uk www.johnsonandstarley.co.uk

Johnson Controls Systems Ltd Johnson House Randalls Way Leatherhead Surrey KT22 7TS tel:(01372) 376111 fax:(01372) 376823 products.uk@jci.com www.johnson-controls

Johnson Matthey Pigments & Dispersions Liverpool Road East Kidsgrove Stoke-on-Trent ST7 3AA tel:(01782) 794400 fax:(01782) 787338 pd@matthey.com www.cmd.matthey.com

Johnson Matthey PLC - Metal Joining York Way Royston Herts. SG8 5HJ tel:(01763) 253200 fax:(01763) 253168 mj@matthey.com www.jm-metal-joining.com

Johnson V B LLP

St John's House 304-310 St Albans Road Watford Herts. WD24 6PW tel:(01923) 227236 fax:(01923) 231134 watford@johnson.co.uk www.vbjohnson.co.uk

Johnson Wax Professional Frimley Green Camberley Surrey GU16 7AJ tel:(01276) 852852 fax:(01276) 852800

Johnston Pipes Ltd Doseley Telford Shropshire TF4 3BX tel:(01952) 630300 fax:(01952) 501537 lou@johnston-pipes.co.uk www.johnston-pipes.co.uk

Johnston Roadstone Ltd Leaton Quarry Leaton Nr Wellington Telford TF6 5HA tel:(01952) 777910 fax:(01952) 777919 www.johnstonroadstone.com

Joint Council For The Building & Civil Engineering Industry (N. Ireland) 143 Marlone Road Belfast BT9 6SU tel:(028) 90 877143 fax:(028) 90 877155 mail@cefni.co.uk www.cefni.co.uk

Joint Industry Board for the Electrical Contracting Industry Kingswood House 47/51 Sidcup Hill Sidcup Kent DA14 6HP 020 8302 0031 020 8309 1103 administration@jib.org.uk www.jib.org.uk

Jolec Electrical Supplies Unit 2, The Empire Centre Imperial Way Watford Herts. WD24 4YH tel:(01923) 243656 fax:(01923) 230301

Jones & Attwood Ltd Titan Works Stourbridge DY8 4LR tel:(01384) 392181 fax:(01384) 371937

Jones & Thomas (Wireworks) Ltd Unit 5, Burtonwood Industial Centre Burtonwood Warrington Cheshire WA5 4HX tel:(01925) 229795 fax:(01925) 228069 sales@joneandthomas.co.uk www.graepel.ie.

Jones of Oswestry Ltd Whittington Road Oswestry Shropshire SY11 1HZ tel:(01691) 653251 fax:(01691) 658222 www.jonesofoswestry.co.uk

Joseph Ash Galvanizing PO Box 16 Charles Henry Street Birmingham B12 OSP tel:(0121) 622 4661 fax:(0121) 622 4317 sales@josephash.co.uk www.josephash.co.uk

Joseph Tipper Ltd Century Works Moat Street Willenhall W. Midlands WV13 1FZ tel:(01902) 608444 fax:(01902) 608445

Jotun Henry Clark Ltd Decorative Division Stather Road Flixborough Lincolnshire DN15 8RR tel:(01724) 400123 fax:(01724) 400100

Joy Statford Joists Ltd 1 Whiting Way London E6 6LR tel:(020) 7474 0550 fax:(020) 7473 0158 stratfordjoist@dial.pipex.com www.stratfordjoists.co.uk

JSB Electrical PLC Manor Lane Holmes Chapel, Crewe Cheshire CW4 8AF tel:(01477) 537773 fax:(01477) 535722 marketng@jsb-electrical.com www.jsb-electrical.com

JT Contract Marketing Ltd 29 Store Street London WC1E 7BS tel:(020) 7580 8535 fax:(020) 7436 2620

Junckers Limited Unit 1A Wheaton Road Witham Essex CM8 3UJ tel:(01376) 534700 fax:(01376) 514401 sales@junkers.co.uk www.junckers.com

Juno Roplasto Ltd Wellington House Wellington Crescent New Malden Surrey KT3 3NE tel:(020) 8942 9323 fax:(020) 8949 5661 info@juno-roplasto.ltd.uk www.juno-roplasto.ltd.uk

KAB Seating Ltd Round Spinney Northampton NN3 8RS tel:(01604) 790500 fax:(01604) 790155 marketing@kabseating.com www.kabseating.com

Kaba (UK) Ltd Lower Moor Way Tiverton Business Park Tiverton Devon EX16 6SS tel:(01884) 256464 fax:(01884) 234415 info@kaba.co.uk www.kaba.co.uk

Kaba Door Systems Ltd Halesfield 4 Telford Shropshire TF7 4AP tel:(0870) 000 5243 fax:(0870) 000 5253 autodoors@adt.kaba.com www.kaba.co.uk

Kaba Garog 14 Leacroft Road Birchwood Warrington WA3 6GG tel:(0870) 000 0600 fax:(0870) 0642 sales@garog.co.uk www.garog.co.uk

KAC Alarm Company Ltd Unit 15 - 19 Trescott Road Trafford Park Worchestershire B49 6EP tel:(01789) 763338 fax:(01789) 400027 marketing@kac.co.uk www.kac.co.uk

Kair ventilation Ltd Unit 6, Chiltonran Industrial Estate 203 Manton Lane Lee London SE12 0TX tel:(08451) 662240 fax:(08451) 662250 info@kair.co.uk www.kair.co.uk

Kalon Decorative Products Huddersfield Road Birstall, Batley W. Yorks. WF17 9XA tel:(01924) 354000 fax:(01924) 354001 www.kalon.com

Kaloric Heater Co Ltd 31-33 Beethoven Street London W10 4LJ tel:(020) 8969 1367 fax:(020) 8968 8913 admin@kaloricheater.co.uk www.kaloricheaters.co.uk

Kasenit Furnaces Ltd Bourne Works Collingbourne, Ducis Marlborough Wilts. SN8 3EH tel:(01264) 850105 fax:(01264) 850445

Kawneer UK Ltd Astmoor Road Astmoor Runcorn Cheshire WA7 1QQ tel:(01928) 563732 fax:(01928) 569297

Kayanson Engineers Ltd Unit 34, Lydney Industrial Estate Harbour Road Lydney Glos. GL15 4EJ tel:(01594) 843494 fax:(01594) 843501

Kaye Aluminium plc Shaw Lane Industrial Estate Ogden Road Wheatley Hills, Doncaster S. Yorks. DN2 4SG tel:(01302) 762500 fax:(01302) 360307 sales@kayealuminium.co.uk www.kayealuminium.co.uk

KCW Windows 25A Shuttleworth Road Goldington Bedford Bedfordshire MK41 0HS tel:(01234) 269911 fax:(01234) 325034 sales@kcwwindows.com www.kcwwindows.com

Keddy-Poujoulat (UK) Ltd 105-109 Oyster Lane Byfleet Surrey KT14 7HJ tel:(01932) 343934 fax:(01932) 343222 sales@keddy.co.uk

Kee Klamp Ltd 1 Boulton Road Reading Berks. RG2 0NH tel:(0118) 931 1022 fax:(0118) 931 1146 sales@keeklamp.com www.keeklamp.com

Keller Ground Engineering Oxford Road Ryton-on-Dunsmore Coventry CV8 3EG tel:(024) 76 511266 fax:(024) 76 305230 martyn.singleton@Keller-ge.co.uk www.keller-ge.co.uk

Kelsey Roofing Industries Ltd Kelsey House, Paper Mill Drive Church Hill South Redditch Worcestershire B98 8QJ tel:(01527) 594400 fax:(01527) 594444 sales@kelseyroofing.co.uk www.kelseyroofing.co.uk

Kem Edwards Ltd Longwood Business Park Fordbridge Road Sunbury-on-Thames Middx TW16 6AZ tel:(01932) 754700 fax:(01932) 754754 sales@kemedwards.co.uk www.kemedwards.co.uk

Kemira Coatings Warth Road, off Radcliffe Road Bury Gtr. Manchester BL9 9NB tel:(0161) 764 6030 fax:(0161) 764 6102

Kemlite Ltd 25 Caker Stream Road Alton Hants. GU34 2QF tel:(01420) 54 10 66 fax:(01420) 541124 info@lampo.co.uk www.Kemlite.com

Kenilworth Daylight Centre Ltd Princes Drive Industrial Estate Kenilworth Warks. CV8 2FD tel:(01926) 511411 fax:(01926) 854155 daylightcentre@btinternet.com

Kenley Manufacturing Ltd Hadida House Carter Lane Worshop Vale Notts NG20 8XF tel:(01782) 597700 fax:(01782) 597702 enquiries@hadida.co.uk www.hadida.co.uk

Kenrick Archibald & Sons Ltd PO Box 9 Union Street West Bromwich W. Midlands B70 6BD tel:(0121) 553 2741 fax:(0121) 500 6332 sales@kenrick.prestel.co.uk

Kensal CMS Kensal House President Way Luton Beds LU2 9NR tel:(01582) 425777 fax:(01582) 425776 sales@kensal.co.uk www.kensal.co.uk

Kent & Co (Twines) Ltd Hartley Trading Estate Long Lane Liverpool L9 7DE tel:(0151) 525 1601 fax:(0151) 523 1410

Kenyon Group Ltd Regent House Regent Street Oldham Lancashire OL1 3TZ tel:(0161) 633 6328 fax:(0161) 627 5072 sales@kenyon-group.co.uk www.kenyon-group.co.uk

Kenyon Performance Adhesives Regent House Regent Street Oldham Lancs OL1 3TZ tel:(0161) 633 6328 fax:(0161) 627 5072 sales@kenyon-group.co.uk www.4adhesives.com

Kermi (UK) Ltd 7 Brunel Road, Earlstrees Industrial Estate Corby Northants NN17 4JW tel:(01536) 400004 fax:(01536) 446614 info@kermi.co.uk

Kern (Design Products) J T Ltd The Grange Budworth Road Noctorum, Birkenhead Cheshire CH43 9TL tel:(0151) 647 9323 fax:(0151) 647 9324

Kershaw Mechanical Services Ltd The Beadle Trading Estate Ditton Walk Cambridge CB5 8PD tel:(01223) 410600 fax:(01223) 411061 k.m.s@dial.pipex.com

Kestner Building Products Ltd Station Road Greenhithe Kent DA9 9NG tel:(01322) 383281 fax:(01322) 386606

Ketley Brick Co Ltd , The Dreadnought Road Pensnett Brierley Hill W. Midlands DY5 4TH tel:(01384) 78361 fax:(01384) 74553 sales@ketleybrick.freeuk.com www.ketley-brick.co.uk

Keymer Hand Made Clay Tiles Nye Road Burgess Hill W. Sussex RH15 0LZ tel:(01444) 232931 fax:(01444) 871852 info@keymer.co.uk. www.keymer.co.uk

Keys W H Ltd Wall End Works Church Lane West Bromwich W. Midlands B71 1BN tel:(0121) 553 0206 fax:(0121) 500 5820

Kidde Fire Protection Ltd Belvue Road Northolt Middx. UB5 5QW tel:(020) 8839 0700 fax:(020) 8845 4304 info@kfp.co.uk www.kfp.co.uk

Kidde Fire Protection Services Ltd 400 Dallows Road Luton Beds. LU1 1UR tel:(01582) 413694 fax:(01582) 402339

Kiltox Damp Free Solutions Unit 6 Chiltonian Industrial Estate 203 Manor Lane Lee London SE12 0TX tel:(08451) 662040 fax:(018451)662050 info@kiltox.co.uk www.kiltox.co.uk

Kilwaughter Chemical Co Ltd Kilwaughter Lime Works 9 Starbog Road Larne Co. Antrim BT40 2JT tel:(028) 28260766 fax:(028) 28260766 www.killwaughter.co.uk

Kimberly-Clark Ltd Cobdown House Larkfield Aylesford Kent ME20 9JS tel:(01622) 615000 fax:(01622) 615070

Kind, J B, Ltd Shobnall Street Burton-on-Trent Staffs. DE14 2HP tel:(01283) 564631 fax:(01283) 511132 sales@jbkind.com www.jbkind.com

Kinder-Janes Engineering Ltd Porters Wood St Albans Herts. AL3 6HU tel:(01727) 844441 fax:(01727) 844247

Kinetico Water Softeners 11 The Maltings Thorney Peterborough Cambs. PE6 0QF tel:(01733) 270463

Kingfisher Building Products Ltd. Cooper Lane Bardsea, Ulverston Cumbria LA12 9RA tel:(01229) 869100 fax:(01229) 869101 kingfisherchem@btopenworld.com www.kingfisherchem.com

Kingsforth Security Fencing Limited Mangham Way Barbot Hall Industrial Estate Rotherham S. Yorks. S61 4RL tel:(01709) 378977 fax:(01709) 838992 info@kingsforthfencing.com.uk

Kingsnorth Bitumen Products Ltd Kingsnorth, Hoo, Rochester Kent ME3 9ND tel:(01634) 250722 fax:(01634) 253601

Kingspan Ltd Greenfield Business Park 2 Greenfield Holywell Flintshire CH8 7GJ tel:(01352) 716110 fax:(01352) 716161 info@kingspanpanels.co.uk www.kingspanpanels.com

Kingspan Access Floors Ltd Burma Drive Marfleet Hull HU9 5SG tel:(01482) 781701 fax:(01482) 799276 enquiries@kingspanaccessfloors.co.uk www.kingspanaccessfloors.co.uk

Kingspan Industrial Ins. Ltd PO Box 3 Charlestown Glossop Derbys. SK13 8LE tel:(01457) 861611 fax:(01457) 852319 enquities@kooltherm.kingspan.co.uk www.kingspan.kooltherm.co.uk

Kingspan Insulation Ltd Pembridge Leominster Herefordshire HR6 9LA tel:(01544) 388601 fax:(01544) 388888

Kingston Craftsmen Structural Timber Engineers Ltd Cannon Street Hull HU2 0AD tel:(01482) 225171 fax:(01482) 217032 info@kingston-craftsmen.co.uk

Kingstonian Point Ltd Mayfield House Sculcoates Lane Hull E. Yorks. HU5 1DR tel:(01482) 342216 fax:(01482) 493096 info@kpaints.co.uk www.kpaints.co.uk

Kinnings Marlow Ltd Franchise Street Wednesbury W. Midlands WS10 9RF tel:(0121) 526 6692 fax:(0121) 526 7166 keithowen-km@garton.co.uk www.garton.co.uk

Kirkpatrick Ltd Frederick Street Walsall Staffs. WS2 9NF tel:(01922) 620026 fax:(01922) 722525

Kirkstone Skelwith Bridge Nr. Ambleside Cumbria LA22 9NN tel:(01539) 433296 fax:(01539) 434006 info@kirkstone.com www.kirkstone.com

Kitchen Specialists Association 12 Top Barn Business Center Holt Heath, Worcester Worcestershire WR6 6NH tel:(01905) 621787 fax:(01905) 621887 info@ksa.co.uk www.ksa.co.uk

Kitsons Insulation Products Ltd Kitsons House Centurion Way Meridian Business Park Leicester LE3 2WH tel:(0116) 201 4499 fax:(0116) 201 4498

Klargester Environmental Ltd College Road Aston Clinton Aylesbury Bucks. HP22 5EW tel:(01296) 633000 fax:(01296) 633001 uksales@klargester.co.uk www.klargester.co.uk.

Klaxon Signal Ltd Warwick Road Tyseley Birmingham B11 2HB tel:(0121) 706 1654 fax:(0121) 708 1220

Klaxon Signals Ltd 502 Honey Pot Lane Stanmore Middx. HA71BE tel:(020) 8952 5566 fax:(020) 8952 6983

Kleeneze Sealtech Ltd Ansteys Road Hanham Bristol BS15 3SS tel:(0117) 958 2450 fax:(0117) 960 0141 sales@ksl.uk.com www.ksltd.com

Kleen-Tex Industries Ltd Causeway Mill, Express Trading Estsate Stone Hill Road Farnworth Nr. Bolton BL4 9TP tel:(01204) 863000 fax:(01204) 863001 sales@kleentexuk.com www.kleentexuk.com

Klick Technology Ltd Claverton Road Wythenshawe Manchester M23 9FT tel:(0161) 998 9726 fax:(0161) 946 0419 sales@klicktechnology.co.uk www.klicktechnology.co.uk

Klinger Fluid Instrumentation Ltd Edgington Way Sidcup Kent DA14 5AG tel:(020) 8300 7777 fax:(020) 8302 8145

Klober Ltd Pear Tree Industrial Estate Upper Langford Bristol N. Somerset BS18 7DJ tel:(01934) 853224 fax:(01934) 853221 support@klober.co.uk www.klober.co.uk

Klockner-Moeller Ltd PO Box 35 Gatehouse Close Aylesbury Bucks. HP19 3DH tel:(01296) 393322 fax:(01296) 21854

Knauf PO Box 133 Sittingbourne Kent ME10 3HW tel:(01795) 424499 fax:(01795) 428651 knauf@knauf.co.uk www.knauf.co.uk

Knauf Insulation Ltd Hunter House Industrial Estate Brenda Road Hartlepool Cleveland TS25 2BE tel:(01429) 855100 fax:(01429) 855138

Knauf Insulation Ltd PO Box 10 Stafford Road St Helens Merseyside WA10 3NS tel:(01270) 824024 fax:(01270) 824025 knauf@tricobyte.co.uk www.knaufinsulation.co.uk

Knight & Smith Ltd, Electrical Contractors 3 High Elms Cottages Woodside Road Watford Herts. WD25 0LA tel:(01923) 894744 fax:(01923) 679073 jknight@btconnect.com

Knightbridge Furniture Productions Ltd 191 Thornton Road Bradford W. Yorks. BD1 2JT tel:(01274) 731442 fax:(01274) 736641 sales@knightsbridge-furniture.co.uk www.knightsbridge-furniture.co.uk

Knowles Ltd R E Knowles Industrial Estate Furness Vale High Peak SK23 7PJ tel:(01663) 44127 fax:(01663) 741 562

Knowles W T & Sons Ltd Ash Grove Sanitary Works Elland W. Yorks. HX5 9JA tel:(01422) 372833 fax:(01422) 370900

Knurr (UK) Ltd Burrel Road St. Ives Cambs. PE27 3LE tel:(01480) 496125 fax:(01480) 496373 knuerr.uk@knuerr.com www.kneurr.com

Kobi Ltd Unit 1B, St. Marks Industrial Estate Oriental Road Silvertown London E16 2BS tel:(020) 7474 3464 fax:(020) 7511 2306 cradles@kobi.co.uk www.kobi.co.uk

Kohler Mira Ltd Cromwell Road Cheltenham Glos. GL52 5EP tel:(01242) 221221 fax:(01242) 221925 mira_technical@mirashowers.com www.mirashowers.com

Komfort Office Environments Plc Unit 1-10 Whittle Way Crawley W. Sussex RH10 9RT tel:(01293) 592500 fax:(01293) 553271 general@komfort.com www.komfort.com

Kompan Ltd 20 Denbigh Hall Bletchley Milton Keynes Buckinghamshire MK3 7QT tel:(01908) 642466 fax:(01908) 270137 customerservice@uk.kompan.com www.kompan.com

Konaflex Ltd Unit 2 Northcote Road Stechford Birmingham B33 9EB tel:(0121) 783 9778

Kone Escalators Ltd Worth Bridge Road Keighley W. Yorks. BD21 4YA tel:(01535) 662841 fax:(01535) 680498

Kone PLC Global House Station Place, Fox Lane North Chertsey Surrey KT16 9HW tel:(0870) 770 1122 fax:(0870) 770 1144 sales&marketing.uk@kone.com www.kone.com

Konematic Bolton Brady 405 Hillington Road Glasgow G52 4BL tel:(0141) 883 2131 fax:(0141) 883 4502 glasgow@boltonbrady.co.uk www.boltonbrady.co.uk

Kronospan Ltd Chirk Wrexham LL14 5NT tel:(01691) 773361 fax:(01691) 773292 sales@kronospan.co.uk www.kronospan.co.uk

KSB Ltd 2 Cotton Way Loughborough Leics. LE11 5TF tel:(01509) 231872 fax:(01509) 215228 sales@ksbgroup.com www.ksbuk.com

Kufa Plastics Ltd 2 Lyon Close Chantry Woburn Road Estate Kempston Beds. MK42 7SB tel:(01234) 854464

Kush UK Ltd 48/50 St Johns Street London EC1M 4DG tel:(01684) 850787 fax:(01684) 850758 admin@kusch.co.uk www.kusch.de

Lab Furnishings (UK) Ltd No.2 Malmo Park Stockholm Road Sutton Fields Industrial Estate, Hull East Yorkshire HU7 0XW tel:(01482) 827999 fax:(01482) 827995 sales@labfurnishings.co.uk www.labfurnishings.co.uk

Lab Systems Furniture Ltd Rotary House Bontoft Avenue Hull Yorkshire HU5 4HF tel:(01482) 444650 fax:(01482) 444730 office@labsystemsfurniture.com

Lab-Craft Ltd Church Road Harold Wood Romford Essex RM3 OHT tel:(01708) 349320 fax:(01708) 376394 sales@labcraft.co.uk

Lafarge Aluminates Ltd 730 London Road Grays Essex RM20 3NJ tel:(01708) 863333 fax:(01708) 861033 www.lafarge.com

Lafarge Plasterboard Ltd Easton-in-Gordano Marsh Lane Bristol BS20 ONF tel:(01275) 377773 fax:(01275) 377737 marketinginfo@lafarge.com www.lafargeplasterboard.co.uk

Lafarge Roofing Ltd Regent House Station Approach Dorking Surrey RH4 1TG tel:(01306) 872 000 fax:(01306) 872 111 roofing@lafarge-roofing.co.uk www.lafarge-roofing.co.uk

Lafaroe Redland Precast Six Hills Melton Mowbray Leics. LE14 3PD tel:(01509) 882121 fax:(01509) 880799

Laidlaw Architectural Hardware The Building Centre 26 Store Street London WC1E 7BT tel:(020) 7733 2101 fax:(020) 7737 2743 london.sales@laidlaw.net www.laidlaw.net

Laidlaw Architectural Hardware Ltd Pennine House Dakota avenue Ssauford Manchester MS 2PU tel:(0161) 848 1700 fax:(0161) 848 1700

Laird Security Hardware Western Road Silver End, Witham Essex CM8 3QB 01367 507507 01376 584687 sales@lairdsecurity.co.uk www.lairdsecurity.co.uk

Lakeside Group Bruce Grove Forest Fach Industrial Estate Swansea SA5 4HS tel:(01792) 561117 fax:(01792) 587046 sales@lakesidesercurity.co.uk www.lakesidesercurity.co.uk

Lamb Macintosh 415-416 Montrose Avenue Slough Trading Estate Slough Berks. SL1 4TP tel:(01753) 522369 fax:(01753) 517216

Lami Doors UK Ltd Station House Stamford New Road Altrincham Cheshire WA14 1EP tel:(0161) 9242217 fax:(0161) 9242218 lamidoors@lamidoors.com www.lamidoors.com

Laminated Wood Ltd Unit 7, PHC Trading Estate Stibb Cross Torrington N. Devon EX38 8LH tel:(01805) 601596 fax:(01805) 601313 lamwood@clara.net www.ascwebindex.com/lamwood

Lampost Construction Co Ltd Greenore Co. Louth tel:(00 353) 42 93 73554 fax:(00 353) 42 93 73378 lampost@iol.ie

Lamps and Tubes 83 Tudor Drive Watford Hertfordshire WD24 7NS tel:(01923) 256800 fax:(01923) 256900 info@lampsandtubes.co.uk www.lampsandtubes.co.uk

Lampways Ltd Allenby House Knowles Lane, Wakfield Road Bradford W. Yorks. BD4 9AB tel:(01274) 686666 fax:(01274) 680157

Lancashire Fittings Ltd The Science Village Claro Road Harrogate N. Yorks. HG1 4AF tel:(01423) 522355 fax:(01423) 506111 kenidle@lancashirefittings.com www.lancashirefittings.com

Lancashire Steel Fabrication Co Ltd 5 Gardeners Place West Gillibrands Skelmersdale WN8 9SP tel:(01695) 727011 fax:(01695) 727611 martin.davies@lansteerfab.co.uk www.lancstedfab.co.uk

Lancaster & Co (Bow) Ltd Hancock Road Bromley by Bow London E3 3DA tel:(020) 8980 2827 fax:(020) 8981 7815

Land Wood & Water Co. Ltd The Barn High Street Hartfield East Sussex TN7 4AE tel:(01892) 770470 fax:(01892) 770760

Landscape Institute, The 6-8 Barnard Mews London SW11 1QU tel:(020) 7305 5200 fax:(020) 7350 5201 mail@l-i.org.uk www.l-i.org.uk

Langford Joinery Ltd 4 Leylands Way Warwick Warwickshire CV34 6DQ tel:(028) 94452787 fax:(028) 94422415

Langley Systems Ltd 12-14 Magdelen Street London SE1 2EW tel:(020) 7407 6271 fax:(020) 7378 1754

Lappset UK Ltd Lappset House, Henson Way Telford Industrial Estate Kettering Northants NN16 8PX tel:(01536) 412612 fax:(01536) 521703 play@lappset.co.uk www.lappset.co.uk

Lastite Building Products Ltd Tenax Road Trafford Park Manchester M17 1JT tel:(0161) 876 7559 fax:(0161) 877 0117

Latchways plc Hopton Park Devizes Wilts. SN10 2JP tel:(01380) 732700 fax:(01380) 732701 info@latchways.com www.latchways.com

Laticrete UK laticrete House Unit 18 roddinglaw buisness park roddinglaw Edinburgh Midlothian EH12 9DW tel:(0131) 333 3777 fax:(0131) 335 3953 laticreteuk@aol.com www.laticrete.com

Laurier M & Sons 18 Marshgate Lane London E15 2NH tel:(020) 8534 7211 fax:(020) 8555 8991

Lawrence Surfaces Plc, Charles Brunel House Jessop Way Newark Notts NG24 2ER tel:(01636) 610777 fax:(01636) 610222 surfaces@clgplc.co.uk www.clgplc.co.uk

Lawton Tube Co Ltd, The Torrington Avenue Coventry CV4 9AB tel:(024) 7646 6203 Fax:(024)7669 4183 iawtontubes@cs.com

Laybond Products Ltd Riverside Saltney Chester CH4 8RS tel:(01244) 674774 fax:(01244) 680215 sales_info@laybond.co.uk www.laybond.com

Laytrad Ltd 30 Harmsworth Way London N20 8JU tel(020) 8445 1114 fax:(020) 8446 5944 sales@laytrad.co.uk www.laytrad.co.uk

Lazawood Ltd Crew Bank Works 79 Farleigh Road Warlingham Surrey CR6 9EJ tel:(01883) 622151 fax:(01883) 624533

LB Lighting Ltd 14 West Burrowfield Burrowfield Industrial Estate Welwyn Garden City Herts. AL7 4TW tel:(01707) 880066 fax:(01707) 880077 sales@lblighting.co.uk www.lblighting.co.uk

LB Plastics Ltd Firs Works Nether Heage Belper Derbys. DE56 2JJ tel:(01773) 852311 fax:(01773) 857080 gary.doxey@ibplastics.co.uk www.litchfield-group.co.uk

LB Structures Ltd Spring Road Ettingshall Wolverhampton WV4 6JT tel:(01902) 353311 fax:(01902) 353049 enquiries@lbstructures.co.uk www.lbstructures.co.uk

Lead Development Association International 42 Weymouth Street London W1G 6NP tel:(020) 7499 8422 fax:(020) 7493 1555 enq@ldaint.org www.ldaint.org

Leaderflush + Shapland Ltd Po Box 5404 Nottingham NG16 4BU tel:(0870) 240 0666 fax:(0870) 240 0777 marketing@leaderflushshapland.co.uk www.leaderflushshapland.co.uk

Leakale Taurus Visual Aids Bargate Manchester Road Linthwaite Huddersfield HD7 5QX tel:(01484) 844016 fax:(01484) 846352 sales@visualdirect.co.uk www.visualdirect.co.uk

Leander Architectural Fletcher Foundry, Hallsteads Close Dove Holes Buxton Derbys. SK17 8BP tel:(01298) 814941 fax:(01298) 814970 sales@leanderarch.demon.co.uk www.leanderarch.demon.co.uk

Leatherhead Food Research Association Randalls Rd Leatherhead Surrey KT22 7RY 01372 376761 01372 386228 help@lfra.co.uk www.lfra.co.uk

Leavlite Electropaint Ltd Linchfield Industrial Estate Tamworth Staffs. B79 7TA tel:(01827) 52083 fax:(01827) 66938

Ledite Glass (Southend) Ltd 168 London Road Southend on Sea Essex SS1 1PH tel:(01702) 345893 fax:(01702) 435099

Legge Thompson F & Co Ltd 1 Norfolk Street Liverpool L1 0BE tel:(0151) 709 7494 fax:(0151) 709 3774 www.livergrease.co.uk

Legrand Electric Ltd Great King Street North Birmingham B19 2LF tel:(0121) 515 0515 fax:(0121) 515 0516 legrand.sales@legrand.co.uk www.legrand.co.uk

Legrand Power Centre Brookside Wednesbury W. Midlands WS10 0QF tel:(0121) 506 4506 fax:(0121) 506 4507

Legrant Electric Ltd Dial Lane West Bromwich West midlands B70 0EB 0845 6054333 0845 605 4334

LEIA 33 Devonshire Street London W1G 6PY tel:(020) 7935 3013 fax:(020) 7935 3321 enquiries@leia.co.uk www.leia.co.uk

Leicester Barfitting Co Ltd West Avenue Wigston Leics. LE18 2FB tel:(0116) 288 4897 fax:(0116) 281 3122

Leicester Fencing Contracts Limited Station Road Great Glen Leics. LE8 0FP tel:(0116) 2592112 fax:(0116) 2592090

Leigh's Paints Tower Works Kestor Street Bolton Lancs. BL2 2AL tel:(01204) 521771 fax:(01204) 382115 enquiries@leighpaints.co.uk www.leighspaints.co.uk

Lenesco Ltd Grove Road Northfleet Kent DA11 9AX tel:(01474) 564692 fax:(01474) 327329

Lesco Products Ltd Wincheap Industrial Estate Canterbury Kent CT1 3RH tel:(01227) 763637 fax:(01227) 762239 sales@lesco.co.uk ww.lesco.co.uk

Letchworth Roofing Company Ltd The Roofing Centre Works Road Letchworth Herts SG6 1JY tel:(01462) 755755 fax:(01462) 755750 sales@letchworthroofing.co.uk www.letchworthroofing.co.uk

Lever James & Sons Ltd Orient Mill Brandwood Street Bolton Lancs. BL3 4BH tel:(01204) 61121 fax:(01204) 658154

Levolux A.T. Limited 24 Easville Close Eastern Avenue Gloucester GL4 3SJ tel:(01452) 500007 fax:(01452) 527496 sales@levoluxat.co.uk www.levolux.com

Lewden Electrical Industries Argall Avenue Leyton London E10 7QD tel:(020) 8539 0237 fax:(020) 8558 2718 hqgen@lewden.co.uk www.lewden.net

Lewes Design Contracts Ltd The Mill Glynde Lewes E. Sussex BN8 6SS tel:(01273) 858341 fax:(01273) 858200 info@spiralstairs.co.uk www.spiralstairs.co.uk

Lewis Spring Products Ltd Studley Road Redditch Worcs. B98 7HJ tel:(01527) 510535 fax:(01527) 500868 sales@lewis-spring.com www.lewis-spring.com

Leyland Paint Company, The Huddersfield Road Birstall Batley W. Yorks. WF17 9XA tel:(01924) 420202 fax:(01924) 420403

LHC 464 Uxbridge Road Hayes Middlesex UB4 0PT tel:(020 8561 4700 fax:(020) 8561 1130 mail@lhc.gov.uk www.lhc.gov.uk

Libraco The Cowbarn Filston Farm, Shoreham Stevenoaks Kent TN14 5JU tel:(01959)524074 fax:(01959) 525218 libraco@aol.com www.libraco-furniture.com

Library Furnishing Consultants Phoenix House 54 Denington Road Wellingborough Northamptonshire NN8 2QH tel:(01933) 442777 fax:(01933 442764 projetcs@ifc-itd.co.uk

Liebert Swindon Ltd Elgin Drive Swindon SN2 6DX tel:(01793) 553355 fax:(01793) 553400

Lift and Escalator Industry Association,The 33/34 devonshire Street London W1G 6PY 020 7935 3013 020 7935 3321 enquiries@leia.co.uk www.leia.co.uk

Lift Boy Ltd Bessemer Close Ebblake Industrial Estate Verwood Dorset tel:(01202) 827388 fax:(01202) 823242

Light (Multiforms), H A Woods Lane Cradley Heath Warley W. Midlands B64 7AL tel:(01384) 569283 fax:(01384) 633712

Light Alloy Ltd Dales Road Ipswich Suffolk IP1 4JR tel:(01473) 740445 fax:(01473) 240002 sales@lightalloy.co.uk

Lightfoot Charles Ltd 42 Higher Ardwick Manchester M12 6DA tel:(0161) 273 1134 fax:(0161) 273 6417

Lighting Association, The Stafford Park 7 Telford Shropshire TF3 3BQ tel:(01952) 290905 fax:(01952) 290906 enquiries@lightingassociation.com www.lightingassociation.com

Lighting Industry Federation Swan House 207, Balham High Road London SW17 7BQ tel:(020) 8765 5432 fax:(020) 8673 5880 keith@lif.co.uk www.lif.co.uk

Lighting Systems UK, Division of Phosco Ltd Charles House Furlong Way Gt Amwell, Ware Herts. SG12 9TA tel:(01920) 462272 fax:(01920) 485915

Lignacite (North London) Ltd Meadgate Works Nazeing Waltham Abbey Essex EN9 2PD tel:(01992) 464661 fax:(01992) 445713

Lignacite Ltd Norfolk House High Street Brandon Suffolk IP27 0AX tel:(01842) 810678 fax:(01842) 814602 info@lignacite.co.uk www.lignacite.co.uk

Linatex Ltd Wilkinson House Galway Road, Blackbush Business Park Yateley Hants. GU46 6GE tel:(01252) 743000 fax:(01252) 743030 linatex.sales@linatex.co.uk www.linatex.net

Lindapter International Lindsay House Brackenbeck Road Bradford W. Yorks. BD7 2NF tel:(01274) 521444 fax:(01274) 521130 enquiries@lindapter.com www.lindapter.com

Lindley George & Sons Ltd Sovereign Quarries Carr Lane Shepley Nr. Huddersfield HD8 8BH tel:(01484) 606203 fax:(01484) 605274

Linear Ltd Coatham Avenue Newton Aycliffe Industrial Estate Aycliffe Co. Durham DL5 6DB tel:(01325) 310151 fax:(01325) 310801

Link 51 (Storage Products) Link House Halesfield 6 Telford Shropshire TF7 4LN tel:(01952) 682221 fax:(01952) 682402 sales@link51.co.uk www.link51.co.uk

Link Build Products Ltd Link House Tag Lane Hare hatch Reading Berkshire RG10 9SH tel:(01189) 404965 fax:(01189) 401186 linkbuild2001@hotmail.com www.linkbuild.co.uk

Lionweld Kennedy Ltd Marsh Road Middlesbrough Cleveland TS1 5JS tel:(01642) 245151 fax:(01642) 224710 sales@lionweldkennedy.co.uk www.lionweldkennedy.co.uk

Liquid Plastics Ltd Iotech House. P.O.Box 7 Miller Street Preston Lancs. PR1 1EA tel:(01772) 255022 fax:(01772) 882016 info@liquidplastics.co.uk www.liquidplastics.co.uk

Liquid Technology Ltd Unit 27, Tatton Court Kingsland Grange, Woolston Warrington Cheshire WA1 4RR tel:(01925) 850324 fax:(01925) 850325 info@liquid-technologyltd.com www.fengroup.com

Liver Grease Oil & Chemical Co Ltd 11 Norfolk Street Liverpool L1 0BE tel:(0151) 709 7494 fax:(0151) 709 3774 www.livergrease.co.uk

Llewellyn Homes Ltd Courtlands Road Eastbourne E. Sussex BN22 8TS tel:(01323) 735271 fax:(01323) 410246

Llewellyn Stonecare Limited Bleak Hall Milton Keynes Bucks. MK6 1LA tel:(01908) 679222 fax:(01908) 235250 b.gardiner@llewellyn.co.uk www.llewellynstonecare.co.uk

Loblite Ltd Third Avenue Team Valley Gateshead Tyne & Wear NE11 0QQ tel:(0191) 487 8103 fax:(0191) 482 0270 info@loblite.co.uk www.loblite.com

Lockhart Catering Equipment Lockhart House Arrowhead Road Theale, Reading Berks. RG7 4AH tel:(0118) 9303900 fax:(0118) 9303931 lockhart@sodexho-uk.com www.lockharteqpt.com

Locking Systems Ltd Unit 2 Windsor Bisiness Park Hanson Lane Halifax Yorks. HX1 4SD tel:(01422) 330312 fax:(01422) 349987 susie@clockrooms.co.uk

Loctite UK Ltd Watchmead Welwyn Garden City Herts. AL7 1JB tel:(01707) 358800 fax:(01707) 358900

Loft Centre Products Thicket Lane Halnaker Nr. Chichester W. Sussex PO18 OQS tel:(01243) 785246/785229 fax:(01243) 533184 sales@loftcentreproducts.co.uk www.leftcentreproducts.co.uk

Loft Ladders Ltd Unit 2 83 Copers Cope Road Beckenham Kent BR3 1NR tel:(020) 8663 1973 fax:(020) 8663 1974

Logic Office Contracts Vestry Estate Otford Road Sevenoaks Kent TN14 5EL tel:(01732) 457636 fax:(01732)740706 contracts.admin@logic-office.co.uk www.logic-office.co.uk

Logic Office Group 748 London Road Hounslow Middx. TW3 1PD tel:(020) 8572 7474 fax:(020) 8570 9643 info@logic-office.co.uk www.logic-office.co.uk

London Fan Company Ltd 75-81 Stirling Road Acton London W3 8DJ tel:(020) 8992 6923 fax:(020) 8992 6928 sales@londonfan.co.uk www.londonfan.co.uk

London Lamplighting Co 21 Old Newtown Road Newbury Berks. RG14 7DP tel:(01635) 48131 fax:(01635) 35675

Longbottom J & J W Ltd Bridge Foundry Holmfirth Huddersfield W. Yorks. HD9 7AW tel:(01484) 682141 fax:(01484) 681513

Longden Doors Ltd Parkwood Road Sheffield S3 8AH tel: 0870 240 0666 fax: enquiries@lslongden.co.uk www.lslongden.co.uk

Longley James & Co Ltd East Park Crawley W. Sussex RH10 6AP tel:(01293) 561212 fax:(01293) 564564 mailroom@longleyco.uk

Lonsdale Metal Company Millmead Industrial Centre Mill Mead Road London N17 9QU tel:(020) 8801 4221 fax:(020) 8801 1287 info@patentglazing.com www.patentglazing.com

Lord Isaac Ltd Desborough Road High Wycombe Bucks. HP11 2QN tel:(01494) 459191 fax:(01494) 461376 info@isaaclord.co.uk www.isaaclord.co.uk

Lotus Water Garden Products Ltd Junction Street Burnley Lancs. BB12 0NA tel:(01282) 420771 fax:(01282) 412719

Louver-Lite Ltd Ashton Road Hyde Cheshire SK14 4BG tel:(0161) 882 5000 fax:(0161) 882 5009 www.louvolite.com

Lowe & Fletcher Ltd Moorcroft Drive Wednesbury W. Midlands WS10 7DE tel:(0121) 5050400 fax:(0121) 5050420 sales@lowe-and-fletcher.co.uk www.lowe-and-fletcher.co.uk

LPA Industries PLC PO Box 15, Tudor Works Debden Road Saffron Walden Essex CB11 4AN tel:(01779) 512802 fax:(01779) 512828 sales@lpa-ind.demon.co.uk www.lpa-industries.co.uk

LSA Projects Ltd The Barn White Horse Lane Witham Essex CM8 2BU tel:(01376) 501199 fax:(01376) 502027 sales@lsaprojects.demon.co.uk www.lsaprojects.demon.co.uk

Lumitron Ltd Park House 15 - 23 Greenhill Crescent Watford Business Park Watford WD18 8PH tel:(020) 8965 0211 fax:(020) 8965 8629 www.lumitron.co.uk

Luxcrete Ltd Premier House Disraeli Road Park Royal London NW10 7BT tel:(020) 8965 7292 fax:(020) 8961 6337 enquiries@luxcrete.co.uk www.luxcrete.co.uk

Luxo UK Ltd Connect House 21 Willow Lane Mitcham Surrey CR4 4NA tel:(020) 8698 7238 fax:(020) 8698 6134 office@luxo.co.uk www.luxo.co.uk

Lycetts Industrial Doors Unit 11, Parkhouse Industrial Estate Winpenny Road Newcastle-U-Lyme Staffs. ST5 7RH tel:(0870) 242 8899 fax:(0870) 241 0917 www.lycettdoors.co.uk

Lydney Containers Ltd Harbour Road Lydney Glos. GL15 4EJ tel:(01594) 842378 fax:(01594) 843213 alan-willette@lydneycontainers.co.uk www.lydneycontainers.com

Lyktan Lighting (UK) Ltd 31 a Bruton Place Mayfair London W1J 6NN tel:(020) 7409 3454 fax:(020) 7409 3414 info@lyktan-lighting.co.uk www.lyktan-lighting.co.uk

Lyncrete Burkle Services Ltd 20 Burns Avenue Church Crookham Fleet Hants. GU52 6BN tel:(01252) 816227 fax:(01252) 653935 email@lyncreteburkle.co.uk www.lyncreteburkle.co.uk

Lytag Ltd Hensall East Yorkshire DN14 0QD tel:(01977 661661 fax:(01977 662221 sales@lytag.co.uk www.lytag.co.uk

M & G Brickcutters (specialist works) Hockley Works Hooley Lane Redhill Surrey RH1 6JE tel:(01737) 771171 fax:(01737) 771171 sales@mgbrickcutters.co.uk www.mgbrickcutters.co.uk

M & M Fencing Limited 8 Lightborne Road Sale Cheshire M33 5EA tel:(0161) 973 0816 fax:(0161) 962 7322

Maby Hire Ltd 1 Railway Street Scout Hill Ravensthorpe, Dewsbury W. Yorks. WF13 3EJ tel:(01924) 460601 fax:(01924) 457932 info@maby-hire.co.uk www.mabey-hire.co.uk

MacAlloy Limited PO Box 71 Hawke Street Sheffield S9 2LN tel:(0114) 242 6704 fax:(0114) 243 1324 sales@macalloy.com www.macalloy.com

MacBee Manufacturing Ltd Unit 1A Building 11 Stanmore Industrial Estate Bridgnorth Shropshire WV15 5HR tel:(01746) 763103/763123 fax:(01746) 764659 sales@macbee.co.uk www.macbee.co.uk

Macc Stone Ltd 2 Robin Hill Biddulph Moor Stoke-on-Trent ST8 7NN tel:(01782) 514353 fax:(01782) 516783

Macclesfield Stone Quarries Ltd 194/198 New Street Budduiph Moor Stoke-on-Trent ST8 7NW tel:(01782) 514353 fax:(01782) 516783 sales@macclesfieldstone.com

Maceman + Amstad Ltd Boyle Road Willowbrook Industrial Estate Corby Northamptonshire NN17 5XU tel:(01536) 401331 fax:(01536) 401298 sales@macemainamstad.com www.macemainamstad.com

Mach & Garbarino Unit K5 Cooper House 2 Michael Road London SW6 7JU tel:(020) 77318555 fax:(020) 77318827 info@machgarbarino.co.uk www.edelmanleather.com

Mackay Carpets, Hugh PO Box 1 Durham Durham DH7 8YQ tel:(0191) 378 4444 fax:(0191) 378 4141 sales@hughmackay.com www.hughmackay.com

Mackay Engineering Unit 2 Mackay Business Park 120 Church End Cherry Hinton Cambridge CB1 3LB tel:(01223) 508222 fax:(01223) 510222 engineering@mackay.co.uk www.mackay.co.uk

Mackwell Electronics Ltd Virgo Place Aldridge W. Midlands WS9 8UG tel:(01922) 58255 fax:(01922) 51263

MacLellan Rubber Ltd Bettys Lane Norton Canes Staffordshire WS11 3UH tel:(0141) 946 5111 fax:(0151) 946 5222 macrubber@compuserve.com

MacMarney Refrigeration & Air Conditioning Ltd Eurolink Ind Est Castle Road Sittingbourne Kent ME10 3LY tel:(01449) 760560 fax:(01449) 760590 sales@macmarney.co.uk www.macmarney.co.uk

Maco Door & Window Hardware (UK) Ltd New Marlborough House Arnolde Close Medway City East, Rochester Kent ME2 4QW tel:(01634) 294555 fax:(01634) 294345 craig.maco@btinternet.com www.maco-europe.com

Macpuarsa (UK) Ltd Unit 43 Walton Business Centre 44/46 Terrace Road Walton-on-Thames Surrey KT12 2SD tel:(01932) 223161 fax:(01932) 223181

Macron Safety Systems (UK) Ltd Fireater House South Denes Road Great Yarmouth Norfolk NR30 3PJ tel:(01493) 859822 fax:(01493) 858374 macron-info@tycoint.com www.macron-safety.com

Made & Glazed For Trade Ltd 2b Hatchpond Road Nuffield Trading Estate Poole BH17 7LQ tel:(01202) 659975 fax:(01202) 659976

Madico Film Products Division Inc Madico House 98 Bolton Road Atherton Lancashire M46 9LD tel:(01942) 891790 fax:(01942) 891432 madicouk@amserve.net

Maestro International Ltd Powerscroft Road Sidcup Kent DA14 5NH tel:(020) 8302 4035 fax:(020) 8302 8933

MAGHansen Unit 6a Spence Mills Mill Lane Bramley, Leeds West Yorkshire LS13 3HE tel:(0113) 255 5111 fax:(0113) 255 1946

Magiboards Ltd Unit B3, Stafford Park 11 Telford Shropshire TF3 3AY tel:(01952) 292111 fax:(01952) 292280 sales@magiboards.co.uk www.magiboards.co.uk

Magiglo Ltd Lysander Close Broadstairs Kent CT10 2YJ tel:(01843) 602863 fax:(01843) 860108 info@magiglo.co.uk www.magiglo.co.uk

Magnet Applications Ltd Northbridge Road Berkhamptsead Herts. HP4 1EH tel:(01442) 875081 fax:(01442) 875009

Magnet Ltd Allington Way Yarm Road Ind. Estate Darlington Co. Durham DR1 4XT tel:(01325) 469441 fax:(01325) 468876 info@magnet.co.uk www.magnet.co.uk

Magnum Ltd T/A Magnum Scaffolding Yard Brook Estate Stockwood Vale Keynsham Bristol BS31 2AL tel:(0117) 986 0123 fax:(0117) 986 2123 swp@globalnomad.co.uk www.magnumltd.co.uk

Magpie Furniture Regulus Works Lynch Lane Weymouth Dorset DT4 9DW tel;(01305) 206000 fax:(01305) 760598 info@magpiefurniture.co.uk www.magpiefurniture.co.uk

Magrini Ltd Unit 5, Maybrook Industrial Estate Brownhills Walsall West Midlands WS8 7DG tel:(01543) 375311 fax:(01543) 361172 sales@magrini.co.uk www.magrini.co.uk

Majestic Shower Co Ltd 1 North Place Edinburgh Way Harlow Essex CM20 2SL tel:(01279) 443644 fax:(01279) 635074 enquiries@majesticshowers.com www.majesticshowers.com

Maldon Fencing Co. Burnham Road Latchingdon Nr Chelmsford Essex CM3 6HA tel:(01621) 740415 fax:(01621) 740769

Manders Paint Ltd PO Box 9 Old Heath Road Wolverhampton WV1 2XG tel:(01902) 871028 fax:(01902) 452435

Mandor Engineering Ltd Turner Street Works Turner Street Ashton Under Lyne Gtr. Manchester OL6 8LU tel:(0161) 330 6837 fax:(0161) 308 3336 sales@mandor.co.uk www.mandor.co.uk

Manitou (Site Lift) Ltd Ebblake Industrial Estate Verwood Wimborne Dorset BH31 6BB tel:(01202) 825331 fax:(01202) 813027 p.bidwell@manitou.co.uk www.manitou.com

Mann McGowan Group Intumescent House, 9 Springlakes Estate Deadbrook Lane Aldershot Hants. GU12 9UH tel:(01252) 333601 fax:(01252) 322724 sales@mannmcgowan.co.uk www.mannmcgowan.co.uk

Mansfield Brick Co Ltd Sanhurst Avenue Mansfield Notts. NG18 4BE tel:(01623) 622441 fax:(01623) 420904

Mansfield Keith Illustration 3 Johns Place Romiley Stockport Cheshire SK6 4BP tel:(0161) 494 9912

Manton Industrial Seals Ltd Unit 5B Bydand Industrial Estate Little Paxton Cambs. PE19 4ES tel:(01480) 214300 fax:(01480) 218987 gculley251@aol.com www.mantonindustrial.com

Mapei (UK) Ltd Mapei House Steel Park Road Halesowen West Midlands B62 8HD tel:(0121) 508 6970 fax:(0121) 508 6960 info@mapei.co.uk www.mapei.co.uk

Mar.Com Systems Ltd 1 MarCom House Heath Gardens Twickenham Middx. TW1 4BP tel:(020) 8891 5061 fax:(020) 8892 9028 kmccarney@mar-com.co.uk www.mar-com.co.uk

Marble Flooring Spec. Ltd Verona house Filwood Road Bristol BS16 3RY tel:(0117) 9656565 fax:(0117) 9656573 www.marbleflooring.uk.com

Marble Mosaic Co Ltd, The Winterstoke Road Weston Super Mare N. Somerset BS23 3YE tel:(01934) 419941 fax:(01934) 625479 sales@marble-mosaic.co.uk www.marble-mosaic.co.uk

Marcrist International Ltd Kirk Sandall Industrial Estate Doncaster South Yorks DN3 1QR tel:(01302) 890888 fax:(01302)883864 info@marchrist.com www.marcrist.com

Marcus M Ltd Narrowboatn Way Peartree Lane Dudley W. Midlands DY2 0XW tel:(01384) 457900 fax:(01384) 457903 www.m-marcus.com

Marflex Chimney Systems Unit 40 Vale Business Park Cowbridge S. Glam. CF71 7PF tel:(01446) 775551 fax:(01446) 772468 www.marflexchimneys.internet2.co.uk.

Marflow Eng Ltd Britannia House Austin Way Hamstead Industrial Estate Birmingham B42 1DU tel:(0121) 358 1555 fax:(0121) 358 1444 sales@marflow.co.uk www.marflow.co.uk

Margolis Office Interiors Ltd 341 Euston Road London NW1 3AD tel(020) 7387 8217 fax:(020) 7388 0625 sales@margolis2.u-net.com wwwmargolisfurniture.co.uk

Marine & Commercial Rigging Systems The Old Farmhouse, 1 Brickwall Close Burnham-on -Crouch Essex CM0 8HB tel:(01621) 782284 fax:(01621) 782263 www.industrialrigging.co.uk

Marketing & Promotions Ltd 6 Fairmead Rise Kings Norton Birmingham B38 8BS tel:(0121) 433 3555 fax:(0121) 459 5175 mapltd@clara.net www.map-ltd.com

Marking Machines International Sportsmark House Lansbury Estate Lower Guildford Road Woking Surrey GU21 2EP 01483 488310 01483 488310 sales@sportsmark.co.uk www.sportsmark.net

Marleton Cross Ltd Unit 2 Alpha Close Delta Drive, Tewkesbury Industrial Estate Tewklesbury Glos. GL20 8JF tel:(01684) 293311 fax:(01684) 293900 www.mx-group.com

Marley Building Materials Ltd Station Road Coleshill Birmingham B46 1HP tel:(01675) 468400 fax:(01675) 468455 mbm@mbm-marley.co.uk www.marley.co.uk

Marley Plumbing and Drainage Dickley Lane Lenham, Maidstone Kent ME17 2DE tel:(01622) 858888 fax:(01622) 858725 marketing@marleyext.com www.marley.co.uk

Marley Waterproofing Dickley Lane Lanham Maidstone Kent ME17 2QX tel:(01622) 854000 fax:(01622) 854300 info@marley.com www.marley.com

Marlow and Company Ltd Marlow House Hollow Road Bury St Edmonds Suffolk IP32 7AP tel:(01284) 772700 fax:(01284) 755567 trusses@marlowsteng.co.uk

Marlow Ropes Ltd Diplocks Way Hailsham E. Sussex BN27 3JS tel:(01323) 847234 fax:(01323) 440093 industrial@marlowropes.com www.marlowropes.com

Marples William & Sons Ltd Oscar Works Meadow Street Sheffield S. Yorks. S3 7BQ tel:(0114) 272 6662 fax:(0114) 273 0257

Marriott & Price Ltd Station House Banstead Road Banstead Surrey SM7 1PZ tel:(01737) 352735 fax:(01737) 359192

Marshall Contracts Ltd Unit 1 Ace Business Park Mackadown Lane, Kitts Green Birmingham B33 0LD tel:(0121) 783 5777 fax:(0121) 783 1117 info@marshallcontract.co.uk www.marshallcontracts.co.uk.

Marshall J A & Co (Brushes) Ltd Parkside House Bent Lane Prestwich Manchester M25 1DL tel:(0161) 773 2526 fax:(0161) 773 1545 jamar75@hotmail.com

Marshalls Clay Products Ltd Robin Hood Wakefield W. Yorks. WF3 3BH tel:(01532) 822141 fax:(01532) 829213

Marshalls Mono Ltd Southowram Halifax Yorks. HX3 9SY tel:(01422) 306000 fax:(01422) 330185 name@marshalls.co.uk www.marshalls.co.uk

Marshall-Tufflex Ltd Churchfields Ind.Estate Hastings E. Sussex TN38 9PU tel:(0870) 2403200 fax:(0870) 2403201 sales@marshall-tufflex.com www.marshall-tufflex.com

Marsland & Co Ltd Station Road Edenbridge Kent TN8 6EE tel:(01732) 862501 fax:(01732) 866737

Martella PLC President Drive Rooksley Milton Keynes MK13 8PD tel:(01908) 667418 fax:(01908) 604799

Martello Plastics Ltd Unit 11, Ross Way Shorncliffe Industrial Estate Folkestone Kent CT20 3UJ tel:(01303) 256848 fax:(01303) 246301 mail@martelloplastics.co.uk www.martelloplastics.co.uk

Martin & Co Ltd 119 Camden Street Birmingham B1 3DJ tel:(0121) 233 2111 fax:(0121) 236 0488

Martin (Marine) Ltd Alec 59 Norman Street Birkenhead Merseyside L41 0AS tel:(0151) 652 1663 fax:(0151) 653 3835

Martin Childs Limited Wimbledon Avenue Brandon IP27 0NZ tel:(01842) 812 882 fax:(01842) 812 002 sales@martinchilds.com www.martinchilds.com

Martin Orgee Ltd 135 Oak Drive Hartlebury Trade Estate Kidderminster Worcs. DY10 4JB tel:(01299) 253 739 fax:(01299) 271188 sales@martinorgee.co.uk www.martinorgee.co.uk

Martin Roberts Grimrod Place East Gillibrands Skelmersdale Lancs. WN8 9UU tel:(01695) 733068 fax:(01695) 50227 sales@martin-robert.co.uk www.martin-roberts.co.uk

Masfield Pollard & Co Ltd Crown Works Parry Lane Bradford BD4 8TL tel:(01274) 724466 fax:(01274) 390175

Mason FW & Sons Ltd Colwick Industrial Estate Colwick Nottingham NG4 2EQ tel:(0115) 911 3500 fax:(0115) 911 35444 mail@masons-timber.co.uk www.masons-timber.co.uk

Masonary Cleaning Services 1A Alpines Road Calaw S44 5AU tel:(01246) 209926 fax:(01246) 221620 maxblastuk@btinternet.com

Masonite CP Ltd Jason House Kerry Hill Horsforth Leeds LS18 4JR tel:(01532) 587689 fax:(01532) 590015

Massey Furguson Tractors Ltd PO Box 62 Banner Lane Coventry CV4 9GF tel:(024) 76 694400 fax:(024) 76 852495

Masson Seeley & Co Ltd Howdale Downham Market Norfolk PE38 9AL tel:(01366) 388000 fax:(01366) 385222 admin@masson-seeley.co.uk www.masson-seeley.co.uk

Masterbill Micro Systems Ltd Woodland Court Soothouse Spring, Valley Road St. Albans Herts. AL3 6NR tel:(01727) 855563 fax:(01727) 854626 sales@masterbill.com www.masterbill.com

Masterframe Windows Ltd 4 Crittall Road Witham Essex CM8 3DR tel:(01376) 510410 fax:(01376) 510400 sales@masterframe.co.uk www.masterframe.co.uk

Mastic Ashphalt Council 8 North Strret Ashford Kent TN24 8JN tel:(01233) 634411 fax:(01233) 634466

Maswell Engineering Ltd 33 Burlington Road Isleworth Middx. TW7 4LU tel:(020) 8568 3630 fax:(0870) 1641217 mxwell@maswell.demon.co.uk

Matki plc Churchward Road Yate Bristol BS37 5PL tel:(01454) 322888 fax:(01454) 315284 helpline@matki.co.uk www.matki.co.uk

Mattersons Cranes PO Box 31, Healey Works Shawclough Road Shawclough, Rochdale Lancs. OL12 6LP tel:(01706) 49321 fax:(01706) 57452

Matthew Hebden 123 Lonsdale Drive Oakwood Enfield Middx. EN2 7LS tel:(020) 8367 6463 fax:(020) 8367 0166 sales@matthewhebden.co.uk www.matthewhebden.co.uk

Matthews & Yates Ltd Peartree Road Stanway Colchester Essex CO3 5LD tel:(01206) 543311 fax:(01206) 760497 sales@matthewyates.co.uk www.matthewyates.co.uk

Matthews Office Furniture Ltd PO Box 81 61-63 Dale Street Liverpool Merseyside L69 2DN tel:(0151) 236 9851 fax:(0151) 236 8096 sales@efgmatthews.co.uk www.efgmatthews.co.uk

Mauser Interiors (UK) Ltd 1 Canterbury Court, Kennington Park 1-3 Brixton Road London SW9 6DE tel:(020) 7735 4565 fax:(020) 7735 4989 info@mauser.co.uk www.mauser.co.uk

Mawdsley's Ltd The Perry Centre, Water Walls Business Park Quedgeley Glos. GL2 4AD tel:(01453) 544131 fax:(01453) 546571

Max Appliances Ltd Wheel Park Farm Ins Est Wheel Lane Hastings East Sussex TN35 4SE tel:(01424) 854 444 fax:(01424) 853862 sales@max-appliances.co.uk www.max-appliances.co.uk

Maxwell Hart Ltd Winnersh Berks. RG11 5HF tel:(01734) 785655 fax:(01734) 785805

May Gurney (Technical Services) Ltd Fencing Division Ringland Lane, Costessey Norwich Norfolk NR8 5BG tel:(01603) 744440 fax:(01603) 747310

May Gurney (Technical Services) Ltd - Fencing Division Trowse Norwich Norfolk NR14 8SZ tel:(01603) 744440 fax:(01603) 747310 piling@maygurney.co.uk www.maygurney.co.uk

May William (Ashton) Ltd Cavendish Street Ashton Under Lyne Lancs. OL6 7BR tel:(0161) 330 3838/4879 fax:(0161) 339 1097 parmet@quista.net www.william-may.com

Mayplas Ltd Butterworth Street Little Borough Lancs. OL15 8JS tel:(01706) 374321 fax:(01706) 374473

McAlpine & Co Ltd Hillington Industrial Estate Glasgow G52 4LF tel:(0141) 882 3213

McAlpine Slate Ltd, Alfred Penrhyn Quarry Bethesda Bangor Gwynedd LL57 4YG tel:(01248) 600656 fax:(01248) 601171 pensales@alfred_mcalpine.co.uk www.amslate.com

McAthur Group Limited Foundry Lane Bristol BS5 7UE tel:(0117) 965 6242 fax:(0117) 958 3536

McCarthy & Stone Construction Services 3 Queensway New Milton Hants. BH25 5PB tel:(01425) 638855 fax:(01425) 638343

McKenzie-Martin Ltd Eton Hill Works Eton Hill Road Radcliffe Manchester M26 2US tel:(0161) 723 2234 fax:(0161) 725 9531 hilda@mckenziemartin.co.uk

McLoughlin Wood Ltd 5 Upper Street South New Ash Green Longfield Kent DA3 8JJ tel:(01474) 872019 fax:(01474) 873343 enquiries@mcwood.co.uk www.mcwood.co.uk

McMullen Architectural Systems 66 Lurgan Road Moira Craigavon County Armagh BT67 0LX tel:(028) 9261 9688 fax:(028) 9261 9711 info@mcmullensystems.co.uk

McNeill-McManus Ltd Hydepark Mallusk, Newtownabbey Co Antrim Northern Ireland BT36 4PX tel:(028) 9083 2025 fax:(028) 9034 2317 mailbox@mcneill-mcmanus.com www.mcneill-mcmanus.com

McQuay UK Ltd Bassington lane Cramlington Northumberland NE23 8AF tel:(01670) 566159 fax:(01670) 566206 info@mcquay.co.uk www.mcquay.com

MDH Ltd Walworth Road Andover Hampshire SP10 5AA tel:(01264) 362111 fax:(01264) 356452

Mea UK Ltd Rectors Lane Pentre Deeside Flintshire CH5 2DH tel:(01244) 534455 fax:(01244) 534477 uk.technical@mea.de www.mea.uk.com

Medina Gimson 60 Churchill Square Kingshill West Malling Kent ME19 4YU tel:(01732) 849440 fax:(01732) 847099

Meggitt Armour Systems Bestobell Ave Slough Berkshire SL1 4UY tel:(01384) 357799 fax:(01384) 357700 mas@meggitt.co.uk www.meggit.com

Mells Roofing Ltd Beehive Works Beehive Lane Chelmsford Essex CM2 9TE tel:(01245) 262621 fax:(01245) 260060 mellsroofing@aol.com www.mellsroofing.co.uk

Mentha & Hessal (Shopfitters) Ltd 95a Linaker Street Southport Merseyside PR8 5BU tel:(01704) 530800 fax:(01704) 500601

Menvier Ltd Southam Road Banbury Oxon. OX16 7RX tel:(01295) 755200 fax:(01295) 270102 enquiries@menvier.co.uk www.menvier.co.uk

Mercian Industrial Doors Pearsall Drive Oldbury W. Midlands B69 2RA tel:(0121) 544 6124 fax:(0121) 552 6793 info@merciandoors.co.uk www.merciandoors.co.uk

Mermet U.K Ryeford Hall Ryeford Nr Ross on Wye Herefordshire HR9 7PU tel:(01989) 750910 fax:(01989) 750768 mermet@btinternet.com www.mermet.co.uk

Merronbrook Ltd Hazeley Bottom Hartley Wintney Hook Hampshire RG27 8LX tel:(01252) 844747 fax:(01252) 845304 sales@merronbrook.co.uk

Metalcraft (Tottenham) Ltd 6-40 Durnford Street Seven Sisters Road Tottenham London N15 5NQ tel:(020) 8802 1715 fax:(020) 8802 1258 sales@makingmetalwork.com www.makingmetalwork.com

Metalliform Products Plc Chambers Road Hoyland Barnsley S. Yorks. S74 0EZ tel:(01226) 350555 fax:(01226) 350112 sales@metalliform-plc.demon.co.uk

Metalline Signs Ltd Unit 18 Barton Hill Trading Estate Bristol BS5 9TE tel:(01179) 555291 fax:(01179) 557518 info@metalline-signs.co.uk www.metalline-signs.co.uk

Metalrax Ltd Bordesley Green Road Birmingham W Midlands B9 4TP tel:(0121) 772 8151 fax:(0121) 772 6135 sales@metalrax-storage.co.uk www.metalrax-storage-co.uk

Metalwood Fencing (Contracts) Limited Soothouse Spring Valley Road Industrial Estate St. Albans Herts. AL3 6PG tel:(01727) 861141/2 fax:(01727) 846018 info@metalwoodfencinguk.co.uk www.metalwoodfencing.co.uk
Metcraft Ltd Harwood Industrial Estate Harwood Road Littlehampton E. Sussex BN17 7BB tel:(01903) 714226 fax:(01903) 723206 sales@metcraft.co.uk www.metcraft.co.uk
Metlex Shepperton Park Caldwell Road Nuneaton Warwks. CV11 4NR tel:(024) 76 324470 fax:(024) 76 324539 metlex@triton.plc.uk
Metra Non Ferrous Metals Ltd Pindar Road Hoddesdon Herts. EN11 0DE tel:(01992) 460455 fax:(01992) 451207 enquiries@metra-metals.co.uk www.metra-metals.co.uk
Metreel Ltd Cossall Industrial Estate Coronation Road Ilkeston Derbyshire DE7 5UA tel:(0115) 932 7010 fax:(0115) 930 6263 sales@metreel.co.uk www.metreel.co.uk
Metsec Building Products Ltd Broadwell Road Oldbury Warley W. Midlands B69 4HE tel:(0121) 601 6000 fax:(0121) 601 6109 lattice_joists@metsec.com www.metsec.com
Metway Electrical Industries Ltd Barrie House 18 North Street Portslade Brighton BN41 1DG tel:(01273) 431600 fax:(01273) 439288 sales@metway.co.uk www.metway.co.uk
Meyer Drenham Ltd Westfield Road Toftwood, Drenham Norfolk NR19 1JA tel:(01362) 697951 fax:(01362) 698980
MGL Carpet Finishers Ltd Unit L2, Melham Mills Knavie Lon Meltham W. Yorks WF13 3JY tel:(01924) 454566 fax:(01924) 468742 mglcarpets@compuserve.com www.mgl-carpets.com
MICE Kaymar Limited Brookhill Road, Brookhill Industrial Estate Pinxton Nottingham NG16 6NS tel:(01773) 864300 fax:(01773) 580286 micekaymar@compuserve.com www.micekaymar.com
Michelmersh Brick & Tile Co Ltd Hillview Road Michelmarsh, Romsey Hants. SO51 0NN tel:(01794) 368506 fax:(01794) 368845 www.michelmersh.co.uk
Mid Essex Trading Co Ltd Montrose Road Dukes Park Industrial Estate Springfield, Chelmsford Essex CM2 6TH tel:(01245) 469922 fax:(01245) 450755
Midgley C T & Co Ltd Scotts Road Hawkhead, Paisley Strathclyde PA2 7AN tel:(0141) 889 3118 fax:(0141) 889 2823
Midland Alloy Ltd Stafford Park 17 Telford Shropshire TF3 3DG tel:(01952) 290961 fax:(01952) 290441
Midland Stom Ltd Chem Road Millfields Wolverhampton W. Midlands WV4 6JJ tel:(01902) 353021 fax:(01902) 353635
Midland Tube and Fabrications Corngreave Works, Unit 4 Corngreave Road Cradley Heath, Warley W. Midlands B64 7DA tel:(01384) 566364 fax:(01384) 566365 Keithcadman@btconnect.com
Midland Wire Cordage Co Ltd Wire Rope House Eagle Rd North Moons Moat Redditch Worcs. B98 9HF tel:(01527) 594150 fax:(01527) 64322 mwc@mid-cord.co.uk www.midlandwirecordage.co.uk
Midven Ltd Hayseech Road Halesowen W. Midlands B63 3PE tel:(0121) 550 6441 fax:(0121) 585 5045
Miele Co Ltd Fairacres Marcham Road Abingdon Oxon. OX14 1TW tel:(01235) 554455 fax:(01235) 554477 info@miele.co.uk www.miele.co.uk
Milesahead Graphic Design/Consultant 11 Aylsham Close, The Willows Widnes Cheshire WA8 4FF tel:(0151) 422 0004 fax:(0151) 422 0004 Milesahead@talk21.com
Miles-Stone, Natural Stone Merchants Quarry Yard Woodside Avenue Boyattwood Industrial Estate Eastleigh, Hants SO50 9ES tel:(023) 80613178
Millar Dennis (1998) Ltd Victoria Works 142 Thornton Road Bradford West Yorkshire BD1 2EB tel:(01274) 722118 fax:(01274) 370977
Millenium Commission, The 2 Little Smith Street London SW1P 3DH tel:(020) 7340 2001
Miller Construction (UK) Ltd Metic House Ripley Drive, Normanton Ind Est Normanton West Yorkshire WF6 1QT tel:(01924) 224370 fax:(01924) 224380 mcofficeatnormanton@miller.co.uk www.construction.miller.co.uk
Milliken Carpet Beech Hill Plant Gidlow Lane Wigan Lancs. WN6 8RN tel:(01942) 826073 fax:(01942) 826570 carpetenquiries@milliken.com www.millikencarpeteurope.com
Mills John Ltd 72 Albert Street London NW1 7NR tel:(020) 7388 7212 fax:(020) 7388 3454
Milner Delvaux Ltd Eastrea Road Whittlesey Peterborough PE7 2AG tel:(01733) 202566 fax:(01733) 205405 mail@blancdebierges.com www.blancdebierges.com
Milton Pipes Ltd Milton Regis Sittingbourne Kent ME10 2QF tel:(01795) 425191 fax:(01795) 420360 sales@miltonpipes.com www.miltonpipes.co.uk
Minsterstone Ltd Pondhayes Farm Dinnington Hinton St. George Somerset TA17 8SU tel:(01460) 52277 fax:(01460) 57865 varyl@minsterstone.ltd.uk www.minsterstone.ltd.uk
Mita (UK) Ltd Bodelwyddan Business Park Bodelwyddan Denbighshire LL18 5SX tel:(01745) 586010 fax:(01475) 586015 sales@mita.co.uk www.mita.co.uk
MiTek Industries Ltd MiTek House, Grazebrook Industrial Park Pear Tree Lane Dudley W. Midlands DY2 OXW tel:(01384) 451400 fax:(01384) 451411 www.mitek.co.uk
Mitel Telecom Ltd Mitel Business Park Caldicot Monmouthshire NP26 5YR tel:(0870) 909 3030 fax:(0870) 909 5050 www.mitel.com
Mitsui Babcock Energy Ltd Porterfield Road Renfrew PA4 8DJ tel:(0141) 886 4141 fax:(0141) 885 3204 enquire@mitsuibabcock.com www.mitsuibabcock.com
MJ Electronics Services (International) Ltd Unit 8, Axe Road Colley Lane Industrial Estate Bridgewater Somerset TA6 5LJ tel:(01278) 422882 fax:(01278) 453331 sales@mjelectronics.freeserve.co.uk
MK (MK Electric Ltd) The Arnold Centre Paycocke Road Basildon Essex SS14 3EA tel:(01268) 563000 fax:(01268) 563563 more@mkelectric.co.uk www.mkelectric.co.uk
MMP International Wakefield Road Swillington Harrogate, Leeds W.Yorkshire LS26 8BN tel:(0113) 232 2880 fax:(0113) 232 2879 info@mmp-international.co.uk www.mmp-international.co.uk
Mockridge Labels & Nameplates Ltd Cavendish Street Ashton-under-Lyne Lancs. OL6 7QL tel:(0161) 308 2331 fax:(0161) 343 1958 sales@mockridge.com www.mockridge.com
Mode Lighting (UK) Ltd The Maltings 63 High Street Ware Herts. SG12 9AD tel:(01920) 462121 fax:(01920) 466881 sales@mode-lighting.co.uk www.mode-lighting.co.uk
Modelscape Adams House Dickerage Lane New Malden Surrey KT3 3SF tel:(020) 8949 9286 fax:(020) 8949 7418 modelscape@btconnect.com www.modelscape.co.uk
Modern Engineering (Bristol) Ltd 456 Badminton Road Yate Bristol BS17 5HX tel:(01454) 318181 fax:(01454) 318231
Moelven Laminated Timber Structures Ltd Unit 10, Vicarage Farm Winchester Road Fair Oak, Eastleigh Hants. SO50 7HD tel:(023) 80695566 fax:(023) 80695577 moelvenlts@aol.com www.moelven.co.uk
Moffat Ltd, E & R Seabegs Road Bonnybridge Stirlingshire FK4 2BS tel:(01324) 812272 fax:(01324) 814107
Moffett Thallon & Co Ltd 143 Northumberland Street Belfast BT13 2JF tel:(028) 90 322802 fax:(028) 90 241428 info@moffet-thallon.co.uk www.moffett-thallon.co.uk
Molyneux G (Products) Ltd 27 Blue Cap Road Stratford-upon-Avon Warwickshire CV37 6TQ tel:(01789) 299601 fax:(01789) 267867
Monarflex Anderson Ltd Barton Rock Rd Stretford Manchester M32 0YC tel:(0161) 865 4444 fax:(0161) 864 1178 enq@monarflex.co.uk www.monarflex.co.uk
Monk Metal Windows Ltd Hansons Bridge Road Erdington Birmingham B24 0QP tel:(0121) 351 4411 fax:(0121) 351 3673 sales@monkmetalwindows.co.uk www.monkmetalwindows.co.uk
Mono Pumps Ltd Martin Street Audenshaw Manchester M34 5DQ tel:(0161) 339 9000 fax:(0161) 344 0727
Monodraught Flues Ltd Halifax House Cressex Business Park High Wycombe Bucks. HP12 3SE tel:(01494) 897700 fax:(01494) 532465
Monowa Manufacturing (UK) Ltd Unit 6 Llyn-yr-Eos, Parc Menter Cross Hands Industrial Estate Llanelli Carmarthenshire SA14 6RA tel:(01269) 845554 fax:(01269) 845515 Monowa.manufacturing@business.ntl.com www.monowa.com
Montague L Meyer Bromford Gate Bromford Lane Heartlands Brimingham B24 8DW 0121 326 3200 0121 326 3230 sales.birmingham@mlmuk.com www.mlmuk.com
Montrose Safety Fixings 6 Bowers Road Shoreham, Sevenoaks Kent TN14 7SS tel:(020) 8304 6615 fax:(020) 8854 1680
Monument Tools Ltd Restmor Way Hackbridge Road Hackbridge, Wallington Surrey SM6 7AH tel:(020) 8288 1100 fax:(020) 8288 1108 info@monument-tools.com www.monument-tools.com
Moor, Andrew Associates 14 Chamberlain Street London NW1 8XB tel:(020) 7586 8181 fax:(020) 7586 8484
Moores Furniture Group Ltd Thorp Arch Trading Estate Wetherby West Yorkshire LS23 7DD tel:(01937) 842394 fax:(01937) 845396
Moray Timber Ltd 11 Perimeter Road Pinefield Ind. Estate Elgin Morayshire IV30 6AP tel:(01343) 545151 fax:(01343) 549518
Moresecure Ltd Haldane House Halesfield 1 Telford TF7 4EH tel:(01952) 683900 fax:(01952) 683982 sales@moresecure.co.uk www.moresecure.co.uk
Morris Material Handling Ltd PO Box 7 North Road Loughborough Leics. LE11 1RL tel:(01509) 643200 fax:(01509) 610666 info@morriscranes.co.uk www.morriscranes.co.uk
Morrison Chemicals Ltd 331-337 Derby Road Liverpool L20 8LQ tel:(0151) 933 0044
Mortar Industry Association Gillingham House 38-44 Gillingham Street London SW1V 1HU 020 7963 8000 020 7730 4355 james@qpa.org www.mortar.org.uk
Moseley GRP Product (Division of Moseley Rubber Co Ltd) Hoyle Street Mancunian Way Manchester M12 6HL tel:(0161) 273 3341 fax:(0161) 274 3743 sales@moseleyrubber.com www.moseleyrubber.com
Moss WM & Sons (Stove Anamellers Ripon) Ltd PO Box 4 Ure Bank Ripon N. Yorks. HG4 1JE tel:(01765) 604351 fax:(01765) 690065
Mott Aluminium Services Ltd Unit C, Watlington Industrial Estate Cuxham Road Watlington Oxfordshire OX9 5LT tel:(01491) 613232 fax:(01491) 613721
Mould J H & Co Ltd Bearmore Road Cradley Heath W. Midlands B60 6DW tel:(01384) 568241 fax:(01384) 411104 sales.jhmould@btconnect.com
Mountford Rubber & Plastics Ltd 44 Bracebridge Street Aston Birmingham B6 4PE tel:(0121) 359 0135/6 fax:(0121) 333 3204
Movement Joints (UK) P.O.Box 10 Wishbech Cambridgeshire PE13 4AN tel: (01945) 450736 fax: (01945) 450819 www.mjuk.co.uk
MR Limited White House Works Bold Road Sutton, St Helens Merseyside WA9 4JG tel:(01744) 820103 fax:(01744) 818997
MSA (Britain) Ltd East Shawhead Coatbridge N. Lanarkshire ML5 4TD tel:(01236) 424966 fax:(01236) 440881 info@msabritain.co.uk www.msabritain.co.uk/msa
Multi Marque Production Engineering Ltd 33 Monkton Road Industrial Estate Wakefield W. Yorks. WF2 7AL tel:(01924) 290231 fax:(01924) 382241 enquiries@multi-marque.co.uk www.multi-marque.co.uk
Multi Mesh Ltd Eurolink House Lea Green Industrial Estate St Helens Merseyside WA9 4QU tel:(01744) 820666 fax:(01744) 821417
Multibeton Ltd 15 Oban Court Wickford Business Park Wickford Essex SS11 8YB tel:(01268) 561688 fax:(01268) 561690
Multikwik Ltd 37 High Street Totton Hampshire SO40 9HL tel:(02380) 663777 fax:(02380) 869996 info@multikwik.com www.multikwik.com
Multitex GRP & GRG Products Unit 5, Dolphin Industrial Estate Southampton Road Salisbury Wilts. SP1 2NB tel:(01722) 332139 fax:(01722) 338458 www.multitex.co.uk
Muraspec 74-78 Wood Lane End Hemel Hempstead Herts. HP2 4RF tel:(08705) 117118 fax:(08705) 329020 customerservices@muraspec.com www.muraspec.com
Muswell Manufacturing Co Ltd Unit D1, Lower Park Road New Southgate Industrial Estate London N11 1QD tel:(020) 8368 8738 fax:(020) 8368 4726
Myddleton Hall Lighting, Division of Peerless Design Ltd Unit 9 Brunswick Industrial Park, Brunswick Way London N11 1JL tel:(020) 8362 8500 fax:(020) 8362 8525 peerlessuk@aol.com www.peerlessdisplay.com
Myson Radiators Ltd Eastern Avenue Team Valley Gateshead Tyne & Wear NE11 0PG tel:(0191) 491 7530 fax:(0191) 491 7568 sales@myson.co.uk www.myson.co.uk
National Association of Shopfitters NAS House 411 Limpsfield Road Warlingham Surrey CR6 9HA 01883 624961 01883 626841 nas@clara.net www.shopfitters.org
National Britannia Ltd Britannia House, Caerphilly Business Park Van Road Caerphilly Mid Glamorgan CF83 3GG tel:(029) 20 852852 fax:(029) 20 867738 enquiries@nb-group.com www.nb-group.com
National Federation of Builders Construction House 56-64 Leonard Street London EC2A 4JX Tel. (020) 7608 5150 Fax: (020) 7608 5151 info@builders.org.uk www.builders.org.uk
National federation of Demolition Contractors Resurgam House 1A New Road, The Causway Staines Middlesex TW18 3DH 01784 456799 01784 461118 info@demolition-nfdc.com www.demolition-nfdc.com
National Federation of Master Steeplejacks & Lightning Conductor Engineers 4d St Mary's Place The Lace Market Nottingham Nottinghamshire NG1 1PH tel: (0115) 955 8818 fax: (0115) 941 2238 info@nfmslce.co.uk www.nfmslce.co.uk
National Federation of Roofing Contractors, The 24 Weymouth Street London W1G 7LX 020 7436 0387 www.nfrc.co.uk

National Federation of Terrazzo Marble & Mosaic Specialists

PO.Box 2843 London W1A 5PG tel:(0845) 609 0050 Fax (0845) 607 8610 donaldslade@nftmms.org.uk www.nftmms.org.uk

National House Building Council Buildmark House Chiltern Avenue Amersham Bucks HP6 5AP 0870 241 4302 01494 735201 www.nhbc.co.uk
National Inspection Council for Electrical Installation Contracting Vintage House 37 Albert Embankment London SE1 7UJ 020 7564 2360 020 7564 2370 enquiries@niceic.org.uk www.niceic.org.uk
National Insulation Association Ltd PO Box 12 Haslemere Surrey GU27 3AH tel:(01428) 654011 fax:(01428) 651401 insulationassociation@aol.com www.insulationAssociation.org.uk
National Starch & Chemical Ltd Wexham Road Slough Berks. SL2 5DS tel:(01753) 533494 fax:(01753) 501241
Nationwide Filter Co Ltd Unit 5, Rufus Business Centre Ravensbury Terrace London SW18 4RL tel:(020) 8944 8877 fax:(020) 8944 8833 info@nationwidefilters.com www.fengroup.com
Naturestone Silwood Road Sunninghill Berkshire SL5 0PZ 01344 627 617 01344 873 185 info@stone.co.uk www.stone.co.uk
Naylor Drainage Ltd Clough Green Cawthorne Nr Barnsley S. Yorks. S75 4AD tel:(01226) 790591 fax:(01226) 790531 info@naylor.co.uk www.naylor.co.uk
NBS The Old Post Office St. Nicholas Street Newcastle-upon-Tyne NE1 1RH tel:(0191) 232 9594 fax:(0191)232 5714 info@thenbs.com www.thenbs.com
NCMP Ltd Falcon Way Feltham Middx. TW14 0XJ tel:(020) 8751 0986 fax:(020) 8751 2446
Nederman UK PO Box 503 91 Walton Summit Bamber Bridge Preston PR5 8AF tel:(01772) 334721 fax:(01772) 315273 info@nederman.co.uk www.nederman.co.uk
NEI Reyrolle Ltd Hebburn Tyne & Wear NE31 1UP tel:(0191) 483 2451 fax:(0191) 483 2446
Nendle Acoustic Company Ltd Tacitus House 153a High Street Aldershot Hants. GU11 1TT tel:(01252) 344222 fax:(01252) 333782 nendle@tacitus.co.uk www.nendle.co.uk
Nenplas Airfield Industrial Estate Ashbourne Derbys. DE6 1HA tel:(01335) 347300 fax:(01335) 340271 enquiries@nenplas.co.uk www.nenplas.co.uk
Neocrylic Signs Ltd Unit 7 Mather Road Eccles Manchester M30 0WQ tel:(0161) 707 8933 fax:(0161) 707 8934
Neopost Ltd Neopost House South Street Romford Essex RM1 2AR tel:(01708) 746000 fax:(01708) 714050 mktg@neopost.co.uk www.neopost.co.uk
Neptune Outdoor Furniture Ltd Thompsons Lane Marwell Nr Winchester Hampshire SO21 1JH tel:(01962) 777799 fax:(01962) 777723 sales@nofl.co.uk www.nofl.co.uk
Neptune Showers Ltd Alpha Close Delton Drive, Tewkesbury Ind Est Tewkesbury Gols GL20 8JF tel:(020) 8960 3185 fax:(020) 8968 5895
Neslo Partitioning & Interiors 1 Westgate Center Westgate Aldridge West Midlands WS9 8EY tel:(01922) 743390 fax:(01922) 743268 gmason@carillionplc.com www.neslo.co.uk
Ness Furniture Ltd Croxdale Durham DH6 5HT tel:(01388) 816109 fax:(01388) 812416 sales@nessfurniture.co.uk www.nessfurniture.co.uk
Nettlefolds Ltd Heath Street Smethwick W. Midlands B66 2SA tel:(0121) 626 0600 fax:(0121) 626 0585 info@nettlefolds.com
Network Flooring Cardiff Earl Road Lane Penarth South Glamorgan CF64 3UR tel:(029) 20 360800 fax:(029) 20 360007
Nevill Long Limited Chartwell Drive West Avenue Wigston Leicester LE18 2FL tel:(0116) 2570670 fax:(0116) 2570044 leicester@nevilllong.co.uk www.nevilllong.co.uk
Neville, Scales & Co Ltd 34 Priests Bridge London SW14 8TA tel:(020) 7254 1033 fax:(020) 8392 2026
New Centurian Industry Services Ltd Sovereign House, 2 Sovereign Park Coronation Road Park Royal London NW10 7 QP tel:(020) 8965 7300 fax:(020) 8965 2615
New Forest Ceilings Ltd 61-65 High Street Totton Southampton SO40 9HL tel:(023) 80869510 fax:(023) 80862244 www.newforestceilings.co.uk
New Haden Pumps Ltd Draycott Cross Road Cheadle, Stoke on Trent Staffs. ST10 2NW tel:(01538) 757900 fax:(01538) 757999 info@nhpumps.com www.nhpumps.com
Neway Doors Ltd 89-91 Rolfe Street Smethwick Warley W.Midlands B66 2AY tel:(0121) 558 6406 fax:(0121) 558 7140 sales@priory-group.com
Newdawn & Sun Ltd Springfield Business Park Arden Forest Inustrial Estate Alcester Warwks. B49 6EY tel:(01789) 764444 fax:(01789) 400164 sales@newdawn-sun.co.uk www.newdawn-sun.co.uk
Newey Ceilings 68 Florence Road Wylde Green Sutton Coldfield W. Midlands B73 5NG tel:(0121) 382 2098 fax:(0121) 382 1764 mike@neweyceiling.fsnet.co.uk
Newington Bricks Ltd London Road Newington Nr Sittingbourne Kent ME9 7NU tel:(01795) 842338
Newlay Concrete Ltd Thornhill Works Calder Road Dewsbury W. Yorks. WF12 9HY tel:(01924) 456416 fax:(01924) 430697
NewTeam Ltd Brunel Road Earlstrees Industrial Estate Corby Northants NN17 4JW tel:(01536) 409222 fax:(01536) 400144 www.newteamshowers.com

Newton & Frost Downsview Yard North Corner, Horam Heathfield E. Sussex TN21 9HJ tel:(01435) 813535 fax:(01435) 813687 enquiries@newtonandfrost.co.uk www.newtonandfrost.co.uk
Newton John & Co Ltd 12 Verney Road London SE16 3DH tel:(020) 7237 1217 fax:(020) 7252 2769 enquiries@newton-membranes.co.uk www.newton-membranes.co.uk
Newtons Computer Services Ltd Wandle House Riverside Drive Mitcham Surrey CR4 4BU tel:(020) 8648 8707 fax:(020) 8648 2240
NFC Specialist Products Ltd NFC House, Vickges Industrial Estate Mellishaw Lane Morecambe Lancashire LA3 3EN tel:(01524) 847600 fax:(01524) 847800 info@nfcspecialistproducts.com www.fengroup.com
Nicholls & Clarke Ltd Niclar House Shoreditch High Street London E1 6PE tel:(020) 7247 5432 fax:(020) 7247 7738
Nicholson Plastics Ltd Riverside Road Kirkfieldbank Lanarkshire ML11 9JS tel:(01555) 664316 fax:(01555) 663056 sales@nicholsonplastics.co.uk www.nicholsonplastics.co.uk
Nico Manufacturing Ltd 109 Oxford Road Clacton on Sea Essex CO15 3TJ tel:(01255) 422333 fax:(01255) 432909 sales@nico.co.uk www.nico.co.uk
Nilfisk Advance Ltd Newmarket Road Bury St Edmunds Suffolk IP33 3SR tel:(01284) 763163 fax:(01284) 750562
Nimlok Ltd Booth Drive Park Farm Wellingborough Northants NN8 6NL Tel: (01933) 409409 Fax: (01933) 409451 info@nimlok.co.uk www.nimlok.co.uk
Nittan (UK) Ltd Hipley Street Old Woking Surrey GU22 9LQ tel:(01483) 769555 fax:(01483) 756686 sale@nittan.co.uk www.nittan.co.uk
nmc (UK) Ltd Prospect Road Beechburn Industrial Estate Crook Co. Durham DL15 8JN tel:(01388) 761300 fax:(01388) 761400 enquiries@nmc.uk.wm www.svrgb2/nmc-uk/
Nobo Heating (UK) Ltd Unit 15 Gravelly Ind Est Park Birmingham W. Midlands B24 8HZ tel:(01323) 641521 fax:(01323 410328 info@nobo.co.uk
Nordson (UK) Ltd Wenman Road Thame Industrial Estate Thame Oxon. OX9 3XB tel:(01844) 264500 fax:(01844) 245358
Norlandia Ltd Unit 14 & 18 Hazel Road Four Marks, Alton Hants. GU34 5EY tel:(01420) 561720 fax:(01420) 561730 info@norlandia.co.uk
Norske Interiors (UK) Ltd Estate Road One South Humberside Industrial Estate Grimsby Humberside DN31 2TA tel:(01472) 240832 fax:(01472) 240304 sales@norske-int.co.uk www.norske-int.co.uk
North Enfield Precast Concrete Co Ltd BR Goods Depot Whitehart Lane London N17 8DP tel:(020) 8801 7217 fax:(020) 8808 0854
North Herts Asphalte Ltd Unit 15, The Cuttings Station Approach Hitchin Herts. SG4 9UW tel:(01462) 434877/454119 fax:(01462) 421539
North Wales Slate Roofing Co, The PO Box 2 Pormadoc Gwynedd tel:(01766) 831532 fax:(01766) 830711
North Yorkshire Artstone Ltd Showfield Lane Malton Yorks. YO17 6BT tel:(01653) 697714 fax:(01653) 692427 artstone@ukonline.co.uk www.northyorkshireartstone.com
Northcot Brick Ltd Blockley, Nr Moreton in Marsh Glos. GL56 9LH tel:(01386) 700551 fax:(01386) 700852 info@northcotbrick.co.uk www.northcotbrick.co.uk
Northern Fencing Contractors Limited Parkhill Walton Road Wetherby W. Yorks. LS22 5DZ tel:(01937) 545155 fax:(01937) 580034 tmorgan@holroydconstruction.co.uk
Northern Incinerators (GB) Ltd Old Grammar School House School Gardens Shrewsbury Shropshire SY1 2AJ tel:(01743) 368134
Northern Joinery Ltd Daniel Street Whitworth, Rochdale Lancs. OL12 8DA tel:(01706) 852345 fax:(01706) 853114 office@northernjoinery.co.uk www.northernjoinery.co.uk
Northgate Solar Controls PO Box 200 Barnet Herts. EN4 9EW tel:(020) 8441 4545
Northwest Pre-cast Ltd Holmefield Works Garstang Road Pilling Nr. Preston PR3 6AN tel:(01253) 790444 fax:(01253) 790085 nwprecast@talk21.com
Norton Abrasives Ltd Bridge Road East Welwyn Garden City Herts. AL7 1HZ tel:(01707) 323484 fax:(01707) 325231
Norton Diamond Products Unit 2, Meridian West Meridian Business Park Leicester LE3 2WX tel:(0116) 263 2302 fax:(0116) 282 7292
Norton Engineering Alloys Co Ltd Norton Grove Industrial Estate Norton Malton N. Yorks. YO17 9HQ tel:(01653) 695721 fax:(01653) 600418 www.neaco.co.uk
Norwood Partition Systems Ltd Oxleasow Road East Moons Moat Redditch Worcs B98 ORE tel:(01527) 510015 fax:(01527) 517222 marketing@norwood.co.uk www.norwood.co.uk
Notcutt Ltd Homeward Farm Newark Lane Ripley Surrey GU23 6DJ tel:(020) 8977 2252 fax:(020) 8977 6423 sales@notcutt.com www.notcutt.com
Nova Garden Furniture Ltd The Faversham Group Graveney Road Faversham Kent ME13 8UN tel:(01795) 535511 fax:(01795) 539215 contract@novagardenfurniture.co.uk
Nova Group Ltd Norman Road Board Heath Altrincham Cheshire WA14 4EN tel:(0161) 941 5174 fax:(0161) 926 8405
Novellini Bathroom Products Ltd Unit 1, Redwood House Orchard Trading Estate Toddington, Cheltenham Glos. GL54 5EB tel:(01242) 620061 fax:(01684) 850912 sales@novellini.freeserve.co.uk
NRG Fabrications 10, St. James Mill Road Northampton Northants NN5 5JW tel:(01604) 580022 fax:(01604) 580033 nrg@cemfencing.co.uk
NT Architectural Hardware Straight Road Short Heath Willenhall W. Midlands WS1 4NA tel:(01922) 401606 fax:(01922) 495344
NT Worcester Parsons Ltd Lifford Lane Kings Norton Birmingham B30 3RJ tel:(0121) 486 2211 fax:(0121) 459 6405
NuAir Ltd Western Industrial Estate Caerphilly M. Glam. CF83 1XH tel:(029) 20 885911 fax:(029) 20 887033
NuLite Ltd 41 Hutton Close Crownther Industrial Estate District 3 Washigton, Tyne & Wear NE38 0AH tel:(0191) 419 111 fax:(0191) 419 1123 sales@nulite-ltd.co.uk www.nulite-ltd.co.uk
Nullifire Ltd Torrington Avenue Coventry CV4 9TJ tel:(+44) 0 24-7685 5000 fax:(+44) 0 24-7646 9547 protect@nullifire.com www.carboline-eu.com
Nuralite (UK) Ltd Nuralite House, Canal Road Higham Rochester Kent ME3 7JA tel:(01474) 823451 fax:(01474) 823961
Nu-Swift International Ltd Elland W. Yorks. HX5 9DS tel:(01422) 372852 fax:(01422) 379569 customer.service@nu-swift.co.uk www.nu-swift.co.uk
Nuttall, Henry (Viscount Catering Ltd) Green Lane Ecclesfield Sheffield S. Yorks. S35 9ZY tel:(0114) 257 0100 fax:(0114) 257 0251 sales@viscount-catering.co.uk www.viscount-catering.co.uk
Nu-Way Ltd PO Box 1 Vines lane Droitwich Worcs. WR9 8AN tel:(01905) 794331 fax:(01905) 794017 info@nu-way.co.uk
Oakdale (Contracts) Ltd Walkerville Industrial Estate Catterick Garrison N. Yorks. DL9 4SA tel:(01748) 834184 fax:(01748) 833003
Oakland Elevators Ltd Mandervell Road Oadby Leics. LE2 5LL tel:(0116) 272 0800 fax:(0116) 272 0904 oakland@elevator.co.uk
Oakleaf Reproductions Ltd Main Street Wilsden Bradford W. Yorks. BD15 0JP tel:(01535) 272878 fax:(01535) 275748 sales@oakleaf.co.uk www.oakleaf.co.uk
O'Brian Manufacturing Ltd Robian Way Swadlincote Derbys. DE11 9DH tel:(01283) 217588 fax:(01283) 215613
Oce (UK) Ltd Langston Road Loughton Essex IG10 3SL tel:(020) 8508 5544 fax:(020) 8508 6689
O'Connor Fencing Limited Station Yard Moor Row Whitehaven Cumbria CA24 3JR tel:(01946) 810602 fax:(01946) 810505 enquiries@conorfencing.co.uk
OEP Furniture Group PLC 7-9 Cartersfield Road Waltham Abbey Essex EN9 1JD tel:(01992) 767014 fax:(01992) 762968
Ogilvie Construction Ltd Pirnhall Works 200 Glasgow Road Whins of Milton Stirling FK7 8ES tel:(01786) 812273 fax:(01786) 816287 enq@ogilvie.co.uk
Ogley Bros Ltd Allen Street Smithfield Sheffield S3 7AS tel:(0114) 276 8948 fax:(0114) 275 8948
Olby H E & Co Ltd 229-313 Lewisham High Street London SE13 6NW tel:(020) 8690 3401 fax:(020) 8690 1408
Oldham Lighting Ltd Claudgen House Eastwick Road Leatherhead Surrey KT23 4DT tel:(020) 8946 5555 fax:(020) 8946 5522 sales@oldhamlighting.co.uk www.oldhamlighting.com
Olivand Metal Windows Ltd Chesley House 43a Chesley Gardens London E6 3LN tel:(020) 8471 8111 fax:(020) 8552 7015
Olley & Sons Ltd, C Iberia House Finchley Avenue Mildenhall Suffolk IP28 7BJ tel:(01638) 712076 fax:(01638) 717304 olley-cork@btconnect.com www.corkproducts.co.uk
Omeg Ltd Imberhorne East Grinstead West Sussex RH19 1RJ tel:(01342) 410420 fax:(01342) 316253 sales@omeg.com www.catelco.co.uk
Onduline Building Products Ltd Eardley House 182-184 Campden Hill Road Kensington London W8 7AS tel:(020) 7727 0533 fax:(020) 7792 1390 enquiries@onduline.net www.onduline.net
Opals (Mirror-Flex) Co Ltd unit 3b Seaden Court Stephenson Road Grose Lane ind. Est Clacton-on-Sea Essex CO15 4XN tel:(01225) 423927 fax:(01225) 221117 enquiries@mirrorflex.co.uk www.mirrorflex.co.uk
Opella Ltd Twyford Road Rotherwas Industrial Estate Hereford Herefordshire HR2 6JR tel:(01432) 357331 fax:(01432) 264014 sales@opella.co.uk www.opella.co.uk
Optelma Lighting Ltd 14 Napier Court The Science Park Abingdon Oxon. OX14 3NB tel:(01235) 553769 fax:(01235) 523005 sales@optelma.co.uk www.optelma.co.uk
Optex (Europe) Ltd Clivemont Road Cordwallis Park Maidenhead Berks. SL6 7BU tel:(01628) 631000 fax:(01628) 636311 sales@optex-europe.com www.optex-europe.com
Optiroc Ltd Adamson House Pomona Strand Manchester M16 0BA tel:(0161) 873 7701 fax:(0161) 872 4385
Opto International Ltd Tower Mill Park Road Dukinfield Cheshire SK16 5LN tel:(0161) 330 9136 fax (0161) 343 7332 enquiry@optoint.co.uk www.optoint.co.uk
Opus 4 Ltd Orchard House Orchard Business Centre, North Farm Road Tunbridge Wells Kent TN2 3DY tel:(01892) 515157 fax:(01892) 515417
Orbik Electrics Ltd Orbik House Northgate Way Aldridge Walsall WS9 8TX tel:(01922) 743515 fax:(01922) 743173
Orchard Street Furniture 119 The Street Crowmarsh & Gifford Nr. Waringford Oxon. OX10 8EF tel:(01491)642123 fax:(01491) 642126 sales@orchardstreet.co.uk www.orchardstreet.co.uk
Ordnance Survey Romsey Road Maybush Southampton SO16 4GU tel:(08456) 050505 www.ordnancesurvey.co.uk
Orsogril Spa UK Distributor Geddings Road Hoddesdon Herts. EN11 0NZ tel:(0870) 606 2070 fax:((0131) 441 4161
OSF Ltd Unit 6, The Four Ashes Industrial Estate Station Road Four Ashes Wolverhampton WV10 7DB tel:(01902) 798080 fax:(01902) 794750 osfsales@aol.com www.wolverhampton.co.uk/osf
OSMO UK Ltd Osmo House Unit 2 Pembroke Road Stocklake Industrial Estate Aylesbury Bucks HP20 1DB tel:(01296) 481220 fax:(01296) 424090 info@osmouk.com www.osmouk.com
Osram Ltd Customer Service Centre Neills Road Bold Industrial Park St. Helens WA9 4XG tel:(01744) 812221 fax:(01744) 831900 csc@osram.co.uk www.osram.co.uk.
Oswestry Reinforced Plastics Ltd The Old Hall Cheswardine Market Drayton Shropshire TF9 2RN tel:(01630) 661410 fax:(01630) 661410
Otis Ltd The Otis Building 187 Twyford Abbey Road London NW10 7DG tel:(020) 8955 3000 fax:(020) 8955 3001 www.otis.com
Outdoor Power Products Dolmer House Clare Street Denton Manchester M34 3LQ tel:(0161) 320 8100 fax:(0161) 335 0114
OWA UK Ltd Elm House 23-25 Elmshott Lane Cippenham Slough SL1 5QS tel:(01628) 663797 fax:(01270) 824025
Owens Corning Automotive (uk) Ltd Lune Ind Est Lancaster Lancastershire LA1 5QP tel:(01524) 591700
Owlett - Jaton Opal Way Stone Business Park Stone Staffs. ST15 0SW tel:(01785) 811300 fax:(01785) 819699 info@owlett-jaton.com
Oxford Double Glazing Ltd Ferry Hinksey Road Oxford OX2 0BY tel:(01865) 248287 fax:(01865) 251070
P & L Systems Ltd Pals House Halfpenny Road Knaresborough N. Yorks. HG5 0PS tel:(01423) 861226 fax:(01423) 863497 info@pandlsystems.co.uk www.pandlsystems.com
P & R Electrical (London) Ltd 11 Hanger Lane Ealing Common London W5 3HH tel:(020) 8992 0174
P & R Laboratory Group Ltd Unit 5 Brindley Road St Helens Merseyside WA9 4HY tel:(01744) 831800 fax:(01744) 831888 sales@pandr.co.uk www.pandr.co.uk
P R A Coating Technology PRA Coatings Technology Centre 8 Waldegrave Road Teddington Middlesex TW11 8LD 020 8614 4800 020 8943 4705 www.pra.org.uk
P.C.I Construction Systems Ltd Albany House Swimton Hall Rd Swimton Manchester M27 4DT tel:(0845) 4589105 fax:(0845) 4589106 pciconstruction@compuserve.com www.pci.uk.com
P4 Limited Unit 7B St Francis Way Shefford Bedfordshire SG17 5DZ tel:(01462) 851144 fax:(01462) 851123 info@p4fastel.co.uk
Pac International Ltd 1 Park Gate Close Bredbury, Stockport Gtr. Manchester SK6 2SZ tel:(0161) 494 1331 fax:(0161) 430 9658 www.pac.co.uk
Package Products Ltd Little Green Works Collyhurst Road Manchester M40 7RT tel:(0161) 205 4181 fax:(0161) 203 4678 sales@packagingproducts.co.uk www.packagingproducts.co.uk
Paddock Fabrications Ltd Fryer Road Walsall, Bloxwich W. Midlands WS2 7NF tel:(01922) 711722 fax:(01922) 476021 sales@paddockfabrication.com www.paddockfabrications.com
Padipa Ltd Sales Office 65 Pennywell Gdns New Milton Hampshire BH25 5YB 01425 610336 01425 616955 enquiries@padipa.co.uk www.padipa.co.uk
Painting & Decorating association 32 Coton Road Nuneaton Warwickshire. CV11 5TW 024 7635 3776 024 7635 4513 info@paintingdecoratingassociation.co.uk www.paintingdecoratingassociation.co.uk
Palgrave Brown Timber & Fireproofing Co Station Road Market Bosworth Nuneaton Warwks. CV13 0PQ tel:(01455) 290917 fax:(01455) 290918 a.bond@palgravebrown.co.uk www.palgravebrown.co.uk
Palgrave Brown UK Ltd Unit C2, C4 and C5 Boston Industrial Centre, Norfolk Street Boston Lincs. PE21 9HG tel:(01205) 362468 fax:(01205) 350982 www.palgravebrown.co.uk
Palmers Scaffolding Ltd Penhall Road Charlton SE7 tel:(020) 8858 6511
Palmyra Ltd Tavistock Works Glasson Estate Maryport Cumbria CA15 8NX tel:(01900) 812796 fax:(01900) 815509 info@palmyra.ltd.com www.palmyro.ltd.com
Panasonic UK Ltd Panasonic House Willoughby Road Bracknell Berks. RG12 8FP tel:(01344) 862444 fax:(01344) 853704
Panel & Louvre Co Ltd 1A Bridle Close Finedon Road Industrial Estate Wellingborough Northants NN8 4RN tel:(01933) 228500 fax:(01933) 228580 palco@pipemedia.co.uk www.ascwebindex.com/palco
Panel Systems Ltd 1-9 Welland Close Parkwood Industrial Estate Rutland Road, Sheffield S. Yorks. S3 9QY tel:(0114) 275 2881 fax:(0114) 278 6840 sales@panelsystems.co.uk www.panelsystems.co.uk
Panelite Green Lane Garforth Leeds LS25 2AE tel:(0113) 2875111 fax:(0113) 286 3377
Pannell Signs Ltd Chelsea Street Nottingham NG7 7HN tel:(0115) 970 0371 fax:(0115) 942 2452 sales@pannellsigns.co.uk
Paragon Oakhurst Drive Cheadle Heath, Stockport Cheshire SK3 0XT tel:(0161) 428 0622 fax:(0161) 428 1239 sales@riwa.co.uk www.insect-o-cutor.co.uk
Paragon Business Furniture York house Empire way 100 Queensbury Rd. Wembley Middlesex HA9 0PA tel:(020) 8900 8800 fax:(020) 8900 8801 sale@para-furn.co.uk www.paragon-businessfurn.com
Paragon by Heckmondwike Farfield Park Manvers Wath on Dearne Rotherham, South Yorkshire S63 5DB tel:(01709) 763800 fax:(01709) 763818 info@pbyh.co.uk www.pbyh.co.uk
Park Products Swan Lane Hinley Green Wigan Lancaster WN2 4HD tel:(01254) 614000 fax:(01254) 614001 sales@parkproducts.co.uk www.parkproducts.co.uk
Parker AG Engineering Ltd Burrowfield Welwyn Garden City Herts. AL7 4SS tel:(01707) 330215 fax:(01707) 372441
Parker Bath Limited Queensway Stem Lane New Milton Hants. BH25 5NN tel:(01425) 624000 fax:(01425) 624039 customer.services@parkerbath.co.uk www.sunriseccg.co.uk
Parker Kislingbury Ltd Wotton Road Brill Bucks. HP18 9UB tel:(0800) 220142 fax:(01844) 238016
Parker Plant Ltd PO Box 146 Canon Street Leicester LE4 6HD tel:(0116) 266 5999 fax:(0116) 268 1254 sales@parkerplant.com www.parkerplant.com
Parkes Josiah & Sons Ltd Union Works Willenhall W. Midlands WV13 1JX tel:(01902) 366931 fax:(01902) 366888
Parkinson Cowan and Tricity Bendix Co, The Cornwall House 55-57 High Street Slough Berks. SL1 1BG tel:(0990) 146146 fax:(01953) 872381
Parklines Building Ltd Gala House 3 Raglan Road Edgbaston Birmingham B5 7RA tel:(0121) 446 6030 fax:(0121) 446 5991
Parsons Brothers Steelwork Griffiths Road West Bromwich W. Midlands B71 2EL tel:(0121) 588 4871 fax:(0121) 588 7513
Parsons C B Co Ltd, The Ashby de la Zouch Leics. LE6 5DR tel:(01530) 412885 fax:(01530) 417315
Partitioning & Interiors Association Jago House 692 Warwick Road Solihull W. Midlands B91 3DX tel:(0121) 705 9270 fax:(0121) 711 2892
Parton Fibreglass Ltd PFG House Claymore, Tame Valley Industrial Estate Tamworth Staffs. B77 5DQ tel:(01827) 251899 fax:(01827) 261390 sales@pfg-tanks.com www.pfg-tanks.com
Passivent Ltd 2 Brooklands Road Sale Cheshire Notts. M33 3SS tel:(0161) 962 7113 fax:(0161) 905 2085 info@passivent.com www.passivent.com
Pavigres-wich Ltd Units E-F Aerial Business park Lambourn Woodlands Hungerford, Berkshire RG17 7RZ tel:(01488) 674 500 fax:(01488) 674 505
PCT Group Sales Ltd Milton Chain Works Shirland Lane Sheffield S. Yorks. S9 3FG tel:(0114 2449761 fax:(0114)2441282 sheffield@pctgroup.co.uk
Peal furniture (Durham) Ltd Littleburn Industrial Estate Langley Moor Co. Durham DH7 8HE tel:(0191) 378 0232 fax:(0191) 378 1660
Pearce Construction(Barnstaple)Ltd Great Western House Old Station Road Barnstaple N. Devon EX32 8GW tel:(01271) 345261 fax:(01271) 322164 barnstaple@pearce.co.uk
Pearce Security Systems Ensign House Green Lane Felling, Gateshead Tyne &Wear NE10 0QH tel:(0191) 438 1177 fax:(0191) 495 0227 admin@pearcesecurity.co.uk
Pearce Signs Ltd 12 Poverest Rd Orpington Kent BR5 2TP tel:(01689) 892500 fax:(01689) 892518 signs@pearcegroup.com www.pearcegroup.com

Pearl Paints Ltd Severn Road Teforest Industrial Estate Pontypridd M. Glam. CF37 5SR tel:(01443) 843231 fax:(01443) 843705 pearladmin@pearlpaints.co.uk www.pearlpaints.co.uk

Pearman Fencing Greystone Yard, Notting Hill Way Lower Weare Axbridge Somerset BS26 2JU tel:(01934) 733380 fax:(01934) 733398

Pearsons J R (Birmingham) Ltd 212/213 New John Street West, Hockley Birmingham West Midlands B19 3UD tel:(0121) 551 9411 fax:(0121) 551 9412 jrpearson@btinternet.com www.drawmet.com

Pedley Furniture International Ltd Shirehill Works Saffron Walden Essex CB11 3AL tel:(01799) 522461 fax:(01799) 513403 sales@pedley.com www.pedley.com

Pegler Hattersley St Cathrines Avenue Doncaster S Yorks. DN4 8DF tel:(01302) 560560 fax:(01302) 367661 uk.sale@pegler.co.uk www.pegler.co.uk

Pegson Ltd Coalville Leics. LE67 3GN tel:(01530) 510051 fax:(01530) 510041

Pel Plc Oldbury Warley W. Midlands B69 4HN tel:(0121) 552 3377 fax:(0121) 552 6067

PEL Services Limited Belvue Business Centre Belvue Road Northolt Middx. UB5 5QQ tel:(020) 8839 2100 fax:(020) 8841 1948 pel@pel.co.uk www.pel.co.uk

Pellfold Parthos Ltd. 1 The Quadrant Howarth Road Maidenhead Berks. SL6 1AP tel:(01628) 773353 fax:(01628) 773363 sales@pellfoldparthos.co.uk www.designs4space.com

Pembury Fencing Limited Leys Industrial Park Maidstone Road Paddock Wood Kent TN12 6PX tel:(01892) 838833 fax:(01892) 838834 pemburyfencing@tiscali.co.uk www.pemburygroup.co.uk

Pentas Office Furniture Asher Lane Ripley Derbys. DE5 3RE tel:(01773) 570 700 fax:(01773) 570 760

PERI Ltd Market Harborough Road Clifton Upon Dunsmore Rugby Warwickshire CV23 0AN tel:(01788) 861600 fax:(01788) 861610 info@peri.ltd.uk www.peri.ltd.uk

Permali Dehoplast Ltd Unit 4 Permali Park Bristol Road Gloucester Glos. GL1 5SR tel:(01452) 411607 fax:(01452) 411617 simon@permalidehoplast.co.uk www.permalidehoplast.co.uk

Perman Briggs Ltd 224 Cheltenham Road Longlevens Gloucester GL2 0JW tel:(01452) 524192 fax:(01452) 309879

Permanoid Ltd Hulme Hall Road Manchester M10 8HH tel:(0161) 205 6161 fax:(0161) 205 9325

PermaRock Products Ltd Jubilee Drive Loughborough Leics. LE11 5TW tel:(01509) 262924 fax:(01509) 230063 permarock@permarock.com www.permarock.com

Permaseal Roofing 4 Fivehouse Tonedale Wallington TA21 0AH tel:(01823) 662262 fax:(01823) 662262

Permclose Door Services Ltd Unit 2, Conqueror Court 505a Kingsland Road Spilsby Road, Harold Hill Romford, essex RM3 8SB tel:(020) 7254 9767 fax:(020) 7923 1447

Permic Emergency Lighting Ltd PO Box 3 Chesterfield Derbys. S40 1EX tel:(01246) 277776 fax:(01246) 275879

Peta Construction Co Ltd 55-69 Willenhall Road Darlaston W. Midlands WS10 8JG tel:(0121) 526 2983 fax:(0121) 526 3531

Petal Postforming Limited Business Link Rochdale Dane Road Rochdale Greater Manchester OL12 6XB 01706 750228 01344 893930 info@petal.co.uk

Peter Cox Ltd Heritage House 234 High Street Sutton Surrey SM1 1NX tel:(020) 8661 6600 fax:(020) 8642 0677 petercox@ecolab.com www.petercox.co.uk

Petit Roque Ltd 5a New Road Croxley Green Rickmansworth Herts. WD3 3EJ tel:(01923) 799291 fax:(01923) 896728

Petrochem Carless Cedar Court Guildford Road Leathead Surrey KT22 9RX tel:(01708) 716600 fax:(01708) 716716

PFP Electrical Products Ltd Fortnum Close Mackadown Lane Kitts Green Birmingham B33 0LB tel:(0121) 783 7161 fax:(0121) 783 5717 slaes@pfp-elec.co.uk www.pfp-elec.co.uk

PH Products Ltd Baird Road Willowbrook Corby Northants NN17 5ZA tel:(01933) 402212 fax:(01933) 679478 sales@ph-products.com www.ph-products.com

PHI Group Ltd Harcourt House 13 Royal Crescent Cheltenham Gloucestershire GL50 3DA tel:(0870) 333 4120 fax:(0870) 333 4121 info@phigroup.co.uk www.phigroup.co.uk

Philips Lighting The Philips Centre 420-430 London Road Croydon CR9 3QR tel:(020) 8665 6655 fax:(020) 8781 8929 www.lighting.philips.com

Philips Lighting Solutions Philips Centre Guildford Business Park Guildford Surrey CR9 3QR tel:(020) 8751 6514 fax:(020) 8890 438 mail@ecs-control.co.uk www.ecs-control.co.uk

Phillip Ader Ltd 9 Green Lane Shepperton Middx. TW17 8DP tel:(01932) 221474 fax:(01932) 224789 www.seranet.com/philliader/

Phoenix Engineering Co Ltd, The Phoenix Works Combe Street Chard Somerset TA20 1JE tel:(01460) 63531 fax:(01460) 67388 sales@phoenixeng.co.uk www.phoenixeng.co.uk

Phoenix Metal Products Ltd Stanmore Industrial Estate Bridgnorth Shropshire WV15 5JA tel:(01746) 765444 fax:(01746) 765440 www.phoenix-metal.com

Phoenix Scales Ltd 7 Church Lane West Bromwich W. Midlands B71 1DD tel:(0121) 553 1855 fax:(0121) 553 0494 sales@salter-wtx.co.uk www.salter-wtx.co.uk

Phoenix Timber Engineering Ltd 44 Ferry Lane Rainham Essex RM13 9DD tel:(01402) 755311 fax:(01402) 750122

Photain Controls Plc Rudford Estate Ford Aerodrome Arundel W. Sussex BN18 0BE tel:(01903) 721531 fax:(01903) 726795 enquiry@photaincontrols.plc.uk www.photaincontrols.plc.uk

Photo Fabrications Services Ltd 14 Cromwell Road St Neots, Huntingdon Cambs. PE19 2HP tel:(01480) 475831 fax:(01480) 475801

PhotoScan Ltd Dolphin Estate Windmill Road Sunbury on Thames Middx. TW16 7HG tel:(01932) 789741 fax:(01932) 787067

Pickard Group Fagley Lane Eccleshill Bradford W.Yorks BD2 3NT tel:(01274) 637307 fax:(01274) 626146 mick@thepickardgroup www.thepickardgroup.co.uk

Pickerings Ltd Globe Elevator Works PO Box 19 Stoke-on-Tees Cleveland TS20 2AD tel:(01642) 607161 fax:(01642) 677638 info@pickering.co.uk www.pickerings.co.uk

Pickersgill Kaye Ltd Pepper Road Leeds LS10 2PP tel:(0113) 277 5531 fax:(0113) 276 0221 bglsales@pkaye.co.uk www.pkaye.co.uk

Pifco Ltd Failsworth Manchester M35 0HS tel:(0161) 681 8321 fax:(0161) 682 1708

Pike Signals Ltd Equipment Works Alma Street Birmingham West Midlands B19 2RS tel:(+44 0121) 359 4034 fax:(+44 0121) 333 3167 enquiries@pikesignals.com www.pikesignals.com

Pilkington Birmingham Nechells Park Road Nechells Birmingham B7 5NQ tel:(0121) 326 5300 fax:(0121) 328 4277 john.hawkins@pilkington.com www.Pilkington.com

Pilkington Plyglass Cotes Park Somercotes Derbys. DE55 4PL tel:(01773) 520000 fax:(01773) 520052 garry.smith@pilkington.com www.pilkington.com

Pilkington UK Ltd Prescot Road St Helens Merseyside WA10 3TT tel:(01744) 692000 fax:(01744) 613049 stephen.lipscombe@pilkington.com www.pilkington.co.uk

Pilkington's Tiles Ltd PO Box 4 Clifton Junction Manchester M27 8LP tel:(0161) 727 1127 fax:(0161) 727 1066

Pillinger G C & Co (Engineers) Ltd Pillinger House 4 Wells Place Merstham,Redhill Surrey RH1 3DR tel:(01737) 733456 fax:(01737) 645036 sales@pillingerengineers.co.uk

Pims Pumps Ltd 22 Invincible Rd Ind Est Farnborough Hampshire GU14 7QU tel:(01252) 513366 fax:(01252) 516404

Pioneer Concrete Holdings PLC Pioneer House 56-60 Northolt Road South Harrow HA2 0EY tel:(020) 8423 3066 fax:(020) 8864 1975

Piper Double Glazing Ltd Hamelin Court 140 Newington Road Ramsgate Kent CT12 6PP tel:(01843) 850500 fax:(01843) 852626 piper@piperwindows.co.uk www.piperwindows.com

Pitchmastic Pmb Ltd Royds Works Attercliffe Road Sheffield South Yorks S4 7WZ tel:(0114) 2700100 fax:(0114) 2768782 buildingproducts@dewpitcnmastic.co.uk www.dewpitchmastic.co.uk

Pittsburg Corning (UK) Ltd Southcourt 29 South Street Reading Berks. RG1 4QU tel:(01189) 500655 fax:(01189) 509019 info@foamglas.co.uk www.Foamglas.co.uk

PJP plc Dixon Hill Road Wellham Green Hatfield Herts. AL9 7JE tel:(01707) 266726 fax:(01707) 263614

Pland Stainless Ltd Lower Wortley Ring Road Leeds W. Yorks. LS12 6AA tel:(0113) 263 4184 fax:(0113) 231 0560 sales@plandstainless.co.uk www.plandstainless.co.uk

Planet Architectural Glazing Ltd Stansdale House Cliftonville Road Northampton NN1 5BU tel:(01604) 29721 fax:(01604) 29726

Planet Communications Group Ltd Partland Tower Partland Street Manchester Lancashire M1 3LD tel:(01753) 807777 fax:(01753) 807700 info@planet.co.uk www.planet.co.uk

Plansee Tizit (UK) Ltd Grappenhall Warrington Cheshire WA4 3JX tel:(01925) 261161 fax:(01925) 267933

Plascoat Systems Ltd Trading Estate Farnham Surrey GU90 0NY tel:(01252) 733777 fax:(01252) 721250 sales@plascoat.com www.plascoat.com

Plastal (SBP Ltd) Alders Way Paignton Devon TQ4 7QE tel:(01803) 697111 fax:(01803) 559619 sales@plastal.co.uk www.plastal.co.uk

Plastic Coatings Ltd Woodbridge Meadow Guildford Surrey GU1 1BG tel:(01483) 531155 fax:(01483) 533534 enquiries@plastic-coatings.com www.plasticcoatings.co.uk

Plastic Construction Fabrications Ltd Evelyn Road Sparkhill Birmingham B11 3JJ tel:(0121) 773 4951 fax:(0121) 772 3588

Plastic Extruders Ltd Russell Gardens Wickford Essex SS11 8DN tel:(01268) 735231 fax:(01268) 560027 sales@plastex.co.uk www.plastex.co.uk

Plasticable Ltd Unit 3 Riverwey Ind Est Newmen Lane Alton Hampshire GU34 2QL tel:(01252) 541385 fax:(01252) 373816 sales@plasticable.co.uk www.plasticable.co.uk

Plastics Design & Fabrications (Glasgow) Ltd Plade Works Gladstone Avenue Barrhead Glasgow G78 1QT tel:(0141) 881 8241 fax:(0141) 881 8143

Platform Lift Company Ltd, The 19 Walworth Enterprise Centre Dukes Close West Way Andover ST10 5AP tel:(01264) 369 469 fax:(01264) 369 669 info@platformlift.co.uk

Platonoff & Harris Plc 206 Mill Studio Crane Mead Ware Hertfordshire SG12 9PY tel:(01920) 444255 fax:(01920) 487673 tony.ph@lineone.net

PLC Hunwick Ltd Harrision Works Kings Road Halstead Essex CO9 1HD tel:(01787) 474547 fax:(01787) 475741

Pollards Fyrespan Haslemere Business Centre Lincoln Way Enfield Middlesex EN1 1AY tel:(020) 8443 5511 fax:(020) 8443 3804

Polycell Products Ltd Wexham Road Slough Berkshire SL2 5DS tel:(01753) 550000 fax:(01753) 578218 www.polycell.co.uk

Polyflor Ltd PO Box 3 Radcliffe New Road Whitefield Manchester M45 7NR tel:(0161) 767 1122 fax:(0161) 767 1128 info@polyflor.com www.polyflor.com

Polypipe Bathroom & Kitchen Products Ltd Warmsworth Halt Industrial Estate Warmsworth, Doncaster S. Yorks. DN4 9SL tel:(01302) 310666 fax:(01302) 856421 sales@polypipe.co.uk www.polypipe.com/bk

Polypipe Civils Ltd Union Works Bishop Meadow Road Loughborough Leics. LE11 5RE tel:(01509) 615100 fax:(01509) 265945 enquiries@polypipecivils.co.uk www.polypipecivils.co.uk

Porn Dunwoody ERS (Lifts & Escalators) Ltd Unit 6, Bermondsey Trading Estate New Rotherhithe Street London SE16 3LL tel:(020) 79199800 fax:(020) 79199803 sales@pders-lifts.co.uk

Portacel Winnall Valley Road Winchester Hampshire SO23 0LL tel:(01962) 705200 fax:(01962) 866084 portacel_sales@tmproducts.com www.portacel.co.uk

Portakabin Ltd Huntington York YO32 9PT tel:(01904) 611655 fax:(01904) 611644 solutions@portakabin.com www.portakabin.com

Portasilo Ltd New Lane Huntington York YO32 9PR tel:(01904) 624872 fax:(01904) 611760 bulk@portasilo.co.uk www.portasilo.co.uk

Portia Engineering Ltd PO Box 9 Star Lane Ipswich IP4 1JJ tel:(01473) 252334 fax:(01473) 233863 portia@btinternet.com

Poselco Lighting 1 The Metropolitan Centre Bristol Road Greenford Middx. UB6 8UW tel:(020) 8813 0101 fax:(020) 8813 0099

Posners the floorstore 35a-37 Fairfax Road Swiss Cottage London NW6 4EW 020 7625 8899 020 7625 8866 sales@posnersfloorstore.co.uk www.posnersfloorstore.co.uk

Potter & Soar Ltd Beaumont Road Banbury Oxon. OX16 1SD tel:(01295) 253344 fax:(01295) 272132 potter.soar@btinternet.com www.wiremesh.co.uk

Potters Insulations (Sales) Ltd Tameside Mills Park Road Dukinfield Cheshire SK16 5LS tel:(0161) 308 3535 fax:(0161) 343 1706 sales@potters-insulations.co.uk

Pouliot Designs by Floralsilk Ltd Unit 6, Marley Industrial Estate Southam Road Banbury Oxon. OX16 2RB tel:(01295) 270848 fax:(01295) 270965 sales@florasilk.com

Power Breaker Temple Fields, South Rd Harlow Essex CM20 2BG tel:(01279) 434561 fax:(01279) 635285 gbe@greenbrook.co.uk

Power Plastics Ltd Station Road Thirsk Yorks. YO7 1PZ tel:(01845) 525503 fax:(01845) 525485 info@powerplastics.co.uk www.powerplastics.com

Precast Flooring Federation 60 Charles Street Leicester Leicestershire LE1 1FB 0116 253 6161 0116 251 4568 info@precastfloors.info www.precastfloors.info

Precolor Sales Ltd Newport Road Market Drayton Shropshire TF9 2AA tel:(01630) 657281 fax:(01630) 655545 precolor@aol.com www.precolor.co.uk

Preformed Components Ltd Davis Road Chessington Surrey KT9 1TU tel:(020) 8391 0533 fax:(020) 8391 2723 sales@pfc-corofil.co.uk www.pfc-corofil.co.uk

Premdor Gemini House Hargreaves Road Swindon Wilts. SN25 5AJ 0870 990 7998 fax:(01793) 708280 ukmarketing@premdor.com www.premdor.co.uk

Premier Group Catwick Lane Brandsburton Nr Driffield East Yorkshire YO25 8RW tel:(01964) 542131 fax:(01964) 542971 sales@premier.group.net www.premier-group.net

President Blinds Ltd Unit 13 Forest Hill Buisness Centre Clyde Vale Forest Hill London SE23 3JF tel:(020) 8699 8885 fax:(020) 8699 8005 president@flyscreens-uk.co.uk www.flyscreens-uk.co.uk

President Office Furniture Ltd The Arenson Centre Arenson Way Dunstable Beds. LU5 5UL tel:(01582) 678200 fax:(01582) 678222 enquiries@president.co.uk www.president.co.uk

Pressalit Care Ltd Riverside Business Park Dansk Way, Leeds Road Ilkley W. Yorks. LS29 8JT tel:(01943) 607651 fax:(01943) 607214 uk@pressalitcare.com www.pressalitcare.com

Pressmain 130 Princess Road Manchester M16 7BY tel:(01422) 349560 fax:(01274) 483459

Presto Engineers Cutting Tools Ltd Penistone Road Sheffield S6 2FN tel:(01742) 349361 fax:(01742) 347446

Preston & Thomas Ltd Woodville Engineering Works Heron Road, off Greenway Road Rumney Cardiff CF3 8YF tel:(029) 20 793331 fax:(029) 20 779195

Prestoplan Four Oaks Road Walton Summit Centre Preston PR5 8AP tel:(01772) 627373 fax:(01772) 627575

Prestoplan Purposebuilt Ltd Spring Lane South Malvern Link Worcester WR14 1AQ tel:(01684) 892602 fax:(01684) 892801

Prestressed Concrete Association 60 Charles Street Leicester Leicestershire LE1 1FB 0116 253 6161 0116 251 4568 pca@britishprecast.org www.bridgebeanns.org.uk

Preussag Fire Protection Ltd Field Way Greenford Middx. UB6 8UZ tel:(020) 8832 2000 fax:(020) 8832 2200

Prima Security & Fencing Products Redwell Wood Farm Ridge Hill Potters Bar Herts. EN6 3NA 01707 663 400 01707 661112 sales@sampsonfencing.co.uk www.sampsonfencing.co.uk

Prima Systems (South East) Ltd The Old Malt House Easole Street Nonington, Dover Kent CT15 4HF tel:(01304) 842888 fax:(01304) 842840

Priory Castor & Engineering Co Ltd Aston Hall Road Aston Birmingham B6 7LA tel:(0121) 327 0832 fax:(0121) 322 2123

Priory Shutter & Door Co Ltd 89-91 Rolfe Street Smethwick Warley W. Midlands B66 2AY tel:(0121) 558 6406 fax:(0121) 558 7140

Prismo Ltd Gleneagles Court Brighton Road Crawley W. Sussex RH10 6UE tel:(01293) 530356 fax:(01293) 530362 roadsafe@prismo.co.uk www.prismo.co.uk

Pro Display Display Technologies Centre 3 Apex Business Centre Apex Way Leeds LS11 5LN tel: (0113) 2420707 fax: (0113) 245757 sales@prodisplay.co.uk www.prodisplay.co.uk

Procter Cast stone & Concrete Products Newhold Aberford Rd Garforth, Leeds Yorkshire LS25 2HG tel:(01132) 863 329 fax:(01132) 867376 sales@proctor-garforth.co.uk www.caststoneuk.co.uk

Procter Bros Limited 4 Beaconsfield Court Garfirth Leeds W. Yorks. LS25 1QH tel:(0113) 287 2777 fax:(0113) 287 1177 sales@procters.demon.co.uk www.fencing-systems.co.uk

Procter Johnson & Co Ltd Excelsior Works Castle Park, Evans Street Flint Clwyd CH6 5NT tel:(01352) 732157 fax:(01352) 735530

Producta Ltd Mount Pleasant Stillingfleet Yorks. YO4 6HS tel:(01904) 728453 fax:(01904) 728140

Profab Access Ltd Unit 52 Fourways Carlyon Road Ind Est Atherstone Warwickshire CV9 1LH 01827 718222 01827 721092 enquiries@profilex.com www.profilex.com

Pro-Fence (Midlands) 3 Barley Croft Perton Wolverhampton W. Midlands WV6 7XX tel:(01902) 894747 fax:(01902) 897007

Profile 22 Systems Ltd Stafford Park 6 Telford Shropshire TF3 3AT tel:(01952) 290910 fax:(01952) 290460

Profile Lighting Services Ltd Links Business Centre Raynham Road Bishops Stortford Herts. CM23 5NZ tel:(01279) 757595 fax:(01279) 755599 mailbox@profile-lighting.co.uk www.profile-lighting.co.uk

Profolia Ltd Unit 3, Plymouth Avenue Brookhill Industrial Estate Pinxton Notts. NG16 6NS tel:(01773) 510505 fax:(01773) 510534

Progress Furnishings Systems Ltd 14 Sextant Park Neptune Close, Medway City Estate Rochester Kent ME2 4LU tel:(01634) 290988 fax:(01634) 291028

Progressive Product Developments Ltd 24 Beacon Bottom Swanwick South Hampton Hampshire SO31 7GQ tel:(01489) 576787 fax:(01489) 578463 sales@ppd-ltd.com www.ppd-ltd.com

Project Aluminium Ltd 418-420 Limpsfield Road Warlington Surrey CR6 9LA tel:(01883) 624004 fax:(01883) 627201 lesley@projects.co.uk

Project Office Furniture PLC Hamlet Green Haverhill Suffolk CB9 8QJ tel:(01440) 705411 fax:(01440) 703376 enquiries@project.co.uk www.project.co.uk

Promat Fire Protection Ltd Victoria Works, Bonsall Street Mill Hill Blackburn Lancs BB2 4DD tel:(01254) 267267 fax:(01254) 680521 promat@promat.co.uk www.promat.co.uk

Promat UK Ltd Iver Lane Uxbridge Middx UB8 2JQ tel:(01895) 463000 fax:(01895) 259262 www.promat-international.com
Property Mechanical Products Ltd Stanton Square Stanton Way Sydenham London SE26 5AB tel:(020) 8676 0911 fax:(020) 8659 1017 info@pmp-walney.co.uk
Protecter Lamp & Lighting Co Ltd Lansdowne Road Eccles Nr. Manchester M30 9PH tel:(0161) 789 5680 fax:(0161) 787 8257
Protim Services Ltd Unit B Mill Mead Stains Middlesex TW18 4UQ tel:(020) 8606 1000 fax:(020) 8569 1743
Protim Solignum Ltd Fieldhouse Lane Marlow Bucks. SL7 1LS tel:(01628) 486644 fax:(01628) 476757
Provend Services Ltd 19 Aintree Road Perivale, Greenford Middx. UB6 7LG tel:(020) 8998 2828 fax:(020) 8998 0704
Provex Products Ltd Unit 1-2 Redwood House Ordland Trading Est Toddington Cheltenham GL54 5EB tel:(01684) 297570 fax:(01684) 292874
Pryorsign Field View Brinsworth Lane Brinsworth, Rotherham S. Yorks. S60 5DG tel:(01709) 839559 fax:(01709) 837659 phil@pryorsign.com www.buyersguide.co.uk/document/pryor-sign/
Pulsford Associates Ltd, John Unit 4 Sphere Industrial Estate Campfield Road St Albans Herts. AL1 5HT tel:(01727) 840800 fax:(01727) 840083 info@jpa-furniture.com www.jpa-furniture.com
Pump International Ltd Trevoole Praze Camborne Cornwall TR14 0PJ tel:(01209) 831937 fax:(01209) 831939 sales@pumpinternational.com www.pumpinternational.com
Pump Technical Services Ltd Beco Works, Cricket Lane off Kent House Lane Beckenham Kent BR3 1LA tel:(020) 8778 4271 fax:(020) 8659 3576 sales@ptsjung.co.uk www.pts-jung.co.uk
Purdie Elcock Ltd Frankfort Street Works Birmingham B19 2YL tel:(0121) 359 5898 fax:(0121) 359 5510
Purewell Timber Works Stony Lane Christchurch Dorset BH23 1EY tel:(01202) 484422 fax:(01202) 490151
Purified Air Ltd Lyon House Lyon Road Romford Essex RM1 2BG tel:(01708) 755414 fax:(01708) 721488 enq@purifieldair.co.uk www.purifield.co.uk
Purimachos Ltd 14 Waterloo Road St Phillips Bristol BS2 0PG tel:(0117) 995 4361 fax:(0117) 935 0844
Pyramid Building Systems Cranswick Ind. Estate Cranswick Nr Driffield East Yorks Y025 9PF tel:(01377) 275100 fax:(01377)275101 info@modularuk.com www.modularuk.com
Pyramid Plastics Ltd Britannia Works Beaumont Street Gainsborough Lincs. DN21 2EN tel:(01427) 677990 fax:(01427) 612204
Q.M.P. Timmis Road Lye, Stourbridge W. Midlands DY9 7BQ tel:(01384) 424181 fax:(01384) 892042 qmp@qmpgroup.com
QualCraft Ltd Presteigne Industrial Estate Presteigne Powys LD8 2UF tel:(01544) 260 033 fax:(01544) 260 044
Qualitas Bathrooms Hartshorne Road Woodville Swadlincote Derbys. DE11 7JD tel:(01283) 550550 fax:(01283) 550314
Quantum Profile Systems Ltd Salmon Fields Royton, Oldham Lancs. OL2 6JG tel:(0161) 627 4222 fax:(0161) 627 4333 qps@quantum-ps.co.uk www.quantum-ps.co.uk
Quarry Products Association 156 Buckingham Palace Rd London SW1W 9TR 020 7730 8194 020 7730 4355 info@qpa.org www.qpa.org
Quartet-GBC UK Ltd Rutherford Road Basingstoke Hampshire RG24 8PD tel:(01256) 842828 fax:(01256) 476001 sales@quartetmfg.co.uk
Quelfire PO Box 35 Altrincham Cheshire WA14 5QA tel:(0161) 928 7308
Quentplass Ltd Thorp Arch Trading Estate Wetherby W. Yorks. LS23 7BZ tel:(01937) 843388 fax:(01937) 541458
Quiligotti Access Flooring Limited PO Box 4 Clifton Junction Manchester M27 8LP tel:(0161) 727 1070 fax:(0161) 727 1080 quiligotti.access@pilkingtons.com www.pilkingtons.com
Quinton & Kaines (Holdings) Ltd Creeting Road Stowmarket Suffolk IP14 5AS tel:(01449) 612145 fax:(01449) 677604 sales@quintonkaines.co.uk www.bbaduiven.demon.nl
Quip Lighting, Div of Bruce & Macintyre Ltd 71 Tenison Road Cambridge Cambs. CB1 2DG tel:(01223) 321277 fax:(01223) 321277 ygj36@dial.pipex.com
R & C Plant Sales (Scotland) Carseview Road Forfar Tayside DD8 3BT tel:(01307) 468200 fax:(01307) 467770
R A International Kingshill Business Park Darlaston Road Wednesbury W. Midlands WS10 7SH tel:(0121) 556 4271 fax:(0121) 502 2562
R.M. Easdale & Co Ltd 67 Washington Street Glasgow G3 8BB tel:(0141) 204 2708 fax:(0141) 204 3159 david@rmeasdale.com www.rmeasdale.com
Rabone Chesterman Ltd Woodside Sheffield S3 9PD tel:(01742) 768888 fax:(01742) 739038
Race Furniture Ltd Bourton Industrial Park Bourton-on-the-Water Glos. GL54 2HQ tel:(01451) 821446 fax:(01451) 821686 sales@racefurniture.com www.racefurniture.com
Rackline Systems Storage Ltd Oaktree Lane Talke Newcastle-under-Lyme Staffs. ST7 1RX tel:(01782) 777666 fax:(01782) 777444 now@rackline.com www.rackline.com
Rada Lighting Unit 30, Wenta Business Centre Colne Way Watford Herts WD24 7ND tel:(01923) 211800 fax:(01923)211688
Radcliffe & Sons (London) Ventura House 5-7 St Leonards Road Horsham W. Sussex RH13 6EH tel:(01403) 251444 fax:(01403) 217174
Radflex Contract Services Ltd Unit 35, Wilks Avenue Questor Dartford Kent DA1 1JS tel:(01322) 276363 fax:(01322) 270606 expjoint@radflex.co.uk www.radflex.co.uk
Radial & Axial Fans Radial House Coldharbour Lane Harpenden Herts. AL5 4UN tel:(01582) 460624 fax:(01582) 460629
Radiant Services Ltd Barrett House 111 Millfields Road Ettingshall Wolverhampton WV4 6JQ tel:(01902) 494266 fax:(01902) 494153 radser@btconnect.co.uk www.radiantservices.co.uk
Radiodetection Ltd Western Drive Bristol BS14 0AZ tel:(0117) 976 7776 fax:(0117) 976 7775 sales.uk@radiodetection.com www.radiodetection.co.uk/
Radius Lighting Ltd 15 South Street Havant Hampshire PO9 1BU tel:(023) 92455525 fax:(023) 92451130
Radway, Door and Windows Ltd. Merse Road North Moons Moat Redditch Worcs. B98 9HL tel:(01527) 585558 fax:(01527) 63965 sales@radways.co.uk www.radways.co.uk
RAF Quedgley Unit 5 Site 1 Quedgley Gloucester GL2 5ZZ tel:(01452) 417575 fax:(01452) 524446
Railex Systems Ltd The Wilson Building 1 Curtain Road London EC2A 3JX tel:(020) 7377 1777 fax:(020) 7247 6181 hwilson@railex.co.uk www.railex.co.uk
Railex Systems Ltd, Elite Division Elite Works Station Road Manningtree Essex CO11 1DZ tel:(01206) 392171 fax:(01206) 391465
Rainham Steel Co Ltd Kathryn House Manor Way Rainham Essex RM13 8RE tel:(01708) 522311 fax:(01708) 559024 sales@rainham-steel.co.uk www.rainham-steel.co.uk
Raised Floor Systems Ltd Peak House Works Road Letchworth Garden City Hertfordshire SG6 1GB 01462 685898 01462 685 353 rfs@raisedfloorsystems.co.uk www.raisefloorsystems.co.uk.
Ramsay & Sons (Forfar) Ltd 61 West High Street Forfar Angus DD8 1BG tel:(01307) 462255 fax:(01307) 466956 enquiries@ramsayladders.co.uk www.ramsayladders.co.uk
Ramset Fasteners Ltd Ramset House Galleymead Road Colnbrook Slough SL3 0EN tel:(01753) 682277 fax:(01753) 680187 pwalsh@ramsetuk.com www.ramsetuk.com
Ranalah Gates Ltd Gloucester Road Industrial Estate Malmesbury Wilts. SN16 9JT tel:(01666) 823001 fax:(01666) 824175
Randalls Fabrications Ltd Hoyle Mill Road Kinsley Pontefract W. Yorks. WF9 5JB tel:(01977) 615132 fax:(01977) 610059 sales@randallsfabrications.co.uk www.randallsfabrications.co.uk
Range Cylinders Tadman Street Wakefield West Yorkshire WF1 5QU tel:(01924) 376026 fax:(01924) 385015 info@range-cylinders.co.uk www.range-cylinders.co.uk
Rangemaster Meadow Lane Long Eaton Notts. NG10 2AT tel:(0115) 946 4000 fax:(0115) 946 0374 sales@rangemaster.co.uk www.rangemaster.co.uk
RAP Group Ltd RAP House Harrison Steet Briercliffe, Burnley Lancs. BB10 2HP tel:(01282) 410678 fax:(01282) 451160 harwood.welpac.@virgin.net
Rapaway Energy Ltd 35 Park Avenue Solihull W. Midlands B91 3EJ tel:(0121) 246 0441 fax:(0121) 246 0442 rapaway@btinternet
Rapid Metal Developments Stubbers Green Road Aldridge Walsall Staffs. WS9 8BW tel:(01922) 743743 fax:(01922) 743400
Rare Stone Group, The 184, Nottingham Road Mansfield Notts NG18 5AP tel:(01623) 623092 fax:(01623) 622509 enquiries@rarestonegroup.co.uk www.rarestonegroup.co.uk
Rawson Carpets Ltd Castle Bank Mills Portobello Road Wakefield W. Yorks. WF1 5PS tel:(01924) 382860 fax:(01924) 366204 sales@rawsoncarpets.co.uk www.rawsoncarpets.co.uk
Rawson Fillings Ltd Castle Bank Mills Portobello Road Wakefield West Yorkshire WF1 5PS Tel:(01924) 373421 Fax:(01924) 290334 sales@rawsonfillings.co.uk
Rawtex Ltd Castle Bank Mills Portobella Road Wakefield West Yorkshirre WF1 5PS Tel:(01924) 373421 Fax:(01924) 290334
R-B International Ltd Beevor Street Lincoln Lincs. LN6 7DJ tel:(01522) 525261 fax:(01522) 512230 sales@rbcranes.com www.rbcranes.com
RB Farquhar Manufacturing Ltd Deveronside Works Huntly Aberdeenshire AB54 4PS tel:(01466) 793231 fax:(01466) 793098 manufacturing.info@rbfarquhar.co.uk www.rbfarquhar.co.uk
RB UK Ltd Element House Napier Road Bedford Beds MK41 0QS tel:(01234) 272717 fax:(01234) 270202 info@rbuk.co.uk www.rbuk.co.uk
RCC, Division of Tarmac Precast Concrete Ltd Barholm Road Tallington Stamford Lincolnshire PE9 4RL tel:(01778) 344460 fax:(01778) 345949 rcc@tarmac.co.uk www.tarmacprecast.com
RDP-Howden Ltd Althorpe Street Leamington Spa Warwks. CV31 2BA tel:(01926) 427782 fax:(01926) 429873
Rea Metal Windows Ltd 126-136 Green Lane Liverpool L13 7ED tel:(0151) 228 6373 fax:(0151) 254 1828 all@reametal.co.uk www.reametal.co.uk
Readymix Limited CEMEX House Coldharbour Lane Thorpe, Egham Surrey TW20 8TD tel:(01932) 583600 fax:(01932) 583611 info.readymix@rmc.co.uk www.readymix.co.uk
Real Door Company The Unit 5 Cadwell Lane Hitchin Herts. SG4 OSA tel:(01462) 451230 fax:(01462) 448459 sales@realdoor.co.uk www.realdoor.co.uk
Realstone Ltd Wingerworth, Chesterfield Derbys. S42 6RG tel:(01246) 270244 fax:(01246) 220095 sales@realstone.co.uk www.realstone.co.uk
Reclaim.UK.Com Maywood House Woodford New Road Woodford Green Essex IG8 9DX tel:(020) 8559 1938 fax:(020) 8559 1932 sarah@reclaim.uk.com www.reclaim.uk.com
Record & Parkes Ltd Crown Wharf Iron Works Dace Road Old Ford London E3 2NL tel:(020) 8985 5383 fax:(020) 8986 8212
Record Electrical Atlantic Street Altrincham Cheshire WA14 5DB tel:(0161) 928 6211 fax:(0161) 926 9750
Record Playground Equipment Ltd Waterfront Complex, Shipyard Industrial Estate Carr Street Selby N. Yorks. YO8 8AP tel:(01757) 703620 fax:(01757) 705158 sales@records.co.uk www.recordplaygrounds.co.uk
Record Power Ltd Parkway Works Sheffield S9 3BL tel:(01742) 434370 fax:(01742) 617141
Record Tools Ltd Parkway Works Sheffield S9 3BL tel:(01742) 449066 fax:(01742) 434302
Red Bank Maufacturing Co Ltd Atherstone Road Measham, Swandlincote Derbys. DE12 7EL tel:(01530) 270333 fax:(01530) 273667
Reddiglaze Ltd The Furlong Droitwich Worcs. WR9 9BG tel:(01905) 795432 fax:(01905) 795757 www.reddiglaze.co.uk
Redditch Plastic Products Pipers Road Park Farm Industrial Estate Redditch Worcs. B98 0HU tel:(01527) 528024 fax:(01527) 520236
Redifix Ltd Queenslie Industrial Estate 44 Coltness Lane Glasgow G33 4DR tel:(0141) 774 2020 fax:(0141) 774 3080 sales@redifix.co.uk www.redifix.co.uk
Rediweld Rubber & Plastics Ltd 6/9 Newman Lane Alton Hampshire GU34 2QR tel:(01420) 543007 fax:(01420) 544090 info@rediweld.co.uk www.rediweld.co.uk
Redman Fisher Engineering Birmingham New Road Tipton W. Midlands DY4 9AA tel:(01902) 880880 fax:(01902) 880446 flooring@redmanfisher.co.uk www.redmanfisher.co.uk
Reed Harris Riverside House 27 Carnwath Road Fulham/London SW6 3HR tel:(020) 7736 7511 fax:(020) 7736 2988 architectural@reed-harris.co.uk www/reedharris.co.uk
Reflecting Roadstuds Ltd Boothtown Halifax W. Yorks. HX3 6TR tel:(01422) 360208 fax:(01422) 349075
Reflex-Rol (UK) Ryeford Hall Ryeford Nr Ross-on-Wye Herefordshire HR9 7PU tel:(01989) 750704 fax:(01989) 750768 reflexrol@btinternet.com www.reflex-rol.co.uk
Rega Ductex Ltd Station Road Sandy Beds. SG19 1BH tel:(01767) 691291 fax:(01767) 692451 sales@rega-uk.com www.regaductex.co.uk
Regency Garage Door Services Ltd Unit 8, Central Trading Estate Signal Way Swindon Wilts. SN3 1PD tel:(01793) 611688 fax:(01793) 495144
Regency Swimming Pools Ltd Regency House 88a Brick Kiln Street Graisley Wolverhampton WV3 0PU tel:(01902) 427709 fax:(01902) 422632 info@jwgswimming.co.uk www.jwgswimming.co.uk
Regon Ltd Station Lane Trading Estate Avenue Two Witney Oxon. OX8 4YE tel:(01993) 771441 fax:(01993) 774105 sales@regon.co.uk www.regon.co.uk
REHAU Ltd Hill Court Walford Ross-on-Wye Herefordshire HR9 5QN tel: (01989) 762600 fax:(01989) 762601 windows.enquiries@rehau.com www.rehau.co.uk
Reid John & Sons (Strucsteel) Ltd Strucsteel House 264-266 Reid Street Christchurch Dorset BH23 2BT tel:(01202) 483333 fax:(01202) 470103 sales@reidsteel.co.uk www.reidsteel.com
Reiner Fixing Devices (Ex: Hardo Fixing Devices) Hall Farm Church Lane North Ockendon, Upminster Essex RM14 3QH tel:(01708) 856601 fax:(01708) 852293 sales@reinerfixings.co.uk www.reinerfixings.co.uk
Relcross Ltd Hambleton Avenue Devizes Wilts. SN10 2RT tel:(01380) 729600 fax:(013080) 729888 sales@relcross.co.uk www.relcross.co.uk
Remploy Ltd Unit 2, Brocastle Avenue Waterton Industrial Estate Bridgend M. Glam. CF31 3YN tel:(01656) 653982 fax:(01656) 767499
Remtox Silexine 14 Spring Road Smethwick Warley W. Midlands B66 1PE tel:(0121) 525 2299 fax:(0121) 525 1740
Renderplas Ltd Stourport Road Bewdley Worcestershire DY12 1BD tel:(01299) 400340 fax:(01299) 400195 info@renderplas.co.uk
Rentokil Healthcare Service Felcourt East Grinstead W. Sussex RH19 2JY tel:(01342) 833022 fax:(01342) 326229
Rentokil Initial U.K. Ltd. Rentokil Initial Products Felcourt East Grinstead West Sussex RH19 2JY tel:(01342) 833022 fax:(01342) 326229 jrowe@rentokil.com
Rentokil Ltd Timber Preserving Products Division Felcourt East Grinstead W. Sussex RH19 2JY tel:(01342) 833022 fax:(01342) 326229
Resdev Ltd Pumaflor House Ainleys Industrial Estate Elland W. Yorks. HX5 9JP tel:(01422) 379131 fax:(01422) 370943 info@resdev.co.uk www.resdev.co.uk
Resiblock Ltd Resiblock House Archers Fields Close Basildon Essex SS13 1DW tel:(01268) 273344 fax:(01268) 273355 mail@resiblock.com www.resiblock.com
Rettig Heating Ltd Hartlebury Trading Estate Nr Kidderminster Worcs. DY10 4JB tel:(01299) 250700 fax:(01299) 251192 uk@rettigheating.com
Revol Products Ltd Samson Close Killingworth Newcastle-upon-Tyne NE12 0DZ tel:(0191) 268 4555 fax:(0191) 216 0004
Reynolds John & Son (Birmingham) Ltd Church Lane West Bromwich W. Midlands B71 1DJ tel:(0121) 553 2754/1287 fax:(0121) 500 5460
Reyven (Sportsfields) Ltd Greenhill Farm Tilsworth Leighton Buzzard Beds. LU7 9PU tel:(01525) 210714 fax:(01525) 211130 venn@reyven.fsnet.co.uk
Reznor UK Ltd Park Farm Road Park Farm Industrial Estate Folkestone Kent CT19 5DR tel:(01303) 259141 fax:(01303) 850002 sales@reznor.co.uk www.reznor.co.uk
RFA Group Ltd Bullhouse Works Bullhouse Bridge Millhouse Green, Sheffield S. Yorks. S36 9NF tel:(01226) 764425 fax:(01226) 767562
Rheinzink UK Cedar House Cedar Lane Frimley, Camberley Surrey GU16 5HY tel:(01276) 686725 fax:(01276) 64480
Rhodar Limited Astra Park Parkside Lane Leeds LS11 5SZ tel:(0113) 270 0775 fax:(0113) 270 4124 info@rhodar.co.uk www.Rhodar.co.uk
Rhone-Poulenc Chemicals Ltd Liverpool Road Barton Moss Site Eccles Manchester M30 7RT tel:(0161) 707 9779 fax:(0161) 910 8812
Rib (UK) Ltd Ellard House Baxter Road, Foxhill Industrial Estate Wadesley Bridge, Sheffield S. Yorks. S6 1JF tel:(0114) 285 2030 fax:(0114) 285 2060
Richard Hose Ltd Unit 7, Roman Way Centre Longridge Road Ribbleton, Preston Lancs. PR2 5BB tel:(01772) 651550 fax:(01772) 651325 richards.fire@btinternet.com www.richardsfire.co.uk
Richards H S Ltd King Street Smethwick Warley W. Midlands B66 2JW tel:(0121) 558 2261 fax:(01785) 840 603
Righton Ltd Righton House Brookvale Road Witton Birmingham B6 7EY tel:(0121) 356 1141 fax:(0121) 331 1347 righton@righton.co.uk www.righton.co.uk
Rigidal Industries Ltd. Blackpole Trading Estate West Worcester WR3 8ZJ tel:(01905) 750500 fax:(01905) 750555 sales@rigidal.co.uk www.rigidal-industries.com
Rigidal Systems LTD Unit 62 Blackpole Trading Estate West Worcester WR3 8ZJ tel:(01905) 750500 fax:(01905) 750555 sales@rigidal.co.uk www.rigidal.co.uk
Rigized Metals Ltd Aden Road Ponders End Enfield Middx. EN3 7SU tel:(020) 8804 0633 fax:(020) 8804 7275
Ring-Gard (UK) Ltd Unit 30, Navagation Drive Hurst Business Park Brierley Hill W. Midlands DY5 1UT tel:(01384) 74849 fax:(01384) 74348
Rite-Vent Ltd Crowrther Estate Washington Tyne & Wear NE38 0AQ tel:(0191) 416 1150 fax:(0191) 415 1263 sales@rite-vent.co.uk www.rite-vent.co.uk
Ritherdon & Co Ltd Lorne Street Darwen Lancs. BB3 1QW tel:(01254) 819100 fax:(01254) 819101 sales@ritherdon.co.uk www.ritherdon.co.uk
Rivermeade Signs Ltd Roslin Road South Acton Industrial Estate London W3 8BW tel:(020) 8896 6900 fax:(020) 8752 1691 sales@rivermeade.com www.rivermeade.com
Rixonway Kitchens Ltd Shaw Cross Busness Park Dewsbury West Yorkshire WF12 7RD tel:(01924) 431300 fax:(01924) 431301 info@rixonway.co.uk www.rixonway.co.uk

RNB Industrial Doors Services Unit 6, Davenport Centre Renwick Road Barking Essex IG11 0SP tel:(020) 8595 1242 fax:(020) 8595 3849

Roadways & Car Parks Ltd 174 Twickenham Road Isleworth Middx. TW7 7DW tel:(020) 8560 7211 fax:(020) 8560 1894

Robert Bosch Ltd PO Box 98, Broadwater Park North Orbital Road Denham, Uxbridge Middx. UB9 5HJ tel:(01895) 834466 fax:(01895) 838388

Roberts-Gordon Oxford Street Bilston W. Midlands WV14 7EG tel:(01902) 494425 fax:(01902) 403200 enquiry@combat.co.uk www.combat.co.uk

Robinson J J & Son Ltd Eden Lane Gainford Darlington DL2 3BN tel:(01325) 730241 fax:(01325) 730596

Robinson Willey Ltd Mill Lane Old Swan Liverpool L134AJ tel:(0151) 228 9111 fax:(0151) 228 6661 info@robinson-willey.com www.robinson-willey.com

Robot (UK) Ltd Wincanton House Wincanton Close Derby Derbys. DE24 8NB tel:(01332) 758000 fax:(01332) 758202

Rockfon Ltd Pencoed Bridgend County Borough CF35 6NY tel:(01656) 864696 fax:(01656) 864549 www.rockfon.co.uk

Rockwell Sheet Sales Ltd Rockwell House Birmingham Road Millisons Wood Coventry CV5 9AZ tel:(01676) 523386 fax:(01676) 523630 info@rockwellsheet.com www.rockwellsheet.com

Rockwool Ltd Pencoed/Bridgend M. Glam. CF35 6NY tel:(01656) 862621 fax:(01656) 862302

Rogers Concrete Ltd Puddletown Road Wareham Dorset BH20 6AU tel:(01929) 462373 fax:(01929) 405326

Rohm & Hass (UK) Ltd Lennig House 2 Mason's Avenue Croydon Surrey CR9 3NB tel:(020) 8686 8844 fax:(020) 8686 8329

Rok-Crete Units Co Ltd 112 Oxford Road Clacton on Sea Essex CO15 3TN tel:(01255) 424884 fax:(01255) 221155

Rolflex Doors Ltd Bldg 63 The Pensnett Estate Kingswinford W. Midlands DY6 7PP tel:(01384) 401555 fax:(01384) 401556

Rollalong Ltd Woolsbridge Industrial Estate Three Legged Cross Wimborne Dorset BH21 6SF tel:(01202) 824541 fax:(01202) 826525 enquiries@rollalong.co.uk www.rollalong.co.uk

Rollins & Sons (London) Ltd Rollins House Murray Road Hertford Herts. SG14 1NR tel:(01992) 587555 fax:(01992) 500159

Rom Ltd Eastern Avenue Trent Valley Lichfield Staffs. WS13 6RN tel:(01543) 414111 fax:(01543) 421672 sales@rom.co.uk www.rom.co.uk

Roman Mosaic Contracts Ltd Bloomfield Road Tipton W. Midlands DY4 9ES tel:(0121) 557 2267 fax:(0121) 557 0975 sales@romanmosaic.co.uk www.romanmosaic.co.uk

Ronacrete Ltd Ronac House Selinas Lane Dagenham Essex RM8 1QL tel:(020) 8593 7621 fax:(020) 8595 6969 sales@ronacrete.co.uk www.ronacrete.co.uk

Roof Units Group Peartree House Pear Tree Lane Dudley W. Midlands DY2 0QU tel:(01384) 74062 fax:(01384) 70435 ru@airmovement.co.uk

Roofing Contractors (Cambridge) Ltd 5 Winship Road Milton Cambridge CB4 6BQ tel:(01223) 423059 fax:(01223) 420737

Roofrite (East Anglia) Ltd The Street Sheering Bishop's Stortford Herts. CM22 7LY tel:(01279) 734515 fax:(01279) 734568 canonburyasphalt@aol.com

Rosa Paula Kitchens Water Lane Trading Estate Storrington W. Sussex RH20 3DS tel:(01903) 746666 fax:(01903) 742410 info@paularose.com www.paularose.com

Rose Emergency Equipment Co Unit 1A, North Orbital Trading Estate Napsbury Lane St Albans Herts. AL1 1XB tel:(01727) 840311 fax:(01727) 844143

Rosengerns Tann Ltd Rosengerns House Whillbury Way Hitchen Herts. SG4 0TY tel:(01462) 471 200

Rotafix Resins Rotafix House Abercraf Swansea SA9 1UX tel:(01639) 730481 fax:(01639) 730858

Rother Co Ltd Meadow Bank Works Meadowbank Road Rotherham S. Yorks S61 2ND tel:(01709) 555416 fax:(01709) 564341 sales@rother.co.uk www.rother.co.uk

Rothley Burn Ltd Macrome Road Wolverhampton West Midlands WV6 9HG tel:(01902) 756461 fax:(01902) 745554 sales@rothley.com www.rothley.co.uk

Roto Frank Ltd Swift Point Rugby Warwks. CV21 1QH tel:(01788) 558600 fax:(01788) 558605 uk-sales@roto-frank.com www.rotofrank.com

Rouse Security Service 345-347 Harrow Road London W9 3RA tel:(020) 7286 1184 fax:(020) 7266 4146 info@rousesecurity.co.uk

Rovacabin Priest End Works Rycote Lane Thame Oxon OX9 2HD tel:(01844)267200 fax:(01844)267201 Rovasales@sgb.co.uk

Rowberry Technical Services 15 Copse Cross Street Ross-on-Wye Herefordshire HR9 5PD tel:(01989) 563850 fax:(01989) 563854

Royair Ltd Heathpark Honiton Devon EX14 8SE tel:(01404) 41651 fax:(01404) 46227 royairltd@aol.com www.royair.co.uk

Royal Dutch Sphinx Ltd Units 1-2 Pipers Court Thatcham Berks. RG19 4ER tel:(01635) 865475 fax:(01635) 861189

Royal Incorporation of Architects in Scotland 15 Rutland Square Edinburgh EH1 2BE tel:(0131) 229 7545 fax:(0131) 228 2188 rias@rias.org.uk www.rias.org.uk

Royal Institute of British Architects (RIBA) 66 Portland Place London W1B 1AD tel:(020) 7580 5533 fax:(020) 7255 1541 info@inst.riba.org www.architecture.com

Royal Institution of Chartered Surveyors (RICS) 12 Great George Street Parliament Square London SW1P 3AD tel:(0870) 333 1600 fax:(020) 7334 3811 contactrics@rics.org www.rics.org

Royal Town Planning Institute 26 Portland Place London W1N 9BE tel:(020) 7636 9107

Royde & Tucker Unit 15-16 High Cross Centre Fountayne Road Tottenham London N15 4QN tel:(020) 8801 7717 fax:(020) 8801 5747

Royston Lead Ltd Pogmoor Works Stocks Lane Barnsley S. Yorks. S75 2DS tel:(01226) 770110 fax:(01226) 730359 info@roystonlead.co.uk www.roystonlead.co.uk

Royston Steel Fencing Limited Tadlow Road Tadlow Royston Herts. SG8 0EP tel:(01767) 631721 fax:(01767) 631721

RPS Data Products (UK) Ltd Unit 2 The Old Bakery Complex South Road Reigate Surrey RH2 7LB tel: (01737) 221228 fax: (01737) 223206 sales@rpsdataproducts.co.uk www.rpsdataproducts.co.uk

RT Display Systems Ltd 212 New Kings Road London SW6 4NZ tel:(020) 7731 4181 fax:(020) 7731 1907

Ruabon Dennis Ltd Hafod Tileries Ruabon, Wrexham Clwyd LL14 6ET tel:(01978) 843484 fax:(01978) 843276

Ruberoid Architectural Patent Glazing/Cutain Wall Contractor St Mungo Street Bishopbriggs Glasgow G64 1QX tel:(0141) 772 1117 fax:(0141) 762 2217

Ruberoid Building Products Ltd 14 Tewin Road Welwyn Garden City Herts. AL7 1BP tel:(01707) 822222 fax:(01707) 375060

Rubert & Co Ltd Acru Works Demmings Road Cheadle Cheshire SK8 2PG tel:(0161) 428 6058 fax:(0161) 428 1146

Rudgwick Brickworks Co Ltd Lynwick Street Rudgwick Horsham W. Sussex RH12 3DH tel:(01403) 822212 fax:(01403) 823352

Rugby Cement Rugby Works Lawford Road Rugby Warwks. CV21 2UZ tel:(01788) 564345 fax:(01788) 564452 tsd@rugbycement.co.uk www.rugbycement.co.uk

Rural design and Building Association ATSS House Station Road East Stowmarket Suffolk IP14 1RQ 01449 676049 01449 770028 secretary@rdba.org.uk www.rdba.org.uk

Ruskin Air Management Wilson Industrial Estate South Street Whistable Kent CT5 3DU tel:(01227) 276100 fax:(01227) 264262 sales@actionair.co.uk www.actionair.co.uk

Rustin Allen Ltd. Mill House and Works Columbia Avenue Edgware Middx. HA8 5DQ tel:(020) 8951 4177 fax:(020) 8951 0801 extrusion@rustinallen.co.uk www.rustinallen.co.uk

Rustins Ltd Waterloo Road Cricklewood London NW2 7TX tel:(020) 8450 4666 fax:(020) 8452 2008 rustins@rustins.co.uk www.rustins.co.uk

Ruthin Precast Concrete Ltd 1st. Floor 11 Paul Street Taunton Somerset TA1 3PF tel:(01823) 274232 fax:(01823) 274231 enquiries@rpcltd.co.uk www.rpcltd.co.uk

Ryall & Edwards Limited Green Lane Sawmills Outwood Redhill Surrey RH1 5QP tel:(01342) 842288 fax:(01342) 843312

Rycroft Ltd Duncombe Road Bradford West Yorkshire BD8 9TB tel:(01274) 490911 fax:(01274) 498580 sales@rycroft.com www.rycroft.com

Ryton's Building Products Ltd Design House, Orion Way Kettering Business Park Kettering Northants NN15 6NL tel:(01536) 511874 fax:(01536) 310455 lit@rytons.com www.vents.co.uk

S & B Ltd Labtec Street Swinton Manchester M27 8SE tel:(0161) 793 9333 fax:(0161) 728 2233

S & P Coil Products Ltd SPC House Evington Valley Road Leicester LE5 5LU tel:(0116) 2490044 fax:(0116) 2490033 spc@spcoils.co.uk www.spcoils.co.uk

S.I. Concrete Systems Hayfield House Devonshire Street Chesterfield Derbyshire S41 7ST tel:(01246) 564200 fax:(01246) 564201 fibermesh@sind.co.uk www.fibermesh.com

Saacke Ltd Marshland Spur Farlington Portsmouth PO6 1RX tel:(023) 92383111 fax:(023) 92327120 admin@saacke.co.uk www.saacke.co.uk

Sabre Access Control Systems Ltd Unit 4, Lismirrane Industrial Park Elstree Road Elstree Herts. WD6 3EE tel:(020) 8953 6724 fax:(020) 8207 4575

Sadolin UK Ltd Sadolin House Meadow Lane St Ives Cambs. PE27 4UV tel:(01480) 496868 fax:(01480) 496801 woodcare@sis.akzonabel.com www.sadolin.co.uk

Safetell Ltd Unit 46, Fawkes Avenue Dartford Kent DA1 1JQ tel:(01322) 223233 fax:(01322) 277751 secure@safetell.co.uk www.safetell.co.uk

Safety Stairways Ltd 45 Owen Road Industrial Estate Owen Road Willenhall W. Midlands WV13 2PX tel:(0121) 526 3133 fax:(0121) 526 2833 sales@safety-stairways.com www.safety-stairways.com

Sage Brent Ltd Alperton Lane Wembley Middx. HA0 1JX tel:(020) 8998 4791 fax:(020) 8991 0673

Saint Gobain pipelines PLC Lows Lane Stanton-By-Dale Ilkeston Derbyshire DE7 4QU tel:(0115) 930 5000 fax:(0115) 932 9513 sales.valves@saint-gobain-pipelines.co.uk www.saint-gobain-pipelines.co.uk

Saint-Gobain Ecophon Ltd Old Brick Kiln Ramsdell Tadley Hants. RG26 5PP tel:(01256) 850977 fax:(01256) 850600 info@ecophon.co.uk www.ecophon.co.uk

Saint-Gobain Glass UK Ltd Waterside Drive Langley Business Park Langley, Slough Berks. SL3 6EZ tel:(01753) 731400 fax:(01753) 731501

Saint-Gobain Pipelines PLC Sinclair Works, PO Box 3 Ketley Telford Shropshire TF1 5AD tel:(01952) 262500 fax:(01952) 262555 innovationssoildrain.uk.pipelines@saint-gobain.com www.saint-gobain-pipelines.co.uk

Sala Ltd. Old Mill Road Portishead Bristol BS20 9BX tel:(01275) 846119 fax:(01275) 849114

Salamandre Davis Hunts Rise South Marston Park South Marston Swindon SN3 4RE tel:(01793) 828000 fax:(01793) 828597

Salex Interiors Ltd Bradbury Drive Springwood Industrial Estate Braintree Essex CM7 2ET tel:(01376) 555200 fax:(080) 555080

Samas Roneo Ltd Maiden Lane Crayford Kent DA1 4AX tel:(01322) 223477 fax:(01322) 222890 enquiries@samasromeo.com www.samasromeo.com

Sampson & Partners Fencing Aubrey Works, 15 Aubrey Avenue London Colney St.Albans Herts. AL2 1NE tel:(01727) 822222 fax:(01727) 826307 primasampson@compuserve.com www.sampsonandpartners.com

Sampson T F Ltd Creeting Road Stowmarket Suffolk IP14 5BA tel:(01449) 613535 fax:(01449) 678381 sales@t-f-sampson.co.uk www.sampsonsblinds.co.uk

Sampson Windows Ltd Maitland Road, Lion Barn Business Park Needham Market Ipswich Suffolk IP6 8NZ tel:(01449) 722922 fax:(01449) 722911

Samual Heath Cobden Works Leopold Street Birmingham B12 0UJ tel:(0121) 772 2303 fax:(0121) 772 3334 mail@samvel-heath.com www.samvel-heath.com

Sanderson Sanderson House Oxford Road Denham Bucks UB9 4DX tel:(01895) 830000 fax:(01895) 830031 contractsales@a-sanderson.co.uk www.sanderson-uk.com

Sandhurst MFG Co Roman Bank Cherry Holt Road Bourne Lincs. PE10 9LQ tel:(01778) 426426 fax:(01778) 393007 info@sandhurst-mfg.com www.sandhurst-mfg.com

Sandler Seating Ltd 58-64 Three Colts Lane London E2 6JR tel:(020) 7799 3000 fax:(020) 7729 2843 sales@sandlerseating.com www.sandlerseating.com

Sandoms (Batley) Ltd 491 Bradford Road Batley Yorks. WF17 8LQ tel:(01924) 474108 fax:(01924) 475409 sandoms@hotmail.com

Sandtoft Roof Tiles Ltd Sandtoft Doncaster South Yorkshire DN8 5SY tel:(01427) 871200 fax:(01427) 871222 info@sandtoft.co.uk www.sandtoft.com

Sangwin Concrete Products Dansom Lane Hull HU8 7LN tel:(01964) 622339 fax:(01964) 624787 info@sangwin.co.uk www.sangwin.co.uk

Sanhall Limited Victoria Stables South Road Bourne Lincs. PE10 9JZ tel:(01778) 426655 fax:(01778) 426899 sales@sanhall.co.uk

Sanitary Appliances Ltd 3 Sandiford Road Kimpton Road Industrial Estate Sutton Surrey SM3 9RN tel:(020) 8641 0310 fax:(020) 8641 6426

Sapa Building Systems (monarch and glostal) Alexandra Way Ashchurch Tewkesbury Glos. GL20 8NB tel:(01684) 297073 fax:(01684) 293904 info.buildingsystems.uk@sapagroup.com www.sapagroup.com/uk/buildingsystems.

Sarena MFG Ltd Sarena House Beechings Way Ind. Estate Gillingham Kent ME8 6PL tel:(01634) 370887 fax:(01634) 370915 info@sarena.co.uk www.sarena.co.uk

Sarnafil Ltd Robberds Way Bowthorpe Industrial Estate Norwich Norfolk NR5 9Jf tel:(01603) 748985 fax:(01603) 743054 roofing@sarnafil.co.uk www.sarnafill.co.uk

Sashless Window Co Ltd Standard Way Northallerton N. Yorks. DL6 2XA tel:(01609) 780202 fax:(01609) 779820

Satchwell Control Systems Ltd Farnham Road Slough Berkshire SL1 4UH tel:(01753) 611000 fax:(01753) 611001 www.ibs.invensys.com

Satec Ltd The Street Engelfield Reading Berkshire RG7 5ES tel:(01189) 649006 fax:(01189) 640074 36421 SATEC G info@satec.co.uk www.satec.co.uk

SAV United Kingdom Ltd Scandia House 131 Armfield Close Molesey Surrey KT8 2RJ tel:(020) 8941 4153 fax:(020) 8783 1132

Sawyer & Fisher Unit B, Aviary Court 138 Miles Road Epsom Surrey KT19 9AB tel:(01372) 742815 fax:(01372) 729710 qs@sawyerfisher.co.uk

SC Johnson Wax Frimley Green Camberley Surrey GU16 5AJ tel:(01276) 852000 fax:(01276) 852308

Scandanavian Timber 30a Roxborough Park Harrow on the Hill Middx. HA1 3AY tel:(020) 8864 0131 fax:(020) 8426 9151 finn@scandinaviantimber.com

Scandinavian Window Systems Ltd 10 Eldon Street Tuxford Newark Notts. Ng22 0LH tel:(01777) 871847 fax:(01777) 872650 enquiries@scandinavianwindows.co.uk www.scandinavian-windows.co.uk

Scapa Tapes UK Ltd. Gordleton Industrial Park Hannah Way Lymington Hampshire SO41 8JD tel:(01590) 684400 fax:(01590) 683728 jon.kitcher@Scapatapes.com www.sellotape.co.uk

Schauman (UK) Ltd Stags End House Gaddesden Row Hemel Hempstead Herts. HP2 6HN tel:(01582) 794661 fax:(01582) 794661

Schindler Ltd Schindler House 3 Plane Tree Crescent Feltham Middx. TW13 7HX tel:(020) 8818 7900 fax:(020) 8818 7999

Schmidlin (UK) Ltd White Lion Court Swan Street Old Isleworth Middx. TW7 6RN tel:(020) 8560 9944 fax:(020) 8568 7081

Schneider (UK) Ten Acres Chobham Road Ottershaw Surrey KT16 0QB tel:(01932) 872137 fax:(01932) 875366

Schneider Electric Ltd Stafford Park 5 Telford Shropshire TF3 3BL tel:(01952) 290029 fax:(01952) 290534 david_jellicoe@schneider.co.uk www.schneider.co.uk

Scholes Windows Ltd St Thomas Road Longroyd Bridge Huddersfield W. Yorks. HD1 3LF tel:(01484) 435116 fax:(01484) 435360

SCHOTT UK Ltd Sales Office Drummond Road Stafford ST16 3EL tel:(01785) 223166 fax:(01785) 223522 info.uk@schott.com www.schott.com/uk

SCHUCO International KG Whitehall Avenue Kingston Milton Keynes MK10 0AL tel:(01908) 282111 fax:(01908) 282124 info@scheveco.co.uk www.schveco.co.uk

Scotia Rigging Services (Industrial Stainless Steel Rigging Specialists) 68 Bridge Street Linwood Paisley Renfrewshire PA3 3DR tel:(01505) 321127 fax:(01505) 321333 enquiries@scotia-handling-rigging.co.uk

Scott Appleton Furniture Ltd Unit 2 New Rock Industrial Estate Chilcompton Somerset BA3 4JE tel:(01761) 232997 fax:(01761) 233526

Scott Fencing Limited Brunswick Industrial Estate Newcastle-upon-Tyne Tyne & Wear NE13 7BA tel:(0191) 236 5314 fax:(0191) 217 0193

Scottish & Northern Ireland Plumbing Employers' Federation 2 Walker Street Edinburgh EH3 7LB 0131 225 2255 0131 226 7638 info@snipef.org www.snipef.org

SCP Concrete Sealing Technology Ltd Ver Houser London Road Markyate Hertfordshire AL3 8JP tel:(01582) 842802 fax:(01582) 842803 info@scp-consealtech.co.uk www.scp-consealtech.co.uk/ver

Screen Systems Ltd PO Box 237, The West Site Britannia Works Bewsey Road Warrington WA5 5JZ tel:(01925) 653306 fax:(01925) 571060

Screenbase Ltd Reith Way West Portway Andover Hampshire SP10 3LF tel:(01264) 332166 fax:(01264) 333560

SE Controls Crossfield Road Industrial Estate Trent valley Lichfield Staffs WS13 6RJ tel:(01543) 415750 fax:(01543) 415747 sales@secontrols.co.uk www.secontrols.com

Sealine Business Products Media House 10 Greencoats Hertford Herts SG13 8AA tel:(01582) 455115 fax:(01582) 453569

Sealmaster Brewery Road Pampisford Cambridge CB2 4HG tel:(01223) 832851 fax:(01223) 837215 sales@sealmaster.co.uk www.sealmaster.co.uk

Sealocrete Ltd Greenfield Lane Rochdale OL11 2LD tel:(01706) 352 255 fax:(01706) 860 8800 bestproducts@sealcrete.co.uk www.sealocrete.co.uk

Seaque Technologies Ltd Church Road Portadown Co. Armagh BT63 5HU tel:(028) 38333131 fax:(028) 38333042

Seco Aluminium Ltd Crittall Road Witham Essex CM8 3AW tel:(01376) 515141 fax:(01376) 500542 seco.aluminium@btinternet.com www.secoaluminium.co.uk

Securikey Ltd PO Box 18 Aldershot Hampshire GU12 4SL tel:(01252) 311888 fax:(01252) 343950 enquires@securikey.co.uk www.securikey.co.uk

Securistyle Ltd Kingsmead Industrial Estate Princess Elizabeth Way Cheltenham Glos. GL51 7RE tel:(01242) 221200 fax:(01242) 520828 info@securistyle.co.uk www.securistyle.co.uk

Selbix Joinery Ltd Unit D, Frenbury Estate Drayton High Road Norwich Norfolk NR6 5DP tel:(01603) 789000 fax:(01603) 405476

Select Heating Ltd 2 Thomas Court East Street Colchester Essex CO1 2CR tel: (01206) 799787 fax:(01206) 799717

Selkin W E Ltd Selray House Ludlow Hill Road West Bridgford Nottingham NG2 6HF tel:(01159) 232286 fax:(01159) 233816
Sellite Blocks Ltd Old Quarry Long Lane Great Heck N. Yorks. DN14 0BT tel:(01977) 661631 fax:(01977) 662155 sales@sellite.co.uk www.sellite.co.uk
Seltek Instruments Ltd Leeside Works Lawrence Avenue Stanstead Abbots Herts SG12 8DL tel:(01920) 871094 fax:(01920) 871853 sales@selteklTd.co.uk www.selteklTd.co.uk
Senior Air Systems Oldfields Business Park Birrell Street Fenton Stoke-on-Trent ST4 3ES tel:(01782) 599995 fax:(01782) 559220
Servitherm Vale House 100 Vale Road Windsor Berks. SL4 5JL tel:(01753) 829449 fax:(01753) 829448
Servowarm Unit 9 The Gateway Centre, Coronation Road High Wycombe Bucks. HP12 3SU tel:(01494) 474474 fax:(01494) 472906 sales@servowarm.co.uk www.servowarm.co.uk
Setsquare Valley Industries Hadlow Road Tonbridge Kent TN11 0BR tel:(01732) 851 888 01732 851 853 sales@setsquare.co.uk www.setaquare.co.uk
SFC (Midlands) Ltd Brooklands 1655 Melton Road East Goscote Leicester LE7 6YQ tel:(0116) 260 1688 fax:(0116) 264 0209
SFS Intec LTD 153 Kirkstall Road Leeds West Yorkshire LS4 2AT tel:(0113) 2085 500 fax:(0133) 2085573 gb.leeds@sfsintec.biz www.sfsintec.biz/uk
SG System Products Ltd Unit 22 Wharfedale Road Ipswich Suffolk IP1 4JP tel:(01473) 240055 fax:(01473) 461616 Sales@sgsystems.co.uk www.handrailsuk.co.uk
SGB Hire Plus Holmwood, Broadlands Business Campus Langhurstwood Road Horsham West Sussex RH12 4SU tel:(08705) 288 388 hireplus@sgb.co.uk www.sgb.co.uk
SGB Youngman 'Holmwood', Broadlands Business Campus Langhurstwood Road Horsham W. Sussex RH12 4QE tel:(01403) 220000 fax:(01403) 321815 youngmandirect@sgb.co.uk
SGL Acotec Ltd Millbuck way Ettiley Heath Sandbach Cheshire CW11 3AB Tel:(01270) 761720 Fax:(01270) 761697 sgl.acotec.ltd@sglcarbon.com www.sglacotec.co.uk
Shackerley (Holdings) Group Ltd incorporating Designer ceramics Ceramics House 139 Wigan Road Euxton, Chorley Lancs. PR7 6JJ tel:(01257) 273114 fax:(01257) 262386 enquiries@shackerley.co.uk www.shackerley.com
Shadbolt F R & Sons Ltd North Circular Road South Chingford London E4 8PZ tel:(020) 8527 6441 fax:(020) 8523 2774 sales@shadbolt.co.uk www.shadbolt.co.uk
Shap Concrete Products Ltd Shap Penrith Cumbria CA10 3QQ tel:(01931) 716444 fax:(01931) 716617
Sharman Fencing & Gate Supplies Robinswood Farm Bere Regis Wareham Dorset BH20 7JJ tel:(01929) 472181 fax:(01929) 472182
Sharpe & Fisher Kelston Road Southmead Bristol Avon BS10 5ER tel:(0117) 950 4700 fax:(0117) 950 4500
Shavrin Levatap Co Ltd 32 Waterside Kings Langley Herts. WD4 8HH tel:(01923) 267678 fax:(01923) 265050 shavrin@levatap.freeserve.co.uk
Shaws of Darwen Waterside Darwen Lancs. BB3 3NX tel:(01254) 775111 fax:(01254) 873462
Sheardown Engineering Ltd 15-17 South Road Templefields Harlow Essex CM20 2AP tel:(01279) 421788 fax:(01279) 435642 sales@sheardown.co.uk www.sheardown.co.uk
shearflow LTD Pharaoh House Arnolde close, Medway City Estate Rochester Kent ME2 4SP tel:(01634)735 020 fax:(01634) 735 019 sales@shearflow.co.uk www.shearflow.co.uk
Shell Chemical Ireland Ltd Dublin Ireland tel:(0110) 3531 785177 fax:(0110) 3531 767489
Shepherd Engineering Services Ltd Mill Mount York N. Yorks. YO24 1GH tel:(01904) 629151 fax:(01904) 610175 skay@ses-ltd.co.uk www.ses-ltd.co.uk
Shipley Fan Co Ltd The Propeller Works 106 Dockfield Road Shipley Yorks. BD17 7BA tel:(01274) 581337 fax:(01274) 531337 shipleyfan@aol.com
Shires Bathrooms Beckside Road Bradford W. Yorks. BD7 2JE tel:(01274) 521199 fax:(01274) 521583 marketing@shires-bathrooms.co.uk www.shires-bathrooms.co.uk
Shower Products Ltd Unit 1, High Cross Centre Fountayne Road Tottenham London N15 4QW tel:(020) 8801 4465 fax:(020) 8801 9936
Showerlux UK Ltd Sibree Road Coventry W. Midlands CV3 4EL tel:(024) 76 639400 fax:(024) 76 305457
ShowerSport Ltd 1 North Place Edinburgh Way Harlow Essex CM20 2SL tel:(01279) 451450 fax:(01279) 635074
Shutter Door Repair & Maintenance Ltd 89-91 Rolfe Street Smethwick Warley W.Midlands B66 2AY tel:(0121) 558 6406 fax:(0121) 558 7140
Sidhil Care Boothtown Halifax W. Yorks. HX3 6NT tel:(01422) 363447 fax:(01422) 344270 info@sidhil-care.ltd.uk www.sidhil-care.ltd.uk
Siemens Communication Ltd Brickhill Street Willen Lake Milton Keynes MK15 0DJ tel:(01908) 855000 fax:(01908) 855001 www.siemenscomms.co.uk
Siemens Building Technologies Ltd Hawthorne Road Staines Middx. TW18 3AY tel:(01784) 461616 fax:(01784) 464646 sales@gb.sibt.com www.sbt.siemens.com
Siemens Building Technologies Ltd, Field Way Woosehill Greenford Middlx UB6 8UZ tel:(0118) 989 4488 fax:(0118) 979 5750
Sierra Windows Yaberton Industrial Estate Alders Way Paignton Devon TQ4 7QE tel:(01803) 697000 fax:(01803) 697071
Sigma Coatings Ltd Back Church Lane London E1 1LU tel:(01280) 812081 fax:(01280) 815656 pcuk@sigmacoatings.com www.sigmacoatings.com
Signature Ltd Signature House Hainge Road Tividale Oldbury West Midlands BL9 2NF tel:(0121) 557 0995 fax:(0121) 557 0995 signature_sales@lineone.net
Signs & Labels Ltd Douglas Bruce House Corrie Way Stockport Cheshire SK6 2RR tel:(0800) 132323 fax:(0161) 430 8514 sales@signsandlabels.co.uk www.safetyshop.com
Signs International PO Box 25 Kingswood Bristol BS30 6LU tel:(0117) 932 4049 fax:(0117) 932 4924 safetysigns@signsinternational.co.uk www.signsinternational.co.uk/signs/
Signs Of The Times Ltd Wingfield Road Tebworth Leighton Buzzard Beds. LU7 9QG tel:(01525) 874185 fax:(01525) 875746 enquiries@soot.co.uk www.sott.co.uk
Sika Ltd Watchmead Welwyn Garden City Herts. AL7 1BQ tel:(01707) 394444 fax:(01707) 393296 technical@uk.sika.com www.sika.com
Silavent Ltd 60 High Street Sandhurst Berkshire GU47 8DY tel:(01252) 878282 fax:(01252) 871212 silavent@btinternet.com www.silavent.co.uk
Silent Gliss Ltd Star Lane Margate Kent CT9 4EF tel:(01843) 863571 fax:(01843) 864503 info@silentgliss.co.uk www.silentgliss.co.uk
Silkolene Lubricants PLC Silkolene Oil Refinery Belper Derby DE5 1WF tel:(01773) 824151 fax:(01773) 823659
Silverwood Fencing Co. Limited 56 Leeds Road Mirfield W. Yorks. WF14 0DE tel:(01924) 498689 fax:(01924) 498689
Simon R W Ltd System Works Hatchmoor Industrial Estate Torrington Devon EX38 7HP tel:(01805) 623721 fax:(01805) 624578
Simon-Hartley Garner Street Stoke-on-Trent Staffs. ST4 7BH tel:(01782) 202300 fax:(01782) 260534 www.hnashe.com/shl.htm
Simons Construction Ltd 401 Monks Road Lincoln Lincs. LN3 4NU tel:(01522) 510000 fax:(01522) 521812 webmaster@simonsgroup.com www.simonsgroup.com
Simplex Lighting Ltd Groveland Road Tipton W. Midlands DY4 7XB tel:(0121) 557 2828 fax:(0121) 557 8900
Simplex Signs Limited Unit C, Peter Road Lancing Business Park Lancing West Sussex BN15 8TH tel:(01903)750333 fax:(1903) 750444 signsystems@simplexltd.com www.simplexltd.com
Simpson Strong-Tie® Winchester Road Cardinal Point Tamworth Staffs. B78 3HG tel:(01827) 255600 fax:(01827) 255616 www.strongtie.com
Single Ply Roofing Association The Building Centre 26 Store Street London WC1E 7BT 0115 914 4445 0115 974 9827 enquiries@spra.co.uk www.spra.co.uk
Sissons W & G Ltd Carrwood Road Sheepbridge Chesterfield Derbys. S41 9QB tel:(01246) 450 255 fax:(01246) 451 276 wgs@sissons.co.uk www.sissons.co.uk
Skanda Broadway House 21 Broadway Maidenhead Berkshire SL6 1NJ tel: (01628) 784330 fax:(01628) 633080
Skarsten Tools Units 16-18 Joseph House Eismann Way Corby Northants NN17 5ZB tel:(01536) 200976 fax:(01536) 400656
Skil Controls Ltd Greenhey Place East Gillibrands Skelmersdale Lancs. WN8 9SB tel:(01695) 714600 fax:(01695) 714611 sales@skilcontrols.com www.skilcontrols.com
Slipstop 9 The Courtyard Whitewick Business Park Coalville LE67 3RZ Tel: (01530) 813 500 Fax:(01530) 813 400
Slottseal Plc Flemming Road Earlstrees Industrial Estate Corby Northants NN17 2TY tel:(01536) 200555 fax:(01536) 204217
Smart F & G (Shopfittings)Ltd The Shopfitting Centre, Tyseley Industrial Estate Seeleys Road Greet Birmingham B11 2LA tel:(0121) 772 5634 fax:(0121) 766 8995 www.smartsshopfittings.co.uk
Smart Systems Ltd North End Road Yatton, nr Bristol BS49 4AW tel:(01934) 876100 fax:(01934) 835169 sales@smartsystems.co.uk www.smartsystems.co.uk
Smith & Choyce Ltd 280 Barton Street Gloucester Gloucestershire GL1 4JJ tel:(01452) 523531 fax:(01452) 310032 martin.choyce@btclick.com
Smith & Rodger Ltd 24-36 Elliot Street Glasgow G3 8EA tel:(0141) 248 6341 fax:(0141) 248 6475 info@smithandrodger.co.uk www.frenchpolishes.com
Smithbrook Building Products Ltd 18 Smithbrook Kilns Cranleigh Surrey GU6 8JJ tel:(01483) 268808 fax:(01483) 278258
Smoke Control Services Ltd The Innovation Centre Tredomen Business Park Hengoed Mid Glamorgan CF82 7FN tel:(0870) 240 64 60 fax:(01443) 819 119 enquiry@smokecontrol.co.uk www.smokecontrol.co.uk
Society for the Protection of Ancient Buildings (SPAB) 37 Spital Square London E1 6DY tel:(020) 7377 1644 fax:(020) 7247 5296 info@spab.org.uk www.spab.org.uk
Solaglas Herald Way Binley Coventry CV3 2ND tel:(024) 76 458844 fax:(024) 76 636473 enquiries@solaglas.co.uk www.solaglas.co.uk
Solaglas Laminated Saffron Way Sittingbourne Kent ME10 2PD tel:(01795) 421534 fax:(01795) 473651
Solair Ltd Pennington Close Albion Road West Bromwich W. Midlands B70 8BA tel:(0121) 525 2722 fax:(0121) 525 6786
Solair GRP Architectural Products Smeaton Road Churchfields Industrial Estate Sailsbury Wilts. SP2 7NQ tel:(01722) 323036 fax:(01722) 337546 sales@solair.co.uk
Solar Trade Association, The The National Energy Centre Davy Avenue Knowlhill Milton Keynes MK5 8NG 01908 442290 0870 0529194 enquiries@solartradeassociation.org.uk www.greenenergy.org.uk/sta/
Sommerfeld Flexboard Ltd Frame Lane Doseley Telford Shropshire TF4 3BY tel:(01952) 503737 fax:(01952) 630132 info@sommerfeld.co.uk www.sommerfeld.co.uk
Sonneborn & Rieck Ltd Jaxa Works 91-95 Peregrine Road Hainault Ilford IG6 3XH tel:(020) 8500 0251 fax:(020) 8500 3696
Sound Service (Oxford) Ltd 55 West End Witney Oxon. OX28 1NJ tel:(01993) 704981 fax:(01993) 779569 soundservice@btconnect.com www.soundservice.com
South Durham Structures Limited Dovecot Hill South Church Enterprise Park Bishop Auckland Co. Durham DL14 6XR tel:(01388) 777350 fax:(01388) 775225 brianhope@sdsltd.demon.co.uk
South Western Tar Distilleries Totton Works High Street Totton Southampton SO40 9TN tel:(023) 80861169 fax:(023) 80864274
Southern Brick Federation Limited Woodside House Winkfield Windsor Berkshire SL4 2DX 01344 885651 01344 890139 melvynhayward@brick.org.uk
Sovereign Brush Co Ltd, The 29-43 Sydney Road Watford Herts. WD18 7P2 tel:(01923) 227301 fax:(01923) 817121 sales@sovereignbrush.com www.sovereignbrush.com
Sovereign Fencing Limited Sovereign House 261 Monton Road, Monton Eccles Manchester M30 9LF tel:(0161) 789 5479
Space - Ray UK Chapel Lane Claydon Ipswich Suffolk IP6 0JL tel:(01473) 830551 fax:(01473) 832055 www.spaceray.com
Sparkes K & L The Forge Claverdon Nr Warwick Warwks. CV35 8P tel:(01926) 842545 fax:(01926) 842559
Spartan Tiles Ltd Slough Lane Ardleigh Colchester Essex CO7 7RU tel:(01206) 230553 fax:(01206) 230516 david@spartantiles.com www.spartantiles.com
Spear & Jackson Atlas Way Atlas North Sheffield S. Yorks. S4 7QQ tel:(0114) 281 4242 fax:(0114) 281 4252 sales@spear-and-jackson.com www.spear-and-jackson.com
Spear & Jackson Interntional Ltd. Atlas Way, Atlasnorth. Sheffield S4 7QQ tel:(0114) 256 1133 fax:(0114) 243 1360
Specialist Building Products PO Box 12422 Holland Park Avenue Holland Park London W11 4GX tel:(020) 8456 8212 fax:(020) 8458 4116 specialistbuildingproducts@yahoo.co.uk
Spectra Glaze Services Unit 7 Rosehill Court St Helier Avenue Morden Surrey SM4 6JT tel:(020) 8640 6737 Fax: (0208) 648 4692 sales@spectraglaze.co.uk www.spectraglaze.co.uk
Spectus Systems Snape Road Macclesfield Cheshire SK10 2NZ tel:(01625) 420400 fax:(01625) 433946 www.spectussystems.com
Speed Saws Ltd. Speed Works Lancaster Street Sheffield S3 8AQ tel:(0114) 275 3333 fax:(0114) 279 5511
Sperrin Metal Products Ltd Cahore Road Draperstown Co. Derry BT45 7AP tel:(028) 79628362 fax:(028) 79628972 raymond@sperrin-metal.com www.sperrin-metal.com
Spindlewood 8 Edward Street Bridgewater Somerset TA6 5ET tel:(01278) 453665 fax:(01278) 453666
Spital Tile Co Ltd Spital Mills Spital Lane Chesterfield Derbys. S41 0EX tel:(01246) 273657 fax:(01246) 220313
Spontex Ltd Berkley Business Park Wainwright Road Worcester WR4 9ZS tel;(01305)450300 Fax;01905 450350
Sportmark™ Ltd Sportsmark House Lansbury Estate Lower Guildford Road Woking Surrey GU21 2EP 01483 488310 01483 487919 sales@sportsmark.net www.sportsmark.net
Sprayed Concrete Association Association House 99 West Street Farnham Surrey GU9 7EN 01252 739153 01252 739140 info@associationhouse.org.uk www.sca.org.uk
Springvale E P S 75 Springvale Road Ballyclare Co. Antrim BT39 0SS tel:(028) 93340203 fax:(028) 93341159 sales@springvale.com www.springvale.com
Spur Shelving Ltd Spur House Otterspool Way Watford Herts WD25 8HT tel:(01923) 830000 fax:(01923) 241040 info@spuruk.com www.spurshelving.com
SPX Process Equipment Ironstone Way Brixworth Northamptonshire NN6 9UD tel:(01604) 880751 fax:(01604) 880145 info.uk@processequipment.spx.com www.spxprocessequipment.com
Square D, Electrical Distribution & Building applicators Division Fordhouse Road Wolverhampton WV10 9ED tel:(01902) 303370 fax:(01902) 303371 www.squared.co.uk
Squire Henry & Sons Linchfield Road New Invention Willenhall W. Midlands WV12 5BD tel:(01922) 476711 fax:(01922) 493490 info@henry-squire.co.uk www.squirelocks.co.uk
Squires Metal Fabrications Ltd Burgess Road Ivyhouse Lane industrial Estate Hastings Sussex TN35 4NR tel:(01424) 428794 fax:(01424) 431567 squires@squiresmetal.freeserve.co.uk
SSAB Dobel Coated Steel Ltd Narrow Boat Way Hurst Business Park Brierley Hill W. Midlands DY5 1UF tel:(01384) 74660 fax:(01384) 77575 sales@dobel.co.uk www.dobel.co.uk
SSI Schaefer Ltd 83-84 Livingstone Road Walworth Industrial Estate Andover Hampshire SP10 5QZ tel:(01264) 386600 fax:(01264) 386611 solutions@ssi-schaefer.co.uk www.ssi-schaefer.co.uk
Stadium Ltd Unit 10, Riverside Industrial Estate Morson Road Ponders End, Enfield Middx. EN3 4TU tel:(020) 8804 4343 fax:(020) 8805 7336
StadiumBuilding Products 11 Jubilee Avenue Highams Park London E4 9UD tel:(020) 8531 4433 fax:(020) 531 4646
Staines Steel Gate Co 20 Ruskin Road Staines Middx. TW18 2PX tel:(01784) 454456 fax:(01784) 466668 www.steelgates@aol.com
Stainton Metal Co Ltd Dukesway Teeside Industrial Estate Thornaby Cleveland TS17 9LT tel:(01642) 766242 fax:(01642) 765509 genenq@stainton-metal.co.uk
Standard Flat Roofing Co Ltd The Street, Sheering Bishops Stortford Herts. CM22 7LY tel:(01279) 734521 fax:(01279) 734568
Standard Patent Glazing Co Ltd Flagship House Forge Lane Dewsbury W. Yorks. WF12 9EL tel:(01924) 461213 fax:(01924) 458083 sales@patent-glazing.com www.patent-glazing.com
Standby Power Company 311 Lidgett Lane Leeds 17 W. Yorks. LS17 6PD tel:(0113) 2682682 fax:(0113) 2682682
Stanhope Chemical Products Ltd 96 Bridge Road East Welwyn Garden City Herts. AL7 1JW tel:(01707) 324373 fax:(01707) 372191
Stannah Lifts Ltd Anton Mill Andover Hampshire SP10 2NX tel:(01264) 339090 fax:(01264) 337942 passengerliftsales@stannah.co.uk www.stannah.co.uk
Stanton plc Lows Lane Stanton- By- Dale Ilkeston Derbyshire DE7 4QU tel:(0115) 930 5000 fax:(0115) 932 9513 sales@stanton.plc.uk www.stanton.plc.uk
Stanton-Thompson (Agencies) Ltd 44 watcombe Road Watlington Oxon. OX49 5QJ tel:(01491) 613515 fax:(01491) 613516
Staytite Ltd Staylite House Halifax Road Cressex Business Park High Wycombe HP12 3SN tel:(01494) 462322 fax:(01494) 464747
Steel Line Ltd 415 Petre Street Sheffield S4 8LL tel:(0114) 231 7330 fax:(0114) 256 0330 sales@steeline.co.uk www.steelline.co.uk
Steelcase Strafor Plc Newlands Drive Poyle Berks. SL3 0DX tel:(01753) 680200 fax:(01753) 680345 www.steelcase.com
Steel-Crete Ltd Stuart House Kingsbury Link Trinity Road Staffordshire B87 2EX 01827 871140 01827 871144
Steelway Fensecure Ltd Queensgate Works Bilston Road Wolverhampton West Midlands WV2 2NJ tel:(01902) 451733 fax:(01902) 452256 sales@steelway.co.uk www.steelway.co.uk
Steinel (UK) Ltd 36-37 Manasty Road Orton Southgate Peterborough PE2 6UP tel:(01733) 238265 fax:(01733) 238270
Stenoak Associated Services PLC Highlands Lane Henley-on-Thames Oxon. RG9 4PS tel:(01491) 575921 fax:(01491) 579713 marketing@associatedasphalt.co.uk www.stenoak.co.uk
Stenoak Fencing Limited Stenoak House The Down Lamberhurst Kent TN3 8ER tel:(01892) 890000 fax:(01892) 891136 info@stenoak.co.uk www.stenoak.co.uk

Stent Foundations Ltd Pavilion C2, Ashwood Park Ashwood Way Basingstoke Hampshire RG23 8BG tel:(01256) 400200 fax:(01256) 400201 foundations@stent.co.uk www.stent.co.uk
Stevenson & Cheyne Ltd 60 Newhaven Road Leith Edinburgh EH6 5QT tel:(0131) 554 5211 fax:(0131) 554 1272 f.smith@stevche.dialnet.com www.platerolling.co.uk
Stewart Wales, Somerville Ltd Glenburn Road College Milton East Kilbride Glasgow G74 5BA tel:(013552) 22101 fax:(013552) 33847 info@swsinternational.com
Stiebel Eltron Ltd Lyveden Road Brackmills Northampton NN4 7ED tel:(01604) 766421 fax:(01604) 765283 info@stiebel-eltron.co.uk www.Stiebel-eltron.co.uk
STO Ltd Unit 3 Lyon Road Lynwood Industrial Estate Paisley Scotland PA3 3BQ tel:(01505) 324262 fax:(01505) 323618 info@sto.co.uk www.sto.co.uk
Stoakes Systems Ltd 1 Banstead Road Purley Surrey CR8 3EB tel:(020) 86607667 fax:(020) 8660 5707 mailbox@stoakes.co.uk www.stoakes.co.uk
Stockbrands LLP Oaklands Yard Flixton Rd Bungay Suffolk NR35 1PD tel:(01986) 893232 fax:(01986) 892552 sales@stockbrands.co.uk www.stockbrands.co.uk
Stocksigns Ltd 43 Ormside Way Redhill Surrey RH1 2 LG tel:(01737 764764 fax:(01737) 763763 info@stocksigns.co.uk www.stocksigns.co.uk
Stockten Heath Forge (Caldwells) Ltd Forge House Dallam Lane Warrington Cheshire WA2 7PZ tel:(01925) 636387 fax:(01925) 234158 sales@caldwells.eu.com www.caldwels.eu.com
Stoke Hall Quarry(Stone Sales)Ltd Eyam Road Grindleford Hope Valley S32 2HW tel:(01433) 630313 fax:(01433) 631353
Stokes R J & Co Ltd 12 Moore Street Sheffield S3 7UQ tel:(01142) 737211 fax:(01142) 755656 info@stokestiles.co.uk
Stokvis R S & Sons Ltd Pool Road West Molesey Surrey KT8 2HN tel:(020) 8941 1212 fax:(020) 8941 4136 info@stokvis.co.uk www.stokvis.co.uk
Stone Fasterners Ltd Woolwich Road London SE7 8SL tel:(020) 8293 5080 fax:(020) 8293 4935
Stone Federation Great Britain Construction House 56-64 Leonard Street London EC2A 4JX 020 7608 5094 020 7608 5081
Stoneham PLc Powerscroft Road Foots Cray Sidecup Kent DA14 5Z tel:(020) 83008181 fax:(020) 8300 8183
Stonewest Ltd Lambert 's Place St James's Road Croydon Surrey CR9 2HX tel:(020) 8684 6646 fax:(020) 8684 9323 info@stonewest.co.uk www.stonewest.co.uk
Stoves PLC Stoney Lane Prescot Liverpool Merseyside L35 2XW tel:(0151) 426 6551 fax:(0151) 426 3261 www.stoves.co.uk
Stowell Concrete Ltd Arnolds Way Yatton, Bristol North Sumerset BS49 4QN tel:(01934) 834000 fax:(01934) 835474 www.stowellconcrete.co.uk
Stramit Industries Ltd Yaxley Eye Suffolk IP23 8BW tel:(01379) 783465 fax:(01379) 783659
Strand Hardware Ltd Strand House, Premier Business Park Long Street Wallsall W. Midlands WS2 9DY tel:(01922) 639111 fax:(01922) 626025 strandhw@aol.com www.strandhardware.co.uk
Streamline Resin Systems Ltd Unit 12 Fowlswick Business Centre Fowlswick Lane Allington Chippenham Wiltshire SN14 6QE tel:(01249) 782963 fax:(01249) 782185
Streamtec Limited 2 Kirkgate House Baden-Powell Road Arbroath Angus DD11 3LS 01241 436862 01241 436787 info@streamtec.com www.streamtec.com
Structural Soils Ltd Chevet House A1 Great North Road Knottingley W. Yorks. WF11 0BS tel:(01977) 674461 fax:(01977) 674475 admin@north-soils.co.uk www.soils.co.uk
Structural Timbers Ltd Whitchurch Lane Whitchurch Bristol BS14 0BH tel:(01275) 832724 fax:(01275) 892593
Stuart Turner Ltd Market Place Henley on Thames Oxon. RG9 2AD tel:(01491) 572655 fax:(01491) 573704 pumps@stuart-turner.co.uk www.stuart-turner.co.uk
Stuarts Industrial Flooring Ltd Stuarts House Church Hill Coleshill Birmingham B46 3AF tel:(01675) 463555 fax:(01675) 467265
Sulzer Infra (UK) Ltd Westmead Farnborough Hampshire GU14 7LP tel:(01252) 544400 fax:(01252)378988
Sumlock J.H.P. Sutherland House Matlock Road Foleshill Coventry CV1 4JQ tel:(024) 76 667891 fax:(024) 76 668075
Sundeala Division of Celotex Ltd Warwick House 27 St Marys Road Ealing London W5 5PR tel:(020) 8579 0811 fax:(020) 8579 0106 sundeala@celotex.co.uk www.celotex.co.uk
Sundwel Solar Ltd 1 Tower Road Glover Washington NE37 2SH tel:(0191) 416 3001 fax:(0191) 415 4297 solar@sundwel.com www.sundwel.com
Sunpipe @ Monodraught Ltd Halifax House Cressex Business Park High Wycombe Bucks HP12 3SE tel:(01494) 897700 fax:(01494) 532465 sunpipe@monodraught.com www.monodraught.com
Sunray Engineering Ltd Kingsnorth Industrial Estate Wotton Road Ashford Kent TN23 6LL tel:(01233) 639039 fax:(01233) 625137 sales@sunraydoors.co.uk www.sunraydoors.co.uk
Sunrite Blinds Ltd Blackchapel Road Edinburgh EH15 3HL tel:(0131) 669 2345 fax:(0131) 657 3595 info@sunrite.co.uk www.sunrite.co.uk
Sunway UK Ltd Mersey Industrial Estate Heaton Mersey Stockport Cheshire SK4 3EQ tel:(0161) 432 5303 fax:(0161) 431 5087
Super Cement Ltd 108 High Street Uckfield E. Sussex TN22 1DX tel:(01825) 767125 fax:(01825) 767125
Supra Acoustics Ltd Hainge Road Tividale, Warley W. Midlands B69 2NF tel:(0121) 557 6884 fax:(0121) 557 6884
Surmak Products Ltd 99 Mabgate Leeds LS9 7DR tel:(01532) 450371 fax:(01532) 428701
Surrey Sackholders Ltd. Trinity Road Richmond Surrey TW9 2LG tel:(020) 8878 7878 fax:(020) 8948 1998
Swanlux Horticultural Services Ltd Unit 10 Courtyard Aurelia Croydon CR0 3BF tel:(020) 8764 6217 fax:(020) 8764 6227
Swedecor Ltd Manchester Street Hull E. Yorks. HU3 4TX tel:(01482) 329691 fax:(01482) 212988 info@swedecor.com www.swedecor.com
Swedhouse Ltd PO Box 10 East Horsley Surrey KT24 6TU tel:(01483) 284004 fax:(01483) 285315
Swift Hitch Ltd Trinity Street Kingsmand Trading Estate St Phillips Bristol BS2 0NT tel:(0117) 941 2121 fax:(0117) 941 3093
Swifts Of Scarborough Ltd Cayton Low Road Eastfield Scarborough N. Yorks. YO11 3BY tel:(01723) 583131 fax:(01723) 584625
Swintex Ltd Derby Works Manchester Road Bury Lancs. BL9 9NX tel:(0161) 761 4933 fax:(0161) 797 1146 alan.armitage@swintex.co.uk www.swintex.com
Swish Building Products Pioneer House Lichfield Rd Industrial Estate Mariner Tamworth Staffs. B79 7TF tel:(01827) 317200 fax:(01827) 317201 marketing@swishbp.co.uk
Switch2 Energy Solutions Ltd High Mill Cullingworth Bradford West Yorkshire BD13 5HA tel:(0118) 956 0060 fax:(0118) 956 7610
SWJ Group Ltd Goodridge Industrial Estate Goodridge Avenue Gloucester Glos. GL2 5EB tel:(01452) 306181 fax:(01452) 300806
Sylvania Lighting International INC. Otley Road Charlestown, Shipley W. Yorks. BD17 7SN tel:(01274) 532 552 fax:(01274) 531 672 sales.office@sylvannia-lighting.co.uk www.sylvania-lighting.com
Symphony Group PLC, The Gelderd Lane Leeds LS12 6AL tel:(0113) 230 8000 fax:(0113) 230 8134 marketing@symphonygroup.co.uk www.symphonygroup.co.uk
Symplex Systems Ltd 418-420 Limpsfield Road Warlingham Surrey CR6 9LA tel:(01883) 624004 fax:(01883) 627201
Syston Rolling Shutters Ltd 33 Albert Street Syston Leicester LE7 2JB tel:(0116) 2608841 fax:(0116) 2640846 www.syston.com
Szerelmey Ltd 369 Kennington Lane Vauxhall London SE11 5QY tel:(020) 7735 9995 fax:(020) 7793 9800 info@szerelmey.com www.szerelmey.com
T & D Packaging Ltd Bowling House Bowling Iron Works Bradford W. Yorks. BD4 8SX tel:(01274) 707050 fax:(01274) 707060
T B S Fabrications Siemens Road, Northbank Ind. Estate Irlam Manchester Greater Manchester M44 5AH tel:(0161) 775 7915 fax:(0161) 775 8929 tbs.fabrications@virgin.net
T. A. Harris Ltd Hargreen House 134 New Kent Road London SE1 6TY 020 7703 3842 020 7708 2596
TA Lucas Sails Porchester House Hospital Lane Portchester Hampshire PO16 9QP tel:(02392) 373699 fax:(02392) 373656 info@lucas-sails.com www.lucas-sails.com
Talbot Winnal Valley Road Winchester Hampshire SO23 0LL tel:(01962) 705200 fax:(01962) 841344 talbot_sales@tmproducts.com www.talbot.co.uk
Tank Storage & Services Ltd Lilley Farm Thurlow Road Withersfield Suffolk CB9 7SA tel:(01440) 712614 fax:(01440) 712615
Tankard York Mills York Street, Fairweather Green Bradford W. Yorks. BD8 0HR tel:(01274) 777666 fax:(01274) 773336 sales@tankard.com www.tankard.com
Tann Synchronome Ltd Unit 50 D,Symondscliffe Way Severn Bridge Industrial Estate Portskewett Monmouthshire NP26 5PW tel:(01291) 431910 fax:(01291) 431911 tann@synchronome.net http://tann.synchronome.net
Tapworks Water Softeners Mill Road Stockenchurch Bucks HP14 3TP 01494 480621 01494 484396 info@tapworks.co.uk www.tapworks.co.uk
Tarkett Sommer Ltd Century House Bridgwater Road Worcester Worcestershire WR4 9FA tel:(01905) 342 700 fax:(01905) 342 777 uksales@tarsom.com
Tarkett-Marley Floors Ltd Lenham Maidstone Kent ME17 2QX tel:(01622) 854000 fax:(01622) 854500 info@marley.com www.marleyfloors.com
Tarmac Ltd Holborough House Holborough Rad Snodland Kent ME6 5PJ tel:(01634) 242514 fax:(01634) 243927 www.tarmac.co.uk
Tarmac Northern Ltd Lingerfield Scotton Knaresborough N. Yorks. HG5 9JN tel:(01423) 796800 fax:(01423) 796808 www.tarmac.co.uk
Tarmac Pellite Ltd Tarmac Wharf Teesport Grangetown Middlesborough TS6 6UG tel:(01642) 452691 fax:(01642) 457793
Tarmac Precast Concrete Limited Tallington Factory Tallington Stamford Lincs. PE9 4RL tel:(01778)381000 fax:(01778)348041 tarmacprecast enquiries@tarmac.co.uk www.tarmacprecast.co.uk
Tarmac Top Floor Weston Underwood Derby DE6 4PH 01332 868 400 01332 868 401 topfloor@tarmac.co.uk www.tarmactopfloor.co.uk
Tarmac Topblock Ltd Millfields Road Ettingshall Wolverhampton WV4 6JP tel:(01902) 382884 fax:(01902) 382219 wendy.hinett@tarmac.co.uk www.topblock.co.uk
Taskworthy Ltd The Old Brickyard Pontrilas Herefordshire HR2 0DJ tel:(01981) 242900 fax:(01981) 242901 sales@taskworthy.co.uk www.taskworthy.co.uk
Tate Fencing Yellowcoat Sawmill Hastings Road Flimwell East Sussex TN5 7PR tel:(01580) 879900 fax:(01580) 879677 sales@tate-fencing.co.uk www.tate-fencing.co.uk
Tavac Insulation Ltd Mitchell Hey Mills College Road Rochdale Lancs. OL12 6AE tel:(01706) 715500 fax:(01706) 715511
Taylor & Atkinson (Leeds) Ltd Swinnow Lane Leeds West Yorkshire LS13 4NB tel:(0113) 257 4374 fax:(0113) 239 3194
Taylor & Portway Ltd 52 Broton Drive Trading Estate off Rosemary Lane Halstead Essex CO9 1HB tel:(01787) 472551 fax:(01787) 476589 sales@portway.demon.co.uk
TBA Sealing Materials Ltd PO Box 21 Rochdale Lancs. OL12 7EQ tel:(01706) 715000 fax:(01706) 42284
TBR Joinery Station Road Lenwade Norwich NR9 5LY tel:(01603) 872485 fax:(01603) 871762 tbr-joinery@ukgateway.net www-tbr-joinery.co.uk
TCW Services (Control) Ltd Bradshaw Works Honley Huddersfield Yorks. HD9 6DT tel:(01484) 662865 fax:(01484) 667574 sales@tcw-services.co.uk www.tcw-services.co.uk
TDI (UK) Ltd Unit 3, Unity Complex Dale Road North Darley Dale, Matlock Derbys. DE4 2HX tel:(01629) 733177 fax:(01629) 732779 gpy40@dial.pipex.com www.polypipe.plc.uk
Teal & Mackrill Ltd Ellenshaw Works Lockwood Street Hull E. Yorkshire HU2 0HN tel:(01482) 320194 fax:(01482) 219266 divid.young@coo-var.co.uk www.teamac.co.uk
Tebrax Ltd International House Cray Avenue Orpington Kent BR5 3RY tel:(01689) 897766 fax:(01689) 896789 brackets@tebrax.co.uk www.tebrax.co.uk
Technical Control Systems Ltd Treefield Industrial Estate Gildersome Leeds LS27 7JU tel:(0113) 252 5977 fax:(0113) 238 0095
Technical Indexes Ltd Willoughby Road Bracknell Berks. RG12 8DW tel:(01344) 426311 fax:(01344) 424971 systems@techindex.co.uk www.tionestop.com
Technology Telford Ltd Halesfield 19 Telford Shropshire TF7 4QT tel:(01952) 585580 fax:(01952) 680240
Technostone Ltd Litterton Lane Shepperton Middx. TW17 0NF tel:(01932) 566999 fax:(01932) 564365
Techrete (UK) Ltd Station Road Scawby Brigg N. Lincolnshire DN20 9AA tel:(01652) 659454 fax:(01652) 659458 ok41@dial.pipex.com
Techspan Systems PLC Church Lane Chalfont St Peter Bucks. SL9 9RF tel:(01753) 889911 fax:(01753) 887496 techspan@jarvis-uk.com
Teddington Controls Ltd Daniels Lane Holmbush St Austell Cornwall PL25 3HG tel:(01726) 74400 fax:(01726) 67953 info@tedcon.com www.tedcon.com
Tefcote Surface Systems Central House 4 Christchurch Road Bournemouth Dorset BH1 3LT tel:(01202) 551212 fax:(01202) 559090 sales@tefcote.co.uk www.tefcote.co.uk
Tektura Plc 1 Heron Quay Docklands London E14 4JA tel:(020) 7536 3300 fax:(020) 7536 3322 enquiries@tekturaonline.com www.tektura.com
Teleflex - Morse Christopher Martin Road Basildon Essex SS14 3ES tel:(01268) 522861 fax:(01268) 282994 p.pinnock@morse-controls.co.uk www.teleflex-morse-architectural.com
Telelarm care Ltd 1 Centurion Business Park Bessemer Way Rotherham S Yorkshire S60 1FB tel:(01709) 389324 fax:(01709) 389344 info@telelarmcare.co.uk
Telford Doors and Barriers Ltd Unit 11, The Bridges Business park Bridge Road Horsham , Telford Shropshire TF4 3EE tel:(01952) 503050 fax:(01952) 5050 sales@telforddoors.co.uk www.telforddoors.co.uk
Tensar International Group LTD New Wellington Street Blackburn Lancs. BB2 4PJ tel:(01254) 262431 fax:(01254) 266869 info@tensar.co.uk www.tensar-international.co.uk
Tensator Ltd Danbury Court Linford Wood Milton Keynes MK14 6TS tel:(01908) 271153 fax:(01908) 274572 dtuppin@tensator.demon.co.uk www.tensator.co.uk
Terminix Property Services Oakhurst Drive Cheadle Heath Stockport Cheshire SK3 0XT tel:(0161) 491 3181 fax:(0161) 428 8138
Termstall Limited 50 Burman Street Droylsden Manchester M43 6TE tel:(0161) 370 2835 fax:(0161) 370 9264 enquiries@termstallfencing.co.uk www.termstallfencing.co.uk
Terrapin Ltd Bond Avenue Bletchley Milton Keynes MK1 1JJ tel:(01908) 270900 fax:(01908) 270052 sales@terrapin-ltd.co.uk www.terrapin-ltd.co.uk
Terry of Redditch Ltd Millsbro Road Redditch Worcs. B98 7BU tel:(01527) 64261 fax:(01527) 62268
Tex Steel Tubes Ltd Unit 35, Claydon Ind Park Gipping Road Great Blakenham, Ipswich Suffolk IP6 0NL tel:(01473) 830030 fax:(01473) 831664
Textile Bonding Ltd Midland Road Higham Ferrers Northants NN10 8ER tel:(01933) 410100 fax:(01933) 410200
Textron Fastening Systems Ltd Mundells Welwyn Garden City Herts. AL7 1EZ tel:(01707) 668668 fax:(01707) 338828 www.advel.textron.com
Textured European Finishes Ltd 1 Dunedin Road Ilford Essex IG1 4LW tel:(020) 8533 1091 fax:(020) 8478 7870 tef.uk@virgin.net www.ascwebindex.com/tef
TFL International Ltd Rivers Way Leather Head Bridge Guildford Road Surrey KT22 9AD 01372 383330 01372 383331 bradleycox@tflinternational.co.uk www.tflinternational.co.uk
TH Kenyon & Sons Ltd Kenyon House 14a Hockerill Street Bishops Stortford Herts CM23 2DW tel:(01279) 661800 fax:(01279) 661807 jonathon@thkenyonplc.co.uk www.thkenyonplc.co.uk
Thakeham Tiles Ltd Rock Road Heath Common Storrington W. Sussex RH20 3AD tel:(01903) 742381 fax:(01903) 746341 stevethompson@mistral.co.uk www.Thakeham.co.uk
Thanet Ware Ltd Ellington Works Princes Road Ramsgate Kent CT11 7RZ tel:(01843) 591076) fax:(01843) 586198
Thatching Advisory Services Ltd Faircross Offices Stratfield Saye Reading Berks. RG7 2BT tel:(01256) 880828 fax:(01256) 880866 cervantes.north@virgin.net www.thatchingadvisoryservices.co.uk
The Abbseal Group incorporating Everseal (Thermovitrine Ltd) P O Box 7 Broadway Hyde Cheshire SK14 4QW tel:(0161) 368 5711 fax:(0161) 366 8155
The Betta Group Pilthers Logde, unio Road Bradford Reading Berkshire RG7 6AB tel:(01189) 502282 fax:(01189) 588259
The Bollom Group Croydon Road Beckenham Kent BR3 4BL tel:(020) 8658 2299 fax:(020) 8658 8672 sales@bollom.com www.bollom.com
The BSS Group PLC Fleet house Lee Circle Leicester LE1 3QQ tel:(0116) 262 3232 fax:(0116) 253 1343 reception@bssgroup.com www.bssgroup.co.uk
The Building Centre Bookshop 26 Store Street Camden London WC1E 7BT tel:(020) 7692 4040 fax:(020) 7636 3628 bookshop@buildingcentre.co.uk www.buildingcentre.co.uk
The Building Centre Group Ltd 26 Store Street Camden London WC1E 7BT tel:(020) 7692 4000 fax:(020) 7580 9641 information@buildingcentre.co.uk www.buildingcentre.co.uk
The Carpet Foundation MCF Complex, 60 New Road P. O. Box 1155 Kidderminster Worcs. DY10 1AQ tel:(01562) 755568 fax:(01562) 865405 info@carpetfounation.com www.carpetfoundation.com
The Cotswold Casement Company Fosse Way Industrial Estate Stratford Road Moreton in Marsh Glos. GL56 9NQ tel:(01608) 650568 fax:(01608) 651699 info@cotswold-casements.co.uk www.cotswold-casements.co.uk
The Cube 14 Holland Street London W8 4LT tel:(020) 7938 2244 fax:(020) 7938 1920 thefrancesdiblasi@ukbusiness.com www.ukbusiness.com/thefrancesdiblasi
The Delabole Slate Co. Ltd Delabole Slate Quarry Pengelly Delabole Cornwall PL33 9AZ tel:(01840) 212242 fax:(01840) 212948 sales@delaboleslate.co.uk www.delaboleslate.co.uk
The Haigh Tweeny Co Ltd Haigh Industrial Estate Alton Road Ross-on-Wye Herefordshire HR9 5LA tel:(01989) 566222 fax:(01989) 767498 sales@tweeny.co.uk www.tweeny.co.uk
The HB Sign Company Ltd 120 The Kings Road Chelsea London SW3 4TR tel:(020) 7851 8044 fax:(020) 7584 1024 hbsign@dial.pipex.com www.hbsugn.co.uk

The Kenco Coffee Company Cheltenham Glos GL50 3AE tel:(0800) 242000 fax:(01295) 264602
The Safety Tread Ltd Crown Wharf Iron Works Dace Road Old Ford London E3 2NL tel:(020) 8985 4407 fax:(020) 8986 8212
The Salex Group Ltd. Broadway Drive Springwood Ind Est Braintree Essex CM7 2ET tel:(01206) 866911 fax:(01206) 865987 sales@salexgroup.co.uk www.salexgroup.co.uk/salex
The Saw Centre Group 650 Eglinton Street Glasgow G5 9RP tel:(0141) 429 4444 fax:(0141) 429 5609 sales@thesawcentre.co.uk www.thesawcentre.co.uk
The Silk Forest Main Street Bagworth Leicestershire LE67 1DW tel:(01530) 231241 fax:(01530) 231240 info@thesilkforest.com www.thesilkforest.com
The Tile Association Forum Court 83 Copers Cope Road Beckenham Kent BR3 1NR 020 8663 0946 020 8663 0949 info@tiles.org.uk www.tiles.org.uk
Thermal Insulation Contractors Association TICA House, Allington Way Yarm Road Business Park Darlington Co. Durham DL1 4QB 01325 466704 01325 487691 ralphbradley@tica-acad.co.uk www.tica-acad.co.uk/tica.htm
Thermalite Marley Building Materials Station Road Coleshill Birmingham B46 1HP tel:(01675) 468400 fax:(01675) 468455 Thermalite@mbm-marley.co.uk www.thermalite.co.uk
Thermax Interglass Ltd Greenfield Industrial Estate Tindale Cresent Bishop Auckland Co. Durham DL14 9TF tel:(01388) 603667 fax:(01388) 609744
Thermica Ltd Vulcan Street Clough Road Hull HU6 7PS tel:(01482) 348771 fax:(01482) 441873 sales@thermica.co.uk www.thermica.co.uk
Therminal Technology (sales) Ltd Bridge House, Station Road Westbury Wilts. BA13 4HR tel:(01373) 865454 fax:(01373)864425 sales@thermaltechnology.com www.thermaltechnology.co.uk
Thermoscreens Ltd Chandlers Ford Industrial Estate Avenger Close Eastleigh Hampshire SO53 4DQ tel:(023) 80254731 fax:(023) 80266563 info@thermoscreens.ltd.co.uk www.thermoscreens.com
Thomas & Betts Manufacturing Services Ltd Bruntcliffe Lane Morley, Leeds West Yorkshire LS27 9LL tel:(0113) 281 0600 fax:(0113) 281 0601 mailbox@emergi-lite.co.uk
Thomas and Betts 4th Floor Regent Court Laporte Way Luton Beds. LU4 8AE tel:(01582) 589000 fax:(01582) 589001 www.tnb.com
Thomas Lowe Joinery Coppiceside Ind Est Brownshill Walsall W Midlands WS8 7EX tel:(01543) 466966 fax:(01543)574884 info@benwin.co.uk
Thompson Kennicott (Incorporating Howl) PO Box 100 Ettingshall Wolverhampton WV4 6JY tel:(01902) 353353 fax:(01902) 354619
Thor Hammer Co Ltd Highlands Road Shirley Birmingham B90 4NJ tel:(0121) 705 4695 fax:(0121) 705 4727 thorhammer@btinternet.com www.thorhammer.com
Thorlux Lighting Merse Road North Moons Moat Redditch Worcs. B98 9HH tel:(01527) 583200 fax:(01527) 584177 thorlux@thorlux.co.uk www.thorlux.com
Thorn Lighting Ltd 3 King George Close Eastern Avenue West Romford Essex RM7 7PP tel:(01708) 766033 fax:(01708) 776303
Thorn Security Ltd Security House, The Summit Hanworth Road Sunbury on Thames Middx. TW16 5DB tel:(01932) 743333 fax:(01932) 743155
Thoro Systems Products Ltd 19 Broad Ground Road Lakeside, Redditch Worcs. B98 8YP tel:(01527) 517989 fax:(01527) 510299
Threshold Flooring Vorda Works Highworth Swindon Wilts SN6 7AJ tel:(01793) 764301 fax:(01793) 765319 sales@thresholdflr.co.uk www.thresholdfl.co.uk
Thurston Building Systems Quarry Hill Industrial Estate Hawkingcroft Road Horbury, Wakefield W. Yorks. WF4 6AJ tel:(01924) 265461 fax:(01924) 280246 sales@thurstongroup.co.uk www.thurstongroup.co.uk
Thwaites Engineering Co Ltd Welsh Road Works Cubbington Leamington Spa Warwks. tel:(01926) 422471 fax:(01926) 337155
Thyssenkrupp Accessibility LTD 72 Jay Avenue Teesside Industrial Estate Stockton - On-Tees Cleveland TS17 9LZ tel:(01642) 750707 fax:(01642) 767145 publicsales@tke-accuk.com www.thyssenkruppaccessibility.com
Tidmarsh & Sons Unit 14 Ash In Est Flex Meadow Wewyn Garden City Pinnacles, West Harlow , Essex CM19 5TJ tel:(01707) 886226 fax:(01707) 886227 blinds@tidmarsh.co.uk
Tiflex Ltd Lion House Oriental Road Woking Surrey GU22 8AP tel:(01483) 757757 fax:(01483) 757715 shall@tiflex.co.uk www.tiflex.co.uk
Tilcon South Ltd Buxton Derbys. SK17 8TG tel:(01298) 768555 fax:(01298) 768454
Till & Whitehead Ltd, t/a L. Birkinshaw Park House Ings Road, Osmondthorpe Lane Leeds W. Yorks. LS9 9HG tel:(0113) 249 6641 fax:(0113) 248 8968 dunn@tillwite.com www.tillwite.com.uk
Tilley International Ltd 30-32 High Street Frimley Surrey GU16 5JD tel:(01276) 691996 fax:(01276) 27282 tilleyuk@globalnet.co.uk www.tilleylamp.co.uk
Tillicoutry Quarries Ltd Tulliallan Tulliallan, Kincardine By Alloa FK10 4DT tel:(01259) 730481 fax:(01259) 731201 wjmtqlta@aol.com www.tillicoultryquarries.com
Timber Components (UK) Ltd Redding Road Redding by Falkirk Stirlingshire Scotland FK2 9TX tel:(01324) 717171 fax:(01324) 713231
Timber Supplies 60-62 Pretoria Road Patchway Bristol Avon BS12 5PX tel:(0117) 969 1356 fax:(0117) 969 4519
Timber Trade Federation Clareville House 26/27 Oxendon Street London SW1Y 4EL 020 7839 1891 ttf@ttf.co.uk www.ttf.co.uk
Timberwise (UK) Plc Caladonian House Tatton Street Knutsford Cheshire WA16 6AG tel:(01565) 621100 fax:(01565) 621000 hq@timberwise.co.uk www.timberwise.co.uk
Timbmet Ltd PO Box 39 Chawley Works Cumnor Hill Oxford OX2 9PP tel:(01865) 862223 fax:(01865) 864367 sales@timbmet.demon.co.uk www.gabriel.co.uk/timbmet
Time and Data Systems International Ltd (TDSI) Sentinel House Nuffield Road Poole Dorset BH17 0RE tel:(01202) 666222 fax:(01202) 679730 info@tdsi.co.uk www.tdsi.co.uk
Timeguard Ltd Apsley Way London NW2 7UR tel:(020) 8450 8944 fax:(020) 8452 5143 csc@timeguard.com www.timeguard.com
Timloc Expamet Building Products Rawcliffe Road Goole E. Yorks. DN14 6UQ tel:(01405) 765567 fax:(01405) 720479
Tinsley & Co Ltd, Eliza Reddal Hill Road Cradley Heath W. Midlands B64 5JF tel:(01384) 566066 fax:(01384) 639156 marketing@elizatinsley.co.uk www.elizatinsley.co.uk
Titan Ladders Ltd 191-201 Mendip Road Yatton Bristol BS19 4ET tel:(01934) 832161 fax:(01934) 876180 sales@titanladders.co.uk www.titanladders.co.uk
Titan plastech Barbot Hall Ind. Est Mangham Rd Rotherham S. Yorks. S61 4RJ tel:(0870) 362448 fax:(0870) 538301 sales@titanplastech.co.uk www.titanplastech.co.uk
Titan Pollution Control Ltd West Portway Andover Hants. SP10 3LF tel:(01264) 357666 fax:(01264) 366446 info@titanpc.co.uk www.titanpc.co.uk
TLC Southern Ltd Unit 10, Chelsea Fields Industrial Estate 278 Western Road Merton London SW19 2QA tel:(020) 8646 6866 fax:(020) 8646 6750
Today Interiors Ltd Hollis Road Grantham Lincs. NG31 7QH tel:(01476) 574401 fax:(01476) 590208 info@today-interiors.co.uk www.todayinteriors.com
Toffolo Jackson(UK) Ltd Burnfield Road Thornliebank Glasgow G46 7TQ tel:(0141) 649 5601 fax:(0141) 632 9314
Tonbridge Fencing Limited Court Lane Farm Court Lane Harlow Tonbridge Kent TN11 0DP tel:(01732) 852596 fax:(01732) 852593 tonfencing@aol.com
Tony Team Ltd Unit 5 Station Road Bakewell Derbys. DE45 1GE tel:(01629) 813859 fax:(01629) 814334 sales@tonyteam.co.uk www.tonyteam.co.uk
Top Office Equipment Ltd 7 Heron Industrial Estate Cooks Road London E15 2PW tel:(020) 8519 3330 fax:(020) 8519 5142 admin@topoffice.co.uk www.topoffice.co.uk
Topcon Corporation Boundary Road Newbury RG1 45RR
Topravit (UK) Ltd Unit B2, Hubert Road Hubert Road Industrial Estate Brentwood Essex CM14 4JY tel:(01277) 218282 fax:(01277) 218284
Tor Coatings Ltd Portobello Industrial Estate Birtley Chester-le-Street Co. Durham DH3 2RE tel:(0191) 410 6611 fax:(0191) 492 0125 enquiries@tor-coatings.com www.tor-coatings.com
Torit DCE Humberstone Lane Thumaston Leicester Leicestershire LE4 8HP tel:+44 (0)116 269 6161 fax:+44 (0)116 269 3028 toritdce.uk@mail.donaldson.com www.toritdce.com
Tormo Ltd Units 4-5, Hurricane Trading Estate Grahame Park Way London NW9 5QY tel:(020) 8205 5533 fax:(020) 8201 3656
Torvale Building Products, A Division of Stadium Group PLC Pembridge Leominster Herefordshire HR6 9LA tel:(01544) 388262 fax:(01544) 388568 office@torvale.co.uk www.torvale.co.uk
Toshiba Carrier UK United Technologies House Guilford Road Leatherhead Surrey KT22 9UT tel:(01372) 220240 fax:(01372) 220241 sarah.hool@carrier.utc.com www.toshiba-aircon.co.uk
Tower Flue Componants Ltd Tower House Vale Rise Tonbridge Kent TN9 1TB tel:(01732) 351555 fax:(01732) 354445 tjt@tfc.org.uk www.Tfc.ukco.com
Tower Manufacturing Navigation Road Worcester WR5 3DE tel:(01905) 763012 fax:(01905) 763610 clips@towerman.co.uk www.towerman.co.uk
Town & Country Landscapes Ltd Burnsall House 6 Birk Dale Bexhill-On-Sea E. Sussex TN39 3TR tel:(01424) 773834 fax:(01424) 774633
Town and Country Planning Association 17 Carlton House Terrace London SW1Y 5AS tel:(020) 7930 8903 Fax: (020) 7930 3280 tcpa@tcpa.org.uk www.tcpa.org.uk
Townhive Ltd Tropical House Charlwoods Road East Grinstead W. Sussex RH19 2HJ tel:(01342) 305420 / 317171 / 317722 fax:(01342) 315362 townhiveltd@clara.co.uk www.faralradiators.co.uk
Townscape Products Ltd Fulwood Road South Sutton in Ashfield Notts. NG17 2JZ tel:(01623) 513355 fax:(01623) 440267 sales@townscape-products.co.uk www.townscape24.com
Trada Technology Ltd Stocking Lane Hughenden Valley High Wycombe HP14 4ND tel:(01494) 569600 fax:(01494) 565487 information@trada.co.uk www.ashtrada.co.uk
Trade Fireseal Systems Ltd Unit 4 127a Reading Road Wokingham Berks. RG11 1HD tel:(01734) 894614 fax:(01734) 890107
Trade Sealants 34 Aston Road Waterlooville Hampshire PO7 7XQ tel:(023) 92251321 fax:(023) 92264307 tsl@adslmail.newnet.co.uk
Trans World Enterprises Ltd 101 Stephenson Street Canning Town London E16 4SA tel:(020) 7511 2288 fax:(020) 7511 1466 trt@sealandair.com
Transtar New York Way New York Ind Park New castle upon tyne Tyne & Wear NE27 0QF tel:(0191) 483 2797 fax:(0191) 428 0262
Travis Perkins (Southern) 149 Harrow Road London W2 6NA tel:(020) 7262 6602 fax:(020) 7724 7485
TRC (Midlands) Ltd 1 Mount Pleasant Street West Bromwich W. Midlands B70 7DL tel:(0121) 5006181 fax:(0121) 5005075 info@totalroofcontrol.co.uk www.totalroofcontrol.co.uk
TRC Service Group Ltd Unit 36, Acorn Industrial Park Crayford Road Crayford Kent DA1 4AL tel:(01322) 555979 fax:(01322) 524804 sales@trcgroup.co.uk
Treetex Acoustics Ltd 23 Kelvin Way Trading Estate West Bromich W. Midlands B70 7TW tel:(0121) 525 1111 fax:(0121) 500 5406
Trelleborg Ltd 9 Somers Road Rugby Warwks. CV22 7DB tel:(01788) 532400 fax:(01788) 579959
Tremco Ltd 86-88 Bestobell Road Slough Berks. SL1 4SZ tel:(01753) 691696 fax:(01753) 822640
Trend Control Systems Ltd PO Box 34 Horsham W. Sussex RH12 2YF tel:(01403) 211888 fax:(01403) 241608 www.trend-controls.com
Trent Bathrooms PO Box 290 Hanley Stoke-on-Trent ST1 3RR tel:(01782) 202334 fax:(01782) 285474 @trent-bathrooms.co.uk www.trent-bathrooms.co.uk
Trespa UK Ltd Grosvenor House Hollinswood, Central Park Telford Shropshire TF2 9TW tel:(01952) 290707 fax:(01952) 290101 info@trespa.co.uk www.trespa.com
Trianco Redfyre Ltd Thorncliffe, Chapeltown Sheffield S. Yorks. S35 2PZ tel:(0114) 257 2300 fax:(0114) 245 3021 info@trianco.co.uk www.trianco.co.uk
Trim Acoustics Unit 4, Leaside Industrial Estate Stockingswater Lane Enfield London EN3 7PH tel:(020) 8443 0099 fax:(020) 8443 1919 sales@trimacoustics.co.uk www.trimacoustics.co.uk
Trimite Ltd Arundel Road Uxbridge Middx. UB8 2SD tel:(01895) 951234 fax:(01895) 256789 info@trimite.com www.trimite.com
Trimite Scotland Ltd 38 Welbeck Road Darnley Industrial Estate Glasgow G53 7RG tel:(0141) 881 9595 fax:(0141) 881 9333
Triton PLC Shepperton Park Caldwell Road Nuneaton Warwks. CV11 4NR tel:(024) 76 344441 fax:(024) 76 324424
Troika Architectural Mouldings Ltd 850 Herries Road Sheffield S. Yorks. S6 1QW tel:(01142) 327000 fax:(01142) 344885 sales@troikaam.co.uk www.plasterware.com
Trox (UK) Ltd Caxton Way Thetford Norfolk IP24 3SQ tel:(01842) 754545 fax:(01842) 763051 trox@troxuk.co.uk www.troxuk.co.uk
TSE Brownson Valmar Trading Estate Valmar Road Camberwell, London SE5 9NP tel:(020) 7274 4577 fax:(020) 7978 8141 www.tsetrolleys.com
T-T Pumps & T-T controls Onneley Works Newcastle Road Woore CW3 9RU tel:(01630) 647200 fax:(01630) 642100 response@ttpumps.com www.ttpumps.com
Tubeclamps Ltd PO Box 41 Petford Street Cradley Heath, Warley W. Midlands B64 6EJ tel:(01384) 565241 fax:(01384) 410490
Tubela Engineering Co Ltd 11 A/B Hoblongs Ind Estate Chelmsford Road GT Dunmow Essex CM6 1JA 01371 859100 01371 859101 tubebending@tubela.com www.tubela.com
Tucker Fasteners Ltd Emhart Fastening Teknologies Walsall Road Birmingham W. Midlands B42 1BP tel:(0121) 356 4811 fax:(0121) 356 1598 www.emhart.com
Tudor Roof Tile Co Ltd Denge Marsh Road Lydd Kent TN29 9JH tel:(01797) 320202 fax:(01797) 320700 info@todorrooftiles.co.uk www.tudorrooftiles.co.uk
Tufton Ltd High Holborn Road Ripley Derbys. DE5 3NT tel:(01773) 740620 fax:(01773) 740629
Tuke & Bell Ltd Lombard house No 1 Cross Keys Lichfield Staffs. WS13 6DN tel:(01543) 414161 fax:(01543) 250462 sales@tukeandbell.co.uk www.tukeandbell.co.uk
Tully Tully Security Division PES House 17 Deer Park Road London SW19 3XJ tel:(020) 8543 3281 fax:(020) 8543 5344 sales@pesparking.com www.pesparking.com
Tunnel Building Products Ltd PO Box 1 Westerham Kent TN16 2BY tel:(01959) 561776 fax:(01959) 561776
Tunstal Telecom Ltd Whitley Lodge Whitley Bridge N. Yorks. DN14 0HR tel:(01977) 661234 fax:(01977) 661993 sales@tunstall.co.uk www.tunstallgroup.com
Turnbull & Scott (Engineers) Ltd 20 Commercial Road Hawick Roxburghshire TD9 7AQ tel:(01450) 372053 fax:(01450) 377800 info@turnbull-scott.co.uk www.turnbull-scott.co.uk
Turner Access Ltd 65 Craigton Road Glasgow G51 3EQ tel:(0141) 309 5555 fax:(0141) 309 5436 enquiries@turner-access.co.uk www.turner-access.co.uk
Turner Plus Eight Ltd 65 Craigton Road Glasgow G51 3EQ tel:(0141) 440 0663 fax:(0141) 425 1073 pluseight@turner.co.uk www.turner-pluseight.co.uk
Turney Turbines (pumps) Ltd High Mead Works 73 Station Road Harrow Middx. HA1 2TZ tel:(020) 8427 1355 fax:(020) 8863 4215
Turnquest UK Ltd Scimitar Close Gillingham Business Park Gillingham Kent ME8 0RN tel:(01634) 377688 fax:(01634) 370612 marketing@turnquest.co.uk www.turnquest.co.uk
TWI Abington Hall Abington Cambridge CB1 6AL tel:(01223) 891162 fax:(10223) 892588
Twide-Paragon Ltd Glasshouse Fields Stepney London E1W 3JA tel:(020) 7790 2333 fax:(020) 7790 0201 sales@twigroup.co.uk www.twigroup.co.uk
Twiflex Ltd 104 The Green Twickenham Middx. TW2 5AQ tel:(020) 8894 1161 fax:(020) 8894 6056
Twin County Fan Company Ltd Hirwaun Industrial Estate Hirwaun Aberdare M. Glam. CF44 9YA tel:(01685) 813454 fax:(01685) 813325
Twin Technology Lighting PO Box 207 Brixworth Northampton NN6 9HU tel:(01604) 881234 fax:(01604) 882217
Twyford Bushboard 9-29 Rixon Road Wellingborough Northants NN8 4BA tel:(01933) 232200 fax:(01933) 232283 sales@bushboard.co.uk www.twyfordbushboard.co.uk
Tyco European Tubes PO Box 13 Popes Lane Oldbury Warley B69 4PF tel:(0121) 5435700 fax:(0121) 543 5750 sales@tycotube.co.uk www.tycotube.co.uk
Tyeo Thermal Control 3 Rotherford Road Stevenson Ind Esd Washington NE37 3HX 0191 419 8200 fax:(0191) 483 4127 sales@biccthermoheat.co.uk www.biccthermoheat.co.uk
Tyrrell Tanks Ltd 37 Seagoe Industrial Estate Portadown Co. Armagh BT63 5QD tel:(028) 38330668 fax:(028) 38350171
Ubbink (UK) Ltd Borough Road Brackley Northants NN13 7TB tel:(01280) 700211 fax:(01280) 705332 csd@ubbink.co.uk www.ubbink.co.uk
UK Cabin Co Kings Court Kingsway TVTE Gateshead NE11 0SH tel:(0191) 461 1000 fax:(0191) 461 1001
UK Fasteners Ltd C1 Liddington Trading Estate Leckhampton Road Cheltenham Glos. GL53 0DL tel:(01242) 577077 fax:(01242) 577078 sales@ukfastners.co.uk www.ukfasteners.co.uk
UK Fire International Ltd The Safety Centre Mountergate Norwich Norfolk NR1 1PY tel:(01603) 727000 fax:(01603) 727072 norwich@ukfire.co.uk www.ukfire.co.uk
UK Mineral Wool Association,The PO Box 35084 London NW1 4XE 020 7935 8532 020 7935 8532 info@eurisol.com www.eurisol.com
Ulster Carpet Mills Ltd Castleisland Factory Craigavon Co. Armagh BT62 1EE tel:(028) 38334433 fax:(028) 38333142 marketing@ulstercarpets.com www.ulstercarpets.com
Ultraframe PLC Salthill Road Clitheroe Lancs. BB7 1PE tel:(01200) 443311 fax:(01200) 425455 marketing@ultraframe.co.uk www.ultraframe.co.uk/ultra
Unicorn Abrasives (UK) Ltd Doxey road Stafford ST16 1EA tel:(01785) 223281 fax:(01785) 213487 info@unicorn-abrasives.co.uk www.unicorn-abrasives.co.uk
Unicorn Containers Ltd 5 Ferguson Drive Knockmore Hill Industrial Park Lisburn BT282EX tel:(02892) 667264 fax:(02892) 625616 info@unicorn-containers.com www.unicorn-containers.com
Unifix Ltd Bridge House Grove Lane Smethwick W. Midlands B66 2SA tel:0800961110 fax:0800961115 info@unifix.com www.unifix-online.co.uk

Unilock Ltd Chichester Road St Leonards-on-Sea E. Sussex TN38 9YS tel:(01424) 712195 fax:(01424) 457330 info@unilock.co.uk www.unilock.co.uk
Unistrut Limited Edison Road Elms Industrial Estate Bedford Beds. MK41 0HU tel:(01234) 211331 fax:(01234) 270819
United Flexibles (A Division of Senior Flex) Abercanaid Merthyr Tydfil M. Glam. CF48 1UX tel:(01685) 385641 fax:(01685) 389683
United Safety Products UK Ltd Unit 15 Bramley Business Centre Station Road Bramley Surrey GU6 0AZ tel:(01483) 898 666 fax:(01483) 894 666 uspuk@btinternet.com
United Wire Ltd Granton Edinburgh EH5 1HT tel:(0131) 552 6241 fax:(0131) 552 8462 info@unitedwire.com www.unitedwire.com
Universal Components Ltd Universal House Pennywell Road Bristol Avon BS5 0ER tel:(0117) 955 9091 fax:(0117) 955 6091 sales@universal-aluminium.co.uk www.universal-aluminium.co.uk
Universal Filters Ltd Unfil House, Unit 12, The Woodford Trading Estate Southend Road Woodford Green Essex IG8 8HF tel:(020) 8551 4447 fax:(020) 8550 9101
Uponor Housing Solutions Snapethorpe House Rugby Road Lutterworth Leicestershire LE17 4HN 01455 550300 01455 550366 hsenquireis@uponor.com www.uponorhousingsolutions.co.uk
Uponor Ltd Hillcote Plant, PO Box 1 Berristow Lane Blackwell, Nr Alfreton Derbys. DE55 5JD tel:(01773) 811112 fax:(01773) 812343 uponor.sales@uponor.co.uk www.uponor.co.uk
UpRight UK Limited Access House Halesfield 17 Telford Shropshire TF7 4PW Tel:(01952) 685200 fax:(01952) 685255 access@upright.co.uk www.upright.co.uk
Urban Planters 202 Pasture Lane Bradford W. Yorks. BD7 2SE tel:(01274) 579331 fax:(01274) 521150 sales@urbanplanters.co.uk www.urbanplanters.co.uk
Urbis Lighting Ltd Telford Road Houndmills, Basingstoke Hampshire RG21 6YW tel:(01256) 354446 fax:(01256) 841314 sales@urbislighting.com www.urbislighting.com
USF Permutit/Ionpure Harforde Court John Tate Road Hertford Herts. SG13 7NW tel:(01992) 823 300 fax:(01992) 501528
USF Surface Preparation PO Box 60 Broadheath Altrincham Cheshire WA14 5EP tel:(0161) 928 6388 fax:(0161) 929 0381 usfsp@usfinternational.com
USG (UK) Ltd 1 Swan Road South West Industrial Estate Peterlee Co. Durham SR8 2HS tel:(0191) 586 1121 fax:(0191) 586 0097 customer-service@usg.com www.usg.com

V B Johnson LLP

St John's House 304-310 St Albans Road Watford Herts. WD24 6PW tel:(01923) 227236 fax:(01923) 231134 watford@vbjohnson.co.uk www.vbjohnson.co.uk

Vallance Muddensfield Road Birstall, Batley W. Yorkshire WF17 9XA tel:(01924) 354000 fax:(01924) 354001 richard.beene@kalon.co.uk www.kalon.co.uk
Valor Wood Lane Erdington Birmingham B24 9QP tel:(0121) 373 8111 fax:(0121) 373 8181 marketing@valor.co.uk www.valor.co.uk
Valspar UK Corporation Ltd Avenue One Station Lane Witeney Oxfordshire OX28 4XR tel:(01993) 707400 fax:(01993) 775579
Valtti Specialist Coatings Ltd Unit 3B South Gyle Cresent Lane Edinburgh EH12 9EG tel:(0131) 334 4999 fax:(0131) 334 3987
Vandex (UK) Ltd PO Box 200 Guilford Surrey GU2 4WD tel:(01372) 363040 fax:(01372) 363373 tech@vandex.co.uk www.vandex.co.uk
Vantrunk Engineering Ltd Goodard Road Astmoor Runcorn Cheshire WA7 1QF tel:(01928) 564211 fax:(01928) 580157 sales@vantrunk.co.uk www.Vantrunk.com
Varley & Gulliver Ltd 57-70 Alfred Street Sparkbrook Birmingham B12 8JR tel:(0121) 773 2441 fax:(0121) 776 6875 sales@v-and-g.co.uk www.v-and-g.co.uk
Veka PLC Farrington Road Rossendale Road Indstrial Estate Burnley Lancs. BB11 5DA tel:(01282) 416611 fax:(01282) 439260
Velfac Ltd The Old livery Hildersham Cambridge CB1 6DR tel:(01223) 897100 fax:(01223) 897101 post@velfac.co.uk www.velfac.co.uk
VELUX Co Ltd, The Woodside Way Glenrothes E. Fife KY7 4ND tel:(01592) 772211 fax:(01592) 771839 enquires@velux.co.uk www.velux.co.uk
Vencel Resil Ltd Infinity House Anderson Way Belvedere Kent DA17 6BG tel:(020) 83209100 fax:(020) 83209110 technical@vencel.co.uk www.vencel.co.uk
Vent-Axia Flemming Way Crawley W. Sussex RH10 2NN tel:(01293) 526062 fax:(01293) 551188 info@vent-axia.com www.vent-axia.com
Ventlane Ltd Betha Street Bolton Gtr. Manchester BL1 8AH tel:(01204) 365858 fax:(01204) 364440
Verco Office Furniture Ltd Chapel lane Sands High Wycombe Bucks. HP12 4BG tel:(01494) 448000 fax:(01494) 464216 info@verco.co.uk www.verco.co.uk
Verine Ltd Folly Faunts House Goldhanger Maldon Essex CM9 8AP tel:(01621) 788611 fax:(01621) 788754 sales@verine.co.uk www.verine.co.uk/verine/
Verna Ltd Folds Road Bolton Lancs. BL1 2TX tel:(01204) 529494 fax:(01204) 521862
Versatile Fittings Ltd Bicester Road Aylesbury Bucks. HP19 3AU tel:(01296) 483481 fax:(01296) 437596 general@versatile-fittings.co.uk
Vertika International Ltd 10 Fountain Crescent Inchinnan Business Park Inchinnan Renfrewshire PA4 9RE tel:(01625) 611622 fax:(01625) 501510
Ves Andover Ltd Eagle Close Chandlers Ford Industrial Estate Chandlers Ford Eastleigh Hants SO53 4NF tel: 08702 40 43 40 fax: 08702 40 45 50 vesltd@ver.co.uk www.ves.co.uk
Vescom (UK) Ltd Canada Close Bradbury Oxfordshir OX16 2RT tel:(01908) 696222 fax:(01908) 696111
Vianen Kitchen Ventilation (UK) Ltd. Welton Road Warwick CV34 5PZ tel:(01926) 496644 fax:(01926) 493977 vainen@ao.com
Vicaima Ltd Marlowe Avenue Greenbridge Industrial Estate Swindon Wilts. SN3 3JF tel:(01793) 532333 fax:(01793) 530193 info@vicaima.com www.vicaima.com
Vicon Industries Ltd Brunel Way Fareham Hampshire PO15 5TX tel:(01489) 566300 fax:(01489) 566322 sales@vicon.co.uk www.vicon-cctv.com
Victaulic Systems 46-48 Wibury Way Hitchin Herts. SG4 0UD tel:(01462) 443322 fax:(01462) 443148 info@victaulic.co.uk www.victaulic.co.uk
Vidionics Security Systems Ltd Systems House, Desborough Industrial Park Desborough Park Road High Wycombe Bucks. HP12 3BG tel:(01494) 459606 fax:(01494) 461936 sales@videonics.co.uk www.videonics.co.uk
Viking Johnson 46-48 Wilbury Way Hichen Herts SG4 0UD tel:(0121) 700 1000 fax:(121) 700 1001 enquiries@glywedpipesystems-uk.com www.glynwedpipesystems-uk.com
Viking Laminates Ltd Spanbourne Avenue Chippenham Wilts. SN15 1LQ tel:(01249) 656353 fax:(01294) 659633 www.viklam.co.uk
Viking Laundry Equipment Ltd 1 Carlisle Road London NW9 0HZ tel:(020) 8205 7285 fax:(020) 8200 3741 mail@viking1959.co.uk www.viking1959.co.uk
Viking Security Systems Ltd Security House, 4 Lismirrane Industrial Park Elstree Road Elstree Herts. WD6 3EE tel:(020) 8207 3838 fax:(020) 82074575 sales@albangroup.co.uk www.albangroup.co.uk
Visco Ltd Stafford Road Croydon Surrey CR9 4DT tel:(020) 8686 3861 fax:(020) 8681 5930 info@visco.co.uk www.visco.co.uk
Viscount Catering Ltd PO Box 16 Green Lane Ecclesfield, Sheffield S. Yorks. S35 9ZY tel:(0114) 257 0100 fax:(0114) 257 0251 sales@viscount-catering.co.uk www.viscount-catering.co.uk
Vision Neon & Display Ltd 2 Altenburg Gardens Battersea London SW11 1JJ tel:(020) 7228 6511 fax:(020) 72230386 signs@visionneon.fsnet.co.uk
Vision Sytems (Europe) Ltd Vision House, Focus 31 Mark Road Hemel Hempstead Herts. HP2 7BW tel:(01442) 242330 fax:(01442) 240327 adpro.sales@vse.co.uk www.adprp.com.au
Visqueen Building Products Zedcor Business Park Bridge Street Witney Oxon. OX8 6LJ tel:(01993) 776346 fax:(01995) 776233 enquiries@visqueenbuilding.co.uk www.visqueenbuilding.co.uk
Vista Engineering Ltd Carr Brook Works Shallcross Mill Road Whaley Bridge High Peak SK23 7JL tel:(01663) 736700 fax:(01663) 736710 sales@vistaeng.co.uk www.vistaeng.co.uk
Vista Plan International PLC High March Daventry Northants NN11 4QE tel:(01327) 704767 fax:(01327) 300243
Visua Ltd 7 Skyport Drive Harmondsworth Middx UB7 0LB tel:(020) 8476 2000 fax:(020) 8476 2111 peter.woodard@itm-group.co.uk www.itm-group.co.uk
Vita Liquid Polymers Ltd Harling Road Wythenshaw Manchester M22 4SZ tel:(0161) 998 3226 fax:(0161) 946 0118 info@vitaliquid.co.uk www.vita-liquid.co.uk
Vitalighting Ltd Unit 4, Sutherland Court Moor Park Industrial Estate, Tolpits Lane Watford Herts. WD18 9NA tel:(01923) 896476 fax:(018923) 897741 sales@vitalighting.com www.vitalighting.com
Vitra (UK) Ltd Park 34 Collet Way Didcot Oxon. OX11 7WB tel:(01235) 750990 fax:(01235) 750980 info@vitrauk.com www.vitrauk.com
Vitra UK Ltd 30 Clerkenwell Road London EC1M 5PQ tel:(020) 7608 6200 fax:(020) 7608 6201 info-uk@vitra.com www.vitra.com
Vodafone Paging Ltd 21-22 Park Way Newbury Berks. RG14 1EE tel:(01635) 521800 fax:(01635) 523016 www.vodafone.co.uk
Vokes Henley Park Guildford Surrey GU3 2AF tel:(01483) 569971 fax:(01483) 235384 vokes@btrinc.com www.vokes.com
Voko (UK) Ltd Fitz Eylwin House 25Holborn Viaduct London EC1A 2BP tel:020) 7583 7000 fax:(020) 7583 1700
Vola Unit 12, Ampthill Business Park Station Road Ampthill Bedfordshire MK45 2QW tel:(01525) 841155 fax:(01525) 841177 sales@vola.co.uk www.vola.com
Volex Group PLC Dornoch House Kelvin Close Birchwood Warrington WA3 7JX tel:(01925) 830101 fax:(01925) 830141
Vortice Ltd Beeches House Eastern Avenue Burton-on-Trent Staffordshire DE13 0BB (01283) 492949 (01283) 544121 sales@vortice.ltd.uk www.vortice.ltd.uk
VT Plastics Ltd Shoreham House Shoreham Street Sheffield S. Yorks. S1 4SR tel:(01742) 757161 fax:(01742) 758920
Vulcan Plastics Ltd Hosey Hill Westerham Kent TN16 1TZ tel:(01959) 562304
Vulcanite Limited Barton Dock Road Mancheater Wakefield M32 0YL tel:(0161) 865 4444 fax:(01924) 252006/240435
Vulcascot Ltd Gatwick Gate Estate Lowfield Heath Crawley W. Sussex RH11 0TG tel:(01293) 560130 fax:(01293) 537743
W Ward Fencing Contractors K & M Hauliers Indust. Estate The Aerodrome, Watnall Road Hucknall Notts. NG15 6EQ tel:(0115) 963 6948 fax:(0115) 963 0864
W. Hawley & Son Ltd Colour Works Lichfield Road, Branston Burton-on-trent Staffordshire DE14 3WH tel:(01283) 714200 fax:(01283) 714201 info@hawley.co.uk www.hawley.co.uk
WA Skinner & Co Limited Dorset Way, off Abbot Close Byfleet Surrey KT14 7LB tel:(01932) 344228 fax:(01932) 348517
Wade Building Services Ltd Groveland Road Tipton W. Midlands DY4 7TN tel:(0121) 520 8121 fax:(0121) 557 7061 sales@wade-bs.co.uk www.wade-bs.co.uk
Wade Ceramics Ltd Westport Road Burslem Stoke-on-Trent ST6 4AP tel:(01782) 577321 fax:(01782) 575195 www.wade.co.uk
Wade International Ltd Third Avenue Halstead Essex CO9 2SX tel:(01787) 475151 fax:(01787) 475579 tech@wadedrainage.co.uk www.wadedrainage.co.uk
Wagner (GB) Ltd VBH House Bailey Drive, Gillingham Business Park Gillingham Kent ME8 0WG tel:(01634) 263300 fax:(01634) 263504 hq@wagnergb.com www.wagnergb.com
Wales & Wales The Long Barn Workshop Muddles Green, Chiddingly Lewes East Sussex BN8 6HW tel:(01825) 872764 fax:(01825) 873197 rod@walesandwales.com www.walesandwales.com
Walker James & Co Ltd Lion House Woking Surrey GU22 8AP tel:(01483) 757575 fax:(01483) 755711
Walker Management Services PO Box 525 Leigh on Sea Essex SS9 5RN tel:(01702) 523628 fax:(01702) 77161
Walker Timber Ltd Carriden Sawmills Bo'ness W. Lothian EH51 9SQ tel:(01506) 823331 fax:(01506) 822590 mail@walkertimber.com
Wall Tie Installers Federation Heald House Heald Street Liverpool L19 2LY 0151 494 2503 0151 494 2511 hughbanks@lineone.net www.wtif.org.uk
Wallbank Wallbank Farm, Wallbank Road, Bramhall, Cheshire SK7 3AP tel:(0161) 439 0908 fax:(0161) 439 0908
Wallcovering Manufacturers Association James House Bridge Street Leatherhead Surrey KT22 7EP 01372 360660 01372 376069 enquiry@bcf.co.uk www.coatings.org.uk
Wallis Office Furniture Ltd 8-18 Fowler Road Hainault Industrial Estate Ilford Essex IG6 3UP tel:(020) 8500 9991 fax:(020) 8500 1949
Walls & Ceilings International Ltd Tything Road Arden Forest Industrial Estate Alcester Warwickshire B49 6EP tel:(01789) 763727 fax:(01789) 400312 sales@walls-and-ceilings.co.uk www.walls-and-ceilings.co.uk
Wandsworth Elecrtrical Ltd Albert Drive Sheerwater Woking Surrey GU21 5SE tel:(01483) 740740 fax:(01483) 740384 info@wandsworth-electrical.com www.wandsworth-electrical.com
Ward Building Components Ltd Sherburn Malton N. Yorks. YO17 8PQ tel:(01944) 712000 fax:(01944) 710555 wbc@wards.co.uk www.wards.co.uk
Wardle Storeys Grove Mill Earby via Colne Lancs. BB8 6UT tel:(01282) 842511 fax:(01282) 843170
Ward's Flexible Rod Company Ltd 22 James Carter Road Mildenhall Suffolk IP28 7DE tel:(01638) 713800 fax:(01638) 716863 sales@wardsflex.co..uk www.wardsflex.co.uk
Warefence Limited Clare Terrace Carterton South Industrial Estate Carterton Oxon. OX18 3ES tel:(01993)845279 fax:(01993)840551
Warner H & Son Ltd Arclion House Hadleigh Road Ipswich Suffolk IP2 0EQ tel:(01473) 253702 fax:(01473) 233330 www.warners-engineers.co.uk
Warner Howard Group Ltd Warner Howard House 2 Woodgrange Avenue Harrow Middlesex HA3 0XD tel:(020) 8206 2900 fax:(020) 8206 1313 info@warnerhoward.co.uk www.warnerhoward.w.uk
Warsop-Metrix Hever Road Edenbridge Kent TN8 5DL tel:(01732) 863081 fax:(01732) 865680
Watco UK Ltd Watco House Filmer Grove Godalming Surrey GU7 3AL tel:(01483) 425050 fax:(01483) 428888 sales@watco.co.uk
Watkins Lichfield Auction Center Fradley Staffs. WS13 8NF tel:(01543) 250088 fax:(01543) 415254
Wavin Plastics Ltd Parsonage Way Chippenham Wilts. SN15 5PN tel:(01249) 766600 fax:(01249) 443286 infi@wavin.co.uk www.wavin.co.uk
Wearparts Ltd Oaks Industrial Estate Gilmorton Road Lutterworth Leics. LE17 4HA tel:(01455) 553551 fax:(01455) 550907 sales@wearparts.com www.wearparts.com
Weatherley Fencing Contractors Limited The Orchard 135 North Cray Road Sidcup Kent DA14 5HE tel:(020) 8308 6421 fax:(020) 8308 1317
Weber Building Solutions Dickens House Enterprise Way Flitwick Bedford MK45 5BY tel:(08703) 33 00 70 fax:(01525) 718988 mail@weberbuildingsolutions.co.uk www.weberbuildingsolutions.co.uk
Wednesbury Tube Oxford Street Bilston W. Midlands WV14 7DS tel:(01902) 491133 fax:(01902) 405838
Welconstruct Co Ltd Woodgate Business Park Kettles Wood Drive Birmingham B32 3GH tel:(0121) 421 9000 fax:(0121) 421 9888 mail@welconstruct.co.uk www.welconstruct.co.uk
Wellman Robey Ltd Newfield Road off Dudley Road East Oldbury, Warley W. Midlands B69 3ET tel:(0121) 552 3311 fax:(0121) 552 4571 sales@wellmanrobey.com www.wellmanrobey.com
Wenmore Rooflights, A Division of W.F.Joy & Co. Ltd Unit 1A Parnall Industrial Estate Parnall Road Fishponds Bristol BS16 3JF tel:(01179) 585865 fax:(01179) 585865
Wentworth Sawmills Ltd, Barrowfield Lane, Wentworth, Nr. Rotherham S. Yorks. S62 7TP tel:(01226) 742206 fax:(01226) 742484
Wernick S & Sons Ltd Molineux House Russell Gardens Wickford Essex SS1 8BL tel:(01268) 735544 fax:(01268) 560026
Wescol Glosford Westercroft Lane Northowram Halifax HX3 7TY tel:(01422) 203522 fax:(01422) 205365
Wessenden Products Prospect Road Alresford Hampshire SO24 9QQ tel:(01962) 732867 fax:(01962) 735331
Wessex Medical Budds Lane Romsey Hampshire SO51 0HA tel:(01794) 830303 fax:(01794) 512621 info@wessexmedical.co.uk www.wessexmedical.co.uk
Wessex Resins & Adhesives Ltd Cuperham House Cuperham Lane Romsey Hants. SO51 7LF tel:(01794) 521111 fax:(01794) 521271 info@wessex-resins.com www.wessex-resins.com
West London Security Security House 19 Stannary Street London SE11 4AA tel:(020) 7735 7733 fax:(020) 7582 6266 www.westlondonsecurity.com
Westbrick Tarmac Bricks & Tiles Ltd Pinhoe Exeter Devon EX4 8JT tel:(01392) 66561 fax:(01392) 68188
Westbury Filters Limited Hall Farm Estate Gadbrook Road Betchworth Surrey RH3 7AH tel:(01306) 611611 fax:(01306) 611613 sales@westburyfilters.co.uk www.westburyfilters.co.uk
Westco Group Ltd Penarth Road Cardiff CF11 8YN tel:(029) 20 376700 fax:(029) 20 383573 westco@westcodiy.co.uk www.westcodiy.co.uk
Western Cork Ltd Penarth Road Cardiff CF11 8YN tel:(029) 20 376700 fax:(029) 20 383573 customerservices@westcodiy.co.uk www.westcofloors.co.uk
Western landscapes Ltd t/a Wyevale Landscapes Upper Buckover Farm Buckover Wotton-under-Edge Glos. GL12 8DZ tel:(01454) 419175 fax:(01454) 412901 wyevale@wyevale-landscapes.co.uk www.wyevale-landscapes.co.uk
Westminster Scaffolding Ltd Westminster House Pensbury Place London SW8 4TP tel:(020) 7720 3404 fax:(020) 7622 5846
Westpile Ltd Dolphin Bridge House Rockingham Road Uxbridge Middx. UB8 2UB tel:(01895) 258266 fax:(01895) 271805 estimating@westpile.co.uk www.westpile.co.uk
WF Electrical Distributors 313-333 Rainham Road South Dagenham Essex RM10 8SX tel:(020) 8515 7000 fax:(020) 8595 0519 www.wf-online.com
WH Wesson Fencing Limited 126 Connaught Road Brookwood Woking Surrey GU24 0AS tel:(01483) 472124 fax:(01483) 472115
Whipp & Bourne Switchgear Works Manchester Road Castleton, Rochdale Lancashire OL11 2SS tel:(01706) 632051 fax:(01706) 674236 sales@st.fki-et.com www.fki-et.com
White Gates Westcoe Hill Lane Weeton Nr. Leeds LS17 0AS tel:(01423) 734466
White Seal Stairways Ltd 88 Hopewel Drive Chatham Kent tel:(01634) 812371 fax:(01634) 812663
Whitecroft Lighting Limited Burlington Street Ashton-Under-Lyne Lancs. OL7 0AX tel:(0870) 5087087 fax:(0870) 5084210 email@whitecroftlight.com www.whitecroftlighting.com

Whitehill Spindle Tools Ltd 6 Union Street Luton Beds. LU1 3AN tel:(01582) 736881 fax:(01582) 488987 david@Whitehill.tools.com www.whitehill-tools.com

Whitmore's Timber Co Ltd Main Road Claybrooke Magna, Nr Lutterworth Leics. LE17 5AQ tel:(01455) 209121 fax:(01455) 209041

Wilec Ltd 17 West Dover Drive Bournemouth BH1 2BY tel:(01202) 290370

Willamette Europe Ltd 10th Floor, Maitland House Warrior Square Southend-on-Sea Essex SS1 2JY tel:(01702) 619044 fax:(01702) 617162 sales@willamette-europe.com www.willamette-europe.com

William Freeman Ltd Wakefield Road Staincross Barnsley S. Yorks. S75 6DH tel:(0870) 4421250 fax:(01226) 731832 sales@williamfreeman.com www.williamfreeman.com

Williams & Farmer Monk Meadow Hempstead Lane Gloucester GL2 5JJ tel:(01452) 300300 fax:(01452) 300791

Williamson T & R Ltd 36 Stonebridge Gate Ripon N. Yorks. HG4 1TP tel:(01765) 607711 fax:(01765) 607908 enquiries@trwilliamson.co.uk trwilliamson.co.uk

Wilo Samson Pumps Ltd 2ND Avenue Centrum 100 Burton on Trent Staffordshire DE14 2WJ tel:(01332)385181 fax:(01332)344423 sales@wilo.co.uk www.wilo.co.uk

Wilson & Garden Ltd 17-21 Newtown Street Kilsyth Glasgow G65 0JX tel:(01236) 823291 fax:(01236) 825356 info@wilsonandgarden.com www.wgltd.com

Wilsonart Ltd 2 Norfolk Court Norfolk Road Rickmansworth Herts WD3 1LA tel:(01923) 712846 fax:(01923) 774642 www.wilsonart.com

Wilsons Fibreglass Ltd Fitzherbert Road Farlington Industrial; Estate Portsmouth PO6 1RU tel:(023) 92384921 fax:(023) 92210716 wilsonsgrp@aol.com

Wiltshire G R & Co Main Road Claybrooke Magna, Lutterworth Leics. LE17 5AQ tel:(01455) 202666 fax:(01455) 209041

Wincro Metal Industires Ltd Fife Street Wincobank Sheffield S. Yorks. S9 1NJ tel:(0114) 242 2171 fax:(0114) 243 4306 sales@wincro.com www.wincro.com

Window Care Systems Limited Unit E, Sawtry Business Park Glatton Road Sawtry, Huntingdon Cambs PE28 5GQ 01487 830311 01487 832876 sales@window-care.net

Windowbuild 55-56 Lewis Road East Moors Cardiff CF24 5EB tel:(029) 20 307200 fax:(029) 20 480030 sales@windowbuild.co.uk www.windowbuild.co.uk

Winn & Coales (Denso) Ltd Denso House Chapel Road London SE27 0TR tel:(020) 8670 7511 fax:(020) 8761 2456 mail@denso.net www.denso.net

Winther Browne & co Ltd 3 Nobel Road Eley Estate Edmonton London N18 3DX tel:(020) 8884 6000 fax:(020) 8884 6001 sales@wintherbrowne.co.uk www.wintherbrowne.co.uk

Witham Oil & Paint Ltd Stanley Road Oultoh Broad Lowestoft, Suffolk HR33 9ND 01502 563434 01502 500010 kathy@withamoil-lowestoft.co.uk www.withamoil.co.uk

Withey Contracts Ltd Woburn House, 3 Adams Close Woburn Road Industrial Estate Kempstone Bedfordshire MK42 7JE tel:(01234) 844600 fax:(01234) 844601 sales@contracts.withey.co.uk www.witheygroup.co.uk

WL West & Son Limited Selham Petworth W. Sussex GU28 0PJ tel:(01798) 861611 fax:(01798) 861633 sales@wlwest.co.uk www.wlwest.co.uk

Wm. Bain Fencing Ltd Lochrin Works Waverley Street Coatbridge Lanarkshire ML5 2BB tel:(01236) 423471 fax:(01236) 435097 sales@lochrin-bain.co.uk www.lochrin-ban.co.uk

Wood E Ltd Standard Way North Allerton N. Yorks. DL6 2XA tel:(01609) 780170 fax:(01609) 780438 grant@ewood.co.uk

Wood International Agency Ltd Wood House 16 King Edward Road Brentwood Essex CM14 4HL tel:(01277) 232991 fax:(01277) 222108 woodia@msn.com www.ourworld.compuserve.com/homepages/wood_international_agency

Wood Wool Slab Manufacturers Association 26 Store St. London WC1E 7BT 020 7323 3770 020 7323 0307

Woodburn Engineering Ltd Rosganna Works Trailcock Road Carrickfergus Co. Antrim BT38 7NU tel:(028) 93366404 fax:(028) 93367539 Terri@woodburnengineeringltd.co.uk

Woodhouse UK Plc Spa Park Lemington Spa Warwickshire CV31 3HL tel:(01926) 314313 fax:(01926) 883778 enquiries@woodhouse.co.uk www.woodhouse.co.uk

Woodscape Ltd Church Works Church Street Church Lancshire BB5 4JT 01254 383322 01254 381166 sales@woodscape.co.uk www.woodscape.co.uk

Woolliscroft Tiles Ltd Melville Street Hanley Stoke-on-Trent ST1 3ND tel:(01782) 208082 fax:(01782) 202631 wtiles@btinternet.com www.wtiles.com

Worcester Heat Systems Ltd Cotswold Way Warndon Worcester WR4 9SW tel:(01905) 754624 fax:(01905) 754619 www.worcester-bosch.co.uk

Wormald Ansul UK Ltd Wormald Park Grimshaw Lane Newton Heath Manchester M40 2WL tel:(0161) 205 2321 fax:(0161) 455 4459

WP Metals Ltd Westgate Aldridge W. Midlands WS9 8DJ tel:(01922) 743111 fax:(01922) 743344 info@wpmetals.co.uk

Wragg Bros (Aluminium Equipment) Ltd Robert Way Wickford Business Park Wickford Essex S11 8DQ tel:(01268) 732607 fax:(01268) 768499

Wright Security incorporating Madewel Electronic Systems Ltd 220 Higher Road Urmston Manchester M41 1BH tel:(0161) 748 1115 fax:(0161) 746 8415

WT Henley Ltd Crete Hall Road Gravesend Kent DA11 9DA tel:(01474) 564466 fax:(01474) 566703 www.wt-henley.com

Wybone Ltd Mason Way, Platts Common Industrial Estate Hoyland Nr Barnsley S. Yorks. S74 9TF tel:(01226) 744010 fax:(01226) 350105 sales@wybone.co.uk www.wybone.co.uk

Wyckham Blackwell Ltd Old Station Road Hampton in Arden, Solihull W. Midlands B92 0HB tel:(01675) 442233 fax:(01675) 442227 wyckham.blackwell@zetnet.co.uk www.wyckham-blackwell.co.uk

Wylex Ltd Wylex Works Wythenshaw Manchester M22 4RA tel:(0161) 998 5454 fax:(0161) 945 1587

Wymark Ltd Runnings Road Industrial Estate Cheltenham Glos. G51 9NQ tel:(01242) 520966 fax:(01242) 519925

Wyndawaye Systems Ltd 9/10 Kelvin Way Trading Estate Kelvin Way West Bromwich W. Midlands B70 9LL tel:(0121) 553 0989 fax:(0121) 553 3280

Wyvern Marlborough Ltd Grove Trading Estate Dorchester Dorset DT1 1SU tel:(01305) 264716 fax:(01305) 264717

Xpelair Ltd PO Box 279 Morley Way Peterborough PE2 9JJ tel:(01733) 456789 fax:(01733) 310606 www.xpelair.co.uk

Yaght Parts (London) 99 Fulham Palace Road London W6 8JA tel:(020) 8741 9803 fax:(020) 8741 5727

Yale Security Products Ltd Wood Street Willenhall W. Midlands WV13 1LA tel:(01902) 366911 fax:(01902) 368535

Yonder Hill Ltd Yonder Hill Chard Junction Chard Somerset TA20 4QR tel:(01460) 220751 fax:(01460) 221392

Yorian Garden Products 311 Lidgett Lane Leeds 17 LS17 6PD tel:(0113) 268 2682 fax:(0113) 268 2682

York Handmade Brick Co Ltd Winchester House Forest Lane Alne, York N. Yorkshire Y061 1TU tel:(01347) 838881 fax:(01347) 838885 sales@yorkhandmade.co.uk www.yorkhandmade.co.uk

Yorkon Ltd New Lane Huntington York YO32 9PT tel:(01904) 610990 fax:(01904) 610880 contact@yorkon.com www.yorkon.com

Yorkshire Copper Tube Ltd East Lancashire Road Kirkby Liverpool L33 7TU tel:(0151) 545 5075 fax:(0151) 549 2139 sales@yorkshirecoppertube.co.uk www.yorkshirecoppertube.co.uk

Yorkstone Products Ltd Britannia Quarries Morley Leeds LS7 0SW tel:(0113) 253 0464 fax:(0113) 252 7520

Yougman S G B Ltd Stane Street Slinford, Horsesham W. Sussex RH13 7RD tel:(01403) 790456 fax:(01403) 790881

Youngs Doors Ltd City Road Works Norwich Norfolk NR1 3AN tel:(01603) 629889 fax:(01603) 764650 mail@youngs-doors.co.uk www.youngs-doors.co.uk

Zehnder Ltd B15 Armstrong Mall Southwood Business Park Farnborough Hampshire GU14 0NR tel:(01252) 515151 fax:(01252) 522528 sales@zhnder.co.uk www.zehnder.co.uk

Zellweger Analytic Ltd, Sieger Division Hatch Pond House 4 Stinsford Road Poole Dorset BH17 0RZ tel:(01202) 676161 fax:(01202) 678 011 post@zellweger-analytics.co.uk www.zelana.com

Zephyr The Visual Communicators Midland Road Thrapston Northants NN14 4LX tel:(01832) 734 484 fax:(01832) 733 064 sales@zephyr-tvc.com www.zephyr-tvc.com

Zinc Alloy Ltd Shakespeare Street Wolverhampton WV1 3LR tel:(01902) 452915 fax:(01902) 352917

Zled (DPC) Co Ltd Unit 9, Tonbridge Chambers Pembury Road Tonbridge Kent TN9 2HZ tel:(01732) 363 443 fax:(01732) 363 553 info@z-led.com www.z-led.com

Zon International PO Box 329 Edgware Middx. HA8 6NH tel:(020) 8381 1222 fax:(020) 8381 1333 sales@zon.co.uk www.zon.co.uk

PRODUCTS AND SERVICES

A01 Institutions/ Associations/ Trade Organisations

Accrediting body and learned society - Chartered Institution of Building Services Engineers (CIBSE)
Aluminium trade association - Aluminium Federation (Alfed) Ltd
Aluminium window association - Council for Aluminium in Building
Architectural aluminium association - Council for Aluminium in Building
Asbestos training courses for management and operatives - Asbestos Removal Contractors' Association
Association and institution - Joint Council For The Building & Civil Engineering Industry (N. Ireland)
Association of concrete technologists - Institute of Concrete Technology The
BCIS guide to dayworks rates - Building Cost Information Service
BCIS guide to house rebuilding costs - Building Cost Information Service
BCIS online - Building Cost Information Service
BCIS review of building prices - Building Cost Information Service
BMI - Building Maintenance Information
Bookshops - Royal Incorporation of Architects in Scotland
Brick information service - Brick Development Association
Builders federation - Federation of Master Builders
Builders federation - National Federation of Builders
Building cost information service - Building Cost Information Service
Building industry joint council - Construction Industry Joint Council
Cable management products trade association - British Electrical Systems Association
Campaigns in Britain and abroad - Society for the Protection of Ancient Buildings (SPAB)
Carpet manufacturers trade association - The Carpet Foundation
Carpet Industry Trade Association - The Carpet Foundation
Carpet promotion - The Carpet Foundation
Casework to save old buildings at risk from damage or demolition - Society for the Protection of Ancient Buildings (SPAB)
Charity for the built environment - Civic Trust
Clay pipe trade asssociation - Clay Pipe Development Association Ltd
Client services and advice - Royal Incorporation of Architects in Scotland
Conferences and events - Royal Incorporation of Architects in Scotland
Construction confederation - Construction Confederation
Construction employers federation - Construction Employers Federation
Construction information on CD-Rom. - Technical Indexes Ltd
Construction information service online - Technical Indexes Ltd
Consultancy services on procurement - Royal Incorporation of Architects in Scotland
Contract flooring journal - Contract Flooring Association
Copper Development Association - Copper Development Association
Cost engineers' association - Association of Cost Engineers
Decorators trade association - Painting & Decorating association
Design competitions administration - Royal Incorporation of Architects in Scotland
Distributor of the Arts' share of the National Lottery - Arts Council of England
Draught proofing trade association - Draught Proofing Advisory Association
Dry stone walling association - Dry Stone Walling Association of Great Britain
Education and training for building professionals - Brick Development Association
Electrical association - Electrical Contractors Association, The
Estimators - V B Johnson LLP
Event organizing/book seller/information provider - Chartered Institution of Building Services Engineers (CIBSE)
Events including lectures and regional group visits - Society for the Protection of Ancient Buildings (SPAB)
Fire protection association - British Fire Protection Systems Asssociation
Free technical advice on the maintenance of old buildings - Society for the Protection of Ancient Buildings (SPAB)
Gallery - Royal Incorporation of Architects in Scotland
Galvanizers trade association - Galvanizers Association
Glazing trade association - Glass and Glazing Federation The
Heating and ventilating trade association - Heating & Ventilating Contractors Association
Housing association - Island Cottages Ltd
Housing associations / registered social landlords - Housing Corporation
Information & advice on craft of dry stone walling. - Dry Stone Walling Association of Great Britain
Information service to businesses - British Library Business Information Service
Institution of Incorporated Engineers in electronic, electrical and mechanical engineering (IIE) - Institution of Incorporated Engineers (IIE)
Institutions and association - Architectural Association
Institutions and association - Asbestos Removal Contractors' Association
Institutions and association - Association of Interior Specialists
Institutions and association - British Ceramic Research Ltd.
Institutions and association - British Precast Concrete Federation
Institutions and association - British Standards Institution
Institutions and association - British Wood Preserving & Damp Proofing Association
Institutions and association - Catering Equipment Suppliers Association
Institutions and association - Cement Admixtures Association
Institutions and association - Commission For Architecture and the Built Environment
Institutions and association - Concrete Pipe Association
Institutions and association - Cookery & Food Association
Institutions and association - Department of Environment
Institutions and association - English Heritage
Institutions and association - Friends of the Earth
Institutions and association - HMSO Publications Office
Institutions and association - Institution of Electrical Engineers
Institutions and association - Institution of Mechanical Engineers
Institutions and association - Institution of Structural Engineers
Institutions and association - Millenium Commission, The
Institutions and association - Mortar Industry Association
Institutions and association - Ordnance Survey
Institutions and association - Quarry Products Association
Institutions and association - Royal Institute of British Architects (RIBA)
Institutions and association - Royal Town Planning Institute
Institutions and associations - British Association for Chemical Specialities
Institutions and associations - Civic Trust
Institutions and associations - Concrete Society
Institutions and associations - Institution of Civil Engineers
Institutions and associations - Interpave
Institutions and associations - Kitchen Specialists Association
Institutions and associations - Lead Development Association International
Institutions and associations - National Federation of Master Steeplejacks & Lightning Conductor Engineers
Institutions and associations - Precast Flooring Federation
Institutions and associations - Prestressed Concrete Association
Institutions and associations - Society for the Protection of Ancient Buildings (SPAB)
Institutions and associations - Southern Brick Federation Limited
Institutions/Associations - British Slate Association
Institutions/Associations - Ferfa
Institutions/Associations - Institute of Quarrying
Institutions/Associations - Leatherhead Food Research Association
Institutions/Associations - Sprayed Concrete Association
Institutions/Associations - Stone Federation Great Britain
Institutions/Associations - Wood Wool Slab Manufacturers Association
Insulated Render & Cladding Association Ltd - Insulated Render & Cladding Association
Journals - Contract Journal
Lift and Escalator Industry Association - LEIA
Lighting association - Lighting Association, The
Lighting industry federation - Lighting Industry Federation
National funding body for arts in England - Arts Council of England
National organisation for the steel construction industry - British Constructional Steelwork Association Ltd
Painting contractors trade association - Painting & Decorating association
Partitioning and interiors association - Partitioning & Interiors Association
Patent glazing contractors association - Council for Aluminium in Building
Perspective artist - Green Gerald
Plant Managers' Journal - Contract Journal
Product and material information and literature - The Building Centre Group Ltd
Product information service, technical publications, library service. - Concrete Information Ltd
Professional & trade association - Forestry and Timber Association
Professional body - Chartered Institute of Building
Professional Body - Royal Institution of Chartered Surveyors (RICS)
Professional body for wastes managment - Chartered Institute of Wastes Management
Professional body registered as an educational charity - Institute of Plumbing The
Professional institution - Chartered Building Company Scheme
Professional institution - Chartered Building Consultancy Scheme
Professional recognition and development for Incorporated Engineers and Engineering Technicians in electronic, electrical and mechanical engineering - Institution of Incorporated Engineers (IIE)
Publications on care and repair of old buildings - Society for the Protection of Ancient Buildings (SPAB)
Publications: contract flooring association members handbook - Contract Flooring Association
Publishing house (rutland press), guides etc - Royal Incorporation of Architects in Scotland
Represent RIBA in Scotland - Royal Incorporation of Architects in Scotland
Scottish architectural trade association - Royal Incorporation of Architects in Scotland
Separate mills section - Society for the Protection of Ancient Buildings (SPAB)
Services to architects including PI insurance - Royal Incorporation of Architects in Scotland
Special events run by SPAB in Scotland - Society for the Protection of Ancient Buildings (SPAB)
TCPA Publications - Town and Country Planning Association
Telephone help line for building professionals - Barbour Enquiry Service
Terrazzo, marble and mosaic federation - National Federation of Terrazzo Marble & Mosaic Specialists
Testing, certification and and technical approval - British Board of Agrement
Thatched roof advice, training and insurance - Thatching Advisory Services Ltd
The welding institute - TWI
Town and Country Planing Journal - Town and Country Planning Association
Trade Association - Biotank Ltd
Trade Association - Bitumen Waterproofing Association
Trade association - Brick Development Association
Trade Association - British Coatings Federation
Trade association - British Constructional Steelwork Association Ltd
Trade Association - British Electrotechnical and Allied Manufacturers Association
Trade Association - British Fenestration Rating Council
Trade Association - British Flat Roofing Council (BFRC)
Trade Association - British Flue and Chimney Manufacturers Association
Trade Association - British Glass Manufacturers Confederation
Trade Association - British Hardware and Housewares Manufacturers Association
Trade Association - British Plastics Federation, The
Trade Association - British Pump Manufacturers Association
Trade Association - British Security Industry Association
Trade Association - British Stainless Steel Association
Trade Association - British Structural Waterproofing Association, The
Trade Association - British Woodworking Federation
Trade Association - Builders Merchants Federation
Trade Association - Building Reserch Establishment
Trade Association - Chartered Institute of Arbitrators
Trade Association - Civil Engineering Contractors Association
Trade Association - Concrete Block Association
Trade Association - Construction Industry Research & Information Association
Trade Association - Construction Industry Training Board
Trade Association - Construction Plant-hire Association
Trade Association - Constructionline
Trade Association - Council for Registered Gas Installers, The
Trade Association - Door and Shutter Manufacturers Association
Trade Association - Drilling & Sawing Association, The
Trade Association - Emporium Brazil Ltd (EBL)

A01 Institutions/ Associations/ Trade Organisations (Cont'd)

Trade Association - Federation of Piling Specialists
Trade Association - Federation of Plastering and Drywall Contractors
Trade Association - Garage Equipment Association Ltd
Trade Association - Guild of Architectural Ironmongers
Trade Association - Guild of Builders & Contractors, The
Trade Association - Heating Ventilating and Air Conditioning Manufacturers Association
Trade Association - Institute of Ashphalt Technology, the
Trade Association - Institute of Remedial Treatment Surveyors, The
Trade Association - Institute of Vitreous Enamellers, The
Trade Association - Joint Industry Board for the Electrical Contracting Industry
Trade Association - Lift and Escalator Industry Association,The
Trade Association - National Association of Shopfitters
Trade Association - National federation of Demolition Contractors
Trade Association - National Federation of Roofing Contractors, The
Trade Association - National House Building Council
Trade Association - National Inspection Council for Electrical Installation Contracting
Trade Association - P R A Coating Technology
Trade Association - Painting & Decorating association
Trade association - Quarry Products Association
Trade Association - Rural design and Building Association
Trade Association - Scottish & Northern Ireland Plumbing Employers' Federation
Trade Association - Single Ply Roofing Association
Trade Association - Solar Trade Association, The
Trade Association - The Tile Association
Trade Association - Thermal Insulation Contractors Association
Trade Association - Timber Trade Federation
Trade Association - UK Mineral Wool Association,The
Trade Association - Wall Tie Installers Federation
Trade Association - Wallcovering Manufacturers Association
Trade association for the ceramic tile industry - The Tile Association
Trade association for the factory produced mortar industry - Mortar Industry Association
Trade association for the lift & escalator industry - LEIA
Trade Association Service - Mastic Ashphalt Council
Trade associations of electrical and electronics sector - Beama Ltd
Trade Body - British Adhesives and Sealants Association
Trade Organisation - Advisory Council and Arbitration Service
Trade Organisation - Association of Cunsulting Engineers
Trade Organisation - Association of Plumbing and Heating Contractors
Trade Organisation - Bathroom Manufacturers Association, The
Trade Organisation - British Cement Association
Trade Organisation - British Ceramic Confederation
Training programs and courses for professionals, craftsmen and owners - Society for the Protection of Ancient Buildings (SPAB)
Urethane foam industry trade association - BUFCA Ltd
Wall ties trade association - British Wall Tie Association
Water and Wastewater Association - British Water

A02 Building contractors

Asbestos removal - Rhodar Limited
Builders - Ogilvie Construction Ltd
Building and civil enginering contractors - Dew Construction Limited
Building contractor - TH Kenyon & Sons Ltd
Building contractor incorporating partnering, guaranteed max price, PFI, lease back, facilities management, site finding - Miller Construction (UK) Ltd
Building contractors - Jarvis Construction (UK) Ltd
Building contractors - Stenoak Associated Services PLC
Civil engineering - Bardon (England) Ltd
Civil engineering - Freyssinet Ltd
Civil engineering and coastal protection works - Wiltshire G R & Co
Design & build - David Mclean Group
Design and build contractor - Longley James & Co Ltd
Design and build fencing contractors - Baker Fencing Limited
Design and build surfacings contractors - Easi-Fall International Ltd
Fencing contractors - Burn Fencing Limited
Insulation and re-roofing contractors - Stockbrands LLP
Interior contractors - Gifford Grant Ltd
Landscape contracting - Land Wood & Water Co. Ltd
Landscape contractors - Gearing F T Landscape Services Ltd
Mechcanical services Contractor - Kershaw Mechanical Services Ltd
Minor civil engineering - Chestnut Products Limited
New build and refurbishment contractor - Longley James & Co Ltd
Pest control - Paragon
Plumbers - Ogilvie Construction Ltd
Retail, Leisure, Ditribution, manufacturing, Pharmaceutical, heathcare, education, offices and food processing sector of the industry - Simons Construction Ltd
Roofing contractor - Evans Howard Roofing Ltd
Roofing contractor - Roofrite (East Anglia) Ltd
Sports field contractor - Bernhards Landscapes Ltd
Sports ground construction - AMK Fence-In Limited
Surfacing contractor - Camas Aggregates Ltd - Aggregate Industries

A10 Project particulars

Architectural and Exhibition model makers - Modelscape
Architectural presentation planning design - Comber Models
ATM enclosures - Pearce Security Systems
Chartered quantity surveyors - Sawyer & Fisher
Chartered quantity surveyors - V B Johnson LLP
Consultancy - Carter Retail Equipment Ltd
Cost Engineers - V B Johnson LLP
Electrical services - H Pickup Mechanical & Electrical Services Ltd
Estimators - V B Johnson LLP
General engineering - Turnbull & Scott (Engineers) Ltd
Mechanical services - H Pickup Mechanical & Electrical Services Ltd
Mechanical, electrical and process services engineers - Shepherd Engineering Services Ltd
Models - Comber Models
Modular cash rooms - Pearce Security Systems
Project management - Carter Retail Equipment Ltd
Project Managers - V B Johnson LLP

Review of consultants fees on construction projects - Building Cost Information Service
Rising armour screens - Pearce Security Systems
Rotary transfer units - Pearce Security Systems
Security lobbies - Pearce Security Systems
Structural Steelwork Design - Billington Structures Ltd
Trade Association - Cast Metals Federation

A20 The Contract/Sub-contract

Advice given to Contractors - National Federation of Builders
Guide to sustainable procurement in the water industry - British Water

A31 Provision, content and use of documents

Books - Town and Country Planning Association
CPD seminars - DW Windsor Lighting
Dettarant design service for Buildings - Birdscarer Products & Net Supplies Ltd
Information Service - NBS
Online bookshop - BRE
Riba. Tl Construction - NBS

A32 Management of the Works

Building regulations approved documents England & Wales - NBS
Civil engineering - David Mclean Group
Partnering - David Mclean Group
Project management - DW Windsor Lighting
Publishers of the National Building Specification. - NBS
Sub Contract Engineers - Multi Marque Production Engineering Ltd
Windows based Q.S, and civils contract administration and estimating software - CSSP (Construction Software Services Partnership)

A33 Quality standards/control

Alarm monitoring & routine maintenance - ADT Fire and Security Plc
ATaC asbestos testing and consulting - a division of Arca - Asbestos Removal Contractors' Association
Cabling installations inspection - Allen P W & Co Ltd
Carpet quality control - The Carpet Foundation
Cavity wall inspection - Allen P W & Co Ltd
Civil engineering test equipment - Capco Test Equipment Ltd
Daylight simulation - Full Spectrum Lighting Ltd
Food safety - National Britannia Ltd
Health & safety - National Britannia Ltd
Heating and ventilation systems inspection - Allen P W & Co Ltd
In-house laboratory - DW Windsor Lighting
Service & Inspection Schemes for mechanical Plant - Multi Marque Production Engineering Ltd
Servicing and maintenance - Planet Communications Group Ltd
Structural inspection - Allen P W & Co Ltd
Systems maintenance - Visua Ltd
Technical support & customer service - Hilmor
Training - Mitel Telecom Ltd
Training - National Britannia Ltd
Training Services for Gas and water and drainage installation - Uponor Ltd

A34 Security/ Safety/ Protection

Access - Magnum Ltd T/A Magnum Scaffolding
Access systems - Clow Group Ltd
Anti bandit & bullet resistant counters - Pearce Security Systems
Asbestos Removal - Kitsons Insulation Products Ltd
Battery operated smoke & fire alarms - Dicon Safety Products (UK) Ltd
Cable fall arrest system - Latchways plc
Carbon monoxide - detectors - Dicon Safety Products (UK) Ltd
Contractors tools - Brettell & Shaw Ltd
Door and window Locks - Securikey Ltd
Energy conservation in built environment - BRE
Fall arrest systems - Latchways plc
Fencing and barriers - SGB Hire Plus
Fire research/consultancy - BRE
Fire safety - National Britannia Ltd
Fire training - Chubb Fire Ltd
Fire training - Griffin and General Fire Services Ltd
Floor protection - Floor Protection Services Ltd
Footwear - industrial - Briggs Industrial Footwear Ltd
Gaseous extinguishing - ADT Fire and Security Plc
Gloves - Builders Range - RAP Group Ltd
Hand and electric tools - BRB Industrial Services Ltd
Hardwire smoke & fire alarms - Dicon Safety Products (UK) Ltd
High security locking systems - Chubb Safe Equipment Company
Industrial footwear - Briggs Industrial Footwear Ltd
Installation and testing of chemical anchors - Montrose Safety Fixings
Ladder safety systems - Latchways plc
Manufacturer of secure steel anti-vandal storage and accomodation buildings - Lydney Containers Ltd
Manufacturers of Fast Tow Mixers - Multi Marque Production Engineering Ltd
Natural gas alarms - Dicon Safety Products (UK) Ltd
Occupational health - National Britannia Ltd
Pedestrian barrier - Wade Building Services Ltd
Personal Protective Equipment - Kem Edwards Ltd
Planning Supervisors - V B Johnson LLP
Protection sheets - Cordek Ltd
Respiratory protection and safety helmets - MSA (Britain) Ltd
Rolled protection - Cordek Ltd
Roof safety systems - Latchways plc
Safes - Chubb Safe Equipment Company
Safety access systems - Kobi Ltd
Safety Equipment - HSS Hire Service Group
Safety eyebolts - Latchways plc
Safety netting - Deborah Services Limited
Safety systems - Unistrut Limited
Safety/ security/ protection - Pilkington UK Ltd
Security / safety protection - Specialist Building Products
Security / Safety Protections - UK Fire International Ltd
Security units - Parklines Building Ltd
Security/safety - Intersolar Group Ltd
Security/safety/protection - BPT Automation Ltd
Site Safety Equipment - B&C Roadmarkings Ltd
Steel security units - Ideal Building Systems Ltd
Strongrooms - Chubb Safe Equipment Company
Superbolt - Furmanite International Ltd
Temporary security fences - Wade Building Services Ltd
Temporary site hoarding - Wade Building Services Ltd
Vaults - Chubb Safe Equipment Company

A36 Facilities/ Temporary works/ services

Bespoke esign services - Camlok Lifting Clamps Ltd
Builders bags - Hilton Banks Ltd
Cabins, site accommodation and event toilets - Rollalong Ltd
Demountable Flood Defense Systems - Bauer Foundations Limited
Facilities/temporary works/services - Interface Europe Ltd
Graphic design services - Milesahead Graphic Design/Consultant
Non-mechanical access and building equipment (hire and sales - Deborah Services Limited
Palletwrap - Hilton Banks Ltd
Repairs and service of industrial doors - Avon Industrial Doors Ltd
Spill Kits - Furmanite International Ltd
Tarpaulins - Hilton Banks Ltd
Temporary fencing and site guard panels - Guardian Wire Ltd
Temporary roadways - Eve Trakway
Temporary works - Frosts Landscape Construction Ltd

A37 Operation/ Maintenance of the finished building

Access audit price guide - Building Cost Information Service
Anti-finger trap devices - Dorma Door Services Ltd
Door safety products - Dorma Door Services Ltd
Engineers tools - Spear & Jackson
Hand tools - Hilmor
Life expectancy of building components - Building Maintenance Information
Maintenance and service - Cooper Group Ltd
Operational and maintenance contracts and technical advice - Mitsui Babcock Energy Ltd

A40 Management and staff

Tralning - Airedale International Air Conditioning Ltd

A41 Site accommodation

Cabin hire - Premier Group
Cabin sales - Premier Group
Doulble Store and Mobile Tool Control - Railex Systems Ltd
Modular buildings - RB Farquhar Manufacturing Ltd
Modular buildings - Wernick S & Sons Ltd
Modular buildings hire - Premier Group
Modular buildings sale - Premier Group
Portable accommodation - Rovacabin
Portacabins, steel cabins and portaloos - UK Cabin Co
Portal frame buildings - Elliott Group Ltd
Rovacabin temporary secure buildings, offices, toilets - SGB Hire Plus
Schools, Key Worker & Student Accomadation - Rollalong Ltd
Secure accommodation - Premier Group
Site accommodation - Premier Group
Site Accomodation - Pyramid Building Systems
Site Accomodation - Thurston Building Systems
Site cabins - RB Farquhar Manufacturing Ltd
Steel frame modular buildings - manufacturer - Lydney Containers Ltd
Temporary buildings - C & R Building Systems Ltd
Timber sectional buildings - Ableson L & D Ltd

A42 Services and facilities

24 hour service - Servowarm
24 hour service call out - Chubb Safe Equipment Company
Abrasives tungsten carbide burs - PCT Group Sales Ltd
Access Equipment hire - HSS Hire Service Group
Advisory service - Birdscarer Products & Net Supplies Ltd
Air quality - National Britannia Ltd
Alloy tower and aerial platform training - Turner Access Ltd
Aluminium towers, aerial platforms, scissors and booms - Turner Access Ltd
Annual Services - Servowarm
Application Tooling - Hellermann Tyton
Asbestos abatement and consultancy services - Rhodar Limited
Automative tools - Spear & Jackson
Bonded abrasives - Marcrist International Ltd
Book & publications - building, architecture, design, contracts - The Building Centre Bookshop
Bookshop - The Building Centre Group Ltd
Breakdown repairs - Servowarm
Bricklifter handlifting tools (bricks) - Edward Benton & Co Ltd
Brooms and paint, industrial, specialist brushes - Davis Burrow & Sons
Buckets - Airflow (Nicoll Ventilators) Ltd
Builders Buckets - Brettell & Shaw Ltd
Builders cleans and one offs - CSL (Services) Limited
Builders tools - Spear & Jackson
Building Project Information - The Building Centre Group Ltd
Cartridge tools - ITW Spit
Coal tar B.P - Legge Thompson F & Co Ltd
Conference & Seminar Facilities - The Building Centre Group Ltd
Conferences and seminars - Town and Country Planning Association
Construction equipment - Burlington Engineers Ltd

A42 Services and facilities (Cont'd)

Construction products - GE Bayer Silicones
Construction research and development - BRE
Construction tools - Ingersoll-Rand European Sales Ltd
Core drills - Norton Diamond Products
Cutting machines - Norton Diamond Products
Decontamination units - Rollalong Ltd
Decorators Tools - Halls Beeline Ltd
Diamond blades - Norton Diamond Products
Disabled access door operators - Dorma Door Services Ltd
Drainage industry and plumbers tools - Monument Tools Ltd
Electric Tools - ITW Spit
Engineering and production tools - Spear & Jackson
Engineering Stockist - Thanet Ware Ltd
Exhibitions - building materials and products - The Building Centre Group Ltd
Exhibitions, conferences, and meeting rooms - The Building Centre Group Ltd
Formliners - Grayhill Trading Ltd
Generators - Standby Power Company
Guideline - The Building Centre Group Ltd
Hand & power tools - Kem Edwards Ltd
Hand and garden tools - Spear & Jackson
Hand tools - Footprint Tools Ltd
Hand tools, power tool accessories and abrasives - Hyde Brian Ltd
Handtools, shovels, spades and hammers - Stockten Heath Forge (Caldwells) Ltd
Highway maintenance contractors - Tarmac Ltd
Hot dip galvanising - Jones of Oswestry Ltd
Importers - WL West & Son Limited
Installation - Servowarm
Kermaster handlifting tools (Kerbs) - Edward Benton & Co Ltd
Ladders - Sarena MFG Ltd
Lifting Equipment hire - HSS Hire Service Group
Local engineers (corgi regestered) - Servowarm
Lubricants - Witham Oil & Paint Ltd
Market & research schemes - The Building Centre Group Ltd
Masking tape - Halls Beeline Ltd
Metal cutting tools - Spear & Jackson
Oak specialist - WL West & Son Limited
Open and enclosed skips, cargo containers and custom built units - Randalls Fabrications Ltd
Paint brushes and rollers, brooms, brushes and sundry items - Sovereign Brush Co Ltd, The
Paper sacks for cementious products and building materials - Assi Doman Sacks (UK)
Pipe tools - Hilmor
Plant - Linatex Ltd
Power tools-pneumatic/electric/hydraulic - PCT Group Sales Ltd
Publications and technical information on timber - Trada Technology Ltd
Recoverable anchor screws - Grayhill Trading Ltd
Scaffold hire - HSS Hire Service Group
Services & facilities - Certifield Laboratories (Div or CPS Industries Ltd)
Services and facilities - Barlow Group
Services and facilities - Frosts Landscape Construction Ltd
Servocare economic insurance - Servowarm
Servocare priority insurance - Servowarm
Site services - Magnum Ltd T/A Magnum Scaffolding
Skin care products - Deb Ltd
Soft faced hammers and mallets - Thor Hammer Co Ltd
Soft soap - Liver Grease Oil & Chemical Co Ltd
Special works / maintenance - Durabella System Floors Ltd
Spindle tooling - Whitehill Spindle Tools Ltd
Suppliers of Bituminous Binders Road Surface Dressing Contractors - South Western Tar Distilleries
Supply of Electrical products to the trade & industry - Imperial Electrical Wholesalers
Surveying Equipment - HSS Hire Service Group
Tar Buckets - Brettell & Shaw Ltd
Timber merchants - WL West & Son Limited
Tool & Equipment Repair - HSS Hire Service Group
Tool hire - HSS Hire Service Group
Tools - Skarsten Tools
Tools - Tinsley & Co Ltd, Eliza
Tools - Walls & Ceilings International Ltd
Tools & Ironmongery - Spital Tile Co Ltd
Trade in non mechanical plant - Grayhill Trading Ltd
Transportation of Brituminous Products, Liquid and Oils - South Western Tar Distilleries
Upgrade to existing system - Servowarm
Wall protection - Moffett Thallon & Co Ltd
Wall tie condition surveys - HCL Contracts Ltd
Wallpaper pasting machines - Halls Beeline Ltd
Wheelborrows - Brettell & Shaw Ltd

A43 Mechanical plant

Access platforms - Manitou (Site Lift) Ltd
Backhoe loaders - Fermec Holdings Ltd
Bandsaw blades, hacksaw blades, mini blades and frames - Speed Saws Ltd.
Boring machines, boring tools and trenchers - B-Trac Equipment Ltd
Builders tools - Benson Industries Ltd
Burnishers and sanders - Alto Cleaning Systems (UK) Ltd
C-hooks/coil hooks - Camlok Lifting Clamps Ltd
Compactors - Sandhurst MFG Co
Compressor crankcase heaters - Flexelec(UK) Ltd
Construction equipment - Watkins
Construction equipment, compressors, rock drills and accessories, compaction and paving equipment - Ingersoll-Rand European Sales Ltd
Core drills - Harbro Supplies Ltd
Crane forks - Camlok Lifting Clamps Ltd
Creteangle bulkbag systems - Edward Benton & Co Ltd
Creteangle hopper vibratory feeder systems - Edward Benton & Co Ltd
Creteangle moulds for precast concrete products - Edward Benton & Co Ltd
Creteangle pan type mixers - Edward Benton & Co Ltd
Demolition grapples - Sandhurst MFG Co
Diamond tipped glasscutters and emery wheel dressers - Radcliffe & Sons (London)
Diesel power generators - Sandhurst MFG Co
Directional drill rigs - B-Trac Equipment Ltd
Double vacuum impregnation plant - Protim Solignum Ltd
Drill bits - Artur Fischer (UK) Ltd
Drum handling Equipment - Camlok Lifting Clamps Ltd
Dumper - HE Services (Plant Hire) Ltd
Dumptruck hire - HE Services (Plant Hire) Ltd
Excavating and earth moving plant - J C Bamford (Excavators) Ltd
Excavator hire - HE Services (Plant Hire) Ltd
Excavators, crawler mounted cranes and draglines - R-B International Ltd
Floor Sanders, overhead & rebuilt - Eureka Products Ltd
Forklift truck attachments - Camlok Lifting Clamps Ltd
Gas nailer - ITW Construction Products
Gas power tools - ITW Spit
General Site Equipment - Kem Edwards Ltd
Grinding machines, surface and double ended grindes - Beacon Machine Tools Ltd
Grinding wheels and abrasive products - Carborundum Abrasives G.B. Ltd
Hand engraving tools - Elliott John H (Monostamp) Ltd
Hight Safety, Controlled Access & Retrieval Systems - Sala Ltd.
Horizontal grinders - B-Trac Equipment Ltd
Horticultural and agricultural tools - Spear & Jackson
Hydraulic attachments - Sandhurst MFG Co
Hydraulic hammers - Sandhurst MFG Co
Inspection & measurement equipment - Allen P W & Co Ltd
Lifting clamps and grabs - Camlok Lifting Clamps Ltd
Loading bay equipment - Amber Doors Ltd
Machine tools for the manufacture of building products - BSA Machine Tools Ltd
Machinery - The Saw Centre Group
Manual and powered hoists, winches and jacks - PCT Group Sales Ltd
Manufacturer of Conveyors - Multi Marque Production Engineering Ltd
Manufacturers of Roller Mixer - Multi Marque Production Engineering Ltd
Material handling equipment - Harbro Supplies Ltd
Mechanical plant spares - O'Brian Manufacturing Ltd
Mechcanical services Contractor - Kershaw Mechanical Services Ltd
Metal detector - Allen P W & Co Ltd
Mini excavators - Sandhurst MFG Co
Mini excavators, dumper trucks, rollers and cement mixers - Chippindale Plant Ltd
Mobile site lighting - Sandhurst MFG Co
Moisture meters - Allen P W & Co Ltd
Moles - B-Trac Equipment Ltd
Pan type mixing machinery - Benton Co Ltd Edward
Plant - Certifield Laboratories (Div or CPS Industries Ltd)
Plant hire - HE Services (Plant Hire) Ltd
Plant hire - HSS Hire Service Group
Plant hire - Signature Ltd
Plate compactors - Chippindale Plant Ltd
Pocket tapes and folding rules - Rabone Chesterman Ltd
Powder actuated tools - ITW Construction Products
Power tools and accessories - Robert Bosch Ltd
Powered barrows - Bitmen Products Ltd
Pulverizers - Sandhurst MFG Co
Rail handling Equipment - Camlok Lifting Clamps Ltd
Road Maintenance Equipment - Phoenix Engineering Co Ltd, The
Rollers - HE Services (Plant Hire) Ltd
Rough terrain forklifts - Manitou (Site Lift) Ltd
Sand blast equipment - Harbro Supplies Ltd
Schedule of basic plant changes - Building Cost Information Service
Screwlok beam clamps - Camlok Lifting Clamps Ltd
Scrubber driers - BVC
Site dumpers, concrete and mortar mixers, sawbenches and telescopic handlers - Multi Marque Production Engineering Ltd
Sorbents - Furmanite International Ltd
Special rigging structures - Scotia Rigging Services (Industrial Stainless Steel Rigging Specialists)
Stone processing machinery - Harbro Supplies Ltd
Telehandler Hire - HE Services (Plant Hire) Ltd
Telehandlers - Chippindale Plant Ltd
Telescopic loaders - Manitou (Site Lift) Ltd
Timber machining service - Brewer T. & Co Ltd
Tipping skips - Woodburn Engineering Ltd
Tools - Ashmead Buidling Supplies Ltd
Tools and plant - Vulcanite Limited
Tractor loaders - Fermec Holdings Ltd
Tree equipment, woodchippers, stump cutters and tub grinders - B-Trac Equipment Ltd
Trenchers - B-Trac Equipment Ltd
Truck mounted forklift - Manitou (Site Lift) Ltd
Tube Bending Machines - Tubela Engineering Co Ltd
Vehicle turntables - Hovair Systems Ltd
Welding Equipment - HSS Hire Service Group
Welding equipment - PCT Group Sales Ltd
Wheeled loader hire - HE Services (Plant Hire) Ltd
Wood working tools - Record Tools Ltd

A44 Temporary works

Abnormal lead support - Eve Trakway
Abrasives - Unicorn Abrasives (UK) Ltd
Access platforms - Chippindale Plant Ltd
Adhesive, solid, metallic, plastic tapes - Marking Machines International
All forms of line (sports and commercial) marking machines - Marking Machines International
Aluminium towers - SGB Youngman
Anchor testing & installation - HCL Contracts Ltd
Bridges - Eve Trakway
Cable grips - Albion Manufacturing
Contract scaffolding - Deborah Services Limited
Contract scaffolding - Magnum Ltd T/A Magnum Scaffolding
Ground protection - Eve Trakway
Hoarding panels - SGB Hire Plus
Ladders - Titan Ladders Ltd
Ladders & steps - SGB Youngman
Loftladders - Titan Ladders Ltd
Measuring tapes and equipment - Marking Machines International
Omni temporary roofing - Deborah Services Limited
Pest control - humane live catch and trap - Albion Manufacturing
Plaited and braided ropes, sash and builders lines - Blakey John H Ltd
Roadway and flooring - temporary - SGB Hire Plus
Scaffold security services - Alpine Autimation Ltd.
Scaffold sheeting - Monarflex Anderson Ltd
Scaffold towers - Titan Ladders Ltd
Scaffolding - Schauman (UK) Ltd
Scaffolding - SGB Hire Plus
Scaffolding - Wade Building Services Ltd
Scaffolding and ladders - SGB Youngman
Scaffolding restraints - HCL Contracts Ltd
Scaffolding service - Turner Plus Eight Ltd
Scaffolding systems - SGB Youngman
Scaffolding Systems - Turner Access Ltd
Stagings - Clow Group Ltd
Steps - Titan Ladders Ltd
Temopary works - Freyssinet Ltd
Temporary fencing - Rom Ltd
Temporary Road Markings - B&C Roadmarkings Ltd
Temporary roadway - Sommerfeld Flexboard Ltd
Tempory protection - British Sisalkraft Ltd
Towers - Clow Group Ltd

B10 Prefabricated buildings/structures

Aluminium green houses, chalets, garages, workshops and sheds - Compton Buildings Ltd
Aluminium special structures - Alifabs (Woking) Ltd
Architectural fabrications - Clow Group Ltd
Bespoke aluminum structures - Clear Span Ltd
Bridges - Structural Timbers Ltd
Bridges in timber or steel, boardwalks, decking, structures pergolas, screens and planters - CTS Ltd
Builders merchant - Jewson Ltd
Builders merchant - Nicholls & Clarke Ltd
Building and development - Pearce Construction(Barnstaple)Ltd
Building systems to buy or hire, self contained buildings and modular buildings - Portakabin Ltd
Cantilever and pigeon hole racking - Hi-Store Ltd
Cedar green houses, chalets, garages, workshops and sheds - Compton Buildings Ltd
Chemical stores - Parklines Building Ltd
Columbariums - Milner Delvaux Ltd
Complete buildings - Elliott Construction Division
Composite housing and containers and portable buildings - Cobham Composites Ltd
Concrete workshops & sheds - Phillip Ader Ltd
Concrete Garages - Parklines Building Ltd
Concrete garages - Phillip Ader Ltd
Conservatories - Alframes Ltd
Conservatories - Caradon Duraflex Systems Ltd
Conservatories - Conservatory Roof Systems Ltd
Conservatories - Hill Leigh Ltd
Conservatories - Solair Ltd
Conservatories - Spectra Glaze Services
Conservatories, garden buildings - Amdega Ltd
Covered Walkways - Fibaform Products Ltd.
Cycle Shelters - Fibaform Products Ltd.
Design and build of timber frame components and floor joist system - Prestoplan
Electrical engineering - Park Products
Environmental trade products - Furmanite International Ltd
Erection and design service - Llewellyn Homes Ltd
Export Buildings - Wernick S & Sons Ltd
Extruded conservatories - Smart Systems Ltd
Fabric tensile structures - Clyde Canvas Fabric Structures Ltd
Flat Pack Buildings - Wernick S & Sons Ltd
Garden buildings - Anthony de Grey Trellises
Gatehouses - Fibaform Products Ltd.
Grp modular buildings - Dewey Waters Ltd
Health Care - Elliott Group Ltd
Hire of buildings - Wernick S & Sons Ltd
Housing - Pearce Construction(Barnstaple)Ltd
Industrial housings - Fibaform Products Ltd.
Jack leg cabins - Ableson L & D Ltd
Mechcanical services - Kershaw Mechanical Services Ltd
Modular and volumetric buildings - Rollalong Ltd
Modular buildings - Ableson L & D Ltd
Modular buildings - Bell & Webster Concrete Ltd
Modular buildings - Rovacabin
Modular buildings - Terrapin Ltd
Modular buildings - Wernick S & Sons Ltd
Modular buildings and garages - Panelite
Modular buildings and relocatable buildings - GE Capital Modular Space
Modular portable and prefabricated buildings - Ideal Building Systems Ltd
Modular solutions - Composite Structures
Module based buildings - Thurston Building Systems
Multi block garages and storage units - Panelite
Office Space - Elliott Group Ltd
Overhead monorail systems - Down & Francis Ltd, Division of Metalrax PLC
Passenger shelters - Signature Ltd
Portable sectional buildings - UK Cabin Co
Pre engineered buildings - Terrapin Ltd
Pre engineered metal buildings - Atlas Ward Structures Ltd
Pre fabricated buildings - Brimar Plastics Ltd
Prefabricated Building Structures - Composite Structures
Prefabricated buildings - C & R Building Systems Ltd
Prefabricated buildings - Freyssinet Ltd
Prefabricated buildings - Lydney Containers Ltd
Prefabricated buildings - Parklines Building Ltd
Prefabricated buildings - Premier Group
Prefabricated buildings / structures - Pyramid Building Systems
Prefabricated buildings /structures - Rom Ltd
Prefabricated buildings/structures - Butler Europe Limited
Prefabricated Buildings/Structures - Dartford Portable Buildings Ltd
Proprietary buildings - Leander Architectural
PVCu & Aluminium Conservatories - KCW Windows
Railway S&T buildings - Elliott Group Ltd
Relocateable buildings - C & R Building Systems Ltd
Sale and hire of relocatable buildings - Initial Plant Services Ltd, Redispace and Johnson
Schools and colleges - Elliott Group Ltd
Silos - CST Vulcan
Smartroof panels - Wyckham Blackwell Ltd
Smoking shelters - Fibaform Products Ltd.
Sports halls - Elliott Group Ltd
Squash and racquet ball courts - Bicester Products Ltd
Steel bus shelters - Planet Architectural Glazing Ltd
Steel frame buildings - Terrapin Ltd
Steel framed buildings - Yorkon Ltd
Steel portal frame buildings - Accomodex
System Buildings - Wernick S & Sons Ltd
Tailor made steel buildings - Astron Buildings Parker Hannifin PLC
Technical consultancy - BFRC Services Ltd
Timber building systems - Bullock & Driffill Ltd
Timber framed buildings - Wyckham Blackwell Ltd
Timber framed houses and structures - Colt W H & Co Ltd
Timber roof design consultancy - Alpine Autimation Ltd.
Timber sheds/ workshops/ gazebos - Phillip Ader Ltd
Timber site offices - Ableson L & D Ltd
Timber Storage Buildings - Parklines Building Ltd

B10 Prefabricated buildings/structures (Cont'd)

Toilets - Inter Public Urban Systems Uk Ltd
Traditional build - Elliott Group Ltd
Units available for hire - Fibaform Products Ltd.

B11 Prefabricated building units

Dock Leveller Pits - Ebor Concretes Ltd
Horizontal panel concrete buildings - Accomodex
Hotels, Timber or Steel Framed Buildings - Rollalong Ltd
Mobile classrooms - C & R Building Systems Ltd
Modular building systems - Elliott Group Ltd
Modular buildings - C & R Building Systems Ltd
Modular buildings - Rovacabin
Pay to enter systems - Inter Public Urban Systems Uk Ltd
Pre fabricated building units - Brimar Plastics Ltd
Prefabricated Building Units - Composite Structures
Prefabricated building units - Ideal Building Systems Ltd
Prefabricated building units - Initial Plant Services Ltd, Redispace and Johnson
Prefabricated building units - Pyramid Building Systems
Public toilets - Inter Public Urban Systems Uk Ltd
Stadium Terracing - Ebor Concretes Ltd
Steel container toilets - Ableson L & D Ltd
Storage and utility products - Elliott Group Ltd
Temporary buildings - Premier Group
Vertical panel concrete buildings - Accomodex

C10 Site survey

Asbestos removal - Hawkins Insulation Ltd
Concrete repair systems - PermaRock Products Ltd
Disability access audits - Peter Cox Ltd
Site survey - Landscape Institute, The
Site surveys - Focal Signs Ltd
Survey anchors and markers - Earth Anchors Ltd

C11 Ground investigation

Asbestos consultancy - National Britannia Ltd
Contaminated land - National Britannia Ltd
Geotechnical consultancy and contamination assessment - Structural Soils Ltd
Ground improvement - Keller Ground Engineering
Ground investigation - Apex Piling Ltd
Ground investigation - May Gurney (Technical Services) Ltd - Fencing Division
Groundworks - Rom Ltd
Site investigation - Haydon & Hagan Construction Ltd
Site investigation,soils and material testing - Structural Soils Ltd
Surveying instruments - Topcon Corporation

C12 Underground services survey

Underground services survey - Radiodetection Ltd

C13 Building fabric survey

Asbestos surveys - Peter Cox Ltd
Building fabric survey - Colebrand Ltd
Building Fabric Survey - Concrete Repairs Ltd
CCTV Surveys - Cico Chimney Linings Ltd
Compresssive/shear tests - Ceram Building Technology
Glass assessment surveys - Durable
Large scale structural laboratory - Ceram Building Technology
on site testing and consultancy - Ceram Building Technology
Testing to standard or bespoke needs - Ceram Building Technology
Toxic mould deection - Peter Cox Ltd

C14 Building services survey

Energy efficiency in lighting advisory service - Poselco Lighting
Performance of masonry/steel/concrete/timber - Ceram Building Technology
Preparation of boilers for insurance inspection - Hodgkinson Bennis Ltd
Preparation of boilers for ultrsonic testing - Hodgkinson Bennis Ltd

C20 Demolition

Alterations - Haydon & Hagan Construction Ltd
Architectural salvage - Reclaim.UK.Com
Demolition - HE Services (Plant Hire) Ltd
Demolition - Parker Plant Ltd
Office refurbishment - Cumberland Construction

C21 Toxic/Hazardous material removal

Asbestos removal - Deborah Services Limited
Toxic / Hazardous material removal - Hertel Services

C30 Shoring/Facade retention

Lifting and moving whole buildings - Apex Piling Ltd
Proping and needling - Maby Hire Ltd
Shoring - SGB Youngman
Shoring and propping - Apex Piling Ltd
Trench strutting equipment - Maby Hire Ltd

C40 Cleaning masonry/concrete

Brick and stone cleaning - Gun-Point Ltd
Brick replacement - Gun-Point Ltd
Building, cleaning and restoration - Cleanstone Co (Bradford) Ltd
Cavity wall ties - Fix Fast (Fasteners) Wales Ltd
Cleaning &restoration of building facades - Szerelmey Ltd
Cleaning &restoration of building stonework - Szerelmey Ltd
Cleaning masonary/concrete - TH Kenyon & Sons Ltd
Cleaning masonry/ concrete - Colebrand Ltd
Cleaning masonry/concrete - Llewellyn Stonecare Limited
Concrete maintenance and repair products - HMG Paints
Concrete repair - Rok-Crete Units Co Ltd
Concrete repair - STO cretec - STO Ltd
Concrete repair mortars - Hines & Sons, P E, Ltd
Concrete repair mortars, crack injection resins - Ronacrete Ltd
Concrete repair systems - Exchem Mining and Construction
Concrete repairs - Llewellyn Stonecare Limited
Designed grout anchoring system - Cintec International Ltd
Epoxy resin repairs - Gun-Point Ltd
Floor seals and concrete repair kits - Tank Storage & Services Ltd
Grouting - Gun-Point Ltd
Physical dpc insertion - Dampcoursing Limited
Pressure pointing - Gun-Point Ltd
Repairing concrete and masonry structures - Vandex (UK) Ltd
Repairing concrete/ brick/ block/ stone - Concrete Repairs Ltd
Repairs and restorations - Alpha M & T Ltd
Reparing and renovating - Canonbury Asphalte Co. Ltd.
Repointing brickwall and stonework - Gun-Point Ltd
Resin and precision grout - Armorex Ltd
Restoration of facades in stone, brick & terracotta - Stonewest Ltd
Stone and brickwork repair and restoration - CSL (Services) Limited
Wall tie replacement - Timberwise (UK) Plc

C41 Repairing/Renovating/Conserving masonry

Cavity wall ties - Fix Fast (Fasteners) Wales Ltd
Chemical DPC and remedial treatments - Protim Solignum Ltd
Chemical dpc installers - Dampcoursing Limited
Chemicals - Baxenden Chemicals Ltd
Conservation works - Szerelmey Ltd
Damp-profing - Timberwise (UK) Plc
External refurbishment works - Szerelmey Ltd
Faience blocks - Shaws of Darwen
In-Situ Chimney & Flue Lining Systems - Cico Chimney Linings Ltd
Lintel replacement - HCL Contracts Ltd
Polyurethane and acylic resins - Baxenden Chemicals Ltd
Remedial wall ties - Peter Cox Ltd
Repairing/ renovating conserving masonry - Colebrand Ltd
Repairing/ renovating/ conserving masonry - Civil Engineered Products Ldt
Repairing/renovating/ conserving masonry - Llewellyn Stonecare Limited
Repairing/renovating/conserving masonry - G. Miccoli & Son Ltd
Stainless steel remedial wall ties - Helifix Ltd
Stainless steel replacement - Fix Fast (Fasteners) Wales Ltd
Structural repairs - Szerelmey Ltd
Technical publications and publication brick bulletin - Brick Development Association
Terracotta blocks - Shaws of Darwen
Underground Waterproffing - Sika Ltd

C42 Repairing/Renovating/Conserving concrete

Commercial Conncrete repair - HCL Contracts Ltd
Concrete repair - Degussa Construction Chemicals (UK)
Concrete repair - Feb MBT
Concrete repair - Fullstop Technology Ltd
Concrete repair - Sealocrete Ltd
Concrete repair products - Flexcrete Ltd
Concrete repair systems - Beton Construction Ltd
Conrete repair - Fosroc Ltd
Corrosion Control - Fosroc Ltd
Corrosion protection, building maintenance coatings and composition - Wood E Ltd
Repairing / renovation / conserving concrete - Freyssinet Ltd
Repairing and conserving concrete - Anti Corrosion Services Ltd
Repairing concrete - Don Construction Products Ltd
Repairing/ renovating conserving Concrete - Colebrand Ltd
Repairing/ renovating/ conserving concrete - Civil Engineered Products Ldt
Repairing/renovating concrete - Anglo Building Products Ltd
Repairing/renovating/ conserving concrete - Llewellyn Stonecare Limited
Repairing/Renovating/Conserving Concrete - Concrete Repairs Ltd
Repairing/Renovating/Conserving Concrete - PermaRock Products Ltd
Repairing/renovating/conserving concrete - TH Kenyon & Sons Ltd

C45 Damp proof course renewal/insertion

Chemical dpcs to existing walls - Peter Cox Ltd
Condensation control and DPC - Kair ventilation Ltd
Damp coursing - Kiltox Damp Free Solutions
Damp proof course renewal/ insertion - Terminix Property Services

C50 Repairing/ Renovating/Conserving metal

Architectural metalwork - renovation and restoration - Albion Manufacturing
Corrosion protection for metal fastenings - Interlux Ltd
Rail infrastructure coatings - Corroless Corrosion Control
Repairing/ renovating conserving Metal - Colebrand Ltd
Repairing/ renovating metal - Certifield Laboratories (Div or CPS Industries Ltd)
Repairing/ renovating/ conserving metal - Wood E Ltd
Repairs and maintenance to doors and shutters - Andersons Ltd
Restoration work in metal - Ballantine, Bo Ness Iron Co Ltd

C51 Repairing/ Renovating/Conserving timber

Repairing/ renewing/ conserving timber - Terminix Property Services
Repairing/ renovating/ conserving timber - Wood E Ltd
Solvent based and water based preservatives - Protim Solignum Ltd
Timber resin repairs - Timberwise (UK) Plc
Timber treatment - Rentokil Ltd
Timber treatment and dry rot eradication - Dampcoursing Limited
Window repair - Window Care Systems Limited
Wood decay repair products - Window Care Systems Limited
Wood pressure - Cuprinol Ltd
Woodworm - Kiltox Damp Free Solutions

C52 Fungus/ Beetle eradication

Dry rot and woodworm treatment - Kair ventilation Ltd
Fungus and beetle eradication - Peter Cox Ltd
Fungus/ beetle eradication - Terminix Property Services
Fungus/beetle eradication - Specialist Building Products
Timber preservation - Kiltox Damp Free Solutions
Timber preservatives - Rentokil Ltd
Timber treatment - Timberwise (UK) Plc
Woodworm killer - Cuprinol Ltd

C90 Alterations - spot items

Refurbishments & replicas - DW Windsor Lighting
Sash renovation - Durable
Wooden floors - new and reclaimed - Reclaim.UK.Com

D11 Soil stabilisation

Bespoke Solutions using natural fibres - Rawtex Ltd
CFA Piling - Stent Foundations Ltd
Chemicals - Baxenden Chemicals Ltd
Civil engineering systems - Sarnafil Ltd
Dynamic Compaction - Bauer Foundations Limited
Erosion control mats - ABG Ltd
Geotextile mesh - Boddingtons Ltd
Geotextiles - Don & Low Ltd
Green roof systems - ABG Ltd
Ground improvement - Keller Ground Engineering
Ground movement products - Cordek Ltd
Lime columns - Stent Foundations Ltd
Load bearing anchors - Earth Anchors Ltd
Retaining walls - ABG Ltd
Retaining walls - PHI Group Ltd
Soil stabilisation - Cementation Foundations SKANSKA Ltd
Soil stabilisation and errosion control - Land Wood & Water Co. Ltd
Soil stabilization - Grass Concrete Ltd
Soil stabilization - May Gurney (Technical Services) Ltd - Fencing Division
Soil stabilization - Stenoak Associated Services PLC
Soil stabilization - Tensar International Group LTD
Soil stabilization - Vencel Resil Ltd
Soil stabilization/ grand anchors - Freyssinet Ltd
Vibro and dynamic compaction - Bauer Foundations Limited
Vibro Concrete Columns - Bauer Foundations Limited
Vibro Stone Columns - Bauer Foundations Limited

D12 Site dewatering

De watering - Keller Ground Engineering
Dewatering - Simon-Hartley
Pumping systems for ground water control - Andrews Sykes Hire Ltd
Water drainage geocomposites - Geosynthetic Technology Ltd

D20 Excavating and filling

Aggregates - Barker Bros Aggregates Ltd
Aggregates - Brikenden (Builders Merchants) Ltd
Aggregates - Camas Aggregates Ltd - Aggregate Industries
Aggregates - Halls Group Ltd
Construction materials - Hanson Building Materials Europe
Crushed rock - Hanson Building Materials Europe
Crushed rock, sands and gravels - Hanson Aggregates
Excavating and filling - HE Services (Plant Hire) Ltd
Excavating and filling - Vencel Resil Ltd
Filling rock - Bardon (England) Ltd
Land reclamation - Bernhards Landscapes Ltd
Sand and gravels - Hanson Building Materials Europe
Sands and gravel - Camas Aggregates Ltd - Aggregate Industries
Specialist sands - Camas Aggregates Ltd - Aggregate Industries
Tunnel linings - Tarmac Precast Concrete Limited

D21 Ground gas venting

Gas drainage geocomposites - Geosynthetic Technology Ltd
Gas venting - Cordek Ltd
Ground gas venting - Keller Ground Engineering
ground gas venting - Vencel Resil Ltd
Methane gas venting - ABG Ltd
Radon Barriers - BRC Ltd
Radon membranes - Visqueen Building Products

D30 Cast in place concrete piling

Bored Piling - Stent Foundations Ltd
Cased flight auger piling - Westpile Ltd
Cast in place concrete piling - Apex Piling Ltd
Cast in place concrete piling - Bauer Foundations Limited
Cast in place concrete piling - Cementation Foundations SKANSKA Ltd
Cast in place concrete piling - Expanded Piling Co Ltd, The
Cast in place concrete piling - Freyssinet Ltd
Cast in place concrete piling - May Gurney (Technical Services) Ltd - Fencing Division
Cast in place concrete piling - Stent Foundations Ltd
Concrete piling - Access Piling Ltd
Continuous flight auger bored piling - Westpile Ltd
Ductile Piling - Bauer Foundations Limited
Filing Accessories - Cordek Ltd
Ground engineering, Piling - Keller Ground Engineering
Large diameter bored piling - Westpile Ltd
Piling - Bauer Foundations Limited
Piling and ground engineering - Stent Foundations Ltd
Rotary angered piling - Apex Piling Ltd
Rotary bored piling - Westpile Ltd
Tripod bored piling - Westpile Ltd

D31 Preformed concrete piling

Bored, CFA, driven precast piling - Stent Foundations Ltd
Driven precast concrete piling, steel sheet piling and steel bearing - DEL Piling Contractors
Driven precast - Westpile Ltd
Ground engineering, piling - Keller Ground Engineering
Preformed concrete piling - Cementation Foundations SKANSKA Ltd
Preformed concrete piling - Expanded Piling Co Ltd, The
Preformed concrete piling - May Gurney (Technical Services) Ltd - Fencing Division
Preformed concrete piling - Stent Foundations Ltd

D32 Steel piling

Driven Piling - Apex Piling Ltd
Driven Tube Piling - Westpile Ltd
Ground engineering, piling - Keller Ground Engineering
Steel piling - Access Piling Ltd
Steel piling - Deepdale Engineering Co Ltd
Steel piling - Stent Foundations Ltd
Steel sheet piling - Rainham Steel Co Ltd
Trench strutting equipment - Maby Hire Ltd
Tubular steel piles - DEL Piling Contractors

D40 Embedded retaining walls

Diaphragm walling - Cementation Foundations SKANSKA Ltd
Diaphragm walling - Keller Ground Engineering
Diaphragm Walling - Stent Foundations Ltd
Precast concrete retaining wall units safety barriers - RCC, Division of Tarmac Precast Concrete Ltd

D41 Crib walls/Gabions/Reinforced earth

Crib walls/Gabions/Reinforced earth - Cementation Foundations SKANSKA Ltd
Crib walls/Gabions/Reinforced earth - Keller Ground Engineering
Cribs walls - Stenoak Associated Services PLC
Cribwalls/Gabions/Reinforced Earth - Grass Concrete Ltd
Geotextiles - Don & Low Ltd
Ground engineering - PHI Group Ltd
Ground slabilization - PHI Group Ltd
Reinforced earth retaining walls/ bridge abutments/ precast arch system - Freyssinet Ltd
Retaining walls - ABG Ltd

D50 Underpinning

Piled raft underpinning - Apex Piling Ltd
Underpinning - Freyssinet Ltd
Underpinning - Haydon & Hagan Construction Ltd
Underpinning - Rok-Crete Units Co Ltd
Underpinning including jacking - Apex Piling Ltd

E05 In situ concrete construction generally

Castle brickbond SR cement - Castle Cement Ltd
Castle high alumina cement - Castle Cement Ltd
Castle hydrated lime - Castle Cement Ltd
Castle natural hydraulic lime - Castle Cement Ltd
Castle OPC cement - Castle Cement Ltd
Castle rapid hardening portland cement - Castle Cement Ltd
Castle sulfate-resisting portland cement - Castle Cement Ltd
Castle white portland cement - Castle Cement Ltd
Cement - Brikenden (Builders Merchants) Ltd
Concretes - Hanson Building Materials Europe
Construction accessories - Creteco Sales Ltd
In-situ concrete - Concrete Society
Lightweight aggregate - Lytag Ltd
Polypropylene & steel fibers for concrete & mortar - S.I. Concrete Systems
Spacer blocks for reinforcement - Creteco Sales Ltd
Synthetic ironoxide colouring agents - Bayer Chemicals Inoegate Pigment Group

E10 Mixing/Casting/Curing in situ concrete

Additives - Mapei (UK) Ltd
Admixtures - Sealocrete Ltd
Aggregates - Pioneer Concrete Holdings PLC
Cavity fill - Knauf Insulation Ltd
Cement - Rugby Cement
Cement - Super Cement Ltd
Cement, high alumina cement, calcium aluminate cement - Lafarge Aluminates Ltd
Coated road stone - Hanson Aggregates
Columns - Broadmead Cast Stone
Concrete admixtures - Beton Construction Ltd
Concrete chemicals - Grace Construction Products Ltd
Concrete plasticizers - Sika Ltd
Concrete products - Hanson Building Materials Europe
Concrete repair materials - Optiroc Ltd
Controlled permeability formliner - Creteco Sales Ltd
Fire cement - Purimachos Ltd
Lightweight and dense aggregates - CAEC Howard (Holdings) Ltd
Pier caps - Broadmead Cast Stone
Porticos - Broadmead Cast Stone
Ready mixed concrete - Fyfe John Ltd
Ready mixed concrete - Pioneer Concrete Holdings PLC
Ready mixed concrete - Tarmac Northern Ltd
Readymix - Tillicoutry Quarries Ltd
Sand and aggregate - Ashmead Buidling Supplies Ltd
Slag cement - Civil & Marine Slag Cement Ltd
Structural concrete - Lytag Ltd
Underwater concrete - Armorex Ltd

E11 Sprayed in situ concrete

Concrete admixtures - Beton Construction Ltd
Polypropylene & steel fibers for concrete & mortar - S.I. Concrete Systems
Shotcrete systems - Clarke UK Ltd
Sprayed insitu concrete - Colebrand Ltd
Tunnel concrete pumps - Clarke UK Ltd

E20 Formwork for in situ concrete

Adjustible steel pipes - Wade Building Services Ltd
Column formers - circular, square, rectangular - Creteco Sales Ltd
Concrete formwork plywood - Wood International Agency Ltd
Film faced formwork ply - Wood International Agency Ltd
Formwork - SGB Youngman
Formwork - Vencel Resil Ltd
Formwork and falsework products - includes design service - SGB Hire Plus
Formwork and Scaffolding - PERI Ltd
Formwork coatings - Hines & Sons, P E, Ltd
Formwork equipment - Maby Hire Ltd
Formwork for in situ concrete - Rom Ltd
Hy-rib permanent formwork - Expamet Building Products
Metal deck flooring - Caunton Engineering Ltd
Permanent shuttering - Cordek Ltd
Specialist formwork 3D - Cordek Ltd
Trough moulds - Cordek Ltd
Void formers & formwork - Cordek Ltd

E30 Reinforcement for in situ concrete

Contractor Detailing Services - BRC Ltd
Polypropylene & steel fibers for concrete & mortar - S.I. Concrete Systems
Prefabrication - BRC Ltd
Re-bar coating - Plastic Coatings Ltd
Reinforcement - Expamet Building Products
Reinforcement accessories - Rom Ltd
Reinforcement accessories and spacers - Hines & Sons, P E, Ltd
Reinforcement for in situ concrete - Civil Engineered Products Ldt
Reinforcement for insitu concrete - Freyssinet Ltd
Reinforcement for Insitu Concrete - Halfen Ltd
Reinforcement mesh and bar - BRC Ltd
Reinforcing accessories - Rainham Steel Co Ltd
Reinforcing bar and mesh - Hy-Ten Reinforcement Co Ltd
Reinforcing Bar Couplers - Ancon Building Products
Reinforcing fabric, bars and accessories - Bromford Iron & Steel Co Ltd
Reinforcing mesh - Rainham Steel Co Ltd
Stainless steel reinforcement - Helifix Ltd
Stainless steel reinforcing rods - Helifix Ltd
Steel reinforcement, fibre, mesh and industrial wire - Rom Ltd
Steel wire fibres for concrete reinforcement - Bekaert Building Products

E31 Post tensioned reinforcement for insitu concrete

Steel wire, cables, rods and bars - MacAlloy Limited

E40 Designed joints in insitu concrete

Concrete floor jointing system - Compriband Ltd
Designed joints - Stenoak Associated Services PLC
Designed joints in concrete - Ancon Building Products
Hydrophilic joint seals - Charcon Tunnels (Division of Tarmac plc)
Structural expansion joints - Radflex Contract Services Ltd
Waterstops - Beton Construction Ltd
Waterstops - RFA Group Ltd

E41 Worked finishes/Cutting into in situ concrete

Anti slip surfacing - Addagrip Surface Treatments Ltd
Concrete airfield pavement protection - Addagrip Surface Treatments Ltd
Dustproofing / anti-skidding concrete floors - Ferguson G A & Co Ltd
Floor ventilators - Airflow (Nicoll Ventilators) Ltd
Formliner for textured concrete - Creteco Sales Ltd
Grouts - Beton Construction Ltd

E42 Accessories cast into insitu concrete

Cast in dovetail slots and ties - Abbey Building Supplies Co
Concrete related accessories - BRC Ltd
Corner protection - Fixatrad Ltd
Grout Equipment - Parker AG Engineering Ltd
Hydrophilic joint seals - Charcon Tunnels (Division of Tarmac plc)
Reinforcement couplers - Halfen Ltd
Ties to bind reinforcing rods together - Huntley & Sparks
Tying wire, spacers and chemicals for concrete - Hy-Ten Reinforcement Co Ltd
Water stops - Creteco Sales Ltd

E50 Precast concrete frame structures

Bespoke concrete products - Hanson Concrete Products
Bridge beams - Tarmac Precast Concrete Limited
Car parks - precast concrete - Bison Concrete Products Ltd
Coffers - Finlay Breton Concrete Products
Composite floors - Finlay Breton Concrete Products
Concrete products - Bell & Webster Concrete Ltd
Engineering castings - Ballantine, Bo Ness Iron Co Ltd
Factory engineered precast concrete components - Tarmac Precast Concrete Limited
Framed structures/components - precast concrete - Bison Concrete Products Ltd
General precast - Finlay Breton Concrete Products
Ground Beams - Border Concrete Products Ltd
Ground Beams - Ebor Concretes Ltd
Ground Beams - Finlay Breton Concrete Products
Lattice girder - Finlay Breton Concrete Products
Polypropylene & steel fibers for concrete & mortar - S.I. Concrete Systems
Precast concrete and cast stone products - Sangwin Concrete Products
Precast concrete fixings - Halfen Ltd
Precast concrete floor beams, staircases, sills and lintols - ACP (Concrete) Ltd
Precast Concrete Frame Structures - Composite Structures
Precast concrete framing - Composite Structures
Precast concrete joists - supply & fix or supply only - C.R. Longley & Co. Ltd
Precast concrete lifting - Halfen Ltd
Precast concrete products - Lafaroe Redland Precast
Precast concrete products - Ruthin Precast Concrete Ltd
Precast concrete staircases - Bison Concrete Products Ltd
Precast Concrete Stairs & Landings - Ebor Concretes Ltd
Precast concrete structural components - Hanson Concrete Products
Precast concrete tunnels,shafts, and cover slabs and bespoke products - Charcon Tunnels (Division of Tarmac plc)
Precast concrete units - Evans Concrete Products Ltd
Precast concrete walling - Rogers Concrete Ltd
Railway crossing bearers - Tarmac Precast Concrete Limited
Railway sleepers - Tarmac Precast Concrete Limited
Retaining panels - Blanc de Bierges
Retaining Walls - Ebor Concretes Ltd
Retaining walls, tee walls - Bell & Webster Concrete Ltd
Spiral stairways - Blanc de Bierges
Stadia - precast concrete - Bison Concrete Products Ltd
Staircases, stadia terracing and retaining wall units - Hanson Building Products
Steps - Blanc de Bierges
Stone faced precast cladding - Marble Mosaic Co Ltd, The
Structural frames - Bison Structures Ltd
Synthetic iron oxide pigments and coloured pigments - Elementis Pigments
T-Beams - precast concrete - Lyncrete Burkle Services Ltd
Timberframed multi storey modular buildings - Accomodex
Wall Panels - Finlay Breton Concrete Products

E60 Precast/ Composite concrete decking

Beam and block floors - precast concrete - Bison Concrete Products Ltd
Concrete flooring - Hanson Concrete Products
Concrete Floors - CEMEX
Double tees - Finlay Breton Concrete Products
Floor beams - Stowell Concrete Ltd
Floors - Finlay Breton Concrete Products
Hollow core - Finlay Breton Concrete Products
Hollowcore floor units - precast concrete - Bison Concrete Products Ltd
Housefloor - precast concrete - Lyncrete Burkle Services Ltd
Interior flooring - Blanc de Bierges
Modular precast concrete stairs - Lyncrete Burkle Services Ltd
Plank floor units - precast concrete - Bison Concrete Products Ltd
Polypropylene, steel fibers for concrete & mortar - S.I. Concrete Systems
Precast composite concrete decking - ACP (Concrete) Ltd
Pre-cast Concrete - Halls Group Ltd
Precast concrete decking - Freyssinet Ltd
Precast concrete flooring - Marshalls Mono Ltd
Precast concrete staircases - Lyncrete Burkle Services Ltd
Precast concrete units - Caunton Engineering Ltd
Precast concrete; stairs, terraced seating, lintols etc - Border Concrete Products Ltd
Precast Prestressed Concrete Flooring - Tarmac Top Floor
Prestressed flooring slabs - Coltman Precast Concrete Ltd
Prestressed floors - Carter Concrete Ltd
Staircases, stadium terraces, columns, wall and ground beams - Coltman Precast Concrete Ltd
Structual components - Tarmac Precast Concrete Limited
Various prestressed precast concrete floorings - Hanson Building Products

F10 Brick/ Block walling

Aircrete Blocks - Tarmac Topblock Ltd
Aireared concrete building blocks - Thermalite - Marley Building Materials Ltd
Arches - Bovingdon Brickworks Ltd
Architectural facing masonry, medium and dense concrete blocks - Lignacite Ltd
Architectural masonry - Broadmead Cast Stone
Autoclaved Aerated Concrete Building Blocks - Thermalite
Block walling - Cheshire Concrete Products
Blocks - Brikenden (Builders Merchants) Ltd
Brick and block walling, cut and bond specials - Castle Brick (Wales) Ltd
Brick cutting specialists - M & G Brickcutters (specialist works)
Brick specials - Michelmersh Brick & Tile Co Ltd
Brick trade association - Brick Development Association
Brick/ block walling - Collier W H Ltd
Brick/ Blockwork - Roadways & Car Parks Ltd
Brick/block walling - Broome Bros (Doncaster) Ltd
Brick/block walling - Mansfield Brick Co Ltd
Bricks - Baggeridge Brick PLC
Bricks - Brikenden (Builders Merchants) Ltd
Bricks - Carlton Main Brickworks Ltd
Bricks - Chelwood Brick Ltd
Bricks - Country Manor Bricks
Bricks - Cranleigh Brick & Tile Co Ltd
Bricks - Freshfield Lane Brickworks Ltd
Bricks - Giscol Bricks
Bricks - Ibstock Scotish Brick
Bricks - Reclaim.UK.Com
Bricks - Rudgwick Brickworks Co Ltd
Bricks - York Handmade Brick Co Ltd
Bricks and blocks - Camas Building Materials
Bricks and blocks - Ashmead Buidling Supplies Ltd
Bricks and blocks - Blockleys Brick Ltd
Buff Clay - Carlton Main Brickworks Ltd
Building blocks - Aggregate Industries UK Ltd
Calcium silicate brick manufacturer - Esk Building Products
Chamfered Facing Block - Esk Building Products
Chemical Resistant Brick Walling - Accrington Brick Tile Co Ltd
Chimney systems - Keddy-Poujoulat (UK) Ltd
Chimney systems - Marflex Chimney Systems
Clay commons, facing bricks and engineering bricks - Hanson Brick
Clay facing bricks - Ibstock Brick Ltd
Clay facing bricks - Ibstock Building Products Ltd
Clay handmade facing bricks - Charnwood Brick Group Ltd
Commons - Northcot Brick Ltd
Concrete blocks - CEMEX
Concrete blocks - Forticrete Ltd
Concrete Blocks - Hanson Concrete Products
Concrete blocks - Sellite Blocks Ltd
Concrete blocks - Stowell Concrete Ltd
Concrete building blocks - Lignacite (North London) Ltd
Concretes - Readymix Limited
Coping stones - Chiltern Concrete and Stone Ltd
Dense and lightweight aggregate building blocks - CAEC Howard (Holdings) Ltd
Dense, lightweight, insulating solid, cellular and hollow concrete Blocks - Armstrong (Concrete Blocks), Thomas, Ltd
Engineering and facing bricks - Carlton Main Brickworks Ltd
Engineering bricks - Northcot Brick Ltd
Facing Brick - Carlton Main Brickworks Ltd
Facing bricks - Baggeridge Brick PLC
Facing Bricks - Hanson Brick Ltd, Northern Regonal Sales Office
Faience - Ibstock Building Products Ltd
Faience blocks - Shaws of Darwen
Fair face blocks - Thakeham Tiles Ltd
Fast Track Calium Silicate Bloca Systems - Esk Building Products
Fireplace briquettes - Michelmersh Brick & Tile Co Ltd
Glazed bricks - Smithbrook Building Products Ltd
Handmade bricks - Bovingdon Brickworks Ltd
Handmade facing bricks - Michelmersh Brick & Tile Co Ltd
Handmade facing bricks - Northcot Brick Ltd
Insulation - TDI (UK) Ltd
interior & exterior - Padipa Ltd
Iron cement - Legge Thompson F & Co Ltd
Large format blocks - Tarmac Topblock Ltd
Library and matching service - Mid Essex Trading Co Ltd
Lime/ sand mortars - Tarmac Northern Ltd
Machine made bricks - Michelmersh Brick & Tile Co Ltd

F10 Brick/ Block walling (Cont'd)

Machine made stock bricks - Bovingdon Brickworks Ltd
Ready to use cemenititious mortars - Tarmac Northern Ltd
Reclaimed bricks - Optiroc Ltd
reclaimed facing bricks - Northcot Brick Ltd
Red Clay - Carlton Main Brickworks Ltd
Refractory brickwork - SGL Acotec Ltd
Rustic wirecut facing bricks - Northcot Brick Ltd
Slab and block - Lindley George & Sons Ltd
Special shaped bricks - Baggeridge Brick PLC
Special Shaped Bricks - Carlton Main Brickworks Ltd
Specials - Bovingdon Brickworks Ltd
Specials - Northcot Brick Ltd
Staffordshire blue bricks and pavers - Ketley Brick Co Ltd , The
Synthetic ironoxide colouring agents - Bayer Chemicals Inoegate Pigment Group
Terracotta - Ibstock Building Products Ltd
Terracotta blocks - Shaws of Darwen
Terracotta wall facings - Smithbrook Building Products Ltd
traditional & thin joint systems - Tarmac Topblock Ltd

F11 Glass block walling

Bullet restraint/ proof glass block panels - Luxcrete Ltd
Glass block systems - Swedecor Ltd
Glass blocks - Cooper H W & Co Ltd
Glass blocks - Luxcrete Ltd
Glass blocks - Shackerley (Holdings) Group Ltd incorporating Designer ceramics
Polycarbonate clip blocks - Rockwell Sheet Sales Ltd

F20 Natural stone rubble walling

Architectural masonry - Farmington Natural Stone Ltd
Customised sheet metal in all metals - Azimex Fabrications Ltd
Dimension stone - Hanson Building Materials Europe
Dressed stone - Forest of Dean Stone Firms Ltd
Granite and natural stone - Bardon (England) Ltd
Heads - Bridge Quarry
Natural limestone quarried blocks - Hanson Bath & Portland Stone
Natural stone - Hanson Aggregates
Natural stone - Miles-Stone, Natural Stone Merchants
Natural stone - Realstone Ltd
Natural stone rubble walling - Macclesfield Stone Quarries Ltd
Natural stone rubble walling - Rare Stone Group, The
Paving and walling stone - The Delabole Slate Co. Ltd
Portland Stone aggregates - Hanson Bath & Portland Stone
Rock face walling - Thakeham Tiles Ltd
Stone masonry specialists - Macc Stone Ltd

F21 Natural stone ashlar walling/ dressings

Architectural terracotta and faience - Ibstock Hathernware Ltd
Ashlar masonry - Macc Stone Ltd
Cills - Bridge Quarry
Cleft stone and screen walling - Thakeham Tiles Ltd
Concrete products - Hanson Aggregates
Dimensional English Stone - Realstone Ltd
Facades in stone, brick & terracotta - Stonewest Ltd
Granite and marble - design, procurement - Szerelmey Ltd
Granite and natural stone - Bardon (England) Ltd
Limestones and sandstones - Rare Stone Group, The
Monumental - Forest of Dean Stone Firms Ltd
Natural derbyshire sandstone - Stoke Hall Quarry(Stone Sales)Ltd
Natural stone - Miles-Stone, Natural Stone Merchants
Natural stone ashlar walling - Macclesfield Stone Quarries Ltd
Natural stone ashlar walling/dressing - G. Miccoli & Son Ltd
Rapid Set Façade StopBonds - 6 min - Laticrete UK
Sawn ashlar - Lindley George & Sons Ltd
Traditional stone - Chelsea Artisans Ltd

F22 Cast stone walling/ dressings

Architectural cast stone - Hampton Stone Ltd
Architectural masonry - Forticrete Ltd
Architectural stonework - Haddonstone Ltd
Artificial stone - Allen (Concrete) Ltd
Artificial stone - Procter Cast stone & Concrete Products
Artificial stone units - Northwest Pre-cast Ltd
Bespoke cast stone service - Procter Cast stone & Concrete Products
Cast stone - Border Concrete Products Ltd
Cast stone - Ibstock Building Products Ltd
Cast stone - Lignacite Ltd
Cast stone - Procter Bros Limited
Cast stone dressings - Forticrete Ltd
Cast stone walling - Chiltern Concrete and Stone Ltd
Cast stone walling - Macclesfield Stone Quarries Ltd
Cast stone walling - Marble Mosaic Co Ltd, The
Cast stone walling - Technostone Ltd
Cast stone walling/ dressings - Chilstone
Cast stone; sills, lintols, copings, features etc - Border Concrete Products Ltd
Decorative Commercial - Aggregate Industries UK Ltd
Fencing - Procter Bros Limited
Precast walling - Oakdale (Contracts) Ltd
Reconstructed cast stone architectural units - Broadmead Cast Stone
Reconstructed stone fireplaces - Taylor & Portway Ltd
Reconstructed stone walling - Atlas Stone Products
Stone - Ibstock Scotish Brick
Stone walling - reconstructed - Forticrete Ltd

F30 Accessories/Sundry items for brick/ block/ stone walling

Abrasive discs and blocks - Harbro Supplies Ltd
Accessories/ sundry items for brick/ block/ stone - Civil Engineered Products Ldt
Accessories/sundry items for brick/block/stone - Haddonstone Ltd
Accessories/undry items for brick/ block/stone - Mansfield Brick Co Ltd
Accessoriesfor brick and block walling - Ensor Building Products
Acoustic ventilators - Ryton's Building Products Ltd
Aluminium copings - Alifabs (Woking) Ltd
Arch lintels - Catnic
Arches - Cavity Trays Ltd
Arches - Northcot Brick Ltd
Brick reinforcement - RFA Group Ltd
Brickties and windposts - Halfen Ltd
Brickwork reinforcement - BRC Special Products
Brickwork support - Halfen Ltd
Building products - cavity closers - Quantum Profile Systems Ltd
Cast in dovetail slots and ties - Abbey Building Supplies Co
Cast stone, sills, lintels, copings, features etc - Border Concrete Products Ltd
Cavity and through wall ventilators - Ryton's Building Products Ltd
Cavity fixings - Artur Fischer (UK) Ltd
Cavity floor boxes - Salamandre Davis
Cavity tanking tiles - Atlas Stone Products
Cavity trays - Cavity Trays Ltd
Cavity Trays - Zled (DPC) Co Ltd
Cavity trays, hight performance dpc - Timloc Expamet Building Products
Chimney flues and linings - Rite-Vent Ltd
Chimney service - Cico Chimney Linings Ltd
Chimney systems - Marflex Chimney Systems
Chimney terminals - Dunbrik (Yorks) Ltd
Clay Chimney Pots - Knowles W T & Sons Ltd
Closers - Cavity Trays Ltd
Coadestone castings - Clifton Nurseries
Columns, canopies, cornices, porticos, bay canopies - Solair GRP Architectural Products
Concrete padstones - Procter Cast stone & Concrete Products
Copings and cills - Burlington Slate Ltd
Cut nails - Glasgow Steel Nail Co Ltd
Damp proof courses - Anderson D & Son
Damp proof courses - Marley Waterproofing
Dense and lightweight blocks - Newlay Concrete Ltd
Diamond blades - Harbro Supplies Ltd
Diamond Blades - Marcrist International Ltd
Diamond core drills - Marcrist International Ltd
Enforce - composite strengthening - Weber Building Solutions
External wall insulation and cladding - Axter Ltd
External wall insulation systems - PermaRock Products Ltd
Faience blocks - Shaws of Darwen
Fireplaces - Farmington Natural Stone Ltd
Flue linings - Dunbrik (Yorks) Ltd
Flues and chimneys - Docherty New Vent Chimney Group
Foundation blocks - Newlay Concrete Ltd
Garden Products - Aggregate Industries UK Ltd
Gas flue blocks - Dunbrik (Yorks) Ltd
Gold leaf - Harbro Supplies Ltd
Granite and natural stone - Bardon (England) Ltd
GRP dummy chimneys - Wilsons Fibreglass Ltd
High performance DPC and cavity trays - Visqueen Building Products
Insulating cavity closers and insulated DPC - TDI (UK) Ltd
Insulation blocks - Newlay Concrete Ltd
Insulation fixings - Artur Fischer (UK) Ltd
Internal room ventilators - Airflow (Nicoll Ventilators) Ltd
Isolation Sleves Remedial - Redifix Ltd
Lateral restraint - HCL Contracts Ltd
Light weight insulating concrete flue system - Taylor & Portway Ltd
Lintels - Ancon Building Products
Lintels - Cavity Trays Ltd
Lintels - CEMEX
Lintels and other hand made shapes - Lignacite Ltd
Lintels, cils and padstones - Chiltern Concrete and Stone Ltd
Masonry coatings - Jotun Henry Clark Ltd Decorative Division
Masonry Design Development - Hanson Bath & Portland Stone
Masonry reinforcement - Bekaert Building Products
Masonry support systems - Ancon Building Products
Masonry support systems - Metsec Building Products Ltd
Masonry to masonry connectors - Simpson Strong-Tie®
Masonry tools - tungsten tipped chisels - Harbro Supplies Ltd
Movement joint - Shackerley (Holdings) Group Ltd incorporating Designer ceramics
Movement joints - Allmat (East Surrey) Ltd
Mullions - Bridge Quarry
Natural stone fixing - Halfen Ltd
Non standard precast units in special finishes - Chiltern Concrete and Stone Ltd
Polythene damp-proof courses - Visqueen Building Products
Power tools - Harbro Supplies Ltd
Pumice stone chimney systems - Sparkes K & L
Remedial wall ties - ITW Construction Products
Remedial wall ties - Redifix Ltd
Repair mortars - Armorex Ltd
Retaining wall products - Forticrete Ltd
Roof ventilation, underfloor ventilation, through the wall ventilation - Timloc Expamet Building Products
Special shaped clay bricks - Ibstock Brick Ltd
Stainless steel remedial wall ties - Helifix Ltd
Steel lintels - Catnic
Steel lintels - Expamet Building Products
Steel lintels - Wade Building Services Ltd
Steel wall connectors - Allmat (East Surrey) Ltd
Stone and brickwork, cleaning - CSL (Services) Limited
Straps - Expamet Building Products
Structural Products - Aggregate Industries UK Ltd
Terracotta blocks - Shaws of Darwen
Through wall ventilators - Airflow (Nicoll Ventilators) Ltd
Timber to masonry connectors - Simpson Strong-Tie®
Ventilators - Cavity Trays Ltd
Wall connectors - Catnic
Wall insulation - Callenders Ltd
Wall insulation - Dow Construction Products
Wall starters - Expamet Building Products
Wall tie replacement - HCL Contracts Ltd
Wall ties - Catnic
Wall ties - Ensor Building Products
Wall ties - Vista Engineering Ltd
Wall ties and reinforcements - Ancon Building Products
Water repelents, antigraffiti and bridge deck membrane system - Degussa Ltd
Weepholes - LB Plastics Ltd
Weepholes for cavity walls - Molyneux G (Products) Ltd
Windpots Masonry support systems reinforcement for masonry - Vista Engineering Ltd
Wire tray - Salamandre Davis

F31 Precast concrete sills/ lintels/ copings/ features

Architectural Masonary - Aggregate Industries UK Ltd
Bespoke - Allen (Concrete) Ltd
Cast stone, sills, lintels, copings, features etc - Border Concrete Products Ltd
Chimney systems - Marflex Chimney Systems
Cills - Allen (Concrete) Ltd
Coping and caps - Rogers Concrete Ltd
Copings - Allen (Concrete) Ltd
Copings Lintels - Stowell Concrete Ltd
General precast items - Brierkrete Ltd
Hydrant boxes for gas and water - Brierkrete Ltd
Lintels - Allen (Concrete) Ltd
Lintels - Allmat (East Surrey) Ltd
Lintels - Anderton Concrete Products Ltd
Lintels - Procter Cast stone & Concrete Products
Plastic cavity closers - LB Plastics Ltd
Polypropylene, steel fibers for concrete & mortar - S.I. Concrete Systems
Precast concrete - Border Concrete Products Ltd
Precast concrete and reconstructed stone manufacturers - Rok-Crete Units Co Ltd
Precast concrete products - Allen (Concrete) Ltd
Precast concrete sills - Hampton Stone Ltd
Precast concrete sills/ lintels/ copings/ features - Chilstone
Precast concrete sills/ lintels/ copings/ features - Evans Concrete Products Ltd
Precast concrete sills/lintels/copings/features - Broadmead Cast Stone
Precast concrete sills/lintels/copings/features - Haddonstone Ltd
Precast concrete sills/lintels/copings/features - Technostone Ltd
Precast deenits - Northwest Pre-cast Ltd
Precast lintels and manhole covers - Caradon Jones Ltd
Precast lintels and prestressed lintels - Procter Bros Limited
Precast Prestressed Concrete Lintels - Tarmac Top Floor
Precast sills/lintels/ coping and features - Chiltern Concrete and Stone Ltd
Precast terrazo columns - Roman Mosaic Contracts Ltd
Prestressed wide slab flooring - ACP (Concrete) Ltd
PVCu products - Howarth Timber (Elland) Ltd
Reconstituted Artstone and Precast Concrete Products - North Yorkshire Artstone Ltd
Retaining walls - Concrete Products (Lincoln) 1980 Ltd
Steel cavity fixings - Lindapter International
Supergalv steel lintels - Birtley Building Products Ltd

G10 Structural steel framing

Aluminium and steel fabricators - Ramsay & Sons (Forfar) Ltd
Aluminium fabrications - Midland Alloy Ltd
Architectural & General Metalwork - Thanet Ware Ltd
Blasting - Bristol Metal Spraying & Protective Coatings Ltd
Brazing alloys, solder and fluxes - Johnson Matthey PLC - Metal Joining
Bridges in steel, boardwalks, decking, lake edging and bespoke landscape structures - CTS Ltd
Coated steel - SSAB Dobel Coated Steel Ltd
Cold rolled purlins and rails - Ward Building Components Ltd
Cold rolled steel sections - Hadley Industries Plc
Collled rolled sections - Walls & Ceilings International Ltd
Copper wire and rods - United Wire Ltd
Cromp - arc studwelding - CSW Coldform Ltd
Domestic and industrial tongue and grooved flooring - Egger (UK) Ltd
Erection of structural steelwork - Billington Structures Ltd
Fabric tensile structures - Clyde Canvas Fabric Structures Ltd
Fabrication of structural steelwork - Billington Structures Ltd
Flat to pitch conversions - Alumasc Exterior Building Products Ltd
Gypframe (steel framed houses) - Metsec Building Products Ltd
Heavy duty protective coatings - International Protective Coatings,
Hot and cold pressings - Barton B C & Son Ltd
Hot dip galvanisers - Lancaster & Co (Bow) Ltd
Industrial Coating Specialists - Bristol Metal Spraying & Protective Coatings Ltd
Meshes - Cadisch MDA Ltd Ltd
Metal fabrication - OEP Furniture Group PLC
Metal framing - Unistrut Limited
Metsec framing - panelised light steel buildings - Metsec Building Products Ltd
Mezzanine floor channels and CEE section floor bea - Metsec Building Products Ltd
Mezzanine Floors - Barton Storage Systems Ltd
Open grill flooring - Norton Engineering Alloys Co Ltd
Rail fixings - Lindapter International
Roof and soffit ventilators - Ryton's Building Products Ltd
Roof truss system - MiTek Industries Ltd
Roof trusses - Llewellyn Homes Ltd
Spacejoist V webs - Gang-Nail Systems Ltd
Special steel sections - British Steel Special Sections
Stainless steel fabrications - Midland Alloy Ltd
Steel fabrication - WA Skinner & Co Limited
Steel fabrication - Woodburn Engineering Ltd
Steel framed buildings complete with fixtures, fittings and cladding - Reid John & Sons (Strucsteel) Ltd
Steel framing - Vantrunk Engineering Ltd
Steel tube stockholders, fabrications, welding and fittings - Midland Tube and Fabrications
Steel wire rope - Bridon International Ltd
Steel wire rope rigging - Marine & Commercial Rigging Systems
Steel wires, cables, rods and bars - Light (Multiforms), H A
Steelwork fabrication - Gilmer Limited
Structural alterations - Haydon & Hagan Construction Ltd
Structural frames - Bison Structures Ltd
Structural steel - Hy-Ten Reinforcement Co Ltd
Structural steel - Woodburn Engineering Ltd
Structural steel framing - Angle Ring Co Ltd
Structural steel framing - Astron Buildings Parker Hannifin PLC
Structural steel framing - Caunton Engineering Ltd
Structural steel framing - Hubbard Architectural Metalwork Ltd
Structural steel framing - Joy Statford Joists Ltd
Structural steel framing - LB Structures Ltd
Structural steel framing and general fabrication - Camtwix Engineering (Division of Terrapan Ltd)
Structural Steel Protection - Abesco Ltd

G10 Structural steel framing (Cont'd)

Structural steel towers - Franklin Hodge Industries Ltd
Structural steelwork - Atlas Ward Structures Ltd
Structural steelwork - Billington Structures Ltd
Structural steelwork - CAEC Howard (Holdings) Ltd
Structural steelwork - Ermine Engineering Co. Ltd
Structural steelwork - South Durham Structures Limited
Structural steelwork - Wescol Glosford
Structural steelwork - supply, delivery and erection - Bourne Steel Ltd
Structural steelwork and associated products - Bourne Steel Ltd
Structural steelwork, portal frames and general steelwork - Bromag Structures Ltd
Structural, mezzanine floors, conveyor support steelwork and architectural structures - Down & Francis Ltd, Division of Metalrax PLC
Support systems - Lindapter International
Timbers - Power Breaker
Universal beams & Columns - Rainham Steel Co Ltd
Wire rope assemblies and fittings - Midland Wire Cordage Co Ltd
Z' purlins - Haze Cladding Systems
Zed and CEE purlins - Metsec Building Products Ltd

G11 Structural aluminium framing

Aluminium extrusions/ sections - Seco Aluminium Ltd
Aluminium Structural Framing - Hartley Clear Span Ltd
Curved canopies, barrel vaults and fabrications - Midland Alloy Ltd
Ground floor framing - Comar Architectural Aluminuim Systems
Structural aluminium framing - Exitex Ltd

G12 Isolated structural metal members

Aluminium isolated structural members - Alcoa Europe-Extrusions & Endproducts (UK)
Blasting - Bristol Metal Spraying & Protective Coatings Ltd
Cold roll sections/forming - Metsec Building Products Ltd
Curved steel sections, tubes, roof and cambered beams - Angle Ring Co Ltd
Industrial Coating Specialists - Bristol Metal Spraying & Protective Coatings Ltd
Isolated structural members - Hubbard Architectural Metalwork Ltd
Isolated structural members - Hy-Ten Reinforcement Co Ltd
Isolated structural members - LB Structures Ltd
Lattice joist and trusses - Metsec Building Products Ltd
Metal web floor beams - Wyckham Blackwell Ltd
Steel chimneys and associated structural steelwork - Beaumont Ltd F E
Structural secondary steelwork - Down & Francis Ltd, Division of Metalrax PLC
Subcontract fabrication - Mackay Engineering
Subcontract machining - Mackay Engineering
Tubes, fittings & flanges - The BSS Group PLC
Welding and fabrication work - Griff Chains Ltd

G20 Carpentry/ Timber framing/ First fixing

Aluminuim Fascia / Soffit - Guttercrest
Associated joinery - ARC Partitioning
Bespoke joinery - Howard Bros Joinery Ltd
Boat nails - Glasgow Steel Nail Co Ltd
Bridges in timber boardwalks, decking, lake edging and bespoke landscape structures - CTS Ltd
British and European hardwoods - Henry Venables Ltd
Carpenters Metalwork - Allmat (East Surrey) Ltd
Carpentry/ Timber framing/ First fixing - Medina Gimson
Composite floor joints - Bullock & Driffill Ltd
Construction & joniery softwoods - Cox Long Ltd
Constructional timber - Brewer T. & Co Ltd
Design, manufacture and supply only of timber engineered floor structures - Marlow and Company Ltd
Design, manufacture and supply only of trussed rafter and associated roof structures - Marlow and Company Ltd
Eaves fillers - Corofil Woodall
Fibreglass products - Yonder Hill Ltd
Fire resisting joinery - Decorfix Ltd
Fixings to trusses - MiTek Industries Ltd
Flooring - Finn Forest
Glulam beams - Laminated Wood Ltd
Hardwood - Brooks Belfast
Hardwood - Finn Forest
Hardwood - James Latham plc
Hardwood timber - Jakem Timbers Ltd
Hardwoods - Brewer T. & Co Ltd
High class joinery - Platonoff & Harris Plc
Industrial pre-treatment preservatives for timber - Protim Solignum Ltd
Intumescent coatings for timber - Quelfire
Joist hangers - Expamet Building Products
Laminated beams - Bullock & Driffill Ltd
Laminated timber beams and structures - Moelven Laminated Timber Structures Ltd
Lightweight lattice joists and trusses - Metsec Building Products Ltd
Load bearing and non LB site fixed panels - Metsec Building Products Ltd
MDF, plywood, and engineered wood products - Willamette Europe Ltd
Nail plates - Gang-Nail Systems Ltd
Paneling (wood) - OSMO UK Ltd
Planed and sawn timber - Mason FW & Sons Ltd
Pneumatic nailers, staples and fastenings - Duo-Fast (UK) Ltd
Portal frames and cranked beams - Laminated Wood Ltd
Rail spikes - Glasgow Steel Nail Co Ltd
Roof trusses - Armstrong (Timber), Thomas, Ltd
Roof trusses - J. Scott (Thrapston) Ltd
Roof trusses - Merronbrook Ltd
Roof trusses - Meyer Drenham Ltd
Roof Trusses - Moray Timber Ltd
Roofing contractor - Evans Howard Roofing Ltd
Saw millers and timber merchants - Whitmore's Timber Co Ltd
Sawmilling - Moray Timber Ltd
Softwood - Finn Forest
Stainless steel timber to masonry fixings - Helifix Ltd
Stainless steel warm roof fixings - Helifix Ltd
Steel decking - Tarmac Top Floor
Stress grading - Cox Long Ltd
Stress Grading - Moray Timber Ltd
Structural timber engineers - Kingston Craftsmen Structural Timber Engineers Ltd
Timber - Ashmead Buidling Supplies Ltd
Timber - Brooks Belfast
Timber - Moray Timber Ltd
Timber and log building specialists - Finwood Leisure Ltd
Timber for structural applications - Wiltshire G R & Co
Timber frame buildings - Fleming Buildings Ltd
Timber frame design and manufacture - Prestoplan Purposebuilt Ltd
Timber frame design supply & fix - Cox Long Ltd
Timber frame housing - Fleming Homes Ltd
Timber framed houses - Cambridge Structures Ltd
Timber frames and roof trusses - Walker Timber Ltd
Timber importers & sawmillers - Walker Timber Ltd
Timber importers, swamillers and merchants - Brewer T. & Co Ltd
Timber merchants - Derby Timber Supplies
Timber merchants - Timber Supplies
Timber merchants - Timbmet Ltd
Timber pallets - Yonder Hill Ltd
Timber panels - Challenge Fencing Ltd
Timber portal frames engineered products, laminated beams, trussed rafters and bolted trusses - Structural Timbers Ltd
Timber preservative treatments - Kingfisher Building Products Ltd.
Timber roof trusses - Dover Trussed Roof Co Ltd
Timber strength grading - Walker Timber Ltd
Timber supplies - Bullock & Driffill Ltd
Timber to masonry connectors - Simpson Strong-Tie®
Timber to timber connectors - Simpson Strong-Tie®
Timber trussed rafters - Bullock & Driffill Ltd
Treatment - Moray Timber Ltd
Trussed rafter design & manufacture - Cox Long Ltd
Trussed rafter nailplate system, builders fixings and beams - Alpine Automation Ltd.
Trussed rafters - Palgrave Brown UK Ltd
Trussed rafters - Wyckham Blackwell Ltd
Trussed rafters and timber frame buildings - Covers Timber Structures Ltd
Trusses, beams, frames, panels , timber engineering - Cowley Structural Timberwork Ltd

G30 Metal profiled sheet decking

Aluminium metal profiled sheet decking - Alcoa Europe-Extrusions & Endproducts (UK)
Composite decking - Broderick Structures Limited
Insulated Roof & Wall Systems - Kingspan Ltd
Metal profile sheet decking - Briggs Roofing & Cladding Ltd, Roofing Contract
Open steel industrial flooring - Eurogrid Ltd
Profiled metal floor decking - Ward Building Components Ltd
PVC decking - LB Plastics Ltd
Sheet metal work - Barton B C & Son Ltd
Structural steelwork - supply, delivery and erection - Bourne Steel Ltd

G31 Prefabricated timber unit decking

Decking - Anthony de Grey Trellises
Sheet materials - Finn Forest
Structural tongue and groove decking - Structural Timbers Ltd
Timber frame kits - Merronbrook Ltd

G32 Edge supported/ Reinforced woodwool/rock fibre decking

Flat and pitched decking products - Torvale Building Products, A Division of Stadium Group PLC

H10 Patent glazing

Aluminium and lead clothed steel patent glazing systems - Standard Patent Glazing Co Ltd
Aluminium conservatory roofs - Glostal - Monarch
Anti-vandal glazing - Amari Plastics Plc
Cintain Walling - Duplus Architectural Systems Ltd
Cladding/ covering - Pilkington UK Ltd
Conservatory components - Exitex Ltd
Curtain walling - Astrofade Ltd
Curtain Walling - Freeman T R
Curtain walling - Heywood Williams Architectural
Curved Glass - Gardner & Newton
Firerated screens - Baydale Architectural Systems Ltd
Glazed barrell vaults - Baydale Architectural Systems Ltd
Glazed steel frame, covered walkways and canopies - Planet Architectural Glazing Ltd
Patent glazing - Cantifix of London Limited
Patent glazing - Duplus Domes Ltd
Patent glazing - EBS Services Ltd
Patent glazing - Lonsdale Metal Company
Patent glazing - Twide-Paragon Ltd
Patent glazing, prorietary glazing systems - Kelsey Roofing Industries Ltd
Rooflights, lantern lights and pyramid lights - Standard Patent Glazing Co Ltd

H11 Curtain walling

Aluminium curtain walling - Glostal - Monarch
Curtain wall systems - Comar Architectural Aluminuim Systems
Curtain walling - Alcover UK Ltd
Curtain walling - Barlow Group
Curtain walling - BLW Associates Ltd
Curtain walling - Cantifix of London Limited
Curtain Walling - CEP Claddings Ltd
Curtain walling - Colt International
Curtain walling - Crittall Steel Windows Ltd
Curtain walling - Deceuninck Ltd
Curtain walling - Duplus Domes Ltd
Curtain walling - English Architectural Glazing Ltd
Curtain Walling - McMullen Architectural Systems
Curtain walling - Panel Systems Ltd
Curtain Walling - SCHUCO International KG
Curtain walling - Stoakes Systems Ltd
Curtain walling - Twide-Paragon Ltd
Curtain walling and cladding - Schmidlin (UK) Ltd
Extruded curtain walling - Smart Systems Ltd
Fibre cement slates - Cembrit Blunn Ltd
Firerated curtain walling - Baydale Architectural Systems Ltd
General extrusions - LB Plastics Ltd
Palent Glazing - Duplus Architectural Systems Ltd
Patent glazing - Astrofade Ltd
Patent glazing - Heywood Williams Architectural
Patent glazing - Solaglas
Structural wall solutions - CGL Cometec Ltd

H12 Plastic glazed vaulting/ walling

Multiwall polycarbonate - Kenilworth Daylight Centre Ltd
Opaque chemical resistant cladding - Filion Products Ltd
Plastic glazed vaulting - Duplus Domes Ltd
Plastic glazed vaulting/ walling - Astrofade Ltd
Plastic glazed vaulting/walling - Amari Plastics Plc
Polycarbonate glazing for flat and industrial roofs - Cox Building Products Ltd
Polycarbonate glazing systems - EBC UK Ltd
Profiled plastic sheets, conservatory roofs and flat plastic sheets - Kenilworth Daylight Centre Ltd

H13 Structural glass assemblies

Atria Glazing - Fenlock-Hansen Ltd
Barrelvaults, ridgelights, pyramids in glass or polycarbonate - Cox Building Products Ltd
Glass arches - Luxcrete Ltd
Glazed security products - Baydale Architectural Systems Ltd
Steel wire, cables, rods and bars - MacAlloy Limited
Structural glass assemblers - Astrofade Ltd
Structural glass assemblies - Cantifix of London Limited
Structural glass assemblies - Colt International
Structural glass assemblies - Pilkington UK Ltd
Structural Glass Assemblies - SCHUCO International KG
Structural glazing - Alcover UK Ltd
Structural Glazing - McNeill-McManus Ltd
Structural silicone glazing - Stoakes Systems Ltd
Stuctural Glazing - Twide-Paragon Ltd

H14 Concrete rooflights/pavement lights

Cell windows and observation units - Luxcrete Ltd
Pavement lights - Cooper H W & Co Ltd

H20 Rigid sheet cladding

Armour sheeting - Meggitt Armour Systems
B.R.E. Structural Render Panels - Gradient Insulations (UK) Ltd
Balcony panels - Trespa UK Ltd
Cladding - Burlington Slate Ltd
Cladding - Eternit Building Materials
Cladding - wall and ceiling - Flexitallic Ltd
Cladding panels - Blaze Neon Ltd
Composite panels - EBC UK Ltd
Composite panels - Gradient Insulations (UK) Ltd
Curtain Walling - Solaglas
Exterior cladding panels - Trespa UK Ltd
Exterior grade high pressure laminate - Abet Ltd
Facade systems/specialist wallcladding - Kelsey Roofing Industries Ltd
Fire Protection boards - Promat UK Ltd
GRP flatroof systems - Filion Products Ltd
High performance building panel - Trespa UK Ltd
High performance claddings - Cembrit Blunn Ltd
Hygenic wall + ceiling services - Altro Floors
Hygienic cladding - Amari Plastics Plc
Kerto LVL (laminated veneer lumber) - Finn Forest
L.P.C.B Roof systems - Gradient Insulations (UK) Ltd
L.P.C.B Wall panels - Gradient Insulations (UK) Ltd
Metal cladding - Coverite Ltd
Metal roofing and cladding - Ash & Lacy Building Products Ltd
PVC claddings and fascias - Rockwell Sheet Sales Ltd
Riggid sheet cladding - Kemlite Ltd
Rigid Sheet Cladding - CEP Claddings Ltd
Rihid sheet cladding - Panel Systems Ltd
Sheet materials - Mason FW & Sons Ltd
Structural Insulated Panels - EBC UK Ltd
Wall cladding - Dunhams of Norwich
Wall Cladding, roof cladding, insulated wall and roof cladding - Kingspan Ltd
Wall lining systems - Formica Ltd
Wood wool boards - Skanda

H21 Timber weatherboarding

Cedar shakes for roofing and cladding - Brash John & Co Ltd
Timber cladding - LSA Projects Ltd
Timber weatherboarding - Brash John & Co Ltd
Timber Weatherboarding - CEP Claddings Ltd

H30 Fibre cement profiled sheet cladding/ covering/ siding

Corrugated sheeting - Eternit Building Materials
Fastenings - Cooper & Turner
Fibre cement profiled sheet cladding / covering - TRC (Midlands) Ltd
Fibre cement sheeting - Gleno Industries Ltd
Profiled Sheets - Cembrit Blunn Ltd

H31 Metal profiled/ flat sheet cladding/ covering/ siding

Aluminium & steel Profiled cladding - Rigidal Systems LTD
Aluminium & steel standing seam systems - Rigidal Systems LTD
Aluminium cladding and roofing - Rigidal Industries Ltd.
CA Profiles - CA Group Ltd - MR-24 Division
Cladding - Astron Buildings Parker Hannifin PLC
Cladding - Caunton Engineering Ltd
Cladding - Letchworth Roofing Company Ltd
Cladding - Orsogril Spa UK Distributor
Cladding and siding - Bromag Structures Ltd
Cladding Systems - Cadisch MDA Ltd Ltd
Column Cladding - Cadisch MDA Ltd Ltd
Composite aluminium and steel flat cladding systems - Booth Muirie
Composite aluminium and steel rain screens - Booth Muirie
Composite panels - Composite Panels Ltd
Composite Panels - Haze Cladding Systems
Expanded Mesh - Cadisch MDA Ltd Ltd
Fastenings - Cooper & Turner
Flat cladding panels - Technology Telford Ltd
Galvanised Corrugated sheet - Rainham Steel Co Ltd
Industrial and commercial roofing - Letchworth Roofing Company Ltd
Insulated Roof & Wall Systems - Kingspan Ltd
Insulated roof panels - Composite Panels Ltd
Insulated wall and roof panels and cladding sheets - Ward Building Components Ltd
Lightweight sheet and panel roofing systems - Britmet Tileform Limited
Metal cladding - Decra Roofing Systems
Metal cladding - LSA Projects Ltd
Metal fabrication, ancillary cladding components - Interlink Group
Metal profiled / flat sheet cladding / covering - TRC (Midlands) Ltd

H31 Metal profiled/ flat sheet cladding/ covering/ siding (Cont'd)

Metal profiled and flat sheet cladding, covering and siding - Haze Cladding Systems
Metal profiled/ flat sheet cladding/ covering - Hadley Industries Plc
Metal profiled/ flat sheet cladding/ covering - Reid John & Sons (Strucsteel) Ltd
Metal profiled/flat sheet cladding/covering - Amari Plastics Plc
Metal Profiled/flat sheet cladding/covering - Britmet Tileform Limited
Metal roof tiles - Catnic
Metal roof, wall and decking - Haironville TAC Ltd
Metal roofing - Coverite Ltd
Metal Roofing - Lafarge Roofing Ltd
Metal roofing and cladding - Ash & Lacy Building Products Ltd
Metal roofing and cladding - Briggs Roofing & Cladding Ltd, Roofing Contract
Metal tiles - EBC UK Ltd
Perforated Sheet - Cadisch MDA Ltd Ltd
Pre Crump Meshes - Cadisch MDA Ltd Ltd
Prebonded claddding - Broderick Structures Limited
Prefabricated bathroom structures - IPPEC Sytsems
Profiled cladding - European Profiles
Profiled metal sheeting - Gleno Industries Ltd
Rainscreen cladding - E. H. S. Roofing Ltd
Rainscreen cladding systems - Booth Muirie
Sheet Metal Work - Thanet Ware Ltd
Sheeting - Durable Contracts Ltd
Sheeting - Letchworth Roofing Company Ltd
Sheeting and cladding - TRC Service Group Ltd
Single ply flat roof system - Intergrated Polymer Systems (UK) Ltd
Standing Seam Roof Decking - Gradient Insulations (UK) Ltd
Steel and aluminium profiled roofing - Kelsey Roofing Industries Ltd
Steel tile effect roof sheeting - Link Build Products Ltd
Traditional flat roofing systems - Ruberoid Building Products Ltd
Vitreous enameled steel cladding panels - Escol Products Ltd
Wall cladding - Drawn Metal Ltd
Waterproofing (inc. standing seam roofing) - Alumasc Exterior Building Products Ltd
Waven Ware Mesh - Cadisch MDA Ltd Ltd

H32 Plastic profiled sheet cladding/ covering/ siding

Cellular PVC fascias and soffits - Celuform
Cellular Pvc fascias, soffits, barge boards and claddings - Swish Building Products
Cladding and roof glazing - Vulcan Plastics Ltd
Corrugated or flat PVC and polycarbonate clear or coloured plastic sheets - Rockwell Sheet Sales Ltd
Covers for new and existing pools of almost any shape - Grando (UK) Ltd
Decorative internal cladding system - Swish Building Products
Fascia, soffit and cladding systems - FloPlast Ltd
Fastenings - Cooper & Turner
Flat roof outlets - Hunter Plastics Ltd
Glass reinforced polyesther profiled and flat sheeting - Filion Products Ltd
GRP cladding - Wilsons Fibreglass Ltd
Internal & external cladding - Amari Plastics Plc
Multiwall polycarbonate sheets - Rockwell Sheet Sales Ltd
Over roofing - Filion Products Ltd
Plastic coated steel flashings - Azimex Fabrications Ltd
Plastic fabrication - Colman Greeves
Plastic mouldings - Invicita Plastics Ltd
Plastic profile extrusion - HIS Ltd
Plastic profiled sheet cladding /covering - TRC (Midlands) Ltd
Plastic profiled sheet cladding and roofing - Ariel Plastics Ltd
Plastic profiled sheet cladding, facias, soffits, interior wall and ceiling panelling - Deceuninck Ltd
Plastic profiled sheets - Leicester Barfitting Co Ltd
Plastic sheets and laminates - Vulcascot Ltd
PVC cladding - LB Plastics Ltd
PVCu sills, trims, cladding and roof line products - Caradon Duraflex Systems Ltd
PVC-V facia & Soffit - Allmat (East Surrey) Ltd
Roof sheets - profiled and structured in GRP, PVC and polycarbonate - Brett Martin Ltd
Soffits and fascias - Spectra Glaze Services
Thermoplastic roofing and cladding - EBC UK Ltd
Vinyl cladding system for fascias soffits and full elevations - Cox Building Products Ltd
Wall cladding - E. H. S. Roofing Ltd
Wall cladding - Opto International Ltd
Woodgrain weather boarding - Vulcan Plastics Ltd

H33 Bitumen & fibre profile sheet cladding/ covering

Bitumen + fibre profile sheet cladding - Onduline Building Products Ltd
Bitumious felt roofing systems - Ruberoid Building Products Ltd
Foundation protection - Onduline Building Products Ltd
Plastic profiled sheet cladding and roofing - Ariel Plastics Ltd

H40 Glassfibre reinforced cement panel cladding/ features

Glass reinforced cement cladding - Techrete (UK) Ltd

H41 Glassfibre reinforced plastics panel cladding/ features

Architectural structures in grp - Design & Display (Sales) Ltd
Bespoke Products Made - Sarena MFG Ltd
G. R. P . Porches - Solair Ltd
Glass fibre reinforced cladding - Brimar Plastics Ltd
Glass fibre reinforced plastics cladding features - Amari Plastics Plc
Glass reinforced plastic cladding - Kemlite Ltd
Glass reinforced plastic features - Troika Architectural Mouldings Ltd
Glass reinforced resins and plastics - BLP (Hamble) Ltd
Glassfibre plastics reinforced cladding/features - Martello Plastics Ltd
Glassfibre reinforced plastics cladding/architectural features - Duroy Fibreglass Mouldings Ltd
Glassfibre reinforced plastics cladding/features - CEP Claddings Ltd
GRG internal cladding - Multitex GRP & GRG Products
GRP - Geosynthetic Technology Ltd
GRP - Hodkin Jones (Sheffield) Ltd
GRP building products - B & M Laminaites
GRP canopies - Halmark Panels Ltd
GRP cladding - Adams- Hydraulics Ltd
GRP cladding and moldings - Armfibre Ltd
GRP cladding, specialist moulding, door canopies and columns - Multitex GRP & GRG Products
GRP Coloumns - Alpha Glass Fibres Ltd
GRP conservatory roofs - Halmark Panels Ltd
GRP door canopies and features - MR Limited
GRP flashings - Hambleside Danelaw Ltd
GRP housings - Precolor Sales Ltd
GRP materials - Wilsons Fibreglass Ltd
GRP Pilasters - Alpha Glass Fibres Ltd
GRP porches - Halmark Panels Ltd
GRP roof sheets - Wilsons Fibreglass Ltd
GRP sheet - Corofil Woodall
GRP structures & plant rooms - APA
GRP/APC roof lights and cladding products - Hambleside Danelaw Ltd
GRP/APC roofing systems for flat, low pitched and barreled applications - Hambleside Danelaw Ltd
Interior and exterior laminated glassfibre components - Pyramid Plastics Ltd
Modular Housings - Sarena MFG Ltd
Plastic profiled sheet cladding and roofing - Ariel Plastics Ltd
Polycarbonate sheeting - Areco Roofing Supplies Company
Road sign plate - Filion Products Ltd
Roadside waymarkers - Filion Products Ltd
RST Cladding Beams - Oakleaf Reproductions Ltd

H43 Metal Panel Cladding/ features

Metal Cladding Pannel - McMullen Architectural Systems

H50 Precast concrete slab cladding/ features

Architectural precast cladding - Marble Mosaic Co Ltd, The
Concrete products - Fyfe John Ltd
Precast concrete cladding - Techrete (UK) Ltd
Precast concrete panels - Anderton Concrete Products Ltd
Precast concrete slab cladding/ features - Evans Concrete Products Ltd
Precast concrete slab cladding/features - Technostone Ltd
Prestressed concrete wall panels - ACP (Concrete) Ltd
Retaining walls - Grass Concrete Ltd

H51 Natural stone slab cladding/ features

Architectural granite - Fyfe John Ltd
Limestone, slate, quartzite, granite, sandstone, and marble - Naturestone
Masonry - Bridge Quarry
Natural granite & marble lightweight panels - Chelsea Artisans Ltd
Natural Limestone Masonry - Hanson Bath & Portland Stone
Natural stone cladding - Kirkstone
Natural stone slab cladding - Macclesfield Stone Quarries Ltd
Natural stone slab cladding - Marble Mosaic Co Ltd, The
Natural stone slab cladding / features - Toffolo Jackson(UK) Ltd
Natural stone slab cladding and features - Rare Stone Group, The
Natural stone slab cladding features - PermaRock Products Ltd
Natural Stone Slab Cladding/ Features - Realstone Ltd
Natural stone slab cladding/features - CEP Claddings Ltd
Natural stone slab cladding/features - G. Miccoli & Son Ltd
Natural stone slab cladding/features - Marriott & Price Ltd
Stone cladding - Alcover UK Ltd
Stone cladding - natural stone - Naturestone
Stonework - design, procurement and installation of - Szerelmey Ltd
Terracotta rain screen cladding - Smithbrook Building Products Ltd
Yorkstone - Pickard Group

H52 Cast stone slab cladding/ features

Architectural cast stone - Hampton Stone Ltd
Architectural cast stone cladding - Marble Mosaic Co Ltd, The
Cast stone cladding/features - G. Miccoli & Son Ltd
Cast stone slab cladding / features - Toffolo Jackson(UK) Ltd
Cast stone slab cladding /features - Haddonstone Ltd
Cast stone slab cladding/ features - Chilstone
Cast stone slab cladding/features - Technostone Ltd
Cast stone slab/ cladding/ features - Evans Concrete Products Ltd
Cladding - Milner Delvaux Ltd
Mullion windows, quoins, cornices and doorways - Minsterstone Ltd

H60 Plain roof tiling

Acrylic glass tiles - Klober Ltd
Clay and concrete tiles - Sandtoft Roof Tiles Ltd
Clay and concrete tiling - Standard Flat Roofing Co Ltd
Clay hanging tiles - clayton - Keymer Hand Made Clay Tiles
Clay peg tiles - traditional handmade - Keymer Hand Made Clay Tiles
Clay plain tiles - traditional handmade - Keymer Hand Made Clay Tiles
Clay roof tile manufacturer - Blyth William
Clay roof tiles - Hinton Perry & Davenhill Ltd
Clay roof tiles and glazed clay roof tiles - Smithbrook Building Products Ltd
Clay roof tiling - HF (GB) Ltd
Clay roofing tiles - Michelmersh Brick & Tile Co Ltd
Concrete roof tiles - Forticrete Ltd
Concrete roof tiles - Forticrete Roofing Products
Concrete roof tiles - Marley Building Materials Ltd
Flat roofing - Mells Roofing Ltd
Handcrafted clay tiles - Eternit Building Materials
Handmade Clay, roof tiles plus under tile vent - Tudor Roof Tile Co Ltd
Machine made clay tiles - Eternit Building Materials
Ornamental Tiles and Fittings - Hinton Perry & Davenhill Ltd
Pitched roofs and accessories - Anderson D & Son
Plain Clay Roof Tiles - Hinton Perry & Davenhill Ltd
Plain Roof Tiling - Britmet Tileform Limited
Plain roof tiling - Decra Roofing Systems
Plain tiles - Lafarge Roofing Ltd
Roof Coverings - Howard Evans Roofing Limited
Roof space ventilating tiles - Klober Ltd
Roof ventalation - Zled (DPC) Co Ltd
Roof ventilation - Hambleside Danelaw Ltd
Roof ventilation - HV Air Conditioning Ltd
Roofing - InstaGroup Ltd
Roofing Accessories - Lafarge Roofing Ltd
Roofing claywork - Red Bank Maufacturing Co Ltd
Roofing fittings - Lafarge Roofing Ltd
Roofing underlay felt - Newton John & Co Ltd
Technical Solutions - Lafarge Roofing Ltd
Terracotta ridges and finals - Keymer Hand Made Clay Tiles
Tiling - Durable Contracts Ltd
Tiling - Letchworth Roofing Company Ltd
Tiling - Roofrite (East Anglia) Ltd
Ventilated ridge system - Klober Ltd
Ventilation system - Keymer Hand Made Clay Tiles

H61 Fibre cement slating

Fibre Cement Slates - Eternit Building Materials
Polypropylene, steel fibers for concrete & mortar - S.I. Concrete Systems
Roof space ventilating slates - Klober Ltd
Roofing slate - Allan Harris & Sons Ltd
Roofing slates, ventilation products and breather membranes - Cembrit Blunn Ltd
Slates - Reclaim.UK.Com
Slating - Durable Contracts Ltd
Synthetic Slates and Tiles - EBC UK Ltd

H62 Natural slating

Copings and sills - Kirkstone
Flooring Slate - Allan Harris & Sons Ltd
Natural slate cladding - McAlpine Slate Ltd, Alfred
Natural slate copings - McAlpine Slate Ltd, Alfred
Natural slate counters - McAlpine Slate Ltd, Alfred
Natural slate flooring - McAlpine Slate Ltd, Alfred
Natural slate landscaping - McAlpine Slate Ltd, Alfred
Natural slate roofing - Burlington Slate Ltd
Natural slate roofing - McAlpine Slate Ltd, Alfred
Natural slate sills - McAlpine Slate Ltd, Alfred
Natural slate walling - McAlpine Slate Ltd, Alfred
Natural slate worktops - McAlpine Slate Ltd, Alfred
Natural slates - Brooks Belfast
Natural Slates - Eternit Building Materials
Natural Stone - Bridge Quarry
Polythene under slating sheet - Visqueen Building Products
Roof ventilation/ accessories - Zled (DPC) Co Ltd
Roofing - E. H. S. Roofing Ltd
Roofing slate - Allan Harris & Sons Ltd
Roofing slate - The Delabole Slate Co. Ltd
Slating - Letchworth Roofing Company Ltd
Slating - Roofrite (East Anglia) Ltd
Slating - Standard Flat Roofing Co Ltd
Welsh roofing slates - Greaves Welsh Slate Co Ltd
Welsh, Spanish and Canadian slates - Cembrit Blunn Ltd

H63 Reconstructed stone slating/ tiling

Green roofs - Alumasc Exterior Building Products Ltd
Recon: stone - Chiltern Concrete and Stone Ltd
Reconstructed stone roofing tiles - Atlas Stone Products
Reconstructed stone slating/tiling - CEP Claddings Ltd
Slates - Lafarge Roofing Ltd
Slates - Sandtoft Roof Tiles Ltd
Terracotta cladding - Alcover UK Ltd
Tiles and specialist works - conglomerate marble, limestone & assimilated granite - Marble Flooring Spec. Ltd

H64 Timber shingling

Cedar shingles for roofing and cladding - Brash John & Co Ltd
Timber shingling - Brash John & Co Ltd

H65 Single lap roof tiling

Finials - Hinton Perry & Davenhill Ltd
Profile tiles - Lafarge Roofing Ltd
Roof Ventalation/Accessories - Zled (DPC) Co Ltd
Single lap roof tiling - Sandtoft Roof Tiles Ltd
Single ply roofing - Roofrite (East Anglia) Ltd
Single roof tiling - Decra Roofing Systems
Thatching, thatching materials and fire retardants - Thatching Advisory Services Ltd

H66 Bituminous felt shingling

Bituminous felt slungies - Onduline Building Products Ltd
Foundation protection - Onduline Building Products Ltd
Tegola asphalt shingles - Matthew Hebden

H70 Malleable metal sheet prebonded covering/ cladding

Fully supported metal roofing - Kelsey Roofing Industries Ltd
Fully supported roofing and cladding - Broderick Structures Limited
Malleable metal sheet prebonded covering/ cladding - Reid John & Sons (Strucsteel) Ltd
Metal cladding coatings - Jotun Henry Clark Ltd Decorative Division

H71 Lead sheet coverings/ flashings

Aluminium flashing - George Gilmour (Metals) Ltd
Cloaks - Cavity Trays Ltd
Flashings - Cavity Trays Ltd
Hyclad matt finish - AvestaPolarit Ltd
Lead and sheet flashings - Royston Lead Ltd
Lead clad steel and lead clad stainless steel cladding - Calder Industrial Materials Ltd
Lead roofing, sand cast lead sheet - Anglia Lead Ltd
Lead sheet - R.M. Easdale & Co Ltd
Lead sheet and flashing - Calder Industrial Materials Ltd
Lead sheet covering and flashings - Broderick Structures Limited
Lead sheet coverings/ flashings - Freeman T R
Lead sheet coverings/ flashings - Harris & Bailey Ltd

H71 Lead sheet coverings/ flashings (Cont'd)

Lead works - Roofrite (East Anglia) Ltd
Leadwork - TRC Service Group Ltd
Low reflective satin finish - AvestaPolarit Ltd
Ternie coated lead finish - AvestaPolarit Ltd
White and red lead - Liver Grease Oil & Chemical Co Ltd

H72 Aluminium strip/ sheet coverings/ flashings

Alucobond and aluminum cladding - Alcover UK Ltd
Aluminium covers - Franklin Hodge Industries Ltd
Aluminium & Steel Composite Roof & Wall Cladding - Rigidal Industries Ltd.
Aluminium and GRP roof edge trim - Areco Roofing Supplies Company
Aluminium cappings & flashings - Aluminium RW Supplies Ltd
Aluminium coping systems - Dales Fabrications Ltd
Aluminium fascias and soffits - Aluminium RW Supplies Ltd
Aluminium flashings and trims - Azimex Fabrications Ltd
Aluminium rain screens cladding - Axter Ltd
Aluminium roof outlets - Marley Plumbing and Drainage
Aluminium sheet coverings/ flashings - Freeman T R
Aluminium sheet coverings/ flashings - Heywood Williams Architectural
Aluminium sheet coverings/ flashings - Reid John & Sons (Strucsteel) Ltd
Aluminium sheet coverings/flashings - Alcoa Europe-Extrusions & Endproducts (UK)
Aluminium Sheet coverings/flashings - Exitex Ltd
Aluminum sheet covering and flashings - Broderick Structures Limited
Anodised Aluminium sheet - Alanod Ltd
Flashings - Corofil Woodall
Insulated Roof & Wall Systems - Kingspan Ltd
Metal facia systems - Alumasc Exterior Building Products Ltd
Metal roofing and cladding - Ash & Lacy Building Products Ltd
Perforated panels in aluminium - Winther Browne & co Ltd
Powder coated aluminium - Dales Fabrications Ltd
Powdercoat aluminium - Leavlite Electropaint Ltd
Standing seam aluminium roofing - Kelsey Roofing Industries Ltd

H73 Copper strip/ sheet coverings/ flashings

Copper flashings and trims - Azimex Fabrications Ltd
Copper sheet coverings and flashings - Broderick Structures Limited
Copper sheet coverings/ flashings - Freeman T R
Copper strip - Metra Non Ferrous Metals Ltd

H74 Zinc strip/ sheet coverings/ flashings

Zinc and titanium alloy sheets and strip - Metra Non Ferrous Metals Ltd
Zinc flashings and trims - Azimex Fabrications Ltd
Zinc roofing - Rheinzink UK
Zinc sheet coverings and flashings - Broderick Structures Limited
Zinc sheet coverings/ flashings - Freeman T R

H75 Stainless steel strip/sheet coverings/ flashings

Insulated Roof & Wall Systems - Kingspan Ltd
Stainless steel flashings and trims - Azimex Fabrications Ltd
Stainless steel sheet and strip - Metra Non Ferrous Metals Ltd
Stainless steel sheet coverings/ flashings - Freeman T R
Stainless steel strip/ sheet coverings/ fashings - Barlow Group
Stainless steel strip/sheet coverings and flashings - Broderick Structures Limited
Stainless steel strip/sheet coverings/flashings - Britmet Tileform Limited
Stainless steel traditional longstrip roofing and cladding material - AvestaPolarit Ltd

H76 Fibre bitumen thermoplastic sheet coverings/ flashings

Roofing membranes - EBC UK Ltd

H90 Tensile fabric coverings

Architectural Fabric Structures - TA Lucas Sails
Banners - TA Lucas Sails
Breathable membranes - Fillcrete Ltd
Covers - TA Lucas Sails
Rigging - TA Lucas Sails
Sails - TA Lucas Sails
Tensile structure material - Mermet U.K
Vapour barriers - Fillcrete Ltd

H92 Rainscreen cladding

Cellular PVC cladding - Celuform
Cladding & Rainscreening Panels Systems - Technology Telford Ltd
Faience blocks - Shaws of Darwen
Metal roofing and cladding - Ash & Lacy Building Products Ltd
PVC Sheeting - Power Plastics Ltd
Rainscreen cladding - Baggeridge Brick PLC
Rainscreen Cladding - CEP Claddings Ltd
Rainscreen cladding - Eternit Building Materials
Rainscreen cladding - Marble Mosaic Co Ltd, The
Rainscreen Cladding - McMullen Architectural Systems
Rainscreen cladding - Panel Systems Ltd
Rainscreen cladding - Promat Fire Protection Ltd
Rainscreen Cladding - SCHUCO International KG
Rainscreen facade systems - CGL Cometec Ltd
Rainscreen Panels - Rheinzink UK
Screening - Power Plastics Ltd
Terracotta and faience - design, procurement and installation of - Szerelmey Ltd
Terracotta blocks - Shaws of Darwen

J10 Specialist waterproof rendering

Cementitious and dry line membrane tanking - Dampcoursing Limited
Conductive coatings - PermaRock Products Ltd
External rendering systems - PermaRock Products Ltd
Lliquid applied dpm - P.C.I Construction Systems Ltd
Potable water coatings - Pearl Paints Ltd
Specialist waterproof rendering - Concrete Repairs Ltd
Specialist waterproof rendering - Don Construction Products Ltd
Specialist waterproof rendering - Mapei (UK) Ltd
Specialist waterproof rendering - Peter Cox Ltd
Specialist waterproof rendering - Ronacrete Ltd
Specialist waterproof rendering - Specialist Building Products
Water proofing systems - Beton Construction Ltd
Waterproof renders(30 colours) - Laticrete UK
Waterproofing - Degussa Construction Chemicals (UK)
Waterproofing - Feb MBT
Waterproofing - Fullstop Technology Ltd
Waterproofing - Kiltox Damp Free Solutions
Waterproofing products - Ruberoid Building Products Ltd
Waterproofing Products - Sealocrete Ltd

J20 Mastic asphalt tanking/ damp proofing

Asphalt - Tillicoutry Quarries Ltd
Asphalt reinforcement - Bekaert Building Products
Basement tanking - Vandex (UK) Ltd
Damp proof membranes - Anderson D & Son
Damp proof membranes - Monarflex Anderson Ltd
Mastic asphalt tanking / damp proof membranes - Durable Contracts Ltd
Mastic tanking and damp proof membranes - Canonbury Asphalte Co. Ltd.
Tanking - Dyke Chemicals Ltd
Tanking - Newton John & Co Ltd
Tanking - Roofing Contractors (Cambridge) Ltd
Tanking and damp proof membranes - Briggs Roofing & Cladding Ltd, Roofing Contract
Waterproofing membranes - Fosroc Ltd

J21 Mastic asphalt roofing/ insulation/ finishes

Asphalt work - TRC Service Group Ltd
Damp-proofing systems - Callenders Ltd
Emergency Repair Compounds - Dyke Chemicals Ltd
GRP and aluminium roof edge trims and flashings - Pitchmastic Pmb Ltd
Insulation - Rockwool Ltd
Mastic asphalt - Coverite Ltd
Mastic asphalt mixers - Bitmen Products Ltd
Mastic asphalt roofing - Standard Flat Roofing Co Ltd
Mastic asphalt roofing and car parks - Briggs Roofing & Cladding Ltd, Roofing Contract
Mastic asphalt roofing/ insulation/ finishes - BFRC Services Ltd
Mastic asphalt roofing/ insulation/ finishes - Thermica Ltd
Mastic roofing - Canonbury Asphalte Co. Ltd.
Porous concrete promenade tiles - Spartan Tiles Ltd
Recycled glass tiles - Spear & Jackson Interntional Ltd.
Roof breather vents - Areco Roofing Supplies Company
Roof waterproofing systems, colour solar reflective roof finishes - Dyke Chemicals Ltd
Roofing - North Herts Asphalte Ltd
Roofing - Roofing Contractors (Cambridge) Ltd
Roofing contractors - Durable Contracts Ltd
Roofing systems - Callenders Ltd
Tarpauling - Power Plastics Ltd
Tarpaulins - Monarflex Anderson Ltd
Walkway tiles - Harris Road Roofing & Flooring Ltd

J22 Proprietary roof decking with asphalt finish

Bitumen boilers - Bitmen Products Ltd
Flat roof decking - Isocrete Floor Screeds Ltd
Flat roofing - Coverite Ltd
Pre coated chipspreaders - Bitmen Products Ltd
Proprietary roof decking with asphalt finish - Durable Contracts Ltd
Roof decking with asphalt finish - Briggs Roofing & Cladding Ltd, Roofing Contract
Roof tile underlays - Monarflex Anderson Ltd
Roofing contractor - Evans Howard Roofing Ltd
Roofing membranes - Onduline Building Products Ltd
Torch on roofing - Ruberoid Building Products Ltd

J30 Liquid applied tanking/ damp proofing

Adhesives - Laticrete UK
Basement tanking - Vandex (UK) Ltd
Basement water proofing - Timberwise (UK) Plc
Damp proof membrane - Ardex UK Ltd
DPC tanking systems - Ruberoid Building Products Ltd
Flashings - Dyke Chemicals Ltd
Grouts and sealers for tiles and surface of floors - P.C.I Construction Systems Ltd
Heldite Jointing Compound - Heldite Ltd
High performance floor coatings - Intergrated Polymer Systems (UK) Ltd
Internal tanking of plant/ tank rooms & suspended floors - Ferguson G A & Co Ltd
Liquid applied damp proof membranes - Kingfisher Building Products Ltd.
Liquid applied tanking/ damp proof membrane - Colebrand Ltd
Liquid applied tanking/ damp proof membranes - Don Construction Products Ltd
Liquid applied tanking/ damp proof membranes - Specialist Building Products
Liquid applied tanking/ damp proof membranes - Wood E Ltd
Liquid applied tanking/damp proof membranes - Trade Sealants
Liquid applied waterproof membranes - Kelsey Roofing Industries Ltd
Structural waterproofing - Alumasc Exterior Building Products Ltd
Structural waterproofing - Grace Construction Products Ltd
Wall coatings - Intergrated Polymer Systems (UK) Ltd
Wall coatings - Lastite Building Products Ltd
Waterproof membranes - Laticrete UK
Waterproofing membranes - Kenyon Group Ltd
Waterproofing products - Kingsnorth Bitumen Products Ltd

J31 Liquid applied waterproof roof coatings

Bitumen based flat roofing membranes - Kelsey Roofing Industries Ltd
Bituminous roofing - TRC (Midlands) Ltd
Bitumious coatings - Protim Services Ltd
Damp-proof membranes - Dyke Chemicals Ltd
Epoxy Resins & Adhesives - Wessex Resins & Adhesives Ltd
Fluid applied roof coatings - Liquid Plastics Ltd
High performance roof coatings - Intergrated Polymer Systems (UK) Ltd
Liquid applied waterproof coating - Bondaglass Voss Ltd
Liquid applied waterproof coating - Brimar Plastics Ltd
Liquid applied waterproof coatings - BFRC Services Ltd
Liquid applied waterproof coatings - Brewer C & Sons Ltd
Liquid applied waterproof coatings - Briggs Roofing & Cladding Ltd, Roofing Contract
Liquid applied waterproof coatings - Certifield Laboratories (Div or CPS Industries Ltd)
Liquid applied waterproof coatings - Colebrand Ltd
Liquid applied waterproof coatings - Concrete Repairs Ltd
Liquid applied waterproof coatings - Don Construction Products Ltd
Liquid applied waterproof coatings - Harris Road Roofing & Flooring Ltd
Liquid applied waterproof coatings - Precolor Sales Ltd
Liquid applied waterproof coatings - Ronacrete Ltd
Liquid applied waterproof coatings - Tor Coatings Ltd
liquid applied waterproof coatings - Trade Sealants
Liquid applied waterproof coatings - Wood E Ltd
Liquid applied waterproof roof coatings - Hyflex Roofing
Liquid membranes - TRC Service Group Ltd
Liquid-applied roofing and waterproofing - Granflex Roofing Ltd
Masonry paints - MR Limited
PmB Structural Waterproofing Systems - Pitchmastic Pmb Ltd

Roof coatings - Crosbie Coatings
Roof coatings - Lastite Building Products Ltd
Roof coatings - Pearl Paints Ltd
Roof repair and protection - Sealocrete Ltd
Roof repair systems - Williamson T & R Ltd
Roof weather proofing - Infil Ltd
Sealing/ dust and oil proofing multi-storey car park decks - Ferguson G A & Co Ltd
Structural roof waterproofing - Alumasc Exterior Building Products Ltd
Structural waterproofing - Coverite Ltd
Waterproofing membranes - Conren Ltd
Waterproofing products - Kingsnorth Bitumen Products Ltd
Waterproofing roofs of multi-storey car parks - Ferguson G A & Co Ltd

J32 Sprayed vapour control layers

Radon and methane gas barriers - Monarflex Anderson Ltd
Sprayer vapuor barriers - Trade Sealants
Vapour barriers and breather membranes - Monarflex Anderson Ltd

J33 In situ glassfibre reinforced plastics

Fibertex geotextiles - Tex Steel Tubes Ltd
In situ glass reinforced plastic - Specialist Building Products
In situ GRP - Precolor Sales Ltd
In-situ glass reinforced plastics - Brimar Plastics Ltd
In-situ GRP - Briggs Roofing & Cladding Ltd, Roofing Contract

J40 Flexible sheet tanking/ damp proofing

Aggregate sacks plain and printed - Anaplast (Greenock)
Basement tanking - Vandex (UK) Ltd
Bitumen membranes - Vulcanite Limited
Brickgrip DPC - Anaplast (Greenock)
Condensation control - Peter Cox Ltd
Damp proof courses - Cavity Trays Ltd
Damp proof membranes - Isocrete Floor Screeds Ltd
Damp proof membranes - Marley Waterproofing
Damp proof tanking - Callenders Ltd
Damproof membranes - Newton John & Co Ltd
Flexible sheet membrane - Anaplast (Greenock)
Flexible sheet tanking / damp proof membranes - Durable Contracts Ltd
Flexible sheet tanking and damp proof membranes - Harris Road Roofing & Flooring Ltd
Flexible sheet tanking, damp proof membanes - Terminix Property Services
Flexible sheet tanking/ damp proof membranes - BFRC Services Ltd
Flexible sheet tanking/ damp proof membranes - John Davidson (Pipes) Ltd
Flexible sheet tanking/ damp proof membranes - Specialist Building Products
Floor coatings - Lastite Building Products Ltd
Gas barrier & ancilliary products - Anaplast (Greenock)
Gas barrier membranes - Visqueen Building Products
Gas barrier membranes and pond liners - Geosynthetic Technology Ltd
Membranes - Cavity Trays Ltd
PIFA standard DPM - Anaplast (Greenock)
Polythene dampproof membranes - Visqueen Building Products
Single Ply Roofing Systems - Sarnafil Ltd
Temporary protective sheeting - Anaplast (Greenock)
Tilene 200P - Anaplast (Greenock)
Vapor barrier - Anaplast (Greenock)
Visqueen handypacks - Anaplast (Greenock)
Waterproof expansion joints - Radflex Contract Services Ltd
Waterproof membranes - Grace Construction Products Ltd
Waterproofing membranes - SCP Concrete Sealing Technology Ltd
Waterproofing sheets - ABG Ltd
Waterstops - Fosroc Ltd

J41 Built up felt roof coverings

Apex roofing - Apex Roofing Contractors
Bituminous roofing felts - Marley Waterproofing
Bituminous roofing system - Vulcanite Limited
Built up felt roof covering - Brewer C & Sons Ltd
Built up felt roof coverings - Canonbury Asphalte Co. Ltd.
Built up felt roof coverings - Durable Contracts Ltd
Built up felt roof coverings - Harris Road Roofing & Flooring Ltd
Built up felt roof coverings - Roofrite (East Anglia) Ltd
Built up felt roof coverings - Standard Flat Roofing Co Ltd
Built up felt roof coverings - TRC (Midlands) Ltd
Built up felt roofing - Briggs Roofing & Cladding Ltd, Roofing Contract
Built Up Felt Roofing - North Herts Asphalte Ltd

J41 Built up felt roof coverings (Cont'd)

Built up felt roofing - Roofing Contractors (Cambridge) Ltd
Built up roofing - Anderson D & Son
Built-up felt roofing - Granflex Roofing Ltd
Felt roofing - TRC Service Group Ltd
Felting - Letchworth Roofing Company Ltd
Flat roofing - Anderson D & Son
Flat roofing built up felt roofing - Chesterfelt Ltd
High performance felt - Coverite Ltd
High performance roofing - Alumasc Exterior Building Products Ltd
Membrane waterproofing - BFRC Services Ltd
Pitched roofing - Mells Roofing Ltd
Porous concrete promenade tiles - Spartan Tiles Ltd
Roof coverings - Howard Evans Roofing Limited
Roof membranes - Axter Ltd
Roofing Accessories - Ubbink (UK) Ltd
Roofing compound - Bartoline Ltd
Roofing contractor - Evans Howard Roofing Ltd
Roofing membrane - Nuralite (UK) Ltd
Roofing products - Cavity Trays Ltd
Roofing products - Klober Ltd
Single ply - Letchworth Roofing Company Ltd
Vapour membranes - Klober Ltd
Weather protection sheeting and sheeting - Power Plastics Ltd

J42 Single layer polymeric roof coverings

Conservatory roof systems, vents and glazing bars - Newdawn & Sun Ltd
Flat roofing - E. H. S. Roofing Ltd
Polymeric roof coatings - MMP International
Single layer plastic roof coverings - BFRC Services Ltd
Single layer plastic roof coverings - Marley Waterproofing
Single layer plastic roof coverings - Standard Flat Roofing Co Ltd
Single layer polymeric roof coverings - Durable Contracts Ltd
Single layer PVC and CPE membrane roofing systems - Alkor Draka Ltd
Single layer roofing - TRC Service Group Ltd
Single ply polymetric membranes - Roofing Contractors (Cambridge) Ltd
Single ply roofing - Canonbury Asphalte Co. Ltd.
Single ply roofing - Coverite Ltd
Single ply roofing - ICB (International Construction Bureau) Ltd
Single Ply Roofing - Sika Ltd
Single ply roofing and roof accessories - Anderson D & Son
Single ply roofing membrane - Kelsey Roofing Industries Ltd
Single ply systems - Briggs Roofing & Cladding Ltd, Roofing Contract
Water repellent solutions - Dyke Chemicals Ltd
Waterproofing System to Waterproof & Refurbish Multi-storey Car Parks & Roofs - Dundec Contracts Ltd
Waterproofing systems - Beton Construction Ltd

J43 Proprietary roof decking with felt finish

Proprietary roof decking with felt finish - Durable Contracts Ltd
Roof decking with felt finish - Briggs Roofing & Cladding Ltd, Roofing Contract
Roofing ventilators - Airflow (Nicoll Ventilators) Ltd
Standing seam roofing - E. H. S. Roofing Ltd

J44 Sheet linings for pools/lakes/waterways

Gas barrier membranes and pond liners - Geosynthetic Technology Ltd
Pond / landfill lining systems - ABG Ltd
Single Ply Roofing Systems - Sarnafil Ltd
Waterproofing systems - Beton Construction Ltd

K10 Plasterboard dry linings/partitions/ceilings

Access panels ceilings - Panel & Louvre Co Ltd
Associated drylining - ARC Partitioning
Custom rolled sections to client specification - Hill Top Sections Ltd
Dry lining systems - British Gypsum Ltd
Drywall systems - Lafarge Plasterboard Ltd
Drywalling collated screws, nails and tools - Duo-Fast (UK) Ltd
Drywalling collated screws, nails and tools - Duo-Fast (UK) Ltd
Fire barrier angle & strap - Hill Top Sections Ltd
Fire stopping and barriers - Firebarrier Services Ltd
GRG (glass reinforced gypsum) products - design, manufacture and installation - Gillespie (UK) Ltd
Gysum fiberboards - Fels UK
Interior wall linings - Trespa UK Ltd
Mineral fibre and vermiculite board - Cryotherm Insulation Ltd
Moveable walls - Becker (SLIDING PARTITIONS) Ltd
Partitioning - Hatmet Ltd
Partitions - Fastwall Ltd
Plaster boards - Ashmead Buidling Supplies Ltd
Plasterboard - British Gypsum Ltd
Plasterboard accessories - British Gypsum Ltd
Plasterboard acoustic wall products - Trim Acoustics
Plasterboard dry lining - Harris & Bailey Ltd
Plasterboard dry lining/ partitions/ ceilings - Logic Office Contracts
Plasterboard dry lining/partitions/ceilings - Unilock Ltd
Plasterboard/dry lining/partitions/ceilings - Promat Fire Protection Ltd
Plasterboards and drywall Accessories - Knauf
Solid, decorated plasterboard partitions - Neslo Partitioning & Interiors
Stud & track partitioning sections - Hill Top Sections Ltd
Supplier of Ceilings, Partitioning, Dry lining and Interior Building products. - Nevill Long Limited
Toilet cubicles - Flexi-Plan Partitions Ltd
Wall liner system - Hill Top Sections Ltd

K11 Rigid sheet flooring/ sheathing/ decking/sarking/ linings/casings

Access flooring - Nevill Long Limited
Acoustic insulation boards - Fels UK
Blockboard - Wood International Agency Ltd
Column and beam encasement systems - Knauf
Concrete Stuctures - Promat UK Ltd
Cutting, drilling and machining - Boardcraft Ltd
Doors - Promat UK Ltd
Dry flooring elements - Fels UK
External Walls - Promat UK Ltd
Film Faced plywood - Wood International Agency Ltd
Fire protection boards - Fels UK
Fire Rated Boards - Abesco Ltd
Floating floors - Designed for Sound Ltd
Flooring accessories - Atkinson & Kirby Ltd
Floors - Threshold Flooring
Fsc certified plywood - Wood International Agency Ltd
Hardboard - Wood International Agency Ltd
Hardwood and softwood - Boardcraft Ltd
Hardwood plywood - Wood International Agency Ltd
Impact resistant boards - Fels UK
Insulating laminates and plasters - Knauf
Internal Walls and Partitions - Promat UK Ltd
Laminate joinery - Decra Ltd
Laminates and MDF's - Kronospan Ltd
Lightweight composite boards and panels - Quinton & Kaines (Holdings) Ltd
Linigs and castings - Cox Long Ltd
MDF sheets - Willamette Europe Ltd
Medium Density Fibrboard (MDF) - Montague L Meyer
Moisture resistant boards - Fels UK
Office flooring - Raised Floor Systems Ltd
Oriented Strand Board (osb) - Montague L Meyer
OSB - Kronospan Ltd
OSB - Wood International Agency Ltd
Panel products - James Latham plc
Perforated MDF - Associated Perforators & Weavers Ltd
Perforated Perforations - Associated Perforators & Weavers Ltd
Pipe and column casings - Alumasc Interior Building Products Limited
Plain and melamine faced chipboard - Kronospan Ltd
Plain and melamine faced flooring - Kronospan Ltd
Plain, melamine faced and lacquered MDF - Kronospan Ltd
Plywood-structure + non-structure - Montague L Meyer
Polyisocyanurate insulation board manufacture - Celotex Ltd
Protaction to Stuctural Steelwork - Promat UK Ltd
Riggid sheet flooring/sheathing/decking/sarking/casting/lining - Kemlite Ltd
rigid sheet flooring, sheathing, decking, sarking - Cox Long Ltd
Rigid sheet flooring/sheathing/decking/sarking - Promat Fire Protection Ltd
Roofs - Promat UK Ltd
Semi finished plastics and building products - Amari Plastics Plc
Sheet materials - Cox Long Ltd
Sheet Materials - Walker Timber Ltd
Softwood plywood - Wood International Agency Ltd
Sprays fire proof - Firebarrier Services Ltd
Thermal insulated panels - Panel Systems Ltd
Undercarrige Systems - Atkinson & Kirby Ltd
Underlay - Atkinson & Kirby Ltd
Veneered Boards - Montague L Meyer
Wet wall and showerwall panels, wet area wall lining and ceiling panels - Norske Interiors (UK) Ltd
Wood Partical Board - Montague L Meyer
Woodbased panels - Egger (UK) Ltd

K12 Under purlin/Inside rail panel linings

Plywood, hard, fibre, chip and particle boards - Boardcraft Ltd

K13 Rigid sheet fine linings/ panelling

Access panels walls - Panel & Louvre Co Ltd
Acoustic flooring and acoustic wall products - Trim Acoustics
Bespoke infil panels - Halmark Panels Ltd
Bespoke joinery - Howard Bros Joinery Ltd
Broadcast and Audio Studios - IAC
Building Boards - Cembrit Blunn Ltd
Carved cornice, frieze and embellishments - Oakleaf Reproductions Ltd
Cladding panel - Panel Systems Ltd
Cladding panels - CEP Claddings Ltd
Cubicle systems - Formica Ltd
Ducting systems - Decra Ltd
Facing laminates - Wilsonart Ltd
Faux book spines, mirrors and frames - Oakleaf Reproductions Ltd
High pressure decorative laminates - Abet Ltd
Hydrotherapy spa baths - Norske Interiors (UK) Ltd
Hygienic and railway arch lining materials - Rockwell Sheet Sales Ltd
Hygienic GRP Cladding - Kemlite Ltd
Hygienic wall linings - Filion Products Ltd
Laminates - Montague L Meyer
Linenfold wall panelling, riven oak wall panelling - Oakleaf Reproductions Ltd
Melamine faced chipboard - Deralam Laminates Ltd
Metal + melamine laminates - Deralam Laminates Ltd
Multi-wall roofing products - Amari Plastics Plc
Panel saws for site work - JKO Ltd
Panelled interiors - Hallidays UK Ltd
Partitions - Unilock Ltd
Plywood, hard, fibre, chip and particle boards - Boardcraft Ltd
Plywood, hard, fibre, chip and particle boards - Boardcraft Ltd
Real wood veneer laminates - Deralam Laminates Ltd
Rigid sheet paneling - Advanced Panels & Products
Rigid sheet wall paneling high performance cladding - Combat Polystyrene Group Ltd
Sheet materials - Brooks Belfast
Slat board - Smart F & G (Shopfittings)Ltd
Softwood - James Latham plc
Structural wall panels - Marshalls Mono Ltd
Veneered boards - Atkinson & Kirby Ltd
Wall lining systems - Knauf
Wall paneling - Viking Laminates Ltd
Wall systems - Neslo Partitioning & Interiors
Wood Based Pannel Products - Montague L Meyer

K20 Timber board flooring/decking/sarking/linings/ casings

Acoustic Flooring - Quiligotti Access Flooring Limited
Alumasc - Brooks Belfast
Cane woven panels - Winther Browne & co Ltd
Decking - Anthony de Grey Trellises
Decorative laminated flooring - Egger (UK) Ltd
Densified wood and industrial flooring - Permali Dehoplast Ltd
Duct walling - LSA Projects Ltd
Fire protection boards - Knauf
First and second fix timber - Brewer T. & Co Ltd
Floor ducting - Alumasc Interior Building Products Limited
Glazed aluminium system with internal blinds - Neslo Partitioning & Interiors
Hardwood flooring - WL West & Son Limited
Hardwood Timber Importers - Parker Kislingbury Ltd
MDF - Finn Forest
Non-Slip Decking - CTS Ltd
Perimeter casing - Alumasc Interior Building Products Limited
Pipe boxing - Alumasc Interior Building Products Limited
Plywood, hard, fibre, chip and particle boards - Boardcraft Ltd
Sanding and Sealing wood floors - Sportmark™ Ltd
Sheet Materials - Moray Timber Ltd
Silent floors - Brooks Belfast
Solid timber flooring - McLoughlin Wood Ltd
Specialised timber panels - McLoughlin Wood Ltd
Timber board flooring/decking - Magnet Ltd
Timber cladding, mouldings and flooring - James Latham plc
Timber decking - Wyckham Blackwell Ltd
Timber Suppliers - Crowthorne Fencing

K21 Timber strip/board fine flooring/ linings

Dance floor - Felix Design Ltd
Flooring - Brooks Belfast
Hardwood flooring - Atkinson & Kirby Ltd
Hardwood flooring - Campbell Marson & Co
Hardwood strip Flooring - Quiligotti Access Flooring Limited
Laminates - Egger (UK) Ltd
Perforated panels in hardboard and MDF - Winther Browne & co Ltd
Saws for shadow joints - JKO Ltd
Sheet materials - Brewer T. & Co Ltd
Solid hardwood flooring - Junckers Limited
Sports Floors - Quiligotti Access Flooring Limited
Strip floor - McLoughlin Wood Ltd
Teak Specialist - Parker Kislingbury Ltd
Timber floors - Norlandia Ltd
Timber strip/board flooring/lining - Magnet Ltd
Timber strips/ board fine flooring / linings - Barlow Group
Timber Suppliers - Crowthorne Fencing

K30 Panel partitions

Access panels - manufacture and installation - Foster W H & Sons Ltd
Acoustic folding partitions - Pellfold Parthos Ltd.
Acoustic movable walls - Hillaldam Coburn Ltd
Acoustic Moveable walls/partitions - Hufcor (Partitions) Ltd
Brushstrip sealing for cable management - Kleeneze Sealtech Ltd
Changing cubicles - unframed - Helmsman
Concertina Partitions - Dividers Modernfold Ltd
Demountable partitioning - Internal Partitions Systems
Demountable partitioning systems - Brown F plc
Demountable partitions - Flexi-Plan Partitions Ltd
Demountable partitions - HEM Interiors Group Ltd
Demountable partitions - Komfort Office Environments Plc
Demountable partitions - Norwood Partition Systems Ltd
Demountable partitions and associated electrical works - Opus 4 Ltd
Demountable suspended ceilings - Clestra Hauserman Ltd
Industrial partitioning - Barton Storage Systems Ltd
Insulated wall panels - Composite Panels Ltd
Interior contractors - Gifford Grant Ltd
Invisible Panel joining - JKO Ltd
Laminate wall claddings - T B S Fabrications
Mesh partitioning - Barton Storage Systems Ltd
Mobile partitions - Dividers Modernfold Ltd
Movable partitions - Alcover UK Ltd
Movable walls - Pellfold Parthos Ltd.
Office partitioning - Dexion Ltd
Operable wall systems - Monowa Manufacturing (UK) Ltd
Panel partitions - Logic Office Contracts
Panel partitions - Unilock Ltd
Partitioning - Hatmet Ltd
Partitioning - New Forest Ceilings Ltd
Partitioning - Newey Ceilings
Partitioning - Promat Fire Protection Ltd
Partitioning - Unilock Ltd
Partitioning - Welconstruct Co Ltd
Partitioning and wall storage systems - Fastwall Ltd
Partitioning systems - Knauf
Partitioning systems - Nevill Long Limited
Partitions - Adex Interiors for industry Ltd
Partitions - Davies John Interiors Ltd
Partitions - Estel Limited UK
Partitions - Firco Construction Ltd
Partitions - ITS Ceilings & Partitions
Plywood, hard, fibre, chip and particle boards - Boardcraft Ltd
Relocatable aluminium trimmed system - Neslo Partitioning & Interiors
Relocatable partitioning and ceilings - Clestra Hauserman Ltd
Relocatable timber and glazed partition systems - Neslo Partitioning & Interiors
Relocateable and demountable partitions - ARC Partitioning
Relocateable partitions - Fastwall Ltd
Sliding folding partition manufacturer - Building Additions Ltd
Sliding partitions - Becker (SLIDING PARTITIONS) Ltd
Solid surface products - Foster W H & Sons Ltd
Spacestor storage wall / partitions - Envoplan
Steel caging/ partitioning - Bradbury Group Ltd
Trimless decorated partition systems - Neslo Partitioning & Interiors
Wall panels - Foster W H & Sons Ltd
Wash hand basin units - manufacture and installation - Foster W H & Sons Ltd

K32 Panel cubicles

Changing cubicles - Detlef Muller Ltd
Changing cubicles - framed - Helmsman
Cubicle partitions - Architectural & Building Products Ltd
Cubicle systems - Panel Systems Ltd
Cubicles - Nevill Long Limited
Cubicles - Twyford Bushboard
Cubicles - Viking Laminates Ltd
Cubicles - manufacture and installation - Foster W H & Sons Ltd
Cubicles/lockers - Trespa UK Ltd
Framed panel cubical partitions - HEM Interiors Group Ltd
Framed panel cubicles - Flexi-Plan Partitions Ltd
Framed panel cubicles - T B S Fabrications
Invisible Panel joining - JKO Ltd
Panel cubicals - Norske Interiors (UK) Ltd
Plywood, hard, fibre, chip and particle boards - Boardcraft Ltd
Purpose made doors, toilet cubicles and washroom systems - Moffett Thallon & Co Ltd

K32 Panel cubicles (Cont'd)

Steel caging/ partitioning - Bradbury Group Ltd
Toilet and shower cubicles - Decra Ltd
Toilet cubicles - Combat Polystyrene Group Ltd
Toilet, shower & changing cubicles - Aaztec Cubicles
Trimless double glazed partition systems - Neslo Partitioning & Interiors
Trimless single glazed partition systems - Neslo Partitioning & Interiors
Washroom cubicle systems-Twyford Bushboard - Twyford Bushboard
Washroom cubicles - Dale Joinery Ltd
WC cubicles - Detlef Muller Ltd
WC cubicles and washroom systems - Industrial Building Componants Ltd

K33 Concrete/ Terrazzo partitions

Specialist contractors in mosaic and terrazzo - Alpha M & T Ltd
Terrazzo partitions - Roman Mosaic Contracts Ltd

K40 Demountable suspended ceilings

Acoustic ceilings - STOsilent panel - STO Ltd
Cavity Barriers - Promat UK Ltd
Ceiling systems - OWA UK Ltd
Ceiling systems - USG (UK) Ltd
Ceiling tiles - Cep Ceilings Ltd
Ceilings - Adex Interiors for industry Ltd
Ceilings (suspended) - Firco Construction Ltd
Celling furring system - Hill Top Sections Ltd
Celling suspendsion components - Hill Top Sections Ltd
Demountable suspended ceilings - Clestra Hauserman Ltd
Designer ceiling range systems - USG (UK) Ltd
Drylining - Neslo Partitioning & Interiors
Fire rated ceiling systems - Fire Protection Ltd
Fire retardent and acoustic suspended ceilings - Howard Home Improvements Ltd
Floors and ceilings - Promat UK Ltd
Grid, Mineral, metal, wood and soft fibre suspended ceilings - Armstrong World Industries Ltd
Kitchen extraction systems - Heydal Ltd
Metal and 3D ceiling systems - USG (UK) Ltd
Metal panels/tiles - Burgess Architectural Products Ltd
Relocatable partitioning and ceilings - Clestra Hauserman Ltd
Specialist in cold formed sections - Hill Top Sections Ltd
Supply and installation of suspended ceilings - ITS Ceilings & Partitions
Suspended ceiling brackets and accessories - Lewis Spring Products Ltd
Suspended ceiling systems - Rockfon Ltd
Suspended ceilings - Brown F plc
Suspended ceilings - Clark & Fenn Ltd
Suspended ceilings - Dampa (UK) Ltd
Suspended ceilings - Davies John Interiors Ltd
Suspended ceilings - Flexi-Plan Partitions Ltd
Suspended ceilings - Hatmet Ltd
Suspended ceilings - HEM Interiors Group Ltd
Suspended ceilings - HT Martingale Ltd
Suspended ceilings - Internal Partitions Systems
Suspended ceilings - Nevill Long Limited
Suspended ceilings - New Forest Ceilings Ltd
Suspended ceilings - Newey Ceilings
Suspended ceilings - Opus 4 Ltd
Suspended ceilings - Walls & Ceilings International Ltd
Suspended ceilings, shaftwall Systems - Knauf

K41 Raised access floors

Access flooring - Moseley GRP Product (Division of Moseley Rubber Co Ltd)
Access panels floors - Panel & Louvre Co Ltd
Cable management - Interface Europe Ltd
Computer Floors - Quiligotti Access Flooring Limited
Flooring and access - Wincro Metal Industires Ltd
Full fit out interiors service - Neslo Partitioning & Interiors
Partial access timber flooring - Durabella System Floors Ltd
Plywood, hard, fibre, chip and particle boards - Boardcraft Ltd
Raised access flooring - Brown F plc
Raised access flooring - Mauser Interiors (UK) Ltd
Raised Access flooring - USG (UK) Ltd
Raised access flooring testing - Ceram Building Technology
Raised Access Floors - Davies John Interiors Ltd
Raised Access Floors - Hatmet Ltd
Raised access floors - HEM Interiors Group Ltd
Raised access floors - Interface Europe Ltd
Raised access floors - Kingspan Access Floors Ltd
Raised access floors - New Forest Ceilings Ltd
Raised Access Floors - Quiligotti Access Flooring Limited
Raised access floors - Raised Floor Systems Ltd
Raised access floors - Redman Fisher Engineering
Raised access floors - Trim Acoustics
Steel raised access flooring - Durabella System Floors Ltd
Timber raised access flooring - Durabella System Floors Ltd

L10 Windows/Rooflights/ Screens/ Louvres

Acoustic Louvres - IAC
Aluminiu clad windows - Broxwood (scotland Ltd)
Aluminium - Glostal - Monarch
Aluminium & Steel Windows - McNeill-McManus Ltd
Aluminium conservatories - Ducana Surrey Ltd
Aluminium windows - Comar Architectural Aluminuim Systems
Aluminium windows - Fabricated Aluminium Services Ltd
Aluminium windows - Glostal - Monarch
Aluminium windows - KCW Windows
Aluminium windows - Midland Alloy Ltd
Aluminium windows - Planet Architectural Glazing Ltd
Aluminium windows - Prima Systems (South East) Ltd
Aluminium/ wood composit windows - Scandinavian Window Systems Ltd
Aluminum windows - Alcover UK Ltd
Anti grafiti films - Madico Film Products Division Inc
Anti UV films - Madico Film Products Division Inc
Anti-Reflective Glass - SCHOTT UK Ltd
Armour plate dorr assemblers - Twide-Paragon Ltd
Atria - Lonsdale Metal Company
Atrium - Clear Span Ltd
Atrium glazing - EBS Services Ltd
Bespoke joinery - AS Newbould Ltd
Bespoke joinery - Howard Bros Joinery Ltd
Bird screen - Timberwise (UK) Plc
Brush strip window seals - Kleeneze Sealtech Ltd
Building hinges - Nico Manufacturing Ltd
Capping bars - Exitex Ltd
Casement - Blair Joinery
Casement windows - Radway, Door and Windows Ltd.
Clad windows - Swedhouse Ltd
Commercial windows - KCW Windows
Composite aluminium windows - Scandanavian Timber
Composite aluminum/wood windows - Velfac Ltd
Composite windows - Crittall Steel Windows Ltd
Composite windows - Crittall Steel Windows Ltd
Conservatories - Smith & Choyce Ltd
Conservatories - TBR Joinery
Conservatory roofing, atria, canopies and rooflights - Ultraframe PLC
Dormer Windows/Louvres - Duroy Fibreglass Mouldings Ltd
Double glazed roof windows - Klober Ltd
Double glazed units - Ledite Glass (Southend) Ltd
Double glazed windows - Alframes Ltd
Entrance Canopies - Duroy Fibreglass Mouldings Ltd
Entrance screens, shopfronts - Pearsons J R (Birmingham) Ltd
Extruded aluminium composite windows - Smart Systems Ltd
Extruded aluminum windows - Smart Systems Ltd
Extruded shopfronts and entrance screens - Smart Systems Ltd
Extruded UPVC windows - Halo, W H S Division Of Bowater Industries
Extruded UPVC windows - Smart Systems Ltd
Filters/Screening - Associated Perforators & Weavers Ltd
Firerated windows - Baydale Architectural Systems Ltd
French opening door bar - Baddeley
Fully reversible windows - Radway, Door and Windows Ltd.
G. R. P. Canopies - Solair Ltd
Glass & window supplies - Spectra Glaze Services
Glazed malls - Lonsdale Metal Company
Grillers/Panels - Associated Perforators & Weavers Ltd
GRP Canopies - Alpha Glass Fibres Ltd
GRP dormer windows - Wilsons Fibreglass Ltd
GRP louvers - APA
Handsaws for removing window frames - JKO Ltd
High Performance Timber Windows - Bennett Windows & Doors
High performance timber windows - Henry Venables Interiors Ltd
High performance timber windows - Henry Venables Ltd
High performance windows - Sashless Window Co Ltd
Hi-spec 22 metric windows - Dale Joinery Ltd
Historic Rooflights - Clement Steel Windows
Industrial grade plastic framed roof windows for slate/tile roofs - Cox Building Products Ltd
Industrial rooflights - Kenilworth Daylight Centre Ltd
Insect screens - President Blinds Ltd
Internal and external doors - Jeld-Wen UK Ltd
Lantern lights - Lonsdale Metal Company
Louvre blind material - Mermet U.K
Louvres - McKenzie-Martin Ltd
Manual control window opening systems - Teleflex - Morse
Metal and glass louvers grilles, diffusers and dampers - Air Diffusion Ltd
Metal framed double glazed windows - Nova Group Ltd
Metal windows - Anglian Home Improvments
Metal windows - Olivand Metal Windows Ltd
Metal Windows - Thanet Ware Ltd
Metal windows - The Cotswold Casement Company
Metal windows/ rooflights/ screens/ louvers - Drawn Metal Ltd
Metal windows/ rooflights/ screens/ louvers - Heywood Williams Architectural
Metal windows/ rooflights/ screens/ louvres - Reid John & Sons (Strucsteel) Ltd
Natural ventilation - Sunpipe @ Monodraught Ltd
New build windows - Radway, Door and Windows Ltd.
North lights - Lonsdale Metal Company
Patio doors - Radway, Door and Windows Ltd.
Perforated Metal - Associated Perforators & Weavers Ltd
Perforated Products - Associated Perforators & Weavers Ltd
Pivot windows - Radway, Door and Windows Ltd.
Plastic framed double glazed windows - Nova Group Ltd
Plastic framed double glazed windows - Radway, Door and Windows Ltd.
Plastic profiled sheet cladding and roofing - Ariel Plastics Ltd
Plastic vertical sliding windows - Masterframe Windows Ltd
Plastic windows - Andersen/ Black Millwork
Plastic windows - Anglian Home Improvments
Plastic windows - Caradon Everest Ltd
PVC extrusions - Nenplas
PVC roofline products - LB Plastics Ltd
PVC window and door systems - LB Plastics Ltd
PVC window profiles - Profile 22 Systems Ltd
PVC windows - Prima Systems (South East) Ltd
PVC windows - Scholes Windows Ltd
PVCU extrusions for window systems - Spectus Systems
PVCU window fabrication - HIS Ltd
PVCu window systems - HW Systems Ltd
PVCu windows - BLW Associates Ltd
PVCu windows - Dale Joinery Ltd
PVCu windows - Interframe Ltd
PVCu windows - KCW Windows
PVCu windows - Plastal (SBP Ltd)
PVC-u windows - IsoSystems
PVC-U Windows - Magnet Ltd
PVCu Windows, Doors & Conservatories - Piper Double Glazing Ltd
PVC-U windows, doors and conservatories fabricated and installed - FineLine
PVC-U windows, patio doors, french doors, steel doors, PVC-U conservatories - Cestrum Building Products Ltd
Reflective films - Madico Film Products Division Inc
Replacement window service - Spectra Glaze Services
Residential doors - Radway, Door and Windows Ltd.
Roller blind material - Mermet U.K
Roof lights - Lonsdale Metal Company
Roof lights - Luxcrete Ltd
Roof vents - Lonsdale Metal Company
Roof Windows - Andersen/ Black Millwork
Roof windows - VELUX Co Ltd, The
Roofglazing - EBS Services Ltd
Rooflights - Alcover UK Ltd
Rooflights - Astrofade Ltd
Rooflights - Astrofade Ltd
Rooflights - Astrofade Ltd
Rooflights - Axter Ltd
Rooflights - Clear Span Ltd
Rooflights - Kenilworth Daylight Centre Ltd
Rooflights - McKenzie-Martin Ltd
Rooflights - Stoakes Systems Ltd
Rooflights - Torvale Building Products, A Division of Stadium Group PLC
Rooflights - Twide-Paragon Ltd
Rooflights - Ubbink (UK) Ltd
Rooflights - profiled and structured in GRP, PVC and polycarbonate - Brett Martin Ltd
Rooflights (Newdome) - Gardner & Newton
Rooflights and skylights - Wenmore Rooflights, A Division of W.F.Joy & Co. Ltd
Rooflights, skylights, atria, covered walkways, barrel vaults, deadlights, aluminium windows - Duplus Architectural Systems Ltd
Rooflights\screen\louvres - Zled (DPC) Co Ltd
Ropes and sashcords - Marlow Ropes Ltd
Sadh & case - Blair Joinery
Sarnalite rooflights - Sarnafil Ltd
Sash balances - Garador Ltd
Sash Windows - Alframes Ltd
Sash Windows - J. Scott (Thrapston) Ltd
Sash windows - Masterframe Windows Ltd
Screening Pannels - Winther Browne & co Ltd
Screens - Enfield Speciality Doors
Screens - Komfort Office Environments Plc
Screens - Project Aluminium Ltd
Screens/louvers - Allaway Acoustics Ltd
Screens/louvers - Levolux A.T. Limited
Security plus windows - Dale Joinery Ltd
Security screens, counters and partitioning - CSI Security LTD
Shop Fronts - Mandor Engineering Ltd
Shopfront systems - Aardee Security Shutters Ltd
Shopfronts - Bassett & Findley Ltd
Sky Lights - Andersen/ Black Millwork
Skyline roof system - Dale Joinery Ltd
Sliding patio door bar - Baddeley
Solar blinds material - Reflex-Rol (UK)
Solar control window films - Madico Film Products Division Inc
Solar shading solutions - Faber blinds UK ltd - Faber Blinds UK Ltd
Solar, safety, security, UV control, aesthetic and architectural films - Northgate Solar Controls
Steel caging - Bradbury Group Ltd
Steel doors - Baddeley
Steel screens - Fitzpatrick Doors Ltd
Steel windows - Crittall Steel Windows Ltd
Steel windows - Crittall Steel Windows Ltd
Steel windows - Monk Metal Windows Ltd
Steel windows and doors - Clement Steel Windows
Steel Windows Installation + glazing - Rea Metal Windows Ltd
Steel Windows Manufacture - Rea Metal Windows Ltd
Structured Steel framing - Phoenix Metal Products Ltd
Suspended glass smoke screens - Fenlock-Hansen Ltd
Tilt and turn windows - Radway, Door and Windows Ltd.
Timber conservatories - Fairmitre Ltd
Timber framed double glazed windows - Nova Group Ltd
Timber porches - Fairmitre Ltd
Timber windows - Andersen/ Black Millwork
Timber windows - Anglian Home Improvments
Timber windows - Broxwood (scotland Ltd)
Timber windows - Fairmitre Ltd
Timber windows - Hill Leigh Ltd
Timber windows - Jeld-Wen Ltd
Timber windows - Langford Joinery Ltd
Timber Windows - Magnet Ltd
Timber windows - Norlandia Ltd
Timber windows - Premdor
Timber windows - Sampson Windows Ltd
Timber windows - Scandanavian Timber
Timber windows - Scandinavian Window Systems Ltd
Timber windows - Swedhouse Ltd
Timber windows - Timber Components (UK) Ltd
Timber windows - Walker Timber Ltd
Timber windows and doorframes - Bennett Windows & Doors
Timber windows and screens - Hansen Windows Ltd
Timber windows mauufacture - Blair Joinery
Tinted films - Madico Film Products Division Inc
Translucent rooflights - Filion Products Ltd
Translucent wall and roof systems - Stoakes Systems Ltd
Unique view control films - Northgate Solar Controls
UPVC conservatories - Ducana Surrey Ltd
UPVC conservatories - Windowbuild
UPVC double glazed windows - Sierra Windows
UPVC window casements - Windowbuild
UPVC window fabrication and fixing supplies - Fab & Fix Ltd
UPVC window systems - Profolia Ltd
UPVC window systems - REHAU Ltd
UPVC windows - Fabricated Aluminium Services Ltd
UPVC windows - Juno Roplasto Ltd
Vertical sliding windows - Radway, Door and Windows Ltd.
Window control gear - Cooper H W & Co Ltd
Window hinges - Nico Manufacturing Ltd
Window Protector - Hilton Banks Ltd
Window repair - Window Care Systems Limited
Window systems - Deceuninck Ltd
Window tinting - Durable
Window ventilator manufactures - Simon R W Ltd
Window/ rooflights/ screens/ Lorrles - McMullen Architectural Systems
Windows - Bowater Containers Ltd
Windows - English Architectural Glazing Ltd
Windows - Smith & Choyce Ltd
Windows - Solair Ltd
Windows - TBR Joinery
Windows (alum) rooflights - Duplus Domes Ltd
windows /rooflight/ screens/ louvres - Universal Components Ltd
Windows/ rooflights/ screens/ louvers - Colt International
Windows/ Rooflights/ Screens/ Louvres - SCHUCO International KG
Windows/rooflights/ screens/louvers - Cantifix of London Limited
Windows/Rooflights/Louvres/Screens - Promat Fire Protection Ltd
Windows/Rooflights/Screens/Louvers - Gardner & Newton
Windows/rooflights/screens/louvres - Smoke Control Services Ltd

L20 Doors/Shutters/hatches

ABS Doors - High Impact - Regency Garage Door Services Ltd
Acoustic Doors - IAC
Acoustic doors - Jewers Doors Ltd
Acoustic seals - Sealmaster
Aircraft hangar doors - Jewers Doors Ltd
All types of fencing & gates - WA Skinner & Co Limited
Aluminium - Project Aluminium Ltd
Aluminium Entrances and patio doors - KCW Windows
Aluminium doors - Comar Architectural Aluminuim Systems
Aluminium doors - Ducana Surrey Ltd
Aluminium doors - Duplus Architectural Systems Ltd
Aluminium doors - Fabricated Aluminium Services Ltd
Aluminium doors - Glostal - Monarch
Aluminium Doors - Mandor Engineering Ltd
Aluminium doors - Planet Architectural Glazing Ltd
Aluminium doors - Prima Systems (South East) Ltd
Aluminium doors - Sapa Building Systems (monarch and glostal)
Apartment Entrance Doors - Vicaima Ltd
Architectural firecheck doors - Dale Joinery Ltd
Armour doors - Meggitt Armour Systems
Automatic & manual door repair & maintenance - Dorma Door Services Ltd
Automatic door operating equipment - Kaba Garog
Automatic door operators - Hillaldam Coburn Ltd
Automatic Doors - Kone PLC
Automatic doors - swing, slide and folding - Kaba Door Systems Ltd
Automatic Entrance Doors - Mandor Engineering Ltd
Ballistic doors - Jewers Doors Ltd
Bespoke joinery - AS Newbould Ltd
Bespoke joinery - Howard Bros Joinery Ltd
Blast doors - Jewers Doors Ltd
Brush strip door seals - Kleeneze Sealtech Ltd
Cedarwood doors - Regency Garage Door Services Ltd
Certified five doors - Bennett Windows & Doors
Clad doors - Swedhouse Ltd
Clean room doors - Clark Door Ltd
Cleanroom & contamination control doorsets - Leaderflush + Shapland Ltd
Cold store doors - Ascot Industrial Doors Ltd
Coloured GRP doors and frames - Regency Garage Door Services Ltd
Composite doors - Halmark Panels Ltd
Composite doors - Laird Security Hardware
Composite doors - Spectra Glaze Services
Conservatories - Hill Leigh Ltd
Conservatories - Solair Ltd
Curtain walling - LB Plastics Ltd
Dock seals and shelters - Konematic Bolton Brady
Door - purpose made flush - Humphrey & Stretton PLC
Door frames - Enfield Speciality Doors
Door frames - Real Door Company The
Door frames - TBR Joinery
Door kits - Vicaima Ltd
Door manufactures - Edmonds A & Co Ltd
Door panels - Door Panels PLC
Door panels - Halmark Panels Ltd
Door protection - Moffett Thallon & Co Ltd
Door sets - Hill Leigh Ltd
Door systems - Deceuninck Ltd
Doors - Altro Floors
Doors - BLP Group
Doors - Cantifix of London Limited
Doors - Crawford Amber Ltd
Doors - Finn Forest
Doors - J. Scott (Thrapston) Ltd
Doors - Komfort Office Environments Plc
Doors - Magnet Ltd
DOORS - McMullen Architectural Systems
Doors - Project Aluminium Ltd
Doors - Scandinavian Window Systems Ltd
Doors - Smith & Choyce Ltd
Doors - TBR Joinery
Doors - Turnquest UK Ltd
Doors - Walker Timber Ltd
Doors & sets - Cox Long Ltd
Doors (alum) - Duplus Domes Ltd
Doors and entrances - Bassett & Findley Ltd
Doors and revolving doors - Pearsons J R (Birmingham) Ltd
Doors architectural - Timbmet Ltd
Doors panelled - Kern (Design Products) J T Ltd
Doors/ Shutters/ Hatches - SCHUCO International KG
Doors/hatches - Allaway Acoustics Ltd
Doors/shutters/hatches - Hadley Industries Plc
Doors/shutters/hatches - Hart Door Systems Ltd
Double glazed doors - Alframes Ltd
Electric openers/ spare parts - Cardale Doors Ltd
Electric openers/ spare parts - Cardale Doors Ltd
Electric openers/ spare parts - Cardale Doors Ltd
Entrance systems - Doors & Hardware Ltd
Extruded aluminium composite doors - Smart Systems Ltd
Extruded aluminium doors - Smart Systems Ltd
Extruded UPVC doors - Halo, W H S Division Of Bowater Industries
Extruded UPVC doors - Smart Systems Ltd
Fast action doors - Neway Doors Ltd
Fd30 and Fd60 fire doors - Real Door Company The
Fibreglass doors - Premdor
Fire and security doors - Andersons Ltd
Fire doors - Enfield Speciality Doors
Fire doors - Fitzpatrick Doors Ltd
Fire doors - J. Scott (Thrapston) Ltd
Fire doors - Rolflex Doors Ltd
Fire Doors - Sunray Engineering Ltd
Fire doors - Vicaima Ltd
Fire doors and door sets - Shadbolt F R & Sons Ltd
Fire doors and roller shutters - Wormald Ansul UK Ltd
Fire escape doors - Fitzpatrick Doors Ltd
Fire rated door sets - Lami Doors UK Ltd
Fire rated doorsets - Leaderflush + Shapland Ltd
fire resistant Glazed screens and doors - Fenlock-Hansen Ltd
Fire resisting doors - Mann McGowan Group
Fire resisting doors and joinery - Decorfix Ltd
Fire shutters - Ascot Industrial Doors Ltd
Fire shutters - Lakeside Group
Fire shutters - Lycetts Industrial Doors
Fire shutters/ fire doors - Stokvis R S & Sons Ltd
Fire, folding, roller shutters and grilles - Bolton Brady Ltd
Fire, folding, roller shutters, grilles and overhead doors - Konematic Bolton Brady
Firerated doors - Baydale Architectural Systems Ltd
Fire-Resistant Glass - SCHOTT UK Ltd
Flat composite panel systems - European Profiles
Flexible doors - Stokvis R S & Sons Ltd
Flush and feature doors - Vicaima Ltd
Flush doors - Enfield Speciality Doors
Foil faced doos - Vicaima Ltd
Folding doors - Dividers Modernfold Ltd
Folding doors - Hillaldam Coburn Ltd
Folding doors and partitions - Pellfold Parthos Ltd.
Folding doors, partitions and screens - Kern (Design Products) J T Ltd
Food industry doors - Ascot Industrial Doors Ltd
Framed and panelled doors, frames, skirtings and architraves. - Real Door Company The
Frames and covers - Bilco UK Ltd
G. R. P . Porches - Solair Ltd
G.R.P Doors - Regency Garage Door Services Ltd
Garage and industrial doors - Hormann (UK) Ltd
Garage door manufacturer - Avon Manufacturing Ltd
Garage doors - Crimeguard Systems Ltd
Garage doors - Doorfit Products Ltd
Garage doors - Roller - Moffett Thallon & Co Ltd
Garage doors - Sectional - Moffett Thallon & Co Ltd
Garage doors - Up and Over - Moffett Thallon & Co Ltd
Gas/ oil fired boilers - Stokvis R S & Sons Ltd
Gates and barriers - Hart Door Systems Ltd
Glazing maintenance - Spectra Glaze Services
Grilles - Bradbury Group Ltd
Grilles and shutter - CSI Security LTD
GRP Door surrounds - Alpha Glass Fibres Ltd
GRP Doors - Piper Double Glazing Ltd - Piper Double Glazing Ltd
GRP hatches - Group Four Glassfibre Ltd
Hardwood doors - Chindwell Co Ltd
Hardwood doorsets to match windows - Blair Joinery
Hatches - metal - McKenzie-Martin Ltd
High integrity doors windows blast and fire protection systems - Booth Industries Ltd
High performance doors - Sashless Window Co Ltd
High speed doors - Clark Door Ltd
High speed doors - Crawford Amber Ltd
High speed overhead doors - Stokvis R S & Sons Ltd
High speed rapid roll doors - Stokvis R S & Sons Ltd
High speed vertical folding and roll -up doors - Mandor Engineering Ltd
Hinged access panels for ceilings and walls - Profab Access Ltd
Hinged steel dorrs - Clark Door Ltd
Industrial and cold storage strip curtain systems - Flexible Reinforcements Ltd
Industrial & commercial doors, shutters, grilles and partitions - Kaba Door Systems Ltd
Industrial & commercial doors, shutters, grilles and partitions - Kaba Door Systems Ltd
Industrial & commercial security & fire resisting metal doors & shutters - Bolton Gate Co. Ltd
Industrial doors - RNB Industrial Doors Services
Industrial doors - Telford Doors and Barriers Ltd
Industrial doors and loading bay equipment - Envirodoor Markus Ltd
Industrial doors rolling, folding and sliding - Lycetts Industrial Doors
Industrial doors, roller, folding and insulated - Avon Industrial Doors Ltd
Industrial sliding and folding doors - European Profiles
Inlay Doors - Vicaima Ltd
Insulated bifolding doors - Jewers Doors Ltd
Insulated overhead doors - Stokvis R S & Sons Ltd
Insulated side folding/ sliding doors - Stokvis R S & Sons Ltd
Insulated sliding doors - Jewers Doors Ltd
Insulated sliding folding doors - Jewers Doors Ltd
Laminated Door sets - T B S Fabrications
Loading bay equipment - Lycetts Industrial Doors
Loft access doors - Timloc Expamet Building Products
Loft Access Hatches - Ryton's Building Products Ltd
Maintenance of industrial doors, shutters and shopfronts - Shutter Door Repair & Maintenance Ltd
Matching Side Doors - Regency Garage Door Services Ltd
Megadoors - Crawford Amber Ltd
Metal doors - EDS Shutter Door Systems Ltd
Metal doors - Fitzpatrick Doors Ltd
Metal doors - Leaderflush + Shapland Ltd
Metal doors - Premdor
Metal doors - The Cotswold Casement Company
Metal doors, shutters - Amber Doors Ltd
Metal doors/ shutters/ hatches - Heywood Williams Architectural
Metal doors/ shutters/ hatches - Reid John & Sons (Strucsteel) Ltd
Metal frames - Fitzpatrick Doors Ltd
Metal meter boxes and replacement architrave units - Ritherdon & Co Ltd
Metal roller shutters - Lakeside Group
Mobility door frames - Dale Joinery Ltd
P.V.C. strip curtains - Dudley Factory Doors Limited
Patio Doors - Alframes Ltd
Patio doors - Andersen/ Black Millwork
Patio doors - Andersen/ Black Millwork
Patio Doors - Solair Ltd
Patio Doors - Solair Ltd
Performance Doors and Doorsets - Leaderflush + Shapland Ltd
Pine doors - Wood International Agency Ltd
Pine furniture doors - Mason FW & Sons Ltd
Plastic door sets - Lami Doors UK Ltd
Plastic doors - Advanced Panels & Products
Plastic doors - Leaderflush + Shapland Ltd
Prehung Door Sets - Bennett Windows & Doors
Profiled composite panel systems - European Profiles
Purpose made flush timber doors - Youngs Doors Ltd
Purpose made panel doors - Youngs Doors Ltd
PVC clad doors & frames - Platonoff & Harris Plc
PVC crash and strip doors - Avon Industrial Doors Ltd
PVC door profiles - Profile 22 Systems Ltd
PVC doors - Prima Systems (South East) Ltd
PVC doors and conservatories - Scholes Windows Ltd
PVC doors and strip curtains - Bolton Brady Ltd
PVC doors and strip curtains - Konematic Bolton Brady
PVC strip curtains - Stokvis R S & Sons Ltd
PVCU door fabrication - HIS Ltd
PVCu doors - BLW Associates Ltd
PVCu doors - Dale Joinery Ltd
PVCu doors - Interframe Ltd
PVCu doors - Plastal (SBP Ltd)
PVCu Entrances and patio doors - KCW Windows
PVCU extrusions for door systems - Spectus Systems
Rapid roll doors - Ascot Industrial Doors Ltd
Rapid roll doors - Konematic Bolton Brady
Rapid roll doors - Lycetts Industrial Doors
Remote control garage doors - Regency Garage Door Services Ltd
Repairs - Syston Rolling Shutters Ltd
Residential garage doors - Garador Ltd
Residential garage doors - Garador Ltd
Residential garage doors - Garador Ltd
Revolving doors - Grothkarst UK Ltd
Revolving doors - Kaba Door Systems Ltd
Roller doors (including insulation) - Stokvis R S & Sons Ltd
Roller doors/ side-hinged doors - Cardale Doors Ltd
Roller doors/ side-hinged doors - Cardale Doors Ltd
Roller doors/ side-hinged doors - Cardale Doors Ltd
Roller Garage Doors - Lakeside Group
Roller shutter and grilles - James & Bloom Ltd
Roller shutter and grilles - James & Bloom Ltd
Roller shutter and grilles - James & Bloom Ltd
Roller shutter doors - Ring-Gard (UK) Ltd
Roller shutter doors, brickbond grilles - Dudley Factory Doors Limited
Roller shutter doors, fast action doors - Priory Shutter & Door Co Ltd
Roller shutters - Ascot Industrial Doors Ltd
Roller shutters - Dallas & Forrest Ltd
Roller shutters - Fitzpatrick Doors Ltd
Roller shutters and industrial doors - Regency Garage Door Services Ltd
Rolling shutters and individual doors in steel, aluminum and timber. Service and maintenance available nationwide - Syston Rolling Shutters Ltd
Ropes and sashcords - Marlow Ropes Ltd
Rubber and PVC flexible crash doors, PVC strip curtains - Neway Doors Ltd
Rubber and PVC flexible doors - Mandor Engineering Ltd
Sectional Doors - Regency Garage Door Services Ltd
Sectional overhead doors - Ascot Industrial Doors Ltd
Sectional overhead doors - Crawford Amber Ltd
Security door systems - Industrial Building Componants Ltd
Security door systems - Vicaima Ltd
Security doors - CSI Security LTD
Security doors - Fitzpatrick Doors Ltd
Security doors - Jewers Doors Ltd
Security doors - Kaba Door Systems Ltd
Security doors - Leaderflush + Shapland Ltd
Security Doors (GRP) - Oswestry Reinforced Plastics Ltd
Security doors, grilles with built in fly screens - United Safety Products UK Ltd
Security gates, Bars and Roller shutters - Marshall Contracts Ltd
Security grilles - Crimeguard Systems Ltd
Security metal doors and shutters - Ascot Industrial Doors Ltd
Security roller shutters - Lakeside Group
Security shutters - Aardee Security Shutters Ltd
Shutters and grilles - Permclose Door Services Ltd
Side Hung Doors - Regency Garage Door Services Ltd
Sliding and folding doors - Hillaldam Coburn Ltd
Sliding and folding partitions - Building Additions Ltd
Sliding closures - Hillaldam Coburn Ltd
Sliding doors - Clark Door Ltd
Sliding doors - Kaba Garog
Sliding doors, partitions and screens - Kern (Design Products) J T Ltd
Sliding folding doors - Ascot Industrial Doors Ltd
Special timber flush doors and door sets - Industrial Building Components Ltd
Specialist steel doorsets - Accent Hansen
Specialized timber doors - AS Newbould Ltd
Stainless steel doors - Fitzpatrick Doors Ltd
Stainless steel, acoustic and insulated freezer doors - Clark Door Ltd
Stainless steel, bronze and aluminium panel doors, entrance screens, shopfronts and window frames - Drawn Metal Ltd
Steel and aluminium security shutters - Lycetts Industrial Doors
Steel Door Installation +glazing - Rea Metal Windows Ltd
Steel Door Manufacture - Rea Metal Windows Ltd
Steel door sets - Ascot Industrial Doors Ltd
Steel doors - Bradbury Group Ltd
Steel doors - Crittall Steel Windows Ltd
Steel doors - Crittall Steel Windows Ltd
Steel doors - Kaba Door Systems Ltd
Steel doors - Monk Metal Windows Ltd
Steel Doors - Regency Garage Door Services Ltd
Steel doors and frames - Fitzpatrick Doors Ltd
Steel doors, Ceilings - Norwood Partition Systems Ltd
Steel doors, frames and entrance systems - Doors & Hardware Ltd
Steel faced residential doors - BLW Associates Ltd
Steel Firescreen Installation + glazing - Rea Metal Windows Ltd
Steel Firescreen Manufacture - Rea Metal Windows Ltd
Steel hinged doors - Lycetts Industrial Doors
Steel hinged Doorsets - Accent Hansen
Steel industrial doors - Mercian Industrial Doors
Steel louvred doorsets - Sunray Engineering Ltd
Steel personal doors - Lakeside Group
Steel security doorsets - Sunray Engineering Ltd
Steel windows and doors - Clement Steel Windows
Synthetic rubber moulding components and ironmongery - Notcutt Ltd
Syston doors - Syston Rolling Shutters Ltd
Teak Specialist - Parker Kislingbury Ltd
Timber door - Leaderflush + Shapland Ltd
Timber doors - Broxwood (scotland Ltd)
Timber doors - Hansen Windows Ltd
Timber doors - Henry Venables Interiors Ltd
Timber doors - Jeld-Wen Ltd
Timber doors - Kind, J B, Ltd
Timber Doors - Langford Joinery Ltd
Timber doors - Norlandia Ltd
Timber doors - Premdor
Timber doors - Swedhouse Ltd
Timber patio doors - Fairmitre Ltd
Traditional oak doors - WL West & Son Limited
Transformer chamber doorsets - Sunray Engineering Ltd
Turnstiles - full and half height - Kaba Door Systems Ltd
Up and over doors/ sectional doors - Cardale Doors Ltd
Up and over doors/ sectional doors - Cardale Doors Ltd

L20 Doors/Shutters/hatches (Cont'd)

Up and over doors/ sectional doors - Cardale Doors Ltd
UPVC door systems - Profolia Ltd
UPVC Doors - Ducana Surrey Ltd
UPVC doors - Fabricated Aluminium Services Ltd
UPVC doors - Spectra Glaze Services
UPVC Doors and conservatories - Juno Roplasto Ltd
UPVC french doors - Windowbuild
veneered doors - Vicaima Ltd
Vision see-through security shutters - Cooper Group Ltd
Wall access hatches - Howe-Green Ltd
Wall panelling - Kern (Design Products) J T Ltd
Wedge wire screens - Screen Systems Ltd
Windows - Jeld-Wen UK Ltd

L30 Stairs/Walkways/Balustrades

Acrylic, glass, steel and timber stairs - Lewes Design Contracts Ltd
Aluminium Bridge parapets - Varley & Gulliver Ltd
Aluminium floor outlets - Marley Plumbing and Drainage
Aluminium handrails/balustrades - Norton Engineering Alloys Co Ltd
Aluminium ladders and step ladders - Ramsay & Sons (Forfar) Ltd
Aluminium stairways - Light Alloy Ltd
Balconys - Woodburn Engineering Ltd
Ballstrades - Strand Hardware Ltd
Balustrade and handrail coating - Plastic Coatings Ltd
Balustrades - Ermine Engineering Co. Ltd
Balustrades - LB Plastics Ltd
Balustrades - Steel Line Ltd
Balustrades and Handrails - Gabriel & Co Ltd
Balustrades and wallrails - Turnquest UK Ltd
Balustrades, staircases and fire escapes - Staines Steel Gate Co
Balustrading - Marine & Commercial Rigging Systems
Bespoke joinery - AS Newbould Ltd
Cast Iron Stairways - Safety Stairways Ltd
Cast Spiral Stairs - Cottage Craft Spirals
Commercial grilles and shutters - Andersons Ltd
Curved glass - Darby Glass Ltd
Decking - B & M Fencing Limited
Design - staircases - Lyncrete Burkle Services Ltd
Domestic and comercial - Loft Centre Products
Double ball industrial handrail systems - Eurogrid Ltd
Energy saving doors - Andersons Ltd
Escape and electric stairway ladders, roof exit systems, spiral, spacesaver and traditional stairs - Loft Centre Products
Fabricators of open metal flooring, handrails and standards - OSF Ltd
Fibreglass ladders and step ladders - Ramsay & Sons (Forfar) Ltd
Fire escape staircases - Crescent of Cambridge Ltd
Fire escapes - Consculpt Spiral Stairs Ltd
Fire Escapes - Safety Stairways Ltd
Fire escapes, gates, railings, balustrades, guardrails and spiral staircases - Metalcraft (Tottenham) Ltd
Fire shutters - Andersons Ltd
Flooring - Orsogril Spa UK Distributor
Glass Balustrade - SG System Products Ltd
GRP Handrails - Brighton (Handrails), W
GRP String course - Alpha Glass Fibres Ltd
Guardrail - Kee Klamp Ltd
Hand rails - RB UK Ltd
Handrail & balustrades - Wenmore Rooflights, A Division of W.F.Joy & Co. Ltd
Handrail systems in steel, stainless steel, aluminium and GRP - Lionweld Kennedy Ltd
Handrailing - Steel Line Ltd
Handrailing & Balustrading - Laidlaw Architectural Hardware Ltd
Handrailing, balustrading and fire escapes - Cameron & Moore (Gridplex) Ltd
Handrails and balustrades - Drawn Metal Ltd
Handrails and balustrades - Pearsons J R (Birmingham) Ltd
Handrails/balustrades - stainless - Norton Engineering Alloys Co Ltd
Helical Staircases - Safety Stairways Ltd
Helical stairs - Cambridge Structures Ltd
Helical stairs - Crescent of Cambridge Ltd
Industrial doors - Andersons Ltd
Industrial flooring, handrailing systems and stairtreads - Redman Fisher Engineering
Insulated Door panels - Ward Building Components Ltd
Internal and external stairs - Camtwix Engineering (Division of Terrapan Ltd)
Ladders, walkways and hand railing - Moseley GRP Product (Division of Moseley Rubber Co Ltd)
Loft Access Ladders - Loft Ladders Ltd
Metal railings - Varley & Gulliver Ltd
Metal Staircases - Albion Manufacturing
Metal Staircases - Safety Stairways Ltd
Metal stairs - Cottage Craft Spirals
Metal stairs / walkways / balustrades - Steelway Fensecure Ltd
Metal stairs/ walkways/ balustrades - Hubbard Architectural Metalwork Ltd
Metal stairs/balustrades - NRG Fabrications
Open steel grating in steel, stainless steel and GRP - Lionweld Kennedy Ltd
Pedstrian walkways - Eve Trakway
Perforated sheets and balustrading - Associated Perforators & Weavers Ltd
Plywood, hard, fibre, chip and particle boards - Boardcraft Ltd
Precast Prestressed Concrete Satirs - Tarmac Top Floor
Precast terrazo treads & riser - Roman Mosaic Contracts Ltd
PVC handrails - Architectural Plastics (Handrail) Ltd
PVC handrails - Brighton (Handrails), W
PVC-u Balcony System - Radway, Door and Windows Ltd.
Rails for special needs - Gabriel & Co Ltd
Roof walkway/ fall arrest system - McKenzie-Martin Ltd
Ropes and sashcords - Marlow Ropes Ltd
Spiral and helical p.c. stairs - Consculpt Spiral Stairs Ltd
Spiral and straight staircases - Mackay Engineering
Spiral Staircase Systems - Lewes Design Contracts Ltd
Spiral Staircases - Safety Stairways Ltd
Spiral Staircases - Titan Ladders Ltd
Spiral staircases and special precast stairs. - Consculpt Spiral Stairs Ltd
Spiral stairs - Gabriel & Co Ltd
Spiral stairs - Pedley Furniture International Ltd
Spiral stairways - Milner Delvaux Ltd
Stainless steel flooring - Ancon Building Products
Stainless Steel Staircases - Safety Stairways Ltd
Stainless steel, aluminium, PVC covered, galvanized and brass balustrades - SG System Products Ltd
Stair and ladder treads - Eurogrid Ltd
Stair edgings - Gradus Ltd
Stair/ walkways - Redman Fisher Engineering
Staircases - Hill Leigh Ltd
Staircases - Lyncrete Burkle Services Ltd
Staircases - TBR Joinery
Staircases - metal, spiral, straight - Crescent of Cambridge Ltd
Staircases - precast and modular concrete - Lyncrete Burkle Services Ltd
Staircases Manufacturers - Safety Stairways Ltd
Staircases, fire escapes, ladders, handrails and balustrades - Down & Francis Ltd, Division of Metalrax PLC
Stairs - Kirkstone
Stairs - Magnet Ltd
Stairs - Smith & Choyce Ltd
Stairs walkways and Balustrades - Midland Wire Cordage Co Ltd
Stairs/ walkways - Steelway Fensecure Ltd
Stairs/ walkways/ balustrades - Broxap Ltd
Stairs/ walkways/ balustrades - Glazzard (Dudley) Ltd
Stairs/ walkways/ balustrades - Quantum Profile Systems Ltd
Stairs/walkways - Ancon Building Products
Stairs/Walkways - Portia Engineering Ltd
Stairs/walkways/ balustrades - Cantifix of London Limited
Stairs/Walkways/Balustrades - Grant & Livingston Ltd
Stairways/Walkways/balustrades - Jones & Thomas (Wireworks) Ltd
Steel bridge parapets - Varley & Gulliver Ltd
Steel fabrication railings, gates, fire escapes - SFC (Midlands) Ltd
Steel spiral stairs - Higginson Staircases Ltd
Steel staircases - Ermine Engineering Co. Ltd
Steel staircases - Gabriel & Co Ltd
Steel stairs - Fire Escapes & Fabrications (UK) Ltd
Straight Flights - Cottage Craft Spirals
Straight helical and spiral stairs - Lewes Design Contracts Ltd
Straight, Winder & geometric timber stairs - Higginson Staircases Ltd
Teak Specialist - Parker Kislingbury Ltd
Timber loft and escape ladders, roof exit systems, spiral, spacesaver and traditional stairs - Loft Centre Products
Timber Spiral Stairs - Cottage Craft Spirals
Timber spiral stairs - Higginson Staircases Ltd
Timber staircases - Jeld-Wen Ltd
Timber staircases - Northern Joinery Ltd
Timber staircases - Thomas Lowe Joinery
Timber stairs - Cottage Craft Spirals
Timber stairs - Jeld-Wen UK Ltd
Timber stairs/balustrades - Timber Components (UK) Ltd
Ultra Modern Staircases - Safety Stairways Ltd
Victorian Spiral Staircases - Safety Stairways Ltd
Walkways - Clow Group Ltd
Walkways - Gabriel & Co Ltd
Walkways and platforms - Woodburn Engineering Ltd
Walkways/balustrades - Levolux A.T. Limited
Wall corner & Door Protection - Gradus Ltd
Wallrails and balustrades - Turnquest UK Ltd
Wooden ladders and step ladders - Ramsay & Sons (Forfar) Ltd
Woodturning - Spindlewood

L40 General glazing

Annealed glass - Darby Glass Ltd
Architectural aluminium systems for glazing - Universal Components Ltd
Architectural glass - Cambridge Structures Ltd
Atrium glazing - English Architectural Glazing Ltd
Bomb blast - Hi-Tec Window Films
Canopies - Clear Span Ltd
Coloured Glass - SCHOTT UK Ltd
Curved & Decorative Glass - McNeill-McManus Ltd
Curved and flat toughened glass - Thermax Interglass Ltd
Decorative glass service - Andy Thornton Ltd
Display Enhancement Glass - SCHOTT UK Ltd
Double glazed units - Ledite Glass (Southend) Ltd
Double glazing - FineLine
Dry glazing systems and ventilated glazing beads etc - Exitex Ltd
Etched and sand blasted glass - Lightfoot Charles Ltd
Films for glass - Northgate Solar Controls
Fire rated glazing - English Architectural Glazing Ltd
Fire resistant and safety glasses - C. G. I. International Ltd
Fire resistant glazing - Mann McGowan Group
Fire resistant glazing - Twide-Paragon Ltd
Ganeral glazing - Saint-Gobain Glass UK Ltd
General glazing - FTW Windows
General glazing - Glaverbel (UK) Ltd
General glazing - Pilkington UK Ltd
General glazing - Solaglas
Glass - Fenlock-Hansen Ltd
Glass - Pilkington Birmingham
Glass - Solaglas
Glass - Spital Tile Co Ltd
Glass & Glazing - Clement Steel Windows
Glass & Glazing - McNeill-McManus Ltd
Glass and sheet plastic processing - Cooper H W & Co Ltd
Glass cutters - Stanton-Thompson (Agencies) Ltd
Glazed anti blast products - Baydale Architectural Systems Ltd
Glazed canopies - Lonsdale Metal Company
Glaziers and glassmerchants - Ledite Glass (Southend) Ltd
Glazing - The Abbseal Group incorporating Everseal (Thermovitrine Ltd)
Glazing Accessories - Reddiglaze Ltd
Glazing Accessories - Simon R W Ltd
Glazing and factory finishing - Bennett Windows & Doors
Glazing bars - Kenilworth Daylight Centre Ltd
Glazing bars for glass sealed units - Newdawn & Sun Ltd
Glazing bars for timber roofs - Newdawn & Sun Ltd
Glazing gaskets to windows and doors - DJ Profiles
Glazing maintenance - Spectra Glaze Services
High performance glass - Darby Glass Ltd
High performance window films - Banafix Limited
Insulated Fire Resistant Glass - SCHOTT UK Ltd
Insulating glass units - Darby Glass Ltd
Laminated - Pilkington Birmingham
Laminated glass - Darby Glass Ltd
Laminated glass - Float Glass Industries
Laminated glass - Solaglas Laminated
Low - emissivity - Pilkington Birmingham
Manifestation graphics - Invicta Window Films Ltd
Mirror - Hi-Tec Window Films
Opaque bullet resistant glass - Attwater & Sons Ltd
Patterned - Pilkington Birmingham
Patterned glass - Darby Glass Ltd
Patterned glass - Float Glass Industries
Plastic profiled sheet cladding and roofing - Ariel Plastics Ltd
Polycarbonate glazing - Cox Building Products Ltd
Privacy - Hi-Tec Window Films
Privacy window films - Invicta Window Films Ltd
Processed glass - Darby Glass Ltd
Putty - Adshead Ratcliffe & Co Ltd
Restoration Glass - SCHOTT UK Ltd
Roof glazing - English Architectural Glazing Ltd
Safety and security window films - Durable Berkeley Co Ltd
Safety/ Security - Hi-Tec Window Films
Secondary glazing - Alframes Ltd
Secondary glazing manufactures - Simon R W Ltd
Security/ safety window films - Invicta Window Films Ltd
Self support glazing beams - Newdawn & Sun Ltd
Selfcleaning - Pilkington Birmingham
Shopfronts - Comar Architectural Aluminuim Systems
Silicone jointed glazing - ARC Partitioning
Silvered - Pilkington Birmingham
Solar Control - Hi-Tec Window Films
Solar control glass - Float Glass Industries
Solar control window films - Invicta Window Films Ltd
Solar, anti shatter glass films - Cooper H W & Co Ltd
Stock glass - Float Glass Industries
Structural bolted glazing - English Architectural Glazing Ltd
Tinted - Pilkington Birmingham
Tinting - Hi-Tec Window Films
Toughened - Pilkington Birmingham
Toughened glass - Darby Glass Ltd
Toughened glass - Float Glass Industries
Toughened, laminate, ceramic, acoustic glass and screen printing on glass - Pilkington Plyglass
Toughenged & Laminated Glass - McNeill-McManus Ltd
White Flashed Opal Glass - SCHOTT UK Ltd
Window films - Bonwyke Ltd
Window films - Durable Berkeley Co Ltd
Window films to existing glass - Hi-Tec Window Films

L41 Lead light glazing

Glazing maintenance - Spectra Glaze Services
Impact resistant safety mirrors and glass - Chelsea Artisans Ltd
Leaded lights - Ledite Glass (Southend) Ltd
Leaded lights - Lightfoot Charles Ltd
Leaded lights - The Cotswold Casement Company
Leaded, decorative, Georgian designs - Darby Glass Ltd
Restoration/ conservation/ design and production of stained glass - Chapel Studio
Specialist glazing - Lightfoot Charles Ltd
Stained glass - Lightfoot Charles Ltd

L42 Infill panels/ sheets

Anti vandal cladding and glazing products - Oswestry Reinforced Plastics Ltd
Clear heat control films - Northgate Solar Controls
Doors - BLP Group
Glazed anti ballistic products - Baydale Architectural Systems Ltd
Glazing maintenance - Spectra Glaze Services
Infill panels/sheets - Kemlite Ltd
Infill panels/sheets - Panel Systems Ltd
Mirrors - Float Glass Industries
Privacy, patterned, decorative and manifestation films - Northgate Solar Controls
Wall & ceiling panels - McLoughlin Wood Ltd

M10 Cement: sand/ Concrete screeds/ toppings

Calcium carbonate - Imerys Minerals Ltd
Cement - Rugby Cement
Cement based screeds - Flowcrete Plc
Cementitious screed - IFT Supplies Ltd
Chemical resistant screed - Addagrip Surface Treatments Ltd
Concrete screeds and toppings - Anti Corrosion Services Ltd
Concrete topping - Ardex UK Ltd
Cristobalite flour - Hepworth Minerals & Chemicals Ltd
Cristobalite sand - Hepworth Minerals & Chemicals Ltd
Floor screed - Lytag Ltd
Floor screeds and floor repair mortars - Ronacrete Ltd
Floor screeds and roof screeds - Isocrete Floor Screeds Ltd
Flooring - Degussa Construction Chemicals (UK)
Flooring - Feb MBT
Flooring - Hanson Concrete Products Ltd
Flooring - Thoro Systems Products Ltd
Flooring surface hardeners - Armorex Ltd
Floors - precast concrete - Lyncrete Burkle Services Ltd
Industrial flooring - Exchem Mining and Construction
Industrial flooring - Stuarts Industrial Flooring Ltd
Mezzanine floors and allied equipment - Bromag Structures Ltd
Norina screed cavity flooring - Durabella System Floors Ltd
Permascreed - DPC Screeding Ltd
Polymer/cementitions screeds - Conren Ltd
Polypropylene & steel fibers for concrete & mortar - S.I. Concrete Systems
Quick drying cementicious screeds - P.C.I Construction Systems Ltd
Repair mortars - Conren Ltd
Sand cement/ Concrete/ Granolithic screeds/ flooring - Don Construction Products Ltd
Sand/cement screed - Tarmac Northern Ltd
Screeds - Lastite Building Products Ltd
Screeds - Readymix Limited
Screeds and floor coatings - Addagrip Surface Treatments Ltd
Self levelling cementicious screeds - P.C.I Construction Systems Ltd
Silica Flour - Hepworth Minerals & Chemicals Ltd
Silica Sand - Hepworth Minerals & Chemicals Ltd
Specialist screeds - Laticrete UK
Steel framed multi storey volumetric buildings - Accomodex
Subfloor preparation products - Ardex UK Ltd

M10 Cement: sand/ Concrete screeds/ toppings (Cont'd)

Tanking - Kiltox Damp Free Solutions
Wilsonart solid surfacing - Wilsonart Ltd

M11 Mastic asphalt flooring/floor underlays

Paving and flooring - Roofing Contractors (Cambridge) Ltd
Underlay - McLoughlin Wood Ltd

M12 Trowelled bitumen/ resin/ rubber-latex flooring

Anti slip surfacing - Addagrip Surface Treatments Ltd
Car Park Decking - Sika Ltd
Celicote resin coatings - SGL Acotec Ltd
Epoxy floor coatings - Ronacrete Ltd
Epoxy flooring systems - Cross-Guard International Ltd
Epoxy resin coatings - IFT Supplies Ltd
Epoxy resin coatings for floors - Watco UK Ltd
Epoxy resin repair materials - Anglo Building Products Ltd
Floor coatings - Jotun Henry Clark Ltd Decorative Division
Flooring systems - Cross-Guard International Ltd
Industrial Flooring - Sika Ltd
Resin based floor & wall finishes - Flowcrete Plc
Resin coatings - Armorex Ltd
Resin Flooring - Altro Floors
Self Leveling Moread Beds Latex - Laticrete UK
Trowelled bitumen/ resins/ rubber latex flooring - Don Construction Products Ltd

M13 Calcium sulphate based screeds

Architechural Mouldings - Hodkin Jones (Sheffield) Ltd
Calcium sulphate based screeds - Flowcrete Plc
Calcium sulphate based screeds - Readymix Limited
Synthascreed - DPC Screeding Ltd

M20 Plastered/ Rendered/ Roughcast coatings

Angle beads and metal arch frames - Easy Arches Ltd
Building plasters - British Gypsum Ltd
Cement based floor screeds, external render systems - Optiroc Ltd
Coil mesh - Catnic
Coloured limestone, dolomite and white sand internal and external renders - Kilwaughter Chemical Co Ltd
Decorative coatings - Textured European Finishes Ltd
Drywall decorating products - Lafarge Plasterboard Ltd
Drywall finishing systems - Lafarge Plasterboard Ltd
Epoxy resin coatings for walls - Watco UK Ltd
Expanded metal products - Ensor Building Products
High performance building boards - Knauf
Metal lath - Catnic
Plaster arris protector - Areco Roofing Supplies Company
Plaster beads - Walls & Ceilings International Ltd
Plasterboard accessories - Walls & Ceilings International Ltd
Plasterboards - Brikenden (Builders Merchants) Ltd
Plastered coatings - Readymix Limited
Plastered/Rendered/Roughcast fittings - PermaRock Products Ltd
Plasterers beads - Vista Engineering Ltd
Plasterers bends - Catnic
Plasters - Kair ventilation Ltd
plastic rendering plaster and dry wall beads - Nenplas
Renders - Weber Building Solutions
Renders and plasters - Thermica Ltd
Roughcast coatings - Andrews Coatings Ltd
Synthetic resin renders - STOrend systems - STO Ltd
Textured coatings & grout - Halls Beeline Ltd

M21 Insulation with rendered finish

Expanded polystyrene insulation - Vencel Resil Ltd
External wall insulation - STOtherm classic - STO Ltd
Floor insulation - Callenders Ltd
Insulated external render - Optiroc Ltd
Insulated external renders/systems - Alumasc Exterior Building Products Ltd
Insulation - Rockwool Ltd
Insulation with rendered finish - Concrete Repairs Ltd
Insulation with rendered finish - Insulated Render & Cladding Association
Insulation with rendered finish - PermaRock Products Ltd
Insulation with rendered finish - Thermica Ltd
Polymer cement renders - MR Limited
Render nail type M for EML lath - Reiner Fixing Devices (Ex: Hardo Fixing Devices)

M22 Sprayed monolithic coatings

Grout pumps - Clarke UK Ltd
Intimescent paint - Firebarrier Services Ltd
Specialist coatings - Valtti Specialist Coatings Ltd
Specialist wall coatings - MR Limited
Spray coating and fire protection - Cafco International
Sprayed & trowelled on resin coatings - SGL Acotec Ltd
Sprayed monolithic coatings - Specialist Building Products
Sprayed monolithic coatings - Wood E Ltd

M23 Resin bound mineral coatings

Anti skid surfacing products - Clare R S & Co Ltd
Coloured sufacing products - Clare R S & Co Ltd
Epoxy Floor Paint - Everlac (GB) Ltd
Industrial Flooring Products - Ardex UK Ltd
Manufacture of Industrial Resin Flooring in Polyurethane or Epoxy materials - Resdev Ltd
Mineral Coating T.E.F. - Textured European Finishes Ltd
Resin bound mineral coatings - Addagrip Surface Treatments Ltd
Resin bound mineral coatings - Certifield Laboratories (Div or CPS Industries Ltd)
Resin bound mineral coatings - Don Construction Products Ltd
Resin sealing coats - Armorex Ltd
Resins - DSM UK Ltd
Seamless resin floors - SGL Acotec Ltd
Specialist coatings - Valtti Specialist Coatings Ltd
Surface finishes - Fels UK
Traffic coatings - Coverite Ltd

M30 Metal mesh lathing/ Anchored reinforcement for plastered coatings

Anchored reinforcement for plastered coatings - Thermica Ltd
Arch formers - Catnic
Arch formers - Expamet Building Products
Cladding systems - BRC Special Products
Mesh arches - Allmat (East Surrey) Ltd
Metal lathing - Expamet Building Products
Metal beads - Expamet Building Products
Metal lathing and mesh - Simpson Strong-Tie®
PVCu plasters beads for drylining, plastering and rendering - Renderplas Ltd
Reinforcement mesh for rendering - EBC UK Ltd
Render nail for Riblath - Reiner Fixing Devices (Ex: Hardo Fixing Devices)
Resin bound mineral coatings - Specialist Building Products
Stainless steel surface protection - Component Developments
Welded wire mesh - Albion Manufacturing

M31 Fibrous plaster

Decorative mouldings, cornices, dados, ceiling roses and coving - Copley Decor Ltd
Decorative plaster coving and ceiling roses - Hilton Banks Ltd
Fibrous plaster - Clark & Fenn Ltd
Fibrous plaster moldings - Troika Architectural Mouldings Ltd
Fibrous plasterware, cornices and ceiling roses - ABC Studios (Plaster Mouldings) Ltd.
Glass fiber reinforced gypsum - Troika Architectural Mouldings Ltd
Ornamental fibrous plasterwork - manufacture and restoration of - Allied Guilds
Plaster cornices and archways - Artistic Plastercraft Ltd
Polypropylene, steel fibers for concrete & mortar - S.I. Concrete Systems

M40 Stone/ Concrete/ Quarry/ Ceramic tiling/ Mosaic

Acid resistant brickwork - SGL Acotec Ltd
Air tools - Marcrist International Ltd
Anit-Fracture Membranes for Natural Stone - Laticrete UK
Brickwall Cladding (interior & exterior) - Padipa Ltd
Ceramic fiber products - Combustion Linings Ltd
Ceramic floor and wall tiles - Stokes R J & Co Ltd
Ceramic floor, wall ties and terracotta floor tiles - Smithbrook Building Products Ltd
Ceramic marble and granite tiles - Reed Harris
Ceramic tile and stone installation systems - Laticrete UK
Ceramic tiles - Bell & Co Ltd
Ceramic tiles - British Ceramic Tile Ltd
Ceramic tiles - Swedecor Ltd
Ceramic, marble and glass mosaics, swimming pool tiles and porcelain stoneware - Domus Tiles Ltd
Ceramic, marble, granite tiles - Reed Harris
Ceramic, porcelain and mosaic tiles - Arnull Bernard J & Co Ltd
Ceramics - Wade Ceramics Ltd
Ceramics and mosaics - Shackerley (Holdings) Group Ltd incorporating Designer ceramics
Chemical Resistant Quarry Tile - Accrington Brick Tile Co Ltd
Concrete floor screeds - Armorex Ltd
Diamond Sawn -Yorkstone - Pickard Group
Faience blocks - Shaws of Darwen
Floor slabs and floor tiles - The Delabole Slate Co. Ltd
Floor systems - MiTek Industries Ltd
Floor tile separators - Compriband Ltd
Floor tiles - ceramic - H & R Johnson Tiles Ltd
Floor tiles - porcelain - H & R Johnson Tiles Ltd
Floor tiling - Fired Earth
Flooring - Bridge Quarry
Flooring - Burlington Slate Ltd
Flooring - Farmington Natural Stone Ltd
Flooring - Kirkstone
Flooring Slate - Allan Harris & Sons Ltd
Glass mosaic -sandstone - Reed Harris
Glazed wall tiles, unglazed floor tiles,swimming pool tiles and mosaics - Pilkington's Tiles Ltd
Grouts - Fosroc Ltd
Industrial floor tiles - Swedecor Ltd
Industrial tiling - Anti Corrosion Services Ltd
Interior flooring - Milner Delvaux Ltd
Keope Tiles - Italian Porecellian - Ruabon Dennis Ltd
Limestone quarry tiles - Reed Harris
Limestone, slate, quartzite, granite, sandstone, and marble - Naturestone
Marble and ceramic tiling - Marriott & Price Ltd
Marble and granite flooring - Diespeker Marble & Terrazzo Ltd
Marble tiles - Emsworth Fireplaces Ltd
Marble, granite, limestone, precast and insitu terrazzo mosaic - Fieldmount (Terrazzo) Ltd
Marble, granite, slate and limestone - Toffolo Jackson(UK) Ltd
Marble, granite, slate, limestone & terrazzo - Diespeker Marble & Terrazzo Ltd
Mosaic - Roman Mosaic Contracts Ltd
New Staining Silicon Joint for String Work - Laticrete UK
Quality tile adhesives - Ardex UK Ltd
Quality Tile Grout - Ardex UK Ltd
Quarry tiling - Ruabon Dennis Ltd
Rapid set high strength bedding mortars - Ronacrete Ltd
recycled glass promenade tiles - Spartan Tiles Ltd
Slate - Reed Harris
Specialist contractors in marble and granite mosaics - Alpha M & T Ltd
Sphinx Ttiles - Glazed Wall and Floor - Ruabon Dennis Ltd
Stone faced raised access flooring - Chelsea Artisans Ltd
Stone flooring - natural stone - Naturestone
Stone quarry - Rare Stone Group, The
Stone tiles - Kirkstone
Stone/ Concrete/ Quarry/ Ceramic tiling/ Mosaic - Network Flooring Cardiff
Stone/concrete/quarry/ceramic tiling/mosaic - Chilstone
Stone/concrete/quarry/ceramic tiling/mosaic - G. Miccoli & Son Ltd
Stone/concrete/quarry/ceramic tiling/mosaic - Spartan Tiles Ltd
Surface Preparation Equipment - Marcrist International Ltd
Swimming pool tiles - Swedecor Ltd
Terracotta - Reed Harris
Terracotta blocks - Shaws of Darwen
Terrazzo - Reed Harris
Tile fixing - Vallance
Tiles - Spital Tile Co Ltd
Tiling adhesives - Axter Ltd
Tiling cement for metal substrates - P.C.I Construction Systems Ltd
Tiling cements (adhesive) for natural stone, marble and ceramic tiles - P.C.I Construction Systems Ltd
Tiling Products - Sealocrete Ltd
Tiling Trims - Imagetrim (UK) Ltd
Unglazed ceramic floor tiles - Woolliscroft Tiles Ltd
Vessel & tank linings - SGL Acotec Ltd
Vibration tile system - SGL Acotec Ltd
Walkway tiles - Harris Road Roofing & Flooring Ltd
Wall and floor tiles - Royal Dutch Sphinx Ltd
Wall tiles - Fired Earth
Wall tiles - ceramic - H & R Johnson Tiles Ltd
Waqll tile fittings - ceramic - H & R Johnson Tiles Ltd
Wire Brushes - Marcrist International Ltd

M41 Terrazzo tiling/ In situ terrazzo

Marble, granite, limestone, precast and insitu terrazzo mosaic - Fieldmount (Terrazzo) Ltd
Precast terrazo - Roman Mosaic Contracts Ltd
Specialist contractors in terrazzo - Alpha M & T Ltd
Stainless Grout - Laticrete UK
Terracotta floor tiles - York Handmade Brick Co Ltd
Terrazzo - Marriott & Price Ltd
Terrazzo flooring - Diespeker Marble & Terrazzo Ltd
Terrazzo polishing and grinding - SGL Acotec Ltd
Terrazzo tile manufacturers - Toffolo Jackson(UK) Ltd
Terrazzo tiles (rubber) - William Freeman Ltd
Terrazzo tiling/ In situ terrazzo - Network Flooring Cardiff
Terrazzo, marble and mosiac fedaration - National Federation of Terrazzo Marble & Mosaic Specialists

M42 Wood block/ Composition block/ Parquet flooring

Commercial Wood Flooring - Tarkett Sommer Ltd
Dance floor - Felix Design Ltd
Floor sanding machines - Eureka Products Ltd
Flooring - Tarmac Top Floor
Flooring (leather) - Mach & Garbarino
HPL flooring - Abet Ltd
Laminated flooring - Kronospan Ltd
Parquet and laminate flooring - Western Cork Ltd
Solid & engineered parquet - Western Cork Ltd
Solid wood flooring - Fired Earth
Sports flooring systems and sprung flooring - Granwood Flooring Ltd
Wood block flooring - Campbell Marson & Co
Wood block/ Composition block/ Parquet flooring - Network Flooring Cardiff
Wood block/ composition block/ parquet flooring - Westco Group Ltd
Wood Flooring - Posners the floorstore
Wood flooring, - DLW Flooring Ltd
Wooden flooring - OSMO UK Ltd

M50 Rubber/ Plastics/ Cork/ Lino/ Carpet tiling/ sheeting

Accessories - Interfloor Limited
Aluminium edge trim for vinyl flooring - Howe-Green Ltd
Anti-Fatique Matting - Kleen-Tex Industries Ltd
Anti-slip surfaces - Redman Fisher Engineering
Bonded tiles - Carpets of Worth Ltd
Brush strips - Manton Industrial Seals Ltd
Carpet and tiles - Adam Carpets Ltd
Carpet fitting tools and accessories - Interfloor Limited
Carpet Protector - Hilton Banks Ltd
Carpet tiles - Brintons Ltd
Carpet tiles - Firth Interface Europe Ltd
Carpet tiles - Gaskell Carpet Tiles
Carpet tiles - Mackay Carpets, Hugh
Carpet tiles - Milliken Carpet
Carpet tiles - Rawson Carpets Ltd
Carpet tiles - Tankard
Carpet, cork and vinyl tiles and laminate flooring - Western Cork Ltd
Carpet, Vinyl and Laminate flooring - Posners the floorstore
Carpets - CFS Carpets
Carpets - HEM Interiors Group Ltd
Carpets - Opus 4 Ltd
Cellular rubber products - E+R Polymers Ltd
Commercial carpet tiles, broadloom and entrance clean off zones - Checkmate Industries Ltd
Commercial Resilient Flooring - Tarkett Sommer Ltd
Commercial Textile Flooring - Tarkett Sommer Ltd
Commercial, barrier, hospital carpet and tiles - MGL Carpet Finishers Ltd
Contamination control flooring - Dycem Ltd
Contamination control mats - Dycem Ltd
Contract carpet and carpet tile manufacturer - Burmatex Ltd
Cork - Olley & Sons Ltd, C
Cork flooring - Amorim (UK) Ltd
Cork, rubber and non slip surfaces - Tiflex Ltd
Custom mixing - Interfloor Limited
Data installations - HEM Interiors Group Ltd
Decoration and home improvements - Cement Glaze Decorators Ltd
Desk chair mats - Jaymart Rubber & Plastics Ltd
Discipative matting soloutions - Kleen-Tex Industries Ltd
Duckboard matting - Jaymart Rubber & Plastics Ltd
Dust control matting soloutions - Kleen-Tex Industries Ltd
Edging - Egger (UK) Ltd
Electrical Safety Matting - Jaymart Rubber & Plastics Ltd
Entrance barrier systems - Kleen-Tex Industries Ltd
Entrance matting - Gradus Ltd
entrance matting systems - Jaymart Rubber & Plastics Ltd
Entrance matting systems and fire retardant flooring - Tufton Ltd
Entrance matting, flooring and roof walkway matting - Plastic Extruders Ltd
Flameproof sheeting - Monarflex Anderson Ltd
Floor coverings - Hometex Trading Ltd
Floor protection coatings - Glixtone Ltd
Floor smoothing underlay - Ball F & Co Ltd
Floor trims - Gradus Ltd
Floorcoverings - Interface Europe Ltd
Flooring - HEM Interiors Group Ltd

M50 Rubber/ Plastics/ Cork/ Lino/ Carpet tiling/ sheeting (Cont'd)

Flooring accessories, stairnosings and PVC accessories - Quantum Profile Systems Ltd
Flooring adhesives - Ball F & Co Ltd
Flooring coatings - Williamson T & R Ltd
Flooring products - Weber Building Solutions
Heavy duty carpet tiles - Paragon by Heckmondwike
High pressure decorative laminates - Abet Ltd
Hygiene matting - Kleen-Tex Industries Ltd
Hygienic walls and ceiling cladding systems - Advanced Hygienic Contracting Ltd
Interlocking tiles - Jaymart Rubber & Plastics Ltd
Laminate flooring - Westco Group Ltd
Lettered & logo mats - Jaymart Rubber & Plastics Ltd
Lino and PVC flooring - DLW Flooring Ltd
Linoleum - Tarkett Sommer Ltd
Linoleum and vinyl floorcoverings - Forbo - Nairn Ltd
Natural floor covering - Fired Earth
Non-slip floor coatings - Andura Textured Masonry Coatings Ltd
Non-slip matting soloutions - Kleen-Tex Industries Ltd
Nylon reinfoced PVC sheeting - Flexible Reinforcements Ltd
Plastic coatings - Plastic Coatings Ltd
Plastic roofing, cladding, and glazing - Amari Plastics Plc
Plastics/ cork/ lino/ carpet tilling/ sheeting - Westco Group Ltd
Polyurethane re-bond underlays - Interfloor Limited
PTFE coatings - Plastic Coatings Ltd
Publications: contract flooring association guide to contract flooring - Contract Flooring Association
PVC floorcoverings and accessories - Polyflor Ltd
PVC foam strip - Manton Industrial Seals Ltd
PVC sheeting - Flexible Reinforcements Ltd
Realwood HPL Decorative Laminates - Abet Ltd
Recycled flooring products - Jaymart Rubber & Plastics Ltd
Room Improvement Products - nmc (UK) Ltd
Rubber flooring - Altro Floors
Rubber flooring - DLW Flooring Ltd
Rubber flooring - William Freeman Ltd
Rubber flooring systems - Jaymart Rubber & Plastics Ltd
Rubber foam strip - Manton Industrial Seals Ltd
Rubber linings - SGL Acotec Ltd
Rubber sheeting and expansion joints & bellows - MacLellan Rubber Ltd
Rubber sheeting and flooring - Linatex Ltd
Rubber sheets - Manton Industrial Seals Ltd
Rubber stud tiles - Polyflor Ltd
Rubber tiles for floors and other surfaces in a huge range of colours and textures - Dalsouple Direct Ltd
Rubber/ plastics/ cork/ lino/ carpet tiling/ sheeting - Bonar Floors Ltd
Rubber/ Plastics/ Cork/ Lino/ Carpet tiling/ sheeting - Network Flooring Cardiff
Rubber/plastics/cork/lino/carpet tiling/sheeting - Kemlite Ltd
Rugs - Brintons Ltd
Safety floorings - Jaymart Rubber & Plastics Ltd
Self-adhesive surface coverings - Rustin Allen Ltd.
Sheeting and matting - Interfloor Limited
Slip resistant flooring - Altro Floors
Smooth vinyl flooring - Altro Floors
Specialised flloring materials, nosing and adhesives - Tiflex Ltd
Sponge rubber underlays (carpets/wood & laminate) - Interfloor Limited
Sports flooring - Jaymart Rubber & Plastics Ltd
Sports Surfaces - Altro Floors
Stair treads, nosing and matting - The Safety Tread Ltd
Underlay and flooring - Interfloor Limited
Vinyl floor coverings - Dermide Ltd
Vinyl floorcovering sheets, tiles and accessories - Tarkett-Marley Floors Ltd
Vinyl flooring - Gerflor Ltd
Vinyl sheet and tile flooring - DLW Flooring
Wall furnishings - Vescom (UK) Ltd
Wetwall - Tarkett Sommer Ltd

M51 Edge fixed carpeting

Broadloom - Tankard
Carpet - Egetaepper (UK) Ltd
Carpet - Hometex Trading Ltd
Carpet (office/Domestic) - Evergreens Uk
Carpet for domestic and contract use - Carpets of Worth Ltd
Carpeting - Davies John Interiors Ltd
Carpets - CFS Carpets
Carpets - Foster C G & Sons, (Contract Soft Furnishers)
Carpets - Ulster Carpet Mills Ltd
Commercial carpet tiles, broadloom and entrance clean off zones - Checkmate Industries Ltd
Commercial, barrier and hospital carpets and tiles - MGL Carpet Finishers Ltd
Contract - Tankard
Contract flooring manufacturers - Rawson Carpets Ltd
Custom carpets - Ulster Carpet Mills Ltd
Decoration and home improvements - Cement Glaze Decorators Ltd
Domestic and contract locations - Adam Carpets Ltd
Edge fixed carpeting - Bonar Floors Ltd
Edge fixed carpeting - Brockway Carpets Ltd
Edge fixed carpeting - Carpets International (UK) Ltd
Edge fixed carpeting - Firth Interface Europe Ltd
Edge fixed carpeting - Mackay Carpets, Hugh
Edge fixed carpeting - Network Flooring Cardiff
Edge fixed carpeting - Quantum Profile Systems Ltd
Edge fixed floorcoverings - Interface Europe Ltd
Fibre bonded floor coverings - Rawson Carpets Ltd
Fitted Carpets - Posners the floorstore
Floorcovering accessories - Gripperrods Ltd
Heavy duty carpets - Paragon by Heckmondwike
Surface membranes and release systems for carpet - Laybond Products Ltd
Wool/Nylon - Tankard
Wool-rich Natural products - Ulster Carpet Mills Ltd
Woven and tufted carpet - Brintons Ltd
Woven axminster carpets - Ulster Carpet Mills Ltd
Woven Wilton and Axminster carpet manufacturers - Tankard
Woven wilton carpets - Ulster Carpet Mills Ltd

M52 Decorative papers/ fabrics

130cm wide fabric backed vinyls - Dixon Turner Wallcoverings
135cm wide fabric backed vinyls - Dixon Turner Wallcoverings
53cm x 10m paper backed vinyls - Dixon Turner Wallcoverings
Abrasive papers & decorating sundry products - Acton & Borman
Acrylic coatd paper - Package Products Ltd
Authentic tribal rugs - Fired Earth
Blown vinyl wallcoverings - Dermide Ltd
Contract wall coverings - Muraspec
Contract wallcoverings - Dixon Turner Wallcoverings
Decoration and home improvements - Cement Glaze Decorators Ltd
Decorative finishes - Glixtone Ltd
Decorative finishes - Weber Building Solutions
Decorative Laminates - Viking Laminates Ltd
Decorative papers/ fabrics - Brewer C & Sons Ltd
Decorative wallcoverings, fabrics and borders - Dixon Turner Wallcoverings
Duplex wallcoverings - Dermide Ltd
Fabric backed wall coverings - Muraspec
Fabrics - Tektura Plc
Fabrics and papers - Interface Europe Ltd
Fabrics and wall papers - Today Interiors Ltd
Fabrics and wallpapers - Sanderson
Glass fibre - Muraspec
Hessian - Muraspec
Interior finishes - Muraspec
Lining papers - Dermide Ltd
Linolem finish material for desks - Forbo - Nairn Ltd
Metallic finishes - Muraspec
Notice board material - Forbo - Nairn Ltd
Paint alternatives - Muraspec
Paint stripper, wall paper stripping machines - Halls Beeline Ltd
Paints and wallcoverings - Kalon Decorative Products
Paperbacked wall coverings - Muraspec
Primers - Tektura Plc
PVC coated fabrics - Flexible Reinforcements Ltd
Simplex wallcoverings - Dermide Ltd
Textile wallcoverings - Dixon Turner Wallcoverings
Textiles - Muraspec
Textured coatings - Glixtone Ltd
Timeless fabrics - Fired Earth
Vinyl wallcoverings - Forbo (Lancaster) Ltd
Wallcoverings - Muraspec
Wallcoverings - Tektura Plc
Wallcoverings and adhesives - Architectural Textiles Ltd
Wholesalers of furnishing fabrics & wallpapers - CrowsonFrabrics Ltd
Woodchip papers - Dermide Ltd

M60 Painting/ Clear finishing

Agricultural paints - Teal & Mackrill Ltd
Air drying enamels - Crosbie Coatings
Anti - static coatings - Kenyon Group Ltd
Anti carbonation coatings - Ronacrete Ltd
Anti Condensation Paints - Witham Oil & Paint Ltd
Anti corrosion coatings - Corroless Corrosion Control
Anti corrosion coatings - Kenyon Group Ltd
Anti corrosive primers - Crosbie Coatings
Anticlimb paint - Coo-Var Ltd
Anti-condensation paints - Pearl Paints Ltd
Antifouling paints - Teal & Mackrill Ltd
Antigraffiti coatings - Coo-Var Ltd
Anti-vandal paints - Pearl Paints Ltd
Bitumen and aluminium bit paint - Liver Grease Oil & Chemical Co Ltd
Bitumen paints - Pearl Paints Ltd
Black varnish - Liver Grease Oil & Chemical Co Ltd
Blast cleaning - Gilmer Limited
Blasting - Bristol Metal Spraying & Protective Coatings Ltd
Brushing tar, creosote, tallow and linseed oil - Legge Thompson F & Co Ltd
Cadmium pigments - Johnson Matthey Pigments & Dispersions
Cement paint - Hunting Industrial Coatings Ltd
Cleaners for paving - Resiblock Ltd
Coal tar creosote & pitch - Liver Grease Oil & Chemical Co Ltd
Coatings for concrete floors - Everlac (GB) Ltd
Colour Schemes - Kalon Decorative Products
Complex inorganic pigments - Johnson Matthey Pigments & Dispersions
Decorating - Davies John Interiors Ltd
Decorating materials - Brewer C & Sons Ltd
Decoration - HEM Interiors Group Ltd
Decoration, painting and home improvements - Cement Glaze Decorators Ltd
Decorative ancillary products - Bartoline Ltd
Decorative coatings for timber - Protim Solignum Ltd
Decorative finishes - ICI Paints
Decorative paints - Akzo Coatings Plc
Decorative paints, opaque and translucent wood stains - Jotun Henry Clark Ltd Decorative Division
Decorative Sundries - Kalon Decorative Products
Distributors of painters' overalls, abrasive papers, masking tape and painters' hand tools - Marshall J A & Co (Brushes) Ltd
Durable, decorative interior coatings - Andura Textured Masonry Coatings Ltd
Elastomeric, anticarbonation coatings - Andura Textured Masonry Coatings Ltd
Electroplating, powder coating and painting - Ritherdon & Co Ltd
Epoxy floor/wall coatings - Conren Ltd
Epoxy paint systems - Pearl Paints Ltd
Epoxy paints - Trade Sealants
Epoxy resin coatings - Anglo Building Products Ltd
Epoxy resins - Addagrip Surface Treatments Ltd
Epoxy systems - Teal & Mackrill Ltd
Exterior woodcare - Blackfriar Paints Ltd
Fillers - ICI Paints
Fillers - Sadolin UK Ltd
Fine and coarse textured coatings - Andura Textured Masonry Coatings Ltd
Finishing - Taskworthy Ltd
Fire Rated Structural Paint - Abesco Ltd
Fire retardant treatment - Palgrave Brown Timber & Fireproofing Co
Flame retardant paints - Bollom Fire Protection Ltd
Flame retardant paints - The Bollom Group
Flame retardent coatings - Pearl Paints Ltd
Flame retardent coatings - Rentokil Initial U.K. Ltd.
Floor - road line paints - Pearl Paints Ltd
Floor and line paint - Coo-Var Ltd
Floor coating systems - Corroless Corrosion Control
Floor coatings - Crosbie Coatings
Floor coatings - Firwood Paints Ltd
Floor coatings, self levellers - Kenyon Group Ltd
Floor Paint s - Witham Oil & Paint Ltd
Floor paints - Kingstonian Point Ltd
Floor paints - Teal & Mackrill Ltd
Floor seals - Evans Vanodine International PLC
Fluorescent Systems - Witham Oil & Paint Ltd
Glass flake reinforced coatings - Corroless Corrosion Control
High performance coatings - Leigh's Paints
High performance paints and specialist coatings - Stewart Wales, Somerville Ltd
Hygiene coatings - Kenyon Group Ltd
Hygienic coatings - Corroless Corrosion Control
Hygienic coatings for walls & ceilings - Tefcote Surface Systems
Industrial Coating Specialists - Bristol Metal Spraying & Protective Coatings Ltd
Industrial coatings - Crosbie Coatings
Industrial paints - Firwood Paints Ltd
Industrial paits - Teal & Mackrill Ltd
Interior coatings - Jotun Henry Clark Ltd Decorative Division
Interior varnishes - Sadolin UK Ltd
Intumescent coatings - Bollom Fire Protection Ltd
Lacquers and seals - Junckers Limited
Lime putty - Legge Thompson F & Co Ltd
Long life exterior wall coating - Everlac (GB) Ltd
Luminous & fluorescent paints - Coo-Var Ltd
Lump and french chalk - Legge Thompson F & Co Ltd
Maintanance Painting service - International Protective Coatings,
Maintenance paints - Firwood Paints Ltd
Manufacturers of paint brushes, paint rollers and dust sheets - Marshall J A & Co (Brushes) Ltd
Marine & antifouling coatings - Coo-Var Ltd
Marine paints - Witham Oil & Paint Ltd
Marine paints and coatings - Teal & Mackrill Ltd
Marking equipment - Elliott John H (Monostamp) Ltd
Masonry paint - Silco color - STO Ltd
Metal - Bristol Metal Spraying & Protective Coatings Ltd
Opaque finishes - Sadolin UK Ltd
Paint - Fired Earth
Paint - Sanderson
Paint and powder manufacturer - Howse Thomas Ltd
Paint brushes, rollers and decorating tools - Hamilton Acorn Ltd
Paint effects - Kingstonian Point Ltd
Paint finishes for timber doors and windows - Wagner (GB) Ltd
Paint spraying equipment - Dawson McDonald & Dawson Ltd
Painting / clear finishing - Tor Coatings Ltd
Painting /Clear Finishing - Wessex Resins & Adhesives Ltd
Painting aids - Polycell Products Ltd
Painting/ clear finishing - Concrete Repairs Ltd
Painting/ clear finishing - Coo-Var Ltd
Painting/ Clear finishing - Don Construction Products Ltd
Painting/ clear finishing - Liquid Technology Ltd
Painting/ clear finishing - Wood E Ltd
Painting/clear finishes - TH Kenyon & Sons Ltd
Painting/Clear Finishing - International Protective Coatings,
Painting/clear finishing - Tefcote Surface Systems
Paints - Crown Trade Paints
Paints - Dexter Paints Ltd
Paints - Emusol Products (LEICS) Ltd
Paints - Herberts, Division of HPG
Paints - Hird Hastie Paints Ltd
Paints - HMG Paints
Paints - ICI Dulux Trade Paints
Paints - Manders Paint Ltd
Paints - Surmak Products Ltd
Paints - The Bollom Group
Paints - Weber Building Solutions
Paints & coatings - Sigma Coatings Ltd
Paints and coatings for tennis courts and other sports surfaces - Everlac (GB) Ltd
Paints and high temperature resistant coatings - Fortafix Ltd
Paints and surface coatings - Dacrylate Paints Ltd
Paints primers and specialized coatings - Coo-Var Ltd
Paints, wood finishes, solvents and chemicals - Anderson Gibb & Wilson
Partnering - Kalon Decorative Products
Pigments for the surface coatings - Johnson Matthey Pigments & Dispersions
Polish and sealers - British Nova Works Ltd
Polyurethane floor paint - IFT Supplies Ltd
Powder and ready mixed fillers - Halls Beeline Ltd
Preparation products - Blackfriar Paints Ltd
Protective and decorative masonry coatings - Andura Textured Masonry Coatings Ltd
Protective coating for steelwork - International Protective Coatings,
Protective coatings - Andrews Coatings Ltd
Protective epoxy resin coatings to concrete - Anti Corrosion Services Ltd
Protective maintenance coatings - Corroless Corrosion Control
Rust convertors - Pearl Paints Ltd
Rust stabilising primers - Corroless Corrosion Control
Seals and coatings for wood floors - Everlac (GB) Ltd
Slurry Granules - Bayer Chemicals Inoegate Pigment Group
Solvent bourne coatings - Sadolin UK Ltd
Specialist Primers - Witham Oil & Paint Ltd
Specialised paints - Teal & Mackrill Ltd
Specialist paint systems - Williamson T & R Ltd
Specialist Paints Varnishes and Woodcare Products - Blackfriar Paints Ltd
Specialist Water Based Coattings - Witham Oil & Paint Ltd
Standard, paint quality and fair face finishes - Tarmac Topblock Ltd
Stencils, branding tools and ink marking - Elliott John H (Monostamp) Ltd
Structural painting and spraying - Masonary Cleaning Services
Surface dressings - Vallance
Surface dressings - Decorating products - Sealocrete Ltd
Swimming Poll Paints - Witham Oil & Paint Ltd
Technical Specifications - Kalon Decorative Products
Tennis court spraying - Sportmark™ Ltd
Timber preservatives, fire retardants - decorative finishes - Arch Timber Protection
Translucent finishes - Sadolin UK Ltd
Transparent ironoxide pigments - Johnson Matthey Pigments & Dispersions
Twin Pack Paints - Witham Oil & Paint Ltd
Two Pack Floor Paints - Coo-Var Ltd
Vapour corrosion inhibitors - Corroless Corrosion Control
Wall and floor coatings - Sealocrete Ltd
Waterbourne coatings - Sadolin UK Ltd
Weatherproof decorative wall coatings and hygiene finishes - Liquid Plastics Ltd

M60 Painting/ Clear finishing (Cont'd)

Wet paint - Plastic Coatings Ltd
Wood finishes - The Bollom Group
Wood finishes, speciality paints, paint removers and decorating sundries - Rustins Ltd
Wood protection products - Sadolin UK Ltd
Woodfinishes - Smith & Rodger Ltd
Woodstain - Cuprinol Ltd

M61 Intumescent coatings for fire protection of steelwork

Fire protection coatings for steelwork - Thermica Ltd
Fire Retart and Coatings - Bristol Metal Spraying & Protective Coatings Ltd
Hyclad matt finish - AvestaPolarit Ltd
Industrial Coating Specialists - Bristol Metal Spraying & Protective Coatings Ltd
Intermescant coatings for fire protection of steelwork - International Protective Coatings,
Intumescant coatings - Andrews Coatings Ltd
Intumescant coatings for fire protection of steelwork - Promat Fire Protection Ltd
Intumescent coating, protecting steel from fire - Smith & Rodger Ltd
Intumescent coatings for steel - Quelfire
Nullifire intumescent range - Nullifire Ltd
Passive fire protection products - Leigh's Paints
Structural steel fire protection - Grace Construction Products Ltd

N10 General fixtures/ furnishings/ equipment

Accommodation furniture and electronic workbenches - Selbix Joinery Ltd
Acoustic telephone enclosure - Burgess Architectural Products Ltd
Aluminum shelving - RB UK Ltd
Anti fatigue mats - Dycem Ltd
Architectural antiques - Andy Thornton Ltd
Archive research - Sanderson
Autofiler - Rackline Systems Storage Ltd
Awnings - Tidmarsh & Sons
Bamer Carpeting/Matting - Floor Protection Services Ltd
Banks of post boxes - Signs Of The Times Ltd
Basket units - Smart F & G (Shopfittings)Ltd
Bathroom cabinets and tube fittings - MacBee Manufacturing Ltd
Bathroom mirrored cabinets and mirrors with lighting - Schneider (UK)
Bathrooms - Rosa Paula Kitchens
Baulstrading - Spindlewood
Bedroom & home study furniture - Ball William Ltd
Bedroom fittings - Bernstein Group PLC
Bedroom Furniture - Lakeside Group
Bedroom furniture - Magnet Ltd
Bedroom furniture - OEP Furniture Group PLC
Bedspreads - Sanderson
Benches - Detlef Muller Ltd
Benching - Envopak Group Ltd
Bespoke Mirrors - Opals (Mirror-Flex) Co Ltd
Bespoke museum display cases - Benbow Interiors
BioCote - Allermuir Contract Furniture Ltd
Blackout/ sunscreen blinds - Tidmarsh & Sons
Blind and awning fabrics - Flexible Reinforcements Ltd
Blinds - Design Line
Blinds - Durable
Blinds - Durable Berkeley Co Ltd
Blinds - Eclipse Blind Systems Ltd
Blinds - Foster C G & Sons, (Contract Soft Furnishers)
Blinds - Hallmark Blinds Ltd
Blinds - Invicta Window Films Ltd
Blinds - Opus 4 Ltd
Blinds - VELUX Co Ltd, The
Blinds and curtains - Contravent Regal Ltd
Blinds and shutters - Andersons Ltd
Blinds in sealed units - Pilkington Plyglass
Blinds, curtains, tracks - Sunrite Blinds Ltd
Block & sheet polyester, bonded products - Rawson Fillings Ltd
Bomb curtains and blankets - Meggitt Armour Systems
Bracketry - Vantrunk Engineering Ltd
Broadcast and audio visual products services - Streamtec Limited
Brooms, brushes, replacement handles - Wessenden Products
Brush strip cable management - Kleeneze Sealtech Ltd
Brushes - Industrial Brushware Ltd
Builders merchant - First Stop Builders Merchants Ltd
Building materials and DIY products - First Stop Builders Merchants Ltd
Cabinet Hinge Manufacturers - Daro Factors Ltd
Canteen and office furniture for cabins - Ableson L & D Ltd
Caravan security boxes - Ritherdon & Co Ltd
Card/ magazine racks - Smart F & G (Shopfittings)Ltd
Carpet Tile - Floor Protection Services Ltd
Cash and security boxes - Securikey Ltd
Cast aliminium mail boxes - Signs Of The Times Ltd
Castors and wheels - Flexello Limited
Changing room furniture - Helmsman
Chemicals and plastics - producer and distributor - Atofina UK Ltd
Chimney sweeping equipment - Wessenden Products
Choir risers - Felix Design Ltd
Chrome shelving - RB UK Ltd
Cinemas fittings - Withey Contracts Ltd
Clothes Posts - Hills Industries Ltd
Coal and log effect gas fires - Verine Ltd
Coir matting & logo - Floor Protection Services Ltd
Conservatory Blinds - Amdega Ltd
Conservatory blinds - Tidmarsh & Sons
Contract fruniture - Taskworthy Ltd
Contract furniture - Metalliform Products Plc
Contract furniture - Ness Furniture Ltd
Contract hardwood seating - Branson Leisure
Contract office and leisure furniture - Zon International
Contract Residential Furniture - Ellis J T & Co Ltd
Contract seat street/park furiture - Branson Leisure
Contract seating - Laytrad Ltd
Contract seating & tables - Knightbridge Furniture Productions Ltd
Contract tables - Laytrad Ltd
Contract/letail beds - Blindcraft Industries
Cosmetic mirrors - Opals (Mirror-Flex) Co Ltd
Cotton & wool felt - Rawson Fillings Ltd
Curtain fabrics - Eclipse Blind Systems Ltd
Curtain poles - Rothley Burn Ltd
Curtain poles - Silent Gliss Ltd
Curtain tracks, cubicle rails and window blinds - electric & manual operation - Silent Gliss Ltd
Curtain tracks, curtains and blinds - Marshall Contracts Ltd
Curtains - Foster C G & Sons, (Contract Soft Furnishers)
Curtains - Sanderson
Cushion pads - Elizabethan Textiles
Custom design staging - Felix Design Ltd
Custom mouldings - Harrison Thompson & Co Ltd
Customer guidence and queue management systems - Tensator Ltd
Customised metal trims in all metals - Azimex Fabrications Ltd
Customized panel work - Ritherdon & Co Ltd
Decorative grilles - Taskworthy Ltd
Decorative melamine faced chipboard - Egger (UK) Ltd
Decorative mouldings - Rustin Allen Ltd.
Desk and door plates - Vision Neon & Display Ltd
Desk lighting - Anglepoise Lighting Ltd
Desks and furniture in metalwork, glass, marble and granite - Howard Bros Joinery Ltd
Desks and seating - Mauser Interiors (UK) Ltd
Domestic bedroom furniture - Ball William Ltd
Doors - BLP Group
Drop down grab rails - Cefndy Healthcare
Electronic fly killers - P & L Systems Ltd
Entrance matting - Bonar Floors Ltd
Entrance Matting - EMS Entrance Matting Systems Ltd
Entrance matting - Norton Engineering Alloys Co Ltd
Entrance matting - Shackerley (Holdings) Group Ltd incorporating Designer ceramics
Enviropol products - Glasdon UK Ltd
Equestrian arenas and products pvc-u - Duralock (UK) Ltd
Equipment cabinets - Arbory Group Limited
External blinds - Tidmarsh & Sons
Fabrication service - MICE Kaymar Limited
Filing and storage cabinets - Railex Systems Ltd, Elite Division
Fine surfaced furniture chip board - Egger (UK) Ltd
Finger posts - Signs Of The Times Ltd
Fire alarm glasses - Opals (Mirror-Flex) Co Ltd
Fire blankets - Rentokil Initial U.K. Ltd.
Fire extinguishers - Asco Extinguishers Co Ltd T/A The Asco Fire Group
Fire extinguishers - Chubb Fire Ltd
Fire extinguishers - Fife Fire Engineers & Consultants Ltd
Fire extinguishers - Firetecnics Systems Ltd
Fire extinguishers - Griffin and General Fire Services Ltd
Fire extinguishers - Kidde Fire Protection Services Ltd
Fire extinguishers - Richard Hose Ltd
Fire fighting equipment - Eagles, William, Ltd
Fire resistant fabrics - Mermet U.K
Fireplace frames and canopies - James Smellie Fabrications Ltd
Fireplace service - Cico Chimney Linings Ltd
Fireplaces - Bridge Quarry
Fireplaces - Emsworth Fireplaces Ltd
Fireplaces - Wyvern Marlborough Ltd
Fireplaces and fires - Bell & Co Ltd
Fireplaces, surrounds, suites and hearths - Winther Browne & co Ltd
Fitted Bedrooms - Hammonds - New Home Division
Fitted Bedrooms - Symphony Group PLC, The
Fitting & Installation - Sanderson
Fittings - Kee Klamp Ltd
Flags, flagstaffs, banners and bunting - Zephyr The Visual Communicators
Flat end grab rails - Cefndy Healthcare
Fragrance dispensers - P & L Systems Ltd
Frames Historical Photographs - Francis Frith Collection
Free standing notice boards - Vision Neon & Display Ltd
Furniture - Andy Thornton Ltd
Furniture - Bevan Funnell Ltd
Furniture - Chase James & Son (Furnishing) Ltd
Furniture - Ducal Ltd
Furniture - Fired Earth
Furniture - Moores Furniture Group Ltd
Furniture - Reclaim.UK.Com
Furniture - Sanderson
Furniture - Welconstruct Co Ltd
Furniture fittings - Nico Manufacturing Ltd
Furniture fittings and accessories - Haffele UK Ltd
Furniture for restuarants and catering facilities - Sandler Seating Ltd
Furniture for schools, public buildings, hotels and restaurants - AEL Furniture Group
Furniture lifting device - Interface Europe Ltd
Furniture manufacturers and contract furnishers - Pulsford Associates Ltd, John
Furniture manufacturers and contract furnishers - Pulsford Associates Ltd, John
Furniture, fittings and hardware - manufacturers and importers - Daro Factors Ltd
Galvanized dustbins - Brettell & Shaw Ltd
Garden canes, ironmongery, woodware, dowels, mouldings, garden sundries - Wessenden Products
Garden sprayers - Hills Industries Ltd
Garment rails/shoe fittings - Smart F & G (Shopfittings)Ltd
Gas fires - Flavel - Leisure Glynwed Consumer & Construction Products Ltd
Geeral fixture/ furnishings and equipment - Hipkiss, H, & Co Ltd
General fixtures - Anders + Kern UK Ltd
General fixtures and furnishings - Be-Modern Ltd
General fixtures and furnishings - Elite Trade Kitchens Ltd
General fixtures/ furnishings/ equipment - Barlow Group
General fixtures/ furnishings/ Equipment - Laidlaw Architectural Hardware
General fixtures/furnishings/ equipment - Libraco
general fixtures/furnishings/equipment - Paragon Business Furniture
General furniture - Ellis J T & Co Ltd
General furniture and perspex furniture - The Cube
General purpose fittings and furnishings - Hettich UK Ltd
Generalfixtures / furnishings/ equipment - Architectural Components Ltd
Glass & Mirror Processors - Opals (Mirror-Flex) Co Ltd
Glass fibre UPVC fabrics - Mermet U.K
GRP customized moldings - Wilsons Fibreglass Ltd
GRP enclosures and kiosks - Dewey Waters Ltd
Handling equipment - Welconstruct Co Ltd
Hardwood worktops & bartops - WL West & Son Limited
Hinges - Cooke Brothers Ltd
Home furniture - Vitra UK Ltd
Hotel bedroom Furniture - Ellis J T & Co Ltd
Hotel bedroom furniture and contract furniture - Pedley Furniture International Ltd
Hotel fittings - Withey Contracts Ltd
Hotel furniture - Laytrad Ltd
Hotel furniture - TFL International Ltd
Hotel Industry - TFL International Ltd
Hotel security boxes - Ritherdon & Co Ltd
Import furniture - Kush UK Ltd
Instrument cabinet - Ruthin Precast Concrete Ltd
Interchangable notice boards - Vision Neon & Display Ltd
Interior and exterior blinds and awnings - Deans Blinds & Awnings (UK) Ltd
Interior Design Mail Order Catalogue - Cube Collection Ltd
Interior Design Service - Cube Collection Ltd
Interior fit outs - OEP Furniture Group PLC
Internal litter bins - Glasdon UK Ltd
Iron and aluminium castings - Cradley Castings Ltd
Iso-design venetian blinds - Eclipse Blind Systems Ltd
Key filing, security cabinets and safes - Securikey Ltd
Kiosks - Fibaform Products Ltd.
Kitchen furniture - OEP Furniture Group PLC
Leaflet holders - Smart F & G (Shopfittings)Ltd
Leather for walling/upholstry - Mach & Garbarino
Letter Boxes - Signs Of The Times Ltd
Lifestyle user friendly shelving kits - RB UK Ltd
Lightweight fire barriers - Rentokil Initial U.K. Ltd.
Local history books - Francis Frith Collection
Lockers - Decra Ltd
Lockers and cubicles - Shackerley (Holdings) Group Ltd incorporating Designer ceramics
Loft hatches - Zled (DPC) Co Ltd
Made-to-measure furnishings - Sanderson
Manufacturers of display & exhibition systems - Nimlok Ltd
Marble vanity tops - Emsworth Fireplaces Ltd
Material handling equipment - Moresecure Ltd
Mats and matting - Floor Protection Services Ltd
Matwell frames - Floor Protection Services Ltd
Matwell frames and entrance matting - Clean Step Ltd
Mechanical extract units - Greenwood Air Management
Metal furniture - H&H Electronics Ltd
Mirror wall covering - flexible material - Opals (Mirror-Flex) Co Ltd
Mirrors - Ledite Glass (Southend) Ltd
Mirrors - Lightfoot Charles Ltd
Mobile benching - Lab Furnishings (UK) Ltd
Modular shelving systems - Screen Systems Ltd
Moulded grab rails - Cefndy Healthcare
Nameplates and fireplaces - The Delabole Slate Co. Ltd
Natural & polyester fillings for bedding & upholstery - Rawson Fillings Ltd
Natural ventilation products - Greenwood Air Management
Non slip bench mat - Dycem Ltd
Nostalgic gift products - Francis Frith Collection
Notice boards - Arbory Group Limited
Odour control products - P & L Systems Ltd
Officefitting - Withey Contracts Ltd
P.O.S support - Eclipse Blind Systems Ltd
Pallet racking, shelving systems and lockers - Sperrin Metal Products Ltd
Panel glides - Silent Gliss Ltd
Parafin pressure lamp - Tilley International Ltd
Personnel equipment - Welconstruct Co Ltd
Pillows: natural & synthetic - Elizabethan Textiles
Pinboards - Vision Neon & Display Ltd
Pine beds, bunk beds and headboards - Airsprung Beds Ltd
Pine room paneling, bookcases and display cupboards - Hallidays UK Ltd
Plastic wheeled bins - Brettell & Shaw Ltd
Portable auditorium seating - Sandler Seating Ltd
Portable gas cookers - Tilley International Ltd
Poster frames - Vision Neon & Display Ltd
Press tooling - Cooke Brothers Ltd
Pressings - Cooke Brothers Ltd
Product finishing - Cooke Brothers Ltd
Promotional products - Marketing & Promotions Ltd
Public area bins, ashtrays and fire safe bins - Lesco Products Ltd
Public Area Furnishings - TFL International Ltd
PVC strip curtains - Mandor Engineering Ltd
Radiator casings - T B S Fabrications
Ready made curtain - Eclipse Blind Systems Ltd
Reconstructed stone fireplaces - Minsterstone Ltd
Recycled paper and board for the building and packaging industries - BPB Paperboard
Refuse sack holders and fire retardant bins - Unicorn Containers Ltd
Roller blinds - Reflex-Rol (UK)
Roller blinds - Tidmarsh & Sons
Roller/roman/venetian/vertical blinds - Silent Gliss Ltd
Ropes and twines - Kent & Co (Twines) Ltd
Rose end grab rails - Cefndy Healthcare
sackholders - Surrey Sackholders Ltd.
Sanibins - Surrey Sackholders Ltd.
Screens & decrative panels - Jali Ltd
Sculptures - Hynes Aden K Sculpture Studios
Sealer units - Pilkington Plyglass
Seating systems - Harrison Thompson & Co Ltd
Security counters & screens - Platonoff & Harris Plc
Security mirrors - Securikey Ltd
Self retracting key reels - Securikey Ltd
Service desking - Grant Westfield Ltd
Shatterproof safety mirrors - Deralam Laminates Ltd
Sheet metal enclosures - Ritherdon & Co Ltd
Shelves, panels - Anderson GEC Ltd
Shelving - Barton Storage Systems Ltd
Shelving - RB UK Ltd
Shelving - Smart F & G (Shopfittings)Ltd
Shelving system - Lab Furnishings (UK) Ltd
Shelving systems - Moresecure Ltd
Shoot bolt Locking Mechanisms - Kenrick Archibald & Sons Ltd
Showroom design - Eclipse Blind Systems Ltd
Slotted angle plastic containers and boxes - Moresecure Ltd
Slotvents - Greenwood Air Management
Soft fabrics - Eclipse Blind Systems Ltd
Soft Furnishing - TFL International Ltd
Solar blinds and tension systems - EBS Services Ltd
Solar blinds material - Reflex-Rol (UK)
Solar reflective fabrics - Mermet U.K
Solar shading - Levolux A.T. Limited
Solid surface fabrication - Grant Westfield Ltd
Solid surfaces - Viking Laminates Ltd
Special Stainless Fabrications - Pland Stainless Ltd
Sprung mattresses, divan bases, pillows and accessories - Airsprung Beds Ltd

N10 General fixtures/ furnishings/ equipment (Cont'd)

Square tube construction systems - Moresecure Ltd
Stainless steel general fixings - Associated Metal (Stainless) Ltd
Stainless steel post boxes - Signs Of The Times Ltd
Stainless steel, aluminium, PVC covered, galvanized and brass handrails - SG System Products Ltd
Stainless steel, cabinets, worktops, mirrors - Anderson GEC Ltd
Steel cabinets - Baddeley
Steel cabinets - H&H Electronics Ltd
Steel shelving uprights and bracketry - RB UK Ltd
Steel tube and fittings - Rothley Burn Ltd
Storage - Rothley Burn Ltd
Storage cabinets - Richard Hose Ltd
Storage wall equipment - ARC Partitioning
Study, lounge, bedroom, office, reception fitted furniture - Hyperion Wall Furniture Ltd
Sunbeds, lockers and cubicles - Dalesauna Ltd
Sunblinds - Sampson T F Ltd
Support rails and systems - Amilake Southern Ltd
Synthetic grass for display - Palmyra Ltd
Table protector - Elizabethan Textiles
Themed features (any material) - Design & Display (Sales) Ltd
Thin processed mirrors and clear glass - Opals (Mirror-Flex) Co Ltd
Track systems and rigging equipment for theatre applications - Harkness Hall Ltd
Training rooms installations - Planet Communications Group Ltd
Trolley - Unicorn Containers Ltd
TV audio brackets - Spur Shelving Ltd
Vanity bathroom units - Topravit (UK) Ltd
Vanity units - Architectural & Building Products Ltd
Vending machines - Unicorn Containers Ltd
Venetian blinds - Komfort Office Environments Plc
Vertical and horizontal carousels - Dexion Ltd
Waiting area seating for train stations, airports, ferry terminals, hospitals and surgeries - Hille Ltd
Wall Murals - Francis Frith Collection
Wall rails - Fixatrad Ltd
Waste bins - Surrey Sackholders Ltd.
Weighing scales and weigh bridges - Phoenix Scales Ltd
Window blind systems, fabrics - vertical, roller, venetian, pleated, woven wood, contract/perofrmance - Louver-Lite Ltd
Window blinds - manufacturers and installers for the construction industry - Hunter Douglas Ltd (Contracts Division) Luxaflex Blinds ®
Window blinds and sunscreening systems - Faber Blinds UK Ltd
Window blinds and sunshades - Dallas & Forrest Ltd
Window furnishings, curtain tracks, poles and valences - Integra Products
Wooden Shutters - Design Line
Wooden Venetian blinds - Design Line
Wooden venetian blinds - Eclipse Blind Systems Ltd
Wooden venetian blinds, timbershades - Tidmarsh & Sons
Work benches - Fine H & Son Ltd
Work benches and cupboards - Emmerich (Berlon) Ltd
Work surfaces - McLoughlin Wood Ltd
Worktops - The Delabole Slate Co. Ltd
Worktops - Viking Laminates Ltd
Worktops and vanity units - Domus Tiles Ltd
Woven fabrics - Mermet U.K
Write-on, wipe-off products - Tektura Plc

N11 Domestic kitchen fittings

Built -in cookers - Stoves PLC
Built-in dishwashers - Stoves PLC
Built-in refrigeerators - Stoves PLC
Cookers - Flavel - Leisure Glynwed Consumer & Construction Products Ltd
Cookers - Stoves PLC
Cookers and electrical appliances - Electrolux Domestic Appliances
Cookers and electrical appliances - Parkinson Cowan and Tricity Bendix Co, The
Customised work tops: stainless steel - Azimex Fabrications Ltd
Domestic kitchen fitting - Johnson & Johnson PLC
Domestic kitchen Fittings - Aga-Rayburn
Domestic kitchen fittings - Be-Modern Ltd
Domestic kitchen fittings - Commodore Kitchens Ltd
Domestic kitchen fittings - Elite Trade Kitchens Ltd
Domestic kitchen fittings - Hipkiss, H, & Co Ltd
Domestic kitchen fittings - Jenkins Newell Dunford Ltd
Domestic kitchen fittings - Sissons W & G Ltd
Domestic kitchen fittings, Lever action - Shavrin Levatap Co Ltd
Domestic kitchen furniture - Alno (United Kingdom) Ltd
Domestic kitchen furniture - Ball William Ltd
Domestic kitchen sinks, taps and accessories - Carron Phoenix Ltd
Electrical appliances - Pifco Ltd
Fitted kitchens - Miele Co Ltd
Flat pack kitchen furniture - Adam Furniture Group PLC
Free standing cookers - Stoves PLC
Granite kitchen work top - Emsworth Fireplaces Ltd
Kitchen and bathroom furniture - Dennis & Robinson Ltd
Kitchen brassware - Midland Stom Ltd
Kitchen Brassware - Polypipe Bathroom & Kitchen Products Ltd
Kitchen fittings - Bernstein Group PLC
Kitchen fittings - Magnet Ltd
Kitchen fittings - Symphony Group PLC, The
Kitchen furniture - Allied Manufacturing Co (London) Ltd
Kitchen furniture - Ellis J T & Co Ltd
Kitchen furniture - Gower Furniture Ltd
Kitchen sinks - Polypipe Bathroom & Kitchen Products Ltd
Kitchen units - Ball William Ltd
Kitchen worksurfaces - Twyford Bushboard
Kitchen worktops - Kronospan Ltd
Kitchens - Bell & Co Ltd
Kitchens - Crickhowell Furniture Co. Ltd
Kitchens - Smith & Choyce Ltd
Kitchens - Taskworthy Ltd
Kitchens - TBR Joinery
Kitchens and bathrooms - Jewson Ltd
Laminator + Solid Hardwood Worktops - Montague L Meyer
Oil stoves - Gazco Ltd
Range Cookers - Gazco Ltd
sackholders - Surrey Sackholders Ltd.
Sanibins - Surrey Sackholders Ltd.
Self assembley kitchens - Ideal Furniture Enterprises Ltd
Sinks - Astracast PLC
Sinks, taps, waste disposal and water filters - Berglen Group Ltd
Solid hardwood worktops - Junckers Limited
Solid wood + melamine worktops - Deralam Laminates Ltd
Stoves inset and free standing - Emsworth Fireplaces Ltd
Taps - kitchen - Hansgrohe
Traditional kitchen taps and Mixers - Shavrin Levatap Co Ltd
Vanitories worktops - Burlington Slate Ltd
Waste bins - Surrey Sackholders Ltd.
Work surfaces - Kirkstone
Worktops - Formica Ltd
Worktops - Twyford Bushboard

N12 Catering Equipment

Bacteria Grease traps - Progressive Product Developments Ltd
Bar Tender Sinks - Pland Stainless Ltd
Bars - Smith & Choyce Ltd
Cast iron multifuel stoves - Valor
Catering equipment - Barlow Group
Catering equipment - BRB Industrial Services Ltd
Catering equipment - Electrolux Foodservice
Catering equipment - Falcon Food Services Equipment
Catering equipment - Grundy Catering
Catering equipment - Miele Co Ltd
Catering equipment - Sissons W & G Ltd
Catering grease and oil traps - Conder Products Ltd
Catering sinks and tables - Pland Stainless Ltd
Catering storage - Hispack Limited
Catering water boilers - Calomax Ltd
Coffee percolators - TSE Brownson
Commercial catering and bar equipment - Imperial Machine Co Ltd
Commercial catering facilities - design, supply, and installation of UK wide - Lockhart Catering Equipment
Counters and Serveries - Viscount Catering Ltd
Dishwashers, glasswashers - Clenaware Systems Ltd
Display carts - TSE Brownson
Domestic and commercial waste disposal units - Max Appliances Ltd
Domestic kitchen foodwaste disposers - The Haigh Tweeny Co Ltd
Drinks vending machines - Craft Jacobs Suchard Ltd
Filtered water dispensers - Airstream Products Ltd
Fish and chip frying ranges - Preston & Thomas Ltd
Fish and Chip Ranges & Catering equipment - Nuttall, Henry (Viscount Catering Ltd)
Food and drink vending machines - Hillday
Gas stoves - Valor
Grease traps, oil and grease water separators - Progressive Product Developments Ltd
Grease treatment plants - Progressive Product Developments Ltd
Heater service counter tops - TSE Brownson
Heavy duty catering equipment - Viscount Catering Ltd
High performance kitchen furniture - Rixonway Kitchens Ltd
Insulated Urns - TSE Brownson
Kitchen equipment - Magnet Ltd
Light duty catering equipment - Viscount Catering Ltd
Medium duty cateing equipment - Viscount Catering Ltd
Milk heaters - TSE Brownson
Oilfired stoves - Don Heating Products Ltd
Range cookers - Don Heating Products Ltd
Rifrgeration - Viscount Catering Ltd
sackholders - Surrey Sackholders Ltd.
Sanibins - Surrey Sackholders Ltd.
Show cases - TSE Brownson
Stainless steel catering equipment - Associated Metal (Stainless) Ltd
Stainless steel catering equipment - Moffat Ltd, E & R
Stainless steel commercial catering equipment - Bartlett Catering Equipment Ltd
Stainless steel shelving systems for kitchens and coldrooms - Beford & Soar Ltd
Tray/plate dispensers - TSE Brownson
Trolleys - TSE Brownson
Vending machines - H&H Electronics Ltd
Vending machines - Provend Services Ltd
Vending systems - The Kenco Coffee Company
Ventilated ceiling system for commercial cooking areas - Heydal Ltd
Waste bins - Surrey Sackholders Ltd.
Waste disposal equipment, balers, shredders, drinks can crushers - Tony Team Ltd
Waste disposal units and catering cupboards - Pland Stainless Ltd

N13 Sanitary appliances/ fittings

Accessories - Caradon Plumbing Solutions
Accessories - Deva Tap Co Ltd
Accessories - Ideal-Standard Ltd
Accessories for showers - Jendico
Actuators for concealed W.C. cisterns - Sheardown Engineering Ltd
Aluminium shower outlets - Marley Plumbing and Drainage
Anti-ligature system for hospital cubicle/shower curtain tracks - President Blinds Ltd
Anti-vandal showers - Shavrin Levatap Co Ltd
Architechtural hardware - Samual Heath
Automatic self Cleaning Toilets - Healthmatic Ltd
Bath and shower mats - William Freeman Ltd
Bath screens - Majestic Shower Co Ltd
Bath systems - whirlpool and spa - Jendico
Bathing equipment for elderly and disabled - Chiltern Invadex
Bathroom Accessories - Marflow Eng Ltd
Bathroom accessories - Metlex
Bathroom accessories - Neptune Showers Ltd
Bathroom accessories - Samual Heath
Bathroom accessories - Turnquest UK Ltd
Bathroom accessories, shower enclosures, overbath screens, bath seats - Provex Products Ltd
Bathroom and shower accessories - Marleton Cross Ltd
Bathroom and showering equipment - Amilake Southern Ltd
Bathroom Brassware - Polypipe Bathroom & Kitchen Products Ltd
Bathroom cabinets, accessories, toilet seats and vanityware - Allibert (UK) Ltd
Bathroom fittings - MacBee Manufacturing Ltd
Bathroom fittings - Rothley Burn Ltd
Bathroom furniture - Ellis J T & Co Ltd
Bathroom Pods - RB Farquhar Manufacturing Ltd
Bathroom products - Qualitas Bathrooms
Bathroom products - Trent Bathrooms
Bathroom systems - Multikwik Ltd
Bathrooms - Bell & Co Ltd
Bathrooms - Fired Earth
Bathrooms - Spital Tile Co Ltd
Baths - Ideal-Standard Ltd
Baths and taps - Midland Stom Ltd
Bedroom furniture - Bell & Co Ltd
Brassware, bath and toilet seats - Topravit (UK) Ltd
Brassware, showers and accessories - Hansgrohe
Cabinet furniture - Turnquest UK Ltd
Ceramic and acrylic sanitaryware - Armitage Shanks Ltd
Ceramic bathroom accessories - Balmoral Porcelain Limited/Longmead Ceramics
Chrome bathroom accessories - MacBee Manufacturing Ltd
Cisterns - Multikwik Ltd
Commercial sanitary ware - Trent Bathrooms
Commercial spa shelves - Golden Coast Ltd
Cubicle rails - Harrison Drape, Divsion of Newell Ltd
Cubicle tracks - Marshall Contracts Ltd
Curved glass bathroom bowls - Gardner & Newton
Domestic kitchen sinks and taps - Fordham
Domestic showering products - Kohler Mira Ltd
Drinking Fountains - Harris T A Ltd
Drinking fountains and water coolers - Maestro International Ltd
Duct panels - Dunhams of Norwich
Electric mixer and power showers, hand wash units and water heaters - Triton PLC
Electric showers - Dolphin Showers Ltd
Electric showers - Heatrae Sadia Heating Ltd
Emergency showers - Rose Emergency Equipment Co
Enclosures - Ideal-Standard Ltd
Ensuite shower pods - RB Farquhar Manufacturing Ltd
Feminine hygiene disposal - Airstream Products Ltd
Fill valves - Multikwik Ltd
Filter water taps-springflow - Fordham
Fitted Bathrooms - Symphony Group PLC, The
Fittings for disabled and aged - Turnquest UK Ltd
Flexible plumbing fittings - Multikwik Ltd
Flush pipes - Airflow (Nicoll Ventilators) Ltd
Flushvalves - Multikwik Ltd
Foot and knee operated self closing taps - Sheardown Engineering Ltd
Frameless shower screens - ShowerSport Ltd
Grab rails - Rothley Burn Ltd
Hand driers - Airstream Products Ltd
Hand dryers - Warner Howard Group Ltd
Hydromassage baths - Novellini Bathroom Products Ltd
Hygiene equipment - Imperial Machine Co Ltd
Hygiene equipment - P & L Systems Ltd
In-Build Automatic Toilets - Healthmatic Ltd
Incinerators - Combustion Linings Ltd
Jacuzzi, baths and showers - Jacuzzi UK Ltd, Division of Jacuzzi Inc
Kits - Ideal-Standard Ltd
Lever taps for disabled people - Sheardown Engineering Ltd
Macerators - Multikwik Ltd
Mirrors, shower seats, and safety grab rails - Provex Products Ltd
Modern bathroom taps and mixers - Shavrin Levatap Co Ltd
Modular multi-User Toilets - Healthmatic Ltd
Nappy changing units - Magrini Ltd
Odor control solutions - Airstream Products Ltd
Pan connectors - Multikwik Ltd
Pan connectors - Polypipe Bathroom & Kitchen Products Ltd
Pillar taps - Booth Samuel & Co Ltd
Plumbing fittings - Flowflex Componants Ltd
Power, mixer valve and electric showers - NewTeam Ltd
Precast terrazo shower trays - Roman Mosaic Contracts Ltd
Push taps - Caradon Plumbing Solutions
Ready Plumbed Modules - Twyford Bushboard
Rigid bathroom furniture - Adam Furniture Group PLC
Sanitary appliances - Arjo Ltd
Sanitary appliances and fittings - Barflow-LCA
Sanitary appliances and fittings - Cefndy Healthcare
Sanitary appliances and fittings - Guest (Speedfit) Ltd, John,
Sanitary appliances and fittings - Gummers A & J Ltd
Sanitary appliances and fittings - Kenley Manufacturing Ltd
Sanitary appliances/ fittings - Harris & Bailey Ltd
Sanitary appliances/fittings - Allgood plc
Sanitary appliances/fittings - FC Frost Ltd
Sanitary appliances/fittings - The Haigh Tweeny Co Ltd
Sanitary Appliances/Fittings - Wybone Ltd
Sanitary bins - Unicorn Containers Ltd
Sanitary disposal units, washroom and baby room products - Cannon Hygiene Ltd
Sanitary equipment and accessories - Ideal-Standard Ltd
Sanitary fittings - Sissons W & G Ltd
Sanitary fixings - Artur Fischer (UK) Ltd
Sanitary manifolds - SAV United Kingdom Ltd
Sanitary towel incinerator and towel rails - Consort Equipment Products Ltd
Sanitary ware and appliances - Diamond Merchants
Sanitary ware and fittings - Caradon Bathrooms Ltd
Sanitaryware, associated fittings and accessories - Shires Bathrooms
Sanitory units - Max Appliances Ltd
Sauna - Golden Coast Ltd
Saunas - Finnish Sauna Baths (Sales) Ltd
Saunas, steam rooms and spa baths - Dalesauna Ltd
Self closing non concussive press lever taps - Sheardown Engineering Ltd
Sharps disposal units - Inter Public Urban Systems Uk Ltd
Shower accessories - Douglas Industrial Ltd
Shower accessories - Gummers A & J Ltd
Shower accessories - NewTeam Ltd
Shower booster pumps - Baresford Pumps Ltd
Shower controls - Douglas Industrial Ltd
Shower curtains - Neptune Showers Ltd
Shower doors - Majestic Shower Co Ltd
Shower doors and carer screens - Norton Engineering Alloys Co Ltd
Shower doors, enclosures, bath screen and shower trays - Daryl Industries Ltd
Shower doors, trays bath screens and shower mixers - Matki plc
Shower enclosers - Majestic Shower Co Ltd
Shower enclosures - Contour Showers Ltd
Shower enclosures and towel rails - Kermi (UK) Ltd
Shower enclosures, bathscreens, shower trays, baths - Novellini Bathroom Products Ltd
Shower grilles - Norton Engineering Alloys Co Ltd

N13 Sanitary appliances/ fittings (Cont'd)

Shower products - Aqualisa Products
Shower pumps - NewTeam Ltd
Shower rails - Harrison Drape, Divsion of Newell Ltd
Shower rails - Neptune Showers Ltd
Shower screens - Majestic Shower Co Ltd
Shower trays, cubicles and bath screens - Coram Showers Ltd
Shower valves - Ideal-Standard Ltd
Showering systems - Daryl Industries Ltd
Showers - Caradon Plumbing Solutions
Showers - Jendico
Sinks - Franke UK Ltd
Spa - Golden Coast Ltd
Specialist bathing etc. - Parker Bath Limited
Spray mixing taps - Sheardown Engineering Ltd
Stainless Steel Faucets - Vola
Stainless steel sanitary appliances and fittings - Blucher UK Ltd
Stainless steel sanitaryware - Pland Stainless Ltd
Stainless steel sanitaryware and washroom equipment - Associated Metal (Stainless) Ltd
Stainless steel sinks, wc's, urinals, sluices, baths and foot baths - Anderson GEC Ltd
Stainless steel, ceramic, synthetic sinks and shower cubicles - Rangemaster
Taps - Astracast PLC
Taps - Deva Tap Co Ltd
Taps - Douglas Industrial Ltd
Taps - Franke UK Ltd
Taps - Shavrin Levatap Co Ltd
Taps & Showers - Samual Heath
Taps and mixers - Ideal-Standard Ltd
Taps and valves - Opella Ltd
Taps for the disabled and elderly - Shavrin Levatap Co Ltd
Taps, mixers, and showers - Pegler Hattersley
Taps, showers and thermostatic mixers - Grohe Ltd
Thermostatic mixing valves - Caradon Plumbing Solutions
Thermostatic mixing valves - Douglas Industrial Ltd
Thermostatic shower panels - Horne Engineering Ltd
Thermostatic shower valves - Shavrin Levatap Co Ltd
Time delay closing taps and valves - Sheardown Engineering Ltd
Toilet Cubicales - Relcross Ltd
Toilet cubicles - Grant Westfield Ltd
Toilet cubicles and washroom systems - Moffett Thallon & Co Ltd
Toilet seats - Polypipe Bathroom & Kitchen Products Ltd
Toilet seats - Pressalit Care Ltd
Toilet, shower and changing cubicles - Dunhams of Norwich
Tool hire - Jewson Ltd
Towel and soap dispensers - Kimberly-Clark Ltd
Towel warmers - Townhive Ltd
Traps - Polypipe Bathroom & Kitchen Products Ltd
Trays - Ideal-Standard Ltd
Urinal filters - Cistermiser Ltd
Urinal fittings - Polypipe Bathroom & Kitchen Products Ltd
Urinal flush controls - Cistermiser Ltd
Valves - Horne Engineering Ltd
Vandal resistant taps and fittings - Sheardown Engineering Ltd
Vandel resistant toilets - Inter Public Urban Systems Uk Ltd
Vanity units - Dunhams of Norwich
Vitreous china ceramic sanitaryware and brassware - Vitra (UK) Ltd
Vitrious Sanitryware - Relcross Ltd
W. C.'s - Ideal-Standard Ltd
Wall mounted basin, bath and shower valves - Sheardown Engineering Ltd
Wall mounted hand wash valves and spouts - Sheardown Engineering Ltd
Warm air hand and face dryers - Airdri Ltd
Warm air hand dryers - Wandsworth Elecrtrical Ltd
Wash basins - Ideal-Standard Ltd
Wash room hygiene products - Airstream Products Ltd
Washroom control systems - Cistermiser Ltd
Washroom Equipment - Relcross Ltd
Washroom Services - Floor Protection Services Ltd
Waste fittings - Polypipe Bathroom & Kitchen Products Ltd
Wastes - Deva Tap Co Ltd
Water dosing system - Airstream Products Ltd
Water management system - Airstream Products Ltd
Water taps and mixers - Barber Wilsons & Co Ltd
Waterless Urinals - Relcross Ltd
WC cisterns - Polypipe Bathroom & Kitchen Products Ltd
Whirlpools, steam showers - Hansgrohe

N15 Signs/ Notices

3D. Domed labels - Mockridge Labels & Nameplates Ltd
All health & safety signs - Masson Seeley & Co Ltd
Anodised aluminium labels - Mockridge Labels & Nameplates Ltd
Architectual signing specialists - The HB Sign Company Ltd
Architectural aluminium systems for signs - Universal Components Ltd
Architectural signage - Signs & Labels Ltd
Badges - Mockridge Labels & Nameplates Ltd
Banners - Green Brothers Ltd
Bespoke Signs - The Delabole Slate Co. Ltd
Blaques - Green Brothers Ltd
Braille signs - Signs & Labels Ltd
Bus stops - Burnham Signs
Cast polyester resin molded letters - Metalline Signs Ltd
Cast signs - Signs Of The Times Ltd
CCTV mandatory sinage - Hoyles Electronic Developments Ltd
Chalk boards, pin boards and display boards - Leakale Taurus Visual Aids
Chemical Etching - Mockridge Labels & Nameplates Ltd
Coats of arms - Signs Of The Times Ltd
Commemorative plaques, coats of arms and architectural lettering in cast bronze & cast aluminum - Metalline Signs Ltd
Commercial signs - Dee-Organ Ltd
Company logos - Signs Of The Times Ltd
Contract enameling - Enamelled Signs (UK) Ltd
Corporate signs - Pryorsign
Custom Made signage - Stocksigns Ltd
Decorative etching - Mockridge Labels & Nameplates Ltd
Design service - Pryorsign
Design, manufacture, installation, & maintenance of internal and external signage - Pannell Signs Ltd
Direction signs - Simplex Signs Limited
Directory systems - Simplex Signs Limited
Doors - Pearce Security Systems
Electric, illuminated and neon signs - Pearce Signs Ltd
Enameling cladding panels - Enamelled Signs (UK) Ltd
Engraved metals - Signs Of The Times Ltd
Engraved signs - Signs & Labels Ltd
Engraved slate - Signs Of The Times Ltd
Engravers - Abbey Craftsmen Ltd
Fabricated signs - Simplex Signs Limited
Fiber optic signs - Burnham Signs
Financial services signs & structures - Pryorsign
Fire signs - Chubb Fire Ltd
GRP structures - Pryorsign
Hand engraving tools and metal etching pens - Elliott John H (Monostamp) Ltd
Health and Safety Signage - Stocksigns Ltd
Heritage and conservation signs - Pryorsign
Housesigns - The Delabole Slate Co. Ltd
Illuminated signs - Burnham Signs
Illuminated signs - Signs & Labels Ltd
Instrumentation panels - Pryorsign
Interactive signs - Signature Ltd
Internal & external signage - Vision Neon & Display Ltd
Internal and external illuminated signs - Interlux Ltd
Internally illminated signs - Signature Ltd
Large format digital printing - Signs & Labels Ltd
Large format printing - Zephyr The Visual Communicators
Large letter signs - Pannell Signs Ltd
Litter bins - Burnham Signs
Marking inks and products - Eyre & Baxter (Stampcraft) Ltd
Metal lettering - Simplex Signs Limited
Nameplates - Eyre & Baxter (Stampcraft) Ltd
Nameplates - Mockridge Labels & Nameplates Ltd
Nameplates & Memorial Tablets in Metal & Plastic - Abbey Craftsmen Ltd
Neon signs & fascias - Vision Neon & Display Ltd
Notice board material - Forbo - Nairn Ltd
Notice panels - Opto International Ltd
Noticeboards, pinboards - Magiboards Ltd
Pinboards - Sundeala Division of Celotex Ltd
Plaques - stainless steel, brass, engraved - Mockridge Labels & Nameplates Ltd
Poster displays - Fairfield Displays & Lighting Ltd
Printed banner material - Mermet U.K
Professional signs - Abacus Signs
RICS & RIBA Site Boards - Abacus Signs
Road traffic signs - Dee-Organ Ltd
Safety signs - Asco Extinguishers Co Ltd T/A The Asco Fire Group
Safety signs - Burnham Signs
Safety signs - Eyre & Baxter (Stampcraft) Ltd
Safety signs - Fife Fire Engineers & Consultants Ltd
Sign manufacturers - Blaze Neon Ltd
Sign posts - Fabrikat (Nottingham) Ltd
Sign systems - Signs & Labels Ltd
Signage - Fairfield Displays & Lighting Ltd
Signage - Furnitubes International Ltd
Signage - Mockridge Labels & Nameplates Ltd
Signboards - Duroy Fibreglass Mouldings Ltd
Signs - Allgood plc
Signs - Dee-Organ Ltd
Signs - Drakard & Humble Ltd
Signs - Green Brothers Ltd
Signs - Hoyles Electronic Developments Ltd
Signs - Neocrylic Signs Ltd
Signs - Vision Neon & Display Ltd
Signs - design - Masson Seeley & Co Ltd
Signs - manufacture - Masson Seeley & Co Ltd
Signs - site surveying - Masson Seeley & Co Ltd
Signs & Notices - BRB Industrial Services Ltd
Signs and notices - Arbory Group Limited
Signs and Notices - Brissco Equipment Ltd
Signs and notices - Enamelled Signs (UK) Ltd
Signs and notices - Focal Signs Ltd
Signs and sign systems - Simplex Signs Limited
Signs installation - Masson Seeley & Co Ltd
Signs/ notices - Laidlaw Architectural Hardware
Signs/ Notices - Leander Architectural
Signs/ notices - Stocksigns Ltd
Signs/notices - Duroy Fibreglass Mouldings Ltd
Signs/notices - Rivermeade Signs Ltd
Specialist signs - Signs International
Stencils - Eyre & Baxter (Stampcraft) Ltd
Traffic management systems - Marshalls Mono Ltd
Traffic signs - Signature Ltd
Transport signage and wayfinding - Burnham Signs
Viewpoint window display signs - Pryorsign
Vinyl signs - Invicta Window Films Ltd
Visual communication equipment - Wilson & Garden Ltd
Vitreous enamel signs - Burnham Signs
Vitreous enameled steel cladding signs - Escol Products Ltd
Wall plaques - Rogers Concrete Ltd
Waterproof, vandal resistant illuminated, health and safety and shop signs, nameplates and fascias labels - Pryorsign
Wayfinding schemes - Masson Seeley & Co Ltd
Whiteboards - Magiboards Ltd

N16 Bird/Vermin control

Bird control - Birdscarer Products & Net Supplies Ltd
Bird/ vermin control - Terminix Property Services
Brushstrip for pest and bird control - Kleeneze Sealtech Ltd
Electric flykillers - Paragon
Flyscreens - Paragon
Glueboard flykillers - Paragon

N20 Safety Equipment

Anti-slip sighting bars - Redman Fisher Engineering
Body armour - Meggitt Armour Systems
Breathing apparatus - Draeger Safety UK Ltd
Burglary resistant equipment - Chubb Safe Equipment Company
Carbon monoxide detectors - Firemaster Extinguisher Ltd
Confined space equipment - MSA (Britain) Ltd
Disabled toilet alarms - Hoyles Electronic Developments Ltd
Domestic safes - Chubb Safe Equipment Company
Domestic smoke alarms - Firemaster Extinguisher Ltd
Emergency evacuation equipment for the mobility impaired - Evac+chair international Ltd
Emergency safety showers and eye and face wash units - Maestro International Ltd
Explosion containment tubes - Attwater & Sons Ltd
Fire blankets - Firemaster Extinguisher Ltd
Fire extinguishers - Macron Safety Systems (UK) Ltd
Fire hose and fittings - Macron Safety Systems (UK) Ltd
Fire hose reels - Kidde Fire Protection Services Ltd
Firemaster powder, water, AFFF foam, and carbon dioxide fire extinguishers - Firemaster Extinguisher Ltd
Gas detection equipment - Draeger Safety UK Ltd
Gas detection systems - Blakell Europlacer Ltd
Harnesses - HCL Verta
Height Safety - Barrow Hepburn Sala Ltd
Industrial safety clothing and equipment - BRB Industrial Services Ltd
Ladder safety post - Bilco UK Ltd
Ladders and safety cages - Redman Fisher Engineering
Personal Protective equipment - MSA (Britain) Ltd
Respiratory filter masks - Draeger Safety UK Ltd
Ropes and sashcords - Marlow Ropes Ltd
Safety & Security - HC Slingsby PLC
Safety and security equipment - Jones & Thomas (Wireworks) Ltd
Safety equipment - Barrow Hepburn Sala Ltd
Safety equipment - Focal Signs Ltd
Safety equipment - Hoyles Electronic Developments Ltd
Safety equipment - Interface Europe Ltd
Safety equipment - Kee Klamp Ltd
Safety equipment - Metreel Ltd
Safety equipment - MSA (Britain) Ltd
Safety equipment - Radiodetection Ltd
Safety films - Madico Film Products Division Inc
Safety harnesses etc - Sala Ltd.
Safety interlocking systems - Castell Safety International
Safety signs - Vision Neon & Display Ltd
Safety treads - Record & Parkes Ltd
Safety, drench, decontamination, emergency showers and eyebaths - Hughes Safety Showers Limited
Security grilles and collapsible gates - Baddeley
Special purpose safety equipment - Hazard Safety Products Ltd
Sports Markings - B&C Roadmarkings Ltd
Steel security doorsets - Sunray Engineering Ltd
Vertical access fall safe access ladders - HCL Verta
Water safety products - Arbory Group Limited
Welding curtains and hangers - Nederman UK

N21 Storage Equipment

Adjustable Pallet Racking - Barton Storage Systems Ltd
Adjustable pallet racking - Link 51 (Storage Products)
Adjustable shelving - Spur Shelving Ltd
Aluminium shelving brackets - Tebrax Ltd
Architectural products - Arbory Group Limited
Archive storage - Dexion Ltd
Coldstores - Carter Retail Equipment Ltd
Construction products - Link 51 (Storage Products)
Cupboards - Brown N C (Storage Equipment) Ltd
Cupboards - Crickhowell Furniture Co. Ltd
Desk-side storage units - Rackline Systems Storage Ltd
Document storage shelving systems - Link 51 (Storage Products)
Drawer units - Brown N C (Storage Equipment) Ltd
Electrical cabinets - H&H Electronics Ltd
Filing systems - Dexion Ltd
Freestanding shelf systems - Spur Shelving Ltd
Garment collectors - Brown N C (Storage Equipment) Ltd
Garment dispensers - Brown N C (Storage Equipment) Ltd
Heavy duty shelving systems - Link 51 (Storage Products)
Library shelving - Sperrin Metal Products Ltd
Lockers - Barton Storage Systems Ltd
Lockers - Brown N C (Storage Equipment) Ltd
Lockers - Detlef Muller Ltd
Lockers - Grant Westfield Ltd
Lockers - electronic lock - Helmsman
Lockers - laminate - Helmsman
Lockers - mild steel - Helmsman
Lockers - wooden - Helmsman
Lockers and cubicles - LSA Projects Ltd
Lockers and personal storage - Moresecure Ltd
Manufactureres of quality filing and storage systems - Railex Systems Ltd
Megyanine Floors - Hi-Store Ltd
Mobile and static shelving equipment - Railex Systems Ltd, Elite Division
Mobile shelving - Dexion Ltd
Mobile shelving - Link 51 (Storage Products)
Mobile Storage Systems - Railex Systems Ltd
Moblie shelving systems - Moresecure Ltd
Pallet parking systems - Moresecure Ltd
Pallet Racking - Dexion Ltd
Picking systems - Dexion Ltd
Radiator cabinets - Taskworthy Ltd
Rails & wardrobe fittings - Hills Industries Ltd
Shelving - Lesco Products Ltd
Shelving / Racking - Smart F & G (Shopfittings)Ltd
Shelving and furniture for libraries and archives - Sealine Business Products
Shelving and storage racking - Dexion Ltd
Shelving systems and safes - Rosengerns Tann Ltd
Single shelf brackets - Spur Shelving Ltd
Small Parts Storage - Barton Storage Systems Ltd
Small parts storage - Link 51 (Storage Products)
Small storage - Moresecure Ltd
Spacestor storage wall / partitions - Envoplan
Stainless steel cabinets IP65 - Kensal CMS
Steel storage and office shelving systems - Sealine Business Products
Storage - Envoplan
Storage - President Office Furniture Ltd
Storage - Rothley Burn Ltd
Storage cabinets - Unicorn Containers Ltd
Storage equipment - Barton Storage Systems Ltd
storage equipment - Emmerich (Berlon) Ltd
Storage equipment - Envopak Group Ltd
Storage equipment - Garran Lockers Ltd
Storage equipment - Hispack Limited
Storage Equipment - Hi-Store Ltd
Storage equipment - Kee Klamp Ltd
Storage equipment - Libraco
Storage equipment - Logic Office Contracts
Storage equipment - Project Office Furniture PLC
Storage equipment - Rackline Systems Storage Ltd
Storage Equipment - SSI Schaefer Ltd
Storage equipment - Welconstruct Co Ltd
Storage Equipment - Wybone Ltd
Storage euipment - Paragon Business Furniture
Storage furniture - Pulsford Associates Ltd, John
Storage racks and shelving - Metalrax Ltd
Storage systems - Komfort Office Environments Plc
Storage systems - Moresecure Ltd
Task lighting - Anglepoise Lighting Ltd
Toughened glass shelves - Spur Shelving Ltd

N21 Storage Equipment (Cont'd)

Vanity units - LSA Projects Ltd
Wallstorage systems - Fastwall Ltd
Waste containment - Randalls Fabrications Ltd

N22 Office Equipment

Binders - Quartet-GBC UK Ltd
Boardroom installations - Planet Communications Group Ltd
Cabinets - Railex Systems Ltd
Carousels - Railex Systems Ltd
Church seating & accessories - Knightbridge Furniture Productions Ltd
Computer cabinets - H&H Electronics Ltd
Computer document paper handling systems - Neopost Ltd
Conference and training room systems - track based - Magiboards Ltd
Conference room installations - Planet Communications Group Ltd
Contract school/ office furniture - Blindcraft Industries
Contract seating for leisure and office industries - FG Contracts Limited
Cupboards - Brown N C (Storage Equipment) Ltd
Desk top products - Lesco Products Ltd
Domestic and commercial office furniture - Ball William Ltd
Electronic file tracking - Rackline Systems Storage Ltd
Ergonomic products - Lesco Products Ltd
Fire resistant cabinets for media and document protection - Chubb Safe Equipment Company
Fire safes - INMAC (a division of microwarehouse Ltd)
Flip top training room tables - Envoplan
Flipcharts - Magiboards Ltd
Flipcharts, hanging rail systems and visual aids - Leakale Taurus Visual Aids
Floor, fire vents, ceiling and roof hatches - Bilco UK Ltd
Folder-inserter systems - Neopost Ltd
Franking machines - Neopost Ltd
Furniture - Deanes Furniture Ltd
Furniture - HEM Interiors Group Ltd
Hand stamps - Elliott John H (Monostamp) Ltd
Home office to the housebuilder market - Hammonds - New Home Division
Interactive whiteboards - Quartet-GBC UK Ltd
IT Furniture - MICE Kaymar Limited
Laminators - Quartet-GBC UK Ltd
Letter openers & extractors - Neopost Ltd
Mailing systems - Neopost Ltd
Mailroom & packaging - HC Slingsby PLC
Mailroom equipment & supplies - Neopost Ltd
Office and industrial cleaning - CSL (Services) Limited
Office and reception fitted furniture - Hyperion Wall Furniture Ltd
Office chairs - KAB Seating Ltd
Office equipment - Antocks Lairn Ltd
Office Equipment - Brettell & Shaw Ltd
Office equipment - Carleton Furniture Group Ltd
Office equipment - Envopak Group Ltd
Office equipment - Hago Products Ltd
Office equipment - HC Slingsby PLC
Office equipment - Intercraft Designs Ltd
Office equipment - Lamb Macintosh
Office equipment - Libraco
Office equipment - Martella PLC
Office equipment - Matthews Office Furniture Ltd
Office equipment - RPS Data Products (UK) Ltd
Office Equipment - SSI Schaefer Ltd
Office equipment - Vista Plan International PLC
Office filing systems - Railex Systems Ltd
Office fit out - Logic Office Group
Office furinture - Vitra UK Ltd
Office furniture - EFG Business Furniture (UK) Ltd
Office furniture - Envoplan
Office furniture - Fray Design Ltd
Office furniture - Girsberger London
Office furniture - Haworth UK Ltd
Office furniture - INMAC (a division of microwarehouse Ltd)
Office furniture - Komfort Office Environments Plc
Office furniture - Laytrad Ltd
Office furniture - Logic Office Group
Office furniture - OEP Furniture Group PLC
Office furniture - OEP Furniture Group PLC
Office furniture - President Office Furniture Ltd
Office furniture - Progress Furnishings Systems Ltd
Office furniture - Project Office Furniture PLC
Office furniture - Pulsford Associates Ltd, John
Office furniture - Sarnas Roneo Ltd
Office furniture - Scott Appleton Furniture Ltd
Office furniture - Top Office Equipment Ltd
Office furniture and accessories - Lesco Products Ltd
Office furniture including storage cabinets and desks - Flexiform Business Furniture Ltd
Office furniture, desking, storage, boardroom and conference furniture - Hands of Wycombe
Office furniture, screens and seating - Intercraft Designs Ltd
Office furniture, seating and board room tables - Steelcase Strafor Plc
Office furniture, seating and storage - Desking Systems Ltd
Office racking, storage cabinets, mobile shelving and filing systems - Rackline Systems Storage Ltd
Office seating - Dauphin PLC
Office seating - Evertaut Ltd
Office seating - Laytrad Ltd
Office seating and desking - Metalliform Products Plc
Office seating, steel filing, storage products, tables and chairs - Harvey Furniture
Office storage equipment and desking - Bisley Office Equipment
Offices - Taskworthy Ltd
Paragon Business Furniture - Paragon Business Furniture
Photocopiers - Oce (UK) Ltd
Planning boards - Magiboards Ltd
Post room furniture - Lesco Products Ltd
Postroom equipment & supplies - Neopost Ltd
Printroom equipment - Envopak Group Ltd
PVC printers - Zephyr The Visual Communicators
Rail presentation systems - Quartet-GBC UK Ltd
Reception - Evertaut Ltd
Reception Desks - Crickhowell Furniture Co. Ltd
Reception desks/counters - Smith & Choyce Ltd
Residential furniture - Carleton Furniture Group Ltd
Rubber stamps - Eyre & Baxter (Stampcraft) Ltd
Scales, electronic postal - Neopost Ltd
Screens - President Office Furniture Ltd
Seating and executive furniture - Verco Office Furniture Ltd
Seating equipment - Evertaut Ltd
Shredders - Quartet-GBC UK Ltd
Side-opening tambour cabinets - Rackline Systems Storage Ltd
Software, laserprinting management - Neopost Ltd
Software, mail & despatch management - Neopost Ltd
Staff locators - Magiboards Ltd
Storage and shelving - HC Slingsby PLC
System desking & sorage - Verco Office Furniture Ltd
Table collection - Envoplan
Tables - offices - Verco Office Furniture Ltd
Task lights - Lesco Products Ltd
Type - Elliott John H (Monostamp) Ltd
Visual display products - Lesco Products Ltd
Waiting area seating for train stations, airports, ferry terminals, hospitals and surgeries - Hille Ltd
Work stations & work benches - Moresecure Ltd
Workstation furniture and mobile filing systems - Mauser Interiors (UK) Ltd
Writing boards, Rollerboards, Notice boards, Rail systems - Wilson & Garden Ltd

N23 Special purpose fixtures/furnishings/equipment

Auditorium, theatre, multi purpose, public area and courtroom seating - Race Furniture Ltd
Bar fittings - TFL International Ltd
Bar Stools - Sandler Seating Ltd
Barstool systems - Laytrad Ltd
Bespoke cabinet makers and fitted furniture - Crickhowell Furniture Co. Ltd
Binders - Quartet-GBC UK Ltd
Canteen Furniture - Sandler Seating Ltd
Computer furniture (education) - Klick Technology Ltd
Conference - Evertaut Ltd
Conference and training room systems - track based - Magiboards Ltd
Corporate seating & tables - Knightbridge Furniture Productions Ltd
Dance floor - Felix Design Ltd
Domestic Furniture - Crickhowell Furniture Co. Ltd
Education equipment - Evertaut Ltd
Educational equipment - Invicita Plastics Ltd
Educational equipment - RPS Data Products (UK) Ltd
Educational furniture - AEL Furniture Group
Educational furniture - Ellis J T & Co Ltd
Educational furniture - MICE Kaymar Limited
Educational furniture for science and technology - S & B Ltd
Educational seating and desking - Metalliform Products Plc
Educational workshop equipment - Emmerich (Berlon) Ltd
Exhibition and display - Wilson & Garden Ltd
Food technology furniture (education) - Klick Technology Ltd
Furniture design and production - Wales & Wales
Interactive whiteboards - Quartet-GBC UK Ltd
Laminators - Quartet-GBC UK Ltd
Libraies etc. - Crickhowell Furniture Co. Ltd
Massage Beds - Lakeside Group
Massage Chairs - Lakeside Group
Public area seating - Evertaut Ltd
Rail presentation systems - Quartet-GBC UK Ltd
Raised slating rostra - Felix Design Ltd
Restaurant Furniture - Sandler Seating Ltd
Roller boards - Leakale Taurus Visual Aids
Sanctuary stages - Felix Design Ltd
Screen printing on whiteboards - Magiboards Ltd
Seating - auditorium, theater, lecture theater, courtroom, conference - Race Furniture Ltd
Special Equipment (mixing Machines for wet-pour Rubber Crumb) - Edward Benton & Co Ltd
Stacking chairs - Laytrad Ltd
Staging - carpet, hard deck, custom build, raised seating - Felix Design Ltd
Staging rostra - Felix Design Ltd
Swivel and fixed frame boards - Leakale Taurus Visual Aids
Technical workstations - MICE Kaymar Limited
Technology furniture (education) - Klick Technology Ltd
Thermostatic mixing valves - Horne Engineering Ltd
Toys & games - Invicita Plastics Ltd
Waiting area seating for train stations, airports, ferry terminals, hospitals and surgeries - Hille Ltd
White boards - Leakale Taurus Visual Aids
Whiteboard accessories - Magiboards Ltd

N24 Shopfitting

Ambient shelving - Carter Retail Equipment Ltd
Bespoke shopfitting - Pearce Security Systems
Cash Management systems and lobby shutters - Safetell Ltd
Checkouts - Carter Retail Equipment Ltd
Commercial shelving - Dexion Ltd
Contour showcases - Opto International Ltd
Contract fit outs - J. Scott (Thrapston) Ltd
Counters and showcases - Smart F & G (Shopfittings)Ltd
Entrances - Comar Architectural Aluminuim Systems
Extrusions - Opto International Ltd
Fabrics - Mermet U.K
Fast security screen counters - Safetell Ltd
Illuminated and non illuminated show cases - Vision Neon & Display Ltd
Interior design - Eaton Elizabeth
Interior fit outs - Welconstruct Co Ltd
Maintenance of industrial doors, shutters and shopfronts - Shutter Door Repair & Maintenance Ltd
Modular shelving systems - Versatile Fittings Ltd
Panel connectors - Opto International Ltd
Point-of-sale, promotional and incentive products - Invicita Plastics Ltd
Premises equipment - HC Slingsby PLC
Purpose made joinery - Opto International Ltd
Reception counters - Sandoms (Batley) Ltd
Reception desks - Opto International Ltd
Refridgerated cabinets - Carter Retail Equipment Ltd
Retail furniture - Selbix Joinery Ltd
Retractable glazed screen counters and secure access systems - Safetell Ltd
Rotary filing cabinets - Rackline Systems Storage Ltd
Screens - Pearce Security Systems
Shelving brackets - Tebrax Ltd
Shop, office, bank and bar fitters - Beasley Joiners Ltd
Shopfit shelf systems - Spur Shelving Ltd
Shopfitters - Cumberland Construction
Shopfitters, display cabinet manufacturers - Sandoms (Batley) Ltd
Shopfitting - Andy Thornton Ltd
Shopfitting - Benbow Interiors
Shopfitting - Blaze Neon Ltd
Shopfitting - Carter Retail Equipment Ltd
Shopfitting - CIL International Ltd
Shopfitting - Kee Klamp Ltd
Shopfitting - Smart F & G (Shopfittings)Ltd
Shopfitting - Steel Line Ltd
Shopfitting - Taskworthy Ltd
Shopfitting - Withey Contracts Ltd
Shopfitting and specialist joinery - Barlow Group
Shopfitting and specialist joinery - Barnwood Shopfitting Ltd
Shopfitting products - Interlux Ltd
Shopfitting products and accessories - RB UK Ltd
Shopfitting profiles and clip-in inserts - Rustin Allen Ltd.
Shopfronts - Comar Architectural Aluminuim Systems
Shopfronts - Project Aluminium Ltd
Specialist shopfitting - TFL International Ltd
Tube connectors - Opto International Ltd

N25 Hospital/ Health Equipment

Anechoic Rooms - IAC
Audiology Rooms - IAC
Beds, mattresses, accessories, cauches, commodes, overbed tables, living aids - Sidhil Care
Clean rooms and Containment Suites - Envair Ltd
Conduit systems - Marshall-Tufflex Ltd
Data management systems - Marshall-Tufflex Ltd
Disabled and elderly equipment - Nicholls & Clarke Ltd
Disposable hospital products - Verna Ltd
Eye and face wash equipment - Rose Emergency Equipment Co
Fireproff snaitary waste chutes - Inter Public Urban Systems Uk Ltd
Health care products - William Freeman Ltd
Hospital and health equipment - Brandon Medical
Hospital and laboratory equipment - Associated Metal (Stainless) Ltd
Hospital Curtains - Marshall Contracts Ltd
Hospital Equipment, Elbow control - Shavrin Levatap Co Ltd
Hospital furniture - Sidhil Care
Hospital Sinks - Pland Stainless Ltd
Hospital/ health equipment - Arjo Ltd
Hospital/ Health Equipment - Sissons W & G Ltd
Hospital/ Health Equipment - The Kenco Coffee Company
Hospital/health equipment - FC Frost Ltd
Hospital/health equipment - Miele Co Ltd
Hospital/Health Equipment - Wybone Ltd
Hospitol Health equipment - Hospital Metalcraft Ltd
Hotel & Leisure seating & tables - Knightbridge Furniture Productions Ltd
Laboratory furniture - Decra Ltd
Lifting, moving and handling products for disabled - Chiltern Invadex
Medical and first aid furniture - Hoskins Medical Equipment Ltd
Medical and first aid furniture - Hospital Metalcraft Ltd
Medical examination lamps - Wandsworth Elecrtrical Ltd
Medical trolleys - H&H Electronics Ltd
Pedal bins - Unicorn Containers Ltd
Polymer pipecover - Coverad Ltd
Polymer radiator covers - Coverad Ltd
Shower enclosures - Contour Showers Ltd
Slophers - Pland Stainless Ltd
Specialist bathing, transfer, hygiene and showering systems - Parker Bath Limited
Sprung mattresses, divan bases, pillows and accessories - Airsprung Beds Ltd
Stainless steel shelving systems for sterile areas - Beford & Soar Ltd
Steel radiator covers (LST) - Coverad Ltd
Steel Trunking systems - Marshall-Tufflex Ltd
Support systems for elderly/disabled - Norton Engineering Alloys Co Ltd
Supportive bathroom products for disabled people - Pressalit Care Ltd
Surgeons scrub up trough - Pland Stainless Ltd
Waiting area seating for train stations, airports, ferry terminals, hospitals and surgeries - Hille Ltd

N26 Gymnastic/ Sport/ Play Equipment

Activity Nets - Record Playground Equipment Ltd
Automatic slatted pvc swimming pool - Grando (UK) Ltd
Children's play areas and playground equipment - Roadways & Car Parks Ltd
Cloakroom equipment - Locking Systems Ltd
Cloakroom equipment including security locks - Locking Systems Ltd
Cloakroom rails - Turnquest UK Ltd
Dojo Matting - Fighting Films
Dojo matting - Geemat
Drama staging - Felix Design Ltd
Furniture, leisure - Allermuir Contract Furniture Ltd
Goals - Sportmark™ Ltd
Golf tee mats & frames - Evergreens Uk
Groundmans Supplies - Sportmark™ Ltd
GRP and plaster castings and molds - Regency Swimming Pools Ltd
Gymnastic matting - Continental Sports Ltd
Gymnastic, sport and play equipment - Carr Gymnasium Equipment Ltd
Gymnastic/sport/play equipment - Kee Klamp Ltd
Hydrotherapy pools - Aqua-Blue Ltd
Judo Matts - Fighting Films
Judo matts - Geemat
Leisure - Withey Contracts Ltd
Maintenance service - Continental Sports Ltd
Medical and first aid furniture - Doherty Medical Ltd
Nappy changers - LSA Projects Ltd
Nets - Sportmark™ Ltd
Panel pool kits - Golden Coast Ltd
Physical education equipment - Continental Sports Ltd
Play equipment - En-tout-cas plc (Scotland)
Playground equipment - Lappset UK Ltd
Playground equipment - Record Playground Equipment Ltd
Playground equipment and associated works - Kompan Ltd
Safety surfacings - En-tout-cas plc (Scotland)
Saunas, steam rooms and jacazzi spa pools - Aqua-Blue Ltd
Seating - cinema, bleacher, sport, multi-purpose, public area, telecopic - Race Furniture Ltd
Skateboard Equipment - Record Playground Equipment Ltd
Spare parts - Continental Sports Ltd
Sport nets - Birdscarer Products & Net Supplies Ltd
Sports arenas products - pvc-u - Duralock (UK) Ltd
Sports benches/Play equipment - Hispack Limited

N26 Gymnastic/ Sport/ Play Equipment (Cont'd)

Sports Covers - Power Plastics Ltd
Sports equipment - Continental Sports Ltd
Sports equipment - En-tout-cas plc (Scotland)
Sports Equipment - Sportmark™ Ltd
Stadia and auditorium seating - Metalliform Products Plc
Swimming Pool Designs - Aqua-Blue Ltd
Swimming pools and equipment - Golden Coast Ltd
Trampoline equipment - Continental Sports Ltd
Vanity units - Helmsman
Waterslides/flumes - Design & Display (Sales) Ltd

N27 Cleaning Equipment

Brooms - Cottam Bros Ltd
Cleaning & Waste - HC Slingsby PLC
Cleaning equipment - Arjo Ltd
Cleaning equipment - Interface Europe Ltd
Cleaning equipment - Monument Tools Ltd
Cleaning equipment - Torit DCE
Cleaning Equipment & Sundries - BRB Industrial Services Ltd
Cleaning maintenance equipment - Cento Engineering Company Ltd
Cleaning materials suppliers - Spontex Ltd
Cleaning Products - SC Johnson Wax
Hand cleaner - Halls Beeline Ltd
Hygiene products - Applied Chemicals Ltd
Industrial and commercial scrubbers, vacuum cleaners and floor maintenance equipment - Nilfisk Advance Ltd
Industrial and commercial vacuum cleaners - BVC
Maintenance, polish and cleaning chemicals - British Nova Works Ltd
Portable vacuum cleaners - Barloworld Vacuum Technology Plc
Power washers, scrubber driers, sweepers, polishers and vacuums - Alto Cleaning Systems (UK) Ltd
Specialist brushes - Cottam Bros Ltd
Specialist cleaning - New Centurian Industry Services Ltd
Wire Brushes - Cottam Bros Ltd

N28 Laboratory furniture/ equipment

Bench top extraction - Nederman UK
CAD Design facility - Lab Furnishings (UK) Ltd
Ceramic and acrylic sanitaryware - Armitage Shanks Ltd
Creteangle vibrating tables - Edward Benton & Co Ltd
Decontamination booths - Rose Emergency Equipment Co
Design, manufacture and installation of laboratory furniture and fume cupboards - Lab Systems Furniture Ltd
Enclosures and technical furniture - Knurr (UK) Ltd
Industrial laboratory furniture - S & B Ltd
Laboratorty furniture - Lab Furnishings (UK) Ltd
Laboratory design and build project management - Fisher Scientific (UK) Ltd
Laboratory Equipment - Edward Benton & Co Ltd
Laboratory Eqyiptment - USF Permutit/Ionpure
Laboratory Furniture - Complete Laboratory Installations
Laboratory furniture - Grant Westfield Ltd
Laboratory furniture - MICE Kaymar Limited
Laboratory furniture - Pulsford Associates Ltd, John
Laboratory furniture - Trespa UK Ltd
Laboratory furniture (education) - Klick Technology Ltd
Laboratory furniture and fume cupboards - P & R Laboratory Group Ltd
Laboratory furniture/equipment - FC Frost Ltd
Laboratory furniture/equipment - Miele Co Ltd
Laboratory furniture/equipment - Sissons W & G Ltd
Laboratory Sinks - Pland Stainless Ltd
Primary & Secondary Pharaceutical Solutions - Envair Ltd
Reprographic room equipment - Envopak Group Ltd

N29 Industrial Equipment

Access Equipment - Barton Storage Systems Ltd
Access equipment - HC Slingsby PLC
Aero-Engine Test Control System - IAC
Bowl Pressings - Pland Stainless Ltd
Commercial and industrial equipment - HC Slingsby PLC
Gatehouse - Arbory Group Limited
Handling & lifting - HC Slingsby PLC
Hopper feeder systems - Benton Co Ltd Edward
Industrial batteries for standby applications - Hawker Ltd
Industrial doors - Telford Doors and Barriers Ltd
Industrial equipment - Benton Co Ltd Edward
Industrial equipment - BRB Industrial Services Ltd
Industrial equipment - Edward Benton & Co Ltd
Industrial Equipment - Emmerich (Berlon) Ltd
Industrial equipment - Kee Klamp Ltd
Industrial equipment - Monument Tools Ltd
Industrial equipment - bins - Unicorn Containers Ltd
Industrial hoses & flexibles - Capper Plastics
Industrial instrumentation - ABB Instrumentation Ltd.
Industrial materials handling equipment - Barton B C & Son Ltd
Industrial shelving - Dexion Ltd
Industrial workbenches and storage drawer cabinets - Q.M.P.
Maintenance of industrial doors, shutters and shopfronts - Shutter Door Repair & Maintenance Ltd
Manual handling equipment - Hymo Ltd
Office and industrial cleaning - CSL (Services) Limited
Paint brushes, rollers and brooms - Cottam Bros Ltd
Power plant noise control systems - IAC
Ropes and sashcords - Marlow Ropes Ltd
Specialist brushes - Cottam Bros Ltd
Ultra violet chambers, lamps and control panels - Hanovia Ltd
Vehicle livery - Signs & Labels Ltd
Vibratory equipment and moulds - Benton Co Ltd Edward
Workshop equipment - HC Slingsby PLC

N30 Laundry Equipment

Clothes lines - Lever James & Sons Ltd
Domestic appliances - Miele Co Ltd
Finishing equipment - Electrolux Laundry Systems
Garment collectors - Brown N C (Storage Equipment) Ltd
Garment dispensers - Brown N C (Storage Equipment) Ltd
Hydro extractors - Electrolux Laundry Systems
Ironing tables - Electrolux Laundry Systems
Laundry equipment - Viking Laundry Equipment Ltd
Lockers - Brown N C (Storage Equipment) Ltd
Roller ironers - Electrolux Laundry Systems
Rotary clothes dryers - Hills Industries Ltd
Soft storage - Hills Industries Ltd
Tumble dryers - Electrolux Laundry Systems
Washers, dryers and ironers - Warner Howard Group Ltd
Washing machines - Electrolux Laundry Systems

N31 Entertainment/ Auditorium equipment

Auditorium & retractable seating - AEL Furniture Group
Auditorium seating - Evertaut Ltd
Conference Equipment - Wilson & Garden Ltd
Entertainment/auditorium euipment - Paragon Business Furniture
Lecture theatre seating - Evertaut Ltd

P10 Sundry insulation/ proofing work/ fire stops

Acoustic door seals - AET.gb Ltd
Acoustic flooring - Interfloor Limited
Acoustic insulation - InstaGroup Ltd
Acoustic insulation - Knauf Insulation Ltd
Acoustic panels - AET.gb Ltd
Automatic fire barrier-curtains - Cooper Group Ltd
Automatic fire shutters - Cooper Group Ltd
Bitumen building papers, film/paper laminated papers - Package Products Ltd
Cellular glass insulation, tapered roof systems - Pittsburg Corning (UK) Ltd
cold seal paper - Package Products Ltd
Expanded polystyrene insulation - Springvale E P S
Expanded polystyrene insulation - Vencel Resil Ltd
External wall insulation systems - MR Limited
Fiberglass fabrics - Fothergill Engineered Fabrics Ltd
Fibreglass loft insulation - Davant Products Ltd
Fire Collars - Abesco Ltd
Fire collars - FloPlast Ltd
Fire doors and doorsets - Industrial Building Componants Ltd
Fire proof cladding - Firebarrier Services Ltd
Fire Protection - Knauf Insulation Ltd
Fire protection and acoustic insulation - Ainsworth B R (Southern) Ltd
Fire protection casings - Cryotherm Insulation Ltd
Fire protection insulation spray - Cryotherm Insulation Ltd
Fire Rated Mastics - Abesco Ltd
Fire resisting partitions - Mann McGowan Group
Fire sound and thermal soloutions - Lafarge Plasterboard Ltd
Fire stopping systems - Cryotherm Insulation Ltd
Fire stops - Décor Ireland
Fire stops - FGF (Ashton) Ltd
Fire stops - Fire Protection Ltd
Fire stops - Quelfire
Fire stops around pipes - Industrial Hangers Ltd
Fire stops for cables pipes and ductwork - Quelfire
Fire stops for joints - Quelfire
Fire walls and barriers - Cryotherm Insulation Ltd
Fire Wraps - Abesco Ltd
Fireproofing and insulation - Thermica Ltd
Fireside Accessories - Brettell & Shaw Ltd
Firestopping materials - Preformed Components Ltd
Flame Retardant Boards - Montague L Meyer
Flame retardant sheeting - Anaplast (Greenock)
Flat roof insulation - Isocrete Floor Screeds Ltd
Floor insulation - Springvale E P S
Foam and rubber products - R A International
Full-Fire Certificate - Firebarrier Services Ltd
Hot and cold water cylinder jackets - Davant Products Ltd
Insuation panels - Cryotherm Insulation Ltd
Insulaion for Heating and Ventilation - Kingspan Industrial Ins. Ltd
Insulation - British Gypsum Ltd
Insulation - Callenders Ltd
Insulation - Deborah Services Limited
Insulation - Nevill Long Limited
Insulation - Rockwool Ltd
Insulation accessories - Astron Buildings Parker Hannifin PLC
Insulation for Process industries - Kingspan Industrial Ins. Ltd
Insulation Products - Allmat (East Surrey) Ltd
Insulation, fire protection and fire stopping - Kitsons Insulation Products Ltd
Intermescant fire and smoke seals - Mann McGowan Group
Intumecsent mastics - Promat Fire Protection Ltd
Intumescent seals for PVC pipes - Décor Ireland
Nullifire fire stopping range - Nullifire Ltd
O-rings - Manton Industrial Seals Ltd
Passive fire prevention - Potters Insulations (Sales) Ltd
Passive fire protection seals - Abesco Ltd
Plasterboard Dry Lining - Phoenix Metal Products Ltd
Plastic extrusions - Manton Industrial Seals Ltd
Plastic extrusions - Plasticable Ltd
Polyethylene and PVC pipe insulation - Davant Products Ltd
Polythene vapor barrier - Visqueen Building Products
Polyurethane and polystyrene insulation panels - Hemsec Manufacturing Ltd
Polyurethane foam (sprayed) - specialists - Stockbrands LLP
Rigid insulation boards - Kingspan Insulation Ltd
Rockarmour - Camas Aggregates Ltd - Aggregate Industries
Roof insulation - Dow Construction Products
Roof insulation - Springvale E P S
Roof Intallation/Tappered Cork - Westco Group Ltd
Rubber extrusions and strips - Manton Industrial Seals Ltd
Seals - Mann McGowan Group
Seals and draught strips - Manton Industrial Seals Ltd
Seals and intumescents - Laidlaw Architectural Hardware Ltd
Self sealing test plugs - Industrial Hangers Ltd
Seperating Floors - Sound Service (Oxford) Ltd
Servery fire curtain-industrial/commercial/residential - Cooper Group Ltd
Sleeving and tubing - Plasticable Ltd
Sound absorbtion panels - Henkel Industrial Adhesives
Sound insulation - Sound Service (Oxford) Ltd
Sound Proofing - InstaGroup Ltd
Spray Foam Insulation - Foamseal Ltd
Standard corkboard - Gradient Insulations (UK) Ltd
Sundry installation/proofing work/fire stops - Reiner Fixing Devices (Ex: Hardo Fixing Devices)
Sundry Insulation - Strand Hardware Ltd
Sundry insulation/proofing work/ fire stops - Knauf Insulation Ltd
Tapered Rockwool - Gradient Insulations (UK) Ltd
Tapered, corkboard, cork/PUR composite, polystyrene and PUR - Gradient Insulations (UK) Ltd
Thermal insulation - Axter Ltd
Vessel,Tank and Pipe insulation - nmc (UK) Ltd
Wall insulation - Springvale E P S
Washers and spacers - Manton Industrial Seals Ltd
Wax coated paper - Package Products Ltd

P11 Foamed/ Fibre/ Bead cavity wall insulation

Acoustic - Thermica Ltd
British wall cavity insulation - Kleeneze Sealtech Ltd
Cavity wall insulation - Callenders Ltd
Cavity wall insulation and loft insulation - InstaGroup Ltd
Expanded polystyrene insulation - Vencel Resil Ltd
External wall insulation - Weber Building Solutions
Extruded polystyrene - Knauf Insulation Ltd
Fibre cavity wall insulation - Knauf Insulation Ltd
Insulation - FGF (Ashton) Ltd
Insulation Products - Allmat (East Surrey) Ltd
Polyisocyanurate insulation board manufacture - Celotex Ltd

P20 Unframed isolated trims/ skirtings/ sundry items

Architectural and decorative mouldings - Mason FW & Sons Ltd
Architraue - Enfield Speciality Doors
Cellular PVC window trims, skirtings & architraves - celuform - Celuform
Doors and joinery - Brewer T. & Co Ltd
Flooring trims and accessories - McLoughlin Wood Ltd
GRP Dentil cornice - Alpha Glass Fibres Ltd
GRP Window architrave - Alpha Glass Fibres Ltd
Hardwood mouldings - Atkinson & Kirby Ltd
Linear grilles - Taskworthy Ltd
MDF moldings - Fibercill
Metal Trims - Homelux Nenplas
Moldings - OSMO UK Ltd
Plaster and wooden traditional molding - Reclaim.UK.Com
Plastic and Foam Components - nmc (UK) Ltd
Plastic handrails - Brighton (Handrails), W
Precast terrazo skirtings - Roman Mosaic Contracts Ltd
Protection rails and corner angles - Harrison Thompson & Co Ltd
PVC trims, angles & edges - Homelux Nenplas
Rubber coving - William Freeman Ltd
Sashcords, clothes lines and building - Kent & Co (Twines) Ltd
Simulated wood mouldings, reproduction oak beams - Oakleaf Reproductions Ltd
Skirting - Enfield Speciality Doors
Skirtings and architraves - Real Door Company The
Trims/ skirtings - Jali Ltd
Wall and Corner protection - Marshall Contracts Ltd
Window board sets - Thomas Lowe Joinery
Window films to existing glass - Hi-Tec Window Films
Wood finishes - OSMO UK Ltd

P21 Door/Window ironmongery

Architectural door furniture and security hardware - Allgood plc
Architectural door furniture, hangers, screws and hinges - RAP Group Ltd
Architectural hardware - Laidlaw Architectural Hardware
Architectural hardware including hinges, locks, levers, flush bolts and other ancilliary items including ceiling fans - James Gibbons Format Ltd
Architectural hinges, door closers and thresholds - Relcross Ltd
Architectural ironmomgery - Samual Heath
Architectural ironmongerers - Moffett Thallon & Co Ltd
Architectural ironmongery - Doorfit Products Ltd
Architectural Ironmongery - Haffele UK Ltd
Architectural ironmongery - Industrial Devices Ltd
Architectural ironmongery, cabinet hardware and fittings - Lord Isaac Ltd
Architectural ironmongery, security locks and panic equipment - Till & Whitehead Ltd, t/a L. Birkinshaw
Automatic door operations - Hillaldam Coburn Ltd
Black Antique Ironmongery - Kirkpatrick Ltd
Brass furniture fittings - Martin & Co Ltd
Brass hardware - Marcus M Ltd
Brass, bronze door and window hardware - Allart, Frank, & Co Ltd
Brush strip door and window seals - Kleeneze Sealtech Ltd
Brushstrip door and letterbox seals - Kleeneze Sealtech Ltd
Butt hinges - Caldwell Hardware (UK) Ltd
Cash transfer units - CSI Security LTD
Castors and wheels - Flexello Limited
Catches & bolts - Caldwell Hardware (UK) Ltd
Computer lockdown units - Baddeley
Concealed door closers, - Samual Heath
Cotton twine - Lever James & Sons Ltd
Door / window ironmongery - Brass Art Ltd
Door closers - Forson Design & Engineering Ltd
Door closers - Permclose Door Services Ltd
Door Controls - Owlett - Jaton
Door fixings - Fairmitre Ltd
Door furniture - ASSA Ltd
Door furniture - Turnquest UK Ltd
Door furniture and hinges - Laidlaw Architectural Hardware Ltd
Door hardware - ASSA Ltd
Door Hardware - Paddock Fabrications Ltd
Door ironmongery - Dorma Door Services Ltd
Door Locks - Bramah Security Equipment Ltd
Door Operators/Closers & Accessories - Abloy Security Ltd
Door/ window ironmongery - Architectural Components Ltd
door/window iironmongery - UK Fasteners Ltd
Door/Window Ironmongery - Crowthorne Fencing
Door/Window Ironmongery - Daro Factors Ltd
Door/window Ironmongery - Draught Proofing Advisory Association
Door/window ironmongery - Hart Door Systems Ltd
Door/window ironmongery - Magnet Ltd

P21 Door/Window ironmongery (Cont'd)

Door/Window ironmongery - Roto Frank Ltd
Dor plates - Fixatrad Ltd
Draught, weather, intumescent, fire and smoke seals - Sealmaster
Electric Door Operator - Relcross Ltd
Emergency Exit hardware - Jebron Ltd
Emergency exit locks, break glass locks - Pickersgill Kaye Ltd
Fasteners - Glasgow Steel Nail Co Ltd
Fire door motors and controls - Kaba Garog
Fire door operating equipment - Kaba Garog
Fixings - Glasgow Steel Nail Co Ltd
Floor springs - Forson Design & Engineering Ltd
Floor springs - Jebron Ltd
Folding openers - Caldwell Hardware (UK) Ltd
Friction stays - Securistyle Ltd
Furniture castors and hardware - Kenrick Archibald & Sons Ltd
Galvanised water bar - Mid Essex Trading Co Ltd
Gas springs - Arvin Motion Control Ltd
Gate locks - Clarke Instruments Ltd
GRP Decorative brackets - Alpha Glass Fibres Ltd
Handles - Securistyle Ltd
Hardware - Joseph Tipper Ltd
Hardware - Reddiglaze Ltd
Hardware for windows & doors - Fab & Fix Ltd
Hardware for windows and doors - Avocet Hardware Ltd
Hardware for windows and doors - Roto Frank Ltd
Heavy duty hinges - Securistyle Ltd
High security locks and locking systems - Abloy Security Ltd
Hinged steel security door sets - Konematic Bolton Brady
Hinges - Cooke Brothers Ltd
Industrial locks - ASSA Ltd
Intumescent fire stopping products - Sealmaster
Ironmongery - Ashmead Buidling Supplies Ltd
Ironmongery - Brewer T. & Co Ltd
Ironmongery - Caldwell Hardware (UK) Ltd
Ironmongery - Challenge Fencing Ltd
Ironmongery - Clayton-Munroe Ltd
Ironmongery - Grorud Industries Ltd
Ironmongery - Groupco Ltd
Ironmongery - Harris & Bailey Ltd
Ironmongery - Hipkiss, H, & Co Ltd
Ironmongery - IR Security and Safety
Ironmongery - Maco Door & Window Hardware (UK) Ltd
Ironmongery - Sampson T F Ltd
Ironmongery - Strand Hardware Ltd
Ironmongery - Tinsley & Co Ltd, Eliza
Ironmongery - Owlett - Jaton
Leading manufacturer for door and window furnitures - HOPPE (UK) Ltd
Locking Systems - Securistyle Ltd
Locks - Aardee Security Shutters Ltd
Locks - Guardian Lock and Engineering Co Ltd
Locks - Ingersol Locks Ltd
Locks - Lowe & Fletcher Ltd
Locks - Parkes Josiah & Sons Ltd
Locksmiths - Doorfit Products Ltd
Mail boxes, door closers, door furniture, door locks and cylinders - JB Architectural Inronmongery Ltd
Malleable Iron - Regon Ltd
Manual and electrical systems for opening windows - Dyer Environmental Controls
Multi point locking systems - Paddock Fabrications Ltd
Overhead door closers - Jebron Ltd
Padlockable bars - Baddeley
Padlocks - Securikey Ltd
Padlocks and locks - Dom-Nemef-Corbin
Padlocks security products, cyclelocks and motorcycle locks - Squire Henry & Sons
Panic Exit Hardware - Abloy Security Ltd
Panic hardware - ASSA Ltd
Patio hardware - Caldwell Hardware (UK) Ltd
Pivot hinges - Caldwell Hardware (UK) Ltd
Porcelain door furniture - Balmoral Porcelain Limited/Longmead Ceramics
Porcelain finger plates - Balmoral Porcelain Limited/Longmead Ceramics
Porcelain furniture knobs - Balmoral Porcelain Limited/Longmead Ceramics
Porcelain mortice sets - Balmoral Porcelain Limited/Longmead Ceramics
Push biutton and cobination locks - Relcross Ltd
Restrictors - Caldwell Hardware (UK) Ltd
Rim, mortice, padlocks, cylinders and panic escape locks - Yale Security Products Ltd
Rim-lock boxes - Baddeley
Roller shutter motors and controls - Kaba Garog
Ropes and sashcords - Marlow Ropes Ltd
Sah window hardware and accessories - Garador Ltd
Sash balances - Caldwell Hardware (UK) Ltd
Sash cord - Lever James & Sons Ltd
Sectional door motors and controls - Kaba Garog
Security cylinders/ padlocks - Dom-Nemef-Corbin
Security inronmongery - CSI Security LTD
Security products - Marcus M Ltd
Security products, black ironmongery and hardware - ERA Products
Sercurity Products - Owlett - Jaton
Shelving brackets - decorative - Winther Browne & co Ltd
Shootbolt - Securistyle Ltd
Shootbolt Locking Systems - Nico Manufacturing Ltd
Sliding door gear - Avon Industrial Doors Ltd
Sliding Door Gear - Lakeside Group
Smoke ventilation systems - Dyer Environmental Controls
Solenoid locks - Clarke Instruments Ltd
Specialist ironmongery - The Cotswold Casement Company
Tills, drawers and paystations - CSI Security LTD
Transom closers - Jebron Ltd
Trickle Ventilations - Ryton's Building Products Ltd
Weatherstriping joinery seal, roof cappings, glazing bars and ridge systems - Exitex Ltd
Weatherstripping - Reddiglaze Ltd
Window and door hardware - PVCu, timber and aluminum - Wagner (GB) Ltd
Window and door ironmongery - Laird Security Hardware
Window hardware - Paddock Fabrications Ltd
Window ironmongery - Securistyle Ltd
Window Locks - Bramah Security Equipment Ltd

P22 Sealant joints

Acoustic seals - Mann McGowan Group
Bath and kitchen seals - Homelux Nenplas
Expansion joints - Fillcrete Ltd
Expansion joints - Pitchmastic Pmb Ltd
Expansion joints - William Freeman Ltd
Fire resistant expansion joint seals - Mann McGowan Group
Joint fillers - Fosroc Ltd
Joint sealants - Fosroc Ltd
Jointing & tooling - The BSS Group PLC
Jointing compounds - The BSS Group PLC
Jointing materials - Furmanite International Ltd
Mechanical movement joints and fire rated sealants - Compriband Ltd
Profiled sealants - Henkel Industrial Adhesives
Sealant joints - Firebarrier Services Ltd
Sealants, solvents, wood preservatives and creosote - Bartoline Ltd
Sealent joints - Canonbury Asphalte Co. Ltd.
Silicones for industry - Ambersil Ltd
Weather draft seals - Mann McGowan Group

P30 Trenches/Pipeways/Pits for buried engineering services

Cable and pipeline locators - Boddingtons Ltd
Cable Duct chambers - Brierkrete Ltd
Cable warning - Boddingtons Ltd
Chambers - Talbot
Detection and protection products - Boddingtons Ltd
Leak location service - Boddingtons Ltd
Meter Boxes - Talbot
Underground warning tape - Boddingtons Ltd

P31 Holes/ Chases/ Covers/ Supports for services

Access covers - Bilco UK Ltd
Access floor service outlet boxes and grommet outlets - Electrak International Ltd
Aluminium and brass recessed floor access covers - Howe-Green Ltd
Brass brackets + tubes - Regon Ltd
Brass bushes + brackets - Regon Ltd
Builders cast iron products - Dudley Thomas Ltd
Channel Supports - Kem Edwards Ltd
Concealed access panels for tiled access - Padipa Ltd
Copper rods + stips - Regon Ltd
Gear Trays - Jerrards Bros. PLC Division
Heavy duty access hatches - Padipa Ltd
Low Surface temperature Covers - Zehnder Ltd
Malleable iron brackets + rings - Regon Ltd
Mechanical Pipe Supports - Kem Edwards Ltd
Pipe clip manufacturers - Bolivar Stamping Ltd
Pipe support componants - Industrial Hangers Ltd
Plastic, steel + rubber clips - Regon Ltd
Pressed grating - Barton B C & Son Ltd
Radiator cabinets / covers - Winther Browne & co Ltd
Roll grooving machine - Hilmor
Services (cables&Pipes), Ducting - Promat UK Ltd
Stainless steel access covers - Howe-Green Ltd
T30/T90 fireproof inspection - Padipa Ltd
Ventilated access panels - Padipa Ltd

Q10 Kerbs/Edgings/Channels/Paving accessories

Bowland stone paving - Concrete Products (Lincoln) 1980 Ltd
Bowling Green Ditches - Brierkrete Ltd
Carpark Marking - B&C Roadmarkings Ltd
Cills, lintols, steps - Forest of Dean Stone Firms Ltd
Combined kerb/drainage units - ACO Technologies Plc
Concrete gutter blocks - Finlock Gutters
Concrete paving, blocks and kerbs - Camas Building Materials
Edging - Rogers Concrete Ltd
Edgings and copings - Oakdale (Contracts) Ltd
Garden edgings - York Handmade Brick Co Ltd
Granite kerbs - CED Ltd
Grass concrete blocks - Concrete Products (Lincoln) 1980 Ltd
Grease traps - ACO Technologies Plc
Handcut & Masoned - Yorkstone - Pickard Group
Hard landscaping - Bridge Quarry
High performance thin wearing courses - Camas Aggregates Ltd - Aggregate Industries
Kerb and channels - Bardon (England) Ltd
Kerbing - Rediweld Rubber & Plastics Ltd
Kerbing & Paving - Gearing F T Landscape Services Ltd
Kerbs - Harris & Bailey Ltd
Kerbs / Edgings / Channels / Paving Accessories - Formpave Ltd
Kerbs and accessories - Brett Paving
Kerbs edgings - Stowell Concrete Ltd
Kerbs, edgings, Channels, Paving accessavial - Grass Concrete Ltd
Kerbs/edgings/ Channels/ Paving - Macclesfield Stone Quarries Ltd
Kerbs/Edgings/channels/pavind accessories - Townscape Products Ltd
Kerbs/Paving Accessories - Realstone Ltd
Landscape Suppliers - Crowthorne Fencing
Path edging - Atlas Stone Products
Pavers - Baggeridge Brick PLC
Paving slab supports - Compriband Ltd
Precast concrete and GRC surface water channels - Allpipe Ltd
Protective skirting - ACO Technologies Plc
Sleeping PoliceMen - B&C Roadmarkings Ltd
Stone kerbs, edgings, and channels - Rare Stone Group, The
Synthetic ironoxide colouring agents - Bayer Chemicals Inoegate Pigment Group
White Lining - B&C Roadmarkings Ltd

Q20 Granular sub-bases to roads/ pavings

Clay Pavers - Hanson Brick Ltd, Northern Regonal Sales Office
Crushed & graded stone - Aggregates - Pickard Group
Granular sub-bases - Stenoak Associated Services PLC
Surfacing and construction - Halls Group Ltd
Suspended ceilings - Phoenix Metal Products Ltd

Q21 Insitu concrete roads/ pavings/ bases

Insitu concrete roads/pavings/bases - Grass Concrete Ltd

Q22 Coated macadam/ Asphalt roads/ pavings

Anti Slip Surface - B&C Roadmarkings Ltd
Asphalt - Pioneer Concrete Holdings PLC
Asphalt Roads - Gearing F T Landscape Services Ltd
Asphalts - Hanson Building Materials Europe
Bitumen boilers - Bitmen Products Ltd
Bitumen macadam, hot rolled asphalt, and sub base aggregates - Johnston Roadstone Ltd
Bituminous coated stone products - Fyfe John Ltd
Civil engineering and road surfacing - Hanson Building Materials Europe
Coated macadam and asphalts - Camas Aggregates Ltd - Aggregate Industries
Coated macadam/ asphalt roads/ pavings - pavement technology - Stenoak Associated Services PLC
Cold emulsion sprayers - Bitmen Products Ltd
Infra ray heaters - Bitmen Products Ltd
Macadam - Bardon (England) Ltd
Macadam and asphalt - Canonbury Asphalte Co. Ltd.
Mastic asphalt mixers - Bitmen Products Ltd
Mastic asphalt road contractors - Tarmac Ltd
Paver hire - Johnston Roadstone Ltd
Pre coated chipspreaders - Bitmen Products Ltd
Red macadam - Johnston Roadstone Ltd
Reflective tarffic markings - Reflecting Roadstuds Ltd
Road surfacing - Fyfe John Ltd
Surfacing contractors - Tarmac Ltd
Synthetic ironoxide colouring agents - Bayer Chemicals Inoegate Pigment Group

Q23 Gravel/ Hoggin/ Woodchip roads/ pavings

Aggregate & Stone - Aggregate Industries UK Ltd
Aggregates - Bardon (England) Ltd
Bowland stone decorative aggregates - Concrete Products (Lincoln) 1980 Ltd
Decorative aggregates - CED Ltd
Footpath gravels & hoggin - CED Ltd
Loose aggregates - B & M Fencing Limited
Slate chippings - The Delabole Slate Co. Ltd

Q24 Interlocking brick/ block roads/ pavings

Block Paving - B & M Fencing Limited
Block paving - Thakeham Tiles Ltd
Blockpaving contractors - Gearing F T Landscape Services Ltd
Clay block paving - Oakdale (Contracts) Ltd
Clay pavers - Marshalls Mono Ltd
Concrete block paving - CEMEX
Interlocking block pavings - Formpave Ltd
Interlocking brick/ block roads/ pavings - Grass Concrete Ltd
Interlocking brick/ block roads/ pavings - Stenoak Associated Services PLC
Landscaping products - Aggregate Industries UK Ltd
Pavers - Baggeridge Brick PLC
Pavers - Cranleigh Brick & Tile Co Ltd
Pavers - Freshfield Lane Brickworks Ltd
Pavers - Ibstock Building Products Ltd
Pavers - York Handmade Brick Co Ltd
Paving - Milner Delvaux Ltd
Paving - Townscape Products Ltd
Paving - block - Brett Paving
Paving/ hard landscaping - Blanc de Bierges
Pavings - Prismo Ltd
Pavings - York Handmade Brick Co Ltd
Synthetic ironoxide colouring agents - Bayer Chemicals Inoegate Pigment Group

Q25 Slab/ Brick/ Sett/ Cobble pavings

Boulders, cobbles and pebbles - CED Ltd
Building materials - block paving - Camas Aggregates Ltd - Aggregate Industries
Clay pavers - Castle Brick (Wales) Ltd
Clay pavers - Hanson Brick
Clay Pavers - Ruabon Dennis Ltd
Cobbles - York Handmade Brick Co Ltd
Concrete slab, block paving - Marshalls Mono Ltd
Decorative precast paving - Oakdale (Contracts) Ltd
Granite paving - CED Ltd
Granite setts - CED Ltd
Grass reinforcement systems - Grass Concrete Ltd
Heavyside paving - Oakdale (Contracts) Ltd
Landscape Suppliers - Crowthorne Fencing
Limestone in lumps - Bennett Jr Ltd Ben
Limestone paving - CED Ltd
Natural stone paving - Marshalls Mono Ltd
Pavac - Park Products
Pavers - Baggeridge Brick PLC
Paving - Atlas Stone Products
Paving - Bardon (England) Ltd
Paving - Burlington Slate Ltd
Paving - Forest of Dean Stone Firms Ltd
Paving - sett - Brett Paving
Paving slabs - Stowell Concrete Ltd
Pavings - Prismo Ltd
Pavings - Ruthin Precast Concrete Ltd
Paviors - Country Manor Bricks
Porphyry setts and edging - CED Ltd
Precast concrete paving - Rogers Concrete Ltd
Regency paving and flagstone paving - Thakeham Tiles Ltd
Roadstone - Pickard Group
Roofing and paving - Bridge Quarry
Slab / Brick / Sett / Cobble pavings - Formpave Ltd
Slab / brick/ sett/ cobble pavings - Harris & Bailey Ltd
Slab/ brick/ sett/ cobble pavings - Chilstone
Slab/ Brick/ Sett/ Cobble pavings - Macclesfield Stone Quarries Ltd
Slab/ brick/ sett/ cobble pavings - Stenoak Associated Services PLC
Slab/ Brick/stett/cobble pavings - Townscape Products Ltd
Slab/Brick/Cobble Pavings - Realstone Ltd
Slab/sett cobble pavings - Rare Stone Group, The
Slabs - B & M Fencing Limited
Stone paving - Macc Stone Ltd
Swimming pool surrounds, hard landscaping and steps - Milner Delvaux Ltd
Synthetic ironoxide colouring agents - Bayer Chemicals Inoegate Pigment Group
Yorkstone paving - CED Ltd

Q26 Special surfacings/ pavings for sport/general amenity

Aesthetic surfacing - colours - Camas Aggregates Ltd - Aggregate Industries
Anti skid surfacing products - Clare R S & Co Ltd
Anti slip surfacing - Addagrip Surface Treatments Ltd
Artificial Grass - Evergreens Uk
Artificial sports surfaces and safety surfacing - En-tout-cas plc (Scotland)
Athletic tracks and synthetic grass sports pitches - Bernhards Landscapes Ltd
Bowling green equipment and channels - Sportmark™ Ltd
Children's play areas and sports pitches - Easi-Fall International Ltd
Clay pavers - Michelmersh Brick & Tile Co Ltd
Coloured sufacing products - Clare R S & Co Ltd
Construction, refurbishing and all types of tennis courts - En-tout-cas plc
Exterior Pitch Markings - B&C Roadmarkings Ltd
Gauged arches - Michelmersh Brick & Tile Co Ltd

Q26 Special surfacings/ pavings for sport/general amenity (Cont'd)

Hardwood decking - WL West & Son Limited
Impact absorbing play surfaces - Lawrence Surfaces Plc, Charles
Indoor/outdoor wet areas - Evergreens Uk
Marking products and materials - Marking Machines International
Outdoor matting soloutions - Kleen-Tex Industries Ltd
Permeable paving - CEMEX
Playground surfacing - Record Playground Equipment Ltd
Poolsurrounds - Evergreens Uk
Resin bonded porous surfacing - Addagrip Surface Treatments Ltd
Resin bonded surface dressing - Addagrip Surface Treatments Ltd
Road markings - Prismo Ltd
Rubber paving - Linatex Ltd
Rubber surfaces for play areas, paddling and swimming pools surrounds, "kick about" play areas and lockers rooms in golf clubs - Easi-Fall International Ltd
Seeding Turfing - Jack Moody Ltd
Special surfacings/pavings for span/general amenity - Grass Concrete Ltd
Specialist surfaces for raods - Conren Ltd
Specialist surfacing contractors - Tarmac Ltd
Sports court marking - Sportmark™ Ltd
Sports Hall Markings - B&C Roadmarkings Ltd
Sports paving - Bardon (England) Ltd
Sports Surfaces - Evergreens Uk
Sports surfacing and pavings - Lawrence Surfaces Plc, Charles
Squash court play surfaces - Bicester Products Ltd
Synthetic grass for residential use - Palmyra Ltd
Synthetic grass for sport - Palmyra Ltd
Synthetic pitches - Balsam (UK) Ltd
Tactile surfaces for visually impaired - Rediweld Rubber & Plastics Ltd
Tennis courts - Roadways & Car Parks Ltd

Q30 Seeding/ Turfing

Aquatic landscaping services - Land Wood & Water Co. Ltd
External planting - Jack Moody Ltd
Grass seed and turf - Sportmark™ Ltd
Hydraulic seedings - Hydraseeders Ltd
Landscape contracting - Land Wood & Water Co. Ltd
Landscaping - Cila Landscapes
Seeding/ turfing - Clifton Nurseries
Seeding/ turfing - Coblands Landscapes Ltd
Seeding/ turfing - Frosts Landscape Construction Ltd
Seeding/ Turfing - Reyven (Sportsfields) Ltd
Seeding/Turfing - Gearing F T Landscape Services Ltd
Soft and hard landscaping - Western landscapes Ltd t/a Wyevale Landscapes
Turfing - Town & Country Landscapes Ltd

Q31 External planting

Aquatic landscaping services - Land Wood & Water Co. Ltd
Artificial plants - Desney Products Ltd
Artificial trees - Desney Products Ltd
Bespoke landscape - CTS Ltd
External planting - Landscape Institute, The
External planting - Urban Planters
Garden Labdscape - Titan plastech
General exterior works contractors - Gearing F T Landscape Services Ltd
Landscape contracting - Land Wood & Water Co. Ltd
Landscape contractors - Gearing F T Landscape Services Ltd
Landscape mulcher - ABG Ltd
Landscape ornaments - Haddonstone Ltd
Landscapes - Ogilvie Construction Ltd
Landscaping - Cila Landscapes
Landscaping and grounds maintenance - Reyven (Sportsfields) Ltd
Planters - Blanc de Bierges
Planting - Bernhards Landscapes Ltd
Planting - Clifton Nurseries
Planting - Coblands Landscapes Ltd
Planting - Frosts Landscape Construction Ltd
Soft and hard landscaping - Western landscapes Ltd t/a Wyevale Landscapes
Tree and shrub planting - Town & Country Landscapes Ltd

Q32 Internal planting

Artificial plants - Desney Products Ltd
Artificial plants and trees - Pouliot Designs by Floralsilk Ltd
Artificial trees - Desney Products Ltd
Interior design (hotels) - TFL International Ltd
Interior landscape - Rare Stone Group, The
Interior planting - Urban Planters
Internal planting - Frosts Landscape Construction Ltd
Internal planting - Haddonstone Ltd
Internal planting - Landscape Institute, The
internal planting - Paragon Business Furniture
Office/ interior planting displays - Clifton Nurseries
Plastic Hardware - Titan plastech
Thermoplastic road marking - Clare R S & Co Ltd

Q35 Landscape maintenance

Garden furniture - Clifton Nurseries
Garden sheds - Ableson L & D Ltd
Garden timbers - Finn Forest
Ground cover materials - Rawtex Ltd
Horticultural mesh - Boddingtons Ltd
Horticultural Products made of natural fibres - Rawtex Ltd
Landscape design - Clifton Nurseries
Landscape featurestone - The Delabole Slate Co. Ltd
Landscape furniture - Clifton Nurseries
Landscape maintenance - Coblands Landscapes Ltd
Landscape maintenance - Frosts Landscape Construction Ltd
Landscape maintenance - Gearing F T Landscape Services Ltd
Landscape maintenance - Landscape Institute, The
Landscape maitenance - Jack Moody Ltd
Landscaping - Bernhards Landscapes Ltd
Light landscaping installation and maintenance - Sportmark™ Ltd
Manufacturer of artificial trees and plant displays - The Silk Forest
Plant pot tops - Rawtex Ltd
Soft and hard landscaping - Western landscapes Ltd t/a Wyevale Landscapes
Soft/ Hard Landscaping - Roadways & Car Parks Ltd
Special surfaces etc - Jack Moody Ltd
Sports Sands and Rootzones - Hepworth Minerals & Chemicals Ltd
Summer houses - Ableson L & D Ltd
Tree guards - Boddingtons Ltd
Tree surrounds - Jones of Oswestry Ltd
Walling and rockery - Forest of Dean Stone Firms Ltd
Wood pole protectors - Anti-Climb Guards Ltd

Q40 Fencing

Acoustic barriers - PHI Group Ltd
Agricultural fencing - Chestnut Products Limited
Alarm fencing - Allen (Fencing) Ltd
All Types - Havering Fencing Co.
All types of fencing - Cardinal Fencing (Northern) Limited
All types of fencing - Dinnington Fencing Co. Limited
Architectural trellises and planters - Anthony de Grey Trellises
Argricultural & Equestrian - Metalwood Fencing (Contracts) Limited
Automated & manual gates - Metalwood Fencing (Contracts) Limited
Automated gates - Dinnington Fencing Co. Limited
Automatic Gates - Tate Fencing
Barbed wire - Anti-Climb Guards Ltd
Barbican steel fencing - Jackson H S & Son (Fencing) Ltd
BFS Pro-net Licensed - Burn Fencing Limited
Bowtop fencing - Jackson H S & Son (Fencing) Ltd
BS1722 parts 1-16 inclusive - Burn Fencing Limited
Builders merchants - Halls Group Ltd
Chain - open linked for fencing or lifting - Griff Chains Ltd
Chain Link - Alltype Fencing Specialists
Chain link - Fencing Design & Development Co Limited
Chain link fencing, concrete fence posts - Procter Cast stone & Concrete Products
Chainlink - B & M Fencing Limited
Chainlink fencing - Hy-Ten Reinforcement Co Ltd
Chestnut steel palisade, chain link, railings, weld mesh, barbed and razor wire - Guardian Wire Ltd
Closeboard - B & M Fencing Limited
Closeboard - Danbury Fencing PLC
Closeboard fencing - Warefence Limited
Commercial fencing and gates - Tonbridge Fencing Limited
Concrete, wire fencing and gates - Challenge Fencing Ltd
Design of security fences - Fencing Design & Development Co Limited
Diamond trellis - Danbury Fencing PLC
Domestic - Metalwood Fencing (Contracts) Limited
Domestic fencing and gates - Tonbridge Fencing Limited
Enclosures - Fibaform Products Ltd.
Environmental barrier/ acoustic fencing - Newton & Frost
Erectors for private, domestic and commercial - Danbury Fencing PLC
Euroguard mesh fencing - Jackson H S & Son (Fencing) Ltd
Fence panel manufacturers - Derby Timber Supplies
Fencepost Support System - Catnic
Fencing - Advanced Fencing Systems Ltd
Fencing - AGN Fencing Limited
Fencing - Allen (Concrete) Ltd
Fencing - Alltype Fencing Specialists
Fencing - Ambec Fencing
Fencing - AMK Fence-In Limited
Fencing - Auckland Construction Ltd
Fencing - B & M Fencing Limited
Fencing - Barker & Geary Limited
Fencing - Bedford Fencing Co. Limited
Fencing - Benfield Fencing Limited
Fencing - Berry Systems
Fencing - Betafence Limited
Fencing - Bowden Fencing Ltd
Fencing - Brads Fencing Co. Limited
Fencing - Burn Fencing Limited
Fencing - Butler Davies (Fencing) Limited
Fencing - Capel Fencing Contractors Ltd.
Fencing - Challenge Fencing Ltd
Fencing - Chestnut Products Limited
Fencing - Cila Landscapes
Fencing - Claydon Architectural Metalwork Ltd
Fencing - Coates Fencing Limited
Fencing - Danbury Fencing PLC
Fencing - Dinnington Fencing Co. Limited
Fencing - Ebor Concretes Ltd
Fencing - Eve Trakway
Fencing - Fabrikat (Nottingham) Ltd
Fencing - Faversham Fencing Ltd
Fencing - FCY Plc
Fencing - Fence Hire Limited
Fencing - Fencing Contractors Association
fencing - Fencing Design & Development Co Limited
Fencing - Fernden Construction (South Western) Limited
Fencing - Fernden Construction Limited
Fencing - Fleetwood Fengate Co Ltd
Fencing - Frosts Landscape Construction Ltd
Fencing - GBG Fences Limited
Fencing - Gilmer Limited
Fencing - GJ Durafencing Limited
Fencing - Goodacres Fencing
Fencing - Gravesend Fencing Limited
Fencing - Hadley Industries Plc
Fencing - Halls Group Ltd
Fencing - Havering Fencing Co.
Fencing - Huntree Fencing Ltd
Fencing - J Pugh-Lewis Limited
Fencing - J. B. Corrie & Co Limited
Fencing - Jack Moody Ltd
Fencing - Kingsforth Security Fencing Limited
Fencing - Leicester Fencing Contracts Limited
Fencing - Maldon Fencing Co.
Fencing - May Gurney (Technical Services) Ltd - Fencing Division
Fencing - Metalwood Fencing (Contracts) Limited
Fencing - Newton & Frost
Fencing - Northern Fencing Contractors Limited
Fencing - Pearman Fencing
Fencing - Pembury Fencing Limited
Fencing - Portia Engineering Ltd
Fencing - Pro-Fence (Midlands)
Fencing - Record Playground Equipment Ltd
Fencing - Reyven (Sportsfields) Ltd
Fencing - Royston Steel Fencing Limited
Fencing - Ryall & Edwards Limited
Fencing - Sanhall Limited
Fencing - Scott Fencing Limited
Fencing - SFC (Midlands) Ltd
Fencing - Silverwood Fencing Co. Limited
Fencing - Sovereign Fencing Limited
Fencing - Stenoak Associated Services PLC
Fencing - Termstall Limited
Fencing - The Betta Group
Fencing - Town & Country Landscapes Ltd
Fencing - W Ward Fencing Contractors
Fencing - Wallbank
Fencing - Warefence Limited
Fencing - Weatherley Fencing Contractors Limited
Fencing - Wentworth Sawmills Ltd,
Fencing - WH Wesson Fencing Limited
Fencing - White Gates
Fencing - WL West & Son Limited
Fencing - all types - Anti-Climb Guards Ltd
Fencing (industrial) - Darfen Durafencing
Fencing and barriers - George James Partnership
Fencing and gates - Farefence Limited
Fencing and gates - May Gurney (Technical Services) Ltd
Fencing and gates, security fencing - Binns Fencing Ltd
Fencing and Sectional Buildings - Howe Fencing & Sectional Buildings
Fencing and security fencing - Jackson H S & Son (Fencing) Ltd
Fencing contractors - Gearing F T Landscape Services Ltd
Fencing contractors - Grange Fencing Erectors Limited
Fencing contractors - James D Cocks (Enterprises)
Fencing gates and environmental barriers - Baker Fencing Limited
Fencing including steel palisade, power operated gates and metal railings - Charnwood Fencing Limited
Fencing mesh - Boddingtons Ltd
Fencing suppliers - Crowthorne Fencing
Fencing Supply - Havering Fencing Co.
Fencing, gates and barriers to ISO 9002 - Albion Fencing Ltd
Fencing, gates, automatic gates and barriers - Tate Fencing
Fencing, gates, railings, pedestrian guard rail and crash barrier - Billericay Fencing Ltd
Fencing, gates, railings, traffic barriers and turnstiles - Elkosta; (UK) Ltd
Fencino (metal) - Steelway Fensecure Ltd
Flow plates - Autopa Ltd
Gate Furniture - RAP Group Ltd
Gates - Alltype Fencing Specialists
Gates - Claydon Architectural Metalwork Ltd
Gates - Coates Fencing Limited
Gates - Fernden Construction (South Western) Limited
Gates - WA Skinner & Co Limited
Gates & railings - Crimeguard Systems Ltd
Gates and access control - Chestnut Products Limited
Gates, barriers, shutters, fencing and parking post - DP Sercurity
Gates, coloured concrete posts, ornamental, vertical bar and bow top railings - Procter Bros Limited
General fencing, safety and noise barrier - Chris Wheeler Construction Limited
Grating fencing - Orsogril Spa UK Distributor
Guard rail and barrier fencing - Stainton Metal Co Ltd
Heavy duty gates and railways - Ermine Engineering Co. Ltd
High Security - Alltype Fencing Specialists
High Visibility Fencing - Polypipe Civils Ltd
Highway safety fencing - Chestnut Products Limited
Hinged poles - Stainton Metal Co Ltd
Industrial and commercial fencing and gates - M & M Fencing Limited
Industrial and domestic fencing - ACP Concrete (Contracts) Limited
Jakcure treated timber fencing - Jackson H S & Son (Fencing) Ltd
Local authority suppliers of fencing - Duralock (UK) Ltd
Lochrin classic rivetless palisade - Wm. Bain Fencing Ltd
Manufacturers of palisade fencing, temporary fencing and systems for all applications - security, commercial, industrial, sports and playgrounds - Fairmile Fencing Ltd
Metal fencing, railings and barriers - Barker W H Ltd
NRG fabrications - NRG Fabrications
On site welding - C & W Fencing Ltd
Paladin - Alltype Fencing Specialists
Palisade - Alltype Fencing Specialists
Palisade - Procter Cast stone & Concrete Products
Palisade fencing and gates - Wm. Bain Fencing Ltd
Panels and posts - B & M Fencing Limited
Perimeter Protection - Guardall Ltd
Perimeter security - Geoquip Ltd
Plastic battens and fencing - Rockwell Sheet Sales Ltd
Post & Rail - B & M Fencing Limited
Post and rail fencing - Warefence Limited
Power fencing - Allen (Fencing) Ltd
Precast concrete fencing - Anderton Concrete Products Ltd
PVC fencing - LB Plastics Ltd
Railings - Alltype Fencing Specialists
Railings - Fernden Construction (South Western) Limited
Railings, gates - Albion Manufacturing
Recycling green waste - Jack Moody Ltd
Retaining wall systems - Fernden Construction (South Western) Limited
Rotating anti climb guards - Anti-Climb Guards Ltd
Safety fencing - Newton & Frost
Security - Metalwood Fencing (Contracts) Limited
Security and industrial fencing - Sampson & Partners Fencing
Security fencing - Allen (Fencing) Ltd
Security fencing - C & W Fencing Ltd
Security fencing - Chestnut Products Limited
Security fencing - Dinnington Fencing Co. Limited
Security fencing - Fencing Design & Development Co Limited
Security Fencing - Sharman Fencing & Gate Supplies
Security fencing - WA Skinner & Co Limited
Security fencing and gates - Stenoak Fencing Limited
Security fencing and gates - Warefence Limited
Security fencing, gates and access systems - O'Connor Fencing Limited
Security grilles & guards - WA Skinner & Co Limited
Security palisade fencing - GJ Durafencing Limited
Security steel palisade fencing and gates - Prima Security & Fencing Products
Security, industrial fencing, railings, crash barriers, gates, palisade, posts and rails - J & M Fencing Services
Security, industrial, commercial new housing and highways - Ambec Fencing

Q40 Fencing (Cont'd)

Sentry bar steel fencing - Jackson H S & Son (Fencing) Ltd
Speed ramps - Autopa Ltd
Sport fencing - Cardinal Fencing (Northern) Limited
Sports - Metalwood Fencing (Contracts) Limited
Sports pitch fencing - Chestnut Products Limited
Steel aluminium grating - Orsogril Spa UK Distributor
Steel fencing and gates - C & W Fencing Ltd
Steel gates - Baddeley
Steel gates - Staines Steel Gate Co
Steel mesh system fences - Warefence Limited
Steel palisade - Fencing Design & Development Co Limited
Steel palisading, welded mesh, chainlink fencing - Fencelines Ltd
Supply & Fix - Havering Fencing Co.
Timber fences - Fencing Design & Development Co Limited
Timber fencing and gates - Tonbridge Fencing Limited
Timber palisade - Danbury Fencing PLC
Timber Panels - Huntree Fencing Ltd
Timber, chainlink and weldmesh fencing systems - Prima Security & Fencing Products
Traffic control barriers - Fernden Construction (South Western) Limited
Trellis - Danbury Fencing PLC
Tubular construction, barriers and rials - Tubeclamps Ltd
Turnstiles - Fernden Construction (South Western) Limited
Vehicle safety fencing, environmental safety barriers,pedestrian guardrail - Stenoak Fencing Limited
Vertical tube fencing - Orsogril Spa UK Distributor
Waney Lap - Danbury Fencing PLC
Weld mesh - Alltype Fencing Specialists
Welded mesh fencing - Orsogril Spa UK Distributor
Weldmesh - Procter Cast stone & Concrete Products
Windposts - Ancon Building Products
Wire - Betafence Limited
Wire mesh - Betafence Limited
Wood - Alltype Fencing Specialists

Q41 Barriers/Guard-rails

Acoustic barriers - Chestnut Products Limited
Architectural railings - Chestnut Products Limited
Automated gates and barriers - Sampson & Partners Fencing
Automated traffic barriers - Dinnington Fencing Co. Limited
Automatic Barriers - Kone PLC
Balustrade - Coates Fencing Limited
Barrier tapes - Boddingtons Ltd
Barriers - Autopa Ltd
Barriers - Cardinal Fencing (Northern) Limited
Barriers - May Gurney (Technical Services) Ltd - Fencing Division
Barriers - Metalwood Fencing (Contracts) Limited
Barriers - Rediweld Rubber & Plastics Ltd
Barriers - Wenmore Rooflights, A Division of W.F.Joy & Co. Ltd
Barriers - temporary steel - Asset International Ltd
Barriers and guard rails - Stenoak Associated Services PLC
Barriers and guardrails - Hart Door Systems Ltd
Barriers and Guardrails - Hubbard Architectural Metalwork Ltd
Barriers and guardrails - J. B. Corrie & Co Limited
Barriers Guardrails - Steelway Fensecure Ltd
Barriers/ guardrails - Claydon Architectural Metalwork Ltd
Barriers/Gaurdrails - Crowthorne Fencing
Barriers/Guard Rails - Gearing F T Landscape Services Ltd
Barriers/guard rails - NRG Fabrications
Barriers/guardrails - Metalcraft (Tottenham) Ltd
Barriers/guard-rails - Berry Systems
Barriers/Guard-rails - BPT Automation Ltd
Barriers/Guard-rails - Grant & Livingston Ltd
Barriers/Guard-rails - Pembury Fencing Limited
Barriers/Guard-rails - Portia Engineering Ltd
Barriers/sGuard-rails - Newton & Frost
Cantilever gates - Jackson H S & Son (Fencing) Ltd
Car park access/egress barriers - Tully
Car parking equipment - Wenmore Rooflights, A Division of W.F.Joy & Co. Ltd
Decorative railings - Claydon Architectural Metalwork Ltd
Emergency breakdowns / call outs - Tully
Endless cords - Lever James & Sons Ltd
Flood prevention gates - Fernden Construction (South Western) Limited
Garden twines - Lever James & Sons Ltd
Gates - May Gurney (Technical Services) Ltd - Fencing Division
Gates and barriers - Hart Door Systems Ltd
Gates, both manual and automatic - Cardinal Fencing (Northern) Limited
Guardrails - Asset International Ltd
Handrail - Kee Klamp Ltd
Housing projects post and rail pvc-u - Duralock (UK) Ltd
Industrial and Commercial - Metalwood Fencing (Contracts) Limited
Industrial gates - Guardian Wire Ltd
Mass multi-applicational safety system - Asset International Ltd
Metal security fencing - Town & Country Landscapes Ltd
Ornamental Rails + Gates - Procter Cast stone & Concrete Products
Palisade - Coates Fencing Limited
Pedguard rail - Jackson H S & Son (Fencing) Ltd
Pedstrian turnstiles - Tully
Perimeter protection - APT Controls Ltd
Planed preventative maintenance - Tully
Post & Rail - Furnitubes International Ltd
Post and rail products -schools pvc-u - Duralock (UK) Ltd
Precast Complete - RCC, Division of Tarmac Precast Concrete Ltd
Racecourse rails and products pvc-u - Duralock (UK) Ltd
Railings - Cardinal Fencing (Northern) Limited
Railings - Coates Fencing Limited
Railings and gates - installers - GJ Durafencing Limited
Ropes and sashcords - Marlow Ropes Ltd
Security barriers - APT Controls Ltd
Security equipment - Anti-Climb Guards Ltd
Security fencing and gates - Tonbridge Fencing Limited
Security point access /egress barriers - Tully
Sliding gates - Tully
Steel security palisade fencing and gates - Wm. Bain Fencing Ltd
Subsea protection systems - Halls Group Ltd
Swing gates - Jackson H S & Son (Fencing) Ltd
Timber gate manufacturers - WL West & Son Limited
Traffic barriers - Crimeguard Systems Ltd
Traffic barriers - Kaba Garog
Traffic barriers - Konematic Bolton Brady

Q50 Site/ Street furniture/ equipment

Access covers and frames - Jones of Oswestry Ltd
Architectural trellises and planters - Anthony de Grey Trellises
Arm chairs - Gloster Furniture Ltd
Automatic Bollards - APT Controls Ltd
Automatic Gates - Crimeguard Systems Ltd
Automatic gates and barriers - BPT Automation Ltd
Balustrade - Kee Klamp Ltd
Barrier posts - Aremco Products
Barriers - Telford Doors and Barriers Ltd
Bench seating - LSA Projects Ltd
Benches - Dee-Organ Ltd
Benches - Erlau AG
Benches - Gloster Furniture Ltd
Bi - foloding Gates - APT Controls Ltd
Bird deterent measures - Birdscarer Products & Net Supplies Ltd
Bollard shells - Signature Ltd
Bollards - Autopa Ltd
Bollards - Dee-Organ Ltd
Bollards - Erlau AG
Bollards - Furnitubes International Ltd
Bollards - Glasdon UK Ltd
Bollards - Rediweld Rubber & Plastics Ltd
Bollards, cable covers, troughs and bespoke products - Anderton Concrete Products Ltd
Bridges in steel, boardwalks and decking - CTS Ltd
Building and site equipment - SGB Hire Plus
Bus & rail shelters - Abacus Lighting Ltd
Bus and glazed shelters - Midland Alloy Ltd
Car Park Posts - Furnitubes International Ltd
Car parks in pvc-u - Duralock (UK) Ltd
Cigarette ash bins - Swintex Ltd
Concrete bollards - Procter Cast stone & Concrete Products
Cycle Lockers - Autopa Ltd
Cycle parking - Furnitubes International Ltd
Cycle parking equipment - Broxap Ltd
Cycle shelters - Dee-Organ Ltd
Cycle Shelters - Fibaform Products Ltd.
Cycle Shelters - Parklines Building Ltd
Cycle stands - Dee-Organ Ltd
Cycle stands - Erlau AG
Cycle stands - Rediweld Rubber & Plastics Ltd
Cycle stands and shelters - Autopa Ltd
Demarcation studs - brass, aluminum, stainless steel - Reflecting Roadstuds Ltd
Dining chairs - Gloster Furniture Ltd
Dog bins - Neptune Outdoor Furniture Ltd
Dog waste bins - Glasdon UK Ltd
Drainage gratings - Jones of Oswestry Ltd
Electrical termination/ service pillar - Ritherdon & Co Ltd
Electrically operated gates - Warefence Limited
Engraving service - Gloster Furniture Ltd
Exterior lighting - Abacus Lighting Ltd
External litter bins - Glasdon UK Ltd
Folding chairs - Gloster Furniture Ltd
Furniture - Barlow Tyrie Ltd
Garden and Entrance Gates - Danbury Fencing PLC
Garden furniture - Nova Garden Furniture Ltd
Garden sheds - Ableson L & D Ltd
Gates and barriers - Rib (UK) Ltd
Gates timber & metal - B & M Fencing Limited
Glass fibre and aluminium flag poles - Harrision Flagpoles
Grit bins - Amberol Ltd
Grit/ salt bins - Balmoral Tanks
Ground and wall mounted banner poles - Harrision Flagpoles
Landscape components and street furniture - Woodscape Ltd
Landscape furniture, seatings, shelters, bridges, litter bins, and trim trails - Lappset UK Ltd
Landscape Suppliers - Crowthorne Fencing
Litter bins - Amberol Ltd
Litter bins - Erlau AG
Litter Bins - Furnitubes International Ltd
Litter bins - Swintex Ltd
Loungers - Erlau AG
Manufacturers - Autopa Ltd
Motorway crash barriers and pedestrian guard rails - Lionweld Kennedy Ltd
Occasional furniture - Gloster Furniture Ltd
Ornamental statues, pots and urns - Rogers Concrete Ltd
Outdoor furniture - Andy Thornton Ltd
Outdoor litter bins - Unicorn Containers Ltd
Parking posts - Autopa Ltd
Parking posts - Rediweld Rubber & Plastics Ltd
Pavior infill covers and frames - Jones of Oswestry Ltd
Pedestrian crossing equipment - Signature Ltd
Pedestrian guard rails - Fabrikat (Nottingham) Ltd
Pergolas, screens and planters - CTS Ltd
Planters - Erlau AG
Planters - Sarena MFG Ltd
Public seats, litter bins, benches, bollards, picnic tables and planters - Neptune Outdoor Furniture Ltd
Purpose made furniture - Orchard Street Furniture
Reconstructed stone balustrading, paving, garden ornaments and architectural dressings - Minsterstone Ltd
Recycling initiatives - Camas Aggregates Ltd - Aggregate Industries
Restaurant chairs - Laytrad Ltd
Retaining panels, planters and street furniture - Milner Delvaux Ltd
Road Blockers - APT Controls Ltd
Road signs and road markers - Glasdon UK Ltd
Rockery stone - The Delabole Slate Co. Ltd
Scaffolding protectors - Anti-Climb Guards Ltd
Seating - Erlau AG
Seating - Furnitubes International Ltd
Seating - Glasdon UK Ltd
Seats - Record Playground Equipment Ltd
Security Bollards - Furnitubes International Ltd
Security gates, barriers, road blockers and turnstiles - Broughton Controls Ltd
Self watering baskets - Amberol Ltd
Shelters - Broxap Ltd
Sign posts - Stainton Metal Co Ltd
Site equipment - Frosts Landscape Construction Ltd
Site furniture, car park furniture and equipment - APT Controls Ltd
Site furniture/ equipment - Broxap Ltd
Site/ street furniture/ equipment - Chilstone
Site/ street furniture/ equipment - Haddonstone Ltd
Site/ Street furniture/ equipment - Leander Architectural
Site/Street funiture/ Equipment - Berry Systems
Site/Street furniture/Equipment - Pembury Fencing Limited
Site/Street Furniture/Equipment - Wybone Ltd
Sleeping policemen and traffic calming products - Rediweld Rubber & Plastics Ltd
Smoking disposable units - Glasdon UK Ltd
Smoking shelters - Autopa Ltd
Speed ramp - Swintex Ltd
Speed ramps - Berry Systems
Steel Storage Buildings - Parklines Building Ltd
Stone street furniture - CED Ltd
Street furniture - Abacus Lighting Ltd
Street furniture - Ballantine, Bo Ness Iron Co Ltd
Street furniture - Blanc de Bierges
Street furniture - Broxap Ltd
Street furniture - Claydon Architectural Metalwork Ltd
Street furniture - Dee-Organ Ltd
Street furniture - DW Windsor Lighting
Street furniture - Earth Anchors Ltd
Street Furniture - Ebor Concretes Ltd
Street Furniture - Falco
Street furniture - Interlux Ltd
Street furniture - Marshalls Mono Ltd
Street furniture - Orsogril Spa UK Distributor
Street Furniture - Townscape Products Ltd
Street furniture - Urbis Lighting Ltd
Street furniture - Wales & Wales
Street furniture - Woodhouse UK Plc
Street furniture, shelters and walkways - design and manufacture - Maceman + Amstad Ltd
Summer houses - Ableson L & D Ltd
Tables - Erlau AG
Tables - Gloster Furniture Ltd
Teen shelters - Record Playground Equipment Ltd
Timber garden products - OSMO UK Ltd
Timber, steel, cast iron street and garden furniture - Orchard Street Furniture
Tower protectors - Anti-Climb Guards Ltd
Traffic barriers - Electro Mechanical Systems Ltd
Traffic products - Swintex Ltd
Tree grilles and guards - Furnitubes International Ltd
Waste containers and systems - Iles Waste Systems
Water gardens and water features - Lotus Water Garden Products Ltd
Wheeled bins - Titan plastech
Winter grit spreaders - Glasdon UK Ltd
Winter grit storage bins - Glasdon UK Ltd
Wrought iron furniture - Desney Products Ltd

R10 Rainwater pipework/ gutters

Accessories - Corofil Woodall
Aluminium flashings - Gilmour Ecometal
Aluminium gutter and rainwater pipe products - Marley Plumbing and Drainage
Aluminium gutters - George Gilmour (Metals) Ltd
Aluminium gutters - Gilmour Ecometal
Aluminium pipework and gutters - Aluminium RW Supplies Ltd
Aluminium rainwater good - Alifabs (Woking) Ltd
Aluminuim rainwater goods - Guttercrest
Building Products incl window & guttering systems - Marshall-Tufflex Ltd
Cast iron rainwater and soil goods - Longbottom J & J W Ltd
Chamber rings - Johnston Pipes Ltd
Channel Drainage - Wavin Plastics Ltd
Clay, Plastic and concrete drainage systems - Hepworth Building Products Ltd
Copper rainwater system - Klober Ltd
Custom fabricated in metals - Azimex Fabrications Ltd
Fascia and soffit systems in powder coated aluminium and steel - Dales Fabrications Ltd
GRP gutters - Wilsons Fibreglass Ltd
GRP valley troughs - Timloc Expamet Building Products
Gutter heating - Findlay Irvine Ltd
Gutter leaf protection system - FloPlast Ltd
Guttering - Letchworth Roofing Company Ltd
Gutters - Corofil Woodall
Insulated gutters - CGL Cometec Ltd
Insulation boards - Vulcanite Limited
Plastic drainage systems - Hunter Plastics Ltd
Plumbing & Heating - Spital Tile Co Ltd
Polypropylene pipe systems - Capper Plastics
PVCu rainwater systems - Marley Plumbing and Drainage
Rain water management systems - Fullflow Group Ltd
Rainwater containers and accessories - T & D Packaging Ltd
Rainwater diverter - FloPlast Ltd
Rainwater drainage products - Drainage Systems
Rainwater goods - Broderick Structures Limited
Rainwater goods - Eternit Building Materials
Rainwater Goods - Rheinzink UK
Rainwater goods - Rockwell Sheet Sales Ltd
Rainwater outlets/ gutters - McKenzie-Martin Ltd
Rainwater pipework and gutters - Geberit UK
Rainwater pipework/ gutters - Harris & Bailey Ltd
Rainwater pipework/ gutters - Interlink Group
Rainwater pipework/ gutters - John Davidson (Pipes) Ltd
Rainwater recirculation systems - Titan Pollution Control Ltd
Rainwater systems - Allmat (East Surrey) Ltd
Rainwater systems - FloPlast Ltd
Rainwater systems - Harrison Thompson & Co Ltd
Rainwater systems in powder coated aluminium and steel - Dales Fabrications Ltd
Rainwater systems, sacias and soffits - Alumasc Exterior Building Products Ltd
Rainwater, soil and wate systems - Brett Martin Ltd
Relining of concrete gutters - Finlock Gutters
Roof outlets - ACO Technologies Plc
Siphonic rainwater drainage systems - Dales Fabrications Ltd
Stainless steel drainage - Blucher UK Ltd
Surface water drainage - Hodkin Jones (Sheffield) Ltd
Syphonic drainage systems - Fullflow Group Ltd
Syphonic roof drainage - Fullflow Group Ltd
T-Pren-gutter expansion material - Matthew Hebden
Water storage containers and accessories - Harcostar Drums
Zinc rainwater goods - Metra Non Ferrous Metals Ltd

R11 Foul drainage above ground

Above ground drainage - Ballantine, Bo Ness Iron Co Ltd
Above ground drainage - FloPlast Ltd
Air admittance valves - FloPlast Ltd
Cast iron, above + below ground System to BSEN877 - Saint-Gobain Pipelines PLC
Drainage above ground - Brett Martin Ltd
Drainage systems - Alumasc Exterior Building Products Ltd
Filter beds - Milton Pipes Ltd
Foul drainage above ground - Drainage Systems

R11 Foul drainage above ground (Cont'd)

Foul drainage above ground - Durey Casting Ltd
Foul drainage above ground - FC Frost Ltd
Foul drainage above ground - Polypipe Civils Ltd
Foul drainage above ground - Wavin Plastics Ltd
Gutters, drainage pipes and fittings - Longbottom J & J W Ltd
MUPVC waste system - Davant Products Ltd
Plastic drainage - Capper Plastics
Plastic drainage systems - Hunter Plastics Ltd
Plastic soil pipes, urinal pipework and traps - Midland Stom Ltd
PVCu soil and waste systems - Marley Plumbing and Drainage
Rainwater pipework/gutters - Blucher UK Ltd
Retro fit above ground & built in submerged covers - Grando (UK) Ltd
Seamless cold drawn copper tube - Lawton Tube Co Ltd, The
Soil and drain down systems - Stanton plc
Soil and vent terminals - Klober Ltd
Stainless steel drainage channels, gratings and gullies - Component Developments
Tanks (retention) - Milton Pipes Ltd
Traps and waste fittings - Opella Ltd
Waste Traps - FloPlast Ltd

R12 Drainage below ground

Access covers and drainage products - Alumasc Construction Products Ltd
Access covers and frames linear surface water drainage system and special fabrication - Clarksteel Ltd
Bacteria Grease traps - Progressive Product Developments Ltd
Barner pipe for contasminated land - Wavin Plastics Ltd
Below ground drainage and maholes - Ballantine, Bo Ness Iron Co Ltd
Box culverts - Brierkrete Ltd
Box culverts - Milton Pipes Ltd
Caisson shafts - Milton Pipes Ltd
Cesspools - Conder Products Ltd
Cesspools and septic tanks - Klargester Environmental Ltd
Channel grating - Shackerley (Holdings) Group Ltd incorporating Designer ceramics
Clay Drainage - Knowles W T & Sons Ltd
Concrete drainage systems - Hanson Concrete Products
Concrete inspection chambers - Brierkrete Ltd
Concrete manholes and soakaways - Milton Pipes Ltd
Concrete pipes - Johnston Pipes Ltd
Corrugated steel buried structures and pipes - Asset International Ltd
Cover slabs - Chiltern Concrete and Stone Ltd
Design, installation and maintenance of building services - Sulzer Infra (UK) Ltd
Drain and pipework cleaning - Dyno Rod PLC
Drain and pipework inspection - Dyno Rod PLC
Drain and pipework installation - Dyno Rod PLC
Drain and pipework repairs - Dyno Rod PLC
Drainage - Marshalls Mono Ltd
Drainage - Polypipe Civils Ltd
Drainage below ground - Certifield Laboratories (Div or CPS Industries Ltd)
Drainage below ground - FC Frost Ltd
Drainage below ground - Formpave Ltd
Drainage below ground - Geberit UK
Drainage below ground - Hanson Building Products
Drainage below ground - Hanson Concrete Products
Drainage below ground - Harris & Bailey Ltd
Drainage below ground - John Davidson (Pipes) Ltd
Drainage below ground - Wavin Plastics Ltd
Drainage clay & plastic - Brikenden (Builders Merchants) Ltd
Drainage media - Lytag Ltd
Drainage products - Ashmead Buidling Supplies Ltd
Drainage products - Drainage Systems
Egg-shpae liners - Johnston Pipes Ltd
Flexible couplings - Flex-Seal Couplings Ltd
Floor gratings - Screen Systems Ltd
Floor gullies, roof outlets, access covers, grease converters and linear drainage - Wade International Ltd
G-flex pipes - Johnston Pipes Ltd
Gravity drain systems - Uponor Ltd
Grease traps, oil and grease water separators - Progressive Product Developments Ltd
Grease treatment plants - Progressive Product Developments Ltd
GRP covers - Armfibre Ltd
GRP Kiosks - Adams- Hydraulics Ltd
GRP pipes, concrete pipes - Johnston Pipes Ltd
Gullies, gratings, channels and traps - ACO Technologies Plc
Gully grating and access covers - Brickhouse
Hic units - Johnston Pipes Ltd
Installation services - Uponor Ltd
Jetters cold and hot, winches, rods and accessories - Ward's Flexible Rod Company Ltd
Large tanks and sewage treatment plant - Conder Products Ltd
Manhole covers - Duct Access Covers Ltd
Manhole covers - Stanton plc
Manhole covers and gratings - Durey Casting Ltd
Manhole covers and reservoir lid - Ruthin Precast Concrete Ltd
Manholes - Brierkrete Ltd
Monolithic road gullies - Milton Pipes Ltd
Non entry inspection chambers - Wavin Plastics Ltd
Package sewage pumping stations - KSB Ltd
Plastic building products - Haymills (Contractors) Ltd
Plastic drainage - Capper Plastics
Plastic drainage systems - Hunter Plastics Ltd
Pumping stations for storm water/Foul Water - Pims Pumps Ltd
PVC & ABS Pipe systems - Capper Plastics
PVCu underground drainage systems - Marley Plumbing and Drainage
Rectangular inspection chambers - Milton Pipes Ltd
Septic tanks - Balmoral Tanks
Septic tanks - Brierkrete Ltd
Septic tanks - Conder Products Ltd
Settlement tanks and components - Armfibre Ltd
Soil and drain down systems - Stanton plc
Stainless steel channels, gratings, gullies, access covers and grease separators - Blucher UK Ltd
Stainless steel manhole covers - Component Developments
Storm water attenuation tanks - Brierkrete Ltd
StormWater Management - Wavin Plastics Ltd
Structured wall HDPE pipes - Asset International Ltd
Surface water drainage systems - Mea UK Ltd
Ultra rib sewer systems - Uponor Ltd
Underground Drainage - FloPlast Ltd
Underground drainage - Naylor Drainage Ltd
Underground drainage systems (PVC) - Brett Martin Ltd

R13 Land drainage

Box Culverts - Tarmac Precast Concrete Limited
Box culvets - Johnston Pipes Ltd
Cover slabs - Johnston Pipes Ltd
GRP soakers - Wilsons Fibreglass Ltd
Inspection chambers - Hunter Plastics Ltd
Land drainage - Drainage Systems
Land drainage - Durey Casting Ltd
Land drainage - Flex-Seal Couplings Ltd
Land drainage - J Pugh-Lewis Limited
Land drainage - John Davidson (Pipes) Ltd
Land Drainage - Naylor Drainage Ltd
Land drainage - Polypipe Civils Ltd
Land Drainage - T-T Pumps & T-T controls
Petrol/oil inceptors, catering grease/oil traps, balancing tanks - Conder Products Ltd
Plumbing and Drainning - Wavin Plastics Ltd
Preformed sheets, cavity and channel drainage - ABG Ltd
PVCu quantum highway and sewer drainage - Marley Plumbing and Drainage
Stormfox - Johnston Pipes Ltd
Vertical Band Drains - Bauer Foundations Limited

R14 Laboratory/ Industrial waste drainage

Bacteria Grease traps - Progressive Product Developments Ltd
Grease traps, oil and grease water separators - Progressive Product Developments Ltd
Grease treatment plants - Progressive Product Developments Ltd
Laboratory and industrial waste - Drainage Systems
Laboratory drainage systems - CPV Ltd
Laboratory Ind Waste - George Fischer Sales Ltd
Laboratory/ Industrial Waste - Naylor Drainage Ltd
Laboratory/industrial waste drainage - Satec Ltd
Stainless steel channels, gratings, gullies, access covers and grease separators - Blucher UK Ltd
Stainless steel drainage systems - ACO Technologies Plc
Thermoplastic pipes - CPV Ltd

R20 Sewage pumping

Centrifugal pumps - Calpeda Limited
Drainage and sewage package systems - Pump Technical Services Ltd
Fuel water seperators - Klargester Environmental Ltd
Maintenance and repair of pumping equipment - Pims Pumps Ltd
Package pump stations - Tuke & Bell Ltd
Pumps and ejectors - Tuke & Bell Ltd
Sewage pumping - Adams- Hydraulics Ltd
Sewage pumping - Certifield Laboratories (Div or CPS Industries Ltd)
Sewage pumping - Conder Products Ltd
Sewage pumping - Polypipe Civils Ltd
Sewage pumping - Satec Ltd
Sewage Pumping - T-T Pumps & T-T controls
Sewage pumps and pumping systems - New Haden Pumps Ltd
Sewage, drainage, pumps and systems - KSB Ltd

R21 Sewage treatment/ sterilisation

Cesspools - Titan Pollution Control Ltd
Cop - Flows & Loads - British Water
Drain clearing rods - Wessenden Products
Effluent treatment - Armfibre Ltd
Grease removal system - Tuke & Bell Ltd
Grease traps - Titan Pollution Control Ltd
Grit removal systems - Tuke & Bell Ltd
GRP Equipment - Adams- Hydraulics Ltd
Package sewage treatment plants - Titan Pollution Control Ltd
Package sewage treatment plants - Tuke & Bell Ltd
Pump stations - Titan Pollution Control Ltd
Rotating biological contactors - Tuke & Bell Ltd
Sample chambers - Titan Pollution Control Ltd
Separators - Titan Pollution Control Ltd
Septic tanks - Titan Pollution Control Ltd
Settling tank scrapers - Tuke & Bell Ltd
Sewage purification equipment - Tuke & Bell Ltd
Sewage storage tanks - Franklin Hodge Industries Ltd
Sewage treatment - Balmoral Tanks
Sewage treatment - Klargester Environmental Ltd
Sewage Treatment - Simon-Hartley
Sewage treatment equipment - Adams- Hydraulics Ltd
Sewage treatment/sterilization - Conder Products Ltd
Sewage treatment/sterilization - Polypipe Civils Ltd
Sewage treatment/sterilization - Satec Ltd
Surface aerators - Tuke & Bell Ltd
Tanks (retention) - Milton Pipes Ltd
Trickling filter distributors - Tuke & Bell Ltd
Ultra violet chambers, lamps and control panels - Hanovia Ltd

R30 Centralised vacuum cleaning

Central Vacuum Systems - Barloworld Vacuum Technology Plc
Centralised vacuum cleaning - Torit DCE
Centralized vacuum cleaning systems - BVC

R31 Refuse chutes

Refuse chutes - Watkins

R32 Compactors/ Macerators

Bag compactors - Tony Team Ltd
Bin compactors - Tony Team Ltd
Compactors/ Macerators - Sissons W & G Ltd
Demountable compactors - Randalls Fabrications Ltd
Sluiceroom macerators - Verna Ltd
Static & portable compaction units - Randalls Fabrications Ltd
Waste management equipment, compactors/ macerators - Imperial Machine Co Ltd

R33 Incineration plant

Incinerators - Northern Incinerators (GB) Ltd

S10 Cold Water

Cold water - Essex
Cold Water - George Fischer Sales Ltd
Cold Water - Holden Brooke Pullen
Cold water - John Davidson (Pipes) Ltd
Cold Water - Polypipe Civils Ltd
Design, installation and maintenance of building services - Sulzer Infra (UK) Ltd
Ductwork clearing - Aquastat Ltd
Level gauges and transmitters - Klinger Fluid Instrumentation Ltd
Manufacturer & supplier of materials & liners for cured-in-place pipe rehabilitation - Applied Felts Ltd
MDPE pipe and fittings - FloPlast Ltd
MDPE water pipes and fittings - Brett Martin Ltd
Non portable water systems - Uponor Ltd
Packaged Cold Water Storage/Pump - Harton Heating Appliances
Pipe bending equipment - Hilmor
Pipe freezing kits - Hilmor
Piped supply systems cold water - Precolor Sales Ltd
Piping systems - George Fischer Sales Ltd
Plumbing products - Drainage Systems
Polyethylene pipes for water and gas - Capper Plastics
Polyethylene systems - Uponor Ltd
Portable water pipe fittings - Uponor Ltd
Radiators - Brikenden (Builders Merchants) Ltd
Seamless cold drawn copper tube - Lawton Tube Co Ltd, The
Submersible pumps - Pillinger G C & Co (Engineers) Ltd
Threading equipment - Hilmor
Water boundary control boxes - Barber Edward & Co Ltd
Water fittings and controls - Flamco UK Ltd
Water meters - SAV United Kingdom Ltd
Water meters - Switch2 Energy Solutions Ltd
Water meters and oil meters - Elster Metering Ltd
Water storage tanks - Sarena MFG Ltd

S11 Hot Water

Boiling water products - Heatrae Sadia Heating Ltd
Cistern & Cistern - Opella Ltd
Cistern & Cistern Fittings - Opella Ltd
Control panels and controls - Pillinger G C & Co (Engineers) Ltd
Copper tube and fittings - Brikenden (Builders Merchants) Ltd
Copper tube for water - Yorkshire Copper Tube Ltd
Design, installation and maintenance of building services - Sulzer Infra (UK) Ltd
Hot water - Essex
Hot Water - George Fischer Sales Ltd
Hot Water - Holden Brooke Pullen
Plumbing products - Drainage Systems
Pre-commisson clearing of heating & chilled water systems - Aquastat Ltd
Stainless Steel Cylinders - AEL
Switches, valves and manifolds - Klinger Fluid Instrumentation Ltd
Temperature blending valves - Deva Tap Co Ltd
Unvented Combination Units with Unvented Hot Water Storage & Boilers - Harton Heating Appliances

S12 Hot & cold water (self-contained specification)

Commercial water control systems - Kohler Mira Ltd
Copper Unvented Units - Harton Heating Appliances
Cylinder Packs With Integral Controls - Harton Heating Appliances
Drinking Water System - Tapworks Water Softeners
Drinking water systems - Ecowater Systems Ltd
Enamelled Steel Unvented Units - Harton Heating Appliances
Filtration equipment - Ecowater Systems Ltd
Fittings - Pegler Hattersley
Flexible plastic plumbing - IPPEC Sytsems
Hot & Cold Water - George Fischer Sales Ltd
Hot & cold water - Hertel Services
Hot & Cold Water (Small Scale) - Essex
Hot and cold water - Guest (Speedfit) Ltd, John,
Hot and cold water and central heating, polybutylene pipework - Brett Martin Ltd
Hot and cold water cylinder jackets - Davant Products Ltd
Hot and cold water systems - Equator PEX - Marley Plumbing and Drainage
Miners - SPX Process Equipment
Mixers - Marflow Eng Ltd
Plumbing - Marflow Eng Ltd
Plumbing accessories - BSS Zenith
Plumbing products - Drainage Systems
Plumbing systems - REHAU Ltd
Pre- insulated pipes - IPPEC Sytsems
Prefabricated Plumbing Units - Harton Heating Appliances
PTFE sealant - Klinger Fluid Instrumentation Ltd
Sealed system equipment - Flamco UK Ltd
Shower Valves - Marflow Eng Ltd
Showers - Deva Tap Co Ltd
Swivel ferrule straps, stop taps, leadpacks and boundary boxes - Booth Samuel & Co Ltd
Taps - Marflow Eng Ltd
Tube Plugs - Furmanite International Ltd
Water heaters - Beeston Heating Group Ltd
Water pumps - Stuart Turner Ltd
Water purifiers - Ecowater Systems Ltd
Water Purifiers - Tapworks Water Softeners
Water saving products - Deva Tap Co Ltd
Water softeners - Ecowater Systems Ltd
Water softeners - tapworks water softeners - Tapworks Water Softeners

S13 Pressurised water

Control valves - Auld Valves Ltd
Design, installation and maintenance of building services - Sulzer Infra (UK) Ltd
Expansion vessels for portable water - Flamco UK Ltd
Pressure booster pumps - Stuart Turner Ltd
Pressure boosting systems - Armstrong Pumps Ltd
Pressure sprayers - Hozelock Ltd
Pressure units - Pressmain
Pressure washers - Fastclean Scotland Ltd
Pressurised water - Armstrong Pumps Ltd
Pressurised water - Danfoss Ltd
Pressurised Water - Holden Brooke Pullen
Pressurised water - Polypipe Civils Ltd
Pressurization - Beeston Heating Group Ltd
Pressurized water - Calpeda Limited
Pressurized water - John Davidson (Pipes) Ltd
Pressurized water - New Haden Pumps Ltd
PVC pressure pipes - Uponor Ltd
Rainwater utilization pump systems - KSB Ltd
Shower booster pumps - Stuart Turner Ltd
Shower pumps - Deva Tap Co Ltd
Valves for high pressure pneumatic and hydraulic appliances - Hale Hamilton Ltd
Water Booster - Pressmain
Water boosters and pressurisation units - Pillinger G C & Co (Engineers) Ltd
Water management - Verna Ltd

S14 Irrigation

Irrigation - Calpeda Limited
Irrigation - Holden Brooke Pullen
Irrigation - Polypipe Civils Ltd
Stainless steel Drainage pipework - Blucher UK Ltd

S15 Fountains/Water features

Fountain display pumps - Stuart Turner Ltd
Fountains/Water features - Holden Brooke Pullen

S20 Treated/ Deionised/ Distilled water

Cooling tower water treatment, cleaning and refurb - Aquastat Ltd
Demineralized, deionized, softened water and filters - Dewplan Ltd
Evaporative water cooling tower - Visco Ltd
Filteration Equipment - Tapworks Water Softeners
Flexible tubes - United Flexibles (A Division of Senior Flex)
Limescale inhibitors - Hydropath (UK) Ltd
Treated/ deionized/distilled water - Satec Ltd
Treated/Deionised/Distilled water - Hertel Services
Treated/Deionised/Distilled Water - USF Permutit/Ionpure
Ultra violet chambers, lamps and control panels - Hanovia Ltd
Waste water treatment - Clearwater Environmental Engineering Ltd
Water chlorination - Portacel
Water conditioners - Cistermiser Ltd
Water filtration (industrial) - Visco Ltd
Water filtration and treatment - Barr & Wray Ltd
Water filtration equipment - Hydropure Group Ltd
Water filtration media - Lytag Ltd
Water purification - USF Permutit/Ionpure
Water purification systems - Elga Ltd
Water treatment - BetzDearborn Ltd
Water treatment - Estec Environmental Ltd
Water treatment - Feedwater Ltd

S21 Swimming pool water treatment

Automatic chemical control and dosing - Golden Coast Ltd
Calorifiors - Beeston Heating Group Ltd
Disinfection systems - Portacel
Modular precoated stainless steel swimming pools - Aqua Design Ltd
Pool covers - Amdega Ltd
Swimming pool surrounds - Blanc de Bierges
Swimming pool installations - Barr & Wray Ltd
Swimming pool kits and accessories-domestic - Fox Pool (UK) Ltd
Swimming pool water treatment - Precolor Sales Ltd
Swimming pools and accessories - Aqua-Blue Ltd
Swimming pools and equipment, spas, heat pumps, water features and chemicals. - Certikin International Ltd
Swimming pools, spa baths, sauna and steam rooms - Buckingham Swimming Pools Ltd
Swimming pools, spas, saunas, steam rooms, baptisteries and mosaic murals - Regency Swimming Pools Ltd
Ultra violet chambers, lamps and control panels - Hanovia Ltd
Water features - Aqua-Blue Ltd
Watertank refurbishment - Aquastat Ltd

S30 Compressed air

Air compressors - F. W. Harris & Co Ltd
Compressed air - Danfoss Ltd
Compressed air - Guest (Speedfit) Ltd, John,
Compressed air - Hertel Services
Compressed air supply systems - Fluidair International Ltd
Compressed air systems - Denco Ltd
Compressors - Chippindale Plant Ltd
Hand and actuated valves - George Fischer Sales Ltd
Oil free air compressors, vacuum pumps and gas circulators. - Dawson McDonald & Dawson Ltd
Pipework and Installation - Fluidair International Ltd

S32 Natural gas

Decorative gas fires - Taylor & Portway Ltd
Gas and sanitation applications - Yorkshire Copper Tube Ltd
Gas detection - Portacel
Gas Detection Equipment - TCW Services (Control) Ltd
Gas detection systems - Blakell Europlacer Ltd
Gas fires - natural - Valor
Gas isolation devices - TCW Services (Control) Ltd
Gas Isolation Equipment - TCW Services (Control) Ltd
Gas meters - Switch2 Energy Solutions Ltd
Gas pipe fittings - Uponor Ltd
Natural gas - John Davidson (Pipes) Ltd
Natural Gas - Saacke Ltd
Seamless cold drawn copper tube - Lawton Tube Co Ltd, The

S33 Liquefied petroleum gas

Calor gas - Bell & Co Ltd
Cylinder & bulk LPG - Calor Gas Ltd
Decorative gas fires - Taylor & Portway Ltd
Gas Detection Equipment - TCW Services (Control) Ltd
Gas Isolation Equipment - TCW Services (Control) Ltd
Heating appliances - Bell & Co Ltd
Liquified petroleum gas storage - Grant & Livingston Ltd
LPG gas fires - Valor
LPG vessels - Flamco UK Ltd
Portable gas heaters - Valor

S40 Petrol/Diesel storage/distribution

Bulk liquid/ chemical storage tanks - Balmoral Tanks
Fill Point cabinets - Metcraft Ltd
Fuel tanks - Balmoral Tanks
Insulation for Petrochemical Industries - Kingspan Industrial Ins. Ltd
Lubricants - MMP International
Oil pollution monitoring and control systems - design, manufacture, installation and commissioning of - Aquasentry
Petrol/diesel storage - Grant & Livingston Ltd

S41 Fuel oil storage/ distribution

Bunded storage tanks - Cookson and Zinn (PTL) Ltd
Fuel oil storage - Grant & Livingston Ltd
Fuel oil storage/ distribution - John Davidson (Pipes) Ltd
Fuel oil storage/ distribution - Metcraft Ltd
Fuel storage tanks - T & D Packaging Ltd
Liquid storage tanks - Franklin Hodge Industries Ltd
Oil pollution monitoring and control systems - design, manufacture, installation and commissioning of - Aquasentry
Oil refinery - Petrochem Carless
Oil storage tanks - Davant Products Ltd
Oil tanks and storage bunkers - Titan plastech

S50 Vacuum

Vacuum - Barloworld Vacuum Technology Plc
Vacuum controls - Danfoss Ltd
Vacuum pumps - F. W. Harris & Co Ltd

S51 Steam

Steam - Golden Coast Ltd
Steam design packages - AEL
Steam Traps - Flowserve Flow Control

S60 Fire hose reels

Fire fighting hoses, nozzles, adaptors and standpipes - Richard Hose Ltd
Fire hose reels - Armstrong Pumps Ltd
Fire hose reels - Fire Security (Sprinkler Installations) Ltd
Fire hose reels - Nu-Swift International Ltd
Fire hose reels - UK Fire International Ltd
Firefighting equipment - The BSS Group PLC
Hose reels - Chubb Fire Ltd
Hosereels - fixed, automatic and swinging - Macron Safety Systems (UK) Ltd

S61 Dry risers

Dry riser - Eagles, William, Ltd
Dry riser equipment - Kidde Fire Protection Services Ltd
Dry Risers - Armstrong Pumps Ltd
Dry risers - Firetecnics Systems Ltd
Dry risers - Macron Safety Systems (UK) Ltd
Flooring installation - Tiflex Ltd
Hose reels for fire fighting - Richard Hose Ltd
Safety and fire resistant flooring - Tiflex Ltd
Sprinkler Systems - Kidde Fire Protection Services Ltd

S62 Wet risers

Wet Risers - Armstrong Pumps Ltd
Wet risers - Holden Brooke Pullen

S63 Sprinklers

Fire sprinkler tanks - Franklin Hodge Industries Ltd
GRP sprinkler and pump suction tanks - Dewey Waters Ltd
Sprinkler systems - Firetecnics Systems Ltd
Sprinklers - Fire Security (Sprinkler Installations) Ltd
Sprinklers - Holden Brooke Pullen
Sprinklers - Richard Hose Ltd
Sprinklers - UK Fire International Ltd
Sprinklers - Wormald Ansul UK Ltd

S64 Deluge

Deluge - Fire Security (Sprinkler Installations) Ltd
Deluge - UK Fire International Ltd
Deluge - Wormald Ansul UK Ltd
Fire hydrant valves including dry riser equipment - Richard Hose Ltd

S65 Fire hydrants

Fire extinguishers supply and maintenance - Nu-Swift International Ltd
Fire hydrants - Bailey Birkett Sales Division

S70 Gas fire fighting

Explosion suppression systems - Kidde Fire Protection Ltd
Fire extinguishers supply and maintenance - Nu-Swift International Ltd
Fire suppression equipment - Kidde Fire Protection Ltd
Fire suppression systems - Macron Safety Systems (UK) Ltd
Gas detection equipment - MSA (Britain) Ltd
Gas detection equipment - Zellweger Analytic Ltd, Sieger Division
Gas detection systems - Blakell Europlacer Ltd
Gas fire fighting - Hygood Ltd
Gas fire fighting - UK Fire International Ltd
Gaseous fire extinguishing system - Wormald Ansul UK Ltd
Marine fire protection systems - Kidde Fire Protection Ltd
Polyethylene pipes for water and gas - Capper Plastics
Rail fire protection systems - Kidde Fire Protection Ltd
Vehicle fire protection systems - Kidde Fire Protection Ltd

S71 Foam fire fighting

Extinguishers - Channel Safety Systems Ltd
Fire extinguisher systems - Fife Fire Engineers & Consultants Ltd
Fire extinguishers - Richard Hose Ltd
Fire extinguishers supply and maintenance - Nu-Swift International Ltd
Foam equipment - Eagles, William, Ltd
Foam equipment and concentrate - Macron Safety Systems (UK) Ltd
Foam fire fighting - Fire Security (Sprinkler Installations) Ltd
Foam fire fighting - UK Fire International Ltd
Foam systems - Wormald Ansul UK Ltd
Hose assemblies for fire fighting, industrial and drinking purposes - Richard Hose Ltd

T10 Gas/ Oil fired boilers

Boiler maintenance - H Pickup Mechanical & Electrical Services Ltd
Boiler Point Instrumentation - Skil Controls Ltd
Boilers - Alpha Therm Ltd
Boilers (commercial condensing) - AEL
Boilers oil, gas & solid fuel - Brikenden (Builders Merchants) Ltd
Combustion equipment - Nu-Way Ltd
Condensing boilers - Broag Ltd
Condensing Boilers - Clyde Combustions Ltd
Flexible liners - Ventlane Ltd
Flue and chimney systems - Flamco UK Ltd
Flue systems - Marflex Chimney Systems
Flue terminals - Tower Flue Componants Ltd
Gas and solid fuel heating appliances - Hepworth Heating Ltd
Gas boilers - AEL
Gas boilers - Halstead Boilers Ltd
Gas central heating boilers - Burco Maxol
Gas fired boilers - Fulton Boiler Works (Great Britain) Ltd
Gas Fired Hot water heaters - HRS Hevac Ltd
Gas fired radiant heating equipment - Horizon International Ltd
Gas flue systems - Ubbink (UK) Ltd
Gas vents - Ventlane Ltd
Gas, oil fired boilers - Broag Ltd
Gas, oil fired boilers - Don Heating Products Ltd
Gas, oil fired boilers - Hartley & Sugden
Gas, oil fired boilers - Kayanson Engineers Ltd
Gas, oil, dual fuel - GP Burners (CIB) Ltd
Gas/ Oil fired boilers - Clyde Combustions Ltd
Gas/oil fired boilers - Docherty New Vent Chimney Group
Gas/oil fired boilers - Hamworthy Heating Ltd
Gas/Oil Fired Boilers - service maintenance - Hodgkinson Bennis Ltd
Gas/oil fired burners - Worcester Heat Systems Ltd
Greenhouse heaters - Brettell & Shaw Ltd
Heat transfer thermal fluid systems and heaters - Babcock Wanson Uk Ltd
Heating - Dantherm Ltd
Heating - HV Air Conditioning Ltd
Heating manifolds - SAV United Kingdom Ltd
Heating products - The BSS Group PLC
Instantaneous hot water systems - HRS Hevac Ltd
Oil and gas fired boilers and ancillaries - Boulter Boilers Ltd
Oil boilers: kitchen/utility, combination, external, wall mounted - Trianco Redfyre Ltd
Plumbing & Heating - Spital Tile Co Ltd
Prefabricated boilers and cylinder sets - IPPEC Sytsems
Refractory boiler linings, bricks, castables, ceramic fibre products & incinerators - Combustion Linings Ltd
Reno- High efficiency Boiler Range - Johnson & Starley
Sectional boilers - Beeston Heating Group Ltd
Service Terminals - Ubbink (UK) Ltd
Steam and hot water boilers and pressure vessels - Wellman Robey Ltd
Steam generators & fire tube boilers - Babcock Wanson Uk Ltd
Steel boilers - Beeston Heating Group Ltd
Thermal fluid heaters, thermal oxidizers for solids, liquids & gaseous waste. - General Combustion Ltd
Water boilers - TSE Brownson

T11 Coal fired boilers

Back boilers for solid fuel open fires - Interoven Ltd
Coal fired boilers - Docherty New Vent Chimney Group
Coal fired boilers - Mitsui Babcock Energy Ltd
Combustion equipment - Saacke Ltd
Fireplaces and accessories - Petit Roque Ltd
Flexible liners for solid fuel - Ventlane Ltd
Flue and chimney systems - Keddy-Poujoulat (UK) Ltd
Flue liners - ISOKERN (UK) Ltd
Flue systems - Marflex Chimney Systems
Plastic coal bunkers - Tyrrell Tanks Ltd
Refractory boiler linings, bricks, castables, ceramic fibre products & incinerators - Combustion Linings Ltd
Solid fuel central HT boiler - Dunsley Heat Ltd
Steam and hot water boilers and pressure vessels - Wellman Robey Ltd
Stokers and associated equipment - service and maintenance - Hodgkinson Bennis Ltd

T12 Electrode/ Direct electric boilers

Electric boilers - Eltron Chromalox
Electric boilers - Fulton Boiler Works (Great Britain) Ltd
Electric boilers - Technical Control Systems Ltd
Electric calorifiers - Eltron Chromalox
Electrical boilers:Aztec Boilers 2-12KW - Trianco Redfyre Ltd
Electrode/direct electric boilers - Collins Walker Ltd, Elec Steam & Hot Water Boilers
Flue systems - Marflex Chimney Systems
Refractory boiler linings, bricks, castables, ceramic fibre products & incinerators - Combustion Linings Ltd

T13 Packaged steam generators

Bird/vermin control - Cannon Hygiene Ltd
Packaged steam generators - Collins Walker Ltd, Elec Steam & Hot Water Boilers
Packaged steam generators - Mitsui Babcock Energy Ltd
Packaged steam generators - Rycroft Ltd
Steam and hot water boilers and pressure vessels - Wellman Robey Ltd
Steam boilers - Fulton Boiler Works (Great Britain) Ltd

T14 Heat pumps

Heat pumps - Eaton-Williams Group Ltd
Heat pumps - Holden Brooke Pullen
Heat pumps - MacMarney Refrigeration & Air Conditioning Ltd
Heat pumps - Rycroft Ltd
Heat Pumps - Stiebel Eltron Ltd
Pumps - Biral Pumps
Reclaim pump packs - S & P Coil Products Ltd
Water chillers - Therminal Technology (sales) Ltd

T15 Solar collectors

Solar - Stiebel Eltron Ltd
Solar collectors - Ubbink (UK) Ltd
Solar collectors, solar electric powered products - Intersolar Group Ltd
Solar panels - CPV Ltd
Solar panelsand cylinders - IPPEC Sytsems
Solar slats - Grando (UK) Ltd
Solar Thermal Collectors - SCHOTT UK Ltd
Solar water heating systems - Sundwel Solar Ltd

T16 Alternative fuel boilers

Alternative fuel boilers - Hartley & Sugden
Combustion burners - GP Burners (CIB) Ltd
Energy saving equipment - Saacke Ltd
Flue and chimney systems - Keddy-Poujoulat (UK) Ltd
Flue systems - Marflex Chimney Systems
Refractory boiler linings, bricks, castables, ceramic fibre products & incinerators - Combustion Linings Ltd

T20 Primary heat distribution

Custom made fires - Valor
District heating - ABB Industrial Systems Ltd
Drum heaters - Flexelec(UK) Ltd
Flue systems - Marflex Chimney Systems
Heater cables - Flexelec(UK) Ltd

T20 Primary heat distribution (Cont'd)

Heater mats - Flexelec(UK) Ltd
Heating - Birdsall Services Ltd
Heating - BSS Zenith
Heating Accessories - Townhive Ltd
Industrial and commercial heating systems - Ambi-Rad Ltd
Industrial heating equipment Manufacturers - Turnbull & Scott (Engineers) Ltd
Primary Heat distribution - Saacke Ltd
Radiators - Townhive Ltd
Redfyre cookers: central heating, traditional, flueless - Trianco Redfyre Ltd
Temperature controls - Flexelec(UK) Ltd
Thermal fluid heaters - Fulton Boiler Works (Great Britain) Ltd

T30 Medium temperature hot water heating

Control equipment - The BSS Group PLC
Extruded PVC profiles - Rustin Allen Ltd.
Heating - Dunham Bush Ltd
Immersion heaters and thermostats - Heatrae Sadia Heating Ltd
LST Radiators - Myson Radiators Ltd
Medium temperature hot water heating - Clyde Combustions Ltd
Medium temperature hot water heating - Hamworthy Heating Ltd
MTHW Heating - Rycroft Ltd
Steam and hot water boilers and pressure vessels - Wellman Robey Ltd
Valves - The BSS Group PLC

T31 Low temperature hot water heating

Boiler and calorifier descaling - Aquastat Ltd
Boilers - HRS Hevac Ltd
Buffer and storage vessels - Flamco UK Ltd
Butterfly valves - Martin Orgee Ltd
Central heating - Property Mechanical Products Ltd
Central heating equipment - Hudevad Britain
Central heating programmers, water heating controllers, motorised valves, room stats and thermostats - Horstmann Controls Ltd
Cistern fed water heaters - Heatrae Sadia Heating Ltd
Compact Radiators - Barlo Radiators Limited
Cylinder Stats - Horstmann Controls Ltd
Design Radiators - Barlo Radiators Limited
Domestic and commercial central heating controls - Danfoss Randall Ltd
Electric ventilated heating system - Elvaco UK Ltd
Gas convector heaters - Burco Maxol
Gate valves - Martin Orgee Ltd
Heating - BSS Zenith
Heating - Dunham Bush Ltd
Heating - Myson Radiators - Myson Radiators Ltd
Heating and plumbing - Diamond Merchants
Heating to pool areas - Buckingham Swimming Pools Ltd
Hot water heating - Jaga Heating Poducts (UK) Ltd
Large unvented water heaters - Heatrae Sadia Heating Ltd
Low surface temperature radiators - Jaga Heating Poducts (UK) Ltd
Low Surface Tttemperature Radiators - Barlo Radiators Limited
Low temperature hot water - Hamworthy Heating Ltd
Low temperature hot water heating - Clyde Combustions Ltd
Low temperature hot water heating - Sundwel Solar Ltd
Low water content heating soloutions - Jaga Heating Poducts (UK) Ltd
LPHW - Turnbull & Scott (Engineers) Ltd
LTHW Heating - Rycroft Ltd
Manifolds - Martin Orgee Ltd
Panel Radiators - Myson Radiators Ltd
Perimeter heating - Unilock Ltd
Point of use water heaters - Heatrae Sadia Heating Ltd
Pressure reducing valves - Martin Orgee Ltd
Radiant Ceiling Panels - Zehnder Ltd
Radiator gaurds (low surface temperature) - Alumasc Interior Building Products Limited
Radiator valves - Martin Orgee Ltd
Radiators - De Longhi Radiators srl
Radiators - Zehnder Ltd
Round top Radiators - Barlo Radiators Limited
Steam and hot water boilers and pressure vessels - Wellman Robey Ltd
Swing check valves - Martin Orgee Ltd
Thermostatic mixinet valves - Martin Orgee Ltd
Towel Radiators - Barlo Radiators Limited
Underfloor heating - FloRad Heating Systems
Underfloor Heating - Wavin Plastics Ltd
Valves - Martin Orgee Ltd
Warm water underfloor heating - IPPEC Sytsems
Water Heating - Stiebel Eltron Ltd

T32 Low temperature hot water heating (self-contained specification)

Aluminum radiators - AEL
Cast Iron Radiators - AEL
Commissioning Valves - Marflow Eng Ltd
Corgi Appilasers Engineers - AEL
Decorative Radiators - Myson Radiators Ltd
Domestic central heating - Boulter Boilers Ltd
Domestic radiators - Barlo Radiators Limited
Electric radiators - Loblite Ltd
Electric water heating - Kaloric Heater Co Ltd
Gas Central Heating - Servowarm
Gas fires - Robinson Willey Ltd
Heating - ACR Heat Transfer Manufacturing Ltd
Heating - Servitherm
Heating and plumbing - Diamond Merchants
Heating systems - Marflow Eng Ltd
Low surface temperature radiator gaurds - Alumasc Interior Building Products Limited
Low Temperature Hot Water Heating - Essex
Low temperature hot water heating - Multibeton Ltd
Low temperature hot water(small scale) - Hamworthy Heating Ltd
LTHW Heating (self contained) - Rycroft Ltd
Manifolds - Marflow Eng Ltd
Radiator valves - Pegler Hattersley
Radiators - Clyde Combustions Ltd
Radiators - Kermi (UK) Ltd
Radiators - Novellini Bathroom Products Ltd
Small unvented water heaters - Heatrae Sadia Heating Ltd
Steel Panel Radiators - AEL
Storage water heating - Johnson & Starley
Underfloor heating - Multibeton Ltd
Water heaters - P & R Electrical (London) Ltd
Water treatment of heating and chilled water systems - Aquastat Ltd

T33 Steam heating

Air conditioning - Dunham Bush Ltd
Steam - Turnbull & Scott (Engineers) Ltd
Steam heating - Rycroft Ltd
Stream heating - Saacke Ltd
Thermostatic shower valves - Horne Engineering Ltd

T40 Warm air heating

Air conditioning - Dunham Bush Ltd
Air Curtains - Robinson Willey Ltd
Air warmers - Seaque Technologies Ltd
Design radiators - Jaga Heating Poducts (UK) Ltd
Electric warm air heating - Kaloric Heater Co Ltd
Fan convectors - S & P Coil Products Ltd
Finned tubing & trench heaters - Turnbull & Scott (Engineers) Ltd
Gas and oil fired indirect warm air heaters and systems - Dragonair Ltd
Gas fired air heaters - Reznor UK Ltd
Gas-fired warm aire unit heaters - Radiant Services Ltd
Heat exchangers gasketted plate - HRS Hevac Ltd
Heating - Encon Air Systems Ltd
Heating equipment - NuAir Ltd
HPHW - Turnbull & Scott (Engineers) Ltd
Industrial & commercial, gas or oil indirect fired warm air heating systems - Elan-Dragonair
Industrial and commercial heating systems - Ambi-Rad Ltd
Industrial heating - Hoval Ltd
Radiant ceiling panels - S & P Coil Products Ltd
Thermal Fluid - Turnbull & Scott (Engineers) Ltd
Trench Heating - S & P Coil Products Ltd
Unit heater - Dravo Environmental Services
Warm air heaters - Saacke Ltd
Warm air heating - Consort Equipment Products Ltd
Warm air heating - Docherty New Vent Chimney Group
Warm air heating - May William (Ashton) Ltd
Warm Air heating - Nu-Way Ltd
Warm air heating - Roberts-Gordon

T41 Warm air heating (self-contained specification)

Basket fires - Valor
Electric warm air heating - Kaloric Heater Co Ltd
Natural and fan assisted floor convectors - Multibeton Ltd
Oil & gas fired warm air cabinet - Radiant Services Ltd
Oil fired warm air unit heaters - Radiant Services Ltd
Unit Heaters - Turnbull & Scott (Engineers) Ltd
Warm Air Heating - Vortice Ltd
Warm air heating (small scale) - Consort Equipment Products Ltd
Warm air heating (small scale) - May William (Ashton) Ltd

T42 Local heating units

Anti condensation heaters - Eltron Chromalox
Central heating equipment - Hudevad Britain
Decorative gas fires & stoves - Magiglo Ltd
Electric fires - Gazco Ltd
Electric fires - Valor
Electric fuel effect fires - Verine Ltd
Flameproof heaters, radiant panel heaters - Seaque Technologies Ltd
Gas fired air heaters - Reznor UK Ltd
Gas fired space heating - domestic/commercial - Johnson & Starley
Gas fires - Gazco Ltd
Gas fires/ Gas conductors - Robinson Willey Ltd
Gas stoves - Gazco Ltd
Gas wall heaters - Valor
Gas-fired radiant tube heater - Radiant Services Ltd
Heaters - Hayward Air
Heating - Andrews Sykes Hire Ltd
Heating products - Pressmain
Immersion heaters - Heating Equipment Ltd
Infra ray heaters - Bitmen Products Ltd
Local Heating - Kiltox Damp Free Solutions
Local heating units - Aga-Rayburn
Local heating units - Consort Equipment Products Ltd
Local heating units - Harvey Fabrication Ltd
Products for commercial and industrial applications - Bush Nelson PLC
Radiant gas heaters - Infraglo (Sheffield) Ltd
radiant heaters - shearflow LTD
Radiant infra-red plaque gas heaters - Space - Ray UK
Radiant infra-red poultry gas brooders - Space - Ray UK
Radiant infra-red tube gas heaters - Space - Ray UK
Radiant tube and radiant heaters - Radiant Services Ltd
Radiators - Rettig Heating Ltd
Radiators - cast iron - Reclaim.UK.Com
Refractory boiler linings, bricks, castables, ceramic fibre products & incinerators - Combustion Linings Ltd
Solid Fuel Suites - Winther Browne & co Ltd
Space Heating - Stiebel Eltron Ltd
Storage and panel heaters - Seaque Technologies Ltd
Storage gas water heaters, focal point fan convector - Burco Maxol
Unit heaters - Biddle Air Systems Ltd
Unit heaters - S & P Coil Products Ltd
Unit heaters - Therminal Technology (sales) Ltd

T50 Heat recovery

Brazed plate heat exchangers - HRS Hevac Ltd
Condensation control - Kiltox Damp Free Solutions
Heat pipes - S & P Coil Products Ltd
Heat recovery - Brook Airchanger
Heat recovery - Dantherm Ltd
Heat recovery - Eaton-Williams Group Ltd
Heat recovery - Greenwood Air Management
Heat recovery - Kair ventilation Ltd
Heat recovery - May William (Ashton) Ltd
Heat recovery - Rycroft Ltd
Heat Recovery - Ubbink (UK) Ltd
Heat Recovery - Vent-Axia
Heat recovery products - Silavent Ltd
Heat recovery unit - Dunsley Heat Ltd
Heat recovery units - Therminal Technology (sales) Ltd
Heat recovery ventilation units - Ves Andover Ltd
Heating and heat recovery equipment thermal wheels - AEL
Run around coils - S & P Coil Products Ltd
Steam and hot water boilers and pressure vessels - Wellman Robey Ltd

T60 Central refrigeration plant

Air conditioning - Clivet UK Ltd
Central refridgeration plants - Eaton-Williams Group Ltd
Central refrigeration plant - MacMarney Refrigeration & Air Conditioning Ltd
Industrial refridgeration equipment and systems - J & E Hall Ltd
Refrigerant dryers - Fluidair International Ltd
Refrigeration - Birdsall Services Ltd
Refrigeration - McQuay UK Ltd
Refrigeration plant - Copeland Corporation Ltd

T61 Chilled water

Chilled water - J & E Hall Ltd
Chilled water - Rycroft Ltd
Dry coolers - Visco Ltd

T70 Local cooling units

Absorption chillers - Birdsall Services Ltd
Chilled beams - ABB Control Ltd
Close cooling units - Clivet UK Ltd
Local cooling units - Eaton-Williams Group Ltd
multi service chilled beam - Trox (UK) Ltd
Portable coolers - London Fan Company Ltd
Portable heating - Andrews Sykes Hire Ltd

T71 Cold rooms

Cold rooms - Clark Door Ltd
Cold rooms - Isowall UK) Ltd
Cold rooms - J & E Hall Ltd
Cold rooms - MacMarney Refrigeration & Air Conditioning Ltd
Cold Rooms - SCHOTT UK Ltd
Modular coldrooms - Hemsec Manufacturing Ltd

T72 Ice pads

Ice pads - J & E Hall Ltd

U10 General ventilation

Access doors - Hotchkiss Air Supply
Air conditioners, dampers, air ductline - Advanced Air (UK) Ltd
Air conditioning - Birdsall Services Ltd
Air conditioning - Purified Air Ltd
Air conditioning - Trox (UK) Ltd
Air conditioning systems - Elvaco UK Ltd
Air distribution equipment - Designed for Sound Ltd
Air seperators - Martin Orgee Ltd
Air transfer grilles - Mann McGowan Group
Air-conditioning - Andrews Sykes Hire Ltd
Airflow instrumentation - Airflow Developments Ltd
Anti-condensation for motors heater tapes - Flexelec(UK) Ltd
Dehumidifiers - Hayward Air
Domestic and commercial exract fans - Greenwood Air Management
Domestic ventilation equipment - Silavent Ltd
Domestic ventilation systems - Airflow (Nicoll Ventilators) Ltd
Exhauster blowers - BVC
Fan convectors - Biddle Air Systems Ltd
Fire resistant air transfer grilles - Décor Ireland
Flanging systems - Hotchkiss Air Supply
Flatoval tube - Hotchkiss Air Supply
Flexible ducting - Hotchkiss Air Supply
Flue systems - Marflex Chimney Systems
Full freshair systems - Dantherm Ltd
General supply/ extract - Emsworth Fireplaces Ltd
General supply/ extract - Torit DCE
General ventilation - Colt International
General ventilation - Dyer Environmental Controls
General ventilation - Intersolar Group Ltd
General ventilation - London Fan Company Ltd
General ventilation - May William (Ashton) Ltd
General ventilation - NFC Specialist Products Ltd
Housing ventilation - ABB Control Ltd
Humidifiers - Hayward Air
Humidity control - Kiltox Damp Free Solutions
Mechanical extract terminals - Klober Ltd
Modular ventilation ducting - Domus Ventilation Ltd.
Natural ventalation Systems - Passivent Ltd
Roof ventilators - (ridge & slope mounted) - McKenzie-Martin Ltd
Sealed system equipment and air separation and dirt removal equipment - Flamco UK Ltd
Spiral tube - Hotchkiss Air Supply
Splits, VRF, Packaged & Chillers - Toshiba Carrier UK
Support systems - Hotchkiss Air Supply
Trade suppliers of commercial and decorative ceiling fans - Fantasia Distribution Ltd
Underfloor ventilators - Ryton's Building Products Ltd
Ventelated Ceilings - Vianen Kitchen Ventilation (UK) Ltd.
Ventilation - Dravo Environmental Services
Ventilation - Encon Air Systems Ltd
Ventilation - HV Air Conditioning Ltd
Ventilation - NuAir Ltd
Ventilation - Therminal Technology (sales) Ltd
Ventilation - Ubbink (UK) Ltd
Ventilation - domestic, commercial and industrial - Vent-Axia
Ventilation and ductwork components - CCL Veloduct Ltd
Ventilation equipment - Johnson & Starley
Ventilation equipment - Roof Units Group
Ventilation of pool areas - Buckingham Swimming Pools Ltd
Ventilation Products - Hinton Perry & Davenhill Ltd
Ventilators - Kair ventilation Ltd
Window ventilator manufactures - Simon R W Ltd
Window Ventilators - Passivent Ltd

U11 Toilet ventilation

Domestic extractor fans - Domus Ventilation Ltd.
Toilet ventilation - Kair ventilation Ltd

U12 Kitchen ventilation

Design, Supply and Installation of Commercial Kitchen Ventelation - Vianen Kitchen Ventilation (UK) Ltd.
Domestic extractor fans - Domus Ventilation Ltd.
Extract fan units - Ves Andover Ltd
Extraction systems and odour control - Purified Air Ltd
Grease eliminators - Nationwide Filter Co Ltd
Grease filters - Farr Filtration Ltd
Grease filters - Westbury Filters Limited
Kitchen Canopies - Vianen Kitchen Ventilation (UK) Ltd.
Kitchen extract - Fire Protection Ltd
Kitchen extraction systems - Heydal Ltd
Kitchen ventilation - Kair ventilation Ltd
Kitchen ventilation - London Fan Company Ltd
Kitchen ventilation - Ryton's Building Products Ltd
Ventilation - domestic, commercial and industrial - Vent-Axia

U13 Car park ventilation

Car park extract - Fire Protection Ltd
car park ventilation - Smoke Control Services Ltd
Carpark ventilation - London Fan Company Ltd

U14 Smoke extract/ Smoke control

Air filtration units - Airstream Products Ltd
Ancillaries - Advanced Air (UK) Ltd
Automatic smoke barrier-curtains / blinds - Cooper Group Ltd
Birdguards - Brewer Metalcraft
Chimney cappers - Brewer Metalcraft
Chimney cowls - Brewer Metalcraft
Chimneys and flues - Metcraft Ltd
Commercial and domestic electrostatic air filters - Robinson Willey Ltd
Emission monitors - Skil Controls Ltd
Powered cowls - Brewer Metalcraft
Rain caps - Brewer Metalcraft
Smoke and natural ventilation systems - SE Controls
Smoke control systems - Smoke Control Services Ltd
Smoke extract - Consort Equipment Products Ltd
Smoke extract - McKenzie-Martin Ltd
Smoke extract fans - London Fan Company Ltd
Smoke extract/ smoke control - Colt International
Smoke extract/ smoke control - Fire Protection Ltd
Smoke extract/ smoke control - Photain Controls Plc
Smoke extract/ Smoke control - Promat Fire Protection Ltd
Smoke extract/control - Trox (UK) Ltd
Smoke management systems - Advanced Air (UK) Ltd
smoke outlets - Luxcrete Ltd
Spark asbestos - Brewer Metalcraft
Specials - Brewer Metalcraft
Spinning cowls - Brewer Metalcraft
Terminals - Brewer Metalcraft
Vehicle exhaust extraction - Nederman UK
Ventilation - domestic, commercial and industrial - Vent-Axia
Ventilation and smoke extract equipment - Twin County Fan Company Ltd

U15 Safety cabinet/Fume cupboard extract

Domestic and commercial flue and ventilation products - Rega Ductex Ltd
Downflow Booths - Envair Ltd
Fume Cupboards - Complete Laboratory Installations
Fume cupboards - S & B Ltd
Fume extract equipment - HV Air Conditioning Ltd
Fume Extraction - Complete Laboratory Installations
Laminar Flow Cabinets - Envair Ltd
Pharmaceutical Isolators - Envair Ltd
Safety cabinet / fume cupboards - Envair Ltd
Safety cabinet/ fume cupboard extract - Active Carbon Filters Ltd
Safety cabinet/fume cupboard - Trox (UK) Ltd
Safety extract/ smoke control - Fire Protection Ltd

U16 Fume extract

Air pollution control plant - Plastic Construction Fabrications Ltd
Dust and fume extraction - Encon Air Systems Ltd
Extractor fans - Tower Flue Componants Ltd
Flue systems - Marflex Chimney Systems
Fume Cupboards - Lab Furnishings (UK) Ltd
Fume extract - Active Carbon Filters Ltd
Fume extract - Emsworth Fireplaces Ltd
Fume extract - Torit DCE
Fume extract - Trox (UK) Ltd
Fume extraction - London Fan Company Ltd
Fume extraction - Nederman UK
Fume extraction - PCT Group Sales Ltd
Nox reduction combustion equipment and technical advice - Mitsui Babcock Energy Ltd
Portable heating - Andrews Sykes Hire Ltd

U20 Dust collection

Dust Collection - Barloworld Vacuum Technology Plc
Dust collection - BVC
Dust collection - Farr Filtration Ltd
Dust collection - Torit DCE
Dust control equipment - Airmaster Engineering Ltd
Dust extraction - Nederman UK

U30 Low velocity air conditioning

Air conditioning - Airedale International Air Conditioning Ltd
Air conditioning - Dalair Ltd
Air conditioning - Eaton-Williams Group Ltd
Air conditioning - Eberle GmbH
Air conditioning - Edenaire Ltd
Air conditioning - HV Air Conditioning Ltd
Air conditioning - Toshiba Carrier UK
Air handling units - ABB Control Ltd
Comfort - Airedale International Air Conditioning Ltd
Condensing units - Ves Andover Ltd
Domestic and commercial flue and ventilation products - Rega Ductex Ltd
Low velocity air conditioning - MacMarney Refrigeration & Air Conditioning Ltd
Portable heating - Andrews Sykes Hire Ltd
Spares - Airedale International Air Conditioning Ltd

U31 VAV air conditioning

VAV air conditioning - MacMarney Refrigeration & Air Conditioning Ltd
VAV air conditioning - May William (Ashton) Ltd
VAV air conditioning - Senior Air Systems
VAV Air conditioning - Trox (UK) Ltd
Vav boxes - ABB Control Ltd
VAV terminal units - Advanced Air (UK) Ltd

U32 Dual-duct air conditioning

Dual-duct air conditioning - Advanced Air (UK) Ltd
Dual-duct air conditioning - MacMarney Refrigeration & Air Conditioning Ltd

U33 Multi-zone air conditioning

Multi-zone air conditioning - MacMarney Refrigeration & Air Conditioning Ltd

U41 Fan-coil air conditioning

Air conditioning - Encon Air Systems Ltd
Dampers - Ruskin Air Management
Fan coil air conditioning - Colt International
Fan coil air conditioning - Trox (UK) Ltd
Fan coil heater casings - Alumasc Interior Building Products Limited
Fan coil units - Therminal Technology (sales) Ltd
Fan coil valves - Martin Orgee Ltd
Fan coils - ABB Control Ltd
Fan coils - McQuay UK Ltd
Fan coils units - Biddle Air Systems Ltd
Fan-coil air conditioning - Advanced Air (UK) Ltd
Fan-coil air conditioning - Eaton-Williams Group Ltd
Fan-coil air conditioning - MacMarney Refrigeration & Air Conditioning Ltd
Fan-coil air conditioning - Senior Air Systems
Legionella prevention and control - Aquastat Ltd
Refrigeration equipment - Dunham Bush Ltd
Service - Airedale International Air Conditioning Ltd

U42 Terminal re-heat air conditioning

Air terminal devices - ABB Control Ltd
Heat exchangers gasketted plate - HRS Hevac Ltd
Terminal re-heat air - Senior Air Systems
Terminal re-heat air conditioning - Advanced Air (UK) Ltd

U43 Terminal heat pump air conditioning

Air conditioning - Clivet UK Ltd
Terminal heat pump air conditioning - MacMarney Refrigeration & Air Conditioning Ltd

U50 Hybrid system air conditioning

Hybrid system air conditioning - Eaton-Williams Group Ltd

U60 Air conditioning units

Air conditioning - Hayward Air
Air conditioning - SGB Youngman
Air conditioning units - Amcor (Appliances) Ltd
Air conditioning units - Colt International
Air conditioning units - Denco Ltd
Air conditioning units - MacMarney Refrigeration & Air Conditioning Ltd
Air conditioning units - Toshiba Carrier UK
Condensing units - Airedale International Air Conditioning Ltd
Free standing air conditioning - Senior Air Systems
Hygrometers - Air Improvement Centre Ltd
Mobile air conditioners, humidifiers, dehumidifiers and air purifiers - Air Improvement Centre Ltd
Portable air conditioners - Thermoscreens Ltd
Portable heating - Andrews Sykes Hire Ltd
Refrigeration equipment - Dunham Bush Ltd
Thermometers - Air Improvement Centre Ltd

U70 Air curtains

Air curtains - Biddle Air Systems Ltd
Air curtains - shearflow LTD
Air curtains - Thermoscreens Ltd
Air curtains / Convectors - S & P Coil Products Ltd
Aircurtains - Ambi-Rad Ltd

V10 Electricity generation plant

Automatic start generators - Cummins Power Generation Ltd
Automatic synchronising generators - Cummins Power Generation Ltd
Baseload diesel generator - Broadcrown Ltd
Building Intergrated Photovoltaics - SCHOTT UK Ltd
Central battery systems - Channel Safety Systems Ltd
Continuous and emergency standby power generator sets from 5.5kVA to 2500kVA power range - FG Wilson
Diesel engined generation plant - Dawson-Keith Ltd
Electrical components - Thomas and Betts
General electrical products - WF Electrical Distributors
General electrical products - wholesale - Jolec Electrical Supplies
Generation plant - Broadcrown Ltd
Generation plant, diesel, and gas turbine - Dale Power Systems
generators - Chippindale Plant Ltd
Mobile generator sets for on-set power requirements - FG Wilson
Mobile generators - Cummins Power Generation Ltd
Peak loppingd diesel generators - Broadcrown Ltd
Power generators - Cummins Power Generation Ltd
Power plants - Cummins Power Generation Ltd
Silenced generators - Cummins Power Generation Ltd
Standby diesel generators - Broadcrown Ltd
Super silenced generators - Cummins Power Generation Ltd
Transformers and packaged substations - Schneider Electric Ltd
Ventilation Products - Allmat (East Surrey) Ltd

V11 HV supply/ distribution/ public utility supply

Brass sections - Boilden MKM Ltd
Electrical cables - AEI Cables Limited
Electrical distribution - Schneider Electric Ltd
Extra high voltage - Schneider Electric Ltd
HV Supply/Distribution/Public Utility Supply - Elequip Projects Ltd
Public distribution - Schneider Electric Ltd

V12 LV supply/ public utility supply

CT Chambers - Blakley Electrics Ltd
Electrical cables - AEI Cables Limited
Electricity meters - Switch2 Energy Solutions Ltd
Final distribution - Schneider Electric Ltd
LV Supply/Public Utility Supply - Elequip Projects Ltd
Retail Lighting - Fitzgerald Lighting Ltd

V20 LV distribution

Accessories for electrical services - Lewden Electrical Industries
Circuit protection and control - Bill Switchgear
Consumer units - Square D, Electrical Distribution & Building applicators Division
Electrical cables - AEI Cables Limited
Electrical components - Thomas and Betts
Electrical distribution equipment - Square D, Electrical Distribution & Building applicators Division
Electrical installations - HEM Interiors Group Ltd
Low voltage distribution - Schneider Electric Ltd
LV Distribution - Elequip Projects Ltd
Modular desk & Screen Power, Telecom & Data Units - Electrak International Ltd
Plug and socket to EN 60309 - Blakley Electrics Ltd
Power Distribution-110v - Blakley Electrics Ltd
Red Projected Electricl Equipment - Blakley Electrics Ltd
Small power distribution - Ackermann Ltd
Surge protection systems - Erico Europa (GB) Ltd
Tranformers 0-1 50 KVQA - Blakley Electrics Ltd

V21 General lighting

110v Lighting - Blakley Electrics Ltd
Access panels to ceiling void - Saint-Gobain Ecophon Ltd
Accessories for electrical services - Lewden Electrical Industries
Aisle and step lighting - Gradus Ltd
Architectural dimming - Home Automation Ltd
Architectural lighting - Optelma Lighting Ltd
Architectural pendants - Beta Lighting Ltd
Asymmetric lighting - Crescent Lighting
Cabinet and display lighting - Crescent Lighting
Cold cathode - Oldham Lighting Ltd
Commercial & architectural lighting - Beta Lighting Ltd
Commercial lighting - Radius Lighting Ltd
Commercial wiring systems - Metway Electrical Industries Ltd
Compact Fluorecent Downlights - Profile Lighting Services Ltd
Decorative and display lighting - Myddleton Hall Lighting, Division of Peerless Design Ltd
Decorative Lighting - Forma Lighting Ltd
Design, supply and install - Oldham Lighting Ltd
Diffusers louvres and glare control products - Interlux Ltd
Display lighting - Cooper Lighting and Security Ltd
Display Lighting - Fitzgerald Lighting Ltd
Display lighting - Illuma Lighting Ltd
Display lighting - Sylvania Lighting International INC.
Domestic lighting - Sylvania Lighting International INC.
Downlights - Illuma Lighting Ltd
Electrical Accessories - Focus SB
Electrical cables - AEI Cables Limited
Electrical components - Thomas and Betts
Emergency batteries - Bernlite Ltd
Emergency modules - Bernlite Ltd
Energy efficiency in lighting - Poselco Lighting
Energy efficient lighting for commercial, retail and industrial markets - IMP Lighting Ltd
Energy saving lighting - Anglepoise Lighting Ltd
Energy saving lighting - Sylvania Lighting International INC.
Exterior lighting - DW Windsor Lighting
Exterior lighting - Lyktan Lighting (UK) Ltd
Exterior lighting - Woodhouse UK Plc
Fibre optic systems - LB Lighting Ltd
Fibreoptic lighting - Crescent Lighting
Flourescent Lighting - Vitalighting Ltd
Genaral Lighting - Commercial Lighting Systems
General and bespoke lighting systems - Concord Lighting
General lighting - Arden Manufacturing Birmingham Ltd
General lighting - Brandon Medical
General Lighting - Consult Lighting
General lighting - Cooper Lighting and Security Ltd
General Lighting - Elequip Projects Ltd
General lighting - Glamox Electric (UK) Ltd
General lighting - IBL Lighting
General lighting - Indirect Lighting Ltd
General lighting - Jerrards Bros. PLC Division
General lighting - Lampways Ltd
General lighting - Oldham Lighting Ltd
General lighting - Sylvania Lighting International INC.
General lighting - The Cube
General lighting and bulbs - GE Lighting Ltd
General lighting and purpose built lighting - Quip Lighting, Div of Bruce & Macintyre Ltd
General lighting products - Whitecroft Lighting Limited
General lighting, magnetic and electronic ballasts - Helvar Electrosonic
General, modular, compact Fluorescent - Lumitron Ltd
Gimbal spots - Profile Lighting Services Ltd
Handlamps - Ensto Briticent Ltd
Hazardous area lighting - Glamox Electric (UK) Ltd
Hazardous area lighting - PFP Electrical Products Ltd
High Bay Lighting - Setsquare
High frequency electronic ballasts - Bernlite Ltd
Highbay lighting - CU Phosco Lighting
Highbay lighting - Lighting Systems UK, Division of Phosco Ltd
Hospital lighting - Thorlux Lighting
Ignitors - Bernlite Ltd
Industrial lighting - Cooper Lighting and Security Ltd
Industrial Lighting - Fitzgerald Lighting Ltd
Intelligent lighting - Cooper Lighting and Security Ltd
Interior and exterior lighting - Andy Thornton Ltd
Lamps - Philips Lighting
Lamps - ceramic - BLV Licht-und Vakuumtecnik
Large lights - Signature Ltd
LED luminaires - Crescent Lighting
LED luminaires - LB Lighting Ltd
Light bulbs, tungsten halogen and metal halide - BLV Licht-und Vakuumtecnik
Light fittings - Profile Lighting Services Ltd
Light fittings - Rada Lighting
Light fittings - Thorn Lighting Ltd
Light fittings - Vitalighting Ltd
Lighting - Davies John Interiors Ltd
Lighting - Designplan Lighting Ltd
Lighting - Full Spectrum Lighting Ltd
Lighting - HT Martingale Ltd
Lighting - Optelma Lighting Ltd
Lighting - Power Breaker
Lighting - Twin Technology Lighting
Lighting busbar - Salamandre Davis
Lighting control equipment - Philips Lighting Solutions
Lighting control gear inc. ballasts, chokes - Bernlite Ltd
Lighting Control Products - Eaton Electric Limited
Lighting control systems - Hamilton R & Co Ltd
Lighting Control systems - Home Automation Ltd
Lighting Controls - Fitzgerald Lighting Ltd
Lighting controls - Philips Lighting
Lighting Design - Concord: Marlin
Lighting equipment - Thorlux Lighting
Lighting equipment - Transtar
Lighting management - Home Automation Ltd
Lighting Manufacturer and supply - Whitecroft Lighting Limited
Lighting products - Friedland Ltd

V21 General lighting (Cont'd)

Lighting support systems - Vantrunk Engineering Ltd
Lighting trunking - Salamandre Davis
Lighting, switches, accessories and weatherproof sockets - Loblite Ltd
Ligthing Controls - Setsquare
Low energy compact fluorescent lighting - Beta Lighting Ltd
low energy downlighter - Vitalighting Ltd
Low energy lighting systems - Connect Lighting Systems (UK) Ltd
Low voltage - Profile Lighting Services Ltd
low voltage downlighter - Vitalighting Ltd
Low-voltage lighting - Fairfield Displays & Lighting Ltd
Luminaires - Philips Lighting
Luminaires - office, retail, industrial, amenity, pharmaceutical, hospital etc. - Whitecroft Lighting Limited
Mains voltage halogen lighting - Sylvania Lighting International INC.
Marine lighting - Chalmit Lighting
Marshalling Boxes - Setsquare
Natural daylighting systems - Sunpipe @ Monodraught Ltd
Office lighting - Cooper Lighting and Security Ltd
Office lighting - Sylvania Lighting International INC.
Office, medical, architectural, industrial and workstation lighting - Luxo UK Ltd
Period and decorative lighting - Denmans Montrose Ltd
Purpose made luminaires - LB Lighting Ltd
Recessed Lighting - Fitzgerald Lighting Ltd
Recessed and Surface LG3 Luminaires - Profile Lighting Services Ltd
Retail Display Luminaires - Profile Lighting Services Ltd
Security PIR lighting - Timeguard Ltd
Specials - Profile Lighting Services Ltd
Spotlights - Illuma Lighting Ltd
Suspended acoustic ceiling systems - Saint-Gobain Ecophon Ltd
Suspended ceiling luminaires - Burgess Architectural Products Ltd
Track - Illuma Lighting Ltd
Uplights - Illuma Lighting Ltd

V22 General LV power

Cable tray - Mita (UK) Ltd
Dimmers, switches, sockets, on plates of brass, bronze,steel, chrome, wood polycarbonate - Hamilton R & Co Ltd
Electrical cables - AEI Cables Limited
Floor boxes - Mita (UK) Ltd
Power - Davies John Interiors Ltd
Powerpoles and powerposts - Mita (UK) Ltd
Support channel - Mita (UK) Ltd

V30 Extra low voltage supply

Accessories for electrical services - Lewden Electrical Industries
Electrical components - Thomas and Betts
Entry sound communoication products - Friedland Ltd
Exrta low voltage supply - Brandon Medical
General electrical products - WF Electrical Distributors
General electrical products - wholesale - Jolec Electrical Supplies
Low-Voltage Garden Lighting - Anthony de Grey Trellises
Miniature precision switches, push button and cord pull etc - Omeg Ltd
Transformers for low voltage lighting - Mode Lighting (UK) Ltd

V31 DC supply

Batteries - MJ Electronics Services (International) Ltd
Battery chargers - MJ Electronics Services (International) Ltd
DC Supplies - Elequip Projects Ltd
Direct current supply - Brandon Medical

V32 Uninterruptible power supply

Central battery systems - Emergi-Lite Safety Systems
Electrical accessories - Electrak International Ltd
Industrial batteries for standby applications - Hawker Ltd
Power electronic repair and service - MJ Electronics Services (International) Ltd
Uninterrupted power supply - Liebert Swindon Ltd
Uniterupted power supply - Channel Safety Systems Ltd
UPS services - MJ Electronics Services (International) Ltd
UPS Systems - Elequip Projects Ltd
UPS systems and inverters - MJ Electronics Services (International) Ltd

V40 Emergency lighting

Bespoke light fitting manufacturers - Vitalighting Ltd
Dimming controls - Jerrards Bros. PLC Division
Electronic starter switches - Jerrards Bros. PLC Division
Emergency lighting - ADT Fire and Security Plc
Emergency lighting - Channel Safety Systems Ltd
Emergency lighting - Connect Lighting Systems (UK) Ltd
Emergency lighting - Cooper Lighting and Security Ltd
Emergency lighting - Designplan Lighting Ltd
Emergency lighting - Emergi-Lite Safety Systems
Emergency Lighting - Fitzgerald Lighting Ltd
Emergency lighting - Gent Limited
Emergency lighting - Glamox Electric (UK) Ltd
Emergency lighting - Lab-Craft Ltd
Emergency lighting - Menvier Ltd
Emergency lighting - Metway Electrical Industries Ltd
Emergency lighting - MJ Electronics Services (International) Ltd
Emergency lighting - Orbik Electrics Ltd
Emergency lighting - PFP Electrical Products Ltd
Emergency lighting - Poselco Lighting
Emergency Lighting - Rada Lighting
Emergency lighting - Thorlux Lighting
Emergency lighting - Vitalighting Ltd
Emergency lighting - Wright Security incorporating Madewel Electronic Systems Ltd
Emergency lighting systems - JSB Electrical PLC
Emergency lighting systems - Mackwell Electronics Ltd
Emergency lighting testing systems - Emergi-Lite Safety Systems
Emergency lighting, linear lighting systems, up and downlighters - Beta Lighting Ltd
Emergency lighting, switch tripping and battery chargers - Elequip Projects Ltd
Emergency lighting/exit lights - Thomas & Betts Manufacturing Services Ltd
Emergency lights - Firetecnics Systems Ltd
Emergency Lights - Whitecroft Lighting Limited
General electrical products - WF Electrical Distributors
General electrical products - wholesale - Jolec Electrical Supplies
Hazardous area lighting - Chalmit Lighting
High frequency ballast - Jerrards Bros. PLC Division
Industrial batteries for standby applications - Hawker Ltd
Light fittings - Thorn Lighting Ltd
Lighting components - Jerrards Bros. PLC Division
Magnetic ballast - Jerrards Bros. PLC Division
Powdered way guidance - Hoyles Electronic Developments Ltd
Security lighting - Asco Extinguishers Co Ltd T/A The Asco Fire Group

V41 Street/ Area/ Flood lighting

Accessories for electrical services - Lewden Electrical Industries
Amenity & area lighting - DW Windsor Lighting
Amentity Lighting - Fitzgerald Lighting Ltd
Architectural interior lighting for commercial applications - Lyktan Lighting (UK) Ltd
Architectural lighting - Thorlux Lighting
Bollard baselights - Signature Ltd
Coloured metal hacide floodlamps - BLV Licht-und Vakuumtecnik
Emergency lighting - Chalmit Lighting
Exterior lighting - Vitalighting Ltd
External Lighting - Guardall Ltd
Flood Lighting - Blakley Electrics Ltd
Flood Lighting - Fitzgerald Lighting Ltd
Floodlighting - Thorlux Lighting
Floodlighting poles - Stainton Metal Co Ltd
Hazardous area lighting - Thorlux Lighting
High Mast Lighting - CU Phosco Lighting
Low-Voltage Garden Lighting - Anthony de Grey Trellises
Outdoor lighting - Holophane Europe Ltd
Pavement lights - Luxcrete Ltd
Security lighting systems - Yorian Garden Products
Sign lights - Signature Ltd
Street lighting - DW Windsor Lighting
Street lighting - Urbis Lighting Ltd
Street lighting - Woodhouse UK Plc
Street lighting and floodlighting - Lighting Systems UK, Division of Phosco Ltd
Street lighting columns and accessories - Lampost Construction Co Ltd
Street lighting columns and high masts - Stainton Metal Co Ltd
Street lighting columns and sign posts - Fabrikat (Nottingham) Ltd
Street lighting systems - Industrial Lighting Ltd
Street lighting, floodlights - CU Phosco Lighting
Street/ area & flood lighting - Sylvania Lighting International INC.
Street/ Area/ Flood lighting - Consult Lighting
Street/ area/ flood lighting - Cooper Lighting and Security Ltd
Street/ area/ flood lighting - Designplan Lighting Ltd
Street/ Area/ Flood lighting - Rada Lighting
Street/Area/Flood Lighting - Commercial Lighting Systems
Street/Area/Flood Lighting - Whitecroft Lighting Limited
Tunnel and underpass lighting - Thorlux Lighting

V42 Studio/ Auditorium/ Arena lighting

Dimming Systems - Setsquare
Display Lighting - Jerrards Bros. PLC Division
Industrial lighting - Chalmit Lighting
Industrial lighting - Glamox Electric (UK) Ltd
Lamps - capsule - BLV Licht-und Vakuumtecnik
Lamps - dichroic - BLV Licht-und Vakuumtecnik
Lighting project design - DW Windsor Lighting
Lux lift - Park Products
Scene Setting - Setsquare
Specialist industrial and commercial lighting - Martin Roberts
Studio/ Auditorium/ Arena Lighting - Consult Lighting
Studio/ auditorium/ arena lighting - Sylvania Lighting International INC.
Studio/Auditorium/Arena Lighting - Elequip Projects Ltd
Theatrical lighting - Multitex GRP & GRG Products

V50 Electric underfloor/ceiling heating

Ceiling Heating - Flexel International Ltd
Electric matting - MacLellan Rubber Ltd
Electric underfloor heating - Bush Nelson PLC
Electric underfloor heating - Heat Trace Ltd
Electric underfloor, road and ramp heating - Tyeo Thermal Control
Floor heating - Flexel International Ltd
Heated raised access floors - Raised Floor Systems Ltd
Ramp heating - Findlay Irvine Ltd
Underfloor heating - H Pickup Mechanical & Electrical Services Ltd
Underfloor heating - Raised Floor Systems Ltd
Underfloor heating - Rettig Heating Ltd
Underfloor heating - Uponor Housing Solutions
Underfloor heating systems - REHAU Ltd

V51 Local electric heatng units

Electric air duct heaters - Eltron Chromalox
Electric cartridge heaters - Eltron Chromalox
Electric convector heaters - Eltron Chromalox
Electric fan heaters - Eltron Chromalox
Electric fire manufacturer - Glen Dimplex UK Ltd
Electric Fires - Winther Browne & co Ltd
Electric heater batteries - Ves Andover Ltd
Electric heaters - Eltron Chromalox
Electric heating - Kaloric Heater Co Ltd
Electric heating - Robinson Willey Ltd
Electric heating system controls - Heat Trace Ltd
Electric panel heaters - Valor
Electric Stoves - Gazco Ltd
Electric stripring heaters - Eltron Chromalox
Electrical heating units - Consort Equipment Products Ltd
General electrical products - WF Electrical Distributors
General electrical products - wholesale - Jolec Electrical Supplies
Infra red heating - Robinson Willey Ltd
Local Electric Heating - Kiltox Damp Free Solutions
Local electric heating units - Bush Nelson PLC
Low energy electric panel heaters - Ecolec

V90 Electrical installation (self-contained spec)

Accessories for electrical services - Lewden Electrical Industries
Design of electrical services - Aercon Wiring Systems Ltd
Domestic wiring accessories and circuit protection equipment - Contactum Ltd
Electrical accessories - Marcus M Ltd
Electrical cables - AEI Cables Limited
Electrical installation equipment - Legrand Power Centre
Electrical installation products - Barton Engineering
Electrical Installations - Knight & Smith Ltd, Electrical Contractors
Electrical wiring accessories - Eaton Electric Limited
Electrical works - Davies John Interiors Ltd
General electrical products - WF Electrical Distributors
General electrical products - wholesale - Jolec Electrical Supplies
Mains lighting - JSB Electrical PLC
Power pole.posts - Salamandre Davis
Ropes and sashcords - Marlow Ropes Ltd
Wire accessories - Hellermann Tyton
Workplace systems - Ackermann Ltd

W10 Telecommunications

Audio visual hire - Planet Communications Group Ltd
Audio visual solutions & sales - Planet Communications Group Ltd
Cables - Alcatel Cablenet
Call centers - Mitel Telecom Ltd
Communications - Ackermann Ltd
Communications - Ogilvie Construction Ltd
Dualtecs - C & K Intellisense
Flexgurds - C & K Intellisense
Hospital communications systems - Tunstal Telecom Ltd
Industrial batteries for standby applications - Hawker Ltd
Least cost routing - Mitel Telecom Ltd
PBX - Mitel Telecom Ltd
Structured IT cabling systems - Visua Ltd
Telecommunication Systems - Siemens Communication Ltd
Telecommunications - Connought Communication Systems Ltd
Telecommunications - Mar.Com Systems Ltd
Telecommunications - Mitel Telecom Ltd
Telecommunications - Vodafone Paging Ltd
Telemetry control - Baxall Ltd
Telephone enclosures - Midland Alloy Ltd
TV/ Telephone systems for hospitals - Wandsworth Elecrtrical Ltd
Video conferencing - Planet Communications Group Ltd
Video conferencing - Visua Ltd
Voice recording, processing and training enhancement software - Computertel Ltd

W11 Paging/ Emergency call

Call systems - C-TEC (Computionics) Ltd
Care communications - ADT Fire and Security Plc
Data paging systems - Channel Safety Systems Ltd
Dispersed alarms - Telelarm care Ltd
Doctor call systems - Channel Safety Systems Ltd
Intercoms - PEL Services Limited
Messaging - Mitel Telecom Ltd
Nurse call and on-site paging systems - Blick Communication Systems Ltd
Nurse call systems - Channel Safety Systems Ltd
Nurse call systems - Wright Security incorporating Madewel Electronic Systems Ltd
Paging - Wandsworth Elecrtrical Ltd
Signalling and warning equipment - Castle Care-Tech
Staff paging/ location - Johnson Controls Systems Ltd
Warden call systems - Telelarm care Ltd

W12 Public address/ Conference audio facilities

Background music - PEL Services Limited
Conference equipment - Anders + Kern UK Ltd
Public address - Blick Communication Systems Ltd
Public address - Blick Communication Systems Ltd
Public address - PEL Services Limited
Public address system - AET.gb Ltd
Public address/ conference audio facilities - RPS Data Products (UK) Ltd
Public address/conference audio - Mar.Com Systems Ltd
Sound masking systems - AET.gb Ltd
Sounders - external - C & K Intellisense
Sounders - internal - C & K Intellisense

W20 Radio/ TV/ CCTV

advanced digital video recording - Vision Sytems (Europe) Ltd
Audio visual - Anders + Kern UK Ltd
Audio visual equipment - Visua Ltd
Audio visual system design and installation - Visua Ltd
C.C.T.V - Sampson & Partners Fencing
CCTV - Abel Alarm Co Ltd
CCTV - ADT Fire and Security Plc
CCTV - Allen (Fencing) Ltd
CCTV - Allgood plc
CCTV - Guardall Ltd
CCTV - Mar.Com Systems Ltd
CCTV - PEL Services Limited
CCTV - Prima Security & Fencing Products
CCTV - UK Fire International Ltd
CCTV - Ward's Flexible Rod Company Ltd
CCTV - White Gates
CCTV cameras - Baxall Ltd
CCTV cameras, switchers, multiplexers and telemetry control video transmission systems - Baxall Ltd
CCTV equipment - Hunting Engineering Ltd
CCTV monitoring - Crimeguard Systems Ltd
CCTV systems - Interphone Security Group PLC
CCTV systems - Viking Security Systems Ltd
CCTV's - Initial Electronic Security Systems Ltd
CCTV's - Wright Security incorporating Madewel Electronic Systems Ltd
Closed circuit television - Crimeguard Systems Ltd
Closed circuit television - Vicon Industries Ltd
Closed circuit television - Vidionics Security Systems Ltd
Design, manufacture and installation of audio visual systems - Wilson & Garden Ltd
Ethernet video surveillance - Vidionics Security Systems Ltd
Induction loop amplifiers - C-TEC (Computionics) Ltd
Monitoring - Mar.Com Systems Ltd
Presentation and conference equipment - Anders + Kern UK Ltd

W20 Radio/ TV/ CCTV (Cont'd)

Radio telemetry - Cooper Security Ltd
Radio/ TV/ CCTV - Johnson Controls Systems Ltd
remote video transmition - Vision Sytems (Europe) Ltd
Social care alarms - Cooper Security Ltd
Transmission poles - Stainton Metal Co Ltd
TV distribution - Blick Communication Systems Ltd
Video entry - Crimeguard Systems Ltd
Video transmission systems - Baxall Ltd
Visual communications systems - Quartet-GBC UK Ltd

W21 Projection

ATM enclosures - CSI Security LTD
Big screens (indoor/outdoor) - Pro Display
Electronic displays - Pro Display
Plasma monitors/TVs - Pro Display
Projection - Mar.Com Systems Ltd
Projection - RPS Data Products (UK) Ltd
Projection equipment - Wilson & Garden Ltd
Projection screen systems - Harkness Hall Ltd
Projection screens - Leakale Taurus Visual Aids
Projection screens - Magiboards Ltd
Projectors - Pro Display
Space management systems for power on manual, vertical or horizontal movement - Harkness Hall Ltd
Video walls - Pro Display

W22 Information/Advertising display

Banner installations - Marine & Commercial Rigging Systems
Information cases - Green Brothers Ltd
Information kiosks - Pryorsign
Information/ advertising display - Focal Signs Ltd
Information/ advertizing display - Stocksigns Ltd
Information/advertising display - RPS Data Products (UK) Ltd
Internet kiosks - H&H Electronics Ltd
LCD displays/kiosks - Pro Display
Light boxes - Simplex Signs Limited
Locakable glazed notice boards - Leakale Taurus Visual Aids
Moving message displays - Pro Display
Multi-line text displays - Pro Display
Poster/ menu cases/cabinets - Simplex Signs Limited
Showcases - Fairfield Displays & Lighting Ltd
Software - sales and database - Golden Coast Ltd
Suspend display systems - Fairfield Displays & Lighting Ltd

W23 Clocks

Clocks - General Time Europe
Clocks - Lesco Products Ltd
Clocks and counters - Pro Display
Time/temperature clocks - Pro Display
Timer switches for central heating and controls for central heating - Tower Flue Componants Ltd
Timers - Elkay Electrical Manufacturing Co Ltd

W30 Data transmission

Cables - Alcatel Cablenet
Construction software - Computer Foundations Ltd
Data cabling - Davies John Interiors Ltd
Data entry and calculating of bills of quantities or other volume information - Sumlock J.H.P.
Data Transmission - Mar.Com Systems Ltd
Datacoms - MK (MK Electric Ltd)
Distance learning systems - Visua Ltd
Electric information systems - Ferrograph Limited
Electronic information display systems - Techspan Systems PLC
Job costing, accounts computer software - Arena Software Ltd
Messaging systems - Visua Ltd
Software for the construction industry - Masterbill Micro Systems Ltd
Take off, bills of quantities and estimating for the PQS or builder - EITE & Associates

W40 Access control

Access control - Abel Alarm Co Ltd
Access control - ADT Fire and Security Plc
Access control - Allen (Fencing) Ltd
Access control - Allgood plc
Access control - BPT Automation Ltd
Access Control - Cooper Security Ltd
Access control - Crimeguard Systems Ltd
Access control - Dorma Door Services Ltd
Access Control - Focus SB
Access Control - Guardall Ltd
Access control - Honeywell Control Systems Ltd
Access control - Hoyles Electronic Developments Ltd
Access control - Initial Electronic Security Systems Ltd
Access control - Johnson Controls Systems Ltd
Access control - Kaba Garog
Access control - Laidlaw Architectural Hardware
Access control - PEL Services Limited
Access control - Permclose Door Services Ltd
Access control - Siemens Building Technologies Ltd
Access Control - Tate Fencing
Access control - Till & Whitehead Ltd, t/a L. Birkinshaw
Access control - Trend Control Systems Ltd
Access control - Turnquest UK Ltd
Access control - UK Fire International Ltd
Access control - Vidionics Security Systems Ltd
Access control - Viking Security Systems Ltd
Access control - Wright Security incorporating Madewel Electronic Systems Ltd
Access control & security products - Clarke Instruments Ltd
Access control equipment - Baydale Architectural Systems Ltd
Access control equipment - Haffele UK Ltd
Access control personal barriers - Electro Mechanical Systems Ltd
Access control products - ASSA Ltd
Access Control Security Solutions - Laidlaw Architectural Hardware Ltd
Access control systems - CEM Systems Ltd
Access control systems - Interphone Security Group PLC
Access control systems - Pac International Ltd
Access control systems - Sabre Access Control Systems Ltd
Access control systems - Sampson & Partners Fencing
Access control systems - Securikey Ltd
Access control systems - Time and Data Systems International Ltd (TDSI)
Access control systems - Wandsworth Elecrtrical Ltd
Access control systems - Yale Security Products Ltd
Access control systems / equipment - APT Controls Ltd
Access control systems, ID and access card production and registered key systems - Kaba (UK) Ltd
Access control, exit devices, electromagnetic locks and electric strikes - Relcross Ltd
Auto sliding gates - Allen (Fencing) Ltd
Automatic doors - Permclose Door Services Ltd
Automatic gates, barriers and railings - Parsons Brothers Steelwork
Automatic Rising Bollards - BPT Automation Ltd
Automatic traffic barriers, gates and doors - Crimeguard Systems Ltd
Barriers and fully automated equipment - Prima Security & Fencing Products
Bird control - Timberwise (UK) Plc
Call systems - Wandsworth Elecrtrical Ltd
Car park access control - Tully
Car park barriers - Maswell Engineering Ltd
Car park payment systems - Tully
Car park revenue systems - Tully
Controlled access - Barrow Hepburn Sala Ltd
Door & window control equipment - Geze UK
Door and gate automation - Automatic Doors & Gates
Door controls and locking devices - Laidlaw Architectural Hardware Ltd
Door entry systems - Aardee Security Shutters Ltd
Electric openers/ spare parts - Cardale Doors Ltd
Electric releases - Clarke Instruments Ltd
Electrical garage door accessories - Garador Ltd
Electrical remote control window opening systems - Teleflex - Morse
Electromagnetic closers - Jebron Ltd
High speed vertical folding and roll -up doors - Mandor Engineering Ltd
ID badging systems - CEM Systems Ltd
Intercom - Wandsworth Elecrtrical Ltd
Linear actuaters - Electro Mechanical Systems Ltd
Mechanical and electrical locks - ASSA Ltd
People counting systems - CEM Systems Ltd
Proximity access control systems - BPT Security Systems (UK) Ltd
Remote access solutions - Mitel Telecom Ltd
Security access and control - Hart Door Systems Ltd
Security detection and alarm control equipment - Castle Care-Tech
Security roller shutters - Lakeside Group
Sliding Door gear - Haffele UK Ltd
Standalone Access control panels - Abloy Security Ltd
Strongrooms and strongroom doors - Rosengerns Tann Ltd
Turnstiles - Allen (Fencing) Ltd
Vehicle access control - Autopa Ltd
Video and audio door entry systems - BPT Security Systems (UK) Ltd
Video Entry systems - Tate Fencing

W41 Security detection and alarm

Access Control Systems - Abacus Integrated Systems Ltd
Anti-terrorist devices - Castle Care-Tech
Burglar alarms - Initial Electronic Security Systems Ltd
CCTV Systems - Abacus Integrated Systems Ltd
Digital image recorders - Optex (Europe) Ltd
Disabled toilet alarm systems - Channel Safety Systems Ltd
Door security locks - Haffele UK Ltd
Electric fence system for commercial & industrial sites - Advanced Perimeter Systems Ltd
Electronic security services - West London Security
Fence Intrusion Detection System - Advanced Perimeter Systems Ltd
Gates and barriers - Hart Door Systems Ltd
Installation and maintenance of intruder alarms - Rouse Security Service
Integrated security systems - Vidionics Security Systems Ltd
Intruder alarm and CCTV - PEL Services Limited
Intruder Alarms - Abacus Integrated Systems Ltd
Intruder alarms - ADT Fire and Security Plc
Intruder alarms - Cooper Security Ltd
Intruder alarms - UK Fire International Ltd
Intruder Detection - Guardall Ltd
passive infared detectors - Vision Sytems (Europe) Ltd
Passive infra red detectors - C & K Intellisense
Passive infra red detectors and photoelectric beams - Optex (Europe) Ltd
PC based security control and monitoring system - Advanced Perimeter Systems Ltd
Perimeter protection - Vidionics Security Systems Ltd
Personal security system - Wandsworth Elecrtrical Ltd
Security - Wright Security incorporating Madewel Electronic Systems Ltd
Security systems - Pac International Ltd
security and surveillance systemas - Vision Sytems (Europe) Ltd
Security ans surveillance - Telford Doors and Barriers Ltd
Security detection - Friedland Ltd
Security detection - Honeywell Control Systems Ltd
Security detection and alarm - Baxall Ltd
Security detection and alarm - Johnson Controls Systems Ltd
Security detection and alarm - Tower Manufacturing
Security detection and alarm control equipment - Castle Care-Tech
Security detection and alarm systems - Abel Alarm Co Ltd
Security detection and alarm systems - ADT Modern Security Systems Ltd
Security detection and alarm systems - C & K Intellisense
Security detection and alarm systems - C-TEC (Computionics) Ltd
Security detection and alarms - Geoquip Ltd
Security detection equipment - Hoyles Electronic Developments Ltd
Security detection screening - Specialist Building Products
Security doors - Jewers Doors Ltd
Security equipment and alarms - ADT Fire and Security Plc
Security equipment and alarms - Advanced Perimeter Systems Ltd
Security equipment, cameras, multiplexes and phoneline video transmission systems - Robot (UK) Ltd
Security fencing - C & W Fencing Ltd
Security grilles - Ring-Gard (UK) Ltd
Security management systems and high security locks - Relcross Ltd
Security markers - Elliott John H (Monostamp) Ltd
Security mirrors - Smart F & G (Shopfittings)Ltd
Security products - Viking Security Systems Ltd
Security Systems - JSB Electrical PLC
Sounders and strobes - KAC Alarm Company Ltd
Vibration sensors - Optex (Europe) Ltd
video intrusion detection - Vision Sytems (Europe) Ltd
Video networks CCTV systems - Baxall Ltd
Voice alarms - PEL Services Limited
Warehouse intrusion detection system - Advanced Perimeter Systems Ltd

W50 Fire detection and alarm

Break glass and call points - KAC Alarm Company Ltd
Emergency lighting - P4 Limited
Fire alarm equipment - Hoyles Electronic Developments Ltd
Fire alarm equipment - Lab-Craft Ltd
Fire Alarm Radio Systems - Tann Synchronome Ltd
Fire alarm system maintenance - Tann Synchronome Ltd
Fire Alarm Systems - Abacus Integrated Systems Ltd
Fire alarms - Fife Fire Engineers & Consultants Ltd
Fire alarms - Griffin and General Fire Services Ltd
Fire alarms - Initial Electronic Security Systems Ltd
Fire alarms and smoke detectors - Asco Extinguishers Co Ltd T/A The Asco Fire Group
Fire and gas control systems - Zellweger Analytic Ltd, Sieger Division
Fire detection - Abel Alarm Co Ltd
Fire detection - ADT Fire and Security Plc
Fire detection - Honeywell Control Systems Ltd
Fire detection - Kidde Fire Protection Ltd
Fire detection - Thomas & Betts Manufacturing Services Ltd
Fire detection and alarm - Clinicon Fire
Fire detection and alarm - Cooper Lighting and Security Ltd
Fire detection and alarm - Johnson Controls Systems Ltd
Fire detection and alarm - PEL Services Limited
Fire detection and alarm - Photain Controls Plc
Fire detection and alarm systems - Blick Communication Systems Ltd
Fire detection and alarm systems - Channel Safety Systems Ltd
Fire detection and alarm systems - Dicon Safety Products (UK) Ltd
Fire detection and alarm systems - Firetecnics Systems Ltd
Fire detection and alarm systems - Gent Limited
Fire detection and alarm systems - Hochiki Europe (UK) Ltd
Fire detection and alarm systems - Kidde Fire Protection Services Ltd
Fire detection and alarm systems - Siemens Building Technologies Ltd,
Fire detection and alarm systems - Tann Synchronome Ltd
Fire detection and alarm systems - Wormald Ansul UK Ltd
Fire detection and alarms - C-TEC (Computionics) Ltd
Fire detection and alarms - Hygood Ltd
Fire detection equipment - Nittan (UK) Ltd
Fire detection systems - Chubb Fire Ltd
Fire detection systems - Emergi-Lite Safety Systems
Fire detection Systems - JSB Electrical PLC
Fire detection systems - Menvier Ltd
Fire protection - Wright Security incorporating Madewel Electronic Systems Ltd
Fire protection and alarm - Siemens Building Technologies Ltd
Fire protection equipment - Preussag Fire Protection Ltd
Fire protection products for doors and building openings - Moffett Thallon & Co Ltd
Fire safety products and alarms - UK Fire International Ltd
Gas detection - Gas Measurement Instruments Ltd
Gates and barriers - Hart Door Systems Ltd
Radio fire alarm systems - Channel Safety Systems Ltd
Smoke and heat detectors - Apollo Fire Detectors Ltd
Smoke detection, carbon monoxide and heat alarms - BRK Brands Europe Ltd
Vesda aspirating smoke detection equipment - Vision Sytems (Europe) Ltd

W51 Earthing and bonding

Contacts - C & K Intellisense
Earth rods and earthing systems - Erico Europa (GB) Ltd
Earthing & Bonding - Tower Manufacturing
Earthing and bonding - Cutting R C & Co
Earthing systems - Furse
Lighting protection engineers - Gray J. W. & Son Ltd
Semiconductors - Mitel Telecom Ltd

W52 Lightning protection

Earthing & bonding installations - Gray J. W. & Son Ld
Lightning conductor systems - Erico Europa (GB) Ltd
Lightning protection - Cutting R C & Co
Lightning protection - Furse
Lightning protection, earthing and electronic protection - Bacon WM & RW Ltd
Transient overvoltage protection - Furse

W53 Electromagnetic screening

Patient tagging system - Wandsworth Elecrtrical Ltd

W54 Liquid detection alarm

Liquid detection alarm - Specialist Building Products
Oil pollution monitoring and control systems - design, manufacture, installation and commissioning of - Aquasentry
Water leak detection equipment - Aquasentry

W60 Central control/Building management

Automation and control - Schneider Electric Ltd
Building and energy management - ADT Fire and Security Plc
Building automation - Guthrie Douglas Ltd
Building Automation - Johnson Controls Systems Ltd
Building control and management systems - Siemens Building Technologies Ltd
Building management systems - Honeywell Control Systems Ltd
Building management systems - Trend Control Systems Ltd

W60 Central control/Building management (Cont'd)

Central control - Johnson Controls Systems Ltd
Central control software - Telelarm care Ltd
Central control systems - Honeywell Control Systems Ltd
Community care alarm and monitering equipment - Tunstal Telecom Ltd
Control equipment - George Fischer Sales Ltd
Control panels - C & K Intellisense
Control systems - Eltron Chromalox
Control systems - Technical Control Systems Ltd
Electric radiator controls - Loblite Ltd
Energy monitoring & targeting - Trend Control Systems Ltd
Energy saving controls - Caradon Plumbing Solutions
Home control products - Friedland Ltd
Indicator panels - Hoyles Electronic Developments Ltd
Industrial controls, air compressor controls and building management systems - Rapaway Energy Ltd
InfoProx stand alone card reader - CEM Systems Ltd
Instrumentation for boiler plant, incinerators and furnaces - Skil Controls Ltd
Lighting control systems - Electrak International Ltd
Monitoring - Initial Electronic Security Systems Ltd
Portable moisture, temperature and noise level meters - Channel Electronics
Remote management and monitoring - Guardall Ltd
Voice recording, processing and training enhancement software - Computertel Ltd
Wireless Nursecall system - Telelarm care Ltd

X10 Lifts

Access, hydraulic and towable lifts - Denka International
Brush sealing for lifts and glass doors - Kleeneze Sealtech Ltd
Disabled access lifts - Hymo Ltd
Disabled access lifts and ramps - Chase Equipment Ltd
Dissabld Access Lifts - Thyssenkrupp Accessibility LTD
Good lifts - Stannah Lifts Ltd
Goods lifts - Hymo Ltd
Goods, passenger, car, freight lifts, hydraulic or electric traction - Porn Dunwoody ERS (Lifts & Escalators) Ltd
Home lifts - Wessex Medical
Hydraulic lifting equipment - Hymo Ltd
Hydraulic passenger lifts, access wheelchair platforms lifts - Stannah Lifts Ltd
Installation, reburibshment and repair of lifts - Cable Lift Installations Ltd
Lift maintenance - Express Evans Lift
Lift maintenance, lift modernization - Express Evans Lifts
Lift manufacturers, installers service and maintenance - Pickerings Ltd
Lifting equipment - Hymo Ltd
Lifting platforms - Wessex Medical
Lifts - Ellis & McDougall Lifts Ltd
Lifts - Express Evans Lift
Lifts - Kone PLC
Lifts - Otis Ltd
Lifts - Schindler Ltd
Lifts, ramps and loading bay equipment - specialist manufacturer - Chase Equipment Ltd
Mechanical screws, electro hydralic goods and lift platforms - Engineering & Lift Ltd
Mobile scissors lifts - Hymo Ltd
New lifts 5kg service two ton goods lifts - Express Evans Lifts
Pallet lifts - Hymo Ltd
Passenger, scenic, goods and vehicle lifts - Oakland Elevators Ltd
Platform lifts - Thyssenkrupp Accessibility LTD
Portable material lifts - Genie Europe
Portable personal lifts - Genie Europe
Public access wheelchair - Thyssenkrupp Accessibility LTD
Pushbuttons and controls for lifts - Dewhurst plc
Scissor lift tables - Hymo Ltd
Scissors lifts - Hymo Ltd
Service lifts, maintenance - Oakland Elevators Ltd
Step lifts - Wessex Medical
Vertical and Stair - Brooks Stairlifts Ltd
Wheelchair stairlifts - Wessex Medical

X11 Escalators

Brush deflection for escalators - Kleeneze Sealtech Ltd
Escalators - Kone PLC
Escalators - Otis Ltd
Escalators - Schindler Ltd
Escalators and passenger conveyers, service and maintenance - Kone Escalators Ltd

X12 Moving pavements

Moving pavements - Schindler Ltd
Passenger conveyors - Kone PLC
Travolators - Otis Ltd

X13 Powered stairlifts

Domestic stairlifts - Thyssenkrupp Accessibility LTD
Stairlifts - Thyssenkrupp Accessibility LTD
Vertical Lifts, Domestic, Straight and Curved Stairlifts - Brooks Stairlifts Ltd

X14 Fire escape chutes/slings

Fire escape chutes/slings - NRG Fabrications
Ropes and sashcords - Marlow Ropes Ltd

X20 Hoists

Access lifting equipment - Genie Europe
Access platforms - Kobi Ltd
Hoists - Metreel Ltd
Hoists - Morris Material Handling Ltd
Hoists - PCT Group Sales Ltd
Hoists - Wessex Medical
Hoists & handling equipment - Randalls Fabrications Ltd
Hoists, davits, tripods, confined space access equipment and manhole cover lifter - Didsbury Engineering Co Ltd
Hydraulic movement systems - Hydratight Sweeney Limited
Lifting equipment, systems for stage and studio applications - Harkness Hall Ltd
Permanent access cradles - Kobi Ltd
Ropes and sashcords - Marlow Ropes Ltd
Slings - Wessex Medical
Wheelchair Platform Lifts - Thyssenkrupp Accessibility LTD
Woven lifting slings - Albion Manufacturing

X21 Cranes

Boom lifts (Telescopic articulating) - Genie Europe
Crane support - Eve Trakway
Cranes - Metreel Ltd
Cranes - Morris Material Handling Ltd
Cranes - PCT Group Sales Ltd
Grade 8 chain slings - PCT Group Sales Ltd
Mini Crane Hire - HE Services (Plant Hire) Ltd
Webbing slings - PCT Group Sales Ltd

X22 Travelling cradles/gantries/ladders

Access cradles - Cento Engineering Company Ltd
Access ladders - Clow Group Ltd
Cradle runways - Cento Engineering Company Ltd
Cradles - Clow Group Ltd
Dock levelers - Crawford Amber Ltd
Dock shelters - Crawford Amber Ltd
Façade access equipment - Cento Engineering Company Ltd
Façade hoists - Cento Engineering Company Ltd
Gantries - Clow Group Ltd
Gantry - Cento Engineering Company Ltd
Latchways - Cento Engineering Company Ltd
Steps - Clow Group Ltd
Suspended access equipment - Cento Engineering Company Ltd
Travelling gantries - Kobi Ltd
Travelling ladders - Cento Engineering Company Ltd
Travelling ladders - Kobi Ltd
Trestles - Clow Group Ltd
Window cleaning cradles - Cento Engineering Company Ltd

X23 Goods distribution/ Mechanised warehousing

Conveyors - Metalrax Ltd
Dock levelers - Chase Equipment Ltd
Dock shelters - Chase Equipment Ltd
Loading bay equipment - Crawford Amber Ltd
Material handling equipment - Barton Storage Systems Ltd
Mobile food conveyors - TSE Brownson
Mobile yard ramps - Chase Equipment Ltd
Modular loading dock - Chase Equipment Ltd
Road haulage - Eve Trakway
Scissor lifts - Genie Europe
Steel conyor plates for sugar industry - manufacturer - Wm. Bain Fencing Ltd

X30 Mechanical document conveying

Despatch equipment - Envopak Group Ltd

X31 Pneumatic document conveying

Air tube conveying systems - Barloworld Vacuum Technology Plc
Mailroom equipment - Envopak Group Ltd
Pneumatic air tube conveying systems:- cash, sample, hospital and document - Barloworld Vacuum Technology Plc
Pneumatic conveying systems - BVC

X32 Automatic document filing and retrieval

Retrieval systems - Barrow Hepburn Sala Ltd
Sorting equipment - Envopak Group Ltd

Y10 Pipelines

DuraFuse - Viking Johnson
Durapipe ABS - Viking Johnson
Friatherm - Viking Johnson
GRE GRP pipes, tubes and fabrications - Fibaflo Ltd
Gullies - Johnston Pipes Ltd
Mechanical joints - Victaulic Systems
Philmac - Viking Johnson
Pipe clips - Unifix Ltd
Pipe couplings - Viking Johnson
Pipelines - Deepdale Engineering Co Ltd
Pipelines - Grant & Livingston Ltd
Pipelines - Hanson Concrete Products
Pipelines - Insituform Permaline Ltd
Pipelines - John Davidson (Pipes) Ltd
Pipelines - Yorkshire Copper Tube Ltd
Pipes ductile iron - Biwater Industries Ltd
Pipework - Woodburn Engineering Ltd
Pipework Fabrication - Turnbull & Scott (Engineers) Ltd
Plastic pipelines - IPPEC Sytsems
Plumbing accessories - BSS Zenith
Polybutylene pipe and fittings - Davant Products Ltd
Polyethylene plastic pipes - Uponor Housing Solutions
Posiflex - Viking Johnson
Preinsulated Piping - Uponor Housing Solutions
Pressure Pipes - Wavin Plastics Ltd
Proctect-line - Viking Johnson
Service fittings - Talbot
Spiral weld steel tube - Tex Steel Tubes Ltd
Stainless steel fittings, flanges and fabrications - Lancashire Fittings Ltd
Stainless steel tubes and sections - Avesta Sheffield Ltd
Stainless steel valves - Lancashire Fittings Ltd
Victaulic - Viking Johnson
Wask - Viking Johnson

Y11 Pipeline ancillaries

Accessories - Talbot
Air valves - Saint Gobain pipelines PLC
Automatic Flow Balancing valves - Hattersley Newman Hender Ltd
Balancing valves - Crane Fluid Systems
Ball valves - Crane Fluid Systems
Ballvalves - Warner H & Son Ltd
Boiler House Equipment - Flowserve Flow Control
Brass Flanges - Essex
Bronze & Cast Iron double regulating valves and commissioning sets - Hattersley Newman Hender Ltd
Bronze thermostatic, wheel head & lockshield radiator valves - Hattersley Newman Hender Ltd
Bronze, cast iron and ductile gate globe, check and butterfly valves - Hattersley Newman Hender Ltd
Bursting (rupture) discs - Bailey Birkett Sales Division
Butterfly valves - Crane Fluid Systems
Check valves - Crane Fluid Systems
Check Valves - Flowserve Flow Control
Control Valves - Flowserve Flow Control
DZR & Bronze ball valves - Hattersley Newman Hender Ltd
Fire and gas safety valves - Teddington Controls Ltd
Fitttings, valves and tanks - CPV Ltd
Flame arresters - Bailey Birkett Sales Division
Gate valves - Crane Fluid Systems
Gate valves - Saint Gobain pipelines PLC
Globe valves - Crane Fluid Systems
Hand- Hydraulic -Electric - Tube & Pipe Manuipulation Service - Tubela Engineering Co Ltd
Hydrant valves and pressure valves - Eagles, William, Ltd
Industrial Flow Control Equipment - Flowserve Flow Control
Isolation Ball Valves - Flowserve Flow Control
Isolation valves and expansion vessels - SAV United Kingdom Ltd
Main Fittings - Talbot
Malleable iron pipe fittings - Crane Fluid Systems
Mixer valves - Gummers A & J Ltd
Pipe clips and brackets - Regon Ltd
Pipe fixings - Ellis Patents
Pipeline ancillaries - Deepdale Engineering Co Ltd
Pipeline ancillaries - Flex-Seal Couplings Ltd
Pipeline ancillaries - Hamworthy Heating Ltd
Pipeline ancillaries - Isolated Systems Ltd
Pipeline ancillaries - John Davidson (Pipes) Ltd
Plug valves, control valves, gate valves, butterfly valves, check valves and consistency transmitters - Dezurik International Ltd
Plumbing accessories - Ashmead Buidling Supplies Ltd
Plumbing accessories - BSS Zenith
Plumbing fittings, pipes, tools and brassware - Hipkin F W Ltd
Plumbing products and pipe fixings - IBP Ltd
Pressure reducing valves - Auld Valves Ltd
Pressure surplus valves - Auld Valves Ltd
Pushfit systems - Davant Products Ltd
Radiator valves - Crane Fluid Systems
Relief valves - Crane Fluid Systems
Safety & relief valves - Auld Valves Ltd
Sealed system equipment, air vents, pressure reducers, mixing valves, flow switches etc. - Altecnic Ltd
Strainers - Crane Fluid Systems
Tap Kits - Essex
Temperature control valves - Horne Engineering Ltd
Thermostatic and manual shower controls - Gummers A & J Ltd
Valves - Pegler Hattersley
Valves - Saint Gobain pipelines PLC
Valves - Townhive Ltd
Waste fittings - Midland Stom Ltd
Water valves and plumbing equipment - Barber Edward & Co Ltd
White petroleum jelly - Legge Thompson F & Co Ltd

Y20 Pumps

AP1610 - Girdlestone Pumps Ltd
Centrifugal pumps - Girdlestone Pumps Ltd
Chemical resistant pumps - Plastic Construction Fabrications Ltd
Circulating, heating and ventilating pumps - Biral Pumps Ltd
Diaphragm hand pumps - Pump International Ltd
Fixing systems for building services - Erico Europa (GB) Ltd
Gear and centrifugal pumps - Albany Standard Pumps
Grout pumps - Clarke UK Ltd
GRP packaged pumping stations - Pims Pumps Ltd
H.V Pumps - Pressmain
Industrial, chemical, oil & gas, and petrochemical pumps - Girdlestone Pumps Ltd
Metering pumps - SPX Process Equipment
Oil pollution monitoring and control systems - design, manufacture, installation and commissioning of - Aquasentry
Packaged pump sets - Armstrong Pumps Ltd
Process pumps - Clarke UK Ltd
Process pumps - SPX Process Equipment
Pump repair - Turney Turbines (pumps) Ltd
Pumps - Albany Engineering Co Ltd
Pumps - Andrews Sykes Hire Ltd
Pumps - Armstrong Pumps Ltd
Pumps - Barr & Wray Ltd
Pumps - Beeston Heating Group Ltd
Pumps - Calpeda Limited
Pumps - Goodwin HJ Ltd
Pumps - Grosvenor Pumps Ltd
Pumps - Ingersoll-Rand European Sales Ltd
Pumps - JLC Pumps & Engineering Co Ltd
Pumps - Mono Pumps Ltd
Pumps - New Haden Pumps Ltd
Pumps - SPX Process Equipment
Pumps - The BSS Group PLC
Pumps - T-T Pumps & T-T controls
Pumps - Turney Turbines (pumps) Ltd
Pumps - Wilo Samson Pumps Ltd
Pumps, fountains, pressure sets and waste water pumping sets - Fordwater Pumping Supplies
Pumps, fountains, pressure sets and waste water pumping sets - Fordwater Pumping Supplies Ltd
Pumps, mixers and aerators - ABS Pumps Ltd
Submersible electric pumps and mixers - ITT Flygt Ltd
Submersible pumps, servicing, maintenance and technical back up - Pump Technical Services Ltd
Sump drainage pumps - Stuart Turner Ltd
Tunnel concrete pumps - Clarke UK Ltd
Variable pumpdrives - Armstrong Pumps Ltd
Water leak detection equipment - Aquasentry
Water pumps - Stuart Turner Ltd
Water, chemical, submersible, magnetic - Baresford Pumps Ltd

Y21 Water tanks/ cisterns

Carbon and stainless steel storage tanks - Cookson and Zinn (PTL) Ltd
Cistern floats - Airflow (Nicoll Ventilators) Ltd
Cold water storage - Polypipe Bathroom & Kitchen Products Ltd
Cold watertank upgrading, refurbishment & replacement Y21 - Nicholson Plastics Ltd
Custom moulded GRP products - Nicholson Plastics Ltd
Domestic hot water tanks - Elsy & Gibbons Ltd
Elson mains pressure hot water units (vented) - Elsy & Gibbons Ltd
Elson plumbing packages - Elsy & Gibbons Ltd
Elson portable emergency tank - Elsy & Gibbons Ltd
Epoxy Tanks - CST Vulcan
Fibreglass water storage tanks - Parton Fibreglass Ltd
G R P Tanks - Pressmain
General GRP molding - Boatman Plastics Ltd
Glass-Fused-to-Steel-Tanks - CST Vulcan
GRP conical settlement tanks, covers and effluent treatment - Armfibre Ltd
GRP housings - Boatman Plastics Ltd

Y21 Water tanks/ cisterns (Cont'd)

GRP one piece and sectional water storage tanks - Dewey Waters Ltd
GRP one piece, semi sectional and sectional water tanks. - Boatman Plastics Ltd
GRP storage cisterns - Precolor Sales Ltd
GRP storage tanks - APA
GRP tank housings - Group Four Glassfibre Ltd
GRP tanks - Fibre Reinforced Products Ltd
GRP water storage tanks - Group Four Glassfibre Ltd
Hot water cylinders and tanks - Gledhill Water Storage Ltd
Hot water storage tanks - Q.M.P.
Liquid Storage Tanks - CST Vulcan
One piece/ semi sectional/ sectional GRP water storage tanks - Aquatanks Ltd
Plastic cisterns - Dudley Thomas Ltd
Plastic water storage tanks - Titan plastech
Plastic water tanks, plastic oil tanks and plastic septic tanks - Tyrrell Tanks Ltd
Rust stabilisation coatings - Corroless Corrosion Control
Sectional and one piece GRP cold water tanks - Nicholson Plastics Ltd
Sectional Steel Tanks and foundations - Goodwin Tanks Ltd
Sprinkler Tanks - CST Vulcan
Storage tanks - Franklin Hodge Industries Ltd
Storage tanks and vessels - Metcraft Ltd
Tanks - Arbory Group Limited
Tanks - Balmoral Tanks
Tanks - Braithaite Engineers Ltd
Tanks - Fibaform Products Ltd.
Water storage containers and accessories - Harcostar Drums
Water storage systems and cylinders - T & D Packaging Ltd
Water storage tanks - Balmoral Tanks
Water storage tanks - Brimar Plastics Ltd
Water storage tanks - CST Vulcan
Water tanks - Denco Ltd
Water tanks - Harris & Bailey Ltd
Water tanks and cisterns - Estec Environmental Ltd
Water tanks/ cisterns - Grant & Livingston Ltd
Water tanks/ cisterns - Harvey Fabrication Ltd

Y22 Heat exchangers

Coil heat exchangers - S & P Coil Products Ltd
Corrugated shell and tube heat exchanger - HRS Hevac Ltd
Heat exchangers - Armstrong Pumps Ltd
Heat exchangers - Grant & Livingston Ltd
Heat exchangers - Rycroft Ltd
Heat exchangers - Sundwel Solar Ltd
Heat exchangers - Turnbull & Scott (Engineers) Ltd
Heat recovery - Kiltox Damp Free Solutions
Manufacturer of heating elements - Electric Elements Co, The
Plate heat exchangers - AEL
Plate heat exchangers - Beeston Heating Group Ltd

Y23 Storage cylinders/ Calorifiers

Carbon and stainless steel pressure vessels - Cookson and Zinn (PTL) Ltd
Cold water storage cisterns - Davant Products Ltd
Corrugated shell and tube heat exchanger - HRS Hevac Ltd
Mains pressure thermal storage - Gledhill Water Storage Ltd
Storage cylinders / calorifiers - Sundwel Solar Ltd
Storage cylinders/ calorifiers - Grant & Livingston Ltd
Storage cylinders/ calorifiers - Harris & Bailey Ltd
Storage cylinders/ calorifiers - Harvey Fabrication Ltd
Storage cylinders/ Calorifiers - Rycroft Ltd
Storage cylinders/Calorifiers - Clyde Combustions Ltd
Storage cylinders/calorifires - Hamworthy Heating Ltd
Unvented cylinders - Titan plastech
Vented hot water storage cylinders - Range Cylinders
Water heaters/calorifiers - Flamco UK Ltd

Y24 Trace heating

Electric space heating - Bush Nelson PLC
Electric surface heating systems manufacturer - Isopad Ltd
Heating jackets and panels manufacturer - Isopad Ltd
Resistance heating equipment manufacturer - Isopad Ltd
Self- regulating cut-to-length heater tapes - Flexelec(UK) Ltd
Trace heating - Heat Trace Ltd
Zone parallel constant wattage heater tapes - Flexelec(UK) Ltd

Y25 Cleaning and chemical treatment

Building, cleaning and restoration - Cleanstone Co (Bradford) Ltd
Central heating system corrosion protectors, cleansers + leak sealers - Fernox Alpha Fry Technologies Europe
Central heating system corrosion protectors, cleansers and leak sealers - Cookson Electronics
Chemical photo etchings - Photo Fabrications Services Ltd
Chemical, solvents, greases and lubricants - Ambersil Ltd
Chemicals - Baxenden Chemicals Ltd
Chemicals - DSM UK Ltd
Chemicals - P & R Laboratory Group Ltd
Chemicals - Rhone-Poulenc Chemicals Ltd
Chlorination of water services - Aquastat Ltd
Cleaning & chemical treatment - Precolor Sales Ltd
Cleaning and chemical treatment - Estec Environmental Ltd
Cleaning and chemical treatment - Frosts Landscape Construction Ltd
Cleaning and chemical treatment - Hertel Services
Cleaning and chemical treatments - Deb Ltd
Cleaning chemicals - Certifield Laboratories (Div or CPS Industries Ltd)
Cleaning materials - Evans Vanodine International PLC
Construction chemicals - Artur Fischer (UK) Ltd
Detergents - Evans Vanodine International PLC
Detergents & Disinfectants - Deb Ltd
Disinfectants - Evans Vanodine International PLC
Disinfection systems - Portacel
Grit blasting, chemical cleaning, high pressure water jetting - Masonary Cleaning Services
Industrial aerosols - Ambersil Ltd
Industrial chemical supplies - IFT Supplies Ltd
Limescale remedies, jointing compounds + quick repair products - Fernox Alpha Fry Technologies Europe
Limescale remedies, jointing compounds, quick repair products and solder & flux - Cookson Electronics
Office and industrial cleaning - CSL (Services) Limited
On site services - flue cleaning, pipework and fabrications - Hodgkinson Bennis Ltd
Open blast cleaning equipment for building cleaning. Spare parts for blast equipment - USF Surface Preparation
Polishes - Evans Vanodine International PLC
Polishes and floor cleaners - Johnson Wax Professional
Portable and mobile blast cleaning machines for concrete preparation and enhancement - USF Surface Preparation
Solvent Degreasers - Deb Ltd
Specialist pigments : artist colours - W. Hawley & Son Ltd
Specialist pigments : asphalt - W. Hawley & Son Ltd
Specialist pigments : concrete paving - W. Hawley & Son Ltd
Specialist pigments : concrete roofing - W. Hawley & Son Ltd
Specialist pigments : concrete walling - W. Hawley & Son Ltd
Specialist pigments : mortar - W. Hawley & Son Ltd
Specialist pigments : paint - W. Hawley & Son Ltd
Specialist pigments : plastic - W. Hawley & Son Ltd
Stone and brickwork cleaning - CSL (Services) Limited
UV treatment - Titan Pollution Control Ltd
Vehicle Cleaners - Deb Ltd
Water treatment chemicals - Tower Flue Componants Ltd
Water treatment products and services - Aldous & Stamp Services Ltd

Y30 Air ductlines/Ancillaries

Air ductlines/Ancillaries - Isolated Systems Ltd
Aluminium air ductwork - Flowline Grilles
Connection and sleeves for air ducts - Westbury Filters Limited
Ducting - Alumasc Interior Building Products Limited
Ducting - MacLellan Rubber Ltd
Ducting and precision sheet metalwork - Ductform Engineering Ltd
Ductwork cleaning - Nationwide Filter Co Ltd
Fire rated ductwork systems - Fire Protection Ltd
Flexible ducting - Flexible Ducting Ltd
Rubber mouldings and elastomeric bearings - MacLellan Rubber Ltd
Sheet metal ductwork - Hi-Vee Ltd

Y40 Air handling units

Air conditioning - Dantherm Ltd
Air handling unit - McQuay UK Ltd
Air handling units - Colt International
Air handling units - Dunham Bush Ltd
Air handling units - Eaton-Williams Group Ltd
Air handling units - Ellis Miller Ltd
Air handling units - Hi-Vee Ltd
Air Handling Units - IAC
Air handling units - May William (Ashton) Ltd
Air Handling Units - Senior Air Systems
Air handling units - Therminal Technology (sales) Ltd
Air handling units - Ves Andover Ltd
Air movement & heating systems - Babcock Wanson Uk Ltd
Connect grid and air handling - Saint-Gobain Ecophon Ltd
Curve and coffers - Saint-Gobain Ecophon Ltd
Displacement Ventelation - Senior Air Systems

Y41 Fans

Air movers - Hayward Air
Axial Fan - Ebm-papst (UK) Limited
Axial, centrifugal, mixed flow fans and dust extraction units - Howden Buffalo
Axial, roof extract and centrifugal fans - Matthews & Yates Ltd
Backward Curve Fan - Ebm-papst (UK) Limited
Chimney extractor fans - Keddy-Poujoulat (UK) Ltd
Crossflow fans - Consort Equipment Products Ltd
Diagonal Fans - Ebm-papst (UK) Limited
Domestic and commercial and industrial fans - Xpelair Ltd
Domestic and industrial fans - Airflow Developments Ltd
EC Fans - Ebm-papst (UK) Limited
Evaporators - Portacel
Fans - BOB Stevenson Ltd
Fans - Colt International
Fans - Elta Fans Ltd
Fans - Gebhardt Kiloheat
Fans - Hotchkiss Air Supply
Fans - Klaxon Signals Ltd
Fans - Mawdsley's Ltd
Fans - McKenzie-Martin Ltd
Fans - Power Breaker
Fans - Roof Units Group
Fans - Shipley Fan Co Ltd The
fans - Smoke Control Services Ltd
Fans - Vortice Ltd
Forward Curve Fan - Ebm-papst (UK) Limited
Industrial axial fans and impellers - London Fan Company Ltd
Industrial fans - ABB Control Ltd
Industrial fans units - Radial & Axial Fans
Industrial, bifurcated, centrifugal and axial fans - Twin County Fan Company Ltd
Jet tunnel fans - Matthews & Yates Ltd
Mixed Flow Fans - Ebm-papst (UK) Limited
Plastic fans - Plastic Construction Fabrications Ltd
Road fans - Matthews & Yates Ltd
Road tunnel fans - Matthews & Yates Ltd
Tangential Fans - Ebm-papst (UK) Limited
Twin fans - Ves Andover Ltd

Y42 Air filtration

10 Year guarantees - Fluidair International Ltd
Air filtration - Air Improvement Centre Ltd
Air ailtration - Bassaire Ltd
Air cleaners - Warner Howard Group Ltd
Air cleaning equipment - Horizon International Ltd
Air filter frames & housings - Auchard Development Co Ltd
Air filters - Nationwide Filter Co Ltd
Air filters - Universal Filters Ltd
Air filters - Vokes
Air filters - absolute - Auchard Development Co Ltd
Air filters - panel & bags - Auchard Development Co Ltd
Air filters - synthetic - Auchard Development Co Ltd
Air filtration - Active Carbon Filters Ltd
Air filtration - Estec Environmental Ltd
Air filtration - Farr Filtration Ltd
Air filtration - McQuay UK Ltd
Air filtration - NFC Specialist Products Ltd
Air filtration - Torit DCE
Air Filtration - Vortice Ltd
Air filtration (aircleaners) - Aircare Europe Ltd
Air purifiers - Amcor (Appliances) Ltd
Air purifiers - Purified Air Ltd
Air quality consultancy - Nationwide Filter Co Ltd
Air quality testing - Westbury Filters Limited
Antimicrobial filters - Nationwide Filter Co Ltd
Carbon filters - Nationwide Filter Co Ltd
Carbon filters - Westbury Filters Limited
Filter loss gauges - Westbury Filters Limited
Filtration - Fluidair International Ltd
Filtration - Trox (UK) Ltd
Fit & disposal service - Nationwide Filter Co Ltd
Odour control - Heydal Ltd
Panel filters - Westbury Filters Limited
Pleateed filters - Westbury Filters Limited
Replacement elements - Fluidair International Ltd
Roll filters - Westbury Filters Limited
Rotary screw - Fluidair International Ltd
Rotary sliding vane - Fluidair International Ltd
Slimline filters - Westbury Filters Limited
Washable filters - Westbury Filters Limited

Y43 Heating/ Cooling coils

Chillers - Airedale International Air Conditioning Ltd
Condensers - Airedale International Air Conditioning Ltd
Fan Coils Units - Ruskin Air Management
Heat exchangers gasketted plate - HRS Hevac Ltd
Heaters - Klaxon Signals Ltd
Heating and cooling radiant panel manufacturer - Comyn Ching & Co Ltd (SOLRAY)

Y44 Air treatment

"HVAC" - Eaton-Williams Group Ltd
Air cleaners - Nationwide Filter Co Ltd
Air dryers - Fluidair International Ltd
Air treament - Air Improvement Centre Ltd
Air treatment - Active Carbon Filters Ltd
Air treatment - Consort Equipment Products Ltd
Air treatment - Danfoss Ltd
Air treatment - Hertel Services
Air treatment - IMI Air Conditioning Ltd
Condensate treatment - Fluidair International Ltd
Dehumidifier and air treatment - Amcor (Appliances) Ltd
Dehumidifiers - S & P Coil Products Ltd
Dehumification - Dantherm Ltd
Disinfection system - Heydal Ltd
Humidifiers - Eaton-Williams Group Ltd
Moisture eliminators - Plastic Construction Fabrications Ltd
Roof extract units - Ves Andover Ltd
Ultra violet chambers, lamps and control panels - Hanovia Ltd

Y45 Silencers/ Acoustic treatment

Accoustic flooring - Monarflex Anderson Ltd
Acoustic Baffles - Acousticabs Industrial Noise Control Ltd
Acoustic Barriers - IAC
Acoustic correction & noise control - Brown F plc
Acoustic curtains, attenuators, enclosures, panels and louvers - Acousticabs Industrial Noise Control Ltd
Acoustic doors and frames - Fitzpatrick Doors Ltd
Acoustic enclosures - Designed for Sound Ltd
Acoustic Flooring Products - Hillaldam Coburn Ltd
Acoustic flooring solutions - Durabella System Floors Ltd
Acoustic inserts - Hotchkiss Air Supply
Acoustic insulation - Potters Insulations (Sales) Ltd
Acoustic insulation,wall ceiling panels,stretchwall systems, - Salex Interiors Ltd
Acoustic linings - Henry Venables Interiors Ltd
Acoustic panels and blocks in terracotta - Smithbrook Building Products Ltd
Acoustic products - Nendle Acoustic Company Ltd
Acoustic roofing & cladding - Axter Ltd
Acoustic tiles - Treetex Acoustics Ltd
Acoustic Walling Products - Hillaldam Coburn Ltd
Acoutic wall panels - Acousticabs Industrial Noise Control Ltd
Anechoic rooms - The Salex Group Ltd.
Attenuators/ silencers - Allaway Acoustics Ltd
Audiology rooms - The Salex Group Ltd.
Ductwork, silencers, acoustic products - Isolated Systems Ltd
Floating floors - The Salex Group Ltd.
Floors - noise resistant - The Salex Group Ltd.
Hinged acoustic doorsets - Scandinavian Window Systems Ltd
Louvres - acoustic - The Salex Group Ltd.
Music practise rooms - The Salex Group Ltd.
Noise absorbing panels - The Salex Group Ltd.
Noise control systems - Acousticabs Industrial Noise Control Ltd
Noise resistant doors - The Salex Group Ltd.
Silencers - Hotchkiss Air Supply
Silencers - Ves Andover Ltd
Silencers - Acoustic - Sound Service (Oxford) Ltd
Silencers, acoustic enclosures and telephone hoods - Environmental Acoustics Ltd
Studios - The Salex Group Ltd.
Timber acoustic panels and doors - Salex Interiors Ltd

Y46 Grilles/Diffusers/Louvres

Acoustic products - Nendle Acoustic Company Ltd
Air diffusers - HT Martingale Ltd
Anodised Aluminium coils - Alanod Ltd
Design and CAD support - Saint-Gobain Ecophon Ltd
Diffusion accessories - Panel & Louvre Co Ltd
Flange Covers - Furmanite International Ltd
Grilles - Panel & Louvre Co Ltd
Grilles & diffusers - Hotchkiss Air Supply
Grilles and diffusers - Trox (UK) Ltd
Grilles and Diffusers and Louvres - Senior Air Systems

Y46 Grilles/Diffusers/Louvres (Cont'd)

Grilles, diffusers and louvres - Grille Diffuser & Louvre Co Ltd The
Grilles, diffusers and louvres - Royair Ltd
Grilles/Diffusers/Louvres - Isolated Systems Ltd
Grills /diffusers/ louvres - Smoke Control Services Ltd
Grills and diffusers - ABB Control Ltd
Integrated lighting systems - Saint-Gobain Ecophon Ltd
Louvers - McKenzie-Martin Ltd
Louvre, hit and miss ventilators - Ryton's Building Products Ltd
Security grilles - Cooper Group Ltd
Smoke ventilation louvres - Matthews & Yates Ltd
Solar shading systems - Dales Fabrications Ltd

Y50 Thermal insulation

Aluzink insulated jacketing - SSAB Dobel Coated Steel Ltd
Cork/ rubber - Olley & Sons Ltd, C
Dobelshield - SSAB Dobel Coated Steel Ltd
Expanded polystyrene insulation - Vencel Resil Ltd
Floor insulation - Springvale E P S
Garden Products - nmc (UK) Ltd
Insulation - Rockwool Ltd
Insulation cork - Olley & Sons Ltd, C
Insulation materials - Gradient Insulations (UK) Ltd
Pre-insulated pipework - CPV Ltd
Roof insulation - Springvale E P S
Sarnatherm insulation - Sarnafil Ltd
Thermal insulation - Ainsworth B R (Southern) Ltd
Thermal Insulation - Draught Proofing Advisory Association
Thermal insulation - Firebarrier Services Ltd
Thermal Insulation - Hawkins Insulation Ltd
Thermal insulation - Hertel Services
Thermal Insulation - Insulated Render & Cladding Association
Thermal Insulation - Isolated Systems Ltd
Thermal insulation - Knauf Insulation Ltd
Thermal insulation - Potters Insulations (Sales) Ltd
Thermal Insulation - Reiner Fixing Devices (Ex: Hardo Fixing Devices)
Thermal insulation - Tavac Insulation Ltd
Thermal insulation - Thermica Ltd
Thermal insulation - Wood E Ltd
Wall insulation - Springvale E P S

Y51 Testing and commissioning of mechanical services

Chlorine residual analysers - Portacel
Contract Smoke Pellets, Matches & Detector Sprays - PH Products Ltd
Corbon Monoxide Detectors - PH Products Ltd
Gas Engineers Record Pads, Lables & Tape - PH Products Ltd
Gas Engineers Tools & Accessories - PH Products Ltd
Gas Leak Etector Spray & Fluids - PH Products Ltd
Portable moisture, temperature and noise level meters - Channel Electronics
Silicone grease & spray WRI Approved - PH Products Ltd
Sleep Safe home Products - PH Products Ltd
Smoke Matches - PH Products Ltd
Smoke Pellets - PH Products Ltd
Testing and commissioning of mechanical - Denco Ltd
Testing facilities for GRP tanks - APA

Y52 Vibration isolation mountings

Acoustic products - Nendle Acoustic Company Ltd
Anti - vibration mountings - MacLellan Rubber Ltd
Anti vibration mounting - Designed for Sound Ltd
Anti-vibration mounts - Isolated Systems Ltd
Anti-vibration pads - Olley & Sons Ltd, C
Machine mounting, anti vibration materials, structural and sliding bearings and resilient seatings - Tiflex Ltd
Noise and vibration control products for mechanical service engineers - The Salex Group Ltd.
Structural. Bridge & siesmic bearings - Tiflex Ltd
Vibration isolation - Allaway Acoustics Ltd
Vibration Isolation - Sound Service (Oxford) Ltd
Vibration isolation mountings - Christie & Grey Ltd

Y53 Control components - mechanical

Ancilleries - Dawson-Keith Ltd
Ancilliary security device - Securistyle Ltd
BMS system support - Satchwell Control Systems Ltd
Building controls - Satchwell Control Systems Ltd
Building management systems - Satchwell Control Systems Ltd
Close Control - Airedale International Air Conditioning Ltd
Control - Airedale International Air Conditioning Ltd
Control - T-T Pumps & T-T controls
Control components - Danfoss Ltd
Control components - Johnson Controls Systems Ltd
Control components - mechanical - Guthrie Douglas Ltd
Control equipment, timers and relays - Thomas & Betts Manufacturing Services Ltd
Control Systems - Ebm-papst (UK) Limited
Control systems - Skil Controls Ltd
Controls - Ves Andover Ltd
Customer training - Satchwell Control Systems Ltd
Data Loggers - GE Protimeter
Diaphragm, butterfly, ball valves and actuators - Alfa Laval Saunders Ltd
Embeded control systems - MJ Electronics Services (International) Ltd
Enclosures - Knurr (UK) Ltd
Energy management systems - Satchwell Control Systems Ltd
Hazardous area control gear - PFP Electrical Products Ltd
Heating controls - SAV United Kingdom Ltd
Heating programmers - Timeguard Ltd
Lighting controls - Thorlux Lighting
Mechanical & electrical connections - Lab Furnishings (UK) Ltd
Millennium management - Satchwell Control Systems Ltd
Moisture meters, hygrometers - GE Protimeter
Nexus bureau - Satchwell Control Systems Ltd
Performance contracting - Satchwell Control Systems Ltd
Process systems and analytical equipment - SPX Process Equipment
Pump selection software - Armstrong Pumps Ltd
Thermometers, hygrometers, hydrometers, pressure and altitude gauges - Brannan S & Sons Ltd
Thermometers, pressure gauges and associated instruments - Brannan Thermometers & Gauges
Thermostats - Teddington Controls Ltd
Valve Actuators - Flowserve Flow Control
Valve Refurbishment - Hodgkinson Bennis Ltd
Water Controls - Setsquare

Y59 Sundry common mechanical items

Air receivers - Flamco UK Ltd
Ancillary components - Llewellyn Homes Ltd
Balancing valves - Ellis Miller Ltd
Diesel engines - Martin (Marine) Ltd Alec
Exterior surface heating - Uponor Housing Solutions
Flexible plumbing - Uponor Housing Solutions
Fry's solder and flux - Fernox Alpha Fry Technologies Europe
Gas Fires - Spital Tile Co Ltd
Gaskets - Olley & Sons Ltd, C
GRP Kiosks & modular enclosures - Group Four Glassfibre Ltd
Lockshield radiator valves - Ellis Miller Ltd
Machines for Plastic Pipe Joinings - George Fischer Sales Ltd
Manual radiator valves - Ellis Miller Ltd
Manufacturerers of: Metal Cutting Circular Saws - Spear & Jackson Interntional Ltd.
O Rings, bonded seals, grease guns, grease nipples and bucket pumps - Ashton Seals Ltd
Parallel slide valves - Bailey Birkett Sales Division
Pipe cutting machines - George Fischer Sales Ltd
Pressure reducing valves - Bailey Birkett Sales Division
Pressure regulating valves - Bailey Birkett Sales Division
Pulley lines - Lever James & Sons Ltd
Resharpners of Circular saw blades - Spear & Jackson Interntional Ltd.
Safety relief valves - Bailey Birkett Sales Division
Seals - Atkinson & Kirby Ltd
Smoke and natural ventilation systems - SE Controls
System 21 - the latest in enrgy efficiency - Servowarm
Thermostatic radiator valves - Ellis Miller Ltd
Valve Manufacturer - Millar Dennis (1998) Ltd
Wear, heat and impact resistanting components - Wearparts Ltd

Y60 Conduit and Cable trunking

Adaptable boxes - Vantrunk Engineering Ltd
All insulated plastic enclosures - Ensto Briticent Ltd
Aluminium Trunking Systems - Marshall-Tufflex Ltd
Cable management - Ackermann Ltd
Cable management - Arena-Walsall Ltd
Cable management - Hager Ltd
Cable management - Kem Edwards Ltd
Cable management systems - MK (MK Electric Ltd)
Cable management systems - REHAU Ltd
Cable routing - Hellermann Tyton
Cable Support System - Marshall-Tufflex Ltd
Cable tray - Legrant Electric Ltd
Cable trunking - Legrant Electric Ltd
Conduit and trunking - Mita (UK) Ltd
Dado trunking - Salamandre Davis
Electrical cable management systems - Marshall-Tufflex Ltd
Electrical Conduit & Accessories - Kem Edwards Ltd
Flexible conduit systems - Adaptaflex Ltd
Floor trunking - Salamandre Davis
Flooring systems - Legrant Electric Ltd
G R P systems - Marshall-Tufflex Ltd
gator shield, flo-coat, plasticoat, Aluminiumised tube, sculptured tube, engineered tube, NHT3,5,7 ERW 1-5 - Tyco European Tubes
L.S.F. Conduit - Aercon Wiring Systems Ltd
Ladder rack - Legrant Electric Ltd
Lighting trunking - Legrant Electric Ltd
Metal cable trunking - Legrand Power Centre
Perimeter trunking systems - Mita (UK) Ltd
Pipe supports - Unistrut Limited
Pre wired PVC conduit - Aercon Wiring Systems Ltd
Skirting trunking - Salamandre Davis
Specailised Applications - Marshall-Tufflex Ltd
Stainless steel basket tray - Electrix International Ltd
Stainless steel conduit - Kensal CMS
Stainless steel enclosures - Electrix International Ltd
Stainless steel trunking and conduit - Electrix International Ltd
Steel and stainless steel enclosures - Ensto Briticent Ltd
Trunking - Vantrunk Engineering Ltd

Y61 HV/ LV cables and wiring

Cable glands and connectors - Elkay Electrical Manufacturing Co Ltd
Cable protection - Hellermann Tyton
Cables - Alcatel Cablenet
Cables and wiring accessories - Ellis Patents
Cabling Components - Hubbell Premise Wiring
HV and LV cables and wiring - Concordia Electric Wire & Cable Co. Ltd
HV/ LV cables & wiring - Metway Electrical Industries Ltd
Mineral insulated cables - Tyeo Thermal Control
Wiring accessories - MK (MK Electric Ltd)

Y62 Busbar trunking

Busbar - Marshall-Tufflex Ltd
Busbar power distribution systems - Electrak International Ltd
Busbar trunking - Arena-Walsall Ltd
Busbar trunking systems - Barduct Ltd
Busbar trunking, low power - Salamandre Davis
Busbar, cable management systems for floors - Legrand Power Centre
Inceiling busbar trunking - Electrak International Ltd
Stainless steel enclosures IP65 - Kensal CMS
Stainless steel trunking - Kensal CMS
Underfloor busbar trunking - Electrak International Ltd

Y63 Support components-cables

Cable ladders - Vantrunk Engineering Ltd
Cable routing - Hellermann Tyton
Cable support systems - Elwell Sections Ltd
Cable supports - Unistrut Limited
Cable ties and cable clips - Unifix Ltd
Cable tray - Ductile Sections
Cable tray - Vantrunk Engineering Ltd
Cable tray/ladder - Unistrut Limited
Cable trays, cable ladder, channel support systems - Swifts Of Scarborough Ltd
Cleats and cable ties - Vantrunk Engineering Ltd
GRP ladders and trays - Mita (UK) Ltd
Stainless steel cable tray - Electrix International Ltd
Stainless steel cable tray - Kensal CMS
Stainless steel supports & fasteners - Electrix International Ltd
Support Components Cables - Ellis Patents
Support systems - Legrant Electric Ltd
Wire basket cable tray - Vantrunk Engineering Ltd

Y70 HV switchgear

Electrical services circuit breakers - Power Breaker
HV Switchgear - Elequip Projects Ltd
HV switchgear - Hager Ltd
HV switchgear - Hubbell Premise Wiring
Residual current devices - Power Breaker
Switchboards - Technical Control Systems Ltd
Switchgear - Power Breaker

Y71 LV switchgear and distribution boards

Circuit protection - MK (MK Electric Ltd)
Contral gear - Square D, Electrical Distribution & Building applicators Division
Custom built distribution panelboards - Square D, Electrical Distribution & Building applicators Division
Domestic switch gear - Eaton Electric Limited
Earth leakage electrical protection equipment - FDB Electrical Ltd
Earth monitoring - FDB Electrical Ltd
Electrical control panels and systems - Warner H & Son Ltd
Electrical switchgear - Dorman Smith Switchgear Ltd
Enclosed isolators - Ensto Briticent Ltd
Enclosures - Elkay Electrical Manufacturing Co Ltd
Feeder pillars - Blakley Electrics Ltd
LV electronic switchgear - Entrelec MTE Ltd
LV switchgear - Klockner-Moeller Ltd
LV switchgear - T-T Pumps & T-T controls
LV Switchgear & Distribution Boards - Elequip Projects Ltd
LV switchgear and distribution boards - Hager Ltd
Mains Distribution Assemblies 100A to 3000 A - Blakley Electrics Ltd
Panel boards - Square D, Electrical Distribution & Building applicators Division
potentiometers - Omeg Ltd
Pre-wired distribution units - Mita (UK) Ltd
RCDs - Power Breaker
Residual current circuit breakers - FDB Electrical Ltd
Residual current monitors - FDB Electrical Ltd
Switch and fusegear - Square D, Electrical Distribution & Building applicators Division
Switch boards - Square D, Electrical Distribution & Building applicators Division
Switchers - Baxall Ltd
Switches - Ensto Briticent Ltd
Switchgear - Power Breaker
Switchgear - Whipp & Bourne
Three phase distribution board - Square D, Electrical Distribution & Building applicators Division
Transformers - Babcock Transformers

Y72 Contactors and starters

Contracts and starters - Danfoss Ltd
Electronic motor protection - Entrelec MTE Ltd
LV motor control equipment - Entrelec MTE Ltd
Mechanical & electrical contracting services - Warner H & Son Ltd
Softstarters - Entrelec MTE Ltd

Y73 Luminaires and lamps

Associated accessories - Bernlite Ltd
Cable management systems - Vantrunk Engineering Ltd
Coloured and clear covers for fluorescent lamps and fittings - Encapsulite International Ltd
Columns and brackets - DW Windsor Lighting
Converters for cold cathode and ballasts - Mode Lighting (UK) Ltd
Desk lamps - Full Spectrum Lighting Ltd
Equipment and luminaires, control gear and accessories - Poselco Lighting
Flameproof EExd fluorescents - PFP Electrical Products Ltd
Fluorescent and discharge lighting - Fitzgerald Lighting Ltd
General fluorescent lighting - Fluorel Ltd
Incandesant light bulbs and other domestic lighting - J.F. Poynter Ltd
Increased safety fluorescents - PFP Electrical Products Ltd
Lampholders - Bernlite Ltd
Lamps and electronic control gear - Osram Ltd
Light bulbs and fittings - British Electric Lamps Ltd
Light fittings - Concord: Marlin
Light sensors - Timeguard Ltd
Lighting components - Astrolighting Ltd
Lighting designers - Thorlux Lighting
Luminaires - Ensto Briticent Ltd
Luminaires and lamps - Brandon Medical
Luminaires and lamps - Connect Lighting Systems (UK) Ltd
Luminaires and Lamps - Consult Lighting
Luminaires and lamps - Focal Signs Ltd
Luminaires and Lamps - Futimis Ltd
Luminares & Lamps - Glamox Electric (UK) Ltd
Project management services - Saint-Gobain Ecophon Ltd
Refurbishment and upgrading service for existing luminaires - Poselco Lighting
Stair components - Spindlewood
Zone 1 Bulkheads - PFP Electrical Products Ltd
Zone 1 floodlights - PFP Electrical Products Ltd
Zone 1 wellglasses - PFP Electrical Products Ltd

Y74 Accessories for electrical services

Accessories for electrical services - Brass Art Ltd
Accessories for electrical services - Tower Manufacturing
Dimming switches - Mode Lighting (UK) Ltd
Din-rail time switches and surge protectors - Timeguard Ltd
Domestic electrical accessories and dry cell batteries and torches - J.F. Poynter Ltd
Electrical Accessories - Arena-Walsall Ltd
Electrical fittings and wire accessories - Legrand Electric Ltd
Electrical products - Panasonic UK Ltd

Y74 Accessories for electrical services (Cont'd)

Electrical swithces - Wandsworth Elecrtrical Ltd
General Electrical Products - WF Electrical Distributors
General electrical products - wholesale - Jolec Electrical Supplies
Invisible light switches and sockets - Forbes & Lomax Ltd
Light fittings - Concord: Marlin
Light switches - Redditch Plastic Products
Modular wiring systems - Thorlux Lighting
Painted, nickel silver, brass and frosted acrylic stainless steel switches and sockets - Forbes & Lomax Ltd
Residual current protected skt outlets - FDB Electrical Ltd
Secure connection systems - Hager Ltd
Switch and socket boxes - Legrand Power Centre
Switches - Teddington Controls Ltd
Thermocouples & RTDs - Tyeo Thermal Control
Time controllers, security light switches and RCD switched sockets - Timeguard Ltd
Track support Materials - Tiflex Ltd
Ultrasonic sensors - Thomas & Betts Manufacturing Services Ltd
Wire accessories - Hellermann Tyton
Wiring accessories - Hager Ltd

Y80 Earthing and bonding components

Earth leakage electrical protection equipment - FDB Electrical Ltd
Earth monitoring - FDB Electrical Ltd
Earth Monitoring Equipment - Blakley Electrics Ltd

Y81 Testing & commissioning of electrical services

Electrical safety testing - National Britannia Ltd
Energy management - National Britannia Ltd
Hygrometers - GE Protimeter
Moisture meters - GE Protimeter

Y82 Identification - electrical

Breezair Evaporative Coolers - Radiant Services Ltd
Cable identification - Hellermann Tyton
Identification labels - Mockridge Labels & Nameplates Ltd

Y89 Sundry common electrical items

Automatic Pool covers - Golden Coast Ltd
Clip in conduit fittings - PFP Electrical Products Ltd
Electrical Distribution Equipment - WT Henley Ltd
Electronic accessories - VELUX Co Ltd, The
Filter and poulton - PFP Electrical Products Ltd
Installation accessories - Mita (UK) Ltd
MV primary and secondary products - Schneider Electric Ltd
Stainless steel isolators - Kensal CMS
Sundry common electrical items - Metway Electrical Industries Ltd
Sundry common electrical items - Tower Manufacturing
Wiring accessories - Mita (UK) Ltd

Y90 Fixing to building fabric

Cable tie and fixing assemblies - Hellermann Tyton
Fixings and fastenings - Tower Manufacturing
Frame cramps, restraint straps, sliding anchor stems - Vista Engineering Ltd
High load anchors - Exchem Mining and Construction
Insulation fixings and stainless steel helexial nails - Reiner Fixing Devices (Ex: Hardo Fixing Devices)
Mild steel and non-ferrous nails, tacks and pins - John Reynolds & Sons (Birmingham) Ltd
Nails - Glasgow Steel Nail Co Ltd
Nails - Stone Fasterners Ltd
Nuts and bolts - Owlett - Jaton
Remedial fixing systems - Redifix Ltd
Rotary impact and masonry drills - Unifix Ltd
Screws, nuts and bolts - Nettlefolds Ltd
Stainless steel building components - Wincro Metal Industires Ltd
Through bolt fixuings - Tucker Fasteners Ltd
Wall ties - remedial - Redifix Ltd
Wood, coach, self tapping, fastbrolley screws, frame anchors, hammer and expansion plugs - Unifix Ltd

Y91 Off-site painting/ Anti-corrosion treatments

Anti-corrosion and sealing systems - Winn & Coales (Denso) Ltd
Anti-rust coating - Bondaglass Voss Ltd
Carboline anti-corrosion products - Nullifire Ltd
Coatings - Thoro Systems Products Ltd
Corrosion protection for metal fastenings - Interlux Ltd
Flame retardant treatment - Gamma Resources Ltd
Galvanized buckets and cans - Brettell & Shaw Ltd
Galvanizers - Joseph Ash Galvanizing
Hot dip galvanizing - Arkinstall Galvanizing Ltd
Hot dip galvinizing - Ash Joseph Galvanizing
Inks and metal coatings - Valspar UK Corporation Ltd
Paint and powder coatings - Trimite Ltd
Potable water coatings - Corroless Corrosion Control
Powder coatings - HB Fuller Coatings Ltd
Printing solutions - Hellermann Tyton
Protective coatings - MMP International
Thermoplastic coating powders - Plascoat Systems Ltd

Y92 Motor drives - electric

Automatic gates - Jackson H S & Son (Fencing) Ltd
Drive Systems - Ebm-papst (UK) Limited
Electric motor repairs - Turney Turbines (pumps) Ltd
Electrical motors - Klockner-Moeller Ltd
Motor drives (electric) - T-T Pumps & T-T controls
Motor drives-electric - Danfoss Ltd
Motors - Electro Mechanical Systems Ltd
Tube motors - Kaba Garog

Z10 Purpose made joinery

Aluminium extrusions, standard (angle, bars, tubes, channels etc) and custom profiles available - Extrusions Direct
Architectural joinery manufacturers - Edmonds A & Co Ltd
Architectural joinery restoration oak, European hardwoods - Henry Venables Interiors Ltd
Archtectural joinery - Henry Venables Ltd
Bespoke - Evertaut Ltd
Bespoke joinery - Howard Bros Joinery Ltd
Bespoke Joinery - Taskworthy Ltd
Bespoke joinery products - Alumasc Interior Building Products Limited
Bespoke laminated components - T B S Fabrications
Bespoke museum display cases - Benbow Interiors
Brass - Aalco Hull Amari Metal Ltd
Bronze - Aalco Hull Amari Metal Ltd
Copper - Aalco Hull Amari Metal Ltd
Decorative wood moulding and carvings, cornice, picture rail, dado rail, skirting, architrave and pediments, radiator cabinet grilles - Winther Browne & co Ltd
Door joinery manufacturer - Humphrey & Stretton PLC
Fire resisting joinery - Decorfix Ltd
Fire surrounds - Reclaim.UK.Com
Fireplaces - Thomas Lowe Joinery
Hardwood mouldings - WL West & Son Limited
Joinery - Firco Construction Ltd
Joinery solutions - Taskworthy Ltd
Machined Hardwoods - Moray Timber Ltd
Machined softwood and machined MDF mouldings - Finn Forest
Machined softwoods - Moray Timber Ltd
Made to measure joinery - Dale Joinery Ltd
Pine room paneling, bookcases and display cupboards - Hallidays UK Ltd
Propose Made Jointery - AS Newbould Ltd
Purpose made joinery - Benbow Interiors
Purpose made joinery - Babcock Joinery
Purpose made joinery - Channel Woodcraft Ltd
Purpose made joinery - Decorfix Ltd
Purpose made joinery - Enfield Speciality Doors
purpose made joinery - Jali Ltd
Purpose made joinery - Jeld-Wen Ltd
Purpose made joinery - Smith & Choyce Ltd
Purpose made joinery - Spindlewood
Purpose made joinery - SWJ Group Ltd
Radiator cabinets - Jali Ltd
Specialist Bespoke Joinery - J. Scott (Thrapston) Ltd
Specialist joinery - Andy Thornton Ltd
Specialist joinery - Barlow Group
Specialist joinery - Barnwood Shopfitting Ltd
Specialist joinery - Cumberland Construction
Specialist joinery manufacturers - Sandoms (Batley) Ltd
Stainless steel - Aalco Hull Amari Metal Ltd
Stockholders of aluminium - Aalco Hull Amari Metal Ltd
Study, lounge, bedroom, office and reception fitted furniture - Hyperion Wall Furniture Ltd
Teak Specialist - Parker Kislingbury Ltd
Timber engineering services - Wyckham Blackwell Ltd
Timber specialist - Challenge Fencing Ltd
Window & door Specials - Blair Joinery

Z11 Purpose made metalwork

Aluminium and steel fabricators - Ramsay & Sons (Forfar) Ltd
Aluminium extrusions, powder coating, anodising and fabrication - Kaye Aluminium plc
Aluminium fabrication - Panel Systems Ltd
Aluminium special structures - Alifabs (Woking) Ltd
Aluminium, bronzes, resistance welding alloys and cupro nickels - Cerro (Manganese Bronze) Ltd
Aluminum and steel standing seam roofing system - CA Group Ltd - MR-24 Division
Architectral metalwork - Broxap Ltd
architectural metalwork - Ermine Engineering Co. Ltd
Architectural metalwork - Mackay Engineering
Architectural metalwork - Multitex GRP & GRG Products
Architectural metalwork - Pearsons J R (Birmingham) Ltd
Architectural metalwork - Woodburn Engineering Ltd
Architectural metalworkers - Edmonds A & Co Ltd
Architectural metalworks - Glazzard (Dudley) Ltd
Bespoke metal products - Alumasc Interior Building Products Limited
Bespoke metalwork - Andy Thornton Ltd
Bespoke museum display cases - Benbow Interiors
Brass and bronze heritage products - Brass Art Ltd
Brass fabrications - Andy Thornton Ltd
Brass Pressure Diecasting - Regon Ltd
Brick bond grilles - Priory Shutter & Door Co Ltd
Cast stanchions - Gabriel & Co Ltd
Cast steelwork - Gabriel & Co Ltd
Cattle grids, duct covers - Cameron & Moore (Gridplex) Ltd
Ceiling tiles & Systems - British Gypsum Ltd
CNC milling and turning of specialist ports in metal and plastic - Griff Chains Ltd
Copper tubing - Wednesbury Tube
Custom sheet metal - Azimex Fabrications Ltd
Decorative castings - Andy Thornton Ltd
Engineering castings - Ballantine, Bo Ness Iron Co Ltd
Fabrications for building/ construction, rail and agriculture - Avon Manufacturing Ltd
Fast action doors, rubber & pvc crash doors - Dudley Factory Doors Limited
GRP, aluminum, mild and stainless steel - Redman Fisher Engineering
Guards for corner protection on walls and columns - Huntley & Sparks
Hot dip galvanizing - Corbett & Co (Galvanising) W Ltd
Iron and aluminium castings - Cradley Castings Ltd
Laser Cutting - Lancashire Steel Fabrication Co Ltd
Lead ingots - R.M. Easdale & Co Ltd
Manufacturers of aluminum extrusion - Alcoa Europe-Extrusions & Endproducts (UK)
Marine Craft - Thanet Ware Ltd
Mesh Screens - Cadisch MDA Ltd Ltd
Metal fabrications - Barton B C & Son Ltd
Metal fabrications - Squires Metal Fabrications Ltd
Metal finishers - Rigized Metals Ltd
Metal framing - Unistrut Limited
Metal parts - H&H Electronics Ltd
Metal roofing and cladding - Ash & Lacy Building Products Ltd
Metal Spraying - Thanet Ware Ltd
On site welding - C & W Fencing Ltd
Pre-engineered metal building systems - Butler Europe Limited
Punch & fold machine - Lancashire Steel Fabrication Co Ltd
Purpose made joinery - Barlow Group
purpose made metal work - Metalcraft (Tottenham) Ltd
Purpose made metalwork - Benbow Interiors
Purpose made metalwork - Civil Engineered Products Ldt
Purpose made metalwork - Claydon Architectural Metalwork Ltd
Purpose made metalwork - Consort Equipment Products Ltd
Purpose Made Metalwork - Deepdale Engineering Co Ltd
Purpose made metalwork - Grant & Livingston Ltd
Purpose made metalwork - Hickton & Co Ltd, John
Purpose made metalwork - Hipkiss, H, & Co Ltd
Purpose made metalwork - Hubbard Architectural Metalwork Ltd
Purpose made metalwork - Leander Architectural
Purpose made metalwork - NRG Fabrications
Remelt Facility - Alcoa Europe-Extrusions & Endproducts (UK)
S.F.S. building systems (Metframe - Metsec Building Products Ltd
Seamless cold drawn copper tube - Lawton Tube Co Ltd, The
Security grilles - Wenmore Rooflights, A Division of W.F.Joy & Co. Ltd
Special arichitectural metalwork - Bassett & Findley Ltd
Special castings in brass and bronze - Brass Art Ltd
Stainless steel and sheet metal - Kensal CMS
Stainless steel brackets - Kensal CMS
Stainless steel cleaning products - Kleen-Tex Industries Ltd
Stainless steel columns - Stainton Metal Co Ltd
Stainless steel Components - Birdscarer Products & Net Supplies Ltd
Stainless steel custom fabrication - Kensal CMS
Stainless steel fabrications - Component Developments
Stainless steel wire mesh - Multi Mesh Ltd
Stainless steel wire supplier - Birdscarer Products & Net Supplies Ltd
Stainless steel, angles, beams, tees, channels, flats and bar - IMS UK
Steel canopies - Gabriel & Co Ltd
Steel fabrication - Lancashire Steel Fabrication Co Ltd
Steel fabrication, plate bending & rolling, welding, drilling, turning, milling etc. - Stevenson & Cheyne Ltd
Steel fabrications - Modern Engineering (Bristol) Ltd
Steel lintels - Jones of Oswestry Ltd
Steel plates, angles, RSJ's, Flats, Tibing - Rainham Steel Co Ltd
Structural fixings - Expamet Building Products
Structures - Kee Klamp Ltd
Welding engineers - Woodburn Engineering Ltd
Wire Products - Owlett - Jaton
Wirecloth, screens, filters, vent mesh and welded mesh - United Wire Ltd
Woven wirecloth, welded wiremesh, perforated metal, quarry screening, wire products - Potter & Soar Ltd

Z12 Preservative/ Fire retardant treatments for timber

Creosote - Legge Thompson F & Co Ltd
Passive fire protection - Deborah Services Limited
Preservation - Cox Long Ltd
Preservative / fire retardent treatment - Brewer C & Sons Ltd
Preservative and fire retardant treatments for timber - Dover Trussed Roof Co Ltd
Preservative/fire retardant treatments for timber - Brash John & Co Ltd
Preservative/fire retardent treatments - Arch Timber Protection
Preservative/Fireretardant treatments for timber - Smith & Rodger Ltd
Stockholm tar - Legge Thompson F & Co Ltd
Timber treatment - Walker Timber Ltd

Z20 Fixings/ Adhesives

3M main distributor tapes & adhesives - Brissco Equipment Ltd
Adhesive - Loctite UK Ltd
Adhesive - Tremco Ltd
Adhesive products - Henkel Consumer Adhesives
Adhesive tapes - Advance Tapes International Ltd
Adhesive Tapes - Hilton Banks Ltd
Adhesives - Atkinson & Kirby Ltd
Adhesives - Brissco Equipment Ltd
Adhesives - Building Adhesives Ltd
Adhesives - Dunlop Adhesives
Adhesives - Evode Ltd
Adhesives - Fortafix Ltd
Adhesives - Henkel Industrial Adhesives
Adhesives - Henkel Industrial Adhesives (AIV)
Adhesives - Kenyon Group Ltd
Adhesives - Kenyon Performance Adhesives
Adhesives - Mapei (UK) Ltd
Adhesives - Sealocrete Ltd
Adhesives - Tarkett-Marley Floors Ltd
Adhesives - Tektura Plc
Adhesives - Viking Laminates Ltd
Adhesives - manufacturer - National Starch & Chemical Ltd
Adhesives and fillers - Polycell Products Ltd
Adhesives and grouts - Shackerley (Holdings) Group Ltd incorporating Designer ceramics
Adhesives and pastes - Clam-Brummer Ltd
Adhesives for wall covering and wood - Halls Beeline Ltd
Adhesives, wood filler and glue - Humbrol Ltd
Admixtures - Degussa Construction Chemicals (UK)
Advance Premier Link Distributor - tapes - Brissco Equipment Ltd
Ahesives - Combustion Linings Ltd
Airtight adhesives and tapes - Fillcrete Ltd
Automated assembly - Textron Fastening Systems Ltd
Blind Bolt - Henry Venables Interiors Ltd
Blind fastening systems - Textron Fastening Systems Ltd
Blind fastening systems - Tucker Fasteners Ltd
Building adhesives - Vallance
Cable clips - Artur Fischer (UK) Ltd
Cast-in fixing channel - Halfen Ltd
Cavity fixings - Ramset Fasteners Ltd
Chemical anchors - ITW Spit
Chemical anchors - Ramset Fasteners Ltd
Clips - Airflow (Nicoll Ventilators) Ltd
Coal, barn, wood and stockholm tar - Liver Grease Oil & Chemical Co Ltd
Cold formed components - Textron Fastening Systems Ltd
Connectors - Simpson Strong-Tie®
Contractors tapes - Airflow (Nicoll Ventilators) Ltd
Corrosion resistant alloys fasteners - Cerro (Manganese Bronze) Ltd
Cut steel and leak-proof nails, tacks, pins and screws - Reynolds John & Son (Birmingham) Ltd

Z20 Fixings/ Adhesives (Cont'd)

Cyanoacrylate adhesives - Kenyon Group Ltd
Cyanoarylate adhesives - Kenyon Performance Adhesives
Drill bits - Ramset Fasteners Ltd
Drop in anchors - Ramset Fasteners Ltd
Elastic Bonding - Sika Ltd
Engineered assemblies - Textron Fastening Systems Ltd
Engineering and construction fasteners - Allman Fasteners Ltd
Epoxy resins - Hatfield Ltd, Roy
Expanding fre rated foam - Ramset Fasteners Ltd
Eye bolts - Marine & Commercial Rigging Systems
Fastener systems for the roofing , cladding and construction industry - EJOT Ecofast Ltd
Fasteners - UK Fasteners Ltd
Fasteners and fastening systems - Duo-Fast (UK) Ltd
Fasteners, fixings, tools, engineering supplies - Arrow Supply Company Ltd
Fixing - Owlett - Jaton
Fixing and adhesives - Degussa Construction Chemicals (UK)
Fixing for masonry - Vista Engineering Ltd
Fixing systems - remedial - Redifix Ltd
Fixings - Airflow (Nicoll Ventilators) Ltd
Fixings - British Gypsum Ltd
Fixings - Corofil Woodall
Fixings - Exchem Mining and Construction
Fixings - Hipkiss, H, & Co Ltd
Fixings - ITW Construction Products
Fixings - ITW Spit
Fixings - Power Plastics Ltd
Fixings - RFA Group Ltd
Fixings / Adhesives - Interface Europe Ltd
Fixings / adhesives - Till & Whitehead Ltd, t/a L. Birkinshaw
Fixings adhesives - Campbell Marson & Co
Fixings and adhesives - Caswell & Co Ltd
Fixings and adhesives - Feb MBT
Fixings and fastenings - Kem Edwards Ltd
Fixings/ adhesives - Bondaglass Voss Ltd
Fixings/ Adhesives - Nettlefolds Ltd
Fixings/ Adhesives - Tower Manufacturing
Fixings/Adhesives - Crowthorne Fencing
Fixings/adhesives - Honeywell Speciality Wax and Additives Ltd
Flat roofing fasteners - SFS Intec LTD
Floor Covering Adhesives - Ardex UK Ltd
Flooring adhesives - Ball F & Co Ltd
Flooring adhesives - Lastite Building Products Ltd
Flooring adhesives, smoothing compounds - Laybond Products Ltd
Frame fixings - Artur Fischer (UK) Ltd
Freyssinet Asphaltic plug joints - Pitchmastic Pmb Ltd
General fixings - Artur Fischer (UK) Ltd
Glue guns - Kenyon Performance Adhesives
Glues, cleaners for PVCu window construction - Wagner (GB) Ltd
Heavy duty anchors - Ramset Fasteners Ltd
Heavy Duty fixings - Artur Fischer (UK) Ltd
Hot melt adhesives - Kenyon Group Ltd
Hot melt ahesives - Kenyon Performance Adhesives
Hot melt guns - Kenyon Performance Adhesives
Installation tools - Textron Fastening Systems Ltd
Large perforated headed, steel and stainless steel anchorages - BigHead Bonding Fasteners Ltd
Lateral Restraint Fixings - Redifix Ltd
Light & heavy weight fixings - ITW Construction Products
Lightweight fixings - Artur Fischer (UK) Ltd
Lindapter - Owlett - Jaton
Lintel Reinforcing - Redifix Ltd
Manilla rope - Marine & Commercial Rigging Systems
Metal repair compounds - MMP International
Metal roofing and cladding - Ash & Lacy Building Products Ltd
Nailing Machines and Nails - Atkinson & Kirby Ltd
Nylon devices up to 300mm - Reiner Fixing Devices (Ex: Hardo Fixing Devices)
Packaging and security fasteners - Hellermann Tyton
Plastic components - Textron Fastening Systems Ltd
Powder acurated cartridge tools. - Ramset Fasteners Ltd
Resin Bonded Steel Female Sockets - Redifix Ltd
Resins and grouts - Helifix Ltd
Rigging screws - Marine & Commercial Rigging Systems
Screws - Owlett - Jaton
Sealants - Kenyon Performance Adhesives
Security fasteners - Kinnings Marlow Ltd
Self drilling fastener - SFS Intec LTD
Self tapping fasteners - SFS Intec LTD
Shackles - Marine & Commercial Rigging Systems
Shear load connectors - Ancon Building Products
Shield anchors - Ramset Fasteners Ltd
Sleeve anchors - Ramset Fasteners Ltd
Special fasteners, solid and hollow cold formed components - Kinnings Marlow Ltd
Special length studs BZP and S/Steel - Redifix Ltd
Splicing - Marine & Commercial Rigging Systems
Spring Steel & other Fixings & Fasteners for Electrical, Mechanical, Construction & Rail Industries - Emlux Holdings Ltd
Stainless steel chain - Marine & Commercial Rigging Systems
Stainless steel channels and bolt fixings - Ancon Building Products
Stainless steel fasteners - SFS Intec LTD
Stainless steel timber to masonry fixings - Helifix Ltd
Stainless steel warm roof fixings - Helifix Ltd
Stamped & formed metals - Textron Fastening Systems Ltd
Steelwork fixings - Lindapter International
Structural adhesives - Kenyon Group Ltd
Structural steel connectors - Unistrut Limited
Stud anchors - Ramset Fasteners Ltd
Styreen Free Resin - Redifix Ltd
Technical adhesive tapes - Beiersdorf (UK) Ltd
Threaded fasteners - Arnold Wragg Ltd.
Threaded fasteners - Textron Fastening Systems Ltd
Tile adhesives - Ruberoid Building Products Ltd
Tiling adhesives and grout - Weber Building Solutions
Tying wire, spacers and chemicals for concrete - Hy-Ten Reinforcement Co Ltd
U bolts - Marine & Commercial Rigging Systems
U bolts, overhead line fittings and wire rope grips - Mould J H & Co Ltd
Various types of metal and plastic fasteners - Righton Ltd
VMI programs - Textron Fastening Systems Ltd
Wall Paper Paste - Halls Beeline Ltd
Wallpaper Adhesives - Bartoline Ltd
Water based adhesives - Kenyon Group Ltd
Wood glue and resins - Wessex Resins & Adhesives Ltd

Z21 Mortars

Admixtures - Building Adhesives Ltd
Admixtures - Feb MBT
Admixtures - Fosroc Ltd
Admixtures - Vallance
Bagged mortar - Tarmac Northern Ltd
Building chemicals - Ashmead Buidling Supplies Ltd
Building solutions and compounds - Ruberoid Building Products Ltd
Cement - Ashmead Buidling Supplies Ltd
Cement - Combustion Linings Ltd
Cement - Rugby Cement
Cement - Super Cement Ltd
Cement admixtures - Ruberoid Building Products Ltd
Cements - Fortafix Ltd
Chemical Resistant Mortar - Accrington Brick Tile Co Ltd
Colour hardeners - Hatfield Ltd, Roy
Coloured mortars - Tarmac Northern Ltd
Colouring admixtures - Hatfield Ltd, Roy
Fire cement - Purimachos Ltd
Fire Rated Mortar - Abesco Ltd
Grouts and mortars - Exchem Mining and Construction
Grouts and resins - Helifix Ltd
Industrial hardeners - Hatfield Ltd, Roy
Lime mortars and aggregates - Tilcon South Ltd
Manufacturers of mortars & Resin systems - SGL Acotec Ltd
Mortars - Degussa Construction Chemicals (UK)
Mortars - Drainage Systems
Mortars - Promat Fire Protection Ltd
Mortars - Readymix Limited
Powder - Bayer Chemicals Inoegate Pigment Group
Ready mixed concrete and mortars - Hanson Aggregates
Ready mixed fillers - Bartoline Ltd
Slag cement - Civil & Marine Slag Cement Ltd
Woodfillers - Bondaglass Voss Ltd

Z22 Sealants

Bonding tape for georgian bars - Alfas Industries Ltd
Butyl sealants and foam fillers - Scapa Tapes UK Ltd.
Ceramic tiling tools & accessories - Building Adhesives Ltd
Concrete products - SFS Intec LTD
Concrete Repair - Sika Ltd
Construction sealents - GE Bayer Silicones
Ducting sealants: ductseal, flangeseal - Alfas Industries Ltd
Façade Sealants - Sika Ltd
Fillers - Loctite UK Ltd
Fire rated sealants - Compriband Ltd
Floor fixings - Lindapter International
Fuel resistant sealants - Honeywell Speciality Wax and Additives Ltd
Glazing sealants, silicones and tapes - Alfas Industries Ltd
Grouting - Feb MBT
Grouts - Building Adhesives Ltd
Grouts - Degussa Construction Chemicals (UK)
Heldite Jointing Compound - Heldite Ltd
Impregnated foam sealants - Compriband Ltd
Intumescent sealant - Mann McGowan Group
Linseed oils - Liver Grease Oil & Chemical Co Ltd
Mastic sealant - Trade Sealants
Metal roofing and cladding - Ash & Lacy Building Products Ltd
Movement joint materials - Building Adhesives Ltd
Natural stone & facade sealant - Dow Corning Ltd
Oakum - Liver Grease Oil & Chemical Co Ltd
Passive fire protection: fire rated and intumescent sealants - Alfas Industries Ltd
Plaster Accessories - British Gypsum Ltd
Polyurathane sealers - Hatfield Ltd, Roy
Pre-compressed impregnated expanding foam tape - Alfas Industries Ltd
Roofing sealants: butyl, PU foam, flashing tape - Alfas Industries Ltd
Sealant applicator quality silicones - Alfas Industries Ltd
Sealants - Adshead Ratcliffe & Co Ltd
Sealants - Amari Plastics Plc
Sealants - Bondaglass Voss Ltd
Sealants - Brewer C & Sons Ltd
Sealants - Campbell Marson & Co
Sealants - Degussa Construction Chemicals (UK)
Sealants - Drainage Systems
Sealants - Evode Ltd
Sealants - Exitex Ltd
Sealants - Feb MBT
Sealants - Firebarrier Services Ltd
Sealants - Fortafix Ltd
Sealants - Fosroc Ltd
Sealants - Henkel Industrial Adhesives
Sealants - Lastite Building Products Ltd
Sealants - Loctite UK Ltd
Sealants - Mapei (UK) Ltd
Sealants - Ogilvie Construction Ltd
Sealants - Promat Fire Protection Ltd
Sealants - Ruberoid Building Products Ltd
Sealants - Sealocrete Ltd
Sealants - Smith & Rodger Ltd
Sealants - Thermica Ltd
sealants - Till & Whitehead Ltd, t/a L. Birkinshaw
Sealants - Tower Manufacturing
Sealants - Tremco Ltd
Sealants - Vallance
Sealants & Primers - Halls Beeline Ltd
Sealants, adhesives and building chemicals - Kalon Decorative Products
Sealents - Canonbury Asphalte Co. Ltd.
Sealers - Hatfield Ltd, Roy
Security glazing tapes - Alfas Industries Ltd
Self-smoothing compounds & surface treatments - Building Adhesives Ltd
Silicone - Unifix Ltd
Silicone products - Jamak Fabrications (Europe) Ltd
Silicone sealants - Compriband Ltd
Silicone weatherseal sealant - Dow Corning Ltd
Trim mounting tapes - Alfas Industries Ltd
Waterproof expansion joints - Radflex Contract Services Ltd
weatherproofing silicone sealant - Dow Corning Ltd
Woodfillers and woodcare - Clam-Brummer Ltd

Z31 Powder coatings

Powder coating - Consort Equipment Products Ltd
Powder coatings - H. B. Fuller Coatings Ltd
Powder Coatings - Howse Thomas Ltd
Powder coatings - Plastic Coatings Ltd
Power Coatings - Lancashire Steel Fabrication Co Ltd

Z32 Liquid coatings

Agricultural - construction enamels - Pearl Paints Ltd
Anti-carbonation coatings - Glixtone Ltd
Anti-graffiti coatings - Andura Textured Masonry Coatings Ltd
Anti-graffiti coatings - Glixtone Ltd
Anti-microbial liquid coatings - Liquid Technology Ltd
Automotive lubricants and industrial paints - Witham Oil & Paint Ltd
Chemical stains - Hatfield Ltd, Roy
Clear protective coatings - Andura Textured Masonry Coatings Ltd
Fungicidal coatings - Glixtone Ltd
Hi-tech Paint Finishes - Bristol Metal Spraying & Protective Coatings Ltd
Liquid coatings - Andrews Coatings Ltd
Liquid coatings - Conren Ltd
Liquid coatings - Smith & Rodger Ltd
Liquid coatings - Tor Coatings Ltd
Liquid Coatings - Wessex Resins & Adhesives Ltd
Nylon coating - Plastic Coatings Ltd
Paint - Fired Earth
Protecting concrete floors - Ferguson G A & Co Ltd
Protective coating - Deborah Services Limited
Specialist decorative protective surface paints and coatings - Glixtone Ltd
Wet Coatings - Lancashire Steel Fabrication Co Ltd

Z33 Anodising

Anodised metal - Alanod Ltd
Hot dip galvanizing - Lionweld Kennedy Ltd

Products and Services
Alphabetical Index

Alphabetical Index (Cont'd)